I C [redacted]
24 0119726 0

AF745632

Lyon Playfair Library

IMPERIAL COLLEGE, LONDON SW7 2AZ

Term Loan
to be returned by last date stamped.
The loan period will be reduced to THREE WEEKS if the book is reserved by another reader.

15 JAN 1988
24 JUN 1988
24 JUN 1988
-7 OCT 1988
11 JAN 1989
28 APR 1989
30 JUN 1989

26 APR 1991
26 APR 1991
4 OCT 1991
19 OCT 1992
-8 OCT 1993
14 JAN 1994
27 APR 1994
13 JAN 1995

26 APR 1995
26 JUN 1998
30 SEP 1998
-5 OCT 2000
11/1/2001
11 JAN 2002
21 JUN 2002
28 APR 2006

Bawl
10-11-86
SD

The chemistry of

organic selenium and tellurium compounds

Volume 1

THE CHEMISTRY OF FUNCTIONAL GROUPS

A series of advanced treatises under the general editorship of Professor Saul Patai

The chemistry of alkenes (2 volumes)
The chemistry of the carbonyl group (2 volumes)
The chemistry of the ether linkage
The chemistry of the amino group
The chemistry of the nitro and nitroso groups (2 parts)
The chemistry of carboxylic acids and esters
The chemistry of the carbon–nitrogen double bond
The chemistry of amides
The chemistry of the cyano group
The chemistry of the hydroxyl group (2 parts)
The chemistry of the azido group
The chemistry of acyl halides
The chemistry of the carbon–halogen bond (2 parts)
The chemistry of the quinonoid compounds (2 parts)
The chemistry of the thiol group (2 parts)
The chemistry of the hydrazo, azo and azoxy groups (2 parts)
The chemistry of amidines and imidates
The chemistry of cyanates and their thio derivatives (2 parts)
The chemistry of diazonium and diazo groups (2 parts)
The chemistry of the carbon–carbon triple bond (2 parts)
The chemistry of ketenes, allenes and related compounds (2 parts)
The chemistry of the sulphonium group (2 parts)
Supplement A: The chemistry of double-bonded functional groups (2 parts)
Supplement B: The chemistry of acid derivatives (2 parts)
Supplement C: The chemistry of triple-bonded functional groups (2 parts)
Supplement D: The chemistry of halides, pseudo-halides and azides (2 parts)
Supplement E: The chemistry of ethers, crown ethers, hydroxyl groups and their sulphur analogues (2 parts)
Supplement F: The chemistry of amino, nitroso and nitro compounds and their derivatives (2 parts)
The chemistry of the metal–carbon bond (3 volumes)
The chemistry of peroxides

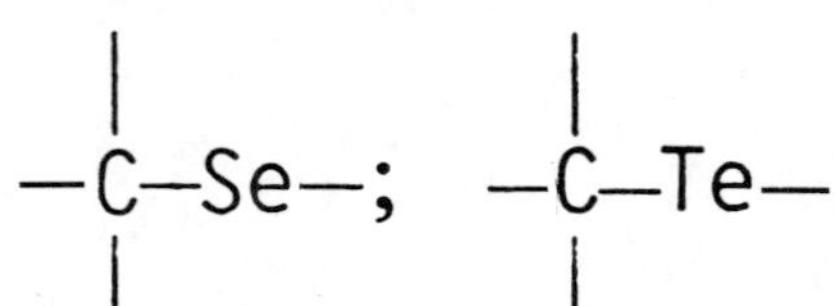

The chemistry of
organic selenium and tellurium compounds
Volume 1

Edited by

SAUL PATAI

and

ZVI RAPPOPORT

The Hebrew University, Jerusalem

1986

JOHN WILEY & SONS

CHICHESTER – NEW YORK – BRISBANE – TORONTO – SINGAPORE

An Interscience ® Publication

547.269

Copyright © 1986 by John Wiley & Sons Ltd.

All rights reserved.

No part of this book may be reproduced by any means, or transmitted, or translated into a machine language without the written permission of the publisher.

Library of Congress Cataloging-in-Publication Data:
Main entry under title:

The Chemistry of organic selenium and tellurium compounds.

(The Chemistry of functional groups)
'An Interscience publication.'
Includes index.
1. Organoselenium compounds. 2. Organotellurium compounds. I. Patai, Saul. II. Rappoport, Zvi. III. Series.
QD412.S5C53 1986 547'.0572 85-17868

ISBN 0 471 90425 2

British Library Cataloguing in Publication Data:

The Chemistry of Organic Selenium and tellurium compounds.—(The Chemistry of functional groups)
Vol. 1
1. Organoselenium compounds 2. Organotellurium compounds
I. Patai, Saul II. Rappoport, Zvi. III. Series
547'.05724 QD412.S5

ISBN 0 471 90425 2

Printed and bound in Great Britain

Contributing Authors

R. Badiello	C. N. R., Servizio Regionale di Sicurezza del Lavoro e Protezione Sanitaria, Facoltà di Farmacia dell'Università di Bologna, Italy
L. Batt	Aberdeen University, Aberdeen, Scotland
J. Bergman	Department of Organic Chemistry, Royal Institute of Technology, S-100 44 Stockholm, Sweden
D. Chakraborti	Department of Chemistry, Texas A&M University, College Station, Texas 77843, USA
I. G. Csizmadia	Department of Chemistry, University of Toronto, Toronto, M5S 1A1, Canada
L. Engman	Department of Organic Chemistry, Royal Institute of Technology, S-100 44 Stockholm, Sweden
F. Fringuelli	Dipartimento di Chimica, Università di Perugia, Perugia, Italy
K. Fujimori	Department of Chemistry, The University of Tsukuba, Sakura-mura Niihari-gun, Ibaraki-ken, 305 Japan
M. L. Gross	Department of Chemistry, University of Nebraska, Lincoln, Nebraska 68588-0304, USA
H. J. Gysling	Research Laboratories, Eastman Kodak Company, Rochester, New York 14650, USA
I. Hargittai	Hungarian Academy of Sciences, Research Laboratory for Inorganic Chemistry, Department of Structural Studies, P.O. Box 117, Budapest, H-1431, Hungary
L. Hevesi	Department of Chemistry, Facultés Universitaires Notre-Dame de la Paix, 61, Rue de Bruxelles, B-5000 Namur, Belgium
K. J. Irgolic	Department of Chemistry, Texas A&M University, College Station, Texas 77843, USA
K. A. Jensen	University of Copenhagen, Chemical Laboratory II, The H. C. Ørsted Institute, Copenhagen, Denmark
A. Kjær	Department of Organic Chemistry, The Technical University of Denmark, Lyngby, Denmark
N. P. Luthra	Department of Chemistry, University of South Carolina, Columbia, South Carolina 29208, USA
S. Oae	Okayama University of Science, Ridai-cho 1-1, Okayama, 700 Japan
J. D. Odom	Department of Chemistry, University of South Carolina, Columbia, South Carolina 29208, USA

A. Ogawa	Department of Applied Chemistry, Osaka University, Suita, Osaka, Japan
Y. Okamoto	Department of Chemistry, Polytechnic Institute of New York, 333 Jay Street, Brooklyn, New York 11201, USA
R. A. Poirier	Department of Chemistry, Memorial University of Newfoundland, St. John's Newfoundland, A1B 3X7, Canada
M. Renson	Université de Liège, Institut de Chimie, Sart-Tilman, 4000 Liège, Belgium
B. Rozsondai	Hungarian Academy of Sciences, Research Laboratory for Inorganic Chemistry, Department of Structural Studies, P.O. Box 117, Budapest, H-1431, Hungary
J. Sidén	Department of Organic Chemistry, Royal Institute of Technology, S-100 44 Stockholm, Sweden
G. Snatzke	Lehrstuhl für Strukturchemie, Ruhruniversität Bochum, POB 10 21 48, D-4630 Bochum, FRG
N. Sonoda	Department of Applied Chemistry, Osaka University, Suita, Osaka, Japan
G. D. Sturgeon	Department of Chemistry, University of Nebraska, Lincoln, Nebraska 68588-0304, USA
A. Taticchi	Dipartimento di Chimica, Università di Perugia, Perugia, Italy
R. A. Zingaro	Department of Chemistry, Texas A&M University, College Station, Texas 77843, USA

Foreword

The present volume in 'The Chemistry of Functional Groups' series deals with organic compounds containing selenium or tellurium atoms. This material falls outside the scope of the set of four volumes in the same series, entitled 'The Chemistry of the Metal–Carbon Bond' now in the process of publication.

The authors have been requested, whenever possible, to make comparisons between analogous compounds containing the three chalcogen atoms sulphur, selenium and tellurium.

Originally we intended to publish all chapters of the present volume simultaneously. However, various technical problems forced us to change this plan and to publish eighteen chapters separately and with separate author and subject indices for this volume. The literature coverage of most chapters is up to the end of 1983, with occasional references from 1984.

A second volume (edited by one of us, S.P.) is now already under active preparation and will hopefully be published towards the end of 1986. The chapters it contains include: PES, Mossbauer, UV, visible and Raman spectroscopy; synthetic methods; preparative uses; seleno and telluro carbonyl derivatives; photochemistry; electrochemistry; H-bonding, acidity and complex formation; biochemistry and pharmacology; insertion and extrusion reactions; organo Se/Te halides; Se—N and Te—N bonds; Se—P, Se—As, Te—P and Te—As bonds; semiconductors, metals and superconductors; Se/Te analogues of ethers; SeCN and TeCN derivatives and Se/Te free radicals. Thus we hope that these two volumes will cover all important aspects of the organic chemistry of the derivatives of selenium and tellurium.

We will be very grateful to readers who would communicate to us mistakes, omissions and proposals relating to this volume as well as to other volumes in the Functional Groups series.

Jerusalem
July 1985

SAUL PATAI
ZVI RAPPOPORT

The Chemistry of Functional Groups
Preface to the Series

The series 'The Chemistry of Functional Groups' is planned to cover in each volume all aspects of the chemistry of one of the important functional groups in organic chemistry. The emphasis is laid on the functional group treated and on the effects which it exerts on the chemical and physical properties, primarily in the immediate vicinity of the group in question, and secondarily on the behaviour of the whole molecule. For instance, the volume *The Chemistry of the Ether Linkage* deals with reactions in which the C—O—C group is involved, as well as with the effects of the C—O—C group on the reactions of alkyl or aryl groups connected to the ether oxygen. It is the purpose of the volume to give a complete coverage of all properties and reactions of ethers in as far as these depend on the presence of the ether group but the primary subject matter is not the whole molecule, but the C—O—C functional group.

A further restriction in the treatment of the various functional groups in these volumes is that material included in easily and generally available secondary or tertiary sources, such as Chemical Reviews, Quarterly Reviews, Organic Reactions, various 'Advances' and 'Progress' series as well as textbooks (i.e. in books which are usually found in the chemical libraries of universities and research institutes) should not, as a rule, be repeated in detail, unless it is necessary for the balanced treatment of the subject. Therefore each of the authors is asked *not* to give an encyclopaedic coverage of his subject, but to concentrate on the most important recent developments and mainly on material that has not been adequately covered by reviews or other secondary sources by the time of writing of the chapter, and to address himself to a reader who is assumed to be at a fairly advanced post-graduate level.

With these restrictions, it is realized that no plan can be devised for a volume that would give a *complete* coverage of the subject with *no* overlap between chapters, while at the same time preserving the readability of the text. The Editor set himself the goal of attaining *reasonable* coverage with *moderate* overlap, with a minimum of cross-references between the chapters of each volume. In this manner, sufficient freedom is given to each author to produce readable quasi-monographic chapters.

The general plan of each volume includes the following main sections:

(a) An introductory chapter dealing with the general and theoretical aspects of the group.

(b) One or more chapters dealing with the formation of the functional group in question, either from groups present in the molecule, or by introducing the new group directly or indirectly.

(c) Chapters describing the characterization and characteristics of the functional groups, i.e. a chapter dealing with qualitative and quantitative methods of determination including chemical and physical methods, ultraviolet, infrared, nuclear magnetic resonance and mass spectra: a chapter dealing with activating and directive effects exerted by the group and/or a chapter on the basicity, acidity or complex-forming ability of the group (if applicable).

(d) Chapters on the reactions, transformations and rearrangements which the functional group can undergo, either alone or in conjunction with other reagents.

(e) Special topics which do not fit any of the above sections, such as photochemistry, radiation chemistry, biochemical formations and reactions. Depending on the nature of each functional group treated, these special topics may include short monographs on related functional groups on which no separate volume is planned (e.g. a chapter on 'Thioketones' is included in the volume *The Chemistry of the Carbonyl Group*, and a chapter on 'Ketenes' is included in the volume *The Chemistry of Alkenes*). In other cases certain compounds, though containing only the functional group of the title, may have special features so as to be best treated in a separate chapter, as e.g. 'Polyethers' in *The Chemistry of the Ether Linkage*, or 'Tetraaminoethylenes' in *The Chemistry of the Amino Group*.

This plan entails that the breadth, depth and thought-provoking nature of each chapter will differ with the views and inclinations of the author and the presentation will necessarily be somewhat uneven. Moreover, a serious problem is caused by authors who deliver their manuscript late or not at all. In order to overcome this problem at least to some extent, it was decided to publish certain volumes in several parts, without giving consideration to the originally planned logical order of the chapters. If after the appearance of the originally planned parts of a volume it is found that either owing to non-delivery of chapters, or to new developments in the subject, sufficient material has accumulated for publication of a supplementary volume, containing material on related functional groups, this will be done as soon as possible.

The overall plan of the volumes in the series 'The Chemistry of Functional Groups' includes the titles listed below:

The Chemistry of Alkenes (*two volumes*)
The Chemistry of the Carbonyl Group (*two volumes*)
The Chemistry of the Ether Linkage
The Chemistry of the Amino Group
The Chemistry of the Nitro and Nitroso Groups (*two parts*)
The Chemistry of Carboxylic Acids and Esters
The Chemistry of the Carbon–Nitrogen Double Bond
The Chemistry of the Cyano Group
The Chemistry of Amides
The Chemistry of the Hydroxyl Group (*two parts*)
The Chemistry of the Azido Group
The Chemistry of Acyl Halides
The Chemistry of the Carbon– Halogen Bond (*two parts*)
The Chemistry of the Quinonoid Compounds (*two parts*)
The Chemistry of the Thiol Group (*two parts*)
The Chemistry of Amidines and Imidates
The Chemistry of the Hydrazo, Azo and Azoxy Groups (*two parts*)
The Chemistry of Cyanates and their Thio Derivatives (*two parts*)
The Chemistry of Diazonium and Diazo Groups (*two parts*)
The Chemistry of the Carbon–Carbon Triple Bond (*two parts*)

Supplement A: The Chemistry of Double-bonded Functional Groups (*two parts*)
The Chemistry of Ketenes, Allenes and Related Compounds (*two parts*)
Supplement B: The Chemistry of Acid Derivatives (*two parts*)
Supplement C: The Chemistry of Triple-Bonded Functional Groups (*two parts*)
Supplement D: The Chemistry of Halides, Pseudo-halides and Azides (*two parts*)
Supplement E: The Chemistry of Ethers, Crown Ethers, Hydroxyl Groups and their Sulphur Analogues (*two parts*)
The Chemistry of the Sulphonium Group (*two parts*)
Supplement F: The Chemistry of Amino, Nitroso and Nitro Groups and their Derivatives (*two parts*)
The Chemistry of the Metal–Carbon Bond (*three volumes*)
The Chemistry of Peroxides
The Chemistry of Organic Se and Te Compounds Vol. 1

Titles in press:

The Chemistry of Cyclopropanes
The Chemistry of Organic Se and Te Compounds Vol. 2

Advice or criticism regarding the plan and execution of this series will be welcomed by the Editor.

The publication of this series would never have started, let alone continued, without the support of many persons. First and foremost among these is Dr Arnold Weissberger, whose reassurance and trust encouraged me to tackle this task. The efficient and patient cooperation of several staff-members of the Publisher also rendered me invaluable aid (but unfortunately their code of ethics does not allow me to thank them by name). Many of my friends and colleagues in Israel and overseas helped me in the solution of various major and minor matters, and my thanks are due to all of them, especially to Professor Z. Rappoport. Carrying out such a long-range project would be quite impossible without the non-professional but none the less essential participation and partnership of my wife.

The Hebrew University
Jerusalem, ISRAEL

SAUL PATAI

Contents

1. Organic derivatives of sulphur, selenium and tellurium—an overview 1
K. A. Jensen and A. Kjaer

2. General and theoretical aspects of organic compounds containing selenium or tellurium 21
R. A. Poirier and I. G. Csizmadia

3. Structural chemistry of organic compounds containing selenium or tellurium 63
I. Hargittai and B. Rozsondai

4. Thermochemistry of selenium and tellurium compounds 157
L. Batt

5. Detection and determination of organic selenium and tellurium compounds 161
K. J. Irgolic and D. Chakraborti

6. Nuclear magnetic resonance and electron spin resonance studies of organic selenium and tellurium compounds 189
N. P. Luthra and J. D. Odom

7. Mass spectrometry of organic selenium and tellurium compounds 243
G. D. Sturgeon and M. L. Gross

8. Radiation chemistry of organic selenium and tellurium compounds 287
R. Badiello

9. Selenium-stabilized carbenium ions and free radicals 307
L. Hevesi

10. Selenium- and tellurium-containing organic polymers 331
Y. Okamoto

11. Organometallic compounds with selenium and tellurium atoms bonded to main group elements of Groups IIIa, IVa and Va 343
R. A. Zingaro

12. Synthesis and uses of isotopically labelled selenium and tellurium compounds 369
K. Fujimori and S. Oae

13. Selenium and tellurium heterocycles 399
M. Renson

14. Tetra- and higher-valent (hypervalent) derivatives of selenium and tellurium 517
J. Bergman, L. Engman and J. Sidén

15. Directing and activating effects involving selenium and tellurium 559
F. Fringuelli and A. Taticchi

16. Functional groups containing selenium and tellurium in various oxidation states 619
N. Sonoda and A. Ogawa

17. Stereochemistry and chiroptical properties of organic selenium and tellurium compounds 667
G. Snatzke

18. Ligand properties of organic selenium and tellurium compounds 679
H. J. Gysling

Author index 857

Subject index 925

The Chemistry of Organic
Selenium and Tellurium
Compounds Volume 1
Edited by S. Patai and Z. Rappoport
© 1986 John Wiley & Sons Ltd.

CHAPTER **1**

Organic derivatives of sulphur, selenium and tellurium—an overview

KAI ARNE JENSEN

University of Copenhagen, Chemical Laboratory II, The H. C. Ørsted Institute, Copenhagen, Denmark

ANDERS KJÆR

Department of Organic Chemistry, The Technical University of Denmark, Lyngby, Denmark

I. INTRODUCTION . . . 2
II. NOMENCLATURE . . . 3
III. HISTORY . . . 5
IV. ANALOGUES OF ALCOHOLS AND ETHERS . . . 5
A. Alcohol Analogues . . . 5
B. Ether Analogues . . . 6
V. ONIUM SALTS AND YLIDES . . . 7
A. Onium Salts . . . 7
B. Ylides . . . 8
VI. INSERTION COMPOUNDS . . . 8
VII. ANALOGUES OF SULPHOXIDES, SULPHONES AND RELATED COMPOUNDS . . . 9
A. Selenoxides and Telluroxides . . . 9
B. Selenones and Tellurones . . . 10
VIII. ANALOGUES OF CARBONYL COMPOUNDS . . . 10
A. Analogues of Aldehydes and Ketones . . . 11
B. Carboxylic and Carbonic Acid Analogues . . . 12
IX. OXO ACIDS OF SULPHUR, SELENIUM AND TELLURIUM . . . 13
A. Valency State Six . . . 13
B. Valency State Four . . . 14
C. Valency State Two . . . 14

X. HALOGEN COMPOUNDS 15
A. Monohalides 15
B. Dihalides 15
C. Trihalides 15
XI. CHALCOGEN DERIVATIVES OF GROUP V ELEMENTS 16
A. Nitrogen Compounds 16
B. Phosphorus and Arsenic Compounds. 16
XII. HETEROCYCLIC COMPOUNDS 17
XIII. REFERENCES 17

I. INTRODUCTION

The chalcogens, constituting Group VI in the Periodic Table, exhibit differences in their chemical properties that run parallel to those observed in other non-metals within Groups IV–VII (Table 1). Thus, O, like other second period members of the family, has unique characteristics rooted in its high electronegativity and lack of d orbitals whereas S and Se share with pairs of third and fourth period elements from other Groups a striking similarity in their chemical properties. By passing on to the fifth period, i.e. Te within Group VI, we note another jump in properties. This overall pattern reflects the increase in atomic radii, and hence in coordination numbers, as clearly brought out by considering the oxo anions derived from the various elements in their highest oxidation states (Table 2): S and Se are comparable, but distinctly different from Te. In their divalent states, however, the Group VI elements exhibit a more gradual change, moving towards lower electronegativity with increase in atomic weight. Consequently, hydride stability decreases. On substitution of H with organic radicals more stable molecules are formed so that R_4Pb, R_3Bi and R_2Po are still species of reasonable stability.

In the present context our attention will be limited to *organic* derivatives of the chalcogens. Based on common knowledge, set out in Tables 1 and 2, we shall enquire into the degree of similarity existing within the organic chemistry of S, Se and Te. By necessity, such a venture must be selective and inevitably biased by the personal interests of the authors. We shall draw on numerous sources including established monographic treatises on organoselenium[1,2] and organotellurium[3] chemistry, assorted reviews on more restricted topics, original articles, and the useful, current awareness publication, *Organic Compounds of Sulphur, Selenium and Tellurium*, issued within the Specialist Periodical Report series by The Royal Society of Chemistry and thus far covering the literature published until March 1980. With a view to ordering the discussion we shall briefly dwell on both historical aspects and nomenclature rules before proceeding to discuss various classes of compound according to functionality, as well as certain aspects of interest to synthetic chemistry. No attention will be given in the present 'overview' to biological aspects which will be discussed in a different chapter of this volume.

TABLE 1. Periodic Table of Group IV–VII elements

	Group			
Period	IV	V	VI	VII
2	C	N	O	F
3	Si	P	S	Cl
4	Ge	As	Se	Br
5	Sn	Sb	Te	I
6	Pb	Bi	Po	At

TABLE 2. Oxo-anions of Group IV–VII elements in their highest oxidation state

Coordination number	Group IV	V	VI	VII
3	CO_3^{2-}	NO_3^-		
4	SiO_4^{4-}	PO_4^{3-}	SO_4^{2-}	ClO_4^-
4	GeO_4^{4-}	AsO_4^{3-}	SeO_4^{2-}	BrO_4^-
6	$Sn(OH)_6^{2-}$	$Sb(OH)_6^-$	TeO_6^{6-}	IO_6^{5-}
6	$Pb(OH)_6^{2-}$			

The present chapter has as its chief objective to introduce the general subject of this volume, to place it in a broader context, but above all to whet the appetite for additional and more detailed information.

II. NOMENCLATURE

Since organoselenium compounds were as a rule discovered later than the corresponding sulphur compounds they have frequently been named by adding the prefix seleno- to the name of the corresponding sulphur compound, e.g. selenocystine, selenoglutathione, selenouracil, selenoxanthate, selenomercaptan. Similarly, a Te analogue of methionine has been called telluromethionine. In the rules formulated by the Commission on Nomenclature of Organic Chemistry of the International Union of Pure and Applied Chemistry (IUPAC)[4] this practice is not accepted. Nevertheless, it is being widely followed in the literature when the sulphur compound is a natural product with an accepted trivial name, as for example selenocysteine. In other cases systematic names should be used. According to the IUPAC rules a 'selenoxanthate' is an *O*-alkyl diselenocarbonate, and a 'selenomercaptan' is a selenol. The use of the seleno- prefix to indicate replacement of S by Se in new compounds is to be strongly discouraged, *inter alia* because the *Chemical Abstracts* indexes enter systematically correct terms without cross-references to new trivial names.

The prefix seleno- has, however, traditionally also been used to indicate replacement of O by Se. The IUPAC Commissions of both organic[4] and inorganic[5] chemistry have adopted this rule if the corresponding oxygen compound has an accepted functional class ending or if an oxygen-containing radical has an accepted prefix. Consequently, selenocyanate, selenourea, selenosemicarbazide, selenoketones, selenobenzamide, etc., and the prefixes selenocyanato- and selenocarbonyl-, are all recommended IUPAC names.

In analogy with the suffix name-thione for $>C{=}S$, the name -selenone has repeatedly been used in the literature to designate $>Se{=}O$. This is, however, confusing since -selenone is the suffix also for an isologue of a sulphone, R_2SeO_2. In sulphur chemistry we have at our disposal the prefixes sulph- and thio-, derived from Latin and Greek, to distinguish, for example, between disulphane, H_2S_2, and the heterocyclic compound dithiane. An analogous opportunity to use Latin luna and tellus along with Greek selene and gea was neglected long ago and is now unrealistic. The IUPAC nomenclature commissions therefore decided to introduce the prefixes sel- and tell- to be used along with selen- and tellur-. In this way, diselane, H_2Se_2, and ditellane, H_2Te_2, can be distinguished from the heterocyclic species diselenane and ditellurane, and the suffix for $>C{=}Se$ becomes -selone (cyclohexaneselone, 4-thiazoline-2-selone, etc.), and for $>C{=}Te$, -tellone.

The last edition of *Nomenclature of Organic Chemistry*[4] contains detailed rules for

TABLE 3. Suffixes and prefixes for selected groups of Se- and Te-containing species

Suffix or functional class name			Prefix
—SH thiol	—SeH selenol	—TeH tellurol	Mercapto, hydroseleno, hydrotelluro
—S—	—Se—	—Te—	Thia, selena, tellura sulfa[a], sela[a], tella[a]
—SR sulfide	—SeR selenide	—TeR telluride	R-thio, R-seleno, R-telluro[b], or R-sulfenyl, R-selenyl, R-tellurenyl
>C=S thione	>C=Se selone	>C=Te tellone	Thiooxo, selenoxo, telluroxo
—CSSH carbodithioic acid	—CSSeH carboselenothioic acid	—CSTeH carbotellurothioic acid	Dithiocarboxy, selenothiocarboxy, tellurothiocarboxy
>SO sulfoxide	>SeO selenoxide	>TeO telluroxide	Sulfinyl, seleninyl, tellurinyl
$>SO_2$ sulfone	$>SeO_2$ selenone	$>TeO_2$ tellurone	Sulfonyl, selenonyl, telluronyl
$—SO_3H$ sulfonic acid	$—SeO_3H$ selenonic acid	—[c]	Sulfo, selenono
$—SO_2H$ sulfinic acid	$—SeO_2H$ seleninic acid	—[c]	Sulfino, selenino
—SOH	—SeOH	—[c]	Sulfeno, seleneno
R_3S^+ sulfonium	R_3Se^+ selenonium	R_3Te^+ telluronium	Sulfonio, selenonio, telluronio

[a]In homogeneous chains (cf. Ref. 4, Rule D–4).
[b]R-tellurio in organometallic compounds
[c]Extension of the system to cover the Te analogues should be carefully considered since such compounds may be polymers.

naming characteristic groups containing Se or Te (Rules C–10.1, C–10.42, C–22.1, C–82.1, C–701, C–833, C–974, C–985 and D–1.54) and guidelines for naming heterocyclic rings with Se or Te (Rules B–1, B–3, B–4 and Appendix, Table IV). A comprehensive discussion of the nomenclature of organoselenium compounds is also available[6].

In Table 3 a list is presented of suffixes and prefixes recommended by IUPAC. A few comments are made as footnotes to the table.

Throughout the present chapter the following abbreviations have been used: X = halogen, Y = chalcogen, R = organyl radical.

III. HISTORY

The organic chemistry of sulphur dates from the discovery by W. C. Zeise of the xanthates in 1822 and 'mercaptan' (ethanethiol) in 1831. His work initiated extensive studies in organic sulphur chemistry with the result that most of the characteristic sulphur-containing organic groups were known by about 1865[7].

Without much delay organic compounds of Se and Te also became known. Berzelius, the discoverer of Se (1818), found that alkali metal selenides and tellurides resemble sulphides, and in 1840 F. Wöhler[8] prepared the first organic Te compound, diethyl telluride, in a similar way to the sulphide. Löwig[9] had already prepared diethyl selenide, mixed with the diselenide, in 1836, but the pure compounds were not isolated until 1869[10]. Several other organic Se and Te compounds were synthesized about the same time, most of them in Wöhler's laboratory[11]: ethaneselenol (1847), diethyl telluroxide (1851), dimethyl telluride, dimethyl diselenide (1856), and others. Selenonium and telluronium salts[12] were discovered in 1865, but selenoxides[13] not until 1893. From SeO_2, inorganic selenocyanate, selenourea (discovered[14] in 1884), and some heterocyclic compounds, containing N and Se in the same ring, were prepared in 1889–90[15–17]. 2,5-Dimethylselenophene[18] was synthesized in 1885, selenophene itself not until 1927, and tellurophene only in 1972.

After the pioneering period organic Se and Te chemistry developed only slowly. The investigations were hampered by the compounds often being evil-smelling, toxic and sensitive to air and light. Many of the compounds prepared were of low purity and several reports on the isolation of new compounds were unwarranted.

Since about 1950, modern methods and equipment have made it possible to prepare organic Se and Te compounds in higher yields and purity and to isolate compounds of low stability. The development has been catalysed by technical and biological interest in Se and Te compounds, and the study of their chemical reactions has resulted in important new methods in organic synthesis.

IV. ANALOGUES OF ALCOHOLS AND ETHERS

A. Alcohol Analogues

Thiols (**1**), selenols (**2**) and tellurols (**3**) are synthetically accessible from salts of hydrogen sulphide (**4**), hydrogen selenide (**5**) and hydrogen telluride (**6**). The acidity of the hydrides increases dramatically from **4** to **6**, the respective pK_{a1} values being 7.0, 3.8 and 2.6.

RYH	H_2Y	R^1YR^2
(**1**) Y = S	(**4**) Y = S	(**7**) Y = S
(**2**) Y = Se	(**5**) Y = Se	(**8**) Y = Se
(**3**) Y = Te	(**6**) Y = Te	(**9**) Y = Te

In analogy with alcohols normally being weaker acids than water, **1**, **2** and **3** may be expected to be weaker acids than **4**, **5** and **6**, respectively, with acidity constants increasing in the order **1**, **2**, **3**. Consequently, the corresponding bases, RS^-, RSe^- and RTe^-, must be weaker bases than RO^-, with the basicity decreasing in the order given. Mainly due to their high polarizability, however, the anions of **2** and **3** are better nucleophiles than the anion derived from **1**, and much better than the RO^- ion.

Selenols (**2**) undergo oxidation to diselenides even more easily than do thiols (**1**) to disulphides. Tellurols (**3**) are so sensitive to oxidation, leading to elemental Te and other products, that ditellurides have not been isolated from this process. Tellurols as such are poorly known; older reports on the characterization of simple alkanetellurols appear highly dubious. Lately, however, arenetellurolates have become available in solution by subjecting diaryl ditellurides to reduction (with $NaBH_4$, or Na in liq. NH_3), or by base-induced disproportionation, yielding tellurinates as the oxidized products[19]. Insertion of metallic Te into the carbon–metal bond of various organolithium compounds, notably of the heteroaromatic series, provides a convenient route to lithium arenetellurolates[20].

Weaker hydrogen bonding accounts for the lower boiling points of the lower members of the classes **1**, **2** and **3** when compared with the analogous alcohols.

Chemically, **1**, **2** and **3** possess reducing properties increasing in that order; thus, arenetellurolates have lately proven useful in the reduction of *vic*-dibromides to alkenes[21], and of α-halocarbonyl compounds to the reduced halogen-free counterparts[22]. Similarly, sodium hydrogen telluride serves as an efficient, selective reagent for reducing α,β-unsaturated carbonyl compounds (aldehydes, ketones and esters) to the saturated analogues[23].

B. Ether Analogues

Organic sulphides (**7**), selenides (**8**) and tellurides (**9**) are generally more stable than the corresponding hydrides, (**1**), (**2**) and (**3**). Numerous diorganyl chalcogenides are known, varying widely in the nature of the radicals R^1 and R^2; thus, besides the more common alkyl and aryl radicals, R^1 and R^2 may also represent, for example, metalorganic radicals (R_3Sn, R_3Ge, etc.). Several selena compounds, in which one or more methylene groups have been substituted by Se, have been synthesized, notably in connection with biological studies.

Generally, selenides (**8**) are light-sensitive, colourless compounds with an obnoxious odour, prepared by methods analogous to those used for making sulphides (**7**). The selenides exhibit great stability towards alkali and reducing reagents but can be oxidized to the synthetically important selenoxides (cf. Section VII of this chapter). Other features of interest in the present context are the ability to undergo alkylation to selenonium salts, and to lose an α-proton to give Se-stabilized carbanions. The C—Se bond is readily cleaved with alkyllithium reagents or by lithium dissolved in amines. These features provide the background for a rapidly expanding synthetic chemistry utilizing Se-containing intermediates[24].

Contrary to the generally inaccessible tellurols, the diorganyl tellurides (**9**) have been the subject of rather detailed studies involving crystal structure determination, spectroscopical characterization, dipole moment measurements, etc. Tellurides share with selenides the ability to form diorganyl chalcogen dihalides on treatment with halogens, and chalcogenonium salts on alkylation, but differ from the selenides in undergoing C—Te fission on oxidation. The yellow or red diorganyl tellurides are stable compounds when aromatic, but far less so in the aliphatic series. Sulphide reduction of organyltellurium trihalides provides an easy entrance into the series of organic ditellurides. In all of the Groups IV–VI, the heaviest element forms a weak bond to carbon. Thus, extensive studies within the class of tetraalkylleads, containing different radicals, have revealed that redistribution reactions easily occur[25]. This seems also to be the case with diorganyl tellurides[26]. Hence, reported syntheses of homogeneous, non-symmetrical diorganyl tellurides must be regarded with scepticism.

Only a few triselenides, and no tritellurides, are known, though unstable species have been recognized containing arrangements such as —SeSSe—, —STeS—, etc. All of these are, as expected, highly sensitive to nucleophilic reagents such as hydroxide ions. Tertiary phosphines have been used to abstract Se or Te from diorganyl dichalcogenides.

All compounds of the types RYH, R_2Y and R_2Y_2 (Y = S,Se,Te) have a pronounced ability to combine with transition metals thus producing a large number of coordination compounds[27].

In summary, the overall chemistry of selenides and tellurides is strongly reminiscent of that of sulphides, with the proviso that the Se, and more notably the Te, analogues are of lower stability and often of greater complexity in their chemical behaviour.

V. ONIUM SALTS AND YLIDES

A. Onium Salts

Triorganylsulphonium ions (**10**) and the isologous selenonium (**11**) and telluronium (**12**) ions were discovered in 1865[12]. Since then a considerable number of their salts have been prepared, either by the classical method (dialkyl chalcogenide and alkyl halide) or by various other methods which have made it possible to prepare also onium salts with different radicals, aromatic radicals, etc. Salts with anions other than those derived from halogens can be prepared by anion exchange or by precipitation with complex anions which often form sparingly soluble salts.

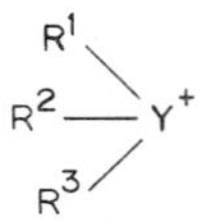

(**10**) Y = S
(**11**) Y = Se
(**12**) Y = Te

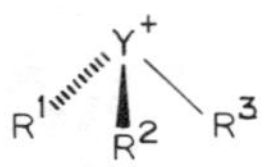

(**13**)

$R^1_2\overset{+}{Y}\overset{-}{C}HR^2_2$

(**14**) Y = S
(**15**) Y = Se
(**16**) Y = Te

Little is known about the Se and Te analogues of the many heterosulphonium salts (**10**) in which one or more of the radicals R^1, R^2 and R^3 represent O, N, S or halogen. Protonated selenols, $RSeH_2^+$, and tellurides, R_2TeH^+, have been observed by NMR spectroscopy in superacid solutions of selenols and tellurides but not been isolated[28].

The chalcogenium ions **10**–**12** form salts which are solid, salt-like compounds, usually insoluble in non-polar solvents but soluble in water. Their aqueous solutions exhibit electrolytic conductivity and give the qualitative reactions of the anions. With silver oxide in water the halides form strongly alkaline solutions. The hydroxides thus formed cannot usually be isolated but their aqueous solutions can be neutralized with HX to form new salts. On heating, the onium salts decompose, usually into R_2Y and RX. Their thermal stability increases from S to Te and generally with the size of the anion.

Sulphonium and selenonium ions adopt the geometry of stable trigonal pyramids (**13**) as evident from their chirality, documented through resolution into enantiomers (for $R^1 \neq R^2 \neq R^3$) which was achieved for **10** in 1900[29,30] and for **11** in 1902[31]. The reported resolution of a telluronium salt[32] could not be confirmed[33].

According to their properties the onium salts would be expected to be strong electrolytes. Recent investigations indicate, however, that this is an oversimplification. In the solid state onium salts of 'hard', complex anions have the expected ionic structure with non-coordinating anions. This has been proved by an X-ray structure analysis of the telluronium salt $[Me_3Te]$ $[BPh_4]$. Onium salts of the 'soft' halide and pseudohalide anions tend to possess a more complicated structure, the tendency increasing from S to Te, and from Cl to I. Therefore, deviations from a purely ionic structure are especially prominent among telluronium halides[34]. Distances are here often less than the sum of the respective van der Waals' radii, which may be characterized as secondary bonding. The intermolecular interactions appear to be the result of directed forces rather than of electrostatic or van der Waals' forces. As a consequence the crystal units may be described as oligomeric, resulting in a distorted octahedral geometry.

Te also forms compounds of the type R_4Te. The most stable representatives contain two 2,2′-biphenyldiyl radicals. An analogous but very unstable Se compound has also been obtained[35]. Attempts to prepare analogous S compounds of the type R_4S were unsuccessful. These compounds are possibly oligomeric in the solid state but no structure determinations have been reported.

B. Ylides

Onium ions may be deprotonated to form ylides, i.e. zwitterions or chalcogen-stabilized carbanions. Stable ylides, **14**, **15** and **16**, e.g. cyclopentadienides, have been prepared from both sulphonium, selenonium and telluronium ions.

More attention has lately been accorded to moderately stable and unstable ylides[36] which undergo reaction with non-enolizable carbonyl compounds to give oxirans in both the Se[37] and the Te[38] series (path a, equation 1).

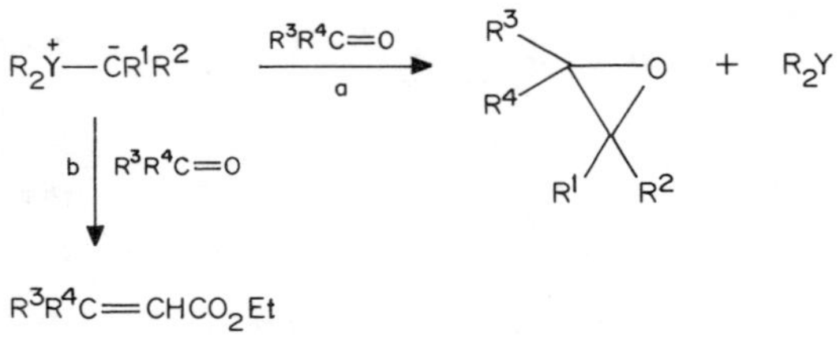

(1)

If $R^1 = CO_2Et, R^2 = H$ and $Y = Te$, however, the reaction proceeds differently, the stabilized tellurium ylide giving an α,β-unsaturated ester[39] (path b, equation 1) in a reaction unprecedented in the sulphur and selenium ylide series. It provides another illustration of qualitatively different reaction paths operating with S/Se compounds on the one hand, and Te compounds on the other.

VI. INSERTION COMPOUNDS

In a characteristic reaction the chalcogens, both as elemental substances (S_8, Se_x, Te_x) and as reactive derivatives, can be inserted into chains, rings and clusters of other atoms, even under mild conditions. A few examples shall serve to illustrate that S, Se and Te behave similarly in such reactions.

S and Se have been found to insert into the Si—Si bond of decamethylcyclopentasilane with the formation of the six-membered ring selenapentasilacyclohexane[40]. Similarly, Te inserts into the Si—P bond of the phosphine $Me_3SiPBu^t_2$ rather than producing a phosphine telluride (cf. Section XI), to give a product containing the — SiTeP group[41]. Other examples comprise insertion of selenium dioxide into stannoxanes, $R_3SnOSnR_3$, to give distannyl selenites, $R_3SnOSe(O)OSnR_3$[42] and of sulphur dioxide, selenium dioxide or tellurium dioxide into the Mo—C bond of the cycloheptatrienylmolybdenum compounds η-$C_7H_7Mo(CO)_2CH_3$, to give products of the type η-$C_7H_7Mo(CO)_2Y(O_2)CH_3$[43]. Often, however, reduction takes place at the same time. Thus, SeO_2 reacts with ditellurides to form tellurenyl tellurinyl selenides, RTeSeTe(O)R[44].

Insertion reactions are quite common with organometallic cyclopentadienyl and carbonyl compounds. As an example, elemental S, Se or Te insert into the Co—Co bond of $(\eta$-$C_5H_5)_2(Me_2P)_2Co_2$ to give products containing a CoYCo structure[45]. Often, however, reactions with carbonyl compounds are complicated by replacement of CO and reduction.

More related to organic chemistry is the reaction of carbon diselenide with tetraalkylmethylenediamines, $R_2NCH_2NR_2$, to give diselenocarbamate esters, $R_2NCH_2SeC(Se)NR_2$[46].

VII. ANALOGUES OF SULPHOXIDES, SULPHONES AND RELATED COMPOUNDS

The pyramidal sulphoxides (**17**) and the tetrahedral sulphones (**20**) have their counterparts in the selenoxides (**18**) or telluroxides (**19**) and the selenones (**21**) or tellurones (**22**). Double bonds are used throughout the present discussion subject to the proviso that varying degrees of polarization and d orbital participation may be involved, as evident from dipole moment measurements and spectroscopic data[47].

(**17**) Y = S (**20**) Y = S
(**18**) Y = Se (**21**) Y = Se
(**19**) Y = Te (**22**) Y = Te

A. Selenoxides and Telluroxides

Although stable, optically active sulphoxides have been known for more than 50 years, the first report on the synthesis of structurally simple, monochiral, optically active selenoxides (**18**) reached the literature only quite recently[48]. Enantiomerically enriched telluroxides (**19**) are unknown[33]. The basic properties of the diorganyl chalcogen oxides increase from **17** to **19** as does the ability to form tetravalent, symmetrical hydrates. With acids, both **18** and **19** form salts of the type $R_2Y(OH)Z$ (Z = halogen, carboxylate, nitrate, etc.), and selenoxides (**18**) form coordination compounds with many metal salts.

Selenoxides (**18**) are reasonably stable species provided that they do not contain β-positioned hydrogen atoms. If so, they undergo a remarkably facile, stereospecific *E* elimination, often at temperatures well below 20 °C (equation 2).

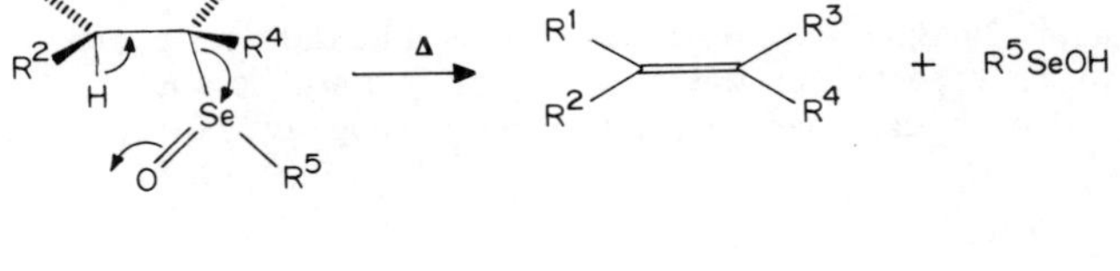

(2)

This reaction, discovered less than a decade ago, has rapidly been put to good use in modern organic synthesis[24]. Selenoxides (**18**) are generally prepared by oxidation of selenides, or by hydrolysis of the mostly easily accessible selenide dihalides.

Telluroxides (**19**), accessible through the same routes[49,50] are more basic compounds exhibiting distinctive alkaline reaction in aqueous solution. They share with the selenoxides (**18**) the ability to undergo thermal elimination to olefins though more drastic conditions are occasionally required[51]. In recent years aromatic telluroxides have attracted interest as mild oxidizing reagents[52].

B. Selenones and Tellurones

Selenones (**21**), like sulphones (**20**), are stable compounds of moderate reactivity, accessible through appropriate oxidation of selenides or selenoxides. An interesting variant of the 1,4-Grob-type elimination, induced by base treatment of vinylic phenylselenones, utilizes the $PhSeO_2$ group as an efficient nucleofuge[53]. The properties and chemistry of selenones deserve further exploration.

The first, fully characterized tellurone, **23**, was described only in 1982[54]. It was prepared by oxidation of the corresponding telluroxide. Previously reported representatives of **22** were obviously assigned erroneous structures. The aromatic tellurone **23** has mildly oxidizing properties of potential synthetic interest[54].

$(4\text{-}MeOC_6H_4)_2TeO_2$

(**23**)

$Ar_2Y{=}NSO_2R$

(**24**) Y = Se

(**25**) Y = Te

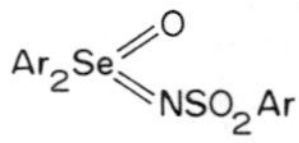

(**26**)

The tri- and tetra-coordinate isologues of *N*-sulphonylated sulphimides and sulphoximides, **24**, **25** and **26**, are known compounds, readily prepared by methods well known from the chemistry of sulphur (cf. Section XI).

Se and Te analogues of the sulphines and sulphenes, **27** and **28**, have yet to be produced and characterized.

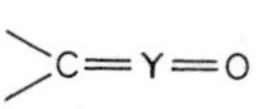

(**27**) Y = S, Se, Te

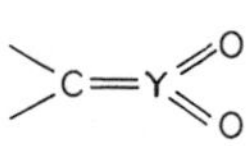

(**28**) Y = S, Se, Te

In general, the chemistry of selenoxides and telluroxides is similar to that of the sulphoxides, though with minor, but synthetically useful, differences. On closer inspection, however, we note once again a greater similarity in chemistry between members of the S and Se series on one hand, and the Te isologues on the other, the latter standing apart notably by their marked ability to attain higher coordination numbers as evident from X-ray structure analyses of compounds such as Ph_2TeO and $Ph_2Te(OH)NO_3$[55].

VIII. ANALOGUES OF CARBONYL COMPOUNDS

Within the series **29**–**32** of carbonyl compounds and their analogues, a diminishing stability is to be expected in the order given, mainly because of the decreasing electronegativity of the chalcogen element (the electronegativity of Te is almost the same as that of C).

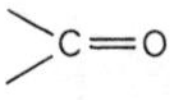
(29)

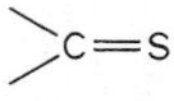
(30)

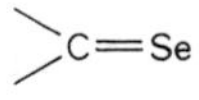
(31)

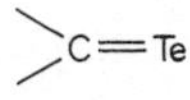
(32)

A. Analogues of Aldehydes and Ketones

A great variety of thials and thiones can be found in the literature[56], many of them stabilized by charge dislocation or tautomerism. Until recently, simpler aliphatic thials, thiones and their Se analogues were known only as polymers. When it was recognized some years ago that the polymers dissociate by pyrolysis, it became feasible to study the microwave spectra of the monomers in a flow system at low temperature. Thus, monomeric thioformaldehyde[57], thioacetaldehyde[58], thioacetone[59], thioketene[59] and selenoacetaldehyde[60] were identified but none of these species had half-lives long enough to permit their isolation. Since 1976, however, it has become evident that bulky substituents or hindered structures may provide enough protection against polymerization to allow the preparation of monomeric species of much higher stability. Moreover, the recognition that the instability of the Se and Te compounds is due both to their high electrophilic reactivity and susceptibility to catalytic influence has made it possible to design methods and equipment suitable for the isolation of these sensitive compounds. A selection of compound types, all formally containing doubly bonded chalcogen atoms, shall serve to illustrate these trends.

The counterparts of the perfectly stable carbonyl sulphide (**33**) and carbon disulphide (**35**), viz. carbonyl selenide (**34**), thiocarbonyl selenide (**36**) and carbon diselenide (**38**), are known compounds, though less stable and more cumbersome to prepare than **33** and **35**. Their general reactivity does not differ significantly from that of the S isologues. As for the Te analogues, thiocarbonyl telluride (**37**) decomposes at temperatures above its melting point, – 54 °C, and carbon ditelluride (**39**) is as yet unknown. A similar trend is noted in the halogen-substituted series: thiocarbonyl chloride (**40**) is a perfectly stable red liquid and selenocarbonyl chloride (**41**) a blue compound only recently prepared by pyrolysis of 2,2,4,4-tetrachloro-1,3-diselenetane (**42**) and decomposing at temperatures above – 130 °C[61].

O=C=Y   S=C=Y   Se=C=Se   Te=C=Te

(**33**) Y = S   (**35**) Y = S   (**38**)   (**39**)

(**34**) Y = Se   (**36**) Y = Se

(**37**) Y = Te

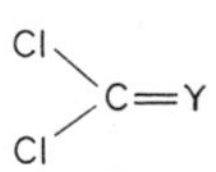
(**40**) Y = S

(**41**) Y = Se

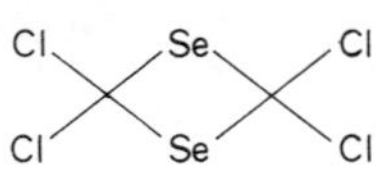
(**42**)

Thials (**43**) and thiones (**46**) no longer constitute chemical curiosities. The recent synthesis of 2,2-dimethylpropanethial (**48**) as a distillable pink monomeric compound[62] raises doubt as to the validity of regarding the non-stabilized thials as species of only transient existence. Monomeric selenals (**44**) and tellurals (**45**), on the other hand, have so far eluded isolation.

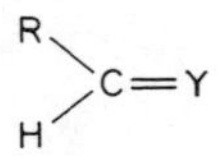

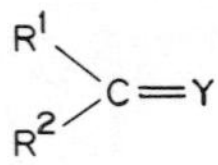

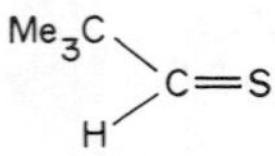

(43) Y = S **(46)** Y = S **(48)**

(44) Y = Se **(47)** Y = Se

(45) Y = Te

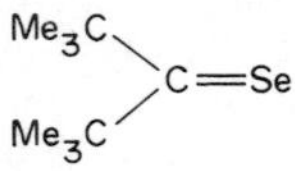

(49)

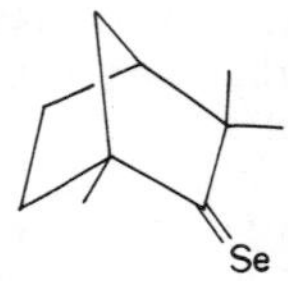

(50)

Aromatic thiones have been known for more than 60 years whereas stable monomeric aliphatic species are of more recent date and notably encountered within the class of polycyclic structures (cf. Ref. 56). Dipole moment measurements reveal a much smaller, perhaps even reversed, polarity when compared with the ketones. Until recently, monomeric authentic selones (**47**) had eluded isolation, but the preparation a few years ago of the blue selones **49** and **50** altered the situation, although the method employed for their synthesis is not of general utility[63]. Convincing evidence for the existence of non-stabilized tellones seems to be lacking.

Stable thioketenes (**51**), known since 1966, became available after 1975, including the parent compound **51** ($R^1 = R^2 = H$), through the remarkably general and efficient flash thermolysis of 1,2,3-thiadiazoles[64]. An analogous approach resulted in the synthesis and characterization of the first selenoketenes (**52**), including the parent compound **52** ($R^1 = R^2 = H$)[65]. In both series, the nature of the substituents, R^1 and R^2, defines the stability. With bulky radicals, selenoketenes can be isolated and stored in the cold[66]. Telluroketenes have not yet been prepared.

$$R^1R^2C{=}C{=}S \qquad R^1R^2C{=}C{=}Se$$

(51) (52)

B. Carboxylic and Carbonic Acid Analogues

Whereas selenocarboxylic and selenocarbonic acids are very unstable species, several esters derived from them are known compounds. Those containing doubly bonded Se, i.e. **53** and **54**, are intensely coloured species, very sensitive to O_2 and light. Se shares with Te the ability to form derivatives of dithioic and diselenoic acids with a central Se or Te atom bound in a planar arrangement to four S or Se atoms, e.g. **55**[67].

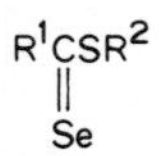

(53)

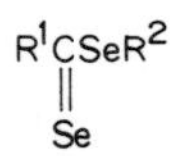

(54)

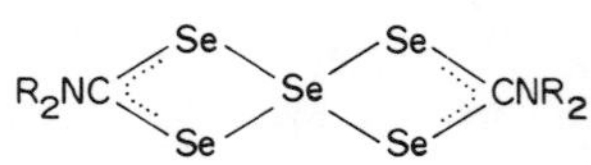

(55)

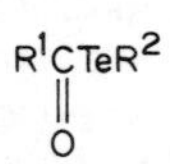

(56)

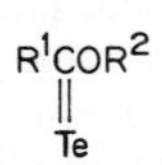

(57)

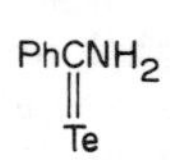

(58)

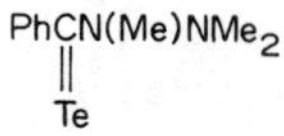

(59)

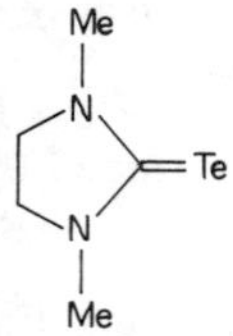

(60)

RTeCN

(61)

Derivatives of tellurocarboxylic acids are of a more recent data. Tellurol esters (**56**) have been prepared by acylation of tellurols[68], and the first species with double-bonded Te (**57**) were prepared from steroid alcohols, *t*-butyl(chloromethylene)dimethylammonium chloride and sodium hydrogen telluride[69].

Selenoamides, selenosemicarbazides and selenoureas, as well as several heterocyclic derivatives of these (selenouracil etc.) are fairly stable compounds. Recently, also telluroamides (**58**)[70,71], tellurohydrazides (**59**)[70] and a derivative of tellurourea (**60**)[72] have become synthetically available.

Attempts to prepare organic tellurocyanates (**61**) only became successful when it was recognized that the tellurocyanate ion is decomposed instantaneously by water[73]. In solvents like dimethylformamide or acetonitrile, however, Te readily reacts with onium cyanates to form onium tellurocyanates, unsuited for alkylation. Organic tellurocyanates have been prepared, however, by alkylation of potassium tellurocyanate, formed *in situ* in dimethyl sulphoxide. The crystal structure of the very stable 4-nitrobenzyl tellurocyanate has recently been determined[74]. The lability of the tellurocyanate ion is attributable to the weak C=Te bond. Association of the nitrogen end of the ion with protic solvents or hard Lewis acids results in further bond weakening. Together, the above observations call for a judicious choice of solvents and cations in the synthesis of tellurocyanates.

The combined experience from the syntheses of tellurocyanates, telluroamides and telluro esters reveals no fundamental difference between selenocarbonyl and tellurocarbonyl derivatives; with due precautions in synthetic methodology the preparation of many additional tellurocarbonyl compounds seems feasible and hence to be expected.

IX. OXO ACIDS OF SULPHUR, SELENIUM AND TELLURIUM

Oxygen-containing isologous acids, with the chalcogens in the valency states 2,4 and 6 (**62**, **63** and **64**), are known for Y = S, Se, Te, yet not without exceptions. Dramatic changes are encountered, however, within the formally analogous series, both with regard to properties and stability. We shall elaborate on this in the following.

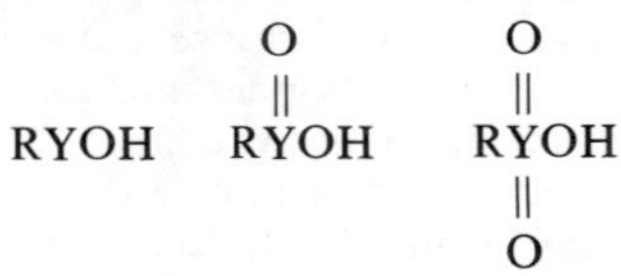

(62) (63) (64)

A. Valency State Six

In the highest valency state, sulphuric and selenic acid, H_2YO_4(Y = S, Se), are very similar strong acids whereas telluric acid, $Te(OH)_6$, is a very weak acid forming salts either with the formal composition M_2TeO_4, or, in the case of certain cations, M_6TeO_6 (e.g. Ag_6TeO_6). However, the TeO_4^{2-} ion is a polymer, containing hexacoordinate Te with oxygen bridges. The only known organic derivative of telluric acid appears to be the ester $Te(OMe)_6$. No derivative of the hypothetical telluronic acid (**64**; Y = Te) is known and probably never will be, because any potential candidate most likely would prove to be an amphoteric polymer and not a true analogue of the well-authenticated sulphonic and selenonic acids (**64**; Y = S, Se); these resemble sulphuric acid in being strong acids. As selenic acid, the selenonic acids possess oxidizing properties, to the extent of becoming explosive (like, for example, organic nitrates). Care must be exercised, however, in accepting structures previously reported or assigned to new selenonic acids; thus, early reports of the preparation of benzeneselenonic acid by selenation of benzene was later revised, the reaction product being a salt formed between protonated benzeneseleninic acid and the benzeneselenonate anion. A few authentic areneselenonic acids are known whereas only salts of the aliphatic counterparts seem stable.

B. Valency State Four

In the tetravalent series, selenious acid differs from sulphurous acid in being weaker but also in exhibiting predominantly oxidizing properties. Selenium dioxide, the anhydride of selenious acid, is a well-established, specific oxidation reagent in synthetic organic chemistry. Again, seleninic acids (**63**; Y = Se) are weaker not only than selenonic acids (**64**; Y = Se) but also than sulphinic acids (**63**; Y = S); they may, in fact, behave as bases forming cations of the type $RSe(OH)_2{}^+$. Unlike sulphinic acids, selininic acids are moderately oxidizing species existing in aqueous solution as hydrates with the structure $RSe(OH)_3$. A great variety of seleninic acid derivatives is known including chlorides, amides, esters and anhydrides. Benzeneseleninic acid anhydride has lately drawn interest as a mild and remarkably specific reagent in organic chemistry[75–78]. Organic derivatives of tetravalent Te behave rather differently. Although TeO_2, like SeO_2, is an oxidizing amphoteric compound, its basic properties are more pronounced; thus, it readily affords Te(IV) salts with strong acids. Organyltellurium trihalides, $RTeX_3$, are hydrolysed to what formally are tellurinic acids. However, the insolubility in water and organic solvents of the latter, as well as their very high and often ill-defined melting points, strongly suggest that one is here dealing with polymers containing oxygen bridges, derived from $RTe(OH)_3$. Apart from a few halogenated derivatives, such as $RTeX_3$ and RTe(O)X, no authentic tellurinic acid seems to have been prepared.

C. Valency State Two

Divalent S and Se compounds containing the radicals RY— are numerous and several of them important. They encompass halides, pseudohalides, acetates, amides, etc. For systematic reasons, the hydroxides, RYOH, are named sulphenic acids, selenenic acids and tellurenic acids. They are amphoteric in nature and only a few are known as stable compounds because they easily disproportionate into compounds of lower and higher oxidation states. Characteristic for several derivatives of **62**, the chalcogen atom behaves as the electrophile towards nucleophilic attack, with displacement of X^-. Thus, species such as RSeSR, RSeCN, RSeSCN and RSeSeCN can be efficiently prepared from selenenyl halides, RSeX, and the appropriate nucleophiles. Though far less well explored, the aromatic tellurenyl halides behave similarly. Thus, 2-formylbenzenetellurenyl bromide reacted with AgCN to form the first organic tellurocyanate[79]. The first known tellurenyl compound, 4-methoxybenzenetellurenyl methanesulphonate, reacted with *O*-methyl dithiocarbonate to form the corresponding tellurenyl derivative[80]. In certain respects, however, tellurenyl halides react differently from selenenyl halides (cf. Section X of this chapter). Of considerable interest in a synthetic context is the application of selenenyl halides for α-arylselenenation of enolizable carbonyl compounds and for the stereospecific *trans* addition to alkenes (cf., for example, Ref. 81).

X. HALOGEN COMPOUNDS

From the many types of known organic compounds containing one or more chalcogen–halogen bonds, three major classes are here singled out for a brief discussion, viz. the monohalides (**65**), the dihalides (**66**) and the trihalides (**67**).

RYX (**65**) R_2YX_2 (**66**) RYX_3 (**67**)

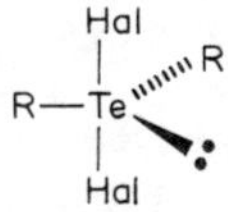

(**68**)

A. Monohalides

Within group **65**, sulphenyl (Y = S) and selenenyl (Y = Se) halides are well-known species, generally available on halogenolysis of disulphides or diselenides. Several, notably aromatic derivatives, are rather stable compounds often serving as important reagents in modern synthetic chemistry. In view of the known stability of other types of organic compounds with Te–halogen bonds, it seems surprising that 2-naphthalenetellurenyl iodide until recently remained the sole well-characterized tellurenyl halide. Additional, though rather unstable, aromatic analogues were prepared in 1975[82]. They combine with halide ions to ions of the type $RTeBrCl^-$ (isolated as for example, tetraphenylarsonium salts). X-ray studies have revealed a T-shaped geometry, with a nearly linear arrangement of the X—Te—X′ group for such ions[83]. S and Se do not form analogous compounds.

B. Dihalides

As known from the beginning of this century, diorganyl chalcogenides, R_2Y, react with halogens to give compounds of type **66**. The stability of these increases from S to Te, and from I to F; thus, the least stable species, such as R_2SI_2, behave as charge-transfer complexes. In general, the bromides and iodides dissociate into the components on heating or dissolution. Non-aqueous solutions exhibit no signs of dissociation whereas the observed conductivity in aqueous solution is attributable to aqua ions, $RYX(H_2O)^+$, and hydrolysis to $R_2YX(OH)$. Diorganyltellurium dihalides (**66**; Y = Te) are reasonably stable, crystalline compounds possessing, according to X-ray analysis, the geometry of trigonal bipyramids (**68**); they are frequently used as intermediates in preparing pure tellurides. Fluorides (**66**; X = F) have been prepared by metathesis with Ag_2F or Na_2F and by direct fluorination at low temperature[84]. On fluorination, Ph_2S forms Ph_2SF_4; analogous Se and Te compounds have not been encountered.

C. Trihalides

The organyl chalcogen trihalides **67** constitute a large and well-studied class of compounds. They may be prepared by halogenation of diselenides or ditellurides; in the aromatic series by electrophilic substitution in activated nuclei by means of selenium and tellurium tetrachloride. The stability of type **67** compounds is fair to good, highest in the Te series in keeping with the tendency of Te to attain tetravalency. The trihalides (**67**) are widely used in the synthesis of other types of products; thus, **67** (X = S, Se) on hydrolysis affords sulphinic or seleninic acids, whereas the Te analogue (**67**; X = Te) gives a formal tellurinyl chloride, RTe(O)Cl. which, however, behaves more like an inorganic oxide chloride. The structure of $PhTeCl_3$ is polymeric with bridging Cl atoms so that each Te atom is surrounded by four Cl atoms[85]. On the other hand, 8-ethoxy-4-cyclooctenyltellurium trichloride has been established as monomeric[86], undoubtedly because the bulky organic radical shields the $TeCl_3$ group.

Once again, the halogen compounds of S and Se display obvious similarities in their properties and reactivities whereas the more 'metallic' character of Te is reflected in a somewhat deviating behaviour of its organic derivatives.

XI. CHALCOGEN DERIVATIVES OF GROUP V ELEMENTS

A. Nitrogen Compounds

The sulphenamides (**69**; Y = S) constitute a large and well-examined class of compounds, whereas only a limited number of selenenyl amides (selenenamides) (**69**; Y = Se), or other compounds with a Se—N bond, have been prepared thus far, mostly by reaction of selenenyl halides or alkoxides (selenenic esters) with ammonia or amines. They seem to be less stable than the sulphenamides, a conclusion which may well require revision, however, when additional members of the class become known.

RYNH$_2$	RN=Se=NR	R$_3$PY
(69)	**(70)**	**(71)**

Trifluromethylselenyl bromide reacts with ammonia to form CF_3SeNH_2, $(CF_3Se)_2NH$ and $(CF_3Se)_3N$[87], all distillable liquids resembling ordinary amines in their reactions. Thus, CF_3SeNH_2, on reaction with an isocyanate, yields a CF_3Se-substituted urea.

Tellurenamides (**69**; Y = Te) have apparently not yet been reported.

The selenenyl amides react with nucleophiles in the same way as selenenyl halides. Thus, *N*-phenylselenenylphthalimide, prepared from potassium phthalimide and phenylselenenyl chloride[88,89], is a useful substitute for selenenyl halides in a great variety of reactions (cf. the reactions of similar *N*-sulphenylphthalimides[90]).

Other chalcogen–nitrogen compounds are known, such as *N*-4-methylbenzenesulphonylated selenimides and tellurimides, **24** and **25**, both stable compound types which have recently been subjected to spectroscopy and X-ray diffraction studies[91,92]. The known selenoximides (**26**) are stable analogues of the extensively studied sulphoximides[93].

The easily prepared selenium diimides (**70**; R = *t*-Bu, 4-$MeC_6H_4SO_2$) have been utilized as efficient reagents for allylic amination of alkenes[94].

B. Phosphorus and Arsenic Compounds

Tertiary phosphines react with elemental S or Se to form phosphine sulphides (**71**; Y = S) and phosphine selenides (**71**; Y = Se). Tertiary arsines and stibines form similar compounds. More recently, elemental Te has been found to react similarly to give phosphine tellurides (**71**; Y = Te)[95].

Diphosphanes react with ditellurides to yield telluradiphosphanes, R_2PTePR_2, by insertion of Te abstracted from the ditelluride[41]. The analogous reaction in the Se series takes a different course, giving a substituted 3-membered ring compound, derived from selenadiphosphirane[96].

Phosphorus trichloride and arsenic trichloride react with $(CF_3Se)_2Hg$ to form the substituted phosphines and arsines, $(CF_3Se)_3P$ and $(CF_3Se)_3As$[87].

Numerous derivatives of Se isologues of phosphorus acids are known[97], whereas very few analogous Te compounds have been investigated, probably because of their extreme sensitivity to air. Several decades ago, an alkaline solution of a dialkyl phosphite was shown to react smoothly with S, Se or Te to form a chalcogenophosphate (phosphorochalcogenate)[98]. The Te compound $(EtO)_2P(O)TeNa$ has recently been utilized as an efficient reagent for the deoxygenation of epoxides[99].

XII. HETEROCYCLIC COMPOUNDS

Thiophene (**72**) represents the classical sulphur analogue of benzene. By the same token, selenophene (**73**) and tellurophene (**74**) assume similar key compound roles in the heterocyclic chalcogen chemistry.

(**72**) Y = S
(**73**) Y = Se
(**74**) Y = Te

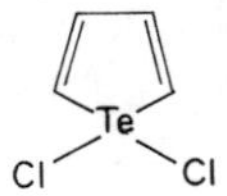

(**75**)

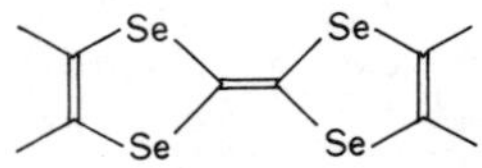

(**76**)

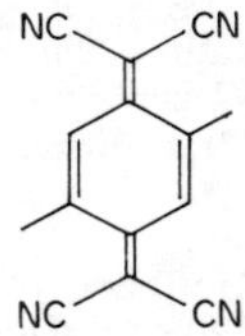

(**77**)

Derivatives of selenophene have been known for about 100 years whereas the parent compound (**73**) was only synthesized, from acetylene and Se, in 1927. Detailed studies of its properties and chemistry have all confirmed its striking similarity to thiophene (**72**)[100].

Tellurophenes are less well explored, and tellurophene (**74**) itself remained unknown until 1972 when its rather straightforward synthesis disproved the suspected instability of **74**. In fact, its properties are surprisingly similar to those of **72** and **73** with the notable exception that **74**, contrary to **72** and **73**, produces a stable 1,1-dichloro compound, (**75**), on reaction with Cl_2. Otherwise, the similarity persists into numerous other Te-containing heterocyclic systems[101].

Much interest has recently been accorded to heterocyclic Se compounds, analogous to the S-containing 1,2-dithioles and 1,3-dithioles, as well as the derived systems, trithiapentalene and tetrathiafulvalene, in connection with the studies of organic compounds with metallic conductivity. Thus tetramethyl tetraselenafulvalene (TMTSF) (**76**) forms a charge-transfer complex with 2,5-dimethyl-7,7,8,8-tetracyano-*p*-quinodimethane (**77**) possessing a conductivity that surpasses that of a closely related sulphur-analogue by a factor of ten[102]. More recently, several salts of the type $TMTSF_2X$, where X represents various inorganic anions, have been found to exhibit superconductivity[103]. Only recently has the synthesis of an analogous Te compound been reported[104].

Saturated Te heterocycles such as 1,4-oxatellurane, 1,4-selenatellurane, tellurolane and others, seem to be very similar to their Se analogues. Many Se-containing nitrogen heterocycles are known, e.g. selenazoles, selenadiazoles, and condensed ring systems derived from these. Surprisingly, analogous Te compounds are rare. Benzo[*d*]-1,2-tellurazole exhibits abnormal physical properties, attributed to very short intermolecular Te—N bonds[105]. To what extent the shortage of such compounds is caused by their properties or rather by the lack of appropriate synthetic methods remains to be established.

XIII. REFERENCES

1. D. L. Klayman and W. H. H. Günther (Eds.), *Organic Selenium Compounds: Their Chemistry and Biology*, John Wiley and Sons, New York–London–Sydney–Toronto, 1973.
2. R A. Zingaro and W. C. Cooper (Eds.), *Selenium*, Van Nostrand–Reinhold, New York, 1974.
3. K. J. Irgolic, *The Organic Chemistry of Tellurium*, Gordon and Breach, New York–London–Paris, 1974; K. J. Irgolic, *J. Organomet. Chem.*, **103**, 91 (1975); **130**, 411 (1977); **158**, 235 (1978).
4. *International Union of Pure and Applied Chemistry: Nomenclature of Organic Chemistry*, Sections A, B, C, D, E, F and H, 1979 Ed., Pergamon Press, Oxford, 1979.

5. *International Union of Pure and Applied Chemistry: Nomenclature of Inorganic Chemistry*, 2nd ed. Butterworths, London, 1971.
6. W. H. H. Günther, Ref. 1, pp. 1–12.
7. K. A. Jensen, *The Early History of Organic Sulfur Chemistry*, to be published.
8. F. Wöhler, *Ann. Chem.*, **35**, 111 (1840).
9. C. J. Löwig, *Pogg. Ann.*, **37**, 552 (1836).
10. B. Rathke, *Ann. Chem.*, **152**, 211 (1869).
11. F. Wöhler, *Ann. Chem.*, **84**, 69 (1852); F. Wöhler and J. Dean, *Ann. Chem.*, **93**, 233 (1855); **97**, 1 (1857); C. Siemens, *Ann. Chem.*, **61**, 360 (1847); J. W. Mallet, *Ann. Chem.*, **79**, 233 (1851); C. A. Joy, *Ann. Chem.*, **86**, 35 (1853).
12. A. Cahours, *Ann. Chem.*, **135**, 356 (1865).
13. F. Krafft and W. Vorster, *Ber. Dtsch. Chem. Ges.*, **26**, 2821 (1893).
14. A. Verneuil, *Bull. Soc. Chim. Fr. [2]*, **41**, 599 (1884).
15. W. Baringer, *Ber. Dtsch. Chem. Ges.*, **23**, 1003 (1890).
16. G. Hofmann, *Ann. Chem.*, **250**, 294 (1889).
17. O. Hinsberg, *Ber. Dtsch. Chem. Ges.*, **22**, 862, 2895 (1889).
18. C. Paal, *Ber. Dtsch. Chem. Ges.*, **18**, 2255 (1885).
19. J. V. Camasseto, J. T. B. Ferreira, J. A. F. Val and A. R. M. de Oliveira, *Abstracts of Fourth International Conference on the Organic Chemistry of Selenium and Tellurium*, Birmingham, 1983.
20. L. Engman and M. P. Cava, *Organometallics*, **1**, 470 (1982).
21. L. Engman, *Tetrahedron Letters*, **23**, 3601 (1982).
22. L. Engman and M. P. Cava, *J. Org. Chem.*, **47**, 3946 (1982).
23. M. Yamashita, Y. Kato, and R. Suemitsu, *Chem. Letters*, 847 (1980).
24. D.L.J. Clive, *Tetrahedron*, **34**, 1049 (1978); H. J. Reich, *Acc. Chem. Res.*, **12**, 22 (1979); D. Liotta, *Acc. Chem. Res.*, **17**, 28 (1984).
25. G. Calingaert and H. A. Beatty, *J. Amer. Chem. Soc.*, **61**, 2748 (1939).
26. P. Granger, S. Chapelle, W. R. McWhinnie and A. Al-Rubaie, *J. Organomet. Chem.*, **220**, 149 (1981); P. Laur, I. Karup Nielsen and M. Scheiter, *Abstracts of Fourth International Conference on the Organic Chemistry of Selenium and Tellurium*, Birmingham, 1983.
27. K. A. Jensen and C. K. Jorgensen, Ref. 1, pp. 1017–1048; V. Krishnan and R. A. Zingaro, Ref. 2, p. 337; A. J. Gysling, *Coord. Chem. Rev.*, **42**, 133 (1982).
28. G. A. Olah, J. J. Svoboda and A. T. Ku, *J. Org. Chem.*, **38**, 4447 (1973).
29. S. Smiles, *J. Chem. Soc.*, **77**, 1174 (1900).
30. W. J. Pope and S. J. Peachey, *J. Chem. Soc.*, **77**, 1072 (1900).
31. W. J. Pope and A. Neville, *J. Chem. Soc.*, **81**, 1552 (1902).
32. T. M. Lowry and F. L. Gilbert, *J. Chem. Soc.*, 2867 (1929).
33. P. Laur, in *Proceedings of Third International Symposium on Organic Selenium and Tellurium Compounds* (Eds. D. Cagniant and G. Kirsch), University of Metz, 1979, pp. 217–301.
34. R. F. Ziolo and D. D. Titus, in *Proceedings of Third International Symposium on Organic Selenium and Tellurium Compounds* (Eds. D. Cagniant and G. Kirsch), University of Metz, 1979, pp. 317–347.
35. D. Hellwinkel, *Ann. N. Y. Acad. Sci.*, **192**, 158 (1972).
36. For a review see D. L. J. Clive, in *Organic Compounds of Sulphur, Selenium and Tellurium*, Vol. 6, Specialist Periodical Reports, Royal Society of Chemistry, 1981, p. 112.
37. W. Dumont, P. Bayet and A. Krief, *Angew. Chem.*, **86**, 308 (1974).
38. A. Osuka and H. Suzuki, *Tetrahedron Letters*, **24**, 5109 (1983).
39. A. Osuka, Y. Mori, H. Shimizu and H. Suzuki, *Tetrahedron Letters*, **24**, 2599 (1983).
40. M. Wojnowska, W. Wojnowski and R. West, *J. Organomet. Chem.*, **199**, C 1 (1980).
41. W. W. duMont and H. J. Kroth, *Z. Naturforsch.*, **36b**, 332 (1981).
42. E. Linder and U. Ansorge, *Chem. Ber.*, **111**, 2455 (1978).
43. W. Dell and M. L. Ziegler, *Angew. Chem.*, **93**, 480 (1981).
44. R. A. Zingaro, E. A. Meyers and M. Renson, *Organometallics*, **1**, 111 (1982).
45. W. Hofmann and H. Werner, *Angew. Chem.*, **93**, 1088 (1981).
46. K. A. Jensen and L. Henriksen, *Acta Chem. Scand.*, **B29**, 871 (1975).
47. H. H. Szmant, in *Sulphur in Inorganic and Organic Chemistry*, Vol. 1 (Ed. A. Senning), Marcel Dekker, New York, 1971, pp. 107–152.

48. F. A. Davis, J. M. Billmers and O. D. Stringer, *Tetrahedron Letters*, **24**, 3191 (1983).
49. M. R. Detty, *J. Org. Chem.*, **45**, 274 (1980).
50. H. Lee and M. P. Cava, *J. Chem. Soc., Chem. Commun.*, 277 (1981).
51. S. Uemura and S. Fukuzawa, *J. Amer. Chem. Soc.*, **105**, 2748 (1983).
52. S. V. Ley, C. A. Meerholz and D. H. R. Barton, *Tetrahedron*, **37**, Suppl. 9, 213 (1981).
53. M. Shimizu, R. Ando and I. Kuwajima, *J. Org. Chem.*, **46**, 5246 (1981).
54. L. Engman and M. P. Cava, *J. Chem. Soc., Chem. Commun.*, 164 (1982).
55. N. W. Alcock and W. D. Harrison, *J. Chem. Soc., Dalton Trans.*, 709, 1421 (1982).
56. F. Duus, in *Comprehensive Organic Chemistry*, Vol. 3 (Ed. D. N. Jones), Pergamon Press, Oxford, 1979, pp. 373–487.
57. E. Block, E. R. Corey, R. E. Penn, T. L. Renken, P. F. Sherwin, H. Bock, T. Hirabayashi, S. Mohmand and B. Solouki, *J. Amer. Chem. Soc.*, **104**, 3119 (1982).
58. H. W. Kroto, B. M. Landsberg, R. J. Suffolk and A. Vodden, *Chem. Phys. Letters*, **29**, 265 (1974).
59. K. Georgiou, H. W. Kroto and B. M. Landsberg, *J. Chem. Soc., Chem. Commun.*, 739 (1974).
60. M. Hutchinson and H. W. Kroto, *J. Mol. Spectry*, **70**, 347 (1978).
61. A. Darmadi, A. Haas, H. Willner and H. Schnöckel, *Z. Naturforsch.*, **36b**, 1261 (1981).
62. E. Vedejs and D. A. Perry, *J. Amer. Chem. Soc.*, **105**, 1683 (1983).
63. T. G. Back, D. H. R. Barton, M. R. Britten-Kelly and F. S. Guziec, Jr., *J. Chem. Soc., Perkin Trans. 1*, 2079 (1976).
64. G. Seybold and C. Heibl, *Angew. Chem.*, **87**, 171 (1975); *Chem. Ber.*, **110**, 1225 (1977).
65. B. Bak, O. J. Nielsen, H. Svanholt and A. Holm, *Chem. Phys. Letters*, **83**, 374 (1978); A. Holm, C. Berg, C. Bjerre, B. Bak and H. Svanholt, *J. Chem. Soc., Chem. Commun.*, 99 (1979).
66. E. Schaumann and F. F. Grabley, *Tetrahedron Letters*, **21**, 4251 (1980).
67. S. Esperås, S. Husebye and Å. Rolandsen, *Acta Chem. Scand.*, **A29**, 608 (1975); L. Henriksen, *Synthesis*, 771 (1982).
68. F. Wudl and D. Nalewajek, *J. Organomet. Chem.*, **217**, 329 (1981); J. L. Piette, D. Denbigh, M. Baiwir and G. L. Llabres, *Spectrochim. Acta*, **36A**, 769 (1980).
69. A. G. M. Barrett, D. H. R. Barton and R. W. Read, *J. Chem. Soc., Chem. Commun.*, 645 (1979).
70. K. A. Lerstrup and L. Henriksen, *J. Chem. Soc., Chem. Commun.*, 1102 (1979).
71. U. Berg, L. Henriksen, K. A. Lerstrup and J. Sandström, *Acta Chem. Scand.*, **B36**, 19 (1982).
72. M. F. Lappert, T. R. Martin and G. M. McLaughlin, *J. Chem. Soc., Chem. Commun.*, 635 (1980).
73. T. Austad, J. Songstad and K. Åse, *Acta Chem. Scand.*, **25**, 331 (1971).
74. K. Maartmann-Moe, K. A. Sanderud and J. Songstad, *Acta Chem. Scand.*, **A35**, 151 (1981).
75. D. H. R. Barton and S. V. Ley, *Ciba Found. Symp.*, No. 53, 53 (1978).
76. D. H. R. Barton, D. J. Lester and S. V. Ley, *J. Chem. Soc., Perkin Trans. 1*, 1212 (1980).
77. T. G. Back, S. Collins and R. G. Kerr, *J. Org. Chem.*, **46**, 1564 (1981).
78. D. H. R. Barton, A. H. F. H. Raymond and S. V. Ley, *J. Chem. Soc., Perkin Trans. 1*, 2179 (1982).
79. J. L. Piette, R. Lysy and M. Renson. *Bull. Soc. Chim. Fr.*, 3559 (1972).
80. O. Foss, *Acta Chem. Scand.*, **6**, 306 (1952).
81. P. D. Magnus, in *Comprehensive Organic Chemistry*, Vol. 3 (Ed. D. N. Jones), Pergamon Press, Oxford, 1979, pp. 491–538.
82. P. Schultz and G. Klar, *Z. Naturforsch.*, **30b**, 40 (1975).
83. P. Klæboe, C. J. Nielsen, R. Suchi and O. Vikane, *Acta Chem. Scand.*, **A32**, 565 (1978).
84. I. Ruppert, *Chem. Ber.*, **112**, 3023 (1979).
85. F. B. W. Einstein and T. Jones, *Acta Cryst.*, **B38**, 617 (1982).
86. A. Bergman and L. Engman, *J. Organomet. Chem.*, **181**, 335 (1979).
87. A. Darmadi, A. Haas and M. Kaschani-Motlach, *Z. Anorg. Allgem. Chem.*, **448**, 35 (1979).
88. K. C. Nicolaou, D. A. Claremon, W. E. Barnette and S. P. Seitz, *J. Amer. Chem. Soc.*, **101**, 3704 (1979).
89. P. A. Grieco, J. Y. Jaw, D. A. Claremon and K. C. Nicolaou *J. Org. Chem.*, **46**, 1215 (1981).
90. D. H. R. Barton, G. Page and D. A. Widdowson, *J. Chem. Soc., Chem. Commun.*, 1466 (1970).
91. V. I. Nadaka, V. P. Krasnov, L. L. Popova and V. L. Menkin, *Zh. Org. Khim.*, **18**, 259 (1982).
92. N. G. Furmanova, V. I. Nadaka and V. P. Krasnov, *Zh. Strukt. Khim.*, **22**, 118 (1981).

93. N. Y. Derkach, T. V. Lyapina and N. A. Pasmurtseva, *Zh. Org. Khim.*, **10**, 807 (1974); Y. A. Nuzhdina, Y. P. Egorov, L. V. Mironyuk and N. Y. Derkach, *Teor. Eksp. Khim.*, **11**, 605 (1975).
94. K. B. Sharpless, T. Hori, L. K. Truesdale and C. O. Dietrich, *J. Amer. Chem. Soc.*, **98**, 269 (1976).
95. R. A. Zingaro, *J. Organomet. Chem.*, **1**, 200 (1963); R. A. Zingaro, B. H. Steeves and K. Irgolic, *J. Organomet. Chem.*, **4**, 320 (1965).
96. M. Baudler, H. Suchomel, G. Fürstenberg and U. Schings, *Angew. Chem.*, **93**, 1087 (1981).
97. J. Michalski and A. Markowska, Ref. 1, pp. 325–377.
98. O. Foss, *Acta Chem. Scand.*, **1**, 8 (1947); **4**, 1245 (1950).
99. D. L. J. Clive and S. M. Mencken, *J. Org. Chem.*, **45**, 2347 (1980).
100. S. Gronowitz, in *Organic Compounds of Sulphur, Selenium and Tellurium*, Vol. 5, Specialist Periodical Reports, The Chemical Society, 1979, p. 247; cf. also Vols. 1–4 and 6 of the same series.
101. L. Christiaens, in *Proceeding of Third International Symposium on Organic Selenium and Tellurium Compounds* (Eds. D. Cagniant and G. Kirsch), University of Metz, 1979, pp. 69–112.
102. K. Bechgaard and D. Jérome, *Sc. Amer.*, **247**, 50 (1982).
103. K. Bechgaard, *Mol. Cryst. Liq. Cryst.*, **79**, 3319 (1982).
104. K. Lerstrup, D. Talham, A. Bloch, T. Poehler and D. Cowan, *J. Chem. Soc., Chem., Commun.*, 336 (1982).
105. R. Weber, J. L. Piette and M. Renson, *J. Heterocycl. Chem.*, **15**, 865 (1978).

The Chemistry of Organic
Selenium and Tellurium
Compounds Volume 1
Edited by S. Patai and Z. Rappoport
© 1986 John Wiley & Sons Ltd.

CHAPTER **2**

General and theoretical aspects of organic compounds containing selenium or tellurium

RAYMOND A. POIRIER

Department of Chemistry, Memorial University of Newfoundland, St. John's, Newfoundland, AIB 3X7, Canada

IMRE G. CSIZMADIA

Department of Chemistry, University of Toronto, Toronto, M5S 1A1, Canada

I. INTRODUCTION 21
II. THEORETICAL BACKGROUND 23
A. Atom Orbital Basis Sets 24
B. *Ab Initio* Computational Methods 30
III. RESULTS 33
IV. DISCUSSION 45
A. Energy-related Properties 45
1. Proton affinities 45
2. Energies of hydrogenation 52
3. Ionization potentials 54
4. Relative stabilities of conformers 55
B. Non-energy-related Properties 55
1. Charge distribution 55
2. Geometries 56
V. CONCLUSIONS 57
VI. ACKNOWLEDGEMENTS 60
VII. REFERENCES 60

I. INTRODUCTION

Even before Mendeleev introduced the concept of periodicity of the chemical elements, certain atoms and their compounds were studied in relation to one another. One of these

TABLE 1. A comparison of experimental first ionization potentials (eV) of Group VI elements and their dihydrides

Species	X = O	S	Se	Te
X^1	13.614	10.357	9.750	9.01
H_2X^{2-7}	12.61	10.47	9.881	9.138

families of elements was O, S and Se, which is now augmented by the inclusion of Te. It is now well known that the basis for the chemical similarities of a family is electronic, that is, the valence electron shells of these elements (O to Te) are all of ns^2np^4 configuration, where n is the principal quantum number. Therefore the four atoms O to Te all have ^3P atomic ground electronic states. The trends one finds for the physical and chemical properties of these elements and their compounds truly justify the original notion that they indeed represent a family and therefore may be studied comparatively. For example, the first ionization potentials (*IP*) of the atoms (X),

$$X(ns^2np^4) \rightarrow X^+(ns^2np^3) + e^-$$

and the first *IP* of their dihydrides,

$$H_2X \rightarrow H_2X^+ + e^-$$

which is normally associated with the removal of an electron from one of the higher occupied orbitals (lone pair), underline the above notion as seen from Table 1.

However, as well as similarities, the differences that exist between these elements also need to be examined. Considering the theoretical aspects only, a number of theoretical computable components of the total electronic energies of these elements can be compared, such as the Hartree–Fock energy (E_{HF}), the correlation energy (E_{cor}) and the relativistic energies (E_{rel}). Two points should be made concerning the energy values quoted in Table 2. Firstly, the energies are quoted in Hartree atomic units (1 hartree = 2625.5 kJ mol^{-1} = 627.51 kcal mol^{-1} = 27.232 eV particle^{-1}). Secondly, it should be noted that E_{cor} and E_{rel} are increasing rapidly as one goes from O to Te. The consequence or implication of this is that experimental quantities such as *IP* are insignificant with respect to E_{cor} and E_{rel}. The ionization potentials of H_2Se and H_2Te for example, are 0.36 and 0.34 hartree, respectively, which compared to their respective relativistic energies of − 26.92 and − 165.63 hartree[8] are certainly insignificant. However, it should be noted that the relativistic effects will be much more important for the core electrons than for the valence electrons and therefore, the relativistic contribution to the first *IP* is not expected to increase significantly from O to Te. Consequently, certain atomic and molecular

TABLE 2. Total experimental electronic energies of Group VI elements and their theoretical computable components[8,9]

	Energy (hartree)			
Component	O	S	Se	Te
E_{HF}	− 74.8095	− 397.5050	− 2399.8669	− 6611.785
E_{cor}	− 0.2575	− 0.6400	—	—
E_{rel}	− 0.0490	− 1.0335	− 26.9247	− 165.6293
E_{total}	− 75.1160	− 399.1785	—	—

quantities cannot be determined theoretically to any desired level of accuracy with the theoretical methods available at the present time. However, it should be pointed out that for certain non-energy-related properties the correlation and relativistic contributions are very small and that for certain energy related properties (ΔEs), the correlation (ΔE_{cor}) and relativistic (ΔE_{rel}) contributions effectively cancel, that is, $\Delta E_{cor} \simeq 0$ and $\Delta E_{rel} \simeq 0$.

II. THEORETICAL BACKGROUND

In any exact science, experiment and theory play an equally important, but complementary, role. This dynamic interaction is illustrated in Figure 1. From this point of view it is necessary to briefly elaborate on some of the important aspects of theoretical chemistry and its application to the Group VI elements, in particular to Se and Te, as well as to their chemistry. In considering the theoretical background necessary for such applications, we shall focus our attention on two factors, namely the atomic orbitals (basis sets) used in the theory and the theoretical method or level of sophistication adopted for the computations. The interdependence of these two components is illustrated in Figure 2. Therefore, two sections covering these methodological aspects will be given followed by a separate section concerning the applications of these methods to the study of organic compounds containing Se and Te.

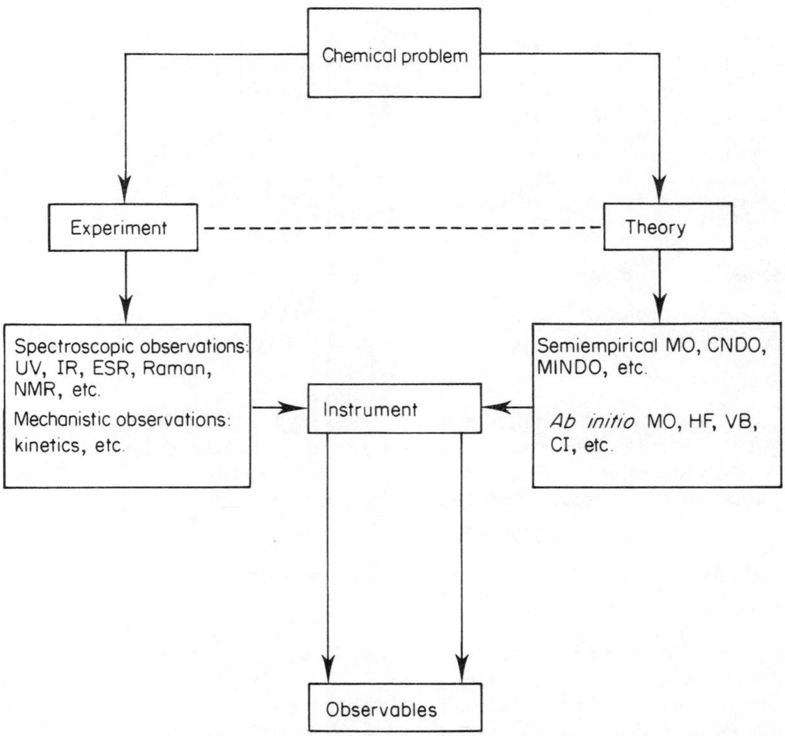

FIGURE 1. An illustration of the dynamic interaction between experiment and theory

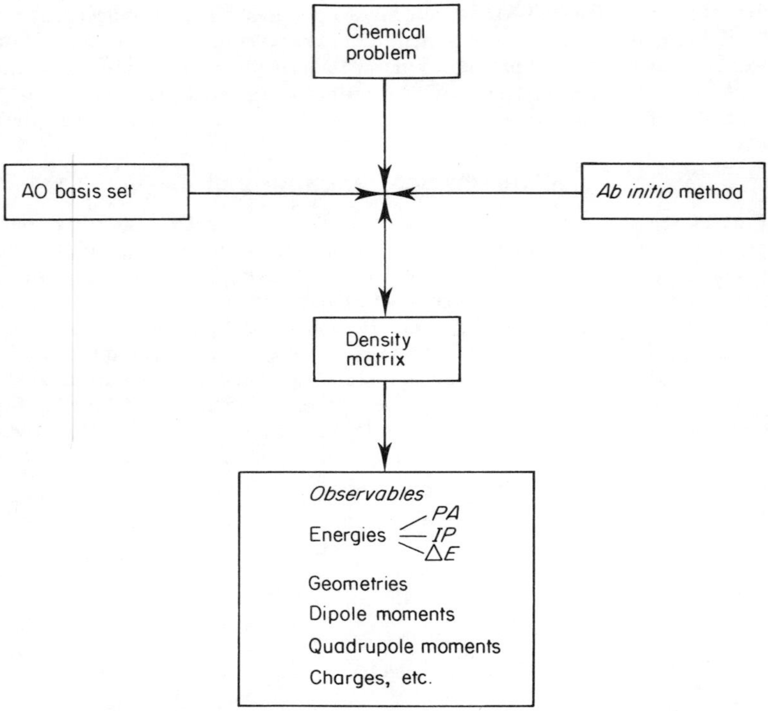

FIGURE 2. An illustration of the factors to be considered for the application of theoretical methods, namely the basis set or atomic orbital (AO) and the *ab initio* method or level of sophistication (e.g. SCF, MC-SCF, CI)

A. Atomic Orbital Basis Sets

In molecular orbital theory[10], molecular orbitals (MO), ϕ, are expressed as a linear combination of a set of functions, η, or atomic orbitals (LCAO),

$$\phi_j = \Sigma_i C_{ij}\eta_i$$

Therefore a fundamental question involved in these types of calculations is choosing the type of functions for η. Two types of functions are widely used, depending on the size of the system: firstly, the exponential-type functions (ETF), frequently called Slater-type orbitals (STO) and which are closely related to the hydrogen orbitals,

$$\text{STO: } \eta(\xi) = N_\xi r^{(n-1)} e^{-\xi r} S_{l,m}(\theta, \phi)$$

and secondly the Gaussian-type orbitals (GTO or GTF),

$$\text{GTO: } \eta(\alpha) = N_\alpha r^{2(n-1)} e^{-\alpha r^2} S_{l,m}(\theta, \phi)$$

where only the lowest angular functions are usually used, that is, 1s, 2p, 3d, 4f,... and no 2s, 3s,... or 3p, 4p,... are used explicitly.

Although nowadays the GTO are more popular for molecular computations, due to the tremendous simplification in the integral evaluations, we cannot, however, ignore Slater-type orbitals altogether for two important reasons. One of the reasons is the conceptual ease in analysing results in terms of STOs, in particular for minimal basis set size (see below). The second reason is due to the fact that in some cases, a linear combination of GTO (contraction) is tailored to fit the shape of STOs (STO-NG basis sets).

In general basis sets can be characterized within one of the categories of these three groups: (*i*) minimal, double-zeta, triple-zeta,... basis sets, (*ii*) split valence (which are minimal in core and can be double-zeta, or triple-zeta... in valence) or (*iii*) general contracted basis sets which do not clearly fall into any of the (*i*) or (*ii*) schemes. These are specified in the following way:

Minimal basis

H	1s
C,N,O,F	1s2s2p
Si,P,S,Cl	1s...3s3p
Ge,As,Se,Br	1s...3d4s4p
Sn,Sb,Te,I	1s...4d5s5p

Double-zeta basis

H	1s1s′
C,N,O,F	1s1s′2s2s′2p2p′
Si,P,S,Cl	1s1s′...3p3p′
Ge,As,Se,Br	1s1s′...3d3d′4s4s′4p4p′
Sn,Sb,Te,I	1s1s′...4d4d′5s5s′5p5p′

Split Valence (*double-zeta in valence*)

H	1s1s′
C,N,O,F	1s2s2s′2p2p′
Si,P,S,Cl	1s2s2p3s3s′3p3p′
Ge, As, Se, Br	1s...3s3p3d4s4s′4p4p′
Sn, Sb, Te, I	1s...4s4p4d5s5s′5p5p′

where the notation 2p, 3p,... stand collectively for $2p_x$, $2p_y$, $2p_z$ and $3p_x$, $3p_y$, $3p_z$... and 3d... stand collectively, in the case of GTOs, for $3d_{xx}$, $3d_{yy}$, $3d_{zz}$, $3d_{xy}$, $3d_{xz}$, $3d_{yz}$....

Among these basis sets just described, certain basis set exponents are optimized with the constraint that the 2s exponent is equal to the 2p exponent (2s = 2p), and similarly for 3s, 3p and 3d exponents (3s = 3p = 3d) and so on. In these cases the functions are refered to as 2sp and 3spd, respectively.

Finding a set of suitable orbital exponents via careful optimization methods is an elaborate, but now routine, process. In the early days of quantum chemistry, in the absence of such sophisticated methods, theoreticians relied on simple rules to obtain acceptable orbital exponents for an AO basis set. The rules are commonly known as Slater's Rules[11], and were originally proposed by Slater and subsequently have been elaborated on by others[12,13]. Several basis sets, many more for the lighter elements and fewer for the heavier elements, exist for most of the elements of the periodic table[14]. Table 3 summarizes the available STO basis sets (single-zeta and double-zeta) for Se and Te. Whenever possible, the compatible O and S basis sets were included. Table 4 gives the corresponding total energies computed at the RHF level. The Hartree–Fock Limit (HFL) of the atoms are also included for the sake of comparison[8].

In the case of GTO-type basis sets, which have been tabulated elsewhere[14], it must be recognized that there are a great many GTO basis sets for O[20–51], a relatively large number for S[24,31,41,42,45,52–64] and only a few for Se[65–69] and Te[70–72]. However, the GTO basis sets which have been used in the present study for Se and Te are reproduced here in Tables 5 and 6. These basis sets will be discussed further in the results section. Descriptions and atomic energies for the basis sets of Tables 5 and 6, along with those of other available Se and Te basis sets are given in Table 7.

TABLE 3. STO basis sets for elements of Group VI

Orbital	O	S	Se	Te
Single-zeta basis sets				
1s	7.6579	15.5253	33.2068	
2s	2.2458	5.2721	12.3157	
3s		1.5309[a]	4.7764[a]	
4s			1.4041[a]	
5s				
2p	2.2266	5.9719	14.8916	
3p		1.2655[b]	4.4567[b]	
4p			1.1248[b]	
3d			6.1697	
Ref.	12	15	15	
1s	7.6579	15.5409	33.2622	
2s	2.2458	5.3144	12.4442	
3s		2.1223	6.4678	
4s			2.4394	
2p	2.2266	5.9885	15.0326	
3p		1.8273	6.2350	
4p			2.0718	
3d			6.1590	
Ref.	12	12	12	
1s			33.255	
2s			12.448	
3s			6.466	
4s			2.569[c]	
4s			2.589[d]	
2p			15.033	
3p			6.235	
4p			2.280[c]	
4p			2.221[d]	
3d			6.159	
Ref.			16	
Double-zeta basis sets				
1s	9.46635	17.07720	35.03650	53.41410
1s′	6.83768	12.69440	24.36140	36.66320
2s	2.68801	6.72875	16.58670	27.70620
2s′	1.67543	5.24284	13.73710	23.28810
3s		2.66221	7.95809	17.29340
3s′		1.68771	5.66700	11.83340
4s			3.13870	6.98359
4s′			1.88996	5.10844
5s				3.14692
5s′				1.90779
2p	3.69445	9.51251	22.43360	33.70530
2p′	1.65864	5.12050	13.83180	22.38020

TABLE 3. (*Contd.*)

Orbital	O	S	Se	Te
3p		2.33793	7.27814	12.65670
3p′		1.33331	4.68101	11.13750
4p			2.71504	6.95455
4p′			1.51140	4.73195
5p				2.73670
5p′				1.56177
3d			9.29756	18.76030
3d′			4.53759	10.90960
4d				6.33002
4d′				3.68881
Ref.	19	19	19	19
1s	10.1085	17.6913	35.0365	
1s′	7.0623	13.7174	24.3614	
2s	2.6216	5.7486	16.5867	
2s′	1.6271	3.0757	13.7371	
3s		3.1596	7.95809	
3s′		1.8151	5.66700	
4s			3.188	
4s′			1.918	
2p	3.6813	8.9026	22.4336	
2p′	1.6537	4.9073	13.8318	
3p		2.3336	7.27814	
3p′		1.3217	4.68101	
4p			2.699	
4p′			1.503	
3d			12.018	
3d′			6.611	
4d			3.658	
4d′			2.440[e]	
Ref.	17	17	16[e]	
1s	9.55070	17.0249		
1s′	6.87575	12.6622		
2s	2.67094	6.28905		
2s′	1.66028	4.88212		
3s		2.74325		
3s′		1.71063		
2p	3.68560	9.50066		
2p′	1.65546	5.11766		
3p		2.33450		
3p′		1.33110		
Ref.	18	18		

[a]These are for 2s-type STOs

[b]These are for 2p-type STOs.

[c]Optimized for the C_{3v} structure of SeH_3^+.

[d]Optimized for the D_{3h} structure of SeH_3^+.

[e]Inner core from Ref. 19.

TABLE 4. Total energies for the STO basis sets of Table 3

Basis	O	S	Se	Te
Single-zeta	– 74.5404	– 396.6411	– 2392.4892	—
	– 74.5404	– 396.6276	– 2392.7274	—
Double-zeta	– 74.8043	– 397.5023	– 2399.756	– 6611.762
	– 74.8042	– 397.4990	—	—
	– 74.8043	– 397.5023	—	—
HF[8]	– 74.8095	– 397.5049	– 2399.867	– 6611.785

TABLE 5. GTO basis sets for Se and Te

		Contraction coefficients		
Function	Exponents	s-coeff.	p-coeff.	d-coeff.
Extended [5*s*4*p*2*d*][67]				
S1	1.10166(+ 5)[a]	1.66(– 3)		
	1.6454(+ 4)	1.280(– 2)		
	3.7725(+ 3)	6.176(– 2)		
	1.0960(+ 3)	2.0702(– 1)		
	3.69448(+ 2)	4.2560(– 1)		
	1.35783(+ 2)	3.9446(– 1)		
S2	3.447(+ 1)	4.0930(– 1)		
	1.51604(+ 1)	6.7456(– 1)		
S3	4.40667	4.8930(– 1)		
	1.92114	7.5122(– 1)		
S4	3.8505(– 1)	1.0		
S5	1.5(– 1)	1.0		
P1	7.872(+ 2)		2.310(– 2)	
	1.86702(+ 2)		1.5197(– 1)	
	5.93369(+ 1)		4.4110(– 1)	
	2.13502(+ 1)		4.8726(– 1)	
P2	6.61516		4.5975(– 1)	
	2.31882		6.5091(– 1)	
P3	3.86413(– 1)		1.0	
P4	1.2(– 1)		1.0	
D1	4.9208(+ 1)			7.258(– 2)
	1.37737(+ 1)			3.1279(– 1)
	4.45			5.2895(– 1)
	1.4			3.6502(– 1)
D2	3.3(– 1)			1.0
Minimal [4*s*3*p*1*d*][66]				
S1	5.7153831(+ 3)	6.291(– 2)		
	8.6651303(+ 2)	3.7333(– 1)		
	1.9030059(+ 2)	6.8429(– 1)		

TABLE 5. (*Contd.*)

Function	Exponents	Contraction coefficients		
		s-coeff.	p-coeff.	d-coeff.
S2	2.5523194(+2)	−1.0632(−1)		
	2.411743(+1)	7.1498(−1)		
	9.81016	3.5219(−1)		
S3	2.057072(+1)	−2.3096(−1)		
	3.26003	7.6658(−1)		
	1.34144	3.6426(−1)		
S4	1.99401	−2.1281(−1)		
	3.4561(−1)	6.5746(−1)		
	1.3017(−1)	4.7783(−1)		
P1	3.3471076(+2)		9.028(−2)	
	7.757597(+1)		4.4372(−1)	
	2.261235(+1)		6.184(−1)	
P2	5.864604(+1)		−2.808(−2)	
	6.58902		4.693(−1)	
	2.20366		6.1892(−1)	
P3	4.50393		−4.306(−2)	
	4.1935(−1)		5.1696(−1)	
	1.2751(−1)		5.8734(−1)	
D1	3.033647(+1)			1.5381(−1)
	7.90696			5.1289(−1)
	2.12105			5.7605(−1)
Minimal STO-3G fits				
1s[27]	2.22766	1.54329(−1)		
	4.05771(−1)	5.35328(−1)		
	1.09818(−1)	4.44635(−1)		
2sp[27]	9.94203(−1)	−9.99672(−2)	1.55916(−1)	
	2.31031(−1)	3.99513(−1)	6.07684(−1)	
	7.51386(−2)	7.00115(−1)	3.91957(−1)	
3spd[68]	4.55950(−1)	−2.27764(−1)	4.95151(−3)	2.19768(−1)
	1.39079(−1)	2.17544(−1)	5.77766(−1)	6.55547(−1
	5.36612(−2)	9.16677(−1)	4.84646(−1)	2.86573(−1)
4sp[68]	2.46458(−1)	−3.08844(−1)	−1.21547(−1)	
	9.09586(−2)	1.96064(−2)	5.71523(−1)	
	4.01683(−2)	1.13103	5.49895(−1)	
4spd[71]	2.33486(−1)	−3.30610(−1)	−1.28393(−1)	1.25066(−1)
	9.09182(−2)	5.76110(−2)	5.85205(−1)	6.68679(−1)
	4.00224(−2)	1.11558	5.43944(−1)	3.05247(−1)
5sp[71]	1.34901(−1)	−3.84264(−1)	−3.48169(−1)	
	7.26361(−2)	−1.97257(−1)	6.29032(−1)	
	3.20846(−2)	1.37550	6.66283(−1)	

[a]This paper contained a printing error: the basis set is listed as [5s4p2d], but the table showed only 4s, the contraction was therefore assumed to be (6, 2, 2, 1, 1/4, 2, 1, 1/4, 1) instead of the (6, 4, 1, 1/4, 2, 1, 1/4, 1).

TABLE 6. Exponents (scale factors) for the Se and Te STO-3G fits of Table 5[68,71]

Orbital	Exponents[a] Se	Te
1s	33.37	51.07
2sp	14.40	22.71
3spd	6.22	12.03
4sp	2.22	—
4spd	—	5.36
5sp	—	2.28

[a]The valence exponents are averages of exponents optimized on a small number of molecules. The atom values are Se: 4sp = 2.19 and Te: 5sp = 2.25.

TABLE 7. Total energies for the Se and Te GTO basis sets

(s/p/d)	[s/p/d]	Contraction scheme	Energy (Ref.)
Se			
(14/11/5)	—	—	− 2399.786915(61)[a]
(14/11/5)	—	—	− 2399.7717(62)[a]
(13/9/5)	—	—	− 2399.703348(61)[a]
(12/8/5)	[5/4/2]	(62211/4211/41)	− 2395.9455(63)[b]
(12/9/3)	[4/3/1]	(3333/333/3)	− 2390.0932(65)[c]
(12/9/3)	[4/3/1]	(3333/333/3)	− 2373.52734(64)[d]
Te			
(18/14/8)	[16/12/8]	(3111.../3111.../111...)	− 6611.6648(66)
(15/12/3)	[5/4/2]	(33333/3333/33)	− 6547.12236(67)[d]
(15/11/6)	[10/8/4]	52111.../4111.../3111...)	− 6611.0593(68)[a,e]

[a]Energy for uncontracted basis set.
[b]First basis set in Table 5.
[c]Second basis set in Table 5.
[d]STO-3G basis set, see Tables 5 and 6.
[e]Other contraction schemes were also considered in Ref. 72.

Although no comparative studies are available on the quality of these Se and Te basis sets (Tables 5–7), earlier studies do exist[47,48,62,73–75] on the quality of basis sets for first-row[47,48,73] and second-row[62,74,75] elements. These studies reveal that care must be taken in choosing a basis set, since poorly balanced basis sets can give drastically different results even when compared with basis sets of similar sizes.

B. *Ab Initio* Computational Methods

All theories of molecular quantum chemistry aim to obtain a solution for the Schrödinger equation, $H\Psi = E\Psi$, of the molecule in question. There are two broad categories, the semiempirical and non-empirical or *ab initio* methods. The semiempirical methods neglect a great many details of the calculations, but try to compensate for this with the use of experimental parameters. In contrast to these, the *ab initio* methods carry out all computations rigorously, without any mathematical neglect. However, even in *ab initio* methods, the explicit form of the wave function Ψ has to be assumed initially[10]. The simplest rigorous wave function that can be used in an *ab initio* calculation is a single Slater determinant (an antisymmetrized spin-orbital product), which will lead to the computation, in a limiting sense, of an energy value normally referred to as the Hartree–Fock Limit (HFL). However, the more general the wave function used in the computation, the more accurate is the resulting energy (E) along with the computed molecular properties which are also expected to improve, but not necessarily monotonically.

The most general wave function that can be used is a linear combination of a very large number of Slater determinants, where the use of such a wave function will lead, in the limiting sense, to an energy value that is normally referred to as the Non-Relativistic Limit (NRL). Traditionally, the molecular Hamiltonian is formulated within the non-relativistic quantum theory. This means that even the most general wave function can only lead to the non-relativistic energy limit. In order to be able to compute the total energy of the atomic

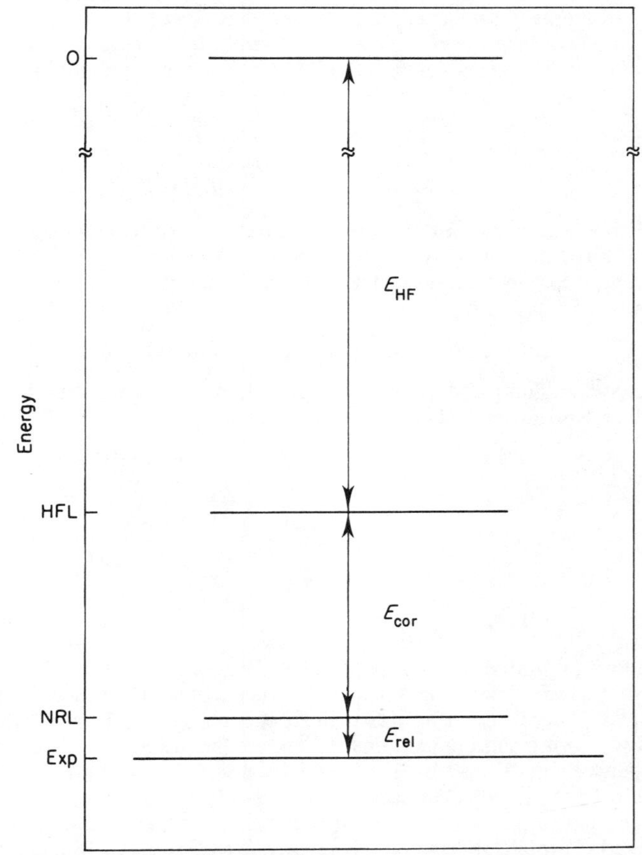

FIGURE 3. The theoretical components of the total experimental energy; E_{HF} at the Hartree–Fock Limit (HFL), $E_{HF} + E_{cor}$ at the Non-Relativistic Limit (NRL) and $E_{exp} = E_{HF} + E_{cor} + E_{rel}$

or molecular system a relativistic Hamiltonian is necessary, so that the relativistic correction can be included in the total non-relativistic energy (Figure 3).

For relatively large chemical systems, and that could mean a medium-size organic molecule containing only H, C, N and O or systems containing one heavy atom with an appropriate number of small ligands (e.g. CH_3), one may hope to use the non-relativistic Hamiltonian with only the simplest possible rigorous wave function. These single determinantal or SCF calculations always yield energies that are above the HFL, sometimes by a substantial amount. Consequently, all the results that are summarized in the next chapter are of this latter level of sophistication. It is perhaps not unreasonable to assume that in the forseeable future, more sophisticated calculations will be performed on organic compounds containing Se and Te. These more sophisticated calculations could include methods such as: Multi Configuration-Self Consistent Field (MC-SCF) theory, Generalized Valence Bond (GVB) theory, Configuration Interaction (CI) theory and the like for which programs exist[76].

In addition to the computation of the total energy with the previous explicitly assumed form of the wave function, that traditionally underline the overall theoretical problem, we also need nowadays the gradients of the energy or forces[77] (i.e. the first partial derivatives of the energy with respect to the geometrical parameters, q_i). For a molecule with N atoms, the energy will be multidimensional function of $3N-6$ independent internal coordinates:

$$E = (q_1, q_2, q_3, \ldots, q_{3N-6})$$

with $3N-6$ partial derivatives:

$$\partial E/\partial q = (\partial E/\partial q_1, \partial E/\partial q_2, \ldots, \partial E/\partial q_{3N-6})$$

These gradients are necessary for the efficient search for geometries of stable molecular structures and for transition states, both of which correspond to critical points on the energy hypersurface, where the gradient of E with respect to each and every internal coordinate is zero:

$$(\partial E/\partial q_1, \partial E/\partial q_2, \ldots, \partial E/\partial q_{3N-6}) = (0, 0, \ldots, 0)$$

Stable molecules (minima) and transition states (first-order saddle-points) are distinguished by their Hessian matrix H:

$$H = \begin{bmatrix} \partial^2 E/\partial q_1^2 & \partial^2 E/\partial q_1 \partial q_2 & \cdots & \partial^2 E/\partial q_1 \partial q_{3N-6} \\ \partial^2 E/\partial q_2 \partial q_1 & \partial^2 E/\partial q_2^2 & \cdots & \partial^2 E/\partial q_2 \partial q_{3N-6} \\ \cdot & \cdot & & \cdot \\ \cdot & \cdot & & \cdot \\ \cdot & \cdot & & \cdot \\ \partial^2 E/\partial q_{3N-6} \partial q_1 & \cdot & \cdots & \partial^2 E/\partial q_{3N-6}^2 \end{bmatrix}$$

For example, for a minimum, all the eigenvalues of H are positive (0th order) and for a first-order saddle-point, one eigenvalue is negative. Other critical points are also possible, but these are the two types of critical points which are chemically significant. Examples of three efficient gradient optimization methods[78–80] used by the authors are, two variable-metric methods, one developed by Broyden, Fletcher, Goldfarb and Shanno (BFGS)[78] and the other is the Optimally Conditioned method of Davidon (OC)[79]. The third method is a minimization of sum of squares (gradients) technique[80] referred to as VA05AD. The gradient method of optimization has not been used in the results we found in the literature on organic compounds containing Se and Te. However, we do include some of our own calculations in which the OC[79] gradient method of geometry optimization has been used for a selected few compounds.

III. RESULTS

Relatively few organic molecules containing Se or Te have been investigated until now. Even those that have been reported in the literature were studied with basis sets of different quality which would not lend themselves to easy comparison. Furthermore, the majority of the studies involved fixed geometries. Results for compounds of Se and Te that have been studied using *ab initio* methods and GTO-type basis sets are summarized in Table 8. We have recently undertaken the computation of a number of organic compounds containing Se[83]. These provide some energy differences as well as molecular properties other than energies. The basis sets used in these calculations are of two types: minimal and 'split valence'. Two minimal basis sets were used; an STO-3G[27,49,68,71] basis set which is

TABLE 8. Previous calculations on organic Se and Te compounds

Species/Geometry[a]		Energy (hartree)	Basis set	Ref.
H_2Se (C_{2v})				
H—Se	1.454		(12s8p5d)→[5s4p2d]	67
HSeH	93.8		(LWD)	
H—Se	1.52		Pseudopotential calculations	81
HSeH	92.0			
H—Se	1.42	−2097.403	FSGO	82
HSeH	92.9			
H—Se	1.439	−2374.69243	STO-3G	68
HSeH	92.4			
CH_3SeH (C_s)				
C—Se	1.931	−2413.27744	STO-3G	68
C—H	1.085			
H—Se	1.441			
CSeH	94.9			
HCH	109.0[b]			
Se=C=O ($C_{\infty v}$)				
Se—C	1.662	−2484.79450	STO-3G	68
C—O	1.168			
H_2Te (C_{2v})				
H—Te	1.624	−6548.26782	STO-3G	71
HTeH	92.4			
Te=C=S ($C_{\infty v}$)				
Te—C	1.859	−6977.67673	STO-3G	71
C—S	1.517			
$TeBr_2$(C_{2v})				
Te—Br	2.512	−11636.51567	STO-3G	71
BrTeBr	98.0			

[a]Bond lengths in Å, bond angles in deg.
[b]Average value.

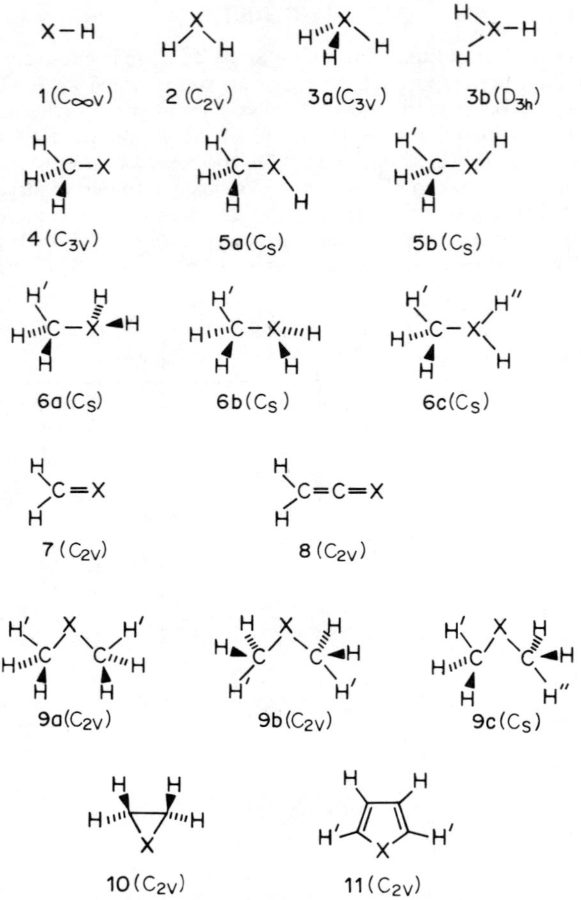

FIGURE 4. The structures which have been studied; X = O, S, Se or Te

available for O, S, Se and Te, and MINI-1[47,61,69] which is available only for O, S and Se. The 'split valence' basis set used for Se is due to Lehn, Wipff and Demuynck (LWD)[67] and a 3-21G[48,49,62] basis set of similar size was used for O and S. The corresponding STO-3G, MINI-1 or 3-21G C and H basis sets were used for the calculations. Therefore the three basis sets which will be referred to here are STO-3G, MINI-1 and 3-21G/LWD basis sets. The total molecular energies for the compounds that have been studied (Figure 4) with the STO-3G, MINI-1 and 3-21G/LWD basis sets are summarized in Tables 9, 10 and 11, respectively. The corresponding geometries for these three basis sets are given in Tables 12, 13 and 14, respectively. Some of the energies given in Tables 9, 10 and 11 (due to the similarities of the basis sets used) can be taken as components in calculating energy differences. The most obvious energy difference is that associated with the protonation and deprotonation process. These differences are in fact the measure of gas-phase basicity and acidity, respectively. One may construct isodesmic (same number of bonds) reactions and with the aid of the total energies summarized in Tables 9, 10 and 11 and of the

TABLE 9. Total STO-3G energies (hartree) of organic Se compounds and their O and S analogues

Molecule	X = O	S	Se
1	− 74.065017	− 393.505782	− 2373.971037
2	− 74.965901	− 394.311630	− 2374.692428
3a	− 75.330440	− 394.666717	− 2375.027599
3b	− 75.328392	− 394.610381	− 2374.976386
4	− 112.706365	− 432.105884	− 2412.558568
5a	− 113.549190	− 432.896073	− 2413.277422
5b	− 113.545983	− 432.893752	− 2413.275780
6a	− 113.929589	− 433.272428	− 2413.635563
6b	− 113.927448	− 433.270066	− 2413.633824
6c	− 113.926562	− 433.220118	− 2413.587693
7	− 112.354347	− 431.671367	− 2412.053568
8	− 149.726105	− 469.047854	− 2449.431475
9a	− 152.133873	− 471.482795	− 2451.863997
9b	− 152.125018	− 471.476644	− 2451.859916
9c	− 152.129285	− 471.479771	− 2451.861986
10	− 150.928501	− 470.276760	− 2450.659590
11	− 225.751253	− 545.092313	− 2525.464839

other molecules involved, calculate the energies of the reactions. The energies of isodesmic reactions may in fact be used as measure of relative stabilities of analogous compounds containing O, S, Se and Te. An example is the energy of hydrogenation, in which the O, S and Se compounds are hydrogenated to methane and their corresponding dihydrides (H_2X).

With the aid of Koopman's theorem[84] we can also estimate the ionization potentials (*IP*) of the closed-shell molecules shown in Figure 4, where the various *IP*s are associated with the removal of an electron from one of the high-lying occupied molecular orbitals and the *IP* is taken as being equal to the negative of the orbital energy. This model has basically two defects in that it does not allow for relaxation of the remaining electrons and does not incorporate correlation effects.

Finally, from these energies conformational and isomeric stabilities of a given compound may be obtained by taking energy differences of different conformations and isomers, respectively. Table 15–19 summarize these four properties based on energy differences for the compounds containing O, S and Se.

Other than the energy-related properties just described, non-energy-related properties can also be calculated; these include geometries and one-electron properties, which are simple expectation values of various one-electron operators. These properties are also of interest and help in making chemical conclusions. One of the key questions is the polarization of the charge distribution that a molecule may have. The most popular way to look at charge distribution is by using Mulliken Population Analysis[85]. However, care

TABLE 10. Total MINI-1 energies (hartree) of organic Se compounds and their O and S analogues

Molecule	X = O	S	Se
1	− 74.700722	− 396.084524	− 2390.654254
2	− 75.495566	− 396.675544	− 2391.220955
3a	—	− 396.897095	− 2391.444659
3b	− 75.796876	− 396.868963	− 2391.400034
4	− 113.520089	− 434.857815	− 2429.425041
5a	− 114.256527	− 435.449603	− 2429.997472
5b	− 114.254128	− 435.448080	− 2429.996165
6a	− 114.573429	− 435.701197	− 2430.249507
6b	—	− 435.699716	− 2430.248149
6c	− 114.573139	− 435.676153	− 2430.210330[a]
7	− 113.058942	− 434.244980	− 2428.794502
8	− 150.638673	− 471.820396	− 2466.371803
9a	− 153.019023	− 474.224778	− 2468.774726
9b	− 153.012318	− 474.220794	− 2468.771489
9c	− 153.015450	− 474.222862	
10	− 151.805715	− 473.024873	− 2467.578289
11	− 227.044415	− 548.229798	—

[a]Energy converged, but the gradient length is 2×10^{-3}.

TABLE 11. Total 3-21G/LWD energies (hartree) of organic Se compounds and their O and S analogues

Molecule	X = O	S	Se
1	− 74.868630	− 396.148616	− 2395.378373
2	− 75.585960	− 396.704666	− 2395.940727
3a	—	− 396.952162	− 2396.218306
3b	− 75.891228	− 396.921097	− 2396.158421
4a	− 113.724796	− 434.958881	− 2434.186645
5a	− 114.398017	− 435.526291	− 2434.761960
5b	− 114.395660	− 435.524560	− 2434.760266
6a	− 114.724919	− 435.802889	− 2435.065389
6b	− 114.713811[a]	− 435.800979	− 2435.063072
6c	− 114.724838	− 435.774001	− 2435.010508
7	− 113.221819	− 434.336245	− 2433.557808
8	− 150.876524	− 471.974929	− 2471.196862
9a	− 153.213209	− 474.348581	− 2472.988686[b]
9b	− 153.206554	− 474.343681	− 2473.579401[b]
9c	− 153.209029	—	− 2473.581864[b]
10	− 152.000703	− 473.157121	− 2471.885849[c]
11	− 227.350082	− 548.473505	− 2547.710125

[a]This structure has a similar conformation to **6a**, except that the OH_2 group is much flatter.
[b]Note that the relative stabilities are reversed.
[c]This structure, although a critical point, is much higher in energy than expected and is perhaps a first-order saddle-point.

TABLE 12. Geometries[a] of organic Se compounds and their O and S analogues calculated with the STO-3G basis set

Molecule/	parameter	X = O	S	Se
1	H—X	1.0681	1.3457	1.4341
2	H—X	0.9893	1.3288	1.4401
	HXH	100.02	92.50	92.52
3a	H—X	0.9903	1.3537	1.4737
	HXH	113.80	96.60	96.34
3b	H—X	0.9831	1.3410	1.4607
4	C—H	1.1324	1.0998	1.0925
	C—X	1.3682	1.7987	1.9278
	XCH	117.13	114.89	113.54
5a	C—H	1.0951	1.0867	1.0852
	C—H′	1.0915	1.0850	1.0838
	C—X	1.4330	1.7975	1.9304
	H—X	0.9911	1.3306	1.4407
	XCH	112.37	110.00	111.42
	XCH′	107.64	108.51	108.44
	CXH	103.86	95.42	94.90
	HCXH′	118.91	119.00	119.15
5b	C—H	1.0942	1.0864	1.0847
	C—H′	1.0920	1.0843	1.0838
	C—X	1.4387	1.8039	1.9359
	H—X	0.9890	1.3293	1.4397
	XCH	110.00	111.02	110.69
	XCH′	113.11	111.32	110.19
	CXH	104.69	95.96	95.27
	HCXH	120.70	120.14	119.92
6a	C—H	1.0954	1.0888	1.0880
	C—H′	1.0968	1.0889	1.0882
	C—X	1.4848	1.8411	1.9787
	H—X	0.9899	1.3491	1.4674
	XCH	105.85	107.04	106.96
	XCH′	108.82	109.61	108.93
	CXH	115.46	100.07	99.28
	HCXH′	120.81	120.79	120.68
	HXCH′	65.73	49.01	48.68
6b	C—H	1.0951	1.0883	1.0878
	C—H′	1.0965	1.0886	1.0876
	C—X	1.4874	1.8487	1.9859
	H—X	0.9879	1.3485	1.4669
	XCH	107.34	108.58	108.02
	XCH′	106.49	106.95	107.04
	CXH	116.97	100.61	99.79
	HCXH′	119.91	119.53	119.69
	HXCH′	111.81	131.04	131.35
6c	C—H	1.0967	1.0914	1.0907
	C—H′	1.0949	1.0914	1.0911
	C—X	1.4757	1.8459	1.9984
	H″—X	0.9827	1.3359	1.4535
	H—X	0.9830	1.3360	1.4535
	XCH	107.45	107.75	106.97
	XCH′	106.15	105.33	104.77
	CXH″	122.25	122.85	123.08
	CXH	121.48	123.17	123.26
	HCXH′	119.58	119.29	119.32

(Contd.)

TABLE 12. (*Contd.*)

Molecule/	parameter	X = O	S	Se
7	C—H	1.1014	1.0900	1.0897
	C—X	1.2167	1.5741	1.6891
	XCH	122.75	124.00	124.07
8	C—H	1.0750	1.0801	1.0820
	C—C	1.2996	1.2953	1.2924
	C—X	1.1828	1.5370	1.6477
	CCH	120.25	121.48	121.53
9a	C—H	1.0949	1.0870	1.0854
	C—H′	1.0915	1.0852	1.0840
	C—X	1.4334	1.7964	1.9277
	CXC	109.49	98.23	96.79
	XCH	112.05	112.08	111.39
	XCH′	107.43	108.60	108.67
	HCXH′	119.10	119.17	119.28
9b	C—H	1.0945	1.0867	1.0850
	C—H′	1.0913	1.0843	1.0840
	C—X	1.4372	1.8040	1.9344
	CXC	113.56	101.26	99.30
	XCH	109.52	110.78	110.64
	XCH′	113.65	111.89	110.58
	HCXH′	120.91	120.34	120.02
9c	C—H	1.0939	1.0865	1.0848
	C′—H	1.0949	1.0869	1.0854
	C′—H′	1.0918	1.0854	1.0842
	C—H″	1.0913	1.0846	1.0842
	C—X	1.4406	1.0840	1.9342
	C′—X	1.4332	1.7971	1.9285
	C′XC	111.00	99.41	97.78
	XCH	109.54	110.96	110.77
	XC′H	112.25	112.22	111.50
	XC′H′	107.19	108.31	108.42
	XCH″	113.39	111.43	110.19
	HCXH″	120.79	120.16	119.87
	HC′XH′	118.91	118.99	119.17
10	C—H	1.0876	1.0833	1.0832
	C—C	1.4828	1.5067	1.5030
	C—X	1.4325	1.7736	1.9079
	CCH	119.47	117.59	117.39
	CCX	58.83	64.86	66.81
	CXC	62.34	50.27	46.39
	HCCX	105.04	110.00	109.92
11	C—H	1.0774	1.0804	1.0815
	C—H′	1.0819	1.0790	1.0793
	C—C	1.4443	1.4542	1.4615
	C═C	1.3397	1.3344	1.3309
	C—X	1.3755	1.7317	1.8640
	CCH	126.71	123.18	123.87
	XCH′	116.54	120.42	121.37
	C—C═C	106.13	112.06	114.20
	C═C—X	111.15	112.76	112.58
	CXC	105.43	90.35	86.43

[a]Bond lengths in Å, bond angles in deg.

TABLE 13. Geometries[a] of organic Se compounds and their O and S analogues calculated with the MINI-1 basis set

Molecule/	parameter	X = O	S	Se
1	H—X	1.0540	1.3895	1.4957
2	H—X	1.0045	1.4123	1.5187
	HXH	103.36	94.59	93.14
3a	H—X	—	1.4695	1.5733
	HXH	—	99.32	96.14
3b	H—X	1.0215	1.4548	1.5592
4	C—H	1.1419	1.1133	1.1107
	C—X	1.4647	1.9288	2.0503
	XCH	116.42	112.28	111.91
5a	C—H	1.1261	1.1157	1.1137
	C—H′	1.1228	1.1156	1.1142
	C—X	1.4773	1.9220	2.0331
	H—X	1.0056	1.4108	1.5169
	XCH	112.30	110.27	110.38
	XCH′	107.26	106.21	106.68
	CXH	106.36	97.20	95.58
	HCXH′	118.70	118.80	118.87
5b	C—H	1.1251	1.1155	1.1136
	C—H′	1.1237	1.1150	1.1134
	C—X	1.4816	1.9280	2.0390
	H—X	1.0037	1.4099	1.5160
	XCH	109.72	108.59	108.97
	XCH′	113.02	109.93	109.75
	CXH	107.06	97.59	95.88
	HCXH′	120.69	120.14	120.05
6a	C—H	1.1355	1.1304	1.1269
	C—H′	1.1366	1.1293	1.1258
	C—X	1.5392	2.0097	2.1010
	H—X	1.0169	1.4529	1.5587
	XCH	105.78	103.39	104.51
	XCH′	107.85	107.51	108.02
	CXH	120.35	102.91	99.58
	HCXH′	120.78	121.10	120.97
	HXCH′	79.07	50.75	48.60
6b	C—H	—	1.1305	1.1262
	C—H′	—	1.1295	1.1269
	C—X	—	2.0192	2.1097
	H—X	—	1.4522	1.5581
	XCH	—	105.93	106.70
	XCH′	—	102.56	103.78
	CXH	—	103.46	100.11
	HCXH′	—	119.33	119.39
	HXCH′	—	129.24	131.45
6c	C—H	1.1363	1.1346	1.1339[b]
	C—H′	1.1348	1.1352	1.1348
	C—X	1.5380	2.0313	2.1457
	H″—X	1.0153	1.4363	1.5413
	H—X	1.0156	1.4364	1.5412
	XCH	106.83	103.25	103.14
	XCH′	105.90	101.61	101.43
	CXH″	122.00	124.07	124.44
	CXH	121.17	124.23	124.44
	HCXH′	119.48	119.30	119.31

(*Contd.*)

TABLE 13. (*Contd.*)

Molecule/	parameter	X=O	S	Se
7	C—H	1.1353	1.1238	1.1215
	C—X	1.2678	1.6970	1.8067
	XCH	122.09	122.22	122.73
8	C—H	1.1084	1.1174	1.1180
	C—C	1.3464	1.3342	1.3333
	C—X	1.2277	1.6722	1.7853
	CCH	119.60	120.67	120.92
9a	C—H	1.1256	1.1151	1.1131
	C—H′	1.1229	1.1155	1.1140
	C—X	1.4753	1.9167	2.0282
	CXC	110.67	98.85	96.84
	XCH	111.69	110.14	110.31
	XCH′	107.26	106.89	107.42
	HCXH′	119.13	119.09	119.13
9b	C—H	1.1257	1.1155	1.1135
	C—H′	1.1225	1.1140	1.1126
	C—X	1.4775	1.9241	2.0353
	CXC	114.65	101.64	99.06
	XCH	109.34	108.80	109.33
	XCH′	112.98	110.00	109.70
	HCXH′	120.78	120.17	119.96
9c	C—H	1.1251	1.1152	—
	C′—H	1.1257	1.1151	—
	C′—H′	1.1233	1.1157	—
	C—H″	1.1225	1.1144	—

Molecule/	parameter	X=O	S	Se
	C—X	1.4811	1.9242	—
	C′—X	1.4746	1.9173	—
	C′XC	112.19	99.84	—
	XCH	109.33	108.96	—
	XC′H	111.90	110.22	—
	XC′H′	107.07	106.65	—
	XCH″	112.79	109.57	—
	HCXH″	120.72	120.06	—
	HC′XH′	118.94	118.95	—
10	C—H	1.1187	1.1140	1.1129
	C—C	1.5383	1.5244	1.5179
	C—X	1.5022	1.9521	2.0688
	CCH	119.08	118.30	118.12
	CCX	59.20	67.02	68.48
	CXC	61.60	45 97	43.04
	HCCX	104.49	106.12	106.72
11	C—H	1.1061	1.1102	
	C—H′	1.1140	1.1100	
	C—C	1.4822	1.4945	
	C=C	1.3785	1.3657	
	C—X	1.4152	1.8495	
	CCH	126.64	122.59	
	XCH′	116.80	119.45	
	C—C=C	106.41	113.70	
	C=C—X	110.58	111.80	
	CXC	106.03	88.99	

[a]Bond lengths in Å, bond angles in deg.
[b]Converged to a gradient length of only 2×10^{-3}.

TABLE 14. Geometries[a] of organic Se compounds and their O and S analogues calculated with the 3-21G/LWD basis set

Molecule/	parameter	X = O	S	Se
1	H—X	1.0289	1.3726	1.4701
2	H—X	0.9666	1.3506	1.4557
	HXH	107.65	95.82	92.58
3a	H—X	—	1.3516	1.4611
	HXH	—	99.55	93.91
3b	H—X	0.9788	1.3341	1.4458
4	C—H	1.1340	1.0861	1.0849
	C—X	1.3482	1.9083	1.9981
	XCH	117.27	110.62	110.34
5a	C—H	1.0853	1.0774	1.0780
	C—H′	1.0786	1.0780	1.0801
	C—X	1.4403	1.8941	1.9701
	H—X	0.9658	1.3523	1.4560
	XCH	112.28	109.69	109.69
	XCH′	106.25	105.49	106.59
	CXH	110.42	97.89	95.33
	HCXH	118.51	118.64	118.80
5b	C—H	1.0814	1.0766	1.0781
	C—H′	1.0848	1.0778	1.0786
	C—X	1.4450	1.9022	1.9777
	H—X	0.9637	1.3507	1.4543
	XCH	109.29	107.84	108.74
	XCH′	112.58	109.46	108.76
	CXH	111.11	98.34	95.86
	HCXH	120.45	120.14	119.74
6a	C—H	1.0744	1.0768	1.0793
	C—H′	1.0740	1.0758	1.0773
	C—X	1.5404	1.9439	1.9774
	H—X	0.9733	1.3500	1.4574
	XCH	104.69	104.50	106.56
	XCH′	106.36	107.91	108.69
	CXH	121.45	101.43	97.17
	HCXH′	120.75	121.05	120.88
	HXCH′	85.69	50.47	47.20
6b	C—H	1.0744	1.0762	1.0780
	C—H′	1.0739	1.0751	1.0777
	C—X	1.5401	1.9588	1.9900
	H—X	0.9732	1.3493	1.4567
	XCH	104.71	106.50	107.76
	XCH′	106.36	103.87	106.54
	CXH	106.59	101.94	97.82
	HCXH′	120.73	119.62	119.78
	HXCH′	62.57	129.51	132.90
6c	C—H	1.0738	1.0732	1.0746
	C—H′	1.0742	1.0762	1.0787
	C—X	1.5398	1.9958	2.0261
	H″—X	0.9727	1.3358	1.4455
	H—X	0.9732	1.3358	1.4458
	XCH	105.58	103.49	105.60
	XCH′	104.55	101.73	103.12
	CXH″	121.93	123.72	123.17
	CXH	121.30	123.95	123.70
	HCXH′	119.36	119.15	118.87

(*Contd.*)

TABLE 14. (*Contd.*)

Molecule/	parameter	X = O	S	Se
7	C—H	1.0832	1.0734	1.0734
	C—X	1.2068	1.6381	1.7392
	XCH	122.53	121.77	121.73
8	C—H	1.0695	1.0736	1.0740
	C—C	1.2960	1.2853	1.2877
	C—X	1.1620	1.6025	1.6937
	CCH	120.06	120.69	120.85
9a	C—H	1.0863	1.0789	1.0599
	C—H′	1.0793	1.0788	1.0633
	C—X	1.4326	1.8851	2.0480
	CXC	114.00	99.06	99.88
	XCH	111.47	109.59	105.13
	XCH′	106.86	106.48	102.61
	HCXH′	119.28	119.03	118.81
9b	C—H	1.0822	1.0775	1.0788
	C—H′	1.0854	1.0786	1.0796
	C—X	1.4345	1.8975	1.9738
	CXC	118.76	102.45	99.99
	XCH	109.35	108.21	109.02
	XCH′	112.01	109.42	108.98
	HCXH′	120.37	120.01	119.75
9c	C—H	1.0819	1.0776	1.0789
	C′—H	1.0856	1.0785	1.0793
	C′—H′	1.0795	1.0789	1.0808
	C—H″	1.0853	1.0792	1.0802
	C—X	1.4392	1.8960	1.9732
	C′—X	1.4326	1.8872	1.9641
	C′XC	116.01	100.38	97.99
	XCH	109.15	108.40	109.20
	XC′H	111.75	109.65	109.69
	XC′H′	106.58	106.19	107.08
	XCH″	112.15	109.08	108.55
	HCXH″	120.38	119.91	119.61
	HC′XH′	119.00	118.84	118.91
10	C—H	1.0708	1.0697	1.0619
	C—C	1.4742	1.4633	1.5495
	C—X	1.4695	1.9337	2.0603
	CCH	119.24	118.65	116.22
	CCX	59.89	67.77	67.91
	CXC	60.21	44.47	44.18
	HCCX	103.30	104.30	99.79
11	C—H	1.0647	1.0692	1.0702
	C—H′	1.0618	1.0653	1.0671
	C—C	1.4501	1.4481	1.4423
	C═C	1.3397	1.3354	1.3416
	C—X	1.3802	1.7970	1.8763
	CCH	126.59	122.87	122.84
	XCH′	116.51	120.34	121.71
	C—C═C	106.67	113.74	115.09
	C═C—X	109.83	111.67	111.47
	CXC	106.99	89.19	86.87

[a]Bond lengths in Å, bond angles in deg.

TABLE 15. Proton affinities ($kJ\,mol^{-1} \times 10^{-2}$) of organic Se compounds and their O and S analogues calculated with the results given in Tables 8, 9 and 10

	X = O			X = S			X = Se		
Molecule	A[a]	B[a]	C[a]	A	B	C	A	B	C
1	23.7	20.9	18.8	21.2	15.5	14.6	18.9	14.9	14.8
2	9.57	7.91[b]	8.01[b]	9.32	5.82	6.50	8.80	5.87	7.29
4	22.1	19.3	17.7	20.7	15.5	14.9	18.9	15.0	15.1
5a	9.99	8.32	8.58	9.88	6.61	7.26	9.40	6.62	7.97
Deviation[c]	3.1	1.0	—	4.5	0.7	—	2.7	+0.7	—

[a]A = STO-3G, B = MINI-1, C = 3-21G/LWD.
[b]Calculated using structure **3b**.
[c]Mean absolute deviation from the 3-21G/LWD basis set.

TABLE 16. Energies of hydrogenation ($kJ\,mol^{-1}$) for a series of isodesmic reactions of organic Se compounds and their O and S analogues

		Energies of hydrogenation[b]		
Reaction[a]	X	A	B	C
$CH_3XH + H_2 \rightarrow CH_4 + H_2X$	O	−68.4	−120	−110
	S	−65.4	−85.1	−84.8
	Se	−64.0	−78.7	−85.8
$H_2C{=}X + 2H_2 \rightarrow CH_4 + H_2X$	O	−271	−318	−250
	S	−347	−302	−261
	Se	−343	−291	−299
$CH_3XCH_3 + 2H_2 \rightarrow 2CH_4 + H_2X$	O	−133	−235	−212
	S	−125	−167	−168
	Se	−124	−155	−175
$H_2C{=}C{=}X + 4H_2 \rightarrow 2CH_4 + H_2X$	O	−587	−593	−450
	S	−650	−588	−503
	Se	−642	−572	−540
$C_2H_4X + 3H_2 \rightarrow 2CH_4 + H_2X$	O	−364	−475	−447
	S	−357	−372	−348
	Se	−352	−351	—
$C_4H_4X + 7H_2 \rightarrow 4CH_4 + H_2X$	O	−787	−816	−742
	S	−799	−802	−730
	Se	−820	—	−728
Mean absolute deviation	O	+67.8	+57.7	—
	S	+62.2	+37.2	—
	Se	+62.2	+16.8	—

[a]Energies (hartree) for H_2 and CH_4 are: −1.117506 and −39.726864 for STO-3G, −1.122073 and −39.928559 for MINI-1 and −1.122960 and −39.976878 for 3-21G/LWD, respectively.
[b]A = STO-3G, B = MINI-1, C = 3-21G/LWD.

TABLE 17. First ionization potentials (eV) (Koopman's theorem) for organic Se compounds and their O and S analogues

	X = O			X = S			X = Se		
Molecule	A[a]	B[a]	C[a]	A	B	C	A	B	C
1	—	—	—	—	—	1.76	—	0.31	1.74
2	10.68	12.64	12.99	7.57	10.76	10.68	7.59	10.04	9.84
3a	24.65	—	—	20.37	21.40	20.97	19.06	20.50	20.28
3b	24.21	25.36	25.22	17.88	19.69	19.20	16.85	18.39	17.85
4	—	—	1.16	—	0.25	1.57	—	0.42	1.50
5a	9.78	11.74	11.82	7.08	10.07	9.91	7.07	9.42	9.17
5b	9.70	11.66	11.74	7.06	10.05	9.88	7.05	9.40	9.16
6a	20.82	21.45	21.34	18.66	19.63	19.25	17.66	18.91	18.73
6b	20.66	—	—	18.59	19.57	19.18	17.61	18.86	18.68
6c	20.40	21.37	21.32	16.40	18.04	17.55	15.53	16.88	16.53
7	9.64	11.89	11.78	6.59	9.79	9.65	6.60	9.07	9.03
8	7.56	10.19	9.78	6.10	9.49	9.07	6.29	8.92	8.65
9a	9.24	11.22	11.15	6.70	9.53	9.30	6.64	8.91	7.41
9b	8.93	10.94	10.88	6.60	9.46	9.22	6.57	8.87	8.57
9c	9.07	11.06	10.99	6.65	9.49	9.25	6.60	—	8.59
10	9.99	12.10	12.12	6.94	9.73	9.51	6.79	9.00	7.89
11	7.39	9.64	9.01	7.22	10.07	9.30	7.14	—	8.98

[a]A = STO-3G, B = MINI-1, C = 3-21G/LWD.

TABLE 18. Experimental ionization potentials (eV) and percentage errors of the calculated first ionization potentials

			% Error[a]		
Molecular	X	Exp.[2]	A	B	C
2	O	12.61	− 15	+ 0.2	+ 3.0
	S	10.47	− 28	+ 2.8	+ 2.0
	Se	9.881	− 23	+ 1.6	− 0.4
5a	O	10.85	− 9.9	+ 8.2	+ 8.9
	S	9.43	− 25	+ 6.8	+ 5.1
7	O	10.88	− 11	+ 9.3	+ 8.3
8	O	9.60	− 21	+ 6.1	+ 1.9
9a	O	9.96	− 7.2	+ 1.3	+ 1.2
	S	8.7	− 23	+ 9.5	+ 6.9
10	O	10.56	− 0.5	+ 1.4	+ 1.4
11	Se	8.86[b]	− 19	—	+ 1.4
Average			− 17	+ 4.7	+ 3.6

[a]A = STO-3G, B = MINI-1, C = 3-21G/LWD.
[b]Average of the two reported values in Refs. 86 and 87.

TABLE 19. Conformational stabilities ΔE (kJ mol^{-1}) for organic Se compounds and their O and S analogues

	X = O			X = S			X = Se		
Conformers	A[a]	B[a]	C[a]	A	B	C	A	B	C
3b–3a	5.38	—	—	148	73.9	81.6	134	117	157
5b–5a	8.42	6.30	6.19	6.09	4.00	4.54	4.31	3.43	4.45
6b–6a	5.62	—	29.2	6.20	3.89	5.01	4.57	3.57	6.08
6c–6a	7.95	0.76	0.21	137	65.8	75.8	126	103	144
9b–9a	23.2	17.6	17.5	16.1	10.4	12.9	10.7	8.50	– 1551[b]
9c–9a	12.0	9.38	11.0	7.94	5.03	—	5.28	—	– 1557[b]

[a]A = STO-3G, B = MINI-1, C = 3-21G/LWD.
[b]This reflects the instability of **9a**.

should be exercised in using this method as the results of Mulliken Population Analysis are somewhat arbitrary and basis-set-dependent to a large degree. Thus only trends may have any significance. A more rigorous measure of the charge distribution, although not so easy to interpret, is the molecular dipole moment. The reason it is rigorous is due to the fact that the dipole moment is a proper observable in the Dirac sense, implying that it is the expectation value of a bonafide quantum-mechanical operator. The reason it is more difficult to interpret a dipole moment is that it is a single vector that is the resultant of the detailed charge distribution. The net charges calculated with the STO-3G, MINI-1 and 3-21G basis sets are given in Tables 20, 21 and 22, respectively. The magnitude of the dipole moments and related results are tabulated in Tables 23 and 24.

IV. DISCUSSION

As far as the applicability of the theory to organic compounds containing Se and Te is concerned, two factors can be considered at this time. First, given the existing comparable basis sets, it is possible to produce as many interesting chemical conclusions as the data will permit. This approach is chemically the most pleasing but has the shortcoming that the reliability of the chemical conclusions cannot be ascertained. Therefore, the second factor addresses itself to this question of basis set reliability. Ultimately it would be desirable to attach a yardstick to each basis set as far as its numerical reliability is concerned in calculation of a given molecular property.

Each property calculated for the selected organoselenium compounds will be discussed in the following sections with special attention being paid to the point concerning basis set reliability.

A. Energy-related Properties

1. Proton affinities

Proton affinities (*PA*) have been calculated for HX^-, H_2X, CH_3X^- and CH_3XH for X = O, S and Se and are tabulated in Table 15. A plot of *PA* as a function of atomic number is shown in Figure 5. The STO-3G basis set consistently gives results of *PA*s greater than the MINI-1 and 3-21G basis set values. The MINI-1 basis set predicts *PA*s which are generally in good agreement with the 3-21G results. Therefore the STO-3G gives the larger mean absolute deviation from the 3-21G basis set, 3.1, 4.5 and 2.7 for O, S and Se respectively, whereas the MINI-1 mean absolute deviations are 1.0, 0.7 and 0.7 for O, S and

TABLE 20. Net charges (from Mulliken populations) of organic Se compounds and their O and S analogues calculated with the STO-3G basis set

Molecule		X = O	S	Se
1	X	−0.742	−0.741	−0.823
	H	−0.258	−0.259	−0.177
2	X	−0.331	+0.072	+0.028
	H	+0.165	−0.036	−0.014
3a	X	−0.256	+0.524	+0.565
	H	+0.419	+0.159	+0.145
3b	X	−0.287	+0.331	+0.357
	H	+0.429	+0.223	+0.214
4	X	−0.640	−0.610	−0.708
	C	−0.046	−0.265	−0.255
	H	−0.105	−0.042	−0.012
5a	X	−0.280	+0.118	+0.061
	C	−0.071	−0.272	−0.264
	H(X)	+0.174	−0.042	−0.025
	H(C)	+0.054	+0.063	+0.074
	H′(C)	+0.070	+0.070	+0.080
5b	X	−0.284	+0.114	+0.058
	C	−0.073	−0.272	−0.264
	H(X)	+0.178	−0.041	−0.023
	H(C)	+0.062	+0.066	+0.077
	H′(C)	+0.056	+0.067	+0.075
6a	X	−0.218	+0.517	+0.537
	C	−0.021	−0.226	−0.221
	H(X)	+0.386	+0.130	+0.118
	H(C)	+0.160	+0.153	+0.153
	H′(C)	+0.147	+0.143	+0.143
6c	X	−0.248	+0.321	+0.311
	C	−0.008	−0.159	−0.143
	H(X)	+0.395	+0.179	+0.172
	H″(X)	+0.397	+0.180	+0.173
	H(C)	+0.154	+0.159	+0.163
	H′(C)	+0.155	+0.159	+0.161
7	X	−0.188	+0.119	+0.058
	C	+0.075	−0.236	−0.215
	H	+0.056	+0.058	+0.078
8	X	−0.186	+0.147	+0.088
	C(X)	+0.260	−0.088	−0.076
	C	−0.246	−0.216	−0.198
	H	+0.087	+0.079	+0.093
9a	X	−0.236	+0.165	+0.101
	C	−0.065	−0.275	−0.269
	H	+0.056	+0.062	+0.070
	H′	+0.071	+0.069	+0.077
9b	X	−0.246	+0.155	+0.090
	C	−0.064	−0.273	−0.266
	H	+0.063	+0.065	+0.075
	H′	+0.060	+0.066	+0.072
9c	X	−0.240	+0.160	+0.096
	C	−0.069	−0.276	−0.269
	C′	−0.062	−0.273	−0.267
	H(C)	+0.063	+0.064	+0.074
	H(C′)	+0.056	+0.062	+0.071
	H′	+0.071	+0.069	+0.078
	H″	+0.061	+0.067	+0.073

6b	X	−0.226	+0.513	+0.533	**10**	X	−0.208	+0.127	+0.041
	C	−0.018	−0.224	−0.219		C	−0.045	−0.205	−0.187
	H(X)	+0.388	+0.130	+0.118		H	+0.074	+0.071	+0.083
	H(C)	+0.152	+0.147	+0.147					
	H′(C)	+0.163	+0.156	+0.156	**11**	X	−0.201	+0.254	+0.197
						C	−0.101	−0.084	−0.075
						C(X)	+0.047	−0.178	−0.168
						H	+0.072	+0.063	+0.067
						H′	+0.082	+0.071	+0.077

TABLE 21. Net charges (from Mulliken populations) of organic Se compounds and their O and S analogues calculated with the MINI-1 basis set

Molecule		X = O	S	Se	Molecule		X = O	S	Se
1	X	−0.855	−0.949	−0.940	**5a**	X	−0.421	−0.156	−0.053
	H	−0.145	−0.051	−0.060		C	−0.253	−0.454	−0.502
						H(X)	+0.267	+0.124	+0.078
2	X	−0.507	−0.277	−0.186		H(C)	+0.129	+0.167	+0.157
	H	+0.253	+0.138	+0.093		H′(C)	+0.148	+0.159	+0.162
3a	X	—	+0.033	+0.243	**5b**	X	−0.424	−0.159	−0.055
	H	—	+0.322	+0.252		C	−0.254	−0.453	−0.501
						H(X)	+0.270	+0.125	+0.078
3b	X	−0.497	−0.172	−0.025		H(C)	+0.140	+0.165	+0.161
	H	+0.499	+0.391	+0.342		H′(C)	+0.128	+0.158	+0.157
4	X	−0.675	−0.727	−0.714	**6a**	X	−0.431	+0.071	+0.293
	C	−0.284	−0.478	−0.514		C	−0.190	−0.356	−0.423
	H	−0.014	+0.068	+0.076		H(X)	+0.463	+0.282	+0.217

(Contd.)

TABLE 21. (*Contd.*)

Molecule		X=O	S	Se
	H(C)	+0.234	+0.244	+0.235
	H′(C)	+0.227	+0.233	+0.226
6b	X	—	+0.066	+0.288
	C	—	−0.351	−0.419
	H(X)	—	+0.281	+0.217
	H(C)	—	+0.236	+0.299
	H′(C)	—	+0.250	+0.239
6c	X	−0.436	−0.139	−0.006
	C	−0.188	−0.277	−0.310
	H(X)	+0.464	+0.325	+0.276
	H″(X)	+0.466	+0.326	+0.277
	H(C)	+0.231	+0.256	+0.256
	H′(C)	+0.231	+0.253	+0.252
7	X	−0.264	−0.055	+0.013
	C	−0.024	−0.301	−0.360
	H	+0.144	+0.178	+0.173
8	X	−0.266	−0.037	+0.026
	C(X)	+0.329	+0.001	−0.062
	C	−0.422	−0.356	−0.353
	H	+0.180	+0.196	+0.194
9a	X	−0.344	−0.023	+0.097
	C	−0.240	−0.458	−0.508
	H	+0.132	+0.154	+0.152
	H′	+0.148	+0.162	+0.156
9b	X	−0.352	−0.034	+0.087
	C	−0.237	−0.455	−0.505
	H	+0.141	+0.160	+0.156
	H′	+0.132	+0.152	+0.150
9c	X	−0.347	−0.029	—
	C	−0.243	−0.458	—
	C′	−0.237	−0.457	—
	H(C)	+0.141	+0.159	—
	H(C′)	+0.132	+0.154	—
	H′	+0.148	+0.162	—
	H″	+0.133	+0.154	—
10	X	−0.297	−0.089	−0.020
	C	−0.170	−0.313	−0.345
	H	+0.159	+0.179	+0.177
11	X	−0.312	+0.076	—
	C	−0.202	−0.172	—
	C(X)	+0.018	−0.213	—
	H	+0.164	+0.167	—
	H′	+0.176	+0.180	—

TABLE 22. Net charges (from Mulliken populations) of organic Se compounds and their O and S analogues calculated with the 3-21G basis set

Molecule		X = O	S	Se
1	X	− 1.025	− 0.911	− 1.042
	H	+ 0.025	− 0.089	+ 0.042
2	X	− 0.733	− 0.170	− 0.373
	H	+ 0.366	+ 0.085	+ 0.187
3a	X	—	+ 0.202	+ 0.037
	H	—	+ 0.266	+ 0.321
3b	X	− 0.714	+ 0.038	− 0.107
	H	+ 0.571	+ 0.321	+ 0.369
4	X	− 0.866	− 0.666	− 0.809
	C	− 0.038	− 0.758	− 0.629
	H	− 0.032	+ 0.141	+ 0.146
5a	X	− 0.677	+ 0.017	− 0.194
	C	− 0.270	− 0.834	− 0.681
	H(X)	+ 0.373	+ 0.069	+ 0.166
	H(C)	+ 0.177	+ 0.246	+ 0.236
	H′(C)	+ 0.220	+ 0.256	+ 0.235
5b	X	− 0.683	+ 0.011	− 0.195
	C	− 0.267	− 0.835	− 0.686
	H(X)	+ 0.377	+ 0.070	+ 0.167
	H(C)	+ 0.201	+ 0.254	+ 0.239
	H′(C)	+ 0.172	+ 0.245	+ 0.236
6a	X	− 0.708	+ 0.293	+ 0.166
	C	− 0.322	− 0.790	− 0.709
	H(X)	+ 0.538	+ 0.240	+ 0.296
	H(C)	+ 0.319	+ 0.343	+ 0.319
	H′(C)	+ 0.317	+ 0.330	+ 0.315
6b	X	− 0.642	+ 0.289	+ 0.167
	C	− 0.389	− 0.784	− 0.714
	H(X)	+ 0.528	+ 0.238	+ 0.294
	H(C)	+ 0.335	+ 0.334	+ 0.318
	H′(C)	+ 0.306	+ 0.350	+ 0.323
6c	X	− 0.710	+ 0.044	− 0.024
	C	− 0.321	− 0.635	− 0.646
	H(X)	+ 0.539	+ 0.275	+ 0.331
	H″(X)	+ 0.537	+ 0.276	+ 0.333
	H(C)	+ 0.320	+ 0.349	+ 0.338
	H′(C)	+ 0.315	+ 0.341	+ 0.330
7	X	− 0.482	+ 0.168	− 0.134
	C	+ 0.131	− 0.703	− 0.374
	H	+ 0.175	+ 0.267	+ 0.254
8	C	− 0.548	+ 0.334	− 0.148
	C(X)	+ 0.544	− 0.288	+ 0.129
	C	− 0.555	− 0.611	− 0.543
	H	+ 0.280	+ 0.283	+ 0.281

(*Contd.*)

TABLE 22. (*Contd.*)

Molecule		X = O	S	Se
9a	X	− 0.643	+ 0.236	− 0.088
	C	− 0.250	− 0.837	− 1.649
	H	+ 0.177	+ 0.236	+ 0.563
	H′	+ 0.217	+ 0.248	+ 0.566
9b	X	− 0.662	+ 0.219	+ 0.003
	C	− 0.237	− 0.837	− 0.693
	H	+ 0.198	+ 0.247	+ 0.234
	H′	+ 0.171	+ 0.234	+ 0.224
9c	X	− 0.650	+ 0.229	+ 0.003
	C	− 0.248	− 0.845	− 0.696
	C′	− 0.246	− 0.837	− 0.685
	H(C)	+ 0.201	+ 0.247	+ 0.233
	H(C′)	+ 0.176	+ 0.236	+ 0.226
	H′	+ 0.217	+ 0.248	+ 0.231
	H″	+ 0.173	+ 0.238	+ 0.227
10	X	− 0.552	+ 0.086	− 0.237
	C	− 0.191	− 0.604	− 1.033
	H	+ 0.234	+ 0.280	+ 0.576
11	X	− 0.645	+ 0.541	+ 0.127
	C	− 0.340	− 0.202	− 0.257
	C(X)	+ 0.141	− 0.598	− 0.315
	H	+ 0.250	+ 0.254	+ 0.249
	H′	+ 0.271	+ 0.276	+ 0.260

TABLE 23. Dipole moments (D)[a] for organic Se compounds and their O and S analogues calculated with the STO-3G, MINI-1 and 3-21G/LWD basis sets

	X = O			X = S			X = Se		
Molecule	A[b]	B[b]	C[b]	A	B	C	A	B	C
2	1.71	2.04	2.39	1.03	2.00	1.83	1.47	1.66	1.19
5a	1.51	1.81	2.12	0.96	2.14	2.12	1.53	1.89	1.81
5b	1.58	1.87	2.23	1.04	2.20	2.19	1.60	1.94	1.86
7	1.54	2.06	2.66	0.82	2.49	2.42	1.69	2.27	2.13
8	0.84	1.48	1.82	0.30	2.47	2.46	1.44	2.28	1.97
9a	1.33	1.64	1.85	0.87	2.11	2.07	1.49	1.88	3.82
9b	1.43	1.70	1.97	1.05	2.20	2.21	1.64	1.98	2.04
9c	1.39	1.69	1.93	0.97	2.17	2.16	1.57	—	2.00
10	1.46	1.83	2.78	0.75	2.54	2.81	1.67	2.39	4.66
11	0.53	0.73	1.10	0.11	1.48	1.35	0.57	—	0.96

[a]1 Hartree/angstrom^2 = 2.5417655 D.
[b]A = STO-3G, B = MINI-1, C = 3-21G/LWD.

TABLE 24. Experimental dipole moments (D) and percentage errors of the calculated dipole moments

				% Error[a]		
Molecule	X	Exp.	Ref.	A	B	C
2	O	1.85	88	− 7.6	+ 10	+ 29
	S	0.97	88	+ 6.2	+ 106	+ 89
	Se	0.62	89	+ 137	+ 168	+ 92
5a	O	1.70	88	− 11.2	+ 6.5	+ 25
	S	1.52	88	− 37	+ 41	+ 39
7	O	2.33	88	− 34	− 12	+ 14
8	O	1.414	90	− 41	+ 4.7	+ 29
9a	O	1.30	91	+ 2.3	+ 26	+ 42
11	O	0.67	94a	− 21	+ 9.0	+ 64
	S	0.55	94a	− 80	+ 169	+ 145
	Se	0.398	94a	+ 43	—	+ 141
MAD[b]	O	—	—	0.31	0.17	0.45
	S	—	—	0.35	0.86	0.75
	Se	—	—	0.51	1.04	0.57

[a]A = STO-3G, B = MINI-1, C = 3-21G/LWD.
[b]MAD = Mean absolute deviation.

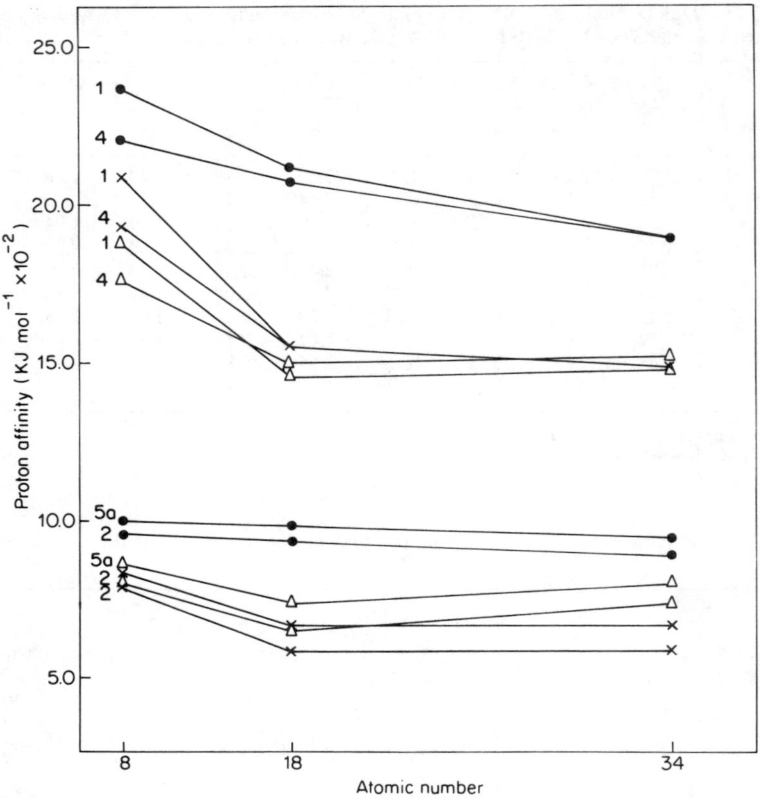

FIGURE 5. Proton affinities (kJ mol^{-1} $\times$ 10^{-2}) as a function of atomic number; ● : STO-3G values, $\times$: MINI-1 values and △: 3-21G/LWD values

Se, respectively. Where the STO-3G basis set has the largest deviation for the S compounds, the MINI-1 basis set has the largest deviation for the O compounds.

The trend on going from O to Se varies slightly with basis set. However, the overall trend is that the *PA*s decrease on going from O to Se compounds with $PA(O) < PA(S) < PA(Se)$ and the *PA*s of S and Se are similar. The 3-21G actually consistently predicts an increase in *PA* on going from S to Se.

2. Energies of hydrogenation

Energies of hydrogenation for isodesmic reactions of the type in which the compounds are hydrogenated to CH_4 and H_2X have been calculated (Table 16). The energies of hydrogenations have also been plotted against the atomic number in Figure 6. Mean absolute deviations from the 3-21G energies of hydrogenation are given at the bottom of Table 16 for the STO-3G and MINI-1 basis sets. The MINI-1 results are in better agreement with the 3-21G results than are the STO-3G results. Both the STO-3G and MINI-1 basis sets do progressively better overall in going from O compounds to Se compounds. Generally the STO-3G basis set is known to do poorly at predicting energies

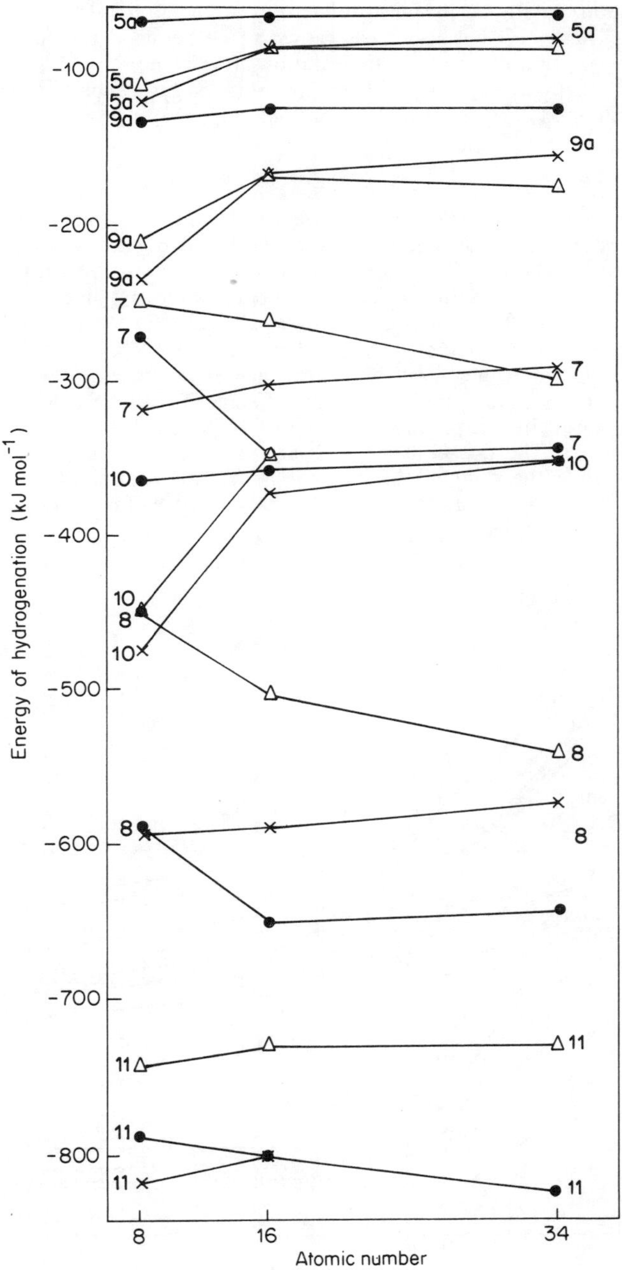

FIGURE 6. Energies of hydrogenation (kJ mol^{-1}) as a function of atomic number; ● : STO-3G values, × : MINI-1 values and △: 3-21G/LWD values

of hydrogenation, giving a mean absolute deviation of 93 kJ mol^{-1} for first-row compounds[48]. The 3-21G basis sets are known to be better at predicting energies of hydrogenation, giving a mean absolute deviation of 41 kJ mol^{-1} for the same set of 18 comparisons[48]. Therefore the results of Table 16 should in general be in similar agreement with the experimental energies of hydrogenation.

3. *Ionization potentials*

First ionization potentials (*IP*s) have been obtained using Koopman's theorem for most of the compounds and are tabulated in Table 17. These can be compared directly with the available experimental values[2,86,87]. Ionization potentials are available for the series of O, S, Se and Te dihydrides (H_2X), for which experimental values are given in Table 1. These are also included in Table 18 along with other available experimental *IP*s and with percentage errors of the *IP* predicted by each basis set. The 3-21G and MINI-1 basis sets are both in excellent agreement with the experimental values, with the 3-21G basis set giving slightly better results overall. The larger difference between STO-3G and the other two basis sets is not that surprising. Both MINI-1 and the 3-21G basis set have been carefully optimized. The 3-21G basis set was optimized for good valence-shell description[48,62], while the minimal MINI-1 basis set was optimized for good orbital energies[47,61,69], thus explaining the excellent values predicted for the *IP*s. In general the

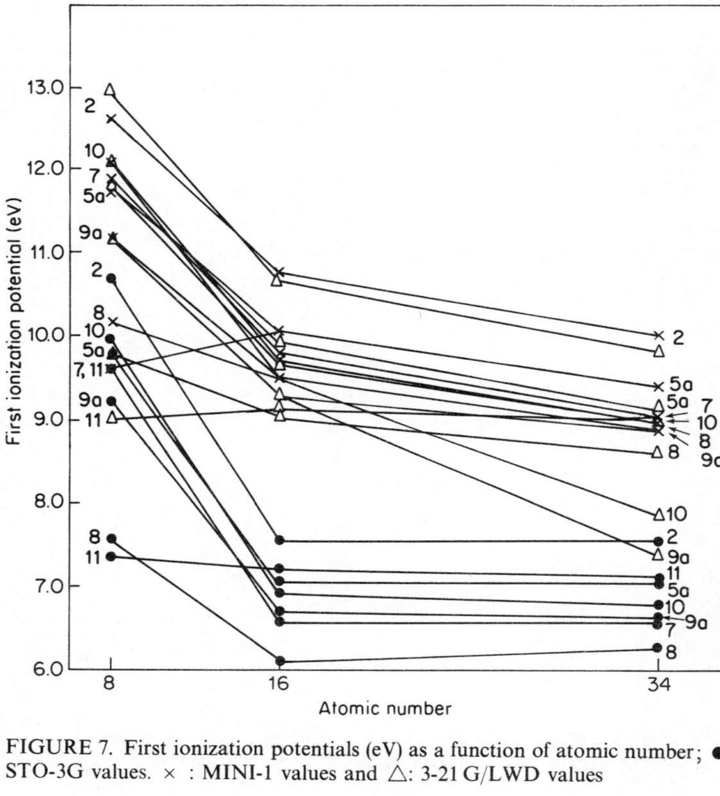

FIGURE 7. First ionization potentials (eV) as a function of atomic number; ● : STO-3G values. × : MINI-1 values and △: 3-21 G/LWD values

calculated values using the 3-21G and MINI-1 basis sets are larger than the experimental values by about 4%, whereas the STO-3G results are lower by about 17%. A modified STO-NG basis set[92] gave a better *IP* for H_2Se, 9.12 eV compared to 7.59 eV with the current standard STO-3G basis set. Normally good-quality orbital energies predict *IP*s which are too high compared to the true values. A systematic study using the DZ basis set gives values which are 8% too high[93].

The *IP*s for the lowest energy conformations of the neutral molecules are plotted as a function of atomic number in Figure 7. The trends of the *IP*s from O to Se are generally consistent for all three basis sets. However, three problems show up in Figure 7. The change in *IP* in going from O to S for both the STO-3G and MINI-1 and from S to Se for the 3-21G basis set do not follow the trends.

4. Relative stabilities of conformers

For structures **3, 5, 6** and **9**, a number of conformations were calculated. The relative stabilities of these conformations are given in Table 19.

The first energy difference in Table 19 (**3b**–**3a**) represents the inversion barriers for the protonated dihydrides. These inversion barriers are predicted to generally increase in going from O to Se, indicating a preference for pyramidality in going from O to Se. Only the STO-3G basis set predicts a non-planar C_{3v} structure for H_3O^+, where the other two basis sets predict only the planar D_{3h} structures.

The stability of conformer **5b** relative to **5a** increases in going from O to Se at all three basis set levels. This is expected since the H–H repulsion in **5b** should decrease due to the increase in C—X distance in going from O to Se.

The relative stabilities of **6a** and **6b** would also be expected to decrease in going from O to Se due to the decreased repulsion between the hydrogen atoms on the C and X atoms. However, the predicted trends are not consistent with this expectation. In the case of **6c**–**6a**, this is again a measure of the increased preference for pyramidality in going from O to Se, whereas the stability of **6c** increases with respect to **6a** in going from O to Se.

For structure **9**, the results indicated a general increase in stability of **9b** and **9c** relative to **9a**. However, the 3-21G basis set predicts **9b** and **9c** to be much more stable than **9a** for the Se compound, indicating that structure **9a** for Se is probably not a minimum (perhaps a saddle-point).

In general the results indicate the large changes in preferred conformation for the O, S and Se compounds. The results are also very sensitive to the basis set.

B. Non-energy-related Properties

1. Charge distribution

Net charges calculated using Mulliken Population Analysis are tabulated in Tables 20, 21 and 22 for the STO-3G, MINI-1 and 3-21G/LWD basis sets, respectively. The dipole moments are tabulated in Table 23. Net charges at O, S and Se are plotted as a function of atomic number in Figure 8 and dipole moment are plotted as a function of atomic number in Figure 9.

As can be seen from Figure 8, the net charge at S and Se is in general more positive than at O and the net charge at Se is predicted to be in some cases more positive and in other cases more negative (in some cases substantially so) than at S (e.g. + 0.09 for S and − 0.24 for Se in structure **10**).

For the dipole moments it is evident that the trends predicted by STO-3G and the other two basis sets are different. The major difference appears to be for the sulphur compounds. From some of the available experimental values the trend, as is expected, is that the

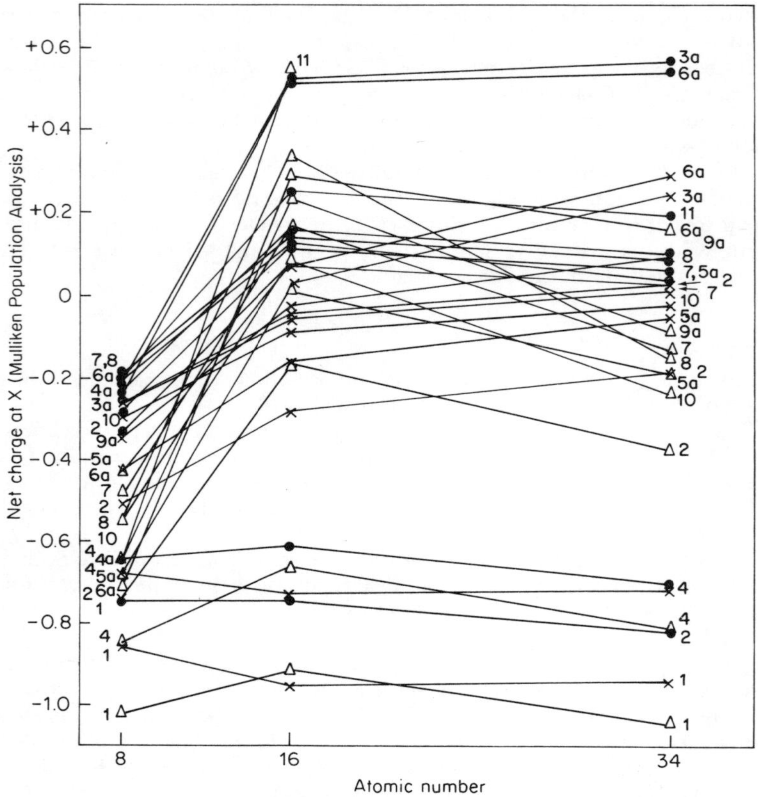

FIGURE 8. Net charges at X(X = O, S and Se) calculated from Mulliken Population Analysis as a function of atomic number; ● : STO-3G values, × : MINI-1 values and △: 3-21G/LWD values

magnitude of the dipole moment decreases from O to Se, e.g., H_2O ($\mu = 1.85$), H_2S ($\mu = 0.97$) and H_2Se ($\mu = 0.62$) and similarly for CH_3XH and C_4H_4X (Table 24). The calculated trends as seen from Figure 9 do not follow the expected or experimentally observed trends. The problems appear to be mainly in going from O to S for the MINI-1 and 3-21G basis sets and from S to Se for the STO-3G basis set. This may be due to the lack of d-polarization functions on S. Table 24 contains the percentage errors for the three basis sets compared to experimental values. Normally, predicted dipole moments are too large except for STO-3G which frequently underestimates the magnitude[48,68]. This is consistent with the calculated errors in Table 24. The MINI-1 minimal basis set, unlike the STO-3G basis set, gives values which are generally too large and are more similar to the 3-21G basis set. Mean absolute deviations are also included in Table 24 for the O, S and Se compounds. These indicate that the STO-3G results are better than the MINI-1 and 3-21G results overall and that MINI-1 does poorly for the S and Se compounds.

2. *Geometries*

The gradient optimized geometries are given in Tables 12, 13 and 14 for the STO-3G, MINI-1 and 3-21G/LWD basis sets, respectively. Some of the available experimental

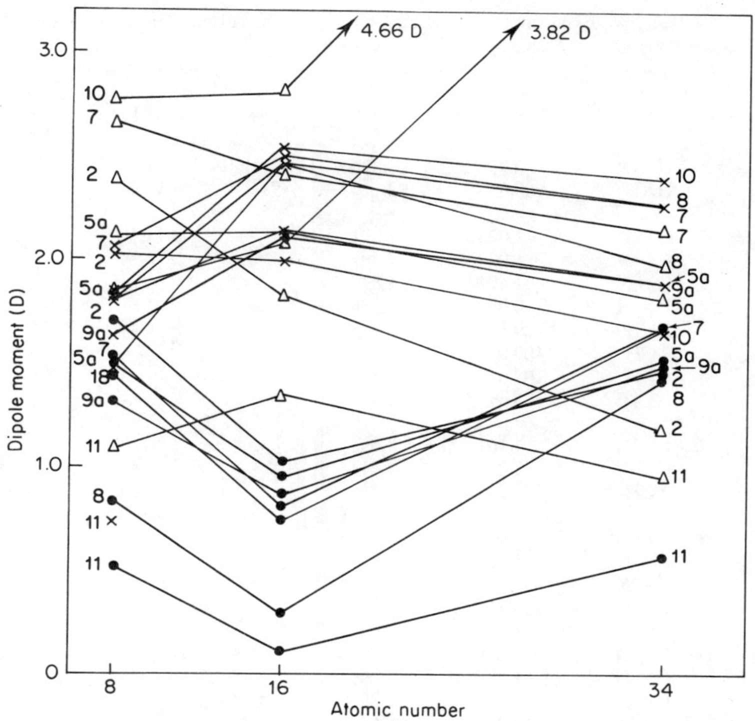

FIGURE 9. Dipole moments (D) as a function of atomic number; ● : STO-3G values, × : MINI-1 values and △: 3-21G/LWD values

geometries[94] are given in Table 25. The deviations from experiment of some of these geometrical parameters and mean absolute deviations for bond lengths and bond angles are given in Table 26. Overall, there is no clear preference. MINI-1 does poorly on bond lengths for the S compounds but very well on the bond angles of compounds containing O, The largest error being for the C—S bonds, which are predicted to be too long by about 1 Å. Previous studies[48] give a mean absolute deviation from experiment for AB bond lengths of 0.028 Å and 0.01 Å (45 comparisons) and mean absolute deviations from experiment for AH bond lengths of 0.035 Å and 0.016 Å (6 comparisons) for the STO-3G and 3-21G basis sets, respectively.

V. CONCLUSIONS

Ab initio MO computations on Se compounds are well within the realms of possibility, but similar computations on Te compounds are more difficult to come by at this time.

The comparison of the available results tend to suggest that there is a general trend on going through a series of analogous compounds containing O, S, Se and Te. The trends indicate a gradual quantitative change rather than a quantum jump that could manifest itself in qualitative differences as one compares S with Se or, in the few cases available, Se with Te. The fundamental cause for the observed trends is most likely related to the change in size of these elements. Since the trends in some cases are rather subtle, this increases the difficulty of predicting these trends reliably at these low-level calculations.

TABLE 25. Experimental geometries[a] of organic Se and Te compounds and their O and S analogues[94]

Molecule/ parameter		X = O	S	Se	Te
2	H—X	0.959	1.336	1.460	1.653
	HXH	103.9	92.1	91	90.2
5a	C—H	1.093	1.091	—	—
	C—H′	1.093	1.091	—	—
	C—X	1.421	1.819	—	—
	H—X	0.963	1.336	—	—
	XCH	129.8	—	—	—
	XCH′	107.0	—	—	—
	CXH	108.0	96.5	—	—
	HCH	108.5	109.8	—	—
7	C—H	1.101	1.093	—	—
	C—X	1.203	1.611	—	—
	XCH	127.0	126.2	—	—
8	C—H	1.071	—	—	—
	C—C	1.329	—	—	—
	C—X	1.150	—	—	—
	CCH	118.35	—	—	—
9a	C—H	1.096	1.091	1.096	—
	C—H′	1.096	1.091	1.088	—
	C—X	1.410	1.802	1.945	—
	CXC	111.7	98.9	96.3	—
	XCH	—	—	110.3	—
	XCH′	—	—	105.0	—
	HCH	109.5	109.6	109.9	—
10	C—H	1.082	—	—	—
	C—C	1.472	—	—	—
	C—X	1.436	—	—	—
	CCH	—	—	—	—
	CCX	—	—	—	—
	CXC	61.4	—	—	—
	HCH	116.7	—	—	—
11	C—H	1.077	1.081	1.081	—
	C—H′	1.075	1.078	1.078	—
	C—C	1.431	1.423	1.423	—
	C=C	1.361	1.370	1.370	—
	C—X	1.362	1.714	1.863	—
	CCH	127.94	124.27	122.54	—
	XCH′	115.92	119.85	124.39	—
	C—C=C	106.06	112.45	114.92	—
	C=C—X	110.68	112.47	111.22	—
	CXC	106.55	92.17	87.72	—

[a]Bond lengths in Å, bond angles in deg.

TABLE 26. Deviation from experiment of some calculated geometrical parameters[a]

Molecule/	parameter	X	A	B	C
Bond lengths					
2	H—X	O	+0.030	+0.046	+0.008
		S	+0.018	+0.076	+0.015
		Se	+0.014	+0.059	−0.004
		Te	−0.029	—	—
5a	H—X	O	+0.028	+0.043	+0.003
		S	−0.005	+0.075	+0.016
	C—X	O	+0.012	+0.056	+0.019
		S	−0.022	+0.103	+0.075
7	C—X	O	+0.014	+0.065	+0.004
		S	−0.037	+0.086	+0.027
8	C—X	O	+0.033	+0.078	+0.012
9a	C—X	O	+0.023	+0.065	+0.023
		S	−0.006	+0.115	+0.083
		Se	−0.017	+0.083	+0.103
10	C—X	O	−0.004	+0.066	+0.034
11	C—X	O	+0.014	+0.053	+0.018
		S	+0.018	+0.136	+0.083
		Se	+0.001	—	+0.013
	MAD[b]	O	0.02	0.06	0.02
		S	0.02	0.10	0.05
		Se	0.01	0.07	0.04
Bond angles					
2	HXH	O	−3.9	−0.5	+3.8
		S	+0.4	+2.5	+3.7
		Se	+1.5	+2.1	+1.6
		Te	+2.2	—	—
5a	CXH	O	−4.1	−1.6	+2.4
		S	−1.1	+0.7	+1.4
9a	CXC	O	−2.2	−1.0	+2.3
		S	−0.7	−0.1	+0.2
		Se	+0.5	+0.5	+3.6
10	CXC	O	+0.9	+0.2	−1.2
11	CXC	O	−1.1	−0.5	+0.4
		S	−1.8	−3.2	−3.0
		Se	−1.3	—	−0.9
	MAD	O	2.4	0.8	2.0
		S	1.0	1.6	2.1
		Se	1.1	1.3	2.0

[a]A = STO-3G, B = MINI-1, C = 3-21G/LWD.
[b]MAD = Mean absolute deviation.

Of the three basis sets studied, the split valence 3-21G/LWD basis set is the most reasonable basis set for studying Se compounds. The minimal MINI-1 basis set, however, gives results which are comparable to the 3-21G/LWD basis set. For this reason, the other MINI-N ($N = 2$–4) or MIDI-N ($N = 1$–4) basis sets[69] are expected to give results which are better than the basis sets compared in the present study. However, comparable studies on Te compounds must wait for the development of such basis sets. At present the only economical basis set available for Te is the STO-3G.

VI. ACKNOWLEDGEMENTS

The continued financial support of the Natural Sciences and Engineering Research Council of Canada is gratefully acknowledged.

VII. REFERENCES

1. C. E. Moore, *Atomic Energy Levels*, NBS (U.S.) Circular No. 467, U.S. Government Printing Office, Washington D.C., 1949.
2. G. Herzberg, *Molecular Spectra and Molecular Structure. III. Electronic Spectra and Electronic Structure of Polyatomic Molecules*, Van Nostrand, Princeton, New Jersey, 1966.
3. L. S. Cederbaum, G. Hohlneicher and W. von Niessen, *Mol. Phys.*, **26**, 1405 (1973).
4. W. C. Price, *Bull. Amer. Phys. Soc.*, **10**, 9 (1935).
5. A. W. Potts and W. C. Price, *Proc. Roy. Soc.* (*London*), **326A**, 165 (1972).
6. A. W. Potts and W. C. Price, *Proc. Roy. Soc.* (*London*), **326A**, 181 (1972).
7. J. Delwicke, P. Natalis and J. E. Collin, *Intern. J. Mass Spectrom. Ion Phys.*, **5**, 443 (1970).
8. S. Fraga, J. Karwowski and K. M. S. Saxena, *Handbook of Atomic Data*, Elsevier, Amsterdam, 1976.
9. W. C. Mackrodt, *Mol. Phys.*, **18**, 697 (1970).
10. See previous volumes in the series *The Chemistry of Functional Groups*: I. G. Csizmadia, in *The Chemistry of the Thiol Group* (Ed. S. Patai), 1974, pp. 44–75; I. G. Csizmadia, M. R. Peterson, C. Kozmutza and M. A. Robb, in *Supplement B; The Chemistry of Acid Derivatives* (Ed. S. Patai), 1979, pp. 4–20.
11. J. C. Slater, *Phys. Rev.*, **36**, 51 (1930).
12. E. Clementi and D. L. Raimondi, *J. Chem. Phys.*, **38**, 2686 (1963).
13. G. Burns, *J. Chem. Phys.*, **41**, 1521 (1964).
14. R. A. Poirier, R. E. Kari and I. G. Csizmadia, *Handbook of Gaussian Basis Sets*, in preparation.
15. D. B. Cook, *Theoret. Chim. Acta* (*Berl.*), **46**, 331 (1977).
16. D. A. Dixon and D. S. Marynick, *J. Amer. Chem. Soc.*, **99**, 6101 (1977).
17. E. Clementi, *J. Chem. Phys.*, **40**, 1944 (1964).
18. S. Huzinaga and C. Arnau, *J. Chem. Phys.*, **53**, 451 (1970).
19. C. Roetti and E. Clementi, *J. Chem. Phys.*, **60**, 4725 (1974).
20. S. Huzinaga, *J. Chem. Phys.*, **42**, 1293 (1965).
21. C. D. Ritchie and H. F. King, *J. Chem. Phys.*, **47**, 564 (1967).
22. H. Basch, M. B. Robin and N. A. Kuebler, *J. Chem. Phys.*, **47**, 1201 (1967).
23. C. Salez and A. Veillard, *Theoret. Chim. Acta* (*Berl.*), **11**, 441 (1968).
24. S. Huzinaga and Y. Sakai, *J. Chem. Phys.*, **50**, 3 (1969).
25. R. F. Stewart, *J. Chem. Phys.*, **50**, 2485 (1969).
26. D. R. Whitman and C. J. Hornback, *J. Chem. Phys.*, **51**, 398 (1969).
27. W. J. Hehre, R. F. Stewart and J. A. Pople, *J. Chem. Phys.*, **51**, 2657 (1969).
28. S. Huzinaga and C. Arnau, *J. Chem. Phys.*, **52**, 2224 (1970).
29. R. Ditchfield, W. J. Hehre and J. A. Pople, *J. Chem. Phys.*, **52**, 5001 (1970).
30. T. H. Dunning, Jr., *J. Chem. Phys.*, **53**, 2823 (1970).
31. B. Roos and B. Siegbahn, *Theoret. Chim. Acta* (*Berl.*), **17**, 209 (1970); **50**, 365 (1979).
32. R. Ditchfield, W. J. Hehre and J. A. Pople, *J. Chem. Phys.*, **54**, 724 (1971).
33. T. H. Dunning, Jr., *J. Chem. Phys.*, **55**, 716 (1971).
34. F. B. Van Duijneveldt, *IBM Res. J.*, 945 (§16437) (1971).

35. W. J. Hehre, R. Ditchfield and J. A. Pople, *J. Chem. Phys.*, **56**, 2257 (1972).
36. A. P. Mortola, H. Basch and J. W. Moskowitz, *Intern. J. Quantum Chem.*, **7**, 725 (1973).
37. P. G. Mezey, R. E. Kari and I. G. Csizmadia, *J. Chem. Phys.*, **66**, 964 (1977).
38. P. G. Mezey, I. G. Csizmadia and R. E. Kari, *J. Chem. Phys.*, **67**, 2927 (1977).
39. P. G. Mezey and I. G. Csizmadia, *Can. J. Chem.*, **55**, 1181 (1977).
40. A. D. McLean, G. H. Loew and D. S. Berkowitz, *J. Mol. Spectry.*, **64**, 184 (1977).
41. L. Gianolio, R. Pavani and E. Clementi, *Gazz. Chim. Ital.*, **108**, 181 (1978).
42. E. L. Mehler and C. H. Paul, *Chem. Phys. Lett.*, **63**, 145 (1979).
43. P. Pulay, G. Fogarasi, F. Pang and J. E. Boggs, *J. Amer. Chem. Soc.*, **101**, 2550 (1979).
44. R. Krishnan, J. S. Binkley, R. Seeger and J. A. Pople, *J. Chem. Phys.*, **72**, 650 (1980).
45. A. N. Tavouktsoglou and S. Huzinaga, *J. Chem. Phys.*, **72**, 1385 (1980).
46. K. Ohta, H. K. Nakatsuji and T. Yonezawa, *J. Chem. Phys.*, **73**, 1770 (1980).
47. H. Tatewaki and S. Huzinaga, *J. Comput. Chem.*, **1**, 205 (1980).
48. J. S. Binkley, J. A. Pople and W. J. Hehre, *J. Amer. Chem. Soc.*, **102**, 939 (1980).
49. GAUSSIAN 80: J. S. Binkley, R. A. Whiteside, R. Krishnan, R. Seeger, D. J. Defrees, H. B. Schlegel, S. Topiol, L. R. Kahn and J. A. Pople, *Q.C.P.E.*, §406, Department of Chemistry, Indiana University, Bloomington, Indiana 47405, USA.
50. R. A. Poirier, R. Daudel and I. G. Csizmadia, *Intern. J. Quantum Chem.*, **19**, 693 (1981).
51. G. L. Lie and E. Clementi, *J. Chem. Phys.*, **60**, 1275 (1974).
52. A. Veillard, *Theoret. Chim. Acta* (*Berl.*), **12**, 405 (1968).
53. S. Huzinaga and C. Arnau, *J. Chem. Phys.*, **53**, 348 (1970).
54. T. H. Dunning, Jr., *Chem. Phys. Letters*, **7**, 423 (1970).
55. T. A. Claxton and N. A. Smith, *Theoret. Chim. Acta* (*Berl.*), **22**, 378 (1971).
56. W. J. Hehre and W. A. Lathan, *J. Chem. Phys.*, **56**, 5255 (1972).
57. W. Von Niessen, L. S. Cederbaum, W. Domcke and G. H. F. Diercksen, *J. Chem. Phys.*, **66**, 4893 (1977).
58. A. D. McLean and G. S. Chandler, *J. Chem. Phys.*, **72**, 5639 (1980).
59. L. Gianolio and E. Clementi, *Gazz. Chim. Ital.*, **110**, 179 (1980).
60. P. N. Skancke, G. Fogarasi and J. E. Boggs, *J. Mol. Struct.*, **62**, 259 (1980).
61. Y. Sakai, H. Tatewaki and S. Huzinaga, *J. Comput. Chem.*, **2**, 100 (1981).
62. M. S. Gordon, J. S. Binkley, J. A. Pople, W. J. Pietro and W. J. Hehre, *J. Amer. Chem. Soc.*, **104**, 2797 (1982).
63. R. A. Poirier, R. Daudel, P. G. Mezey and I. G. Csizmadia, *Intern. J. Quantum Chem.*, **21**, 799 (1982).
64. M. M. Francl, W. J. Pietro, W. J. Hehre, J. S. Binkley, M. S. Gordon, D. J. Defrees and J. A. Pople, *J. Chem. Phys.*, **77**, 3654 (1982).
65. T. H. Dunning, Jr., *J. Chem. Phys.*, **66**, 1382 (1977).
66. S. Huzinaga, *J. Chem. Phys.*, **66**, 4245 (1977).
67. J.-M. Lehn, G. Wipff and J. Demuynck, *Helv. Chim. Acta,* **60**, 1239 (1977).
68. W. J. Pietro, B. A. Levi, W. J. Hehre and R. F. Stewart, *Inorg. Chem.*, **19**, 2225 (1980).
69. Y. Sakai, H. Tatewaki and S. Huzinaga, *J. Comput. Chem.*, **3**, 6 (1982).
70. S. Huzinaga, *J. Chem. Phys.*, **71**, 1980 (1979).
71. W. J. Pietro, E. S. Blurock, R. F. Hout, Jr., W. J. Hehre, D. J. Defrees and R. F. Stewart, *Inorg. Chem.*, **20**, 3650 (1981).
72. A. Stromberg, O. Gropen and U. Wahlgren, *J. Comput. Chem.*, **4**, 181 (1983).
73. R. A. Poirier, R. Kari, R. Daudel and I. G. Csizmadia, in preparation.
74. R. A. Poirier, R. Daudel, P. G. Mezey and I. G. Csizmadia, *Intern. J. Quantum Chem.*, **21**, 799 (1982).
75. R. A. Poirier, R. Daudel, R. E. Kari and I. G. Csizmadia, *Intern. J. Quantum Chem.*, in press.
76. MONSTERGAUSS: M. R. Peterson and R. A. Poirier, University of Toronto, Ontario, M5S 1A1, Canada.
77. H. B. Schlegel, program FORCE, *Ph.D. Thesis,* Queen's University, Kingston, Ontario, Canada, 1975.
78. M. J. D. Powell, Subroutine VA13AD, Harwell Fortran Subroutine Library, Atomic Energy Research Establishment, Harwell, UK. The algorithm is described in R. Fletcher, *Computer J.*, **13**, 317 (1970).
79. W. C. Davidon, *Mathematical Programming*, **25**, 1 (1975). The routine is described in technical

Memos 303 and 306, W. C. Davidon and L. Nazareth, Applied Mathematics Division, Argonne National Laboratories, Argonne, Illinois 60439, USA.

80. M. J. D. Powell, Subroutine VA05AD, Aere Subroutine Library Harwell Didcot, Berkshire, UK. The algorithm is very similar to one described by M. J. D. Powell in *Numerical Methods for Non-linear Algebraic Equations* (Ed. P. Robinowitz), Gordon and Breach, London, 1970, Chaps. 6 and 7.
81. J. C. Barthelat, P. Durand and A. Serafini, *Mol. Phys.*, **33**, 159 (1977).
82. G. Simons and E. R. Talaty, *J. Chem. Phys.*, **66**, 2457 (1977).
83. R. A. Poirier, in preparation.
84. T. Koopmans, *Phys. Fenn.*, **1**, 104 (1934).
85. R. S. Mulliken, *J. Chem. Phys.*, **23**, 1833 (1955).
86. G. Distetano, S. Pignataro, G. Innorta, F. Fringuelli, G. Marino and A. Taticchi, *Chem. Phys. Letters*, **22**, 132 (1973).
87. W. Schaefer, A. Schweig, S. Gronowitz, A. Taticchi and F. Fringuelli, *Chem. Commun.*, 541 (1973).
88. R. D. Nelson, D. R. Lide and A. A. Maryott, *Selected Values of Electric Dipole Moments for Molecules in the Gas Phase.*, NBS Ref. Data Ser., No. 10, National Bureau of Standards, Washington, D. C., 1967.
89. A. L. McLellan, *Table of Experimental Dipole Moments*, Freeman, San Franscisco, 1963.
90. H. R. Johnson and M. W. P. Strandberg, *J. Chem. Phys.*, **20**, 687 (1952).
91. L. G. Groves and S. Sugden, *J. Chem. Soc.*, 1779 (1937).
92. R. H. Findlay, *J. Chem. Soc., Faraday Trans. 2*, **72**, 388 (1976).
93. L. C. Snyder and H. Basch, *Molecular Wavefunctions and Their Properties: Tabulated from SCF Calculations in a Gaussian Basis Set*, Wiley–Interscience, New York, 1972.
94. Experimental geometries are from Ref. 2 or (a) R. D. Brown, F. R. Burden and P. D. Godfrey, *J. Mol. Spectry.*, **25**, 415 (1968); (b) J. H. Callomon, E. Hirota, K. Kuchitsu, W. J. Lafferty, A. G. Maki and C. S. Pote, *Structure Data on Free Polyatomic Molecules*, Vol. 7 of *Landolt–Bornstein: Numerical Data and Function Relationships in Science and Technology*, New Series (Ed. K. H. Hellwege), Springer-Verlag, West Berlin, 1976; (c) G. K. Pandey and H. Dreizler, *Z. Naturforsch.*, **32a**, 482 (1977).

The Chemistry of Organic
Selenium and Tellurium
Compounds Volume 1
Edited by S. Patai and Z. Rappoport
© 1986 John Wiley & Sons Ltd.

CHAPTER **3**

Structural chemistry of organic compounds containing selenium or tellurium

ISTVÁN HARGITTAI† and BÉLA ROZSONDAI

Hungarian Academy of Sciences, Research Laboratory for Inorganic Chemistry, Department of Structural Studies, P.O. Box 117, Budapest, H-1431, Hungary

I. INTRODUCTION 63
II. ONE-COORDINATED SELENIUM AND TELLURIUM 64
III. TWO-COORDINATED SELENIUM AND TELLURIUM 70
A. Acyclic Selenides and Tellurides 70
B. Acyclic Diselenides and Ditellurides 82
C. Heterocycles 86
D. Organometallic Complexes 103
IV. THREE-COORDINATED SELENIUM AND TELLURIUM 106
V. FOUR-COORDINATED SELENIUM AND TELLURIUM 125
VI. FIVE- AND HIGHER-COORDINATED SELENIUM AND TELLURIUM 143
VII. ACKNOWLEDGEMENTS 147
VIII. REFERENCES 148

I. INTRODUCTION

The structural chemistry of selenium and tellurium is somewhat in the shadow of the structural chemistry of sulphur. With the development of the chemistry of these elements, progress in their structural chemistry is inevitable. This progress is expected primarily in experimental studies where their position in the periodic system is not a disadvantage, whereas for theoretical calculations it may be.

The present review is concerned with the metrical aspects of the structural chemistry of Se and Te. Although a comprehensive literature search preceded writing, completeness is not aimed at, except perhaps for gas-phase studies which are not numerous anyway but usually deal with the most fundamental molecules.

†Visiting Professor of Physics (1983/84) and Chemistry (1984/85) at The University of Connecticut, Storrs, CT 06268, U.S.A.

The criterion for a structure to be included in this review is that at least one of the Se or Te bonds is to carbon. As earlier periods have been at least partially covered in other works (e.g. References 1 and 2) our attention has focused on the past decade.

A convenient guiding principle in systematizing the structure is the coordination number of Se or Te. Further subdivision is applied only to the two-coordinated selenides and tellurides.

Although gas-phase and crystal-phase structures are discussed together, the available material does not as yet allow a meaningful discussion of gas/crystal structural differences.

Attempts have been made to systematize the collected structural information and to search for regularities in the structural variations. Substituent effects and characteristic differences as compared with sulphur have been looked for. For possible interpretation of such observations only qualitative considerations have been employed. Electron-pair repulsions and non-bonded interactions are useful in discussing sulphur stereochemistry[3]. Of these two the relative importance of the non-bonded interactions diminishes for Se and Te as compared with S due to the longer bonds of the former two elements.

Very little attention is devoted in our review to the consequences of molecular vibrations and librations. The well-defined physical meaning of the parameters is important especially in discussing subtle structural effects and those structural variations which are on the borderline of the observed differences. Furthermore, in the case of large-amplitude vibrations, the determined structures may differ considerably from what they would be in the equilibrium structure (cf. Reference 4). As neither subtle structural differences nor large-amplitude motion have special importance in the accumulated structural chemical information for Se or Te *for the time being*, a detailed discussion of the physical meaning of the determined parameters has been avoided. The different representations of the molecular geometry and the meaning of the parameters r_e, r_0, r_s, r_a, r_g, r_α, r_z etc. are summarized and references are given in the review[4] cited above.

The experimental errors and uncertainties have been cited as communicated in the original reports; in some cases, though, they have been rounded to one digit. Generally they correspond to the usual requirements of the respective techniques used at the time of the particular studies. Throughout this review, the uncertainties are given in parentheses following the parameter value and they refer to the last digit of the parameter. The most common experimental techniques are denoted as follows: MW, microwave spectroscopy; ED, gas electron diffraction; XD, X-ray diffraction crystallography.

II. ONE-COORDINATED SELENIUM AND TELLURIUM

The only structure belonging here and involving tellurium is S=C=Te. It was determined three decades ago by means of microwave spectroscopy (MW)[5]. The length of the C=Te bond is 1.904 Å. The carbon atom is formally in sp hybridization. Our discussion will first deal with C=Se bonds with sp carbon, and then with C=Se bonds with sp^2 carbon. All the available geometrical data for the C(sp)=Se bond refer to the vapour phase and mostly originate from MW.

Table 1 presents the bond distances in the analogues of CO_2 and CO. The selenium–

TABLE 1. C—X bond distances (Å)[a] with X = O, S or Se

X	OCX	SCX	SeCX	CX
O	1.1598[6]	1.1543(3)[7]	1.1535(1)[8]	1.12834(1)[9]
S	1.5628(4)[7]	1.5526[10]	1.553(2)[11]	1.53496(3)[9]
Se	1.7098(1)[8]	1.695(2)[11]	1.6917(15)[12]	1.67620(7)[9]

[a]All distances are r_e except those for SCSe.

TABLE 2. Bond lengths and bond angles of isocyanic acid, isothiocyanic acid and isoselenocyanic acids, HN=C=X, from MW

X =	O[16]	S[17]	Se[15]
H—N (Å)	0.986	0.989	0.99 (assumed)
∠H—N=C(deg.)	128.0	135.0	143
N=C(Å)	1.209	1.216	1.195
C=X(Å)	1.166	1.560	1.717

carbon bond distance is 1.73(1) Å in crystalline (1-6-η-methylbenzoate)dicarbonylselenocarbonylchromium[13], [Cr(PhCOOMe)$(CO)_2$(CSe)]. The Cr—C(Se) bond is 1.786(11) Å, shorter than the Cr—C(O) bonds, 1.862 Å (average). This is in accord with the observation, based on spectroscopic, structural and theoretical evidence, that the strength of the linkages to CSe, CS and CO groups decreases in this order[13,14].

A new molecular species, isoselenocyanic acid, HN=C=Se, was detected in the microwave spectrum[15] as HBr gas was passed through dry AgNCSe. The molecular parameters are given in Table 2 together with those of isocyanic acid[16] and isothiocyanic acid[17].

Selenoketene, H_2C=C=Se, was obtained by pyrolysis of 1,2,3-selenadiazole[18,19]. Bak and coworkers[19] suggested the reaction path shown in Scheme 1. MW allowed the elucidation of a complete substitution geometry (r_s)[19]. It is presented in Table 3 together with the structural parameters of ketene[20] and thioketene[21].

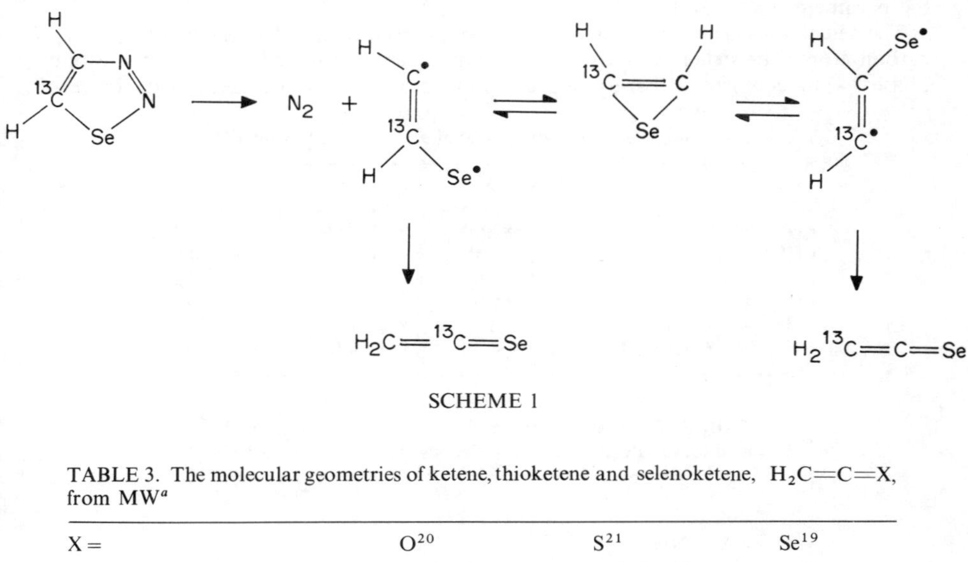

SCHEME 1

TABLE 3. The molecular geometries of ketene, thioketene and selenoketene, H_2C=C=X, from MW[a]

X =	O[20]	S[21]	Se[19]
C=X(Å)	1.1614 (r_s)	1.554 (r_s)	1.698 (r_s)
C=C(Å)	1.3142 (r_s)	1.314 (r_s)	1.311 (r_s)
C—H (Å)	1.0768 (r_0)	1.090 (r_s)	1.090 (r_s)
∠H—C—H (deg.)	122.24 (r_0)	120.3 (r_s)	119.7 (r_s)

[a] r_0 is an effective parameter derived directly from ground-state rotational constants; r_s is the substitution parameter obtained from an appropriate set of isotopically substituted species.

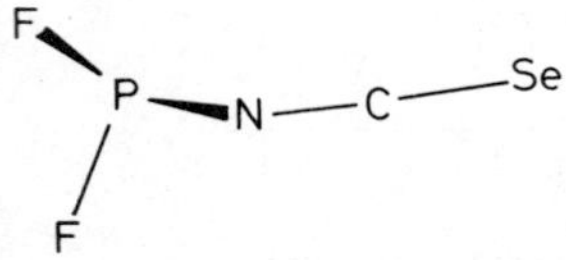

FIGURE 1. The molecular model* of difluoro(isoselenocyanato)phosphine (after Reference 22)

The molecular structure of difluoro(isoselenocyanato)phosphine, F_2PNCSe, has been determined by electron diffraction (ED)[22] (Figure 1). The effects of perpendicular vibrations on the molecular configurations of isothiocyanates and analogues have been discussed in detail[4]. Due to the low-frequency bending modes, the effective bond angle P—N=C as determined from ED (r_a parameter) may be considerably smaller than that which would correspond to the average structure (r_α). The average structure may be obtained from the effective parameters by applying harmonic corrections based on the vibrational spectra and normal coordinate analysis. In the ED investigation of F_2PNCSe the correction procedure was part of the structure refinement and both r_a *and* r_α parameters have been directly obtained from the analysis. The results are shown in Table 4 together with those for F_2PNCS and F_2PNCO. Noteworthy is the gradual increase in the P—N=C bond angles from O to Se.

The consequences of the low-frequency, large-amplitude bending vibrations are particularly striking for the bond angles Si—N=C of isoselenocyanatosilane[24], $H_3SiNCSe$, and isothiocyanatosilane[25], H_3SiNCS. Unfortunately the strong correlation among the parameters prevented the determination of an unambiguous structure for the selenium derivative from the ED data[24a]. Two somewhat differing parameter sets have been obtained and their mean values are presented for orientation in Table 5 together with the parameters of H_3SiNCS.

The selenium–carbon bond in selenocarbonyl difluoride, $F_2C{=}Se$, involves an sp^2 carbon atom. The structure of this molecule has been determined by ED in the vapour phase[26]. The geometrical parameters are presented in Table 6 together with those of

TABLE 4. Bond lengths and bond angles of F_2PNCX from ED

	X = O[23]	S[23]	Se[22]
r_a(P—N) (Å)	1.683(6)	1.686(6)	1.670(12)
r_a(P—F) (Å)	1.563(3)	1.566(3)	1.547(4)
r_a(N=C) (Å)	1.256(6)	1.221(6)	1.220(8)
r_a(C=X) (Å)	1.165(6)	1.553	1.700(10)
∠ P—N=C(r_α)(deg.)	134.8(8)	144.0(7)	149.0(15)
∠ P—N=C(r_a)(deg.)	130.6(8)	140.5(7)	143.9(13)

TABLE 5. Bond lengths and bond angles of H_3SiNCX, isothiocyanatosilane and isoselenocyanatosilane, from ED

X =	S[25]	Se[24]
r_a(Si—N) (Å)	1.704(6)	1.716
r_a(N=C) (Å)	1.197(7)	1.183
r_a(C=X) (Å)	1.563(6)	1.754
∠ Si—N=C(r_α)(deg.)	180	180
∠ Si—N=C(r_a)(deg.)	163.8(26)	159.5

*This figure and many others are simplified projections of three-dimensional structures. Bonds that make an angle with the plane of the drawing are shown as wedges, in order to illustrate them in an exaggerated perspective. The broader end of the wedge is considered nearer to the observer.

TABLE 6. The molecular geometries of $F_2C{=}X$

X =	O[a]	S[b]	Se[c]
C=X(Å)	1.170(3)	1.589(2)	1.743(3)
C—F (Å)	1.317(2)	1.316(2)	1.314(2)
∠F—C—F (deg.)	107.6(2)	107.1(2)	107.5(4)

[a]MW[27], r_z parameters.
[b]ED + MW[26], r_{av} parameters.
[c]ED[26], r_α^0 parameters.

thiocarbonyl difluoride[26] and carbonyl difluoride[27]. A characteristic feature of these structures is that the F—C—F bond angle is smaller than 120°, and accordingly, the F—C=X angles are invariably larger than 120°. This is in complete agreement with the valence-shell electron-pair repulsion (VSEPR) model[28] according to which multiple bonds exercise greater repulsions towards the neighbouring electron pairs than do single bonds.

The molecular structure of selenoacetaldehyde, $CH_3CH{=}Se$, has been determined by microwave spectroscopy[29]. The geometry of the CH_3CH fragment was assumed to be the same as in thioacetaldehyde[30], $CH_3CH{=}S$. The parameters for the thio and seleno derivatives are as follows:

$CH_3CH{=}S$		$CH_3CH{=}Se$	
C=S	1.610(9) Å	C=Se	1.758(10) Å
∠C—C=S	125.3(11)°	∠C—C=Se	125.7(3)°

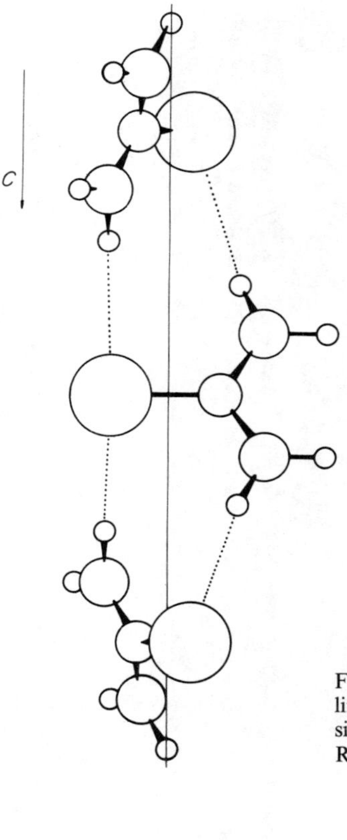

FIGURE 2. A chain of selenourea molecules in the crystal, linked by N—H···Se hydrogen bonds. Reproduced by permission of the Akademische Verlagsgesellschaft, Wiesbaden from Reference 33

These bond angles, as well as the comparison of $F_2C{=}S$ and $F_2C{=}Se$ geometries, clearly indicate that, in accord with the above assumption, the S/Se substitution has no appreciable effect on the structure of the rest of the molecule.

On the other hand, Anthoni and coworkers[31], based on CNDO/2 calculations, have recently suggested that 'selenation' may involve considerable changes in the electronic structure. The term selenation[32] is used to compare the spectra of compounds containing a thioamide group with the corresponding selenium analogues. It has been stated that the method works 'almost like an isotopic substitution'[31]. Anthoni and coworkers[31] have examined in detail the foundation and limitations of this method.

The crystal and molecular structure of selenourea, $(H_2N)_2C{=}Se$ (**1**), and several of its derivatives has been determined by X-ray diffraction (XD). The crystal structure of selenourea[33] contains N—H···Se hydrogen bonds similar to the N—H···O and N—H···S hydrogen bonds of urea and thiourea, respectively. There are hydrogen bonds forming chains of molecules (Figure 2), which are then also connected to each other by N—H···Se hydrogen bonds. Rutherford and Calvo[33] found the crystal structure of selenourea to be very similar to that of urea and thiourea when they form inclusion compounds with certain hydrocarbons, rather than to the crystal structure of urea and thiourea themselves. In the inclusion compounds the urea or thiourea molecules form a honeycomb structure and the hydrocarbon molecules form the honey. Analogues of the inclusion compounds have not been observed for selenourea in which both the honeycomb frame and the honey in it consist of selenourea molecules (Figure 3). The unit cell contains 27 molecules, of which 18 molecules are involved in building the frame and the remaining 9 are distributed in the channels, three to each. The 27 molecules of the unit

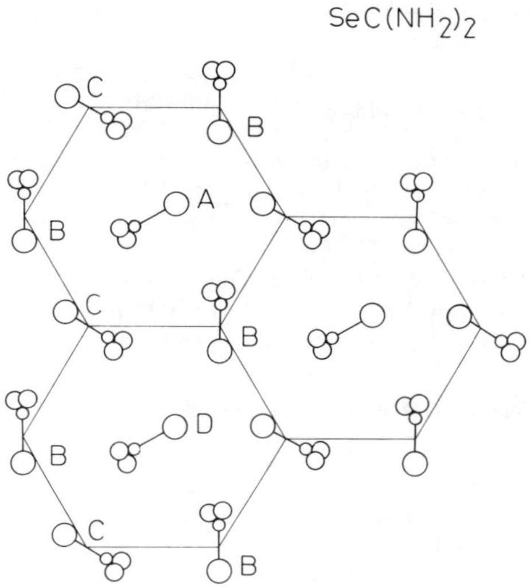

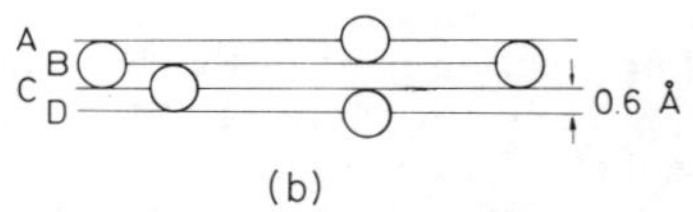

FIGURE 3.(a) Part of a honeycomb to illustrate the selenourea crystal structure. Photograph reproduced by courtesy of Professor Pál Zoltán Örösi. (b) The crystal structure of selenourea projected onto the *x*,*y* plane. Displacements of molecules along the *z* axis are indicated below. Reproduced by permission of the Akademische Verlagsgesellschaft, Wiesbaden from Reference 33

cell indeed form nine spiral chains one of which is represented by Figure 2. The dimensions of the selenourea molecule are reported from a low-temperature study (– 100 °C). The weighted averages of the bond lengths referring to nine molecules per asymmetric unit are *r*(C=Se) = 1.86(3) Å and *r*(C—N) = 1.37(2) Å. The C=Se bond is much longer than in selenoacetaldehyde[29]. The average Se=C—N bond angle is barely larger than 120°, viz. 120.5°. In fact it was reported to be 120°. One would expect the Se=C—N bond angle to be larger than the N—C—N bond angle. However, the Se=C—N bond angles in other derivatives have also been seen to only slightly exceed 120°. In tris(acetylacetonato)cobalt(III)–selenourea(1/2)[34], $Co(acac)_3 \cdot 2\,SeC(NH_2)_2$, for example, the Se=N—C angle is 120.8(4)°. The other molecular parameters for the selenourea moiety are: C=Se = 1.853(5) Å, C—N = 1.320(12) Å and ∠N—C—N = 118.3(4)°. The

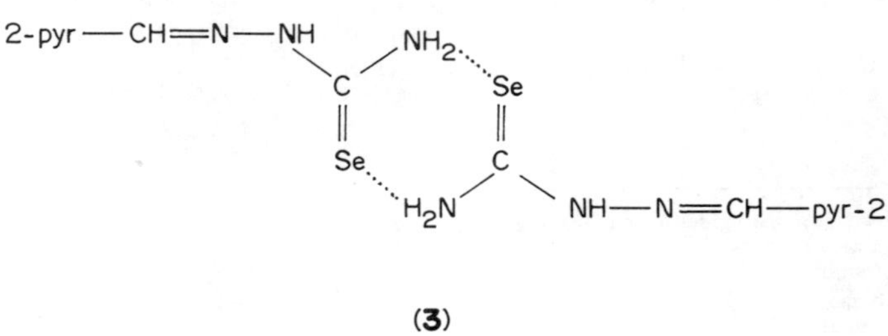

(3)

selenourea moieties are brought into chains by N—H···Se intermolecular hydrogen bonds of 3.49 Å length. The selenourea moiety of the selenourea solvate of tris(selenourea)sulphate–selenourea–water(1/1/2)[35], $[SeC(NH_2)_2]_3SO_4{\cdot}SeC(NH_2)_2{\cdot}2H_2O$ (**2**) has the following geometrical parameters: C=Se = 1.867(4) Å, C—N = 1.316(5) Å, ∠Se=C—N = 120.6(3)° and ∠N—C—N = 118.8(4)°. The bonding arrangement in the N_2CSe part of 2-formylpyridine selenosemicarbazone[36] (**3**) is somewhat asymmetric, with two different Se=C—N bond angles, 122.9(24) and 119.8(23)°, whose difference is not however significant. The C=Se and C—N bonds are 1.83(2) and 1.37(2) Å long, respectively, and the N—C—N angle is 117.2(28)°. There are weak N—H···Se intermolecular hydrogen bonds forming dimer-like molecules, which in turn, are linked into a three-dimensional network by strong intermolecular N—H···N hydrogen bonds. Other selenosemicarbazones have also been investigated.

III. TWO-COORDINATED SELENIUM AND TELLURIUM

A. Acyclic Selenides and Tellurides

The carbon–selenium bond lengths and the selenium bond angles of *free* organoselenide molecules are listed in Table 7 for quick reference.

The molecular geometry of dimethyl selenide, Me_2Se, (Figure 4) was determined by MW[37,38]. Both Me groups are staggered with respect to the other Se—C bond. The rotational axes (axes of quasi-C_{3v} symmetry) of the Me groups do not coincide with the Se—C bonds but are tilted by 2.6° toward the lone electron pairs of the Se atom. The substitution structure shows a slight asymmetry in the structure of the Me group characterized by the parameters given in Figure 4. The structure is similar to that of Me_2S[50,51]. The C—Se—C skeleton is highly bent (cf. Table 7). A suggestion that it is

TABLE 7. C—Se bond lengths and Se bond angles in acyclic organoselenides from MW and ED

Molecule	C—Se (Å)		∠C—Se—X (deg.)	Reference
Me_2Se	r_s 1.945(0.4)		96.3(1)	37
	r_s 1.943(1)		96.2(2)	38
$(CF_3)_2Se$	r_a 1.980(9)[a]		94.1(20)[a]	39
	r_a 1.975(9)[b]		97.0(20)[b]	
MeSeH	r_0 1.959(10)		95.45(174)	40
EtSeH, *anti*	r_s 1.962(2)		93.5(6)	41
gauche	r_s 1.957(4)		93.1(3)	
anti	r_0 1.958(5)		93.3(10)	42
gauche	1.958[c]		93.9(10)	
MeSeCN	r_s 1.956[d]	r_0 1.837[e]	96.1	43
	r_0 1.950(25)[d]	1.851[e]	95.6	44
F_3CSeCN	r_a 1.984(20)[a,d]	1.854(16)[a,e]	92.2(20)[a]	45
	r_a 1.984(20)[b,d]	1.851(20)[b,e]	103.0(25)[b]	
PhSeMe	r_a 1.957(19)[d]	1.912(13)[f]	99.6(25)	46
PhSeBr		r_a 1.899(6)[f]	99.8(12)	47
$(F_3CSe)_3N$	r_g 1.973(5)		104.3(9)	48
$(MeSe)_3B$	r_a 1.954(4)		102.5(5)	49

[a,b]Two possible solutions: model A and B (see text).
[c]Assumed parameter.
[d]$C(sp^3)$—Se bond.
[e]C(sp)—Se bond.
[f]$C(sp^2)$—Se bond.

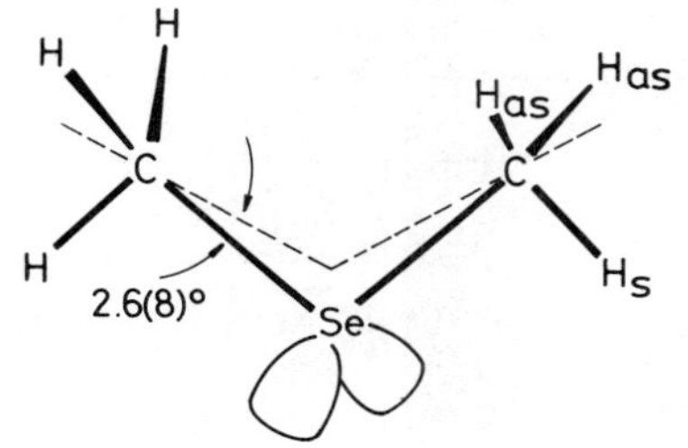

		C–Hs	1.088(13) Å
		C–Has	1.096(4) Å
Se–C–Hs	105.0(7)°	Has–C–Has	109.9(5)°
Se–C–Has	110.3(3)°	Has–C–Hs	110.6(5)°

FIGURE 4. The molecular model of dimethyl selenide and the geometrical parameters (r_s) of the Me group[37]

linear[52] has been strongly criticized[53] on the basis of extended spectroscopic evidence in addition to the unambiguous MW structure determination. Spectroscopically calculated mean amplitudes of vibration of Me_2Se have been reported[54] together with those of Me_2S and Me_2O.

Two solutions (A and B forms in Figure 5) have been reported as to the conformational properties of bis(trifluoromethyl) selenide, $(CF_3)_2Se$ from an ED investigation[39]. Both forms have C_2 symmetry although form A has essentially C_{2v} symmetry. On the basis of ED data it was not decided whether both forms were present in the vapour or whether one was merely a mathematically possible solution. This uncertainty did not hinder the determination of the geometry of the $SeCF_3$ part of the molecule. However, the result for the bond angle C—Se—C was sensitive to the choice of the conformational model: 94.1(20)° and 97.0(20)° were determined for models A and B, respectively. It is in fact this sensitivity that may facilitate our preferring one of the two conformational models on the basis of observed geometrical variations in analogous molecules. The C—S—C bond angle decreases from 98.9(2)° in Me_2S[50] to 97.3(8)° in $(CF_3)_2S$[55]. This change is in complete agreement with the valence-shell electron-pair repulsion (VSEPR) model[28]. According to this model, increasing ligand electronegativity results in increasing electron-withdrawing

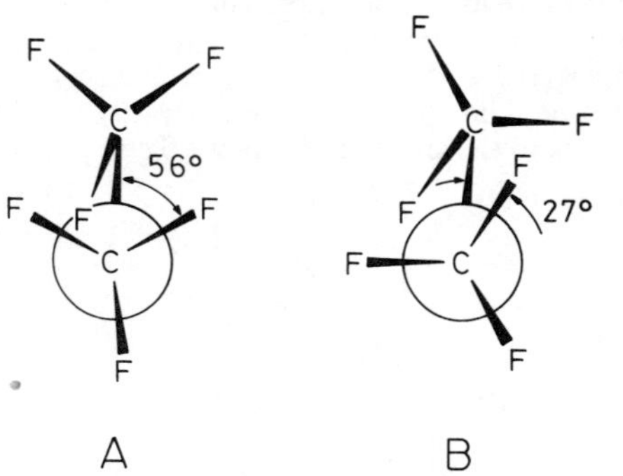

FIGURE 5. The two models of bis(trifluoromethyl) selenide[39] viewed along the C—Se bond

TABLE 8. Bond angles and non-bonded distances in X_2S and X_2Se

X_2Y	X_2S			X_2Se		
	∠ X—S—X (deg.)	X···X (Å)	Ref.	∠ X—Se—X (deg.)	X···X (Å)	Ref.
H_2Y	92.12	1.923[a]	56	90.57	2.075[a]	59
Me_2Y	99.05	2.746[a]	51	96.3	2.898[a]	37
$(CF_3)_2Y$	97.3	2.731[a]	55	94.1	2.90	39
$(SiH_3)_2Y$	97.4	3.210	57	96.6	3.396	60
$(GeH_3)_2Y$	98.9	3.358	58	94.6	3.445	61

[a]Calculated from the bond distance and bond angle given in the reference.

ability, which in turn lessens the electron-pair repulsion in the vicinity of the central atom. Thus the respective bond angles may close somewhat. Atom–atom non-bonded interactions might be expected to cause an opening of the C—S—C bond angle upon CH_3/CF_3 substitution. That the reverse occurs is witness to the prevailing importance of the electron-pair repulsions in these structures. The relative importance of the atom–atom interactions is certainly smaller in the Se analogues as compared with the S derivatives. Accordingly, the C—Se—C bond angle in $(CF_3)_2Se$ is predicted to be smaller than that in Me_2Se. This favours model A for $(CF_3)_2Se$. Marsden and Sheldrick noted[39] that model B may be fortuitous as it involves a very short F···F distance (2.51 Å) between the two different trifluoromethyl groups.

Comparison of the analogous X_2S and X_2Se molecules shows increasing non-bonded distances and slightly decreasing bond angles from S to Se (Table 8). All the cited non-bonded distances are greater than twice the postulated 1,3 non-bonded radii, viz. $r_{1,3}$ (H) = 0.92, $r_{1,3}$(C) = 1.25, $r_{1,3}$(Si) = 1.55 and $r_{1,3}$(Ge) = 1.58 Å[62,63]. H_2S and H_2Se are included in Table 8 for comparison. The angle H—Te—H is 90.25° in H_2Te[64] with a 2.350 Å H···H distance. A recent ED investigation of dimethyl telluride, Me_2Te yielded Te—C = 2.142(5) Å, C···C = 3.14 Å, ∠ C—Te—C = 94(2)°[65]. The analogous O bond angles are considerably greater with smaller non-bonded distances: Me_2O 111.7°, 2.334 Å[66]; $(SiH_3)_2O$ 144.1°, 3.107 Å[67]; and $(GeH_3)_2O$ 126.5°, 3.154 Å[58].

The diminishing bond angles in the series of analogous O, S and Se derivatives have been noted also by Thomas[40]. He pointed out the importance of this geometrical variation in relation to the changes in the methyl rotational barrier. This barrier is 4.2(2) kJ mol^{-1} in methaneselenol, MeSeH, as determined by MW[40]. The conformation (viewed along the Se—C bond in **4**) is staggered and the Me group is tilted by 1.5(10)° towards the lone electron pairs of the Se atom. The Me tilt in the Se derivatives occurs in the same direction and is similar in magnitude to those in the analogous S compounds, viz. Me_2S 2.8°[50] and MeSH 2.2°[68].

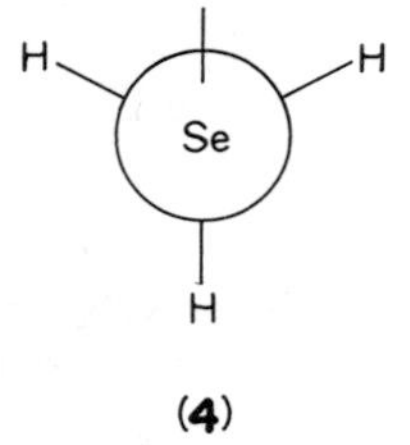

(**4**)

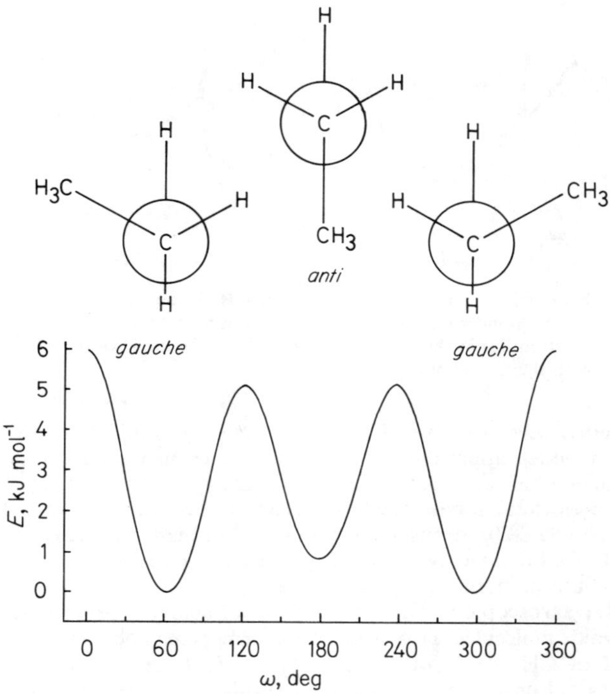

FIGURE 6. The potential energy (*E*) of ethaneselenol as a function of the dihedral angle ω(CCSeH), the angle of rotation about the C—Se bond. Above: projections of the *gauche* and *anti* forms along the C—Se bond. Drawn after Reference 42

Both *gauche* and *anti* conformers, with respect to rotation about the C—Se bond, were found in the vapour of ethaneselenol, EtSeH, by MW[41,42]. The potential energy curve describing internal rotation about the C—Se bond is shown in Figure 6 after Durig and Bucy[42]. The Et group is assumed to have a staggered conformation. The complete structures of both forms have been determined from the rotational spectra[41]. The dihedral angle CCSeH of the *gauche* form is 61.6(8)° [41]. Notable differences between the two conformers occur in the bond angles of the methylene carbon. This was attributed to a tilt δ of the axis of internal rotation from the C—Se bond (Figure 7). The bond angles around the methylene carbon were then used in both studies[41,42] to calculate the tilt angle and the angle β between the rotation axis and the C—C bond. Values obtained for ethaneselenol, $\beta = 111.9°$ and $\delta = 3.2°$ [41], are very similar to those for ethanethiol[41,69].

The MW of methyl selenocyanate, MeSeCN, has been analysed[43,44]. The available data allowed an unambiguous determination of the Se—CH_3 bond length. A tentative refinement based on several assumed parameters yielded *r*(Se—CN) and the bond angle C—Se—C (Table 7). The observed difference between the two types of selenium–carbon bond length is 0.12 Å (0.14 Å is given in the original paper[43]), nearly the same as found between the analogous sulphur–carbon bond lengths in MeSCN[70]. The barrier to internal rotation of the Me group was determined from MW to be 5.2(2) kJ mol^{-1} in methyl selenocyanate[43,44] and 6.6 kJ mol^{-1} in methyl thiocyanate[71]. The rotational barrier and the Me tilt in MeSeCN are similar to those in MeSeH.

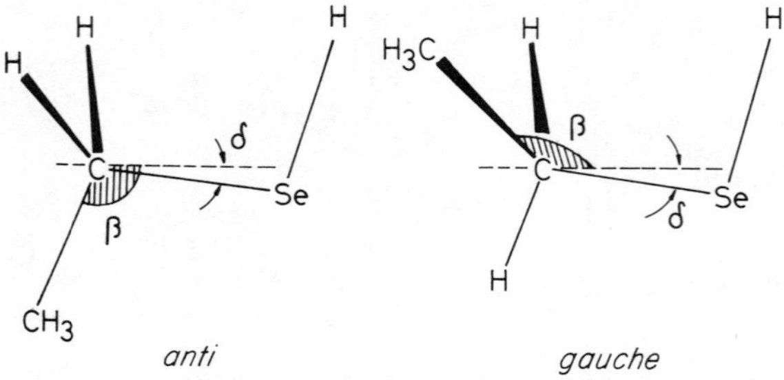

FIGURE 7. Tilt (δ) of the rotational axis (dashed line) from the C—Se bond in ethaneselenol. It is assumed that the axis of rotation lies in the C—Se—H plane. β is the angle of the axis with the C—C bond

Trifluoromethyl selenocyanate, F_3CSeCN, has been studied by ED[45]. Again two conformational models approximated well to the experimental data (Figure 8). Model A may be considered as an effective structure arising from a staggered equilibrium conformation performing torsional vibrations about the F_3C—Se bond. The bond angle ∠C—Se—CN = 92.2(20)° seems to be more plausible than the much larger one obtained for model B (103°). Fortunately, the other geometrical parameters are unaffected by the choice of the conformational model (cf. Table 7).

Vibrational spectroscopic studies indicated the variability of conformational properties of organic selenide molecules. Two or more conformers may be present in the liquid state and solutions of CH_2=CHSePh and CH_2=$CHSeC_6H_4NO_2$-*p* according to their infrared spectra[72]. For diisopropyl selenide the solid state spectra were interpreted by a single conformer (C_2 symmetry), while three forms were detected in the liquid state (presumably C_2, C_s and C_1) with the C_2 form being the most stable[73].

The presence of a single conformation was assumed in the ED structure analysis of selenoanisole, PhSeMe[46]. The Se—CH_3 bond is rotated from the plane of the benzene ring by about 40°. A similar conformation has been reported for thioanisole[74] (Figure 9).

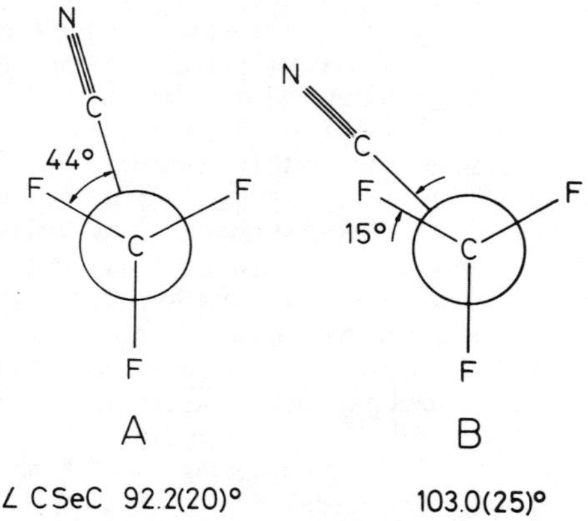

FIGURE 8. The two models of trifluoromethyl selenocyanate[45] projected along the F_3C—Se bond

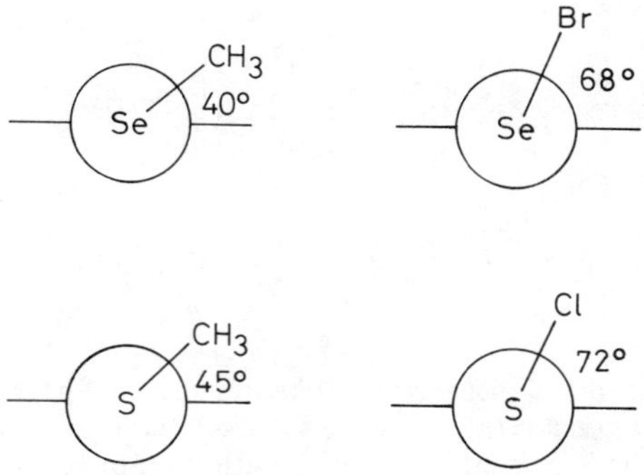

FIGURE 9. Conformation of selenoanisole[46], thioanisole[74], benzeneselenenyl bromide[47] and benzenesulphenyl chloride[75]. Projections along the Se—C(phenyl) or S—C(phenyl) bond. The horizontal line represents the plane of the Ph ring

The structure of benzeneselenenyl bromide, PhSeBr, has also been determined by ED[47]. It has a greater deviation from planarity than selenoanisole. A similar change is observed in benzenesulphenyl chloride[75] as compared with thioanisole (cf. Figure 9).

It is especially interesting to compare the gas-phase structure of benzeneselenenyl bromide with the crystal XD molecular structure of 2-formylbenzeneselenenyl bromide[76] (Figure 10). The crystals are orthorhombic, $Pc2_1n$. The molecule has a planar *syn* conformation (**5**) similar to the analogous Te derivative[77]; the Br—Se···O and Br—Te···O chains are roughly linear. The planarity of the crystal molecular conformation is strikingly different from the non-planar structure of free benzeneselenenyl bromide. The coplanarity may be a consequence of the presence of the aldehyde group in the *ortho* position and/or the intermolecular forces in the crystal. Generally speaking, the most common of the possible gas/crystal structural differences are the conformational changes. Thus it is tempting to ascribe this change to crystal field effects. On the other hand, similarly drastic conformational change is observed in the free molecules of benzenesulphenyl chloride[75] and 2-nitrobenzenesulphenyl chloride[78a], both studied by ED. The flattening here may be a consequence of interaction between the substituents in the *ortho* positions (Figure 11). The S—Cl bond lengths are the same in the two molecules and much larger than in free sulphur dichloride itself viz. 2.015 Å[79]. The considerable lengthening of

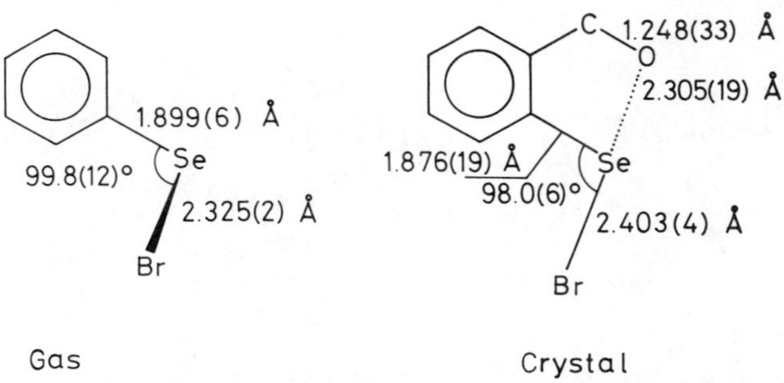

FIGURE 10. Geometrical parameters of benzeneselenenyl bromide[47] and 2-formylbenzeneselenenyl bromide[76]

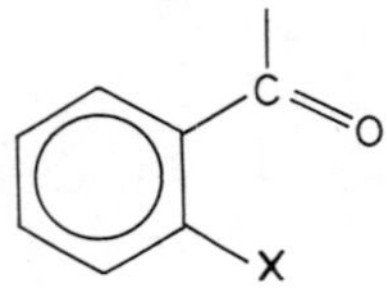

(**5**)
syn

the S—Cl bond thus appears to be a consequence of Ph/Cl substitution, and no appreciable change is observed that could be ascribed to the presence of the nitro group in the *ortho* position. This is in contrast with the situation of the Se—Br bond as seen in Figure 10. Unfortunately, there is no reliable information on the structure of selenium dibromide (cf. References 3 and 80). If, however, the presence of the *ortho* aldehyde group were assumed to produce no change in the Se—Br bond length, the considerable lengthening of the Se—Br bond in 2-formylbenzeneselenenyl bromide as compared with benzeneselenenyl bromide should be ascribed to the crystal field effect. This situation could be clarified by information on structural changes induced by aldehyde versus nitro groups in the *ortho* position, and from gas/crystal structure determinations of the same molecules. According to a recent ED study of 2-nitrobenzeneselenenyl bromide, this molecule is practically planar, and the Se bond configuration is characterized by C—Se 1.917(11) Se—Br 2.354(3) Å, ∠ C—Se—Br 98.5(7)° and Se···O 2.36 Å[78b].

The most interesting structural feature of the *ortho*-substituted benzeneselenenyl and benzenesulphenyl halides is undoubtedly the interaction between the O and the Se (or S) atom. The relatively short Se···O distance indicates partial bonding. A similar situation was observed in 2-formylbenzenetellurenyl bromide by XD[77]. The Se···O and Te···O interaction has been discussed by Baiwir and coworkers[76] in relation to C=O bond lengths and vibrational frequencies and proton magnetic resonance data in analogous molecules. Table 9 contains further available structural data on such bonding situations. The Se···O and Te···O distances invariably fall between the sums of the covalent radii and 1,3 intramolecular non-bonded radii[62,86] (Table 10).

Zaripov and coworkers[46] discussed the conformational properties of the anisole analogues PhXMe with X = O, S, Se and Te. Of the many effects influencing internal rotation two were singled out for closer examination, viz. the p, π conjugation tending to

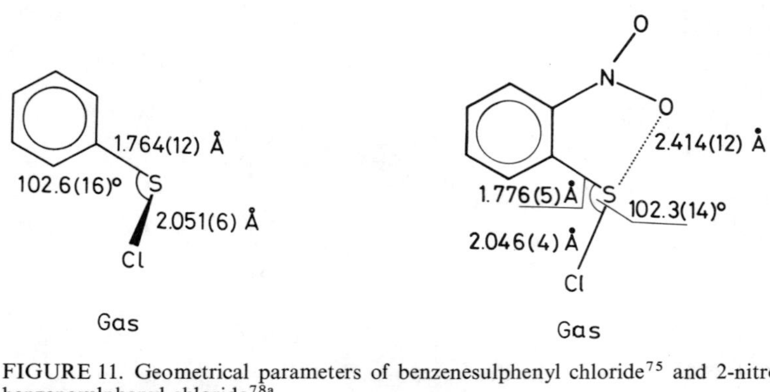

FIGURE 11. Geometrical parameters of benzenesulphenyl chloride[75] and 2-nitrobenzenesulphenyl chloride[78a]

TABLE 9. Se and Te bond configurations in *ortho*-substituted benzeneselenenyl and benzenetellurenyl derivatives from XD

Molecule	Bond lengths (Å)		Angle (deg.)	Distance (Å)	Reference
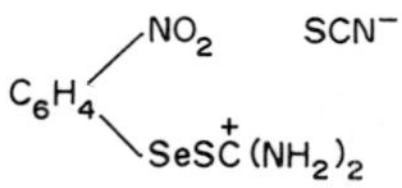	C—Se 1.939(8)	Se—S 2.189(3)	C—Se—S 98.9(3)	Se···O 2.505(8)	81
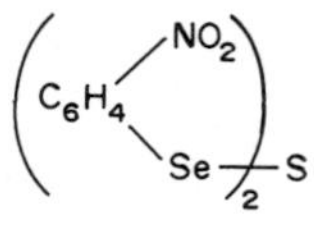	C—Se 1.918(8)	Se—S 2.202(2)	C—Se—S 102.0(3)	Se···O 2.574(8)	82
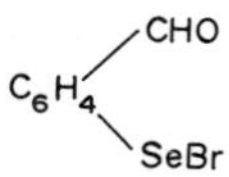	C—Se 1.876(9)	Se—Br 2.403(4)	C—Se—Br 98.0(6)	Se···O 2.305(19)	76
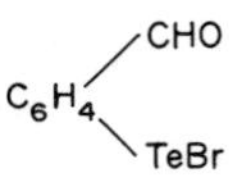	C—Te 2.081(21)	Te—Br 2.618(3)	C—Te—Br 94.2(6)	Te···O 2.31	77
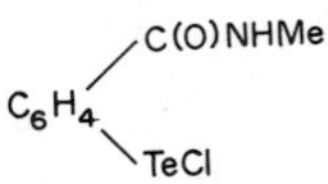	C—Te 2.133(10)	Te—Cl 2.516(3)	C—Te—Cl 92.7(3)	Te···O 2.250(7)	83
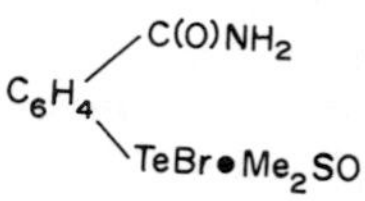	C—Te 2.105(9)	Te—Br 2.646(1)	C—Te—Br 95.1(2)	Te···O 2.237(8)	83
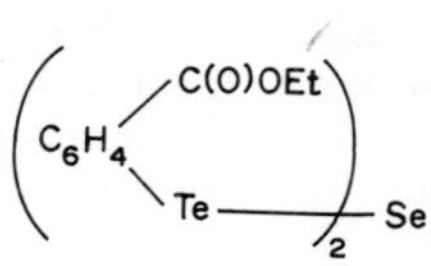	C—Te 2.123(8)	Te—Se 2.536(1)	C—Te—Se 97.9(2)	Te···O 2.658(6)	84
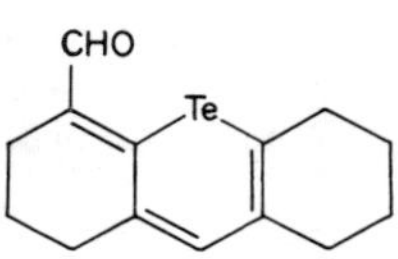	C—Te 2.076[a]	Te—C 2.055[a]	C—Te—C 94.2[a]	Te···O 2.575[a]	85

[a] Average from two crystallographically independent molecules.

TABLE 10. Atomic radii (Å) and calculated internuclear distances (Å): covalent, non-bonded 1,3 and van der Waals'

	r_{cov}^{87}	$r_{1,3}^{62,86}$	$r_{vdW}^{88,89}$
C	0.767	1.25	1.70
O	0.745	1.12	1.40
S	1.020	1.45	1.85
Se	1.163	1.60	2.00
Te	1.356	1.75	2.20
	Σr_{cov}	$\Sigma r_{1,3}$	Σr_{vdW}
SC	1.787	2.70	3.55
SeC	1.930	2.85	3.70
TeC	2.123	3.00	3.90
SO	1.765	2.57	3.25
SeO	1.908	2.72	3.40
TeO	2.101	2.87	3.60

stabilize the planar form, and the Me/Ph hydrogen–hydrogen interaction tending to displace the system from coplanarity. Tschmutowa and Bock[90] concluded from photoelectron spectra that the p, π conjugation diminishes in the order of $O \gg S > Se > Te$ and accordingly the probability of planar conformation should also decrease in the same order. The non-planarity of the structures may be the consequence of d, π or σ, π interaction as well as that of steric hindrance[46]. All this refers to the intramolecular interactions governing the conformational choice of the free molecules. In the crystalline phase, effects from intermolecular interaction have to be considered in addition.

Some results of X-ray crystallographic studies will be cited here for comparison with the gas-phase data of Table 7.

The C—Se—C bond angle in crystalline acetylselenocholine iodide, $MeC(O)SeCH_2CH_2\overset{+}{N}Me_3{\cdot}I^-$ (**6**), is 97(1)°[91], the same as in Me_2Se (Table 7). In analogous molecules of biological interest the $\overset{+}{N}$—C—C—O chain prefers the *gauche* conformation while $\overset{+}{N}$—C—C—S and $\overset{+}{N}$—C—C—Se adopt the *anti* conformation[91,92]. The *anti* form has been found in **6** and in (2-(dimethylamino)ethyl) selenobenzoate hydrochloride, $PhC(O)SeCH_2CH_2\overset{+}{N}HMe_2{\cdot}Cl^-$ (**7**)[92]. The C—Se—C angle in **7** is 96.4(2)°, and both C—Se bond lengths have been determined as 1.945(5) Å[92]. In the Se analogue of a tetrapeptide derivative, *S*-benzyl—L-Cys—L-Pro—L-Leu—Gly—amide, the Se—C bonds are 1.92(4)–1.99(2) Å long and the C—Se—C angles are 100(1) and 102(1)° in two crystallographically independent molecules[93]. In the crystal of **8**,

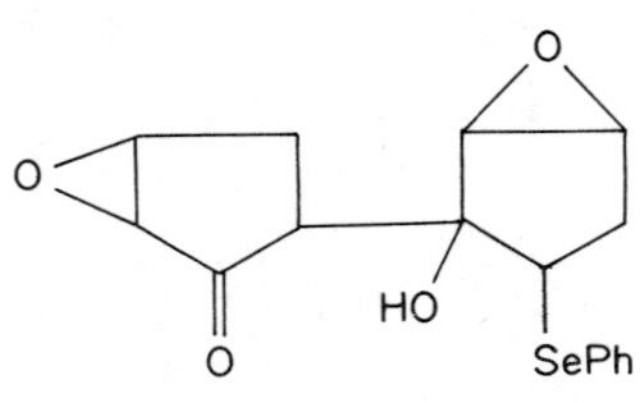

(**8**)

disorder in the position of the Ph ring hinders the accurate determination of the parameters, viz. Se—C(sp^3) = 1.97(1) Å, and for the two positions of the Ph ring, Se—C(sp^2) = 1.97(2) and 1.83(2) Å, and ∠ C—Se—C = 97.0(7) and 99.7(7)°[94].

The structures of crystalline bis(6-hydroxy-4,4-dimethyl-2-oxo-6-cyclohexenyl) selenide (**9**)[95] and bis(2-hydroxy-4,4,6,6-tetramethyl-3-oxo-1-cyclohexenyl) selenide (**10**)[96]

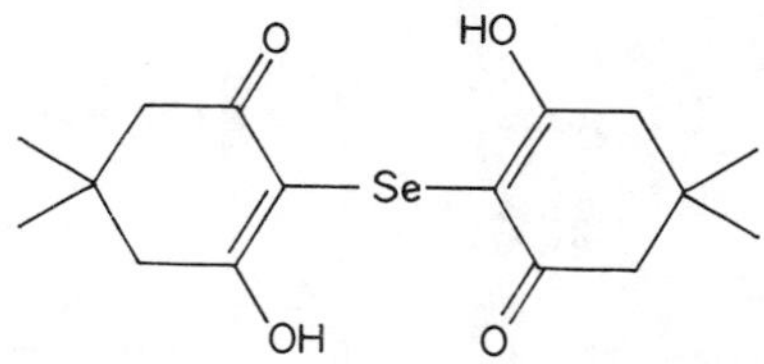

(**9**)

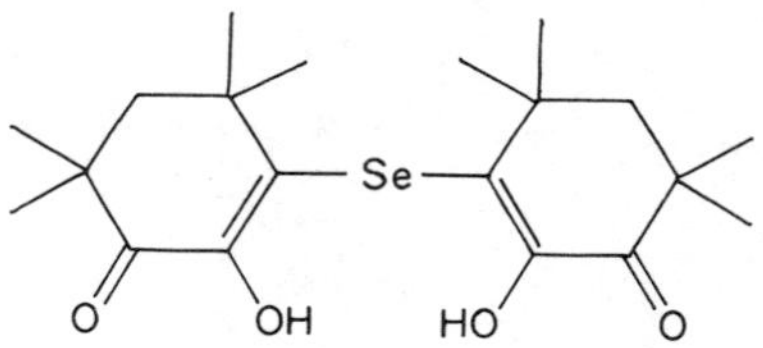

(**10**)

are different in that **9** is intramolecularly hydrogen-bonded while **10** is intermolecularly hydrogen-bonded. The molecular packing of **10** is illustrated in Figure 12 after Kivekäs and Laitalainen[95]. The Se—C bonds are of the same length (1.916 Å) in the two molecules whereas the Se bond angle opens somewhat in **10** as compared with **9**, viz. 103.9(2) vs. 100.9(4)°. The Se bond configurations show no unusual features. The conjugated bond

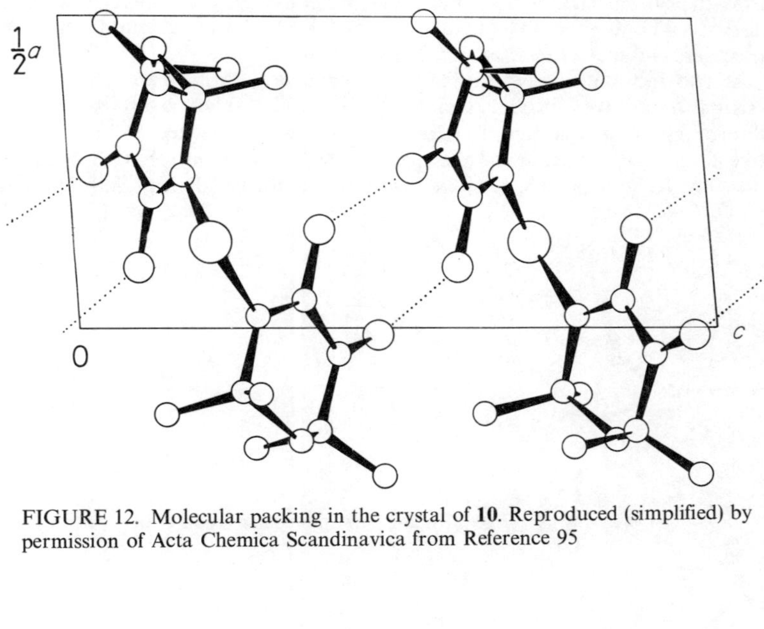

FIGURE 12. Molecular packing in the crystal of **10**. Reproduced (simplified) by permission of Acta Chemica Scandinavica from Reference 95

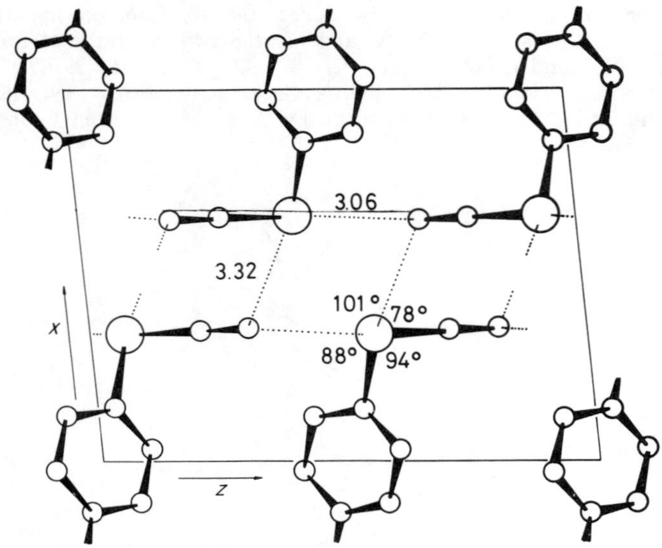

FIGURE 13. Molecular packing in the crystal of *p*-diselenocyanatobenzene (**11**): projection along the *b* axis. Reproduced by permission of the Royal Society of Chemistry from Reference 97

system in **10** is coplanar and the rings are in half-chair conformation. The rings of **9** take a half-boat form.

The crystal structure of *p*-diselenocyanatobenzene, *p*-$NCSeC_6H_4SeCN$ (**11**), has been determined by XD[97]. The two types of selenium–carbon bonds, Se—CN = 1.837(23) Å and Se—C_6H_4 = 1.916(19) Å, are similar in length to the corresponding single bonds to sp- and sp^2-hybridized carbon in free molecules of methyl selenocyanate and selenoanisole (see Table 7). The C—Se—C angle in **11** is 94°, and the selenium obviously has intermolecular contacts as seen in Figure 13.

The last two molecules listed in Table 7 possess practically planar Se_3X skeletons. Both structures have been elucidated by ED. The overall structure of tris(trifluoromethylseleno)amine, $(F_3CSe)_3N$, is shown in Figure 14. The C—Se—N planes are nearly perpendicular to the Se_3N plane[48]; the deviations have been reported to be 10–14°. The CF_3 groups are staggered with respect to the adjacent Se—N bonds. The

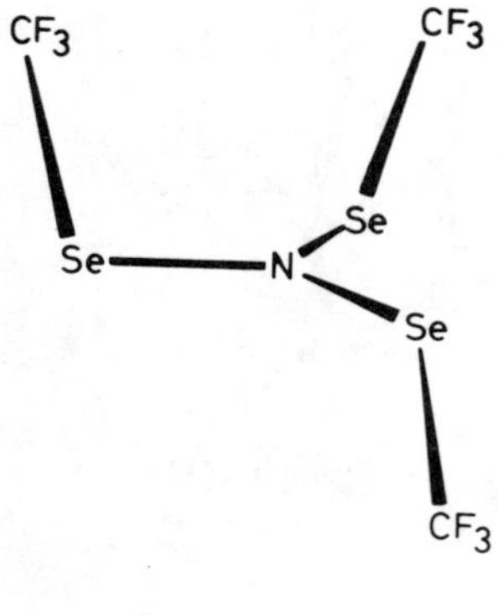

FIGURE 14. The molecular model of tris(trifluoromethylseleno)amine

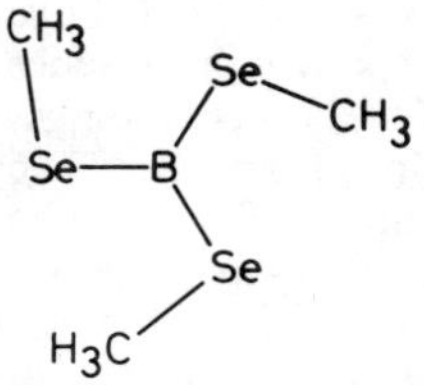

FIGURE 15. The planar skeleton of tris(methylseleno)borane with C_3 symmetry

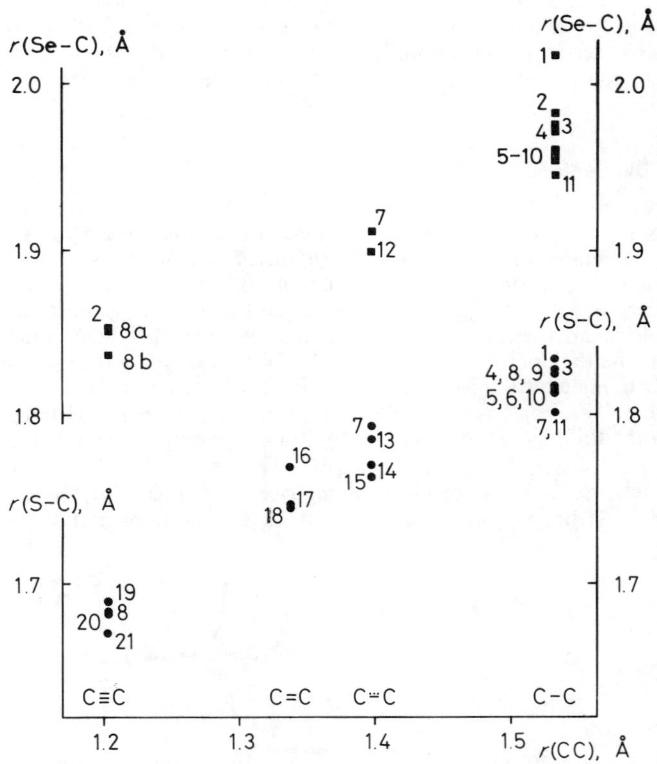

FIGURE 16. The lengths of Se—C and S—C single bonds with sp, sp^2 (non-aromatic and aromatic) and sp^3 carbon atoms. The abscissa represents the carbon–carbon bond distance in acetylene, ethylene, benzene and ethane. *Se compounds.* 1: $(CF_3)_2Se_2$, 2: F_3CSeCN, 3: $(CF_3)_2Se$ (mean of two models[39]), 4: $(F_3CSe)_3N$, 5: MeSeH, 6: EtSeH (mean of *anti* and *gauche* forms[41]), 7: PhSeMe, 8: MeSeCN (a: Reference 44, b: Reference 43), 9: $(MeSe)_3B$, 10: Me_2Se_2, 11: Me_2Se, 12: PhSeBr. See Tables 7 and 11 for references. *S compounds.* 1, 3–11: see the Se analogues, 13: Pyr_2S, 14: Ph_2S, 15: PhSCl, 16: HC(O)SH, 17: CH_2=CH—SMe, 18: CH_2=C=CH—SMe. 19: EtS—CN. 20: HC≡C—SMe, 21: MeS—C≡C—SMe. See References 3, 20 and 100

most interesting feature of this structure is that two of the CF_3 groups are situated above the Se_3N plane and the third CF_3 group below this plane. Thus there is a striking absence of threefold symmetry. The structure is very much the same as that of the corresponding S compound, tris(trifluoromethylthio)amine, $(F_3CS)_3N$, as also determined from ED[98]. The barrier to internal rotation about the S—N bond was estimated from NMR data to be 25(4) kJ mol^{-1} [98], which is probably greater than the as yet unestimated barrier about the longer Se—N bond.

An essentially planar heavy-atom skeleton characterizes the geometry of tris(methylseleno)borane, $(MeSe)_3B$[49] (Figure 15). Tris(methylthio)borane has a similar structure[99].

In concluding this section it is of interest to examine the variation in lengths of Se—C single bonds in different environments. A characteristic sample from the available data is presented in Figure 16. There is an appreciable change in the Se—C bond lengths upon changing carbon hybridization. The variation nicely parallels the tendency observed for the S—C bond lengths in sulphides[3,100], which are also presented in Figure 16 for comparison.

B. Acyclic Diselenides and Ditellurides

The structure and conformation of dimethyl diselenide, Me_2Se_2[101] and bis(trifluoromethyl) diselenide, $(CF_3)_2Se_2$[102] have been determined by ED. The experimental data are consistent with C_2 overall symmetry, which has been assumed in both studies. The spatial geometry is illustrated in Figure 17 while the conformational properties are depicted in Figure 18 for both molecules. The geometrical data of the CSeSeC skeleton are given in Table 11, and are compared with those of the analogous S compounds. The angle of rotation about the C—Se bonds (Figure 18) is primarily deduced from the scattering contribution of atomic pairs involving the ligands of carbon and the other Se atom. This angle is better determined in the F than in the H derivative. The deviation from the ideal staggered form is not surprising in view of the asymmetric environment of the $SeCX_3$ moiety. Thus this deviation is believed not to be entirely an apparent one as a consequence of averaging over intramolecular torsional vibrations about the C—Se bonds. On the other hand, these vibrations certainly influence the

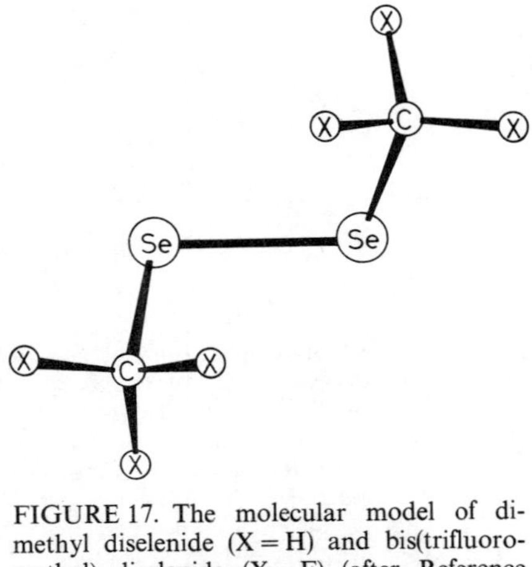

FIGURE 17. The molecular model of dimethyl diselenide (X = H) and bis(trifluoromethyl) diselenide (X = F) (after Reference 101 b)

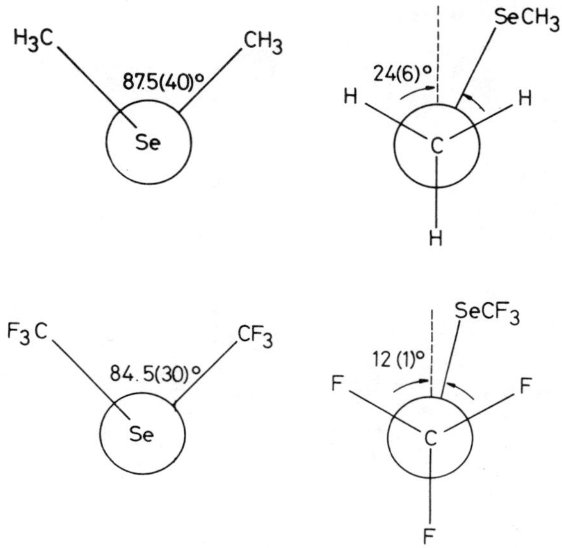

FIGURE 18. Conformation of dimethyl diselenide[101] (above) and bis(trifluoromethyl) diselenide[102] (below): projections along the Se—Se (left) and the C—Se (right) bonds

effective angle of rotation as determined from ED. This may be of importance, especially for the methyl derivative, as relatively large-amplitude, low-barrier torsional vibrations are anticipated.

For a series of XSSX disulphanes it has been observed that the central bond considerably shortens with increasing ligand electronegativity[3]. The Se—Se bond shortening in the diselenides upon CH_3/CF_3 substitution is consistent with this

TABLE 11. Parameters of Me_2S_2, Me_2Se_2 and perfluoro derivatives from gas electron diffraction

	Me_2Y_2	
	Me_2S_2[103]	Me_2Se_2[101]
Y—Y (Å)	2.029(3)	2.326(4)
Y—C (Å)	1.816(3)	1.954(5)
∠ Y—Y—C (deg.)	103.2(2)	98.9(2)
ω (CYYC) (deg.)	85.3(37)	87.5(40)
	$(CF_3)_2Y_2$	
	$(CF_3)_2S_2$[104]	$(CF_3)_2Se_2$[102]
Y—Y (Å)	2.030(5)	2.292(10)
Y—C (Å)	1.835(5)	2.018(20)
∠ Y—Y—C (deg.)	101.6(6)	98.0(5)
ω(CYYC) (deg.)	104.4(40)	84.5(30)

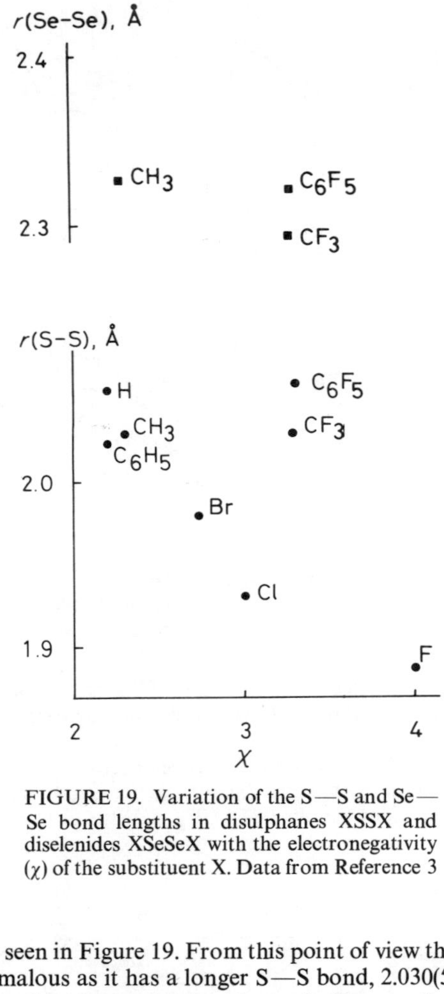

FIGURE 19. Variation of the S—S and Se—Se bond lengths in disulphanes XSSX and diselenides XSeSeX with the electronegativity (χ) of the substituent X. Data from Reference 3

observation as can be seen in Figure 19. From this point of view the structure of $(CF_3)_2S_2$ seems somewhat anomalous as it has a longer S—S bond, 2.030(5) Å [104], than would be expected according to the above mentioned trend. Noteworthy is the greater CSSC dihedral angle ($104 \pm 4°$) determined for this molecule as compared with the rest of the disulphane series—except for $Me_2BSSBMe_2$—or the diselenide molecules, for that matter[3].

Two examples from among crystalline organic diselenides are cited here. Cyclic molecules with a Se—Se fragment will be discussed in Section III.C.

The structure of α,α′-diselenobis(formamidinium) dichloride, $[SeC(NH_2)_2]_2Cl_2$, has been determined by XD[105]. The planes of the two selenourea groups in the bis(selenourea) cation, $[SeC(NH_2)_2]_2^{2+}$ (**12**), are nearly perpendicular to the Se—Se bond. The heavy-atom skeleton is shown in Figure 20. On the other hand, the orientation of the thiourea groups in two salts of the S analogue of **12** has been found to be essentially parallel to the S—S bond[106] (Figure 20). It has been argued[105] that the conformational differences are due to intramolecular spatial interactions rather than packing requirements. Some of the parameters of the S and Se derivatives are listed in Table 12. The Se—Se bond is longer

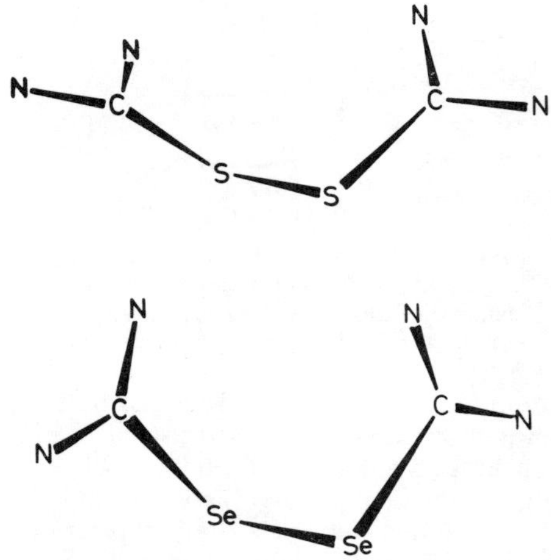

FIGURE 20. Conformation of the $[SC(NH_2)_2]_2{}^{2+}$ and $[SeCNH_2)_2]_2{}^{2+}$ cation after Reference 105)

than in Me_2Se_2 (Table 11). The Se—C bond of **12** in its diiodide salt has been determined as 1.943(4) Å [108].

The other example is bis(pentafluorophenyl) diselenide, $(C_6F_5)_2Se_2$ and its S analogue[107], which are isostructural in the crystal, orthorhombic $P2_12_12_1$. Their main parameters are also given in Table 12. An unexpected finding was[107] that the crystal

TABLE 12. Parameters of diselenides and analogous disulphides from X-ray crystallographic studies

	$[YC(NH_2)_2]_2X_2$		
	$[SC(NH_2)_2]_2X_2\cdot H_2O$[106]		
	X = Br	X = I	$[SeC(NH_2)_2]_2Cl_2$[105]
Y—Y (Å)	2.044(10)	2.044(20)	2.380(6)
Y—C (Å)	1.78(3)	1.75(4)	1.94(1)
∠ Y—Y—C (deg.)	104.0	98.9	95.5(6)
ω(CYYC) (deg.)	89.2	104.8	89.5

	$(C_6F_5)_2Y_2$	
	$(C_6F_5)_2S_2$[107]	$(C_6F_5)_2Se_2$[107]
Y—Y (Å)	2.059(4)	2.319(4)
Y—C (Å)	1.770(7)	1.910[a]
Y—Y—C (deg.)	101.2[a]	98.8[a]
ω(CYYC) (deg.)	76.5	75.3
ϕ^b (deg.)	39.3	34.0

[a]Mean value.
[b]ϕ is the dihedral angle between the planes of the rings.

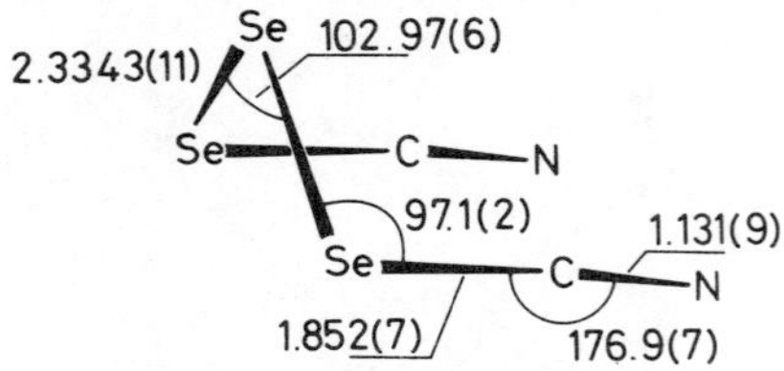

FIGURE 21. The molecular model, bond lengths (Å) and bond angles (deg.) of selenium diselenocyanate[121]. Drawn from the atomic coordinates given in Reference 121

packing and conformation of these molecules proved to be different from those of the corresponding unsubstituted Ph derivatives. Also the chalcogen bond angles seem to be significantly larger in the latter, viz. about 106°, both in Ph_2S_2[109] and Ph_2Se_2[110].

The Se—Se bond lengths in these molecules (Tables 11 and 12) can be compared with those in dimorpholino-di-, -tri- and -tetra-selane and dipiperidinotetraselane[111], where they lie in the range 2.327(2)–2.356(2) Å, and in the cyclic Se_8 molecule of γ-selenium (monoclinic), 2.326(3)–2.344(3) Å [112].

The Te—Te bond length is 2.712(2) Å in crystalline diphenyl ditelluride, Ph_2Te_2, from XD[113]. The mean of the observed Te—C bond lengths is 2.115 Å. In a subsequent study the crystal and molecular structure of p,p'-ditolyl ditelluride has also been determined[114]: Te—Te = 2.697(3) Å and Te—C = 2.13(1) Å. The authors[114] noted a great dissimilarity between the molecular conformation in the crystals of Ph_2Te_2[113], $(p\text{-}MeC_6H_4)_2Te_2$[114] and $(p\text{-}ClC_6H_4)_2Te_2$[115]. It has been suggested that these differences are due to steric hindrances arising from packing peculiarities[114].

Conformational properties and dynamic behaviour of organic dichalcogenide molecules and, among them, diaryl dichalcogenides, have been studied by experimental and theoretical methods (see, for example, References 116–120 and references therein). Apart from intermolecular interactions, the adopted conformation of diaryl dichalcogenides is a result of steric effects on the one hand, and interaction of chalcogen lone pairs with the π electrons of the aromatic rings on the other[116]. The steric effects predominate in the case of ditellurides[116].

The crystal and molecular structures of selenium diselenocyanate, $Se(SeCN)_2$, and also of the isomorphous selenium dithiocyanate, $Se(SCN)_2$, have been redetermined[121], based on XD data, since they were discussed in another volume of this series[4]. Bond lengths and angles of $Se(SeCN)_2$ (Figure 21) are in agreement with the data in Tables 7, 11 and 12 and the parameters of p-diselenocyanatobenzene (**11**)[97]. Notable deviations from the results of the earlier investigations have been found in the geometry of the —SeCN group. The molecule lies on a crystallographic symmetry plane and has a *syn* conformation. The SeSeSeC torsional angle is 93.2° (cf. the data on diselenides in Tables 11 and 12). The coordination around each Se atom is completed by N atoms of neighbouring molecules to a roughly square planar arrangement, which, at the same time, gives rise to a network of nearly linear N···Se(C)–Se···N and C–Se···N sequences with intermolecular contacts of 3.253(6), 3.142(6) and 3.085(6) Å, respectively[121] (cf. the stereoscopic view of crystal packing in Reference 4).

C. Heterocycles

Structural data on free non-aromatic ring molecules with Se are scarce. Selenetane, $(CH_2)_3Se$, has been studied by microwave spectroscopy[122]. The geometry has not been determined except for the ring puckering. It is found to be essentially the same as

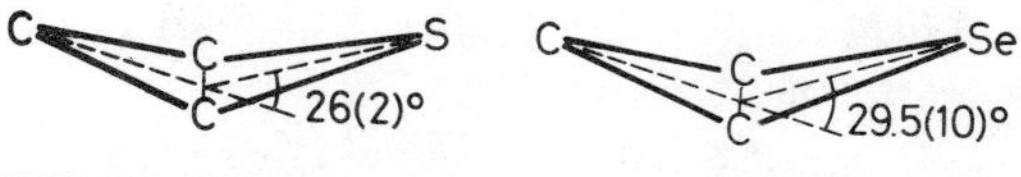

FIGURE 22. Ring puckering in thietane[123] and selenetane[122]

TABLE 13. Puckering angles and inversion barriers of four-membered rings $(CH_2)_3X$

X	Puckering angle CXC/CCC (deg.)	Barrier ($kJ\ mol^{-1}$)	Technique	Reference
CH_2	35	6.02(2)	ED	124
	26(3)		ED	125
NH	33.1(24)	5.27	ED	126
O	0	0.1856(6)	MW	127
SiH_2	33.6(21)	5.26	ED	128
S	26(2)	3.28(2)	ED	123
Se	29.5(10)	4.58(4)	MW	122

that in thietane shown in Figure 22. The puckering angle and barrier to inversion of some simple four-membered rings are collected in Table 13.

The four-membered ring of tetrafluoro-1,3-diselenetane, $(F_2CSe)_2$, is planar with D_{2h} symmetry[129] (Figure 23) similar to that of tetrafluoro-1,3-dithietane, $(F_2CS)_2$[130,131]. Both structures have been determined by ED. Ring planarity in $(F_2CSe)_2$ has been confirmed by CNDO/2 calculations, which indicated some degree of bonding interaction between the two Se atoms[129]. As two parallel ED studies of $(F_2CS)_2$[130,131] showed some discrepancy, the information on the Se derivative proved to be helpful in resolving the controversy. The two sets of ED results on $(F_2CS)_2$ were in agreement concerning the shape of the molecule. The discrepancy occurred as regards its size. The final arguments in deciding between the two parameter sets were based on the geminal F ··· F distances. This distance was reported to be 2.15(1) Å in $(F_2CSe)_2$[129]. Further, the mean value of F ··· F distances calculated from C—F bond lengths and F—C—F bond angles of 40(!) molecules containing the CF_3 group is 2.162 Å with a standard deviation of 0.008 Å! The F ··· F distance was never found to be smaller than 2.14 Å in those 40 molecules[3,132]. The striking stability of the F ··· F distances points to the importance of the non-bonded interactions between F ligands separated by one bond angle. The mean F ··· F distance is in excellent agreement with the 1,3 non-bonded radius of F, viz. 1.08 Å postulated a long time ago[62]. On this basis one of the two sets of results[130] for $(F_2CS)_2$ could be preferred[3,132]. The most important feature of $(F_2CSe)_2$ and $(F_2CS)_2$ is the planar four-membered ring. In this they are markedly different from selenetane and thietane, or from cyclobutane itself. A matrix isolation

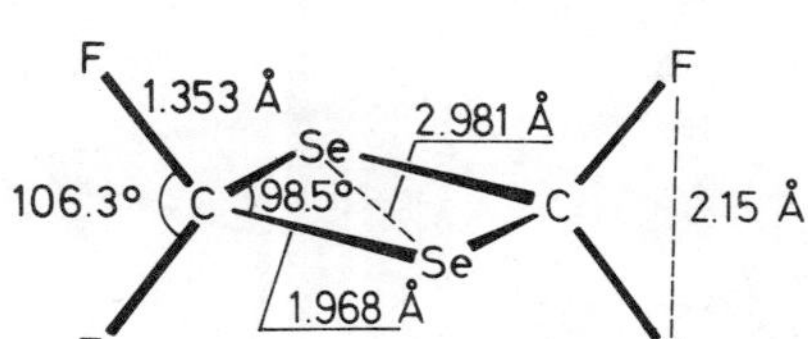

FIGURE 23. Model and geometrical parameters of tetrafluoro-1,3-diselenetane[129]

TABLE 14. The ring geometry of tetrahydroselenophene (C_2 symmetry) from ED[134] and MW[135] studies

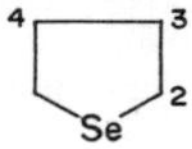

	ED (r_a)	MW (r_s)
Bond lengths (Å)		
Se—C	1.975(3)	1.963(2)
C(2)—C(3)	1.538(4)[a]	1.549(3)
C(3)—C(4)		1.527(2)
Bond angles (deg.)		
C—Se—C	89.1(5)	90.73(12)
Se—C—C	105.8(3)	104.97(17)
C—C—C	106.0(7)	106.87(20)
Dihedral angles (deg.)		
CSeCC	15.4(5)	
SeCCC	− 42.7(14)[b]	
CCCC	56.9(17)	
θ[c]		29.73(23)

[a]Mean C—C bond distance.
[b]Relative sign of the dihedral angle according to the convention by Klyne and Prelog[136].
[c]θ is the angle between the projections of the bonds Se—C and C(3)—C(4) on a plane perpendicular to the C_2 symmetry axis.

vibrational spectroscopic investigation of 1,3-dithietane[133] concluded C_{2v} symmetry for this molecule with a considerably puckered ring. On the other hand, this molecule was found to have D_{2h} symmetry in the solid state[133]. 1,3-Dithietane may in fact have a quasi-planar ring with low-frequency, large-amplitude deformation motion governed by a double-minimum potential.

Table 14 presents the geometrical parameters of tetrahydroselenophene from both ED[134] and MW[135]. The rotational spectroscopic information permitted distinction between the two different carbon–carbon bond lengths. Tetrahydroselenophene has a well-defined conformation, viz. the one with C_2 symmetry as shown in Figure 24, in accord with spectroscopic results[137,138]. The experimental findings were in complete agreement with the results of molecular mechanics calculations[134]. Tetrahydrothiophene has the same well-defined C_2 conformation as determined by ED[139] and molecular mechanics

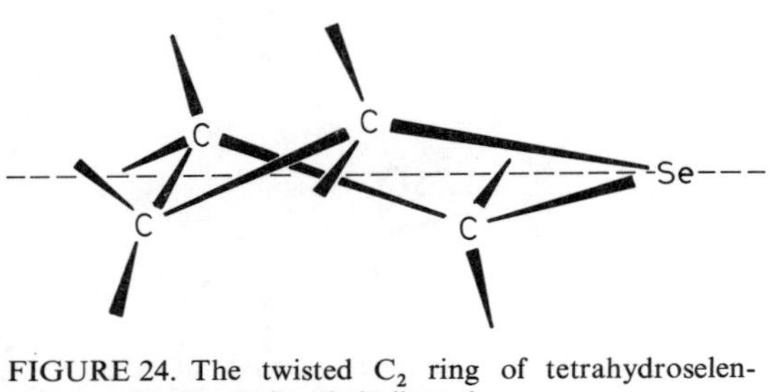

FIGURE 24. The twisted C_2 ring of tetrahydroselenophene with C—H bonds indicated

TABLE 15. Geometries of the planar rings of furan[146], thiophene[147], selenophene[143] and tellurophene[145] from MW

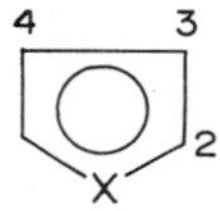

	X = O[a]	S[a]	Se[a]	Te
Bond lengths (Å)				
X—C	1.3621(10)	1.7140(14)	1.8547(9)	2.055
C(2)—C(3)	1.3609(10)	1.3696(17)	1.3695(12)	1.375
C(3)—C(4)	1.4309(20)	1.4232(23)	1.4332(30)	1.423[b]
Bond angles (deg.)				
C—X—C	106.55(7)	92.17(10)	87.77(7)	82.53
X—C—C	110.68(7)	112.47(23)	111.57(13)	110.81
C—C—C	106.05(7)	112.45(18)	114.55(10)	117.93

[a]Substitution structure (r_s).
[b]Assumed parameters.

calculations[139]. On the other hand, the structures of cyclopentane[140] and tetrahydrofuran[141,142] are characterized by pseudorotation. An interesting feature of tetrahydroselenophene is that its Se—C bond is longer than that of open-chain Me_2Se (cf. Table 7).

The complete molecular geometry of selenophene has been derived[143] from MW studies[143,144]. In a study of tellurophene[145] some parameters had to be assumed from related molecules. Ring bond lengths and bond angles are shown in Table 15. The most important structural feature of these molecules and their O and S analogues is their planarity. Their aromatic character has been studied and discussed widely. In the chemist's shorthand they are depicted in essentially two forms:

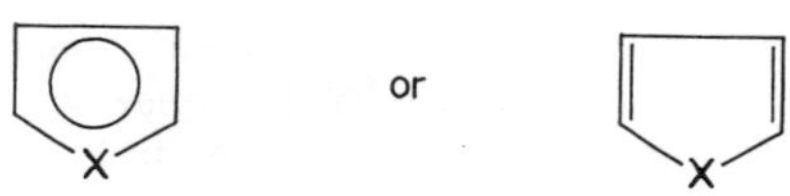

representing either the delocalized system of six π electrons or the two double bonds. The nice geometric and electronic symmetry of benzene is of course disturbed when a heteroatom with lone electron pairs replaces the —CH=CH— grouping. The middle carbon–carbon bond in selenophene is longer than the others and they are very similar in length to the corresponding bonds in furan[146] and thiophene[147]—a justification for the assumption made in the study of tellurophene. This bond pattern resembles the conjugated system in 1,3-butadiene[20]:

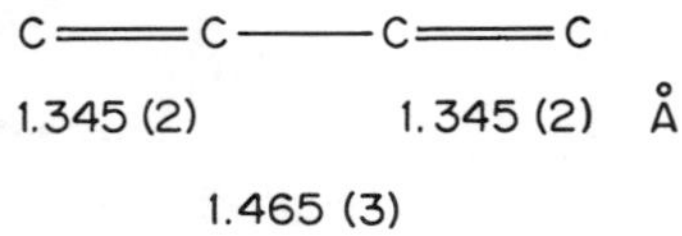

The bond length in benzene[148], 1.399(1) Å, lies just between these values.

The bond angle of the heteroatom decreases from furan to tellurophene (Table 16). However, due to the bond lengthening, the C(2)···C(5) distance increases in this order. The ring accommodates itself to this change first of all by opening the angles at C(3) and C(4) (see Table 15). The bond angles of O, S, Se and Te in the ring are smaller and their

TABLE 16. Parameters of dimethyl chalcogenides and non-aromatic and aromatic five-membered heterocycles

	X = O	S	Se	Te
Me_2X				
C—X (Å)	1.410	1.805	1.945	2.142
C···C[a] (Å)	2.334	2.746	2.898	3.14
C—X—C (deg.)	111.7	99.05	96.3	94
Reference	66	51	37	65

C—X (Å)	1.428	1.839	1.975	
C(2)···C(5)[a] (Å)	2.3[b]	2.677	2.771	
C—X—C (deg.)	106–111[b]	93.4	89.1	
Reference	142	139	134	

C—X (Å)	1.362	1.714	1.855	2.055
C(2)···C(5)[a] (Å)	2.184	2.470	2.572	2.710
C—X—C (deg.)	106.6	92.2	87.8	82.5
Reference	146	147	143	145
Δ[c] (deg.)	5.1	6.8	8.5	12

[a]Calculated from the other parameters.
[b]The ring cannot be characterized by constant values because of pseudorotation.
[c]The difference between angle C—X—C in Me_2X and C_4H_4X.

bonds are shorter than in the corresponding dimethyl chalcogenides, and the difference in bond angles (Δ) increases from O to Te (Table 16). The angles C—S—C and C—Se—C in thiophene and selenophene are even smaller than in the corresponding saturated heterocycles.

Microwave studies of 1,3,4-selenadiazole (**13**)[149] and 1,2,5-selenadiazole (**14**)[150] confirmed the planarity and C_{2v} symmetry of these molecules and with many assumptions yielded C—Se = 1.868 Å and the surprisingly small C—Se—C = 81.8° for **13** and N—Se = 1.80 Å and N—Se—N = 94.3° for **14**.

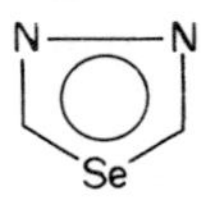

(13)

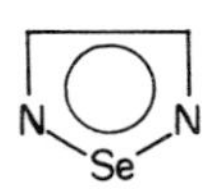

(14)

The crystal and molecular structures of diphenyl-substituted[151] and ring-fused derivatives[152] of **14** have been determined by XD and are discussed in relation to the structures of their O and S analogues.

In a $\Delta^4-1,4,2\lambda^5$-selenazaphospholine derivative (**15**) the two rings occupy axial–equatorial positions at the distorted trigonal bipyramidal phosphorus atom[153]. The selenazaphospholine ring is nearly planar with single bonds P—Se = 2.273(2) Å, Se—C = 1.972 Å and angle P—Se—C = 89.9°.

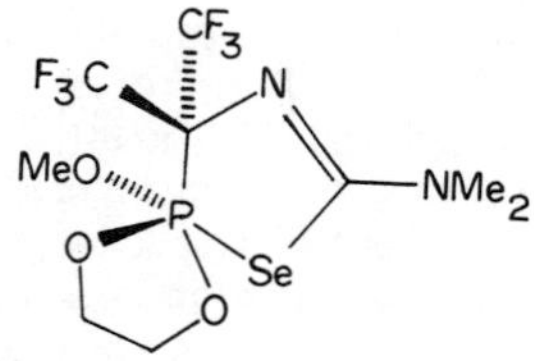

(15)

The different crystal structures of 1, 2-benzisoselenazole (**16**) and 1, 2-benzisotellurazole (**17**) explain why the latter compound has an unexpectedly high melting point and low solubility[154]. There are only van der Waals' intermolecular contacts in the crystals of **16**, whereas short (2.5 Å) Te···N interactions link the molecules of **17** into chains. Mean bond lengths and angles from crystallographically different molecules are C—Se = 1.86 Å, Se—N = 1.86 Å, C—Se—N = 88°, C—Te = 2.08 Å, Te—N = 2.11 Å and C—Te—N = 80°.

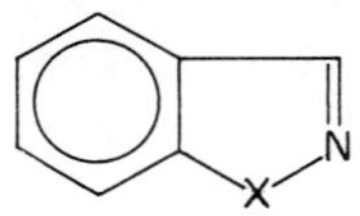

(16) X = Se

(17) X = Te

In a 3-substituted derivative of **16** (**18**) C—Se = 1.845(11) Å, Se—N = 1.833(7) Å and C—Se—N = 91.0(4)° were determined[155].

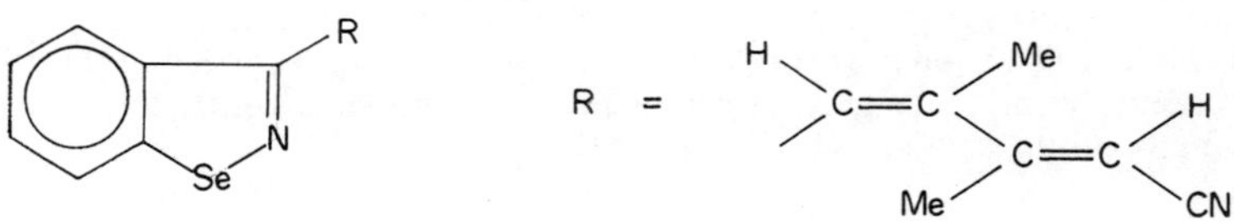

(18)

5-Acetoxy-6, 7-dichloro-3, 4-benzobicyclo[3.2.0]-2-selenaheptene (**19**) is a photo-addition product[156]. The bond angle of the Se atom, 87.4(4)°, is equal to that in selenophene (Table 15). Its bond lengths are the normal values for single bonds, Se—C(sp^3) = 1.949(9) and Se—C(sp^2) = 1.905(9) Å [156].

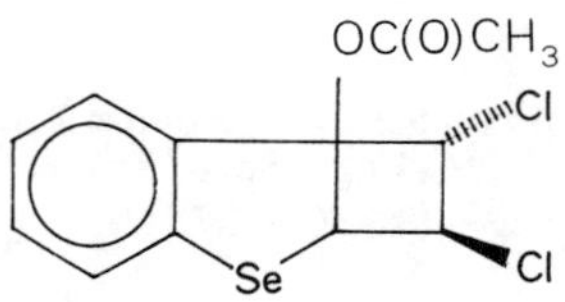

(19)

The molecular structure and conformation of some Se compounds, which are formed in the oxidation of diketones by SeO_2, have been elucidated recently by XD. Two structures, **9** and **10**, have been treated in Section III.A. 5,6-Dihydro-4,4,4′,4′,6,6,6′,6′-

octamethylspiro[1,3-benzoxaselenole-2, 1′-cyclohexane]-2′,3′,7(4*H*)-trione (**20**)[157] and 6,7-dihydro-4′,4′,6,6-tetramethylspiro[1,3-benzoxaselenole-2,1′-cyclohexane]-2′,4,6′(5*H*)-trione (**21**)[158], as well as 1,5,5-trimethyl-7-selenabicyclo[2.2.1]heptane-2,3-dione (**22**)[159], contain the Se atom in five-membered rings. The bond angles of Se are 83.6(2), 82.5 and 78.6(3)° in **20**,**21** and **22**, respectively, and therefore very small in the Se-bridged compound **22**. The selenium–carbon bonds are, in part, longer than in acyclic selenides (cf. Table 7): Se—C(sp^2) = 1.885(5) and 1.903 Å, Se—C(spiro) = 2.012(5) and 2.007 Å in **20** and **21**, respectively (values for **21** are averages from two crystallographically independent molecules), while in **22** the bond to the quaternary carbon is 2.021(7) Å, and to the other bridgehead atom 1.977(7) Å.

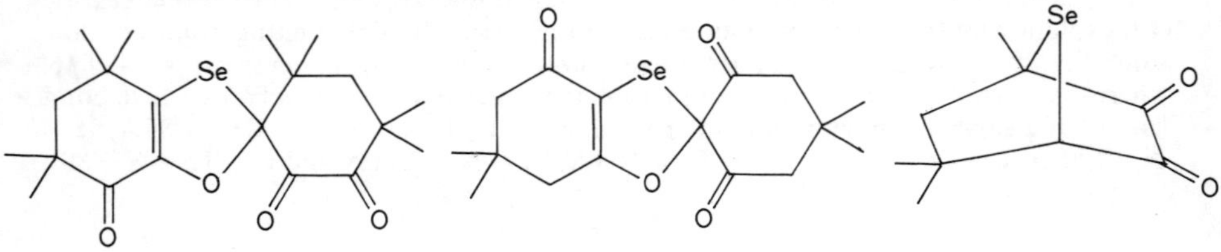

(20) **(21)** **(22)**

Dibenzoselenophene (**23**)[160] and dibenzotellurophene (**24**)[161] have been studied by XD. The molecules are practically planar, with only small dihedral angles between the best planes of the individual rings. Bond lengths and bond angles within the five-membered rings are given in Table 17, together with those in dibenzofuran (X = O) and dibenzothiophene (X = S). McCullough[161] has discussed the structural changes in this series of molecules and compared chalcogen bond angles with those in dimethyl chalcogenides. The angle pattern is very similar to that found in the respective isolated five-membered rings (Table 15). The bond angle of the heteroatom is somewhat smaller in the fused ring system. Bonds seem to lengthen by ring fusion. Caution is called for, however, in comparing bond distances from MW and XD.

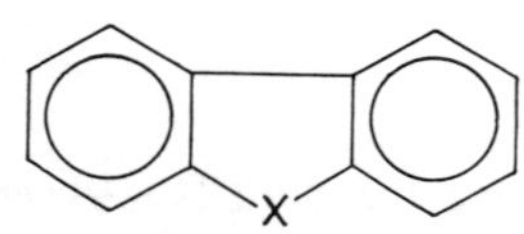

(23) X = Se

(24) X = Te

The bond lengths and bond angles of heteroatoms in selenolo[2,3-*b*] benzothiophene (**25**)[164] are: Se—C(2) = 1.896(17), Se—C(8a) = 1.859(13), S—C(7a) = 1.741(15), S—C(8a) = 1.687(15) Å, ∠C—Se—C = 87.7(7), ∠C—S—C = 91.4(7)°; and in

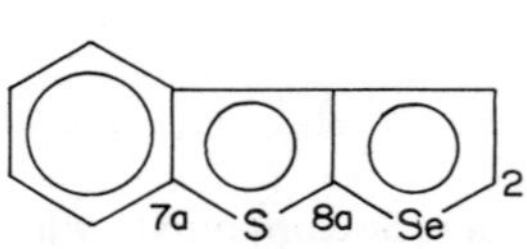

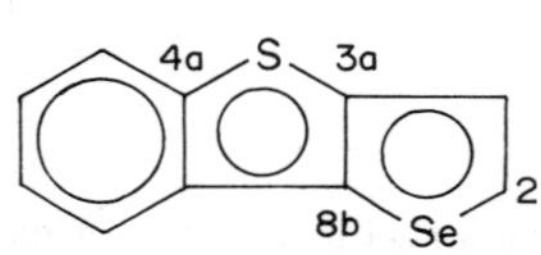

(25) *cis* **(26)** *trans*

TABLE 17. Geometries of the planar five-membered rings of dibenzo-furan[162], -thiophene[163],-selenophene[160] and -tellurophene[161] from XD[a]

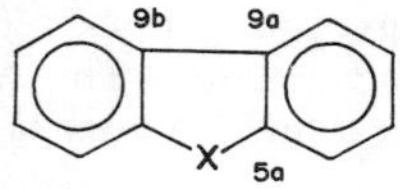

	X = O	S	Se	Te
Bond lengths (Å)				
X—C	1.418(6)	1.740(8)	1.899(5)	2.087(5)
C(5a)—C(9a)	1.382(7)	1.409(11)	1.398(7)	1.394(6)
C(9a)—C(9b)	1.480(6)	1.441(1)	1.453(7)	1.460(7)
Bond angles (deg.)				
C—X—C	104.4(4)	91.5(4)	86.7(2)	81.7(2)
X—C(5a)—C(9a)	112.9(4)	112.3(6)	112.3(4)	112.1(4)
C(5a)—C(9a)—C(9b)	105.6(4)	111.9(7)	114.3(5)	117.1(5)

[a]Mean values from chemically equivalent distances and angles as well as their standard deviations are presented after McCullough[161].

selenolo [3,2-*b*] benzothiophene (**26**)[164]: Se—C(2) = 1.861(10), Se—C(8b) = 1.863(9), S—C(3a) = 1.751(8), S—C(4a) = 1.734(8) Å, ∠C—Se—C = 87.0(4), ∠C—S—C = 91.0(4)°. The shapes of the heterocyclic rings are similar to the corresponding ones in other molecules.

Tetrathiafulvalene (TTF, $\Delta^{2,2'}$-bis-1,3-dithiole) (**27**) derivatives and Se analogues have

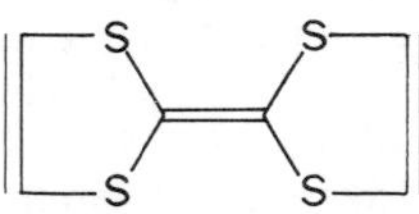

(**27**)

been widely studied because of their ability to form charge-transfer salts with electron acceptors like TCNQ (**28**). The structures of a few Se compounds of this type will be

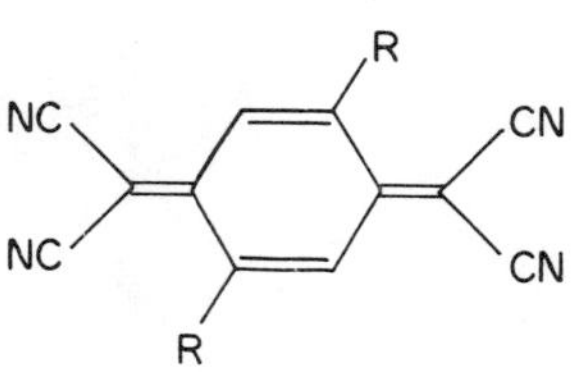

(**28**) R = H

(**29**) R = Me

touched on here. The pure electron donor 4,4′,5,5′-tetramethyl-$\Delta^{2,2'}$-bis-1,3-diselenole (tetramethyltetraselenafulvalene, TMTSF) (**30**) crystallizes in the triclinic space group P$\bar{1}$ with one molecule in the unit cell[165]. The molecule has thus a centre of symmetry (C_i). The rings are slightly puckered at the Se ··· Se lines, taking envelope forms with 6.1° dihedral angle. The inner Se—C bonds are 1.892(7), the outer ones 1.906(7) Å on the average, while the C—Se—C angles are 93.9°. The molecular packing is determined essentially by van der Waals' contacts. The crystal of the 1:1 salt of **30** with 2,5-dimethyl-7,7,8,8-tetracyano-*p*-quinodimethane (**29**) consists of separate stacks of the radical anions of **29** and of the radical cations of **30**[166]. The rings of **30** are puckered by only 2.3° and the centrosymmetric molecule has an overall chair conformation. The two kinds of Se—C bonds, 1.879(5) and 1.896(5) Å, are somewhat shorter than in the neutral compound, and the C—Se—C angle is 94.1°.

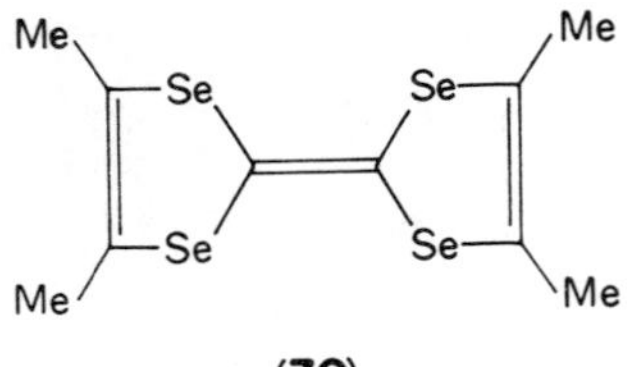

(30)

The salt of TMTSF with TCNQ has two crystalline forms with interesting properties. One is a highly conducting crystal (black form) built of segregated columns of the donor and acceptor molecules[167]. The other is a semiconducting modification (red form) which consists of mixed stacks of alternating donor and acceptor molecules[168]. The overlap of molecules within a stack is shown in Figure 25. The distance between molecular planes is 3.6 Å in the stack of cations (black form) and 3.5 Å within the mixed stack (red form), about the same as in the TCNQ crystal itself[169]. The stacks of TCNQ anions in the black form of the salt are characterized, on the other hand, by interplanar spacings of 3.26 Å [167]. The Se bond lengths and angles are (in the above order) 1.88(1), 1.90(1) Å and 94.5° in the black form[167,168] and 1.908(4), 1.900(4) Å and 94.3° in the red form[168]. It is noteworthy that S has the same bond angle, 94.4°, in pure crystalline TTF (**27**)[170], and the S—C bonds to the bridgehead carbons and longer than the other S—C bonds, viz. 1.757(2) and 1.730(2) Å. This is in accord with calculated electron populations[170].

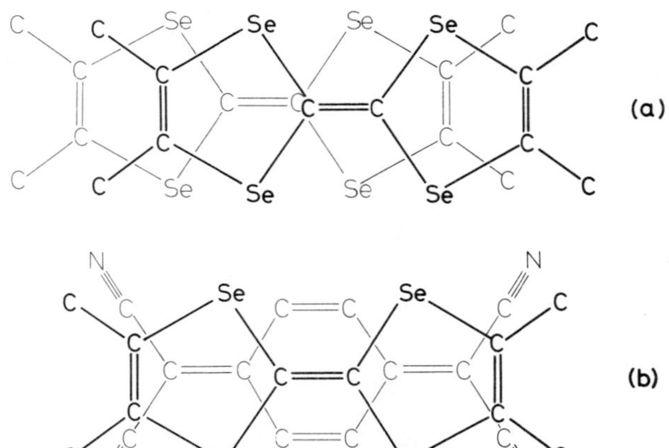

FIGURE 25. Overlap of molecules, as seen along the normal to the mean molecular plane, (a) in the stack of cations in the black form of TMTSF·TCNQ (after Reference 167) and (b) in the mixed stack of cations and anions in the red form of TMTSF·TCNQ (after Reference 168)

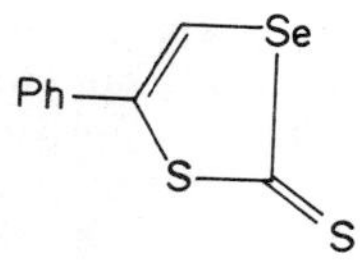

(**31**)

5-Phenyl-1,3-thiaselenole-2-thione (**31**) is the donor in a charge-transfer complex with TCNQ. **31** is planar in the complex and has bond lengths Se—C(=C) = 1.858 Å, Se—C(=S) = 1.878 Å and angle C—Se—C = 92.0°[171].

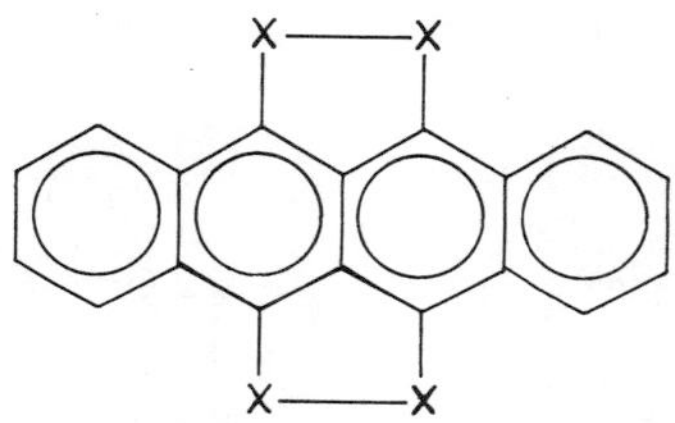

(**32**) X = S

(**33**) X = Se

(**34**) X = Te

Chalcogen–chalcogen bonds are present in the planar fused ring system of tetrathiotetracene (TTT) (**32**) and its derivatives and analogues, which also form charge-transfer salts with different anionic species. The crystal structures of 2:1 complexes of tetraselenotetracene (**33**) with chloride[172] and thiocyanate[173] ions have been determined. In both crystals the molecules of **33** lie in centres of symmetry and are nearly planar: the atoms of the Se—Se group are on opposite sides of the mean plane of carbons, destroying complete planarity. The molecules form columns along the *c* axis and the anions reside in the canals between these stacks. The stacks, the interstack spaces and the positions of the anions explain why the chloride salt behaves as a metal and has an electric conductivity about a hundred times higher than has the thiocyanate complex[172]. The geometry of the diselenide group in cation **33** is characterized by the following mean parameters:

Anion	Se—Se(Å)	Se—C(Å)	C—Se—Se(deg.)
Cl^-	2.323(1)	1.901(6)	91.6
SCN^-	2.320(7)	1.86	90

The molecules form stacks in the crystals of neutral tetratellurotetracene (**34**)[174] and tetrathiotetracene (**32**)[175]. The corresponding structural data are:

Molecule	X—X(Å)	X—C(Å)	C—X—X(deg.)
34(X = Te)	2.680(7)	2.135(15)	87.7(8)
32((X = S)	2.100(3)	1.781	95.9

Chalcogen–chalcogen bond distances are comparable with those in acyclic dichalcogenides (Section III.B). Chalcogen–carbon bond lengths are normal for single bonds to an aromatic carbon. Bond angles in these rings are smaller than in acyclic molecules (Tables 7 and 9) and larger than in the rings listed in Table 16. The increasing X—X distance from S to Te gives rise to an opening of carbon bond angles in the five-membered rings of **32**–**34**.

The structures of six-membered rings with alternating carbon and chalcogen atoms have been determined from three-dimensional XD photographic microdensitometer data. (Figure 26). All these rings have the chair conformation in the crystal, and the Me substituents are in equatorial positions. The intraring bond angles of carbon exceed the regular tetrahedral value (see Figure 26 for references). The ring shapes in (a), (b), (c) and (d) of Figure 26 seem to be very similar. The Se and S atoms have practically the same bond angles. The introduction of O atoms in the ring, however, causes appreciable changes. The sum of the bond angles, which is 720° in a planar hexagon and may be a measure for the non-planarity of the ring, is 641–645° in the trithiane and triselenane rings of (a), (b), (c) and (d), 647° in (e) and 633° in(f) of Figure 26, while it is 660° in free 1,3,5-trioxane, with angle C—O—C 108.9(8)°, from an ED study[182]. The average Se—C bond distances in (c), (d), (e) and (f), 1.94–1.95 Å, agree with single-bond lengths in non-cyclic compounds (Table 7).

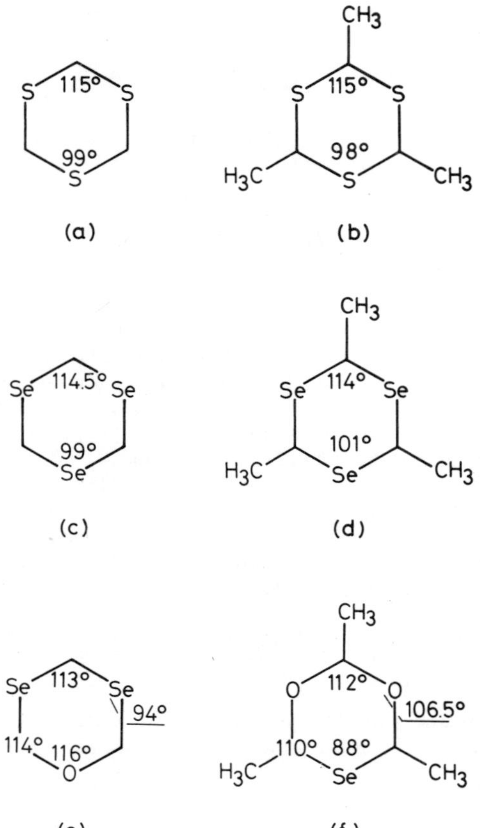

FIGURE 26. Mean endocyclic bond angles in chair-form six-membered rings: (a) 1,3,5-trithiane[176], (b) *cis*-2, 4, 6-trimethyl-1, 3, 5-trithiane[177], (c) 1, 3, 5-triselenane[178], (d) *cis*-2, 4, 6-trimethyl-1, 3, 5-triselenane[179], (e) 1, 3, 5-oxadiselenane[180], (f) *cis*-2, 4, 6-trimethyl-1, 3, 5-dioxaselenane[181]

TABLE 18. Structural data of 1-telluracyclohexane-3,5-dione derivatives from XD

Molecule	Te—C (Å)	C—Te—C (deg.)	Reference
35	2.15(1), 2.18(1)[a]	89.5(4)	186
	2.17(1), 2.15(1)[a]	90.8(4)	183
36	2.202(8)[b], 2.157(8)[c]	88.4(3)	184
37	2.184(6), 2.206(7)[a]	89.7(3)	187
38	2.168(5)	86.4(2)	185
39	2.18(1), 2.18(1)[a]	89.3(6)	188

[a]The two Te—C bonds are crystallographically different.
[b]The Te—C(Me) bond.
[c]The Te—CH_2 bond.

Cyclic complexes of Se(II) or Te(II) are formed in the reaction of selenium or tellurium tetrachloride with 1,3-diketones. Acetylacetone or its derivatives function in these complexes as bivalent bidentate ligands, bonded to the chalcogen atom through the carbon atoms next to the carbonyl group. This type of bonding is completely different from the bonding of acetylacetone with other central atoms (see References 183–185 for references). The crystal and molecular structures of some 1-telluracyclohexane-3,5-dione derivatives **35**–**39** (Table 18) were determined in the 1970s. In some of these studies by Dewan and Silver, some museum pieces of crystals obtained about fifty years before were used (see Table 18 for references).

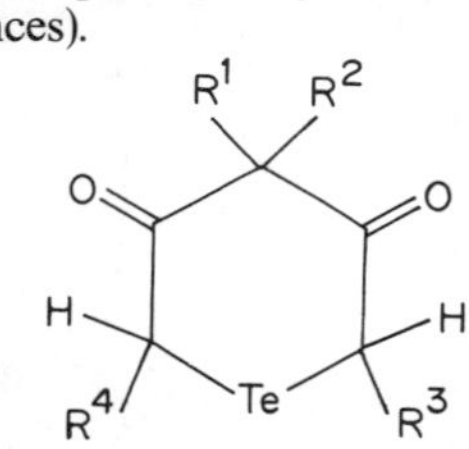

(**35**) $R^1 = R^2 = R^3 = R^4 = H$
(**36**) $R^1 = R^4 = H$; $R^2 = R^3 = Me$
(**37**) $R^1 = R^2 = H$; $R^3 = R^4 = Me$
(**38**) $R^1 = R^2 = Me$; $R^3 = R^4 = H$
(**39**) $R^1 = H$; $R^2 = R^3 = R^4 = Me$

The telluracyclohexane rings have a chair form and the Me groups occupy equatorial positions—except in 4,4-dimethyl-1-telluracyclohexane-3,5-dione (**38**) where this is not possible. The parent compound, 1-telluracyclohexane-3, 5-dione (**35**) itself has been the subject of three XD studies[183,186,189]. The molecules are aligned in piles along axis *a* (Figure 27). Weak association is formed[183] between four of these piles via Te lone pairs at Te···Te distances of 3.95(1), 3.95(1), 3.97(1), 3.97(1) and 4.18(1) Å. Each Te atom has five contacts to two molecules each in neighbouring piles shifted by $\pm a/2$ and to one molecule in the diagonally opposite stack[183]. The above listed distances are all shorter than twice the Te van der Waals' distance (Table 10). Crystals of **37** and **39** are isomorphous, monoclinic, $P2_1$; **36** is also monoclinic, $B2_1/c$ (see Table 18 for references). There are zig-zag Te···Te···Te chains in these structures along axis *b* with Te···Te distances of 4.068(7) Å (**37**), 4.138(7) Å (**39**) and 4.042(5) Å (**36**). In crystalline **38**, which is orthorhombic, Pmnb, the shortest Te···Te distances are 5.05 Å, much longer than in the other crystals, probably due to the space requirement of the axial methyl group in **38**. The schematic projection of this structure in Figure 28 demonstrates how the lone pairs of tetrahedrally sp^3-hybridized Te(II) atoms are oriented into available space in the structure[185]. Another possible hybridization of Te(II) would be the planar trigonal sp^2 hybrid with the remaining lone pair on a p orbital perpendicular to the plane[183,184]. Some physical properties of these crystals such as Mössbauer spectra[190], colour, crystal growth and variations of the C—Te—C bond angle have been correlated with the existence of one-dimensional chains of molecules in the crystal and with geometric and electronic characteristics of intermolecular Te···Te interactions (see References 183–185, 187, 188 and 190 and references therein). The relatively small angle in **38** has been related[185] to weaker intermolecular Te···Te interactions in this crystal (see above).

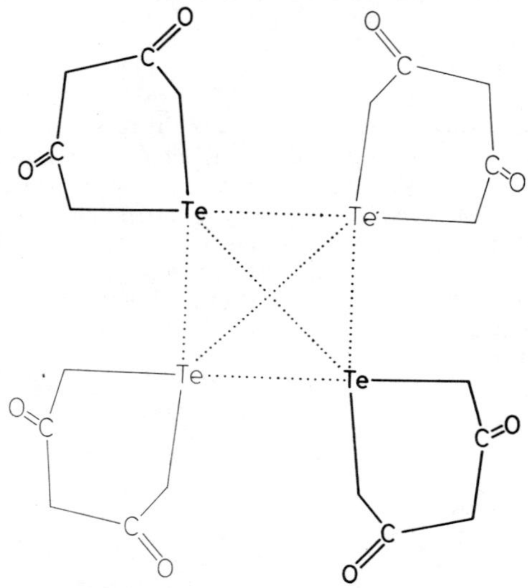

FIGURE 27. Molecular packing in the crystal of 1-telluracyclohexane-3, 5-dione (**35**): projection of four piles of molecules along the *a* axis (after References 183 and 186)

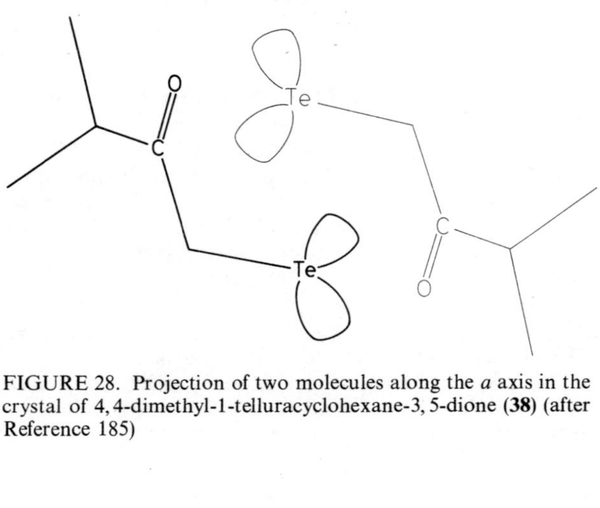

FIGURE 28. Projection of two molecules along the *a* axis in the crystal of 4, 4-dimethyl-1-telluracyclohexane-3, 5-dione (**38**) (after Reference 185)

The non-planar heterocyclic ring of 3-formyl-5,8-dimethyl-1,2-benzothiaselenin-1,1-dioxide (**40**) has bond lengths Se—C = 1.888(7) and S—Se = 2.205(1) Å and bond angle S—Se—C = 92.8(2)°[191].

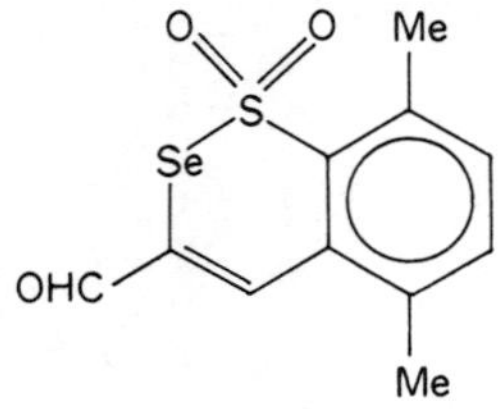

(**40**)

9,10-Dihydroanthracene[192a] and its heterocyclic analogues are folded at atoms 9 and 10, having a boat-form central ring, and are referred to epithetically as 'butterfly' molecules (Figure 29). An exception is dibenzo-*p*-dioxin (**41**), which is, at least in the crystal phase, practically planar[192b]. Geometrical parameters of such molecules containing Se or Te atoms are listed in Table 19, together with those of O and S analogues for comparison,

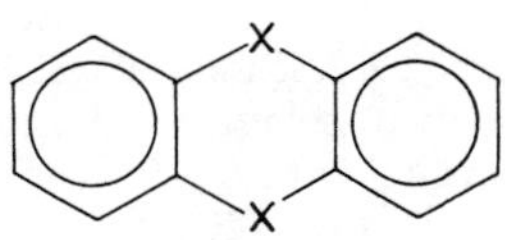

(**41**) X = O
(**42**) X = S
(**43**) X = Te

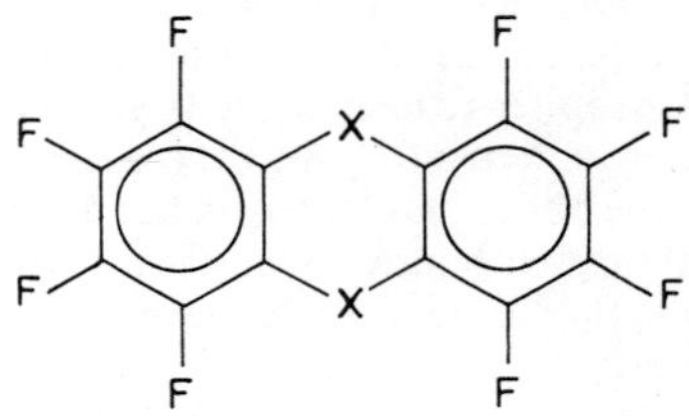

(**44**) X = O
(**45**) X = S
(**46**) X = Se
(**47**) X = Te

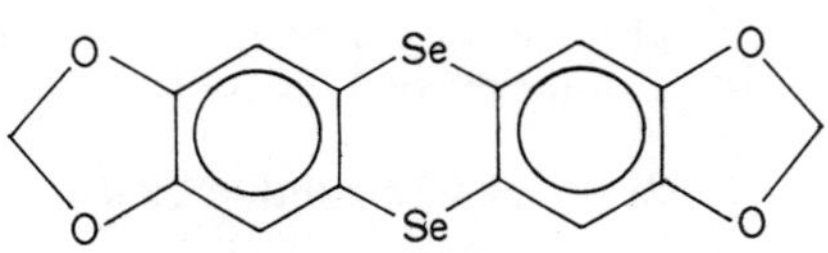

(**48**)

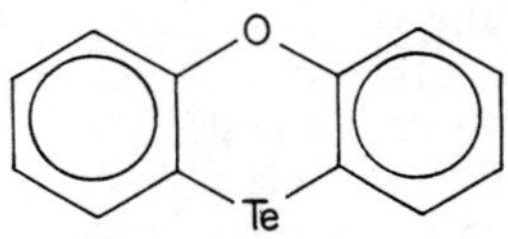

(**49**)

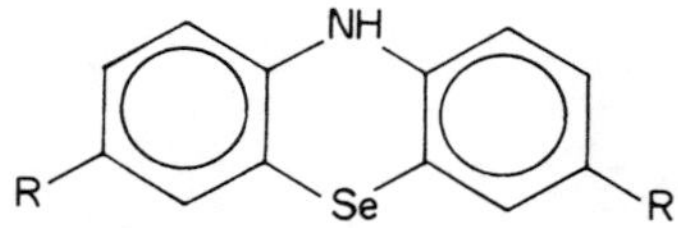

(**50**) R = H
(**51**) R = Cl

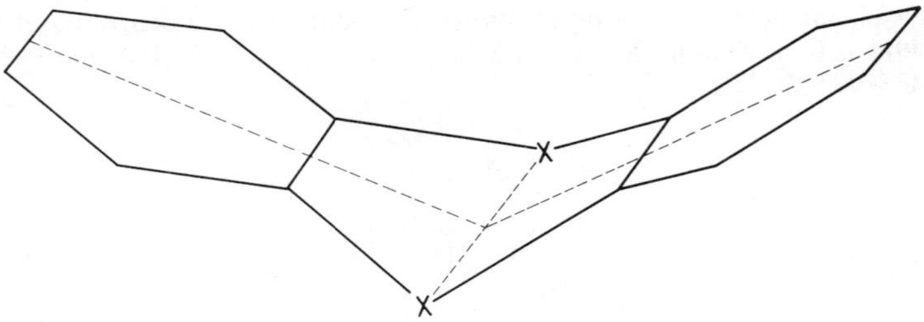

FIGURE 29. The shape of a 'butterfly' molecule

mainly from XD investigations. Thianthrene (**42**) has been studied in the gas phase by ED[194]. Chalcogen–carbon bond distances are normal for single bonds to sp^2-hybridized carbon (cf. Table 7 and diphenyl ditellurides in Section III.B). Bond angles show the usual trend down the chalcogen group. The bond lengths in the perfluoro derivatives **44**–**47** are practically the same as in the other molecules; chalcogen bond angles in **44**,**45** and **47** are smaller than in the corresponding non-fluorinated molecules **41**,**42** and **43** (Table 19). It is instructive to follow the changes[201] in the dihedral angle θ_1 of the two phenylene planes and the fold angle θ_2 of the central ring in the perfluoro compounds (Figure 30). Both angles diminish as the size of the chalcogen atom increases and thus the repulsion of the tetrafluorophenylene rings gets smaller. This is seen for example, from the 1,3 C···C distances in the central rings of the four molecules **44**–**47**: 2.31, 2.69, 2.85 and 3.06 Å. At the same time, the usual chalcogen bond angles are approached only by stronger folding of the central ring, without substantially distorting the carbon bond angles from 120°. The latter angles in the central ring are, nevertheless, systematically larger[195] than 120°, viz. O—C—C 122.8(2), S—C—C 121.3(2), Se—C—C 121.7(9) and Te—C—C 122.5(3)°. Another interesting observation[201] is that, except for perfluorotelluranthrene (**47**), $\theta_1 \geq \theta_2$ is valid in these and similar molecules. This is demonstrated by the data in Figure 30 as well as by θ_1, θ_2 124.0, 120.1° in telluranthrene (**43**), and 145, 138° in phenoxatellurin (**49**), while $\theta_1 = \theta_2$ was found to be 131.4° in thianthrene (**42**), 150° in phenoselenazine (**50**) and 146° in 3,7-dichlorophenoselenazine (**51**) (see Table 19 for references). Repulsion and attraction between the phenylene 'wings' may give an explanation for this phenomenon[201].

The eight-membered ring of 1,3,5,7-tetraselenocane, $(CH_2Se)_4$, has an asymmetric twist-chair conformation in the crystal (Figure 31) with Se—C bond distances of 1.91–1.98 Å and C—Se—C angles of 98.0–101.5°[203]. The relatively large Se—C—Se angles between 114 and 119°, accompanied by Se···Se distances of 3.26–3.38 Å, may be a consequence of the size of the Se atoms[203].

The crystal molecular structure of a dimer (**52**) of 1*H*,4*H*-naphtho[1,8-*d*,*e*] [1,2]-diselenepin has been determined[204]. The molecule possesses C_2 symmetry; the angle between the two naphthalene planes is 87.6°. The CSeSeC dihedral angle is 88.1°, similar to this angle in acyclic diselenides. The Se—Se bond is 2.315(2) Å long and the mean Se—C distance and Se—Se—C bond angle are 1.991(11) Å and 101.8°.

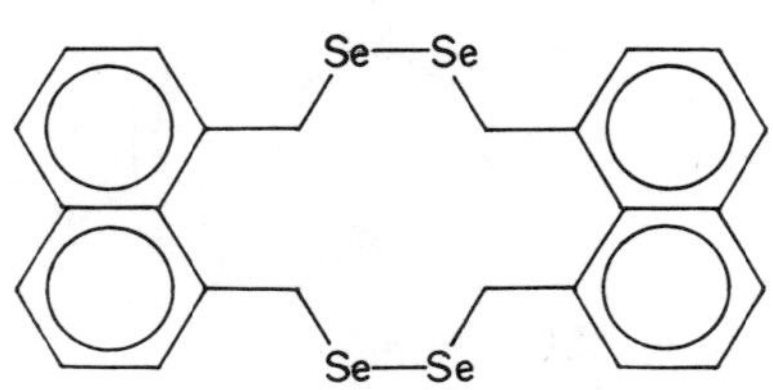

(**52**)

TABLE 19. Chalcogen bond lengths and bond angles in 9,10-diheteroanthracene molecules **41**–**51**

Molecule	Bond length (Å)	Bond angle (deg.)	Reference
	O—C	C—O—C	
41	1.383(8)	116.4(5)	192b
44	1.374(2)	114.3(2)	193
	S—C	C—S—C	
42	1.770(3)	104.1(1)	194
45	1.765(2)	99.3(1)	195
	Se—C	C—Se—C	
46	1.910(8)	96.5(10)	196
48	1.920(8)	98.2(4)	197
50	1.89(2)	97(1)	198
51	1.921(5)	95.4(2)	199
	Te—C	C—Te—C	
43	2.112(4)	95.6(3)	200
47	2.114	92.9(1)	201
49	2.098	89.4(3)	202

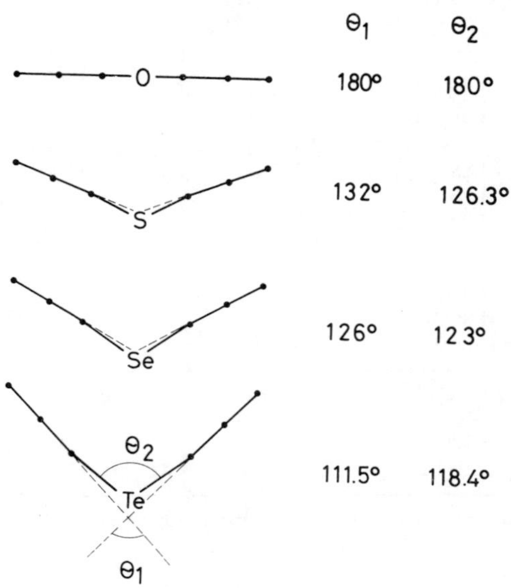

FIGURE 30. Schematic projections of the ring planes, along the chalcogen···chalcogen line, in perfluoro-dibenzo-*p*-dioxin (**44**) -thianthrene (**45**), -selenanthrene (**46**) and -telluranthrene (**47**). Deviations are slightly exaggerated in the drawing. θ_1 is the dihedral angle between the mean planes of the phenylene rings, θ_2 is the fold angle of the central ring. See Table 19 for references

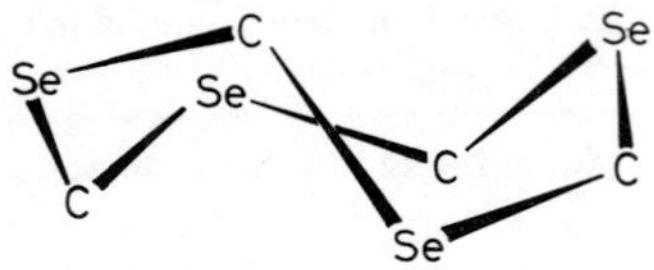

FIGURE 31. The ring conformation in crystalline 1,3,5,7-tetraselenocane. Drawn from the atomic coordinates given in Reference 203

TABLE 20. Bond configuration of Se and Te in organometallic complexes

Molecule	M	Se—C(Å)	M—Se (Å)	M—Se—C (deg.)	Reference
53	Cu	1.86	2.43	84	205
59	Cu	1.82	2.50	81.2	206
54	Zn	1.88	2.48	77–95	205
60	Hg	1.89(3)	2.477(3)	104.4(8)	207
57	Se	1.902	2.462	89.1	208
		1.839	2.816	80.1	
58	Se	1.896	2.450	89.5	209
		1.849	2.867	78.5	
61	Te	1.897(7)	2.7229(16)	101.1(2)	210
		1.893(7)	2.8895(17)	99.8(2)	
62	Te	1.884	2.812	91.7(7)	211
				98.0(8)	
63	Mo	1.86(1)	2.492(2)	105.1(4)	212
64	Co	1.952(3)	2.378(1)	92.1(1)	213
65	Co	1.949(10)	2.355(1)	92.7(3)	214
55	Ni	1.88	2.317	85.2	205
56	Ni	1.86	2.31	85.2	215
66	Ni	1.906(10)	2.280(3)	102.8(3)	216
67	Ni	1.86(1)	2.387(1)	87.3[a]	217
68	Ni	1.78(3)	2.391(5)	85[a]	218
69	Rh	1.926(7)	2.527	106.3	219
70	Pt	1.89(2)	2.462(4)	[b]	220
		1.86(3)	2.491(3)	[b]	
		Te—C (Å)	M—Te (Å)	M—Te—C (deg.)	
71	Cr	2.12(2)	2.765(4)	96.1(15)	221
72	Hg	2.10(1.2)	2.697(10)	99.9(16)	222

[a]Calculated from the atomic coordinates given in the original paper; mean value.
[b]Not given.

D. Organometallic Complexes

Bond lengths and bond angles of Se and Te in some organometallic complexes with a two-coordinated chalcogen atom directly bonded to a carbon and a metal atom (**53–72**) are summarized in Table 20. Included are some compounds where the central atom is also Se or Te. Complexes with pseudohalide ligands have not been considered. The selenocyanate ligand may be coordinated to metals through the Se or N atom or both, it may form a bridge between two metal atoms, and the choice of binding mode depends rather on steric than on electronic factors[223].

The diselenocarbamate complexes of Cu(II), Zn(II) and Ni(II) (**53–56**) are isomorphous with the corresponding dithiocarbamate compounds. Average differences of Se—C and S—C (0.152 Å) and Se–metal and S–metal bond lengths (0.113 Å), compared with the difference (0.14 Å) of Se and S covalent radii (Table 10) may indicate that Se–metal bonds are of higher order than S–metal bonds[205]. The diselenocarbamate complexes (**53–58, 68, 70**) all contain practically planar Se_2CNC_2 skeletons in their ligands but the environment of the Se atoms and the coordination of the central atom is different in their crystals. The Ni atom has a distorted square planar coordination and monomeric molecules are present in the crystals of **55**[205] and **56**[215] while a pair of centrosymmetrically related molecules is linked by a pair of Se–metal bonds in **53** and **54**[205]. In **53** and **54** the Cu and Zn atoms obtain five close neighbours in different ways (Figure 32) so that the geometry is a distorted tetragonal pyramid around Cu and something between this and a trigonal bipyramid around Zn[205]. Two diselenocarbamate ligands are coordinated to a Se atom in bis(*N,N*-diethyldiselenocarbamato)selenium(II) (**57**)[208] and selenium bis(1-pyrrolidinecarbodiselenoate) (**58**)[209], forming two approximately linear three-centre four-electron (3c–4e) bonding systems in a nearly planar trapezoid $SeSe_4$ structure. There are two shorter Se—Se bonds and, in *trans* positions to them, two longer ones (Table 20) and they are all longer than the Se—Se bond in acyclic or cyclic diselenides (Sections III.B and III.C). The adjacent Se—C bonds are, in the same order, longer and shorter and the corresponding Se—Se—C bond angles are also systematically different (see Table 20).

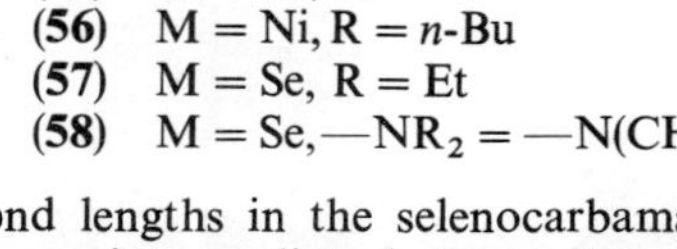

(**53**) M = Cu, R = Et
(**54**) M = Zn, R = Et
(**55**) M = Ni, R = Et
(**56**) M = Ni, R = *n*-Bu
(**57**) M = Se, R = Et
(**58**) M = Se, —NR_2 = —$N(CH_2)_4$

The selenium–carbon bond lengths in the selenocarbamate (**53–59, 68, 70**) and selenourea (**60–62**) complexes are intermediate between the expected length of a single (1.94 Å) and a double bond (1.73 Å)[209] (see also data in Sections II and III.A for comparison), and are comparable with the bond length in crystalline selenourea and related systems (see Section II). The Se—C bonds are remarkably short in the cation of **68**, which is an octahedrally coordinated Ni(IV) complex. The Se bond angles in the bidentate ligands are influenced by strain in the four-membered ring.

The central metal atom is octahedrally coordinated in **63–65**, **67–69** and **71**. In the cobalt complexes (**64, 65**) a lengthening of the Co—N bond *trans* to the Se atom has been observed[213,214] and compared to the '*trans* effect' in analogous ethanethiolamine complexes. The Se bond configuration in the selenol complexes (**64–66, 69**) is similar to that found in acyclic selenides (cf. Table 7).

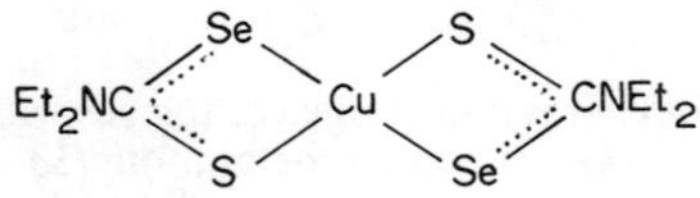

(**59**)

$[MeHgSeC(NH_2)_2]^+ NO_3^-$
(**60**)

$PhTe[SeC(NH_2)_2]_2{}^+ Cl^-$
(**61**)

$Te[SeC(NH_2)_2]_4{}^{2+}$ 2 Cl^-
(**62**)

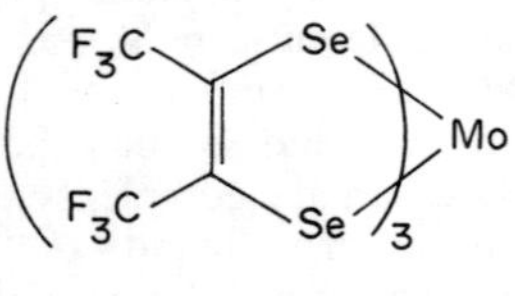

(**63**)

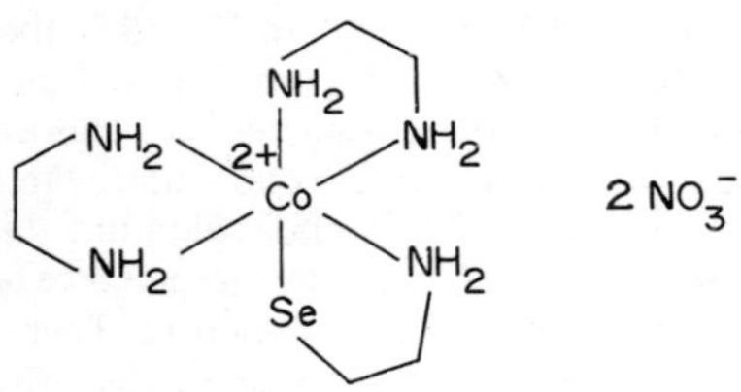

(**64**)

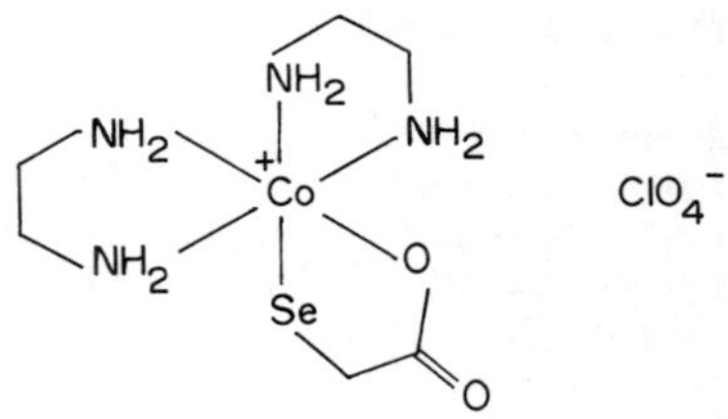

(**65**)

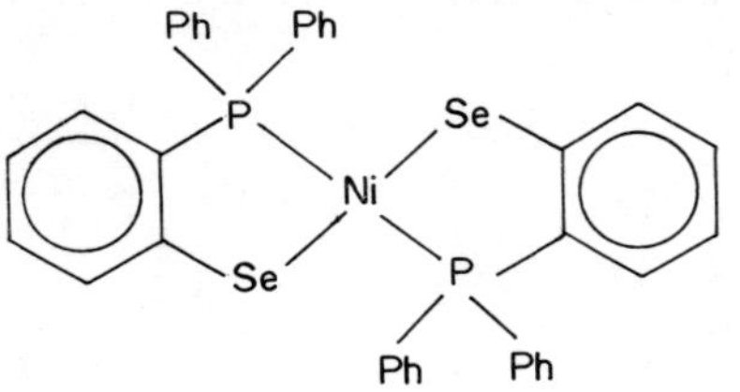

(**66**)

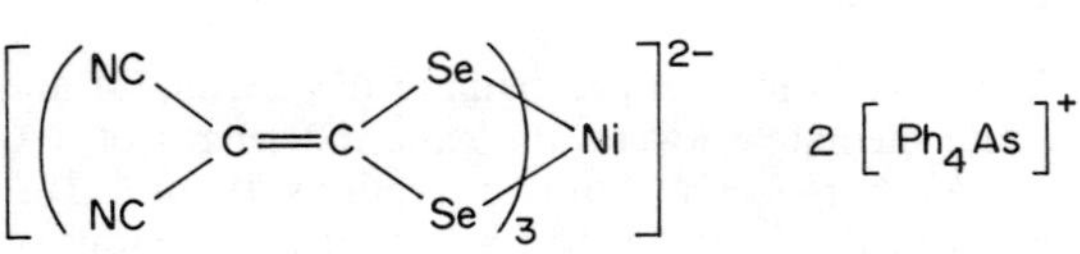

(**67**)

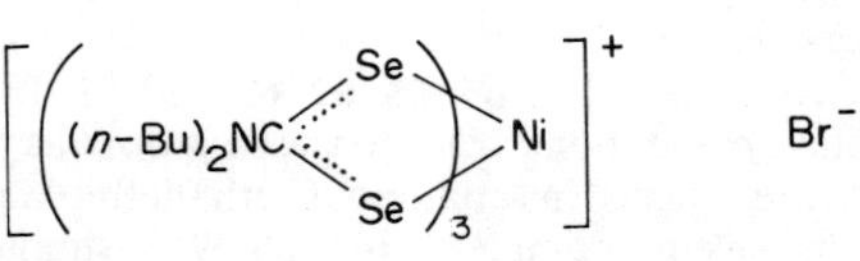

(**68**)

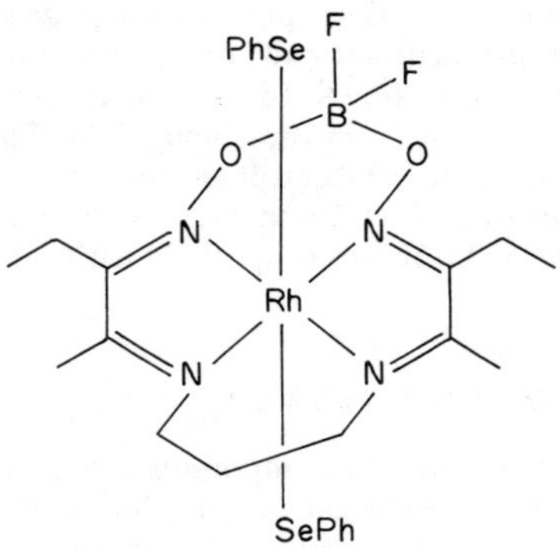

(69)

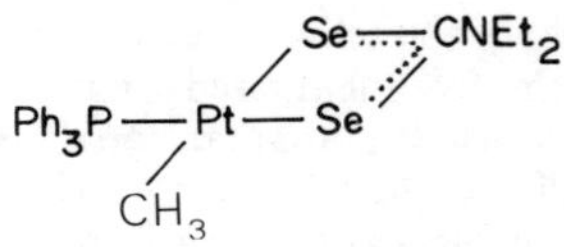

(70)

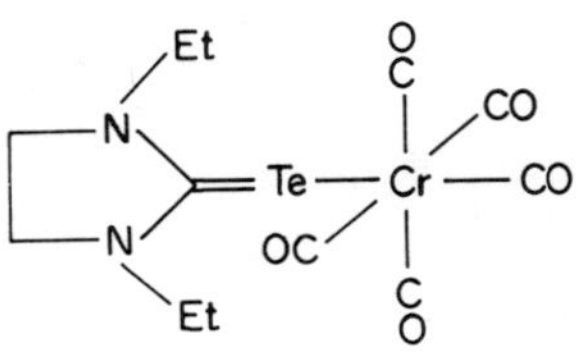

(71)

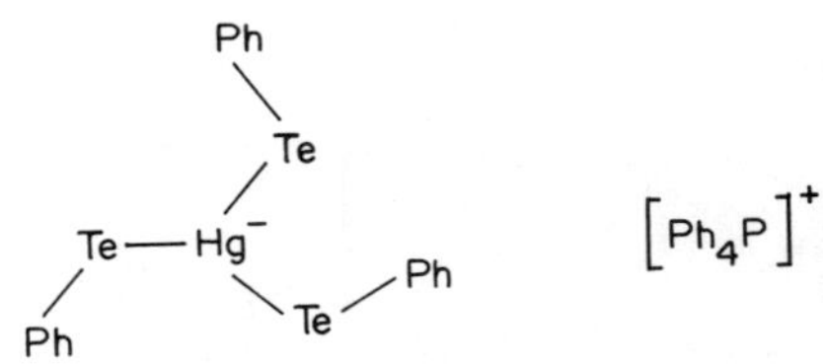

(72)

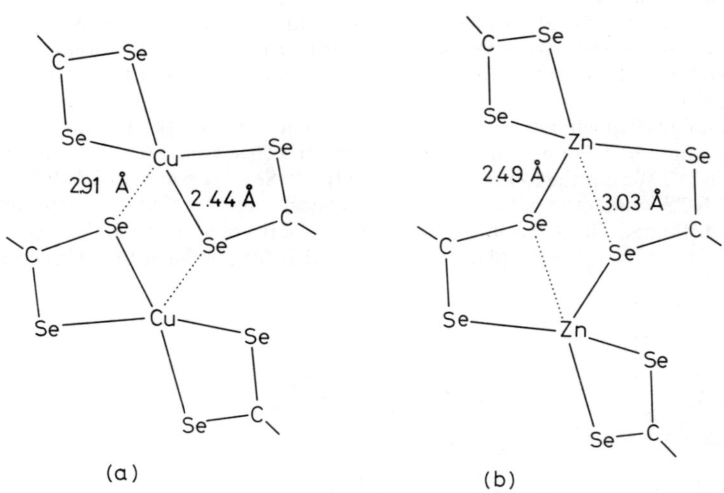

FIGURE 32. Schematic projection of the dimeric molecule in the crystal of (a) bis(*N,N*-diethyldiselenocarbamato)copper (II) (**53**) and (b) bis(*N,N*-diethyldiselenocarbamato)zinc (II) (**54**) (after Reference 205). The diethylamino groups are not shown

The molecule of the tellurourea complex (**71**) lies on a crystallographic mirror plane, which bisects the five-membered ring[221]. The tris(tellurophenolato)mercurate(II) anion (**72**) has a distorted planar trigonal $HgTe_3$ core, which holds the propeller-like arrangement of the phenyl rings[222]. The Hg atom lies 0.09 Å from the plane of the Te atoms, and the distances of *ipso* carbon atoms of the rings from the same plane are − 0.59, − 0.21 and 0.12 Å. The Ph rings form dihedral angles of 34.6, 71.2 and 11.3° with that plane[222]. The Te—C bond lengths in both compounds (**71, 72**) are comparable with the sum of the covalent radii (Table 10).

IV. THREE-COORDINATED SELENIUM AND TELLURIUM

Molecular geometry data are relatively scarce for organic Se and Te compounds in which the chalcogen atom is three-coordinate. The structure of no such organic Se or Te compound has been determined in the vapour phase. There are though a few inorganic compounds whose vapour-phase molecular geometry has been elucidated by ED: $SeOF_2$[224], $SeOCl_2$[225], ethylene selenite (**73**)[226] and trimethylene selenite (**74**)[227].

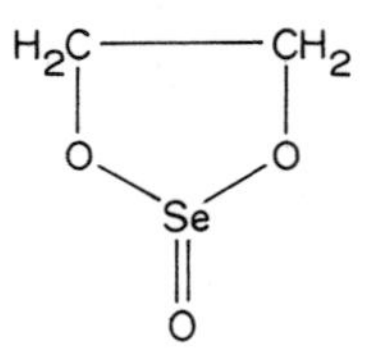

(**73**)

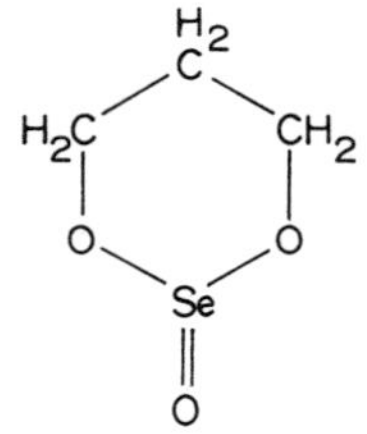

(**74**)

XD has been used to elucidate the molecular structure of several organic Se and Te compounds in the crystalline phase. The Ph group often occurs as ligand both for Se and Te. Especially noteworthy is the interest in structures in which the organic Se moiety is linked to a transition metal. The fluxional behaviour of some of these structures has also attracted interest.

Crystalline triphenylselenonium isothiocyanate, $Ph_3Se(NCS)$, consists of discrete Ph_3Se^+ and NCS^- ion pairs separated from other ion pairs by van der Waals' distances[228]. Within the ion pairs (Figure 33), the Se ··· N contact of 3.197(4) Å and the Se ··· C(NCS) contact of 3.260(5) Å are considerably shorter than the respective van der Waals' distances. The Se bond configuration has a trigonal pyramidal shape with the Se atom lying 0.87 Å out of the plane of the three adjacent carbon atoms. This configuration

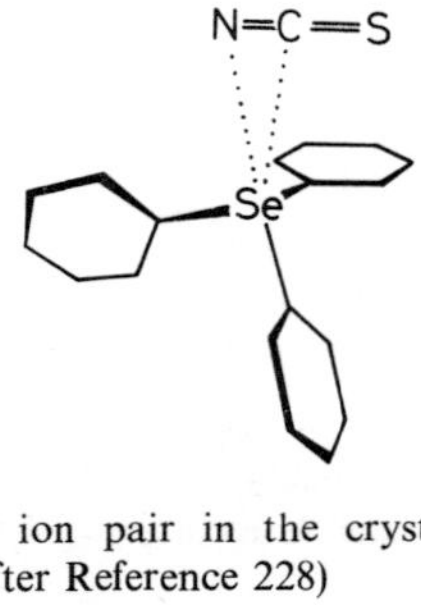

FIGURE 33. An ion pair in the crystal of triphenylselenonium isothiocyanate (after Reference 228)

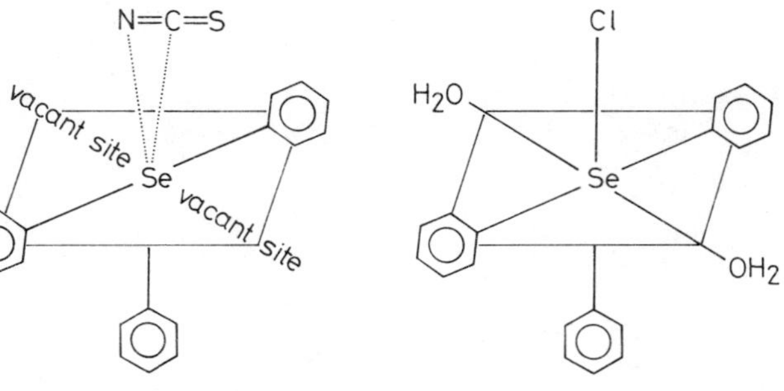

FIGURE 34. Schematic Se coordination in $Ph_3Se(NCS)$ and $Ph_3SeCl\cdot 2\,H_2O$

may also be considered to be distorted tetrahedral with a lone pair of electrons occupying the fourth position.

The Se bond configurations are the same in $Ph_3Se(NCS)$ and in $Ph_3SeCl\cdot 2H_2O$[229]. The mean values of the Se—C bond lengths and C—Se—C bond angles are as follows:

	r(Se—C)(Å)	C—Se—C(deg.)
$Ph_3Se(NCS)$	1.923(8)	100.9(3)
$Ph_3SeCl\cdot 2H_2O$	1.93(1)	100(1)

The Se bond angles are slightly larger than the corresponding mean Te bond angles in the Ph_3Te^+ ion, viz. 95.7(8)° in $Ph_3Te(NCO)\cdot\frac{1}{2}CHCl_3$ [230] and 97.3(6)° in $Ph_3Te(NCS)$[231]. The mean Te—C bond distance in both crystals is 2.13(2) Å [230,231]. The mean C—Te—C bond angle and Te—C bond length in triphenyltelluronium chloride, Ph_3TeCl, is 95.6° and 2.130(4) Å [232].

In the two above-mentioned Se compounds the coordination of the Se atoms is different in spite of the structural similarity of the two cations. The Se is actually four-coordinated in $Ph_3Se(NCS)$ and six-coordinated in $Ph_3SeCl\cdot 2H_2O$. Considering the magnitude of the Se bond angles, viz. N···Se—C and C—Se—C, the Se coordination even in $Ph_3Se(NCS)$ can be visualized as octahedral with three bonds, one coordination linkage and two vacant sites[228]. In $Ph_3SeCl\cdot 2H_2O$ the octahedral coordination of Se comprises three bonds and three coordination linkages. The two different Se coordinations are depicted schematically in Figure 34. The differences between the structures of $Ph_3Se(NCS)$ and its Te analogue are more marked according to Ash and coworkers[228]. They are neither isomorphous, nor isostructural[231,233]. Dimers and polycyclic tetramers coexist in the unit cell of $Ph_3Te(NCS)$[231]. There are tetramers in $Ph_3Te(NCO)\cdot\frac{1}{2}CHCl_3$ [230]. The thiocyanate and cyanate anions occur both in end-to-end and terminally bridging positions in these structures. Te is present in both crystals as five- and six-coordinated atoms in distorted square-pyramidal and octahedral environments, respectively. Ph_3TeCl is dimeric in the crystal, with five-coordinated Te atoms[232].

It is of interest to examine the effect of Se substitution on the benzene ring deformation. In $Ph_3Se(NCS)$ the mean value of the *ipso* endocyclic C—C—C bond angles from the three independently determined Ph geometries is 121.4° (see Table 22). This indicates an appreciable amount of ring deformation in the direction usually characteristic for markedly electronegative substituents.

The ylide resonance structures (**75**) of diacetylmethylenediphenylselenurane gained support from the XD determination of its molecular structure in the crystal[234]. The Se bond configuration is pyramidal with the Se atom lying 0.78 Å out of the plane of the three

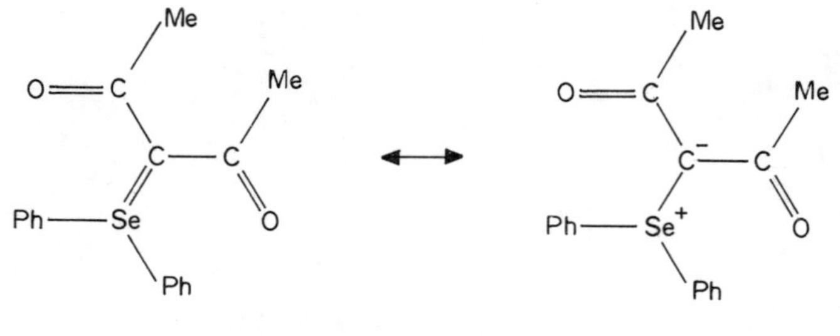

(**75a**) (**75b**)

adjacent carbon atoms. The mean value of the three C—Se—C bond angles is 104.4°, but the three angles are considerably different. The smallest is the Ph—Se—Ph angle, viz. 100.8(3)°, not far from the C—Se—C bond angle in Me_2S (cf. Table 7). Noteworthy is the Se—C(alkyl) bond length, 1.906(8) Å, which is the same as the mean of the two Se—C(phenyl) bond lengths (1.912 Å). This is indeed the main indication that there is a large contribution from the ylide resonance structure **75b**. Similarly, and even more so, in **76** the Se—C bonds are longer and the Se atom lies further out of the adjacent CCC plane (by 0.91 Å). This structure was determined by X-ray crystallography by Saatsazov and coworkers[235]. Incidentally, the five-membered ring is non-planar and its geometrical parameters are consistent with those of the free tetrahydroselenophene molecule determined by ED (see Table 14).

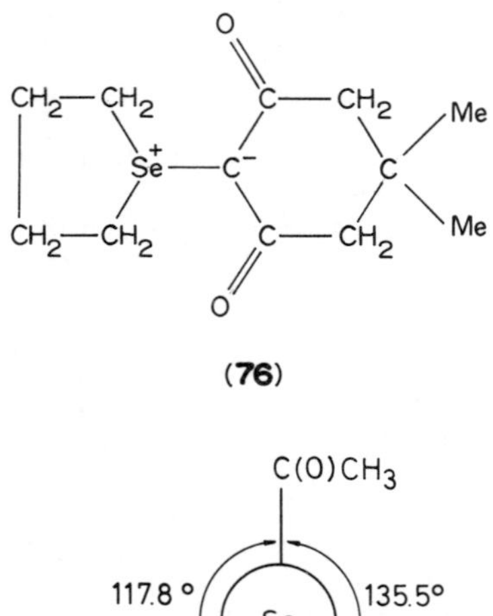

(**76**)

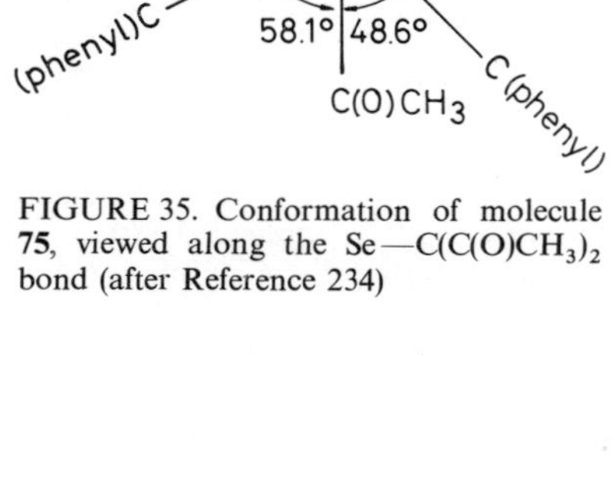

FIGURE 35. Conformation of molecule **75**, viewed along the Se—$C(C(O)CH_3)_2$ bond (after Reference 234)

The molecule **75** does not have exact C_s symmetry in the crystal[234]. The orientation of the two Ph groups is markedly different with respect to what would be the symmetry plane containing the Se atom and the acetyl acetonate moiety. Even the rotational form of the $C_2Se—CC_2$ skeleton about the Se—C bond is asymmetrical (Figure 35).

The two independently determined Ph geometries differ appreciably. One of them shows an elongation of the ring usually characteristic of substitution with an electropositive ligand. Possible Se and Te substituent effects upon the benzene ring deformation are discussed in the conclusion of this section.

There are two modifications of tetrakis (diphenylseleno) dimercury(I) diperchlorate $(Ph_2Se)_2HgHg(SePh_2)_2(ClO_4)_2$ (77), yellow and red. Both have been studied by X-ray crystallography[236,237]. The structural differences of the two modifications are pro-

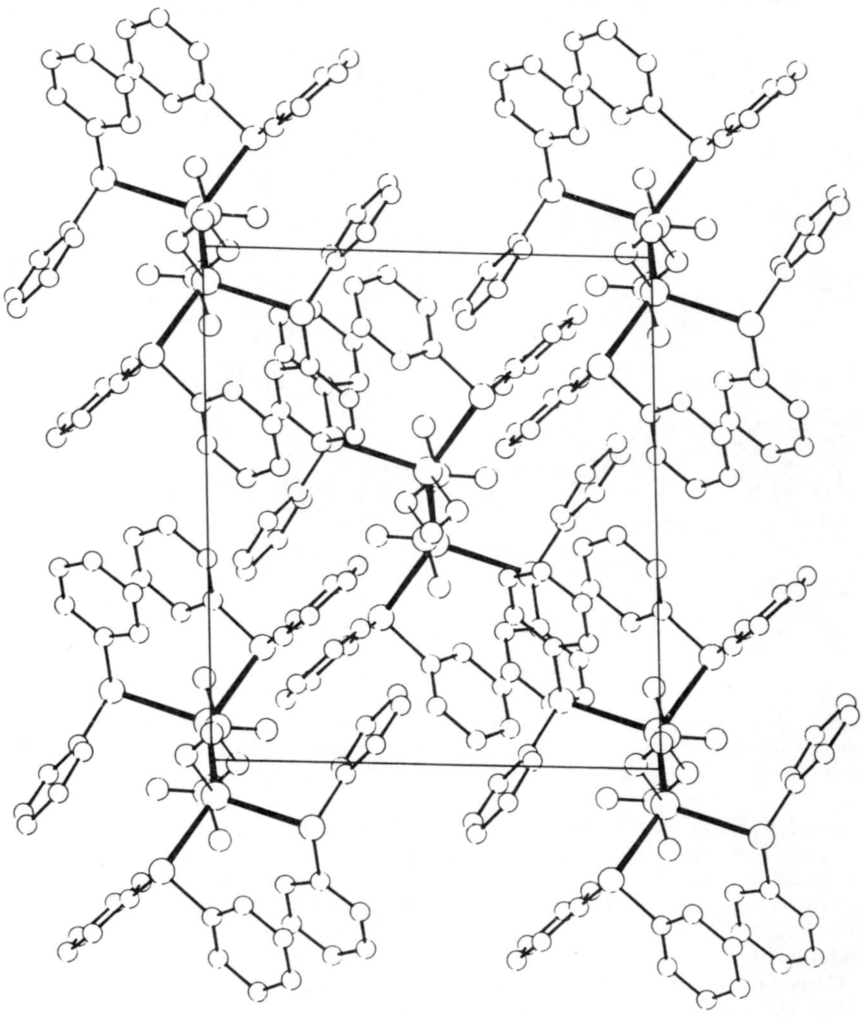

FIGURE 36(a)

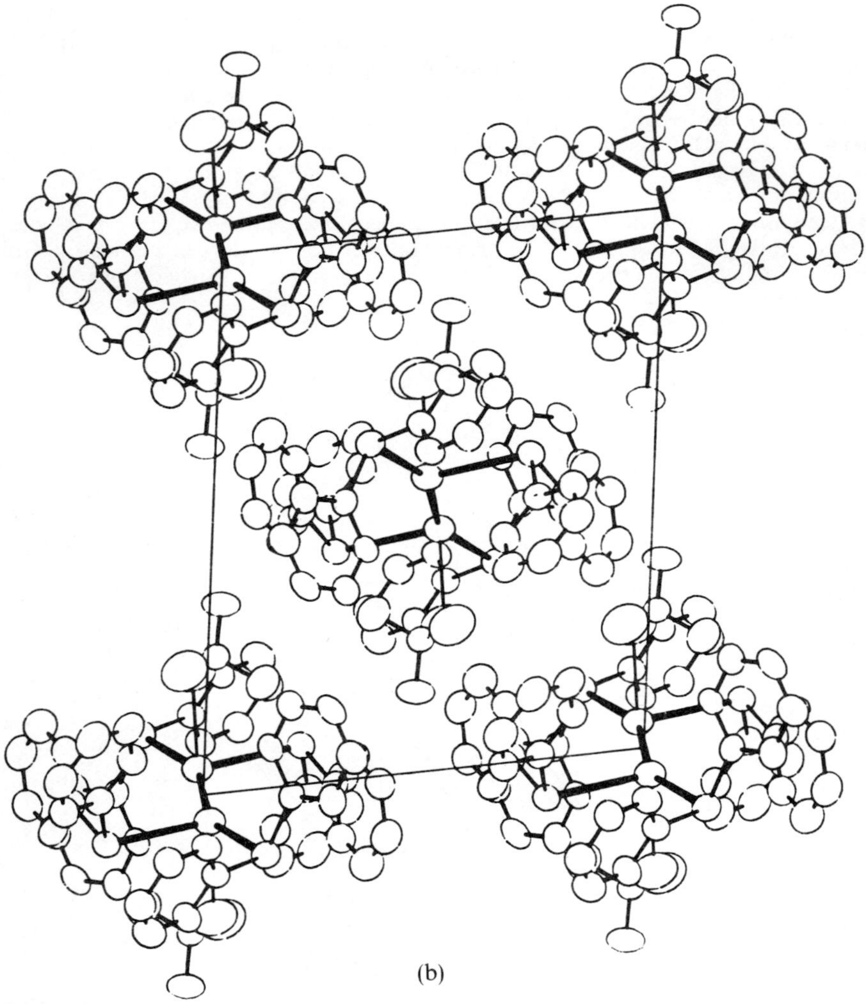

(b)

FIGURE 36. Packing of molecules in the crystals of **77**, viewed along direction (010): (a) yellow modification, (b) red modification. Reproduced by permission of the Verlag der Zeitschrift für Naturforschung from Reference 237

nounced in the Hg bond configuration. Although the $Se_2HgHgSe_2$ skeleton is coplanar in both, the Hg bond angles and the Hg—Se bond lengths change considerably. The Se bond configurations may also be somewhat different in the two modifications but large deviations of crystallographically non-equivalent bonds and angles prevent a meaningful discussion of the differences. The configurations are peaked pyramidal, the mean Se bond angles being about 101° in both structures.

The two modifications of **77** have markedly different molecular packing (Figure 36). The 'red' modification has nearly spherical molecular units, while the Ph groups are *sticking out* in the packing of the 'yellow' modification. Thus the most striking differences might be expected in the structures of the Ph groups. The ring sizes appear to be appreciably

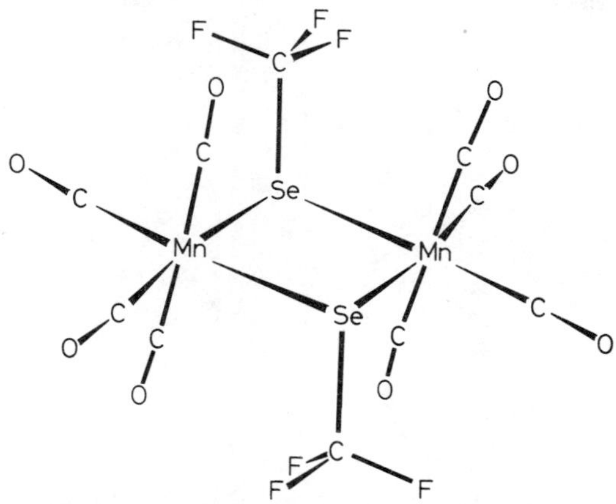

FIGURE 37. The molecular model of **78** (after Reference 238)

different, although the experimental errors are rather large. There are four independent pairs of Ph groups in each molecule, and the four mean C—C ring bond lengths have been reported for both modifications. The mean values of the respective four mean bond lengths are 1.403 Å (yellow) and 1.375 Å (red). Although the angular parameters for the benzene rings were not communicated, it is expected on this basis that considerable differences in the angular deformation also occur in the two modifications.

The structure of the $SeCF_3$ moiety of bis{μ[(trifluoromethyl)seleno]-manganesetetracarbonyl}, $[(CF_3Se)Mn(CO)_4]_2$ (**78**), (Figure 37) is of interest for our discussion. This molecular structure has been determined by XD[238]. The Se—C bond is 1.97 Å long. Noteworthy is that the mean of the F···F dis-

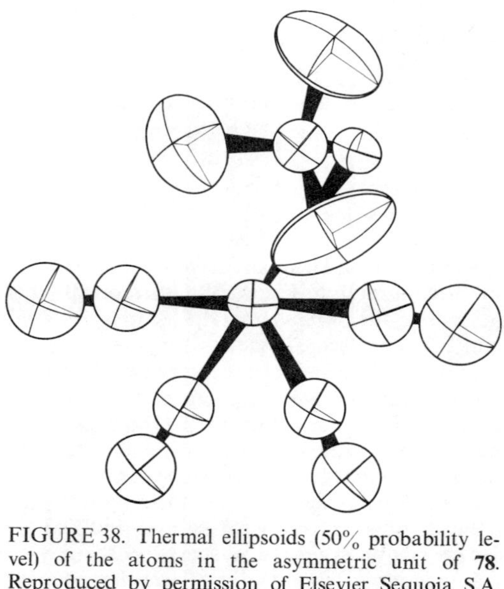

FIGURE 38. Thermal ellipsoids (50% probability level) of the atoms in the asymmetric unit of **78**. Reproduced by permission of Elsevier Sequoia S.A. from Reference 238

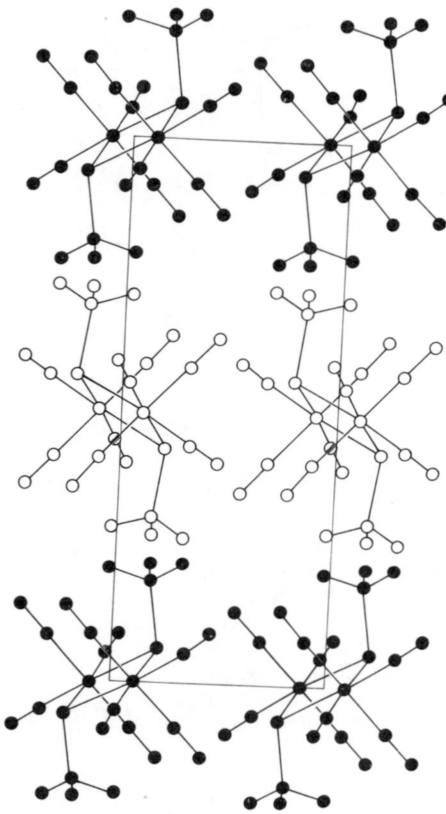

FIGURE 39. Crystal structure of **78** projected along the *z* axis. Reproduced by permission of Elsevier Sequoia S.A. from Reference 238

tances, referring to fluorines separated by one bond angle, is only 2.10 Å. They are markedly shorter than the mean value of 2.162 ± 0.008 Å found from a large series of free molecules containing CF_3 groups as was discussed in Section III.C. It is not obvious what may be the origin of this discrepancy. An apparent shortening of the C—F bonds may be a consequence of the librational motion. Figure 38 illustrates the thermal motion of the atoms in the asymmetric unit after Marsden and Sheldrick[238]. Packing considerations may also be of importance. The molecule possesses a crystallographic centre of symmetry, the asymmetric unit consisting of one half of one molecule. The molecules are arranged in layers. As for intramolecular interactions, the conformation adopted by the CF_3 groups appears to minimize the repulsion from the two CO groups on the same side of the $(MnSe)_2$ plane, cf. Figure 37. A fluorine may be almost equidistant from two carbons and two oxygens, thus non-bonded interactions seem to be of importance. There may also be non-bonded interactions between molecules. The shortest intermolecular contacts occur between oxygens and fluorines, viz. 2.84 Å within a layer and 3.03 Å between layers. The shortest intermolecular F $\cdots$ F non-bonded distances are 3.14 Å between layers. Figure 39 shows the crystal structure in projection down the *z* axis after Marsden and Sheldrick[238].

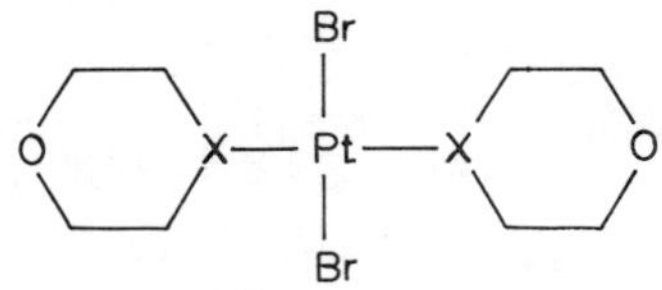

(79) X = S
(80) X = Se

The molecular structure of *trans*-dibromobis(1,4-oxaselenane) platinum(II) (**80**) has been determined by XD[239] along with that of *trans*-dibromobis(1,4-oxathiane) platinum (II)(**79**). In addition to the bond length changes, the largest angular change occurs in the C—Se—C versus C—S—C endocyclic bond angles. Some of the geometrical parameters in the two analogous molecules are:

	79		**80**
S—C	1.81(1) Å	Se—C	1.96(2) Å
	1.82(2) Å		1.86(4) Å
C—S—Pt	113.2(5)°	C—Se—Pt	110.5(7)°
	104.0(4)°		107.2(10)°
C—S—C	96.4(7)°	C—Se—C	90.8(13)°

The C—S—C bond angle in the thio derivative agrees well with the analogous angle in *free* 1,4-oxathiane, viz. 97.1(20)° as determined by ED[240]. 1,4-oxaselenane itself has not been investigated. The most interesting difference between the structures of **79** and **80** occurs in their conformations. Whereas the Pt—S bond is equatorial to the ring, the Pt—Se bond adopts an axial position (Figure 40). According to Barnes and coworkers[239] the difference cannot be explained by intermolecular packing considerations. It was also noted[239] that similar conformational differences have been observed.

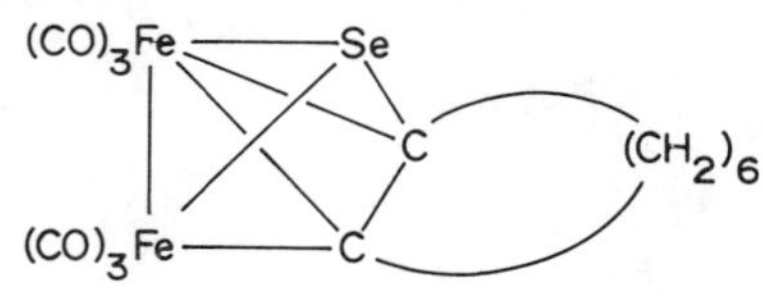

(81)

The hexacarbonyl [μ-[1,2-η:2-η)-1-cyclooctene-1-selenolato(2–)-*Se*:*Se*]]diiron(*Fe–Fe*) molecule (**81**)[241] is interesting in that it contains iron–selenium bonds. The selenoketocarbene moiety establishes asymmetric bonding to the diironhexacarbonyl group. The Se

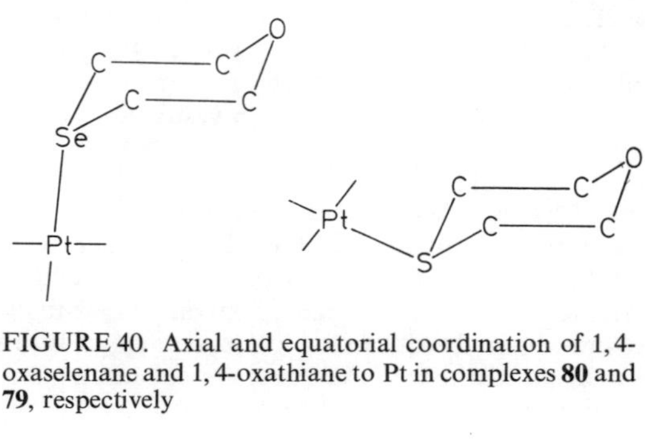

FIGURE 40. Axial and equatorial coordination of 1,4-oxaselenane and 1,4-oxathiane to Pt in complexes **80** and **79**, respectively

bond configuration is characterized by the following parameters:

r(Se—C) 1.917(3) Å ∠Fe—Se—Fe 64.1(0)°,
∠Fe—Se—C 77.1(1) and 57.4(1)°

The lengths of the iron–selenium bonds were determined to be 2.398(1) and 2.364(1) Å. Similarly, unequal iron–selenium bonds were found in another compound containing a selenaferrole ring[242].

The trimethyltellurium ion, Me_3Te^+, is a discrete entity in the trimethyltelluronium tetraphenylborate, Me_3TeBPh_4(**82**), structure as has been determined by X-ray crystallography[243]. The Me_3Te^+ ion has a trigonal pyramidal geometry with approximate C_{3v} symmetry. It was presumed that the fourth tetrahedral site is occupied by the Te lone pair of electrons, similarly to Me_3TeCl, in which the electron density distribution has also been determined[244]. The geometry of the TeC_3 skeleton in **82** is characterized by the following mean parameters:

r(Te—C) 2.14(2) Å and ∠C—Te—C 92(1)°

As there are no appreciable coordination linkages the lone pair of electrons may be expected to exercise its full stereochemical activity. The absence of secondary interactions has been ascribed to the non-coordinating character of the BPh_4^- ion. The mean value of the crystallographically independent Te—C bond lengths corresponds to the calculated single bond, and agrees well with Te—C bond lengths determined in other substances as well (for references see Reference 243). The C—Te—C bond angles appear slightly smaller than generally observed. This may be a consequence of the repulsion effect of the stereochemically fully active lone pair of electrons of Te. The lone pair may have somewhat diminished activity in other molecules due to secondary coordination.

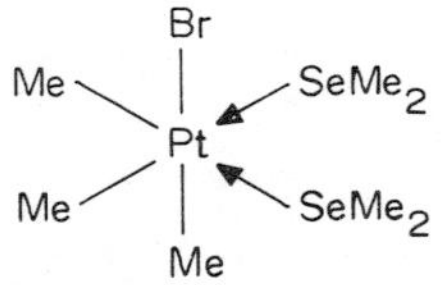

(**83**)

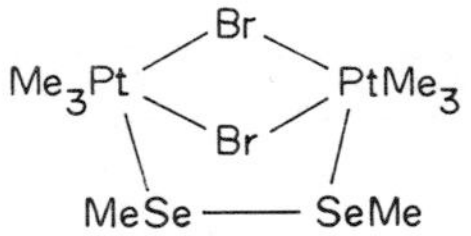

(**84**)

Interesting structures, **83** and **84**, are produced when tetrameric bromotrimethylplatinum reacts with Me_2Se or Me_2Se_2, respectively[245]. **83** is supposed to have an octahedral bond configuration about the central Pt atom. It has been investigated by ^{1}H-NMR spectroscopy. The structure of **84** was determined by XD[245]. The Se—Se bond length is 2.36(1) Å, slightly larger than in open-chain diselenides discussed in Section III.B. The mean of the C—Se—Se and C—Se—Pt bond angles is 99(2)°. The Pt atoms retain an approximate octahedral coordination. The Pt_2Br_2 four-membered ring is puckered, the dihedral angle BrPtBr/BrPtBr is 29°. An ^{1}H-NMR investigation of **84** revealed an atomic inversion process in which the two Se atoms were exchanged. The activation energy was determined to be 64.5(62) kJ mol^{-1} [245].

Abel and coworkers[246] have studied the various structural factors governing the energy barriers to inversion of three-coordinated Se and S. In a typical investigation, a complex, $[ReI(CO)_3\{MeSe(CH_2)_2SeMe\}]$ (**85**), was prepared, its crystal and molecular structure (Figure 41) was determined by XD and the energy barrier associated with the pyramidal inversion at the three-coordinated Se atom was calculated by means of total band-shape dynamic NMR spectroscopic methods[246]. The geometrical parameters characterizing the selenium–carbon bonding systems are listed below:

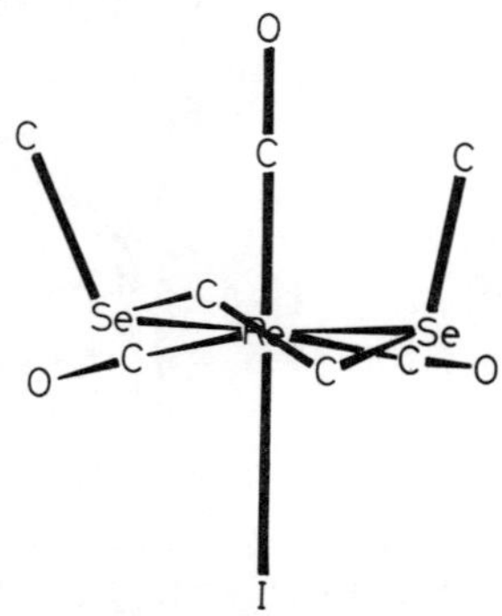

FIGURE 41. The molecular model of **85** (after Reference 246)

Se—CH_3	1.946(11) Å	CH_2—Se—CH_3	96.8(5)°
	1.991(11) Å		97.5(5)°
Se—CH_2	1.970(9) Å	Re—Se—CH_2	102.3(3)°
	1.989(9) Å		100.9(3)°
		Re—Se—CH_3	107.7(4)°
			108.9(4)°

The mean Se—C bond length in the ring is slightly greater than the mean exocyclic Se—C bond length. Concerning the inversion barriers, the following statements can be made[246]: (1) The inversion barriers at Se are higher than those at analogous S. (2) Changing the halogen in the *cis* position to the inverting centre has negligible effect on the overall barrier height. (3) The nature of the metal–Se (S) bond is of great importance for the barrier heights. (4) Substituting the aliphatic backbone of the ligand by an unsaturated backbone lowers the inversion barrier. The lowering is smaller for S inversion than for Se inversion. This change was ascribed to (p–p)π conjugation between the chalcogen lone pair of electrons and the ligand backbone during the inversion. It was suggested that the (3p–2p)π sulphur–carbon conjugation was more effective than the (4p–2p)π selenium–carbon conjugation.

Three-coordinated Se and Te may have a planar T-shaped bond configuration. Such cases will be reviewed next.

Structures of crystals containing the tris(selenourea), $[SeC(NH_2)_2]_3{}^{2+}$, ion have been determined from visually estimated photographic[247] and diffractometer[35] X-ray data. Crystals of $[SeC(NH_2)_2]_3Cl_2 \cdot H_2O$(**86**) and $[SeC(NH_2)_2]_3Br_2 \cdot H_2O$(**87**) are isomorphous, orthorhombic[247], Pbca; the sulphate, $[SeC(NH_2)_2]_3SO_4 \cdot SeC(NH_2)_2 \cdot 2H_2O$ (**88**), is triclinic[35], P$\bar{1}$ and contains a solvate selenourea molecule. The cation geometry is similar in the three salts (Figure 42). All selenourea groups, including the uncomplexed solvate molecule in **88**, are nearly planar. The selenourea planes of the cation are nearly perpendicular to the approximately linear Se—Se—Se sequence, and are thus nearly parallel to each other: the central plane makes dihedral angles of 5–18° with the terminal planes. The two terminal Se—C bonds are nearly coplanar, and the dihedral angle between the least-squares planes of the atoms

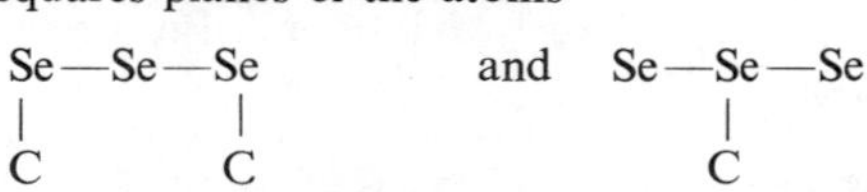

respectively, is about 75°. This conformation of the tris(selenourea) cation (Figure 42)

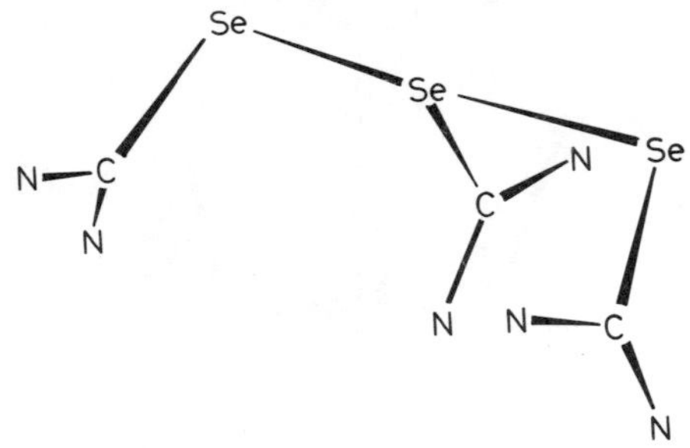

FIGURE 42. The 'easel' shape of the tris(selenourea) cation. Drawn from the atomic coordinates given in Reference 35

resembles the 'easel' shape of the triselenocyanate, $(SeCN)_3^-$ anion, whose dihedral angle is smaller, viz. 44, 57 and 65° in the crystals of the Cs[248], K[249] and Rb salts[250] (see also Reference 4). As to the Se—Se—Se moiety, it is only slightly bent in the triselenocyanate ions with angles of 176–178° while it is more bent in the tris(selenourea) ions, which have Se—Se—Se angles of 173.8° in the dichloride (**86**)[247] and dibromide (**87**)[247] and 168.3° in the sulphate (**88**)[35]. The latter differs from **86** and **87** also in the direction of bending.

The average bond angles (deg.) at the Se atoms of the tris(selenourea) group are as follows:

	86[247]	**87**[247]	**88**[35]
Se—Se—C(terminal)	96.4	96.4	98.0
Se—Se—C(central)	89.1	89.2	86.2

They are similar to those found in the triselenocyanate ion[4,248–250]. The crystallographically different Se—Se bond lengths (Å) are the following:

	86	**87**	**88**
Se—Se	2.597(2)	2.624(2)	2.6336(15)
Se—Se	2.717(2)	2.712(2)	2.6639(15)

and the lengths (Å) of the Se—C bonds in the same order:

Se—C(terminal)	1.921(8)	1.921(12)	1.903(5)
Se—C(central)	1.940(8)	1.947(12)	1.925(4)
Se—C(terminal)	1.905(8)	1.906(12)	1.903(5)

The central Se—C bond in the tris(selenourea) ion is longer than the terminal Se—C bonds. A similar observation was made for the chalcogen–carbon bonds in trithiapentalenes and analogues[251] (see below). A comparison with triselenocyanates is not feasible because of large estimated errors in the parameters of the latter. There is asymmetry in the Se—Se and terminal Se—C bonds of the tris(selenourea) ion. The longer Se—C bond belongs to the shorter Se—Se bond, although differences in terminal Se—C lengths are hardly significant. Selenium–carbon bonds seem to have a certain amount of double-bond character (cf. Table 7) and are longer than this bond, 1.867(4) Å, in the solvate selenourea molecule in the crystals of **88**. The selenium–selenium bonds are longer than single bonds (see Tables 11 and 12), similarly to those in triselenocyanates[4,248–250] and triselenapentalenes.

Trithiapentalenes and related systems contain a three-coordinated, formally tetravalent chalcogen atom Y (**89a**), which is involved in a three-centre four-electron bond in the roughly linear chain of atoms X—Y—Z. Form **89b** reflects the delocalized system of ten π

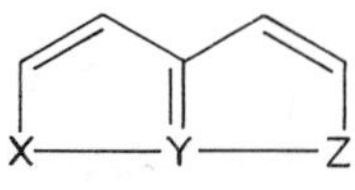
(89a)

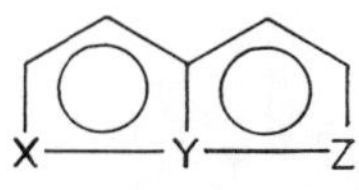
(89b)

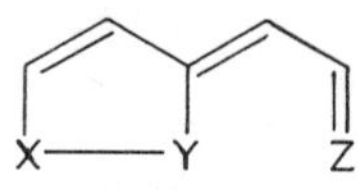
(89c)

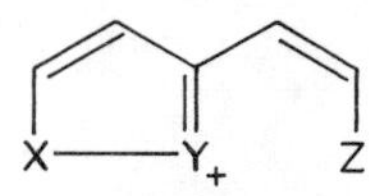
(89d)

electrons in the nearly planar bicyclic molecule, while asymmetric forms like **89c** or **89d** indicate its dynamic properties. Compounds of this type have been extensively studied since their discovery 25 years ago. (See Reference 252 for references to reviews.) Bond length data of Se and Te, as well as O and S atoms in such heteropentalene molecules are compiled in Table 21 from X-ray crystallographic investigations. Some S analogues are also included for comparison. The molecular structures of two compounds have been determined in the gas phase: 1,6,6a-trithiapentalene (**90**) was studied by electron diffraction[266] and 1,6-dioxa-6a-thiapentalene by microwave spectroscopy[267].

Chalcogen–chalcogen bonds in these systems are longer than normal covalent single bonds, and the relative lengthening diminishes from S to Te[263]. This is in accord with observed ESCA line widths[268,269] and results of CNDO/2 calculations[269]; namely, the total energy of **90** as a function of the displacement of the central S atom (6a) between its fixed neighbours S(1) and S(6) has a rather flat minimum, roughly 0.3 Å wide, about the symmetric position[269,270]. This potential well is much narrower[269] when a Se atom takes

TABLE 21. Bond lengths (Å) of chalcogen atoms in trithiapentalenes and analogues from XD

Molecule		Bond (5)–(6)	(6)–(6a)	(3a)–(6a)	(6a)–(1)	(2)–(1)	Reference
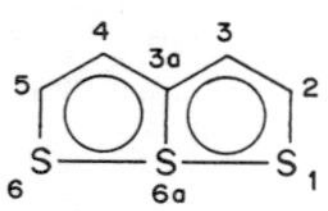	(**90**)	1.684(3)	2.363(1)	1.748(3)	2.363(1)	1.684(3)	251, 253
	(**91**)[a]	1.703(6)	2.304(3)	1.753(6)	2.362(3)	1.712(6)	254
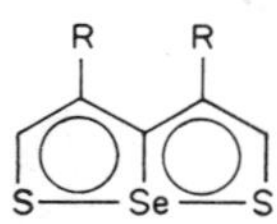	(**92**) R = H	1.66(2)	2.446(5)	1.95(2)	2.446(5)	1.66(2)	255
	(**93**) R = Me	1.691(3)	2.414(1)	1.917(3)	2.414(1)	1.691(3)	256
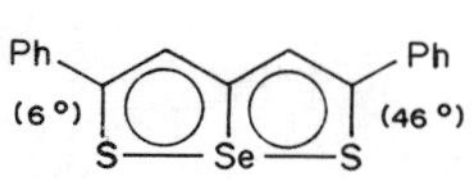	(**94**)[a]	1.72(1)	2.419(3)	1.87(1)	2.433(3)	1.71(1)	257
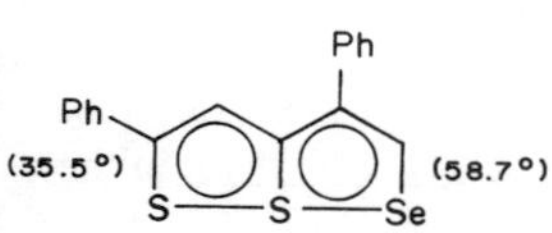	(**95**)[a]	1.69(1)	2.492(3)	1.71(1)	2.333(3)	1.82(1)	258

(*Contd.*)

TABLE 21. (*Contd.*)

	Bond					
Molecule	(5)–(6)	(6)–(6a)	(3a)–(6a)	(6a)–(1)	(2)–(1)	Reference
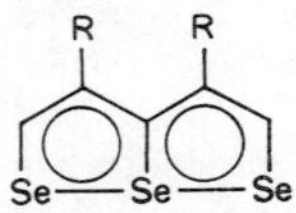 **(96)** R = H	1.86(3)	2.586(3)	1.90(2)	2.579(3)	1.81(3)	259
(97) R = Me		2.542(6)				260
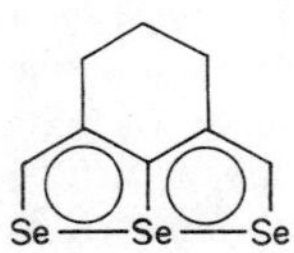 **(98)**	1.78(2)	2.568(3)	1.91(2)	2.554(3)	1.82(2)	261
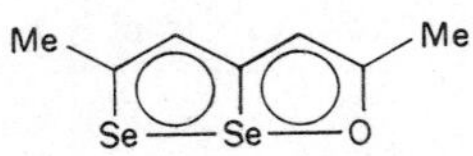 **(99)**	1.848(15)	2.384(3)	1.881(15)	2.336	1.22	262
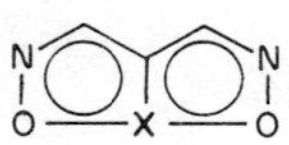 **(100)** X = S	1.353(5)	1.853(3)	1.683(4)	1.850(3)	1.347(5)	263
(101) X = Se	1.32(1)	1.987(7)	1.827(8)	1.997(6)	1.34(1)	263
(102) X = Te	1.36	2.094	1.980	2.080	1.39	263
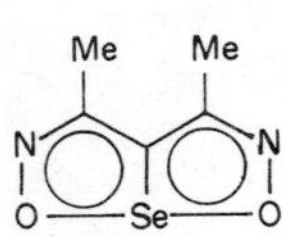 **(103)**		1.942(6)				264
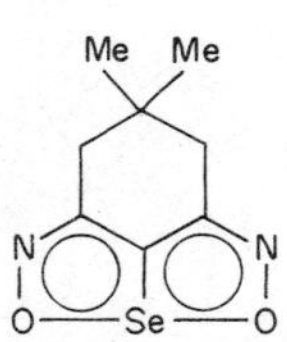 **(104)**	1.337(14)	2.017(9)	1.802(11)	2.030(9)	1.346(14)	265

[a]Dihedral angles of the Ph rings with the heterocyclic plane are given.

the role of the 'bell-clapper'—as this kind of vibrational motion is called. A double-minimum potential was used, on the other hand, to interpret the electronic spectrum[271], and was also obtained form EHT calculations without using sulphur 3d orbitals[272]. Inclusion of d orbitals yields, however, a broad single-minimum potential[272]. Unfortunately neither XD[273] nor ED could distinguish the two cases of (1) a flat potential well with a single minimum (symmetric equilibrium structure) and (2) a potential with a small hump in the middle (asymmetric equilibrium structure). The MW study of 1,6-dioxa-6a-thiapentalene unambiguously gave a symmetric model[267].

These weak bonds are of course sensitive to changes in their intra- and inter-molecular environment. CNDO/2 calculations have shown that a Me substituent on C(2) causes a lengthening of the distance S(1)—S(6a) but a shortening arises from a 3-Me substitution[270,274]. The influence of a Ph group depends also on its dihedral angle with the plane of the heterocyclic rings[270]. These results are in agreement with structural data on trithiapentalenes[270] and with bond lengths given in Table 21. The scarcity of data and diversity of heteroatoms in Se- or Te-containing pentalenes (Table 21) make it difficult, however, to observe clear trends in the geometrical changes in these molecules. It seems from a comparison of pairs **90** and **91** versus **92** and **94** that Ph substituents have a smaller effect on S—Se bonds, in accord with the narrower potential well, than on S—S bonds. Similar statements cannot be made about other substituents or other sequences of heteroatoms.

The central carbon–sulphur bonds are longer than the terminal ones in a number of trithiapentalenes[251], and this is the case in **90**,**91**,**96** and **98** (Table 21). Bond angles of chalcogen atoms do not seem to be very characteristic and seem to be determined mainly by the geometry of the rest of the ring, e.g. by the carbon–chalcogen bond distances. Just a few bond angles at terminal and central chalcogen atoms should be listed here for rough orientation (mean values when there is asymmetry):

	terminal (deg.)		central (deg.)	
90	C—S—S	92.0(1)	C—S—S	89.1(1)
96	C—Se—Se	88.5	C—Se—Se	88.0
100	N—O—S	113.3	C—S—O	85.9
101	N—O—Se	113.2	C—Se—O	82.2
102	N—O—Te	115	C—Te—O	77.0

It has been noted that the central C—C bonds in trithiapentalenes are longer than the terminal C—C bonds, a feature that resembles the structure of naphthalene, which is also a fused ring system with ten π electrons[251]. A few examples of C—C bond lengths are:

	terminal (Å)	central (Å)
90	1.354(3)	1.409(2)
92	1.35(3)	1.41(3)
96	1.37(mean)	1.41(mean)

In 2,5-dimethyl-1-oxa-6, 6a-diselenapentalene (**99**), on the other hand, C—C bond lengths alternate as in a conjugated system, viz. 1.36(3), 1.42(3), 1.34(3) and 1.43(3) Å, continued with the C—O bond of 1.22 Å length[262].

The molecular structures of a series of phenyltellurium derivatives with three-coordinated Te have been determined by Vikane and associates[277–280] using XD, following earlier work by Foss and coworkers[275]. In both phenylbis(thiourea)tellurium (**105**) chloride[210] and phenylbis(selenourea)tellurium (**106**) chloride[210] (cf. **61**) the Te is three-coordinated being bonded to a Ph group and two chalcogen atoms. The S(Se)—Te—S(Se) bonding system is essentially linear and the Te—Ph bond nearly bisects this three-centre arrangement[276]. There is an important difference in the Te coordination in the two structures[210]. In the Se derivative the Te is strictly three-coordinated as the chloride ion is removed into a position that may be considered independent from it. On the other hand, in the thio derivative the chloride approaches the fourth coordination site in what may be considered as a square planar arrangement about Te.

$$(H_2N)_2CY—\overset{+}{Te}—YC(NH_2)_2$$
$$|$$
$$Ph$$

(**105**) Y = S
(**106**) Y = Se

In ethylenethiourea(iodo)phenyltellurium (**107**)[277] and ethyleneselenourea(iodo)-phenyltellurium (**108**)[277] each Te atom is strictly three-coordinated with nearly planar bond configuration.

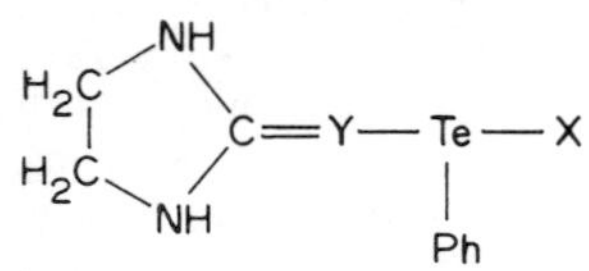

(**107**) Y = S, X = I
(**108**) Y = Se, X = I
(**109**) Y = S, X = Br
(**110**) Y = Se, X = Br
(**111**) Y = S, X = Cl

Crystals with two different space groups may occur when the 1:1 complex of benzenetellurenyl bromide and ethylenethiourea crystallizes yielding bromo(ethylenethiourea)phenyltellurium (**109**). The crystal and molecular structures of both modifications have been determined and a series of interesting differences in packing and conformation have been noted[278] (Figures 43 and 44). Most of the bond lengths and bond angles in the two modifications cannot be considered to be significantly different with the striking exception of the Te—Br bond being 2.8348(10) and 2.9694(10) Å in the two crystals, respectively.

Bromo(ethyleneselenourea)phenyltellurium (**110**) has only one (known) crystal modification[279] being isomorphous with one ($P2_1/c$) of the two above mentioned forms of its thio analogue. Incidentally, the Te—Br bond in this compound is considerably longer, viz. 3.0537(16) Å, than in either of the two modifications of **109**. While this bond shows great sensitivity, the Te—C(phenyl) bond appears to be remarkably constant (see below). The structure of **110** is characterized by three-coordinated Te, with the Te—C bond nearly perpendicular to the Br—Te—Se chain.

The structure of chloro(ethylenethiourea)phenyltellurium (**111**)[279] is analogous in all essential respects to that of **110**.

The crystal molecular structures of tetramethylammonium phenyl(dithiocyanato)tellurate (**112**) and phenyl(diselenocyanato)tellurate (**113**) have been determined by Hauge and Vikane[280]. The crystals are isomorphous, monoclinic (space group C2/c). The Te bond configuration is regarded as square-planar with one position, *trans* to the Ph ligand, vacant. The Te—C(phenyl) is perpendicular to the S(Se)—Te—S(Se) chain. The latter itself is nearly linear. This description quite generally characterizes the above-described three-coordinated complexes of divalent Te.

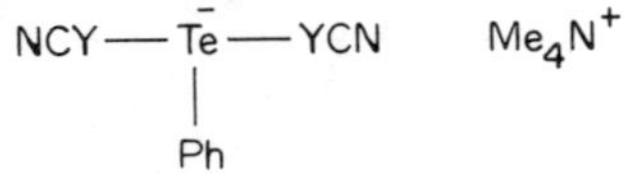

(**112**) Y = S
(**113**) Y = Se

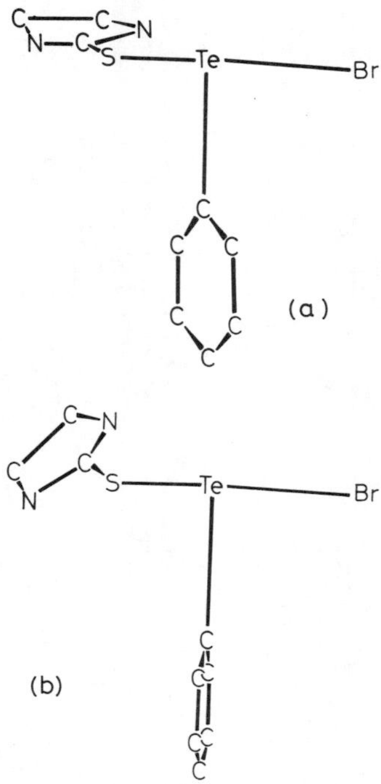

FIGURE 43. Molecular models of **109** in the two crystallographic modifications (a) space group C2/c, (b) space group $P2_1/c$ (drawn after Reference 278)

Among the structures discussed in this section, three Se derivatives and several Te derivatives contain bonding to a Ph group. The lengths of the Se—Ph and Te—Ph bonds are collected in Tables 22 and 23, respectively. Parameters characterizing the deformation of the benzene rings are also given, viz. the *ipso* endocyclic C—C—C angle (α), the mean value of the ring C—C bond distances, and the difference of bond lengths, C(2)—C(3) minus C(1)—C(2), where C(1) is the Ph carbon adjacent to the substituent (Se or Te). Using the covalent radii of two-coordinated Se and Te given in Table 10, the estimated single bond lengths are r(Se—C) = 1.93 Å and r(Te—C) = 2.12 Å, without using electronegativity corrections. According to Tables 22 and 23 the experimentally determined Se—C and Te—C bond lengths are remarkably stable from compound to compound and stay around the values estimated for the single bonds. Where there are crystallographically independent Ph groups in the same structure, and this notably occurs in the Se derivatives, considerable variations are observed. Here again, though, the mean values fall very much in line, viz. 1.923, 1.927 and 1.912 Å for the three Se compounds listed in Table 22. As there is a relatively large series of Te derivatives, it seems worth mentioning that the mean Te—C bond length for the nine compounds listed in Table 23 is 2.114 Å with a standard deviation of 0.010 Å.

Considering all the crystallographically independent Ph groups in the Se derivatives, there seems to be some indication for a possible correlation between the bond length of the substituent to the Ph group and the *ipso* endocyclic C—C—C angles which is known to be the most characteristic of all the parameters related to the benzene ring deformation[281]. Figure 45 shows that with increasing Se—C(phenyl) bond length there may be an increase in the *ipso* endocyclic C—C—C bond angle of the Ph group. The sample is too small to

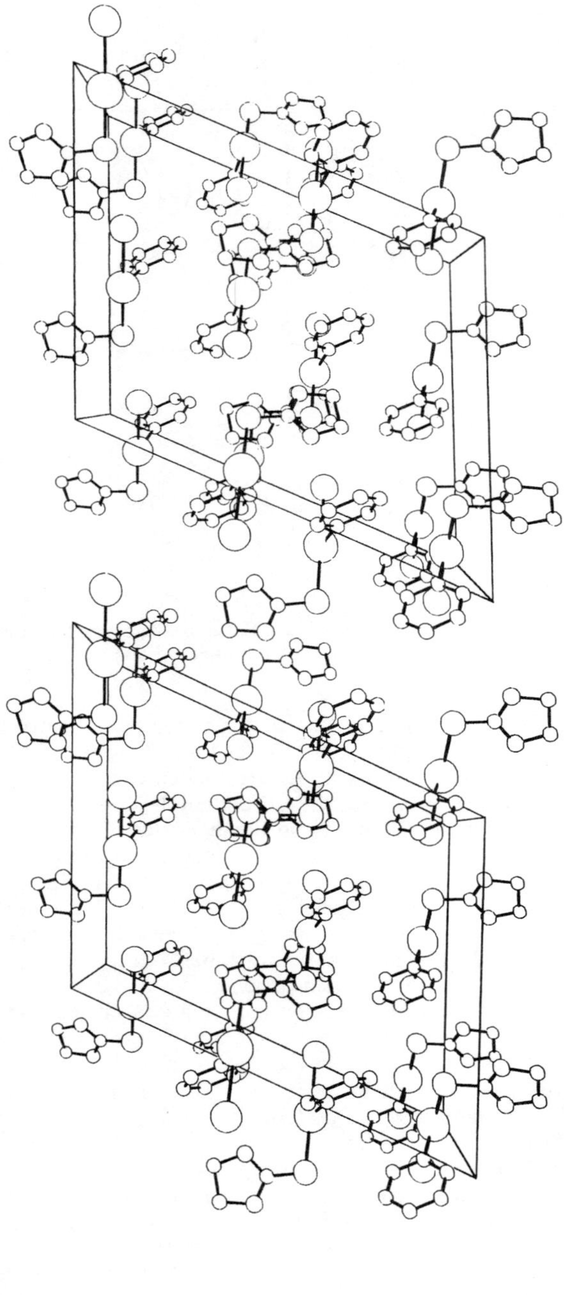

(a)

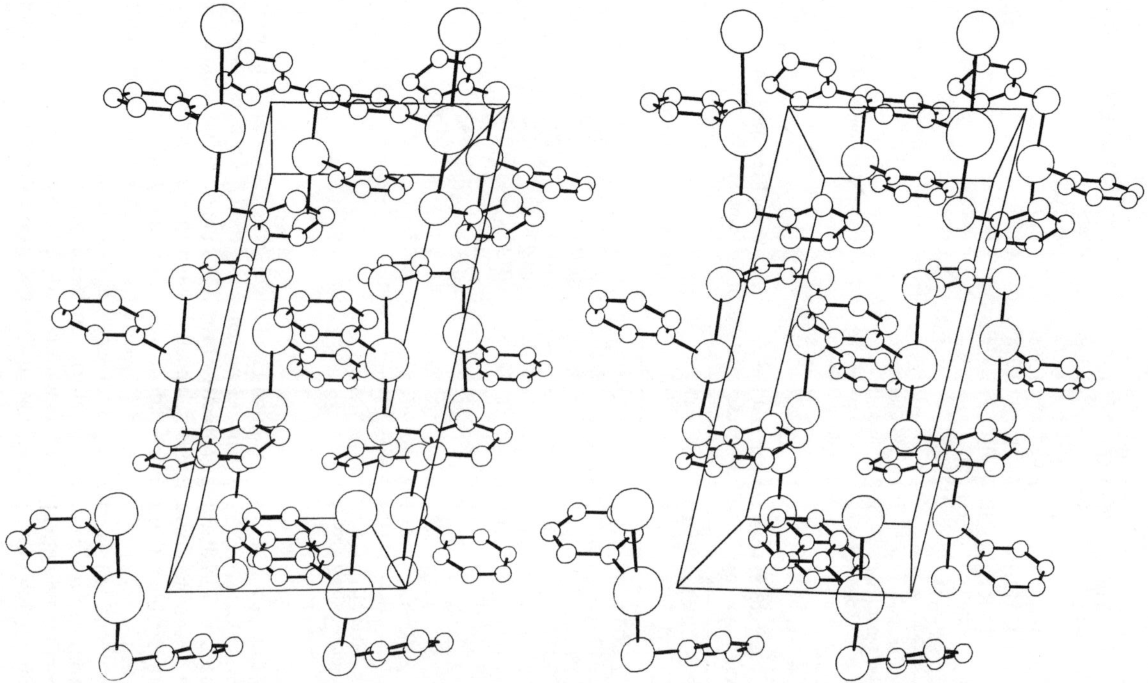

(b)

FIGURE 44. Stereoscopic view of crystal packing, seen along the *b* axis, in the two modifications of **109**: (a) space group C2/c, (b) space group $P2_1/c$. Reproduced by permission of Acta Chemica Scandinavica from Reference 278.

TABLE 22. Se—C(phenyl) bond lengths and the characterization of the benzene ring deformation in three-coordinated complexes of divalent Se from XD

Compound	Se—C(Ph) (Å)	α^a (deg.)	$\langle CC \rangle^b$ (Å)	ΔCC^c (Å)	Reference
$Ph_3Se(NCS)$	1.936(4)	122.5(4)	1.370	0.020	228
	1.904(5)	119.8(5)	1.369	−0.005	
	1.928(4)	121.8(4)	1.377	0.002	
$Ph_3SeCl \cdot 2H_2O$	1.911(10)	121.6(9)	1.386	0.019	229
	1.936(12)	121.6(11)	1.378	0.038	
	1.934(11)	121.3(10)	1.378	0.000	
75	1.898(9)	118.5(8)	1.382	−0.002	234
	1.926(9)	120.7(8)	1.378	0.002	

[a]Endocyclic C—C—C bond angle adjacent to the substituent.
[b]Mean value of the C—C bond lengths in the Ph group.
[c]Mean value of the differences of bond lengths C(2)—C(3) minus C(1)—C(2) and C(5)—C(6) minus C(6)—C(1), where C(1) is the Ph carbon atom adjacent to the substituent.

TABLE 23. Te—C(phenyl) bond lengths and the characterization of the benzene ring deformation in three-coordinated complexes of divalent Te from XD

Compound	Te—C(Ph) (Å)	α^a (deg.)	$\langle CC \rangle^b$ (Å)	ΔCC^c (Å)	Reference
105	2.102(7)	120.2(8)	1.400	0.022	210
106	2.129(6)	122.0(6)	1.397	0.020	210
107	2.124(6)	122.4(5)	1.398	0.017	277
108	2.112(7)	120.1(6)	1.408	−0.020	277
109[d]	2.116(3)	119.6(3)	1.384	0.000	278
	2.123(4)	120.0(4)	1.379	0.011	278
110	2.118(7)	119.2(7)	1.382	0.018	279
111	2.120(2)	118.6(2)	1.373	−0.001	279
112	2.104(5)	119.0(5)	1.378	−0.020	280
113	2.100(6)	120.2(6)	1.381	0.033	280

[a,b,c]See footnotes to Table 22.
[d]Crystallizes in two different space groups.

give much credibility to this correlation, although in a formal sense it is consistent with expectation: If we have a $C_6H_5—XY_n$ system, a lengthening of the C—X bond may be associated with the electron-withdrawing character of the XY_n ligand. At the same time, the presence of a more electronegative ligand is expected to increase the *ipso* endocyclic C—C—C angle. The available data may not be adequate for examining such subtle effects as the case is here. On the other hand, due to various circumstances some important systematic errors may cancel in this series of parameters. The changes depicted in Figure 45 cannot originate from electronic effects, but rather from steric ones as various C_6H_5Se moieties occur in different environments. Even if the correlation is not spurious, in any case this is as far as we can stretch our desire to find substituent effects upon benzene ring deformation in these compound series, relying upon the available data. Even that much indication for correlation cannot be established for the larger sample of the Te derivatives. This in itself is not surprising as the changes in the Te ligands in the series examined are expected to influence appreciably the nature of the Te as a benzene substituent.

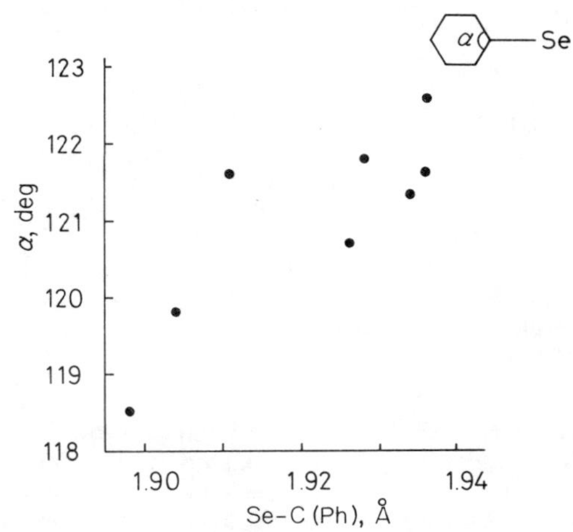

FIGURE 45. Variation of the Se—C(phenyl) bond length and the *ipso* endocyclic C—C—C angle (α) of the benzene ring. See Table 22 for data and references

V. FOUR-COORDINATED SELENIUM AND TELLURIUM

With four-coordinated Se or Te only crystal-phase molecular structures have been determined. The typical bond configuration of the metal is trigonal bipyramidal, made up from four bonds to various ligands and one lone pair of electrons. The systems are of the AX_4E type where A is the central atom, Se or Te, X are ligands and E is a lone pair of electrons. The lone pair invariably occupies one of the equatorial sites (Figure 46). This configuration is in complete agreement with the VSEPR model[28]. It will be of interest to examine structural variations from the point of view of the applicability of the VSEPR model. It has been demonstrated recently[282] that while testing the applicability of the

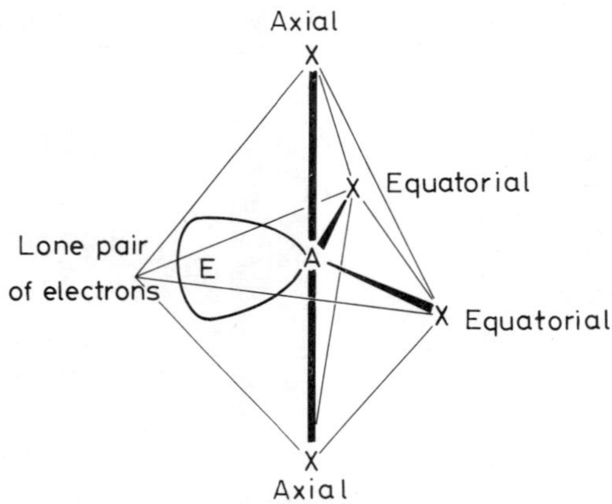

FIGURE 46. The trigonal bipyramidal configuration of an AX_4E-type molecule

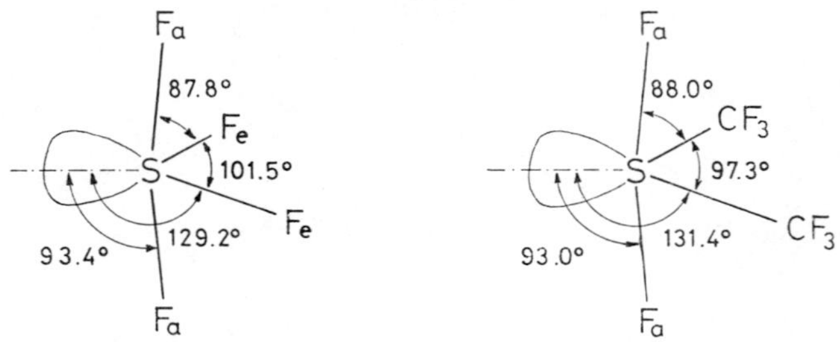

FIGURE 47. Bond angles of the central S atom in SF_4[283] and $(CF_3)_2SF_2$[284]

VSEPR model, all angles of all electron pairs about the central atom have to be examined rather than the bond angles only. Thus, for example in the equatorial plane of the AX_4E bond configuration, there are two E—A—X angles and one bond angle X—A—X. Obviously the two E—A/A—X lone-pair/bond interactions are more important than the single A—X/A—X bond/bond interaction. Fortunately, in this system the angles made by the lone pair can easily be calculated from the bond angles by virtue of symmetry. Comparison of the SF_4 and $(CF_3)_2SF_2$ structures demonstrates the utility of the approach suggested above (Figure 47). As the fluorines in the equatorial positions are substituted by the less electronegative CF_3 groups, the change in the bond angles in themselves would indicate incompatibility with the VSEPR model. The decisive factor, however, is the lone-pair/bonding-pair interactions in the equatorial plane in complete agreement with the VSEPR model.

It has been suggested[285] that the quadruple average angle of the lone pair, α_4^E, be considered as a measure of its generalized space requirement. This angle is the mean of the four angles made by the lone pair in the AX_4E configuration. It has been observed to be fairly constant in such structures in spite of the sometimes considerably varying bond angles. An example is shown by the series of SF_4, OSF_4 and $H_2C{=}SF_4$ molecules in Figure 48. The quadruple average angle will be especially useful in our discussion as in many structures the non-equality of the equatorial ligands will prevent the calculation of the individual E—A—X angles.

In an ideal trigonal bipyramidal structure the two axial bonds are colinear and the equatorial bond angles are 120°. Due to the relatively large space requirement of the lone

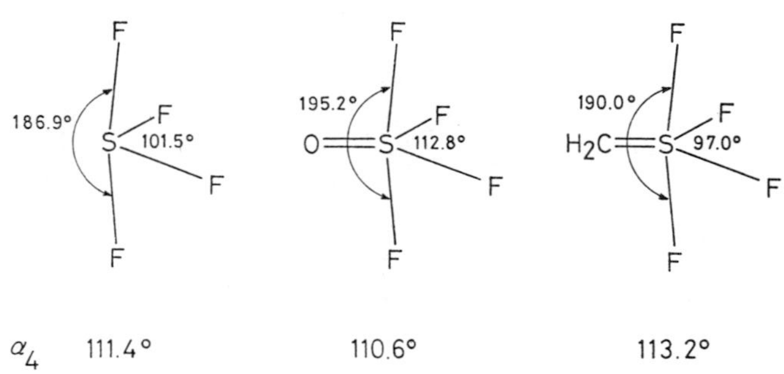

FIGURE 48. Bond angles and the quadruple average angle[285] (α_4) of the lone pair, the O=S and C=S bonds in SF_4[283], OSF_4[286] and $H_2C{=}SF_4$[287], respectively

pair of electrons the angle between the two axial bonds is usually smaller than 180° and the equatorial bond angle is smaller than 120°. For the ideal arrangement the quadruple average angle of an equatorial bond would be 105°. The lone pair of electrons has usually considerably larger quadruple average angles.

Relatively low accuracy was achieved in the X-ray diffraction investigation[288] of dichloro(2-chloropropyl)-*p*-tolylselenium(IV) (**114**). The equatorial

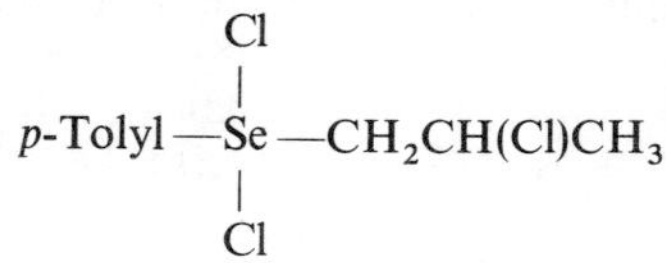

(**114**)

C—Se—C angle was found to be 102° while the two axial Se—Cl bonds were reported to be colinear. This does not agree with general experience. The quadruple average angle of the Se lone pair, α_4^E(Se) is 109.5° according to the reported bond angles, which is much too low as compared with similar structures where a typical value would be about 112°. This may indicate that as much as even 10° deviation from linearity might be expected for the Cl—Se—Cl skeleton.

The crystalline molecular structure of *o*-carboxyphenyl methyl selenoxide (**115**) was studied together with its S analogue (**116**)[289]. The two compounds show very similar atomic positions, but there are important differences in their structures as is already apparent in the structural formulae below. There is a ring closure in the Se derivative (see

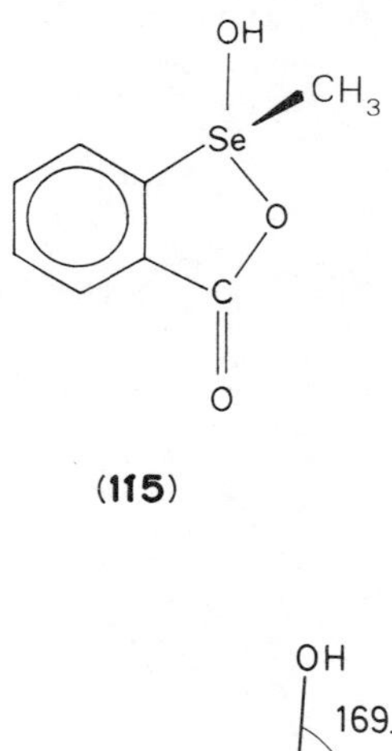

(**115**)

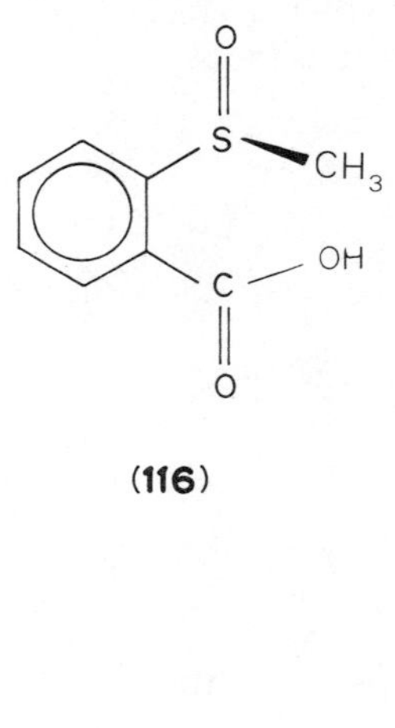

(**116**)

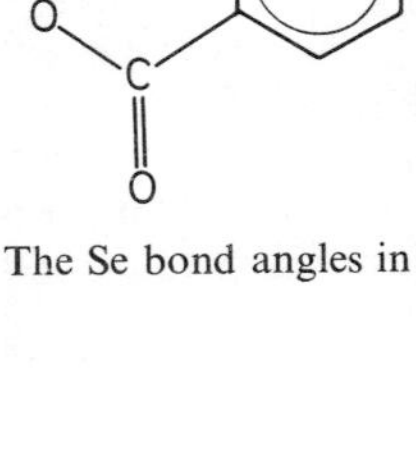

FIGURE 49. The Se bond angles in **115**[289]

Figure 49) with an Se—O linkage of 2.378 Å length. The selenium–carbon(Me) bond, 1.941(6) Å, is only slightly longer than the selenium–carbon(Ph) bond, 1.928(3) Å. For the S analogue the difference in the two analogous bonds is of the opposite sign, but considering the experimental errors they cannot really be distinguished. The bond configuration about S in **116** is very much the same as in dimethyl sulphoxide[290]. The selenium–oxygen bond, 1.774(3) Å, is nearer to a single bond than to a double bond. The calculated single Se—O bond would be 1.82 Å whereas the Se=O bond in $SeOCl_2$ is 1.614(5) Å[225]. The quadruple average angle of the Se lone pair, α_4^E(Se), is 112.8°.

The quadruple average angle, α_4^E(Se), is 110.0°, somewhat small, for 1-thia-4-selenacyclohexane 4,4-dibromide (**117**)[291] with Br—Se—Br 175.1(1)° and C—Se—C 105(1)°.

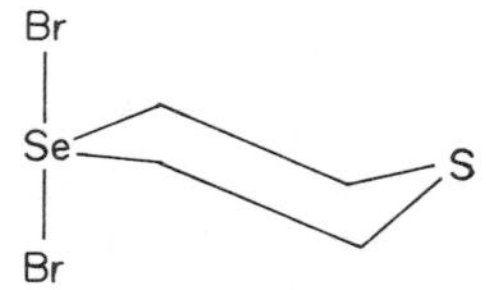

(**117**)

The Se bond configurations are essentially the same in 4,4′-spirobi(4-selena-4-butanolide)(**118**)[292] and 3,3′-spirobi(3-selenaphthalide)(**119**)[293]. In both compounds the two halves of the molecules are related by a two-fold axis. The angles about the Se atom in **118** and **119** are shown in Table 24. The Se–methylene bond in **118** is somewhat longer than the Se–phenylene bond in **119**: Se—C(methylene) = 1.959(3) Å, Se—C(phenylene) = 1.930(1) Å, whereas the Se—O bonds are the same, viz. 1.974(3) Å in **118** and 1.968(7) Å in **119**.

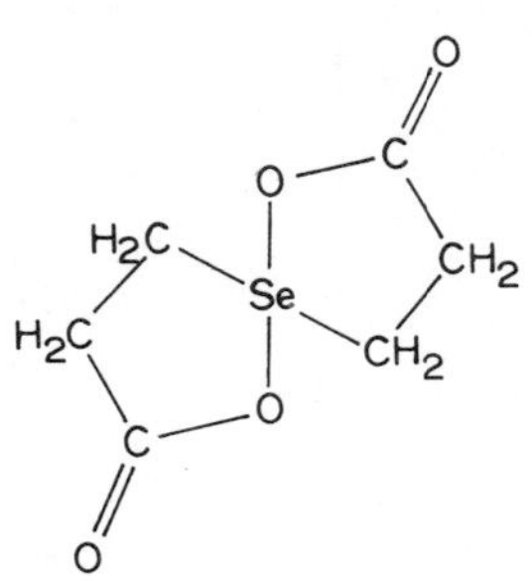

(**118**)

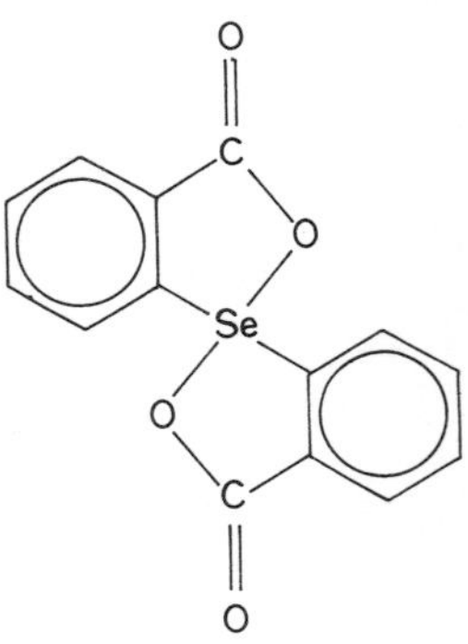

(**119**)

In 1,1-dichloro-2,5-bis[*N*-(chlorothio)imino]-3,4-dicyano-1,1,2,5-tetrahydroselenophene (**120**)[294] the relatively small endocyclic Se bond angle C—Se—C, 86.4(2)°, is the origin of the relatively large E—Se—C angle and consequently of the large quadruple average angle α_4^E(Se), 115.0°. In free selenophene[20,143] the C—Se—C angle is slightly larger, 87.8(2)°, as determined by MW. The Se—C bond of the four-coordinated arrangement is much longer, viz. 1.970(3) Å, than in selenophene itself, 1.855(2) Å. Wudl and Zellers[294] noted some unusual features of the molecular packing in the crystal. Whereas the related **121** forms head-to-tail sheets[295] (Figure 50a) the head-to-head ordering (Figure 50b) of **120** results in rather short intermolecular Cl···Cl contacts.

TABLE 24. Angles (deg.) about the Se atom in spiro compounds **118** and **119** from XD[a]

Angle	**118**[292]	**119**[293]
E—Se—O	93.8	93.8
E—Se—C	129.2	129.2
O—Se—O	172.4	172.4
C—Se—C	101.5	101.0
α_4^E(Se)	111.5	111.6

[a]Angles involving the lone pair of electrons (E) have been calculated from bond angles by virtue of symmetry. The estimated standard deviations for the bond angles are 0.3°.

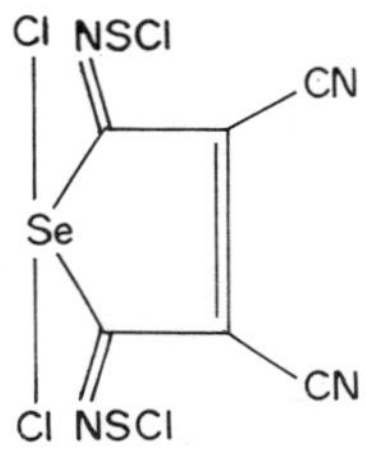

(120)

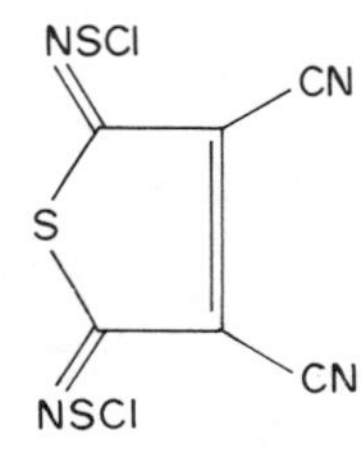

(121)

There is a relative abundance of data on four-coordinated organic Te derivatives. The Te bonds again show the typical features of the trigonal bipyramidal configuration, taking also the Te lone pair of electrons into consideration.

Dimethyltellurium tetraiodide, Me_2TeI_4 (**122**), is in fact dimethyltellurium diiodide–diiodine(1/1), an adduct of Me_2TeI_2 with I_2, linked by I···I bonds[296]. One of the I atoms of a Me_2TeI_2 molecule participates in this bonding and it is connected to two I_2 molecules (Figure 51). An I_2 molecule, on the other hand, forms a linear I···I—I···I bridge between two Me_2TeI_2 molecules as part of a zig-zag chain of I atoms. There are also two weak intermolecular Te···I contacts and, accordingly, the Te coordination may also be described as distorted octahedral. Figure 52 gives the bond angles about Te and the angles made by the lone pair of electrons. The distortion from the ideal trigonal bipyramidal arrangement is very well interpreted by the VSEPR model[28,282,296]. An alternative description of the bonding of Te has also been advanced by Pritzkow[296] by considering only s and p orbitals. Huheey[297] has recently shown the consistency of the VSEPR model and Bent's rule involving the p and s character in bonding for simple trigonal bipyramidal systems, even taking the directional effects[298] into consideration. Incidentally, the quadruple average angle of the Te lone pair of electrons, α_4^E(Te), is 112.1°, which seems to be in the region of typical values for well determined structures. The two axial Te—I bonds have strikingly different lengths in **122**. The bond to I that has intermolecular contacts with other I_2 molecules (cf. Figure 51) is much longer, 3.082(2) Å, than the other bond, 2.809(2) Å. Even this latter is considerably longer than what would correspond to the sum of the covalent radii (2.70 Å).

A similar bond configuration was found in an adduct (**123**) of tellurium tetrachloride and propylene by Kobelt and Paulus[299] (Figure 53). The Te—Cl bonds are again longer,

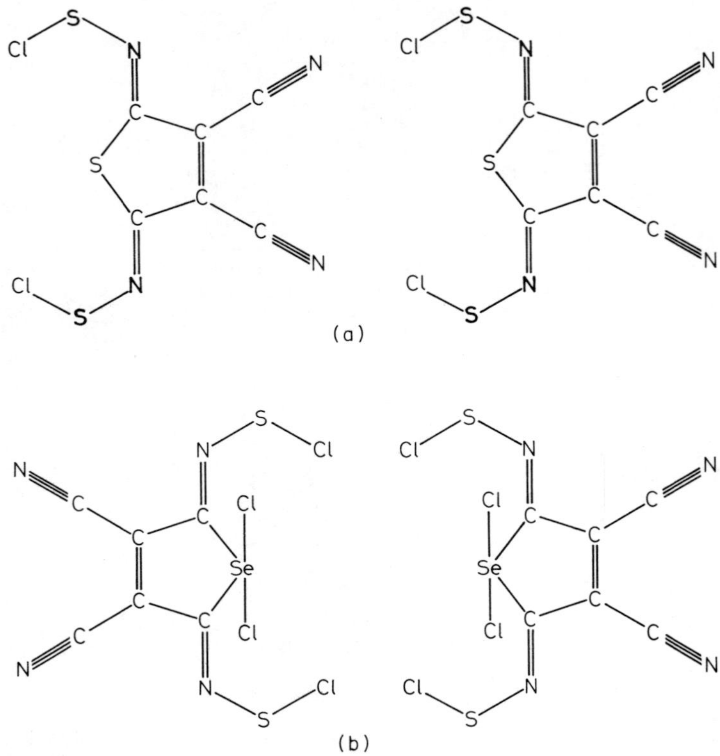

FIGURE 50. Schematic drawings of (a) the head-to-tail ordering of molecules in the crystal of **121** (after Reference 295) and (b) the head-to-head ordering in **120** (after Reference 294)

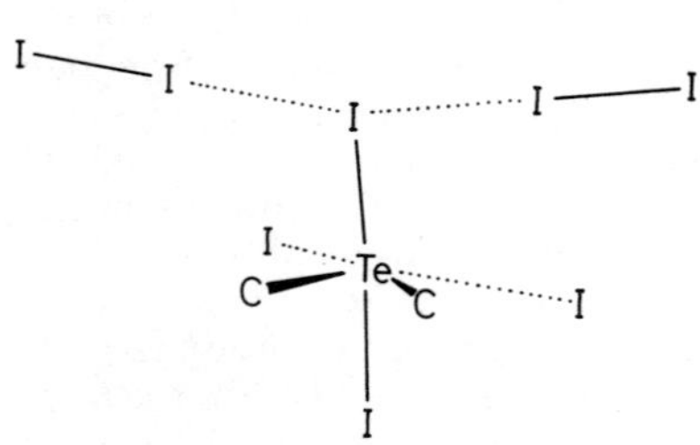

FIGURE 51. The molecular geometry and intermolecular I···I and Te···I contacts in crystalline dimethyltellurium tetraiodide (**122**) (after Reference 296)

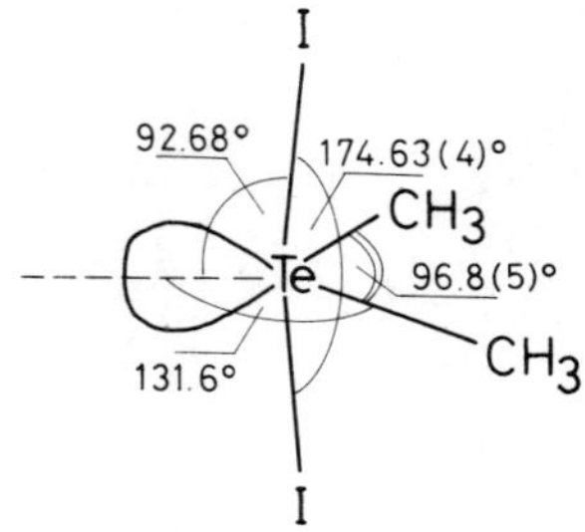

FIGURE 52. Angles around the Te atom in **122**[296]

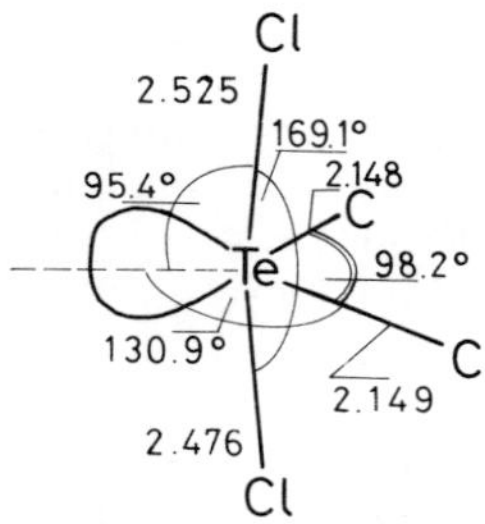

FIGURE 53. Angles around and bond lengths of the Te atom in **123**[299]

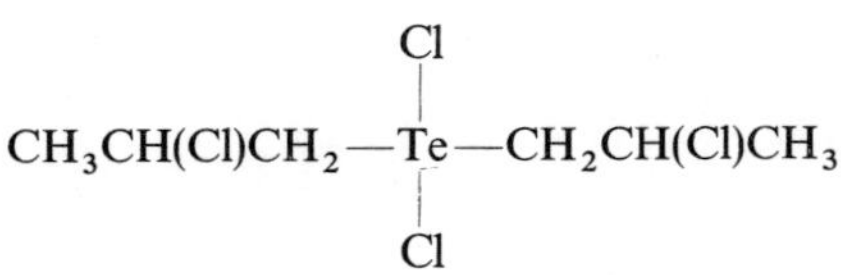

(123)

2.525(15) and 2.476(15) Å than what would correspond to the sum of the covalent radii (2.35 Å), but they are not significantly different.

In dichlorobis(2-chlorocyclohexyl)tellurium(IV) (**124**)[300] the quadruple angle, α_4^E(Te), is markedly small, 109.5°. It would be tempting to ascribe this effect to the bulkiness of the 2-chlorocyclohexyl groups. However, even this would not explain why even the E—Te—Cl angles are smaller (Figure 54) than those in **123**. In both cyclohexyl rings the Te—C and the adjacent C—Cl bonds are equatorial.

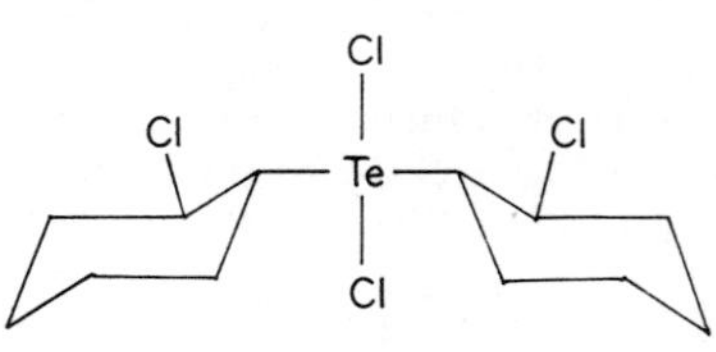

(124)

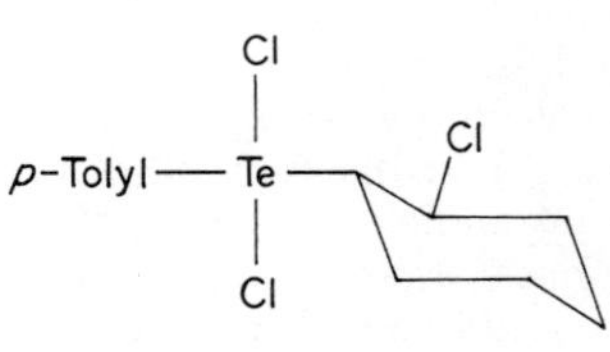

(125)

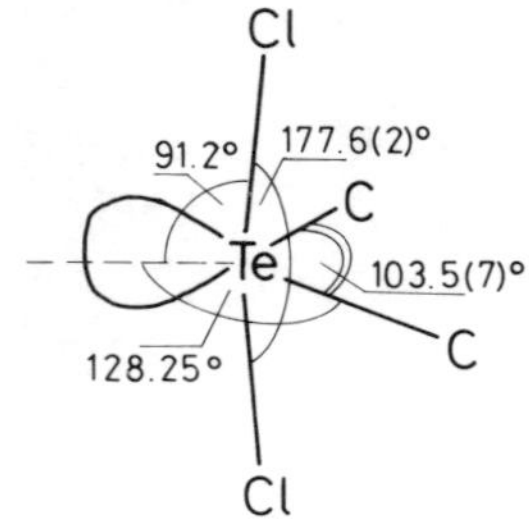

FIGURE 54. Angles around the Te atom in **124**[300]

The molecular geometry of dichloro(2-chlorocyclohexyl)-*p*-tolyltellurium(IV)(**125**)[301] is similar to that of **124**. The Cl—Te—Cl and C—Te—C bond angles are 176.4(3) and 101.0(6)°, respectively, and α_4^E(Te) is 110.6°. The Te—Cl bond lengths are also similar to those in **124**.

Another trigonal bipyramidal Te configuration appears in the crystal of 1,1-dichloro-1-telluracyclohexane-3,5-dione (**126**)[186]. The Cl ligands are in axial positions with Te—Cl 2.49 Å and Cl—Te—Cl 171.8°. The endocyclic C—Te—C angle is somewhat larger, viz. 95.5°, than this angle in 1-telluracyclohexane-3,5-dione (**35**) itself[183,186] (cf. Table 18).

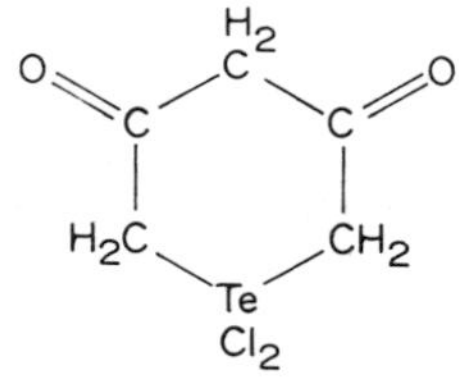

(**126**)

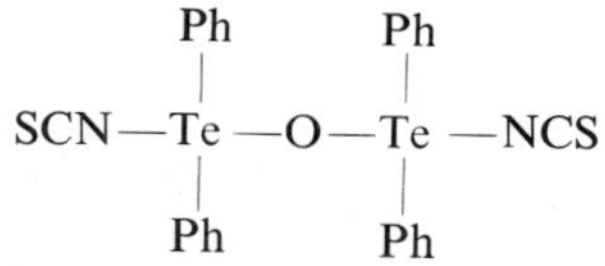

(**127**)

The molecular configuration of bis(isothiocyanatodiphenyltellurium(IV)) oxide (**127**)[302] is shown in Figure 55. The angular arrangement about the Te atom shows no unusual features, N—Te—O 172.2(3)° and C—Te—C 97.2(4)° with α_4^E(Te) 112.6°. There are relatively short intermolecular Te···S contacts, viz. 3.416(3) Å, serving to link the **127** molecules into molecular chains as shown in Figure 56. Mancinelli and coworkers[302] suggested that the Te bond configuration may be considered to be square pyramidal. Notwithstanding this, the angular parameters of the Te bond configuration also indicate very typical trigonal bipyramidal arrangement.

Smith and coworkers[303] recently determined the crystal and molecular structure of tetraphenyltellurium–benzene(8/1), $Ph_4Te \cdot \frac{1}{8}C_6H_6$ (**128**). There are four independent molecules in the unit cell containing altogether eight molecules. The space group is $P\bar{1}$. The bond configuration about Te is essentially the same in the four independent molecules

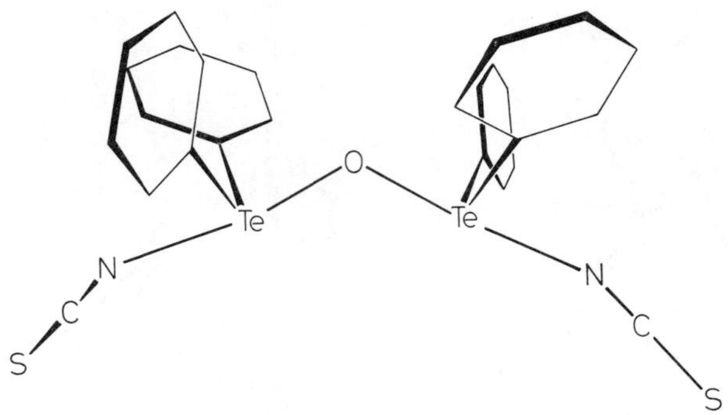

FIGURE 55. The molecular configuration of **127** (after Reference 302)

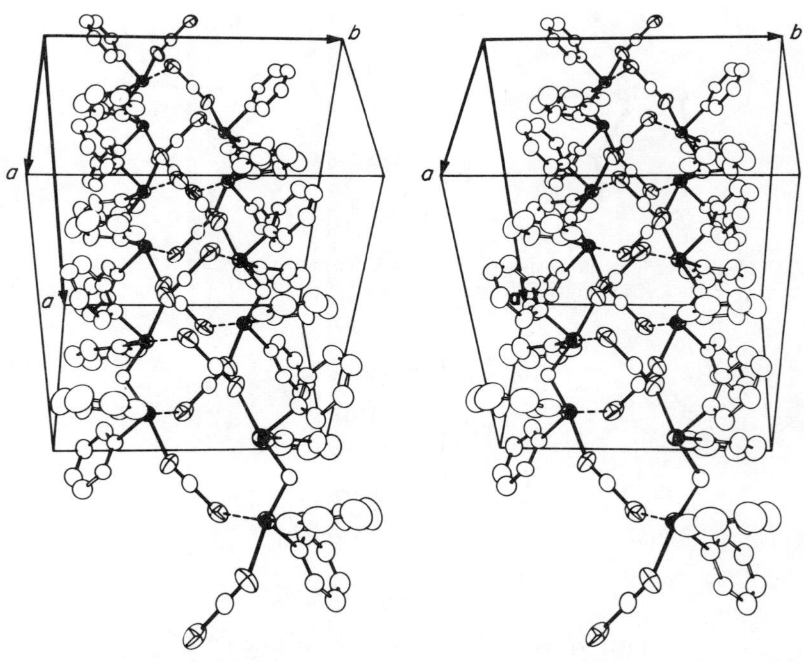

FIGURE 56. Chains of molecules along the *c* axis in the crystal of **127**, a stereoscopic view. Reproduced by permission of Elsevier Sequoia S.A. from Reference 302

which are, however, conformationally different due to differences in the Ph group orientation (Figure 57). The angles about Te are shown in Figure 58. The $\alpha_4^E(Te) = 110.7°$ quadruple average angle suggests that weaker repulsion interactions are in effect between the lone pair and the bonding pairs in **128** than $\alpha_4^E(Te) = 113.2°$ in **123**. The axial bond angles Cl—Te—Cl in **123** and C—Te—C in **128** are the same, 169°. Thus the difference can conveniently be reduced to that in the interactions in the equatorial plane. The difference in the bond angles C—Te—C of **123** and **128** is 10°! Considering the ligand

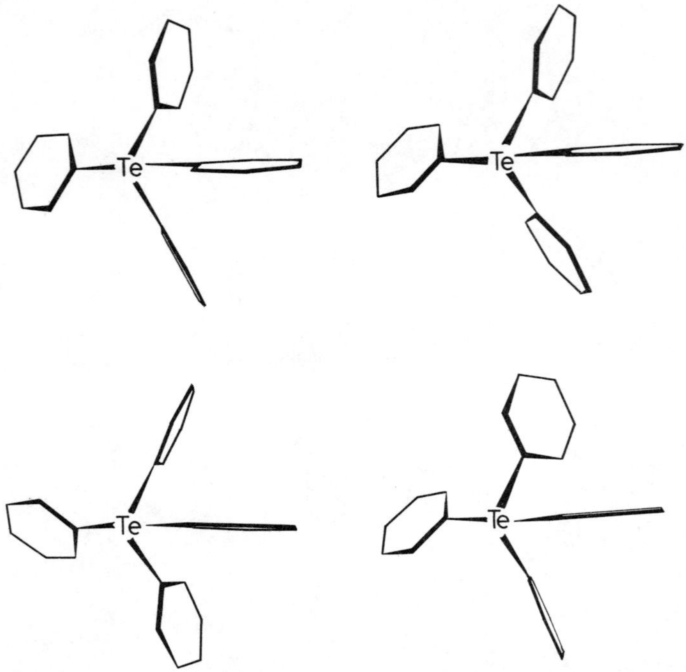

FIGURE 57. Conformations of the four independent molecules of Ph_4Te in crystals of **128** (after Reference 303)

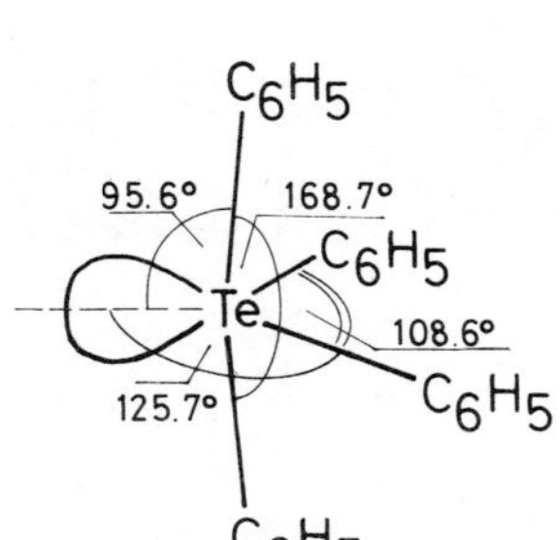

FIGURE 58. Angles around the Te atom in **128**[303]

electronegativities, the Ph group would be expected to draw more electron density from the vicinity of the central atom than would the alkyl group. Thus from a superficial approach in the application of the VSEPR model, but one which is widely used, the C_{eq}—Te—C_{eq} of **128** would be predicted smaller. However, the observed difference in the bond angles is in complete agreement with the VSEPR model if consideration also includes the importance of lone-pair/bonding-pair interactions of which there are two, versus the less important and only *one* bond/bond interaction. Of course, this model may predict the direction of the change but not its magnitude. The amount of the angular opening in the equatorial plane might suggest that steric effects cause at least part of it. On the other hand

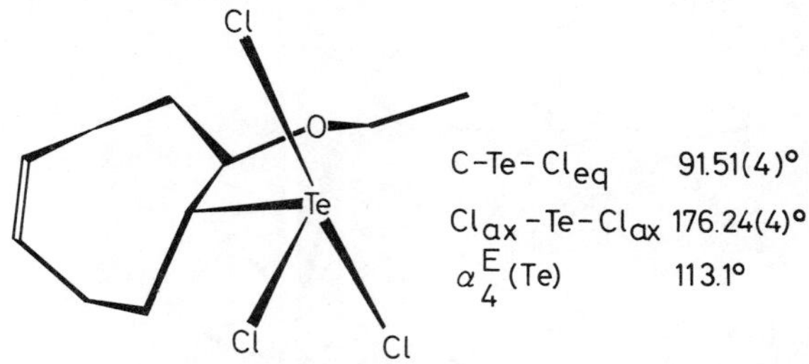

FIGURE 59. The molecular model of **129** (after Reference 304)

the C(phenyl)—Te—C(phenyl) bond angle involving an axial and an equatorial position may be as small as 85°! Indeed the Ph arrangements in Figure 57 do not indicate strong steric hindrances. The axial and equatorial Te—C bond lengths are markedly different, again in agreement with the VSEPR model.

The molecular structure of 8-ethoxy-4-cyclooctenyltellurium trichloride (**129**)[304] has no unusual features in its Te bond configuration as shown in Figure 59.

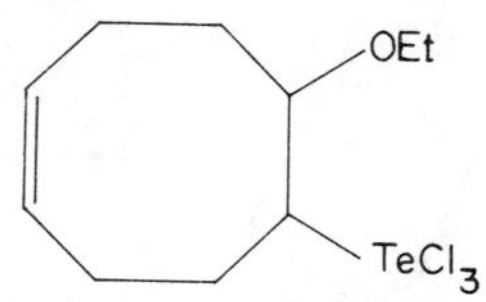

(**129**)

McCullough and Knobler[305–307] have studied some 2-biphenylyltellurium trihalides

o-$PhC_6H_4TeBr_3$	*o*-$PhC_6H_4TeI_3$
(**130**)	(**131**) α-modification
	(**132**) β-modification

(**130–132**) by XD. For all three structures, again, the trigonal bipyramidal configuration is characteristic with two halogens in axial positions, and the third halogen and a carbon atom plus the lone pair of electrons of Te in the equatorial positions. The parameters are given in Table 25. Important and interesting are the differences in intermolecular contacts in the three structures. In the Br derivative (**130**) relatively long Te···Br interactions (3.713 Å) join pairs of molecules into *dimers* across the symmetry centre (Figure 60). In the α-modification of the I derivative (**131**) two different systems of I···I intermolecular

TABLE 25. Angles (deg.) about the Te atom in 2-biphenylyltellurium trihalides from XD

Compound	X_{ax}—Te—X_{ax}	X_{eq}—Te—C	α_4^E(Te)[a]	Reference
130 (X = Br)	178.46(4)	97.1(3)	111.1	305
131 (X = I)	176.54(4)	100.4(3)	110.8	306
132 (X = I)	176.02(5)	98.1(3)	111.5	307

[a]Calculated from the bond angles.

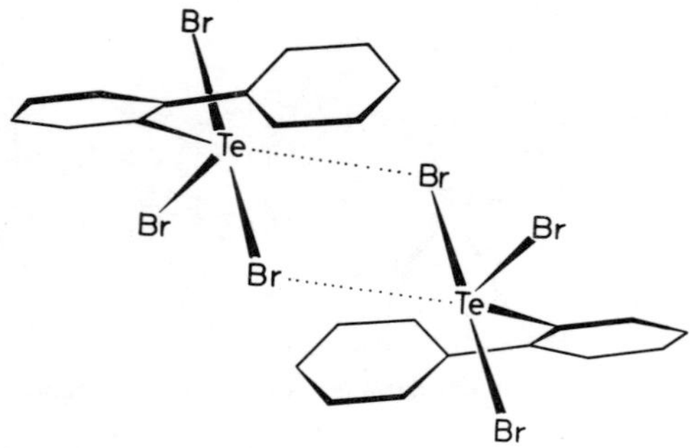

FIGURE 60. A dimer in the crystal of **130** (after Reference 305)

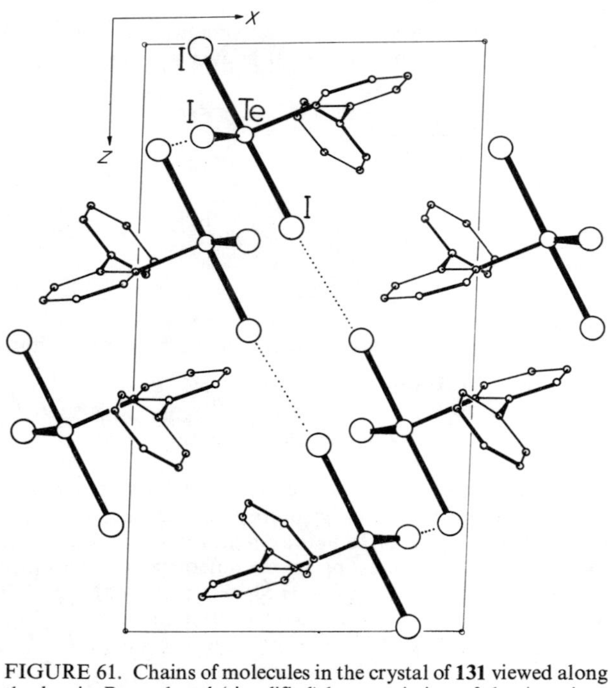

FIGURE 61. Chains of molecules in the crystal of **131** viewed along the *b* axis. Reproduced (simplified) by permission of the American Chemical Society from Reference 306

linkages (of 3.239 and 3.772 Å, respectively) connect the molecules into *chains* (Figure 61). Finally, in the β-modification of the I derivative (**132**), again molecular *chains* are formed by intermolecular contacts. Here, however, they are between Te and I atoms. A selected portion of the structure is shown in Figure 62. McCullough discusses possible correlations between the colour of **131** and **132** and the peculiarities of molecular packing. He notes a relatively short intramolecular Te $\cdots$ C distance[307] as shown in Figure 63. As to the benzene ring angular deformations in **130**, **131** and **132**, the mean values of the endocyclic

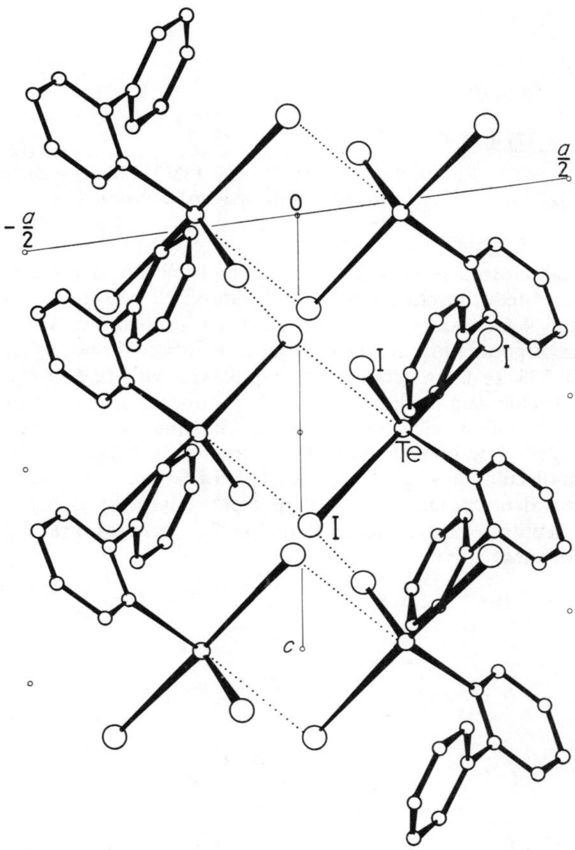

FIGURE 62. Chains of molecules in the crystal of **132** viewed along the *b* axis. Reproduced (simplified) by permission of the American Chemical Society from Reference 307

bond angles are given in Figure 64. The bond angle adjacent to the Te substituent is markedly larger than 120° and the angle adjacent to the second benzene ring is markedly smaller. Then the angular deviations gradually diminish. The mean of the *ipso* angles of the second benzene ring in the three structures is 118.8(7)° in agreement with the deformation observed in biphenyl itself[308].

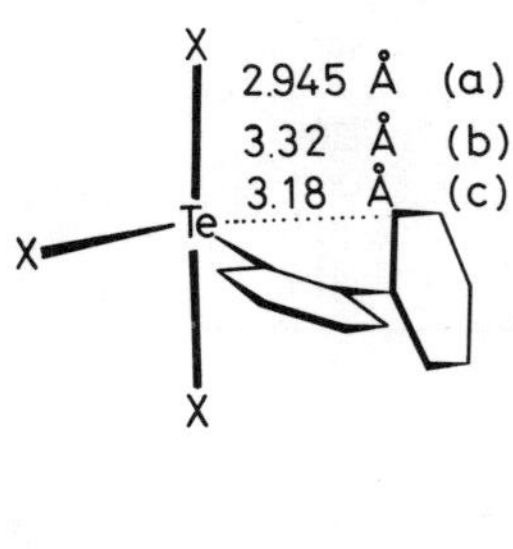

FIGURE 63. Short intramolecular Te···C contacts[307] in the crystals of (a) **130**, (b) **131** and (c) **132**, and the molecular model of **132** (after Reference 307)

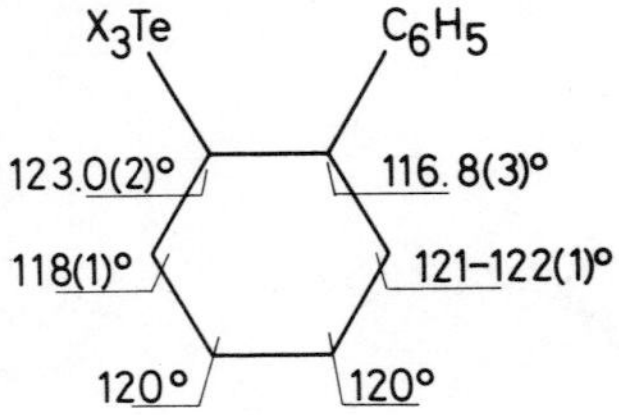

FIGURE 64. Endocyclic bond angles in the phenylene ring of **130**[305], **131**[306] and **132**[307]. Mean values from the three structures with standard deviations

The crystal and molecular structures of 1-thia-4-telluracyclohexane 4,4-dibromide (**133**)[309], 1-oxa-4-telluracyclohexane 4,4-diiodide (**134**)[310] and 1-thia-4-telluracyclohexane 4,4-diiodide (**135**)[311] have been studied by XD. The parameters characterizing the trigonal bipyramidal configuration about Te are collected in Table 26. In the crystal of **133** Te forms coordination linkages with one of the Br atoms of a neighbouring molecule and with the S atom of another neighbouring molecule. An octahedral Te bond configuration is thus formed. The same situation was observed for the Se analogue (**117**)[291]. The bond lengths are compared in Figure 65. It is seen that the analogous intermolecular linkages involving Te and Se are approximately of the same length. As the covalent radius of Te is considerably larger (1.356 Å) than that of Se (1.163 Å), these lengths may indicate that the Te linkages are stronger than the corresponding Se ones.

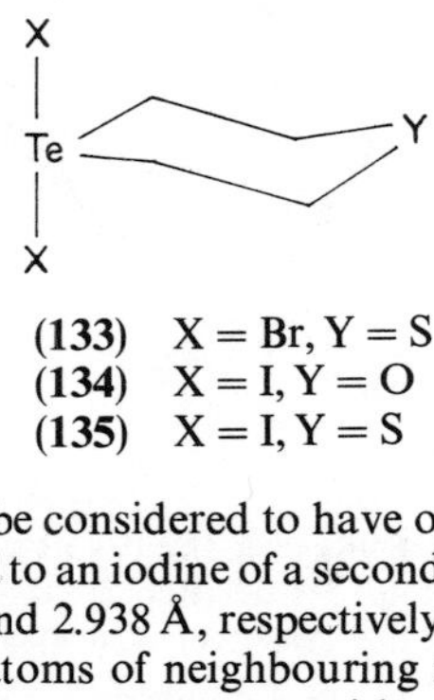

(**133**) X = Br, Y = S
(**134**) X = I, Y = O
(**135**) X = I, Y = S

The Te atom in **134** may also be considered to have octahedral bond configuration, as coordination linkages are formed to an iodine of a second molecule and to one more iodine of a third molecule, at 2.886(1) and 2.938 Å, respectively. The Te of **135** again establishes intermolecular contacts with I atoms of neighbouring molecules.

The ring puckering in the three structures with *chair* conformation discussed is demonstrated by the average torsional angles in Figure 66. The puckering is the same in the three cases and the molecules seem to be somewhat flatter than free 1,4-dioxane or 1,4-

TABLE 26. Angles (deg.) about the Te atom in dihalides **133**, **134** and **135** from XD

Compound	X—Te—X	C—Te—C	α_4^E(Te)[a]	Reference
133	176.63(6)	99.4(6)	111.0	309
134	177.08(4)	94.1(4)	112.2	310
135[b]	174.9(1)	100(1)	111.3	311
	178.1(1)	100(1)	110.5	

[a]Calculated from the bond angles.
[b]There are two molecules in the asymmetric unit.

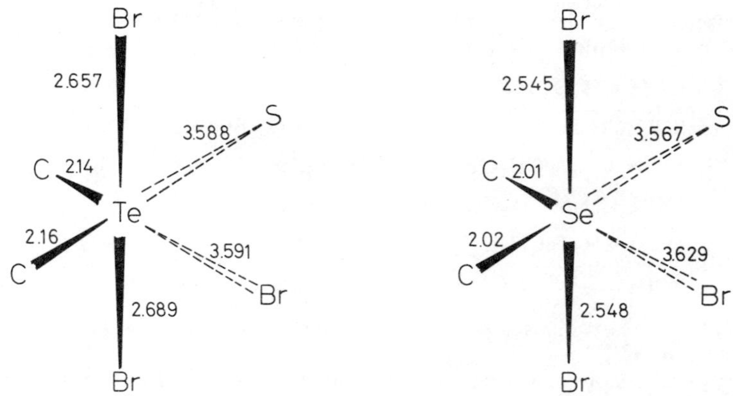

FIGURE 65. Bond lengths and intermolecular contacts (Å) of Te in **133**[309] and of Se in **117**[291]

oxathiane or especially 1,4-dithiane. For comparison, data on *trans*-2,3-dichlorides of analogous molecules in the crystal are also presented in Figure 66.

A normal coordinate analysis by Hagen and coworkers[315] indicated that the six-membered rings of 1,4-dioxane, 1,4-dithiane, 1,4-diselenane, 1,4-oxaselenane, 1,4-oxatellurane and 1,4-thiaselenane are relatively rigid. The relative rigidity of the rings is

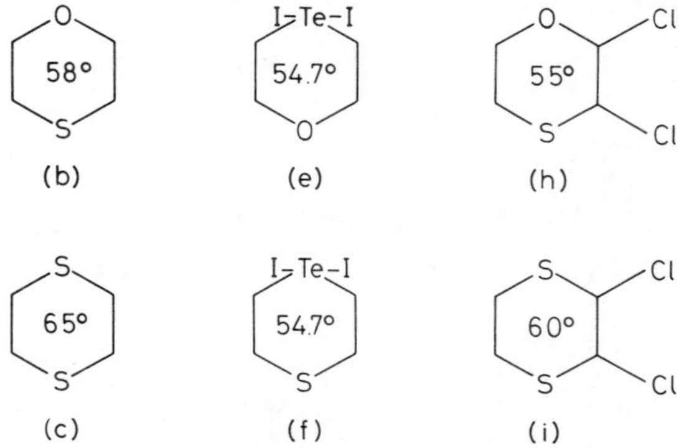

FIGURE 66. 1, 4-Diheterocyclohexane derivatives: the mean torsional angles as characterization of ring puckering; (a)[312], (b)[240] and (c)[313], free molecules, (d) **133**[309], (e) **134**[310], (f) **135**[311], (g), (h) and (i)[314], crystals

well characterized by the relatively small mean parallel vibrational amplitudes (l)[316], referring to the 1,4-interactions between the heteroatoms:

$\overline{OCH_2CH_2OCH_2CH_2}$	l(O···O)	0.071 Å
$\overline{SCH_2CH_2SCH_2CH_2}$	l(S···S)	0.062 Å
$\overline{SeCH_2CH_2SeCH_2CH_2}$	l(Se···Se)	0.076 Å
$\overline{OCH_2CH_2SeCH_2CH_2}$	l(O···Se)	0.078Å
$\overline{OCH_2CH_2TeCH_2CH_2}$	l(O···Te)	0.080 Å
$\overline{SCH_2CH_2SeCH_2CH_2}$	l(S···Se)	0.078 Å

The calculated effects of the perpendicular vibrations[316] for the same interactions practically vanish[315].

The crystal and molecular structure of the α-modification of 1,1-diiodo-3,4-benzo-1λ^4-telluracyclopentane (**136**) was determined by XD[317]. The trigonal bipyramidal arrangement about Te is characterized by the axial I—Te—I and equatorial C—Te—C bond angles of 176.53(4) and 86.0(5)°, respectively. The quadruple average angle of the lone pair is relatively large, α_4^E(Te) = 114.4°. The Te···I intermolecular contacts, however, create a distorted octahedral Te bond configuration (Figure 67).

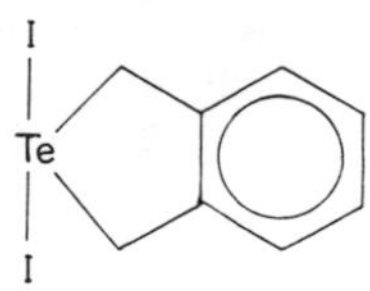

(**136**)

Te also forms intermolecular linkages with I atoms in each of the two neighbouring molecules in the crystal of dibenzotellurophene diiodide (**137**)[318]. The intermolecular Te···I links which are 3.717 and 3.696 Å long bring the molecules into infinite chains (Figure 68). The intermolecular contacts make the Te bond configuration distorted octahedral. The axial I—Te—I and equatorial C—Te—C bond angles are 178.47(1) and

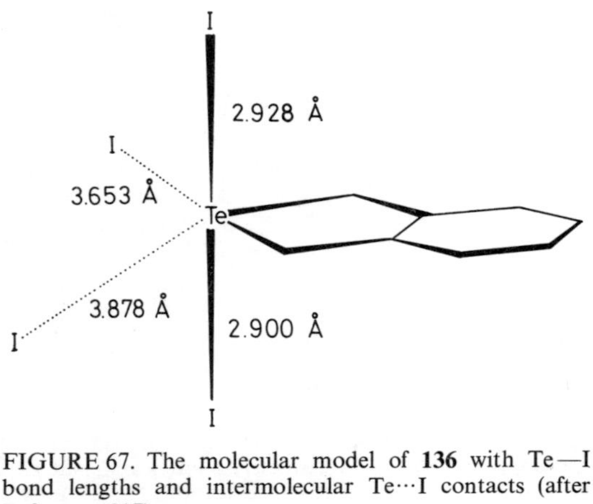

FIGURE 67. The molecular model of **136** with Te—I bond lengths and intermolecular Te···I contacts (after Reference 317)

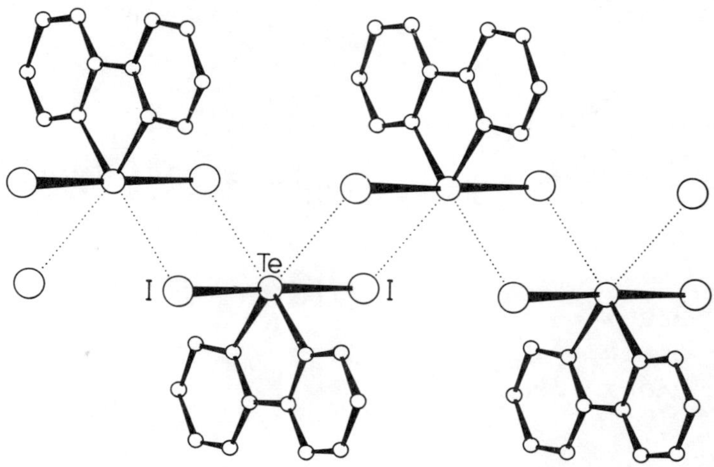

FIGURE 68. Chains of molecules in the crystal of **137**. Reproduced (simplified) by permission of the American Chemical Society from Reference 318

81.8°, respectively. The α_4^E(Te) is 114.9°. This relatively large quadruple average angle of the Te lone pair, i.e. about 115°, seems to be characteristic for the distorted octahedral Te bond configuration; thus it may indicate an increased participation of coordination bonding for the lone pair of electrons.

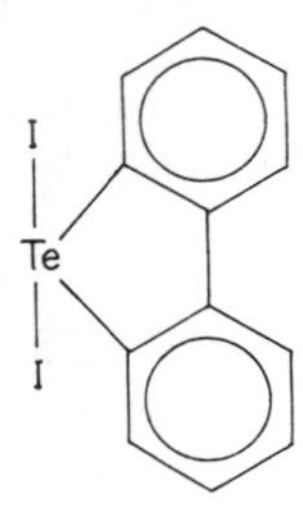

(**137**)

The crystal and molecular structures of several phenoxatellurin derivatives have been determined, i.e. the 10,10-diiodide (**138**)[319], bis(trifluoroacetate) (**139**)[320] and dinitrate (**140**)[321] and 10,10′-oxybis [10,10-dihydro-10-(nitrooxy)phenoxatellurin] (**141**)[322]. One of the characteristic features of the phenoxatellurin structures is the fold angle θ_2 along the Te···O axis where the two planar or nearly planar halves of the molecules meet (Figure 30). These angles are the following: **138** 164°; **139** 152°; **140** 175°; **141** 147°, 163°. The structure of phenoxatellurin (**49**)[202] itself has been discussed in Section III.C dealing with two-coordinated Te. Mangion, Smith and Meyers[321] have interpreted the differences in the fold angles of phenoxatellurin with Te(II) and its derivatives with Te(IV) on the basis of molecular orbital theory. Of the Te(IV) derivatives only the dinitrate (**140**) has a nearly planar ring structure.

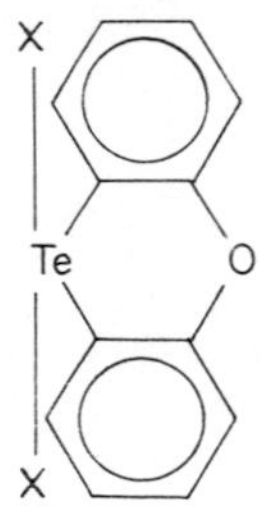

(**138**) X = I
(**139**) X = $-OC(O)CF_3$
(**140**) X = ONO_2

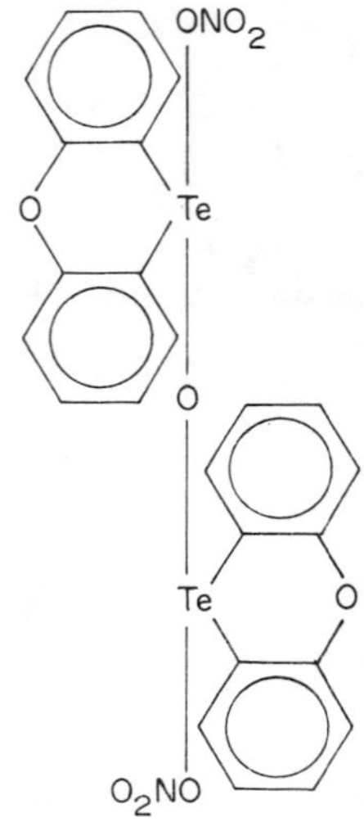

(**141**)

Table 27 presents data on the trigonal bipyramidal configuration about Te. Noteworthy are the relatively large quadruple average angles of the Te lone pair of electrons. In **138** there are again Te ··· I intermolecular contacts with two neighbouring molecules at 3.739 and 3.788 Å. These contacts bring the molecules into infinite chains, similar to those depicted for **137** in Figure 68. For **139** the possible intermolecular contacts have not been discussed. The shortest intermolecular contacts in **140** are O ··· O at 3.14 Å, excluding H atoms. No noteworthy intermolecular contacts have been observed in **141** either.

In conclusion, the various Te—C bond lengths and C—Te—C bond angles for the molecules **122** to **141** are summarized in Table 28. The general tendency for Te—C(alkyl) to be longer than Te—C(phenyl) seems to be present, although the data scatter and overlap. Both types, at least in their mean values, appear to be somewhat longer than what would correspond to the sum of the covalent radii, 2.12 Å. The latter occurs as the mean of the Te—C bond lengths for endocyclic systems involving saturated rings or moieties. The bond angles referring to these systems are obviously smaller than those occurring in open moieties.

TABLE 27. Angles (deg.) about the four-coordinated Te atom in phenoxatellurin derivatives

Compound	ax—Te—ax	C—Te—C	α_4^E(Te)[a]	Reference
138	176.44(6)	91.5(6)	113.0	319
139	167.5(2)	91.5(3)	115.2	320
140	168.0(2)	93.5(2)	114.6	321
141[b]	168.6(1)	90.2(2)	115.3	322
	171.8(2)	91.3(3)	114.2	

[a] Calculated from the bond angles.
[b] The two Te atoms are crystallographically different.

TABLE 28. Te—C bond lengths (Å) and C—Te—C bond angles (deg.) with four-coordinated Te

Compound	Te—C(alkyl)	Te—C(ring)[a]	Te—C(phenyl)	C—Te—C	Reference
122	2.14(1)			96.8(5)	296
123	2.148(5)			98.2	299
124	2.18(1)			103.0(5)	300
125	2.24(2)		2.08(2)	101.0(6)	301
127			2.112(9)	97.2(4)	302
128			2.13(1)eq[b] 2.29(2)ax[b]	109(2)[c]	303
129	2.172(3)				304
130			2.136(8)		305
131			2.152(12)		306
132			2.153(12)		307
126		2.16[d]		95.5[d]	186
133		2.14(1)		99.4(6)	309
134		2.17(1)		94.1(4)	310
135		2.16[d]		100[d]	311
136		2.139(12) 2.145(12)		86.0(5)	317
137		2.111(4)		81.8(2)	318
138		2.10(2)[d]		91.5(6)	319
139		2.066[d]		91.5(3)	320
140		2.068(4)		93.5(2)	321
141		2.091[d]		90.7[d]	322
Mean (σ)	2.18(4)	2.12(4)	2.15(6)	96(6)	
Mean of **122–125**, **127**, **128**				101(4)	
Mean of **126**, **133–141**				92(5)	

[a]Endocyclic Te—C bonds.
[b]Mean of eight values.
[c]Mean of four C(eq)—Te—C(eq) values.
[d]Mean value.

VI. FIVE- AND HIGHER-COORDINATED SELENIUM AND TELLURIUM

Trifluoromethylselenium trichloride, CF_3SeCl_3, is the only Se compound belonging to this section. Its molecular structure was determined in the solid state by Marsden and coworkers[323] using XD. The crystal is orthorhombic, Pbca and built from discrete dimeric molecules (Figure 69). The Se atoms are *five-coordinated* and have a distorted octahedral bond configuration with a Se—C bond and Se lone pair of electrons in the axial positions. The C—Se—Cl bond angles, at least in their mean, do not deviate from 90° significantly and this may suggest relatively weak repulsions from the lone pair of electrons. The reduced stereochemical activity of the lone pair may be related to intermolecular Se···Cl contacts. Although the Se—Cl bond lengths scatter, the bridging bonds are distinctly longer than the terminal ones, viz. 2.51–2.75 vs. 2.13–2.25 Å. The Se—C bonds are 2.01(2) and 2.05(2) Å long. The X-ray scattering by the CF_3 groups was smeared out by the effects of thermal motion, hence assumed C—F bond lengths and F—C—F bond angles were used in the structure refinement. The selected values (1.33 Å and 108.5°) correspond to 2.00 Å F···F non-bonded distances which are much shorter than twice the postulated

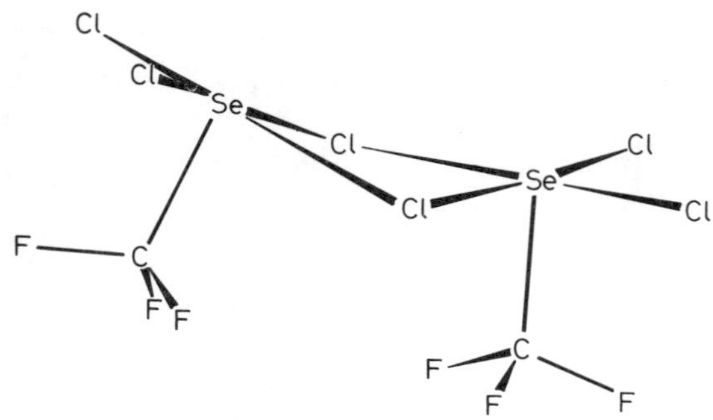

FIGURE 69. The dimeric molecule in the crystal of CF_3SeCl_3. Drawn from the atomic coordinates given in Reference 323

fluorine 1,3 intramolecular non-bonded radius, viz. 1.08 Å (cf. the discussion of the structure of tetrafluoro-1,3-diselenetane in Section III.C). The four-membered ring is puckered and the deviation from planarity may be characterized by the ClSeCl/ClSeCl angle which is about 26°, the same as the analogous feature of the free trimethylene sulphide structure.

The Te bond configuration in di-μ-bromo-μ-1,2-cyclohexylenetetrabromoditellurium, $C_6H_{10}Br_6Te_2$ (**142**), is similar to the Se bond configuration of CF_3SeCl_3. The crystal of **142** is orthorhombic, Pnma or $Pn2_1a$[324]. The Te atoms are somewhat displaced with respect to the plane of the Br atoms and away from the C atom (Figure 70). This indicates that the Te lone pair of electrons may have greater influence in determining the bond configuration here than was the case for Se in CF_3SeCl_3. The Te—C bond lengths are 2.19(4) and 2.26(5) Å. The four-membered ring is considerably puckered and the BrTeBr/BrTeBr angle is 48°. The molecule has a symmetry plane which contains all its six C atoms.

Another octahedral Te bond configuration occurs in phenyltellurium trihalide, $PhTeX_3$, i.e. $PhTeBr_{1.3}Cl_{1.7}$, whose molecular structure was determined by XD[325] (Figure 71 with X being a mixture of Cl and Br). The $PhTeX_3$ units are connected into infinite chains by halogen bridges. The C—Te—X bond angles do not deviate significantly from 90° in their mean. The Te—C bond lengths are 2.122(5) and 2.133(5) Å.

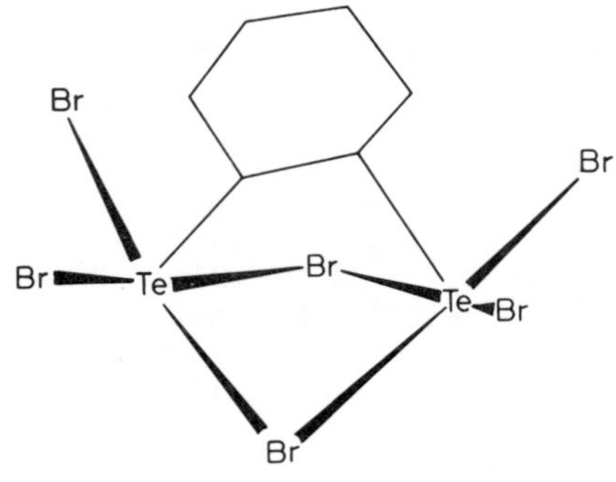

FIGURE 70. The molecular model of **142** (after Reference 324)

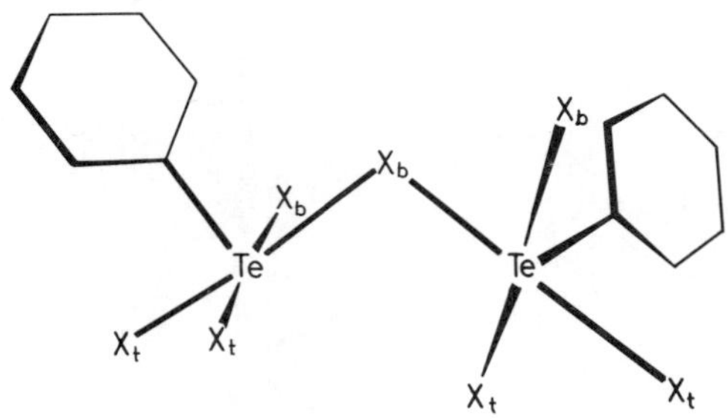

FIGURE 71. A chain molecules in crystalline $PhTeBr_{1.3}Cl_{1.7}$, with bridging (X_b) and terminal (X_t) halogen atoms (after Reference 325)

The ring deformations in the two independent Ph groups show a characteristic pattern. The mean values of the endocyclic bond angles are the following, with α being the *ipso* angle: $\alpha = 122.0$, $\beta = 118.6$, $\gamma = 120.2$ and $\delta = 120.2°$. The error limits of the individual bond angle values are about 0.5–0.6°.

The product of the reaction of $TeBr_2$ with cycloheptene in ethanol was identified by XD as *cis*-2-ethoxycycloheptyltribromotellurium(IV) (**143**)[326]. There are two molecules in the asymmetric unit, the crystal is triclinic, $P\bar{1}$. The bond configuration about Te is distorted octahedral with the Te lone pair of electrons occupying one of the 'equatorial' positions. The 'axial' Te—Br bonds are considerably longer than the 'equatorial' one, viz. 2.66 vs. 2.50 Å (mean values). The Te—C bonds are 2.29(4) and 2.24(4) Å long.

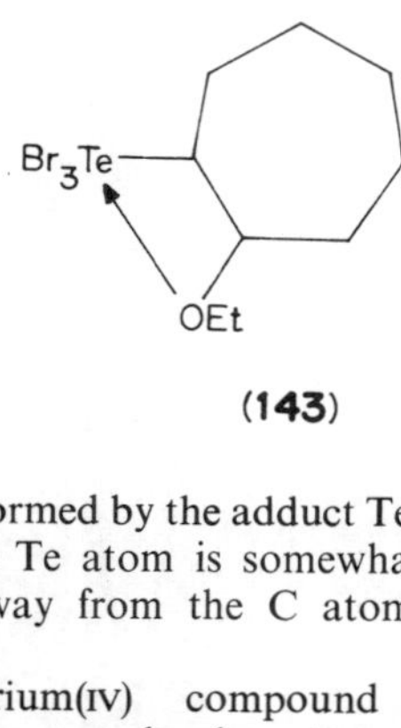

(**143**)

Infinite polymeric chains are formed by the adduct $TeCl_4{\cdot}C_2H_4$ (Figure 72). Its crystal is orthorhombic, Pcma[327]. The Te atom is somewhat displaced with respect to the plane of the Cl atoms and away from the C atoms. The Te—C bond length is 2.164(13) Å.

Another new organotellurium(IV) compound is [2-(6-acetyl-2-pyridinyl)-2-oxoethyl]trichlorotellurium, whose molecular structure was determined by XD[328] (Figure 73). The Te atom is *six-coordinated* and its bond configuration is distorted pentagonal bipyramidal with the Te lone pair of electrons occupying one of the equatorial positions. The Te—C bond length is 2.129(3) Å. This is a normal bond whereas the other Te linkages are appreciably longer than the corresponding sums of the covalent radii.

Esperås and Husebye[329] determined the crystal molecular structure of tris-(diethyldithiocarbamato)phenyltellurium(IV), $(Et_2NCS_2)_3TePh$ (**144**), by XD. The crystal is monoclinic, $P2_1/c$. The Te atom is *seven-coordinated* with six linkages to S atoms and the seventh to a Ph group. The bond configuration is distorted pentagonal bipyramidal with the Te—C bond in an axial position. The structure is depicted in the simplified model of

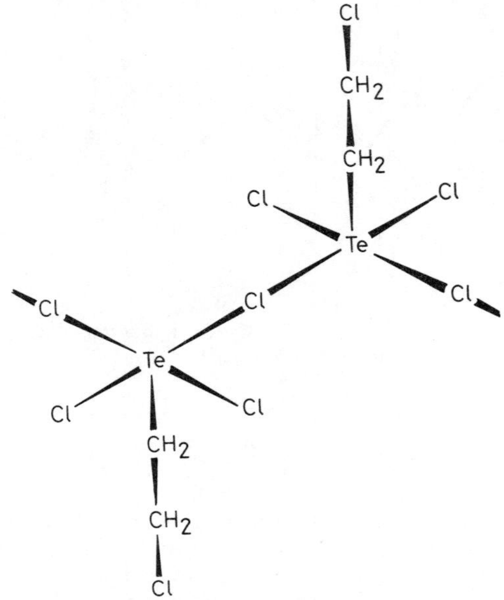

FIGURE 72. A chain of molecules in the crystal of $TeCl_4 \cdot C_2H_4$ (after Reference 327)

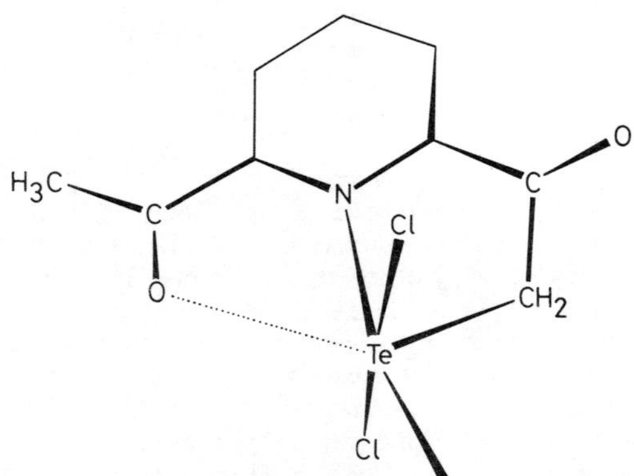

FIGURE 73. The molecular model of [2-(6-acetyl-2-pyridinyl)-2-oxoethyl]trichlorotellurium (after Reference 328). H atoms on the pyridine ring and double bonds are not shown. Wedges indicate a perspective view. The dotted line is a long Te···O linkage

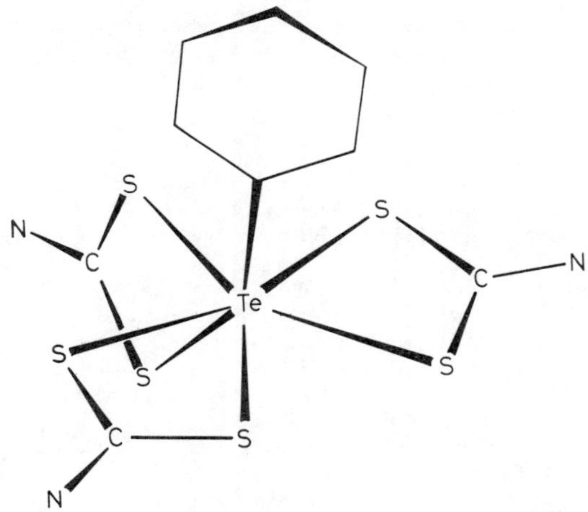

FIGURE 74. The molecular model of **144**. Drawn from the atomic coordinates given in Reference 329

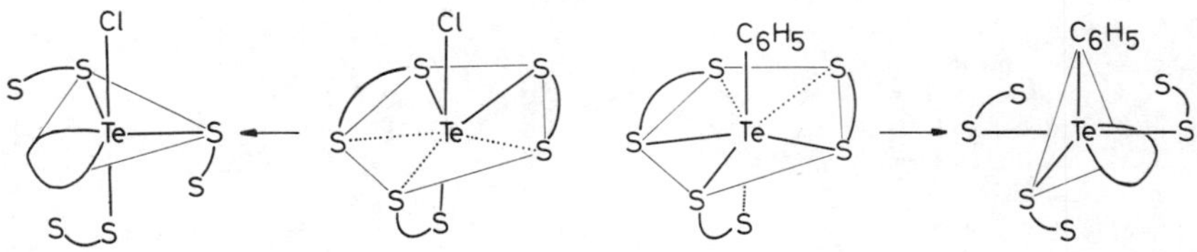

FIGURE 75. The effect of the Te lone pair on a distorted pentagonal bipyramidal arrangement with a Cl ligand (**145**) or with a Ph ligand (**144**) (after Reference 330)

Figure 74. The molecular structure of its Cl analogue $(Et_2NCS_2)_3TeCl \cdot C_4H_8O_2$ (**145**), was recently determined by XD[330]. The Te—S bonds are not all equivalent. In both compounds there are three shorter and three longer ones. The Te—C bond is a normal single bond in the Ph derivative, viz. 2.124(11) Å, but the Te—Cl bond is strikingly long in the Cl analogue, viz. 2.686(4) Å. The relative orientation of the Te—S bonds is also different in the two compounds. Von Deuten and coworkers[330] descriptively characterized this difference invoking the VSEPR model. Assuming the limiting case in which only the shorter Te—S linkages are considered to be bonds, both structures may be described by a trigonal bipyramidal configuration with the Te lone pair of electrons in an equatorial position. The more electronegative ligand (Cl) will be in an axial position and the Ph group will take an equational one[330] (Figure 75).

There are no organic eight- or higher-coordinated Se or Te derivatives containing Se—C or Te—C bonds whose structure has been elucidated. Several Te derivatives with purely inorganic Te environment have been investigated by XD[331–333]. It is anticipated that research will expand to include compounds in which one and possibly more organic ligands will find their way into the Te bonding sphere.

VII. ACKNOWLEDGEMENTS

Many colleagues kindly sent us the reprints of their papers. Dr. J. E. Boggs read the text of this chapter and made valuable comments. Ms. Márta Szentgyörgyi typed the manuscript and drew the figures with care and competence. We also appreciate the kindness and assistance from the authors and copyright holders for granting permission to reproduce figures from their publications.

VIII. REFERENCES

1. (a) I. L. Karle and J. Karle in *Organic Selenium Compounds: Their Chemistry and Biology* (Eds. D. L. Klayman and W. H. H. Günther), Wiley–Interscience, New York, 1973, Chap. XV.H, pp. 989–1015.
   (b) A. W. Cordes, in *Selenium* (Eds. R. A. Zingaro and W. C. Cooper), Van Nostrand Reinhold, New York, 1974, Chap. 6, pp. 298–336.
2. (a) M. M. Mangion, R. A. Zingaro and E. A. Meyers, *Chem. Scr.*, **8A**, 45 (1975).
   (b) H. J. Gysling, *Coord. Chem. Rev.*, **42**, 133 (1982).
3. I. Hargittai, *The Structure of Volatile Sulphur Compounds*, Akadémiai Kiadó, Budapest and D. Reidel, Dordrecht–Boston–London, 1985.
4. I. Hargittai and I. C. Paul in *The Chemistry of Cyanates and Their Thio Derivatives* (Ed. S. Patai), John Wiley and Sons, Chichester, 1977, Part 1, Chap. 2, pp. 69–129.
5. W. A. Hardy and G. Silvey, *Phys. Rev.*, **95**, 385 (1954).
6. C. P. Courtoy, *Ann. Soc. Sci. Bruxelles, Ser. I*, **73**, 5 (1959).
7. A. G. Maki and D. R. Johnson, *J. Mol. Spectry*, **47**, 226 (1973).
8. A. G. Maki, R. L. Sams and R. Pearson, *J. Mol. Spectry*, **64**, 452 (1977).
9. F. J. Lovas and E. Tiemann, *J. Phys. Chem. Ref. Data*, **3**, 609 (1974).
10. G. Blanquet, J. Walrand and C. P. Courtoy, *Ann. Soc. Sci. Bruxelles, Ser. I*, **88**, 87 (1974).
11. C. Hirose and R. F. Curl, *J. Chem. Phys.*, **55**, 5120 (1971).
12. A. G. Maki and R. L. Sams, *J. Mol. Spectry*, **90**, 215 (1981).
13. J.-Y. Saillard and D. Grandjean, *Acta Cryst.*, **B34**, 3772 (1978).
14. J.-Y. Saillard, D. Grandjean, P. Caillet and A. Le Beuze, *J. Organomet. Chem.*, **190**, 371 (1980).
15. B. M. Landsberg, *Chem. Phys. Letters*, **60**, 265 (1979).
16. W. H. Hocking, M. C. L. Gerry and G. Winnewisser, *Can. J. Phys.*, **53**, 1869 (1975).
17. R. Kewley, K. V. L. N. Sastry and M. Winnewisser, *J. Mol. Spectry*, **10**, 418 (1963).
18. B. Bak, O. J. Nielsen, H. Svanholt and A. Holm, *Chem. Phys. Letters*, **53**, 374 (1978); **55**, 36 (1978).
19. B. Bak, N. A. Kristiansen, H. Svanholt, A. Holm and S. Rosenkilde, *Chem. Phys. Letters*, **78**, 301 (1981).
20. Landolt–Börnstein, New Series (Ed.-in-Chief K.-H. Hellwege), Group II, Vol. 7: *Structure Data of Free Polyatomic Molecules*, Springer Verlag, Berlin, 1976.
21. K. Georgiou, H. W. Kroto and B. M. Landsberg, *J. Mol. Spectry*, **77**, 365 (1979).
22. S. Cradock, G. S. Laurenson and D. W. H. Rankin, *J. Chem. Soc., Dalton Trans.*, 187 (1981).
23. D. W. H. Rankin and S. J. Cyvin, *J. Chem. Soc., Dalton Trans.*, 1277 (1972).
24. (a) A. Fraser, G. S. Laurenson and D. W. H. Rankin, unpublished result, cited in Ref. 22.
    (b) D. W. H. Rankin, personal communication, 1982.
25. C. Glidewell, A. G. Robiette and G. M. Sheldrick, *Chem. Phys. Letters*, **16**, 526 (1972).
26. D. Christen, H. Oberhammer, W. Zeil, A. Haas and A. Darmadi, *J. Mol. Struct.*, **66**, 203 (1980).
27. J. H. Carpenter, *J. Mol. Spectry*, **50**, 182 (1974).
28. R. J. Gillespie, *Molecular Geometry*, Van Nostrand Reinhold, London, 1972.
29. M. Hutchinson and H. W. Kroto, *J. Mol. Spectry*, **70**, 347 (1978).
30. H. W. Kroto and B. M. Landsberg, *J. Mol. Spectry*, **62**, 346 (1976).
31. U. Anthoni, G. Borch, P. Klaeboe and P. H. Nielsen, *Acta Chem. Scand.*, **A36**, 69 (1982).
32. K. A. Jensen and P. H. Nielsen, *Acta Chem. Scand.*, **20**, 597 (1966).
33. J. S. Rutherford and C. Calvo, *Z. Krist.*, **128**, 229 (1969).
34. W. B. Wright and E. A. Meyers, *Cryst. Struct. Commun.*, **9**, 1173 (1980).
35. S. Hauge, *Acta Chem. Scand.*, **A33**, 317 (1979).
36. A. Conde, A. López-Castro and R. Márquez, *Cryst. Struct. Commun.*, **1**, 155 (1972); *Acta Cryst.*, **B28**, 3464 (1972).
37. G. K. Pandey and H. Dreizler, *Z. Naturforsch.*, **32a**, 482 (1977).
38. J. F. Beecher, *J. Mol. Spectry.*, **21**, 414 (1966).

39. C. J. Marsden and G. M. Sheldrick, *J. Mol. Struct.*, **10**, 405 (1971).
40. C. H. Thomas, *J. Chem. Phys.*, **59**, 70 (1973).
41. J. Nakagawa, H. Okutani and M. Hayashi, *J. Mol. Spectry*, **94**, 410 (1982).
42. J. R. Durig and W. E. Bucy, *J. Mol. Spectry*, **64**, 474 (1977).
43. B. M. Landsberg, *Z. Naturforsch.*, **33a**, 1333 (1978).
44. T. Sakaizumi, Y. Kohri, O. Ohashi and I. Yamaguchi, *Bull. Chem. Soc. Japan*, **51**, 3411 (1978).
45. C. J. Marsden and G. M. Sheldrick, *J. Mol. Struct.*, **10**, 413 (1971).
46. N. M. Zaripov, A. V. Golubinskii, G. A. Chmutova and L. V. Vilkov, *Zh. Strukt. Khim.*, **19**, 894 (1978).
47. N. M. Zaripov, M. V. Popik, L. V. Vilkov and T. G. Mannafov, *Zh. Strukt. Khim*, **21**, (2) 37 (1980).
48. M. von Carlowitz, *Diplomarbeit*, Universität Tübingen, Tübingen, BRD, 1979; H. Oberhammer, personal communication, 1983.
49. S. Lindøy, H. M. Seip and W. Siebert, *Acta Chem. Scand.*, **A29**, 265 (1975).
50. L. Pierce and M. Hayashi, *J. Chem. Phys.*, **35**, 479 (1961).
51. T. Iijima, S. Tsuchiya and M. Kimura, *Bull. Chem. Soc. Japan*, **50**, 2564 (1977).
52. K. Hamada and H. Morishita, *J. Mol. Struct.*, **44**, 119 (1978).
53. D. C. McKean and J. L. Duncan, *J. Mol. Struct.*, **51**, 149 (1979).
54. O. Gebhardt and S. J. Cyvin, *J. Mol. Struct.*, **12**, 205 (1972).
55. H. Oberhammer, W. Gombler and H. Willner, *J. Mol Struct.*, **70**, 273 (1981).
56. T. H. Edwards, N. K. Moncur and L. E. Snyder, *J. Chem. Phys.*, **46**, 2139 (1967).
57. A. Almenningen, K. Hedberg and R. Seip, *Acta Chem. Scand.*, **17**, 2264 (1963).
58. C. Glidewell, D. W. H. Rankin, A. G. Robiette, G. M. Sheldrick, B. Beagley and S. Cradock, *J. Chem. Soc.* (*A*), 315 (1970).
59. R. A. Hill and T. H. Edwards, *J. Chem. Phys.*, **42**, 1391 (1965).
60. A. Almenningen, L. Fernholt and H. M. Seip, *Acta Chem Scand.*, **22**, 51 (1968).
61. J. D. Murdoch, D. W. H. Rankin and C. Glidewell, *J. Mol. Struct.*, **9**, 17 (1971).
62. L. S. Bartell, *J. Chem. Phys.*, **32**, 827 (1960).
63. C. Glidewell, *Inorg. Chim. Acta*, **12**, 219 (1975).
64. N. K. Moncur, P. D. Willson and T. H. Edwards, *J. Mol. Spectry*, **52**, 380 (1974).
65. R. Blom, A. Haaland and R. Seip, *Acta Chem. Scand.*, **A37**, 595 (1983).
66. U. Blukis, P. H. Kasai and R. J. Myers, *J. Chem. Phys.*, **38**, 2753 (1963).
67. A. Almenningen, O. Bastiansen, V. Ewing, K. Hedberg and M. Traetteberg, *Acta Chem. Scand.*, **17**, 2455 (1963).
68. T. Kojima and T. Nishikawa, *J. Phys. Soc. Japan*, **12**, 680 (1957); T. Kojima, *J. Phys. Soc. Japan*, **15**, 1284 (1960).
69. J. Nakagawa, K. Kuwada and M. Hayashi, *Bull. Chem. Soc. Japan*, **49**, 3420 (1976).
70. H. Dreizler, H. D. Rudolph and H. Schleser, *Z. Naturforsch.*, **25a**, 1643 (1970).
71. H. Dreizler and A. M. Mirri, *Z. Naturforsch.*, **23a**, 1313 (1968).
72. A. B. Remizov and T. G. Mannafov, *Zh. Prikl. Spektrosk.*, **26**, 765 (1977).
73. K. Ohno, A. Mitsui and H. Murata, *Bull. Chem. Soc. Japan*, **52**, 2178 (1979).
74. N. M. Zaripov, *Zh. Strukt. Khim.*, **17**, 741 (1976).
75. N. M. Zaripov, M. V. Popik and L. V. Vilkov, *Zh. Strukt. Khim.*, **21**, (6) 38 (1980).
76. M. Baiwir, G. Llabres, O. Dideberg, L. Dupont and J. L. Piette, *Acta Cryst.*, **B31**, 2188 (1975).
77. M. Baiwir, G. Llabres, O. Dideberg, L. Dupont and J. L. Piette, *Acta Cryst.*, **B30**, 139 (1974).
78. (a) G. Schultz, I. Hargittai, I. Kapovits and Á. Kucsman, *J. Chem. Soc., Faraday Trans. 2*, **80**, 1273 (1984).
(b) N. M. Zaripov, S. V. Sokolov, A. V. Golubinskii, L. V. Vilkov and T. G. Mannafov, *Zh. Strukt. Khim.*, in press.
79. R. W. Davis and M. C. L. Gerry, *J. Mol. Spectry*, **65**, 455 (1977).
80. P. A. Akishin, V. P. Spiridonov and R. A. Mishulina, *Vestn. Mosk. Univ., Ser. II, Khim.*, **17**, (2) 23 (1962).
81. R. Eriksen and S. Hauge, *Acta Chem. Scand.*, **26**, 3153 (1972).
82. R. Eriksen, *Acta Chem. Scand.*, **A29**, 517 (1975).
83. L. Dupont, O. Dideberg, J. Lamotte and J. L. Piette, *Acta Cryst.*, **B35**, 849 (1979).
84. N. L. M. Dereu, R. A. Zingaro and E. A. Meyers, *Cryst. Struct. Commun.*, **10**, 1353 (1981).
85. J. Lamotte, H. Campsteyn, L. Dupont and M. Vermeire, *Cryst. Struct. Commun.*, **6**, 749 (1977).
86. C. Glidewell, *Inorg. Chim. Acta*, **36**, 135 (1979).

87. C. Glidewell, *Inorg. Chim. Acta*, **20**, 113 (1976).
88. L. Pauling, *The Nature of the Chemical Bond*, 3rd ed., Cornell University Press, Ithaca, N.Y., 1960.
89. A. Bondi, *J. Phys. Chem.*, **68**, 441 (1964).
90. G. Tschmutowa (= G. A. Chmutova) and H. Bock, *Z. Naturforsch.*, **31b**, 1611 (1976).
91. E. Shefter and O. Kennard, *Science*, **153**, 1389 (1966).
92. D. D. Dexter, *Acta Cryst.*, **B28**, 49 (1972).
93. A. D. Rudko and B. W. Low, *Acta Cryst.*, **B31**, 713 (1975).
94. C. Rømming and P. E. Hansen, *Acta Chem. Scand.*, **A33**, 265 (1979).
95. R. Kivekäs and T. Laitalainen, *Acta Chem. Scand.*, **B37**, 61 (1983).
96. M. Klinga, R. Kivekäs, T. Simonen, T. Laitalainen, R. Uggla and M. Sundberg, *Cryst. Struct. Commun.*, **7**, 531 (1978).
97. W. S. McDonald and L. D. Pettit, *J. Chem. Soc. (A)*, 2044 (1970).
98. C. J. Marsden and L. S. Bartell, *J. Chem. Soc., Dalton Trans.*, 1582 (1977).
99. R. Johansen, E. Wisløff Nilssen, H. M. Seip and W. Siebert, *Acta Chem. Scand.*, **27**, 3015 (1973).
100. B. Rozsondai and I. Hargittai, *Kém. Közlem.*, **54**, 268 (1980).
101. (a) P. D'Antonio, C. George, A. H. Lowrey and J. Karle, *J. Chem. Phys.*, **55**, 1071 (1971). (b) J. Karle, *Proc. Natl. Acad. Sci. U.S.A.*, **74**, 4707 (1977).
102. C. J. Marsden and G. M. Sheldrick, *J. Mol. Struct.*, **10**, 419 (1971).
103. A. Yokozeki and S. H. Bauer, *J. Phys. Chem.*, **80**, 618 (1976).
104. C. J. Marsden and B. Beagley, *J. Chem. Soc., Faraday Trans. 2*, **77**, 2213 (1981).
105. A. Chiesi Villa, M. Nardelli and M. E. Vidoni Tani, *Acta Cryst.*, **B26**, 1504 (1970).
106. O. Foss, J. Johnsen and O. Tvedten, *Acta Chem. Scand.*, **12**, 1782 (1958).
107. C. M. Woodard, D. S. Brown, J. D. Lee and A. G. Massey, *J. Organomet. Chem.*, **121**, 333 (1976).
108. S. Hauge, unpublished result, cited in Reference 35.
109. J. D. Lee and M. W. R. Bryant, *Acta Cryst.*, **B25**, 2094 (1969).
110. R. E. Marsh, *Acta Cryst.*, **5**, 458 (1952).
111. O. Foss and V. Janickis, *J. Chem. Soc., Dalton Trans.*, 628 (1980); 620 (1980).
112. O. Foss and V. Janickis, *J. Chem. Soc., Dalton Trans.*, 624 (1980).
113. G. Llabres, O. Dideberg and L. Dupont, *Acta Cryst.*, **B28**, 2438 (1972).
114. M. R. Spirlet, G. Van den Bossche, O. Dideberg and L. Dupont, *Acta Cryst.*, **B35**, 1727 (1979).
115. F. H. Kruse, R. E. Marsh and J. D. McCullough, *Acta Cryst.*, **10**, 201 (1957).
116. G. Llabres, M. Baiwir, L. Christiaens and J.-L. Piette, *J. Mol. Struct.*, **70**, 231 (1981).
117. M. Baldo, A. Forchioni, K. J. Irgolic and G. C. Pappalardo, *J. Amer. Chem. Soc.*, **100**, 97 (1978).
118. G. Tschmutowa (= G. A. Chmutova) and H. Bock, *Z. Naturforsch.*, **31b**, 1616 (1976).
119. R. Laitinen, *J. Mol. Struct.*, **91**, 337 (1983).
120. V. Renugopalakrishnan, H. R. Wyssbrod, R. Walter and M. E. Druyan, *J. Chem. Phys.*, **77**, 1362 (1982).
121. S. Hauge, *Acta Chem. Scand.*, **A33**, 313 (1979).
122. M. G. Petit, J. S. Gibson and D. O. Harris, *J. Chem. Phys.*, **53**, 3408 (1970).
123. K. Karakida and K. Kuchitsu, *Bull. Chem. Soc. Japan*, **48**, 1691 (1975).
124. A. Almenningen, O. Bastiansen and P. N. Skancke, *Acta Chem. Scand.*, **15**, 711 (1961).
125. F. Takabayashi, H. Kambara and K. Kuchitsu, *Seventh Austin Symposium on Gas Phase Molecular Structure*, Austin, Texas, 1978, p. 63.
126. V. S. Mastryukov, O. V. Dorofeeva, L. V. Vilkov and I. Hargittai, *J. Mol. Struct.*, **34**, 99 (1976).
127. S. I. Chan, J. Zinn, J. Fernandez and W. D. Gwinn, *J. Chem. Phys.*, **33**, 1643 (1960).
128. V. S. Mastryukov, O. V. Dorofeeva, L. V. Vilkov, B. N. Sivin (= B. N. Cyvin) and S. I. Sivin (= S. J. Cyvin), *Zh. Strukt. Khim.*, **16**, 473 (1975).
129. T. Wehrung, H. Oberhammer, A. Haas, B. Koch and N. Welcman, *J. Mol. Struct.*, **35**, 253 (1976).
130. Z. Smith and R. Seip, *Acta Chem. Scand.*, **A30**, 759 (1976).
131. J. F. Chiang and K. C. Lu, *J. Phys. Chem.*, **81**, 1682 (1977).
132. I. Hargittai, *J. Mol. Struct.*, **54** 287 (1979).
133. V. F. Kalasinsky, E. Block, D. E. Powers and W. C. Harris, *Appl. Spectry*, **33**, 361 (1979).
134. Z. Náhlovská, B. Náhlovský and H. M. Seip, *Acta Chem. Scand.*, **24**, 1903 (1970).

135. A. H. (= Kh.) Mamleev, N. M. Pozdeev and N. N. Magdesieva, *J. Mol. Struct.*, **33**, 211 (1976).
136. W. Klyne and V. Prelog, *Experientia*, **16**, 521 (1960).
137. Ya. M. Kimel'fel'd, A. Usmanov, G. N. Zhizhin, V. A. Vdovin and N. N. Magdesieva, *Zh. Strukt. Khim.*, **11**, 1126 (1970).
138. W. H. Green, A. B. Harvey and J. A. Greenhouse, *J. Chem. Phys.*, **54**, 850 (1971).
139. Z. Náhlovská, B. Náhlovský and H. M. Seip, *Acta Chem. Scand.*, **23**, 3534 (1969).
140. W. J. Adams, H. J. Geise and L. S. Bartell, *J. Amer. Chem. Soc.*, **92**, 5013 (1970).
141. H. J. Geise, W. J. Adams and L. S. Bartell, *Tetrahedron*, **25**, 3045 (1969).
142. A. Almenningen, H. M. Seip and T. Willadsen, *Acta Chem. Scand.*, **23**, 2748 (1969).
143. N. M. Pozdeev, O. B. Akulinin, A. A. Shapkin and N. N. Magdesieva, *Dokl. Akad. Nauk SSSR*, **185**, 384 (1969); *Zh. Strukt. Khim.*, **11**, 869 (1970).
144. R. D. Brown, F. R. Burden and P. D. Godfrey, *J. Mol. Spectry*, **25**, 415 (1968).
145. R. D. Brown and J. G. Crofts, *Chem. Phys.*, **1**, 217 (1973).
146. B. Bak, D. Christensen, W. B. Dixon, L. Hansen-Nygaard, J. Rastrup-Andersen and M. Schottländer, *J. Mol. Spectry*, **9**, 124 (1962).
147. B. Bak, D. Christensen, L. Hansen-Nygaard and J. Rastrup-Andersen, *J. Mol. Spectry*, **7**, 58 (1961).
148. K. Tamagawa, T. Iijima and M. Kimura, *J. Mol. Struct.*, **30**, 243 (1976).
149. D. M. Levine, W. D. Krugh and L. P. Gold, *J. Mol. Spectry*, **30**, 459 (1969).
150. G. L. Blackman, R. D. Brown, F. R. Burden and J. E. Kent, *Chem. Phys. Letters*, **1**, 379 (1967).
151. M. Mellini and S. Merlino, *Acta Cryst.*, **B32**, 1074 (1976); **B32**, 1079 (1976).
152. R. Neidlein, D. Tran-Viet, A. Gieren, M. Kokkinidis, R. Wilckens, H.-P. Geserich and W. Ruppel, *Chem. Ber.*, **115**, 2898 (1982); A. Gieren and V. Lamm, *Acta Cryst.*, **B38**, 2605 (1982).
153. K. Burger, R. Ottlinger, A. Frank and U. Schubert, *Angew. Chem.*, **90**, 824 (1978).
154. H. Campsteyn, L. Dupont, J. Lamotte-Brasseur and M. Vermeire, *J. Heterocycl. Chem.*, **15**, 745 (1978).
155. M. R. Bryce, C. D. Reynolds, P. Hanson and J. M. Vernon, *J. Chem. Soc., Perkin Trans.* 1, 607 (1981).
156. O. Dideberg, L. Dupont and M. Baiwir, *Acta Cryst.*, **B33**, 1 (1977).
157. R. Kivekäs, M. Klinga, R. Uggla, M. Sundberg, T. Laitalainen and T. Simonen, *Cryst. Struct. Commun.*, **7**, 441 (1978).
158. T. Laitalainen, T. Simonen, R. Kivekäs and M. Klinga, *J. Chem. Soc., Perkin Trans.* 1, 333 (1983).
159. R. Kivekäs, T. Laitalainen and T. Simonen, *Cryst. Struct. Commun.*, **8**, 477 (1979).
160. H. Hope, C. Knobler and J. D. McCullough, *Acta Cryst.*, **B26**, 628 (1970).
161. J. D. McCullough, *Inorg. Chem.*, **14**, 2639 (1975).
162. A. Banerjee, *Acta Cryst.*, **B29**, 2070 (1973).
163. R. M. Schaffrin and J. Trotter, *J. Chem. Soc. (A)*, 1561 (1970).
164. H. Campsteyn, L. Dupont, J. Lamotte and M. Vermeire, *Cryst. Struct. Commun.*, **6**, 639 (1977).
165. T. J. Kistenmacher, T. J. Emge, P. Shu and D. O. Cowan, *Acta Cryst.*, **B35**, 772 (1979).
166. J. R. Andersen, K. Bechgaard, C. S. Jacobsen, G. Rindorf, H. Soling and N. Thorup, *Acta Cryst.*, **B34**, 1901 (1978).
167. K. Bechgaard, T. J. Kistenmacher, A. N. Bloch and D. O. Cowan, *Acta Cryst.*, **B33**, 417 (1977).
168. T. J. Kistenmacher, T. J. Emge, A. N. Bloch and D. O. Cowan, *Acta Cryst.*, **B38**, 1193 (1982).
169. R. E. Long, R. A. Sparks and K. N. Trueblood, *Acta Cryst.*, **18**, 932 (1965).
170. F. W. Cooper, N. C. Kenny, J. W. Edmonds, A. Nagel, F. Wudl and P. Coppens, *Chem. Commun.*, 889 (1971).
171. V. F. Kaminskii, R. P. Shibaeva, M. Z. Aldoshina, P. N. Lyubovskaya and M. L. Khidekel', *Zh. Strukt. Khim.*, **20**, 157 (1979).
172. R. P. Shibaeva and V. F. Kaminskii, *Kristallografiya*, **23**, 1183 (1978).
173. R. P. Shibaeva, V. F. Kaminskii, A. I. Kotov, É. B. Yagubskii and M. L. Khidekel', *Kristallografiya*, **24**, 271 (1979).
174. R. P. Shibaeva and V. F. Kaminskii, *Cryst. Struct. Commun.*, **10**, 663 (1981).
175. O. Dideberg and J. Toussaint, *Acta Cryst.*, **B30**, 2481 (1974).
176. G. Valle, V. Busetti, M. Mammi and G. Carazzolo, *Acta Cryst.*, **B25**, 1432 (1969).
177. G. Valle, V. Busetti, M. Mammi and G. Carazzolo, *Acta Cryst.*, **B25**, 1631 (1969).
178. M. Mammi, G. Carazzolo, G. Valle and A. Del Pra, *Z. Krist.*, **127**, 401 (1968).

179. G. Valle, A. Del Pra and M. Mammi, *Cryst. Struct. Commun.*, **2**, 169 (1973).
180. G. Valle, G. Zanotti and A. Del Pra, *Cryst. Struct. Commun.*, **4**, 349 (1975).
181. G. Valle, V. Busetti and M. Mammi, *Cryst. Struct. Commun.*, **2**, 167 (1973).
182. A. H. Clark and T. G. Hewitt, *J. Mol. Struct.*, **9**, 33 (1971).
183. J. C. Dewan and J. Silver, *Acta Cryst.*, **B33**, 1469 (1977).
184. J. C. Dewan and J. Silver, *J. Organomet. Chem.*, **125**, 125 (1977).
185. J. C. Dewan and J. Silver, *J. Chem. Soc., Dalton Trans.*, 644 (1977).
186. C. L. Raston, R. J. Secomb and A. H. White, *J. Chem. Soc., Dalton Trans.*, 2307 (1976).
187. J. C. Dewan and J. Silver, *Australian J. Chem.*, **30**, 487 (1977).
188. J. C. Dewan and J. Silver, *Acta Cryst.*, **B33**, 2671 (1977).
189. G. L. Parks, *Dissertation*, Clemson University, Clemson, South Carolina, U.S.A., 1968; *Diss. Abstr.*, **B29**, 3450 (1969).
190. F. J. Berry and J. Silver, *J. Organomet. Chem.*, **188**, 255 (1980).
191. A. Michel and F. Durant, *Cryst. Struct. Commun.*, **9**, 219 (1980).
192. (a) W. G. Ferrier and J. Iball, *Chem. Ind.* (*London*), 1296 (1954); M. C. Malmstrom, *Dissertation*, University of Arkansas, Fayetteville, U.S.A., 1972; *Diss. Abstr.*, **B33**, 1978 (1972). (b) M. Senma, Z. Taira, T. Taga and K. Osaki, *Cryst. Struct. Commun.*, **2**, 311 (1973); P. Singh and J. D. McKinney, *Acta Cryst.*, **B34**, 2956 (1978).
193. D. Rainville, R. A. Zingaro and E. A. Meyers, *Cryst. Struct. Commun.*, **9**, 771 (1980).
194. K. L. Gallaher and S. H. Bauer, *J. Chem. Soc., Faraday Trans.*, 2, **71**, 1173 (1975).
195. D. Rainville, R. A. Zingaro and E. A. Meyers, *Cryst. Struct. Commun.*, **9**, 909 (1980).
196. D. Rainville, R. A. Zingaro and E. A. Meyers, *Cryst. Struct. Commun.*, **9**, 291 (1980).
197. K. von Deuten, W. Hinrichs and G. Klar, *Cryst. Struct. Commun.*, **10**, 327 (1981).
198. P. Villares, R. Jiménez-Garay, A. Conde and R. Márquez, *Acta Cryst.*, **B32**, 2293 (1976).
199. F. Bernier, A. Conde and R. Márquez, *Acta Cryst.*, **B30**, 1332 (1974).
200. N. L. M. Dereu, R. A. Zingaro and E. A. Meyers, *Cryst. Struct. Commun.*, **10**, 1359 (1981).
201. D. P. Rainville, R. A. Zingaro and E. A. Meyers, *J. Fluorine Chem.*, **16**, 245 (1980).
202. M. R. Smith, M. M. Mangion, R. A. Zingaro and E. A. Meyers, *J. Heterocycl. Chem.*, **10**, 527 (1973).
203. G. Valle, G. Zanotti and M. Mammi, *Acta Cryst.*, **B34**, 2634 (1978).
204. S. Aleby, *Acta Cryst.*, **B28**, 1509 (1972).
205. M. Bonamico and G. Dessy, *J. Chem. Soc.* (*A*), 264 (1971).
206. J. Kaiser, R. Richter and J. Sieler, *Acta Cryst.*, **B31**, 2600 (1975).
207. A. J. Carty, S. F. Malone and N. J. Taylor, *J. Organomet. Chem.*, **172**, 201 (1979).
208. R. O. Gould, C. L. Jones, W. J. Savage and T. A. Stephenson, *J. Chem. Soc., Dalton Trans.*, 908 (1976).
209. S. Esperås, S. Husebye and Å. Rolandsen, *Acta Chem. Scand.*, **A29**, 608 (1975).
210. S. Hauge, Ø. Johannesen and O. Vikane, *Acta Chem. Scand.*, **A32**, 901 (1978).
211. S. Hauge and M. Tysseland, *Acta Chem. Scand.*, **25**, 3072 (1971).
212. C. G. Pierpont and R. Eisenberg, *J. Chem. Soc.* (*A*), 2285 (1971).
213. C. A. Stein, P. E. Ellis, R. C. Elder and E. Deutsch, *Inorg. Chem.*, **15**, 1618 (1976).
214. R. C. Elder, K. Burkett and E. Deutsch, *Acta Cryst.*, **B35**, 164 (1979).
215. J. H. Noordik and J. M. M. Smits, *Cryst. Struct. Commun.*, **3**, 253 (1974).
216. R. Curran, J. A. Cunningham and R. Eisenberg, *Inorg. Chem.*, **9**, 2749 (1970).
217. J. Kaiser, W. Dietzsch, R. Richter, L. Golič and J. Siftar, *Acta Cryst.*, **B36**, 147 (1980).
218. P. T. Beurskens and J. A. Cras, *J. Cryst. Mol. Struct.*, **1**, 63 (1971).
219. J. P. Collman, R. K. Rothrock, J. P. Sen, T. D. Tullius and K. O. Hodgson, *Inorg. Chem.*, **15**, 2947 (1976).
220. H.-W. Chen, J. P. Fackler, A. F. Masters and W.-H. Pan, *Inorg. Chim. Acta*, **35**, L333 (1979).
221. M. F. Lappert, T. R. Martin and G. M. McLaughlin, *J. Chem. Soc., Chem. Commun.*, 635 (1980).
222. U. Behrens, K. Hoffmann and G. Klar, *Chem. Ber.*, **110**, 3672 (1977).
223. W. Beck and W. P. Fehlhammer in *MTP International Review of Science*, *Inorganic Chemistry*, *Series One* (Consultant Ed. H. J. Emeléus), Vol. 2, Butterworths, London and University Park Press, Baltimore, 1972, Chap. 9, pp. 253–300.
224. I. C. Bowater, R. D. Brown and F. R. Burden, *J. Mol. Spectry*, **23**, 272 (1967); **28**, 461 (1968).
225. D. C. Gregory, I. Hargittai and M. Kolonits, *J. Mol. Struct.*, **31**, 261 (1976).

226. B. A. Arbuzov, V. A. Naumov, N. M. Zaripov and L. D.Pronicheva, *Dokl. Akad. Nauk SSSR*, **195**, 1333 (1970).
227. B. A. Arbuzov, V. A. Naumov and I. V. Anonimova, *Dokl. Akad. Nauk SSSR*, **192**, 327 (1970).
228. P. Ash, J.-S. Lee, D. D. Titus, K. B. Mertes and R. F. Ziolo, *J. Organomet. Chem.*, **135**, 91 (1977).
229. J.-S. Lee and D. D. Titus, *J. Cryst. Mol. Struct.*, **6**, 279 (1976).
230. D. D. Titus, J.-S. Lee and R. F. Ziolo, *J. Organomet. Chem.*, **120**, 381 (1976).
231. J. -S. Lee, D. D. Titus and R. F. Ziolo, *J. Chem. Soc., Chem. Commun.*, 501 (1976); *Inorg. Chem.*, **16**, 2487 (1977).
232. R. F. Ziolo and M. Extine, *Inorg. Chem.*, **19**, 2964 (1980).
233. R. F. Ziolo and D. D. Titus, *J. Appl. Cryst.*, **9**, 506 (1976).
234. K.-T. H. Wei, I. C. Paul, M.-M. Y. Chang and J. I. Musher, *J. Amer. Chem. Soc*, **96**, 4099 (1974).
235. V. V. Saatsazov, R. A. Kyandzhetsian, S. I. Kuznetsov, N. N. Magdesieva and T. L. Khotsyanova, *Dokl. Akad. Nauk SSSR*, **206**, 1130 (1972).
236. K. Brodersen, G. Liehr and M. Rosenthal, *Chem. Ber.*, **110**, 3291 (1977).
237. K. Brodersen, G. Liehr, M. Rosenthal and G. Thiele, *Z. Naturforsch.*, **33b**, 1227 (1978).
238. C. J. Marsden and G. M. Sheldrick, *J. Organomet. Chem.*, **40**, 175 (1972).
239. J. C. Barnes, G. Hunter and M. W. Lown, *J. Chem. Soc., Dalton Trans.*, 458 (1977).
240. G. Schultz, I. Hargittai and L. Hermann, *J. Mol. Struct.*, **14**, 353 (1972).
241. R. C. Pettersen, K. H. Pannell and A. J. Mayr, *Cryst. Struct. Commun.*, **9**, 643 (1980).
242. R. C. Pettersen, K. H. Pannell and A. J. Mayr, *Acta Cryst.*, **B36**, 2434 (1980).
243. R. F. Ziolo and J. M. Troup, *Inorg. Chem.*, **18**, 2271 (1979).
244. J. M. Troup, M. W. Extine and R. F. Ziolo in *Electron Distributions and the Chemical Bond* (Eds. P. Coppens and M. B. Hall), Proc. Symp. 1981, Plenum Press, New York, 1982, pp. 285–296; see *Chem. Abstr.*, **97**, 203382b (1982); cf. also R. F. Ziolo and J. M. Troup, *J. Amer. Chem. Soc.*, **105**, 229 (1983).
245. E. W. Abel, A. R. Khan, K. Kite, K. G. Orrell, V. Šik, T. S. Cameron and R. Cordes, *J Chem. Soc., Chem. Commun.*, 713 (1979).
246. E. W. Abel, S. K. Bhargava, M. M. Bhatti, K. Kite, M. A. Mazid, K. G. Orrell, V. Šik, B. L. Williams, M. B. Hursthouse and K. M. A. Malik, *J. Chem. Soc., Dalton Trans.*, 2065 (1982).
247. S. Hauge, D. Opedal and J. Aarskog, *Acta Chem. Scand.*, **24**, 1107 (1970); **A29**, 225 (1975).
248. S. Hauge, *Acta Chem. Scand.*, **25**, 1134 (1971).
249. S. Hauge and J. Sletten, *Acta Chem. Scand.*, **25**, 3094 (1971).
250. S. Hauge, *Acta Chem. Scand.*, **25**, 3103 (1971).
251. L. K. Hansen and A. Hordvik, *Acta Chem. Scand.*, **27**, 411 (1973).
252. B. Birknes and A. Hordvik, *Acta Chem. Scand.*, **A36**, 813 (1982).
253. L. K. Hansen and A. Hordvik, *Acta Chem. Scand.*, **24**, 2246 (1970).
254. A. Hordvik, *Acta Chem. Scand.*, **25**, 1583 (1971).
255. A. Hordvik and K. Julshamn, *Acta Chem. Scand.*, **25**, 1895 (1971).
256. A. Hordvik, T. S. Rimala and L. J. Saethre, *Acta Chem. Scand.*, **27**, 360 (1973).
257. A. Hordvik, T. S. Rimala and L. J. Saethre, *Acta Chem. Scand.*, **26**, 2139 (1972).
258. J. H. van den Hende and E. Klingsberg, *J. Amer. Chem. Soc.*, **88**, 5045 (1966).
259. A. Hordvik and K. Julshamn, *Acta Chem. Scand.*, **25**, 2507 (1971).
260. K. Jynge, *Cand. Real Thesis*, University of Tromsø, Tromsø, Norway, 1976, cited in Reference 263.
261. A. Hordvik and J. A. Porten, *Acta Chem. Scand.*, **27**, 485 (1973).
262. V. Busetti, G. Valle and R. Bardi, *Acta Cryst.*, **B34**, 691 (1978).
263. F. A. Amundsen, L. K. Hansen and A. Hordvik, *Acta Chem. Scand.*, **A36**, 673 (1982).
264. E. Stevens, unpublished result, cited in Reference 263.
265. R. J. S. Beer, J. R. Hatton, E. C. Llaguno and I. C. Paul, *Chem. Commun.*, 594 (1971); E. C. Llaguno and I. C. Paul, *J. Chem. Soc., Perkin Trans.* 2, 2001 (1972).
266. Q. Shen and K. Hedberg, *J. Amer. Chem. Soc.*, **96**, 289 (1974).
267. T. Pedersen, S. V. Skaarup and C. T. Pedersen, *Acta Chem. Scand.*, **B31**, 711 (1977).
268. L. J. Saethre, N. Martensson, S. Svensson, P. Å. Malmquist, U. Gelius and K. Siegbahn, *J. Amer. Chem. Soc.*, **102**, 1783 (1980).
269. L. J. Saethre, P. A. Malmquist, N. Martensson, S. Svensson, U. Gelius and K. Siegbahn, *Inorg. Chem.*, **20**, 399 (1981).

270. L. K. Hansen, A. Hordvik and L. J. Saethre, *J. Chem. Soc., Chem. Commun.*, 222 (1972).
271. R. Gleiter, D. Schmidt and H. Behringer, *Chem. Commun.*, 525 (1971).
272. R. Gleiter and R. Hoffmann, *Tetrahedron*, **24**, 5899 (1968).
273. A. Hordvik, E. Sletten and J. Sletten, *Acta Chem. Scand.*, **23**, 1852 (1969).
274. L. J. Saethre and A. Hordvik, *Acta Chem. Scand.*, **A29**, 136 (1975).
275. O. Foss and K. Marøy, *Acta Chem. Scand.*, **20**, 123 (1966); O. Foss and S. Husebye, *Acta Chem. Scand.*, **20**, 132 (1966).
276. O. Foss, *Pure Appl. Chem.*, **24**, 31 (1970).
277. O. Vikane, *Acta Chem. Scand.*, **A29**, 787 (1975).
278. O. Vikane, *Acta Chem. Scand.*, **A29**, 738 (1975).
279. O. Vikane, *Acta Chem. Scand.*, **A29**, 763 (1975).
280. S. Hauge and O. Vikane, *Acta Chem. Scand.*, **A29**, 755 (1975).
281. A. Domenicano and P. Murray-Rust, *Tetrahedron Letters*, 2283 (1979).
282. I. Hargittai, *Inorg. Chem.*, **21**, 4334 (1982).
283. W. M. Tolles and W. D. Gwinn, *J. Chem. Phys.*, **36**, 1119 (1962).
284. H. Oberhammer, R. C. Kumar, G. D. Knerr and J. M. Shreeve, *Inorg. Chem.*, **20**, 3871 (1981).
285. I. Hargittai, *J. Mol. Struct.*, **56**, 301 (1979).
286. L. Hedberg and K. Hedberg, *J. Phys. Chem.*, **86**, 598 (1982).
287. H. Oberhammer and J. E. Boggs, *J. Mol. Struct.*, **56**, 107 (1979).
288. D. G. Garratt, P.-T. Cheng, S. C. Nyburg and G. H. Schmid, *Cryst. Struct. Commun.*, **3**, 361 (1974).
289. B. Dahlén, *Acta Cryst.*, **B29**, 595 (1973).
290. V. Typke, *Z. Naturforsch.*, **33a**, 842 (1978).
291. L. Battelle, C. Knobler and J. D. McCullough, *Inorg. Chem.*, **6**, 958 (1967).
292. B. Dahlén and B. Lindgren, *Acta Chem. Scand.*, **A33**, 403 (1979).
293. B. Dahlén and B. Lindgren, *Acta Chem. Scand.*, **27**, 2218 (1973); B. Dahlén, *Acta Cryst.*, **B30**, 647 (1974).
294. F. Wudl and E. T. Zellers, *J. Amer. Chem. Soc.*, **102**, 5430 (1980).
295. F. Wudl and E. T. Zellers, *J. Amer. Chem. Soc.*, **102**, 4283 (1980).
296. H. Pritzkow, *Inorg. Chem.*, **18**, 311 (1979).
297. J. E. Huheey, *Inorg. Chem.*, **20**, 4033 (1981).
298. K. O. Christe and H. Oberhammer, *Inorg. Chem.*, **20**, 296 (1981).
299. D. Kobelt and E. F. Paulus, *J. Organomet. Chem.*, **27**, C63 (1971).
300. T. S. Cameron, R. B. Amero and R. E. Cordes, *Cryst. Struct. Commun.*, **9**, 533 (1980).
301. T. S. Cameron, R. B. Amero and R. E. Cordes, *Cryst. Struct. Commun.*, **9**, 539 (1980).
302. C. S. Mancinelli, D. D. Titus and R. F. Ziolo, *J. Organomet. Chem.*, **140**, 113 (1977).
303. C. S. Smith, J.-S. Lee, D. D. Titus and R. F. Ziolo, *Organometallics*, **1**, 350 (1982).
304. J. Bergman and L. Engman, *J. Organomet. Chem.*, **181**, 335 (1979).
305. C. Knobler and J. D. McCullough, *Inorg. Chem.*, **16**, 612 (1977).
306. J. D. McCullough and C. Knobler, *Inorg. Chem.*, **15**, 2728 (1976).
307. J. D. McCullough, *Inorg. Chem.*, **16**, 2318 (1977).
308. L. Fernholt, personal communication, 1981.
309. C. Knobler and J. D. McCullough, *Inorg. Chem.*, **11**, 3026 (1972).
310. H. Hope, C. Knobler and J. D. McCullough, *Inorg. Chem.*, **12**, 2665 (1973).
311. C. Knobler, J. D. McCullough and H. Hope, *Inorg. Chem.*, **9**, 797 (1970).
312. O. Bastiansen, H. M. Seip and J. E. Boggs, in *Perspectives in Structural Chemistry*, Vol. 4 (Eds. J. D. Dunitz and J. A. Ibers), John Wiley and Sons, New York, 1971.
313. O. Hassel and H. Viervoll, *Acta Chem. Scand.*, **1**, 149 (1947).
314. C. Romers, C. Altona, H. R. Buys and E. Havinga, in *Topics in Stereochemistry*, Vol. 4 (Eds. E. L. Eliel and N. L. Allinger), Interscience, New York, 1969, pp. 39–97.
315. G. Hagen, I. Hargittai and G. Schultz, *Acta Chim. Acad. Sci. Hung.*, **86**, 219 (1975).
316. S. J. Cyvin, *Molecular Vibrations and Mean Square Amplitudes*, Universitetsforlaget, Oslo and Elsevier, Amsterdam, 1968.
317. C. Knobler and R. F. Ziolo, *J. Organomet. Chem.*, **178**, 423 (1979).
318. J. D. McCullough, *Inorg. Chem.*, **14**, 1142 (1975).
319. J. D. McCullough, *Inorg. Chem.*, **12**, 2669 (1973).
320. M. M. Mangion and E. A. Meyers, *Cryst. Struct. Commun.*, **2**, 629 (1973).

321. M. M. Mangion, M. R. Smith and E. A. Meyers, *J. Heterocycl. Chem.*, **10**, 533 (1973).
322. M. M. Mangion, M. R. Smith and E. A. Meyers, *J. Heterocycl. Chem.*, **10**, 543 (1973).
323. C. J. Marsden, G. M. Sheldrick and R. Taylor, *Acta Cryst.*, **B33**, 139 (1977).
324. A. C. Hazell, *Acta Chem. Scand.*, **26**, 1510 (1972).
325. D. Rainville, R. A. Zingaro and E. A. Meyers, *Cryst. Struct. Commun.*, **9**, 77 (1980).
326. T. S. Cameron, R. B. Amero, C. Chan and R. E. Cordes, *Cryst. Struct. Commun.*, **9**, 543 (1980).
327. D. Kobelt and E. F. Paulus, *Angew. Chem.*, **83**, 81 (1971).
328. H. J. Gysling, H. R. Luss and S. A. Gardner, *J. Organomet. Chem.*, **184**, 417 (1980).
329. S. Esperas and S. Husebye, *Acta Chem. Scand.*, **26**, 3293 (1972).
330. K. von Deuten, W. Schnabel and G. Klar, *Phosphorus Sulfur*, **9**, 93 (1980).
331. S. Husebye and S. E. Svaeren, *Acta Chem. Scand.*, **27**, 763 (1973).
332. S. Esperas and S. Husebye, *Acta Chem. Scand.*, **A29**, 185 (1975).
333. S. Husebye, *Acta Chem. Scand.*, **A33**, 485 (1979).

The Chemistry of Organic
Selenium and Tellurium
Compounds Volume 1
Edited by S. Patai and Z. Rappoport
© 1986 John Wiley & Sons Ltd.

CHAPTER **4**

Thermochemistry of selenium and tellurium compounds

L. BATT

Aberdeen University, Aberdeen, Scotland

I. INTRODUCTION . . . 157
II. ORGANIC SELENIUM COMPOUNDS . . . 157
A. Heats of Formation . . . 157
B. Bond Dissociation Energies . . . 158
III. CONCLUSIONS . . . 160
IV. REFERENCES . . . 160

I. INTRODUCTION

In addition to three comprehensive reviews on thermochemistry[1–3], *Organometallic Chemistry Reviews* and *Advances in Organometallic Chemistry* were consulted. Also the following periodicals from 1970 were searched: *Journal of Organometallic Compounds, Thermochimica Acta, Journal of Chemical Thermodynamics* and *Journal of Inorganic and Nuclear Chemistry*. This search revealed very little information on the thermochemistry of organic selenium compounds and no information whatsoever on organic tellurium compounds. Accordingly this review is confined to selenium compounds and the production of group properties for the prediction of the heats of formation of other compounds. Some indications are made for further desirable studies.

II. ORGANIC SELENIUM COMPOUNDS

A. Heats of Formation

All of the heats of formation of organoselenium compounds have been determined via their heats of combustion. In one sense this is quite a straightforward procedure because the combustion of selenium and its compounds produce only one metallic oxide, selenium dioxide, SeO_2. Since SeO_2 is soluble in water, it is essential to use rotating bomb methods to ensure complete dissolution of the oxide[4]. Early determinations using static bomb methods are therefore suspect for this reason[5].

Although not strictly relevant to this chapter, it is first important to establish the heat of formation of crystalline SeO_2 for the reasons outlined above. Barnes and Mortimer[6]

TABLE 1. Values for ΔH_f^0 (SeO_2)(c)

Bomb calorimetry Method	ΔH_f^0 (SeO_2, c) (kcal mol^{-1})	Ref.
Static	− 53.86	7, 8
Static	− 53.35	9
Static	− 53.95	10
Rotating	− 53.9 ± 0.5	6

determined the heat of combustion of Se in the presence of benzoic acid as a combustion acid using a rotating bomb calorimeter. The combustion process refers to reaction (1):

$$Se(c) + O_2(g) + 401\ H_2O(l) \rightarrow SeO_2.401\ H_2O(l) \quad (1)$$

By making use of the enthalpy of solution ΔH^0 (solution) $= 0.92 \pm 0.02$ kcal mol^{-1}, the heat of formation of the pure crystalline form was found to be -53.90 ± 0.47 kcal mol^{-1}. Table 1 compares this value with previous static bomb determinations. In fact the agreement is very good which suggests that the static bomb experiments *may* not be seriously in error. However Skinner[4] quotes a value of -56.5 ± 1 kcal mol^{-1}.

Merten and Schlüter[5] determined the heat of combustion of diethyl selenide/paraffin oil/benzoic acid mixtures in a static bomb calorimeter. Using ΔH_f^0 (SeO_2, c) $= -53.9 \pm 0.5$ kcal mol^{-1}, this gives ΔH_f^0 (Et_2Se, l) $= -21.7 \pm 0.9$ kcal mol^{-1}. Skinner's[4] value for the latent heat of vaporization of $+9.3$ kcal mol^{-1} gives ΔH_f^0 (Et_2Se, g) $= -12.4$ kcal mol^{-1}.

Barnes and Mortimer[6] determined the heat of combustion of diphenyl selenide using a rotating bomb calorimeter. This yielded a value for ΔH_f^0 (Ph_2Se, l) $= 54.1 \pm 1.4$ kcal mol^{-1}. They calculated a value for the latent heat of vaporization of 15.2 ± 0.6 kcal mol^{-1} from the boiling points at 4, 42 and 760 Torr, which gives ΔH_f^0 (Ph_2Se, g) $= 69.3 \pm 1.5$ kcal mol^{-1}.

Merten and Schlüter[5] determined the heat of combustion of dibenzyl selenide using a static bomb calorimeter. The already cited value for ΔH_f^0 (SeO_2, c) leads to ΔH_f^0 ($(PhCH_2)_2Se$, c) $= 0.5 \pm 5.0$ kcal mol^{-1} as previously calculated by Cox and Pilcher[2]. No information is available on the heat of sublimation.

Mortimer and Waterhouse[11] determined the heat of combustion of diphenyl diselenide using a rotating bomb calorimeter. This leads to ΔH_f^0 (Ph_2Se_2, c) $= (28.8 \pm 0.5)$ kcal mol^{-1}. A Knudsen-effusion technique was used to determine the enthalpy of sublimation $\Delta H_{subl} = (27.9 \pm 0.6)$ kcal mol^{-1} and thus ΔH_f^0 (Ph_2Se_2, g) $= 56.7 \pm 0.8$ kcal mol^{-1}.

Arshadi and Shabang[12] determined the heat of combustion of dibenzyl diselenide using an adiabatic static bomb calorimeter. They found that ΔH_f^0 ($(PhCH_2)_2Se_2$, c) $= 40.8 \pm 0.8$ kcal mol^{-1}. They used a transpiration method to determine the heat of sublimation $\Delta H_s^0 = 31.2 \pm 0.2$ kcal mol^{-1} and thus ΔH_f^0 (Ph_2Se_2, g) $= 72.0 \pm 0.8$ kcal mol^{-1}.

Arshadi and Shabang[12] also determined the heat of combustion of 4-phenyl-1,2,3-selenadiazole using the same technique. They found that ΔH_f^0 ($C_8H_7N_2Se$, c) $= 85.6 \pm 2$ kcal mol^{-1}, $\Delta H_s^0 = 22.5 \pm 0.2$ kcal mol^{-1} and thus ΔH_f^0 ($C_8H_7N_2Se$, g) $= 108.1 \pm 2$ kcal mol^{-1}.

The results for these organoselenium compounds are summarized in Table 2.

B. Bond Dissociation Energies

The R—X bond dissociation energy is defined as the energy required to break the R—X bond. No information of this type is available for organoselenium compounds. However,

TABLE 2. Heats of formation for organoselenium compounds

Compound	Bomb calorimetry method	ΔH_f^0 (g) (kcal mol^{-1})	Ref.
Et_2Se	Static	-12.4 ± 0.9	6
Ph_2Se	Rotating	69.3 ± 1.5	7
$(PhCH_2)_2Se$	Static	0.5 ± 5.0 [a]	6
$C_8H_7N_2Se$ [b]	Static	108.1 ± 2	13
Ph_2Se_2	Rotating	56.7 ± 0.8	12
$(PhCH_2)_2Se_2$	Static	72.0 ± 0.8	13

[a] ΔH_f^0(c) only.
[b] 4-Phenyl-1,2,3-selenadiazole.

mean bond dissociation energies may be determined. This is based upon the heats of formation of the compounds in the gaseous state together with that for the radicals formed and ΔH_f^0 (Se, g) = 54.3 kcal mol^{-1} [7]. For diethyl selenide the mean Se—C bond energy is defined as $\Delta H_2^0/2$:

$$Et_2Se \rightarrow 2\ \dot{E}t + Se \tag{2}$$

Using a value of ΔH_f^0 ($\dot{E}t$) = 26.5 ± 1 kcal mol^{-1} [13], $\bar{D}$(Se—C) = 59.9 ± 1.7 kcal mol^{-1} [15]. Similarly for diphenyl selenide, with ΔH_f^0 ($\dot{P}h$) = 78.5 ± 1 kcal mol^{-1} [13], $\bar{D}$(Se—C) = 71 kcal mol^{-1}. It appears that the effect of the benzene ring is to strengthen the Se—C bond by some 11 kcal mol^{-1}. Unfortunately a similar calculation cannot be made for dibenzyl selenide.

Assuming that $\bar{D}$(Se—Ph) is the same in diphenyl selenide, a consideration of the process:

$$Ph{-}Se{-}Se{-}Ph \rightarrow 2\ \dot{P}h + 2\ Se \tag{3}$$

would conclude that $\Delta H_3^0 = 2\ \bar{D}(Se{-}Ph) + D(Se{-}Se) = 66.9 \pm 2.9$ kcal mol^{-1} close to Gaydon's[14] value for D(Se—Se) of 64.6 kcal mol^{-1}. For comparison D(Te—Se) = 57.6 kcal mol^{-1} and D(Te—Te) = 52 ± 2 kcal mol^{-1} [15]. A similar calculation for dibenzyl diselenide, assuming that $\bar{D}$($PhCH_2$—Se) here is equal to $\bar{D}$(Et—Se) gives D(Se—Se) = 63.3 kcal mol^{-1}. This leads to an average value for D(Se—Se) = 64.9 kcal mol^{-1} with a spread of ± 1.2 kcal mol^{-1}.

Returning to the monoselenide compounds the mean bond dissociation energies may be compared with those for similar compounds involving other elements in Group VI of the Periodic Table as shown in Table 3. In the two sets of examples, $\bar{D}$(X—R) decreases with the downward movement in Group VI suggesting that D(R—Te) < $\bar{D}$(R—Se).

In the manner of Benson and coworkers[3], Cox and Pilcher[2] derived group additivity values for ΔH_f^0. Using their assignment for $Se(C_2) = 0$ makes $C(H_2)(C)(Se)$ =

TABLE 3. Mean bond dissociation energies for derivatives of Group VI elements

Compound	$\bar{D}$(X—R) (kcal mol^{-1})	Ref.
Et_2O	86.5	2
Et_2S	69.7	2
Et_2Se	59.9	6
Ph_2O	102.4	2
Ph_2S	84.2	2
Ph_2Se	71	7

3.9 kcal mol^{-1}. This allows heats of formation for other aliphatic monoselenides to be generated. It is emphasized that the group values depend upon only one experimental datum and therefore the results should be treated with caution. No other group values may be derived at this time.

III. CONCLUSIONS

It is obvious that far more data are required to make a comprehensive review of the thermochemistry of Se and Te compounds in terms of both combustion studies and bond dissociation energy determinations. For the latter data only mean bond energy information is available. Probably the best technique here is the VLPP method (very low pressure pyrolysis)[16]. It would be foolish to come to any conclusion other than that Te—R bonds are weaker than the Se—R bonds for the corresponding Se compounds.

IV. REFERENCES

1. D. R. Stall, M. Prophet and coworkers, *JANAF Thermochemical Tables*, NSRDS-NBS37, Washington, D.C., 1971.
2. J. D. Cox and G. Pilcher, *Thermochemistry of Organic and Organometallic Compounds*, Academic Press, London, 1970.
3. S. W. Benson, F. R. Cruickshank, D. M. Golden, G. R. Haugen, H. E. O'Neal, A. S. Rodgers, R. Shaw and R. Walsh, *Chem. Rev.*, **69**, 279 (1969).
4. H. A. Skinner, *Advances in Organometallic Chemistry*, Vol. 2 (Eds. F. G. A. Stone and R. West), Academic Press, London, 1964.
5. H. Merten and H. Schlüter, *Ber. Deut. Chem. Ges.*, **69**, 1364 (1936).
6. D. S. Barnes and C. T. Mortimer, *J. Chem. Thermodynamics*, **5**, 371 (1973).
7. *N.B.S. Tech. Note*, 270 (1968).
8. G. Gattow, *Z. Anorg. Allgem. Chem.*, **317**, 245 (1962).
9. S. N. Gadzhiev, *Zh. Fiz. Khim.*, **40**, 241 (1966).
10. CODATA, *J. Chem. Thermodynamics*, **4**, 331 (1972).
11. C. T. Mortimer and J. Waterhouse, *J. Chem. Thermodynamics*, **12**, 961 (1980).
12. M. R. Arshadi and M. Shabang, *J. Chem. Soc., Perkin Trans. 2*, 1732 (1973).
13. S. W. Benson, *Thermochemical Kinetics*, 2nd ed., John Wiley and Sons, New York, 1976.
14. A. G. Gaydon, *Dissociation Energies and Spectra of Diatomic Molecules*, Chapman and Hall, London, 1953.
15. B. de B. Darwent, *Bond Dissociation Energies in Simple Molecules*, NSRDS-NBS31, Washington, D.C., 1970.
16. G. P. Smith and R. Patrick, *Int. J. Chem. Kinet.*, **15**, 167 (1983).

The Chemistry of Organic
Selenium and Tellurium
Compounds Volume 1
Edited by S. Patai and Z. Rappoport
© 1986 John Wiley & Sons Ltd.

CHAPTER **5**

Detection and determination of organic selenium and tellurium compounds

KURT J. IRGOLIC† and DIPANKAR CHAKRABORTI‡

Department of Chemistry, Texas A&M University, College Station, Texas 77843, USA

I. INTRODUCTION . . . 162
II. DETERMINATION OF Se/Te INORGANIC COMPOUNDS . . . 163
A. Qualitative Tests for Se/Te . . . 163
B. Determination of Se/Te in Organic Compounds . . . 163
1. Determination of selenium . . . 163
2. Determination of tellurium . . . 165
III. IDENTIFICATION AND DETERMINATION OF Se/Te COMPOUNDS 165
A. Selenols and Tellurols, RYH . . . 165
B. Selenenyl Compounds, RSeX . . . 166
C. Aryl-selenium and -tellurium Trihalides, RYX_3 . . . 166
D. Seleninic and Selenonic Acids, $RSeO_2H$ and $RSeO_3H$ . . . 166
E. Diorganyl Diselenides, R_2Se_2 . . . 166
F. Diorganyl Selenides and Tellurides, R_2Y . . . 167
G. Diorganyl-selenium and -tellurium Dihalides, R_2YX_2 . . . 171
H. Triorganyl-selenonium and telluronium Salts . . . 171
I. Selenoamino Acids and Related Compounds . . . 171
J. Diorganyl Selenoxides, R_2SeO . . . 177
K. Selenourea . . . 177
L. Selenophene, Tellurophene and Related Compounds. . . . 177
M. Piazselenols . . . 183
N. Selenium and Tellurium Diethyldithiocarbamates, $(Et_2NCSS)_2Y$ . . . 183
O. Trialkylsilyl Selenides/Tellurides and Related Compounds . . . 183
IV. POLAROGRAPHIC AND OTHER ELECTROCHEMICAL METHODS 183
V. REFERENCES . . . 185

†Author to whom all correspondence should be addressed.
‡On leave from Jadavpur University, Calcutta, India.

I. INTRODUCTION

The need to ascertain the composition of organic Se/Te compounds, to qualitatively prove the presence of these chalcogens and to quantitatively determine their percentages has existed since the first organic Se/Te compounds were prepared approximately 150 years ago. During the past one and a half centuries many organic Se compounds and not quite as many organic Te compounds were prepared. The pertinent literature is well summarized in several books[1,2], chapters[3–5] and review articles[6]. Analytical techniques and physicochemical methods were used as they became available for the characterization of organic Se/Te compounds. However, the necessity of determining specific organic Se/Te derivatives in mixtures of such compounds or other matrices did not arise frequently until the effects of Se and its compounds on living organisms became the concern of environmental scientists and the medical profession. Se is now known to be—depending on concentrations—an essential and a toxic element[7–11]. Of special interest is its anticarcinogenic effect against experimentally induced cancer in several animals. Se, chemically related to S, replaces S in many biologically important molecules and is subject to metabolic transformations[8]. There can be no doubt that beneficial and inimical effects of Se are attributable in most cases not to elemental Se but to specific compounds of Se. The identification and determination of organic Se compounds in environmental samples is, therefore, of great importance.

Most of the efforts in the area of analytical Se chemistry were directed toward accurate and precise determinations of Se in inorganic, organic and biological matrices. Few methods were developed which allowed the determination of Se compounds. The sections on analytical chemistry in the books devoted to the biochemistry[8–11] and the medical and biological effects of Se[7] provide little if any information on the methods for the determination of Se compounds, although techniques exist for the gas chromatographic determination of volatile Se compounds[12] and for the element-specific detection of Se compounds in eluents from high-pressure liquid[13] and ion chromatographs[14]. The technique of element-specific detection[15] using graphite furnace atomic absorption spectrometers[14] or plasma emission spectrometers[13] does not seem to have been applied to organic Se compounds.

Future work in analytical Se/Te chemistry will have to concentrate on speciation to provide the techniques required for routine application in research efforts to eludicate the biochemical, medical and biogeochemical roles of these chalcogens.

This chapter discusses the determination of Se/Te in organic compounds and presents the analytically useful methods for the identification and determination of organic Se/Te compounds. The potentially very useful polarographic and other electrochemical techniques have not yet been used for the determination of these compounds. Most of the work in this area, which is summarized in the last section of this chapter, explored the mechanisms of the electrochemical reduction of organic Se compounds.

Infrared spectroscopy, UV–visible spectroscopy, and Mössbauer spectroscopy, which are frequently used to characterize Se/Te compounds, were reviewed recently[1,2,6]. The few photoelectron spectra reported for organic Se/Te compounds are at present not very useful for the characterization of these compounds. Dipole moments, although available for many Se/Te compounds, are often influenced more by functional groups other than Se/Te[1,6], and are, therefore, of rather limited use for the characterization of organic Se/Te compounds. The determination of the conformations of organic Se/Te compounds in solution by means of dipole moment measurements is more in the domain of physical chemistry than analytical chemistry. For these reasons, UV–visible, infrared, Mössbauer and photoelectron spectroscopic techniques, and dipole moment measurements are not covered in this chapter. Other physicochemical techniques such as nuclear magnetic resonance and mass spectrometry are the topics of separate chapters in this book.

II. DETERMINATION OF Se/Te IN ORGANIC COMPOUNDS

The qualitative detection and quantitative determination of Se/Te are the first steps in the characterization of organic compounds of these elements. Although some, rather unstable, Se/Te compounds deposit red Se or black Te on standing and thus reveal that they contain these chalcogen atoms, the decomposition of the organic compounds with concomitant conversion of Se/Te to inorganic chalcogen derivatives is in most cases necessary for the detection and determination of Se/Te.

A. Qualitative Tests for Se/Te

Sensitive, reliable, quick and convenient tests for checking whether organic compounds contain Se or Te are very helpful in synthetic and analytical work.

A spot test for the detection of Se in organic compounds was developed by Feigl[16]. A small quantity of the sample and one drop of 70% perchloric acid are gradually heated in a glycerol bath to 205 °C. After three minutes at 205 °C the reaction mixture is cooled and then treated with one drop of a saturated solution of hydrazine sulphate. Upon warming this mixture on a boiling water bath, a pink colouration or a red precipitation appears indicating the presence of Se. Another method[17] decomposes the sample with potassium chlorate and precipitates selenate with barium ion in the presence of potassium permanganate. The formation of violet crystals is the sign for the presence of Se.

No qualitative tests seem to have been developed specifically for the detection of Te in organic compounds. However, mineralization to an inorganic Te compound and its reduction with sulphur dioxide or hydrazine to black elemental Te, in a procedure similar to the one used for Se, should be applicable.

S, Se and Te may be present together in an organic molecule[1,6]. After mineralization and conversion of selenate to selenite in boiling HCl, hydroxylamine hydrochloride in strongly acidic solution reduces only selenite. Te can then be precipitated by hydroxylamine in ammoniacal solution[18]. S should not interfere with these tests.

Many other reactions potentially useful for the detection of Se/Te in organic compounds[4,18,19] and in biological samples[20,21] are the subject of reviews, which give references to the pertinent literature.

B. Determination of Se/Te in Organic Compounds

Many methods are available for the determination of Se in organic compounds and in biological matrices. The biological samples contain in many cases unidentified organic Se compounds at low concentrations. Te, an element much less abundant than Se, does not have an organic chemistry as extensive as that of Se, and does not have its environmental importance. Therefore, much less attention has been given to the development of methods for the determination of Te.

Unless a non-destructive method is used, the first step in the determination of Se/Te in organic compounds and biological matrices is the mineralization of the organic matter and the conversion of the chalcogen to an inorganic compound. The methods for the mineralization, the subsequent determination of Se[18–28] and Te[6,18,19,26,27,29,30] and the separation of Se and Te[31] have been summarized and evaluated in several reviews. Therefore, only a brief survey of these methods is presented to provide information about the available options.

1. Determination of selenium

Non-destructive methods for the analysis of Se are X-ray fluorescence[19,21,23–25,32,33], neutron activation[19–25,32] and proton-induced X-ray[34,35] or γ-ray[35] emission. These methods have been used almost exclusively to determine Se in biological matrices or to detect its presence in chromatographic fractions, although the chalcogen content of organic Se/Te compounds were claimed to have been obtained within $\pm 0.1\%$ by X-ray fluorescence of KCl or borax disks[36]. An X-ray fluorescence procedure for the determination of Se in solid and liquid samples in the range of 2–40 $\pm$ 1 mg/kg or mg/L Se was developed using a seleno steroid and diphenyl selenide as the standards[37]. These results suggest that non-destructive techniques are useful for the determination of Se/Te in organic compounds.

Destructive methods for the analysis of Se mineralize the organic compounds by wet digestion, dry ashing or by combustion in an oxygen atmosphere. The inorganic Se compounds can then be determined by gravimetric, titrimetric, spectrophotometric, fluorometric, electrochemical, gas-chromatographic, mass spectrometric or neutron activation methods.

Wet ashing procedures employ mixtures of acids (sulphuric/nitric, sulphuric/perchloric, sulphuric/nitric/perchloric), mixtures of acids and salts (sulphuric acid/potassium permanaganate, nitric/perchloric acid/ammonium vanadate or sodium molybdate) or a mixture of nitric/perchloric acid and hydrogen peroxide. Several of these wet digestion methods for the determination of Se have been critically reviewed[38]. Dry ashing involves heating the organic compound with magnesium nitrate or with sodium peroxide in a Parr bomb. Combustion methods oxidize the organic compounds in an oxygen-filled flask, in a combustion tube with oxygen flowing over the sample[39] or at low temperature in an oxygen plasma[40].

The solutions obtained after mineralization may contain selenite, selenate or a mixture of these two compounds. Boiling HCl will reduce all selenate to selenite. Se can then be determined gravimetrically by reduction of selenite to selenium with sulphur dioxide or hydrazine sulphate, or titrimetrically through titration of selenite with thiosulphate or of iodine liberated in the reaction between potassium iodide and selenite. Potassium permanganate oxidizes selenite to selenate and excess permanganate can be back-titrated. Selenite can also be determined by titration with silver nitrate or lead nitrate[18,19,27] using a lead-sensitive electrode[41]. A method for the microdetermination of Se in organic compounds employs oxygen flask combustion or digestion with sulphuric/nitric acid and argentometric titration of selenite. Halides can often be determined simultaneously[42]. The simultaneous microdetermination of Se and S in organic compounds uses oxygen-flask combustion, titration of S with barium perchlorate and iodometric determination of Se[43].

Selenite reacts with aromatic *ortho*-diamines to form yellow- to red-coloured piazselenols, which have absorption maxima in the range 270–500 nm and molar extinction coefficients of approximately 20,000. The piazselenols can be determined spectrophotometrically in aqueous solution or after extraction into an organic solvent such as toluene[19,24]. The spectrophotometric determination of Se can be similarly carried out using dithizone, 2-mercaptobenzoic acid, phenyl thiosemicarbazide or similar S-containing reagents[19,24].

Piazselenols extracted into hydrocarbon solvents fluoresce at 580 nm after excitation at 450 nm. This fluorescence allows the very sensitive determination of Se[19,24]. Piazselenols after extraction into toluene can be quantitated by gas chromatography using electron-capture detection[24,44]. This method has the advantage of removing Se from interfering ions.

Electrochemical techniques for the determination of selenite include polarography, anodic and cathodic stripping voltammetry, and amperometric, coulometric or potentiometric titration[19,24]. A coulometric method for the simultaneous determination of C, H, Se and C, Se, S in organic compounds after combustion in an oxygen current has been described[45].

Flame atomic absorption[46,47] or emission[47] spectrometry and the much more sensitive flameless atomic absorption methods[47–49] can be used to determine Se in the aqueous digests or in organic extracts[50–53]. The absolute sensitivities may reach 1 pg Se[19,24]. An inductively coupled plasma emission spectrometer was used to determine Se in organic compounds after their digestion with HNO_3[54].

2. Determination of tellurium

The choice of methods for the determination of Te is rather limited. In principle, many of the methods proven to be applicable to the determination of Se should be useful for Te also. However, tellurite does not appear to react with aromatic *ortho*-diamines to form piaztellurols.

Organic Te compounds were mineralized with fuming nitric acid, with mixtures of nitric/perchloric acid or nitric/sulphuric acid, with sodium peroxide/potassium chlorate in a Parr bomb, by the oxygen flask method[42,55] and in a stream of oxygen[56,57]. The tellurium dioxide or tellurite formed during mineralization is then determined gravimetrically[56], by titration with potassium permanganate[58], potassium dichromate[59] or silver nitrate[44], or iodometrically with amperometric indication of the end-point[56]. A method for the simultaneous microdeterminations of C, H and Te employs the coulometric titration of tellurite with sodium thiosulphate[57]. Te in aqueous solution was quantitated by flame atomic absorption spectrometry in an air/acetylene flame at 214.2 nm[55,60]. The general analytical chemistry of Te is summarized in several reviews[18,19,30]. General aspects of the flame atomic absorption spectroscopy of Te[61], the thermal stabilization of inorganic and organically bound Te[62] and interferences by cations in the determination of Te[63] using electrothermal atomic absorption spectrometry are discussed in recent reports.

The formation of complexes of Te with S-containing ligands and the extraction of these complexes has been summarized[64]. The spectrophotometry of these complexes is a sensitive method for the determination of Te, which has not yet been applied to the estimation of Te in organic compounds.

III. IDENTIFICATION AND DETERMINATION OF Se/Te COMPOUNDS

The techniques most often used for the identification and determination of organic Se/Te compounds are paper chromatography, thin-layer chromatography, gas chromatography with electron capture, flame ionization, flame photometric or thermal conductivity detection, ion exchange chromatography, high-pressure liquid chromatography and iodometry. The identification and determination of organic Se/Te compounds becomes easier when element-specific detectors are coupled to chromatographs. Atomic absorption spectrometers with silica tube[65–69] or graphite furnaces[70,71] and microwave-excited emission spectrometers were used as Se-specific detectors[72,73] for gas chromatographs. High-pressure liquid chromatographs were interfaced with a graphite furnace atomic absorption spectrometer[74] and an inductively coupled argon plasma emission spectrometer[13] to determine organic and inorganic Se compounds, respectively. Neutron activation analysis was also suggested as a method for Se-specific detection[75]. The combination of chromatography with element-specific detection systems provides analytical capabilities especially useful when Se/Te compounds need to be determined in complex matrices.

A. Selenols and Tellurols, RYH

Benzeneselenol, PhSeH, was separated from several diorganyl sulphides, diorganyl selenides and benzenethiol by gas chromatography on 20% squalene–80/100-mesh Celite-545 using an argon ionization detector[76]. The selenols corresponding to pantheteine, 4′-phosphopantheteine, dephosphocoenzyme A and coenzyme A were separated from the thiols, disulphides and diselenides of these compounds (see Table 6, items 47–49, 51). The spots were visualized by spraying with an aqueous solution of starch/sodium hydrogen carbonate/0.1 N iodine. The blue background turned brown on drying. Spots caused by selenols, diselenides and selenides returned to the blue background colour upon spraying with dilute phosphoric acid. Spots caused by S compounds remained white. The Se compounds on inorganic thin-layer materials can be determined through oxidation with starch/iodine on the TLC support, extraction of Se compounds from the TLC support into an aqueous solution followed by acidification and colorimetric measurement of the starch–iodine complex. Reproducible results were obtained at concentrations as low as 10^{-8} mole/mL[77].

B. Selenenyl Compounds, RSeX

Benzeneselenenyl bromide was determined by addition of aqueous 0.1 M KI and 6 M H_2SO_4 solution to the solution of the Se compound in CCl_4. The mixture was titrated with sodium thiosulphate solution using starch for the end-point determination[78].

2-Nitrobenzene- and 2,4-dinitrobenzene-selenenyl compounds, RSeX (X = Br, SCN, OMe, OEt, NH_2) reacted in a mixed solvent consisting of ethyl acetate, ethanol and glacial acetic acid with 0.01 N sodium thiosulphate to form $RSeS_2O_3^-$. The excess thiosulphate was back-titrated with an iodine solution[79].

Ethyl selenocyanate, EtSeCN, was separated by gas chromatography from dialkyl selenides and diselenides[80] (see Table 1, item 5).

C. Aryl-selenium and -tellurium Trihalides, RYX_3

Phenylselenium tribromide reacted with KI in H_2SO_4 medium. The mixture was titrated with sodium thiosulphate solution to the starch end-point[78].

Phenyltellurium trichloride was separated by TLC on alumina from tellurium tetrachloride, triphenyltelluronium chloride, diphenyltellurium dichloride and diphenyl telluride[81] (see Table 3, item 6).

D. Seleninic and Selenonic Acids, $RSeO_2H$ and $RSeO_3H$

Benzeneseleninic acids in CCl_4 shaken with an aqueous H_2SO_4 solution of KI were converted to diselenides with liberation of iodine. Iodine was titrated with sodium thiosulphate[78]. The iodometric equivalent masses of *para*-substituted benzeneseleninic acids ($4\text{-}RC_6H_4SeO_2H$, R = H, Me, Cl, Br) are one third of the molecular masses. These iodometric titrations were also carried out in aqueous NaOH solutions with a visual or biamperometric end-point determination[82]. The equivalent masses of the seleninic acids were also determined by titration with NaOH[82].

Benzeneseleninic acid was separated from benzeneselenonic acid by descending chromatography on glass-fibre paper using amyl alcohol/pyridine/ammonium hydroxide (6:14:20). The R_f values were 0.47 ($RSeO_2H$) and 0.21 ($RSeO_3H$)[83].

2-Aminoethaneseleninic acid was separated from 3-aminopropaneseleninic acid and the corresponding selenonic, sulphinic and sulphonic acids on Whatman No. 1 paper with water-saturated 2,4,6-trimethylpyridine/2,6-dimethylpyridine (1:1)[83]. Ion-exchange chromatography with Aminex A-6 and A-5 resins using an amino acid analyser separated the four Se compounds and the sulphinic acids. The selenonic and sulphonic acids eluted together[84].

E. Diorganyl Diselenides, R_2Se_2

Diselenides were detected with the iodine–azide reaction. Heating a diselenide in ethanolic HCl with Raney alloy generated a selenol which liberated nitrogen from a solution containing sodium azide and iodine. Other compounds which gave this test are disulphides, thiols, selenols, thio ketones, thio acids and their derivatives, and mustard oils. In the absence of these substances one microgram of dibenzyl diselenide was detectable[85]. Aromatic diselenides, $(RC_6H_4)_2Se_2$ (R = H, 4-Me, 2-Ph), were titrated iodometrically. The diselenide was added to a solution of iodine monochloride in 12 M HCl. After shaking, the mixture was titrated with standard potassium iodate solution. The diselenide was converted during the titration to arylselenium trichloride[78].

Several dialkyl diselenides were successfully separated from each other, from dialkyl selenides and ethyl selenocyanate by gas chromatography using flame ionization and electron capture detectors. The hydrogen flame ionization detector was much more sensitive to the selenides than the diselenides and ethyl selenocyanate, whereas the electron capture detector had higher sensitivity for the diselenides and ethyl selenocyanate[80].

Dimethyl diselenide and dimethyl selenide are formed by plants and animals from inorganic Se compounds. Several methods were developed to separate these two compounds. The air samples were passed through a column (Alusil[86], silicone oil DC-550[68]) or traps at low temperature[65] to concentrate the Se compounds. After desorption, dimethyl diselenide was separated from dimethyl selenide[65–68,86], dimethyl and diethyl selenide[71] and unidentified Se compounds[68] by gas chromatography using atomic absorption spectrometers as Se-specific detectors[65–68,71].

The methods for the identification, separation and determination of diorganyl diselenides are summarized in Table 1. Diselenides derived from amino acids, and diselenides corresponding to oxytocin, panthetine and coenzyme A are discussed in Section III.I (Table 6).

F. Diorganyl Selenides and Tellurides, R_2Y

Diphenyl chalcogenides, R_2Y, were chromatographed on alumina thin layers. The spots corresponding to Se/Te compounds were visualized with iodine vapour[87].

Attempts were made to separate diphenyl selenide, diphenyl sulphide, diphenyl telluride, diphenyl chalcogen dichlorides, R_2YCl_2 (Y = S, Se, Te) and triphenylselenonium and -telluronium tetrafluoroborates on Silufol UV-254 thin layers. Seven of these compounds were successfully separated. A 0.02% solution of bromocresol green was used for visualization[88].

The separation of dimethyl selenide from dimethyl diselenide and the determination of dimethyl selenide was achieved by gas chromatography using flame ionization or Se-specific atomic absorption spectrometric detectors[65–68,71,80,86]. Several other dialkyl selenides, phenyl alkyl selenides and diphenyl selenide were determined by gas chromatography in mixtures containing selenides, diselenides[80], sulphides, disulphides[76], heterocyclic Se compounds[89] or trimethylarsine[76,90], trimethylstibine[90] and tetramethyltin[70,80].

Dibenzyl selenide and 1,1-dimethylselenourea were separated by high-pressure liquid chromatography on Partisil-PXS-ODS with methanol/water (2:1) as the mobile phase. A graphite furnace atomic absorption spectrometer served as Se-specific detector. Nickel nitrate was added as coanalyte to reduce the volatility of Se in the furnace and enhance the intensity of the Se signal[74]. More details about this and the other separation procedures are given in Table 2.

The applicability of microwave emission systems for the detection of Se in gas chromatographic effluents was checked with dimethyl selenide[72] and diethyl selenide[73]. Absolute detection limits of 12 pg[72] and 62 pg[73] were obtained. The atomic absorption system had a detection limit of 7 ng[70].

The flame photometric detector can distinguish between S and Se compounds. Doping the gas stream with carbon disulphide generated negative Se peaks[92]. Doping with methane reduced the S signal much more than the Se signal[89,92]. The flame photometric detector had an exponential response to Se and Te. The response was made linear by

TABLE 1. Separation, identification and determination of diorganyl diselenides by thin-layer and gas chromatography[a]

Item No.	Diselenide	Separated from	Method	Remarks	Ref.
1	Me_2Se_2	Me_2Se	GC-FID, TP, 15% Si-polymer Se-30 or 10% Apiezon on 80/100 Gas-Chrom-P	Analysis of air exhaled by rats injected with Na_2SeO_3; elution sequence: Me_2Se, Me_2Se_2	86
2	Me_2Se_2	Me_2Se	GC-AA (quartz furnace), TP, 196 nm, 3% OV1 on 80/100 Chromosorb W, DL 0.1 ng Se	Determination of Me_2Se_n in air after preconcentration on GC column at −80°C	65 66 67
3	Me_2Se_2 (17)[b]	Me_2Se (3), unknown Se compounds	GC-AA (quartz furnace), TP 196 nm, 20% PMPE on 60/80 Chromosorb W	Analysis of transpiration gases of *Astragalus racemosus*	68
4	Me_2Se_2 (1)	Me_2Se (3.2), Et_2Se (2)	GC-AA (graphite furnace), 196 nm, 10% PMPE on 80/100 Chromosorb W, DL 0.1 ng Se	Determination of Se compounds in air after preconcentration	71
5	Me_2Se_2 (5.2)	Et_2Se_2 (12.6), Pr_2Se_2 (35.5), Me_2Se (0.6), Et_2Se (1.8), Pr_2Se (4.3), EtSeCN (7.0), MeSeSeEt (4.0), EtSeSePr (8.5)	GC-FID or ECD, TP 20% PMPE on 60/80 Chromosorb W coated with hexamethyldisilazane	Retention times for column temperature of 125°C and injector temperature of 180°C	80
6	Et_2Se_2	See item 5			80
7	Pr_2Se_2	See item 5			80
8	MeSeSeEt	See item 5			80
9	EtSeSePr	See item 5			80
10	Diselenides of amino acids, oxytocin, panthetine, coenzyme A		TLC	See Table 6, items 47–52	77

[a] AA: Atomic Absorption Spectrometry; DL: Detection Limit; ECD: Electron Capture Detector; FID: Flame Ionization Detector; GC: Gas Chromatography; PMPE: PolyMetaPhenyl Ether; TP: Temperature Programmed. [b] Numbers in parentheses represent retention times in minutes.

TABLE 2. Separation, identification and determination of diorganyl selenides by thin-layer and gas chromatography[a]

Item no.	Selenide	Separated from	Method	Remarks	Ref.
1	Me_2Se	Me_2Se_2	GC-FID, GC-AA	See Table 1, items 1 – 5	65 – 68, 71, 80, 86
2	Me_2Se	Et_2Se, Pr_2Se, Bu_2Se, PhSeH, PhSeMe, PhSeEt, corresponding S compounds	GC-Sr-90 ID, 90°C, 20% squalene on 80/100 Celite 545	Pr_2Se not sepd. from PhSH Me_2Se not sepd. from Me_2S	76
3	Me_2Se (1.5)[b]	Me_3As (1.5), Me_4Sn (1.8)	GC-AA (graphite furnace) 196 nm, 5% methylsilicone SP-2100, 3% fluoropropylsilicon SP-2401 on 80/100 Supelcon AWDMCS	Detection limit 12 ng Se	70
4	Me_2Se (9.8)	Me_3As (1.5), Me_3Sb (12), Me_4Sn (8.4), CH_3I (7.6)	GC-TCD, 70°C, 20% DBS on $NaCl/SO_4$/phosphate		90
5	Me_2Se	Et_2Se, Pr_2Se, Me_2Se_2, Et_2Se_2, Pr_2Se_2, MeSeSeEt, EtSeSePr, EtSeCN	GC-FID or ECD	See Table 1, item 5	80
6	Et_2Se	See item 2	GC-Si-90 ID		76
7	Et_2Se (5)	Et_2O (1), EtI (2)	GC-TCD, 20% silicone rubber on Chromaton N-AW-DMCS		91
8	Et_2Se	See item 5	GC-FID or ECD	See Table 1, item 5	80

(*Contd.*)

TABLE 2. (*Contd.*)

Item no.	Selenide	Separated from	Method	Remarks	Ref.
9	Et_2Se	Ph_2Se, piazselenol, 2,3-dihydro-3-azabenzoselenophene	GC-FPD, TP, 5% PEGA on 80/100 Chromosorb W AW		89
10	Pr_2Se	See item 2	GC-Sr-90 ID		76
11	Pr_2Se	See item 5	GC-FID or ECD	See Table 1, item 5	80
12	Bu_2Se	See item 2	GC-Sr-90 ID		76
13	$(PhCH_2)_2Se$ (26)	1,1-Dimethylselenourea (10)	HPLC-AA (graphite furnace), 196 nm, Partisil-PXS-ODS, $MeOH/H_2O$ (2:1)		72
14	PhSeMe	See item 2	GC-Sr-90 ID		76
15	PhSeEt	See item 2	GC-Sr-90 ID		76
16	Ph_2Se	Ph_2O, Ph_2S, other Se and Te compounds	TLC-alumina	See Table 3, items 2, 3	87 88
17	Ph_2Se	See item 9	GC-FID		89

[a]AA: Atomic Absorption Spectrometry; DBS: DodecylBenzeneSulphonate; ECD: Electron Capture Detector; FID: Flame Ionization Detector; GC: Gas Chromatography; Sr-90 ID: Sr-90 Ionization Detector.
[b]Numbers in parentheses represent retention times in minutes.

providing a high sulphur background with carbon disulphide[93].

The paper and thin-layer chromatographic behaviour of several aromatic ditellurides was investigated with the goal to separate reaction mixtures, identify products and isolate products. Diaryl tellurides were efficiently separated from tellurium tetrachloride, diaryltellurium dihalides, aryltellurium trihalides and triorganyltelluronium halides. The separation of organic Po compounds from the corresponding Te compounds was generally not possible with these methods. The systems investigated and pertinent references are listed in Table 3.

Diethyl telluride was separated from diethyl ether and ethyl iodide by gas chromatography on silicone rubber/Chromaton. A thermal conductivity detector was used[91].

G. Diorganyl-selenium and -tellurium Dihalides, R_2YX_2

Diarylselenium dichlorides and dibromides were determined by a volumetric procedure. Upon shaking the dihalides, R_2SeX_2 (R, X: Ph, Cl; Ph, Br; 4-Tol, Br)[78,101], $RR'SeX_2$ (R, R′, X: Ph, 4-Tol, Cl; Ph, 4-BrC_6H_4, Br; 4-Tol, Ph, Br) or dibenzoselenophene dibromide[102] with aqueous KI, iodine was liberated. The iodine was titrated with sodium thiosulphate solution to the starch end-point.

To determine Se, 9 M HCl solutions of selenite were reacted with acetophenone to form an organic Se compound, which might have been bis(benzoylmethyl)selenium dichloride. The organic Se compound was extracted and the extract analysed by gas chromatography employing an electron-capture detector[102].

The thin-layer and paper chromatographic separations of diaryltellurium dichlorides from organic S, Se and Po compounds, other organic Te compounds and tellurium tetrachloride are summarized in Table 4.

H. Triorganyl-selenonium and -telluronium Salts

Milligram amounts of triarylselenonium salts, $(4\text{-}RC_6H_4)_3Se^+X^-$ (R, X: H, Cl; Me, Cl; Me, HSO_4) were determined in aqueous solutions of pH 1–13 by spectrophotometry at 227 nm (Ph derivative, $\varepsilon = 18{,}400$) or 232 nm (Tol derivatives, $\varepsilon = 30{,}900$)[103].

Data relating the thin-layer[81,87,88] and paper chromatographic[95] separation of triphenylselenonium tetrafluoroborate[88] and of triaryltelluronium salts from other organic chalcogen compounds are summarized in Table 5.

I. Selenoamino Acids and Related Compounds

Se with properties similar to S replaces S in S-containing amino acids forming selenoamino acids such as selenocystine and selenomethionine. Selenoamino acids occur naturally. The separation of the seleno- from the thio-amino acids and other amino acids was achieved by ion-exchange chromatography using amino acid analysers. Paper chromatography generally did not separate selenoamino acids from the corresponding S compounds. However, selenomethionine was successfully separated from methionine by thin-layer chromatography on silica gel[119]. Trimethylsilylated selenocystine and selenomethionine had gas chromatographic retention times different from those of the corresponding S compounds[115].

Several aminoalkyl selenides and diselenides related to selenoamino acids [seleno(homo)cystamine, seleno(homo)lanthionamine, selenocystathionamine, carboxymethylselenocysteamine] were separated from each other and the corresponding S derivatives by ion-exchange chromatography[116,117]. A dipeptide, glutamyl-*Se*-methylselenocysteine[123], had a significantly longer retention time than the S-peptide on a Dowex 1-X4 column. The paper chromatographic behaviour of the selenols, thiols,

TABLE 3. Separation, identification and determination of diorganyl tellurides by thin-layer, paper and gas chromatography[a]

Item no.	Telluride	Separated from	Method	Remarks	Ref.
1	Et_2Te (8)[b]	EtI (3.5), Et_2O (1)[b]	GC-TCD		91
2	Ph_2Te	(*o*-Tol)$_2$Te (0.61), (*m*-Tol)$_2$Te (0.54), (*p*-Tol)$_2$Te (0.54), (2,5-Xyl)$_2$Te (0.62), (*p*-An)$_2$Te (0.54), (1-Naph)$_2$Te (0.33), Ph_2O (0.59), Ph_2S (0.60)	TLC, alumina, pet. ether	Visualization with I_2 vapour	87
3	Ph_2Te	Ph_2S, Ph_2Se, Ph_2SCl_2, Ph_2SeCl_2, Ph_2TeCl_2, $Ph_3Y^+BF_4^-$ (Y = S, Se, Te)	TLC, Silufol UV-254, C_6H_6/EtOAc/H_2CO(?)(4:4:1)	Seven compounds sepd., visualization with 0.02% bromocresol green in EtOH	88
4	Ph_2Te (0.55)[c]	$TeCl_4$ (0.0)	TLC, alumina, pet. ether	Separation of ^{127m}Te compounds	94
5	Ph_2Te (0.20)	Ph_2TeCl_2 (0.67), Ph_3TeCl (0.96), Ph_3TeI (0.90)	PC, Me_2CO/MeOH/H_2O (4:3:2), several other solvent systems investigated	Po compounds have same R_f as Te compounds	95
6	Ph_2Te (0.95)	$TeCl_4$ (0.0), $PhTeCl_3$ (0.08), Ph_2TeCl_2 (0.54), Ph_3TeCl (0.32)	TLC, alumina, C_6H_6/EtOH/MeCOOH/H_2O (44:11:3.5:1)	Several other solvent systems investigated	81
7	(*p*-Tol)$_2$Te (0.64)	(*o*-Tol)$_2$Te (0.71), (*m*-Tol)$_2$Te (0.67)	TLC, alumina, solvent not reported	Visualized by spraying with aq. $KMnO_4$	96
8	(*p*-Tol)$_2$Te (0.0)	R_2TeCl_2 (0.40), $RTeCl_3$ (0.60), R_3TeCl (0.90)	PC, MeOH/H_2O (3:1)	Paper treated with 2% butyl sebacate/MeOH	97
9	(*p*-Tol)$_2$Te (1.0)	$TeCl_4$ (0.0), R_2TeCl_2, $RTeCl_3$, R_3TeCl (1.0)	PC, $CHCl_3$	Without sebacate	97

10	$(p\text{-Tol})_2Te$ (1.0)	$TeCl_4$, R_2TeCl_2, $RTeCl_3$, R_3TeCl (0.0)	PC, pet. ether	Without sebacate	97
11	$(p\text{-Tol})_2Te$ (0.76)	R_2TeF_2 (0.40), R_2TeCl_2 (0.58), R_2TeBr_2 (0.49), R_2TeI_2 (0.43), R_3TeCl (0.08), R_3TeI (0.21)	PC, EtOAc, paper treated with DMF	Po derivatives have R_f similar to Te compounds	98
12	$(m\text{-Tol})_2Te$	See item 7	TLC, alumina		96
13	$(m\text{-Tol})_2Te$	See item 2	TLC, alumina, pet. ether		87
14	$(o\text{-Tol})_2Te$	See item 7	TLC, alumina		96
15	$(o\text{-Tol})_2Te$	See item 2	TLC, alumina, pet. ether		87
16	$(p\text{-An})_2Te$ (0.66)	R_2TeF_2 (0.42), R_2TeCl_2 (0.59), R_3TeCl (0.12), R_3TeBr (0.13), R_3TeI (0.17)	PC, EtOAc, paper treated with DMF	Po compounds have R_f similar to Te compounds	99
17	$(p\text{-An})_2Te$	See item 2	TLC, alumina, pet. ether		87
18	$(2,5\text{-Xyl})_2Te$	See item 2	TLC, alumina, pet. ether		87
19	$(1\text{-Naph})_2Te$ (0.63)	R_2TeF_2 (0.50), R_2TeCl_2 (0.45), R_2TeBr_2 (0.50), R_2TeI_2 (0.69), R_3TeCl (0.09), R_3TeBr (0.02), R_3TeI (0.010), $R_3TeCl\cdot HgCl_2$ (0.0)	PC, EtOAc, paper treated with DMF	Po compounds have R_f similar to Te compounds	100
20	$(1\text{-Naph})_2Te$	See item 2	TLC, alumina, pet. ether		87

[a] GC: Gas Chromatography; PC: Paper Chromatography; TCD: Thermal Conductivity Detector; TLC: Thin-Layer Chromatography.
[b] Retention time.
[c] Numbers (O.XY) represent R_f values.

TABLE 4. Separation, identification and determination of diorganyl-selenium/tellurium dihalides by paper and thin-layer chromatography[a]

Item no.	Dihalide	Separated from	Method	Remarks	Ref.
1	Ph_2SeCl_2	Ph_2Y, $Ph_3Y^+BF_4^-$ (Y = S, Se, Te) Ph_2SCl_2, Ph_2TeCl_2	TLC, Silufol UV-254, C_6H_6/EtOAc/H_2CO(?) (4:4:1)	Seven compounds separated, visualization with 0.02% bromocresol green in EtOH	88
2	Ph_2TeCl_2	See item 1	See item 1	See item 1	88
3	Ph_2TeCl_2	Ph_2Te, Ph_3TeX[b]	PC, Me_2CO/MeOH/H_2O (4:3:2)	See Table 3, item 5	95
4	Ph_2TeCl_2	$TeCl_4$, $PhTeCl_3$, Ph_3TeCl	TLC, alumina	See Table 3, item 6	81
5	Ph_2TeCl_2 (0.49)[c]	(*o*-Tol)$_2TeCl_2$ (0.25), (*m*-Tol)$_2TeCl_2$ (0.67), (*p*-Tol)$_2TeCl_2$ (0.49), (*p*-Tol)$_3TeCl$ (0.80)	TLC, alumina, EtOAc/MeOH (3:1)		87
6	Ph_2TeCl_2 (0.68)	(*p*-Tol)$_2TeCl_2$ (0.73), (*p*-Tol)$_3TeCl$ (0.22), Ph_3TeCl (0.27)	TLC, alumina, C_6H_6/EtOH (4:1)	C_6H_6/EtOH (9:1) also used	87
7	(*p*-Tol)$_2TeF_2$	R_2Te, R_2TeX_2, R_3TeX	PC, EtOAc	See Table 3, item 11	98
8	(*p*-Tol)$_2TeCl_2$	R_2Te, $RTeCl_3$, R_3TeCl	PC, MeOH/H_2O (3:1)	See Table 3, item 8	97
9	(*p*-Tol)$_2TeCl_2$ (1.0)	$TeCl_4$ (0.0)	PC, $CHCl_3$	See Table 3, item 9	97
10	(*p*-Tol)$_2TeCl_2$ (0.0)	R_2Te (1.0)	PC, pet. ether	See Table 3, item 10	97

11	(p-Tol)$_2$TeCl$_2$	R$_2$Te, R$_2$TeX$_2$, R$_3$TeX	PC, EtOAc	See Table 3, item 11	98
12	(p-Tol)$_2$TeCl$_2$	See items 5, 6	TLC, alumina		87
13	(p-Tol)$_2$TeBr$_2$	R$_2$Te, R$_2$TeX$_2$, R$_3$TeX	PC, EtOAc	See Table 3, item 11	98
14	(p-Tol)$_2$TeI$_2$	R$_2$Te, R$_2$TeX$_2$, R$_3$TeX	PC, EtOAc	See Table 3, item 11	98
15	(m-Tol)$_2$TeCl$_2$	See item 5	PC, alumina		87
16	(o-Tol)$_2$TeCl$_2$	See item 5	PC, alumina		87
17	(p-An)$_2$TeF$_2$	R$_2$TeCl$_2$, R$_3$TeX	PC, EtOAc	See Table 3, item 16	99
18	(p-An)$_2$TeCl$_2$	R$_2$TeF$_2$, R$_3$TeX	PC, EtOAc	See Table 3, item 16	99
19	(1-Naph)$_2$TeF$_2$	R$_2$TeX$_2$, R$_3$TeX	PC, EtOAc	See Table 3, item 19	100
20	(1-Naph)$_2$TeCl$_2$	R$_2$TeX$_2$, R$_3$TeX	PC, EtOAc	See Table 3, item 19	100
21	(1-Naph)$_2$TeBr$_2$	R$_2$TeX$_2$, R$_3$TeX	PC, EtOAc	See Table 3, item 19	100
22	(1-Naph)$_2$TeI$_2$	R$_2$TeX$_2$, R$_3$TeX	PC, EtOAc	See Table 3, item 19	100

[a]PC: Paper Chromatography; TLC: Thin-Layer Chromatography.
[b]X = other halides. Numbers represent R_f values.

TABLE 5. Separation and identification of triaryl-telluronium and -selenonium compounds[a]

Item no.	$R_3Te^+X^-$	Separated from	Method	Remarks	Ref.
1	Ph_3TeCl	R_2TeCl_2, R_3TeI	PC, Me_2CO/MeOH	See Table 3, item 5	95
2	Ph_3TeCl	$TeCl_4$, $RTeCl_3$, R_2TeCl_2	TLC, alumina	See Table 3, item 6	81
3	Ph_3TeI	R_2TeCl_2, R_3TeCl	PC, Me_2CO/MeOH	See Table 3, item 5	95
4	Ph_3TeBF_4	Ph_2Y, Ph_2YCl_2, Ph_3YBF_4 (Y = S, Se, Te)	TLC, Silufol UV-254	See Table 3, item 3	88
5	Ph_3SeBF_4	Ph_2Y, Ph_2YCl_2, Ph_3YBF_4 (Y = S, Se, Te)	TLC, Silufol UV-254	See Table 3, item 3	88
6	*p*-Tol_3TeCl	R_2Te, R_2TeCl_2, $RTeCl_3$, $TeCl_4$	PC, MeOH/H_2O, $CHCl_3$, pet. ether, EtOAc	See Table 3, items 8, 9, 10, 11	97 98
7	*p*-Tol_3TeI	R_2TeX_2, R_3TeCl	PC, EtOAc	See Table 3, item 1	98
8	*p*-An_3TeX	R_2TeF_2, R_2TeCl_2, R_3TeX	PC, EtOAc	See Table 3, item 16	99
9	1-$Naph_3TeX$	R_2TeX_2, R_3TeX, $R_3TeCl \cdot HgCl_2$	PC, EtOAc	See Table 3, item 19	100

[a]PC: Paper Chromatography; TLC: Thin-Layer Chromatography.

selenides, sulphides, diselenides and disulphides related to pantethine, coenzyme A, 4-phosphopantethine and dephosphocoenzyme A was investigated[77].

The results of these investigations of selenoamino acids and related compounds are summarized in Table 6.

J. Diorganyl Selenoxides, R_2SeO

Diaryl selenoxides reacted with aq. KI in 6 M H_2SO_4 solution. The liberated iodine was titrated with sodium thiosulphate to the starch end-point. This reaction converts the selenoxides to selenides. The equivalent mass for selenoxides is half the molecular mass[78]. Although only bis(4-ethoxyphenyl) selenoxide was determined, the method should be applicable to other selenoxides and perhaps to telluroxides.

Four methods for the microdetermination of selenoxides were investigated. Potentiometric titrations of selenoxides dissolved in acetic anhydride with perchloric acid gave acceptable results[125,126]. The iodometric method[78] was found to be erratic and unreliable in the presence of compounds with other functional groups[125]. Selenoxides oxidized iron(II) to iron(III) in acidic methanolic ferroammonium thiocyanate solution. The intensely coloured iron(III) thiocyanate complex was determined photometrically. The instability of the iron(II) thiocyanate solution proved to be a serious disadvantage for routine application of this method[125]. The degree of inhibition of the hydrolysis of urea by urease was found to be proportional to the concentration of selenoxides. The inhibition constants are compound-specific. It might be possible to determine an inhibitory compound in the presence of others[125]. These methods were used to determine diphenyl selenoxide and 2-(phthalimido)ethyl methyl selenoxide[125].

K. Selenourea

Selenourea was determined in sulphite solutions buffered with $NaHCO_3$ by addition of 0.1 N iodine solution to oxidize selenium to selenite. After addition of acetic acid the excess iodine was titrated with thiosulphate solution. The selenite was then determined iodometrically in HCl solution. An appropriate modification of this procedure allowed the analysis of mixtures containing selenourea and selenosulphate[127].

1,1-Dimethylselenourea was separated from dibenzyl selenide by high-pressure liquid chromatography using a graphite furnace atomic absorption spectrometer as a Se-specific detector[74] (Section III. F, Table 2, item 13).

L. Selenophene, Tellurophene and Related Compounds

The molar responses relative to benzene and the effective carbon numbers of selenophene, tellurophene and their 2-acetyl derivatives were determined with a gas chromatograph – flame ionization detector system. The molar responses were approximately half that of benzene[128].

The equivalent mass of benzoselenophene dibromide was determined by treating the dibromide with a H_2SO_4 solution of KI and titrating the liberated iodine with thiosulphate solution[78].

The gas chromatographic peak caused by 2,5-dimethyl-3-azabenzoselenophene was identified by doping the gas stream with carbon disulphide. Se peaks turned negative, but S peaks remained positive[92].

2,3-Dihydro-3-azabenzoselenophene was separated by gas chromatography from diethyl selenide, diphenyl selenide and piazselenol[89] (Section III. F; Table 2, item 9).

TABLE 6. Separation, identification and determination of selenoamino acids and related compounds[a]

Item no.	Compound	Separated from	Method	Ref.
1	*Se*-Methylselenocysteine (0.48)[b] $MeSeCH_2CH(NH_2)COOH$	Selenocystine (0.14), selenomethionine (0.55), cystine (0.14), methionine (0.55); S compounds not separated from corresponding Se compounds	DPC, BuOH/pyr./H_2O (1:1:1); several other solvent systems investigated	104
2	*Se*-Methylselenocysteine	Cystine, homocystine, lanthionine, cystathionine, djenkolic acid, methionine, methionine sulphoxide, *S*-methylcysteine sulphoxide; not separated from *S*-methylcysteine	PC (2 dim.), HCO_2H/*t*-BuOH/H_2O (3:14:3), EtOH/BuOH/H_2O/(*c*-Hex)$_2$NH 10:10:5:2)	105
3	*Se*-Methylselenocysteine (115)[c]	Selenocystine (201), selenomethionine (210), selenocystathionine (203), selenohomocystine (305), corresponding S compounds, other amino acids	AAA, Aminex A-4, NaC, 56°C	106
4	*Se*-Methylselenocysteine (121)[c]	*S*-Methylcysteine (99), selenocystine (157), selenomethionine (170), selenoethionine (186), *Se*-benzylselenocysteine (282), and corresponding S compounds	AAA, Beckman P-28 resin, NaC with Brij, pentachlorophenol, thiodiglycolic reagent, 55°C	107
5	*Se*-Carboxymethylselenocysteine (78)[d] $HOOCCH_2SeCH_2CH(NH_2)COOH$	*S*-Carboxymethylcysteine (60), aspartic acid (68), threonine (72)	AAA, Aminex A-6, NaC, 30°C	108
6	*Se*-Benzylselenocysteine $PhCH_2SeCH_2CH(NH_2)COOH$	See item 4	AAA	107
7	*Se*-(*N*-Ethyl)succinimidoselenocysteine (0.66)[b]	Selenocystine (0.17), cystine (0.16), cysteic acid (0.20), *N*-ethylsuccinimidocysteine (0.66)	PC, EtOH/*t*-BuOH/HCO_2H/H_2O (12:4:1:3)	109
8	Selenocystine $[HOOCCH(NH_2)CH_2]_2Se_2$	See item 7	PC	109

9	Selenocystine	See item 1	DPC	104
10	Selenocystine	See item 3	AAA	106
11	Selenocystine (157)[c]	Cystine (134) and compounds in item 4	AAA	107
12	Selenocystine (410)[c]	Selenomethionine (455), cystine (368), methionine (420), isoleucine (423), leucine (442)	AAA, 8% divinylbenzene–styrene copolymer, NaC, 33°C	110
13	Selenocystine (20)[d]	Methionine (20), leucine (84), Selenomethionine (76) (time of buffer change: 0)	AAA, 8% divinylbenzene–styrene copolymer, NaC	111
14	Selenocystine	R_f values determined in various solvents	TLC, cellulose, PC	112
15	Selenocystine (45)[d]	Valine (36), methionine (62), isoleucine (65), leucine (82), selenomethionine (90)	AAA, Beckman PA-35, NaC, 30°C	113
16	Selenocystine (0.08)[b] oxidized with hydrogen peroxide	Selenomethionine (ox.) (0.22), methionine (0.55), cystine (0.04)	PC, $BuOH/AcOH/H_2O$ (4:1:1); other solvents used also	114
17	$[Me_3SiOOCCH(NH_2)CH_2]_2Se_2$ silylated selenocystine (15.8)[d]	$[Me_3SiOOCCH(NH_2)CH_2]_2S_2$ (9.8) $Me_3SiOOCCH(NH_2)CH_2$—SSe—$CH_2CH(NH_2)COOSiMe_3$ (15.8)	GC-FID, 170°C, 2% SE-30 on 90/100 Anakrom SD	115
18	$[HOOCCH(NH_2)CH_2]_2SeS$ (137)[c]	Selenocystine (143), cystine (130)	AAA, see item 4, without thiodiglycolic reagent	107
19	$HOOCCH(NH_2)CH_2SeSeCH_2CH_2COOH$ (146)[c]	Selenocystine (139)	AAA, see item 4, without thiodiglycolic reagent	107
20	*Se*-Carboxymethylselenocysteamine $HOOCCH_2$—Se—$CH_2CH_2NH_2$	Corresponding S compound	AAA, Aminex A-6, NaC, 50°C PC, $BuOH/AcOH/H_2O$ (4:1:5)	116

(*Contd.*)

TABLE 6. (*Contd.*)

Item no.	Compound	Separated from	Method	Ref.
21	Selenocystamine (163, 97*)[d] $(H_2NCH_2CH_2)_2Se_2$	Selenohomocystamine (58, 45*), selenolanthionamine (124, 93*), selenocystathionamine (82, 63*), selenohomolanthionamine (47, 44*) and corresponding S compounds (ret. time of S compounds marked*)	AAA, Aminex A-5, 2.35 M pot. citrate pH 5.6, then 0.2M NaOH; underlined ret. times relative to time of buffer change	117
22	Selenohomocystamine $(H_2NCH_2CH_2CH_2)_2Se_2$	See item 21	AAA, Aminex A-5	117
23	Selenalanthionamine $(H_2NCH_2CH_2)_2Se_2$	See item 21	AAA, Aminex A-5	117
24	Selenocystathionamine $H_2NCH_2CH_2—Se—CH_2CH_2CH_2NH_2$	See item 21	AAA, Aminex A-5	117
25	Selenohomolanthionamine $(H_2NCH_2CH_2CH_2)_2Se$	See item 21	AAA, Aminex A-5	117
26	Selenohomocystine $[HOOCCH(NH_2)CH_2CH_2]_2Se$	See item 3	AAA, Aminex A-4, NaC	106
27	Selenomethionine $MeSeCH_2CH_2CH(NH_2)COOH$	See item 3	AAA, Aminex A-4, NaC	106
28	Selenomethionine	See item 1; not separated from S compound	DPC	104
29	Selenomethionine	See item 4	AAA, Beckman P-28, NaC	107
30	Selenomethionine	See item 12	AAA	110
31	Selenomethionine	See item 13	AAA	111
32	Selenomethionine	See item 15	AAA, P-35 resin, NaC	113
33	Selenomethionine	Not separated from methonine	DPC, $BuOH/pyr./H_2O$ (1:1:1)	118

34	Selenomethionine (0.41)[b]	Methionine (0.32)	TLC, silica gel, *i*-PrOH/BuOH/H_2O (1:3:1)	119
35	Selenomethionine (0.22) oxidized	Methionine (0.55), see item 16	PC	114
36	$Me_3SiOOCCH(NH_2)CH_2CH_2SeCH_3$ (3.8)[d] silylated selenomethionine	Sil. methionine (2.8), sil. methionine sulphoxide (10.5), sil. methionine sulphone (14.3)	GC-FID, SE-30 on 80/100 Anakrom SD, 117°C	115
37	Selenoethionine $EtSeCH_2CH_2CH(NH_2)COOH$	See item 4	AAA, Beckman P-28, NaC	107
38	Selenocystathionine $HOOCCH(NH_2)CH_2SeCH_2CH_2CH(NH_2)COOH$	See item 3	AAA, Aminex A-4, NaC	106
39	Selenalysine (51) $H_2NCH_2CH_2SeCH_2CH(NH_2)COOH$	Lysine (43), thialysine (46), histidine (53)	AAA, Aminex A-5, 30°C, 0.35M Na citrate, pH 5.28	120
40	Selenalysine	Lysine, thialysine: only one spot, no separation; distinguishable with various spray reagents	PC, 10 mobile phases investigated	120
41	3-Selenaproline (34)[d]	4-Selenaproline (37), 3-thiaproline (30), 4-thiaproline (34), proline (42)	AAA, Durum DC-2A, 54°C, NaC, pH 3.25	121
42	3-Selenaproline (0.66)[b]	4-Selenaproline (0.36), 3-thiaproline (0.63), 4-thiaproline (0.59)	PC, $EtOH/NH_3(25\%)/H_2O$ (20:2:4)	121
43	4-Selenaproline	See items 41, 42	AAA, PC	121
44	4-Selenaproline (62)[d]	4-Thiaproline (53), proline (68)	AAA, Aminex A-6, 50°C, 0.2N Na citrate, pH 3.25	122
45	4-Selenaproline (0.20)[b]	4-Thiaproline (0.45)	PC, collidine/lutidine (1:1), water-saturated	122

(Contd.)

TABLE 6. (*Contd.*)

Item no.	Compound	Separated from	Method	Ref.
46	γ-L-Glutamyl-*Se*-methylseleno L-cysteine (163)[d]	Corresponding S-peptide (134), *Se*-methylselenocysteine (310)	AAA, Dowex 1-X4 (acetate form) pyr./acetate buffer pH 4 (0.1M acetate)	123
47	Panteth(e)ine	—SeH (0.61)[b], —SH (0.63), Se_2 (0.86), S_2 (0.72)	PC, BuOH/AcOH/H_2O (5:2:1)	77
48	4′-Phosphopanteth(e)ine	—SeH (0.72)[b], —SH (0.67), Se_2 (0.50), S_2 (0.50)	PC, see item 47	77
49	Coenzyme A	—SeH (0.34)[b], —SH (0.36), Se_2 (0.10), S_2 (0.10)	PC, EtOH/0.5M NH_4OAc (3:2), pH 4	77
50	Benzoylcoenzyme A	Se (0.59)[b], S (0.63)	PC, EtOH/0.5M NaOAc (3:2), pH 4	77
51	Dephosphocoenzyme A	—SeH (—), —SH (0.48), Se_2 (0.25), S_2 (0.23)	PC, see item 49	77
52	Selenoxytocin	Oxytocin; no separation	TLC, silica gel; Gel Filtration Sephadex G-25, 0.2 M AcOH	124

[a]AAA: Amino Acid Analyser; DPC: Descending Paper Chromatography; NaC: Na Citrate buffer, 0.2 M, pH 3.25, then 4.25; PC: Paper Chromatography; TLC: Thin-Layer Chromatography.
[b]R_f values.
[c]Elution volume (mL).
[d]Retention time (min).

M. Piazselenols

Piazselenols are five-membered heterocyclic compounds which form when aromatic *ortho*-diamines condense with selenites. The absorption and fluorescence properties of piazselenols were exploited for the determination of Se. Piazselenols in contrast to selenite are soluble in non-polar organic solvents and can be extracted, for instance with hexane, toluene, chloroform or cyclohexane, from aqueous solutions. The extracts can then be analysed by gas chromatography employing electron capture detectors[129–134], flame photometric detectors[89,135] or a microwave emission detector[136]. The preconcentration achieved by the extraction, the separation of the piazselenol from interfering compounds by gas chromatography and the sensitivity of the detectors make it possible to determine Se at concentrations of 0.1 μg/L. The absolute detection limits are in the low nanogram range. The sensitivity of the methods is influenced by the substituents present in the piazselenol. The relative sensitivities of piazselenols derived from diaminonaphthalene and substituted diaminobenzenes were investigated by Shimoishi[132].

Piazselenol was separated by gas chromatography from diethyl selenide, diphenyl selenide and 2,3-dihydro-3-azabenzoselenophene[89] (Section III. F; Table 2, item 9).

The piazselenol formed from selenite and diaminonaphthalene was extracted with cyclohexane. The extract was chromatographed on C-18 reverse-phase TLC plates (EtOH/H_2O/AcOH 65:35:1) and silica gel plates (ethyl acetate/toluene 1:4). Se at nanogram levels was quantitated by densitometry[137].

Piazselenol, 5-chloro, 5-nitro- and benz-piazselenol (from 2,3-diaminonaphthalene[138,139]) were separated from excess reagents by reverse-phase (Nucleosil C-18[138], Bondapak C-18[139]) high-pressure liquid chromatography with methanol/water, acetonitrile/water[138] or ethanol/water[139] as mobile phases. With ultraviolet or fluorometric detection Se was determined in the nano- to pico-gram range[138].

N. Selenium and Tellurium Diethyldithiocarbamates, $(Et_2NCSS)_2Y$

Inorganic Se/Te compounds react with diethyldithiocarbamates to form compounds extractable with chloroform. Dithiocarbamates are formed by many cations. High-pressure liquid chromatography using reverse-phase columns allowed the separation of the Se[140,141] and Te[141] compounds from other dithiocarbamates. With ultraviolet spectroscopic detectors one microgram of Se/Te was detected[141].

Tellurium diethyldithiocarbamate was separated on silica gel thin layers, the spot extracted with methanol and Te determined in the extract by graphite furnace atomic absorption spectrometry[142].

O. Trialkylsilyl Selenides/Tellurides and Related Compounds

The compounds $(Et_3M)_2Y$ (Y = Te, Se, S) were separated by gas chromatography with a detector based on heat conductivity. The column material was 20% Apiezon L on silanized Chromosorb W. The retention time increased with increasing boiling points of the compounds[143]. Bis(trimethylsilyl) selenone $[(Me_3Si)_2SeO_2]$ formed from selenite and *N,O*-bis(trimethylsilyl)acetamide, is a volatile, stable compound, which was used to determine Se by gas chromatography with flame ionization detection[144].

IV. POLAROGRAPHIC AND OTHER ELECTROCHEMICAL METHODS

Polarography has been applied extensively for the determination of inorganic Se[18–25]. Although several organic Se compounds have been the subject of polarographic studies, few methods were developed which are suitable for quantitative determinations. Most of the investigations explored the mechanisms of reduction of the Se compounds and compared their reduction behaviour with that of the corresponding S compounds.

Benzeneselenols gave under various conditions a single reduction wave. The half-wave potential was dependent on the pH of the medium and the substituents in the benzene ring[145]. Nygard investigated the polarographic reduction of bis(carboxyalkyl) diselenides, $[HOOC(CH_2)_n]_2Se_2$ ($n = 1$[146,147], 2–4[148,149]), diphenyl diselenide[150], 1,2-diselenacyclopentane[151] and its 4,4-bis(hydroxymethyl)[152,154], 4-carboxy[153] and 3-(4′-carboxybutyl)[152] derivatives, 4-carboxy-1-thia-2-selenacyclopentane[147,153], 3,6-dicarboxy-1,2-diselenacyclohexane[152], 4,5-dicarboxy-1,2-diselenacyclohexane[152] and 2,2-bis(hydroxymethyl)-1,3-propanediseleninic acid[147,154]. Information about the electrode processes was gained from direct-current polarographic curves, electrocapillary curves and oscillopolarographic investigations. The diselenides, R_2Se_2, form initially the mercury compounds $(RSe)_2Hg$ or RSeHg which are subsequently reduced to selenols, RSeH, by transfer of one electron per Se atom[147]. The diseleninic acid is first irreversibly reduced to the selenium–mercury compound which in a subsequent reversible step is transformed to the diselenol[147,154] The polarographic reduction of 2-aminoethylselenosulphuric acid is preceded by the formation of RSeHg and sulphite and finally yields 2-aminoethaneselenol[155].

Selenocystine and selenocysteine were investigated by several polarographic methods and the properties of the Se compounds compared with those of the corresponding S derivatives[148,156,157]. Selenocystine can be determined by cathodic stripping voltammetry in dilute aqueous acid. The detection limit for selenocystine is 5×10^{-10} M in the presence of 100-fold amounts of cystine and cysteine[158]. Selenocystine causes catalytic prewaves in the polarographic reduction of cobalt[159] and nickel[160]. The prewaves are probably produced by complexes formed between the metal ions and selenocysteine[159,160]. Selenocystamine [bis(2-aminoethyl) diselenide] gave polarographic waves more positive than cystamine. The diselenide can be distinguished from the disulphide in the same solution[161].

An oscillopolarographic investigation of selenourea proved that the reduction of selenourea yielded selenide ion which reacted with mercury. Selenourea is reduced at a potential 140 mV more negative than thiourea. The detection limit for selenourea is 5×10^{-6} M[162,163].

Substituted 2-nitroselenophenes can be determined polarographically at concentrations of 10^{-5} M[164]. The general features of the polarographic reduction of substituted 2-nitroselenophenes are similar to those of thiophenes and furans[165]. AC polarographic studies of heteroacene quinones in acetonitrile containing the 2,5-diphenylselenophene ring showed that the reduction proceeded in two one-electron steps[166].

Polarographic investigations of 2,1,3-selenadiazole[167], benzoselenadiazole[168–174], benzoselenadiazoles substituted in the 4- or 5-position[174], pyridinoselenadiazole[175] and dioxopyrimidinoselenadiazole[175] in aqueous medium and in dimethylformamide established that a radical anion is first formed which is subsequently reduced to the *ortho*-diamine and selenide.

The selenadiazoles obtained from 3,3′,4,4′-tetraaminobiphenyl and selenite were used for the polarographic determination of Se[176–178]. A sensitivity of one microgram Se per litre was achieved with single sweep polarography[178]. With 4-chlorobenzoselenadiazole and differential pulse polarography the detection limit was 0.4 microgram Se per litre[179].

The polarographic reduction of bis(4-methoxyphenyl) ditelluride produced 4-methoxybenzenetellurol[180]. Bis(4-methoxyphenyl)tellurium oxide yielded under conditions of classical and oscillographic polarography bis(4-methoxyphenyl) telluride in an irreversible reaction[181].

V. REFERENCES

1. K. J. Irgolic, *The Organic Chemistry of Tellurium*, Gordon and Breach, New York, 1974.
2. D. L. Klayman and W. H. H. Günther (Eds.), *Organic Selenium Compounds: Their Chemistry and Biology*, John Wiley and Sons, New York, 1973.
3. H. Rheinboldt, in *Houben–Weyl: Methoden der Organischen Chemie*, Vol. IX (Ed. E. Müller), Georg Thieme Verlag, Stuttgart, 1955, pp. 917–1209.
4. R. A. Zingaro and K. J. Irgolic, in *Tellurium* (Ed. W. C. Cooper), Van Nostrand Reinhold, New York, 1971, pp. 184–280.
5. K. J. Irgolic and M. Kudchadker, in *Selenium* (Eds. R. A. Zingaro and W. C. Cooper), Van Nostrand Reinhold, New York, 1974, pp. 408–545.
6. K. J. Irgolic, *J. Organomet. Chem.*, **103**, 91 (1975); **130**, 411 (1977); **158**, 235, 267 (1978); **189**, 65 (1980); **203**, 367 (1980).
7. *Selenium*, National Academy of Sciences, Washington, D.C., 1976.
8. R. J. Shamberger, *Biochemistry of Selenium*, Plenum Press, New York, 1983.
9. O. H. Muth, J. E. Oldfield and P. H. Weswig (Eds.), *Selenium in Biomedicine*, Air Publishing Co., Westport, Connecticut, 1967.
10. 'Selenium–Tellurium in the Environment', *Proceedings of Symposium, University of Notre Dame, Notre Dame, Indiana, May 11–13, 1976*, Industrial Health Foundation, Inc., Pittsburgh, 1976.
11. J. E. Spallholz, J. L. Martin and H. E. Ganther (Eds.), *Selenium in Biology and Medicine*, Air Publishing Co., Westport, Connecticut, 1981.
12. T. R. Crompton, *Gas Chromatography of Organometallic Compounds*, Plenum Press, New York, 1982, pp. 519–525.
13. K. J. Irgolic, R.A. Stockton, D. Chakraborti and W. Beyer, *Spectrochim. Acta*, **38B**, 437 (1983).
14. D. Chakraborti, D. C. Hillman, K. J. Irgolic and R. A. Zingaro, *J. Chromatogr.*, **249**, 81 (1982).
15. K. J. Jewett and F. E. Brinckman, in *Detectors in Liquid Chromatography* (Ed. T. M. Vickrey), Marcel Dekker, New York, 1983, pp. 205–241.
16. F. Feigl, *Anal. Chim. Acta*, **24**, 501 (1961).
17. A. Del'Acqua and D. F. Gastmans, *Ecletica Quim.*, **1**, 43 (1976).
18. T. E. Green and M. Turley, in *Treatise on Analytical Chemistry*, Part II, Vol. 7 (Eds. I. M. Kolthoff and P. J. Elwing), Interscience, New York, 1961, p. 170.
19. I. I. Nazarenko and A. N. Ermakov, *Analytical Chemistry of Selenium and Tellurium*, Halsted Press, New York, 1972.
20. J. H. Watkinson, in *Selenium in Biomedicine* (Eds. O. H. Muth, J. E. Oldfield and P. H. Weswig), Air Publishing Co., Westport, Connecticut, 1967, pp. 97–117.
21. O. E. Olson in Ref. 10, pp. 67–84.
22. W. C. Cooper, in *Selenium* (Eds. R. A. Zingaro and W. C. Cooper), Van Nostrand Reinhold Co., New York, 1974, pp. 615–653.
23. J. F. Alicino, and J. A. Kowald, in Ref. 2, pp. 1049–1081.
24. Ref. 8, pp. 311–327.
25. O. E. Olson, I. S. Palmer and E. I. Whitehead, in *Methods of Biochemical Analysis*, Vol. 21 (Ed. D. Glick), John Wiley and Sons, New York, 1973, pp. 39–78.
26. T. R. Crompton, *Chemical Analysis of Organometallic Compounds*, Vol. 5, Academic Press, New York, 1977, pp. 6–20.
27. M. R. Masson, *Mikrochim. Acta*, 419 (1976 I).
28. E. W. Bem, *Environ. Health Perspect.*, **37**, 183 (1981).
29. Ref. 1, pp. 366–368.
30. W. C. Cooper (Ed.), in *Tellurium*, Van Nostrand Reinhold, New York, 1971, pp. 281–312.
31. K. L. Cheng and R. A. Johnson, *Chem. Anal.* (*N.Y.*), **8**, 371 (1978).
32. Ref. 7, pp. 136–138.
33. S. E. Raptis, W. Wegscheider, G. Knapp and G. Toelg, *Anal. Chem.*, **52**, 1292 (1980).
34. M. Berti, G. Buso, P. Colantti, G. Moschini, B. M. Stievano and C. Tregnoghi, *Anal. Chem.*, **49**, 1313 (1977).
35. F. Bodart and G. Deconnick, *Int. J. Nondestruct. Test.*, **4**, 171 (1972).
36. S. J. Anderson, D. S. Brown and A. H. Norbury, *J. Organomet. Chem.*, **64**, 301 (1974).
37. E. C. Olson and J. W. Shell, *Anal. Chim. Acta*, **23**, 219 (1960).

38. J. Neve, M. Hanocq and L. Molle, *Mikrochim. Acta*, 259 (1980 I).
39. A. A. Tsurkan, T. S. Tsurkan, I. B. Levshin, G. F. Anisimova, E. B. Kurbatova, and S. D. Kopylov, *Khim. Prom-st., Ser.: Metody Anal. Kontrolya Kach. Prod. Khim. Prom-sti.*, 25 (1980); *Chem. Abstr.*, **94**, 131698 (1981).
40. M. A. Volodina and S. Y. Medvedev, *Dokl. Akad. Nauk SSSR*, **240**, 605 (1978).
41. A. Campiglio, *Mikrochim. Acta*, 245 (1979).
42. M. R. Masson, *Mikrochim. Acta*, 399 (1976 I).
43. V. Nuti and D. Bertini, *Farmaco. Ed. Prat.*, **35**, 544 (1980).
44. H. Uchida, Y. Shimoishi and K. Toei, *Analyst*, **106**, 757 (1981).
45. G. F. Anisimova and V. A. Klimova, *Zh. Anal. Khim.*, **37**, 1100 (1982).
46. C. L. Chakraborti, *Anal. Chim. Acta*, **42**, 379 (1968).
47. M. Verlinden, H. Deelstra and E. Adriaenssens, *Talanta*, **28**, 637 (1981).
48. J. Alexander, K. Saeed and Y. Thomassen, *Anal. Chim. Acta*, **120**, 377 (1980).
49. T. Kamada and Y. Yamamoto, *Talanta*, **27**, 473 (1980).
50. K. S. Subramanian and J. C. Meranger, *Anal. Chim. Acta*, **124**, 131 (1981).
51. T. Kamada, T. Shiraishi and Y. Yamamoto, *Talanta*, **25**, 15 (1978).
52. J. C. Chambers and B. E. McClellan, *Anal. Chem.*, **48**, 2061 (1976).
53. N. Iordanov, K. Daskalova, I. Khaverzov and V. Toncheva, *Usp. Anal. Khim.*, 54 (1974).
54. E. Koyama, S. Katsura, M. Fukumura and K. Nagashima, *Bunseki Kagaku*, **29**, 211 (1980).
55. E. R. Clark and M. A. Al-Turaihi, *J. Organomet. Chem.*, **118**, 55 (1976).
56. G. F. Anisimova and V. A. Klimova, *Zh. Anal. Khim.*, **35**, 607 (1980).
57. G. F. Anisimova and V. A. Klimova, *Zh. Anal. Khim.*, **36**, 1137 (1981).
58. F. H. Kruse, R. W. Saftner and J. F. Suttle, *Anal. Chem.*, **25**, 500 (1953).
59. T. S. Ma and W. G. Zoellner, *Mikrochim. Acta*, 329 (1971).
60. P. Thavornyutikarn, *J. Organomet. Chem.*, **51**, 237 (1973).
61. C. L. Chakraborti, *Anal. Chim. Acta*, **39**, 293 (1967).
62. G. Weibust, F. J. Langmyhr and Y. Thomassen, *Anal. Chim. Acta*, **128**, 23 (1981).
63. R. V. D. Robert and G. Balaes, *Report No. 2053*, National Institute for Metallurgy, Randburg, 2125 South Africa, March 31, 1980; *Chem. Abstr.*, **93**, 160586 (1980).
64. I. Havezov and N. Jordanov, *Talanta*, **21**, 1013 (1974).
65. Y. K. Chan, P. T. S. Wong and P. D. Goulden, *Anal. Chem.*, **47**, 2279 (1975).
66. Y. K. Chan and P. T. S. Wong, *NBS Special Publ.*, **464**, 485 (1977).
67. G. A. Cutter, *Anal. Chim. Acta*, **98**, 59 (1978).
68. B. Radziuk and J. van Loon, *Sci. Total Environ.*, **6**, 251 (1976).
69. J. C. van Loon, B. Radziuk, N. Kahn, J. Lichwa, F. J. Fernandez and J. D. Kerber, *At. Absorpt. Newsletter*, **16**, 79 (1977).
70. G. E. Parris, W. R. Blair and F. E. Brinckman, *Anal. Chem.*, **49**, 378 (1977).
71. S. Jiang, W. de Jonghe and F. Adams, *Anal. Chim. Acta*, **136**, 183 (1982).
72. D. C. Reamer, T. C. O'Haver and W. H. Zoller, *NBS Special Publ.*, **464**, 609 (1977).
73. S. A. Estes, P. C. Uden and R. M. Barnes, *Anal. Chem.*, **53**, 1829 (1981).
74. T. M. Vickrey, M. S. Buren and H. E. Howell, *Anal. Lett., Part A*, **11**, 1075 (1978).
75. E. Orvini and M. Gallorini, *J. Radioanal. Chem.*, **71**, 75 (1982).
76. J. Benes and V. Prochazkova, *J. Chromatogr.*, **29**, 239 (1967).
77. W. H. H. Günther and H. G. Mautner, *J. Amer. Chem. Soc.*, **87**, 2708 (1965).
78. J. D. McCullough, T. W. Campbell and N. J. Krilanovich, *Ind. Eng. Chem., Anal. Ed.*, **18**, 638 (1946).
79. O. Foss and S. R. Svendsen, *Acta Chem. Scand.*, **8**, 1351 (1954).
80. C. S. Evans and C. M. Johnson, *J. Chromatogr.*, **21**, 202 (1966).
81. V. D. Nefedov, E. N. Sinotova, S. A. Timofeev and M. I. Kzakov, *Zh. Obshch. Khim.*, **40**, 1103 (1970).
82. D. De Filippo and C. Preti, *Ric. Sci.*, **39**, 231 (1969).
83. O. H. Wheeler, C. L. Gonzalez and H. Lopez-Alonso, *Int. J. Appl. Radiat. Isotop.*, **21**, 244 (1970).
84. C. DeMarco, P. Cossu, S. Dernini and A. Rinaldi, *J. Chromatogr.*, **129**, 369 (1976).
85. V. Anger and G. Fischer, *Mikrochim. Acta*, 501 (1962).
86. V. Vlasakova, J. Benes and J. Parizek, *Radiochem. Radioanal. Lett.*, **10**, 251 (1972).
87. M. Vobetsky, V. D. Nefedov and E. N. Sinotova, *Zh. Obshch. Khim.*, **35**, 1684 (1965).
88. N. E. Schepina, S. G. Sukhov, V. D. Nefedov and M. A. Toropova, *Sint. Osn. Magnii Tsinkorg. Soedin.*, 113 (1980); *Chem. Abstr.*, **95**, 143586 (1981).

89. C. G. Flinn and W. A. Aue, *J. Chromatogr.*, **153**, 49 (1978).
90. V G. Makarenko, V. A. Fedorov, E. A. Efremov, V. Y. Orlov and E. P. Filippov, *Zh. Anal. Khim.*, **30**, 1183 (1975).
91. E. A. Efremov, V. G. Makarenko, E. P. Filippov and V. A. Fedorov, *Zh. Anal. Khim.*, **32**, 1041 (1977).
92. J. R. Hancock, C. G. Flinn and W. A. Aue, *Anal. Chim. Acta*, **116**, 195 (1980).
93. W. A. Aue and C. G. Flinn, *J. Chromatogr.*, **158**, 161 (1978).
94. E. N. Sinotova and S. A. Timofeev, *Radiokhimiya*, **10**, 756 (1968).
95. V. D. Nefedov, V. E. Zhuravlev and M. A. Toropova, *Zh. Obshch. Khim.*, **34**, 3719 (1964).
96. M. Vobetsky, V. D. Nefedov and E. N. Sinotova, *Zh. Obshch. Khim.*, **33**, 4023 (1963).
97. V. D. Nefedov, S. A. Grachev and Z. A. Grant, *Zh. Obshch. Khim.*, **32**, 1179 (1962).
98. V. D. Nefedov, V. E. Zhuravlev, M. A. Toropova, S. A. Grachev and A. V. Levchenko, *Zh. Obshch. Khim.*, **35**, 1436 (1965).
99. V. D. Nefedov, V. E. Zhuravlev, M. A. Toropova, L. N. Gracheva and A. V. Levchenko, *Radiokhimiya*, **7**, 245 (1965).
100. V. D. Nefedov, M. A. Toropova, V. E. Zhuravlev and A. V. Levchenko, *Radiokhimiya*, **7**, 203 (1965).
101. J. D. McCullough, *J. Amer. Chem. Soc.*, **64**, 2672 (1942).
102. N. Jordanov, K. Daskalova and N. Rizov, *Talanta*, **21**, 1217 (1974).
103. V. S. Epifanov, L. V. Konyakhina and V. P. Syskova, *Zh. Anal. Khim.*, **35**, 2250 (1980).
104. R. J. Peterson and G. W. Butler, *J. Chromatogr.*, **8**, 70 (1962).
105. A. A. Di Somma and A. L. Jacobs, *Science*, **132**, 618 (1960).
106. J. L. Martin and M. L. Gerlach, *Anal. Biochem.*, **29**, 257 (1969).
107. R. Walter, D. H. Schlesinger and I. L. Schwartz, *Anal. Biochem.*, **27**, 231 (1969).
108. A. Rinaldi, P. Cossu and C. De Marco, *J. Chromatogr.*, **120**, 221 (1976).
109. J. P. Portanova and A. Shrift, *J. Chromatogr.*, **139**, 391 (1977).
110. J. L. Martin and L. M. Cummins, *Anal. Biochem.*, **15**, 530 (1966).
111. K. P. McConnel and C. H. Wabnitz, *Biochim. Biophys. Acta*, **86**, 182 (1964).
112. R. P. Spencer, K. R. Brady, W. H. H. Günther and H. G. Mautner, *J. Chromatogr.*, **21**, 342 (1966).
113. J. V. Benson, Jr. and J. A. Patterson, *Anal. Biochem.*, **29**, 130 (1969).
114. J. Scala and H. H. Williams, *J. Chromatogr.*, **15**, 546 (1964).
115. K. A. Caldwell and A. I. Tappel, *J. Chromatogr.*, **32**, 635 (1968).
116. A. Rinaldi, S. Dernini, M. R. Dessy and C. De Marco, *Anal. Biochem.*, **69**, 289 (1975).
117. C. De Marco, P. Cossu, A. Rinaldi and S. Dernini, *J. Chromatogr.*, **115**, 621 (1975).
118. A. J. Barak and S. C. Swanberg, *J. Chromatogr.*, **31**, 282 (1967).
119. K. R. Millar, *J. Chromatogr.*, **14**, 344 (1966).
120. C. De Marco, A. Rinaldi, S. Dernini, P. Cossu and D. Cavallini, *J. Chromatogr.*, **114**, 291 (1975).
121. C. De Marco, C. Cini, R. Coccia and C. Blarzino, *Ital. J. Biochem.*, **28**, 104 (1979).
122. C. De Marco, R. Coccia, A. Rinaldi and D. Cavallini, *Ital. J. Biochem.*, **26**, 51 (1977).
123. S. N. Nigam and W. B. McConnell, *Biophys. Acta*, **192**, 185 (1969).
124. R. Walter and V. Du Vigneaud, *J. Amer. Chem. Soc.*, **87**, 4192 (1965).
125. Z. Stefanac, M. Tomaskovic and I. Bregovec, *Microchem. J.*, **16**, 226 (1971).
126. Z. Stefanac and M. Tomaskovic, *Bull. Sci. Conseil Acad. RSF Yougoslavie*, **10**, 317 (1965); *Chem. Abstr.*, **65**, 4640 (1966).
127. Z. E. Mel'chekova and G. A. Kitaev, *Izv. Vyssh. Ucheb. Zaved., Khim. Khim. Tekhnol.*, **17**, 805 (1974); *Anal. Abstr.*, **30**, 2B–115 (1976).
128. S. Clementi, G. Savelli and M. Vergoni, *Chromatographia*, **5**, 413 (1972).
129. B. M. Bycroft and D. E. Clegg, *J. Assoc. Off. Anal. Chem.*, **61**, 923 (1978).
130. S. Nakashima and K. Toei, *Talanta*, **15**, 1475 (1968).
131. Y. Shimoishi and K. Toei, *Talanta*, **17**, 165 (1970).
132. Y. Shimoishi, *J. Chromatogr.*, **136**, 85 (1977).
133. Y. Shimoishi and K. Toei, *Anal. Chim. Acta*, **100**, 65 (1978).
134. T. Stijve and E. Cardinale, *J. Chromatogr.*, **109**, 239 (1975).
135. W. Aue and C. R. Hastings, *J. Chromatogr.*, **87**, 232 (1973).
136. Y. Talmi and A. W. Andren, *Anal. Chem.*, **46**, 2122 (1974).
137. H. M. Stahr, J. Kinker, D. Nicholson and W. Hyde, *J. Liq. Chromatogr.*, **5**, 1191 (1982).
138. G. Schwedt, *Fresenius Z. Anal. Chem.*, **288**, 50 (1977).

139. G. L. Wheeler and P. F. Lott, *Microchem. J.*, **19**, 390 (1974).
140. G. Schwedt, *Chromatographia*, **11**, 145 (1978).
141. S. R. Hutchins, P. R. Haddad and S. Dilli, *J. Chromatogr.*, **252**, 185 (1982).
142. H. J. Issaq and E. W. Barr, *Anal. Chem.*, **49**, 189 (1977).
143. G. N. Bortnikov, M. N. Bochkarev, N. S. Vyazankin, S. K. Ratushnaya, and Y. I. Yashin, *Izv. Akad. Nauk SSSR, Ser. Khim.*, **20**, 851 (1971).
144. C. A. Burgett, *Anal. Lett.*, **7**, 799 (1974).
145. M. A. Shakhgel'diev, B. R. Gasanov and F. A. Kuliev, *Nov. Polyarogr., Tezisy Dokl. Vses. Soveshch. Polyarogr., 6th*, 130 (1975).
146. B. Nygard, *Acta Chem. Scand.*, **15**, 1039 (1961).
147. R. Danielson, B. L. Johansson, B. Nygard and B. Persson, *Chem. Scr.*, **20**, 19 (1982).
148. B. Nygard, *Chem. Zvesti*, **16**, 320 (1962).
149. B. Nygard, *Arkiv Kemi*, **27**, 325 (1967).
150. B. Nygard, *Acta Chem. Scand.*, **20**, 1710 (1966).
151. B. Nygard, E. Johansson and J. Olofsson, *J. Electroanal. Chem.*, **12**, 564 (1966).
152. B. Nygard, *Arkiv Kemi*, **27**, 405 (1967).
153. B. Nygard, J. Olofsson and G. Bergson, *Arkiv Kemi*, **28**, 41 (1968).
154. B. Nygard, *Arkiv Kemi*, **27**, 425 (1967).
155. W. Stricks and R. G. Mueller, *Anal. Chem.*, **36**, 40 (1964).
156. B. Nygard, *Arkiv Kemi*, **27**, 341 (1967).
157. R. E. Huber and R. S. Criddle, *Arch. Biochem. Biophys.*, **122**, 164 (1967).
158. R. A. Grier and R. W. Andrews, *Anal. Chim. Acta*, **124**, 333 (1981).
159. A. Calusaru and V. Voicu, *Electroanal. Chem.*, **20**, 463 (1969).
160. A. Calusaru and V. Voicu, *Electroanal. Chem.*, **32**, 427 (1971).
161. P. van Caneghem, *Bull. Cl. Sci., Acad. Roy. Belg.*, **52**, 770 (1966).
162. L. I. Klimova and V. I. Gorokhovskaya, *Nov. Polyarogr., Tezisy Dokl. Vses. Soveshch. Polyarogr., 6th*, 114 (1975).
163. L. I. Klimova and V. I. Gorokhovskaya, *Zh. Obshch. Khim.*, **47**, 650 (1977).
164. J. Stradins, S. Hillers and Y. K. Yu'rev, *Teoriya i Proktika Polyarogr. Analiza, Akad. Nauk Moldavsk, SSSR, Materialy Pervogo Vses. Sovesch.*, 420 (1962); *Chem. Abstr.*, **59**, 7463 (1963).
165. Y. P. Stradyn', R. A. Gavar, L. K. Baumane, L. M. Baider, B. A. Medynya and M. A. Trushule, *Khim. Geterotsikl. Soedin.*, **16**, 1606 (1980).
166. E. Mueller and W. Dilger, *Chem. Ber.*, **106**, 1643 (1973).
167. V. S. Tsveniashvili, V. N. Gaprindashvili, L. A. Tskalobadze and V. A. Sergeev, *Zh. Obshch. Khim.*, **43**, 2122 (1973).
168. S. I. Zhdanov and V. S. Tsveniashvili, *Elektrokimiya*, **4**, 1383 (1968).
169. V. S. Tsveniashvili, V. N. Gaprindashvili and N. S. Khavtasi, *Zh. Obshch. Khim.*, **42**, 2049 (1972).
170. V. S. Tsveniashvili, S. I. Zhdanov and Z. V. Todres, *Khim. Geterotsikl. Soedin.*, **4**, 712 (1968).
171. V. S. Tsveniashvili, S. I. Zhdanov and Z. V. Todres, *Z. Anal. Chem.*, **224**, 389 (1967).
172. V. S. Tsveniashvili, Z. V. Todres and S. I. Zhdanov, *Zh. Obshch. Khim.*, **38**, 1888 (1968).
173. V. S. Tsveniashvili, Z. V. Todres and S. I. Zhdanov, *Zh. Obshch. Khim.*, **38**, 1894 (1968).
174. Z. V. Todres, S. I. Zhdanov and V. S. Tsveniashvili, *Izv. Akad. Nauk SSSR, Ser. Khim.*, 975 (1967).
175. V. S. Tsveniashvili, L. A. Tskalobadze and V. N. Gaprindashvili, *Soobshch. Akad. Nauk Gruz. SSR*, **71**, 625 (1973).
176. C. LePeintre, *Compt. Rend.*, **252**, 1968 (1961).
177. G. D. Christian, E. C. Knoblock and W. C. Purdy, *J. Assoc. Off. Agr. Chem.*, **48**, 877 (1965).
178. D. A. Griffin, *Anal. Chem.*, **41**, 462 (1969).
179. A. G. Howard, M. R. Gray and A. J. Waters, *Anal. Chim. Acta*, **118**, 87 (1980).
180. B. N. Bukharin, E. G. Ignatenko and K. N. Bagdasarov, *Izv. Vyssh. Uchebn. Zaved., Khim. Khim. Tekhnol.*, **23**, 241 (1980); *Chem. Abstr.*, **93**, 56690 (1980).
181. B. N. Bukharin, E. G. Ignatenko, K. N. Bagdasarov and I. D. Sadekov, *Izv. Vyssh. Uchebn. Zaved., Khim. Khim. Tekhnol.*, **22**, 1414 (1979); *Chem. Abstr.*, **93**, 83545 (1980).

The Chemistry of Organic
Selenium and Tellurium
Compounds Volume 1
Edited by S. Patai and Z. Rappoport
© 1986 John Wiley & Sons Ltd.

CHAPTER **6**

Nuclear magnetic resonance and electron spin resonance studies of organic selenium and tellurium compounds

NARENDER P. LUTHRA and JEROME D. ODOM

Department of Chemistry, University of South Carolina, Columbia, South Carolina 29208, USA

I. INTRODUCTION . . . 190
II. NMR STUDIES . . . 191
A. Organoselenium Compounds . . . 191
1. Relaxation times . . . 191
2. Chemical shifts . . . 193
a. Medium effects . . . 194
i. Solvent effects . . . 194
ii. Concentration effects . . . 195
iii. pH effects . . . 195
iv. Temperature effects . . . 195
b. ^{77}Se chemical shift referencing . . . 196
c. Classes of compounds . . . 196
i. Selenols and selenolates . . . 196
ii. Dialkyl selenides . . . 198
iii. Alkyl aryl selenides . . . 202
iv. Diaryl selenides. . . . 204
v. Alkyl diselenides . . . 205
vi. Diaryl diselenides . . . 206
vii. Selenenyl sulphides . . . 206
viii. Seleno esters . . . 207
ix. Compounds containing C=Se double bonds . . . 208
x. Selenocyanates . . . 209
xi. Selenenyl halides . . . 211
xii. Heterocyclic selenium compounds . . . 212
xiii. Selenium–oxygen compounds . . . 212

xiv. Miscellaneous selenium compounds 217
d. Isotope effects 217
3. Coupling constants 220
B. Organotellurium Compounds 221
1. Relaxation times 221
2. Chemical shifts. 221
a. Tellurides 222
b. Ditellurides 224
c. Tellurium halides 228
d. Heterocyclic tellurium compounds 232
e. Miscellaneous tellurium compounds 232
3. Coupling constants 233
III. ESR STUDIES 234
A. Organoselenium Compounds 234
B. Organotellurium Compounds 238
IV. ACKNOWLEDGEMENTS 238
V. REFERENCES 238

I. INTRODUCTION

Every element in Group VI in the Periodic Table possesses at least one magnetically active isotope[1]. (Table 1 lists the NMR properties of these nuclei and compares them to those of the ^{1}H and ^{13}C nuclei.) However, the two lighter elements, O and S, both possess only quadrupolar nuclei (Table 1) and the dominant isotopes of the elements, ^{16}O and ^{32}S, have zero spin. Although the number of studies using ^{17}O-NMR spectroscopy has increased considerably with the advent of pulsed Fourier transform (FT) techniques, ^{33}S-NMR spectroscopy continues to be difficult experimentally, due to the quadrupolar spin (I = 3/2) and the low natural abundance (0.76%). On the other hand, Se and Te both possess spin-$\frac{1}{2}$ nuclei with adequate natural abundance (^{77}Se and ^{125}Te) such that NMR experiments using pulsed FT methods to observe these nuclei are relatively routine. However, only in the last five years have researchers begun to extensively investigate the NMR parameters of organoselenium and organotellurium compounds using pulsed FT ^{77}Se- and ^{125}Te-NMR spectroscopy.

Because the material has not been reviewed previously, this article will concentrate heavily on NMR studies utilizing the ^{77}Se and ^{125}Te nuclei. ^{1}H-NMR studies have been reviewed previously[2] and, because of the small chemical shift range, they have not been found to be very informative. ^{13}C-NMR investigations of organoselenium and organotel-

TABLE 1. NMR properties of the Group VI nuclei[a]

Isotope	Spin	Natural abundance (%)	NMR frequency (MHz)	Relative receptivity[b] R^{P}	R^{C}
^{1}H	1/2	99.985	100.1	1.000	5.68×10^{3}
^{13}C	1/2	1.108	25.1	1.76×10^{-4}	1.00
^{17}O	5/2	0.037	13.6	1.08×10^{-5}	6.11×10^{-2}
^{33}S	3/2	0.76	7.7	1.71×10^{-5}	9.73×10^{-2}
^{77}Se	1/2	7.58	19.1	5.26×10^{-4}	2.98
^{123}Te	1/2	0.87	26.2	1.57×10^{-4}	0.88
^{125}Te	1/2	6.99	31.5	2.21×10^{-3}	12.5

[a]Data taken from Ref. 1.
[b]Receptivity relative to ^{1}H is given as R^{P} and relative to ^{13}C is given as R^{C}.

lurium compounds are fairly numerous and will be dealt with where they complement the ^{77}Se and ^{125}Te studies or are in some other way especially noteworthy. A discussion and comprehensive listing of ^{13}C chemical shifts and other carbon parameters in organoselenium and organotellurium compounds is beyond the scope of this article.

In searching the literature, every attempt has been made to review publications through June 1983. Not every paper may be referenced, but certainly the large majority are discussed or referenced in some way. This article is divided into two parts. The first and largest part of the paper deals with NMR studies and the second part deals with ESR studies. As mentioned previously, material reviewing ^{77}Se- and ^{125}Te-NMR studies has not appeared since the large explosion of pulsed FT data has begun. To demonstrate how recent most of the data are, it is interesting to note that a 1978 book[1] covering NMR studies of all nuclei in a very comprehensive manner devoted only ten pages to ^{77}Se-NMR spectroscopy and only five papers were referenced[3-7] which had utilized pulsed FT methods. Only two pages described ^{125}Te-NMR spectroscopy and there were no publications describing direct observation of ^{125}Te resonances. Accordingly, it was felt that efficient use could be made of tables of ^{77}Se and ^{125}Te chemical shifts classified according to type of compound. The chemical shift ranges of both of these nuclei are very large ($\sim$ 3000 ppm for ^{77}Se and $\sim$ 7000 ppm for ^{125}Te) and it is hoped that, in addition to the discussion in this chapter, the reader will find it instructive to refer to the chemical shift values given for various classes of organoselenium and organotellurium compounds. Because the chemical shift range is so large for these nuclei and because these shifts are sensitive to various factors, in some cases more than one value has been reported for the same molecule. Every attempt has been made to report all values in the tables and to comment on any discrepancies. Also, where possible, a comparison of Se and Te values with corresponding O and/or S parameters has been made.

II. NMR STUDIES

A. Organoselenium Compounds

1. Relaxation times

The problems of the relatively low natural abundance and relatively low NMR receptivity of ^{77}Se with respect to the proton have been minimized with the advent of FT NMR spectrometers and techniques, through which substantial gains in the signal-to-noise ratio over conventional continuous wave NMR spectroscopy may be realized. To take full advantage of the FT method, a knowledge of the inherent spin-lattice relaxation time, T_1, of the Se nucleus is desirable since it influences the time duration of the experiments via the recycle time between pulses[8,9]. The T_1 values can also provide valuable information concerning molecular dynamics and interactions, molecular structure, conformation and composition. It is always advantageous, when studying any particular nucleus, to be forearmed with a knowledge of the range of T_1 values of the nucleus in a number of functional forms and under a variety of conditions (e.g. temperature, solvent, concentration). Furthermore, it is also very important to determine which of the various mechanisms contribute to the overall spin-lattice relaxation of the nucleus[8,9].

In 1977–79 several studies[10-14] appeared which reported the first investigations of ^{77}Se spin-lattice relaxation times. Values for T_1 of representative compounds are shown in Table 2. Several important conclusions emerged from these and later studies[15,16]. First, the dipole–dipole (DD) relaxation mechanism which is so important in ^{13}C-NMR spectroscopy[17,18] is almost non-existent in Se compounds. For ^{77}Se, a maximum Nuclear Overhauser Enhancement (NOE)[19] of 2.6 is possible, but for only one compound has any

TABLE 2. ^{77}Se spin-lattice relaxation times, T_1, for representative compounds

Compound	T_1(s)	Temp. (°C)	Conditions	Ref.
H_2Se	0.7	34	1.0M, D_2O	11
	0.34	35	40% (v/v), $CDCl_3$	11
MeSeH	1.3	40	20% (v/v), $(CD_3)_2CO$	11
EtSeH	1.7	40	20% (v/v), $CDCl_3$	11
$C_{10}H_{21}SeH$	1.9	42	20% (v/v), $CDCl_3$	11
$H_2N(CH_2)_2SeH$	7.1	32	0.5M, D_2O, pD 8.3	11
PhSeH	1.8	27	80% (v/v), C_6F_6	14
Me_2Se	7.5	32	20% (v/v), $CDCl_3$	11
	4.5	27	7.5M, CH_2Cl_2, 20% (v/v)	14
Bu_2Se	19.1	40	20% (v/v), $CDCl_3$	11
$(C_8H_{17})_2Se$	10.4	41	20% (v/v), $CDCl_3$	11
i-Pr_2Se	8.7	41	20% (v/v), $CDCl_3$	11
$MeSe(CH_2)_2NH_2$	8.4	43	0.5M, $CDCl_3$	11
D,L-$MeSe(CH_2)_2CH(NH_2)CO_2H$	13.5	34	0.1M, D_2O, pD 4	11
$MeSeCH_2SeMe$	13.9	29	30% (v/v), $CDCl_3$	16
$MeSe(CH_2)_2SeMe$	14.4	29	30% (v/v), $CDCl_3$	16
$MeSe(CH_2)_3SeMe$	12.9	29	30% (v/v), $CDCl_3$	16
$(MeSe—)_2$	9.0	45	0.5M, $CDCl_3$	11
$(PhSe—)_2$	20.0	45	0.5M, $CDCl_3$	11
	16.0	*a*	3.0M, 20% (v/v), C_6F_6	14
$(PhCH_2Se—)_2$	27.0	55	0.5M, $CDCl_3$	11
$(C_{10}H_{21}Se—)_2$	21.0	43	0.5M, $CDCl_3$	11
Me_3SeI	13.7	32	1.0M, D_2O, pD 7	11
$Me_2Se(Br)CH_2CO_2Et$	1.8	32	0.5M, $CDCl_3$	11
$Me_2NC(Se)NH_2$	8.6	55	0.5M, D_2O, pD 4	11
Me_2SeO	8.9	30	0.5M, D_2O, pH 7	11
$(Me_2CHO)_2SeO$	11.9	*a*	*a*	15
$(CF_3CH_2O)_2SeO$	13.1	*a*	*a*	15
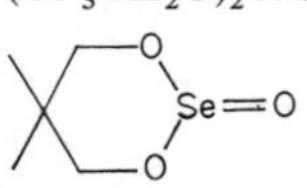	7.4	*a*	*a*	15
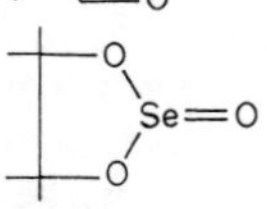	6.7	*a*	*a*	15
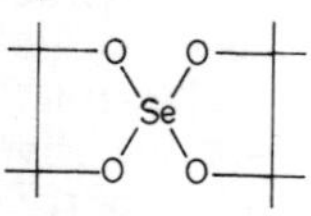	10.8	*a*	*a*	15
NaSeMe	16.3	43	1.0M, D_2O	11
Na_2SeO_4	16.8	15	0.5M, D_2O, pH 6.6	11
	10.2	*a*	0.5m, H_2O	13
Na_2SeO_3	10.7	*a*	1.0m, H_2O	13
H_2SeO_3	2.1	12	1.0M, D_2O, pH 9.6	11
	8.5	10	1.0M, D_2O, pH 1.5	11
	1.4	*a*	4.0m, D_2O	13
$NaHSeO_3$	0.34	*a*	4.0m, H_2O	13
$Zn(Se_2CNEt_2)_2$	4.4	24	1.0M, $CDCl_3$	12
$Pd[Se_2CN(Bu\text{-}i)_2]_2$	1.4	27	0.43M, $CHCl_3$	12
$Et_2NH_2Se_2CNEt_2$	2.7	−8	0.87M, $CDCl_3$	12

[a]Value not given.

significant enhancement been reported (4 M aqueous Na_2SeO_3, NOE = 0.4[13]). Second, for small Se-containing molecules, the spin-rotation (SR) relaxation mechanism dominates the spin-lattice relaxation and as the size of the molecule increases, the chemical shift anisotropy mechanism (CSA) becomes increasingly important[11,12,16].

These findings have important implications concerning the future of ^{77}Se-NMR spectroscopy, particularly if this nucleus is studied in large, macromolecular, biological systems as well as higher molecular weight organic molecules. The relaxation times which have been reported thus far are not extremely long ($\sim$ 1–30 s) and thus ^{77}Se-NMR spectroscopy should be able to be utilized as a routine tool for characterization. Also, in higher molecular weight systems where CSA is the dominant relaxation rate ($1/T_1$) the use of high-field spectrometers will be very beneficial since the CSA mechanism is proportional to the square of the external magnetic field[20]. As yet, no variable field T_1 studies have been reported for ^{77}Se to determine quantitatively the contributions of the SR and CSA mechanisms and experiments of this type are needed.

2. Chemical shifts

It is not within the purview of this article to thoroughly discuss the theory of chemical shifts. However, it is useful to discuss chemical shift theory in a rather specific way as it relates to ^{77}Se. The chemical shift reflects the distribution of electrons surrounding the observed nucleus and is, in general, a sensitive probe for characterization of the bonding in a molecule. Although considerable progress has been made in the recent past towards a refined theory of nuclear shielding, the *ab initio* calculation of chemical shifts still constitutes a formidable problem, and satisfactory correlation with experimental data remains limited to relatively small molecules containing light atoms, which unfortunately are of little practical importance to most chemists. This is certainly the case for Se where no published attempts toward a unified chemical shift theory are available. Thus we will endeavour to discuss chemical shifts in chemically relevant terms and interpret these shifts, where possible, using empirical substituent effects.

For any magnetically active nucleus in the presence of an external field, H_0, the effective field, H'_0, is given by

$$H'_0 = H_0 - H_0\sigma = H_0(1 - \sigma) \tag{1}$$

$$\sigma = \frac{H_0 - H'_0}{H_0} \tag{2}$$

The factor σ is called the magnetic shielding constant for the nucleus under observation and characterizes the electronic and chemical environment of that nucleus. The magnitude of the chemical shift is determined by this parameter which can be further subdivided into three different types of shielding as shown in equation (3).

$$\sigma = \sigma^{\text{dia}} + \sigma^{\text{para}} + \sigma^{\text{N}} \tag{3}$$

Here σ^{dia} represents contributions from local diamagnetic effects, σ^{para} is the shielding term arising from electronic circulations on the observed atom and σ^{N} represents neighbouring-group anisotropy effects, i.e. fields arising from electronic circulations around atoms surrounding the observed nucleus. The diamagnetic term decreases with the distance r between the nucleus and circulating electrons and is therefore dominant for atoms with s electrons only. The σ^{N} term is dependent on the nature and geometry of neighbouring atoms and is independent of the nature of the observed nucleus. The paramagnetic term σ^{para} is zero for nuclei with spherical distributions of electrons. However, for Se, which has a valence-shell electron configuration of $4s^2 4p^4$, a non-spherical distribution of electrons exists and variations in paramagnetic terms are expected to be the dominant contributor to Se chemical shifts. The paramagnetic term σ^{para} for Se is related to several factors as shown in expression (4)[21]

$$\sigma^{\text{para}} \propto -\frac{\langle r^{-3}\rangle_{4p}}{\Delta E}\Sigma Q \quad (4)$$

where $\langle r^{-3}\rangle_{4p}$ is the mean inverse cube of the radius of the 4p orbitals, the Q terms (obtained from the charge density-bond order matrix) denote the imbalance of charge in the valence shell of Se and ΔE is an effective excitation energy. The above is a somewhat drastic approximation of a more rigorous formulation but will suffice for our purposes.

From expression (4), it is clear that the *deshielding* of Se will increase (*a*) as the paramagnetic circulation gets closer to the nucleus, i.e. the larger the radial factor $\langle r^{-3}\rangle_{4p}$, (*b*) as the asymmetry of the valence electron cloud increases, i.e. the greater the ΣQ term and (*c*) as ΔE becomes smaller and excitation becomes easier.

An increase in the electron-withdrawing ability of the groups attached to Se should decrease the value of *r*, thereby increasing $\langle r^{-3}\rangle$ and increasing the deshielding. Such a correlation is indeed shown by the ^{77}Se chemical shifts for a series of Me derivatives[22] in which the Se shielding increases in the order MeSeOOH < $MeSeCl_3$ < Me_2SeO < Me_2SeCl_2 < Me_2SeBr_2 < Me_2Se_2 < Me_3Se^+ < Me_2Se < MeSeH < $MeSe^-$. This is the order of decreasing electronegativity of the groups attached to Se. A similar order is obtained when the chemical shifts of a series of phenyl selenenyl derivatives is compared[23]. For example, the shielding increases in the order PhSeOOH < PhSeCl < PhSeBr < PhSeSPh < PhSeSePh < PhSeMe < PhSeH < $PhSe^-$.

Deviations from spherical symmetry of the electron cloud occurs in molecules having low-lying electronic excited states which mix with the ground state to yield non-zero matrix elements of the magnetic moment operator. The ΔE term tends to decrease as the asymmetry of the molecules increases so that the ΣQ and ΔE^{-1} terms tend to act in the same way. In general, whenever the chemical shift of a diselenide is compared to that of the corresponding selenide, a deshielding of Se is observed[22]. This has been attributed to the influence of the low-energy electronic band associated with the diselenide moiety, i.e. to a small value of ΔE.

In this regard, it has been shown that in some organoselenium compounds, particularly selones (which contain carbon–selenium double bonds, >C=Se), there is a correlation of the ^{77}Se chemical shift with the wavelength λ of the band in the visible spectrum[24]. The major contribution to the chemical shifts comes from the deshielding of Se by the $n \rightarrow \pi^*$ circulation of lone-pair electrons in the π^* orbitals of the C=Se group. In fact, the deshielding of Se in selones increases as the $n \rightarrow \pi^*$ band moves to longer wavelength. This can be understood by expression (4) since the larger the value of λ, the smaller the excitation energy ΔE and thus the greater the deshielding of the Se nucleus.

a. Medium effects. (*i*) Solvent effects. ^{77}Se-NMR chemical shifts are very susceptible to solvent effects. The solvent dependence of ^{77}Se-NMR shifts can vary by up to 50 ppm in many organoselenium compounds[23,25]. (Although not as high as several hundred ppm as indicated by Wong and coworkers[26] in reference to the ^{77}Se-NMR shifts of selenophosphorus compounds studied by Dean[27]; in this case the solvent (SO_2) forms a complex with the selenophosphorus compound.) While this property may be exploited for studies of many molecular properties, comparison of chemical shift data from different sources is made correspondingly more difficult. The solvent shifts[28,29] of the ^{77}Se resonance of 5% solutions of Me_2Se in 20 solvents are presented in Table 3. These shifts cover a range of approximately 22 ppm. Deshielding of Se resonances is observed in going from polar solvents such as dimethyl sulphoxide to non-polar solvents such as cyclohexane and carbon tetrachloride. Employing a linear solvation energy relationship and the data of Table 3, Taft recently demonstrated that in non-chlorinated solvents, ^{77}Se solvent shifts in Me_2Se are influenced primarily by the dipolarity of the solvent[30]. Solvent shifts have been

TABLE 3. ^{77}Se solvent shifts of Me_2Se

Solvent	Chemical shift (ppm)	Standard	Ref.
Cyclohexane	4.5; 7.3	*a*; *b*	28; 29
Carbon tetrachloride	2.1; 5.46	*a*; *b*	28; 29
Triethylamine	0.4	*a*	28
Mesitylene	0.3	*a*	28
Diethyl ether	− 2.8	*a*	28
Benzene	− 3.2; − 0.58	*a*; *b*	28; 29
N,*N*-Dimethylaniline	− 4.2	*a*	28
Nitrobenzene	− 5.4	*a*	28
Pyridine	− 5.5	*a*	28
Acetic acid	− 6.7	*a*	28
Phenylacetonitrile	− 7.0	*a*	28
Methylene chloride	− 7.2; − 4.50	*a*; *b*	28; 29
Chloroform	− 7.6; − 4.86	*a*; *b*	28; 29
Methanol	− 8.3; − 4.34	*a*; *b*	28; 29
Acetonitrile	− 8.5	*b*	29
Hexamethylphosphoramide	− 9.4	*a*	28
Acetone	− 9.8; − 6.30	*a*; *b*	28; 29
Dimethylformamide	− 11.9	*a*	28
Nitromethane	− 13.4	*a*	28
Dimethyl sulphoxide	− 14.9; − 14.33	*a*; *b*	28; 29

[a]Relative to neat Me_2Se (shifts are reported for 5% mol solution).
[b]Relative to Me_2Se in $CDCl_3$ (60% v/v); all samples 5% (v/v) in the specified solvent.

observed for alkyl and aryl selenides[23,28], diselenides[23] and methyl selenocyanate[28]. From this discussion it is clear that solvent effects are significant in ^{77}Se-NMR studies, making it imperative to specify the solvent in which ^{77}Se spectra are obtained.

(*ii*) Concentration effects. Dilution shifts may reach a magnitude of several ppm. The ^{77}Se-NMR resonance of neat Me_2Se is shielded by 9 ppm upon dilution in $CDCl_3$ (1.25% v/v)[29]. Shifts of similar magnitude have been observed for other alkyl selenides and diselenides[23] whereas the shifts of diaryl diselenides show much smaller concentration dependence. The concentration dependence of the chemical shift probably accounts, at least in part, for discrepancies of reported chemical shifts which have appeared in the literature when Me_2Se has been used as the reference. For example, reported ^{77}Se chemical shifts for dimethyl diselenide are 275 and 281 ppm, and for dibenzyl diselenide, $(PhCH_2Se)_2$, 402 and 412 ppm. In these two studies[11,22] neat Me_2Se and 1.0 M Me_2Se in $CDCl_3$, respectively, were used as reference standards. A constant shift independent of concentration can be attained by using very dilute samples, which, in some cases, could be time-consuming because of the large number of repetitive scans which would be required to obtain a spectrum. Again, for reproducibility between laboratories, it is important to specify the concentrations of the sample used in the study.

(*iii*) pH effects. Selenols, RSeH, in general, are relatively acidic and can be easily deprotonated. For example, 2-aminoethaneselenol has been found to be completely in the zwitterionic form at pH 7–10, in contrast to 2-aminoethanethiol, which is only 60% zwitterionic at pH 10[31]. When the selenolate anion is formed, the ^{77}Se resonance of the selenol is shielded by approximately 200 ppm. A typical weak acid titration plot has been obtained for the change in ^{77}Se-NMR chemical shifts of selenocysteamine, $H_2NCH_2CH_2SeH$, as a function of pH[11].

(*iv*) Temperature effects. The temperature dependence of ^{77}Se chemical shifts of Se compounds was first noted by Lardon[32] in 1972 and by Odom, Dawson and Ellis[11] in 1979. Subsequently it has been reported for selenides[23,29], diselenides[23] and selenocarbonyl[26] ($R_2C{=}Se$) compounds. The temperature dependence of the ^{77}Se chemical shifts of $(CH_3)_2Se$ in $CDCl_3$ is approximately 2.5 ppm over the range of 222–323 K[29]. The temperature dependence of selones, $R_2C{=}Se$, is found to be 0.34–0.48 ppm K^{-1}. Over the temperature range 287–313 K, the variation of ^{77}Se shifts with temperature is reported to be linear[26]. Aryl diselenides[32] are reported to have a temperature dependence of 0.4 ppm K^{-1}. In all cases a deshielding of Se is observed. Diselenides and selones are highly coloured compounds which would be expected to have small excitation energies, ΔE, the value of which would decrease with increasing temperature. As noted previously, a smaller value of ΔE would reduce the shielding of Se.

b. *^{77}Se chemical shift referencing.* This area has been a somewhat vexing problem. There has been no universal agreement on a reference standard. The earliest studies employed either neat dimethyl selenide[22], Me_2Se, or seleninyl chloride[33], $SeOCl_2$, as a reference. More recently, ^{77}Se FT NMR studies have been reported using aqueous selenous acid[34,35], H_2SeO_3, selenophene[3–6] and 4,4′-dimethyldiphenyldiselenide[36], $(4\text{-}CH_3C_6H_4Se)_2$, as well as the two standards mentioned above. Since, as noted, Se chemical shifts are susceptible to solvent effects, reproducibility is best attained by using an external standard which appears to be a general consensus. A proposal that a solution of Me_2Se in $CDCl_3$ (60% v/v) be accepted as universal reference standard for ^{77}Se chemical shifts has appeared in the literature[29]. The rationale for doing so is that (1) Me_2Se resonates at a frequency which is close to one extreme of the chemical shift range, (2) Me_2Se is inexpensive and commercially available and (3) a spectrum of Me_2Se can be obtained with a good signal-to-noise ratio in one pulse on any of the commercially available FT instruments. The magnetic field of a magnet of an FT instrument is generally 'shimmed' to high homogeneity by locking to a ^2H resonance signal and adjusting the current in the shim coils until maximum signal and resolution are achieved. This process requires repetitive acquisition of the spectrum of the observed nucleus. Thus, the 60% Me_2Se in $CDCl_3$ provides both the ^2H lock signal and sufficient signal-to-noise to achieve maximum homogeneity of the magnetic field in a series of one-pulse acquisitions.

c. *Classes of compounds.* The tables are organized according to chemical shift values, i.e. from the most shielded resonance to the most deshielded resonance for that particular class of compounds. In cases where there are substituted phenyl rings the parent compound is given first and all derivatives of that parent are then presented in the order stated above. To standardize ^{77}Se chemical shift values in this chapter, they are tabulated with respect to the reported value of Me_2Se in $CDCl_3$ (60% v/v). In the tables we have attempted to present values which may be compared in as consistent a manner as possible. However, in many cases experimental conditions have not been specified clearly, or more commonly, investigators were probably not aware that the resonance positions of their reference standards were so sensitive to variable conditions. Thus, to avoid confusion, in those investigations where values are reported with respect to Me_2Se as a standard reference, we have tabulated the original values irrespective of the conditions employed, whereas, for the values which are referenced to standards other than Me_2Se, relevant conversion factors are employed with respect to Me_2Se in $CDCl_3$ (60% v/v). For the reader's convenience, the conversion constants employed for other reference materials are given as footnotes in each table.

(*i*) Selenols and selenolates. The chemical shift data for selenols and selenolates are presented in Table 4. The total chemical shift range is approximately 600 ppm with the most shielded ^{77}Se resonance being that in the selenolate $MeSe^-\ Na^+$ and the most deshielded resonance occurring in CF_3SeH. As expected, selenolates are more shielded than the corresponding selenol, usually by ~200 ppm. For example, the ^{77}Se chemical shifts of MeSeH and $MeSe^-\ Na^+$ are −130 and −330 ppm, respectively. The increased shielding in selenolates occurs as a result of the increased electron density around the Se

TABLE 4. ^{77}Se chemical shifts of selenols and selenolates

Compound	Chemical shift (ppm)	Standard	Solvent	Ref.
$MeSe^- Na^+$	−332; −330	*a*;*b*	H_2O	22;11
H_2Se	−288	*b*	D_2O	11
$HSeCH_2CH_2NH_2$	−212	*b*	D_2O, pD = 8.3	11
$EtSe^- Na^+$	−150	*a*	H_2O	22
DL-$HSeCH_2CH(NH_2)COOH$	−141	*b*	D_2O, pD = 5	11
MeSeH	−116; −130	*a*;*b*	Neat; $CDCl_3$	22;11
C_6F_5SeH	−16	*a*	Not specified	37
$C_{10}H_{21}SeH$	−7	*b*	$CDCl_3$	11
i-$PrSe^- Na^+$	8.7	*a*	H_2O	22
EtSeH	39;41	*b*	$CDCl_3$; Me_2CO	11
$PhCH_2SeH$	107	*a*	CH_2Cl_2	22
t-$BuSe^- Na^+$	129	*a*	H_2O	22
PhSeH	145	*a*	Neat	22
o-AnSeH	84	*a*	Neat	22
o-TolSeH	112	*a*	Neat	22
o-ClC_6H_4SeH	158	*a*	Neat	22
o-FC_6H_4SeH	191	*a*	Neat	22
m-TolSeH	144	*a*	Neat	22
m-AnSeH	153	*a*	Neat	22
m-$CF_3C_6H_4SeH$	159	*a*	Neat	22
m-FC_6H_4SeH	164	*a*	Neat	22
m-ClC_6H_4SeH	167	*a*	Neat	22
p-AnSeH	122	*a*	Neat	22
p-TolSeH	128	*a*	Neat	22
p-FC_6H_4SeH	141	*a*	Neat	22
p-ClC_6H_4SeH	142	*a*	Neat	22
i-PrSeH	159	*a*	Neat	22
t-BuSeH	278	*a*	Neat	22
CF_3SeH	287	*c*	C_6F_6	41

[a]Relative to neat Me_2Se.
[b]Relative to 1.0M Me_2Se in $CDCl_3$.
[c]Relative to Me_2Se in C_6D_6.

nucleus causing an increase in the 4p radial distance. Thus the $\langle r^{-3}\rangle_{4p}$ factor of expression (4) decreases. A comparison of the data for MeSeH, $PhCH_2SeH$ and PhSeH shows that the replacement of a Me group by a $PhCH_2$ group deshields the Se nucleus by ~220 ppm whereas the deshielding caused by substitution of a Ph group is ~280 ppm, presumably due to the increased electronegativity of the sp^2-hybridized C atom in the Ph group. Interestingly, introduction of a C_6F_5 group shields the Se nucleus as evidenced by a 168 ppm chemical shifts difference between selenophenol, PhSeH, and pentafluoroselenophenol, C_6F_5SeH. In this case the greater electronegativity of the ring F atoms should lead to a positive charge on the sigma framework of the polyfluoroaromatic ring, making it a stronger inductive acceptor than the Ph group. However, an increased nuclear shielding has been found in polyfluoroaromatic compounds with respect to their hydrocarbon analogues. Thus, the CF_3 group which exhibits a strong inductive effect deshields Se, whereas the C_6F_5 group shields it. This shielding has been discussed[37] in terms of a decreased conjugation between the unshared electron pair of the Se and the π system of the perfluorinated benzene ring. Reduced conjugation of the unshared electron pair of S with the π system of the polyfluoroaryl ring is confirmed by X-ray fluorescent spectral data for organic S compounds[38]. A similar dependence may also be expected for the Se derivatives.

McFarlane and Wood[22], who were the first to report ^{77}Se chemical shift data for a variety of organoselenium compounds, observed that, in general, Se resonances are deshielded as the electronegativity of the substituent attached to Se increases when the overall range of substituent electronegativities is relatively large. However, when substituents had similar electronegativity values, as, for example, when alkyl groups were varied, the relationship broke down. Thus, in the alkyl selenols and selenolates (RSeH and RSe^-), the shielding effect of the group increases in the order t-Bu < n-Pr < Et < Me which is opposite to that expected if the Se chemical shifts were dominated by the inductive effect of the alkyl groups. As will be seen later in this chapter, the same trend is observed with dialkyl selenides, R_2Se, aryl alkyl selenides, ArSeR, dialkyl diselenides, R_2Se_2, and similar behaviour has been observed in ^{31}P chemical shifts of trialkylphosphines[39]. An explanation for this reverse shielding phenomenon has been proposed[40] and will be discussed in the section concerning Se chemical shifts of selenides.

Due to the large chemical shift range of ^{77}Se, NMR studies of this nucleus should be a powerful technique to study the transmission of electronic substituent effects. McFarlane and Wood[22] carried out such a study for *meta*- and *para*-substituted phenyl selenols and reported a poor correlation of ^{77}Se chemical shifts with Hammett σ constants of the substituents. The lack of a good fit was attributed to H-bond formation in selenols which these investigators felt could affect the Se shielding by more than 20 ppm. This is significant considering that the total chemical shift range covered in the compounds studied was ~45 ppm. It should also be noted that, as yet, no additional definitive studies demonstrating significant H-bonding to Se have appeared.

(*ii*) Dialkyl selenides. ^{77}Se chemical shifts of dialkyl selenides are reported in Table 5. When hydrogen atoms of the methyl groups of Me_2Se are successively replaced by Me groups, a deshielding of approximately 100 ppm per H atom replaced is observed (e.g., Me_2Se (0 ppm), MeSeEt (108 ppm), Et_2Se (217 ppm), MeSeBu-t (294 ppm), (i-Pr)$_2$Se (432 ppm) and (t-Bu)$_2$Se (601 ppm)). The replacement of the methyl H by a Ph group causes a deshielding of ~170 ppm per replacement, e.g. Me_2Se (0 ppm), $PhCH_2SeMe$ (173 ppm), $(PhCH_2)_2Se$ (333 ppm). Although replacement of H by a more electronegative F atom causes a considerable deshielding, the replacement by a Cl (less electronegative than F) causes even greater deshielding. For example, the ^{77}Se chemical shifts of Se in CF_3SeCH_3, CF_3SeCF_3, CF_3SeCF_2Cl, $CF_3SeCFCl_2$ and CF_3SeCCl_3 are 370, 717, 815, 927 and 953 ppm, respectively.

The ^{77}Se-NMR studies of a large number of substituted *Se*-benzylselenoalkanoic acids ($XC_6H_4CH_2Se(CH_2)_nCOOH$, where $n = 1-10$) are presented in Table 6. The shifts caused by *meta* and *para* substituents have been correlated with Hammett's σ values; however, in our opinion there were not enough substituents studied to obtain a meaningful correlation. It is interesting to note that a maximum ^{77}Se shielding occurs for the butyric acid derivatives ($n = 3$) and it has been proposed that this could perhaps be due to the interaction between the free electron pairs on the carbonyl group and empty Se valence-shell d orbitals[5].

Previously, the abnormal behaviour of the ^{77}Se chemicals shifts in alkyl selenols, selenolates, selenides and diselenides was mentioned (*vide supra*). Briefly, for selenols, a deshielding of ~120 ppm is observed as H is replaced by Me in the series MeSeH, EtSeH, i-PrSeH, t-BuSeH and a similar trend is observed for selenolates. In alkyl selenides, the analogous substitution deshields the Se resonance by ~110 ppm per replacement by a Me group for the series Me_2Se, MeSeEt, Et_2Se, MeSeBu-t, (i-Pr)$_2$Se, (t-Bu)$_2$Se. In dialkyl diselenides a similar, but smaller (~65 ppm per replacement) deshielding is observed in the ^{77}Se chemical shifts of MeSeSeMe, EtSeSeEt, i-PrSeSe-Pr-i, t-BuSeSeBu-t. In all cases, each Me group contributes an additive deshielding to the ^{77}Se resonance which is surprisingly large for a relatively small change in the electronic nature and electronegativity of these groups.

A clearer understanding of this effect and its origin can be obtained by more closely

TABLE 5. ^{77}Se chemical shifts of dialkyl selenides

Compound	Chemical shift (ppm)	Standard	Solvent	Ref.
Me_2Se	0	*a*	$CDCl_3$	27
$MeSeCH_2CH_2NH_2$	44	*c*	$CDCl_3$	11
DL-$MeSeCH_2CH_2CH(NH_2)COOH$	75	*c*	D_2O(pD 4)	11
MeSeEt	108	*b*	CH_2Cl_2	22
Bu_2Se	167	*c*	$CDCl_3$	11
$(C_8H_{17})_2Se$	168	*c*	$CDCl_3$	11
$PhCH_2SeMe$	173	*b*	$CDCl_3$	7
Et_2Se	217; 236.6	*d*; *b*	Neat; neat	34; 22
$CF_3SeHgCl$	267.9	*e*	CH_3OH	33
MeSeBu-*t*	294	*b*	Neat	22
$(PhCH_2)_2Se$	328.5; 333	*f*; *b*	$(CD_3)_2CO$; $CDCl_3$	5; 7
$(o\text{-}ClC_6H_4CH_2)_2Se$	302.4	*f*	$(CD_3)_2CO$	5
$(m\text{-}ClC_6H_4CH_2)_2Se$	330.2	*f*	$(CD_3)_2CO$	5
$(m\text{-}BrC_6H_4CH_2)_2Se$	331.5	*f*	$(CD_3)_2CO$	5
$(p\text{-}ClC_6H_4CH_2)_2Se$	331.9	*f*	$(CD_3)_2CO$	5
$(p\text{-}BrC_6H_4CH_2)_2Se$	338.0	*f*	$(CD_3)_2CO$	5
$(o,p\text{-}Cl_2C_6H_3CH_2)_2Se$	309.2	*f*	$(CD_3)_2CO$	5
$(CF_3Se)_2Hg$	337.4	*e*	MeOH	33
CF_3SeMe	370	*g*	C_6D_6	41
$(i\text{-}Pr)_2Se$	432; 436	*c*; *d*	$CDCl_3$; $CDCl_3$	11; 22
$(CF_3)_2Se$	697.9; 717; 724.6	*e*; *g*; *e*	Neat; C_6D_6; not specified	33; 41; 52
CF_3SeCF_2Cl	799.6; 815	*e*; *g*	Not specified; C_6D_6	52; 41
$CF_3SeCFCl_2$	910.6; 927	*e*; *g*	Not specified; C_6D_6	52; 41
CF_3SeCCl_3	953	*g*	C_6D_6	41

[a]Relative to Me_2Se in $CDCl_3$ (60% v/v).
[b]Relative to neat Me_2Se.
[c]Relative to 1.0M Me_2Se in $CDCl_3$.
[d]Aq. H_2SeO_3 was used as a reference; chemical shift values have been converted relative to Me_2Se in $CDCl_3$ (60% v/v) using the expression $\delta[Me_2Se] = \delta[H_2SeO_3\ (aq.)] - 1285.6$.
[e]$SeOCl_2$ was used as a reference; chemical shift values have been converted relative to Me_2Se in $CDCl_3$ (60% v/v) using the expression $\delta[Me_2Se] = \delta[SeOCl_2] - 1482.6$.
[f]Selenophene in $(CD_3)_2CO$ (20%) was used as a reference; chemical shift values have been converted relative to Me_2Se in $CDCl_3$ (60% v/v) using the expression $\delta(Me_2Se) = \delta(\text{selenophene}) - 608.6$.
[g]Relative to Me_2Se in C_6D_6.

examining solvent effects on ^{77}Se chemical shifts[40]. As pointed out earlier in this chapter, substantial solvent-induced ^{77}Se chemical shifts have been observed. The shifts have been reported to range up to 40–50 ppm for alkyl selenides and diselenides and diamagnetic susceptibility contributions are small compared to these shifts. It should be noted that the extreme values for these solvent shifts for low molecular weight alkyl selenides and diselenides occur for the solvents trifluoroacetic acid (maximum shielding) and diiodomethane (maximum deshielding)[40].

In an attempt to understand the ^{77}Se solvent-induced chemical shifts with respect to some particular property of the solvent, the shifts have been correlated with the following parameters: (1) the refractive index of the solvent, (2) the molar polarizability of the solvent and (3) the ratio of the molar polarizability of the solvent to its molar volume[40].

TABLE 6. ^{77}Se chemical shifts of *Se*-benzylselenoalkanoic acids[a,b]

Compound	Chemical shift (ppm)	Compound	Chemical shift (ppm)
$PhCH_2Se(CH_2)_3COOH$	243.3	$4\text{-}BrC_6H_4CH_2Se(CH_2)_9COOH$	253.2
$4\text{-}BrC_6H_4CH_2Se(CH_2)_3COOH$	250.1	$4\text{-}NO_2C_6H_4CH_2Se(CH_2)_9COOH$	265.5
$4\text{-}NO_2C_6H_4CH_2Se(CH_2)_3COOH$	263.4	$2,4\text{-}Cl_2C_6H_3CH_2Se(CH_2)_9COOH$	245.9
$PhCH_2Se(CH_2)_7COOH$	247.2	$PhCH_2Se(CH_2)_4COOH$	264.4
$4\text{-}MeC_6H_4CH_2Se(CH_2)_7COOH$	245.9	$4\text{-}MeC_6H_4CH_2Se(CH_2)_4COOH$	245.1
$4\text{-}BrC_6H_4CH_2Se(CH_2)_7COOH$	253.5	$4\text{-}BrC_6H_4CH_2Se(CH_2)_4COOH$	252.4
$PhCH_2Se(CH_2)_{10}COOH$	247.2	$4\text{-}NO_2C_6H_4CH_2Se(CH_2)_4COOH$	264.7
$2\text{-}ClC_6H_4CH_2Se(CH_2)_{10}COOH$	242.0	$PhCH_2Se(CH_2)_2COOH$	266.8
$2\text{-}BrC_6H_4CH_2Se(CH_2)_{10}COOH$	242.8	$2\text{-}ClC_6H_4CH_2Se(CH_2)_2COOH$	261.1
$2\text{-}FC_6H_4CH_2Se(CH_2)_{10}COOH$	247.7	$2\text{-}BrC_6H_4CH_2Se(CH_2)_2COOH$	262.1
$2\text{-}NO_2C_6H_4CH_2Se(CH_2)_{10}COOH$	261.6	$2\text{-}FC_6H_4CH_2Se(CH_2)_2COOH$	266.8
$3\text{-}FC_6H_4CH_2Se(CH_2)_{10}COOH$	251.7	$2\text{-}NO_2C_6H_4CH_2Se(CH_2)_2COOH$	280.9
$3\text{-}ClC_6H_4CH_2Se(CH_2)_{10}COOH$	253.5	$3\text{-}FC_6H_4CH_2Se(CH_2)_2COOH$	271.8
$3\text{-}BrC_6H_4CH_2Se(CH_2)_{10}COOH$	254.0	$3\text{-}ClC_6H_4CH_2Se(CH_2)_2COOH$	272.8
$3\text{-}NO_2C_6H_4CH_2Se(CH_2)_{10}COOH$	259.8	$3\text{-}BrC_6H_4CH_2Se(CH_2)_2COOH$	273.4
$4\text{-}FC_6H_4CH_2Se(CH_2)_{10}COOH$	249.6	$3\text{-}NO_2C_6H_4CH_2Se(CH_2)_2COOH$	278.8
$4\text{-}ClC_6H_4CH_2Se(CH_2)_{10}COOH$	251.0	$4\text{-}FC_6H_4CH_2Se(CH_2)_2COOH$	269.6
$4\text{-}NO_2C_6H_4CH_2Se(CH_2)_{10}COOH$	265.8	$4\text{-}ClC_6H_4CH_2Se(CH_2)_2COOH$	272.6
$2,4\text{-}Cl_2C_6H_3CH_2Se(CH_2)_{10}COOH$	246.2	$4\text{-}BrC_6H_4CH_2Se(CH_2)_2COOH$	273.1
$PhCH_2Se(CH_2)_5COOH$	247.5	$4\text{-}NO_2C_6H_4CH_2Se(CH_2)_2COOH$	285.1
$4\text{-}BrC_6H_4CH_2Se(CH_2)_5COOH$	253.5	$2,4\text{-}Cl_2C_6H_3CH_2Se(CH_2)_2COOH$	265.0
$4\text{-}NO_2C_6H_4CH_2Se(CH_2)_5COOH$	265.3	$PhCH_2SeCH_2COOH$	291.1
$PhCH_2Se(CH_2)_6COOH$	247.5	$2\text{-}ClC_6H_4CH_2SeCH_2COOH$	277.8
$4\text{-}BrC_6H_4CH_2Se(CH_2)_6COOH$	253.5	$2\text{-}BrC_6H_4CH_2SeCH_2COOH$	279.9
$4\text{-}NO_2C_6H_4CH_2Se(CH_2)_6COOH$	265.3	$2\text{-}FC_6H_4CH_2SeCH_2COOH$	283.0
$PhCH_2Se(CH_2)_8COOH$	247.5	$2\text{-}NO_2C_6H_4CH_2SeCH_2COOH$	300.8
$4\text{-}NO_2C_6H_4CH_2Se(CH_2)_8COOH$	265.3	$3\text{-}FC_6H_4CH_2SeCH_2COOH$	296.0
$PhCH_2Se(CH_2)_9COOH$	247.7	$3\text{-}BrC_6H_4CH_2SeCH_2COOH$	296.9
$2\text{-}ClC_6H_4CH_2Se(CH_2)_9COOH$	242.0	$3\text{-}ClC_6H_4CH_2SeCH_2COOH$	297.4
$2\text{-}BrC_6H_4CH_2Se(CH_2)_9COOH$	242.8	$3\text{-}NO_2C_6H_4CH_2SeCH_2COOH$	302.9
$2\text{-}FC_6H_4CH_2Se(CH_2)_9COOH$	248.3	$4\text{-}MeC_6H_4CH_2SeCH_2COOH$	288.5
$3\text{-}FC_6H_4CH_2Se(CH_2)_9COOH$	251.4	$4\text{-}FC_6H_4CH_2SeCH_2COOH$	294.5
$3\text{-}ClC_6H_4CH_2Se(CH_2)_9COOH$	253.5	$4\text{-}ClC_6H_4CH_2SeCH_2COOH$	297.7
$3\text{-}BrC_6H_4CH_2Se(CH_2)_9COOH$	254.0	$4\text{-}BrC_6H_4CH_2SeCH_2COOH$	297.7
$4\text{-}FC_6H_4CH_2Se(CH_2)_9COOH$	249.3	$4\text{-}NO_2C_6H_4CH_2SeCH_2COOH$	306.5
$4\text{-}ClC_6H_4CH_2Se(CH_2)_9COOH$	252.7	$2,4\text{-}Cl_2C_6H_3CH_2SeCH_2COOH$	282.8

[a]Solvent C_6D_6 (20% solutions). Chemical shifts were reported relative to selenophene; converted to Me_2Se in $CDCl_3$ (60% v/v) using the relationship $\delta(Me_2Se) = \delta(\text{selenophene}) - 608.6$.
[b]Ref. 5.

Although none of these parameters individually yields an excellent correlation, all three reproduce the overall trend of the data with the refractive index being the parameter which most successfully correlates with the solvent shifts. For example, CF_3COOH has the lowest value of refractive index and CH_2I_2 has the highest value of the solvents studied[42]. Thus polarizability and dispersion forces clearly figure predominantly (although perhaps not exclusively) in the intermolecular interactions causing these shifts. Se, with its relatively large size and two unshared pairs of electrons should be highly polarizable and should interact significantly with highly polarizable solvents.

With such an intermolecular polarizability concept in mind, it is rather easy to extrapolate to an *intramolecular* situation where dispersion forces within a molecule could be exerted by neighbouring alkyl groups and influence Se shielding. Alkyl groups have been shown to be very polarizable in the gas phase in that they stabilize a negative or a positive charge[43]. This polarizability effect is based on the charge-induced dipole interaction and attenuates rapidly with distance (proportional to r^{-4}) between the centres of polarizability and of charge. Since the polarizability of a molecule is an additive property[44], successive additions of a Me group at a fixed distance should exert nearly the same influence. Thus, in going from MeSeMe → MeSeEt → EtSeEt → MeSeBu-*t* → *i*-PrSePr-*i*, successive replacement of a Me group at a fixed distance disperses the electron density at the highly polarizable Se atom by about the same magnitude which in turn increases the $\langle r^{-3} \rangle$ factor (r decreases) as well as the asymmetry factor, ΣQ, in expression (4). The net result is an increased deshielding of about equal magnitude in each case. The deshielding observed in alkyl selenols and selenolates can be explained in a similar manner as the methyl hydrogens in methyl selenol (or selenolate) are successively replaced by Me groups.

In molecules where the alkyl group is *n*-Pr, a *shielding* of Se is observed relative to the case where the alkyl group is Et. For example, this is seen in the compounds PhSeMe (202 ppm), PhSeEt (322 ppm), PhSePr-*i* (424 ppm), PhSeBu-*t* (521 ppm) and PhSePr-*n* (285 ppm). This shielding in *n*-Pr derivatives can be understood due to the fact that a Me group is now introduced two C atoms away from the Se and thus the dispersion effect introduced by an extra Me group is very small compared to that in PhSe-Pr-*i*. In addition, introduction of a Me group in the γ-position with respect to the Se atom causes shielding by the 'γ-effect', which is well known and accepted in ^{13}C-NMR spectroscopy[45]. Confirmation of this γ-effect in Se-NMR spectroscopy is further established by comparing the ^{77}Se chemical shifts of (1) PhSeBu-*i* (264 ppm) and PhSeBu-*n* (288 ppm) and of (2) PhSeBu-*s* (394 ppm) and PhSePr-*i* (424 ppm). In the first case the *i*-Bu derivative has two γ-carbons with respect to Se whereas the *n*-Bu derivative has only one such carbon; thus an increased shielding of 24 ppm is observed with the *i*-Bu derivative. In the second case introduction of a γ-carbon in the *i*-Pr compound yields a *s*-Bu derivative and causes a shielding of 30 ppm. The replacement of a methyl H of the *n*-Pr group by an alkyl group has virtually no effect on the Se shielding, as e.g. in PhSePr-*n* (285 ppm) and PhSeBu-*n* (288 ppm) and in *n*-Bu_2Se (167 ppm) and (n-$C_8H_{17})_2Se$ (168 ppm). Thus, intramolecular dispersion effects of neighbouring alkyl groups have a great influence on the shielding of the Se nucleus. Such an effect is also observed in the gas-phase acidities (also an intrinsic property) of alcohols[43]. The observed order of gas-phase acidities of alkyl alcohols is *t*-BuOH > *i*-PrOH > EtOH > MeOH > H_2O which is reversed from that found in solution. In fact, it has been suggested that the solution order is an artifact and does not represent any intrinsic property of these molecules.

Another example where a group which is less electronegative (i.e. has lower electron-withdrawing ability) influences the shielding of ^{77}Se in the reverse manner is the observed increased deshielding in going from $CF_3SeCH_3 \rightarrow CF_3SeCF_2Cl \rightarrow CF_3SeCFCl_2 \rightarrow CF_3SeCCl_3$. Gombler explained[41] this phenomenon by considering C—F and C—Cl hyperconjugation so that mesomeric structures **A** and **B** contribute to the shielding of Se.

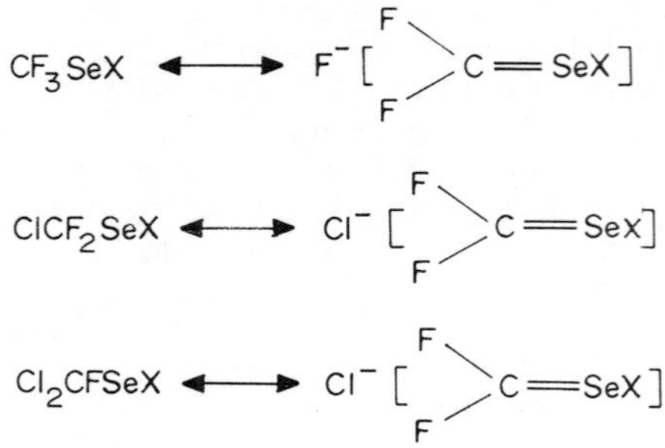

(A) (B)

Although hyperconjugation involving the $—CF_3$ group was proposed[46] as early as 1950, serious reservations have been expressed and reviewed[47–49] concerning this concept and conclusions have also been advanced that C—F hyperconjugation does *not* play a significant role in the stability or reactivity of aliphatic organofluorine compounds or aromatic compounds with perfluoroalkyl substituents. In this regard, it should be noted that in halogen-substituted acetic acid molecules, gas-phase acidities increase in the order $FCH_2COOH < ClCH_2COOH < BrCH_2COOH$ which is the reverse of the order in aqueous solution[50]. This acidity order can be explained by comparing the polarizability of the halogen substituents. The atomic polarizabilities of F, Cl and Br are 0.53, 2.61 and 3.79 $Å^3$, respectively[51]. Thus it was proposed that the more polarizable atom can better stabilize a charge on the O in the acetate anion. In the Se compounds above, the more polarizable Cl atom can polarize the electron density on the polarizable Se atom more effectively than F. Polarization of the Se electron density causes an increase in the ΣQ and $\langle r^{-3} \rangle$ factors in expression (4), thereby increasing the deshielding at the Se nucleus.

(*iii*) Alkyl aryl selenides. ^{77}Se chemical shifts of alkyl aryl selenides are shown in Table 7. The ^{77}Se resonance in MeSePh is deshielded by ~200 ppm when compared to that in Me_2Se. Again, deshielding of Se is incremented by ~120 ppm when the Me group is replaced by Et, *i*-Pr and *t*-Bu groups. Table 7 also provides the ^{77}Se chemical shifts for a range of substituted selenoanisoles, XC_6H_4SeMe. The shifts in the *para* series[53] cover a range of ~50 ppm as the substituent is varied from $—NMe_2$ to $—NO_2$. The direction of the shifts is normal, i.e. electron donors cause upfield shifts and electron acceptors cause downfield shifts. The ^{77}Se chemical shifts of these substituted selenoanisoles have been examined rather carefully in an attempt to understand their origin. Although single-parameter equations of the Hammett or Brown–Okamoto type have been used extensively to derive relationships between Substituent Chemical Shift (SCS) values and substituent parameters (σ constants) in order to obtain one or more transmission coefficients (commonly denoted by ρ), Dual Substituent Parameter (DSP) equations have also been successfully used[54,55]. In the DSP method, the SCS values are related to a linear combination of previously defined polar and resonance substituent parameters (σ_I and $\bar{\sigma}_R$, respectively) as shown in equation (5):

$$SCS = \rho_I \sigma_I + \rho_R \bar{\sigma}_R \qquad (5)$$

The symbol $\bar{\bar{\sigma}}_R$ denotes the fact that any one of four resonance scales (σ_R^-, σ_R^0, σ_R^{BA}, σ_R^+) may be used for a given correlation. It is usual practice to perform four separate correlations, each with a different resonance scale, and then to utilize the one which yields the best fit to the experimental data. The main advantage of the DSP method over single-parameter treatments is that it allows the calculation of separate transmission coefficients for resonance and polar effects. The 'goodness of fit' of a DSP correlation is judged from the 'f'

TABLE 7. ^{77}Se chemical shifts of alkyl aryl selenides[a]

Compound	Chemical shift (ppm)	Solvent	Ref.
1-NaphSeMe	155	$CDCl_3$	7
2-NaphSeMe	202	$CDCl_3$	7
PhSeMe	197	$CDCl_3$	57
	202	CH_2Cl_2	22
	202	$CDCl_3$	58
	202	Neat	53
	203	$CDCl_3$	7
o-AnSeMe	150	CH_2Cl_2	22
o-TolSeMe	162	CH_2Cl_2	22
o-ClC_6H_4SeMe	201	CH_2Cl_2	22
o-$NaOOCC_6H_4Me$	225	50% aq. MeOH	58
o-$MeOOCC_6H_4SeMe$	259; 265	$CDCl_3$; $CDCl_3$	59; 58
o-$CH(O)C_6H_4SeMe$	267	$CDCl_3$	59
o-$HOOCC_6H_4SeMe$	272	$CDCl_3$	58
o-AcC_6H_4SeMe	282	$CDCl_3$	59
o-$NO_2C_6H_4SeMe$	292.1	Neat	53
m-TolSeMe	199; 200	CH_2Cl_2; neat	22; 53
m-AnSeMe	207	Neat	53
m-FC_6H_4SeMe	214	Neat	53
m-$CF_3C_6H_4SeMe$	217	Neat	53
m-ClC_6H_4SeMe	215; 217.7	CH_2Cl_2; neat	22; 53
m-BrC_6H_4SeMe	219.8	Neat	53
m-$NO_2C_6H_4SeMe$	225.6	Neat	53
p-$Me_2NC_6H_4SeMe$	181.2	Neat	53
p-$NH_2C_6H_4SeMe$	182.0	Neat	53
p-AnSeMe	189.5; 191	Neat; $CDCl_3$	53; 22
p-TolSeMe	196.1; 197	Neat; neat	53; 22
p-FC_6H_4SeMe	200; 200.0	CH_2Cl_2; neat	22; 53
p-ClC_6H_4SeMe	203; 203.6	CH_2Cl_2; neat	22; 53
p-$MeOOCC_6H_4SeMe$	218.1	Neat	53
p-$NO_2C_6H_4SeMe$	233.4	Neat	53
PhSeBu-*i*	264	$CDCl_3$	57
PhSePr-*n*	285	$CDCl_3$	57
PhSeBu-*n*	288	$CDCl_3$	57
PhSeEt	322	$CDCl_3$	57
	326	$CDCl_3$	7
	327	CH_2Cl_2	22
o-TolSeEt	276	CH_2Cl_2	22
o-AnSeEt	362	CH_2Cl_2	22
m-AnSeEt	331	Neat	22
m-ClC_6H_4SeEt	336	CH_2Cl_2	22
m-$CF_3C_6H_4SeEt$	340	Neat	22
p-AnSeEt	318	CH_2Cl_2	22
p-TolSeEt	323	Neat	22
p-FC_6H_4SeEt	324	CH_2Cl_2	22
PhSeBu-*s*	394	$CDCl_3$	57
PhSePr-*i*	424	$CDCl_3$	57
o-AnSePr-*i*	355	CH_2Cl_2	22
PhSeBu-*t*	521	$CDCl_3$	57

[a]All chemical shifts are relative to neat Me_2Se.

value where

$$f = SD/RMS \tag{6}$$

Here, SD is the standard deviation of the fit and RMS represents the root mean square value of the experimental data. In practice, $f = 0.0$–0.1 represents excellent correlation, $f = 0.1$–0.2 is felt to be moderately good correlation and f values of 0.3 and greater represent only crude trends[56].

In the substituted selenoanisoles, a DSP analysis of the ^{77}Se chemical shifts as the substituent is varied indicates that the shifts are precisely related to substituent electronic effects and the following equations are obtained[56]:

$$\textit{ortho}: \mathrm{SCS} = 47\sigma_{\mathrm{I}} + 137\sigma_{\mathrm{R}}^{-} \qquad f = 0.21 \tag{7}$$

$$\textit{meta}: \mathrm{SCS} = 34\sigma_{\mathrm{I}} + 8\sigma_{\mathrm{R}}^{\mathrm{BA}} \qquad f = 0.14 \tag{8}$$

$$\textit{para}: \mathrm{SCS} = 32\sigma_{\mathrm{I}} + 52\sigma_{\mathrm{R}}^{0} \qquad f = 0.17 \tag{9}$$

It is interesting to note that, while 'good' correlations are obtained, the ρ_{I} values for the *meta* and *para* transmission coefficients are similar but the ρ_{R} values differ substantially.

(*iv*) Diaryl selenides. Gronowitz and coworkers carried out a systematic ^{77}Se-NMR study[6] of the effect of substituents on the Se chemical shift in 4,4′-disubstituted diphenyl selenides (Table 8). The ^{77}Se shifts vary in a regular way with the character of the substituents, i.e. electron-donating groups cause upfield shifts and electron-attracting groups induce downfield shifts. A good linear correlation was obtained when the ^{77}Se shifts were plotted against the σ_p^+ values of Swain and Lupton or with the σ_p values of Hammett. A linear correlation was also found between ^{77}Se chemical shifts and ^{19}F

TABLE 8. ^{77}Se chemical shifts of diaryl selenides

Compound	Chemical shift (ppm)	Standard	Solvent	Ref.
$(C_6F_5)_2Se$	110	*a*	Not specified	37
$(4\text{-}CF_3C_6F_4)_2Se$	179	*a*	Not specified	37
$PhSeC_6F_5$	265	*a*	Not specified	37
Ph_2Se	402; 414.5	*a*; *b*	$CDCl_3$; $CDCl_3$: DMSO	7; 6
$(o\text{-}NaOOCC_6H_4)_2Se$	435	*c*	50% aq. MeOH	58
$(o\text{-}MeOOCC_6H_4)_2Se$	469	*c*	$CDCl_3$	58
$(o\text{-}HOOCC_6H_4)_2Se$	479	*c*	$CDCl_3$	58
$(p\text{-}H_2NC_6H_4)_2Se$	373.4	*b*	$CDCl_3$: DMSO (1:1)	6
$p\text{-}An_2Se$	387.5	*b*	$CDCl_3$: DMSO (1:1)	6
$(p\text{-}PhOC_6H_4)_2Se$	393.8	*b*	$CDCl_3$: DMSO (1:1)	6
$p\text{-}Tol_2Se$	399.3	*b*	$CDCl_3$: DMSO (1:1)	6
$(p\text{-}IC_6H_4)_2Se$	402.2	*b*	$CDCl_3$: DMSO (1:1)	6
$(p\text{-}MeSC_6H_4)_2Se$	404.0	*b*	$CDCl_3$: DMSO (1:1)	6
$(p\text{-}FC_6H_4)_2Se$	404.8	*b*	$CDCl_3$: DMSO (1:1)	6
$(p\text{-}BrC_6H_4)_2Se$	408.4	*b*	$CDCl_3$: DMSO (1:1)	6
$(p\text{-}ClC_6H_4)_2Se$	411.8	*b*	$CDCl_3$: DMSO (1:1)	6
$(p\text{-}HOOCC_6H_4)_2Se$	428.6	*b*	$CDCl_3$: DMSO (1:1)	6
$(p\text{-}AcC_6H_4)_2Se$	429.9	*b*	$CDCl_3$: DMSO (1:1)	6
$(p\text{-}NO_2C_6H_4)_2Se$	439.3	*b*	$CDCl_3$: DMSO (1:1)	6

[a]Relative to neat Me_2Se.

[b]Reported relative to selenophene; converted to Me_2Se in $CDCl_3$ (60% v/v) using the relationship $\delta(Me_2Se) = \delta(\text{selenophene}) - 608.6$.

chemical shifts of 4-substituted fluorobenzenes with the ^{77}Se shifts being approximately 2.5 times more sensitive than the ^{19}F shifts. Again it can be seen from the data in Table 8 that the introduction of a perfluorinated benzene ring causes a shielding of the Se nucleus compared to the parent hydrogen compound. For example, compare the ^{77}Se chemical shifts of Ph_2Se (402 ppm), $PhSeC_6F_5$ (265 ppm) and $(C_6F_5)_2Se$ (110 ppm).

(*v*) Alkyl diselenides. ^{77}Se chemical shifts of dialkyl and alkyl aryl diselenides are provided in Table 9. As mentioned previously, a deshielding of approximately 65 ppm is observed in going from $Me_2Se_2 \rightarrow Et_2Se_2 \rightarrow i\text{-}Pr_2Se_2 \rightarrow t\text{-}Bu_2Se_2$. The magnitude of the deshielding is considerably less than that observed in alkyl selenides, as, for example, Me_2Se (0 ppm) and Et_2Se (217 ppm). This smaller deshielding in diselenides has been discussed in terms of the difference in the site where substitution takes place[40]. In going from Me_2Se to Et_2Se, the substitution takes place at each of the α-carbons, whereas in going from Me_2Se_2 to Et_2Se_2, each of the Me groups is introduced at the α-position with respect to one Se atom and at the β-position with respect to the second Se atom. Thus, in diethyl diselenide each of the Se atoms has a Me group in the γ-position, which induces a shielding of Se by the γ-effect (*vide supra*). Therefore, in going from $Me_2Se_2 \rightarrow Et_2Se_2 \rightarrow i\text{-}Pr_2Se_2 \rightarrow t\text{-}Bu_2Se_2$, each new Me group induces deshielding of the β-Se by a dispersion effect and shielding of the distant Se by the γ-effect, resulting in inducing a smaller

TABLE 9. ^{77}Se chemical shifts of alkyl diselenides

Compound	Chemical shift (ppm)	Standard	Solvent	Ref.
Dialkyl				
$(MeSe)_2$	268.3; 270; 275; 275; 281	*a*; *b*; *c*; *d*; *e*	$CDCl_3$; neat; neat; $CDCl_3$; $CDCl_3$	40; 32; 22; 7; 11
$(C_{10}H_{21}Se)_2$	316	*e*	$CDCl_3$	11
$(EtSe)_2$	319; 333.5; 339	*b*; *a*; *c*	Neat; $CDCl_3$; neat	32; 40: 22
$(PhCH_2Se)_2$	401.4; 402; 412	*f*; *d*; *e*	$CDCl_3$: DMSO (1:1); $CDCl_3$; $CDCl_3$	5; 7; 11
$(2\text{-}ClC_6H_4CH_2Se)_2$	395.1	*f*	$CDCl_3$: DMSO (1:1)	5
$(2\text{-}BrC_6H_4CH_2Se)_2$	395.9	*f*	$CDCl_3$: DMSO (1:1)	5
$(4\text{-}FC_6H_4CH_2Se)_2$	403.0	*f*	$CDCl_3$: DMSO (1:1)	5
$(4\text{-}ClC_6H_4CH_2Se)_2$	407.9	*f*	$CDCl_3$: DMSO (1:1)	5
$(4\text{-}BrC_6H_4CH_2Se)_2$	408.2	*f*	$CDCl_3$: DMSO (1:1)	5
$(2,4\text{-}Cl_2C_6H_3CH_2Se)_2$	398.8	*f*	$CDCl_3$: DMSO (1:1)	5
$(i\text{-}PrSe)_2$	401.9; 407	*c*	$CDCl_3$; neat	40, 22
$(t\text{-}BuSe)_2$	493	*c*	Neat	22
$(CF_3Se)_2$	528; 531; 550	*c*; *g*; *h*	Not specified; neat; C_6D_6	37, 33, 41
Alkyl aryl				
PhSe*SeMe	445(*), 294	*c*	Neat	22

[a]Relative to Me_2Se in $CDCl_3$ (60% v/v).
[b]Aq. H_2SeO_3 was used as a reference; chemical shift values have been converted relative to Me_2Se in $CDCl_3$ (60% v/v) using the expression $\delta[Me_2Se] = \delta[H_2SeO_3(aq.)] - 1285.6$.
[c]Relative to neat Me_2Se.
[d]Relative to Me_2Se in $CDCl_3$.
[e]Relative to 1.0M Me_2Se in $CDCl_3$.
[f]Selenophene in $(CD_3)_2CO$ (20%) was used as a reference; chemical shift values have been converted relative to Me_2Se in $CDCl_3$ (60% v/v) using the expression $\delta(Me_2Se) = \delta(\text{selenophene}) - 608.6$.
[g]$SeOCl_2$ was used as a reference; chemical shift values have been converted relative to Me_2Se in $CDCl_3$ (60% v/v) using the expression $\delta(Me_2Se) = \delta(SeOCl_2) - 1482.6$.
[h]Relative to Me_2Se in C_6D_6.

TABLE 10. ^{77}Se chemical shifts of diaryl diselenides, $R^1C_6H_4Se^*SeC_6H_4R^2$[a]

R^1	R^2	Chemical shift (ppm)
H	H	464.1
2-NO_2	2-NO_2	485.0
4-OMe	2-NO_2	352.2(*), 618.6
4-Me	2-NO_2	353.1(*), 596.4
4-H	2-NO_2	356.9(*), 583.6; 357(*), 584[b]
4-NO_2	2-NO_2	372.9(*), 565.7
4-COOMe	4-COOMe	451.4
4-NO_2	4-NO_2	455.8
4-CN	4-CN	457.5
4-OMe	4-H	462.0(*), 505.9
4-Me	4-H	462.0(*), 478.2
4-I	4-I	470.5
4-Me	4-Me	475.7
4-Br	4-Br	475.9
4-Cl	4-Cl	478.2
4-F	4-F	493.5
4-OMe	4-OMe	505.8

[a]Relative to Me_2Se in $CDCl_3$ (60% v/v); solvent $CDCl_3$; Ref. 60.
[b]Relative to Me_2Se; solvent $CDCl_3$; Ref. 61.

magnitude of deshielding compared to that observed in the alkyl selenide series. A similar effect has also been seen in ditellurides, which will be discussed in more detail later in this chapter.

(*vi*) Diaryl diselenides. ^{77}Se chemical shifts of symmetrical and unsymmetrical diaryl diselenides are presented in Table 10. For the symmetrical diselenides, ^{77}Se shifts of 4-substituted derivatives containing strongly electron-accepting groups, strongly electron-donating groups as well as substituents of intermediate character such as Me and halogens have been reported[60]. The total chemical shift range covered is ~55 ppm. Correlation of ^{77}Se shifts with substituent constants of the Hammett and Brown–Okamoto type is poor and a negative slope is obtained. Thus, electron-withdrawing groups shield the ^{77}Se resonance whereas electron-donating groups deshield the ^{77}Se resonance. From the ^{77}Se chemical shifts of unsymmetrical diselenides it can be seen that electron-withdrawing substituents induce deshielding of the proximate Se atom but induce shielding of the distant Se atom. For example, compare *p*-AnSe*SePh (462(*) ppm, 505.9 ppm) and PhSeSePh (464.1 ppm), in which introduction of an OMe group in the 4-position of diphenyl diselenide deshields the distant Se by 41.8 ppm whereas the proximate Se resonance is *shielded*, but only by ~2 ppm. Thus the main effect of a substituent is to affect the distant Se. The reason(s) for this rather unusual chemical shift behaviour certainly require and deserve further study.

(*vii*) Selenenyl sulphides. The large majority of selenenyl sulphides which have been studied by ^{77}Se-NMR spectroscopy (Table 11) were investigated as model compounds for biological systems which had Se covalently bound to thio groups by using the reactant 6,6′-diselenobis-(3-nitrobenzoic acid)[62]. Even though the compounds are relatively similar, the total chemical shift range is over 320 ppm with the most shielded system being 4-$NO_2C_6H_4$SeSBu-*t* and the most deshielded being 2-$NO_2C_6H_4$SeSCN. The Se resonance in these compounds is generally deshielded with respect to the corresponding diselenides presumably due to the slightly greater electronegativity of S compared to Se. For example, the ^{77}Se chemical shift of Ph_2Se_2 (464 ppm) is 62 ppm shielded from

TABLE 11. ^{77}Se chemical shifts of selenenyl sulphides

Compound	Chemical shift (ppm)	Standard	Solvent	Ref.
$4-NO_2C_6H_4SeSBu$-*t*	416.6	*a*	$CDCl_3$	62
$PhSeSCH_2Ph$	475.0	*a*	$CDCl_3$	62
$2-NO_2,4-BrC_6H_3SeSBu$-*t*	514.6	*a*	$CDCl_3$	62
PhSeSPh	526.0	*a*	$CDCl_3$	62
$2-NO_2C_6H_4SeSEt$	563.7	*a*	$CDCl_3$	62
$2-NO_2C_6H_4SeSCH_2CH_2COOH$	565.7	*a*	$CDCl_3$	62
$2-NO_2C_6H_4SeSBu$-*n*	573.1	*a*	$CDCl_3$	62
PhSeSMe	574.0	*b*	Not specified	22
$2-NO_2C_6H_4SeSCH_2CH_2OH$	574.6	*a*	$CDCl_3$	62
$2-NO_2C_6H_4SeSC_{16}H_{33}$	574.7	*a*	$CDCl_3$	62
$2-NO_2C_6H_4SeSCH_2Ph$	577.0	*a*	$CDCl_3$	62
$2-NO_2,4-BrC_6H_3SeSCH_2Ph$	584.6	*a*	$CDCl_3$	62
CF_3SeSCF_3	590	*c*	C_6D_6	41
$2,4-(NO_2)_2C_6H_3SeSEt$	593.9	*a*	$CDCl_3$	62
$2,4-(NO_2)_2C_6H_3SeSCH_2Ph$	607.5	*a*	$CDCl_3$	62
$2-NO_2C_6H_4SeSCH_2COOH$	609.8	*a*	$CDCl_3$	62
$2-NO_2C_6H_4SeSPh$	618.8	*a*	$CDCl_3$	62
$2-NO_2C_6H_4SeS(4-NO_2C_6H_4)$	588.9	*a*	$CDCl_3$	62
$2-NO_2C_6H_4SeS$(4-Tol)	632.3	*a*	$CDCl_3$	62
$2-NO_2C_6H_4SeS(4-FC_6H_4)$	641.7	*a*	$CDCl_3$	62
$2-NO_2C_6H_4SeS$(4-An)	660.9	*a*	$CDCl_3$	62
$2-NO_2,4-BrC_6H_3SeS$(4-Tol)	639.9	*a*	$CDCl_3$	62
$2,4-(NO_2)_2C_6H_3SeS$(4-Tol)	668.9	*a*	$CDCl_3$	62
CF_3SeSCF_2Cl	620	*c*	C_6D_6	41
$CF_3SeSCFCl_2$	656	*c*	C_6D_6	41
$CF_3SeSCCl_3$	680	*c*	C_6D_6	41
$2-NO_2C_6H_4SeSCN$	737.7	*a*	$CDCl_3$	62

[a]Relative to Me_2Se in $CDCl_3$ (60% v/v).
[b]Relative to neat Me_2Se.
[c]Relative to Me_2Se in C_6D_6.

PhSeSPh (526 ppm). Since many of these compounds were studied as model systems, the Se had a Ph group attached. Introduction of an electron-withdrawing group (e.g. —NO_2) in the *ortho* or *para* position deshielded the Se resonance. Interestingly, when a *para* substituent was introduced on a Ph ring bound to S, behaviour of the ^{77}Se chemical shift was opposite to that expected. For example, a NO_2 group on the sulphur Ph ring caused a shielding of the Se while electron donors (in a mesomeric sense) effected a deshielding of the Se resonance. This same effect has been noticed in unsymmetrical diselenides (Table 10).

(*viii*) Seleno esters. The ^{77}Se chemical shifts for this class of compounds are shown in Table 12. Replacement of a Me group in dimethyl selenide by a benzoyl group, —COPh, leads to a deshielding of the ^{77}Se resonance by 445 ppm. Baiwir and coworkers found that the ^{77}Se chemical shifts of *ortho*-substituted butyl selenobenzoates varied in a linear manner with the electronegativities of the substituents if the substituents were halogens[63]. The ^{77}Se chemical shifts of 4-substituted phenyl selenobenzoates, $4-XC_6H_4SeCOPh$, were found to vary in a regular manner with the electronic properties of the substituent[64]. The total chemical shift range for this series covers approximately 28 ppm, only half as much as that observed for 4-substituted selenoanisoles. The observed ^{77}Se chemical shifts and the ^{13}C chemical shifts of the *ipso* and carbonyl carbon were evaluated with the Dual Substituent Parameter analysis (Taft) and all the correlations were found to be

TABLE 12. ^{77}Se chemical shifts of seleno esters, ArCOSeR

Compound	Chemical shift (ppm)	Standard	Ref.
BzSeMe	445.0	*a*	63
BzSeBu	517.0	*a*	63
2-IC_6H_4COSeBu	562.3	*a*	63
2-BrC_6H_4COSeBu	566.2	*a*	63
2-ClC_6H_4COSeBu	569.4	*a*	63
2-FC_6H_4COSeBu(*trans*)	576.2	*a*	63
2-FC_6H_4COSeBu (*cis*)	582.3	*a*	63
2-MeC_6H_4COSeBu	590.6	*a*	63
BzSePh	627.6; 641.5	*a*; *b*	63; 64
BzSe(2-An)	549.5	*a*	64
BzSe(2-Tol)	593.5	*a*	63
BzSe(4-$NMe_2C_6H_4$)	622.9	*b*	64
BzSe(4-An)	628.9	*b*	64
BzSe(4-Tol)	633.1; 634.4	*c*; *b*	36; 64
BzSe(4-FC_6H_4)	634.4	*b*	64
BzSe(4-ClC_6H_4)	637.0	*b*	64
BzSe(4-BrC_6H_4)	637.4	*b*	64
BzSe(4-$EtOOCC_6H_4$)	642.3	*b*	64
BzSe(4-AcC_6H_4)	642.7	*b*	64
BzSe(4-$NO_2C_6H_4$)	645.7	*b*	64
BzSe(4-CNC_6H_4)	650.4	*b*	64
2-IC_6H_4COSe(2-An)	595.4	*a*	63
2-BrC_6H_4COSe(2-Tol)	636.6	*a*	63
2-$MeSeC_6H_4$COSePh	640.8	*a*	63
2-$MeSC_6H_4$COSePh	650.7	*a*	63
2-IC_6H_4COSePh	668.0	*a*	63
2-ClC_6H_4COSePh	674.1	*a*	63
2-AnCOSePh	700.2	*a*	63
4-$OctOC_6H_4$COSe(4-$PenC_6H_4$)	618.9	*c*	36
4-AnCOSe(4-Tol)	620.5	*c*	36
AcSePh	660.0	*a*	63

[a]Relative to Me_2Se in $CDCl_3$; solvent $CDCl_3$.
[b]Relative to Me_2Se in $CDCl_3$ (60% v/v); solvent $CDCl_3$.
[c]$(4\text{-TolSe})_2$ in $CDCl_3$ (~10%) was used as a reference; chemical shifts values have been converted relative to Me_2Se in $CDCl_3$ (60% v/v) using the expression $\delta[Me_2Se] = \delta[(4\text{-TolSe})_2] - 475.7$; solvent $CDCl_3$.

discriminatory towards the σ_R^+ resonance scales. Mesomeric donors were found to interact well with the Ph ring. Thus it is proposed that the selenobenzoato moiety, —SeCOPh, is acting as an electron-deficient group and that the delocalization of electron density of Se onto carbonyl O takes place via an $n_{Se}-\pi^*_{Se}$ interaction.

(*ix*) Compounds containing C═Se double bonds. ^{77}Se chemical shifts of compounds containing C═Se double bonds are spread over a very broad range (Table 13). The ^{77}Se chemical shift of COSe is most shielded at −447 ppm and the Se resonance in $(t\text{-Bu})_2CSe$ is the most deshielded at 2131 ppm. In fact, the chemical shifts of compounds containing the C═Se moiety appear as a class to have a larger chemical shift range than any other type of Se compound. Cullen and coworkers[24], Gombler[65] and Wong, Guziec and Moustakis[26] found a linear correlation, with two exceptions, between the ^{77}Se shifts and λ_{max} of the $n\rightarrow\pi^*$ transition. As the ΔE term appears well correlated with the energies of the $n\rightarrow\pi^*$ transition, the correlation between ^{77}Se chemical shifts and λ_{max} $(n\rightarrow\pi^*)$ clearly shows that the ^{77}Se chemical shift is dominated by the local paramagnetic screening term, σ^{para}. Wong, Guziec and Moustakis[26] have also determined ^{17}O chemical shifts of the corresponding carbonyl compounds and found that the correlation of ^{77}Se chemical shifts

TABLE 13. ^{77}Se chemical shifts of C=Se compounds

Compound	Chemical shift (ppm)	Standard	Solvent	Ref.
Se=C=O	– 447	*a*	C_6D_6	65
Se=C=S	102	*a*	C_6D_6	65
Se=C(NH$_2$)(NMe$_2$)	147	*b*	D_2O, pD 4	11
Se=C(NH$_2$)$_2$	195	*a*	C_6D_6	65
Se=C=Se	331	*a*	C_6D_6	65
Se=C(*t*-Bu)NMe$_2$	640	c	$CDCl_3$	24
Se=CF$_2$	688	*a*	C_6D_6	65
Se=C(Ph)NMe$_2$	733	*c*	$CDCl_3$	24
Se=C(Ph)OEt	915	*c*	$CDCl_3$	24
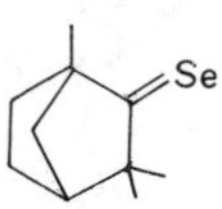	1613	*c*	$CDCl_3$	24
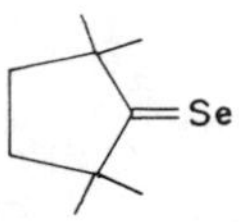	1737	*c*	$CDCl_3$	26
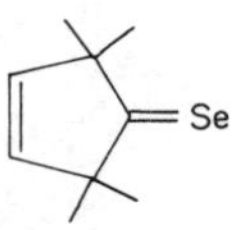	1803	*c*	$CDCl_3$	24
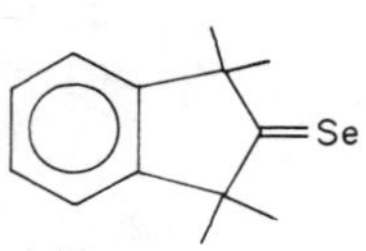	1844; 1849	*c*; *a*	$CDCl_3$; C_6D_6	24; 65
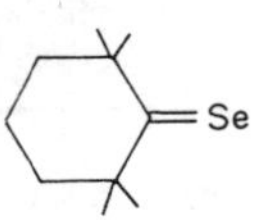	2034	*c*	$CDCl_3$	26
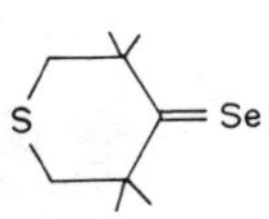	2135	*c*	$CDCl_3$	24
Se=C(Bu-*t*)$_2$	2131; 2162	*c*; *a*	$CDCl_3$; C_6D_6	24; 65

[a]Relative to Me_2Se in C_6D_6.
[b]Relative to 1.0M Me_2Se in $CDCl_3$.
[c]Internal standard Ph_3PSe; converted to Me_2Se in $CDCl_3$ (60% v/v) using the relationship $\delta(Me_2Se) = \delta(Ph_3PSe) + 263$.

of selones and ^{17}O chemical shifts of the corresponding ketones is excellent.

(*x*) Selenocyanates. Se resonances of selenocyanates fall in the range of 125–509 ppm (Table 14). The CN group exhibits a surprisingly small deshielding effect on the Se chemical shift in these compounds. In fact, it is much smaller than might be expected on the

TABLE 14. ^{77}Se chemical shifts of selenocyanates

Compound	Chemical shift (ppm)	Standard	Solvent	Ref.
Alkyl derivatives				
MeSeCN	125	*a*	$CDCl_3$	7
$PhCH_2SeCN$	291; 299.9	*a*; *b*	$CDCl_3$; $(CD_3)_2CO$	7, 5
$2\text{-}ClC_6H_4CH_2SeCN$	284.7	*b*	$(CD_3)_2CO$	5
$2\text{-}BrC_6H_4CH_2SeCN$	285.0	*b*	$(CD_3)_2CO$	5
$2\text{-}FC_6H_4CH_2SeCN$	295.7	*b*	$(CD_3)_2CO$	5
$3\text{-}FC_6H_4CH_2SeCN$	305.8	*b*	$(CD_3)_2CO$	5
$3\text{-}ClC_6H_4CH_2SeCN$	309.0	*b*	$(CD_3)_2CO$	5
$3\text{-}BrC_6H_4CH_2SeCN$	309.8	*b*	$(CD_3)_2CO$	5
$4\text{-}FC_6H_4CH_2SeCN$	306.4	*b*	$(CD_3)_2CO$	5
$4\text{-}ClC_6H_4CH_2SeCN$	309.3	*b*	$(CD_3)_2CO$	5
$4\text{-}BrC_6H_4CH_2SeCN$	309.3	*b*	$(CD_3)_2CO$	5
$2,4\text{-}Cl_2C_6H_3CH_2SeCN$	291.8	*b*	$(CD_3)_2CO$	5
CF_3SeCN	509	*a*	C_6D_6	41
Aryl derivatives				
PhSeCN	320.8; 322.3	*c*; *a*	$CDCl_3$; neat	60, 67
$2\text{-}MeOC(O)C_6H_4SeCN$	394	*a*	$CDCl_3$	59
$2\text{-}NO_2C_6H_4SeCN$	413; 417	*d*; *c*	$CDCl_3$; $CDCl_3$	61, 68
$2\text{-}HC(O)C_6H_4SeCN$	423	*a*	$CDCl_3$	59
$2\text{-}AcC_6H_4SeCN$	434	*a*	$CDCl_3$	59
4-AnSeCN	305.0; 308.8	*a*; *c*	$CDCl_3$; $CDCl_3$	67, 60
4-TolSeCN	313.0	*c*	$CDCl_3$	60
$4\text{-}FC_6H_4SeCN$	318.3	*c*	$CDCl_3$	60
$4\text{-}ClC_6H_4SeCN$	321.0	*c*	$CDCl_3$	60
$4\text{-}BrC_6H_4SeCN$	321.7	*c*	$CDCl_3$	60
$4\text{-}EtOC(O)C_6H_4SeCN$	329.4	*c*	$CDCl_3$	60
$4\text{-}AcC_6H_4SeCN$	329.9	*c*	$CDCl_3$	60
$4\text{-}MeOC(O)C_6H_4SeCN$	330; 330.5	*a*; *c*	$CDCl_3$; $CDCl_3$	7; 60
$4\text{-}NO_2C_6H_4SeCN$	338.8; 340	*c*; *a*	$CDCl_3$; $CDCl_3$	60; 67
$4\text{-}CNC_6H_4SeCN$	341.8	*c*	$CDCl_3$	60
$2,4\text{-}(NO_2)_2C_6H_3SeCN$	449	*c*	$CDCl_3$	68

[a]Relative to neat Me_2Se.
[b]Selenophene in $(CD_3)_2CO$ (20%) was used as a reference; chemical shift values have been converted relative to Me_2Se in $CDCl_3$ (60% v/v) using the expression $\delta(Me_2Se) = \delta(\text{selenophene}) - 608.6$.
[c]Relative to Me_2Se in $CDCl_3$ (60% v/v).
[d]Relative to Me_2Se in $CDCl_3$ (concentration not specified).

basis of its electron-withdrawing power (i.e. its electronegativity) as well as by comparison with the ^{77}Se chemical shifts of selenoisocyanates and selenoisothiocyanates (see Table 22). The deshielding induced by the CN group on the ^{77}Se chemical shift is approximately of the same magnitude as that induced by an acetylenic group[28]. Although the CN group and the C≡C bond are isoelectronic, the more polar CN group would be expected to induce a larger deshielding if inductive polarization dominates. Thus other factors must contribute substantially in the case of triply bonded groupings. Similar behaviour of the CN group is observed in the α-carbon chemical shifts of alkyl cyanides[66].

The ^{77}Se chemical shifts of *para*-substituted aryl selenocyanates vary in a regular way with the character of the substituent[60], i.e. electron-donating groups cause shielding and electron-withdrawing groups cause deshielding. The shifts cover a range of approximately 33 ppm (becoming more deshielded from —Me to —CN), which is similar to the range observed for the substituent effect on the Se chemical shifts of *para*-substituted phenyl selenobenzoates (27.5 ppm). A Dual Substituent Parameter analysis of the shift data support the concept that the level of π interaction between the —SeCN group and the Ph ring is approximately 1.5 times the level of polar effects ($\rho_R/\rho_I = 1.49$)[60].

(*xi*) Selenenyl halides. This class of compounds includes mono-, di- and tri-halides and thus a relatively broad range of chemical shifts is observed (Table 15). Although, as expected, replacement of a hydrogen by a halogen induces a large deshielding, further introduction of halogens will reverse the trend. This can be seen when comparing the ^{77}Se chemical shifts of CF_3SeH (287 ppm), CF_3SeCl (1077 ppm) and CF_3SeCl_3 (890 ppm). It should be noted that this behaviour is also observed in the ^{119}Sn chemical shifts of alkyltin halides and in the ^{13}C chemical shifts of alkyl halides when the halogens are Br and I.

In a recent study[41] of trifluoromethyl selenenyl halides, a plot of the ^{77}Se chemical shift of the selenenyl derivative CF_3SeX (where X = Cl, Br, CN, H, Ag) vs. the electronegativity of the substituent X yielded a relatively straight line although there is some question concerning the scientific validity of this plot. All substituent electronegativity values were chosen from one scale, except that for the CN group which was taken from a separate scale. Electronegativity scales are internally consistent but values from different scales should not be mixed as has been done in this case. Thus, with respect to a correlation with electronegativity, there is reason to believe that the ^{77}Se chemical shift for CF_3SeCN may be anomalous.

TABLE 15. ^{77}Se chemical shifts of selenenyl halides[a]

Compound	Chemical shift (ppm)	Solvent	Ref.
Alkyl derivatives			
Me_2SeBr_2	389	CH_2Cl_2	22
Me_2SeCl_2	448	CH_2Cl_2	22
Et_2SeBr_2	540	CCl_4	22
i-Pr_2SeBr_2	742	CCl_4	22
$(CF_3)_2SeF_2$	830	C_6D_6	41
CF_3SeBr	886	C_6D_6	41
MeSeCl	890	CH_2Cl_2	22
CF_3SeCl_3	953	C_6D_6	41
$EtSeCl_3$	995	CH_2Cl_2	22
CF_3SeCl	1077	C_6D_6	41
Aryl derivatives			
Ph_2SeCl_2	586	$CDCl_3$	22
C_6F_5SeCl	812	Not specified	39
PhSeBr	869.0	$CDCl_3$	68
$2\text{-}NO_2C_6H_4SeBr$	908;912	$CDCl_3$;$CDCl_3$	61;68
$2\text{-}CH(O)C_6H_4SeBr$	1019	$CDCl_3$	59
$2\text{-}AcC_6H_4SeBr$	1029	$CDCl_3$	59
$2\text{-}MeOC(O)C_6H_4SeBr$	1042	$CDCl_3$	59
$4\text{-}NO_2C_6H_4SeBr$	823.0	$CDCl_3$	68
4-TolSeBr	876.9	$CDCl_3$	68
4-AnSeBr	887.7	$CDCl_3$	68
$2,4\text{-}(NO_2)_2C_6H_3SeBr$	897.6	$CDCl_3$	68
PhSeCl	1042	$CDCl_3$	68
$2\text{-}NO_2C_6H_4SeCl$	999	$CDCl_3$	68
$2\text{-}MeOC(O)C_6H_4SeCl$	1017	$CDCl_3$	59
$2\text{-}AcC_6H_4SeCl$	1087	$CDCl_3$	59
$2\text{-}HC(O)C_6H_4SeCl$	1097	$CDCl_3$	59

[a]Chemical shifts are reported relative to Me_2Se.

Similar to the above study, a plot of ^{77}Se chemical shifts of *o*-nitrophenyl selenenyl compounds, *o*-$NO_2C_6H_4SeX$ (where the first element of the X group is O, Cl, Br, N, S, Se, C), vs. the electronegativities of X yields two straight lines. One line represents the substituents X = C, N and O and has a different slope than the line where X = Cl, Br, Se and S.

(*xii*) Heterocyclic selenium compounds. ^{77}Se NMR parameters for a series of 2- and 3-substituted selenophenes are presented in Table 16. Acceptable linear correlations were observed between the substituent-caused ^{77}Se shifts of 2-substituted selenophenes and similarly *ortho*-positioned ^{13}C shifts when carbonyl-containing derivatives and 2-nitroselenophene were excluded[4]. The substituent-caused ^{77}Se shifts in the 3-substituted derivatives behaved as if the substituents were in the para position rather than in the *meta* position and these ^{77}Se shifts were approximately six times larger than ^{13}C shifts of similarly positioned carbons. The ^{77}Se shifts are much smaller than expected for the 2-CHO group and furthermore an upfield shift was observed even though the 2-CHO group is electron-withdrawing. Although the difference between the ^{77}Se shift caused by the CN group (712.9 ppm) and the CHO group (602.2 ppm) is substantial, the ^{13}C chemical shifts of the 3-carbon are almost the same (11.5 ppm and 10.5 ppm from the chemical shift of the 3-carbon of selenophene, respectively) The anomalous ^{77}Se shifts of 2-carbonyl derivatives were explained by invoking through-space bonding between empty Se d orbitals and the carbonyl O lone pair in the *cis* conformation of the 2-carbonyl derivatives. A similar through-space d orbital interaction was also suggested to occur for 2-nitroselenophene, while this interaction is absent in 2-cyanoselenophene.

In addition to the substituted selenophenes discussed above, a series of mono- and di-substituted benzo[b]selenophenes have been studied and their ^{77}Se chemical shifts are listed in Table 17 and 18, respectively[69]. The general trends observed in the monosubstituted benzo[*b*]selenophenes are the same as those observed in the corresponding selenophenes. As observed in the selenophenes, nitro- and carbonyl-containing derivatives show relatively lower deshielding of Se resonances. No correlation has been established between the ^{77}Se chemical shifts in 2- and 3-substituted benzo[*b*]selenophenes. In disubstituted derivatives, additivity rules do not usually hold; however, a linear correlation has been established between the ^{77}Se chemical shifts of 2-substituted benzo[*b*]selenophenes and those of the corresponding 2,3-disubstituted derivatives when the 3-substituent is Me. In these compounds, the observed shifts are in good agreement with the shifts calculated by additivity.

Additional heterocyclic compounds of Se and their ^{77}Se chemical shifts are listed in Tables 19 and 20.

(*xiii*) Selenium–oxygen compounds. ^{77}Se chemical shifts of Se compounds containing a Se–O–C moiety are presented in Table 21. In general, the presence of a neighbouring O causes a large deshielding of Se and thus this class of compounds exhibits the most deshielded Se resonances among the organoselenium compounds with certain exceptions (e.g. some C=Se compounds, *vide supra*). It is interesting to note that the replacement of a Me group by a Ph group in methylseleninic acid, MeSeOOH, leads to a shielding (−40 ppm) rather than the expected deshielding of the Se nucleus; e.g. MeSeOOH (1216 ppm), PhSeOOH (1173 ppm), 3-TolSeOOH (1176 ppm). Actually the Se chemical shift in this class of compounds appears to be less sensitive to the changes in the structure of the neighbouring groups than most of the classes of compounds which we have discussed. For example, the ^{77}Se chemical shift difference between $PhCH_2SeOOH$ and MeSeOOH is only 19 ppm and that between $(MeO)_2SeO_2$ and $(EtO)_2SeO_2$ is only 6 ppm. In the former series, RSeOOH, the Se atom has one non-bonded pair of electrons and in the latter series, R_2SeO_2, the Se atom has no lone pairs. Moreover, for the transition $(MeO)_2SeO \rightarrow (MeO)_2SeO_2$ an increased shielding of Se of approximately 290 ppm is observed. Therefore, it appears that Se shielding is less sensitive to changes in substituents

TABLE 16. ^{77}Se chemical shifts of substituted selenophenes[a,b]

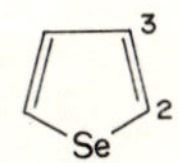

Substituent	Chemical shift (ppm)	Substituent	Chemical shift (ppm)
2-F	513.2	2-Br	674.6
2-OMe	517.3	2-CN	712.9
2-OAc	576.2	2-I	721.2
2-CH(Me)OAc	590.5	3-OMe	527.9
2-CH_2OH	600	3-Me	593.4
2-CHO	602.2	3-Cl	622.2
2-H	608.6	3-SMe	628.1
2-Me	612.4	3-Br	646.7
2-NO_2	614.2	3-COOH	652.1
2-Ac	620.3	3-Ac	653.4
2-COOMe	633.0	3-NO_2	654.4
2-COOH	634.8	3-CHO	663.9
2-Cl	649.8	3-CN	667.0
2-SMe	650.3	3-I	681.0
2-$CONMe_2$	650.9		

[a]Solvent C_6D_6 (20% solutions). Chemical shifts are reported relative to selenophene; converted to Me_2Se in $CDCl_3$ (60% v/v) using the relationship $\delta(Me_2Se) = \delta(\text{selenophene}) - 608.6$.

[b]Ref. 4.

TABLE 17. ^{77}Se chemical shifts of substituted benzo[*b*]selenophenes[a,b]

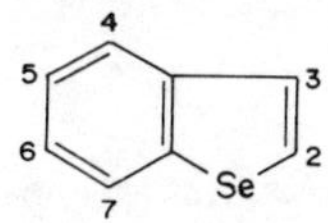

Compound	Chemical shift (ppm)	Compound	Chemical shift (ppm)
2-CHO	509.5	2-CN	604.6
2-CH_2OH	517	2-I	639.1
2-CH_2CN	517.1	3-OMe	425.7
2-NO_2	517.8	3-NHAc	454.9
2-Ac	527.4	3-Me	493.5
2-H	529.6	3-CH_2CN	497.3
2-CH_2SeCH_2COOH	531.3	3-Cl	508.6
2-CONHMe	536	3-CH_2SeCH_2COOH	510.2
2-COOEt	536.6	3-Ac	518.6
2-COOMe	537.7	3-SMe	526.7
2-COSMe	537.7	3-NO_2	531.4
2-Me	538.6	3-Br	533.7
2-CH_2Ph	538.8	3-SeMe	542.3
2-Bz	544.8	3-CHO	544.8
2-$CONMe_2$	564.1	3-Ph	555.1
2-SMe	569.2	3-CN	559.6
2-Cl	577.1	4-NO_2	553.9
2-SeMe	580.3	5-NO_2	562.0
2-$Se(CH_2)_2COOH$	587.6	7-NO_2	618.1
2-Br	601.1		

[a]Solvent $CDCl_3$. Chemical shifts are reported relative to benzo[*b*]selenophene; converted to Me_2Se in $CDCl_3$ (60% v/v) using the relationship $\delta(Me_2Se) = \delta$(benzo[*b*]selenophene) − 529.6.

[b]Ref. 69.

TABLE 18. ^{77}Se chemical shifts of disubstituted benzo[*b*]selenophenes[a,b]

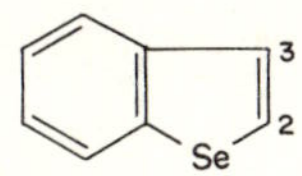

Substituents	Chemical shift (ppm)	Substituent	Chemical shift (ppm)
2-COOEt, 3-OH	465.1	2-Bz, 3-Me	549.8
2-Ac, 3-OAc	481.6	2-SeMe, 3-Me	552.8
2-Me, 3-OAc	484.0	2-Cl, 3-Me	553.4
2-NO_2, 3-Br	484.1	2-CH_2Ph, 3-CHO	560.9
2-COOEt, 3-OAc	487.8	2-Me, 3-Br	561.5
2-NHCOMe, 3-Me	488.7	2-Me, 3-Ac	564.0
2-CHO, 3-Me	497.2	2-SMe, 3-Ac	566.3
2-COOEt, 3-OMe	498.2	2-SeMe, 3-Ac	569.7
2-CH_2OH, 3-Me	500.7	2-Me, 3-COOMe	570.3
2-OAc, 3-Ac	515.4	2-Br, 3-Me	573.9
2-Me, 3-Me	520.3	2-COOEt, 3-Br	573.9
2-Ac, 3-Me	521.2	2-CN, 3-Me	575.8
2-COOMe, 3-Me	526.8	2-Ac, 3-SeMe	582.1
2-CONMe_2, 3-Me	527.7	2-Me, 3-CN	582.4
2-CONH_2, 3-Me	534.9	2-SeMe, 3-CHO	591.6
2-COOMe, 3-NO_2	536.9	2-Me, 3-I	599.8
2-CHO, 3-SPh	537.7	2-Ac, 3-OH	601.6
2-Me, 3-CONH_2	539.7	2-SPh, 3-CHO	602.7
2-SMe, 3-Me	540.7	2-I, 3-Me	614.8
2-OH, 3-Ac	540.8	2-Br, 3-Br	615.9
2-CHO, 3-Br	544.3	2-Br, 3-CHO	622.9
2-CHO, 3-SeMe	549.6		

[a]Solvent $CDCl_3$. Chemical shifts are reported relative to benzo[*b*]selenophene; converted to Me_2Se in $CDCl_3$ (60% v/v) using the relationship $\delta(Me_2Se) = \delta$(benzo[*b*]selenophene) − 529.6.
[b]Ref. 69.

TABLE 19. ^{77}Se chemical shifts of S–Se and Se–Se isosteres of benzopentalene[a]

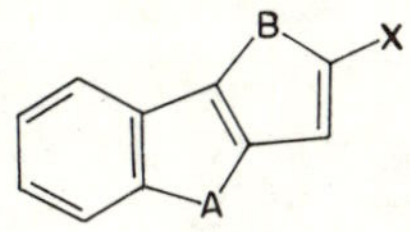

trans

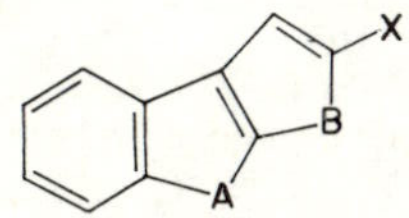

cis

A–B	X = H		X = Ac		X = CHO		X = Br		X = Me		$X = NO_2$	
	A	B	A	B	A	B	A	B	A	B	A	B
Se–Se (*cis*)	499.7	583.4	521.5	597.1	525.0	583.8	515.6	655.6	498.2	595.8	550.2	599.6
Se–Se (*trans*)	486.7	549.6	494.9	588.8	486.6	549.4	508.5	634.0	488.6	555.3	524.4	563.2
Se–S (*cis*)	475.4	—	491.4	—	496.1	—	495.8	—	475.2	—	526.0	—
Se–S (*trans*)	459.4	—	465.2	—	468.8	—	479.7	—	457.2	—	504.5	—
S–Se (*cis*)	—	555.1	—	565.7	—	553.1	—	628.9	—	567.3	—	573.7
S–Se (*trans*)	—	525.8	—	533.9	—	519.8	—	611.0	—	525.5	—	541.3

[a]Relative to neat Me_2Se; solvent $CDCl_3$; Ref. 70.

TABLE 20. ^{77}Se chemical shifts of other heterocycles containing selenium

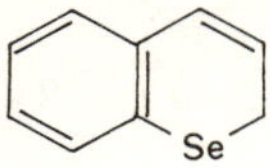

(**1**)

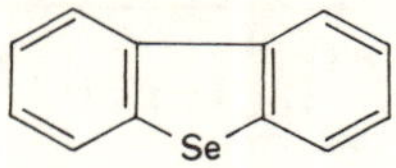

(**2**)

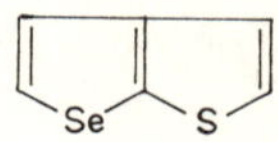

(**3**)

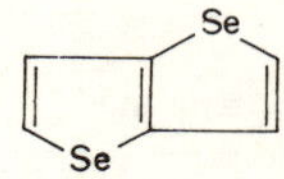

(**4**)

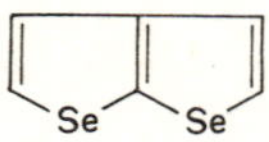

(**5**)

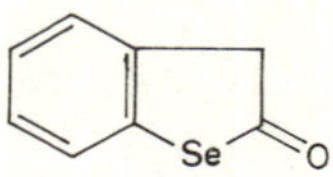

(**6**)

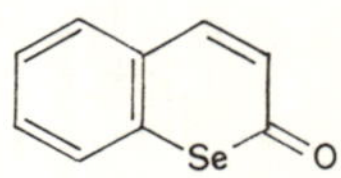

(**7**)

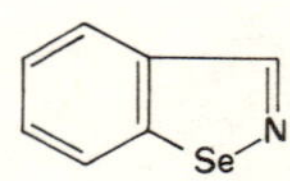

(**8**)

Compound	Chemical shift (ppm)	Standard	Solvent	Ref.
1	157	*a*	$CDCl_3$	7
2	451	*a*	$CDCl_3$	7
3	545	*a*	$CDCl_3$	7
4	552.6	*b*	$(CD_3)_2CO$	71
5	581	*a*	$CDCl_3$	7
6	584	*a*	$CDCl_3$	7
7	654	*a*	$CDCl_3$	7
8	1013	*a*	$CDCl_3$	7

[a]Relative to neat Me_2Se.
[b]Selenophene in $(CD_3)_2CO$ was used as a reference; chemical shift values have been converted relative to Me_2Se in $CDCl_3$ (60% v/v) using the expression $\delta(Me_2Se) = \delta(\text{selenophene}) - 608.6$.

in compounds in which Se does not have lone pairs, like R_2SeO_2, and much more sensitive in compounds in which Se has two non-bonded electron pairs, like alkyl selenides and dialkyl diselenides. This, of course, fits well with the polarizability concept discussed earlier in this chapter.

(*xiv*) Miscellaneous selenium compounds. A substantial deshielding of Se is observed in going from selenides to selenonium salts. The magnitude of this deshielding depends both upon the nature of the added alkyl group and the groups already present. Thus the formation of Me_3Se^+ from Me_2Se is accompanied by a deshielding of 253 ppm, whereas when Me_2EtSe^+ is formed from MeSeEt the deshielding is only 183 ppm (Table 22). Similarly the transformation $Et_2Se \rightarrow Et_3Se^+I^-$ and $Me_2Se \rightarrow Me_2EtSe^+I^-$ are accompanied by deshieldings of 144 and 291 ppm.

d. Isotope effects. Several groups have studied the effect of isotopes on ^{77}Se nuclear shielding. Actually, the first isotope effect involving Se was observed on the F chemical shift in ^{19}F-NMR spectroscopy in several Se–F compounds[74]. The first example of Se isotope effects on ^{77}Se shielding was that of Jakobsen and Hansen in a study[75] of the ^{77}Se-NMR spectrum of Me_2Se_2 in which four well-resolved lines due to the naturally occurring isotopic species $Me^{77}Se—^{n}SeMe$, where $n = 76$, 78, 80 and 82, were observed (natural abundances of selenium isotopes are: ^{74}Se, 0.96%; ^{76}Se, 9.12%; ^{77}Se, 7.5%; ^{78}Se, 23.61%; ^{80}Se, 49.61%; ^{82}Se, 8.85%). An increase by two mass units in Se mass was found to induce a

TABLE 21. ^{77}Se chemical shifts of selenium–oxygen compounds

Compound	Chemical shift (ppm)	Standard	Solvent	Ref.
$PhSe(OCH_2Ph)_2Me$	672	*a*	$CDCl_3$	58
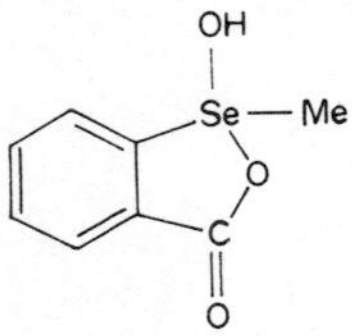	799	*c*	MeOH	58
Me_2SeO	812	*a*	H_2O	22
$2\text{-}HOOCC_6H_4Se(ONa)Me$	830	*a*	MeOH	58
PhSe(O)Me	832	*a*	$CDCl_3$	58
$2\text{-}MeOC(O)C_6H_4Se(O)Me$	852	*a*	$CDCl_3$	58
Ph_2SeO	863	*a*	$CDCl_3$	58
$(EtO)_2SeO_2$	1047	*a*	CH_2Cl_2	22
$(MeO)_2SeO_2$	1053	*a*	CH_2Cl_2	22
$(CF_3)_2SeO$	1095	*a*	C_6D_6	41
MeSeOOH	1216	*a*	Neat	22
$PhCH_2SeOOH$	1235	*b*	$(CD_3)_2SO$	72
$2\text{-}ClC_6H_4CH_2SeOOH$	1228.5	*b*	$(CD_3)_2SO$	72
$2\text{-}FC_6H_4CH_2SeOOH$	1228.7	*b*	$(CD_3)_2SO$	72
$2\text{-}BrC_6H_4CH_2SeOOH$	1230.8	*b*	$(CD_3)_2SO$	72
$2\text{-}TolCH_2SeOOH$	1240.5	*b*	$(CD_3)_2SO$	72
$3\text{-}BrC_6H_4CH_2SeOOH$	1228.2	*b*	$(CD_3)_2SO$	72
$3\text{-}FC_6H_4CH_2SeOOH$	1231.1	*b*	$(CD_3)_2SO$	72
$3\text{-}ClC_6H_4CH_2SeOOH$	1232.6	*b*	$(CD_3)_2SO$	72
$3\text{-}TolCH_2SeOOH$	1234.1	*b*	$(CD_3)_2SO$	72
$4\text{-}ClC_6H_4CH_2SeOOH$	1226.8	*b*	$(CD_3)_2SO$	72
$4\text{-}BrC_6H_4CH_2SeOOH$	1228.9	*b*	$(CD_3)_2SO$	72
$4\text{-}TolCH_2SeOOH$	1229.8	*b*	$(CD_3)_2SO$	72
$4\text{-}FC_6H_4CH_2SeOOH$	1231.7	*b*	$(CD_3)_2SO$	72
$2,4\text{-}Cl_2C_6H_3CH_2SeOOH$	1227.1	*b*	$(CD_3)_2SO$	72
PhSeOONa	1173	*c*	D_2O	23
3-TolSeOOH	1176.7	*b*	$(CD_3)_2SO$	72
4-TolSeOOH	1175.9	*b*	$(CD_3)_2SO$	72
$(MeO)_2SeO$	1339	*a*	Neat	22

[a]Relative to neat Me_2Se.
[b]Selenophene in $(CD_3)_2CO$ (20%) was used as a reference; chemical shift values have been converted relative to Me_2Se in $CDCl_3$ (60% v/v) using the expression $\delta(Me_2Se) = \delta(\text{selenophene}) - 608.6$.
[c]Relative to Me_2Se in $CDCl_3$ (60% v/v).

shift of 2.62×10^{-2} ppm (0.50 Hz at 19.09 MHz). In this same study[75], 2-deuteroselenophene exhibited a ^{77}Se chemical shift 32.1 Hz to lower frequency from that of the hydrogen compound which was a two-bond isotope shift of 1.682 ppm. Also, in the same sample of selenophene a one-bond ^{13}C (^{12}C) isotope effect of 0.257 ppm (4.9 Hz) on the ^{77}Se chemical shift was determined from the ^{13}C satellite doublet.

The largest secondary isotope shift ever observed in high-resolution NMR spectroscopy was reported for the ^{77}Se spectrum of a liquid mixture of H_2Se, HDSe and D_2Se[76]. The isotope shift per D atom is 7.02 ppm (133 Hz at 18.95 MHz) to lower frequency and in the

TABLE 22. ^{77}Se chemical shifts of miscellaneous organoselenium compounds

Compound	Chemical shift (ppm)	Standard	Solvent	Ref.
CF_3SeAg	26	*a*	Pyridine	41
$Me_3Se^+I^-$	253	*b*	H_2O	22
$Et_2MeSe^+I^-$	325	*b*	H_2O	22
$PhSeCH_2CH{=}CH_2$	321.9	*b*	Not specified	73
$4\text{-}MeOC_6H_4SeCH_2CH{=}CH_2$	313.5	*b*	Not specified	73
$4\text{-}NO_2C_6H_4SeCH_2CH{=}CH_2$	344.1	*b*	Not specified	73
$4\text{-}AnSe^+Me_2I^-$	325	*b*	H_2O	22
PhSeCH=CHCl (*trans*)	368.6	*b*	Not specified	73
4-AnSeCH=CHCl (*trans*)	361.8	*b*	Not specified	73
4-TolSeCH=CHCl (*trans*)	364.6	*b*	Not specified	73
$4\text{-}ClC_6H_4SeCH{=}CHCl$ (*trans*)	366.3	*b*	Not specified	73
$4\text{-}BrC_6H_4SeCH{=}CHCl$ (*trans*)	366.5	*b*	Not specified	73
$4\text{-}NO_2C_6H_4SeCH{=}CHCl$ (*trans*)	377.0	*b*	Not specified	73
$PhSeC_3H_3Cl_2$ (cyclo)	370.2	*b*	Not specified	73
$4\text{-}AnSeC_3H_3Cl_2$ (cyclo)	356.8	*b*	Not specified	73
$4\text{-}TolSeC_3H_3Cl_2$ (cyclo)	362.9	*b*	Not specified	73
$4\text{-}ClC_6H_4SeC_3H_3Cl_2$ (cyclo)	370.5	*b*	Not specified	73
$4\text{-}BrC_6H_4SeC_3H_3Cl_2$ (cyclo)	371.5	*b*	Not specified	73
$4\text{-}NO_2C_6H_4SeC_3H_3Cl_2$ (cyclo)	396.8	*b*	Not specified	73
$Et_3Se^+I^-$	377	*b*	H_2O	22
$PhSeCH{=}CH_2$	395.5	*b*	Not specified	73
$4\text{-}AnSeCH{=}CH_2$	386.7	*b*	Not specified	73
$4\text{-}TolSeCH{=}CH_2$	390.9	*b*	Not specified	73
$4\text{-}ClC_6H_4SeCH{=}CH_2$	395.4	*b*	Not specified	73
$4\text{-}BrC_6H_4SeCH{=}CH_2$	396.0	*b*	Not specified	73
$4\text{-}NO_2C_6H_4SeCH{=}CH_2$	404.9	*b*	Not specified	73
$(CF_3Se)_3As$	505	*a*	C_6D_6	41
$(CF_3Se)_3P$	541	*a*	C_6D_6	41
$CF_3SeSO_2CF_2Cl$	975	*a*	C_6D_6	41
$CF_3SeSO_2CFCl_2$	976	*a*	C_6D_6	41
$CF_3SeSO_2CF_3$	984	*a*	C_6D_6	41
CF_3SeNSO	1091	*a*	C_6D_6	41
CF_3SeNCO	1104	*a*	C_6D_6	41
$CF_3SeNPPh_3$	1112	*a*	C_6D_6	41
$(CF_3Se)_3N$	1617	*a*	C_6D_6	41

[a]Relative to Me_2Se in C_6D_6.
[b]Relative to neat Me_2Se.

^1H-coupled ^{77}Se spectrum each resonance is fully separated from resonances of the other isotopomers. In this study primary and secondary isotope effects on the $^{77}Se{-}^1H$ coupling constants were also observed.

In two other recent reports, Gombler reported further studies on C and Se isotope effects on ^{77}Se nuclear shielding[77,78]. ^{13}C isotope effects were measured for 30 organoselenium compounds and were found to range from 0.012 ppm in CF_3SeCl to 1.099 ppm in Se=C $(Bu\text{-}t)_2$[77]. Several important findings were (1) the magnitude of the isotope shift could be correlated with C—Se bond distance, (2) appreciable substituent effects were observed in closely related compounds, (3) there was no general correlation between the electronegativity of the substituent and the isotope shift and (4) the isotope shift could be temperature-dependent. In the study of the effect of Se isotopes on ^{77}Se shielding[78], the compounds $(CF_3)_2Se_2$, Me_2Se_2 and $CF_3SeSeMe$ were investigated.

Jakobsen's earlier work on Me_2Se_2 was essentially confirmed, although instead of a constant isotope shift it was found that the isotope shift per unit mass difference decreased with increasing mass of the Se isotopes. A very interesting observation was that the ^{77}Se spectrum of the MeSe group of $CF_3SeSeMe$ exhibited a fine splitting for each isotopomer which was attributed to the occurrence of two conformers. This suggests a relatively high barrier to rotation around the Se—Se bond in this molecule and why this should be so is not clear. This molecule and its ^{77}Se-NMR spectrum deserve further study.

3. Coupling constants

Spin–spin coupling constants involving ^{77}Se have been measured for a relatively large number of nuclei and this area has been reviewed relatively recently[79]. Of primary importance in this review are $^{77}Se-^{13}C$ and $^{77}Se-^{77}Se$ coupling constants in organoselenium compounds although $^{77}Se-^{1}H$ coupling constants deserve mention. A large number of $^{77}Se-^{1}H$ coupling constants have been reported and the area has not been reviewed since 1973[2]. However, in the intervening years, nothing more of substance has been uncovered. Of particular interest is the possible use of coupling constants to probe structure and bonding and $^{77}Se-^{1}H$ couplings appear to be sensitive to stereochemistry[80–82].

Although fewer $^{77}Se-^{13}C$ coupling constants have been reported, they are increasingly being investigated with regard to structure and bonding in organoselenium compounds. McFarlane and coworkers[83] were the first to study $^{77}Se-^{13}C$ coupling constants in a variety of organoselenium compounds in which the Se atom had from zero to three unshared electron pairs. These coupling constants ranged from -13.0 Hz in $MeSeO_3^-$ K^+ to -123 Hz in $PhSe^-$ K^+ and the sign was always negative. It was suggested that, in general, a coupling constant of more than 45 Hz was indicative of a direct Se—C bond but the variety of factors affecting the magnitudes of the couplings would make it difficult to use them as structural probes. However, Reich demonstrated[82] that two-bond Se–C coupling constants in a selenide, a selenoxide, a selenium salt and a selenonium ylide in which the Se is part of a dihydrobenzoselenophene system are stereospecific. More recently one-bond Se–C coupling constants were found[58] to be useful in distinguishing between selenuranes (61–65 Hz) and selenoxides (77–81 Hz) and, in an attempt to further understand the factors influencing the Se–C spin–spin coupling, a series of diorganyl selenides, RSeR′, was investigated[84]. The conclusions of this study[84] were that the Fermi contact term is the principal contributor to one-bond and two-bond Se–C couplings and that $^{1}J(^{77}Se-^{13}C)$ can be used to estimate the bond order of the Se—C bond as well as the conformation of the selenide. It remains to be seen if this will be the case in a large number of systems.

That the Fermi contact contribution is important in Se–C spin–spin coupling in compounds containing a C=Se bond was demonstrated in 1981 by Gombler[65] as the couplings exhibited a strong dependence on the s character of the C atom. As expected, coupling constants between Se and C in these doubly bonded molecules are larger (>200 Hz) with the largest coupling constant being 286.9 Hz for COSe. Wong and coworkers[24,26,85] have also studied a series of selenocarbonyl compounds and confirm that the Fermi contact interaction is dominant in determining $^{1}J(^{77}Se-^{13}C)$. They also reported[26] that where the selenocarbonyl moiety is part of a ring system, due to a change in the s character of the C atom in the C=Se group, the smaller the CC(Se)C angle, the larger the one-bond Se–C coupling constant.

Se–Se coupling constants have understandably received even less attention than Se–C coupling constants. First, the particular compound must possess chemically non-equivalent Se atoms (i.e. $J(^{77}Se-^{77}Se)$ cannot be determined in a symmetrical diselenide) and second, the number of molecules in any diselenide sample possessing two ^{77}Se nuclei will be small (0.56%) and thus, the $^{77}Se-^{77}Se$ coupling constant determination must be from satellites and will be time-consuming. The first measurement of the sign and magnitude of a $^{77}Se-^{77}Se$ spin coupling constant was reported by McFarlane[86] in PhSeSeMe ($+22 \pm 4$ Hz). With the exception of $^{77}Se-^{77}Se$ coupling constants which have been measured in polyatomic inorganic cations[87] containing Se and Te, only one report has appeared[61] in the literature describing the determination of these couplings. The compounds were unsymmetrical diselenides and, in agreement with McFarlane's report, the values were less than 20 Hz. Attempts were also made to determine two- and three-bond $^{77}Se-^{77}Se$ coupling constants but these were unsuccessful and the values were said to be less than 2 Hz. Recent work in this laboratory[88] has resulted in the measurement of several $^{1}J(^{77}Se-^{77}Se)$ values in unsymmetrical diorganyl diselenides. These values range from 2.7 Hz in *t*-BuSeSeMe to 36.3 Hz in *n*-BuSeSeMe. Using enriched ^{77}Se we have also obtained a value of 64 Hz in *o*-$NO_2C_6H_4$SeSeCN for the one-bond $^{77}Se-^{77}Se$ coupling constant[68]. Also, Johannsen has recently observed $^{3}J(^{77}Se-^{77}Se)$ values in a series of substituted tetraselenafulvalenes and used the magnitude of this homonuclear coupling constant to distinguish between *cis* and *trans* isomers[89]. The *cis* orientation resulted in relatively large values of the coupling constant (80–100 Hz) while the *trans* geometry yielded much smaller values (10–25 Hz). Clearly, more extensive studies of this parameter and its relationship to molecular and electronic structure should be rewarding.

B. Organotellurium Compounds

1. Relaxation times

In general, progress in Te-NMR spectroscopy has lagged somewhat behind that in ^{77}Se-NMR spectroscopy. However, in the last 2–3 years publications involving ^{125}Te-NMR studies have increased dramatically, although ^{125}Te relaxation times have not been studied nearly so thoroughly. One reason for this may be that researchers have felt that what has been found for Se will also hold true for Te, but the few preliminary studies which have appeared[15,90–93,114] demonstrate that spin-lattice relaxation times for ^{125}Te are approximately 6–7 times shorter than those of ^{77}Se in analogous compounds (Table 23). It has been shown that the spin rotation mechanism is the dominant mechanism for small Te-containing molecules. It has not yet been demonstrated that the chemical shift anisotropy mechanism may be important in larger organotellurium molecules as has been shown for ^{77}Se.

2. Chemical shifts

Like Se, the chemical shift pattern of Te compounds is considered to be dominated by the paramagnetic term of the chemical shift equation. The paramagnetic term for Te is given by the following proportionality:

$$\sigma_{\mathrm{Te}}^{\mathrm{para}} \propto \frac{-\langle r^{-3}\rangle_{5\mathrm{p}}}{\Delta E}\Sigma Q \tag{10}$$

Expressions (4) and (10) predict that if the ΣQ terms for Se and Te are the same in analogous compounds, a plot of $\delta_{125\mathrm{Te}}$ vs. $\delta_{77\mathrm{Se}}$ should be a straight line with a slope of $\langle r^{-3}\rangle_{5\mathrm{p(Te)}}\cdot\Delta E_{\mathrm{Se}}/\langle r^{-3}\rangle_{4\mathrm{p(Se)}}\cdot\Delta E_{\mathrm{Te}}$. Such a plot has been found to be linear with a slope of 1.7–1.8[57,94]. Whereas a part of this can be attributed to the ratio of $\langle r^{-3}\rangle_{5\mathrm{p(Te)}}/\langle r^{-3}\rangle_{4\mathrm{p(Se)}} \simeq 1.25$[94], there is still an unexplained factor which must come from the difference in the ΔE values. It is difficult to obtain a reasonable estimate of ΔE for Se and Te; however, changes in these ΔE values should roughly parallel those found

TABLE 23. ^{125}Te spin-lattice relaxation times, T_1

Compound	T_1 (s)	Temp. (°C)	Conditions	Ref.
Me_2Te	0.58	26	90%, C_6D_6	92
Et_2Te	1.1	30	2M, $CDCl_3$	91
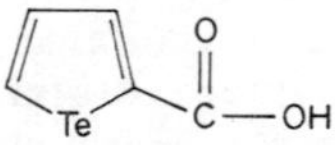	~3	*a*	$(CD_3)_2CO$	114
$Te(OCHMe_2)_4$	2.23	*a*	*a*	15
$Te(OCH_2CF_3)_4$	1.57	*a*	*a*	15
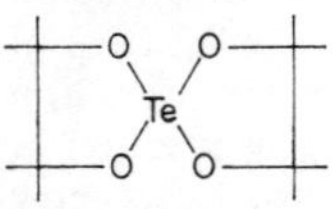	2.99	*a*	*a*	15
p-Tol_2Te_2	1.4	26	$CDCl_3$	93
$(p\text{-}EtOC_6H_4)_2Te_2$	1.7	26	$CDCl_3$	93
Te	0.04	26	Oleum	92
$TeCl_4$	0.8	26	12N, HCl	92
K_2TeO_3	2.5[b]	27	4N, D_2O	90
Na_2TeO_4	1.95	26	10%, HNO_3	92

[a]Not specified.
[b]$T_2 = 0.31$ s.

for absorption maxima, λ_{max}, in the UV–visible spectrum. Therefore, it seems reasonable to take $\Delta E_{Te}/\Delta E_{Se} \simeq \lambda_{Te}/\lambda_{Se}$. The value of the latter ratio has been found to be ~1.25[94]. A calculated value of the slope of the plot of $\delta_{^{125}Te}$ vs. $\delta_{^{77}Se}$ is 1.5–1.6; hence the agreement is reasonable in view of the approximations involved in the preceding treatment. In general, trends in ^{125}Te shielding are very similar to those found for analogous Se compounds and thus, a separate detailed discussion for each class of compounds, as for Se compounds, is scarcely warranted.

Dimethyl telluride appears to be the reference of choice for reported ^{125}Te chemical shifts although 4,4′-dimethyldiphenyl ditelluride[36,95–97] and bis[diethyldithiocarbamato]tellurium(II)[98,99] have also been used. To standardize ^{125}Te chemical shift values in this chapter, they are reported with respect to the literature value of Me_2Te (neat). This creates some discrepancies between our values and some values found in the literature[57,105] since there is a definite concentration effect on the chemical shift of Me_2Te in solution (e.g. 20 ppm difference between neat Me_2Te and dilute solutions of Me_2Te in $CDCl_3$[102]). As was pointed out previously for Me_2Se, the chemical shift value of the Te reference standard, Me_2Te, is also solvent-dependent. The magnitude of the solvent shift of Me_2Te is approximately twice that of the Se standard[100]. These solvent shifts are provided in Table 24.

^{125}Te chemical shifts have also been found to be concentration-dependent. For example, a 2M solution of Me_2Te_2 in benzene gave $\delta_{^{125}Te} = 303$ ppm and on dilution changed monotonically to 293 ppm at a concentration of 0.02M[94]. It has been found that λ_{max} for Me_2Te_2 decreases from 410 to 385 nm on dilution in hexane and other solvents, and other tellurides behave similarly[94]. Interestingly, the value of λ_{max} for Me_2Se_2 was unaffected by dilution.

a. Tellurides. ^{125}Te chemical shifts of dialkyl, alkyl aryl and diaryl tellurides are given in

TABLE 24. ^{125}Te solvent shifts of Me_2Te[a]

Solvent	Chemical shift (ppm)	Solvent	Chemical shift (ppm)
Et_3N	5.9	Me_2CO	−12.7
CCl_4	0.9	CH_2Cl_2	−14.9
Et_2O	−0.2	Me_2NCHO	−18.6
C_6H_6	−3.1	$CHCl_3$	−18.8
$PhNO_2$	−7.1	$MeNO_2$	−20.6
C_5H_5N	−9.4	Me_2SO	−26.4
MeOH	−10.4		

[a]Relative to neat Me_2Te; solutions are 5 mole %; Ref. 100.

TABLE 25. ^{125}Te chemical shifts of dialkyl tellurides[a]

Compound	Chemical shift (ppm)	Solvent	Ref.
$(Me_3Sn)_2Te$	−1214	CH_2Cl_2	101
$(Me_3Si)_2Te$	−43	$CDCl_3$	97
MeTePen-*neo*	5	$CDCl_3$	57
(*neo*-Pen)$_2$Te	25	$CDCl_3$	57
MeTeBu-*i*	44	$CDCl_3$	57
neo-PenTeBu-*i*	92	$CDCl_3$	57
MeTePr-*n*	95	$CDCl_3$	57
MeTeBu-*n*	103	$CDCl_3$	57
(*i*-Bu)$_2$Te	113	$CDCl_3$	57
neo-PenTePr-*n*	120	$CDCl_3$	57
neo-PenTeBu-*n*	128	$CDCl_3$	57
n-PrTeBu-*i*	163	$CDCl_3$	57
MeTeEt	165	$CDCl_3$	57
n-BuTeBu-*i*	171	$CDCl_3$	57
EtTePen-*neo*	191	$CDCl_3$	57
n-Pr$_2$Te	213	$CDCl_3$	57
$(Me_3Si(CH_2)_3)_2Te$[b]	221	Not specified	98
n-PrTeBu-*n*	222	$CDCl_3$	57
n-Bu$_2$Te	228	$CDCl_3$	57
n-Pen$_2$Te	230	$CDCl_3$	57
$(C_{16}H_{33})_2Te$[b]	232	$CDCl_3$	102
EtTeBu-*i*	234	$CDCl_3$	57
$(Me_2CHCH_2CH_2)_2Te$	241	$CDCl_3$	57
MeTeBu-*s*	246	$CDCl_3$	57
$(PhCH_2CH_2)_2Te$	277	CH_2Cl_2	99
EtTePr-*n*	285	$CDCl_3$	57
s-BuTePen-*neo*	287	$CDCl_3$	57
EtTeBu-*n*	292	$CDCl_3$	57
MeTePr-*i*	322	$CDCl_3$	57
i-BuTeBu-*s*	332	$CDCl_3$	57
i-PrTePen-*neo*	351	$CDCl_3$	57
EtTeEt	356; 380; 392	$CDCl_3$; C_6H_6; $CDCl_3$	57; 94; 91
n-PrTeBu-*s*	381	$CDCl_3$	57
n-BuTeBu-*s*	388	$CDCl_3$	57
i-PrTeBu-*i*	394	$CDCl_3$	57
n-PrTePr-*i*	445	$CDCl_3$	57

(*Contd.*)

TABLE 25. (*Contd.*)

Compound	Chemical shift (ppm)	Solvent	Ref.
EtTeBu-*s*	449	$CDCl_3$	57
i-PrTeBu-*n*	451	$CDCl_3$	57
MeTeBu-*t*	479	$CDCl_3$	57
t-BuTePen-*neo*	504	$CDCl_3$	57
EtTePr-*i*	512	$CDCl_3$	57
i-BuTeBu-*t*	554	$CDCl_3$	57
$(s\text{-Bu})_2Te$	558	$CDCl_3$	57
$(s\text{-Pen})_2Te$	564	$CDCl_3$	57
n-PrTeBu-*t*	602	$CDCl_3$	57
n-BuTeBu-*t*	608	$CDCl_3$	57
i-PrTeBu-*s*	618	$CDCl_3$	57
EtTeBu-*t*	670	$CDCl_3$	57
$(i\text{-Pr})_2Te$	676; 707	$CDCl_3$; CH_2Cl_2	57; 94
s-BuTeBu-*t*	786	$CDCl_3$	57
i-PrTeBu-*t*	846	$CDCl_3$	57
$(t\text{-Bu})_2Te$	979	$CDCl_3$	57
$(CF_3)_2Te$	1368	CH_3CN	41
CF_3TeCF_2Cl	1566	CH_3CN	41

[a]Chemical shifts reported relative to neat Me_2Te.
[b]Chemical shifts reported relative to bis(diethyldithiocarbamato)tellurium(II); converted relative to neat Me_2Te using the expression $\delta[Me_2Te] = \delta[\text{bis(diethyldithiocarbamato)tellurium(II)}] - 833.6$.

Tables 25, 26 and 27, respectively. The effect of replacing α-hydrogens by alkyl groups on the ^{125}Te chemical shifts parallels that observed in the ^{77}Se chemical shifts for analogous Se compounds. A deshielding of Te is observed when hydrogens in Me_2Te are replaced by Me groups as shown, for example, in the following series: (1) MeTeMe (0 ppm), MeTeEt (165 ppm), MeTePr-*i* (322 ppm), MeTeBu-*t* (477 ppm); (2) EtTeMe (165 ppm), EtTeEt (356 ppm), EtSePr-*i* (512 ppm), EtTeBu-*t* (670 ppm); (3) *i*-PrTePr-*i* (676), *i*-PrTeBu-*t* (846) and (4) *t*-BuTeBu-*t* (979 ppm). The chemical shift values of all possible Me derivatives of Me_2Te are considered above and in each case the deshielding introduced per replacement of a H atom by a Me group is approximately 165 ppm. Furthermore, the shielding caused by the introduction of a γ-carbon in tellurides can be clearly seen in the ^{125}Te chemical shifts if the following compounds are compared, i.e. MeTeEt (156 ppm), MeTePr (95), MeTeBu-*i* (44 ppm), MeTePen-*neo* (5 ppm). Thus, the replacement of one hydrogen atom on the β-carbon of the Et group in MeTeEt by a Me group introduces a shielding of 70 ppm. Further replacement introduces additional shielding although the magnitude of the shielding decreases. A small deshielding is observed if alkyl substitution takes place at the γ-carbon and alkyl substitution more remote to Te than the γ-carbon has little or no effect on the ^{125}Te chemical shifts, e.g. (1) MeTePr (95 ppm), MeTeBu (103 ppm); (2) Bu_2Te (228 ppm), $(C_{16}H_{33})_2Te$ (232 ppm).

The effect of alkyl substitution on the ^{125}Te chemical shifts for methyl phenyl telluride is consistent with that observed with Me_2Te. Thus, for alkyl phenyl tellurides, the most shielded Te resonance is found in methyl phenyl telluride and the most deshielded resonance is observed for *t*-butyl phenyl telluride. ^{125}Te chemical shifts for *ortho*-, *meta*- and *para*-substituted telluroanisoles are provided in Table 26. There have not been enough substituents studied to meaningfully evaluate the data.

b. Ditellurides. ^{125}Te chemical shifts of dialkyl, alkyl aryl and diaryl ditellurides are

TABLE 26. ^{125}Te chemical shifts of alkyl aryl tellurides[a]

Compound	Chemical shift (ppm)	Solvent	Ref.
PhTeMe	329; 330	$CDCl_3$; $CHCl_3$	57; 100
2-AnTeMe	205	$CDCl_3$	102
2-TolTeMe	221	$CDCl_3$	102
3-TolTeMe	324	$CDCl_3$	102
3-AnTeMe	341	$CDCl_3$	102
3-FC_6H_4TeMe	355	$CDCl_3$	102
4-$EtOC_6H_4$TeMe	317	$CDCl_3$	102
4-AnTeMe	318; 321	$CDCl_3$; CCl_4	102; 100
4-TolTeMe	318; 329	$CDCl_3$; CCl_4	102; 100
4-$PhOC_6H_4$TeMe	326	$CDCl_3$	102
4-FC_6H_4TeMe	336	$CDCl_3$	102
4-ClC_6H_4TeMe	338	$CDCl_3$	102
PhTeBu-*i*	409	$CDCl_3$	57
PhTePr-*n*	460	$CDCl_3$	57
PhTeBu-*n*	468	$CDCl_3$	57
2-$OHCC_6H_4$TeBu-*n*	517	$CDCl_3$	104
PhTeEt	532; 549	$CDCl_3$; CCl_4	102; 100
PhTeBu-*s*	640	$CDCl_3$	102
PhTePr-*i*	702; 719	$CDCl_3$; CCl_4	102; 100

[a]Relative to neat Me_2Te.

TABLE 27. ^{125}Te chemical shifts of diaryl tellurides[a]

Compound	Chemical shift (ppm)	Solvent	Ref.
Ph_2Te	688; 688	CH_2Cl_2; $CDCl_3$	94; 102
2-An_2Te	445	$CDCl_3$	102
2-AnTePh	579	$CDCl_3$	102
4-Tol_2Te	660; 663	$CDCl_3$; CH_2Cl_2	102; 103
4-AnTePh	664; 668	$CDCl_3$; CH_2Cl_2	102; 104
4-BrC_6H_4TePh	699	CH_2Cl_2	104
(2,4,6-$Me_3C_6H_2$)$_2$Te	612	$CDCl_3$	91
(2-Naph)$_2$Te	692	$CDCl_3$	102

[a]Relative to neat Me_2Te.

given in Tables 28, 29 and 30, respectively. ^{125}Te magnetic shielding of dialkyl ditellurides closely parallels that of Se in analogous diselenides, but the sensitivity to changes in alkyl groups is even greater. For example, compare Me_2Te_2 (49 ppm), Et_2Te_2 (166 ppm), *i*-Pr_2Te_2 (293 ppm) and *t*-Bu_2Te_2 (477 ppm). The deshielding effect per replacement of a H atom by a Me group is approximately 120 ppm compared to 65 ppm in the ^{77}Se chemical shifts of analogous Se compounds. As previously discussed in the ^{77}Se-NMR section using Et_2Se_2 as an example, the influence of each Me group is two-fold, (1) to induce deshielding of the proximate Se by the dispersion (or polarizability) effect and (2) to cause shielding of the distant Se by the γ-effect. The same effects are found for Te shielding in dialkyl ditellurides. It can be seen further if the shifts of the following ditellurides are compared: MeTeTeMe (49 ppm), MeTe*TeEt (− 48 ppm(*), 264 ppm), MeTe*TePr-*i* (− 94 ppm(*), 435 ppm), MeTe*TeBu-*t* (− 68 ppm(*), 597 ppm). In these compounds the Te bonded to

TABLE 28. ^{125}Te chemical shifts of dialkyl ditellurides[a]

Compound	Chemical shift (ppm)	Compound	Chemical shift (ppm)
MeTe*TeBu-*s*	– 108(*), 130	*n*-BuTe*TeBu-*t*	71(*), 522
MeTe*TePr-*i*	– 94(*), 435	(*i*-BuTe)$_2$	79
s-BuTeTe*Pen-*neo*	– 79(*), 353	EtTeTe*Pr-*n*	88(*), 182
MeTe*TeBu-*t*	– 68(*), 597	EtTeTe*Bu-*n*	97(*), 183
i-PrTeTe*Pen-*neo*	– 65(*), 415	(*n*-PrTe)$_2$	105
MeTe*TeEt	– 48(*), 264	EtTe*TeBu-*s*	105(*), 276
t-BuTeTe*Pen-*neo*	– 45(*), 577	*n*-PrTe*TeBu-*n*	195(*), 112
MeTe*TePr-*n*	– 31(*), 185	(*n*-BuTe)$_2$	113
MeTe*TeBu-*n*	– 31(*), 193	($C_{10}H_{21}$Te)$_2$	113
EtTeTe*Pen-*neo*	– 25(*), 249	(C_9H_{19}Te)$_2$	114
i-BuTe*TeBu-*s*	– 19(*), 315	($C_{18}H_{37}$Te)$_2$	114
n-PrTeTe*Pen-*neo*	– 9(*), 172	(C_5H_{11}Te)$_2$	115
n-BuTeTe*Pen-*neo*	– 8(*), 181	(C_6H_{13}Te)$_2$	117
i-PrTeTe*Bu-*i*	– 6(*), 378	($C_{11}H_{23}$Te)$_2$	117
MeTeTe*Bu-*i*	– 4(*), 130	EtTe*TePr-*i*	118(*), 337
i-BuTe*TeBu-*t*	16(*), 545	(C_8H_{17}Te)$_2$	118
i-BuTeTe*Pen-*neo*	17(*), 122	($Me_2CHCH_2CH_2$Te)$_2$	120
n-PrTe*TeBu-*s*	30(*), 291	(C_7H_{15}Te)$_2$	120
EtTeTe*Bu-*i*	37(*), 208	EtTe*TeBu-*t*	135(*), 503
n-BuTe*TeBu-*s*	37(*), 292	(EtTe)$_2$	166; 188[b]
MeTe*TePen-*neo*	42(*), 72	(*t*-BuTe)$_2$	223
n-PrTe*TePr-*i*	44(*), 354	*i*-PrTeTe*Bu-*s*	236(*), 281
(MeTe)$_2$	49; 63[b]	*s*-BuTe*TeBu-*t*	249(*), 460
i-PrTeTe*Bu-*n*	51(*), 355	(*i*-PrTe)$_2$	293; 303[b]
n-PrTeTe*Bu-*i*	53(*), 131	*i*-PrTe*TeBu-*t*	305(*), 470
i-BuTe*TeBu-*n*	53(*), 138	(*s*-BuTe)$_2$	477
(*neo*-PenTe)$_2$	60	($HOOCCH_2$Te)$_2$	538[c]
n-PrTe*TeBu-*t*	63(*), 520	(CF_3Te)$_2$	686[d]

[a]Relative to neat Me_2Te; solvent $CDCl_3$ unless otherwise indicated; Ref. 105.
[b]Solvent CH_2Cl_2; Ref. 94.
[c]Solvent Me_2SO; Ref. 103.
[d]Solvent C_6D_6; Ref. 41.

the Me group is shielded by successive methylation of the distant Me group, whereas the shielding of the remote Te decreases progressively. The Te atom remote to the Me group does not have any γ-atom and thus experiences only the deshielding caused by the dispersion effect of the alkyl group in going from $Me_2Te_2 \rightarrow \rightarrow \rightarrow$ MeTeTeBu-*t*. On the other hand, the Me-bonded Te experiences a net shielding from the two opposing effects (shielding from the γ-effect and deshielding from the dispersion effect). The magnitude of the net shielding decreases with additional alkylation (actually, a deshielding is observed in going from MeTeTePr-*i* $\rightarrow$ MeTeTeBu-*t*).

Unlike diselenides, dialkyl[105] and diaryl[93] ditellurides apparently undergo exchange reactions between two symmetrical ditellurides. The exchange is slow on the ^{125}Te-NMR time-scale and, consequently, four ^{125}Te resonances were observed in the spectra of solutions initially containing two symmetrical ditellurides, one resonance corresponding to the Te in each symmetric ditelluride and two resonances representing the non-equivalent Te atoms of the unsymmetric ditelluride. Thus, O'Brien and coworkers[105] were able to report the ^{125}Te chemical shifts of 36 unsymmetrical dialkyl ditellurides by initially preparing nine symmetrical dialkyl ditellurides. Such an exchange is also shown

TABLE 29. ^{125}Te chemical shifts of alkyl aryl ditellurides[a]

Compound	Chemical shift (ppm)	Ref.
MeTe*TePh	95(*), 384	105
i-BuTe*TePh	170(*), 331	105
n-PrTe*TePh	223(*), 305	102
n-BuTe*TePh	226(*), 304	102
EtTe*TePh	295(*), 289	102

[a]Relative to neat Me_2Te, solvent $CDCl_3$.

TABLE 30. ^{125}Te chemical shift of diaryl ditellurides, $XC_6H_4TeTeC_6H_4X'$[a]

X = X'	Chemical shift (ppm)	Solvent	Ref.
H	420; 421	$CDCl_3$; $CDCl_3$	102; 93
2-OMe	168	$CDCl_3$	106
2-F	255	$CDCl_3$	102
2-Cl	312	$CDCl_3$	102
2-SPh	325	$CDCl_3$	102
2-COOEt	411; 414	$CDCl_3$; $CDCl_3$	102; 106
2-CHO	423	$CDCl_3$	106
2-Ac	473	$CDCl_3$	102
2-Me	495	$CDCl_3$	102
3-Me	416	$CDCl_3$	102
3-F	444	$CDCl_3$	102
3-OMe	435	$CDCl_3$	102
4-Me	427; 432.2; 433	$CDCl_3$; $CDCl_3$; $CDCl_3$	102; 93; 106
4-SPh	434	$CDCl_3$	102
4-Br	442	$CDCl_3$	102
4-Cl	447; 452	$CDCl_3$; $CDCl_3$	102; 93
4-OPh	449	$CDCl_3$	102
4-OEt	455; 456	$CDCl_3$; $CDCl_3$	102; 93
4-F	457; 458	$CDCl_3$; $CDCl_3$	102; 106
4-Me_2N	466	$CDCl_3$	102

X	X'	Chemical shift (ppm)	Solvent	Ref.
4-Me*	4-Cl	433.9(*); 456.3	$CDCl_3$	93
4-Me*	4-OEt	436.8(*); 446.9	$CDCl_3$	93
4-Cl*	4-OEt	444.3(*); 475.2	$CDCl_3$	93

Compound	Chemical shift (ppm)	Solvent	Ref.
(2-Thiophenyl)$_2$Te$_2$	264	$CDCl_3$	93
(1-Naph)$_2$Te$_2$	336	$CDCl_3$	102
(2-Naph)$_2$Te$_2$	438	$CDCl_3$	102

[a]Relative to neat Me_2Te.

to occur in diaryl ditellurides and between dialkyl and diaryl ditellurides. Granger and coworkers[93] have shown that the exchange is a thermal process. The substituent effect on the ^{125}Te chemical shifts of *para*-substituted diaryl ditellurides has been shown to be consistent with that observed on the ^{77}Se chemical shifts for *para*-substituted diaryl diselenides. Thus electron-donating groups (in a mesomeric sense) deshield the Te resonance in diaryl ditellurides. However, mesomeric withdrawing groups have not been studied in sufficient number to enable a good comparison.

Finally, it should be noted that ^{125}Te chemical shifts of diaryl ditellurides have been found to be concentration-dependent[93], more so than the ^{77}Se chemical shifts for analogous diaryl diselenides. Also, the assignment[93] of the Te resonances for unsymmetrical diaryl ditellurides is not consistent with the reported values of ^{77}Se resonances for the analogous unsymmetrical diaryl diselenides[60]. Thus, saturation transfer or labelling experiments should be performed to assign the Te resonances unequivocally in these unsymmetrical ditellurides.

c. Tellurium halides. ^{125}Te chemical shifts of alkyl and aryl tellurium halides (mono-, di- and tri-) are reported in Tables 31, 32 and 33. From the chemical shift data it is apparent that introduction of halogens bonded to Te causes a deshielding of the Te nucleus. For example, a deshielding of 293 ppm is observed in going from Ph_2Te (688 ppm) to Ph_2TeCl_2 (981 ppm) and deshielding of 749 ppm is seen in going from Me_2Te to Me_2TeCl_2. Thus, Cl has a relatively greater effect on the shielding of Te in the alkyl derivative than the aryl derivative. Large solvent shifts of the order of 400 ppm have been reported in this class of compounds. The observed chemical shift values of $PhTeCl_3$ and $PhTeBr_3$ in a benzene/toluene mixture are 900 and 892 ppm, respectively[107], whereas in CH_2Cl_2 : Me_2SO (1:1) the corresponding values are 1229 and 1193 ppm, respectively[104]. Similarly, the effect of solvent on the shielding of Te is also observed in dihalides, e.g. chemical shifts of Me_2TeCl_2 and Me_2TeBr_2 in CH_2Cl_2 are 749 and 669 ppm[94], respectively, and in toluene these are 1218 and 858 ppm[107], respectively. Although there is

TABLE 31. ^{125}Te chemical shifts of aryl- and alkyl-tellurium trihalides[a]

Compound	Chemical shift (ppm)	Solvent	Ref.
$MeTeBr_3$	647	$C_6H_6/PhCH_3$	107
$MeTeCl_3$	758	$C_6H_6/PhCH_3$	107
$EtTeBr_3$	849	$C_6H_6/PhCH_3$	107
$(Cl_3TeCH_2CO)_2O$	864	Me_2SO	103
$PhTeBr_3$	892; 1193	$C_6H_6/PhCH_3$; $CH_2Cl_2:Me_2SO(2:1)$	107; 104
$EtTeCl_3$	900	$C_6H_6/PhCH_3$	107
$PhTeCl_3$	917; 1229	$C_6H_6/PhCH_3$; $CH_2Cl_2:Me_2SO(2:1)$	107; 104
$PhTeI_3$	1101	$CH_2Cl_2:Me_2SO(2:1)$	104
$4\text{-}AnTeI_3$	1078	$CH_2Cl_2:Me_2SO(1:1)$	104
$4\text{-}AnTeBr_3$	1204	$CH_2Cl_2:Me_2SO(2:1)$	104
$(Cl_3Te)_2CH_2$	1198	Me_2SO	103
$4\text{-}BrC_6H_4TeCl_3$	1208	$CH_2Cl_2:Me_2SO(1:1)$	104
$4\text{-}ClC_6H_4TeCl_3$	1208	$CH_2Cl_2:Me_2SO(1:1)$	104
$4\text{-}PhOC_6H_4TeCl_3$	1226	$CH_2Cl_2:Me_2SO(1:1)$	104
$4\text{-}TolTeCl_3$	1234	$CDCl_3:Me_2SO(1:1)$	104
$4\text{-}AnTeCl_3$	1246	Me_2SO	104

[a]Relative to neat Me_2Te.

TABLE 32. ^{125}Te chemical shifts of diorganyl tellurium dihalides, R_2TeX_2[a]

Compound	Chemical shift (ppm)	Solvent	Ref.
Me_2TeBr_2	669; 858	CH_2Cl_2; $PhCH_3$	94; 107
$PhTeMeI_2$	698	CH_2Cl_2	103
4-$TolTeMeI_2$	663	CH_2Cl_2	103
4-$AnTeMeI_2$	664	CH_2Cl_2	103
4-$EtOC_6H_4TeMeI_2$	670	CH_2Cl_2	103
Me_2TeCl_2	749; 1218	CH_2Cl_2; $PhCH_3$	94; 107
Et_2TeBr_2	879; 1153	CH_2Cl_2; $PhCH_3$	94; 107
Ph_2TeCl_2	981; 917	CH_2Cl_2; $PhCH_3$	103
4-$TolTe(Ph)Cl_2$	1070.6	CH_2Cl_2	104
4-$BrC_6H_4Te(Ph)Cl_2$	1079.6	CH_2Cl_2	104
i-Pr_2TeBr_2	1105	CH_2Cl_2	94
$(CF_3)_2TeCl_2$	1114	MeCN	41
$(CF_3)_2TeBr_2$	1180	MeCN	41
$(CF_3)_2TeF_2$	1187	MeCN	41
Ph_2TeBr_2	1481; 1508; 890	THF; $CDCl_3$; $PhCH_3$	91; 102; 107
4-Tol_2TeBr_2	905	$CDCl_3$	102

[a]Relative to neat Me_2Te.

TABLE 33. ^{125}Te chemical shifts of triorganyl tellurium halides, R_3TeX[a]

Compound	Chemical Shifts (ppm)	Solvent	Ref.
Me_3TeI	443	Me_2SO-d_6	94
$Et_2TeMeBr$	470	D_2O	94
$PhTeMe_2I$	550	Me_2SO	103
4-$EtC_6H_4TeMe_2I$	542	Me_2SO	103
4-$AnTeMe_2I$	550	Me_2SO	103
Et_3TeBr	573	D_2O	94
Ph_2TeMeI	595	Me_2SO	103
2-Tol_2TeMeI	580	Me_2SO	103
$(Et_3Te^*)_2(TeCl_4Br_2)$	598.4(*), 1365.4	CH_2Cl_2 : Me_2SO(1:2)	104
$(Et_3Te^*)_2(TeBr_6)$	603.5(*), 1347.7	CH_2Cl_2 : Me_2SO(1:2)	104
i-Pr_2TeMeI	630	D_2O	94

[a]Relative to neat Me_2Te.

no explanation provided for this very substantial solvent effect observed for the ^{125}Te chemical shift of these halides, we find it hard to rationalize the ^{125}Te shifts of this class of compounds based on simple inductive and solvent effects.

From the results of many cryoscopic studies[108] it has been suggested that $MeTeCl_3$ exists as a mixture of monomers and dimers in benzene and that $MeTeBr_3$ is more associated than $MeTeCl_3$. Thus, equilibria involving higher aggregates are more important for $MeTeBr_3$. Perhaps due to the different degree of oligomerization in various solvents, the state of aggregation may play an important role in determining the shielding of Te in these compounds. Even the ^1H-NMR spectrum of $MeTeCl_3$ exhibited a pronounced solvent dependence[108]. A peak was found at 2.83 ppm in benzene, while this resonance occurred at 3.70 ppm in CH_2Cl_2. Clearly large solvent effects will make it rather difficult to interpret the shielding behaviour of Te in the organic tellurium halides. A consideration of shifts in the same solvent should enable a more valid comparison of various effects on ^{125}Te chemical shifts. The four diiodides, $XC_6H_4TeMeI_2$, have chemical shifts in the region 663–698 ppm (Table 32). The shielding of Te in the two diarylmethyl tellurium iodides (Table 33) increases (580 and 595 ppm), indicating that the Te nucleus becomes more shielded when I is replaced by an aryl group. The replacement of aryl by Me causes further shielding. The chemical shifts of the three aryl dimethyltellurium iodides are

TABLE 34. ^{125}Te chemical shifts of substituted tellurophenes[a]

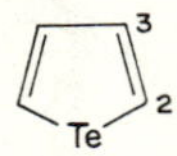

Compound	Chemical shift (ppm)
2-CH_2OH	764
2-H	782; 782[b]
2-CHO	789
2-Ac	813
2-COOEt	820
2-COOH	848; 861[c]
2-Cl	905
2-Br	949

[a]Tellurophene in $(CD_3)_2CO$ was used as a reference; chemical shift values have been converted relative to neat Me_2Te using the expression $\delta(Me_2Te) = \delta(\text{tellurophene}) - 782$; solvent $(CD_3)_2CO$; Ref. 109.
[b]Relative to neat Me_2Te; solvent $(CD_3)_2CO$; Ref. 100.
[c]Ref. 102.

TABLE 35. ^{125}Te chemical shifts of substituted benzo[*b*]tellurophenes[a]

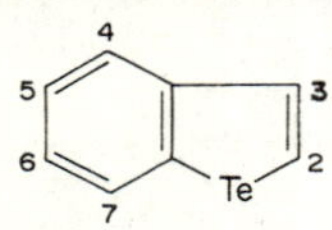

Compound	Chemical shift (ppm)	Compound	Chemical shift (ppm)
2-CH_2OH	721	2-SMe	839
2-H	727	2-Cl	868
2-CHO	727	2-Br	911
2-Me	750	2-CN	952
2-Ac	751	3-Me	659
2-COOH	764	3-CH_2COOH	680
2-$CONH_2$	764	3-CH_2OH	683
2-CONHMe	765	3-CH_2CN	709
2-COCl	769	3-Cl	721
2-COOEt	774	3-CHO	750
2-COOMe	775	3-Br	762
2-COPh	786	3-CN	787
2-$CONMe_2$	830		

[a]Relative to neat Me_2Te; Ref. 110.

TABLE 36. ^{125}Te chemical shifts of other heterocycles containing tellurium[a]

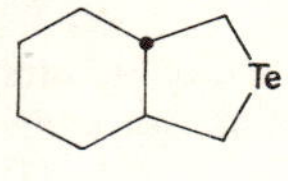

(9)

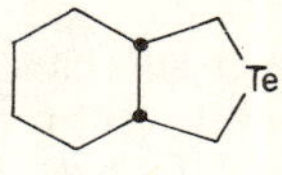

(10)

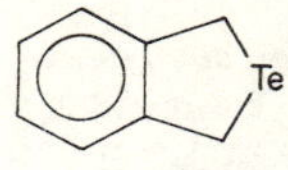

(11)

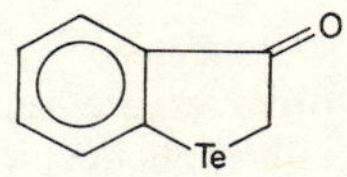

(12)

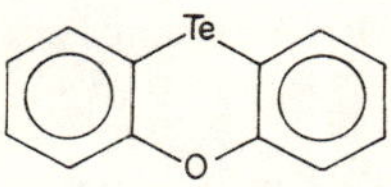

(13)

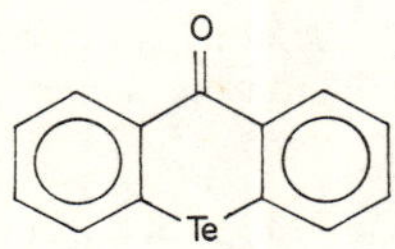

(14)

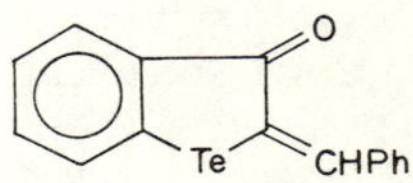

(15)

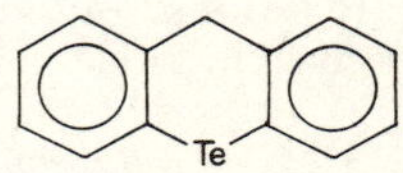

(16)

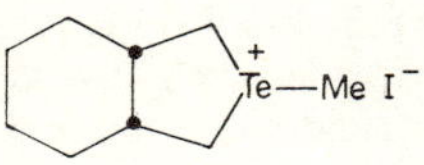

(17)

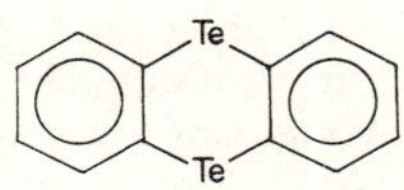

(18)

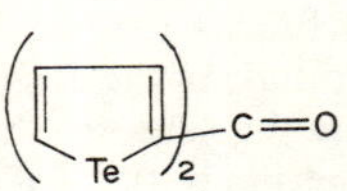

(19)

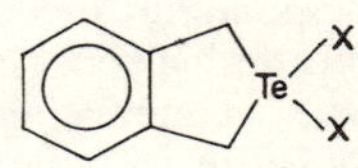

(20)

(a) X = Cl, Cl
(b) X = Br, Cl
(c) X = Br, Br
(d) X = I, I

Compound	Chemical shift (ppm)	Solvent	Ref.
9	175	—	111
10	209	—	111
11	269[c]	$(CD_3)_2NCHO$	98
12	383[b]	$CDCl_3$	95
13	424[b]	$CDCl_3$	95
14	468[b]	$CDCl_3$	95
15	501	$CDCl_3$	95
16	512[b]	$CDCl_3$	95
17	630	—	111
18	888	Not specified	112
19	901[b]	$CDCl_3$	96
20a	994[c]	$(CD_3)_2NCHO$	98
20b	972[c]	$(CD_3)_2NCHO$	98
20c	940[c]	$(CD_3)_2NCHO$	98
20d	831[c]	$(CD_3)_2NCHO$	98

[a]Relative to neat Me_2Te.

[b]$(4\text{-TolTe})_2$ in $CDCl_3$ was used as a reference; chemical shift values have been converted relative to neat Me_2Te using the expression $\delta(Me_2Te) = \delta(4\text{-TolTe})_2 - 427$.

[c]Chemical shifts reported relative to bis(diethyldithiocarbamato)tellurium(II); converted relative to neat Me_2Te using the expression $\delta[Me_2Te] = \delta[\text{bis(diethyldithiocarbamato)tellurium(II)}] - 833.6$.

in the region 542–550 ppm whereas the value for trimethyl tellurium iodide is 443 ppm. As expected, a shielding of Te is observed in going from $PhTeCl_3 \rightarrow PhTeBr_3 \rightarrow PhTeI_3$.

d. Heterocyclic tellurium compounds. ^{125}Te chemical shifts of 2-substituted tellurophenes are reported in Table 34. When the substituent-caused shifts of the eight tellurophenes (substituent: 2-Cl, 2-Br, 2-CH_2OH, 2-CHO, 2-Ac, 2-COOH, 2-COOMe, H) were plotted[109] against the ^{77}Se chemical shifts of the same set of selenophene derivatives, a good linear correlation was observed. This should be an indication that the transmission of substituent effects follows the same mechanism in the two analogous heterocyclic systems. Furthermore, the slope of the regression line showed that Te is 2.44 times more sensitive to substituent effects than Se in this type of organic compound[109]. Te chemical shifts for substituted benzo[*b*]tellurophenes[110] and other heterocyclic Te compounds are given in Tables 35 and 36, respectively.

e. Miscellaneous tellurium compounds. The ^{125}Te chemical shifts of miscellaneous compounds are shown in Table 37, whereas Table 38 provides the ^{125}Te chemical shifts of alkenyl and alkynyl tellurides. The presence of two chemically different Te atoms in tellurenyl tellurinyl selenides, RTeSeTe(O)R, was indicated[106] by two widely separated ^{125}Te-NMR resonances (Table 37). One resonance was the same as that observed for the corresponding ditelluride whereas the second was observed approximately 600 ppm further deshielded. The magnitude of the deshielding effect would be expected for a Te—O bond. It is interesting to note that even in these compounds mesomeric donors deshield both Te nuclei, although the magnitude of the deshielding is smaller for the telurinyl Te(Te—O). It has been reported [106] that when a solution containing two aryl tellurenyl

TABLE 37. ^{125}Te chemical shifts of miscellaneous compounds containing tellurium[a]

Compound	Chemical shift (ppm)	Solvent	Ref.
(PhTe*)Se(TeBz)	431(*), 1014	$CDCl_3$	106
(2-AnTe*)Se(2-AnTe(O))	159(*), 931	$CDCl_3$	106
(4-TolTe*)Se(4-TolTe(O))	424(*), 1010	$CDCl_3$	106
(4-AnTe*)Se(4-AnTe(O))	455(*), 1012	$CDCl_3$	106
(4-FC_6H_4Te*)Se(4-FC_6H_4Te(O))	456(*), 1027	$CDCl_3$	106
MeTeSeMe	512	C_6H_6	94
i-PrTeSeMe	816	CH_2Cl_2	94
(4-OctC_6H_4C(O))Te(4-PenC_6H_4)	899.9[b]	$CDCl_3$	36
(4-AnC(O))Te(4-Tol)	904.7[b]	$CDCl_3$	36
(2-EtOOCC_6H_4Te)$_2$Se	907	$CDCl_3$	106
BzTe(4-Tol)	929.4[b]	$CDCl_3$	36
(2-CHOC_6H_4Te)$_2$Se	1120	$CDCl_3$	106
$(OCH_2CMe_2CH_2O)_2Te$	1355[c]	Not specified	15
$((CF_3)_2CHO)_4Te$	1394[c]	Not specified	15
$(CF_3CH_2O)_4Te$	1463[c]	Not specified	15
$(EtO)_4Te$	1503[c]	Not specified	15
$(MeO)_4Te$	1510[c]	Not specified	15
(*i*-PrO)$_4$Te	1523[c]	Not specified	15
(*i*-BuO)$_4$Te	1525[c]	Not specified	15
$(OCMe_2CMe_2O)_2Te$	1526	Not specified	15
$(OCH_2CH_2O)_2Te$	1601[c]	Not specified	15

[a]Relative to neat Me_2Te unless otherwise specified.
[b]$(4\text{-TolTe})_2$ in $CDCl_3$ was used as a reference; chemical shift values have been converted relative to neat Me_2Te using the expression $\delta[Me_2Te] = \delta[(4\text{-TolTe})_2] - 427$.
[c]Relative to 2.0M Me_2Te in C_6D_6.

TABLE 38. ^{125}Te chemical shifts of alkenyl and alkynyl[a] tellurides

Compound	Chemical shift (ppm)	Solvent	Ref.
MeTeC≡CBu	159.1	Neat	100
MeTeC≡CCH=CH$_2$	174	Neat	100
PhTeC≡CPh	475	$CDCl_3$	102
4-TolTeC≡CH	483.7	CCl_4	100
4-FC$_6$H$_4$TeC≡CH	499.0	CCl_4	100
4-ClC$_6$H$_4$TeC≡CH	503.1	CCl_4	100
(CH$_2$=CH)$_2$Te	529.9	CCl_4	100
PhTeCH=CH$_2$	615.3	CCl_4	100
4-AnTeCH=CH$_2$	596.3	CCl_4	100
4-BrC$_6$H$_4$TeCH=CH$_2$	621.0	CCl_4	100
4-TolTeCH=CHOCOEt	743.9	CCl_4	100
4-TolTeC(Ph)=CHCHO	918.0	CCl_4	100

[a]Relative to neat Me_2Te.

aryltellurinyl selenides was examined by ^{125}Te-NMR, a rapid exchange similar to the exchange reported for diaryl ditellurides appeared to occur. It can also be noted that the replacement of a Te in Me_2Te_2 by a Se (i.e. MeTeSeMe) caused a deshielding of Te by 449 ppm[94]. A similar deshielding (425 ppm) was observed[94] in going from MeTeTePr-*i* → MeSeTePr-*i*. A deshielding of similar magnitude was observed[106] in the aryl derivatives, e.g. (*o*-CHOC$_6$H$_4$Te)$_2$ (423 ppm), (*o*-CHOC$_6$H$_4$Te)$_2$Se (1120 ppm).

To date there have been no published reports dealing with isotope effects in ^{125}Te- or ^{123}Te-NMR spectroscopy.

3. Coupling constants

As a general rule, spin–spin coupling constants involving Te are about 2–3 times greater than the same coupling constants involving Se. Although coupling constants of Te to other magnetically active nuclei have not been studied in any great detail, coupling to ^{1}H, ^{13}C, ^{19}F, ^{31}P, ^{77}Se, ^{119}Sn, ^{129}Xe and ^{195}Pt have been reported and these were recently reviewed[113]. Analogously to Se, our primary concern will be Te–C and Te–Te spin–spin coupling, although the fact that the ^{125}Te nucleus couples strongly with ^{1}H allowed McFarlane to obtain Te chemical shifts of organotellurium compounds by heteronuclear double resonance over a decade ago[94,103]. At the same time two- and three-bond ^{125}Te–^{1}H coupling constants were obtained and published[94,103]. These values were generally between 20 and 30 Hz.

More recently, by use of selective population transfer experiments, the sign of ^2J(Te–H) and ^3J(Te–H) in 2-tellurophenecarboxylic acid were shown[114] to be negative while the one-bond J(Te–C) coupling constant was positive. In this same report[114], a personal communication from Jakobsen was cited in which the J(Te–C) one-bond coupling in the parent tellurophene was also found to be positive. Also, relatively recently, reports of Te–C coupling constants have appeared for various organotellurium compounds although the value for Me_2Te had appeared[115,116] as early as 1967. The mechanism which contributes to Te–C coupling has not been thoroughly studied although in a follow-up study to the sign determinations discussed above, Martin and coworkers[117] determined ^{125}Te–^{13}C coupling constants in 2-substituted tellurophenes. Based on a comparison of Se–C and Te–C coupling constants in analogous compounds, it was postulated[117] that, in addition to the Fermi contact term, orbital and spin-dipolar terms also contributed significantly to these couplings. Interestingly, in perhaps the most complete study to date of ^{13}C parameters of organotellurium compounds, Chadha and Miller[104] assume that Te–C coupling is dominated by the Fermi contact term. In their study, a linear relationship between $\Delta\delta(^{13}C_1)$ and ^1J(Te–C) is observed and different straight lines are found for phenyl- and *p*-methoxyphenyl-tellurium compounds. On this basis, and the fact that there is an approximate linear relationship between ^1J(Te–C) and ^{125}Te Mossbauer isomer shift values, it was postulated[104] that one-bond Te–C coupling constants depend primarily on the s electron density at the Te nucleus. No correlation between the magnitude of one-bond and two-bond Te–C couplings was evident, leading to the postulate[104] that the two-bond couplings are dependent on complex steric and electronic factors.

With the exception of the determination of Te–Te coupling constants in polyatomic cations[87,118] only one report has appeared describing the determination of this parameter in organotellurium compounds. Granger and coworkers[93] reported $J(^{125}\text{Te}-^{125}\text{Te})$ values for two unsymmetrical ditellurides and found that, because of exchange, the magnitude of this coupling may change with solvent and concentration.

Several Se–Te coupling constants have been reported although most of these are from polyatomic cations[87,118]. Two of these coupling constants are published for organic compounds[106].

III. ESR STUDIES

A. Organoselenium Compounds

In 1964, Shimazu and Tappel[119] first found that Se-amino acids are much better protectors *in vitro* against ionizing radiations than the analogous S-amino acids and this was later confirmed *in vivo* by Breccia and coworkers[120]. To explain these results, it was suggested that Se-containing compounds were able to stabilize free radicals better than the thio analogues. Even though the hypothesis that Se-amino acids are able to stabilize free radicals was advanced some time ago, to date there have been only a few reports concerned with electron spin resonance (ESR) studies of organoselenium radicals and, with the exception of a Se–Te compound, we are not aware of a single study describing the ESR of organotellurium compounds.

No radicals could be detected by ESR spectroscopy when Me_2Se_2 was photolysed in hydrocarbon solvents nor when mixtures of MeSeH and di-*t*-butyl peroxide were photolysed[121]. This is hardly unexpected since alkoxy radicals[122] and alkylthiyl radicals[123,124] in solution are also undetected by the ESR technique[125]. This is due to the fact that these radicals have orbitally degenerate (or nearly degenerate) ground states which lead, by spin-orbit coupling, to a markedly anisotropic *g* factor. As a consequence, these radicals have very short relaxation times and hence extremely broad lines[122–124]. However, a radical that appeared to be MeSe· was trapped when Me_2Se was photolysed in the presence of *t*-butyl phenyl nitrone. The adduct had a well-resolved ESR spectrum[121],

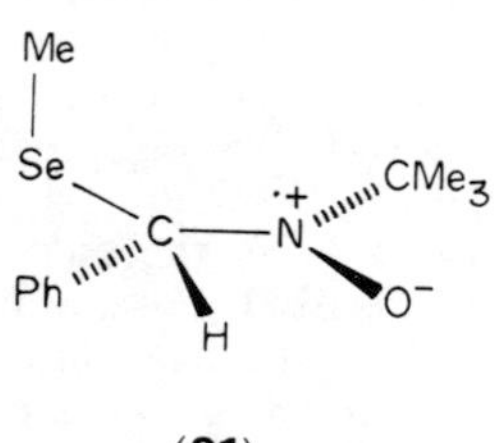

(**21**)

the parameters for which are $g = 2.0097$, $a^{N} = 13.5$G, $a^{H}(1\,H) = 2.06$G and $a^{H}(3H) = 0.85$G at -50°C. On the basis of these parameters structure **21** was assigned to the adduct. No adduct was identified when 2-methyl-2-nitrosopropane or 1,1-di-*t*-butylethylene were used as spin traps. However, di-*t*-butyl selenoketone gave a short-lived adduct ($t_{1/2} = 0.5$ s at -80°C) having $g = 2.0026$ to which the structure $MeSeSe\dot{C}(t\text{-}Bu)_2$ was assigned. With di-*t*-butyl thioketone at -105°C, a weak signal due to $MeS\dot{C}(t\text{-}Bu)_2$ could be obtained together with an even weaker signal at $g = 2.0056$ which was attributed to $MeSeS\dot{C}(t\text{-}Bu)_2$[121].

The ESR spectrum[126] of a γ-irradiated single crystal of Me_2Se at -196°C consisted of lines from more than one radical. The main spectrum (^{even}Se) showed hyperfine coupling to twelve equivalent protons. The analysis is what would be expected for a radical cation dimer of Me_2Se, i.e. $Me_2SeSeMe_2{}^{+\cdot}$. Dimer radical cations of the congeneric species $Me_2SSMe_2^{+\cdot}$ have been detected by ESR studies during chemical oxidation of Me_2S in aqueous solution[127]. The identification of this radical cation dimer of Me_2Se was completed by the observation of ^{77}Se satellite spectra at several different orientations of the crystal. The relative intensities in the ^{77}Se spectrum were in agreement with a hyperfine interaction involving two anisotropically equivalent Se atoms ($a^{H}(12H) = 4.9$ G). From the principal values and the direction cosines of the ^{77}Se hyperfine and g tensors, the structure of the dimer radical cation was interpreted as a centrosymmetrical dimer, with the unpaired electron occupying an antibonding (σ^*) orbital formed almost entirely from the p_σ orbitals of the two Se atoms. The results seemed to rule out the possibility of any significant d orbital participation in the bonding[126].

That dimerization of the dimethyl selenide radical cation can be avoided was shown by using a γ-irradiation technique which generated the monomer radical cations by positive charge transfer to the parent molecules dispersed in a Freon matrix[128]. Thus, the ESR spectrum of the dimethyl selenide radical cation, generated in a Freon solution[129], consisted of a seven-line pattern with a binomial intensity distribution with the parameter $a^{H}(6H) = 15.6$ G. The ^1H coupling constant for the monomer species is about three times greater than that for the dimer species. Assuming that the spin population is shared between the Se orbitals in going from monomer to dimer without any significant change of orbital hybridization, a ratio of 2 would be expected if the ^1H coupling in the Me groups were determined only by the local spin densities in the orbitals of the adjacent Se atom. It was proposed that the admixture of a Me group orbital into the heavy atom orbitals of a dimeric radical species is determined by the character of the molecular orbital (MO) formed between the heavy atoms. If the MO containing the unpaired electron is antibonding, the Me group admixture coefficients into the two heavy atom orbitals will have opposite signs resulting in considerable cancellation and a much lower spin density at the Me hydrogens in the dimer than half the value for the monomeric species. Since the MO which contains the odd electron in $Me_2SeSeMe_2^{+\cdot}$ is antibonding (σ^*) the results are clearly in accord with this proposal[129].

All attempts at the direct observation by ESR spectroscopy of unhindered β-selenoalkyl radicals formed by hydrogen atom abstraction from Me_2Se, Et_2Se, *n*-BuSeBu-*t*, MeSePh and Me_2Se_2 were unsuccessful[121]. In contrast to the unhindered β-selenoalkyls, well-resolved ESR spectra of hindered β-selenoalkyl radical species $R_nXSe\dot{C}(t\text{-}Bu)_2$ (where X = C,S,Se) were obtained when transient radicals $R_nX\cdot$ were added to di-*t*-butyl selenoketone[130,131].

The ESR parameters for a series of $R_nXSe\dot{C}(t\text{-}Bu)_2$ radicals are listed in Table 39. The failure to detect unhindered β-selenoalkyl radicals (which were shown to be produced by spin-trapping experiments) was attributed to the conformation adopted by these radicals. It was proposed that the hindered radicals adopt a conformation in which the unpaired electron interacts only weakly with the Se nucleus because a Se unshared pair of electrons lies in the C_α-$2p_z$ nodal plane. The unhindered radicals presumably adopt a conformation

TABLE 39. ESR parameters for $R_nXSeC_\alpha(t\text{-Bu})_2$ radicals[a]

R_nX	g	$a(^{13}C_\alpha)$(G)	$a(^{77}Se)$(G)	a^X(G)	Temp. range (°C)
Me	2.0021	49.5	23.7	19.00	30 to −100
t-Bu	2.0019	48.5	21.3	*b*	20 to −150
F_3C	2.0005	46.5	36.6	*b*	30 to −130
Cl_3C	2.0020	45.5	46.5	*b*	−20 to −70
Ph	2.0030	49.5	18.7	*b*	30 to −50
C_6F_5	2.0026	*b*	*b*	*b*	−20 only
Me_3CO	2.0024	38.4	66.3	*b*	0 to −100
$(Me_2CHCH_2)_2(Me)CO$	2.0024	*b*	64.2	*b*	−70 only
t-BuS	2.0014	41.1	59.4	*b*	20 to −80
n-BuS	2.0018	*b*	65.8	*b*	−110 only
CF_3S	2.0024	*b*	*b*	*b*	−10 to −80
$(t\text{-Bu})_2CHSe$	2.0022	*b*	39.22	65.06	−110 only
MeSe	2.0026	*b*	*b*	*b*	−120 only

[a]Unless otherwise noted, parameters are reported in *n*-pentane or isopentane at −50°C and they are essentially invariant over the range of temperatures studied.
[b]Poorly resolved.

in which the Se lone pair is eclipsed by the C_α-$2p_z$ orbital and the resultant strong interaction with the Se nucleus leads to a line-broadening both by the spin-rotation and the spin-orbit mechanisms.

The formation of γ-selenoalkyl radicals presented no major problems provided the intervening β-atom was not carbon[121]. Thus, the $MeSeSe\dot{C}(t\text{-Bu})_2$ and $MeSeS\dot{C}(t\text{-Bu})_2$ radicals could be formed by the addition of the MeSe· radical to the selenoketone and thioketone (for ESR parameters, see Table 39). However, no γ-selenoalkyl radicals having a β-carbon could be detected by ESR spectroscopy[121]. The failure to detect these radicals was attributed to a facile and probably exothermic β-scission reaction, i.e.

$$R_nXSeCH_2\dot{C}H_2 \rightarrow R_nXSe\cdot + CH_2{=}CH_2 \qquad (11)$$

From chemical observation, Chu and coworkers[132] proposed that both Se—Se and Se—C scissions occur in dibenzyl diselenide photolysis. Brown and coworkers[133] had considered that for the same compound the cleavage of the Se—Se bond was more important than that of the Se—C bond, and a similar result had been reported for Et_2Se_2[134]. Franzi and Geoffroy[135] concluded from their results that diphenyl diselenide and dibenzyl diselenide, when subjected to UV irradiation in the presence of the spin-trapping reagent nitrosodurene, probably reacted by different mechanisms.

$$PhCH_2SeSeCH_2Ph \xrightarrow[O=NC_{10}H_{13}]{h\nu} [(PhCH_2)_2] \rightarrow PhCH_2N(\dot{O})C_{10}H_{13} \qquad (12)$$

$$PhSeSePh \xrightarrow[O=NC_{10}H_{13}]{h\nu} [PhSe] \rightarrow PhSeN(\dot{O})C_{10}H_{13} \qquad (13)$$

Equation (12) involves C—Se scission and a subsequent formation of $\dot{S}eSeCH_2Ph$, which was not detected by ESR spectroscopy. The spectrum obtained for $PhSeN(O)C_{10}H_{13}$ was characterized by a high g value, which indicated some spin delocalization into the Se atom ($g = 2.0098$, $a(^{14}N) = 18.8$ G, $a(^{77}Se) = 8.7$ G). An ESR spectrum was also observed for o-$NO_2C_6H_4SeN(O)C_{10}H_{13}$ during photolysis of $(o\text{-}NO_2C_6H_4Se)_2$ in the presence of nitrosodurene. The electron-attracting group increased the nitrogen participation in the $NO(\pi)$ orbital and thus decreased the nitrogen contribution in the $NO(\pi^*)$ orbital which contained the unpaired electron. The diminution of the ^{14}N-$2p_z$ spin density involved a smaller spin polarization of the ^{77}Se s electrons, resulting in a smaller value of $a(^{77}Se)$ ($g = 2.0075$), $a(^{14}N) = 16.5$ G, $a(^{77}Se) = 7.8$ G)[135].

It has been shown that irradiation of Ph_3SeX (X = Cl, Br)[136,137] and of R_2SeO_2[138] lead respectively to the formation of $Ph_2\dot{S}eX$ and $R\dot{S}eO_2$. The X-irradiation of a single crystal of Ph_3SeCl produced a radical showing hyperfine interaction with ^{35}Cl and ^{77}Se. Through the principal values of g and the hyperfine coupling tensors, it was proposed that the totality of the spin was shared between the Cl and Se atoms and that, for these two atoms, the unpaired electron was confined in a p orbital. The calculated spin distribution for $Ph_2\dot{S}eCl$ trapped in Ph_3SeCl was $\rho_z(^{35}Cl) = 0.24$, $\rho_x(^{77}Se) = -0.09$ and $\rho_z(^{77}Se) = 0.72$. The X-irradiation of single crystals of Ph_3SeBr leads to the formation of a radical identified as $Ph_2\dot{S}eBr$. Using the magnetic hyperfine interaction or the quadrupolar interaction, the value calculated for ρ_z(Br) was found to be the same (0.27). Thus, when $Ph_2\dot{S}eCl$ trapped in Ph_3SeCl was compared with $Ph_2\dot{S}eBr$ trapped in Ph_3SeBr, it appeared that substitution of Cl by Br led to a slight increase in the spin density of the halogen. This difference was further evident when $Ph_2\dot{S}eCl$ trapped in Ph_3SeCl was compared with $Ph_2\dot{S}eBr$ trapped in Ph_2SeBr_2 [ρ_z(Br) = 0.47][137]. This behaviour is in accordance with an antibonding character of the orbital containing the unpaired electron. From the analysis of the data it was concluded that the σ^* orbital was oriented perpendicular to the CSeC plane. The ESR parameters for the $Ph_2\dot{S}eCl$ radical trapped in Ph_2SeBr_2 have also been provided and the spin distribution was again found to be very sensitive to the nature of the host matrix[137].

The first ESR study of an organic seleninyl radical trapped in a single-crystal matrix was reported by Franzi and coworkers[139], when they produced the radical $Ph\dot{S}eO$ by irradiating diphenyl selenoxide, Ph_2SeO. From the analysis of the hyperfine interaction with ^{77}Se it was concluded that the unpaired electron was mainly localized ($\sim 80\%$) in a Se $4p_x$ orbital. The reported ESR parameters are very similar to those obtained for $Ph_2\dot{A}s$ and from that it was estimated that the p_x orbital was probably perpendicular to the CSeO plane. The spin density on the Se atom in the phenylseleninyl radical, $Ph\dot{S}eO$, was calculated to be twice as much as that reported in the phenylselenenoyl radical, $Ph\dot{S}eO_2$[139].

Radical anions of 2,2,4,4-tetramethyl-3-pentaneselone, t-$Bu_2C{=}Se$, and 1,1,3,3-tetramethyl-2-indaneselone were produced by one-electron electroreduction and were studied by ESR spectroscopy[140]. From the values of the ^{13}C hyperfine coupling constants, it was concluded that the spin density was located at the central carbon atom. The ESR parameters were similar to those obtained for the corresponding thioketyl radical anions. For the thioketyl molecules a temperature dependence of the ^{13}C hyperfine coupling constants was observed which indicated the presence of a planar geometry for these radical anions.

The ESR spectra of cation radicals (**22**) and nitroxide radicals (**23**) derived from phenoselenazine have been reported[141]. Hyperfine splittings from ^{77}Se have been observed and these are employed in the evaluation of the unpaired electron density

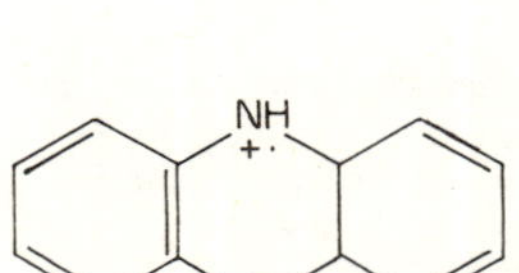

(**22**)

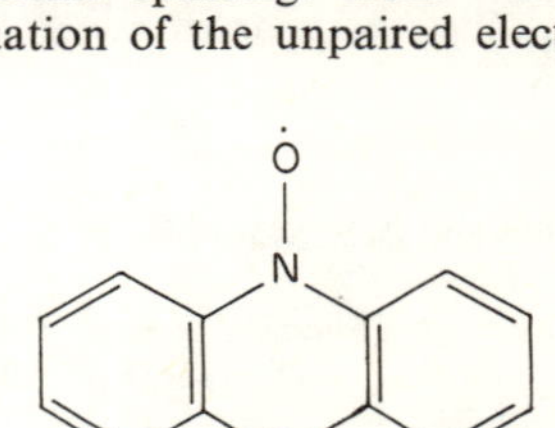

(**23**)

distributions. For the cation radical it has been concluded that spin-rotation relaxation caused an appreciable contribution to the line-width of the ESR spectrum in solution at room temperature. The ESR spectra of several Se-containing anion radicals from 2,1,3-benzoselanediazole and the corresponding perfluorocompound have been reported. No ^{77}Se satellite lines have been reported for these compounds.

The radical cations generated from the peri-bridged naphthalenes $C_{10}H_6XY$, where XY = SS, SSe, STe, SeSe, SeTe are found to be stable at room temperature except when XY = TeTe for which no ESR signal could be detected even at 200 K[142]. The ESR spectra of other compounds showed line-widths increasing in the sequence S < Se < Te. From the 17 lines observed for the disulphide radical cation, only a quintet remained in the selenothiol derivative. It was still recognizable in the ESR spectrum of $C_{10}H_6SeSe^{+\cdot}$ but could no longer be detected for $C_{10}H_6SeTe^{+\cdot}$ (only one broad line was observed). The *g* values of the radical cations generated increased from 2.0086 (XY = SS) to 2.0409 (SY = SeTe).

The formation and stabilization of free radicals in X-irradiated S- and Se-amino acids have been investigated by ESR spectroscopy[143]. It was difficult to identify radical species from powder spectra; therefore, speculative interpretations have been made based on studies on the related compounds. Se-Cystine and Se-methionine at 295 K showed very similar ESR spectra, whereas Se-ethionine exhibited a very different one. The spectra of Se-cystine and Se-methionine were poorly resolved whereas Se-ethionine showed a well-resolved spectrum. It was speculated that the main radicals in Se-cystine and Se-methionine were RSeSeĊHR′ and MeSeĊHR″. For Se-cystine the spectrum at 100 K showed a signal in the low-field region and was attributed to the RSė radical. On warming the low-field signal disappeared very quickly and the spectrum changed to that observed at 295 K.

Finally, with regard to ESR studies of organoselenium compounds, brief mention should be made of studies of biological systems which normally contain S but which have had Se substituted for labile S. ESR studies of such seleno biomolecules include ^{77}Se- and ^{80}Se-replaced putidaredoxin and adrenodoxin[144–146], Se-containing parsley ferredoxin[147] and *N,N*-dimethylselenocysteamine iron(III) complexes of bleomycin and hemoglobin[148].

B. Organotellurium Compounds

To our knowledge, there have been no published reports concerning ESR studies of organotellurium compounds.

IV. ACKNOWLEDGEMENTS

The authors are especially grateful to Ms. Sue Collins for her painstaking care in the preparation of this manuscript. Se nuclear magnetic resonance research at the University of South Carolina is supported by the National Institutes of Health (Grant ESO-2836).

V. REFERENCES

1. R. K. Harris and B. E. Mann (Eds.), *NMR and the Periodic Table*, Academic Press, New York, 1978.
2. U. Svanholm, in *Organic Selenium Compounds: Their Chemistry and Biology*, (Eds. D. L. Klayman and W. H. H. Günther), Wiley-Interscience, New York, 1973, Chap. XV D, pp. 903–932.
3. S. Gronowitz, I. Johnson and A.-B. Hörnfeldt, *Chem. Scr.*, **3**, 94 (1973).
4. S. Gronowitz, I. Johnson and A.-B. Hornfeldt, *Chem. Scr.*, **8**, 8 (1975).
5. A. Fredga, S. Gronowitz and A.-B. Hörnfeldt, *Chem. Scr.*, **8**, 15 (1975).

6. S. Gronowitz, A. Konar and A.-B. Hörnfeldt, *Org. Magn. Reson.*, **9**, 213 (1977).
7. L. Christiaens, J.-L. Piette, L. Laitem, M. Baiwir, J. Denoel and G. Llabres, *Org. Magn. Reson.*, **8**, 354 (1976).
8. R. Ernst, *Adv. Magn. Reson.*, **2**, 1 (1966).
9. T. C. Farrar and E. D. Becker, *Pulse and Fourier Transform NMR*, Academic Press, New York, 1971.
10. W. H. Dawson and J. D. Odom, *J. Amer. Chem. Soc.*, **99**, 8352 (1977).
11. J. D. Odom, W. H. Dawson and P. D. Ellis, *J. Amer. Chem. Soc.*, **101**, 5815 (1979).
12. W.-H. Pan and J. P. Fackler, *J. Amer. Chem. Soc.*, **100**, 5383 (1978).
13. W. Koch, O. Lutz and A. Nolle, *Z. Naturforsch.*, **33a**, 1025 (1978).
14. O. A. Gansow, W. D. Vernon and J. J. Dechter, *J. Magn. Reson.*, **32**, 19 (1978).
15. D. B. Denney, D. Z. Denney, P. J. Hammond and Y. F. Hsu, *J. Amer. Chem. Soc.*, **103**, 2340 (1981).
16. E. W. Abel, K. G. Orrell and A. W. G. Platt, *Org. Magn. Reson.*, **21**, 196 (1983).
17. J. R. Lyerla and D. M. Grant, *Int. Rev. Sci., Phys. Chem. Ser.*, **1**, 1 (1972).
18. G. C. Levy, *Acc. Chem. Res.*, **6**, 161 (1973).
19. J. H. Noggle and R. E. Schirmer, *The Nuclear Overhauser Effect: Chemical Applications*, Academic Press, New York, 1971.
20. A. Abragam, *The Principles of Nuclear Magnetism*, Oxford University Press, London, 1961.
21. N. F. Ramsey, *Phys. Rev.*, **78**, 699 (1950); **86**, 243 (1952).
22. W. McFarlane and R. J. Wood, *J. Chem. Soc. (A)*, 1397 (1972).
23. N. P. Luthra, R. B. Dunlap and J. D. Odom, unpublished results.
24. E. R. Cullen, F. S. Guziec, Jr., C. J. Murphy, T. C. Wong and K. K. Andersen, *J. Amer. Chem. Soc.*, **103**, 7055 (1981).
25. S. W. Carr and R. Colton, *Aust. J. Chem.*, **34**, 35 (1981).
26. T. C. Wong, F. S. Guziec and C. A. Moustakis, *J. Chem. Soc., Perkin Trans. 2*, 1471 (1983).
27. P. A. W. Dean, *Can. J. Chem.*, **57**, 754 (1979).
28. R. B. Valeev, G. A. Kalabin and D. F. Kushnarev, *Zh. Org. Khim.*, **16**, 2482 (1980).
29. N. P. Luthra, R. B. Dunlap and J. D. Odom, *J. Magn. Reson.*, **52**, 318 (1983).
30. R. W. Taft and M. J. Kamlet, *Inorg. Chem.*, **22**, 250 (1983).
31. H. Tanaka, H. Sakurai and A. Yokoyama, *Chem. Pharm. Bull. (Tokyo)*, **18**, 1015 (1970).
32. M. Lardon, *Ann. N.Y. Acad. Sci.*, **192**, 132 (1972).
33. T. Birchall, J. R. Gillespie and S. L. Vekris, *Can. J. Chem.*, **43**, 1672 (1965).
34. M. Lardon, *J. Amer. Chem. Soc.*, **92**, 5063 (1970).
35. H. Kolshorn and H. Meir, *J. Chem. Res. (S)*, 338 (1977).
36. B. Kohne, W. Lohner, K. Praefcke, H. J. Jakobsen and B. Villadsen, *J. Organomet. Chem.*, **166**, 373 (1979).
37. G. G. Furin, A. I. Rezvukhin, M. A. Fedotov and G. G. Yakobson, *J. Fluorine Chem.*, **22**, 231 (1982).
38. G. N. Dolenko, S. A. Krupoder, L. I. Nasonova, L. N. Mazalov, G. G. Furin and G. G. Yakobson, *Izv. Akad. Nauk SSSR, Ser. Khim.*, 343 (1979).
39. S. O. Grim, W. McFarlane and E. F. Davidoff, *J. Org. Chem.*, **32**, 781 (1967).
40. N. P. Luthra and J. D. Odom, manuscript in preparation.
41. W. Gombler, *Z. Naturforsch.*, **36b**, 535 (1981).
42. R. C. Weast (Ed.), *CRC Handbook of Chemistry and Physics*, CRC Press, Florida, 1978.
43. J. I. Braumann and L. K. Blair, *J. Amer. Chem. Soc.*, **92**, 5986 (1970).
44. J. D. Hirschfelder, C. F. Curtiss and R. B. Bird, *Molecular Theory of Gases and Liquids*, Wiley, New York, 1964, p. 947.
45. (a) G. E. Maciel, in *Topics in Carbon-13 NMR Spectroscopy* (Ed. G. C. Levy), Interscience, New York, 1974, p. 53.
(b) J. B. Stothers, *Carbon-13 NMR Spectroscopy*, Academic Press, New York, 1972.
46. J. D. Roberts, R. L. Webb and E. A. McElhill, *J. Amer. Chem. Soc.*, **72**, 408 (1950).
47. W. A. Sheppard and C. M. Sharts, *Organic Fluorine Chemistry*, Benjamin, New York, 1969, Chap. 2.
48. D. Holtz, *Chem. Rev.*, **71**, 139 (1971).
49. D. Holtz, *Progr. Phys. Org. Chem.*, **8**, 1 (1971).
50. K. Hiraoka, R. Yamidagni and P. Kebarle, *J. Amer. Chem. Soc.*, **95**, 6833 (1973).
51. S. Fraga, K. M. Saxena and B. W. N. Lo, *Atomic Data*, **3**, 323 (1971).

52. A. Darmadi, A. Haas and K. Tebbe, *Z. Naturforsch.*, **36b**, 426 (1981).
53. G. A. Kalabin, D. F. Kushnarev, V. M. Bzesovsky and G. A. Tschmutova, *Org. Magn. Reson.*, **12**, 598 (1979).
54. (a) R. R. Wells, S. Ehrenson and R. W. Taft, *Progr. Phys. Org. Chem.*, **6**, 147 (1968).
(b) S. Ehrenson, R. T. C. Brownlee and R. W. Taft, *Progr. Phys. Org. Chem.*, **10**, 1 (1973).
55. M. Charton, *Progr. Phys. Org. Chem.*, **13**, 119 (1981).
56. D. J. Craik and R. T. C. Brownlee, *Progr. Phys. Org. Chem.*, **14**, 1 (1981).
57. D. H. O'Brien, N. Dereu, C.-K. Huang, K. J. Irgolic and F. F. Knapp, Jr., *Organometallics*, **2**, 305 (1983).
58. W. Nakanishi, Y. Ikeda and H. Iwamura, *Org. Magn. Reson.*, **20**, 117 (1982).
59. G. Llabres, M. Baiwir, J.-L. Piette and L. Christiaens, *Org. Magn. Reson.*, **15**, 152 (1981).
60. N. P. Luthra, J. D. Odom and R. B. Dunlap, manuscript submitted to *J. Org. Chem.*
61. P. Granger, S. Chapelle and C. Paulmier, *Org. Magn. Reson.*, **14**, 240 (1980).
62. N. P. Luthra, R. B. Dunlap and J. D. Odom, *J. Magn. Reson.*, **46**, 152 (1982).
63. M. Baiwir, G. Llabres, J.-L. Piette and L. Christiaens, *Spectrochim. Acta*, **38A**, 575 (1982).
64. G. P. Mullen, N. P. Luthra, J. D. Odom and R. B. Dunlap, *J. Org. Chem.*, in press.
65. W. Gombler, *Z. Naturforsch.*, **36b**, 1561 (1981).
66. Ref. 45b, p. 157.
67. G. A. Kalabin, D. F. Kushnarev and G. A. Tschmutova, *Zh. Org. Khim.*, **15**, 24 (1979).
68. N. P. Luthra and J. D. Odom, manuscript submitted to *Org. Magn. Reson.*
69. M. Baiwir, G. Llabres, L. Christiaens and J. L. Piette, *Org. Magn. Reson.*, **16**, 14 (1981).
70. M. Baiwir, G. Llabres, J.-L. Piette, and L. Christiaens, *Org. Magn. Reson.*, **14**, 293 (1980).
71. S. Gronowitz, A. Fredga and A.-B. Hörnfeldt, *Chem. Scr.*, **5**, 236 (1974).
72. A. Fredga, S. Gronowitz, and A.-B. Hörnfeldt, *Chem. Scr.*, **11**, 37 (1977).
73. G. A. Kalabin, D. F. Kushnarev and T. G. Mannafov, *Zh. Org. Khim.*, **16**, 505 (1980).
74. T. Birchall, S. L. Crossley and R. J. Gillespie, *J. Chem. Phys.*, **41**, 2760 (1964).
75. H. J. Jakobsen and R. S. Hansen, *J. Magn. Reson.*, **30**, 397 (1978).
76. H. J. Jakobsen, A. J. Zozulin, P. D. Ellis and J. D. Odom, *J. Magn. Reson.*, **38**, 219 (1980).
77. W. Gombler, *J. Amer. Chem. Soc.*, **104**, 6616 (1982).
78. W. Gombler, *J. Magn. Reson.*, **53**, 69 (1983).
79. Ref. 1, Chap. 12, Sect. C3.
80. K. Olsson and S. O. Almquist, *Acta Chem. Scand.*, **23**, 3271 (1969).
81. W. McFarlane and D. S. Rycroft, *J. Chem. Soc., Chem. Commun.*, 10 (1973).
82. H. J. Reich and J. E. Trend, *J. Chem. Soc., Chem. Commun.*, 310 (1976).
83. W. McFarlane, D. S. Rycroft and C. J. Turner, *Bull. Soc. Chim. Belg.*, **86**, 457 (1977).
84. G. A. Kalabin and D. F. Kushnarev, *Zh. Strukt. Khim.*, **20**, 617 (1979).
85. E. R. Cullen, F. S. Guziec, Jr., C. J. Murphy, T. C. Wong and K. K. Andersen, *J. Chem. Soc., Perkin Trans. 2*, 473 (1982).
86. W. McFarlane, *J. Chem. Soc. (A)*, 670 (1969).
87. G. J. Schrobilgen, R. C. Burns and P. Granger, *J. Chem. Soc., Chem. Commun.*, 957 (1978).
88. J. A. Anderson, A. J. Zozulin and J. D. Odom, *Organometallics*, **3**, 1458 (1984).
89. I. Johannsen and H. Eggert, *J. Amer. Chem. Soc.*, **106**, 1240 (1984).
90. K. U. Buckler, J. Kronenbitter, O. Lutz and A. Nolle, *Z. Naturforsch.*, **32a**, 1263 (1977).
91. G. V. Fazakerley and M. Celotti, *J. Magn. Reson.*, **33**, 219 (1979).
92. P. Granger and S. Chapelle, *J. Magn. Reason.*, **39**, 329 (1980).
93. P. Granger, S. Chapelle, W. R. McWhinnie and A. Al-Rubaic, *J. Organomet. Chem.*, **220**, 149 (1981).
94. H. C. E. McFarlane and W. McFarlane, *J. Chem. Soc., Dalton Trans.*, 2416 (1973).
95. W. Lohner and K. Praefcke, *J. Organomet. Chem.*, **208**, 39 (1981).
96. W. Lohner and K. Praefcke, *J. Organomet. Chem.*, **208**, 43 (1981).
97. W.-W. duMont and H.-J. Kroth, *Z. Naturforsch*, **36b**, 332 (1981).
98. N. Zumbulyadis and H. J. Gysling, *J. Organomet. Chem.*, **192**, 183 (1980).
99. H. J. Gysling, N. Zumbulyadis and J. A. Robertson, *J. Organomet. Chem.*, **209**, C41 (1981).
100. G. A. Kalabin, R. B. Valeev and D. F. Kushnarev, *Zh. Org. Khim.*, **17**, 947 (1981).
101. Ref. 1, pp. 383–419.
102. D. H. O'Brien, personal communication.

103. W. McFarlane, F. J. Berry and B. C. Smith, *J. Organomet. Chem.*, **113**, 193 (1973).
104. R. K. Chadha and J. M. Miller, *Can. J. Chem.*, **60**, 2256 (1982).
105. D. H. O'Brien, N. Dereu, R. A. Grigsby, K. J. Irgolic and F. F. Knapp, Jr., *Organometallics*, **1**, 513 (1982).
106. N. L. M. Dereu, R. A. Zingaro and E. A. Meyers, *Organometallics*, **1**, 111 (1982).
107. H. Schumann and M. Magerstadt, *J. Organomet. Chem.*, **232**, 147 (1982).
108. K. J. Wynne and P. S. Pearson, *Inorg. Chem.*, **9**, 106 (1970).
109. T. Drakenberg, F. Fringuelli, S. Gronowitz, A.-B. Hörnfeldt, I. Johnson and A. Taticchi, *Chem. Scr.*, **10**, 139 (1976).
110. T. Drakenberg, A.-B. Hörnfeldt, S. Gronowitz, J.-M. Talbot and J.-L. Piette, *Chem. Scr.*, **13**, 152 (1978–1979).
111. P. H. Laur, in *Proceedings of the Third International Symposium on Organic Selenium and Tellurium Compounds, Metz, France, July 9–12*, 1979, p. 219.
112. N. L. M. Dereu and R. A. Zingaro, *J. Organomet. Chem.*, **212**, 141 (1981).
113. Ref. 1, Chap. 12, Sect. 12D.
114. M. L. Martin, M. Trierweiler, V. Galasso, F. Fringuelli and A. Taticchi, *J. Magn. Reson.*, **42**, 155 (1981).
115. W. McFarlane, *Mol. Phys.*, **12**, 243 (1967).
116. H. Dreeskamp and G. Pfisterer, *Mol. Phys.*, **14**, 295 (1968).
117. M. L. Martin, M. Trierweiler, V. Galasso, F. Fringuelli and A. Taticchi, *J. Magn. Reson.*, **47**, 504 (1982).
118. C. R. Lassigne and E. J. Wells, *J. Chem. Soc., Chem. Commun.*, 956 (1978).
119. F. Shimazu and A. L. Tappel, *Radiat. Res.*, **23**, 210 (1964).
120. A. Breccia, R. Badiello, A. Trenta and M. Mattii, *Radiat. Res.*, **38**, 483 (1969).
121. J. C. Scaiano and K. U. Ingold, *J. Amer. Chem. Soc.*, **99**, 2079 (1977)..
122. M. C. R. Symons, *Advan. Phys. Org. Chem.*, **1**, 283 (1963); *J. Amer. Chem. Soc.*, **91**, 5924 (1969).
123. M. C. R. Symons, *J. Chem. Soc., Perkin Trans. 2*, 1618 (1974).
124. B. C. Gilbert, H. A. H. Laue, R. O. C. Norman and R. C. Sealy, *J. Chem. Soc., Perkin Trans. 2*, 892 (1975).
125. D. Nelson and M. C. R. Symons, *Chem. Phys. Letters*, **36**, 340 (1975).
126. K. Nishikida and F. Williams, *Chem. Phys. Letters*, **34**, 302 (1975).
127. B. C. Gilbert, D. K. C. Hodgeman and R. O. C. Norman. *J. Chem. Soc., Perkin Trans. 2*, 1748 (1973).
128. J. T. Wang and F. Williams, *J. Phys. Chem.*, **84**, 3156 (1980).
129. J. T. Wang and F. Williams, *J. Chem. Soc., Chem. Commun.*, 1184 (1981).
130. J. C. Scaiano and K. U. Ingold, *J. Phys. Chem.*, **17**, 80, 1901 (1976).
131. J. C. Scaiano and K. U. Ingold, *J. Chem. Soc., Chem. Commun.*, 205 (1976).
132. J. Y. C. Chu, D. G. Marsh and W. H. H. Günther, *J. Amer. Chem. Soc.*, **97**, 4905 (1975).
133. D. H. Brown, R. J. Cross and D. Millington, *J. Chem. Soc., Dalton Trans.*, 159 (1977).
134. R. J. Cross and D. Millington, *J. Chem. Soc., Chem. Commun.*, 455 (1975).
135. R. Franzi and M. Geoffroy, *J. Organomet. Chem.*, **218**, 321 (1981).
136. M. Geoffroy, *J. Chem. Phys.*, **70**, 1497 (1979).
137. M. Franzi, M. Geoffroy, L. Ginet and N. Leray, *J. Phys. Chem.*, **83**, 2898 (1979).
138. M. Geoffroy and N. Leray, *J. Chem. Phys.*, **72**, 775 (1980).
139. R. Franzi, M. Geoffroy and N. Leray, *J. Organomet. Chem.*, **193**, 315 (1980).
140. C.-P. Klages and J. Voss, *Chem. Ber.*, **113**, 2255 (1980).
141. M. F. Chiu, B. C. Gilbert and P. Hanson, *J. Chem. Soc. (B)*, 1700 (1970).
142. H. Bock, G. Brahler, D. Dauplaise and J. Meinwald, *Chem. Ber.*, **114**, 2622 (1981).
143. G. Colombetti and S. Monti, *Intern. J. Radiat. Biol.*, **21**, 321 (1972).
144. J. C. M. Tsibris, M. J. Namtuedt and I. C. Gunsalus, *Biochem. Biophys. Res. Commun.*, **30**, 323 (1968).
145. W. H. Orme-Johnson, R. E. Hansen, H. Beinert, J. C. M. Tsibris, R. C. Bartholomaus and I. C. Gunsalus, *Proc. Natl. Acad. Sci. USA*, **60**, 368 (1968).
146. K. Mukai, J. J. Huang and T. Kimura, *Biochem. Biophys. Res. Commun.*, **50**, 105 (1973).
147. J. A. Fee and G. Palmer, *Biochim. Biophys. Acta*, **245**, 175 (1971).
148. Y. Sugiura, J. Kuwahara and T. Suzuki, *J. Chem. Soc., Chem. Commun.*, 908 (1982).

The Chemistry of Organic
Selenium and Tellurium
Compounds Volume 1
Edited by S. Patai and Z. Rappoport
© 1986 John Wiley & Sons Ltd.

CHAPTER **7**

Mass spectrometry of organic selenium and tellurium compounds

GEORGE D. STURGEON and MICHAEL L. GROSS

Department of Chemistry, University of Nebraska, Lincoln, Nebraska 68588-0304, USA

I. INTRODUCTION . . . 244
A. Ionization Energies . . . 245
B. Role of Charge Localization in Fragmentation . . . 247
II. MASS SPECTRA . . . 249
A. Alkyl Selenides and Tellurides . . . 249
B. Alkyl Aryl Selenides and Tellurides . . . 249
C. Diaryl Selenides and Tellurides . . . 252
D. Cyclic Selenides and Tellurides . . . 252
E. Diselenides and Ditellurides . . . 253
F. Other Selenides . . . 255
G. Dialkyl Selenoxides . . . 255
H. Cyclic Selenoxides . . . 256
I. Alkyl Aryl Selenoxides . . . 257
J. Diaryl Selenoxides . . . 257
K. Oxygen Atom Transfer Reactions: Rearrangements to Selenoxides . . . 257
L. Dialkyl Selenones ($RSeO_2R$) . . . 258
M. Diaryl Selenones (Ar_2SeO_2) . . . 260
N. Organo-selenium and -tellurium Dihalides . . . 260
O. Selenium Ylides . . . 261
P. Diorganotellurium Dicarboxylates . . . 261
Q. Seleninic Acids and Selenocyanates . . . 262
R. Carbonate Analogues . . . 263
S. Urea Analogues . . . 264
T. Selenophenes and Tellurophenes . . . 264
U. Benzoselenophenes and Related Compounds . . . 265
V. Fulvalene and Pentalene analogues . . . 269
W. Complex Heterocyclic Compounds . . . 272
X. Organometallic Compounds Involving Transition Metals . . . 281
III. SUMMARY . . . 282
IV. ACKNOWLEDGEMENT . . . 283
V. REFERENCES . . . 283

I. INTRODUCTION

The number of scientific papers concerning the mass spectra of organic Se and Te compounds has grown in parallel with the increasing interest in such compounds in the last decade. Among the reasons for the growing interest are the use of Se and Te and related species in chemical synthesis[1], the potential of these compounds for antitumour and antileptospire activity[2], the development of molecules in the organic conductor class[3], and the study of the mass spectral behaviour of compounds which contain the Group VIa metals in a rational series.

Because of the decline in activity in 'main-group element' chemistry and the concentration of inorganic chemists upon transition-metal complexes and organometallic species, most of the work in this area has been performed from the organic chemistry perspective. But the general organic chemistry view of the organo-selenium and -tellurium class is likely to consist of the opinion that these compounds by-and-large emulate their S analogues except insofar as the heavier species are less stable and hence more difficult to prepare. Although this is generally true, we shall see examples in this review where large differences in the gas-phase ion chemistry of Se- and Te-containing compounds occur in comparison to their S analogues.

The rich isotopic mixture for both Se and Te provides advantages and drawbacks. 'Fingerprinting' is more detailed in the mass spectra, and the discernment of the presence and number of atoms of Se or Te present in a species is usually quite direct. Ideal mass distributions for combinations of two or more chalcogens in the same species, even including a halogen or two, have been reported by several authors[4–6]. On the other hand, the presence of species containing one less or one additional H atom and the contributions by ^{13}C etc. mixed into the isotopic distributions can complicate the spectra. Therefore, most authors thoughtfully convert their data to a basis consistent with the presence of only ^{80}Se or ^{130}Te. Deconvolution techniques are useful and often essential for this purpose. A recent example of the analytical use of isotopic ratios is the quantitation of selenocysteine in glutathione peroxidase[7].

The generally reduced stability of Se- and Te-containing compounds, by comparison to the O- and S-containing analogues, has two effects. First, use of a direct inlet method becomes desirable since decompositions and rearrangements can result during attempts at chromatographic separation and purification. Secondly, low mass ions, corresponding to such species as CH_3Se or C_3HSe, which may be produced prior to ionization as a result of attempts to vaporize samples, are rather typically observed in the spectra. (Metastable ion studies can rule out these species as resulting from the fragmentation of heavier ions.)

A few other features of the mass spectra rather commonly occur for Se- and Te-containing organics: (1) combinations of molecular groupings leading to m/z ratios greater than that of the parent ion, very often at $[M + 80]^+$ when Se is involved; (2) the relatively larger tendency of ions to lose chalcogen, even to 'extrude' chalcogen from the middle of a group or from a ring system; and (3) the subsequent propensity of chalcogen atoms to combine to give species of the general formula $Se^{+\bullet}$ or $Te^{+\bullet}$.

The natural expectation that Se—C and, even more so, Te—C bonds would be more easily broken than their C—O and C—S counterparts is most often borne out in the fragmentation behaviours upon electron ionization. But some instances of surprising resilience of C—Se are encountered. Cases where the parent ion is too unstable to be detected are rare but do occur: e.g. alkyl selenides and tellurides.

Nearly all workers who have reported mass spectral studies of organic Se and Te compounds have made use of electron ionization (usually at high energy, 70 eV) and low-resolution mass analysis. In those cases, the proposed fragmentation mechanisms and ion structures should be viewed as entirely speculative. However, studies are beginning to appear which have involved high mass resolution to confirm elemental compositions, metastable ion studies to verify pathways, and isotopic labelling to shore-up mechanisms. Most metastable ion studies are somewhat haphazard; that is, only those metastable ions which appear in the normal mass spectra are used. With the advent of tandem mass spectrometry instruments[8] including reverse geometry and triple sector quadrupoles, more detailed investigations of decomposition pathways are urged and should appear. Moreover, standard high-resolution instruments of forward geometry can be modified with linked scanning capabilities[9] to permit studies of metastable ions unencumbered by normal ions which usually overlap severely in conventional spectra.

Other ionization modes may be useful for certain organic Se and Te compounds. Complex acyclic materials such as alkyl selenides and tellurides give low abundance molecular ions. These substances may be profitably ionized by using chemical ionization, field ionization or field desorption. Very few examples of the use of these more gentle methods of ionization have been published to date. The utility of fast atom bombardment[10] and field desorption will no doubt be realized for thermally sensitive or ionic compounds in the Se and Te series. One example compound is benzeneseleninic acid which decomposes under normal solid sample handling. Finally, negative ion mass spectra may be very useful for structure studies and for determining molecular masses. This approach has also been slighted as we were able to find only one report of organoselenium compounds in which negative ions were investigated. Nevertheless, it is expected that the recent advances in technique in mass spectrometry will have a salutary effect on future studies of organic-Se and Te compounds.

In the review which follows, we begin with a discussion of the ionization of and the charge localization in organic compounds containing Se and Te. This is intended to serve as an introduction to the chemistry of gas-phase ions of these compounds. These sections are followed by a review of the mass spectra of compound types, beginning with acyclic selenides and tellurides and moving toward complex heterocycles. The review includes papers published up to early 1984.

A. Ionization Energies

The first event in producing a mass spectrum is ionization, usually to form a radical cation. The initially formed radical cation may exist in various excited states, but internal conversion to a vibrationally excited ground state should occur very rapidly. Fragmentation is then viewed to involve the ground state of the ion. Therefore, it is of interest to consider the nature of the lowest energy ionization of Se- and Te-containing organic molecules before discussing their mass spectral fragmentations.

The lowest energy ionization of various dimethyl chalcogenides has been measured by using photoelectron spectroscopy (PES), and assigned to ionization from the chalcogen lone pair[11,12]. The trend of ionization potentials for Me_2X is 10.04, 8.71, 8.40 and 7.89 eV for X = O, S, Se and Te, respectively. The decreasing electronegativity of X as we go from O to Te destabilizes the n molecular orbital and is responsible for the trend. Similar variations were found for a series of compounds $(MH_3)_2X$ where M = C, Si, Ge, and X = S, Se, Te[13].

The lowest energy ionization of the series PhXMe also smoothly decreases as X is varied from O to Te[14]. Again, the most facile ionization is that of the n electrons of X except for PhOMe where π electron ionization is more favoured.

For a series of disubstituted benzenes RC_6H_4SeMe, where R = NH_2, OMe, SMe and SeMe and is substituted *ortho*, *meta* and *para* to the SeMe, little difference can be seen in the lowest ionization energies[15]. Again, electrons in the n orbital of the Se are most readily removed. Little interaction occurs between the two substituents, apparently because of the non-planar structures of these disubstituted molecules. On the other hand, the lowest ionization potential of p-$MeOC_6H_4OMe$ is determined by π_s orbitals of the benzene ring and not by the chalcogen.

Simple cyclic molecules, $(CH_2)_4X$, also show a smooth decrease in the ionizing energy as we go from O to Te[16]. The ionization is thought to involve an n_π molecular orbital which is perpendicular to the C—X—C plane and thus reflects the electronegativity of X. The trend of values 9.53, 8.42, 8.14 and 7.73 eV for X = O, S, Se and Te, respectively, parallels nicely that observed for the Me_2X series.

The picture is more complex for the series furan, thiophene, selenophene and tellurophene: $IE = 8.99$, 9.12, 9.01 and 8.60 eV, respectively, as determined by using electron ionization[17]. It was concluded by using PES that the highest occupied molecular orbital (HOMO) of tellurophene is π_2 whereas the other analogues have π_3 as the HOMO. The π_2 MO involves bonding between carbon atoms and antibonding between the heteroatom and carbons whereas π_3 is almost exclusively C—C bonding and, thus, its energy is nearly independent of the heteroatom. Low-energy ionization of selenophene, as of furan and thiophene, does not involve the n electrons of the chalcogen.

This inversion of π_2 and π_3 molecular orbitals in tellurophene was verified by a PES study of a large series of substituted furans, thiophenes, selenophenes and tellurophenes[18]. The π_2 is inevitably the HOMO for all tellurophenes, but π_3 is of highest energy for all others.

The PES of benzo[*b*]selenophene and benzo[*b*]tellurophene have also been measured and compared with those of benzofuran, benzothiophene, indole and indene[19]. The HOMO for benzo[*b*]tellurophene is highly localized on the heteroatom and has an ionization energy of 7.76 eV whereas the HOMO ionization energy for the benzo[*b*]selenophene is 8.03 eV.

Ionization energies determined by electron ionization have been used to determine the extent of tautomerism of various substituted selenophenes (see equation 1) in the gas phase[20–22]. The strategy involved comparisons of the ionization energy of the unknown with various model or reference compounds which exist in either a keto or enol form. For both 2- and 3-hydroxyselenophenes, the keto form is preferred whereas 3-thio-substituted selenophenes exist preferentially in the thiol (enol) form.

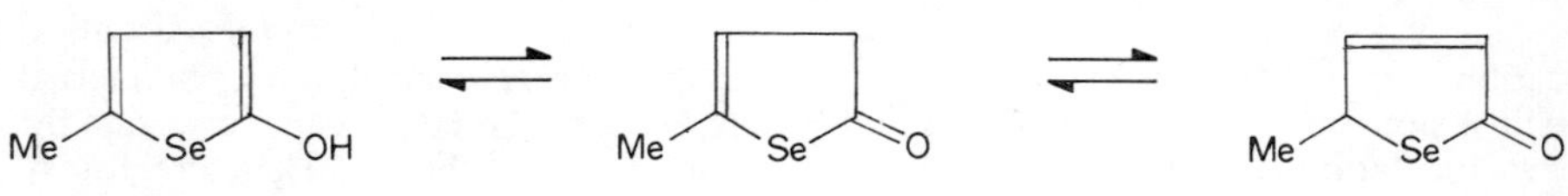

(1) **(2)** **(3)** (1)

Ionization energies from mass spectrometric measurements have also been used as correlates for understanding electrophilic substitution reactions on selenophene[23]. It was concluded that variation of reactivity with substituent effects is similar to that of thiophene.

The lowest ionization energies for a series of phenoxachalcogenins (**4**), where X = O, S, Se and Te, were determined by using PES and found to be nearly identical whatever the

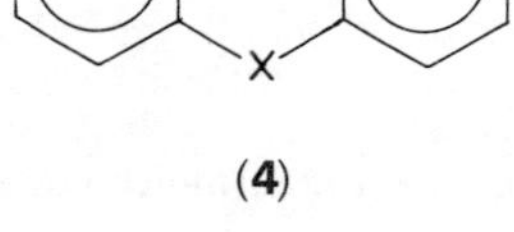

(4)

identity of X[24]. If the molecules were planar, the π MOs would become progressively destabilized as the mass of X was increased. Since this was not the case, it was postulated that the centre ring is not planar, and π interactions with the chalcogen and the aromatic rings are reduced. This conclusion is in agreement with interpretations drawn from dipole moment data[24].

The lowest ionization energies of various non-benzofused analogues (**5**), where X = O, S, Se and Te, also do not vary significantly for various identities of X, which indicates that the favoured ionization does not principally involve the chalcogen atom[25]. This is not true for various $(MeO)_3PX$ compounds where X = O, S and Se[26]. The lowest ionization energy drops in the order 10.82, 9.16 and 8.67 eV for X = O, S and Se, respectively, which can be interpreted in terms of favoured ionization of the n electrons of the X atom.

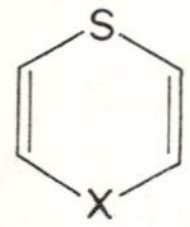

(**5**)

A unique application of PES has been the proof of the structure of 6-fulveneselone, a product of the pyrolysis of 1,2,3-benzoselenadiazole[27]. The first ionization potential of 8.34 eV is lower than for the S analogue and the lowest of known fulvene-like analogues. However, a reversal is found for tetramethyltetraselenafulvalenewhose HOMO is stabilized ($IE = 6.58$ eV) with respect to the tetrathia compound ($IE = 6.40$ eV)[28].

In summary, caution should be exercised in viewing the positive charge localized on Se or Te in ionized organic molecules containing these atoms. This seems to be an accurate picture for selenides and tellurides, but it clearly is not for more complex molecules. The role of localization of charge in triggering fragmentation reactions is discussed in the next section.

B. Role of Charge Localization in Fragmentation

The replacement of an O or S atom by Se in an organic molecule may result in a lowered ionization energy. For example, the ionization energies for Me_2O, Me_2S, Me_2Se and Me_2Te are 10.0 eV, 8.71 eV, 8.40 eV and 7.93 eV[12]. This has presented theorists in mass spectrometry with an opportunity to examine the widely held concept of 'charge localization'. It has been hypothesized that the unimolecular dissociations of gas-phase radical cations are triggered or initiated by the nearby presence of a charged site. For example, the loss of R in the series of compounds $MeXCH_2R$, where X is a chalcogen, is due to the ionization of the chalcogen (equation 2).

$$Me\overset{+\cdot}{X}CH_2R \rightarrow Me\overset{+}{X}{=}CH_2 + R\cdot \quad (2)$$

A now classic example of the use of charge localization theory is the explanation of the mass spectra of methionine and selenomethionine. Svec and Junk[29] postulated that the reason for the higher intensity of $C_2H_5X^+$ (X = Se, S) is the preferred charge localization on S or Se rather than N (equations 3 and 4). The ionization energy of Se non-bonding electrons in molecules such as selenomethionine is ca. 8.6 eV compared to ca. 10 eV for nitrogen non-bonding electrons.

$$Me\overset{+\cdot}{X}CH_2CH_2\underset{\displaystyle NH_2}{\underset{|}{C}}HCOOH \rightarrow Me\overset{+}{X}=CH_2 \quad (3)$$

$$\mathrm{MeXCH_2CH_2\underset{\overset{|}{\overset{+\cdot}{N}H_2}}{C}HCOOH \rightarrow MeXCH_2CH_2CH = \overset{+}{N}H_2} \quad (4)$$

However, Bentley and coworkers[30] have shown conclusively by means of metastable ion measurements and low energy spectra that $C_2H_5X^+$ is not produced by direct decomposition of the molecular ion but rather by some circuitous routes such as loss of H_2O followed by expulsion of methyl radical, CO and finally HCN. The loss of water has an appearance energy lower than the ionization energy of the non-bonding electrons of —NH_2 or —COOH, and, thus, the charge triggering or localization of charge on the NH_2 or COOH moieties is not a necessary prerequisite for the decomposition. In other words, the decomposition takes place 'thermally' from the vibrationally excited ground electronic state of methionine or selenomethionine, an explanation fully in accord with the quasiequilibrium theory of mass spectra.

Van den Heuvel and Nibbering[31] have shown that the corresponding multistep process leading to $C_2H_5S^+$ from ionized methionine contributes little to the intensity of this ion actually observed in the mass spectrum. This serves as an appropriate caution that metastable-denoted fragmentation routes may 'refer to processes which contribute to the intensities of peaks to only a minor extent…'. Furthermore, they were able to rationalize the decomposition route of the methionine radical cation by invoking charge localization on nitrogen. Bentley and coworkers[30] anticipated this type of rationalization by noting that the triggering approach does have the notable properties of 'flexibility and vagueness'. The criticism is undeserved, in part, because there is no doubt that charge localization is a valuable predictive and didactic tool in the hands of the practicing mass spectroscopist. Nevertheless, it is now clear that the Svec and Junk explanation of the relatively intense $C_2H_5Se^+$ ion in the mass spectrum of selenomethionine is incorrect.

Budzikiewicz and Pesch[32] have also addressed this problem of charge localization by a study of a well-chosen set of homologues $CH_3X(CH_2)_nNH_2$ where X = O, S or Se. Here, as for methionine and selenomethionine, little $C_2H_5X^+$ is obtained; the major ions produced in the mass spectral decompositions contain N, viz. $CH_2{=}NH_2^+$. The direct cleavage to give $CH_3\overset{+}{X}{=}CH_2$ is probably not favoured energetically when X = Se because of the difficulty in forming a C—Se double bond. The ionization energy and appearance energy for production of $CH_2{=}NH_2^+$ were measured as a function of the number of intervening methylene groups between Se and NH_2 (see Table 1). As can be seen from the table, the *IE* is nearly the same for all homologues and is principally determined by the orbital energy of the chalcogen. The appearance energy for $CH_2{=}NH_2^+$ is between 9 and 10 eV, which is above the ionization energy of *n*-propylamine (*IE* = 9.1 eV) but suspiciously, and perhaps fortuitously, close to the *AE* of $CH_2{=}NH_2^+$ from $CH_3CH_2CH_2NH_2$ (*AE* = 9.7 eV).

There are two notable exceptions; the homologues for which $n = 3$ have appearance energies significantly lower and nearly equal to the ionization energies. One explanation, favoured by Budzikiewicz and Pesch, is that charge exchange from Se to N can now occur

TABLE 1. Ionization (*IE*) and appearance energies (*AE*) for the homologous series of compounds: $CH_3X(CH_2)_nCH_2NH_2$

	X = S		X = Se	
n	*IE*	*AE*	*IE*	*AE*
1	8.4	8.9	8.4	9.8
2	8.5	9.7	8.3	9.5
3	8.4	8.4	8.3	8.3
4	8.5	10.1	8.3	10.0

via a six-membered ring, a transition state of lower energy than those which would be possible for the other homologues. Thus, transfer and localization of charge on the N triggers the fragmentation to produce $CH_2{=}NH_2^+$.

An alternate explanation rests upon the unusual stability of the neutrals formed along with $CH_2{=}NH_2^+$ when $n = 3$; e.g. $CH_3XCH_2^{\cdot}$ and $CH_2{=}CH_2$ or a cyclic structure. If this is so, then the question must be addressed as to why no unusual stability pertains when Se or S is replaced by oxygen. Thus, the question of charge localization remains an open one. In our view, because it provides a means of conveniently interpreting spectra of Se-containing molecules, the concept should be retained on a utilitarian basis until more solid evidence to require its rejection is generated.

II. MASS SPECTRA

A. Alkyl Selenides and Tellurides

Agenas, in a 1973 review of the mass spectrometry of Se compounds[33], discussed the Se analogues of ethers. The mass spectra of several prototypic compounds, e.g. diethylselenium, dibenzylselenium, diphenylselenium as well as more complex species such as selenodialkanoic acids, selenomethionines, 3,3′-diindolyl selenide, 3,3′-di(l-methylindolyl)selenide and mixed side-chain compounds such as selenocyanatobenzene were presented. Typical decomposition reactions of the selenides included progressive loss of alkyl groups, H migration and alkene (e.g. C_2H_4) loss, and loss of RSe, HSe or Se itself.

For cyclic analogues such as selenolane and tellurolane with five-membered saturated rings various eliminations leading to three-membered rings were most commonly reported, although cyclic structures had to be inferred.

More recently, Rebane[34] used deuterium substitution to demonstrate supplemental skeletal rearrangements which implied that Se extrusion and carbon–carbon bond formation took place. Thus, the one-step loss of C_2H_4 from dimethylselenium was confirmed by metastable ion studies. The base peak in the mass spectrum of this compound corresponded to $MeSe^+$; successive losses of hydrogens from $MeSe^+$, leading finally to $CSe^{+\cdot}$, were posited as well. The most common decomposition reaction for higher molecular weight dialkyl selenides was the cleavage of a C—Se bond accompanied by migration of H to Se to give $RSeH^{+\cdot}$. This process is rare for O- or S-containing analogues. Other ions resulted from fragmentations involving the alkyl groups.

Budzikiewicz and Pesch[32] reported the spectra of several dialkylselenium compounds, all possessing one Me group on the Se. They, along with others, observed the disfavour of the Se compounds for α-cleavage. However, their assertion that this is due to the impossibility of a Se=C double bond (to form, say, $MeSe{=}CH_2^+$) is, at best, of qualitative value, since the appearance of such a species has been reported by other workers. They also studied the decomposition of $MeSe(CH_2)_nNH_2$ as was discussed in the previous section.

The mass spectra of $(C_2F_5)_2Te$, $MeTeC_2F_5$, $(CF_3)_2Te$ and of CF_3TeMe were reported by Denniston and Martin[35]. They noticed no interchange of H and F in the mixed fluorine–hydrogen compounds. For the perfluorinated bis-ethyltellurium, Paige and Passmore[36] have similarly reported only the parent ion and an expected fragmentation pattern.

B. Alkyl Aryl Selenides and Tellurides

Rebane[37,38], Greiner and coworkers[39] and Busse and coworkers[40] have added to our knowledge of the mixed alkyl aryl ether analogues. Thus, it is now possible to make comparisons of behaviour descending the Group VI column. For phenyl methyl ether and its analogues, the parent ion is invariably the base peak in the mass spectra. For the ether proper, the familiar losses of formaldehyde and $CH_3 + CO$ are equally dominant (ca. 60%). For the sulphide, loss of SCH_2 (ca. 50%) still occurs, but new processes involving loss of Me radical (ca. 50%) and of HS expulsion (ca. 35%) enter in. With the selenide, the processes insinuated into the scheme by replacement of O by S grow in importance, and, by the time the telluride is reached, Me radical loss is entirely dominant. These results are summarized in Scheme 1 for comparison.

A similar comparison of the phenyl ethyl ether and its analogues (see Scheme 2) first shows, for the ether itself, the familiar expulsion of ethene and production of the $PhOH^{+\cdot}$ ion (via the so-called the phenetole rearrangement) which provides the base peak in the spectrum. While the same sort of process occurs for the sulphide, the $PhS{=}CH_2^+$ ion, by loss of a Me radical, is nearly as important. With the replacement of S by Se, yet another

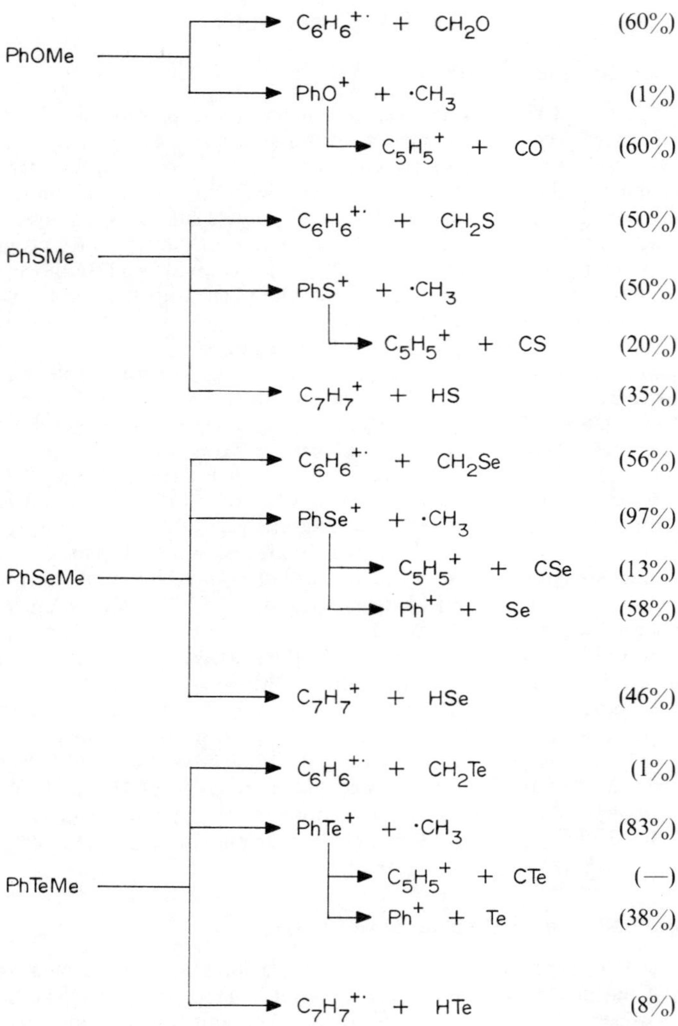

SCHEME 1

process, loss of a Et radical, becomes significant. Finally, for phenyl ethyl tellurium, the important ions (parent ion = 100%) are: Ph^+ (108%, perhaps by progressive loss of Et and Te), $PhTe^+$ (83% loss of Et radical), $C_6H_6^{+\cdot}$ (1% loss of ethene and Te), $Te^{+\cdot}$ (46.0%), HTe^+ (8.6%) and $C_7H_7^+$ (17.2%). This latter species may be viewed as a product of reductive elimination of Te from $PhTeCH_2^+$. The rearrangement to give $PhTeH^{+\cdot}$ still occurs, but it is relatively unimportant.

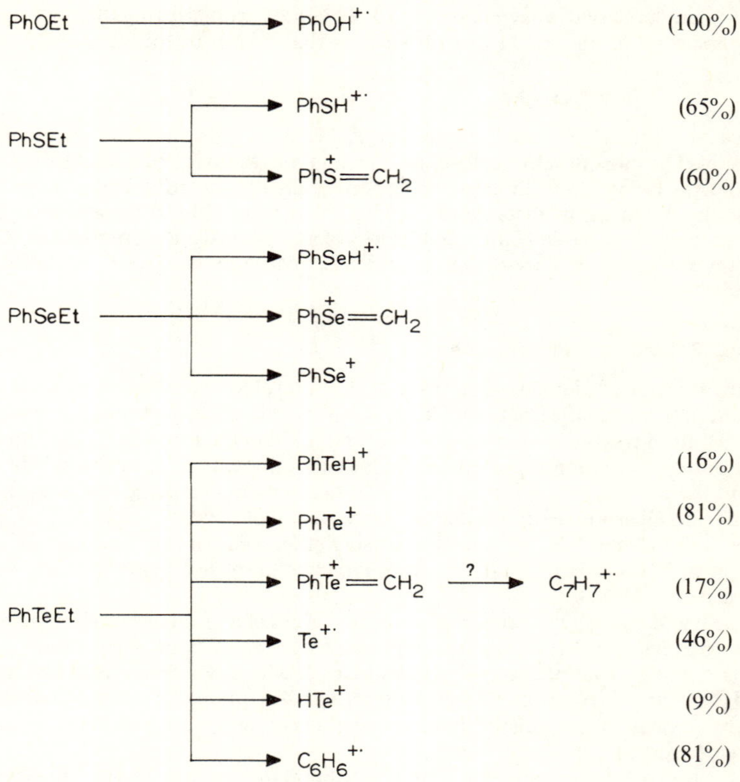

SCHEME 2

Greiner and coworkers[39] reported the fragmentations of ionized ether analogues involving Ph and Et groups substituted by CN or CO_2Me groups. α-Cleavage to produce Ar—X=CH_2^+ (X = O, S, Se, Te) was encouraged by the substituent groups; $PhXH^{+\cdot}$ fragment ions lost CX to give $C_5H_6^{+\cdot}$ when X = O or S. Production of PhX^+ became important for X = Se or Te as for the PhXMe and PhXEt series discussed above.

Busse and coworkers[40] also noted the presence of low molecular weight species of the class TeC_pH_q ($p = 1$–4, $q = 1,2$) in the mass spectra of alkyl phenyl tellurides. These fragment ions were found to derive their hydrogens from the Ph ring as was proved by studies of deuterium-labelled analogues. The $MeTe^+$ ion from methyl phenyl telluride contained exclusively the Me group on the Te of the parent compound with no H/D exchange reactions preempting this simple cleavage reaction.

C. Diaryl Selenides and Tellurides

Fewer studies have been carried out on diaryl selenides and tellurides, although some information has arisen as a result of studies upon the dichalcogenides (*vide infra*). Rebane[41] has supplied data concerning bis(substituted phenyl)selenium. He has reported that the main fragmentations involve scission at the Se atom although loss of the substituents on the Ph ring is also typical. Most interesting are the strong tendency to extrude the Se entirely, presumably giving rise to a substituted biphenyl species, the tendency to undergo rearrangements towards expanded aromatic rings (tropylium ion, for example) or even to undergo a process such as that shown in equation (5).

$$(MeC_6H_4)_2Se^{+\cdot} \xrightarrow{-2\,Me} (C_6H_4)_2Se^{+\cdot} \xrightarrow{-Se} (C_6H_4)_2^{+\cdot} \qquad (5)$$

Albeck and Shaik[4] reported studies on Ph_2Te, *p*-An_2Te and their dichlorides and dibromides. The most notable features for the dihalides (halogen bounded to Te) were the losses of the halogens leading to the species already noted for the simple diaryl compounds. Based on metastable ion studies, they were able to demonstrate that the dihalides were also capable of the stepwise loss of aryl groups giving rise to $ArTeCl^{+\cdot}$ and $TeCl^+$, for example, but since the initial step was invariably loss of $Cl^\cdot$, no $ArTeCl_2^+$ was observed.

D. Cyclic Selenides and Tellurides

Rebane[34] reported the mass spectrum of the saturated cyclic selenane, $C_5H_{10}Se$. An important process was the fracture of both Se—C bonds, allowing the elimination of HSe after an H atom migration so that the resulting ion has the formula $C_5H_9^+$. An entirely parallel pattern has been reported for the selenacyclopentane[42], but this behaviour is foreign to the corresponding S analogues. Additional modes involved elimination of one-, two-, and three-carbon atom fragments giving rise to $C_4H_7Se^+$, $C_3H_6Se^{+\cdot}$, $C_3H_3Se^+$, $C_2H_4Se^{+\cdot}$, $C_2H_5^+$ and $CH_2Se^{+\cdot}$, the last arising at least in part via a succession of C_2H_4 eliminations. $CHSe^+$, HSe^+ and $Se^{+\cdot}$ ions, typical of aliphatic Se compounds, were also observed here.

Kulkarni and coworkers[43] reported detailed studies of a series of saturated heterocyclic hexanes analogous to 1,4-dioxane. Important decompositions resulted in cyclic intermediate ions by elimination of C_2H_4 and CH_2X where X = chalcogen. The formation of $C_2H_4X^{+\cdot}$ and $C_2H_5^+$ ions was also important. As would be anticipated, there was considerable variation among the different members of the series and a fairly regular trend was often observed.

In the case of 1,4-dioxane itself, the most abundant ion was $C_2H_4^{+\cdot}$, the parent ion lagging at 77% and the ion resulting from loss of formaldehyde at 59%. For all other members of the series studied, the parent ion was the most abundant and the occurrence of $C_2H_4^{+\cdot}$ was at less than 10% except for the 1,4-dithiane (80%). Loss of formaldehyde dropped off in importance as one O atom was successively replaced with S, Se or Te; of course, loss of formaldehyde was impossible in the absence of O, but no formaldehyde analogue figured significantly among the species ejected from the parent ions. The authors suggested that a difficulty in forming the resultant cyclic intermediates militates against these eliminations for those species bearing the heavier chalcogens.

Especially in the cases involving the heavier chalcogens, the fragment ions retained the chalcogen atom because, apparently, of the capacity of the chalcogen for stabilizing positive charge through a greater ability to share electrons or to be polarized.

Formation of $H_2C{=}X^{+\cdot}$ via C_3H_6O loss assumed the most importance for OC_4H_8S (99%) and for OC_4H_8Se (85%), dropping off thereafter for OC_4H_8Te (13%). The dithiane (49%) and to a lesser extent the diselenane (17%) produced a similar $CH_2X^{+\cdot}$ ion.

Production of $C_2H_4X^{+\cdot}$ was of little importance for the low molecular weight members of the series, but their direct production by loss of C_2H_4O became very important when X = Se (90%) or Te (44%). Hydrogen migration and loss of C_2H_3O led to $C_2H_5X^+$ [X = S (66%), Se (51%) and Te (0.7%)]. For 1,4-dithiane, the $C_2H_2^{+}$ peak was again fairly large at 53%.

Nanjappan and coworkers[44] have published recently a study of an extensive series of 4-selenanones (**6**) with both alkyl and aryl substituents on the ring carbon atoms. In addition to cleavage reactions involving ring carbons, loss of CO occurred to form presumably a five-membered cyclic selenide, a process not found for 4-thianones.

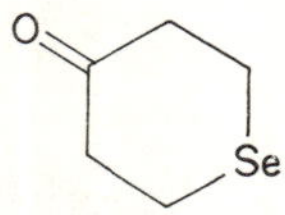

(**6**)

E. Diselenides and Ditellurides

Reactions of tetrafluoroethylene with $Se_8(AsF_6)_2$ and with $Se_8(Sb_2F_{11})$ were carried out by Desjardin and Passmore[45]. Among the unusual perfluorinated alkylselenium compounds which were discovered and studied using mass spectrometry were $(F_5C_2Se)_2$, $(F_5C_2Se)_2Se$, their Me analogues and some oxyfluoro species tracing from chemical contaminants.

The triselenium species gave rise to $F_4C_2Se_3C_2F_5^{+\cdot}$, and $F_4C_2Se_3^{+\cdot}$ along with species typical of the diselenium version of the compound (*vide infra*). In light of the occurrence of triselenium species in mass spectra of samples where only diselenium species are supposed to exist, it is not clear whether real quantities of a neutral triselenium compound were actually prepared, although the excess of Se present would certainly allow for this. The reports of the mass spectrometry of the triselenium species are sketchy, but the spectra do show that F atom loss and the fracture of F—C as well as C—Se bonds, while the Se_3 group remains intact, both occur.

In the spectrum arising from $(F_5C_2Se)_2$, the most abundant ion is CF_3^+ (100%); other important fluorocarbon fragments are $C_2F_5^+$(37%) and $C_2F_4^{+\cdot}$(23%). As for Se-containing species, the parent ion (26%), $[M - C_2F_5]^+$ (36%), $SeC_2F_5^+$ (5%), $F_4C_2Se^{+\cdot}$ (7%), $Se_2C_2F_4^{+\cdot}$ (8%), $SeCF^+$ (17%) and $Se_2^{+\cdot}$ (39%) appear with notable abundances. Many less abundant species appear to have arisen possibly by loss of a fluorine atom, breaking a C—F bond. The ion at *m/z* 379 (1%) certainly corresponds to $[M - F]^+$. This information together with the albeit slight presence of species such as $F_2CSe_2^{+\cdot}$, $FCSe_2^{+\cdot}$, and $SeC^{+\cdot}$ serve as a warning that fragmentations depend upon factors aside from inherent or relative strengths of bonds such as Se—C or C—F.

Irgolic and Haller[5] have presented and discussed the mass spectra of symmetrical diaryl ditellurides where the aryl groups were XC_6H_4(X = 4-Me, 3-F, 4-F, 4-Br, 4-Ph). In most cases, the parent ion was most abundant, but for the bis(fluorophenyl) and bis(bromophenyl) compounds, the $ArTe^{+\cdot}$ species were the most favoured; for bis(4-biphenylyl) ditelluride the parent ion was of low abundance and the dominant species was $Ar^{+\cdot}$. All of the spectra exhibited the foreseeable species $Ar_2Te^{+\cdot}$, $ArTe_2^+$, $ArTe^+$, $Ar_2^{+\cdot}$, Ar^+, $Te_2^{+\cdot}$ and $Te^{+\cdot}$ (cf. Scheme 3). In addition, the species $Ar_2Te_3^{+\cdot}$ was noted in several of the spectra. Whether this arose from a tritelluride impurity in the initial compound or from thermolytic rearrangement reactions upon evaporation into the ionization region or upon electron ionization is not known. Again, low intensity peaks corresponding to

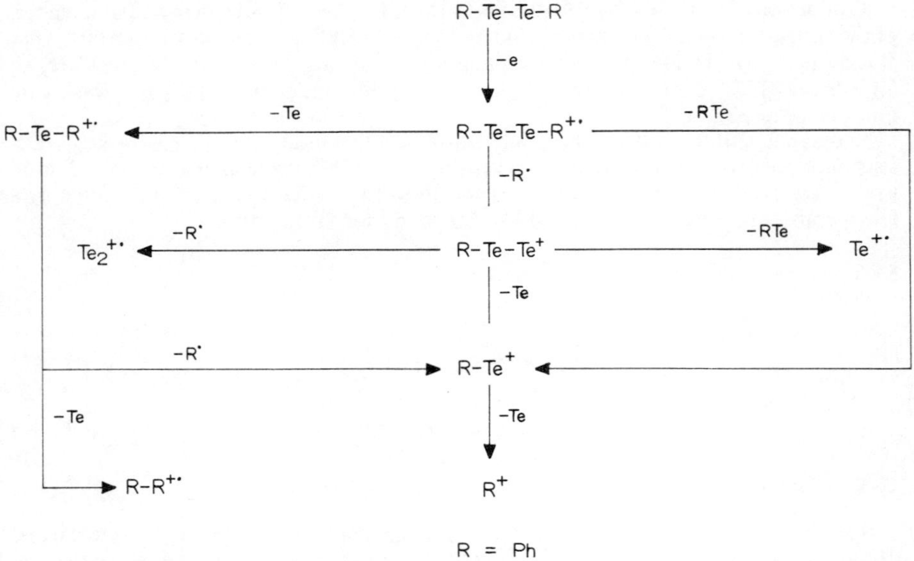

SCHEME 3

TeC_nH ($n = 2$–5) were observed. Also, typically, species such as $ArTe_2H^{+\cdot}$, $ArTeH^{+\cdot}$, Te_2H^+ and TeH^+ were present. Finally, smaller hydrocarbon fragments, attributable to decomposition and possible rearrangements of Ar^+ and $Ar_2^{+\cdot}$ were observed. Therefore, the authors noted that, with the exception of HS loss from disulphides, which is not found for Se and Te cases, the spectra of the diaryl disulphides, diselenides and ditellurides are largely the same.

For the decomposition of ionized bis(4-bromophenyl) ditelluride, the presence of a $C_7H_7Te^+$ species was noted. Although unexplained, it is reminiscent of $C_7H_7Te^+$ observed in the spectrum of PhTeEt and must necessarily result from some rearrangement process or processes.

Dance and McWhinnie[46] studied a similar group of R_2Te_2 compounds but with R = Ph, *p*-Tol, *p*-An, *p*-$EtOC_6H_4$, *p*-$PhOC_6H_4$ or C_6F_5. These spectra presumably exhibited the features as described by Irgolic and Haller[5]. Dance and McWhinnie performed appearance energy measurements which permitted them to argue for the production of $R_2Te^{+\cdot}$ by the fragmentation of $R_2Te_2^{+\cdot}$ rather than via thermolysis on the sample introduction probe. The curvature of the ionizing energy vs. intensity curves for RTe^+ was cited as an argument in favour of two routes for production of this ion from the parent $R_2Te_2^{+\cdot}$—the first by rupture of the Te—Te bond and the second by progressive loss of Te, then R. They suggested that both the Te—C and Te—Te bonds have bond strengths of the order of 290 kJ mol^{-1}, thus supporting the low-energy pathway represented by Te extrusion.

Another interesting feature of this work involved the physical mixing of two different diaryl ditelluride species prior to admission to the sample introduction probe and ionization. In addition to the spectral features anticipated from each component of the mixture, evidence for mixed species such as $PhTeTeC_6H_4OPh^{+\cdot}$ and $PhTeC_6H_4OPh^{+\cdot}$ was noted, in spite of their lack of success in preparing the corresponding unsymmetrical neutral ditelluride molecules.

Finally, these workers also prepared and studied the series Ph_3SnTeR (R = Ph, *p*-An, *p*-$EtOC_6H_4$, *p*-$PhOC_6H_4$) and also observed a parent ion for $(n\text{-}Bu)_3SnTePh$, all of which might be considered 'pseudo-ditellurides'. Appearance energies for some ionic species and the note that $Ph_6Sn_2^{+\cdot}$ was observed from thermolysis were the limits of information provided.

Continuing their pursuits of the 1960s, Buu-Hoi and coworkers[47] studied the mass spectra of RTeTeR′ compounds, most of them with R = R′ (R = Ph, *p*-Tol, α,α′-Naph, β,β′-Naph). Parent ions were obtained for these compounds, presumably because the aryl groups provided sufficient stabilization for the positive charge. Many of the features of the spectra were expected, e.g. the stepwise extrusion of Te atoms leading to $RTeR'^{+\cdot}$ and $RR'^{+\cdot}$ species. However, there was no evidence presented for the symmetric fracture of the parent ion at the Te—Te bond. The spectra of these compounds qualitatively, if not quantitatively, emulated those of the analogous Se compounds. However, the counterparts of species such as C_3HSe^+ and CHSe, species which are typical in the Se systems, were not reported here.

F. Other Selenides

Shafiee and coworkers[48–57] have prepared a number of Se-containing compounds and have used mass spectrometry to characterize them, chiefly by observing the parent ion and noting the intensity pattern appropriate for the isotopic distribution of Se. In one study[48], the mass spectra of a group of related compounds were studied. They can be interpreted as discussed below although no demonstration of the fragmentations or supporting data are available.

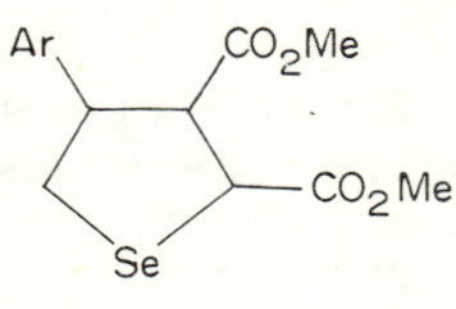

(**7**)

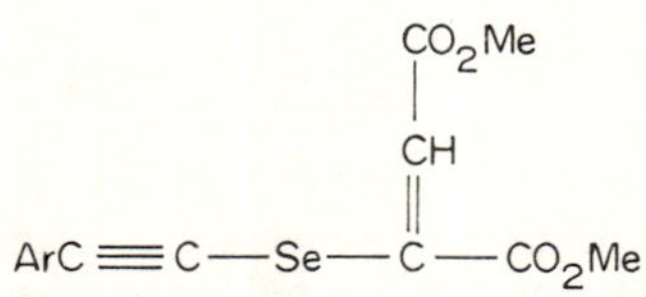

(**8**)

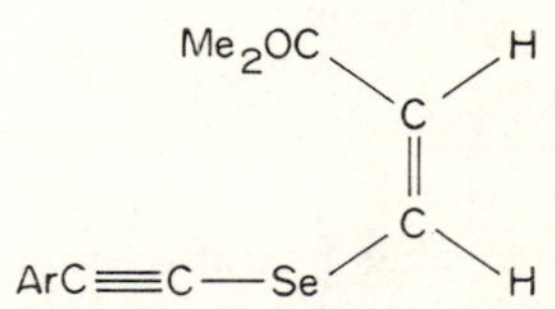

(**9**)

For all of these compounds Ar = *p*-Tol. Progressive losses from the parent ion at *m/z* 338 for **7** gave rise to daughter peaks at *m/z* 307 (loss of MeO), *m/z* 275 (loss of CH_4O from *m/z* 307), *m/z* 219 (loss of two CO) and *m/z* 139 (loss of Se). The isomer, **8**, appeared to lose Me (giving *m/z* 323) and $(CCOOMe)_2$ followed by Se (giving *m/z* 195 and 115), and also to undergo a rearrangement yielding *m/z* 146 which corresponds to $ArC{\equiv}COMe^{+\cdot}$. Compound **9** with parent ion at *m/z* 280 apparently lost OMe (giving *m/z* 249), $SeCHCHCO_2Me$ (giving *m/z* 115) and $CHCHCO_2Me$ (giving *m/z* 195). In addition, the ArC≡COMe radical cation (*m/z* 146) is again produced along with ions at *m/z* 209 corresponding to $ArC{\equiv}C{-}Se{-}CH_2^{+\cdot}$ and at *m/z* 129 corresponding to $ArC{\equiv}C{-}CH_2^{+\cdot}$.

G. Dialkyl Selenoxides

Dialkyl selenoxides are the Se analogues of sulphoxides, and their mass spectra have been investigated by Rebane[38]. Dimethyl selenoxide gives an intense molecular ion upon ionization, but the molecular ion intensity drops precipitously for higher homologues such as diethyl, dipropyl, etc. The mass spectra of these latter compounds are dominated by alkyl and alkenyl ions. For example, the most abundant fragments produced in the decomposition of dibutyl selenoxide are $C_4H_9^+$ and its decomposition products $C_3H_5^+$ and $C_2H_5^+$. Unlike $Me_2Se{=}O$ and Ph(Me)Se=O which give abundant fragments corresponding to $MeSeO^+$ and $PhSeO^+$, the higher dialkyl selenoxides give very low-abundance ions corresponding to $C_nH_{2n+1}SeO^+$.

This is also true of the dialkyl selenides themselves. The relative abundances of RSe^+ are surprisingly low compared to R^+ (R = Et, Pr, Bu). From this we may conclude that the ionization potentials of the alkyl selenide and alkyl selenoxide radicals are higher than those of the alkyl radical, i.e.:

$$Et^{\cdot}\ (IE = 8.4\,\mathrm{eV}) < EtSe^{\cdot} \text{ and } EtSeO^{\cdot} \quad (6)$$

$$Pr^{\cdot}\ (IE = 8.1\,\mathrm{eV}) < PrSe^{\cdot} \text{ and } PrSeO^{\cdot} \quad (7)$$

There exist interesting rearrangement reactions of the dialkyl selenoxide radical cations which compete only weakly with the decomposition to produce alkyl ions. The rearrangement processes are most noteworthy for MeSeOMe. In addition to the rearrangement of H followed by loss of OH, losses of MeO and CH_2O to give $MeSe^+$ and $MeSeH^{+\cdot}$, respectively, have been observed. These latter ions must originate in a rearrangement (termed a selenoxide–selenate rearrangement) (see equation 8). The relative abundances of $MeSe^+$ and MeO^+ are comparable, which may mean that the ionization potential of MeSe is approximately equal to that of MeO. However, it is possible that these ions are produced by rearrangement to give CH_2OH^+, for example.

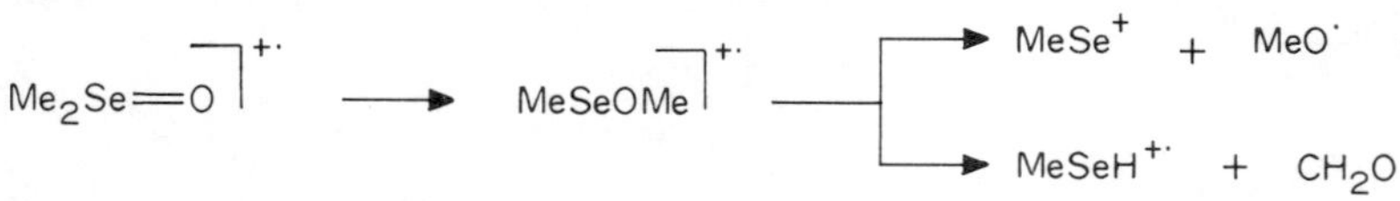

(8)

The corresponding selenoxide–selenate rearrangement occurs for diethyl, dipropyl and dibutyl selenoxides, but it is of little consequence. The mass spectrum of $Pr_2Se{=}O$ shows a weak molecular ion, loss of O, OH and C_3H_6. The losses of OH and of the olefin are also found in the decompositions of the corresponding sulphoxides.

The dibenzyl compound shows the loss of C_7H_6 (which corresponds to C_3H_6 loss for dipropyl), loss of O and a low abundance selenoxide–selenate rearrangement. Specifically, losses of $PhCH_2Se$ and $PhCH_2O$ constitute less than 5% of the total ion current.

H. Cyclic Selenoxides

Compared to the acyclic compounds, the five- and six-membered ring selenoxides show considerably more abundant molecular ions[38]. Losses of O, OH and C_2H_4 all imply no rearrangement. However, the low intensity losses of SeH and C_2H_4O require a prior rearrangement, presumably selenoxide-to cyclic selenate (equation 9). The corresponding process is less important for the six-membered ring, probably because expansions to seven-membered rings are less likely.

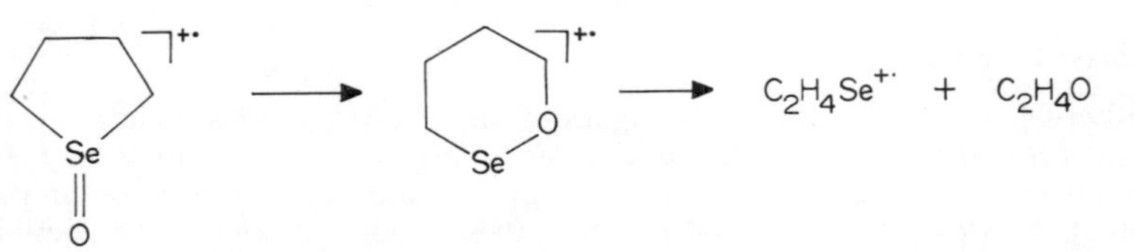

(9)

I. Alkyl Aryl Selenoxides

Methyl phenyl selenoxide undergoes a very facile rearrangement to form m/z 94 (base peak) by expulsion of neutral CH_2Se[38]. Again, this can be accounted for by the selenoxide-selenate rearrangement involving preferential Ph migration (equation 10). The corresponding rearrangement also occurs from methyl phenyl sulphoxide, but it is considerably less facile. The other decompositions of Ph(Me)Se=O radical cations are the expected ones: loss of Me and O. Following loss of O, the expulsion of SeH to give $C_7H_7^+$ is found, just as for PhSeMe.

$$Ph(Me)Se{=}O^{+\cdot} \rightarrow PhOSeMe^{+\cdot} \rightarrow PhOH^{+\cdot} \quad (10)$$

J. Diaryl Selenoxides

Here skeletal rearrangements totally dominate, which is consistent with the high aptitude of Ph groups for migration to electron-deficient centres[38]. First of all, one sees abundant loss of Se (reminiscent of the diaryl selenides) to give, in the case of diphenyl selenoxide, $PhOPh^{+\cdot}$. the expected loss of O is observed and expulsion of Se again gives $Ph{-}Ph^{+\cdot}$. However, the most intense ion is $PhSe^+$ (equation 11). Weakly competitive is the formation of PhO^+ as is expected in view of the greater ability of PhSe to accommodate a

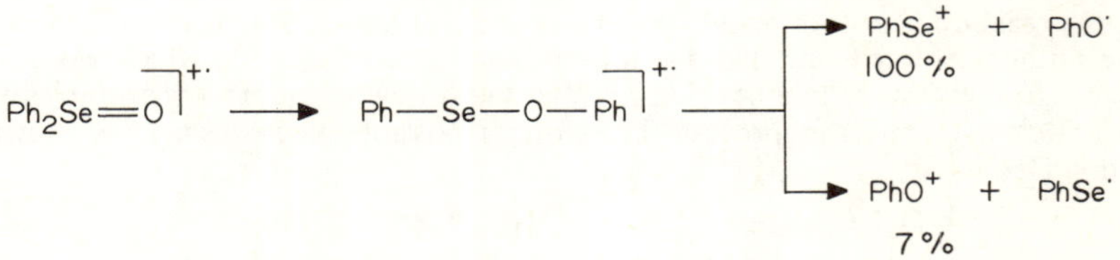

(11)

positive charge. The loss of H_2O is reasonably facile and must be accompanied by an interesting rearrangement; Rebane has suggested the pathway as shown in equation (12). The rearrangement to selenate finds precedent in $Ph_2S{=}O$ with PhS^+ as one of the most abundant fragments formed in its decomposition. The comparable rearrangement also occurs for substituted diphenyl selenoxides and is probably general for all diaryl selenoxides.

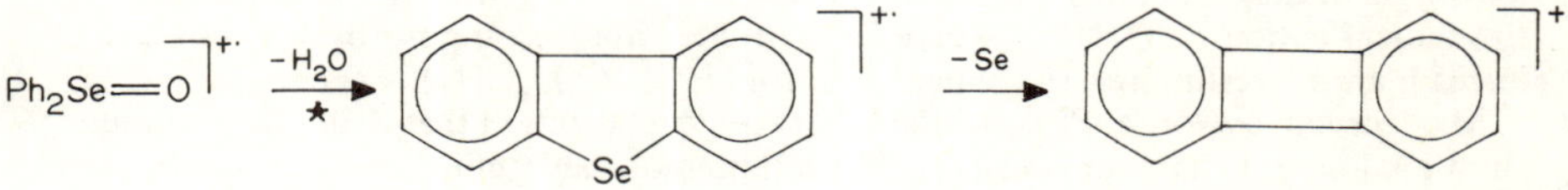

(12)

K. Oxygen Atom Transfer Reactions: Rearrangements to Selenoxides

Certain molecules containing S and o-$NO_2C_6H_4$ groups undergo novel oxygen-transfer reactions upon electron ionization (see equation 13). Martens and coworkers[58] have recently investigated the scope of this reaction by replacing the S atom with Se. For example, the radical cation of the Se analogue of the compound pictured decomposes to two ions, **10** and **11**, which must arise via an oxygen transfer. The most abundant ion (100% rel. abund.) is $\overset{+}{S}eC_6H_4Me$, which is formed by simple α-cleavage.

$$o\text{-}O_2NC_6H_4C({=}O)SC_6H_4Me^{+\cdot} \longrightarrow O{=}\overset{+}{S}{-}C_6H_4Me \quad (13)$$

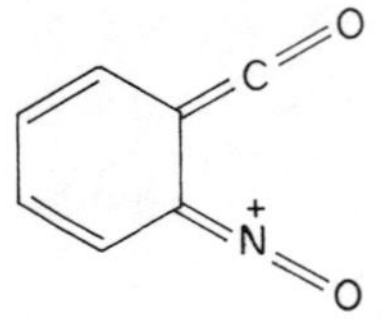

$O{=}SeC_6H_4Me$

(**10**)
21% rel. abund.

(**11**)
95% rel. abund.

A more stringent test of the generality of the reaction is the mass spectral decomposition of *o*-nitrophenyl *p*-tolyl selenide (equation 14). The indicated fragmentation is very facile which shows

$o\text{-}O_2NC_6H_4SeTol\text{-}p]^{+\cdot} \longrightarrow$ 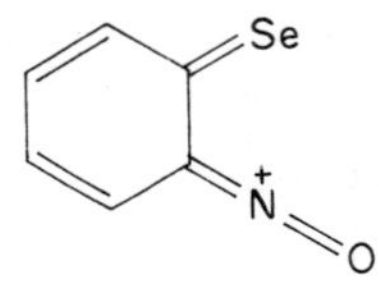 (14)

that oxygen transfer can occur via a five-centred transition state. When the Se is replaced by S, the reaction is not observed; rather the unusual loss of SO_2 is found instead.

The authors pointed out that the intermediacy of a selenoxide which would then rearrange to a selenate cannot be ruled out. We have reviewed a number of examples which implicate clearly the latter process. Therefore, a possible mechanism is as shown by equation (15).

$$o\text{-}O_2NC_6H_4SeTol\text{-}p^{+\cdot} \rightarrow o\text{-}ONC_6H_4Se({=}O)Tol\text{-}p^{+\cdot} \rightarrow o\text{-}ONC_6H_4SeOTol\text{-}p^{+\cdot} \quad (15)$$

If the Se is replaced by SeO, again the rearrangement of O occurs with expulsion of C_7H_7O, to give the most intense ion in the mass spectrum. The corresponding sulphoxide radical cation does not undergo this loss of C_7H_7O, but oxygen-transfer reactions must still pertain as losses of SO_2, SO_3 and HSO_3 (the last gives rise to the most abundant ion) are conspicuous.

The aromatic ring might be regarded as necessary to orient the *o*-nitro group properly for oxygen transfer, but this is not so for acyclic S-containing nitro compounds. However, the radical cation of $O_2NCH_2CH_2COSeTol$ does not decompose by loss of C_7H_7O; instead, an interesting rearrangement to form $[HOSeC_6H_4CH_3]^{+\cdot}$ occurs.

Martens and coworkers[58] concluded that the generalization that S and Se compounds show analogous behaviour is suspect. The compounds they studied are the most dramatic examples which support their conclusion, but additional examples are reviewed here that show that mass spectral decompositions of Se-containing compounds are related but not identical to those of S analogues.

L. Dialkyl Selenones (RSe(=O)(=O)R)

The mass spectra of this class of compounds have been principally investigated by Rebane[59]. Only dimethyl selenone gives an abundant molecular ion; for the diethyl homologue, the molecular ion is barely visible, and for higher dialkyl homologues, the molecular ion is so low in abundance as to preclude inclusion in the bar graph spectrum.

Clearly alternative methods of ionization, such as chemical or field ionization, will be required for obtaining unambiguous molecular weight information of even moderately complex dialkyl selenones. This precipitous decline in molecular ion abundance and the higher propensity to produce alkyl ions is also a property of sulphones, selenoxides and selenides (*vide supra*).

The principal decomposition reactions of the diethyl selenones and higher homologues is to produce alkyl ions. Just as for the selenoxides, the fragmentation to ions carrying the Se atom is overshadowed by the high abundance of alkyl cations. Nevertheless, some diagnostic ions of higher mass are produced by loss of an alkyl group or by rearrangement of H followed by loss of an alkene (see equation 16). These decomposition reactions are characteristic of the selenides and the selenoxides as well.

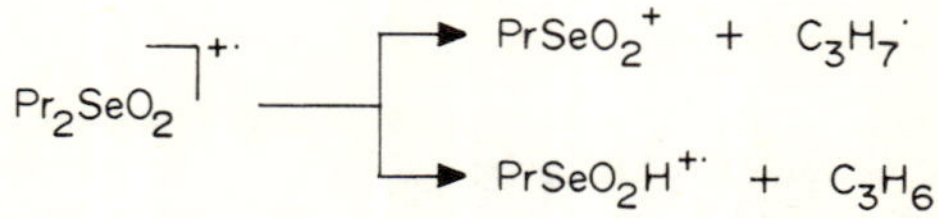

(16)

Isomerization of dialkyl selenones to dialkyl seleninates must occur for these compounds as evidenced by losses of CH_2O, CH_3O and formation of CH_3O^+ from the molecular ion of dimethyl selenone (equation 17). A second Me migration either before, concurrent with, or following loss of MeO and Me was also postulated to account for the consecutive loss of CH_2O (equations 18 and 19) and was substantiated by observation of the appropriate metastable ions.

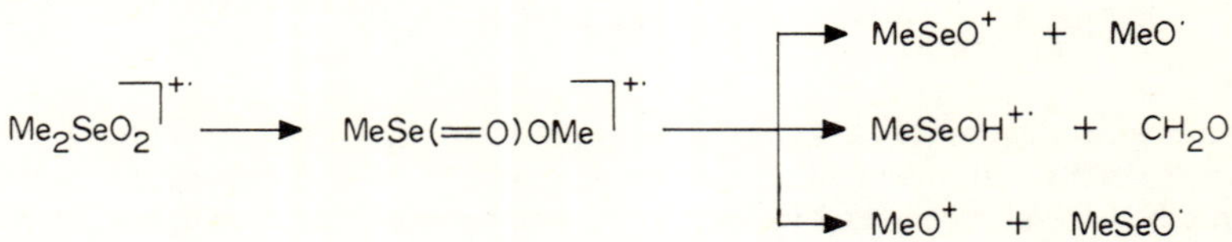

(17)

$$\overset{O}{\overset{\|}{MeSe^+}} \rightarrow MeOSe^+ \rightarrow HSe^+ + CH_2O \quad (18)$$

$$MeSeO_2^+ \rightarrow MeOSeO^+ \rightarrow HSeO^+ + CH_2O \quad (19)$$

The mass spectrum of dimethyl sulphone serves as an interesting contrast to that of dimethyl selenone. The sulphone decomposes principally by loss of Me, and the losses of MeO, CH_2O and formation of MeO^+ are only weakly competitive. Replacement of S by Se apparently lowers the activation energy for the selenone–seleninate rearrangement such that this process is highly competitive with the simple cleavage to expel a methyl group.

The higher molecular weight homologues also undergo the selenone to seleninate rearrangement, but the principal fragmentations of the isomerized ion no longer include expulsion of the alkoxy radical. For example, dipropyl selenone decomposes to give roughly comparable amounts of PrO^+ and $PrSeOH^{+\cdot}$ (equation 20). Apparent loss of O_2

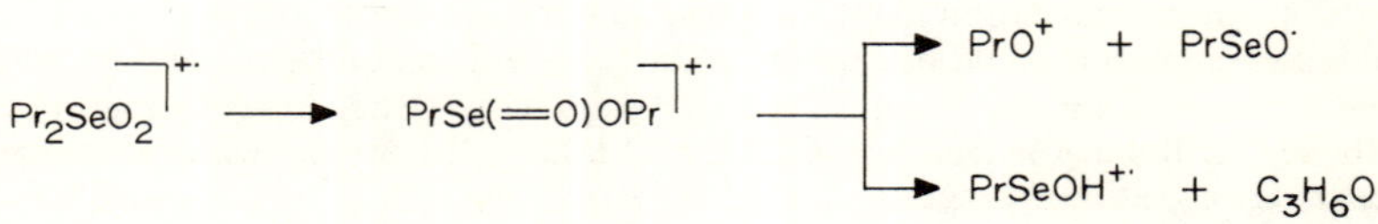

(20)

either as such or via a two-step process is also characteristic of higher homologues, but it is not possible to rule out trace quantities of the dialkyl selenides being present initially or formed thermally on the sample introduction probe of the mass spectrometer.

A unique rearrangement occurs for the dibutyl selenone. Apparently a double hydrogen rearrangement occurs with expulsion of $C_4H_7^{\cdot}$, which is followed by loss of $C_4H_9^{\cdot}$ (equation 21).

$$Bu_2SeO_2^{+\cdot} \longrightarrow BuSe(OH)_2CH_2CH{=}CHCH_3^{+\cdot} \longrightarrow BuSe(OH)_2^{+} \longrightarrow Se(OH)_2^{+\cdot} \quad (21)$$

M. Diaryl Selenones (Ar_2SeO_2)

Diaryl selenones undergo extensive rearrangement reactions, similar to diaryl selenoxides, an observation not unexpected given the greater migratory aptitude of aryl groups. These rearrangements find precedent in diaryl sulphone mass spectral reactions as well. Thus, the base peak in the mass spectrum of diphenyl selenone arises by elimination of $PhO^{\cdot}$. Furthermore, PhO^{+} formation presumably occurs in competition with the elimination of $PhO^{\cdot}$ (equation 22). Both of these fragment ions undergo consecutive decarbonylation reactions to give the abundant $C_5H_5Se^{+}$ and $C_5H_5^{+}$ daughters, respectively.

$$Ph_2SeO_2^{+\cdot} \longrightarrow PhOSe({=}O)Ph^{+\cdot} \longrightarrow PhSeO \text{ and } PhO^{+} \quad (22)$$

Reductive elimination of SeO_2, of Se following loss of O_2 (or O + O) and of SeO are other noteworthy features of the decomposition reactions of the diaryl selenones and appear to be general processes. Finally, all of the diaryl selenones investigated by Rebane show losses of $ArSe^{\cdot}$. This interesting rearrangement may be a route to protonated *o*-quinoid structures as pictured for the diphenyl compound in equation (23). Evidence for this hypothesis is the stepwise elimination of two CO molecules from the presumed *o*-quinoid ion, as confirmed by detection of appropriate metastable ions.

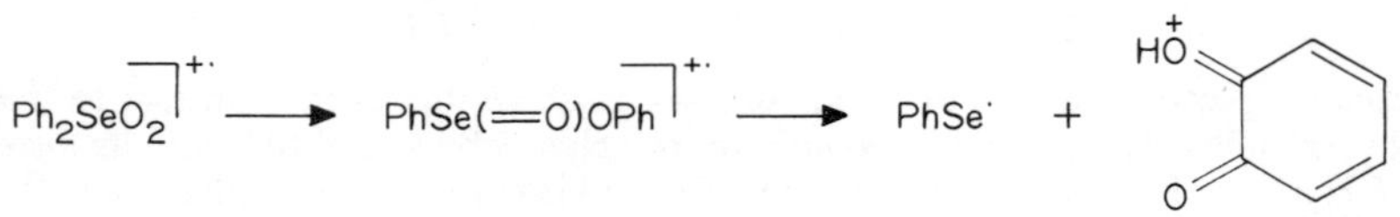

(23)

N. Organo-selenium and -tellurium Dihalides

Several organoselenium dihalides, R_2SeX_2 (R = Me, Pr, Ph; X = Cl, Br), as reported by Rebane[37], provided quite untrammelled mass spectra. Loss of a halogen was the first process in all cases studied, and that reaction was so facile that a parent ion was often not observed; $X_2^{+\cdot}$ was also very weak or not observed. The $[M-X]^{+}$ ion, in one common mode of fragmentation, eliminated the second halogen, thus presumably providing an organic selenide radical cation identical to that directly available by ionizing the neutral selenide; it presumably would be capable of the same fragmentations. The other mode of fragmentation, in which the remaining halogen stays with the Se, involved fracture of one or both Se—C bonds, or cleavage of a C—C bond α to the Se, with or without an associated H migration.

Mass spectra of dihalides of cyclic selenides have also been investigated by Rebane[37] and by Kulkarni and coworkers[43].

Futekov and coworkers[60] reported in passing that several compounds of the type R_2SeCl_2, in which R represents a monochloro, saturated alkyl group, cyclic or not, presented mass spectra exhibiting the parent ion as well as successive losses of Cl and of HCl. Their suggestion that the departing halogen and hydrohalogen species arose from the hydrocarbon portions and were not attached to the Se is probably an untenable speculation in light of the work of Rebane just mentioned.

Dimethyltellurium diiodides and tetraiodides were studied by Smith and Thayer[61]. The highest mass ions they observed in both cases were those of $Me_2TeI^{+\cdot}$ and a trace of TeI_2^{+}. The major ion corresponded to $I_2^{+\cdot}$ and other ions corresponding to $MeTeI^+$, TeI^+, Me_2Te^+, $MeTe^+$, MeI^+, $Te^{+\cdot}$ and I^+ testified to the rather straightforward decompositions induced by electron ionization of these molecules.

O. Selenium Ylides

A brief description of the mass spectrum of (diacetylmethylene) diphenylselenurane was reported by Wei and coworkers[62]. Other than loss of Me group and loss of the complete diacetylmethylene groups, the spectrum exhibited the features of diphenyl selenurane; viz.: $Ar_2Se^{+\cdot}$, $ArSe^+$, $Ar_2^{+\cdot}$ and Ar^+.

A more complete study of the mass spectra of selenium ylides was made by Terent'ev and coworkers[63]. They observed the parent ion for all seven of the ylides which they studied; in fact, the molecular ion was quite abundant possibly because of ready ionization at the anionic site (equation 24).

$$R_2\overset{+}{Se}—\overset{-}{C}R_2 \rightarrow R_2\overset{+}{Se}—\dot{C}R_2 + e^- \quad (24)$$

Invariably, the main fragment ions corresponded to the rupture of the ylide bond ($R^1R^2Se—C$) accompanied by H transfer from R^1R^2 when $R^1R^2 = —C_4H_8—$. Thus, species such as $R^1R^2Se^{+\cdot}$, $R^1Se = R^2H^+$, R^1Se^+ etc. could be explained.

More striking was the transfer of Ph from Se to the O of the anion moiety which was postulated to explain the losses of PhSe and PhO and the formation of $PhSe^+$ (equation 25). Various explicable purely organic species were also observed, but $SeH_2^{+\cdot}$, $RSeO^+$, $RSeOH^{+\cdot}$ and $RSeH^{+\cdot}$ ions were not observed even though these are characteristic of selenides and selenoxides.

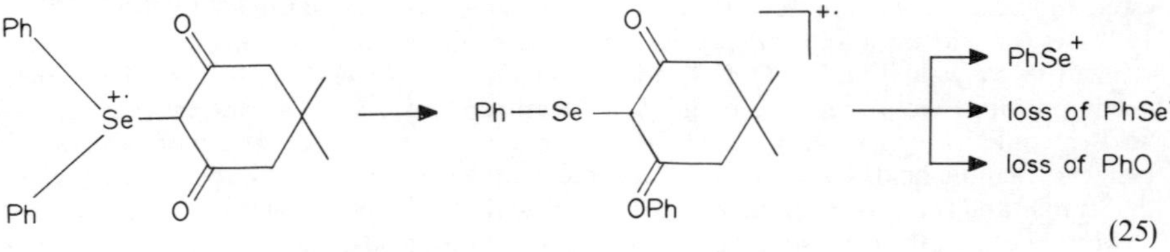

(25)

P. Diorganotellurium Dicarboxylates

A fairly thorough study of the mass spectra of some diorganotellurium dicarboxylates was reported by Adley and coworkers[64] (see Scheme 4). The carboxylate groups were either acetate or benzoate; the organo groups were: Ph, *p*-An, or *p*-$EtOC_6H_4$. Among the general features were: (1) The parent ions never exceeded 0.05% Σ_{40} and mostly were not observed. Alternate ionization methods are recommended for these compounds. (2) Purely organic ions were important in the spectra, particularly species such as the biphenyl ion. (3) Loss of one complete carboxylate group from the parent ion always occurred and the resulting ion generally showed loss of the second carboxylate group as part of its fragmentation pattern. (4) By contrast, loss of an aryl group was not favoured, particularly when the aryl group was Ph. (5) By shedding of various groups and Te extrusion, some arylcarboxylate ions (e.g. $PhOAc^+$) were observed. (6) Small fragments such as $TeCH_2^+$ turned up, but were not important in the spectra.

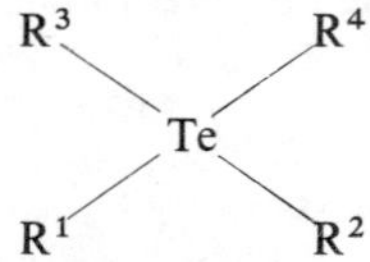

$R^1 = R^2 = Ph$ $R^3 = R^4 = CO_2Me$	$R^1 = R^2 = Ph$ $R^3 = R^4 = CO_2Ph$	$R^1 = R^2 = —C_4H_8—$ $R^3 = R^4 = CO_2Me$
$R^1 = R^2 = p$-An $R^3 = R^4 = CO_2Me$	$R^1 = Ph$, $R^2 = p$-$PhOC_6H_4$ $R^3 = R^4 = CO_2Me$	$R^1R^2 = —C_4H_8—$ $R^3 = R^4 = CO_2Ph$

SCHEME 4

Additionally, Adley and coworkers ordered the groups according to their facility to detach from the central Te atom: acetate $\approx$ benzoate $>$ p-$EtOC_6H_4 \approx p$-An $>$ Ph. Disubstituted tellurium ions, observed in the course of decomposition of the tetrasubstituted species or by direct ionization of the disubstituted neutrals, behaved similarly. However, the trisubstituted species, again arising in the fragmentation of the tetrasubstituted species, behaved somewhat differently. Most interestingly, the loss of acetate, for example, was not a single-step process but rather required loss of ketene followed by loss of OH, whereas the loss of benzoate was direct.

Necessarily, some rearrangements must occur to account for the formation of biphenyl and other ions with two originally ligated groups conjoined: the conversion of $PhCOOTeAr^{+\cdot}$ to $PhTeAr^{+\cdot}$ with loss of CO_2, loss of RCO_2 from $C_4H_8TeOCOR^+$, as well as the loss of C_2H_4 from biaryl$^{+\cdot}$ ions.

Q. Seleninic Acids and Selenocyanates

The first report of the mass spectra of methaneseleninic acid, CH_3SeO_2H, its methyl ester, and benzeneseleninic acid, $PhSeO_2H$, and its methyl ester was made by Rebane[65] in 1974. As for selenones and selenoxides, methaneseleninic acid rearranged after electron ionization to yield $[MeOSeOH]^{+\cdot}$ and MeO and CH_2O were then lost. The other decompositions were predictable. Similar rearrangements occurred for benzeneseleninic acid en route to formation of $[C_6H_6O]^{+\cdot}$, for example, and for a variety of substituted benzeneseleninic acids. Although the molecular ions were of low abundance, they were detectable, and the mass spectra were consistent with the structures of the acids and esters.

More recently, the mass spectra of the benzeneseleninic acid class were reported again in two papers by Benedetti and coworkers[66,67]. No mention was made of the earlier work of Rebane. By far the most notable feature of the mass spectra of these compounds was that they were dominated by peaks corresponding to diphenyl diselenide ions and their daughter fragments. The spectra were totally different to those reported by Rebane[65]. The fragment ions were presumed to arise via a probe reaction prior to ionization. After ionization, the fragmentation behaviour of the diphenyl diselenide ion was that already discussed. Of lesser importance was the occurrence and the fragmentation of $Ph_2SeO^{+\cdot}$-type species, giving $PhSeO^+$ and $PhSeOH^{+\cdot}$, for example. Again, these were analogous to the selenoxide–selenate rearrangements discussed earlier.

Regardless of the presence of a substituent (X = m-Cl, p-Cl, m-Br, p-Br, m-NO_2, p-NO_2, *p*-Me) on the Ph ring or of the position of the substituent, the principal fragmentation mechanisms remained the same. The authors cited several species such as $XC_6H_4Se^+$, $XC_6H_4^+$, $C_6H_4^{+\cdot}$ as possible daughters of an originally ionized acid molecule. However, peaks corresponding to these ions had slight intensities in the spectrum and the ions may have originated by other decomposition routes. Low intensity peaks in appropriate multiplets were observed at *m/z* ratios ascribable to $XC_6H_4SeOC_6H_4X$ moieties. This species surely arises as part of the thermally induced probe reactions.

The appearance of ions with masses appropriate to $C_6H_4OH^+$, $XC_6H_5^{+\cdot}$ and $XC_6H_4OH^{+\cdot}$ argued for a possible parent acid molecular ion (not seen in the spectrum) which rapidly eliminated SeO and then lost either CO or X. The loss of SeO was supported by metastable ion studies.

In the case of the nitro-substituted benzeneseleninic acids only, the steps of formation of selenoxide ion via a 1,2-shift of $NO_2C_6H_4$ group to O, then loss of NO_2 groups and Se to give $C_6H_4OC_6H_4^{+\cdot}$ (possibly a dibenzofuran radical cation) was supported by a metastable ion. This ion was not observed by Rebane in his earlier study[65]. The dramatic differences in spectra reported by these two authors point to the requirement of careful sample handling during sample introduction into the mass spectrometer. We suggest that a more reliable way of mass spectral analysis of these seleninic acids would be to employ fast atom bombardment.

Finally, Benedetti and coworkers[67] presented a study of the mass spectrum of *m*-nitrophenyl selenocyanate. This compound presented a very abundant molecular ion and subsequent fragments. Here too, the diselenide and its fragment ions were present, but at much reduced abundance when compared to the seleninic acids. The selenocyanate parent ion itself fragmented by rather expected routes: initial loss of CN, NO_2 or Se followed by subsequent losses such as CN, NO_2, Se, CO, NO, HCN. A variety of small-mass hydrocarbon ions as well as the species C_3HSe and HCSe were explained in the authors' proposed decomposition reaction schemes.

R. Carbonate Analogues

Dimethyl carbonate analogues with S and/or Se were studied by Drager and Gattow[6]. It is worth recalling that, after ionization, MeOCOOMe itself primarily ejects H and neutral CO_2 to yield $CH_2{=}OMe^+$. Although the charged species $CS_2^{+\cdot}$ was observed in the spectrum of MeSCSSMe, it is unlikely that the $CS_2^{+\cdot}$ arose directly from the molecular ion. Since the ionization potential of neutral Me_2S is about 8.7 eV, and that of neutral CS_2 is about 10 eV, the charge of a parent ion should be retained by the Me_2S fragment rather than by the CS_2 portion. Rather, we propose that the origin of CS_2^+ was by loss of Me from $MeSC{=}S^+$.

For the case of MeSeCSSMe, the expected preferential cleavage of the Se—C bond was observed, giving $S{=}CSMe^+$ as the base peak in the spectrum. Again this species ejected $Me^{\cdot}$ to give $CS_2^{+\cdot}$. Because the alternate route involving initial loss of MeS by cleavage of a C—S bond to produce $S{=}C{=}\overset{+}{S}eMe$ is less significant, only a small peak due to $SeCS^{+\cdot}$ was expected and was observed. There is no reason to believe that the Se atom in the original molecule does not bear a Me group.

The compound Me_2CSSe_2, according to expectation and to the NMR spectrum, shows both Se atoms bearing Me groups. The high abundance of $SeCS^{+\cdot}$ in the decompositions is explicable regardless of S and Se positions, but were the structure of the parent ion $MeSeCSeSMe^+$, there should be at least some fracture of a C—S bond giving rise to $MeSeCSe^+$ which in turn ought to yield $CSe_2^{+\cdot}$. However, no $MeSeCSe^+$ was observed. With two Se atoms present in the parent, the occurrence of the species $Se_2^{+\cdot}$, Se_2Me^+ and $(MeSe)_2^{+\cdot}$ is not startling.

When the compound Me_2CSe_3 is ionized, the parent ion is not so abundant as in the previous cases; CSe_2^+ becomes the base peak. In all of these cases, the lower mass region contains several of the by-now-familiar fragments such as $SeCH_x^+ (x = 1-3)$.

S. Urea Analogues

Kirkien and coworkers[68] extended their study of the mass spectra of numerous *N*-substituted ureas and thioureas to include five selenoureas, and two related selenothiocarbamic esters. In addition to graphical representations of the actual spectra, they presented a tabular summary of the features they considered important in comparable $R^1R^2NCXNR^3R^4$ compounds (X = O, Se; R = H, Me, Ph). There appears to be some disagreement between the numbers shown in their Table 2 and the data found in their references (in particular, the *m/z* and relative intensity data for $H_2NCSNHPh$ and for $Me_2NCSNHPh$). Nonetheless, the reader may find the article useful in identifying spectra or in rationalizing spectra of related compounds.

A few generalizations can be drawn from their table: there is a very strong tendency to produce a $R^1R^2NH^{+\cdot}$ species when at least one of the R groups is Ph. By contrast, evidence for the same sort of species when the R groups are H or Me is minimal or vanishingly small. The S analogue species exhibit the greatest tendency to form intact molecular ions. Falling outside the authors' standard set of fragmentation behaviors is the loss of SePh from $Me_2NCSeNHPh$ to give the base peak at *m/z* 71. Expulsion of HSe is also notable. On the other hand, the 1,1-diphenyl analogue, $H_2NCSeNPh_2$, shows loss of HSe, H_2Se and SePh, but derives its base peak from the *m/z* 51 ion, a standard fragment of the Ph group.

The other 'standard fragmentation behaviours' must involve cleavage at the carbonyl C—N bond, with the charge apportioned to either of the resulting fragments and with or without H migration in either direction. Needless to say, most of the observed fragments can be fitted to such a set of behaviour patterns, especially when migration of H or R groups to the O or chalcogen atom is added to the list of possibilities. Some of the same patterns were observed, in part, in the fragmentation patterns of the selenothiocarbamates although some rather more complicated mechanisms would have to be invoked to explain the prominent occurrence of $[M - SeMe]^+$ for $Me_2NCSeSMe$ or of $PhCN^+$ from PhHNCSeSMe.

The authors concluded that the Se compounds generally emulate the S analogues more closely than the O analogues, particularly as R group migration to Se is involved, and that the charge of the molecular ion must be 'largely centred on the selenium atom'.

T. Selenophenes and Tellurophenes

When one turns to the mass spectrometric behaviour of this aromatic class of organochalcogen compounds, it is found that Fringuelli and Taticchi[69] have provided a

TABLE 2. Percent total ionization between *m/z* 29 and $M^{+\cdot}$

	C_4H_4O	C_4H_4S	C_4H_4Se	C_4H_4Te
$M^{+\cdot}$	26.8	30.4	44.8	44.4
$[M - CHX]^+$	43.2	8.3	3.3	0.4
$[M - C_3H_3]^+$	6.8	17.5	5.2	1.1
$[M - C_2H_2]^{+\cdot}$	3.0	19.7	16.6	4.1
$[M - HX]^{+\cdot}$		1.1	5.7	6.6
X^+			4.8	21.6
$[M - X]^+$			0.8	3.8
$[M - H]^+$				2.0

useful comparison by a study of the furan to tellurophene series of model compounds. The principal fragmentations noted for the unsubstituted members of the series are summarized in Table 2.

That S and Se exhibit the most closely similar behaviours within this family is aptly illustrated by these data. Fairly regular variation in a single direction is exhibited for the most part. Only Se and Te atoms show any propensity to be eliminated as monatomic neutrals or by retaining the charge as monatomic cations.

This authors were able to suggest a few decay routes which rationalize the observed spectra (Scheme 5).

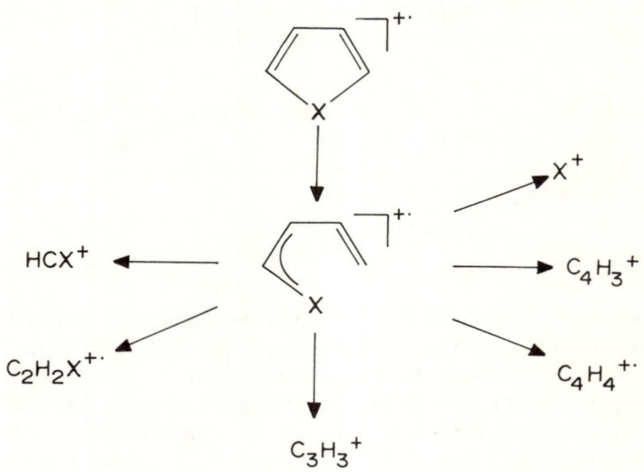

SCHEME 5

A parallel statement by Barton and coworkers[70] regarding the furan class with chalcogen replacement reads: 'Thus there is a great variety in the modes of decomposition of excited Group VI heterocyclopentadienes, but at the same time there is a high degree of continuity in going from furans to tellurophenes with selenophenes providing a key link in that continuity at the interface of the non-metallic elements of the group with the metallic members of the group.'

Further, Barton's group reported the mass spectra of selenophenes and tellurophenes with a Ph group substituted in either the 2- or 3-position, and mono- or tri-substituted with D. Of two important observations, one was that a compound substituted with a Ph group, regardless of location, gives a mass spectrum with the same fragmentations as those of the unsubstituted heterocycles except for the straightforward offset of some peaks by 77 units due to the presence of Ph; the other was that the photolytic decompositions of these parent molecules are also dominated by the loss of Se, or of Se and HCSe.

U. Benzoselenophenes and Related Compounds

A set of 18 disubstituted benzo[*b*]selenophenes (**12**) substituted in the 2- and 3-positions was studied by Croisy and coworkers[71]. When either R^1 or R^2 = Me, the molecular ions were the most abundant ions seen. Where the substitution was appropriate, rearrangements in the category of '*ortho* effects' occurred just as for other 1,2-disubstituted aromatic compounds. For example when R^2 = Me and $R^1 = CO_2H$ or CO_2Me, losses of H_2O or MeOH were found. In a consecutive reaction, loss of CO occurred to give **13** which was postulated to undergo expansion to give the interesting radical cation **14** (equation 26).

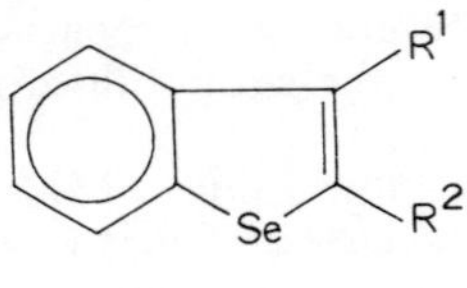

(**12**)

When $R^1 = OH$ and $R^2 = CO_2Et$, the '*ortho* effect' rearrangement led to elimination of EtOH which was followed by expulsion of two molecules of CO to yield **15**. The structures are purely speculative as little supporting evidence involving isotopic labelling, metastable or collisional activation spectra of the ions was obtained.

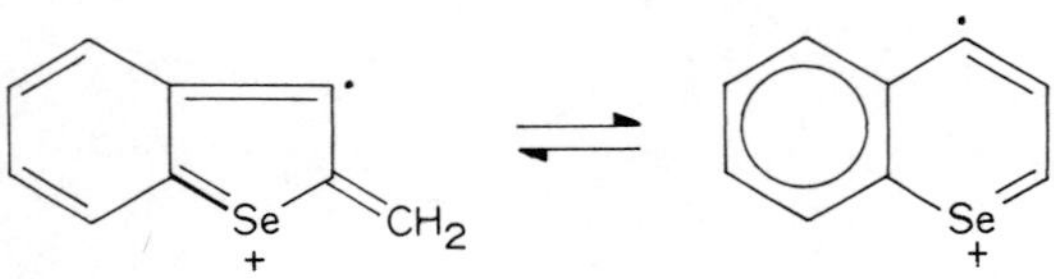

(26)

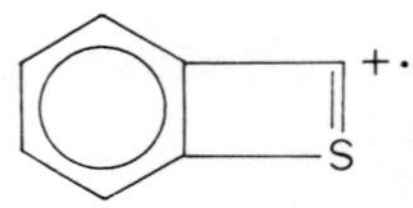

(**13**) (**14**) (**15**)

Closed-shell ions possibly of similar structure were found as products in many of the mass spectral decompositions of substituted selenophenes. For example, when $R^1 = COR$ and $R^2 = Me$, the expected losses of R followed by CO occurred to give ion **16** which might undergo ring-expansion to **17**, a rearrangement entirely analogous to the tolyl–tropylium ion interconversion. Losses of H from the Me group of Me-substituted selenophenes might also be accompanied or followed by a similar ring-expansion to give substituted selenopyrylium ions. Additional evidence from metastable ions or collisional activation spectroscopy and from isotopic labelling should be gathered before the ring-expansion hypothesis is accepted.

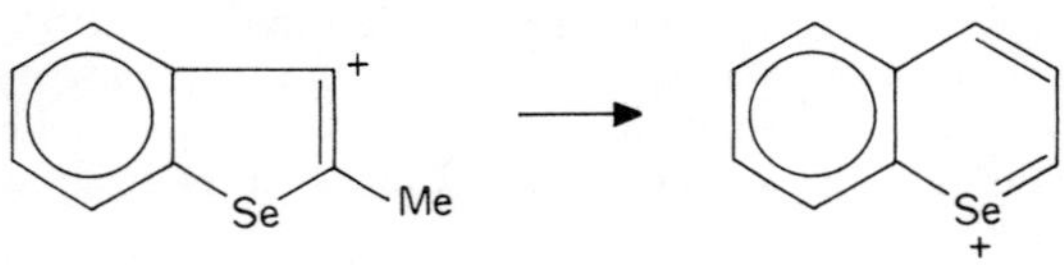

(27)

(**16**) (**17**)

Isomers which involve interchange of the substituents R^1 and R^2 could be detected in some cases by comparison of quantitative peak intensities in the mass spectrum. The consecutive losses of R and CO were more prevalent when COR was R^2 than when it is R^1 (in **12**).

It is noteworthy that loss of Se from these disubstituted selenophenes was *not* observed except when a second Se atom was introduced into a side-chain of the molecule. This contrasts with the mass spectrum of the parent or unsubstituted benzoselenophene. The chemistry of disubstituted benzoselenophenes is entirely analogous to other 1,2-disubstituted aromatics. The decomposition reaction to expel Se is preempted by other, more standard, fragmentation processes such as *ortho* effects, loss of substituents, etc.

Croisy and coworkers[2] have also reported the mass spectra of 13 selenolo[2,3-b]pyridines (**18**) including the parent compound and mono- and di-substituted analogues. The unsubstituted compound, like selenophene, gave the molecular ion as the most abundant species. For **18** and selenophene, ca. 50% of the total ion current was carried by

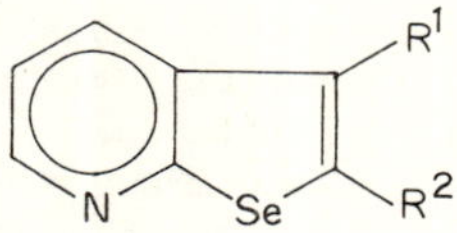

(**18**)

the molecular ion. Fragmentation of **18** was a composite of that of selenophene and quinoline; i.e. losses of C_2H_2, HCN and Se were observed, the latter being in 36% relative abundance.

Me-substituted **18** showed more extensive fragmentation when ionized. The expected loss of H might be followed by ring-expansion. Losses of Se and HSe (possibly in two steps) were still observed. Substitution with COR, where R = H, Me or OEt, led to fragmentation by loss of R to give abundant $ArCO^+$ ions which then decarbonylated to form Ar^+. No loss of Se was observed, presumably because the decomposition reactions involving COR were of lower critical energy.

The compound with an OEt substituent gave an unexpected mass spectrum. Unlike ethyl aryl ethers, no loss of C_2H_4 to give ArOH was found. Instead losses of Et and OC_2H_4 occurred. No explanation has been put forth.

For disubstituted compounds such as those with $R^1 = OH$ and $R^2 = CO_2Et$, the expected loss of EtOH ('*ortho* effect') occurred, just as for the corresponding benzoselenophene, to give the most abundant ion in the spectrum. This type of rearrangement was not found when $R^1 = Me$ and $R^2 = CO_2Et$, however.

The decomposition reactions of these substituted compounds are also typical of mono- and di-substituted aromatic molecules, and they have low energy barriers which preempt the loss of Se which is so prevalent for the unsubstituted prototype.

The mass spectra of some tricyclic analogues **19** and **20** where X and Y = S or Se have been studied by Jacquignon and coworkers[72]. The molecular ions were relatively stable yet losses of Se were facile processes. When X = S and Y = Se, the greatest propensity to lose Se was found ($[M - Se]^{+\cdot}$ carried ca. 20% of the total ion current). When Se and S were interchanged (X = Se, Y = S), loss of Se was somewhat attenuated, which may suggest that the selenophene ring is more stabilized when fused to two other aromatic rings.

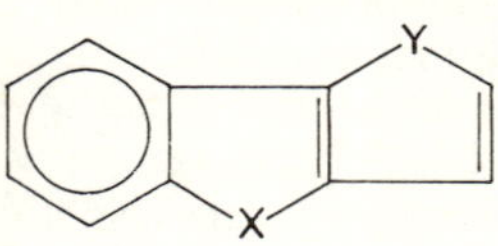

(**19**)

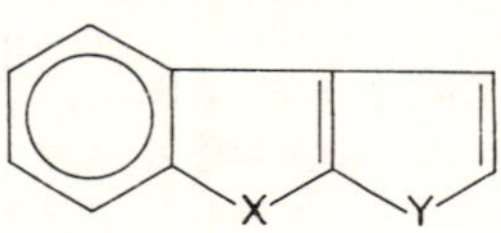

(**20**)

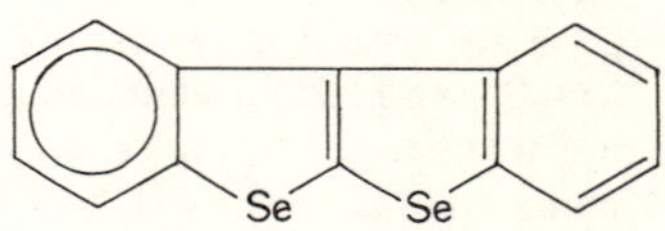

(**21**)

When X = Y = Se, consecutive losses of both Se atoms were observed, and this was the major decomposition route for this ionized compound. Losses of two Se atoms were also characteristic of **21**[73]. The structural possibilities for the $[M - 2\,Se]^{+\cdot}$ are intriguing. Could it be **22**, a benzyne analogue of naphthalene (**23**) or diethynyl benzene (**24**)?

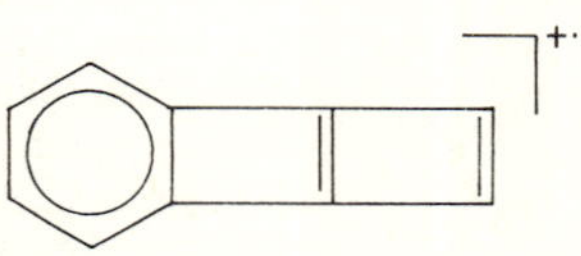

(**22**)

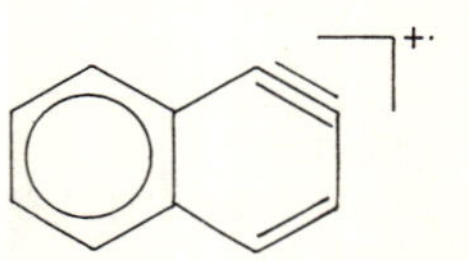

(**23**)

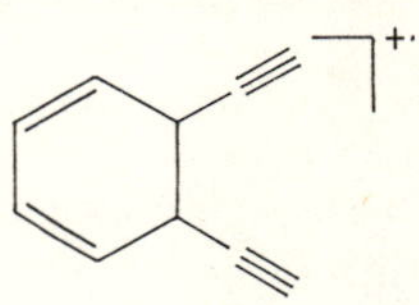

(**24**)

Other fragmentations such as losses of C_2H_2, CS or CSe compete only weakly for this series of compounds. Once again the ability of unsubstituted selenophene radical cations to undergo reductive elimination of the Se is seen.

Shafiee and Behnam[51] prepared a series of seleno[3,4-*b*]benzofurans (**25**), but the mass spectrum of only one member of the series (R = OMe) was reported. The molecular ion was the most abundant.

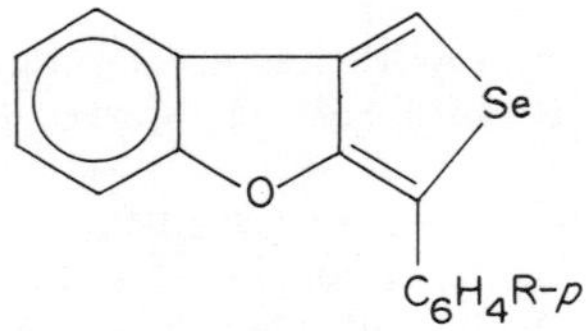

(**25**)

Evers and coworkers[74] reported the preparation of a number of benzochalcogen-ophenones, and mass spectra were obtained for several compounds. In the case of the benzo[*b*]selenophene-2(3*H*)-ones (**26**), a selenolactone model, and closely related structures, the mass spectral behaviour was characterized by the strong presence of the parent ion, the loss of the monocarbon species, either CO or HCO, and of Se.

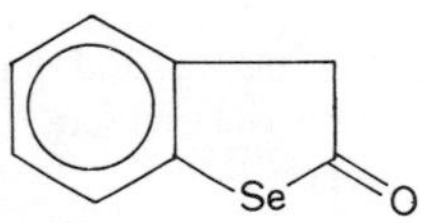

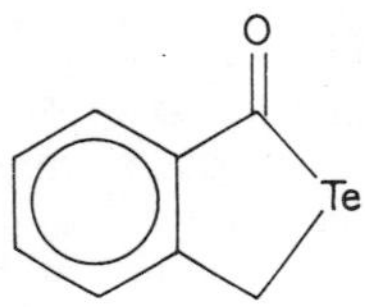

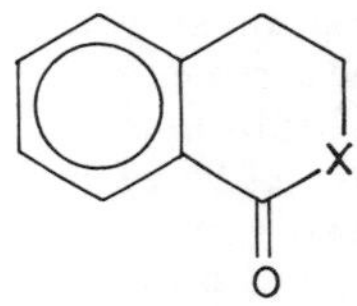

(**26**) (**27**) (**28**)

In the case of $^1H,^3H$-benzo[*c*]tellurophen-1-one (2-tellurophthalide) (**27**) Loth-Compere and coworkers[75] were able to ascertain that the expulsion of monatomic Te generally preceded the loss of CO or HCO. Their comparison of the fragmentations of the Te-, Se- and S-containing analogues of **28** illustrated anew the pivotal position of Se. Se approximated the behaviour of S in participating in a retro-Diels–Alder reaction via emission of CH_2S or CH_2Se (like the familiar formaldehyde loss), or in suffering the loss of CO or of HCO while the Te compound showed one-fourth to one-half the propensity for these decay routes. On the other hand, both Se and Te species were capable of producing $C_8H_8^{+\cdot}$ and $C_9H_4O^{+\cdot}$. Indeed, $C_8H_8^{+\cdot}$, presumably the cyclooctatetraene radical cation, was the most abundant in the decomposition of the Te compound. As has been often observed, the chalcogens can be viewed to exhibit metallic behaviour by exiting from the species to which they had been bound. For example, this process of reductive elimination is also exhibited by Se- and Te-containing heteroaromatics.

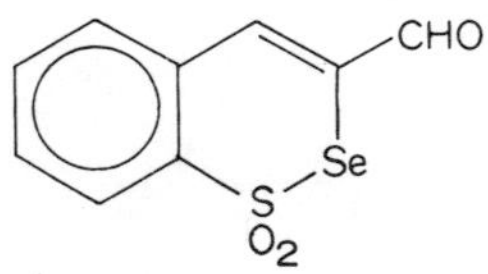

(**29**)

Van Coppenolle and Renson[76] reported a molecular ion for a selenosulphone (**29**) which lost an SO_2 molecule and gave an ion corresponding in formula and fragmentation to that obtained directly from formyl-2-benzo[*b*]selenophene.

Cohen and coworkers[77] presented considerable detail concerning the mass spectra of two classes of compounds, represented by **30** and **31**, which are related in their mass spectra even more than they may appear to be according to their structures.

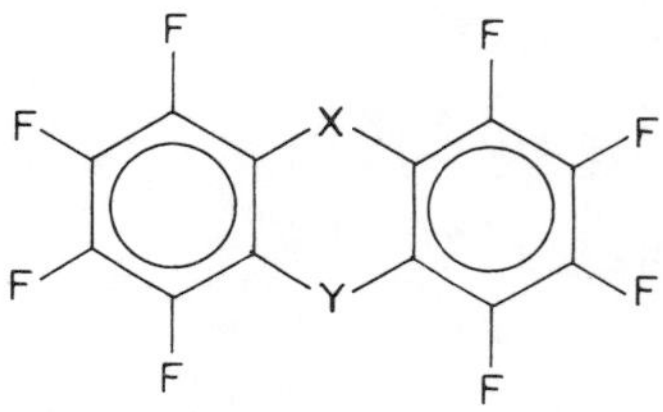

(30)
X = Y = S
X = Y = Se
X = S, Y = Se

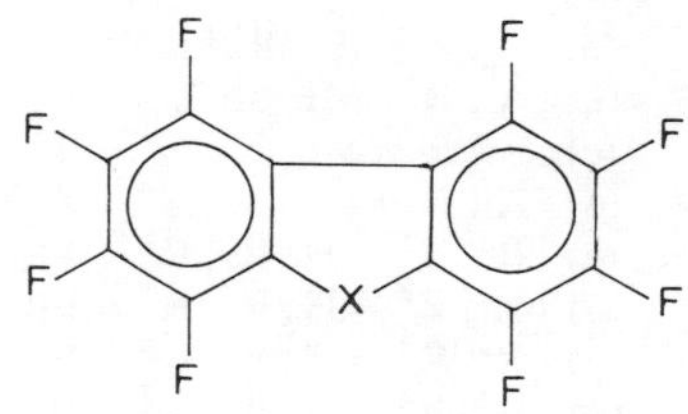

(31)
X = S, Se, Te

Through the use particularly of metastable studies, the authors were able to discern many implications in the spectra of these species. For compound **30** with one or two S atoms, the ion current was distributed broadly among various ions, but these need not concern us here. When two or even one atom of Se was present, the spectra were more straightforward. For example, with X = Y = Se, the major ion resulted from loss of Se from the parent ion. The parent ion itself, $[M - Se]^{+\cdot}$, $[M - 2\,Se]^{+\cdot}$ and $Se_2^{+\cdot}$ accounted for 73% of the total ion current. When X = S and Y = Se, the parent ion was relatively abundant, but the ion $[M - Se]^{+\cdot}$ provided the base peak and these two peaks alone accounted for nearly 75% of the total ion current.

For compounds of structure **31** the parent ions were so dominant as to account for over half of the ion current. The S compound showed the greatest stability of its parent ion, but the Se and Te analogues demonstrated increasing tendencies to lose neutral Se or Te atoms (10.2% and 17.1%, respectively, of the total ion current). Other ion current was attributable to a myriad of species which had diminished numbers of C and F atoms with or without chalcogens.

The substantial degree of fragmentation via loss of Se is once again cited by these authors as a reflection or indicator of a relatively weak C—Se bond strength. This must be tempered by the observation that loss of Se is more favoured when there is an S atom across from it than when two Se atoms are present. In part, this probably reflects the greater stability of the tricyclic product when the S atom is a part of the central ring. It is also bemusing to note the authors' comment: 'Another common reaction is the loss of neutral F' among the many metastable-ion-supported fragmentations of fluorocarbon groupings, with or without a chalcogen. What, if anything, does this suggest regarding strength of the C—F bond?

V. Fulvalene and Pentalene Analogues

The mass spectra of Se analogues of tetrathiofulvalenes (TTF) (**32**) were reported by Andersen and coworkers[3]. In these compounds, all of, or two of, the S atoms, symmetrically or not, across the inversion centre of the molecular core, were replaced by Se. Various substituent groups were attached to the periphery of the molecules.

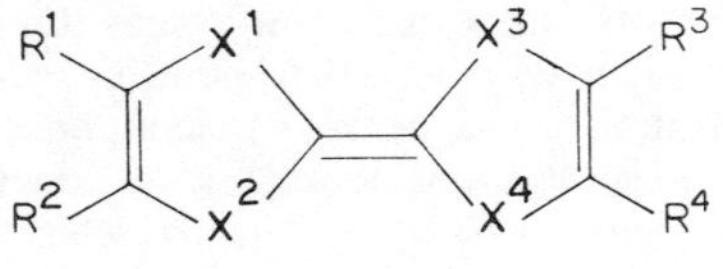

(**32**)

(**a**) All R = H; all X = Se
(**b**) All R = Me; all X = Se
(**c**) All R = Me; $X^1, X^3 = S; X^2, X^4 = Se$
(**d**) All R = Me; $X^1, X^4 = S; X^2, X^3 = Se$
(**e**) $R^1, R^3 = H; R^2, R^4 = Me$; all X = Se
(**f**) $R^1, R^4 = H; R^2, R^3 = Me$; all X = Se
(**g**) $R^1, R^3 = H; R^2, R^4 = Ph$; all X = Se
(**h**) $R^1, R^4 = H; R^2, R^3 = Ph$; all X = Se
(**i**) $R^1, R^3 = H; R^2, R^4 = Me; X^1, X^3 = S; X^2, X^4 = Se$
(**j**) $R^1, R^3 = H; R^2, R^4 = Me; X^1, X^4 = S; X^2, X^3 = Se$
(**k**) $R^1R^2 = —(CH_2)_3—; R^3R^4 = —(CH_2)_3—$; all X = Se
(**l**) $R^1R^2 = —(CH_2)_3—; R^3R^4 = —(CH_2)_3—; X^1, X^3 = S; X^2, X^4 = Se$
(**m**) $R^1R^2 = —(CH_2)_3—; R^3R^4 = —(CH_2)_3—; X^1, X^4 = S; X^2, X^3 = Se$

This thorough study made good use of field ionization (FI); the observation of only the molecular ions in the FI spectra demonstrated the high purity of the samples. Thus, the electron impact spectra were open to interpretation without fear of being misled by peaks arising from impurities or from prior thermolysis.

A number of fundamental patterns were observed. As is the case for many other compounds, the ease of elimination of Se was greater than for S. But this Se loss, commonly observed in Se organics, was only of 1–6% relative abundance for this class. The loss of C_2R_2 was most important for the tetraseleno class, less so for the dithiadiseleno class, and least important for the tetrathia class, in support of the idea that the C—Se bond is relatively fragile. Although the loss of SCR had a relative abundance up to 43%, no loss of SeCR occurred detectably.

The authors pictured the primary products as originating either directly from the parent ion or via parallel routes, all substantiated by metastable ion studies.

One particularly noteworthy rearrangement reaction takes place (see equation 28) starting with the $[M - C_2R_2]^{+\cdot}$ fragment **33**. When the diselenodithia compound is

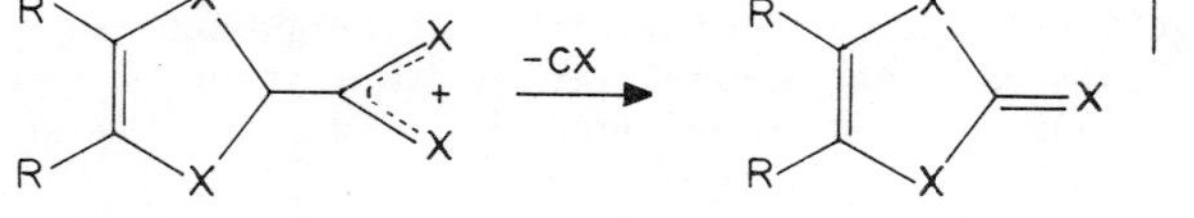

(**33**) (**34**) (28)

involved, the only product is **35**. The intermediate species can be viewed as reacting with a preferred migration of a five-membered ring moiety onto the Se atom, followed by elimination of CS. A second intermediate **36** is formed from the parent ion by initial loss of XCR. It subsequently loses C_2R. Since no corresponding loss of SeCR was observed, this route must be blocked for the tetraseleno analogues.

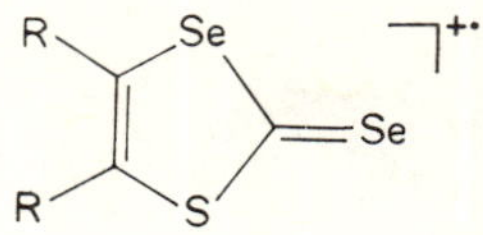

(35)

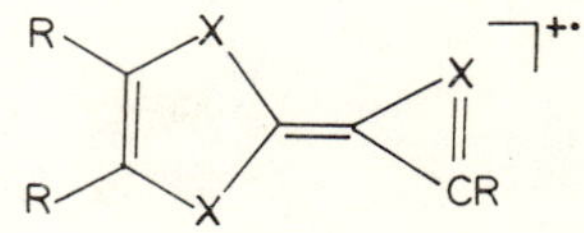

(36)

Undoubtedly, an important factor in the fragmentation of the TTF-type compounds is the production of species **37**, the stability of which has already been demonstrated by the mass spectra of the corresponding neutrals, which invariably show the molecular ion as the most abundant.

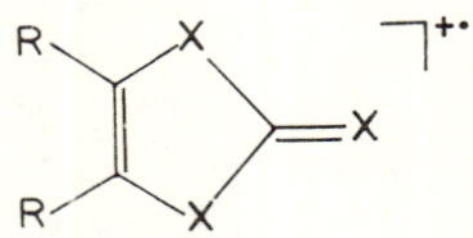

(37)

A small group of rather unusual, conjugated fused ring structure species was studied by Moller and coworkers[78]. The two basic structures are shown as **38** and **39**. Part of the incentive for studying this group of compounds arose from the linear arrangement of the chalcogen atoms and the implied three-centre, four-electron bonding. Some possible reflection of these unusual features was anticipated in the mass spectra. What was observed for **38** and derivatives was a very abundant molecular ion, often providing the base peak of the spectrum; the inevitable loss of Se_2, confirmed by a metastable ion, occurred in parallel with or followed by losses of Se or HSe. Note that there was no loss of S_2 by the S analogues for which, rather, losses of SH, S_2H and S_2H_2 were observed.

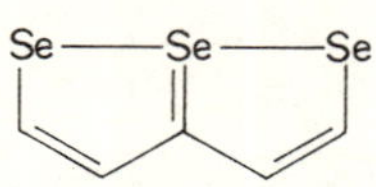

(38)

1,6,6aλ^4-Triselenapentalene

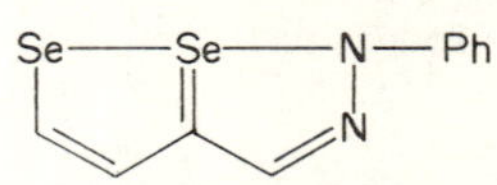

(39)

Phenyl-1,6aλ^4-diselena-5,6-diazapentalene

For the class of compounds represented by **39**, about 50% of the molecular ions fragmented to give Ph^+ when the N atom bore a Ph group. For compound **40**, losses of Se_2 and PhN_2 led to the formation of $C_5H_5^+$. However, for compound **41**, only a trace of $C_6H_7^+$ was observed.

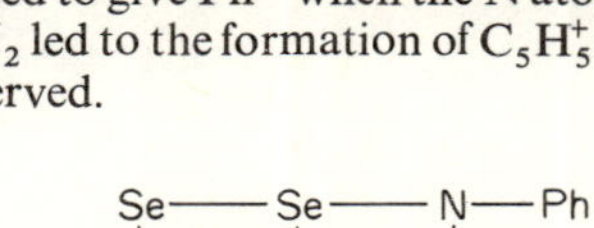

(40)

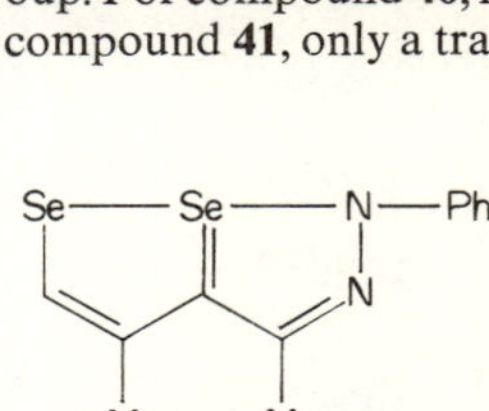

(41)

A rather interesting class of compounds (**42**) was studied by Perrier and coworkers[79]. By analogy with the symmetric trithiapentalenes, this 2,5-diaza-1,6-dioxa-6aλ^4-thiapentalene and its Se and Te analogues are presumed to be bicyclic with the chalcogen atom assuming a tetravalent role.

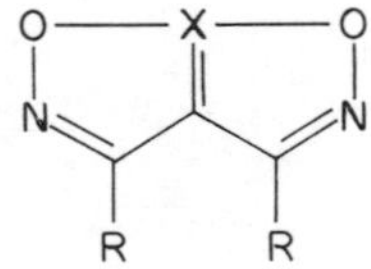

(42)
R = H, Me
X = S, Se, Te

The stability of these species upon electron ionization is reflected in the high abundance of the molecular ion, which generally, gives the base peak. The formulae for the parent compounds are thereby demonstrated.

All species, after ionization, are capable of loss of first one and then a second NO molecule, leading to a cyclopropenone-type stoichiometry and, perhaps, structure. It is this $HC_3HTe^{+\cdot}$ species which provides the most abundant ion in the spectrum arising when R = H and X = Te.

A possible but less favoured alternative to the loss of the second NO is the loss of XO although it is about equally likely that the charge will remain with the XO as with the organic moiety.

Loss of HCNO is observed when R = H, but there is no observed loss of MeCNO.

Finally, a few unusual decomposition reactions are observed: loss of HO when R = Me, X = Se and the appearance of $HC{\equiv}S^+$ or $HC{\equiv}Se^+$, but not $HC{\equiv}Te^+$, and of $H_3CC{\equiv}Se^+$. All in all, the spectra of this group are very similar to each other.

W. Complex Heterocyclic Compounds

In the category of the more complex monocyclic compounds, Jham and coworkers[80] have discussed some substituted thiazoles and the analogous selenazoles. They reported no major differences in the electron ionization mass spectral behaviour relating to the substitution of Se for S.

In addition to the predictable fragmentation behaviour such as release of Ph˙, Ph^+, PhCO or $PhCO^+$, etc., the following were presented, based on metastable ion evidence and deuterium replacements of O—H and N—H hydrogens (equations 29–32).

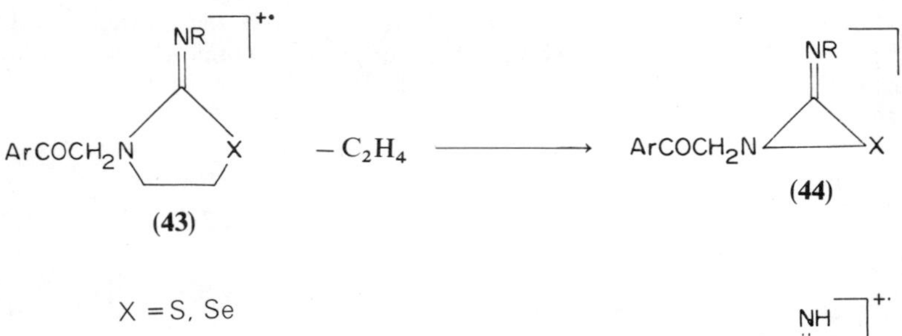

(29)

(43) **(44)**

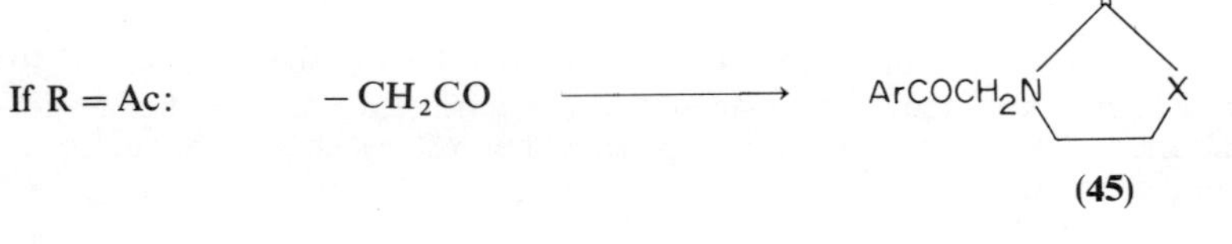

(30)

(45)

If R = H: $- H_2O$ ⟶

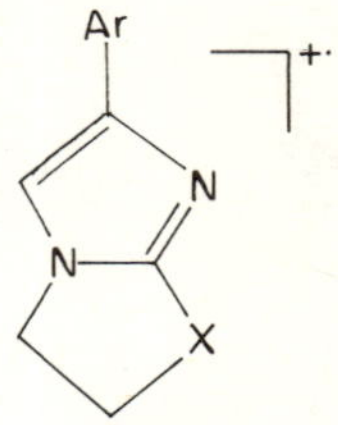

(46) (31)

If R = H: $-[H_2O, C_2H_4, HCN] \longrightarrow Ar{-}C{\equiv}C{-}N{=}X$ (32)

(47)

The rather startling loss of water from **49** was proposed to proceed via the mechanism in equation (33) leading to a highly conjugated, stabilized species. Replacement of an imine hydrogen by an alkyl group caused the water elimination to shut down, an observation in accord with the proposed mechanism. It is interesting to note that, in the case where R = Ac, a fragmentation specific for that species, elimination of CH_2CO with an H migration, should produce a species identical to the parent ion where R = H originally, except for possible differences in energy. Nonetheless, the authors did not report evidence for the novel water elimination from the $[M - CH_2CO]^{+\cdot}$ which was reported for the parent compound where R = H. Whether this process indeed does not occur when R originally equals Ac or the data relating to this process have been omitted by the authors is not clear.

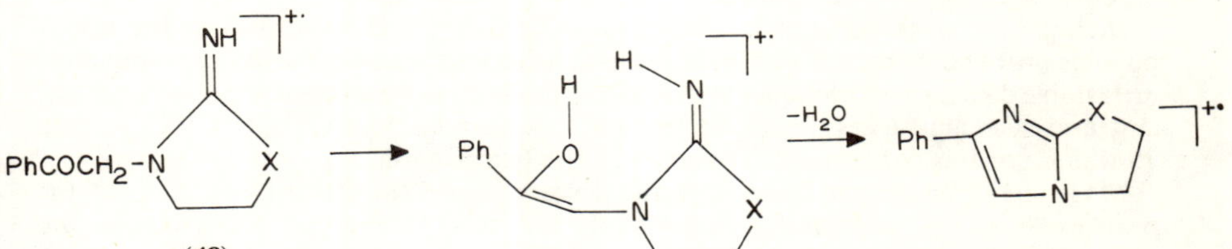

(48) (49) (50)

(33)

Both the Se and S compounds with R = H have, as the base peak in their respective mass spectra, ions which correspond in fact to an approximately 50/50 mixture of isobars arrived at through separate pathways: the first involved progressive loss of aldehyde and ethene (see equation 34); the second loss of PhCHOH and hydrogen cyanide (see equation 35).

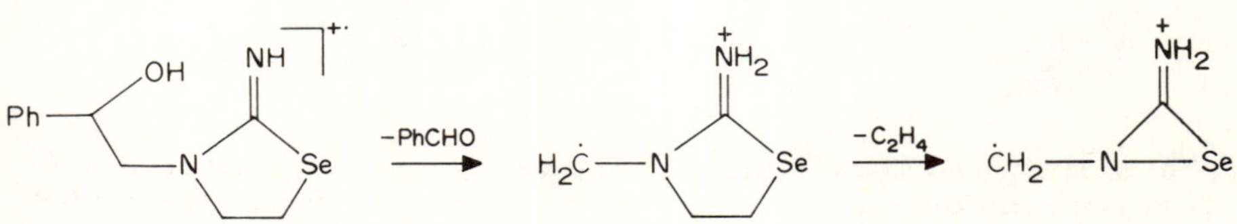

(51) (52) (53)

(34)

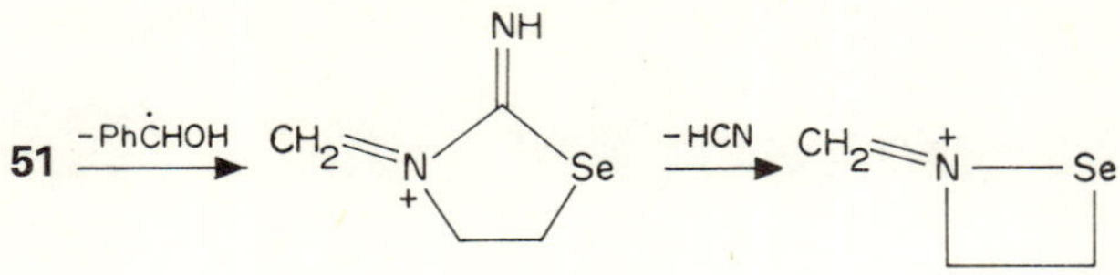

(54) (55) (35)

For the class of compounds of type **56** it was postulated that the initial fragmentations, confined to the ring adjoining that containing the chalcogen, consisted of the ordinary eliminations of isocyanate-type species. These all appeared to originate in the cleavage of the C—N bond between the carbonyl carbon and the adjoining tertiary nitrogen.

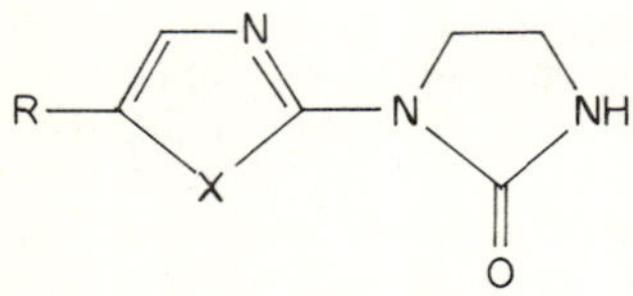

(56)

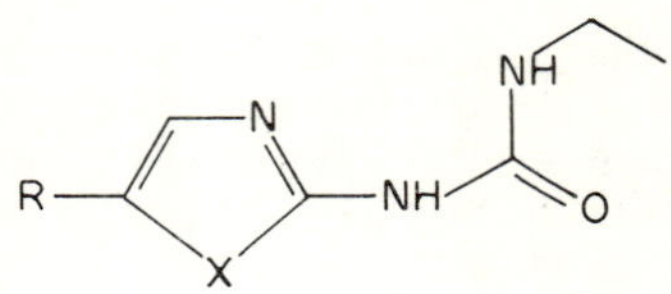

(57)

Similarly, compounds typified by **57** kept the chalcogen-containing ring system intact in all of the initial fragmentations of the parent ions so that no Se-specific chemistry was noted.

A thorough study of the mass spectra of 1,2,5-selenadiazoles, together with their O and S analogues, was carried out by Pedersen and Moller[81]. They worked at the lowest possible probe temperatures to minimize thermal decomposition of the samples, employed metastable defocusing techniques to support virtually all of their fragmentation pathways, and used exact mass measurements to ascertain the chemical formula of $C_5H_2N^{+\cdot}$ for m/z 76 which previously had been posited to be the benzyne ion.

Most of the important features of the observed spectra with the exception of the propensity for the ionized O-containing compounds to undergo NO emission are summarized by equation (36). In comparison, NS emission was observed only barely, and NSe not at all, from their respective analogous species.

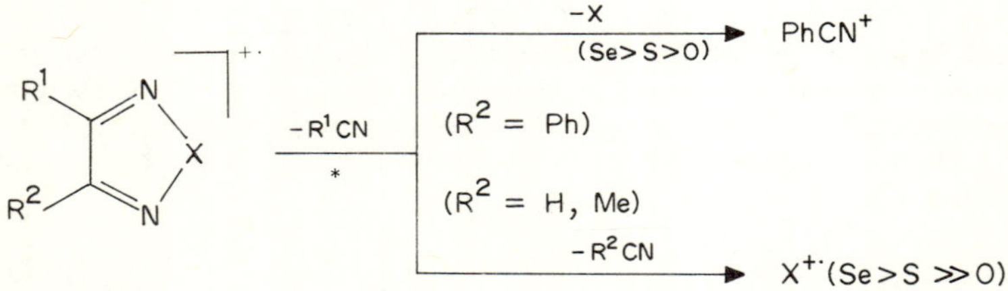

(36)

(58)

The molecular ions always appeared quite strongly, sometimes providing the most abundant species. To be noted is the consistent first fragmentation step of the loss of RCN. The chemical formula for the resulting species is $RCNX^{+\cdot}$ (X = O, S, Se). The authors raised the question of the structure of this species. The loss of X, which was strongest for Se of course, would appear to call for a simple cleavage of the N—X bond. However, other fragmentation products such as losses of CX and HCN neutrals argued for the rearrangement of $RCNX^{+\cdot}$. Comparison with spectra from PhNCX molecules did not completely answer the question, which remains open at this point.

In the presence of a Ph group, the charge of the ions tended to be retained on the organic moiety producing high abundances of species such as $PhCN^{+\cdot}$ and Ph^{+}. In the absence of a Ph substituent, the tendency for the charge to reside on those species containing S, and even more so, Se, was enhanced. Thus, when $R^1 = R^2 = H$, the dominant fragmentation was by progressive emission of HCN groups. Similarly, when $R^1R^2 = —C_4H_4—$, loss of HCN was by far the most important fragmentation process.

Arshadi[82] also studied the mass spectra of the 1,3-benzodiazoles with O, S and Se in the 2-position and found results parallel to those of Pedersen and Moller[81]. However, he also investigated the negative ions of these species—an unusual approach when it comes to organoselenium compounds although the author points out the propensity of these and similar compounds for electron capture and formation of molecular anions.

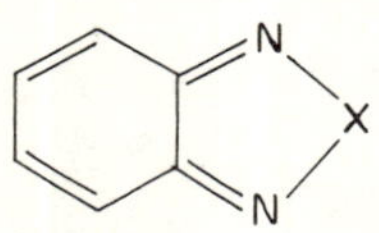

(59)
X = O, S, Se

All of the various species **59** exhibited formation of molecular anions at relatively high pressures (9×10^{-5} Torr) where there is an abundance of secondary electrons with approximately thermal energy. However, even at low pressure (1×10^{-6} Torr), the Se compound was able to form molecular anions in significant quantities (38% relative intensity).

Dissociative electron capture processes produced very abundant amounts of CN^- ion for all species under all circumstances, generally providing the base peak in the spectrum. In addition, the oxygen compound showed loss of NO from the parent anion, formation of CNO^-, and of very small amounts of O^-; the Se analogue predominantly gave CN^- and Se^- ions; and the S analogue was intermediate between the O- and Se-containing species in its behaviour.

Following the study of 1,2,5-selenadiazoles[81], Pedersen[83] reported the synthesis and mass spectra of some substituted 1,2,5-selenadiazole *N*-oxides. He found that the position of *N*-oxide could be detected; for example, **60** underwent a facile loss of MeCNO and little MeCN elimination was seen.

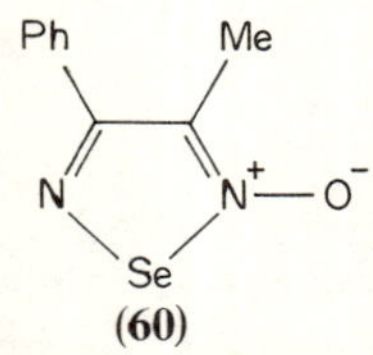

(60)

A particulaly straightforward fragmentation scheme largely appears to be followed by the molecular ion of **61** as reported by Fitjer and Luttke[84]. Although the base peak was provided by $C_3H_5^+$, the molecular ion showed a 79% relative abundance. The molecular ion mainly underwent a cycloreversion, expelling C_4H_8 (isobutene), followed by loss of CO. The second ring then emulated the first leading to formation of $Se_2C_2^{+\cdot}$, of undetermined structure. Besides a peak for $C_4H_8Se^{+\cdot}$, only hydrocarbon fragments, including that yielding the base peak, were to be found in the decompositions.

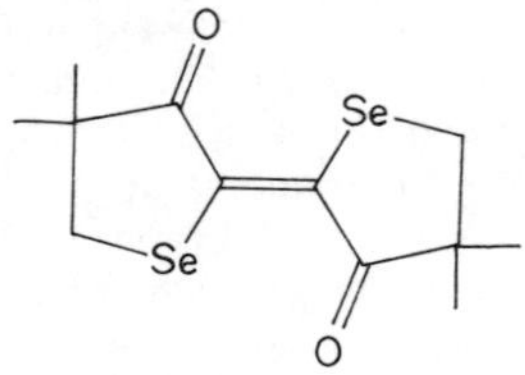

(61)

trans-3,3′-Dioxo-4,4,4′,4′-tetramethyl-2,2′-biselenolanylidene

The mass spectra of phthalide, together with some S and Se analogues (**62**) were reported and discussed by McMurray and coworkers[85]. They observed that when both X and Y = O, the abundant parent ion gave rise to relatively abundant species (40–50% compared to the parent ion) corresponding to the loss of CHO and subsequent loss of CO, forming Ph^+ ion.

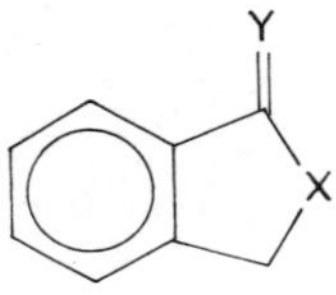

(**62**)

When both X and Y = S, the predominant decomposition of the abundant parent ion was via loss of CS and CHS, mimicking the O prototype. However, loss of monatomic S became significant here, along with some loss of CS_2, which introduced a route not observed when X = Y = O.

Compounds with mixed chalcogens, corresponding to X = O, Y = S, or X = S, Y = O, gave nearly identical spectra when ionized. Loss of CO, loss of CHO or CO then H, some loss of COS and a trace of S loss were all observed. The authors explained the near-identity of the spectra on the basis of a facile interchange of the X and Y atoms (see equation 37).

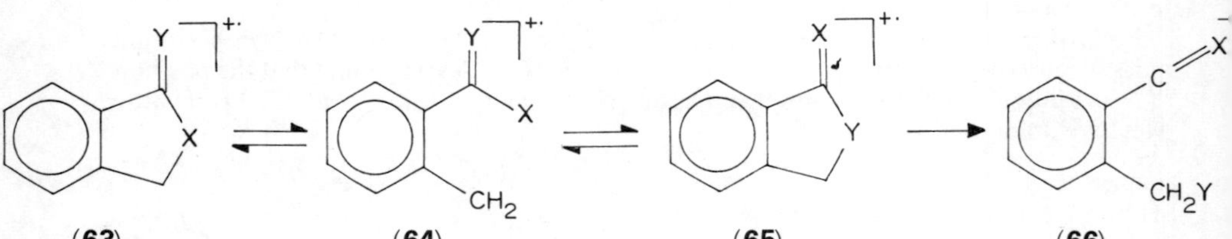

(**63**) (**64**) (**65**) (**66**) (37)

A similar interchange occurred when X = Se and Y = S or X = S and Y = Se; the spectra were nearly identical. Here, loss of Se was so pronounced as to give rise to the most abundant ion with the parent ion at only 40% relative abundance.

All of the fragmentations of this group of species can be explained by the authors' fragmentation schemes which invoke no unusual processes beyond the isomerization of the parent ion prior to decompositions.

The electron ionization induced decompositions of some model benzoselenazoles (**67**) and benzoisoselenazole (**68**) have been investigated by Croisy and coworkers[86]. It is informative to compare their spectra with benzothiazoles and benzoisothiazole (also reported in this paper) as well as with benzoselenophene itself (*vide supra*).

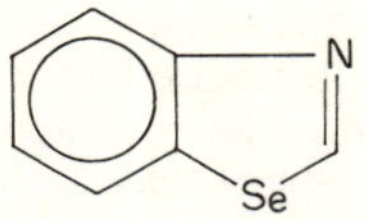
(**67**)

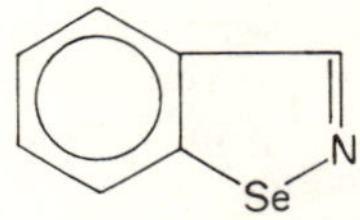
(**68**)

Both of the parent compounds underwent loss of HCN after ionization to yield $C_6H_4Se^{+\cdot}$ which then eliminated Se to give $C_6H_4^{+\cdot}$. In competition with this process was the expected reductive elimination of Se, which was considerably more favoured (by a factor of seven times) for the isoselenazole. Benzoselenophene also showed elimination of Se, but the corresponding benzoisothiazole and benzothiazole give barely detectable losses of S.

The loss of HCN shifted to an elimination of MeCN for both 2-methylbenzoselenadiazole and 3-methylbenzoisoselenazole. The loss of Se, CSe and H were competitive but the differences in the abundances in $[M - Se]^+$ could no longer be observed.

Replacing the Me groups with Ph groups led to expulsion of PhCN instead of MeCN. The isoselenazole now showed considerably more loss of Se and HSe (presumably in two steps) which permitted ready distinction of the two isomers.

The mass spectrometric fragmentations of several 2-acylmethylbenzo-thia- and -selenazoles (**69**) were reported by Ciurdaru and coworkers[87] whose work was distinguished by the use of high-resolution as well as metastable ion studies. For these compounds very little of the fragmentation was determined by the presence of S or Se. Beyond the loss of PhCO, exhibited by all of the analogues studied, the Se-containing compound showed loss of Se which was not reflective of the other two compounds.

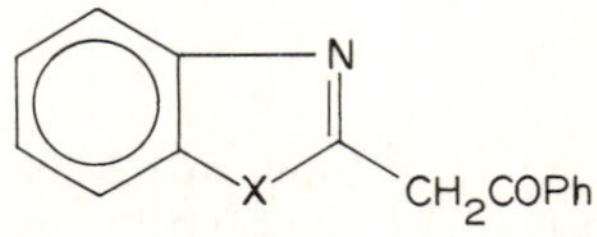
(**69**)
X = O, S, Se, NH (imidazoles)

Very similar fragmentations were found by Bologa and coworkers[88] when they attempted to introduce quaternary nitrogen salts of the structure **70** into the mass spectrometer. The salts lost HI and then evaporated and were ionized, presumably to give **71**. A preferred method for these compounds would be fast atom bombardment. Once again, with X = O, the principal decomposition was loss of CO, when X = S the loss was of SH and when X = Se the loss was of Se itself. Exact masses and the study of metastable species were used to advantage again in this study.

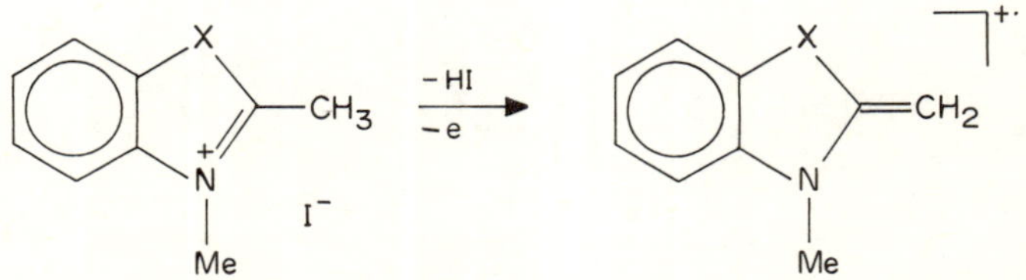
(**70**) X = O, S, Se

(**71**)

(38)

Stackhouse and coworkers[89], in a report of the first example of a selenabenzene (**72**), presented its mass spectrum and those of related synthetic products (e.g. **73**). The mass spectrum was used to provide a fingerprint and to confirm the mass of the parent molecule through the parent ion. They presented no analysis of the mass spectra but it would appear that among the fragmentations were found some familiar and anticipated routes such as the loss of Se, and of H_xSe.

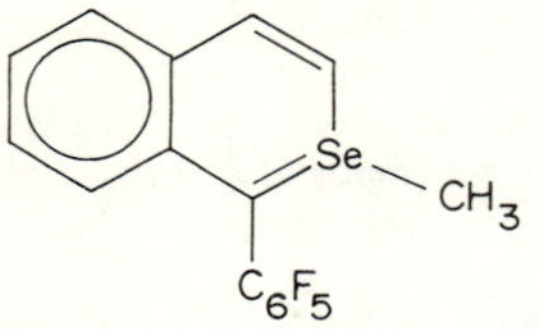

(**72**)

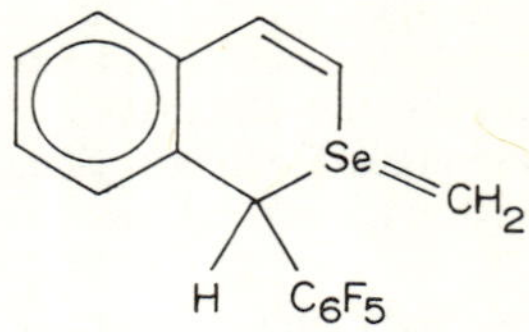

(**73**)

A number of complex heterocyclic organic molecules containing Se have been subjected to mass spectrometric examination by Shafiee working with a number of different colleagues[48–57]. This work was inspired chiefly by the pharmacological potentialities of the compounds, and so the mass spectral information was sought primarily to provide a molecular mass and a 'fingerprint' of each compound. Therefore, little more than the m/z of the parent ion, perhaps along with masses of a few other prominent ions, devoid of interpretation, was offered. The compounds which were reported have structures **74–97**.

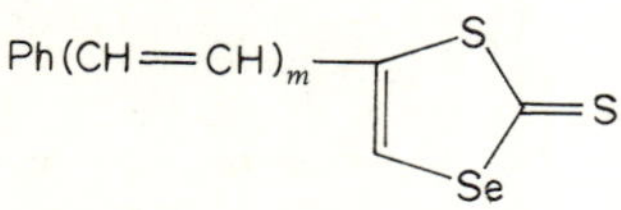

(**74**) $m = 1$: 5-Styryl-2-thioxo-1,3-thiaselenole[55]
(**75**) $m = 2$: 5-(4-Phenyl-1,3-butadienyl)-2-thioxo-1,3-thiaselenole[55]

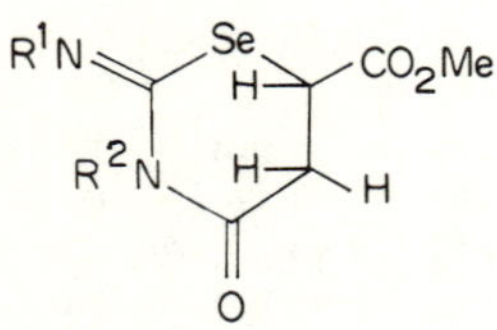

(**76**)
Methyl 2-amino-3,4,5,6-tetrahydro-4-oxo-2*H*-1,3-selenazine-6-carboxylate[49]

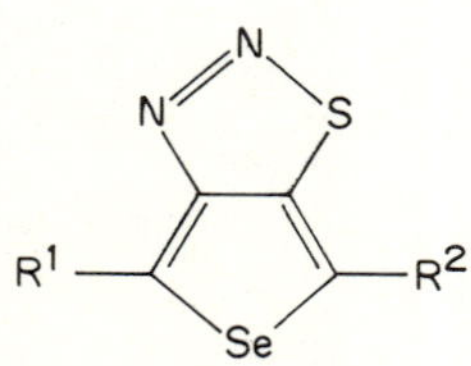

(**77**) $R^1 = H$, $R^2 = Ph$: 6-Phenylselenolo[3,4-*d*] [1,2,3]thiadiazole[50]
(**78**) $R^1 = Ph$, $R^2 = H$: 4-Phenylselenolo[3,4-*d*] [1,2,3]thiadiazole[50]

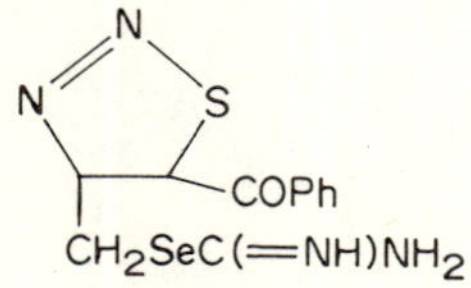

(79)
4-Isoselenoureidomethyl-5-benzoyl-[1,2,3]thiadiazole[50]

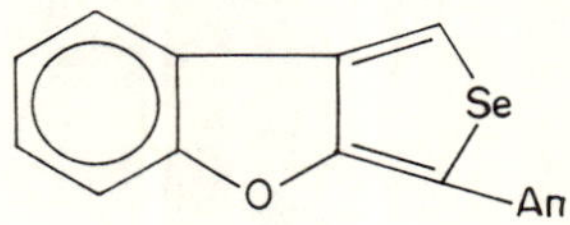

(80)
4-*p*-Methoxyphenylselenolo[3,4-*b*]benzofuran[51]

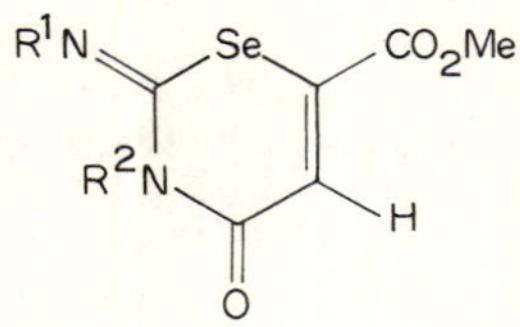

(81)
$R^1 = R^2 = H$: Methyl 2-imino-3,4-dihydro-4-oxo-2*H*-1,3-selenazine-6-carboxylate[49]
$R^1 = R^2 = Ph$: Methyl 2-phenylimino-3-phenyl-3,4-dihydro-4-oxo-2*H*-1,3-selenazine-6-carboxylate[49]

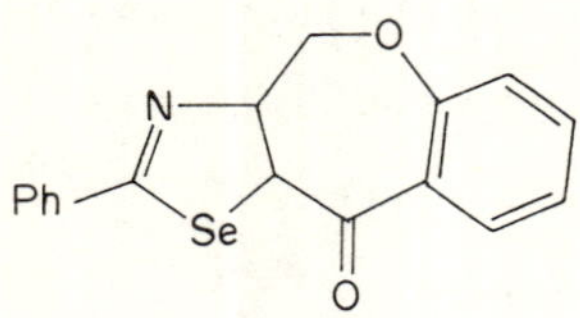

(82)
2-Phenyl-4,10-dihydro-10-oxo[1] benzoxepino[3,4-*d*]selenazole[57]

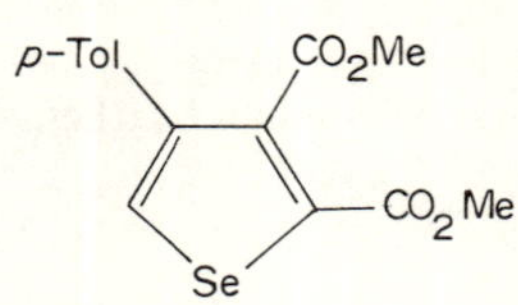

(83)
Dimethyl 4-*p*-tolylselenophene-2,3-dicarboxylate[48]

p-TolC≡C—Se—C(CO_2Me)=CHCO_2Me

(84)

Dimethyl 2-(2-*p*-tolylethynylselenomercapto)-1,2-ethenedicarboxylate[48]

p-TolC≡C—Se—CH=CHCO_2Me

(85)

Methyl *cis*-3-(2-tolylethynylselenomercapto)acrylate[48]

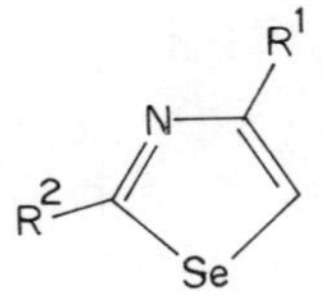

(86) $R^1 = CH_2Cl, CH_2OH, HCO; R^2 = Ph$: 2-Phenyl-4-chloromethylselenazole[53]
(87) $R^1 = CH_2OH, R^2 = Ph$: 2-Phenyl-4-hydroxymethylselenazole[53]
(88) $R^1 = CHO, R^2 = H$: 2-Phenyl-4-formylselenazole[53]

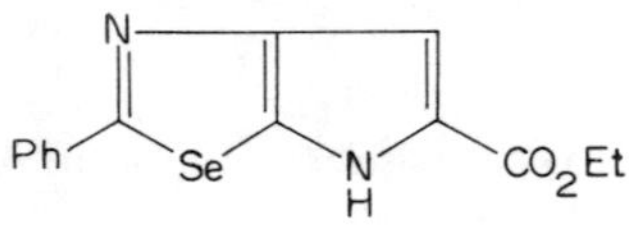

(89)
Ethyl 2-phenylpyrrolo[3,2-*d*]selenazole-5-carboxylate[53]

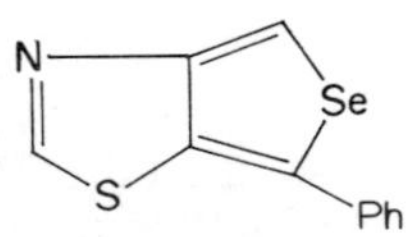

(90)
6-Phenylselenolo[3,4-*d*]thiazole[52]

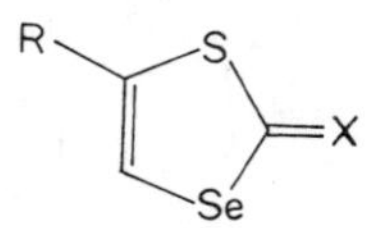

(91) $R = Ph, X = CHCO_2Et$: 2-Phenyl-ω-carbethoxy-1,4-thiaselenafulvene[54]
(92) $R = CHMe_2, X = NPh$: 5-Isopropyl-2-phenylimino-1,3-thiaselenole[54]
(93) $R = Me, X = NCO_2Et$: 5-Methyl-2-carbethoxyimino-1,3-thiaselenole[54]

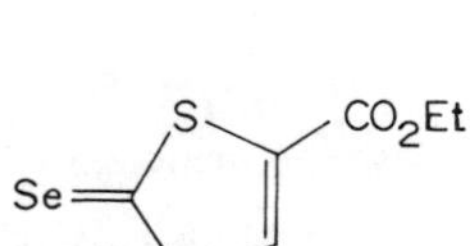

(94)
4-Carbethoxy-2-selenoxo-1,3-dithiole[56]

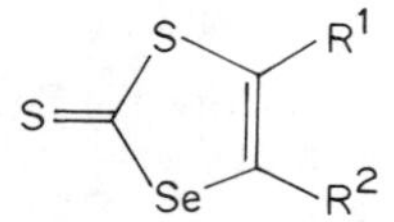

(95) $R^1 = CO_2Et$, $R^2 = H$: 5-Carbethoxy-2-thioxo-1,3-thiaselenole[56]
(96) $R^1 = H$, $R^2 = CO_2Et$: 4-Carbethoxy-2-thioxo-1,3-thiaselenole[56]

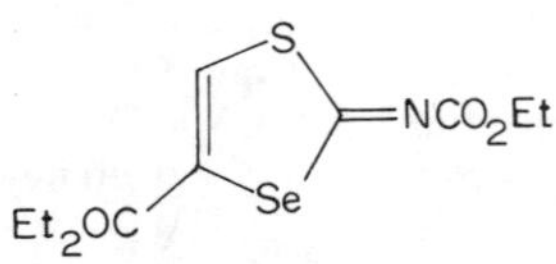

(97)
4-Carbethoxy-2-carbethoximino-1,3-thiaselenole[56]

X. Organometallic Compounds Involving Transition Metals

Chaudhuri and coworkers[90–92] carried out mass spectrometric studies of a class of compounds which involve chalcogen together with manganese and iron carbonyl species. In the case of species of the general formulae $Fe_2(CO)_6X_2$($X = S$, Se) and $Fe_3(CO)_9X'_2$ ($X' = S$, Se, Te), the parent ion appeared in all cases; its abundance increased with the atomic weight of X or X′ but was much greater for the diiron species than for the trisiron ones. In all cases, the base peak was provided by the ions resulting from the complete shedding of CO molecules and with iron and chalcogen atoms retained. Thereafter, progressive losses occurred, first of X, then of Fe for $Fe_2X_2^{+}$, or of Fe, then both X and another Fe ultimately resulting in monatomic Fe^{+}[90]. These results were well supported by studies of metastable species.

In the case of the species $[Mn(CO)_4SeCF_3]_2$ and $[Fe(CO)_3SeCF_3]_2$[91] the molecular ion was observed and successive losses of six carbonyl groups led to $(CF_2Se)_2Fe_2$. An ion corresponding to $CF_3SeFe_2F^+$ indicated occurrence of a transfer of F from C to metal; this turned out to be standard in the decomposition of these compounds. A different placement of the charge led to $SeCF_2^{+\cdot}$ which appeared as 'one of the strongest ions' in the mass spectra of these two metal carbonyl species.

For $CF_3SeFe(CO)_2C_5H_5$, a similar pattern was followed. Beginning with the parent ion, progressive losses were observed as follows: $-CO$, $-CO$, $-SeCF_2$, $-F$, $-C_2H_2$ and $-C_3H_3$. Chaudhuri and his coworkers[92] presented a thorough mass spectrometric study of compounds of the class $[CF_3XMn(CO)_4]_2$, $[CF_3XFe(CO)_3]_2$ ($X = S$, Se), $CF_3SeFe(CO)_2C_5H_5$ and $CF_3SCr(NO)_2C_5H_5$. Low-abundance molecular ions were observed for both the $[CF_3XMn(CO)_4]_2$ compounds. The molecular ions lost three CO groups in a single step, another CO, and then XCF_2 with an accompanying F atom shift to the Mn. The resultant fragment, $[CF_3XMn_2(CO)_4F]^+$ lost the remaining CO groups stepwise, then another XCF_2 unit which gave $[Mn_2F_2]^{+\cdot}$. This lost its F atoms, one by one, then an Mn atom leading to the Mn^+ ion. Little distinction between the S and Se cases was to be noted.

For the iron compounds $[CF_3SFe(CO)_3]_2$ and $[CF_3SeFe(CO)_3]_2$, the molecular ions were again present but sparse. The preferred decompositions proceeded via the progressive loss of all six CO groups and progressive loss of XCF_2 groups, leading to species containing only Fe and F atoms: $[Fe_2F_2]^{+\cdot} \rightarrow [Fe_2F]^+ \rightarrow [Fe]^+$.

III. SUMMARY

An examination of the mass spectra of organic Se and Te compounds presents the opportunity to test whether their chemical properties can be understood in analogy with those of corresponding O and S analogues. In general, smooth variations occur for the reactivity of gas-phase radical cations. Starting with alkyl selenide radical cations, $RSeR^{+\cdot}$, losses of R to give RSe^+ and hydrogen rearrangements to yield $RSeH^{+\cdot}$ are foreshadowed by the chemistry of S analogues, but these reactions are not seen when S or Se is replaced by O. Diselenides and ditellurides react similarly to disulphides. The aryl alkyl chalcogenides ArXR, constitute a useful system for making detailed comparisons. For PhXMe, the propensity for loss of CH_2X smoothly drops whereas production of PhX^+ smoothly increases as one goes from O to Te. For PhXEt, the common rearrangement to expel C_2H_4 is dominant for X = O, but is barely detectable for X = Te. Replacement of O with Se or Te presumably opens up new low-energy pathways, such as formation of PhX^+.

The ability to expel X increases smoothly from X = O to X = Te for ArXAr, a process which may be viewed as a reductive elimination and which occurs commonly in the fragmentations of Se and Te compounds. As expected, the elimination is only important for compounds containing the more metal-like chalcogens.

Periodic variations are also seen for cyclic chalcogenides. Notable is the smooth decrease in the facility of **98** to undergo cycloreversion reactions to expel CH_2O or C_3H_6O as X is varied from O to Te.

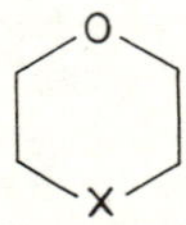

(**98**)

Various oxidized organoselenides have been extensively studied, and examples include selenoxides and selenones. Noteworthy is a facile rearrangement which usually involves

$$PhSe(=O)Me \rightarrow Ph—O—SeMe \tag{39}$$

aryl transfer from Se to O (equation 39), which is termed as a selenoxide–selenate rearrangement. Hints of this rearrangement are found by examining the decompositions of the corresponding sulphoxides and sulphones. Oxygen transfer from a nitro group to Se followed by the selenoxide–selenate rearrangement may be the explanation for the abundant loss of C_7H_7O from **99**. For **99** and related compounds, the variation of chemical properties is abrupt rather than smooth as the S-containing analogues show absolutely no sign of this interesting rearrangement. This abrupt change, however, appears to be an exception, not the rule.

o-$O_2NC_6H_4SeTol$-p (**99**)

Examination of the mass spectra of various heteroaromatics also reveals smooth variations of decomposition reactions. Loss of CHX smoothly drops off as one proceeds from furan to tellurophene whereas the reductive elimination of Te from tellurophene is foreshadowed by the fragmentation of selenophene. For benzoselenophene, benzoselenazole, benzoisoselenazole and selenolo[2,3-*b*]pyridines, reductive elimination of Se is always important. The reactions of substituted benzoselenophenes, however, involve decompositions which are determined by the nature of substituents rather than the presence of the Se. The mass spectra are entirely similar to the corresponding benzofurans and benzothiophenes. Apparently, the energy requirement for substituent-driven processes is lower than that of the reductive elimination. Systematic variations can also be found for the mass spectral decompositions of selenadiazoles, 2,5-diazoles substituted with O, S or Se in the 1-position, and other heterocycles.

Besides the selenone/selenate-type rearrangement, two other rearrangements are noteworthy for Se-containing compounds, and they merit further study. The exchange of heteroatoms in the S and Se analogues of phthalide (see equation 37) and the C—Se interchange which occurs for various selenofulvalenes (see equation 28) are characterized by transfer of organic moieties to the Se. An interesting question is whether these processes pertain when these species are in solution under one-electron oxidation conditions.

Final remarks are focused on the mass spectrometric techniques used in the studies reviewed here. Most compounds have been ionized by using electron beams and that works reasonably well. It is recommended that materials which give low-abundance molecular ions be reinvestigated by using newer methods such as field ionization, field desorption and fast atom bombardment, or by electron attachment to give radical anions. Two reports show clearly the advantages of field ionization and negative ions[3,82]. Many, but not all, fragmentation pathways and ion structures are speculative. Future studies should include metastable ion measurements, collisional activation and tandem mass spectrometry (MS/MS) to remove some of the uncertainties associated with the mechanisms given in this review.

IV. ACKNOWLEDGEMENT

The preparation of this review was supported in part by the Midwest Center for Mass Spectrometry, a US National Science Foundation Instrumentation Facility (Grant No. CHE-8211164).

V. REFERENCES

1. J. March, *Advanced Organic Chemistry*, 2nd ed., McGraw-Hill, New York, 1977, pp. 829, 934.
2. A. Croisy, P. Jacquignon, P. Pirson and L. Christiaens, *Org. Mass Spectrom.*, **9**, 970 (1974).
3. J. Andersen, H. Egsgaard, E. Larsen, K. Bechgaard and E. Engler, *Org. Mass Spectrom.*, **13**, 121 (1978).
4. M. Albeck and S. Shaik, *J. Organomet. Chem.*, **91**, 307 (1975).
5. K. J. Irgolic and W. S. Haller, *Intern. J. Sulfur Chem. (A)*, **2**, 267 (1972).
6. M. Drager and G. Gattow, *Spectrochim. Acta*, **28A**, 425 (1972).
7. R. J. Kraus, S. J. Foster and H. E. Ganther, *Biochemistry*, **22**, 5853 (1983).
8. F. W. McLafferty (Ed.), *Tandem Mass Spectrometry*, John Wiley and Sons, New York, 1983.
9. K. R. Jennings and R. S. Mason, Ref. 8, pp. 197–222.
10. M. Barber, R. S. Bordeli, R. D. Sedgwick and A. N. Tyler, *J. Chem. Soc., Chem. Commun.*, 325 (1981).
11. R. D. Levin and S. G. Lias, 'Ionization Potential and Appearance Potential Measurements, 1971–81', *Natl. Bur. Stand. Ref. Data Ser.*, **71**, 179, 330, 459, 506 (1982).
12. J. D. Scott, G. C. Causley and B. R. Russell, *J. Chem. Phys.*, **59**, 6577 (1973).
13. S. Cradock and R. A. Whiteford, *J. Chem. Soc., Faraday Trans. 2*, **68**, 281 (1972).
14. G. Tschmutowa and H. Bock, *Z. Naturforsch.*, **31b**, 1611 (1976).
15. G. A. Chmutova, N. N. Vtyurina, T. V. Komina, I. G. Gazizov and H. Bock, *Zh. Obshch. Khim.*, **49**, 192 (1979).
16. S. Pignataro and G. Distefano, *Chem. Phys. Letters*, **26**, 356 (1974).
17. G. Distefano, S. Pignataro, G. Innorta, F. Fringuelli, G. Marino and A. Taticchi, *Chem. Phys. Letters*, **22**, 132 (1973).
18. F. Fringuelli, G. Marino, A. Taticchi, G. Distefano, F. P. Colonna and S. Pignataro, *J. Chem. Soc., Perkin Trans. 2*, 276 (1976).
19. J.-F. Muller, *Helv. Chim. Acta*, **58**, 2646 (1975).
20. O. Thorstad, K. Undheim, B. Cederlund and A.-B. Hörnfeldt, *Acta Chem. Scand.*, **29B**, 647 (1975).

21. O. Thorstad, K. Undheim, R. Lantz and A.-B. Hörnfeldt, *Acta Chem. Scand.*, **29B**, 652 (1975).
22. B. Cederlund, R. Lantz, A.-B. Hörnfeldt, O. Thorstad and K. Undheim, *Acta Chem. Scand.*, **31B**, 198 (1977).
23. P. Linda, G. Marino and S. Pignataro, *J. Chem. Soc.* (*B*), 1585 (1971).
24. F. P. Colonna, G. Distefano, V. Galasso, K. J. Irgolic, C. E. King and G. C. Pappalardo, *J. Organomet. Chem.*, **146**, 235 (1978).
25. F. P. Colonna, G. Distefano and V. Galasso, *J. Electron Spectry Relat. Phenom.*, **18**, 75 (1980).
26. A. H. Cowley, M. Lattman, R. A. Montag and J. G. Verkade, *Inorg. Chim. Acta*, **25**, L151 (1977).
27. R. Schulz and A. Schweig, *Angew. Chem. Intern. Ed. Engl.*, **19**, 69 (1980).
28. R. Gleiter, M. Kobayashi, J. Spanget-Larsen, J. P. Ferraris, A. N. Bloch, K. Bechgaard and D. O. Cowan, *Ber. Bunsenges.*, **79**, 1218 (1975).
29. H. J. Svec and G. A. Junk, *J. Amer. Chem. Soc.*, **89**, 790 (1967).
30. T. W. Bentley, R. A. W. Johnstone and F. A. Mellon, *J. Chem. Soc.*, **13**, 1800 (1971).
31. C. van den Heuvel and N. Nibbering, *Org. Mass Spectrom.*, **10**, 250 (1975).
32. H. Budzikiewicz and R. Pesch, *Org. Mass Spectrom.*, **9**, 861 (1974).
33. L. B. Agenäs, in *Organic Selenium Compounds: Their Chemistry and Biology*, (Eds. W. H. H. Günther and D. L. Klayman), John Wiley and Sons, New York, 1973, pp. 963–988.
34. E. Rebane, *Acta Chem. Scand.*, **27**, 2861 (1973).
35. M. L. Denniston and D. R. Martin, *J. Inorg. Nucl. Chem.*, **37**, 1871 (1975).
36. H. L. Paige and J. Passmore, *Inorg. Nucl. Chem. Letters*, **9**, 277 (1973).
37. E. Rebane, *Acta Chem. Scand.*, **27**, 2870 (1973).
38. E. Rebane, *Chem. Scri.*, **4**, 219 (1973).
39. J. Greiner, E. Rouvier and A. Cambon, *Org. Mass Spectrom.*, **13**, 197 (1978).
40. P. J. Busse, R. A. Grigsby, F. L. Kolar and K. J. Irgolic, *Org. Mass Spectrom.*, **11**, 640 (1976).
41. E. Rebane, *Acta Chem. Scand.*, **27**, 2861 (1973).
42. A. M. Duffield, H. Budzikiewicz and C. Djerassi, *J. Amer. Chem. Soc.*, **87**, 2920 (1965).
43. P. S. Kulkarni, M. M. Khan, G. D. Sturgeon and M. L. Gross, *J. Heterocycl. Chem.*, **17**, 929 (1980).
44. P. Nanjappam, K. Ramalingam and K. D. Berlin, *Indian J. Chem.*, **22B**, 640 (1983).
45. C. D. Desjardins and J. Passmore, *J. Chem. Soc., Dalton Trans.*, 2314 (1973).
46. N. S. Dance and W. R. McWhinnie, *J. Organomet. Chem.*, **125**, 291 (1977).
47. N. P. Buu-Hoi, M. Mangane, P. Jacquignon, J. L. Piette and M. Renson, *Bull. Soc. Chim. Fr.*, **3**, 925 (1971).
48. A. Shafiee, I. Lalezari and F. Savabi, *Synthesis*, 765 (1977).
49. A. Shafiee, F. Assadi and V. I. Cohen, *J. Heterocycl. Chem.*, **15**, 39 (1978).
50. A. Shafiee, *J. Heterocycl. Chem.*, **15**, 473 (1978).
51. A. Shafiee and E. Behnam, *J. Heterocycl. Chem.*, **15**, 589 (1978).
52. A. Shafiee and A. Mazloumi, *J. Heterocycl. Chem.*, **15**, 1455 (1978).
53. A. Shafiee, A. Mazloumi and V. I. Cohen, *J. Heterocycl. Chem.*, **16**, 1563 (1979).
54. A. Shafiee, M. Vosooghi and R. Asgharian, *J. Heterocycl. Chem.*, **17**, 117 (1980).
55. A. Shafiee, M. Vosooghi and I. Lalezri, *J. Heterocycl. Chem.*, **17**, 545 (1980).
56. A. Shafiee and F. Assadi, *J. Heterocycl. Chem.*, **17**, 549 (1980).
57. A. Shafiee, G. Kiaeay and M. Vosooghi, *J. Heterocycl. Chem.*, **18**, 789 (1981).
58. J. Martens, K. Praefcke, U. Schulze, H. Schwarz and H. Simon, *Tetrahedron*, **32**, 2467 (1976).
59. E. Rebane, *Chem. Scr.*, **5**, 65 (1974).
60. L. Futekov, S. Stojanov and H. Specker, *Fresenius Z. Anal. Chem.*, **295**, 7 (1979).
61. K. V. Smith and J. S. Thayer, *Inorg. Chem.*, **13**, 3021 (1974).
62. K. H. Wei, I. C. Paul, M. Y. Chang and J. I. Musher, *J. Amer. Chem. Soc.*, **96**, 4099 (1974).
63. P. B. Terent'ev, R. A. Kyandzhetsian and N. N. Magdesieva, *Zh. Org. Khim.*, **12**, 2504 (1976) [*J. Org. Chem. USSR*, **12**, 2425 (1976)].
64. T. J. Adley, B. C. Pant and R. T. Rye, *Can. J. Spectry.*, **20**, 71 (1975).
65. E. Rebane, *Chem. Scr.*, **5**, 5 (1974).
66. A. Benedetti, C. Preti, G. Tosi and P. Zannini, *J. Chem. Soc., Dalton Trans.*, 1467 (1980).
67. A. Benedetti, C. Preti, L. Tassi and G. Tosi, *Austr. J. Chem.*, **35**, 1365 (1982).
68. A. M. Kirkien, R. J. Shine and J. R. Plimmer, *Org. Mass Spectrom.*, **7**, 233 (1973).
69. F. Fringuelli and A. Taticchi, *J. Heterocycl. Chem.*, **15**, 137 (1978).

70. T. J. Barton, C. R. Tully and R. W. Roth, *J. Organomet. Chem.*, **108**, 183 (1976).
71. A. Croisy, P. Jacquignon and L. Christiaens, *Org. Mass Spectrom.*, **10**, 1084 (1975).
72. P. Jacquignon, A. Croisy, M. Renson, E. Iteke and L. Christiaens, *Org. Mass Spectrom.*, **7**, 1235 (1973).
73. D. Elmaleh, S. Patai and Z. Rappoport, *J. Chem. Soc (C)*, 939 (1970).
74. M. Evers, R. Weber, Ph. Thibaut, M. Renson, L. Christiaens, A. Croisy and P. Jacquignon, *J. Chem. Soc., Perkin Trans. 1*, 2452 (1976).
75. M. Loth-Compere, A. Luxen, Ph. Thibaut, L. Christiaens, M. Guillaume and M. Renson, *J. Heterocycl. Chem.*, **18**, 343 (1981).
76. J. van Coppenolle and M. Renson, *Tetrahedron*, **31**, 2099 (1975).
77. S. C. Cohen, A. G. Massey, G. F. Lanthier and J. M. Miller, *Org. Mass Spectrom.*, **6**, 373 (1972).
78. J. Moller, R. Christie, C. Pedersen and D. Reid, *Acta Chem. Scand.*, **30B**, 600 (1976).
79. M. Perrier, R. Pinel and J. Vialle, *J. Heterocycl. Chem.*, **12**, 639 (1975).
80. G. N. Jham, R. N. Hanson, R. W. Giese and P. Vouros, *J. Heterocycl. Chem.*, **18**, 1335 (1981).
81. C. L. Pedersen and J. Moller, *Acta Chem. Scand.*, **29B**, 483 (1975).
82. M. R. Arshadi, *Org. Mass Spectrom.*, **13**, 379 (1978).
83. C. Pedersen, *Acta Chem. Scand.*, **30B**, 675 (1976).
84. L. Fitjer and W. Luttke, *Chem. Ber.*, **105**, 919 (1972).
85. W. J. McMurray, S. R. Lipsky, R. J. Cushley and H. G. Mautner, *J. Heterocycl. Chem.*, **9**, 1093 (1972).
86. A. Croisy, P. Jacquignon, R. Weber and M. Renson, *Org. Mass Spectrom.*, **6**, 1321 (1972).
87. Gh. Ciurdaru, Z. Moldovan and I. Oprean, *Monatsh. Chem.*, **109**, 379 (1978).
88. M. Bologa, N. Palibroda, Z. Moldovan and Gh. Ciurdaru, *Org. Mass Spectrom.*, **12**, 562 (1977).
89. J. Stackhouse, G. H. Senkler, Jr., B. E. Maryanoff and K. Mislow, *J. Amer. Chem. Soc.*, **96**, 7835 (1974).
90. M. K. Chaudhuri, A. Haas, M. Rosenberg, M. Velicescu and N. Welcman, *J. Organomet. Chem.*, **124**, 37 (1977).
91. M. K. Chaudhuri, A. Haas and A. Wensky, *J. Organomet. Chem.*, **116**, 323 (1976).
92. M. K. Chaudhuri, H. S. Dasgupta, N. Roy and D. T. Khathing, *Org. Mass Spectrom.*, **16**, 534 (1981).

The Chemistry of Organic
Selenium and Tellurium
Compounds Volume 1
Edited by S. Patai and Z. Rappoport
© 1986 John Wiley & Sons Ltd.

CHAPTER **8**

Radiation chemistry of organic selenium and tellurium compounds

ROBERTO BADIELLO

C. N. R., Servizio Regionale di Sicurezza del Lavoro e Protezione Sanitaria, Facoltà di Farmacia dell'Università di Bologna, Italy

I.	INTRODUCTION	287
II.	SELENIUM COMPOUNDS	287
	A. Antiradiation Effects in Chemical and Biological Systems	287
	B. Radiation Chemistry of Selenium Compounds	292
III.	TELLURIUM COMPOUNDS	302
IV.	CONCLUDING REMARKS	303
V.	REFERENCES	304

I. INTRODUCTION

The heaviest elements of the sulphur subgroups—Se, Te and Po—have been less intensively studied than either O or S; consequently the radiation chemistry of Se- and Te-containing compounds has been less exploited. Nevertheless, the interest in radiation effects, particularly in Se-containing compounds has increased in recent years, as, on the one hand, the identification of radiation damage is helpful in understanding the part played by Se in radioprotection, and on the other, such studies enable a direct comparison with the more important element S. The organic Te-containing compounds could be important in relation to hot-atom chemistry.

II. SELENIUM COMPOUNDS

A. Antiradiation Effects in Chemical and Biological Systems

Radiation chemical studies on Se-containing compounds were recently performed in connection with the role played by these compounds as radioprotectors.

Many chemical substances can act as modifiers of radiation damage (i.e. radioprotectors and radiosensitizers) in chemical and biological systems and it is well known that the most important class of radioprotective agents is the S-containing compounds[1].

Compounds in which Se replaces S are also of interest in relation to the phenomenon of radiobiological protection. Some chemical and biological features of Se- and S-containing compounds appear to be similar, but some differences are observed[2–4]. The atomic rays and the solvation abilities of S and Se analogues are rather similar. S is a typical metalloid and can expand its electron shell, but does so with some difficulty; Se presents some metallic properties and the expansion of the electron shells is easier. Their compounds are therefore quite different in terms of electronic distribution and chemical reactivity. For example, in conjugated systems, the C—Se bond is more polarized than the C—S bond. The differences in the bond polarization are reflected in the different reactivities of the systems in analogous series of compounds.

Se-containing compounds are generally very reactive, especially in redox processes. The Se—Se bond is less stable than the S—S bond and it is easily cleaved in various reversible reactions. In the last few years the biochemistry of Se has also developed extensively due to the role of Se as an essential micronutrient element. Recently it has been shown that some enzyme-catalysed redox reactions require the participation of Se-containing proteins, and a detailed review on this subject has been published[5].

Considerations concerning mainly ionization potentials, bond strengths and electropositivities suggest that compounds in which Se replaces S should be good radioprotective agents. Some attempts made in the past to test this assumption have not always met with success. The first Se-containing compound tested for radioprotective ability was selenophenol. When administered to mice i.p. (0.2 or 0.4 mg per kg body weight) in olive oil 10 minutes before X-irradiation, this compound was found by Bacq and coworkers to be ineffective[6].

Shimazu and Tappel[7,8] have compared the radioprotection, in model systems, offered by the Se-containing amino acids, selenocystine and selenomethionine, with that of the known radioprotectors, cystine, 2-aminoethylthiopseudourea dihydrochloride and 2-aminoethanethiol hydrochloride. The selenoamino acids were found to protect different amino acids and the enzymes ribonuclease and yeast-alcohol deydrogenase against ionizing radiation better than the S-containing compounds. The strong radioresistance of selenomethionine was attributed to its ability to form stable radical intermediates which can combine with free H atoms or electrons to restore the molecule to its original state.

This behaviour could be linked to the biological role of Se and in particular to its antioxidant action. However, in these and in other *in vitro* experiments, the Se levels were very much higher than those commonly present in normal tissues and cells and often the pH values were outside the physiological range[9].

In other related experiments, Dickson and Tappel[10] explored the effects of the selenoamino acids, as compared to the analogous thioamino acids, on the activation and the activity of the sulphydryl enzymes papain and glyceraldeyde-3-phosphate dehydrogenase. They observed that the selenoamino acids bind reversibly to the thiol groups of the enzymes to form substrate-displaceable complexes which protect the enzymes against oxidative inactivation.

The synthesis of some potential antiradiation Se compounds, analogues of the well-known radioprotective drugs, are reported in the literature: 2-aminoethaneselenol (selenocysteamine) hydrochloride[11], bis(2-aminoethyl) diselenide(selenocystamine) dihydrochloride[11,12], 2-aminoethaneselenosulphuric acid[11] and 2-aminoethyl selenopseudourea dihydrobromide[13]. None of these compounds showed any protective activity[3].

The radioprotective activities of selenourea[14] and of selenocystine, selenomethionine, selenoxanthene, selenoxanthone, selenochromone and colloidal Se[15] in rats exposed at sublethal (600 R), lethal (750 R) and superlethal doses (900 R) have been reported. All these substances show activity similar to that of cysteine and, in some instances, superior to it. This results in reduction in mortality, of some clinical symptoms of radiation injury and of the severity of haematological syndrome (depression of white cell count and of neutrophiles in blood). Moreover, other experiments have shown that the preadministration of selenourea in rats modifies favourably other biochemical changes induced by ionizing radiation[16].

The parameters investigated in the last paper were the total protein content and the protein pattern of serum and some changes in the activity of plasma enzymes like alkaline phosphatase (AP), glutamate oxalacetate transaminase (GOT) and glutamate pyruvate transaminase (GPT).

The distribution of ^{75}Se after 6 hours to 14 days from labelled selenourea in rat organs has been determined after i.p. administration[17]. The distribution of ^{75}Se varied depending upon the amount of Se and was strictly related to the physiological activity of the organs and to the catabolism of selenourea. At the highest levels of administered selenourea, chemical and radiochemical toxicity phenomena were observed[18].

The rates of incorporation of seleno analogues of cystine, methionine, cystamine and taurine, as well as their toxicities have also been investigated in rats[19,20]. Generally, the highest values of incorporation of all the tested compounds occurred in the liver with a fast metabolic exchange, while the blood values remained practicallly constant from three hours to one week. The toxic effects were estimated from the survival, the general clinical conditions and the qualitative changes in the components of the red and white blood-cell series.

With selenourea, selenocystine and selenomethionine the retention of Se in blood constituents and its distribution at the subcellular levels of the liver have been studied[15]. For all compounds the blood radioactivity was concentrated in the red cells, while that in the plasma was low. In the subcellular fractions of the liver, the ^{75}Se was distributed mainly in the non-protein-soluble fraction, in the microsomes and in the nuclei. There was practically none in the protein and in the mitochondria. The microsomes should be the initial point of Se incorporation into proteins and should therefore be the first to utilize the organoselenium compounds and the elementary Se. The high activity of the nuclei was probably due to elementary Se formed by decomposition of the compound and precipitated with the nuclear fraction. The radioactivity of the soluble fraction was due to the presence of intact organoselenium compounds, as tested by chromatography.

Inorganic seleno compounds have also been reported to exhibit radioprotection in animals. Sodium selenate, when administered to rats one hour postirradiation (800 R) at 4.2–4.6 mg/kg i.p., followed by an additional 80% subcutaneously, resulted in the survival of all the animals[21]. Sodium selenite, administered *per os* to male Swiss albino mice, using a dose of 0.05 mg/kg day for ten consecutive days prior to their 500 R irradiation, increased their survival rates in the 15th day after irradiation[22]. Other experiments have demonstrated the protective effects of Se against teratogenic and carcinogenic effects[23].

The radioprotective characteristics of sodium selenite on rats have also been investigated[24]. It was ascertained that the sodium selenite in combination with vitamins A and E has radioprotective ability. The greatest protective effect was achieved by introduction of the combination of sodium selenite with the vitamins A and E half an hour before irradiation. The possible mechanism of the radioprotective effect has been discussed on the base of the antioxidizing characteristics of Se and vitamin E and on their influence on the subcellular particles, specifically on lysozomes.

The radioprotective properties of Se in 2-amino ethylisoselenouronium bromide–hydrobromide, 2-aminoselenoazoline and the Se analogue of mercaptoethylguanidine were investigated in white mice[25]. A radioprotective effectiveness, comparable with classical thiol radioprotectors, was manifested by 2-aminoselenoazoline. In the other two compounds, on the contrary, a synergism of toxicity and radiation was observed.

The time dependence of the distribution of ^{75}Se-labelled 2-aminoselenoazoline in the organisms of rats and mice was followed[26]. By comparison of the results obtained with the previous study of radioprotective properties of this compound, its efficiency was shown to be dependent on the actual concentration in tissues.

Se-containing compounds have also been reported to act as radioprotectors in cellular and molecular systems.

The ability of various Se heterocycles to protect ATP from losing orthophosphate on irradiation has been studied by Brucker and Bulka[27]. Only 2-amino-4,5-dimethylselenazole hydrochloride acted as radioprotector. Other 2-aminoselenazoles, as well as selenocarbazide and acetoselenosemicarbazone, did not show any radioprotective ability for *Phycomyces blakesleeanus*, but on the contrary showed some sensitizing effect[28].

Selenourea has been reported to protect amino acids from radiation damage in aqueous solution and in the solid state[29] by serving as an efficient free-radical scavenger[30]. Similar behaviour in selenourea, as well as in selenocystine and selenomethionine, was observed for enzyme systems, i.e. ribonuclease[31] and yeast-alcohol dehydrogenase[32].

The inactivation of ribonuclease was strongly reduced by the presence of the Se-containing additive. In view of the high reactivity of selenourea, selenocystine and selenomethionine with primary water radicals, as deduced from the radiation chemical data, the radioprotective effect could be explained on the basis of a simple competition between organic Se compounds and ribonuclease for the radicals themselves.

In contrast with the results obtained from multicellular systems, when cultures of *E. coli* B/r were irradiated in the presence of selenourea, they show an increased survival when irradiated in air, and the opposite effect in nitrogen[33]. The anoxic sensitization was explained by the presence of colloidal Se which is the main product of selenourea radiolysis, while selenourea itself is a most active protector in the presence of O_2, confirming the general trend found for Se-containing compounds in mammalian systems.

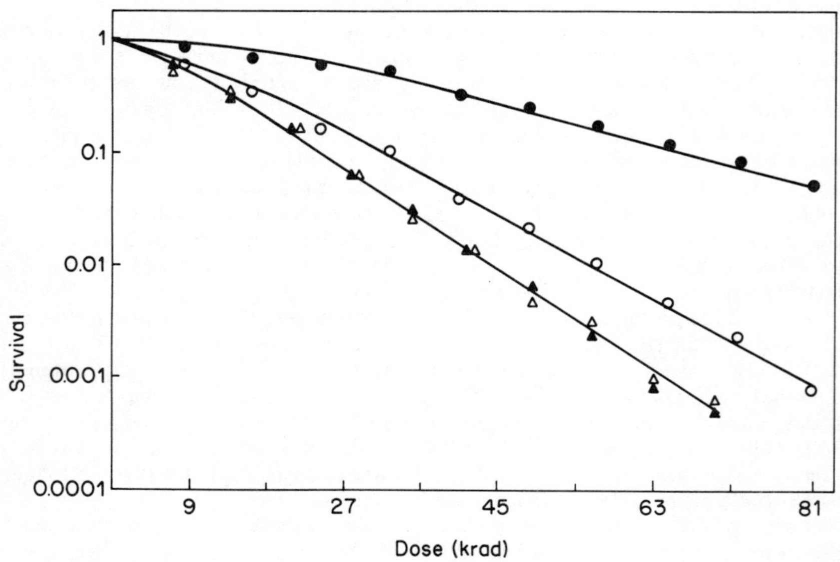

FIGURE 1. Survival of *E. Coli* B/r irradiated with X-rays in the presence of colloidal Se: ●—● control and ○—○ colloidal Se in N_2, ▲—▲ control and △—△ colloidal Se in air. Reproduced with permission from Ref. 33

TABLE 1. Radioprotective effects of Se-containing compounds on various biological systems

Compound	Test system	End-point	Ref.
	Molecular		
Selenocystine, Selenomethionine	Amino acids · Enzymes (ribonuclease, yeast-alcohol-dehydrogenase)	Chemical degradation Loss of catalytic activity	7,8
Selenourea	Amino acids	Chemical degradation and inhibition of ESR signals	29
Selenourea	Yeast-alcohol dehydrogenase	Loss of catalytic activity	32
Selenourea, Selenocystine, Selenomethionine	Ribonuclease	Loss of catalytic activity	31
Selenocyanate	Enzymes (ribonuclease, lysozime, α-chymotrypsin, alcohol dehydrogenases)	Loss of catalytic activity	35
Selenium dioxide	Enzymes (ribonuclease, lysozime, α-chymotrypsin, alcohol dehydrogenases)	Loss of catalytic activity	37
Colloidal Se	Yeast-alcohol dehydrogenase	Loss of catalytic activity	32
2-Amino-4,5-dimethyl-selenazole; HCl	ATP	Loss of ortophosphate	27
	Cellular		
Selenourea	*E. coli* B/r	Cell survival	32,33
Colloidal Se	*E. coli* B/r	Cell survival	32,33
2-Aminoselenazoles, Selenocarbazide, Acetoselenosemicarbazone	*Phycomyces blakesleeamus*	Spore survival	28
	Multicellular organisms		
Selenourea	Rats	30 days survival, haematological syndrome	14
Selenourea	Rats	Protein and enzyme parameters	16
Selenocystine, Seleno-methionine, Selenoxanthene, Selenoxanthone, Selenochromone	Rats	30 days survival, haematological syndrome	15
Sodium selenate	Rats	30 days survival	21
Sodium selenite	Mice	15 days survival	22
Sodium selenite (+ vitamins A and E)	Rats	30 days survival	24
Selenophenol	Mice	30 days survival	6
2-Aminoselenoazoline, Selenium-2-aminoethyl-isoselenouronium bromide–HBr, Selenoethylguanidine	Mice	30 days survival	25
2-Aminoethaneselenol (selenocysteamine)·HCl	Mice	30 days survival	39
Bis(2-aminoethyl) diselenide (selenocystamine)·2 HCl	Mice	30 days survival	3
2-Aminoethaneselenosulphuric acid	Mice	30 days survival	39
2-Aminoethylselenopseudourea· 2 HBr	Mice	30 days survival	3

Figure 1 shows the sensitizing activity of colloidal Se on bacterial systems. In particular the data emphasize that colloidal Se has no effect in air but it shows appreciable sensitization in the absence of O_2. Although colloidal Se does not appear to have any possible practical application, mainly because of its toxicity and instability, it exhibits a type of radiosensitization which could contribute to an understanding of mechanisms of anoxic sensitization[34]. Colloidal Se forms as the result of chemical and physical actions on aqueous solutions of selenourea. The hydrosols of elementary Se are made up by micellae which may contain different numbers of atoms. The micellae tend to grow sedimenting amorphous red Se which then goes over into the grey form. The elementary Se particles are rather unlikely to be involved in an intracellular sensitization mechanisms just because of their size, and therefore one is inclined to postulate an extracellular mechanism in which the free-radical cleavage and reformation of —Se—Se— bonds is likely to play a role. The sensitization is probably due to the formation, from irradiated colloidal Se, of short-lived species acting in processes occurring extracellularly.

It should be noted that the two inorganic secondary radicals $(CNSe)_2^{-}$[35] and SeO_3^{-}[36], derived from the radiation-induced oxidation of $CNSe^-$ and SeO_2, show on the inactivation of some enzymes a dose-modifying effect which is related to the specificity of attack of the secondary radical towards certain amino acids and to the structure of the protein itself[31,35,37] (see also Section II.B).

A series of 2-substituted selenoazolidines have been synthesized as potential radioprotective agents[38].

A comprehensive survey of the radioprotective activity reported so far for a variety of systems of different biological organization and end-groups is given in Table 1. With multicellular organisms only a few compounds have apparently been tested. However, one can try a tentative extrapolation to such organisms, based on the information concerning mechanisms of action obtained with molecular and single-cell systems.

B. Radiation Chemistry of Selenium Compounds

Understanding of the radiation chemistry of Se-containing compounds is especially important in order to explain their role in radioprotection and to permit a comparison with the S analogues. Identification of the transient species and radiolysis products arising from Se compounds and knowledge of the nature of their reactions with biologically interesting compounds should help to explain the radioprotection offered by these compounds.

All these studies are far from being exhaustive since the instability of many Se compounds with consequent analytical difficulties does not permit accurate experiments of steady-state radiolysis.

The most suitable techniques for studying such systems are pulse radiolysis and electron spin resonance spectroscopy and most papers deal with these subjects. In particular, pulse radiolysis uses a short intense pulse of radiation to induce the initial physicochemical damage and fast recording techniques (i.e. absorption kinetic spectrophotometry with oscillographic output) to investigate the short-lived chemical species produced and to follow their subsequent reaction pathway[40].

Unlike photochemistry, where the incident energy is absorbed by a particular chromophore, the absorption of ionizing radiation energy occurs non-selectively and, in dilute solution, exclusively by the solvent. The most important transient species formed in water or in aqueous solutions are OH radicals, H atoms and the hydrated electron (e_{aq}^-), which are the precursors of the subsequent chemical damage.

When aqueous solutions of Se-containing compounds are pulse-irradiated, an intense transient absorption with a maximum from 380 to 450 nm is produced (Figure 2). In all cases this absorption is attributed to radicals where Se atom (or atoms) is (or are) involved.

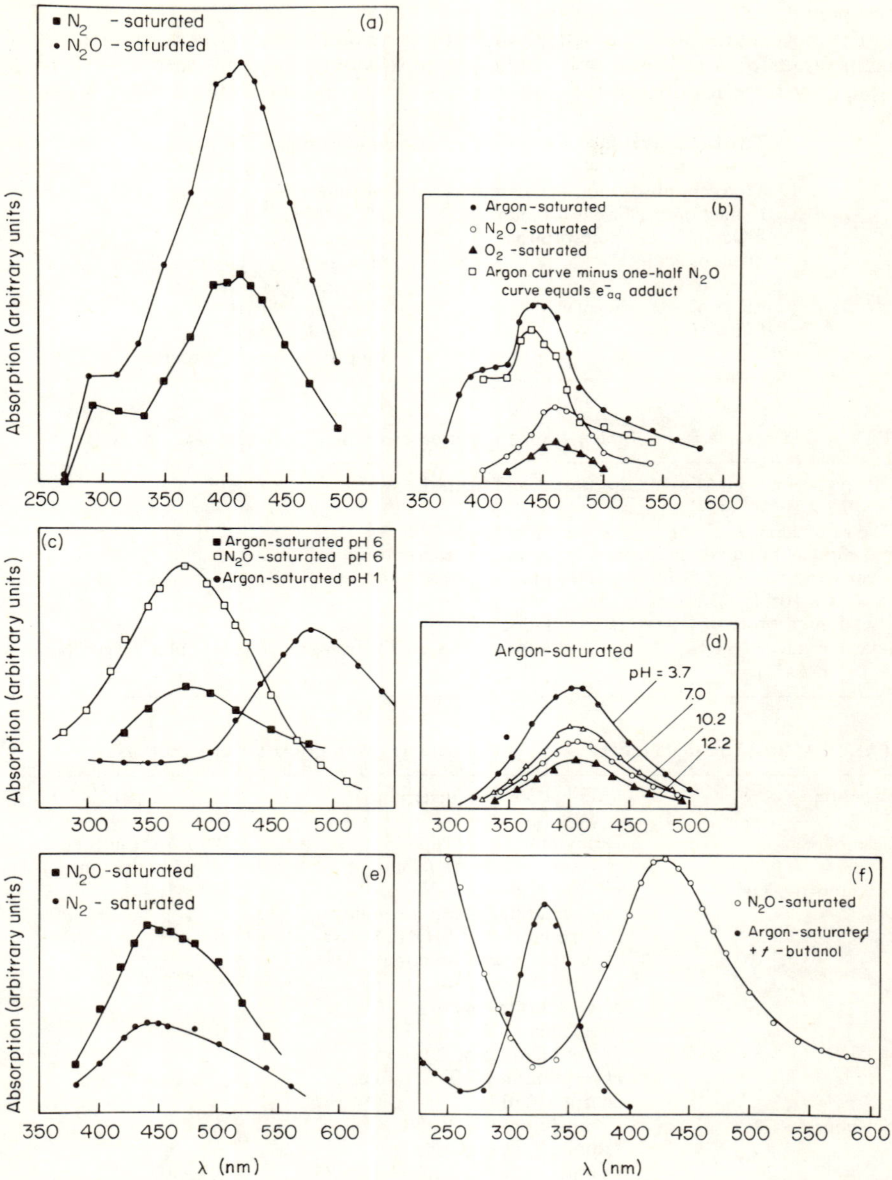

FIGURE 2. Transient absorption spectra from aqueous solutions containing seleno derivatives. (a) selenourea[41] (b) selenocystine[42] (c) selenomethionine[43] (d) selenidric acid[44] (e) selenocyanate[35] and (f) selenium dioxide[36]. A spectrum similar to (c) was obtained in the case of selenoethionine[45]. Reproduced with permission from the above references

Radicals which absorb in a similar spectral region have been identified for S-containing compounds[46].

The most studied and best understood is the radiation chemistry of selenourea[41]. The main results obtained from pulse radiolysis of selenourea are summarized in Table 2, which gives the features of the transient absorption spectra, Table 3 which gives the

TABLE 2. Transient spectra on irradiation of selenourea solutions

Transient absorption maximum	410 nm
Extinction coefficient (410 nm)	8.7×10^3
Effect of N_2O on absorption	Increased by a factor of 2
Effect of acid pH	Increased by a factor of 2
Effect of O_2	Reduced by a factor of 0.2
Effect of OH scavenger	Strongly reduced
Effect of solute	Not in accordance with a radical scavenging process

TABLE 3. Characteristics of the 410 nm absorption decay in selenourea pulse radiolysis

Always second order in N_2O-saturated and neutral solution ($k = 5.7 \times 10^9 \mathrm{M^{-1} s^{-1}}$)
The rate increases as the ionic strength increases
No effect of hundredfold increase in selenourea concentration
Nearly second-order in N_2-saturated and neutral solution ($k = 1.4 \times 10^{10} \mathrm{M^{-1} s^{-1}}$)
Pseudo-first-order in O_2-saturated solution. (average value of the bimolecular rate constant between the transient and molecular oxygen, ($k = 1.2 \times 10^8 \mathrm{M^{-1} s^{-1}}$)

TABLE 4. Rate constants for the reaction of selenourea with primary water radicals[a]

Reaction	Method used to determine k	$k(\mathrm{M^{-1} s^{-1}})$
Selenourea + e^-_{aq}	Decay of e^-_{aq} at 700 nm, in N_2-saturated, neutral solution	4.0×10^9 (independent of pH in the range 6–11)
Selenourea + OH	Competition with CNS^-, in N_2O-saturated, neutral solution	7.2×10^9
	Competition with EtOH in N_2O-saturated, neutral solution	6.8×10^9
	Competition with MeOH in N_2O-saturated, neutral solution	6.5×10^9
	Bleaching at 250 nm	6.9×10^9
	Direct build up of 410 nm absorption in N_2O-saturated, neutral solution	5.5×10^9
Selenourea + H	Competition with EtOH in N_2O-saturated, neutral solution	7.5×10^8
	Competition with MeOH in N_2O-saturated, neutral solution	6.4×10^8
	Direct build up of 410 nm absorption in N_2-saturated, acid and concentrated solution	6.3×10^8

[a]Reproduced with permission from Ref. 41.

characteristics of the transient absorption decay and Table 4 which gives the rate constants for the reaction of selenourea with primary water radicals.

From the above data and from the data in the tables both the OH radical and the hydrogen atom give rise to the same radical species when reacting with selenourea. The radical formed by this reaction with a strong absorption band at 410 nm is shown to be a charged complex involving two selenourea molecules.

The reaction involves the abstraction of one H atom from the —SeH group of the pseudoenol form of the compound, followed by the association with another molecule of selenourea in an equilibrium reaction (Scheme 1).

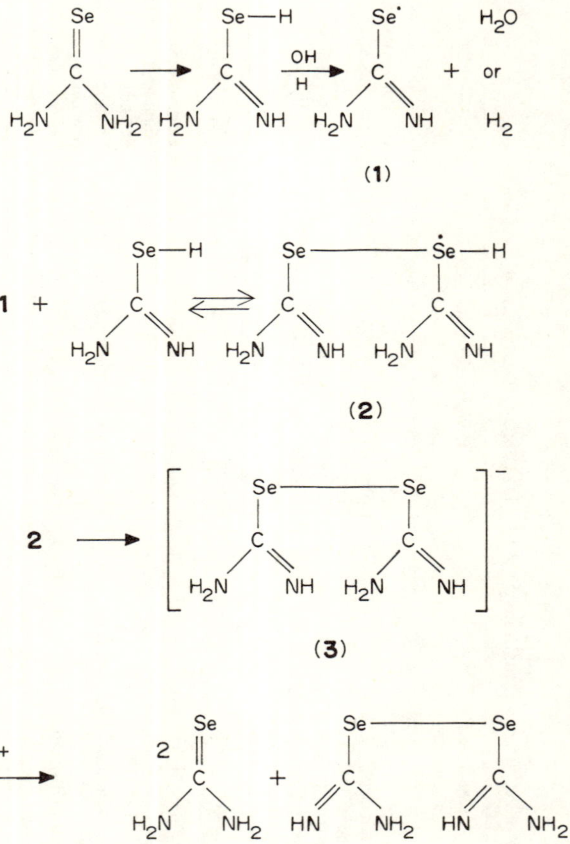

SCHEME 1

The radical species decays with the reformation of two selenourea molecules and the formation of the condensed dimer. The latter is very unstable and serves as a precursor of the red Se found under all experimental conditions.

The structure of α,α′-diselenobisformamidinium cation, derived from selenourea oxidation product has been determined by X-ray structural analysis[47].

Table 4 shows the high reactivity of selenourea towards the primary radicals of water radiolysis. Selenourea is superior to urea and thiourea in competing for these free radicals.

TABLE 5. Radiolytic products from γ-irradiation of 3×10^{-4} M selenocystine in aqueous solutions[a]

Dose (rads)	Deaerated solutions	Deaerated solutions + 0.1 M CH_3OH	N_2O-saturated solutions
1×10^5	G(—Se—cystine) = 0.65 Colloidal Se($G = 0.40$) Alanine ($G < 0.1$) Serine ($G < 0.01$) Selenocysteine seleninic acid (traces) Unidentified product containing Se	G(—Se—cystine) = 0.38 Colloidal Se($G = 0.20$) Alanine ($G = 0.24$) Selenocysteine seleninic acid (traces) Unidentified product containing Se	G(—Se—cystine) = 0.87 Colloidal Se($G = 0.70$) Alanine ($G < 0.1$) Serine ($G < 0.1$) Selenocysteine seleninic acid (traces) Three unidentified products containing Se
2×20^5	G(—Se—cystine) = 0.55 Colloidal Se ($G = 0.50$) Alanine ($G < 0.1$) Serine ($G < 0.01$) Selenocysteine seleninic acid (traces) Unidentified product containing Se	G(—Se—cystine) = 0.40 Colloidal Se ($G = 0.30$) Alanine ($G = 0.24$) — Selenocysteine seleninic acid (traces) Unidentified product containing Se	G(—Se—cystine) = 0.84 Colloidal Se ($G = 0.70$) Alanine ($G < 0.1$) Serine ($G < 0.1$) Selenocysteine seleninic acid (traces) Three unidentified products containing Se

[a] Reproduced with permission from Ref. 42.

The order of reactivity with radicals in the series of ureas seems to parallel their known radioprotective activity in biological systems. The ability of a secondary radical to form a relatively inert complex may also have some bearing on the protective efficiency of the parent molecule.

The high reactivity of selenourea with free radicals is also found in the irradiation of organic compounds in the solid state, where the presence of selenourea suppresses the formation of stable free radicals by an amount proportional to its concentration[29]. The fact that neutron activation has proved a good method for labelling selenourea with ^{75}Se with high yield[48], although the recoil of the (n, γ) reaction has sufficient energy to break the Se—C bond, reflects the efficiency of Se as a radical scavenger.

Gamma and pulse radiolysis of selenoamino acids have been reported and mechanisms of the radiation–degradation processes can be suggested.

Table 5 shows some radiolytic products derived from γ-irradiation of selenocystine[42,49]. The main product is elementary Se, which is produced with a yield proportional to selenocystine decomposition. Some amino acids not containing Se are present, as well as the oxidation product selenocysteine seleninic acid (**4**). Deamination and decarboxylation reactions occur but they have not been followed quantitatively.

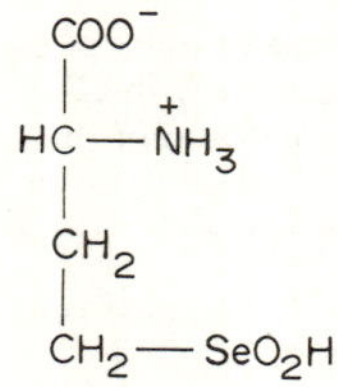

(**4**)

Tables 6 and 7 show some radiolytic products derived from γ-radiolysis of selenomethionine[50,51]. The main radiation–degradation reactions of selenomethionine include, in addition to decarboxylation and deamination, reactions leading to cleavage of the Se bonds accompanied by formation of volatile compounds, as well as recombination reactions, which prevail in a nitrogen atmosphere. In oxygen, oxidation reactions take place on Se, on the carbon skeleton and on the cleaved groups.

Some of these degradation products have been found, although to a lower extent, during the storage of radiopharmaceutical selenomethionine labelled with ^{75}Se, due to self-radiolytic processes[51,52].

The pulse radiolysis results demonstrate the high reactivity of selenocystine and selenomethionine with the hydrated electron and the hydroxyl radical. The rate constants (Table 8) are higher than for other aliphatic amino acids[53] and are of the same order of

TABLE 6. Radiolytic products from γ-irradiation of selenomethionine[50]

Deaerated solution	N_2O-saturated solution	O_2-saturated solution
G(—selenomethionine) = 1.4	G(—selenomethionine) = 5.2	G(—selenomethionine) = 5.7
Ammonia ($G = 0.52$)	Ammonia ($G = 1.2$)	Ammonia ($G = 1.5$)
Selenohomolanthionine	(No other product was tested)	Selenomethionine oxide
Selenohomocysteine		Homoserine
Me_2Se		Aspartic acid
Me_2Se_2		Methylselenous acids
		Selenous acids

TABLE 7. Radiolytic products from γ-irradiation of selenomethionine (1 Mrad)[51]

Deaerated solution	Air solution
Undestroyed selenomethionine (30%)	Undestroyed selenomethionine (30%)
Inorganic Se and highly oxidized selenomethionine (7.0%)	Inorganic Se and highly oxidized selenomethionine (15%)
Selenomethionine oxide (43%)	Selenomethionine oxide (11%)
Other decomposition products (5.0%)	Other decomposition products (negligible at this dose and about 10% at 2 Mrad)

TABLE 8. Rate constants for the reaction of e_{aq}^- and OH with selenoamino acids

Reaction	pH	Method used	$k(\text{M}^{-1}\,\text{s}^{-1})$	Ref.
Selenocystine + e_{aq}^-	6.0	Decay of e_{aq}^- at 700 nm	7.6×10^9	42
	11.5		3.9×10^9	42
Selenomethionine + e_{aq}^-	7.0	Decay of e_{aq}^- at 700 nm	1.8×10^9	43
Selenocystine + OH	7.0	Direct build-up of the transient absorption in N_2O	1.0×10^{10}	42
	7.0	Competition with CNS^- in N_2O	1.7×10^{10}	42
Selenomethionine + OH	7.0	Direct build-up of the transient absorption in N_2O	1.3×10^{10}	43

magnitude of those of thioamino acids[54]. The compound selenodicysteine shows similarly a high rate constant for scavenging OH radicals[55].

The product of the reaction of e_{aq}^- with selenomethionine has no significant absorption in the ultraviolet and visible region. The most likely reaction of e_{aq}^- with selenocystine occurs at the —Se—Se— group and gives rise to a transient absorption with a maximum at 400 nm. The reaction scheme is shown in Scheme 2[42].

$$\text{RSeSeR} + e_{eq}^- \longrightarrow \overset{-}{\text{RSeSeR}} \xrightarrow{\text{1st order}} \text{RSe}^{\bullet} + \text{RSe}^-$$

$$2\ \text{RSe}^{\bullet} \longrightarrow \text{RSeSeR} \xrightarrow{\text{OH}} \text{RSeOH} + \text{RSe}^{\bullet}$$

SCHEME 2

The electron adduct decays in an equilibrium reaction and the radical $\text{RSe}^{\bullet}$ is responsible for the absorption peak at 460 nm. A back-reaction should occur with reformation of selenocystine and this would explain the low values of G(—RSeSeR) in γ-radiolysis experiments.

As OH radicals also generate a transient absorption at 460 nm, a possible mechanism is indicated in Scheme 2. Such mechanism is in conformity with the γ-radiolysis experiments, as RSeOH could be the precursor of the Se analogue of cysteine sulphinic acid.

In the case of selenomethionine, the OH attack occurs at the Se atom with the formation of transient absorption at 380 nm.

The decay of this radical is complex due to the existence of concurrent reactions, and it requires a special kinetic treatment[56]. The exponential decay observed at low doses may be due to a unimolecular decomposition or to an internal rearrangement; the bimolecular reaction, observed at high doses, probably involves neutral radicals.

In conclusion, from the steady-state and pulse radiolysis data, the degradation of both selenoamino acids is mainly due to OH attack; meanwhile the e_{aq} are partially scavenged by Se atoms with less damage to the molecule.

TABLE 9. Radical yields by X-ray irradiation of thio- and seleno-amino acids at 295 K and 100 K[a]

Compound	295 K	100 K
Cysteine	0.3	1.1
Cystine	2.5	2.1
Methionine	1.0	1.7
Ethionine	1.4	0.8
Se-cystine	0.02	0.1
Se-methionine	0.3	0.2
Se-ethionine	0.4	0.1

[a]Reproduced with permission from Ref. 58.

The relative radiation stability of both selenoamino acids may be ascribed to a repair mechanism and to back-reactions during the irradiation process. Likewise in the case of selenourea steady-state radiolysis the dose dependence is initially linear with a sharp kink at the higher doses, attributed to a self-repair mechanism[29].

Of particular interest in this context are the comparative ESR studies by Colombetti and coworkers on the radiation resistance of thio- and seleno-amino acids in the solid state[57,58]. The formation and the stabilization of free radicals in X-ray irradiated thio- and seleno-amino acids have been investigated by ESR in an attempt to identify the main radical species present. A low free-radical yield was found in irradiated Se compounds, indicating the presence of repair mechanisms (Table 9). The radicals formed in these compounds probably react readily with H atoms or electrons, returning to the original molecule or forming non-radical species. Another possibility is that the absorbed energy is utilized to break bonds to Se, which are readily reformed.

In support of this hypothesis are some polarographic results[59] which demonstrate that the dissociation–formation rate of bonds to Se is higher than for the S bonds.

Comparison of some spectroscopic properties of selenoamino acids[60] with those of their thio analogues show important structural modifications which could be related to the different reactivity and radiation response of the compounds.

The substitution of S by Se results in a bathochromic shift of the UV absorption band and this is due to the higher excitability of the unbound electron pairs of the Se atom with respect to S. This should favour a greater chemical reactivity for selenoamino acids, as was indeed observed in pulse radiolysis for the process concerning the reactivity with water free radicals.

From IR spectra, the substitution of Se for S in methionine/selenomethionine results in a weakening of the C—C bonds of the side-chain and of the bond between the heteroatom and the Me group. The different radiation response of the two amino acids[50,51] may be explained in the light of these results.

The role of some Se-containing compounds in the transfer of excitation energy produced by ionizing radiation has also been investigated in order to give a more complete picture of radiation effects on such compounds.

In particular, the technique of pulse radiolysis has been applied to the study of luminescence phenomena of the radioprotective compound selenoxanthene in the crystalline form[61]. The spectral distributions of the radiation-induced luminescence have been studied as a function of time and temperature giving information about the disposition and the fate of excitation energy produced by ionizing radiation. Selenoxanthene is an extremely efficient luminescent material as may be expected from its aromatic structure. No fluorescent emission was found following pulse irradiation. Only a single sharp phosphorescence peak at 500 nm was observed. An oscillogram of the decay of

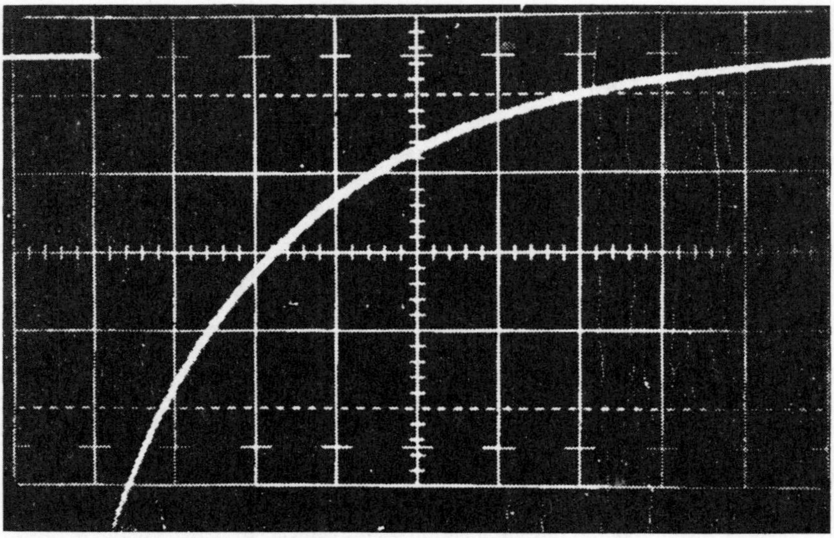

FIGURE 3. Decay of the phosphorescence from irradiated selenoxanthene at 490 nm. Sample temperature, 95 K, time-scale 50 $\mu s\,cm^{-1}$. Reproduced with permission from Ref. 61

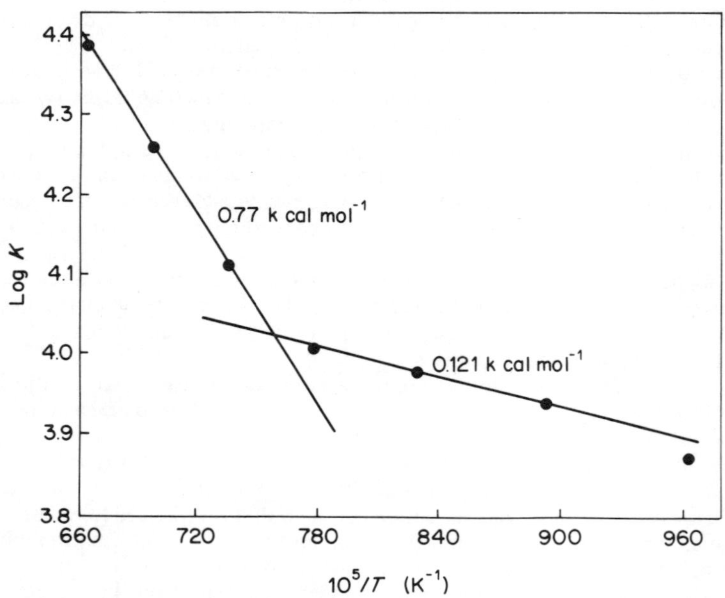

FIGURE 4. Arrhenius plot for the temperature-sensitive phosphorescence decay of selenoxanthene emission. Reproduced with permission from Ref. 61

this phosphorescence is shown in Figure 3. Below 150 K this decay is exponential but the half-life is strongly dependent on the temperature of the sample. The emission decay rate as a function of temperature can be plotted in the form of an Arrhenius plot as shown in Figure 4.

It is obvious from these data that there must be at least two temperature-sensitive quenching reactions in competition with the phosphorescence decay. The activation energies for these two quenching reactions were calculated to be 0.77 and 0.12 kcal mol^{-1}.

The fluorescent efficiencies of compounds excited by ionizing radiation are in the order xanthene > thioxanthene > selenoxanthene, with marked differences between them, whereas the phosphorescent efficiences are in the reverse order. The luminescence yields, defined as the proportion of the absorbed energy which is reemitted, are shown in Table 10.

The enhanced intersystem crossing within the molecule, promoted by the high atomic number of Se, could explain why at low temperature the total luminescence yields of selenoxanthene and xanthene are within an order of magnitude but the emission from xanthene is 99% fluorescence. At room temperature, the total luminescence yields of xanthene and thioxanthene are reduced slightly, while the phosphorescence yield of selenoxanthene is dramatically decreased. As intersystem crossing is very fast compared with the lifetime of the triplet state the transfer of energy to the triplet state probably still occurs but it is followed by dissipation of the energy before light emission. At room temperature the triplet energy level of selenoxanthene is also lower than that of thioxanthene and many other organic molecules of biological interest (i.e. nucleic acids). It is possible that the action of Se in radioprotection is to promote the transfer of electronic excitation energy away from the initial site of energy absorption to the n,π^* triplet states associated with the Se atom. On the basis of this hypothesis, the molecule containing the Se could act as an innocuous 'energy sink' in some vital part of the cell.

The ESR technique has contributed to the study of radiation damage in Se-containing compounds, giving information on the structure of radicals trapped in irradiated molecules.

Geoffroy studied the irradiation and ESR analysis of single crystals of Ph_3SeCl[62] and of diphenylselenone (Ph_2SeO_2)[63]. The identification of the radicals was made by comparison with the homologous thio radicals, and the radiation mechanism aspects were discussed.

Some radiation-induced Se-containing free radicals have found applications in biochemical problems. Radiation-induced oxidizing free radicals have been introduced by Adams and coworkers[64,65] and are now used extensively as selective probes in identifying sites of radical attack at essential amino acid residues in proteins[66].

The primary radicals formed in water radiolysis, e^-_{aq}, OH and H, damage biomolecules, and enzymes in particular, in a non-specific way, because these species can attack many sites of the molecule. In contrast, secondary radicals, derived from the introduction of

TABLE 10. Luminescence yields at 95 and 295 K[a]

Compound	Fluorescence		Phosphorescence	
	95 K	295 K	95 K	295 K
Xanthene	6×10^{-3} (2×10^{-5})	2×10^{-3}	3.5×10^{-5}	—
Thioxanthene	2.5×10^{-3} (2.8×10^{-4})	1.8×10^{-3} (1.5×10^{-4})	4.4×10^{-3}	4.1×10^{-5}
Selenoxanthene	—	—	3.0×10^{-2}	1.0×10^{-5}

[a]Reproduced with permission from Ref. 61.

TABLE 11. Absorption maxima and bimolecular rate constants, k, with amino acids for Se-containing secondary radicals

		$k(\mathrm{M}^{-1}\mathrm{s}^{-1})$					
Radical anion	λ_{max}(nm)	Cysteine	Methionine	Tyrosine	Tryptophan	Histidine	Ref.
$(CNSe)_2^{-}$	440	6.8×10^7	$< 1 \times 10^7$	1×10^7	1×10^7	$< 1 \times 10^7$	35
$SeO_3^{\cdot-}$	430	—	1.2×10^8	1.1×10^9	3.3×10^9	4.3×10^7	36,37

another solute into the solution which competes as a radical scavenger, can attack specific functional groups and could be, therefore, useful reagents for the identification of residues related to the biological activity.

The secondary radicals of different redox properties derived from halide or pseudo-halide ions, or from carbonate and selenate ions, have been shown to be of potential interest as selective reagents. These ions react with OH quantitatively and rapidly according to the reactions:

$$X^- + OH^{\bullet} \longrightarrow X^{\bullet} + OH^- \qquad X^- = Cl^-, Br^-, I^-, CNS^-, CNSe^-$$

$$X^{\bullet} + X^- \rightleftharpoons X_2^-$$

$$Y^{2-} + OH^{\bullet} \longrightarrow Y^- + OH^- \qquad Y^{2-} = CO_3^{2-}, SeO_3^{2-}$$

The secondary radicals derived from the pulse radiolysis of selenocyanate[35] and selenium dioxide[36] answer the need for a good selective secondary radical. Table 11 shows the absorption maximum for each radical and the bimolecular rate constants with some amino acids. Both radicals present strong transient absorption spectra, so their reactions can be observed directly. They react rapidly with amino acids containing residues involved in the composition of the active site of enzymes.

Some of the rate constants are pH-dependent due to the ionic equilibria in the particular amino acids and the solute system[36]. Furthermore, the absorption spectra of the reaction products between the secondary radicals and the amino acids studied by means of pulse radiolysis are typical and characteristic of the precise amino acid involved in the reaction.

The combination of pulse radiolysis data, i.e. the rate constants and the spectral characteristics, and the data derived from the assay of biological activity measured under comparable conditions, can give useful information on the composition of the active site of simple and complex enzymes.

The effects of $SeCN^{\cdot-}$ [35] and $SeO_3^{\cdot-}$ [37] on the inactivation of some enzymes (ribonuclease, lysozime, α-chymotrypsin, alcohol dehydrogenases, etc.) are consistent with the present knowledge of the structure of the catalytic site and of the crucial role played by the amino acid residues.

III. TELLURIUM COMPOUNDS

The radiation chemistry of inorganic Te compounds has mainly been studied by Haissinsky and coworkers[67–69]. They have described some results on the γ-radiolysis of acid solutions of Te(IV) and Te(VI). The apparent existence of a 'quasiequilibrium' between these two valency states is pointed out and a mechanism of radiolytic oxidation of Te(IV) and reduction of Te(VI) is proposed.

The rate constants of the reactions: (1) Te(IV) + OH and (2) Te(VI) + H have been determined by competition with the $H_2O_2 + OH$ and $H_2O_2 + H$ reactions: $k_1 = 4.7 \times$

$10^6 M^{-1}s^{-1}$ and $k_2 = 1.05 \times 10^8 M^{-1}s^{-1}$, respectively. In deaerated solutions neither the oxidation nor the reduction is complete, but after a sufficient irradiation a 'quasiequilibrium' state is apparently established. It is shown that this is due to competition between the Te compounds and the stable radiolytic products (H_2O_2, O_2,H_2) for the free OH radicals and the H atoms.

Other detailed studies on the radiation-induced reactions of Te compounds have been performed by a Japanese group[70,71]. γ-ray-induced reactions of Te(IV) and Te(VI) in sulphuric acid solutions have been investigated by using double tracers for the element[71]. The results are interpreted on the basis of reactions of solutes with H, OH, H_2 and H_2O_2 which are the primary species in acid solutions. The direct absorption of γ-rays in H_2SO_4 is neglected as the concentration of H_2SO_4 is much lower than that of water. The experimental data support a sequence in which the unstable valence states Te(III) and Te(V) play an important role in the reaction mechanisms.

The radiation damage of solid $TeCl_4$ irradiated by fast protons (600 MeV) was investigated by the use of the combined methods of microfiltration and neutron activation analysis[72]. The results indicate that the radiolytic decomposition is negligible compared to other types of decomposition (i.e. thermal effects). The final products of any decomposition of solid $TeCl_4$ may be $TeCl_2$, elementary Cl and elementary Te. As the dichloride probably does not exist in the solid state[73], the amount of elementary Te in the tetrachloride is a measure of the degree of decomposition of this compound. An estimation of the pure radiolytic decomposition shows good agreement with the experimental data.

The organotellurium compounds are of potential importance in the chemistry of hot atoms. Te has several radioisotopes, which can be prepared artificially and can be utilized to study the chemical effects associated with nuclear transformations such as (n,γ) reactions on ^{126}Te, ^{128}Te and ^{130}Te, isomeric transitions (^{129}Te m $\rightarrow$ ^{129}Te) and β-disintegration of ^{131}Te m, ^{131}Te and ^{132}Te.

It is well known that nuclear processes are associated with radiolysis of the target during irradiation inside the reactor or with the effect of rays emitted during the nuclear disintegration or the nuclear deexcitation.

These facts imply the necessity of a good knowledge of the radiation resistance of the organotellurium compounds.

Studies of this type have been carried out by Llabador [74,75].

The role of the energy and charge-transfer processes which occur during the reactions of ^{131}I formed by β–γ decay of ^{132}Te in mixtures of ^{132}Te-labelled and unlabelled diphenyltellurium (Ph_2Te) and dibutyltellurium (Bu_2Te) has been determined[74]. The results show that the radiolysis of the organotellurium molecules in the autoradiation zone is mainly due to energy-transfer processes involving highly excited states.

The products formed during the γ-radiolysis of Ph_2Te, in the pure state and in benzene solution, have been analysed by gas chromatography[75]. In the pure compound, the main products are benzene, biphenyl and diphenylditelluride. The yield of the latter decreases markedly in the presence of oxygen. In no case is elementary Te found. The proposed primary process involves the rupture of a C—Te bond.

IV. CONCLUDING REMARKS

The radiation chemistry of Se-containing compounds has been much more investigated than that of Te-containing compounds. This fact is explained by the possible role played by Se compounds in chemical radioprotection, and by the interest in development of drugs that, when given before exposure to lethal ionizing radiation, can prolong the life of the irradiated organism. Se-containing compounds are certainly not the best and most versatile radioprotective agents. Their toxicity and chemical instability do not permit their practical application. The best and most effective antiradiation drugs are probably those containing S in their molecules.

However, Se compounds do show a type of radioprotection which could contribute to an understanding of phenomena of radiobiological interest. These compounds exhibit radioprotective ability in many chemical and biological systems, but it is not known whether they act by the same mechanism at all levels of organization.

Since free-radical mechanisms involving transient species from Se compounds acting on chemical or biological structures are important in the radiobiological phenomena, the radiochemical investigations have contributed to the elucidation of these mechanisms. In particular the studies on the pulse radiolysis of some Se-containing organic molecules have shown the chemical nature and reactivity of free radicals which could be active in the radioprotective events. However, our knowledge of the radiobiology and radiochemistry of Se compounds is quite incomplete and most of the different aspects of the field are still wide open to investigation.

V. REFERENCES

1. A. Pihl and T. Sanner, *Radiation Protection and Sensitization* (Eds. H. L. Morson and M. Quintiliani), Taylor and Francis, London, 1970, p. 43.
2. I. Rosenfeld and O. A. Beath, *Selenium Geobotany, Biochemistry, Toxicity and Nutrition*, Academic Press, New York, 1965.
3. D. L. Klayman and W. H. H. Günther (Eds.), *Organic Selenium Compounds: Their Chemistry and Biology*, John Wiley and Sons, New York, 1973.
4. M. Schmidt, W. Siebert and K. W. Bagnall, *The Chemistry of Sulphur, Selenium, Tellurium and Polonium*, Pergamon Press, Oxford, 1973.
5. T. C. Stadman, *Science*, **183**, 915 (1974).
6. Z. M. Bacq, C. Onkelinx and G. Barac, *Compt. Rend. Soc. Biol.*, **157**, 899 (1963).
7. F. Shimazu and A. L. Tappel, *Experientia*, **143**, 369 (1964).
8. F. Shimazu and A. L. Tappel, *Radiat. Res.*, **23**, 210 (1964).
9. K. A. Caldwell and A. L. Tappel, *Arch. Biochem. Biphys.*, **112**, 196 (1965).
10. R. C. Dickson and A. L. Tappel, *Arch. Biochem. Biophys.*, **131**, 100 (1969).
11. D. L. Klayman, *J. Org. Chem.*, **30**, 2454 (1965).
12. W. H. H. Günther and H. G. Mautner, *J. Amer. Chem. Soc.*, **82**, 2762 (1960).
13. S. H. Chu and H. G. Mautner, *J. Org. Chem.*, **27**, 2899 (1962).
14. R. Badiello, A. Trenta, M. Mattii and S. Moretti, *Med. Nucl. Radiobiol. Latina*, **10**, 1, (1967).
15. A. Breccia, R. Badiello, A. Trenta and M. Mattii, *Radiat. Res.*, **38**, 483 (1969).
16. R. Badiello, E. Gattavecchia, M. Mattii and M. Tamba, *Z. Naturforsch.*, **29c**, 647 (1974).
17. A. Breccia, A. Trenta, R. Badiello, S. Moretti and M. Mattii, *Experientia*, **22**, 475 (1966).
18. R. Badiello A. Trenta, M. Mattii and S. Moretti, *Ric. Sci.*, **37**, 452 (1967).
19. R. Badiello, M. Mattii and E. Cecchetti, *Med. Nucl. Radiobiol. Latina*, **10**, 95 (1967).
20. M. Mattii, R. Badiello and E. Cecchettii, *Radiobiol. Radioter. Fis. Med.*, **22**, 56 (1967).
21. Z. M. Hollo and S. Zlatarov, *Naturwissenschaften*, **47**, 328 (1960).
22. J. Aleksandrowicz and M. Wazewsda-Czyzewska, personal communication.
23. J. Aleksandrowicz, J. Dobrowolski, J. Lisiewicz, B. Smyk and A. Skotnicki, *Pol. Arch. Med. Wewn.*, **53**, 209 (1975).
24. G. B. Abdullaev, G. G. Gasanov, M. A. Mekthiev, S. T. Sadykhov and A. I. Dzafarov, *Selen v biologii* (Ed. G. G. Gasanov) Baku, Elm, 1974.
25. I. Kozak, L. Kronrad and Z. Dienstbier, *Strahlentherapie*, **150**, 539 (1975).
26. I. Kozak, L. Kronrad and Z. Dienstbier, *Strahlentherapie*, **150**, 590 (1975).
27. W. Brucker and E. Bulka, *Studia Biophys.*, **1**, 253 (1966).
28. W. Brucker and H. G. Rohde, *Pharmazie*, **23**, 310 (1968).
29. R. Badiello and A. Breccia, *Radiation Protection and Sensitization* (Eds. H. L. Moroson and M. Quintiliani), Taylor and Franics, London, 1970, p. 103.
30. R. Badiello and E. M. Fielden, Ref. 29, p. 109.
31. R. Badiello, M. D. Solenghi and M. Tamba, *Ital. J. Biochem.*, **26**, 255 (1977).

32. R. Badiello, O. Sapora, G. Simone and M. Tamba, *Ann. Ist. Super. San.*, **10**, 147 (1974).
33. R. Badiello, D. Di Maggio, M. Quintiliani and O. Sapora, *Int. J. Radiat. Biol.*, **20**, 61 (1971).
34. A. Breccia, C. Rimondi and G. E. Adams (Eds.), *Radiosensitizers of Hypoxic Cells*, Elsevier/North Holland Biomedical Press, Amsterdam, 1979.
35. R. Badiello and M. Tamba, *Radiochem. Radioanal. Letters*, **37**, 165 (1979).
36. M. Tamba and R. Badiello, *Int. J. Radiat. Phys. Chem.*, **10**, 238 (1977).
37. R. Badiello and M. Tamba, 12th Annual Meeting of the European Society for Radiation Biology, Budapest, 19–22 October, 1976.
38. T. Sakesh and G. Shtacher, IV Congrès Intern. de Radiobiologie et de Physicochemie des Rayonnements, Evian (France) 29 June-4 July, 1970, Abstract No. 733.
39. D. L. Klayman, M. M. Grenan and D. P. Jacobus, *J. Med. Chem.*, **12**, 510 (1969).
40. R. Badiello and A. Breccia, *Chim. Ind.*, **57**, 525 (1975).
41. R. Badiello and E. M. Fielden, *Int. J. Radiat. Biol.*, **17**, 1 (1970).
42. M. Tamba and R. Badiello, *Int. J. Radiat. Biol.*, **23**, 435 (1973).
43. M. Tamba, S. Bonora and R. Badiello, *Z. Naturforsch.*, **29b**, 571 (1974).
44. M. Schöneshöfer, W. Karmann and A. Henglein, *Int. J. Radiat. Phys. Chem.*, **1**, 407 (1969).
45. R. Badiello and M. Tamba, unpublished data.
46. G. E. Adams, *Radiation Chemistry of Aqueous Solutions* (Ed. G. Stein), The Weizmann Science Press of Israel, Jerusalem, 1968, p. 244.
47. A. Chiesi, G. Grossoni, M. Nardelli and M. E. Vidoni, *Chem. Commun.*, **261**, 404 (1969).
48. R. Badiello and A. Breccia, *Ric. Sci.*, **36**, 335 (1966).
49. R. Badiello and M. Tamba, *Studia Biophysica*, **29**, 193 (1971).
50. J. Kopoldova and L. Nemcova, *Collect. Czech. Chem. Commun.*, **37**, 1089 (1972).
51. J. Galateanu, V. V. Lungu and D. Viorel, *J. Radioanal. Chem.*, **33**, 273 (1976).
52. M. Besnard, P. Frapart and Y. Cohen, *Int. J. Appl. Radiat. Isotop.*, **22**, 469 (1971).
53. P. Neta, M. Simic and E. Hayon, *J. Phys. Chem.*, **74**, 1214 (1970).
54. G. E. Adams, *Quad. Area Ric. Emilia-Romagna*, **2**, 1 (1972).
55. M. Czauderna and K. Samochocka, *Radiochem. Radioanal. Letters*, **42**, 129 (1980).
56. S. Bonora, R. Badiello and M. Tamba, IV National Congress of Radiochemistry, Radiation Chemistry, Nuclear Chemistry, Padova, 14–17 June, 1982.
57. G. Colombetti and G. Falconi, *Nuovo Cimento Lett.*, **2**, 127 (1969).
58. G. Colombetti and S. Monti, *Int. J. Radiat. Biol.*, **21**, 321 (1972).
59. B. Nygärd, *Arch. Kem.*, **27**, 341 (1967).
60. M. Tamba, B. Righetti and R. Badiello, *Atti Istit. Ven. Sci. Lett. Arti*, **31**, 105 (1973).
61. R. Badiello, E. M. Fielden and S. C. Lillicrap, *Int. J. Radiat. Phys. Chem.*, **5**, 173 (1973).
62. M. Geoffroy, *J. Chem. Phys.*, **70**, 1497 (1979).
63. M. Geoffroy and N. Leray, *J. Chem. Phys.*, **72**, 775 (1980).
64. G. E. Adams, J. E. Aldrich, R. H. Bisby, R. B. Cundall, J. L. Redpath and R. L. Willson, *Radiat. Res.*, **49**, 278 (1972).
65. G. E. Adams, R. H. Bisby, R. B. Cundall, J. L. Redpath and R. L. Willson, *Radiat. Res.*, **49**, 290 (1972).
66. R. Badiello and M. Tamba, *LAB J. Res. Lab. Med.*, **5**, 597 (1978).
67. M. Haissinsky, G. Lunde and P. Patigny, *J. Chem. Phys.*, **63**, 1093 (1966).
68. M. Haissinsky, *The Chemistry of Ionization and Excitation* (Eds. G. R. A. Johnson and G. Scholes), Taylor and Francis, London, 1967, p. 277.
69. M. Haissinsky, *J. Chem. Phys.*, **65**, 1386 (1968).
70. T. Nishi, I. Fujiwara and H. Moriyama, *Bull. Inst. Chem. Res., Ktoto Univ.*, **55**, 1 (1977).
71. H. Moriyama, I. Fujiwara and T. Nishi, *J. Phys. Chem.*, **84**, 1801 (1980).
72. G. C. Christensen and J. Alstad, *Radiochem. Radioanal. Letters*, **13**, 235 (1973).
73. G. C. Christensen and J. Alstad, *Radiochem. Radioanal. Letters*, **13**, 227 (1973).
74. Y. Llabador, *Radiochim. Acta*, **14**, 100 (1970).
75. Y. Llabador, *Radiochim. Acta*, **17**, 202 (1972).

The Chemistry of Organic
Selenium and Tellurium
Compounds Volume 1
Edited by S. Patai and Z. Rappoport
© 1986 John Wiley & Sons Ltd.

CHAPTER **9**

Selenium-stabilized carbenium ions and free radicals

LÁSZLÓ HEVESI

Department of Chemistry, Facultés Universitaires Notre-Dame de la Paix, 61, Rue de Bruxelles, B-5000 Namur, Belgium

I. INTRODUCTION. 307
II. SELENIUM-STABILIZED CARBENIUM IONS 308
A. Cyclic Selenocarbenium Ions 308
1. 1,2-Diselenolylium ions 308
2. 1,3-Diselenol-2-ylium ions 309
3. 1,3-Diselenolan-2-ylium and 1,3-thiaselenolan-2-ylium ions. . . . 312
4. Miscellaneous cationic systems. 314
B. Aliphatic Selenocarbenium Ions 315
1. Monoselenocarbenium ions 315
2. Bis(seleno)carbenium ions 320
3. 1-Seleno- and 2-seleno-allyl cations 321
III. α-SELENO FREE RADICALS 321
IV. REFERENCES 328

I. INTRODUCTION

Since the discovery of selenium by Berzelius some 150 years ago its organic chemistry has developed considerably, especially in the years 1920–1950 and during the last decade[1,2]. It has been recognized only recently that divalent Se is able to stabilize positively charged, negatively charged and radical centres located at the α-position. Therefore, selenium closely resembles in this respect its congener sulphur. While α-selenocarbanions have already proved to be extremely useful synthetic intermediates, the use of their cationic and radical counterparts is currently under investigation. The aim of this chapter is to give an overview of what is presently known about Se-stabilized carbenium ions and radicals.

II. SELENIUM-STABILIZED CARBENIUM IONS

A. Cyclic Selenocarbenium Ions

Besides being cyclic all these cations have the common feature of containing additional structural features which contribute to their stabilization. Some of them are aromatic in the sense that they formally possess a delocalized π sextet; others bear substituents such as $—NH_2$ or $—NMe_2$.

1. 1,2-Diselenolylium ions

The 3,5-diamino-1,2-diselenolylium ion (**1**) was described first[3] in this series. It can be prepared in 76% yield by iodine or iron (III) chloride oxidation of diselenomalonamide[4] in ethanol solution (equation 1) and isolated as a yellow crystalline solid. Both the chloride and iodide salts are so stable that they decompose (without melting) only when heated to 200 °C. 3,5-Dimethyl-1,2-diselenolylium salts (**2**) have also been prepared by a completely different method[5,6] (equation 2).

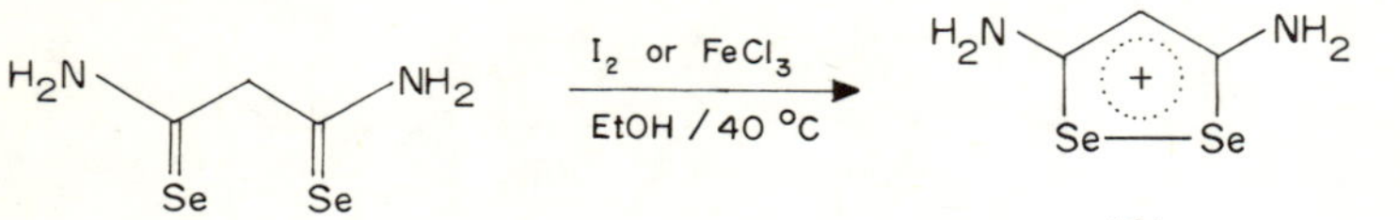

(1)

(**1**)

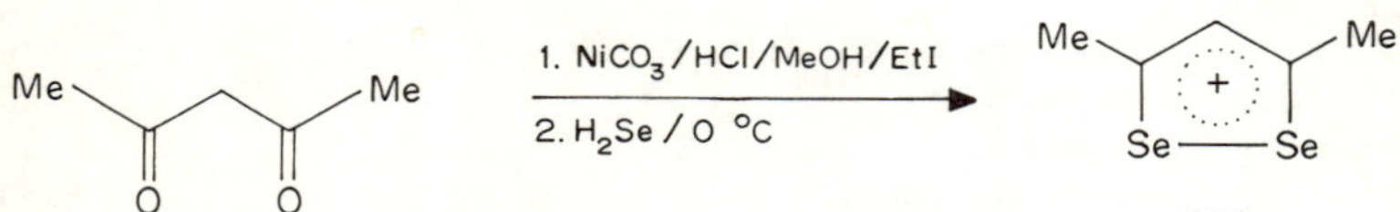

(2)

(**2**)

The close similarity of the spectral properties of cations **1** and **2** to those of the corresponding 1,2-dithiolylium systems[3–7] led the authors to the conclusion that the two types of cations must also have similar structures. Table 1 illustrates UV and a few of the IR absorptions of **2** as compared to those of the 3,5-dimethyl-1,2-dithiolylium ion (**3**).

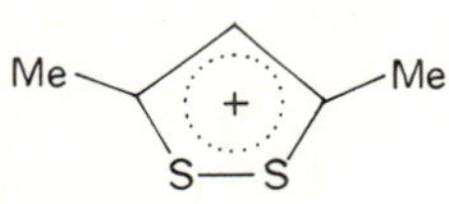

(**3**)

The crystal structures of several differently substituted 1,2-dithiolylium cations have been determined [8–11]. These ions all appear to be planar with all the ring-forming bonds being intermediate between single and double bond lengths[11,12]. The close structural similarity between **2** and **3** suggests therefore that 1,2-diselenolylium ions also have delocalized 6 π electron structure. Presumably, the relatively high thermal stability of **1** and **2** as well as their inertness towards halide and alcohol nucleophiles are also due to this aromatic character. It is worth noting in this connection that non-empirical calculations using linear combinations of gaussian orbitals on the parent 1,2-dithiolylium cation[13] have shown that almost all the positive charge is shared by the S and H atoms and quite

TABLE 1. IR and UV absorptions of **2** and **3**

IR[5,7]				UV[a 6]			
3		**2**		**3**		**2**	
$\bar{\nu}$(cm^{-1})	Assignment	$\bar{\nu}$(cm^{-1})	Assignment	λ_{max}	log ε	λ_{max}	log ε
1476	ν(C—C)	1465	ν(C—C)				
1318	ν(C—C)	1340	ν(C—C)	265	3.89	300	3.75
1232	δ(C—H)	1240	δ(C—H)				
1204	ν(C—CH_3)	1170	ν(C—CH_3)	288	4.00	320	3.81
1092	ν(C—CH_3) + ν(C—S)	1000	ν(C—CH_3) + ν(C—Se)				
697	ν(C—S)	600	ν(C—Se)				

[a] In 1M HCl solution.

surprisingly, all the ring C atoms bear net negative charges. Again, if a similar situation holds for 1,2-diselenolylium systems, it explains their lack of reactivity towards common nucleophiles.

2. 1,3-Diselenol-2-ylium ions

Two methods are now available for the preparation of 1,3-diselenolylium salts (**4**). Meier and Menzel[14] obtained a precursor of **4** by photolysis of 4-carboethoxy-1,2,3-

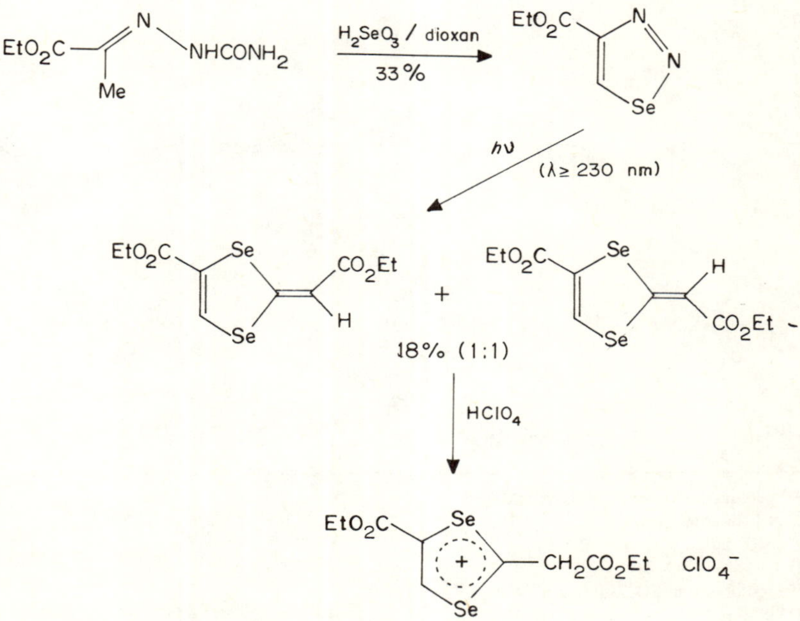

(**4**)

SCHEME 1

TABLE 2. Selected ^{1}H-NMR data for 1,3-dichalcogenolylium ions

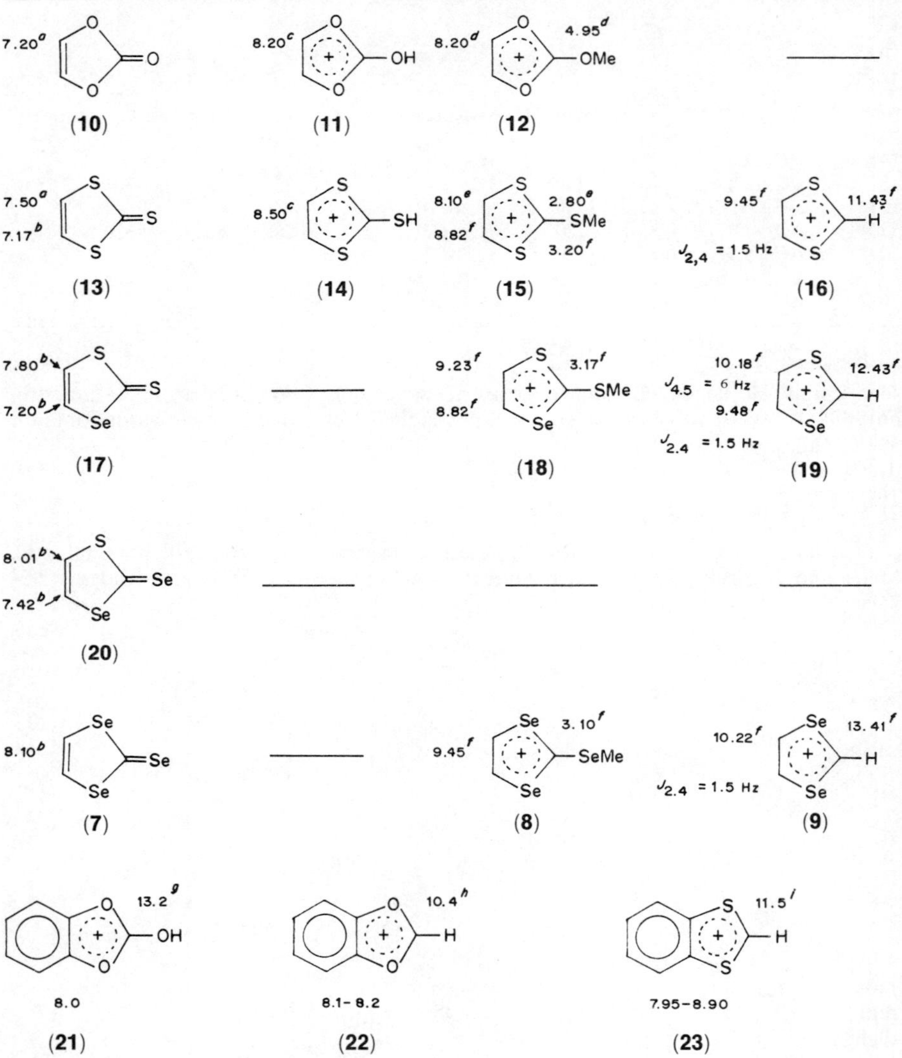

[a] In SO_2 solution, external (capillary) $SiMe_4$.[18]
[b] In $CDCl_3$ internal $SiMe_4$[16].
[c] In FSO_3H–SbF_5–SO_2 at −60 °C[18].
[d] In H_3CF–SbF_5–SO_2 at −40 °C[18].
[e] Stable iodide salt in SO_2 solution at −40 °C[18].
[f] In DMSO-d_6[16]
[g] In FSO_3H–SbF_5–SO_2ClF at −80 °C[18].
[h] In FSO_3H–SO_2 at −70 °C[18].
[i] In CD_3CN at room temperature[18].

selenadiazole; the sequence is outlined in Scheme 1. However, perchlorate **4** decomposed during NMR measurements. Changing the protonation medium ($CDCl_3$/CF_3COOH instead of $HClO_4$), as well as the ring substituents, has the effect of slowing down the decomposition rate sufficiently to permit NMR measurements. Thus the trifluoroacetate salts of 2-cyclopentyl-4,5-tetramethylene-1,3-diselenolylium (**5**) and 2-benzyl-4-phenyl-1,3-diselenolylium (**6**) ions could be observed[15] in solution.

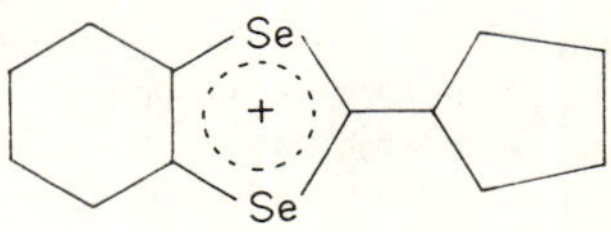

(**5**)

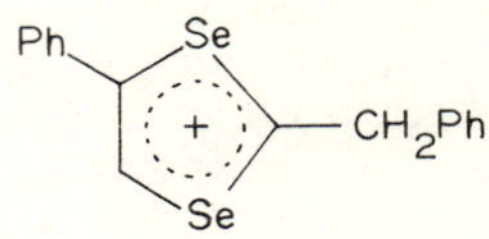

(**6**)

The method of Engler and Patel[16] is more versatile with regard to the nature of the heteroatoms (S, Se or Se, Se) present in the five-membered ring. However, it is less flexible with regard to the substituents on C(4) or C(5) (Scheme 2). Starting from 1,3-diselenole-2-selone (**7**) or from 1,3-thiaselenole-2-selone (or from 2-thione)[17], 1,3-diselenolylium (**9**) or 1,3-thiaselenolylium salts can be obtained. These compounds appear to be more stable thermally than compounds **4**, **5** and **6**.

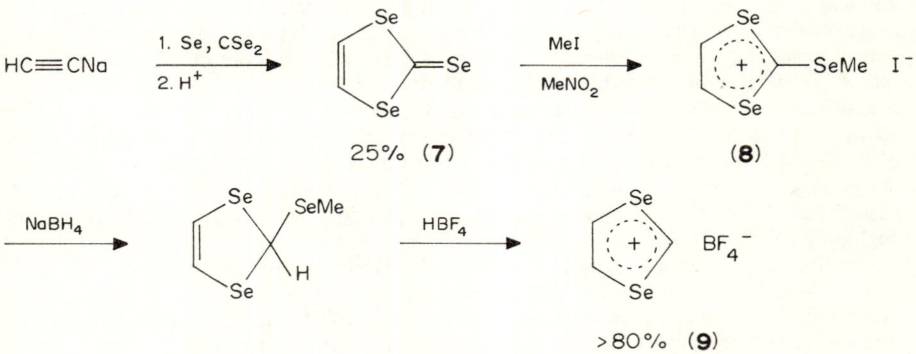

SCHEME 2

Although UV and IR absorptions have been provided[16] for some of the compounds, the most relevant information about their electronic structure has been obtained from ^1H- and ^{13}C-NMR data[15,18]. Proton chemical shifts are available for a fair number of 1,3-dichalcogenolylium systems (Table 2). Direct comparisons are however difficult, not only because data for 1,3-dioxol-2-ylium ions are missing, but also because the solvent effects are in many cases too large to be ignored. The most striking examples of this are the 0.33 ppm difference for the vinyl protons of **13** in SO_2 or $CDCl_3$ solution and the 0.72 ppm difference for the vinyl protons of **15** in SO_2 at $-40\,°C$ or in DMSO-d_6 at room temperature. Nevertheless, **16** and **9** have been measured in the same solvent and the vinyl and methine protons are shifted downfield by 0.77 ppm and 2 ppm, respectively on going from **16** to **9**. It has been concluded[16] that the greater displacement to low field of the methine proton is a consequence of more positive charge being on C(2) in **9** than in **16**, and that resonance structures **B** and **C** are less important than **A** in the case of **9** compared with **16**. The picture is less clear however, if one compares other pairs of data in Table 2.

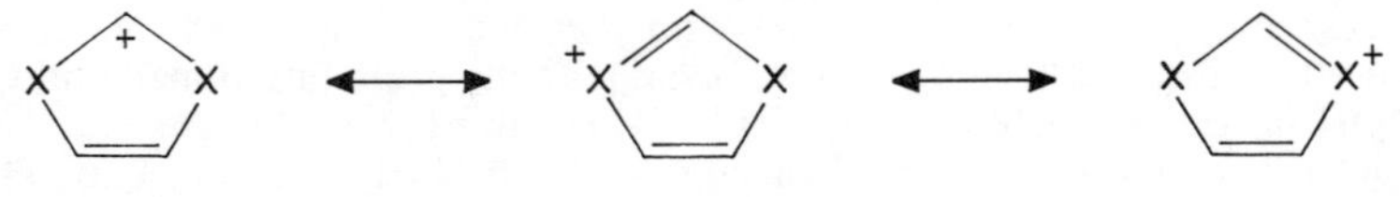

A B C

Conversion of vinylene carbonate (**10**) to the cation **12** brings about a downfield shift of 1.0 ppm for the vinyl protons. The same transformation results in a downfield shift of ca. 1.6 ppm for the change from **13** to **15** and of ca. 1.35 ppm for the change from **7** to **8**. Even taking into account the important solvent effects, it seems difficult to rationalize the same sort of downfield shift for **8** in terms of dominant resonance structure **A**. Also, comparing the methine proton chemical shifts of **22** and **23** and those of **16**, **19** and **9** it is intriguing to see that replacement of two O atoms by two S atoms brings about the same difference of ca. 1 ppm in the chemical shift as does the replacement of one S atom by Se.

It is not our purpose to deny the better positive-charge-stablizing ability of S compared with Se, but we feel that the existing data do not exclude the simultaneous operation of rather significant ring current effects through the O, S and Se analogues.

On the other hand, Table 3 presents some pertinent ^{13}C-NMR data for 1,3-dioxol-2-ylium and 1,3-diselenol-2-ylium ions. With the exception of **11**, **12**, **14** and **15** there is a general trend of increasing chemical shifts for the C(2) carbon atoms on going from O to S to Se derivatives. Thus one finds a $\Delta\delta$ of 14.5 ppm between **22** and **23**, 26.3 ppm between **25** and **5** and 24.2 ppm between **28** and **6**. Increasing C(2) chemical shifts point to more and more localized and increasing positive charge on that carbon and one could therefore conclude not only that the aromatic character decreases in the order O > S > Se, but also that it is eventually negligible for the thio and seleno cations[15]. The same sort of conclusion is arrived at when one considers $\Delta\delta$ values of C(4) carbon atoms in couples like **10** and **11** or **13** and **14**. There is only a small change in C(4) chemical shift on protonation of vinylene carbonate (**10**) and vinylene trithiocarbonate (**13**). Actually, the largest part of the change may be due to a solvent effect[15,18], which indicates that in the case of trisubstituted ions like **11** and **14** charge delocalization onto C(4) and C(5) is indeed negligibly small.

The situation appears even more complicated in the case of disubstituted ions **25, 28, 5** and **6**, where $\Delta\delta \sim 30$ ppm (free of solvent effect) for C(4) of the ion and its corresponding precursor. Taking into account the small differences in C(4) chemical shifts between the thio and seleno precursors (**24, 26, 27** and **29**) it is difficult to explain the above relatively large values solely in terms of localized positive charge on C(2), especially for the Se derivatives **5** and **6**.

In summary, the question of charge delocalization and aromatic character of 1,3-diselenol-2-ylium ions remains open to dispute until X-ray structure data are available for these species.

3. *1,3-Diselenolan-2-ylium and 1,3-thiaselenolan-2-ylium ions*

A number of saturated cyclic Se-stabilized carbenium ions bearing *N,N'*-dialkylamino or *N,N'*-arylalkylamino substituents on the positively charged C atom have also been prepared as their dimethyltin (IV) tribromide salts (Scheme 3). These compounds can also be regarded as imminium salts. In agreement with this view compounds **30**–**34** exhibit the ν(C$\cdots$N) absorptions in the IR at relatively high frequencies (1520–1588 cm^{-1}), suggesting thereby high percentages of double-bond character of the C—N bonds[19]. Also, the *N*-Me protons of 2-(dimethylamino)-1,3-thiaselenolan-2-ylium (**30**) and 2-(dimethylamino)-1,3-thiaselenan-2-ylium (**31**) ions appear as two singlets in the ^{1}H-NMR spectrum at room

TABLE 3. Selected ^{13}C chemical shift of 1,3-dichalcogenolylium ions

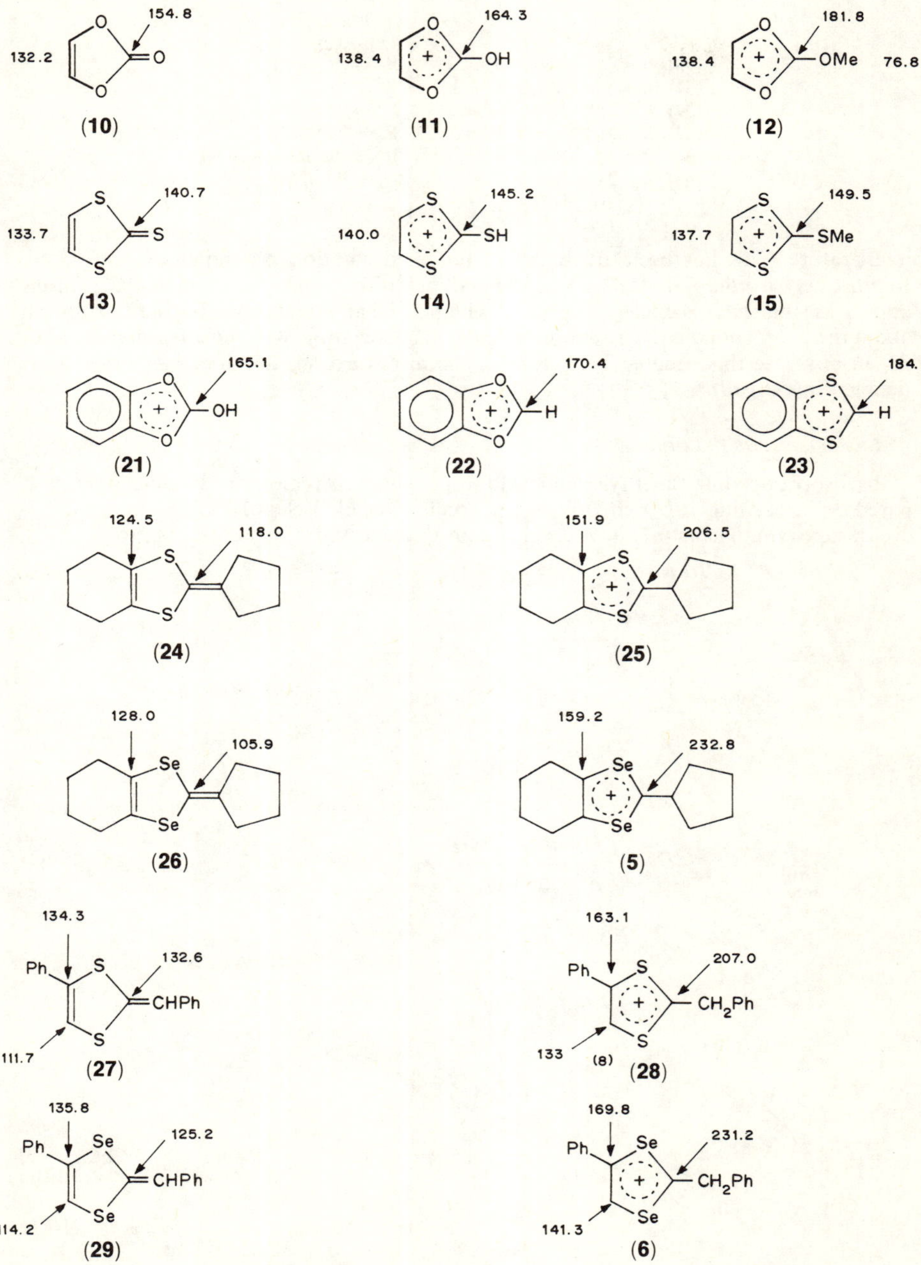

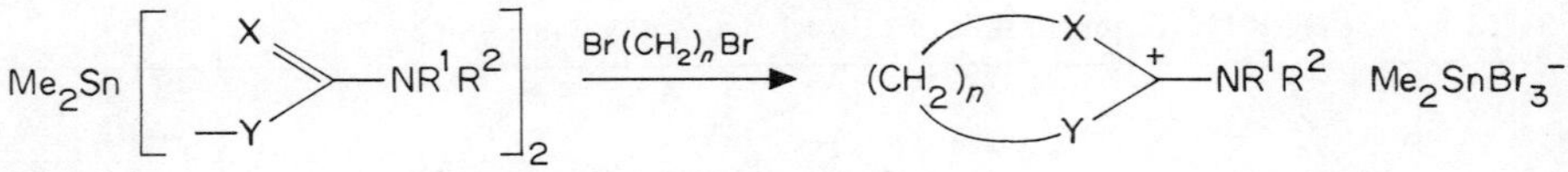

(**30**) $n = 2$; X = S, Y = Se; $R^1 = R^2 =$ Me
(**31**) $n = 3$; X = S, Y = Se; $R^1 = R^2 =$ Me
(**32**) $n = 2$; X = Y = Se; $R^1 = R^2 =$ Me
(**33**) $n = 3$; X = Y = Se; $R^1 = R^2 =$ Me
(**34**) $n = 2$; X = Y = Se; $R^1 =$ Ph, $R^2 =$ Me
(**35**) $n = 2$; X = Y = S; $R^1 = R^2 =$ Me

SCHEME 3

temperature. This has been attributed to hindered rotation around the C—N bond. Interestingly, in compounds **32, 33** and **34** and their dithio analogue **35** the *N*-Me protons appear as singlets. It is unclear at present whether the apparent lower barrier to rotation about the C—N bond is due to a more efficient delocalization of the positive charge in **32**–**35**. In any case this peculiar behaviour of the mixed cations **30** and **31** deserves more detailed investigation.

4. *Miscellaneous cationic systems*

Bis (isopropylamino)methylselenocyclopropenylium perchlorate (**36**) has been prepared[20] according to equation (3) and tris(alkylseleno) and -(arylseleno)-cyclopropenylium salts (**37**) have been obtained according to equations (4)–(6)[21–23]

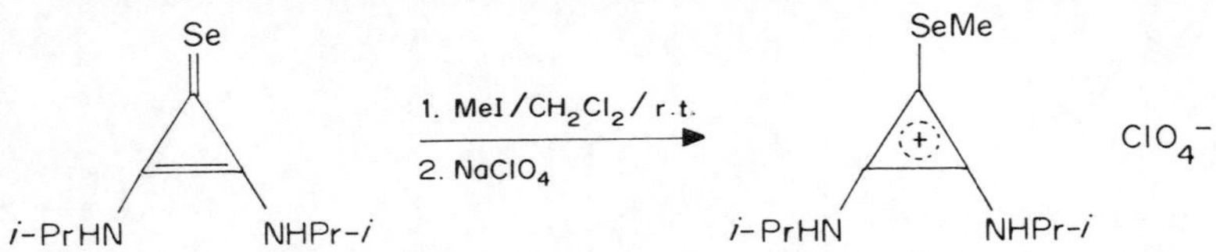

53% (**36**)

(3)

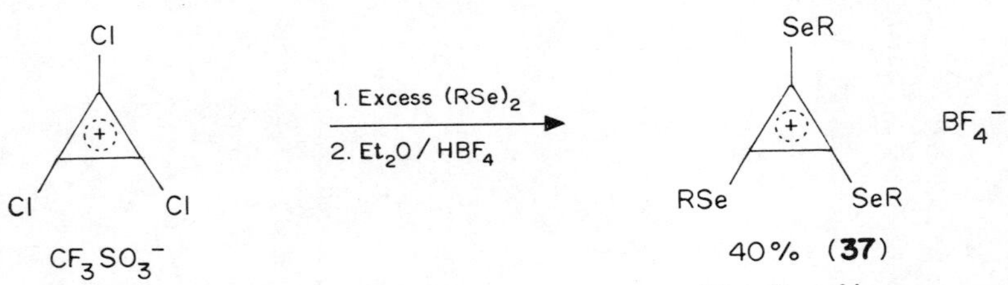

40% (**37**)

(**a**) R = Me

(**b**) R = Ph

(4)

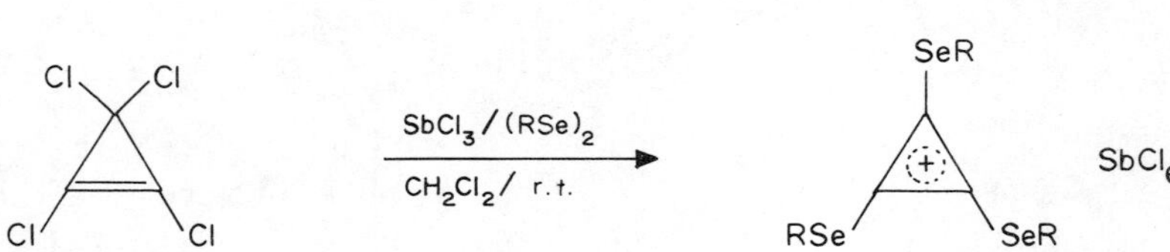

92% (**37a**)

89% (**37b**)

(5)

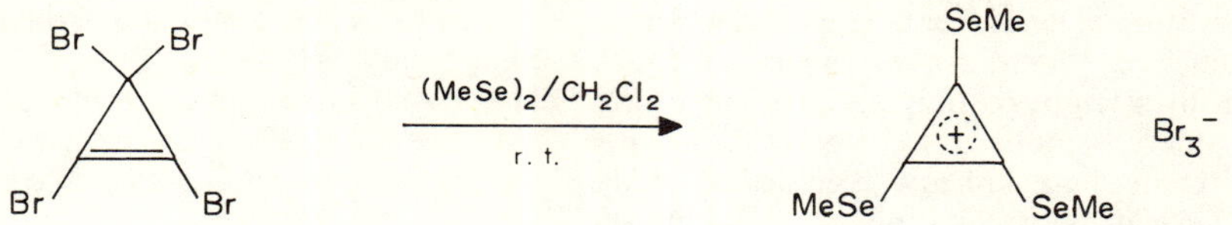

92% (**37a**) (6)

Tris(methyltelluro)cyclopropenylium salts can also be prepared by a method analogous to that of equation (4). While the thio and seleno derivatives very closely resemble each other in stability and in their spectroscopic properties, the telluro analogues appear to be much more labile. Although no attempt has been made to evaluate the contribution of the alkylthio or alkylseleno groups in these systems to the total stabilization, it should not be very high since trihalogenocyclopropenylium ions are equally stable.

A large number of aromatic heterocyclic cationic systems have been described, such as selenocyanine (**38**), selenazinium (**39**) and selenoxanthylium (**40**) dyes[24] or selenopyrylium (**41**) and selenochromylium (**42**) salts[25–27].

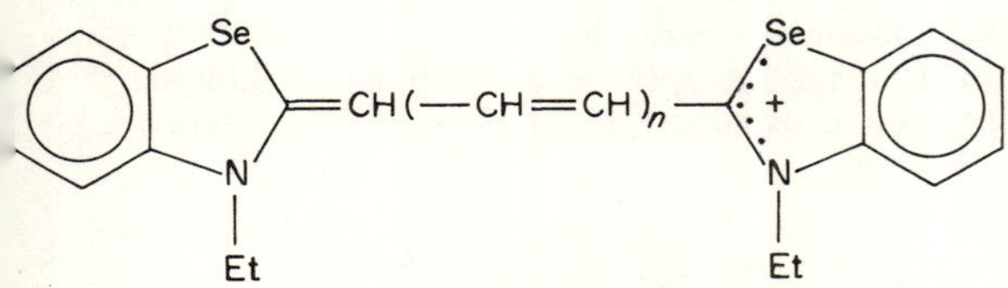

(**38**)

$n = 0, 1, 2, 3$

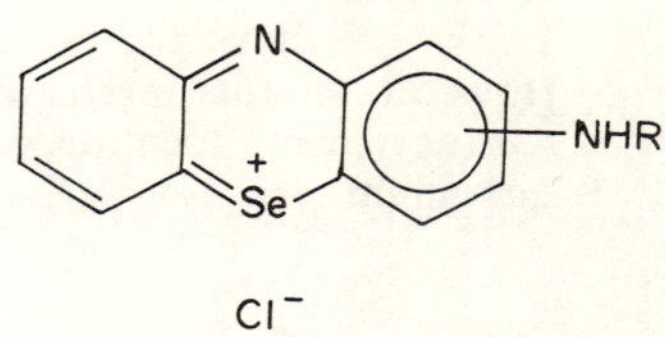

(**39**)

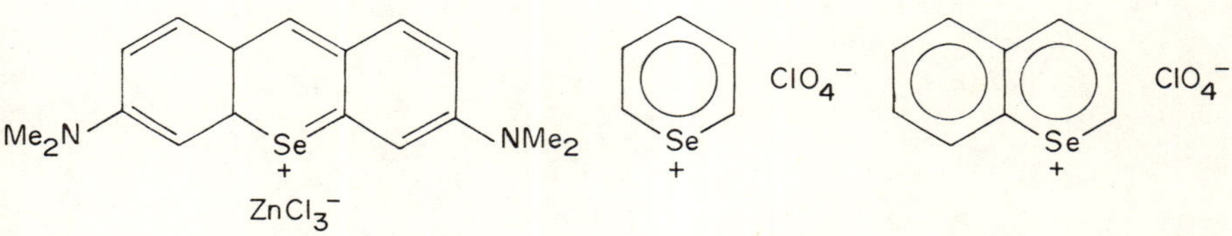

(**40**) (**41**) (**42**)

In all these cases the UV–visible spectra alone provide convincing evidence for a efficient delocalization of the positive charge through the Se atom.

B. Aliphatic Selenocarbenium Ions

1. *Monoselenocarbenium ions*

Although these species have not yet been isolated and characterized structurally, there are good reasons to believe in their existence. The first line of evidence arises from the observation of a number of reactions which appear to be the seleno analogues of the well-documented Pummerer rearrangement for thio compounds[28,29].

In an attempt to polymerize phenyl vinyl selenide (**43**) in the presence of benzoyl peroxide Okamoto and coworkers[30] isolated crystalline **44** in 56% yield (equation 7). Treatment of alkyl phenyl selenides (**45**) under the same reaction conditions yielded (50–92%) α-benzoyloxyalkyl phenyl selenides (**46**)[31].

$$\underset{(\mathbf{43})}{PhSeCH{=}CH_2} + (PhCOO)_2 \xrightarrow[\text{reflux}]{CCl_4} \underset{(\mathbf{44})}{PhSeCH(OCOPh)CH_2OCOPh}$$

Both reactions have been rationalized[31] by postulating the initial formation of a dibenzoyloxyselenurane which decomposes on heating to the corresponding ion pair and then leads to products via a Se-stabilized carbenium ion (**47**) as shown in Scheme 4. Starting from methyl phenyl selenide the corresponding selenurane was isolated in 92% yield[31]. Similarly alkyl phenyl selenides have on treatment with peracids been transformed to α-acyloxyphenyl selenides via Pummerer rearrangement[32]. Other interesting examples of this type of reaction are included in Table 4. It should be noted that in Scheme 4 and Table 4 we have represented the key intermediates as free α-selenocarbenium ions. However, another mechanistic possibility is that at least some of the seleno-Pummerer rearrangements mentioned above proceed in a concerted manner without involving carbenium ions.

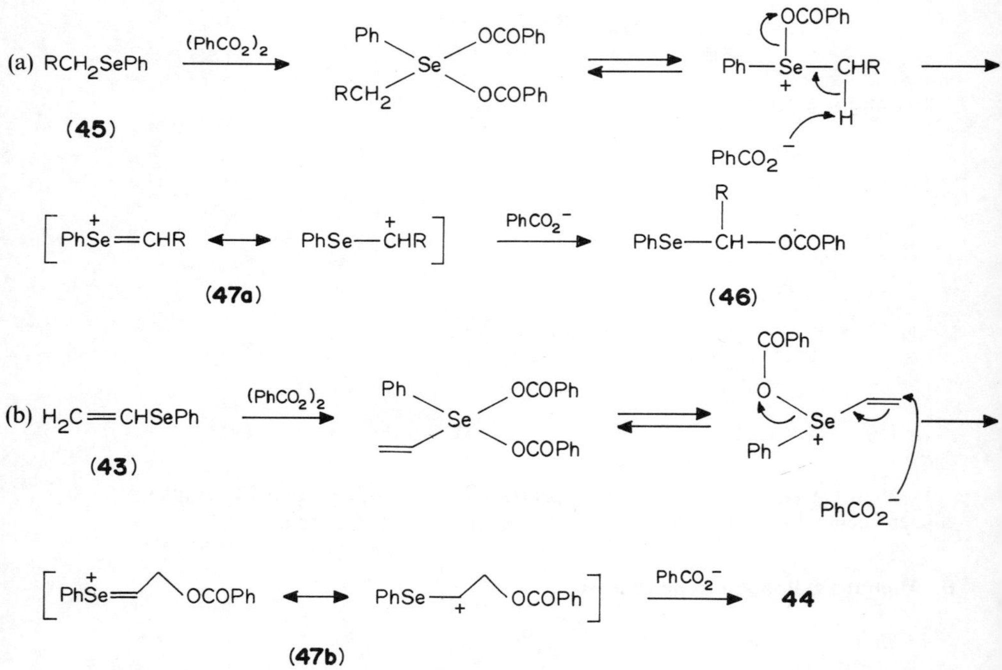

SCHEME 4

TABLE 4. Examples of the seleno-Pummerer rearrangement

Starting material	Conditions	Key intermediate	Products (yield, %)	Ref.
MeSe(O)Me	AcOH/60 °C	$MeSeCH_2^+$	$MeSeCH_2OAc$ (29)	33
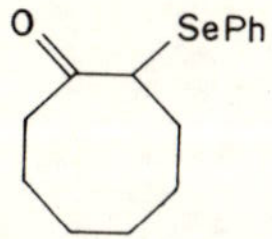	$O_3/CH_2Cl_2/-78\,°C \rightarrow$ r.t.	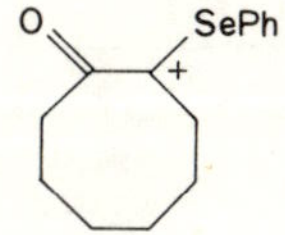	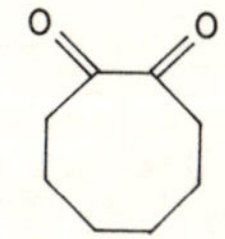 (43)	34
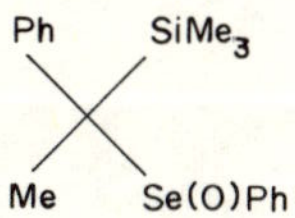	THF/i-Pr_2NH/r.t.	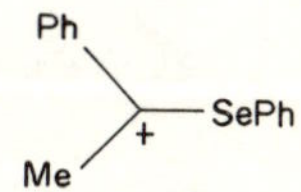	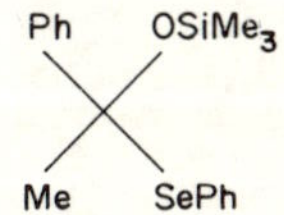 + PhC(O)Me (22)	35
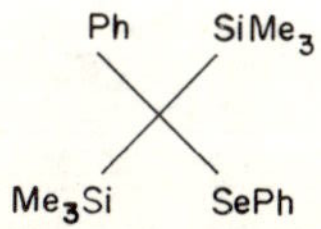	H_2O_2/CH_2Cl_2/r.t.	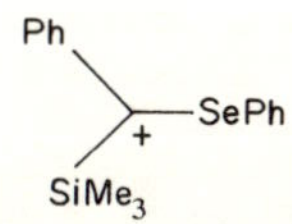	$PhC(O)SiMe_3$ (46)	35
$(Me(CH_2)_4C(O)CH(Me)Se)_2$	O_3/AcOEt/25 °C	$Me(CH_2)_4C(O)\overset{+}{C}(Me)Se—$	$Me(CH_2)_4C(O)C(O)Me$ (34)	36
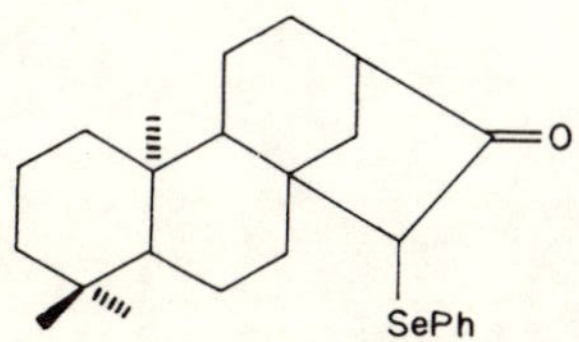	$Tl(NO_3)_3$/MeOH/24 h	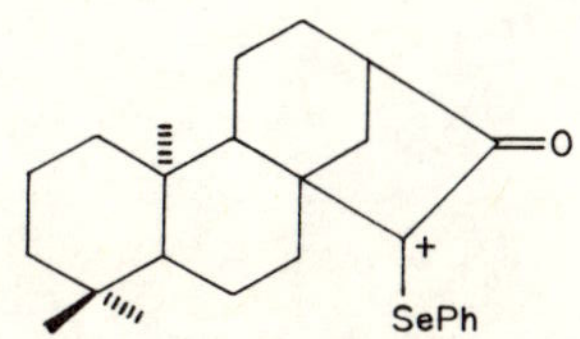	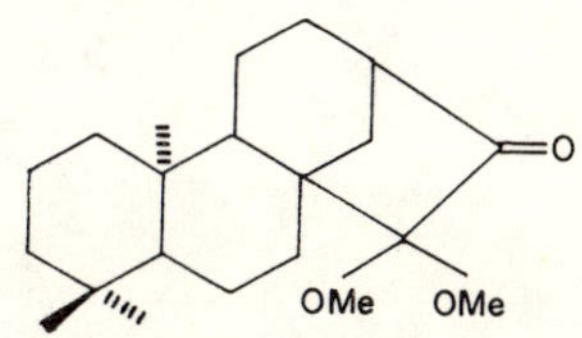	37

(*Contd.*)

TABLE 4 (*Contd.*)

Starting material	Conditions	Key intermediate	Products (yield, %)	Ref.
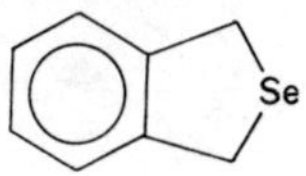	H_2O_2/AcOH	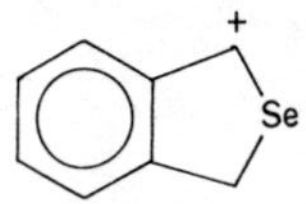	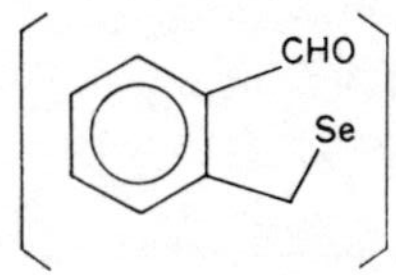 (58)	38
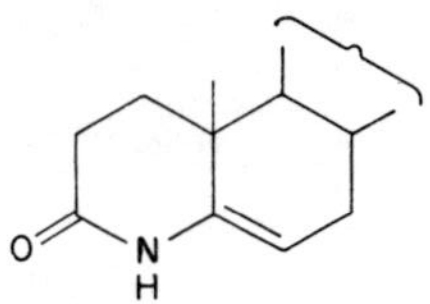	$(PhSeO)_2O/CH_2Cl_2$/r.t.	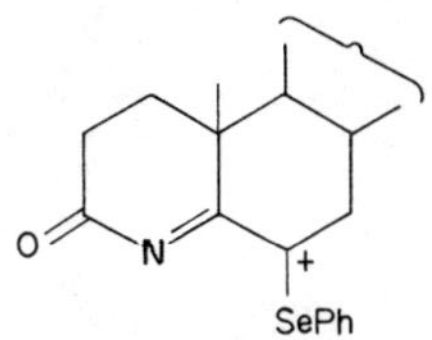	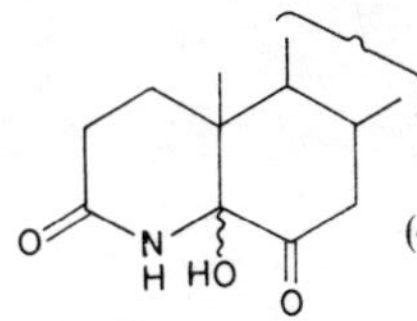 (40)	39, 40

The second line of evidence for substantial stabilization of a positive charge by an adjacent selenyl moiety stems from investigations of the mechanism of acidic hydrolysis of vinyl selenides **48**[41] and **49**[42,43]. It has been established that all the aryl vinyl selenides **48** undergo hydrolysis according to the conventional mechanism characteristic for vinyl ethers and vinyl sulphides. A slow protonation of the olefinic β-carbon atom is followed by hydration of the intermediate arylselenocarbenium ion **50**. By using the correlation equation for the protonation of alkenes[44,45] McClelland and Leung were able to estimate[41] the following σ_p^+ constants: $\sigma_p^+(\text{PhO}) = -0.62$, $\sigma_p^+(\text{PhS}) = -0.54$ and $\sigma_p^+(\text{PhSe}) = -0.47$. The value for the phenylseleno group is comparable to that of the cyclopropyl group[46].

$H_2C{=}CHSeC_6H_4X$

(**48**)

X = 4-MeO, 4-Me, H, 4-F, 4-Br, 4-Cl, 3-Cl, 4-CF_3, 4-NO_2

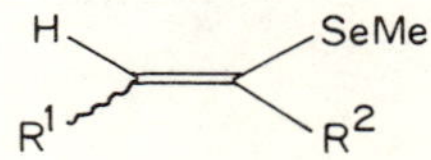

(**49**)

(**a**) $R^1 = n$-Dec; $R^2 = $ Me
(**b**) $R^1 = $ H; $R^2 = 4\text{-}XC_6H_4$; X = MeO, Me, H, Cl, NO_2

In the case of vinyl selenides **49b** the hydrolysis rate constants do not correlate with the σ^+ parameters, but they obey a Yukawa–Tsuno-type equation[43]. This can be interpreted as the result of a cross-conjugative competition between ArX and SeMe for delocalization of the positive charge in **51**.

$Me\overset{+}{C}HSeC_6H_4X$ (**50**)

$R^1CH_2\overset{+}{C}(SeMe)C_6H_4X$ (**51**)

Finally, a few other reactions in which α-selenocarbenium ions are the most likely intermediates have also been carried out in our laboratory (Scheme 5)[47–49]. Reactions (a) and (b) proved to be synthetically useful[47,48].

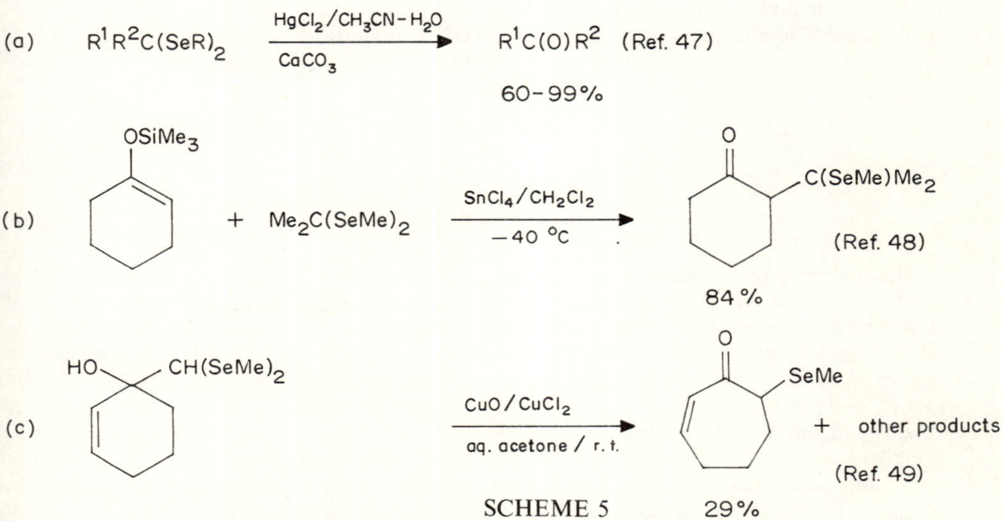

SCHEME 5

2. Bis(seleno)carbenium ions

Acidic hydrolysis of ketene selenoacetals (**52**) (Scheme 6) provided[50] the first examples of bis(seleno)carbenium ions (**53**). From the mechanistic point of view it is of interest to note that there is a progressive changeover from a rate-limiting protonation to a rapid preequilibrium protonation of the double bond on going from **48** to **49** to **52**. Preequilibrium protonation was likewise observed during hydrolysis of ketene thioacetals[51]. Subsequently a few cations of the type **53** were also prepared and isolated[52] as their hexahaloantimonate salts (Scheme 7).

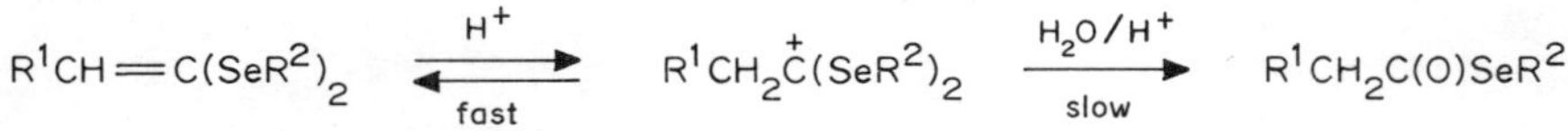

$R^1, R^2 = Me, Ph$

(**a**) $R^1 = R^2 = Me$
(**b**) $R^1 = Ph, R^2 = Me$

SCHEME 6

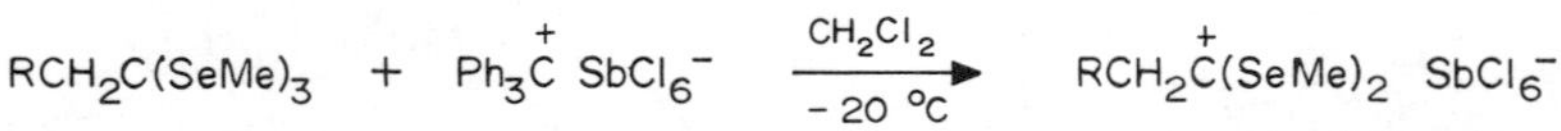

R = Me 85%
R = Ph 80%

SCHEME 7

Bis(methylseleno)benzylcarbenium hexachloroantimonate (**53b**) has been structurally investigated by ^1H- and ^{13}C-NMR as well as by X-ray diffraction and compared to its bis (methylthio) analogue[52]. ^1H-NMR spectra (room temperature, CD_3CN solution) of these two compounds appeared to be identical within a few hundredths of a ppm. No significant conclusion can therefore be drawn from ^1H chemical shifts. However, the shapes of both the SMe and the SeMe signals were temperature-dependent and this was attributed to hindered rotation about the C^+—S and C^+—Se bonds. The corresponding Arrhenius activation energies are 14.3 ± 2 and 12.7 ± 2 kcal mol^{-1} respectively. The available values for the oxo derivatives range from 8 to 15 kcal mol^{-1}. Thus, the rotational barriers appear to be roughly identical for all three types of carbenium ions, suggesting thereby a very large contribution from resonance structures **54**, even in the case of X = Se.

(**54**)

X = O, S, Se

Although ^{13}C resonances for C^+ occur at a rather low field (263.4 ppm for **53b** and 244.5 ppm for its thio analogue), the above idea gains considerable support from X-ray structural data, especially from the bond distances (Scheme 8). The bond angles around the positive carbon atom show that both cations are essentially planar with virtually no distortion of the regular triangular structure. More striking is the finding that the C^+—S and C^+—Se bond distances are considerably reduced with respect to the corresponding single bond lengths of 1.81 Å (C—S) and 1.98 Å (C—Se). (Note that the C═S and C═Se bond lengths are 1.54–1.61 Å and 1.67 Å, respectively.) This undoubtedly shows that Se-containing substituents are able to stabilize a neighbouring positive charge and that the mechanism of this stabilization involves non-negligible conjugative interaction.

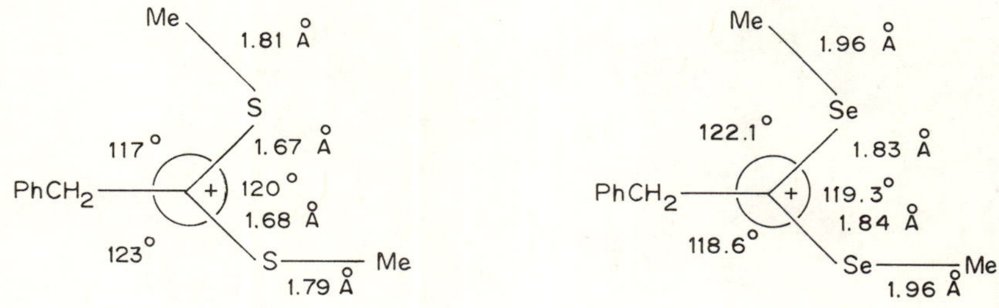

SCHEME 8

The latter results shed some light also on the 1,2-diselenolylium and 1,3-diselenolylium ions as well as on other 'aromatic' Se-containing cationic systems discussed in the preceding sections.

3. 1-Seleno- and 2-seleno-allyl cations

The presence of the allylic delocalization should in principle make these monoselenated species more stable than their saturated analogues. However, the preliminary work so far carried out suggests that both 1-seleno- and 2-seleno-allyl cations **55** and **56** must possess some structural peculiarities which are responsible for the observed reactivities[53,54] illustrated in Scheme 9. The fact that *E*- and *Z*-1,3-bis(phenylseleno)propenes react with furan with a remarkable retention of the initial double-bond configuration, as well as the notable difference in the reaction rates, suggest more complex intermediates such as **55a** and **55b** instead of the fully delocalized free allylcarbenium ion. Similarly, the intermediacy of alkylideneseleniranium ion **56** could explain that 1-bromo-2-(phenylseleno)-2-pentene gives no cycloaddition at all and that C(1) of the starting material is the exclusive site of attachment to the furan ring. However, these structures are proposed only tentatively in order to rationalize the observed results; more work is necessary to fully characterize the nature of the intermediates.

III. α-SELENO FREE RADICALS

The field of Se-containing radicals is much less well documented than that of Se-containing anions or cations. A large part of the existing work has been devoted to radical anions or radical cations, so that at present time only a very limited number of papers deal with neutral α-seleno free radicals. The best method for investigating radicals is by ESR and the following discussion will mainly be centred on the results obtained by this technique.

Radical anions **57** and **58** derived from heteroaromatic Se compounds have been prepared by one-electron transfer to the parent heterocycle from anions, alkali metals or from an electrode[55-59]. Neutral and cation radicals derived from phenoselenazine[56,60] (**59** and **62**), from phenoxselenin[56,61] (**60**), from dibenzodiselenin[61], (**61**), from 9-phenylselenoxanthene[62], (**63**) and phenoselenazine nitroxide[56,60] (**64**) have also been generated in solution and studied by ESR. Hyperfine coupling constants a_i and g values for these species are presented in Table 5. In terms of resonance theory the extent of delocalization of the odd electron of a radical is a measure of its stability. In so far as the hyperfine splitting constants a_i are linearly connected to spin densities, the magnitude of the splitting by ^{77}Se should in principle give a measure of the contribution of Se atoms to the stabilization of radical species. On the other hand, semiempirical theory has been

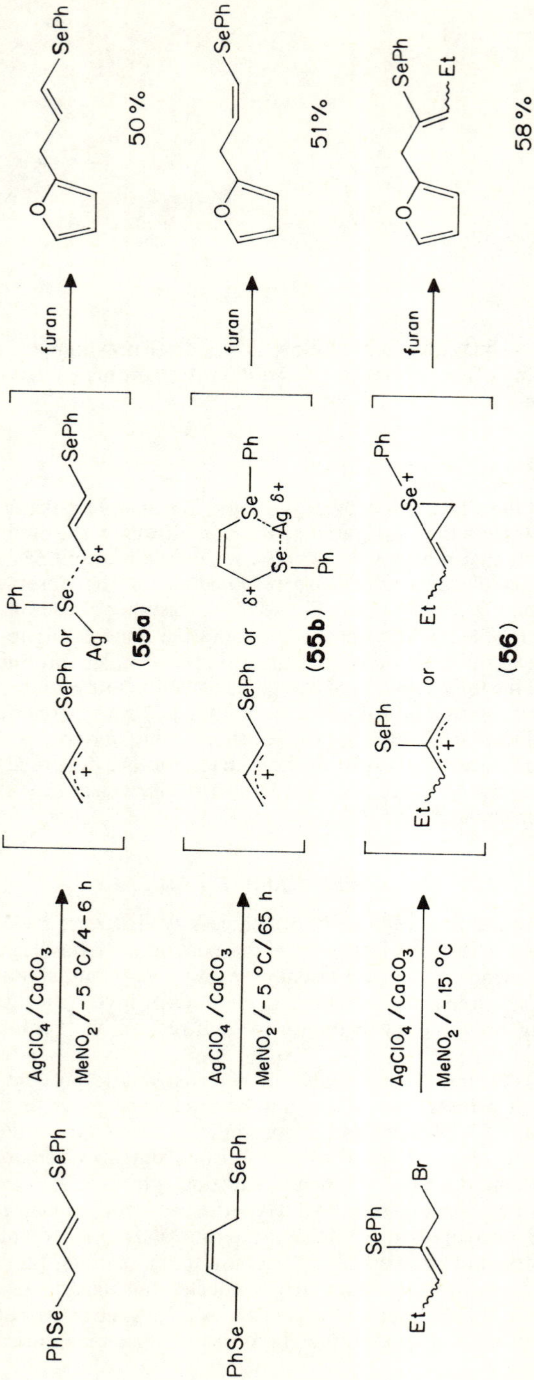
PhSe
SePh
AgClO4 / CaCO3
MeNO2 / −5 °C/4−6 h
or
Ph
Se
Ag
δ+
(55a)
furan
50%
AgClO4 / CaCO3
MeNO2 / −5 °C/ 65 h
Se—Ph
Se··Ag δ+
(55b)
51%
Et
Br
AgClO4 / CaCO3
MeNO2 / −15 °C
Se+
(56)
58%

SCHEME 9

TABLE 5. Hyperfine coupling constants a_i (G) and g values of heteroaromatic seleno radicals

		a_i								
Species	i =	Se	N	C(1)	C(2)	C(3)	C(4)	g	Solvent	Ref.
57		—		4.27	1.02	5.27	1.02	—	DME	55
		5.44		4.21	1.03	5.18	1.03	2.0030	DME	56
58		—	5.97	1.99	1.99			—	DMSO–*t*-BuOH	57
		4.9	5.79	2.48	1.65			2.0063	DME	56
		—	5.79	2.48	1.65			—	DME	58
		—	5.67	1.88	1.88			—	DME	59[a]
59		25.0	6.17	1.18	0.40	2.50	0.40	2.0161	Nitroethane	56,60
60		39.5		—	—	—	—	2.0228	Nitroethane	56,61
61		—		—	—	—	—	2.0315		61
62		10.9	6.96	3.82	1.00	2.86	0.100	2.0104	Benzene	56
63		—		3.70	0.59	3.70	0.59	—	THF	62
64		7.70	9.26	2.15	0.64	2.12	0.64	2.0071	Benzene	60

[a] Variation of a_i with the solvent was also reported.

suggested[63] and successfully applied[56] to correlate isotropic g shifts with spin densities and spin-orbit coupling constants. Therefore, experimentally determined g values can also give valuable information on the stabilization of radicals by a heteroatom. This is especially true for Se whose spin-orbit coupling constant is large compared to that of S (1688 and 382 cm^{-1} for Se and S, respectively). Inspection of Table 5 then shows that a large ^{77}Se coupling constant (a_{Se}) is associated with a large g value (species **59, 60, 62**) and vice-versa. It is interesting to note that a_{Se} and g values are also strongly dependent on the charge of the radical. Anion radicals **57** and **58** exhibit the lowest ^{77}Se coupling, neutral phenoselenazine radical **62** and nitroxide **64** have intermediate values, while cation radicals **59**–**61** have the highest ^{77}Se coupling constants. The same trend is observed for the corresponding g values. Gilbert and coworkers have derived spin densities ρ^{π} for **59** as well as its oxo and thio analogues using HMO and McLachlan calculations[56,60], which in the case of **59** reproduced correctly all the observed coupling constants including a_{Se}. The calculations also appeared quite adequate for the other seleno cation radicals. The calculated spin densities can therefore be considered correct also in the case of phenoxazine and phenothiazine cation radicals, with the result $\rho^{\pi}_{Se} \geqslant \rho^{\pi}_{S} > \rho^{\pi}_{O}$ (0.198, 0.179 and 0.086, respectively, using McLachlan calculations). The method also applies for neutral radicals and yields the same trend with considerably lower spin densities ($\rho^{\pi}_{X} = 0.085$; 0.076 and 0.046 for X = Se, S and O, respectively), but it fails in the case of radical anions **57** and **58**. The reasons for this failure are not clear[56] but it seems quite certain that the spin densities on Se in these anionic species are very low, as confirmed by the coupling constants and g values of Table 5.

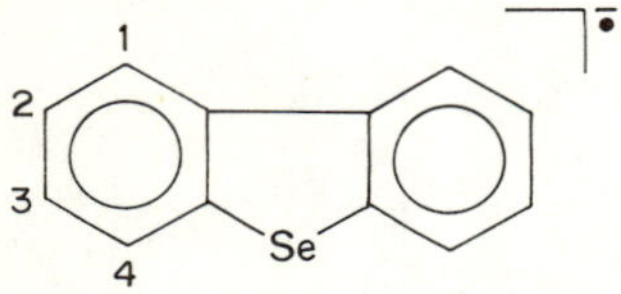

(**57**)

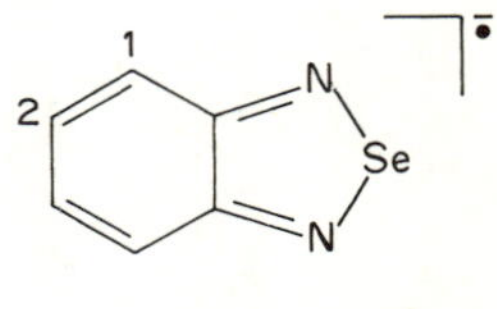

(**58**)

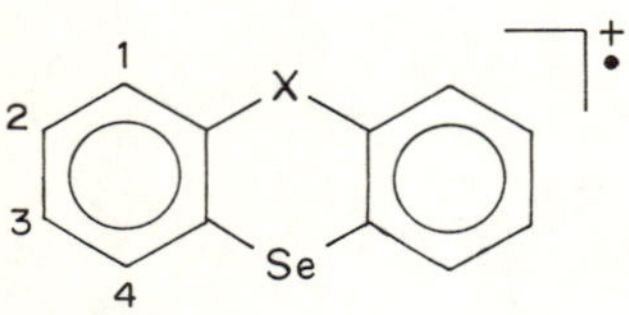

(**59**) X = NH
(**60**) X = O
(**61**) X = Se

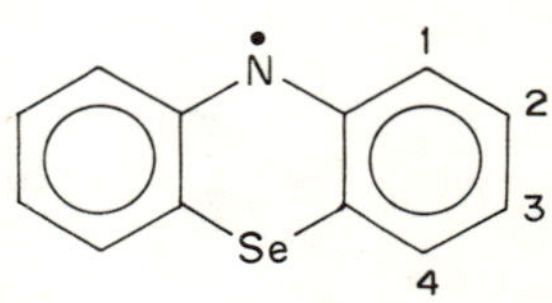

(**62**)

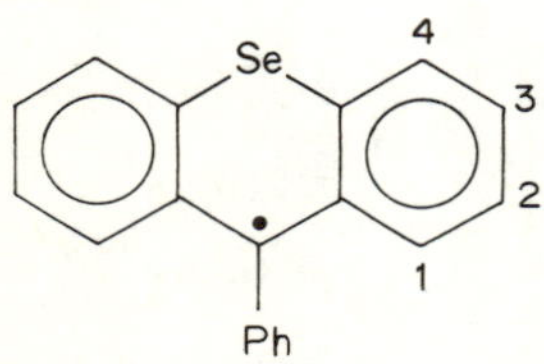

(**63**)

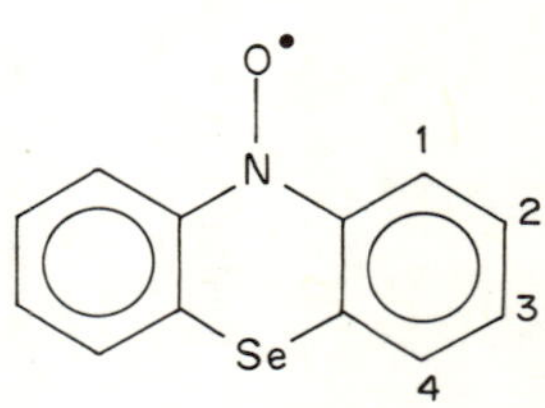

(**64**)

It therefore appears reasonable to conclude that the unpaired electron of neutral and cation radicals derived from heteroaromatic molecules is efficiently delocalized onto Se. However, at present, it is not possible to make a statement about the relative stability of the oxo, thio and seleno species.

Selenuranyl-type radicals are relevent to the problem of radical stabilization by Se because in a way they make the link between the aromatic species discussed above and the carbon radicals bearing α-seleno substituents.

The dimer radical cation of dimethyl selenide (**65**) was obtained by γ-irradiation of a single crystal at − 196 °C and investigated by ESR spectroscopy[64]. Dialkyl selenides have also been found[65] to react with various radicals to produce the neutral selenuranyl radicals **66**. Se-containing heterapentalenes[66] react with tri-*n*-butylstannyl to produce radicals such as **67** and **68**. In the absence of conjugation one expects high spin densities on the Se atoms of **65** and **66**. This is confirmed by both the isotropic *g* values and the hyperfine coupling constants to ^{77}Se ($g = 2.0344$, $a_{Se} = 108$ G for **64**; $g = 2.0206$, $a_{Se} = 162.6$ G for **66** where $R^1 = R^2 =$ 1-adamantyl, X = *t*-BuC(O)S). As can be judged from the *g* values a similar spin distribution holds in radical **68** ($g = 2.0169$), while in the persistent radical **67** a very small part, if any, of the odd electron resides on Se ($g = 2.0047$).

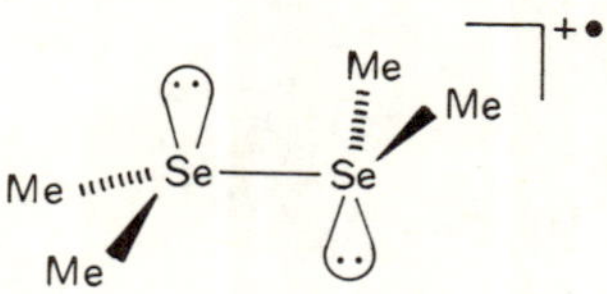

(**65**)

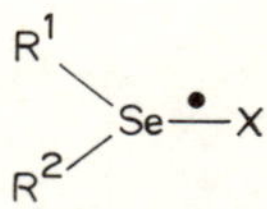

(**66**)

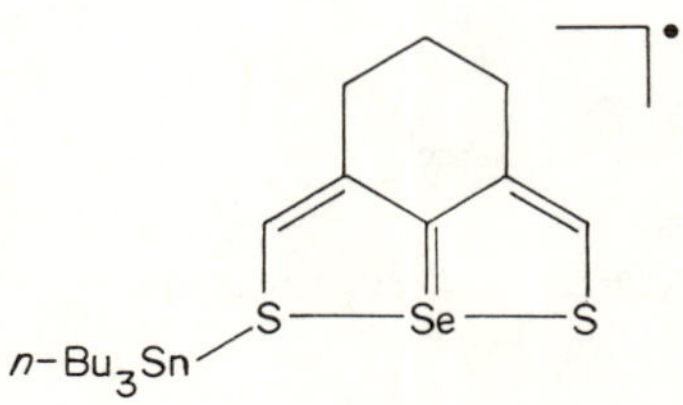

(**67**)

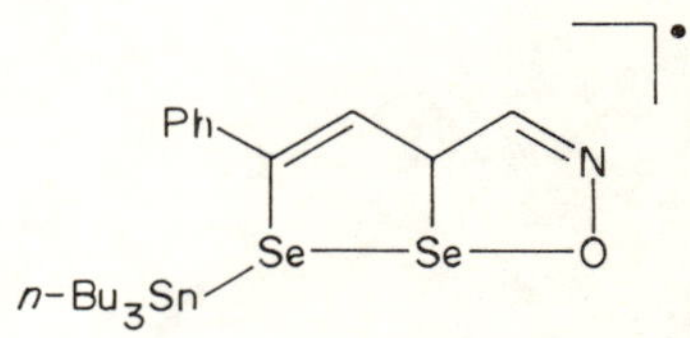

(**68**)

It is of importance to note that the ESR spectrum of **65** at − 196 °C also contained lines at higher field which were attributed[64] to carbon-centred radicals, mainly $MeSeCH_2\cdot$, but it was not possible to analyse this portion of the spectrum in detail. On the other hand Scaiano and Ingold have described[67] the easy generation of hindered α-selenoalkyl radicals **69** by the reaction of radicals X· with di-*t*-butyl selenoketone in hydrocarbon solution.

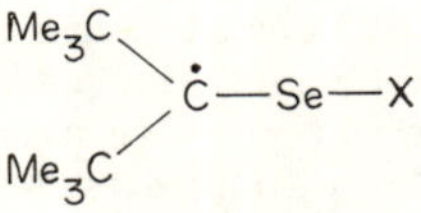

(**69**)

	X	g	$a_{Se}(G)$
(a)	Me	2.0021	24.6
(b)	*t*-Bu	2.0020	21.3
(c)	CF_3	2.0005	36.6
(d)	*t*-BuO	2.0024	65.0
(e)	Me_3Sn	2.0043	10.9

However, all attempts at the direct observation of unhindered α-selenoradicals formed in solution by hydrogen abstraction from dimethyl selenide, diethyl selenide, methyl phenyl selenide or dimethyl diselenide and under a variety of experimental conditions were unsuccessful[68]. Yet, spin trapping experiments using di-*t*-butyl thioketone provided good evidence that methylselenomethyl ($MeSeCH_2\cdot$) and 1-ethyseleno-1-ethyl ($EtSe\dot{C}HMe$) radicals were indeed efficiently formed. The authors suggested that their failure to detect unhindered alkylselenoalkyl radicals was due to the operation of both the spin-rotation and the spin-orbit line broadening mechanisms. Thus it was estimated[68] that the line width for $MeSeCH_2\cdot$ should be ca. 6.6 G arising from spin rotation alone, while some of the radicals **69** have line widths as small as 0.1–0.2 G.

This great difference in line widths between hindered and unhindered α-selenoalkyl radicals has been attributed to conformational effects. The preferred conformation of **69** is given by **70** and that of $MeSeCH_2\cdot$ by **71**. The unpaired electron in **70** cannot be appreciably delocalized onto the Se atom, hence the low *g* values (see **69**) and sharp lines. In contrast, delocalization in **71** can be efficient and causes very important line broadening, and such radicals cannot be observed in solution.

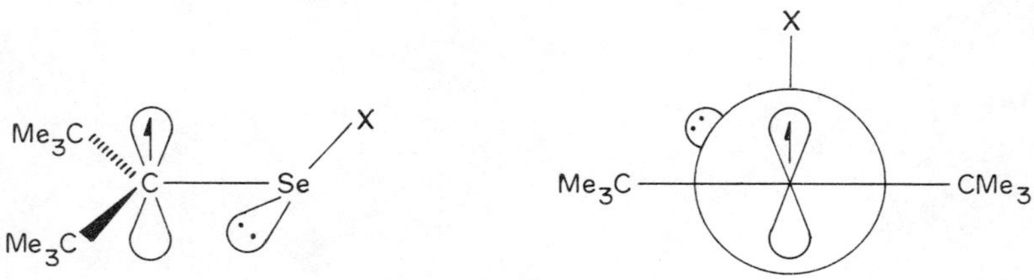

(70)

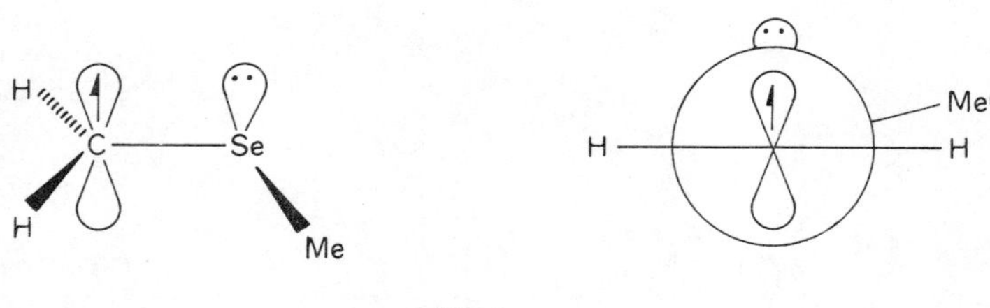

(71)

The last series of examples illustrating the stabilization of carbon radicals by adjacent seleno substituents comes from studies of the addition of radicals to 1, 1-disubstituted olefins (equation 8). It has been found that the formation and stability of the adduct

$$H_2C{=}CXY + R\cdot \rightarrow RCH_2\dot{C}XY \tag{8}$$

radicals are highly dependent on the nature of the geminal substituents of the starting olefin, and that olefins bearing one electron-attracting and one electron-donating group on the same sp^2 carbon react far more easily and yield the most stable adduct radicals. Furthermore, the latter radicals do not undergo polymerization or hydrogen abstraction, but rather they trap another radical or dimerize[69]. Scheme 10 features two examples of such reactions involving 'captodative'-stabilized α-phenylseleno radicals[69–71]. From competition and rate measurements it was also possible to establish the relative rate of

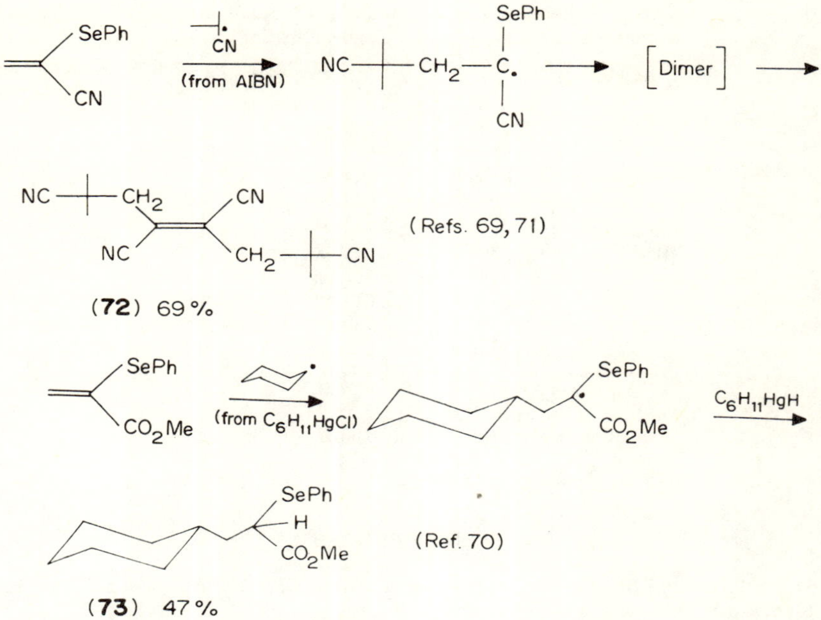

SCHEME 10

formation of adduct radicals. The figures of Table 6 clearly demonstrate the beneficial effect of the phenylseleno group on the rate of reaction which, of course, also depends on the nature of the electron-attracting group and of the reacting radical species. However, recent results suggest that these additions proceed via very early transition states, and the observed rate differences arise primarily from polar effects of the substituents X and Y on the transition state rather than from their radical stabilizing abilities[72].

α-Phenylseleno radicals (**74**) have also been studied[73] by ESR. Isotropic g factors indicate by themselves that in all cases the Se atom carries a significant amount of spin density, and thereby confirm, at least to some extent, the results of kinetic measurements as well as the conclusions arrived at in the cases of aromatic neutral and cation radicals. There is no doubt, however, that additional experimental evidence such as ^{77}Se hyperfine coupling and appropriate theoretical treatment are necessary for a closer and more quantitative evaluation of the radical-stabilizing ability of α-selenyl moieties.

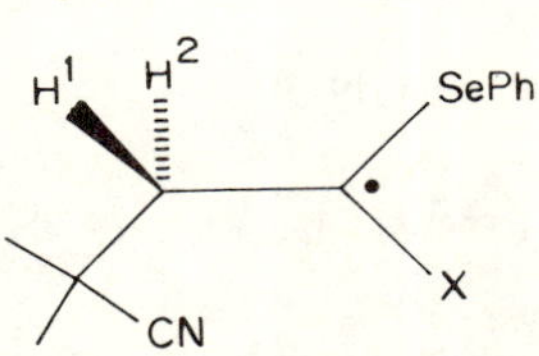

(**74**)

X	g	a_1	a_2	$(a_1 + a_2)$	a_F
CF_3	2.0107	10.75	10.75	—	24.5
S(O)Ph	2.0095	8.55	13.16	—	—
SO_2Et	2.0119	—	—	19.65	—
SO_2Ph	2.0113	—	—	19.19	—

TABLE 6. Relative rates of addition of cyclohexyl and isobutyronitrile radicals to *gem*-disubstituted olefins[a]

R	X	Y	Rel. rate
C_6H_{11}[70 b]	SePh	CO_2Me	62
	Ph	CO_2Et	40
	$SCMe_3$	CO_2Me	35
	Me	CO_2Me	4.7
	OMe	CO_2Me	1 .
IBN[72 c]	SePh	CO_2Me	6.9
	SePh	SO_2Ph	1.5
	OMe	CO_2Me	1
	SePh	S(O)Ph	0.5
	SePh	CF_3	0.4

[a] Cf. equation (8).
[b] At 20 °C in CH_2Cl_2.
[c] At 70 °C in benzene; IBN = isobutyronitrile.

IV. REFERENCES

1. H. Rheinboldt, in *Houben–Weyl: Methoden der Organischen Chemie*, Georg Thieme Verlag, Stuttgart, 1955, p. 917.
2. D. L. Klayman and W. H. H. Günther (Eds), *Organic Selenium Compounds: Their Chemistry and Biology*, Wiley–Interscience, New York, 1973.
3. K. A. Jensen and U. Henriksen, *Acta Chem. Scand.*, **21**, 1991 (1967).
4. K. A. Jensen, H. R. Baccaro and O. Buchardt, *Acta Chem. Scand.*, **17**, 163 (1963). K. A. Jensen and P. H. Nielsen, *Acta Chem. Scand.*, **20**, 597 (1966).
5. G. A. Heath, I. M. Stewart and R. L. Martin, *Inorg. Nucl. Chem. Letters*, **5**, 169 (1969).
6. G. A. Heath, R. L. Martin and I. M. Stewart, *Aust. J. Chem.*, **22**, 83 (1969).
7. O. Siimann and J. Fresco, *Inorg. Chem.*, **9**, 294 (1970).
8. A. Hordvik and H. M. Kjøge, *Acta Chem. Scand.*, **19**, 935 (1965).
9. A. Hordvik, *Acta Chem. Scand.*, **19**, 1039 (1965).
10. A. Hordvik and E. Sletten, *Acta Chem. Scand.*, **20**, 1974 (1966).
11. H. C. Freeman, G. H. W. Milburn, C. E. Nockolds, P. Hemmerich and K. H. Knauer, *J. Chem. Soc., Chem. Commun.*, **55**, (1969).
12. A. Hordvik, *Acta Chem. Scand.*, **20**, 1885 (1966).
13. M. H. Palmer and R. H. Findlay, *Tetrahedron Letters*, 4165 (1972).
14. H. Meier and I. Menzel, *Tetrahedron Letters*, 445 (1972).
15. U. Timm, U. Plücken, H. Petersen and H. Meier, *J. Heterocyclic Chem.*, **16**, 1303 (1979).
16. E. M. Engler and V. V. Patel, *Tetrahedron Letters*, 1259 (1975).
17. E. M. Engler and V. V. Patel, *J. Org. Chem.*, **40**, 387 (1975).
18. G. A. Olah and J. L. Grant, *J. Org. Chem.*, **42**, 2237 (1977).
19. K. Tanaka and T. Tanaka, *Bull. Chem. Soc. Japan*, **45**, 489 (1972).
20. Z. Yoshida, *Japan Kokai*, 75, 117, 753 (Cl C07CD, AOIN) (1974); *Chem. Abstr.*, **84**, 43422r (1974).
21. R. Weiss, C. Schlierf and K. Schloter, *J. Amer. Chem. Soc.*, **98**, 4668 (1976).
22. R. Weiss and P. Marolt, *Synthesis*, 225 (1980).
23. R. Weiss and P. Marolt, *Synthesis*, 227 (1980).
24. L. G. S. Brooker, J. A. Ford and E. J. Van Lare, Ref. 1(b), pp. 509–523 and references cited.
25. R. B. Silverman, Ref. 1(b), p. 537–553 and references cited.
26. M. Renson and P. Pirson, *Bull. Soc. Chim. Belges*, **75**, 456 (1966).
27. A. Tadino, L. Christiaens and M. Renson, *Bull. Soc. Roy. Sci. Liège*, **42**, 146 (1973) and other papers in this series.

28. R. Pummerer, *Chem. Ber.*, **42**, 2282 (1909); *Chem. Ber.*, **43**, 1401 (1910).
29. S. Oae and T. Numata, in *Isotopes in Organic Chemistry* (Eds. E. Buncel and C. C. Lee), Vol. 5, Elsevier, Amsterdam, 1980, pp. 45–102.
30. Y. Okamoto, R. Homsany and T. Yano, *Tetrahedron Letters*, 2529 (1972).
31. Y. Okamoto, K. L. Chellappa and R. Homsany, *J. Org. Chem.*, **38**, 3172 (1973).
32. G. Galambos and V. Simonidesz, *Tetrahedron Letters*, 4371 (1982).
33. N. Miyoshi, S. Murai and N. Sonoda, *Tetrahedron Letters*, 851 (1977).
34. H. J. Reich, J. M. Renga and I. Reich, *J. Amer. Chem. Soc.*, **97**, 5434 (1975).
35. H. J. Reich and S. K. Shah, *J. Org. Chem.*, **42**, 1773 (1977).
36. K. B. Sharpless and K. M. Gordon, *J. Amer. Chem. Soc.*, **98**, 300 (1976).
37. Y. Nagao, M. Ochiai, K. Kaneko, A. Macda, K. Watanabe and E. Fujita, *Tetrahedron Letters*, 1345 (1977).
38. B. E. Norcross, J. M. Lansinger and R. L. Martin, *J. Org. Chem.*, **42**, 369 (1977).
39. T. G. Back and N. Ibrahim, *Tetrahedron Letters*, 4931 (1979).
40. T. G. Back, N. Ibrahim and D. J. McPhee, *J. Org. Chem.*, **47**, 3238 (1982).
41. R. A. McClelland and M. Leung, *J. Org. Chem.*, **45**, 188 (1980).
42. J. L. Piquard and L. Hevesi, *Tetrahedron Letters*, 1901 (1980).
43. L. Hevesi, J. L. Piquard and H. Wautier, *J. Amer. Chem. Soc.*, **103**, 870 (1981).
44. V. J. Nowland and T. T. Tidwell, *Acc. Chem. Res.*, **10**, 252 (1977).
45. K. M. Koshy, D. Ray and T. T. Tidwell, *J. Amer. Chem. Soc.*, **101**, 357 (1979).
46. O. Exner, in *Correlation Analysis in Chemistry* (Eds. N. B. Chapman and J. Shorter), Plenum Press, New York–London, 1978, p. 482.
47. A. Burton, L. Hevesi, W. Dumont, A. Cravador and A. Krief, *Synthesis*, 877 (1979).
48. K. M. Nsunda and L. Hevesi, *Tetrahedron Letters*, 4441 (1984).
49. J. Lucchetti, J. L. Laboureur and A. Krief, private communications of unpublished results.
50. H. Wautier, S. Desauvage and L. Hevesi, *J. Chem. Soc., Chem. Commun.*, 739 (1981).
51. T. Okuyama and T. Fueno, *J. Amer. Chem. Soc.*, **102**, 6590 (1980).
52. L. Hevesi, S. Desauvage, B. Georges, G. Evrard, P. Blanpain, A. Michel, S. Harkema and G. J. van Hummel, *J. Amer. Chem. Soc.*, **106**, 3784 (1984).
53. M. Renard and L. Hevesi, *Tetrahedron Letters*, 3911 (1983).
54. S. Halazy and L. Hevesi, *J. Org. Chem.*, **48**, 5242 (1983).
55. R. Gerdil and E. A. S. Lucken, *J. Amer. Chem. Soc.*, **87**, 213 (1965).
56. M. F. Chin and B. C. Gilbert, *J. Chem. Soc., Perkin Trans. 2*, 258 (1973).
57. E. T. Strom and G. A. Russell, *J. Amer. Chem. Soc.*, **87**, 3326 (1965).
58. N. M. Atherton, J. N. Ockwell and R. Dietz, *J. Chem. Soc.* (*A*), 771 (1967).
59. M. Kamiya and Y. Akahori, *Bull. Chem. Soc. Japan*, **43**, 268 (1970).
60. M. F. Chiu, B. C. Gilbert and P. Hansen, *J. Chem. Soc* (*B*), 1700 (1970).
61. B. Lamotte and G. Berthier, *J. Chim. Phys.*, **63**, 369 (1966).
62. K. Maruyama, M. Yoshida and K. Murakami, *Bull. Chem. Soc. Japan*, **43**, 152 (1970).
63. A. J. Stone, *Mol. Phys.*, **6**, 502 (1966).
64. K. Nishikida and F. Williams, *Chem. Phys. Letters*, **34**, 302 (1975).
65. J. R. H. Giles and B. P. Roberts, *J. Chem. Soc., Chem. Commun.*, 504 (1980).
66. K. U. Ingold, D. H. Reid and J. C. Walton, *J. Chem. Soc., Perkin Trans. 2*, 431 (1980).
67. J. C. Scaiano and K. U. Ingold, *J. Chem. Soc., Chem. Commun.*, 205 (1976).
68. J. C. Scaiano and K. U. Ingold, *J. Amer. Chem. Soc.*, **99**, 2079 (1977).
69. H. G. Viehe, R. Merenyi, L. Stella and Z. Janušek, *Angew. Chem.* (*Intern. Ed. Engl.*), **18**, 917 (1979) and references cited.
70. B. Giese and J. Meixner, *Chem. Ber.*, **114**, 2138 (1981).
71. Z. Janušek, S. Piettre, F. Gorissen-Hervens and H. G. Viehe, *J. Organometal. Chem.*, **250**, 197 (1983).
72. B. Giese, *Angew. Chem.* (*Intern. Ed. Engl.*), **22**, 753 (1983).
73. S. Piettre and H. G. Viehe, personal communication (I.R.S.I.A. Research Report of S.P.).

The Chemistry of Organic
Selenium and Tellurium
Compounds Volume 1
Edited by S. Patai and Z. Rappoport
© 1986 John Wiley & Sons Ltd.

CHAPTER **10**

Selenium- and tellurium-containing organic polymers

Y. OKAMOTO

Department of Chemistry, Polytechnic Institute of New York, 333 Jay Street, Brooklyn, New York 11201, USA

I. INTRODUCTION 331
II. ORGANOSELENIUM POLYMERS CONTAINING A MONOSELENIDE CHAIN 332
A. Poly(selenoalkylene)s 332
1. Poly(selenomethylene) 332
2. Poly(selenoalkylene)s 333
B. Oxygen- and Selenium-containing Polymer 333
C. Poly(seleno-*p*-xylene) 334
D. Poly(seleno-1,4-phenylene) 334
E. Poly(seleno-4,4′-biphenylene) 334
F. Poly(seleno-2,5-thienylene) 335
G. Poly(carbon diselenide) 335
III. ORGANOSELENIUM POLYMERS CONTAINING A DISELENIDE CHAIN 336
A. Poly(diselenoalkylene)s 336
1. Poly(diselenomethylene) 336
2. Poly(diselenoethylene) 336
3. Poly(1,3-diselenopropylene) 337
4. Poly(1,4-diselenotetramethylene) 337
B. Poly(α,α′-diseleno-*m*-xylene) 338
C. Poly(1,4-diselenophenylene) 339
IV. POLYSELENIUM-CONTAINING POLYMERS 340
V. ORGANOTELLURIUM POLYMERS 340
VI. REFERENCES 341

I. INTRODUCTION

Recently, the considerable interest in Se- and Te-containing polymers[1,2] has been greatly enhanced by the observation of the highly conductive properties of modified polyarylyene sulphides. The most promising polymer among the latter is the commercially available poly(thio-1,4-phenylene), PPS (**1**). The polymer becomes highly conducting on exposure to strong oxidants[3]. The strong similarity between S and Se has often been a lead in investigating chalcogen analogues of S compounds.

The C—Se bond is much weaker than the C—S bond and the C—Te bond is even weaker than the C—Se bond. Thus, although Se- and Te-containing polymers may be prepared by using methods similar to those containing S, the synthetic routes for Se- and Te-containing polymers are more limited.

There are three general synthetic routes for producing Se-containing polymers (please note that in this chapter the symbol —R— denotes a divalent alkylene or arylene radical):

(1) *Polycondensation of alkyl or aryl dihalide compounds*:

$$X—R—X \ + \ Na_2Se \longrightarrow —\!\!(R—Se)_n\!—$$

$$X—R—X \ + \ Na_2Se_2 \longrightarrow —\!\!(R—Se_2)_n\!—$$

(2) *Hydrolysis and oxidation of alkyl- or aryl-diselenocyanates in an alkaline medium*:

$$NCSe—R—SeCN \longrightarrow HOSe—R—SeOH \longrightarrow —\!\!(Se—R—Se)_n\!—$$

(3) *Ring-opening polymerization of Se-containing cyclic oligomers*: Methods (1) and (2) are the most generally utilized for synthesis of Se-containing polymers. However, these methods often yield both Se-containing cyclic oligomers as well as linear polymers. The third method involves the ring-opening polymerization of the Se-containing cyclic oligomers obtained by methods (1) and (2).

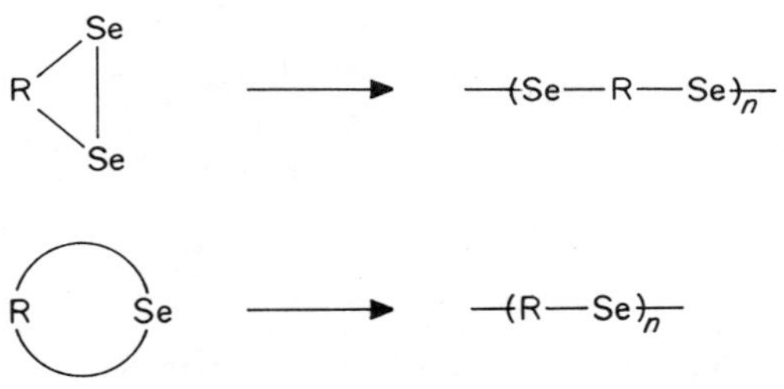

II. ORGANOSELENIUM POLYMERS CONTAINING A MONOSELENIDE CHAIN

A. Poly(selenoalkylene)s, $\!-\!\!\!(C_nH_{2n}Se)\!\!\!-_n$

1. *Poly(selenomethylene),* $-\!\!\!(CH_2Se)\!\!\!-_n$ *(**2**)*

This is one of the most widely investigated organoselenium polymers with respect to both its synthesis and its physical properties.

In 1915 Vanino and Schinner found that on passing H_2Se into an aqueous solution of formaldehyde in dilute HCl, a pasty solid was obtained[4]. The structure of this compound was later determined as cyclic trimer[5]. Various polymers **2** were obtained from the ring-opening polymerization of the trimer with cationic initiation[6,7] or in the solid state by γ-irradiation[8]. Prince and Bremer reported that **2** could also be prepared in 76% yield without the intermediate cyclic oligomers from the reaction of CH_2Br_2 with Na_2Se in ethyl acetate[9].

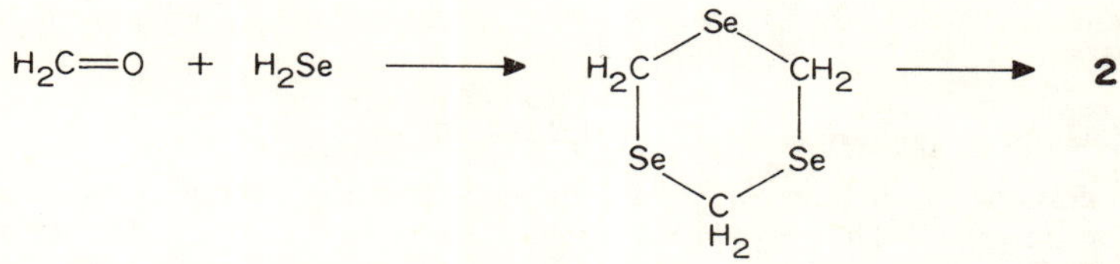

Two crystalline structures of **2**, one hexagonal and one orthorhombic, were obtained and their occurrence was found to be dependent on the polymerization conditions. In particular, the polycondensation method[9] and the ring-opening polymerization[6] gave the hexagonal structure, while solid-state irradiation or solid-state cationic polymerization (BF_3–etherate catalyst at 180 °C) of triselenane yielded an orthorhombic polymer[10]. The two crystalline systems of **2** give rise to a monotropic dimorphic system with an orthorhombic to hexagonal transition temperature near the melting range of the hexagonal polymer (185–190 °C). It is very interesting that the same monotropic dimorphism, orthorhombic–hexagonal, observed in polymer **2**, appears also in polyoxymethylene[11].

Prince and Bremer reported that **2** had an electrical conductivity of 6.72×10^{-4} $\Omega^{-1}\,cm^{-1}$ at 25°C[9]. However, Sandrolini and coworkers who reinvestigated the conductivity of **2** synthesized by various methods found in contrast to Prince and Bremer's data, that none of the samples of **2** studied showed conductivity values exceeding 7.5×10^{13} Ω^{-1} cm^{-1} [10].

2. *Poly(selenoalkylene)s*

The reaction of *vic*-dihaloethanes with Na_2Se in DMSO of DMF did not produce any polymeric compounds, but rather produced ethylene and Se: The elimination reaction was stereospecific, thus *d,l*-2,3-dibromobutane gave 90% *cis*-butene-2[13]. Longer chain α,ω-dibromoalkanes react readily with Na_2Se to produce the corresponding cyclic seleno hydrocarbons. However, Morgan and Burstall found that 1,3-dibromopropane and Na_2Se gave only a small proportion of cycloselenopropane, the main product being a low molecular weight polymer, $\leftarrow\!C_3H_6Se\!\rightarrow_6$, m.p. 38–40 °C[14]. They also obtained poly-(selenohexylene), $\leftarrow\!C_6H_{12}Se\!\rightarrow_n$ ($n \sim 12$), m.p. 36–37 °C, by reacting the corresponding dibromide with Na_2Se[14]. This polymer was thermally stable up to 200 °C, but on further heating to 220 °C, it depolymerized without loss of Se giving 2-methylcycloselenopentane.

$$XCH_2{-}CH_2X \quad + \quad Na_2Se \quad \longrightarrow \quad CH_2{=}CH_2 \quad + \quad 2\,NaX \quad + \quad Se$$

Okamoto and coworkers had prepared these polymers according to the literature procedures and found that the electric conductivities were in the range of $\sim 10^{-14}\,\Omega^{-1}\,cm^{-1}$ [15].

B. Oxygen- and Selenium-containing Polymer

The O- and Se-containing linear copolymer, $\leftarrow\!CH_2O(CH_2Se)_2\!\rightarrow_n$ (**3**), m.p. 195–198 °C, was formed by the cationic (BF_3–etherate) ring-opening polymerization of 1,3,5-

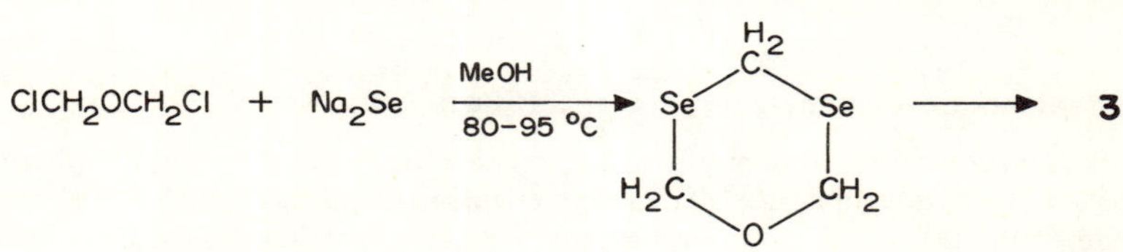

oxadiselenane, prepared by the reaction of α,α′-dichlorodimethyl ether with Na_2Se in methanol[12].

C. Poly(seleno-*p*-xylene), $-(CH_2C_6H_5CH_2Se)_n-$ (4)

The reaction of α,α′-dichloro-*p*-xylene with Na_2Se in refluxing ethyl acetate yielded the pale yellow polymer **4**. The polymer was slightly soluble in boiling pyridine, DMSO and DMF; it was found to have a narrow melting point, 174–175 °C, but the molecular weight was not determined[15].

D. Poly(seleno-1,4-phenylene), $-(C_6H_4Se)_n-$ (PPSe) (5)

The polymer **5**, which is an analogue of poly(thio-1,4-phenylene) (PPS), has recently been investigated with respect to synthetic methods and physical properties, particularly electric properties.

Cava and coworkers reported that **5** was synthesized from bis(4-bromophenyl) diselenide by reaction with electrolytic copper in refluxing *n*-hexanol-pyridine[16]. Sandman and coworkers found that **5** could also be easily obtained in 80% yield from the reaction between *p*-dibromobenzene with Na_2Se in DMF[17].

The polymer obtained was yellow and melted at 220 °C. Elemental analysis showed that the polymer contained bromine and, assuming that both ends of the polymer chain were bromine atoms, the molecular weight of the polymer was calculated to be about 10,000 ($n = 60$). The polymer was partially crystalline and its density was 2.05–2.15 $g\,cm^{-3}$. The structure was found to be isomorphous to that for PPS[13].

Tanaka and coworkers also prepared **5** ($n \sim 25$), m.p. 250 °C, from the reaction of *p*-dibromobenzene with Na_2Se in *N*-methylpyrrolidone in a sealed tube at 180 °C[19]. The polymer obtained was thermally stable up to 350 °C in a N_2 atmosphere and yielded a residue of 46% upon heating to 500 °C. The electric properties of **5** and the modified polymers prepared by different investigators are summarized in Table 1.

The conductivities of the doped **5** varied. However, Sandman and coworkers[18] showed that doping of **5** with AsF_5 led to an insulator–conductor transformation and this was accompanied by structural cross-linking and apparent dibenzoselenophene formation which is analogous to processes occurring for PPS[3].

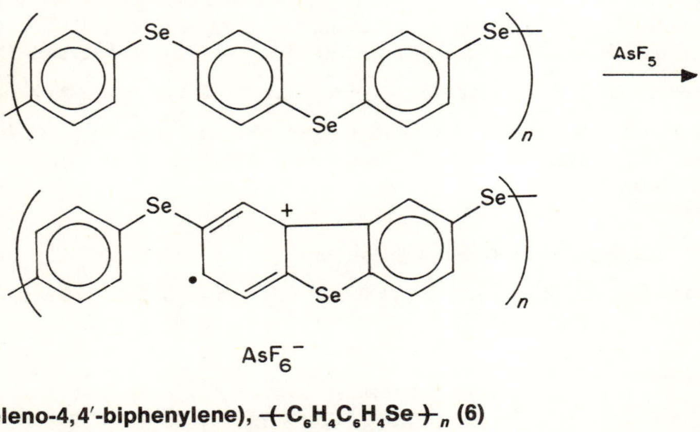

E. Poly(seleno-4,4′-biphenylene), $-(C_6H_4C_6H_4Se)_n-$ (6)

Tanaka and coworkers have prepared **6** in a sealed tube using 4,4′-diiodobiphenyl with Na_2Se in *N*-methylpyrrolidone[19]. The polymerization degree *n* was calculated as $n \sim 24$ by the determination of the iodine end-groups. **6** was also synthesized by the polycondensation reaction of the Grignard reagent of bis(4-bromophenylselenide) with 2,2′-

TABLE 1. Electric conductivities of PPSe

Dopant	Conductivity ($\Omega^{-1}cm^{-1}$)	T°(C)	Ref.
None	8.3×10^{-3}	50	17
AsF_5	10^{-6}	25	14
	10^{-8} (after 24 h)	25	14
AsF_5	10^{-2}–10^{-3} (40–45% weight increase)	25	16
SO_3^-	1.5×10^{-7}	25	17
BF_3	3.4×10^{-8}	25	17
I_2	7.1×10^{-12}	25	17
SbF_5	1.3×10^{10}	25	17

bipyridine nickel dichloride. The molecular weight of the **6** thus prepared was found to be much lower than that of **6** prepared by the former method[19]. Okamoto and coworkers found that **6** was obtained from the reaction of the Grignard reagent of 4,4′-diiodobenzene with Se followed by oxidation[15].

The electrical conductivities of **6** as well as of the doped polymer were in the range of $10^{-2}\ \Omega^{-1}\ cm^{-1}$ [19].

F. Poly(seleno-2,5-thienylene) (7)

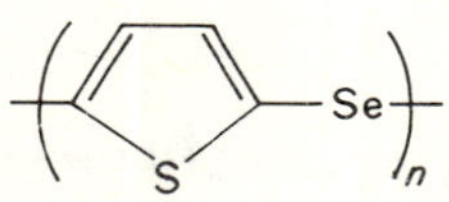

(7)

2,2′-Dithienyl diselenide was treated with SO_2Cl_2 in CH_2Cl_2 in a N_2 atmosphere and the resulting 2-thienylselenyl chloride was then polymerized into **7** in 87% yield. The polymer obtained was a cream-coloured powder and softened at 220–235 °C. A preliminary result showed that the conductivity of **7** was not appreciable and changed after doping with AsF_5[16].

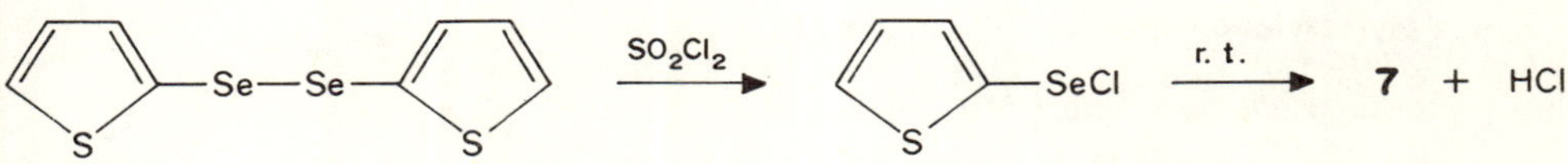

G. Poly(carbon diselenide) (8)

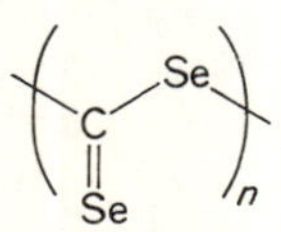

(8)

CSe_2 was reported to yield a black solid at a rate of about 1% per month at room temperature[20,21]. The structure of the solid was investigated by IR over the range of 4000–50 cm^{-1} and found to have a linear polymeric structure similar to that of poly(carbon disulphide)[22].

Jensen and Nielsen have reported that when the complex of CSe_2 and a trialkylphosphine was allowed to stand, it was transformed to a brown–red semisolid from which a dark crystalline material was isolated. The following structure was suggested[23]:

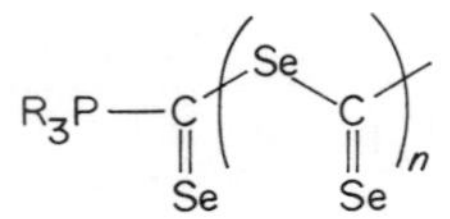

Okamoto and Wojciechowski found that when a CSe_2 solution in CH_2Cl_2 or dioxane was pressurized to about 5000 atm and heated at 100 °C for 20 h a black solid was obtained in 95% yield. When the solid was further heated under reduced pressure, it was transformed into another black solid at around 160 °C. The results of IR and electrical conductivity measurements suggested that CSe_2 was first trimerized into a cyclic compound, which was then polymerized upon heating at 160 °C to yield a linear polymer **8**. The **8** obtained was found to be amorphous by X-ray diffraction, similar to poly(carbon disulphide) and its electrical conductivity was $10^{-3}\,\Omega^{-1}\,cm^{-1}$ at room temperature[24].

The copolymerization of CSe_2 with CS_2 or CSeS was investigated under a higher pressure, ~6000 atm. However, only Se-containing solids were isolated[24].

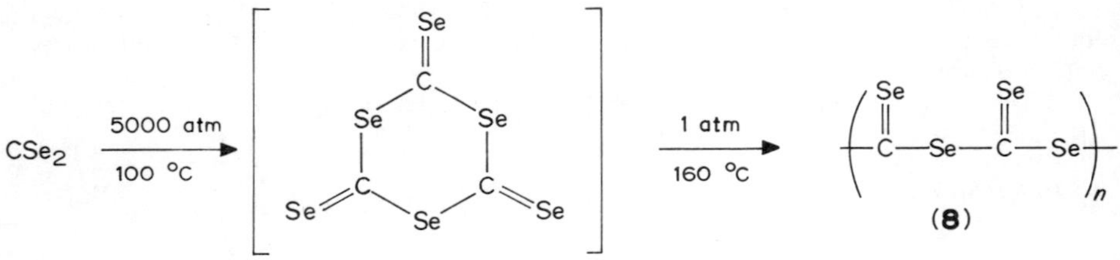

(**8**)

III. ORGANOSELENIUM POLYMERS CONTAINING A DISELENIDE CHAIN

The Se—Se bond is weaker than the S—S bond. The bond energies are 44 and 50.9 kcal mol^{-1}, respectively. Therefore, the Se—Se bond can be easily cleaved and various reversible reactions can be observed. The polymer containing Se—Se bonds may be synthesized by similar reactions to those decribed for the organic monoselenide polymer.

A. Poly(diselenoalkylene)s, $\text{-(}C_nH_{2n}Se_2\text{)-}_n$

*1. Poly(diselenomethylene), $\text{-(}CH_2Se_2\text{)-}_n$ (**9**)*

The reaction of H_2Se with formaldehyde in aqueous solution produced poly(selenomethylene) (**2**). However, Prince and Bremer found that under anhydrous conditions, formaldehyde and Na_2Se reacted to give **9**[25]. Two crystalline forms were found to exist with a transition temperature of about 120 °C. The crystalline red–brown low-temperature solid **9** could be moulded at 70 °C under pressure and the conductivity was found to be $10^{-5}\,\Omega^{-1}\,cm^{-1}$. Paetzold and Knaust also obtained **9** in 98% yield by the reaction of dibromomethane and Na_2Se_2 in aqueous solution[26].

*2. Poly(diselenoethylene), $\text{-(}C_2H_4Se_2\text{)-}_n$ (**10**)*

10 was prepared via a three-step process involving reaction of 1,2-dibromoethane with KSeCN followed by hydrolysis of the diselenocyanate formed with alkali or acid to produce the corresponding diselenol. The diselenol was then oxidized by air in aqueous alcohol to give **10**[15].

$$BrCH_2CH_2Br + 2\ KSeCN \longrightarrow NCSeCH_2CH_2SeCN \xrightarrow[alc.]{KOH} HSeCH_2CH_2SeH \xrightarrow{air} \mathbf{10}$$

The **10** obtained was insoluble in all common organic solvents and had a melting point of 120 °C. The electrical conductivity was $10^{-13}\ \Omega^{-1}\ cm^{-1}$ and no photoconductivity was detected. The polymer was thermally stable at room temperature, but it decomposed into ethylene and Se at around 225 °C.

3. Poly(1,3-diselenopropylene), $\text{-}\!\!(C_3H_6Se_2)\!\!\text{-}_n$ ***(11)***

When 1,3-propylenediselenol, obtained by the method described above, was oxidized with air, a cyclotrimethylene 1,2-diselenide was obtained instead of the polymeric material[27,28]. The cyclic monomer was polymerized upon heating to 80 °C, but when the **11** obtained was further heated to 170 °C, it was depolymerized to the original cyclic monomer[27,28]. The polymer was insoluble in organic solvent.

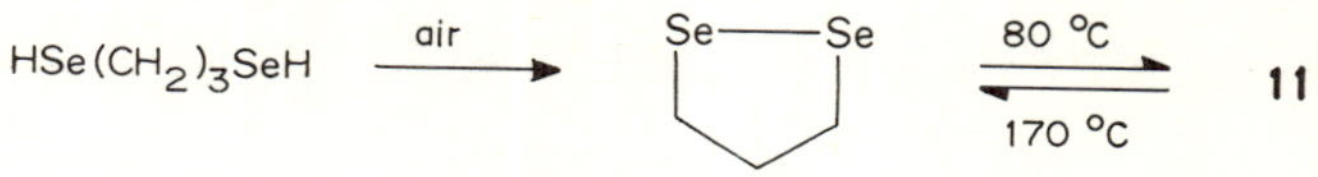

4. Poly(1,4-diselenotetramethylene), $\text{-}\!\!(C_4H_8Se_2)\!\!\text{-}_n$ ***(12)***

The polymer **12** was obtained by air oxidation of 1,4-butanediselenocyanate[29,30]. Brown and Gillman reported in detail the properties of **12**, which was obtained from this precursor[31]. The polymer was purified by dissolving it in warm chloroform and then cooling the solution at 0 °C, and it melted at 38–39 °C with decomposition. The molecular weight was found to be 516 in a freshly prepared $CHBr_3$ solution using a cryoscopic method. However, when the solution was heated at 60 °C in the dark for 24 h, the molecular weight was decreased and ultimately reached a steady value of 209.7, which is close to the value of 214.0 expected for 1,2-diselenotetramethylene. A similar decrease in the molecular weight with time was also detected by viscosity measurements. The UV spectrum of a freshly prepared solution of **12** in $CHCl_3$ gave an absorption maximum at 308 nm which is characteristic of the Se—Se bond in linear diselenide. On allowing the solution to stand at 60 °C in the presence or absence of air, the peak at 308 nm disappeared while a new maximum at 364 nm appeared. The latter absorption is characteristic of the Se—Se bond in the cyclic 1,2-diselenide. These observations suggested that the polymer was depolymerized to give the cyclic monomer. The depolymerization was found to obey first-order kinetics with $k = 2.75 \times 10^{-5}\,s^{-1}$ at 60 °C in $CHCl_3$ and $8.96 \times 10^{-5}\,s^{-1}$ at

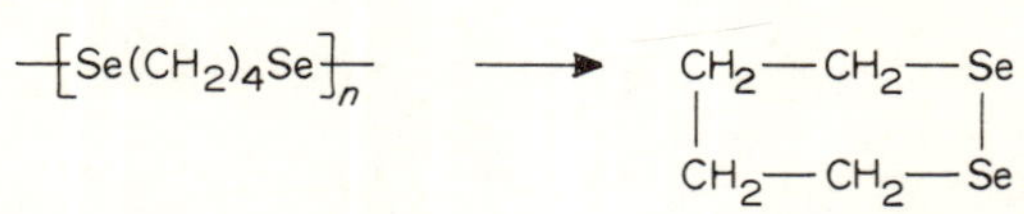

room temperature by irradiation with 366 nm. Since the depolymerization might involve the formation of radicals, the polymerization of various vinyl monomers in the presence of **12** was investigated by Brown and Gillman[31]. The vinyl monomers studied included styrene, methyl methacrylate, vinyl acetate and acrylonitrile. However, the alkylselenium radicals produced thermally did not initiate the polymerization of these vinyl monomers

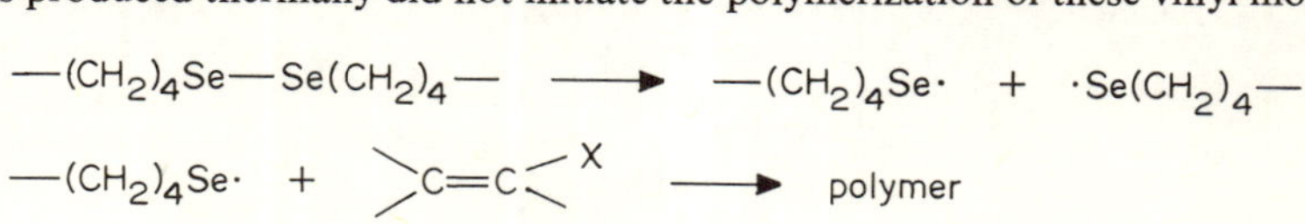

(except possibly for methyl methacrylate). However under photodepolymerization conditions (360 nm at 25 °C), the polymerizations of these vinyl monomers were accelerated in the presence of **12**, and the degree of polymerization was decreased. This behaviour resembles that of linear disulphides which show little activity as thermal initiators of vinyl polymerization but which are excellent photoinitiators[32]. Since the Se end-radicals were not capable of initiating the polymerization of monomers, the results might be explained if the light was absorbed by the Se—Se bond, but the adjacent C—Se bond was broken. The resulting methylene radical might initiate the polymerization of the monomers and then the growing polymer radical could react with the diselenide, with the polymer **12** acting as a chain-transfer agent. Thus, this mechanism (Scheme 1) explains the decrease in the degree of the polymerization in the presence of **12**.

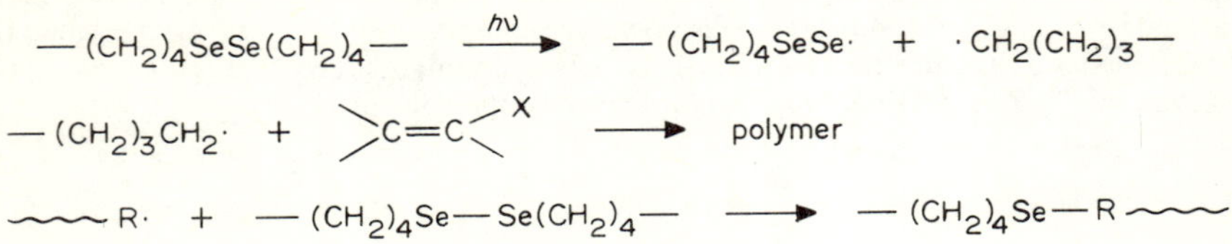

SCHEME 1

A carboxylic-acid-substituted poly(diselenoalkylene) can be prepared by the ring-opening polymerization of the corresponding substituted cyclodiselenide. Fredga reported the preparation of a cyclodiselenide–dicarboxylic acid which was readily converted to a functionalized diselenoalkylene polymer[33].

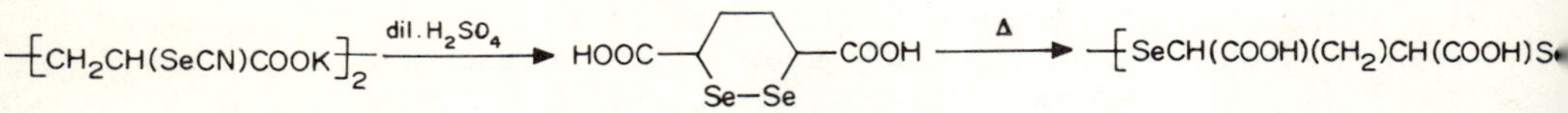

A Se—S bond in a cyclic compound was also found to be polymerized by ring opening. Bergson and Biezais reported that 1-thia-2-selenolane-4-carboxylic acid was polymerized at just above its melting point, to a polymeric material for which two possible alternative structures have been proposed, as shown in Scheme 2[34].

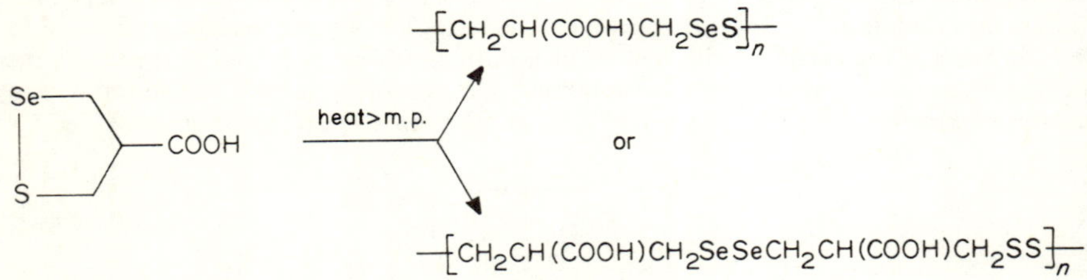

SCHEME 2

B. Poly (α,α′-diseleno-*m*-xylene) (13)

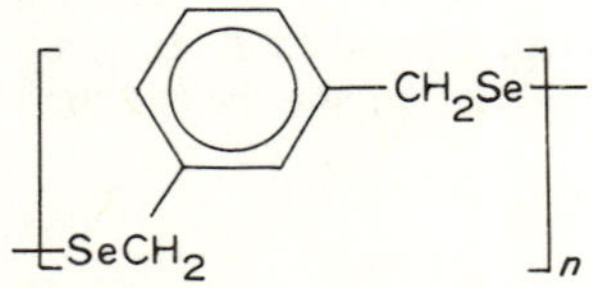

(13)

Günther and Salzman have reported that benzylic polydiselenides can be produced fairly readily via a thermal polymerization of a cyclic dimer[35]. The polymer **13** was obtained as a yellow amorphous powder. When a solution of **13** in 1,1,2,2-tetrachloroethane (TCE) was heated at 80–100 °C for several days it depolymerized into the cyclic dimer (Scheme 3) which was insoluble in organic solvents.

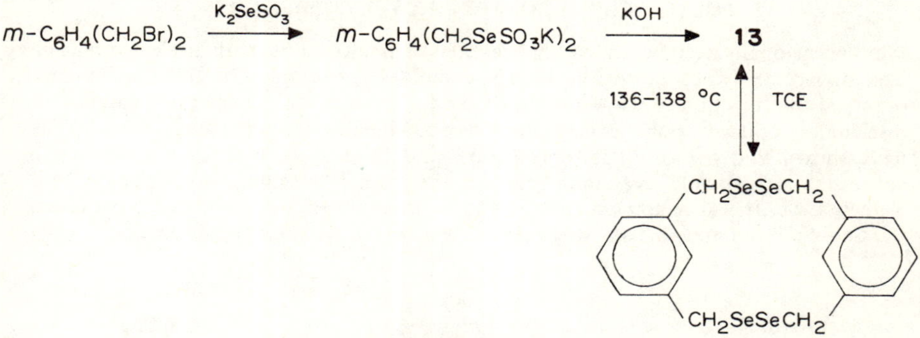

SCHEME 3

C. Poly(1,4-diselenophenylene) (14)

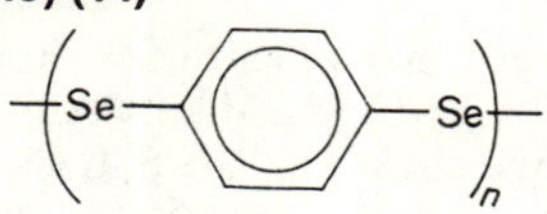

(**14**)

The aromatic Se—C bond is generally more stable than that of the aliphatic Se—C bond. Keimatsu and Satoda[36] first prepared **14** in 1935 and Okamoto and coworkers later modified their method[15]. The synthetic route is shown in Scheme 4. The possible precursor of the polymer, paraselenocyclophane, could not be isolated. The **14** obtained melted at 248 °C and was stable up to 400 °C in a N_2 atmosphere.

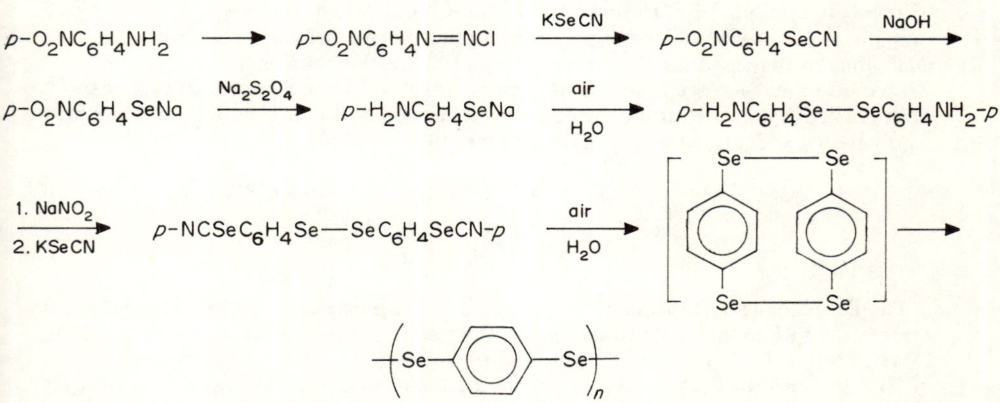

(**14**)

SCHEME 4

Günther and Saltzman also obtained **14** (m.p. 252–256 °C) and showed that it was a crystalline tetramer (mol. wt. 930) of *p*-phenylene diselenide [35]. Okamoto and coworkers also concluded by mass spectrometric analysis that **14** was the cyclic tetramer and found that the oligomer had an electrical conductivity in the range of $10^{-13}\,\Omega^{-1}\,cm^{-1}$ [15].

IV. POLYSELENIUM-CONTAINING POLYMERS, $-(RSe_x)-_n$

It has occasionally been found that the reactions of dihaloalkanes with diselenide reagents yield unexpectedly products with a high Se content. For example, Günther and Saltzman observed that reaction of α,α′-dibromo-(*m*- and *p*-)xylenes with bis(methoxymagnesium) diselenide produced a polymer containing six equivalents of Se per xylene moiety[35]. They have shown that the result is reproducible and have accounted for the product by suggesting a possible reaction between the initial product, a diselenide, with elemental Se. It was suggested that Se can be incorporated into the chain containing the Se—Se bond and this alloying process was investigated on the diselenide.

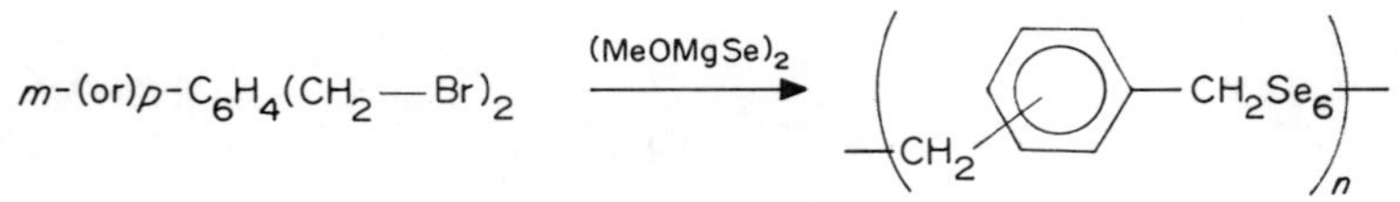

Günther and Saltzman showed that when the cyclic dimer shown in Scheme 3 was heated with elemental Se at 220–230 °C, 15 additional equivalents of Se per monomer unit were readily assimilated into the chain and the polyselenide compound was thermally stable to 140 °C, but on further heating, it disproportionated with deposition of grey Se. However, reheating the mixture to 220 °C reproduced the original alloy[35].

V. ORGANOTELLURIUM POLYMERS

Organotellurium compounds are generally less well known than organoselenium compounds. Only a few organotellurium polymers have been reported in the literature. Organic tellurides and ditellurides may be prepared by the reaction of alkyl or aryl halides with Na_2Te and Na_2Te_2. However, Russo and Credali noted that it was not possible to prepare a cyclic oligomer or a polymer of telluroformaldehyde by a synthetic route analogous to that used for the preparation of poly(selenomethylene)[11].

Livingston and Krosec reported the first polymer containing Te as part of a chain[37]. They reacted diphenyltellurium dichloride with dicarboxylic acid silver salts and obtained a polymer ($R = C_3H_6$) having a molecular weight of ~3860.

$$Ph_2TeCl_2 + AgOOC{-}R{-}COOAg \longrightarrow -[Te(Ph_2)OOC{-}R{-}COOTe(Ph_2)]_n- + AgCl$$

$$R = C_2H_4,\ C_3H_6$$

The Te analogue of PPS and PPSe, poly(telluro-1,4-phenylene), $-(C_6H_4Te)-_n$ (**15**), was synthesized by Sandman and coworkers by the reaction of *p*-diidobenzene with Na_2Te in DMF at 110–120 °C in 70% yield[38]. The infrared spectrum of **15** is superimposable on that of PPS between 4000 and 600 cm^{-1} while the bands at 550 and 475 cm^{-1} in PPS are shifted to 489 and 465 cm^{-1} in **15**. The molecular weight was calculated from the iodine content to be about 8000, and the polymer decomposed at 162–170 °C.

The electrical conductivity of **15** at room temperature was $10^{-11}\,\Omega^{-1}\,cm^{-1}$. When **15** was exposed to iodine vapour, the weight was increased by 150% and the conductivity increased to 10^{-8}–$10^{-6}\,\Omega^{-1}\,cm^{-1}$. The large increase in weight was accounted for by the formation of tetracoordinate Te in the polymer, by attachment of two iodines to each Te atom. This species was transformed back to **15** by reaction with Na_2S. When **15** was treated with AsF_5, the tan colour initially turned to black and then returned to tan. The conductivity did not increase appreciably. The IR spectrum of this material after NH_3 treatment and salt removal revealed that structural modification had occurred in **15**. Polymer **15** was decomposed by $SbCl_5$ in CH_2Cl_2 at 0 °C[18].

VI. REFERENCES

1. Y. Okamoto and W. H. H. Günther (Eds.), 'Organic Selenium and Tellurium Chemistry', *Ann. N.Y. Acad. Sci.*, **192** (1972).
2. L. Mortillaro and M. Russo, in *Organic Selenium Compounds: Their Chemistry and Biology* (Eds. D. L. Klayman and W. H. H. Günther), Wiley–Interscience, New York, 1973.
3. R. H. Baughman, J. L. Bredas, R. R. Chance, R. L. Elsenbaumer and L. W. Shacklette, *Chem. Rev.*, **82**, 209 (1982).
4. J. Vanino and A. Schinner, *J. Prakt. Chem.*, **91**, 116 (1915).
5. H. J. Bridger and R. W. Pitman, *J. Chem. Soc.*, 1371 (1950).
6. L. Mortillaro, L. Credali, M. Russo and C. De Checchi, *J. Polym. Sci. (B)*, **3**, 581 (1965).
7. M. Russo, L. Mortillaro, L. Credali and C. De Checchi, *J. Polym. Sci (A-1)*, **4**, 248 (1966).
8. G. Garazzolo and M. Mammi, *J. Polym. Sci. (C)*, **16**, 1521 (1967).
9. M. Prince and B. Bremer, *J. Polym. Sci. (B)*, **5**, 843 (1967).
10. F. Sandrolini, P. Manaresi, S. Pietra, L. Mortillaro and M. Russo, *J. Polym. Sci. (B)*, **6**, 749 (1970).
11. M. Russo and L. Credali, *J. Macromol. Sci.*, **Al(3)**, 387 (1967).
12. M. Russo, L. Mortillaro, L. Credali and C. De Checchi, *J. Polym. Sci. (B)*, **4**, 167 (1966).
13. M. Prince and B. Bremer, *J. Org. Chem.*, **31**, 4292 (1966); **32**, 1655 (1967).
14. G. T. Morgan and F. H. Burstall, *J. Chem. Soc.*, 173 (1931).
15. Y. Okamoto, T. Yano and R. Homsany, *Ann. N.Y. Acad. Sci.*, **192**, 60 (1972).
16. M. Cava, M. V. Lakshmikanthan, K. Y. Jen, N. Benfaremo, W. S. Huang and A. G. MacDiamid, *Polymer Preprints, ACS*, **24**, 251 (1983).
17. D. J. Sandman, M. Rubnet and L. Samuelson, *J. Chem. Soc., Chem. Commun.*, 1133 (1982).
18. L. A. Acampora, D. L. Dugger, T. Emma, J. Mohammed, M. Rubner, L. Samuelson, D. J. Sandman and S. Tripathy, *A.C.S. Org. Coat. Appl. Polym. Sci. Proc.*, **48**, 556 (1983).
19. S. Tanaka, M. Sato, K. Kaeriyama, H. Kanetsura, M. Kato and Y. Suda, *Makromol. Chem. Rapid. Commun.*, **4**, 231 (1983).
20. H. G. Grimm and H. Metzger, *Ber.*, **69**, 1356 (1936).
21. D. J. G. Ives, R. W. Pittman and W. Wardlaw, *J. Chem. Soc.*, 1080 (1947).
22. A. J. Brown and E. Whalley, *Inorg. Chem.*, **7**, 1254 (1968).
23. K. A. Jensen and P. H. Nielsen, *Acta Chem. Scand.*, **17**, 549 (1963).
24. Y. Okamoto and P. S. Wojciechowski, *J. Chem. Soc., Chem. Commun.*, 386 (1982).
25. M. Prince and B. Bremer, *J. Polym. Sci. (B)*, **5**, 847 (1967).
26. R. Paetzold and D. Knaust, *Z. Chem.*, **10**, 269 (1970).
27. L. Hagelberg, *Ber.*, **23**, 1083 (1890).
28. G. T. Morgan and F. H. Burstall, *J. Chem. Soc.*, 1497 (1930).
29. G. T. Morgan and F. H. Burstall, *J. Chem. Soc.*, 1096 (1929).
30. G. Bergson, *Arkiv Kemi*, **13**, 11 (1958).
31. J. R. Brown and G. P. Gillman, *J. Polym. Sci. (A-1)*, **5**, 903 (1967).
32. K. Tsuda, S. Kobayashi and T. Otsu, *Bull. Chem. Soc. Japan*, **38**, 1517 (1965).
33. A. Fredga, *Arkiv Kemi, Mineral. Geol.*, **11B**, No. 15, 1 (1933); *Uppsala Univ. Arsskrift*, No. 5, 125 (1935).
34. G. Bergson and A. Biezais, *Arkiv Kemi*, **18**, 143 (1961).
35. W. H. H. Günther and M. N. Salzman, *Ann. N.Y. Acad. Sci.*, **192**, 25 (1972).
36. S. Keimatsu and I. Satoda, *J. Pharm. Soc. Japan*, **55**, 58 (1935).
37. H. K. Livingston and R. Krosec, *Polymer Letters*, **9**, 95 (1971).
38. D. J. Sandman, J. C. Stark, L. A. Acampora and P. Gagne, *Organometallics*, **2**, 549 (1983).

The Chemistry of Organic
Selenium and Tellurium
Compounds Volume 1
Edited by S. Patai and Z. Rappoport
© 1986 John Wiley & Sons Ltd.

CHAPTER **11**

Organometallic compounds with selenium and tellurium atoms bonded to main group elements of Groups IIIa, IVa and Va

RALPH A. ZINGARO

Department of Chemistry, Texas A&M University, College Station, Texas 77843, USA

I. INTRODUCTION . . . 343
II. COMPOUNDS CONTAINING AN X—Se OR X—Te BOND WHERE X IS A GROUP IIIa ELEMENT . . . 344
A. Compounds Containing a B—Se or B—Te Bond . . . 344
B. Compounds Containing an Al—Se, Al—Te, Ga—Se or Ga—Te Bond . . . 348
III. COMPOUNDS CONTAINING AN X—Se OR X—Te BOND WHERE X IS A GROUP IVa ELEMENT . . . 348
A. Compounds Containing a Si—Se or Si—Te Bond . . . 348
B. Compounds Containing a Ge—Se or Ge—Te Bond . . . 356
C. Compounds Containing a Sn—Se or Sn—Te Bond . . . 356
D. Compounds Containing the Pb—Se Bond . . . 359
IV. COMPOUNDS CONTAINING AN X—Se OR X—Te BOND WHERE X IS A GROUP Va ELEMENT . . . 359
A. Compounds Containing a P—Se or P—Te Bond . . . 359
B. Compounds Containing As—Se, As—Te and Other P—Se or P—Te Bonds . . . 362
V. ACKNOWLEDGEMENT . . . 365
VI. REFERENCES . . . 365

I. INTRODUCTION

This chapter is concerned with developments which have taken place in recent years and which describe organometallic compounds bearing X—Se or X—Te bonds; specifically, X = B, Al, Ga, In, Tl from Group IIIa, X = Si, Ge, Sn, Pb from Group IVa and X = P, As,

Sb, Bi from Group Va. Compounds containing C—Se or C—Te bonds are not included in this chapter. They form the major portion of the other chapters which make up this volume.

Compounds of the type which form the subject of this chapter are relatively few in number. The literature is not extensive. Among the reasons for the relative paucity of literature references in this field is the fact that many X—Se and X—Te bonds are thermodynamically, and often photochemically, unstable.

The reader should consult the 'Annual Surveys on Tellurium' prepared by Irgolic which appear regularly in the *Journal of Organometallic Chemistry*. Unfortunately, 'Annual Surveys on Selenium' have not appeared as part of this series. A very useful review entitled 'The Ligand Chemistry of Tellurium' by Gysling[1] furnishes the reader with a detailed coverage of Te—M bonded compounds in which M is a transition metal.

II. COMPOUNDS CONTAINING AN X—Se OR X—Te BOND WHERE X IS A GROUP IIIa ELEMENT

A. Compounds Containing a B—Se or B—Te Bond

Alkaneselenols or benzeneselenol undergo reaction with boron trihalides in carbon disulphide to give tris(organoseleno)boranes[2]. The alkyl derivatives, $B(SeR)_3$ (R = Et, *i*-Pr, *n*-Bu), are pale yellow liquids, whereas tris(phenylseleno)borane, $B(SePh)_3$, is a white solid. The compounds can also be prepared from the selenolates.

$$BY_3 + 3\ NaSeR \longrightarrow B(SeR)_3 + 3\ NaY$$

$$Y = Cl, Br, I$$

The B—Se bond is ruptured by nucleophilic protic reagents.

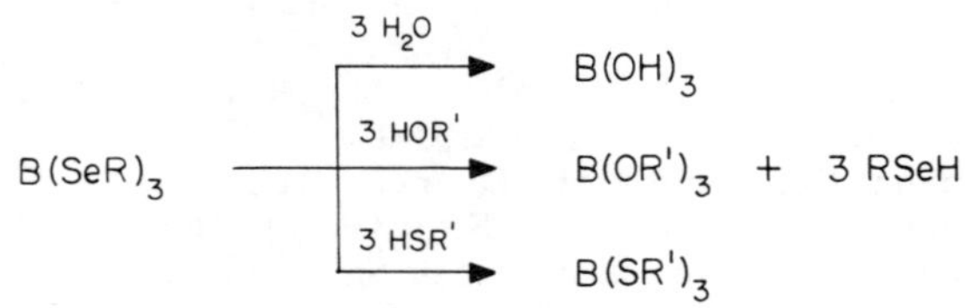

Siebert and Ospici[3] have described the synthesis of alkylseleno(diiodo)boranes, RSe—BI_2, (alkylseleno)alkyliodoboranes, RSe—BR′I, and other derivatives of the type $(MeSe)_nBR'_{3-n}$ where $n = 1,2$ and R′ = Me, Ph.

For the synthesis of the diiodoboranes, the following reactions have been utilized. The

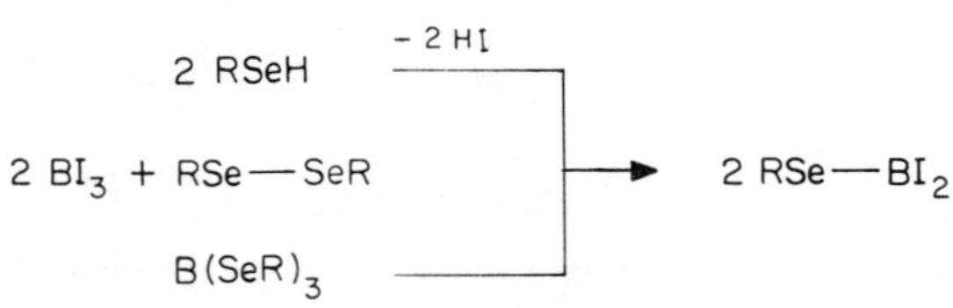

R = Me, Ph

monoiodoborane derivatives are prepared by the reaction between a diiodo(organyl)-borane and a diselenide or a (diorganylseleno)organylborane. The term 'organyl' is used in this chapter to include both alkyl and aryl substituents. The reaction between

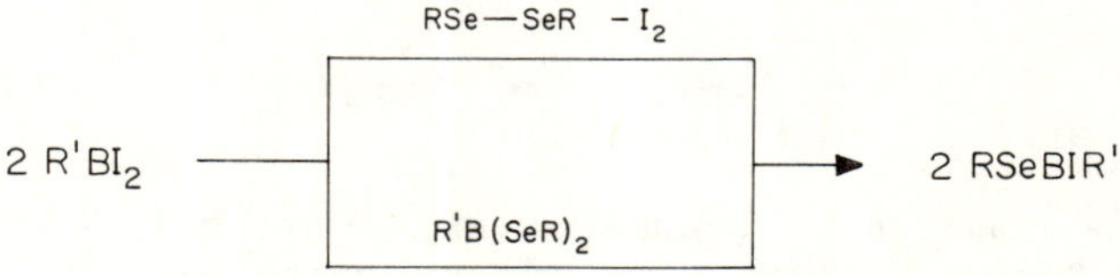

R = Me, 1-Pr; R' = Me, Ph

triiodoborane and a diselenide or tri(organylseleno)borane gives the iododi-

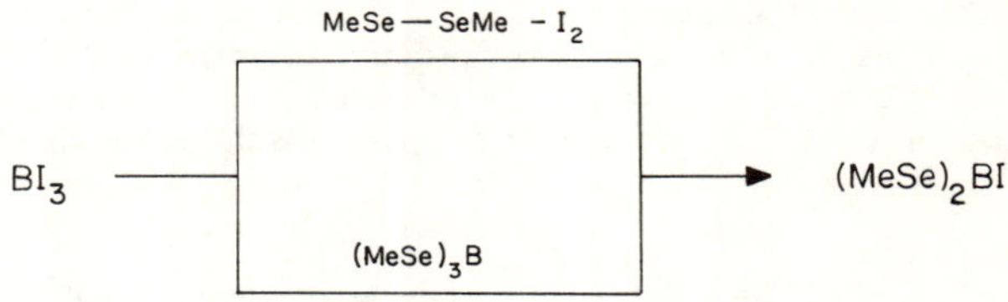

(organylseleno)borane. The reactions are carried out in carbon disulphide under a nitrogen atmosphere. Methylselenoboranes of the composition MeSe—BI_2 and MeSe—B(Me)I form stable trimers. However, $(MeSe)_2BI$, MeSe—B(Ph)I and $(MeSe)_nBR'_{3-n}$ (R' = Me, Ph) are monomeric. The ^{11}B-NMR chemical shift observed for the trimer, $[MeSe—BMeI]_3$, is 11.1 ppm. This is consistent with an sp^3-hybridized boron atom.

The synthesis of 1,2,4,3,5-triselenadiborolanes has been described by Siebert and Riegel[4]. This ring system has been synthesized in a variety of ways. The reaction between tri(*n*-butyl)borane and elemental Se yields the ring system shown below. Elemental Se also

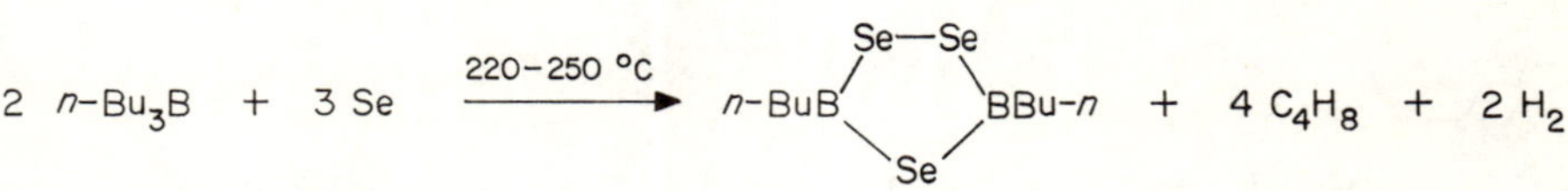

undergoes reactions with various boranes to yield derivatives having the same ring system.

$$2\ RBI_2 + 3\ Se \longrightarrow (RB)_2Se_3 + 2\ I_2$$

R = Me, Ph

$$(c\text{-}HexBH_2)_2 + 3\ Se \xrightarrow{\Delta} (c\text{-}HexB)_2Se_3 + 2\ H_2$$

$$2\ [(n\text{-}Pr)_2BH]_2 + 3\ Se \xrightarrow{\Delta} (n\text{-}PrB)_2Se_3 + 2\ (n\text{-}Pr)_3B + 2\ H_2$$

The B—Se bonded compound, PhSe—BI_2, disproportionates at room temperature to yield the diphenyl derivative. The reaction between boron tribomide and dicyclopen-

$$3\ PhSe{-}BI_2 \longrightarrow (PhB)_2Se_3 + PhBI_2 + 2\ I_2$$

tadienyltitanium pentaselenide (Cp_2TiSe_5) gives the Br-substituted derivative. NMR

chemical shifts for some of the triselenadiborolanes are given in Table 1.

B—Se—B bonds are formed by the reaction between dimethyltin selenide and dimethylboron iodide in carbon disulphide[5]. The product is removed by vacuum

$$2\ R_2BI + Me_2SnMe \longrightarrow R_2B{-}Se{-}BR_2 + Me_2SnI_2$$

distillation from the solvent. Compounds of the same type shown have been isolated for R = *c*-Hex or Ph. When R = Me or *n*-Bu the product polymerizes to $(RBSe)_n$.

Compounds having B—Se—B bonds are prepared in two ways. The first utilizes the reaction between $(Cp_2)TiSe_5$ and Me_2BI in carbon disulphide at room temperature.

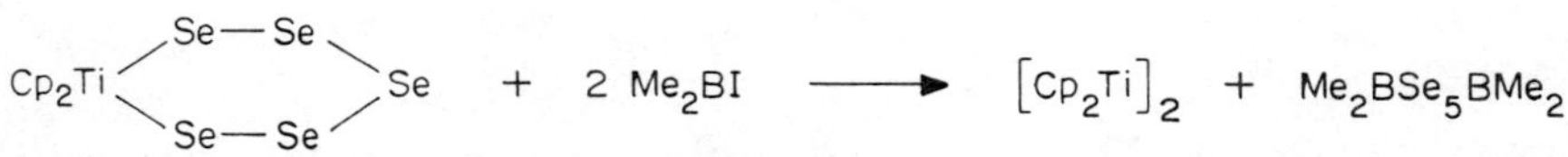

$$Me_2BSe_5BMe_2 \xrightarrow{35\ ^\circ C} Me_2BSe{-}SeBMe_2 + 3/8\ Se_8$$

The second procedure involves the reaction between a borolane and elemental Se. Thus, 3-methylborolane reacts with elemental Se at 120–140 °C to give bis(3-methylboranyl) diselenide.

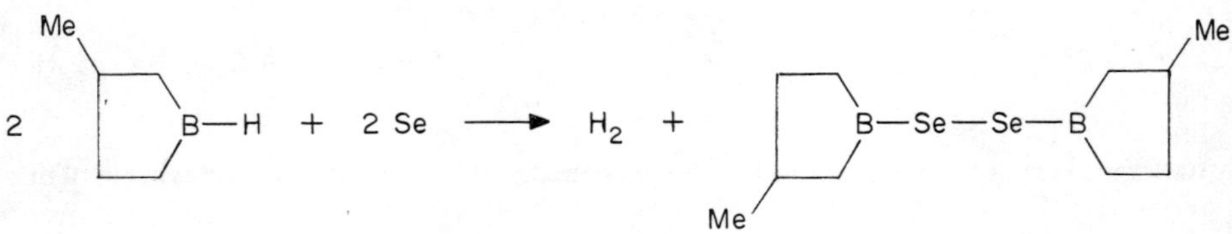

TABLE 1. NMR chemical shifts for triselenadiborolanes $(RB)_2Se_3$

R	NMR chemical shift, (ppm)
Me	– 0.93 (s)
n-Pr	– 1.16 (m); – 1.75 (m)
n-Bu	– 1.2 (m)
Ph	– 7.8 (m)
p-Tol	– 2.45 (6,s); – 7.22 (4d); – 7.83 (4,d), $J = 8$ Hz

The heterocyclic system, 1,2,5-selenadiborolene, has also been reported[6]. Polymeric iodoboron selenide, $\leftarrow\!\!(\mathrm{BISe})\!\!\rightarrow_m$, is first prepared by the reaction between BI_3 and Se at 140 °C. This polymer, when heated with 3, 4-$Me_2C_6H_3BI_2$, gives yellow crystals of xyleno-1, 2, 5-selenadiborolene diiodide.

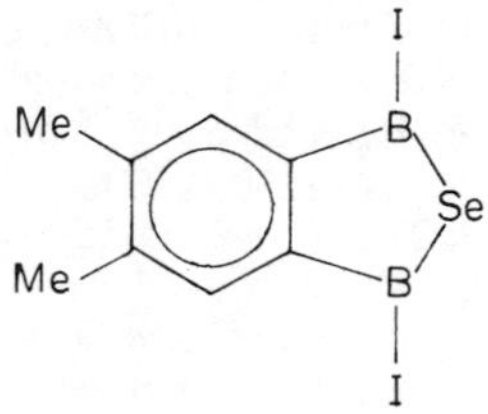

Bregadze and coworkers have reported[7] that B-mercurated and B-thalliated carboranes undergo reaction with elemental Se and Te to give carboranyl derivatives having B—Se or B—Te bonds. The following example illustrates the reaction. The same

$$\mathrm{(C_2H_2B_{10}H_9)_2Hg} + \mathrm{X} \xrightarrow{270\text{–}300\ ^\circ\mathrm{C}} \mathrm{(C_2H_2B_{10}H_9)_2X}_n$$

$$\mathrm{X = Se},\ n = 2\ ;\ \mathrm{X = Te},\ n = 1$$

investigators report that the monotelluride is formed in the reaction between [*m*-carboranyl(9)]thallium bis(trisfluoroacetate) with Te in dimethylformamide in the presence of iodide. The carboranyl derivatives of Se and Te are solids which melt at temperatures > 200 °C.

$$\mathrm{(CF_3CO_2)_2TlB_{10}H_9C_2H_2} + \mathrm{Te} \xrightarrow[\mathrm{DMF}]{\mathrm{I^-}} \mathrm{(C_2H_2B_{10}H_9)_2Te}$$

Clive and Menchen[8–10] have utilized tris(phenylseleno)borane, $(PhSe)_3B$, for organic functional group interconversions. This reagent is a reasonably stable crystalline solid, first reported by Schmidt and Block[11]. It readily undergoes reaction with carbonyl compounds to yield selenoacetals which are easily reduced by organotin hydrides. The reaction sequence utilized by Clive and Menchen is summarized below. Besides the

$$\mathrm{R^1R^2C{=}O} + \mathrm{(PhSe)_3B} \longrightarrow \mathrm{R^1R^2C(SePh)_2} \xrightarrow{\mathrm{Ph_3SnH}} \mathrm{R^1R^2CH_2}$$

deoxygenation of carbonyl compounds, the reagents $(PhSe)_3B$, $MeSe_3B$ and $(n\text{-}BuB)_2Se_3$ deoxygenate sulphoxides efficiently. NMR evidence suggests that the formation of a selenosulphoxide, R_2SSe, as an intermediate is *not* involved. These investigators have suggested the initial formation of the following species:

$$\mathrm{R^1R^2C{=}\overset{+}{O}{-}\overset{-}{B}(SeR)_3} \quad \text{and} \quad \mathrm{R^1R^2C(SeR)OB}\!<\,\}$$

B. Compounds Containing a Al—Se, Al—Te, Ga—Se or Ga—Te Bond

Coates[12] has described the formation of adducts between Me_3Ga or Me_3Al and Me_2Se or Me_2Te. The following liquid adducts were prepared: $Me_3\bar{Ga}$—$\overset{+}{Se}Me_2$, $Me_3\bar{Ga}$—$\overset{+}{Se}Me_2$, $Me_3\bar{Al}$—$\overset{+}{Se}Me_2$ and $Me_3\bar{Al}$—$\overset{+}{Te}Me_2$. Coates found the $Me_3\bar{Ga}$—$\overset{+}{Se}Me_2$ adduct to be more stable than that formed with Me_2S. The greater stability was attributed in part to contributions arising from double-bonded structures of the type $Me_3Ga{=}SeMe_2$.

The Al—Se bond has been the subject of only limited investigations. The air-sensitive compound, $K[MeSe(AlMe_3)_3]\cdot 2\,C_6H_6$ is prepared[13] by the thermal decomposition, in a sealed tube, of $K(Al_2Me_6SeCN)\cdot n\,C_6H_6$ at 80 °C. The crystal structure reveals that the Se atom is tetrahedrally coordinated. The Al—Se bond lengths average 2.578(5) Å and the Se—C bond distance is 1.93(2) Å.

The reaction between Me_3Al and powdered Se in toluene gives dimethylaluminium methylselenolate (**1**)[14]. The compound is prepared *in situ* and reacts with esters to give

$$Me_3Al + Se \xrightarrow[2\ h]{\Delta,\ toluene} Me_2AlSeMe \quad (\mathbf{1})$$

$$PhCOOMe + 1 \longrightarrow PhCOSeMe$$

methyl selenol esters in good yield. With δ-lactones **1** reacts to give the δ-hydroxy selenol

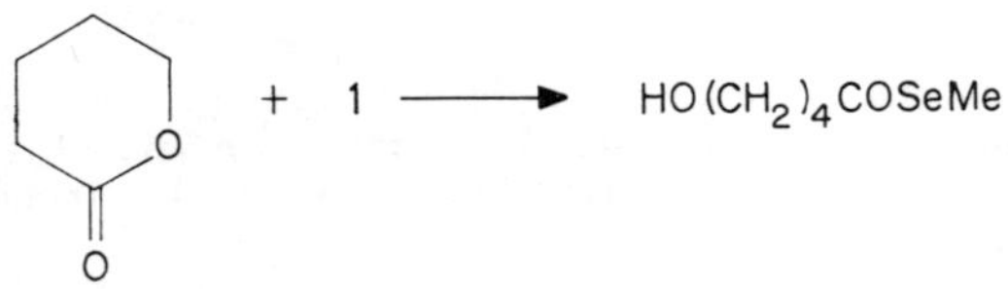

ester. With epoxides **1** gives methylseleno alcohols.

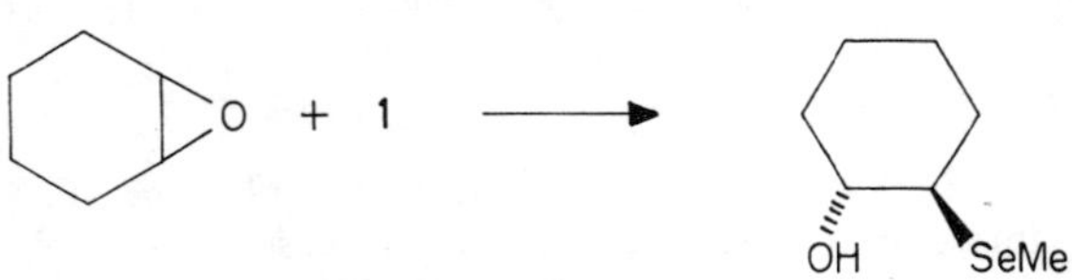

III. COMPOUNDS CONTAINING AN X—Se OR X—Te BOND WHERE X IS A GROUP IVa ELEMENT

A. Compounds Containing a Si—Se or Si—Te Bond

Trisilylamine reacts readily with hydrogen selenide[15] and analagous reactions take place with *N*-methyldisilyamine. If air is eliminated, the salts are stable at room

$$(H_3Si)_3N + 2\,H_2Se \longrightarrow Se(SiH_3)_2 + H_3Si\bar{Se}\ NH_4^+$$

$$2\,(H_3Si)_2NMe + 3\,H_2Se \longrightarrow Se(SiH_3)_2 + 2\,H_3Si\bar{Se}\ NH_3Me^+$$

$$H_3SiNMe_2 + H_2Se \longrightarrow H_3Si\bar{Se}\ \overset{+}{N}H_2Me_2$$

$$H_4\overset{+}{N}\ \overset{-}{Se}SiH_3 + MeI \longrightarrow NH_4I + MeSeSiH_3$$

temperature for about a day. Infrared spectra suggest the presence of H_3SiSe^- anions in the crystal lattice. The salts react slowly with methyl iodide and among the products formed is silyl methyl selenide. With trimethylsilyl chloride, a Si—Se—Si derivative is formed in 50% yield. With F_2PBr the expected $F_2PSeSiH_3$ is not formed. Instead, the

$$H_4\overset{+}{N}\ \overset{-}{Se}SiH_3 + Me_3SiCl \longrightarrow NH_4Cl + Me_3SiSeSiH_3$$

phosphine selenide, $F_2P(Se)SiH_3$, is obtained. The ammonium salts undergo reaction with MeCOCl with the formation of a variety of products among which is $(H_3Si)_2Se$, $(MeCO)_2Se$ and $MeCOSeSiH_3$.

Anderson and coworkers[16] have reported on the synthesis of a number of methylseleno derivatives of the elements of Group IVa. Methaneselenol undergoes reaction with chloromethylsilanes in the presence of Me_3N to give methylselenosilanes in yields up to

$$Me_{4-n}SiCl_n + n\ MeSeH + n\ Me_3N \longrightarrow Me_{4-n}Si(SeMe)_n + n\ Me_3NHCl$$

70%. A second method involves an interesting reagent prepared by these investigators, i.e. lithium tetra(methylseleno)aluminate. The latter undergoes reaction with trimethylstannyl halides to give yields of the desired products. The third synthetic method described by

$$LiAlH_4 + 4\ MeSeH \longrightarrow 4\ H_2 + LiAl(SeMe)_4$$

$$LiAl(SeMe)_4 + 4\ Me_3MX \longrightarrow 4\ Me_3MSeMe + LiX + AlX_3$$

$$M = Si, Ge, Sn;\quad X = Br, Cl$$

these workers involves the reaction between Si—N, Ge—N or Sn—N bonded compounds and methaneselenol.

$$Me_3MMe_2 + MeSeH \longrightarrow Me_3MSeMe + Me_2NH$$

$$M = Si, Ge, Sn$$

Methylseleno(trimethyl)silane can be utilized as a starting material for the synthesis of the Ge and Sn derivatives. The driving force for these reactions is attributed by these

$$n\ Me_3SiSeMe + Me_{4-n}MX_n \longrightarrow Me_{4-n}M(SeMe)_n + n\ Me_3SiX$$

$$M = Ge, Sn;\quad n = 1\text{–}4;\quad X = \text{halogen}$$

investigators to the 'hardness' of Si and the 'softness' of Se, Ge and Sn. The 'harder' halogens form stronger bonds to Si, while the heavier Group IV metals and Se prefer a mutual 'soft–soft' interaction. The exchange reaction does not occur with Me_3PbCl. In this latter case only decomposition products are formed.

NMR studies give evidence of long-range ^1H–^{77}Se NMR couplings. The NMR data and the results of the reaction studies by these investigators are summarized in Table 2, 3 and 4.

Barker and coworkers[17] have utilized the exchange reaction which takes place between the Si—Se and Ge—Cl bonds for the preparation of Ge—Se bonded molecules. A rapid

TABLE 2. The ^{1}H-NMR parameters for the series $Me_{4-n}M(SeMe)_n$ (M = Si, Ge, Sn; $n = 1$–4)[a,b]

Compound	δ(SeMe)	δ(MMe)	$\lvert J(HC^{77}Se)\rvert$	$\lvert J(HCM^{77}Se)$	$\lvert J(^{13}CH)\rvert(M)$	$\lvert J(^{13}CH)\rvert(Se)$	$\lvert J(MCH)\rvert$	$\lvert J(MSeCH)\rvert$
$Me_3SiSeMe$	1.70	0.36	9.8	3.9	122	141	8.1	3.6
$Me_2Si(SeMe)_2$	1.78	0.62	10.5	4.2	122	143	7.8	4.2
$MeSi(SeMe)_3$	1.84	0.93	10.5	3.3	123	143	7.8	5.4
$Si(SeMe)_4$	2.00	—	10.5	—	—	145	—	5.7
$Me_3GeSeMe$	1.78	0.55	10.2	3.6	128	143	—	
$Me_2Ge(SeMe)_2$	1.85	0.85	10.5	2.7	131	142	—	
$MeGe(SeMe)_3$	1.95	1.22	10.8	2.1	131	141	—	
$Ge(SeMe)_4$	2.04	—	10.9	—	—	144	—	
$Me_3SnSeMe$	1.83	0.52	10.5	4.1	131	144	57.8 53.5	33.3 31.5
$Me_2Sn(SeMe)_2$	1.90	0.74	10.8	4.2	132	144	57.2 55.4	37.4 36.2
$MeSn(SeMe)_3$	2.05	1.12	10.8	4.0	132	143	59.7 57.3	46.1 43.9
$Sn(SeMe)_4$	2.09	—	10.8	—	—	141	—	54.0 52.2

[a] All spectra recorded at room temperature; δ in ppm (± 0.02) downfield from internal tetramethylsilane; J in Hz(± 0.2 for $J < 12$; ± 1 for $J > 12$).
[b] Reproduced with permission from Ref. 16.

TABLE 3. The ^{1}H-NMR parameters of some chloro(methyl)seleno derivatives of Si and Ge[a,b]

Compound	δ(SeMe)	δ(MMe)	$\|J(HC^{77}Se)\|$	$\|J(HCM^{77}Se)\|$	$\|J(^{13}CH)\|$(M)	$\|J(^{13}CH)\|$(Se)	$\|J(MCH)\|$
$Me_2SiCl(SeMe)$	1.83	0.66	9.6	3.1	123	145	7.2
$MeSiCl(SeMe)_2$	1.90	0.99	9.9	3.0	122	144	n.o.
$MeSiCl_2(SeMe)$	2.15	1.04	9.7	3.0	122	144	n.o.
$SiCl(SeMe)_3$	2.08	—	10.2	—	—	144	—
$SiCl_2(SeMe)_2$	2.13	—	10.6	—	—	144	—
$SiCl_3(SeMe)$	2.23	—	10.4	—	—	144	—
$Me_2GeCl(SeMe)$	1.99	1.01	9.9	2.4	129	141	—
$MeGeCl(SeMe)_2$	2.03	1.38	9.8	1.8	130	144	—
$MeGeCl_2(SeMe)$	2.19	1.52	10.5	1.8	131	144	—
$GeCl(SeMe)_3$	2.16	—	10.5	—	—	144	—
$GeCl_2(SeMe)_2$	2.26	—	10.5	—	—	145	—
$GeCl_3(SeMe)$	2.44	—	10.0	—	—	144	—

[a] All spectra recorded at room temperature on equilibrium mixtures; δ in ppm ($\pm$ 0.02) downfield from internal tetramethylsilane; J in Hz ($\pm$ 0.1 for $J < 12$; $\pm$ 1 for $J > 12$).

[b] Reproduced with permission from Ref. 16.

TABLE 4. Reactions of methylseleno(trimethyl)silane with halides[d]

Halide	[Halide] (mmol)	[$Me_3SiSeMe$] (mmol)	Reaction time (h)	Product	[Product] (mmol)	[$Me_3SiSeMe$] (mmol)	[Me_3SiCl] (mmol)	
Me_2SiCl_2	0.2	0.5	ca. 170	No reaction		0.5		
$MeSiCl_3$	0.2	0.7	ca. 170	No reaction		0.7	—	
$SiCl_4$	1.0	4.5	72	$Si(SeMe)_4$	0.95	0.5	4.0	
Me_3GeBr	0.6	0.5	0.5	$Me_3GeSeMe$	0.47	Trace	0.49[a]	Me_3GeBr
Me_2GeF_2	0.5	1.1	1.0	$Me_2Ge(SeMe)_2$	0.5	1.0	*ca.*1.0[b]	
$MeGeCl_3$	0.4	1.4	2.0	$MeGe(SeMe)_3$	0.38	ca 0.2	1.2	
$GeCl_4$	0.4	2.0	2.0	$Ge(SeMe)_4$	0.36	0.5	1.5	
Me_3SnCl	0.4	0.5	1.0	$Me_3SnSeMe$	0.38	Trace	0.4	Trace of yellow solid
Me_2SnCl_2	0.9	2.0	1.0	$Me_2Sn(SeMe)_2$	0.86	Trace	1.75	
$MeSnCl_3$	0.8	2.5	1.0	$MeSn(SeMe)_3$	0.75	0.2	2.3	
$SnCl_4$	0.5	2.3	1.0	$Sn(SeMe)_4$	0.45	0.5	2.0	
SiH_3Br	1.0	0.6	1.0	SiH_3SeMe	*c*	—	*c*	SiH_3Br (0.4 mmol)
HBr	0.7	0.5	1.0	HSeMe	0.48	—	49[a]	HBr (0.2 mmol)

[a] Me_3SiBr. [b] Me_3SiF. [c] Inseparable mixture of SiH_3SeMe and Me_3SiBr (approximately equimolar by NMR).
[d] Reproduced with permission from Ref. 16.

$$MeGeH_2Cl + Me_3SiSeMe \longrightarrow MeGeH_2SeMe + Me_3SiCl$$

exchange takes place when an excess of halogenogermane undergoes reaction with $Me_3SiSeMe$. Evidence of the formation of $H_2Ge(SeMe)_2$ was obtained, but the compound was not characterized due to its extreme instability. $MeGeH_2SeMe$ and $Me_2GeHSeMe$ were prepared by the direct combination of the reactants at − 196 °C. The reaction mixture was taken to room temperature for 15 min and the components were separated by trap-to-trap fractionation. The componenets were identified by ^{1}H-NMR as well as IR and Raman spectroscopy. However, the Si—Se and Ge—Se vibrations were not reported.

The preparation of Te—Si and Te—Ge bonded species has been described by Drake and Hemmings[18]. Lithium organotellurolates, LiTeR(R = Me, Ph), are prepared by the addition of elemental Te to the organolithium reagent in THF. The THF is removed by distillation at reduced pressure and the Cl—Si reagent is added to the residual solid. Trimethyl(methyltelluro)silane and trimethyl(phenyltelluro)silane are isolated as pure compounds.

$$Me_3SiCl + LiTeR \xrightarrow{R.T.} Me_3SiTeR + LiCl$$

$$R = Me, Ph$$

They have been characterized by elemental analysis, mass spectrometry, ^{1}H-NMR and vibrational spectroscopy. Both fluorogermanes and chlorostannanes undergo quantitative exchange reactions with the Si—Te derivatives to give tellurogermanes and tellurostannanes.

$$Me_3SiTeMe + H_3GeF \longrightarrow H_3GeTeMe + Me_3SiF$$

$$Me_3SiTePh + Me_3SnCl \longrightarrow Me_3SnTePh + Me_3SiCl$$

The tellurosilanes are sensitive to protolytic cleavage and make possible the synthesis of the otherwise unavailable arenetellurols. On prolonged storage or when subjected to laser

$$2\ Me_3SiTePh + H_2O \longrightarrow (Me_3Si)_2O + PhTeH$$

excitation, the derivatives display a variety of decomposition routes which are summarized below.

$$2\ Me_3SnTeMe \longrightarrow (Me_3Sn)_2Te + Me_2Te$$

$$2\ H_3GeTeMe \longrightarrow (H_3Ge)_2Te + Me_2Te$$

$$2\ H_3SiTeMe \longrightarrow H_2Si(TeMe)_2 + SiH_4$$

The following tentative assignments have been made for the M—Te vibrational frequencies: ν(Si—Te) 327, 227 cm^{-1}; ν(Ge—Te) 257, 238 and 258, 233; ν(Sn—Te) 248, 177.

Charov and coworkers[19] have prepared compounds of the type Et_3ESeLi(E = Si, Ge) as follows: The reaction is run in hexane and the following crystalline compounds are

$$EtLi + [(Et)_3E]_2Se \longrightarrow Et_4E + Et_3ESeLi$$

$$E = Si, Ge$$

obtained: $(Et_3SiSeLi)_4$, $(Et_3SiTeLi)_4$, $(Et_3GeSeLi)_{4.3}$ and $[(i\text{-}Pr)_3GeSeLi]_x$. When treated with haloalkanes the lithium compounds yield E—Se(or Te)—R bonded derivatives.

$$Et_3GeSeLi + n\text{-}BuCl \longrightarrow LiCl + Et_3GeSeBu\text{-}n$$

The compounds $Et_3SiSeBu$-*n*, $Et_3SiTeBu$-*n*, $Et_3GeSeBu$-*n* and $(i\text{-}Pr)_3GeSeBu$-*n* are liquids at room temperature.

Diselenides, following initial treatment with sodium, which presumably converts them to the selenolates, react readily with chlorosilanes[20,21].

$$PhSeSePh + 2\ Na \longrightarrow PhSeNa$$

$$PhSeNa + ClSiMe_3 \longrightarrow PhSeSiMe_3$$

The product, $PhSeSiMe_3$, reacts with alkyl acetates (and also with lactones) in the presence of ZnI_2[22].

$$ROAc + PhSeSiMe_3 \longrightarrow RSePh$$

This reaction allows for the introduction of the phenylseleno group into an organic molecule. The reaction with a lactone, carried out in acetonitrile or toluene, is described as follows:

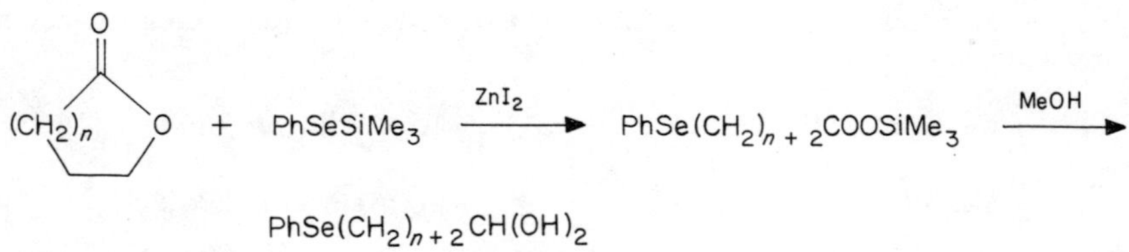

The same reagent also undergoes reaction with epoxides in the presence of ZnI_2 or *n*-BuLi[23].

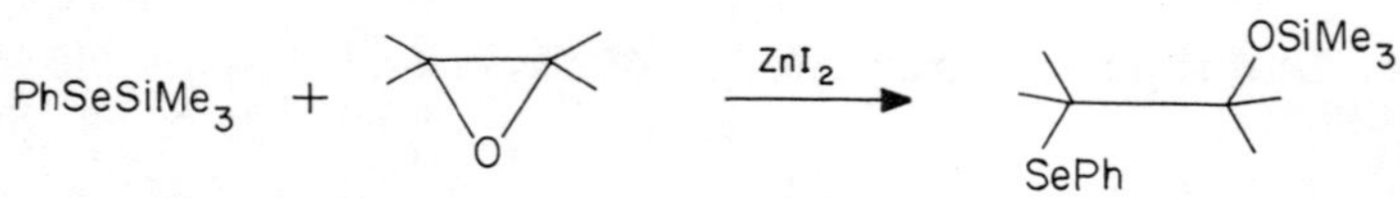

Detty[24] has demonstrated that phenylselenotrimethylsilane is a useful reagent for the gentle reduction (deoxygenation) of sulphoxides, selenoxides and telluroxides. The reaction depicted below is generally applicable. The reaction is successful with a variety of

$$R_2M{\rightarrow}O + 2\ PhSeSiMe_3 \longrightarrow R_2M + (PhSe)_2 + (Me_3Si)_2O$$

$$M = S,\ Se,\ Te$$

functional groups. Detty describes a mechanism in which the oxide oxygen attacks the $SiMe_3$ group to give the onium species together with the phenylselenide anion shown below. Attack by the phenylselenide anion on the onium compound results in the displacement of Me_3SiO^- and formation of the onium species, R_2M^+—SePh. The Me_3SiO^- could attack a second mole of $PhSeSiMe_3$ to give $Me_3SiOSiMe_3$ and $PhSe^-$.

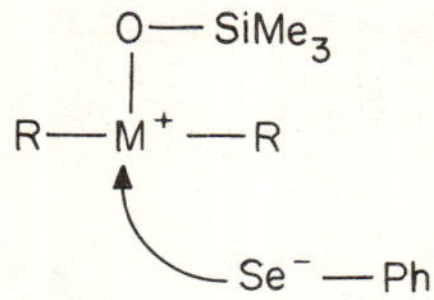

The addition of iodine to phenylselenotrimethylsilane produces trimethylsilyl iodide and diphenyldiselenide[25]. Because of the utility of Me_3SiI in organic synthesis, this has

$$PhSeSiMe_3 + \tfrac{1}{2} I_2 \longrightarrow (PhSe)_2 + Me_3SiI$$

been described as a useful method for its preparation. The reaction between phenylselenotrimethylsilane and alcohols has been recommended for the synthesis of benzeneselenol[26].

$$PhSeSiMe_3 + MeOH \longrightarrow PhSeH + Me_3SiOMe$$

Selenoamides undergo reaction with lithium diisopropylamide in tetrahydrofuran to give the eneselenolates which, on trimethylsilylation, give the vinylselenosilane[27].

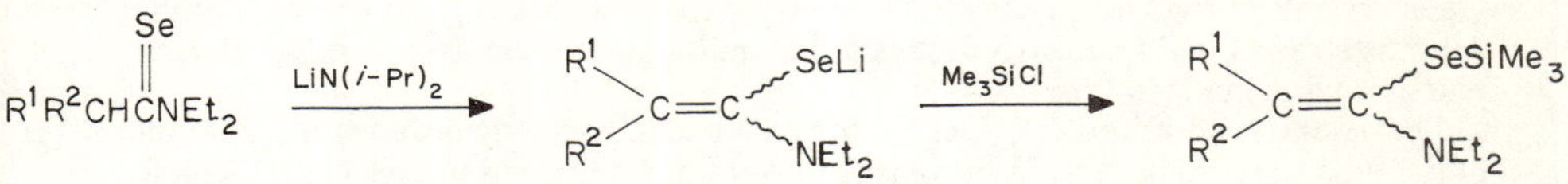

$R^1 = R^2 = H, Me, Et, t\text{-}Bu$

When decamethylcyclopentasilane dissolved in decalin containing a 10% molar excess of Se is heated to 190 °C for 10 h a quantitative yield of the six-membered ring, c-$(Me_2Si)_5Se$, is obtained[28]. The mass spectrum gives a strong parent ion peak and the ^1H-NMR shows three singlets at δ0.496, 0.172 and 0.200 ppm in a 2:2:1 ratio, respectively. Insertion by Se into the Si—Si bond of $(Me_2Si)_6$ does not take place.

A similar study has been reported by Hengge and Schuster[29] Octamethylcyclotetrasilane undergoes a reaction with Se in benzene at 80 °C to give the five-membered ring system, c-$(Me_2Si)_4Se$. The ν(Si—Se) vibration is observed at 368 (Raman) and 348 cm^{-1} (IR).

Praefcke and Weichsel[30] have described the syntheses of 4-methylphenylseleno- and 4-methylphenyltelluro-trimethylsilane. The synthesis if fairly straightforward. The appropriate Grignard reagent is treated with Se or Te to give the arylchalcogenomagnesium

$$p\text{-Tol—Y—MgBr} + SiMe_3Cl \longrightarrow p\text{-Tol—Y—SiMe}_3 + MgBrCl$$

$$Y = Se, Te$$

bromide. The latter, with trimethylsilyl chloride, gives the desired compound. Both the Se and Te compounds are colourless oils.

Bis(trifluorosilyl)tellurium has been prepared by the condensation of Te vapours in an atmosphere of $SiF_3\cdot$ radicals generated from hexafluorodisilane in a low-temperature glow discharge[31]. The compound is unstable at room temperature. It reacts with water and polar organic solvents. It can be vaporized at −30 °C in vacuo without decomposition. The ^{19}F-NMR spectrum at −45 °C gives a singlet at 102.0 ppm from $CFCl_3$ with $J(^{125}Te—F) = 210$ Hz. The $Si_2F_6Te^+$ ion is present in the mass spectrum.

B. Compounds Containing a Ge—Se or Ge—Te Bond

Elemental Se or Te react with trialkylgermanes at high temperatures[32] to form the unstable R_3GeXH compounds (X = Se, Te). The exchange reaction between $(C_6F_5)_3GeBr$

$$R_3GeH + X \xrightarrow{\Delta} R_3GeXH$$

$$X = Se, Te$$

and $(Et_3Ge)_2Se$ is carried out at 150 °C for 35 h. Fractionation of the mixture at reduced pressure gives $(C_6F_5)_3GeSeGeEt_3$ in 66% yield. When $(C_6F_5)_3GeBr$ and $(Et_3Ge)_2$ are heated at 150 °C for 32 h and the products are separated *in vacuo*, $[(C_6F_5)_3Ge]_2$ Se is obtained[33]. In this work, the synthesis of a Ge—Se—Sn bonded compound is also described.

The reaction between $(C_6F_5)_3GeSeGeEt_3$ and Ph_3SnBr in toluene at 100 °C for 2 h gives Et_3GeBr and a solid residue. The residue, recrystallized from hexane, was identified as $(C_6F_5)_3GeSeSn(C_6F_5)_3$.

The four-membered ring systems *c*-2,2,4,4-tetra-(*t*-butyl)-1,3-diselena-2,4-digermane and the analagous 1,3-ditellura compound have been described[34]. The hydride, $[(t\text{-}Bu)_2GeH]_2$ is prepared by the reaction between $GeCl_4$ and excess *t*-BuLi. It reacts with elemental Se to give orange cystals of the four-membered Ge—Se heterocycle or pale green crystals of the Ge—Te heterocycle. Mass spectral data suggest the formation of *c*-3,3,5,5-tetra-(*t*-butyl)-1,2,4-triselena-3,5-digermane as a secondary product during the reaction with elemental Se.

The oxidative addition of $(R_3Ge)_2Cd$ to vanadocene gives organobimetallic compounds of the type $Cp_2V(GeR_3)$. Se inserts easily into the V—Ge bond to give $Cp_2V(SeGeR_3)$[35].

Phenyl(triphenylgermyl)tellurium(II) has been synthesized by Gardner and coworkers[36]. Diphenyl ditelluride in benzene–ethanol is converted to PhTeNa by reduction with sodium borohydride followed by the addition of sodium hydroxide. The addition of triphenylgermanium chloride, followed by work-up, gives a yellow solid, analysed as $PhTeGePh_3$. In a similar manner *p*-$TolTeGePh_3$ and *p*-$AnTeGePh_3$ can also be prepared. In this study, the following M—Te vibrational frequences (cm^{-1}) have been identified by Raman spectroscopy: M = Ge 173, 182; M = Sn 161, 164, 168; M = Pb 147, 154, 153, 156. In derivatives of the type $(Ar_3M)_2Te$, the ν(M—Te) vibrations have been reported as follows: M = Ge 167; M = Sn 153; M = Pb 113. For $(Me_3M)_2Te$, the following assignments were made for ν(M—Te): M = Ge 236; M = Sn 191; M = Pb 161.

C. Compounds Containing a Sn—Se or Sn—Te Bond

Dialkyltin oxides undergo a neutralization reaction with benzeneselenol, and dialkyltin dihalides in the presence of triethylamine react with benzeneselenol to give organotin benzeneselenolates[37], Some reported physical properties of these compounds are the

$$R_2SnO + 2\ PhSeH \longrightarrow R_2Sn(SePh)_2 + H_2O$$

$$R_2SnCl_2 + 2\ PhSeH + 2\ Et_3N \longrightarrow R_2Sn(SePh)_2 + 2\ Et_3NHCl$$

following: *n*-$Bu_2Sn(SePh)_2$, orange oil, ν(Sn—Se) 305 and 275 cm^{-1}; *n*-$Bu_3SnSePh$, red–orange oil, ν(Sn—Se) 295; $Et_2Sn(SePh)_2$, red–orange oil, ν(Sn—Se) 328 and 309; $Ph_3Sn(SePh)\cdot 4\ Me_2CO$, white solid, ν(Sn—Se) 339; $Me_2Sn(SePh)_2$, yellow oil, ν(Sn—Se) 305 and 285.

Japanese workers[38] have described *N,N*-dialkylselenocarbamate complexes of tin(IV) The reaction of a dialkyltin dihalide with the dimethylammonium salt of R_2NCSeO in dry benzene gives derivatives of the type $R^1_2Sn(OSeCNR_2)_2$. The stereochemical configurations of these compounds is uncertain. The fundamental stretching frequencies, ν(Sn—Se), were reported to occur in the range 322–343 cm^{-1}.

A number of heterocyclic systems which have Sn—Se bonds have been described[39–45].

2,4,5-Hexamethyl-1,3-diselena-2,4,5-tristannolane (**2a**) and its Te analogue (**2b**) were prepared[39] by the reaction of dimethyltin dihydride in ether:dimethylformamide (30:1/v/v) with elemental Se or Te under N_2.

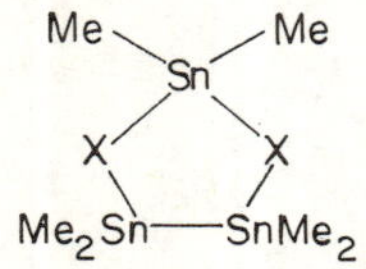

(2)

(a) X = Se

(b) X = Te

The compound $(MeSn)_4Se_6$ has an adamantane-type of structure[41]. It is prepared by the reaction of $MeSnBr_3$ with NaSeH in aqueous media. The crystal structure of the molecule has been determined and its structure is shown below.

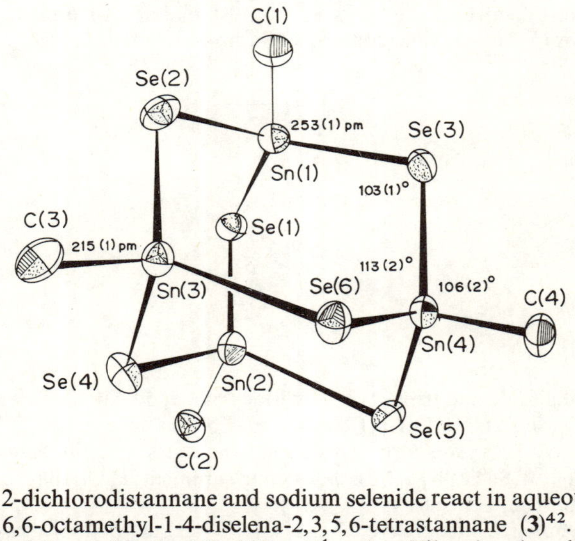

Tetramethyl-1,2-dichlorodistannane and sodium selenide react in aqueous solution, to form 2,2,3,3,5,5,6,6-octamethyl-1-4-diselena-2,3,5,6-tetrastannane (**3**)[42]. The Sn—Se stretching vibration is located in the 217–255 cm^{-1} region. Vibrational analysis excludes a chain conformation for this six-membered ring.

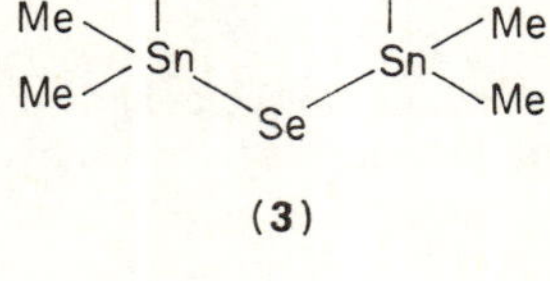

(**3**)

Four-membered cyclic ring systems having two Sn—Se or Sn—Te bonds have been reported[40,43]. The reaction is described by the following equation:

$$2\,(t\text{-Bu})_2SnCl_2 + 2\,Na_2X \longrightarrow 4\,NaCl + [(t\text{-Bu})_2SnX]_2$$

$$X = Se,\ Te$$

The four-membered ring:

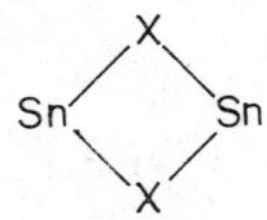

is planar and the Sn—Se distance is 2.55 Å.

The six-membered ring compound, $(Me_2SnSe)_3$, has been prepared and its structure determined[44]. The measured Sn—Se bond distances are reported as 2.51, 2.54 and 2.54 Å. The molecules have a twist-boat conformation.

Sn–Se and Sn–Te five-membered ring compounds (**4**) have been reported[45]. These are prepared by the reaction between $(t\text{-}Bu_2Sn)_4I_2$ and H_2Se or H_2Te. The crystal structures have been determined and the rings are almost planar. An unusual feature of these heterocyclic compounds is their long Sn—Sn bond distances.

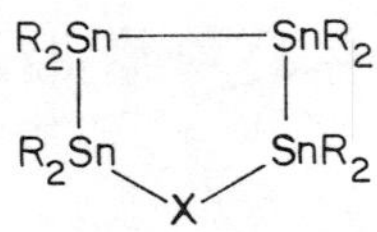

(**4**)

X = Se, Te

R = *t*-Bu

Diphenyl ditelluride and triphenyltin hydride react at 60–70 °C to give $Ph_3SnTePh$. In a similar way $Ph_3SnTe(C_6H_4OR\text{-}p)$ (R = Me, Et, Ph) can be prepared[46]. The reaction of $(t\text{-}Bu)_3SnH$ with $(ArTe)_2$ gives a red oil which, from mass spectral observation, shows the presence of $[(t\text{-}Bu)_3SnTePh]^+$. Mössbauer measurements show that the sign of the ^{119}Sn quadrupole splitting is negative.

The six-membered heterocycle, 2,2,4,4,6,6-hexamethylcyclotristannatellurane, a Sn_3Te_3 ring system, has been prepared by Blecher and Dräger[47], as follows:

$$3\,Me_2SnCl_2 + 3\,NaHTe \xrightarrow{H_2O} [Me_2SnTe]_3 + 3\,NaCl + 3\,HCl$$

Its structure is shown below. Analagous molecules had been synthesized earlier by Blecher and Mathiasch[48] by the reaction between dimethyltin hydride and elemental S, Se or Te.

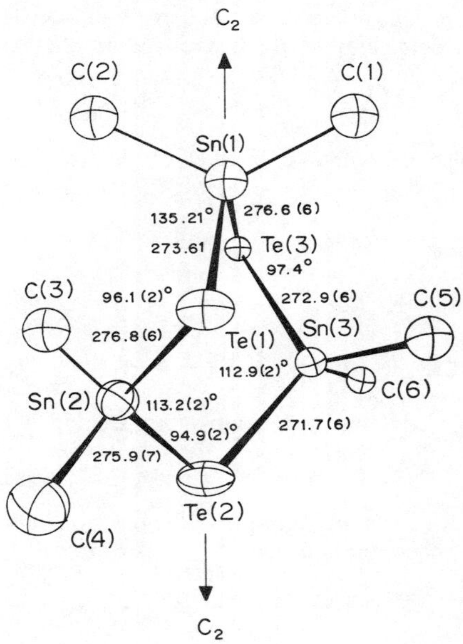

D. Compounds Containing the Pb—Se Bond

Several Pb—X (X = Se, Te) bonded compounds have been described in the preceding sections of this chapter. Only a limited amount of additional work has been carried out.

The hexacarbonyls of Cr, Mo and W undergo a photochemical displacement of CO by

$$M(CO)_6 + C_4H_8O \longrightarrow (CO)_5MOC_4H_8 + CO$$

$$M = Cr, Mo, W$$

tetrahydrofuran, C_4H_8O. The C_4H_8O can be displaced by $(Me_3M')_2Se$, where M′ is Ge, Sn or Pb:

$$(CO)_5MOC_4H_8 + (Me_3M')_2Se \longrightarrow (CO)_5MSe(M'Me_3)_2 + C_4H_8O$$

The compound $(Me_3Pb)_2Se$ is prepared by the reaction of MePbCl with Na_2Se in dry benzene[49].

IV. COMPOUNDS CONTAINING AN X—Se OR X—Te BOND WHERE X IS A GROUP Va ELEMENT

A. Compounds Containing a P—Se or P—Te Bond

A number of reviews have appeared which cover developments in this area up until about 1971. Selenothiophosphates and diselenophosphates and their transition-metal complexes are described in a review by Wasson and coauthors[50].

Kuchen and Knop[51] have described the reaction between tetraethyldiphosphine disulphide, sodium selenide and Se which gives sodium diethylphosphoselenothioate (**5**).

$$Et_2P(S){-}P(S)Et_2 + Na_2Se + Se \longrightarrow 2\ Et_2PSSe^- Na^+ \quad \mathbf{(5)}$$

Compound **5** can also be prepared by the reaction between diethylchlorophosphine sulphide or selenide with NaHSe or NaHs, respectively. Diethylchlorophosphine

$$Et_2P(S)Cl + NaHSe \longrightarrow \mathbf{5}$$

$$Et_2P(Se)Cl + NaHS \longrightarrow \mathbf{5}$$

$$Et_2P(Se)Cl + 2\ NaHSe \longrightarrow \mathbf{5} + NaCl + H_2Se$$

undergoes nucleophilic addition by Se to give the selenide. Trialkylphosphines add Se to

$$Et_2PCl + Se \longrightarrow Et_2P(Se)Cl$$

form phosphine selenides[52,53]. In acetonitrile, the reaction between triarylphosphines and KSeCN gives triarylphosphine selenides[54].

$$R_3P + Se \longrightarrow R_3PSe$$

A review entiled 'The Chemistry of Selenium-bearing Organometallic Derivatives of Group Va Elements' was published in 1972[55]. This review covers the reactions of P_2Se_5 with alcohols and amines, compounds of the type R_3MX (X = Se, Te; M = P, As, Sb), R_2AsSeR' derivatives and the addition of H_2Se and selenols to Vaska's compounds. This report now proceeds to more recent developments.

Austad and coworkers[56] were able to prepare a triphenylphosphine telluride–triphenylphosphine adduct which was found to be quite stable. Previously, tri-alklphosphine tellurides had been found to be quite unstable[57]. These workers found that triphenylphosphine and tetraphenylarsonium tellurocyanate in pure acetonitrile, in the presence of lithium perchlorate, react to give a yellow precipitate identified as Ph_3PTePh. Te—P 'short distances' in this adduct (the crystals are disordered) range from 2.270 to 2.424 Å. Each formula unit is found to contain a linear P—Te—P configuration. Other Scandinavian investigators[58] have prepared tris(morpholino)phosphine telluride by the reaction between tris(morpholino)phosphine and $TeCN^-$. In this compound the P—Te bond distance is 2.356 Å.

Du Mont[59] has found that phosphine tellurides behave as kinetically labile Te complexes in the presence of an excess of the phosphine. On the ^{1}H- and ^{31}P-NMR time-scale, even at room temperature, all PR_3Te/PR'_3 systems which have been investigated display fast Te transfer reactions which have been termed 'fluxional redox systems'. Such an equilibrium is illustrated by the following equation:

$$R_3P{=}Te + R'_3P \rightleftharpoons R'_3P{=}Te + R_3P$$

R, R' = alkyl, dialkylamino, sometimes aryl

However, as found by du Mont, if Te is transferred from tri-*n*-butylphosphine telluride to di-*t*-butyl(trimethylsilyl)phosphine, the transfer of Te to silylphosphine is followed by a rapid insertion of Te into the P—Si bond.

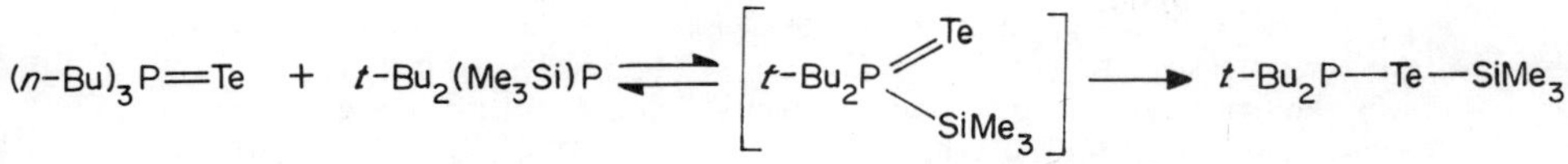

In an analogous manner, elemental Te reacts with P—Si, P—Ge and P—Sn bonded compounds.

$$t\text{-Bu}_2\text{P—XMe}_3 + \text{Te} \longrightarrow t\text{-Bu}_2\text{P—Te—XMe}_3$$

$$\text{X} = \text{Si}, \text{Ge}, \text{Sn}$$

The compounds, t-Bu_2P—Te—XMe_3 are detected in solution by ^{31}P-NMR and ^{119}Sn-NMR spectroscopy. In a few hours, the following redistribution equilibria occur:

$$2\ t\text{-Bu}_2\text{P—Te—XMe}_3 \rightleftharpoons (\text{Me}_3\text{X})_2\text{Te} + (t\text{-Bu})_2\text{P—Te—P(Bu-}t)_2$$

The R_2P—Te—PR_2 type compounds are identified by ^{1}H- and ^{31}P-NMR spectroscopy. They are pale yellow in colour and have been termed 'tellurophosphinous anhydrides'.

^{125}Te-NMR chemical shifts and coupling constants $^1J(^{125}\text{Te}-^{31}\text{P})$ have been reported by du Mont and Kroth[60]. In R_3PTe/PR'_3 mixtures only ^{125}Te singlets are observed. This is taken as evidence of a rapid Te migration. In tellurophosphines such as R—Te—PR'_2 well-resolved doublets arising from $^1J(^{125}\text{Te}-^{31}\text{P})$ coupling is noted, If $R = PR'_2$, such doublets give rise to triplets. The ^{125}Te-NMR data reported by these investigators are as follows: compound, $\delta(^{125}\text{Te})$(ppm), ^1J(Te–P)(Hz); all chemical shifts are relative to (p-$MeC_6H_4Te)_2$ as a standard – (t-Bu)$_3$PTe, – 480(s), 1600; (t-Bu)$_3$PTe/(t-Bu)$_3$P, – 481 (s); (n-Bu)$_3$PTe/(n-Bu)$_3$P, – 512(s); Te[P(Bu-t)$_2$]$_2$, – 574(t), 451; Te[P(Bu-t)$_2$]$_2$·Cr(CO)$_4$, + 142, 324; Te[P(Pr-i)$_2$]$_2$, – 704(t), 324, Me_3Si—Te—P(Bu-t)$_2$, – 772(d), 384; p-TolTe—P(Bu-t)$_2$, – 542(d), 532; p-TolTe—P(Bu-t)$_2$·Ni(CO)$_3$, – 396(d), 651; p-TolTe-P(Pr-i)$_2$, – 588(d), 556; ($Me_3Si)_2$Te, – 460(s).

Tellurium bis(di-t-butyl)phosphine displaces norbornadiene from $C_7H_8Cr(CO)_4$ to give a red crystalline solid, **6**, with a four-membered CrP_2Te cyclic group[61]. A nickel

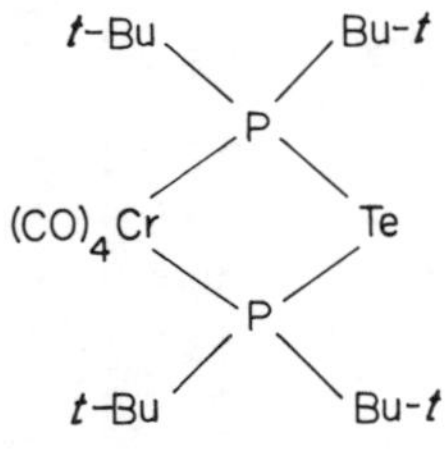

(**6**)

compound p-TolTeP(Bu-t)$_2$Ni(CO)$_3$ can also be prepared. In the chromium compound, the ^{125}Te chemical shift is + 142 ppm(t) and $^1J(^{125}\text{Te}-^{31}\text{P})$ is 324 Hz.

Bergesen and coworkers[62] have prepared a series of compounds of the type $(R_2N)_3P$, $(R_2N)_3PX$(X = O, S, Se, Te) and $(R_2N)_3As$. In these compounds, R_2N is morpholine(Mor), piperidine, pyrrole, Me_2N, Et_2N and (n-Pr)$_2$N. In the vast majority of cases, the ^{13}C-NMR chemical shifts are found to be insensitive to whether the central atom is P or As and relatively insensitive to the nature of X in the pentacovalent series. For example, in the series Mor_3P, Mor_3PO, Mor_3PS, Mor_3PSe and Mor_3PTe, the ^{13}C-NMR shifts of the carbons attached to the nitrogen atom are, respectively, 47.4, 45.4, 46.5 and

47.5 ppm. These values are within 2 ppm of those of the parent amines. In Mor_3P and Mor_3As, the corresponding shifts are 47.1 and 47.4 ppm, respectively.

Among the pentacovalent phosphorus species, a slight, but distinct downfield trend is noted with increasing size of the chalogen atom. To quote the investigators: 'A compensation between sterically induced upfield shifts and varying downfield shifts due to differences in the electron withdrawing power of the chalcogen atom may well be the cause of the small effects observed in the present study'.

Among the tervalent compounds which have been studied, a small, but generally observed downfield shift is noted (1–2 ppm) in going from the phosphines to the arsines. This modest effect is attributed to the fact that the As atom, being less electronegative, deshields the amino carbon atoms more effectively relative to the P atom.

A cyclic three-membered selenadiphosphorane ring system has been synthesized by Baudler and coworkers[63]. The reagents are 1,2-di-*t*-butyl-1,2-dichlorodiphosphine and bis(trimethylstannyl) selenide. The reaction is carried out in boiling THF. Based on NMR spectral evidence, the alkyl groups assume a *trans* configuration.

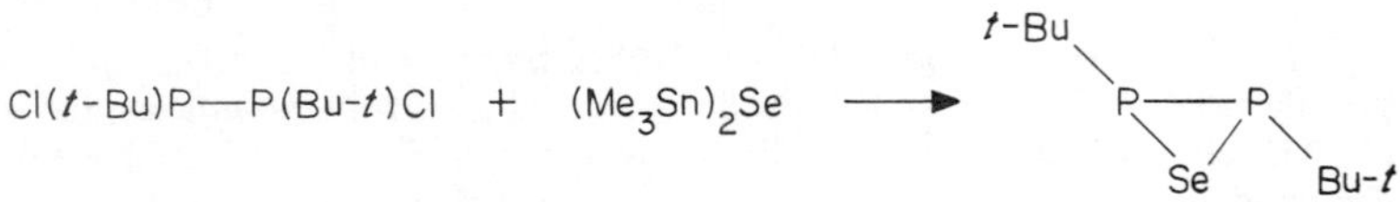

B. Compounds Containing As—Se, As—Te and Other P—Se or P—Te Bonds

A detailed NMR(^{1}H, ^{19}F) spectroscopic investigation of the cleavage of element–element bonds by Me_3SnH in ligands coordinated to $Cr(CO)_5$ and $Mo(CO)_5$ has been published by Grobe and Le Van[64]. The coordinated ligands are of the type $R_2XXR'_2$ and R_2XGR', where X is P or As and G is S, Se or Te. The rate of reaction of the coordinated ligands is much slower than that of the free ligands. In Table 5 are listed the results by these workers for compounds containing an XGR′ bond.

Emeléus and coworkers[65] introduced trifluoromethylselenobis(trifluoromethyl)-phosphines and -arsines. The reactions used for the syntheses of these compounds are given below.

$$2\ (CF_3)_2PI + (CF_3Se)_2Hg \longrightarrow HgI_2 + 2\ (CF_3)_2PSeCF_3$$

$$2\ (CF_3)_2AsI + (CF_3Se)_2Hg \longrightarrow HgI_2 + 2\ (CF_3)_2AsSeCF_3$$

The reáction was carried out in sealed glass ampoules for 72 h. The precipitated red mercuric oxide was separated by filtration and the colourless liquid products were separated by vacuum distillation. Heptafluoropropylselenobis(trifluoromethyl)arsine was also prepared from bis(heptafluoropropylseleno)mercury and iodobis(trifluromethyl)arsine.

The conformations of tris(phenylthio)-, tris(phenylseleno)- and tris(phenyltelluro)-arsines as well as of the $Me_2As(XPh)$ derivatives (X = S, Se, Te) have been studied by ^{1}H-NMR and UPS (ultiaviolet photoelectron spectroscopy)[66]. It was concluded that (in benezene solution) (*a*) in all the compounds studied the As is pyramidal; (*b*) the $As(XPh)_3$ molecules are symmetric; (*c*) the twisting angle about the As—X bond is close to zero; (*d*) the Ph—X bond is directed toward the external of the pyramid; (*e*) the Ph rings are symmetrically twisted at an angle of about 40° with respect to the As—X—C plane in the tris(phenylchalcogeno)arsines and (*f*) the bond moments, As(L)—X, where L is the electron lone pair, have been determined to be 0.35 D, 0.30 D and 0.28 D for X = S, Se and Te, respectively. It is concluded from these measurements that there is a virtual absence of resonance contributions.

TABLE 5. Cleavage reactions of complexes of the type $Cr(CO)_5R_2XGR'$ and $MO(CO)_5R_2XGR'$ by Me_3SnH[64]

Reactant	Reaction temp. (°C)	Products
$Cr(CO)_5Me_2PSeMe$	20	$Cr(CO)_5Me_2PH$, $Me_3SnSeMe$
$Cr(CO)_5Me_2AsSeMe$	20	$Cr(CO)_5Me_2AsSnMe_3$
$Cr(CO)_5(CF_3)_2PSeMe$	<0	$Cr(CO)_5(CF_3)_2PH$, $Me_3SnSeMe$
	>0	$Cr(CO)_5(CF_3)_2SnMe_3$, $Cr(CO)_5(CF_3)_2PH$
$Cr(CO)_5Me_2AsSeCF_3$	20	$Cr(CO)_5Me_2AsH$, $CF_3SeSnMe_3$
$Mo(CO)_5Me_2PSeMe$	20	$Mo(CO)_5Me_2PH$ $Me_3SnSeMe$
$Mo(CO)_5Me_2AsSeMe$	20	$Mo(CO)_5Me_2AsSnMe_3$
$Mo(CO)_5(CF_3)_2PSeMe$	−10–0	$Mo(CO)_5(CF_3)_2PH$, $Me_3SnSeMe$
	>0	$Mo(CO)_5(CF_3)_2PSnMe_3$, Me_3SnF
$Mo(CO)_5(CF_3)_2PTeMe$	−40–0	$Mo(CO)_5(CF_3)_2PH$, $Me_3SnTeMe$
	>0	$Mo(CO)_5(CF_3)_2PSnMe_3$, Me_3SnF
$Mo(CO)_5Me_2AsSeCF_3$	0–20	$Mo(CO)_5Me_2AsH$, $CF_3SeSnMe_3$

The ^{1}H-NMR chemical shifts of the *ortho*, *meta* and *para* protons move slightly downfield with decreasing electronegativity of the X atom. This is in agreement with the decreasing ability of the chalcogen atoms to transfer charge from the lone-pair electrons into the aromatic ring as the atomic number increases.

Herrmann[67] has described the synthesis of the five-membered and six-membered heterocyclic systems, 1,4-diarsa-2,3,5-triselenacyclopentane (**7**) and 1,3,5-triarsa-2,4,6-triselenacyclohexane (**8**). The preparation involves the reaction between elemental Se and either hexaphenylcyclohexylarsine or pentamethylcyclopentaarsine. Compound **7** is prepared by heating the reactants at 220 °C and crystallization from benzene in the form of orange–yellow crystals. Compound **8** is obtained as yellowish crystals by heating the reactants at 180 °C. Details about the geometries have not been obtained.

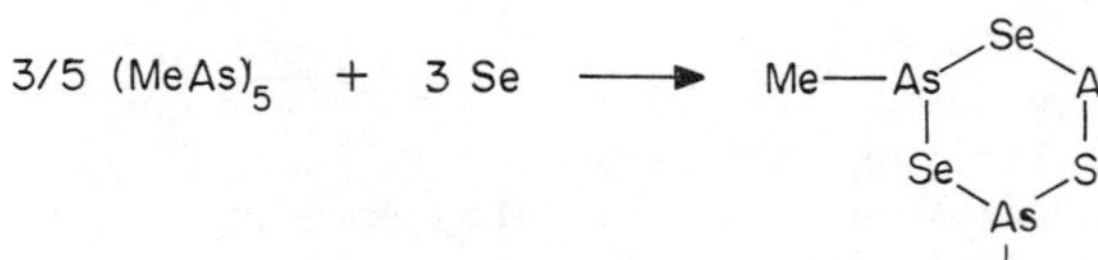

(**7**)

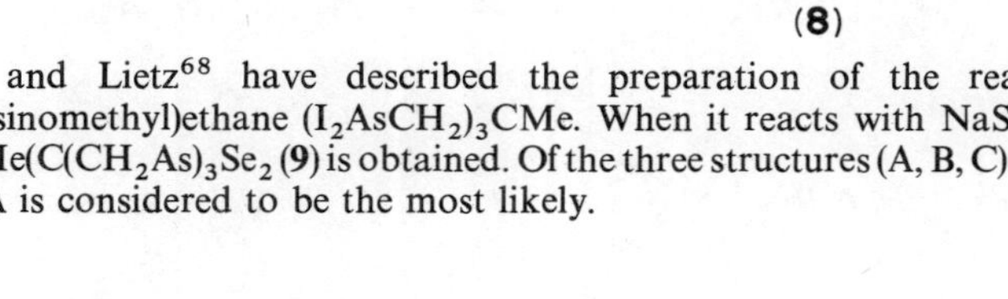

(**8**)

Ellerman and Lietz[68] have described the preparation of the reagent, 1,1,1-tris(diiodoarsinomethyl)ethane $(I_2AsCH_2)_3CMe$. When it reacts with NaSeH, the cage compound $Me(C(CH_2As)_3Se_2$ (**9**) is obtained. Of the three structures (A, B, C) proposed for **9** structure A is considered to be the most likely.

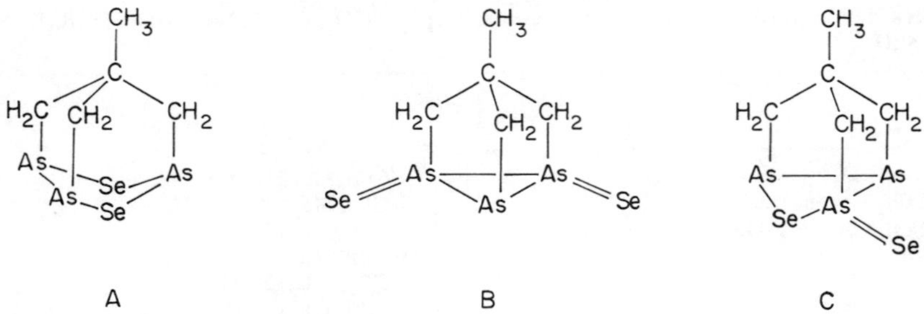

(9)

All four members of the 5,10-epichalcogenodihydroarsanthrene series have been prepared[69]. The epithio, episeleno and epitelluro compounds are prepared by saturating a solution of 5,10-dichloro-5,10-dihydroarsanthrene (**10**) in ethanol with H_2S, H_2Se or H_2Te.

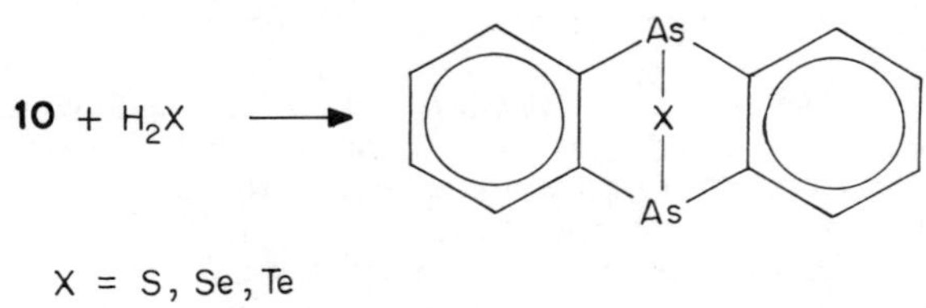

The structure of the epithio compound has been determined crystallographically and is shown below. The 5,10-episeleno and 5,10-epitelluro analogues have closely related cell constants, but they are not isomorphous. The investigators are of the opinion that the Se and Te analogues have structures very similar to that shown for the thio compound.

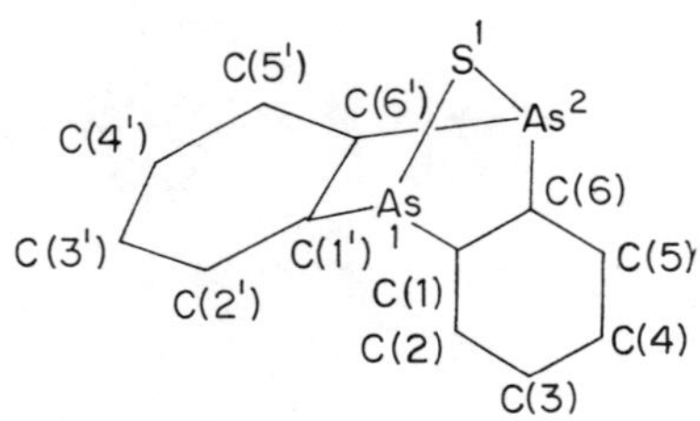

Chi and Kober[70] have prepared, by the oxidation of the parent arsines with Se, di- and tri-tertiary arsine selenides. Typical examples of the types of compounds prepared are shown below.

$$N[CH_2CH_2AsMe_2]_3 \xrightarrow{Se} N[CH_2CH_2As(Se)Me_2]_3$$

$$ClCH_2C(Me)[CH_2AsMe_2]_2 \xrightarrow{Se} ClCH_2C(Me)[CH_2As(Se)Me_2]_2$$

$$(ClCH_2)_2C[CH_2AsMe_2]_2 \xrightarrow{Se} (ClCH_2)_2C[CH_2As(Se)Me_2]_2$$

The coordination compounds shown below were also prepared.

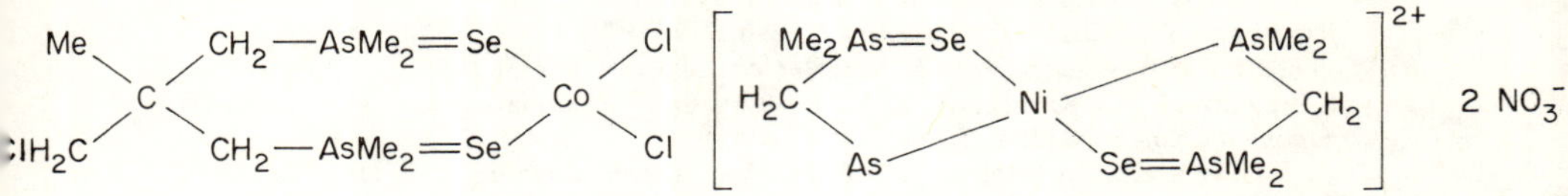

The As—O—C bond is hydrolytically unstable, but the As—S—C and As—Se—C bonds are hydrolytically quite stable. This has led Zingaro to the conclusion that although monosaccharide esters of arsenic acid have never been successfully prepared it should be possible to prepare thio- or seleno-sugar esters of arsinous acids. The synthesis of 1- and 6-*S*- and -*Se*-arsinous acid esters have been reviewed[71]. A typical reaction procedure is that described for the synthesis of 2,3,4,6-tetra-*O*-acetyl-1-*SE*-dimethylarsino-*β*-D-galactopyranose[72].

Acetobromogalactose is treated with selenourea which displaces bromine at C(1) to give the selenoureide. Reduction of the selenoureide by bisulphite or borohydride followed by air oxidation gives the galactosyl-1–1′-diselenide. The diselenide is cleaved by $Me_2AsAsMe_2$ under nitrogen to give the acetylated 1-*Se*-dimethylarsinogalactose compound.

Another example of such a reaction, for the synthesis of derivatives of 2-amino-2-deoxy-D-glucopyranose, is given in Scheme 1[73].

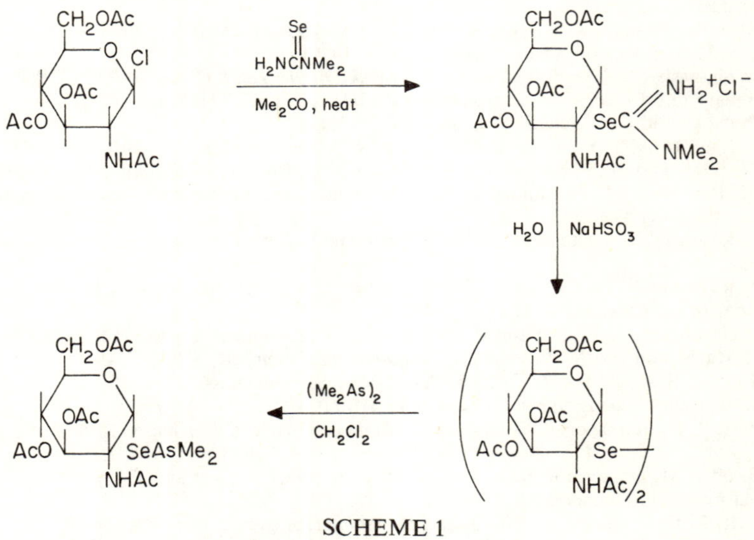

SCHEME 1

V. ACKNOWLEDGEMENT

The financial assistance made available by the Robert A. Welch Foundation during the preparation of this manuscript is appreciated.

VI. REFERENCES

1. H. J. Gysling, *Coord. Chem. Rev.*, **42**, 133 (1982).
2. M. Schmidt and H. D. Block, *J. Organometal. Chem.*, **25**, 17 (1970).

3. W. Siebert and A. Ospici, *Chem. Ber.*, **105**, 454 (1972).
4. W. Siebert and F. Riegel, *Chem. Ber.*, **106**, 1012 (1973).
5. F. Riegel and W. Siebert, *Z. Naturforsch.*, **29b**, 719 (1974).
6. W. Siebert and B. Asgarouladi, *Z. Naturforsch.*, **30b**, 647 (1975).
7. V. I. Bregadze, V. Ts. Kampel, A. Ya. Usiatinsky, O. B. Ponomareva and N. N. Godovikov, *J. Organometal. Chem.*, **233**, C33 (1982).
8. D. L. J. Clive and S. M. Menchen, *J. Chem. Soc., Chem. Commun.*, 356 (1978).
9. D. L. J. Clive and S. M. Menchen, *J. Org. Chem.*, **44**, 4279 (1979).
10. D. L. J. Clive and S. M. Menchen, *J. Chem. Soc., Chem. Commun.*, 168 (1979).
11. M. Schmidt and H. D. Block, *J. Organometal. Chem.*, **25**, 17 (1970).
12. G. E. Coates, *J. Chem. Soc.*, 2003 (1951).
13. J. L. Atwood and S. K. Seale, *J. Organometal. Chem.*, **114**, 107 (1976).
14. A. P. Kozikowski and A. Ames, *J. Org. Chem.*, **43**, 2735 (1978).
15. S. Cardock, E. A. V. Ebsworth and H. F. Jessep, *J. Chem. Soc., Dalton. Trans.*, 359 (1972).
16. J. W. Anderson, G. K. Barker, J. E. Drake and M. Rodger, *J. Chem. Soc., Dalton Trans.*, 1716 (1973).
17. G. K. Barker, J. E. Drake and R. T. Hemmings, *J. Chem. Soc., Dalton Trans.*, 450 (1974).
18. J. E. Drake and R. T. Hemmings, *Inorg. Chem.*, **19**, 1879 (1980).
19. A. I. Charov, M. N. Bochkarev and N. S. Vyazankin, *Zh. Obshch. Khim.*, 772 (1973).
20. M. R. Detty, *Tetrahedron Letters*, 5087 (1978).
21. N. Miyoshi, H. Ishii, K. Kondo, S. Murai and N. Sonoda, *Synthesis*, 300 (1979).
22. N. Miyoshi, H. Ishii, S. Murai and N. Sonoda, *Chem. Letters*, 873 (1979).
23. N. Miyoshi, K. Kondo, S. Murai and N. Sonoda, *Chem. Letters*, 909 (1979).
24. M. R. Detty, *J. Org. Chem.*, **44**, 4528 (1979).
25. M. R. Detty, *Tetrahedron Letters*, 4189 (1979).
26. N. Miyoshi, H. Ishii, K. Kondo, S. Murai and N. Sonoda, *Synthesis*, 300 (1979).
27. R. S. Sukhai and L. Brandsma, *Synthesis*, 455 (1979).
28. M. Wojnowska, W. Wojnowski and R. West, *J. Organometal. Chem.*, **199**, C1 (1980).
29. E. Hengge and G. Schuster, *J. Organometal. Chem.*, **231**, C17 (1982).
30. K. Praefcke and C. Weichsel, *Synthesis*, 216 (1980).
31. T. R. Bierschenk, T. J. Juhlke and R. J. Lagow, *J. Amer. Chem. Soc.*, **103**, 7341 (1981).
32. M. N. Bochkarev, L. P. Sanina and N. S. Xyazankin, *Zh. Obshch. Khim.*, **39**, 135 (1969).
33. M. N. Bochkarev, L. P. Maiorova, N. S. Vzazankin and G. A. Razuvaev, *J. Organometal. Chem.*, **82**, 65 (1974).
34. M. Wojnowska, M. Noltemeyer, H.-J. Füllgrabe and A. Meller, *J. Organometal. Chem.*, **228**, 229 (1982).
35. G. A. Razuvaev, V. N. Latyaeva, E. N. Gladyshev, E. V. Krasilnikova, A. N. Linyova and A. P. Kosina, *Inorg. Chim. Acta*, **31**, 357 (1978).
36. S. A. Gardner, P. J. Trotter and H. J. Gysling, *J. Organometal. Chem.*, **212**, 35 (1981).
37. E. C. MacMullin and M. E. Peach, *J. Organometal. Chem.*, **52**, 355 (1973).
38. K. Tanaka, S. Araki and T. Tanaka, *Bull. Chem. Soc. Japan*, **46**, 2136 (1973).
39. B. Mathiasch, *Z. anorg. allgem. Chem.*, **432**, 269 (1977).
40. H. Puff, R. Gattermayer, R. Hundt and R. Zimmer, *Angew. Chem. (Intern. Ed. Engl.)*, **16**, 547 (1977).
41. A. Blecher, M. Drager and B. Mathiasch, *Z. Naturforsch.*, **36b**, 1361 (1981).
42. B. Mathiasch, *J. Organometal. Chem.*, **141**, 189 (1977).
43. H. Puff, R. Gattermayer, R. Hundt and R. Zimmer, *Angew. Chem.*, **89**, 556 (1977).
44. M. Dräger, A. Blecher, H.-J. Jacobsen and B. Krebs, *J. Organometal. Chem.*, **161**, 319 (1978).
45. P. Heinrich, A. Bongartz, W. Schuh and R. Zimmer, *J. Organometal. Chem.*, **248**, 61 (1983).
46. N. S. Dance, W. R. McWhinnie and C. H. W. Jones, *J. Organometal. Chem.*, **125**, 291 (1977).
47. A. Blecher and M. Dräger, *Angew. Chem. (Intern. Ed. Engl.)*, **18**, 677 (1979).
48. A. Blecher and B. Mathiasch, *Z. Naturforsch.*, **33b**, 246 (1978).
49. H. Schumann, R. Mohtachemi, H.-J. Kroth and U. Frank, *Chem. Ber.*, **106**, 1555 (1973).
50. J. R. Wasson, G. M. Woltermann and H. J. Stoklosa, *Topics Curr. Chem.*, **35**, 65 (1973).
51. W. Kuchen and B. Knop, *Chem. Ber.*, **99**, 1663 (1966).
52. C. Screttas and A. F. Isbell, *J. Org. Chem.*, **27**, 2573 (1962).
53. R. A. Zingaro and R. E. McGlothin, *J. Chem. Eng. Data*, **8**, 226 (1963).

54. P. Nicpon and D. W. Meek, *Inorg. Chem.*, **5**, 1297 (1966).
55. R. A. Zingaro, *Ann. N.Y. Acad. Sci.*, **192**, 72 (1972).
56. T. Austad, T. Rød, K. Åse, J. Songstad and A. H. Norbury, *Acta Chem. Scand.*, **27**, 1939 (1973).
57. R. A. Zingaro, *J. Organometal. Chem.*, **1**, 200 (1963).
58. C. Rømming, A. J. Iversen and J. Songstad, *Acta Chem. Scand.*, **A34**, 333 (1980).
59. W.-W. du Mont, *Angew. Chem (Intern. Ed. Engl.)*, **19**, 554 (1980).
60. W.-W. du Mont and H.-J. Kroth, *Z. Naturforsch.*, **36b**, 332 (1981).
61. W.-W. du Mont and E. Nordhoff, *J. Organometal. Chem.*, **198**, C58 (1980).
62. K. Bergesen, B. Pedersen and J. Songstad, *Acta Chem. Scand.*, **A35**, 147 (1981).
63. M. Baudler, H. Suchomel, G. Fürstenberg and U. Schings, *Angew. Chem.*, **93**, 1087 (1981).
64. J. Grobe and D. Le Van, *Z. Naturforsch.*, **36b**, 666 (1981).
65. H. J. Emeléus, K. J. Packer and N. Welcman, *J. Chem. Soc.*, 2529 (1962).
66. R. A. Pyles, K. J. Irgolic and G. C. Pappalardo, *XII Congr. Naz. Chem. Inorg., Univ. of Trieste, Italy, Sept. 11–15*, 1979, 141–142.
67. D. Herrmann, *Z. anorg. allgem. Chem.*, **416**, 50 (1975).
68. J. Ellermann and M. Lietz, *Z. Naturforsch.*, **35b**, 1514 (1980).
69. D. W. Allen, J. C. Coppola, O. Kennard, F. G. Mann, W. D. S. Motherwell and D. G. Watson, *J. Chem. Soc. (C)*, 810 (1970).
70. P. B. Chi and F. Kober, *Z. anorg. allgem. Chem.*, **466**, 183 (1980).
71. R. A. Zingaro, *Chem. Scripta*, **8A**, 51 (1975).
72. J. R. Daniel and R. A. Zingaro, *Phosphorus and Sulfur*, **4**, 179 (1978).
73. G. C. Chen, C. H. Banks, K. J. Irgolic and R. A. Zingaro, *J. Chem. Soc., Perkin. Trans. 1*, 2287 (1980).

The Chemistry of Organic
Selenium and Tellurium
Compounds Volume 1
Edited by S. Patai and Z. Rappoport
© 1986 John Wiley & Sons Ltd.

CHAPTER **12**

Synthesis and uses of isotopically labelled selenium and tellurium compounds

KEN FUJIMORI

Department of Chemistry, The University of Tsukuba, Sakura-mura Niihari-gun, Ibaraki-ken, 305 Japan

SHIGERU OAE

Okayama University of Science, Ridai-cho 1-1, Okayama, 700 Japan

I. INTRODUCTION . . . 370
II. PREPARATION OF ISOTOPICALLY LABELLED SELENIUM AND TELLURIUM COMPOUNDS . . . 370
A. Inorganic Selenium and Tellurium Compounds Labelled with Radioactive Isotopes . . . 370
B. Selenols, Diselenides, Selenides and Tellurides . . . 371
1. Acyclic compounds . . . 371
2. Cyclic compounds . . . 376
3. Amino acids . . . 378
4. Miscellaneous . . . 382
III. USES OF ISOTOPICALLY LABELLED SELENIUM AND TELLURIUM COMPOUNDS . . . 382
A. Physicochemical Uses . . . 382
1. Infrared and Raman spectroscopy . . . 382
2. Microwave spectroscopy . . . 383
3. Nuclear magnetic resonance spectroscopy . . . 383
B. Mechanistic Studies of the Reactions of Organic Selenium and Tellurium Compounds . . . 385
1. Isotopic exchange . . . 385
2. Photochemical reaction . . . 386
C. Biochemical Uses . . . 387
1. Metabolism of selenium and tellurium . . . 387
a. Selenium . . . 387

b. Tellurium 389
2. Selenoproteins 389
a. Glutathione peroxidase 389
b. Glycine reductase 390
c. Formic acid dehydrogenase 390
d. Other proteins 391
3. Selenium-containing t-RNA 391
4 Non-haemeiron proteins containing acid-labile sulphur 391
C. Medicinal Uses 392
1. Scintiscanning 392
3. Miscellaneous 393
IV. CONCLUSION 393
V. REFERENCES 393

I. INTRODUCTION

Selenium and tellurium have many stable and radioactive isotopes. For example, the stable isotopes of selenium, with their natural abundances, are ^{74}Se (0.87%), ^{76}Se (9.02%), ^{77}Se (7.58%), ^{78}Se (23.52%), ^{80}Se (49.82%) and ^{82}Se (9.19%), while ^{72}Se, ^{73}Se, ^{75}Se and ^{77m}Se are the main radioactive isotopes. Tellurium has 8 different stable isotopes and more than 10 radioisotopes[1]. Among the most often used are ^{75}Se (half-life: 118.5 days), ^{123m}Te (half-life: 117 days), ^{127m}Te (half-life: 109 days) and ^{132}Te (half-life: 78 h)[1]. Positron-emitting ^{73}Se may be useful for imaging human body organs by what is called PETT (positron emission transaxial tomography), but it has not yet been put into use.

Most of the radioactively labelled selenium and tellurium compounds discussed in this chapter have been used for biochemical studies and radiomedical research, whereas the deuteriated compounds have been used for infrared, Raman, microwave and nuclear magnetic resonance spectroscopic studies. Very rarely, these isotopes have also been used for studying organic reaction mechanisms. Accordingly, our attention will be focused on the preparation of these isotopically labelled compounds and their biochemical and medicinal uses.

II. PREPARATION OF ISOTOPICALLY LABELLED SELENIUM AND TELLURIUM COMPOUNDS

A. Inorganic Selenium and Tellurium Compounds Labelled with Radioactive Isotopes

Commercially available radioactive elemental selenium and tellurium, selenium dioxide, alkali selenite and tellurite can be reduced to radioactive Na_2Se_2, Na_2Se, NaSeH, Na_2Te and NaTeH, which are versatile reagents for the synthesis of a variety of radioactive organic selenium and tellurium compounds, since these reagents are extremely strong nucleophiles which by S_N2 processes can readily displace halides and tosyloxy groups attached to carbon. This is mainly due to the softness of these reagents which have donor orbitals (HOMOs) of quite high energy levels[2] and even behave as potential reducing agents[3].

Radioactive selenium can readily be reduced with sodium borohydride to sodium selenide and sodium diselenide (equation 1)[4,5].

$$^{75}Se \xrightarrow{NaBH_4} Na^{75}SeH \xrightarrow{^{75}Se} (Na^{75}Se)_2 \qquad (1)$$

Radioactive elemental selenium is also often reduced by metallic sodium in liquid ammonia to generate sodium diselenide which in turn reacts with various alkylating agents *in situ* to afford dialkyl diselenide labelled with ^{75}Se (equation 2)[6–9].

$$2\ ^{75}Se + 2\ Na \xrightarrow{\text{liq. } NH_3} Na_2Se_2 \qquad (2)$$

Kronrád and coworkers reduced $^{75}SeO_2$ with stannous chloride and phosphoric acid to gaseous hydrogen selenide, $H_2{}^{75}Se$, which can be driven into the reactor containing a suitable electrophilic reagents by a stream of nitrogen (equation 3)[10].

$$^{75}SeO_2 + 3\ Sn^{2+} + 6\ H^+ \longrightarrow H_2{}^{75}Se + 3\ Sn^{4+} + 2\ H_2O \qquad (3)$$

Sadek, Basmadjian and Ice reduced radioactive selenious acid with sodium borohydride in an aqueous buffer solution by adjusting the final pH at 6.5–7.0 to obtain $Na^{75}SeH$ quantitatively. Under these reaction conditions evolution of highly toxic hydrogen selenide and the formation of $Na_2{}^{75}Se$ can be avoided[11].

Radioactive elemental selenium obtained by treatment of $Na_2{}^{75}SeO_3$ with sulphur dioxide was shown to be reduced to potassium diselenide, which can be converted to ^{75}Se-labelled dialkyl diselenides upon treatment with alkyl halides or tosylates *in situ* (equation 4)[12].

$$Na_2{}^{75}SeO_3 \xrightarrow{SO_2/H^+} {}^{75}Se \xrightarrow{K_2SO_3} K_2{}^{75}Se_2 \qquad (4)$$

^{127m}Te-labelled elemental tellurium was prepared by treating radioactive sodium tellurite with hydrogen bromide and subsequent addition of stannous chloride and natural tellurium as an entraining agent. The radioactive tellurium so obtained was purified by sublimation and can be converted to ^{127m}Te-labelled tellurium tetrachloride by direct contact with chlorine gas (equation 5)[13].

$$Na_2{}^{127m}TeO_3 \xrightarrow{HBr,\ SnCl_2} {}^{127m}Te \xrightarrow{Cl_2} {}^{127m}TeCl_4 \qquad (5)$$

B. Selenols, Diselenides, Selenides and Tellurides

1. Acyclic compounds

Methaneselenol-d_4 was prepared by the following sequence of reactions: treatment of dimethyl sulphate-d_6 with sodium diselenide gave dimethyl diselenide-d_6 which was reduced with metallic sodium in liquid ammonia to obtain the desired compound (equation 6)[6,14,15]. Methaneselenol and methaneselenol-d_3 can be readily converted to CH_3SeD and CD_3SeD simply by treating with deuterium oxide[15].

$$Na_2Se_2 \xrightarrow{(CD_3)_2SO_4} (CD_3Se)_2 \xrightarrow[2.\ H^+]{1.\ Na} CD_3SeH \xrightarrow{D_2O/\text{gas phase}} CD_3SeD \qquad (6)$$

Alkyllithium and Grignard reagents react readily with elemental selenium to afford the selenols (equation 7). Isotopically labelled selenols can be obtained if suitably labelled starting materials are used[16,17].

$$RMgX + Se \longrightarrow RSeMgX \xrightarrow{H_2O/H^+} RSeH \qquad (7)$$

Basmadjian, Hetzel and Ice synthesized ^{75}Se-labelled α-tolueneselenol and dibenzyl diselenide by treating benzyl chloride with ^{75}Se-labelled sodium hydrogen selenide or sodium diselenide, respectively (equation 8 and 9)[4]. The first method may be used to afford any non-symmetrical selenide. The same authors claim that ^{75}Se-labelled selenourea

$$Na^{75}SeH \xrightarrow{RX} R^{75}SeH \xrightarrow{R'X/B:} R^{75}SeR' \qquad (8)$$

$$Na_2{}^{75}Se_2 \xrightarrow{2\ RX} R_2{}^{75}Se_2 \qquad (9)$$

derivatives, which can be readily obtained upon heating ^{75}Se-labelled sodium hydrogen selenide with substituted carbodiimides, are versatile reagents for preparation of ^{75}Se-labelled selenols (equation 10)[4].

$$Na^{75}SeH \xrightarrow{RN=C=NR} RNHC(={}^{75}Se)NHR \xrightarrow{R'X} RNH\overset{+}{C}({}^{75}SeR')NHR\ X^- \xrightarrow{H_2O} R'^{75}SeH \qquad (10)$$

Monks and coworkers prepared 6β-(alkylseleno-^{75}Se-methyl)norcholestane[7] (**1**) and ^{75}Se-labelled 8-methylselenoadenosine 3′,5′-cyclic phosphate[18] (**2**) by the reaction sequence shown in equations (11) and (12).

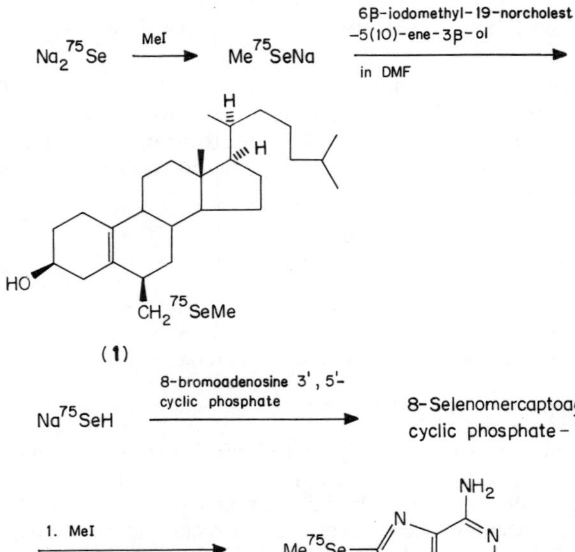

(11)

(**1**)

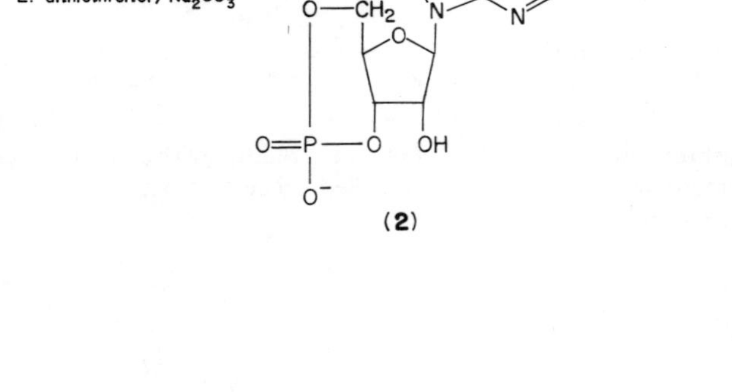

(12)

(**2**)

Chambers and coworkers prepared ^{75}Se-labelled 2-methylselenoprednisolone (**3**) by treating cortisol with ^{75}Se-labelled selenium dioxide as shown in equation (13)[19].

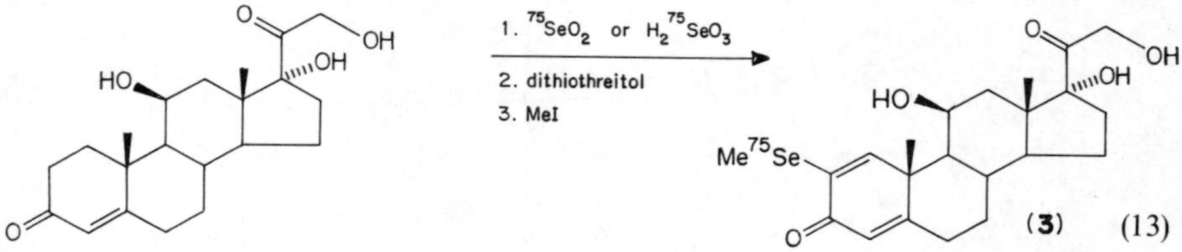

(**3**) (13)

Monks and coworkers studied two routes for the preparation of ^{75}Se-labelled steroids. One route involves conversion of a hydroxyl group in the steroids at the 3, 6, 17, 19 or 21 position to the corresponding methaneselenyl group as shown in equation (14)[20]. The

$$\text{Steroid—OH} \xrightarrow{\text{TsCl}} \text{Steroid—OTs} \xrightarrow{K^{75}SeCN} \text{Steroid—}^{75}\text{SeCN}$$

$$\xrightarrow{\text{1. dithiothreitol 2. MeI or } Me_2SO_4} \text{Steroid—}^{75}\text{SeMe} \quad (\mathbf{4}) \qquad (14)$$

second route involves conversion of steroids bearing a carbonyl group at the 3, 6, 17 or 20 position into oximes having a radioactive selenium-containing group as shown in equation (15)[20].

$$\text{Steroid=O} \xrightarrow{H_2NOCH_2COOH} \text{Steroid=NOCH}_2\text{COOH} \xrightarrow{\text{ClCOOBu-}i\text{ / Et}_3\text{N}}$$

$$\xrightarrow[\text{or } Me^{75}SeCH_2CH_2NH_2]{Me^{75}SeCH_2CH_2CH(NH_2)COOH} \text{Steroid=NOCH}_2\text{CONHCH(COOH)CH}_2\text{CH}_2{}^{75}\text{SeMe} \ (\mathbf{5}) \text{ or Steroid=NOCH}_2\text{CONHCH}_2\text{CH}_2{}^{75}\text{SeMe} \ (\mathbf{5a}) \qquad (15)$$

Otto prepared ^{75}Se-labelled glycerine triethers (**6a** and **6b**) by the method shown for **6a** in equation (16)[21]. They also showed that when ^{123m}Te was used instead of ^{75}Se in the above synthetic procedure, ^{123m}Te-labelled glycerine triethers, e.g. **7**, could be obtained[21].

$$^{75}\text{Se} \xrightarrow{\text{BuLi}} \text{Bu}^{75}\text{SeLi} \xrightarrow{CH_2O(CH_2)_{17}CH_3 - CHO(CH_2)_{17}CH_3 - CH_2Br} CH_2O(CH_2)_{17}CH_3 - CHO(CH_2)_{17}CH_3 - CH_2{}^{75}Se(CH_2)_3CH_3 \qquad (16)$$

(**6a**)

$CH_2O(CH_2)_{17}CH_3$
$CH\,{}^{75}Se(CH_2)_3CH_3$
$CH_2O(CH_2)_{17}CH_3$

(**6b**)

$CH_2O(CH_2)_{17}CH_3$
$CHO(CH_2)_{17}CH_3$
$CH_2{}^{123m}Te(CH_2)_3CH_3$

(**7**)

Kung and Blau synthesized ^{75}Se-labelled tertiary amines (**8**) by reducing ^{75}Se-labelled selenious acid with sodium borohydride and subsequent treatment of the intermediate formed with *N,N*-substituted aminoethyl chloride (equation 17)[22].

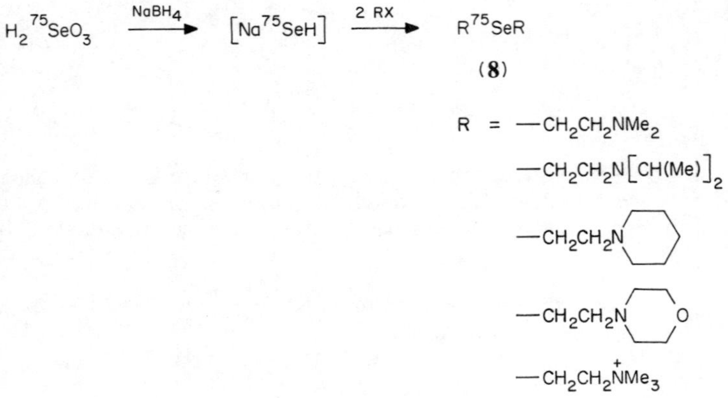

(17)

Carrillo and Nassiff succeeded in the separation of $^{127m}Te/^{127g}Te$ obtained by [n, γ] reaction of a sample enriched with ^{126}Te. The elemental tellurium containing a mixture of radioactive isotopes obtained by the neutron irradiation was converted to dimethyltellurium iodide, which was then dissolved in an organic solvent and kept until radioactive equilibrium was attained. At this stage, ^{127g}Te (half-life: 9.4 h), formed from ^{127m}Te (half-life: 109 days) during the radioactive equilibration, was shown to have changed through recoil reaction to inorganic tellurium which could be separated by coprecipitation with ferric hydroxide from other isotopes of organic form contained in the organic layer (equation 18)[23].

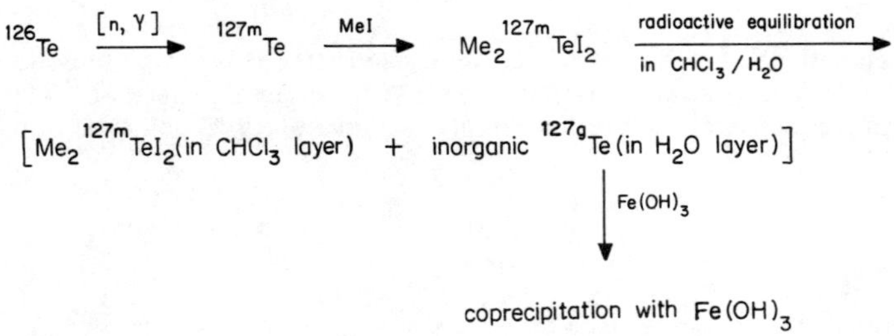

(18)

Adloff and Adloff prepared radioactive ^{132}Te-labelled diphenyltellurium dichloride by displacement of mercury from diphenylmercury by ^{132}Te under nitrogen and subsequent treatment with chlorine[24].

The ^{123m}Te-labelled norchorane derivative **10** was prepared by the reaction of bromonorcholane (**9**) with ^{123m}Te-labelled isopropyl telluride in an alkaline solution (equation 19)[25].

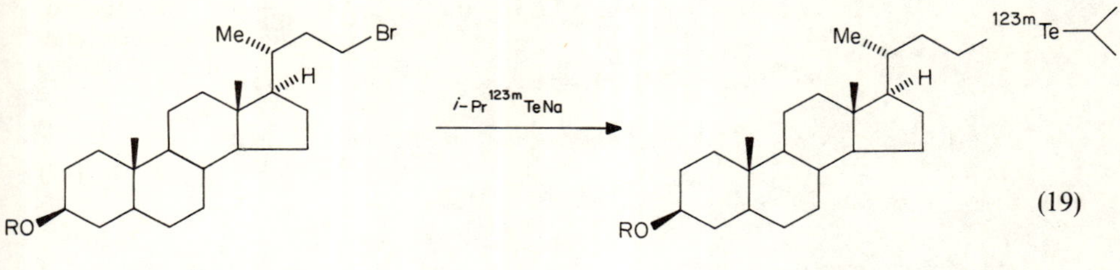

(9) (10)

(19)

Knapp and coworkers prepared ^{123m}Te-labelled isosters of palmitoleic acid and oleic acid, i.e. 9-tellura-^{123m}Te-pentadecanoic acid (**11a**) and 9-tellura-^{123m}Te-heptadecanoic acid (**11b**), respectively, by the synthetic route shown in equation (20)[26]. 17-

$$n\text{-}C_nH_{2n+1}{}^{123m}TeNa \xrightarrow[\text{2. } H_2O/OH^- \quad \text{3. } H^+]{\text{1. } Br(CH_2)_7COOCH_3} n\text{-}C_nH_{2n+1}{}^{123m}Te(CH_2)_7COOH \qquad (20)$$

(**11**)

(**a**) $n = 6$

(**b**) $n = 8$

[^{123m}Te]tellura-9-octadecenoic acid (**13a**) was prepared by Basmadjian and coworkers through the two routes shown in equation (21)[27]. The four ^{123m}Te-labelled hexadecenoic acid analogues **13b–e** were synthesized in the same manner[27].

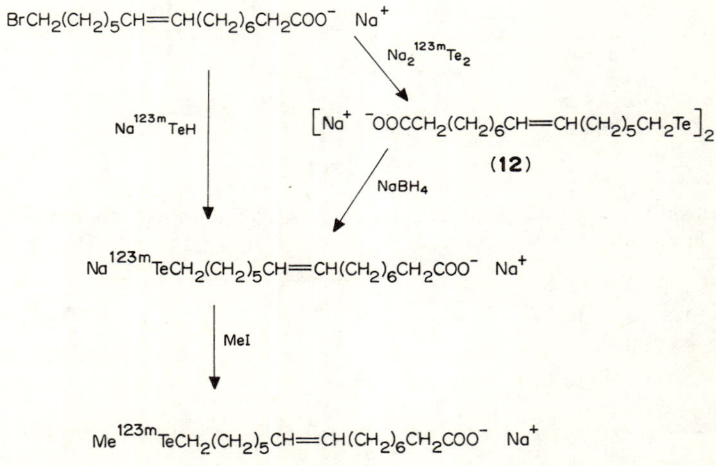

(21)

(**13a**)

$Et\,^{123m}TeCH_2(CH_2)_5CH{=}CH(CH_2)_6CH_2COOH$ **(13b)**

$Pr\,^{123m}TeCH_2(CH_2)_5CH{=}CH(CH_2)_6CH_2COOH$ **(13c)**

$Me_2\,^{123m}\overset{+}{Te}(CH_2)_5CH{=}CH(CH_2)_6CH_2COOH\ Cl^-$ **(13d)**

$Me(CH_2)_{15}\,^{123m}TeCH_2COOH$ **(13e)**

2. Cyclic compounds

Lambert and coworkers synthesized several deuteriated selenanes (**14a–c**)[28–30] by treating suitable deuteriated 1,5-dibromopentanes with sodium selenide or a mixture of sodium formaldehydesulphoxylate and elemental selenium according to the procedure developed by McCullough and Lefohn (equations 22–24)[31]. In the latter process elemental selenium is believed to be reduced to form sodium selenide in the initial step of the reaction[31].

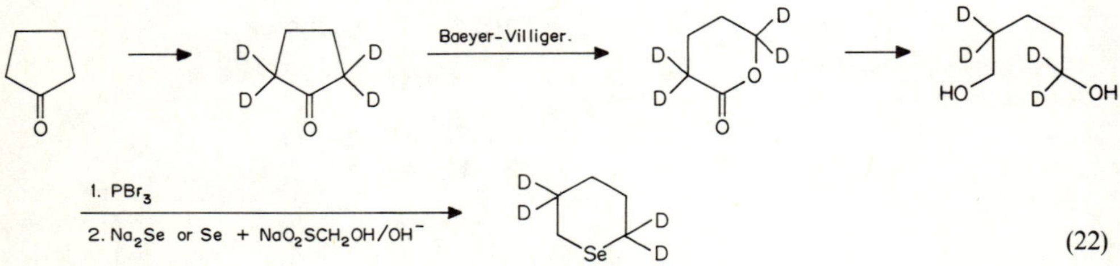

(**14a**) (22)

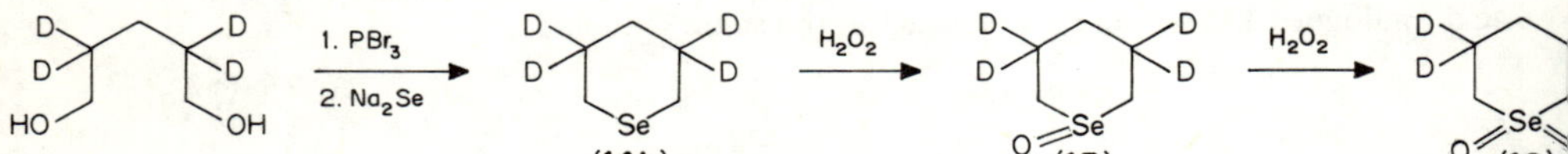

(**14b**) (**15**) (**16**) (23)

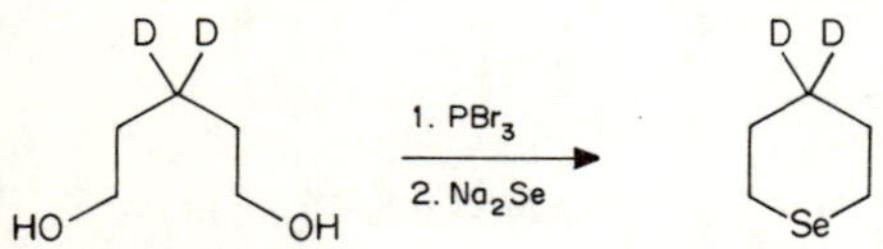

(**14c**) (24)

Meanwhile, Russian chemists prepared 2,2,5,5-tetradeuterioselenolane (**17**) as shown in equation (25)[32].

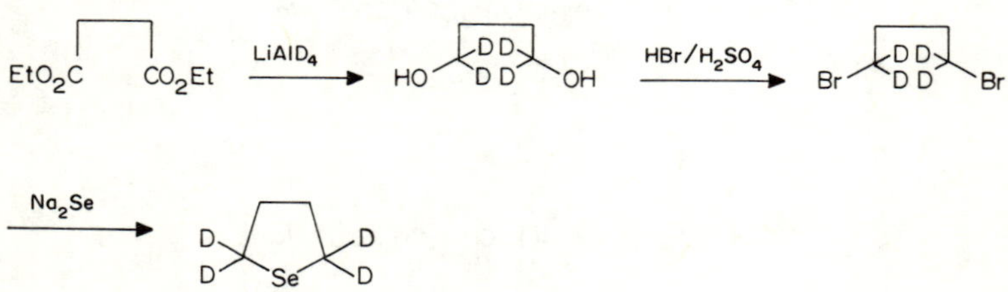

(**17**) (25)

Lambert and coworkers also synthesized 4,4-dideuteriotellurane (**18a**) and 3,3,5,5-tetradeuteriotellurane (**18b**) by the reaction between the corresponding deuteriated 1,5-dibromopentane and sodium telluride (equations 26 and 27)[28].

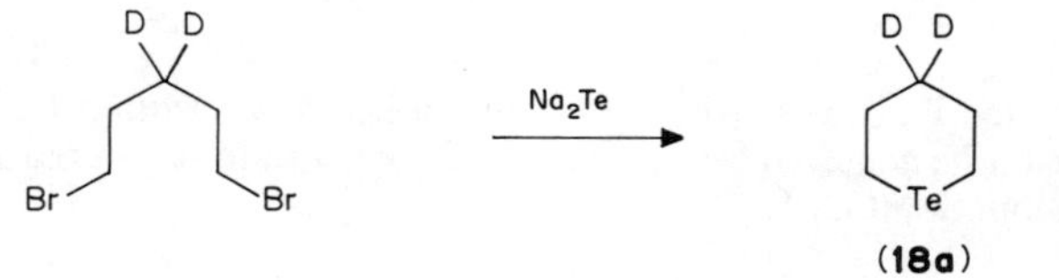

(26)

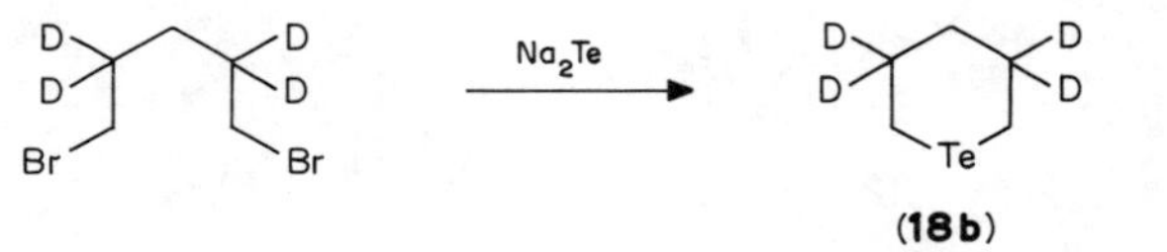

(27)

Several deuteriated selenophenes (**19a–f**) were prepared by Magdesieva and coworkers by treating the corresponding iodoselenophenes with zinc and AcOD (equation 28)[33].

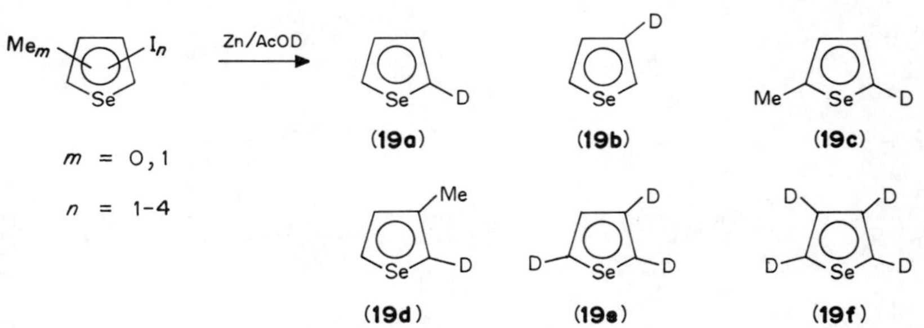

(28)

Martin and coworkers obtained deuteriated selenophenes (**19a,g–i**) by quenching the corresponding Grignard reagents with D_2O (equation 29)[34].

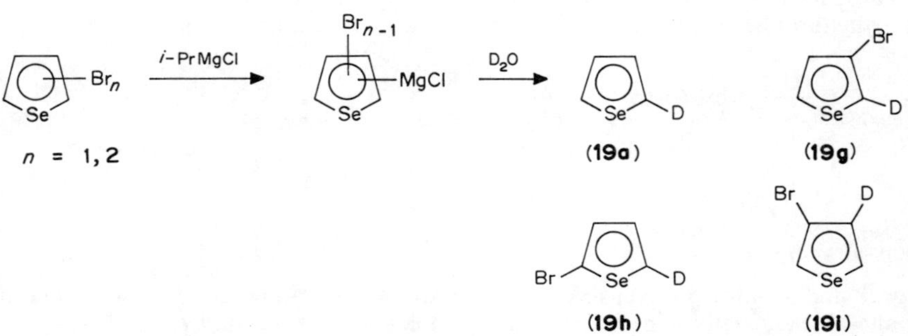

(29)

4-Phenyl-5-deuterio-1,2,3-selenodiazole (**20**) was prepared by the reaction of the semicarbazone of acetophenone-d_3 with selenium dioxide (equation 30)[35]. Hanson and

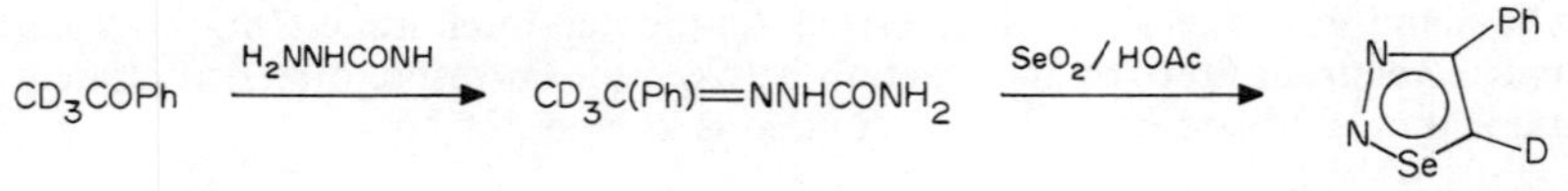

(30)

(20)

Davis reported that the synthesis of ^{75}Se-labelled 4-substituted 1,2,3-selenadiazoles, **21** and **22**, could be achieved by cyclization of appropriate semicarbazones by the method shown in equation (30)[36].

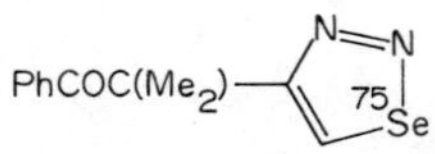

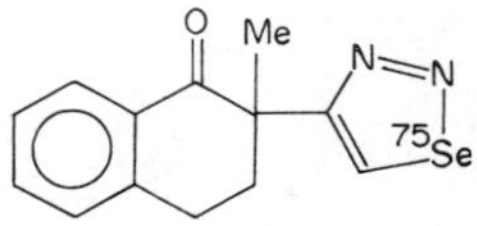

(21) (22)

Paliani and coworkers synthesized tellurophene (**23**) and various deuteriated derivatives (**23a–c**) by the processes shown in equation (31)[37]. Barton and coworkers

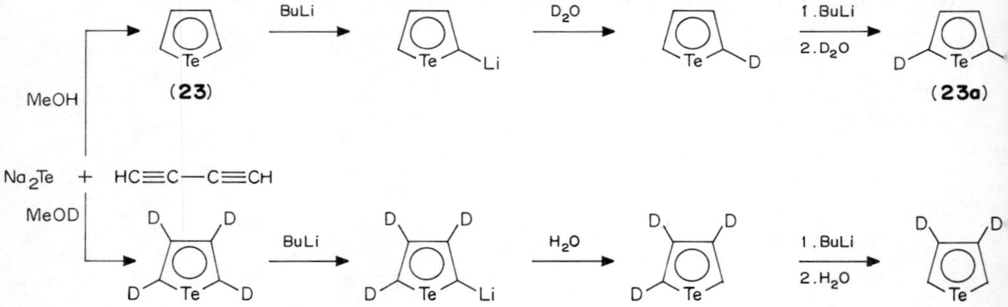

(23) (23a) (23c) (23b)

(31)

prepared 2-phenyltellurophene-5-d_1 (**24a**) by initial lithiation of 2-phenyltellurophene and subsequent quenching of the lithiated compound by deuterium oxide, while 2-phenyltellurophene-3,4,5-d_3 (**24b**) was prepared as shown in equation (32)[38].

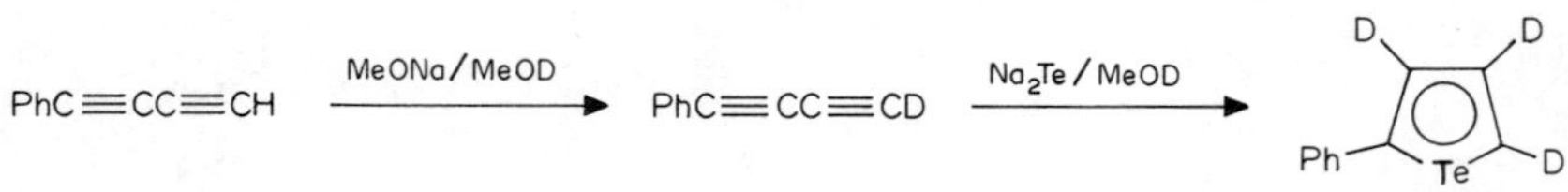

(32)

(24b)

3. Amino acids

Barak and Swanberg established a quantitative analytical method for estimating the amount of selenomethionine in biological samples such as serum, liver and muscle by coupling the techniques of paper chromatography and neutron activation analysis; this involves the irradiation of the sample with neutron flux to convert ^{76}Se to radioactive ^{77m}Se[39].

McConnell and coworkers irradiated selenomethionine, 6-selenoguanine and 6-selenopurine in a water-cooled compartment of a graphite reactor with a neutron flux of 7.5×10^{11} neutron cm^{-2} s^{-1} for 62 h to obtain directly the corresponding ^{75}-Se-labelled compounds. They claimed that no notable degradation of the starting substances was observed[40]. Later, however, Spencer and coworkers found that the neutron irradiation of selenocystine gave ^{75}Se-labelled selenocystine together with some degradation products via recombination of symmetrical halves of the molecule or by exchange with non-radioactive molecules found as cleavage products in the $[n, \gamma]$ reaction as observed in the neutron irradiation of diphenyl selenide which had been known to give a mixture of radioactive diphenyl selenide and labelled diphenyl diselenide. These observations suggest that the $[n,\gamma]$ process involves Se—Se bond rupture–recombination[41].

Dilworth reported a convenient enzymatic synthesis of isotopically labelled selenocysteine and its derivatives. At first serine was converted to *O*-acetylserine which was then incubated with *O*-acetylserine sulphydrylase at pH 7.2 in the presence of sodium hydrogen selenide to obtain selenocysteine which was converted to ^{75}Se-labelled selenocysteine via ^{75}Se-labelled selenocystine as shown in equation (33)[5]. When ^{14}C- or ^{3}H-labelled serine or ^{75}Se-labelled sodium hydrogen selenide was used as the starting material, the corresponding radioactive selenoamino acid was obtained[5].

$$HOCH_2CH(NH_2)COOH \xrightarrow{AcOH/HCl} AcOCH_2CH(NH_2)COOH \xrightarrow[\textit{O}\text{-acetylserine sulphydrylase}]{NaSeH}$$

$$[SeCH_2CH(NH_2)COOH]_2 \xrightarrow{KBH_4} \underset{[^{14}C,\ ^{3}H,\ ^{75}Se]}{HSeCH_2CH(NH_2)COOH} \xrightarrow{HOCH_2CH_2Br} \begin{matrix} SeCH_2CH_2OH \\ | \\ CH_2CH(NH_2)COOH \end{matrix} \quad (33)$$

^{75}Se-labelled selenomethionine was prepared according to equation (34)[12].

$$(Me^{75}Se)_2 \xrightarrow{Na} Me^{75}SeNa \xrightarrow{BrCH_2CH_2CH(NH_2)COOMe}$$

$$Me^{75}SeCH_2CH_2CH(NH_2)COOMe \xrightarrow{H^+/H_2O} Me^{75}SeCH_2CH_2CH(NH_2)COOH \quad (34)$$

Otto obtained ^{75}Se-labelled selenomethionine by treating ^{75}Se-labelled lithium methaneselenolate with bis(chloroethyl)dioxopiperazine and subsequent hydrolysis (equation 35)[42].

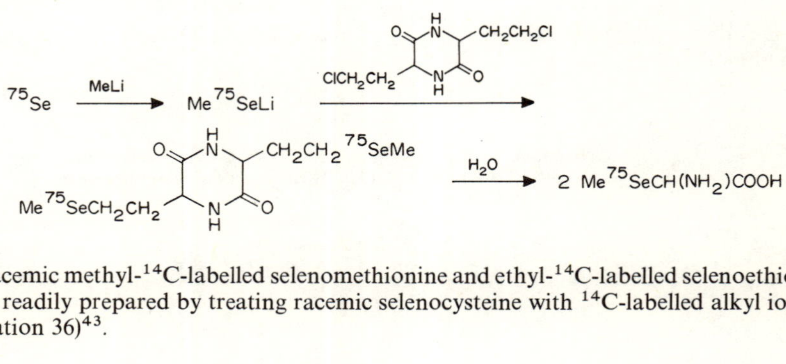

(35)

Racemic methyl-^{14}C-labelled selenomethionine and ethyl-^{14}C-labelled selenoethionine were readily prepared by treating racemic selenocysteine with ^{14}C-labelled alkyl iodides (equation 36)[43].

$$^{14}CH_3I + HSeCH_2CH_2CH(NH_2)COOH \longrightarrow {}^{14}CH_3SeCH_2CH_2CH(NH_2)COOH \quad (36)$$

Bremer and Natori tritiated selenomethionine randomly by simple exposure of the compound to tritium gas [44]according to the procedure reported by Wilzbach[45]. ^{3}H-labelled selenomethionine was then converted to ^{3}H-labelled *Se*-adenosylselenomethionine (**25**) through the procedure developed by Cantoni and Durell (equation 37)[46].

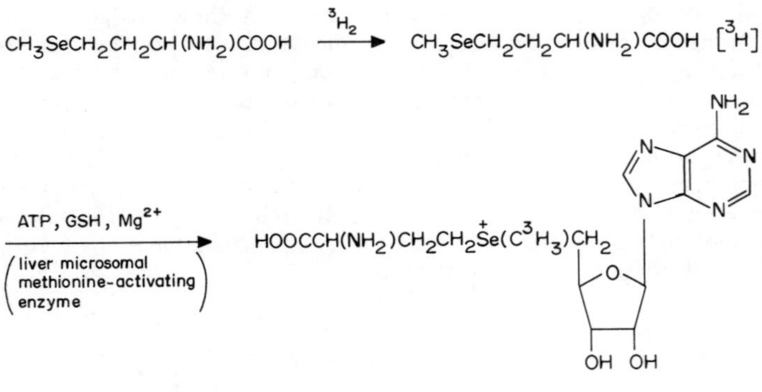

(**25**)

(37)

Blau reported a successful small-scale biosynthesis of radioactive selenocysteine and selenomethionine[47]. At first baker's yeast (*Saccharomyces cerevisa*) was grown in the presence of ^{75}Se-labelled selenous acid. Subsequent hydrolysis and ion exchange fractionation provided the desired ^{75}Se-labelled amino acids.

Wong and coworkers prepared ^{75}Se-labelled selenaproline (4-selenazolidine-1-^{75}Se-carboxylic acid) (**26**) by condensation of ^{75}Se-labelled L-selenocystine with formaldehyde after reduction with sodium borohydride as shown in equation (38)[48].

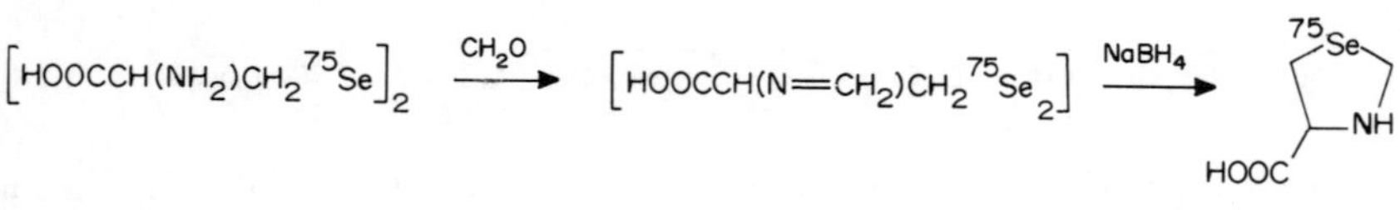

(**26**)

(38)

Monks prepared ^{75}Se-labelled derivatives of folates **28** and **29** as shown in equations (39) and (40)[8,9].

$$Na_2{}^{75}Se_2 \xrightarrow{ClCH_2CH(NH_2)COONa} [{}^{75}SeCH_2CH(NH_2)COONa]_2 \xrightarrow{Na/liq.\ NH_3}$$

$$Na^{75}SeCH_2CH(NH_2)COONa \xrightarrow[\text{2. } H_2O/NaOH \quad \text{3. } H^+]{\text{1. } N^{10}\text{-trifluoroacetylpteroyl isobutyl carbonate } (\mathbf{27})}$$

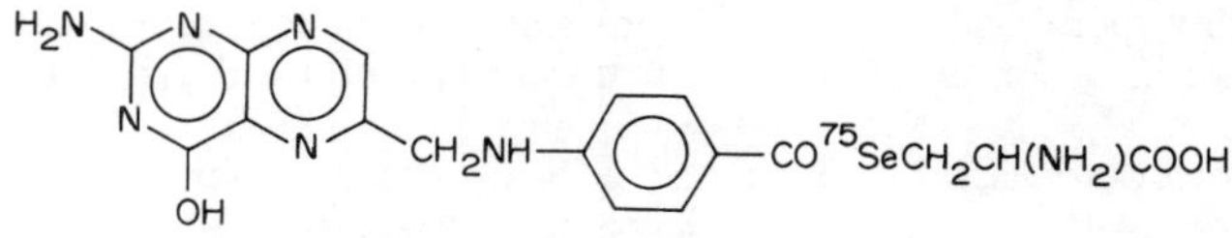 (39)

(28)

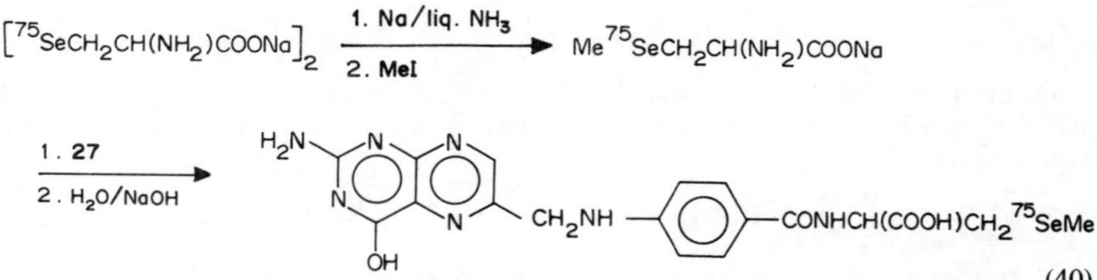 (40)

(29)

^{123m}Te-labelled racemic α-amino-γ-(phenyltelluro)butyric acid (**30**) was synthesized by Knapp and coworkers by alkaline hydrolysis of racemic 5-[β-(phenyltelluro)ethyl]hydantoin-^{123m}Te obtained as shown in equation (41)[49].

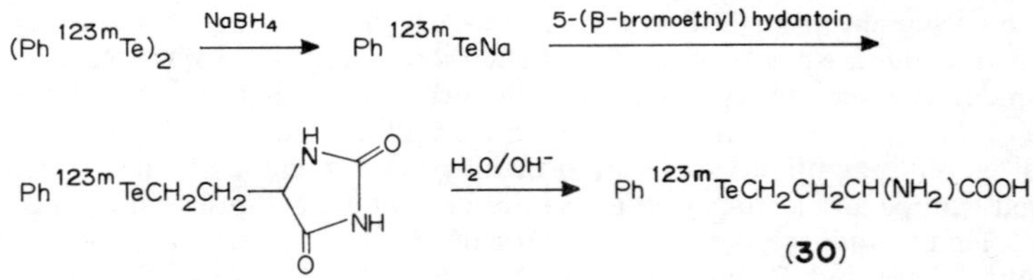 (41)

(30)

Celander and Celander isolated ^{75}Se-labelled urokinase from the urine of dogs after injecting $Na_2{}^{75}SeO_3$[50], while ^{75}Se-labelled fibrinogen was obtained from the plasma of dogs which were administered with $Na_2{}^{75}SeO_3$[51] or ^{75}Se-selenomethionine[52].

Insulin containing radioactive selenocystine was prepared by Tavaarwerk according to the following *in vivo* reaction[53]: Selenocystine labelled with ^{75}Se was injected into the tail vein of a rat. After one day, radioactive insulin was extracted from the rats pancreas and purified.

Leon and coworkers successfully prepared ^{75}Se-labelled plague murine toxins produced by *Pasteurella pestis*. The toxic proteins A and B were isolated from the soluble protein obtained from the cells grown in the presence of ^{75}Se-labelled selenomethionine[54].

Snake venom labelled with ^{75}Se was obtained from *Vipera ammodytes* one week after ^{75}Se-labelled selenomethionine was injected into the snake's stomach. The labelled snake venom exhibited the same electrophoretic behaviour as non-labelled venom[55].

The technique of labelling proteins with ^{75}Se has also been applied to immunological research[56–60]. For example, Engler and coworkers obtained highly radioactive ^{75}Se-labelled glycoproteins, i.e. hepatoglobin and α_2-macroglobulin, formed by acute-phase reaction of the inflammatory process caused by turpentine[57]. Rats were injected with ^{75}Se-selenomethionine one day after treating with turpentine. Five and a half hours later the blood of the animals was collected and the radioactive glycoproteins were isolated from the plasma[57].

In vivo labelling of living lymphocytes with ^{75}Se was performed through incubation of mice or rat lymphocytes with ^{75}Se-selenomethionine[61,62].

Meanwhile, highly specific labelling of viral proteins was achieved by Enzmann, who revealed that viruses incorporated ^{75}Se-selenomethionine in culture, during their protein synthesis[63].

Kishore and coworkers labelled the living cells of *Pseudomonas aeruginosa* and *Staphylococcus aureus* with ^{75}Se by growing the bacteria in trypticase soy in the presence of ^{75}Se-labelled selenomethionine[64].

4. Miscellaneous

Kronrád and Hladik prepared ^{75}Se-labelled selenourea by treating ^{75}Se-labelled hydrogen selenide, obtained from the reaction shown in equation (3), with cyanamide (equation 42)[10b].

$$NCNH_2 + H_2{}^{75}Se \longrightarrow {}^{75}Se{=}C(NH_2)_2 \qquad (42)$$

Martinsen and Songstad obtained bis(triphenylphosphine)imminium selenocyanate-^{13}C by the reaction of ^{13}C-labelled cyanide with elemental selenium (equation 43)[65].

$$Ph_3P\overset{+}{N}PPh_3\ Cl^- \xrightarrow{K^{13}CN} Ph_3P\overset{+}{N}PPh_3\ {}^{13}CN^- \xrightarrow{Se} Ph_3P\overset{+}{N}PPh_3\ {}^-Se^{13}CN \qquad (43)$$

Sulphur atoms in acid-labile sulphur–iron clusters of bacterial origin were found to be replaced by selenium in an ESR study of these enzymes. Putidaredoxin isolated from *Pseudomonas putida* is a protein of mol. wt. 12,000 containing two atoms each of acid-labile sulphur and iron. Selenium-containing homologues of this iron protein labelled with ^{80}Se (97 atom%) and ^{77}Se (87 atom%) containing a small amount of radioactive ^{75}Se were prepared by treating putidaredoxin apoprotein with ammonium selenide enriched with the desired isotope and ferrous iron in the presence of 2-mercaptoethanol under argon. Both the selenium and iron contents of the modified protein were two atoms per mole of the enzyme[66]. Fee and Palmer prepared $^{32}S^{57}Fe$, $^{80}Se^{56}Fe$, $^{77}Se^{56}Fe$ and $^{80}Se^{57}Fe$ homologues of parsley ferredoxin (native enzyme: $^{32}S^{56}Fe$). They treated the apoenzyme with a mixture of ^{80}Se (80 atom%)- or ^{77}Se (88 atom%)-labelled Na_2SeO_3 containing a small amount of $Na_2{}^{75}SeO_3$ and excess dithiothreitol in a buffer solution under helium, and subsequently with ^{56}Fe or ^{57}Fe (90 atom%) citrate. The apoprotein was prepared by precipitation with trichloroacetic acid, while the isotopically labelled Na_2SeO_3 was prepared by the reaction of enriched isotopic elemental selenium with nitric acid[67]. Similarly, Mukai and coworkers prepared ^{80}Se- or ^{77}Se-containing selenium homologues of adrenodoxin isolated from beef adrenal gland[68]. Displacement of labile sulphur atoms in these native non-haemeiron proteins with selenium changed little the nature of the original enzymes, e.g. the enzymes modified with selenium were found to retain about 80% of the native enzyme activity in all three cases above.

III. USES OF ISOTOPICALLY LABELLED SELENIUM AND TELLURIUM COMPOUNDS

A. Physicochemical Uses

1. Infrared and Raman spectroscopy

Harvey and coworkers recorded IR spectra of light and heavy methaneselenols[15] and dimethyl diselenides[6]. They assigned absorption bands at 3027 cm^{-1} (CH_3SeH, CH_3SeD)

and 2270 cm^{-1} (CD_3SeH, CD_3SeD) for antisymmetrical stretching vibrations, while absorption bands at 2955 cm^{-1} (CH_3SeH, CH_3SeD), 2194 cm^{-1} (CD_3SeH) and 2148 cm^{-1} (CD_3SeD) were taken to be due to symmetrical C—H stretching vibrations. The following bands were assigned for Se—H or Se—D stretching vibrations: 2330 cm^{-1} (CH_3SeH), 2323 cm^{-1} (CD_3SeH) and 1680 cm^{-1} (CH_3SeD, CD_3SeD). The absorption bands due to the antisymmetrical stretching vibrations of the C—H or C—D bond for dimethyl diselenide were found to be located at 3027 cm^{-1} for $(CH_3Se)_2$, and at 2273 and 2258 cm^{-1} for $(CD_3Se)_2$. The symmetrical stretching vibration of the C—H bond was at 2940 cm^{-1} for $(CH_3Se)_2$, while that of the C—D bond was at 2137 cm^{-1} for $(CD_3Se)_2$. These authors claimed the complete assignment of both the IR and Raman spectra of these compounds.

Magdesieva and coworkers recorded the IR and Raman spectra of selenophene and its deuteriated derivatives (**19a–f**) in the region of 400–3100 cm^{-1}[69,70]. The observed data were compared with the values of frequencies, force constants and intensities theoretically calculated. Magdesieva summarized C—H(D) stretching and deformation and 9 ring strecthing and deformation vibrations for light and heavy selenophenes together with those for thiophenes[71].

The infrared and Raman spectra of tellurophene (**23**) and the deuteriated tellurophenes (**23a–c**) have been studied by Italian investigators, who reported that the vibrational spectra of thiophene, selenophene and tellurophene are very similar, although somewhat different from that of furan[37]. They proposed that the spectral behaviour of five-membered aromatics containing one heteroatom can be attributed to the mass and the electronegativity of the heteroatom, the delocalization of π electrons, the geometry of the molecule and to the differences in vibrational coupling on the normal mode in these molecules. The increase in frequency of the ν_5 mode due to the symmetric stretching of the double bond in the order of thiophene, selenophene, tellurophene was explained in terms of the same decreasing order of aromaticity which increases the localization of the double bonds in the heterocycles.

2. *Microwave spectroscopy*

Earlier, planarity of selenophene had been a matter of controversy[72]. In order to clarify this problem, Pozdeev and coworkers recorded microwave spectra of selenophene and deuteriated selenophenes at $-40\,^{\circ}C$. 17 Selenophenes containing different isotopes such as ^{12}C, ^{13}C, ^{1}H, ^{2}H, ^{76}Se, ^{78}Se, ^{80}Se and ^{82}Se, were studied. The experimental results verified the planarity of the selenophene molecule[73].

3. *Nuclear magnetic resonance spectroscopy*

Lambert and coworkers achieved conformational characterization of protonated thiane, selenane and tellurane in FSO_3H–SO_2 by measuring NMR spectra of the deuteriated compounds. Vicinal coupling constants between H_X and the axial proton (J_{ax})

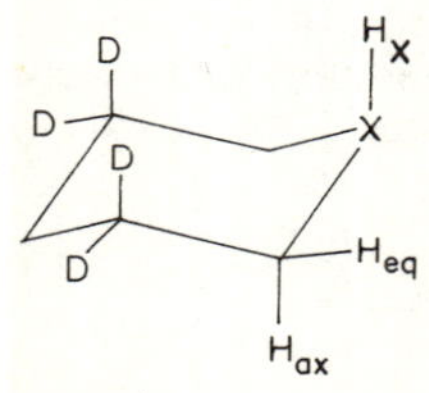

X = S, Se, Te

of 3,3,5,5-tetradeuterio heterocycles were found to be 14.1, 13.0 and 11.2 Hz for X = S, Se and Te, respectively. These values are similar to those found in piperidine (10.5 Hz), *N*-methylpiperidine (10.0 Hz) and *N*-methylpiperidine (11.3 Hz) in which the *N*-bound protons are axial, indicating that the protons attached to the Group IV heteroatoms are also at the axial position. Similarly,the values of J_{eq}, the coupling constant between H_X and H_{eq} for the Group IV heterocycles (2.3, 2.1 and 2.4 Hz for X = S, Se and Te, respectively) were found to be similar to those of the nitrogen heterocycles in accordance with the above results[28].

The same authors also studied the ring deformation of these heterocycles on the basis of the NMR data of 4,4-dideuterio heterocycles, and showed that the following three cases, i.e. **31**, **32** and **33**, can be diagnosed based on an *R* value which is equal to J_{trans}/J_{cis} and is independent of the electronegativity of X.

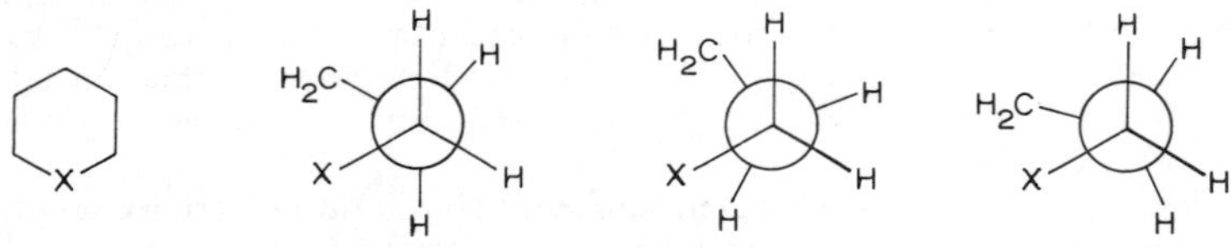

(**31**) (**32**) (**33**)

$R = J_{trans} / J_{cis}$ $R = 2$ $R > 2$ $R < 2$

R values were found to be 2.61, 2.74 and 2.76 for thiane-4-d_2, selenane-4-d_2 (**14c**) and tellurane-4-d_2 (**18a**), respectively. These values appear to indicate that the pseudo-equatorial protons are pushed more closely together as in **32** in these heterocycles[28].

Lambert and coworkers[30] determined the free energy barrier for the conformational reversal for six-membered heterocycyclic rings containing a Group IV heteroatom by analyzing the temperature-dependent proton NMR spectra of these heterocycles tetra-deuteriated at the 3- and 5-positions. The free energy for the ring reversal decreases with the increase of the size of the Group IV atoms, i.e. 10.3 kcal mol^{-1} at -64 °C for tetrahydropyrane, 9.4 kcal mol^{-1} at -81 °C for thiane, 8.2 kcal mol^{-1} at -105 °C for selenane (**14b**) and 7.3 kcal mol^{-1} at -119 °C for tellurane (**18b**). The barriers are well correlated with the torsional properties of the C—X bond. The same trend was observed for the corresponding oxides, i.e. 10.1 kcal mol^{-1} at -70 °C for thiane-1-oxide, 8.3 kcal mol^{-1} at -102 °C for selenane-1-oxide (**15**), 10.3 kcal mol^{-1} at -63 °C for thiane-1,1-dioxide and 6.7 kcal mol^{-1} at -133 °C for selenane-1,1-dioxide (**16**). They also found that in **15** the proportion of axial isomer (84%) is more than that in thiane-1-oxide (62%), since the increase of carbon–heteroatom bond length increases the 1,3- and 1,5-attractive interactions. For the same reason, in the *Se*-methylselenanium iodide, the Me group was found to be nearly exclusively axial, while in the *S*-methylthianium iodide the Me group was nearly completely equatorial[30].

Lambert and coworkers determined the structure of complexes between Group IV heterocycles and halogens by an NMR study using 4,4-dideuterio derivatives and 2,2,5,5-tetradeuterio heterocycles, and also by conductance measurements. In solution, selenane

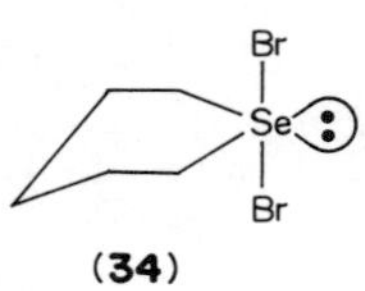

(**34**)

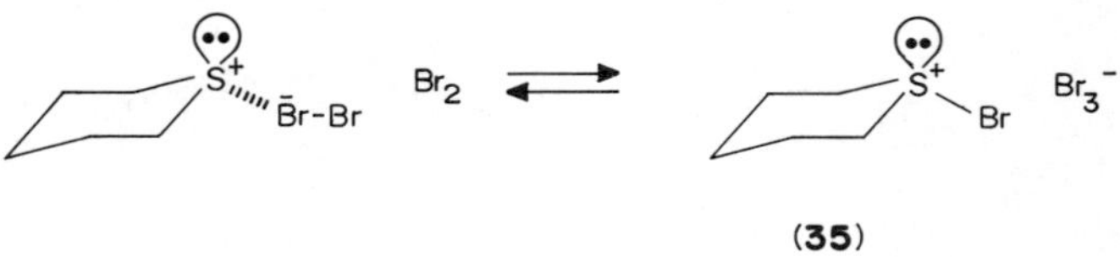

(**35**)

dibromide exists as a trigonal bipyramid (**34**), whereas selenane diiodide, thiane dibromide and thiane diiodide are in the form of a molecular complex like **35**[29].

Gronowitz and coworkers determined the ^{1}H- and ^{13}C-NMR parameters for a series of selenophenes substituted at the 2- and 3-positions. The chemical shifts were linearly correlated with the reactivity parameters of Swain–Lupton's two-parameter equation. Deuteriated selenophenes were used for the chemical shift assignment[74].

Fringuelli and coworkers also used deuteriotellurophenes in an NMR study[75]. They discussed physicochemical properties involving NMR and chemical reactivities of tellurophene in comparison with those of other five-membered heteroaromatics containing a Group IV element[75].

B. Mechanistic Studies of the Reactions of Organic Selenium and Tellurium Compounds

1. *Isotopic exchange*

Magdesieva and coworkers investigated both acid- and base-catalysed D–H exchange reactions of 2- and 3-deuterioselenophene (**19a** and **b**), 2-deuterio-5-methylselenophene (**19c**) and 2-deuterio-3-methylselenophene (**19d**)[33b]. Electrophilic D–H exchange reactions were conducted in 4:1 mixture of acetic and trifluoroacetic acids at 25 °C. Relative rates were reported to be **19a**: **19c**: **19d** = 1 : 107 : 236, while the D–H exchange in **19b** was too slow to compete with the acid-catalysed degradation. The results clearly reveal that the rate of D–H exchange reaction is correlated with the stability of the protonated intermediate **36** (equation 44).

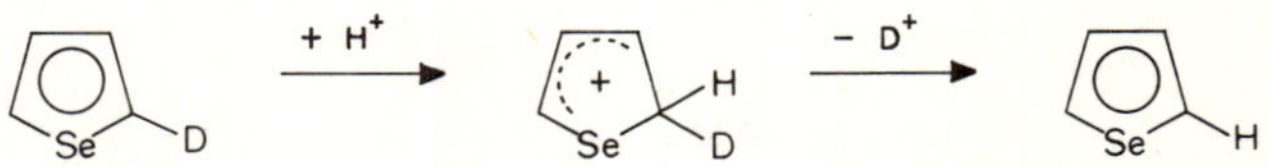

(**19a**) (**36**) (**19**) (44)

Schwetlick and Unverferth determined relative reactivities of different aromatics in the acid-catalysed D–H exchange[76]. Obviously, the rate appears to increase as the aromaticity of the heteroaromatics decreases as shown below. Magdesieva also discussed the relative reactivities of selenophene, thiophene and furan in both acid- and base-catalysed D–H exchange reactions at the α- and β-positions, on the basis of the abilities of electron pair donating effects of the heteroatoms[71].

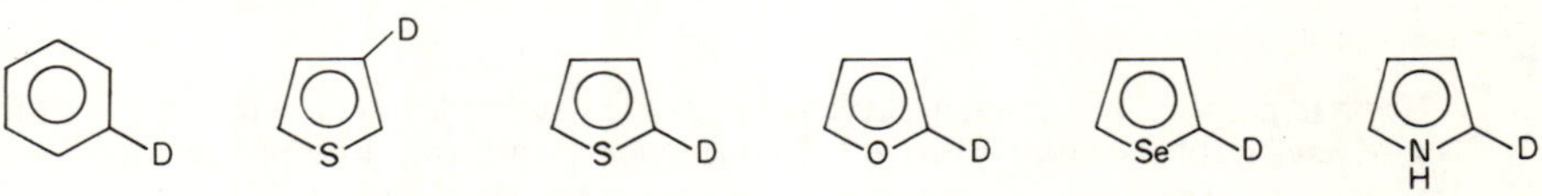

k^{rel}	1	10^4	2×10^7	3×10^7	2×10^8	10^{15}

Magdesieva and coworkers also carried out alkaline D–H exchanges of deuterioselenophenes in *t*-BuOLi/DMSO or *t*-BuOK/in 70% *t*-BuOH and 30% diglyme at 25 °C. In the former system, the relative exchange rates were observed to be **19a**: **19c**: **19d**: **19b** = 1:

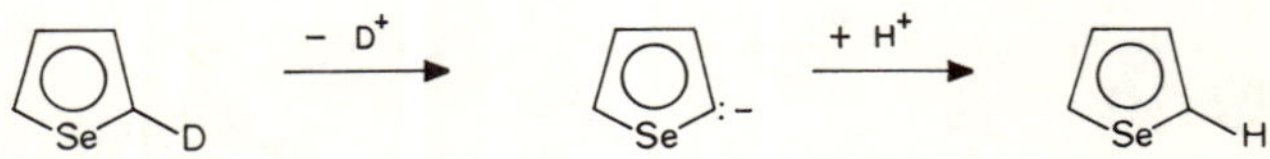

(**37**) (45)

0.12: 0.05: 2.5×10^{-5}. These rates vary with the stability of the intermediary anion (**37**) as shown in equation (45)[33b,71].

Ghandehari and coworkers[35] investigated the mechanism of the base-catalysed decomposition of 4-aryl-substituted 1,2,3-selenodiazoles (**20a**) to give *para*-substituted phenylethynylselenolate (equation 46a)[38b] which eventually affords substituted 1,3-diselenafulvenes (**38**) by the reaction shown in equation (46b)[35]. They observed that an electronegative *para* substituent in the Ph group accelerates the reaction (46a) and the kinetic data were nicely correlated with the Hammett σ constants giving $\rho = +2.37$.

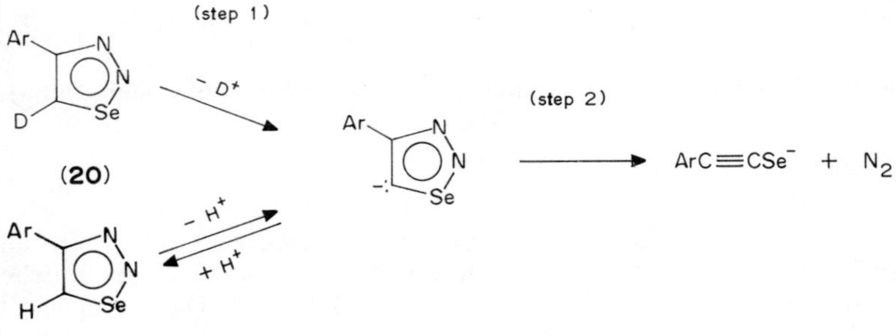

(**20**)

(**20a**)

(46a)

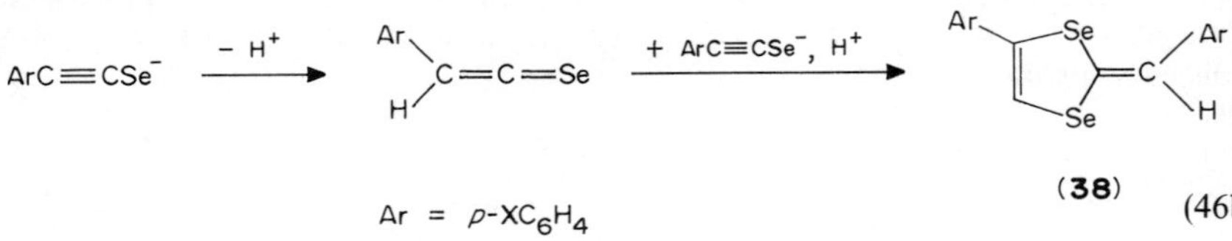

(**38**)

(46b)

However, when 5-deuterio-4-phenyl-1,2,3-selenadiazole (**20**) was subjected to the reaction, the D–H exchange took place readily before the evolution of nitrogen gas was noticed. The value of $k_H/k_D = 2$ was observed for step 1, while no significant D–H kinetic isotope effect was observed in the overall reaction. Thus the authors concluded that step 1 is a fast equilibrium and the positive ρ value is due to the shift of the equilibrium by the electronegative substituents to the right, hence step 2 seems to be facilitated[35].

2. Photochemical reaction

Upon irradiation of 2-phenylselenophene, Barton and coworkers obtained a mixture of 3-phenylselenophene and 1-phenyl-2-vinylacetylene, while the latter product was the sole product in the irradiation of 2-phenyltellurophene (**24**). They studied the mechanism of the formation of phenylvinylacetylene using deuterium as the tracer[38a]. Photolysis of 2-phenyl-5-deuteriotellurophene (**24a**) in ether with a 300 nm lamp gave a 1:1 *syn anti* mixture of (2-deuteriovinyl)phenylacetylenes (equation 47). Similarly, a 1: 1 mixture of *syn-* and *anti-*

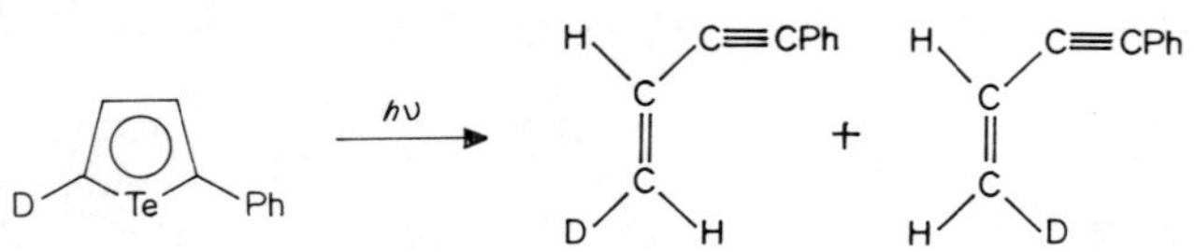

(**24a**)

(47)

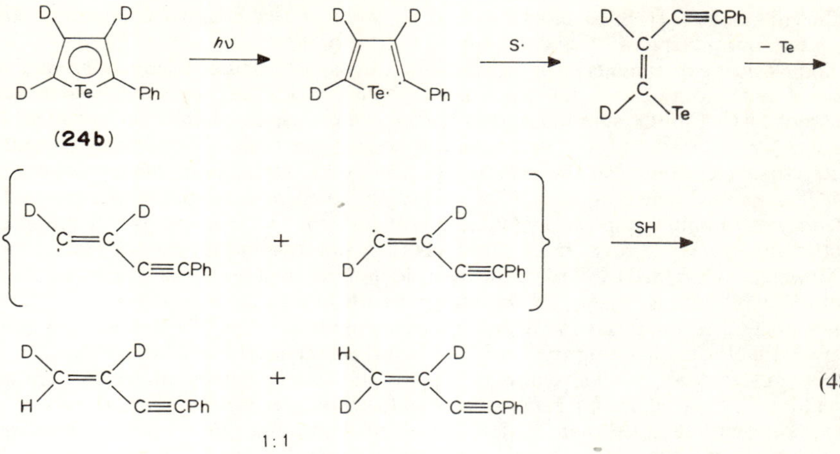

(48)

1,2-dideuteriovinylphenylacetylenes, was afforded in the photolysis of 2-phenyl-3,4,5-trideuteriotellurophene(**24b**). This clearly demonstrates that the reaction involves intermolecular hydrogen transfer. These deuterium tracer results were rationalized by the mechanism shown in equation (48). Actually this is an extension of van Tamelen's general mechanism for the photochemistry of aromatic heterocyclopentadienes. The mechanism involves homolytic cleavage of the weakest C—Te linkage of the excited tellurophene followed by hydrogen abstraction by either a solvent radical or the photoexcited substrate. The subsequent second C—Te homolysis forms a phenylpropargylvinyl radical which then abstracts hydrogen from solvent giving the final products[38].

C. Biochemical Uses

Earlier, selenium compounds were regarded as only toxic to most living organisms. Later, especially by the use of isotopically labelled selenium compounds in biochemical studies, selenium was found to be an essential element.

1. *Metabolism of selenium and tellurium*

a. Selenium. Absorption, retention, distribution and excretion of selenium in living bodies have been tested very extensively with humans[78–81] animals[82–92], chickens[93–98], bacteria[99–100], etc. Usually ^{75}Se, a γ-emitter, is used because of its convenience in monitoring. This radioactive nuclide can be used from macro- to molecular level investigations. For example, by whole-body counting of ^{75}Se it is possible to determine its retention after administration of ^{75}Se-labelled substances[77]. On the other hand, molecular level investigation of a trace amount of ^{75}Se-containing metabolite is also possible as mentioned in the following sections.

When $Na_2{}^{75}SeO_3$ is administered to selenium-deficient animals or chickens, selenium is effectively absorbed in the body and retained for a long time. Generally, ^{75}Se is detected more in the liver, kidney and pancreas than in the muscles, heart, lungs and spleen. When $Na_2{}^{75}SeO_3$ was administered to cocks, the highest ^{75}Se concentration was found in the protein fraction of sperm after 16 days. The ^{75}Se was found to remain in a non-dialysable protein probably as selenocysteine[93]. The same trend was found in rats[84]. Janghorbani and coworkers found that chicken tissue which was fed with $Na_2{}^{74}SeO_3$ contained a sufficient amount of ^{74}Se to allow the use of these chicken for human feeding experiments by faecal monitoring of ^{74}Se with neutron activation analysis[98].

Some tracer experiments have revealed that the metabolisms of selenoamino acids are quite similar to those of the sulphur analogues. According to some evidence the metabolisms of both sulphur and selenium amino acids share the same enzyme systems[5,43,44,99–103]. Mudd and Cantoni found that selenomethionine is utilized better than methionine in the partially purified yeast methionine adenosyltransferase (MAT)-catalysed reaction with ATP, affording *Se*-adenosylselenomethionine and *S*-adenosylmethionine, respectively (equations 49 and 50)[102a]. Pan and Tarver found that methionine works 1.7 times more effectively than selenomethionine in the reaction with ATP which is catalysed by purified rat-liver methionine adenosyltransferase (equations 49 and 50), while the K_m value for the reaction with selenomethionine (equation 50) was found to be less than that for with methionine (equation 49).[102b]. Bremer and Natori showed that the methylating ability of ^3H-labelled *Se*-adenosylselenomethionine (**25**) is 1.4 times greater than *S*-adenosylmethionine in the phosphatidylcholine (**39**) synthesis (equation 51)[44]. This was further evidenced in the choline biosynthesis *in vivo* by Pan and coworkers, who revealed that ^{14}C-labelled selenomethionine (**25a**) is twice as efficient as methionine in the methylation of phosphatidylethanolamine to afford phosphatidylcholine (**39a**) in intact rat liver (equation 51)[43]. They demonstrated further that *Se*-adenosylselenomethionine (**25b**) can donate a Me group to RNA similarly to *S*-adenosylmethionine 43.

$$MeSCH_2CH_2CH(NH_2)COOH + ATP + H_2O \xrightarrow{MAT} P_i + P_iP_i + S\text{-adenosylmethionine} \quad (49)$$

$$MeSeCH_2CH_2CH(NH_2)COOH + ATP + H_2O \xrightarrow{MAT} P_i + P_iP_i + Se\text{-adenosylselenomethionine} \quad (50)$$

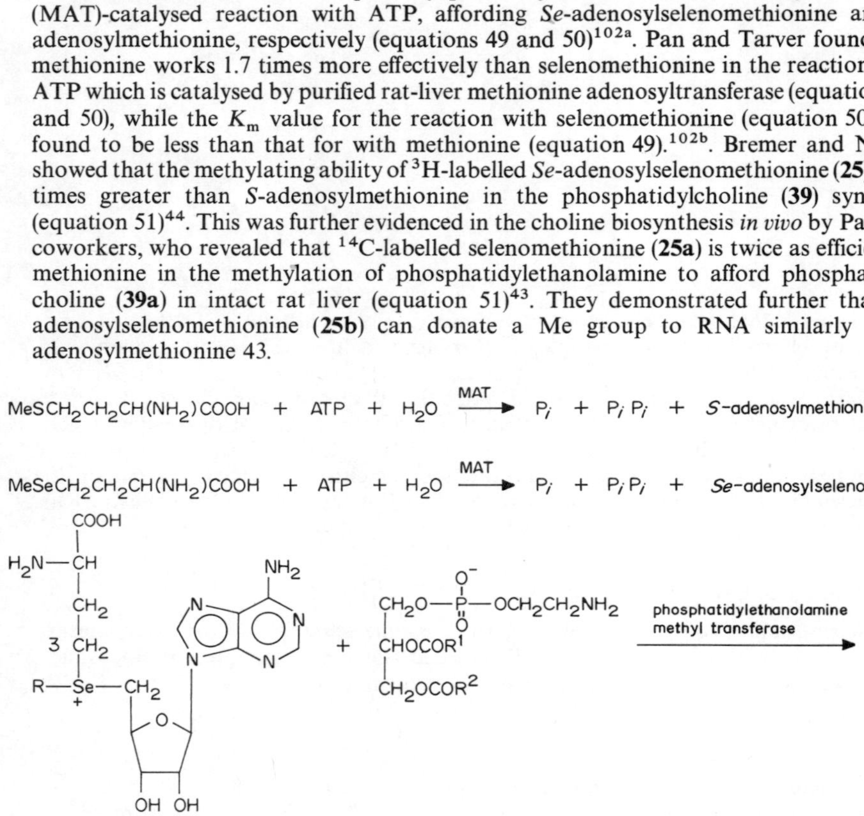

(**25**) R = C^3H_3

(**25a**) R = $^{14}CH_3$

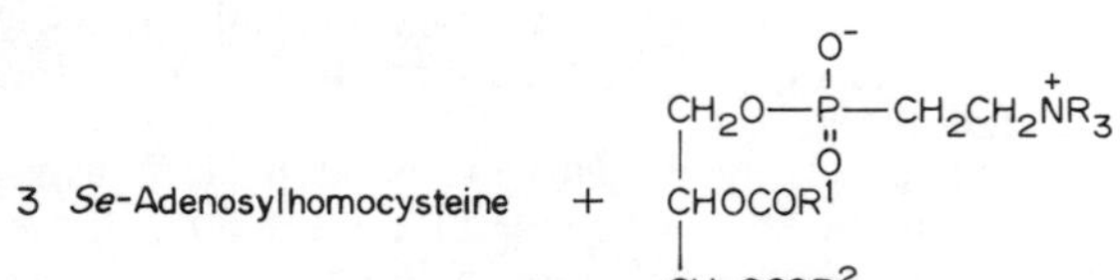

(51)

(**39**) R = C^3H_3

(**39a**) R = $^{14}CH_3$

Wilson and Bandurski suggested that the sulphate-activating enzyme system to form adenosine monophosphate–sulphate from adenosine triphosphate (ATP) and sulphate can also promote the selenate activation in the same manner (equation 52)[103].

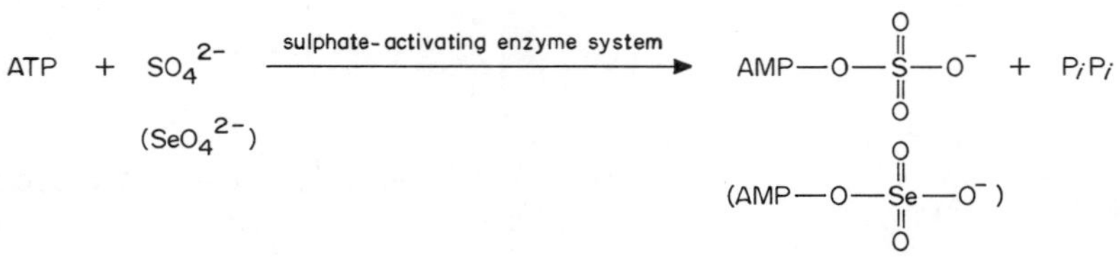

(52)

Spencer and Blau revealed that ^{75}Se-selenomethionine seems to come out from the hamster's intestine through the same route as for methionine[101].

Awwad and coworkers reported that when ^{75}Se-selenomethionine and ^{35}S-cystine are injected into rats, the formation of radioselenate is much less than that of radiosulphate, suggesting that the selenoamino acid is utilized for peptide synthesis better than thioamino acid[91].

b. Tellurium. When $^{132}TeCl_4$ is injected into a rat, the liver contains 20.17% of the ^{132}Te after 6 h, whereas a lesser amount of ^{132}Te is found in the bones (8.21%), kidneys (7.30%), lungs (2.23%), spleen (0.7%) and brain (0.37%)[104]. Russian investigators administered $^{127}TeCl_4$ to rats in order to see the fate of the tellurium and several experimental results have been reported[105–112].

2. Selenoproteins

a. Glutathione peroxidase. Glutathione peroxidase (EC 1.11.1.9) is the only known mammalian enzyme which requires selenium. Glutathione peroxidase reduces poisonous hydrogen peroxide and a variety of lipid hydroperoxides formed by autooxidation to prevent the damage of components in living bodies, while vitamin E prevents the hydroperoxide formation. Rotruck and coworkers revealed by the use of ^{75}Se as tracer that selenium is an essential element of the glutathione peroxidase in erythrocytes. Erythrocytes were obtained from rats after they were injected with $Na_2{}^{75}SeO_3$ for 2 or 4 weeks[113]. On the chromatograph of the supernatant prepared by haemolysis of the erythrocytes both the peaks of the enzyme activity and the selenium content appeared in the same position[113].

$$\left.\begin{matrix} H_2O_2 \\ \text{or} \\ ROOH \end{matrix}\right\} + 2\ GSH \xrightarrow{\text{glutathione peroxidase}} \left.\begin{matrix} H_2O \\ \text{or} \\ ROH \end{matrix}\right\} + GSSG \qquad (53)$$

Flohé and Günzler found that only the reduced form of glutathione peroxidase is inhibited by iodoacetic acid[114]. Thus Tappel and coworkers treated the reduced form of ^{75}Se-labelled glutathione peroxidase (obtained from rat liver pretreated with $Na_2{}^{75}SeO_3$) with ^{14}C-labelled iodoacetic acid. Acid hydrolysis of the modified enzyme gave carboxymethylselenocysteine labelled with ^{75}Se and ^{14}C (equation 54). Since about 60–90% of the original selenium in the enzyme was recovered in the form of carboxymethylselenocysteine by this treatment, it was concluded that all of the selenium in the enzyme is in the form of selenocysteine. This was further confirmed by a similar procedure using ethyleneimine as an alkylating agent of the reduced form of glutathione peroxidase instead of iodoacetic acid[115].

$$\sim\!\sim\text{NHCH}(\text{CH}_2{}^{75}\text{SeH})\text{COO}\sim\!\sim \xrightarrow{\text{I}^{14}\text{CH}_2\text{COOH}} \sim\!\sim\text{NHCH}(\text{CH}_2{}^{75}\text{Se}^{14}\text{CH}_2\text{COOH})\text{COO}\sim\!\sim \xrightarrow{\text{H}_2\text{O}/\text{H}^+} \text{H}_2\text{NCH}(\text{CH}_2{}^{75}\text{Se}^{14}\text{CH}_2\text{COOH})\text{COOH} \quad (54)$$

Recently, in order to test which selenium compound would serve as the best precursor to stimulate glutathione peroxidase, human erthrocytes were cultured in the presence of $Na_2{}^{75}SeO_3$, $Na_2{}^{75}SeO_4$, ^{75}Se-selenocystine, ^{75}Se-selenomethionine and ^{75}Se-labelled foetal calf serum protein in vitro. Among these, selenocystine was found to give the highest increase in the glutathione peroxidase activity (79%). The foetal calf serum protein labelled with ^{75}Se also increased markedly the enzyme activity (47%), while selenite showed a very low stimulation (7%). Hence selenocystine is the compound most effectively utilized for the protein synthesis[116].

b. Glycine reductase. Glycine reductase which catalyses the reductive deamination of glycine (equation 55) has been found in bacteria, e.g. *Clostridium sticklandii.* The presence

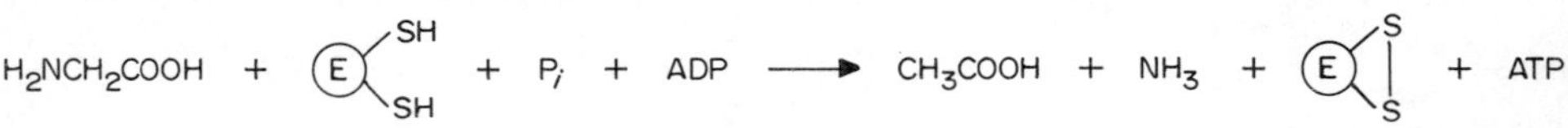

(55)

of selenite in the culture of the bacteria increases the glycine reductase activity of bacteria, but their growth is only increased slightly[117]. In this enzymic reaction, at least three proteins, called A, B, and C, seem to cooperate[117]. Stadtman and coworkers confirmed that in the purification of protein A obtained from bacteria grown in the presence of sodium selenite-^{75}Se, the ^{75}Se content and the activity of the enzyme increased in parallel, showing that protein A is a selenoprotein[118]. Later they treated the purified enzyme labelled with ^{75}Se, obtained above, with various alkylating agents, such as iodoacetamide, and hydrolysed it to amino acids by heating with HCl. Thus, they isolated both ^{75}Se-carboxymethylselenocysteine and *S*-carboxymethylcysteine from the hydrolysis mixture, indicating that both selenocysteine and cysteine are essential components. The enzyme (mol. wt. = 12,000–16,200) was found to contain 1 gram atom of Se per mole. The fully reduced enzyme possesses two sulphydryl groups in addition to the selenol group titrable with 5,5′-dithiobis(2-nitrobenzoic acid), Ellman's reagent[119].

c. Formic acid dehydrogenase. The presence of selenium-dependent formic acid dehydrogenase has been identified in *E. coli* and several anaerobic bacteria. These enzymes are known to contain acid-labile iron, molybdenum and cytochrome b besides selenium[117].

Shum and Murphy suggest the presence of selenium in *E. coli* formic acid dehydrogenase based on parallel increases of the ^{75}Se content and the activity during the purification of the enzyme[120]. Similarly, Andressen and Ljungdahl found that there is a correlation between the bound ^{75}Se and the enzyme activity in the formate dehydrogenase obtained from *Clostridium thermoaceticum* grown in the presence of $Na_2{}^{75}SeO_3$[121].

Jones, Dilworth and Stadtman grew *Methanococcus vannielii* in the presence of $Na_2{}^{75}SeO_3$ and isolated the formate dehydrogenase labelled with ^{75}Se from the bacteria, reduced and alkylated with iodoacetamide (equation 56). The alkylated enzyme was hydrolysed in two different ways. The acid hydrolysis gave ^{75}Se-labelled carboxymethyl selenocysteine, while the enzymic hydrolysis with chymotripsin followed by carboxypeptidase B and carboxypeptidase A-DEP afforded ^{75}Se-labelled carboximidomethylselenoeysteine. Thus, the formate dehydrogenase was firmly shown to contain selenocysteine[122].

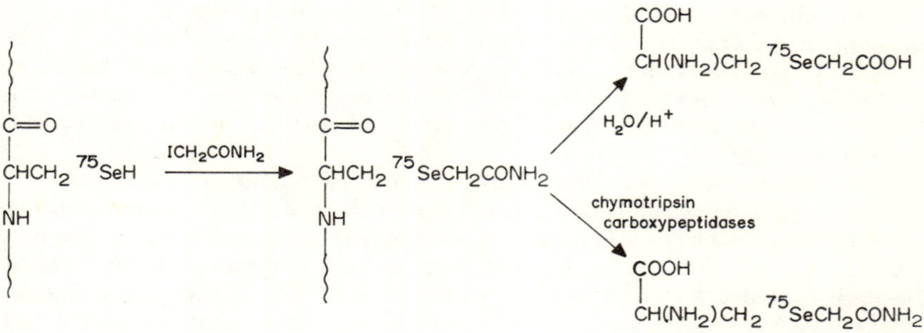

(56)

d. Other proteins. Pedersen and coworkers successfully isolated a selenoprotein (mol. wt. = 10,000) from the muscle of Se-supplemented lambs. They injected lambs with $Na_2{}^{75}SeO_3$ and isolated a selenoprotein labelled with ^{75}Se from the muscle of the lambs. This protein could not be detected in the muscle of white muscle diseased lambs but was present in normal ones[123].

Imhoff and Andressen found that the activity of bacterial nicotinic acid hydroxylase (which catalyses the addition of water to the double bond of nicotinic acid, the initial step in the anaerobic fermentation of nictinic acid to propionate, acetate, NH_3 and CO_2) is increased by the addition of 10^{-7} M sodium selenite to the growth medium of *Clostridium barkeri*, suggesting that the enzyme is selenoprotein[124]. This was evidenced by a parallel increase of both the ^{75}Se radioactivity and the enzyme activity during the purification of the enzyme obtained from the bacteria grown in the presence of $Na_2{}^{75}SeO_3$[117].

Burk and Gregory isolated ^{75}Se-containing proteins from plasma (mol. wt. = 79,000) and liver (mol. wt. = 90,000) of pigs injected with $Na_2{}^{75}SeO_3$; however the biochemical function of these selenoproteins is not yet known[125].

3. *Selenium-containing t-RNA*

Recently, various naturally occurring t-RNAs containing selenium have been isolated from bacteria; for example, Hoffman and coworkers isolated a ^{75}Se-containing t-RNA from *E. coli* cultured in the presence of $Na_2{}^{75}SeO_3$. ^{75}Se-t-RNA was subjected to RNase digestion followed by chromatography. A selenium-containing component remained on the column, suggesting that selenium is not incorporated as a selenoaminoacyl-t-RNA but in the base of the t-RNA. This prediction has been proved by the isolation of ^{75}Se-labelled 4-selenouridine from the ^{75}Se-t-RNA after digestion of the latter with a mixture of bovine pancreatic RNase, snake venom phosphodiesterase and alkaline phosphatase[126].

Chen and Stadtman were able to isolate three varieties of ^{75}Se-containing t-RNA from *Clostridium sticklandii* grown in cultures supplemented with $Na_2{}^{75}SeO_3$ or ^{75}Se-selenocysteine. The selenium was shown to be located in the polynucleotide portion of the t-RNA but not as the alkali-labile selenocysteine[127].

4. *Non-haemeiron proteins containing acid-labile sulphur*

Non-haemeiron proteins containing two irons and two acid labile sulphurs are known to be widely present in a variety of living bodies. ESR has been shown to be a powerful tool to investigate how the odd electron is shared in the enzyme's active site which consists of cysteine sulphur and two acid-labile sulphur and iron atoms according to research groups who utilized selenium isotopes. The odd electron densities can be estimated from the ESR hyperfine structure. However, naturally occurring non-haemeiron proteins possess ^{32}S and ^{56}Fe which have unfortunately both the nuclear spin 0 and do not give any hyperfine splitting. Although the nuclear spin of ^{33}S is 3/2, ^{33}S gives seven hyperfine lines if two nuclei are involved. Therefore the analysis becomes very complicated[66], especially since only 50% enriched ^{33}S is available. Thus, one or two of the labile sulphur atoms have been substituted by ^{77}Se (nuclear spin = 1/2) and ^{80}Se (nuclear spin = 0) up to 80 atom% prior to ESR spectroscopic studies as mentioned in Section II.B.4. ^{57}Fe(nuclear spin = 1/2) is also a suitable isotope for this purpose.

Gunsalus and coworkers recorded ESR spectra of the reduced form of native (^{32}S), ^{80}Se and ^{77}Se containing reduced forms of putidaredoxin obtained from *Pseudomonas putida* and adrenodoxin obtained from pig and beef. A single ESR signal appeared at a low field for the ^{32}S^{32}S and ^{80}Se^{80}Se combinations whereas both enzymes with the ^{77}Se^{77}Se combination exhibited hyperfine splitting in the same region. This suggests that the unpaired electron is shared among the two iron and the two selenium atoms[66].

Fee and Palmer examined by ESR isotopic derivatives of parsley ferredoxin containing ^{32}S^{56}Fe, ^{32}S^{57}Fe, ^{80}Se^{56}Fe, ^{77}Se^{56}Fe and ^{80}Se^{57}Fe. They concluded that two labile sulphur and two iron atoms are present at the active centre of the enzyme and share the unpaired electron in the reduced form[67].

Mukai and coworkers also studied isotopic derivatives of adrenodoxin prepared from beef adrenal gland containing. ^{80}Se^{80}Se, ^{77}Se^{77}Se,^{80}Se^{32}S and ^{77}Se^{32}S combinations. They concluded that the paramagnetic centre of the enzyme with one sulphur and one selenium atom is more rhombohedral than those of the native adrenodoxin and its Se—Se derivative[68].

C. Medicinal Uses

1. Scintiscanning

Much attention has been paid to ^{75}Se-selenomethionine as a pancreas-imaging agent[12,49,128–136] and the method is now used clinically. Seven to eight times more selenomethionine-^{75}Se appeared in the pancreas of mice than in their liver 25–30 min after *in vivo* administration[131]. Cottrall and Taylor tested the influence of the pretreatments with a variety of agents on the uptake of ^{75}Se-selenomethionine by rat pancreas. Only propylthiouracil increased the pancreatic uptake of ^{75}Se-selenomethionine[134]. Selenomethionine-^{75}Se photoscan of the pancrease is useful for the detection of cancer which is difficult to diagnose by ordinary means[133]. Several papers have appeared on the effect on internal irradiation with ^{75}Se-selenomethionine on the endocrine systems[137–143]. Knapp and coworkers found that ^{123m}Te-labelled DL-α-amino-γ-(phenyltelluro)butyric acid (**30**) is a potential pancreatic imaging agent[49].

Kozák and coworkers reported that ^{75}Se-labelled 2-aminoisoselenouronium bromide is accumulated in myocardium and may be useful for scanning of myocardium[144]. Since the normal myocardial tissues utilize fatty acids as the energy source, ^{123m}Te-labelled fatty acid analogues (**11**) are useful as myocardial imaging agents[26]. The ^{123m}Te-labelled unsaturated fatty acid analogues (**13**)[27] and ^{75}Se- or ^{123m}Te-labelled glycerine triethers (**6, 7**)[21] are also useful in the determination of the fat absorption[21,27].

Di(β-dialkylaminoethyl) selenides (**8**) labelled with ^{75}Se[22] and 4-selenazolidine-1-^{75}Se-carboxylic acid (**26**) were examined as brain-imaging agents[48].

^{75}Se-labelled 4-substituted 1,2,3-selenadiazoles (**21, 22**) were tested for the possibility of adrenal gland imaging[36]. 6-Benzylseleno-19-norcholesterol and alkylselenanilide labelled with ^{75}Se are known to be specific for adrenal cortex and adrenal modula, respectively[4].

Parizek and Benes reported that trimethylselenium iodide labelled with ^{75}Se may be useful for nephrography[145]. Blottner and coworkers suggested the possibility of using 8-^{75}Se-substituted purine nucleosides for tumour-specific diagnosis[146]. A scientigraphic measurement of the relative concentration of ^{75}Se-labelled selenomethionine and radioactive colloidal ^{198}Au has been shown to detect primary and secondary liver cancers in patients[147].

2. *Miscellaneous*

^{75}Se-labelled selenomethionine can be used in immunological investigations [57–60,148]. For example, Gutman and coworkers labelled immunoglobins secreted by plasmacytomas with ^{75}Se-selenomethionine which binds to foetal calf serum protein so that it is releaseable by mercaptoethanol treatment[58]. Dosseto and coworkers labelled plasma membrane antigens of human and mouse lymphocytes with ^{75}Se-selenomethionine[59].

^{75}Se-selenomethionine was found to be a good isotopic marker for cell localization studies of lymphocytes[61,62] and erythrocytes[149,150]. ^{75}Se-selenomethionine has also been used for labelling of insulin[53], of platelets to determine their kinetics[151,152], of urokinase[50] and of fibrinogen[51,52,153]. Some muscle tumours were found to accumulate ^{75}Se-selenomethionine more than liver and muscle[154]. The Metabolism of ^{75}Se-selenomethionine has been shown to be useful in the diagnosis of some diseases[155–159].

Celander and coworkers achieved *in vivo* labelling of erythrocytes with ^{75}Se by injecting $Na_2{}^{75}SeO_3$ to a dog. ^{75}Se appeared to be bound to the haemoglobin[160].

Monks synthesized ^{75}Se-containing folate derivatives (**28, 29**) for the saturation analysis of trace folates[8,9]. Steroid derivatives labelled with ^{75}Se (**1, 3–5, 10**) were applied for radioimmunological determination of steroid hormones[9,20], for the investigation of steroid metabolism and for imaging to detect disorders[7,25]. ^{75}Se-labelled cyclic nucleotide (**2**) was used for the saturation analysis of nucleotides[18].

IV. CONCLUSION

As mentioned in the introduction, the major portion of the research activities with isotopically labelled selenium and tellurium compounds is in the fields of biology and medical sciences, tracing the locations and the fates of these labelled compounds in living bodies. Only a few simple synthetic methods have been developed to prepare these labelled compounds. However, these compounds have contributed enormously to clarifying many of the biological functions of selenium compounds which are by now considered quite essential in living bodies. Thus, it is our hope that this review will be of some use for the overview of the various applications of the isotopically labelled selenium and tellurium compounds in biomedical studies.

V. REFERENCES

1. C. M. Lederer and V. S. Shirley, in *Isotopes*, 7th ed., John Wiley and Sons, New York, 1978.
2. (a) R. E. Huber and R. S. Criddle, *Arch. Biochem. Biophys.*, **122**, 164 (1967).
   (b) Tse-Lok Ho, *Hard and Soft Acids and Bases: Principles in Organic Chemistry*, Academic Press, New York, 1977, pp. 4, 126.
3. (a) K. Fujimori, Y. Yoshimoto and S. Oae, *Tetrahedron Letters*, 4397 (1979).
   (b) K. Fujimori, Y. Yoshimoto and S. Oae, *Tetrahedron Letters*, **21**, 3385 (1980).
4. G. P. Basmadjian, K. R. Hetzel and R. D. Ice, *Int. J. Appl. Radiat. Isotopes*, **26**, 695 (1975).
5. G. L. Dilworth, *J. Labelled Compd. Radiopharm.*, **19**, 1197 (1982).
6. W. H. Green and A. B. Harvey, *J. Chem. Phys.*, **49**, 3586 (1968).

7. R. Monks and A. L. M. Riley, *German Patent*, No. 2,629,871 (1977); *Chem. Abstr.*, **87**, 102540q (1977).
8. Radiochemical Centre Ltd., *British Patent*, No. 1,501,119 (1978); *Chem. Abstr.*, **90**, 6422u (1979).
9. R. J. Bayly, V. E. M. Chambers and R. Monks, *U.S. Patent*, No. 4,202,976 (1980); *Chem. Abstr.*, **94**, 4248a (1981).
10. (a) R. Stejskal, L. Kronrád and J. Cífka, *Int. J. Appl. Radiat. Isotopes*, **22**, 254 (1971).
(b) L. Kronrád and V. Hladík, *Int. J. Appl. Radiat. Isotopes*, **23**, 237 (1972).
11. S. A. Sadek, G. P. Basmadjian and R. D. Ice, *Nucl. Med. Commun.*, **3**, 247 (1982).
12. (a) H. Yao and C. Dan, *Zhonghua Heyixue Zazhi*, **2**, 166 (1982); *Chem. Abstr.*, **97**, 211697y (1982).
(b) G. D. Grummon and P. E. Wiegert, *German Patent*, No. 2,047,983 (1971); *Chem. Abstr.*, **75**, 49581 (1971).
13. Y. Llabador and J. P. Adloff, *Radiochim. Acta*, **6**, 49 (1966).
14. A. B. Harvey and M. K. Wilson, *Inorg. Nucl. Chem. Letters*, **1**, 101 (1965).
15. A. B. Harvey and M. K. Wilson, *J. Chem. Phys.*, **45**, 678 (1966).
16. E. P. Painter, *J. Amer. Chem. Soc.*, **69**, 229 (1947).
17. H. H. Sisler and N. K. Kotia, *J. Org. Chem.*, **36**, 1700 (1971).
18. R. Monks and I. L. Thomas, *German Patent*, No. 2,615,151 (1976); *Chem. Abstr.*, **86**, 39824x (1977).
19. V. E. M. Chambers, J. S. Glover and R. Tudor, *Steroid Immunoassay, Proc. Tenovus Workshop, 5th 1974*, 177 (1975); *Chem. Abstr.*, **83**, 203457r (1975).
20. R. J. Bayly, V. E. M. Chambers and R. Monks, *German Patent*, No. 2,458,671 (1975); *Chem. Abstr.*, **83**, 110865g (1975).
21. P. P. H. L. Otto, *German Patent*, No. 2,451,055 (1975); *Chem. Abstr.*, **83**, 175163w (1975).
22. H. Kung and M. Blau, *J. Med. Chem.*, **23**, 1127 (1980).
23. L. Carrillo and S. J. Nassiff, *Radiochim Acta*, **8**, 124 (1967).
24. M. Adloff and J.-P. Adloff, *Bull. Soc. Chim. Fr.*, 3304 (1966).
25. F. F. Knapp, Jr., *U.S. Patent*, No. 4,215, 045 (1980); *Chem. Abstr.*, **94**, 47621j (1981).
26. F. F. Knapp, Jr., K. R. Ambrose, A. P. Callahan, R. A. Grigsby and K. J. Irgolic, *Radiopharm. 2, Proc. Int. Symp. 2nd* (Ed. J. A. Sorenson), Soc. Nucl. Med. Inc., New York, 1979, p. 101; *Chem. Abstr.*, **92**, 106543 (1980).
27. G. P. Basmadjian, S. L. Mills and R. D. Ice, *Nucl. Med. Commun.*, **3**, 150 (1982).
28. J. B. Lambert, R. G. Keske and D. K. Weary, *J. Amer. Chem. Soc.*, **89**, 5921 (1967).
29. J. B. Lambert, D. H. Johnson, R. G. Keske and C. E. Mixan, *J. Amer. Chem. Soc.*, **94**, 8172 (1972).
30. J. B. Lambert, C. E. Mixan and D. H. Johnson, *J. Amer. Chem. Soc.*, **95**, 4634 (1973).
31. L. D. McCullough and A. Lefohn, *Inorg. Chem.*, **5**, 150 (1966).
32. N. N. Magdesieva and A. A. Ibragimov, *Zh. Org. Khim.*, **8**, 885 (1972); *Chem. Abstr.*, **77**, 19479z (1972).
33. (a) Y. K. Yur'ev, N. N. Magdesieva and L. Y. Petrova, *Khim. Geterotsikl. Soedin.*, 910 (1966); *Chem. Abstr.*, **67**, 11384n (1967).
(b) A. I. Shatenshtein, N. N. Magdesieva, Y. I. Ranneva, I. O. Shapiro and A. I. Serebryanskaya, *Teor. Eksp. Khim.*, **3**, 343 (1967); *Chem. Abstr.*, **68**, 58799v (1968).
34. G. J. Martin, B. Mechin, Y. Leroux, C. Paulmier and J. C. Meunier, *J. Organometal. Chem.*, **67**, 327 (1974).
35. M. H. Ghandehari, D. Davalian, M. Yalpani and M. H. Partovi, *J. Org. Chem.*, **39**, 3906 (1974).
36. R. N. Hanson and M. A. Davis, *J. Pharm. Sci.*, **70**, 91 (1981).
37. (a) R. Cataliotti and G. Paliani, *Can. J. Phys.*, **55**, 1541 (1977).
(b) G. Paliani, R. Cataliotti, A. Poletti, F. Fringuelli, A. Taticchi and M. G. Giorgini, *Spectrochim. Acta*, **32A**, 1089 (1976).
38. (a) T. J. Barton, C. R. Tully and R. W. Roth, *J. Organometal. Chem.*, **108**, 183 (1976).
(b) E. E. van Tamelen and T. H. Whitesides, *J. Amer. Chem. Soc.*, **93**, 6129 (1971).
39. A. J. Barak and S. C. Swanberg, *J. Chromatog.*, **31**, 282 (1967).
40. K. P. McConnell, H. G. Mautner and G. W. Leddicotte, *Biochim. Biophys. Acta*, **59**, 217 (1962).
41. R. P. Spencer, K. R. Brody, W. H. H. Günther and H. G. Mautner, *J. Chromatog.*, **21**, 342 (1966).

42. P. P. H. L. Otto, *German Patent*, No. 2,026, 937 (1970); *Chem. Abstr.*, **74**, 126057v (1971).
43. F. Pan, Y. Natori and H. Tarver, *Biochim. Biophys. Acta*, **93**, 521 (1964).
44. J. Bremer and Y. Natori, *Biochim. Biophys. Acta*, **44**, 367 (1960).
45. K. E. Wilzbach, *J. Amer. Chem. Soc.*, **79**, 1013 (1957).
46. G. L. Cantoni and J. Durell, *J. Biol. Chem.*, **225**, 1033 (1957).
47. M. Blau, *Biochim. Biophys. Acta,* **49**, 389 (1961).
48. S. H. Wong, R. P. Spencer and A. Weaver, *Radiophram. 2, Proc. Int. Symp. 2nd* (Ed. J. A. Sorenson), Soc. Nucl. Med. Inc., New York, 1979, p. 109; *Chem. Abstr.*, **92**, 106580d (1980).
49. F. F. Knapp, Jr., K. R. Ambrose and A. P. Callahan, *J. Med. Chem.*, **24**, 794 (1981).
50. D. R. Celander and E. Celander, *Progr. Nucl. Hematol. Allied Fields*, 134 (1967); *Chem. Abstr.*, **72**, 128888z (1970).
51. D. R. Celander and E. Celander, *Texas Rept. Biol. Med.*, **23**, 589 (1965): *Chem. Abstr.* **63**, 18787d (1965).
52. C. A. Owen, Jr. and E. J. W. Bowie, *Proc. Soc. Exp. Biol. Med.*, **146**, 349 (1974).
53. G. J. M. Tavaarwerk, *Diabetes*, **18**, 51 (1969).
54. S. A. Leon, S. Kadis and G. Kollmann, *Nature*, **229**, 120 (1971).
55. D. Lebez, *Radiat. Sensitivity Toxins Anim. Poisons, Proc. Panel, May 1966*, IAEA, Vienna, 1970, pp. 75–86; *Chem. Abstr.*, **74**, 21580g (1971).
56. R. M. Barton, P. J. Higgins and K. P. McConnell, *Biochim. Biophys. Acta*, **493**, 323 (1977).
57. R. Engler, C. Lombart and N. F. Jayle, *Biochim. Biophys. Res. Commun.*, **46**, 1483 (1972).
58. G. A. Gutman, N. L. Warner, A. W. Harris and A. Bowles, *J. Immunol. Methods*, **21**, 101 (1978).
59. M. Dosseto, C. Rohner, M. Pierres and C. Goridis, *J. Immunol. Methods*, **41**, 145 (1981).
60. G. F. W. Smith, R. A. J. Stevens and B. Jacoby, *German Patent*, No. 3,004,382 (1980); *Chem. Abstr.*, **94**, 1796s (1981).
61. D. R. Bainbridge, *Immunol.*, **30**, 135 (1976).
62. M. L. Rose and H. S. Micklem, *J. Immonol. Methods*, **9**, 281 (1976).
63. P. J. Enzmann, *J. Virol. Methods*, **5**, 243 (1982).
64. R. Kishore, A. L. Baltch, M. Hammer, N. T. Sutphen and R. P. Smith, *Radiochem. Radioanal. Letters*, **43**, 55 (1980).
65. A. Martinsen and J. Songstad, *Acta Chem. Scand.*, **A31**, 645 (1977).
66. (a) W. H. Orme-Johnson, R. E. Hansen, H. Beinert, J. C. M. Tsibris, R. C. Bartholomaus and I. C. Gunsalus, *Proc. Natl. Acad. Sci. USA,* **60**, 368 (1968).
(b) J. C. M. Tsibris, M. J. Namtvedt and I. C. Gunsalus, *Biochem. Biophys. Res. Commun.*, **30**, 323 (1968).
67. J. A. Fee and G. Palmer, *Biochim. Biophys. Acta*, **245**, 175 (1971).
68. K. Mukai, J. J. Huang and T. Kimura, *Biochem. Biophys. Res. Commun.*, **50**, 105 (1973).
69. (a) V. T. Aleksanyan, Y. M. Kimel'fel'd, N. N. Magdesieva and Y. K. Yur'ev, *Opt. Spektrosk., Akad. Nauk SSSR, Otd. Fiz-Mat. Nauk, Sb. Statei*, **3**, 178 (1967); *Chem. Abstr.* **67**, 86227x (1967).
(b) V. T. Aleksanyan, Y. M. Kimel'fel'd, N. N. Magdesieva and Y. K. Yur'ev, *Opt. Spektrosk.*, **22**, 216 (1967); *Chem. Abstr.*, **67**, 48641v (1967).
70. V. T. Aleksanyan, Y. M. Kimel'fel'd and N. N. Magdesieva, *Zh. Strukt. Khim.*, **9**, 633 (1968); *Chem. Abstr.*, **70**, 6701w (1969).
71. N. N. Magdesieva, *Advan. Heterocycl. Chem.*, **12**, 1 (1970).
72. (a) G. Milazzo and L. Poloni. *Gazz. Chim. Ital.*, **82**, 576 (1952).
(b) H. Gerding, G. Milazzo, and H. H. Rossmark, *Rec. Trav. Chim.*, **72**, 957 (1953).
(c) M. Nardelli, G. Favo and G. Gilraldi, *Acta Cryst.*, **15**, 732 (1962).
(d) B. Tamamushi, H. Akiyama and S. Umezawa, *Bull. Chem. Soc. Japan*, **14**, 310 (1939).
(e) M. L. Hefferman and A. Humffray, *Mol. Phys.*, **7**, 527 (1964).
73. N. M. Pozdeev, O. B. Akulinin, A. A. Shapkin and N. N. Magdesieva, *Zh. Struct. Khim.*, **11**, 869 (1970); *Chem. Abstr.*, **74**, 47751p (1971).
74. S. Gronowitz, I. Johnson and A. B. Hornfeldt, *Chem. Scripta*, **7**, 111 (1975); *Chem. Abstr.*, **83**, 113162m (1975).
75. F. Fringuelli, G. Marino and A. Taticchi, *Advan. Heterocycl. Chem.*, **21**, 119 (1977).
76. K. Schwetlick and K. Unverferth, *Wiss. Z. Tech. Hochsch. Chem. 'Carl Schorlemmer' Leuna-Merseburg*, **12**, 230 (1970); *Chem. Abstr.*, **74**, 124421k (1971).
77. (a) A. K. Said and D. M. Hegsted, *Proc. Soc. Exp. Biol. Med.*, **133**, 1388 (1970); *Chem. Abstr.*, **74**, 39057b (1971).

(b) R. L. Ropez, R. L. Preston and W. H. Pfander, *J. Nutr.*, **97**, 123 (1969).
78. C. D. Thomson, C. E. Burton and M. F. Robinson, *Brit. J. Nutr.*, **39**, 579 (1978).
79. K. A. Lathrop, P. V. Harper and F. D. Malkinson, *Strahlentherapie Sonderbaende*, **67**, 436 (1968).
80. T. Hirooka and J. T. Galambos, *Biochem. Biophys. Acta*, **130**, 313, 321 (1966).
81. K. A. Lathrop, R. E. Johnson, M. Blau and E. O. Rothchill, *J. Nucl. Med.*, **13**, 7 (1972).
82. R. Schoental, *Nature*, **218**, 294 (1968).
83. C. F. Ehlig, D. E. Hogue, W. H. Allaway and D. J. Hamm, *J. Nutr.*, **92**, 121 (1967).
84. D. G. Brown and R. F. Burk, *Federal Proc.*, **31**, 692 (1972).
85. P. L. Wright and M. C. Bell, *Amer. J. Physiol.*, **211**, 6 (1966).
86. P. L. Wright and M. C. Bell, *J. Nutr.*, **84**, 49 (1964).
87. D. Fijal, *Folia Med. Cracov*, **23**, 269 (1981); *Chem. Abstr.*, **97**, 123030v (1982).
88. S. E. Khait and T. A. Norets, *Radiobiol. Radiother.*, **22**, 697 (1981); *Chem. Abstr.*, **96**, 64935 (1982).
89. S. E. Khait, T. A. Norets and V. S. Deeva, *Med. Radiol.*, **24**, 40 (1979); *Chem. Abstr.*, **92**, 106581e (1980).
90. J. L. Martin and M. L. Gerlach, *Ann. N. Y. Acad. Sci.*, **192**, 193 (1972).
91. H. K. Awwad, A. S. El Sheraki, S. El Shetaiwi and S. Helmi, *Intern. J. Appl. Radiation Isotopes*, **24**, 91 (1973).
92. E. A. Carr, Jr. and B. J. Walker, *Arch. Exptl. Pathol. Pharmakol.*, **248**, 287 (1964).
93. H. Patrick and W. G. Martin, *Radioisotop. Animals Nutr. Physiol. Proc. Symp. Prague, 1964*, 487 (1965); *Chem. Abstr.*, **64**, 14647e (1966).
94. C. A. Swanson, D. C. Reamer, C. Veillon and O. A. Levander, *J. Nutr.*, **113**, 793 (1983).
95. L. S. Jensen, E. D. Walter and J. S. Dunlap, *Proc. Soc. Exptl. Biol. Med.*, **112**, 899 (1963).
96. J. S. Edwardly, *Maj. BATAN*, **14**, 18 (1981).
97. K. J. Jenkins, *Can. J. Biochem.*, **46**, 1417 (1968).
98. M. Janghorbani, M. J. Christensen, F. H. Steinke and V. R. Young, *J. Nutr.*, **111**, 817 (1981).
99. D. B. Cowie and G. N. Cohen, *Biochim. Biophys. Acta*, **26**, 252 (1957).
100. T. W. Tuve and H. H. Williams, *J. Biol. Chem.*, **236**, 597 (1961).
101. R. P. Spencer and M. Blau, *Science*, **136**, 155 (1962).
102. (a) S. H. Mudd and G. L. Cantoni, *Nature*, **180**, 1052 (1957).
(b) F. Pan and H. Tarver, *Arch. Biochem. Biophys.*, **119**, 429 (1967).
103. L. G. Wilson and R. S. Bandurski, *Arch. Biochem. Biophys.*, **62**, 503 (1956).
104. (a) G. K. Korolev, *Med. Radiol.*, **10**, 33 (1965); *Chem. Abstr.*, **64**, 8611g (1966).
(b) G. K. Korolev, *Radioaktiv. Izotopy Organizm*, 145 (1969); *Chem. Abstr.*, **72**, 51563e (1970).
105. Y. I. Moskalev, *Raspredelenie Biol. Deistvie, Uskorenie Vyvedeniya Radioaktivn. Izotopov Sb. Rabot*, 62 (1964); *Chem. Abstr.*, **62**, 9568f (1965).
106. Y. I. Moskalev, *Med. Radiol.*, **5**, 54 (1960); *Chem. Abstr.*, **55**, 27468i (1961).
107. L. A. Sokov, *Radioaktiv. Izotopy Vnesh. Srede Organizme*, 63 (1970); *Chem. Abstr.*, **74**, 121072m (1971).
108. V. Slouka, *Sb. Ved. Pr. VLVDU Hradci Kralove*, **47**, 3 (1970); *Chem. Abstr.*, **75**, 59454c (1971).
109. T. M. Ponikarova and D. K. Popov, *Radiobiologiya*, **19**, 633 (1979); *Chem. Abstr.*, **91**, 170865g (1979).
110. G. V̌. Dobryakova, *Radioaktiv. Izotopy Vnesh. Srede Organizme*, 57 (1970); *Chem. Abstr.*, **74**, 121004r (1971).
111. S. M. Puchkova and L. A. Sokov, *Radioaktiv. Izotopy Vnesh Srede Organizme*, 54 (1970); *Chem. Abstr.*, **74**, 120979a (1971).
112. A. Ando, K. Hisada and I. Ando, *Radioisotopes*, **26**, 696 (1977); *Chem. Abstr.*, **92**, 142438x (1980).
113. J. T. Rotruck, A. L. Pope, H. E. Ganther, A. B. Swanson, D. G. Hafeman and W. G. Hoekstra, *Science*, **179**, 588 (1973).
114. L. Flohé and W. A. Günzler, in *Glutathione* (Eds. L. Flohé, H. C. Benohr, H. Sies, H. D. Waller and A. Wendel), Academic Press, New York, 1974, p. 132.
115. (a) J. W. Forstrom, J. J. Zakowski and A. L. Tappel, *Biochem.*, **17**, 2639 (1978).
(b) J. J. Zakowski, J. W. Forstrom, R. A. Condell and A. L. Tappel, *Biochem. Biophys. Res. Commun.*, **84**, 248 (1978).
116. J. A. Karle, F. J. Kull and A. Shrift, *Biol. Trace Elem. Res.*, **5**, 17 (1983).
117. T. C. Stadtman, *Ann. Rev. Biochem.*, **49**, 93 (1980).

118. D. C. Turner and T. C. Stadtman, *Arch. Biochem. Biophys.*, **154**, 366 (1973).
119. J. E. Cone, R. M. d. Río and T. C. Stadtman, *J. Biol. Chem.*, **252**, 5337 (1977).
120. A. C. Shum and J. C. Murphy, *J. Bacteriol.*, **110**, 447 (1972).
121. J. R. Andressen and L. G. Ljungdahl, *J. Bacteriol.*, **116**, 867 (1973).
122. J. B. Jones, G. L. Dilworth and T. C. Stadtman, *Arch. Biochem. Biophys.*, **195**, 255 (1979).
123. P. D. Whanger, N. D. Pedersen and P. H. Weswig, *Biochem. Biophys. Res. Commun.*, **53**, 1031 (1973).
124. D. Imhoff and J. R. Andressen, *FEMS Microbiol. Letters.*, **5**, 155 (1979).
125. R. F. Burk and P. E. Gregory, *Arch. Biochem. Biophys.*, **213**, 73 (1982).
126. (a) D. A. Saelinger, J. L. Hoffman and K. P. McConnell, *J. Mol. Biol.*, **69**, 9 (1972).
(b) J. L. Hoffman and K. P. McConnell, *Biochim. Biophys. Acta*, **366**, 109 (1974).
127. C.-S. Chen and T. C. Stadtman, *Proc. Natl. Acad. Sci. USA*, **77**, 1403 (1980).
128. A. Laconi, F. Melfi, F. Cataliotti and M. Semilia, *Minerva Med.*, **56**, 3645 (1965); *Chem. Abstr.*, **64**, 7145d (1966).
129. J. Fruhling, *DHEW Publ. (FDA) (US), FDA-76-8044, Radiopharm. Dosim. Symp.*, 432 (1976); *Chem. Abstr.*, **87**, 80483u (1977).
130. D. M. McCarthy and P. Brown, *Gut*, **10**, 913 (1969).
131. C. -C. Chen and K. -C. Wang, *Tai-Wan I Hsueh Hui Tsa Chih*, **68**, 341 (1969); *Chem. Abstr.*, **72**, 1604y (1970).
132. R. N. Melmed, J. E. Agnew and I. A. D. Bouchier, *J. Nucl. Med.*, **10**, 575 (1969).
133. S. Yamagata, *Japan. J. Med.*, **8**, 211 (1969).
134. M. F. Cottrall and D. M. Taylor, *Nuklearmedizin (Stuttgart)*, **20**, 187 (1981).
135. D. M. Taylor, V. R. McCready and D. O. Cosgrove, *Nucl. Med. Commun.*, **2**, 80 (1981).
136. V. I. Dedov and T. A. Norets, *Med. Radiol.*, **26**, 26 (1981); *Chem. Abstr.*, **94**, 135082q (1981).
137. V. I. Dedov and T. A. Norets, *Izv. Akad. Nauk SSSR, Ser. Biol.*, 785 (1981); *Chem. Abstr.*, **95**, 183141a (1981).
138. V. I. Dedov and T. A. Norets, *Radiobiologiya*, **21**, 553 (1981); *Chem. Abstr.*, **95**, 146188a (1981).
139. V. I. Dedov and T. A. Norets, *Med. Radiol.*, **26**, 48 (1981); *Chem. Abstr.*, **96**, 64920q (1982).
140. V. I. Dedov, *Izv. Akad. Nauk SSSR, Ser. Biol.*, 454 (1982); *Chem. Abstr.*, **97**, 19942h (1982).
141. V. I. Dedov and T. A. Norets, *Radiobiologiya*, **22**, 409 (1982); *Chem. Abstr.*, **97**, 68631q (1982).
142. Y. N. Anokhin, T. A. Norets and A. A. Yarilin, *Radiobiologiya*, **23**, 256 (1983); *Chem. Abstr.*, **98**, 212213v (1983).
143. V. I. Dedov and T. A. Norets, *Radiobiol. Radiother.*, **23**, 159 (1982); *Chem. Abstr.*, **97**, 19943j (1982).
144. I. Kozak, L. Kronrad and Z. Dienstbier, *Strahlentherapie*, **146**, 590 (1973).
145. J. Parizek and J. Benes, *Radiopharm. Label. Compounds, Proc. Symp. New Develop. Radiopharm. Label. Compounds 1973*, **2**, 141 (1973); *Chem. Abstr.*, **82**, 25612g (1975).
146. A. Blottner, H. Deckart and H. Spies, *Zentralinst. Kernforsch, Rossendorf Dresden ZfK 340, Jahresbericht*, 58 (1977); *Chem. Abstr.*, **89**, 175772t (1978).
147. J. De Saint-Laurent, P. Hadchouel, J. Caroli and G. Milhaud, *Compt. Rend., Ser. D*, **272**, 3221 (1971).
148. F. H. F. Cheng, R. E. Peterson and T. C. Evans, *J. Nucl. Med.*, **10**, 63 (1969).
149. J. A. Penner, *J. Lab. Clin. Med.*, **67**, 427 (1966).
150. P. A. McIntyre, B. Evatt, B. A. Hodkinson and U. Scheffel, *J. Lab. Clin. Med.*, **75**, 472 (1970).
151. Y. Najean and N. Ardaillou, *Scand. Haematol.*, **6**, 395 (1969).
152. K. Rechowicz, *Ann. Med. Sect. Pol. Acad. Sci.*, **15**, 115 (1970); *Chem. Abstr.*, **77**, 111128q (1972).
153. A. K. Moza, M. Kumar and R. P. Sapru, *Bull. Postgrad. Inst. Med. Educ. Res. Chandigrarh*, **10**, 88 (1976); *Chem. Abstr.*, **90**, 18611t (1979).
154. O. P. Savel'eva, *Eksp. Onkol.*, **2**, 71 (1980); *Chem. Abstr.*, **93**, 64833b (1980).
155. S. Nakayama, T. Mizukami, S. Morita, N. Umezaki, S. Fukae, H. Ohtake, A. Kondo and K. Yano, *Kaku Igaku*, **17**, 579 (1980); *Chem. Abstr.*, **93**, 163584c (1980).
156. I. Szantai, S. Cotul and M. Gidali, *Rev. Int. Hepatol.*, **18**, 663 (1968).
157. W. H. Oldendorf, W. B. Sisson and A. Silverstein, *Arch. Neurol. (Chicago)*, **24**, 524 (1971); *Chem. Abstr.*, **75**, 32758t (1971).
158. G. R. Youngs, J. E. Agnew, G. E. Levin and I. A. D. Bouchier, *Brit. Med. J.*, **2**, 252 (1971).
159. E. J. Potchen and H. G. Watts, *Endocrinology*, **80**, 467 (1967).
160. L. Gatien, E. Celander, G. Gordon, M. Jacquo, Jr. and D. R. Celander, *J. Amer. Osteopath. Assoc.*, **63**, 871 (1964); *Chem. Abstr.*, **61**, 6129a (1964).

The Chemistry of Organic
Selenium and Tellurium
Compounds Volume 1
Edited by S. Patai and Z. Rappoport
© 1986 John Wiley & Sons Ltd.

CHAPTER **13**

Selenium and tellurium heterocycles

MARCEL RENSON

Université de Liège, Institut de Chimie, Sart-Tilman, 4000 Liège, Belgium

I. INTRODUCTION 401
II. Se/Te THREE-MEMBERED RINGS 402
A. One Heteroatom in the Ring 402
1. Seleniranes and telluriranes 402
2. Seleniranium salts 403
3. Seleniren and selenirenium salts 403
B. Two Heteroatoms in the Ring 403
III. Se/Te FOUR-MEMBERED RINGS 404
A. One Heteroatom in the Ring 404
B. Two Heteroatoms in the Ring 404
1. One oxygen and one selenium 404
2. One sulphur and one selenium 404
3. Two selenium atoms 405
a. In the 1,2-position 405
b. In the 1,3-position 405
4. Two tellurium atoms 406
C. Three Heteroatoms in the Ring 406
IV. Se/Te FIVE-MEMBERED RINGS 406
A. One Heteroatom: Selenophenes and Tellurophenes 406
1. Monocyclic selenophenes and tellurophenes 406
a. Synthesis 406
b. Ground-state aromaticity: comparison with furan and thiophene . 410
c. Electrophilic and nucleophilic substitution 410
d. Metalated selenophenes and tellurophenes 414
e. Hydrogenated selenophenes and tellurophenes 418
2. Benzoselenophenes and condensed analogues 419
a. Benzo(*b*)selenophene 419
b. Benzo(*c*)selenophene and derivatives 426
c. Dibenzoselenophene and related compounds 428

3. Hetero-condensed selenophenes 429
a. Selenophenes condensed to five-membered non-nitrogen heterocycles 429
b. Selenophenes condensed to six- or seven-membered non-nitrogen heterocycles 435
c. Selenophenes condensed to nitrogen heterocycles 436
4. Benzo- and dibenzo-tellurophenes 439
a. Benzo(*b*)tellurophene 439
b. Benzo(*c*)tellurophene 442
c. Dibenzotellurophene 442
5. Tellurophene condensed to heterocycles 443
B. Two Heteroatoms. 444
1. Se/Te rings without nitrogen 444
a. Heteroatoms in the 1,2-position 444
b. Heteroatoms in the 1,3-position 447
2. Nitrogen Se/Te rings 452
a. Heteroatoms in the 1,2-position 452
b. Heteroatoms in the 1,3-position 455
C. Three Heteroatoms 462
1. Se/Te rings without nitrogen 462
2. Nitrogen Se/Te rings 462
a. One nitrogen in the ring 462
b. Two nitrogens in the ring: chalcogenadiazoles 463
D. Four Heteroatoms 468
V. Se/Te SIX-MEMBERED RINGS 468
A. One Heteroatom 468
1. Monocyclic systems 468
a. Selenanes and telluranes 468
b. Selenopyrans 469
c. Chalcogenopyrones and tetrahydro derivatives 469
d. Chalcogenopyrylium salts 470
e. Bis-4,4′-chalcogenopyranylidenes 470
2. Benzo-, naphtho- and hetero-condensed systems 470
a. (*b*)Condensed systems: chroman derivatives 470
b. (*c*)Condensed systems: isochroman derivatives 476
3. Dibenzo-condensed systems and analogues 477
a. Selenium derivatives 477
b. Tellurium derivatives 479
B. Two Heteroatoms without Nitrogen 480
1. Monocyclic systems 480
a. Heteroatoms in the 1,2-position: 1,2-dichalcogenanes 480
b. Heteroatoms in the 1,3-position: 1,3-dichalcogenanes 480
c. Heteroatoms in the 1,4-position 480
2. Benzo-condensed systems 481
3. Dibenzo-condensed systems 482
a. Phenoxaselenins and phenoxatellurins 482
b. Phenothiatellurin 483
c. Selenanthrene 483
d. Telluranthrene 484
C. Two Heteroatoms with Nitrogen 484
1. Monocyclic systems 484
a. 1,3-Selenazines 484

b. 1,4-Selenazines 485
2. Benzoselenazines and hetero analogues 485
a. 1,3-Benzoselenazines 485
b. 1,4-Benzoselenazines 486
3. Dibenzochalcogenazines and hetero analogues 486
a. Phenoselenazines 486
b. Phenotellurazines 487
D. Three Heteroatoms without Nitrogen 487
E. Three Heteroatoms with Nitrogen 488
1. One nitrogen 488
2. Two nitrogens: selenadiazines 488
a. 1,2,4-Selenadiazines 488
b. 1,2,6-Selenadiazines 488
c. 1,3,4-Selenadiazines 489
d. 1,3,5-Selenadiazines 489
F. Four Heteroatoms 489
VI. SEVEN-MEMBERED RINGS 489
A. One Heteroatom 489
1. Monocyclic systems 489
2. Benzoselenepins and analogues 489
3. Dibenzoselenepins 490
a. Dibenzo(*b,f*)selenepins 490
b. Dibenzo(*b,e*)selenepins 490
c. Dibenzo(c,e)selenepins 490
4. Condensed telluracycloheptane 491
B. Two Heteroatoms. 491
1. Without nitrogen 491
2. With nitrogen: selenazepines 491
C. Three Heteroatoms 491
VII. LARGER RINGS 492
A. Eight-membered Rings 492
B. Macrocycles 492
VIII. SPIROHETEROCYCLES 493
IX. BRIDGED Se/Te HETEROCYCLES. 493
X. REFERENCES 494

I. INTRODUCTION

The present chapter is devoted to the chemistry of organic Se/Te heterocycles. Generally, it will not take into account heterocycles derived from Se (IV) or Te(IV), to avoid overlap with another chapter of this book. For the same reason, it will discuss neither the physicochemical properties nor the photochemistry of Se/Te heterocycles.

The literature on the theory, preparation and applications of Se/Te heterocycles is too voluminous to enable an extensive coverage of the subject. This chapter will attempt to cover the most important advances in this field, mainly during the last few years, and through 1983. Earlier reviews include small chapters on Se heterocycles [1,2], and a complete coverage of the subject has been given in Klayman and Günther's book[3]. The chemistry of Te heterocycles has been described in detail in Cooper's book[4] as well as in Irgolic's book[5] and his annual reviews[6]. Information about Se/Te heterocycles can also be found in the *Specialist Periodical Reports of the Royal Society of Chemistry* (*London*)[7–10].

We must also note a review covering the years 1975–1979[11]. Special reviews devoted to particular Se or Te ring systems will be mentioned in the relevant chapters.

When information is available, comparison will be made with the corresponding S heterocycles. Differences between S and heavier chalcogen heterocycles are related to the greater size of Se/Te atoms, their greater polarizability, their lower electronegativity and their lower bond energies with other elements. These differences will sometimes lead to lower stability of Se/Te heterocycles, and to specific ring-opening or ring-transformation reactions.

This chapter will be divided according to the size of the ring systems. For a given size, it will successively follow the number of heteroatoms.

II. Se/Te THREE-MEMBERED RINGS

A. One Heteroatom in the Ring

1. Seleniranes and telluriranes

No seleniranes **1a**, (episelenides) have been isolated, but they were tentatively assumed to be formed during flash photolysis of CSe_2–olefin mixtures[12], and their structure was confirmed by mass spectral assignments[13]. Seleniranes were also suggested as intermediates in the following reactions:

(*a*) In the stereospecific conversion of oxiranes to olefins with triphenylphosphine selenide[14] (equation 1), tri-*n*-butylphosphine selenide[15], phosphole and phospholene selenide[16] or potassium selenocyanate[17].

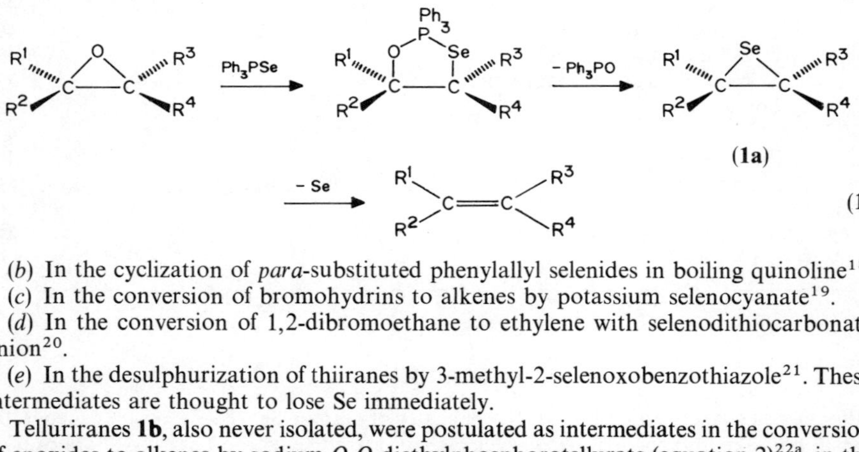

(**1a**)

(1)

(*b*) In the cyclization of *para*-substituted phenylallyl selenides in boiling quinoline[18].

(*c*) In the conversion of bromohydrins to alkenes by potassium selenocyanate[19].

(*d*) In the conversion of 1,2-dibromoethane to ethylene with selenodithiocarbonate anion[20].

(*e*) In the desulphurization of thiiranes by 3-methyl-2-selenoxobenzothiazole[21]. These intermediates are thought to lose Se immediately.

Telluriranes **1b**, also never isolated, were postulated as intermediates in the conversion of epoxides to alkenes by sodium *O,O*-diethylphosphorotellurate (equation 2)[22a], in the

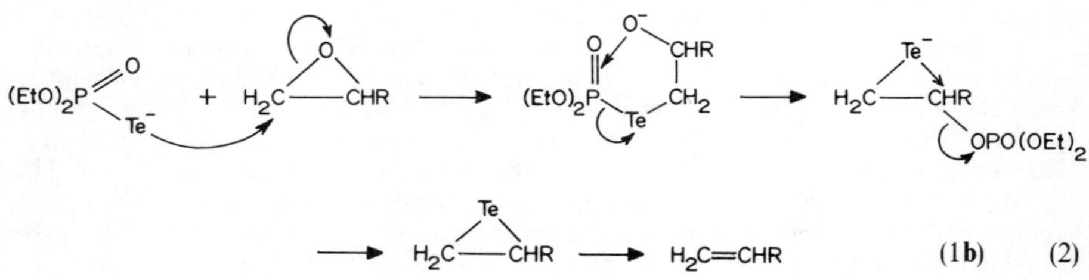

(**1b**) (2)

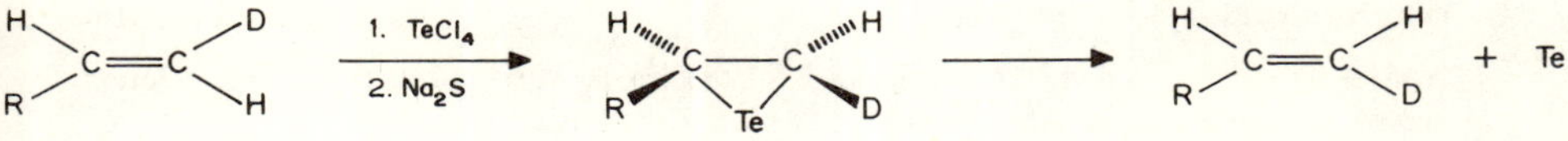

(3)

dehalogenation of α-haloketones by the same reagent[22b], and in the Na_2S reduction with olefin inversion of chloroalkyltellurium trichlorides, the addition products of $TeCl_4$ with alkenes (equation 3)[23,24]. The transient tellurirane **1b** (R = H) and its Me derivative were detected in the flash photolysis of dimethyl telluride in ethylene and propene, respectively[25].

2. *Seleniranium salts*

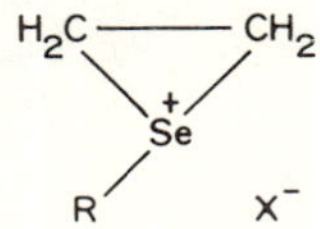

(2)

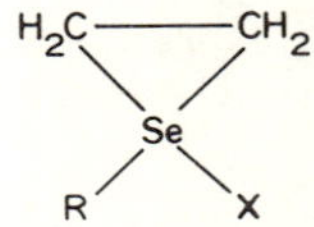

(3)

Thiiranium (episulphonium) ions are known to be very unstable[26]. The first long-lived thiiranium salt, stable for weeks below − 10 °C, was synthezised in 1975[27]. Seleniranium salts were not isolated. Seleniranium intermediates **2** or their selenurane equivalent **3** are, however, invoked in the addition of selenium monochloride[28], *p*-tolueneselenenyl chloride[29], areneselenenyl hexafluorophosphate[30], methaneselenium trichloride[31], benzene and methaneselenenyl bromide[32–35] to alkenes. In one case[29], the isolation of a stable episelenurane was claimed, but this cyclic structure has been shown to be incorrect and is apparently the acyclic compound, **4**[36] (equation 4).

$$p\text{-}MeC_6H_4SeCl + H_2C{=}CH_2 \rightarrow \underset{\mathbf{(4)}}{p\text{-}MeC_6H_4SeCl_2CH_2CH_2Cl} \qquad (4)$$

Seleniranium ions are also proposed as intermediates in the reaction of bis(2-bromoethyl) selenide with various nucleophiles[37].

3. *Seleniren and selenirenium salts*

Similarly to thiirens, seleniren intermediates are considered to be formed from the reaction of 1,2,3-selenadiazole with diiron enneacarbonyl[38,39] which forms complexes with the heterocycles and catalyses the elimination of N_2, and by photolysis[40,41]. Further irradiation results in a cleavage of this small ring to acetylene and selenoketene[40].

Selenirenium ions are believed to be intermediates in the electrophilic addition of areneselenenyl halides to alkynes[30,45,46a], according to spectroscopic, kinetic and stereochemical studies[45]. Not surprisingly, no selenirenium salt has been isolated up to now, since the first stable thiirenium compounds were isolated only recently. These are 1,2,3-trimethylthiirenium tetrafluoroborate, stable below − 10 °C[42], and the 1-methyl-2,3-di-*t*-butyl analogue[43], stable for weeks at room temperature, and whose structure was confirmed by X-ray analysis[44].

Benzoseleniren has been detected by ESCA[46b], by IR spectroscopy[46c], and matrix-isolated as a short-lived intermediate in the pyrolysis and the photolysis of 1,2,3-benzoselenadiazole. It is rapidly rearranged to fulvene-6-selone.

B. Two Heteroatoms in the Ring

Parallel to sulphur chemistry, unsuccessful attempts were made to obtain selenaziridines as intermediates in the treatment of oxaziridine with potassium selenoxanthate[47]. Selenaziridine and oxaselenirane were suggested as intermediates in the decomposition of 1,2-benzodithiole-3-selenoxoimide[48] and in the oxidation of 3-selenoxo-1,2-benzodithiole[49], respectively. This must be regarded with caution, since the structure of the last compound was demonstrated later to be an isomeric structure[511].

III. Se/Te FOUR-MEMBERED RINGS

A. One Heteroatom in the Ring

Selenetane (**5**) itself is obtained in a low yield from 1,3-dibromopropane and alkali selenide[50] (equation 5). It is stable only in the dark when cooled and it polymerizes easily. The more stable 3,3-dimethyl derivative **6**, prepared under similar conditions[51] or with

(5)

(**5**)

KSeCN as the nucleophilic salt[52], is stable up to 140 °C. Its Se dibromo derivative **7** is stable at temperatures below −20 °C. At more elevated temperatures there is a ring-opening to the 1-bromo-3-bromoseleno compound **8** which subsequently gives 1-bromo-3-tribromoseleno-2,2-dimethylpropane **9**[51,53a] (equation 6). Methyl iodide gives also ring-opening to 1-dimethyliodoseleno-2,2-dimethyl-3-iodopropane[52].

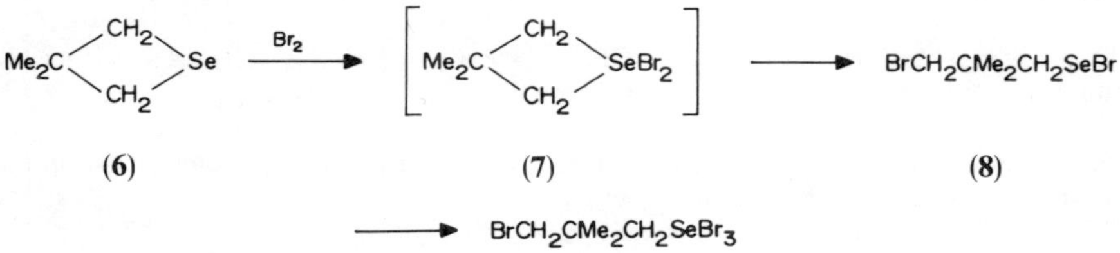

(6)

Compound **6** is oxidized by H_2O_2 to a five-membered lactone of corresponding seleninic acid, rather than to a selenone[53a]. Selenetan-3-ol has also been prepared[53b].

B. Two Heteroatoms in the Ring

1. One oxygen and one selenium

An oxaselenetane is suggested as intermediate in the oxidation of exocyclic alkenes with SeO_2/H_2O_2[54].

2. One sulphur and one selenium

A sulphone of thiaselenetane (**10**) was claimed to be obtained by the reaction of divinyl sulphone with $SeBr_4$, instead of the corresponding six-membered ring[55] (equation 7).

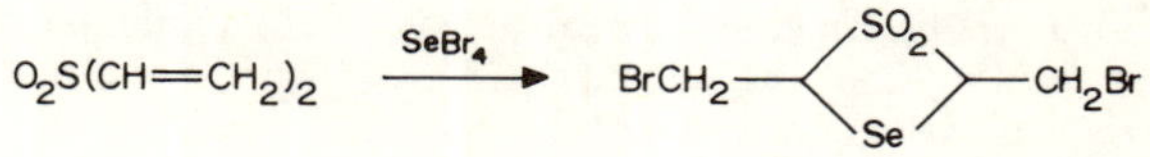

(7)

(10)

3. Two selenium atoms

a. In the 1,2-*position.* 3,4-Bis(trifluoromethyl)-1,2-diselenetene (**11**) has been prepared in 25% yield by pyrolysis of bis(trifluoromethyl)acetylene with Se vapour at 750 °C[56] (equation 8).

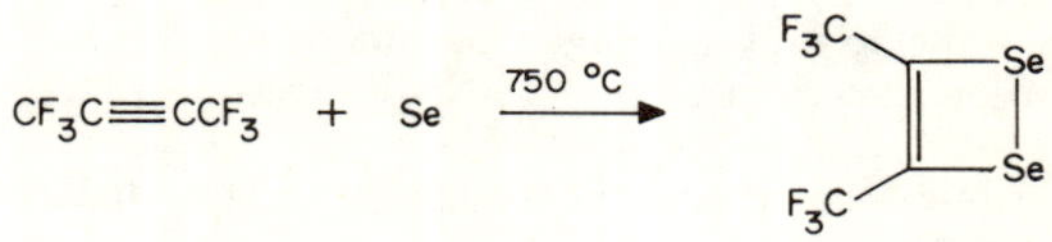

(8)

(11)

b. In the 1,3-*position.* Derivatives **13** and **15** of 2,4-bis(methylene)-1,3-diselenetane have been obtained by two different methods:
(*i*) Pyrolysis of 1,2,3-selenadiazoles at 500–600 °C[57], via a selenoketene **12** (equation 9a).

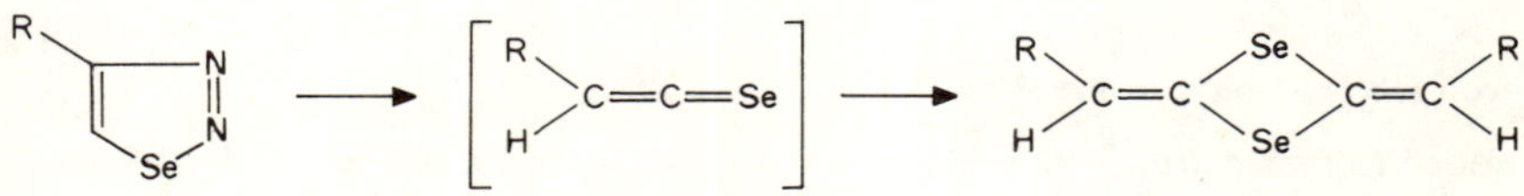

(12) **(13)**

(*ii*) Reaction of CSe_2 with dimethyl malonate[58], giving initially the intermediate **14** by a route parallel to that used in sulphur chemistry (equation 9b).

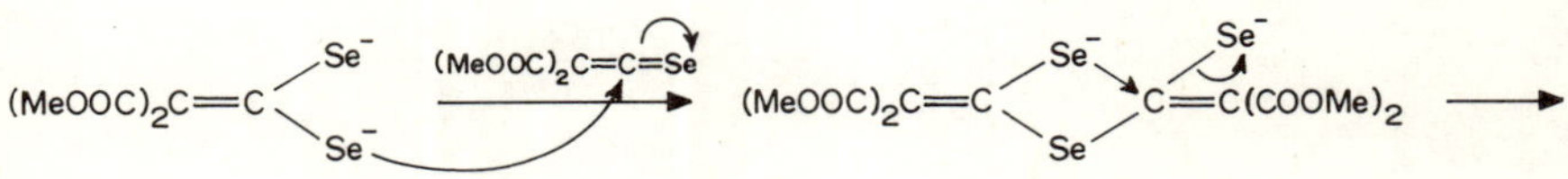

(14)

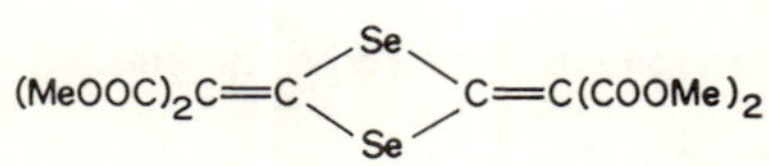

(9b)

(15)

2,2,4,4-Tetraacetyl-1,3-diselenetane (**16**) for which another structure was previously assigned[59] has been prepared from $SeCl_4$ and acetylacetone[60a] (equation 10).

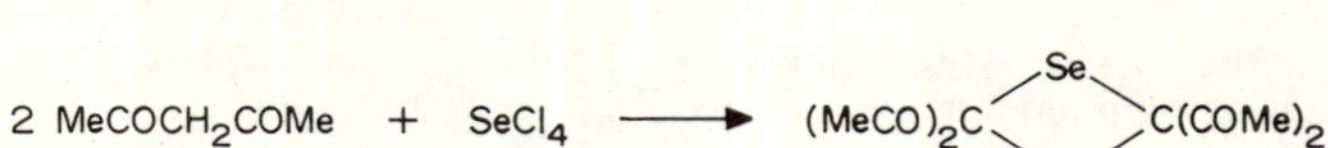

(10)

(16)

Hydrogen cyanide gives ring-opening to 3-cyanoseleno-2,4-pentanedione. 1,3-Diselenetane-2,4-diselenol diacetate is formed by treating acetyl chloride with H_2Se and $AlCl_3$[60b]. Perfluoro-1,3-diselenetane structures have been tentatively assigned to the pyrolysis product at 150 °C of the polymer $(CF_2Se)_n$[61], or to the reaction product of perfluoropropene with selenium and antimony pentafluoride[62a]. 2,2,4,4-Tetrakis(trifluoromethyl)-1,3-diselenetane is obtained from triphenylphosphine selenoxide and hexafluoroacetone[62b].

4. Two tellurium atoms

The first 1,3-ditelluretane system **17** was recently obtained by reaction of phenylacetylene and Te in DMSO[63–65,69]. The ditelluretane structure, which was previously believed to be a 1,4-ditellurafulvene ring[66], by analogy with known S and Se compounds[67,68], has been determined by ^{1}H-NMR, mass spectrometry and X-ray measurements.

Compound **17** and its *cis* isomer[64] may be considered as the cyclodimerization products of a transient telluroketene (equation 11).

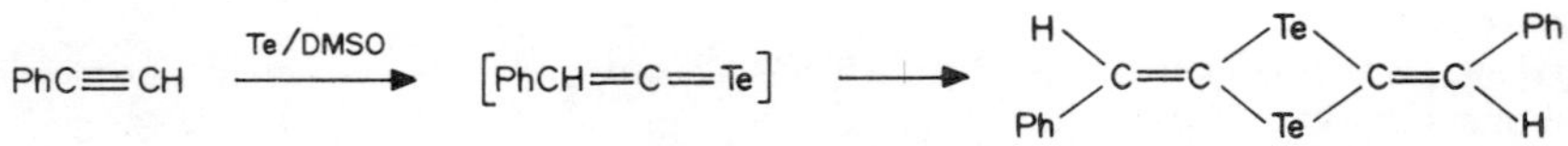

(11)

(17)

C. Three Heteroatoms in the Ring

A four-membered ring with three Se atoms **18**, regarded as a bis(diselenocarbamato)selenium(II) compound, and whose structure was proven by X-ray diffraction, is obtained from CSe_2 and tetrasubstituted methylene diamine[70] (equation 12).

(18) (12)

IV. Se/Te FIVE-MEMBERED RINGS

A. One Heteroatom: Selenophenes and Tellurophenes

1. Monocyclic selenophenes and tellurophenes

The chemistry of selenophene has been reviewed through 1950 by Hartough[71], for the period up to 1970 by Magdesieva[72] and by Magdesieva and Zefirov[73], and for the last decade by Hörnfeldt[74]. The chemistry of tellurophene has been completely covered up to 1975 by Fringuelli and coworkers[75]. A more condensed review was presented by Marino[76]. The current literature on selenophene and tellurophene has been continuously reviewed by Gronowitz in his biennial reviews on thiophene and Se/Te analogues[9] and for tellurophene in the annual surveys of Irgolic[6].

Selenophene and tellurophene complete with furan and thiophene a series of four stable aromatic systems. These systems differ mainly in their degree of reactivity, but generally display similar reactions. However the larger heteroatoms can be a centre of reactivity towards some reagents, and sometimes a centre of ring destruction.

In this chapter, we shall use the terms 'selenienyl' and 'tellurienyl' for the heterocyclic radicals, and for the corresponding condensed derivatives the prefixes 'selenolo' and 'tellurolo' will be added.

a. Synthesis. (*i*) *Selenophenes.* Selenophene or its homologues can be obtained by two main methods: from acyclic compounds, or from other ring systems. The first true synthesis of selenophene, a liquid (b.p. 100 °C, m.p. − 38 °C) of a disagreeable odour, in contrast to thiophene, was realized in 1927 by Mazza and Solazzo from acetylene and Se at high temperature[77]. The same methodology was later applied by Briscoe and coworkers[78,79] and by Umezawa[80,81]. Recently, Gronowitz and coworkers[82a] modified this reaction to a larger scale synthesis (150–200 g per day), in a 58% yield, by mixing Se with alumina under well-determined conditions. In this reaction, about 30 compounds were identified including 2- and 3-alkylselenophenes, biselenienyls and condensed ring systems. The yield of this reaction is now improved to 70%[82b]. Perveev and coworkers[83] used acetylenic epoxides to obtain alkyl-, vinyl- and hydroxyalkyl-selenophenes. Selenophene derivatives, e.g., alkyl-substituted selenophenes, were also obtained from paraffins[87,88], olefins or conjugated dienes with SeO_2[84–86] or metallic Se[87,88].

More elaborated syntheses start from a four-carbon chain. Substituted 1,3-diyne systems including the bis(trimethylsilyl) derivative, reacted with NaSeH or H_2Se in a general synthesis of 2,5-disubstituted selenophenes **19**[89,90] (equation 13). For example, 2,5-diphenyl- and 2,5-dihydroxymethyl-selenophenes were synthesized by this method.

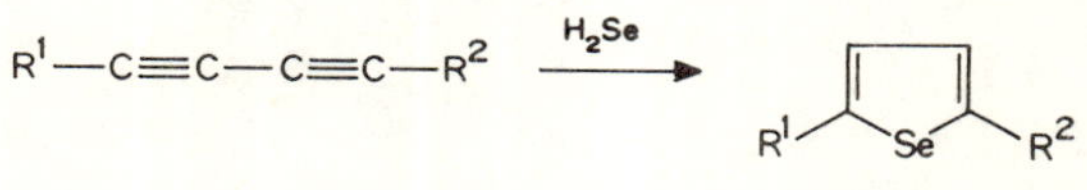

(13)

(19)

Using the bis(trimethylsilyl) analogue ($R^1 = R^2 = SiMe_3$) leads, after hydrolysis, to the unsubstituted selenophene[91]. The monosilylated diyne **20** was used for the synthesis of ^{75}Se-labelled 2-selenienylalanine **21**[92] (equation 14). Vinylacetylene and Se also give selenophene as the main product (24%)[93].

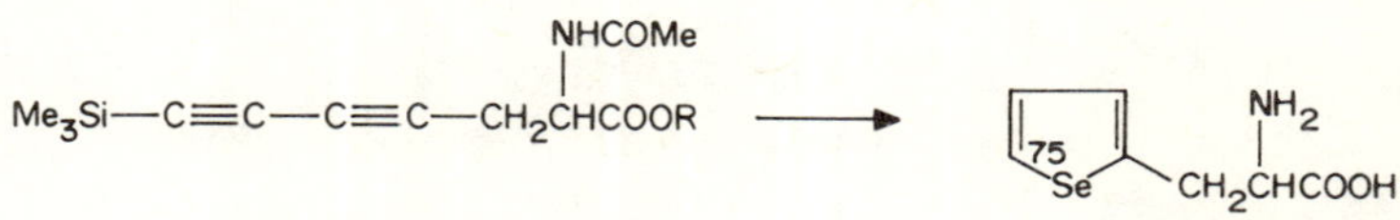

(14)

(20) **(21)**

2,5-Disubstituted selenophenes can also be obtained in a general way by the Paal-Knorr five-membered heterocycle synthesis from γ-diketones and phosphorus pentaselenide[94–96]. The first known derivative of a selenophene (2,5-dimethylselenophene, also named selenoxene, **19**, $R^1 = R^2 = Me$) was prepared by this method[96] (equation 15). The

$$R^1CO(CH_2)_2COR^2 \xrightarrow{P_4Se_{10}} \mathbf{19} \qquad (15)$$

synthetic equivalents of γ-diketones, 1,4-dihalogeno-1,3-dienes, can also be used in reaction with Li_2Se. The reagents of the corresponding opposite polarities, i.e., 1,4-dilithiodienes and Se_2Br_2 were also used.[97a] (equation 16).

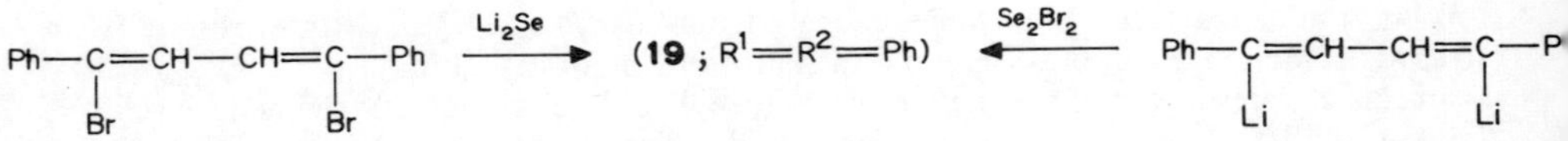

(16)

Starting from a molecule bearing a carbonyl and an alkyne group, low yields of 3-ethoxycarbonyl-2,5-dimethyl- or 5-methyl-2-phenyl-selenophenes are obtained[97b].

Another general ring-closure synthesis of selenophenes developed by Cagniant and coworkers consists of the Fiesselmann reaction involving β-chlorovinylaldehyde (**22**), Na_2Se and ethyl bromoacetate[98–100] (equation 17). Using the appropriate aldehyde **22** and α-bromopropionate gives 2,5-dimethylselenophene in a 40% yield[100].

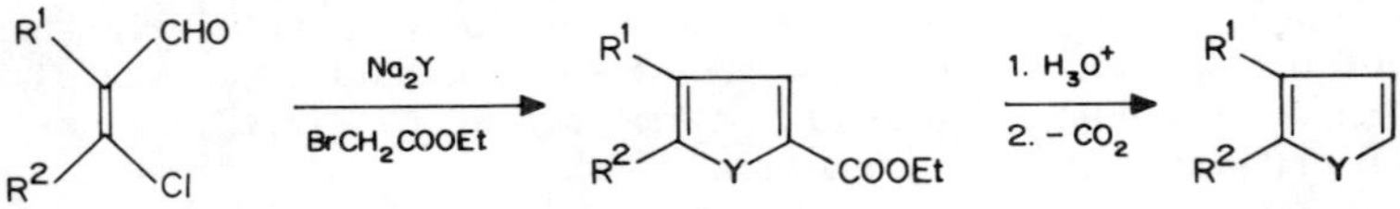

(17)

(**22**)

Liebscher and Hartmann developed a methodology, starting from 2-aminovinyl selenoketones (**23**) for the formation of 2-acylselenophenes **24**[101] (equation 18).

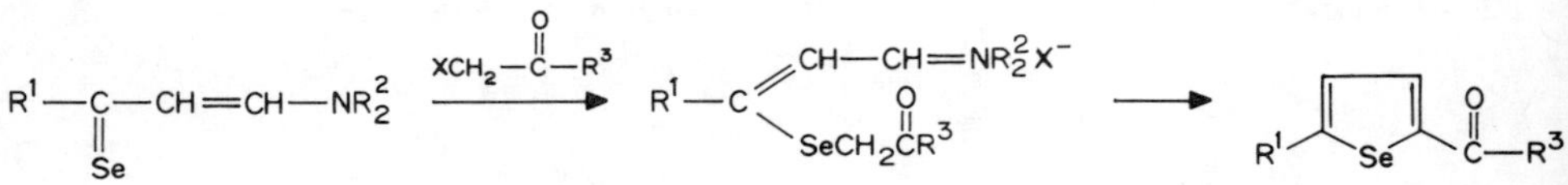

(**23**) (**24**)

(18)

Selenophene derivatives **26** can be obtained from α-diketones by the Hinsberg ring-closure reaction[102], in mixture with mono- and di-decarboxylated heterocycles (equation 19).

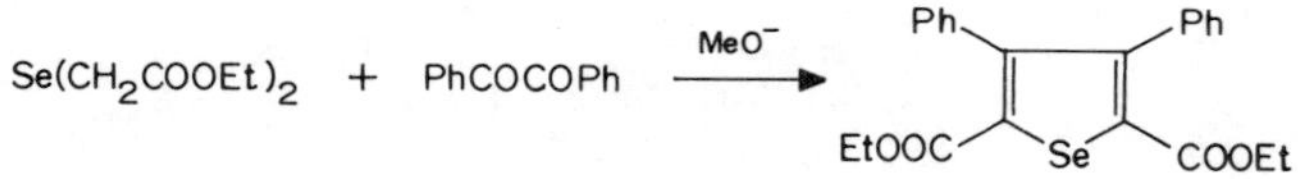

(19)

(**25**) (**26**)

A recent novel method of synthesis of thiophene and selenophene consists of a double nucleophilic substitution of the dichloro precursor **27** to give either thienyl- or selenienyl-3-malonic ester **28**[103] (equation 20).

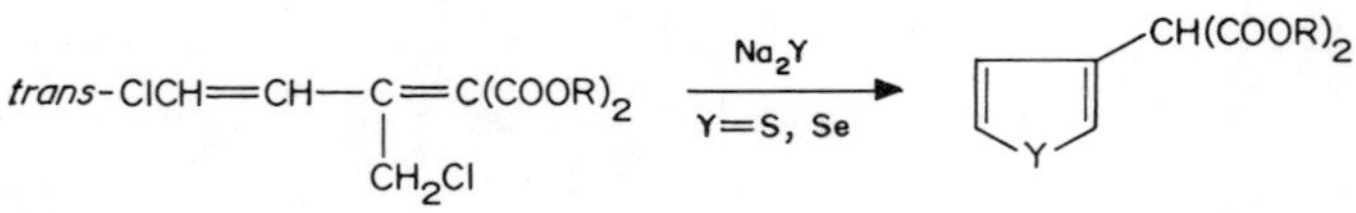

(20)

(**27**) (**28**)

Cycloaromatization of diallenyl chalcogenides **29** through a 6π electrocyclic ring-closure with a 1,5-hydride shift gives 3,4-disubstituted furans, thiophenes and selenophenes **30**[104] (equation 21).

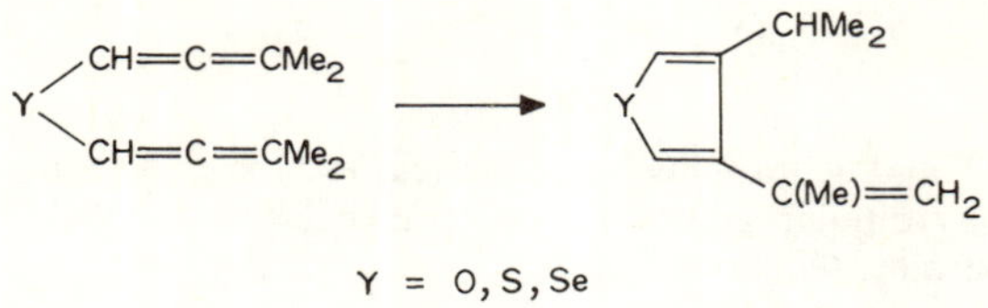

(21)

(**29**) (**30**)

Selenophene derivatives can also be obtained from furans and H_2Se[105,106], from pyrolysis of 4-aryl-1,2,3-selenadiazoles (**31**)[107,108], probably via intermediate monoaryl acetylene and Se, and the subsequent 2-arylethynyl selenol (equation 22), and from displacement of sulphone groups by Se[109].

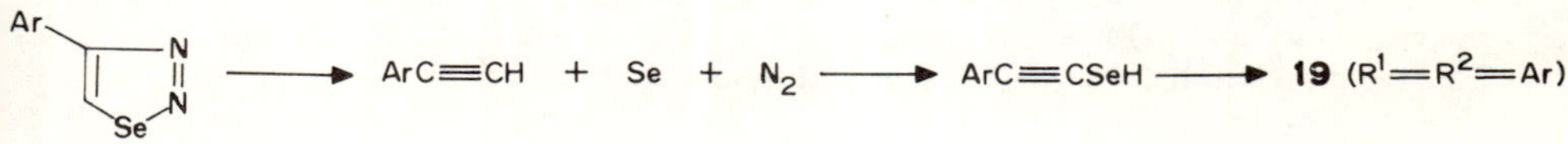

(**31**) (22)

(*ii*) *Tellurophenes.* Unsuccessful attempts have been made to prepare tellurophene from acetylene or sodium succinate and aluminium telluride[110]. The first synthesis of the unsubstituted heterocycle was realized only in 1966 by Mack[111] using butadiyne and Na_2Te (equation 23). However, the procedure was first well described by Fringuelli and Taticchi in 1972[112,113], with yields of 37–50%. Tellurophene (**32**; $R^1 = R^2 = H$) is a light-yellow bad-smelling liquid (b.p. 151 °C, m.p. −36 °C).

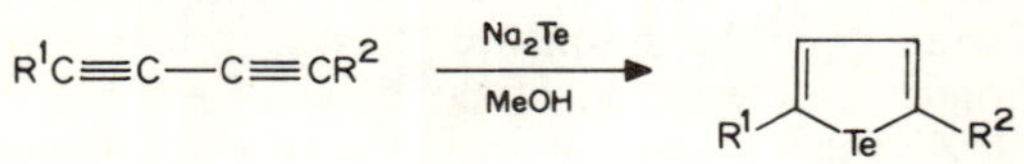

(23)

(**32**)

The use of the bis-silylated diyne seemed first to give a lower yield[114], but when Na_2Te solution was prepared from sodium formaldehyde sulphoxylate (rongalite), the yield improved to 59%[115]. 2-Phenyltellurophene was obtained by the same methodology[116]. Applying reaction 23 when $R^1 = R^2 =$ PhCH(OH) gives a bis(hydroxbenzyl)tellurophene which is the precursor of the synthesis of a tellurathiaporphyrine[117], a reaction parallel to that in the selenophene series[118a]. Recently, reaction of lithium and Se or Te on 3-methyl-3-butene-1-yne gave, after prototropy, a good yield of 3-methylselenophene and 3-methyltellurophene[118b]. The Fiesselmann reaction was also extended by Cagniant and coworkers to the tellurophene series, (equation 17; $R^1 = H$, $R^2 = t$-**Bu**, Y = Te) to give some mono and polycyclic tellurophenes in 35% yield[119]. Applied to arylacetic synthons, this method leads to 3-aryl-thiophenes,-selenophenes and -tellurophenes **33**[120] (equation 24).

Tetraphenyltellurophene was also obtained in 1961, by the two alternative procedures described for selenophenes in equation (16) (right-hand side, $TeBr_4$ as the reagent[97]).

(24)

(33)

Selenophene[69,121] and tellurophene derivatives **34**[122] can also be obtained, among other products, by dipolar addition of acetylenic esters to phenylethynyl metal(M)-selenolate or tellurolate (equation 25).

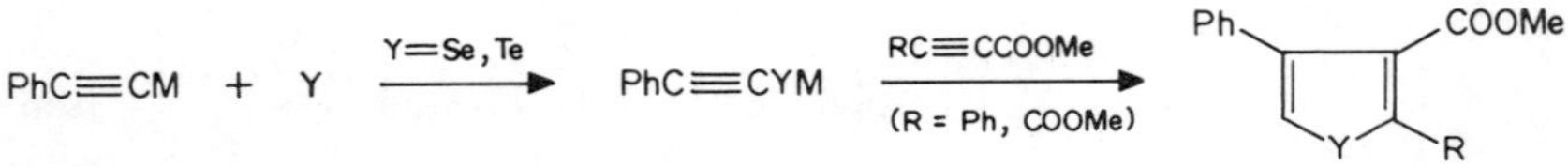

(25)

(34)

2-Benzylidene-3-oxo-2,3-dihydrotellurophenes can sometimes be obtained from dialkynyl ketones and $(t\text{-Bu}_3\text{Si})_2\text{Te}$, instead of the six-membered isomer telluropyrone[123]. This reaction will be discussed in the corresponding section.

b. Ground-state aromaticity: comparison with furan and thiophene. Selenophene and tellurophene show typical aromatic behaviour: they are more stable than the corresponding dienic compounds and tend to react by substitution rather than by addition. Like thiophene, selenophene does not give a normal Diels–Alder reaction, and loses Se when heated with maleic anhydride[124]. The geometry of these rings, determined by microwave spectral measurements[125,126], is planar and similar to that of furan and thiophene. The more characteristic variations for the four congeners are an increase of the Y—C (2) bond from furan to tellurophene (O:1.362 Å; S:1.714; Se:1.855; Te:2.005), and a decrease of the C(2)—Y—C(5) bond angle (O:106.55°; S:92.17; Se:87.76; Te:82.53)[76].

The ground-state aromaticity can be estimated by a study of seven different criteria, the more direct feature, i.e. the resonance energy, being not available: NMR dilution shifts experiments, observation of the effect of a 2-Me substituent on the heterocyclic proton NMR shifts, the difference of chemical shifts between α and β protons, the diamagnetic susceptibility exaltation, the sum of the bond orders, the Julg parameter and the mesomeric dipole moments[127]. These criteria, within their limit of validity, are in good agreement with one another, giving good linear relations from which empirical values of resonance energy can be estimated for selenophene (29 kcal mol^{-1}) and tellurophene (25 kcal mol^{-1}). All these criteria, but one, agree with the following order of ground-state aromaticity:

$$\text{thiophene} > \text{selenophene} > \text{tellurophene} > \text{furan}$$

For example, mesomeric dipole moments, which measure the π delocalization give the following values: O:1.03 D; S:1.35; Se:1.29; Te:1.17[128,129]. This sequence must be explained by taking into account an interplay of two opposite factors: the decrease of electronegativity from O to Te, which should favour the contribution of the lone pair of the heteroatom to the aromatic sextet, and the corresponding increase of the covalent radius, which is unfavourable to good overlap between the p orbitals of C(2) and of the heteroatom.

c. Electrophilic and nucleophilic substitution. (*i*) *Kinetic and physicochemical results.* Selenophene and tellurophene easily give electrophilic substitution and α-metalation. Electrophilic substitution in furan, thiophene, selenophene and pyrrole has been reviewed by Marino[130]. Quantitative studies of electrophilic substitution have been conducted with the four heterocycles in three reactions: formylation by phosgene and dimethylformamide, acetylation with acetic anhydride and $SnCl_4$, and trifluoroacetylation by trifluoroacetic anhydride[131,132]. In all three reactions, the reactivity sequence is: furan > tellurophene > selenophene > thiophene. The relative reactivities in this sequence are 107:36.8:3.64:1.0 for the formylation, 11.9:7.75:2.28:1.0 for the acetylation and 140:46.4:7.33:1.0 for the trifluoroacetylation[132].

Measurements of activation parameters in α-formylation show that the activation entropy $\Delta S^{\ddagger}$ can be regarded as constant, so that the relative rates are controlled by the activation enthalpies $\Delta H^{\ddagger}$. These activation enthalpies give a good linear correlation with the resonance energy of the four congeners, showing that the ground-state energy plays an important role in controlling the sequence of reactivity.

Thus, the more aromatic system should give the lower substitution rate and this is corroborated by the experimental data. The ability of a more polarizable heteroatom to delocalize the positive charge should also play a role in the stabilization of the transition complexes. That selenophene is more reactive than thiophene in electrophilic substitution reactions is confirmed by the determination of the ratios of competitive Vilsmeier–Haack formylation (S:Se = 20:80), and of Friedel–Crafts acetylation (S:Se = 25:75) in selenienyl-thienyl-methane[133]. Solvolysis in side-chain reactions[134] gives results in complete agreement with the electrophilic substitution, except for the inversion between furan and tellurophene. On the other hand, the solvolysis of 1-(2-aryl)ethyl acetate shows that tellurophene is less sensitive to substituents than selenophene, and more sensitive in electrophilic substitutions. By these side-chain reactions, a σ_{α}^{+} parameter has been determined for each heteroatom (O: − 0.93; S: − 0.79; Se: − 0.88; Te: − 0.92). The pK_as of the 2-carboxylic acids give a linear correlation with the electronegativity (O:3.16; S:3.53; Se:3.60; Te:3.97)[135] and with the $\nu_{C=O}$ in the IR spectra. The reaction constant ρ, determined from the pK_a of 2-carboxy-5-substituted heterocycles is equal for tellurophene, selenophene[136] and thiophene, indicating that the electronic effects of substituents are transmitted in the same manner in the three heterocycles.

From the ionization constants, a σ-Hammett-like value can be calculated for the heteroatoms which are considered as endocyclic substituents of the —CH=CH— group in benzene. Their σ_{α} values (O: + 1.04; S: + 0.67; Se: + 0.60; Te: + 0.23) also appear to be a function of the electronegativity. The validity of the Hammett equation in selenophene confirms that 2,4- and 2,5-substituents can be considered as *meta*-like and *para*-like substituents, respectively; σ and σ^{-} constants were also evaluated for the 2-selenienyl and the 2-tellurienyl groups[137].

The increasing polarizability of the heteroatom, which is connected with increased stabilization of a negative charged transition state intermediate, should also play a role in the faster nucleophilic reactions of selenophene compared to thiophene. The nucleophilic substitution of 3-bromo-2-nitroselenophene by thiophenoxide ion is four times faster than that of the corresponding thiophene[138–141]. Some Meisenheimer complexes have been detected in the selenophene series[142–145].

(*ii*) *Synthetic results.* Very few organized studies were known in the selenophene series before Yur'ev and coworkers started their investigations in the fifties. More recent and systematic studies in this field are mainly due to the groups in Lund and in Rouen. Studies on tellurophene were started mainly by the group of Marino and Fringuelli in the seventies. Selenophene and tellurophene, like their congeners, undergo electrophilic substitution in the α- rather than in the β-position.

Selenophenes undergo electrophilic nitration, sulphonation, halogenation, mercuration, acylation, formylation, chloro- and amino-methylation and hydrogen exchange.

Nitration of selenophene with fuming nitric acid in acetic anhydride[146,147] or with nitric acid at − 40 °C[148,149] occurs mainly in the α-position, with a total yield of 15–25%, giving a mixture of 85% 2-nitro and 15% 3-nitro isomer, which is very difficult to separate. When position 2 is occupied by an electron-withdrawing group (—NO_2, —CHO, —COMe), the nitration occurs mainly in position 4 and less (10–15%) in position 5[150,151]. For example, nitration of 2-formylselenophene by fuming HNO_3/H_2SO_4 gives 45% 2-formyl-4-nitroselenophene, 5% of its 5-nitro isomer and nearly 50% 2,4-dinitroselenophene. However, in acetic anhydride with 7% H_2SO_4, the NO_2 group enters only in position 5[150]. The 2- and 3-nitroselenophenes were prepared by decarboxylation of 2-carboxy-5-nitro- and 2-carboxy-4-nitro-selenophenes, the latter being easily obtained by oxidation of the corresponding formyl or acetyl-derivative[151a] (equation 26). 3-Nitroselenophene (**36**) can be reduced by Sn/HCl to 3-aminoselenophene hexachlorostannate (**37**). Isolation of the 3-amino base is impossible, but **37** can be acylated in 30%

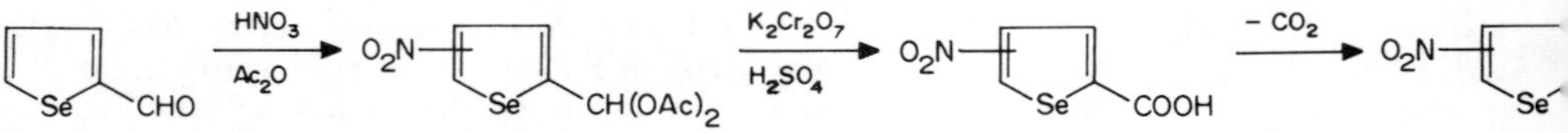

(35)

(26)

yield[148] to compound **38** (equation 27). The hexachlorostannate of the 2-amino isomer can be similarly obtained[148]. Aminoselenophenes are stabilized by an *ortho* electron-attracting substituent[149]. They can be obtained from activated bromo selenophenes via the azido derivative and its subsequent reduction[151b].

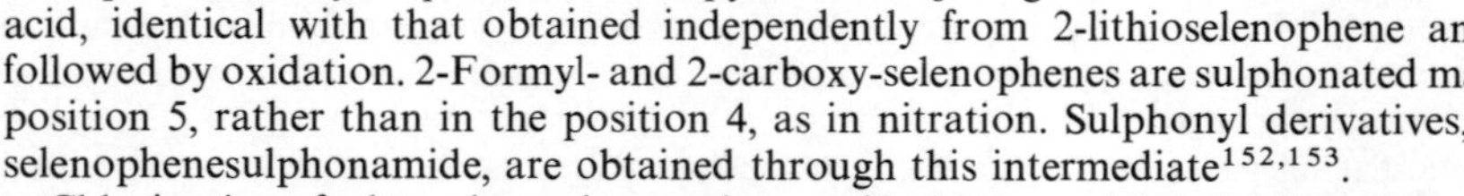

(27)

(36) **(37)** **(38)**

Sulphonation by sulphuric acid or pyridine SO_3[152] gives a 2-substituted sulphonic acid, identical with that obtained independently from 2-lithioselenophene and SO_2 followed by oxidation. 2-Formyl- and 2-carboxy-selenophenes are sulphonated mainly in position 5, rather than in the position 4, as in nitration. Sulphonyl derivatives, e.g. 2-selenophenesulphonamide, are obtained through this intermediate[152,153].

Chlorination of selenophene gives a mixture of 2-chloro- and 2,5-dichloro-selenophene, which, with chlorine in excess, leads to addition and substitution products[154]. Tetrachloroselenophene is obtained directly from Se and hexachlorobutadiene at 250 °C[155]. 2-Chloroselenophene is obtained in 60% yield from selenophene and sulphuryl chloride[154], and a fairly recent patent claims its preparation through dechlorination of tetrachloroselenophene by Na_2Te[156].

Bromination in CS_2 at − 20°C gives 2-bromoselenophene (**39**). Excess of bromine leads to 2,5-dibromo- and 2,3,5-tribromo-selenophene (**40**)[80]. The latter compound is the key intermediate in the preparation of 3-bromoselenophene (**41**) which cannot be obtained by direct monobromination; **41** is obtained from **40** by debromination with zinc in acetic acid[157] (equation 28). A recent simplified synthesis of **41** using the same methodology, but without isolating intermediates, has been described, with an overall yield of 61%[158]. A similar α,α'-debromination is realized starting from tetrabromoselenophene[159–161]. The product, 3,4-dibromoselenophene, is a starting material for 3,4-fusion on selenophene.

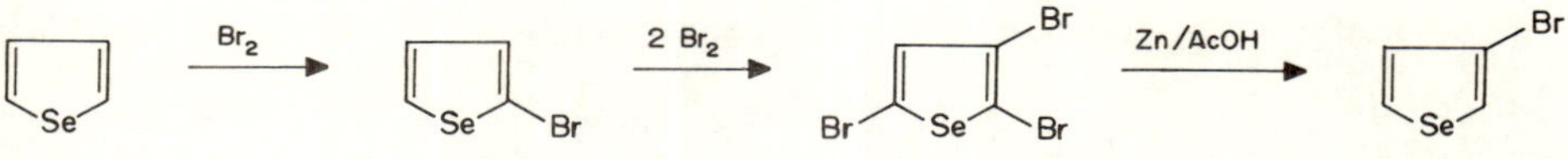

(39) (40) (41) (28)

Iodine can be introduced in the α-position of activated selenophenes with the aid of mercuric oxide[86,162], or through exchange with mercuri salts, which can enter directly by electrophilic substitution to the α-position of selenophene and substituted selenophenes[81c,88,94,163–165]. The HgX group is readily exchanged, not only by iodine, but by various halogens[88] and by the cyano group[94]. For example, from tetramercuriacetate **42**, tetraiodoselenophene (**43**) can be obtained, from which by reduction with aluminium amalgam 3-iodoselenophene (**44**) was obtained[163] (equation 29). From 2,5-dihaloselenophene the selenophene analogue of the electron acceptor TCNQ is prepared[166].

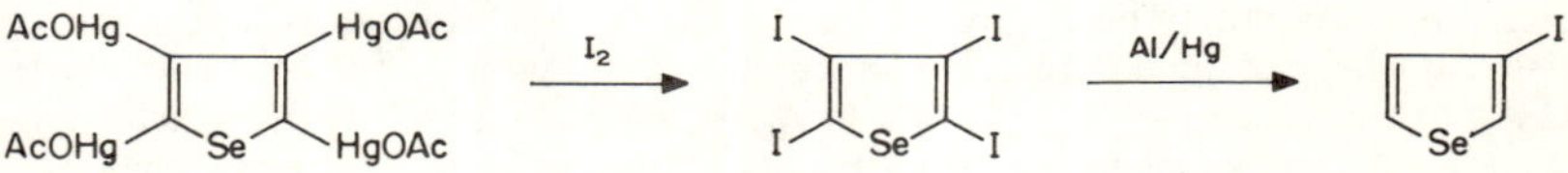

(29)

(42) (43) (44)

Acylation can take place in the α-position under Friedel–Crafts conditions[81,167]. No diacylated selenophenes can be obtained, nor products of direct β-acylation on unsubstituted selenophenes. Hence, 3-acetylselenophene (**46**) is synthesized by standard malonate synthesis from the corresponding acid chloride **45**[157], the acid being obtained through 3-metalated selenophenes (see following section). **46** could be transformed to 3-acetylaminoselenophene (**47**) by the Schmidt reaction with a 75% yield[149] (equation 30).

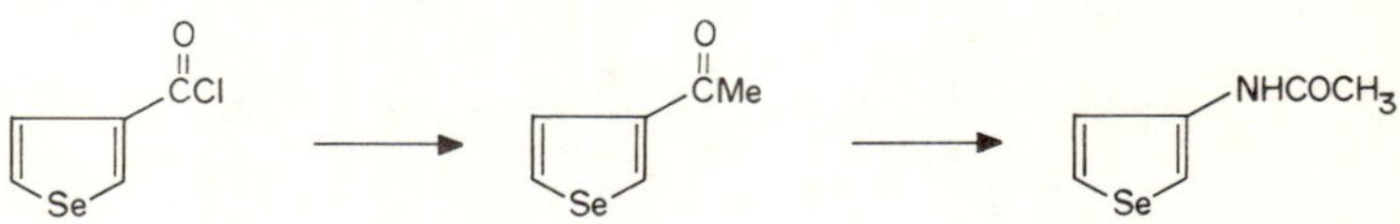

(30)

(45) (46) (47)

α-Formylation is generally realized by the Vilsmeier–Haack reaction. The 2-formylselenophene obtained is the starting material for the 2-carboxy- and 2-hydroxymethyl-selenophenes[168]. 2-Methylselenophene gives formylation in position 5[169] and 3-methylselenophene, like compound **47**[149], gives the 2-formyl derivative[86,170,171]. 3-Formylselenophene is obtained through standard modifications of conveniently available 3-substituted selenophenes, e.g. by Sommelet reaction on 3-bromomethylselenophene[163], by reaction of DMF on 3-lithioselenophene[163] or by reduction of 3-cyanoselenophene[172]. Tri- and tetra-formylselenophenes can be obtained by a suitable combination of electrophilic formylation, halogen–metal exchange, reduction of nitriles, Kröhnke or Sommelet reaction and protection of existing groups[173–176].

Selenophene carboxaldehydes behave like typical aromatic aldehydes, undergoing the Darzens, Perkin, Claisen, Wittig and crotonic type condensations. 2-Vinylselenophene can also be obtained by reaction with MeMgBr[177].

Chloro-[178,179] and amino-methylation[180] also occur in the 2- and 2,5-positions. Direct β-chloromethylation can only be obtained with 2,5-disubstituted selenophenes. For example, 2,5-dimethylselenophene (**48**) which cannot be directly formylated, is chloromethylated in the β-position[181,182a] and the product **49** is transformed to the 3-formyl analogue **50** (equation 31). By aminomalonate synthesis 2- and 3-chloromethylselenophenes lead to the corresponding 2- and 3-selenienyl alanines[182b].

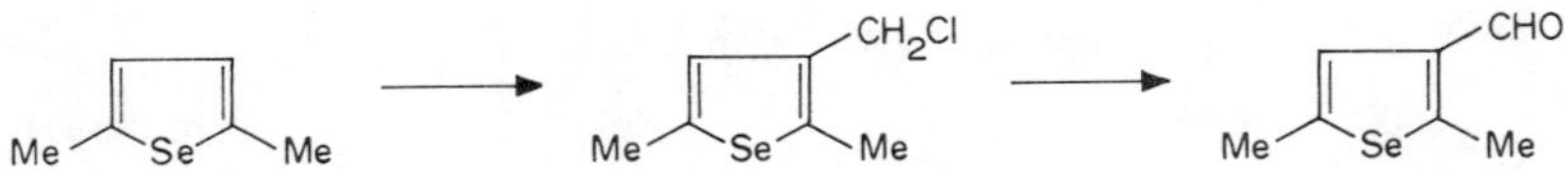

(48) (49) (50) (31)

Finally, with powerful 3-substituted + M donating groups (—$NHCOCH_3$,—OCH_3) the electrophilic Kaufmann reaction (Br_2 + KSCN or KSeCN) can be realized in position 2, giving the corresponding 2-thiocyano- and 2-selenocyano-selenophenes[183,184]. The introduction of the groups —SCN and —SeCN in the β-position can, on the other hand, be achieved through diazotation of 3-amino-substituted selenophenes[185]. The groups directly introduced by electrophilic substitution can be transformed by standard methods to other appropriate functional groups, which can be used further for the synthesis of many selenienyl-substituted heterocycles. For example, 3-(2′-selenienyl)-pyrazoles or -isoxazoles can be obtained from 2-selenienylalkynyl ketones[186]; 2-(2′-selenienyl)-oxazolines and -thiazolidines[150,187] from 2-formylselenophene; 2-(2′-and 3′-selenienyl)-oxazoles, -thiazoles and -selenazoles from the corresponding 2- or 3-selenienylchalcogenoamides[188,189]; 3-selenienyl-1,2,4-oxadiazoles from selenienylamidoximes[190]; 2-selenienylpyrimidine from selenienylamidines[191] and 4-selenienyl-1,2,3,5-tetrazoles from cyanoselenophene or the corresponding imino esters[192a].

Very recently 2-selenienyl isosters of levamisole, an immuno-potentiating agent, were prepared for biological evaluation[192b].

It should be noted that Ullmann reaction on 3-bromo-2-selenienyl and 2- or 4-bromo-3-selenienyl ketones leads to bis-hetero-condensed cyclopentadienones **51**[193] (equation 32).

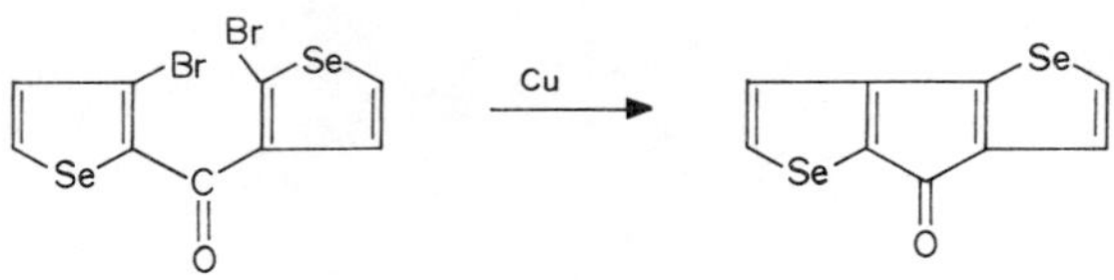

(32)

(**51**)

The chemistry of tellurophene is much less known. Tellurophene is fairly stable to air and light at room temperature; at 0 °C and in the dark, it can be stored for long periods[112]. Since tellurophene is decomposed by strong mineral acids, electrophilic substitution is conducted under moderate conditions. Because of the very high α/β ratio, only 2-substituted or 2,5-disubstituted tellurophenes could be obtained and no 3-monosubstituted tellurophenes are known except the 3-methyl derivative[118b].

Attempts to prepare 2-nitrotellurophene by nitration have been unsuccessful[75,113] and halogens give only 1,1-addition products on the Te atom[97,111,194]. Direct acylation with Ac_2O–$SnCl_4$[113] and with trifluoroacetic anhydride at − 75 °C[75] leads to 2-acetyl- and 2-trifluoroacetyl-tellurophene, respectively. 2,5-Dideuteriotellurophene can also be prepared from D_2SO_4/MeOD[111], and 2,5 bis(acetoxymercuri)tellurophene from mercuric acetate[111]. All other known 2-monosubstituted tellurophenes are obtained from 2-lithiotellurophene. 2-Methoxycarbonyltellurophene (**52**) has been acetylated in position 5 (equation 33), and 2,5-diphenyltellurophene (**54**) has been bis(chloromethylated) in positions 3 and 4, to give **55**. The latter compound was transformed to 3,4-diformyl-2,5-diphenyltellurophene (**56**) (equation 34) via the corresponding diol[195].

(52) (53) (33)

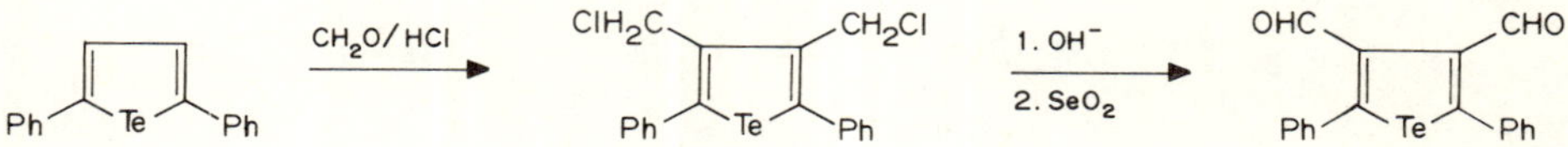

(54) (55) (56)

(34)

d. Metalated selenophenes and tellurophenes. (i) Preparation and standard transformations. Analogously to furan and thiophene, selenophene can be directly metalated in the α-position by organolithium compounds, through hydrogen–metal exchange[157]. α-Metalation can also be realized through halogen–metal exchange, e.g. from 2-iodo-selenophene and phenyl-lithium[162] or -magnesium.

β-Metalation can only be achieved by reacting 3-halogeno(Br, I)selenophene and organolithium compounds at the temperature of dry ice[157,196]. 2-Selenienyl- and 3-selenienyl-lithium are valuable nucleophilic intermediates for the preparation of various 2- and 3-substituted selenophenes. In this way were obtained 2-[157] and 3-carboxyselenophenes[157,196] from CO_2, selenophene-2-sulphinic acid from SO_2[141,152], 3-formylselenophene from DMF[163,175], isomer-free 2- and 3-chloro- and 2- and 3-bromoselenophenes, respectively, from hexachloroethane at $-70\,^{\circ}C$ and ethylene bromide[197], and 2- and 3-selenopheneboronic acids **58** from triethylborate. These boronic acids are converted by oxidation to 2-[198–200] and 3-hydroxyselenophenes[199–201]. Compounds **59** are also accessible from the lithium compounds by reaction with *t*-butyl perbenzoate[198,200,201], followed by hydrolysis (equation 35). 2,5-Dihydroxyselenophene, regarded as the Se analogue of maleic anhydride, is obtained by this method[198].

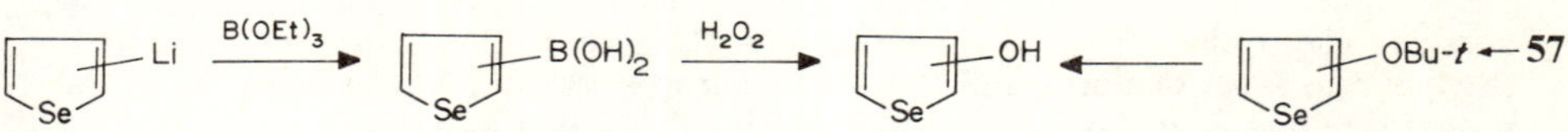

(57) (58) (59) (35)

Reaction of sulphur on 2- and 3-lithioselenophenes gives the corresponding 2- and 3-mercaptoselenophenes[201]. The tautomerism of these hydroxy and mercapto compounds has been studied[200–203]: 2-hydroxyselenophene exists mainly as the α,β-unsaturated selenolactone form, 3-selenolene-2-one, and undergoes Michael addition[202]; 3-hydroxyselenophene and its 2,5-dimethyl analogue are present only in the keto form, 3-oxo-4-selenolene[201]. This is reminiscent of the corresponding furans, whereas the corresponding 2,5-dimethyl-3-hydroxythiophene shows 32% of the enol form[199]. The 2- and 3-mercaptoselenophenes, whose tautomerism was studied in the more stable methyl analogues, exist in the enethiol form, due to difficulties in the formation of thione groups[201,203]. The enol form in hydroxy compounds becomes preponderant in the stabilized *o*-acylated derivatives[204,205].

Reversed reactivity of selenienyllithium is obtained through iodonium salts **60**, by reaction with *trans*-iodosovinyl dichloride. This conversion to an electrophilic reagent is of great synthetic values for condensation with various nucleophiles, in order to obtain substituents which are difficult to prepare by other methods (equation 36). For example, with sodium nitrite in DMF 2- and 3-nitroselenophenes were obtained in 37 and 18% yield, respectively[206]. The nucleophile SCN^- gives 48% of 3-thiocyanoselenophene[207], but the reaction has failed in introducing the SeCN group.

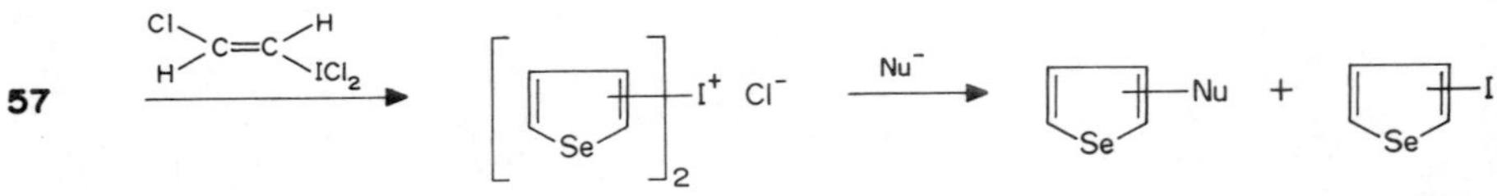

(60) (36)

Direct α-metalation is also preponderant in substituted selenophenes. 3-Monosubstituted selenophenes such as 3-methoxy-[208] 3-cyano-[209] 3-diethoxymethyl-[175,209] and 3-*t*-butoxy-[205] are lithiated in position 2. A 2-substituted selenophene, like 2-cyanoselenophene, is lithiated in the second α-position[209]. To obtain β-metalation it is necessary to start from the corresponding bromo compound. 3-Bromoselenophene itself gives, depending on the reagent and temperature, either 3-lithioselenophene (BuLi, − 50°C), or 3-bromo-2-lithioselenophene[157]. 3,4-Dibromoselenophene gives β-lithiation to 3-bromo-4-lithioselenophene[159,176]. However, with a 2,3-dibromoselenophene, halogen–lithium exchange takes place preferentially on the 2-bromo atom[176]. Easy exchange in the α-position is used to obtain a 2,5-selenienylated polymer from a Grignard reagent and 2,5-dibromoselenophene via a poly-Wurtz reaction[210]. There is thus experimental evidence that selenophene is metalated in the same manner as thiophene.

2-Lithiotellurophene, easily obtained by α-metalation of the parent heterocycle[113], is the starting product for the synthesis of some 2-substituted tellurophenes. In this way were obtained 2-carboxytellurophene[113] (leading by diazomethane to 2-methoxycarbonyl tellurophene), 2-formyltellurophene (leading to 2-methyl, 2-hydroxymethyl- and 2-α-hydroxyethyl-tellurophenes[113]) and 2-dimethylcarboxamido-[211] and 2-methylthio-tellurophenes[212]. 2-Methyltellurophene is lithiated in the remaining α-position[113,213], giving 2,5-disubstituted tellurophenes. It was shown later that by carbonation of 2-lithiotellurophene di-tellurienyl ketone was obtained in 6% yield as a by-product[214].

2-Lithiotellurophene also gives di-α-tellurienyl telluride by reaction with Te[215], but does not lead to the 2-fluoro derivative from $FClO_3$, nor to the 2-nitro derivative from ethyl nitrate. The 2-chloro- and 2-bromo-tellurophenes were obtained by using the reversed methodology shown in equation (36), but the 2-nitro derivative was not obtained in this way[215]. 2-Copper derivatives and iodoarenes give coupling reactions to 2-aryltellurophenes[216]. No β-lithiotellurophene has been obtained so far.

(*ii*) *Ring-opening of lithium derivatives.* The first indication of abnormal behaviour in some selenophene derivatives towards organolithium compounds was the lack of coupling in the reaction of 2,5-dimethyl-3-iodoselenophene with ethyllithium, shown by the failure of subsequent treatment with copper chloride to give the corresponding biselenienyl[217] a reaction working well for the synthesis of bithienyls.

The isolation of the acyclic vinylacetylene **63** in 49% yield when 2,5-dimethyl-3-iodoselenophene (**61**) was treated with ethyllithium at − 70 °C was the evidence of a ring-opening of the selenophene ring (equation 37). The corresponding lithium derivative **62** was trapped in the same reaction at − 100 °C as the carboxylic acid. Consequently, it is very probable that **62** is the key intermediate of this ring-opening[218].

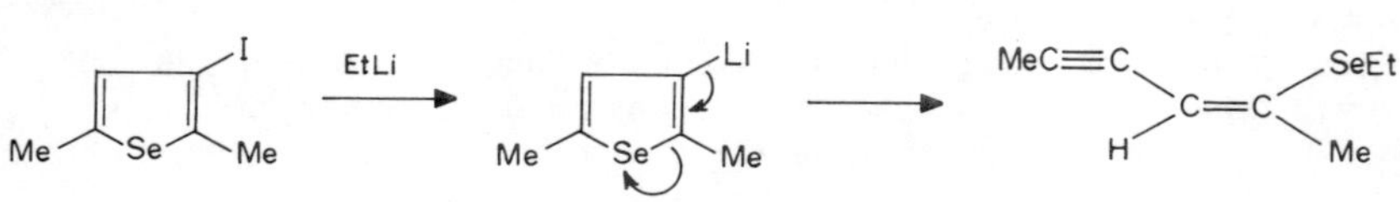

(37)

(61) **(62)** **(63)**

The corresponding thiophene derivative is much more stable and it decomposes only at room temperature, with complicated rearrangements. These facts have led to a careful comparison of the ring-opening of 3-lithiothiophene, -selenophene and -furan derivatives. The previous indication that 3-lithiothiophene gave some unsaturated aliphatic compounds[219] can be explained in this way. A mechanism which has been proposed to explain this ring-opening consists of a retro-Michael reaction from a carbanion situated in a position β to a heteroatom (equation 37). A similar bond-rupture is known for β-carbanions or vinyl ethers[220]. The ring-opening is trans-stereospecific giving the *Z*-compound, where the alkyne and the Se group are in *cis* positions, a conclusion also reached for 3-lithio-2-methylthiophene[221]. Although some 3-lithiofurans give this ring-opening (but not 3-lithiofuran itself which decomposes at -5 °C), no pyrrole derivatives give this reaction.

A general review of the base-induced ring-opening of heterocycles[222] has led to a classification of three types of ring-opening. The preceding reaction belongs to the first group, named 'R.O.I' (Ring Opening Type I) which is extended to all eliminative ring-opening of β-carbanions. That the selenophene ring is less stable than the thiophene ring in this reaction is due to the greater size of the Se atom, rendering it more polarizable and

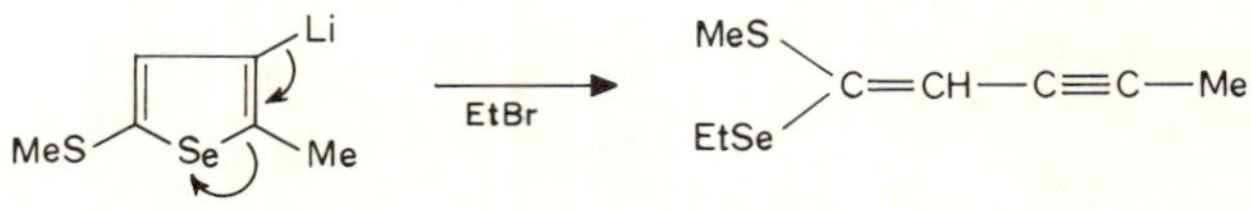

(38)

(64) **(65)**

reducing the bond energy between the heteroatom and carbon. Extension of this reaction to some 3-lithio-5-(methylthio)selenophenes **64** leads to the formation of a mixed thioselenoacetal of acetylenic ketene (**65**)[223], a compound difficult to prepare by other methods (equation 38). The R.O.I is easier with donor groups (for example Me groups) in positions 2 or 5. It was found to be easy, even in the sulphur series and was observed at -70 °C for both 2-trimethylsilyl-3-lithio-thiophene and -selenophene[224]. It could therefore be anticipated that $-$I groups in the α-position should stabilize the 3-lithio intermediates. For example, 2,5-dichloro-3-lithioselenophene (**66**), prepared by abstraction of proton of 2,5-dichloroselenophene by lithium diethylamide is highly stable. However, when one tries to obtain this lithio compound by the classical method from 2,5-dichloro-3-iodoselenophene and ethyllithium, only the unstable ring-opening product (**67**) is formed[225] (equation 39). Its formation can be explained by reaction of the 3-lithio

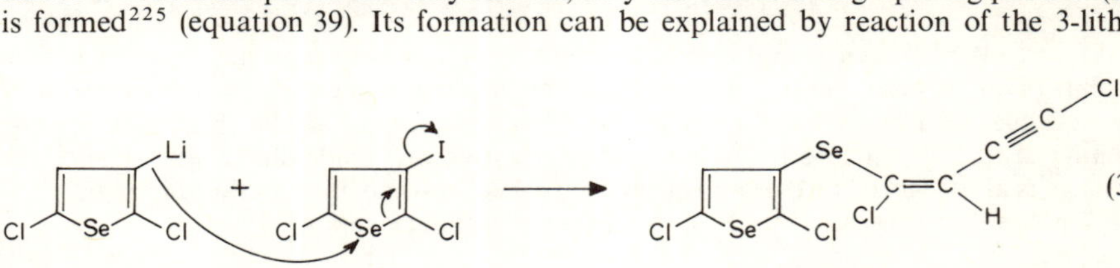

(39)

(66) **(67)**

derivative on the unreacted 3-iodo precursor. It brings evidence of a second type of ring-opening (R.O.II), consisting of a heterophilic reaction (reaction on the heteroatom) of the lithium nucleophile, which can be of two different types. The first one, exemplified in the preceding reaction, consists of an eliminative R.O.II, and requires the presence of a good leaving group β to the heteroatom. To this R.O.II type belongs also the formation of Ph_2Se (in 54% yield) from tetrachloroselenophene and phenyllithium[226]. Due to the greater polarizability of Se, the heterophilic R.O.II reaction is easier in the selenophene than in the thiophene series. But, contrary to the R.O.I, this eliminative R.O.II proves easier with electro-withdrawing groups. Consequently, even in the sulphur series, 3,4-dichloro 2,5-dimethoxythiophene shows this ring-opening reaction[227]. Another type of heterophilic ring-opening was demonstrated on 2,5-diphenyltellurophene (**68**)[228] and on fused selenophenes independently in Liège and in Moscow (see Sections IV.A.2 and IV.A.3). It consists of substitutive R.O.II with cleavage of the heteroatom–carbon bond and formation of a carbanion on this α-carbon (equation 40). The dilithio compound **69** was used to prepare a variety of 1,4-disubstituted butadienes. This reaction is normally easier with the largest and more polarizable heteroatom. A substitutive R.O.II was demonstrated later[229] in 2,5-dimethoxyselenophene, giving 55% yield of dibutyl selenide. This reaction failed in the corresponding thiophene. 2,5-Diphenylselenophene itself (**70**), the Se analogue of **68**, gives a ring-opening, which seems to be of the R.O.I eliminative type, since the isolation of 5,8-diphenyldodeca-5,7-diene (**72**) can only be explained by addition of butyllithium to intermediate 1,4-diphenylbutadiyne, probably formed by preliminary metalation in the β-position (equation 41)[229]. The substitutive R.O.II mechanism, which can also be viewed as a chalcogen–metal interconversion reaction, an

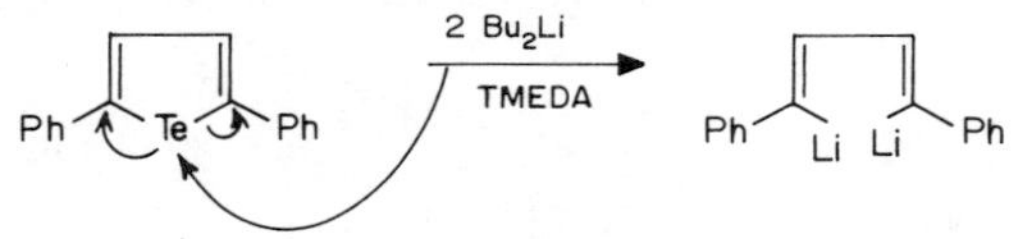

(**68**) (**69**) (40)

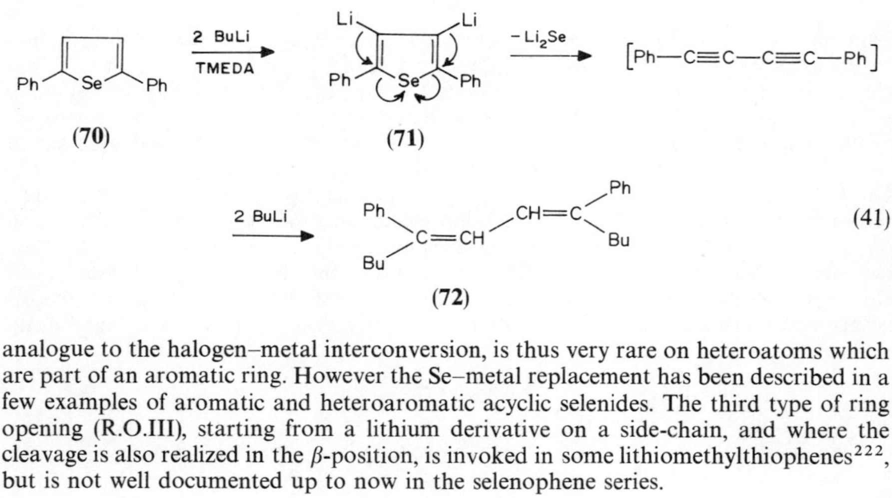

(**70**) (**71**) (**72**) (41)

analogue to the halogen–metal interconversion, is thus very rare on heteroatoms which are part of an aromatic ring. However the Se–metal replacement has been described in a few examples of aromatic and heteroaromatic acyclic selenides. The third type of ring opening (R.O.III), starting from a lithium derivative on a side-chain, and where the cleavage is also realized in the β-position, is invoked in some lithiomethylthiophenes[222], but is not well documented up to now in the selenophene series.

e. Hydrogenated selenophenes and tellurophenes. The chemistry of tetrahydroselenophene (selenolane) (**73a**) and tetrahydrotellurophene (**73b**) is similar to that of acyclic selenides and tellurides. The starting materials for their preparation are generally tetramethylene halides (equation 42). The most reliable method uses sodium chalcogenide:

(**a**) Y = Se
(**b**) Y = Te

(**73**) (42)

Na_2Se[230], Se + rongalite and NaOH[231] (Se + rongalite without NaOH forms a low-molecular-weight polymer, $(Se\text{-}(CH_2)_4\text{-}Se)_n$, which can lead to tetrahydroselenophene by heating[230] or the corresponding Te compounds[232]. Aluminium telluride gives a very low yield[233]), but Te at 130–140 °C can also be used[233,234]. Tetrahydroselenophene has also been prepared from tetrahydrofuran, Al_2O_3 and H_2Se at 400 °C[105] and in 97% yield from the stabilized selenonium ylid **74** (equation 43). The chalcogenide methodology was used

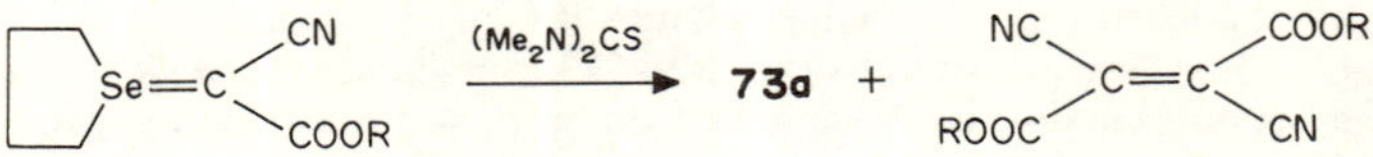

(43)

(**74**)

to prepare 2,5-dicarboxytetrahydroselenophene from α,α′-dibromoadipic acid[235], 4,4-dimethyl-3-oxo-tetrahydroselenophene[236] and 3,3′-bis-tetrahydrotellurienyl[237]. 2,5-Dihalomethyl-1,1-dihalotetrahydrotellurophenes were obtained from 1,5-hexadiene, using the ability of TeX_4 to add to alkenes[238]. The heteroatom of **73a**[230] and **73b**[233] can undergo addition reactions with halogen or MeI; when 1,1-dihalogenotetrahydrotelluro-phene is treated by SO_2, it gives back **73b**. The halogens on Te can be substituted by various nucleophiles (N_3^-, NCS^-, $NCSe^-$)[239]. Compound **73b** can be oxidized by air to the corresponding telluroxide[233]. Selenoxide **76** has been obtained by hydrolysis of selenium-*N*-tosylimide **75**[240] (equation 44).

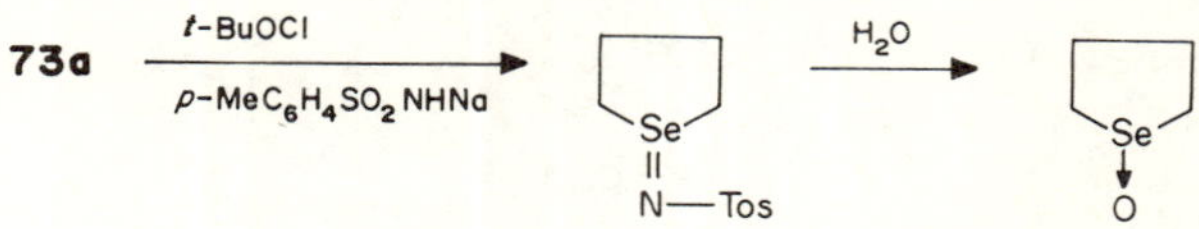

(44)

(**75**) (**76**)

The dipole moments of **73a** and **73b** are lower than that of tetrahydrothiophene[241] (S: 1.89 D; Se: 1.81; Te: 1.63). The Se analogue of biotine has been synthesized, the Se being introduced by NaSeH through a pentacyclic lactone to give a corresponding selenolactone which is then transformed by standard methods to selenobiotine[242,243]. 2,5-Dihydrotellurophene and its 3,4-dimethyl analogue **78** were recently prepared from the corresponding butadienes and $TeCl_4$[244] (equation 45). The same reaction failed with $SeCl_4$ or $SeOCl_2$. **78** is detellurated by butyllithium, and the 1,1-dichloro precursor **77** reacts with arylmagnesium bromide to give a good synthetic micro-method for the preparation of diaryl tellurides.

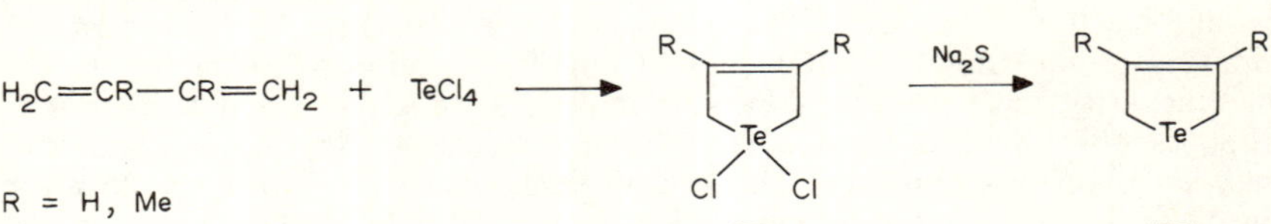

(45)

(77) (**78**)

2. *Benzoselenophenes and condensed analogues*

A review of the chemistry of benzo(*b*)selenophene and benzo(*b*)tellurophene was published in 1975[245].

a. Benzo(b)*selenophene. (i) Synthesis.* Benzo(*b*)selenophene has been obtained, among other products and, often with very small yields, by reaction of Se with acetylene or phenylacetylene[77,246–248], by basic hydrolysis and oxidation of *o*-cyanoselenocinnamic acid[246], by reaction of SeO_2 with styrene[249] and by pyrolysis of *o*-ethylselenophenol[250]. Systematic methods of synthesis consist rarely of benzologation of selenophene. We can only note in this respect some ring-opening–ring-closure reactions of selenolo-(2,3-*c*)- or (3,2-*c*)-pyrylium cations by secondary amines to 5- or 6-dialkylamino- or hydroxy-benzoselenophenes[251,252]. The more general syntheses involve the building of the selenophene ring on the benzenic substrate.

Following the pyrolysis of diacids, unsubstituted[253,254] or substituted[255–260] Se analogues of indoxyl are prepared by the C(2)–C(3) ring-closure. The yields are improved by isolating the intermediate enol-acetate **80** (equation 46). Sodium amalgam[254], sodium borohydride or hypophosphorous acid reduction of **81** gives the corresponding benzo(*b*)selenophene[261]. By this general reaction naptho(2,3-*b*)selenophene was also prepared[262]. The alternative Dieckmann condensation of the corresponding diesters of **79** gives an α-keto ester, which is also directly reduced in high yields[261] to

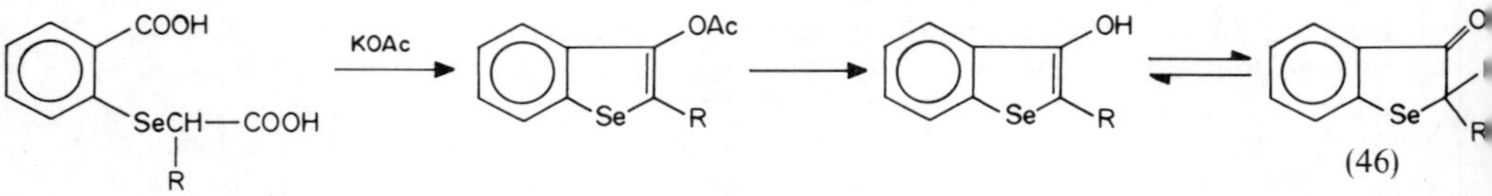

(46)

(79) **(80)** **(81)** **(82)**

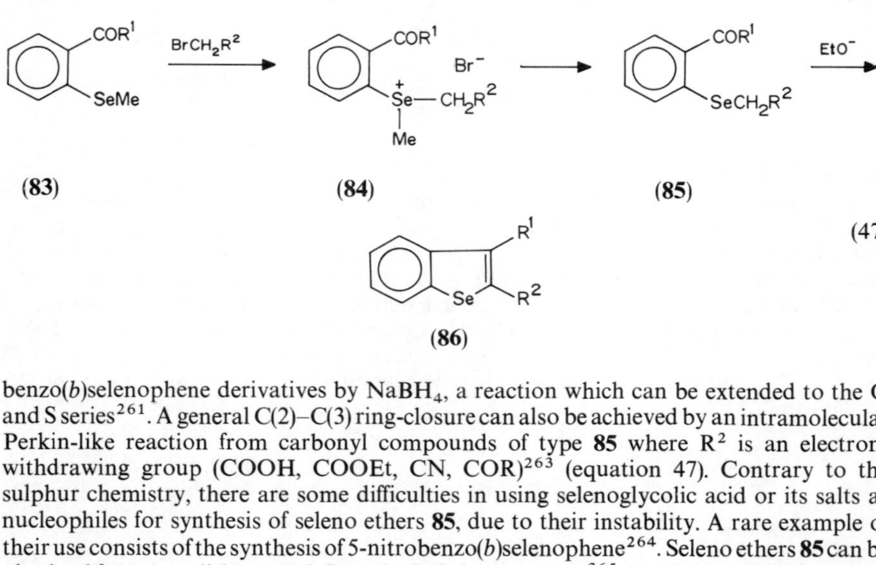

(83) **(84)** **(85)**

(47)

(86)

benzo(*b*)selenophene derivatives by $NaBH_4$, a reaction which can be extended to the O and S series[261]. A general C(2)–C(3) ring-closure can also be achieved by an intramolecular Perkin-like reaction from carbonyl compounds of type **85** where R^2 is an electron-withdrawing group (COOH, COOEt, CN, COR)[263] (equation 47). Contrary to the sulphur chemistry, there are some difficulties in using selenoglycolic acid or its salts as nucleophiles for synthesis of seleno ethers **85**, due to their instability. A rare example of their use consists of the synthesis of 5-nitrobenzo(*b*)selenophene[264]. Seleno ethers **85** can be obtained from an *o*-lithioacetal, Se and ethyl chloroacetate[265], but they are best prepared by alkylation of the selenoethers **83** to **84** and subsequent heating[263,266,267]. The ease of alkylation of the heteroatoms increases with the size of the chalcogen. The reaction works also with an acid cyanide to give 3-cyano-2-substituted benzo(*b*)selenophene[265].

A C(3)–C(aromatic) ring-closure gives one of the most general preparations of benzo(*b*)thiophenes, by using acidic cyclodehydration of phenylthiomethyl ketones, aldehydes or acetals. However, it fails generally in the Se series, due to the cleavage of this type of selenium–aliphatic carbon bond in acidic conditions. A unique example, described in the literature[268], could not be repeated[260]. Electrophilic cyclization of acid chlorides to five-membered Se heterocycles is also very rare, and is only described in the cyclization of α-phenylselenopropionyl chloride (**87**; $R^1 = H$, $R^2 = Me$) to 2-methylselenoindoxyl (**88**)[269], and of oxalyl chloride monophenylseleno ester (**87**; $CR^1R^2 = C{=}O$) to 2,3-dioxo-2,3-dihydrobenzo(*b*)selenophene (selenanaphtenequinone)[270] (equation 48). A third type of C(3)–C(aromatic) ring-closure consists of a cyclization of phenylallyl selenides. It is realized only in quinoline solution and is thus different from the traditional Claisen rearrangement[18,271]. Depending on the nature of the allyl group this cyclization gives 2-methyl- or 2,2-dimethyl-2,3-dihydrobenzo(*b*)selenophene, but not a six-membered ring nor *o*-allylselenophenol. The selenoxide of 2,2-dimethyl-2,3-dihydrobenzo(*b*)selenophene gives, by ring-chain tautomerism, an unstable selenenic acid, trapped by intra- and inter-molecular reactions[272].

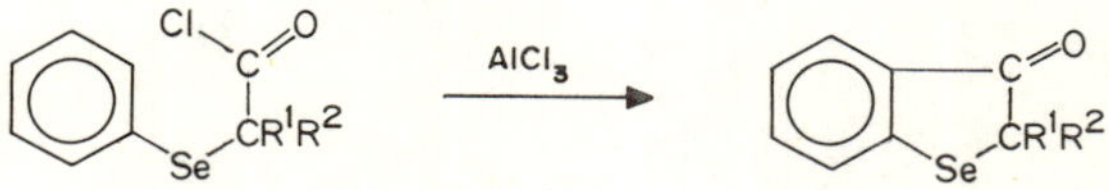

(**87**) (**88**) (48)

Selenoindoxyls can be obtained by base-catalysed C(2)–Se cyclization of *o*-acetylselenenyl halide[260] or selenocyanate[273,274]. This cyclization is easier than in the sulphur series due to the increasing size of the chalcogen and to its decreasing electronegativity (equation 49). Selenoindoxyl can also be similarly obtained from *o*-

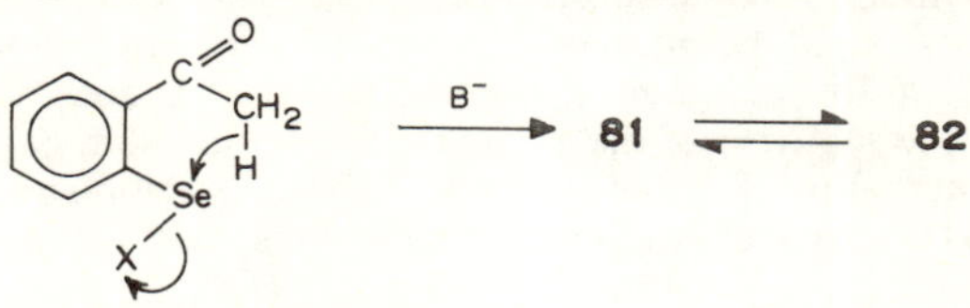

X = Cl, Br, CN

(**89**) (49)

diazoacetylselenenyl bromide[275]. The tautomerism of selenoindoxyls has been the subject of discussion. The chemical behaviour of the compounds is consistent with ketonic behaviour. They form selenoaurones[276] with aromatic aldehydes and anils with nitrosodimethylaniline [followed by hydrolysis to 2,3-dioxo-2,3-dihydrobenzo(*b*)selenophene (selenanaphthene quinone)][253]. They are oxidized to selenoindigo (for a review on selenoindigoid dyes see Ref. 277) and give a positive Brady's test. On the other hand, enolic behaviour is shown by their solubility in NaOH, and by methylation and acetylation to *O*-substituted benzo(*b*)selenophenes[276]. Previous studies indicate that selenoindoxyl exists only in the keto form **82**[256,278]. A thorough investigation has shown by IR and NMR spectra the presence of the ketonic form in freshly distilled selenoindoxyl. However, an equilibrium was determined after more than a hundred hours. The equilibrium mixture consists of 13% of enol form in $CDCl_3$ solution, 21% enol in acetone-d_6 and 50% enol in DMSO-d_6[279]. The corresponding results for thioindoxyl are 2.4% of the enol form in $CDCl_3$, and 30% enol in acetone-d_6[280]. Substituents on the benzene ring modify the position of equilibrium and the rate of its achievement, but freshly prepared solutions are exclusively in the keto form. In DMSO solution, the equilibrium percentages of the enol form are: 4-MeO: 40%; 5-MeO: 80%; 5-NO_2: 100%. In this last example the value is obtained after 8 hours[279].

A C(2)–Se ring closure, by electrophilic substitution of selenenyl halides on styrenes, discovered during attempts to prepare selenocoumarine, was realized in pyridine solution (equation 50)[281]. A selenenyl bromide is an intermediate. This very general method can be

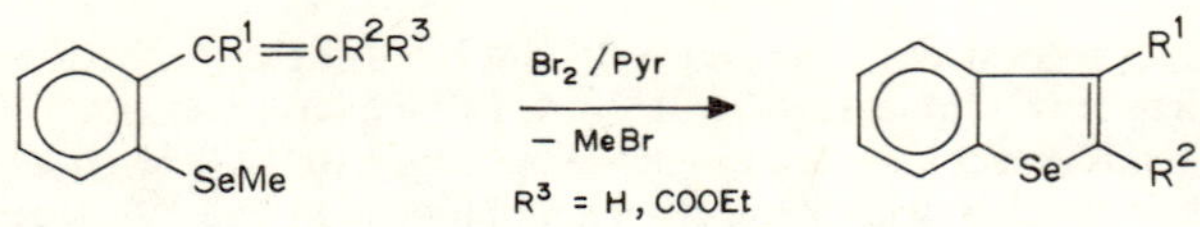

(90) (91)

(50)

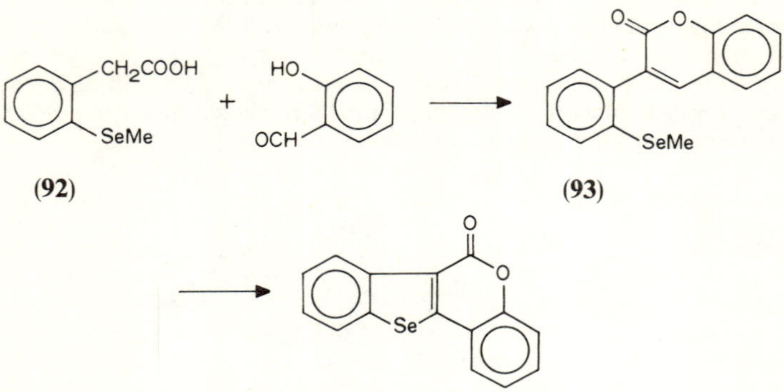

(92) (93)

(94)

(51)

applied to the synthesis of unsubstituted, 2- or 3-, or 2- and 3-substituted benzo(*b*)selenophenes[281], of homocycle-substituted benzo(*b*)selenophenes (4-NO_2, 5-NO_2, 7-NO_2, 6-Cl, 6-Br and their 2-Me derivatives)[282], and of some polycyclic derivatives, e.g. for the synthesis of the Se analogue of coumestan (**94**)[283] (equation 51). This method has been applied to the synthesis of benzo(*b*)thiophene: the intermediate sulphenyl halide was prepared by reaction of sulphuryl chloride on the corresponding thio ether[281].

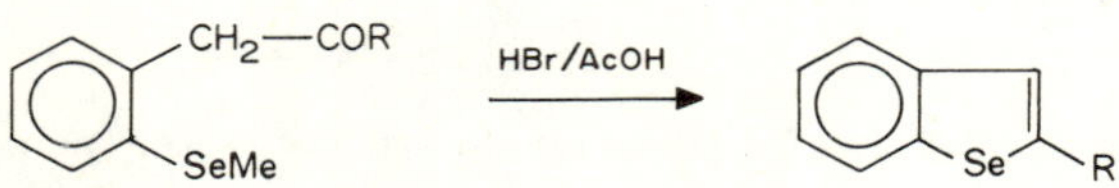

(95) (96)

(52)

Cyclodehydration of *o*-methylselenobenzyl ketones involves demethylation by HBr in acetic acid[284] (equation 52). This method has also been applied to the O and S congeners[284].

Cyclodehydration of demethylated *o*-methylselenophenylacetic acid results in selenolactonization. By this method 2-oxo-2,3-dihydrobenzo(*b*)selenophene (**98**), a lactone isomer of selenoindoxyl (**82**), was obtained in a one-pot reaction from *o*-methylselenophenylacetic acid (**97**) or its nitrile[283] (equation 53). Under more drastic

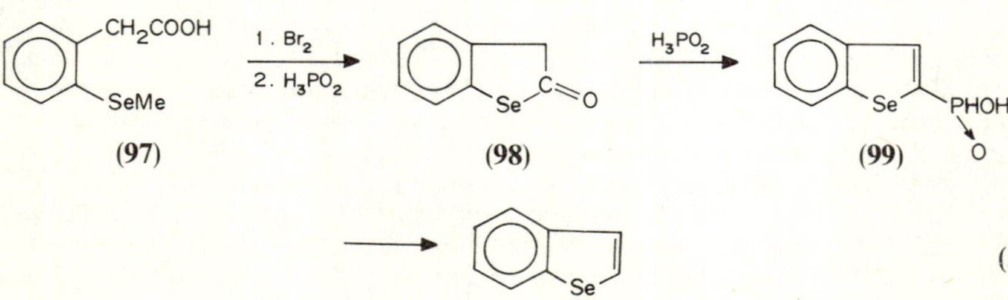

(97) (98) (99)

(53)

conditions, **98** gives a good yield of benzo(*b*)selenophene, through the 2-phosphinic acid **99** as intermediate. This lactonization method is applicable to the corresponding thiolactone derived from sulphenyl chloride. Selenolactone **98** undergoes ring-opening by conventional nucleophilic reagents, and reacts at the CH_2 by aldol condensation with benzaldehyde in ethanol–Et_3N solution. This is followed by ring-opening of the selenolactone and internal Michael addition of the SeH group, to give the ring-rearrangement product 2-phenyl-3-ethoxycarbonyl-2,3-dihydrobenzo(*b*)selenophene[283].

2,3-Dihydrobenzo(*b*)selenophene (**101**) is prepared in 45–50% yield by a one-pot H_3PO_2 reduction of *o*-bromoselenophenethyl alcohol, followed by cyclodehydration of the intermediate selenophenol[285] (equation 54). Compound **101** is difficult to obtain by other methods: electrophilic ring-closure of *β*-phenethylselenenyl chloride (**102**) gives a very low yield (5–10%); hydroboration or catalytic hydrogenation of benzo(*b*)selenophene, or Wolff–Kishner or Clemmensen reduction of selenoindoxyl

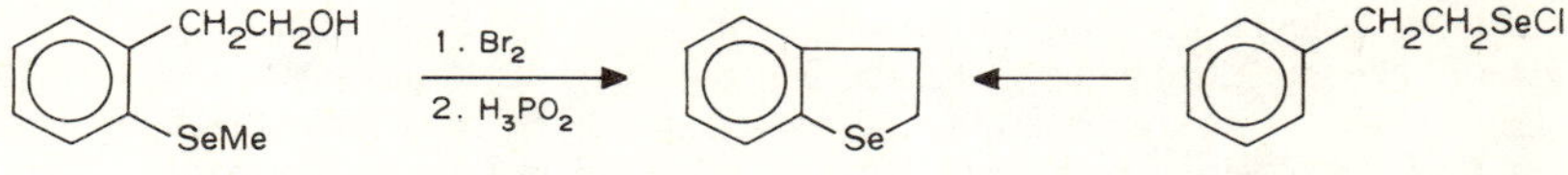

(54)

(**100**) (**101**) (**102**)

failed[285]. Compound **101** can be quantitatively dehydrogenated by S to benzo(*b*)selenophene.

Allylic SeO_2 oxidation of selenochromenes **103** under well-chosen conditions gives rise to a ring-contraction leading to very good yields of 2-formylbenzo(*b*)selenophenes **105**[286], probably through an intermediate of type **104** (equation 55). In other solvents and at other temperatures 1-selenocoumarines, 1-selenochromones and diselenocinnamaldehydes are formed[286]. The ease of ring-contraction increases with the size of the chalcogen; for thiochromenes, the reaction is much more complex, giving also, depending on experimental conditions, dithiocinnamaldehyde-*S*,*S*-dioxide[287], 2-(2′-benzo(*b*)thienyl)-1-thiochromones[288] and a six-membered selenosultone by insertion of SeO_2[289] (Section V.B.2). The yields are lower for the corresponding ketones. MnO_2 oxidative hydrolysis of selenochromylium salts can also give **105**, through intermediates[290a] of a similar type. The intramolecular nucleophilic cyclization of *o*-(ethynyl)selenophenolate, obtained by *o*-lithiation of phenylacetylene and reaction with Se gives directly benzo(*b*)selenophene[290b].

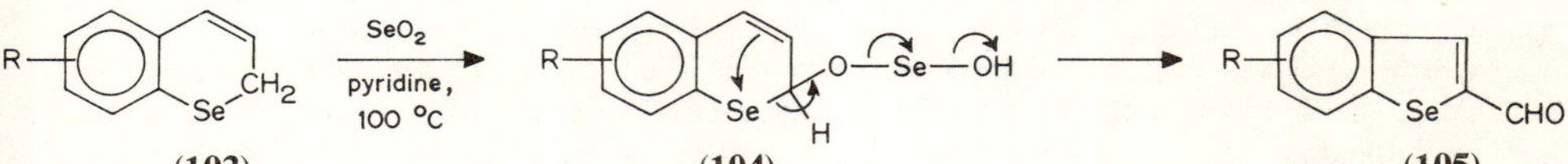

(**103**) (**104**) (**105**)

(55)

$SeOCl_2$ cyclization of cinnamic acids by analogy with the $SOCl_2$–Krubsack cyclization has only been applied on hetarylpropenoic acids, and on diarylethylenes (Section IV.A.3).

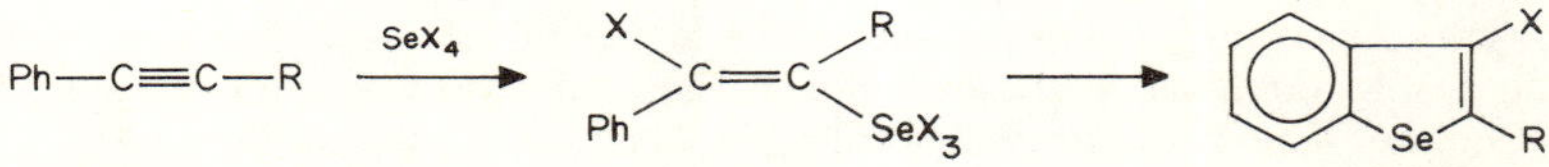

(56)

(**106**) (**107**) (**108**)

C(aromatic)–Se cyclization by addition of $SeCl_4$ or $SeBr_4$ to acetylenic compounds (e.g. **106**) to give a Markownikoff-type addition compound **107** has been realized to give 3-halobenzo(*b*)selenophenes **108** (equation 56). By this method were obtained 2-phenyl-3-chloro-[291], 3-bromo-[292], 2,3-dibromo-[292], 2,3-dichloro-[292] and 3-chloro-2-methyl-benzo(*b*)selenophenes[292]. This reaction was recently realized by phase-transfer catalysis[293]. By a similar reaction with selenium monochloride (Se_2Cl_2) 3-chloro-2-substituted-benzo(*b*)selenophenes (2-Ph, 2-Me, 2-*n*-Bu, 2-COOEt) were obtained, probably via a selenirenium ion[294]. By reaction of SeO_2 and HBr on phenylpropiolic acid or its *m*-methoxy derivative the 3-bromo-2-carboxy and 3-bromo-2-carboxy-5-methoxy-derivatives were obtained in yields of 28%[295] (compared with 63% in the corresponding sulphur series). Similarly, by the same reagent, 3-bromo-2-sulphonamido-benzo(*b*)selenophene was obtained in a 14% yield[296]. $SeBr_4$ can also give by addition to benzal or dibenzalacetone 3-bromo-2-acetyl or 2-cinnamoyl-2,3-dihydrobenzo(*b*)selenophene. The latter is dehydrohalogenated by MeONa to 2-cinnamoyl benzo(*b*)selenophene[297].

(*ii*) *Electrophilic substitution and standard transformations.* In parallel with its sulphur analogue benzo(*b*)selenophene undergoes preferential electrophilic substitution on the heterocycle. However, whereas mononitration gives isolated yields of 5% of the 2-nitro and 35% of the 3-nitro derivatives[298], acylation occurs preferentially in the 2-position[299,300] (2-acylation: 3-acylation = 9). This ratio is the opposite to that in the acetylation of benzo(*b*)thiophene, where the 2-acylation: 3-acylation ratio is 14: 86. The destabilization of the Wheland intermediate resulting from reaction in position 3, by a less efficient + E effect of Se is invoked[299]. Rieche or Vilsmeier formylation leads to high-melting-point compounds; chloromethylation by dichloromethyl methyl ether, followed by nitropropane reaction, leads however to a 7% yield of 3-formyl-benzo(*b*)selenophene[299].

Lewis acids alone can give two isomeric benzoselenophene dimers, 2- and 3-(2′-benzoselenienyl)-2,3-dihydrobenzoselenophene[300]. On the other hand, monobromination of benzo(*b*)selenophene by bromine in CCl_4[301] gives a 3:1 ratio of the 3- to 2-products[292]. Halogenation on Se occurs initially and 1,1,3-trihalogenated intermediates can be isolated. An important isotope effect on the $\beta:\alpha$ ratio is evidenced: for 3-deuterio-benzo(*b*)selenophene the preceding ratio is 3:2; for the 2-deuteriated isomer, it is 86:14. This could indicate that the reaction is not a traditional electrophilic substitution and thus suggests an intermediate bromonium ion as in alkenes, whose formation could be favoured by primary *Se*-dihalogenation which lowers the aromatic character. Excess of bromine gives successively 2,3-dibromo and 2,3,6-tribromo derivatives[292]. Pure 3-bromobenzo(*b*)selenophene can be obtained from the 2,3-dibromo derivative by reaction with Zn and AcOH (95% yield) or with butyllithium at $-80\,^{\circ}C$ (80%). Monomethyl homologues of benzo(*b*)selenophene can be obtained, the 2-methyl isomer by a direct ring-closure of the corresponding α-methylated diacid **79** (R = Me)[256], from cyclodehydration following equation (52) or from Wolff–Kishner reduction of the 2-formyl derivative[302]. The 3-methyl isomer can be obtained by reaction of MeMgBr with selenoindoxyl[256], or best by direct ring-closure followed by decarboxylation of 2-acetylphenylselenoacetic acid, following equation (47). Electrophilic reactions (formylation[303], acylation[304], aroylation[304]), on these monomethyl benzo(*b*)selenophenes proceed easily on the neighbouring free heterocyclic position, leading by standard transformations to a variety of disubstituted benzo(*b*)selenophenes[303], such as the 2,3-dimethyl derivative, the 2,3-dicarboxy derivative and its corresponding 2,3-anhydride. From 2,3- or 3,2-acid alcohols (**109**; Y = Se) the corresponding lactones **110** and **111** (Y = Se) can be easily obtained (equation 57). In contrast, in the benzofuran series the corresponding **109a** or **109b** (Y = O) does not lactonize, and in the benzothiophene series, only the acid **109b** (Y = S) gives lactonization under these conditions. The lactone **110** (Y = S) can only be

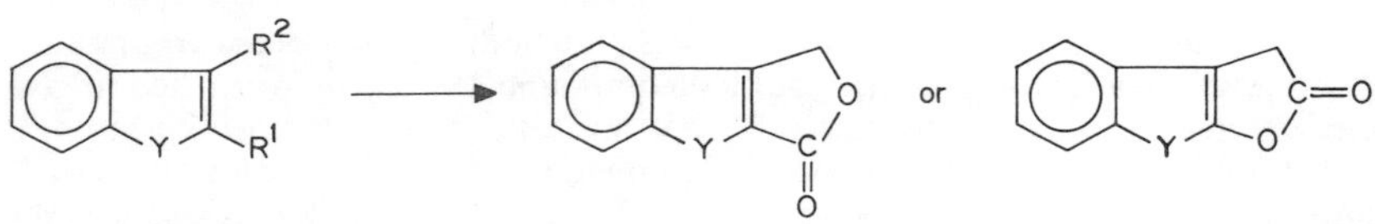

(109) **(110)** **(111)** (57)

(**a**) $R^1 = COOH, R^2 = CH_2OH$
(**b**) $R^1 = CH_2OH, R^2 = COOH$

obtained from the corresponding nitrile alcohol, but this reaction fails in the benzofuran series[305]. Electrophilic reactions can also lead to 2,3-disubstituted bromocarbonyl- or carboxybenzo(*b*)selenophenes.

Nucleophilic substitution of the bromine in the bromocarbonyl-activated compounds by thiolate or selenolate ion can be easily realized, leading to precursors for further cyclization to fused heterocycles[306]. This reaction leads to introduction of a S or Se atom into position 3 of the benzo(*b*)selenophenes. The 2-bromo-3-substituted derivatives can be debrominated by butyllithium, giving rise from the 2-bromo-3-diethoxymethyl compound (**112**; $R = CH(OEt)_2$) to the 3-formyl derivative (**113**; R = CHO), and successively to 3-carboxy-and 3-acetyl-benzo(*b*)selenophenes[299] (equation 58). The Se analogue of gramine (**113**; $R = CH_2NMe_2$) was obtained similarly from **112** ($R = CH_2NMe_2$)[307].

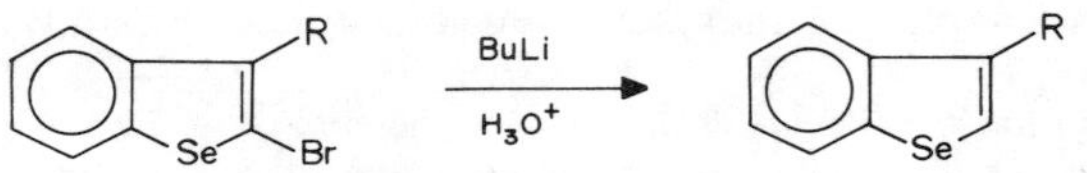

(58)

(**112**) (**113**)

A Se analogue of tryptamine was prepared from the nitrile **113** ($R = CH_2CN$), which was obtained by a Wittig reaction on selenoindoxyl; the Se-analogue of tryptophane (**113**; $R = CH_2CH(NH_2)COOH$) was obtained from the 3-formylbenzo(*b*)selenophene via hippuric acid synthesis[307]. Corresponding isomers in position 2 were also synthesized. From selenotryptamine the Se analogues of desmethoxyharmaline, harmine and hexadehydroyohimbane can be obtained[308]; the Se analogue of serotonin (5-hydroxyselenotryptamine) can be prepared from 5-nitroselenoidoxyl in a similar way[309a]. In this same field, a benzoselenienyl glycine was obtained by direct electrophilic reaction on benzo(*b*)selenophene[309b].

Electrophilic substitution on the benzene ring takes place only when the 2- and 3-positions are occupied. For example, the 2,3-dimethyl-derivative is acylated in position 6[310]. In the same way *β*-(3-benzo(*b*)selenienyl)propionic acids can be cyclized at position 4 of the homocycle only if position 2 is occupied by a Me or halogen substituent as in the transformation **113a** → **113b**[311]. Note that the non-aromatic 2,3-dihydrobenzo(*b*)selenophene is also benzoylated in position 6[285].

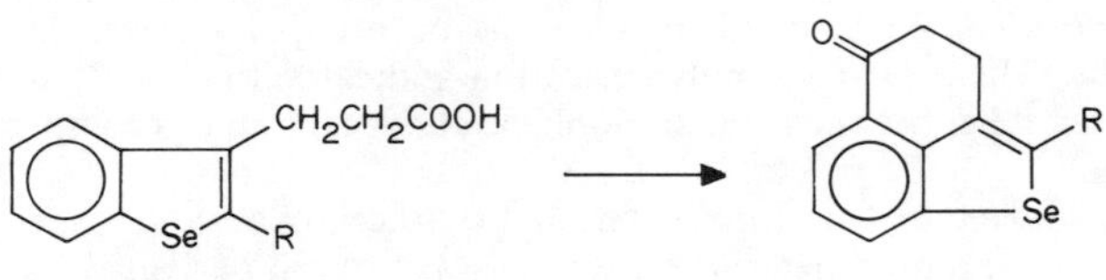

(**113a**) (**113b**)

(*iii*) *Metalation and ring-opening reactions.* 2-Lithio-benzo(*b*)selenophene is easily obtained by direct metalation of the parent heterocycle. It is used for many transformations, e.g. for the synthesis of the pure 2-bromo derivative[299]. The 3-lithio isomer (**115**) is obtained by halogen–lithium interconversion of 3-bromobenzo(*b*)selenophene at $-80\,^{\circ}C$[312] (equation 59) and is stable for a long time at room temperature. It can be transformed with good yield to 3-bromo, 3-carboxy- or 3-methyl-benzo(*b*)selenophenes. No lithium transfer is realized as in 3-lithiobenzo(*b*)thiophene. Treated by a second equivalent of butyllithium at 0 °C, it gives a ring-opening of eliminative R.O.I type. A study of the methylation products is in agreement with the formation of the intermediary 2,3-dilithio derivative (**116**). The latter is responsible for the ring-opening reaction, and its instability confirms the ease of the R.O.I transformation when electron density is increased in position 2 (equation 59). 3-Lithio-2-methyl-benzo(*b*)selenophene is also less stable and begins to ring-open after 4 h at room temperature. The formation of a 70% yield of the 2-carboxy derivative by treatment of **114** with 2.5 equivalents of phenyllithium, followed by carbonation can be explained by a ring-opening of **116**, carbonation of the resulting lithioacetylene and acidic ring-closure. This explanation takes into account the absence of

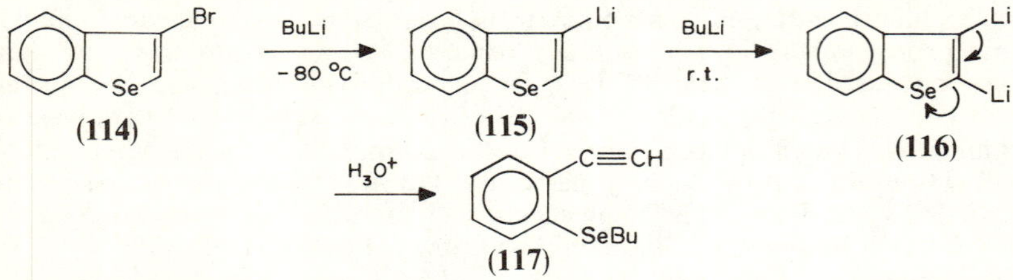

(59)

2-lithio-benzo(*b*)selenophene in the system, as demonstrated by a study of the methylation products[312].

(*iv*) *Transformation and ring fusion on the homocycle.* The 4-, 5- and 7-nitro-benzo(*b*)selenophenes, obtained by direct ring-closure of conveniently nitrated precursors[264,298], give by reduction the corresponding 4-,5- and 7-amino derivatives. The 6-nitro isomer could not be prepared. The corresponding 6-amino derivative is obtained by reaction of potassium amide on the 6-bromo compound. These amino compounds have been used in the Bernthsen synthesis of acridines[313], in the Beyer–Combes and Skraup syntheses of quinolines, in the Ullmann–Fetiadjan cyclization of benzacridine derivatives of selenophene[314], and in the Fischer indole synthesis via the corresponding hydrazines. From 4,5-diamino-benzo(*b*)selenophene, quinoxalines, pyrroloquinoxalines, thiadiazoles and selenadiazoles condensed on the benzene ring were obtained[315].The 5-mercapto derivative, obtained from the 5-amino compound, has been used for the formation of a thiophene ring, giving a selenolobenzothiophene whose linear or angular structure is not firmly established[260].

(*v*) *Base-catalysed H/D exchange.* The stability of the carbanion α to the heteroatom has been tested by base-catalysed H/D exchange[316] in benzo-condensed furan and congeners. The ratios of *k*(exchange) values for the sequence S/Se/O are respectively, 7/1/2, indicating that, surprisingly in these systems, Se is less effective than O or S in stabilizing an α-carbanion. It can therefore be concluded that none of the preceding factors has a decisive contribution to this stabilization.

(*vi*) *Copolymerization.* Benzo(*b*)selenophene has been copolymerized as alkene with maleic anhydride. The ease of copolymerization indicates that, with respect to the aromatic character, it lies between benzothiophene and indole, and is completely different from benzofuran[317].

b. Benzo(c) *selenophene and derivatives.* The first synthesis of benzo(*c*)selenophene (**121**), a highly reactive *o*-quinonoid heterocycle whose substitution derivatives are unknown, was realized in 1976 by Saris and Cava[318] via a cold alkaline destruction of the dibromide **119** of 1,3-dihydrobenzo(*c*)selenophene (**118**) (2-selenaindane) (equation 60).

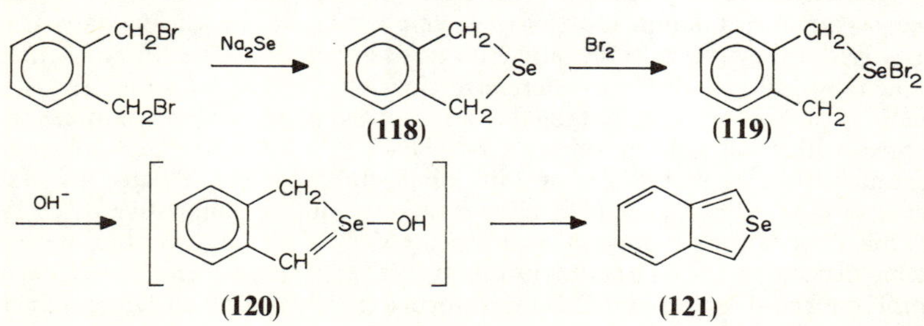

(60)

The selenoxide of **118**, readily available by oxidation of **118** with H_2O_2 in cold methanol, also liberated **121** when treated with 40% NaOH. Dehydration of the selenoxide hydrate, via the ylid **120**, should be a likely mechanism. **121** polymerizes easily

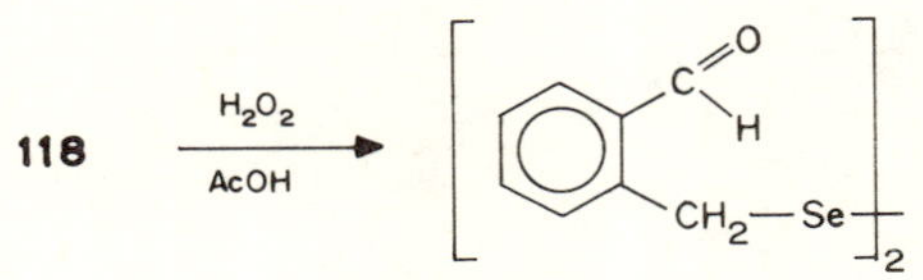

(61)

(122)

in attempts to isolate it in a pure state. It gives the expected crystalline Se-containing adduct with tetracyanoethylene. Compound **118** has been thought to be oxidized to 2-selenophthalide by H_2O_2 in acetic acid[319], but in fact it gives 2,2′-diformyldibenzyl diselenide (**122**)[320] (equation 61). A possible mechanism should be a Pummerer rearrangement of the intermediate **120** followed by ring-chain tautomerism.

Chacko and coworkers[321] used the NMR J_{5-6}/J_{4-5} ratio (J ratio) as a measure of the degree of electronic delocalization in butadiene fragments. This ratio is 0.74 for **121** and is in the same range as for other benzo(*c*) five-membered heterocycles. This value is lower than for the isomeric (*b*) condensed congeners ($J_{ratio} = 0.9$), indicating a residual butadienic character, since $J = 0.52$ for cyclohexadiene. The annelation energies for (*c*) condensed systems are very low, indicating a low contribution of the benzene ring in the delocalization of the heterocycle.

The oxo compound **126** (2-selenophthalide) has been prepared by hydrolysis of the corresponding imino compound, 2-selenophthalimidine (**125**), a cyclic tautomer of **124**[322,323], by heating di(*o*-carboxybenzyl)diselenide (**127**) with 50% aqueous hypophosphorous acid[324] (equation 62) or by $AlCl_3$ cyclization of *o*-methylselenomethyl benzoyl chloride[11]. Compounds of type **123** can be transformed directly to the corresponding selenophthalide in a one-pot reaction with H_3PO_2[325]. The thione corresponding to **126** was obtained by treating **125** with hydrogen sulphide[323]. Dipole moments and relative basicities of the carbonyl group were studied in order to obtain information about the change in the electron density when replacing the O of phthalide by S and Se, in the ring as well as in the carbonyl group[326]. The 1,3-dioxo-1,3-dihydrobenzo(*c*)selenophene (2-selenophthalic anhydride) was obtained from phthaloyl dichloride, either with H_2Se in the presence of $AlCl_3$[327], or in 61% yield with Se and sodium borohydride in phase-transfer catalysis[328].

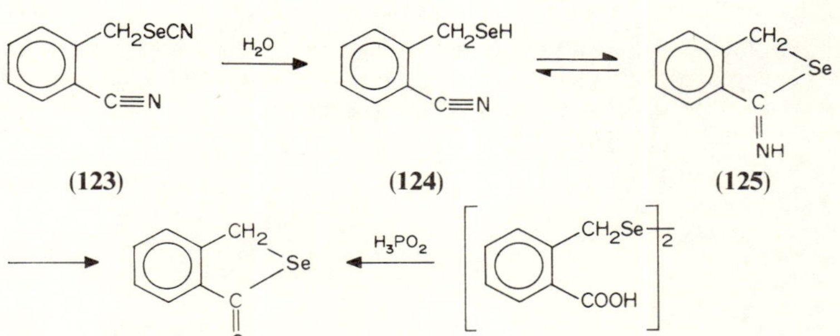

(123) **(124)** **(125)**

(62)

(126) **(127)**

Some isobenzoselenophene quinones (**130**) and their O and S isologues were obtained by a general method starting from the reaction of bis-acetylenic γ-diketones **128** with tris(triphenylphosphine)rhodium(I) chloride and chalcogen[329–332] (equation 63).

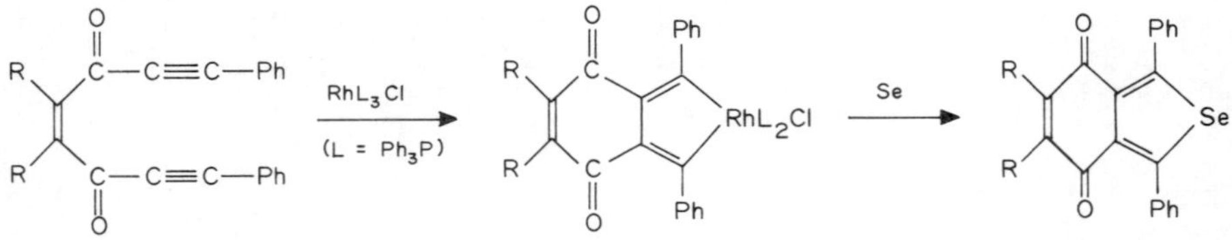

(128) (129) (130) (63)

c. Dibenzoselenophene and related compounds. Dibenzoselenophene (**132**) has been first prepared by cyclodehydration of diphenyl selenoxide with sodium amide[333]. Other syntheses giving low yields are a Pschorr reaction on diazotized diphenyl selenide or the heating treatment of *o*-biphenylselenium trihalide with KOH[334]. Better yields are obtained by cyclodehydrohalogenation of the unstable diphenylselenenyl halide **131**, prepared from *o*-aminobiphenyl[335] (equation 64). Analogously the selenium dichloride of phenyl 2-biphenylyl selenide gives 5-phenyldibenzoselenophenium chloride[336]. Dibenzoselenophene has also been obtained by heating dibenzothiophene *S,S*-dioxide with Se, from selenanthrene and copper-bronze[337], from the 2,2′-biphenylylenmercury tetramer and Se[338], or by Se insertion into biphenylene[339]. The selenanthrene route was also applied to the synthesis of perfluorodibenzoselenophene[340]. Dibenzoselenophene selenium-oxide was prepared from the selenium-dibromide by alkaline hydrolysis[341], or by direct oxidation with peracetic acid[337]. The *N*-tosylimide was also obtained from the parent heterocycle[338]. Dibenzoselenophene-*Se*-oxide can transfer oxygen photochemically to aryl alkyl sulphides[342]. Very few reactions have been tried on dibenzoselenophene. Nitration gives 2-nitro- and by subsequent reduction 2-aminodibenzoselenophene[343,344]. Friedel–Crafts reaction, followed by Haworth and Fischer synthesis, was realized[345,346] on position 2. Metalation was realized at position 4, and followed by carbonation it gave dibenzoselenophene carboxylic acid. Deselenization to biphenyl occurs with Raney nickel[347].

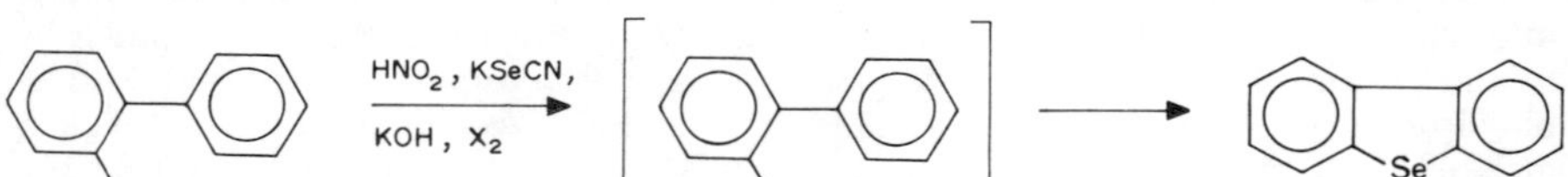

(131) (132) (64)

2-Allylbenzo(*b*)selenophenes **133**, prepared from the 2-lithio- and 2-lithio-5-methyl analogues, are cyclized, following Bradsher reaction with ethyl dichloroethoxy acetate, to 1-ethoxycarbonyldibenzoselenophene (**134**) and its 8-methyl analogue, which can be easily hydrolysed and decarboxylated[348] (equation 65).

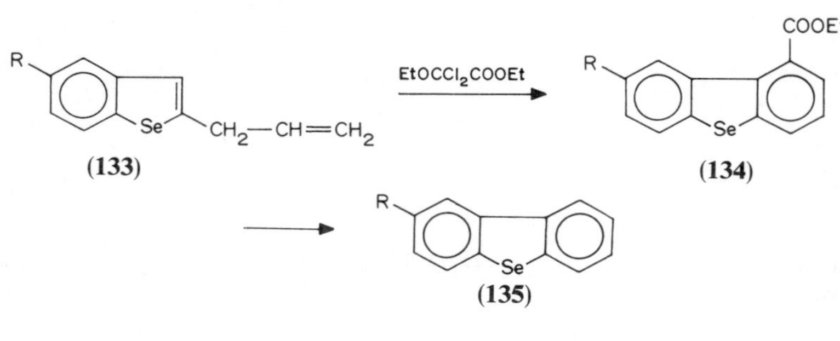

(133) (134) (135) (65)

Esters **134** can be transformed in nine steps to the 1-γ-carboxypropyl derivatives which, after cyclization, Wolff–Kishner reduction and aromatization, lead to the tetracyclic benzo(*b*)naptho(1,2-*d*)selenophene[349].

The 2,3-naptho isomer (**139**), and other Se analogues of benzo(*a*)anthracene and dibenzoanthracenes, were obtained by cyclodehydration of 2-benzyl- or 2-α-naphtylmethyl-3-formyl-benzo(*b*)selenophenes **138**, obtained in turn by Rieche formylation of the corresponding 2-benzyl compounds **137**. Compounds **137** are easily prepared by reduction of selenoaurones **136** (equation 66)[350]. A generalization of this reaction leads also to a Se analogue of cholanthrene[351].

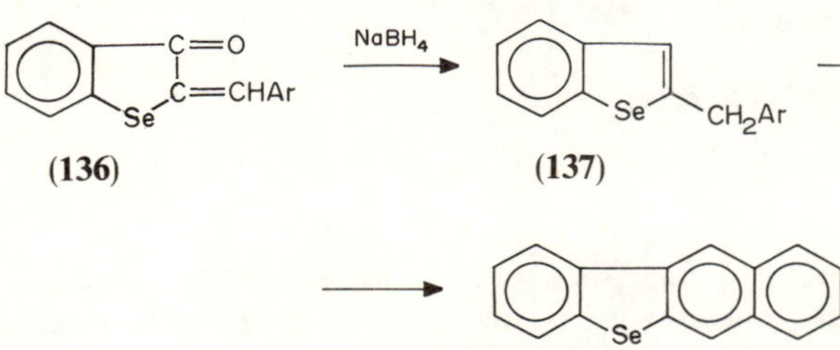

(**136**) (**137**) (**138**)

(**139**) (66)

A Robinson–Stork stereospecific annelation of 3-pyrrolidino-benzo(*b*)selenophene or -benzofuran (**140**)[352] by methyl vinyl ketones gives a 2-oxo-1,2,3,4-tetrahydrodibenzoselenophene or its dibenzofuran analogue (**141**)[353] (equation 67).

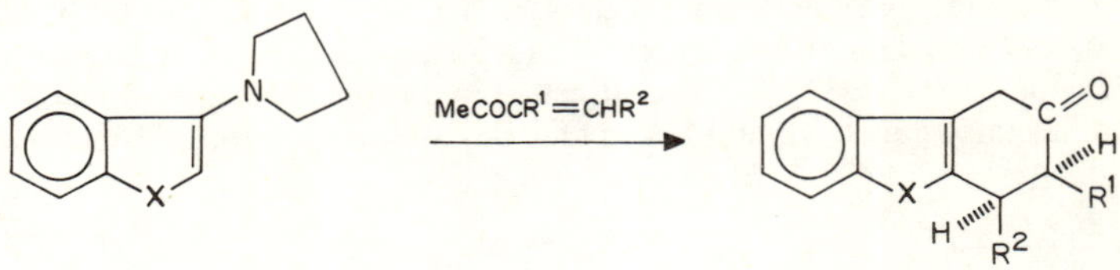

(67)

(**140**) (**141**)

X = O, Se

3. Hetero-condensed selenophenes

a. Selenophenes condensed to five-membered non-nitrogen heterocycles. (*i*) *Selenolofurans.* These are usually obtained from the oxygen heterocycle. For example, selenolo(3,2-*b*)benzofuran **144** (Y = O) was obtained by the general Fiesselmann–Hauptmann reaction of chlorovinyl aldehydes, Na_2Se and ethyl bromoacetate, followed by carbocyclization[354] (equation 68). However, this reaction has failed for the naptho (2,3-*b*)furan analogue of

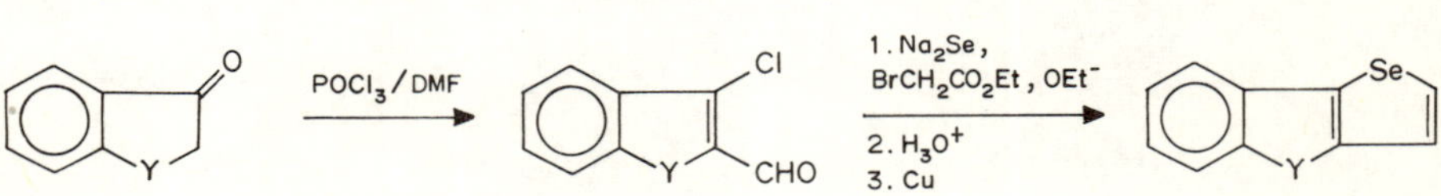

(**142**) (**143**) (**144**) (68)

143[355]. 2-Phenylselenolo(2,3-*b*)furan is obtained similarly, an imino intermediate being used instead of the formyl group[356,357]. 4-Phenylselenolo(3,4-*b*)furan (**146**)[358] and its (*d*)benzologue[359] are obtained from selenoacetamide with 2-bromomethyl-3-benzoylfuran (**145**) or benzofuran (equation 69). 2-Chloromethyl-3-formylbenzofuran gives no corresponding reaction.

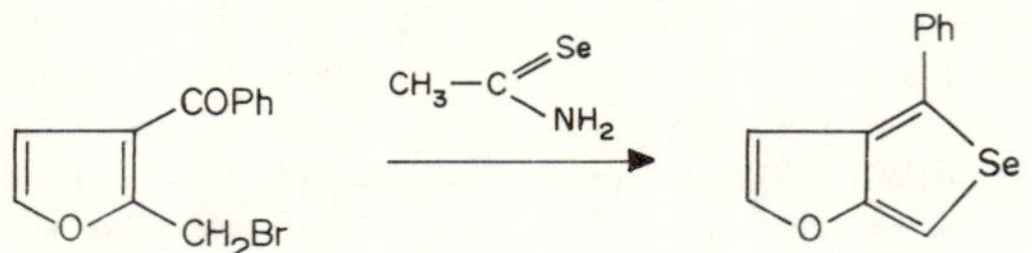

(145) (146) (69)

The 1H,3H-4,6-diphenyl derivative of the non-classical selenolo(3,4-*c*)furan is obtained by a Rhodium method similar to equation (63)[360]. Starting from a selenophene, in one case only a condensed derivative, i.e. the selenolo(3,2-*b*)furan derivative **148** is obtained[361] (equation 70). Decarboxylation of **148** has failed.

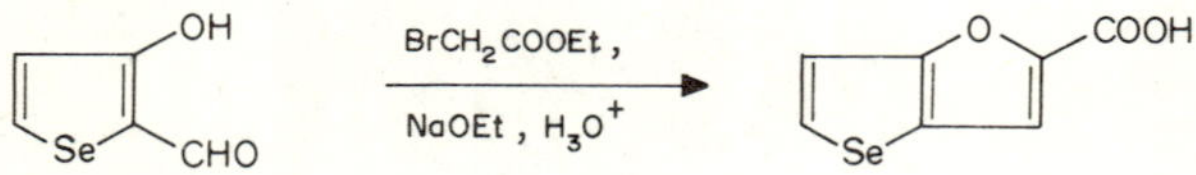

(147) (148) (70)

The three γ-lactones at the 2,3- and 3,4-positions of selenophene, regarded as oxodihydroselenolofurans, are obtained from the corresponding hydroxymethylcarboxyselenophenes[362].

(*ii*) *Selenolothiophenes*. These have been more studied. There are four possible classical selenolothiophenes, where no tetracovalent S or Se atom appears in the noncharged structure: the (2,3-*b*) (**149**), (3,2-*b*) (**150**), (2,3-*c*) (**151**) and (3,4-*b*) (152) fused systems. Generally, their syntheses start from the more accessible thiophene compounds. **149**[363] and **150**[364] were prepared by a Perkin-like cyclization of the appropriate formyl thienylseleno acetates, in a reaction parallel to the last steps of equation (47). The 2-phenyl derivative of **149** was similarly prepared[356]. Two reviews have been published on the chemistry of these systems[365,366]. The direct preparation of its precursor 3-formyl 4-thienylselenoacetate failed, but **151** can be obtained from 3-methylseleno-4-formylthiophene[367].

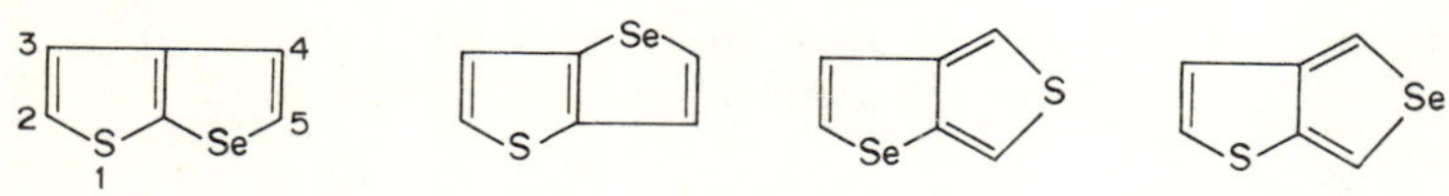

(**149**) (**150**) (**151**) (**152**)

There are only a few recent publications in this field. Application of the $SOCl_2$ Krubsack method of cyclization of arylpropenoic acids to 2-[368,369] and 3-selenienylpropenoic acids[370] gives only the 2-chlorocarbonyl-3,5-dichloroselenolo(3,2-*b*)thiophene (**154**) or the corresponding (2,3-*b*) isomer, in low yields (equation 71). However, by treating the

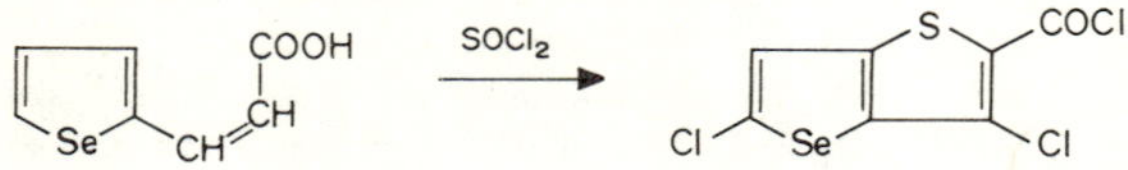

(**153**) (**154**) (71)

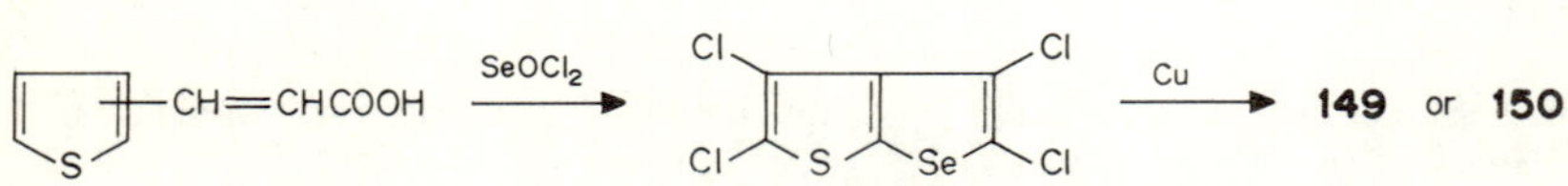

(72)

(**155**) (**156a**) or isomer (**156b**)

corresponding thienylpropenoic acids with $SeOCl_2$, completely chlorinated cyclized compounds **156** can be obtained. Complete dechlorination can be achieved by copper, to give the parent heterocycles **149** and **150** in 60% yields[371] (equation 72). A 6-methoxy derivative of **150**, i.e. (**159**) could be obtained from 2-acetyl-3-aminothiophene (**157**) (equation 73). This method has also been applied to the corresponding methoxyselenoloselenophene[149].

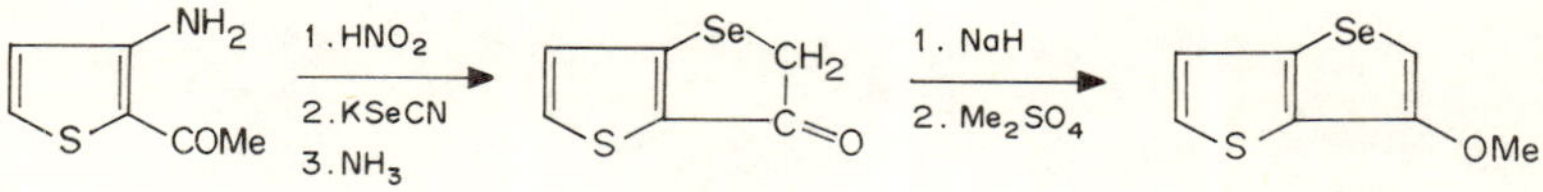

(**157**) (**158**) (**159**) (73)

The second methodology, starting from a selenophene ring, and applying the benzothiophene-like synthesis from cyclodehydration of arylthiomethyl ketones, has led recently to the synthesis of derivatives of **149**, substituted on the selenophene ring[372]. Metalation of **150** is realized non-selectively on the two α-positions. However, the (2,3-*b*) isomer **149** gives an interesting ring-opening characteristic of the Se-containing ring, which was identified as a R.O.II selenophilic substitutive type[373,374] (equation 74). The driving force of this ring-opening is the larger polarizability of Se compared with S, combined with the formation of a stabilized α-carbanion on the thiophene ring.

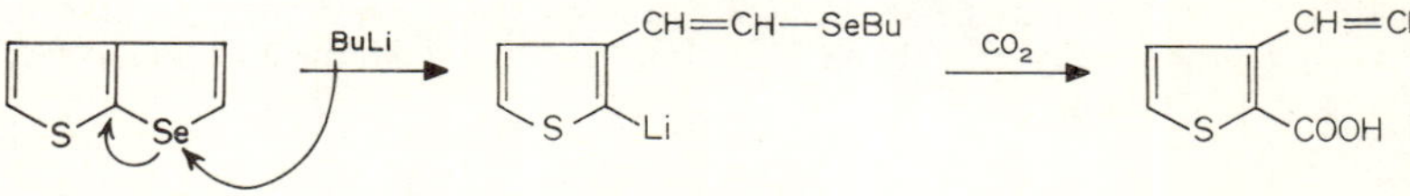

(**149**) (**160**) (**161**) (74)

The R.O.II reaction has been extended to various '*cis*-fused' tricyclic systems **168** such as selenolo(2,3-*b*)benzothiophene and benzoselenolo(2,3-*b*)thiophene. In all cases, regiospecific ring-opening of the Se ring has been realized, regardless of the nature of its central or lateral position[374]. A ring-opening of the selenophene ring of the more strained isomer selenolo(2,3-*c*)thiophene (**151**) led to the formation of a non-stabilized β-carbanion on the thiophene ring[375] (equation 75).

(**151**) (**162**) (75)

Bromination of **150** by bromine gives 70% of the 5-bromo (α to Se) and 25% of the 2,5-dibromo derivatives, but no 2-bromo derivative. Bromination with *N*-bromosuccinimide gives a 65% yield of a 13:87 mixture of the 2-bromo and 5-bromo derivatives[376]. The Vilsmeier–Haack reagent also reacts preferentially on the selenophene ring, giving a 35:65 ratio of attack at the 2- vs. the 5-position. Acetylation of **150** at position 5 is 1.29 times faster than in the 2-position. The corresponding ratio of 1.89 for reactions at position 5 and at the α-position of the corresponding thieno(3,2-*b*)thiophene[377] demonstrates the larger reactivity of this condensed selenophene ring. The Vilsmeier formylation of **151** gives a mixture of 4- and 6-substitution, in a 60:40 ratio[375]. The transmission of substituent effects in thiophene, selenophene and their condensed analogues[378], and a quantum-mechanical analysis of ^{77}Se chemical shifts of these compounds[379] have been recently studied.

The first derivative of the fourth classical isomer **152**, a selenolo(3,4-*b*)thiophene bearing a 2-thienyl group in position 4 and a diselenide in position 6, has been obtained from a 2-thienyl-substituted phosphorane and $NaHSeO_3$[380]. Unsubstituted **152** was obtained in 1982. The methodology used for the preparation of the corresponding thienothiophene was used (equation 76). The critical steps were the formation of **164**, which was obtained with a 15% yield and gave predominantly the dimer, and the hydrolysis of the ester. A

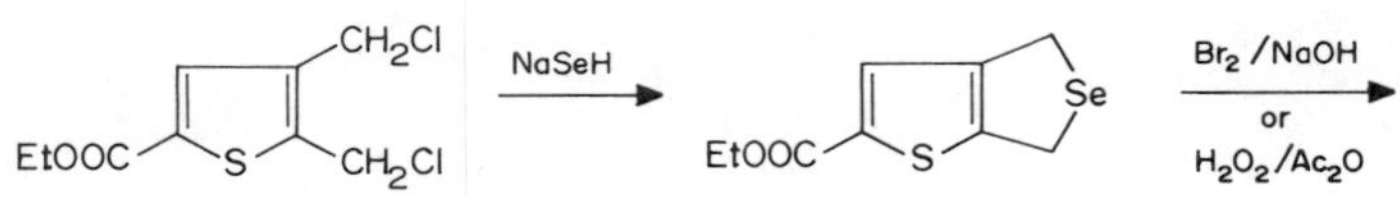

(**163**) (**164**)

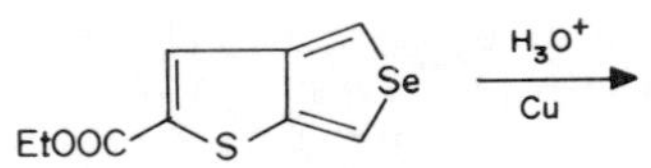

(**165**) (76)

better yield was obtained from 3-formyl-4(methylthio)selenophene, followed by reaction on S with methyl bromoacetate, and cyclization[381].

The non-classical (3,4-*c*) isomer **167** has been prepared by a reaction similar to equation (76) from a 3,4-bis(chloromethyl)thiophene (equation 77). This unstable compound has been trapped by *N*-phenylmaleimide[382].

 (77)

(**166**) (**167**)

The two benzo-condensed *cis*-fused selenolothiophenes (**168a, b**) and the two corresponding *trans*-fused isomers (**169a, b**) have been synthesized by the classical Perkin-like ring-closure reaction and the process extended to the selenolobenzoselenophene analogues **168c** and **169c**[383–385]. Intermediates of the (3,2-*b*) condensation are obtained more easily by Vilsmeier Haack formylation of indoxyl analogues[384,385].

(**168**) (**169**)

(**a**) X = S, Y = Se (**b**) X = Se, Y = S (**c**) X = Y = Se

The electrophilic substitution of **168** and **169** gives substitution only at the 2-position[386]. Tetracyclic (1)benzoselenolo(2,3-*b*)(1)benzothiophene is obtained by cyclo-

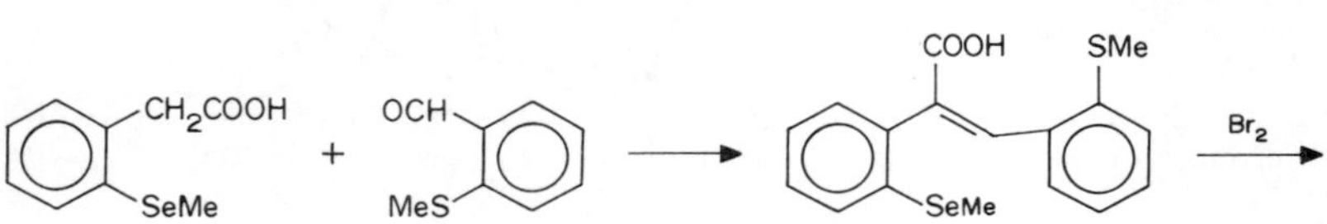

(**170**)

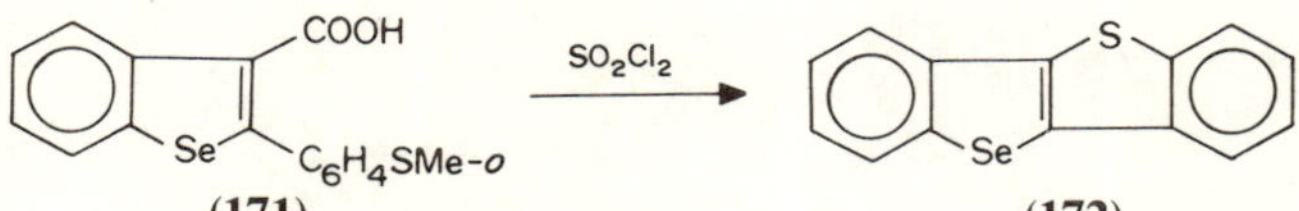

(171) (172)

(78)

dehydration of α-(2-benzoselenienylthio)cyclohexanone followed by aromatization; its (3,2-*b*)-*trans* isomer (**172**) is obtained by applying the new cyclization method of *o*-halochalcogenenyl styrenes (equation 78). The two steps of the cyclization can be easily separated, since Br_2 reacts only on seleno ethers[387].

Thio- and seleno-lactones fused to the heterocycle of benzo(*b*)thiophene and benzo(*b*)selenophene, which can be regarded as oxo derivatives of 1,3-dihydro(3,4-*b*) benzo-condensed selenolothiophenes, are obtained in one step by H_3PO_2 reduction of the *o*-cyanohetaryl chalcogenocyanates[325].

(*iii*) *Selenoloselenophenes*. There are four possible isomeric selenoloselenophenes **173**–**176**, including the 'non-classical' isomer **176**. Umezawa was the first to search for selenoloselenophene among the by-products of the reaction of Se with acetylene, and to suggest the possible existence of **173**–**175**[388] but his structural assignment has been shown to be erroneous. It was later shown by NMR and dipole moment measurements, and by independent synthesis that only the (3,2-*b*) isomer **174** was present among 33 other products[248,389]. The syntheses of the isomers **173**–**176** are parallel to those of the selenolothiophene analogues: the isomers (2,3-*b*)-**173** and (3,2-*b*)-**174** were synthesized in 1974 from the suitable selenienyl-1,3 dioxolanes via *o*-formyl selenienylselenoacetates[248]. The third classical (3,4-*b*) isomer **175** was prepared in 1980[390] by a reaction similar to that presented in equation (76).

(173) (174) (175) (176)

Evidence has been obtained for the transient existence of the non-classical isomer **176**[182]. A particular synthesis of a dihydro derivative, the 1,1,4,4-tetramethyl-1*H*, 4*H*-selenolo(3,4-*c*)selenophene (**178**) and of its dithio analogue, was realized from an allenic selenocyanate or thiocyanate **177**, via the cyclization of diseleno- or dithio-bridged

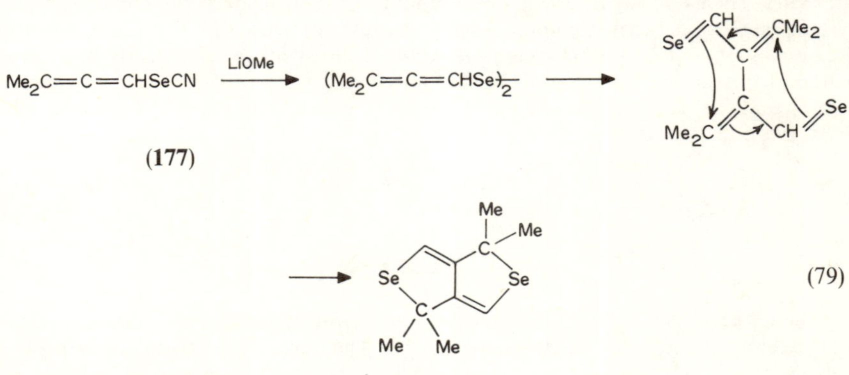

(177)

(178)

(79)

diallenyls[391] (equation 79). Another dihydro derivative, the 1*H*,3*H* system **179**, was obtained by an application of the Paal–Knorr synthesis of monocycles[392] (equation 80). Under the same conditions, P_4S_{10} gives the entirely aromatic ring. The selenoxide **180** has been prepared in order to assign the *cis* or *trans* configuration to **179**.

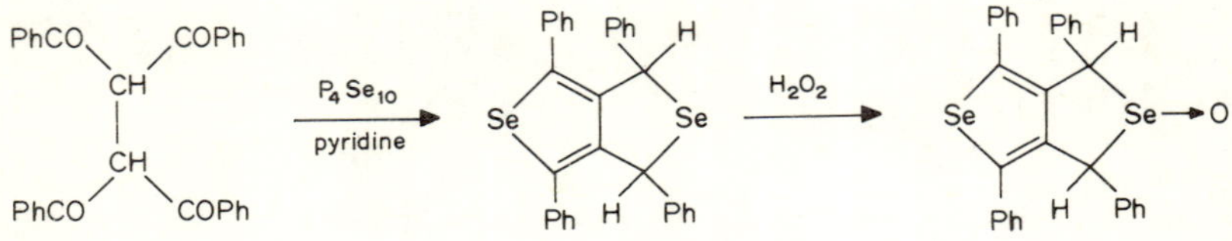

(80)

(**179**) (**180**)

The electrophilic substitution of **173** and **174** (Vilsmeier formylation, chlorination with *N*-bromosuccinimide, iodination by iodine–iodic acid), gives only 2-substituted products in good yields[369,393]. Acetylation of **174** gives the expected 2-acetyl derivative, while **173** gives only the 2,5-diacetyl derivative. Nitration gives a low yield of the 2-isomer, but it is the only electrophilic substitution that gives detectable 3-substituted isomers[394]. The 3-bromo derivative of **173** has been obtained by successive halogen–lithium exchange of 3,4-dibromoselenophene, reaction with dimethylacetamide, metalation of the intermediate in the 2-position, reaction with Se and methyl bromoacetate and cyclization[395]. Vilsmeier formylation of **175** gives a mixture of the 4- and 6-formyl derivatives in a 55:45, ratio, a parallel reaction as regards the sites of electrophilic attack and the ratios of the derivatives, to those on the corresponding thienothiophenes and selenolothiophenes. These results are also consistent with the quantum-chemical analysis of Litvinov and coworkers[396]. Substitutive selenophilic ring-opening of selenolo(2,3-*b*)- and -(3,4-*b*)-selenophenes are also parallel to similar reactions in selenolothiophenes[394] (equations 74 and 75).

Similarly, tricyclic benzo-condensed '*cis*' selenoloselenophene gives the same type of ring-opening where the less aromatic central selenophene ring is always opened regiospecifically (equation 81)[374]. The tricyclic systems themselves **168c** and **169c**, are

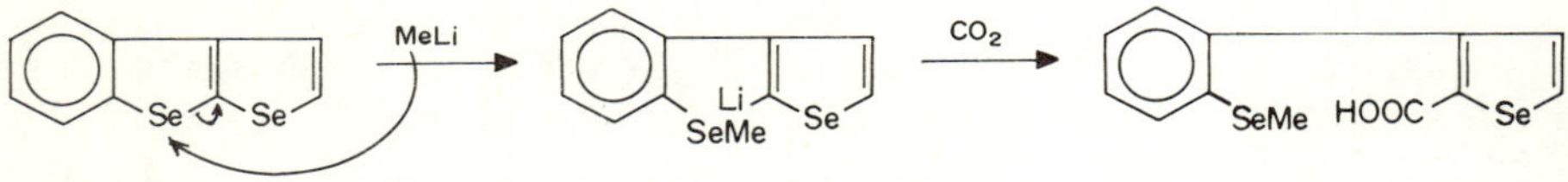

(**168c**) (**181**) (81)

obtained as mentioned before[383] and the (3,2-*b*) isomer is also obtained in fewer steps by a Vilsmeier–Haack formylation of selenoindoxyl[354]. Tetra- and poly-condensed (2,3-*b*) systems, such as (1)benzoselenolo(2,3-*b*)(1)benzoselenophene (**183**; R = H) or substituted derivatives or their naphtho analogues were obtained in 28–87% yields by a general method from diarylethylenes and $SeOCl_2$, specific for the Se series[397–399] (equation 82). $SOCl_2$ and $TeCl_4$ react on the same alkenes to give only acyclic thio and telluro derivatives, respectively.

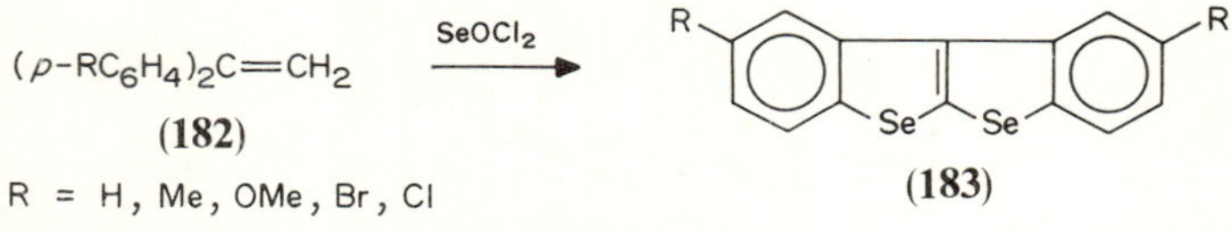

(82)

(**182**)

R = H, Me, OMe, Br, Cl

(**183**)

A heptacyclic system, containing four selenophene rings was obtained by the same methodology from 1,4-di-α-styrylbenzene[399]. The tetracyclic '*trans*' isomer of **183**, (1)benzoselenolo(3,2-*b*) (1)benzoselenophene, was obtained as a by-product from reaction of bromoacetic acid with *o*-methylselenobenzaldehyde, and a mechanism was proposed[267]. Charge-transfer complexes of benzo(*b*)selenophene, selenolo(3,2-*b*)thiophene and selenolo(3,2-*b*)selenophene with tetracyanoethylene have been investigated by UV and IR spectroscopy, and association constants, enthalpies of formation and ionization potentials of the donors determined[400].

b. Selenophenes condensed to six- or seven-membered non-nitrogen heterocycles. A few monocyclic selenophenes condensed to non-nitrogen six-membered or larger rings are known. 2-Oxo-selenolo(3,2-*b*)pyran (**184**), a selenophene analogue of coumarin, has been obtained from 3-hydroxy-2-formylselenophene. The diketo seven-membered oxacycloheptenedione derivative **185**[205] and a selenolopyrylium salt **186** have also been

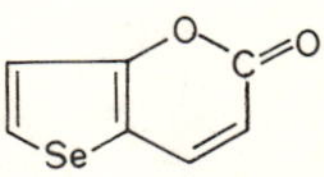

(**184**)

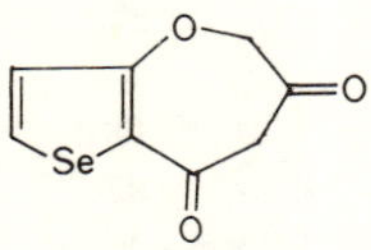

(**185**)

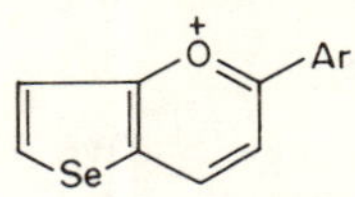

(**186**)

prepared[208,401]. Other selenolopyrylium salts **187**–**189** with isomeric skeletons have been synthesized from disubstituted selenophenes, and transformed, either to hydroxy and dialkylaminobenzoselenophenes or to selenolonitrogen heterocycles[251,402].

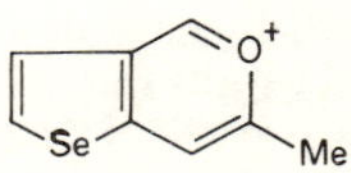

(**187**)

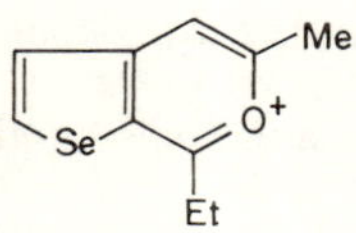

(**188**)

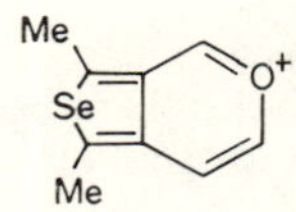

(**189**)

A selenolothiopyranone, 6-methyl-4*H*-selenolo(2,3-*b*)6*H*-benzothiopyran-4-one and several selenoloselenopyranones[403,404] have been obtained by photochemistry of 3-bromo selenophene-2-thio- or -2-selleno carbonic acid -*S*- or -*Se*- aryl esters. 8H-Diselenolo(3,2-*b*:2′,3′-*e*)thiopyran-8-one is obtained by thermolysis of 3*H*-selenolo(2,3-*d*)-1,2-dithiol-3-one, prepared by reaction of 4-mercaptopyridine with 2-chlorocarbonylselenophene[405]. More systematic syntheses are based on benzo(*b*)selenophene and have been realized by the group in Metz. From 2-mercaptobenzo(*b*)selenophene, obtained by sulphurization of the 2-lithio derivative, some benzo(*b*)selenienylmercaptoalkanoic acids **190** are obtained. Cyclization of **190** gives the dihydrothiopyranone **191a** or thiepinone **191b**, and by standard transformation the corresponding very unstable thiopyran analogue is obtained[406] (equation 83). The tetracyclic thiepinone **192** is obtained by a similar reaction[407]. Benzoselenolo(2,3-*c*)condensed thiin and thiepin are obtained by

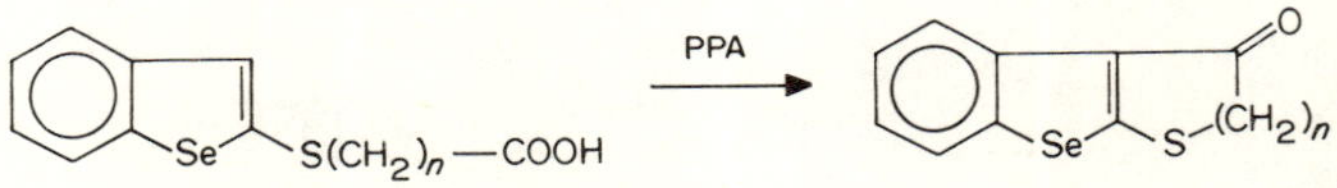

(83)

(**190**) (**a**) $n = 2$ (**191**)

(**b**) $n = 3$

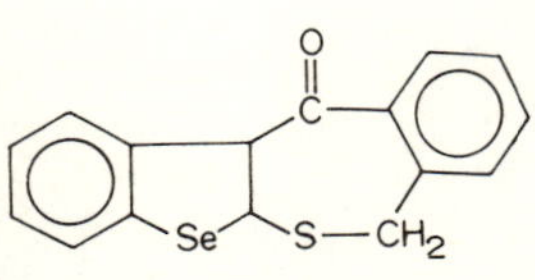

(**192**)

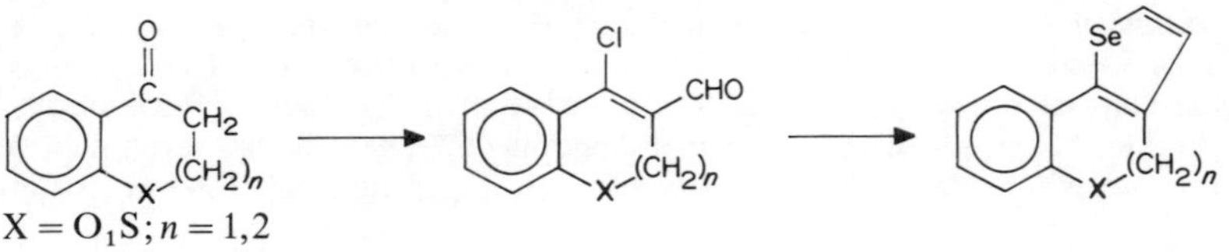

(84)

$X = O_1S; n = 1,2$

(**193**) (**194**) (**195**)

similar intramolecular Friedel–Crafts ring-closure from 2-benzoseleni-enylmethylmercapto-acetic and -propionic acids[408]. The (3,2-*b*) isomers and their Wolff–Kishner reduction products are similarly obtained, from 3-carboxyalkylmercaptobenzo(*b*)selenophenes which are obtained in turn from reaction of thioglycolic or *β*-mercaptopropionic acids with selenoindoxyl[409]. Cagniant and coworkers have applied their selenophene synthesis to the Vilsmeier chloroformylated products obtained from chromanone, thiochromanone and benzothiepinone[355] (equation 84). A similar selenophene-forming condensation gives from dibenzo(*b*, *f*)oxepinone the corresponding selenolo(2,3-*d*) condensed system[410].

c. Selenophenes condensed to nitrogen heterocycles. The first methodology for the synthesis of selenolo nitrogen heterocycles starts from a suitably substituted selenophene ring. The nitrogen heterocycles are built by the classical methods used for the corresponding benzo-condensed nitrogen systems. The principal difficulties arise from the accessibility of the convenient mono- or di-substituted selenophenes, and from the relative unstability of selenophene derivatives towards some cyclization reagents. Selenolo-(2,3-*b*)- (**199**) and -(3,2-*b*)-pyrroles (**198**) have been obtained by the Hemetsberger cyclization of ethyl azidoaryl acrylates **196** (equation 85)[411,412]. The ester **197** has also been

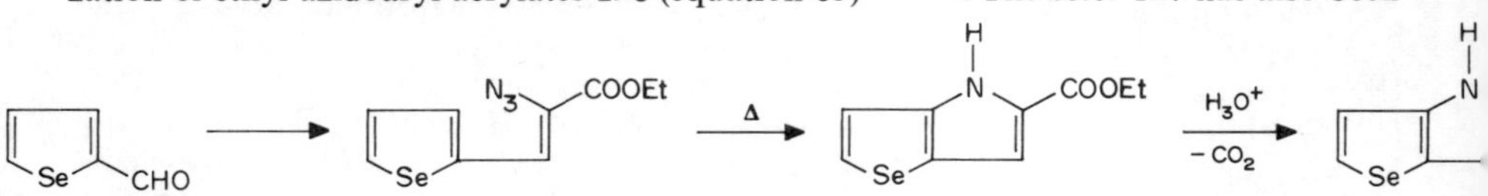

(**196**) (**197**) (**198**)

(85)

obtained by cyclization of a disubstituted selenophene, ethyl 3-(2-formylselenienyl)-*N*-acylaminoacetate[412]. The synthetic utility of the azido group for synthesis of condensed pyrroles is also demonstrated in the thermal cyclization of 2-(3-azidoselenienyl)vinyl alkyl ketones **201** to **202**[413] (equation 86). Compound **200** is also the precursor for obtaining

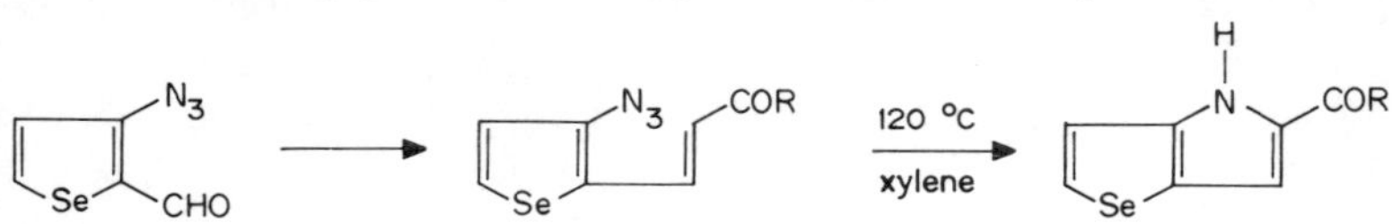

(86)

(**200**) (**201**) (**202**)

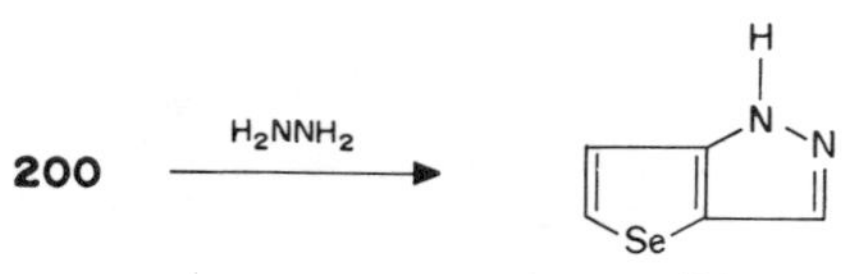

(87)

(**203**)

selenolo(3,2-*c*)pyrazole (**203**) (equation 87)[414] by application of the recent Rees synthesis of indazole, for which a mechanism involving a pentazene intermediate has been postulated. This type of heterocycle has been formylated by Vilsmeier–Haack reaction in the position α to Se, a reaction parallel to that for the analogous thio compound[415]. Compound **200** or its acetyl analogue (**204**; R = Me) can also be thermally cyclized to selenolo(3,2-*c*)isoxazoles **205** (R = H, Me)[416,417] (equation 88). This biheterocycle is acylated in a position β to Se[417].

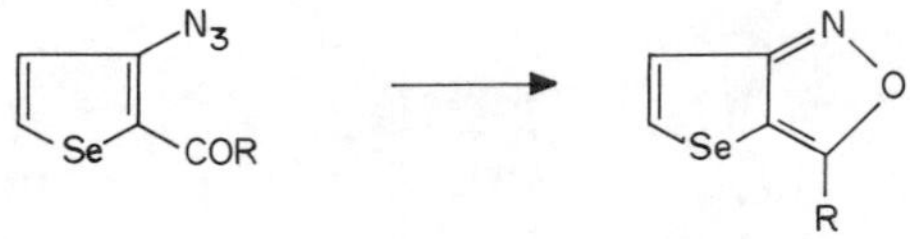

(88)

(**204**) (**205**)

Compound **206** is cyclized by heating at 220 °C to 2-acetylaminoselenolo(3,2-*d*)thiazole (**207**). Contrary to its thiophene analogue, compound **207** cannot be transformed to the parent heterocycle[418,419]. By acidic hydrolysis, compound **206** gives the selenolo-

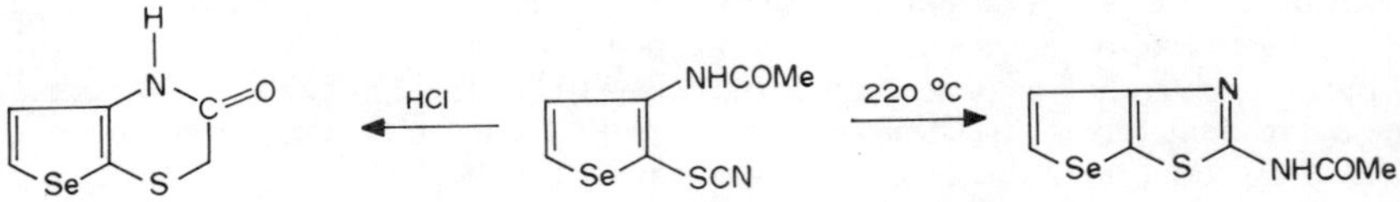

(89)

(**208**) (**206**) (**207**)

thiazinone **208**, whose CO group can be reduced to CH_2 by $LiAlH_4$[419]. The isomeric selenolo(2,3-*d*)thiazole systems, **209a** and **209b** have been obtained by two different syntheses, starting from 2-amino- and 3-bromo-selenophene, respectively[420,421] (equa-

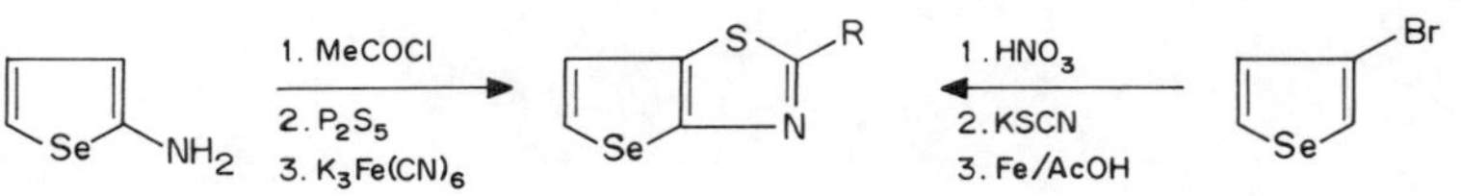

(90)

(**209**)
(**a**) R = Me
(**b**) R = NH_2

tion 90). Deamination of **209b** has failed. A selenothiazinone isomer of **208** can also be obtained[422]. Selenolo(3,2-*d*)isothiazole (**211**) has been obtained in 16% yield by PPA ring-closure of the 2-(methylthio)aldoxime **210**[423] (equation 91). The corresponding isoselen-

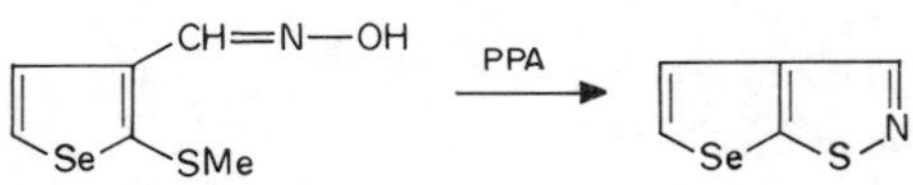

(91)

(**210**) (**211**)

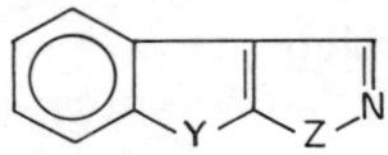

Y = S, Se
Z = S, Se

(**212**)

azole and its (2,3-*d*) isomer are obtained from the suitable methylseleno aldehyde with Br_2/NH_3 under dry ice conditions, with respective yields of 23 and 7%. The method fails for the (3,4-*d*) isomer[424]. Tricyclic benzo-condensed ring systems **212** and isomers are obtained similarly. Ring-opening reactions by methyllithium[425] will be discussed in the section on isoselenazoles. Among the six-membered nitrogen heterocycles, four isomeric selenolopyridines have also been obtained by the methodology starting from selenophene. Ammonolysis of selenolopyrylium salts gives derivatives of selenolo-(2,3-*c*)-[251] and -(3,2-*c*)-pyridines[251,426].

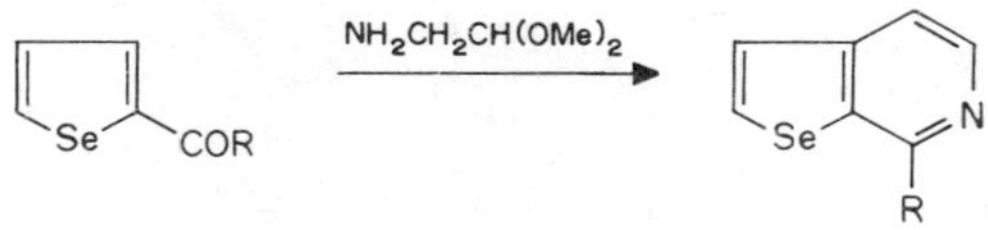

(92)

The corresponding parent heterocycles and their Me derivatives are obtained by a Pomeranz–Fritsch cyclization of 2- and 3-formyl- or -acetyl-selenophenes with the dimethyl acetal of aminoacetaldehyde[427,428] (equation 92). They are nitrated and brominated in a position β to Se. Their two (*b*)condensed isomers are obtained from the hexachlorostannate of 2- and 3-aminoselenophene and malonaldehyde tetraethyl acetal[148,429]; the (3,2-*b*) system is also prepared by the Friedlander method from 3-amino-2-formylselenophene and acetylacetone (equation 93)[430]. Treating **214** by the excess Vilsmeier reagent gives cyclization to 5-chloro-6-formylselenolo(3,2-*b*)pyridine[431]. Deuteriodeprotonation of the selenolo(3,2-*b*) isomer works preferentially in position 3 as in the furo and thieno analogues. The C_3: C_2 reactivity ratio is 10^3 [432]. Compound **214** gives, by reaction with ammonium formate, the selenolo(3,2-*d*)pyrimidine **215**[149,433] (equation 93).

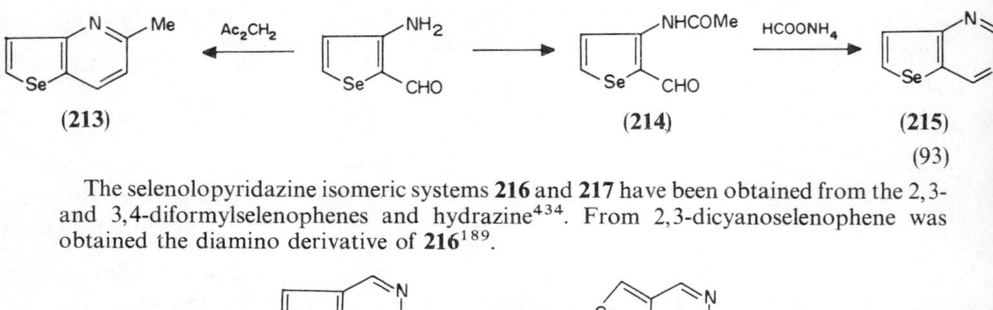

(213) **(214)** **(215)**

(93)

The selenolopyridazine isomeric systems **216** and **217** have been obtained from the 2,3- and 3,4-diformylselenophenes and hydrazine[434]. From 2,3-dicyanoselenophene was obtained the diamino derivative of **216**[189].

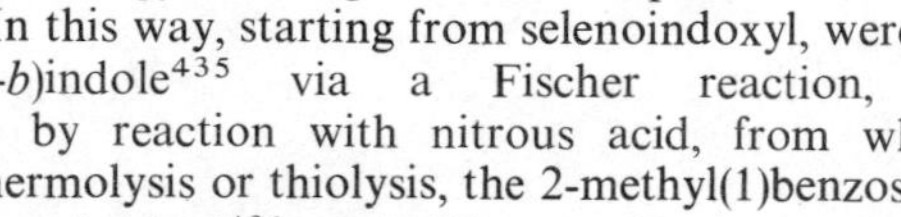

(216) **(217)**

The general methodology starting from selenophene has also been applied to condensed systems. In this way, starting from selenoindoxyl, were obtained: condensed (1)benzoselenolo(3,2-*b*)indole[435] via a Fischer reaction, 2-nitroso-3-hydroxy-benzo(*b*)selenophene by reaction with nitrous acid, from which upon reduction, *N*-acetylation and thermolysis or thiolysis, the 2-methyl(1)benzoselenolo(2,3-*d*)-oxazole and -thiazole were synthesized[436], and (1)benzoselenolo(3,2-*d*)-1,2,3-selenadiazole, in 8% yield[437] from semicarbazone and SeO_2 oxidation. *N*-Methyl(1)benzoselenolo-(2,3-*b*)-(**218**) and -(3,2-*b*)-pyrrole were prepared by reaction of methyl sarcosinate,

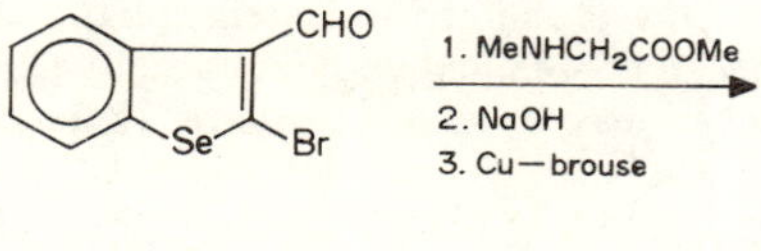

(94)

cyclization and decarboxylation on 2-bromo-3-formylbenzo(*b*)selenolophene, and Vilsmeier chloroformylated selenoindoxyl, respectively[438] (equation 94). In contrast to its isomer, the (2,3-*b*) system gives with methyllithium, a typical substitutive R.O.II of the selenophene '*cis*-condensed' on an aromatic five-membered heterocycle[439] (equation 95).

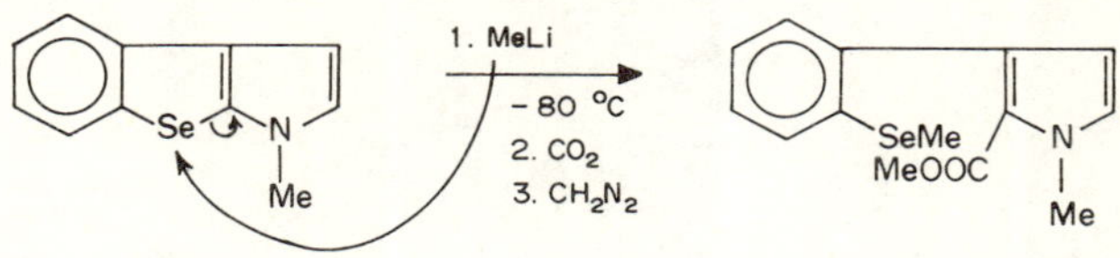

(95)

(218) **(219)**

The second synthetic methodology consists of construction of the selenophene ring by reactions generally used for benzo(*b*)selenophene. For example, from the intramolecular Perkin-like condensation between a formyl group and a selenoacetic derivative, 4-Me-6-substituted selenolo(3,2-*d*)pyrazoles[440], 6-substituted selenolo(3,2-*d*)imidazoles[441] and 5-phenyl-6-methylselenolo(2,3-*b*)pyrrole[442] can be synthesized. By the dehydrohalogenation of 2-selenomercapto-3-bromoacetyl-6-methylpyridine, the 3-hydroxy-6-methylselenolo(2,3-*b*)pyridine can be obtained and reduced to the corresponding 6-methylselenolopyridine[443]. By the bromine–pyridine cyclization of *o*-methylselenostyrenes (equation 50) the first synthesis of the unsubstituted selenolo(2,3-*b*)pyridine was realized[444]. A methyl-substituted (1)benzoselenolo(2,3-*b*)pyridine was synthesized by a Pschorr reaction of the 4-methyl-2-phenylselenopyridine 3-diazonium salt[445]. A Fischer indole synthesis between *o*-(cyanoseleno)benzophenone and phenylhydrazine gives, via dehydrocyanation, the same benzoselenoloindole[446] mentioned previously[435]. Shafiee extended his method of synthesis of selenolo-(3,4)-condensed heterocycles (equation 69) to conveniently substituted *N*-phenyl-indole[358], -thiazole[447] or-1,2,3-thiadiazole[448] precursors. Finally, the synthesis of selenolo(2,3-*b*)quinoline derivatives was achieved in different ways, each starting from a 2-chloroquinoline or its synthetic equivalent. The reactions are summarized in equation (96) ($R^1 = H, Me$). Compounds **221** are obtained,

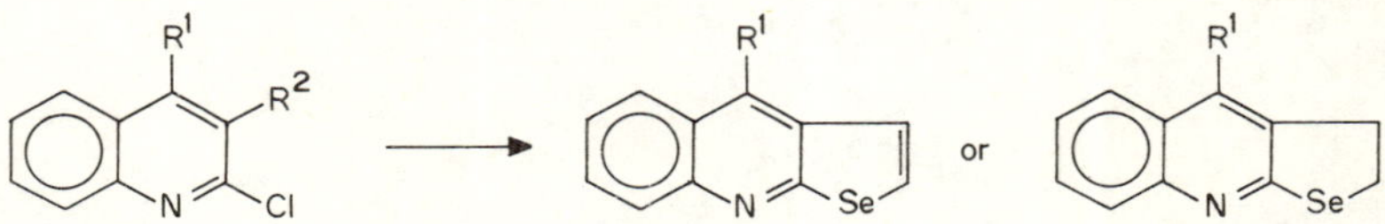

(96)

(220) **(221)** **(222)**

for $R^2 = CH{=}CH_2$ with Na_2Se_2 in ca. 10% yield[449] or for $R^2 = CHBrCH_2Br$, with selenourea/Et_3N[450]. For $R^2 = CH_2CH_2Cl$ with NaSeH[451] or $R^2 = CH_2CH_2OH$, and NaSeH, the necessary 2-chloro compound being prepared in the same pot from the corresponding lactam and $POCl_3$[452], the dihydro system **222** is obtained.

4. Benzo- and dibenzo-tellurophenes

a. Benzo(b) *tellurophene.* The unsubstituted benzo(*b*) tellurophene (**226**) was the first compound of this system to be prepared and it has been known since 1971[453]. The structure of a previous compound, which was claimed to be telluroindoxyl[454], was later demonstrated to be erroneous[453,455]. **226** is obtained with an overall yield of 75% via a C(2)–C(3) ring-closure, following the very easy quaternization of Te (equation 97).

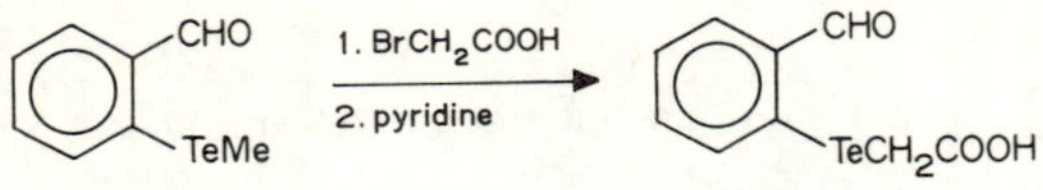

(**223**) (**224**) (**225**)

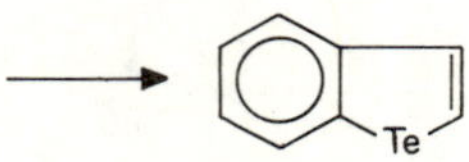

(97)

(**226**)

This method gives directly a very low yield of the 2-acetyl derivative from bromoacetone and is very difficult to apply to *O*-acetyl derivatives from which only 2-carboxy-3-methylbenzo(*b*)tellurophene can be obtained in 10% yield[456]. By standard methods, the acid **225** is transformed to the 2-COCl, 2-$CONH_2$, 2-CONHPh and 2-CH_2OH derivatives[456]. A more general synthesis giving improved yields but generally failing to obtain 3-methyl derivatives (e.g. only 5% yield of 2-carboxy-3-methyl derivative is obtained) uses the intramolecular tellurenyl halide addition–elimination on styrenic compounds, which are obtained in turn by a Wittig reaction (equation 98). Good yields

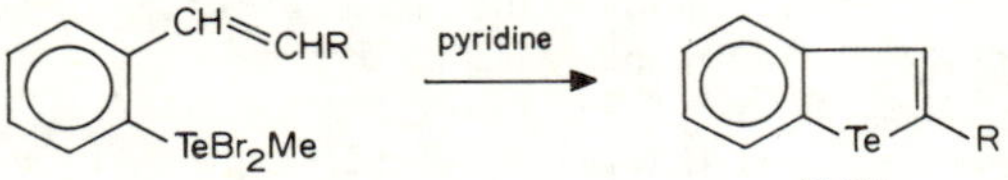

(98)

(**227**)

are obtained for R = COMe, COOEt, CN, COPh, but the reaction fails for R = CHO[456]. Telluroaurones **229** are prepared with 80% yield, by a similar reaction on *o*-cinnamoylbenzenetellurenyl bromide (**228**)[457] (equation 99). Selenoaurones are similarly obtained[457]. 2-Substituted benzo(*b*)tellurophenes (2-NO_2, 2-COOEt, 2-COMe) and

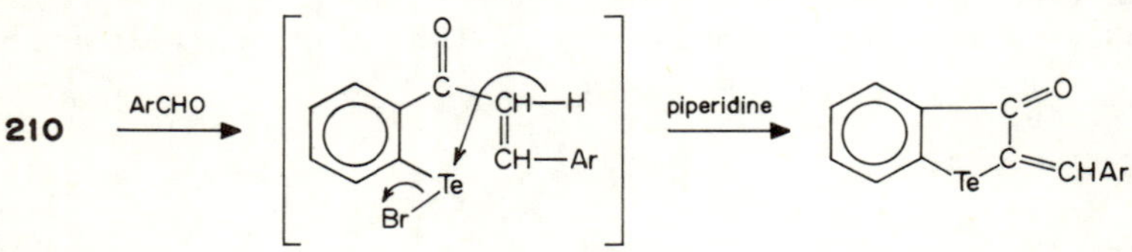

(99)

(**228**) (**229**)

naphtho(1,2-*b*)tellurophene can also be prepared by DDQ aromatization of the tetrahydro or dihydro derivatives, obtained by the general cyclization methods starting from the Vilsmeier products of cyclohexanone, Na_2Te and ethyl bromoacetate[119]. A key intermediate is the Te analogue of indoxyl (**231**), obtained in low yields by C(2)–C(3) ring-closure of the diacid **230**, but more easily by a dehydrohalogenation of *o*-acetylbenzenetellurenyl halides[458] (equation 100) which is easier with larger Te atom than with Se, and also from *o*-diazoacetyl tellurenyl bromide[275]. Telluroindoxyl (**231**) shows only the ketonic form in all solvents by all spectroscopic techniques, and with no structural variation with time. It does not give 3-methoxy benzo(*b*)tellurophene and gives many

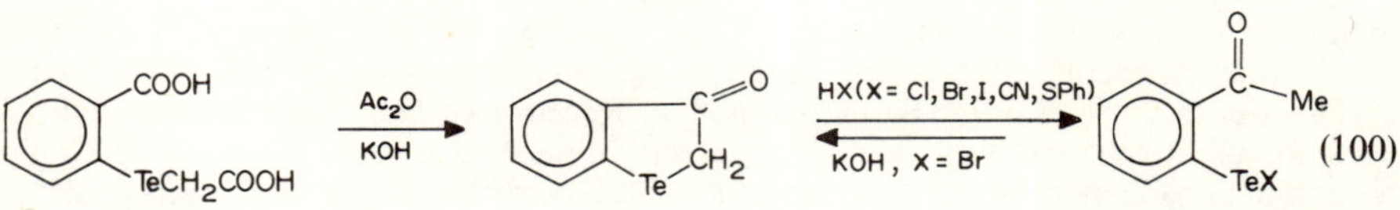

(230) (231) (232) (100)

reactions of the keto form such as formation of oxime, semicarbazone, and phenylhydrazone, followed by Fischer synthesis, reduction to benzo(*b*)tellurophene in low yields[458] and easy oxidation to telluroindigo[458,459], a violet crystalline compound (λ_{max} 637 nm (4.00) compared with λ_{max} 564 nm (4.0) for selenoindigo). It also condenses on the CH_2 group, with aromatic aldehydes to give telluroaurones, with the 3-(C=O) group of isatine, with *p*-nitrosodimethylaniline to give the corresponding anil, and with dimethylacetamide dimethylacetal to give the dimethylaminomethylene derivative[458]. In contrast to selenoindoxyl, **231** is unstable in acidic medium, giving ring-opening reactions by rupture of the Te—C(2) bond (equation 100). The Wittig–Horner reaction gives an interesting rearrangement, characteristic of the Te series, which is the key procedure for obtaining various 3-substituted benzo(*b*)tellurophenes. For example the intermediate **233** has been demonstrated to give a mixture of the prototropic tautomer **235**, and the rearranged product **234**, formed via a ring-opening between Te and C(2) (R = COOEt, CN, COMe)[460] (equation 101).

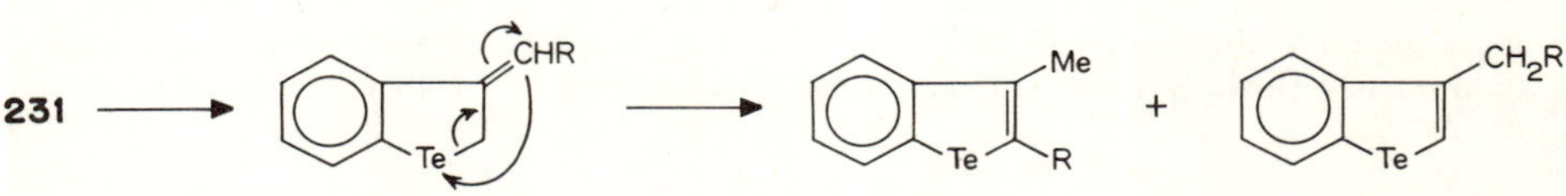

(233) (234) (235) (101)

A **234** derivative (R = COOH), gives 3-methylbenzo(*b*)tellurophene by decarboxylation. The derivative **235** (R = COOH) is transformed with pyridine–*N*-oxide to 3-formylbenzo(*b*)tellurophene, which can be reduced to the 3-CH_2OH, oxidized to the 3-COOH or transformed to the 3-CN derivative, giving access to the 3-$CONH_2$ derivative. 3-Acetylbenzo(*b*)tellurophene cannot be obtained by these standard transformations. As Lewis acids give complexes with benzo(*b*)tellurophene, direct acetylation has been performed with Ac_2O/trifluoroacetic acid. A mixture of 2- and 3-acetyl isomers is obtained in a 76/24 ratio, from which the pure 3-acetyl isomer can be isolated. The corresponding ratios for a similar acetylation of congeners are: O: 73/27; S: 35/65; Se: 65/35. Formylation by hexamethylenetetramine in the same solvent gives a 10% yield of the 2-formyl derivative only. By the same acetylation method, 3-methyl-, 3-bromo- and 2-methyl-benzo(*b*)tellurophenes can be acetylated on the other free heterocyclic position[460]. Halogenation of benzo(*b*)tellurophene gives a very stable dihalogenated Te derivative from which, by successive halogenation and hydrazinolysis, the 2-chloro and 2-bromo derivatives can be isolated. The 2,3-dibromo analogue is obtained by subsequent bromination and hydrazinolysis of the 1,1,3-tribromo derivative. 3-Methylbenzo(*b*)tellurophene gives in the same manner the 2-chloro-3-methyl and 2-bromo-3-

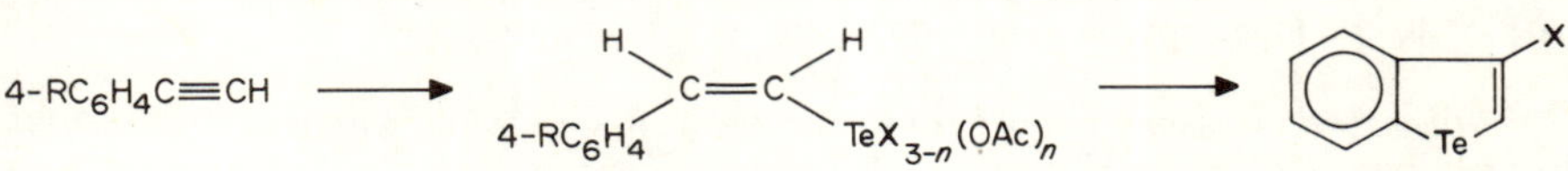

(236) (237)

(102)

methyl derivatives. 3-Monohalogeno derivatives can be obtained, either from **231** and CX_4/triphenylphosphine[460] or by reaction of TeO_2 dissolved in HOAc/LiX with arylacetylenes[461,462] (X = Cl, Br) (equation 102). Reaction of TeX_4 with phenylacetylene does not lead to cyclization products[460] but with diphenylacetylene, 3-chloro-2-phenylbenzo(*b*)tellurophene is obtained after reduction of the dichlorotellurium compound[463].

Compound **237** gives **231** by reflux in CF_3COOH. Whereas **237** gives eliminative R.O.I with butyllithium, even at low temperatures, with formation of a 76% yield of *o*-butyltelluroacetylene[462,464] and only a 10% yield of the 3-carboxy derivative by carbonation, benzo(*b*)tellurophene itself is stable towards butyllithium and is easily lithiated on C(2) like its congeners. From this 2-lithio derivative the following 2-substituted benzo-(*b*)tellurophenes can be obtained[464]: 2-CHO (55%); 2-COOH (40%); 2-Me (28%); 2-Cl (26%); 2-MeS (31%); 2-MeSe (31%). The pK_as of the four 2-carboxybenzo congeners have been determined (O:4.21; S:4.67; Se:4.79; Te:5.12)[465] and compared with the corresponding monoheterocyclic acids. The benzo(*b*) fusion lowers the pK_a by a nearly constant value of 0.35 unit, showing the importance of the $-I$ effect of the heteroatom. The pK_a of the 3-carboxy congeners, on the other hand, give very similar values (O:5.54; S:5.67; Se:5.65; Te:5.79)[460]. The solvolysis of 1-(2-benzoheteroaryl)ethyl acetates, shows the reactivity order Te > Se ~ O > S, with a decrease by a factor 10^2 with respect to the corresponding monocycles[466]. Among aromaticity indices, $\delta_{NMR}(\alpha-\beta)$ give lower values (0.71) than for tellurophene (0.937)[465] but with a very good linear correlation relative to monocycles, with a coefficient of ~0.80. On the other hand, the $J_{5,6}/J_{4,5}$ ratio, proposed by Günther as a measure of the delocalization of benzo-condensed rings[467], gives the following values: O:0.92; S:0.90; Se:0.91; Te:0.91; NH:0.90 which should be compared with the values for naphthalene (0.82) and for a completely localized alkene (0.5)[464].

b. Benzo(c)*tellurophene.* Benzo(*c*)tellurophene itself is unknown. However, tellurolo(*c*)-condensed quinones are obtained by the general method of rhodium intermediates (equation 63)[468,469]. 1,3-Dihydrobenzo(*c*)tellurophene is prepared from α,α'-dibromo-*o*-xylene and Na_2Te[470], or Te and NaI, via the 2,2-diiodo compound[471,472]. The naphtho(2,3) analogue is obtained in a similar manner[470]. These dihydro(*c*)condensed compounds easily extrude Te to give condensed benzocyclobutane[471]. A dioxo-1,3 derivative, tellurophthalic anhydride (**239**; $R_2C=C{=}O$) is obtained by a phase- transfer reaction from phthaloyl dichloride[328] (equation 103). Using

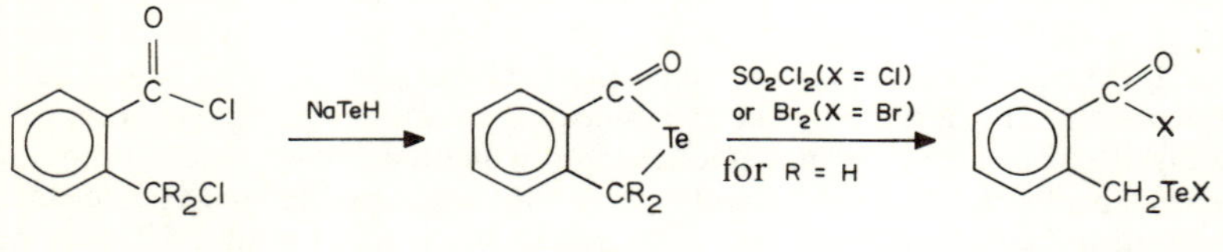

(103)

(**238**) (**239**) (**240**)

the same methodology, 2-tellurophthalid (**239**; R = H), the first example of a five-membered cyclic tellurol ester, is obtained from *o*-bromomethylbenzoyl chloride[473,474]. Its 3-thioxo derivative has been prepared by thionation[474]; 2-tellurophthalid was ring-opened to the first aliphatic *o*-aryl-stabilized tellurenyl halide **240**[474].

c. Dibenzotellurophene. This was obtained for the first time, but with low yield, in 1936 by reaction of TeX_4 with biphenyl[475]. It was also prepared by tellurodesulphonation of dibenzothiophene-*S,S*-dioxide[337] or thianthrene-*S*-tetraoxide[476]. Hellwinkel obtained it and its 3,3-dimethyl derivative in good yields from appropriate biphenylyl mercuric tetramers and Te, or from 2,2′-dilithiobiphenyl and $TeCl_4$[477,478]. Dibenzotellurophene was obtained quantitatively by cyclization of o-(trichlorotelluro)biphenyl at 210 °C[479] and nitrated in position 2. It showed a ring-opening by butyllithium to 2,2′-dilithiobiphenyl and dibutyl telluride[480]. Octafluorodibenzotellurophene was also described[340].

5. *Tellurophene condensed to heterocycles*

Tellurophene has been condensed in some cases to heterocycles. The only monocyclic tellurophene system used as a starting material for fusing another heterocycle is 2,5-diphenyltellurophene. From its 3,4-dichloromethyl derivative and sodium chalcogenide (S, Se) 4,6-diphenyl-1*H*,3*H*-tellurolo(3,4-*c*)-thiophene and -selenophene were obtained[481]. Attempts to prepare the first tellurolotellurophene system for Na_2Te led only to hydrogenolysis of the two dichloromethyl groups, an example of the recently recognized reducing properties of Te^{2-} anions. From 2,5-diphenyl-3,4-diformyltellurophene (**241**) and hydrazine is obtained the tellurolopyridazine **242** (equation 104).

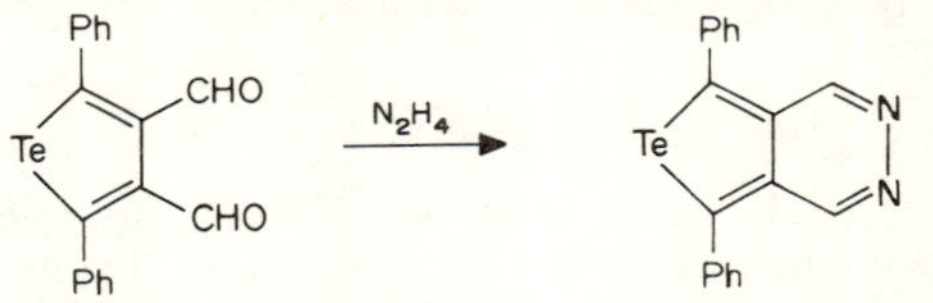

(**241**) (**242**) (104)

By Hinsberg condensation with di(ethoxycarbonylmethyl) sulphide, **241** gives the corresponding tellurolo(3,4-d)thiepin[482]. All other known hetero-condensed tellurophenes are prepared by constructing the tellurophene ring, e.g., the unsubstituted (2,3-*b*)- and (3,2-*b*)-tellurolothiophenes, are prepared similarly to benzo(*b*)tellurophene[245]. The (3,2-*b*) system, in contrast to its Se analogue which is stable under these conditions, shows a tellurophilic substitutive ring-opening of the tellurophene (R.O.II) by organolithium compounds, with the unusual formation of a carbanion at the β-position of the thiophene ring. The two thieno-(3,2-*b*) and -(2,3-*b*) analogues of telluroindoxyl (**243** and **244**) are obtained in good yields by dehydrobromination[464] of the suitable *o*-acetylthio-

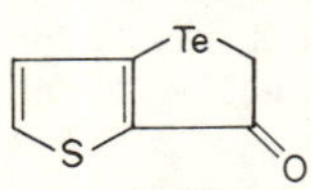

(**243**)

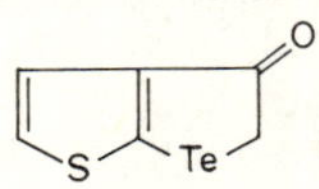

(**244**)

phenetellurenyl bromide[483]. The 5-formyltellurolo(3,2-*b*)thiophene is obtained similary to 2-formyl benzo(*b*)tellurophene by methylene oxidation and ring-contraction of the corresponding 5*H*-thienotellurin[484]. 4,6-Diphenyl-1*H*,3*H*-tellurolo(3,4-*c*)furan is prepared, as its selenophene analogue by the rhodium method[360]. As tricyclic systems, tellurolo(3,2-*b*)-(1)benzofuran and -(1)benzoselenophene are obtained by the general methodology starting from a Vilsmeier–Haack reaction on coumarone and selenoindoxyl[354] (equation 105). 2-Chloro-3-chloroethylquinoline is used as a starting material for tellurolo(2,3-*b*)quinoline[485] or its 2,3-dihydro analogue[486]. A tetracyclic system,

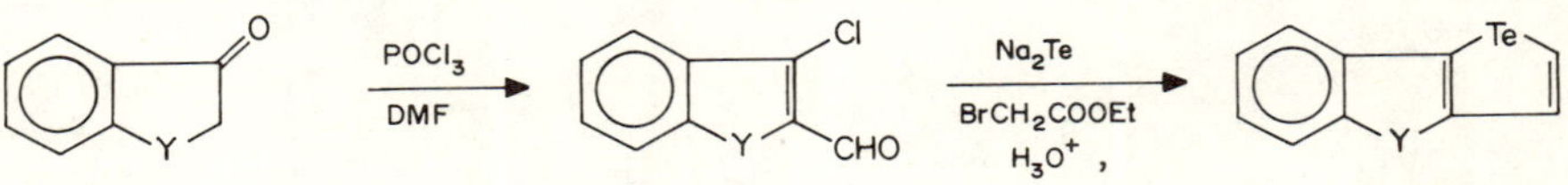

Y = O, Se (105)

(1)benzotellurolo(3,2-*b*)indole is obtained by an appropriate Fischer indole synthesis on telluroindoxyl[458].

B. Two Heteroatoms

1. Se/Te rings without nitrogen

a. Heteroatoms in the 1,2-*position.* (*i*) *Monocyclic systems.* 1,2-Diselenolanes can be considered as cyclic dichalcogenides and can be obtained by oxidative coupling of the intermediate dichalcogenols, prepared by suitable known methods. 1,2-Diselenolane (**246**) has been claimed to be prepared as a yellow powder by hydrolysis of propylene diselenocyanate (**245**)[50,487] (equation 106). Bergson and Claeson[488,489] have suggested that this powder is in fact a dimer or a polymer of **246** the existence of which has been

(106)

(**245**) (**246**)

shown only in solution. The more stable 4,4-dimethyl, as well as the 4-methyl-4-phenyl[490a] and the 3,5-diethoxycarbonyl[489] derivatives are obtained by the same method. The 4,4-dimethyl derivative is oxidized by MCPBA at $-45\,°C$ to the corresponding mono Se oxide, a cyclic selenolseleninate ester, which gives ring opening at $-20\,°C$[490b]. The 1,2-diselenolane system has also been prepared by sodium diselenide nucleophilic substitution of substituted propylene bromides, which give the 4-carboxy and 4,4-bis(hydroxymethyl) derivatives[489,491] or by debenzylation and oxidation of the corresponding dibenzylseleno ethers. A Se analogue of thioctic acid, i.e. a 3-ω-carboxybutyl derivative of **246**, is obtained by these two methods[489,491] and also by using potassium selenite and Se as nucleophiles[492]. The 4-carboxy[493] and 4-aminomethyl mixed 1,2-thiaselenolanes[494] are obtained by the debenzylation method. 3,4-Diphenyl-1,2-diselenol-5-one, recently prepared from diphenylcyclopropenone and selenoamides or NaHSe loses the two Se atoms with hydrazine, giving a corresponding pyrazolinone[495]. Reaction of 2-phenyl-1-diethylaminoethene-1 selenolate salt with CS_2 gives 3-diethylamino-4-phenyl-5-thioxo-1,2-thiaselenole, a ring which loses Se by reaction with dimethyl acetylenedicarboxylate[496]. 4,4-Dimethyl-1,2-oxaselenolane-2-oxide is obtained by H_2O_2 oxidation of 3,3-dimethylselenetane[53].

The first 1,2-diselenolylium system (**248**) was prepared by I_2 oxidation of malonodiselenoamide (**247**)[497] (equation 107). The 3,5-dimethyl analogue of **248** is obtained similarly from diselenoxoacetylacetone and a cobalt(II) salt[498,499]. 1,2-Oxaselenolylium cations are formed by protonation of 1,6-dioxa-6a-selenapentalenes[500]. The first 1,2-oxatel-

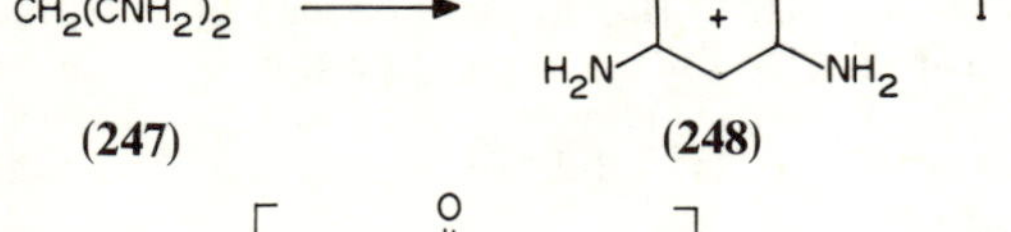

(107)

(**247**) (**248**)

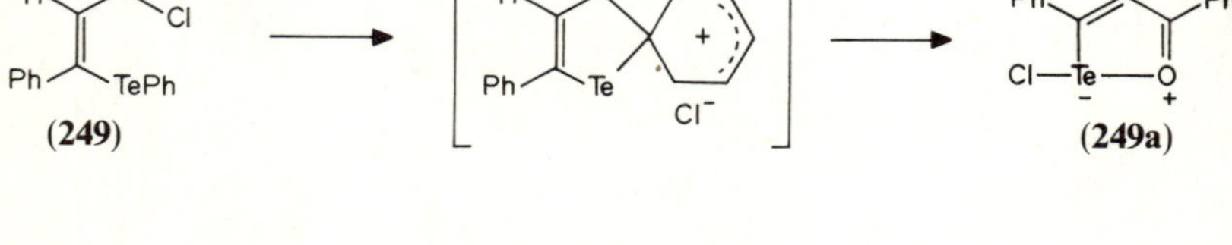

(108)

(**249**) (**249a**)

lurolylium system (**249a**) has been obtained by Lewis acid or thermal catalysed rearrangement of β-(phenyltelluro)cinnamoyl chloride (**249**) via *ipso* attack and Te—C aromatic ring-opening. Evidence for this structure has been obtained by X-ray and spectroscopic analysis[501] (equation 108).

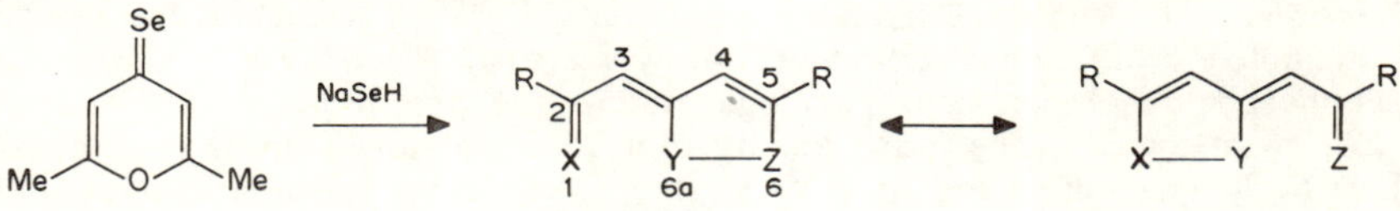

(**250**) (**251**)

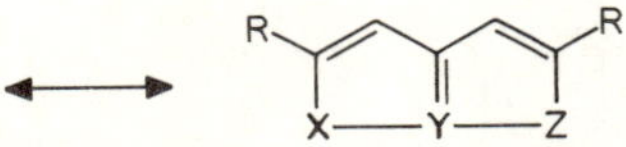

(109)

(*ii*) *Chalcogenapentalenes.* The selenapentalene derivatives **251** can be regarded according to their no-bond resonance hydrids, as derivatives of 1,2-oxaselenoles, 1,2-thiaselenoles or 1,2-diselenoles. An oxadiselenapentalene system (**251**; R = Me, Y = Z = Se, X = O) was obtained by reaction of 2,6-dimethylpyran-4-selone (**250**) with NaHSe[502] (equation 109); it was first incorrectly formulated, but its correct structure was suggested from X-rays[503] and proved[504]. Its formation was proposed to involve a ring-opening of the pyran system. When treated with P_2S_5, it gives a mixture of the corresponding thiadiselenapentalene (**251**; Y = Z = Se, X = S), and as major product the trans-chalcogen exchange product, the 2,5-dimethyl-1,6a, 6-dithiaselenapentalene (**251**; Y = Z = S, X = Se, R = Me). Similarly, thiopyran-4-selone and its 3,5-dimethyl and 2,6-diphenyl derivatives give with Na_2S and subsequent ferricyanide oxidation, a 1,6,6a-dithiaselenapentalene isomeric system (**251**; X = Z = S, Y = Se, R = H, Ph)[505]. 1,6,6a-Dioxaselenapentalene analogues (X = Y = O, Y = Se), precursors of the previously mentioned 1,2-oxaselenolylium cations, are obtained analogously from **250** and thallium (III) halides via a 4-thallioselenopyrylium salt[506a]. Recently, the 1,6,6a-dioxatellurapentalene was similarly synthesized[506b]. Another general route to selenapentalenes consists of the reaction of NaHSe with the Vilsmeier salt formed from the 1,2-dithiolylium salts **252** (equation 110)[507a]. Compound **253** is also obtained by P_2Se_5 selenation of the corresponding oxadithiapentalene[507b].

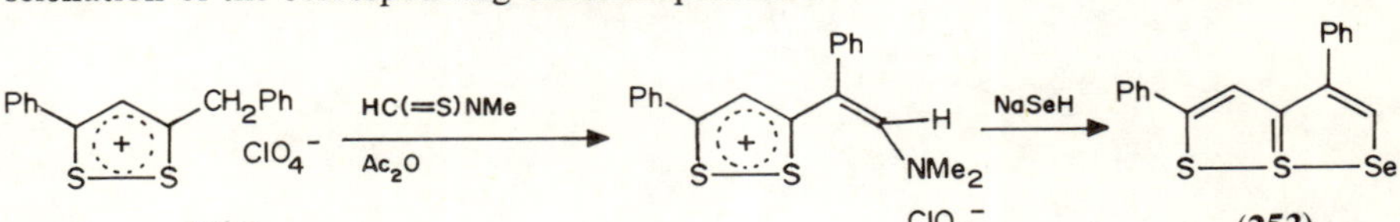

(**252**) (**253**)

(110)

(*iii*) *Condensed systems.* Some oxo derivatives of benzothiaselenole have been studied. 3-Oxo-2,1-benzothiaselenole **255a** and its diselenole analogue **255b** were prepared in 1924

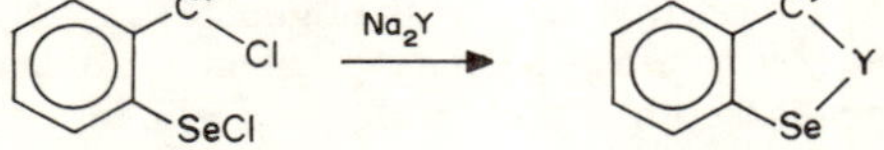

(111)

(**254**) (**255**)

(**a**) Y = S
(**b**) Y = Se

by Lesser's group from *o*-chloroselenobenzoyl chloride (**254**) (equation 111)[508]. The 1,2-benzothiaselenole isomer of **255a** is obtained by P_2Se_5 selenation of dithiosalicylic acid[509]. These systems sometimes give isomerization or chalcogen exchange reactions. Thus, thionation of **255b** gives the anomalous 2,1-benzothiaselenol-3-thione[510]. Its isomer **258** is obtained from **256** following equation (112)[511]; this reaction does not give the attempted selone isomer **259** as previously described[48,49]. However, with diethyl selenophosphate and the 3-chloro analogue of **257**, a mixture of 60% of **259** and 6.5% of **258** is obtained[511]. The 3-oxo analogue of **258** is obtained in 1.2% yield from dithiosalicyloyl dichloride and *p*-tolyl-SeMgBr[512a]. As another rearrangement example, **259** gives 3-diphenylmethylene-1,2-benzothiaselenole with diphenyldiazomethane[512b]. Another example of rearrangement is also given in Ref. 597.

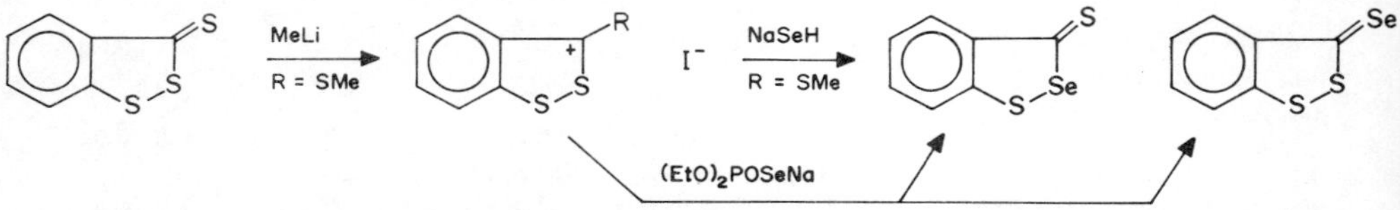

(**256**) (**257**) (**258**) (**259**)

(112)

1-Hydroxy-1-methyl-2,1-benzoxaselenol-3-one has been demonstrated to be a ring tautomer of *O*-methylselenobenzoic acid Se oxide[512c,d].

From 2-chloro-3-vinylquinoline and NaHSe 3-methyl-1,2-diselenolo(4,5-*b*)quinoline and a derivative with a dihydroselenophene ring were obtained[513,514].

Some *peri*-bridged 1,2-diselenoles or 1,2-ditelluroles have been prepared for use as π donor systems for the design of highly conducting organic metals. These systems may be principally obtained, either by sodium dichalcogenide nucleophilic substitution of the corresponding chloro compounds, or via organolithium reagents. By the first method 5,6:11, 12-bis(diseleno)tetracene (**261a**)[515] and 5,6:11, 12-bis(ditelluro)tetracene (**261b**)[516] were obtained from **260** (equation 113); also obtained were 1,9:4,10-bis(diseleno)anthracene[517] and 1,8:4,5-bis(diseleno)naphthalene[518]. **261a** was prepared

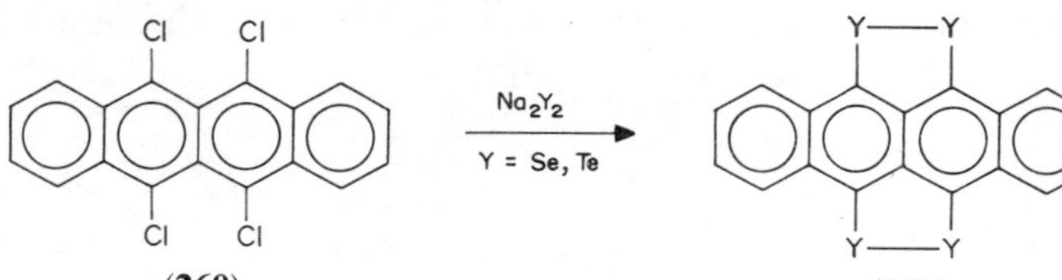

(113)

(**260**) (**261**)

(**a**) Y = Se
(**b**) Y = Te

previously in low yield from 5,11-dichlorotetracene which is difficult to obtain[519]. On the other hand, a patented method from tetracene and Se could not be reproduced[515,520]. **261a** is an organoselenium π donor which gives an ion radical with metallic states below 30 K. Electrochemical data suggest that the Te analogue **261b** is more easily oxidized in solution than **216a**, a situation existing also in naphthalene 1,8-dichalcogenides[521]. These dichalcogenonaphthalenes **263**, both symmetrical (Y = Z = S, Se, Te)[521] or unsymmetrical (Y—Z=S—Se, S—Te, Se—Te)[522], and dichalcogenoacenaphthylene analogues[523] are obtained via organolithium compounds (equation 114). The unsymmetrical derivatives, obtained in very low yields together with the symmetrical impurities, are best isolated, when Y = S, from 1-chloro-8-mercaptonaphthalene, via the butyllithium method. One of the Se atoms in **263** (Y = Z = Se) and the two Te atoms in **263** (Y = Z = Te) are easily exchanged

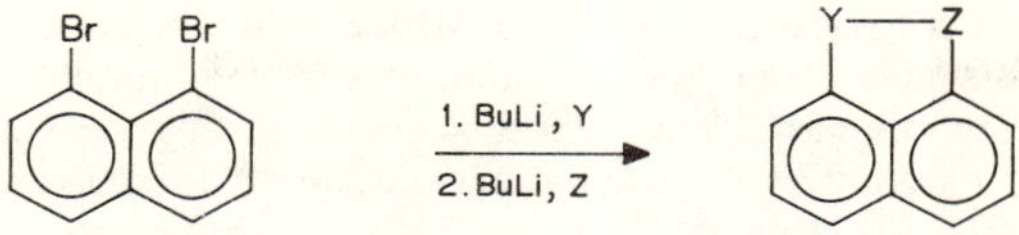

(114)

(262) **(263)**

with organolithium compounds[521]. In this series, S and Se derivatives exhibit similar conducting properties.

b. Heteroatoms in the 1,3-*position.* (*i*) 1,3-*Dichalcogenoles.* The first 1,3-oxaselenole derivatives were reported in 1978: 2,5-di-*p*-methoxyphenyl-2,4-diphenyl-2*H*-1,3-oxaselenole together with the corresponding 2-oxo derivative, was obtained from *p*-anisoyl α,α-dibromobenzyl ketone and potassium ethyl diselenocarbamate[524]; the 2-acetyliminio-5-cyano-*p*-substituted phenyl derivative[525] was obtained from the 2,2-dicyano-3-aryloxirane and potassium selenocyanate, and a tricyclic derivative was obtained by selenium dioxide oxidation of dimedone[526a]. New 1,3-oxaselenoles were recently obtained in this manner[526b]. A general synthesis of 1,3-dichalcogenoles **(264)** starts from alkynylchalcogenolates and bromochloromethane (equation 115) (R = Ph, *t*-Bu; Y = S; Z = S, Se, Te)[527]. This method has been applied to 1,3-ditelluroles (R = H, Me, *t*-Bu, Ph)[528]. Parent 1,3-ditellurole is lithiated in the 4-position by lithium diethylamide (LDA)[529]. On the other hand the 4-phenyl derivative gives products derived from lithiation in the 2-position, results parallel to those in the sulphur series.

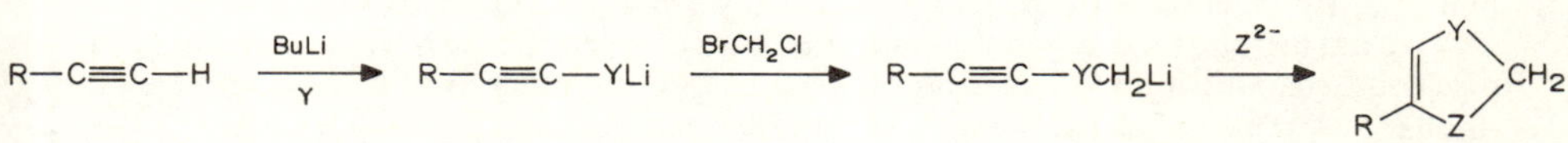

(264)

(115)

(*ii*) 1,4-*Dichalcogenafulvenes.* 1,4-Diselenafulvenes **(266)** are obtained by dimerization of alkynylselenols **(265)**[67] (equation 116). The corresponding 2,6-diphenyl -1,4-

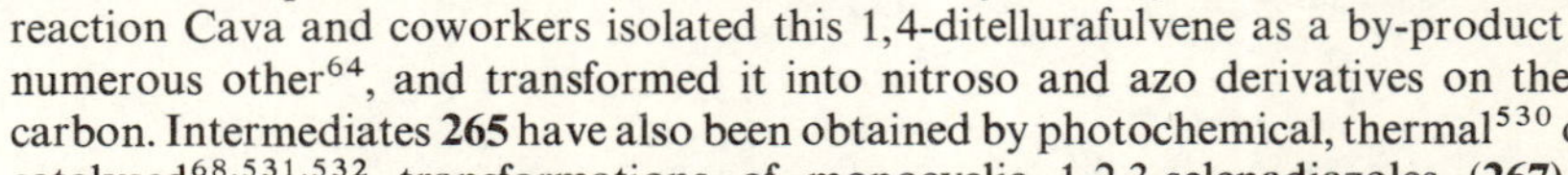

(116)

(265) **(266)**

ditellurafulvene was claimed by Russian workers[66] to be obtained by a similar method, but their compound was in fact a ditelluretane system. By careful examination of the reaction Cava and coworkers isolated this 1,4-ditellurafulvene as a by-product among numerous other[64], and transformed it into nitroso and azo derivatives on the lateral carbon. Intermediates **265** have also been obtained by photochemical, thermal[530] or base-catalysed[68,531,532] transformations of monocyclic 1,2,3-selenadiazoles **(267)** (equation 117), and they are also dimerized to the corresponding fulvenes **266**.

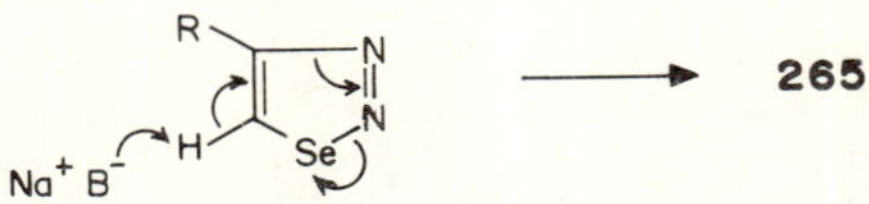

(117)

(267)

6,6-Dicyano- and diethoxycarbonyl-2,3-dihydroselenafulvenes can be obtained from ethylene triselenocarbonate, MeI and malononitrile[533] or diethyl malonate[534]. 1,4-Thiaselenafulvenes are prepared from 1,3-thiaselenole-2-thiones via a phosphonium derivative[535] or by reaction with ethyl diazoacetate[536]. The same thiones give 6-ethoxycarbonyl-6-aza-1,4-thiaselenafulvene with ethyl azidoformate[536]. 6-Phenyl-6-aza-1,4-thiaselenafulvene and 1,4-diselenafulvene are also prepared from **265** and phenylisothiocyanate[537].

(*iii*) *2-Chalcogenoxo-1,3-dichalcogenoles.* 2-Oxo-, 2-thioxo- or 2-selenoxo-1,3-dichalcogenoles have been the subject of many publications, due to the fact that they can lead by coupling reactions to tetrachalcogenafulvalenes, the most popular π donors for designing organic metals. Several types of synthesis have been tested for their preparation. The first one, which leads finally to the parent tetraselenafulvalene (see following section), consists of treating **265** with carbon diselenide and leads to 2-selenoxo-1,3-diselenole

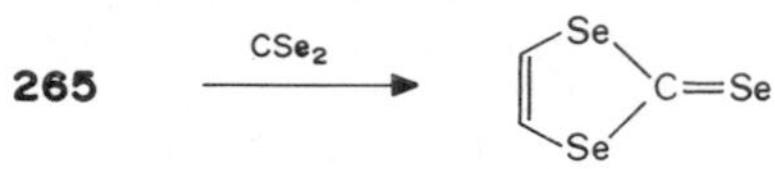

(118)

(268)

(268)[538,539] (equation 118). The reaction has been applied to the 4-phenyl derivative[540], which by P_4S_{10} thionation gives the 2-thioxo and subsequently the 2-oxo analogue. In fact, these reactions are applications of Mayer's synthesis of 4-phenyl-1,3-thiaselenol-3-thione, the first derivative of the series, from phenylacetylene, Se and carbon disulphide[541]. By reaction with mercuric oxide, Mayer obtained the 2-oxo analogue.

A second method for the synthesis of compounds of type **268** consists of the reaction of a diselenocarbamate ion with an α-bromoketone, followed by cyclization and subsequent thionation[542,543] or selenation with H_2Se ($R^1 = Me$, $R^2 = H$[542]; $R^1 = R^2 = Me$; $R^1 = Ph$, $R^2 = H$; $R^1 = R^2 = Ph$[543]; $R^1 = R^2 = CD_3$[544] (equation 119). The cyclization may be effected by H_2SO_4 or best by HBF_4[545]. One of the resonance forms of **270** is a diselenolylium cation. Modifications have been proposed to avoid the use of the very fetid CSe_2 which is a precursor to the diselenocarbamate ion. For example, this anion can be replaced by tetramethylselenourea[546]; when the method is applied to dimethylthiourea and H_2Se it gives trimethylene thiaselenol selone[546]. In another modification the diselenocarbamate ion can also be prepared from *N,N*-dimethylphosgeniminium chloride[547–549]. In the last reference the gaseous and toxic H_2Se is replaced by NaHSe obtained from Se and $NaBH_4$ in DMF, and the synthetic conditions are optimized. Since carbon ditelluride and tellurourea are unkown, and NaTeH or H_2Te reduce the phosgeniminium salt, application of this method to the Te series has been tentatively realized by forming vinyltellurothiocarbamate from a vinylmagnesium reagent, Te and a thiocarbamoyl chloride. The tellurothiocarbamate decomposed in the cyclization essays[550].

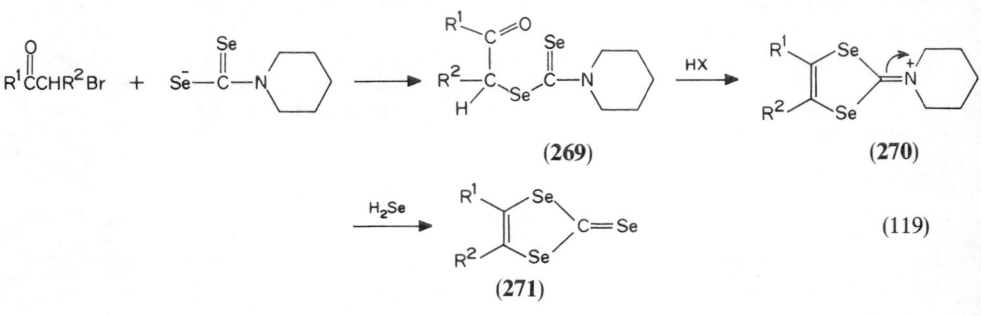

(269) **(270)**

(119)

(271)

A third method for the synthesis of compounds of type **271** consists of the dipolar addition of ethylene triselenocarbonate[551] or thiodiselenocarbonate (**272**) to dimethyl acetylenedicarboxylate[552,553] (equation 120). This reaction requires for a concerted

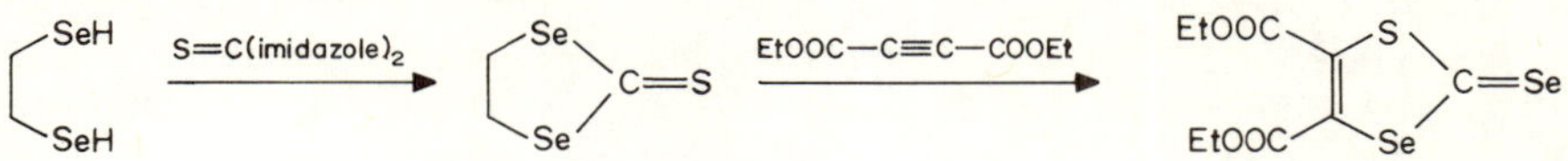

(**272**) (**273**) (120)

mechanism that the selone Se atom of the triselenocarbonate will be incorporated in the ring in the Se analogue of **273**[553]. Substituted **268** may also be formed by electrochemical reduction of CSe_2[554,555] giving after alkylation the 4,5-bis(methylseleno) derivative of **271**[554] and the 4,5-ethylenediseleno derivative[555]. 1,3-Benzothiaselenol-2-thione is obtained by heating 1,2,3-benzoselenadiazole with CS_2[556].

Condensed 2-thioxo-1,3-thiaselenole or diselenole systems can also be obtained from thiocarbonyl diimidazolide and 1,2-benzene diselenolate[557], from thiophosgene and the 2,3- and 3,4-thiophene diselenolate or 3-thiolate 2-selenolate[558]and from thiophosgene and *o*-bis(methylseleno)benzene or *o*-(methylseleno)thioanisole[559]. The last reaction failed for the methyltelluro analogues. The condensed benzo, 5,6-dimethylbenzo-[560] and 4*H*,6*H*-thieno-[561] selenoxo-1,3-selenoles are prepared by heating the corresponding condensed 1,2,3-selenadiazole with CSe_2. 1,3-Ditellurol-2-thione *C*-condensed on the thiophene ring has also been prepared[558b].

Some problems of chalcogen exchange can arise when two different chalcogens, especially S and Se, are used for the synthesis of analogues of **271**. When alkynylselenide anion is reacted with CS_2 five different heterocyclic compounds **274** are obtained, form an unexpected S–Se scrambling process (equation 121). This may arise either from an ambident anion obtained by deprotonation and ring-opening or from a rearrangement of the non-cyclic precursor[562]. After very difficult separations, the required

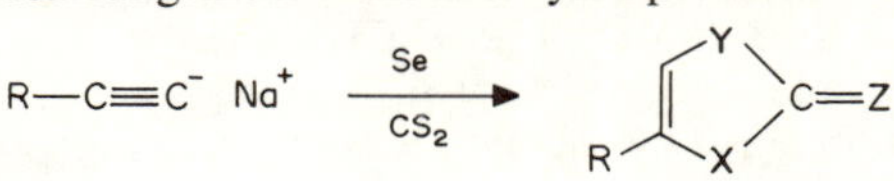

(121)

(**274**)

(**a**) Y = Se; X = Z = S; R = Me, Ph
(**b**) Y = X = S; Z = Se
(**c**) Y = S; X = Z = Se
(**d**) Y = X = Se; Z = S
(**e**) Y = X = Z = Se

274a is obtained in a 0.82% yield. A similar mixture results by using S and CSe_2. Compounds of type **274a** (R = aryl) are also obtained by treatment of the selenolate anion generated by base-catalysed fragmentation of 4-aryl-1,2,3-selenadiazoles with CS_2[563a]. When the reaction is applied to 1,2,3-selenadiazole itself, it gives under the usual conditions a low-yield mixture of the required thione **274a** (R = H) and its isomeric selone **274b**. Non-isomerizing conditions can, however, be found[552]. A fused system, an unstable 2,5-dipiperidinium-1,3,4,6-tetraselenapentalene, has been obtained from piperidinium *N*,*N*-pentamethylene diselenocarbamate and methyl dichloroacetate[563b].

(*iv*) *Tetrachalcogenafulvalenes*. The interest in the chemistry of the tetrachalcogenafulvalenes arose when it was demonstrated in 1972 that charge-transfer complexes of the π donor tetrathiafulvalene (TTF) and the acceptor tetracyanoquinodimethane (TCNQ) possessed remarkable high conducting properties[564,565]. Considerable interest arose in the preparation of a parallel series of tetraselenafulvalene (TSeF) systems. Interesting changes in physical properties were demonstrated by replacing S by Se[566] in TSeF/TCNQ salts. One of these complexes, hexamethylene TSeF/TCNQ, behaves as a semi-metal and becomes an insulator at low temperatures[567,568]. Recently, one salt of these derivatives (tetramethyl-TSeF)$_2$PF$_6$, was the first organic compound found to lose all resistance to

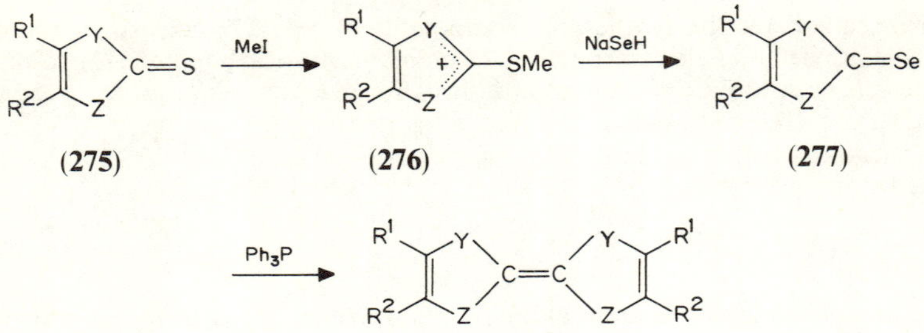

(275) (276) (277)

(278)

(122)

electric current around 1 K, showing superconductivity[569]. Now at least six $(TMSeF)_2X$ salts have been reported to be superconductors ($X^- = TaF_6^-$, SbF_6^-, AsF_6^-, PF_6^-, ReO_4^- and ClO_4^-), this last salt being the only simple pressure organic superconductor at 1.3 K[570]). The most common synthesis of tetrachalcogenafulvalenes is the coupling of the 1,3-dichalcogenol-2-thiones or -2-selones. Peracid oxidation of 1,3-diselenol-2-thione or -selone, a reaction working well in the dithiole chemistry, failed in the Se series[538,571], perhaps because oxidation on Se atoms took place. Coupling reactions were effected by triphenylphosphine or triethyl phosphite, and they work much better with the 2-selone than with the 2-thione[556]. Consequently, the selone will be preferentially prepared when the method used involves intermediary iminium salts (equation 119). When only the thione is available, it is interesting to transform it to the selone, via *S*-methylation to form a 1,3-dichalcogenolylium salt, and subsequent reaction with NaSeH or H_2Se (equation 122). One of the best methylating reagents is methyl fluorosulphonate[553,556]. The coupling reactions lead to the majority of the known tetraselenafulvalene derivatives **278** (Y = Z = Se): $R^1 = R^2 = H$[538]; $R^1 = Me$, $R^2 = H$[543]; $R^1 = R^2 = Me$[542,543,547–549]; $R^1 = Ph$, $R^2 = H$[543]; $R^1 = R^2 = Ph$[543]; $R^1 = R^2 = (CH_2)_3$[542,543,546–548]; $R^1 = R^2 = CD_3$[544]; $R^1 = R^2 = COOMe$[551]; $R^1 = R^2 = SeMe$[554]; $R^1R^2 = CH_2SCH_2$[561]. The method has been applied to dibenzotetraselenafulvalene[560,572a]. Surprisingly, the 4,5-ethylenediseleno derivative of **278** cannot be obtained by triethyl phosphite coupling of the corresponding selone, but it is obtained in 10% yield from the 2-oxo derivative[572b].

The tetramethoxycarbonyl derivative of **278** can be easily transformed to the parent compound by LiBr[551]. An unsymmetrical non-linear condensed diselenoleselone such as the dihydronaphtho (1,2-*b*) derivative, gives by coupling a mixture of *cis*- and *trans*-condensed TSeF[573]. Mixed thiaselenole derivatives **274** nearly always give unseparable mixtures of thermally stable *cis* (*Z*) and *trans* (*E*) geometrical isomers of coupled derivatives[553,556,574,575]. The parent *E,Z*-dithiadiselenafulvalene mixture was obtained by tetradecarboxymethylation. Unexpected S–Se interchanges were discovered by triethylphosphite coupling of 1,3-diselenol-2-thiones. For example **275** ($R^1 = R^2 = COOMe$, CF_3; Y = Z = Se) gives a triselenathiafulvalene, instead of the attempted tetraselenafulvalene. The dimethyl derivative gives the scrambled and unscrambled product[576]. A few other methods were used for the synthesis of tetrachalcogenafulvalenes. So, tetratellurafulvalenes have been recently synthesized. The coupling method was unsuitable for their preparation due to lack of a method for preparing the corresponding

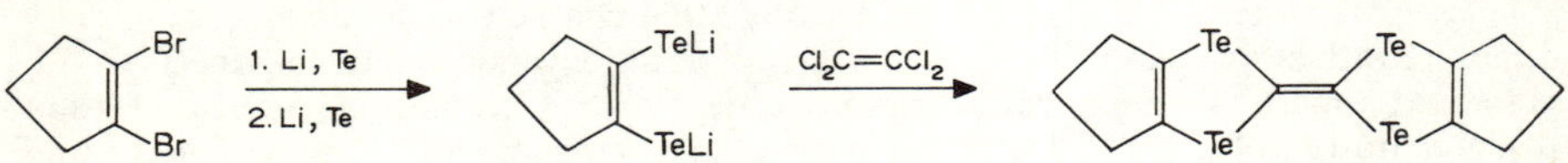

(279) (280)

(123)

thiones or selones. The first two known Te derivatives were obtained from a suitable 1,2-ditellurolate ion and tetrachloroethylene (equation 123)[577]. This reaction could also lead to the bis-six-membered ditellurinoditellurin isomer, but structure **280** was proved by X-ray analysis[578]. The method has also been applied to dibenzotetratellurafulvalene[579]. A very easy preparation of dibenzotetraselenafulvalene (**282**) starts from the easily available *o*-bis(methylseleno)benzene (**281**)[559] and gives an overall 48% yield (equation 124).

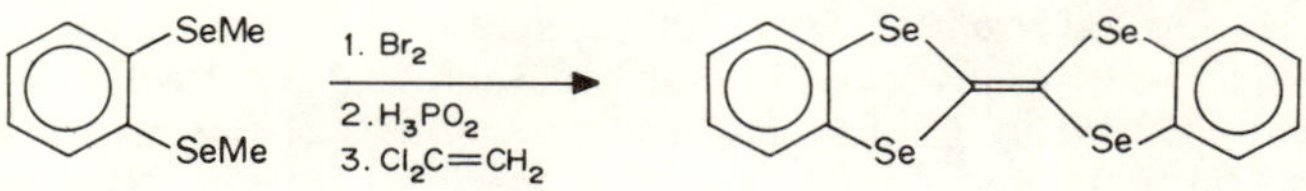

(**281**) (**282**) (124)

The surprising behaviour of vinylidene chloride is also shown in the preparation of tetra(methylseleno)ethylene from sodium methaneselenolate[559]. It seems that the mechanism involves an elimination step to an alkyne. Monosubstituted TSeF can be obtained from the parent compound via the 2-lithio derivative[580]. Recently, a one-step high-pressure synthesis of TSeF was realized from acetylene and a two-fold excess of carbon diselenide at 4500–5000 atm[581,582]. Some thiaselenafulvenes are also prepared from the thiaselenolylium salt **283** (equation 125)[583]. Some corresponding imidazoles can also be employed for this reaction.

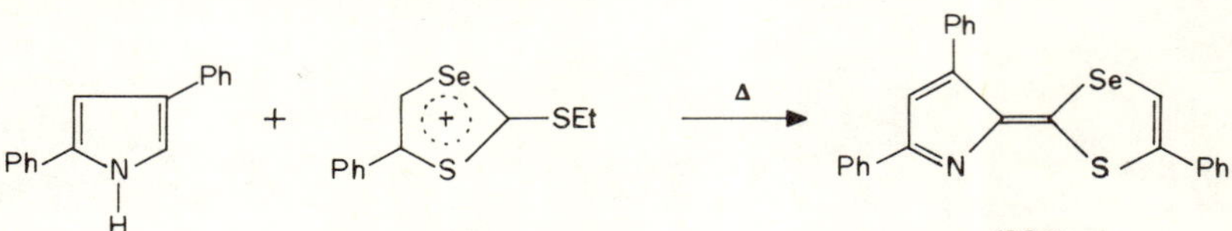

(**283**) (**284**) (125)

(*v*) 1,3-*Dichalcogenolylium salts.* 1,3-Thiaselenolylium and 1,3-diselenolylium salts are sometimes obtained as intermediates in the transformation of a thione to a selone group. These MeS-substituted cations **286** can be transformed to the relatively unstable parent cations **288** by successive $NaBH_4$ reduction and treatment with HBF_4[574] (equation 126). 2-Benzyl-4-phenyl-1,3-dioxolylium, 1,3-dithiolylium and 1,3-diselenolylium cations can also be obtained by trifluoroacetic acid protonation of 2,6-diphenyldichalcogenafulvenes[584]. NMR evidence indicates that the unsubstituted 1,3-ditellurolylium cation is formed by treatment of 1,3-ditellurole with the hydride acceptor triphenylcarbenium tetrafluoroborate. The unsubstituted cation **290** rearranges rapidly to the quite unstable 1,2-ditellurolylium cation, **291** (equation 127), which rapidly deposits Te. Substituted intermediates **290** are not detectable[528].

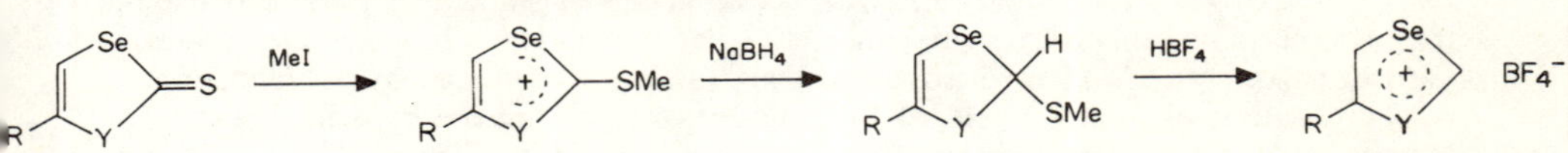

(**285**) (**286**) (**287**) (**288**)

R = H, Me (126)
Y = S, Se

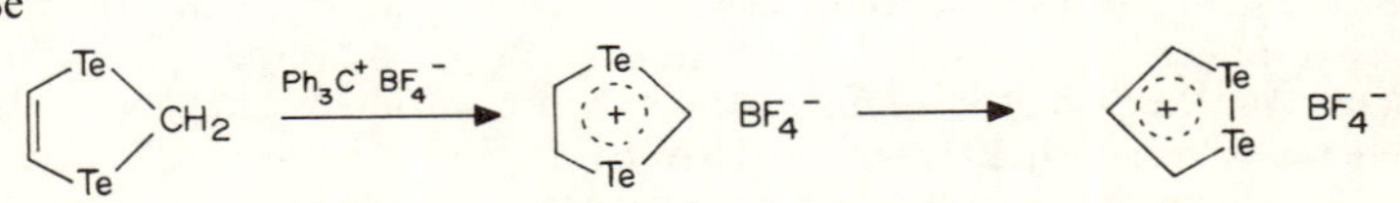

(**289**) (**290**) (**291**) (127)

2. Nitrogen Se/Te rings

Se–N heterocycles have been reviewed three times in the last decade: twice by Bulka through 1970[585] and 1974[586] and once by Lalezari and Shafiee through 1976[587].

a. Heteroatoms in the 1,2-*position.* (*i*) *Isoselenazoles and condensed analogues.* The first known isoselenazole, 10-oxoanthra(9,1-*c*,*d*)isoselenazole, was synthesized in 1936 by reaction of NH_3 and Na_2Se with 1-chloroanthraquinone[588,589]. Preparation of isoselenazole itself, and its 3-methyl derivative from selenocyanic acid and acrolein or 3-butyn-2-one were mentioned briefly[590]. The first general synthesis of this ring system which includes the parent compound was achieved by the reaction of Br_2 and NH_3 under dry-ice conditions with *Z*-β-(methylseleno)acrolein derivatives (equation 128)[591]. These precursors are obtained either by reaction of sodium methaneselenolate on the product of Vilsmeier reaction on ketones or for the parent compound by addition of methaneselenol to propargylaldehyde, avoiding in this manner the use of the unstable selenocyanic acid. The following derivatives **294** were obtained: $R^1 = R^2 = H$; $R^1 = Me$, $R^2 = H$; $R^1 = R^2 = Me$; $R^1R^2 = (CH_2)_3$ or $(CH_2)_4$ (equation 128). Probable intermediate selenenyl bromides could never be isolated. The more stable corresponding benzeneselenenyl bromides **295** are in fact good precursors for the synthesis of 1,2-benzisoselenazole by the same methodology[591] (equation 129) (R^1 = H, Me, Ph, CH=CHPh, R^2 = H; R^1 = H, R^2 = 5-NO_2, 7-NO_2).

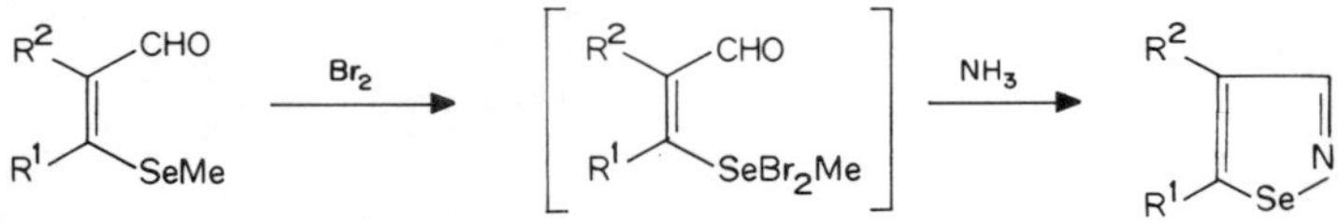

(128)

(**292**) (**293**) (**294**)

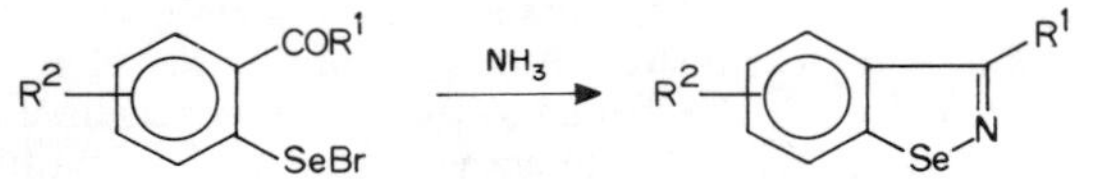

(129)

(**295**) (**296**)

This method, previously used in the sulphur series for the synthesis of 5-nitrobenzisothiazole, has been extended to a general method of preparation of 1,2-benzisothiazoles. It has also been applied to the synthesis of pyrido(3,2-*d*)isoselenazole, of the three classical bicyclic thieno-condensed 1,2-selenazoles, i.e. the (3,2-*d*), (2,3-*d*) and (3,4-*d*) systems[423,424], of selenolo(3,2-*d*)isoselenazole[424], of its (2,3-*d*) isomer (7% yield) and also of the corresponding benzotricyclic systems: the (1) benzothieno- and (1) benzoselenolo-(3,2-*d*)isoselenazoles[425]. Condensed isoselenazoles can also be obtained by polyphosphoric acid cyclization of 2-(methylseleno)-3-oximinobenzo(*b*)-thiophene or -selenophene[425]. Isoselenazole can be nitrated in position 4, and brominated in positions 4 and 5. Thieno (2,3-*d*)isoselenazole is nitrated or brominated at the β-position of the thiophene ring, whereas its (3,2-*d*) isomer is α-brominated and nitrated to give a mixture of α- and β-derivatives. The parent benzo compound **296** ($R^1 = R^2 = H$) is nitrated and brominated in the homocycle in positions 5 and 7, dibrominated in the 4,5- and 4,7-positions and tribrominated in the 4,5,7-positions. No acylation products can be isolated[591]. Chichibabin reaction leads to the 3-amino derivative, whose structure has been confirmed by independent Curtius reaction on the 3-carboxy derivative. This acid has been prepared from the 3-carboxamide which was directly obtained by reaction of Br_2 and NH_3 with 2,3-dioxo-2,3-dihydrobenzo(*b*)selenophene, or from the 3-styryl derivative via ozonolysis to the corresponding 3-formyl derivative and oxidation[591]. Isoselenazoles behave towards organolithium compounds in the same manner as isothiazoles; they give concomitantly an eliminative R.O.I from the 3-carbanion, forming a *o*-(alkylseleno)nitrile, and a substitutive selenophilic R.O.II, with rupture of the Se—N bond and formation, after hydrolysis, of the corresponding aldehyde[591]. Even in fused bicyclic[423] or tricyclic[425] '*cis*-condensed' selenoloisoselenazoles **297**, the isoselenazole ring is regiospecifically opened (equation 130), giving the nitrile **298**, the aldehyde **299** and products of metal–Se interconversion.

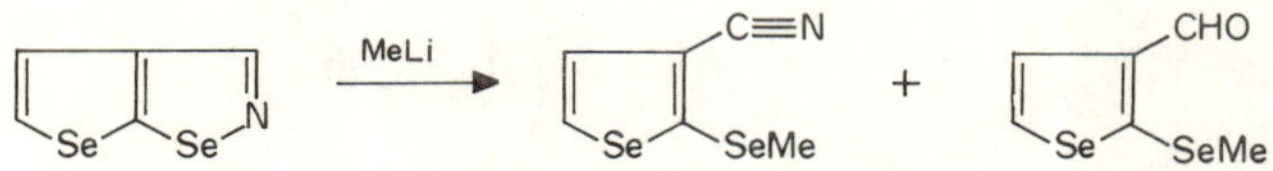

(297) **(298)** **(299)** (130)

Regiospecific ring-opening of the selenophene ring occurs only in selenophenes '*cis*-condensed' to isothiazole (equation 131), as well as in the bicyclic and the tricyclic systems[423,425]. Compound **300** is prepared from the 2-(methylthio)-3-oximinoselenophene[423]. An interesting addition of benzyne across the C=N—Se group of 2,1,3-benzoselenadiazole is accompanied by rearrangement to **304**, whose structure was confirmed by ozonolysis to the known 3-formyl-1,2-benzisoselenazole (**305**)[592] (equation 132).

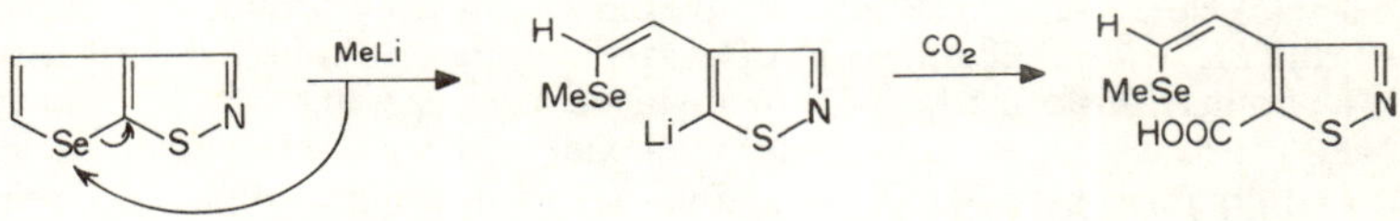

(131)

(300) **(301)** **(302)**

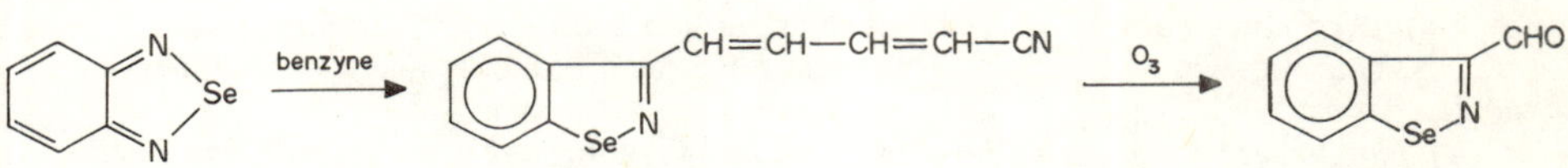

(303) **(304)** **(305)**

(132)

The 5,6-dimethyl analogue of adduct **304** gives a thermal rearrangement to 4,5-dimethyl-2-(*o*-selenocyanophenyl)pyridine[593]. A corresponding adduct of naphthoselenadiazole with dimethyl acetylenedicarboxylate rearranges similarly to 3-(2′,3′-dimethoxycarbonyl-4′-cyano-1′-naphthyl)benzisoselenazole[594]. 2-(Selenocyano)-benzophenone or -acetophenone gives by treatment with $NH_2OH{\cdot}HCl$ the corresponding 3-phenyl- or 3-methyl-*N*-oxide of benzisoselenazole. With phenylhydrazine and a trace of acetic acid, this acetophenone gives, probably via a Fischer cyclization, the known benzoselenolo(3,2-*b*)indole[446].

An isoselenazolo(4,3-*d*)pyrimidine is obtained by SeO_2 oxidation of 5-amino-6-methyl-3-phenyl-4-(3*H*)-pyrimidine[595].

(*ii*) *Isoselenazolium salts*. Liebscher and Hartmann realized a one-pot synthesis of 5-arylisoselenazolium salts from alkali selenocyanates and *N*,*N*-dimethyl-3-chloro-3-aryl-2-chloropropeniminium salts[596]. 3-Chloro-1,2-benzisoselenazolium salts can be prepared from PCl_5 and *o*-(methylseleno)-substituted benzamides or benzisoselenazolinones[597]. They are hydrolysed to benzisoselenazolin-3-ones, thiolysed by thioacetic acid to 3*H*-*N*-methylimino-2,1-benzothiaselenoles and ammonolysed to 3-substituted amino-1,2-benzisoselenazoles. The 3-methyliminobenzothiaselenole is rearranged in DMSO solution to the 2-methylbenzisoselenazolin-3-thione isomer. The *N*-ethoxycarbonylbenzisoselenazolium salt can be obtained by quaternization of benzisoselenazole with ethyl chloroformate[598].

(*iii*) *Benzisoselenazolinones.* The first synthesis of benzisoselenazolin-3-ones (**305a**) was achieved in 1924 from *o*-chloroselenobenzoyl chloride[599] and NH_3 or amines RNH_2(R = Me, Et, Ph) (equation 133). They can also be prepared from 3-chlorobenzisoselena-

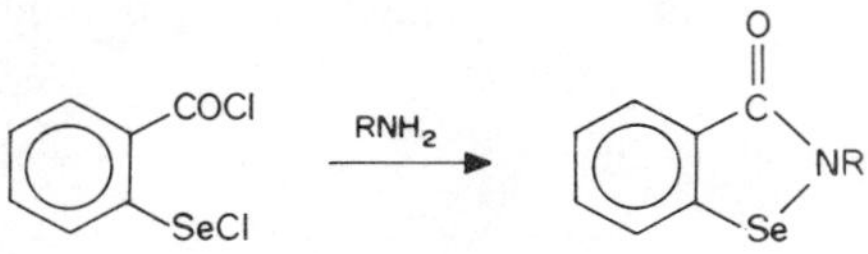

(133)

(**305a**)

zolium salts[597] and by reaction of Br_2 with *o*-methylselenobenzamides. The parent compound is also obtained from *o*-bromoselenobenzamide[597] and by ammonolysis of ethyl *o*-cyanoselenobenzoate[600]. Their modes of preparation, for pharmaceutical purposes, are summarized in a recent patent[601]. The parent compound can be *N*-acetylated by acetic anhydride. Treated by potassium acetate and acetic anhydride this reaction leads to three different types of products: 3-acetylaminobenzo(*b*)selenophene, 3-acetoxybenzo(*b*)selenophene and its transformation products and 2-carbamoyl-3-hydroxybenzo(*b*)selenophene. The last compound is probably formed via the seven-membered ring 2*H*,4*H*-3,5-dioxobenzo(*b*)selenazepine, since its 4-methyl derivative was isolated in a parallel transformation of *N*-methylbenzisoselenazolinone[602]. A Se analogue of saccharin, which can be regarded as a Se-dioxide of benzisoselenazolin-3-one, was claimed to be obtained by $KMnO_4$ oxidation of 2,2′-dicarbamoyldiphenyl diselenide, followed by a crystallization in HCl solution[603]. This compound was in fact *o*-carboxybenzeneseleninic acid, which was also obtained by HCl treatment of an oxidation product of benzisoselenazole. Chemical and X-ray determination of the structure of this oxidation product showed that it is the monohydrate of the ammonium salt of *o*-carboxybenzeneselenonic acid[604]. HCl acts in fact as a reducing agent of the selenonic acid.

(*iv*) *Isotellurazole and derivatives.* The first known compound having this ring system was 1,2-benzisotellurazole, prepared in 1978 from *o*-bromotellurobenzaldehyde, by a reaction analogous to that described for the Se derivative in equation (129)[605]. It was also obtained by polyphosphoric acid ring-closure of *o*-butyltellurobenzaldoxime. Reaction with MeI gives *N*-methyl-1,2-benzisotellurazolium iodide. Compared with the Se analogue, 1,2-benzisotellurazole presents a higher melting point and a lower solubility. These two anomalies are related to particular crystallographic properties which show very short (2.4 Å) Te...N intermolecular contacts[606]. Thienoisotellurazoles could not be obtained by this method[423]. Monocyclic 3-substituted isotellurazoles **307** were recently obtained in 5–10% yields, from alkynyl ketones **306**, hydroxylamine-*O*-sulphonic acid, sodium acetate and K_2Te together with the by-products telluro-bis-alkenyl ketones (equation 134) (R^1 = H, Me, Ph; R^2 = Me, Et, Pr). The method failed for the corresponding aldehydes (R^2 = H), and it gave instead tellurobis(alkenenitriles) via an elimination reaction of the intermediate *O*-sulphonated oximes[607].

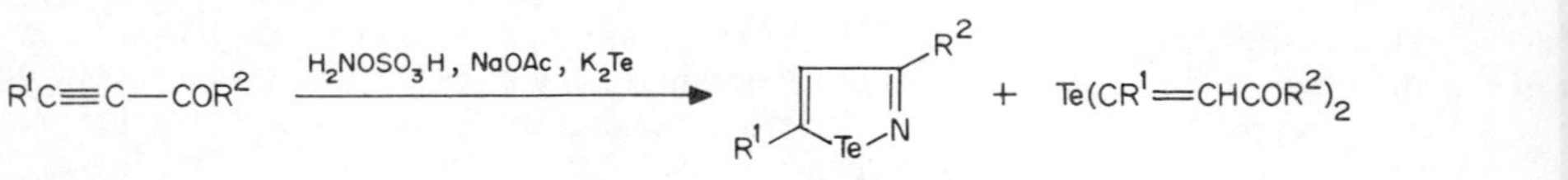

(**306**) (**307**)

(134)

b. Heteroatoms in the 1,3-*position.* (*i*) *Monocyclic selenazoles.* In addition to other reviews mentioned in the beginning of this chapter, a review covering the state of monocyclic selenazole chemistry was published by Bulka in 1963[608]. Monocyclic selenazoles are almost exclusively obtained by an extension of the Hantzsch synthesis of thiazoles, presented in a general form in equation (135).

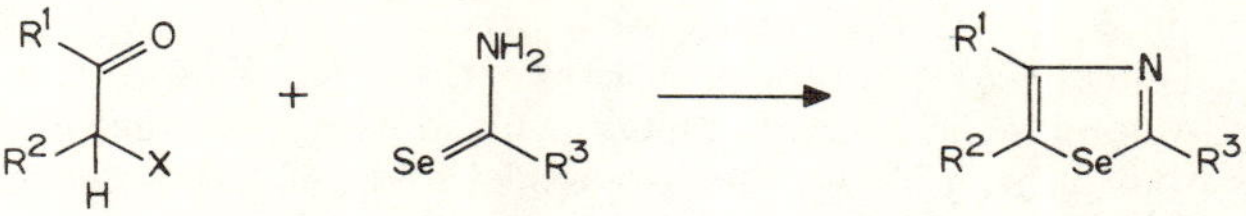

(308) (309) (310) (135)

The selenazole ring was reported for the first time in 1889 by Hofmann[609], a student of Hantzsch. All the variants of this reaction are achieved by a convenient choice of the groups R^1, R^2 and R^3. The choice of R^3 determines the substituent in position 2. Starting from primary aliphatic or aromatic selenoamides, 2-substituted, 2,4-disubstituted and 2,4,5-trisubstituted selenazoles (2-Me[609,610], 2,4-di-Me[610,611], 2,4,5-tri-Me[610], 2-Me-4-Ph[610,614], 2-Et-4-Me[610], 2,4-di-Ph[609,614], 2-Ph-4-Me, 2-Ph-4-Me-5-COOEt, 2-Ph-4-Me-5-COOH[609], 2-(-2′ or -3′-furyl, -thienyl and -selenienyl)[188,612,613] and their 4′-chloromethyl and 4′,5′-diethoxycarbonyl derivatives) were obtained. Some recent 2,4-disubstituted selenazoles with $R^3 = CH_2Cl$, Me, Ph and $R^2 = 4\text{-}XC_6H_4$(X = H, Me, Cl, Br), $3,4\text{-}Me_2C_6H_3$, 2-thienyl[614] were obtained by this method. Since the primary selenoamides are very unstable, a mixture of a nitrile and H_2Se in the presence of condensation catalysts was used[610]. 4-Methylselenazole, one of the sole selenazoles unsubstituted in the 2-position, was obtained from hydrogen cyanide but in a yield of only 2.5%[610]. The parent selenazole still remains unknown. Starting from selenourea ($R^3 = NH_2$), Hantzsch reactions give 2-aminoselenazole derivatives (unsubstituted[609,615,616], 5-Me and 5-Et[615,616], 4-Me[609,616–618], 4-Et, 4,5-di-Me and 4-*t*-Bu[617], 4- and 5-COOEt[619], 4-CH_2Cl, from which was obtained *β*-(2-amino-1,3-selenazol-4-yl)alanine[620] and 4-formylselenazole by Sommelet's reaction[621], 4-Ph[609,616–618,622] and some 4-aryl, 4,5-di-Ph, 4-Ph-5-Me or 4-Me-5-COOEt derivatives[622], 4-aryl-5-carboxymethyl[623]), 2,2′-diamino-4,4′-biselenazolyl[617] and 2′-amino-4-selenazolyl derivatives of cephalosporines[624]. 2-Alkylamino-[625,626] 2-arylamino-[627,628], 2-dialkylamino-[625,626], 2-benzamido- or 2-acylamino-selenazoles[585] can be prepared by using the corresponding *N*-substituted selenoureas. The 2-benzamido derivative can be easily hydrolysed to the 2-aminoselenazole, thus avoiding the use of unsubstituted selenourea and its preparation from the toxic H_2Se. The condensation of *N*′-benzoyl *N*-monosubstituted selenoureas gives the non-tautomerizable 2-(benzoylimino)-3-phenyl selenazolines following Hantzsch reaction, whereas the corresponding *N*-disubstituted selenoureas give selenazoles **312** by ring-closure on the C=O amide, the usual type of ring-closure being hindered by the absence of the necessary hydrogen on the second nitrogen[629] (equation 136). Reaction of α-halonitriles with selenourea leads to 2,4-diaminoselenazoles[630,631]. 2-Hydrazinoselenazoles[632,633] are obtained by the Hantzsch synthesis from acetone or benzaldehyde selenosemicarbazones, after hydrolysis of the 2-

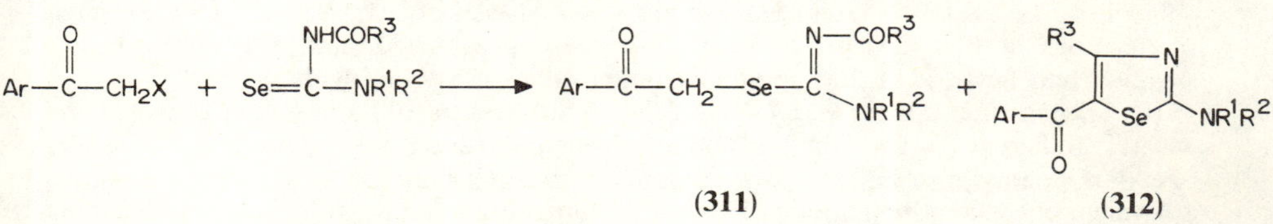

(311) (312)

(136)

selenazolylisopropylidene or benzylidene hydrazones[632,633]. Arylidene hydrazones give 2-selenazolyl formazans by coupling with diazonium salts[634]. 2-Hydrazinoselenazoles show the typical properties of arylhydrazines, by being reducing agents and by giving 2-selenazolyl pyrazolones with keto esters[635]. Coupled with *p*-nitrosodimethylaniline, these pyrazolones give a series of azomethines tested as dyes for photography[635]. Similar Hantzsch reactions with 4-phenylselenosemicarbazones and 4-phenyl-1-acyl-selenosemicarbazides were studied by Polish workers[636–638] and gave substituted 2-imino-Δ^4-selenazolines. Very few other syntheses of the monocyclic selenazole ring have been published. Among them, a recent reaction which formally belongs to the Hantzsch reaction, starts from primary thio- or seleno-amides and 2-chlorooxiranes and gives 4,5-tetramethylene- (equation 137) and 4-isopropyl-selenazole[639]. Bis-2′-(4′-dimethylcarboxamido-5′-dimethylamino)selenazol, whose structure has been determined by X-rays, is obtained by a specific reaction between carbon diselenide and 2*H*-2-dimethylcarboxamido-3-dimethylaminoazirine[640]. Using other azirines, such as 2*H*-2-phenyl-3-dimethylaminoazirine or the 2,2-dimethyl analogue gives, respectively, 5*H*-5-phenyl-4-dimethylaminoselenoazoline-2-selone and 4,4-dimethyl-2-selenoxo-5-selenazolidinone[640].

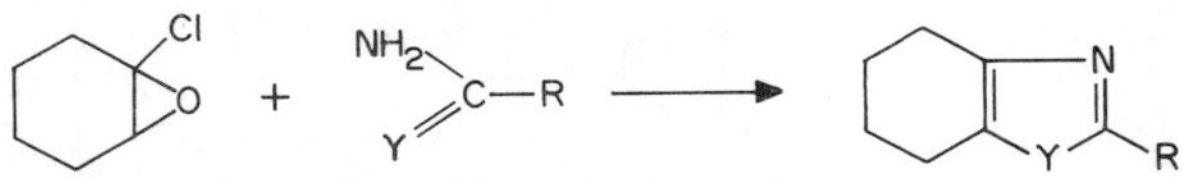

(137)

(313) **(314)**

Electrophilic nitration[641], nitrosation[642], sulphonation[641] and diazonium coupling[625] of selenazoles are directed to the 5-position. Nitration gives, however, additional complex reactions[586]. Halogenation sometimes gives unstable compounds except with the 2-acylamino derivatives, where the 5-bromo derivative is stable[586]. Acylation and Gattermann formylation failed. Nucleophilic amination failed on 4-methylselenazole, which was the only 2-unsubstituted derivative for a long time. Unsuccessful attempts[616,641] were made to obtain 2-unsubstituted derivatives from 2-aminoselenazoles via a diazotation reaction. Diazotation of 2-amino-4,5-diphenylselenazole, gives coupling to an azo compound, which decomposes on standing to a 2-hydroxy derivative, which gives the 2-chloro derivative with phosphorus oxychloride and the 2-hydrazino derivative with hydrazine hydrate. From the 2-chloro derivative and thiourea the corresponding 2-mercapto analogue is obtained, and this gives 4,5-diphenylselenazole with H_2O_2. This 2-unsubstituted derivative is more easily obtained by oxidation of the 2-hydrazino derivative by use of mercuric oxide, copper(II) salts or silver oxide. The corresponding 2,2′-azoselenazole can also be obtained. Recently, the first successful Sandmeyer–Schiemann reaction on a selenazole ring was realized although in a low yield, when 2-amino-5-acetoxyethyl-4-methylselenazole reacted with nitrosonium tetrafluoroborate, NaF and HBF_4[643]. The reactivity of a 2-Me group on selenazole is higher than that of the 4-Me group. Thus, 2,4-dimethylselenazole gives selectively the 2-styryl derivative and is oxidized by SeO_2 to an unstable 2-carboxy-4-methylselenazole. This acid, as in the thiazole series, decarboxylates easily, giving 4-methylselenazole[585]. Selenazoles are easily quaternized on the nitrogen. The formed selenazolium salts add methoxide ion to the 2-position but they lose Se, forming imidazolium salts, on reaction with anilines[644].

(*ii*) *Selenazolines and selenazolidines.* 2-Methyl-Δ^2-selenazoline has been obtained in several different ways: by phosphorus pentachloride ring-closure of bis-(2-acetylaminoethyl)diselenide[645], by reaction of selenoacetamide with **315**[646] and by treatment of acetamidoethanol with P_2Se_5[647]. Other 2-alkylselenazolines are obtained in a similar manner[648]. 2-Amino-Δ^2-selenazoline (**316**) is prepared from **315** and potassium

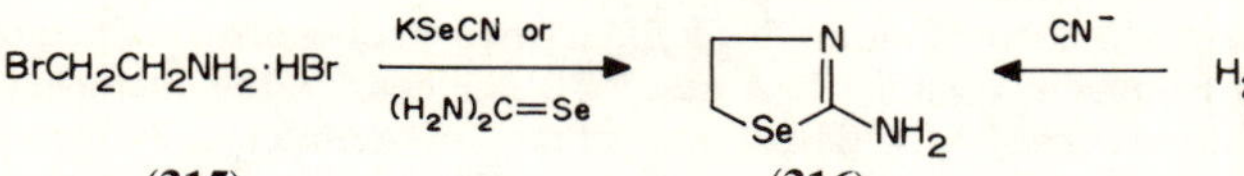

(315) (316) (317) (138)

selenocyanate[649] or selenourea[650], or from **317** and cyanide ion[651] (equation 138). **316** ring-opens easily by alkaline or H_2S medium, giving bis-ureo- and thioureo-ethyl diselenide by air oxidation.

The selenazolidine systems **318**, including the parent compound, are obtained by treating aziridines and aldehydes or ketones with H_2Se (equation 139). In some cases, the intermediary aziridinylmethanols or β-selenomercaptoethylamines ($H_2NCH_2CH_2SeH$) can also be used. In this way various 2-alkyl, dialkyl, aryl and some *N*-substituted derivatives (R^1 = H, Me, Ph) are obtained[652]. Starting from unsymmetrical aziridines, 4-methyl- and 4,4-dimethyl- selenazolidines are prepared[653]; in the same manner 4-carboxy

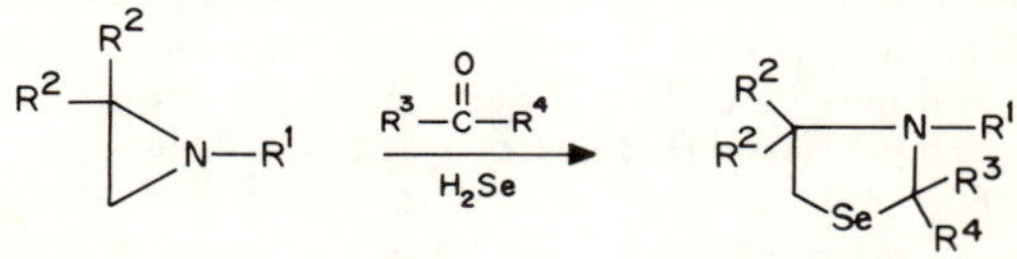

(318) (139)

or 4-carboxy-5,5-dimethyl derivatives are obtained from selenocysteine and selenopenicillamine, respectively[654]. By the same method a selenaproline is also obtained[655]. Carbonyl exchange reactions in this selenoaminal system occur easily. Selenazolidines can be *N*-acylated in pyridine solution. Benzoylation in alkali medium gives hydrolysis of this acetal-like system, forming *Se,N*-dibenzoylselenoethylamine derivatives. Although 2-phenylthiazolidine gives mainly *N*-methyl-2-phenylthiazolidine by reaction with MeI, 2-phenylselenazolidine gives a product of *Se*-methylation of the chain tautomer[656]. Selenazolidine-2-selones can be obtained by reaction of CSe_2 with aziridine[657]. The corresponding 2-thiones have been prepared from β-chloroethylisothiocyanate and NaHSe[658]. The 2-benzyliminoselenazolidine system **321** ($R^3 = PhCH_2$) is obtained by condensation of benzylisoselenocyanate with the appropriate aminoacetylenes **319**, whereas aryl isoselenocyanate gives the 2-arylaminoselenazoline tautomeric structure **320**[659].

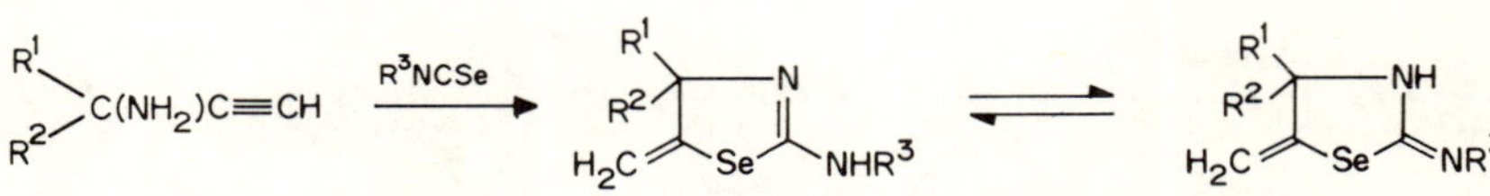

(319) (320) (321) (140)

A similar tautomerism between aminoselenazoline and iminoselenazolidine exists in the 4-oxo derivatives. These are prepared by an adaptation of the Hantzsch reaction between α-halogeno acids and selenourea[626,660–663]. The tautomerism has been investigated by comparison with systems where only one tautomer can exist. The isomers **323** and **322** are obtained by using *N,N*-disubstituted and *N,N'*-disubstituted selenoureas,

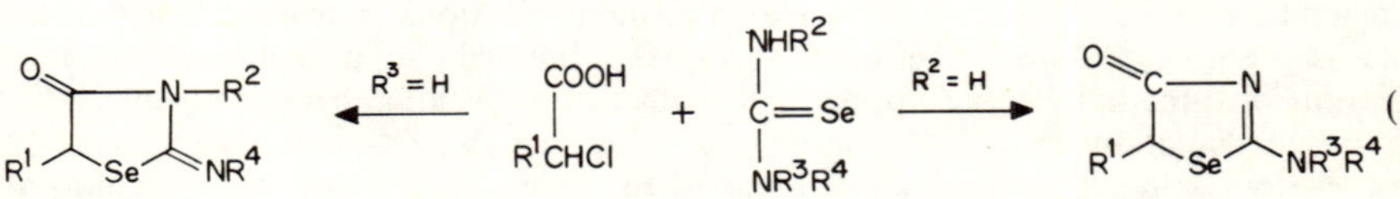

(322) (323) (141)

respectively (equation 141). The corresponding 2-alkylidene hydrazones are obtained by a similar reaction starting from selenosemicarbazones[664,665]. Generally, the chemistry of the tautomeric compounds is consistent with reactions of the 2-iminoselenazolidine form **322**, which gives easy hydrolysis without ring-opening to the 2,4-dioxoselenazolidine system, *N*-iminobenzoylation, condensation on the 5-CH_2 group with aromatic aldehydes[666] and *N*-iminoureation with isocyanates[667]. However, phenyl isothiocyanate gives reaction on the 3-nitrogen. Similarly, 2,4-selenazolidinediones are also stable and give normal Mannich products on N[667] and *N*-methylation[668–670]. In contrast, the 4-thioxo derivative, obtained by thionation of the 2,4-dione, ring-opens with hydrazine to give a 3-selenomercaptomethyl-5-oxo-2-pyrazolinone. Numerous 2-thioxo-4-oxoselenazolidine isomers are described in the patent literature, from selenomercaptoacetic acid and thiocyanates[585]. Among the selenazolidine selones, 4,4-dimethyl-5-oxo-selenazolidine-2-selone is obtained from CSe_2 and 3,3-dimethyl-2-dimethylaminoazirine, via a ring-opening[640].

(*iii*) *Mesoionic selenazoles.* Treating α-halo acid derivatives with *N*-substituted selenobenzamides, in a Hantzsch reaction, gives mesoionic selenazoles. The first derivative of this series (**324**) was obtained by Cava and Saris in 1975[671] (equation 142); **324** acts as a dipolarophile towards methylacetylenedicarboxylate and gives a pyridone with a loss of Se. The same reaction has been applied to a number of derivatives of variable stability[672,673]. The derivative with R = SMe gives 3,5-diphenylselenazolidine-2,4-dione easily[672].

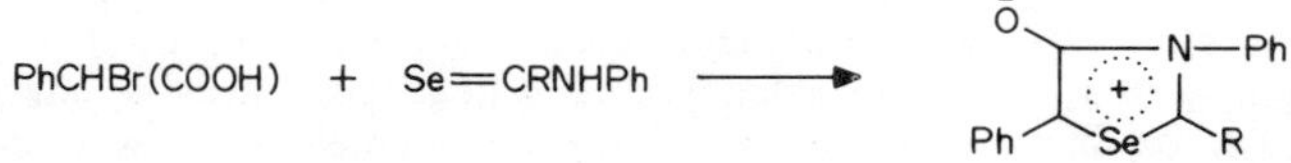

(142)

(**324**)

(*iv*) *Benzoselenazoles.* The most general synthesis of benzoselenazoles uses the reaction of the zinc salt of *o*-aminobenzeneselenolate (**325**) with acid chlorides (equation 143). In this way are obtained 2-alkyl-[672–677], 2-aryl-[676,678,679] (and among them the 2-Ph derivative, the first example of the series in 1913[680]) and 2-hetaryl-benzoselenazoles **326**[679]. Bis(2,2′-benzoselenazolyl) alkanes are formed from acid dichlorides[676] and the parent compound (R = H) is prepared from formic acid[674,676,681].

(143)

(**325**) (**326**)

Cyclic anhydrides react with one equivalent of **325**, and give benzoselenazolyl-alkane arene- or -heteroarene(pyridine)-carboxylic acids[682]. A recent alternative of this method, which uses protected selenide, starts from *o*-methylseleno-acyl- or -aroyl-anilides and phosphorus oxychloride[683] and gives **326** (R = H, Me, Ph, COOEt, COPh). Interestingly, 2-functionalized benzoselenazoles can be obtained in this manner. 2-Substituted benzoselenazoles can also be prepared by rearrangement reactions instead of benzisoselenazole isomers; e.g. 2-methylbenzoselenazole is obtained by polyphosphoric acid cyclization of *o*-methylselenoacetophenone oxime, in a Beckmann rearrangement[605]. From a benzyne addition to 3,4-dimethyl- or 3,4-diphenyl-1,2,5-selenadiazoles, Bryce and coworkers isolated 2-methyl- and 2-phenyl-benzoselenazole, respectively, in yields of

1–2%[684]. 2-Aminobenzoselenazole (**328**; R = H) and substituted derivatives can be obtained by Br_2[685,686] or H_2O_2[687] oxidation of various 1-phenyl-2-selenoureas or from cyclization of *o*-aminoaryl selenocyanates (**327**)[688] (equation 144).

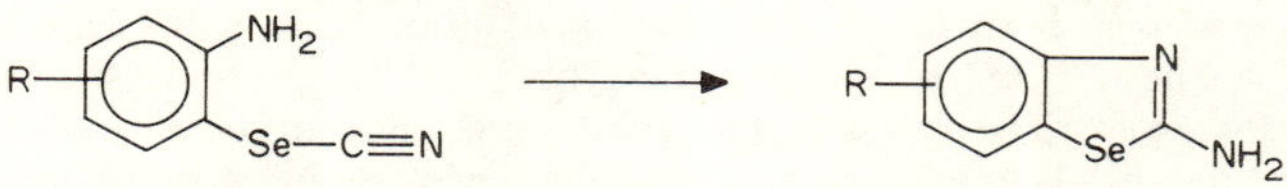

(**327**) (**328**) (144)

2-Hydroxy- and 2-mercapto-benzoselenazoles **329** (Y = O, S) can be prepared directly from **325** and phosgene or thiophosgene[685] or from bis(*o*-aminophenyl) diselenide, by reaction with ethyl chloroformate or CS_2, respectively[685,689] (equation 145). 2-Mercapto-

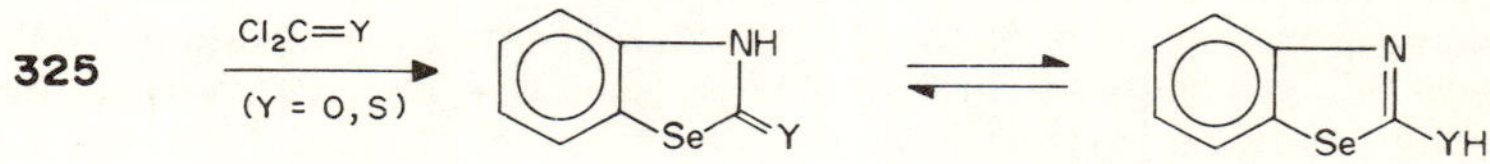

(**329**) (145)

and 2-selenomercapto-benzoselenazole can also be obtained via a diazotation of the 2-amino group, which behaves typically as an aromatic amine. The 2-chloro derivative, obtained from a Sandmeyer reaction, is easily converted by nucleophilic substitution to the 2-OH[685], 2-SH, 2-SeH[690] and 2-hydrazino groups[691]. The SH and SeH groups can be transformed to the corresponding thio or seleno ethers[690,692,693]. These compounds are described in the patent literature as vulcanization accelerators. Direct attachment of a substituent to the free 2-position is not documented. Nucleophilic amination failed[674,676]. Benzoyl chloride in the presence of KCN gives a ring fission[674], as in benzothiazole. The H/D base-catalysed exchange, measured for the three O, S, Se congeners, gives the relative k_h values of 20:1:4, respectively, a sequence different from that for benzoselenophene and its analogues[316], but similar to that of nucleophilic substitution on furan and its congeners. However, this ratio could not be rationalized by the same arguments.

Among the electrophilic reactions, halogenation gives undefined products[676] and although no Friedel–Crafts reactions have been described, mild nitration gives the 6-nitro derivatives[674,678]. Like the 2-chloro group, the 2-Me group is also activated, giving styrenes with aromatic aldehydes[674,676,694], and the 2-formyl derivative by a three-step reaction with chloral[695], by SeO_2 oxidation[696] or by a Kröhnke reaction via the 2-bromomethyl derivative[697]. The 2-carboxylic acid, obtained as a by-product[695] or by further oxidation of the aldehyde, decarboxylates easily upon melting[698]. An interesting ring-contraction with rearrangement was surprisingly demonstrated during the SeO_2 oxidation in pyridine of 3-phenyl-2*H*-1,4-benzothiazine and selenazine, giving 2-benzoylbenzo-thiazole and selenazole[598]. Benzoselenazole gives benzoselenazolium salts easily[699]. The rate of methylation is controlled by the inductive effect of the chalcogen, and the values of $\log k_{rel}$ correlate well with the pK_a of azoles. Methylation of the 2-amino derivative **328** (R = H) gives the 2-imino-3-methyl-2,3-dihydrobenzoselenazole (**330**)[685,690]. Its isomer, 2-(methylamino)benzoselenazole, is obtained directly by oxid-

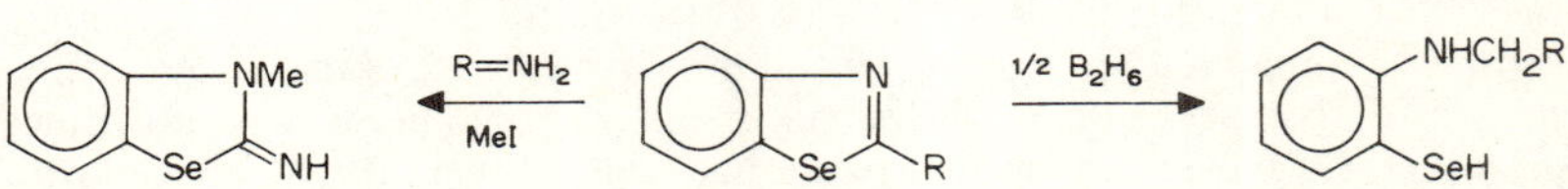

(**330**) (**331**) (**332**) (146)

ation of 1-methyl-3-phenylselenourea[685]; **330** is easily hydrolysed to 2-oxo-3-methylbenzoselenazolinone. The easy conversion of benzoselenazolium salts to benzoselenazolines is important, because these last compounds are key substances for the synthesis of cyanine dyes. They cannot be obtained by direct reduction of benzochalcogenazoles. For example, the reaction of diborane with **331** leads to *N*-alkyl-*o*-selenomercaptoanilines (**332**) via a benzoselenazaborole[700] (equation 146). This reduction is similar to those in the oxygen and sulphur series. Benzothiazole, quaternized by a $-(CH_2)_2Cl$ group, leads in basic medium to ring-expansion in 4-formyl-2,3-dihydro-1,4-benzothiazine[701]. Similar ring-expansions are successful with other halide groups. An example of application for a benzoselenazolium derivative will be described in Section VII.A.

(*v*) *Other condensed selenazoles.* Two naphthoselenazole ring systems are known. 2-Acetamidonaphtho(1,2-*d*)selenazole is obtained by cyclization of 4-phenyl-2-amino-5-(carboxyethyl)selenazole with acetic anhydride[623]. The 2-amino derivative of the (2,1-*d*) analogue is prepared by a reaction similar to equation (144), from 1-cyanoseleno-2-aminonaphthalene[685]. Other condensation derivatives on the benzene ring of benzoselenazole have been described. By heating **333** with S or Se, respectively, a thiazole or a second selenazole ring is fused on the (*f*) bond, giving **334a** and **334b**[702] (equation 147).

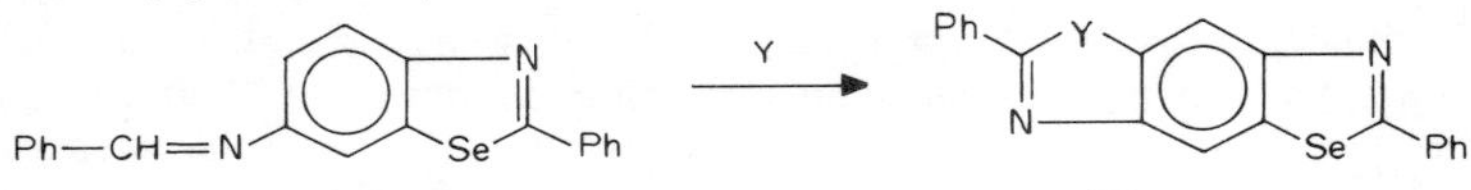

(147)

(**333**) (**334**)

(**a**) Y = S
(**b**) Y = Se

By a Doebner–Miller reaction on 6-amino-2-methylbenzoselenazole, 2,7,9-trimethylselenazolo(5,4-*f*)quinoline was obtained[703]. The corresponding 6-hydrazino-2-methylbenzoselenazole gives by successive condensation with cyclohexanone and dehydrogenation with chloranil 2-methylcyclohepta(4,5)-pyrrolo(3,2-*f*)benzoselenazole, an aza-azulene derivative[704]. Among heterocondensed systems on the selenazole ring some are condensed on the 4,5-bond. (2′-Thienyl and selenienyl)-2-selenazole analogues of phtalazine have been obtained by reaction of hydrazine on the previously mentioned diester[613]. 2-Aryl-5-(ethoxycarbonyl)pyrrolo(3,2-*d*)selenazole (**335**) has been obtained by constructing the pyrrole ring on the selenazole ring via the corresponding 3-chloromethyl- → 3-formyl- → 3-azidovinyl-selenazole[705].

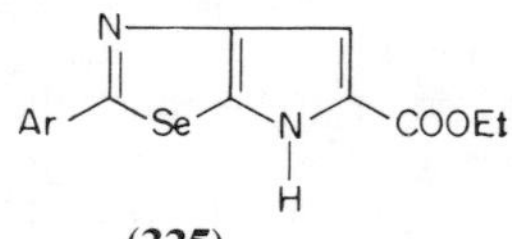

(**335**)

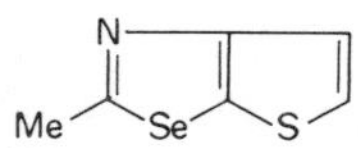

(**336**)

In contrast, the 2-methylthieno(3,2-*d*)selenazole system (**336**) has been synthesized from the thiophene ring, via the ring-closure of 3-acetylamino-2-acetylselenothiophene[419]. However, an attempted cyclization of 2-amino-3-acylselenothiophene failed to give the (2,3-*d*) isomer[412]. On the other hand, some selenazoles are condensed on the 2,3-bond, giving a nitrogen common to two rings. Nearly all these systems are obtained by starting from selenazole derivatives. Some 2-bromoalkyl- or 2-bromomethylaryl-substituted benzoselenazoles (**337**) give intramolecular quaternary salts of type **338**, which lead with bases to isoindolo(1,2-*b*)benzoselenazole (**339**) or its pyridopyrrolo analogue[706] (equation 148).

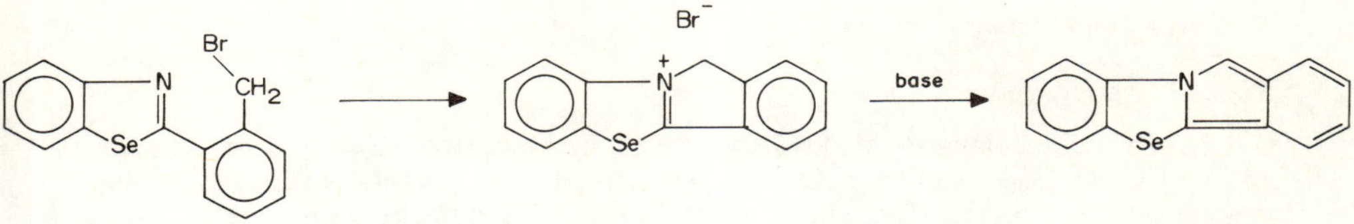

(337) (338) (339)

(148)

The selenazolopyrimidine system **340** is obtained by reaction of 2-amino-4-substituted selenazoles with ethyl propiolate[619] (equation 149). The Se in the ring of the levamisole analogue **341** is part of an imidazo(2,1-*b*)selenazole system[192a,707,708]. Selenazolo-*s*-triazine systems **342** (R = Ph, Y = NPh) are prepared by reaction of phenylisothiocyanate with *N*-phenyl-*N'*-(2-selenazolyl)thiourea. Reaction of 2-aminoselenazoline and ethoxycarbonyl isothiocyanate, gives **342** (R = H, Y = O)[709]. Selenazolo(3,2-*b*)-*as*-triazines **343** can be obtained from 6-substituted 2,3,4,5-tetrahydro-*as*-triazin-5-one-3-selone, ethyl bromoacetate and acetic anhydride[710]. Condensation of 2,4-diamino-5-phenylselenazole with ethyl ethoxymethylenecyanoacetate gives tricyclic polyaza systems bearing a selenazole ring[631]. 2-Methylbenzoselenazole can be condensed with malonic ester to give a tetracyclic benzoselenazolo(3,2-*a*)pyridine[711], and with 2 moles of ethyl acetylenedicarboxylate to give a selenazolo(3,2-*a*)azepine derivative[712]. Benzoxepino- and benzothiepino-(3,4-*d*)selenazoles are prepared from 2-aryl-4-(chloromethyl)selenazoles[713].

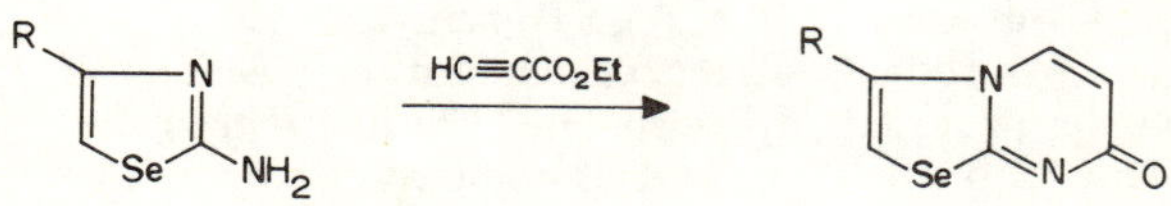

(149)

(340)

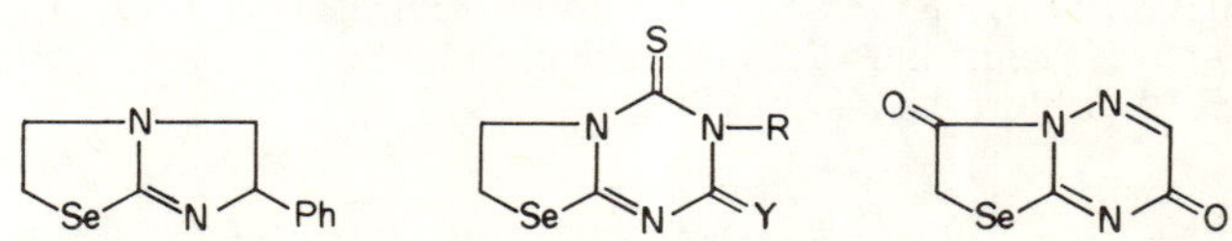

(341) (342) (343)

(*vi*) *Tellurazoles*. No derivative of the tellurazoles ring was known until 1983. The laboratory of Liège succeeded very recently in preparing the first five derivatives of the benzotellurazole ring, including the parent compound itself (**345**; R = H, Me, Ph, COOEt, COPh) (equation 150)[683]. They are obtained in low yields from *o*-(methyltelluro)-anilides **344** by a method which is also applicable to benzoselenazole derivatives.

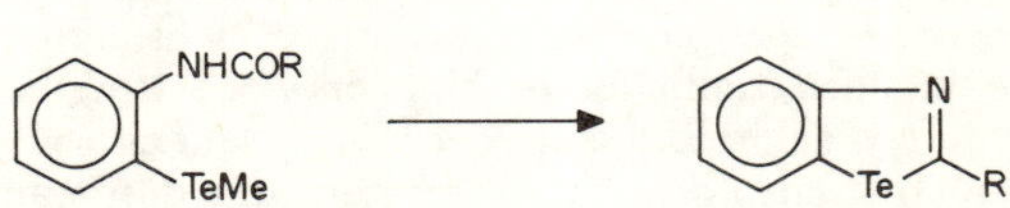

(150)

(344) (345)

C. Three Heteroatoms

1. Se/Te rings without nitrogen

1,3,2-Dioxaselenolan-2-oxide (**346**), which is a cyclic organic selenite, is obtained by reaction of selenous acid or SeO_2 with ethylene glycol[714] (equation 151). A similar reaction has been previously applied to erythritol[715]. A 2,2-diphenylselenurane derivative (**348**) is postulated as intermediate in the oxidation of catechols (e.g. **347**) to *o*-quinones by

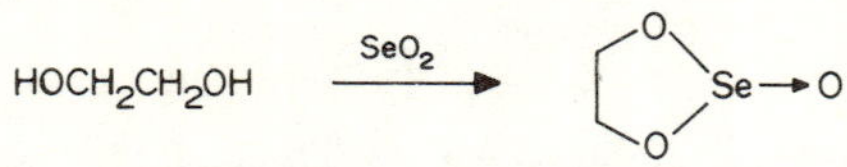

(151)

(**346**)

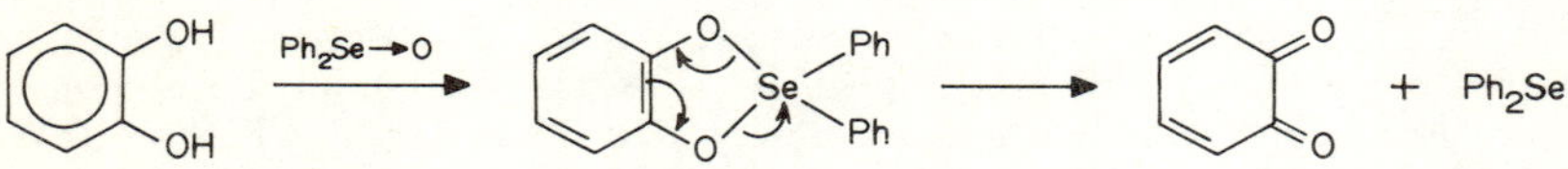

(152)

(**347**) (**348**)

diphenyl selenoxide[716] (equation 152). A diaryl telluroxide gives the same overall reaction, but the intermediate postulated is acyclic[717]. 2,3-Dimercaptoquinoxaline, gives similarly a dithioselenite, i.e. 1,3,2-dithiaselenolo(4,5-*b*)quinoxaline-2-oxide[718]. Unsubstituted 1,2,3-triselenolane is obtained as a by-product in the hydrolysis of diselenocyanoethane[553]. 2,1,3-Oxadiselenane-1,3-dioxide, which may be regarded as a diseleninic anhydride, is obtained by peracid oxidation of ethylene bis(1,2-diselenide), diselenocyanoethane[719] or 1,4-diselenane[720,721]. 3,5-Dimethylene substituted 1,2,4-triselenoles are obtained as by-products of the decomposition reactions of 1,2,3-selenadiazoles[722], probably via phenylethynyl selenolate. The first tritellurole system, namely E,E-1,3-dibenzylidene-1,2,4-tritellurole (**351**) is obtained as a by-product in the HCl treatment of the corresponding phenylethynyltellurolate (**349**), probably by addition of an intermediate ditelluride anion Ph-C≡C-Te-Te⁻ to the transient telluroketene **350**[723a] (equation 153). The structure of **351** is confirmed by X-ray diffraction. 5,10-Episeleno- and 5,10-epitelluro-5,10-dihydroarsanthrene are prepared from the corresponding 5,10-dichloro derivative[723b].

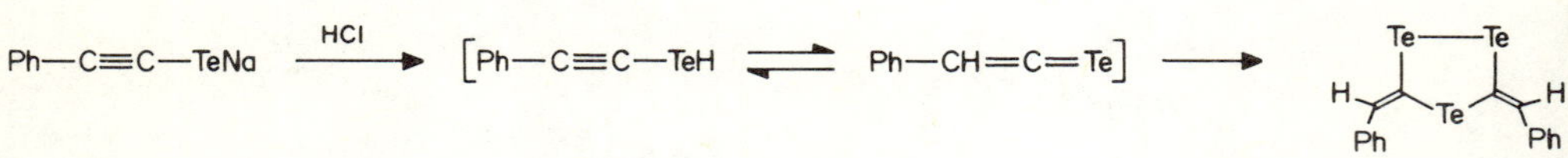

(**349**) (**350**) (**351**)

(153)

2. Nitrogen Se/Te rings

a. One nitrogen in the ring. Two classes of compounds are included in this group. The first involves 1,2,3-benzodichalcogenazolium cations **353** which are the Se analogues of Herz salts. They are obtained from *o*-aminothiophenol or selenophenol hydrochlorides (**352**; Y = S, Se) and thionyl chloride, giving **353a**[724] or selenous acid, giving **353b**[725] and **353c**[726] (equation 154). It is also possible to transform the dithio analogue to **353b**, by reaction with selenous acid, a reaction which is easier for non-electron-donating substituents and in acidic solution. The *S*-oxide, which reacts vigorously with H_2SeO_3, is a

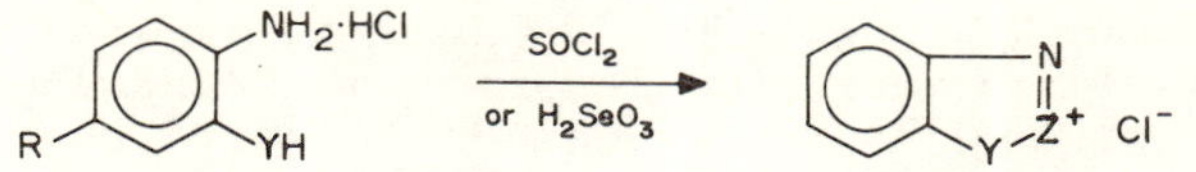

(154)

(352) **(353)**

(**a**) Y = Se, Z = S
(**b**) Y = S, Z = Se
(**c**) Y = Z = Se

possible intermediate[727]. The 2-*Se*-oxide is formed by buffered hydrolysis of the salt **353b**[728]. Two stereoisomers are observed, which are interconverted in acidic medium. Hydrolysis of **353a** gives di(*o*-aminophenyl)diselenide[729]. In reaction with amines[730] and especially with *p*-toluidine in acetic acid, **353c** reacts four times faster than its dithio analogue, in agreement with charge densities based on MO calculations[731].

The second class of compounds, obtained by Burger and coworkers according to equation (155), are of the 3*H*-1,2,4-dichalcogenazole type **355** (**a**: Y = Se, Z = S[732]; **b**: Y = Z = Se[733], also obtained from hexafluoroacetone and selenoamides; **c**: Y = Te, Z = S[734]). Some ring transformations, leading to the first cyclic selenophosphorane[735] or using these heterocycles as synthons for heterodienes, have been developed[736a]. Some 1,2,4-diselenazolium salts are obtained by I_2 oxidation of selenoureas[736b].

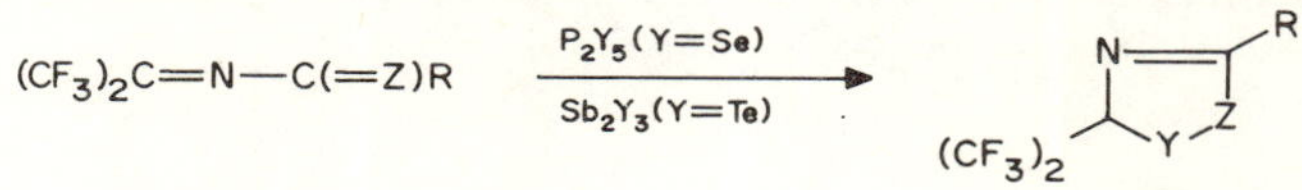

(155)

(354) **(355)**

b. Two nitrogens in the ring: chalcogenadiazoles. Derivatives of all four possible isomers of selenadiazoles are known. In the Te series, only 1,2,5-telluradiazoles have been described.

(*i*) 1,2,3-*Selenadiazoles.* 1,2,3-Benzoselenadiazole derivatives (**357**), namely 5-Me, 4-Me and 5,6-di-Me derivatives, have been known since 1935. They are obtained by diazotation of *o*-aminoselenophenols (**356**)[737,738] (equation 156). No further work was

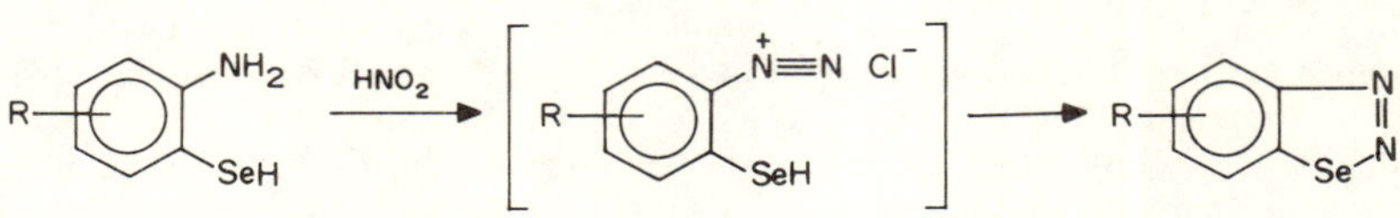

(156)

(356) **(357)**

reported until 1975, when the unsubstituted **357** was prepared by the same method[556]. The non-fused system has been known only since 1969, when Lalezari and coworkers obtained 4-phenyl-1,2,3-selenadiazole by SeO_2 oxidation of phenyl ketone semicarbazones[739] (equation 157). The method has been used for the preparation of various 4-substituted[740] and 4,5-disubstituted[741] derivatives.

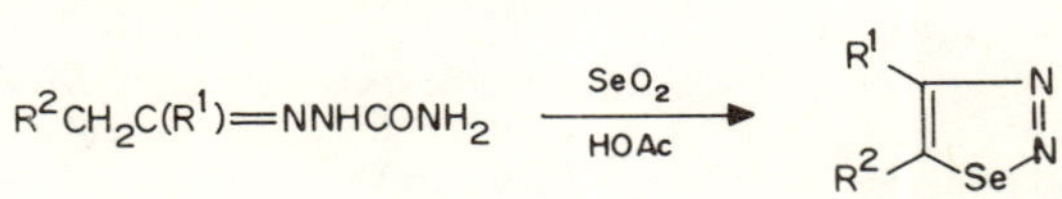

(157)

(358) **(359)**

Starting from aldehyde semicarbazones, the method could be applied to the parent heterocycle, and to various 5-substituted derivatives[741]. The parent compound (**359**;

$R^1 = R^2 = H$) is stable in the dark at 0 °C, but it decomposes slowly under other conditions. The method has been used for obtaining the three 4-pyridyl-[531], 5-cyano-, 5-ethoxycarbonyl-[741], 4,5-cycloocteno-[107,742,743] and other cycloalkeno-selenadiazoles[107]; the 4,4′-alkyl bis-selenadiazolyls[744], 4,5-indeno[745] and other 4,5-benzoalkeno derivatives[746]; the 4-*β*-naphthyl-, phenanthrenyl- and steroidal[747], 5-(1,3,4-oxadiazolyl)-[748], 4-vinyl- and -butadienyl-[749], 4-aroylmethyl-[750] and 5-arylsulphonyl-selenadiazoles[751]. ^{75}Se-labelled[752] and 5-deuterio-1,2,3-selenadiazoles[753] have been synthesized. SeO_2 is therefore an excellent Se-transfer reagent in these syntheses. Fused phenanthro(*d*)-[754] and benzoselenolo(3,2-*d*)-1,2,3-selenadiazoles[755] are among the rare fused aromatic systems obtained by this method. The regioselectivity of the ring-closure has been studied[756]. 1,2,3-Benzoselenadiazole can be quaternized on the 3-nitrogen[757] but the 1,2,3-selenadiazole ring is generally a rather unstable system, which shows versatile utility in organic synthesis. Unlike the 1,2,3-thiadiazole ring system, the corresponding selenadiazoles are easily pyrolysed with loss of N and Se to give high yields of alkynes[107,531,740,741–743,744,746,747,758]. When the selenadiazole ring is fused to a ring which is lower than eight-membered the acetylene formation is unfavoured and the denitrogenated intermediate can dimerize to diselenin derivatives **361**, which can lose in turn a Se atom to give selenophene derivatives[107,746,747,758] (equation 158). In the other cases, the proportions of compounds **360**–**362** depend on the temperature, concentration and medium of the thermolysis. As seen in a previous section, diselenetane derivatives can also be isolated as by-products[57].

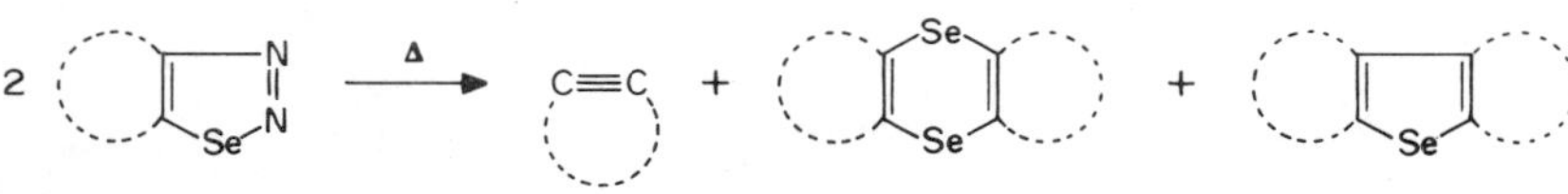

(158)

(360) **(361)** **(362)**

4-Aryl-1,2,3-selenadiazoles can be directly transformed to 2,4- and 2,5-diarylselenophenes without passing through the corresponding selenin, and the mechanism of this reaction has been discussed[108]. The 1,2,3-selenadiazole ring is unstable towards Br_2, and from the bromodediazoniation bis(2-bromovinyl)diselenide was isolated[759]. Reaction with bases (OH^-, EtO^- [563,760,761], BuLi[762]), results initially in proton abstraction from the 5-position (equation 117), and subsequent ring-opening gives an alkyne selenolate **265** which can be dimerized under acid catalysis to diselenafulvenes **266**. It has been demonstrated that selenoketenes are also intermediates in the diselenafulvene formation[532] which involves cycloaddition of alkyne selenolate to the selenoketene species. 1,2,3-Selenadiazoles are also photolysed to alkynes, together with small amounts of diselenafulvenes[763]. Alkyne selenol and selenoketene were identified in these reactions[764]. Intermediate alkyne selenolates can be easily transformed to 1,3-diselenolethiones or -selones by addition of CS_2 or CSe_2, respectively (see above), to selenoamides by addition of secondary amines[761,762] and to seleno esters by reaction with alcohols[741].

(*ii*) 1,2,4-*Selenadiazoles*. The chemistry of 1,2,4-selenadiazoles is poorly known. This ring system can be obtained by two synthetic methods. The first one, which led to the two first derivatives in 1904[765] and which was developed in 1978 by Cohen[766], consists of the oxidative N–Se coupling of aryl- or hetaryl-selenoamides by I_2

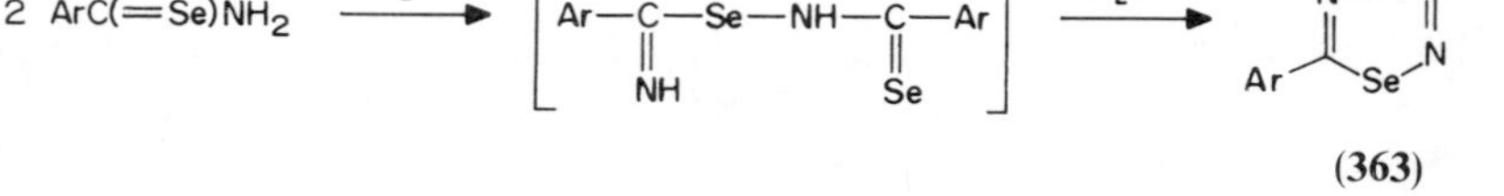

(159)

(363)

(equation 159). In this way the 3,5-disubstituted derivatives **363** (Ar = Ph, *p*-Tol[765], 2-thienyl) are obtained. The second method, developed by Goerdeler in 1963, is analogous to the preparation of 1,2,4-thiadiazoles and consists of the reaction of KCNSe with *N*-haloamidine **364**. It leads to the 5-amino-3-substituted derivatives (R = Ph, Me) **365**[767] (equation 160). A recent synthesis of 1,2,4-selenadiazolidine-3,5-dione consists of dehydrocyanation of allophanoyl selenocyanate, the first well-documented acyl selenocyanate[768].

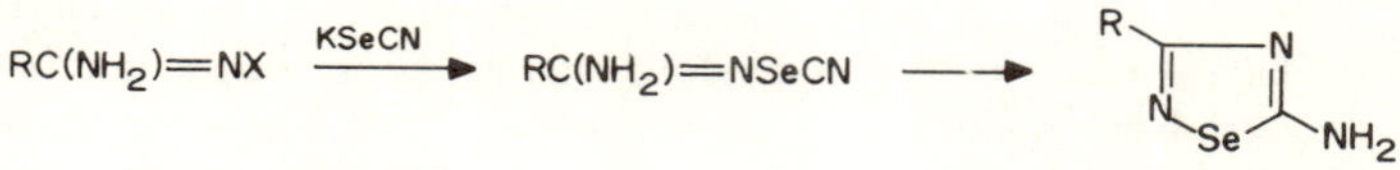

(160)

(**364**) (**365**)

(*iii*) 1,2,5-*Selenadiazoles*. Fused 1,2,5-selenadiazoles were among the first known Se–N heterocycles. The benzo-condensed system, 2,1,3-benzoselenadiazole (**367**) also called piaselenole or piazselenole, was prepared by Hinsberg in 1889 from phenylenediamine **366** and SeO_2[769] (equation 161).

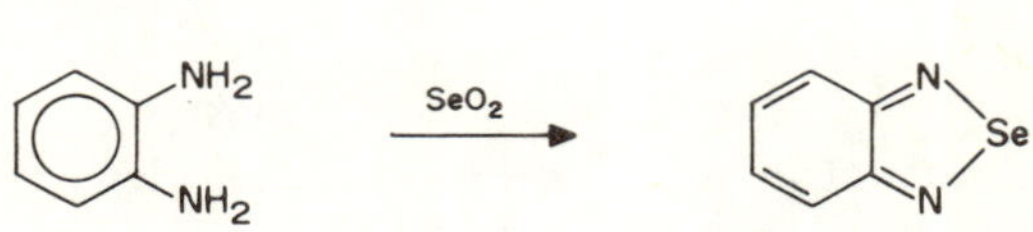

(161)

(**366**) (**367**)

This very reproducible reaction has been used for absorption or fluorescence spectrophotometric detection of small amount of Se in chemical or natural media, for the detection of *o*-diamines and for the coulometric titration of the Pd^{2+} ion. It is the most widely used route to fused 1,2,5-selenadiazoles. The following fused systems[315] are obtained from SeO_2 or $SeOCl_2$: naphtho(1,2-*c*)-[315,769] and naphtho(2,3-*c*)-phenanthro(9,10-*c*)-[315,770], fluoreno(2,3-*c*)-[771] and pyrazolo(3,4-*c*)-1,2,5-selenadiazoles[772]; (1)benzothieno(2,3-*e*)-[773], (1)benzoselenolo-(2,3-*e*)-[774] and -(3,2-*e*)[775], thiazolo(4,5-*g*)-[776,777] and selenolo(3,2-*e*)-2,1,3-benzoselenadiazoles[315]; 1,2,5-selenadiazolo-5*H*-(3,4-*b*)carbazole[778]; 6*H*-(3,4-*c*)carbazole[778]; (3,4-*b*)- and (3,4-*c*)-pyridine[779,780]; (3,4-*d*)pyrimidine (8-selenapurine)[781–783]; (3,4-*c*)-1,2,5-thiadiazole[784]; (3,4-*b*)quinoxaline[784]; (3,4-*b*)- and (3,4-*h*)-quinoline[785] and the quinonoid systems naphtho(2,3-*c*)-1,2,5-selenadiazole-4,9-dione and 4*H*,8*H*-benzo(1,2-*c*:4,5-*c*′)bis-1,2,5-selenadiazole-4,8-dione[786].

From *N*-methyl-*o*-phenylenediamine are obtained similarly 1-methyl-2,1,3-benzoselenadiazolium salts[787] which can also be prepared by methylation of **367**[788a]. This compound, and the corresponding 1,3-dimethyl-bis-quaternized salt, are easily pyrolysed at 70–110 °C to give, respectively, unsubstituted and 1-methyl-benzimidazole[788b]. The *N*-oxide derivative and some analogues are obtained from *o*-benzoquinone dioxime and Se_2Cl_2[789,790]. The *N*-oxide can be thermolysed to **367** and benzofuroxan[790], and photolysed to benzofuroxan and Se, probably via 2-(selenonitroso)nitrosobenzene[791]. The chemical properties of 2,1,3-benzoselenadiazoles are similar to those of the corresponding thio compounds. Halogenation and sulphonation occur mainly in the 4-position[792]. However, **367** gives many ring-scission reactions including reductions to phenylenediamine under Stephen's conditions[792], and ring-opening by phenyllithium to *N*-phenylethylenediamine and diphenyl selenide via a double selenophilic Se—N rupture. Dienophilic behaviour towards arynes gives the 1,2-benzoselenazole ring system[592–594], and towards ethyl acetylenedicarboxylate gives a quinoxaline diester[592]. The polar nature of 2,1,3-benzochalcogenadiazoles is manifested by their dipole moments (O: 4.04 D; S: 1.79; Se: 1.19). The mesomeric charge transfer, which is not extensive for O, increases in importance for S and Se[793].

A review on fused 1,2,5-thiadiazoles and 1,2,5-selenadiazoles was published recently[794]. It must be noted that the first description of a monocyclic 1,2,5-selenadiazole is much more recent. In 1967, Shealy described the formation of the 4-amino-3-carboxy derivatives by degradation of the pyrimidine ring of 1,2,5-selenadiazolo(3,4-*d*)pyrimidin-7(6*H*) one[795]. The first general synthesis was reported nearly simultaneously by two other groups and involves a Se transfer from Se_2Cl_2 or SeO_2 to ethylene diamine[796,797] (equation 162). Several 3- and 3,4-substituted derivatives have been obtained by this method[797,798]. The parent system **368** undergoes nucleophilic attack by butyllithium on the Se centre, which is easier than that on 1,2,5-thiadiazole, giving, as for the benzo derivative, a double Se—N cleavage with formation of an α-diketone and dibutyl selenide[799]. The nucleophilic attack of organometallic compounds on chalcogen have been used to prepare 3-vinyl-1,2,5-selenadiazole by a one-pot ring-transformation of 1,2,5-thiadiazole with vinylmagnesium chloride and Se_2Cl_2[800]. The 2-oxide of **368** has been prepared similarly to its benzo analogue[690]. Photolytic cleavage of derivatives of **368** gives a nitrile selenide as the initial product[801,802].

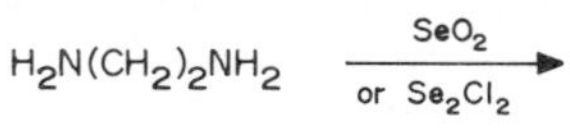

(368)

(162)

(*iv*) 1,3,4-*Selenadiazoles*. The first 1,3,4-selenadiazole derivatives (**370**, $R^1 = R^2 =$ Ph, Me) were obtained in low yield by Stollé and Gutmann in 1904, by heating *N,N*-dibenzoylhydrazine or its diacetyl analogue with phosphorus pentaselenide[803] (equation 163). Unsubstituted 1,3,4-selenadiazole (**370**, $R^1 = R^2 =$ H) is obtained in 25% yield

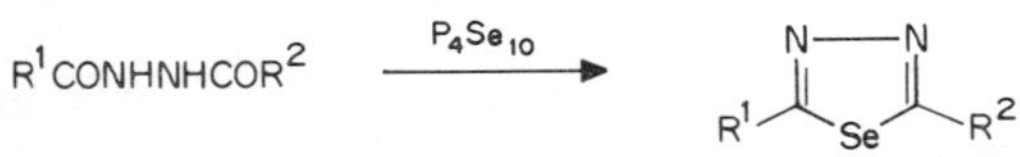

(163)

(369) **(370)**

from the azine of DMF, a synthetic equivalent of **369**, and H_2Se[804], together with *N,N*-dimethylselenoformamide. Its bond angle, of 81.8°, is the smallest angle known for a planar five-membered ring[805]. Via the diselenoxo derivatives of **369**, obtained by hydrazinolysis of two equivalents of selenoamides, Cohen similarly prepared some derivatives of **370** ($R^1 = R^2 =$ Ar,2-thienyl)[806]. Likewise, starting from 1-acyl-4-benzoylselenosemicarbazide (monoselenoxo derivative of **369**, $R^2 =$ NHCOPh), 2-benzoylamino derivatives of **370** ($R^1 =$ Ar, Me, Pr, 2-furyl) were prepared [807]. These last derivatives give by hydrolysis 2-amino-1,3,4-selenadiazoles (**370**; $R^2 = NH_2$)[807], which were previously obtained by the same method in a one-pot reaction. The precursors, 1-acyl- or -aroyl-selenosemicarbazide, were prepared *in situ* from selenosemicarbazide, an acid and phosphorus oxychloride[808]. 2-Amino-1,3,4-selenadiazoles behave as typical aromatic amines, giving azomethines with aromatic aldehydes, and azo compounds via diazotation. They are alkylated on the 3-nitrogen, giving the tautomeric 2-imino-1,3,4-selenadiazoline form[809]. The 3-aryl derivatives of this imino form **373** are obtained from α-halohydrazones (**371**) and either KSeCN[810,811] or better, aroylselenourea or selenosemicarbazide[809] (equation 164).

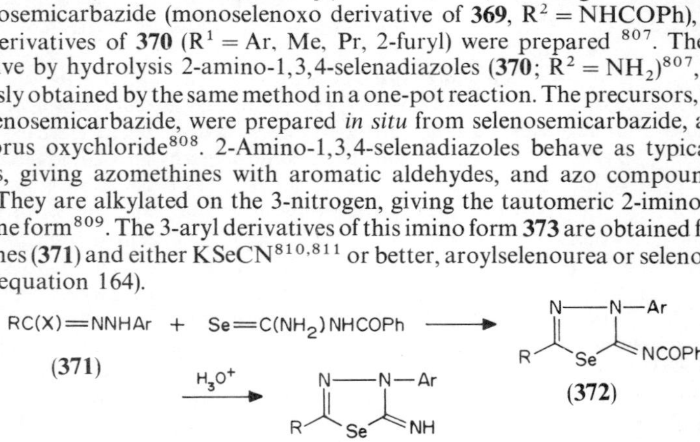

(371) (372) (373)

(164)

373 gives by nitrosation the 2-nitrosoamino derivatives, which are easily hydrolysed to the corresponding 2-oxo-Δ^4-1,3,4-selenadiazolines[810,811]. A Δ^3-1,3,4-selenadiazoline has been obtained as intermediate in the reaction between di-*t*-butylselenoketone and diphenyldiazomethane, which by thermolysis gives olefins with loss of Se and N[812]. 2-Iminoselenadiazoline derivatives or their 2-aminoselenadiazole tautomers, are starting materials for the synthesis of the following fused 1,3,4-selenadiazoles: imidazo (2,1-*b*)-1,3,4-selenadiazoles[808,809]; 1,3,4-selenadiazolo(3,2-*a*)pyrimidin-7-ones, by reaction with methyl acetylenedicarboxylate[619]; (3,2-*a*)pyrimidium perchlorates, from acetylacetone or diethyl malonate[813], and (2,3-*b*)quinazolines[814].

(*v*) 1,2,5-*Telluradiazoles*. The 1,2,5-telluradiazole ring system was the first known monocyclic ring system with a Te—N bond. It was obtained by Bertini in 1982, from the corresponding 1,2,5-thiadiazole or better selenadiazole, via a R.O.II substitutive chalcogenophilic ring-opening, followed by reaction of the dimetalloimine **374** with $TeCl_4$[815] (equation 165). This new Te system (**375**) is hydrolysed by acids to the α-dione, ammonium cation, Te and tellurous acid. As for benzisotellurazole, the melting point is anomalously high (> 185 °C), and the solubility anomalously low. Te has been trapped by *o*-chloromercuridiazobenzene, giving a compound $C_{12}H_9ClN_2Te$, which was reduced to di-(*o*-aminophenyl) ditelluride and whose X-ray crystallography was in agreement with a quasi-1-chloro-1,2,3-telluradiazole ring system[816].

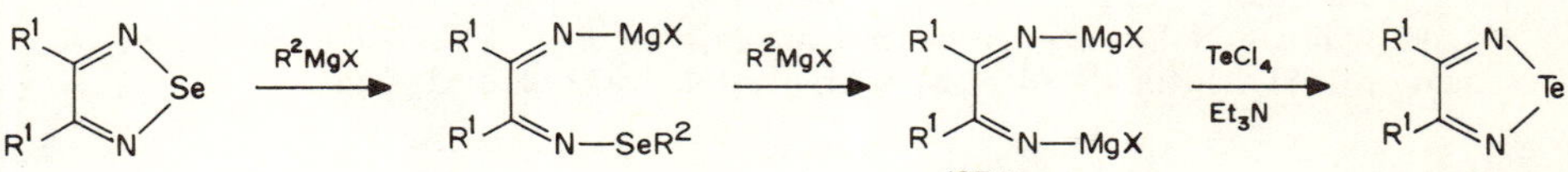

(**374**) (**375**)

(165)

(*vi*) *Chalcogenaazapentalenes*. The reaction products of SeO_2 with 1,3-cyclohexanedione oximes, previously assigned as monoheterocyclic systems[817], were later proved by NMR[818,819] and X-ray analysis[820,821] to possess a no-bond resonance system of type **377**, as in heterapentalenes. They can be considered as aza analogues of chalcogenapentalenes; 2,5-diaza-1,6-dioxa-6*a*-selena-Se (IV) pentalenes, obtained in this manner have a very short Se—O contact of 2.02 Å. The reaction has been extended to the corresponding Te compounds[822,823] and to various 1,3-dioximes (equation 166) giving the parent compound **377** (R = H)[823] and to the S analogue with the aid of SCl_2[824]. Compounds **377** are sensitive to alkali and to reducing agents, giving, for example, the 1,3-dioxime of 1,2,3-cyclohexanetrione, and with hydroxylamine the corresponding trioxime. They give no electrophilic substitution[823]. By replacing the starting dioxime **376** by an arylhydrazone, a 1,2,5,6-tetraaza analogue is obtained, both in the Se and in the Te series[825].

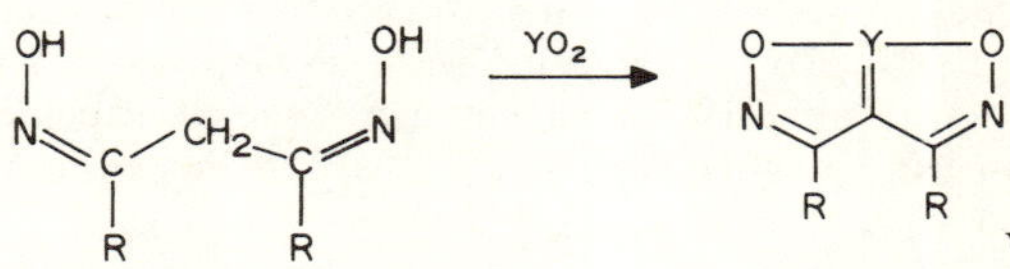

(166)

Y = Se, Te

(**376**) (**377**)

The method using the acidity of the 3-methylene group of 1,2-dichalcogenolylium salts can be applied to the reaction of 3-methyl(ene)-1,2-diselenolylium salts (**378**) with arenediazonium fluoroborates, giving 6,6*a*-diselena-1,2-diazapentalenes (**379**)[826a] (equation 167; $R^1 = R^2 = R^3 = H$, Me). A parallel nitrosation gave a 1-oxa-6,6*a*λ^4-diselena-2-aza analogue[826b]. This last compound can be transformed with Hg(II) acetate into the 1,6-dioxa-6*a*-selena-2-aza analogue[827a]. 3,4-Dimethyl- and 3,4-trimethylene-6*a*λ^4-thia-6-selena-1,2-diazapentalenes are obtained from the corresponding 6-oxa derivative, by reaction with the Vilsmeier reagent and potassium selenosulphate[827b].

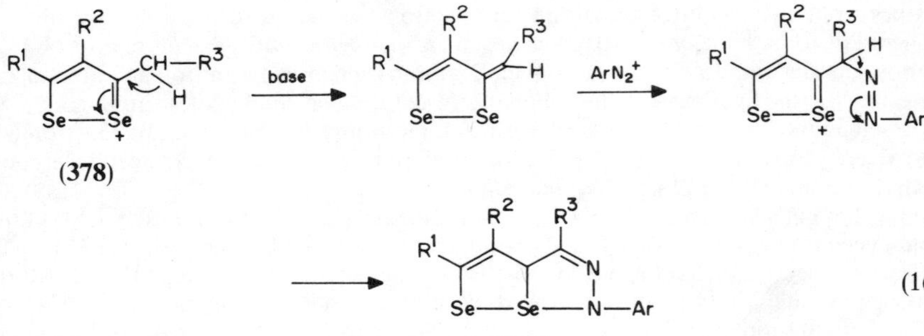

(378)

(379)

(167)

D. Four Heteroatoms

The first selenatriazole ring system has been recently prepared by treating 1,2-diamino-4-phenylimidazole (**380**) and its 4,5-diphenyl analogue with SeO_2, giving 5-phenylimidazolo(1,2-*c*)-1,2,3,5-selenatriazole (**381**)[772] and the corresponding 5,6-diphenyl analogue, following the general reaction of SeO_2 with *o*-diamino derivatives (equation 168). Benzimidazolo(1,2-*c*)selenatriazole was obtained similarly.

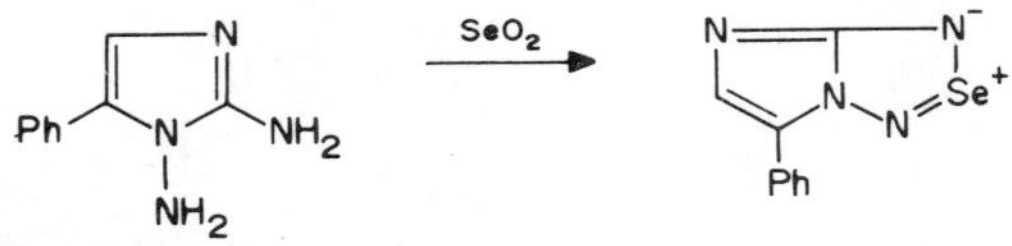

(168)

(**380**) (**381**)

Since then, another ring system, 5-diethylamino-1,2,3,4-selenatriazole, has been prepared from 4,4-diethylselenosemicarbazide and HNO_2. Its half-life time is 180 h in $CHCl_3$ at 20 °C[827c].

V. Se/Te SIX-MEMBERED RINGS

A. One Heteroatom

1. Monocyclic systems

a. Selenanes and telluranes. Selenanes and telluranes **382** are obtained from pentamethylene halides and Na_2Se[828], Se and rongalite in alkaline medium[829], MgTe or Al_2Te_3[830], Te[232] or Na_2Te[831] (equation 169). 2-Methyl-[832] and 2,6-dicarboxyselenanes[833] are obtained similarly. Selenane is also obtained by passing pentamethylene oxide and alumina in a stream of H_2Se and in 57% yield from pentane, SeO_2 and a zeolite catalyst[85].

(169)

(**382**)

Y = Se, Te

The conformations of selenanes and telluranes[834] and their halogen adducts[835] have been studied as a part of a more extended work on the conformational analysis of pentamethylene heterocycles, which has been reviewed[836]. The inversion barrier decreases with the size of the chalcogen (ΔG_{inv}:O:10.5 kcal mol^{-1}; S: 9.4; Se: 8.3; Te: 7.3). Tellurane-1, 1-diiodide has been submitted to nucleophilic substitution on Te by various nucleophiles (F, CN, NCO, NCS, NCSe, CCl_3)[837].

b. Selenopyrans. The parent compound **384** (4*H*-selenin) has been prepared in 16% yield, in the same time as its sulphur analogue, by reaction of H_2Se and HCl on glutaraldehyde[838] (equation 170). The 4-methyl[839], 2,4,6-triphenyl and 3,5-dimethyl-2,6-diphenyl[840a] derivatives are synthesized in a similar way. 6-Phenyl-2-formylmethylene-2*H*-selenopyrans were obtained recently from the selenobenzamide vinylogue and β-chlorocrotonaldehyde[840b]. No monocyclic telluropyrans are so far known.

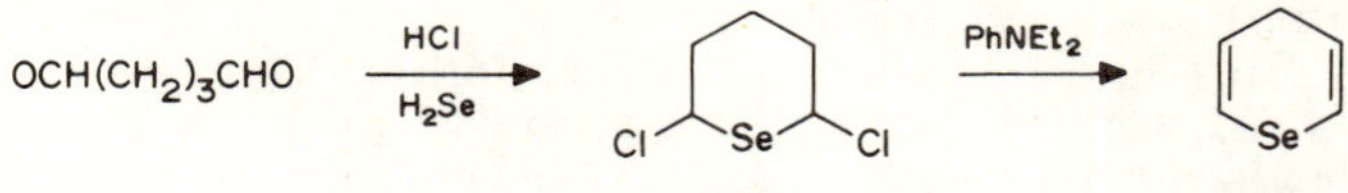

(170)

(383) **(384)**

c. Chalcogenopyrones and tetrahydro derivatives. 2,6-Diarylselenan-4-ones (**386**) with *trans*-aryl groups[843] are obtained by addition of H_2Se to distyryl ketones **385**[841–844] (equation 171). The reaction also works for the keto derivative obtained in one pot from

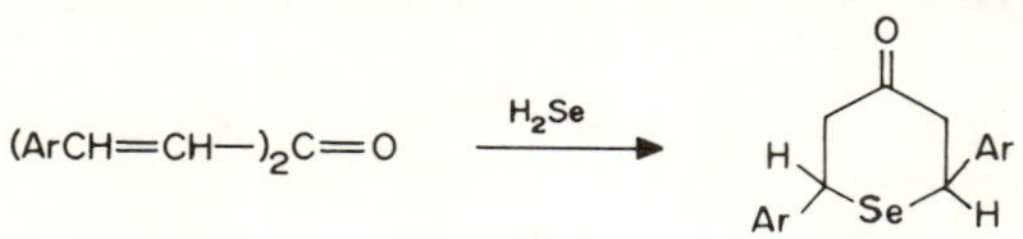

(171)

(385) **(386)**

diethyl acetonedicarboxylate and acetaldehyde, giving 2,6-dimethyl-3,5-diethoxycarbonylselenan-4-one[845]. Conformational studies by NMR, suggest a chair conformation with four equatorial substituents[845]. Conformational studies have also been conducted for **386**, their cyanhydrins[844] and the corresponding selenanols[846]. The same method has been applied to the preparation of 2,6-disubstituted 4*H*-4-selenopyrones and -telluropyrones (**388**), starting from the corresponding dialkynyl ketones (**387**) (equation 172; R = Me, Ph, *t*-Bu)[847,848]. These precursors are prepared directly in one pot from alkynes, ethyl formate and MnO_2[848,849]. Using tellurobis(tri-*t*-

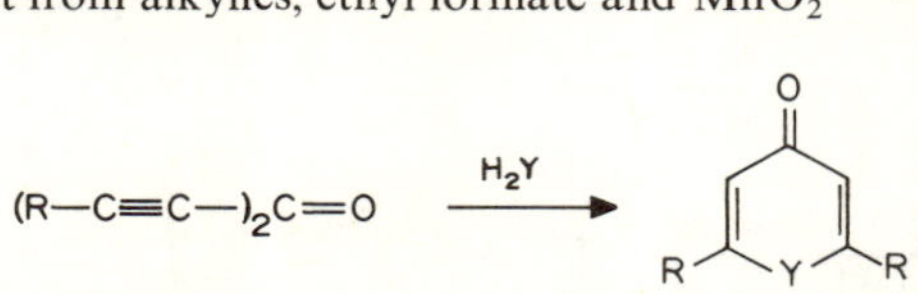

(172)

(387) **(388)**

Y = Se, Te

butylsilane) as a nucleophile, the reaction gives, in addition to **388**, the isomeric five-membered 5-substituted 2-arylidene-3-oxo-2,3-dihydrotellurophene as a major product[848]. A 2-selenopyrone isomer of **388**, i.e. **390**, has been prepared from a selenoamide vinylogue **389** and arylacetyl chloride[850] (equation 173). 3,5-Dioxo-1,1-dichlorotelluranes, erroneously formulated in the earlier publications, were obtained by the Morgan's group in the twenties from 2,4-pentanedione derivatives and $TeCl_4$[851–853]. These structures have been recently confirmed by X-ray diffraction and NMR spectroscopy[854]. The reaction follows a different route in the Se series, giving 1,3-diselenetane derivatives as previously mentioned[60].

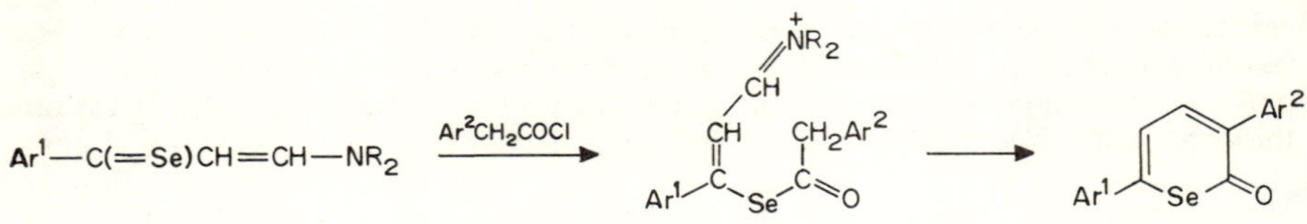

(389) (390)

(173)

d. Chalcogenopyrylium salts. Selenopyrylium (or seleninium) salts **391**, comprising the parent compound, have been prepared either by hydride abstraction from selenopyrans of type **384**[838,839] (equation 174) or directly from δ-diketones[855]. Their relative stability has been studied and compared with that of the thio analogues[856]. A 2-amino-6-aryl-3-substituted selenopyrylium salt has also been obtained from **389** and derivatives of malononitrile[850]. 2,6-Diphenyltelluropyrylium salts have been recently synthesized, as precursors for the preparation of telluropyrylium dyes[857]. The 4-ethoxy, the 4-*p*-dimethylaminophenyl and the 4-methyl derivatives are obtained from 2,6-diphenyltelluropyrone and ethyl fluorosulphate, *p*-dimethylaminophenylmagnesium bromide and Meldrum's acid respectively, followed by reduction and decarboxylation.

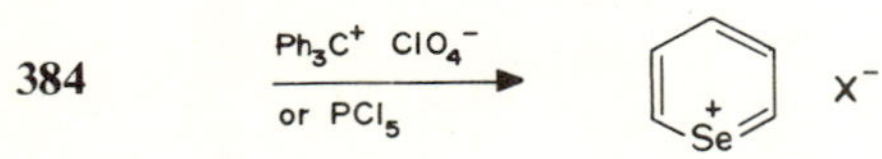

(174)

(391)

e. Bis-4,4′-chalcogenopyranylidenes. Compounds **393** may be good electron-donating heterocycles, since the corresponding cations have an aromatic 6π structure. The O analogue was discovered in 1975 to form electrically conductive complexes with acceptors. The corresponding S[858–860], Se[849] and Te analogues[861] are synthesized by copper coupling of chalcogenopyran-4-thiones (**392**), obtained by thionation of the corresponding pyrones **388** (equation 175; Y = Se, R = Me, Ph; Y = Te, R = Me, Ph, *t*-Bu). Indeed, compounds **393** give with TCNQ complexes of high specific conductivity ($\sim 0.5\ \mathrm{ohm}^{-1}\ \mathrm{cm}^{-1}$). Their first oxidation potential follows the general trend of increasing with the size of the heteroatom.

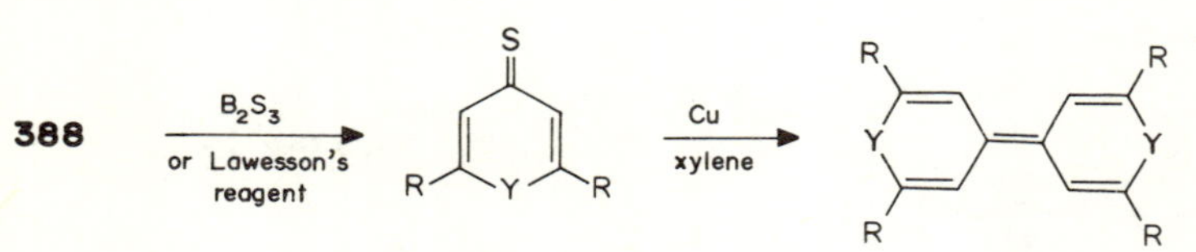

(175)

(392) (393)

2. Benzo-, naphtho- and hetero-condensed systems

a. (b)*Condensed systems: chroman derivatives.* (*i*) *Chalcogeno-chromanones, -chromenes and -chromans.* Benzochalcogenopyrans (chalcogenochromenes) (**397**) are commonly obtained by standard methods from the corresponding chalcogenochromanones **395**. The first known selenochromanone, the parent compound itself (**395a**; $R^1 = H$) was prepared initially in 1958 in a 23% yield by pyrolysis of the corresponding diacid, β-(*o*-

carboxyphenylseleno)propionic acid[862]. A more general synthetic method described in 1964, consists of an electrophilic cyclization of β-phenylselenopropionic acid **394a** with polyphosphoric acid[863] (equation 176). Sulphuric acid, which is used for preparation of thiochromanone, cleaves the Se—C (aliphatic) bond.

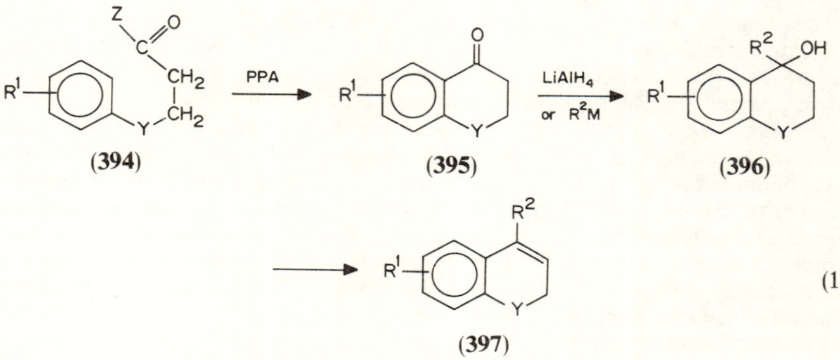

(394) (395) (396)

(397)

(176)

(**a**) Y = Se
(**b**) Y = Te

The method has been widely applied to acids **394a** (Z = OH) or to their acid chlorides (Z = Cl) for the synthesis of 2-Me, 3-Me, 7-Me, 8-Me[864,874], 6-Me[269,864,874] and 6,7-di-Me[874] selenochromanones **395a**. The corresponding condensed selenopyranones: naphtho-(1,2-*b*)- and -(2,1-*b*)-[865–867], (1)benzothieno-(3,2-*b*)- and -(2,3-*b*)-[868] and (1)benzoselenolo(2,3-*b*)-[869,870] were obtained similarly. It has also been applied to various tellurochromanones **395b**[871] and to the naphtho(2,1-*b*) and thieno(3,2-*b*) analogues, starting from the acid chlorides at dry-ice temperature with aluminium trichloride as catalyst. The 2-phenylchalcogenochromanones (selenoflavanone **399a**[872] and telluroflavanone **399b**[873]) are prepared by a Y—C(2) ring-closure of *o*-cinnamoylchalcogenoanisoles (**398**) and hydrobromic acid, with respective yields of 80 and 20% (equation 177). In the Te series, an important by-product was identified as *o*-cinnamoylbenzenetellurenyl bromide. The same ring-closure has been applied to the synthesis of 4*H*-2,3-dihydro-2-phenyl-4-oxo-(1)-benzothieno- and -(1)benzoselenolo-(2,3-*b*)selenopyrans[869].

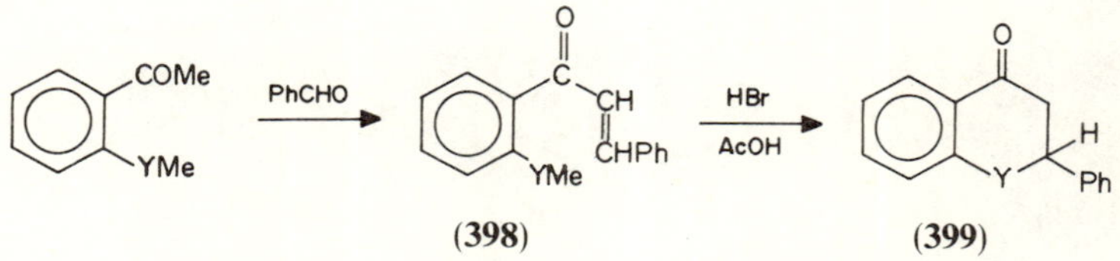

(177)

(398) (399)

(**a**) Y = Se
(**b**) Y = Te

The chromanols **396**, obtained from **395** by reduction or via addition of organometallic reagents, are easily dehydrated to the corresponding chromenes **397**, or their analogues[863–871]. Selenochromanone leads also, via the Vilsmeier reagent, to 4-chloro-3-formylselenochromene, a starting material for the further condensation of a thiophene[874] or a selenophene[100] ring. The known selenopyrans condensed to a nitrogen heterocycle are obtained by a direct ring-closure of the selenopyran ring: in this manner, 7-methyl-2*H*-

selenino(2,3-*b*)pyridine is prepared, as its O and S analogues, from 2-ethylseleno-6-methyl-3-(1′,3′-dihydroxypropyl)pyridine and HBr[875]. A selenino(3,2-*d*)pyrazole, the 6*H*-3-methyl-1-phenyl-5,6-dicarboxy derivative, is prepared by a Diels–Alder reaction between maleic anhydride and 3-methyl-1-phenyl-4-methylaminomethylene-5-selenoxopyrazole[876].

The reduction of **395a** to the corresponding selenochromans has been performed by the Clemmensen[863] and the Wolff–Kishner reductions[877,878]. Whereas the Clemmensen reaction leads to the required product, sometimes mixed with variable quantities of selenochromenes, the Wolff–Kishner reaction on the parent compound leads partially to the ring-contraction product, 2-methyl-2,3-dihydrobenzo(*b*)selenophene, besides the attempted selenochroman. The reaction differs from that in the sulphur series, where only thiochroman is formed, and from that in the oxygen series where the formation of *o*-cyclopropylphenol is established. The isolation of pure selenochroman has been used as evidence that selenochroman is not formed by thermal rearrangement of phenylallyl selenide as previously described[879]. However, the isolated product is 2-methylbenzo(*b*)selenophene[877]. Pure tellurochroman has been prepared by a diimide reduction of tellurochromene[880].

(*ii*) *Benzochalcogenopyrylium salts* (*chalcogenochromylium salts*). Chalcogenochromylium salts are obtained from selenochromenes **397a** by a reaction parallel to that described in equation (174)[864,867–869,881]. Several 2- or 4-phenylselenochromylium salts are prepared in the same manner[869,882]. 4-Methyl-2-phenyl-[883] and 2,4-diphenyl-selenochromylium[884] salts are also prepared from selenoflavone for the synthesis of polymethine dyes. On the other hand, 4-methoxyselenochromylium salts are obtained by methylation of selenochromone and selenoflavone[885]. The pK_{R^+} value, which characterizes the equilibrium constant between the cation and its pseudobase and hence the stability of the cation, has been determined for the chalcogenopyrylium cation[885] (O: -1.96; S: $+3.15$; Se: $+1.20$), and for the naphtho-(2,1-*b*)- and -(1,2-*b*)-selenopyrylium salts ($pK_{R^+} = +2.7$ and $+1.9$, respectively)[867]. The reaction of selenochromylium salts and analogues with nucleophiles has been shown to occur at the 2- and 4-positions. Whereas the monosubstituted 2-aryl- and 4-aryl-selenochromylium salts react exclusively at the other free activated position with all the nucleophiles studied ($NaBH_4$, PhMgBr, PhSH, PhSeH, $(MeCO)_2CH^-$)[882], the unsubstituted selenochromylium cation **400** gives with aryl organometallic compounds a mixture of addition products at positions 2 (**401**) and 4 (**402**) depending on the nature of the metal and the organic radical (equation 178). The isomer

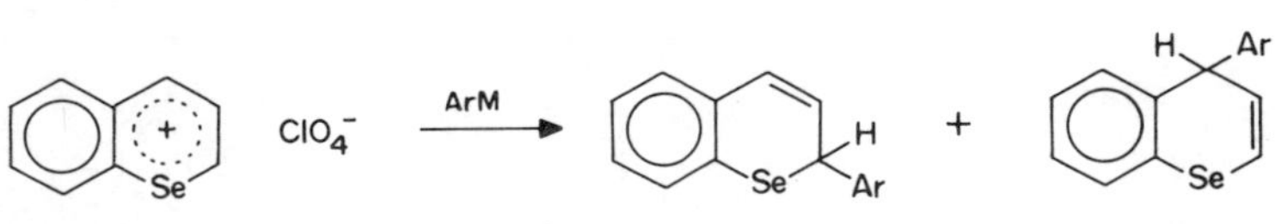

(178)

(400) **(401)** **(402)**

ratio is affected by a secondary oxidation–reduction reaction between 2-arylselenochromenes and the unsubstituted selenochromylium salt[886]. A comparison with the analogues S cation, which gives a 55:45 ratio between positions 2 and 4 (in agreement with the calculated positive charge density), is therefore difficult. On the other hand, other nucleophiles as dimethylaniline, thiophenol and selenophenol react regiospecifically at the 2-position[886]. However, a 2-regiospecificity is demonstrated with all the nucleophiles, including organometallic compounds, for some heterocondensed selenopyrylium cations[887]. The MnO_2 oxidative hydrolysis of the selenochromylium cation differs appreciably from the corresponding reaction of the thiochromylium cation.

Although the size of the ring of the latter is conserved in the oxidation products, and the thiocoumarin : thiochromone ratio corresponding to respective nucleophilic attack on the 2- and 4-positions, is 90:5, the selenochromylium cation gives exclusively a product of ring-contraction, namely 2-formylbenzo(*b*)selenophene[286,888]. A parallel increase in the extent of ring-contraction with the size of the chalcogen has also been demonstrated in the oxidation of thio-, seleno- and telluro-chromenes. Depending on the oxidant, the reaction can be directed to give selenocoumarines (with chromium trioxide in pyridine, 50% yield) or to ring-contraction products (e.g. 2-acylbenzo(*b*)selenophene)[286,287]. The selenochromones are always minor products. In the Te series, only products of ring-contraction can be obtained[484]. Dichalcogenocinnamaldehydes are intermediates in the ring-contraction reactions.

(*iii*) *Chalcogenochromones.* The parent selenochromone (**403**; $R^1 = R^2 = H$), was obtained in 1964 by dehydrogenation of selenochromanone with chloranil[863] or triphenylmethyl cation[881] (equation 179). Later, a photodehydrogenation of chalcogenochromanones gave a 50% yield of thiochromones, a very low yield of selenochromone (5%) and failed for tellurochromanone[889].

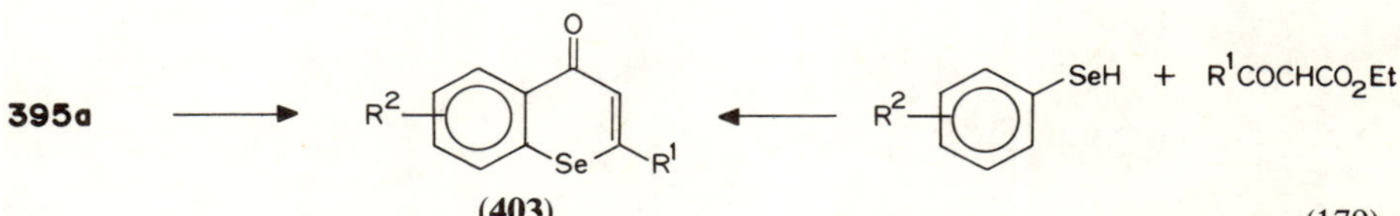

(179)

Application of the Simonis synthesis of thiochromones gives from selenophenols and β-oxo esters in polyphosphoric acid (PPA) medium, selenochromones, selenoflavones (**403**; R^1 = Ph)[890,891] and isoselenoflavone (3-phenylselenochromone)[890]. The yields are poor (7–30%), due to secondary oxidation of the selenophenols. The following selenochromones are obtained: 2-Me; 3-Et-2-Me; 2,3-, 2,6-, 2,7- and 2,8-di-Me; 2,3,6-, 2,3,7- and 2,3,8-tri-Me; and the 6-Me, 7-Me and 8-Me selenoflavones. The parent selenochromone is obtained by this method in a very low yield from malic acid and selenophenol[890]. 2-Methylselenochromone can also be prepared in the same medium from selenophenol and diketene, or from the isolated intermediate phenyl selenoacetoacetate (**404**)[892]. Compound **404**, which possesses the selenol ester group, does not give the isomeric selenocoumarin and is rearranged to selenochromone. The same selenochromone is also obtained from another isolated intermediate **405**[892] (equation 180).

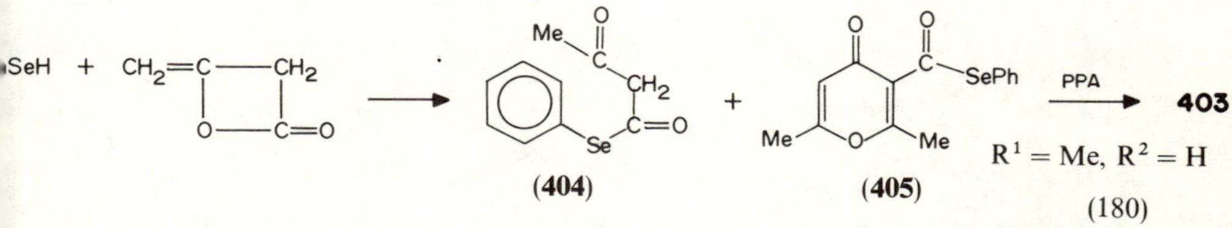

R^1 = Me, R^2 = H

(180)

Under the same conditions, malonic esters give with thio- or seleno-phenols 2-(phenylthio)thiochromone and 2-(phenylseleno)selenochromone derivatives, respectively[893]. In the Simonis reaction, β-arylthio- or β-arylseleno- cinnamates (**406**), could also be possible precursors. The isolation of these compounds has been achieved by a regiospecific control of the β-addition of chalcogenophenols across the triple bond of an ethyl arylpropiolate in basic medium[894] (equation 181). In contrast, a regiospecific α-addition is achieved under uncatalysed, probably radical, conditions. *Z*-β-Aryl-thio- and -

seleno-cinnamic acids are cyclized to the corresponding flavones, whereas the α-arylchalcogeno isomers give the corresponding chalcogenoaurones. With cinnamates substituted by activating substituent in the aryl group of **406** the cyclization can be realized on this aryl ring, following the necessary Z–E rearrangement, to give 3-(arylseleno)inden-1-ones.

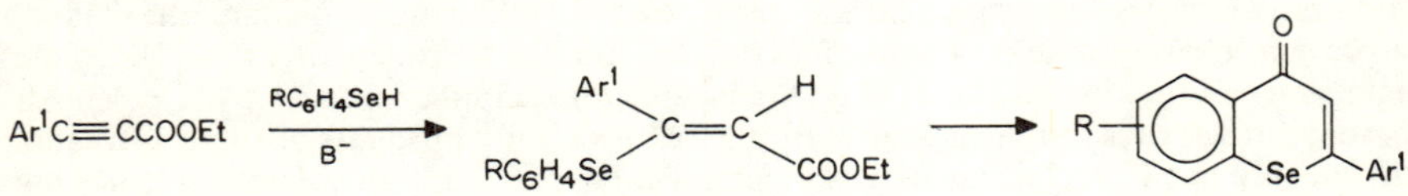

(**406**) (**407**)

(181)

Selenoflavones (**409**; Y = Se) have also been obtained by another general cyclization method, consisting of a hydrobromic acid ring-closure of *o*-methylchalcogenobenzoyl phenylacetylenes (**408**; R^1 = Ph)[895] (equation 182). An access to 2-methyl- and 2-phenyl-selenochromone, characteristic of the Se series, is the ring-expansion of **410**, which

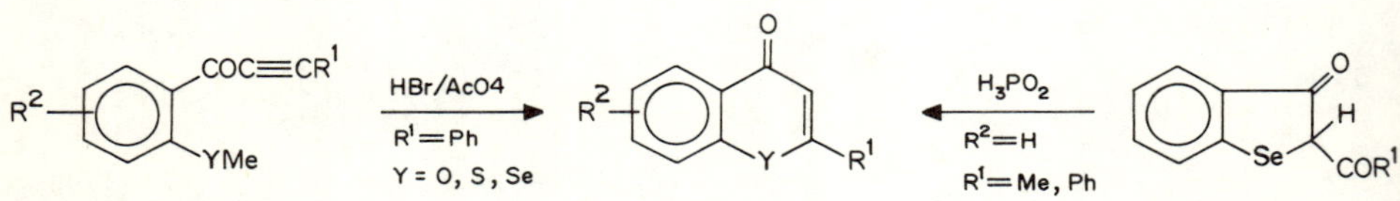

(**408**) (**409**) (**410**)

(182)

gives 40–50% yields (equation 182)[245]. The S analogues of **410** give only reduction to benzo(*b*)thiophene derivatives. This ring-expansion reaction has been applied to the synthesis of benzothieno- or benzoselenolo-(2,3-*b*)- or -(3,2-*b*)-selenochromones, where the exocyclic carbonyl group of **410** belongs to thioindoxyl, selenoindoxyl or its 2-oxo isomer[245]. Nearly all of the preceding methods failed in the Te series, so that tellurochromone (**413**) was prepared as late as 1981 by a new method, starting from *o*-acetylbenzenetellurenyl bromide (**411**) and the acetal of DMF. The condensation product **412** has been cyclized by one equivalent of hypophosphorous acid to **413**[896] (equation 183). A second equivalent of H_3PO_2 reduces **413** to tellurochromanone (**414**). This is a typical reduction of the Te series, since selenochromone is not reduced under these conditions.

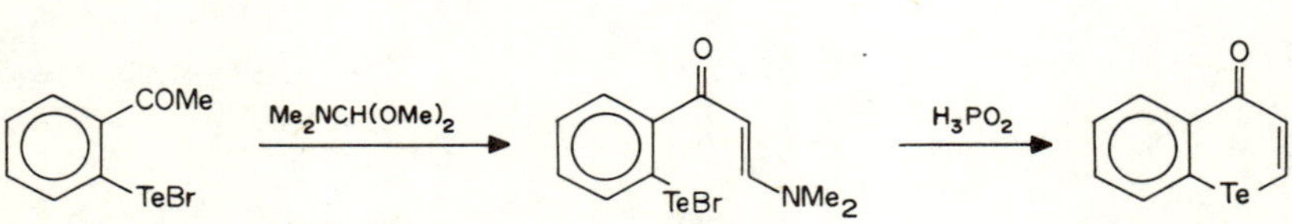

(**411**) (**412**) (**413**)

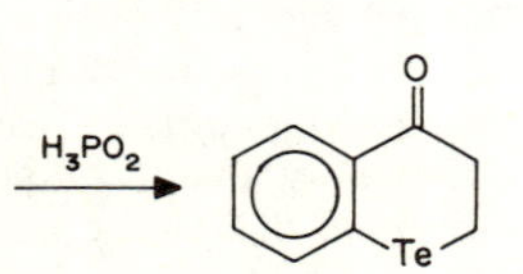

(183)

(**414**)

This synthesis has been extended to the formation of 2-methyltellurochromone, of the thieno-(3,2-*b*)-, -(2,3-*b*)- and -(3,4-*b*)-tellurin analogues and of selenochromone. The major difficulty in the latter case is in the formation of the Se analogue of **412**, since the acidity of the acetyl hydrogen of **411** is lowered[896]. A structural comparison of the four chalcogenochromones (O, S, Se, Te), based on dipole moments, IR, UV and ^{1}H- and ^{13}C-NMR spectra has been conducted[897,898].

Among the preceding synthetic methods for selenochromones, only the ring-closure analogous to that in equation (181) has led recently to tellurochromone derivatives in particular cases. Generally, the attempted electrophilic ring-closure of the acid chloride of the Te analogue of **406** leads to a tellurenyl chloride (**416**) via an *ipso* acylation,[500] due to the polarizability and electronegativity of Te. The chloride **416** can be considered as a resonance hybrid of the no-bond hybrid 1,2-oxatellurolylium halide. The Te—O distance is effectively low. The *ortho* cyclization to telluroflavones **417** can be controlled in selecting *ipso*-deactivating or *ortho*-activating methoxy or fluoro groups in the *meta* position of the aryltellurocinnamoyl chlorides **415**[899]. In this reaction, the thio compounds give preferentially thiochromones, whereas the seleno compounds show competition between the two reactions. 7-Methoxytelluroflavone (**417**) has been coupled via the corresponding thione in order to obtain a donor molecule for studies in the field of organic metals[900].

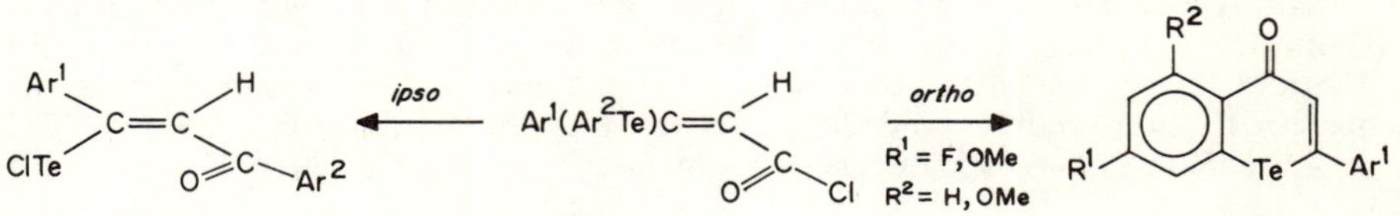

(**416**) (**415**) (**417**)

(184)

(*iv*) 1-*Chalcogenocoumarins*. 4-Hydroxy-1-selenocoumarins (**419**) are synthesized by cyclization of diphenyl selenomalonate with $AlCl_3$[901], by treatment of substituted malonic acids with selenophenols in PPA[902] (whereas malonate esters give 2-(arylseleno)selenochromones[903]) and, with a very good yield, by ring-expansion of 2-(ethoxycarbonyl)selenoindoxyls (**418**) with hypophosphorous acid. This latter reaction probably involves a ring-opening affected by a selenophilic attack[283] (equation 185). The ring-expansion is much more difficult in the sulphur series, giving only 4-hydroxy-thiocoumarin with an electron-withdrawing R group at position 3 in a poor yield (25%).

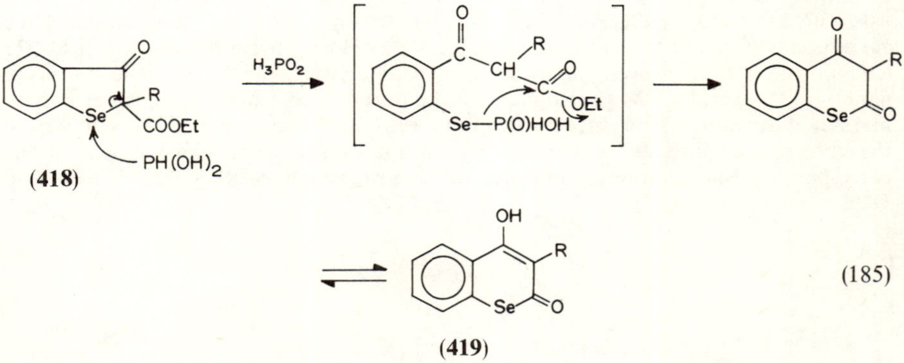

(**418**)

(**419**)

(185)

The parent non-hydroxylated 1-seleno-[904] and 1-telluro-coumarins[905] (**421**) are prepared by cyclization of *o*-methylchalcogeno-cinnamoyl chlorides **420** (equation 186). The unsubstituted; 3-Me, 4-Me, 6-Me, 7-Me and 8-Me; 3,4-, 3,6-, 3,7-, 4,6- and 4,7-di-Me; 3-CN-4-Me, 3-CN-4,6-di-Me and 4,7-di-Me; 3-Ph, 4-Ph, 3-Ph-6-Me, 3-Ph-7-Me and 4-Ph-6-Me derivatives were prepared when Y = Se, and the unsubstituted derivative when Y = Te. Some selenocoumarins are also obtained by CrO_3 oxidation of seleno-chromenes[286,287]. 3,4-Dihydro-1-selenocoumarin has been prepared by a somewhat similar method: treatment of *o*-(methylseleno)dihydrocinnamic acid with Br_2 and H_3PO_2 transforms the seleno ether to the corresponding selenophenol, without cyclization. The selenolactonization is effected by *p*-toluenesulphonic acid[905]. A parallel reaction failed in the Te series.

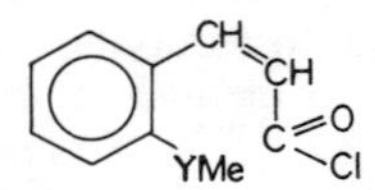

(186)

(**420**) (**421**)

b. (c)*Condensed systems: isochroman derivatives* (*i*) *Isoseleno- and isotelluro-chromans.* Isoseleno- and isotelluro-chroman (**423**; R^n = H) were prepared in 1945 with respective yields of 58 and 50% by a nucleophilic substitution of the corresponding dihalo derivative **422** (R^n = H) with the chalcogenide anion[906] (equation 187). They were transformed to quaternarization products, and the Se derivative to the corresponding *Se*-oxide. The 1*H*,3*H*-naphtho(1,8-*cd*) selenin has been obtained by a similar reaction[907].

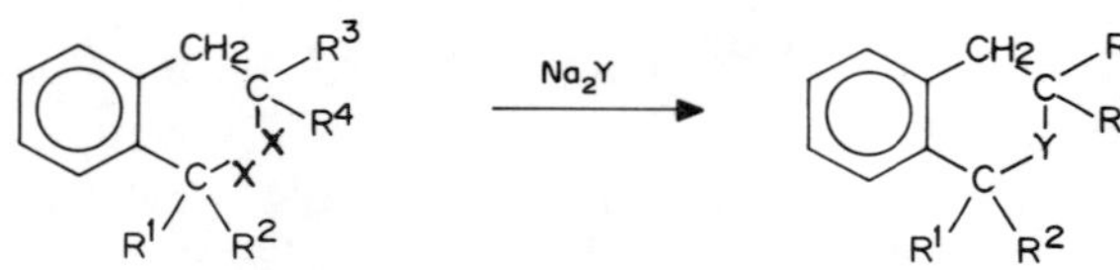

(187)

(**422**) (**423**)

(**a**) Y = Se
(**b**) Y = Te

(*ii*) *Oxo derivatives.* The Se/Te 1-oxo derivatives, i.e. 3,4-dihydro-isoseleno- and -isotelluro-coumarins **423** (CR^1R^2 = C=O)[473], and the isomeric Se/Te 3-oxo derivatives (**423**; CR^3R^4 = C=O) are prepared by the same method, starting from the appropriate acid chloride and using phase-transfer catalysis[908]. The isomeric 4-oxo derivative, isoselenochromanone (**426**), could not be obtained by electrophilic cyclization of the corresponding acid, but it was obtained in 31% yield by pyrolysis of the diacid **425** (equation 188)[909]. However, the direct electrophilic cyclization by P_2O_5 leads to the naphtho-(2,1-*c*) and -(1,2-*c*) analogues[867]. The unsaturated heterocycles, isoselenocoumarin and isotellurocoumarin (**428**) are obtained by a method parallel to the synthesis of the chalcogenocoumarin isomers i.e., by intramolecular electrophilic attack of the corresponding acid chloride on an unsaturated seleno or telluro ether (**427**)[910] (equation 189).

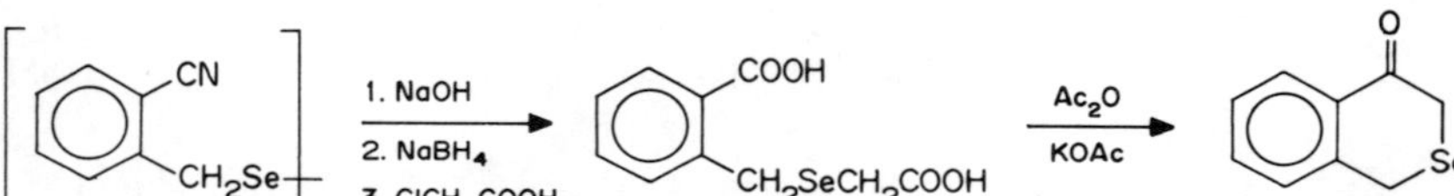

(188)

(**424**) (**425**) (**426**)

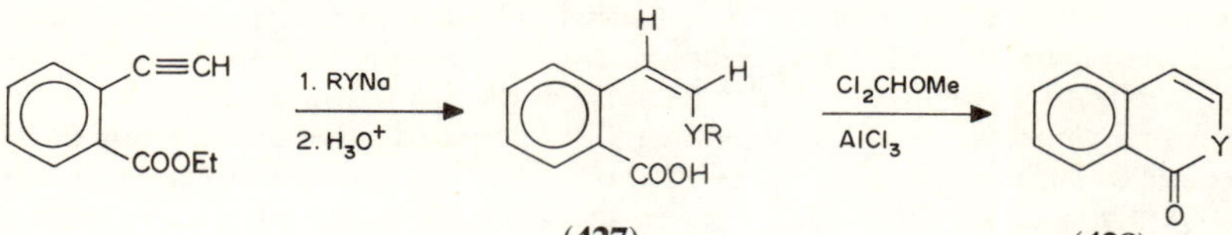

(189)

(427) (428)

(a) Y = Se
(b) Y = Te

(*iii*) *Isoselenochromylium salts.* The parent isoselenochromylium salt **429a** (2-selenonianaphthalene cation)[909] and its 2- and 3-selenoniaphenanthrene analogues[867] are prepared by similar reaction paths to those used for the selenochromylium cations. The pK_{R^+} value for the 2-selenonianaphthalene cation (+0.2) is lower than that for its 1-selenonia isomer (+1.20) and its S analogue 2-thionianaphthalene (+2.17).

(*iv*) *Selenabenzene derivatives.* Reaction of **429a** with perfluorophenylmagnesium bromide furnishes 1-perfluorophenyl-2-seleno-Δ^3-chromene (**429b**), which upon methylation leads to the first example of an authentic selenabenzene **429c**[911] (equation 190). Solution **429c** in DMSO shows intense colour (λ_{max} = 500 nm compared with 490 nm for the S analogue). Its stability is lower than that of its S analogue, and it has a half-life of 47 min at 25 °C. The NMR spectra indicate the ylide nature of this selenabenzene. Previously claimed selenaanthracenes have properties which are inconsistent with those expected for selenabenzenes.

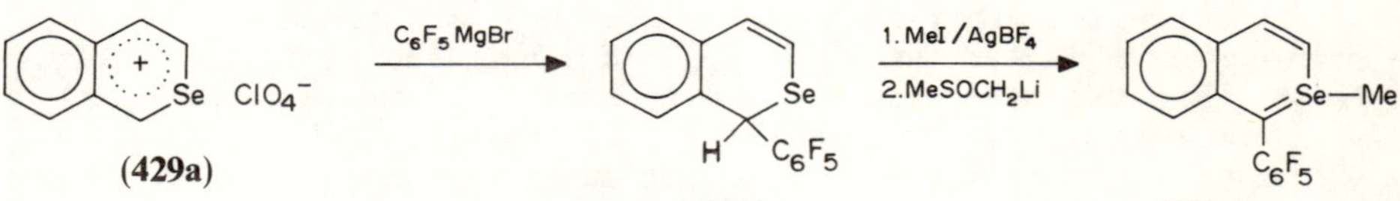

(429a) (429b) (429c)

(190)

3. Dibenzo-condensed systems and analogues

a. Selenium derivatives. (*i*) *Selenoxanthones.* The parent selenoxanthone **431** was synthesized in 1914[912]. Its synthesis can be achieved by an electrophilic ring-closure of *o*-phenylselenobenzoic acid (**430**) in sulphuric acid (45[912]–88%[913] yield), or in polyphosphoric acid (82%)[914] (equation 191). By this method 2-chloro-[913] and 2- and 3-methylselenoxanthones have been prepared[914]. Another mode of formation consists of a double electrophilic cyclization of *o*-chloroselenobenzoyl chloride (**432**) with aromatic compounds[914,915]; e.g. reaction with toluene gives the 2-methyl derivative[914]. The parent compound leads to the *Se*-oxide by reaction with chromium trioxide[915]. Praefcke's group synthesized selenoxanthone and some analogues[916,917] photochemically from aromatic *o*-chloro-, *o*-tosyloxy- or *o*-methylsulphinyl-selenol esters, via cleavage of the acyl–Se bond. In this way 7-methyl-cyclopenta(*b*)(1)benzoselenin-9-one[918], 7-methyl-5*H*-

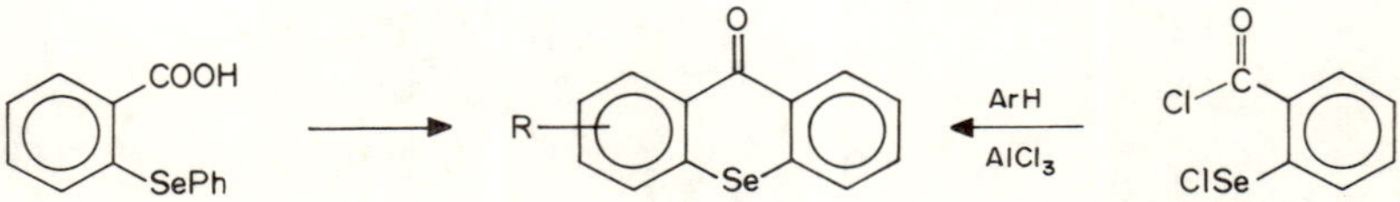

(191)

(430) (431) R = H (432)

benzoselenino(2,3-*b*)pyridine-5-one[919,920], 7-methyl-9*H*-selenolo(3,2-*b*)(1)benzoselenin-9-one and the isomeric 6-methyl-4*H*-selenolo(2,3-*b*)(1)benzoselenin-4-one[921] were obtained. This cyclization reaction is similar to those in the O and the S series.

(*ii*) *Selenoxanthylium salts.* The selenoxanthen-9-ols, prepared either by reduction or by organometallic treatment of selenoxanthones[913,914,922] are the pseudobases of the selenoxanthylium salts (equation 192). They easily give these salts by treatment with a strong acid. In another method, a 1→8-octahydroselenoxanthylium cation has been prepared by reaction of H_2Se with bis(2-oxocyclohexyl)methane[923]. The stability of the selenoxanthylium cation has been compared by Degani and coworkers with those of the O and S analogues. Their pK_{R^+} values are O: -0.83; S: -0.21; Se: -1.67[885]. These cations react exclusively with nucleophiles at the 9-position. For example, the parent cation is transformed by MnO_2 oxidative hydrolysis to selenoxanthone (**431**), in a nearly quantitative yield[888]. On the other hand, other nucleophiles give 9-monosubstituted or disubstituted selenoxanthenes. Some reports regarding a possible reaction on Se in position 10 of the cation to give selenaanthracenes[924] could not be confirmed[911]. A review on chalcogenoxanthylium salts has appeared recently[925].

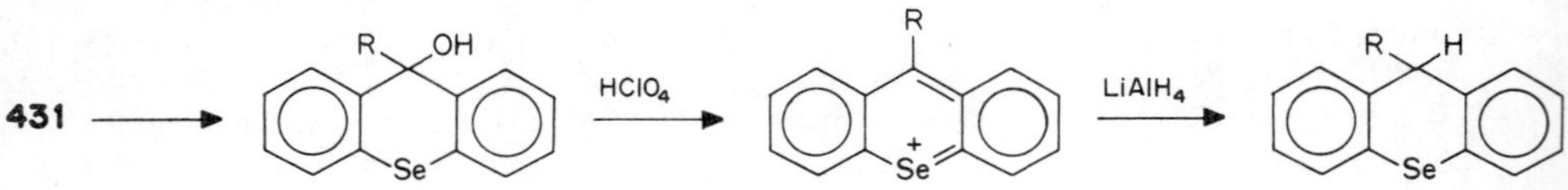

(**433**) (**434**) (**435**)

(192)

(*iii*) *Selenoxanthenes.* The parent selenoxanthene (**435**; R = H), the first derivative of the series, is prepared in a nearly quantitative yield from selenoxanthone, red phosphorus and HI[926]. The other selenoxanthenes are prepared from selenoxanthylium salts. Unsubstituted derivatives in position 9 (e.g. 3,6-dimethylamino-[927] and 3-methylselenoxanthenes[914]) are obtained by reduction of the corresponding cation with complex hydrides. 9-Monosubstituted derivatives are prepared by reduction of the 9-substituted cations, or by reaction of organomagnesium compounds with the parent cation (9-Me[914], 9-Ph[914,924]). The last reaction on the chosen cation leads to 9,9-disubstituted derivatives (9,9-di-Me, 9-Me-9-Ph[914,924]). 3′-Dimethylaminopropylidene and propylselenoxanthenes have been prepared for pharmaceutical screening by dehydration and HI reduction of the corresponding 9-selenoxanthydrols, respectively[913]. 9-Phenylselenoxanthene is transformed by chloramine T to the corresponding *N*-arylsulphonylselenilimine[928]. Other dipole-stabilized Se-substituted selenoxanthenes including *Se*-oxide were prepared recently and transformed to several 10-selenoxanthenium ylids[929].

(*iv*) *Dibenzo*(b,d)*selenin derivatives.* The angular isomer **437** of selenoxanthene and the corresponding selenoniaphenanthrene (**438**) have been prepared by a Pschorr cyclization[930] (equation 193). Hydride abstraction from **437** by trityl perchlorate is only

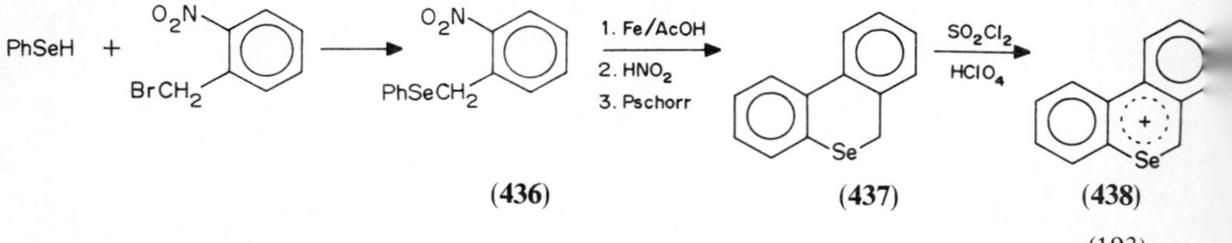

(**436**) (**437**) (**438**)

(193)

partial. The pK_{R^+} values of **438** and of its O and S analogues (O: − 5.96; S: − 1.67; Se: − 4.28)[855] have been compared with the pK_{R^+} values of other similar cations. Indole analogues of the selenin **437**, namely (1)benzoselenino(3,4-*b*)indoles, have been prepared by constructing the indole ring from the phenylhydrazone of selenochromanone **395a**[931].

b. Tellurium derivatives. The chemistry of telluroxanthone and its derivatives is much more recent. Acidic cyclization of the Te analogue of **430** failed, due to destruction of the precursor. From the acid chloride **440**, a non-cyclized rearrangement isomer, *o*-benzoylbenzenetellurenyl chloride (**441**), is obtained in the presence of Lewis acids; the reaction proceeds via an *ipso* attack on the carbon attached to the Te atom followed by cleavage of the aromatic C—Te bond[932] (equation 194). This isomer **441** cannot be cyclized further[932,933].

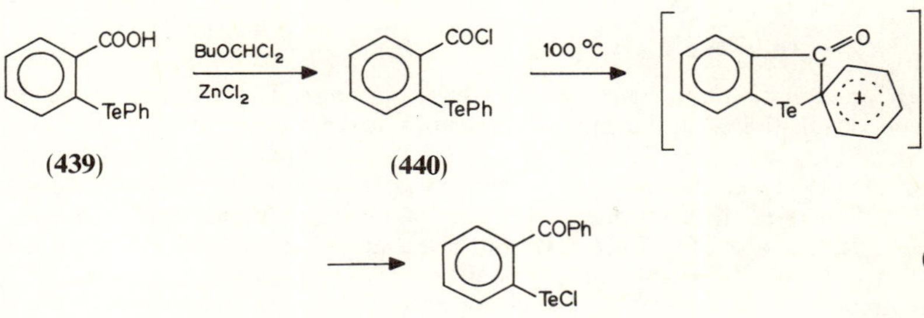

(**439**) (**440**) (**441**)

(194)

A similar rearrangement occurs in the thiophene series[932]. Telluroxanthone (**443**) is therefore obtained by more complicated methods. It first preparation was realized in 1979 with a 2% yield by a double nucleophilic ring-closure of 2,2′-benzophenone bis-diazonium salt (**442**) with Na_2Te[934] (equation 195).

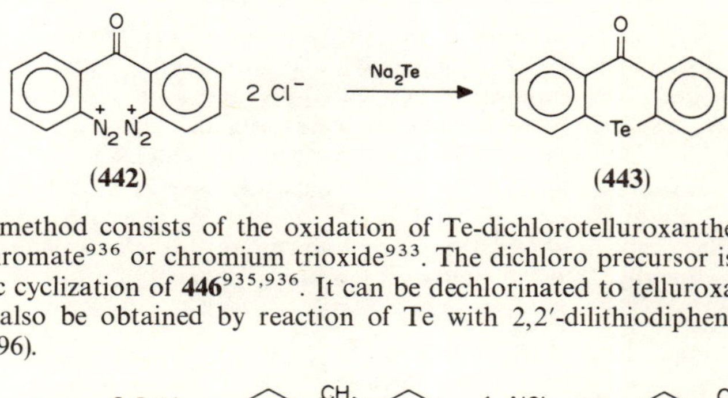

(**442**) (**443**)

(195)

Another method consists of the oxidation of Te-dichlorotelluroxanthene with potassium dichromate[936] or chromium trioxide[933]. The dichloro precursor is prepared by electrophilic cyclization of **446**[935,936]. It can be dechlorinated to telluroxanthene (**445**), which can also be obtained by reaction of Te with 2,2′-dilithiodiphenylmethane[933] (equation 196).

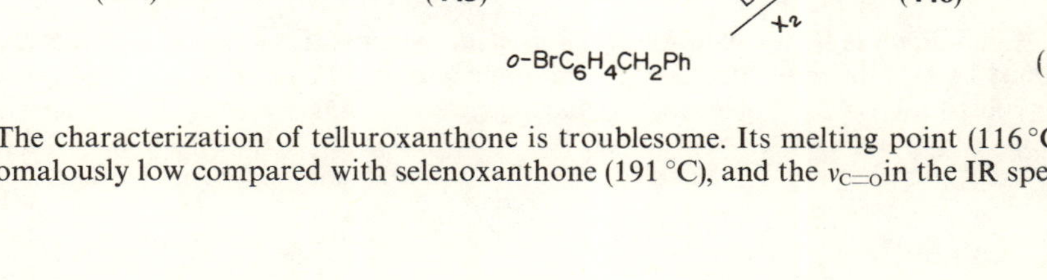

(**444**) (**445**) (**446**)

(196)

The characterization of telluroxanthone is troublesome. Its melting point (116 °C) is anomalously low compared with selenoxanthone (191 °C), and the $\nu_{C=O}$ in the IR spectra measured by two research groups are different (1590 and 1640 cm^{-1}). Furthermore, the more logical melting point (202 °C) given in the first publication[934] was later regarded as erroneous by the same authors. Telluroxanthene is characterized by reactions typical of diaryl tellurides such as Te-dihalogenation. Quaternarization by MeI is achieved only with an equimolar amount of silver perchlorate[936]. Telluroxanthone is transformed to 10-telluroniaanthracene perchlorate, the first known telluropyrylium compound, by the standard methods used for the Se analogues (equation 192). These salts also react at position 9 with various nucleophiles[937]. The 9-Ph derivative gives with zinc a stable radical which can be reversibly dimerized[938]. A 1-formyl-2→8-heptahydrotelluroxanthene has been surprisingly obtained from bis(2-formylcyclohexen-1-yl) telluride[939].

B. Two Heteroatoms without Nitrogen

1. Monocyclic systems

a. Heteroatoms in the 1,2-*position* : 1,2-*dichalcogenanes*. The structure of the reaction products of substituted butadienes with selenous acid (equation 197) at room temperature, previously proposed as selenones of 2*H*,5*H*-selenophene, is actually of a cyclic seleninic ester type, namely 4,5-disubstituted 1-oxa-2-selena-4-cyclohexene-2-oxide (**447**)[940]. However, the structure of the corresponding sulphone was confirmed in a parallel reaction with SO_2. Formation of the 1,2-diselenane **448**[230] leads to a problem similar to that for 1,2-diselenolane. The compound is a low-molecular polymer, present as the monomer only in solution[941]. The 3,6-dicarboxy derivative is prepared by hydrolysis of *rac*-di(selenocyano)adipic acid[942].

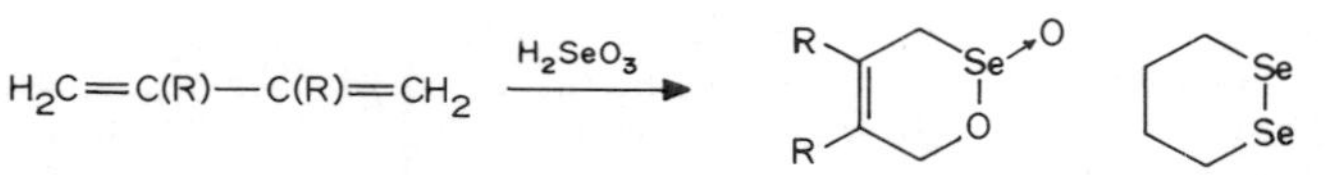

(197)

(**447**) (**448**)

b. Heteroatoms in the 1,3-*position*: 1,3-*dichalcogenanes*. 1,3-Oxaselenanes, 1,3-thiaselenanes and 1,3-diselenanes (**450**) were prepared for conformational studies, from reactions of 1-hydroxy- and 1-mercapto-propaneseleno-3-ols and from 1,3-propanediselenol derivatives (**449**), respectively, with acidic aqueous formaldehyde or dichloromethane[52] (equation 198). 1,3-Diselenane-2-selone (a triselenocarbonate) and 1,3-diselenane-2-dimethoxycarbonylmethylene (a ketene diselenoacetal) have been obtained from the reaction of 1,3-dibromopropane and CSe_2[533] or of CSe_2 and malonate anion[534], respectively. The only 1,3-heteratelluranes known are 1,3-tellurasilane derivatives, obtained from dimethyl(chloromethyl) (γ-chloropropyl)silane[943].

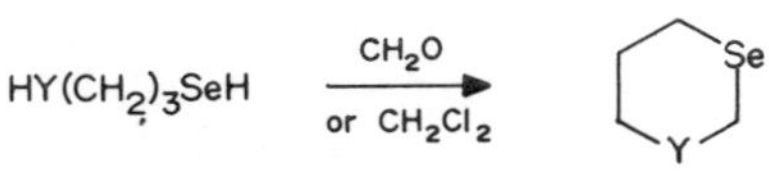

(198)

(**449**) (**450**)

Y = O, S, Se

c. Heteroatoms in the 1,4-*position*. (*i*) 1,4-*Dichalcogenanes*. These compounds (**452**) are obtained by two different methods; the more general one consists of a double nucleophilic substitution of β,β′-dichloroethylchalcogeno ethers (**451**) by the chalcogenide anion (equation 199). 1,4-Oxaselenane (Z = O, Y = Se, 33%)[231,944] 1,4-thiaselenane (Z = S, Y = Se, 44%)[945] (but not Z = Se and Y = S)[946], 1,4-diselenane (0.3% from Li_2Se)[947], 1,4-

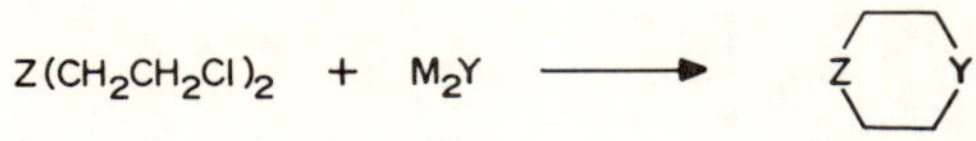

(199)

(451) **(452)**

oxatellurane (Z = O, Y = Te, 46%)[232] and 1,4-thiatellurane (Z = S, Y = Te, 6%)[948] were obtained by this method. The yield of 1,4-diselenane can be improved to 10% starting from ethylene bromide and aluminium selenide[949]. These heterocycles easily form complexes on Se or Te with halogens. 1,4-Diselenane gives easily the bis-selenoxide, and is quaternarized by MeI[947]. The second method consists of an electrophilic addition of selenium tetrahalide[950] or tellurium tetrahalide[238] to the diallyl ethers or thio ethers **453** (equation 200). A methanolic solution of potassium selenocyanate and cupric chloride gives the same type of derivative **454** (X = OMe)[951].

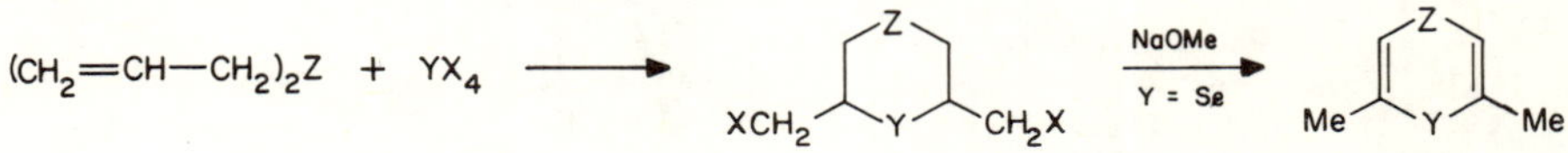

(453) **(454)** **(455)**

(200)

(*ii*) 1,4-*Dichalcogenins.* 3,5-Dimethyl-1,4-thiaselenin (**455**) (Y = Se, Z = S) and the corresponding 1,4-oxaselenin[950] have been prepared by dehydrohalogenation of **454**; formation of 2,5-diphenyl-1,4-diselenin by basic thermolysis of phenylacetylene with Se has been reported[952a], while tetra(trifluoromethyl)-1,4-selenin is prepared from *cis*-1,2-diiodo-1,2-(trifluoromethyl)ethylene and Se[952b,c]. However, the general method of preparation of 1,4-dichalcogenins (**457**) consists of the bis-nucleophilic addition of chalcogenide anion or hydrogen chalcogenide to the bis-acetylenic system **456** (equation 201) (R^1 = H; R^2 = Me, Et, *t*-Bu; Y = S, Se, Te; Z = S[953,954], SO_2[955], or P(O)R[956], in the two

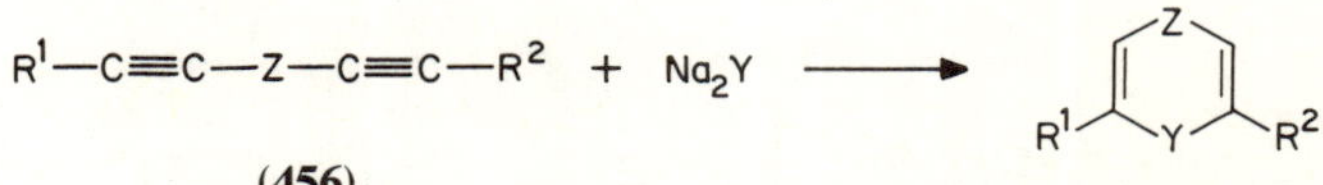

(201)

(456) **(457)**

last cases with $R^1 = R^2$). When this method was applied to alkenyl alkynyl sulphoxides and Na_2S or Na_2Se it gave the *S*-oxides of 2,3-dihydro-1,4-dithiin and 1,4-thiaselenin[957]. The 2,3-dihydro derivatives **459** are prepared by a more versatile method: a mixed ring closure from **458** is based on a nucleophilic substitution of the halogen followed by addition to the alkyne (equation 202)[958]. 1,4-Diselenin can also be obtained by thermolysis of 1,2,3-selenadiazoles[758].

(202)

(458) **(459)**

Y = Se, Te

2. Benzo-condensed systems

Only two publications have appeared relating to this area. The first one reports the formation of 3-formyl benzo(*e*)-1,2-thiaselenin-1,1-dioxide (**460**), from the oxidation of thiochromenes with four moles of SeO_2 in pyridine (equation 203). The unsubstituted, 6-Me, 7-Me and 8-Me and 5,8-, 6,7- and 6,8-di-Me derivatives were prepared[289] with yields

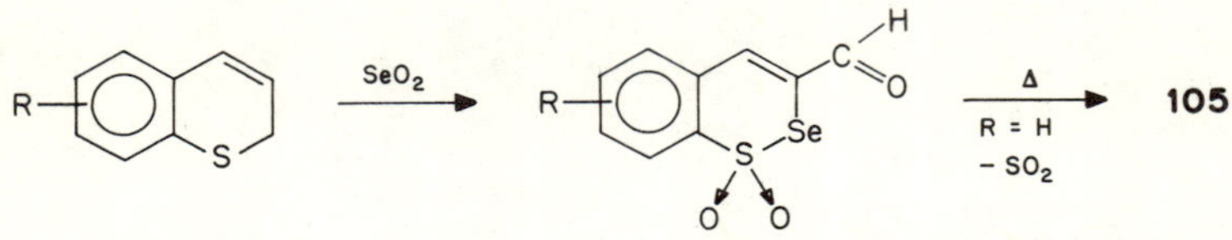

(460) (203)

of 35%. The structure of **460**, which is a rare example of a selenosultone, has been proved by X-ray diffraction. Compound **460** loses quantitatively SO_2 by thermolysis and its 3-formyl group can be transformed to the 3-carboxy and the 3-ethoxycarbonyl functionalities. The second report deals with the formation of **462** by a simultaneous addition and electrophilic substitution of $SeBr_4$ on phenyl allyl ether (**461**)[959] (equation 204).

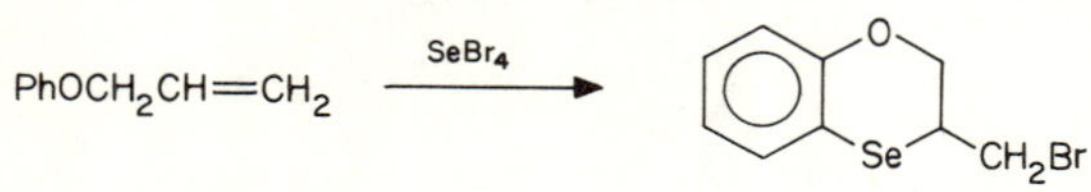

(461) (462) (204)

3. Dibenzo-condensed systems

a. Phenoxaselenins and phenoxatellurins. The main synthesis of phenoxatellurins consists of heating $TeCl_4$ and diphenyl ethers at temperature of ca. 200 °C. In this way the parent compound **466**, which is the first derivative of the series, was prepared in 1926[960,961] (equation 205). Under milder conditions, **463** was initially obtained, but it

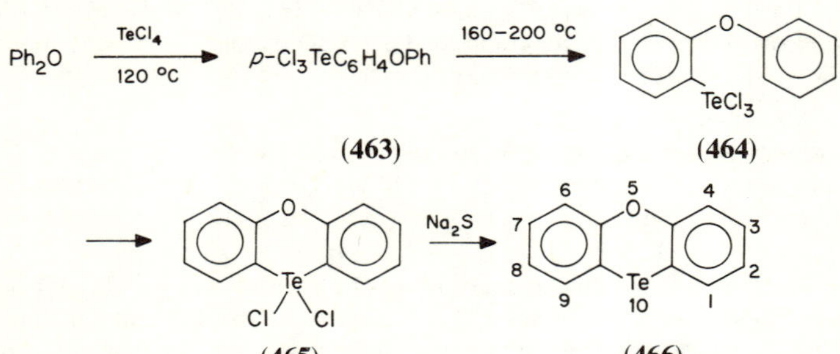

(463) (464)

(465) (466) (205)

isomerizes at higher temperatures to **464**. Dechlorination can be affected by sodium sulphite[960] or sulphide[962]. The method has been applied to the preparation of various 2,8-disubstituted derivatives: 2,8-dimethyl[963], 2,8-difluoro[964], 2,8-dichloro[965,966], 2-chloro-8-methyl[968,969] and 2-fluoro-8-methyl[966]. The 2,8-dibromo derivative cannot be prepared by this method. It is obtained in 86% yield by bromination of the dibromo analogue of **465**[967]. The 4′-substituted analogues of **464** can be conveniently obtained from the corresponding 2-chloromercuri salts and $TeCl_4$. By cyclization they give 2-monosubstituted phenoxatellurins (e.g., 2-Me, 2-COOH)[968]. The 2-nitro derivative cannot be prepared by this reaction but it is obtained by nitration of **466**, via the corresponding 2-nitro-10,10-dinitrate[961,968], together with the 4-nitro and the 2,6- and 2,8-dinitro derivatives. It gives by reduction the 2-amino analogue. The reaction described in equation (205) does not work with $SeCl_4$, whose chlorinating power is too high. The phenoxaselenin was obtained for the first time in 1928[970] from **466** and Se and the mixture of the two chalcogenins was separated via their 10,10-dibromides. The parent phenoxase-

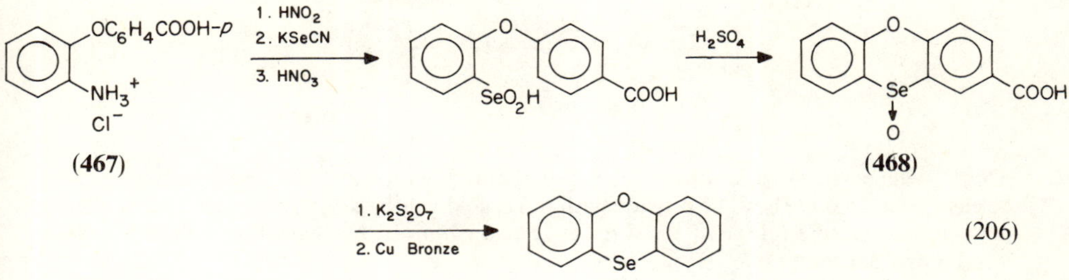

(467) (468) (469)

(206)

lenin 469 was also synthesized ten years later from a suitably substituted aminodiphenyl ether 467[971] (equation 206). Phenoxatellurin is stable in acidic medium, but extruses Te in basic medium, with formation of diphenyl ether[961]. The Te atom can also be lost by replacement with S[966] or Se. The tellurium dichloride **465** gives dibenzofuran by treatment with degassed Raney Ni and Cu[972]. The strong ability of the Te atom to react with halogens can be used to dehalogenate gem-dihalides[973], such as phenoxaselenin dibromide. Analysis of the dipole moments of phenoxachalcogenins (O: 0.55D; S: 1.18; Se: 0.73; Te: 0.38) leads to the conclusion that these molecules adopt a non-planar conformation in solution, and are folded along the axis of the two heteroatoms. UV and photoelectron spectroscopic data support these conclusions[974]. Phenoxachalcogenins form conductive complexes with TCNQ, and the conductivity, of the order of 10^{-7}–10^{8} ohm^{-1} cm^{-1}, shows no significant change with the nature of the chalcogen[975].

b. Phenothiatellurin. This has been prepared with a yield of 40% by the method described in equation (205), starting from $TeCl_4$ and diphenyl thioether, at temperatures of ca. 250 °C[976]. The structure has been confirmed by transforming it to thianthrene by means of S[973].

c. Selenanthrene. The first synthesis of selenanthrene (**471**)[977] which was achieved in 1896 and gave later an improved yield of 44%[337] consists of the reaction of Se with thianthrene tetraoxide **470** (equation 207). Selenanthrene was also prepared in 70% yield

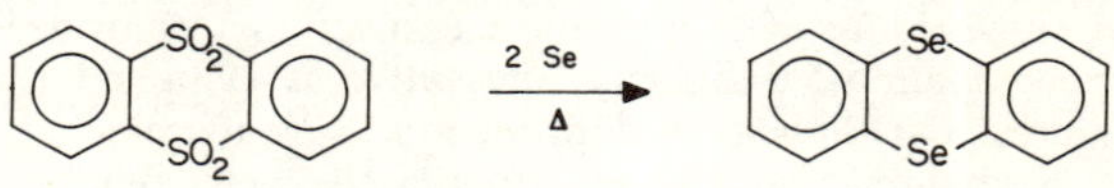

(207)

(470) (471)

by Japanese workers at the same time as the 1-Me, 2-Me and some polymethyl derivatives (**473**), from 2-(methylseleno)diphenyl selenides **472** and H_2SO_4 in the presence of Zn[738,978] (equation 208). Selenanthrenes were also obtained in high yields by pyrolysis of 1,2,3-benzoselenadiazoles, and this was the first example of a new method which led later to important developments (see Section on condensed 1,2,3-selenadiazoles), for the synthesis of many 1,4-diselenin analogues. Selenanthrene has also been prepared in 8% yield by heating benzeneselenenyl chloride with P_2O_5[737], in 50% yield by reaction of potassium amide with diphenyl selenide in liquid ammonia[979], in 9% yield by pyrolysis of tetraphenyltin with Se at 300 °C[980] and in low yield from 1-aminobenzotriazole and SeO_2[981]. 2,3,7,8-Tetramethoxyselenanthrene, obtained by Se insertion by SeO_2 into *o*-dimethoxybenzene[982], has been studied as a charge-transfer π donor[983,984]. Perfluoroselenanthrene can be produced by heating *o*-diiodotetrafluorobenzene with Se at

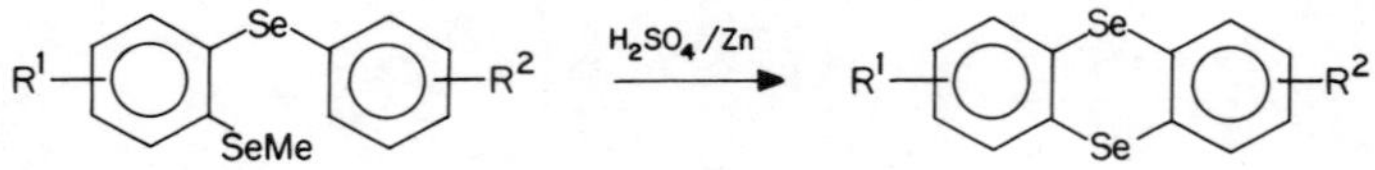

(472) (473)

(208)

320 °C, whereas loss of a Se atom occurs at higher temperatures[340,985]. Selenanthrene can be easily oxidized to the 5, 10-dioxide with nitric acid via the dihydronitrate[986], as well as with hydrogen peroxide, and the dioxide is also obtained by hydrolysis of the *Se, Se*-tetrachloro derivative[987].

d. Telluranthrene. Telluranthrene (**475**), whose production was claimed in 1964 from the reaction between tetraphenyltin and Te metal[988], was actually prepared in 72% yield, and well characterized, in 1981 by Dereu and Zingaro from hexameric *o*-phenylenemercury (**474**) and Te[989] (equation 209). Telluranthrene is convered thermally to dibenzotellurophene. X-ray diffraction confirms the structure and shows a dihedral angle of 124° between phenyl rings[990]. Selenanthrene is also obtained by this method, but in only 5% yield. Perfluorotelluranthrene has been prepared by heating diiodotetrafluorobenzene with Te and Br_2 at 300 °C, and subsequent debromination with Na_2S[991a]

(474) (475)

(209)

C. Two Heteroatoms with Nitrogen

1. Monocyclic systems

a. 1,3-Selenazines. Only one 1,3-selenazine has actually been described: the 4*H* derivative **477** is obtained from the Se heterocycle **476**, which serves as a potential heterodiene towards an ynamine[733,736] (equation 210). However, some oxo derivatives are known: the 4-oxo compound 2-chloro-4-oxo-1,3-selenazine is produced by hydrochloric acid ring-closure of *Z*-*β*-(cyanoseleno)acryloyl chloride[991b]. The 4-oxo-5,6-dihydro 2-diphenylamino-5,6-diphenyl derivative is obtained (38% yield) from 1,1-diphenylselenourea and diphenylcyclopropenone. Selenoamides, contrary to thioamides which give normal reaction, lead only to a diselenole system by Se insertion into the small ring[495]. Reaction of *N,N'*-dimethyl (or diphenyl)selenourea with methyl acetylenedicarboxylate gives 4-oxo-2-methylimino-(or 2-phenylimino)-3-methyl-(or phenyl)-6-ethoxycarbonyl-3,4-dihydro-selenazine[992], whose ring can be dihydrogenated for confirmation of its structure. Some perhydroselenazines **480** ($R^1 = H$, $R^2 =$ Me, 4-XC_6H_4 (X = H, Cl, Br); $R^1 = R^2 =$ Me) have been prepared by a general method from (di(*γ*-aminopropyl) diselenide and a carbonyl compound[993] (equation 211). Compounds **480** can be *N*-benzoylated in pyridine, but not in aqueous

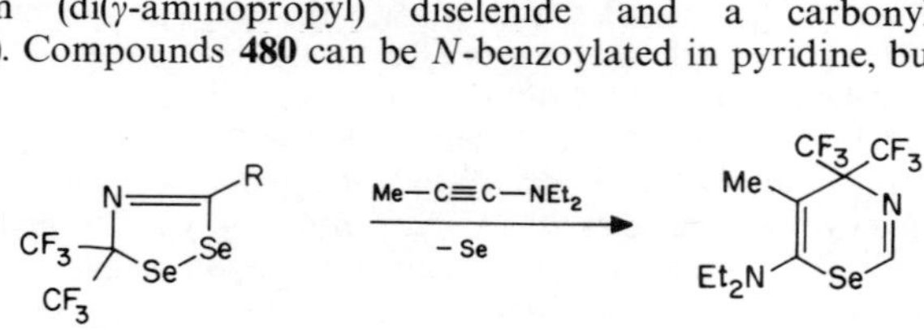

(476) (477)

(210)

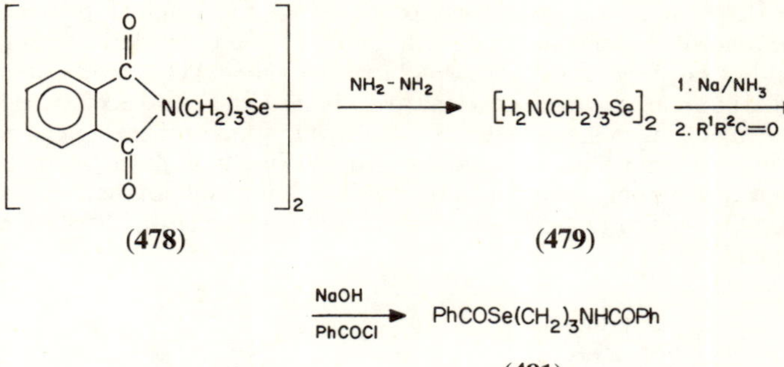

(478) (479) (480)

(211)

(481)

alkaline medium, where they convert to the *N,Se*-dibenzoyl derivative **481**, a reaction parallel to that of the corresponding five-membered ring. The 2-iminoperhydroselenazine, the first derivative of this ring-system, is known since 1890, and obtained by boiling γ-bromopropylamine and potassium selenocyanate in aqueous solution[994]. Some 4-oxo-2-imino and 2,4-dioxo derivatives are prepared from trichloromethylpropiolactone and *Se*-ethylselenopseudourea[995]. Reaction of NaHS with cinnamoyl isoselenocyanate gives the corresponding 6-phenyl-4-oxo-2-thioxo system[996]. 2-Amino-1,3-selenazinium salts are obtained from β-chloropropenylideniminium salts and *N*-substituted selenoureas[997]. Mesoionic derivatives of anhydro-1,3-selenazinium hydroxides are formed from a reaction between a selenoamide and a 1,3-bielectrophile such as chlorocarbonyl phenyl ketene.

b. 1,4-Selenazines. The reaction between aziridine, Se and 3-pentanone results in a direct selenization of the acidic CH group, and forms the 5,6-dihydro-4*H*-1,4-selenazine **482**. By reduction of the ring, **482** leads to 2-methyl-3-ethylselenomorpholine[998] (equation 212). **482** can be normally *N*-acetylated to **484**.

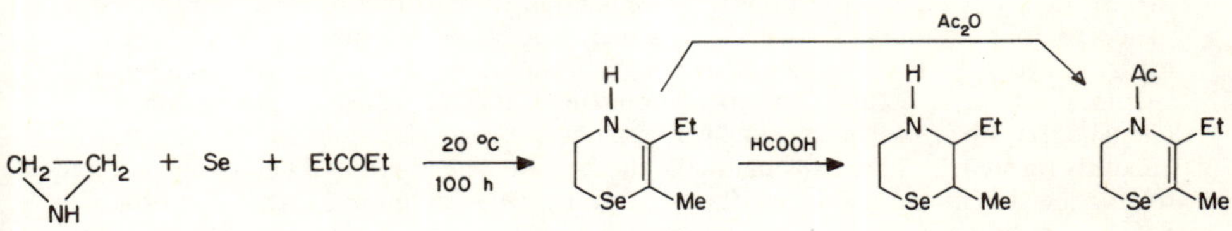

(482) (483) (484)

(212)

2. Benzoselenazines and hetero analogues

a. 1,3-Benzoselenazines. The only described products in this series are 2-chloro-1,3-benzoselenazin-4-one **486**, prepared by cyclization of *o*-(cyanoseleno)benzoyl chloride with HCl, and its 2-dimethylamino derivative**487**[499,1000a] (equation 213). A 4,8-dimethyl-

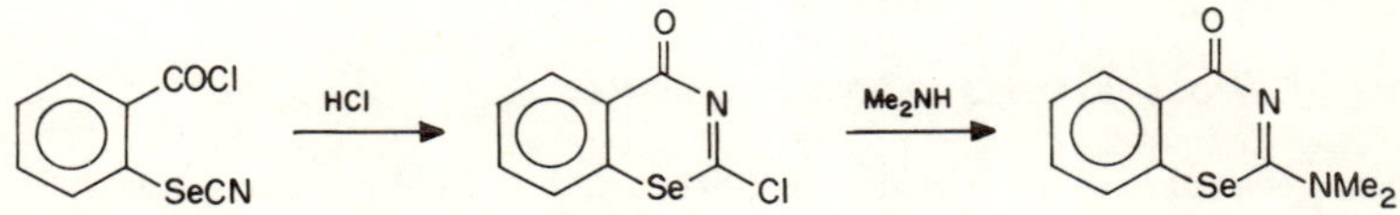

(213)

(485) (486) (487)

2-imino-1,3-benzoselenazine-3-oxide has been proposed as the product between 3-methyl-2-cyanoselenoacetophenone and hydroxylamine[595]. Two other derivatives have also been prepared but not reported[598]: the 2,4-dioxo derivative **488** is synthesized by ring-expansion of 2-ethoxycarbonylbenzisoselenazolin-3-one with Fe and acetic acid. It is exclusively methylated on N with diazomethane. 2-Methyl-2,3-dihydrobenzoselenazin-4-one (**489**) can be formed from 2-acetylbenzoselenazolin-3-one and Zn in acetic acid. Hypophosphorous acid gives only a ring-opening in this reaction, but not the subsequent ring-closure. Pyrido-(3,2-*e*) and (3,4-*e*)-1,3-selenazine-4-one-2-thione or -2-selone have been recently obtained[1000b]

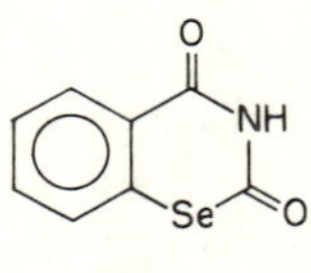

(**488**)

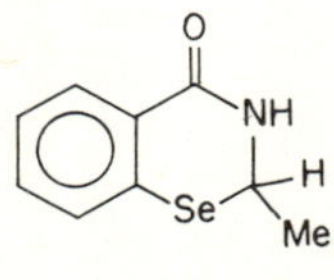

(**489**)

b. 1,4-Benzoselenazines. Behagel and Rollmann[1001] prepared the first derivative of the series (**491**) by reduction of *o*-nitrophenyselenoacetic acid and direct cyclization. It is also obtained from the Zn salt **492** and chloracetyl chloride[1002,1003] (equation 214). By a

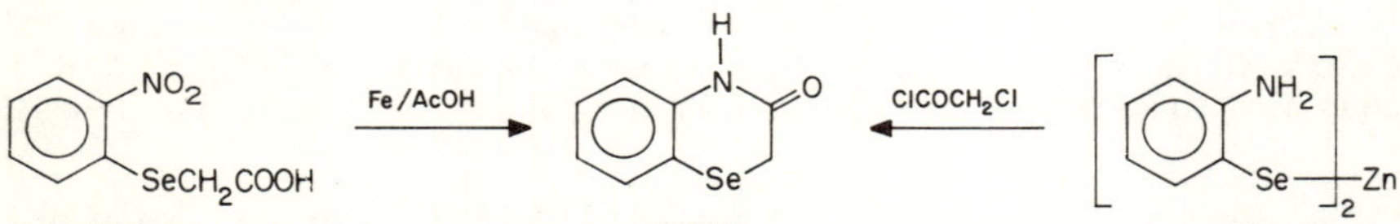

(**490**) (**491**) (**492**)

(214)

similar procedure, reaction of **492** with acetylenedicarboxylic acid results in the 2-carboxymethylene derivative of **491**[1004]. When reated with P_2S_5, **491** is transformed to the corresponding thione[1003] which is further condensed with bromoacetone to give 1-methylthiazolo(2,3-*c*)-1,4-benzoselenazine, a compound used for further synthesis of cyanine dyes. From **492** and ω-bromoacetophenone in DMF a mixture comprising 50% 3-phenyl-2*H*-1,4-benzoselenazine (**493**), 30% 2-phenylbenzoselenazole and 20% benzoselenazole is formed[598]. Like its S analogue, the 2-methylene group of **493** is easily oxidized to give the system **494** and its oxidation derivative **495** (equation 215). By reaction with SeO_2 in pyridine, a product of ring-contraction, 2-benzoylbenzoselenazole has been isolated, besides 2-phenylbenzoselenazole.

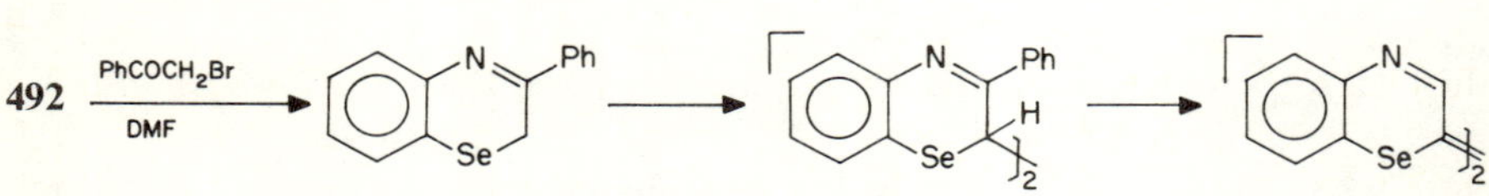

(**493**) (**494**) (**495**)

(215)

3. Dibenzochalcogenazines and hetero analogues

a. Phenoselenazines. Phenoselenazine (**497**) and its 2-chloro derivative have been prepared by the reaction of diarylamines (e.g. **496**) and $SeCl_2$[1005–1008] (equation 216). Bauer has also prepared the 4-nitro and the 2,4-dinitro derivatives[1009] from *o*-

(496) (497) (216)

aminoselenophenol and picryl chloride. Various reactions on the N of **497** have been performed: methylation[1005,1007], cyanoethylation[1007], attachment of aminoalkyl derivatives for pharmacological screening[1010], alkylation[1011] and aroylation[1008]. Friedel–Crafts has been studied[1012] and the *N*-aroyl derivatives have been oxidized to the corresponding selenoxide with phenyl iodosoacetate[1013]. The dimerization of the cation radicals of phenoselenazine and of its benzo(*c*)derivative, prepared by equation (216), has been studied[1014]. Monopyrazine and monoquinoxaline analogues of **497** have been prepared by reaction of **492** with 2,3-dichloropyrazine[1015] and 2,3-dichloroquinoxaline respectively[595,1015]. Se analogues of methylene blue, in particular 3,6-bis(dimethylamino)phenoselenazinium have also been prepared and studied[1016–1018] as selenazine dyes. They are obtained by the general reaction of equation (216) by using H_2Se, or by bromination of **497**, followed by amination.

b. Phenotellurazines. The chemistry of phenotellurazine is very recent. The first publication deals with the preparation of a monoquinoline analogue, the 1*H*-dibenzo(*b,g*)-4-tellura-1,8-naphthyridine[1019], obtained from 2-anilinoquinoline and $TeCl_4$, followed by dechlorination with Na_2SO_3. 2,8-Dimethyl-5-ethylphenotellurazine (**499**: $R^1 = Et$, $R^2 = Me$)[1020] and the 5-methylphenotellurazine (**499**; $R^1 = Me$, $R^2 = H$)[1021] are prepared by reaction of the 2,2′-dilithio derivative **498** with $TeCl_2$ and $TeCl_4$, respectively (equation 217). The direct reaction of $TeCl_4$ on the amine **496**, a reaction working well for the synthesis of phenoxatellurine, failed completely.

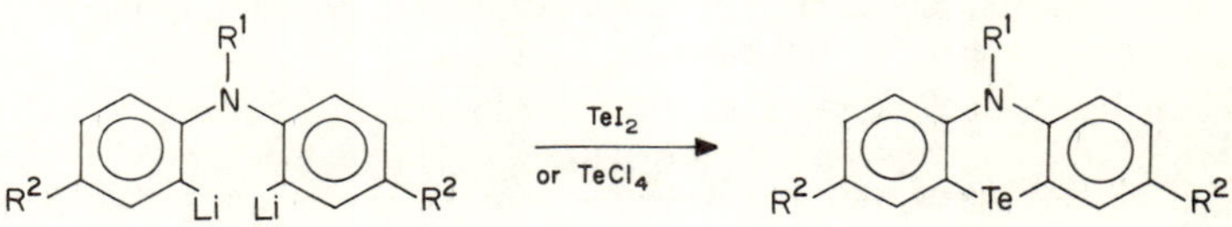

(498) (499) (217)

D. Three Heteroatoms without nitrogen

Nearly all the known derivatives of this series are of the 1,3,5-type. Compound **500**, the first known derivative of the 1,3,5-dioxaselenane type, also known as the monoselenium analogue of paraldehyde, is prepared by reaction of H_2Se with acetaldehyde in aqueous acidic solution, near 0 °C[1022]. Similarly, the parent derivative is obtained from formaldehyde, together with 1,3,5-oxadiselenane[1023]. The parent 1,3,5-oxadiselenane (**501**) is also obtained from α,α'-dichlorodimethyl ether and Na_2Se in methanol[1024]. This compound is easily polymerized. The symmetric 1,3,5-triselenane (**502a**), the trimer of selenoformaldehyde, is the best known of these compounds. It was first obtained in 1915[1025] and again in 1938[1026], but it was incorrectly formulated as selenoformaldehyde. It was prepared in 1950 in 11% yield by reaction of H_2Se with acidic formaldehyde, or by hydrolysis of α,α'-dichloromethyl selenide[1027] and identified as the trimer. The by-products of the former reaction are a linear polymer and a tetramer. The structure was confirmed by X-ray crystallography. **502a** was also later obtained from dichloromethane and Na_2Se[1028,1029], and in 1982 by the first method[1030]. It is decomposed by Cl_2 to

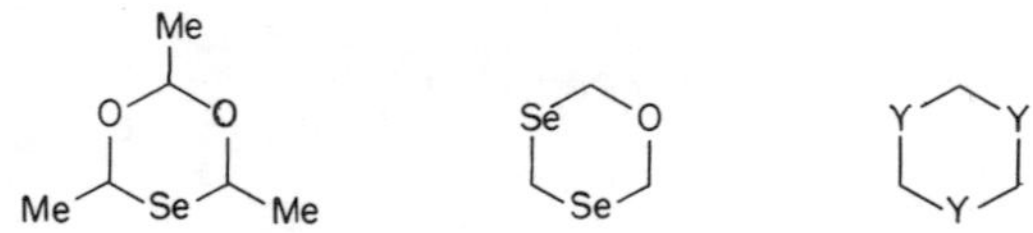

(500) **(501)** **(502)**
(**a**) Y = Se
(**b**) Y = Te

dichloromethyl selenide, diselenide and chloromethylselenenyl chloride[1027]. The trimethyl homologue of **502a** is prepared similarly from acetaldehyde[1022]. The 1,3,5-tritellurane analogue **502b** is also similarly prepared[1031–1033]. More recently it has been obtained by reaction of Te on the methylene produced by thermolysis of diazomethane[1034]. Its chemistry remains unexplored. Among the other Te derivatives, the 1,3,2-benzodioxatellurane has been prepared according to the general synthesis of dialkoxytellurides by trans-esterification of diols, by using *o*-hydroxybenzyl alcohol as the diol[1035]. A 1,2,6-selenadigermane is also known[1036].

E. Three Heteroatoms with Nitrogen

1. One nitrogen

Three 5,6-dihydro-1,4,3-oxaselenazines were obtained in 1979 1,4-cycloaddition of styrene to the heterodiene Se=N=C=O system of *N*-benzoylselenimides[1037] (equation 218). This was the first six-membered system with Se linked to N. The 5-trichloromethyl-3-imino-3*H*-1,4,2-thiaselenazine-1,1-dioxide, a unique member of this ring system is obtained via the general ring-opening of *β*-lactones with *Se*-ethyl selenourea, except that the propiolactone is replaced by a propiosultone[1038]. *N*-Methyl-1,3,5-perhydrodiselenazine is obtained by thermolysis of a eight-membered ring-system, a diselenadiazocane[1039] (cf. equation 225).

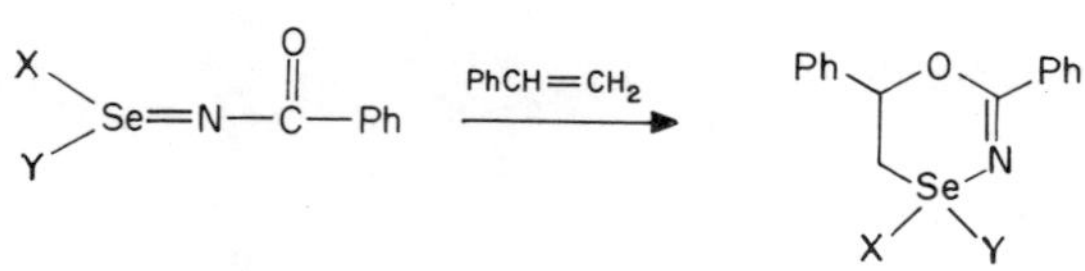

(218)

(503) **(504)**

2. Two nitrogens: selenadiazines

a. 1,2,4-*Selenadiazines.* Reaction of allylthiourea and selenium monochloride leads to 3-thioxo-6-chloromethyl-1,2,4-perhydroselenadiazine[1040a]. 5-Methyl-3,4-diphenyl-1,2,4-benzoselenadiazine has been prepared recently by thermolysis at 180 °C of the ylid form of a diaza analogue of selenabenzene, 5-methyl-1,3-diphenyl-1λ^4, 2,4-benzoselenadiazine[1040b]. This reaction is parallel to that in the sulphur series.

b. 1,2,6-*Selenadiazines.* Se insertion in *o*-diamines by means of SeO_2 has been applied to 1,8-diaminonaphthalene, giving 1*H*,3*H*-naphtho(1,8-*cd*)-1,2,6-selenadiazine[1041]. This reaction could not be reproduced, but with $SeOCl_2$ the corresponding 1*H*,3*H*-2-selenoxide has been isolated in 51% yield. Applied to the tin double salt of 1,4,5,8-tetraaminonaphthalene (**505**), this reaction leads by double Se insertion to naphtho (1,8-*c,d*: 4,5-*c′,d′*) bis-1,2,6-selenadiazine (**506**)[1042] (equation 219).

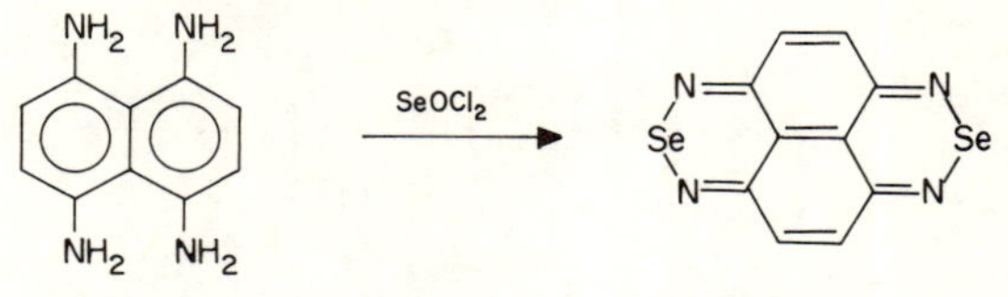

(219)

(505) (506)

c. 1,3,4-*Selenadiazines*. Reaction of 4-arylselenosemicarbazides with α-haloketones does not give five-membered selenazoline derivatives, but 2-arylamino-5-substituted 6*H*-1,3,4-selenadiazine **507**[637,1043] (equation 220). A structural proof is based on the reaction of the 6-methylene group with *p*-nitrobenzaldehyde. The isolated intermediate eliminates Se easily, as in the sulphur series.

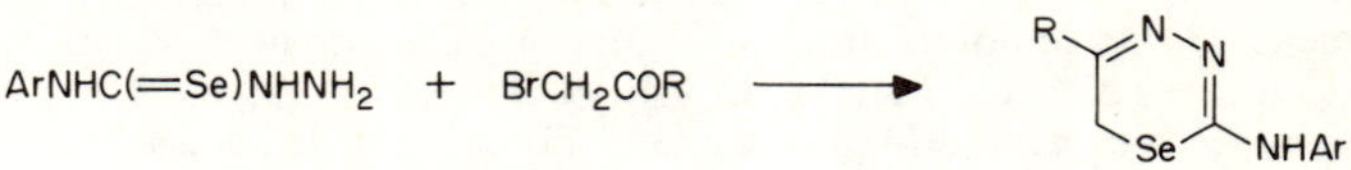

(220)

(507)

d. 1,3,5-*Selenadiazines*. The reaction of primary amines, formaldehyde and CSe_2[1044] or phenyl isoselenocyanate and NaSeH[1045] leads to tetrahydro-1,3,5-selenadiazine-2-selones. By reaction with H_2Se and formaldehyde, benzylamine, contrary to other aliphatic or aromatic amines, gives 3,5-dibenzyl-1,3,5-selenadiazine, a six-membered ring derivative, instead of an eight-membered ring[1039] (see Section VII.A). A tricyclic dibenzoselenadiazinium dibromide, a dipyrido analogue of selenoxanthene, has been synthesized[1046] by a reaction parallel to that in the sulphur series, but which failed in the corresponding oxygen and nitrogen series.

F. Heteroatoms

1-Oxa and 1-thia-4-selena-2,6-disilanes or -digermanes are formed by bis-nucleophilic attack of the appropriate dichlorodisiloxane or digermoxane by Na_2Se[1047,1048]. In the Te series, the 3*H*,6*H*-1,2,4,5-tetratellurin is formed by reaction of K_2SO_3 and methylene bis(tellurium trichloride). The structure has been confirmed by mass spectrometry, NMR and Mössbauer spectroscopies[1049].

VI. SEVEN-MEMBERED RINGS

A. One Heteroatom

1. *Monocyclic systems*

In 1931 Morgan and Burstall[832] prepared selenepane in low yield by the general procedure of ring-closure from Na_2Se and 1,6-dibromohexane. The 1,1,3,6-tetrachloro derivative is obtained by reaction of 1,5-hexadiene with SeO_2 in concentrated HCl[1050]. The 3,6-dimethoxy derivative is also formed in small amounts in the oxyselenation of 1,5-hexadiene with potassium selenocyanate and Cu salts[951].

2. *Benzoselenepins and analogues*

5-Oxo-2,3,4,5-tetrahydrobenzoselenepins (**509**) (homoselenochromanones) can be synthesized by the electrophilic ring-closure of the corresponding γ-arylselenobutyric acids

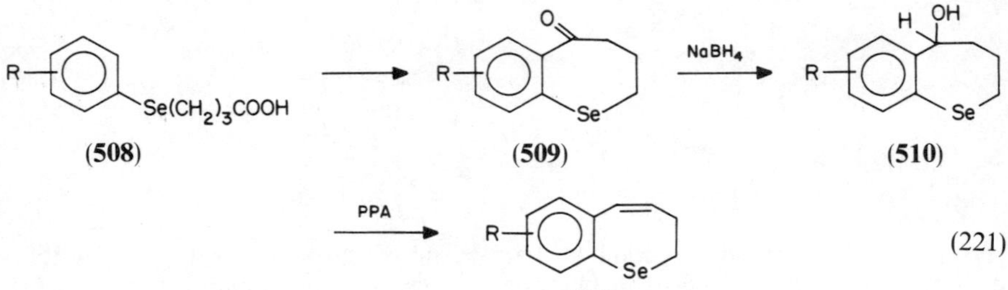

(508) (509) (510) (511) (221)

(**508**) with PPA, or of their chlorides with $SnCl_4$[863,1051]. Compounds **509** have been transformed by standard methods to homoselenochromanols (**510**), the corresponding homochromenes (**511**) and homochromans (equation 221). The method has been applied to various derivatives of **509** (7-Me, 7-t-Bu, 7-Br, 9-Me, 6,9-di-Me and 7,8-di-Me) and to the naphtho[866], benzofuro(2,3-*b*) and benzoselenolo(2,3-*b*)analogues[870].

3. Dibenzoselenepins

a. Dibenzo(b,f)*selenepins*. The 10,11-dihydro-10-keto compound **513** and its 8-chloro derivative have been prepared in 47% yield by the classical electrophilic ring-closure from **512**. Compound **513** is transformed successively to the corresponding 10-hydroxy, 10-chloro and 10-amino derivatives (**514**) required for the synthesis of analogues of psychotropic benzothiepins[1052] (equation 222). By reaction of *N*-methylpiperazine with the ketone **513**, the parent dibenzoselenepin itself (**515**) has been obtained as a by-product together with the corresponding expected enamine[1053].

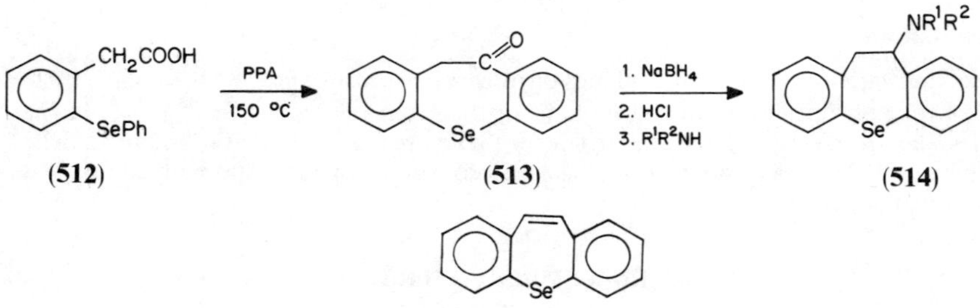

(512) (513) (514) (222)

(515)

b. Dibenzo(b,e)*selenepins*. The isomer **516** of **513** has been similarly prepared and transformed to the amino derivatives[1054].

c. Dibenzo(c,e)*selenepins*. A third system, **517**[1055], and its 3,9-diphenyl derivative[1056] have been synthesized from the 2,2′-dibromomethylbiphenyls and Na_2Se.

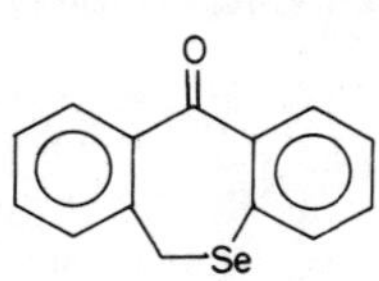

(516)

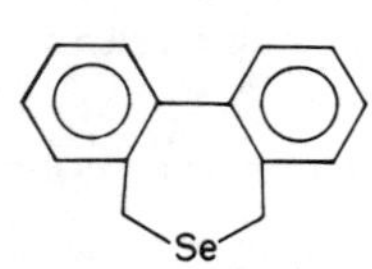

(517)

4. Condensed telluracycloheptane

A steroidal telluracycloheptane has been prepared from a steroidal methanesulphonate and Na_2Te[1057].

B. Two Heteroatoms

1. Without nitrogen

Treating 1,5-pentanediselenocyanate with alcoholic KOH gives a compound regarded as 1,2-diselenepane, but which is probably, as are its lower ring homologues, a low-molecular-weight polymer[828]. The intramolecular oxyselenation of diallyl ether gives 3,6-dimethoxy-1,4-oxaselenepane in isomeric mixture with the six-membered ring[951]. By the general reaction of chloroalkylthioalkynes with alkali metal selenide and telluride, Brandsma and coworkers have prepared 5*H*-6,7-dihydro-1,4-thiaselenepin and -thiatellurepin and their 3-Me, 3-Et, 3-*t*-Bu and 3-Ph derivatives (**518**)[958] (equation 223).

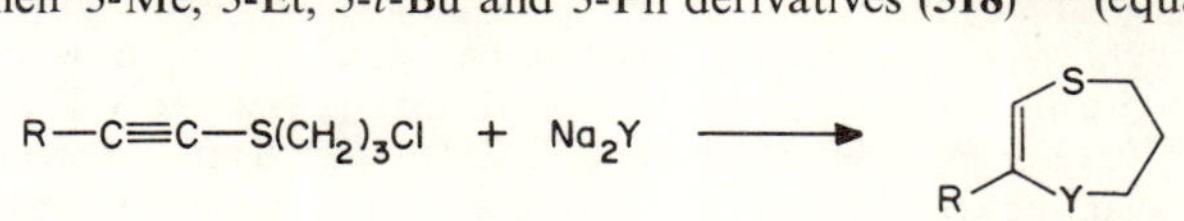

(223)

Y = Se, Te
R = H, Me, Et, *t*-Bu, Ph

(**518**)

2. With nitrogen: selenazepines

Investigations on the condensation of the Zn salt **492** with mesityl oxide[1058], 3-methylhept-3-en-5-one[1059] and crotonic acid[1060] resulted in the preparation of the 2,3-dihydro-1,5-benzoselenazepines **519** (R = Me, Et), and of the lactam **520**. The dioxo-1,4-benzoselenazepine **521** has been isolated in 74% yield as major product in the reaction of *N*-methylbenzisoselenazolin-3-one with acetic anhydride and sodium acetate[602].

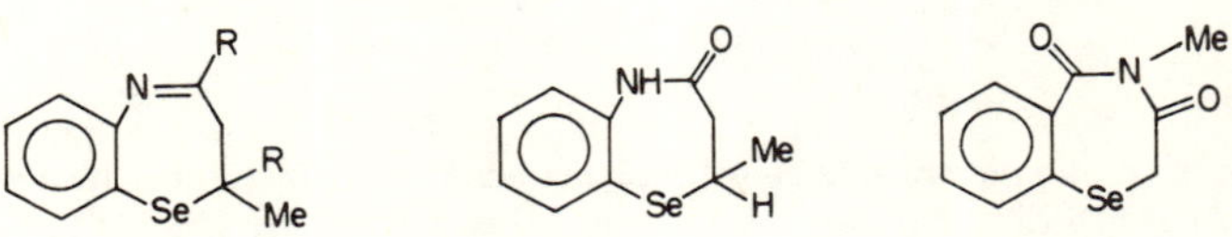

(**519**) (**520**) (**521**)

C. Three Heteroatoms

By the general method of trans-esterification of diols a seven-membered dialkoxytelluride is obtained from 2,2′-dihydroxybiphenyl[1034]. A new heterocyclic system, the 2*H*,7*H*-1,4,5-selenadiazepine ring **523** has been obtained by reaction of equimolecular amounts of hydrazine and selenodiacetophenone **522** (equation 224)[1061a]. This ring undergoes easy thermal ring-contraction to afford the corresponding pyridazine **524**.

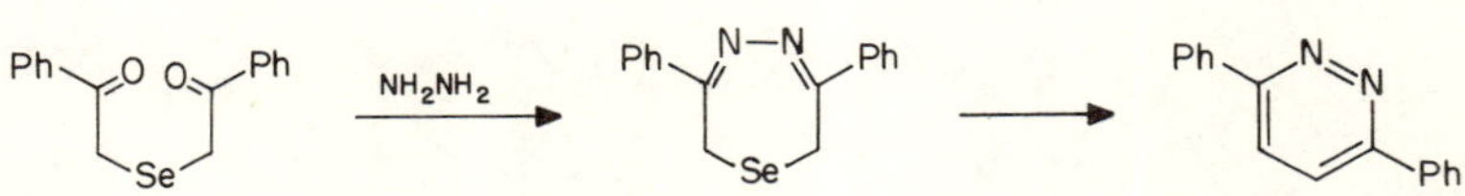

(224)

(**522**) (**523**) (**524**)

VII. LARGER RINGS

A. Eight-membered Rings

Two tetraselenocane isomer rings are known. The 1,2,5,6-tetraselenocane, a bis-diselenide of diselenoethylene glycol, has been obtained by alkaline hydrolysis of 1,2-ethanediselenocyanate[487,719]. Its dibenzo-condensed system is mentioned as an air oxidation derivative of 1,4-benzenediselenol[1061b]. The 1,3,5,7-tetraselenocane, a cyclic tetramer of selenoformaldehyde, is isolated as a by-product from the reaction of

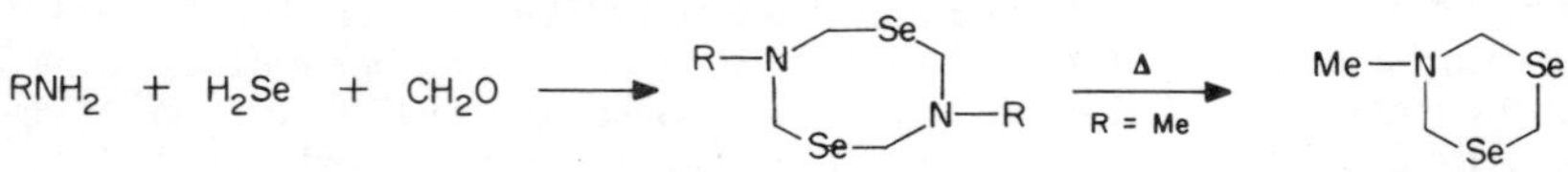

(525) (526) (225)

dichloromethane with Na_2Se[1028] and from H_2Se and formaldehyde, which also give the trimer and the pentamer[1023]. The cyclic pentamer, 1,3,5,7,9-pentaselenecane, is also isolated in the former reaction[1029]. 3,7-disubstituted perhydrodiselenadiazocines **525** (R = Me[1039a,b], Et, *n*-Pr, *n*-Bu, Ph, *n*-Hex, PhCHMe[1039b]) are the products generally formed in the reaction of formaldehyde, H_2Se and a primary amine except for benzylamine. These eight-membered rings are thermally unstable and are decomposed to perhydro-1,3,5-diselenazines (**526**)[1039b] (equation 225). The general ring-enlargement of benzoxazolium and benzothiazolium salts has been extended to the benzoselenazolium salt **527**, giving the dibenzo-1,4-selenazocine system **528** (equation 226)[701].

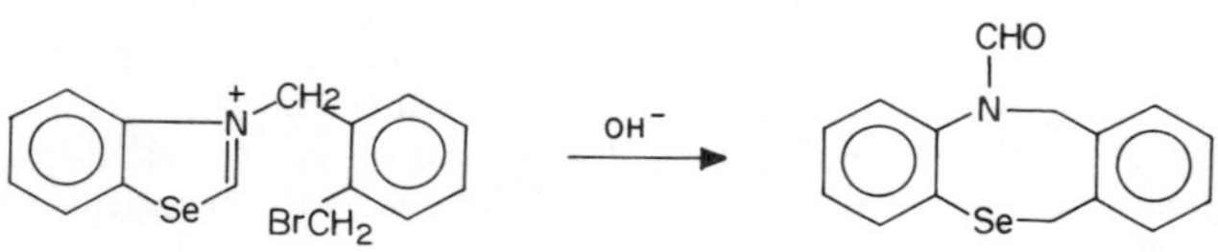

(527) (528) (226)

B. Macrocycles

A few Se-containing peri-bridged naphthalenes are known, including the monoselena 9-membered ring **529**[1062] and the tetraselena 14-membered ring **530**[907]. The latter is obtained by alkaline hydrolysis of 1,8-di(cyanoselenomethyl)naphthalene. A 7-membered cyclic diselenide, the corresponding 'monomer' seems to be the primary product of reaction, and this dimerizes easily to **530**. A dioxaditellura 14-membered ring has been synthesized from sodium orthophthalate and diaryltellurium dichloride[1063]. Among *meta*-substituted benzene macrocycles (metacyclophanes), the 2,11-diselena-(3,3)metacyclophane ring system **531** has been synthesized in 5–10% yield by reaction of *m*-xylylene dibromide with anhydrous Na_2Se[1064a] or KSeCN[1064b]. The corresponding bis-diselenide, namely 2,3,12,13-tetraselena(4,4)metacyclophane (**532**), has been authenti-

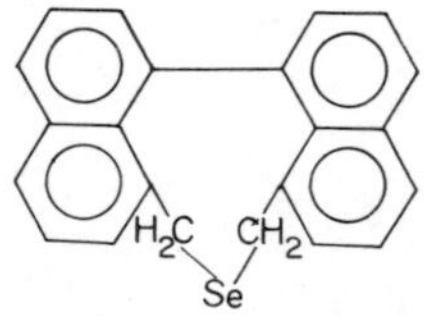

(529)

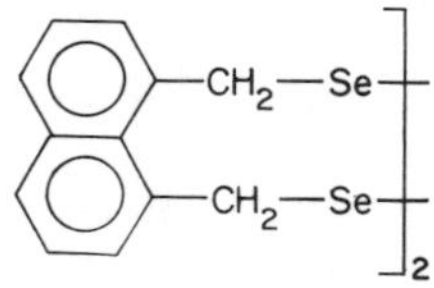

(530)

cally obtained by alkaline hydrolysis of *m*-xylylene diselenocyanate. Its structure has been confirmed by its thermal conversion to **531**[1065], and the dynamic stereochemistry has been studied[1066].

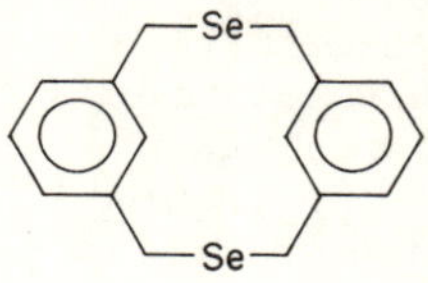

(531)

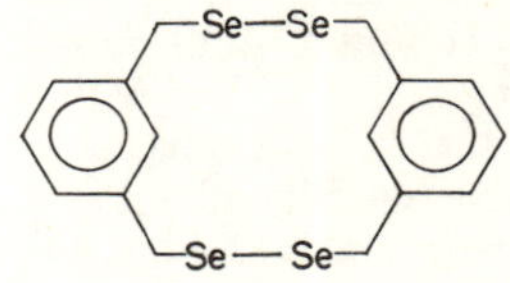

(532)

VIII. SPIROHETEROCYCLES

The known spiroselenium heterocycles are generally saturated ring systems, where the Se function is a selenide (four-membered ring), a diselenide or a mixed five-membered sulphide and selenide ring system. They are prepared according to equation (227) (X = Y = Br). Four-membered ring systems (**534**; $n = 1$), where the first ring may be a cyclohexane (2-selenaspiro(3,5)nonane)[51], an oxetane[1061b] or a second identical selenetane (2,6-diselenaspiro(3,3)heptane)[51], are obtained in this last case by a double ring-closure from sym-tetrabromoneopentane. For the diselenide ring system (reagent KSeCN), the first ring is also either a cyclohexane[1067] or an oxetane[1061b]. The corresponding mixed selenide and sulphide is prepared by reaction of Se with the corresponding four-membered 2-thiaspiro(3,5)nonane[1068a]. The selenone corresponding to the 2-selenaspiro(3,5)nonane is directly obtained by thermal cyclization on the selenate **533** (Y = SeO_2Na) in a sealed tube[51].

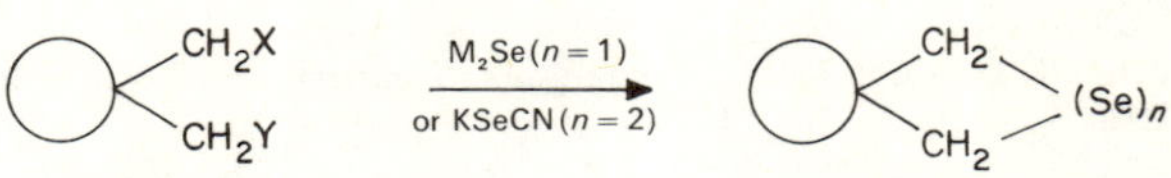

(227)

(533) **(534)**

A 1,6-dioxa-5-selena(5-Se^{IV})spiro(4,4)nonane-2,7-dione has been obtained by another method, consisting of oxidation of 3,3′-selenodipropionic acid[1068b].

IX. BRIDGED Se/Te HETEROCYCLES

1-Vinyl-2-selenabicyclo(3.2.0)hepten-3-one is obtained as a by-product of reaction of Se with vinylacetylene[1068c]. 1,3,5-Trimethyl-7-selenabicyclo(2.2.1)heptane-2,3-dione is prepared by SeO_2 oxidation of a cyclic diketone[526b]. A 8-selena-2-azabicyclo(3.2.1)octane **535** is formed through a second ring-closure of dimethyl 4-chloro-2,7-dimethyl-4*H*-azepine-3,6-dicarboxylate by H_2Se, or by ring-enlargement of the corresponding halomethyldihydropyridine[1069], via the same intermediate and with the same Se reagent. Among the bicyclononane analogues the 9-selenabicyclo(3.3.1)nonane **536** is obtained by oxyselenation of *cis,cis*-1,5-cyclooctadiene[951]. The oxathiaselenabicyclo(3.3.1)nonane **537** is also known and is prepared from a tetraiodomethylene cyclobutane[1070]. Based on

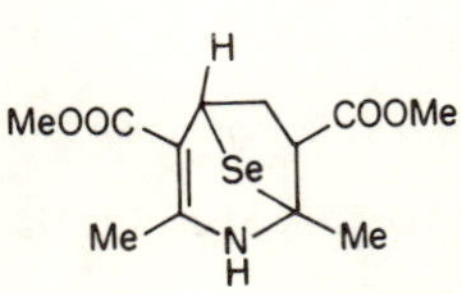

(535)

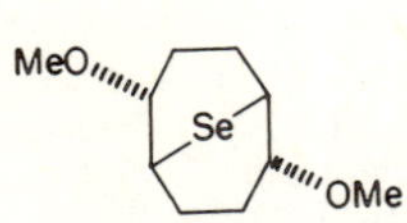

(536)

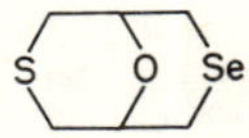

(537)

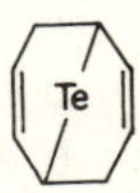

(538)

stereochemical and mechanistic considerations, and episelenonium bridged ion has been suggested as intermediate in the reductive elimination of β-hydroxyselenides of bicyclo(4.2.1)nonatriene in a superacid medium[1071]. A tellura-bridged heterocycle, namely 9-tellurabicyclo(3.3.1)nona-2,6-diene (**538**) has been prepared with 18% yield by reaction of Na_2Te with 3,7-dibromo-1,5-cyclooctadiene. At 175 °C the reaction leads to a bicyclo(5.0.0)octadiene[1072,1073] with bromine elimination induced by telluride anion. A selenatricyclo(2.2.1.1)octane has been obtained by reaction of norbornadiene with $SeBr_4$[1074]. Some selenaadamantanes have also been prepared[1075]. A 2,4,6,8-tetraselenaadamantane and a 2,4,6,8,9,10-hexaselenaadamantane are formed by a route similar to the preparation of the S analogue[60b].

X. REFERENCES

1. E. Painter, *Chem. Rev.*, **28**, 200 (1941).
2. T. Campbell, H. Walker and G. Coppinger, *Chem. Rev.*, **50**, 279 (1952).
3. D. Klayman and W. Günther (Eds.), *Organic Selenium Compounds, their Chemistry and Biology*, Wiley–Interscience, New York, 1973.
4. W. Cooper, *Tellurium*, Van Nostrand, New York, 1977.
5. K. Irgolic, *The Organic Chemistry of Tellurium*, Gordon and Breach, New York, 1974.
6. K. Irgolic, *J. Organometal. Chem.*, **103**, 91 (1975); **130**, 411 (1977); **158**, 235 (1978); **158**, 269 (1978); **189**, 65 (1980); **203**, 367 (1980).
7. G. Pattenden, *Saturated Heterocyclic Chemistry*, The Royal Society of Chemistry, London, 1973–1978.
8. (a) C. Bird and G. Cheeseman, *Aromatic and Heteroaromatic Chemistry*, Vols. 1–5, The Royal Society of Chemistry, London, 1973–1977.
   (b) H. Suschitzky and O. Meth-Cohn, *Aromatic and Heteroaromatic Chemistry*, Vols. 6–7, Royal Society of Chemistry, London, 1978–1979.
9. D. Reid, *Organic Compounds of Sulphur, Selenium and Tellurium*, Vols. 1–6, The Royal Society of Chemistry, London, 1970–1981.
10. H. Suschitzky and O. Meth–Cohn, *Heterocyclic Chemistry*, The Royal Society of Chemistry, London, 1980–.
11. L. Christiaens, in *Proceedings of the Third International Symposium on Organic Se and Te Compounds* (Eds. D. Cagniant and G. Kirsch), University of Metz, 1979, p. 69.
12. A. Callear and W. Tyerman, (a) *Proc. Chem. Soc.*, 296 (1964); (b) *Trans. Faraday Soc.*, **61**, 2395 (1965); **62**, 371 (1966).
13. W. Tyerman, W. O'Callaghan, P. Kebarle, O. Strausz and H. Gunning, *J. Amer. Chem. Soc.*, **88**, 4277 (1966).
14. D. Clive and C. Denyer, *J. Chem. Soc., Chem. Commun.*, 253 (1973).
15. T. Chan and J. Finkenbine, *Tetrahedron Letters*, 2091 (1974).
16. F. Mathey and G. Muller, *Compt. Rend., Ser. C*, **281**, 881 (1975).
17. J. Behan, R. Johnstone and M. Wright, *J. Chem. Soc., Perkin Trans. 1*, 1216 (1975).
18. E. Kataev, G. Chmutova, A. Musina and A. Anastaseva, *Zh. Org. Khim.*, **8**, 1531 (1972).
19. D. Van Ende and A. Krief, *Tetrahedron Letters*, 2709 (1975).
20. M. Lakshmikantham and M. Cava, *J. Org. Chem.*, **41**, 879 (1976).
21. V. Calò, L. Lopez, A. Mincuzzi and G. Pesce, *Synthesis*, 200 (1976).
22. (a) D. Clive and S. Menchen, *J. Chem. Soc., Chem. Commun.*, 658 (1977).
   (b) D. Clive and P. Beaulieu, *J. Org. Chem.*, **47**, 1124 (1982).
23. J. Bergman and L. Engman, *J. Organomet. Chem.*, **181**, 335 (1979).
24. J.-E. Bäckvall and L. Engman, *Tetrahedron Letters*, **22**, 1919 (1981).
25. J. Connor, G. Grey and O. Strausz, *J. Amer. Chem. Soc.*, **91**, 5695 (1969).
26. (a) L. Rasteikiene, D. Greiciute, M. Lun'kova and I. Knunyants, *Russ. Chem. Rev.*, **46**, 548 (1977).
   (b) D. C. Dittmer and B. H. Patwardahom, in *The Chemistry of the Sulphonium Group* (Eds. C. J. M. Stirling and S. Patai), Chap. 13, Wiley–Interscience, Chichester, 1981.
27. G. Capozzi, O. De Lucchi, V. Lucchini and G. Modena, *Tetrahedron Letters*, 2603 (1975).
28. F. Lautenschlager, *J. Org. Chem.*, **34**, 4002 (1969).
29. D. Garratt and G. Schmid, *Can. J. Chem.*, **52**, 1027 (1974).

30. G. Schmid and D. Garratt, *Tetrahedron Letters*, 3991 (1975).
31. D. Garratt and G. Schmid, *J. Org. Chem.*, **42**, 1776 (1977).
32. J.-N. Denis, J. Vicens and A. Krief, *Tetrahedron Letters*, 2697 (1979).
33. A. Toshimitsu, T. Aoai, H. Owada, S. Uemura and M. Okano, *J. Org. Chem.*, **46**, 4727 (1981).
34. S. Raucher, *Tetrahedron Letters*, 2261 (1978).
35. N. Miyoshi, S. Ohno, K. Kondo, S. Murai and N. Sonoda, *Chem. Letters*, 1309 (1979).
36. H. Reich and J. Trend, *Can. J. Chem.*, **53**, 1922 (1975).
37. B. Lindgren, *Acta Chim. Scand.*, **27**, 726 (1973); *Tetrahedron Letters*, 4347 (1974).
38. P. Mente and C. Rees, *J. Chem. Soc., Chem. Commun.*, 418 (1972).
39. T. Gilchrist, P. Mente and C. Rees, *J. Chem. Soc., Perkin Trans. 1*, 2165 (1972).
40. J. Laureni, A. Krantz and R. Hajdu, *J. Amer. Chem. Soc.*, **98**, 7872 (1976).
41. A. Krantz and J. Laureni, *J. Amer. Chem. Soc.*, **99**, 4842 (1977).
42. G. Capozzi, O. De Lucchi, V. Lucchini and G. Modena, *J. Chem. Soc., Chem. Commun.*, 248 (1975).
43. G. Capozzi, V. Lucchini, G. Modena and P. Scrimin, *Tetrahedron Letters*, 911 (1977).
44. R. Destro, T. Pilati and M. Simonetta, *Nouv. J. Chim.*, **3**, 533 (1979).
45. G. Schmid and D. Garratt, *Chem. Scripta*, **10**, 76 (1976).
46. (a) K. Nicolaou, R. Magolda, W. Sipio, W. Barmette, Z. Lysenko and M. Jouille, *J. Amer. Chem. Soc.*, **102**, 3784 (1980).
(b) H. Bock, S. Aygen and B. Solouki, *Z. Naturforsch.*, **38b**, 611 (1983).
(c) R. Schulte and A. Schweig, *Tetrahadron Letters*, **25**, 2337 (1984).
47. D. Black and K. Watson, *Australian J. Chem.*, **26**, 2167 (1973).
48. S. Tamagaki, K. Sakaki and S. Oae, *Heterocycles*, **2**, 39 (1974).
49. S. Tamagaki, K. Sakaki and S. Oae, *Heterocycles*, **2**, 45 (1974).
50. G. Morgan and F. Burstall, *J. Chem. Soc.*, 1497 (1930).
51. H. Backer and H. Winter, *Rec. Trav. Chim.*, **56**, 492 (1937).
52. A. Geens and M. Anteunis, *Bull. Soc. Chim. Belges*, **80**, 639 (1971).
53. (a) B. Lindgren, *Chem. Scripta*, **16**, 1 (1980).
(b) A. Arnold and A. Canty, *Australian J. Chem.*, **36**, 815 (1983).
54. M. Francis, P. Grant, K. Low and R. Weavers, *Tetrahedron*, **32**, 95 (1976).
55. Yu. Migalina, V. Lendel, A. Koz'min and N. Zefirov, *Khim. Geterotsikl. Soedin.*, 708 (1978).
56. A. Davison and E. Shawl, *J. Chem. Soc., Chem. Commun.*, 670 (1967); *Inorg. Chem.*, **9**, 1820 (1970).
57. A. Holm, C. Berg, C. Bjerre, B. Bak and H. Svanholt, *J. Chem. Soc., Chem. Commun.*, 99 (1979).
58. K. Jensen and L. Henriksen, *Acta Chem. Scand.*, **24**, 3213 (1970).
59. G. Morgan, H. Drew and T. Barker, *J. Chem. Soc.*, 2432 (1922).
60. (a) D. Dewar, J. Fergusson, P. Hentschel, C. Wilkins and P. Williams, *J. Chem. Soc.*, 688 (1964).
(b) K. Olsson and S. Almqvist, *Acta Chem. Scand.*, **23**, 3271 (1969).
61. A. Haas, *Chem. Scripta*, **8A**, 75 (1975).
62. (a) Y. Kopaevich, G. Belen'kii, E. Mysov, L. German and I. Knunyants, *Zh. Vses. Khim. Obshchest*, **17**, 226 (1972); *Chem. Abstr.*, **77**, 101427 (1972).
(b) M. Raasch, *J. Org. Chem.*, **45**, 3517 (1980).
63. S. Bender, N. Haley and H. Luss, *Tetrahedron Letters*, **22**, 1495 (1981).
64. M. Lakshmikantham, M. Cava, M. Albeck and L. Engman, *J. Chem. Soc., Chem. Commun.*, 829 (1981).
65. M. Lakshmikantham, M. Cava, M. Albeck, L. Engman, P. Carroll, J. Bergman and F. Wudl, *Tetrahedron Letters*, **22**, 4199 (1981).
66. M. Petrov, V. Laishev and A. Petrov, *J. Org. Chem.* (*USSR*) (*Engl. Transl.*), **15**, 2346 (1979).
67. R. Mayer, B. Hunger, R. Prousa and A. Muller, *J. Prakt. Chem.*, **35**, 294 (1967).
68. I. Lalezari, A. Shafiee and M. Yalpani, *J. Org. Chem.*, **38**, 338 (1973).
69. M. Petrov, V. Laishev and A. Petrov, *Zh. Org. Khim.*, **17**, 667 (1981).
70. K. Jensen and L. Henriksen, *Acta Chem. Scand.*, **29B**, 877 (1975).
71. H. Hartough, in *The Chemistry of Heterocyclic Compounds*, (Ed. A. Weissberger), Chap. XVI, Interscience, New York, 1952.
72. N. Magdesieva, *Advances in Heterocyclic Chemistry* (Eds. A. Katritzky and A. Boulton), Vol. 12, Academic Press, New York, 1970, p. 1.

73. N. Magdesieva and N. Zefirov, in Ref. 3, p. 427.
74. A.-B. Hörnfeldt, in *Advances in Heterocyclic Chemistry* (Eds. A. Katritzky and A. Boulton), Vol. 30, Academic Press, New York, 1982, p. 127.
75. F. Fringuelli, G. Marino and A. Taticchi, in *Advances in Heterocyclic Chemistry* (Eds. A. Katritzky and A. Boulton), Vol. 21, Academic Press, New York, 1977, p. 120.
76. G. Marino, *Chem. Scripta*, **8A**, 23 (1975).
77. F. Mazza and L. Solazzo, *Rend. Acad. Sci. Napoli*, **33**, 236 (1927).
78. H. Briscoe and J. Peel, *J. Chem. Soc.*, 1741 (1928).
79. H. Briscoe, J. Peel and P. Robinson, *J. Chem. Soc.*, 2628 (1928).
80. H. Suginome and S. Umezawa, *Bull. Chem. Soc. Japan*, **11**, 157 (1936).
81. S. Umezawa, *Bull. Chem. Soc. Japan*, (a)**11**, 775 (1936); (b)**12**, 4 (1937); (c)**14**, 155 (1939).
82. (a) S. Gronowitz, T. Frejd, A. Moberg-Ogard and L. Trege, *J. Heterocycl. Chem.*, **13**, 1319 (1976).
(b) S. Mohmand, J. Bargon and R. Waltman, *J. Org. Chem.*, **48**, 3544 (1983).
83. F. F. Perveev, N. Kudryashova and D. Gliborskii, *Zh. Obshch. Khim.*, **26**, 3331 (1956).
84. Yu. K. Yur'ev and L. Khmel'nitskii, *Dokl. Akad. Nauk SSSR*, **94**, 265 (1954).
85. E. Mamedov, S. Kurbanov, R. Mishiev and T. Shakhtakhtinskii, *J. Org. Chem.* (*USSR*), (*Engl. Transl.*), **15**, 1388 (1979).
86. Yu. Yur'ev, N. Sadovaya and M. Gal'bershtam, *Zh. Obshch. Khim.*, **28**, 620 (1958).
87. B. Arbuzov and E. Kataev, *Dokl. Akad. Nauk SSSR*, **96**, 983 (1954).
88. Yu. Yur'ev, N. Magdesieva and L. Petrova, *Khim. Geterotsikl. Soedin.*, 910 (1966).
89. R. Curtis, S. Hasnain and J. Taylor, *J. Chem. Soc., Chem. Commun.*, 365 (1968).
90. L. Lepage and Y. Lepage, *J. Heterocycl. Chem.*, **15**, 1185 (1978).
91. P. Jacobs, M. Davis and H. Norton, *J. Heterocycl. Chem.*, **14**, 1115 (1977).
92. P. Jacobs and M. Davis, *J. Org. Chem.*, **44**, 178 (1979).
93. B. Trofimov, G. Musorin, G. Kalabin and S. Amosova, *J. Org. Chem.* (*USSR*) (*Engl. Transl.*) **16**, 448 (1980).
94. M. Bogert and C. Andersen, *J. Amer. Chem. Soc.*, **48**, 223 (1926).
95. S. Gronowitz and A. Konar, *Chem. Scripta*, **12**, 11 (1977).
96. C. Paal, *Chem. Ber.*, **18**, 2255 (1885).
97. (a) E. Braye, W. Hubel and I. Caplier, *J. Amer. Chem. Soc.*, **83**, 4406 (1961).
(b) K. Schulte, J. Reisch and A. Bergenthal, *Chem. Ber.*, **101**, 1540 (1968).
98. P. Cagniant, P. Perin, G. Kirsch and D. Cagniant, *Compt. Rend., Ser. C*, **277**, 37 (1973).
99. P. Cagniant, P. Perin and G. Kirsch, *Compt. Rend., Ser. C*, **278**, 1201 (1974).
100. P. Cagniant, G. Kirsch and P. Perin, *Compt. Rend., Ser. C*, **279**, 851 (1974).
101. J. Liebscher and H. Hartmann, *Synthesis*, **8**, 521 (1976).
102. H. Backer and W. Stevens, *Rec. Trav. Chim.*, **59**, 423 (1940).
103. P. Clayton, A. Guest, A. Taylor and R. Ramage, *J. Chem. Soc., Chem. Commun.*, 500 (1979).
104. S. Braverman and I. Duar, *Tetrahedron Letters*, 1493 (1978).
105. Yu. Yur'ev, *Zh. Obshch. Khim.*, **16**, 851 (1946).
106. V. Kharchenko, I. Markushina and S. Voronin, *USSR Patent, SU 929,642; Chem. Abstr.*, **97**, 198097f (1982); *Khim. Geterotsikl. Soedin.*, 219 (1983).
107. I. Lalezari, A. Shafiee and M. Yalpani, *J. Heterocycl. Chem.*, **9**, 1411 (1972).
108. I. Lalezari, A. Shafiee, F. Rabet and M. Yalpani, *J. Heterocycl. Chem.*, **10**, 953 (1973).
109. I. Lalezari, A. Shafiee and A. Rashidbaigi, *J. Heterocycl. Chem.*, **13**, 57 (1976).
110. F. McMahon, T. Pearson and P. Robinson, *J. Chem. Soc.*, 1644 (1933).
111. W. Mack, *Angew. Chem.* (*Intern. Ed.*), **5**, 896 (1966).
112. F. Fringuelli and A. Taticchi, *Ann. Chim.* (*Rome*), **62**, 777 (1972).
113. F. Fringuelli and A. Taticchi, *J. Chem. Soc., Perkin Trans. 1*, 199 (1972).
114. T. Barton and R. Roth, *J. Organomet. Chem.*, **39**, C66 (1972).
115. W. Lohner and K. Praefcke, *Chem. Ber.*, **111**, 3745 (1978).
116. T. Barton, C. Tully and R. Roth, *J. Organomet. Chem.*, **108**, 183 (1976).
117. A. Ulman, J. Manassen, F. Frolow and D. Rabinovich, *Tetrahedron Letters*, 1885 (1978).
118. (a) A. Ulman, J. Manassen, F. Frolow and D. Rabinovich, *Tetrahedron Letters*, 167 (1978).
(b) W. Kulik, H. Verkruijsse, R. De Jong, H. Hommes and L. Brandsma, *Tetrahedron Letters*, **24**, 2203 (1983).
119. P. Cagniant, R. Close, G. Kirsch and D. Cagniant, *Compt. Rend., Ser. C*, **281**, 187 (1975).
120. G. Kirsch, P. Cagniant, D. Cagniant and C. Backes, *Phosphorus and Sulfur*, **6**, 161 (1979).

121. (a) V. Laishev, M. Petrov and A. Petrov, *Zh. Org. Khim.*, **18**, 281 (1982).
(b) A. Shafiee, I. Lalezari and F. Savabi, *Synthesis*, **9**, 765 (1977).
122. V. Laishev, M. Petrov and A. Petrov, *Zh. Org. Khim.*, **17**, 2064 (1981).
123. M. Detty, B. Murray and M. Seidler, *J. Org. Chem.*, **47**, 1968 (1982).
124. B. Arbuzov and E. Katar, *Dokl. Akad. Nauk SSSR*, **65**, 493 (1939).
125. N. Pozdeev, O. Akulinin, A. Shapkin and N. Magdesieva, *Dokl. Akad. Nauk SSSR*, **1851**, 384 (1969).
126. R. Brown and J. Crofts, *Chem. Phys.*, **1**, 217 (1973).
127. F. Fringuelli, G. Marino and A. Taticchi, *J. Chem. Soc., Perkin Trans. 2*, 332 (1974).
128. H. Lumbroso, D. Bertin, F. Fringuelli and A. Taticchi, *J. Chem. Soc., Chem. Commun.*, 342 (1973).
129. H. Lumbroso, D. Bertin, F. Fringuelli and A. Taticchi, *Compt. Rend., Ser. C*, **277**, 203 (1973).
130. G. Marino, in *Advances in Heterocyclic Chemistry* (Eds. A. Katriksky and A. Boulton), Vol. 13, Academic Press, New York, 1971, p. 235.
131. F. Fringuelli, G. Marino, G. Savelli and A. Taticchi, *J. Chem. Soc., Chem. Commun.*, 1441 (1971).
132. S. Clementi, F. Fringuelli, P. Linda, G. Marino, G. Savelli and A. Taticchi, *J. Chem. Soc., Perkin Trans. 2*, 2097 (1973).
133. P. Konstantinov, N. Koloskova, R. Shupik and M. Volkov, *Zh. Obsch. Khim.*, **43**, 872 (1973).
134. F. Fringuelli, G. Marino and A. Taticchi, *Gazz. Chim. Ital.*, **102**, 534 (1972).
135. F. Fringuelli, G. Marino and A. Taticchi, *J. Chem. Soc., Perkin Trans. 2*, 1738 (1972).
136. D. Spinelli, G. Guanti and C. Dell'Erba, *La Ricerca Scientifica*, **38**, 1048 (1968).
137. F. Fringuelli, B. Serena and A. Taticchi, *J. Chem. Soc., Perkin Trans. 2*, 971 (1980).
138. G. Guanti, C. Dell'Erba and G. Garbarino, *J. Heterocycl. Chem.*, **7**, 1425 (1970).
139. L. Chierici, C. Dell'Erba, A. Guareschi and D. Spinelli, *Ann. Chim. (Rome)*, **57**, 632 (1967).
140. C. Dell'Erba, A. Guareschi and D. Spinelli, *J. Heterocycl. Chem.*, **4**, 438 (1967).
141. L. Chierici, C. Dell'Erba and D. Spinelli, *Ann. Chim. (Rome)*, **55**, 1069 (1965).
142. F. Terrier, A. Chatrousse, R. Schaal, C. Paulmier and P. Pastour, *Tetrahedron Letters*, 1961 (1972).
143. C. Paulmier, M. Simonnin, A. Chatrousse and F. Terrier, *Tetrahedron Letters*, 1123 (1973).
144. F. Terrier, A. Chatrousse, C. Paulmier and R. Schaal, *J. Org. Chem.*, **40**, 2911 (1975).
145. F. Terrier, A. Chatrousse and C. Paulmier, *J. Org. Chem.*, **44**, 1634 (1979).
146. Yu. Yur'ev and E. Zaitseva, *Zh. Obshch. Khim. (Engl. Transl.)*, **29**, 1057 (1959).
147. Yu. Yur'ev, E. Zaitseva and G. Rozantsev, *Zh. Obshch. Khim., (Engl. Transl.)*, **30**, 2189 (1960).
148. F. Outurquin, G. Ah-Kow and C. Paulmier, *Compt. Rend., Ser. C*, **277**, 29 (1973).
149. G. Ah-Kow, C. Paulmier and P. Pastour, *Bull. Soc. Chim. Fr.*, 151 (1976).
150. Yu. Yur'ev and E. Zaitseva, *Zh. Obshch. Khim.*, **28**, 2164 (1958).
151. (a) Yu. Yur'ev and E. Zaitseva, *Zh. Obshch. Khim.*, **30**, 859 (1960).
(b) S. Gronowitz, C. Westerlund and A.-B. Hörnfeldt, *Acta Chem. Scand.*, **B29**, 224 (1975).
152. Yu. Yur'ev and N. Sadovaya, *Zh. Obshch. Khim.*, **34**, 1803 (1964).
153. A. Arcoria, E. Maccarone and A. Mamo, *J. Chem. Soc., Perkin Trans. 2*, 1347 (1979).
154. Yu. Yur'ev, N. Mezentsova and A. Monakhova, *Zh. Obshch. Khim.*, **30**, 2726 (1960).
155. W. Mack, *Angew. Chem.*, **77**, 260 (1965).
156. *French Patent*, No. 1,568,529; *Chem. Abstr.*, **72**, 100490q (1970).
157. Yu. Yur'ev, N. Sadoyava and E. Grekova, *Zh. Obshch. Khim.*, **34**, 847 (1964).
158. A. Hallberg, S. Liljefors and P. Pedaja, *Synth. Commun.*, **11**, 25 (1981).
159. C. Dell'Erba, D. Spinelli, G. Garbarino and G. Leandri, *J. Heterocycl. Chem.*, **5**, 45 (1968).
160. J. Morel, C. Paulmier and P. Pastour, *Compt. Rend., Ser. C*, **266**, 1300 (1968).
161. C. Paulmier, J. Morel and P. Pastour, *Bull. Soc. Chim. Fr.*, 2511 (1969).
162. Yu. Yur'ev and N. Sadoyava, *Zh. Obshch. Khim.*, **26**, 3154 (1956).
163. C. Paulmier and P. Pastour, *Compt. Rend., Ser. C*, **265**, 926 (1967).
164. Yu. Yur'ev, N. Sadovaya and M. Gal'bershtam, *Zh. Obshch. Khim.*, **29**, 1970 (1959).
165. Yu. Yur'ev, M. Gal'bershtam and I. Kondroi, *Khim. Geterotsikl. Soedin*, 897 (1966).
166. S. Gronowitz and P. Uppström, *Acta Chem. Scand.*, **28B**, 981 (1974).
167. N.-P., Buu-Hoï, P. Demerseman and R. Royer, *Compt. Rend., Ser. C*, **237**, 397 (1953).
168. Yu. Yur'ev and N. Magdesieva, *Zh. Obshch. Khim.*, **27**, 179 (1957).
169. Yu. Yur'ev, N. Mezentsova and V. Vaskowskii, *Zh. Obshch. Khim.*, **27**, 3155 (1957).
170. Yu. Yur'ev, N. Sadovaya and M. Gal'bershtam, *Zh. Obshch. Khim.*, **29**, 1917 (1959).
171. C. Paulmier and P. Pastour, *Bull. Soc. Chim. Fr.*, 4021 (1966).

172. Yu. Yur'ev, N. Magdesieva and A. Monakhova, *Khim. Geterotsikl. Soedin.*, 650 (1968).
173. J. Morel, C. Paulmier and P. Pastour, *Compt. Rend., Ser. C*, **266**, 1300 (1968).
174. J. Morel, C. Paulmier and P. Pastour, *Compt. Rend., Ser. C*, **267**, 1842 (1968).
175. C. Paulmier, J. Morel and P. Pastour, *Bull. Soc. Chim. Fr.*, 2511 (1969).
176. B. Decroix, J. Morel and P. Pastour, *Bull. Soc. Chim. Fr.*, 1848 (1972).
177. Yu. Yur'ev, N. Mezentsova and J. Balashova, *Zh. Obshch. Khim.*, **27**, 2536 (1957).
178. Yu. Yur'ev, N. Sadovaya and E. Lyubimova, *Zh. Obshch. Khim.*, **30**, 2732 (1960).
179. Yu. Yur'ev, N. Sadovaya and M. Gal'bershtam, *Zh. Obshch. Khim.*, **32**, 259 (1962).
180. Yu. Yur'ev, N. Sadovaya and A. Ibragimova, *Zh. Obshch. Khim.*, **29**, 3647 (1959).
181. Yu. Yur'ev, M. Gal'bershtam and N. Sadovaya, *Zh. Obshch. Khim.*, **32**, 1301 (1962).
182. (a) S. Gronowitz and A. Konar, *J. Chem. Soc., Chem. Commun.*, 163 (1977).
(b) T. Frejd, M. Davis, S. Gronowitz and T. Sadeh, *J. Heterocycl. Chem.*, **17**, 759 (1980).
183. C. Paulmier, *Tetrahedron Letters*, 1797 (1978).
184. C. Paulmier, *Bull. Soc. Chim. Fr.*, II-592 (1979).
185. C. Paulmier, *Bull. Soc. Chim. Fr.*, II-237 (1979).
186. A. Nakhmanovich, N. Elokhina and R. Karnaukhova, *Khim. Geterotsikl. Soedin.*, 920 (1971).
187. Yu. Yur'ev, *Zh. Obshch. Khim.*, **29**, 3865 (1959).
188. P. Chauvin, J. Morel and P. Pastour, *Compt. Rend., Ser. C*, **276**, 1453 (1973).
189. P. Chauvin, J. Morel, C. Paulmier and P. Pastour, *Compt. Rend., Ser. C*, **274**, 1347 (1972).
190. P. Dubus, B. Decroix, J. Morel and P. Pastour, *Ann. Chim. (Paris)*, **10**, 333 (1975).
191. J. Pankiewicz, B. Decroix, C. Fugier, J. Morel and P. Pastour, *J. Chem. Res.*, 133 (1978).
192. (a) B. Decroix, P. Dubus, J. Morel and P. Pastour, *Bull,. Soc. Chim. Fr.*, 621 (1976).
(b) G. Jones, R. Hanson and M. Davis, *J. Heterocycl. Chem.*, **20**, 523 (1983).
193. C. Maletras, B. Decroix, J. Morel and C. Paulmier, *Compt. Rend., Ser. C*, **276**, 1305 (1973).
194. W. Mack, *Angew. Chem. (Intern. Ed.)*, **4**, 245 (1965).
195. E. Müller, E. Luppold and W. Winter, *Z. Naturforsch.*, **31b**, 367 (1976).
196. S. Gronowitz and T. Frejd, *Int. J. Sulfur Chem., (A)*, **2**, 165 (1972).
197. S. Gronowitz, A.-B. Hörnfeldt and K. Pettersson, *Synth. Commun.*, **3**, 213 (1973).
198. J. Morel, C. Paulmier, D. Semard and P. Pastour, *Compt. Rend., Ser. C*, **270**, 825 (1970).
199. R. Lantz and A.-B. Hörnfeldt, *Chem. Scripta*, **10**, 126 (1976).
200. B. Cederlund and A.-B. Hörnfeldt, *Acta Chem. Scand.*, **B30**, 101 (1976).
201. J. Morel, C. Paulmier, D. Semard and P. Pastour, *Compt. Rend., Ser. C*, **271**, 1005 (1970).
202. G. Henrio, J. Morel and P. Pastour, *Ann. Chim. (Paris)*, **10**, 37 (1975).
203. B. Cederlund, R. Lantz, A.-B. Hörnfeldt, O. Thorstad and K. Undheim, *Acta Chem. Scand.*, **B31**, 198 (1977).
204. J. Morel, C. Paulmier and P. Pastour, *J. Heterocycl. Chem.*, **9**, 355 (1972).
205. C. Paulmier, J. Morel, D. Semard and P. Pastour, *Bull. Soc. Chim. Fr.*, 2434 (1973).
206. S. Gronowitz and B. Holm, *Synth. Commun.*, **4**, 63 (1974).
207. S. Gronowitz and B. Holm, *J. Heterocycl. Chem.*, **14**, 281 (1977).
208. G. Henrio, J. Morel and P. Pastour, *Bull. Soc. Chim. Fr.*, 265 (1976).
209. P. Dubus, B. Decroix, J. Morel and P. Pastour, *Bull. Soc. Chim. Fr.*, 628 (1976).
210. M. Bezoari, P. Kovacic, S. Gronowitz and A.-B. Hörnfeldt, *J. Polym. Sci., Polym. Lett. Ed.*, **19**, 347 (1981).
211. S. Caccamese, S. Montaudo, A. Recca, F. Fringuelli and A. Taticchi, *Tetrahedron*, **30**, 4129 (1974).
212. G. Distefano, S. Pignataro, G. Innorta, F. Fringuelli, G. Marino and A. Taticchi, *Chem. Phys. Letters*, **22**, 132 (1973).
213. S. Clementi, F. Fringuelli, P. Linda, G. Marino, G. Savelli, A. Taticchi and J.-L. Piette, *Gazz. Chim. Ital.*, **107**, 339 (1977).
214. W. Lohner and K. Praefcke, *J. Organomet. Chem.*, **208**, 43 (1981).
215. F. Fringuelli, S. Gronowitz, A.-B. Hörnfeldt, I. Johnson and A. Taticchi, *Acta Chem. Scand.*, **B30**, 605 (1976).
216. F. Fringuelli, B. Serena and A. Taticchi, *J. Chem. Soc., Perkin Trans. 2*, 971 (1980).
217. S. Gronowitz and T. Frejd, *Acta Chem. Scand.*, **23**, 2540 (1969).
218. S. Gronowitz and T. Frejd, *Acta Chem. Scand.*, **24**, 2656 (1970).
219. S. Gronowitz and P. Moses, *Arkiv Kemi*, **18**, 119 (1961).
220. S. Gronowtiz and T. Frejd, *Int. J. Sulfur Chem.*, **A2**, 165 (1972).
221. H. Jakobsen, *Acta Chem. Scand.*, **24**, 2663 (1970).

222. S. Gronowitz and T. Frejd, *Khim. Geterotsikl. Soedin.* (*Engl. Transl.*), **14**, 353 (1978).
223. S. Gronowitz and T. Frejd, *Acta Chem. Scand.*, **27**, 2242 (1973).
224. S. Gronowitz, T. Frejd, O. Karlsson, K. Lawitz, P. Pedaja and K. Pettersson, *Chem. Scripta*, **18**, 192 (1981).
225. S. Gronowitz and T. Frejd, *Acta Chem. Scand.*, **B30**, 439 (1976).
226. T. Frejd, *Chem. Scripta*, **10**, 133 (1976).
227. A. Hallberg, T. Frejd and S. Gronowitz, *Chem. Scripta*, **13**, 186 (1978).
228. E. Luppold, E. Müller and W. Winter, *Z. Naturforsch.*, **31b**, 1654 (1976).
229. S. Gronowitz, A. Hallberg and T. Frejd, *Tetrahedron*, **35**, 2607 (1979).
230. G. Morgan and F. Burstall, *J. Chem. Soc.*, 1096 (1929).
231. J. McCullough and A. Lefohn, *Inorg. Chem.*, **5**, 150 (1966).
232. W. Farrar and J. Gulland, *J. Chem. Soc.*, 11 (1945).
233. G. Morgan and F. Burstall, *J. Chem. Soc.*, 180 (1931).
234. A. Duffield, H. Budzikiewicz and C. Djerassi, *J. Amer. Chem. Soc.*, **87**, 2920 (1965).
235. A. Fredga, *J. Prakt. Chem.*, **127**, 103 (1930).
236. L. Fitjer and W. Luettke, *Chem. Ber.*, **105**, 907 (1972).
237. E. Buchta and K. Greiner, *Chem. Ber.*, **94**, 1311 (1961).
238. Yu. Migalina, I. Balog, V. Lendel, A. Koz'min and N. Zefirov, *Khim. Geterotsikl. Soedin.*, 1212 (1978).
239. T. Srivastava, R. Srivastava, H. Singh and M. Singh, *Indian J. Chem.*, **18A**, 367 (1979).
240. S. Tamagaki, S. Oae and K. Sakaki, *Tetrahedron Letters*, 649 (1975).
241. H. Lumbroso, D. Bertin, F. Fringuelli and A. Taticchi, *J. Chem. Soc., Chem. Commun.*, 342 (1973).
242. S. Bory and A. Marquet, *Tetrahedron Letters*, 2035 (1976).
243. A. Marquet, *Pure Appl. Chem.*, **49**, 183 (1977).
244. J. Bergman and L. Engman, *J. Amer. Chem. Soc.*, **103**, 2715 (1981).
245. M. Renson, *Chem. Scripta*, **8A**, 29 (1975).
246. S. Umezawa, *Bull. Chem. Soc. Japan*, **14**, 363 (1939).
247. E. Kataev, L. Kataeva and Z. Titova, *Khim. Geterotsikl. Soedin.*, 172 (1968).
248. S. Gronowitz, T. Frejd and A.-B. Hörnfeldt, *Chem. Scripta*, **5**, 236 (1974).
249. Yu. Yur'ev, N. Nezentsova, T. Melenteva and E. Vreshova, *Zh. Obshch. Khim.*, **27**, 1587 (1957).
250. C. Hansch and C. Geiger, *J. Org. Chem.*, **24**, 1025 (1959).
251. V. Dulenko, N. Alekseev and L. Kaplan, *Khim. Geterotsikl. Soedin.*, 1212, 1324 (1973).
252. V. Dulenko and N. Alekseev, *Dokl. Akad. Nauk SSSR*, **206**, 351 (1972).
253. R. Lesser and A. Schoeller, *Chem. Ber.*, **47**, 2292 (1914).
254. G. Komppa and G. Nyman, *J. Prakt. Chem.*, **139**, 229 (1934).
255. B. Muth and I. Kiss. *J. Org. Chem.*, **21**, 576 (1956).
256. N. Magdesieva and V. Vdovin, *Khim. Geterotsikl. Soedin.*, 1475 (1970).
257. N. Magdesieva and V. Vdovin, *Khim. Geterotsikl. Soedin.*, 1640 (1971).
258. P. Cagniant and G. Kirsch, *Compt. Rend., Ser. C*, **274**, 1394 (1972).
259. P. Cagniant and G. Kirsch, *Compt. Rend., Ser. C*, **275**, 1029 (1972).
260. M. Vafai and M. Renson, *Bull. Soc. Chim. Belges*, **75**, 145 (1966).
261. T. Q. Minh, L. Christiaens and M. Renson, *Tetrahedron*, **28**, 5393 (1972).
262. P. Cagniant, G. Kirsch and P. Legendre, *Compt. Rend., Ser. C*, **276**, 1629 (1973).
263. L. Christiaens and M. Renson, *Bull. Soc. Chim. Belges*, **77**, 153 (1968).
264. G. Maréchal, A. Ruwet, M. Renson and N.-P. Buu-Hoï, *Bull. Soc. Chim. Belges*, **79**, 601 (1970).
265. L. Christiaens and M. Renson, *Bull. Soc. Chim. Belges*, **79**, 133 (1970).
266. F. Dallacker, E. Kaiser and P. Uddrich, *Ann. Chem.*, **689**, 179 (1965).
267. P. Faller and F. Mantovani, *Bull. Soc. Chim. Fr.*, 1643 (1972).
268. R. Mitra, K. Rabindran and B. Tilak, *Current Sci.* (*India*), **23**, 263 (1954).
269. E. Kataev, G. Chmutova and Y. Mareev, *Khim. Geterotsikl. Soedin.*, 333 (1971).
270. L. Laitem and L. Christiaens, University of Liège, unpublished results.
271. E. Kataev, G. Chmutova, A. Musina and P. Anastaseva, *Zh. Obshch. Khim.*, **37**, 597 (1967).
272. H. Reich and J. Trend, *J. Org. Chem.*, **41**, 2503 (1976).
273. E. Giesbrecht and I. Mori, *Ann. Acad. Brasil Cienc.*, **30**, 521 (1958).

274. J. Gosselck, *Chem. Ber.*, **91**, 2345 (1958).
275. W. Lohner and K. Praefcke, *J. Organomet. Chem.*, **194**, 173 (1980).
276. P. Cagniant, G. Kirsch and D. Cagniant, *Compt. Rend., Ser, C*, **274**, 711 (1972).
277. L. Brooker, S. Ford and E. Vanhare, in Ref. 3, p. 507.
278. G. Grandolini, A. Ricci, N.-P. Buu-Hoï and F. Perin, *J. Heterocycl. Chem.*, **5**, 133 (1968).
279. L. Laitem and L. Christiaens, unpublished results; L. Laitem, *Doctoral Thesis*, University of Liège, 1975.
280. B. Stridsberg and S. Allenmark, *Chem. Scripta*, **6**, 184 (1974).
281. A. Ruwet and M. Renson, *Bull. Soc. Chim. Belges*, **79**, 593 (1970).
282. G. Maréchal, L. Christiaens, M. Renson, P. Jacquignon and A. Croisy, *Bull. Soc. Chim. Fr.*, 157 (1977).
283. M. Evers, R. Weber, Ph. Thibaut, L. Christiaens, M. Renson, A. Croisy and P. Jacquignon, *J. Chem. Soc., Perkin Trans. 1*, 2452 (1976).
284. L. Christiaens and M. Renson, *Bull. Soc. Chim. Belges*, **97**, 235 (1970).
285. Ph. Thibaut, L. Christiaens and M. Renson, *Compt. Rend., Ser. C*, **281**, 937 (1975).
286. A. Ruwet, J. Meessen and M. Renson, *Bull. Soc. Chim. Belges*, **78**, 459 (1969).
287. J. Van Coppenolle, *Doctoral Thesis*, University of Liège, 1975.
288. J. Van Coppenolle and M. Renson, *Compt. Rend., Ser. C*, **280**, 283 (1975).
289. J. Van Coppenolle and M. Renson, *Tetrahedron*, **31**, 2099 (1975).
290. (a) I. Degani, R. Fochi and C. Vicenzi, *Gazz. Chim. Ital.*, **97**, 397 (1967).
(b) H. Hommes, N. Verkruysse and L. Brandsma, *J. Chem. Soc., Chem. Commun.*, 366 (1981).
291. R. Riley, J. Flato and P. McIntyre, *J. Org. Chem.*, **28**, 1138 (1963).
292. T. Q. Minh, L. Christiaens and M. Renson, *Bull. Soc. Chim. Fr.*, 2239 (1974).
293. Y. Migalina, S. Galla-Bobic, V. Lendel and V. Staninets, *Khim. Geterotsikl. Soedin.*, 1283 (1981).
294. W. Ried and G. Sell, *Synthesis*, 447 (1976).
295. I. Smirnov-Zamkov and I. Zborovskii, *Zh. Org. Khim.*, **13**, 667 (1977).
296. I. Smirnov-Zamkov, I. Zborovskii and V. Staninets, *Zh. Org. Khim.*, **15**, 1782 (1979).
297. V. Lendel, Y. Migalina, S. Galla, A. Koz'min and N. Zefirov, *Khim. Geterotsikl. Soedin.*, 1340 (1977).
298. G. Maréchal, L. Christiaens, M. Renson, P. Jacquignon and A. Croisy, *Bull. Soc. Chim. Fr.*, 157 (1977).
299. T. Q. Minh, F. Mantovani, P. Faller, L. Christiaens and M. Renson, *Bull. Soc. Chim. Fr.*, 3955 (1972).
300. T. Q. Minh, L. Christiaens, M. Renson and P. Faller, *Chem. Scripta*, **9**, 143 (1976).
301. N. Magdesieva and V. Vdovin, *Khim. Geterotsikl. Soedin.*, 20 (1972).
302. L. Christiaens and M. Renson, *Bull. Soc. Chim. Belges*, **77**, 153 (1968).
303. L. Christiaens, R. Dufour and M. Renson, *Bull. Soc. Chim. Belges*, **79**, 143 (1970).
304. F. Mantovani, L. Christiaens and P. Faller, *Bull. Soc. Chim. Fr.*, 1595 (1972).
305. L. Christiaens and M. Renson, *Bull. Soc. Chim. Belges*, **81**, 609 (1972).
306. T. Q. Minh, Ph. Thibaut, L. Christiaens and M. Renson, *Tetrahedron*, **28**, 5393 (1972).
307. L. Laitem and L. Christiaens, *Bull. Soc. Chim. Fr.*, 2294 (1975).
308. L. Laitem, L. Christiaens and A. Welter, *Heterocycles*, **4**, 1107 (1976).
309. (a) L. Laitem, Ph. Thibaut and L. Christiaens, *J. Heterocycl. Chem.*, **13**, 469 (1976).
(b) T. Sadeh, M. Davis, R. Gil and U. Zoller, *J. Heterocycl. Chem.*, **18**, 1605 (1981).
310. P. Cagniant, G. Kirsch and L. Christiaens, *Compt. Rend., Ser. C*, **287**, 333 (1978).
311. L. Laitem, L. Christiaens and G. Llabres, *Bull. Soc. Chim. Fr.*, 681 (1974).
312. T. Q. Minh, L. Christiaens and M. Renson, *Bull. Soc. Chim. Fr.*, 2244 (1974).
313. N.-P. Buu-Hoï, M. Dufour, P. Jacquignon, M. Renson, G. Maréchal and A. Ruwet, *J. Chem. Soc. (C)*, 2308 (1971).
314. G. Maréchal, L. Christiaens, M. Renson and P. Jacquignon, *Collect. Czech. Chem. Commun.*, **43**, 2916 (1978).
315. P. Jacquignon, G. Maréchal, M. Renson, A. Ruwet and Do Phuoc Hien, *Bull. Soc. Chim. Fr.*, 677 (1973).
316. O. Attanasi, P. Battistoni, and G. Fava, *Phosphorus and Sulphur*, **5**, 305 (1979).
317. E. Goethals, G. Hubrecht and C. Verschuere, *Eur. Polymer J.*, **10**, 847 (1974).
318. L. Saris and M. Cava, *J. Amer. Chem. Soc.*, **98**, 867 (1976).
319. N. Magdesieva and V. Vdovin, *Khim. Geterotsikl. Soedin.*, 24 (1972).

320. B. Norcross, J. Lansinger and R. Martin, *J. Org. Chem.*, **42**, 369 (1977).
321. E. Chacko, J. Bornstein and D. Sardella, *J. Amer. Chem. Soc.*, **99**, 8248 (1977).
322. A. Drory, *Chem. Ber.*, **24**, 2564 (1891).
323. M. Renson and R. Collienne, *Bull. Soc. Chim. Belges*, **73**, 491 (1964).
324. W. Günther, *J. Org. Chem.*, **32**, 3929 (1967).
325. L. Christiaens and M. Renson, *Tetrahedron*, **28**, 5405 (1972).
326. I. Wallmark, M. Krackov, S.-H. Chu and H. Mautner, *J. Amer. Chem. Soc.*, **92**, 4447 (1970).
327. L. Szperl and J. Böhm, *Roczn. Chem.*, **12**, 378 (1932); *Chem. Abstr.*, **27**, 3926 (1933).
328. J. Bergman and L. Engman, *Org. Prep. Proceed. Int.*, **10**, 289 (1978).
329. E. Müller and E. Langer, *Tetrahedron Letters*, 735 (1970).
330. E. Müller and W. Dilger, *Chem. Ber.*, **106**, 1643 (1973).
331. J. Hambrecht, H. Straub and E. Müller, *Tetrahedron Letters*, 1790 (1976).
332. J. Hambrecht and E. Müller, *Ann. Chem.*, 387 (1977).
333. C. Courtot and A. Matamedi, *Compt. Rend.*, **199**, 531 (1934).
334. O. Behagel and K. Hofmann, *Chem. Ber.*, **72**, 697 (1939).
335. J. McCullough, T. Campbell and E. Gould, *J. Amer. Chem. Soc.*, **72**, 5753 (1950).
336. B. Greenberg, E. Gould and W. Burlant, *J. Amer. Chem. Soc.*, **78**, 4028 (1956).
337. N. Cullinane, A. Rees and C. Plummer, *J. Chem. Soc.*, 151 (1939).
338. D. Hellwinkel and G. Fahrbach, *Ann. Chem.*, **715**, 68 (1968).
339. J. Gaidis, *J. Org. Chem.*, **35**, 2811 (1970).
340. C. Woodard, G. Hughes and A. Massey, *J. Organomet. Chem.*, **112**, 9 (1976).
341. L. Chierici and R. Passerini, *J. Chem. Soc.*, 3249 (1954).
342. T. Tezuka, H. Suzuki and H. Miyazaki, *Tetrahedron Letters*, 4885 (1978).
343. E. Sawicki and F. Ray, *J. Amer. Chem. Soc.*, **74**, 4120 (1952).
344. G. Wiseman and E. Gould, *J. Amer. Chem. Soc.*, **77**, 1061 (1955).
345. N.-P. Buu-Hoï and N. G. Hoan, *J. Org. Chem.*, **17**, 643 (1952).
346. N.-P. Buu-Hoï and N. G. Hoan, *J. Chem. Soc.*, 3745 (1952).
347. G. Wiseman and E. Gould, *J. Amer. Chem. Soc.*, **76**, 1706 (1954).
348. P. Cagniant, N. Bellinger and D. Cagniant, *Compt. Rend., Ser. C*, **277**, 383 (1973).
349. P. Cagniant, N. Bellinger and D. Cagniant, *Compt. Rend., Ser. C*, **277**, 779 (1973).
350. F. Girardin, L. Christiaens and P. Faller, *Compt. Rend., Ser. C*, **277**, 113 (1973).
351. F. Girardin, P. Faller, L. Christiaens and D. Cagniant, *Bull. Soc. Chim. Fr.*, 2095 (1974).
352. J. Weber and P. Faller, *Compt. Rend., Ser. C*, **281**, 389 (1975); *Bull. Soc. Chim. Fr.*, 783 (1975).
353. M. Schaefer, J. Weber and P. Faller, *Synthesis*, 122 (1979).
354. P. Cagniant, P. Perin and G. Kirsch, *Compt. Rend., Ser. C*, **278**, 1011 (1974).
355. P. Cagniant, G. Kirsch and P. PerIn, *Compt. Rend., Ser. C*, **279**, 851 (1974).
356. I. Kvitko, N. Sokolova and L. Efros, *Khim. Geterotsikl. Soedin.*, 715 (1973).
357. I. Kvitko, N. Sokolova, S. Fadkina and A. El'tsov, *Zh. Org. Khim.*, **12**, 1574 (1976).
358. A. Shafiee and S. Sattari, *J. Heterocycl. Chem.*, **19**, 227 (1982).
359. A. Shafiee and E. Behnam, *J. Heterocycl. Chem.*, **15**, 589 (1978).
360. A. Scheller, W. Winter and E. Müller, *Ann. Chem.*, 1448 (1976).
361. C. Paulmier, J. Morel, D. Semard and P. Pastour, *Bull. Soc. Chim. Fr.*, 2434 (1973).
362. C. Paulmier, J. Bourguignon, J. Morel and P. Pastour, *Compt. Rend., Ser. C*, **270**, 494 (1970).
363. A. Bugge, *Acta Chem. Scand.*, **23**, 1823 (1969).
364. Ya. Gol'dfarb, V. Litvinov and S. Ozolin, *Izv. Akad. Nauk SSSR, Ser. Khim.*, 1419 (1968).
365. V. Litvinov and Ya. Gol'dfarb, *Adv. Heterocycl. Chem.*, **19**, 123 (1976).
366. V. Litvinov, Ya. Gol'dfarb, V. Bogdanov, I. Konjaeva and A. Sukiasjan, *J. Prakt. Chem.*, **315**, 850 (1973).
367. V. Litvinov, A. Sukiasjan, Ya. Gol'dfarb and L. Bochacheva, *Izv. Akad. Nauk SSSR, Ser. Khim.*, 1592 (1971).
368. B. Capron, C. Paulmier and P. Pastour, *Compt. Rend., Ser. C*, **277**, 167 (1973).
369. B. Capron, C. Paulmier and P. Pastour, *Bull. Soc. Chim. Fr.*, 2575 (1975).
370. S. Gronowitz and B. Maltesson, *Acta Chem. Scand.*, **26**, 2982 (1972).
371. B. Capron and C. Paulmier, *Compt. Rend., Ser. C*, **279**, 947 (1974).
372. N. Alekseev and A. Kirilash, *Khim. Geterotsikl. Soedin.*, 1424 (1979).
373. V. Litvinov, I. Konyaeva and Ya. Gol'dfarb, *Izv. Akad. Nauk SSSR*, 477 (1976).
374. F. Iteke, L. Christiaens and M. Renson, *Tetrahedron*, **32**, 689 (1976).
375. S. Gronowitz, A. Konar and V. Litvinov, *Izv. Akad. Nauk SSSR*, 363 (1981).

376. S. Gronowitz, A. Konar and V. Litvinov, *Chem. Scripta*, **15**, 206 (1980).
377. V. Litvinov, Ya. Gol'dfarb and I. Konyaeva, *Izv. Akad. Nauk SSSR*, 372 (1980).
378. I. Abronin, A. Djumanazarova, V. Litvinov and A. Konar, *Chem. Scripta*, **20**, 208 (1982).
379. I. Abronin, A. Djumanazarova, V. Litvinov and A. Konar, *Chem. Scripta*, **19**, 75 (1982).
380. G. Ruban, D. Zobel, G. Kossmehl and I. Sgustov, *Chem. Ber.*, **114**, 818 (1981).
381. A. Konar and V. Litvinov, *Chem. Scripta*, **19**, 176 (1982).
382. L. Saris and M. Cava, *Heterocycles*, **6**, 1349 (1977).
383. E. Iteke, L. Christiaens and M. Renson, *Bull. Soc. Chim. Fr.*, 4767 (1972).
384. G. Kirsch and P. Cagniant, *Compt. Rend., Ser. C*, **274**, 2179 (1972).
385. P. Cagniant, G. Kirsch and L. Perrin, *Compt. Rend., Ser. C*, **276**, 1561 (1973).
386. F. Iteke, L. Christiaens and M. Renson, *Bull. Soc. Roy. Sci. Liège*, **44**, 640 (1975).
387. E. Iteke, L. Christiaens and M. Renson, *Compt. Rend., Ser. C*, **275**, 277 (1972).
388. S. Umezawa, *Bull. Soc. Chem. Japan*, **14**, 363 (1939).
389. S. Gronowitz, A. Konar and A.-B. Hörnfeldt, *Chem. Scripta*, **10**, 159 (1976).
390. A. Konar and S. Gronowitz, *Tetrahedron*, **36**, 3317 (1980).
391. S. Bravermann, M. Freund and I. Goldberg, *Tetrahedron Letters*, **21**, 3617 (1980).
392. S. Gronowitz and A. Konar, *Chem. Scripta*, **12**, 15 (1977).
393. S. Gronowitz, A. Konar and A.-B. Hörnfeldt, *Chem. Scripta*, **19**, 5 (1982).
394. Ya. Gol'dfarb, V. Litvinov and I. Konyaeva, *Khim. Geterotsikl. Soedin.*, 1072 (1979).
395. A. Konar, *Chem. Scripta*, **22**, 177 (1983).
396. I. Abronin, V. Litvinov, G. Zhidomirov, A. Dzhumanazarova and Ya. Gol'dfarb, *Khim. Geterotsikl. Soedin.*, 199 (1980).
397. S. Patai, M. Sokolovsky and A. Friedlander, *Proc. Chem. Soc.*, 181 (1960).
398. S. Patai, K. Muszkat and M. Sokolovsky, *J. Chem. Soc.*, 734 (1962).
399. S. Patai, D. Elmaleh and Z. Rappoport, *J. Chem. Soc.* (*C*), 2637 (1971).
400. A. Dzhumanazarova, V. Litvinov and A. Konar, *Chem. Scripta*, **20**, 214 (1982).
401. G. Henrio, G. Plé and J. Morel, *Compt. Rend., Ser. C*, **278**, 125 (1974).
402. V. Dulenko and N. Alekseev, *Khim. Geterotsikl. Soedin.* (*Engl. Transl.*), 551 (1975).
403. K. Beelitz, K. Praefcke and S. Gronowitz, *Ann. Chem.*, 1597 (1980).
404. K. Beelitz, K. Praefcke and S. Gronowitz, *J. Organomet. Chem.*, **194**, 167 (1980).
405. K. Beelitz and K. Praefcke, *Ann. Chem.*, 1620 (1980).
406. P. Cagniant, G. Kirsch and M. Renson, *Compt. Rend., Ser. C*, **272**, 1363 (1971).
407. P. Cagniant and G. Kirsch, *Compt. Rend., Ser. C*, **283**, 751 (1976).
408. G. Kirsch and P. Cagniant, *Compt. Rend., Ser. C*, **283**, 211 (1976).
409. P. Cagniant, G. Kirsch and D. Cagniant, *Compt. Rend., Ser. C*, **274**, 711 (1972).
410. P. Cagniant and G. Kirsch, *Compt. Rend., Ser. C*, **283**, 683 (1976).
411. K. Nain Java, S. Soth, M. Farnier and C. Paulmier, *Compt. Rend., Ser. C*, **281**, 793 (1975).
412. S. Soth, M. Farnier and C. Paulmier, *Can. J. Chem.*, **56**, 1429 (1978).
413. S. Gronowitz, C. Westerlund and A.-B. Hörnfeldt, *Acta Chem. Scand.*, **B30**, 391 (1976).
414. S. Gronowitz, C. Westerlund and A.-B. Hörnfeldt, *Chem. Scripta*, **12**, 1 (1977).
415. Yu. Koshelev, A. Reznichenko, L. Efros and I. Kvitko, *Zh. Org. Khim.*, **9**, 2201 (1973).
416. C. Paulmier, *Compt. Rend., Ser. C*, **281**, 317 (1975).
417. S. Gronowitz, C. Westerlund and A.-B. Hörnfeldt, *Chem. Scripta*, **10**, 165 (1976).
418. C. Paulmier, *Tetrahedron Letters*, 1797 (1978).
419. C. Paulmier, *Bull. Soc. Chim. Fr., II*, 151 (1980).
420. P. Abramenko and V. Zhiryakov, *Khim. Geterotsikl. Soedin.*, 1495 (1977).
421. V. Ziryakov and I. Levkoev, *Dokl. Akad. Nauk SSSR*, **120**, 1035 (1968).
422. C. Paulmier and F. Outurquin, *J. Heterocycl. Chem.*, **20**, 113 (1983).
423. N. V. Onyamboko, *Doctoral Thesis*, University of Liège, 1983.
424. N. V. Onyamboko, R. Weber, N. Dereu, M. Renson and C. Paulmier, *Bull. Soc. Chim. Belges*, **89**, 773 (1980).
425. N. V. Onyamboko, R. Weber, A. Fauconnier and M. Renson, *Bull. Soc. Chim. Belges*, **92**, 53 (1983).
426. V. Dulenko, N. Alekseev and S. Baranov, *Khim. Geterotsikl. Soedin.*, 997 (1971).
427. F. Outurquin, C. Paulmier, J. Morel and P. Pastour, *Compt. Rend., Ser. C*, **274**, 1696 (1972).
428. F. Outurquin, C. Paulmier, J. Morel and P. Pastour, *Bull. Soc. Chim. Fr.*, 3039 (1974).
429. F. Outurquin, G. Ah-Kow and C. Paulmier, *Bull. Soc. Chim. Fr.*, 883 (1976).
430. S. Gronowitz, C. Westerlund and A.-B. Hörnfeldt, *Acta Chem. Scand.*, **B29**, 233 (1975).
431. C. Paulmier and F. Outurquin, *J. Chem. Res.* (*S*), 318 (1977).

432. S. Clementi, S. Lepri, G. Sebastiani, S. Gronowitz, C. Westerlund and A.-B. Hörnfeldt, *J. Chem. Soc., Perkin Trans. 2*, 861 (1978).
433. G. Ah-Kow, C. Paulmier and P. Pastour, *Compt. Rend., Ser. C*, **278**, 1513 (1974).
434. B. Decroix, J. Morel, C. Paulmier and P. Pastour, *Bull. Soc. Chim. Fr.*, 3453 (1972).
435. N.-P. Buu-Hoï, G. Saint-Ruf, A. Martani, A. Ricci and D. Balucani, *J. Chem. Soc. (C)*, 609 (1968).
436. P. Abramenko and V. Zhiryakov, *Khim. Geterotsikl. Soedin.*, 38 (1971).
437. V. Litvinov and I. Dzhumaev, *Izv. Akad. Nauk SSSR*, 478 (1979).
438. J. M. Webert, D. Cagniant, P. Cagniant, G. Kirsch and J.-V. Weber, *J. Heterocycl. Chem.*, **20**, 49 (1983).
439. J. M. Webert, D. Cagniant, P. Cagniant, G. Kirsch and J.-V. Weber, *J. Heterocycl. Chem.*, **20**, 61 (1983).
440. Y. Koshelev, I. Kvitko and E. Samartseva, *Zh. Org. Khim.*, **8**, 2204 (1972).
441. R. Khozeeva, I. Kvitko, E. Baranov and A. El'tsov, *Zh. Org. Khim.*, **13**, 232 (1977).
442. I. Kvitko and N. Sokolova, *Khim. Geterotsikl. Soedin.*, 565 (1973).
443. P. Abramenko, *Zh. Obshch. Khim.*, **17**, 595 (1972).
444. P. Pirson and L. Christiaens, *Bull. Soc. Chim. Fr.*, 704 (1973).
445. P. Abramenko, T. Balykova and T. Ponomareva, *Khim. Geterotsikl. Soedin.*, 796 (1974).
446. D. Amis, A. Singh and W. Smyth, *Tetrahedron*, **39**, 831 (1983).
447. A. Shafiee and A. Mazloumi, *J. Heterocycl. Chem.*, **15**, 1455 (1978).
448. A. Shafiee, *J. Heterocycl. Chem.*, **15**, 473 (1978).
449. T. Raja, S. Nagarayan and P. Shanmugam, *Chem. Scripta*, **12**, 44 (1977).
450. T. Raja, N. Soundararayan, V. Bakthavachalam and P. Shanmugam, *Z. Naturforsch.*, **33b**, 228 (1978).
451. T. Raja, *Curr. Sci.*, **49**, 147 (1980).
452. P. Shanmugam and T. Raja, *Synthesis*, 117 (1977).
453. J.-L. Piette and M. Renson, *Bull. Soc. Chim. Belges*, **80**, 521 (1971).
454. F. Mazza and L. Melchiona, *Rend. Acad. Sci. Fis. Napoli*, **34**, 54 (1928).
455. W. Farrar, *Research*, **4**, 178 (1951).
456. J.-L. Piette, J.-M. Talbot, J.-C. Genard and M. Renson, *Bull. Soc. Chim. Fr.*, 2468 (1973).
457. J.-L. Piette, A. Petit and M. Renson, *Compt. Rend., Ser. C*, **276**, 1035 (1973).
458. J.-M. Talbot, J.-L. Piette and M. Renson, *Bull. Soc. Chim. Fr.*, 294 (1976).
459. W. Lohner and K. Praefcke, *J. Organomet. Chem.*, **208**, 35 (1981).
460. J.-M. Talbot, J.-L. Piette and M. Renson, *Bull. Soc. Chim. Belges*, **89**, 763 (1980).
461. J. Bergman and L. Engman, *Tetrahedron Letters*, 1509 (1979).
462. J. Bergman and L. Engman, *J. Organomet. Chem.*, **199**, 377 (1980).
463. I. Sadekov and V. Minkin, *Khim. Geterotsikl. Soedin.*, 138 (1971).
464. J.-M. Talbot, *Doctoral Thesis*, University of Liège, 1978.
465. F. Fringuelli and A. Taticchi, *J. Heterocycl. Chem.*, **10**, 89 (1973).
466. F. Fringuelli, P. Linda, G. Marino, G. Savelli, A. Taticchi and J.-L. Piette, *Gazz. Chim. Ital.*, **107**, 339 (1977).
467. H. Günther, *Tetrahedron Letters*, 2967 (1967).
468. E. Müller, E. Luppold and W. Winter, *Synthesis*, 265 (1975).
469. E. Müller, E. Luppold and W. Winter, *Chem. Ber.*, **108**, 237 (1975).
470. E. Cuthbertson and D. McNicol, *Tetrahedron Letters*, 1893 (1975).
471. R. Ziolo and W. Günther, *J. Organomet. Chem.*, **146**, 245 (1978).
472. T. Srivastava, P. Srivastava and R. Kumar, *J. Indian Chem. Soc.*, **59**, 186 (1982).
473. M. Loth-Compère, A. Luxen, Ph. Thibaut, L. Christiaens, M. Guillaume and M. Renson, *J. Heterocycl. Chem.*, **18**, 343 (1981).
474. L. Engman and M. Cava, *J. Org. Chem.*, **46**, 4194 (1981).
475. C. Courtot and M. Bastani, *Compt. Rend., Ser. C*, **203**, 197 (1936).
476. R. Passerini and G. Purello, *Ann. Chim. (Rome)*, **48**, 738 (1958).
477. D. Hellwinkel and G. Fahrbach, *Tetrahedron Letters*, 1823 (1965).
478. D. Hellwinkel and G. Fahrbach, *Ann. Chem.*, **712**, 1 (1968).
479. J. McCullough, *Inorg. Chem.*, **14**, 2285 (1975).
480. D. Hellwinkel and G. Fahrbach, *Chem. Ber.*, **101**, 574 (1968).
481. E. Luppold and W. Winter, *Chem. Ztg.*, **101**, 303 (1977).
482. E. Luppold, W. Winter and E. Müller, *Chem. Ber.*, **109**, 3886 (1976).
483. N. Dereu and J.-L. Piette, *Bull. Soc. Chim. Fr.*, 623 (1979).

484. N. Dereu and M. Renson, *J. Organomet. Chem.*, **258**, 163 (1983).
485. K. Ramasamy and P. Shanmugam, *Synthesis*, 842 (1978).
486. K. Ramasamy and P. Shanmugam, *Z. Naturforsch.*, **32b**, 605 (1977).
487. L. Hagelberg, *Chem. Ber.*, **23**, 1083 (1890).
488. G. Bergson and G. Claeson, *Acta Chem. Scand.*, **11**, 911 (1957).
489. G. Bergson, *Arkiv Kemi*, **19**, 195 (1962).
490. (a) H. Backer and H. Winter, *Rec. Trav. Chim.*, **56**, 691 (1937).
(b) H. Reisch, C. Hoeger and W. Willis, *J. Amer. Chem. Soc.*, **104**, 2936 (1982).
491. G. Bergson, *Acta Chem. Scand.*, **12**, 582 (1958).
492. Y. Deguchi, *Yakugaku Zasshi*, **80**, 931 (1960); *Chem. Abstr.*, **54**, 24384 (1960).
493. G. Bergson and A. Biezais, *Arkiv Kemi*, **18**, 143 (1961).
494. G. Bergson, *Arkiv Kemi*, **19**, 75 (1962).
495. M. Takahashi, S. Watanabe and T. Kasai, *Heterocycles*, **14**, 1921 (1980).
496. M. Petrov, V. Bobylev, V. Paramonov and A. Petrov, *Zh. Org. Khim.*, **18**, 2619 (1982).
497. K. Jensen and U. Henriksen, *Acta Chem. Scand.*, **21**, 1991 (1967).
498. G. Heath, I. Stewart and R. Martin, *Inorg. Nucl. Chem. Letters*, **5**, 169 (1969).
499. G. Heath, R. Martin and I. Stewart, *Australian J. Chem.*, **22**, 83 (1969).
500. D. Reid and R. Webster, *J. Chem. Soc., Perkin Trans. 1*, 2097 (1975).
501. M. Detty, B. Murray, D. Smith and N. Zumbulyadis, *J. Amer. Chem. Soc.*, **105**, 875 1983.
502. G. Traverso, *Ann. Chim.* (*Rome*), **47**, 3, 1244 (1957).
503. S. Bezzi, C. Garbuglio, M. Mammi and G. Traverso, *Gazz. Chim. Ital.*, **88**, 1226 (1958).
504. S. Pietra, C. Garbuglio and M. Mammi, *Gazz. Chim. Ital.*, **94**, 48 (1964).
505. D. Reid, *J. Chem. Soc.* (*C*), 3187 (1971).
506. (a) D. Reid and R. Webster, *J. Chem. Soc., Perkin Trans. 1*, 775 (1975).
(b) M. Detty and H. Luss, *J. Org. Chem.*, **48**, 5149 (1983).
507. (a) J. Dingwall, S. McKenzie and D. Reid, *J. Chem. Soc.* (*C*), 2543 (1968).
(b) J. Van den Hende and E. Klingsberg, *J. Amer. Chem. Soc.*, **88**, 5045 (1966).
508. R. Lesser and R. Weiss, *Chem. Ber.*, **57**, 1077 (1924).
509. E. Klingsberg, *US Patent*, No. 3, 511, 853 (1970); *Chem. Abstr.*, **73**, 66589 (1970).
510. U. Schmidt, *Ann. Chem.*, **635**, 109 (1960).
511. S. Oae, K. Sakaki, M. Fukumura, S. Tamagaki, Y. Matsuura and M. Kahudo, *Heterocycles*, **19**, 657 (1982).
512. (a) B. Kohne, K. Praefcke and C. Weichsel, *Phosphorus and Sulfur*, **7**, 211 (1979).
(b) S. Tamagaki, R. Ichihara and S. Oae, *Bull. Chem. Soc. Japan*, **48**, 355 (1975).
(c) W. Nakanishi, S. Matsumoto, Y. Ikeda, T. Sugawara, Y. Kawada and H. Iwamura, *Chem. Letters*, 1353 (1987).
(d) W. Nakanishi, S. Murata, Y. Ikeda, T. Sugawara, Y. Kanada and H. Iwamura, *Tetrahedron Letters*, **22**, 4241 (1981).
513. T. Raja and P. Shanmugam, *Tetrahedron Letters*, 1531 (1977).
514. T. Raja, *J. Indian Chem. Soc.*, **58**, 174 (1981).
515. K. Balodis, A. Livdane, R. Medne and O. Neiland, *J. Org. Chem.* (*USSR*), **15**, 343 (1979).
516. S. Sandman, J. Stark and B. Foxman, *Organometallics*, **1**, 739 (1982).
517. S. Onishi, T. Nogami and H. Mikawa, *Chem. Letters*, 1841 (1982).
518. A. Yamahira, T. Nogami and H. Mikawa, *J. Chem. Soc., Chem. Commun.*, 904 (1983).
519. C. Marschalk, *Bull. Soc. Chim. Fr.*, 800 (1952).
520. B. Hilti, C. Mayer and G. Rihs, *Helv. Chim. Acta*, **61**, 1462 (1978).
521. J. Meinwald, D. Dauplaise, F. Wudl and J. Hauser, *J. Amer. Chem. Soc.*, **99**, 255 (1977).
522. J. Meinwald, D. Dauplaise and J. Clardy, *J. Amer. Chem. Soc.*, **99**, 7743 (1977).
523. L.-Y. Chiang and J. Meinwald, *Tetrahedron Letters*, **21**, 4565 (1980).
524. C. Bak, K. Praefcke and L. Henriksen, *Chem. Ztg.*, **102**, 361 (1978).
525. A. Robert and A. Le Maréchal, *J. Chem. Soc., Chem. Commun.*, 447 (1978).
526. (a) T. Laitalainen, T. Simonen, M. Klingen and R. Kivekaes, *Finn. Chem. Letters*, 145 (1979); *Chem. Abstr.*, **92**, 163902j (1980).
527. R. Sukhai, W. Verboom, J. Meijer, M. Schoufs and L. Brandsma, *Rec. Trav. Chim.*, **100**, 10 (1981).
528. S. Bender, M. Detty and N. Haley, *Tetrahedron Letters*, **23**, 1531 (1982).
529. S. Bender, M. Detty, M. Fichtner and N. Haley, *Tetrahedron Letters*, **24**, 237 (1983).

530. H. Meier and I. Menzel, *Tetrahedron Letters*, 445 (1972).
531. I. Lalezari, A. Shafiee and S. Yazdany, *J. Pharm. Sci.*, **63**, 628 (1974).
532. M. Ghandehari, D. Davalian, M. Yalpani and M. Partovi, *J. Org. Chem.*, **39**, 3906 (1974).
533. L. Henriksen, *Acta Chem. Scand.*, **B21**, 1981 (1967).
534. K. Jensen and L. Henriksen, *Acta Chem. Scand.*, **B24**, 3213 (1970).
535. M. Lakshmikantham and M. Cava, *Heterocycles*, **14**, 271 (1980).
536. A. Shafiee, M. Vosooghi and R. Asgharian, *J. Heterocycl. Chem.*, **57**, 117 (1980).
537. (a) H. Spies, K. Gewald and R. Mayer, *J. prakt. Chem.*, **313**, 804 (1971).
(b) V. Laishev, M. Petrov and A. Petrov, *Zh. Org. Khim.*, **18**, 514 (1982).
538. E. Engler and V. Patel, *J. Amer. Chem. Soc.*, **96**, 7376 (1974).
539. R. Lyubovskaya, Y. Lipshan, O. Krasochka and L. Atovanyan, *Izv. Akad. Nauk SSSR, Ser. Khim.*, 179 (1976).
540. H. Poleschner and E. Fanghaenel, *J. Prakt. Chem.*, **324**, 691 (1982).
541. R. Mayer and K. Müller, *Z. Chem.*, **4**, 384 (1964).
542. K. Bechgaard, D. Cowan and A. Bloch, *J. Chem. Soc., Chem. Commun.*, 937 (1974).
543. K. Bechgaard, D. Cowan, A. Bloch and L. Henriksen, *J. Org. Chem.*, **40**, 746 (1975).
544. F. Wudl, E. Aharon-Shalom and S. Bertz, *J. Org. Chem.*, **46**, 4612 (1981).
545. J. Andersen and K. Bechgaard, *J. Org. Chem.*, **40**, 2016 (1975).
546. P. Shu, A. Bloch, T. Carruthers and D. Cowan, *J. Chem. Soc., Chem. Commun.*, 505 (1977).
547. F. Wudl and D. Nalewajek, *J. Chem. Soc., Chem. Commun.*, 866 (1980).
548. L.-Y. Chiang, Th. Poehler, A. Bloch and D. Cowan, *J. Chem. Soc., Chem. Commun.*, 866 (1980); *Mol. Cryst. Liq. Cryst.*, **56**, 1767 (1982).
549. A. Moradpour, V. Peyrussan, I. Johannsen and K. Bechgaard, *J. Org. Chem.*, **48**, 388 (1983).
550. F. Wudl, *J. Organomet. Chem.*, **217**, 329 (1981).
551. M. Lakshmikantham and M. Cava, *J. Org. Chem.*, **41**, 882 (1976).
552. M. Lakshmikantham, M. Cava and A. Garito, *J. Chem. Soc., Chem. Commun.*, 383 (1975).
553. M. Lakshmikantham and M. Cava, *J. Org. Chem.*, **45**, 2632 (1980).
554. E. Engler, D. Green and J. Chambers, *J. Chem. Soc., Chem. Commun.*, 148 (1976).
555. V. Lee, E. Engler, R. Schumaker and S. Parkin, *J. Chem. Soc., Chem. Commun.*, 235 (1983).
556. H. Spencer, M. Lakshmikantham, M. Cava and A. Garito, *J. Chem. Soc., Chem. Commun.*, 867 (1975).
557. K. Lerstrup, M. Lee, F. Wiygul, Th. Kistenmacher and D. Cowan, *J. Chem. Soc., Chem. Commun.*, 294 (1983).
558. (a) V. Litvinov and I. Dzhumaev, *Izv. Akad. Nauk SSSR, Ser. Khim.*, 718 (1982).
(b) L. Chiang, P. Shu, D. Holt and D. Cowan, *J. Org. Chem.*, **48,** 4713 (1983).
559. C. Lambert and L. Christiaens, *Tetrahedron Letters*, **25**, 833 (1984).
560. I. Johannsen, K. Bechgaard, K. Mortensen and C. Jacobsen, *J. Chem. Soc., Chem. Commun.*, 295 (1983).
561. C. Berg. K. Bechgaard, K. Andersen and C. Jacobsen, *Tetrahedron Letters*, 1719 (1976).
562. E. Engler and V. Patel, *J. Org. Chem.*, **40**, 387 (1975).
563. (a) A. Shafiee, I. Lalezari and F. Savabi, *Synthesis*, 764 (1977).
(b) R. Schumaker and E. Engler, *J. Amer. Chem. Soc.*, **99**, 5519 (1977).
564. F. Wudl, D. Wobschall and J. Hulfnagel, *J. Amer. Chem. Soc.*, **94**, 670 (1972).
565. J. Ferraris, D. Cowan, V. Walatka and J. Perlstein, *J. Amer. Chem. Soc.*, **95**, 948 (1973).
566. E. Engler, R. Craven, Y. Tomkiewicz, B. Scott, K. Bechgaard and J. Andersen, *J. Chem. Soc., Chem. Commun.*, 337 (1976).
567. A. Bloch, D. Cowan, K. Bechgaard, R. Pyle, R. Banks and T. Poehler, *Phys. Rev. Letters*, **34**, 1561 (1975).
568. J. Cooper, M. Weger, D. Jerome, D. Lefur, K. Bechgaard, A. Bloch and D. Cowan, *Solid State Commun.*, **19**, 749 (1976).
569. D. Jerome, A. Mazaud, M. Ribaud and K. Bechgaard, *J. Phys. Letters*, **41**, L-95 (1980).
570. K. Bechgaard, K. Carneiro, F. Rasmussen, M. Olsen, G. Rindorf, C. Jacobsen, H. Pedersen and J. Scott, *J. Amer. Chem. Soc.*, **103**, 2440 (1981).
571. E. Engler and V. Patel, *Tetrahedron Letters*, 1259 (1975).
572. (a) K. Lerstrup, M. Lee, F. Wiygul, T. Kistenmacher and D. Cowan, *J. Chem. Soc., Chem. Commun.*, 294 (1983).
(b) V. Lee, E. Engler, R. Schumaker and S. Parkin, *J. Chem. Soc., Chem. Commun.*, 235 1983.

573. E. Fanghängel, G. Schukat and J. Schüzendübel, *Z. Chem.*, **21**, 447 (1981).
574. E. Engler and V. Patel, *J. Chem. Soc., Chem. Commun.*, 671 (1975).
575. R. Wheland and J. Gillson, *J. Amer. Chem. Soc.*, **98**, 3916 (1976).
576. E. Engler and V. Patel, *J. Chem. Soc., Chem. Commun.*, 835 (1977).
577. F. Wudl and E. Aharon-Shalom, *J. Amer. Chem. Soc.*, **104**, 1154 (1982).
578. P. Carroll, M. Lakshmikantham, M. Cava, F. Wudl, E. Aharon-Shalom and S. Cox, *J. Chem. Soc., Chem. Commun.*, 1316 (1982).
579. K. Lerstrup, D. Talham, A. Bloch, Th. Poehler and D. Cowan, *J. Chem. Soc., Chem. Commun.*, 336 (1982).
580. D. Green, *J. Chem. Soc., Chem. Commun.*, 161 (1977).
581. Y. Okamato and P. Wojciechowski, *J. Chem. Soc., Chem. Commun.*, 668 (1981).
582. J. Rice, P. Wojciechowski and Y. Okamato, *Heterocycles*, **18**, 191 (1982).
583. R. Weiss and R. Gompper, *Tetrahedron Letters*, 481 (1970).
584. U. Timm, U. Plücken, M. Petersen and H. Meier, *J. Heterocycl. Chem.*, **16**, 1303 (1979).
585. E. Bulka, in Ref. 3, p. 459.
586. E. Bulka, *Chem. Scripta*, **8A**, 39 (1975).
587. I. Lalezari and A. Shafiee, *Advances in Heterocyclic Chemistry* (Eds. A. Katritzky and A. Boulton), Vol. 24, Academic Press, New York, 1979, p. 109.
588. R. Lulek, *US Patent*, No. 2,049,212 (1928); *Chem. Abstr.*, **30**, 6577 (1936).
589. I. G. Farbenindustrie, *French Patent*, No. 803,304 (1936); *Chem. Abstr.*, **31**, 2616 (1937).
590. F. Wille, A. Ascherl, G. Kaupp and C. Capeller, *Angew. Chem.*, **74**, 753 (1962).
591. R. Weber and M. Renson, *J. Heterocycl. Chem.*, **12**, 1091 (1975).
592. C. Campbell, C. Rees, M. Bryce, M. Cooke, P. Hanson and J. Vernon, *J. Chem. Soc., Perkin Trans. 1*, 1006 (1978).
593. M. Bryce, C. Reynolds, P. Hanson and J. Vernon, *J. Chem. Soc., Perkin Trans. 1*, 607 (1981).
594. J. Vernon, M. Bryce and T. Dransfield, *Tetrahedron*, **39**, 835 (1983).
595. T. Ueda. Y. Shibata, J. Sakakibara, M. Inoue and T. Ishida, *Chem. Pharm. Bull.*, **30**, 3424 (1982).
596. J. Liebscher and H. Hartmann, *Synthesis*, 273 (1976).
597. R. Weber and M. Renson, *Bull. Soc. Chim. Fr.*, 1124 (1976).
598. M. M'Buyi, *Doctoral Thesis*, University of Liège, 1983.
599. R. Lesser and R. Weiss, *Chem. Ber.*, **57**, 1079 (1924).
600. G. Wagner and P. Nuhn, *Pharm. Zentralhalle*, **104**, 328 (1965).
601. A. Welter, L. Christiaens and P. Wirtz, *European Patent Appl.*, EP 44, 453; *Chem. Abstr.*, **96**, 199699v (1982).
602. R. Weber and M. Renson, *Bull. Soc. Roy. Sci. Liège*, **48**, 146 (1979).
603. R. Lesser and R. Weiss, *Chem. Ber.*, **45**, 1835 (1912).
604. L. Dupont, O. Dideberg, J. Lamotte, M. Baiwir and R. Weber, *Tetrahedron*, **33**, 3083 (1977).
605. R. Weber, J.-L. Piette and M. Renson, *J. Heterocycl. Chem.*, **15**, 865 (1978).
606. H. Campsteyn, L. Dupont, J. Lamotte-Brasseur and M. Vermeire, *J. Heterocycl. Chem.*, **15**, 745 (1978).
607. F. Lucchesini and V. Bertini, *Synthesis*, 824 (1983).
608. E. Bulka, *Advances in Heterocyclic Chemistry* (Ed. A. Katritzky), Vol. 2, Academic Press, New York, 1963, p. 343.
609. G. Hofmann, *Ann. Chem.*, **250**, 294 (1889).
610. J. Haginiwa, *Yakugaku Zasschi*, **68**, 191 (1948); *Chem. Abstr.*, **47**, 8074f (1953).
611. L. Brooker, G. Keyes and F. White, *J. Amer. Chem. Soc.*, **57**, 2492 (1935).
612. P. Chauvin, J. Morel and P. Pastour, *Compt. Rend., Ser. C*, **276**, 1453 (1973).
613. P. Chauvin, J. Morel, P. Pastour and J. Martinez, *Bull. Soc. Chim. Fr.*, 2099 (1974).
614. V. Cohen, *Synthesis*, 66 (1979).
615. J. Metzger and P. Bailly, *Compt. Rend., Ser. C*, **237**, 906 (1953).
616. J. Metzger and P. Bailly, *Bull. Soc. Chim. Fr.*, 1249 (1955).
617. H. Backer and H. Bos, *Rec. Trav. Chim.*, **62**, 580 (1943).
618. A. Roy and P. Guha, *J. Indian Chem. Soc.*, **22**, 82 (1945).
619. A. Shafiee and I. Lalezari, *J. Heterocycl. Chem.*, **12**, 675 (1975).
620. R. Hanson and M. Davis, *J. Heterocycl. Chem.*, **18**, 205 (1981).
621. A. Tsurkan, *Farm. Zh. (Kiev)*, 51 (1977); *Chem. Abstr.*, **87**, 23163 (1977).
622. L. King and R. Hlavacek, *J. Amer. Chem. Soc.*, **73**, 1864 (1951).
623. E. Knott, *J. Chem. Soc.*, 628 (1945).

624. R. Rainer and V. Weiss, *European Patent Appl.*, EP 49, 855 (1982); *Chem. Abstr.*, **97**, 162684f (1982).
625. J. Haginiwa, *Yakugaku Zasschi*, **69**, 566 (1949); *Chem. Abstr.*, **44**, 4465d (1950).
626. R. Zingaro, F. Bennett and G. Hammar, *J. Org. Chem.*, **18**, 292 (1953).
627. E. Bulka, K. Ahlers, H. Ewert, W. Haase, F. Meier and W. Pilz, *Chem. Ber.*, **101**, 1910 (1968).
628. E. Bulka and R. Oppermann, *Z. Chem.*, **17**, 99 (1977).
629. J. Liebscher and H. Hartmann, *Z. Chem.*, **16**, 18 (1976).
630. W. Davies, J. McLaren and L. Wilkinson, *J. Chem. Soc.*, 3491 (1950).
631. O. Ceder and B. Beijer, *Tetrahedron*, **32**, 173 (1976).
632. E. Bulka, H.-G. Patzwaldt, F.-K. Peper and H. Beyer, *Chem. Ber.*, **94**, 1127 (1961).
633. E. Bulka, M. Morner and H. Beyer, *Chem. Ber.*, **94**, 2763 (1961).
634. E. Bulka, G. Rodekirch and H. Beyer, *Chem. Ber.*, **95**, 658 (1982).
635. E. Bulka, W. Dietz, H.-G. Patzwaldt and H. Beyer, *Chem. Ber.*, **96**, 1996 (1963).
636. S. Bilinski, J. Chmeclewski and D. Misiuna, *Acta Pol. Pharm.*, **32**, 159 (1975).
637. S. Bilinski, J. Chmeclewski and D. Misiuna, *Acta Pol. Pharm.*, **36**, 307 (1979).
638. B. Marcevicziojewski and S. Bilinski, *Acta Pol. Pharm.*, **37**, 169 (1980).
639. J. Gasteiger and C. Herzig, *Tetrahedron*, **37**, 2607 (1981).
640. E. Schaumann, H. Nimmesgern, G. Adiwidjaja and L. Carlsen, *Chem. Ber.*, **115**, 2516 (1982).
641. J. Haginiwa, *Yakugaku Zasshi*, **68**, 195, 197 (1948); *Chem. Abstr.*, **47**, 8074f, 8075b (1953).
642. E. Bulka and D. Ahlers, *Z. Chem.*, **3**, 388 (1963).
643. S. Archer and R. McGarry, *J. Heterocycl. Chem.*, **19**, 1245 (1982).
644. O. Shvaika and V. Lipnitskii, *Zh. Obshch. Khim.*, **51**, 1842 (1981).
645. W. Michels, *Chem. Ber.*, **25**, 3048 (1892).
646. *British Patent*, No. 392, 410 (1932); *Chem. Zentralbl., II*, 1934 (1933).
647. A. Van Dormael, *Bull. Soc. Chim. Belges*, **57**, 341 (1948).
648. Y. Mizuhara, *Bull. Chem. Soc. Japan*, **25**, 263 (1952).
649. W. Baringer, *Chem. Ber.*, **23**, 1003 (1890).
650. S.-H. Chu and H. Mautner, *J. Org. Chem.*, **27**, 2899 (1962).
651. D. Klayman, *J. Org. Chem.*, **30**, 2454 (1965).
652. C. Draguet and M. Renson, *Bull. Soc. Chim. Belges*, **81**, 279 (1972).
653. C. Draguet and M. Renson, *Bull. Soc. Chim. Belges*, **81**, 295 (1972).
654. C. Draguet and M. Renson, *Bull. Soc. Chim. Belges*, **81**, 303 (1972).
655. C. Di Marco, R. Coccia, A. Rinaldi and D. Cavallini, *Ital. J. Biochem.*, **26**, 51 (1977); **28**, 104 (1979).
656. C. Draguet and M. Renson, *Bull. Soc. Chim. Belges*, **81**, 289 (1972).
657. F. Devillanova and L. Kniezo, *Spectrochim Acta*, **36A**, 371 (1980).
658. P. Kristian and G. Verani, *Collect. Czech. Chem. Commun.*, **43**, 2298 (1978).
659. I. Azerbaev, L. Tsoi, S. Cholpankulova, A. Asmanova and V. Artyukhin, *Khim. Geterotsikl. Soedin.*, 917 (1978).
660. J.-F. Giudicelli, J. Menin and H. Najer, *Compt. Rend., Ser. C*, **262**, 285, 778 (1966).
661. J.-F. Giudicelli, J. Menin and H. Najer, *Bull. Soc. Chim. Fr.*, 1099 (1968).
662. A. Comrie, D. Dingwall and J. Stenlake, *J. Chem. Soc.*, 5713 (1963).
663. S. Behrendt, R. Borsdorf, E. Kleinpeter, D. Gründel and A. Hautschmann, *Z. Chem.*, **16**, 405 (1976).
664. A. Comrie, D. Dingwall and J. Stenlake, *J. Pharm. Pharmacol.*, **16**, 268 (1964).
665. A. Tsurkan, V. Groshev, V. Efremenko and E. Troshenko, *Farm. Zh.* (*Kiev*), **27**, 69 (1972); *Chem. Abstr.*, **78**, 72013q (1973).
666. A. Tsurkan and V. Groshev, *Farm. Zh. Kiev*, **28**, 82 (1973); *Chem. Abstr.*, **78**, 159525 (1973).
667, H. Siaglo, S. Andrzejewski, E. Kleczek and D. Prelicz, *Pol. J. Pharmacol. Pharm.*, **27**, 57 (1975); *Chem. Abstr.*, **83**, 79157 (1975).
668. V. Kononenko, B. Zhetar and S. Baranov, *Zh. Org. Khim.*, **9**, 61 (1973).
669. I. Levshin, K. V'yunov, A. Tsurkan, A. Knizhnik and A. Ginak, *Zh. Org. Khim.*, **19**, 654 (1983).
670. O. Shvaika, V. Artemov, V. Kononenko and S. Baranov, *Khim. Geterotsikl. Soedin.*, 930 (1973).
671. M. Cava and I. Saris, *J. Chem. Soc., Chem. Commun.*, 617 (1975).
672. K. Potts, F. Huang and R. Khattack, *J. Org. Chem.*, **42**, 1644 (1977).
673. M. Baudy, A. Robert and C. Guimon, *Tetrahedron*, **38**, 2129 (1982).
674. (a) L. Clark, *J. Chem. Soc.*, 2313 (1928).

(b) E. Ochiai and J. Haginiwa, *Yakugaku Zasshi*, **67**, 138 (1947); *Chem. Abstr.*, **45**, 9539e (1951).
675. E. Ochiai, J. Haginiwa and K. Komatsu, *Yakugaku Zasshi*, **70**, 372 (1950); *Chem. Abstr.*, **45**, 2475d (1951).
676. J. Develotte, *Ann. Chim. (Paris)*, **5**, 215 (1950).
677. H. Barany and M. Pianka, *J. Chem. Soc.*, 2217 (1953).
678. M. Bogert and Y. Chen, *J. Amer. Chem. Soc.*, **44**, 2352 (1922).
679. M. Bogert and A. Stull, *J. Amer. Chem. Soc.*, **49**, 2011 (1927).
680. H. Bauer, *Chem. Ber.*, **46**, 92 (1913).
681. L. Clark, *J. Chem. Soc.*, 2805 (1927).
682. F. Babichev and V. Likhitskaya, *Khim. Geterotsikl. Soedin.*, 508 (1969).
683. M. M'Buyi, M. Evers, G. Tihange, A. Luxen and L. Christiaens, *Tetrahedron Letters*, **24**, 5873 (1983).
684. M. Bryce, P. Hanson and M. Vernon, *J. Chem. Soc., Chem. Commun.*, 299 (1982).
685. G. Hasan and R. Hunter, *J. Chem. Soc.*, 1762 (1935).
686. J. Haginiwa, *Yakugaku Zasshi*, **69**, 565 (1949); *Chem. Abstr.*, **44**, 4465f (1950).
687. S. Treppendahl, *Acta Chem. Scand.*, **B29**, 385 (1975).
688. *German Patent*, No. 852, 392 (1952); *Chem. Abstr.*, **52**, 10206c (1958).
689. O. Gorgon and M. Simek, *Czech. Patent*, No. 127, 919 (1968); *Chem. Abstr.*, **70**, 96799m (1969).
690. *British Patent*, No. 489, 202 (1938); *Chem. Abstr.*, **33**, 435 (1939).
691. G. Reynolds and J. Van Allan, *J. Org. Chem.*, **24**, 1478 (1959).
692. S. Hünig and K. Fritsch, *Ann. Chem.*, **609**, 143 (1957).
693. R. Kada, V. Knoppova and J. Kovac, *Collect. Czech. Chem. Commun.*, **42**, 3333 (1977).
694. B. Lugovkin, *Khim. Geterotsikl. Soedin.*, 571 (1966).
695. F. Hamer, *J. Chem. Soc.*, 3197 (1952).
696. M. Seyhan, *Chem. Ber.*, **86**, 888 (1953).
697. W. Ried and R. Gross, *Chem. Ber.*, **90**, 2646 (1957).
698. T. Ukai and S. Kanahara, *Yakugaku Zasshi*, **75**, 31 (1955); *Chem. Abstr.*, **50**, 961h (1956).
699. L. Deady, *Australian J. Chem.*, **26**, 1949 (1973).
700. K. Knapp, P. Keller and J. Rund, *J. Chem. Soc., Chem. Commun.*, 971 (1978).
701. H. Federsel and J. Bergman, *Tetrahedron Letters*, **21**, 2429 (1980).
702. M. Bogert and H. Hopkins, *J. Amer. Chem. Soc.*, **46**, 1912 (1924).
703. E. Barni and G. di Modica, *J. Heterocycl. Chem.*, **8**, 693 (1971).
704. W. Treibs and E. Poppe, *J. Prakt. Chem.* (*4*), **13**, 330 (1961).
705. A. Shafiee, A. Mazloumi and V. Cohen, *J. Heterocycl. Chem.*, **16**, 1563 (1979).
706. V. Likhitskaya and F. Babichev, *Ukr. Khim. Zh.*, **35**, 746 (1969); *Chem. Abstr.*, **72**, 12654 (1970).
707. R. Hanson, R. Giese, M. Davis and S. Costello, *J. Med. Chem.*, **21**, 496 (1978).
708. D. Klayman and G. Milne, *Tetrahedron*, **25**, 191 (1964).
709. D. Klayman and T. Woods, *J. Org. Chem.*, **39**, 1819 (1979).
710. A. Shafiee and I. Lalezari, *J. Heterocycl. Chem.*, **8**, 1011 (1971).
711. T. Kappe, A. Chirazi and E. Ziegler, *Monatsh. Chem.*, **103**, 234 (1972).
712. R. Acheson, M. Foxton and G. Miller, *J. Chem. Soc.*, 3200 (1965).
713. A. Shafiee, G. Kiaevay and M. Vosooghi, *J. Heterocycl. Chem.*, **18**, 789 (1981).
714. A. Simon and G. Heintz, *Naturwissenschaften*, **47**, 468 (1960).
715. C. Chabrie and R. Jacob, *Compt. Rend., Ser. C*, **134**, 1508 (1902).
716. J.-P. Marino and A. Schwartz, *Tetrahedron Letters*, 3253 (1979).
717. S. Ley, C. Meerholz and D. Barton, *Tetrahedron*, **37**, 213, suppl. (1981).
718. K. Sasse, R. Wegler, G. Unerstenhofer and F. Grewe, *Angew. Chem.*, **72**, 973 (1960).
719. R. Paetzold and D. Lienig, *Z. Anorg. Allgem. Chem.*, **335**, 289 (1965).
720. E. Gould and W. Burlant, *J. Amer. Chem. Soc.*, **78**, 5825 (1956).
721. E. Gould and B. Post, *J. Amer. Chem. Soc.*, **78**, 5161 (1956).
722. H. Meier, T. Molz, U. Merkle, T. Echter and M. Lorch, *Ann. Chem.*, 914 (1982).
723. (a) M. Lakhsmikantham, M. Cava, M. Albeck, L. Engman, P. Carroll, J. Bergman and F. Wudl, *Tetrahedron Letters*, 4199 (1981).
(b) D. Allen, J. Coppola, O. Kennard, F. Mann, W. Motherwell and D. Watson, *J. Chem. Soc. (C)*, 810 (1970).

724. Y. Akulin, B. Strelets and L. Efros, *Khim. Geterotsikl. Soedin.*, 275 (1975).
725. Y. Akulin, B. Strelets and L. Efros, *Khim. Geterotsikl. Soedin.*, 138 (1974).
726. A. Evdokimov, Y. Akulin, B. Strelets and L. Efros, *Khim. Geterotsikl. Soedin.*, 1429 (1975).
727. L. Efros, B. Strelets and Y. Akulin, *Khim. Geterotsikl. Soedin.*, 1361 (1976).
728. B. Strelets, M. Gel'mont, Y. Akulin and L. Efros, *Khim. Geterotsikl. Soedin.*, 1205 (1979).
729. B. Strelets, M. Gel'mont, Y. Akulin and L. Efros, *Khim. Geterotsikl. Soedin.* (*Engl. Transl.*), **15**, 978 (1979).
730. Y. Akulin, B. Strelets and L. Efros, *Khim. Geterotsikl. Soedin.*, 1499 (1977).
731. Y. Akulin, M. Gel'mont, B. Strelets and L. Efros, *Khim. Geterotsikl. Soedin.*, 912 (1978).
732. K. Burger and R. Ottlinger, *Synthesis*, 44 (1978).
733. K. Burger and R. Ottlinger, *Tetrahedron Letters*, 973 (1978).
734. K. Burger, R. Ottlinger, A. Proksch and J. Firl, *J. Chem. Soc., Chem. Commun.*, 80 (1979).
735. K. Burger, R. Ottlinger, A. Frank and U. Schubert, *Angew. Chem.*, **90**, 824 (1978).
736. (a) K. Burger, R. Ottlinger, H. Goth and J. Firl, *Chem. Ber.*, **113**, 2699 (1980).
(b) J. Liebscher, A. Hantschmann and H. Hartmann, *East German Patent*, No. 134093 (1979); *Chem. Abstr.*, **91**, 39497 (1979).
737. S. Keimatsu and I. Satoda, *J. Pharm. Soc. Japan*, **55**, 233 (1935); *Chem. Abstr.*, **31**, 6661 (1937).
738. S. Keimatsu and I. Sotoda, *J. Pharm. Soc. Japan*, **56**, 869 (1936); *Chem. Abstr.*, **33**, 154 (1939).
739. I. Laelezari, A. Shafiee and M. Yalpani, *Tetrahedron Letters*, 5105 (1969).
740. I. Lalezari, A. Shafiee and M. Yalpani, *Angew. Chem.*, **82**, 484 (1970).
741. I. Lalezari, A. Shafiee and M. Yalpani, *J. Org. Chem.*, **36**, 2836 (1971).
742. H. Meier and I. Menzel, *J. Chem. Soc., Chem. Commun.*, 1059 (1971).
743. H. Meier and H. Petersen, *Synthesis*, 596 (1978).
744. I. Lalezari, A. Shafiee and H. Golgolab, *J. Heterocycl. Chem.*, **10**, 655 (1973).
745. H. Meier, S. Schniepp and W. Combrink, *Chem. Ztg.*, **99**, 461 (1975).
746. H. Meier, M. Layer, W. Combrink and S. Schniepp, *Chem. Ber.*, **109**, 1650 (1976).
747. H. Golgolab and I. Lalezari, *J. Heterocycl. Chem.*, **12**, 801 (1975).
748. A. Shafiee, I. Lalezari, M. Mirrashed and D. Nercesian, *J. Heterocycl. Chem.*, **14**, 567 (1977).
749. I. Lalezari, A. Shafiee and S. Sadeghi-Milani, *J. Heterocycl. Chem.*, **16**, 1405 (1979).
750. R. Hanson and M. Davis, *J. Heterocycl. Chem.*, **17**, 1245 (1980).
751. I. Lalezari, A. Shafiee, J. Khorami and A. Soltani, *J. Pharm. Sci.*, **67**, 1336 (1978).
752. (a) R. Hanson and M. Davis, *J. Labeled Compd. Radiopharm.*, **16**, 31 (1979).
(b) S. Dhawan and J. Kagan, *J. Labeled Compd. Radiopharm.*, **19**, 331 (1982).
753. B. Bak, O. Nielsen, H. Svanholt and A. Holm, *Chem. Phys. Letters*, **55**, 36 (1978).
754. I. Lalezari and S. Sadeghi-Milani, *J. Heterocycl. Chem.*, **15**, 501 (1978).
755. V. Litvinov and I. Dzhumaev, *Izv. Akad. Nauk SSSR, Ser. Khim.*, 478 (1979).
756. O. Zimmer and H. Meier, *Chem. Ber.*, **114**, 2938 (1981).
757. G. Jaffari, A. Nunn and J. Ralph, *J. Chem. Soc.* (*C*), 2060 (1970).
758. H. Meier and E. Voigt, *Tetrahedron*, **28**, 187 (1972).
759. H. Meier, J. Zountsas and S. Petersen, *Nouv. J. Chim.*, **6**, 73 (1982).
760. I. Lalezari, A. Shafiee and M. Yalpani, *J. Org. Chem.*, **38**, 338 (1973).
761. F. Malek-Yazdi and M. Yalpani, *Synthesis*, 328 (1977).
762. M. Barth, H. Buhl and H. Meier, *Chem. Ztg.*, **101**, 452 (1977).
763. H. Meier and I. Menzel, *Tetrahedron Letters*, 445 (1972).
764. J. Laureni, A. Krantz and R. Hajdu, *J. Amer. Chem. Soc.*, **98**, 7872 (1976).
765. W. Becker and J. Meyer, *Chem. Ber.*, **37**, 2550 (1904).
766. V. Cohen, *Synthesis*, 768 (1978).
767. J. Goerdeler, D. Gross and H. Klinke, *Chem. Ber.*, **96**, 1289 (1963).
768. H. Werchan and G. Dittrich, *Tetrahedron Letters*, **23**, 2833 (1982).
769. O. Hinsberg, *Chem. Ber.*, **22**, 862, 2895 (1889).
770. N.-P. Buu-Hoi, *J. Chem. Soc.*, 2882 (1949).
771. E. Sawicki, B. Chastain and H. Bryant, *J. Org. Chem.*, **21**, 754 (1956).
772. K. Potts, R. Cody and R. Dennis, *J. Org. Chem.*, **46**, 4065 (1981).
773. E. Sawicki, *J. Org. Chem.*, **18**, 1492 (1953).
774. E. Sawicki, *J. Org. Chem.*, **19**, 1163 (1954).
775. E. Sawicki, *J. Amer. Chem. Soc.*, **77**, 957 (1955).
776. V. Pesin, M. Khaletskii and L. Zolotova-Zolotukhina, *Zh. Obshch. Khim.*, **33**, 1101, 3716 (1963).

777. J. Ralph and C. Marks, *Tetrahedron*, **22**, 2487 (1966).
778. E. Sawicki, *J. Amer. Chem. Soc.*, **76**, 664 (1954).
779. N. Brown and P. Bladon, *Tetrahedron*, **24**, 6577 (1968).
780. G. Harts, K. De Roos and C. Salemink, *Rec. Trav. Chim.*, **89**, 5 (1970).
781. A. Carr, E. Sawicki and F. Ray, *J. Org. Chem.*, **23**, 1940 (1958).
782. V. Pesin, A. Khaletskii and L. Zolotova-Zolotukhina, *Zh. Obshch. Khim.*, **31**, 3000 (1961).
783. F. Sachs and G. Meyerheim, *Chem. Ber.*, **41**, 3957 (1908).
784. A. Komin and M. Carmack, *J. Heterocycl. Chem.*, **13**, 13 (1976).
785. K. Sharma, S. Kumari and R. Singh, *Synthesis*, 316 (1981).
786. R. Neidlein, T. V. Dao, A. Gieren, M. Kokkinidis, R. Wilckens, H. Geserich and W. Ruppel, *Chem. Ber.*, **115**, 2898 (1982).
787. G. Eremeeva, B. Strelets and L. Efros, *Khim. Geterotsikl. Soedin.*, 276 (1975).
788. (a) G. Eremeeva, B. Strelets and L. Efros, *Khim. Geterotsikl. Soedin.*, 340 (1976).
(b) G. Eremeeva, Y. Akulin, T. Timofeeva, B. Strelets and L. Efros, *Khim. Geterotsikl. Soedin.*, 1129 (1982).
789. C. Pedersen, *J. Chem. Soc., Chem. Commun.*, 704 (1974).
790. C. Pedersen, *Acta Chem. Scand.*, **B30**, 675 (1976).
791. C. Pedersen, *Tetrahedron Letters*, 745 (1979).
792. C. Bird, G. Cheeseman and A. Sarsfield, *J. Chem. Soc.*, 4767 (1963).
793. F. Tobiason, L. Huestis, C. Chandler, S. Pedersen and P. Peters, *J. Heterocycl. Chem.*, **10**, 773 (1973).
794. W. Sliwa and A. Thomas, *Heterocycles*, **20**, 71 (1983).
795. Y. Shealy and J. Clayton, *J. Heterocycl. Chem.*, **4**, 96 (1967).
796. L. Weinstock, P. Davis, D. Mulvey and J. Schaeffer, *Angew. Chem. (Intern. Ed.)*, **6**, 364 (1967).
797. V. Bertini, *Angew. Chem. (Intern. Ed.)*, **6**, 563 (1967).
798. V. Bertini, *Gazz. Chim. Ital.*, **97**, 1870 (1967).
799. V. Bertini, A. De Munno, A. Menconi and A. Fissi, *J. Org. Chem.*, **39**, 2294 (1974).
800. V. Bertini, F. Lucchesini and A. de Munno, *Synthesis*, 979 (1979).
801. C. Pedersen, H. Harrit, M. Poliakoff and I. Dunkin, *Acta Chem. Scand.*, **B31**, 848 (1977).
802. C. Pedersen and N. Hacker, *Tetrahedron Letters*, 3981 (1977).
803. R. Stollé and R. Gutmann, *J. Prakt. Chem. (2)*, **69**, 509 (1904).
804. R. Kendall and R. Olofsson, *J. Org. Chem.*, **35**, 806 (1970).
805. D. Levine, N. Krugh and L. Gold, *J. Mol. Spectry.*, **30**, 459 (1969).
806. V. Cohen, *J. Heterocycl. Chem.*, **16**, 365 (1979).
807. E. Bulka, D. Ehlers and H. Storm, *J. Prakt. Chem.*, **315**, 164 (1973).
808. I. Lalezari and A. Shafiee, *J. Heterocycl. Chem.*, **8**, 835 (1971).
809. E. Bulka and D. Ehlers, *J. Prakt. Chem.*, **315**, 510 (1973).
810. R. Fusco and C. Musante, *Gazz. Chim. Ital.*, **68**, 665 (1938).
811. H. Hassaneen, A. Shetta and A. Shawali, *J. Heterocycl. Chem.*, **17**, 1185 (1980).
812. T. Back, D. Barton, M. Britten-Kelly and F. Guziec, *J. Chem. Soc., Perkin Trans. 1*, 2079 (1976).
813. V. Chiuguk, D. Sheiko and V. Glustrakov, *Khim. Geterotsikl. Soedin.*, 1435 (1974).
814. (a) H. Hassaneen, A. Abdelhamid, A. Shetta and A. Shawali, *Gazz. Chim. Ital.*, **112**, 545 (1982).
(b) A. Abdelhamid, H. Hassaneen and A. Shawali, *J. Heterocycl. Chem.*, **20**, 719 (1983).
815. V. Bertini, F. Lucchesini and A. de Munno, *Synthesis*, 681 (1982).
816. R. Cobbledick, F. Enstein, W. McWhinnie and F. Musa, *J. Chem. Res. (S)*, 145 (1979).
817. F. King and D. Felton, *J. Chem. Soc.*, 274 (1949).
818. D. Paquer, M. Perrier and J. Vialle, *Bull. Soc. Chim. Fr.*, 4527 (1970).
819. M. Perrier and J. Vialle, *Bull. Soc. Chim. Fr.*, 4591 (1971).
820. (a) R. Beer, J. Hatton, E. Llaguno and I. Paul, *J. Chem. Soc., Chem. Commun.*, 594 (1971).
(b) E. Lalaguno, I. Paul and W. Noyes, *J. Chem. Soc., Perkin Trans. 2*, 2001 (1972).
821. F. Amundsen, L. Hansen and A. Hordvik, *Acta Chem. Scand.*, **A36**, 673 (1982).
822. M. Perrier, R. Pinel and J. Vialle, *J. Heterocycl. Chem.*, **12**, 639 (1975).
823. M. Perrier and J. Vialle, *Bull. Soc. Chim. Fr., II*, 199 (1979).
824. R. Beer and A. Poole, *Tetrahedron Letters*, 1835 (1972).
825. M. Perrier and J. Vialle, *Bull. Soc. Chim. Fr., II*, 205 (1979).
826. (a) R. Christie and D. Reid, *J. Chem. Soc., Perkin Trans. 1*, 229 (1976).

(b) J. Dingwall, A. Dunn, D. Reid and K. Wade, *J. Chem. Soc., Perkin Trans. 1*, 1360 (1972).
827. (a) D. Reid, R. Walker and D. Webster, *J. Chem. Soc., Perkin Trans. 1*, 1596 (1981).
(b) J. Czyzewski and D. Reid, *J. Chem. Soc., Perkin Trans. 1*, 777 (1983).
(c) M. Jacobsen, L. Henriksen and A. Holm, *Acta Chem. Scand.*, **37B**, 585 (1983).
828. G. Morgan and F. Burstall, *J. Chem. Soc.*, 2197 (1929).
829. I. McCullough and A. Lefohn, *Inorg. Chem.*, **5**, 150 (1966).
830. G. Morgan and H. Burgess, *J. Chem. Soc.*, 321 (1928).
831. U. Anthoni, C. Larsen and P. Nielsen, *Acta Chem. Scand.*, **21**, 2571 (1967).
832. G. Morgan and F. Burstall, *J. Chem. Soc.*, 173 (1931).
833. A. Fredga and K. Styrman, *Arkiv Kemi.*, **14**, 461 (1959).
834. J. Lambert, C. Mixan and D. Johnson, *J. Amer. Chem. Soc.*, **95**, 4634 (1973).
835. J. Lambert, D. Johnson, R. Keske and C. Mixan, *J. Amer. Chem. Soc.*, **94**, 8172 (1972).
836. J. Lambert and S. Featherman, *Chem. Rev.*, **75**, 612 (1975).
837. T. Srivastava, R. Srivastava and H. Singh, *Indian J. Chem.*, **18A**, 71 (1979).
838. I. Degani, R. Fochi and C. Vincenzi, *Gazz. Chim. Ital.*, **94**, 203 (1964).
839. I. Degani and C. Vincenzi, *Boll. Sci. Fac. Chim. Ind. Bologna*, **25**, 51 (1967).
840. (a) V. Karchenko and B. Drevko, *Zh. Org. Khim.*, **18**, 2995 (1982).
(b) M. Müller, A. Hantschmann, M. Puls and M. Weisenfels, *Z. Chem.*, **23**, 145 (1983).
841. J. Gosselck and E. Wolters, *Z. Naturforsch.*, **17b**, 131 (1962).
842. I. Lalezari, A. Ghanbarpour, F. Ghapgaran, M. Niazi and R. Jafari-Namin, *J. Heterocycl. Chem.*, **11**, 469 (1974).
843. W. Ziriakus, W. Hinsel and R. Haller, *Arch. Pharm.*, **304**, 681 (1971).
844. P. Nanjappan, K. Ramalingam, M. Herd, P. Arjunan and K. Berlin, *J. Org. Chem.*, **45**, 4622 (1980).
845. (a) W. Hansel and R. Haller, *Naturwissenschaften*, **5**, 583 (1968).
(b) W. Hänsel and R. Haller, *Arch. Pharm.*, **302**, 147 (1979); **303**, 819 (1970).
(c) T. Raja, *Indian J. Chem.*, **19B**, 812 (1980).
846. P. Nanjappan, K. Ramalingam, N. Satyamurthy and K. Berlin, *J. Org. Chem.*, **46**, 2542 (1981).
847. A. Tolmachev and M. Kudinova, *Khim. Geterotsikl. Soedin.*, 274 (1974).
848. M. Detty, B. Murray and M. Seidler, *J. Org. Chem.*, **47**, 1968 (1982).
849. S. Es-Seddiki, G. Le Coustumer, Y. Mollier and M. Devaud, *Tetrahedron Letters*, **22**, 2771 (1981).
850. J. Liebscher and H. Hartmann, *Tetrahedron*, **33**, 731 (1977).
851. G. Morgan and H. Drew, *J. Chem. Soc.*, 731 (1924).
852. G. Morgan and E. Holmes, *J. Chem. Soc.*, 760 (1924).
853. G. Morgan and A. Taylor, *J. Chem. Soc.*, 797 (1925).
854. (a) J. Dewan, W. Jennings, J. Silver and M. Tolley, *Org. Mag. Res.*, **11**, 449 (1978).
(b) J. Dewan and J. Silver, *J. Organomet. Chem.*, **125**, 125 (1977).
(c) C. Reston, R. Secomb and A. White, *J. Chem. Soc., Dalton Trans.*, 2307 (1976).
855. M. Kudinova, S. Krivun and A. Tolmachev, *Khim. Geterotsikl. Soedin.*, 857 (1973).
856. I. Degani, R. Fochi and C. Vincenzi, *Boll. Sci. Fac. Chim. Ind. Bologna*, **23**, 21 (1965).
857. M. Detty and B. Murray, *J. Org. Chem.*, **47**, 5235 (1982).
858. C. Fabre, R. Fugnitto and H. Strzelecka, *Compt. Rend., Ser. C*, **282**, 175 (1976).
859. D. Sandman, A. Epstein, T. Holmes and A. Fisher, *J. Chem. Soc., Chem. Commun.*, 177 (1977).
860. D. Sandman, A. Epstein, T. Holmes, J. Lee and D. Titus, *J. Chem. Soc., Perkin Trans. 2*, 1578 (1980).
861. M. Detty, B. Murray and J. Perlstein, *Tetrahedron Letters*, **24**, 539 (1983).
862. J. Gosselck, *Chem. Ber.*, **91**, 2345 (1958).
863. M. Renson, *Bull. Soc. Chim. Belges*, **73**, 483 (1964).
864. J.-M. Danze and M. Renson, *Bull. Soc. Chim. Belges*, **75**, 169 (1966).
865. N. Bellinger and P. Cagniant, *Compt. Rend., Ser. C*, **268**, 1385 (1969).
866. P. Cagniant, N. Bellinger and D. Cagniant, *Bull. Soc. Chim. Fr.*, 2699 (1971).
867. I. Degani and R. Fochi, *Ann. Chim. (Roma)*, **63**, 319 (1973).
868. A. Tadino, L. Christiaens, M. Renson and P. Cagniant, *Bull. Soc. Chim. Belges*, **81**, 595 (1972).
869. A. Tadino, L. Christiaens and M. Renson, *Bull. Soc. Roy. Sci. Liège*, **42**, 496 (1973).
870. P. Cagniant and G. Kirsch, *Compt. Rend., Ser. C*, **285**, 327 (1977).

871. N. Dereu, J.-L. Piette, J. Van Coppenolle and M. Renson, *J. Heterocycl. Chem.*, **12**, 423 (1975).
872. J. Gosselck and E. Wolters, *Chem. Ber.*, **95**, 1237 (1962).
873. J.-L. Piette and M. Renson, *Bull. Soc. Chim. Belges*, **80**, 669 (1971).
874. A. Ricci, D. Balucani, C. Rossi and A. Croisy, *Boll. Sci. Fac. Chim. Ind. Bologna*, **27**, 279 (1969).
875. P. Abramenko and V. Zhiryakov, *Khim. Geterotsikl. Soedin.*, 1539 (1972).
876. G. Lebedeva, I. Kvitko and A. El'tsov, *Khim. Geterotsikl. Soedin.* (*Engl. Transl.*), **15**, 429 (1979).
877. N. Bellinger, P. Cagniant, D. Cagniant and M. Renson, *Bull. Soc. Chim. Fr.*, 2689 (1971).
878. N. Bellinger, D. Cagniant and P. Cagniant, *Tetrahedron Letters*, 49 (1971).
879. G. Chmutova, *Sb. Aspir. Rat. Kazan. Cos. Univ. Khim. Geol.*, 70 (1967); *Chem. Abstr.*, **70**, 87444q (1969).
880. N. Dereu, *Doctoral Thesis*, University of Liège, 1978.
881. I. Degani, R. Fochi and C. Vincenzi, *Gazz. Chim. Ital.*, **94**, 451 (1964).
882. A. Tadino, L. Christiaens and M. Renson, *Bull. Soc. Roy. Sci. Liège*, **42**, 129 (1973).
883. A. Tolmachev, M. Kudinova and L. Shulezhko, *Khim. Geterotsikl. Soedin.* 142 (1977).
884. R. Neidlein and J. Koerber, *Arch. Pharm.*, **311**, 170 (1978).
885. I. Degani, R. Fochi and G. Spunta, *Boll. Sci. Fac. Chim. Ind. Bologna*, **23**, 244 (1965).
886. A. Tadino, L. Christiaens and M. Renson, *Bull. Soc. Roy. Sci. Liège*, **42**, 146 (1973).
887. A. Tadino, L. Christiaens and M. Renson, *Bull. Soc. Roy. Sci. Liège*, **42**, 505 (1973).
888. I. Degani and R. Fochi, *Ann. Chim.* (*Roma*), **58**, 251 (1968).
889. A. Couture, A. Lablache-Combier and T.-Q. Minh, *Tetrahedron Letters*, 2873 (1977).
890. A. Ruwet and M. Renson, *Bull. Soc. Chim. Belges*, **75**, 260 (1966).
891. F. Bossert, *Angew. Chem.*, **77**, 913 (1965).
892. A. Ruwet, D. Janne and M. Renson, *Bull. Soc. Chim. Belges*, **79**, 81 (1970).
893. A. Ruwet, A. Pourveur and M. Renson, *Bull. Soc. Chim. Belges*, **79**, 631 (1970).
894. D. Wadsworth and M. Detty, *J. Org. Chem.*, **45**, 4611 (1980).
895. A. Luxen, *Doctoral Thesis*, University of Liège, 1983.
896. N. Dereu and M. Renson, *J. Organomet. Chem.*, **208**, 11 (1981).
897. H. Lumbroso, C. Liegeois, N. Dereu, L. Christiaens and A. Luxen, *J. Mol. Struct.*, **67**, 251 (1980).
898. N. Dereu, M. Renson, Y. Mollier and J. Le Coustumer, *J. Organomet. Chem.*, **208**, 23 (1981).
899. M. Detty and B. Murray, *J. Amer. Chem. Soc.*, **105**, 883 (1983).
900. M. Detty and B. Murray, *J. Org. Chem.*, **47**, 1146 (1982).
901. E. Ziegler and E. Nölken, *Monatsh. Chem.*, **89**, 737 (1958).
902. A. Ruwet, C. Draguet and M. Renson, *Bull. Soc. Chim. Belges*, **79**, 639 (1970).
903. A. Ruwet, A. Pourveur and M. Renson, *Bull. Soc. Chim. Belges*, **79**, 631 (1970).
904. A. Ruwet and M. Renson, *Bull. Soc. Chim. Belges*, **78**, 449 (1969).
905. L. Christiaens, J.-L. Piette, A. Luxen and M. Renson, *J. Heterocycl. Chem.*, **21**, 1281 (1984).
906. R. Holliman and F. Mann, *J. Chem. Soc.*, 37 (1945).
907. A. Biezais-Zirnis and A. Fredga, *Acta Chem. Scand.*, **25**, 1174 (1971).
908. C. Lemaire, A. Luxen, L. Christiaens and M. Guillaume, *J. Heterocycl. Chem.*, **20**, 811 (1983).
909. M. Renson and P. Pirson, *Bull. Soc. Chim. Belges*, **75**, 456 (1966).
910. A. Luxen, L. Christiaens and M. Renson, *J. Org. Chem.*, **48**, 3535 (1980).
911. J. Stackhouse, G. Senkler, B. Maryanoff and K. Mislow, *J. Amer. Chem. Soc.*, **96**, 7835 (1974).
912. R. Lesser and R. Weiss, *Chem. Ber.*, **47**, 2510 (1914).
913. K. Sindelar, E. Svatek, J. Metysova, J. Metys and M. Protiva, *Collect. Czech. Chem. Commun.*, **34**, 3792 (1969).
914. M. Renson and L. Christiaens, *Bull. Soc. Chim. Belges*, **79**, 511 (1970).
915. R. Lesser and R. Weiss, *Chem. Ber.*, **57**, 1077 (1924).
916. J. Martens, K. Praefcke and H. Simon, *Z. Naturforsch.*, **31b**, 1717 (1976).
917. R. Lüdesdorf, J. Martens, B. Pakzad and K. Praefcke, *Ann. Chem.*, 1992 (1977).
918. K. Praefcke and D. Schmidt, *J. Heterocycl. Chem.*, **16**, 47 (1979).
919. B. Pakzad, K. Praefcke and H. Simon, *Angew. Chem.* (*Intern. Ed.*), **16**, 319 (1977).
920. K. Praefcke and U. Schulze, *J. Organomet. Chem.*, **184**, 189 (1980).
921. K. Beelitz, K. Praefcke and S. Gronowitz, *J. Organomet. Chem.*, **194**, 167 (1980).
922. F. François, *Compt. Rend., Ser. C*, **190**, 191 (1930).

923. A. Blinokhvatov, O. Markovtseva, I. Shlaider and V. Karchenko, *Khim. Geterotsikl. Soedin.*, 640 (1981).
924. M. Hori, T. Kataoka, H. Shimizu and C. F. Hsu, *Chem. Letters*, 391 (1973).
925. I. Sadekov, A. Ladatko and E. Sadekova, *Khim. Geterotsikl. Soedin.*, 435 (1983).
926. B. Muth, *Chem. Ber.*, **93**, 283 (1960).
927. R. Nealey and J. Driscoll, *J. Heterocycl. Chem.*, **3**, 228 (1966).
928. M. Hori, T. Kataoka, H. Shimizu and K. Tomimatsu, *Tetrahedron Letters*, **23**, 901 (1982).
929. T. Kataoka, K. Tomimatsu, H. Shimizu and M. Hori, *Tetrahedron Letters*, **24**, 75 (1983).
930. I. Degani, R. Fochi and G. Spunta, *Boll. Sci. Fac. Chim. Ind. Bologna*, **23**, 165 (1965).
931. N.-P. Buu-Hoi, A. Croisy, P. Jacquignon, M. Renson and A. Ruwet, *J. Chem. Soc. (C)*, 1058 (1970).
932. J.-L. Piette, P. Thibaut and M. Renson, *Tetrahedron*, **34**, 655 (1978).
933. W. Lohner and K. Praefcke, *J. Organomet. Chem.*, **205**, 167 (1981).
934. W. Lohner and K. Praefcke, *Chem. Ztg.*, **103**, 265 (1979).
935. I. Sadekov, A. Ladatko and V. Minkin, *Khim. Geterotsikl. Soedin.*, 1567 (1978).
936. I. Sadekov, A. Ladatko and V. Minkin, *Khim. Geterotsikl. Soedin.*, 1342 (1980).
937. I. Sadekov, A. Ladatko, E. Sadekova, G. Dorofeenko and V. Minkin, *Khim. Geterotsikl. Soedin.*, 343 (1981).
938. V. Panov, I. Sadekov, A. Ladatko, O. Okhlobystin and V. Minkin, *Khim. Geterotsikl. Soedin.*, 1421 (1980).
939. R. Close, D. Cagniant, G. Le Coustumer, C. Andrieu and Y. Mollier, *J. Chem. Res. (S)*, 4 (1978).
940. L. Brooker, *US Patent*, No. 2, 189, 599 (1940); *Chem. Abstr.*, **34**, 4003 (1940).
941. G. Bergson, *Arkiv Kemi*, **13**, 11 (1958).
942. A. Fredga, *Arkiv Kemi, Mineral. Geol.*, **11B**, 6 (1933).
943. R. Dedeyne and M. Anteunis, *Bull. Soc. Chim. Belges*, **85**, 319 (1976).
944. C. Gibson and J. Johnson, *J. Chem. Soc.*, 266 (1931).
945. J. McCullough and P. Radlick, *Inorg. Chem.*, **3**, 924 (1964).
946. C. Gibson and J. Johnson, *J. Chem. Soc.*, 1529 (1933).
947. E. Gould and J. McCullough, *J. Amer. Chem. Soc.*, **73**, 1105 (1951).
948. J. McCullough, *Inorg. Chem.*, **4**, 862 (1965).
949. J. McCullough and N. Tideswall, *J. Amer. Chem. Soc.*, **76**, 3091 (1954).
950. Y. Migalina, V. Staninets, V. Lendel, I. Balog, V. Palyulin, A. Koz'min and N. Zefirov, *Khim. Geterotsikl. Soedin.*, 58 (1977).
951. A. Toshimitsu, S. Uemura and M. Okano, *J. Chem. Soc., Perkin Trans. 1*, 1206 (1979).
952. (a) L. Brooker and R. Sprague, *US Patents*, No. 2, 332, 433 (1943); *Chem. Abstr.*, **38**, 1968 (1944).
(b) E. Kataev, L. Kataeva and Z. Titova, *Khim. Geterotsikl. Soedin.*, 172 (1968).
(c) C. Krispan, *J. Amer. Chem. Soc.*, **83**, 3434 (1961).
953. A. Zilverschoon, J. Meijer, P. Vermeer and L. Brandsma, *Rec. Trav. Chim.*, **94**, 163 (1975).
954. J. Meijer, P. Vermeer, H. Verkruijsse and L. Brandsma, *Rec. Trav. Chim.*, **92**, 1326 (1973).
955. W. Verboom, M. Schoufs, J. Meijer, H. Verkruijsse and L. Brandsma, *Rec. Trav. Chim.*, **97**, 244 (1978).
956. A. Naaktgeboren, J. Meijer, P. Vermeer and L. Brandsma, *Rec. Trav. Chim.* , **94**, 92 (1975).
957. W. Verboom, R. Sukhai and J. Meijer, *Synthesis*, 47 (1979).
958. R. Sukhai, R. de Jong, H. Verkruijsse and L. Brandsma, *Rec. Trav. Chim.*, **100**, 369 (1981).
959. Y. Migalina, S. Galla-Bobic, S. Khripak and V. Staninets, *Khim. Geterotsikl. Soedin.*, 911 (1982).
960. H. Drew, *J. Chem. Soc.*, 223 (1926).
961. H. Drew and R. Thomason, *J. Chem. Soc.*, 116 (1927).
962. L. Reichel and E. Kirschbaum, *Chem. Ber.*, **76B**, 1105 (1943).
963. A. Gioaba and G. Vasiliu, *Ann. Univ. Bucaresti, Ser. Strint. Natur. Chim. 15*, 89 (1966); *Chem. Abstr.*, **70**, 68319 (1969).
964. G. Vasiliu and A. Gioaba, *Rev. Chim. (Bucarest)*, **20**, 357 (1969).
965. G. Vasiliu and A. Gioaba, *Rev. Chim. (Bucarest)*, **19**, 253 (1968).
966. A. Gioaba, O. Maior and M. Mitran, *Rev. Roum. Chim.*, **21**, 733 (1976).
967. A. Gioaba and O. Maior, *Ann. Univ. Bucarest Chim.*, **22**, 111 (1973).
968. I. Campbell and E. Turner, *J. Chem. Soc.*, 37 (1938).

969. H. Drew, *J. Chem. Soc.*, 506 (1928).
970. H. Drew, *J. Chem. Soc.*, 511 (1928).
971. M. Thompson and E. Turner, *J. Chem. Soc.*, 29 (1938).
972. J. Bergman, *Chem. Scripta*, **8A**, 116 (1975).
973. A. Gioaba and O. Maior, *Rev. Roum. Chim.*, **15**, 1967 (1970).
974. F. Colonna, G. Distefano, V. Galasso, K. Irgolic, C. King and G. Pappalardo, *J. Organomet. Chem.*, **146**, 235 (1978).
975. D. Rainville, R. Zingaro and J. Ferraris, *Can. J. Chem.*, **58**, 1133 (1980).
976. N. Petragnani, *Tetrahedron*, **11**, 15 (1960).
977. F. Krafft and A. Kaschau, *Chem. Ber.*, **29**, 443 (1896).
978. S. Keimatsu, I. Satoda and T. Kobayashi, *J. Pharm. Soc. Japan*, **57**, 190 (1937).
979. O. Schmitz-Du Mont and B. Ross, *Angew. Chem. (Intern. Ed.)*, **6**, 1071 (1967).
980. M. Schmidt and H. Schumann, *Chem. Ber.*, **96**, 780 (1963).
981. C. Campbell and C. Rees, *J. Chem. Soc. (C)*, 752 (1969).
982. T. Weiss, W. Nitsche, F. Böhnke and G. Klar, *Ann. Chem.*, 1418 (1973).
983. A. Addison, T. Li and T. Weiler, *Can. J. Chem.*, **55**, 766 (1977).
984. W. Hinrichs and G. Klar, *J. Chem. Res. (S)*, 336 (1982).
985. S. Cohen, M. Reddy and A. Massey, *J. Chem. Soc., Chem. Commun.*, 451 (1967).
986. N. Cullinane, *J. Chem. Soc.*, 237 (1951).
987. G. Bergson, *Acta Chem. Scand.*, **11**, 580 (1957).
988. M. Schmidt and H. Schumann, *Z. Naturforsch.*, **19b**, 74 (1964).
989. N. Dereu and R. Zingaro, *J. Organomet. Chem.*, **212**, 141 (1981).
990. N. Dereu, R. Zingaro and E. Meyers, *Cryst. Struct. Commun.*, **10**, 1359 (1981).
991. (a) D. Rainville, R. Zingaro and E. Meyers, *J. Fluorine Chem.*, **16**, 245 (1980).
(b) G. Simchen and G. Entenmann, *Ann. Chem.*, 1249 (1977).
992. A. Shafiee, F. Assadi and V. Cohen, *J. Heterocycl. Chem.*, **15**, 39 (1978).
993. C. Draguet and M. Renson, *Compt. Rend., Ser. C*, **274**, 1637 (1972).
994. W. Baringer, *Chem. Ber.*, **23**, 1003 (1890).
995. F. Luknitskii, D. Taube and B. Vovsi, *Zh. Org. Khim.*, **5**, 1844 (1969).
996. M. Dzurilla and P. Kristian, *Collect. Czech. Chem. Commun.*, **41**, 1388 (1976).
997. J. Liebscher and H. Hartmann, *Tetrahedron Letters*, 2005 (1976).
998. F. Asinger, H. Berding and H. Offermanns, *Monatsh. Chem.*, **99**, 2084 (1968).
999. G. Simchen, *Angew. Chem.*, **80**, 484 (1968).
1000. (a) G. Simchen and J. Wanzelburger, *Chem. Ber.*, **103**, 413 (1970).
(b) P. Kristian, D. Kuscik and J. Gonda, *Collect. Czech. Chem. Commun.*, **48**, 3567 (1983).
1001. O. Behagel and M. Rollmann, *J. Prakt. Chem.*, **123**, 336 (1930).
1002. J. Develotte, *Ann. Chim.*, **5**, 215 (1950).
1003. A. Kiprianov and V. Khilya, *Zh. Org. Khim.*, **3**, 1097 (1967).
1004. L. Mushkako and D. Sheiko, *Ukr. Khim. Zh.*, **30**, 384 (1964); *Chem. Abstr.*, **61**, 4359e (1964).
1005. W. Cornelius, *J. Prakt. Chem.*, **88**, 395 (1913).
1006. P. Karrer, *Chem. Ber.*, **49**, 603 (1916).
1007. P. Müller, N.-P. Buu-Hoï and R. Rips, *J. Org. Chem.*, **24**, 37 (1959).
1008. B. Podolesov and V. Jordanovska, *Croatica Chim. Acta*, **42**, 61 (1970).
1009. H. Bauer, *Chem. Ber.*, **47**, 1873 (1914).
1010. M. Olmsted, P. Craig, J. Lafferty, A. Pavloff and C. Zirkle, *J. Org. Chem.*, **26**, 1901 (1961).
1011. D. Simov and N. Khristova, *God. Sufii. Univ. Khim. Fak.*, **59**, 237 (1864–5); *Chem. Abstr.*, **67**, 108642h (1967).
1012. G. Cordella and F. Sparatore, *Farmaco Ed. Sci.*, **20**, 446 (1965).
1013. B. Podolesov and V. Jordanovska, *Croatica Chim. Acta*, **44**, 411 (1972).
1014. Y. Matsunaga and T. Tanaka, *Bull. Chem. Soc. Japan*, **48**, 1043 (1975).
1015. G. Cheeseman and G. Rishman, *Tetrahedron*, **36**, 2681 (1980).
1016. P. Karrer, *Chem. Ber.*, **51**, 190 (1918).
1017. S. Keimatsu and K. Yokota, *J. Pharm. Soc. Japan*, **52**, 961 (1932).
1018. J. Groves, S. Lindenauer, B. Haywood, J. Knol and J. Schulz, *J. Med. Chem.*, **17**, 902 (1972).
1019. W. Zankowska-Jasinska and M. Burgiel, *Zesz. Nauk Univ. Jagiellon. Pr. Chem.*, **22**, 49 (1977); *Chem. Abstr.*, **88**, 50807f (1978).
1020. I. Sadekov, G. Abakarov, A. Garnovskii and V. Minkin, *Khim. Geterotsikl. Soedin.* 707 (1982).

1021. J. Bergman and L. Engman, *J. Organomet. Chem.*, **251**, 223 (1983).
1022. L. Credali, M. Russo, L. Mortillaro, C. De Checchi, C. Valle and M. Mammi, *J. Chem. Soc. (B)*, 117 (1967).
1023. E. Weissflog, *Z. Anorg. Allgem. Chem.*, **488**, 60 (1982).
1024. M. Russo, L. Mortillaro, I. Credali and C. De Checchi, *J. Polymer Sci. (B)*, **4**, 167 (1966).
1025. L. Vanino and A. Schinner, *J. Prakt. Chem.*, **91**, 116 (1915).
1026. T. Pearson, R. Purcell and G. Saigh, *J. Chem. Soc.*, 409 (1938).
1027. H. Bridger and R. Pittman, *J. Chem. Soc.*, 1371 (1950).
1028. M. Russo, L. Mortillaro, L. Credali and C. De Checchi, *J. Polymer Sci. (A)*, **4**, 248 (1966).
1029. M. Schmidt, K. Blaettner, P. Kochendorfer and H. Ruf, *Z. Naturforsch.*, **21b**, 622 (1966).
1030. E. Weissflog, *Z. Anorg. Allgen. Chem.*, **488**, 99 (1982).
1031. F. Rice and A. Glasebrook, *J. Amer. Chem. Soc.*, **56**, 2381 (1934).
1032. L. Belchetz and E. Rideal, *J. Amer. Chem. Soc.*, **57**, 1168 (1935).
1033. T. Pearson, R. Purcell and G. Saigh, *J. Chem. Soc.*, 409 (1938).
1034. F. Williams and F. Dunbar, *J. Chem. Soc. (D)*, 459 (1968).
1035. M. Wieber and E. Kaunzinger, *J. Organomet. Chem.*, **129**, 339 (1977).
1036. P. Mazerolles, M. Lesbre and M. Joanny, *J. Organomet. Chem.*, **16**, 227 (1969).
1037. N. Derkach, T. Lyapina and G. Barashenkov, *Zh. Org. Khim.*, **15**, 788 (1979).
1038. F. Luknitskii, B. Taube and B. Vovsi, *Dokl. Akad. Nauk SSSR*, **184**, 355 (1969).
1039. (a) J.-M. Lehn and F. Riddel, *J. Chem. Soc., Chem. Commun.*, 803 (1966).
(b) C. Draguet, H. Della Fiorentina and M. Renson, *Compt. Rend. Ser. C*, **274**, 1700 (1972).
1040. (a) M. Apostolescu, *Bull. Inst. Politech. Iasi*, **20**, 9 (1974); *Chem. Abstr.*, **82**, 112049x (1975).
(b) T. Gilchrist, C. Rees and D. Vaughan, *J. Chem. Soc., Perkin Trans. 1*, 55 (1983).
1041. F. Sachs, *Ann. Chem.*, **365**, 142 (1909).
1042. M. Kaplan, R. Haddon, F. Schilling, J. Marshall and F. Bramwell, *J. Amer. Chem. Soc.*, **101**, 3306 (1979).
1043. E. Bulka and K. Ahlers, *Z. Chem.*, **3**, 349 (1963).
1044. G. Suchar and P. Kristian, *Chem. Zvesti*, **28**, 425 (1974); *Chem. Abstr.*, **81**, 120586q (1974).
1045. G. Suchar and R. Stefko, *Chem. Zvesti*, **36**, 419 (1982); *Chem. Abstr.*, **97**, 182370d (1982).
1046. H. Grant and L. Summers, *Z. Naturforsch.*, **33b**, 118 (1978).
1047. M. Schmidt and M. Wieber, *Chem. Ber.*, **94**, 1426 (1961).
1048. M. Wieber and G. Schwarzmann, *Monatsh. Chem.*, **100**, 74 (1969).
1049. F. Berry, B. Smith and C. Jones, *J. Organomet. Chem.*, **110**, 201 (1976).
1050. H. Funk and W. Papenroth, *J. Prakt. Chem.*, **8**, 256 (1959).
1051. N. Bellinger, P. Cagniant, D. Cagniant and M. Renson, *Bull. Soc. Chim. Fr.*, 2689 (1971).
1052. K. Sindelar, J. Metysova and M. Protiva, *Collect. Czech. Chem. Commun.*, **34**, 3801 (1969).
1053. J. Jilek, K. Sindelar, J. Metysova, J. Metys, J. Pomykacek and M. Protiva, *Collect. Czech. Chem. Commun.*, **35**, 3721 (1970).
1054. K. Sindelar, J. Metysova, E. Svatek and M. Protiva, *Collect. Czech. Chem. Commun.*, **34**, 2122 (1969).
1055. W. Truce and D. Emrick, *J. Amer. Chem. Soc.*, **78**, 6130 (1956).
1056. R. Taber, G. Daub, F. Hayes and D. Ott, *J. Heterocycl. Chem.*, **2**, 181 (1965).
1057. G. Zanati and M. Wolff, *J. Med. Chem.*, **15**, 368 (1972).
1058. L. Mushkalo and D. Sheiko, *Zh. Obshch. Khim.*, **33**, 157 (1963).
1059. L. Mushkalo, D. Sheiko and E. Lanovaya, *Ukr. Khim. Zh.*, **30**, 502 (1964); *Chem. Abstr.*, **61**, 5654c (1964).
1060. L. Mushkalo and D. Sheiko, *Ukr. Khim. Zh.*, **30**, 384 (1964); *Chem. Abstr.*, **61**, 4359e (1964).
1061. (a) E. Ajello, *J. Heterocycl. Chem.*, **9**, 1427 (1972).
(b) W. Günther and M. Salzman, *Ann. N. Y. Acad. Sci.*, **192**, 25 (1972).
1062. M. Haris and P. Patel, *J. Chem. Soc., Perkin Trans. 2*, 304 (1978).
1063. N. Dance and W. McWhinnie, *J. Organomet. Chem.*, **104**, 317 (1976).
1064. (a) R. Mitchell, *Tetrahedron Letters*, 1363 (1975).
(b) H. Higuchi and S. Misumi, *Tetrahedron Letters*, **23**, 5571 (1982).
1065. R. Mitchell, *Can. J. Chem.*, **54**, 238 (1976).
1066. K. Dixon and R. Mitchell, *Can. J. Chem.*, **61**, 1598 (1983).
1067. G. Bergson and A. Biezais, *Arkiv Kemi*, **22**, 475 (1964).
1068. (a) A. Biezais and G. Bergson, *Acta Chem. Scand.*, **18**, 815 (1964).
(b) L. Agenas and B. Lindgren, *Acta Chem. Scand.*, **24**, 3301 (1970).

(c) B. Trofimov, G. Musorin, G. Kalabin and S. Amosova, *Zh. Org. Khim.*, **16**, 518 (1980).
1069. J. Ashby, L. Cort, J. Elvidge and U. Eisner, *J. Chem. Soc.* (*C*), 2311 (1968).
1070. N. Zefirov and S. Rogozina, *Strukt. Svoistva Krist.*, **1**, 94 (1974); *Chem. Abstr.*, **87**, 84859u (1977).
1071. H. Gillissen, P. Schipper and H. Buck, *Rec. Trav. Chim.*, **99**, 346 (1980).
1072. E. Cuthbertson and D. MacNicol, *J. Chem. Soc., Chem. Commun.*, 498 (1974).
1073. E. Cuthbertson, J. Gall and D. MacNicol, *Tetrahedron Letters*, 3203 (1977).
1074. Y. Migalina, V. Lendel, I. Balog and V. Staninets, *Ukr. Khim. Zh.*, **47**, 1293 (1981); *Chem. Abstr.*, **96**, 104001z (1982).
1075. Y. Migalina, V. Staninets, V. Lendel, A. Koz'min and N. Zefirov, *Khim. Geterotsikl. Soedin.*, 1633 (1977).

The Chemistry of Organic
Selenium and Tellurium
Compounds Volume 1
Edited by S. Patai and Z. Rappoport
© 1986 John Wiley & Sons Ltd.

CHAPTER **14**

Tetra- and higher-valent (hypervalent) derivatives of selenium and tellurium

J. BERGMAN, L. ENGMAN and J. SIDÉN

Department of Organic Chemistry, Royal Institute of Technology, S-100 44 Stockholm, Sweden

I. INTRODUCTION . . . 517
II. PREPARATION AND PROPERTIES OF RZX_3, R_2ZX_2, AND R_3ZX . . . 519
A. General Aspects and Preparation of $ArZX_3$, Ar_2ZX_2 and Ar_3ZX . . . 519
B. Reactions of ZX_4 and RZX_3 with Carbonyl Compounds . . . 523
C. Reactions of ZX_4 and RZX_3 with Alkenes and Acetylenes . . . 526
1. Reactions with alkenes . . . 526
2. Reactions with acetylenes . . . 532
3. Regio- and stereo-chemistry of the adducts . . . 533
III. PREPARATION AND PROPERTIES OF R_4Z . . . 534
IV. TETRAVALENT ORGANO-SELENIUM AND -TELLURIUM COMPOUNDS WITH Z—O, Z—S OR Z—N BONDS . . . 536
V. TETRAHALOORGANYLTELLURATES (IV) . . . 542
VI. HEXAVALENT SELENIUM AND TELLURIUM COMPOUNDS . . . 544
A. General Aspects . . . 544
B. Selenones . . . 546
C. Tellurones . . . 550
VII. HIGHER-VALENT COMPOUNDS CONTAINING Te—Te OR Se—Se BONDS . . . 552
VIII. REFERENCES . . . 552

I. INTRODUCTION

In discussing tetra- and higher-valent Se and Te compounds it is of interest to consider the following hypothetical molecules, which, following Musher[1], are called selenane (**1**), tellurane (**2**), perselenane (**3**) and pertellurane (**4**).

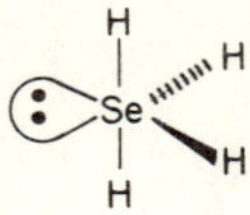

(1)

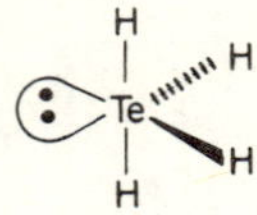

(2)

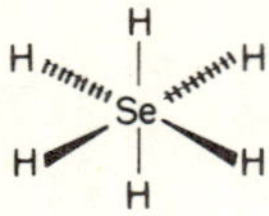

(3)

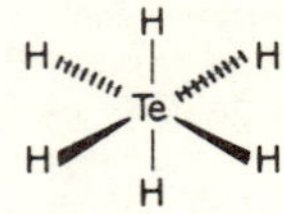

(4)

Simple derivatives of these systems do include $TeCl_4$, SeF_4, SeF_6, TeF_6 and $HOTeF_5$. It should also be noted that the indicated nomenclature is different from that recommended[2,3], but we feel that Musher's system is more suitable for analysis of all the delicate stereochemical problems involved and the system has also been used in a recent comprehensive review[4]. Likewise Martin[5] has recently used this nomenclature when describing a non-dissociative permutational isomerization of a pertellurane. Reich[6] has advocated the name 'selenurane' instead of selenane because the word selenane has been used as a name for selenacyclohexane (**5**). We feel, however, that selenacyclohexane is an excellent name for **5** and 'selenane' a confusing one for that molecule. Hence the name selenane should be reserved for **1**.

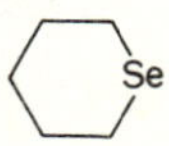

(5)

To illustrate Musher's system, consider diphenyldibromotellurane (diphenyltellurium dibromide, by conventional nomenclature) which has been determined[7] by X-ray analysis to be of structure **6a**, i.e. with the halogen atoms in apical positions. The five (still unknown) topomers **6b–f** are also expected to be stable and spectroscopically observable even though it might not be possible to isolate them, due to the likely existence of intra- or inter-molecular equilibration (topomerization).

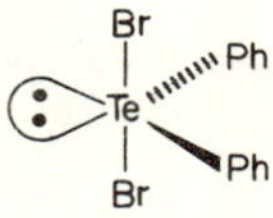

(6a)

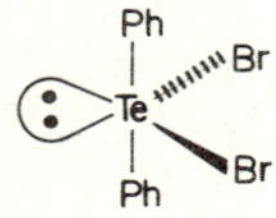

(6b)

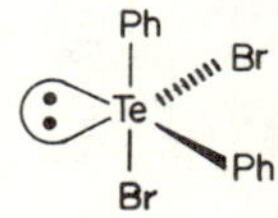

(6c)

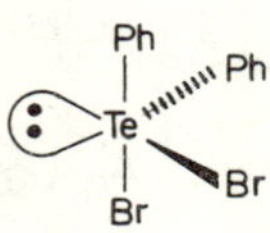

(6d)

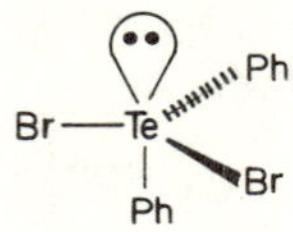

(6e)

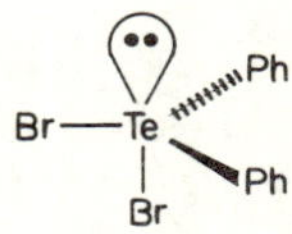

(6f)

In this connection it is also of interest to note that, in agreement with VSEPR theory[8], all telluranes studied so far[9,10] have their electron pairs in equatorial positions, whereas tetravalent organotellurium compounds of the general structure **7** invariably have a pyramidal arrangement around Te. However, with a suitable choice of substituents or ring constructions other isomers might become favourable (cf. Ref. 11).

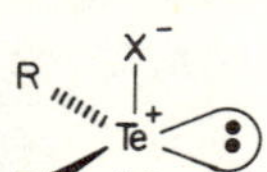

(7)

X = O, NR, CR_2

The classical problem[12] with α- and β-forms of Me_2TeI_2 has been resolved by Einstein[13,14], who found that while the α-form is *e,e*-dimethyl-*a,a*-diiodotellurane (**8**) the β-form has in fact the more complex structure **9**.

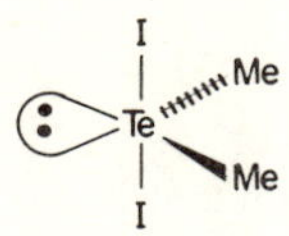

(**8**)

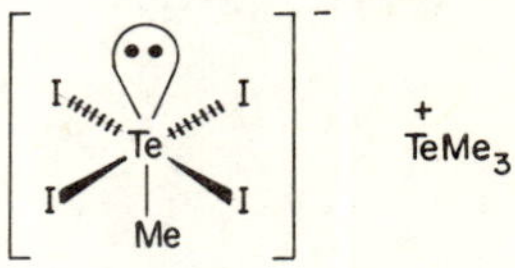

(**9**)

In this connection the report[15] by Ziolo and Günther about the simultaneous formation of at least two crystalline forms of 1,1-diiodo-3,4-benzo-1-telluracyclopentane (**10**) is of interest. It was concluded that the two forms are polymorphs.

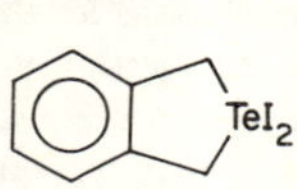

(**10**)

Although all diorganyldihaloselenanes and diorganyldihalotelluranes studied so far have, in the solid state, the halogen atoms in apical positions the situation might be different in solution. In fact there are indications (dipole moment measurements)[16] that diphenyldichloroselenane, which has the chlorine atoms in apical positions in the solid state (i.e. **11a**), might in solution (e.g. benzene) prefer other arrangement(s) around the central atom (e.g. **11b**). Such topomers, with one or both chlorine atoms in the equatorial position, should be stabilized by solvation as indicated.

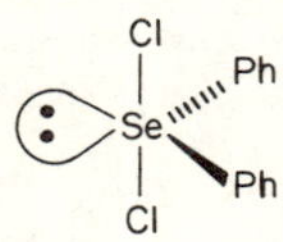

(**11a**)

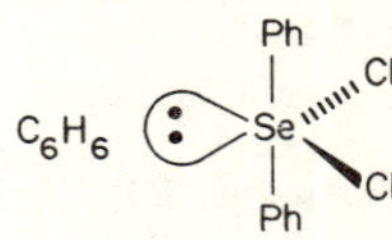

(**11b**)

In solution the situation is also complicated by equilibria of the following type (Z = Se, Te; X = halogen):

$$R_2ZX^+ + X^- \rightleftharpoons R_2ZX_2 \rightleftharpoons R_2Z\ldots X_2 \rightleftharpoons R_2Z + X_2 \qquad (1)$$

Problems involved in these equilibria have been studied with various spectroscopic techniques[17] including Se- and Te-NMR. Particularly suitable for such studies is the analysis of magnetic circular dichroism (MCD) data[17a].

II. PREPARATION AND PROPERTIES OF RZX_3, R_2ZX_2 AND R_3ZX

A. General Aspects and Preparation of $ArZX_3$, Ar_2ZX_2 and Ar_3ZX

Many tetravalent organotellurium and organoselenium compounds are readily available via direct methods as indicated by equations (2)–(6). ZCl_4 (or the related combinations TeO_2/HOAc/LiCl and SeO_2/HCl, respectively) are the reagents of choice. If the substrates are insufficiently reactive, addition of $AlCl_3$ or other suitable Lewis acids might be beneficial[18]. Benzene, for instance, will not react with $TeCl_4$ even at reflux temperature but addition of $AlCl_3$ (1 equivalent or more) will cause a vigorous reaction[19] that might even yield Ph_3TeCl. Benzene and a $SeCl_4$–$AlCl_3$ reagent will not similarly[20,32,67–69] yield Ph_3SeCl although Ph_2SeCl_2 can be arylated to Ph_3SeCl by benzene in the presence of $AlCl_3$.

$$\mathrm{AnH} + \mathrm{TeCl_4} \xrightarrow[\Delta]{\mathrm{CHCl_3}} \mathrm{AnTeCl_3} \xrightarrow[\Delta]{\mathrm{AnH}} \mathrm{An_2TeCl_2} \tag{2}$$

$$\mathrm{An} = p\text{-}\mathrm{MeOC_6H_4}$$

$$\mathrm{AnH} + \mathrm{SeCl_4} \xrightarrow[\mathrm{r.t.}]{26\,\mathrm{h}} \mathrm{An_2SeCl_2} \tag{3}$$

$$\mathrm{PhH} \xrightarrow[\mathrm{AlCl_3}]{\mathrm{TeCl_4}} \mathrm{Ph_3TeCl} \tag{4}$$

As indicated above, $TeCl_4$ is acting as a moderate electrophile (a pseudohalogen with a reactivity somewhere between I_2 and Br_2) with an excellent selectivity for telluration and only highly reactive substrates such as undeactivated indoles[21] will be chlorinated. Actually, more than 150 aromatic substrates have been successfully converted into $ArTeCl_3$ or Ar_2TeCl_2 in high yields. For tabulations see Refs. 3 and 4.

The electrophiles ZCl_4 and $ArZCl_3$ do show a strong preference (equations 2 and 3) for attack in the *para* position and the *ortho/para* ratio is unusually low. Thus An_2TeCl_2 is readily isolated as beautiful crystals in 90% yield[22] by reacting $TeCl_4$ with anisole at reflux temperature.

$ArTeX_3$ Ar_2TeX_2 and Ar_3TeX are generally nice crystalline compounds that are easy to store. The aryltellurium trihalides are sensitive to moisture and are hydrolysed[23] to $(ArTe(O))_2O$. The diaryltellurium dihalides are usually quite insensitive to moisture and An_2TeCl_2, for example, can be nicely crystallized[22] from ethanol. Treatment of diaryltellurium dihalides with NaOH will give diaryl telluroxides, Ar_2TeO. Similar exchange reactions[24–26] have been performed with, for example, AcO^-, SCN^- and MeO^- to yield the expected substitution products. Diaryl selenium dihalides behave similarly[27].

Triaryltellurium halides are often distinctly ionic[28,29] (hence they should be written $Ar_3Te^+X^-$) and can often be recrystallized from water or ethanol/water. Needless to say, exchange reactions are easy to perform. Thus addition of a solution of KI in water to a solution of Ar_3TeCl will cause the precipitation of Ar_3TeI. The low solubility of this iodide can be utilized for isolation, e.g. if the compound is prepared according to equation (5), or the variant (6). As indicated, reaction (6) is suitable for the preparation of unsymmetrical compounds.

Preparations according to equations (5) and (6) are relatively complicated and the yields are often unsatisfactory[3].

$$\mathrm{ArMgX} + \mathrm{TeCl_4} \rightarrow \mathrm{Ar_3TeX} \tag{5}$$

$$\mathrm{Ar_2TeX_2} + \mathrm{RMgX} \rightarrow \mathrm{Ar_2RTeX} \tag{6}$$

Ziolo and Titus[30] and McWhinnie and coworkers[28,29] have made extensive studies of the structural and solution chemistry of this class of compounds. Solution conductivities and molecular weights show triaryltellurium halides and pseudohalides to be associated non-electrolytes in weakly polar to non-polar solvents such as chloroform and benzene and weak to strong electrolytes in polar solvents.

The triphenyltellurium pseudohalides, Ph_3TeN_3, Ph_3TeNCO, Ph_3TeNCS and $Ph_3TeNCSe$, do all show much more complex IR absorptions in the 1900–2000 cm^{-1} region than one *a priori* should expect. These facts are explained in terms of their oligomeric structures[30].

Ziolo and coworkers[31] have also recently found that Me_2TeCl_2 and Ph_2TeCl_2 undergo direct facile (25 °C in H_2O/CH_3OH) phenylation reactions with $NaBPh_4$ to produce $Me_2TePh^+\ BPh_4^-$ and $Ph_3Te^+\ BPh_4^-$ respectively. Tellurium tetrachloride in benzene similarly undergoes triarylation with excess $NaBPh_4$ to yield $Ph_3Te^+\ BPh_4^-$.

Triarylselenium halides have also ionic character and their chemistry is similar to that of Ar_3TeX, although they are more prone to undergo decomposition in solution according to equation (7).

$$R_3ZX \rightarrow R_2Z + RX \tag{7}$$

According to Ogawa[32] $Ph_3Se^+\ Br^-$ is not formed when PhMgBr is allowed to react with Ph_2SeBr_2. Instead Ph_2Se and biphenyl are formed, possibly via the unstable tetraphenylselenane (equation 8). The non-formation of $Ph_3Se^+\ Br^-$ thus provides still another difference (cf. equation 6) between organoselenium and organotellurium chemistry. On the other hand Ogawa[32] could prepare $Ph_3Se^+\ Br^-$ in fair yields using equations (9) or (10).

$$Ph_2SeBr_2 + PhMgBr \rightarrow [Ph_4Se] \rightarrow PhSePh + PhPh \tag{8}$$

$$PhMgBr + SeOCl_2 \rightarrow Ph_3Se^+\ Br^- \tag{9}$$

$$Ph_2SeO + PhMgBr \rightarrow Ph_3Se^+\ Br^- \tag{10}$$

In contrast to the Te series, very few[33] aromatic compounds have been converted into $ArSeCl_3$ or Ar_2SeCl_2 with $SeCl_4$ or $SeOCl_2$ (cf. equations 2 and 3), which are much more reactive than $TeCl_4$. Here chlorination is a much more serious problem. Sometimes dehalogenations occur (cf. equation 1) and selenides are formed, which might react further. Thus 1,2-dimethoxybenzene, when treated[34] with H_2SeO_3 at 130–150 °C, will give **12**.

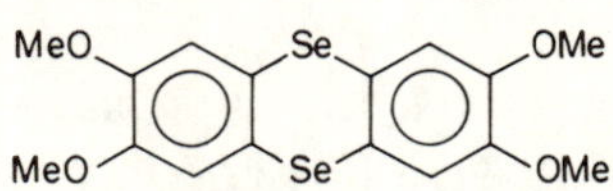

(**12**)

Electrophilic cyclization reactions of suitable diaryl selenium dichlorides do occur readily[35], as exemplified by equation (11).

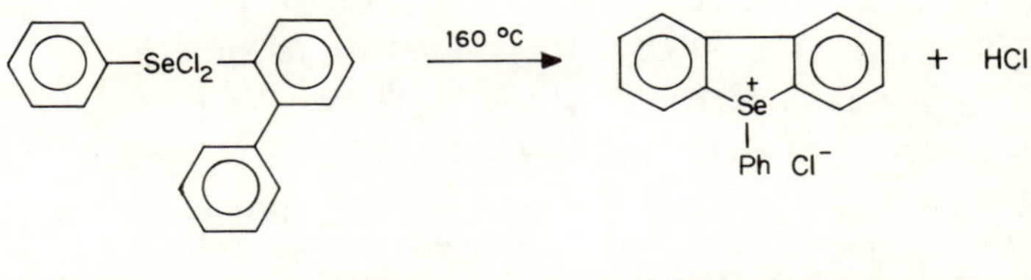

(11)

There have been no similar cyclizations reported in the Te series.

$$RSeX_3 \rightarrow RSeX + X_2 \tag{12a}$$

$$RSeX_2R \rightarrow RSeR + X_2 \tag{12b}$$

$$RSeR \xrightarrow{X_2} RSeX_3 \tag{12c}$$

$$RSeSeR \xrightarrow[\text{or } SO_2Cl_2]{X_2} RSeX + RSeX_3 \tag{12d}$$

$$RSeCN \xrightarrow{X_2} RSeX_3 \tag{12e}$$

Generally speaking, diorganylselenium dihalides and monoorganylselenium trihalides are much less stable than the corresponding Te isologues. The selenium trihalides are generally prepared from[36] diorganyl diselenides or selenocyanates (i.e. by indirect methods) under carefully controlled conditions as indicated by equations (12c–12e). Actually very few organyl selenium trihalides have been isolated using direct procedures and it is tempting to suggest that the often cited stable product **13**[37], obtained by treating methyl salicylate with $SeCl_4$, might in fact be the isomer **14**, which should be stabilized by Se—OH coordination. Certainly this problem should merit a detailed study. In this connection the reported[38] formation of **15** from the reaction of *o*-cresol with SeO_2 in the presence of HCl is of interest, as well as the reported[39] formation of $(4\text{-}HOC_6H_4)_3Se^+ Cl^-$ together with an unidentified isomer from the reaction of phenol with $SeOCl_2$.

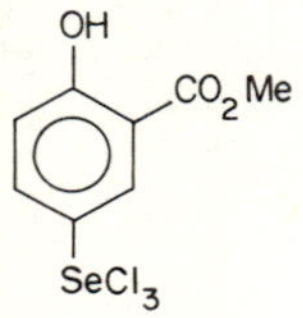

(13)

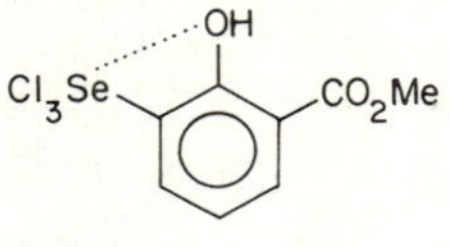

(14)

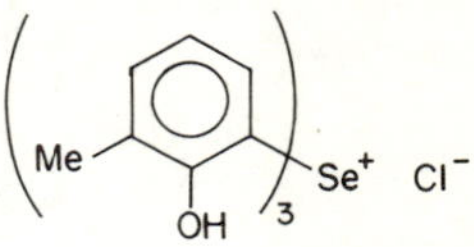

(15)

In addition to the direct methods for the preparation of tetravalent organo-selenium and -tellurium compounds discussed above, other methods, involving, for example, oxidative addition to selenides and tellurides[24,25,40] as well as exchange reactions[24,41–52] of organometallic and organoelemental compounds, are frequently used. The last-mentioned technique is particularly useful for the preparation of unsymmetrical compounds of the type RZX_2R^1.

$$R_2Te + (SCN)_2 \rightarrow R_2Te(SCN)_2 \quad (13)$$

$$R_2TeI_2 + HOSO_2Cl \rightarrow R_2Te(OSO_2Cl)_2 + 2\,HI \quad (14)$$

$$RC_6H_4Si(CH_3)_3 + TeCl_4 \xrightarrow[\text{reflux}]{\text{toluene}} RC_6H_4TeCl_3 \quad (15)$$

$$ArHgCl + Ar^1TeCl_3 \rightarrow ArTeCl_2Ar^1 + HgCl_2 \quad (16)$$

$$AnTeCl_3 + Ph_6Pb_2 \rightarrow AnTeCl_2Ph \quad (17)$$

$$4\text{-}EtOC_6H_4TeCl_3 + 1,3\text{-}(HO)_2C_6H_4 \rightarrow 4\text{-}EtOC_6H_4TeCl_2C_6H_4(OH)_2\text{-}1,3 \quad (18)$$

$$4\text{-}EtOC_6H_4TeCl_3 + AnH \rightarrow An_2TeCl_2 + PhOEt \quad (19)$$

$$ArHgX + PhH \rightarrow ArH + PhHgX \quad (20)$$

Direct methods for the preparation of unsymmetrical diaryltellurium dihalides can sometimes be used, particularly when the aromatic reactant is readily susceptible to electrophilic aromatic substitution (equation 18)[51]. In other cases (equation 19) exchange reactions[18] might be troublesome. Nothing is known about the mechanism of such exchange reactions in the Te series (they are unknown in the Se series) but related exchange reactions in the Hg and Tl series have been studied[53–57] using kinetic methods (equation 20).

$TeBr_4$ and $SeBr_4$ are less reactive than the corresponding chlorine compounds and should be interesting alternatives when sensitive (notably π-excessive heterocyclic) reactants are involved. TeF_4 and SeF_4 are certainly interesting reagents for introduction of Te and Se into organic substrates, albeit that very limited information is available in the literature. Actually no compounds of the types $ArZF_3$ or Ar_2ZF_2 have been prepared using these reagents. Compounds of the type R_2SeF_2 were prepared for the first time by Wynne and Puckett[58,59] by treating a selenide with AgF_2 (equation 21).

$$ArSeAr \xrightarrow{AgF_2} Ar_2SeF_2 \tag{21}$$

By this reaction also, aliphatic derivatives such as dimethylselenium difluoride and diethylselenium difluoride can be obtained. Later other fluorinating agents[60–66] such as SF_4, XeF_2, F_2 and ClF have been used as examplified by equations (22)–(25).

Other methods for the preparation[28,42] of organic Te and Se compounds containing fluorine depend on exchange reactions (equations 26 and 27).

$$Ph_2Te + Ph_2XF_2(X = S,Se) \rightarrow Ph_2TeF_2 + Ph_2X \tag{22}$$

$$(CF_3)_2Te + ClF \xrightarrow{-78\,^{\circ}C} (CF_3)_2TeClF \tag{23a}$$

$$(CF_3)_2TeClF \rightarrow (CF_3)_2TeF_2 + (CF_3)_2TeCl_2 \tag{23b}$$

$$(CF_3)_2TeX_2 + X_2 \rightarrow CF_3TeX_3 + CF_3X \tag{24}$$

$$ArZAr \xrightarrow{SF_4} ArZF_2Ar \tag{25}$$

$$Ar_2TeO \xrightarrow{HF} Ar_2TeF_2 \tag{26}$$

$$Ar_3TeCl \xrightarrow[2.\ HF]{1.\ Ag_2O} Ar_3TeF \tag{27}$$

B. Reactions of ZX_4 and RZX_3 with Carbonyl Compounds

$TeCl_4$ and $SeCl_4$ (and related reagents, e.g. $SeOCl_2$) do readily attack a wide variety of aliphatic and aromatic ketones, as indicated in equations (28)–(34). Early studies in the field were made by Morgan and Elvins[70], Michaelis and Kunckell[71,72] as well as others[73]. Thus, condensation of acetone and of methyl aryl ketones with ZX_4 readily occur in ether or chloroform yielding dichlorides. The products are often unstable and, particularly in the presence of HCl or HBr, α-haloketones as well as other products are formed. Sometimes quite unexpected products are formed (equation 31)[74]. The selenapyrylium salt formed in this reaction undergoes dimerization in the presence of base to give **16**; it is not yet clear which of the two possible isomers (**16a** or **16b**) is formed.

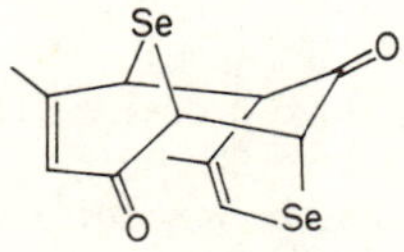

(16a)

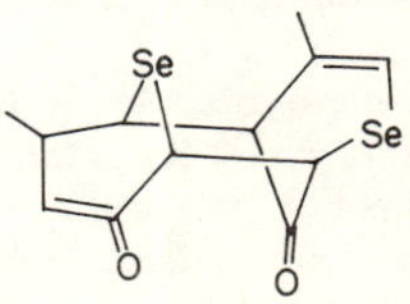

(16b)

Silylated carbonyl compounds do react readily with ZX_4 (equations 32 and 33)[75] and might offer advantages when sensitive materials are involved. The reaction according to equation (32) quickly yielded $(PhCOCH_2)_2TeCl_2$ as colourless silky needles. However, so far, very few interactions between electrophilic Te and Se species and silylated reactants have been studied. Much more research in this area is desirable. It should be added that the activation of the ketone indicated in equation (33) is not always necessary. In many cases $RTeX_3$ will give reactions with unactivated ketones[76].

$$PhCOMe + ZCl_4 \rightarrow PhCOCH_2ZCl_3 \xrightarrow{PhCOMe} (PhCOCH_2)_2ZCl_2 \quad (28)$$

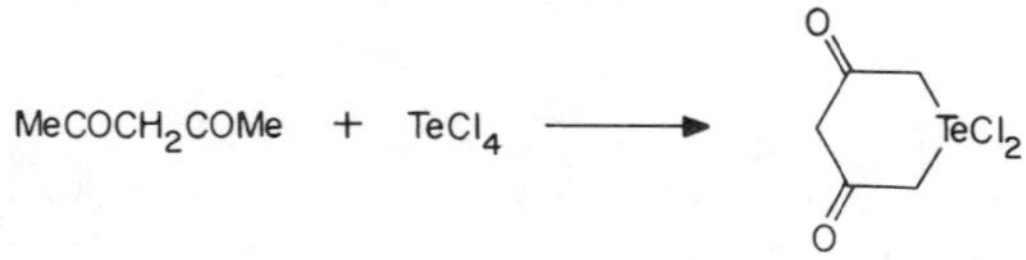
(29)

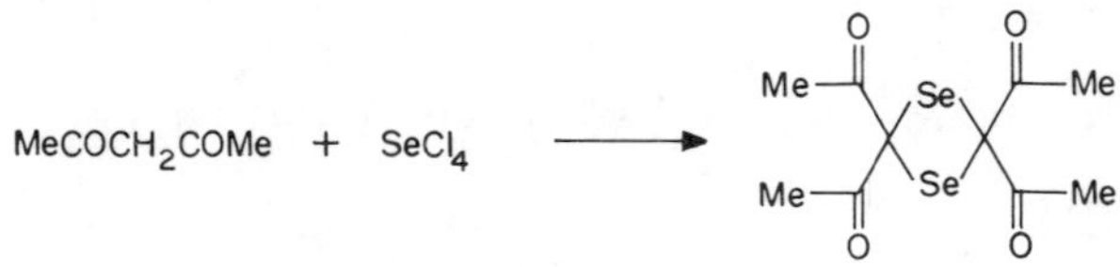
(30)

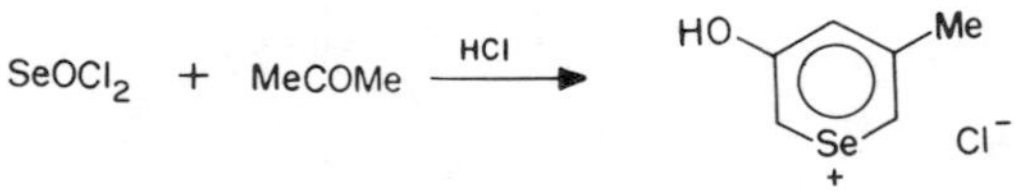
(31)

$$PhC({=}CH_2)OSiMe_3 \xrightarrow{TeCl_4} (PhCOCH_2)_2TeCl_2 + ClSiMe_3 \quad (32)$$

$$PhC({=}CH_2)OSiMe_3 + p\text{-}EtOC_6H_4TeCl_3 \rightarrow p\text{-}EtOC_6H_4TeCl_2CH_2COPh + ClSiMe_3 \quad (33)$$

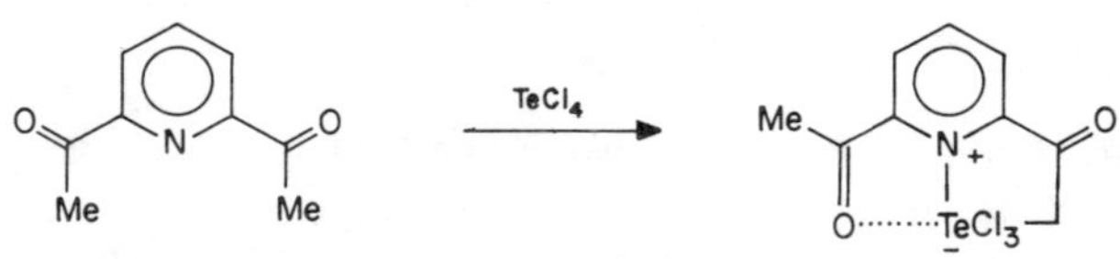
(34)

Morgan and Elvins[70] studied the interaction of $TeCl_4$ with several unsymmetrical ketones and claimed, for example, the formation of $(PrCOCH_2)_2TeCl_2$ from 2-pentanone. This regiochemistry is, however, erroneous and O'Brien[77] has recently shown by NMR studies that the correct structure is $(MeCOCHEt)_2TeCl_2$. O'Brien also isolated the compound **17** from the condensation of acetone with $TeCl_4$. The structure of **17** was verified by an X-ray analysis (the C—Te distance is 2.15 Å and the O—Te distance is 2.44 Å) and it was concluded that acetone did first condense to 4-methyl-3-pentene-2-one(mesityl oxide) (**18**), which is subsequently attacked by $TeCl_4$ at one of the γ-methyl groups.

Bis(benzoylmethyl)tellurium dichloride, $(PhCOCH_2)_2TeCl_2$, is sensitive to light and has been considered[78] for imaging purposes in photographic processes. The products from equations (29) and (34)[79] are also of interest in this respect.

The solution photochemistry of $(PhCOCH_2)_2TeCl_2$ has been studied in detail by Marsh and coworkers[80] who found that the compound undergoes facile β-cleavage of the C—Te bonds and an intramolecular Norrish Type II photoelimination reaction as the sole photochemical processes. Te and acetophenone are the two major photoproducts formed during photolysis of $(PhCOCH_2)_2TeCl_2$ at 313 nm in degassed hydrogen atom-donating solvents, while photolysis in degassed solutions of benzene or acetonitrile yields Te and $PhCOCH_2Cl$ as principal photoproducts. In addition smaller amounts of 1,2-dibenzoylethane and acetophenone are produced.

In view of the interesting properties found for $(PhCOCH_2)_2TeCl_2$ and related compounds it is surprising to find that $(PhCOCH_2)_2Te$ is not reported in the literature. Attempts[70] to prepare this divalent Te compound by sulphite reduction failed, due to C—Te bond cleavage. On the other hand Ajello found[81] that the Se isologue,

$(PhCOCH_2)_2SeCl_2$, can be successfully reduced by zinc powder in CS_2, a method introduced by Kunckell and Zimmermann[72]. Pyrosulphite reductions are also possible[82]. Ajello also reported that $(PhCOCH_2)_2Se$ with hydrazine yielded the seven-membered compound **19**.

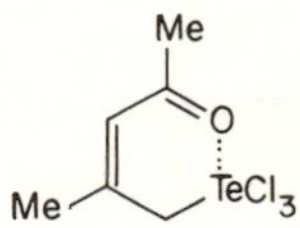

(17)

$Me_2C{=}CHCOMe$

(18)

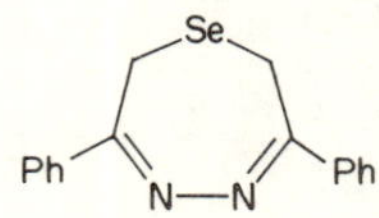

(19)

Very little seems to be known about reactions between aldehydes and tetravalent Se and Te reagents, and defined Se- or Te-containing products have not yet been published in spite of extensive studies[83] of the reactions of SeO_2/HCl with aldehydes in the area of analytical chemistry. However, very recently[218] the formation of 2,6-dialkoxy-3,5-dialkyl-1,4-oxaselanacyclohexane 1,1-dichlorides by reaction of aliphatic aldehydes with SeO_2 in alcohols containing aqueous HCl has been reported.

An interesting distinction between $SeCl_4$ and $TeCl_4$ in their reactions with acetylacetone has already been observed by Morgan and coworkers[84,85]. Thus $SeCl_4$ will attack the CH_2 group to yield what was finally[87] identified as the diselenetane shown in equation (30), while $TeCl_4$ will attack the Me groups yielding the cyclic tellurium dichloride shown in equation (29). The structures given in the original papers for these two products were erroneous and were later corrected (e.g. the diselenetane by NMR studies)[86] and finally verified by X-ray crystallography[87–94].

1,5-Substituted acetylacetones will also give products of type **20**. In fact only when the carbon atoms 1 and 5 do not bear any hydrogen atoms will $TeCl_4$ attack the CH_2 group. Thus, dibenzoylmethane yields the unstable compound **21**.

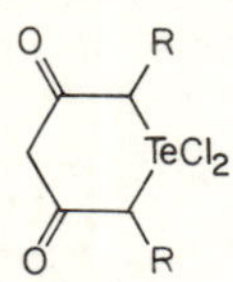

(20)

$PhCOCH(TeCl_3)COPh$

(21)

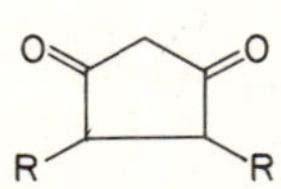

(22)

$R^2CH{=}CHCOCH_2COCHR^1$

(23)

Finally, it might be mentioned that attempts have been made[95] to explore the unique regiochemistry in the product **20b** (from the reactions of derivatives of acetylacetone with $TeCl_4$) for synthetic purposes. However, attempted coupling reactions (e.g. with degassed Raney nickel) gave a disappointingly low yield of the desired cyclopentan-1,3-diones **22** and the major isolated products had the structure **23**. Further experiments in this area, e.g. with silylated reactants, would be of interest.

The outcome of the reactions between Se reagents and 1,3-diketones is strongly dependent on the specific structures as underlined by equations (35) and (36). In none of these reactions were diselenetanes observed[96–98]. The formation of 1-chlorocyclohexan-

1,3-dione probably proceeds via the selenium dichloride indicated in equation (35). This chlorine insertion reaction is closely related to the transformations ($RSeR + Br_2 \rightarrow RBr$ and $RSe(O)R + HBr \rightarrow RBr$) reported by Krief and coworkers[99].

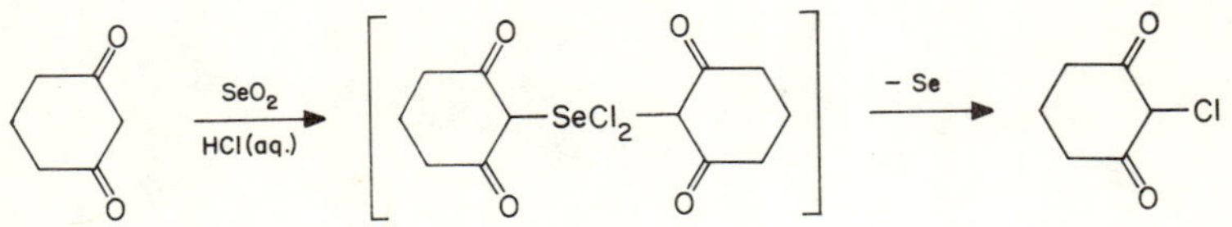

(35)

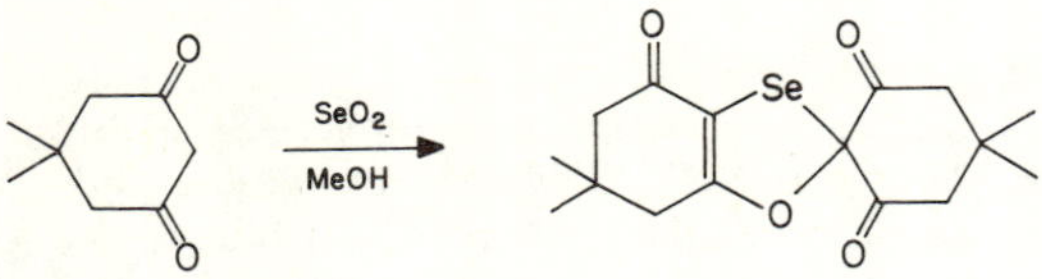

(36)

Acetic anhydride and $TeCl_4$ in a 6:1 molar ratio in chloroform yield bis(carboxymethyl)tellurium dichloride and $CH_2(TeCl_3)_2$ (equation 37), as discovered by Morgan and coworkers[100,101]. The primary condensation products are most likely derivatives of acetic anhydride. Hydrolysis to the acid derivative ($HOOCCH_2TeCl_3$ has also been isolated in some experiments) occurs during isolation of the products.

$$MeCOOCOMe \xrightarrow[2.\ H_2O]{1.\ TeCl_4} Cl_2Te(CH_2COOH)_2 + CH_2(TeCl_3)_2 \qquad (37)$$

The ability of tellurium tetrachloride to react with homologues of acetic anhydride diminishes rapidly with increasing size of the hydrocarbon chain. Whereas trichlorotelluropropionic acid has been obtained in the reaction with propionic anhydride and identified as the ditelluride, the reaction with butyric and isovaleric anhydrides causes reduction of tellurium tetrachloride and formation of unidentified products[101].

Reduction of $CH_2(TeCl_3)_2$ yields 1,2,4,5-tetratelluracyclohexane (**24**)[100,102]. The structure of **24** is supported[102] by mass spectrometry, NMR data and ^{125}Te Mossbauer spectroscopy. The interesting compound $CH_2(TeCl_3)_2$ has recently been prepared in better yields by Wudl[103,104], who has also used this compound for the preparation of some interesting conducting polymers $(CH_2Te_2)_n$. For further details see the chapter on Se/Te-containing polymers.

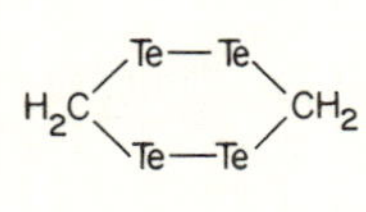

(**24**)

C. Reactions of ZX_4 and RZX_3 with Alkenes and Acetylenes

1. Reactions with alkenes

Selenium and tellurium tri- and tetra-halides undergo addition reactions with alkenes to form three principally different addition compounds **25**–**27** as shown in equation (38), The regio- and stereo-chemistry are not indicated in this simplified scheme but will be discussed separately.

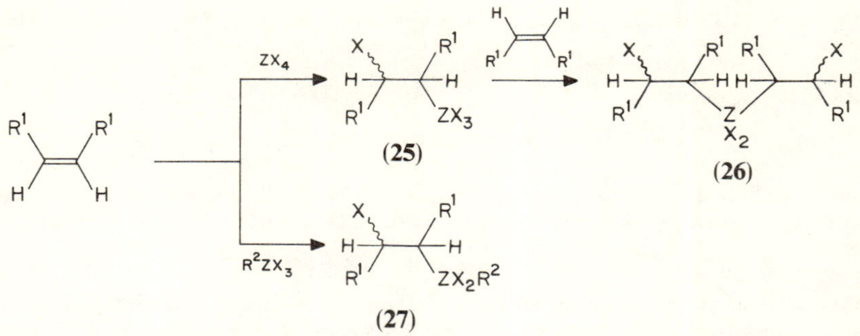

(38)

In the Se series, no examples of compound **25** seem to be isolable from selenium tetrahalides and olefins. However, these compounds are undoubtedly formed as reactive intermediates during the preparation of compounds **26**. The reaction of $SeBr_4$ with benzalacetone[105] affords products where the unstable 1:1 adduct **28** has undergone a cyclization reaction (equation 39). The formation of benzoselenopheno[2,3-*b*] benzoselenophene (**29**) from $SeOCl_2$ and 1,1-diphenylethylene provides a more complicated example of a similar series of reactions[106–108].

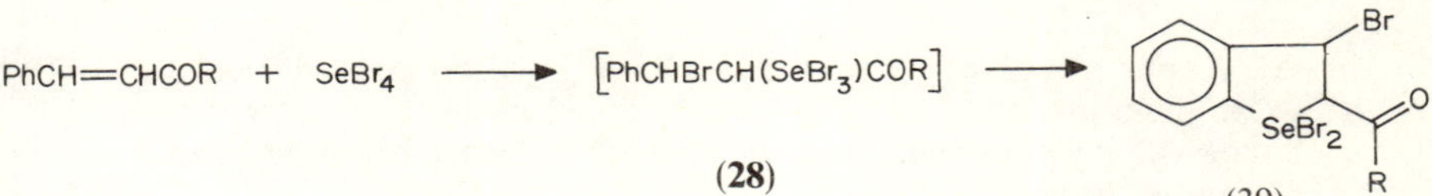

(39)

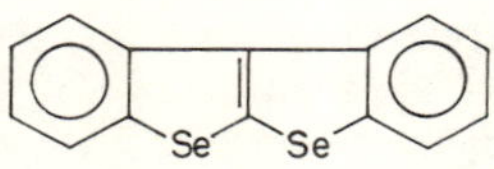

(29)

On the other hand, compounds of the type **25** are quite stable in the Te series. Tellurium tetrachloride and cyclohexene yield the crystalline adduct **30**[109–111]. Similar addition compounds are obtained from ethene[112], propene[113], *E*- and *Z*-2-butene, 1-decene, cyclopentene, cyclooctene[113,114] and 1,5-cyclooctadiene[115]. Tellurium tetrabromide and ethene form[116] the 1:1 adduct $BrCH_2CH_2TeBr_3$ (**31**).

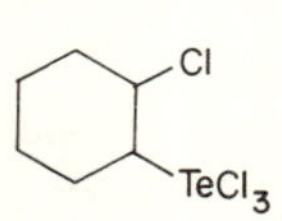

(30)

Compounds **26** in the Se series are readily available via several methods.

Diselenium dichloride, Se_2Cl_2, and ethene produce a bis(β-chloroalkyl)selenium dichloride **32** as shown in equation (40)[117,118]. This process does involve several steps, including the formation of $SeCl_2$ (equation 41), which is probably the active selenating agent. Diselenium dichloride is then functioning as a chlorinating agent to give the observed product (equation 42). The described procedure is useful only for a limited number of olefins and often results in formation of bis(β-chloroalkyl)selenides[119,120], or mixtures with the corresponding selenium dichloride[118].

$$2\,CH_2{=}CH_2 + 2\,Se_2Cl_2 \rightarrow \underset{\mathbf{(32)}}{(ClCH_2CH_2)_2SeCl_2} + 3\,Se \qquad (40)$$

$$Se_2Cl_2 \rightarrow SeCl_2 + Se \qquad (41)$$

$$SeCl_2 + 2\,CH_2{=}CH_2 \rightarrow (ClCH_2CH_2)_2Se \xrightarrow[-2\,Se]{Se_2Cl_2} \mathbf{32} \qquad (42)$$

Selenium oxychloride behaves in a similar way to diselenium dichloride towards olefins. Selenium dioxide is a by-product in this process (equation 43)[121].

$$2\,CH_2{=}CH_2 + 2\,SeOCl_2 \rightarrow \mathbf{32} + SeO_2 \qquad (43)$$

The most general method for the synthesis of compounds **26** uses selenium dioxide in aqueous hydrochloric or hydrobromic acid, respectively, for the generation of $SeCl_4$ and $SeBr_4$[38,122–124]. The synthesis of bis(2-chlorocyclopentyl)selenium dichloride (**33**), exemplifies this process (equation 44).

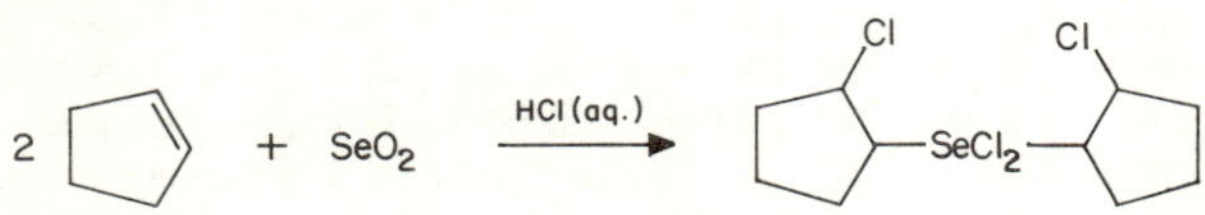

(44)

(**33**)

Ethene, propene, 1-hexene, styrene, allyl chloride, allyl bromide, methallyl chloride, methallyl bromide, vinyl chloride, vinyl acetate, cyclohexene[122–124] as well as methylenecyclobutane, allylbenzene and allyl phenyl ether[125,126] have been treated analogously with $SeCl_4$ and/or $SeBr_4$.

The reaction of $SeCl_4$ and $SeBr_4$, respectively, with olefins under non-aqueous conditions represents another method of obtaining compounds **26**. Thus, styrene and $SeCl_4$ afford compound **34** in high yield (equation 45)[127]. Ethene, propene[128,129], *E*- and *Z*-2-butene[130], 1-hexene, 1-pentene, cyclohexene[127] and vinyl chloride[119] yield similar products with $SeCl_4$. $SeBr_4$ gives addition compounds with ethene[128] and allyl bromide[119].

$$2\,PhCH{=}CH_2 + SeCl_4 \rightarrow \underset{\mathbf{(34)}}{(PhCHClCH_2)_2SeCl_2} \qquad (45)$$

One problem with these reactions is the halogenating properties of selenium tetrahalides. Uemura[131,132] has observed chlorination of cyclohexene, cyclooctene and norbornene with $SeCl_4$ and Migalina[133] has similarly observed halogenation of a series of styrenes using $SeCl_4$ and $SeBr_4$.

Compounds **26** are not so well represented in the Te series, but a few examples are known. Tellurium tetrachloride can be combined with ethene[112,123], propene[112,134], 1-butene[135] and cyclohexene[111,123] to give normal 1:2-addition compounds. A 1,4-addition is observed for butadiene[112] which yields compound **35**. However, under slightly different reaction conditions[136] (refluxing acetonitrile instead of CCl_4 at 30–60 °C), a 2,5-dihydrotellurophene **36** is obtained as the only product. The latter is probably a cyclization product of the former (loss of 1,4-dichloro-2-butene).

$$(ClCH_2CH{=}CHCH_2)_2TeCl_2$$

(**35**)

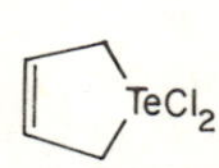

(**36**)

The unsaturated acid **37** undergoes internal lactonization during the reaction with $TeCl_4$ to give a symmetrical bis-adduct **38**[110].

$CH_2{=}CHCH_2CPh_2COOH$

(37)

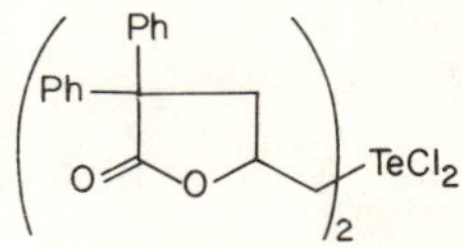

(38)

Similar cyclizations are observed when γ- and δ-hydroxy olefins are treated with tellurium dioxide, TeO_2, in acetic acid containing lithium chloride (this system might be viewed as a source of $TeCl_4$)[137]. The oxidative cyclization of 2-allylphenol is shown in equation (46). These adducts are nice crystalline compounds. The NMR spectra do indicate coordination between O and Te.

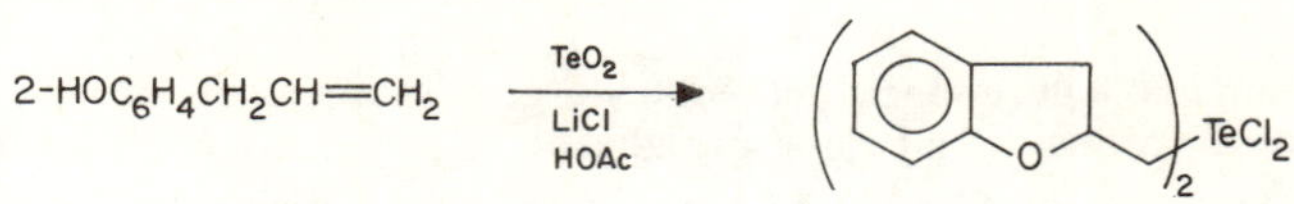

(46)

Tellurium tetrabromide forms the symmetrical adduct $(BrCH_2CH_2)_2TeBr_2$ (**39**) with ethene[116].

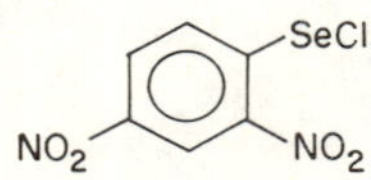

(40)

$MeSeCl_3$

(41)

The few known compounds **27** in the selenium series were all prepared by Garratt and coworkers[138–140]. They added 2,4-dinitrophenylselenium trichloride (**40**) and β-methylselenium trichloride (**41**), respectively, to simple olefins like propene, 2-methylpropene, *E*- and *Z*-1-phenyl-1-propene, *E*- and *Z*-2-butene, 3-methyl-1-butene and 3,3-dimethyl-1-butene. Mixtures of compounds resulting from Markownikoff and *anti*-Markownikoff addition were frequently obtained, as shown for the addition of **40** to *Z*-1-phenyl-1-propene (equation 47).

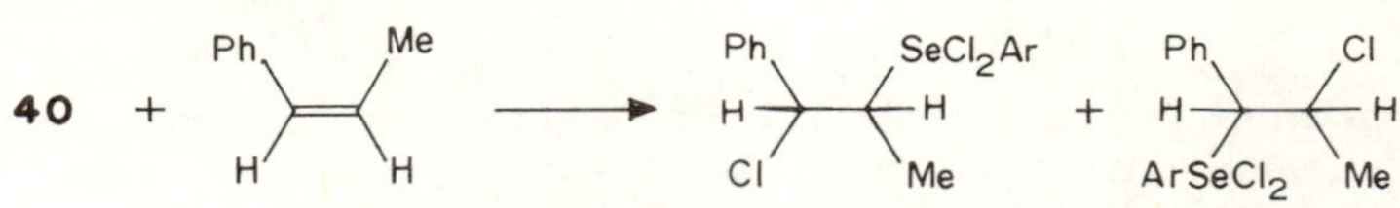

(47)

Ar = 2,4-dinitrophenyl

Various aryltellurium trichlorides were added to olefins to obtain compounds **27** in the Te series. Phenyl-, 4-phenoxyphenyl-, 4-ethoxyphenyl- and 1- and 2-naphthyl-tellurium trichloride, respectively, yielded compounds **42** with cyclohexene[109,110].

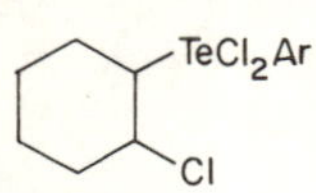

(42)

2-Naphthyltellurium trichloride was similarly allowed to react with propene[141], *E*- and *Z*-2-butene, 1-decene, cyclopentene and cyclohexene[114]. Lactonization was again observed when the unsaturated acid **37** was treated with a series of aryltellurium trichlorides (equation 48)[109,110].

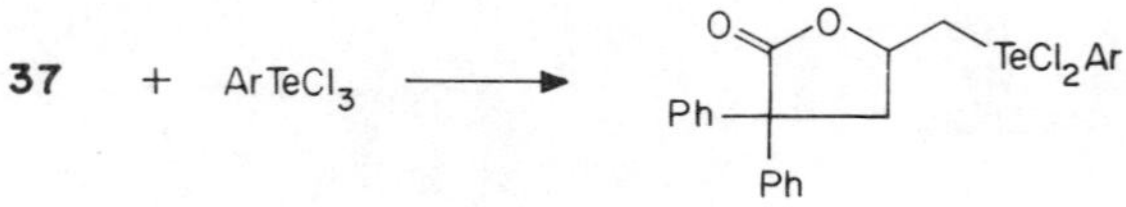

(48)

Uemura[142] allowed aryltellurium tribromides to react with a series of olefins in different alcohols to obtain (β-alkoxyalkyl)aryltellurium dibromides. A typical example is shown in equation (49). When the solvent was changed to aqueous tetrahydrofuran or aqueous *t*-butanol a hydroxy group could be similarly introduced in the β-position.

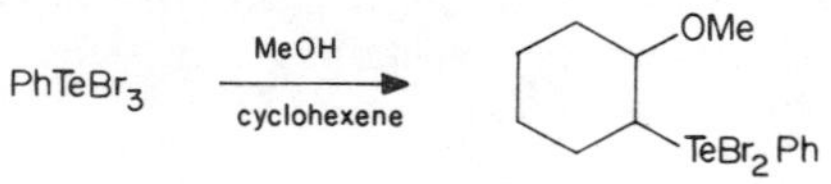

(49)

Both selenium and tellurium tetrahalides undergo addition reactions with diolefins to give heterocyclic products. Thus, norbornadiene[143] gives compounds **43** using either $SeBr_4$ or $TeBr_4$ (equation 50). Bicyclo(3,3,1)nona-2,6-diene (**44**) similarly affords compound **45** with $SeCl_4$ and $SeBr_4$[144] and $TeCl_4$ added to 2,2′-bicyclopentene to give compound **46**[145].

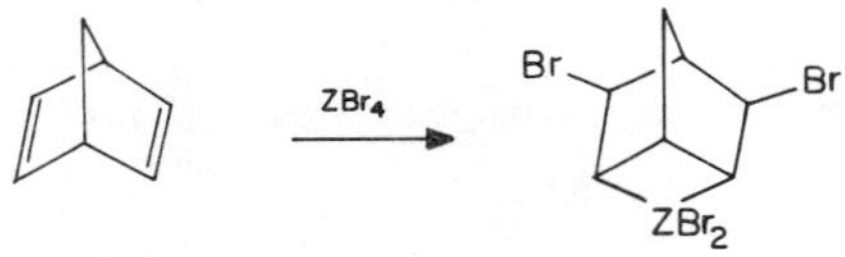

(50)

(43)

Z = Se, Te

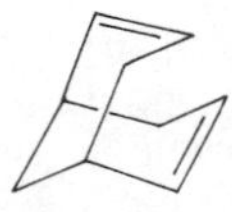

(44)

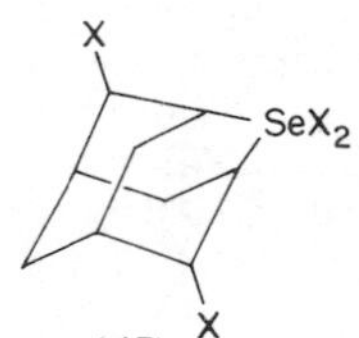

(45)

X = Cl, Br

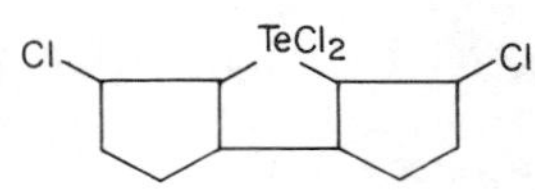

(46)

Selenium tetrachloride generated in aqueous medium gives a selenophene derivative **47** with butadiene[123]. The analogous reaction with $TeCl_4$ (generated from TeO_2 and HCl(aq.)) apparently does not work. As already mentioned, $TeCl_4$ and butadiene form 1,1-dichloro-2,5-dihydrotellurophene (**36**) in acetonitrile[136].

An unusual selenathietan, **48**, has been claimed as the product from divinyl sulphone and $SeBr_4$[146].

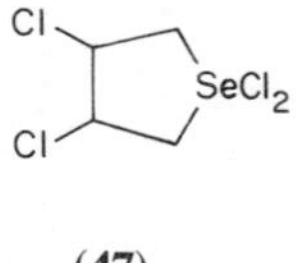

(47)

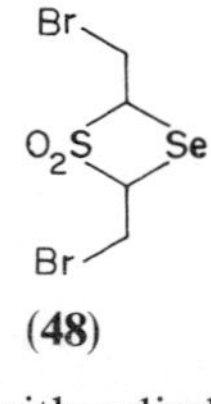

(48)

1,5-Hexadiene forms an addition compound **49** with either diselenium dichloride[120] or selenium tetrachloride[145]. This material was first erroneously formulated as a seven-membered compound **50**[124].

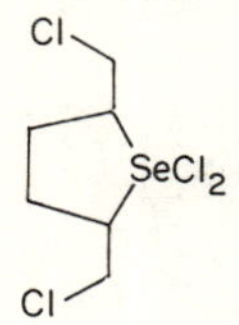

(49)

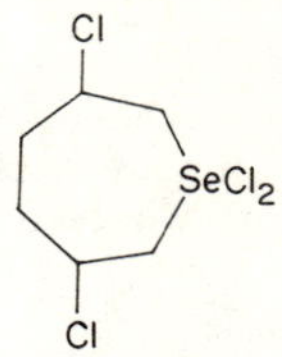

(50)

Migalina and coworkers[148] isolated the Te isologue of compound **49** from the reaction of $TeCl_4$ with 1,5-hexadiene in ethyl ether. However, during a recent reinvestigation of this reaction[149], products **51** and **52** were isolated. This is the first example of a Te-induced

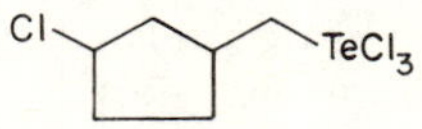

(51)

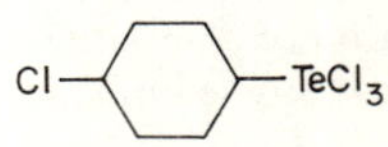

(52)

carbon–carbon bond formation in reactions with olefins. The 1,6-diolefins **53** undergo addition reactions with selenium and tellurium tetrahalides to form six-membered heterocyclic compounds (equation 51)[147,148,150].

(51)

(53)
Y = O, S, NH, NMe

Z = Se, Te
X = Cl, Br

However, when $TeCl_4$ and diallyl sulphide were allowed to interact in acetonitrile, only one double bond was involved in the reaction, yielding the zwitterionic compound **54** (equation 52)[149].

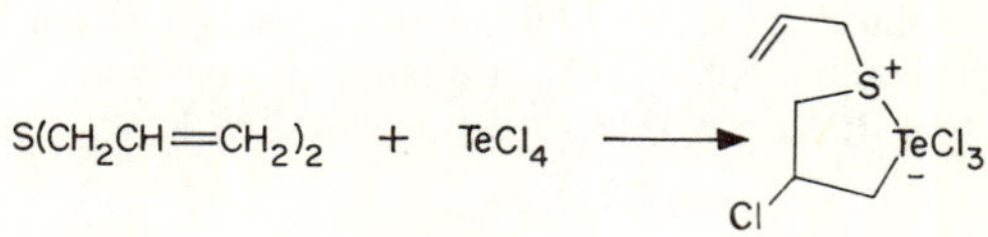

(52)

(54)

A series of *N*-acyldiallylamines behaved similarly[151]. Treatment with $TeCl_4$ and/or $TeBr_4$ afforded a series of zwitterionic oxazolines **55** (equation 53). This is another example of an internal cyclization during the addition of a tellurium tetrahalide to an unsaturated system.

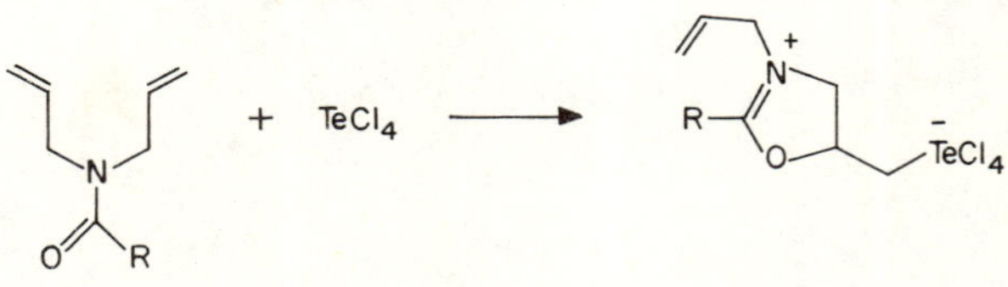

(53)

(55)

2. Reactions with acetylenes

Selenium and tellurium tetrahalides undergo addition reactions with acetylenes to give two different addition compounds **56** and **57** (equation 54); the regio- and stereochemistry of these reactions will be discussed separately.

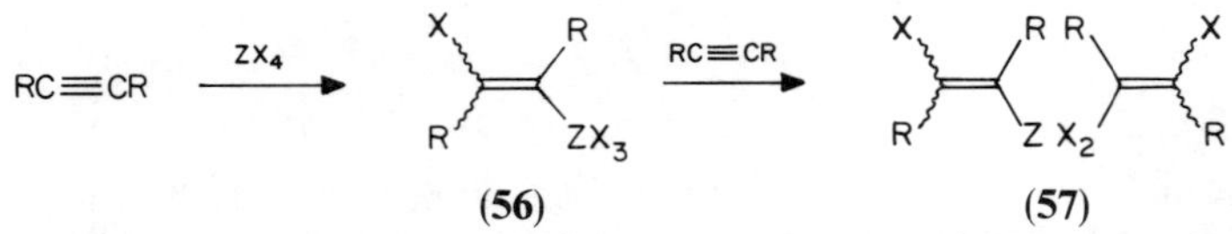

(54)

No compounds of structure **56** seem to have been isolated from the reaction of selenium tetrahalides with acetylenes. However, they were undoubtedly formed as intermediates during the preparation of compounds **57**. They were also postulated as intermediates in the synthesis of benzo(*b*)selenophenes according to equation (55)[152,153].

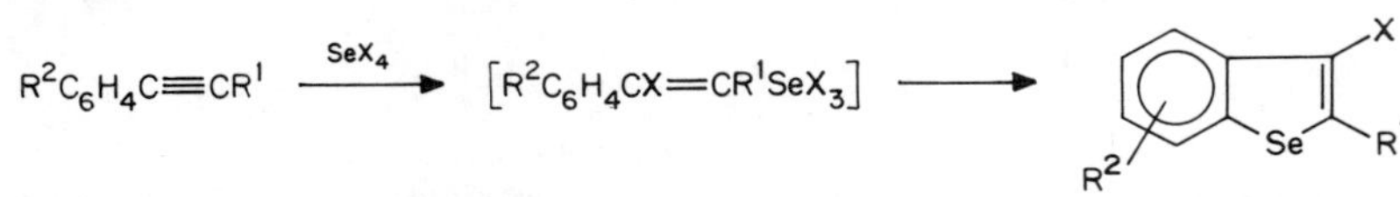

(55)

Several examples of compounds **56** have been isolated in the Te series. Phenyl- and diphenyl-acetylene, respectively, yield compounds **58a** and **58b** when treated with $TeCl_4$[154–156]. A series of alkylphenylacetylenes has given similar results[156].

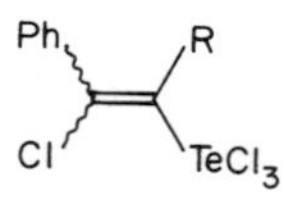

(58)

(a) R = H
(b) R = Ph

Compounds of the principal structure **56** have also been postulated as intermediates in the synthesis of benzo(*b*)tellurophenes from phenylacetylenes and tellurium dioxide in acetic acid containing a lithium halide (equation 56)[157,158].

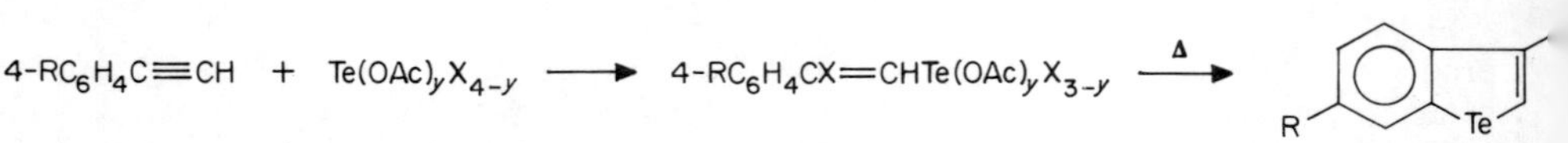

(56)

Examples of compound **57** can only be found in the Se series. Thus, acetylene was combined with $SeCl_4$ and $SeBr_4$, respectively, to afford compounds **59a** and **59b**[123,124,127]. Phenylacetylene gave an analogous product[127].

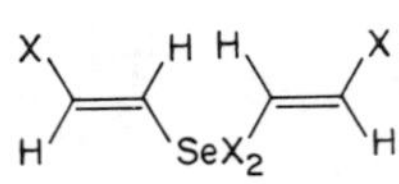

(59)

(a) X = Cl
(b) X = Br

3. Regio- and stereo-chemistry of the adducts

Although a considerable number of compounds of the different structures **25**–**27** and **56**–**57** have been synthesized, surprisingly little is known about the regio- and stereochemistry of the additions. Consequently, the following conclusions have been arrived at from relatively few experimental observations.

Concerning the regiochemistry of the additions of selenium tetrahalides and alkyl- and aryl-selenium trihalides, both Markownikoff and *anti*-Markownikoff addition has been observed. Equation (47) describes the non-regiospecific addition of 2,4-dinitrophenylselenium trichloride to *Z*-1-phenyl-1-propene[138]. Mixtures of isomers were similarly obtained when selenium tetrahalides were allowed to react with allylic halides[159] and allyl benzene[125]. The addition compounds **27** from selenium tetrahalides and terminal olefins have always been formulated as Markownikoff adducts[119,123], e.g. compound **34**[127]. By the use of modern spectroscopic methods, Garratt has shown that terminal monosubstituted olefins give products **60** and **61** of *anti*-Markownikoff addition under kinetic control as shown in equation (57)[160]. Subsequent rearrangement allows the isolation of the thermodynamically more stable Markownikoff adduct **62**.

$$2\ RCH{=}CH_2 + SeCl_4 \longrightarrow Cl_2Se(CHRCH_2Cl)_2\ \textbf{(60)} + RCHClCH_2SeCl_2CHRCH_2Cl\ \textbf{(61)} \longrightarrow Cl_2Se(CH_2CHClR)_2\ \textbf{(62)} \quad (57)$$

The stereochemistry of the addition of $SeCl_4$[130] and alkyl-[139] and aryl-selenium[138] trichlorides, respectively, to different olefins has been studied. A stereospecific *anti* addition was observed in all cases.

Tellurium tetrachloride and organyltellurium trihalides always yield products of Markownikoff addition with olefins. However, the few examples of compounds **25**–**27** include only terminal or symmetric 1,2-disubstituted olefins[110,114,134].

Concerning the stereochemistry Moura Campos and Petragnani[110] have postulated a conventional *anti* addition mechanism as operative for both $TeCl_4$ and aryltellurium trichlorides. This mechanism has recently been confirmed for the addition of 2-naphthyltellurium trichloride to various olefins[114]. The oxytellurations described in equation (49) also occur *anti*-specifically[142].

However, tellurium tetrachloride usually gives mixtures of *syn* and *anti* addition products when the reaction is carried out in chloroform. If a radical inhibitor (*p*-benzoquinone) is added in catalytic amount, the *syn* addition can be highly promoted. A more or less concerted *syn* addition mechanism (equation 58) is therefore postulated for the $TeCl_4$ addition, competing with a non-specific radical process (equations 59–61)[114].

(58)

$$TeCl_4 \longrightarrow {}^{\bullet}TeCl_3 + {}^{\bullet}Cl \quad (59)$$

(60)

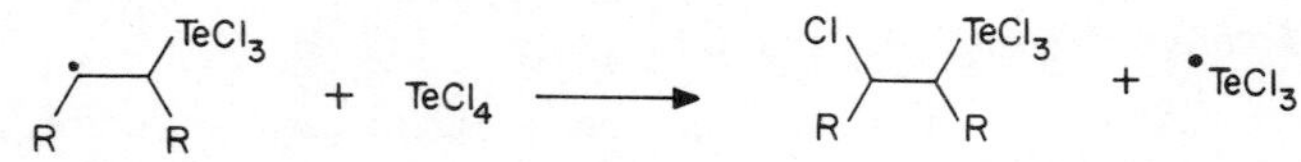 (61)

The few studies of addition reactions of $SeCl_4$ with acetylenes do not indicate the regio- and stereo-chemistry of the reaction. However, the reaction described in equation (55) has necessarily to involve an *anti* addition to allow cyclization to the benzo(*b*)selenophene system[152,153].

Uemura and coworkers[156] have recently shown that the addition of tellurium tetrachloride to acetylenes is a *syn* process. This conclusion is based on a halodetelluration of the primary addition compound **63** which gives predominantly *Z*-dihaloalkenes **64** (equation 62).

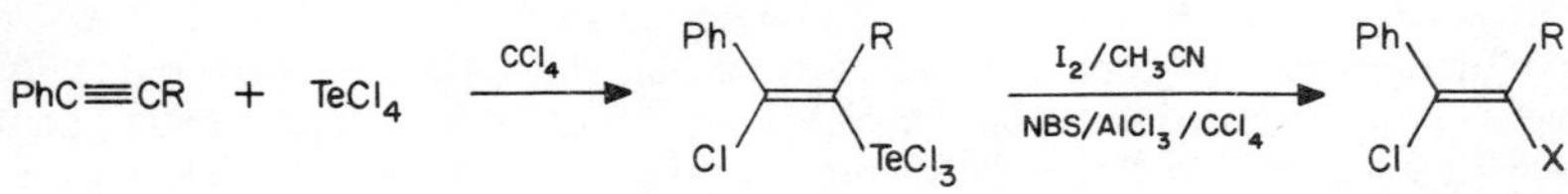 (62)

(63) **(64)**

X = Br, I

On the other hand, equation (56) does indicate that an *anti* addition might also occur under certain reaction conditions, to allow cyclization to a benzo(*b*)tellurophene[157,158]. When Sadekov[155] heated compound **58b** (unknown stereochemistry) in refluxing trichlorobenzene, the benzo(*b*)tellurophene derivative **65** was isolated.

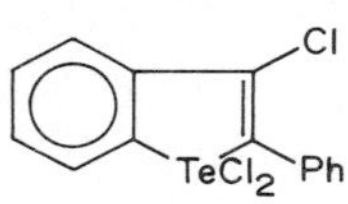

(65)

III. PREPARATION AND PROPERTIES OF R_4Z

As early as 1888 Marquardt and Michaelis[161] made an unsuccessful attempt to prepare a tetraorganyltellurium compound. They treated $TeCl_4$ with Et_2Zn, which yielded Et_3TeCl. This telluronium salt when heated with Et_2Zn gave Et_2Te and butane.

The first tetraorganyltellurium compound, Ph_4Te, was prepared according to equation (63) by Wittig and Fritz[162] in 1952. Tetraphenyl tellurane could also be prepared from Ph_3TeCl and PhLi. An excess of PhLi is necessary in these reactions, which does indicate the pentacoordinated species, $Ph_5Te^- Li^+$, as an intermediate.

$$Ph_2TeCl_2 + 2\,PhLi \rightarrow Ph_4Te \qquad (63)$$

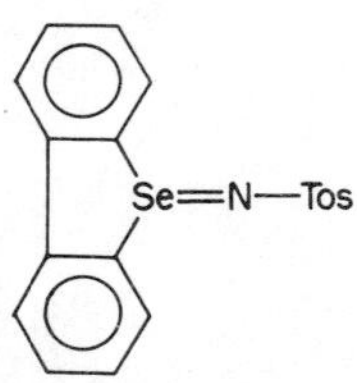

(66)

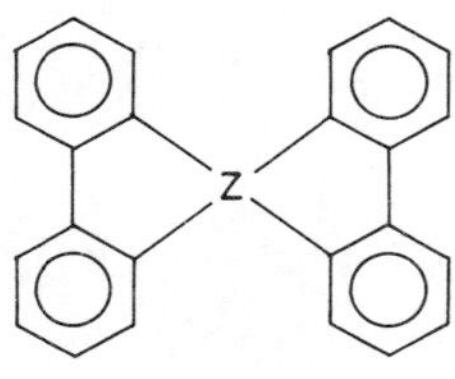

(67)

(a) Z = Se
(b) Z = Te

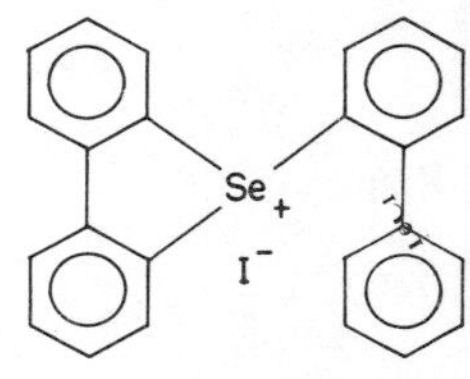

(68)

Ph_4Te melts at 104–106 °C with decomposition, but is apparently much more stable than Ph_4Se because treatment of Ph_3SeBr with PhLi gives Ph_2Se and biphenyl. These facts can, however, be explained if Ph_4Se is postulated as an unstable intermediate (cf. also equation 8). This postulate has further been substantiated by subsequent work by Hellwinkel and Fahrbach[163], who prepared, by treating compound **66** with 2,2′-dilithiobiphenyl, compound **67a**, which underwent ring-opening on treatment with water and KI to give 2,2′-biphenylylene-2-biphenylylselenonium iodide (**68**).

Hellwinkel and Fahrbach[163] also prepared compound **67b** by reacting 2,2′-biphenylylene dilithium with $Te(OCH_3)_6$, $Te(OCH_3)_4$ or $TeCl_4$. Compound **67b**, which crystallized as yellow needles and melted at 214 °C is much more stable than tetraphenyltellurium. However, heating to 260 °C for 30 min leads to decomposition yielding dibenzotellurophene, biphenylene and tetraphenylene (equation 64).

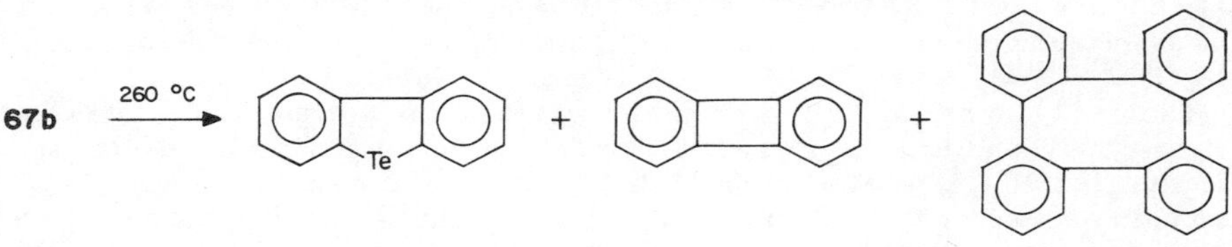

(64)

The chemistry of the less stable Se isologue **67a** parallells that of **67b**. Attempts to introduce, with MeI, a fifth group to the spirocyclic Te compound **67b**, resulted in a ring-opening which was interpreted in terms of a pentacoordinated intermediate or transition state (equation 65)[163].

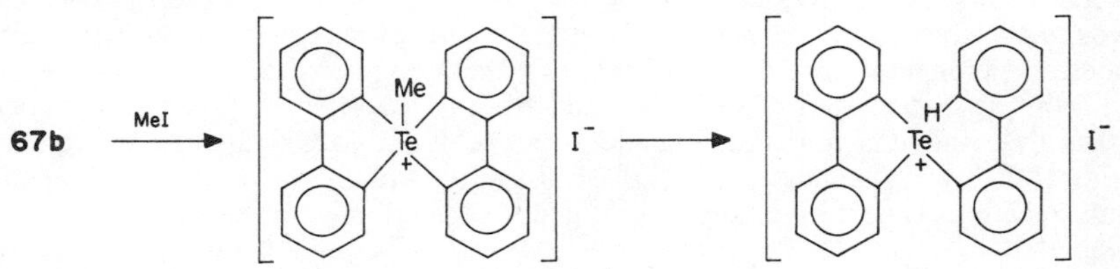

(65)

Another indication of a positively charged pentaorganyltellurium transition state or intermediate was found in the reactions of tetrabutyltellurium with methyl iodide (equation 66). In this reaction a high yield of trimethyltelluronium iodide was obtained and this has been explained by a series of electrophilic exchange processes involving pentacoordination around Te[163].

$$Bu_3Te^+ I^- + BuLi \rightarrow Bu_4Te \xrightarrow{MeI} (Bu_4Te^+Me)I^-$$

$$\xrightarrow{-BuI} Bu_3TeMe \xrightarrow[-3\,BuI]{3\,MeI} Me_4Te \xrightarrow{MeI} Me_3Te^+ I^- + C_2H_6 \qquad (66)$$

Reactions of related interest are given in equations (67), (68) and (69). The first-

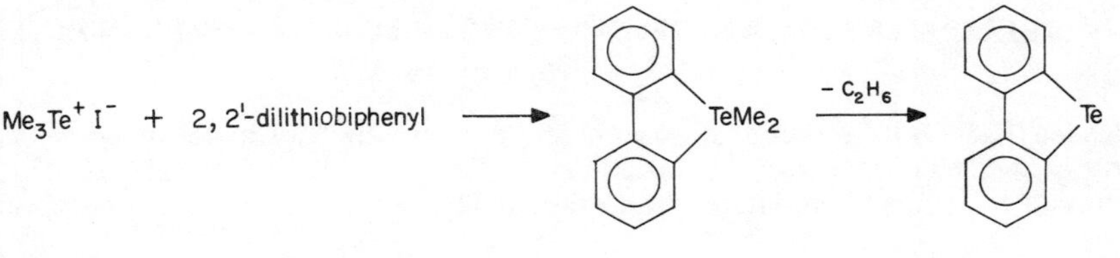

(67)

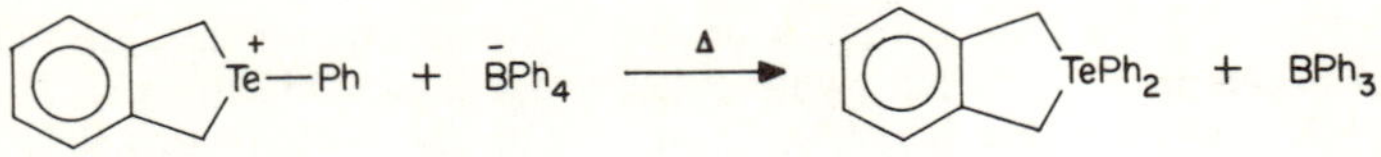
(68)

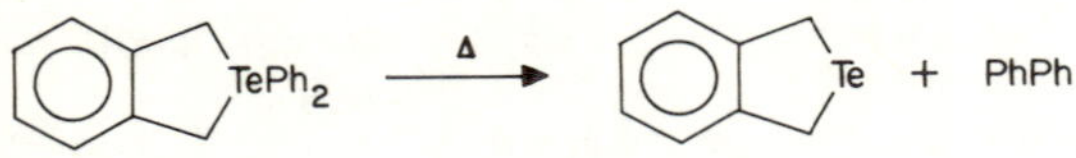
(69)

$$Ph_4Te + t\text{-}BuSH \xrightarrow{r.t.} Ph_2Te(SBu\text{-}t)_2 \longrightarrow Ph_2Te + t\text{-}BuSSBu\text{-}t \quad (70)$$

mentioned is due to Hellwinkel[163] and the others are due to McWhinnie[9]. Mechanistic details of the decomposition of tetraorganyltellurium species have recently been discussed by Glover[164], who also discussed the related reaction (70).

Glover[164] also heated Ar_4Te with known aryl-radical scavengers such as furan and styrene and found that the product pattern did not differ significantly from the neat reaction ($Ar_4Te + \Delta$). It was concluded therefore that the decomposition of R_4Te is a symmetry-allowed concerted process and has an analogy in the reaction of triarylsulphonium salts with optically active alkyllithium reagents (equation 71). This reaction proceeds via the tetravalent Ar_3SR^* intermediate which does decompose with retention of configuration at R.

$$Ar_3S^+ + R^*Li \rightarrow Ar_2S + ArR^* \quad (71)$$

The results of Glover's radical-scavenging experiments contrast with those from cross-over experiments. Thus, thermal decomposition of mixtures of Ph_4Te and p-Tol_4Te *in vacuo* or in toluene under nitrogen gave, together with the expected symmetrical diaryl tellurides and biphenyls, extensive amounts of 4-methylbiphenyl and phenyl 4-methylphenyl telluride. Since no aryl radicals could be trapped in these experiments it was proposed that the unsymmetrical products arose by a rapid random interchange (equation 72) of aryl ligands prior to decomposition. Analogous non-radical ligand-exchange reactions have been reported for organo-lead and -mercury compounds (cf. also equation 19).

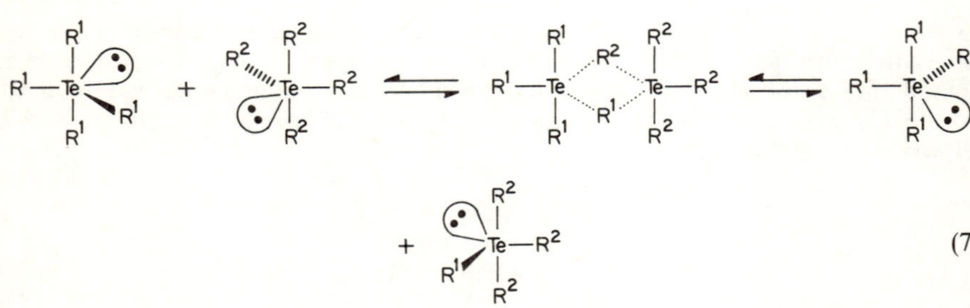
(72)

IV. TETRAVALENT ORGANO-SELENIUM AND -TELLURIUM COMPOUNDS WITH Z—O, Z—S OR Z—N BONDS

Selenoxides (R_2SeO) have been studied intensely due to their importance in synthetic procedures and recent reviews[165,166] are available. This chemistry will therefore not be reproduced here. Some recent interesting entries do include the selective oxidation[167] of

RSH to RSSR and RSR to the corresponding sulphoxides and the introduction of $R_2Se(OCOCF_3)_2$ as an oxidant in organic chemistry by Marino and Larsen[168,169].

Tellurium oxides (R_2TeO), on the other hand, have been relatively neglected. This class of compounds and the closely related compounds, $R_2Te(OR)_2$, can be prepared using methods already established in the Se series as exemplified in equations (73a) and (73b)[65,170,171]. Alcock and Harrison[172] have recently determined the X-ray structure of Ph_2TeO. The compound is monomeric.

$$Ar_2Te \xrightarrow[PhCH_2NEt_3^+, 25\,°C, H_2O_2]{RCOOH, CHCl_3} Ar_2Te(OCOR)_2 \qquad (73a)$$

$$R_2TeX_2 \overset{OH^-}{\qquad\quad} R_2TeO \qquad (73b)$$

All the known telluroxides, R_2TeO, are white solids. The water-soluble compounds give a basic solution, presumably due to formation of $R_2Te(OH)_2$. In this connection it is interesting to note that the basicity[173] of the oxides increases in the series sulphoxide, selenoxide, telluroxide.

Telluroxides have recently been used for synthetic purposes. Engman and Cava[174] found that bis(*p*-methoxyphenyl)telluroxide did function as a mild catalyst for a variety of aldol condensations.

Lee and Cava[175] and later also Uemura and Fukuzawa[176] studied the elimination of alkenes from telluroxides (equation 74).

$$PhTeBr_2CHMeCH_2R \xrightarrow[H_2O]{OH^-} PhTe(O)CHMeCH_2R \rightarrow RCH_2CH{=}CH_2 + RCH{=}CHMe \qquad (74)$$

Sadekov and coworkers[177] have recently found that Ar_2TeO in *i*-PrOH reacts with formic acid at 60 °C and with trichloroacetic acid at room temperature to give $Ar_2Te(OCOR)_2$ in high yields. The product from trichloroacetic acid, $Ar_2Te(OCOCCl_3)_2$, can be used as a source of dichlorocarbene as indicated in equation (75).

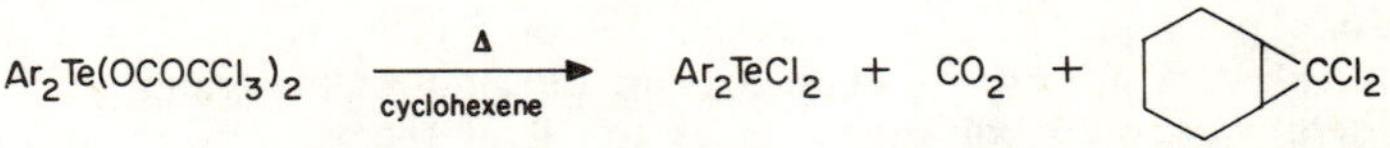

(75)

Telluroxides, notably An_2TeO, have recently been shown to be particularly mild and selective oxidants for a number of substrates[178,179]. Phosphines can be oxidized to their corresponding oxides, while thiols are converted to disulphides. Oxidation of acyl hydrazines produces hydrazides in high yields, whereas aryl hydrazines affords arenes and symmetrical and unsymmetrical tellurides. Of particular significance is the fact that An_2TeO does not react with a number of fairly easily oxidized substrates such as simple phenols, enamines and amines and heterocycles such as pyrrole and indole.

Although a few tetraalkoxytelluranes, e.g. $(MeO)_4Te$, had already been prepared by Meerwein[180] in 1929, their chemistry had not received much attention until the recent work by Denney[181] although Paetzold[182] had studied a number of simple tetraalkoxyselenanes in the early '70s. Generally speaking, $(RO)_4Se$ is much less stable than $(RO)_4Te$ and decomposition according to equation (76) is much more pronounced for the Se compounds. Thus $(EtO)_4Te$ is a distillable colourless liquid, while $(EtO)_4Se$ will readily decompose (above 0 °C) according to equation (76). However, it is well-known that electronegative groups do often stabilize hypervalent molecules and consequently $Se(OCH_2CF_3)_4$ and related compounds are relatively stable distillable liquids[181,183]. Some preparative methods for various $(RO)_4Se$ are given in equations (77)–(80).

$$(RO)_4Z \rightarrow (RO)_2ZO + ROR \qquad (76)$$

$$SeCl_4 + PrOH + 4\,Et_3N \rightarrow Se(OPr)_4 + Et_3NH^+\,Cl^- \quad (77)$$

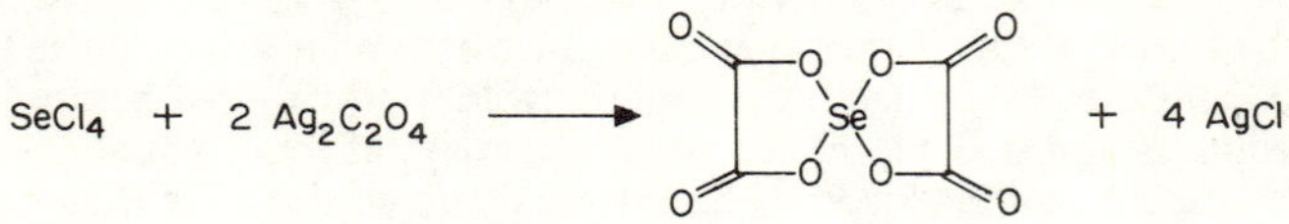

(78)

$$Se(OMe)_4 + 4\,ROH \rightarrow Se(OR)_4 + 4\,MeOH \quad (79)$$

$$PhSeBr_3 + 3\,NaOMe \rightarrow PhSe(OMe)_3 \quad (80)$$

Tetraalkoxyselenanes are stabilized when the central atom is incorporated in a five-membered ring. Thus compound **69a** is a relatively stable crystalline solid[181]. The PMR spectrum of **69a** at room temperature shows[181,184] a single resonance. Upon cooling the spectrum becomes quite complicated and many lines can be observed. Addition of (diethylamino)trimethylsilane, a known acid and water scavenger, led to an extremely complicated but symmetrical ambient spectrum centred at δ 4.07.

Compound **69b** was also prepared as a crystalline solid. The PMR spectrum had resonances at δ 1.20 and 1.25 for hydrogens of non-equivalent pairs of Me groups. The ^{13}C-NMR spectrum of **69b** at room temperature showed two resonances at δ 24.40 and 24.70, assignable to pairs of non-equivalent Me group carbons. A singlet for the quaternary carbons appeared at δ 81.40. At $-112\,°C$, the ^{13}C-NMR spectrum of compound **69b** exhibited two absorptions separated by 76 Hz at δ 79.0 and 82.8 for non-equivalent quaternary carbons.

The low-temperature ^{13}C-NMR spectrum is consistent with a static trigonal–bipyramidal, TBP, or nearly TBP structure. When the compound is warmed, a reorganization process occurs rapidly on the NMR time-scale which renders the quaternary carbons equivalent but not the pairs of Me group carbons. This process most probably involves axial–equatorial switching of the rings with the lone pair remaining in an equatorial position.

The acid-catalysed process (equation 81) which renders all of the groups equivalent undoubtedly involves **70a** and **70b** in equilibrium with **69a** and **69b**, respectively. Switching of an equatorial position of the rings by ionization, rotation and ring-closure renders *trans* groups equivalent. This process, coupled with intramolecular ligand reorganization, which renders *cis* groups equivalent, leads to all of the various groups becoming equivalent.

The crystal structure of **69b** reveals it to be a distorted TBP with the distortion away from the lone pair. The apical O—Se bond length is 1.871 Å and the equatorial O—Se bond length is 1.77 Å. The angle O—Se—O in the rings is 85.7°, while the O—Se—O angle between the two equatorial oxygens is 109.6°.

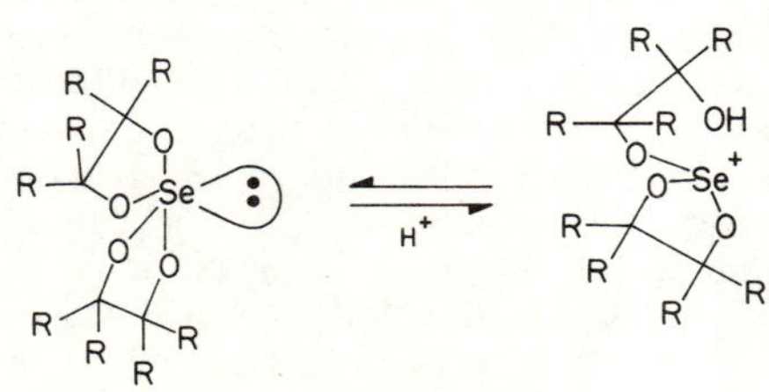

(81)

(69)

(a) R = H
(b) R = Me

(70)

(a) R = H
(b) R = Me

The Te analogues of **69a** and **69b** have also been studied and similar conclusions about ring-switchings and structures have been drawn[181].

Tetravalent organic selenium and tellurium sulphides have not been greatly studied and part of the field is still controversial. The first compound of the general structure R_2ZS, namely **71**, was reported in 1982 by Detty and Murray[185]. Compound **71** was isolated in low yield when compound **72** was treated with two equivalents of the Lawesson reagent **73** in benzene at room temperature for 17 h. The product **71** was a sharp-melting (153.5–155 °C) purple-black solid, whose spectral and analytical data agreed with the formulation **71**. In the transformation **72** → **71** Te has been sulphurated directly by S, presumably from **73**.

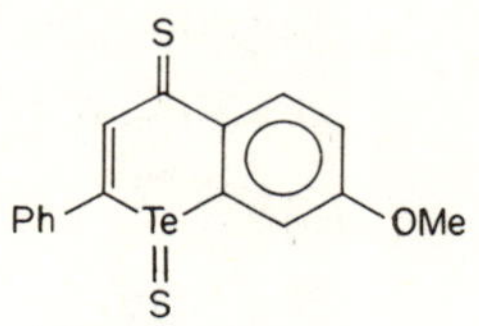

(71a)

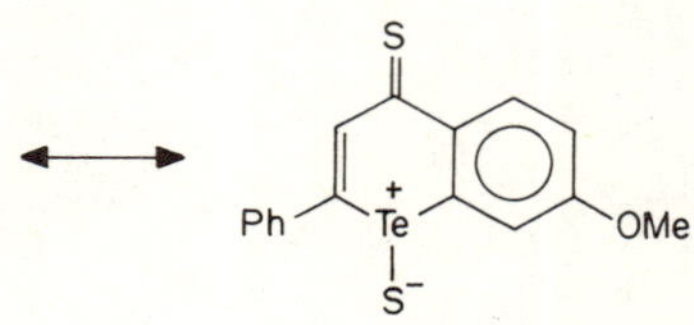

(71b)

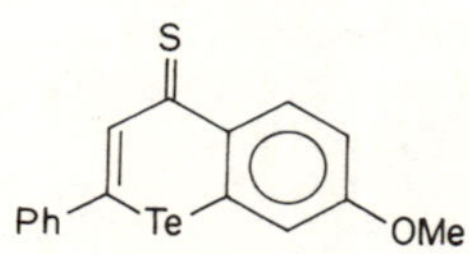

(72)

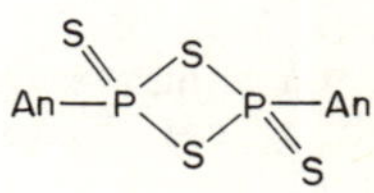

(73)

Compounds of the general structure $Te(SR)_4$ and $Se(SR)_4$ have also been studied[186–188] for some time. Formulae **74**–**77** are some specific examples claimed in the literature. Compound **74** was obtained by oxidation of *meso*-dimercaptosuccinic acid dimethyl ester with SeO_2 in methanol[187]. Similar reactions were later studied[188] by Czauderna and Samochocka; however, the evidence for the structures is only supported by elemental analytical data. Clearly NMR studies would be of interest in this area.

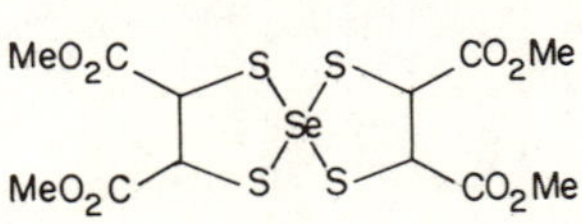

(74)

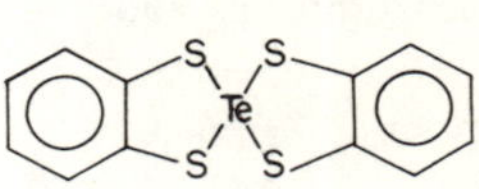

(75)

$Te(SPh)_4$

(76)

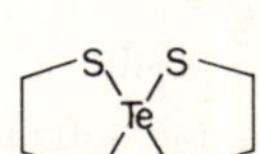

(77)

Compounds **75** and **76** were first claimed by Nakhdjavan and Klar[186] who treated the thiols or dithiols with $TeCl_4$ at low temperatures (−8 °C for **76**). The products were relatively unstable and **76** was claimed to decompose around 100 °C. However, very recently Stukalo and coworkers[189] reported that the products described by Klar[186] are in fact equimolar mixtures of diaryl disulphides and bis(arylthio)tellurium. This decomposition (equation 82) took place even at −60 °C. The compound $(PhS)_2Te$ was obtained as low-melting (64–67 °C) orange prisms. The same type of product mixture was obtained with the tetraalkoxytellurane, $(CHF_2CF_2CH_2O)_4Te$, as reactant[189].

$$TeCl_4 + ArSH \rightarrow [(ArS)_4Te] \rightarrow ArSSAr + (ArS)_2Te \tag{82}$$

Ready formation of disulphides upon treatment of thiols with tetravalent organotellurium compounds was also reported by Wieber and Kaunziger[171] (equation 83).

$$Ar_2Te(OEt)_2 + HSCH_2CH_2SH \rightarrow Ar_2Te + (SCH_2CH_2S)_n \tag{83}$$

Stukalo and coworkers[189] also reported that treatment of $(ArS)_2Te$ with MeI at room temperature yielded ArSSAr and Me_2TeI_2. This was interpreted in terms of decomposition $(ArS)_2Te$ into ArSSAr and elemental Te, which will then add MeI oxidatively to yield the observed product. However, we think that a reaction pathway as outlined in equation (84) is more likely.

(84)

The interaction of TeX_4 with dithio ligands (dithiocarbamates and xanthates) has been studied by Huseby and Klar[190–196]. In many cases the formation of complexes with interesting structures have been observed. Oxidation–reduction (i.e. formation of disulphides) is common in this area too. Interaction of Et_2NCSSH with $TeBr_4$ and $TeCl_4$ gives $Br_2Te(S_2CNEt_2)_2$ and $ClTe(S_2CNEt_2)_3$, respectively. Simplified structures are represented by **78** and **79**.

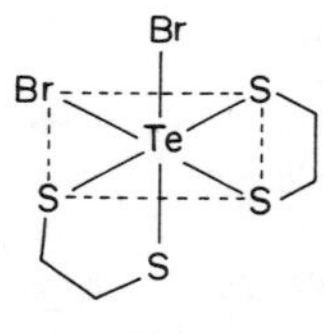

(**78**)

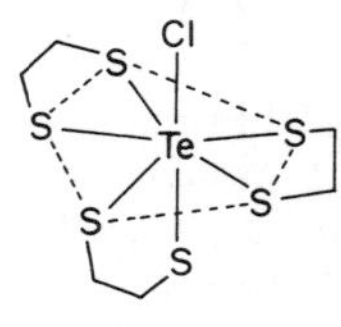

(**79**)

2-Mercaptobenzothiazole and $TeCl_4$, yield[116] the interesting complex **80**.

(**80**)

Organic Te and Se compounds containing Te—N or Se—N bonds had hardly been investigated at all until around 1975. Although the adduct $Me_2TeI_2 . nNH_3$ had been reported in old literature (see Ref. 3, p. 190) the first well-defined compounds with Te—N bonds were reported by Russian workers[197,198], who prepared diaryl tellurimides (equation 85) and tellurium diimide species (equations 86a and 86b, where X = Cl or F).

$$Ar_2TeO + RSO_2NH_2 \rightarrow Ar_2Te{=}NSO_2R \tag{85}$$

$$TeX_4 + RN(SiMe_3)_2 \rightarrow RN{=}TeX_2 \tag{86a}$$

$$RN{=}TeX_2 + RN(SiMe_3)_2 \rightarrow RN{=}Te{=}NR \tag{86b}$$

The compounds $ArSO_2N{=}TeF_2$ react with *N*-trimethylsilylmorpholine in benzene solution to produce[198] the Te derivative **81**. The Te—N bonds in the compounds $Ar_2Te{=}NSO_2R$ and RN=Te=NR are readily hydrolysed by hot water producing TeO_2 and the amide. Chlorolysis, yielding $TeCl_4$, is also possible.

The *N*-trifluoroacetyltellurimides **82** can be isolated in two forms **82a** and **82b** with quite different melting points (80–81 °C for the *syn* form and 168–169 °C for the *anti* form)[199]. An X-ray structure investigation is available for $An_2Te{=}NTos$[200].

$$(\text{Morpholino})_2\text{Te}{=}\text{NTos}$$

(81)

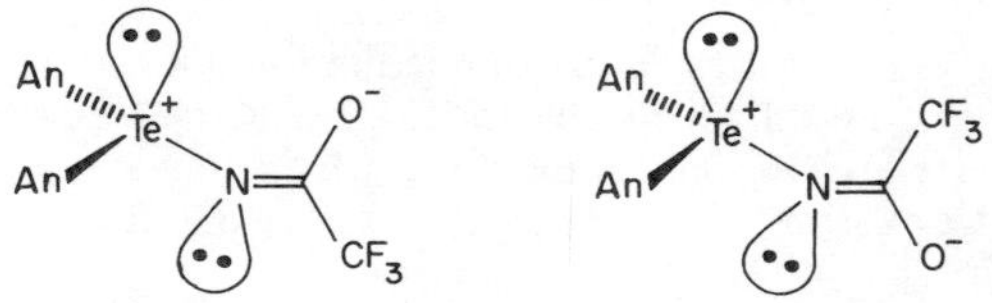

(82a) **(82b)**

Ogura and coworkers[201] prepared diorganyl tellurimides according to equation (87) and used the product to prepare olefins (equation 88), with yields in the range 66–93%.

$$RCH_2CH_2TePh + TosNClNa \xrightarrow{THF} RCH_2CH_2Te({=}NTos)Ph \quad (87)$$

$$RCH_2CH_2Te({=}NTos)Ph \xrightarrow[\text{reflux}]{THF} RCH{=}CH_2 + PhTeNHTos \quad (88)$$

Only a few organoselenium compounds of the type $Ar_2Se{=}NR$ had been prepared[202] before 1974 (cf. compound **66**), but since then the development has been rapid[203–214]. Of particular importance is probably the discovery that TosN=Se=NTos could effect allylic amination[208] of olefins and 1,2-diamination of 1,3-dienes[207].

Derkach and coworkers[211] have studied the interaction of $PhSO_2N{=}Se{=}NSO_2Ph$ with aromatic aldehydes, dimethylformamide, dimethylsulphoxide, Ph_3PO and PH_2SeO. The reactions are exemplified by equation (89). PhCON=Se=NCOPh was found to be a less active reagent and only dimethyl sulphoxide reacted with it (equation 90).

$$PhSO_2N{=}Se{=}NSO_2Ph + 2\,PhCHO \rightarrow 2\,PhCH{=}NSO_2Ph + SeO_2 \quad (89)$$

$$PhCON{=}Se{=}NCOPh + 2\,Me_2SO \rightarrow 2\,Me_2S{=}NCOPh + SeO_2 \quad (90)$$

Treatment of MeCON=Se=NCOMe with aldehydes (ArCHO) resulted in the unexpected formation of $ArCH(NHCOMe)_2$, which seems to require the presence of moisture (equations 91 and 92).

$$MeCON{=}Se{=}NCOMe + ArCHO \rightarrow ArCH{=}NCOMe + SeO_2 \quad (91)$$

$$ArCH{=}NCOMe \xrightarrow[MeCONH_2]{H_2O} ArCH(NHCOMe)_2 \quad (92)$$

Treatment of ArSeSeAr with $ArCONCl_2$ will result in a cleavage reaction (equations 93–95) yielding **84**, possibly via the hypervalent intermediate **83**[212]. *N*-Acylareneseleninimidoyl chlorides (**84**) are readily hydrolysed (e.g. by moisture) to the corresponding amides and areneseleninic acids.

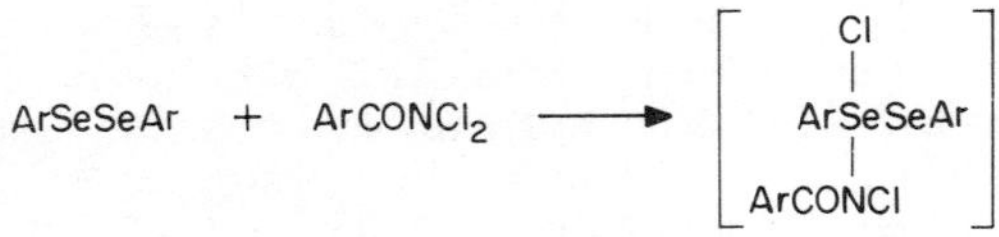 (93)

(83)

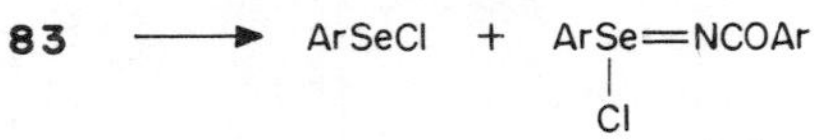 (94)

(84)

$$2\,ArSeCl + ArCONCl_2 \rightarrow \mathbf{84} + ArSeCl_3 \qquad (95)$$

Compound **84** undergoes 1,4-cycloaddition (equation 96) with styrene to yield a new type of Se heterocycle, 5,6-dihydro-4-selena-1,3-oxazine (**85**) as colourless prisms[212a]. Compound **85** shows relatively high conductivity, which indicates that the ionic form **85b** might be preferred. Compounds of the type RN=Se=NCOR give similar adducts with suitable alkenes[212b].

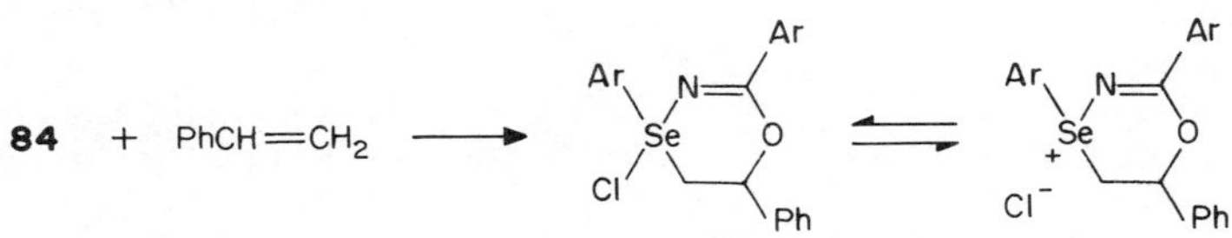 (96)

(85a) **(85b)**

A wide range of arylaminoselenium dichlorides can be prepared[213] by the interaction of ArSeCl with *N*-chloroimides, *N*-chloroimidic esters, *N*-chloroketimines and related compounds (equation 97). The same compounds can also be prepared from $ArSeCl_3$ and R_2NSiMe_3 (equation 98). $ArTeCl_3$ can be similarly used as a reactant[214].

$$ArSeCl + R_2NCl \rightarrow Ar(R_2N)SeCl_2 \qquad (97)$$

$$ArSeCl_3 + R_2NSiMe_3 \rightarrow Ar(R_2N)SeCl_2 \qquad (98)$$

V. TETRAHALOORGANYLTELLURATES (IV)

Tetrahaloaryltellurates(IV) can be readily obtained[23,215–217] by equation (99). The reactants are simply mixed in, for example, dry $CHCl_3$ and the crystalline product collected after a reflux period of 4 h in chloroform. A wide range of cations (e.g. Me_4N^+, Ph_3Te^+, Ph_4As^+, Ph_2I^+) have been employed.

$$ArTeX_3 + Y^+X^- \rightarrow Y^+ + (ArTeX_4)^- \qquad (99)$$

The ionic nature of the tetrahaloaryltellurates is evident from reactions with ion exchange resins and by conductivity measurement[23,215]. The salts are fairly soluble in water albeit not completely stable, as aryltellurium oxohalides start to separate after ca. 20 h (equations 100 and 101). The operation of equation (100) also explains the formation of $PhTeCl_2CH_2COMe$ on attempted recrystallization of $PyH^+(PhTeCl_4)^-$ from acetone[23].

$$ArTeX_4^- \rightleftharpoons ArTeX_3 + X^- \qquad (100)$$

$$ArTeX_3 + H_2O \rightarrow ArTe(O)X + 2\,HX \qquad (101)$$

As would be expected, exchange reactions, such as (102), can be readily performed.

$$(Ph_3Te)^+(PhTeCl_4)^- \underset{Cl^-}{\overset{Br}{\rightleftharpoons}} (Ph_3Te)^+(PhTeBr_4)^- \tag{102}$$

Sometimes tetrahaloaryltellurates(IV) are obtained, due to redox processes, in connection with the preparation of dihaloaryltellurates(II). Thus, addition of bromine to ArTeTeAr in the presence of R_4PBr leads to the formation of $(R_4P)^+(ArTeBr_2)^-$ and $(R_4P)^+(ArTeBr_4)^-$. The results are explained by equations (103)–(106).

$$Br_2 + Ar_2Te_2 \rightarrow 2\,ArTeBr \tag{103}$$

$$ArTeBr + R_4PBr \rightarrow (R_4P)^+(ArTeBr_2)^- \tag{104}$$

$$ArTeBr + Br_2 \rightarrow ArTeBr_3 \tag{105}$$

$$ArTeBr_3 + R_4PBr \rightarrow (R_4P)^+(ArTeBr_4)^- \tag{106}$$

Whether a pure tellurate(II) or tellurate(IV) or a mixture thereof is obtained apparently depends on a delicate balance involving the relative stability and solubility of the Te(II)- versus the Te(IV)-containing salt in a particular system.

The first aliphatic tetrahalotellurates(IV) were prepared[12] more than 50 years ago, although their true nature was not revealed until 1967 by Einstein[13]. No systematic studies of the chemistry have been reported.

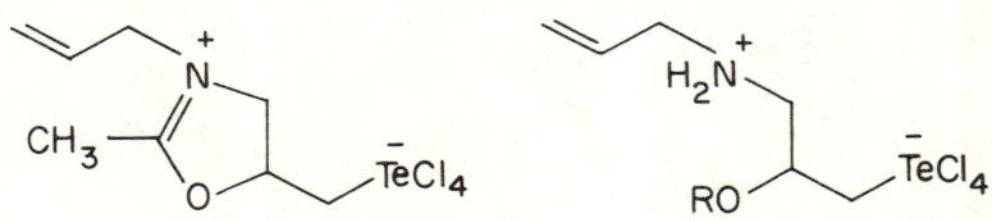

(86) **(87)**

(a) R = H
(b) R = COMe

Recently Bergman and coworkers[151] have reported a new type of tetrahalotellurate readily formed when, for example, $MeCON(CH_2{-}CH{=}CH_2)_2$ in MeCN is treated with $TeCl_4$ (or $TeBr_4$) at 25 °C. The product **86** is quickly formed in an exothermic reaction. The structure of **86** has been elucidated by X-ray analysis. The four Cl atoms are all equatorial and nearly coplanar, in other words the structure is similar to that of **9**. Compound **86** will readily dissolve in water at 25 °C. Concentration of the solution after 1 h will yield crystals of the ring-opened and hydrolysed product **87a**. The surprisingly facile hydrolysis is explained in terms of an equilibrium between a tetrachlorotellurate(IV) and the corresponding tellurium trichloride followed by intramolecularly catalysed hydrolysis as indicated in equation (107).

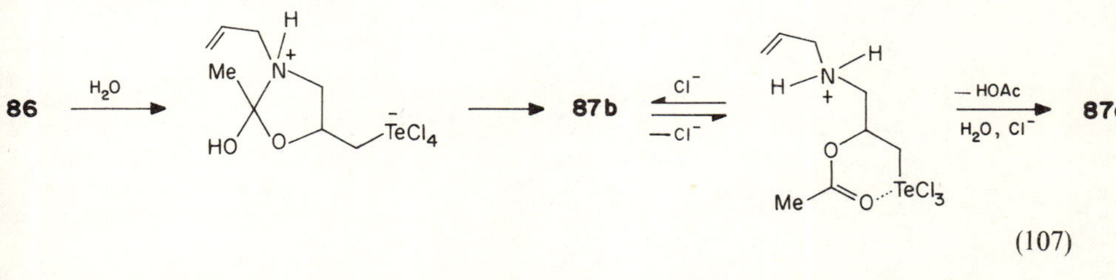

(107)

Considering the ready and exothermic formation of **86** from $TeCl_4$ and $MeCON(CH_2CH{=}CH_2)_2$ it is tempting to assume that tetrahalotellurates might also play a role in the reaction of other alkenes with $TeCl_4$. One example is the interaction between $TeCl_4$ and 1,5-cyclooctadiene which yields an unstable adduct, assigned structure **88a** with a highly reactive C—Cl bond[115]. Attempted recrystallization of the adduct from ethanol resulted in the formation of **88b**, whose structure was proven by X-ray analysis[115].

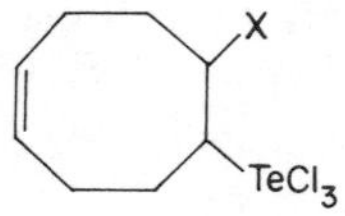

(88)

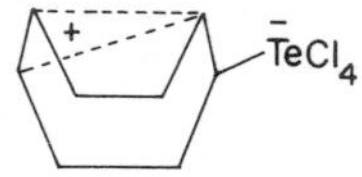

(89)

(**a**) X = Cl
(**b**) X = OEt

It now appears possible that the formulation **89** of the adduct might better explain its properties. Anyhow, further studies of the problem should be rewarding.

Compounds with zwitterionic (**90**) structures related to **89** can be obtained (as already discussed in Section II.C) by interaction of diallyl ether, diallyl sulphide and diallyl selenide with $TeCl_4$ (equation 52)[219]. In this connection a series of compounds, **91**, recently obtained by Detty, is also of interest[220].

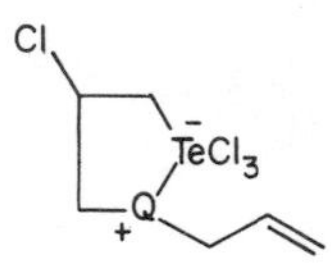

(90)

Q = O, S, Se

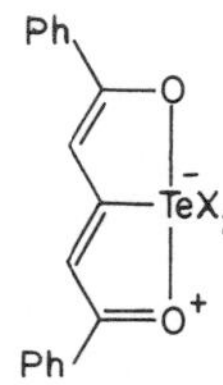

(91)

X = Br, Cl

VI. HEXAVALENT SELENIUM AND TELLURIUM COMPOUNDS

A. General Aspects

Although the pertellurane, $Te(OMe)_6$, readily prepared[221] from $Te(OH)_6$ and diazomethane, has been known since 1916, comparatively little systematic research (except for selenones and tellurones which are treated separately) has been done in the field of hexavalent Se and Te compounds. Attempts[163] to prepare compound **92** by treatment of $Te(OMe)_6$ with 2,2′-dilithiobiphenyl resulted in the tetravalent compound **67b**, and there are no compounds known with six carbon atoms bonded to one Te or Se atom.

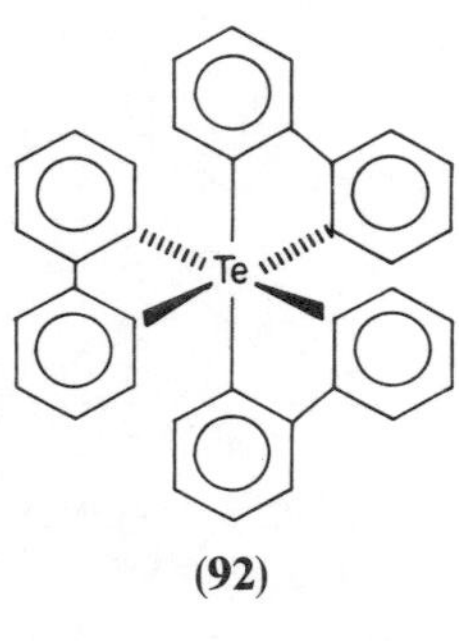

(92)

Recently the interesting acids $HOTeF_5$ (available via equation 108 or 109) and $(HO)_2TeF_4$ have been prepared. Esters (e.g. $MeOTeF_5$) are known but no derivatives (such as $MeTeF_5$ or $PhTeF_5$) with C—Te bonds have yet been described[222]. Some interesting derivatives based on $HOTeF_5$ or $HOSeF_5$ do include $Te(OTeF_5)_6$ and $C(OSeF_5)_4$. It is of interest to note that $Te(OTeF_5)_6$ can be sublimed at 110 °C and will melt at 242 °C as a stable molecule with a molecular weight of 1567.

$$Te(OH)_6 + 5\,FSO_3H \rightarrow 5\,H_2SO_4 + HOTeF_5 \qquad (108)$$

$$3\,SeO_2F_2 + 4\,HF \rightarrow 2\,HOSeF_5 + H_2SeO_4 \qquad (109)$$

$$MeOTeF_5 + C_5H_5N \xrightarrow{-40\,^{\circ}C} C_5H_5NMe^+ \ ^-OTeF_5 \qquad (110)$$

The hydrolysis of TeF_6 as well as the solvolysis of $Te(OH)_6$ in HF have been investigated and practically all the intermediate species $(HO)_xTeF_{6-x}$ have been detected[223–227]. As is evident from the structure of **93**, isomerism should be possible for $(HO)_2TeF_4$ and indeed *cis* and *trans* forms of this composition have been isolated. Derivatives, such as *trans*- and *cis*-$(Me_3SiO)_2TeF_4$ are also known[228–230]. Methanolysis[231–236] and aminolysis of TeF_6 have also been studied[237–239]. During this study Fraser and Meikle[236] found that $MeOTeF_5$ is a fairly strong alkylating agent (equation 110). TeF_6 and $MeCONHCH_2CH_2OH$ similarly give the oxazoline salt **94**. Recently Shack has reported[290] that both TeF_5OCl and TeF_5OF react with fluoro olefins to form TeF_5O-containing fluorocarbons.

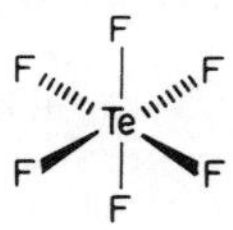

(**93**)

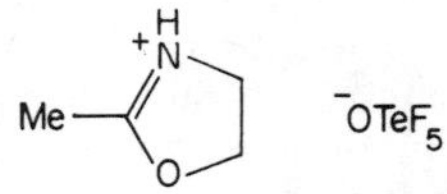

(**94**)

Depending on the conditions, TeF_6 reacts with $(Me_3Si)_2NH$ yielding $Me_3SiNHTeF_5$ or the explosive *cis*-$(Me_3SiNH)_2TeF_4$. The interesting compound $O{=}C{=}NTeF_5$ (m.p. −95 °C, b.p. 39 °C) has been prepared according to equation (111). Another interesting conversion yielding the orange crystalline $Cl_4W{=}NTeF_5$ is given in equation (112).

$$(Me_3SiNH)TeF_5 + COF_2 \rightarrow O{=}C{=}NTeF_5 + Me_3SiF + HF \qquad (111)$$

$$(Me_3SiNH)TeF_5 + WCl_6 \rightarrow Cl_4W{=}NTeF_5 + Me_3SiCl + HCl \qquad (112)$$

Compounds of the general structure R_2TeX_4 are scarce in the literature and the first compound of this sort, *trans*-$(C_2F_5)_2TeF_4$, was obtained together with other compounds, $(C_2F_5)_2TeF_2$ and *trans*-$C_2F_5TeClF_4$, upon treatment of $(C_2F_5)_2Te$ with ClF. The compound *trans*-$(C_2F_5)_2TeF_4$ was not isolated but was identified by its ^{19}F-NMR spectrum. *trans*-$C_2F_5TeClF_4$ was isolated as a colourless liquid. The same research group had earlier also described[240] the Se isologue, *trans*-$C_2F_5SeClF_4$.

It is not known if other fluorinating agents, e.g. XeF_2, can be used to convert say Ar_2TeF_2 into Ar_2TeF_4 (cf. Refs. 60–65). Michalak and Martin[5,241] have recently synthesized the pertellurane **96** as outlined in equations (113a) and (113b).

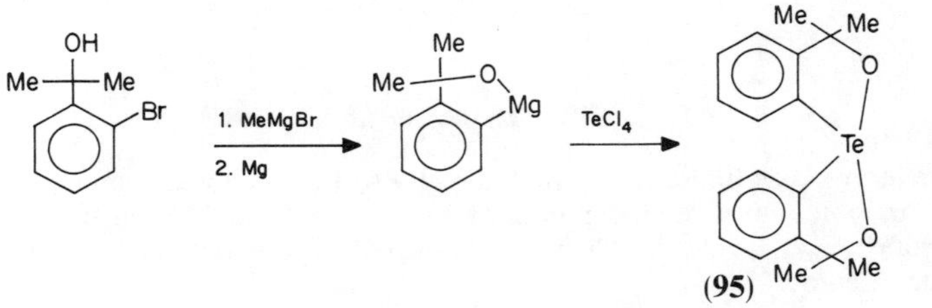

(**95**) (113a)

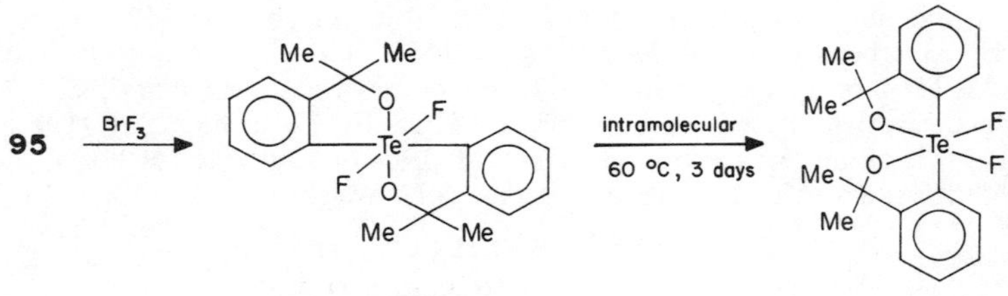

(113b)

(96) (97)

The *trans*-pertellurane **96** is converted to the thermodynamically more stable *cis* isomer **97** by an intramolecular twist mechanism, a polytopal rearrangement. The detailed geometry of **97** has been established by a complete X-ray crystallographic study.

Non-dissociative isomerizations are well established in a few cases for derivatives of transition metals[242,243]. In connection with the now established non-dissociative isomerization **96**→**97**, it is of particular interest to note that the S isologue of **96** will undergo a dissociative, acid-catalysed isomerization to the isologue of **97**. Nothing has yet been published about the Se isologues, although a detailed study[241] of the S isologues is available.

Factors which might lower the activation barrier for non-dissociative isomerizations include the presence of low-lying empty orbitals on the central atom. The Te atom of **96** will provide empty 4 f and 5 d orbitals, not available in the S isologue, which may be low enough in energy to contribute significantly to the bonding, perhaps more in the transition state than in the ground state. Substitution of the much larger Te atom of **96** for S could also lower the energy of activation for the isomerization by introducing ground-state strain resulting from the incorporation of the longer bonds to octahedral Te into the five-membered rings of **96**. The longer bonds to Te may also reduce steric crowding between substituents in the transition state for the transformation **96**→**97** relative to that for a non-dissociative process.

B. Selenones

Diorganyl selenones, R_2SeO_2, have been known for almost 90 years and their chemistry has been reviewed several times[244–246]. Unfortunately, the early literature contains several erroneous structures which have been corrected only recently. Despite this confusion, selenones are, although little studied, a fully respectable class of compounds available via several synthetic routes.

The simple dialkyl selenones **98** are best prepared by ozonation of the corresponding dialkyl selenoxides[247] (equation 114).

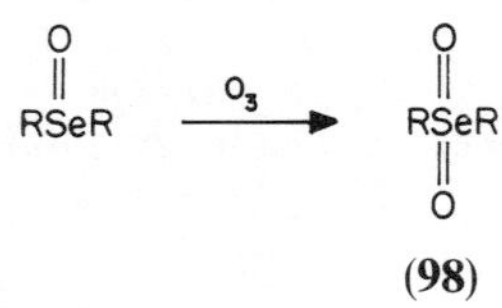

(114)

(98)

R = Me, Et, Pr, Bu

Attempts to oxidize dimethyl selenide directly to the selenone, using either hydrogen peroxide or potassium permanganate, were unsuccessful[248]. On the other hand, selenacyclohexane-1,1-dioxide (**99**) was obtained by perhydrol treatment of the corresponding selenide[249].

(99)

Diaryl selenones have been obtained by oxidation of the corresponding selenide or selenoxdie with a variety of oxidants. Rebane[250] prepared several substituted diphenyl selenones, **100**, using either peracetic acid[251] or hydrogen peroxide as the oxidant. One noteworthy property of this reaction is the fact that perhydrol is only capable of oxidizing the selenides to selenoxides when R = 3-Cl, 4-Cl or 4-Br.

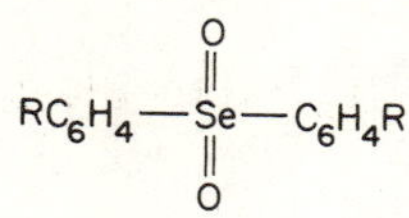

(100)

R = H, 4-Me, 4-OMe, 3-Cl, 4-Cl, 4-Br

Potassium permanganate was similarly used for the preparation of 4,4′-dicarboxydiphenyl selenone[252], 3,3′-dicarboxydiphenyl selenone[253], diphenyl selenone[254], 1,1′-dinaphthyl and 2,2′-dinaphthyl selenone[255].

Dostál[256,257] has prepared diaryl selenones in low yield by direct selenonation of aromatic compounds with selenium trioxide, SeO_3, in liquid sulphur dioxide (equation 115). The main products in these reactions were the arylselenonic acids.

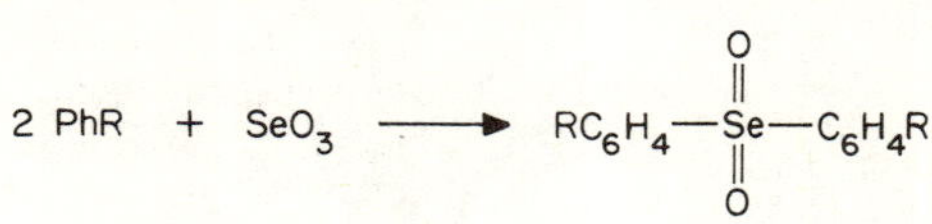

(115)

R = H, Me, Cl, Br

Reich[258] has recently prepared the alkyl aryl selenones **101a** and **101b** by *m*-chloroperbenzoic acid oxidation of the corresponding selenides between 0 °C and room temperature. Yagupolskii[259] has similarly prepared the trifluoromethyl derivative **101c** using trifluoroperacetic acid as the oxidant.

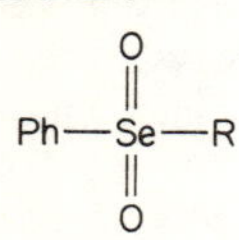

(101)

(a) R = Me

(b) R = *i*-Bu

(c) R = CF_3

The early literature described the preparation of methyl 2-naphthyl selenone (**102**) by two different methods (equation 116)[255].

$$\text{2-NaphSeO}_2\text{Na} \xrightarrow{\text{MeI}} \text{2-NaphSe(=O)}_2\text{Me} \xleftarrow{\text{KMnO}_4} \text{2-NaphSeMe} \qquad (116)$$

(102)

The alkylation of a seleninate represents a rarely used method for the preparation of selenones. It is apparently not always possible to predict if the alkylation is going to occur on Se or on O. Backer and Winter[260] gave a selenone structure **103** for the cyclization product of the sodium seleninate **104**. However, Lindgren[261] has recently shown that alkylation occurs on O instead of Se to give the five-membered compound **105** (equation 117).

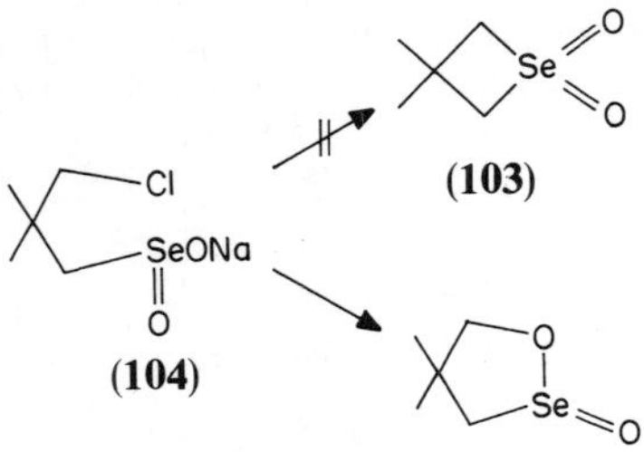

(103) (104) (105) (117)

The addition products formed from selenium dioxide and a series of butadienes were originally formulated as cyclic selenones **106**, by analogy with the well-established sulphur dioxide addition to give sulphones[262]. A reinvestigation by Mock[263] has shown that the products do in fact have a cyclic seleninic ester structure **107** (equation 118).

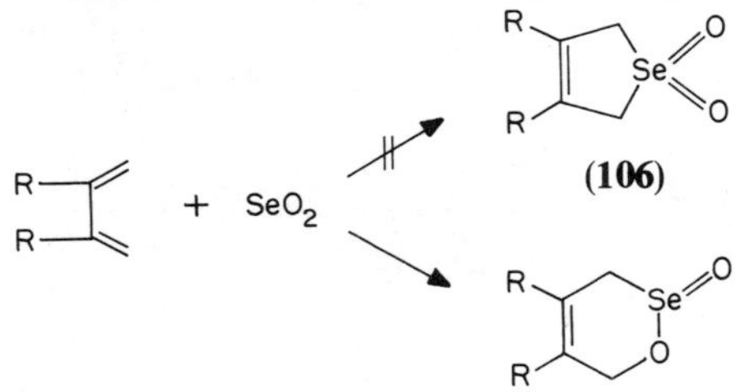

(106) (107) (118)

The IR and Raman spectra of selenones have been studied[247,264,265] as well as their mass spectral behaviour[250]. The tendency of selenones to form mixed crystals with sulphones, sulphoxides and selenoxides was early recognized[266,267]. The behaviour of certain selenones in disulphonic[268] and chlorosulphuric acid[269] has also been investigated.

Until recently, very few chemical reactions have been carried out with selenones. Yagupolskii[259] observed a haloform-type decomposition of a series of aryl trifluoromethyl selenones under very mild basic conditions (equation 119).

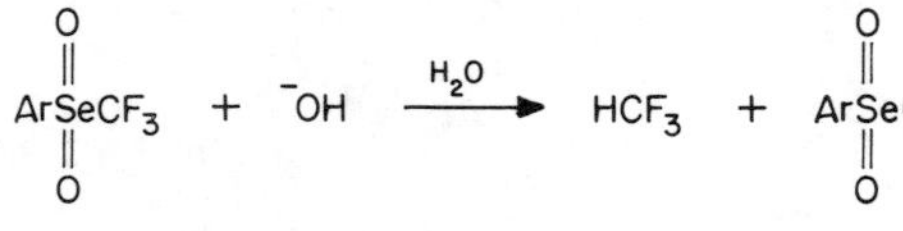

(119)

It has also been found[270,271] that diaryl selenones undergo a reaction with *N*-sulphinylarenesulphonamides to form *Se*,*Se*-diaryl(-*N*-(arylsulphonyl)selenoximines (equation 120).

$$Ar^1_2SeO_2 + Ar^2SO_2NSO \xrightarrow{-SO_2} Ar^1_2Se(=O)=NSO_2Ar^2 \quad (120)$$

Reich[258] has studied the thermal behaviour of simple selenones. Methyl phenyl selenone rearranges at ca. 100 °C to give a methyl seleninate (equation 121).

$$PhMeSeO_2 \xrightarrow[100\,^{\circ}C,\,CDCl_3]{\Delta} PhSe(=O)OMe \quad (121)$$

When β-hydrogens are present, as in the selenone **108**, an elimination occurs to give an olefin (equation 122) (2-methoxypropene is required as a trap for electrophilic Se species).

$$PhSe(=O)_2CH_2CH_2CMe_2Ph \xrightarrow[CH_2=C(Me)OMe]{C_6D_6,\,100\,^{\circ}C} CH_2=CHCMe_2Ph \quad (122)$$

(108)

The stereochemistry is shown to be *syn* by the observation that the *threo* deuterium-labelled selenone **109** gives only the *cis*-dideuterio and *cis*-diprotio olefins (equation 123).

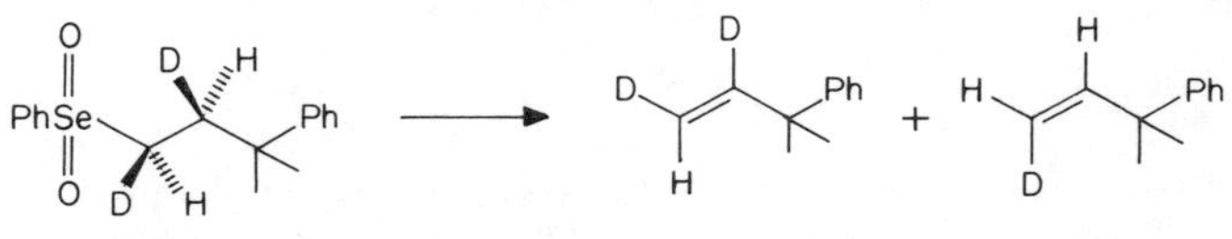

(123)

(109)

The excellent leaving-group properties of selenones in S_N2 displacements have recently been used for some synthetic applications. Towards dimethyl sulphide compound **101a** is approximately three times as reactive as methyl iodide[258] (equation 124). The cleavage of dimethyl selenone by HCl was observed early by Paetzold and Bochmann[247].

$$PhMeSeO_2 \xrightarrow[CD_3OD,\,35\,^{\circ}C]{Me_2S} PhSe(=O)O^- + Me_3S^+ \quad (124)$$

$$Me_2SeO_2 \xrightarrow{HCl(aq.)} MeCl + MeSe(=O)OH$$

Shimizu and Kuwajima[272] have developed a synthetic sequence for the synthesis of oxetanes, involving a selenone as a leaving group. In this process 3-(phenylseleno)-2-propenal (**110**) is attacked by a Grignard reagent and the resulting alcohol oxidized with two equivalents of *m*-chloroperbenzoic acid (MCPBA) to a selenone. On treatment with sodium hydroxide in aqueous methanol (Michael attack by methoxide), good yields of isomeric oxetanes can be isolated as outlined in equation (125).

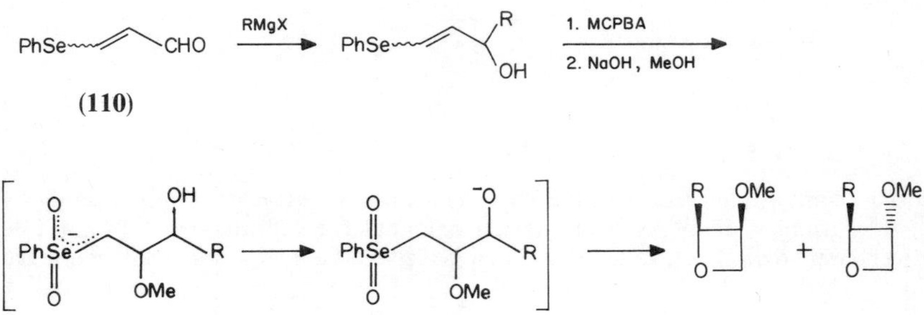

(125)

The vinyl selenones **111** and **112** have been shown to undergo fragmentation on treatment with bases[273]. The former, after Michael attack by alkoxide, gives an ethylenic ketone (equation 126) and the latter (a more hindered tetrasubstituted olefin) directly gives an acetylenic ketone (equation 127).

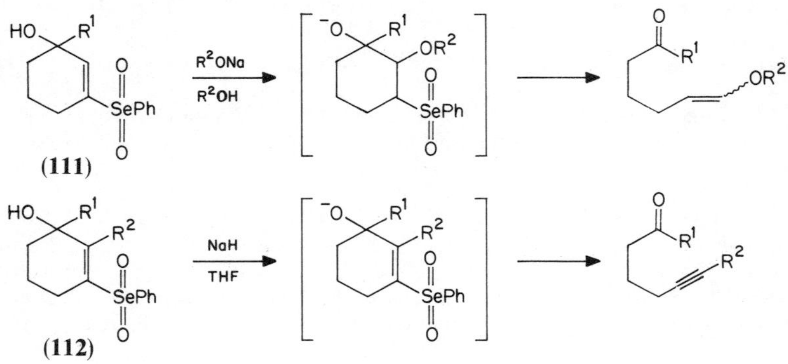

(126)

(127)

C. Tellurones

Diorganyl tellurones, R_2TeO_2, the Te analogues of the well-known sulphones and selenones, are ill-defined substances. Although reported in the literature since 1920, their existence has been questioned in two more recent review articles[274,275].

Vernon[276] oxidized dimethyl telluride with hydrogen peroxide and claimed to have isolated dimethyl tellurone (**113**), an insoluble, white, amorphous powder. This compound possessed all the characteristics of a peroxide; thus it had explosive properties and oxidized halogen acids to the corresponding halogen. It is therefore highly probable that Vernon's compound was a 'hydroxyperhydrate' (**114**) instead of the purported tellurone. This type of structure for the hydrogen peroxide oxidation product was first suggested by Balfe and coworkers[277].

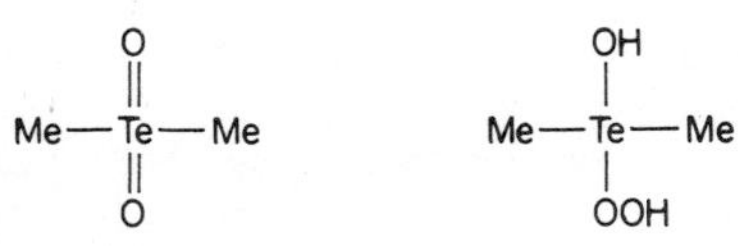

(**113**) (**114**)

Diethyl telluride[278], telluracyclohexane (**115**)[279], phenoxtellurine (**116**)[280] and telluroisochroman (**117**)[281], respectively, have all been submitted to perhydrol treatment in order to obtain the corresponding tellurones. These products were probably also of the 'hydroxyperhydrate' type.

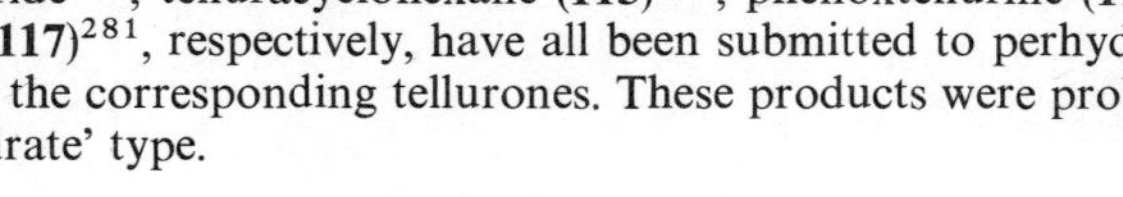

(**115**) (**116**) (**117**)

The first definitely characterized diorganyl tellurones were prepared only recently by Cava[175,282], using sodium periodate as the oxidant. The telluroxides **118** and **119**, respectively, were oxidized in aqueous methanol to give the corresponding tellurones **120** and **121**.

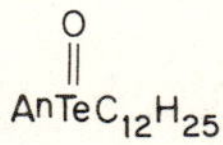

(118)

$An_2Te{=}O$

(119)

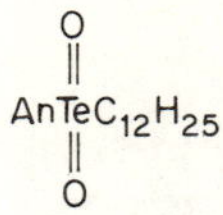

(120)

An_2TeO_2

(121)

An = 4-$MeOC_6H_4$

The role of tellurones in organic synthesis has of course been very little explored. However, the tellurone **120**, when thermolysed in refluxing toluene for 40 h, gives a mixture of 1-dodecene (55%) and 1-dodecanol (15%)[175]. These products are probably formed via the rearrangement product **122**.

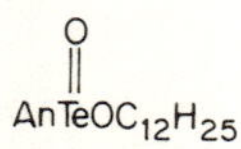

(122)

The mild oxidizing properties of tellurone **121** have recently been investigated[282]. Thus, benzenethiol was oxidized to diphenyl disulphide and hydroquinone converted into *p*-benzoquinone. Benzylic alcohols could be converted to the corresponding carbonyl compounds, e.g. veratryl alcohol (**123**) was oxidized to veratraldehyde in 79% yield. On the other hand, hydrobenzoin (**124**) was cleaved by the tellurone to give benzaldehyde in 79% yield.

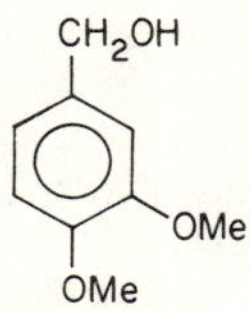

(123)

PhCH(OH)CH(OH)Ph

(124)

Uemura[283,284] has recently discovered an oxidative procedure for cleavage of C—Te bonds with introduction of a methoxy group (equation 128).

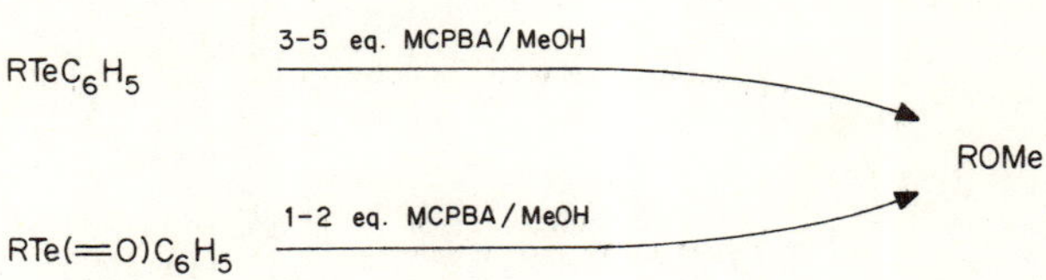

(128)

R = $C_{12}H_{25}$

These reactions probably involve the formation of a tellurone **125** and its *m*-chloroperbenzoic acid adduct **126**. A preformed tellurone **125** required treatment with excess *m*-chloroperbenzoic acid to give a good yield of the ether ROMe[285].

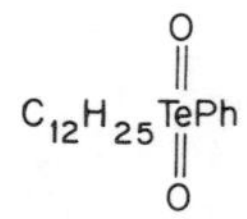

(125)

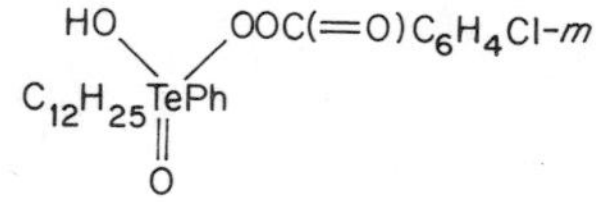

(126)

VII. HIGHER-VALENT COMPOUNDS CONTAINING Te—Te OR Se—Se BONDS

Although ditellurides of the type RTeTeR are well known, related hypervalent compounds such as $RTeX_2TeX_2R$ are rare and the only reasonably well-characterized case has been reported by Schulz and Klar[286], who isolated, by treatment of the corresponding ditelluride with iodine, compound **127**, as a lilac-brown diamagnetic powder which decomposed at 145–150 °C.

$$2\text{-}PhC_6H_4TeI_2{-}TeI_2C_6H_4Ph\text{-}2$$

(127)

Recently Zingaro and coworkers[287] have obtained evidence (^{13}C- and ^{125}Te-NMR data) for the formation of the ditellurone **128** together with **129** and ArTeSeTeAr from the oxidation of ArTeTeAr with SeO_2.

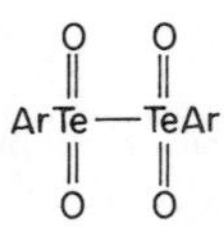

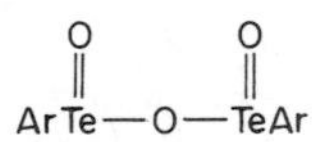

$Ph_3TeTePh_3$

(128) **(129)** **(130)**

No published data are available about hexaaryl- or hexaalkyl-ditelluranes, but in connection with studies[288] of the interaction of PhLi and PhMgBr with elemental Te a compound with elemental analytical and mass spectrometric data in agreement with the formulation **130** could be obtained in low and varying yields. Attempts to reductively couple $Ph_3Te^+Cl^-$ to $Ph_3TeTePh_3$ failed.

Reich and coworkers have recently reported[289] the first example of an observable selenolseleninate ester (**131**). The compound is moderately stable at $-50\,°C$ ($t_{1/2} \approx 1$ h). The cyclic selenolseleninate **132** was also prepared by oxidation of the corresponding cyclic diselenide with *m*-chloroperbenzoic acid. The PMR spectrum of **132** is particularly characteristic, showing *AB* quartets for the CH_2 protons and two singlets for the diastereotopic Me groups.

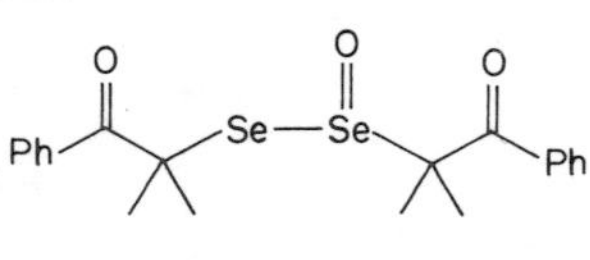

(131)

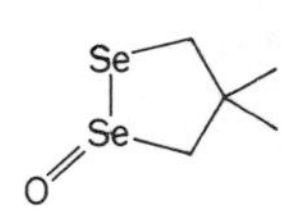

(132)

VIII. REFERENCES

1. J. I. Musher, *Ann. N.Y. Acad. Sci.*, **192**, 52 (1972).
2. D. L. Klayman and W. H. H. Günther, *Organic Selenium Compounds*, Wiley–Interscience, New York, 1973.

3. K. J. Irgolic, *The Organic Chemistry of Tellurium*, Gordon and Breach, New York, 1974.
4. I. D. Sadekov, A. Ya. Bushkov and V. I. Minkin, *Russ. Chem. Rev.*, **48**, 343 (1979).
5. R. S. Michalak, S. R. Wilson and J. C. Martin, *J. Amer. Chem. Soc.*, **106**, 7529 (1984).
6. H. J. Reich, *J. Amer. Chem. Soc.*, **95**, 964 (1973).
7. J. D. McCullough and G. Hamburger, *J. Amer. Chem. Soc.*, **64**, 508 (1942).
8. R. Gillespie and R. Nyholm, *Quart. Rev.*, **9**, 339 (1957).
9. W. R. McWhinnie, *Proceedings of the 4th International Symposium on Organic Selenium and Tellurium Compounds, Birmingham, UK*, 1983.
10. R. F. Ziolo and J. M. Troup, *J. Amer. Chem. Soc.*, **105**, 229 (1983).
11. M. M. Mangion, R. A. Zingaro and E. A. Meyers, *Chem. Scripta*, **8A**, 45 (1975).
12. H. D. K. Drew, *J. Chem. Soc.*, 564 (1929).
13. F. Einstein, J. Trotter and C. Williston, *J. Chem. Soc. (A)*, 2018 (1967).
14. L. Chan and F. Einstein, *J. Chem. Soc., Dalton Trans.*, 316 (1972).
15. R. F. Ziolo and W. H. H. Günther, *J. Organometal. Chem.*, **146**, 245 (1978).
16. C. P. Smyth, S. J. Grossman and S. R. Ginsburg, *J. Amer. Chem. Soc.*, **62**, 192 (1940).
17. (a) D. Bendix, W. Grütjen, P. Laur and I. Karup Nielsen, *Proceedings of the 4th International Symposium on Organic Selenium and Tellurium Compounds, Birmingham, UK*, 1983.
(b) W. Nakanishi, Y. Yamamoto, Y. Kusuyama, Y. Ikeda and H. Iwamura, *Chem. Letters*, 675 (1983).
18. J. Bergman, *Tetrahedron*, **28**, 3323 (1972).
19. W. H. H. Günther, J. Nepywoda and J. Y. C. Chu, *J. Organometal. Chem.*, **74**, 79 (1974).
20. Y. Ishii, Y. Iwama and M. Ogawa, *Synth. Commun.*, **8**, 93 (1978).
21. J. Bergman and N. Eklund, *Tetrahedron*, **36**, 1439 (1980).
22. J. Bergman, R. Carlsson and B. Sjöberg, *Org. Synth.*, **57**, 18 (1977).
23. P. Thavornyutikarn and W. R. McWhinnie, *J. Organometal. Chem.*, **50**, 135 (1973).
24. F. M. Musa and W. R. McWhinnie, *J. Organometal. Chem.*, **159**, 37 (1978).
25. T. N. Srivastava, R. C. Srivastava and M. Singh, *J. Organometal. Chem.*, **160**, 499 (1978).
26. B. C. Pant, W. R. McWhinnie and N. S. Dance, *J. Organometal. Chem.*, **63**, 305 (1973).
27. R. Paetzold and U. Lindner, *Z. Anorg. Allgem. Chem.*, **350**, 295 (1967).
28. W. R. McWhinnie and J. Mallaki, *Polyhedron*, **1**, 13 (1982).
29. N. S. Dance, W. R. McWhinnie, J. Mallaki and Z. Monsef-Mirzai, *J. Organometal. Chem.*, **198**, 131 (1980).
30. R. F. Ziolo and D. D. Titus, *Proceedings of the 3rd International Symposium on Organic Selenium and Tellurium Compounds, Metz 9–12 July*, 1979, pp. 319–343.
31. R. F. Ziolo, C. J. Thornton, A. C. Smith, D. D. Titus, C. S. Smith and N. Buno, Jr., *J. Organometal. Chem.*, **190**, C64 (1980).
32. Y. Iwama, M. Aragai, M. Suiyama, K. Matsui, Y. Ishii and M. Ogawa, *Bull. Chem. Soc. Japan*, **54**, 2065 (1981).
33. E. R. Clark and M. A. Al-Turaihi, *J. Organometal. Chem.*, **96**, 251 (1975).
34. T. Weiss, W. Nitsche, F. Böhnke and G. Klar, *Ann.*, 1418 (1973).
35. B. Greenberg. E. S. Gould and W. Burlont, *J. Amer. Chem. Soc.*, **78**, 4028 (1956).
36. E. R. Clark and M. A. Al-Turaihi, *J. Organometal. Chem.*, **124**, 391 (1977).
37. R. E. Nelson, E. F. Degering and J. A. Bilderback, *J. Amer. Chem. Soc.*, **60**, 1239 (1938).
38. H. Funk and W. Papenroth, *J. Prakt. Chem.*, **11**, 191 (1960).
39. G. T. Morgan and F. H. Burstall, *J. Chem. Soc.*, 3260 (1928).
40. T. N. Srivastava, R. C. Srivastava and M. Singh, *Inorg. Chim. Acta*, **33**, L99 (1979).
41. I. D. Sadekov and A. A. Maksimenko, *Zh. Obshch. Khim.*, **47**, 1918 (1977).
42. I. D. Sadekov, A. Y. Bushkov, V. L. Pavlova, Y. L. Yureva and V. I. Minkin, *Zh. Obshch. Khim.*, **47**, 1305 (1977).
43. R. M. Mingaev, I. D. Sadekov and V. I. Minkin, *Zh. Obshch. Khim.*, **47**, 2011 (1977).
44. I. D. Sadekov, A. I. Usachev, V. A. Bren, M. S. Korolov, I. D. Tseimakh and V. I. Minkin, *Zh. Obshch. Khim.*, **47**, 2332 (1977).
45. R. C. Paul, K. K. Bhasim and R. K. Chadha, *Inorg. Nucl. Chem.*, **37**, 2339 (1975).
46. B. C. Pant, *J. Organometal. Chem.*, **54**, 191 (1973).
47. V. Kumar, P. H. Bird and B. C. Pant, *Synth. React. Inorg. Met.-org. Chem.*, **9**, 203 (1979).
48. T. N. Srivastava, R. C. Srivastava and K. Kapoor, *J. Inorg. Nucl. Chem.*, **41**, 413 (1979).
49. Z. A. Siddiqi, M. Shakir, M. Aslam and S. A. A. Zaidi, *Synth. React. Inorg. Met.-org. Chem.*, **13**, 173 (1983).

50. T. N. Srivastava, P. C. Srivastava and R. Kumar, *J. Ind. Chem. Soc.*, **59**, 186 (1982).
51. N. Petragnani, *Tetrahedron*, **12**, 219 (1961).
52. G. Kirsch, M. M. Goodman and F. F. Knapp, Jr., *Organometallics*, **2**, 357 (1983).
53. E. C. Kooyman, A. V. Huygens, J. P. J. de Lepper, D. de Vos and J. Wolters, *Rec. Trav. Chim.*, **100**, 24 (1981)
54. J. Spierenburg, *Dissertation*, University of Leiden, 1970.
55. A. V. Huygens, *Dissertation*, University of Leiden, 1972.
56. E. C. Kooyman, J. Wolters, J. Spierenburg and J. Reedijk, *J. Organometal. Chem.*, **3**, 487 (1965).
57. J. Verweij, *Dissertation*, University of Leiden, 1969.
58. K. J. Wynne and J. Puckett, *Chem. Commun.*, 1532 (1968).
59. K. J. Wynne, *Inorg. Chem.*, **9**, 299 (1970).
60. I. Ruppert, *Chem. Ber.*, **112**, 3023 (1979).
61. D. Aumann and S. Herberg, *J. Fluorine Chem.*, **19**, 205 (1982).
62. S. Herberg and D. Naumann, *Z. Anorg. Allgem. Chem.*, **492**, 95 (1982).
63. S. Herberg and D. Naumann, *Z. Anorg. Allgem. Chem.*, **494**, 151, 159 (1982).
64. C. D. Desjardins, C. Lau and J. Passmore, *Inorg. Nucl. Chem. Letters*, **10**, 151 (1974).
65. C. D. Desjardins, H. L. Paige, J. Passmore and P. Taylor, *J. Chem. Soc., Dalton Trans.*, 488 (1975).
66. I. D. Sadekov, A. Yu. Bushkov, L. N. Markovskii and V. I. Minkin, *Zh. Obshch. Khim.*, **46**, 1660 (1976).
67. H. M. Leicester and F. W. Bergstrom, *J. Amer. Chem. Soc.*, **51**, 3587 (1929).
68. W. E. Bradt and J. F. Green, *J. Org. Chem.*, **1**, 540 (1937).
69. H. J. Emeleus, *J. Chem. Soc.*, 1126 (1946).
70. G. T. Morgan and O. C. Elvins, *J. Chem. Soc.*, 2625 (1925).
71. A. Michaelis and F. Kunckell, *Ber.*, **30**, 2826 (1897).
72. F. Kunckell and R. Zimmermann, *Ann.*, **314**, 281 (1901).
73. R. E. Nelson and R. N. Jones, *J. Amer. Chem. Soc.*, **52**, 1588 (1930).
74. R. G. Turnbo, *Dissertation*, University of Texas, 1965.
75. I. D. Sadekov, A. A. Maksimenko and B. B. Rivkin, *Zh. Org. Khim.*, **14**, 874 (1978).
76. N. Petragnani, *Tetrahedron*, **11**, 15 (1960).
77. D. H. O'Brien, *Proceedings of the 4th International Symposium on Organic Selenium and Tellurium Compounds, Birmingham, UK*, 1983.
78. H. J. Gysling, *Proceedings of the 3rd International Symposium on Organic Selenium and Tellurium Compounds, Metz 9–12 July*, 1979, pp. 375–390.
79. H. J. Gysling, H. R. Luss and S. Gardner, *J. Organometal. Chem.*, **184**, 417 (1980).
80. D. G. Marsh, J. Y. C. Chu, J. W. Lewicki and J. L. Weaver, *J. Amer. Chem. Soc.*, **98**, 8432 (1976).
81. E. Ajello, *J. Heterocycl. Chem.*, **9**, 1427 (1972).
82. H. Rheinboldt, *Houben–Weyl: Methoden der Organischen Chemie*, Vol. IX (Ed. E. Müller), Georg Thieme Verlag, Stuttgart, 1955, p. 1008.
83. L. Futekov, S. Stojanov and H. Specker, *Fresenius Z. Anal. Chem.*, **294**, 125 (1979).
84. G. T. Morgan and H. D. K. Drew, *J. Chem. Soc.*, 1456 (1920).
85. G. T. Morgan, H. D. K. Drew and T. V. Barker, *J. Chem. Soc.*, 2432 (1922).
86. D. H. Dewar, J. E. Fergusson, P. R. Hebtchek, C. J. Wilkins and P. P. Williams, *J. Chem. Soc.*, 688 (1964).
87. C. L. Raston, R. J. Secomb and A. H. White, *J. Chem. Soc., Dalton Trans.*, 2307 (1976).
88. G. L. Parks, *Dissertation*, Clemson University, 1968.
89. J. C. Dewan and J. Silver, *Australian J. Chem.*, **30**, 487 (1977).
90. J. C. Dewan and J. Silver, *Acta Cryst.*, **33B**, 2671 (1977).
91. J. C. Dewan and J. Silver, *Acta Cryst.*, **33B**, 1469 (1977).
92. J. C. Dewan and J. Silver, *J. Chem. Soc., Dalton Trans.*, 644 (1977).
93. J. C. Dewan and J. Silver, *J. Organometal. Chem.*, **125**, 125 (1977).
94. R. Kivekäs, T. Simonen and T. Laitalainen, to be published.
95. J. Bergman and L. Engman, unpublished results.
96. T. Laitalainen, *Dissertation*, University of Helsinki, 1982; *Ann. Acad. Sci. Fenn (A)*, 195 (1982).

97. T. Laitalainen, *Finn. Chem. Letters*, 15 (1982).
98. T. Laitalainen, T. Simonen, R. Kivekäs and M. Klinga, *J. Chem. Soc., Perkin Trans. 1*, 333 (1982).
99. M. Sevrin, W. Dumont, L. Hevesi and A. Krief, *Tetrahedron Letters*, 2647 (1976).
100. G. T. Morgan and H. D. K. Drew, *J. Chem. Soc.*, 531 (1925).
101. G. T. Morgan and R. E. Kellett, *J. Chem. Soc.*, 1080 (1926).
102. F. J. Berry, B. C. Smith and C. Jones, *J. Organometal. Chem.*, **110**, 201 (1976).
103. F. Wudl, in *Chemistry and Physics of One-dimensional Metals* (Ed. H. J. Keller), Plenum, New York, 1977, p. 233.
104. F. Wudl, *Proceedings of the 4th International Symposium on Organic Selenium and Tellurium Compounds, Birmingham, UK*, 1983.
105. V. L. Lendel, Yu. V. Migalina, S. V. Galla, A. S. Kozmin and N. S. Zefirov, *Khim. Get. Soedin.*, 1340 (1977).
106. S. Patai, K. A. Muszkat and M. Sokolovsky, *J. Chem. Soc.*, 734 (1962).
107. D. Elmaleh, S. Patai and Z. Rappoport, *J. Chem. Soc. (C)*, 939 (1970).
108. D. Elmaleh, S. Patai and Z. Rappoport, *Israel J. Chem.*, **9**, 155 (1971).
109. M. de Moura Campos and N. Petragnani, *Tetrahedron*, **6**, 11 (1959).
110. M. de Moura Campos and N. Petragnani, *Tetrahedron*, **18**, 521 (1962).
111. M. Ogawa and R. Ishioka, *Bull. Chem. Soc. Japan*, **43**, 496 (1970).
112. H. J. Arpe and H. Kuckertz, *Angew. Chem. (Intern. Ed. Engl.)*, **10**, 73 (1971).
113. J. E. Bäckvall and L. Engman, *Tetrahedron Letters*, **22**, 1919 (1981).
114. J. E. Bäckvall, J. Bergman and L. Engman, *J. Org. Chem.*, **48**, 3918 (1983).
115. J. Bergman and L. Engman, *J. Organometal. Chem.*, **181**, 335 (1979).
116. Halcon Research and Development Corp., New York, *US Patent*, No. 4, 271 090 (1979).
117. H. W. Bausor, C. S. Gibson and W. J. Pope, *J. Chem. Soc.*, 1453 (1920).
118. C. E. Boord and F. F. Cope, *J. Amer. Chem. Soc.*, **44**, 395 (1922).
119. H. Brintzinger, K. Pfannstiel and H. Vogel, *Z. Anorg. Allgem. Chem.*, **256**, 75 (1948).
120. F. Lautenschlager, *J. Org. Chem.*, **34**, 4002 (1969).
121. C. E. Frick, *J. Amer. Chem. Soc.*, **45**, 1795 (1923).
122. H. L. Riley and N. A. C. Friend, *J. Chem. Soc.*, 2342 (1932).
123. H. Funk and W. Weiss, *J. prakt. Chem. (4)*, **1**, 33 (1954).
124. H. Funk and W. Papenroth, *J. prakt. Chem. (4)*, **8**, 256 (1959).
125. Yu. V. Migalina, S. V. Galla-Bobik, S. M. Khripak and V. I. Staninets, *Khim. Get. Soedin.*, 911 (1982).
126. Yu. V. Migalina, S. V. Galla-Bobik and V. I. Staninets, *Zh. Org. Khim.*, **18**, 2609 (1982).
127. F. Riley, J. Flato and D. Bengels, *J. Org. Chem.*, **27**, 2651 (1962).
128. H. C. Bell and C. S. Gibson, *J. Chem. Soc.*, 1877 (1925).
129. T. Smedslund, *Finska Kemistsamfundets Medd.*, **41**, 13 (1932).
130. D. Garratt, M. Ujjainwalla and G. H. Schmid, *J. Org. Chem.*, **45**, 1206 (1980).
131. S. Uemura, A. Onoe and M. Okano, *Bull. Chem. Soc. Japan*, **48**, 3702 (1975).
132. S. Uemura, A. Onoe and M. Okano, *Bull. Chem. Soc. Japan*, **50**, 1078 (1977).
133. Yu. V. Migalina, S. V. Galla-Bobik, V. G. Lendel and V. I. Staninets, *Zh. Obshch. Khim.*, **52**, 1563 (1982).
134. D. Kobelt and E. F. Paulus, *J. Organometal. Chem.*, **27**, C63 (1971).
135. M. Ogawa, *Bull. Chem. Soc. Japan*, **41**, 3031 (1968).
136. J. Bergman and L. Engman, *J. Amer. Chem. Soc.*, **103**, 2715 (1981).
137. J. Bergman and L. Engman, *J. Amer. Chem. Soc.*, **103**, 5196 (1981).
138. D. G. Garratt and G. H. Schmid, *Can. J. Chem.*, **52**, 3599 (1974).
139. D. G. Garratt and G. H. Schmid, *J. Org. Chem.*, **42**, 1776 (1977).
140. G. H. Schmid and D. G. Garratt, in *Supplement A: The Chemistry of Double-bonded Functional Groups* (Ed. S. Patai), John Wiley and Sons, London, 1977, p. 866.
141. J. Bergman and L. Engman, *Tetrahedron*, **36**, 1275 (1980).
142. S. Uemura, S.-I. Fukuzawa and A. Toshimitsu, *J. Organometal. Chem.*, **250**, 203 (1983).
143. Yu. V. Migalina, V. G. Lendel, I. M. Balog and V. I. Staninets, *Ukr. Khim. Zh.*, **47**, 1293 (1981); *Chem. Abstr.*, **96**, 104001z (1982).
144. Yu. V. Migalina, V. I. Staninets, V. G. Lendel, A. S. Koz'min and N. S. Zefirov, *Khim. Get. Soedin.*, 1633 (1977).

145. M. Albeck and T. Tamary, *J. Organometal Chem.*, **164**, C23 (1979).
146. Yu. V. Migalina, V. G. Lendel, A. S. Koz'min and N. S. Zefirov, *Khim. Get. Soedin.*, 708 (1978).
147. Yu. V. Migalina, V. I. Staninets, V. G. Lendel, I. M. Balog, V. A. Palyulin, A. S. Koz'min and N. S. Zefirov, *Khim. Get. Soedin.*, 58 (1977).
148. Yu. V. Migalina, I. M. Balog, V. G. Lendel, A. S. Koz'min and N. S. Zefirov, *Khim. Get. Soedin.*, 1212 (1978).
149. J. Bergman, *Proceedings of the 4th International Symposium on Organic Selenium and Tellurium Compounds, Birmingham, UK*, 1983.
150. *USSR Patent*, No. 450, 806 (1974); *Chem. Abstr.*, **82**, 125418 (1975).
151. J. Bergman, J. Sidén and K. Maartmann-Moe, *Tetrahedron*, **40**, 1607 (1984).
152. R. F. Riley, J. Flato and P. McIntyre, *J. Org. Chem.*, **28**, 1138 (1963).
153. Yu. V. Migalina, S. V. Galla-Bobik, V. G. Lendel and V. I. Staninets, *Khim. Get. Soedin.*, 1283 (1981).
154. M. de Moura Campos and N. Petragnani, *Tetrahedron*, **18**, 527 (1962).
155. I. D. Sadekov and V. I. Minkin, *Khim. Get. Soedin.*, 138 (1971).
156. S. Uemura, H., Miyoshi and M. Okano, *Chem. Letters*, 1357 (1979).
157. J. Bergman and L. Engman, *Tetrahedron Letters*, 1509 (1979).
158. J. Bergman and L. Engman, *J. Organometal. Chem.*, **201**, 377 (1980).
159. Yu. V. Migalina, S. V. Galla-Bobik, I. I. Ershova and V. I. Staninets, *Zh. Obshch. Khim.*, **52**, 1559 (1982).
160. See Ref. 140, p. 872.
161. A. Marquardt and A. Michaelis, *Ber.*, **21**, 2042 (1888).
162. G. Wittig and H. Fritz, *Ann.*, **577**, 39 (1952).
163. (a) D. Hellwinkel and G. Fahrbach, *Ann.*, **715**, 68 (1968).
(b) D. Hellwinkel and G. Fahrbach, *Ann.*, **712**, 1 (1968).
(c) D. Hellwinkel and G. Fahrbach, *Chem. Ber.*, **101**, 574 (1968).
(d) D. Hellwinkel, *N.Y. Acad. Sci.*, **192**, 158 (1972).
164. S. A. Glover, *J. Chem. Soc., Perkin Trans. 1*, 1338 (1980).
165. D. L. J. Clive, *Tetrahedron*, **34**, 1049 (1978).
166. H. J. Reich, *Acc. Chem. Res.*, **12**, 22 (1979).
167. F. Ogura, H. Uamaguchi, T. Otsubo and H. Tanaka, *Bull. Chem. Soc. Japan*, **55**, 641 (1982).
168. R. D. Larsen, *Dissertation*, University of Michigan, Ann Arbor, 1982.
169. J. H. Marino and R. D. Larsen, *J. Amer. Chem. Soc.*, **103**, 4642 (1981).
170. I. D. Sadekov and B. B. Rivkin, *Zh. Org. Khim.*, **18**, 227 (1982).
171. M. Wieber and E. Kaunziger, *J. Organometal Chem.*, **129**, 339 (1977).
172. N. W. Alcock and W. D. Harrison, *J. Chem. Soc., Dalton Trans.*, 709 (1982).
173. C. Klofutar, F. Krasovec and M. Kusar, *Croat. Chem. Acta*, **40**, 23 (1968).
174. L. Engman and M. P. Cava, *Tetrahedron Letters*, **22**, 5252 (1981).
175. H. Lee and M. P. Cava, *Chem. Commun.*, 277 (1981).
176. S. Ucmura and S. Fukuzawa, *J. Amer. Chem. Soc.*, **105**, 2748 (1983).
177. I. D. Sadekov, B. B. Rivkin and A. A. Maksimenko, *Izv. Sev. Kauk. Nauchn. Tsentra Vyssh. Shk. Estestv. Nauki.*, 54 (1982).
178. D. H. R. Barton, S. V. Ley and C. A. Meerholz, *Chem. Commun.*, 755 (1979).
179. S. V. Ley, C. A. Meerholz and D. H. R. Barton, *Tetrahedron*, **37**, Suppl. 1, 213 (1981).
180. H. Meerwein and T. Bersin, *Ann.*, **476**, 113 (1929).
181. D. B. Denney, D. Z. Denney, Ph. J. Hammond and Y. F. Hsu, *J. Amer. Chem. Soc.*, **103**, 2340 (1981).
182. R. Paetzold and M. Reichenbacher, *Z. Chem.*, **10**, 307 (1970).
183. L. N. Markovskii, E. A. Stukalo, A. N. Ogloblin and S. V. Iksanova, *Zh. Org. Chim.*, **19**, 72 (1983).
184. D. B. Denney, D. Z. Denney and Y. F. Hsu, *J. Amer. Chem. Soc.*, **100**, 5982 (1978).
185. M. R. Detty and B. J. Murray, *J. Org. Chem.*, **47**, 1148 (1982).
186. B. Nakhdjavan and G. Klar, *Ann.*, 1683 (1977).
187. F. A. H. Friedheim, *US Patent*, No. 3,544,593 (1967); *Chem. Abstr.*, **74**, 87991 (1971).
188. M. Czauderna and K. Samochocka, *Tetrahedron*, **38**, 2421 (1982).
189. E. A. Stukalo, E. M. Yureva and L. N. Markovskii, *Zh. Org. Khim.*, **19**, 343 (1983).
190. S. Huseby, *Acta Chem. Scand.*, **19**, 1045 (1965).

191. S. Huseby and S. E. Svaeren, *Acta Chem. Scand.*, **27**, 763 (1973).
192. D. Huseby, *Proceedings of the 4th International Symposium on Organic Selenium and Tellurium Compounds, Birmingham*, 1983.
193. W. Schnabel, K. von Deuten and G. Klar, *Phosphorus and Sulfur*, **13**, 345 (1982).
194. W. Schnabel, K. von Deuten and G. Klar, *Cryst. Struct. Commun.*, **10**, 1405 (1981).
195. K. von Deuten, W. Schnabel and G. Klar, *Phosphorus and Sulfur*, **9**, 93 (1980).
196. K. von Deuten, W. Schnabel and G. Klar, *Cryst. Struct. Commun.*, **8**, 221 (1974).
197. V. I. Naddaka, V. P. Garkin, I. D. Sadekov and V. I. Minkin, *Zh. Org. Khim.*, **13**, 220 (1977).
198. L. N. Markovskii, E. A. Stukalo, and G. P. Kunitskaya, *Zh. Org. Khim.*, **13**, 2055, 2514 (1977).
199. V. I. Naddaka and V. P. Garkin, *Zh. Org. Khim.*, **17**, 401 (1981).
200. N. G. Furmanova, V. I. Naddaka and V. P. Krasnov, *Zh. Struct. Khim.*, **22**, 118 (1981).
201. T. Otsubo, F. Ogura and H. Yamaguchi, *Chem. Letters*, 447 (1981).
202. D. Spinelli and G. Monako, *Boll. Sci. Fac. Chim. Ind. Bologna*, **20**, 56 (1962).
203. N. Ya. Derkach, T. V. Lyapina and N. A. Tasmurtseva, *Zh. Org. Khim.*, **10**, 807 (1974).
204. S. Tamagaki and K. Sakaki, *Chem. Letters*, 503 (1975).
205. S. Tamagaki, S. Oae, and K. Sakaki, *Tetrahedron Letters*, 649 (1975).
206. S. Oae, M. Fukumura and N. Furukawa, *Phosphorus and Sulfur*, **10**, 153 (1981).
207. K. B. Sharpless and S. P. Singer, *J. Org. Chem.*, **41**, 2504 (1976).
208. K. B. Sharpless, T. Hori, L. K. Truesdale and C. O. Dietrich, *J. Amer. Chem. Soc.*, **98**, 269 (1976).
209. N. Ya. Derkach and G. G. Barashenkov, *Zh. Org. Khim.*, **12**, 100 (1976).
210. N. Ya. Derkach, G. G. Barashenkov and E. S. Levchenko, *Zh. Org. Khim.*, **15**, 1547 (1979).
211. N. Ya. Derkach, G. G. Barashenkov and E. I. Slyusarenko, *Zh. Org. Khim.*, **18**, 70 (1982).
212. (a) N. Ya. Derkach. T. V. Lyapina and E. S. Levchenko, *Zh. Org. Khim.*, **16**, 33 (1980).
(b) N. Ya. Derkach, G. G. Barashenkov and E. S. Levchenko, *Zh. Org. Khim.*, **19**, 1622 (1983).
213. N. Ya. Derkach and T. V. Lyapina, *Zh. Org. Khim.*, **17**, 529 (1981).
214. J. M. Miller and R. K. Chadha, *J. Organometal. Chem.*, **216**, 177 (1981).
215. N. Petragnani, L. T. Castellanos, K. J. Wynne and W. Maxwell, *J. Organometal. Chem.*, **55**, 295 (1973).
216. N. Petragnani, J. V. Comasseto and Y. Kawazsuo, *J. Inorg. Nucl. Chem.*, **38**, 608 (1976).
217. N. Petragnani, L. Torres, K. J. Wynne and D. J. Williams, *J. Organometal. Chem.*, **76**, 241 (1974).
218. J. Bergman and T. Laitalainen, *Heterocycles*, **23**, 37 (1985).
219. J. Bergman, J. Sidén and K. Maartmann-Moe, unpublished results.
220. M. R. Detty and H. R. Luss, *J. Org. Chem.*, **48**, 5149 (1983).
221. G. Pellini, *Gazz. Chim. Ital.*, **46**, 247 (1916).
222. K. Seppelt, *Angew. Chem.*, **94**, 890 (1982), and Refs. therein.
223. G. W. Fraser and G. D. Meikle, *Chem. Commun.*, 625 (1974).
224. U. Elgad and H. Selig, *Inorg. Chem.*, **14**, 140 (1975).
225. W. Tötsch and F. Sladky, *Chem. Commun.*, 927 (1980).
226. W. Tötsch, P. Peringar and F. Sladky, *Chem. Commun.*, 841 (1981).
227. K. Seppelt and D. Nötke, *Inorg. Chem.*, **12**, 2727 (1973).
228. W. Tötsch and F. Sladky, *J. Fluorine Chem.*, **19**, 213 (1982).
229. B. Bildstein, W. Tötsch and F. Sladky, *Z. Naturforsch.*, **36B**, 1542 (1981).
230. W. Tötsch and F. Sladky, *Chem. Ber.*, **115**, 1019 (1982).
231. A. Clouston, R. D. Peacock and G. W. Fraser, *Chem. Commun.*, 1197 (1970).
232. G. W. Fraser and J. B. Millar, *J. Chem. Soc., Dalton Trans.*, 2029 (1974).
233. G. W. Fraser and G. D. Meikle, *J. Chem. Soc., Perkin Trans. 2*, 312 (1975).
234. G. W. Fraser and G. D. Meikle, *J. Chem. Soc., Dalton Trans.*, 1033 (1975).
235. G. W. Fraser and G. D. Meikle, *J. Chem. Soc., Dalton Trans.*, 1985 (1977).
236. G. W. Fraser and G. D. Meikle, *J. Chem. Soc., Perkin Trans. 1*, 2005 (1979).
237. G. W. Fraser and R. D. Peacock and P. M. Watkins, *J. Chem. Soc. (A)*, 1125 (1971).
238. K. Seppelt, *Inorg. Chem.*, **12**, 2837 (1973).
239. H. Hartl, P. Huppmann, D. Lentz and K. Sepplet, *Inorg. Chem.*, **22**, 2183 (1983).
240. C. D. Desjardins, C. Lou and J. Passmore, *Inorg. Nucl. Chem. Letters*, **9**, 1037 (1973).
241. R. S. Michalak and J. C. Martin, *J. Amer. Chem. Soc.*, **104**, 1683 (1982).
242. R. K. Pomeroy, L. Vancea, H. P. Calhoun and W. A. G. Graham, *Inorg. Chem.*, **16**, 1508 (1977).

243. F. N. Tebbe, P. Meakin, J. P. Jesson and E. L. Muetterties, *J. Amer. Chem. Soc.*, **92**, 1068 (1970).
244. H. Rheinboldt, in *Houben–Weyl: Methoden der Organischen Chemie*, Vol. IX, 4th ed. (Ed. E. Müller), Georg Thieme Verlag, Stuttgart, 1955, p. 1030.
245. L.-B. Agenäs, in *Organic Selenium Compounds: Their Chemistry and Biology* (Eds. D. L. Klayman and W. H. H. Günther), John Wiley and Sons, New York, 1973, p. 209.
246. R. A. Zingaro and W. C. Cooper, *Selenium*, Van Nostrand Reinhold Company, New York, 1974, p. 504.
247. R. Paetzold and G. Bochmann, *Z. Anorg. Allgem. Chem.*, **360**, 293 (1968).
248. M. L. Bird and F. Challenger, *J. Chem. Soc.*, 570 (1942).
249. J. B. Lambert, C. E. Mixan and D. H. Johnson, *J. Amer. Chem. Soc.*, **95**, 4634 (1973).
250. E. Rebane, *Chem. Scripta*, **5**, 65 (1974).
251. B. A. Arbusow, *J. Prakt. Chem.* (*2*), **131**, 357 (1931).
252. W. Dilthey, L. Neuhaus, E. Reis and W. Schommer, *J. Prakt. Chem.* (*2*), **124**, 81 (1930).
253. R. Lesser and R. Weiss, *Ber. Dtsch. Chem. Ges.*, **46**, 2640 (1913).
254. F. Krafft and R. E. Lyons, *Ber. Dtsch. Chem. Ges.*, **29**, 424 (1896).
255. J. Loevenich, H. Fremdling and M. Föhr, *Ber. Dtsch. Chem. Ges.*, **62**, 2856 (1929).
256. K. Dostàl, P. Mosna and Z. Zàk, *Z. Chem.*, **6**, 153 (1966).
257. K. Dostàl, Z. Zàk and M. Cernik, *Chem. Ber.*, **104**, 2044 (1971).
258. H. J. Reich, *Proceedings of the 4th International Symposium on Organic Selenium and Tellurium Compounds, Birmingham, UK*, 1983.
259. L. M. Yagupolskii and V. G. Voloshchuk, *Zh. Obshch. Khim.*, **38**, 2509 (1968).
260. H. J. Backer and H. J. Winter, *Rec. Trav. Chim.*, **56**, 493 (1937).
261. (a) H. J. Reich, Plenary Lecture given at the *4th International Conference on the Organic Chemistry of Selenium and Tellurium, Birmingham*, 1983.
(b) B. Lindgren, *Chem. Scripta*, **16**, 24 (1980).
262. H. J. Backer and J. Strating, *Rec. Trav. Chim.*, **53**, 1113 (1934).
263. W. L. Mock and J. H. McCausland, *Tetrahedron Letters*, 391 (1968).
264. R. Paetzold and G. Bochmann, *Spectrochim. Acta*, **26A**, 391 (1970).
265. G. Hopf and R. Paetzold, *J. Mol. Struct.*, **13**, 371 (1972).
266. H. Rheinboldt and E. Giesbrecht, *J. Amer. Chem. Soc.*, **68**, 2671 (1946).
267. H. Rheinboldt and E. Giesbrecht, *J. Amer. Chem. Soc.*, **69**, 644 (1947).
268. A. Bali and K. C. Malhotra, *Bull. Chem. Soc. Japan*, **48**, 1929 (1975).
269. R. C. Paul, D. Konwer, D. S. Dhillon and J. K. Puri, *J. Inorg. Nucl. Chem.*, **41**, 55 (1979).
270. N. Ya. Derkach, N. A. Pasmurtseva and E. S. Levchenko, *Zh. Obshch. Khim.*, **42**, 1652 (1972).
271. N. Ya. Derkach, T. V. Lyapina and N. A. Pasmurtseva, *Zh. Org. Khim.*, **10**, 807 (1974).
272. M. Shimizu and I. Kuwajima, *J. Org. Chem.*, **45**, 4063 (1980).
273. M. Shimizu, R. Ando and I. Kuwajima, *J. Org. Chem.*, **46**, 5246 (1981).
274. H. Rheinboldt, in *Houben-Weyl: Methoden der Organishen Chemie*, Vol. IX, 4th ed. (Ed. E. Müller), Georg Thieme Verlag, Stuttgart, 1955, p. 1074.
275. Ref. 3, p. 196.
276. R. H. Vernon, *J. Chem. Soc.*, 889 (1920).
277. M. P. Balfe, C. A. Chaplin and H. Phillips, *J. Chem. Soc.*, 341 (1938).
278. F. L. Gilbert and T. M. Lowry, *J. Chem. Soc.*, 3179 (1928).
279. G. T. Morgan and H. Burgess, *J. Chem. Soc.*, 321 (1928).
280. H. D. K. Drew, *J. Chem. Soc.*, 3054 (1926).
281. F. G. Holliman and F. G. Mann, *J. Chem. Soc.*, 37 (1945).
282. L. Engman and M. P. Cava, *Chem. Commun.*, 164 (1982).
283. S. Uemura and S. Fukuzawa, *J. Chem. Soc., Perkin Trans. 1*, in press.
284. S. Uemura and S. Fukuzawa, *Tetrahedron Letters*, **24**, 4347 (1983).
285. S. Uemura, Lecture given at the *4th International Conference on the Organic Chemistry of Selenium and Tellurium, Birmingham*, 1983.
286. P. Schulz and G. Klar, *Z. Naturforsch.*, **30B**, 43 (1975).
287. N. L. M. Dereu, R. A. Zingaro, E. A. Meyers and M. Renson, *Organometallics*, **1**, 111 (1982).
288. J. Bergman, N. Eklund, E. E. Engler and N. Martinez, unpublished results.
289. H. J. Reich, C. A. Hoeger and W. W. Willis, Jr., *J. Amer. Chem. Soc.*, **104**, 2936 (1982).
290. C. J. Shack and K. O. Christe, *J. Fluorine Chem.*, **24**, 467 (1984).

The Chemistry of Organic
Selenium and Tellurium
Compounds Volume 1
Edited by S. Patai and Z. Rappoport
© 1986 John Wiley & Sons Ltd.

CHAPTER **15**

Directing and activating effects involving selenium and tellurium

FRANCESCO FRINGUELLI and ALDO TATICCHI

Dipartimento di Chimica, Università di Perugia, Perugia, Italy

---

I. INTRODUCTION AND GENERAL FEATURES 560
II. SUBSTITUENT CONSTANTS FOR SELENIUM- AND TELLURIUM-CONTAINING GROUPS 560
A. The Substituent Constant 560
B. Bivalent Selenium- and Tellurium-containing Groups 563
C. Tetravalent Selenium- and Tellurium-containing Groups 566
D. Heterocyclic Rings as Substituents 567
E. Replacement Substituent Constants 568
F. Special Substituent Constants 569
G. Hydrophobic Substituent Constants 570
III. SUBSTITUENT EFFECTS INVOLVING SELENIUM AND TELLURIUM 570
A. The Reaction Constant 570
B. Electronic Transmission Through Heterocyclic Systems 572
C. Electronic Transmission Through Selenium and Tellurium Atoms . . 574
D. Substituent Effects in Electrophilic and Nucleophilic Aromatic Reactions 577
E. Substituent Effects on Ring Protons and Carbons of Heterocyclic Systems by NMR Spectroscopy 579
F. Miscellaneous 580
IV. REACTIVITY AND DIRECTING EFFECTS IN HETEROCYCLIC SYSTEMS 581
A. Electrophilic Aromatic Substitutions 581
B. Nucleophilic Aromatic Substitutions 589
C. Meisenheimer Complexes 590
D. Addition Reactions 591
E. Side-chain Reactions 593
F. Protophilic Reactions 594
G. Metalation Reactions 595

H. Equilibrium Reactions 598
V. SYSTEMS WITH EXOCYCLIC FUNCTIONAL GROUPS 599
A. Selenium Acids and Selenols 599
B. Onium Salts 601
C. Selenonium and Telluronium Ylides 602
D. Selenides and Tellurides 603
E. Selenoxides and Selenones 609
VI. REFERENCES 611

I. INTRODUCTION AND GENERAL FEATURES

Over the last decade the chemistry of Se/Te organic compounds has been developed greatly and various books, reviews and advances, have appeared[1–6]. The synthetic, theoretical and physicochemical aspects are the subjects mainly reviewed, with the reactivity of these compounds being insufficiently covered leading one to think that little work has been done in this area. Although the synthetic difficulties and instability of the compounds sometimes limit or preclude obtaining quantitative reactivity data, careful examination of the literature reveals that much work has been done.

The reactivity of organic compounds containing Se and Te discussed in this chapter is mainly confined to those reactions in which Se and Te are not the primary centres of the reaction. Those reactions in which the attack occurs on the heteroatom or in which bonds to the heteroatom are broken, are considered only marginally.

In Section II the electronic, steric and biological effects of groups of atoms containing Se and Te are discussed in terms of the so-called substituent constants. The tangled subject of σ and σ-like substituent constants is introduced to facilitate understanding. All substituent constants of Se and Te-containing groups are, to the authors' knowledge, collected for the first time.

Section III reports the ability of Se and Te to transmit electronic effects from a remote substituent to the reaction centre. The electronic transmission is usually quantified in terms of ρ reaction constant, the physicochemical meaning of which has given rise to controversy[7–9]. The present position in this area is briefly and critically summarized. All relevant ρ reaction constants are, to the authors' knowledge, collected for the first time.

The reactivity of heterocyclic compounds containing Se and Te is discussed in Section IV. Because of the variety of heterocyclic systems, the data are discussed according to reaction type.

Section V reports the directing and activating effects of well-known functional groups containing Se and Te.

In all sections particular emphasis is given to the comparison with O and S analogues. A comparative study is fundamental for evaluating the influence of the basic parameters of the chalcogen atom (mass, electronegativity, polarizability, possibility of utilizing d orbitals, etc.) and the molecule (geometry, resonance energy, etc.) on the chemical reactivity.

This work is not encyclopaedic, but most of the pertinent literature up to March 1983 has been covered.

II. SUBSTITUENT CONSTANTS FOR SELENIUM- AND TELLURIUM-CONTAINING GROUPS

A. The Substituent Constant

A variable substituent S attached to a skeletal group G of a SGW compound affects the chemical or physical phenomena of the molecule measured at the reaction site W. The substituent constant is an empirical parameter of substituent S defined relative to a standard substituent and to a standard property under well-defined conditions. This constant, which is usually denoted by the letter σ, in principle depends only on the nature of the substituent and measures the effect of the substituent in the process concerned.

This approach is very interesting because it should permit correlation of a vast amount of data, calculation of chemical data from those reactions for which experimental data are not available and collection of mechanistic information.

The problems arise when one attempts to investigate the nature of the effect of the substituent (field, inductive, resonance, steric, etc.) and to find compounds and reactions in which only one effect is operative.

The result has been a proliferation of substituent constants denoted by subscripts and superscripts of the letter σ and other symbols. To facilitate the reading of the following sections, the problem is briefly examined. More details can be found in specialist reports[10–12].

The substituent constant on a quantitative basis was first stated by Hammett[13,14] as a measure of the electronic effects of *meta* (σ_m) and *para* substituents (σ_p) on the dissociation of benzoic acids in water at 25 °C. Hydrogen was taken as reference with a zero value. The σ_m value is a measure of polar effects (inductive and/or field) of the substituent and σ_p is a measure of the polar effect including no important resonance interaction between the substituent and the reaction centre. A positive or negative value of the σ constant indicates an electron-withdrawing or an electron-releasing capacity of the substituent, respectively. The difference ($\sigma_p - \sigma_m$) has been used as a measure of conjugative interaction of the substituent.

The Hammett σ values are able to correlate many rate and equilibrium data[10,11,15,16] and many physical measurements[10,11], but deviations are observed by *para* substituents which can enter into direct resonance interactions with the reaction site in the transition state. By using the Hammett procedure, two new sets of substituent constants were defined[10,17] (σ_p^+ and σ_p^-) when the substituent interacts with a positive or a negative charge developed in the transition state. Strong electron-donating groups have high negative values of σ_p^+ (-0.5 to -1.5) and strong electron-withdrawing groups have high positive σ_p^- values (0.4 to 1.3). When there is no important resonance interaction between the substituent and the reaction centre, σ^+ and σ^- should in principle have the same value as the Hammett σ constants. The differences ($\sigma_p^+ - \sigma_p$) and ($\sigma_p^- - \sigma_p$) have been used as a measure for the conjugative ability of the substituent.

By observing that in the ionization of benzoic acids a cross-conjugation effect occurs between an electron-donor substituent and the carboxyl group, Taft defined[18] σ_m^o and σ_p^o substituent constants based on the ionization of phenylacetic acids, a reaction in which the reaction centre is insulated from the π electrons of the benzene ring. By using a different procedure Van Bekkum and coworkers[19] defined analogous σ_p^n and σ_m^n constants. σ^o and σ^n should therefore be free of the mesomeric component and should be a measure for the inductive effect of the substituent depending on its *para* or *meta* position on the benzene ring.

Hammett treatment fails for *ortho* substituents probably because of the interplay of steric, polar and mesomeric effects. A separation of the effects was proposed by Taft[20] who defined a set of σ *ortho* values (σ_o, σ_o^*) as a measure of the polar effect of the substituent.

A σ_I scale based on the reactivity of rigid alicyclic and aliphatic systems and reflecting only inductive effects was defined by Taft[21]. The inductive constant σ_I of the substituent is, in principle, independent of the method of determination and of the position of the substituent with respect to the reaction site. In an attempt to divide the inductive and resonance contribution of the substituent, Taft defined[22] the resonance contribution σ_R by equation (1):

$$\sigma_R = \sigma_p - \sigma_I \tag{1}$$

Consequently, σ_m can be expressed by equation (2):

$$\sigma_m = \sigma_I + 0.33\,\sigma_R \quad (2)$$

The two fundamental assumptions in the Taft approach are: (*i*) the σ_I, σ_p and σ_m values are on the same scale and (*ii*) the inductive effect of the substituent is the same from both positions. An additional problem is that σ_R depends on the nature of the reaction. In a first attempt to overcome the problem, new mesomeric constants, called σ_R^o, were defined[22] by using σ_p^o constants (equations 3) and the σ_m^o constants were expressed by equation 4.

$$\sigma_R^o = \sigma_p^o - \sigma_1 \quad (3)$$

$$\sigma_m^o = \sigma_1 + 0.5\,\sigma_R^o \quad (4)$$

Equation (3) was then refined by Exner[23], assuming that σ_I and σ_R^o do not contribute equally to σ_p^o (equation 5).

$$\sigma_R^o = \sigma_p^o - 1.14\,\sigma_I \quad (5)$$

A complete σ_R reaction dependence was then recognized[24] and by using the σ_p^- and σ_p^+ values, σ_R^- and σ_R^+ constants were defined for electron-withdrawing and electron-donating groups which entered into direct resonance interaction with the reaction site in the transition state.

Another attempt to divide the overall electronic contribution of the substituent into inductive and resonance contributions is that by Swain and Lupton[25]. This approach however is not substantially different from Taft's and the symbols $\mathscr{F}$ and $\mathscr{R}$ replace σ_I and σ_R respectively.

Further limitedly successful substituent constants which, like those recorded above, do not take into account the steric effects, are the σ^* constants[26] which have practically the same meaning as σ_I constants and the *F* (field) and *M* (mesomeric) quantities derived by Dewar and Grisdale[27] using a semiempirical approach.

Substituent constants which are claimed to measure only the steric effect of the substituent (E_s) were defined by Taft using acid-catalysed hydrolysis of esters in aqueous acetone[28]. The methyl group is taken as reference ($E_s = 0$) and the E_s values approximate the size of the group. Another set of *steric* constants, called E_s^* has been defined[29] by using the hydroboration reaction.

We designate all the above recorded substituent constants as *normal substituent constants* because they refer to a group which replaces the hydrogen atom of benzene or of a heterocyclic ring, and in order to distinguish them from *replacement substituent constants* and *special substituent constants*[30]. The replacement substituent constants denoted as σ_{ar} or σ_a refer to the electronic effects of a heteroaromatic system considered as a substituted benzene[15] or to a substituent fused on to benzene ring[31]. For instance if the selenophene is viewed as an endocyclic substituted benzene, the Se atom is considered as a substituent replacing the —CH=CH— moiety in the benzene ring and $\sigma_{\alpha\text{-selenienyl}}$ and $\sigma_{\beta\text{-selenienyl}}$ are calculated as depending on the site (α or β) of the reaction. The σ_{ar} depends on the reaction and therefore σ_{ar}^-, σ_{ar}^+, σ_{ar}^0, etc. have also been defined.

A new set of σ constants is defined when the concept of the *replacement* is extended to other aromatic rings besides that of benzene. These are known as special substituent constants[30]. Examples are σ_r calculated[32] for the protonation of arenes taking the α-position of naphthalene as reference and σ_{het} calculated[33] for the ionization reaction of *N*-ethyl-2-aminoheteroarenes by using *N*-ethyl-2-aminopyridinium chloride as reference.

Two methods may be employed to evaluate the substituent constant. In the first procedure a single reaction is chosen as a *standard* and the σ constant is determined from the rate (or equilibrium) constant of this reference reaction. Sometimes chemical difficulties preclude this possibility and other reactions or physical methods (^{1}H—, ^{13}C—

and ^{19}F-NMR, IR, UV, $E_{1/2}$) are used. This and the dependence on solvent and temperature[10,11] further complicate the jungle of substituent constants. In the second procedure the substituent constant is derived from a statistical analysis of the available data for that substituent in a wide spectrum of related reactions. The most suitable statistical treatment is that based on the application of the *extended selectivity relationship*[34] in which the σ constant is obtained from the slope of a plot of $\log k_S/k_H$ against the ρ constants for several similar reactions.

A Hammett-like relationship was developed in biology to correlate physical properties of molecules and their ability to penetrate biological membranes[11,35]. The chosen property is the lipid–aqueous partition coefficient P which represents the relative lipophilicity (now called hydrophobicity) of biologically active organic compounds. Most partition coefficients are obtained in an *n*-octanol–water system which is generally accepted as standard for a biological lipid–water system. Other lipid solvent–water systems are sometimes used. By using $\log P$ as a measure for the hydrophobicity of a whole molecule, the hydrophobic (lipophilic) substituent constant π is defined according to equation (6):

$$\pi = \log P_S - \log P_H \qquad (6)$$

where P_H and P_S are the partition coefficients of the reference and substituted compounds respectively. Positive and negative π values indicate that the substituent facilitates the penetration of the molecule to the *n*-octanol phase or to water, respectively. A dependence of π value on the reaction system was observed, especially for strongly electron-donating and electron withdrawing substituents.

Only a limited number of substituent constants for Se and Te having organic groups are listed in the specialist reports[1–4]. In the following section the constants available in the literature are reported without discrimination as to their origin and reliability. Where possible, a comparison with analogous O and S substituent constants is made.

B. Bivalent Selenium- and Tellurium-containing Groups

The substituent constants of groups containing bivalent Se and Te are reported in Table 1 with the corresponding data for O and S for comparison. The σ value of the group without heteroatom is also reported to estimate the electronic contribution of the heteroatom.

The SeMe and TeMe groups show $-$I and $+$M effects. The electron-withdrawing inductive effect increases from TeMe to OMe by a factor of two, while the electron-donating resonance effect increases about eight times. ^{13}C-NMR investigations[36–38] confirm the weak electron donor character of TeR and SeR groups and show that the degree of p-π interaction decreases as the size of R (Me, Et, *i*-Pr) increases. The increase of the $+$M effect in the order TeMe $<$ SeMe $<$ SMe $<$ OMe was also observed[39] in the cyanohydrin equilibrium with substituted benzaldehydes, in the UV absorption spectra of substituted benzoic acids and more recently in the calculated[40] mesomeric moments of telluroanisole (0.18 D), selenoanisole (0.25 D), thioanisole (0.5 D) and anisole (1.1 D) and in the ^{13}C-NMR chemical shifts of the *para* carbons of chalcogenoanisoles[41]. A theoretical treatment of the connection between the size of the heteroatom and its mesomeric effect has been published[42]. The σ_p and σ_m values show that the $+$M effect prevails over the $-$I effect. A comparison with the Me group shows that the presence of the chalcogen atom increases the electron-withdrawing power of the group according to the electronegativity of the heteroatom and that the Te atom does not change the releasing mesomeric effect of the group.

The $SeCF_3$ group shows a $-$I effect of the same magnitude as the SCF_3 group but lower than that of OCF_3. In contrast to the OCF_3 group, the $SeCF_3$ and SCF_3 show a $-$M effect owing to the involvement of the d orbitals of S and Se atoms in the conjugation. The

TABLE 1. Substituent constants for groups containing bivalent calcogen atoms

Substituent	σ_p	σ_m	σ_p^-	σ_I	σ_R	σ_R^o	$\mathscr{F}$	$\mathscr{R}$
TeMe	0.04[a]			0.12[a]	−0.08[a]	−0.10[b]		
SeMe	−0.15[c]	0.05[c]		0.15[d]	−0.14[d]	−0.15[b]	0.13[e]	−0.12[e]
	0.0[f]	0.09[d]						
	0.07[g]	0.1[f]						
	0.01[d]							
SMe[h]	0.0	0.15	0.04	0.19	−0.24	−0.24	0.20	−0.18
OMe[h]	−0.27	0.12		0.25	−0.63	−0.41	0.26	−0.51
Me[h]	−0.17	−0.07		−0.05	−0.07	−0.10	−0.04	−0.13
$SeCF_3$	0.38[i]	0.32[i]	0.53[j]	0.28[i]	0.04[k]		0.29[e]	0.12[e]
	0.45[k]	0.44[k]	0.58[l]	0.42[k]	0.10[i]			
SCF_3[h]	0.50	0.40	0.64	0.44	0.12[d]	0.00	0.35	0.18
OCF_3[h]	0.35	0.40	0.25[j]	0.55	−0.14	0.17	0.38	0.00
CF_3[h]	0.43	0.54	0.65	0.41	0	0.01	0.38	0.19
SeCN	0.66[m]	0.61[n]		0.58[o]				
SCN[h]	0.52	0.41	0.60	0.55[o]		−0.05	0.36	0.19
CN[h]	0.66	0.56	0.88	0.56	0	0.07	0.51	0.19
TePh	0.34[p]			0.38[p]	−0.04[p]			
SePh	0.13[q]		0.13[q]	0.24[r]		−0.19[s]		
	0.42[g]			0.37[h]				
SPh[h]	0.13	0.17	0.29[t]	0.30		−0.19		
OPh[h]	−0.32	0.25		0.39	−0.31[a]	−0.36[s]	0.34	−0.35
Ph[h]	−0.01	0.06	0.11	0.10	−0.10	−0.10	0.08	−0.08
Se—c-C_6H_{11}				0.38[u]				
S—c-C_6H_{11}				0.31[u]				
O—c-C_6H_{11}				0.26[u]				
c-C_6H_{11}[h]	−0.22	−0.15		0.02			−0.13	−0.10
$SeCH_2CH{=}CH_2$[v]	0.15	0.21						
$SCH_2CH{=}CH_2$[v]	0.12	0.19						
$OCH_2CH{=}CH_2$[v]	−0.25	0.09						
$SeCH{=}CH_2$[v]	0.21	0.26						
$SCH{=}CH_2$[v]	0.20			0.21[h]		0.27[h]		
$CH{=}CH_2$[h]	−0.02	0.00						
$SeCH{=}CHCl$[v]	0.26	0.28						
$SCH{=}CHCl$[v]	0.24	0.31						
$SeC{\equiv}CPh$[w]	0.05							
	−0.03							
$SC{\equiv}CPh$[w]	0.20							
$C{\equiv}CPh$[h]	0.16	0.14	0.39	0.22		0.00	0.12	0.05
$SeC_6H_4NO_2$			0.41[e]					
$P(Se)(4\text{-}FC_6H_4)_2$				0.42[x]	0.10[x]			
$P(Se)(3\text{-}FC_6H_4)_2$				0.47[x]	0.11[x]			
$P(Se)Ph_2$				0.36[x]	0.09[x]	0.13[y]		
$P(S)Ph_2$[h]	0.47	0.29	0.63	0.40	0.09[x]	0.11		
$P(O)Ph_2$[h]	0.53	0.38	0.68	0.27	0.11[x]	0.12	0.31	0.24
PPh_2[h]	0.19	0.11	0.26	0.17	−0.01[x]	−0.01		
NCSe[aa]	0.31							
NCS[aa]	0.20							
Se^-	−0.98[z]							
S^-	−1.21[z]					−0.33[h]		
O^-	−1.86[z]	−0.71[h]		−0.16[h]		−0.60[h]		

weaker electron-accepting conjugation of the $SeCF_3$ compared to the SCF_3 group has been explained by suggesting[43] that the more diffuse 4d orbitals of Se have a smaller tendency towards (p–d)π conjugation than the 3d orbitals of S. The analysis of values of dipole moments of *p*-SeMe- and *p*-SMe-substituted anilines supports this hypothesis[44]. Indeed when the SeMe group is present the interaction dipole moment ($\mu_{int} = 0.43$ D) is lower than when the substituent SMe ($\mu_{int} = 0.68$ D) is present. A comparison with the CF_3 group shows that the contribution of S and Se atoms is small and therefore it is not surprising that the $SeCF_3$ group does not show an electron-donating effect even when a nitro group is present in the *para* position of the benzene ring[44]. Consequently σ_p and σ_m have the same values.

The behaviour of the SeCN group is the same as that of the $SeCF_3$ group but the $-$I and $-$M effects are stronger as a consequence of the electronic effects of the cyano group compared with those of the trifluoromethyl group.

The SePh and TePh groups, like the Ph group, show an electron-withdrawing inductive effect ($-$I) and a push–pull resonance effect (+M and $-$M). Theoreticians[45,46] generally invoke only a polarization effect to explain the stabilization of the incipient carbanion in the transition state by a bivalent S-containing group, but the lower value of σ_p^- of the SePh group with respect to that of the SPh group does not agree with this hypothesis. A larger effect would be expected for the larger and more polarizable Se atom. Therefore a conjugative interaction, in which the d orbitals of the larger Se atom are less effective than those of S in overlapping with the p orbitals of C, must be invoked. From measurements of the hydrolysis rates of aryl vinyl selenides, McClelland and Leung[47] have calculated a σ_p^+ value of -0.47 for the SePh group which is higher in absolute value than that of the Ph group ($\sigma_p^+ = -0.18$)[48] and lower than those of SPh ($\sigma_p^+ = -0.54$)[47] and OPh ($\sigma_p^+ = -0.62$)[47]. The chalcogen atoms stabilize an adjacent carbonium ion centre, but the type of electron donation is supposed to depend on the nature of the heteroatom[47]. Dipole moment data[49] show that when strongly electron-donating or electron-withdrawing substituents are present in the position *para* to SePh and TePh the push–pull resonance effect is markedly exalted.

The cyclohexylseleno group has the expected $-$I effect but in the series the order is the reverse of that expected on the basis of the inductive effect of methylchalcogen groups.

The allyl-, vinyl- and chlorovinyl-seleno groups have a $-$I effect and a weak +M effect ($\sigma_p - \sigma_m$, negative).

The Se atom markedly reduces the electron-withdrawing effect of the phenylethynyl

Footnotes to Table 1.

[a] From ^{19}F-NMR, Ref. 51.
[b] From ^{13}C-NMR, Ref. 52.
[c] From ionization of benzoic acids in 30% EtOH at 25 °C, Ref. 39.
[d] Ref. 53.
[e] Ref. 35.
[f] From ionization of benzoic acids in 50% EtOH at 25 °C, Ref. 54.
[g] From methyl hyperfine splitting of 1-phenyl-1,2-propanesemidones, Ref. 55.
[h] From compilations of Exner (Ref. 10 and 11) and Hansch (Ref. 35).
[i] From ionization of benzoic acids in 50% EtOH, Ref. 53.
[j] From ionization of anilinium ions, Ref. 43.
[k] From ^{19}F-NMR of fluorobenzenes, Ref. 53.
[l] From ^{1}H-NMR chemical shifts of amino group, Ref. 43.
[m] From the ionization of *p*-substituted *p*′-dimethylaminoazobenzenes in 25% EtOH at 25 °C, Refs. 15, 56.
[n] Ref. 57.
[o] Ref. 58.
[p] From ^{19}F-NMR, Ref. 49.
[q] Ref. 59.
[r] Refs. 60, 61.
[s] From integrated IR intensities, Ref. 62.
[t] From ionization of phenols, Ref. 63.
[u] From dissociation of acetic acids in H_2O at 25 °C, Refs. 58, 64.
[v] From the half-wave reduction potential of nitrobenzenes in DMF, Ref. 65.
[w] From polarographic and spectrophotometric data of *p*-nitrophenyl phenylethynyl chalcogenides. Ref. 50.
[x] Ref. 67; these values were incorrectly[66] labelled σ_R^o.
[y] From ^{13}C-NMR, Ref. 66.
[z] From the half-wave reduction potential of nitrophenyldichalcogenides in DMF, Ref. 68.
[aa] From UV data, Ref. 104.

group, probably as a consequence of a larger conjugative interaction with the benzene ring than with the ethynyl bond[50].

On the other hand, the Se atom greatly increases the negligible Ph–P π interaction and the electron acceptor effect of the PPh_2 group. Examination of the sequence of phosphine chalcogenides indicates that the effect is due to a concordant action of – M and – I effects of the heteroatom.

The negatively charged Se atom shows a strong electron donor power in the *para* position, which increases in the order $Se^- < S^- < O^-$ as expected on the basis of the + M effects.

C. Tetravalent Selenium- and Tellurium-containing Groups

In compounds containing Se(IV) and Te(IV) (selenuranes and telluranes), the chalcogen atoms have a valence shell of ten electrons. These compounds are usually classified as σ-selenuranes (σ-telluranes) and π-selenuranes (π-telluranes) depending on whether four of the five valence electron pairs of the heteroatom form four σ bonds (**1**) or one electron pair forms a π bond with a neighbouring element of the second period (**2**). Theoretical[72,73]

(1)

(2)

Z = Te, Se, S

Y = O, C, N

TABLE 2. Substituent constants for Se(IV)- and Te(IV)-containing groups

Substituent	σ_p	σ_m	σ_I	σ_R	σ_r^0
$TeCl_3$[a]	0.66		0.55	0.11	
$TeBr_3$[a]	0.62		0.51	0.11	
$TeMeCl_2$[a]	0.56		0.50	0.06	
$TeMeBr_2$[a]	0.59		0.53	0.06	
$TeMeI_2$[a]	0.62		0.58	0.04	
$Te(Ph)Cl_2$[a]	0.59		0.52	0.07	
	0.73[b]		0.66[b]		
$Te(Ph)Br_2$[a]	0.60		0.54	0.06	
	0.59[b]		0.52[b]	0.07	
$SeOCF_3$[c]	0.86	0.81	0.76	0.10	
	0.63[d]				
$SOCF_3$[e]	0.80	0.74	0.67	0.13	
SeO_2CF_3[c]	1.21	1.08	0.96	0.25	
SO_2CF_3[e]	1.04	0.88	0.73	0.31	
$\overset{+}{Te}MeR$[f]					– 0.11
$\overset{+}{Se}MeR$[f]					0.13
$\overset{+}{S}Me_2$[f]					0.20

[a] From ^{19}F-NMR data, Ref. 51.
[b] From ^{19}F-NMR data, Ref. 69.
[c] From ^{19}F-NMR data, Ref. 70.
[d] From UV data, Ref. 71.
[e] From ^{19}F-NMR data, Ref. 71.
[f] From ^{13}C-NMR data, $R = C^-$—$(CO)_2C_6H_4$ for Te and C^-—$(CO)_2(CH)_2CMe_2$ for Se, Ref. 52.

and experimental data[74–76] show that the configuration of σ-selenuranes and -telluranes is that of a slightly distorted trigonal bipyramid in which the more electronegative ligand occupies the *axial* position. Whereas diorganotellurium dihalides are monomeric in the solid phase and in solution, trihalotelluranes have a dimeric or polymeric structure in the solid state and the Te atom is pentacoordinate[77–79]. π-Selenuranes and -telluranes have the same tetrahedral configuration[80] in oxides (**2**; Y = O) and a trigonal pyramid structure with the heteroatom at the apex in ylides[81] (**2**; Y = C). The σ constants for Se(IV)- and Te(IV)-containing groups are reported in Table 2.

^{19}F-NMR data show[51] that the trihalotelluro groups exhibit a strong $-$I effect in agreement with the strong deshielding effect of the $TeCl_3$ group (almost equal to that of the nitro group) on the ring protons of trichlorophenyltellurium[73]. Dipole moment data of a series of trichlorophenyltelluriums show an absence of mesomeric and interaction moments indicating that the p and d orbitals of Te are not conjugated with the π orbitals of the Ph ring[73]. The calculated value for σ_R should therefore be seen as the result of the arbitrary choice in the separation of inductive and mesomeric effects (equation 1). By using for σ_I a coefficient greater than unity, the σ_R value is reduced. Theoretical calculations[73] explain the electronic properties of the $TeCl_3$ group.

Replacing a halogen atom in the TeX_3 group (X = Cl, Br, I) with Me or Ph, the electron-withdrawing power of the group is reduced by 5–15% and appears to depend on the nature of the halogen.The values $\sigma_p = 0.73$ and $\sigma_I = 0.66$ for the Te(Ph)Cl_2 group are probably printing errors because they conflict with the values of 0.59 and 0.52 reported by the same authors using the same technique[51,69]. ^{19}F-NMR data show that Te(R)X_2 groups (R = Me, Ph; X = Cl, Br, I) do not exhibit electron-donating mesomeric properties. However, experimental and calculated[69] dipole moments of substituted diaryldihalotelluriums show that when a + M substituent (i.e. NMe_2) is present in the *para* position, the Te(R)X_2 group has a $-$M effect probably via d–π conjugation.

The $SeOCF_3$ and SeO_2CF_3 groups exhibit a marked enhancement of the electron-withdrawing inductive effect in respect to the $SeCF_3$ group. The effect is related to the number of O atoms. The SeO_2CF_3 group is one of the most powerful electron acceptors. The presence of O also influences the conjugative effect, indeed the low $\sigma_p - \sigma_m$ value increases on passing from $SeCF_3$ (mean value $\simeq 0.03$) and $SeOCF_3$ (0.05) to SeO_2CF_3 (0.13). $SeOCF_3$ and SeO_2CF_3 exhibit a higher $-$I and lower $-$M effect than the analogous S-containing groups. This has been ascribed to the greater contribution of resonance structures in which a high positive charge is present on the Se atom. The higher values of the group moment[82,83] of $SeOCF_3$ (4.13 D) and SeO_2CF_3 (5.3 D) with respect to $SOCF_3$ (3.88 D) and SO_2CF_3 (4.32 D) and the lower interaction moment of compounds with electron-donating substituents (*p*-NH_2-$C_6H_4SeOCF_3$ = 0.97 D, *p*-$NH_2C_6H_4SOCF_3$ = 1.11 D; *p*-$MeC_6H_4SeO_2CF_3$ = 0.18 D, *p*-$MeC_6H_4SO_2CF_3$ = 0.78 D) support this hypothesis[82,83]

The opposite conjugative electron-withdrawing effect of the selenonium and telluronium groups has been ascribed[52] to the greater difference in energy of the non-occupied Te d orbitals and the benzene 2p orbitals in comparison to Se and S.

The electronic effects of the $TeMeF_2$, Te(Ph)$(NCS)_2$ and Te(Ph)F_2 groups have also been investigated[84] by the ^{19}F-NMR technique. These substituents interact with the π system of the benzene ring by an inductive mechanism[84].

D. Heterocyclic Rings as Substituents

Literature data concerning the electronic effects caused by the substitution of a H atom of the benzene ring with a heterocyclic ring having Se or Te are limited to 2-selenienyl and 2-tellurienyl groups[85–87]. The data are reported in Table 3 with the pertinent data of 2-furyl and 2-thienyl for comparison[86,88,89]. The σ^* constants refer to the polar effect of the

TABLE 3. Substituent constants for heterocyclic groups containing a chalcogen atom

Substituent	σ_m	σ_p	σ_m^-	σ_p^-	σ^*	σ_I	σ_R
2-Tellurienyl	0.06[a]	0.03[a]	0.10[b]	0.25[b]			
			0.12[a]	0.21[a]			
2-Selenienyl	0.06[a]	0.04[a]	0.16[b]	0.22[b]	0.85[d]	0.15[c]	−0.14[c]
	0.09[c]	0.01[c,i]	0.15[a]	0.20[a]			
2-Thienyl[l]	0.09[e]	0.05[e]	0.11[f]	0.19[f]	0.93[d]	0.12[c]	−0.10[c]
2-Furyl[l]	0.06[g]	0.02[g]	0.11[h]	0.21[h]	1.08[d]	0.09[c]	−0.08[c]
5-Me-(2-selenienyl)		−0.03[i]					
5-Et-(2-selenienyl)		−0.02[i]					
5-Cl-(2-selenienyl)		0.13[i]					
5-Br-(2-selenienyl)		0.12[i]					
5-I-(2-selenienyl)		0.11[i]					
5-Ac-(2-selenienyl)		0.18[i]					
5-NO_2-(2-selenienyl)		0.24[i]					

[a] From ^{1}H-NMR, Ref. 85.
[b] From ionization of phenols in 30% EtOH at 25 °C, Ref. 85.
[c] From ionization of benzoic acids in 50% EtOH, Ref. 86.
[d] From the alkaline and acid hydrolysis of 2-ethoxycarbonyl of five-membered heterocycles in 62% acetone at 25 °C, Ref. 90.
[e] From ionization of benzoic acids in 50% EtOH at 25 °C, Ref. 88.
[f] From ionization of phenols in 30% EtOH at 25 °C, Ref. 88.
[g] From ionization of benzoic acids in 30% EtOH at 25 °C, Ref. 89.
[h] From ionization of phenols in 30% EtOH at 25 °C, Ref. 89.
[i] From ionization of biselenienylcarboxylic acids in 50% buthylcellosolve, Ref. 87.
[l] For additional σ values see Ref. 30.

C_4H_3Z group and were calculated[90] from rate constants of $C_4H_3ZCOOEt$ (Z = Se, S, O) hydrolysis taking ethyl acetate as reference according to the Taft procedure[28].

The 2-selenienyl and 2-tellurienyl groups exhibit an inductive electron-withdrawing effect like 2-furyl and 2-thienyl. The σ^* values reflect the order expected on the basis of the electronegativity of the heteroatoms. The surprising opposite order of σ_I values has been explained[86] by invoking the magnitude and direction of dipole moments and conformational effects. In the absence of important resonance interactions with the reaction site an electron-releasing effect is observed from conjugative positions (negative values of $(\sigma_p - \sigma_m)$ and σ_R). The σ_p^- value and the difference $(\sigma_p^- - \sigma_p)$ show that 2-selenienyl and 2-tellurienyl easily delocalize a negative charge and that the latter appears to show a stronger effect than 2-furyl and 2-thienyl. The electronic behaviour of the selenienyl is markedly influenced by a group at C(5). The ability of heteroaromatic rings containing Se and Te to delocalize both positive (+M effect) and negative (−M effect) charges is ascribable to the polarizability of the heteroatom which can release its p electrons or accept electrons into its free d orbitals. The degree of effect is probably related to the resonance energy of the heteroaromatic ring.

E. Replacement Substituent Constants

The so-called replacement substituent constants have been calculated for a large number of heterocyclic groups[11]. However, their use is questionable both from the point of view of nomenclature and the concept of the constant.

The terms $\sigma_{\alpha\text{-selenienyl}}$, $\sigma_{\alpha\text{-tellurienyl}}$ etc. should be used [89] to indicate the effect caused by the substitution of a hydrogen atom of the benzene ring with C_4H_3Se or C_4H_3Te groups (see Section II.D) and not used to measure the effect caused by the replacement of the CH=CH moiety of the benzene ring with Se or Te atoms.

TABLE 4. Replacement substituent constants for chalcogenaryl groups

Substituent[a]	Heterocyclic group[b]		σ_{ar}
Te	Tellurophene	σ_α	0.23[c]
		σ_α^+	−0.92[d]
Se	Selenophene	σ_α	0.62[e] 0.60[c] 0.55[d] 0.28[f] −0.4[g] −0.22[h] −0.24[i] −1.16[h]
		σ_α^+	−1.28[c] −1.25[j] −0.95[k] −0.88[l] −0.43[i]
		σ_β^+	> −0.53[k]
S	Thiophene	σ_α	0.71[q] 0.67[c] 0.36[f] 0.05[f] −0.28[h] −0.21[i] −0.92[h]
		σ_α^+	−1.15[j] −0.79[l] −0.38[i]
O	Furan	σ_α	1.04[c] 0.61[f] 0.28[f] 0.10[m] −0.29[h,g] −0.79[h]
		σ_α^+	−1.26[j] −0.93[l] −0.51[i]
—N=CH—Se—	Benzoselenazole	σ_α^o	1.9[n] 1.2[o]
		σ_α^-	1.8[p]
—N=CH—S—	Benzothiazole	α_α^o	1.6[n] 1.3[o]
		σ_α^-	1.6[p]
—N=CH—O—	Benzoxazole	σ_α^o	1.4[o] 1.2[n]
		σ_α^-	1.7[p]

[a] Substituent which replaces the CH=CH moiety of the benzene ring or which is fused on the benzene ring.
[b] From which the substituent constant is derived.
[c] From the ionization of α-carboxylic acids in H_2O at 25 °C, Ref. 91.
[d] From statistical analysis, Ref. 92.
[e] Ref. 31.
[f] Ref. 99.
[g] Ref. 100, the value has been calculated erroneously[87].
[h] From the ionization of *E*- and *Z*-α,β-diarylacrylic acids in 80% 2-methoxyethanol at 25 °C, Ref. 93.
[i] From the ionization of *para*-substituted chalcones in AcOH–H_2SO_4, Ref. 94.
[j] From a protodemercuriation reaction, Ref. 95.
[k] From acid protodedeuteriation, Ref. 98.
[l] By an extended selectivity relationship, Ref. 96.
[m] Ref. 101.
[n] From ^{1}H-NMR, Ref. 97.
[o] From IR data, Ref. 97.
[p] From alkaline deuteriodeprotonation reaction, Ref. 97.
[q] Ref. 30.

The σ_{ar} values reported in Table 4 question the concept of the constant and its electronic meaning. The polar constant σ_{ar} sometimes has a high degree of variation for the same substituent and appears to exhibit both an electron-releasing and an electron-withdrawing inductive effect. The effect is also strongly dependent on minor changes in polarity of the solvent. Significant examples are the $\sigma_{\alpha\ \mathrm{selenienyl}}$ values of −0.22 and −1.16 calculated[93] from the ionization of *E*- and *Z*-α-phenyl-β-(2-selenienyl)acrylic acids. However, it has been observed[93] that the greater electron-donating effect of the Se in the *Z*-isomer conflicts with stereochemical data indicating that the carboxyl group is twisted out of plane and therefore a minor conjugation is possible.

The σ_{ar}^+ values obtained by the extended selectivity relationship[96] are more reliable, but deviations are observed in this case also.

In the authors' opinion the replacement substituent constants are the result of an incorrect application of the Hammett equation and should be viewed with caution.

F. Special Substituent Constants

The σ_{het} constants for chalcogenazidinium salts are reported in Table 5. Their use is very limited and discrepancies in values are observed[33,102]. In the authors' opinion these

TABLE 5. Special substituent constants for chalcogen heteroarenes

Compound	σ_{het}
2-Amino-3-ethylbenzoselenazolium–BF_4	4.98[a–c]
2-Amino-3-ethylbenzothiazolium–BF_4	4.92[a] 4.89[b,c]
2-Amino-3-ethylbenzoxazolium–BF_4	5.73[a] 4.33[b]

[a] From ionization of azidinium salts in water at 25 °C, Ref. 33.
[b] From the reaction rates of azidinium salts with 1-(*p*-sulphophenyl)-3-methylpyrazolin-5-one in water at 25 °C, Ref. 102.
[c] From the reaction rates of azidinium salts with 1-(*m*-sulphophenyl)-3-methyl-5-aminopyrazole in water at 40 °C, Ref. 102.

values are also the result of an arbitrary extension of the Hammett equation and therefore their physicochemical meaning is questionable.

G. Hydrophobic Substituent Constants

Substituent constants of chalcogen atoms containing groups for structure–activity correlations are reported in Table 6. The hydrophobic substituent constant π of the SeMe group is slightly higher than that of SMe but opposite in sign to that of the OMe group. This indicates that when going from O to Se the molecule's movement into the organic phase is facilitated. A comparison with the Me group shows that the Se atom increases the hydrophobic character. This is also evident from the π constant of the $SeCH_2COOH$ group when compared with that of the S and O analogues. As the chain is lengthened (Me, Et, Pr) the hydrophobic character increases.

TABLE 6. Hydrophobic substituent constants for chalcogen-containing groups[a]

Substituent	π	Substituent	π
SeMe	0.74	SePr	1.82
SMe	0.61	SPr	1.61
OMe	−0.02	OPr	1.05
Me	0.56	Pr	1.55
SeEt	1.28	$SeCH_2COOH$	−0.08[b]
SEt	1.07	SCH_2COOH	−0.25[b]
OEt	0.38	OCH_2COOH	−0.87
Et	1.02	CH_2COOH	−0.72

[a] Values taken from Ref. 35 and 103.
[b] Calculated from the partition coefficient[35].

III. SUBSTITUENT EFFECTS INVOLVING SELENIUM AND TELLURIUM

A. The Reaction Constant

Structure–reactivity relationships are usually expressed by single- or dual-parameter equations. The most popular treatment of the first type is that proposed by Hammett[13] (equation 7) for side-chain equilibria (K) or rate (k) reactions of *para* and *meta*-substituted benzenes. K_o or k_o refer to the unsubstituted parent compound and σ is the substituent constant defined by ionization of *para*- and *meta*-substituted benzoic acids in water at

25 °C (σ_p, σ_m) assuming the reaction constants $\rho = 1$ (see Section II.A).

$$\log K/K_o = \rho\sigma \quad \text{or} \quad \log k/k_o = \rho\sigma \tag{7}$$

The Hammett equation has been extended[10,11] to physical data (^{1}H-, ^{19}F- and ^{13}C-NMR, $E_{1/2}$, IR, UV etc.), to other substrates besides that of benzene and to other reactions determining new types of σ constants (σ^+, σ^-, etc., see Section II.A). A more general Hammett-like equation (equation 8) can be written, where ΔQ is the substituent effect on equilibria, kinetic or physical data and σ is any type of substituent constant (σ_m, σ_p, σ^+, σ^-, etc.) which correlates ΔQ. The ρ constant is the reaction constant which is independent of the substituent, depending only on the reaction or physical measurement and experimental conditions (solvent, catalyst, temperature).

$$\Delta Q = \rho\sigma \tag{8}$$

The ρ constant is a measure of the sensitivity of the reaction to the substituent effects and of the transmission of the electronic effects from the substituent to the reaction centre.

In equilibrium processes ρ is the result of the difference between the ρ values of the direct and inverse reaction and therefore its value is positive or negative depending on the considered direction of the reaction. If in the ionization equilibria the acid is on the left, the ρ value is positive.

For side-chain solvolytic reactions of aromatic substrates in which a positive delocalizable charge is developed in the rate-determining step and in electrophilic aromatic substitutions, the sign of ρ is negative. The value of ρ is positive in aromatic nucleophilic substitution and in other rate processes in which a delocalizable negative charge is developed in the rate-determining step.

The best known dual-parameter equations used for correlating equilibrium, kinetic and other physical data of aromatic substrates are those proposed by Taft[21], Swain–Lupton[25] and Yukawa–Tsuno[104]. The first two (equations 9 and 10) are those almost exclusively used in Se and Te organic compounds. The regression parameters ρ_I and f, ρ_R and r should respectively be a measure of the relative importance of polar and mesomeric effects transmitted from the substituent to the reaction centre.

$$\Delta Q = \rho_I\sigma_I + \rho_R\sigma_R \tag{9}$$

$$\Delta Q = f\mathscr{F} + r\mathscr{R} + i \tag{10}$$

Two serious criticisms have been made about the use of these equations. The first is statistical in nature[105]. The significance of regression parameters of a multiple linear regression depends strongly on the number of points and on the width of the range covered. Therefore when a limited number of substituents are used, or there is multicollinearity between the two substituent constants, the regression parameters ρ_I, f, ρ_R and r have no meaning. This was made clear by Taft himself[24] who suggested a minimal set of substituents; nevertheless many of the correlations reported in literature commit this error.

The second criticism is one of concept[7,9]. Supporting statistics show that within a given type of reaction the regression parameters of the dual parameter equations (9) and (10) are not significantly different from those expected on the basis of the Hammett ρ constant and of the correlation between the proper set of σ constants. Equations (9) and (10) cannot therefore distinguish between a different blend of polar and resonance contributions even when the aromatic system is changed but the reaction remains the same.

The use of both single- and dual-parameter equations in ^{1}H-NMR spectroscopy has been criticised[106–110] because there is no clear theoretical relationship between the shielding mechanism and the chemical reactivity parameters and other effects, in addition to electronic effects, influence the chemical shifts. Wiley and Miller[106] have stated: '^{1}H-NMR correlations which depend solely on the reactivity constants ($\sigma, \mathscr{F}, \mathscr{R}$) are theoretically deficient and we would discourage their use. The introduction of additional terms, e.g. to correct for substituent magnetic anisotropy, appears to be essential, but the merits of such a hybrid approach are doubtful.' This criticism can be extended to other spectroscopic parameters.

Finally it must be recalled that the goodness of a correlation, and therefore of the regression parameters ρ, ρ_I, f, ρ_R and r, cannot be established solely on the basis of the value of the correlation coefficient. The use of appropriate statistic tests is absolutely necessary[105].

B. Electronic Transmission Through Heterocyclic Systems

Transmission of the substituent electronic effects through selenophene (**3**; Z = Se) and tellurophene (**3**; Z = Te) rings has been determined[4,91,111] for the ionization reaction in water at 25 °C of 4- and 5-substituted selenophen-2-carboxylic acids (**4**; Z = Se) and 5-

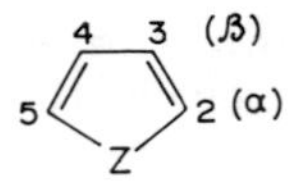

(3)

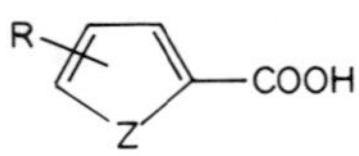

(4)

substituted tellurophen-2-carboxylic acids (**4**; Z = Te). By plotting the pK_a values against the Hammett σ constants excellent linear correlations are obtained. The ρ values are reported in Table 7 with those for the ionization of substituted thiophen-2-carboxylic acids[4,112] (**4**; Z = S) and furan-2-carboxylic acids[4,113] (**4**; Z = O). A comparison shows that the furan ring is the most sensitive to structural change and that selenophene, tellurophene and thiophene behave similarly. Analogous results are obtained[87] in the ionization reaction of 2,2′-biselenienyl and 2,2′-bithienyl-5′-substituted 5-carboxylic acids (**5**; Z = Se, S) in water–butylcellosolve. The ρ values (Table 7) show that

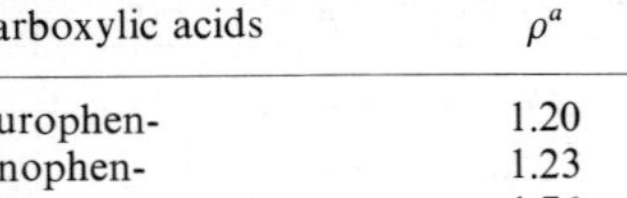

(5)

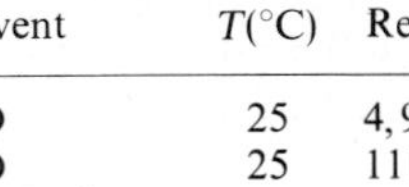

(6)

TABLE 7. Transmission of substituent effects in the ionization of α-carboxylic acids of chalcogenic heterocycles

α-Carboxylic acids	ρ^a	Solvent	T(°C)	Ref.
Tellurophen-	1.20	H_2O	25	4, 91
Selenophen-	1.23	H_2O	25	111
	1.76	H_2O/BC^b	20	87
Thiophen-	1.23	H_2O	25	4, 112
	1.71	H_2O/BC	20	87
Furan-	1.40	H_2O	25	4, 113
2,2′-Biselenienyl-	0.50	H_2O/BC	20	87
2,2′-Bithienyl-	0.56	H_2O/BC	20	87

^{a}Calculated by using the Hammett σ constants and pK_a values determined potentiometrically.
^{b}Butylcellosolve in 1:1 v/v mixture.

the transmission through the biselenienyl and bithienyl systems is the same ($\rho = 0.50$ and 0.56) and very similar to that of the biphenyl system[114] (0.59), but markedly lower than that through the corresponding monocyclic compounds **4** ($\rho = 1.76$ and 1.71).

Quantitative information on the ability of the selenophene ring to transmit the substituent electronic effects was obtained through a study[115] of the protonation of 5-substituted 2-acetylselenophenes (**6**) in an aqueous solution of H_2SO_4 by ^{1}H-NMR spectroscopy and by IR investigation[116] of the proton acceptor power of compounds **6** in the incomplete acid–base interaction with phenol in tetrachloroethylene. In the first study the correlation between the values of $pK = H_0$ (half-protonation) and the electrophilic σ^+ constants gives a ρ value of 3.35 which compared with those of 5-substituted 2-acetylthiophenes[115] ($\rho = 3.34$) and substituted acetophenones[115] ($\rho = 2.49$) indicates that the two heterocyclic rings once again exhibit the same sensitivity to structural changes but transmit the substituent effects more efficiently than the benzene ring. The d orbitals of Se and S are invoked[115] to justify the higher transmission ability. In the second study the association constants K_{as} and the shifts $\Delta\nu(OH)$ of the stretching vibrations of the hydroxy group of phenol occurring in the hydrogen bond formation, were correlated with the σ_p and σ_p^+ constants. Although the statistical analysis does not give clear information[116], the correlation with σ_p^+ constants was preferred because it seems to take better account of the conjugative interactions of the substituent and the carbonyl group. The values of -0.64 (K_{as} vs. σ_p^+) and -0.08 ($\Delta\nu$ vs. σ_p^+) compared with those calculated for the substituted acetophenones under the same experimental conditions (-0.36 and -0.09) seem to confirm the higher sensitivity of the selenophene compared to the benzene ring to the electronic effects of the substituent. However, these results[115,116] should be regarded with suspicion because of the type and limited number of substituents used in the correlations.

The transmission ability of selenophene and tellurophene rings has also been determined[117] in the side-chain solvolyses of 1(2-heteroaryl)ethyl acetates (**7**; Z = Se, Te and **8**; Z = Se, Te) in 30% ethanol by the k(Me)/k(H) reactivity ratios. The values of the ratios are 23 and 12 for the derivatives of selenophene and tellurophene, respectively. If the

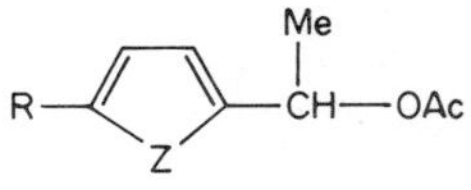

(**7**) R = H
(**8**) R = Me

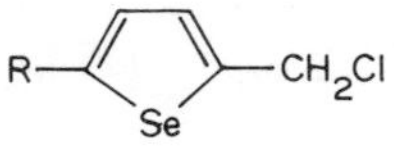

(**9**)

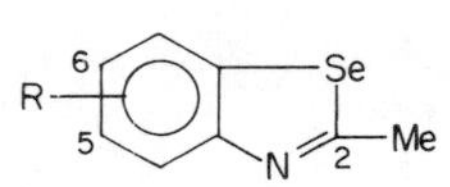

(**10**)

k(Me)/k(H) ratios are considered an estimate of the transmission through the individual ring system, it may be concluded that the selenophene transmits the electronic effects of the substituent better than the tellurophene. A comparison[117] with the k(Me)/k(H) ratios of the analogous derivatives of furan and thiophene (160 and 70, respectively) shows that the transmission ability decreases in the order furan > thiophene > selenophene > tellurophene. The transmission across the selenophene ring has also been determined[118] in the solvolyses of 5-substituted 2-chloromethylselenophenes (**9**). A ρ value of -6.42 can be calculated from the regression analysis between the log of the reaction rate constants and σ^+ values.

The sensitivity of the benzoselenazole system to structural changes has been evaluated[119] by ^{1}H-NMR spectroscopy. The chemical shifts of the 2-Me protons of a number of 5- and 6-substituted 1,3-benzoselenazoles (**10**) have been correlated with Hammett's σ constants using σ_m values for the substituents bonded at C(5) and σ_p values for those bonded at C(6). The value of the ρ constant is 6.72 and is very similar to those obtained[119] for 5- and 6-substituted 2-methyl-1,3-benzoxazoles ($\rho = 6.82$) and benzothiazoles

($\rho = 7.31$). The chemical shifts of the 2-Me protons are strongly dependent[119] on the nature of the chalcogen atom, but the ρ values for the three benzoazoles show that the heteroatom does not play an important role in the transmission of electronic effects, most of which are therefore transmitted across the N atom.

C. Electronic Transmission Through Selenium and Tellurium Atoms

The transmission of electronic effects of the substituents through the —Te— and —TeX_2—(X = Cl, Br) bridges has been investigated by Sadekov and coworkers[120–122] by measuring the thermodynamic basicity constants of *p*- and *m*-aryl [*p*-(dimethylamino)phenyl]tellurides (**11**) and corresponding dihalotelluranes (**13**, **14**) and the ^{19}F-NMR chemical shifts of *p*-aryl [*p*-fluorophenyl]tellurides (**15**) and corresponding dichlorotelluranes (**16**). The results of single-parameter correlations using σ, σ^o and σ^+ constants indicate that the Hammett σ constants generally give the best correlation. Examination of the ρ constants reported in Table 8 show that the Te in tellurides is able to transmit the electronic effect better than the Te in the telluranes by a factor of 2.3–4 and that ^{19}F chemical shifts are more sensitive than pK_a values.

m- or *p*-RC_6H_4—Z—$C_6H_4NMe_2$-*p*
(**11**) Z = Te
(**12**) Z = Se

m- or *p*-$RC_6H_4TeX_2C_6H_4NMe_2$-*p*
(**13**) X = Cl
(**14**) X = Br

p-$RC_6H_4TeC_6H_4F$-*p*
(**15**)

p-$RC_6H_4TeCl_2C_6H_4F$-*p*
(**16**)

Transmission through a —TeX_2— bridge is mainly of the inductive type, but through the —Te— bridge a contribution of polar resonance structures **17** and **18**, with the participation of the vacant d orbitals of Te, is possible. Dipole moment data seem to support this hypothesis[49,69]. The ^{19}F chemical shifts of **15** and **16** were also elaborated by

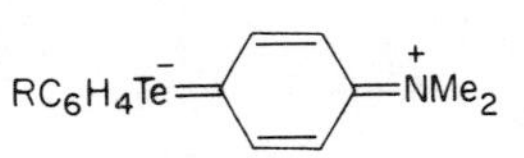

(**17**)

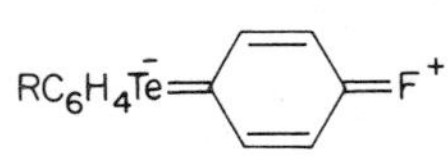

(**18**)

TABLE 8. Hammett ρ constants for the transmission of substituent effects through Se and Te bridges

Compound	Bridge	ρ^a	Ref.
11	—Te—	0.776	120
15		3.83	122
28		− 19.3	136
13	—$TeCl_2$—	0.335	121
16		0.984	122
14	—$TeBr_2$—	0.251	121
19	—Se—	1.47	127
20		0.35	60
21		− 1.55	47
22		-1.05×10^{-4}	133
27		− 18.7	136

^{a} Hammett σ constants were used except for **20** and **22** (see text).

TABLE 9. Taft ρ_I and ρ_R constants for the transmission of substituent effects through chalcogenic bridges[a]

Constant[b]	Bridge				Ref.
	—Te—	—$TeCl_2$—	—O—	—S—	
ρ_I	3.18	1.14	3.84	4.77	122, 123, 124
ρ_R	3.06	1.11	5.07	5.90	122, 123, 124

[a] From ^{19}F-NMR data of **15**, **16** and O and S analogues.
[b] σ_I and σ_R were used in the dual-parameter equation.

the Taft equation and comparison of ρ_I and ρ_R values (see Table 9) with those obtained from *p*-fluorodiphenyloxides[123] and *p*-fluorodiphenylsulphides[124] seems to indicate that Te is less able to transmit the electronic effects of the substituents than O and S.

The transmission through the —Se— bridge has been investigated in a variety of compounds and reactions (Table 8). The first studies were by Litvinenko and coworkers[59,125] who measured the $k(H)/k(NO_2)$ reactivity ratios of selenides (**12**; R = *p*-NO_2, H) in the reactions with picryl chloride and *p*-nitrobenzyl chloride. At 25 °C the $k(H)/k(NO_2)$ ratios for the two reactions are 22.2 and 17, respectively and are higher than those for O (16 and 12.1) and lower than those for S (30.6 and 23.8) analogues. Further studies[126] extended to other bridging groups have pointed out the following decreasing order of transmissivity: NMe > NH > S > Se > O > CH_2.

The log of the rate constants for alkaline hydrolysis of aryl selenoesters (**19**) in 70% aqueous acetone gives a good linear correlation with the Hammett σ constants[127,128]. The value of the reaction constant ($\rho = 1.47$) is the same as that found[127] for O ($\rho = 1.46$) and S ($\rho = 1.46$) analogues. Although the selenoesters (**19**) are more reactive than the analogous esters and thioesters, the Se atom exhibits the same power for transmitting electronic interaction as O and S. Parallel lines are also obtained in the correlations of carbonyl frequencies of **19** and O and S analogues with σ values confirming the previous results[127]. Analogous results are obtained in the ionization reaction of *para*- and *meta*-substituted aryl chalcogenacetic acids (**20**). The pK_a values correlate[60] with Bekkum's σ^n substituent constants[19], where possible, and otherwise with the Hammett σ values, and give a ρ value of 0.35 for substituted phenylselenoacetic acids (**20**; Z = Se) which is similar to that calculated for phenoxyacetic acids[129] (**20**; Z = O) (0.30) and phenylthioacetic acids[60] (**20**; Z = S) (0.32), but lower than that for phenylacetic acids[130,131] (0.49), and higher than that of β- arylpropionic acids[132] (**20**; Z = CH_2) (0.21). Chalcogen atoms have therefore a lower insulating effect than that of a methylene group. The similarity of transmission of electronic effects of chalcogen atoms is probably the result of an interplay between the increase in the polarizability of the heteroatom and the increased distance of the substituent from the reaction centre going from O to Se[60].

p-$RC_6H_4SeC(=O)CH_2CH_2Ph$ **(19)** *m*- or *p*-RC_6H_4—Z—CH_2COOH **(20)**

Differences in the electronic transmission through —Se—, —S— and —O— bridges are also observed in the acid-catalysed hydrolysis of aryl vinyl chalcogenides. The mechanism of the reaction of aryl vinyl selenides (**21**) has been studied by McClelland and Leung[47]. The rate-determining step of the reaction is the proton transfer to the olefinic bond (equation 11) and the reaction products are acetaldehyde and benzoselenols which quickly convert to diselenides under the reaction conditions. The Hammett ρ constant of selenides (− 1.55) is smaller than those of sulphides (− 1.84) and ethers (− 2.00). The ability to transmit the substituent effects to an adjacent positive reaction centre therefore appears to be largely determined by the degree of the electron-donating character of the chalcogen atom.

$$\underset{\textbf{(21)}}{ArSeCH{=}CH_2} \xrightarrow[\text{slow}]{H^+} ArSe\overset{+}{C}HMe \rightarrow \rightarrow ArSH + MeCHO \quad (11)$$

In aryl allyl chalcogenides (**22**; Z = O, S, Se) and aryl cyanoselenides a linear dependence between the $\bar{\nu}_{max}$ (cm^{-1}) of the K-band in UV spectra and σ_n^- has been found[133]. One-parameter regression analysis shows that the power of transmission of a chalcogenic bridge in allyl compounds decreases in the order —Se—($\rho = -1.05 \times 10^{-4}$) > —O—($\rho = -1.03 \times 10^{-4}$) > —S—($\rho = -0.83 \times 10^{-4}$) and is higher than in the cyanoselenides ($\rho = -0.71 \times 10^{-4}$).

$$p\text{-}RC_6H_4{-}Z{-}CH_2CH{=}CH_2$$
(22)

The transmissivity of the —Se— bridge has been evaluated[134] and compared with those of —Se—Se—, —CH_2— and —CH_2—CH_2— bridges by measurements of half-potential of diferrocenyl compounds **23**–**26**. The values $E'_{1/2}$ and $E''_{1/2}$ of the first and second half-wave potentials and their differences $\Delta E_{1/2}$ are reported in Table 10.

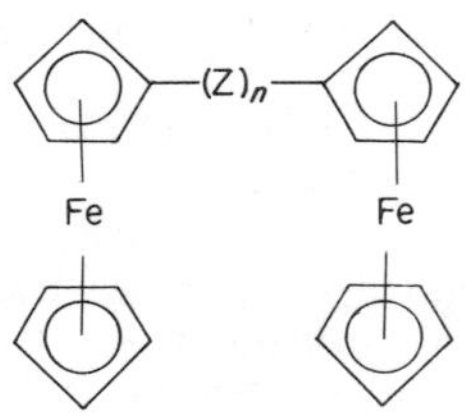

(**23**) Z = Se, $n = 1$
(**24**) Z = Se, $n = 2$
(**25**) Z = CH_2, $n = 1$
(**26**) Z = CH_2, $n = 2$

Assuming[135] the $\Delta E_{1/2}$ value as a quantitative estimate of the electronic interaction between the two ferrocene moieties, the order of decreasing power of transmission is: —Se— > —Se—Se > —CH_2— > —CH_2—CH_2—. A field effect should not be important since the C—Se and Se—Se bonds are much longer than the C—C bond and therefore the increased interaction between the ferrocene moieties in **23** and **24** is probably due to an inductive effect which is transmitted more efficiently from the more polarizable Se atom.

The ability of transmission of —Se— and —Te— bridges has been compared[136] by ^{1}H-NMR investigation of the complexes **27** and **28** of π-cyclopentadienylnickel tri-*n*-butylphosphine with *meta*- and *para*-substituted phenyl selenides and phenyl tellurides. Good linear correlations are observed between the cyclopentadienyl proton chemical

TABLE 10. First and second half-wave potentials of bridged diferrocenes[a]

Compound	$E'_{1/2}$	$E''_{1/2}$	$\Delta E_{1/2}$
23	0.46	0.68	0.22
24	0.53	0.67	0.14
25	0.39	0.56	0.17
26	0.37	0.37	0.00

[a] In acetonitrile except for **25** for which MeCN (90% EtOH) was used. For references see text.

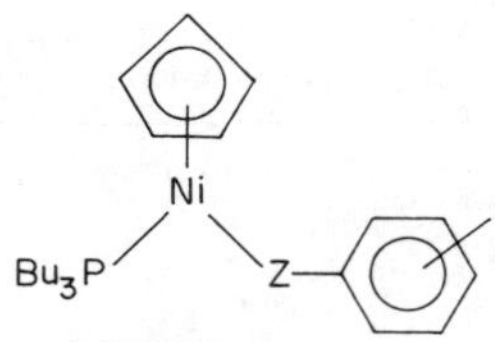

(**27**) Z = Se
(**28**) Z = Te

shifts and the Hammett σ constants. The ρ values for Se and Te ligands are -18.7 and -19.3, respectively, and the comparison with the ρ value of -17.9 for the S ligand[137] shows that the order of transmission of the substituent electronic effects through the heteroatom–Ni bond increases in the order —S— < —Se— < —Te—.

The electronic transmission in π-tellurane system has been evaluated[138] by single- and multi-parameter correlations between the thermodynamic basicity constants of telluronium ylides (**29**) and $\Sigma\sigma$, $\Sigma\sigma^{o}$, $\Sigma\sigma^{+}$, $\Sigma\sigma_{I}$, $\Sigma\sigma_{R}$, etc.

The $\Sigma\sigma^{o}$ constants seem to give the best one-parameter correlation ($\rho = 0.958$) but the results should be used cautiously.

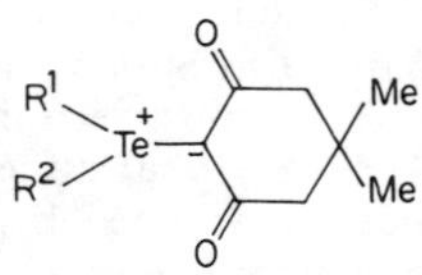

(**29**)

Hammett-like correlation between the basicity constants of *para*-substituted diphenylselenoxides and $\Sigma\sigma^{*}$ has been used[142] to evaluate the transmission in π-selenuranes. The ρ constant (-0.89) is very similar to that calculated for analogous sulphoxides (-0.92), indicating that the two heteroatoms have the same transmission ability.

The sensitivity of the Se–halogen bond to electronic effects has been evaluated[139–141] by determining the dissociation constants of symmetrically and unsymmetrically disubstituted diphenylselenium dibromides (**30**; X = Br) and symmetrically disubstituted diphenylselenium diiodides (**30**; X = I, $R^1 = R^2$). These compounds dissociate in CCl_4 solution to give diphenyl selenides (**31**) and halogen (equation 12). The pK values of the dissociation constants plotted against the Hammett σ constants give ρ values of 2.1 and 0.7 for dibromides and diiodides, respectively, indicating a higher sensitivity of the Se—Br bond as compared with the Se—I bond to electron density.

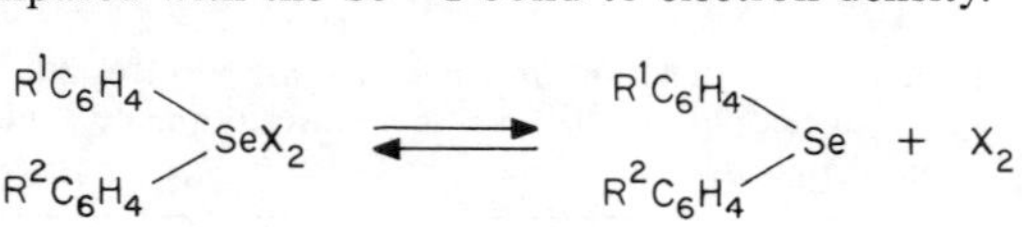

(12)

(**30**) (**31**)

D. Substituent Effects in Electrophilic and Nucleophilic Aromatic Reactions

The k(Me)/k(H) reactivity ratios for all chalcogen heterocycles (**3**; Z = Te, Se, S, O) and their 2-Me derivatives have been determined in our laboratories[117,143] by kinetic and competitive procedures in two electrophilic substitution reactions (see Section IV.A), i.e. formylation and trifluoroacetylation. The values are summarized in Table 11. If the k(Me)/k(H) ratios are considered as estimates of the sensitivity of the heterocyclic ring to

TABLE 11. $k(Me)/k(H)$ reactivity ratios for electrophilic substitution of five-membered chalcogen heterocycles

Ring	Formylation[a]	Trifluoroacetylation[b]
Tellurophene	620	500
Selenophene	300	280
Thiophene	290	380
Furan	880	1700

[a] By $COCl_2$–DMF in $CHCl_3$ at 20 C.
[b] By trifluoroacetic anhydride in dichloroethane at 75 °C. For references see text.

substituent electronic effects the following order is observed: furan > tellurophene > thiophene > selenophene. The order is different from that observed[117] in side-chain reactions in which a delocalizable positive charge is developed in the intermediate carbocation (see Section III.B) indicating that parallelism between the two classes of reactions is not always possible (see Section IV.A). The different order may be the result of concomitant factors such as the charge distribution, the position of transition state along the reaction coordinate and the interaction between the substituent and heteroatom[117].

The sensitivity of selenophene (**3**; Z = Se) to substituent electronic effects in electrophilic and nucleophilic reactions has been determined by mercuriation[144] of 2-substituted selenophenes and by piperidino-debromination[145] of 2-bromo-3-nitro-5-substituted selenophenes, respectively. The reaction constants ρ are reported in Table 12 together with those calculated for the analogous thiophenes[144,146]. The sensitivity of the selenophene ring to structural changes is high (high ρ values) and is the same as that of the thiophene ring.

A theoretical study of the sensitivity of selenophene, selenopheno[2,3-*b*]selenophene and selenopheno[3,2-*b*]selenophene rings to the electronic effect of substituents bonded at C(2) has been reported by Konar and coworkers[147]. Localization energies were chosen as reactivity indices and calculations were performed for the aromatic compounds and their *C*-protonated form (σ complex). The theoretical values correlated with Swain–Lupton $\mathscr{F}$ and $\mathscr{R}$ constants seem to indicate a dominant resonance effect for the α-position of five-membered rings and a more efficient transmission of resonance effect at C(5) through the [3,2-*b*] system than through the [2,3-*b*] system.

TABLE 12. Reaction ρ constants for electrophilic and nucleophilic substitutions at selenophene and thiophene rings

Ring	Mercuriation[a]	Piperidino-debromination[b]
Selenophene	− 5.77 (25 °C)	3.15 (20 °C)
	− 5.80 (35 °C)	
	− 5.14 (50 °C)	
Thiophene	− 5.86 (35 °C)	3.21 (20 °C)
	− 5.60 (50 °C)	

[a] Log of rate constants vs. σ^+, Ref. 144.
[b] Log of rate constants vs. σ^-, Refs. 145, 146.

E. Substituent Effects on Ring Protons and Carbons of Heterocyclic Systems by NMR Spectroscopy

The proton and carbon chemical shifts of substituted compounds relative to parent unsubstituted ones ($\Delta\delta = \Delta H_i^j$ or ΔC_i^j where i indicates the position of proton and carbon and j that of the substituent) are frequently related to the electronic effects of the substituents by σ or σ-like constants[10,11]. In Se/Te organic compounds $\mathscr{F}$ and $\mathscr{R}$ Swain–Lupton constants[25] are the most used. The regression constants f and r (equation 10) have been calculated for 2- and 3-substituted selenophenes[148,149], 2-substituted tellurophenes[150], 2- and 3-substituted benzo[*b*]selenophenes[150], benzo[*b*]tellurophenes[150], 2-substituted selenopheno[3,2-*b*]selenophenes[151], 2-substituted selenopheno[2,3-*b*]selenophenes[147] and *para*-substituted 4-aryl-1,2,3-selenadiazoles[152]. Some examples are summarized in Table 13. One-parameter regressions between $\Delta\delta$ and σ_p^+ have been calculated for 2-substituted selenopheno[3,2-*b*]selenophenes[151]. Theoretical considerations discourage the use of reactivity constants in ^{1}H-NMR correlations and supporting statistics question the meaning of dual-parameter equations (see Section III.A).

Another approach is to correlate the relative chemical shifts of a substrate $S_1[\Delta\delta(S_1)]$, against those of another substrate S_2, $[\Delta\delta(S_2)]$, by equation (13). The straight line of the regression should pass through the origin, but deviations are observed. The regression constant A can be viewed (B should be zero) as the ratio of the transmissions of the substituent effect of the two substrates. The constant A has been determined for the couples selenophene–thiophene[153,154], selenophene–tellurophene[153,154], tellurophene–thiophene[153,154], selenophene–benzo[*b*]selenophene[150], tellurophene–

TABLE 13. Regression constants (f and r) and intercept (i) of the Swain–Lupton equation for substituted heterocyclic rings containing Se and Te[a]

$\Delta\delta$ (Compound)[b]	f	r	i	Ref.
ΔH_3^2 (Te)	0.51	2.97	−0.07	150
ΔH_3^2 (Se)	0.34	2.67	0.00	148, 149
ΔH_3^2 (BTe)	0.33	2.49	0.04	150
ΔH_3^2 (BSe)	0.28	2.13	0.12	150
ΔH_2^3 (Se)	0.57	3.23	0.02	148, 149
ΔH_2^3 (BTe)	0.17	4.20	0.33	150
ΔH_2^3 (BSe)	0.49	3.28	0.22	150
ΔC_3^2 (Se)	−5.7	46.6	4.0	148, 149
ΔC_2^3 (Se)	−3.7	61.7	2.3	148, 149
ΔC_5^2 (2,3-SeSe)	2.0	4.9	−0.3	147
ΔC_5^2 (3,2-SeSe)	7.5	15.2	0.86	147, 151

[a] Data for other positions of proton, carbon and substituent are reported in the cited references.
[b] Te = tellurophene, Se = selenophene, BTe = benzo[*b*]tellurophene, BSe = benzo[*b*]selenophene, 2,3-SeSe = selenopheno[2,3-*b*]selenophene, 3,2-SeSe = selenopheno[3,2-*b*]selenophene.

TABLE 14. Relative shifts of 2-substituted tellurophenes, selenophenes and furans vs. those of 2-substituted thiophenes[a]

Relative proton chemical shift				Relative carbon chemical shift			
$\Delta\delta(S_1)$	$\Delta\delta(S_2)$	*A*	*B*	$\Delta\delta(S_1)$	$\Delta\delta(S_2)$	*A*	*B*
				ΔC_2^2(Te)	ΔC_2^2(S)	1.18	1.73
				ΔC_2^2(Se)	ΔC_2^2(S)	1.11	−0.62
				ΔC_2^2(O)	ΔC_2^2(S)	0.81	−5.05
ΔH_3^2(Te)	ΔH_3^2(S)	1.23	−0.16	ΔC_3^2(Te)	ΔC_3^2(S)	1.10	−0.99
ΔH_3^2(Se)	ΔH_3^2(S)	1.11	−0.08	ΔC_3^2(Se)	ΔC_3^2(S)	1.17	−1.57
ΔH_3^2(O)	ΔH_3^2(S)	1.16	0.16	ΔC_3^2(O)	ΔC_3^2(S)	1.22	−0.11
ΔH_4^2(Te)	ΔH_4^2(S)	1.64	0.04	ΔC_4^2(Te)	ΔC_4^2(S)	1.04	−0.86
ΔH_4^2(Se)	ΔH_4^2(S)	1.06	0.01	ΔC_4^2(Se)	ΔC_4^2(S)	1.18	−0.43
ΔH_4^2(O)	ΔH_4^2(S)	1.07	0.19	ΔC_4^2(O)	ΔC_4^2(S)	1.28	−0.75
ΔH_5^2(Te)	ΔH_5^2(S)	1.16	−0.02	ΔC_5^2(Te)	ΔC_5^2(S)	1.25	−0.45
ΔH_5^2(Se)	ΔH_5^2(S)	1.02	0.00	ΔC_5^2(Se)	ΔC_5^2(S)	1.14	−0.50
ΔH_5^2(O)	ΔH_5^2(S)	0.68	0.10	ΔC_5^2(O)	ΔC_5^2(S)	0.58	0.28

[a] Te = tellurophene, Se = selenophene, S = thiophene, O = furan; *A* and *B* are the regression constant and the intercept, respectively, of equation (13); data taken from Refs. 153 and 154.

benzo[*b*]tellurophene[150], selenopheno[3,2-*b*]selenophene–thieno[3,2-*b*]thiophene[151] and dibenzylchalcogenides–dibenzyldichalcogenides[155]. The data relative to 2-substituted five-membered heterocycles are summarized in Table 14. Except for a few cases no significant differences are observed in the slopes with values near to unity. Therefore the four rings transmit the substituent effects by the same mechanism and with practically the same intensity.

$$\Delta\delta(S_1) = A\cdot\Delta\delta(S_2) + B \qquad (13)$$

The coupling constants *J*(H–H) and *J*(Se–H) of 2- and 3-substituted selenophenes have also been correlated with Swain–Lupton $\mathscr{F}$ and $\mathscr{R}$ constants[156], but generally poor correlations are obtained.

F. Miscellaneous

The sensitivity of the seleninic group (—SeO_2H) to substituent effects has been evaluated by the correlation[157,158] between the ionization constants[157–160] of *p*- and *m*-benzeneseleninic acids determined in water at 25 °C and the Hammett σ constants. The ρ value ($\rho = 1.0$) indicates that in aromatic systems the seleninic group has the same sensitivity as the carboxylic group.

Kristian and Suchar[161] found a linear correlation between the vibration frequencies of the isoselenocyano group (—NCSe) of *p*- and *m*-substituted phenyl isoselenocyanates and Hammett's σ constants. The reaction constant ($\rho = -16.14$ in chloroform) is solvent-dependent and is markedly lower than that for the isothiocyanates ($\rho = -36.60$ in chloroform). The result agrees with the greater mesomeric interaction of the isothiocyano group with the aryl ring.

Correlations with σ constants were also found in the base-catalysed decomposition of 4-aryl-1,2,3-selenadiazoles[162] ($\rho = 2.4$) and in the ionization reaction of 5,6-substituted 2,1,3-benzoselenadiazoles[163].

IV. REACTIVITY AND DIRECTING EFFECTS IN HETEROCYCLIC SYSTEMS

Organic compounds of Se and more particularly Te are sometimes unstable in mineral acids and sensitive to light. They give complexes at the heteroatom and undergo carbon–heteroatom fission with nucleophilic reagents. Hence the study of their reactivity is difficult, and most of the available reactivity data are qualitative. Quantitative data refer mainly to five-membered rings and will be discussed first. When necessary, a comparison with analogous compounds of O and S is made.

A. Electrophilic Aromatic Substitutions

The behaviour of selenophene (**3**; Z = Se) and tellurophene (**3**; Z = Te) in electrophilic substitutions is similar to that of other aromatic systems. The electrophile is directed mainly or exclusively to the 2-position (α-position) as in furan (**3**; Z = O) and thiophene (**3**; Z = S). The $\alpha:\beta$ ratio of selenophene in the acid-catalysed protodedeuteriation reaction[98] ($MeOH–H_2O$, 20 °C) is $> 5 \times 10^3$ and only the α-isomer is observed in sulphonation[168], halogenation[164,169], mercuriation[170], trifluoroacetylation[166], acetylation[164], formylation[165], chloromethylation[144] and aminomethylation[171]. Nitration[172] gives 15% of the 3-nitro derivative (β-isomer) so that the $\alpha:\beta$ ratio is similar to that found for thiophene[173]. Tellurophene is acetylated[143,165], formylated[143,165] and trifluoroacetylated[143,165] exclusively in the α-position. Bromination gives 1,1-dibromotellurophene[174].

An electron-donating or -withdrawing substituent present at the 2- or 3-position does not change the orientation and the electrophilic substitution still proceeds mainly or exclusively at the free α-position[181]. An exception is nitration with fuming HNO_3 and concentrated H_2SO_4 of 2-substituted selenophenes containing strong electron-withdrawing groups (i.e. CHO, COOH, CN, COMe, SO_2Cl). In this case the reaction occurs[175,181] mainly at the 4-position (60–80%). The preference for the α-position may be

TABLE 15. Rates of tellurophene, selenophene and furan relative to thiophene

Electrophilic substitution[a]	$k(Te)/k(S)$	$k(Se)/k(S)$	$k(O)/k(S)$
Bromination, Br_2, AcOH, 25 °C		47.5[b]	120[c]
Chlorination, Cl_2, AcOH, 25 °		6.5[b]	48.7[c]
Bromination, Br^+		4.5[b]	
Trifluoroacetylation, TAA, DCE, 75 °C	46.4[d]	7.33[d]	140[e]
Acetylation, Ac_2O, $SnCl_4$, DCE, 25 °C	7.55[d]	2.28[d]	11.9[b]
Protodemercuriation, HCl, EtOH, 70 °C		2.2[f,i]	2.3[f,i]
Protodedeuteriation, $MeOH–H_2O$, H^+, 20 °C		10[g,h]	1.8[g,j]
Formylation, $COCl_2$, $HCONMe_2$, 30 °C	36.8[d]	3.64[d]	107[d]

[a] Te = tellurophene, Se = selenophene, S = thiophene, O = furan, TAA = trifluoroacetic anhydride, DCE = 1,2-dichloroethane.
[b] Ref. 164. [c] Ref. 96. [d] Ref. 165. [e] Ref. 166.
[f] Ref. 95. [g] Ref. 98.
[h] $k(Se)/k(S) = 6–10$, Ref. 167, p. 24.
[i] It has been assumed[96] that the reaction proceeds via preliminary coordination of Hg with the heteroatom.
[j] Partial cleavage of the furan ring occurs, Ref. 96, p. 267.

rationalized in terms of the relative stabilities of Wheland intermediates as models for the transition states[96].

The α-reactivity data for selenophene and tellurophene relative to thiophene are reported in Table 15. Both compounds are more reactive than thiophene. A comparison with furan was made[165] in three electrophilic reactions: trifluoroacetylation, acetylation and formylation. In all three reactions the reactivity sequence is: furan > tellurophene > selenophene > thiophene. This order is the reverse of that of ground-state aromaticities[176] and suggests that the relative differences of ground-state energies play a more important role than the relative stabilities of intermediate carbocations in determining the reactivity scale. The activation parameters of the formylation of compounds **3** support this observation[165]. The activation entropy has the same value in all four reactions and the activation enthalpies are linearly correlated both with localization energies and empirical values of resonance. The constancy of entropy suggests that the transition states of all chalcogenides **3** lie in a similar position along the reaction coordinate.

Recent[177] theoretical calculations of electron density distribution of the intermediate carbocations show that the π charge on the heteroatom decreases in the sequence selenophene > thiophene > furan both for α- and β-substitution and is greater for β- than for α-substitution (Table 16). This supports the previous hypothesis[165] that in electrophilic substitutions the α:β ratio can be explained by the relative stabilities of carbocation intermediates, and that the relative rates of chalcogenides **3** are not determined only by the relative stabilities of the intermediates.

Reactivity and substituent effects of the 2-selenienyl group in electrophilic substitution reactions have also been studied[87,178] in compounds **32**, **34** and **36**. In all cases α-substitution at the selenienyl group was the prevalent reaction for giving the corresponding **33**, **35** and **37** compounds.

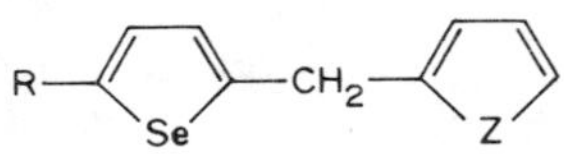

(**32**) R = H ; Z = S

(**33**) R = CHO , OMe ; Z = S

(**34**) R = H ; Z = Se

(**35**) R = Cl , Br , I , NO_2 , COMe ; Z = Se

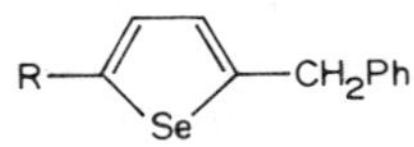

(**36**) R = H

(**37**) R = COMe

TABLE 16. π Charge distribution of Wheland intermediates of electrophilic substitution reactions[a]

Carbocation		Atom				
	Z	Z	C(2)	C(3)	C(4)	C(5)
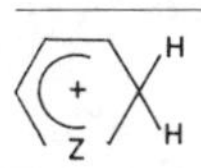	O	0.248	−0.007	0.368	−0.093	0.297
	S	0.277	0.010	0.374	−0.038	0.260
	Se	0.312	0.019	0.345	−0.027	0.250
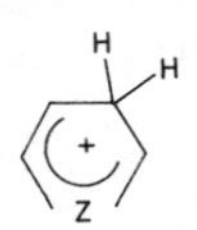	O	0.394	0.399	0.008	−0.017	−0.044
	S	0.600	0.288	0.019	0.146	−0.073
	Se	0.638	0.249	0.024	0.141	−0.058

[a] Data from Ref. 177.

A N atom in the five-membered ring markedly reduces the reactivity of the ring. When it is in the α-position, as for example in 1,2-chalcogenazoles (usually called isochalcogenazoles) (**38**; Z = O, S, Se, Te), the directing effect of the aza group appears to prevail over that of the chalcogen atom and the electrophile attacks in the 4-position[179]. Protodedeuteriation of 3,5-dimethylisotellurazole (**39**; Z = Te) and 3,5-dimethylisoselenazole (**39**; Z = Se) occurs at the free position and the former is more reactive than the

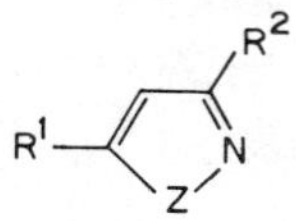

(**38**) $R^1 = R^2 = H$

(**39**) $R^1 = R^2 = Me$

(**40**) $R^1 = C_6H_5$, $R^2 = Me$

latter[179]. Attacks at the 4-position have also been observed in nitration, bromination and chlorosulphonation reactions[179]. The nitration of 3-methyl-5-phenylisoselenazole (**40**; Z = Se) also gives 3-methyl-5(*p*-nitrophenyl)isoselenazole indicating that the α-isoselenazolyl is a *para*-directing group[179]. The attack of the electrophile in 1,3-selenazoles (**41**) occurs in the 5-position as in the O and S analogues as a consequence of convergence of the directing effects of the aza group and the chalcogen atom[204]. The

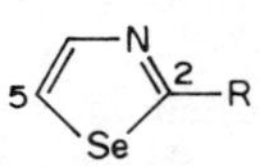

(**41**)

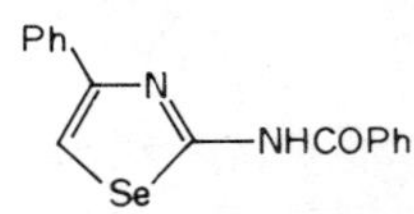

(**42**)

reactivity of the 5-position has been compared[204] to that of the *para* and *meta* positions of the benzene ring in the nitration of 2-benzamido-4-phenylselenazole (**42**), which is first nitrated in the 5-position and then a second NO_2 group enters in the *meta* position of the phenyl ring of the benzoyl group under more drastic conditions. A third NO_2 group is directed to the *para* position of the Ph ring bonded at C(4). The high reactivity of the 5-position in selenazoles has also been observed in bromination and azo-coupling reactions[204].

The reaction products **44**, **45** and **47** of nitration of 2-selenienyl-2-selenazole (**43**; Z = Se), 2-thienyl-2-selenazole (**43**; Z = S) and 3-selenienyl-2-selenazole (**46**) indicate

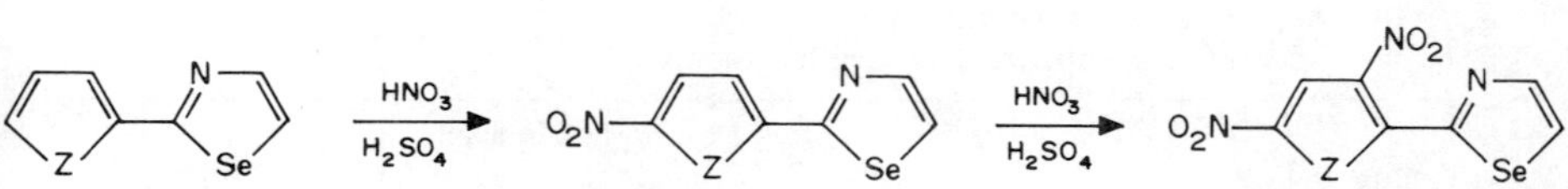

(**43**) (**44**) (**45**)

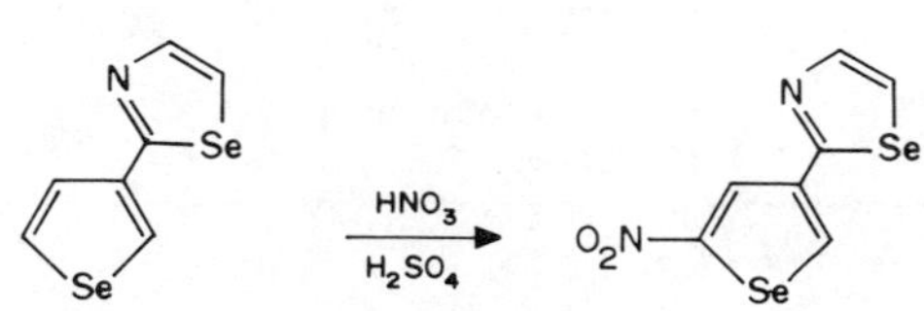

(**46**) (**47**)

that the reactivity of 2-selenazole is even less than that of thiophene[180]. When two N atoms are present, as in 1,2,5-selenadiazole (**48**; Z = Se), the compound is unreactive to hot fuming nitric acid and the 1,2,5-selenadiazolyl group is *ortho–para* directing[182]. Bromination with Br_2 and nitration using 99% HNO_3 of **49** gives **50** and **51**, respectively.

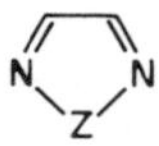

(49)

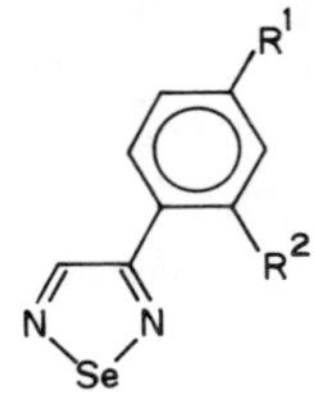

(49) $R^1 = R^2 = H$
(50) $R^1 = Br$, $R^2 = H$
(51) $R^1 = NO_2$, $R^2 = H$
(52) $R^1 = R^2 = NO_2$

By using a 1:2 v/v mixture of concentrated HNO_3 (99%) and H_2SO_4 (94%), **52** is obtained[182]. Competitive nitration (99% HNO_3) and bromination reactions show that **49** is less reactive than S and O analogues [k(S/Se) = 3.5 and k(O/Se) = 1.5 for nitration and k(S/Se) = 7 for bromination]. The higher reactivity of 3-phenyl-1,2,5-thiadiazole has been related to the higher delocalization of π electrons[182]. 1,2,5-Telluradiazole (**48**; Z = Te) has recently been synthesized[206], but easy decomposition by aqueous acid obstructs the study of its electrophilic substitution.

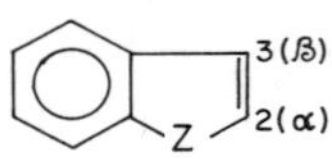

(53)

Benzo[*b*]selenophene (**53**; Z = Se) and benzo[*b*]tellurophene (**53**; Z = Te) are less reactive than the parent compounds and undergo preferential electrophilic substitution on the heterocyclic ring[183–186]. The α:β ratio depends on the nature of the electrophile and the experimental conditions. The isomer distributions in the acetylation and trifluoroacetylation reactions of **53**(Z = Te, Se, S, O) are reported in Table 17. Nitration of benzo[*b*]selenophene with concentrated HNO_3 in AcOH at 20 °C gives a mixture of α- and β-mononitro adducts with a ratio of 1:7. At 70 °C only 2,3-dinitrobenzo[*b*]selenophene is obtained[205]. The monobromination of benzo[*b*]selenophene occurs[185] preferentially in the β-position with a α:β ratio of 25:75. The reaction goes through a tribromide intermediate and an isotopic effect is observed[185]. Further bromination gives 2,3-dibromobenzo[*b*]selenophene. A third Br atom is directed in the benzene ring at C(6). The bromination[185,186] of benzo[*b*]tellurophene (Scheme 1) gives a 1,1-adduct which is then brominated exclusively in the α-position. Bromination in the β-

TABLE 17. Isomer distributions in the acetylation and trifluoroacetylation of benzo[*b*]-fused five-membered chalcogen heterocycles.

Compound	Electrophilic reagent			
	$MeCOCl/AlCl_3$[a]		Ac_2O/CF_3COOH[b]	
	α	β	α	β
Benzo[*b*]tellurophene	100	0	76	24
Benzo[*b*]selenophene	90	10	65	35
Benzo[*b*]thiophene		Predominant	35	65
Benzo[*b*]furan	Predominant		73	27

[a] Refs 96 and 185.
[b] Ref. 186.

position occurs neither in halogenation of **53** (Z = Te) nor of its 2-bromo derivative. 3-Halobenzo[*b*]tellurophenes are obtained[186] by the reaction of telluroindoxyl (**54**) with CCl_4 or CBr_4 in the presence of triphenylphosphine (Scheme 1), and by refluxing phenylacetylene with TeO_2 in acetic acid in the presence of lithium halide[187].

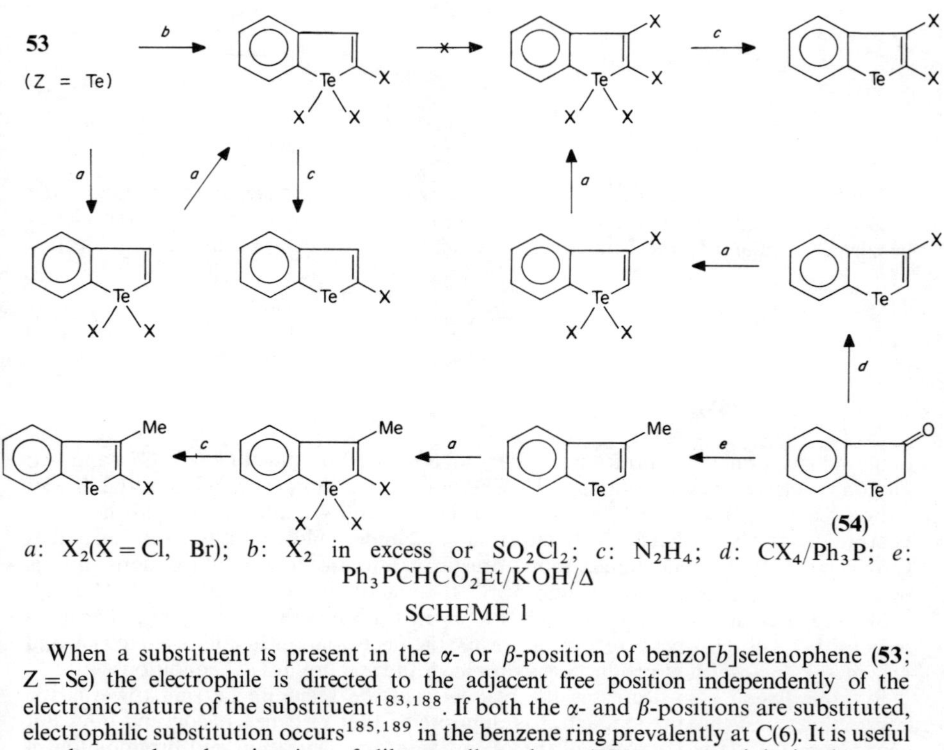

(**54**)

a: X_2(X = Cl, Br); *b*: X_2 in excess or SO_2Cl_2; *c*: N_2H_4; *d*: CX_4/Ph_3P; *e*: $Ph_3PCHCO_2Et/KOH/\Delta$

SCHEME 1

When a substituent is present in the α- or β-position of benzo[*b*]selenophene (**53**; Z = Se) the electrophile is directed to the adjacent free position independently of the electronic nature of the substituent[183,188]. If both the α- and β-positions are substituted, electrophilic substitution occurs[185,189] in the benzene ring prevalently at C(6). It is useful to observe that the nitration of dibenzotellurophene (**55**) occurs mainly in the *para* position with respect to the heteroatom[207].

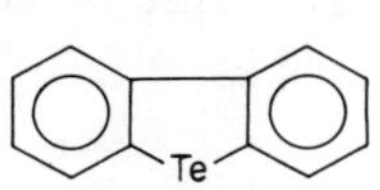

(**55**)

The isosters of phenanthrene **56**–**58** are more reactive than the parent benzo[*b*] compounds and electrophilic substitution occurs in the α-position[190]. The

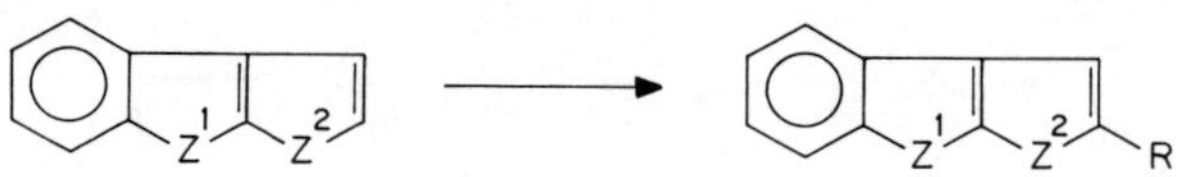

(**56**) $Z^1 = S$, $Z^2 = Se$
(**57**) $Z^1 = Se$, $Z^2 = S$
(**58**) $Z^1 = Se$, $Z^2 = Se$
$R = NO_2$, COMe, Br, CHO

benzo[*b*]seleno system affects the $\alpha:\beta$ reactivity of the fused *N*-methylpyrrole ring in **59** and **60**. The $\alpha:\beta$ ratios of the acetylation reactions of **59** and **60** are 66:34 and 42:58, respectively[191].

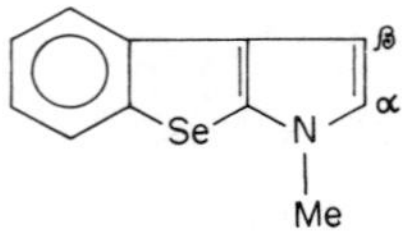

(59)

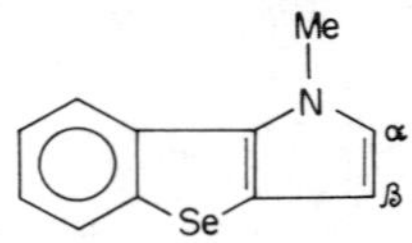

(60)

The benzo fusion further decreases the reactivity of the selenazole and selenadiazole rings and electrophilic substitutions occur on the Ph ring. The 3- and 2-positions of 1,2-benzisoselenazole (**61**) and benzoselenazole (**62**), respectively, are unreactive to electro-

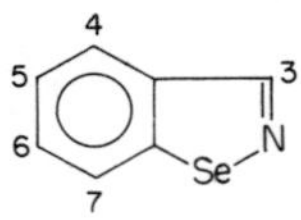

(61)

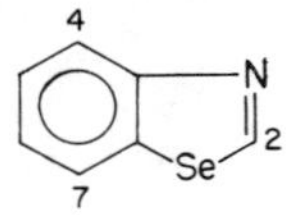

(62)

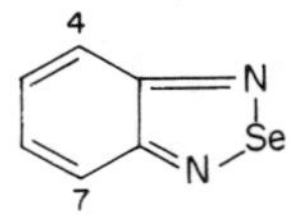

(63)

philic substitutions. The nitration (H_2SO_4–KNO_3) of **61** occurs at C(5) or C(7) and two mononitro derivatives are obtained[208]. The same orientation is observed in the bromination reaction (Br_2–H_2SO_4–Ag_2SO_4) but the 4,5- and 4,7-dibromides and the 4,5,7-tribromide are also isolated[208]. By nitration of **62** under mild conditions substitution at C(6) occurs[209,210] while under more drastic conditions a 4,6-dinitro derivative is obtained. In the detritiation reaction[192,204] the reactivity order of **62** and its O and S analogues decreases in the order $O > S > Se$. In the sulphonation[211,212] and halogenation[212–214] of 2,1,3-benzoselenadiazole (**63**) the electrophile is first directed to C(4) and 4,5- or 4,7-disubstituted products are obtained under more drastic conditions.

Pyridine fusion also decreases the reactivity of the selenophene ring towards the electrophiles and the directing effect is opposite to that of benzo fusion and does not depend on the relative position of the N and Se atoms. Nitration and bromination of selenopheno[2,3-*c*]-, [3,2-*c*]-, [2,3-*b*]- and [3,2-*b*]-pyridines **64**–**67** give β-nitro and β-bromo derivatives[193,194]*. The $\alpha:\beta$ ratio of deuteriodeprotonation[195] of **67** is ca. 10^{-3} as in the S analogue and higher than that for furo [3,2-*b*]pyridine ($\alpha:\beta = 10^{-5}$).

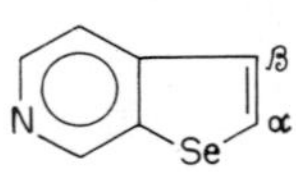

(64) [2,3-*c*]

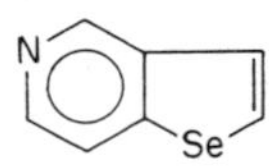

(65) [3,2-*c*]

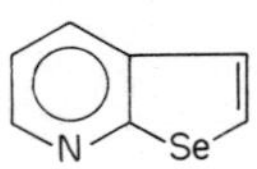

(66) [2,3-*b*]

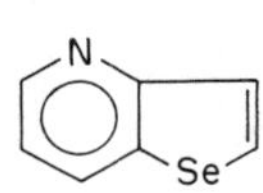

(67) [3,2-*b*]

*Most of the literature uses the prefix 'selenopheno'. *IUPAC Rules of Nomenclature* and *Chemical Abstracts* prefer the prefix 'selenolo'.

The fusion with α-phenyl-*N*-methylpyrrole increases[196] the electrophilic reactivity of the selenophene ring and substitution occurs in the α-position. The rate constants of acetylation of selenopheno[2,3-*b*] compounds **68**–**70** relative to selenophene are 8.1, 4.3 and 0.89, respectively. Compounds **68**–**70** are more reactive than analogous thieno compounds and the heterocyclic fusion increases the reactivity of thiophene more than that of selenophene.

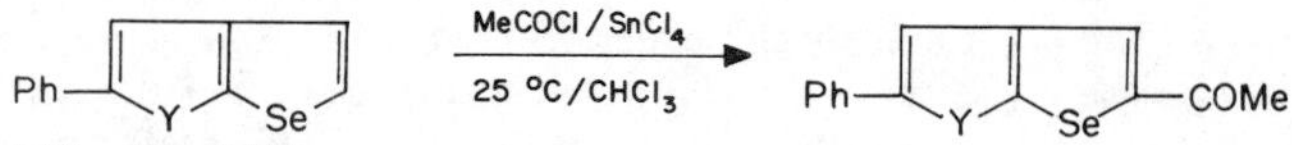

(**68**) Y = NMe
(**69**) Y = O
(**70**) Y = S

The reactivity of selenophene decreases when fused with the pyrazole ring but the directing effect does not change[197]. Vilsmeier formylation and bromination of selenopheno[3,2-*b*]pyrazole (**71**) gives only α-substituted derivatives and the reactivity of **71** is higher than that of **72**.

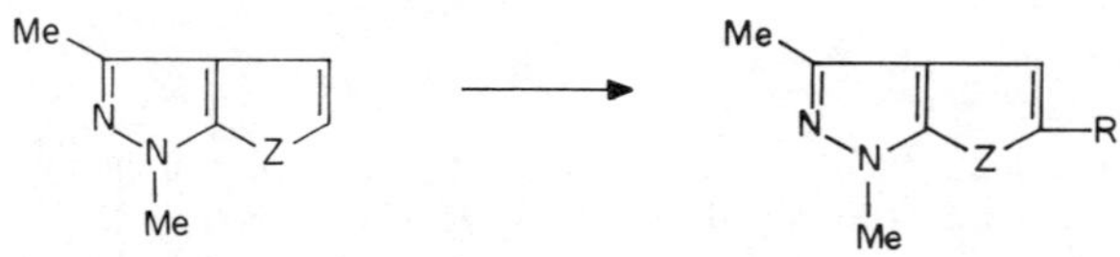

(**71**) Z = Se
(**72**) Z = S

R = COMe, Br

A recent quantitative study on the electrophilic substitution reactions of selenopheno[3,2-*b*]thiophene (**74**), selenopheno[3,2-*b*]selenophene (**75**), selenopheno[2,3-*c*]thiophene (**76**) and selenopheno[2,3-*b*]thiophene (**77**) has been re-

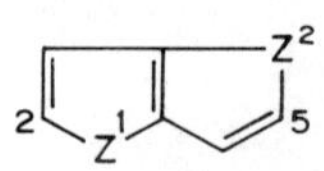

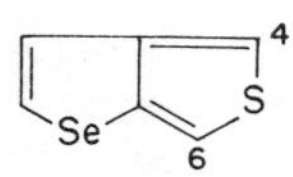

(**76**)

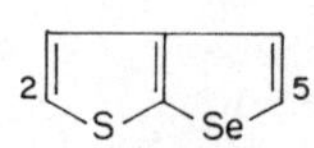

(**77**)

(**73**) $Z^1 = Z^2 = S$
(**74**) $Z^1 = S$, $Z^2 = Se$
(**75**) $Z^1 = Z^2 = Se$

TABLE 18. Isomer distributions (%) in electrophilic substitutions of selenophenothiophenes **74**, **76** and **77**[a]

	Compound					
	74		**76**[c]		**77**	
Reaction	C(2)	C(5)	C(4)	C(6)	C(2)	C(5)
Formylation	27	73	60	40	45	55
Acetylation	44	56				
Chlorination	0	100				
Bromination[b]	0	100			13	87

[a] Ref. 198 and references cited therein.
[b] 20% of the 2,5-dibromo derivative is present.
[c] Ref. 199.

ported by Gronowitz and coworkers[198–200]. The isomer distributions for **74**, **76** and **77** are reported in Table 18. Compounds **74** and **77** give predominant substitution in the selenophene ring while in **76** the formylation occurs exclusively on the thiophene ring. The high selectivity of the halogenation is probably ascribable to a primary attack of the halogen at the Se ring atom. The reactivity of **74** and **75** relative to **73** and to thiophene and selenophene has been studied in three electrophilic substitution reactions, namely acetylation, formylation and chlorination, by means of competitive experiments. The overall reactivity (Table 19) follows the order **75** > **74** > **73** > thiophene and the ratios increase in the series acetylation < chlorination < formylation. Positions 2 and 5 (Table 20) of **74** are more reactive than the α-positions of **73** and the parent five-membered compounds. Quantum-mechanical calculations substantially agree with the experimental data[198–200].

Isoxazole fusion increases the reactivity of the β-position of the selenophene ring[201]. Nitration with KNO_3–H_2SO_4 (conc.) and H–D exchange of **78** occur preferentially at C(3) as in the thieno analogues.

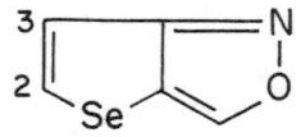

(78)

Hori and coworkers[202], in the course of their studies on the chemistry of selenoxanthylium salts, reported the nitration of 9-phenylselenoxanthylium perchlorate (**79**; Z = Se).

TABLE 19. Relative reactivities of **74** and **75** in electrophilic substitutions[a]

Relative reactivity	Acetylation[b]	Formylation[c]	Chlorination[d]
k(SSe)/*k*(SS)	1.69	4.59	4.19
k(SeSe)/*k*(SS)	2.81	14.3	10.5
k(SeSe)/*k*(SSe)	1.66	3.12	2.50
k(SSe)/*k*(S)	5.02	204	139
k(SeSe)/*k*(S)	8.35	635	347
k(SSe)/*k*(Se)	2.20	39.4	
k(SeSe)/*k*(Se)	3.66	123	

[a] Ref. 198; S = thiophene, Se = selenophene; SS = **73**, SSe = **74**, SeSe = **75**.
[b] $SnCl_4/Ac_2O$ in dichloroethane at 20 °C.
[c] $DMF/POCl_3$ at 20 °C.
[d] *N*-Chlorosuccinimide in acetic acid at 20 °C.

TABLE 20. Partial reactivities of the 2- and 5-positions of **74** in electrophilic substitutions[a]

Partial reactivity	Acetylation[b]	Formylation[c]
k(SSe-2)/*k*(SS-2)	1.46	2.41
k(SSe-5)/*k*(SS-2)	1.89	6.70
k(SSe-5)/*k*(SSe-2)	1.29	2.78
k(SSe-2)/*k*(S-2)	4.34	107
k(SSe-5)/*k*(Se-2)	2.46	57.5

[a–c] See Table 19.

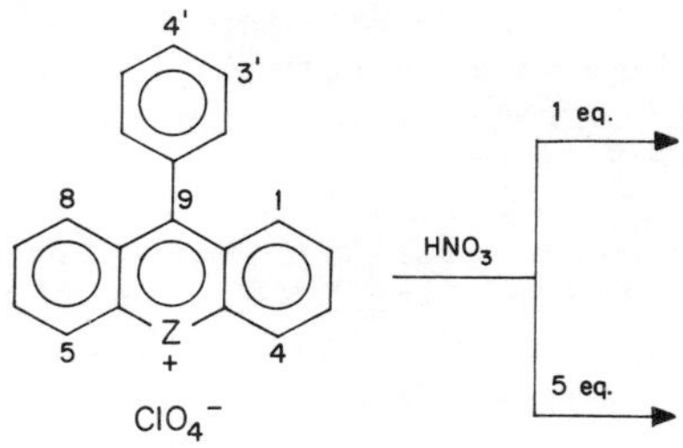

(79)

		Z	
	Se	S	O
4'	40	55.5	18.2
3'	60	44.5	81.8
4',4	44.5	54.5	-
3',4	55.5	45.5	-

SCHEME 2

With HNO_3–H_2SO_4 (1 eq. HNO_3) the electrophile is directed exclusively on the phenyl ring giving a mixture of 4′- and 3′-derivatives in the ratio 1:1.5 (Scheme 2). By using 5 eq. HNO_3, 4′,4- and 3′,4-dinitro compounds are obtained in the ratio 1:1.25. The S analogue (**79**; Z = S) gives similar results but the O analogue (**79**; Z = O) behaves differently. The reactivity decreases in the order $—S^+=\ >—Se^+=\ >—O^+=$. The reaction indices calculated from SCF-MO computation agree with experimental data[203].

B. Nucleophilic Aromatic Substitutions

Quantitative data on nucleophilic substitutions in selenophenes were first reported by Italian workers[216,217]. The kinetic constants of the reactions of compounds **80**

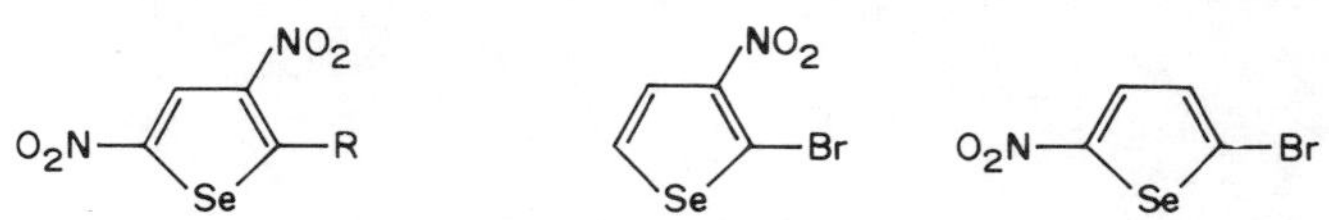

(80) (81) (82)

R = Cl, Br, I, SO_2Ph, $OC_6H_4NO_2$

with piperidine follow the order expected on the basis of the leaving-group type. The relative reactivities of **81** and **82** depend on the nucleophile and solvent. With piperidine **81** is more reactive than **82** [k(**81**)/k(**82**) = 7.5 in dioxane–water and 620 in benzene], but with sodium thiophenoxide in methanol a reverse reactivity is observed [k(**81**)/k(**82**) = 0.89]. Comparison with S analogues shows that selenophenes are about 1.2–10 times more reactive both with piperidine and sodium thiophenoxide[146,218]. The greater reactivity of selenophene compared to thiophene, both in nucleophilic and electrophilic reactions, may be explained by the lower aromatic character and greater ability of the selenophene ring to delocalize both negative and positive charges of intermediate[165,176,216] carbon ions.

The second-order rate constants for the reaction at 130 °C between 4- and 5-chloro-2,1,3-benzoselenadiazoles (**83**; Z = Se) and MeONa have been determined and com-

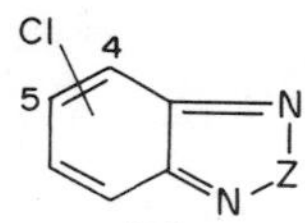

(83)

pared[219] with those of S and O analogues. Compound **83** (5-Cl, Z = Se) is more reactive than **83** (4-Cl, Z = Se) and both are more reactive than chlorobenzene. The activation effect depends on the chalcogen atom and decreases in the order Se(13.7) > S(5.2) > O(3.46). The relative reactivities [k(Z)/k(S); Z = O,S,Se] of 4- and 5-chloro derivatives decrease in the order O > Se ⩾ S and O > Se > S, respectively.

The reactivities of the 2-position of benzoselenazole (**62**) and the 3-position of 1,2-benzisoselenazole (**61**) towards nucleophilic attacks are very different. While **62** does not give Cicibabin amination[209,215] with $NaNH_2$, **61** reacts[208] with KNH_2 in liquid NH_3 to give the corresponding 3-amino derivative in fair yield. A Me group at C(2) in **62** is on the other hand activated and reacts with aromatic aldehydes to give styrene derivatives[209,215].

Nucleophilic substitution data on heteroaromatic compounds having Te atoms are not abundant. An example[187] is the reaction of 2-aryl-3-chlorobenzo[*b*]tellurophenes with EtONa in refluxing EtOH to give the corresponding ethoxy derivatives.

C. Meisenheimer Complexes

Selenophenes, 2-methoxyselenophenes, thiophenes and 2-methoxythiophenes 3,5-disubstituted with NO_2 and CN groups (**84**) give stable Meisenheimer complexes[220–224] (equation 14). The equilibrium and rate constants for the formation and decomposition of complexes **85**–**92** are reported in Table 21.

The complexes **85**, **87** and **89** are thermodynamically more stable than the analogous

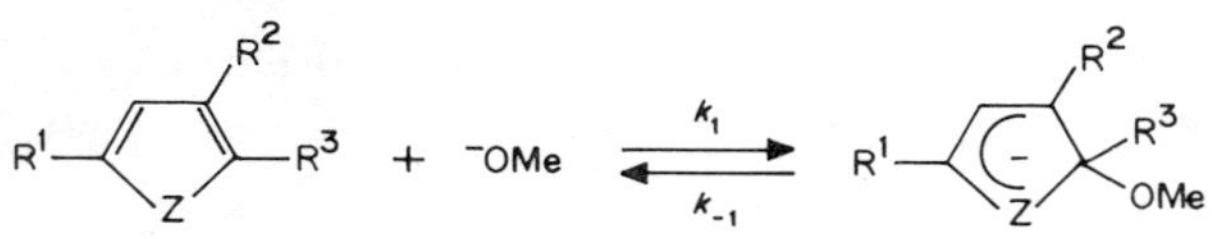

$$k_1 / k_{-1} = K_1 \qquad (14)$$

(**84**) Z = Se, S

(**a**) $R^1 = R^2 = NO_2$, R^3 = H

(**b**) $R^1 = NO_2$, R^2 = CN, R^3 = H

(**c**) R^1 = CN, $R^2 = NO_2$, R^3 = H

(**d**) $R^1 = R^2 = NO_2$, R^3 = OMe

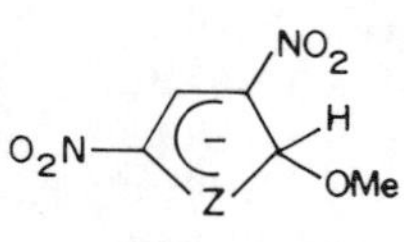

(**85**) Z = Se
(**86**) Z = S

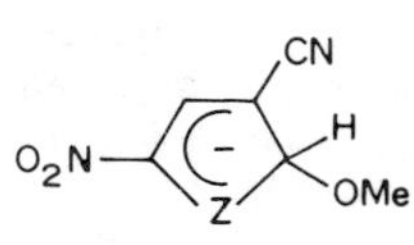

(**87**) Z = Se
(**88**) Z = S

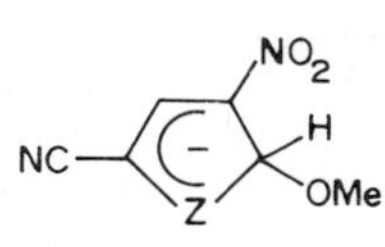

(**89**) Z = Se
(**90**) Z = S

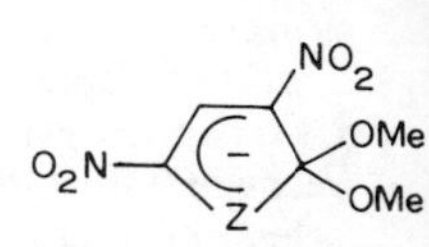

(**91**) Z = Se
(**92**) Z = S

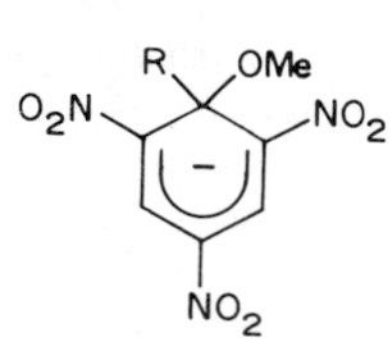

(**93**) R = H
(**94**) R = MeO

TABLE 21. Ratios of equilibrium constants and rates for the formation and decomposition of selenophene and thiophene Meisenheimer complexes with methoxide ion in methanolic solution at 25 °C (equation 14)

Compounds[a] A	B	Equilibrium $K_1(A)/K_1(B)$	Decomposition $k_{-1}(A)/k_{-1}(B)$	Formation $k_1(A)/k_1(B)$
85	**86**	68	2.8×10^{-2}	1.8
87	**88**	113	1.9×10^{-2}	2.1
89	**90**	72	1.5×10^{-2}	1.1
86	**88**	6.7	3.5	20
86	**90**	125	5×10^{-2}	6.2
85	**87**	4	5	20
85	**89**	118	9×10^{-2}	10
91	**92**	20	13.3×10^{-2}	2.5
91	**85**	100	3.6×10^{-b}	3.7
92	**86**	400	$0.6 \times 10^{-2\,b}$	2.7

[a] Data from Ref. 224 and 225; A and B indicate the complexes listed on p. 590; k_{-1} in s^{-1}, k_1 in $M^{-1}s^{-1}$, K_1 in M^{-1}.
[b] At 20 °C.

adducts of thiophene, **86**, **88** and **90**, and can be isolated as stable potassium salts. The presence of two methoxy groups has a stabilizing influence[222,225] and therefore **91** and **92** are more stable than **85** and **86** respectively. The complexes **85** and **91** are about 2500- and 300-times more stable than the benzene adducts **93** and **94**, respectively, and the reason has been discussed in terms of geometrical differences[222]. The stability also depends on the electron-withdrawing power of substituents and on their position in the five-membered ring[225].

The rates of formation of selenophene complexes are only a little higher[225] than those of the corresponding thiophene adducts, confirming the higher reactivity of selenophenes towards nucleophilic reagents[145,216,217]. The presence of a NO_2 group in the *β-ortho*-like position influences the rate of methoxide ion attack more than when the NO_2 group is in the *α-para*-like position[225]. This is in agreement with Spinelli's suggestion that the structure of the transition state coming from an attack at C(2) is affected more by a change in the *ortho*-like position than in the *para*-like position to the reaction centre[226,227].

The rates of decomposition of **85**, **87** and **89** are lower than those of **86**, **88** and **90**, respectively[225]. Surprisingly[225] the decomposition rates of **87** and **88** are lower than those of the corresponding **85** and **86**.

D. Addition Reactions

The Diels–Alder reaction has sometimes been used to estimate the aromaticity of heterocyclic compounds on the assumption that the less the ring is aromatic, the easier is the formation of the Diels–Alder adduct[167].

Selenophene, thiophene and furan give 1:1 adducts with maleic anhydride but the first two compounds require more drastic reaction conditions than the third in agreement with their greater aromatic character[167]. Quantitative data on the rate of cycloaddition of tetracyanoethylene oxide to selenophene, thiophene and furan have been reported by Gronowitz and Uppstrom[228]. The order of reactivity (furan > selenophene > thiophene) is the opposite of the order of ground-state aromaticity[176] and is the same as that observed in the electrophilic substitution reactions[165,228].

Benzo[*c*]selenophene (**95**,**96**) gives Diels–Alder adducts more easily than selenophene. With tetracyanoethylene the reaction occurs in the heterocyclic ring and is very fast[229] in

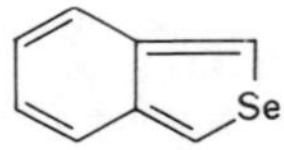

(95)

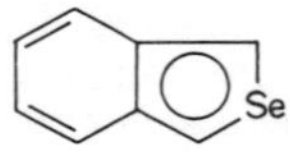

(96)

agreement with the poor aromatic character[230] of the compound. ^{1}H-NMR data and theoretical analysis[230] indicate that the structure **96**, in which 6 π electrons are delocalized on the heterocyclic ring, is more representative than the *ortho* quinoidal one (**95**).

The rates of quaternarization of benzoselenazole (**62**) and 2,1,3-benzoselenadiazole (**63**) have been determined[231,232] by means of a competitive method using Me_2SO_4 in sulpholane and have been compared with those of S and O analogues. Benzo fusion decreases the reactivity of parent five-membered compounds and the order of activation by the chalcogen atom in **62** and **63** is Se = S > O and Se > S > O, respectively.

The 2- and 4-aryl-substituted selenochromylium salts **97** and **98** direct the nucleophile R^- (H^-, Ph^-, PhS^-, Ac_2CH^-, etc.) into *para* and *ortho* positions of the heterocyclic ring like their O and S analogues and give **99** and **100** respectively[233]. The orientation does not depend either on the electronic nature of the aryl substituent or the type of nucleophile[233].

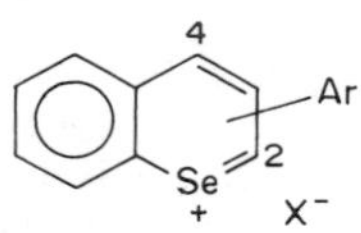

(**97**) X = ClO_4 , BF_4 ; 2-Ar

(**98**) X = ClO_4 , BF_4 ; 4-Ar

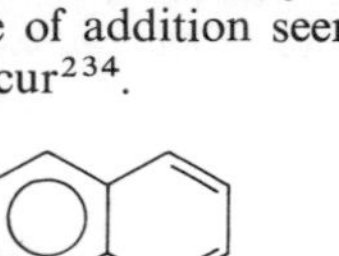

(99)

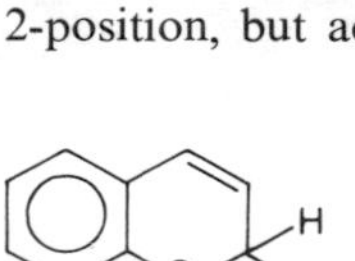

(100)

The unsubstituted selenochromylium cation[101] behaves in a more complex manner. The preferential site of addition seems to be the 2-position, but addition at C(4) and substitution also occur[234].

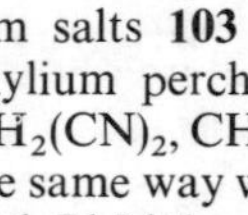

ClO_4^-

(101)

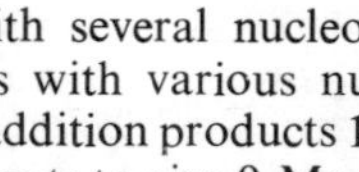

(102)

When the nucleophile is OH^- the first reaction product of **101** is the α-pseudobase **102** which is converted in the opened desmotropic form in the medium of the reaction[238].

Selenoxanthylium salts **103** and **104** react with several nucleophiles[203,235,236]. 9-Phenylselenoxanthylium perchlorate (**104**) reacts with various nucleophiles [NaOH, $LiAlH_4$, MeOH, $CH_2(CN)_2$, CH_2Ac_2] to give the addition products **105**. Compounds **103** and **104** react in the same way with Grignard reagents to give 9-Me and 9-Ph derivatives **106** and **107**, but with PhLi the reaction takes place exclusively or predominantly at the Se atom. The compound **104** gives only[236] **109**, and the compound **103** gives a mixture of **108** (87%) and **110** (13%). The reaction of **104** with organometallic reagents (PhLi, PhMgBr,

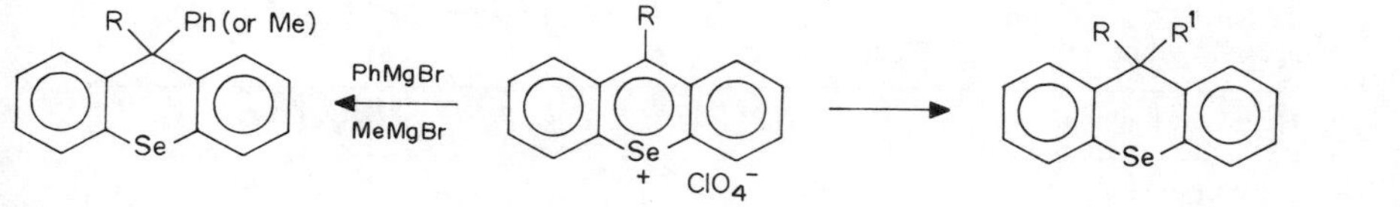

(106) R = H
(107) R = Ph

(103) R = H
(104) R = Ph

(105) R = Ph
R¹ = H, OH, OMe, CH(CN)₂, CH(Ac)₂

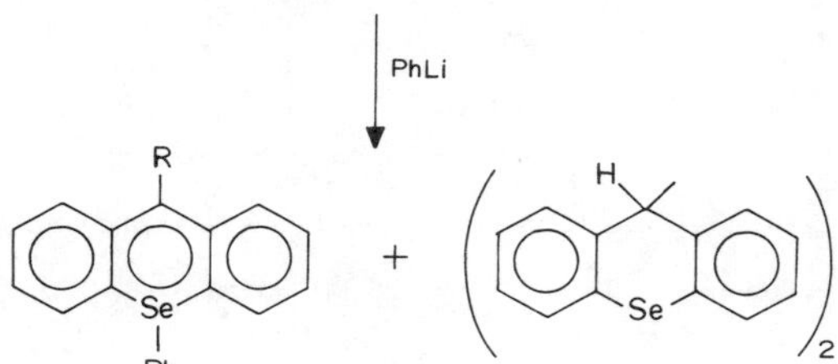

(108) R = H
(109) R = Ph

(110)

MeMgBr) has been investigated by ESR spectroscopy and a radical mechanism has been proposed[203,235,236]. Stackhouse and coworkers[237] report that **108** and **109** are not selenoanthracenes but oligomers of undetermined composition.

E. Side-chain Reactions

It is usual to consider complete parallelism between aromatic electrophilic substitutions and side-chain reactions in which the formation of a carbocation is the rate-determining step[96,239]. The first-order rate constants for the solvolysis of 1-(2-tellurienyl)- and 1-(2-selenienyl)-ethylacetates (**7**; Z = Se, Te) and their S and O analogues **7** (Z = S, O) in 30% ethanol (Table 22) show[117] a reactivity order (Te > O > Se > S) which is different from that (Table 15) found[117] in electrophilic substitutions (O > Te > Se > S).

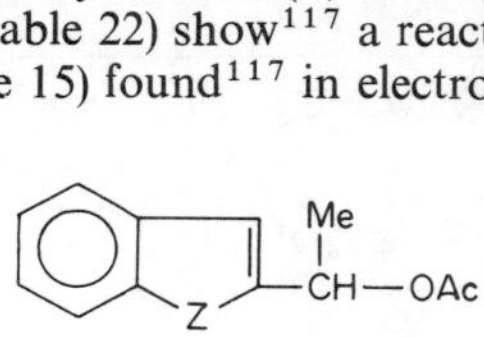

(111)

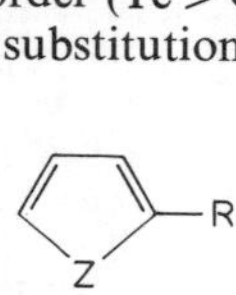

(112)

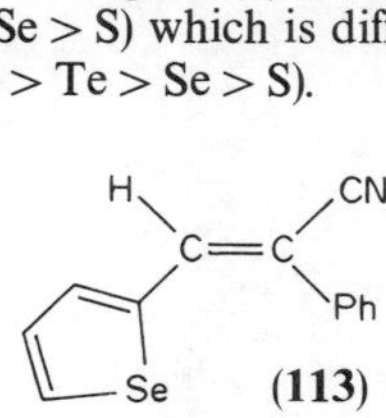

(113)

A different order of sensitivity to the substituent effect is also observed[117] (see Section III.B.D). Tellurophene behaves in a different way in the two types of reaction and the reason has been discussed in terms of solvation effects, charge distribution in the transition state and the position of the transition state along the reaction coordinate in the two types of reactions[117].

Solvolysis of **112** (Z = Se, R = CH_2Cl) and of 3-chloromethyl-2,5-dimethylselenophene proceeds by a S_N1 mechanism and comparison[118] with S and O analogues gives the same reactivity order found in the solvolysis of acetates **7**.

Benzo fusion decreases the reactivity of the esters **7** and the reactivity order of compounds **111** (Table 22) is similar to that observed for the monocyclic compounds. The

TABLE 22. Rate constants and relative rates for the solvolysis of 1-(2-heteroaryl)ethyl acetates **7** and **111** in 30% ethanol at 60 °C[a]

Heteroaryl group	$k \times 10^5 (s^{-1})$	$k(Z)/k(S)^b$	$[k(B)/k(M)] \times 10^{3c}$
2-Tellurienyl	341	5.2	
2-Selenienyl	109	1.7	
2-Thienyl	65	1	
2-Furyl	205	3.1	
2-Benzo[*b*]tellurienyl	3.26	5.2	8.9
2-Benzo[*b*]selenienyl	1.04	1.7	9.5
2-Benzo[*b*]thienyl	0.63	1	9.7
2-Benzo[*b*]furyl	0.95	1.5	4.8

[a] Data from Ref. 117.
[b] Ratios relative to thienyl(S) for five-membered compounds and to benzo[*b*]thienyl for benzo-fused ones; Z = Te, Se, S, O.
[c] Reactivity ratios between bicyclic (B) and monocyclic (M) systems.

annelation effect $k(B)/k(M)$ is larger for **7** (Z = O) whereas it is the same for the other systems[117].

Arcoria and coworkers[90] reported kinetic data for the nucleophilic side-chain reactions of 2-selenophene derivatives **112**. The reactions and the reactivity sequences are reported in Table 23. The reactivities of corresponding Ph derivatives do not occupy the same position in the six sequences. The variability of the reactivity order has been interpreted in terms of different mechanisms of the reactions.

Maccarone and coworkers[240] found that the base-catalysed isomerization of *cis*-1-(2-selenienyl)-2-phenylacrylonitrile (**113**) is faster than that of the acid-catalysed one and the reactivity order is: 2-thienyl > 2-selenienyl > 2-furyl. In the acid-catalysed isomerization the reverse order is found: 2-furyl > 2-selenienyl > 2-thienyl. A mechanism is suggested and the results are discussed in terms of inductive, mesomeric and steric effects of the heterocyclic ring.

TABLE 23. Reactivity of 2-selenienyl, 2-thienyl and 2-furyl derivatives in nucleophilic side-chain reactions[a]

Reaction	Compound **112**	Reactivity order[b]
Acid-catalysed hydrolysis	R = COOEt; Z = Se, S, O	O > Se > S
Base-catalysed hydrolysis	R = COOEt; Z = Se, S, O	O > S > Se
Substitution with aniline	$R = SO_2Cl$; Z = Se, S, O	Se > O > S
Substitution with aniline	$R = CH_2Cl$; Z = Se, S, O	O > S > Se
Condensation with aniline	R = CHO; Z = Se, S	S > Se
Condensation with BMTF[c]	R = CHO; Z = Se, S, O	O > S > Se

[a] Data from Ref. 90.
[b] Se = selenienyl, S = thienyl, O = furyl.
[c] BMTF = benzoylmethylenetriphenylphosphorane.

F. Protophilic Reactions

Hydrogen exchange reactions in strong alkaline media (*t*-BuOK in DMSO, KNH_2 in liquid NH_3, alkoxides in alcoholic solutions) are called 'protophilic reactions' and proceed via a carbanion intermediate[241,242].

Selenophene in the presence of *t*-BuOLi in Me_2SO at 25 °C exchange[167] a D atom in the α-position faster than thiophene (1.5 times) and furan (700 times) and the estimated α:β ratio is 2.5×10^5. Me groups at C(3) or C(5) reduce the rate of the reaction[167].

The 1,2,5-selenadiazole (**48**) in opposition to 1,2,5-thiadiazole does not give H–D exchange in strong basic conditions[243]. On the contrary 4-aryl-substituted 1,2,3-selenadiazoles (**114**), give easily H exchange at C(5), but the intermediate anion quickly decomposes to give an arylethynylselenolate ion[162].

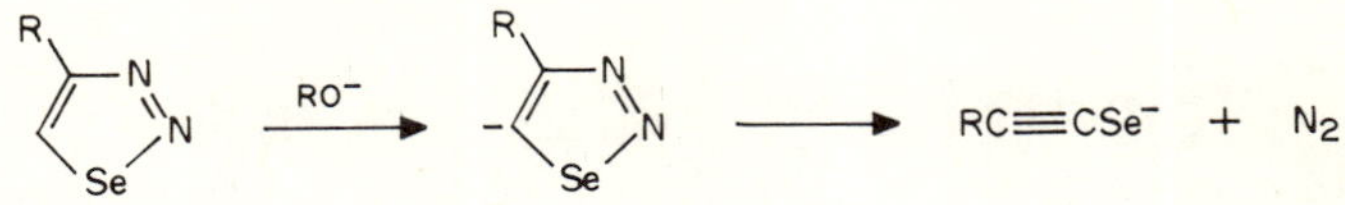

(**114**)

G. Metalation Reactions

Organolithium compounds give two main reactions with heterocyclic compounds of Se and Te: hydrogen– (or halogen–) metal exchange and nucleophilic attack at the heteroatom. The organolithium reagent, type of heteroatom, electronic nature of the substituent, experimental procedure and solvent influence either of the two reactions[244]. Organolithium intermediates from hydrogen– and halogen–metal exchange are used as nucleophiles[3] or converted to electrophilic reagents[245,246] to prepare substituted heterocyclic compounds[245,246]. Sometimes an eliminative ring-fission or a substitutional ring-opening occurs giving organolithium intermediates of synthetic utility[3,244].

Tellurophene[174], selenophene[167,244], 2-iodoselenophene[169], 2-bromoselenophene[247], benzo[*b*]tellurophene[185] and benzo[*b*]selenophene[185] are readily metalated in the α-position with BuLi or PhLi at 25–35 °C. The preferred α-orientation has been related[258] to ^{13}C–H one-bond coupling constants. The most reactive position is that with the greater $J(^{13}C\text{–}H)$. 3-Bromoselenophene[247] is metalated by PhLi in the 2-position at 36 °C whereas BuLi at − 50 °C gives Br–Li exchange. On the contrary 2,5-disubstituted 3-bromo or 3-iodo-selenophenes (**115**) undergo a partial or total ring-opening reaction[248,250]. 2,5-Dichloroselenophene gives a stable 3-lithium derivative with diisopropylaminolithium[250]. Contrary to 2,5-dimethoxythiophene (**116**; Z = S, R =

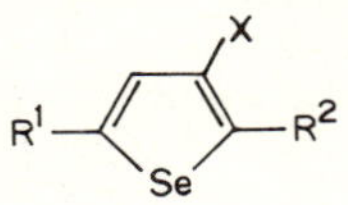

(**115**)

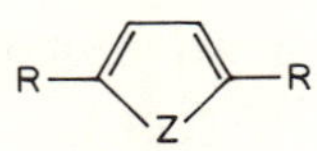

(**116**)

BuPhC═CHCH═CPhBu

(**117**)

$R^1 = R^2 = Cl$
$R^1 = Me$, $R^2 = SMe$
$R^1 = SMe$, $R^2 = Me$
X = Br, I

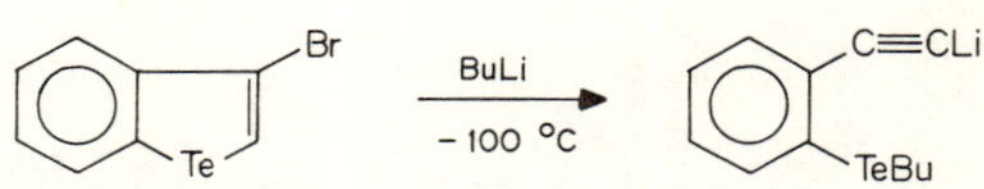

(**118**)

OMe), 2,5-dimethoxyselenophene (**116**; Z = Se, R = OMe) is not metalated in the 3-position by BuLi and PhLi, but a substitutional ring-opening occurs[256] giving dienes. The 2,5-diphenylselenophene (**116**; Z = Se, R = Ph) and 2,5-diphenyltellurophene (**116**; Z = Te, R = Ph) behave differently[255,256] when metalated with the complex BuLi–tetramethylethylenediamine (TMEDA). Both of these give ring-cleavage, but the first gives 5,8-diphenyl-5,7-dodecadiene (**117**) probably via β-lithiation of **116** (Z = Se, R = Ph) followed by eliminative ring-fission and addition of BuLi to 1,4-diphenylbutadiyne and

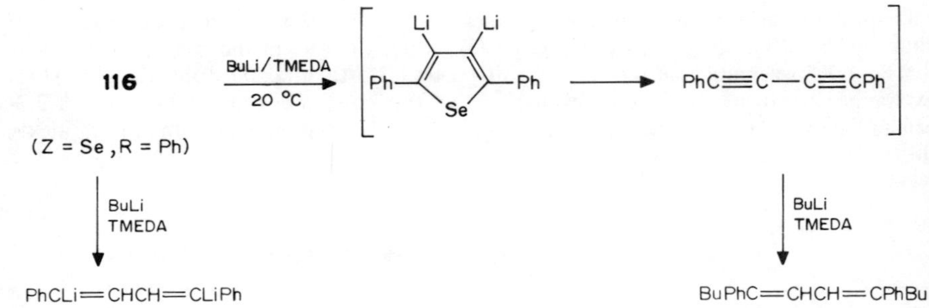

(117)

the second gives 1,4-diphenyl-1,4-dilithiumbutadiene which is trapped by various electrophiles. 3-Bromobenzo[*b*]tellurophene (**118**) undergoes[147] a ring-opening reaction even at − 100 °C, whereas the Se analogue **120** gives[185] the expected 3-lithium derivative **121** at low temperature. The behaviour of 2,3-dibromobenzo[*b*]selenophene (**122**) when is treated with two equivalents of BuLi is interesting[185]. At room temperature halogen–metal exchange at C(2) occurs, but at − 80 °C both Br atoms are exchanged and the intermediate 2,3-dilithiumbenzo[*b*]selenophene undergoes a ring-opening reaction giving **119**.

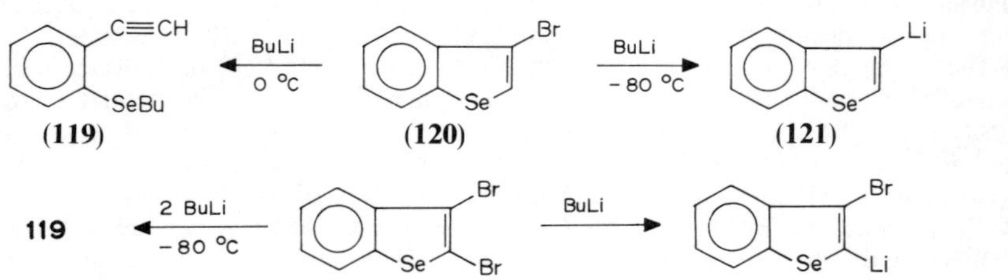

(119) (120) (121)

(122)

The metalation of **123**, **124** and **125** allows evaluation of both the relative sensitivity of selenophene and thiophene rings to the lithium reagent and the importance of the type of ring-fusion[185,187,248–251]. The tendency to ring-opening is greater in the selenophene than in the thiophene series.

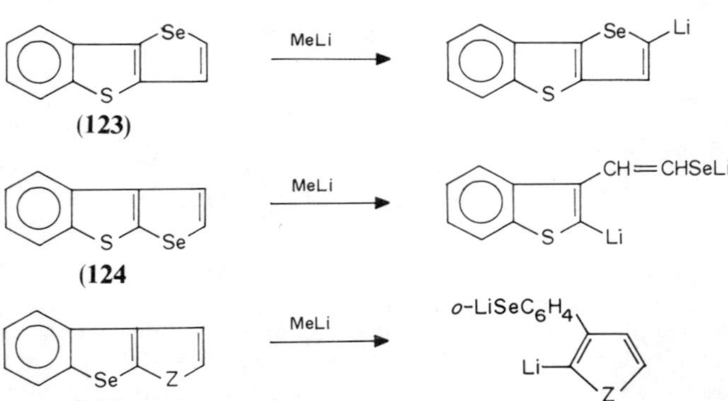

(123)

(124

(125)

Z = Se , S

Gronowitz's[199] recent report on the metalation of selenopheno[2,3-*c*]thiophene (**76**) confirms that the Se atom is more sensitive than the S to nucleophilic attack and shows that there is a parallelism between the reactivity of α- and β-protons in hydrogen–metal exchange and the attack sites in electrophilic substitution reactions. Only the H atoms at C(4) and C(6) (as in the formylation reaction[199]) were exchanged giving **126** and **127** in the ratio 18:82. The main reaction product **128** was the only one obtained by the ring-opening reaction.

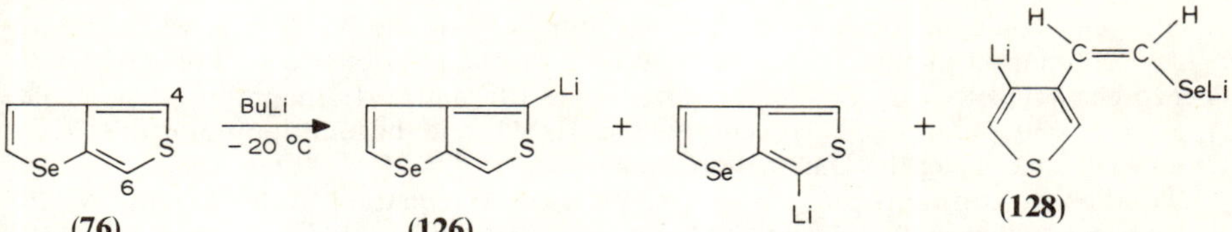

(**76**) (**126**) (**127**) (**128**)

The behaviour of selenopheno[2,3-*b*]thiophene (**77**) is similar to **76** when treated with MeLi[251,257]. Through the reaction of 1,2-benzisoselenazole (**61**) with BuLi under different experimental conditions, both the hydrogen–metal exchange and nucleophilic attack at the heteroatom are observed[208].

Nucleophilic cleavages of 1,2,5-selenadiazoles, 1,2,5-telluradiazoles and isotellurazoles by organolithium and Grignard reagents have recently been reported[252–254].

TABLE 24. pK_a values of carboxy-, hydroxy- and amino-heterocyclic compounds having one chalcogen atom

Compound	pK_a H$_2$O, 25 °C	pK_a H$_2$O–EtOH, 25 °C
2-Tellurophenecarboxylic acid	3.97[a]	5.48[e]
2-Selenophenecarboxylic acid	3.60[b]	5.14[e]
2-Thiophenecarboxylic acid	3.53[c]	5.05[e]
2-Furancarboxylic acid	3.16[d]	4.54[e]
2-Benzo[*b*]tellurophenecarboxylic acid		5.13[e]
2-Benzo[*b*]selenophenecarboxylic acid		4.79[e]
2-Benzo[*b*]thiophenecarboxylic acid		4.67[e]
2-Benzo[*b*]furancarboxylic acid		4.20[e]
4-Hydroxy-2,1,3-benzoselenadiazole	8.16[f]	9.06[g]
4-Hydroxy-2,1,3-benzothiadiazole	7.86[f]	8.80[g]
4-Hydroxy-2,1,3-benzoxadiazole	6.83[f]	7.58[g]
5-Hydroxy-2,1,3-benzoselenadiazole	8.06[f]	8.73[g]
5-Hydroxy-2,1,3-benzothiadiazole	8.16[f]	8.82[g]
5-Hydroxy-2,1,3-benzoxadiazole	7.28[f]	7.84[g]
4-Amino-2,1,3-benzoselenadiazole	2.18[h]	
4-Amino-2,1,3-benzothiadiazole	2.02[h]	
4-Amino-2,1,3-benzoxadiazole	0.78[h]	

[a] Ref. 91.
[b] Ref. 11, in H$_2$O-butylcellosolve 1:1 at 20 °C, pK_a = 5.00 (Ref. 87).
[c] Ref. 112.
[d] Ref. 259
[e] H$_2$O–EtOH 1:1, Ref. 260.
[f] Ref. 261.
[g] 47.5% EtOH, Ref. 261.
[h] Ref. 262.

H. Equilibrium Reactions

Ionization constants of carboxy-, hydroxy- and amino-heterocyclic compounds having one chalcogen atom are reported in Table 24. The main conclusion that may be drawn is that the order of acidity appears to be a function of the inductive effect of the heteroatom: O > S > Se > Te. The difference in pK_a values of 5-hydroxybenzothia- and benzoselenadiazole falls within the experimental error and therefore the observed reverse order is probably not significant. Monoprotonation of 4-amino-2,1,3-benzoselenadiazole and its chalcogenic analogues occurs at the amino group but 5-amino-2,1,3-benzoselenadiazole gives protonation mainly at the aza group[262]. Protonation of two N atoms of 2,1,3-benzoselenadiazole ($pK_{a1} = -1.41$; $pK_{a2} = -8.10$) and 2,1,3-naphthoselenadiazole ($pK_{a1} = -1.30$; $pK_{a2} = -7.78$) occurs in conc. H_2SO_4 and the ionization constants have been calculated using the Hammett H_0 acidity scale[163,263].

Equilibrium constants pK_{R^+} for cation–pseudobase (equation 15) for 1- (**101**) and 2-selenachromylium (**130**), selenaxanthylium (**103**) and dibenzo[*b*,*d*]selenapyrilium (**131**) cations are 1.20, 0.20, −1.67 and −4.28, respectively[264,265]. The benzo fusion decreases the stability of the cation and the benzo[*b*] cation **101** is more stable than its isomer **130**. These results coupled with those from the equilibrium reaction of H exchange (equation 16) show that the selenapyrilium cation **129** is more stable than the corresponding benzo-fused cations and that S-containing cations are more stable than O- and Se-containing cations[266].

$$R^+ + H_2O \rightleftharpoons ROH + H^+ \qquad (15)$$

$$RH + R^{1+} \rightleftharpoons R^+ + R^1H \qquad (16)$$

There is a close structural relationship between the conjugate acid **134** (equation 17) of selenachromone (**132**; Z = Se) and selenaxanthone (**133**; Z = Se) and the oxy derivatives of selenachromylium (**101**) and selenaxanthylium (**103**) salts.

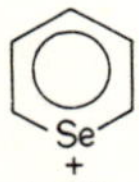

(129)

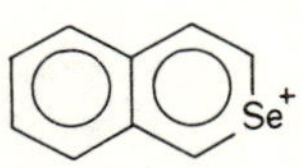

(130)

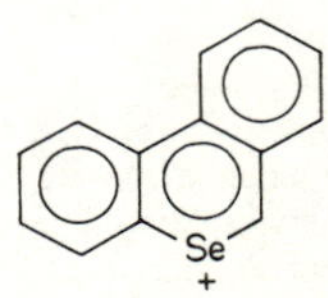

(131)

The pK_{BH^+} values of selenachromone (**132**; Z = Se) and selenaxanthone (**133**; Z = Se) have been determined (Table 25) and compared with those of O and S analogues[267]. The basicity decreases with benzo fusion in the order S > Se > O and S > O > Se in the chromone and xanthone series, respectively.

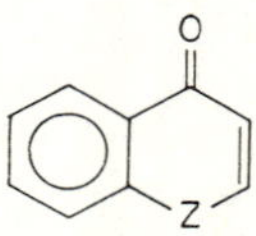

(132)

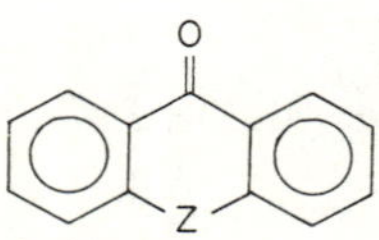

(133)

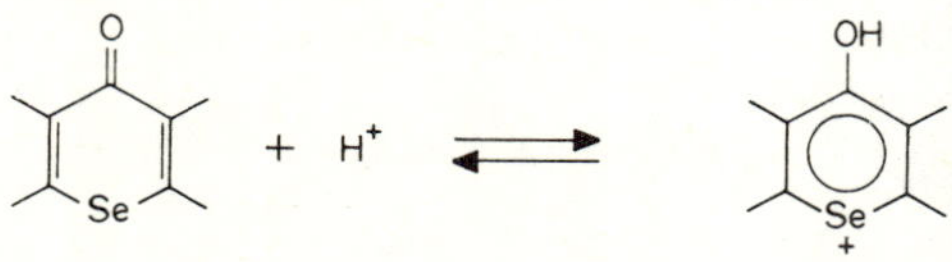

(134) (17)

TABLE 25. pK_{BH^+} of selenachromone, selenaxanthone and their S and O analogues at 25 °C in H_2SO_4 solution[a]

Compound	pK_{BH^+}
Selenachromone	−1.46
Thiachromone	−1.20
Chromone	−2.05
Selenaxanthone	−4.36
Thiaxanthone	−3.95
Xanthone	−4.12

[a] By UV spectrophotometric method, Ref. 267.

2-Hydroxyselenophene and its 5-Me derivative exist in an equilibrium in which the α, β-unsaturated selenalactone (**135**; R = H, Me) is the main isomer. Alkylation using an ion-pair extraction method mainly gives C- or O-alkylation with the 'soft' MeI and 'hard' Me_2SO_4, respectively. 2,5-Dimethyl-3-hydroxyselenophene (**136**) exists in oxo–enol equilibrium and selectively gives *C*- and *O*-alkylation depending on the nucleophilic reagent[268–271]

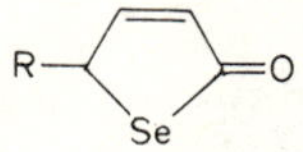

(**135**)

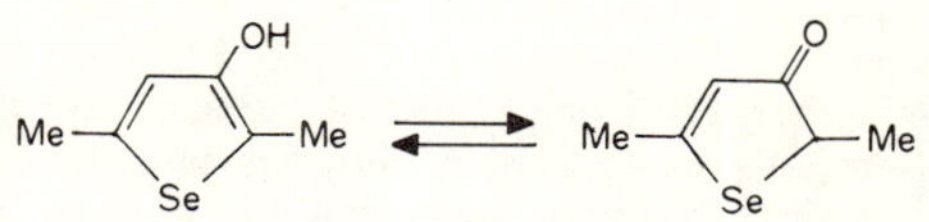

(**136**)

V. SYSTEMS WITH EXOCYCLIC FUNCTIONAL GROUPS

In recent years the chemistry of functional groups containing Te and, more so Se has been developed[6,272] and new reagents and new reactions have been introduced in the organic synthesis. The present discussion is mainly confined to directing effects of typical functional groups and how they affect the remaining part of the molecule.

A. Selenium Acids and Selenols

Selenenic acids (RSeOH) are supposed intermediates in a number of reactions and except for the antraquinone-1,4-diselenenic acid, no selenenic acid is stable in the pure form[273], but they easily disproportionate to diselenides and seleninic acids. Selenonic acids ($RSeO_3H$) are strong oxidizing substances, unstable in heat and light[274]. A limited number of selenonic acids, with proven structures, are known[274]. Quantitative data on the oxidizing power of aromatic selenonic and seleninic acids have recently been reported[275].

Seleninic acids ($RSeO_2H$) represent the most stable class of oxyselenium acids and are more stable than the corresponding sulphinic acids[159]. Alkyl- and aryl-seleninic acids (Table 26) are weaker than the corresponding carboxylic and sulphinic acids. The acidity decreases in the order S > C > Se. Theoretical calculations[159] indicate that conjugation between the Ph ring and the sulphinic group is prevented. The greater acidity of benzenesulphinic acid ($pK_a = 2.76$)[159] has been explained by invoking the solvent effects in the dissociation reaction. To explain the weaker acidity of benzeneseleninic acid a π(d–p) conjugation has been invoked[159]. This is supported by the fact that *p*-methoxybenzenesulphinic acid is stronger than the unsubstituted one ($\Delta pK_a = 0.04$) but

TABLE 26. Dissociation constants of seleninic acids in water at 25 °C

Compound	pK_a	pK_{a1}	pK_{a2}	Ref.
$MeSeO_2H$	5.19			279
$EtSeO_2H$	5.27			279
$PrSeO_2H$	5.25			279
$BuSeO_2H$	5.29			279
$PhSeO_2H$[a]	4.70–4.90			157, 159
$HO_2CCH_2SeO_2H$		2.60	5.43	280
$HO_2CCH(Me)SeO_2H$		2.47	5.48	280
$HO_2CCH(Et)SeO_2H$		2.53	5.48	280
$HO_2CCH(Pr)SeO_2H$		2.56	5.48	280
$HO_2C(CH_2)_2SeO_2H$		3.47	5.99	280

[a] For dissociation constants of substituted benzeneseleninic acids see Refs 157 and 159 and Section III.F.

the *p*-methoxybenzeneseleninic acid is weaker than the corresponding unsubstituted ($\Delta pK_a = -0.26$) acid.

To evaluate the substituent effect of a negatively charged Se atom on the acidity of the carboxyl group the microscopic dissociation constants for the SeH and COOH groups of selenoglycolic acid were determined[276] (Scheme 3) by spectrophotometry at 25 °C. The effect of a negatively charged Se atom in $^-SeCH_2COOH$ is 0.8 pK_a units larger than that produced by a charged N or O atom. A similar effect (0.6 pK units) is observed for a charged S atom. This effect is explained by decrease of the effective dielectric constant as a consequence of the increase of the atomic radius in the series O, S, Se.

The selenols have a relatively high acidity and Table 27 shows that the increased acidity of selenols over thiols is almost constant 3–3.2 pK units); the pK_a of PhSeH is probably not corrected[277]. A consequence is that selenols are able to protonate the amines in organic solvents, whereas thiols do not[277]. The same difference of pK_a is observed[278] between HSH ($pK_a = 7.0$) and HSeH ($pK_a = 3.74$). The pK_a value of HTeH[278] is 3.64 and therefore the tellurols are probably stronger acids than the selenols. The order of acidity of the —ZH groups (Z = O, S, Se, Te) is therefore reversed with respect to the acidity order of the $—ZO_2H$ and $—ZCH_2COOH$ (see Section V.D) groups when going from O to Te.

TABLE 27. Dissociation constants of selenols and thiols in water at 25 °C

Compound	pK_a Z = Se	Z = S	Ref.
PhZH	5.9[a]	6.5	281
$HOOCCH_2ZH$	4.7	8.1	276
$^-OOCCH_2ZH$	7.3	10.58	276
$MeOOCCH_2ZH$	4.70	8.08	276
$HOOCCH(NH_2)CH_2ZH$	5.24	8.25	283
$H_3\overset{+}{N}CH_2CH_2ZH$	5.0	8.3	284
C_9H_6NZH[b]	4.94	7.68	285
$H_2NCH_2CH_2ZH$	6.47	10.21	286

[a] Ref. 282; doubts on this value have been reported[277].
[b] C_9H_6N = 8-quinoline.

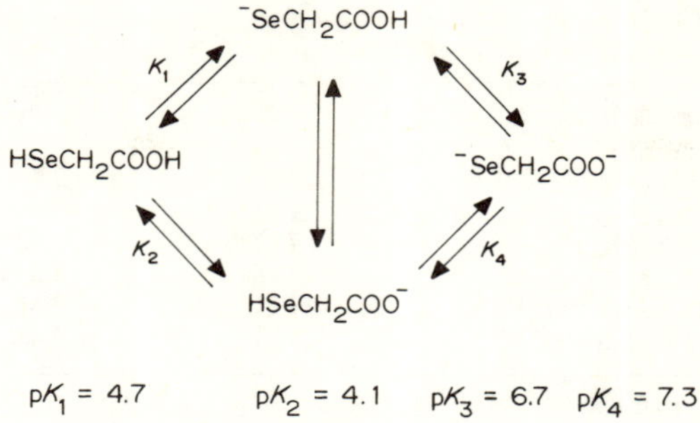

SCHEME 3

B. Onium Salts

Gilov and coworkers[287–289] have investigated the reactivity and orientation of selenonium ions[137–140] in electrophilic substitutions. The isomer distribution in nitration, chlorination and bromination of **137** and **138** is reported in Table 28. These ions exhibit —I, —M and *meta*-directing effects. Bromination of **137** in acetic acid without Ag^+ ions surprisingly gives a *para* derivative. The *meta* substitution decreases and the *para* substitution increases as the positive pole s insulated from the ring by a methylene group. By changing the counterion no significant differences were observed in isomer distribution. The sulphonium ion behaves similarly to the selenonium ion but the oxonium group exhibits a different directing effect. Nitration of triphenyloxonium tetrafluoroborate results in 100% *meta* substitution[290]. Selenonium ions are more reactive than sulphonium ions and bromination is faster than chlorination. Reactivity data of **137**, **139**

$[Ph\overset{+}{S}eMe_2]MeSO_4^-$ (**137**)

$[Ar\overset{+}{S}eMe_2]A^-$
(**138**) Ar = $PhCH_2$, A = picrate
(**139**) Ar = $PhCH_2$, A = ClO_4
(**140**) Ar = $PhCH_2CH_2$, A = ClO_4

TABLE 28. Isomer distribution in nitration, chlorination and bromination of selenonium salts[a]

Compound[b]	Reaction conditions[c]	*ortho*(%)	*meta*(%)	*para*(%)
	Nitration			
$PhSeMe_2MS$	HNO_3–H_2SO_4 conc.	2.6	91.3	6.1
$PhCH_2SeMe_2PI$	HNO_3–H_2SO_4 conc.	18.8	11.9	69.1
$PhCH_2SeMe_2PI$	HNO_3 fuming	12.5	18.0	69.5
	Chlorination			
$PhSeMe_2MS$	Cl_2/Ag_2SO_4; H_2SO_4	17.4	74.9	7.7
$PhSeMe_2MS$	Cl_2/AgO_2CCF_3; CF_3COOH	24.5	71.1	4.4
	Bromination			
$PhSeMe_2MS$	Br_2/Ag_2SO_4; H_2SO_4	6.4	88.1	5.5
$PhSeMe_2MS$	Br_2/AgO_2CCF_3; CF_3COOH	6.4	89.3	4.2
$PhSeMe_2MS$	Br_2; MeCOOH	9.4	0	90.6

[a] Data from Refs 287–290.
[b] MS = methyl sulphate, PI = picrate.
[c] At room temperature except for picrates (0 °C).

TABLE 29. Rate constants for nitration of selenonium and sulphonium salts in aqueous H_2SO_4 at 25 °C[a]

Compound[b]	% H_2SO_4	$10^2 k_2$ ($l\ M^{-1} s^{-1}$)
$PhSeMe_2MS$	98.1	2.56
$PhSMe_2MS$	98.1	0.113
$PhCH_2SeMe_2PC$	75.5	1.62
$PhCH_2SMe_2PC$	75.5	0.558
$PhCH_2CH_2SeMe_2PC$	70.1	3.52
$PhCH_2CH_2SMe_2PC$	70.1	3.44

[a] Data from Refs 287–290.
[b] MS = methyl sulphate, PC = perchlorate.

TABLE 30. Relative rates and partial factors for nitration at 25 °C of selenonium and sulphonium ions in aqueous H_2SO_4[a]

Compound	10^8 relative reactivity	$10^8 f_o$	$10^8 f_m$	$10^8 f_p$
$Ph\overset{+}{Se}Me_2$	9.22	0.719	25.3	3.37
$Ph\overset{+}{S}Me_2$	0.407	0.044	1.10	0.147
$PhCH_2\overset{+}{Se}Me_2$	100,000	56,400	35,000	415,000
$PhCH_2\overset{+}{S}Me_2$	80,000	38,400	93,800	210,000

[a] Data from Ref. 288.

and **140** and those of S analogues for the nitration reaction are reported in Tables 29 and 30. The selenonium ion **137** is 22.7 times more reactive than the S analogue and an insulating methylene group increases the reactivity by about 10^4 times. The results have been explained[228] in terms of π(d–p) overlap for the positive pole bonded to the Ph ring and invoking a hyperconjugative effect when the insulating group is present.

C. Selenonium and Telluronium Ylides

In the recent years selenonium and telluronium ylides have received much attention especially from the synthetic point of view[291–293].

The stability of these ylides is related to the delocalization of the negative charge. Ylides in which the carbanionic centre is part of a delocalized system (i.e. cyclopentadienyl) or is bonded to two electron-withdrawing substituents are stable and can be isolated. However, as the stability increases, the reactivity towards electrophilic reagents decreases. The reactivity also depends on the nature of the electrophilic reagent. Stabilized ylides **141**, **142** and **143** (Z = Se, Te) do not react[294,295] with *p*-nitrobenzaldehyde. Stable ylides **144** do

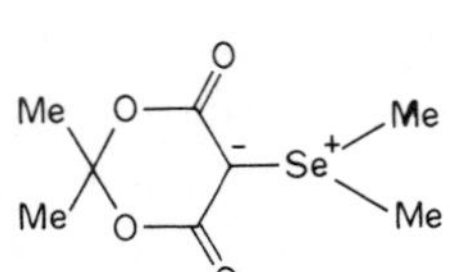

(141)

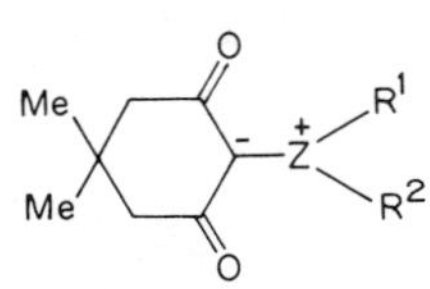

(142)

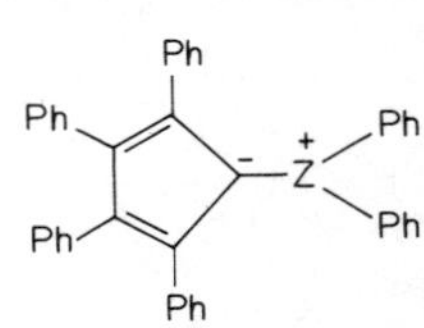

(143)

not react[296] with *o*-nitrobenzaldehyde, acrylonitrile, diethyl methylenemalonate and dimethyl fumarate, but react with dimethyl acetylenedicarboxylate to give intermediates

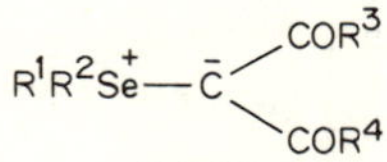

(144)

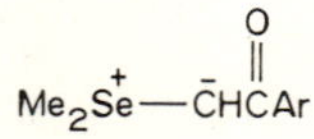

(145)

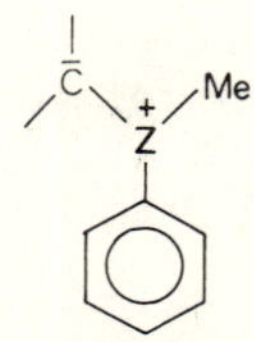

(146)

which then evolve to tetrasubstituted furans. The unstable ylides **145**, prepared *in situ* by deprotonation of the corresponding selenonium salts, give electrophilic additions with benzaldehyde, *p*-nitrobenzaldehyde, diethyl diacetylenedicarboxylate, salicylaldehyde, 2-formylfuran, 2-formylthiophene and 2-formylselenophene at 0 °C. The intermediate addition products cannot be isolated but evolve to trisubstituted furans, epoxides or benzo[*b*]furans according to the electrophiles used[291,292]. Sulphur ylides analogous to **145** show a lower reactivity[297–299].

The thermodynamic basicity constants of ylides **142** (Z = Se, Te) in anhydrous acetonitrile have been determined[138]. Telluronium ylides of this type are fairly strong bases (Z = Te, $R^1 = R^2$ = Ph, $pK_a = 10.59$; Z = Se, $R^1 = R^2$ = Ph, $pK_a = 9.59$; pK_a of aniline under the same conditions[138] is 10.57) and the basicity decreases in the order Te > Se > S. The substituents bonded to the Te atom of **142** interact with the positive pole mainly by an inductive mechanism[138].

A recent spectroscopic study[52] of stabilized ylides **146** (Z = S, Se, Te) reports that sulphonium and selenonium groups show an electron-withdrawing effect on the Ph ring whereas the telluronium group exhibits a π-donating effect; in our opinion these results should be used with caution.

Selenium ylide salts with two onium centres have recently been synthesized[293] and their reactivity towards electrophilic and nucleophilic reagents has been investigated.

D. Selenides and Tellurides

Selenides are important sources of organoselenium reagents and their chemical behaviour is related to the ability of alkyl- and aryl-seleno groups to stabilize both carbanions and carbocations. PhSe and MeSe groups increase the acidity of the hydrogens bonded to the adjacent C atom so that selenides are α-deprotonated by strong bases. Alkyllithium reagents are generally used for the deprotonation of the Me group[300] since in absence of the latter an attack of the lithium reagent on the Se atom occurs[301–303].

The deprotonation of phenyl selenides (**147**) can be conveniently carried out by using lithium amides which have a low tendency to attack the soft electrophilic Se atom.

$$\underset{(\mathbf{147})}{PhSeCH_2R} \rightarrow \underset{(\mathbf{148})}{PhSeCHLiR}$$

$$R = Ph, SiMe_3, CO_2Et, COPh, CH{=}CR_2, CH_2, C{\equiv}CH, SePh$$

Lithium diisopropylamide (LDA) and lithium 2,2,6,6-tetramethylpiperidide (LTMP) have been found to be the most useful reagents and their choice depends on the nature of the group R[304–308]. Lithium diisobutylamide and $LiNEt_2$ have sometimes[300,309,310] been used. Phenylselenomethane (**147**; R = H) is not satisfactorily deprotonated by lithium amide bases. A CF_3 group in the *meta* position on the phenylseleno group greatly increases the acid-strengthening effect and LTMP easily gives deprotonation[304,311]. The α-alkyl substitution of selenide decreases the acidity of the C—H bond and the deprotonation can be carried out only if a strong electron-withdrawing substituent is present[312] as in **149**.

$$\underset{\textbf{(149)}}{PhSeCH(CN)(CH_2)_5Me} \xrightarrow[-78^\circ C]{LDA/THF} \underset{\textbf{(150)}}{PhSeCLi(CN)(CH_2)_5Me}$$

The relative acidity increasing effect of Se and S on adjacent C—H hydrogens is a problem debated on theoretical[45,46,313,314] and experimental grounds[61,315]. A compendium of experimental data (Table 31) shows that carbanions with the negative charge in p or sp^3 orbitals are stabilized by S (probably by conjugative interaction) better than by Se and hence S compounds are more acid[315]. In carbanions coming from vinyl systems the conjugation with the heteroatoms is precluded, nevertheless the Se stabilizes more effectively than S because of its higher polarizability; therefore a higher acid-strengthening effect of Se is observed[315].

The anions **148** coming from the deprotonation reaction are powerful nucleophiles of great interest in organic synthesis[305]. Allyl selenide anions **152** give prevalently α-alkylation[310,319], but sometimes a remarkable amount of γ-alkylation is observed (Table 32). These results are similar to those observed in S systems[320,321].

$$\underset{\textbf{(151)}}{PhSeCH_2CH{=}CR^1} \xrightarrow[-78\ ^\circ C]{LDA} \underset{\textbf{(152)}}{PhSeCHLiCH{=}CR^2}$$

$$R^1, R^2 = H, H;\ Me, Cl;\ Me, H;\ Ph, H$$

In principle, selenoketone enolates can direct alkylating reagents to Se, O and C atoms. Treatment[322] of the α-(phenylseleno)acetophenone enolate (**154**) with prenyl halides (Scheme 4) gives only *C*- and *Se*-alkylation, the preferred site of attack depending on the reaction conditions. *Se*-Alkylation is followed by a 2,3-sigmatropic rearrangement to give compound **158**.

TABLE 31. Relative acid-strengthening effect of S and Se on hydrogens bonded to the adjacent C atom

Reaction (Z = S, Se)	Ratio of constants[c]	Ref.
1. Deprotonation of $(PhZ)_2CH_2$	K (S/Se) ~ 100	300, 309
2. Deprotonation of $PhCOCH_2ZPh$	K (S/Se) = 32	61
3. Isotopic exchange of PhZMe[a]	k (S/Se) = 10	316
4. Deprotonation of $PhZCH_2CH{=}CH_2$	k (S/Se) = 75	310
5. Base-catalysed isomerization of $PhZCH_2C{\equiv}CH$ to $PhZCH{=}C{=}CH_2$	k (S/Se) = 6.14	317, 318
6. Base-catalysed isomerization of $PhZCH{=}C{=}CH_2$ to $PhZC{\equiv}CMe$	k (S/Se) = 4.8	317, 318
7. Deprotonation of m-$CF_3C_6H_4ZMe$	k (S/Se) = 3.8	304
8. Isotopic exchange of C_4H_4Z[b]	k (S/Se) = 0.67	167
9. Deprotonation of m-$CF_3C_6H_4ZCH{=}CH_2$	k (S/Se) = 0.42	315
	K (S/Se) = 0.3	315
10. Deprotonation of $PhZCH{=}CH_2$	k(S/Se) = 0.37	315
	K (S/Se) = 0.21	315

[a] In KNH_2/NH_3.
[b] Thiophene and selenophene in t-BuOLi/DMSO, see Section IV.F.
[c] K and k refer to equilibrium and rate processes, respectively.

TABLE 32. α:γ Ratio in alkylation[a] of allyl selenide anions **152**[b]

Anion (R^1,R^2)	Electrophile	α:γ
H,H	$PhCH_2CH_2CH_2Br$	~80:20
H,H	Me_3SiCl	82:18
H,H	$Me_2PhSiCl$	41:59
H,H	PhCOMe	15:85
Me,H	$PhCH_2CH_2Br$	>90:10
Me,Me	$Me_2PhSiCl$	>90:10
Ph,H	MeCOMe	~50:50

[a] The γ-alkylation products are a 1:1 mixture of *E*- and *Z*-isomers. [b] Data from Ref. 310.

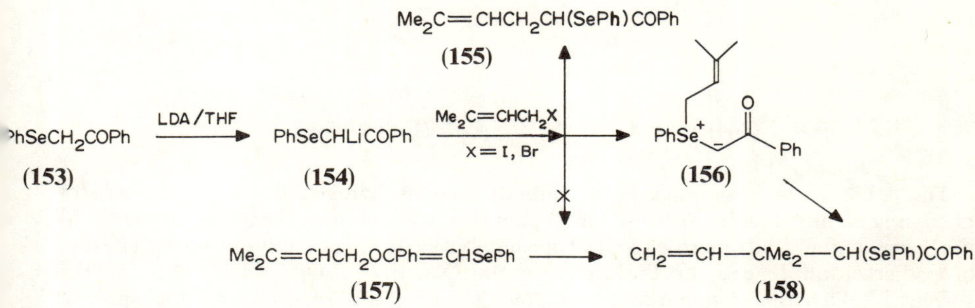

Under proper reaction conditions, alkyllithium reagents can be added to vinyl selenide **159** to give α-lithiumalkyl phenyl selenides (**160**) which can then be trapped by electrophiles[323]. The directing effect of the phenylseleno group is that of an electron-withdrawing substituent.

$$\underset{(\mathbf{159})}{PhSeCH{=}CH_2} \xrightarrow[R = Bu, i\text{-}Pr]{RLi} \underset{(\mathbf{160})}{PhSeCHLiCH_2R}$$

The ability of the phenylseleno group to stabilize carbocations has recently been pointed out by measuring the acid-catalysed hydrolysis rate of vinyl selenides[324,325]. Vinyl selenides **161** bearing an alkyl or aryl substituent at the α-position of the vinyl moiety hydrolyse by an A2-type mechanism in which (Scheme 5) the initial protonation is a slow reversible step and the decomposition of the intermediate hemiselenoacetal **162** is also slow[324]. Unsubstituted aryl vinyl selenides (**161**; $R^1 = R^2 = H$, $R^3 = Ar$) hydrolyse

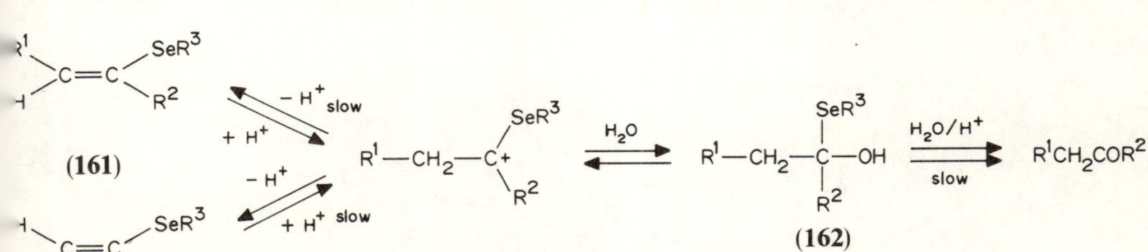

SCHEME 5

according to the classical A2 mechanism[47,325] in analogy with vinyl ethers and vinyl sulphides. The nature of the *para* substituent present in the aryl group of vinyl selenides (**161**; $R^1 = R^2 = H$, $R^3 = Ar$) and in the aryl group of α-(methylseleno)styrenes (**161**; $R^1 = H$, $R^2 = Ar$, $R^3 = Me$) does not affect the reaction mechanism. This is an indication that the ability of the Se moiety to stabilize the carbocation is fundamental in determining the mechanism. Hydrolysis rate ratios show that a methylseleno group stabilizes a positive charge better than a phenylseleno group (MeSe: PhSe = 15) and this effect increases in S (MeS: PhS = 41) and O (MeO: PhO = 133) analogues.

Aryl vinyl selenides involved in addition and elimination reactions have been reported by Chierici and Montanari[326]. *Trans*- and *cis*-β-arylselenoacrylic acids **163** and **165** add Br_2 very easily giving stereoisomers of 1-arylseleno-2,3-dibromopropionic acid which in water easily eliminate CO_2 and HBr giving *trans*- and *cis*-1-bromo-2-arylselenoethylenes **164** and **166**. The *cis* compound **166** always prevails (80%) over the *trans* one (**164**).

(**163**) R = COOH (**165**) R = COOH
(**164**) R = Br (**166**) R = Br

The phenylseleno group affects the stability, dehydrohalogenation and solvolysis of phenylselenoalkyl halides. β-Bromoalkyl phenyl selenides (**167**; X = Br) isomerize[327] to the thermodynamically more stable **169** presumably via the seleniranium ion **168**. The rate of isomerization depends on the leaving group (X = Br, Cl, $OCOCF_3$, OAc) and the solvent[327]. Dehydrobromination[328] of **169** (X = Br) in LDA–Et_2O at 0 °C gives a prevalence of the *E*-isomer **170**. The stereoselectivity of the reaction depends on the reaction conditions and nature of the R group. The first-order rate constant of solvolysis of phenylselenoethyl chloride (**167**; R = H, X = Cl) in MeOH is five times higher than that of the S analogue but in 80% EtOH it is 200 times higher[329]. This extraordinary phenylseleno effect is explained supposing that the reaction is anchimerically assisted and that the solvent could nucleophilically assist the PhSe neighbouring group[329].

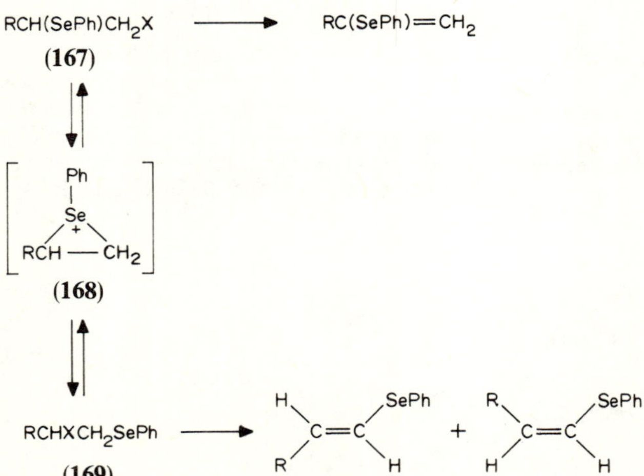

(**167**) (**168**) (**169**) (**170**) (**171**)

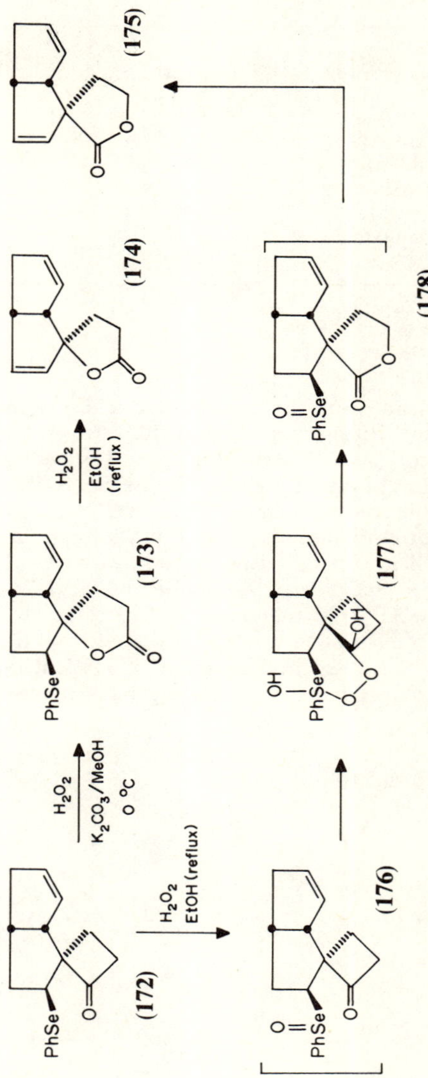
PhSe
(172)
H_2O_2
K_2CO_3/MeOH
0 °C
(173)
H_2O_2
EtOH
(reflux)
(174)
(175)
H_2O_2
EtOH (reflux)
(176)
(177)
(178)
OH

SCHEME 6

TABLE 33. Constants for substituted benzaldehyde–cyanohydrin equilibria and for dissociation of benzoic acids[a]

Substituent	Benzaldehydes[b] $10^3 K$	Benzoic acids[c] pK_a
p-SeMe	35.5	5.00
m-SeMe	3.95	4.74
p-SMe	38.1	5.02
m-SMe	4.07	4.75
p-OMe	42.9	5.11
m-OMe	4.27	4.71
H	4.40	4.81

[a] Data from Ref. 39.
[b] In EtOH at 20 °C.
[c] In 30% EtOH at 25 °C.

The phenylseleno group also increases the acidity of the carboxy group insulated by a methylene group[60]. Phenylselenoacetic acid is stronger ($pK_a = 3.75$) than acetic acid but weaker than S ($pK_a = 3.38$) and O ($pK_a = 3.15$) analogues[60]. The inductive electron-withdrawing effect of the PhZ group (Z = Se, S, O) decreases in the order O > S > Se.

The equilibrium constant of the addition of CN^- ion to benzaldehyde is increased[39] by a methylseleno group in the *para* position and decreased if the group is in the *meta* position (Table 33). Similar electron-releasing and -withdrawing effects are observed in the dissociation of *p*- and *m*-methylselenobenzoic acids[39]. A comparison with S and O analogues (Table 33) shows that the + M effect of the MeZ (Z = Se, S, O) group decreases in the order O > S > Se and it is also operative from the *meta* position.

An example of a regiochemical effect of a selenide group is the influence of the phenylseleno group on the Baeyer–Villiger rearrangement[330]. Through oxidation of

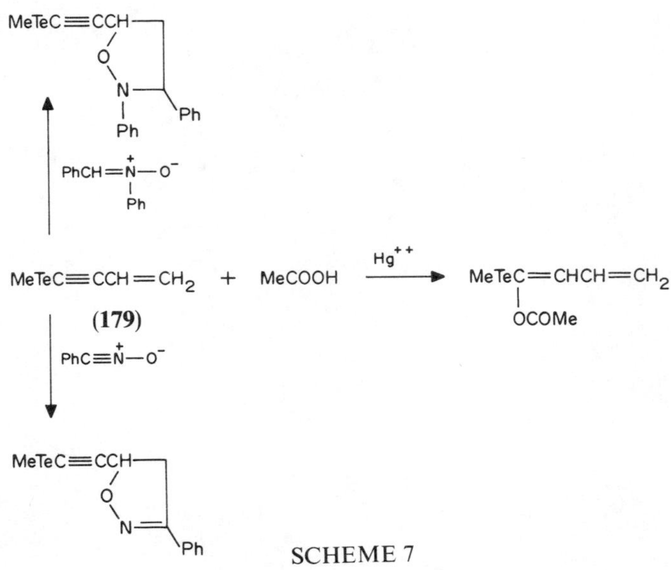

SCHEME 7

selenide **172** (Scheme 6) with H_2O_2 under basic conditions at 0 °C the expected compound **173** is obtained which gives **174** by treating with H_2O_2 in refluxing EtOH. When **172** is treated in absence of the base, **175** is obtained. The different outcome of the rearrangement has been interpreted in terms of different rates in selenide oxidation and in the Baeyer–Villiger reaction. In the presence of the base, selenide oxidation is slower than Baeyer–Villiger oxidation and the migration of quaternary carbon occurs giving the expected **173** which is converted to **174** by selenoxide elimination. In the absence of the base, H_2O_2 reacts with the selenoxide as well as with the carbonyl group of the cyclobutanone system of intermediate **176** giving **177** which in turn gives **175** via migration of the cyclobutanone methylene group and selenoxide elimination from **178**.

Information on the reactivity and directing effects of methyl- and phenyl-telluro groups is very limited. Radchenko and coworkers[331,332] report that the methyltelluro group of methyltellurovinylacetylene (**179**) directs the addition of the acetoxyl part of acetic acid to the triple bond in the α-position and the addition of benzonitrile oxide and α,*N*-diphenylnitrone to the double bond (Scheme 7).

E. Selenoxides and Selenones

Phenylseleninyl (PhSeO—) and phenylselenonyl ($PhSeO_2$—) are strong electronegative groups which increase the acidity of a H atom bonded to an adjacent C atom and activate an olefinic bond in the vinylic position. These characteristics, coupled with the fact that they are good leaving groups, have recently been utilized to develop new syntheses of olefins[304], allyl alcohols[304], dienes[304], cyclopropyl ketones[333], oxetanes[334], ethylenic and acetylenic ketones[335].

Selenoxides are not generally very stable at room temperature. They are hygroscopic substances and exhibit a greater basicity than the analogous sulphoxides[142].

Selenoxides are deprotonated better than selenides and the reaction proceeds quickly at −78 °C with lithium diisopropylamide (LDA)[304]. Alkyllithium reagents (i.e. butyllithium) give partial or total cleavage rather than deprotonation[304]. Lithium selenoxides **181** react with a variety of electrophiles (aldehydes, ketones, acyl and alkyl halides) to give α-substituted selenoxides which are directly converted to olefins by selenoxide *syn* elimination or to selenides by reduction reaction[304].

$$\underset{(\mathbf{180})}{\text{PhSeCHRMe}} \xrightarrow[-78\,^\circ\text{C}]{\text{LDA}} \underset{(\mathbf{181})}{\text{PhS(=O)CLiRMe}} \xrightarrow{\text{Me}_2\text{CO}}$$

$$\text{PhS(=O)CRMeCMe}_2\text{OH} \xrightarrow{\Delta} \text{H}_2\text{C=CRCMe}_2\text{OH}$$

$$\downarrow \text{red.}$$

$$\text{PhSeCRMeCMe}_2\text{OH}$$

Vinyl selenoxides and vinyl selenones give conjugative nucleophilic addition with ketones, ester enolates and alkoxides, followed by displacement of the seleno group. Examples are given in Scheme 8 which also outlines the difference in reactivity between the PhSeO and $PhSeO_2$ groups[333–335].

The proposed mechanism for the formation of cyclopropyl ketone (**184**) and oxetane (**186**) involves three steps: nucleophilic addition, proton transfer and nucleophilic substitution[333,334].

The oxetane formation is not a stereospecific reaction but a prevalence of *cis*-oxetane is observed[334].

The synthesis of ketones **189** shows good stereoselectivity because the attack of methoxide from the *B* side is energetically favoured and the phenylselenonyl group in the intermediate **190** is in a favourable position for 1,4-fragmentation.

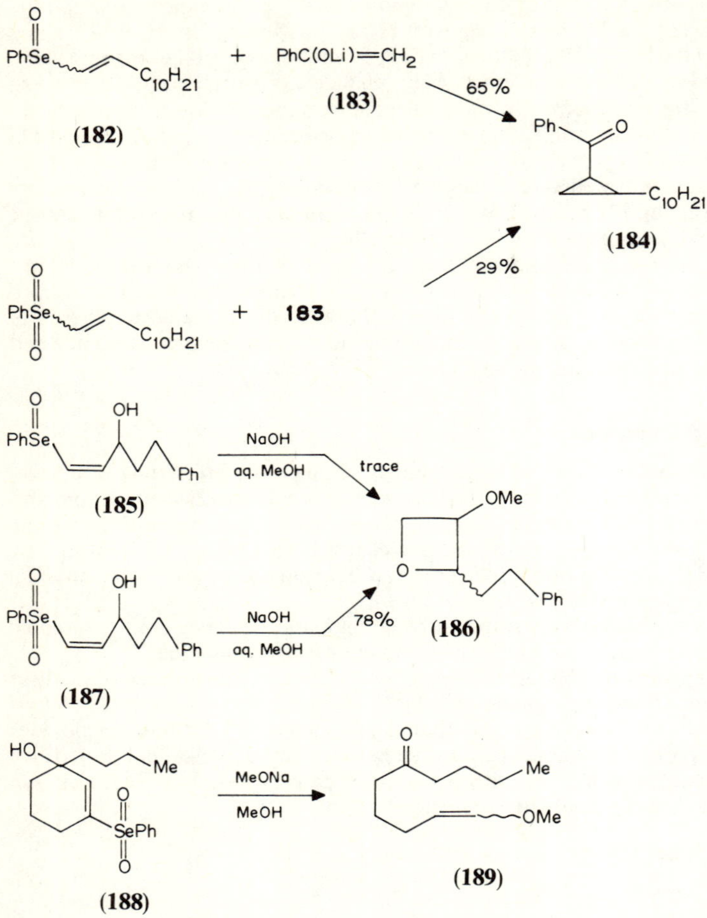
PhC(OLi)=CH2
(182)
(183)
65%
(184)
29%
183
NaOH
aq. MeOH
trace
(185)
78%
(186)
(187)
MeONa
MeOH
(188)
(189)

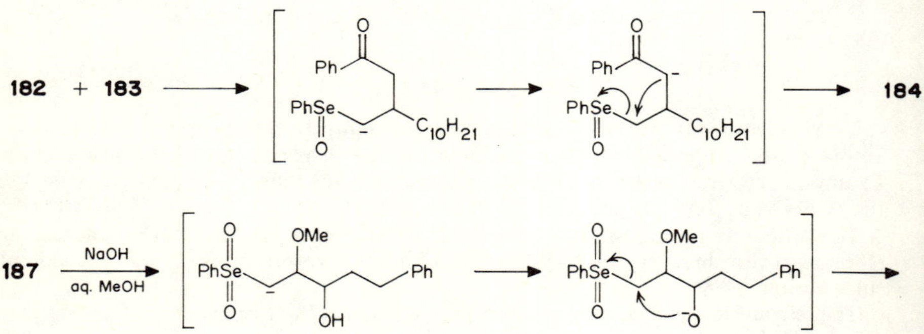
182 + 183
184
187
NaOH
aq. MeOH

(Con

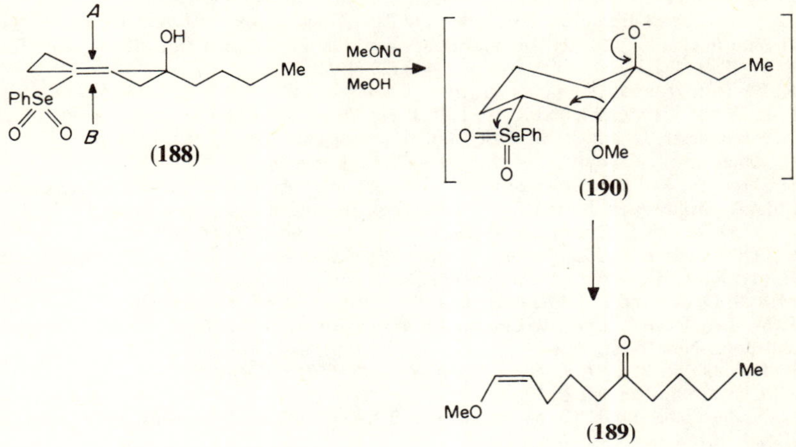

$Z:E$ ratio = 80 : 20

SCHEME 8

Vinyl selenones containing a tetrasubstituted double bond have a very low reactivity towards nucleophilic addition[304] of alkoxide and the fragmentation process which gives an acetylenic ketone, is predominant.

VI. REFERENCES*

1. *Organic Compounds of Sulfur, Selenium and Tellurium*, Vols 1–6, specialist periodical reports of the Royal Society of Chemistry, London, 1970–1981.
2. *Organic Selenium Compounds: Their Chemistry and Biology* (Eds. D. L. Klayman and W. H. H. Günther), Wiley–Interscience, New York, 1973.
3. A. B. Hörnfeldt, in *Advances in Heterocyclic Chemistry*, (Ed. A. R. Katritzy), Vol. 30, Academic Press, New York, 1982, p. 127.
4. F. Fringuelli, G. Marino and A. Taticchi, in *Advances in Heterocyclic Chemistry*, (Ed. A. R. Katritzky), Vol. 21, Academic Press, New York, 1977, p. 119.
5. I. D. Sadekov, A. Ya. Bushkov and V. I. Minkin, *Usp. Khim.*, **48**, 635 (1979); E.E. 343.
6. D. L. Clive, *Tetrahedron*, **34**, 1049 (1978).
7. S. Clementi, F. Fringuelli, P. Linda and G. Savelli, *Gazz. Chim. Ital.*, **105**, 281 (1975).
8. S. Clementi, F. Fringuelli and G. Savelli, *Proceedings of Euchem Conference on Correlation Analysis in Organic Chemistry*, Assisi, 1979, p. 4.
9. S. Clementi and F. Fringuelli, *Anal. Chim. Acta,* **103**, 477 (1978).
10. N. B. Chapman and J. Shorter (Eds.), *Advances in Linear Free Energy Relationships*, Plenum Press, New York, 1972.
11. N. B. Chapman and J. Shorter (Eds.), *Correlation Analysis in Chemistry: Recent Advances*, Plenum Press, New York, 1978.
12. J. Shorter, *Correlation Analysis in Organic Chemistry: an Introduction to Linear Free Energy Relationships*, Oxford Chemistry Series, 1973.
13. L. P. Hammett, *Physical Organic Chemistry*, 2nd ed., McGraw-Hill, New York, 1970.
14. C. D. Johnson, *The Hammett Equation*, Cambridge University Press, Cambridge, 1973.
15. H. H. Jaffé, *Chem. Rev.*, **53**, 191 (1953).
16. P. Tomasik and C. D. Johnson, in *Advances in Heterocyclic Chemistry*, (Ed. A. R. Katritzky), Vol. 20, Academic Press, New York, 1976, p. 1.

*When quoting Russian journals, E.E. means the corresponding English translation, followed by the page number.

17. H. C. Brown and Y. Okamoto, *J. Amer. Chem. Soc.*, **80**, 4979 (1958).
18. R. W. Taft, Jr., S. Ehrenson, I. C. Lewis and R. E. Glick, *J. Amer. Chem. Soc.*, **81**, 5352 (1959).
19. H. Van Bekkum, P. E. Verkade and B. M. Wepster, *Rec. Trav. Chim.*, **78**, 815 (1959).
20. R. W. Taft, Jr., *J. Amer. Chem. Soc.*, **74**, 3120 (1952).
21. R. W. Taft, Jr. and I. C. Lewis, *J. Amer. Chem. Soc.*, **80**, 2436 (1958); **81**, 5343 (1959).
22. P. R. Wells, S. Ehrenson and R. W. Taft, Jr., in *Progress in Physical Organic Chemistry*, (Eds. A. Streitwieser, Jr. and R. W. Taft), Vol. 6, Interscience, New York, 1968, p. 147.
23. O. Exner and J. Lakomy, *Collect. Czech. Chem. Commun.*, **35**, 1371 (1970).
24. S. Ehrenson, R. T. C. Brownlee and R. W. Taft, in *Progress in Physical Organic Chemistry*, (Eds. A. Streitwieser, Jr. and R. W. Taft), Vol. 10, Interscience, New York, 1973p. 1.
25. C. G. Swain and E. C. Lupton, *J. Amer. Chem. Soc.*, **90**, 4328 (1968)
26. C. D. Ritchie and W. F. Sager, in *Progress in Physical Organic Chemistry*, (Eds. A. Streitwieser, Jr. and R. W. Taft), Vol. 2, Interscience, New York, 1964, p. 323.
27. M. J. S. Dewar and P. J. Grisdale, *J. Amer. Chem. Soc.*, **84**, 3548 (1962).
28. R. W. Taft, in *Steric Effects in Organic Chemistry* (Ed. M. S. Newman), Chapt. 13, John Wiley and Sons, New York, 1956.
29. R. Fellous and R. Luft, *J. Amer. Chem. Soc.*, **95**, 5593 (1973).
30. M. Charton, in Ref. 11, p. 176.
31. N. N. Zatsepina, Yu. L. Kaminskii and I. F. Tupitsyn, *Reakts. Sposobnost Org. Soedin.*, **6**, 778 (1969); E.E. 333.
32. A. Streitwieser, *Molecular Orbital Theory for Organic Chemists*, John Wiley and Sons, New York, 1961.
33. H. Balli, B. Hellrung and H. Hinsken, *Helv. Chim. Acta*, **57**, 1174 (1974).
34. H. C. Brown and L. M. Stock, *J. Amer. Chem. Soc.*, **84**, 3298 (1962).
35. C. Hansch and A. J. Leo, *Substituent Constants for Correlation Analysis in Chemistry and Biology*, John Wiley and Sons, New York, 1979.
36. V. V. Bairov, G. A. Kalabin, M. L. Al'pert, V. M. Bzhezovskii, I. D. Sadekov, B. A. Trofimov and V. I. Minkin, *Zh. Org. Khim.*, **14**, 617 (1978); E.E. 622.
37. V. M. Bzhezovskii, G. A. Kalabin, I. A. Aliev, B. A. Trofimov, M. A. Shakhgel'diev and M. Kuliev, *Izv. Akad. Nauk. SSSR, Ser. Khim.*, 1999 (1976).
38. V. M. Bzhezovskii, G. A. Kalabin, G. A. Chmutova and B. A. Trofimov, *Izv. Akad. Nauk. SSSR, Ser. Khim.*, 586 (1977).
39. J. W. Baker, G. F. Barrett and W. T. Tweed, *J. Chem. Soc.*, 2831 (1952).
40. H. Lumbroso, Ch. Liegeois, N. Dereu, L. Christiaens and A. Luxen, *J. Mol. Struct.*, **67**, 251 (1980).
41. G. Llabres, M. Baiwir, L. Christiaens, J. Denoel, L. Laitem and J. L. Piette, *Can. J. Chem.*, **56**, 2008 (1978).
42. G. Baddeley, *J. Chem. Soc.*, 663 (1950).
43. E. V. Titov, N. G. Korzhenevskaya, L. M. Kapkan, L. N. Sedova, L. Z. Gandel'sman, Yu. A. Fialkov and L. M. Yagupol'skii, *Zh. Org. Khim.*, **7**, 2552 (1971); E.E. 2651.
44. A. E. Lutskii, E. M. Obukhova, L. M. Yagupol'skii and V. G. Volashchuk, *Zh. Strukt. Khim.*, **10**, 349 (1969).
45. J. I. Musher, *J. Amer. Chem. Soc.*, **94**, 1370 (1972).
46. J. B. Florey and L. C. Cusachs, *J. Amer. Chem. Soc.*, **94**, 3040 (1972).
47. R. A. McClelland and M. Leung, *J. Org. Chem.*, **45**, 187 (1980).
48. O. Exner, in Ref. 10, p. 1.
49. I. D. Sadekov and A. Ya. Bushkov, *Zh. Obshch. Khim.*, **47**, 631 (1977); E.E. 576.
50. Yu. M. Kargin, V. Z. Latypova, O. G. Yakovleva, A. A. Vafina, A. N. Ustingov and A. V. Il'yasov, *Zh. Obshch. Khim.*, **52**, 655 (1982); E.E. 572.
51. I. D. Sadekov, A. Ya. Bushkov, V. S. Yur'eva and V. I. Minkin, *Zh. Obshch. Khim.*, **47**, 2541 (1977); E.E. 2321.
52. D. F. Kushnarev, G. A. Kalabin, R. A. Kyandzhetsian and N. N. Magdesieva, *Zh. Org. Khim.*, **18**, 119 (1982); E.E. 103.
53. V. G. Boloshchuk, L. M. Yagupol'skii, G. P. Syrova and V. F. Bystrov, *Zh. Obshch. Khim.*, **37**, 118 (1967); E.E. 105.
54. D. H. McDaniel and H. C. Brown, *J. Org. Chem.*, **23**, 420 (1958).
55. E. T. Strom and J. R. Norton, *J. Amer. Chem. Soc.*, **92**, 2327 (1970).
56. M. T. Rogers, T. W. Campbell and R. W. Maatman, *J. Amer. Chem. Soc.*, **73**, 5122 (1951).

57. M. Charton, *J. Org. Chem.*, **30**, 557 (1965).
58. M. Charton, *J. Org. Chem.*, **29**, 1222 (1964).
59. L. M. Litvinenko, *Izv. Akad. Nauk. SSSR, Ser. Khim.*, 1737 (1962); E.E. 1653.
60. L. D. Pettit, A. Royston, C. Sherrington and R. J. Whewell, *J. Chem. Soc. (B)*, 588 (1968).
61. F. G. Bordwell, J. E. Bares, J. E. Bartmess, G. E. Drucker, J. Gerhold, G. J. McCollum, M. Van der Puy, N. R. Vanier and W. S. Matthews, *J. Org. Chem.*, **42**, 326 (1977).
62. J. M. Angelelli, R. T. Brownlee, A. R. Katritzky, R. D. Topsom and L. Yakhoontov, *J. Amer. Chem. Soc.*, **91**, 4500 (1969).
63. C. Y. Meyers, *Gazz. Chim. Ital.*, **93**, 1206 (1963).
64. M. Mousseron and P. Brun, *Bull. Soc. Chim. Fr.*, 616 (1947).
65. Yu. M. Kargin, T. A. Podkovyrina, G. A. Chmutova and T. G. Mannafov, *Zh. Obshch. Khim.*, **46**, 2459 (1976); E.E. 2357.
66. L. F. Wuyts, D. F. Van De Vondel and G. P. Van Der Kelen, *J. Organomet. Chem.*, **129**, 163 (1977).
67. R. F. De Ketelaere and G. P. Van Der Kelen, *J. Mol. Struct.*, **27**, 33 (1975).
68. Yu. M. Kargin, V. Z. Kondranina, L. M. Kataeva, L. A. Sorokina, V. V. Yanilkin and M. M. Khysalnov, *Zh. Obshch. Khim.*, **47**, 672 (1977); E.E. 611.
69. I. D. Sadekov, A. Ya. Bushkov, V. L. Parlova, V. S. Yur'eva and V. I. Minkin, *Zh. Obshch. Khim.*, **47**, 1305 (1977); E.E. 1202.
70. L. M. Yagupol'skii, G. P. Syrova, V. G. Voloshchuk and V. F. Bystrov, *Zh. Obshch. Khim.*, **38**, 2591 (1968).
71. L. M. Yagupol'skii, A. Ya Il'chenko and N. B. Kondratenko, *Usp. Khim.*, **43**, 64 (1974); E.E. 32.
72. V. I. Minkin and R. M. Minyaev, *Zh. Org. Khim.*, **11**, 1993 (1975); E.E. 2023.
73. V. I. Minkin, I. D. Sadekov, L. M. Sayapina and R. M. Minyaev, *Zh. Obshch. Khim.*, **43**, 809 (1973); E.E. 808.
74. G. D. Christofferson, R. A. Sparks and J. D. McCullough, *Acta Cryst.*, **11**, 782 (1958).
75. G. D. Christofferson and J. D. McCullough, *Acta Cryst.*, **11**, 249 (1958).
76. D. Kobelt and E. F. Paulus, *J. Organomet. Chem.*, **27**, C63 (1971).
77. W. R. McWhinnie and P. Thavornyutikarn, *J. Chem. Soc., Dalton Trans.*, 551 (1972).
78. D. Kobelt and E. F. Paulus, *Angew. Chem. (Intern. Ed.)*, **10**, 74 (1971).
79. B. Buss and B. Krebs, *Inorg. Chem.*, **10**, 2795 (1971).
80. H. Rheinboldt and E. Giesbrecht, *J. Amer. Chem. Soc.*, **69**, 2310 (1947).
81. V. V. Saatsazov, R. A. Kyandzhetsian, S. I. Kuznetsov, N. N. Magdesieva and T. L. Khotsyanova, *Izv. Akad. Nauk. SSSR, Ser. Khim.*, 671 (1973).
82. A. E. Lutskii, E. M. Obukhova, V. A. Granzhan, S. A. Volchenok, Z. M. Kanevskaya, L. M. Yagupol'skii and V. G. Voloschuk, *Teor. Exper. Khim.*, **5**, 614 (1969).
83. A. E. Lutskii, L. M. Yagupol'skii and E. M. Obukhova, *Zh. Obshch. Khim.*, **34**, 2641 (1964).
84. I. D. Sadekov, A. Ya. Bushkov and V. V. Metlushenco, *Zh. Obshch. Khim.*, **48**, 1658 (1978).
85. F. Fringuelli, B. Serena and A. Taticchi, *J. Chem. Soc., Perkin Trans. 2*, 971 (1980).
86. G. I. Matyushecheva, A. I. Tolmachev, A. A. Shulezhko, L. M. Shulezhko and L. M. Yagupol'skii, *Zh. Obshch. Khim.*, **46**, 162 (1976); E.E. 161.
87. C. Dell'Erba, D. Spinelli and G. Garbarino, *Gazz. Chim. Ital.*, **100**, 777 (1970).
88. F. Fringuelli, G. Marino and A. Taticchi, *J. Chem. Soc. (B)*, 2302, 2304 (1971).
89. F. Fringuelli, G. Marino and A. Taticchi, *J. Chem. Soc. (B)*, 1595 (1970).
90. A. Arcoria, E. Maccarone and A. Mammo, *J. Chem. Soc., Perkin Trans. 2*, 1347 (1979).
91. F. Fringuelli, G. Marino and A. Taticchi, *J. Chem. Soc., Perkin Trans. 2*, 1738 (1972).
92. F. Fringuelli, G. Marino and A. Taticchi, *Gazz. Chim. Ital.*, **102**, 534 (1972).
93. E. Maccarone, A. Mammo, D. Sciotto and M. Torre, *J. Chem. Soc., Perkin Trans. 2*, 161 (1980).
94. S. V. Tsukerman, L. A. Kutulya, V. F. Lavrushin and N. Min'Tyau, *Russ. J. Phys. Chem.*, **48**, 1015 (1968).
95. R. D. Brown, A. S. Buchanan and A. A. Humffray, *Australian J. Chem.*, **18**, 1513, 1521, 1527 (1965).
96. G. Marino, in *Advances in Heterocyclic Chemistry*, (Ed. A. R. Katritzky), Vol. 13, Academic Press, New York, 1971, p. 235.
97. N. N. Zatsepina, I. F. Tupistin, Y. L. Kaminskii and N. S. Kolodina, *Reakts. Sposobnost. Org. Soedin.*, **6**, 766 (1969).
98. Von K. Schwetlick and K. Unverferth, *J. Prakt. Chem.*, **314**, 603 (1972).

99. H. H. Jaffé and H. L. Jones, in *Advances in Heterocyclic Chemistry*, (Ed. A. R. Katritzky), Vol. 3, Academic Press, New York, 1964, p. 209.
100. S. V. Tsukerman, Yu. N. Surov and V. F. Lavrushin, *Zh. Obshch. Khim.*, **37**, 364 (1967); E.E. 339.
101. Š. Toma, *Collect. Czech. Chem. Commun.*, **34**, 2771 (1969).
102. H. Balli, B. Hellrung and A. Kneubuhler-Hof, *Helv. Chim. Acta*, **57**, 1178 (1974).
103. C. Hansch, A. Leo, S. H. Unger, Kihwan Kim, D. Nikaitani and E. J. Lien, *J. Medicinal Chem.*, **16**, 1207 (1973).
104. Y. Yukawa and Y. Tsuno, *Bull. Chem. Soc. Japan*, **32**, 971 (1959).
105. S. Clementi, F. Fringuelli and G. Savelli, *Chim. Ind.* (*Milano*), **60**, 598 (1978).
106. G. R. Wiley and S. I. Miller, *J. Org. Chem.*, **37**, 767 (1972).
107. M. J. S. Dewar and Y. Takeuky, *J. Amer. Chem. Soc.*, **89**, 390 (1967).
108. K. Bowden, J. G. Irving and M. J. Price, *Can. J. Chem.*, **46**, 3903 (1968).
109. M. D. Bentley and M. J. S. Dewar, *J. Org. Chem.*, **35**, 2707 (1970).
110. R. R. Fraser and R. N. Renaud, *Can. J. Chem.*, **49**, 800 (1971).
111. D. Spinelli, G. Guanti and C. Dell'Erba, *Ricerca Scient.*, **38**, 1048 (1968).
112. A. R. Butler, *J. Chem. Soc.* (*B*), 867 (1970).
113. W. E. Catlin, *Iowa State Coll. J. Sci.*, **10**, 65 (1935).
114. D. J. Byron, G. W. Gray and R. C. Wilson, *J. Chem. Soc* (*C*), 831 (1966).
115. S. V. Tsukerman, L. P. Pivovarevich, L. A. Kutulya, V.G. Gordienko, N. N. Magdesieva and V. F. Lavrushin, *Zh. Obshch, Khim.*, **45**, 1070 (1975) and references cited therein; E.E. 1055.
116. Yu. N. Surov, L. P. Pivovarevich, L. A. Kutulya, Yu. A. Sukach, N. N. Magdesieva and S. V. Tsukerman, *Zh. Obshch. Khim.*, **46**, 2473 (1976); E.E. 2336.
117. S. Clementi, F. Fringuelli, P. Linda, G. Marino, G. Savelli and A. Taticchi, *Gazz. Chim. Ital.*, **107**, 339 (1977).
118. Yu. K. Yur'ev, M. A. Gal'bershtam and A. F. Prokof'eva; *Izv. Vysshikh. Uchebn. Zavedenii SSR, Khim. i Khim. Tekhnol.*, **7**, 419, 598 (1964); **8**, 421 (1965).
119. G. Di Modica, E. Barni and A. Gasco, *J. Heterocycl. Chem.*, **2**, 457 (1965).
120. I. D. Sadekov, M. L. Cherkinskaya, V. L. Pavlova, V. A. Bren' and V. I. Minkin, *Zh. Obshch. Khim.*, **47**, 2774 (1977); E.E. 2521.
121. I. D. Sadekov, M. Cherkinskaya, V. L. Pavlova, V. A. Bren' and V. I. Minkin, *Zh. Obshch. Khim.*, **48**, 390 (1978); E.E. 351.
122. I. D. Sadekov, M. L. Cherkinskaya, V. P. Metlushenko, Yu. E. Chernish and V. I. Minkin, *Zh. Obshch. Khim.*, **51**, 1060 (1981); E.E. 884.
123. S. K. Dayal and R. W. Taft, *J. Amer. Chem. Soc.*, **95**, 5595 (1973).
124. I. R. Ager and L. Phillips, *J. Chem. Soc., Perkin Trans. 2*, 1975 (1972).
125. L. M. Litvinenko and R. S. Cheshko, *Zh. Obshch. Khim.*, **30**, 3682 (1960).
126. L. M. Litvinenko, A. F. Popov, R. S. Popova and L. P. Snagoshenko, *Reakts. Sposobnost Org. Soedin.*, **5**, 744 (1968).
127. J. Gosselck, *Angew. Chem.* (*Intern. Ed.*), **2**, 660 (1963).
128. J. Gosselck, H. Barth and L. Berress, *Ann. Chem.*, **671**, 1 (1964).
129. N. V. Hayes and G. E. K. Branch, *J. Amer. Chem. Soc.*, **65**, 1555 (1943).
130. F. J. Dippy and R. F. Williams, *J. Chem. Soc.*, **161** (1934).
131. A. Fisher, B. R. Mann and J. Vaughan, *J. Chem. Soc.*, 1093 (1961).
132. F. J. Dippy and J. E. Page, *J. Chem. Soc.*, 357 (1938).
133. G. A. Chmutova and V. I. Neonilina, *Zh. Fiz. Khim.*, **46**, 201 (1972).
134. P. Shu, K. Bechgaard and D. O. Cowan, *J. Org. Chem.*, **41**, 1849 (1976).
135. J. E. Gorton, H. L. Lentzener and W. E. Watts, *Tetrahedron*, **27**, 4353 (1971) and references cited therein.
136. M. Sato and T. Yoshida, *J. Organomet. Chem.*, **51**, 231 (1973).
137. M. Sato and T. Yoshida, *J. Organomet. Chem.*, **39**, 389 (1972).
138. I. D. Sadekov, A. I. Usachev, V. A. Bren', M. S. Korobov, I. D. Tseimakh and V. I. Minkin, *Zh. Obshch. Khim.*, **47**, 2232 (1977); E.E. 2037.
139. J. D. McCullough and M. K. Barsh, *J. Amer. Chem. Soc.*, **71**, 3029 (1949).
140. J. D. McCullough and B. A. Eckerson, *J. Amer. Chem. Soc.*, **67**, 707 (1945).
141. J. D. McCullough and B. A. Eckerson, *J. Amer. Chem. Soc.*, **73**, 2954 (1951).

142. E. P. Buchikhin and R. P. Klyshevich, *Zh. Obshch. Khim.*, **47**, 1300 (1977); E.E. 1198.
143. F. Fringuelli, G. Marino, G. Savelli and A. Taticchi, *J. Chem. Soc., Chem. Commun.*, 1441 (1971).
144. Yu. K. Yur'ev, M. A. Gal'bershtam and L. L. Kandror, *Khim. Geterotsikl. Soedin.*, **6**, 897 (1966).
145. C. Dell'Erba, A. Guareschi and D. Spinelli, *J. Heterocycl. Chem.*, **4**, 438 (1967).
146. C. Dell'Erba and D. Spinelli, *Tetrahedron*, **21**, 1061 (1965).
147. I. A. Abronin, A. Z. Djumanazarova, V. P. Litvinov and A. Konar, *Chem. Scripta*, **20**, 208 (1982).
148. S. Gronowitz, I. Johnson and A. B. Hörnfeldt, *Chem. Scripta*, **7**, 111 (1975).
149. S. Gronowitz, I. Johnson and A. B. Hörnfeldt, *Acta Chem. Scand.*, **26**, 1726 (1972).
150. J. M. Talbot, J. L. Piette, L. Christiaens, T. Drakenberg, S. Gronowitz, G. Llabres and M. Baiwir, *Chem. Scripta*, **18**, 147 (1981).
151. S. Gronowitz, A. Konar and A. B. Hörnfeldt, *Chem. Scripta*, **19**, 5 (1982).
152. G. A. Caplin, *Org. Mag. Res.*, **5**, 169 (1973).
153. F. Fringuelli, S. Gronowitz, A. B. Hörnfeldt. I. Johnson and A. Taticchi, *Acta Chem. Scand.*, **B28**, 175 (1974).
154. F. Fringuelli, S. Gronowitz, A. B. Hörnfeldt, I. Johnson and A. Taticchi, *Acta Chem. Scand.*, **B30**, 605 (1976).
155. G. Llabres, M. Baiwir, L. Christiaens and J. L. Piette, *Bull. Soc. Roy. Sci. Liège*, **51**, 365 (1982).
156. M. P. Simmonnin, M. J. Ponet and J. M. Cense, *Org. Mag. Res.*, **8**, 508 (1976).
157. J. D. McCullough and E. S. Gould, *J. Amer. Chem. Soc.*, **71**, 674 (1949).
158. E. S. Gould and J. D. McCullough, *J. Amer. Chem. Soc.*, **73**, 1109 (1951).
159. D. De Filippo and F. Momicchioli, *Tetrahedron*, **25**, 5733 (1969).
160. D. De Fillippo, E. Moretti and C. Preti, *Ann. Chim. (Rome)*, **58**, 603 (1968).
161. P. Kristian and G. Suchar, *Collect. Czech. Chem. Commun.*, **37**, 3066 (1972).
162. M. H. Ghadendehari, D. Davaliani, M. Yalpani and M. N. Partovi, *J. Org. Chem.*, **39**, 3906 (1974).
163. J. Neve, M. Hanocq and L. Molle, *Talanta*, **26**, 1173 (1979).
164. P. Linda and G. Marino, *J. Chem. Soc. (B)*, 43 (1970).
165. S. Clementi, F. Fringuelli, P. Linda, G. Marino, G. Savelli and A. Taticchi, *J. Chem. Soc., Perkin Trans. 2*, 2097 (1973).
166. S. Clementi and G. Marino, *Tetrahedron*, **25**, 4599 (1969).
167. N. N. Magdesieva, in *Advances in Heterocyclic Chemistry*, (Ed. A. R. Katritzky), Vol. 12, Academic Press, New York, 1970, p. 1.
168. Yu. K. Yur'ev and N. K. Sadovaya, *Zh. Obshch. Khim.*, **34**, 1803 (1964).
169. Yu. K. Yur'ev and N. K. Sadovaya, *Zh. Obshch. Khim.*, **26**, 3154 (1956).
170. Yu. K. Yur'ev, N. K. Sadovaya and M. A. Gal'bershtam, *Zh. Obshch. Khim.*, **29**, 1970 (1959).
171. Yu. K. Yur'ev, N. K. Sadovaya and A. B. Ibragimova, *Zh. Obshch. Khim.*, **29**, 3647 (1959).
172. Yu. K. Yur'ev, E. L. Zaitseva and G. G. Rozantsev, *Zh. Obshch. Khim*, **30**, 2207 (1960).
173. B. Ostman, *Arkiv Kemi*, **19**, 499 (1962).
174. F. Fringuelli and A. Taticchi, *J. Chem. Soc., Perkin Trans. 1*, 199 (1972).
175. Yu. K. Yur'ev and E. L. Zaitseva, *Zh. Obshch. Khim.*, **29**, 2245 (1959).
176. F. Fringuelli, G. Marino, A. Taticchi and G. Grandolini, *J. Chem. Soc., Perkin Trans. 2*, 332 (1974).
177. L. I. Belen'Kii and L. Abroin, *Zh. Org. Khim.*, **17**, 1129 (1981); E.E. 997.
178. P. A. Konstantinov, N. N. Koloskova, R. I. Shupik and M. N. Volkov, *Zh. Obshch. Khim.*, **43**, 872 (1973).
179. F. Lucchesini and V. Bertini, *Proceedings XIII Convegno Naz. Chim. Org.*, Milano, 1982.
180. P. Chauvin, J. Morel and P. Pastour, *Bull. Soc. Chim. Fr.*, 2099 (1974).
181. G. Ah-Kow, C. Paulmier and P. Pastour, *Bull. Soc. Chim. Fr.*, 151 (1976).
182. A. De Nunno, V. Bertini, P. Raseo, M. Picci and L. Bonfanti, *Atti Acad. Naz. Lincei Sez. Sci. Fis. Mat. Nat.*, 385 (1978).
183. L. Christiaens, R. Dufour and M. Renson, *Bull. Soc. Chim. Belges*, **79**, 143 (1970).
184. J. L. Piette, J. M. Talbot, J. C. Genard and M. Renson, *Bull. Soc. Chim. Fr.*, 2468 (1973).

185. M. Renson, *Chem. Scripta*, **8A**, 29 (1975).
186. J. M. Talbot, J. L. Piette and M. Renson, *Bull. Soc. Chim. Belges*, **89**, 763 (1980).
187. J. Bergman and L. Engman, *J. Organomet. Chem.*, **201**, 377 (1980).
188. F. Mantovani, L. Christiaens and P. Faller, *Bull. Soc. Chim. Fr.*, 1595 (1972).
189. P. Cagniant, G. Kirsh and L. Christiaens, *Compt. Rend., Ser. C*, **287**, 333 (1978).
190. I. Fefe, L. Christiaens and M. Renson, *Bull. Soc. Roy. Sci. Liège*, **9–10**, 640 (1975).
191. J. M. Webert, D. Cagniant, P. Cagniant, G. Kirsh and J. V. Weber, *J. Heterocycl. Chem.*, **20**, 61 (1983).
192. J. A. Elvidge, J. R. Jones, R. Salih, M. Shandalo and S. E. Taylor, *J. Chem. Res.*, 172(S) 2375(M) (1980).
193. F. Outurquin, C. Paulmier and P. Pastour, *Bull. Soc. Chim. Fr.*, 3039 (1974).
194. F. Outurquin, G. Ah-Kow and C. Paulmier, *Bull. Soc. Chim. Fr.*, 883 (1976).
195. S. Clementi, S. Lepri, G. V. Sebastiani, S. Gronowitz, C. Westerlund and A. B. Hörnfeldt, *J. Chem. Soc., Perkin Trans. 2*, 861 (1978).
196. I. Ya. Kvitko, N. B. Sokolova, S. P. Fradkina and A. V. El'tsov, *Zh. Org. Khim.*, **12**, 1754 (1976); E.E. 1550.
197. Yu. N. Koshelev, A. V. Reznichenko, L. S. Efros and I. Ya. Kvitko, *Zh. Org. Khim.*, **9**, 2201 (1973); E.E. 2216.
198. S. Gronowitz, A. Konar and V. P. Litvinov, *Chem. Scripta*, **15**, 206 (1980).
199. S. Gronowitz, A. Konar and V. P. Litvinov, *Izv. Akad. Nauk SSSR, Ser. Khim.*, 1363 (1981); E.E. 1089.
200. S. Gronowitz, A. Konar, I. A. Abronin and V. P. Litvinov, *Izv. Akad. Nauk. SSSR, Ser. Khim.*, 1285 (1981); E.E. 1019.
201. S. Gronowitz, C. Westerlund and A. B. Hörnfeldt, *Chem. Scripta*, **10**, 165 (1976).
202. M. Hori, T. Kataoka and Chen-Fu Hsu, *Chem. Pharm. Bull.*, **22**, 21 (1974).
203. M. Hori, T. Kataoka, Chen-Fu Hsu, Y. Asahi and E. Mizuta, *Chem. Pharm. Bull.*, **22**, 27 (1974).
204. E. Bulka, *Chem. Scripta*, **8A**, 39 (1975), and references cited therein.
205. G. Marechal. L. Christiaens, M. Renson, P. Jaquignon and A. Croise, *Bull. Soc. Chim. Fr.*, 157 (1977).
206. V. Bertini, F. Lucchesini and A. De Nunno, *Synthesis*, 681 (1982).
207. R. Passerini and G. Purello, *Ann. Chim. (Rome)*, **48**, 738 (1958).
208. R. Weber and M. Renson, *J. Heterocycl. Chem.*, **12**, 1091 (1975).
209. E. Ochiai and J. Haginiwa, *Yakugaku Zasshi*, **67**, 138 (1947); *Chem. Abstr.*, **45**, 9539c (1951).
210. M. T. Bogert and Yu-Gwan Chen, *J. Amer. Chem. Soc.*, **44**, 2352 (1922).
211. L. S. Efros and Z. V. Todres-Selektor, *Zh. Obshch. Khim.*, **27**, 983 (1957).
212. C. W. Bird, G. W. H. Cheeseman and A. A. Sarsfield, *J. Chem. Soc.*, 4767 (1963).
213. V. G. Pesin and R. S. Muravnik, *Latv. PSR Zinat. Akad. Vastis*, **725** (1964); *Chem. Abstr.*, **63**, 4279c (1965).
214. V. G. Pesin, V. A. Sergeev and A. G. Nesterova, *Khim. Geterotsikl. Soedin.*, 95 (1968).
215. E. Ochiai, J. Haginiwa and K. Komatsu, *Yakugaku Zasshi*, **70**, 372 (1950); *Chem. Abstr.*, **45**, 2475 (1951).
216. L. Chierici, C. Dell'Erba, A. Guareschi and D. Spinelli, *Ann. Chim. (Rome)*, **57**, 632 (1967).
217. G. Guanti, C. Dell'Erba and G. Garbarino, *J. Heterocycl. Chem.*, **7**, 1425 (1970).
218. D. Spinelli and C. Dell'Erba, *Ann. Chim. (Rome)*, **54**, 281 (1964).
219. D. Dal Monte and E. Sandri, *Ann. Chim. (Rome)*, **53**, 1697 (1963).
220. G. Doddi, G. Illuminati and F. Stegel, *Chem. Commun.*, 953 (1969).
221. G. Doddi, G. Illuminati and F. Stegel, *J. Org. Chem.*, **36**, 1918 (1971).
222. G. Baldini, G. Doddi, G. Illuminati and F. Stegel, *J. Org. Chem.*, **41**, 2153 (1976).
223. F. Terrier, A. P. Chatrousse, R. Shaal, C. Paulmier and P. Pastour, *Tetrahedron Letters.*, 1961 (1972).
224. F. Terrier, A. P. Chatrousse, C. Paulmier and R. Shaal, *J. Org. Chem.*, **40**, 2911 (1975).
225. F. Terrier, A. P. Chatrousse and C. Paulmier, *J. Org. Chem.*, **44**, 1634 (1979).
226. D. Spinelli and G. Consiglio, *J. Chem. Soc., Perkin Trans. 2*, 989 (1975).
227. D. Spinelli, R. Noto and G. Consiglio, *J. Chem. Soc., Perkin Trans. 2*, 747 (1976).
228. S. Gronowitz and B. Uppstrom, *Acta Chem. Scand.*, **B29**, 441 (1975).
229. L. E. Saris and M. P. Cava, *J. Amer. Chem. Soc.*, **98**, 867 (1976).
230. E. Chacko, J. Bornstein and D. J. Sardella, *J. Amer. Chem. Soc.*, **99**, 8248 (1977).
231. M. Davis, L. W. Deady and E. Homfeld, *Australian J. Chem.*, **27**, 1917 (1974).
232. L. W. Deady, *Australian J. Chem.*, **26**, 1949 (1973).

233. A. Tadino, L. Christiaens and M. Renson, *Bull. Soc. Roy. Sci. Liège*, 129 (1973).
234. A. Tadino, L. Christiaens and M. Renson, *Bull. Soc. Roy. Sci. Liège*, 146 (1973).
235. M. Hori, T. Kataoka and Chen-Fu Hsu, *Chem. Pharm. Bull.*, **22**, 15 (1974).
236. M. Hori, T. Kataoka, H. Shimizu, Chen-Fu Hsu, Y. Asahi and E. Mizuta, *Chem. Pharm. Bull.*, **22**, 32 (1974).
237. J. Stackhouse, G. H. Senkler, Jr., B. E. Maryanoff and K. Mislow, *J. Amer. Chem. Soc.*, **96**, 7835 (1974).
238. I. Degani, R. Fochi and G. Spunta, *Bull. Fac. Chim. Ind. Bologna*, **23**, 151 (1965).
239. L. M. Stock and H. C. Brown, in *Advances in Physical Organic Chemistry* (Ed. V. Gold), Vol. 1, Academic Press, New York, 1963, p. 35.
240. E. Maccarone, A. Mamo, G. Scarlata and M. Torre, *J. Org. Chem.*, **44**, 2896 (1979).
241. A. I. Shatenshtein, in *Advances in Physical Organic Chemistry* (Ed. V. Gold), Vol. 1, Academic Press, New York, 1963, p. 155.
242. A. Streitweiser, D. E. Van Sickle and L. Reif, *J. Amer. Chem. Soc.*, **82**, 1513 (1960).
243. V. Bertini, *Gazz. Chim. Ital.*, **97**, 1870 (1967).
244. S. Gronowitz and F. Frejd, *Chem. Heterocycl. Comp.* (*E.E.*), **14**, 353 (1978).
245. S. Gronowitz and B. Holm, *Synth. Commun.*, **4**, 63 (1974).
246. S. Gronowitz and B. Holm, *J. Heterocycl. Chem.*, **14**, 281 (1977).
247. Yu. K. Yur'ev, N. N. Sadovaya and E. A. Grekova, *Zh. Obshch.Khim.*, **34**, 847 (1964).
248. S. Gronowitz and F. Frejd, *Acta Chem. Scand.*, **23**, 2540 (1969).
249. S. Gronowitz and F. Frejd, *Acta Chem. Scand.*, **27**, 2242 (1973).
250. S. Gronowitz and F. Frejd, *Acta Chem. Scand.*, **B30**, 439 (1976).
251. F. B. Iteke, L. Christiaens and M. Renson, *Tetrahedron*, **32**, 689 (1976).
252. V. Bertini, A. De Nunno, A. Menconi and A. Fissi, *J. Org. Chem.*, **39**, 2294 (1974).
253. F. Lucchesini and V. Bertini, *Proceedings XII Convegno Naz. Chim. Org.*, Ancona, 1980, p. 235.
254. F. Lucchesini and V. Bertini, *Proceedings VIII Intern. Congress Heterocyclic. Chem.*, Graz, 1981, p. 121.
255. E. Luppold, E. Miller and W. Winter, *Z. Naturforsch.*, **31b**, 1654 (1976).
256. S. Gronowitz, A. Hallberg and F. Frejd, *Tetrahedron*, **35**, 2607 (1979).
257. V. P. Litvinov, I. P. Konjajeva and Ya. L. Goldfarb, *Izv. Akad. Nauk SSSR, Ser. Khim.*, **25**, 477 (1976).
258. E. B. Pedersen, *J. Chem. Soc., Perkin Trans. 2*, 473 (1977).
259. P. O. Lumme, *Suomen Khemistil*, **B33**, 87 (1960).
260. F. Fringuelli and A. Taticchi, *J. Heterocycl. Chem.*, **10**, 89 (1973).
261. D. Dal Monte and E. Sandri, *Ann. Chim.* (*Rome*), **54**, 486 (1964).
262. D. Dal Monte and E. Sandri, *Bull. Sci. Fac. Chim. Ind. Bologna*, **22**, 33 (1964).
263. J. Neve, M. Hanocq and L. Molle, *Mikrochimica Acta*, 41 (1980).
264. I. Degani, R. Fochi and G. Spunta, *Bull. Sci. Fac. Chim. Ind. Bologna*, **23**, 243 (1965).
265. M. Renson and P. Pirson, *Bull. Soc. Chim. Belges*, **75**, 456 (1966).
266. I. Degani, R. Fochi and C. Vincenzi, *Bull. Sci. Fac. Chim. Ind. Bologna*, **23**, 21 (1965).
267. I. Degani, R. Fochi and G. Spunta, *Bull. Sci. Fac. Chim. Ind. Bologna*, **26**, 3 (1968).
268. O. Thorstand, K. Undheim, B. Cederlund and A. B. Hörnfeldt, *Acta Chem. Scand.*, **B29**, 647 (1975).
269. B. Cederlund and A. B. Hörnfeldt, *Acta Chem. Scand.*, **B30**, 101 (1976).
270. R. Lantz and A. B. Hörnfeldt, *Chem. Scripta*, **10**, 126 (1976).
271. O. Thorstand, K. Undheim, R. Lantz and A. B. Hörnfeldt, *Acta Chem. Scand.*, **B29**, 652 (1975).
272. A. Krief, *Tetrahedron*, **36**, 2531 (1980).
273. H. J. Reich, W. W. Willis, Jr. and S. Wollowitz, *Tetrahedron Letters*, **23**, 3319 (1982).
274. D. L. Klayman, Ref. 2, p. 67.
275. L. G. Fahel and J. Kice, *J. Org. Chem.*, **44**, 2357 (1979).
276. J. L. Kurz and J. C. Harris, *J. Org. Chem.*, **35**, 3086 (1970).
277. H. J. Reich and M. L. Cohen, *J. Org. Chem.*, **44**, 3148 (1979).
278. A. J. Band (Ed.), *Encyclopedia of Electrochemistry of the Elements*, Vol. IV, Marcel Dekker, New York, 1975.
279. H. J. Backer and W. Van Dam, *Rec. Trav. Chim.*, **49**, 479 (1930); **54**, 531 (1935).
280. H. J. Backer and W. Van Dam, *Rec. Trav. Chim.*, **49**, 482 (1930).
281. C. L. Liotta, E. M. Perdue and H. P. Hopkins, Jr., *J. Amer. Chem. Soc.*, **96**, 7981 (1974).
282. F. Fagioli, F. Pulidori, C. Bighi and A. De Battisti, *Gazz. Chim. Ital.*, **104**, 639 (1974).

283. R. E. Huber and R. S. Criddle, *Arch. Biochem. Biophys.*, **122**, 164 (1967).
284. Y. Sugiura, Y. Hojo, Y. Tamai and H. Tanaka, *J. Amer. Chem. Soc.*, **98**, 2339 (1976).
285. N. Nakamura and E. Sekido, *Talanta*, **17**, 515 (1970).
286. H. Tanaka, H. Sukuari and A. Yokoyama, *Chem. Pharm. Bull.*, **18**, 1015 (1970).
287. H. M. Gilow and G. L. Walker, *J. Org. Chem.*, **32**, 2580 (1967).
288. H. M. Gilow, M. De Shazo and W. C. Van Cleave, *J. Org. Chem.*, **36**, 1745 (1971).
289. H. M. Gilow, R. Bradford Camp, Jr. and E. C. Clifton, *J. Org. Chem.*, **33**, 230 (1968).
290. N. N. Nesmeyanov, T. P. Tolstaya, L. S. Isaeva and A. V. Grid, *Dokl. Akad. Nauk SSSR*, **133**, 602 (1960).
291. N. N. Magdesieva, Le. Nguyen Ngi and N. M. Koloskova, *Zh. Org. Khim.*, **15**, 609 (1979); E.E. 541.
292. N. N. Magdesieva and N. G. Chovnikova, *Zh. Org. Khim.*, **15**, 2402 (1979); E.E. 2174.
293. N. N. Magdesieva, R. A. Kyandzhetsian, N. G. Chovnikova and N. N. Emel'yanova, *Zh. Org. Khim.*, **17**, 340 (1981); E.E. 286.
294. E. E. Ernstbrunner and D. L. Lloyd, *Ann. Chem.*, **753**, 196 (1971).
295. B. H. Freeman, D. L. Lloyd and M. I. C. Singer, *Tetrahedron*, **28**, 343 (1972).
296. N. N. Magdesieva, R. A. Kyandzhetsian and V. A. Dinilenko, *Khim. Geterotsikl. Soedin.*, 1447 (1973).
297. R. W. Rates and A. N. Yao, *J. Org. Chem.*, **31**, 1689 (1966).
298. B. M. Trost, *J. Amer. Chem. Soc.*, **89**, 138 (1967).
299. A. W. Johnson and R. T. Amel. *J. Org. Chem.*, **34**, 1240 (1969).
300. D. Seebach and N. Peleties, *Chem. Ber.*, **105**, 511 (1972).
301. D. Seebach, N. Meyer and A. K. Beck, *Ann. Chem.*, 846 (1977).
302. W. Dumont, P. Bayet and A. Krief, *Angew. Chem.* (*Intern. Ed.*), **13**, 804 (1974).
303. H. J. Reich and S. K. Shah, *J. Amer. Chem. Soc.*, **97**, 3250 (1975).
304. H. J. Reich, F. Chow and S. K. Shah, *J. Amer. Chem. Soc.*, **101**, 6638 (1979).
305. H. J. Reich, *Acc. Chem. Res.*, **12**, 22 (1979).
306. M. W. Rathke and R. Kow, *J. Amer. Chem. Soc.*, **94**, 6854 (1972).
307. R. A. Olofson and C. M. Daugherty, *J. Amer. Chem. Soc.*, **95**, 582 (1973).
308. H. J. Reich, S. K. Shah, P. M. Gold and R. E. Olson, *J. Amer. Chem. Soc.*, **103**, 3112 (1981).
309. D. Seebach and N. Peleties, *Angew. Chem.*, **81**, 465 (1969).
310. H. J. Reich, *J. Org. Chem.*, **40**, 2570 (1975).
311. H. J. Reich and S. K. Shah, *J. Org. Chem.*, **42**, 1773 (1977).
312. P. A. Grieco and Y. Yokoyama, *J. Amer. Chem. Soc.*, **99**, 5210 (1977).
313. J. M. Lehn, G. Wipff and J. Demuynck, *Helv. Chim. Acta*, **60**, 1239 (1977).
314. C. A. Coulson, *Nature* (*London*), **221**, 1106 (1969).
315. H. J. Reich and W. W. Willis, Jr., *J. Org. Chem.*, **45**, 5227 (1980).
316. A. I. Shatenshtein and H. A. Gvozdeva, *Tetrahedron*, **25**, 2749 (1969).
317. G. Pourcelot and C. Georgoulis, *Bull. Soc. Chim. Fr.*, 866 (1964).
318. G. Pourcelot and J. M. Cense, *Bull. Soc. Chim. Fr.*, 1578 (1976).
319. H. J. Reich, M. C. Clark and W. W. Willis, Jr., *J. Org. Chem.*, **47**, 1618 (1982).
320. J. F. Biellmann and J. B. Ducep, *Tetrahedron Letters*, 5629 (1968).
321. D. A. Evans and G. C. Andrews, *Acc. Chem. Res.*, **7**, 147 (1974).
322. H. J. Reich and M. L. Cohen, *J. Amer. Chem. Soc.*, **101**, 1307 (1979).
323. S. Raucher and G. A. Koolpe, *J. Org. Chem.*, **43**, 4252 (1978).
324. L. Hevesi, J. L. Piquard and H. Wautier, *J. Amer. Chem. Soc.*, **103**, 870 (1981).
325. R. A. McClelland, *Can. J. Chem.*, **55**, 548 (1977).
326. L. Chierici and F. Montanari, *Gazz. Chim. Ital.*, **86**, 1269 (1956).
327. S. Raucher, *J. Org. Chem.*, **42**, 2950 (1977).
328. S. Raucher, M. R. Hansen and M. A. Colter, *J. Org. Chem.*, **43**, 4885 (1978).
329. S. P. Mc Manus and D. H. Lam, *J. Org. Chem.*, **43**, 650 (1978).
330. B. M. Trost, P. Buhlmayer and M. Mao, *Tetrahedron Letters.*, **23**, 1443 (1982).
331. S. I. Radchenko, I. G. Savich and A. A. Petrov, *Zh. Org. Khim.*, **10**, 2269 (1974); E.E. 2284.
332. S. I. Radchenko, V. N. Christokletov and A. A. Petrov, *Zh. Obshch. Khim.*, **35**, 1753 (1965).
333. M. Shimizu and I. Kuwajima, *J. Org. Chem.*, **45**, 2921 (1980).
334. M. Shimizu and I. Kuwajima, *J. Org. Chem.*, **45**, 4063 (1980).
335. M. Shimizu, R. Ando and I. Kuwajima, *J. Org. Chem.*, **46**, 5246 (1981).

The Chemistry of Organic
Selenium and Tellurium
Compounds Volume 1
Edited by S. Patai and Z. Rappoport
© 1986 John Wiley & Sons Ltd.

CHAPTER **16**

Functional groups containing selenium and tellurium in various oxidation states

NOBORU SONODA and AKIYA OGAWA

Department of Applied Chemistry, Osaka University, Suita, Osaka, Japan

I. INTRODUCTION . . . 620
II. SELENOLS . . . 620
A. Synthesis . . . 621
B. Reactions . . . 622
1. Reactions with alkylating reagents . . . 622
a. Reaction with various halides . . . 622
b. Reaction with aromatic halides . . . 622
c. Reaction with alkylamines . . . 623
2. Reduction by selenols . . . 623
3. Ring-opening reactions . . . 623
4. Addition reactions . . . 624
III. SELENENIC ACIDS AND THEIR DERIVATIVES . . . 624
A. Selenenic Acids and Their Anhydrides . . . 624
1. Generation . . . 624
a. β-Elimination of alkyl aryl selenoxides . . . 624
b. Oxidation of selenols and diselenides . . . 625
c. Reduction of seleninic acids and their anhydrides . . . 626
d. Hydrolysis of selenenyl halides . . . 626
2. Reactions . . . 626
B. Selenenyl Halides . . . 628
1. Synthesis . . . 629
2. Reactions . . . 629
a. With carbonyl compounds . . . 629
b. With unsaturated compounds . . . 632
c. With nucleophiles . . . 639
3. Selenenyl iodides and fluorides . . . 639
C. Miscellaneous . . . 639
1. Selenenic esters and selenocarboxylates . . . 639
a. Synthesis . . . 639

b. Reactions 640
2. Selenenamides 640
a. Synthesis 640
b. Reactions 640
3. Selenocyanides 641
a. Synthesis 641
b. Reactions 641
IV. SELENINIC ACIDS AND THEIR ANHYDRIDES 642
A. Synthesis 643
B. Reactions 643
1. Oxidation of alcohols 643
2. α-Hydroxylation of ketones 644
3. Oxidation of benzylic hydrocarbons 645
4. Epoxidation of olefins 645
5. Dehydrogenation 646
6. Oxidation of nitrogen compounds 647
7. Reaction with sulphur and phosphorus compounds 648
8. Deprotection 648
V. SELENONIC ACIDS AND THEIR DERIVATIVES 649
A. Synthesis 649
B. Reactions 650
VI. TELLURIUM ANALOGUES 650
A. Tellurols 650
B. Tellurenyl Compounds 651
C. Tellurinyl Compounds 653
1. $RTeX_3$ (X = halogen, OH, etc.) 653
2. R_2TeX_2 654
3. R_3TeX 655
4. RTe(O)X 655
D. Telluronyl Compounds 656
VII. REFERENCES 656

I. INTRODUCTION

There have been many reports of organic molecules having functional groups containing Se and Te in various oxidation states involving higher valencies. The main purpose of this chapter is to describe the chemistry of organic compounds containing Se or Te in various oxidation states such as selenols (RSeH), selenenic acids, seleninic acids, selenonic acids and their derivatives and Te analogues. Attention has been mainly focused on their preparative methods and their characteristic reactions. Since Klayman's excellent review[1] has covered the main literature on Se chemistry up to 1972, we shall deal with the subject mainly on the basis of references published in the last decade.

II. SELENOLS

Selenols (RSeH) are acidic and air-susceptible compounds having an intolerable odour. They are soluble in alkaline aqueous solution due to their acidic nature, but are usually insoluble in water. In general, selenols have greater acidities than those of the corresponding S analogues. For example, the pK_a values of H_2Se and benzeneselenol are 3.7 and 5.9, respectively[2–5], whereas H_2S and benzenethiol show 7.0 and 6.5. Selenols are generally more sensitive to air than thiols, and are oxidized to the corresponding diselenides. Selenolate ions (RSe^-), formed in alkaline solution of selenols, are generally more sensitive to molecular oxygen and are rapidly converted to diselenides (equation 1).

$$\text{RSeH} \xrightarrow{[\text{O}]} \text{RSeSeR} \tag{1}$$

A. Synthesis

In general, alkaneselenols and aromatic selenols can be synthesized by the reaction of Grignard reagents and organolithium compounds with controlled quantities of elemental Se (equation 2)[1,6–9].

$$\text{RLi(MgX)} \xrightarrow{\text{Se}} \text{RSeLi(MgX)} \xrightarrow{\text{H}_2\text{O}} \text{RSeH} \tag{2}$$

The other readily available and convenient starting materials for preparation of selenols

TABLE 1. Preparation of selenols and selenolates

Starting Material	Reagents	Products	Ref.
Se	$RMgX/H_2O/H^+$	RSeH	1
	$ArMgX/H_2O/H^+$	ArSeH	1
	ArLi	ArSeLi	1
	RLi	RSeLi	1, 6–8
	RC≡CNa	RC≡CSeNa	9
	1. Na 2. ArX/$h\nu$,e^-	ArSeNa	12
H_2Se	RX	RSeH	1
	$RCH{=}CH_2$	RCH_2CH_2SeH	1
		$HOCH_2CH_2SeH$	1
RSeSeR	Na/liq. NH_3	RSeNa	1
	$NaBH_4$	RSeNa	1
	H_3PO_2/H_2O	RSeH	13, 14
$(H_2NCH_2CH_2Se)_2$	Na_2S/KCN	$H_2NCH_2CH_2SeH$	15
ArSeSeAr	Na/THF	ArSeNa	10
	$NaBH_4$	ArSeNa	1, 11
	NaOH/phase-transfer catal.	ArSeNa	16
	$Zn/HCl/H_2O$	ArSeH	17
RSeCN	H^+/H_2O	RSeH	1
ArSeCN	H^+/Zn	ArSeH	1, 17
	H_3PO_2	ArSeH	1
	$NaBH_4$	ArSeH	18
ArSe(O)OH	$NaBH_4$	ArSeH	1
$ArSeO_2OH$	H_2S/SO_2/Zn/HCl	ArSeH	1
$ArSeSiMe_3$	MeOH/H^+	ArSeH	19, 20
	KF/18-crown-6	ArSeK	22
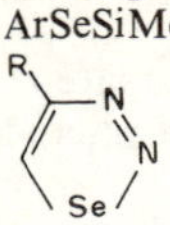	KOH/EtOH/dioxane	RC≡CSeK	23–26
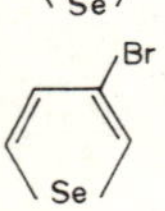	EtLi	Z-LiSeCH=CHC≡CLi	27, 28
t-$Bu_2C{=}Se$	$LiAlH_4$	t-Bu_2CHSeH	29

are alkyl or aryl diselenides which are stable in air, and some of them are commercially available. They are easily reduced to the corresponding selenols by appropriate reducing reagents such as Na[10] or $NaBH_4$[11].

A number of other methods involving the use of other Se compounds have been devised. Table 1 shows a variety of methods for the synthesis of selenols or selenolates[1–29].

B. Reactions

1. Reactions with alkylating reagents

a. Reaction with various halides. Alkylation of alkylselenols and aromatic selenols with alkyl halides in the presence of base gives the corresponding dialkyl selenides in satisfactory yields (equation 3)[16,30–39]. Various types of selenides are synthesized by this method. For example, acetylenic selenols and selenolates are converted to the acetylenic selenides (equations 4 and 5)[9,28,40,41].

$$\mathrm{RSeH} + \mathrm{R^1X} \xrightarrow{\text{base}} \mathrm{RSeR^1} \tag{3}$$

$$\mathrm{RC{\equiv}CSeH} \xrightarrow{\mathrm{R^1X}} \mathrm{RC{\equiv}CSeR^1} \tag{4}$$

$$\mathrm{RC{\equiv}CSeLi(Na)} \xrightarrow{\mathrm{R^1X}} \mathrm{RC{\equiv}CSeR^1} \tag{5}$$

Reich and coworkers have prepared an *E,Z*-mixture of 1,3-bis(phenylseleno)propene by the alkylation of benzeneselenolate with 1,3-dichloropropene. This product is converted to α,β-unsaturated aldehydes as shown in equation (6)[42].

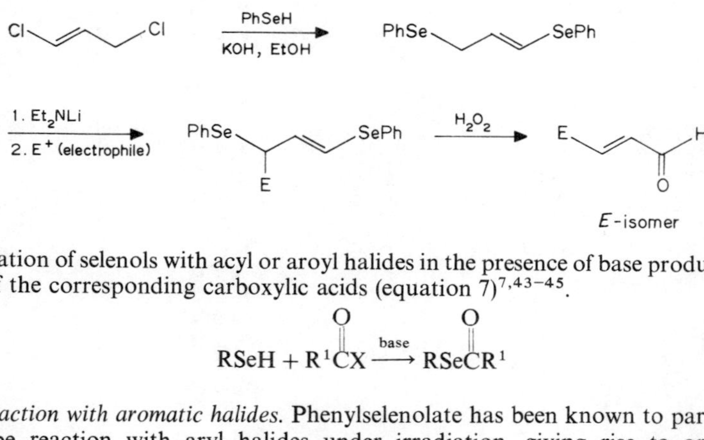

(6)

Alkylation of selenols with acyl or aroyl halides in the presence of base produces seleno esters of the corresponding carboxylic acids (equation 7)[7,43–45].

$$\mathrm{RSeH} + \mathrm{R^1\overset{O}{\overset{\|}{C}}X} \xrightarrow{\text{base}} \mathrm{RSe\overset{O}{\overset{\|}{C}}R^1} \tag{7}$$

b. Reaction with aromatic halides. Phenylselenolate has been known to participate in $S_{RN}1$-type reaction with aryl halides under irradiation, giving rise to aryl phenyl selenides[12,46,47]. The reaction mechanism is depicted in equation (8).

$$\mathrm{ArX} + \mathrm{PhSe^-} \xrightarrow{h\nu} (\mathrm{ArX})^{\cdot-} + \mathrm{PhSe}$$

$$(\mathrm{ArX})^{\cdot-} \longrightarrow \mathrm{Ar^{\cdot}} + \mathrm{X^-}$$

$$\mathrm{Ar^{\cdot}} + \mathrm{PhSe^-} \rightarrow (\mathrm{ArSePh})^{\cdot-} \xrightarrow{\mathrm{ArX}} \mathrm{ArSePh} + (\mathrm{ArX})^{\cdot-} \tag{8}$$

It has been shown that the reaction of arylselenolate ion with aryl iodides is catalysed efficiently by cuprous iodide in HMPA solution (equation 9)[48,49].

$$ArI + Ar^1Se^- \xrightarrow[\text{HMPA}]{\text{CuI}} ArSeAr^1 \tag{9}$$

c. Reaction with alkylamines. In general, selenols form quarternary ammonium salts of the selenolate ion with various amines by ionization of the acidic proton. Ammonium salts of benzeneselenols usually decompose on heating to produce the alkyl phenyl selenide[2]. This pyrolytic reaction offers a method for alkyl group migration from the N atom of amine to the Se atom. The salts formed from tosylated amines[50] or from tertiary alkylamines[2] give alkyl phenyl selenides as shown in equations (10) and (11).

$$RCH_2NH_2 \xrightarrow{\text{TsCl}} RCH_2NHTs \xrightarrow{\text{PhSeH}} [RCH_2NH_2Ts]^+[PhSe]^- \longrightarrow RCH_2SePh + TsNH_2 \tag{10}$$

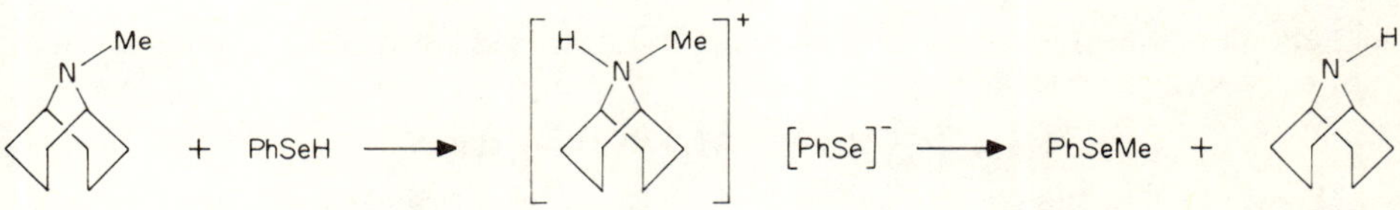

(11)

It has been shown that an alkyl shift from tertiary amines to benzeneselenolates to give alkyl phenyl selenides can be achieved by use of a Ru catalyst (equation 12)[51].

$$PhSeNa + R^1R^2R^3N \xrightarrow[\text{2. } H_2O]{\text{1. Ru catalyst}} PhSeR^1 + R^2R^3NH \tag{12}$$

2. Reduction by selenols

It has been known that selenols are good reducing agents for various classes of organic compounds. Functional groups such as nitroso (—N=O), azo (—N=N) and imino ($>$C=N—) are readily reduced to give amino, hydrazo and amino groups, respectively[52,53]. Organic sulphoxides are generally reduced to the corresponding sulphides[54].

Methylselenol reacts reductively with ketones[55,56] or benzyl halides[57] as shown in equations (13) and (14).

$$R^1R^2C{=}O + MeSeH \xrightarrow[CH_2Cl_2]{ZnCl_2} R^1R^2CHSeMe + R^1R^2C(SeMe)_2 \tag{13}$$

$$PhCH_2X + MeSeH \rightarrow PhCH_2SeMe + PhMe \tag{14}$$

3. Ring-opening reactions

Treatment of epoxides[58–63,70], lactones [6,10,64–69] and cyclopropanes[68,71] with selenols results in ring-opening to selenenylated derivatives. These reactions are due to the high nucleophilicity of the Se atom in the selenols. Examples are given in equations (15)–(17).

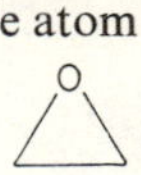

$$+ RSeH \rightarrow HOCH_2CH_2SeR \tag{15}$$

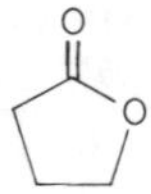

$$+ \mathrm{RSeH} \rightarrow \mathrm{HOOC(CH_2)_3SeR} \tag{16}$$

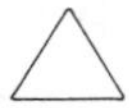

$$+ \mathrm{RSeH} \rightarrow \mathrm{MeCH_2CH_2SeR} \tag{17}$$

4. Addition reactions

Reaction of selenols with conjugated ketones and aldehydes having the C=C—C=O bond system, gives products of addition to the C=C bonds (equation 18)[19,21,72–74].

(18)

Addition of benzeneselenol to mono- and di-substituted acetylenes at room temperature gives vinylic selenides (equation 19)[75–85].

$$\mathrm{RC{\equiv}CH + PhSeH \rightarrow RCH{=}CHSePh} \tag{19}$$

Some other studies concerning Section II.B, i.e. on selenols and their derivatives have been reported[86–111].

III. SELENENIC ACIDS AND THEIR DERIVATIVES

The generalized structure of selenenyl compounds can be represented by R—Se—X, in which R is an alkyl, aryl or heterocyclic moiety. Selenenyl compounds are classified into five general groups: (*i*) selenenic acid (X = OH), (*ii*) selenenic esters (X = OR^1) and selenocarboxylates (X = OC(O)R^1), (*iii*) selenenamides (X = NR$^1_2$), (*iv*) selenocyanates (X = CN) and (*v*) selenenyl halides (X = halogen). All of these types of selenenyl compounds share a common feature, i.e. polarization of the Se$^{\delta+}$—X$^{\delta-}$ bond, resulting in reactions involving positively charged Se species (equation 20).

$$\mathrm{R-Se-X \rightarrow R-Se^+ + X^-} \tag{20}$$

A. Selenenic Acids and Their Anhydrides

Selenenic acids and their anhydrides are usually generated *in situ* and used without isolation, because of their instability. There are several reactions in which selenenic acids and their anhydrides play important roles as transient intermediates.

1. Generation

a. β-Elimination of alkyl aryl selenoxides. β-Elimination of alkyl aryl selenoxides proceeds under mild conditions with high selectivity and has been used as a facile and convenient method for synthesis of olefins[11,112–115]. β-Hydroxyselenides, which may be derived by addition of the eliminated arylselenenic acid (ArSeOH) to the olefinic products, are incidentally formed as by-products (equation 21)[117,131,148].

$$\begin{aligned}\mathrm{RCH_2CH_2SeAr} &\xrightarrow{[O]} \mathrm{RCH_2CH_2\overset{\overset{\displaystyle O}{\|}}{Se}Ar} \xrightarrow{\Delta} \mathrm{RCH=CH_2 + [HOSeAr]}\\ &\longrightarrow \mathrm{RCH(OH)CH_2SeAr}\end{aligned} \tag{21}$$

Decomposition of 3,3-dimethyldihydrobenzoselenophene oxide (**1**) leads to generation of an unstable intermediate selenenic acid (**2**), which is trapped intramolecularly to give the isomeric hydroxyselenides **3** and **4** (equation 22)[116,117].

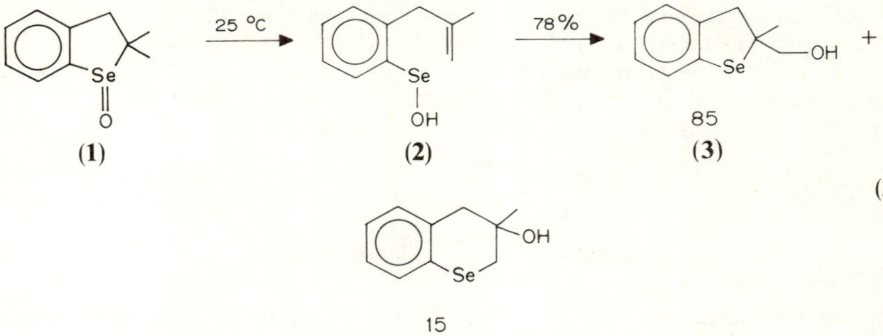

(**1**) (**2**) (**3**) (**4**)

(22)

A ^{77}Se-NMR signal assignable to the selenenic acid **6** has been observed during the *syn* elimination of selenoxide **5** in CD_3OD (equation 23)[118]. The formation of the seleninic ester **7** serves as a chemical evidence supporting the intermediacy of the selenenic acid in this reaction.

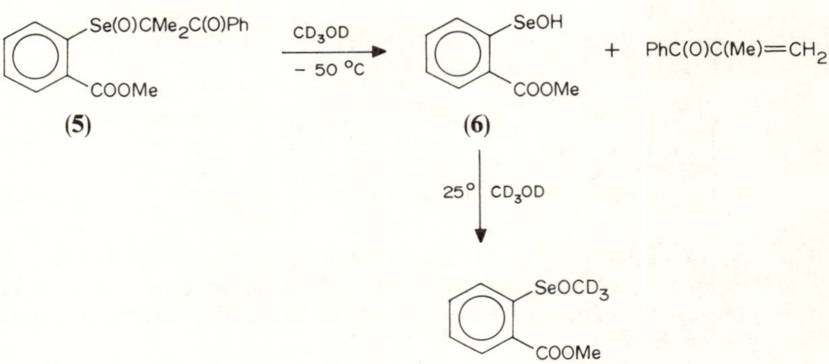

(**5**) (**6**) (**7**)

(23)

b. Oxidation of selenols and diselenides. Selenenic acids are known to disproportionate into the corresponding diselenides and seleninic acids (going right to left in equation 24[11,131]. The reverse processes were sometimes termed 'comproportionation' in the literature[117,131] and benzeneselenenic acid can be generated *in situ* by this comproportionation.

$$PhSeSePh + PhSeO_2H + H_2O \rightleftharpoons 3\,[PhSeOH] \quad (24)$$

Similar comproportionation between diphenyl diselenide and benzeneseleninic anhydride (2:1 molar ratio) has been also reported to give benzeneselenenic anhydride (equation 25)[125–127,129].

$$2\,PhSeSePh + (PhSeO)_2O \rightleftharpoons 3\,[PhSeOSePh] \quad (25)$$

t-Butyl hydroperoxide is employed as an oxidizing agent for the oxidation of diphenyl diselenide to benzeneselenenic anhydride (equation 26)[125,126,128,129]. However, it seems likely that the anhydride is not present in high concentration during the reaction[130].

$$\mathrm{PhSeSePh} + t\text{-}\mathrm{BuOOH} \rightarrow [\mathrm{PhSeOSePh}] + t\text{-}\mathrm{BuOH} \qquad (26)$$

c. Reduction of seleninic acids and their anhydrides. Several reducing reagents, such as hypophosphorous acid (H_3PO_2)[132,137], hydrazines (NH_2NH_2[134], $RNHNH_2$[138,139]) and thiols (RSH)[133], have been used for reducing seleninic acids and their anhydrides to seleninic acids and their anhydrides, respectively (equations 27 and 28).

$$\mathrm{R\overset{\overset{O}{\|}}{Se}OH} \xrightarrow{\text{reducing agent}} [\mathrm{RSeOH}] \qquad (27)$$

$$\mathrm{R\overset{\overset{O}{\|}}{Se}O\overset{\overset{O}{\|}}{Se}R} \xrightarrow{\text{reducing agent}} [\mathrm{RSeOSeR}] \qquad (28)$$

Recent reinvestigations[135,136] of the reduction products of *ortho*-substituted benzeneseleninic acids, which were initially reported[132–134,151,152] to be sufficiently stable for isolation, have shown that the products are not the selenenic acids but the corresponding selenenic anhydrides (equations 29 and 30).

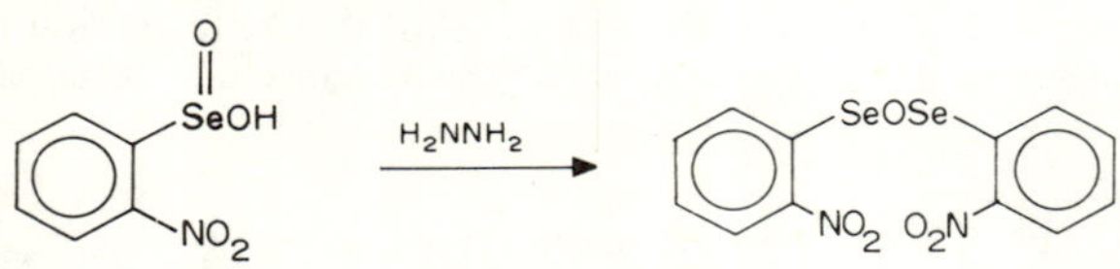

(29)

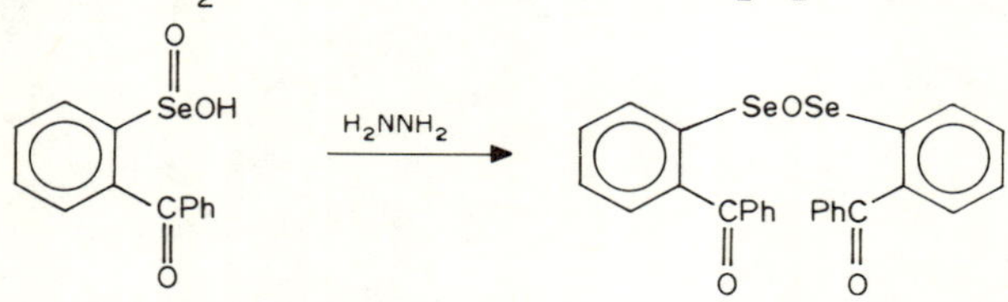

(30)

d. Hydrolysis of selenenyl halides. Hydrolysis of selenenyl halides has been reported to give the corresponding selenenic acids (equation 31)[44]. Anthraquinone-1-seleninic acid is prepared by the reaction of the corresponding selenenyl bromide with moist silver oxide in dioxane (equation 32)[145].

$$o\text{-}\mathrm{O_2NC_6H_4SeBr} + \mathrm{H_2O} \rightarrow o\text{-}\mathrm{O_2NC_6H_4SeOH} + \mathrm{HBr} \qquad (31)$$

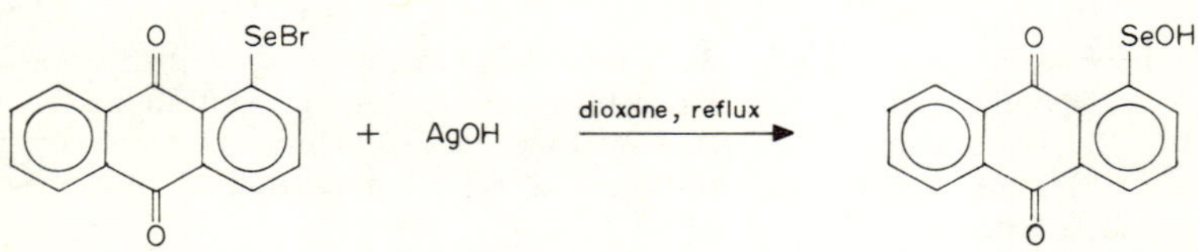

(32)

2. Reactions

In general selenenic acids are unstable and rapidly disproportionate to diselenides and seleninic acids (equation 33)[115].

$$3\,\mathrm{RSeOH} \rightleftharpoons \mathrm{RSeSeR} + \mathrm{RSeO_2H} + \mathrm{H_2O} \qquad (33)$$

Selenenic acid is a good electrophilic reagent and often used for *in situ* oxyselenation of olefins (equation 34)[137]. The resulting β-hydroxyselenides can be easily converted to the

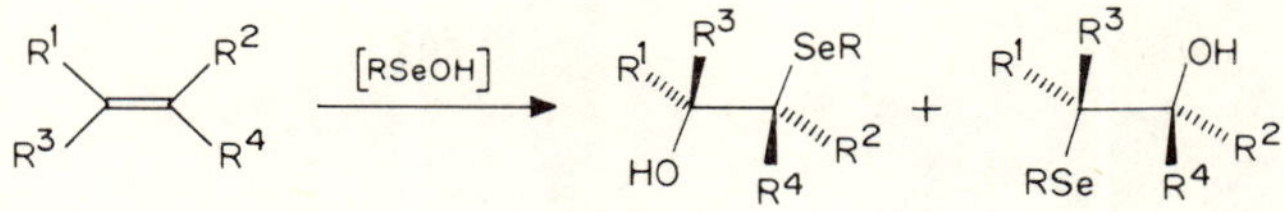

(34)

corresponding allylic alcohols or epoxides, in a one-pot reaction, via oxidation to selenoxides by appropriate oxidizing reagents.

The disproportionation (equation 33) is a reversible reaction, and the reverse process can be used for the *in situ* formation of selenenic acid (equation 24)[117,131]. Hori and

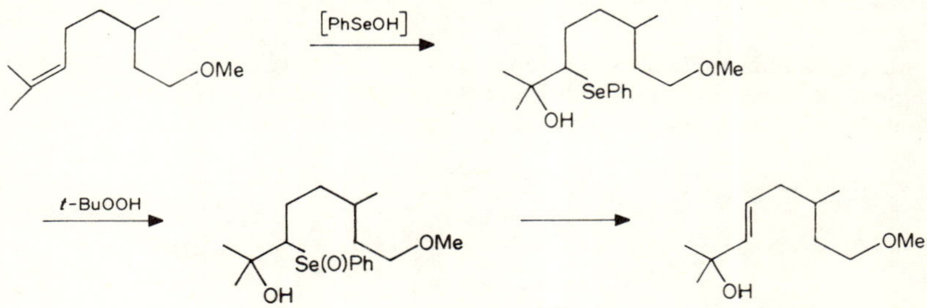

(35)

Sharpless[131] have reported the conversion of olefins to allylic alcohols using this reverse reaction system (equation 35). This method can be applied to the synthesis of various cyclic ether derivatives from dienes (equations 36 and 37)[150].

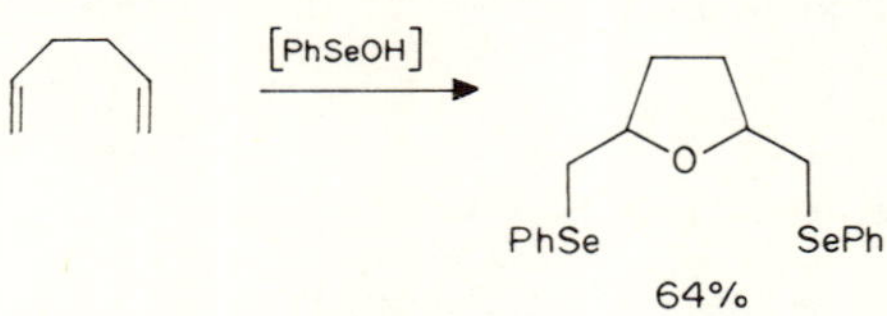

(36)

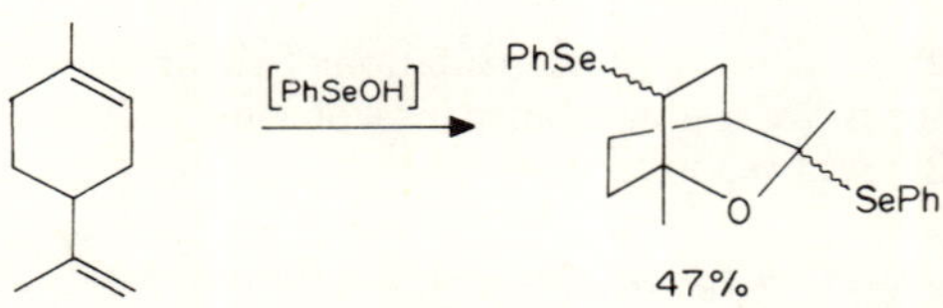

(37)

Electrochemical *in situ* generation of phenylselenenic acids has been undertaken by Torii[149] (equation 38), and this method was used for a one-step synthesis of allylic derivatives (**9**) from isoprenoids (**8**). In this case the selenenylation reagents can be recycled by using a catalytic amount of diphenyl diselenide (equation 38)[149].

$$(PhSe)_2 \xrightarrow[2\ ROH]{-2\,e} 2\,PhSeOR + 2\,H^+\ (R = Me, H) \qquad (38)$$

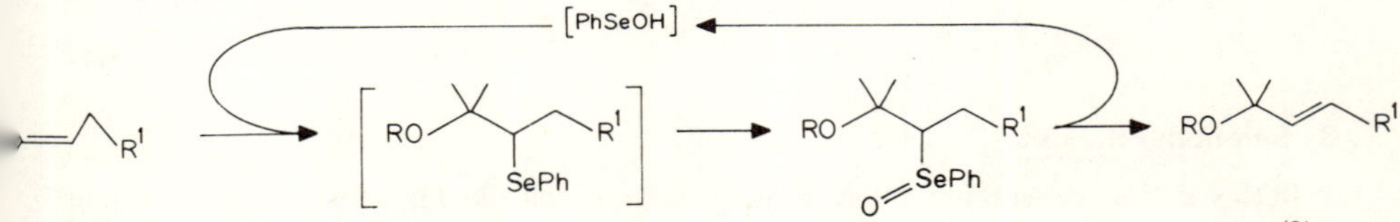

(8) **(9)**

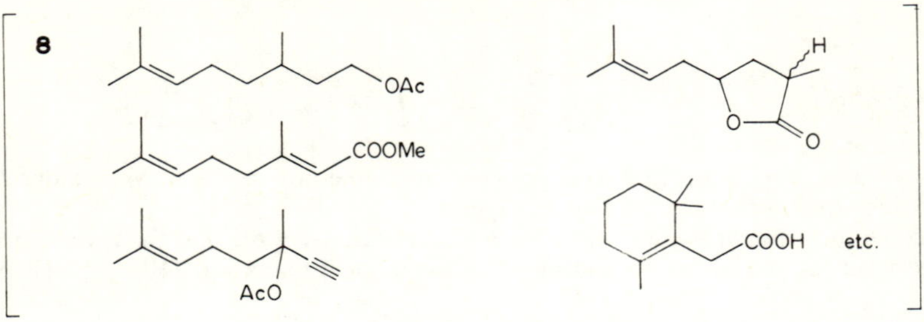

Recently, Kuwajima and coworkers reported a regioselective oxidation of olefins to α-phenylseleno ketones by using $(PhSe)_2$–*t*-BuOOH or $(PhSe)_2$–$(PhSeO)_2O$ systems (equation 39)[126]. The nature of the reactive species in this reaction has not yet been

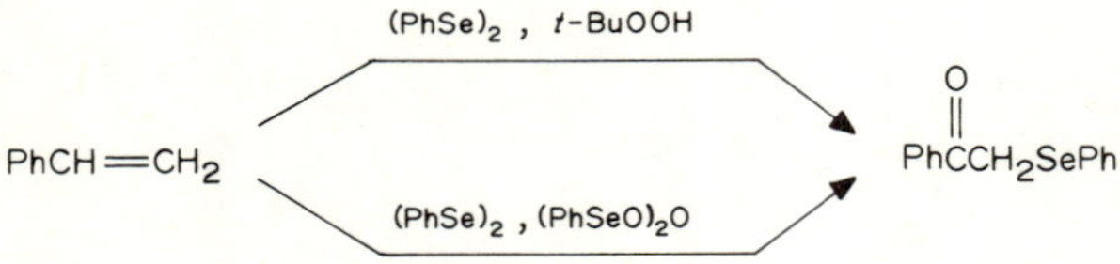

(39)

confirmed. However, benzeneselenenic anhydride (PhSeOSePh) is assumed to be a plausible one as shown in equations (25) and (26). *anti*-Markownikoff-type oxidation is observed with allylic alcohols, while the oxidation of terminal olefins proceeds in Markownikoff manner (equation 40)[125–127,129].

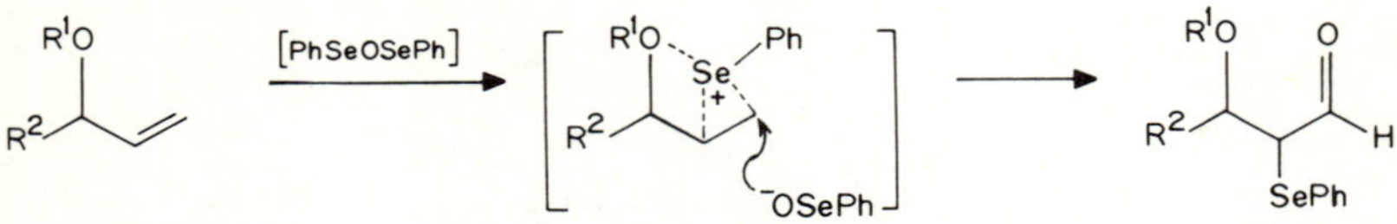

(40)

In connection with this reaction system, oxidation of allylic alcohols to the corresponding aldehydes or ketones by the combined use of *t*-butyl hydroperoxide and diaryl diselenide (equation 26) was reported (equations 41 and 42)[128].

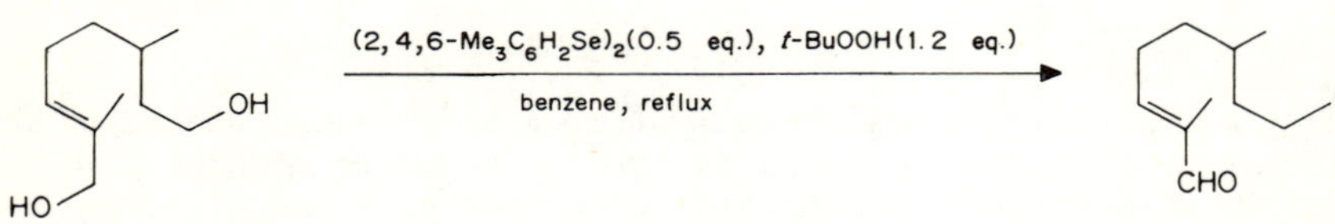

(41)

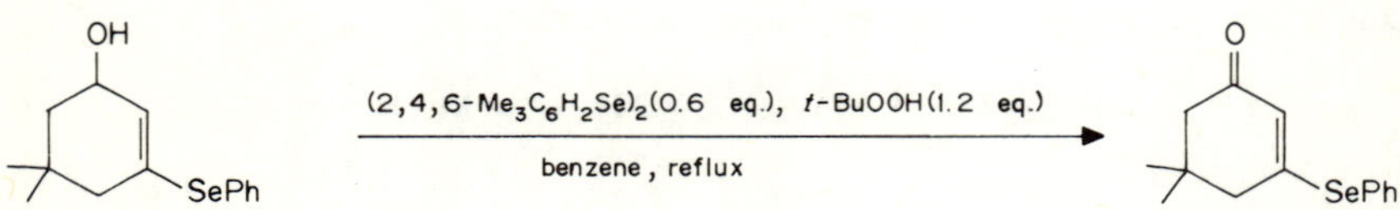

(42)

B. Selenenyl Halides

In this section, general methods for the synthesis of selenenyl halides, and some of their characteristic reactions will be described.

1. Synthesis

The most general preparative method of selenenyl halides is direct halogenation of corresponding diselenides (equation 43)[114,151,154–161].

$$RSeSeR + Cl_2 \text{ (or } Br_2) \rightarrow 2\,RSeCl \text{ (or } 2\,RSeBr) \qquad (43)$$

Benzeneselenenyl chloride has been prepared from bromobenzene as shown in equation (44) and is commercially available[162].

$$PhBr \xrightarrow[2.\ Se]{1.\ Mg} PhSeMgBr \xrightarrow[64-70\%]{Br_2} PhSeSePh \xrightarrow[88-93\%]{Cl_2} PhSeCl \qquad (44)$$

Alternative methods for the synthesis of selenenyl halides are shown in equations (45)–(47), and their details have been described in Klayman's review[1].

$$ArSeCN + Cl_2 \text{ (or } Br_2) \rightarrow ArSeCl \text{ (or } ArSeBr) \quad \text{(Refs. 134, 157, 163, 164)} \qquad (45)$$

$$RSeX_3 \rightleftharpoons RSeX + X_2 \quad \text{(Refs. 153, 154)} \qquad (46)$$

$$Ar\overset{+}{Se}(Br)R\ Br^- \xrightarrow{\Delta} ArSeBr + RBr \quad \text{(Ref. 165)} \qquad (47)$$

2. Reactions

a. With carbonyl compounds. Sharpless and coworkers have reported that benzeneselenenyl chloride reacts with various ketones and aldehydes to give α-phenylselenenyl carbonyl compounds, which on treatment with H_2O_2 or $NaIO_4$ at room temperature yield the corresponding α,β-unsaturated ketones or aldehydes (equations 48 and 49)[114,154].

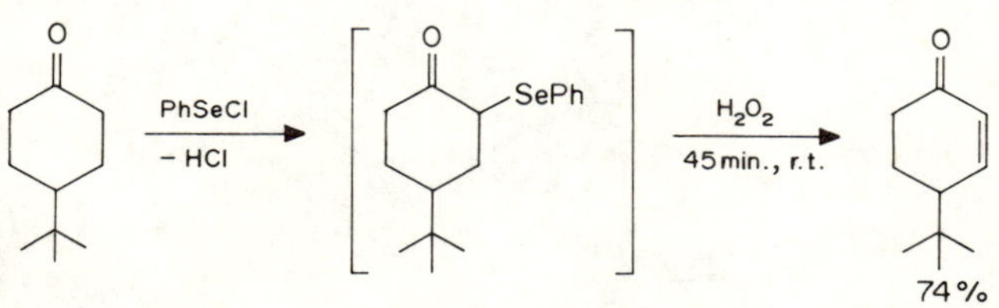

(48)

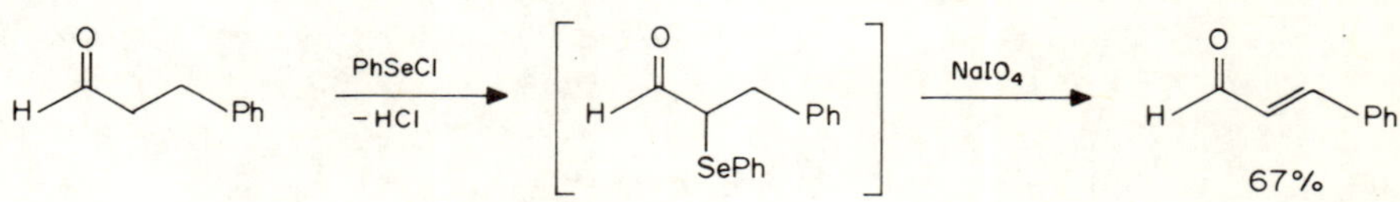

(49)

Reich and coworkers have reported a higher yield method in which benzeneselenenyl halides are allowed to react with lithium enolates prepared *in situ* from lithium diisopropylamide (LDA) and ketones (Table 2, entries 1–4)[115,168]. Similarly, α,β-unsaturated esters[115] and lactones[154] are synthesized from the corresponding saturated compounds as shown in Table 2 (entries 5–7).

A similar method can be applied for the preparation of substituted furans (equation 50)[171].

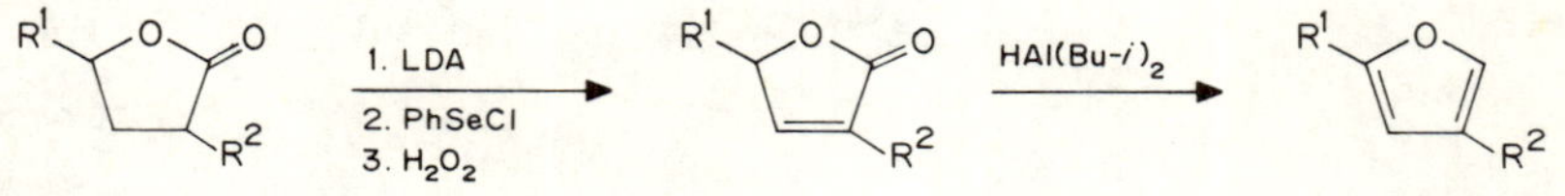

(50)

TABLE 2. Synthesis of α,β-unsaturated carbonyl compounds by reaction of phenylselenenyl halides with lithium enolates[114,115,154,166–170]

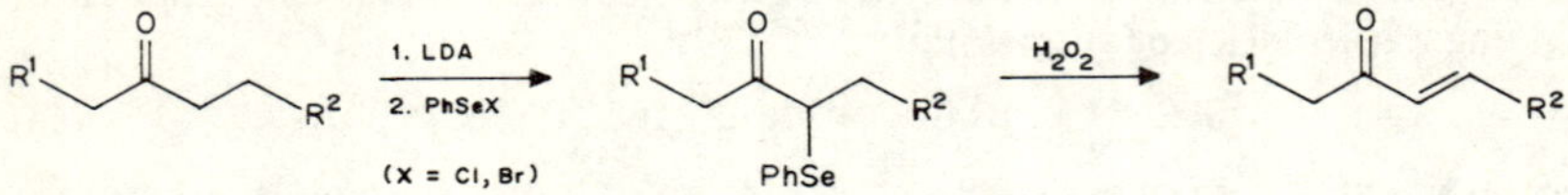

Entry	Starting material	Product	Yield (%)	Ref.
1	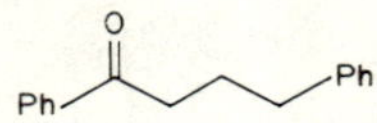	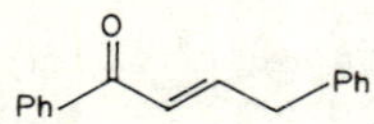	84	115, 168
2	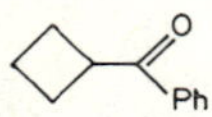	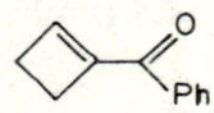	83	115, 168
3	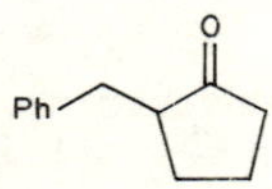	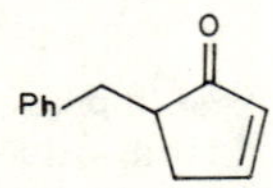	66	115, 168
4	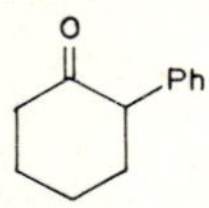	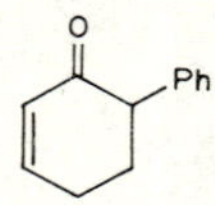	60	115, 168
5	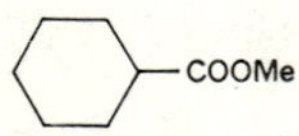	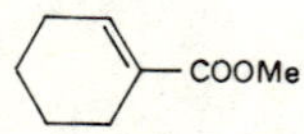	96	115
6	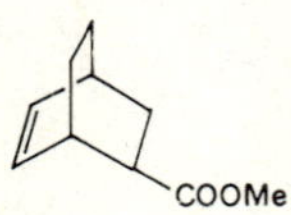	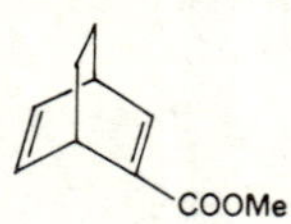	68	115
7	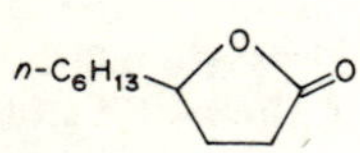	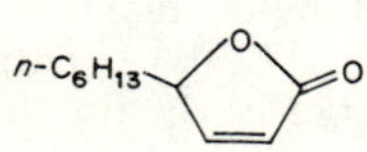	56[a]	154
8	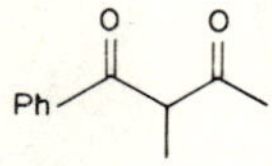	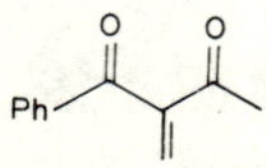	80[b]	115

(*Contd.*)

TABLE 2. (*Contd.*)

Entry	Starting material	Product	Yield (%)	Ref.
9	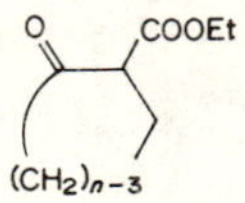	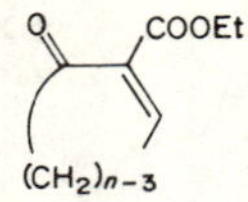	$n = 8$ 93[b] $n = 7$ 93[b] $n = 6$ 89[b] $n = 5$ 81[b]	115
10	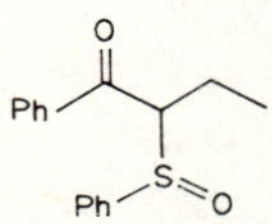	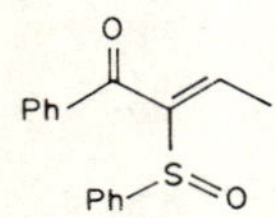	71	115
11	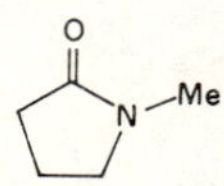	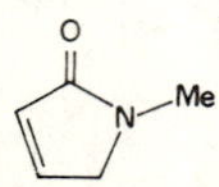	55	170, 113

[a] $MeCO_3H$ is used as the oxidizing agent.
[b] NaH is used instead of LDA.

In the case of 1,3-dicarbonyl compounds, sodium hydride is used as a base, and α,β-unsaturated 1,3-dicarbonyl compounds are obtained regioselectively (Table 2, entries 8–9)[115].

The reaction with a β-ketosulphoxide proceeds through intermediate **10**, and gives only a sulphinoenone by *syn* elimination of a benzeneseleninyl group and a β-hydrogen. Elimination of a benzenesulphinyl group does not take place (Table 2, entry 10)[115].

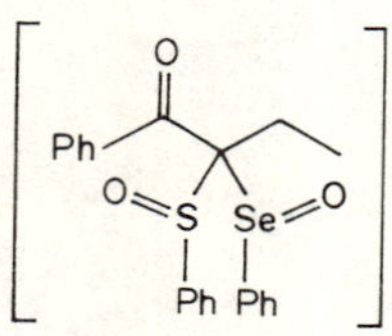
(10)

Two equivalents of LDA are employed for the preparation of α,β-unsaturated lactams (Table 2, entry 11)[170].

Other examples of α,β-unsaturated carbonyl compounds prepared by selenenylation followed by selenoxide β-elimination are summarized in Reich's review[113].

Benzeneselenenyl halides also react similarly with copper enolates[167], aluminium enolates[172] and zirconium enolates[172] (equations 51 and 52).

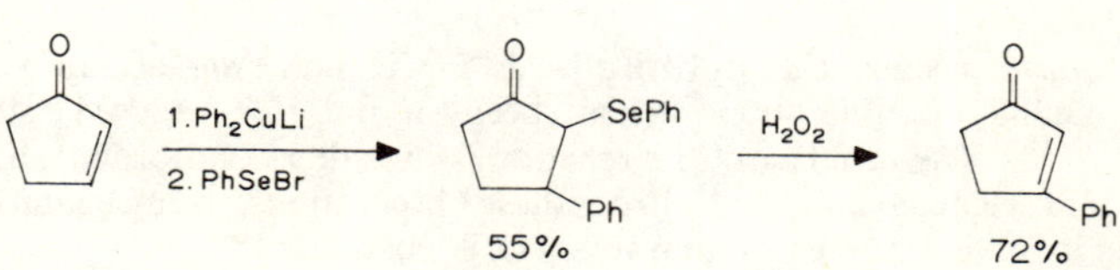

(51)

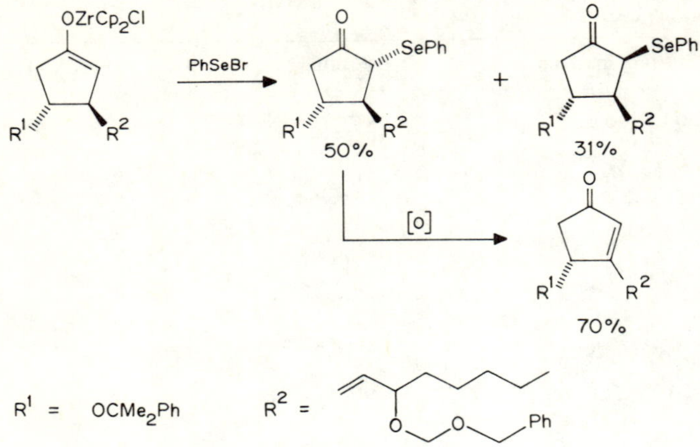

(52)

The reactions of selenenyl halides with α,β-unsaturated carbonyl compounds or α-diazoketones are found to give α-selenenyl α,β-unsaturated carbonyl compounds (equations 53–55).

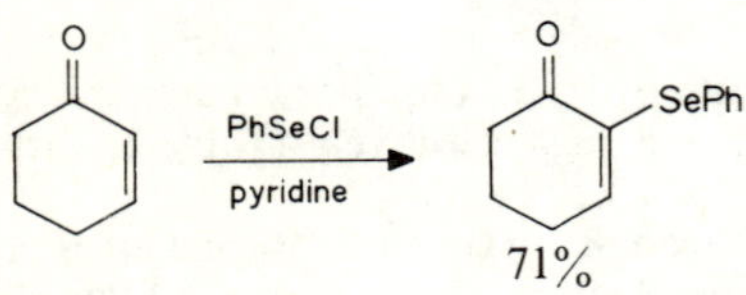

(Ref. 173) (53)

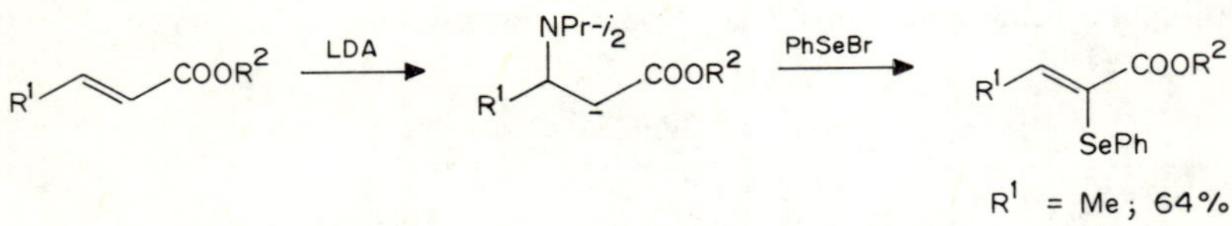

(Ref. 174) (54)

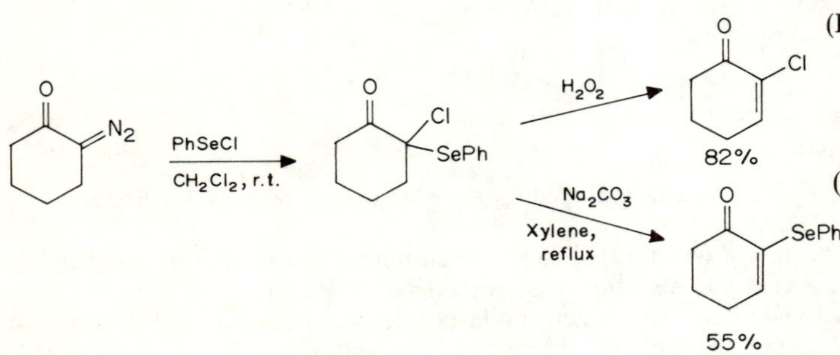

(Ref. 175) (55)

b. With unsaturated compounds. (*i*) *With alkenes.* Electrophilic *anti* addition of selenenyl halides to olefinic double bonds has been found to proceed stereospecifically[114,160,176–191]. The rate law of the reaction is overall second order—first order in both alkenes and selenenyl halides[181]. From these observations, the mechanism shown in equation (56), involving a seleniranium ion (**11**), is postulated[180].

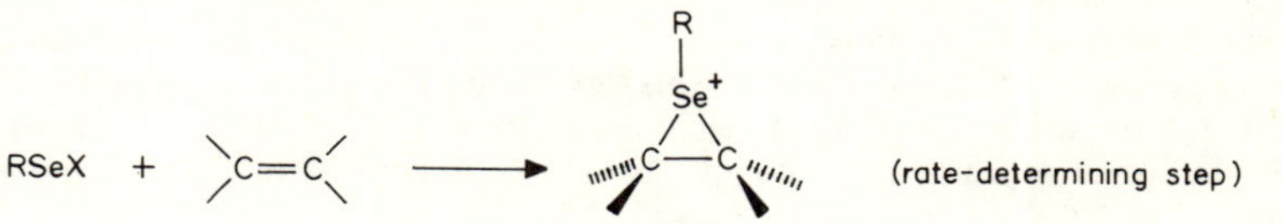

(11)

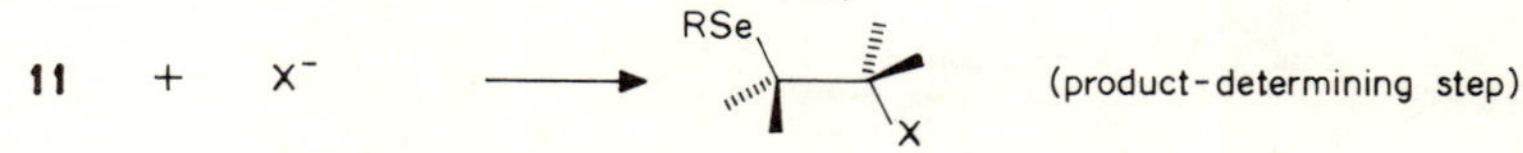

With respect to the regiochemistry of the addition, Raucher[178,179] has shown that the reaction of benzeneselenenyl bromide with terminal olefins gives predominantly *anti*-Markownikoff adducts under kinetically controlled conditions (CCl_4, -20 °C). The adducts isomerize to give predominantly Markownikoff adducts in 48 h at 25 °C in CCl_4 (equation 57).

$$Me(CH_2)_3CH{=}CH_2 \xrightarrow[CCl_4,\,-20\,^{\circ}C]{PhSeBr} Me(CH_2)_3CH(SePh)CH_2Br$$
$$\xrightarrow{CCl_4,\,-25\,^{\circ}C} Me(CH_2)_3CH(Br)CH_2SePh \qquad (57)$$

On the other hand the reaction with internal olefins usually gives a mixture of Markownikoff and *anti*-Markownikoff adducts. Their ratio is influenced by both electronic and steric effects of the substituents[180,181]. Liotta and Zima have synthesized adducts isomerize to give predominantly Markownikoff adducts in 48 h at 25 °C in CCl_4 (equation 57).

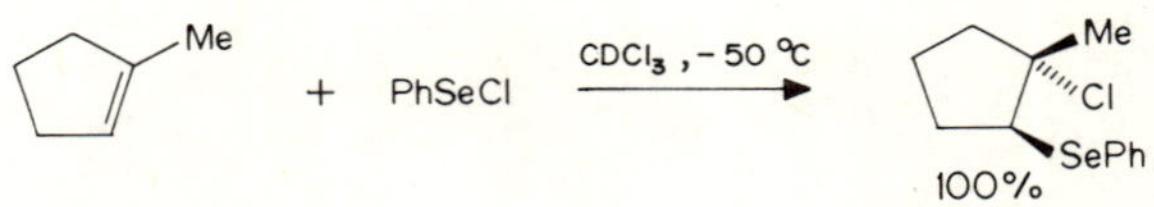

(58)

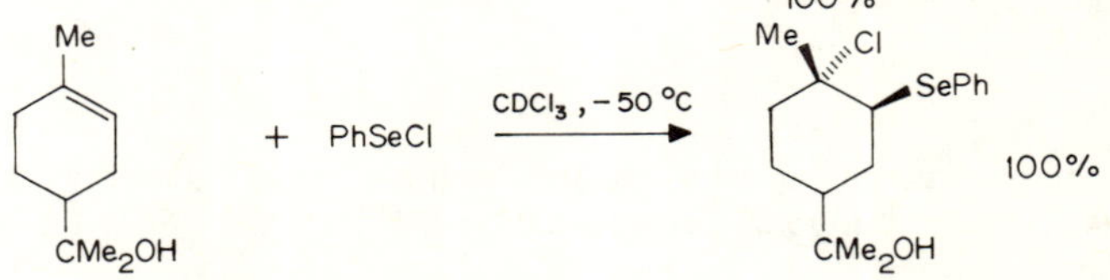

(59)

The adducts of selenenyl halides and olefins are generally unstable both thermally and solvolytically. However, when the reactions are carried out in the presence of various nucleophiles such KOAc[114], $AgOCOCF_3$[185,186], H_2O[187,191], ROH[188], MeCN[189] and $AgNO_2$[190], thermodynamically stable adducts can be isolated in high yields (equation 60).

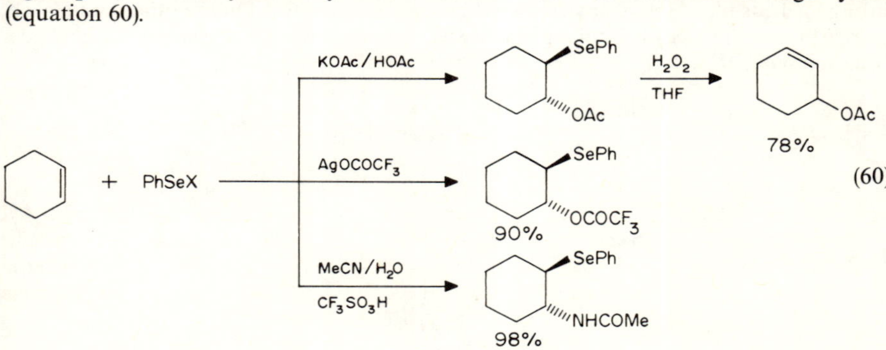

(60)

Benzeneselenenyl chloride adds to allylic alcohols and acetates in a highly regio- and stereo-selective fashion (equation 61) and this type of addition can be used as the key step of a simple 1,3-enone transposition sequence (equation 62)[192,193]. However, the addition to 3,4-dihydro-2*H*-pyran is known to be exceptionally non-stereospecific (equation 63)[177].

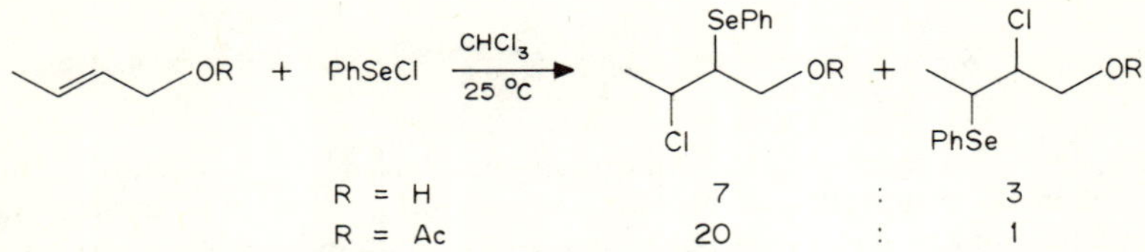

(61)

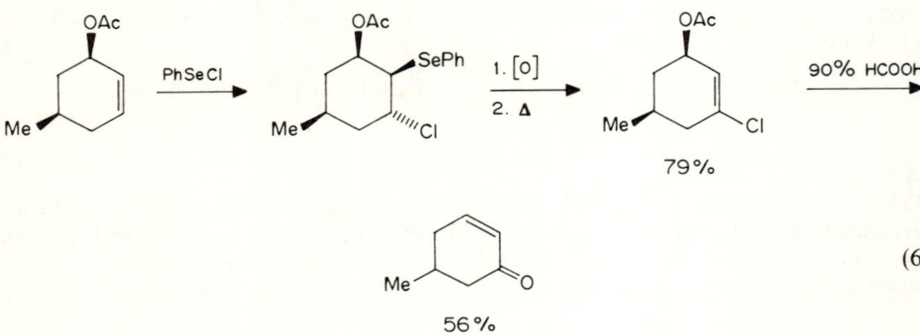

(62)

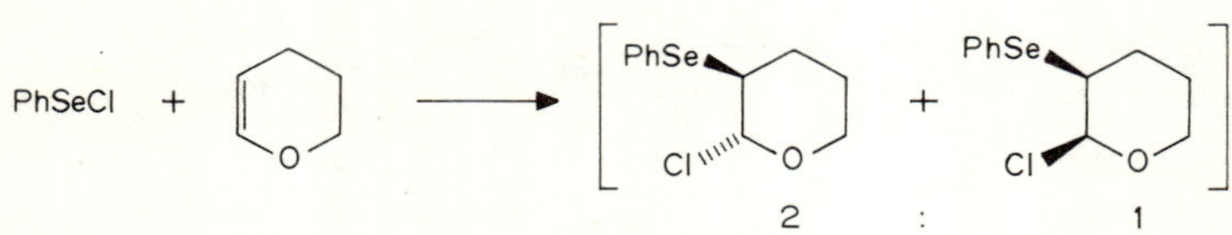

(63)

The reaction of benzeneselenenyl halides with enol ethers gives the corresponding α-phenylseleno ketones in good yields (equation 64)[194–197].

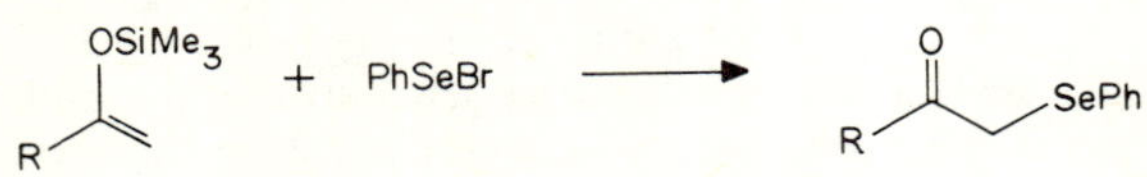

(64)

Benzeneselenenyl chloride reacts regioselectively with allylsilanes to give allyl selenides, through a 1,3-shift of the phenylselenenyl group (equation 65). On the other hand, such 1,3-rearrangement does not take place in the reaction with benzenesulphenyl chloride (equation 66)[198,199].

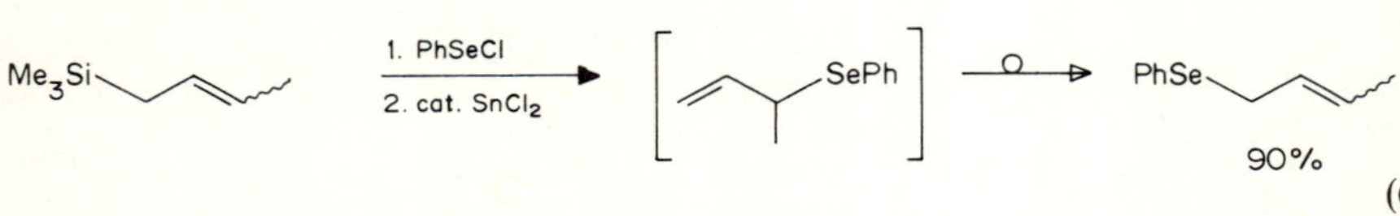

(65)

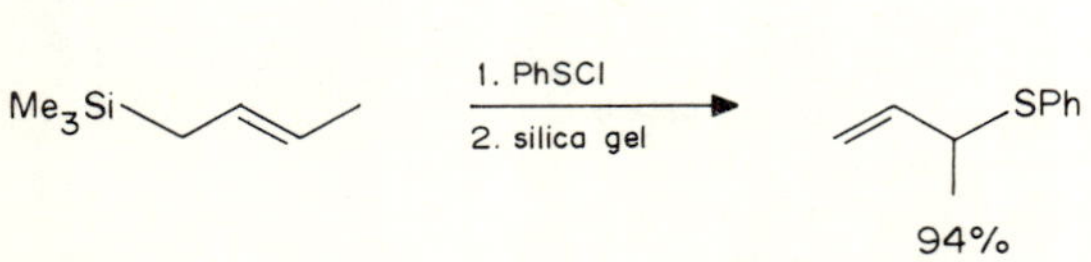

(66)

In addition, the reaction of fluoroalkenes with benzeneselenenyl halides has been also reported[183,184].

(*ii*) *With alkynes.* Little is known about the reaction of selenenyl halides with simple alkynes[200]. The reaction of propargyl alcohols with benzeneselenenyl chloride gives regioselectively the corresponding vinyl selenides (equation 67)[201].

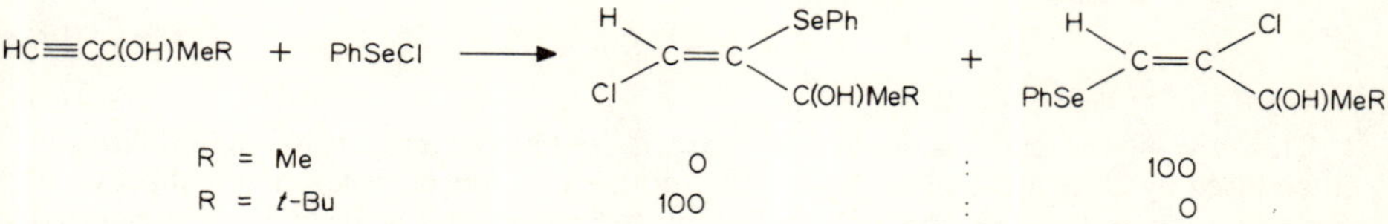

(67)

The reaction of lithium trialkylalkynylborate with benzeneselenenyl chloride, followed by selective oxidation, provides a new synthetic pathway from acetylenes to acyclic *trans*-α,β-unsaturated ketones (equation 68)[202].

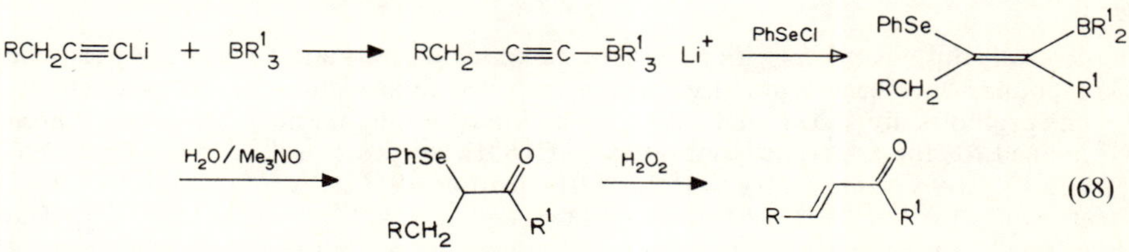

(68)

(*iii*) *With allenes.* Benzeneselenenyl halides also add to allenes, giving 1:1 adducts[203–205] in which the phenylselenenyl group is predominantly located at the central carbon of allenes (equation 69).

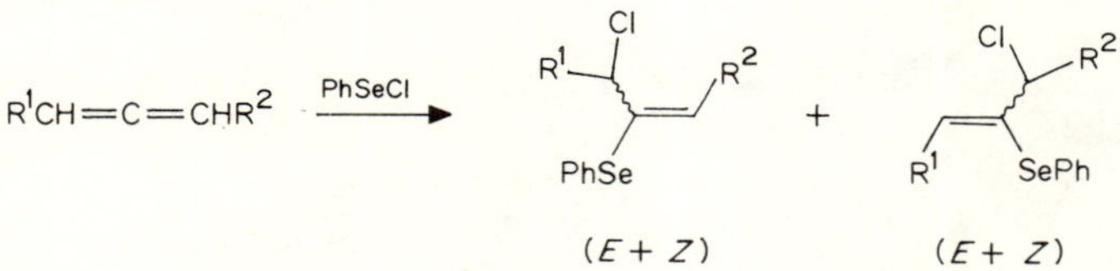

(69)

(*iv*) *With quadricyclene.* The reaction of benzeneselenenyl chloride with tetracyclo[3.2.0.0^{2,7}.0^{4,6}]heptane (quadricyclene) yields adducts of both 1,3- and conjugative 1,6-addition (equation 70)[206].

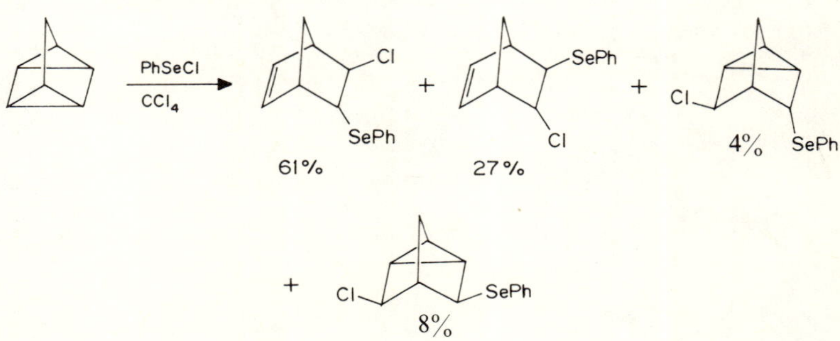

(70)

(*v*) *Cyclofunctionalization.* Electrophilic addition of the phenylselenenyl group to olefins possessing nucleophilic moieties (Nu^-), such as hydroxyl or carboxyl groups, followed by intramolecular ring-closure gives stereospecifically cyclized products, such as cyclic ethers or lactones, respectively (equation 71).

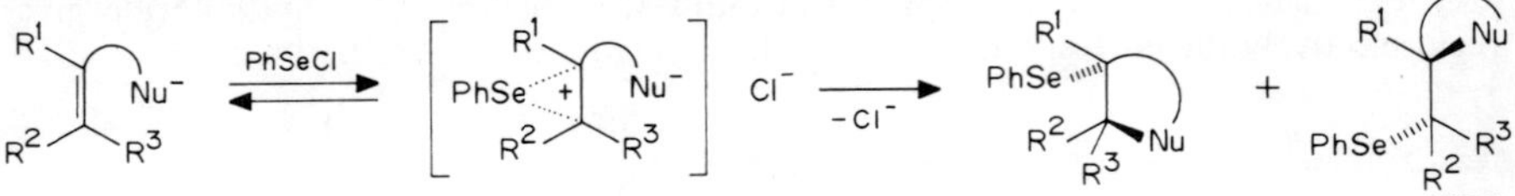

(71)

This type of synthetic reaction is well characterized by the term cyclofunctionalization introduced by Clive and coworkers[207,209]. Another example of cyclofunctionalization in Se chemistry is the reaction of γ,δ-unsaturated carboxylic acids with arylselenenyl bromides (equation 72)[208].

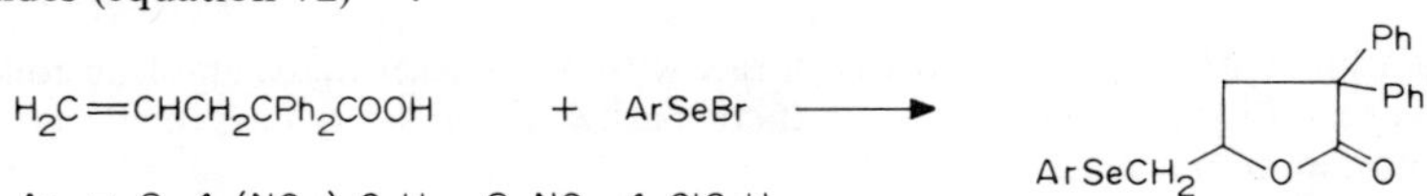

(72)

Ar = 2,4-$(NO_2)_2C_6H_3$, 2-NO_2-4-ClC_6H_3

Recently this type of cyclization has been studied extensively and a wide variety of cyclic compounds have been synthesized with a high degree of regio- and stereo-selectivity. Some examples are shown in Table 3. Many kinds of internal nucleophiles have been examined for the cyclofunctionalization:—COOH (entries 1–6)[207–212],—COOR (entry 7)[213,217] $\geq$C=O (entry 8)[214],—OH (entries 9–11)[215–217,222],—NHCOOEt (entries 12, 13)[218,219,228],—SH (entry 14)[220] and $\geq$C=C$\leq$ (entry 15)[221,229]. The resulting benzeneselenenyl lactones (entry 4) can be treated with H_2O_2 or some reducing agents such as Raney Ni or tri-*n*-butyltin hydride to give unsaturated or saturated lactones (equation 73)[210,211].

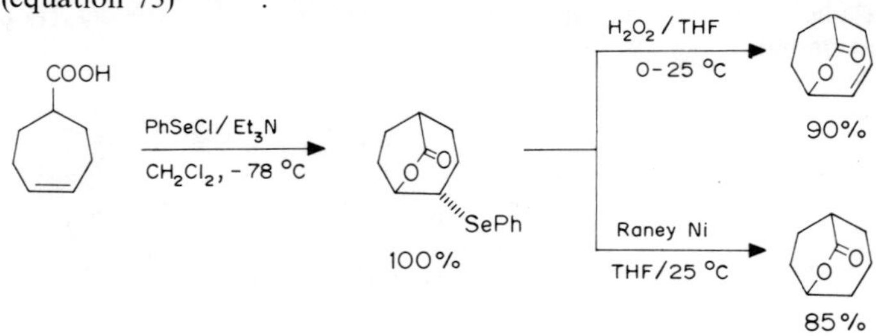

(73)

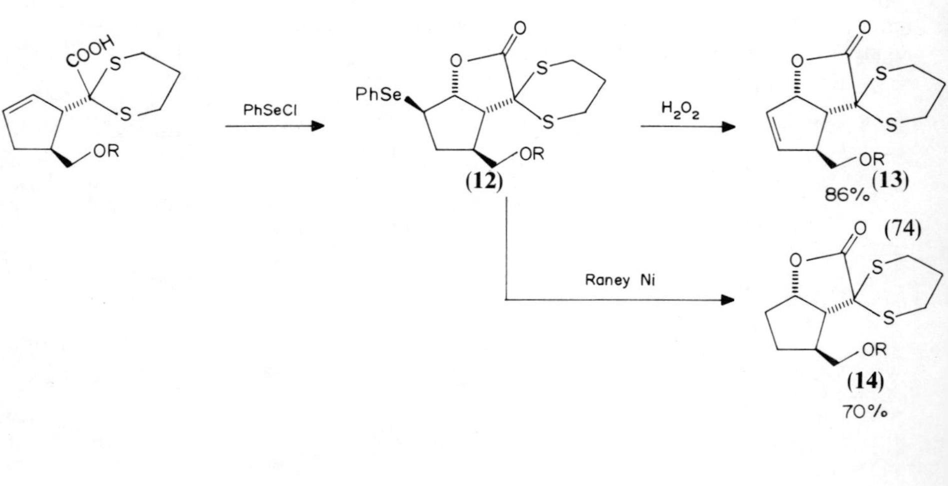

(74)

TABLE 3. Cyclofunctionalization of unsaturated compounds with benzeneselenenyl chloride

Entry	Substrate	Product	Yield (%)		Ref.
1	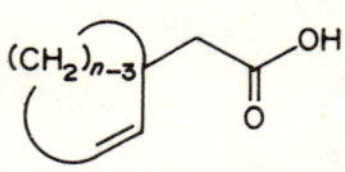	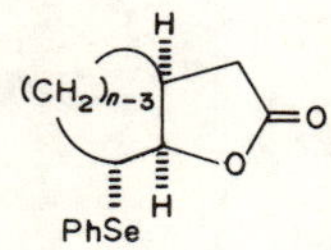	$n = 5$ $n = 6$ $n = 7$	73 82 97	207, 209
2	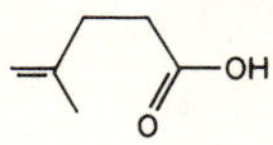	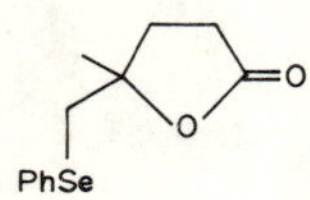		85	209
3	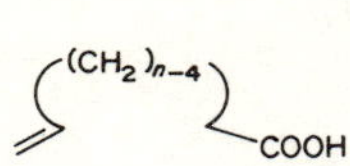	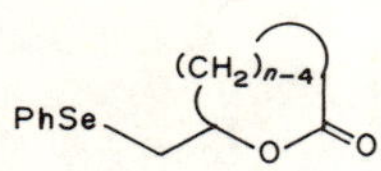	$n = 5$ $n = 6$ $n = 7$	80 74 70	210
4	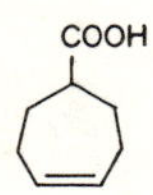	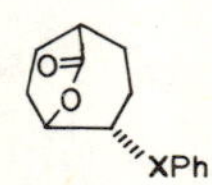	X = Se X = S	100 (82)[a]	210, 211
5		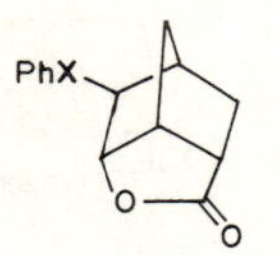	X = Se X = S	95 (95)[a]	210, 211
6	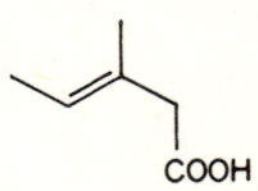	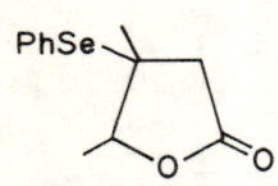		86	212
7	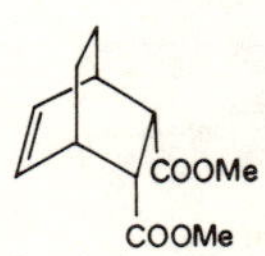	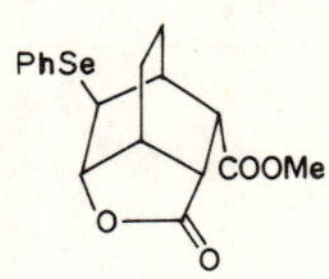 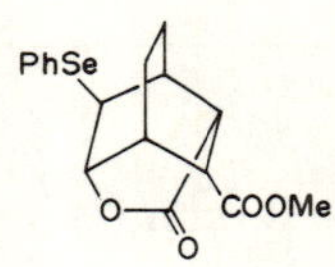		68:25	213
8	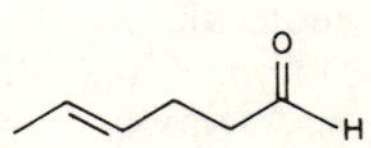	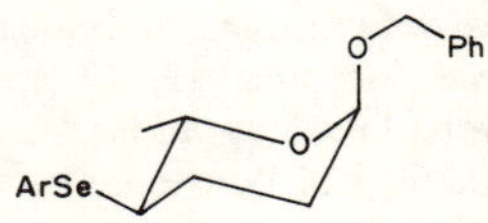		67[b]	214
9	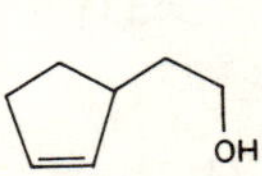	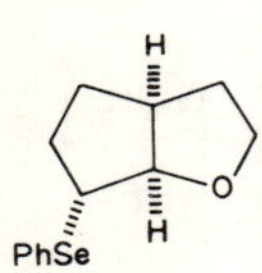		84	215

(*Contd.*)

TABLE 3. (*Contd.*)

Entry	Substrate	Product	Yield (%)	Ref.
10	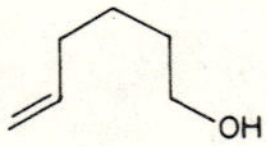	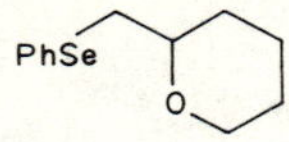	90	216
11	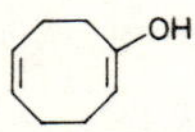	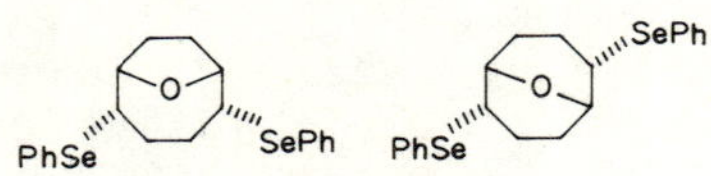 (88 : 12)	92[c]	217
12	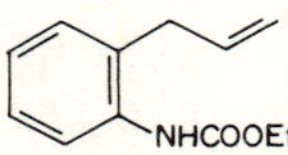	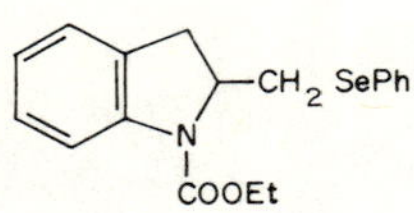	73	218
13	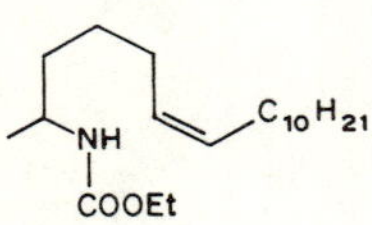	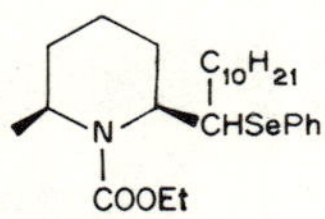	72	219
14	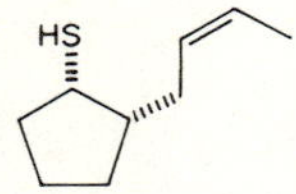	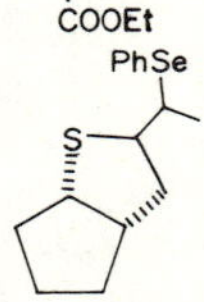	80	220
15	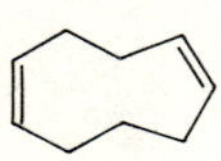	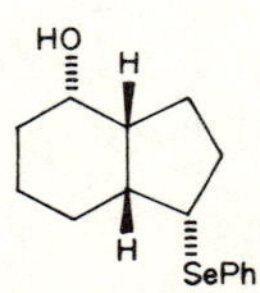	68[d]	221

[a] PhSCl was used instead of PhSeCl.
[b] $p\text{-}ClC_6H_4SeBr/PhCH_2OH$.
[c] PhSeCl/aq. MeCN.
[d] PhSeCl/AcOH/AcONa and then $MeOH/H_2O/K_2CO_3$.

Applications to the synthesis of complex biologically active molecules have also been reported[224–227]. For example, lactones **12**, **13** and **14** are important synthetic intermediates for the construction of prostaglandin A2 and blefeldin A (equation 74)[210,211].

On the other hand, the reaction of benzeneselenenyl chloride with alkynyl alcohols gives only 1,2-adducts (equation 75)[223].

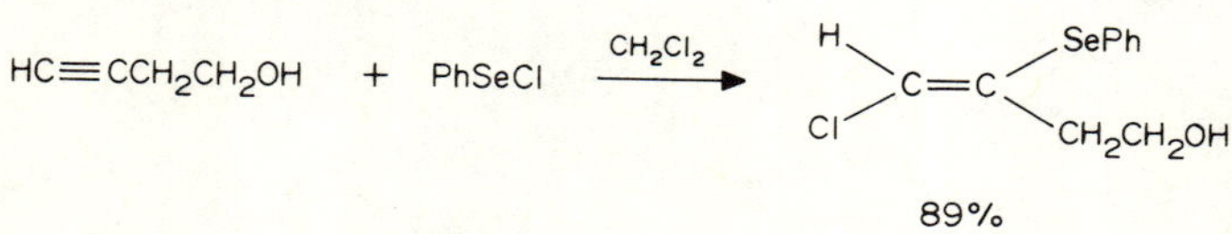

(75)

c. With nucleophiles. A variety of selenenic acid derivatives RSeX (X = OH[144,187], OR[114,185,186], NR_2[216,230], CN[231,236]) have been synthesized by nucleophilic displacement on the Se atom of selenenyl halides. Some examples are shown in equations (76)–(79). SCN^-[232,233] and RSO_2^-[234] have also been used as nucleophiles.

$$2\text{-}O_2NC_6H_4SeX + H_2O \rightarrow 2\text{-}O_2NC_6H_4SeOH \qquad (76)$$

$$RSeX + KOAc \xrightarrow{AcOH} RSeOAc \qquad (77)$$

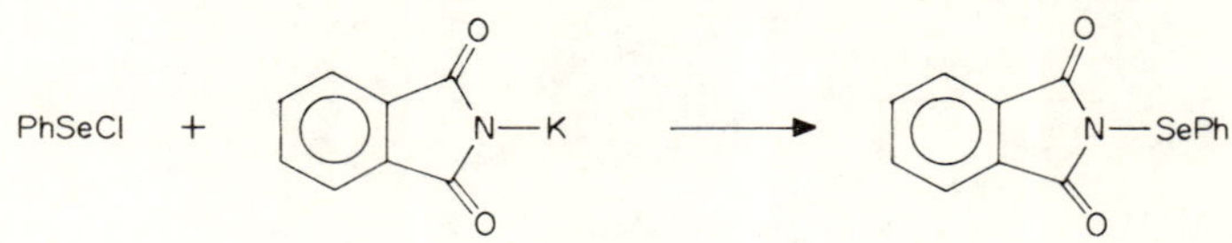

(78)

$$PhSeCl + Me_3SiCN \rightarrow PhSeCN + Me_3SiCl \qquad (79)$$

3. Selenenyl iodides and fluorides

Recently the reaction of diphenyl diselenide with I_2 and 1,5-hexadiene has been reported to give a 68:32 mixture of **15** and **16** in 62% yield (equation 80)[235]. It is assumed that benzeneselenenyl iodide is formed *in situ* and plays an important role as the active reagent.

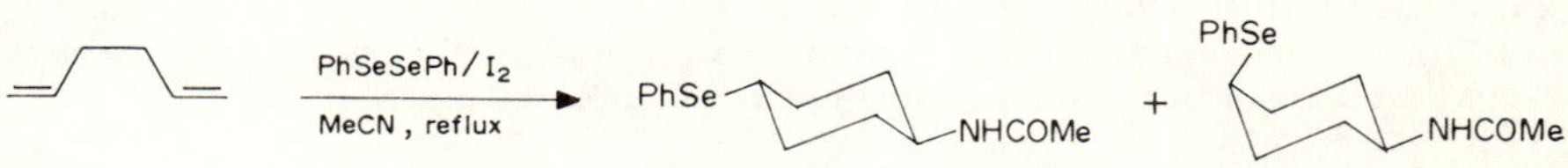

(15) **(16)** (80)

No successful synthesis of selenenyl fluoride has been reported so far.

C. Miscellaneous

1. *Selenenic esters and selenocarboxylates*

a. Synthesis. Selenenic esters (R^1SeOR^2) are usually prepared by a alcoholysis of selenenyl halides or selenocyanides (equations 81 and 82)[151,237–240]. The simple esterification of selenenic acids has not been reported. It is likely that selenenic esters are formed *in situ* in the rearrangement of selenoxides (equations 83 and 84)[119–124,142]. Selenocarboxylates ($R^1SeOCOR^2$) are produced by the reaction of selenenyl halides with some metal carboxylates (equation 85)[114,166].

$$R^1SeBr \xrightarrow{R^2OH} R^1SeOR^2 \qquad (81)$$

$$R^1SeCN \xrightarrow{R^2OH} R^1SeOR^2 \qquad (82)$$

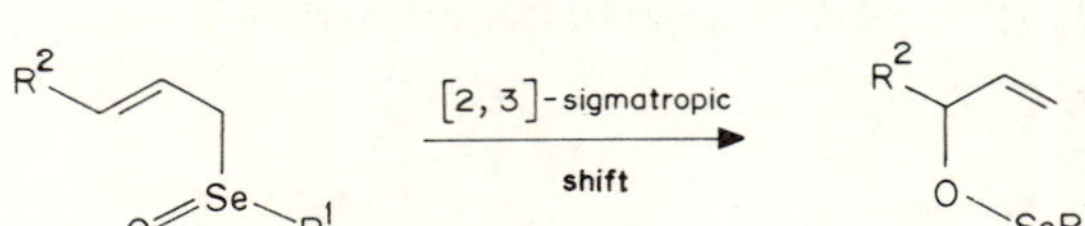

(83)

$$R^1-\overset{\overset{\displaystyle O}{\|}}{Se}-R^2 \xrightarrow{\Delta} R^1SeOR^2 \qquad (84)$$

$$R^1SeX \xrightarrow{\text{KOC(O)}R^2 \text{ or AgOC(O)}R^2} R^1SeOC(O)R^2 \qquad (85)$$

b. Reactions. Selenenic esters and selenocarboxylates are hydrolysed with water to the selenenic acids. The selenenyl group of selenenic esters and selenocarboxylates is electrophilic and can be introduced into organic molecules via oxyselenenylation of various olefins (equation 86)[140,143,166,186].

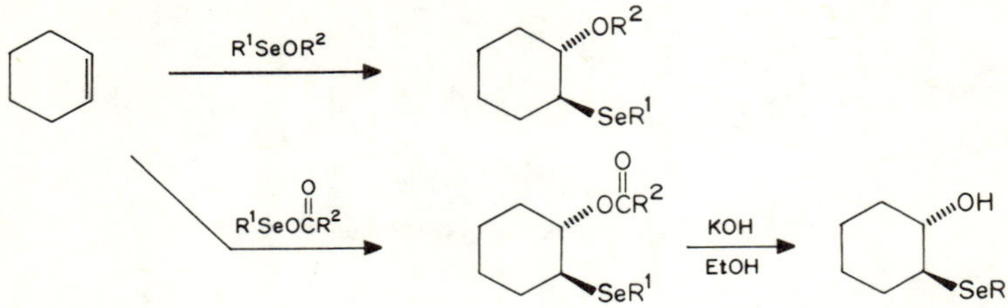

(86)

2. Selenenamides

a. Synthesis. Selenenamides are synthesized by the reaction of selenenyl halides or selenenic acids with amines (equation 87)[232,237,245,248,249]. *N*-Phenylselenosuccinimide and *N*-phenylselenophthalimide are prepared by the reaction of diphenyl diselenide with *N*-chlorosuccinimide or *N*-chlorophthalimide (equations 88 and 89)[216,257,283].

$$R^1SeX \text{ or } R^1SeOH \xrightarrow{R^2_2NH} R^1SeNR^2_2 \qquad (87)$$

$$X = Cl, Br$$

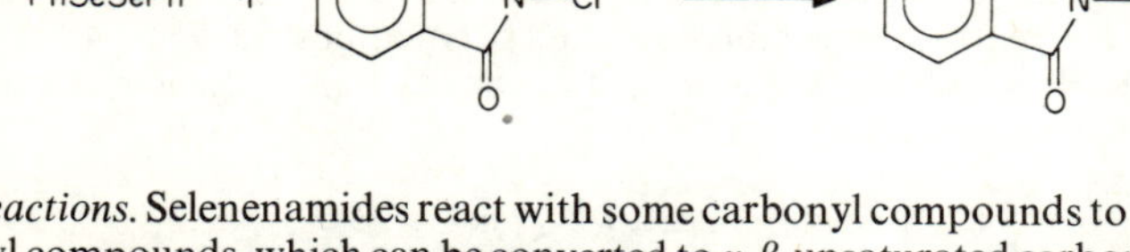

(88)

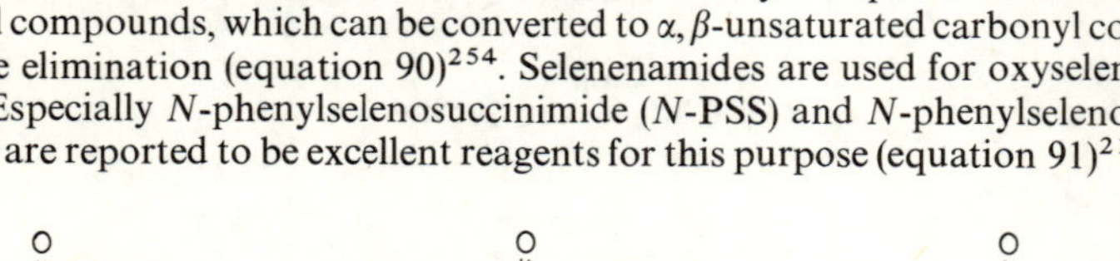

(89)

b. Reactions. Selenenamides react with some carbonyl compounds to afford α-selenenyl carbonyl compounds, which can be converted to α,β-unsaturated carbonyl compounds by oxidative elimination (equation 90)[254]. Selenenamides are used for oxyselenenylation of olefins. Especially *N*-phenylselenosuccinimide (*N*-PSS) and *N*-phenylselenophthalimide (*N*-PSP) are reported to be excellent reagents for this purpose (equation 91)[216,257,283]. *N*-

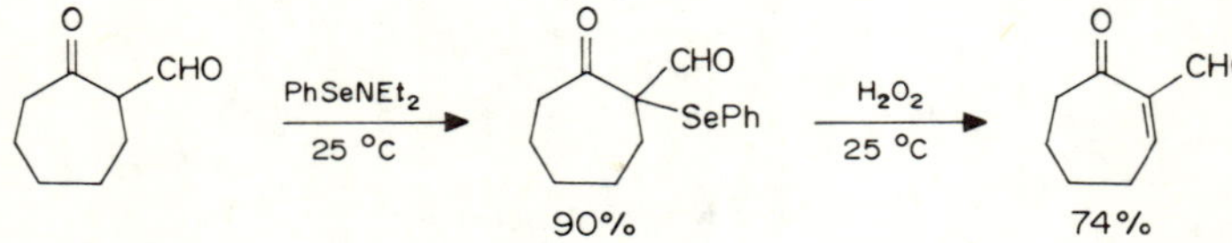

(90)

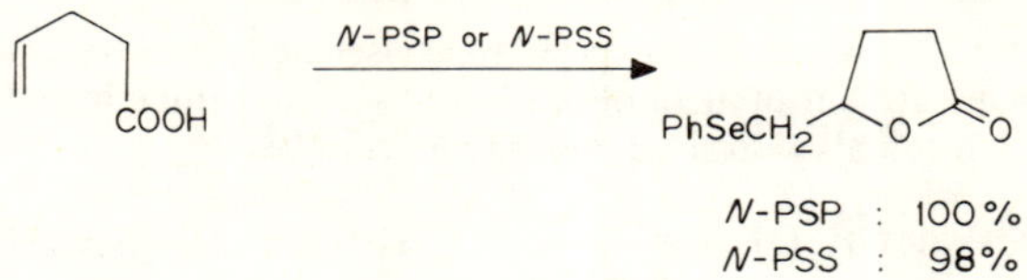

(91)

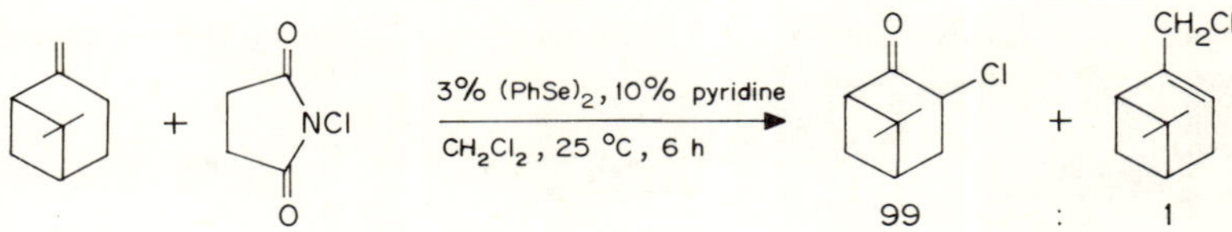

(92)

88%

PSS, formed *in situ* also acts as a catalyst for the allylic chlorination by *N*-chlorosuccinimide (NCS) (equation 92)[257].

3. Selenocyanides

a. Synthesis. Selenocyanides are generally prepared by the reaction of aryl or alkylselenenyl halides (or sulphonates) with cyanide ion (equation 93)[269,270]. Aliphatic selenocyanides can also be produced by the reaction between alkyl halide and $SeCN^-$ (equation 94)[260–268,275,276,295–299]. The use of selenocyanogen as a substitution reagent gives 3-cyanoindole from indole (equation 95)[134,271–274,292,294,300,301].

$$RSeX + CN^- \rightarrow RSeCN + X^- \quad (93)$$
$$X = \text{halogen}, —SO_2R^1$$

$$RX + KSeCN \rightarrow RSeCN + KX \quad (94)$$
$$X = \text{halogen}, —OSO_2R^1, —HgBr$$

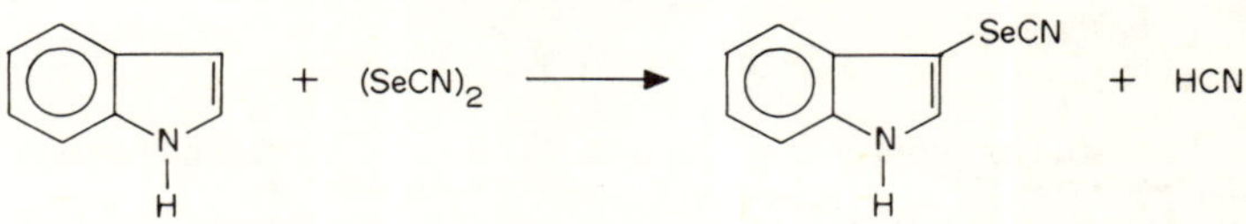

(95)

b. Reactions. Phenyl selenocyanides add to olefinic double bonds in the presence of tetrachlorostannane as catalyst to give 2-phenylselenocyanide (equation 96)[284–287].

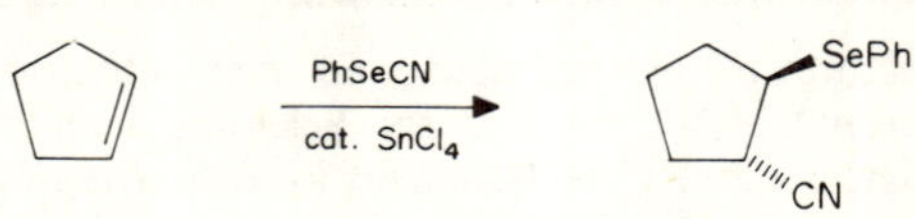

(96)

On the other hand, phenyl selenocyanide causes electrophilic selenenylation in the presence of cupric chloride. An application of this reaction is a facile preparation of cyclic ethers by oxyselenation of diolefins using phenyl selenocyanide. For example, 1,5-cyclooctadiene gives a mixture of cyclic ethers in aqueous tetrahydrofuran solution (equation 97)[296,302,306–308,311].

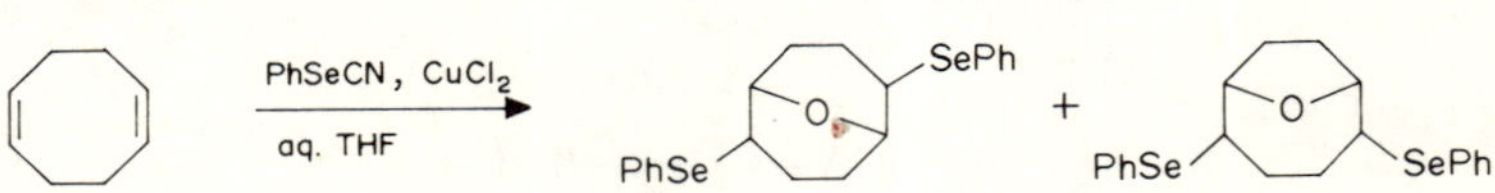

(97)

Hydroxyl groups can be easily converted to arylselenenyl groups using selenocyanides and phosphines. Treatment of alcohols with aryl selenocyanide in the presence of tri-*n*-butylphosphine affords alkyl aryl selenides (equation 98)[312].

$$CH_3(CH_2)_{10}CH_2OH \xrightarrow[Bu_3P,\ THF]{\delta\text{-}NO_2C_6H_4SeCN} CH_3(CH_2)_{10}CH_2SeC_6H_4NO_2\text{-}o \qquad (98)$$

$$94\%$$

In connection with this reaction, a 1,3-rearrangement of primary allylic alcohols can take place as shown in equation (99)[288].

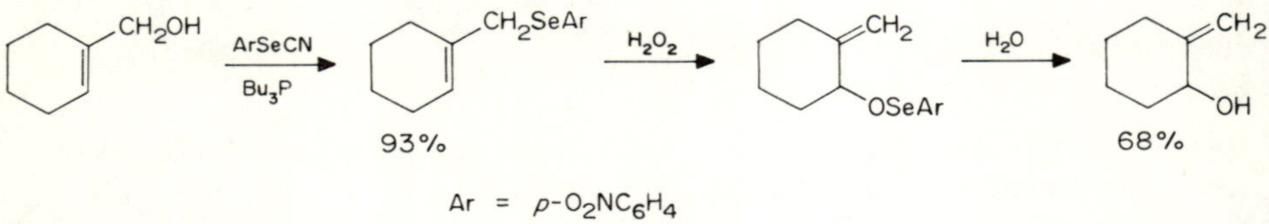

Aryl selenocyanides also react with carboxylic acids and aldehydes, giving rise to seleno esters and cyanoselenides in excellent yields (equations 100 and 101)[289,290].

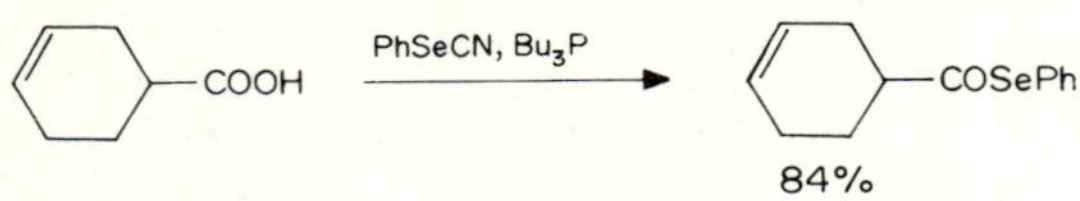
(100)

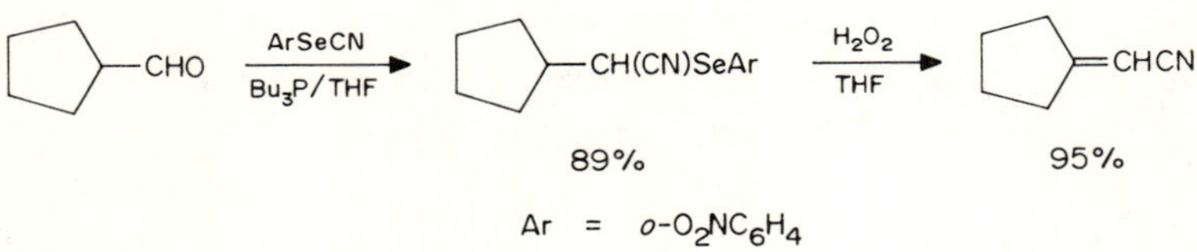
(101)

Some other studies concerning Section III, i.e. on selenenic esters, selenocarboxylates[146,147,241–247], selenenamides[250–253,255,256,258,259] and selenocyanides[153,277–282,288,291,293,303–305,309,310] have been reported.

IV. SELENINIC ACIDS AND THEIR ANHYDRIDES

The generalized structures of seleninic acids and their anhydrides can be described as RSe(O)OH and RSe(O)OSe(O)R, respectively. Seleninic acids are less acidic than the corresponding carboxylic acids RCOOH, and are known to be amphoteric: both base and acid react with seleninic acids to give the corresponding salts (equations 102 and 103). Arylseleninic acids and their anhydrides have been used as versatile and specific oxidizing reagents for various organic compounds.

$$R-\overset{\overset{\displaystyle O}{\|}}{Se}-OH + B \rightarrow R-\overset{\overset{\displaystyle O}{\|}}{Se}-O^- \cdot HB^+ \qquad (102)$$

$$R-\overset{\overset{\displaystyle O}{\|}}{Se}-OH + HA \rightarrow R-\underset{+}{\overset{\overset{\displaystyle OH}{|}}{Se}}-OH \cdot A^- \qquad (103)$$

A. Synthesis

In general, seleninic acids are prepared by oxidation of the corresponding selenols, diselenides and selenocyanides with various oxidizing agents such as HNO_3[264,269,355–357,360], H_2O_2[279,357–359] and AcOOH[279] (equation 104).

$$\left.\begin{array}{l} RSeH \\ (RSe)_2 \\ RSeCN \end{array}\right] \xrightarrow{[o]} RSe(=O)OH \qquad (104)$$

Benzeneseleninic acid is commercially available. Alternative methods for preparation of seleninic acids are shown in equations (105) and (106).

$$RMgBr + SeO_2 \rightarrow RSeO_2MgBr \xrightarrow[H_2O]{H^+} RSeO_2H \qquad \text{(Ref. 1)} \quad (105)$$

$$R = \text{alkyl}$$

$$3\,RSeX + 2\,H_2O \rightarrow RSeO_2H + RSeSeR + 3\,HX \qquad \text{(Refs. 1,269)} \quad (106)$$

$$X = Cl, Br; R = \text{alkyl, aryl}$$

Seleninic anhydrides are synthesized by dehydration of seleninic acids[246,360] or by oxidation of diselenides with oxidizing reagents such as ozone[246,361] (equations 107 and 108). Benzeneseleninic anhydride is commercially available and can be prepared by the oxidation of diphenyl diselenide with nitric acid or ozone[315].

$$2\,RSe(=O)OH \xrightarrow{\Delta} RSe(=O)OSe(=O)R + H_2O \qquad (107)$$

$$RSeSeR \xrightarrow{[O]} RSe(=O)OSe(=O)R \qquad (108)$$

B. Reactions

1. Oxidation of alcohols

Benzeneseleninic anhydride is a mild oxidizing agent for a variety of alcohols, giving high yields of the corresponding carbonyl compounds (equations 109 and 110)[314]. In some cases, the corresponding α,β-unsaturated compounds are obtained by further dehydrogenation (equation 111)[314]. Benzeneseleninic anhydride can be also used to oxidize

$$PhCH_2OH \xrightarrow{(PhSeO)_2O} PhCHO \quad 100\% \qquad (109)$$

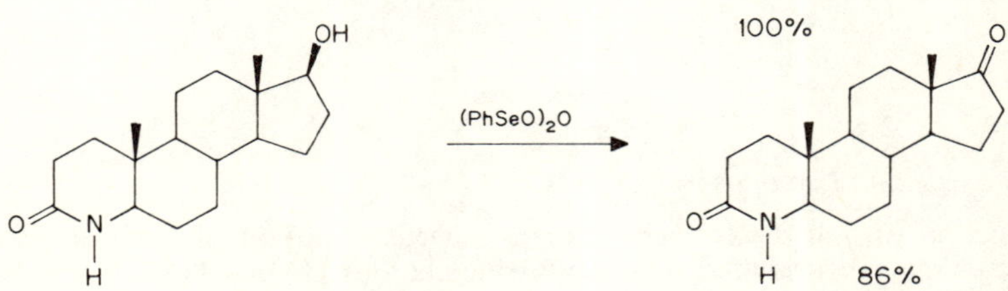

(110)

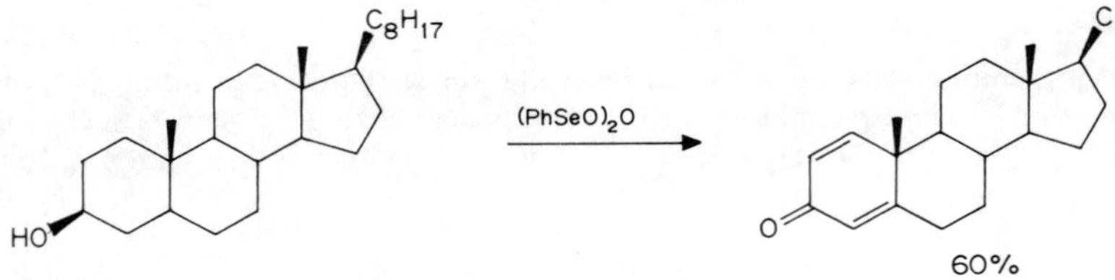

(111)

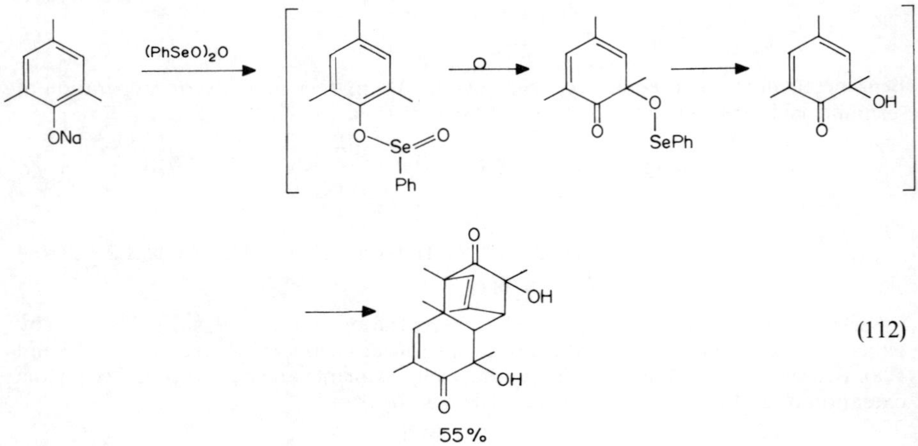

(112)

simple alkylphenols to hydroxydienones (equation 112)[315], and to convert phenols to *o*-quinones[313,316] (equation 113). The oxidation of phenols in the presence of hexamethyldisilazane gives the corresponding phenylselenoimines, which in turn can be readily reduced to aminophenols (equation 114)[317,318].

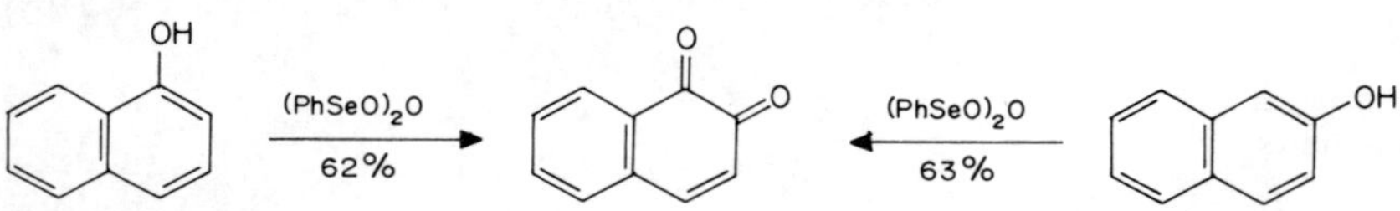

(113)

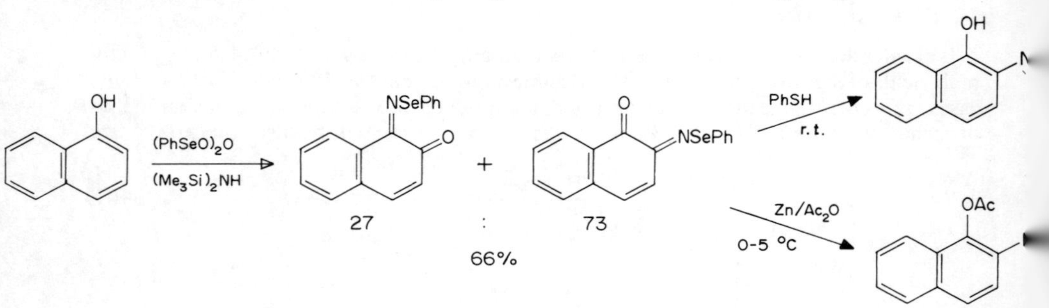

(114)

2. α-Hydroxylation of ketones

Introduction of a hydroxyl group to the α-tertiary carbon of ketones by using benzeneseleninic anhydride has been reported (equations 115 and 116)[323,324,340].

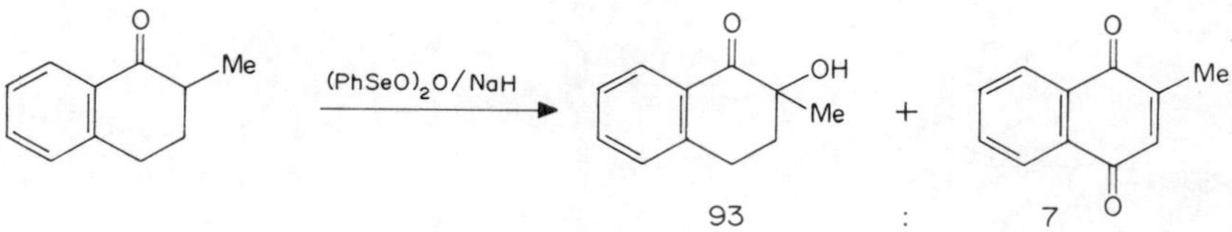

93 : 7

86%

(115)

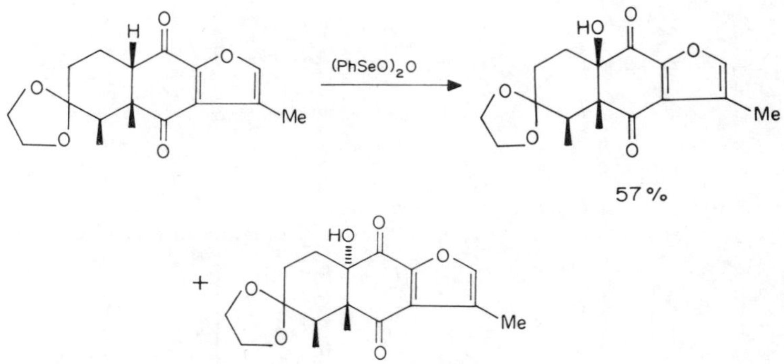

57%

17%

(116)

3. Oxidation of benzylic hydrocarbons

A methyl or methylene group adjacent to an aromatic ring is converted to a carbonyl group by benzeneseleninic anhydride (equations 117 and 118)[319,320].

$$\mathrm{PhCH_2Ph} \xrightarrow{(\mathrm{PhSeO})_2\mathrm{O}} \mathrm{PhC(=O)Ph} \quad (117)$$

90%

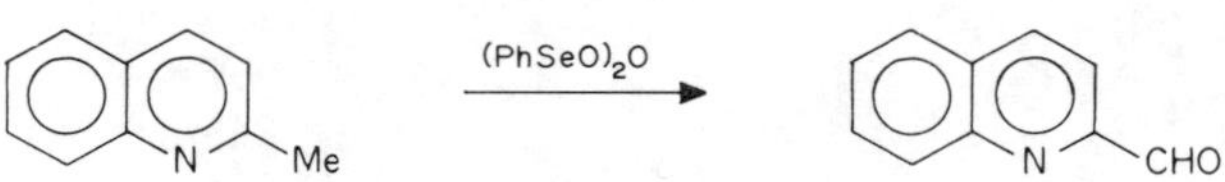

(118)

4. Epoxidation of olefins

Reaction of benzeneseleninic acid with H_2O_2 generates *in situ* benzeneperoxyseleninic acid (equation 119) which epoxidizes olefinic double bonds (equations 120 and 121)[328,329]. Indeed seleninic acids are found to catalyse epoxidations with H_2O_2 (equation 122)[330].

$$\mathrm{Ph-Se(=O)-OH + H_2O_2 \rightleftharpoons Ph-Se(=O)-OOH + H_2O} \quad (119)$$

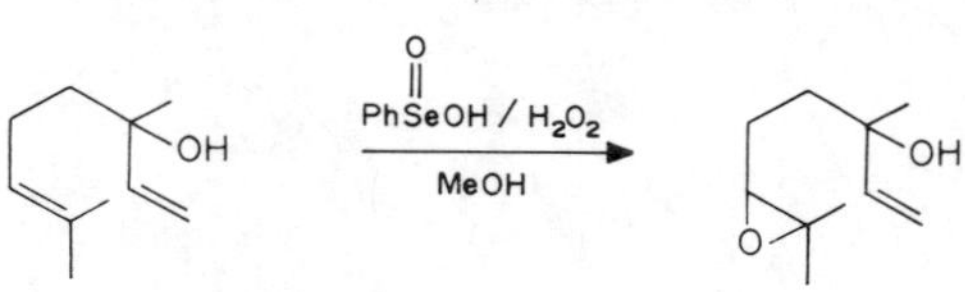

63%

(120)

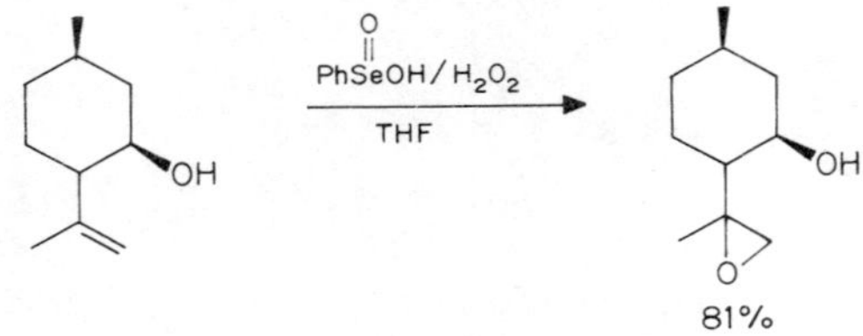

(121)

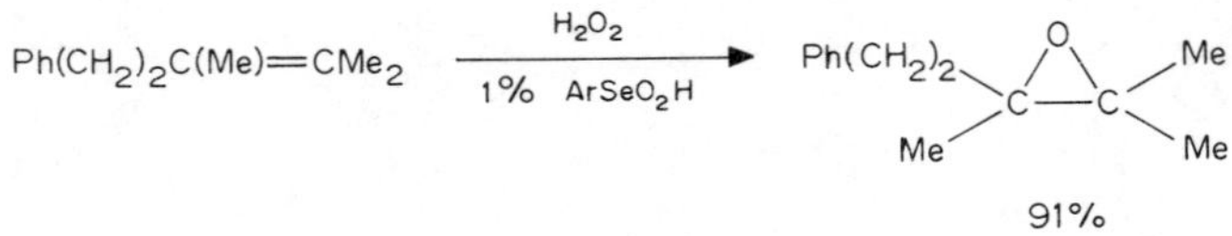

(122)

The *t*-butyl ester of benzeneperoxyseleninic acid is synthesized by reacting several seleninic acid derivatives with *t*-butyl hydroperoxide or its sodium salt (equation 123)[331].

$$
\left.
\begin{array}{l}
(PhSeO)_2O + t\text{-}BuOOH \\
PhSe(O)OEt + t\text{-}BuOOH \\
PhSe(O)Br + t\text{-}BuOONa
\end{array}
\right\} \longrightarrow PhSe(O)OOBu\text{-}t \qquad (123)
$$

5. *Dehydrogenation*

A number of steroidal ketones are dehydrogenated to α,β-unsaturated ketones in excellent yields under mild conditions using benzeneseleninic anhydride. When the

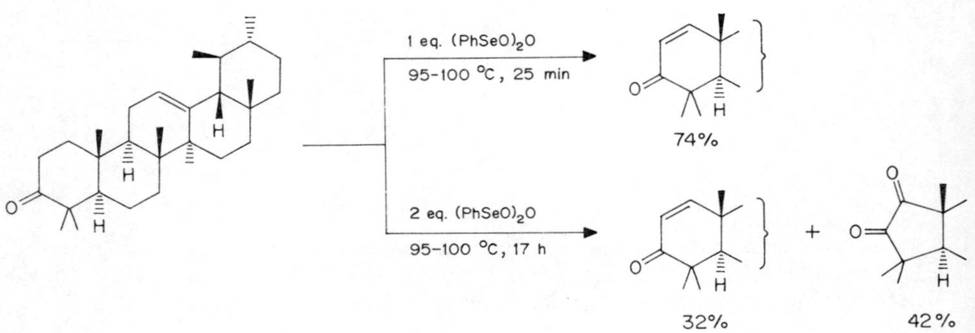

(124)

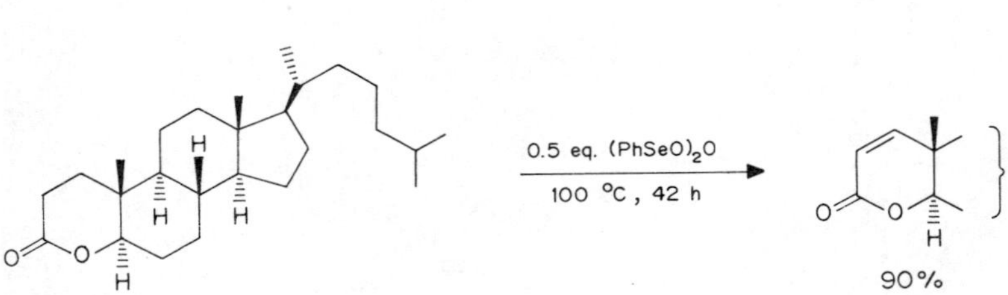

(125)

reaction is conducted with excess anhydride for a prolonged reaction time, ring-*A*-contracted diketones are produced in moderate yields (equation 124)[325,339]. Similarly, δ-lactones[338] and lactams[345] undergo smooth dehydrogenation (equation 125). Transformation of indolines to indoles by dehydrogenation with benzene seleninic anhydride is also possible (equation 126)[337].

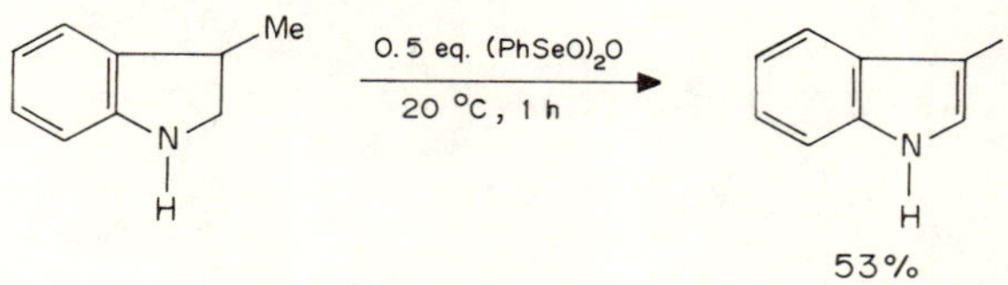

(126)

6. Oxidation of nitrogen compounds

Primary amines are oxidized in high yields to carbonyl compounds through the corresponding imines by benzeneseleninic anhydride (equation 127)[326,327]. Hydroxylamines can be converted into nitroso compounds using benzeneseleninic anhydride (equation 128)[347]. In the case of *N*-arylhydroxamic acids, rearrangement to *p*-hydroxybenzanilides has been reported (equation 129)[332].

$$Ph_2CHNH_2 \xrightarrow{(PhSeO)_2O} [Ph_2CH{-}NH{-}\overset{\overset{O}{\|}}{S}ePh] \xrightarrow[-PhSeOH]{} Ph_2C{=}NH \xrightarrow{H_2O} Ph_2C{=}O \qquad (127)$$

$$Me_3CNHOH \xrightarrow[1\ min]{1\,eq.(PhSeO)_2O} Me_3CN{=}O \quad 96\% \qquad (128)$$

$$Ph\overset{\overset{O}{\|}}{C}{-}\overset{\overset{OH}{|}}{N}Ph \xrightarrow[r.t.]{PhSeO_2H} Ph\overset{\overset{O}{\|}}{C}NHC_6H_4OH\text{-}p \quad 70\% \qquad (129)$$

Back has reported the reaction of a variety of hydrazines with benzeneseleninic acid and anhydride, in which aryl, acyl or sulphonhydrazides are converted to phenyl selenides as shown in equations (130–132). Hydrazones from aldehydes and hydrazo compounds can

$$p\text{-}O_2NC_6H_4NHNH_2 \xrightarrow[or\ PhSeO_2H]{(PhSeO)_2O} \underset{60\%}{p\text{-}O_2NC_6H_4SePh} + \underset{20\%}{PhNO_2} \qquad \text{(Ref. 336) (130)}$$

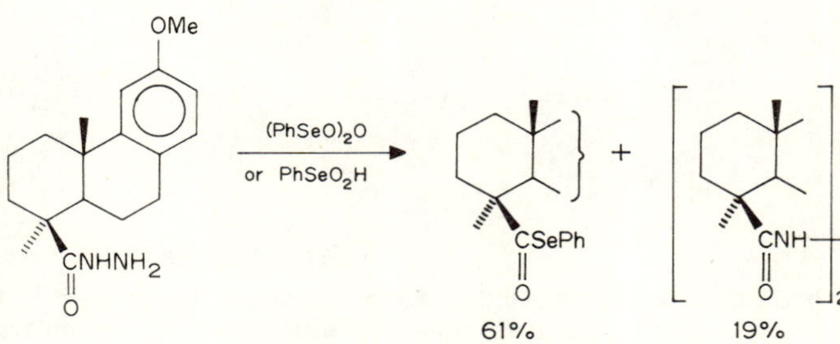

(Refs. 138, 139, 336)

(131)

$$PhSO_2NHNH_2 \xrightarrow[or\ PhSeO_2H]{(PhSeO)_2O} \underset{88\%}{PhSO_2SePh} \qquad \text{(Ref. 333) (132)}$$

be readily oxidized with benzeneseleninic anhydride to afford high yields of azo compounds (equation 133)[347].

$$Ph{-}CH{=}N{-}NHPh \xrightarrow{(PhSeO)_2O} \underset{67\%}{Ph{-}\overset{\displaystyle O}{\overset{\|}{C}}{-}N{=}N{-}Ph} \tag{133}$$

Enamides react with benzeneseleninic anhydride to give Pummerer-type products (equation 134)[344,354].

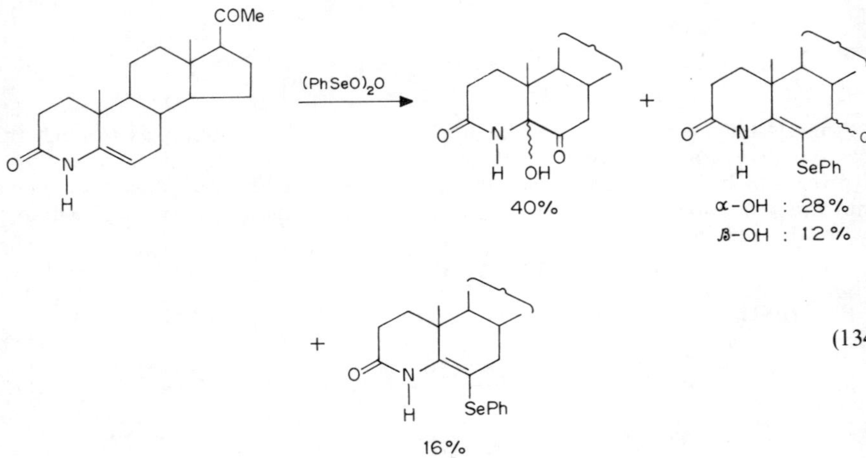

(134)

7. *Reaction with sulphur and phosphorus compounds*

Seleninic acids react with thiols[349,350], sulphides[351], sulphinic acids[352] and phosphines[351] to give the corresponding oxidized products (equation 135–138). The kinetics of these reactions have been studied in detail[349–352].

$$RSH \xrightarrow{ArSeO_2H} RSSR \tag{135}$$

$$RSR' \xrightarrow{ArSeO_2H} R\overset{\displaystyle O}{\overset{\|}{S}}R' \tag{136}$$

$$Ar'SO_2H \xrightarrow{ArSeO_2H} ArSe{-}\underset{\underset{\displaystyle O}{\|}}{\overset{\overset{\displaystyle O}{\|}}{S}}Ar' + Ar'SO_3{}^{-}ArSeO_2H_2{}^{+} \tag{137}$$

$$Ph_3P \xrightarrow{ArSeO_2H} Ph_3P{=}O \tag{138}$$

8. *Deprotection*

Benzeneseleninic anhydride smoothly regenerates parent ketones and aldehydes from their thioacetals in high yields under mild conditions (equation 139)[341–343]. Similarly ketone hydrazones, oximes and semicarbazones are converted to the parent carbonyl compounds by treatment with benzeneseleninic anhydride (equation 140)[346,348]. Conversion of thiocarbonyl compounds to their corresponding oxo derivatives has been performed successfully by using benzeneseleninic anhydrides (equation 141)[321,322].

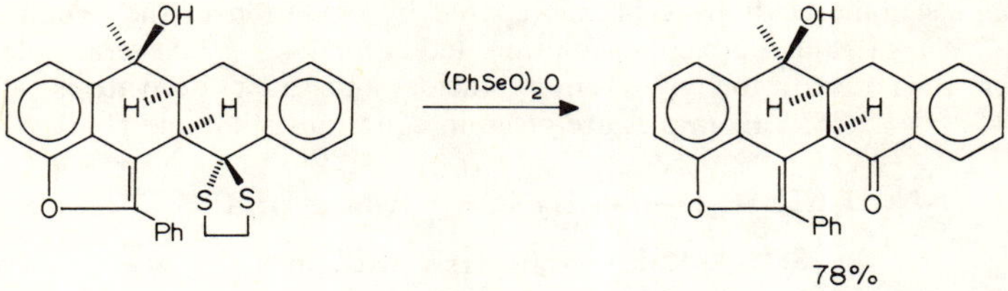

(139)

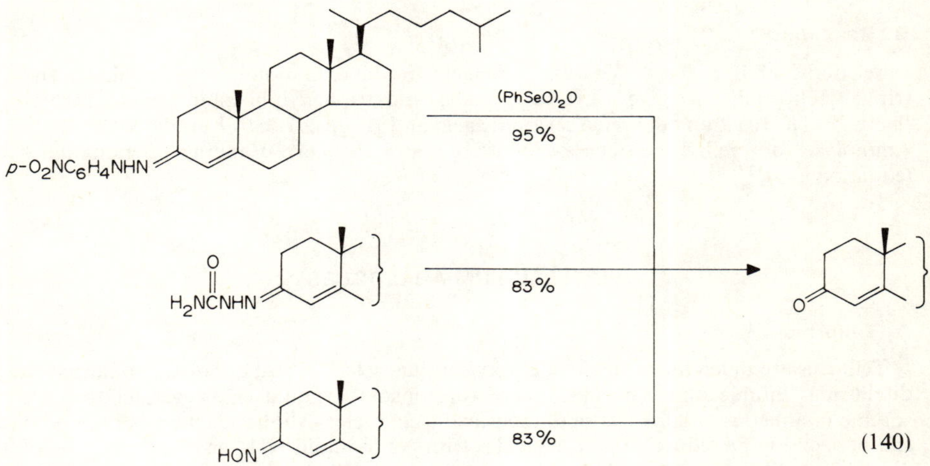

(140)

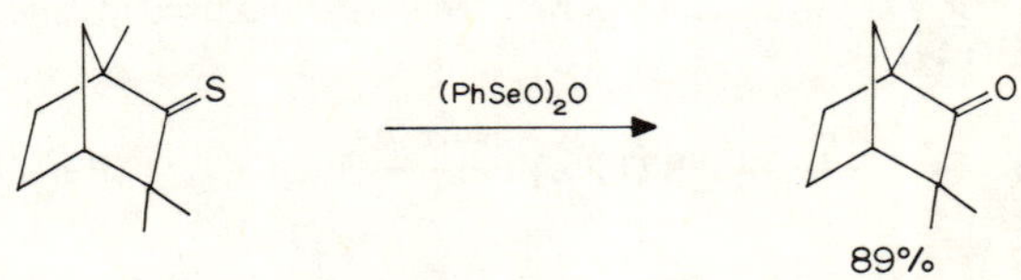

(141)

Some other studies, concerning Section IV, i.e. on seleninic acids and their anhydrides have been reported[281,334,335,353].

V. SELENONIC ACIDS AND THEIR DERIVATIVES

A. Synthesis

Little is known about the chemistry of selenonic acids ($RSeO_3H$) and their derivatives compared with the chemistry of selenenic and seleninic compounds, because of their lower thermal stability.

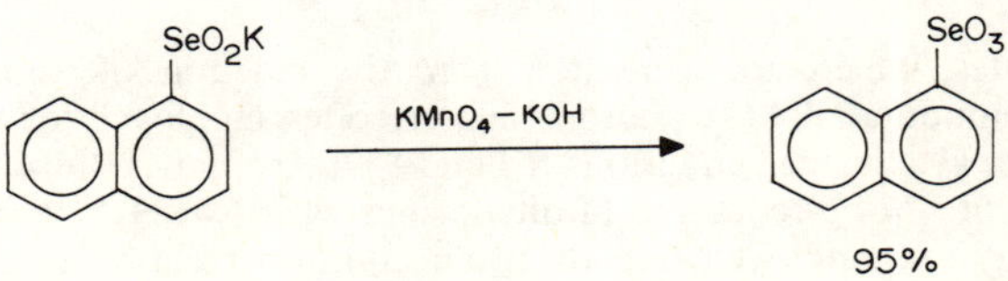

(Ref. 370) (142)

In general, selenonic acids have been prepared by oxidation of the seleninic acids with $KMnO_4$ in alkaline aqueous solution (equation 142)[370]. Several alternative methods for the preparation of selenonic acids and their derivatives are also known[1,141,262,358,362-370]. Examples are given in equations (143) and (144).

$$(H_2NCH_2CH_2Se)_2 \xrightarrow{Cl_2,H_2O} H_2NCH_2CH_2SeO_3H \quad \text{(Ref. 262)} \qquad (143)$$

$$SeO_3 + RSH \rightarrow RSSeO_3H \quad \text{(Ref. 363)} \qquad (144)$$

B. Reactions[362,351,371,372]

Selenonic acids are used as oxidizing reagents in the oxidation of alkyl sulphides and triphenylphosphine to give alkyl sulphoxides and triphenylphosphine oxide, respectively[351]. The reaction between a selenonic acid and HI gives rise to I_2 and diselenide[262]. Aminolysis of methyl benzeneselenonate provides the corresponding selenonamides (equation 145)[371].

$$PhSeO_3Me + HNR^1R^2 \rightarrow PhSeO_2NR^1R^2 \qquad (145)$$

VI. TELLURIUM ANALOGUES

A. Tellurols

Tellurols are much more sensitive to oxygen than selenols, and are readily oxidized to ditellurides. Simple aliphatic tellurides are water-insoluble liquids. However, tellurols are soluble in aqueous alkali because of their acidic character. Aliphatic tellurols have been synthesized by the reduction of ditellurides with Na in liquid NH_3 or by the reaction of aluminium telluride with alcohols (equations 146 and 147). Aromatic tellurols cannot be isolated in a pure form. They are usually generated *in situ* and used for successive reactions (equations 148 and 149).

$$R_2Te_2 \xrightarrow{Na/liq.NH_3} [RTeNa] \xrightarrow{dil.H_2SO_4} RTeH \quad \text{(Ref. 381)} \qquad (146)$$

$$Al_2Te_3 + ROH \xrightarrow[\text{under } H_2]{240-350\,^\circ C} RTeH \quad \text{(Refs. 382–384)} \qquad (147)$$

$$(RTe)_2 + R^1MgX \longrightarrow RTeMgX + RTeR^1 \quad \text{(Ref. 385)} \qquad (148)$$

$$RTeMgX \xrightarrow{H_2O} RTeH + MgX(OH)$$

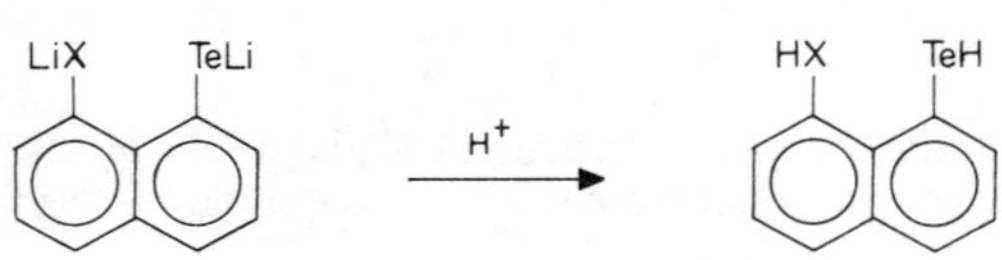

(Refs. 386, 387) (149)

X = S, Se, Te

Metal tellurolates, which are generated from the reaction of elemental Te with organometallic compounds RM (equation 150), are often employed for the synthesis of asymmetric diorganyl tellurides (equations 151 and 152). Aromatic tellurols or tellurolates are also prepared *in situ* by reduction of ditellurides with $NaBH_4$[421,422]. The tellurols formed in this way are employed directly in an addition reaction to alkynes or in a substitution reaction with alkyl halides (equation 153)[421–425].

$$\mathrm{RM + Te \rightarrow RTeM} \qquad (150)$$

$$\mathrm{M = MgX^{388-390},\ Li^{390-393},\ Na^{394-396}}$$

$$\mathrm{PhLi + Te \rightarrow PhTeLi \xrightarrow{RX} PhTeR} \qquad \text{(Refs. 40, 397–402)} \qquad (151)$$

$$\mathrm{ArTeNa + R^1C{\equiv}C{-}\overset{\displaystyle O}{\overset{\|}{C}}{-}R^2 \xrightarrow[\text{r.t.}]{\text{EtOH}} ArTe\overset{\displaystyle R^1}{\overset{|}{C}}{=}CH{-}\overset{\displaystyle O}{\overset{\|}{C}}R^2} \qquad \text{(Refs. 403, 404)} \qquad (152)$$

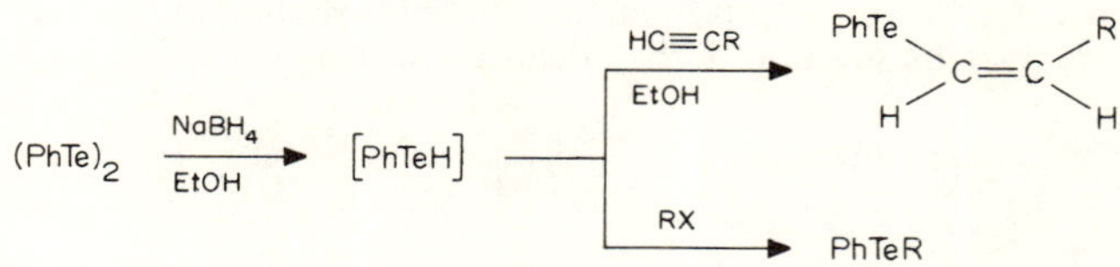

(153)

Recently, biologically important compounds[405–414] such as telluro steroids[405–410] and telluroamino acids[411] and superconductive materials such as tetratellurofulvalenes[415–417] have been synthesized from tellurols or tellurolates which were prepared by these methods (equations 154–156).

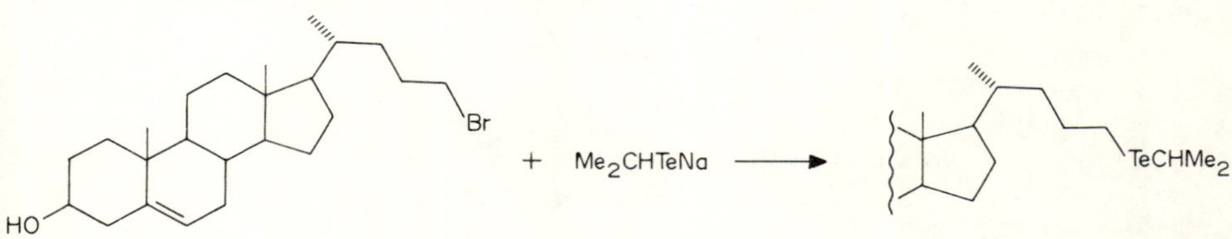

(154)

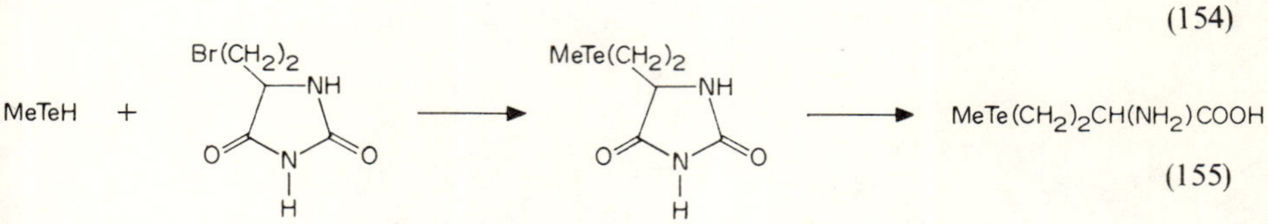

(155)

(156)

Lithium and sodium tellurolates are used for the reductive dehalogenation of α-halocarbonyl compounds (equation 157)[418–420].

$$\mathrm{RCOCH_2X + ArTe^- \rightarrow [RCOCH_2TeAr] \xrightarrow{ArTe^-/H^+} RCOMe + (ArTe)_2} \qquad (157)$$

B. Tellurenyl Compounds

In general tellurenyl derivatives are unstable. Therefore they have been isolated as complexes with donor molecules such as thioureas or selenoureas. An alkyltellurenyl chloride–thiourea complex has been prepared from tellurinyl trichloride and thiourea[426–429] (equation 158).

$$\mathrm{RTeCl_3 + 3\,S{=}C(NH_2)_2 \rightarrow RTeCl{\cdot}S{=}C(NH_2)_2} \qquad (158)$$

The isolation of aryltellurenyl halides having an electron-withdrawing group in the position *ortho* to the Te atom has been successfully achieved (equation 159)[430–433]

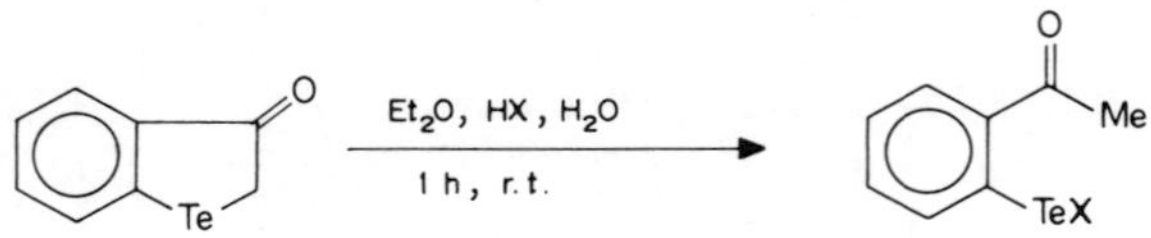

X = Cl, Br, I, CN (159)

Vicentini and coworkers have reported that 2-naphthyltellurenyl iodide **17** can be isolated in a pure form by reaction of the ditelluride with I_2 (equation 160)[158].

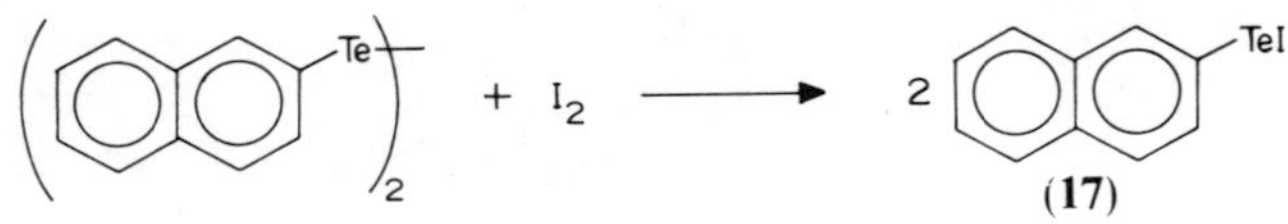

(**17**) (160)

Similarly, several aryltellurenyl halides (ArTeX: Ar = Ph, 4-$MeOC_6H_4$, 3,4-$(MeO)_2C_6H_3$, 4-PhC_6H_4, 2-PhC_6H_4; X = Br, I) have been isolated from the reaction of diaryl ditellurides with halogens[435,436]. 2-Naphthyltellurenyl iodide undergoes various types of reactions as shown in equation (161).

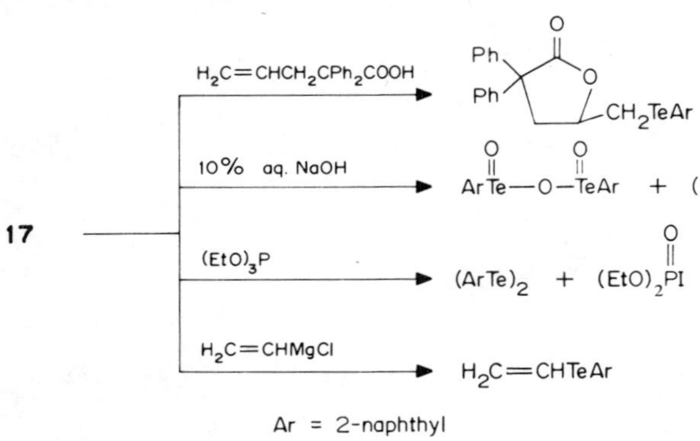

(Ref. 208)

(Ref. 158)

(161)

(Ref. 434)

(Refs. 437, 438)

Several tellurocyanides have been prepared as shown in equations (162) and (163).

$$(PhTe)_2 \xrightarrow[\text{C}_6\text{H}_6-\text{EtOH}]{\text{NaBH}_4} PhTeNa \xrightarrow[\text{EtOH}]{\text{BrCN}} PhTeCN \quad \text{(Ref. 439)} \qquad (162)$$

$$KCN + Te \xrightarrow{\text{Me}_2\text{SO}} KTeCN \xrightarrow{\text{PhCH}_2\text{Cl}} PhCH_2TeCN \quad \text{(Ref 440)} \qquad (163)$$

Benzyl tellurocyanide is reactive towards both oxidizing and reducing reagents (equation 164)[441–444].

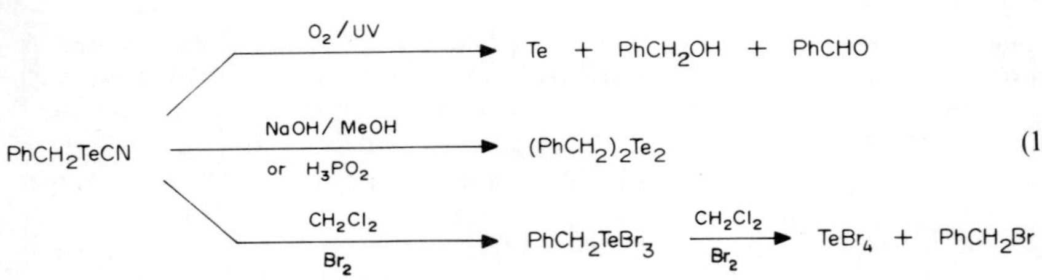

(164)

C. Tellurinyl Compounds

Tellurinyl compounds are classified into four groups ($RTeX_3$, R_2TeX_2, R_3TeX, and RTe(O)X) with respect to their structures.

1. $RTeX_3$ (X = halogen, OH, etc.)

Organotellurinyl trihalides have been prepared by the four methods given in equations (165)–(168).

(*i*) *Reaction of ditellurides with halogens*

$$R_2Te_2 + X_2 \rightarrow RTeX_3 \qquad \text{(Refs. 158, 454, 455)} \qquad (165)$$

(*ii*) *Addition to olefins*

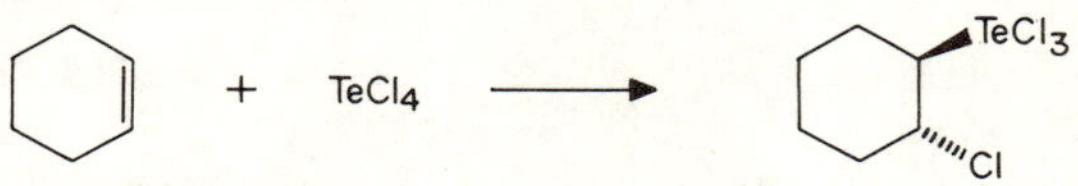

(Refs. 445–451) (166)

(*iii*) *Aromatic substitution*

(Refs. 452, 453) (167)

(*iv*) *Reaction of tellurium tetrachloride with active methylene groups*

$$R_3C\overset{\overset{O}{\|}}{C}CH_2\overset{\overset{O}{\|}}{C}CR_3 + TeCl_4 \rightarrow R_3C\overset{\overset{O}{\|}}{C}\underset{\underset{TeCl_3}{|}}{C}H\overset{\overset{O}{\|}}{C}CR_3 \qquad \text{(Refs. 454, 456, 457)} \qquad (168)$$

vic-Dihalides can be derived from the adducts of $TeCl_4$ to olefins as shown in equations (169) and (170). Combined application of the addition reaction of $TeCl_4$ to olefins and a

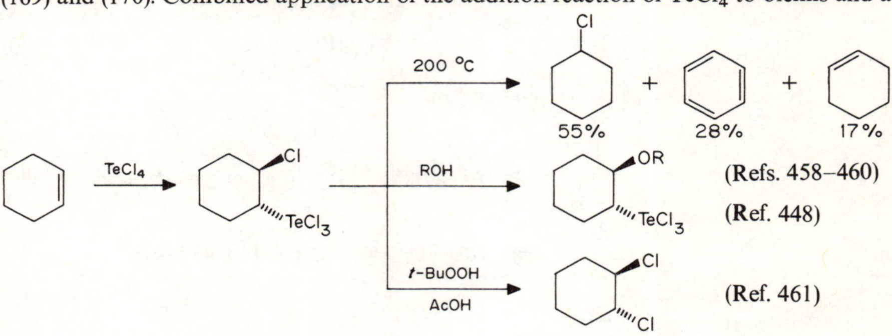

(Refs. 458–460)

(Ref. 448)

(Ref. 461)

(169)

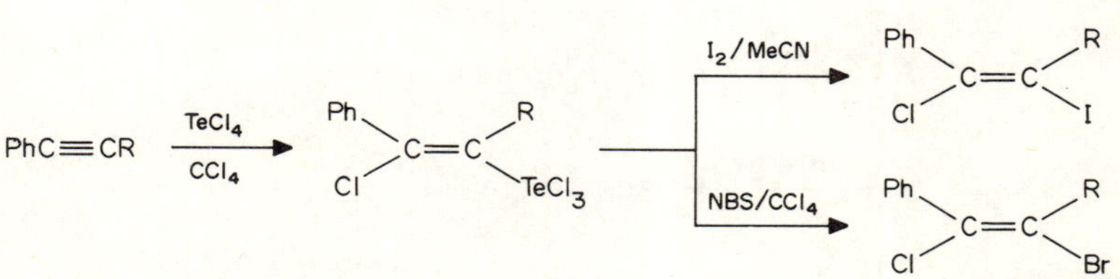

(Ref. 451) (170)

subsequent reduction of the adducts with Na_2S provides a new method for the conversion of *trans*-olefins into *cis* isomers (equation 171)[449].

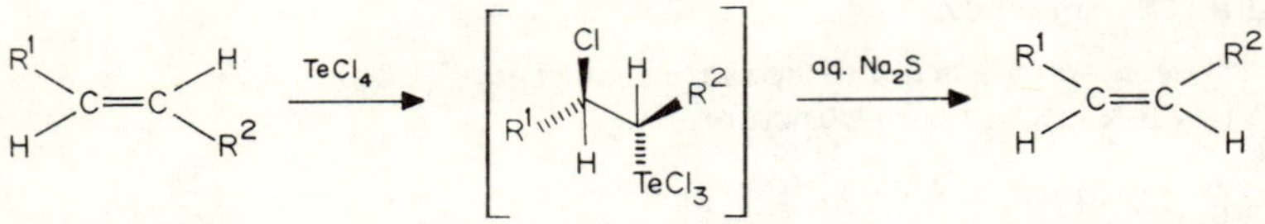 (171)

The C—C bond formations shown in equations (172)–(174) have been successfully achieved with the aid of transition metal derivatives.

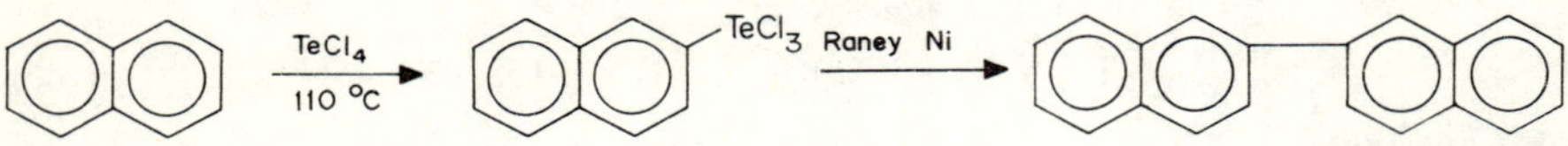

(Ref. 462) (172)

$$ArTeCl_3 \xrightarrow[2.\ H_2O/H^+]{1.\ Ni(CO)_4} ArCOOH$$ (Ref. 463) (173)

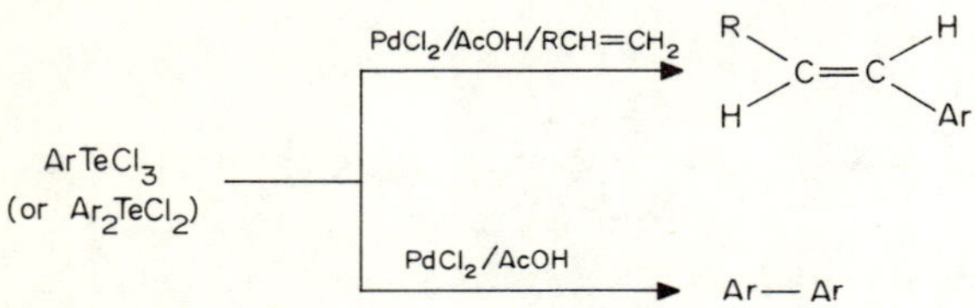

(Refs. 464, 465) (174)

2. R_2TeX_2

Several methods for the preparation of R_2TeX_2 have been reported (equations 175–178).

$$Te \xrightarrow{RI \text{ or } RN_2{}^+Cl^- \text{ or } R_2I^+Cl^-} R_2TeX_2$$ (Refs. 466–470) (175)

$X = Cl, I$

$$RTeR \xrightarrow{X_2 \text{ or } SO_2X_2} R_2TeX_2$$ (Refs. 471–473) (176)

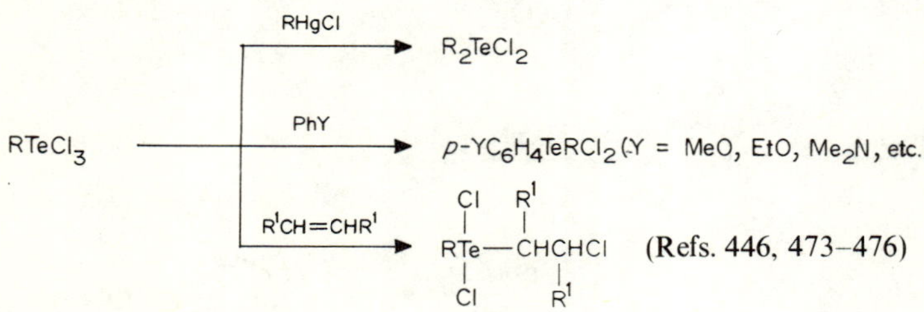

(Refs. 446, 473–476) (177)

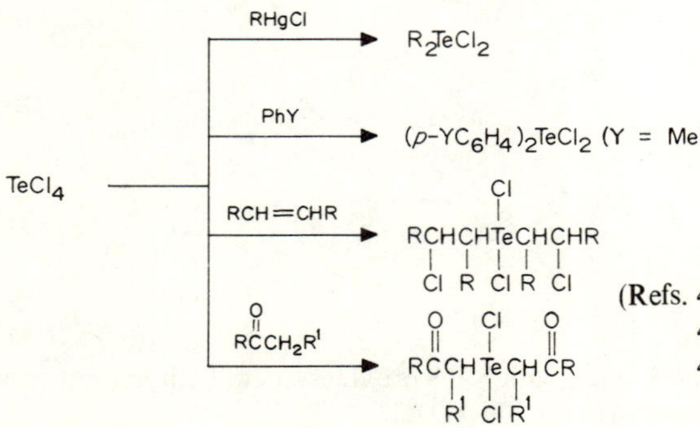

(Refs. 445, 452, 455, 456, 458, 460, 466, 474–477) (178)

Diaryltellurium dichlorides are reduced with Raney Ni to give aromatic coupling products (equation 179)[462,463,478,479]

$$(p\text{-}XC_6H_4)_2TeCl_2 \xrightarrow{\text{Raney Ni}} p\text{-}XC_6H_4{-}C_6H_4X\text{-}p \qquad (179)$$

Recently, Uemura and Fukuzawa have reported that olefins, allylic alcohols and/or allylic ethers are obtained in high yields by treatment of *s*-alkyl(phenyl)tellurium dibromides with aqueous NaOH (equations 180–182)[480].

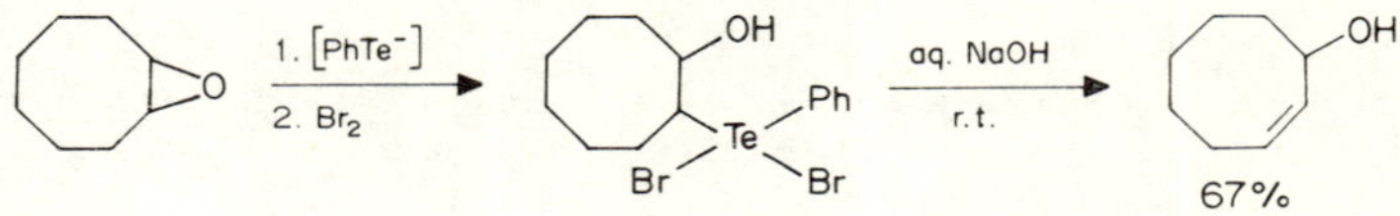
(180)

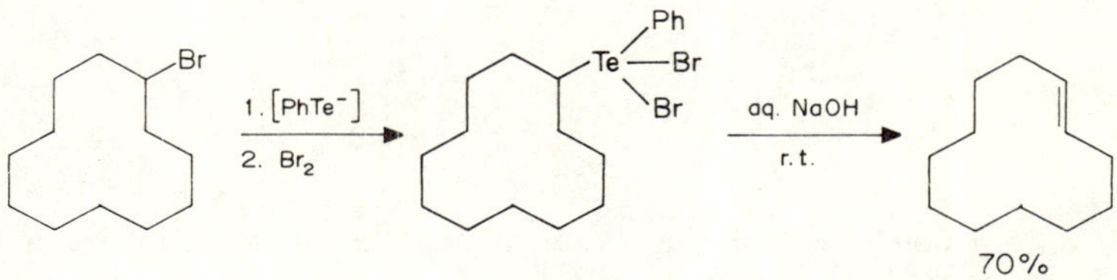
(181)

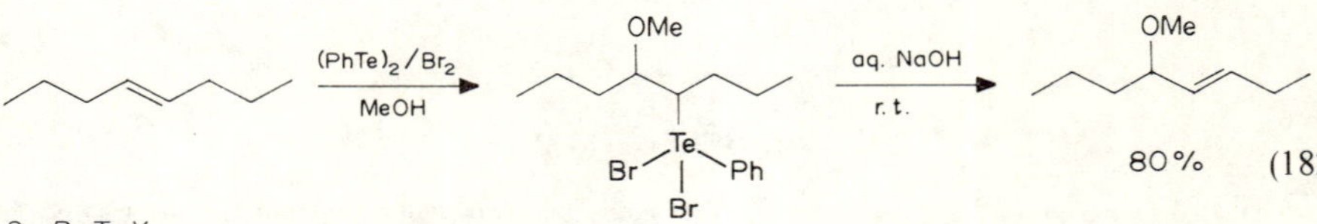
(182)

3. R_3TeX

Triorganyltellurium compounds have been synthesized by the methods shown in equation (183).

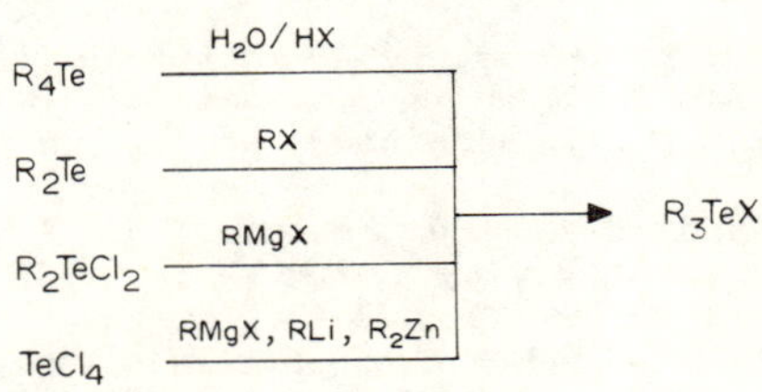
(Refs. 481, 482)
(Refs. 391, 392, 483–485)
(183)
(Refs. 481, 486)
(Refs. 481, 487)

4. RTe(O)X

Several synthetic routes for tellurinic acids (RTe(O)OH) and their derivatives (RTe(O)X; X = halogen, NO_3, OTe(O)R) have been reported, and examples are given in equations (184)–(187).

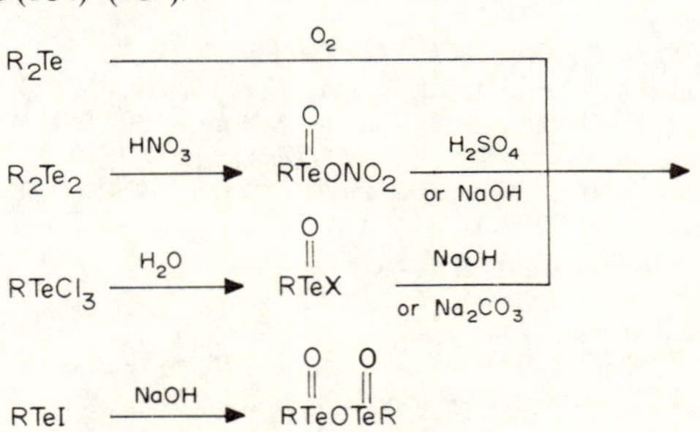
(Refs. 488, 489) (184)
(Ref. 490) (185)
(Refs. 392, 491) (186)
(Ref. 158) (187)

D. Telluronyl Compounds

Little is known about telluronyl compounds, probably because of their instability. Lee and Cava have reported that alkyl aryl telluroxide is oxidized by $NaIO_4$ to the corresponding telluron (or its hydrate) which is then pyrolysed to give a mixture of the olefin, the alcohol and the coupling product. Pyrolysis of the telluroxide gives only the olefinic product (equation 188)[492].

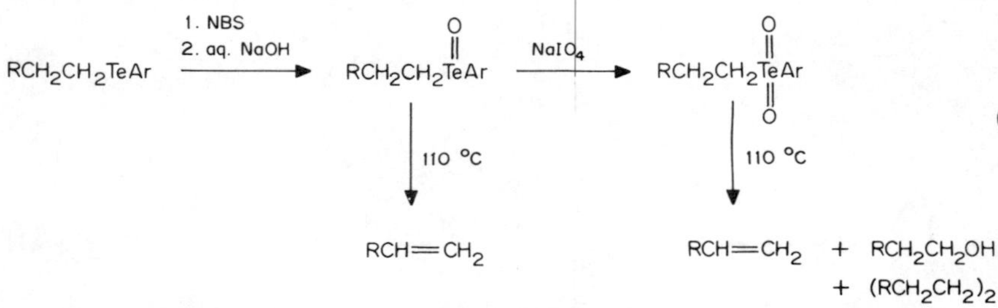

(188)

More details on organotellurium chemistry are summarized in several of Irgolic's books[373,374] and reviews[375–380].

VII. REFERENCES

1. D. L. Klayman and W. H. H. Günther (Eds.), *Organic Selenium Compounds: Their Chemistry and Biology*, John Wiley and Sons, New York, 1973.
2. H. J. Reich and M. L. Cohen, *J. Org. Chem.*, **44**, 3148 (1979).
3. A. J. Bard (Ed.), *Encyclopedia of Electrochemistry of the Elements*, Vol. IV, Marcel Dekker, New York, 1975, pp. 280, 363.
4. C. L. Liotta, E. M. Perdue and H. P. Hopkins, Jr., *J. Amer. Chem. Soc.*, **96**, 7981 (1974).
5. F. Fagioli, F. Pulidori, C. Bighi and A. De Battisti, *Gazz. Chim. Ital.*, **104**, 639 (1974).
6. M. Tiecco, L. Testaferri, M. Tingoli, D. Chianelli and M. Montanucci, *Synth, Commun.*, **13**, 617 (1983).
7. K. Kondo, M. Takarada, S. Murai and N. Sonoda, *Synthesis*, 597 (1979).
8. D. Liotta, G. Zima, C. Barnum and M. Saindane, *Tetrahedron Letters*, **21**, 3643 (1980).
9. S. I. Radchenko and A. A. Petrov, *Zh. Org. Khim.*, **13**, 40 (1977).
10. D. Liotta, W. Markiewicz and H. Santiesteban, *Tetrahedron Letters*, 4365 (1977).
11. K. B. Sharpless and R. F. Rauer, *J. Amer. Chem. Soc.*, **95**, 2697 (1973).
12. R. A. Rossi and A. B. Peñéñory, *J. Org. Chem.*, **46**, 4580 (1981).
13. W. H. H. Günther, *J. Org. Chem.*, **31**, 1202 (1966).
14. M. Schmidt and H. D. Block, *Chem. Ber.*, **103**, 3348 (1970).
15. G. Polidoro, P. De Cicco, S. Di Luzio and G. Federici, *Boll. Soc. Ital. Biol. Sper.*, **55**, 2346 (1979).
16. J. V. Comasseto, J. T. B. Ferreira, C. A. Brandt and N. Petragnani, *J. Chem. Res., Synop.*, 212 (1982).
17. G. Wagner and G. Lehmann, *Pharm. Zentralhalle,* **100**, 160 (1961).
18. N. Esaki, H. Tanaka, S. Uemura, T. Suzuki and K. Soda, *Biochemistry*, **18**, 407 (1979).
19. N. Miyoshi, H. Ishii, K. Kondo, S. Murai and N. Sonoda, *Synthesis*, 300 (1979).
20. J. E. Drake and R. T. Hemmings, *J. Chem. Soc., Dalton Trans.*, 1730 (1976).
21. D. Liotta, P. B. Paty, J. Johnston and G. Zima, *Tetrahedron Letters*, 5091 (1978).
22. M. R. Detty, *Tetrahedron Letters*, 5087 (1978).
23. A. Shafiee, I. Lalezari and F. Savabi, *Synthesis*, 765 (1977).
24. A. Shafiee, I. Lalezari and F. Savabi, *Synthesis*, 764 (1977).
25. F. Malek-Yazdi and M. Yalpani, *J. Org. Chem.*, **41**, 729 (1976).
26. V. Z. Laishev, M. L. Petrov and A. A. Petrov, *Zh. Org. Khim.*, **18**, 514 (1982).
27. Q. M. Tran, L. Christiaens and M. Renson, *Bull. Soc. Chim. Fr.*, 2244 (1974).

28. S. Gronowitz and T. Frejd, *Acta Chem. Scand.*, **B30**, 313 (1976).
29. B. J. McKinnon, P. De Mayo, N. C. Payne and B. Ruge, *Nouv. J. Chim.*, **2**, 91 (1978).
30. J.-L. Fourrey, G. Henry and P. Jouin, *Tetrahedron Letters*, **21**, 455 (1980).
31. L. R. M. Pitombo, E. Flumignan, and Y. Umemura, *Reu. Latinoam. Quim.*, **13**, 108 (1982).
32. A. Rinaldi, S. Dernini, M. R. Dessy and C. De Marco, *Anal. Biochem.*, **69**, 289 (1975).
33. W. Levason, C. A. McAuliffe and S. G. Murray, *J. Chem. Soc., Dalton Trans.*, 269 (1976).
34. V. G. Voloshchuk, V. N. Boiko and L. M. Yagupol'skii, *Zh. Org. Khim.*, **13**, 2008 (1977).
35. A. Fredga and R. Golmohammadi, *Chem. Scr.*, **19**, 234 (1982).
36. C. De Marco, A. Rinaldi, S. Dernini and D. Cavallini, *Gazz. Chim. Ital.*, **105**, 1113 (1975).
37. W. G. Salmond, *German Patent*, No. 2547199 (1976).
38. T. Wakamatsu, K. Akasaka and Y. Ban, *Tennen Yuki Kagobutsu Toronkai Yoshishu, 21st*, 301 (1978).
39. I. Stirling and G. Bruton, *European Patent Application*, EP 80286 (1983).
40. S. I. Radchenko and A. A. Petrov, *Zh. Org. Khim.*, **11**, 1988 (1975).
41. R. S. Sukkai and L. Brandsma, *Rec. Trav. Chim.*, **98**, 55 (1979).
42. H. J. Reich, M. C. Clark and W. W. Willis, Jr., *J. Org. Chem.*, **47**, 1618 (1982).
43. H.-J. Gais, *Angew. Chem.*, **89**, 251 (1977).
44. H.-J. Gais and T. Lied, *Angew. Chem.*, **90**, 283 (1978).
45. C. O. Meese, W. Walter, H. Mrotzek and H. Mirzai, *Chem. Ber.*, **109**, 956 (1976).
46. A. B. Pierini and R. A. Rossi, *J. Org. Chem.*, **44**, 4667 (1979).
47. R. A. Rossi, *Acc. Chem. Res.*, **15**, 164 (1982).
48. H. Suzuki, H. Abe and A. Osuka, *Chem. Lett.*, 151 (1981).
49. H. Suzuki, H. Abe, N. Ohmasa and A. Osuka, *Nippon Kagaku Kaishi*, 445 (1952).
50. P. Mueller and M. P. Nguyen-Thi, *Helv. Chim. Acta*, **63**, 2168 (1980).
51. S. Murahashi and T. Yano, *J. Amer. Chem. Soc.*, **102**, 2456 (1980).
52. K. Fujimori, H. Yoshimoto and S. Oae, *Tetrahedron Letters*, 4397 (1979).
53. K. Fujimori, H. Yoshimoto and S. Oae, *Tetrahedron Letters*, **21**, 3385 (1980).
54. M. J. Perkins, B. V. Smith, B. Terem and E. S. Turner, *J. Chem. Res., Synop.*, 341 (1979).
55. A. Cravador, A. Krief and L. Hevesi, *J. Chem. Soc., Chem. Commun.*, 451 (1980).
56. M. J. Perkins, B. V. Smith and E. S. Turner, *J. Chem. Soc., Chem. Commun.*, 977 (1980).
57. L. Hevesi, *Tetrahedron Letters*, 3025 (1979).
58. J. R. Bull and J. Floor, *J. Chem. Soc., Perkin Trans. 1*, 437 (1981).
59. G. H. Posner and D. Z. Rogers, *J. Amer. Chem. Soc.*, **99**, 8208 (1977).
60. J. Rémion, W. Dumont, and A. Krief, *Tetrahedron Letters*, 1385 (1976).
61. T. Kametani, K. Suzuki, H. Kurobe and H. Nemoto, *J. Chem. Soc., Chem. Commun.*, 1128 (1979).
62. I. M. Akhmedov, F. G. Gasanov, S. B. Kurbanov and M. M. Guseinov, *Zh. Org. Khim.*, **14**, 881 (1978).
63. F. G. Gasanov, I. M. Akhmedov, S. B. Kurbanov and M. M. Guseinov, *Azerb. Khim. Zh.*, 50 (1978).
64. D. Liotta, U. Sunay, H. Santiesteban and W. Markiewicz, *J. Org. Chem.*, **46**, 2605 (1981).
65. R. M. Scarborough, Jr. and A. B. Smith, 3, *Tetrahedron Letters*, 4361 (1977).
66. D. Liotta and H. Santiesteban, *Tetrahedron Letters*, 4369 (1977).
67. T. R. Hoye and A. J. Caruso, *Tetrahedron Letters*, 4611 (1978).
68. R. M. Scarborough, Jr., B. H. Toder and A. B. Smith, 3., *J. Amer. Chem. Soc.*, **102**, 3904 (1980).
69. T. Kometani, Y. Takeuchi and E. Yoshii, *J. Org. Chem.*, **48**, 2630 (1983).
70. M. Evers and L. Christiaens, *Tetrahedron Letters*, **24**, 377 (1983).
71. A. B. Smith, 3. and R. M. Scarborough, Jr., *Tetrahedron Letters*, 1649 (1978).
72. H. Pluim and H. Wynberg, *Tetrahedron Letters*, 1251 (1979).
73. J. Lucchetti and A. Krief, *Compt. Rend., Ser. C*, **289**, 287 (1979).
74. M. Miyashita and A. Yoshikoshi, *Synthesis*, 664 (1980).
75. J. V. Comasseto, J. T. B. Ferreira and N. Petragnani, *J. Organometal. Chem.*, **216**, 287 (1981).
76. D. H. Wadsworth and M. R. Detty, *J. Org. Chem.*, **45**, 4611 (1980).
77. I. N. Azerbaev, L. A. Tsoi and A. K. Patsaev, *Zh. Org. Khim.*, **12**, 751 (1976).
78. I. N. Azerbaev, N. M. Kolesnikova, L. A. Tsoi, A. B. Aswanova and A. K. Patsaev, *Izv. Akad. Nauk Kaz. SSR, Ser. Khim.*, **26**, 72 (1976).
79. I. N. Azerbaev, L. A. Tsoi and A. K. Patsaev, *Tezisy Dokl.-Vses. Konf. Khim. Atsetilena, 5th*, 253 (1975).

80. I. M. Akhmedov, F. G. Gasanov, A. Yu. Aliev and E. G. Mamedov, *Mater Sumgaitskoi Gor. Nauchno-Tekh. Konf. Prool, Org. Khlororg. Sint.*, 12 (1980).
81. A. P. Arnold and A. J. Cauty, *Australian J. Chem.*, **36**, 815 (1983).
82. I. M. Akhmedov, A. Yu. Aliev, G. G. Gadzhiev and M. M. Guseinov, *Dokl. Akad. Nauk Az. SSR*, **38**, 37 (1982).
83. M. L. Petrov, L. S. Rodionova and A. A. Petrov, *Zh. Org. Khim.*, **13**, 1564 (1977).
84. V. Z. Laishev, M. L. Petrov and A. A. Petrov, *Zh. Org. Khim.*, **18**, 281 (1982).
85. M. L. Petrov, V. A. Bobylev, V. A. Paramonov and A. A. Petrov, *Zh. Org. Khim.*, **18**, 2619 (1982).
86. T. Schaefer, W. Danchura, W. Niemczura and W. J. E. Parr, *Can. J. Chem.*, **56**, 1721 (1978).
87. J.-M. Lehn, G. Wipff and J. Demuynck, *Helv. Chim. Acta*, **60**, 1239 (1977).
88. A. E. Lutskii, A. I. Mitichkin, T. N. Marchenko and G. I. Sheremeteva, *Zh. Obshch. Khim.*, **46**, 1103 (1976).
89. V. K. Pogorelyi and T. F. Divnich, *Teor. Eksp. Khim.*, **12**, 822 (1976).
90. J. R. Durig and W. E. Bucy, *J. Mol. Spectry*, **64**, 474 (1977).
91. V. A. Bobylev, M. L. Petrov, V. Ya. Komarov, V. V. Vasilev and B. I. Ionin, *Zh. Org. Khim.*, **19**, 20 (1983).
92. W. Dietzsch, J. Kaisen, R. Richter, L. Golic, J. Siftar and R. Heber, *Z. Anorg. Allgem. Chem.*, **477**, 71 (1981).
93. J. Laureni, A. Krantz and R. A. Hajdu, *J. Amer. Chem. Soc.*, **98**, 7872 (1976).
94. J. Martens, K. Praefcke and H. Simon, *Z. Naturforsch., Anorg. Chem., Org. Chem.*, **31b**, 1717 (1976).
95. K.-Y. Jen and M. P. Cava, *J. Org. Chem.*, **48**, 1449 (1983).
96. D. D. NacNicol and D. R. Wilson, *J. Chem. Soc., Chem. Commun.*, 494 (1976).
97. P. J. Giddings, D. I. John, E. J. Thomas and D. J. Williams, *J. Chem. Soc., Perkin Trans. 1*, 2757 (1982).
98. M. J. Perkins and E. S. Turner, *J. Chem. Soc., Chem. Commun.*, 139 (1981).
99. M. G. Voronkov, V. I. Rakhlin, S. Kh. Khangazheev, R. G. Mirskov and A. I. Albanov, *Zh. Obshch. Khim.*, **50**, 1771 (1980).
100. F. G. James, M. J. Perkins, O. Porta and B. V. Smith, *J. Chem. Soc., Chem. Commun.*, 131 (1977).
101. H. M. J. Gillissen, P. Schipper, P. J. J. M. van Ool and H. M. Buck, *J. Org. Chem.*, **45**, 319 (1980).
102. S. Halazy, J. Lucchetti and A. Krief, *Tetrahedron Letters*, 3971 (1978).
103. M. Sevrin and A. Krief, *Tetrahedron Letters*, 187 (1978).
104. D. Van Ende, W. Dumont and A. Krief, *Angew. Chem.*, **87**, 709 (1975).
105. D. Van Ende and A. Krief, *Tetrahedron Letters*, 457 (1976).
106. A. Burton, L. Hevesi, W. Dumont, A. Cravador and A. Krief, *Synthesis*, 877 (1979).
107. M. Sevrin, J. N. Denis and A. Krief, *Tetrahedron Letters*, **21**, 1877 (1980).
108. R. Seshadri, W. J. Pegg and M. Israel, *J. Org. Chem.*, **46**, 2596 (1981).
109. M. R. Detty, *Tetrahedron Letters*, 4189 (1979).
110. W. Dumont, M. Sevrin and A. Krief, *Angew. Chem.*, **89**, 561 (1977).
111. W. Dumont and A. Krief, *Angew. Chem.*, **89**, 559 (1977).
112. D. L. J. Clive, *Tetrahedron*, **34**, 1049 (1978).
113. H. J. Reich, *Oxidation in Organic Chemistry: Part C* (Ed. W. Trahanovsky), Academic Press, New York, 1978.
114. K. B. Sharpless and R. F. Lauer, *J. Org. Chem.*, **39**, 429 (1974).
115. H. J. Reich, J. M. Renga and I. L. Reich, *J. Amer. Chem. Soc.*, **97**, 5434 (1975).
116. H. J. Rèich and J. E. Trend, *J. Org. Chem.*, **41**, 2503 (1976).
117. H. J. Reich, S. Wollowitz, J. E. Trend, F. Chow and D. F. Wendelborn, *J. Org. Chem.*, **43**, 1697 (1978).
118. H. J. Reich, C. A. Hoeger and W. W. Willis, Jr., *J. Amer. Chem. Soc.*, **104**, 2936 (1982).
119. H. J. Reich, S. K. Shah, P. M. Gold and R. E. Olson, *J. Amer. Chem. Soc.*, **103**, 3112 (1981).
120. K. B. Sharpless and R. F. Lauer, *J. Amer. Chem. Soc.*, **94**, 7154 (1972).
121. H. J. Reich, *J. Org. Chem.*, **40**, 2570 (1975).
122. H. J. Reich and S. K. Shah, *J. Amer. Chem. Soc.*, **99**, 263 (1977).
123. H. P. Jensen and K. B. Sharpless, *J. Org. Chem.*, **40**, 264 (1975).
124. D. Arigoni, A. Vasella, K. B. Sharpless and H. P. Jensen, *J. Amer. Chem. Soc.*, **95**, 7917 (1973).

125. M. Shimizu, R. Takeda and I. Kuwajima, *Bull. Chem. Soc. Japan*, **54**, 3510 (1981).
126. M. Shimizu, R. Takeda and I. Kuwajima, *Tetrahedron Letters*, 419 (1979).
127. M. Shimizu, R. Takeda and I. Kuwajima, *Tetrahedron Letters*, 3461 (1979).
128. I. Kuwajima, M. Shimizu and H. Urabe, *J. Org. Chem.*, **47**, 837 (1982).
129. M. Shimizu and I. Kuwajima, *Bull. Chem. Soc. Japan*, **54**, 3100 (1981).
130. R. A. Gancarz and J. L. Kice, *Tetrahedron Letters*, **22**, 1661 (1981).
131. T. Hori and K. B. Sharpless, *J. Org. Chem.*, **43**, 1689 (1978).
132. H. Rheinboldt and E. Giesbrecht, *Chem. Ber.*, **88**, 1974 (1955).
133. H. Rheinboldt and E. Giesbrecht, *Chem. Ber.*, **88**, 1037 (1955).
134. H. Rheinboldt and E. Giesbrecht, *Chem. Ber.*, **89**, 631 (1956).
135. H. J. Reich, W. W. Willis, Jr. and S. Wollowitz, *Tetrahedron Letters.*, **23**, 3319 (1982).
136. J. L. Kice, F. McAfee and H. Slebocka-Tilk, *Tetrahedron Letters*, **23**, 3323 (1982).
137. D. Labar, A. Krief and L. Hevesi, *Tetrahedron Letters*, 3967 (1978).
138. T. G. Back, S. Collins and R. G. Kerr, *J. Org. Chem.*, **46**, 1564 (1981).
139. T. T. Back and S. Collins, *Tetrahedron Letters*, 2661 (1979).
140. N. Miyoshi, Y. Takai, S. Murai and N. Sonoda, *Bull. Chem. Soc. Japan*, **51**, 1265 (1978).
141. M. L. Bird and F. Challenger, *J. Chem. Soc.*, 570 (1942).
142. I. D. Entwistle, *J. Chem. Soc., Chem. Commun.*, 61 (1976).
143. N. Miyoshi, S. Murai and N. Sonoda, *Tetrahedron Letters*, 851 (1977).
144. O. Behaghel and W. Müller, *Ber.*, **68**, 1540 (1935).
145. W. Jenny, *Helv. Chim. Acta*, **35**, 1591 (1952).
146. H. J. Reich, I. L. Reich and S. Wollowitz, *J. Amer. Chem. Soc.*, **100**, 5981 (1978).
147. I. Kuwajima and M. Shimizu, *Tetrahedron Letters*, 1277 (1978).
148. F. A. Davis, O. D. Stringer and J. M. Billmers, *Tetrahedron Letters*, **24**, 1213 (1983).
149. S. Torii, K. Uneyama, M. Ono and T. Bannou, *J. Amer. Chem. Soc.*, **103**, 4606 (1981).
150. R. M. Scarborough, Jr., A. B. Smith, III, W. E. Barnette and K. C. Nicolaou, *J. Org. Chem.*, **44**, 1742 (1979).
151. W. Jenny, *Helv. Chim. Acta*, **40**, 317 (1958).
152. D. L. Klayman and J. W. Lown, *J. Org. Chem.*, **31**, 3396 (1966).
153. D. D. Lawson and N. Kharasch, *J. Org. Chem.*, **24**, 857 (1959).
154. K. B. Sharpless, R. F. Lauer and A. Y. Teranishi, *J. Amer. Chem. Soc.*, **95**, 6137 (1973).
155. R. Paetzold and E. Wolfram, *Z. Anorg. Allgen. Chem.*, **353**, 167 (1967).
156. H. Rheinboldt and E. Giesbrecht, *Justus Liebigs Ann. Chem.*, **574**, 227 (1951).
157. G. Hölzle and W. Jenny, *Helv. Chim. Acta*, **41**, 356 (1958).
158. G. Vicentini, E. Giesbrecht and L. R. M. Pitombo, *Chem. Ber.*, **92**, 40 (1959).
159. T. W. Campbell and J. D. McCullough, *J. Amer. Chem. Soc.*, **67**, 1965 (1945).
160. N. N. Yorovenko and M. A. Raksha, *Zh. Obshch. Khim.*, **30**, 4064 (1960).
161. J. W. Dale, H. J. Emeléus and R. N. Haszeldine, *J. Chem. Soc.*, 2939 (1958).
162. H. J. Reich, M. L. Cohen and P. S. Clark, *Org. Synth.*, **59**, 141 (1980).
163. M. Renson and J. L. Piette, *Bull. Soc. Chim. Belges*, **73**, 507 (1964).
164. O. Behaghel and H. Seibert, *Ber.*, **65**, 812 (1932).
165. J. Gosselck, *Chem. Ber.*, **91**, 2345 (1958).
166. D. L. J. Clive, *J. Chem. Soc., Chem. Commun.*, 695 (1973).
167. H. J. Reich, J. M. Renga and I. L. Reich, *J. Org. Chem.*, **39**, 2133 (1974).
168. H. J. Reich, I. L. Reich and J. M. Renga, *J. Amer. Chem. Soc.*, **95**, 5813 (1973).
169. K. B. Sharpless and M. W. Young. *J. Org. Chem.*, **40**, 947 (1975).
170. P. A. Zoretic and P. Soja, *J. Org. Chem.*, **41**, 3587 (1976).
171. P. A. Grieco, C. S. Pogonowski and S. Burke, *J. Org. Chem.*, **40**, 542 (1975).
172. J. Schwartz and Y. Hayashi, *Tetrahedron Letters*, **21**, 1497 (1980).
173. G. Zima and D. Liotta, *Synth. Commun.*, **9**, 697 (1979).
174. T. A. Hase and P. Kukkola, *Synth. Commun.*, **10**, 451 (1980).
175. D. J. Buckley, S. Kulkowit and A. McKervey, *J. Chem. Soc., Chem. Commun.*, 506 (1980).
176. N. N. Yorovenko, G. B. Gazieva, V. N. Shemanina and N. A. Fedorova, *Zh. Obshch. Khim.*, **29**, 940 (1959).
177. D. G. Garratt, *Can. J. Chem.*, **56**, 2184 (1978).
178. S. Raucher, *J. Org. Chem.*, **42**, 2950 (1977).
179. S. Raucher, *Tetrahedron Letters*, 3909 (1977).
180. D. G. Garratt and A. Kabo, *Can. J. Chem.*, **58**, 1030 (1980).

181. G. H. Schmid and D. G. Garratt, *Tetrahedron*, **34**, 2869 (1978).
182. D. Liotta and G. Zima, *Tetrahedron Letters*, 4977 (1978).
183. A. E. Feiring, *J. Org. Chem.*, **45**, 1962 (1980).
184. A. E. Feiring, *J. Org. Chem.*, **45**, 1958 (1980).
185. D. L. J. Clive, *J. Chem. Soc., Chem. Commun.*, 100 (1974).
186. H. J. Reich, *J. Org. Chem.*, **39**, 428 (1974).
187. A. Toshimitsu, T. Aoai, H. Owada, S. Uemura and M. Okano, *J. Chem. Soc., Chem. Commun.*, 412 (1980).
188. A. P. Kozikowski, R. J. Schmiesing and K. L. Sorgi, *J. Amer. Chem. Soc.*, **102**, 6577 (1980).
189. A. Toshimitsu, T. Aoai, H. Owada, S. Uemura and M. Okano, *J. Org. Chem.*, **46**, 4727 (1981).
190. T. Hayama, S. Tomoda, Y. Takeuchi and Y. Nomura, *Tetrahedron Letters*, **23**, 4733 (1982).
191. J. N. Denis, J. Vicens and A. Krief, *Tetrahedron Letters*, 2697 (1979).
192. D. Liotta, G. Zima and M. Saindane, *J. Org. Chem.*, **47**, 1258 (1982).
193. D. Liotta and G. Zima, *J. Org. Chem.*, **45**, 2551 (1980).
194. I. Ryu, S. Murai, I. Niwa and N. Sonoda, *Synthesis*, 874 (1977).
195. S. Danishefsky, C. F. Yan, and P. M. McCurry, Jr., *J. Org. Chem.*, **42**, 1819 (1977).
196. M. Petrzilka, *US Patent*, No. 4247717 (1981).
197. K. C. Nicolaou, R. L. Magolda and W. J. Sipio, *Synthesis*, 982 (1979).
198. H. Nishiyama, S. Narimatsu and K. Itoh, *Tetrahedron Letters*, **22**, 5289 (1981).
199. H. Nishiyama, K. Itagaki, K. Sakuta and K. Itoh, *Tetrahedron Letters*, **22**, 5285 (1981).
200. E. G. Kataev, T. G. Mannafov and Yu. Yu. Samitov, *Zh. Org. Khim.*, **11**, 2324 (1975).
201. D. G. Garratt, P. L. Beaulieu and V. M. Morisset, *Can. J. Chem.*, **59**, 927 (1981).
202. J. Hooz and R. D. Mortimer, *Can. J. Chem.*, **56**, 2786 (1978).
203. D. G. Garratt, P. L. Beaulieu, V. M. Morisset and M. Ujjainwalla, *Can. J. Chem.*, **58**, 2745 (1980).
204. D. G. Garratt, P. L. Beaulieu and M. D. Ryan, *Tetrahedron*, **36**, 1507 (1980).
205. S. Halazy and L. Hevesi, *Tetrahedron Letters*, **24**, 2689 (1983).
206. P. L. Beaulieu, V. M. Morisset and D. G. Garratt, *Can. J. Chem.*, **58**, 1005 (1980).
207. D. L. J. Clive and G. Chittattu, *J. Chem. Soc., Chem. Commun.*, 484 (1977).
208. M. de Moura Compos and N. Petragnani, *Ber.*, **93**, 317 (1960).
209. D. L. J. Clive, C. G. Russell, G. Chittattu and A. Singh, *Tetrahedron*, **36**, 1399 (1980).
210. K. C. Nicolaou, S. P. Seitz, W. J. Sipio and J. F. Blount, *J. Amer. Chem. Soc.*, **101**, 3884 (1979).
211. K. C. Nicolaou and Z. Lysenko, *J. Amer. Chem. Soc.*, **99**, 3185 (1977).
212. D. Goldsmith, D. Liotta, C. Lee and G. Zima, *Tetrahedron Letters*, 4801 (1979).
213. D. G. Garratt, M. D. Ryan and P. L. Beaulieu, *J. Org. Chem.*, **45**, 839 (1980).
214. S. Current and K. B. Sharpless, *Tetrahedron Letters*, 5075 (1978).
215. D. L. J. Clive, G. Chittattu and C. K. Wong, *Can. J. Chem.*, **55**, 3894 (1977).
216. K. C. Nicolaou, *Tetrahedron*, **37**, 4097 (1981).
217. S. Uemura, A. Toshimitsu, T. Aoai and M. Okano, *Tetrahedron Letters*, **21**, 1533 (1980).
218. D. L. J. Clive, C. K. Wong, W. A. Kiel and S. M. Menchen, *J. Chem. Soc., Chem. Commun.*, 379 (1978).
219. D. L. J. Clive, V. Farina, A. Singh, C. K. Wong, W. A. Kiel and S. M. Menchen, *J. Org. Chem.*, **45**, 2120 (1980).
220. K. C. Nicolaou, W. E. Barnette and R. L. Magolda, *J. Amer. Chem. Soc.*, **100**, 2567 (1978).
221. D. L. J. Clive, G. Chittatu and C. K. Wong, *J. Chem. Soc., Chem. Commun.*, 441 (1978).
222. D. L. J. Clive, G. Chittatu, N. J. Curtis, W. A. Kiel and C. K. Wong, *J. Chem. Soc., Chem. Commun.*, 725 (1977).
223. C. N. Filer, D. Ahern, R. Fazio and E. J. Shelton, *J. Org. Chem.*, **45**, 1313 (1980).
224. K. C. Nicolaou, G. P. Gasic and W. E. Barnette, *Angew. Chem.*, (*Intern. Ed. Engl.*), **17**, 293 (1978).
225. E. J. Corey, G. E. Keck and I. Székely, *J. Amer. Chem. Soc.*, **99**, 2006 (1977).
226. A. P. Kozikowsky and Y.-Y. Chen, *J. Org. Chem.*, **45**, 2236 (1980).
227. J.-C. Gfeller, A. K. Beck and D. Seebach, *Helv. Chim. Acta*, **63**, 728 (1980).
228. S. R. Wilson and R. A. Sawicki, *J. Org. Chem.*, **44**, 287 (1979).
229. D. G. Garratt, M. D. Ryan and A. Kabo, *Can. J. Chem.*, **58**, 2329 (1980).
230. K. C. Nicolaou, D. A. Claremon, W. E. Barnette and S. P. Seitz, *J. Amer. Chem. Soc.*, **101**, 3704 (1979).

231. H. Rheinboldt and M. Perrier, *Bull. Soc. Chim. Fr.*, 251 (1956).
232. L. R. M. Pitombo, *Chem. Ber.*, **92**, 745 (1959).
233. H. Rheinboldt and M. Perrier, *Bull. Soc. Chim. Fr.*, 245 (1950).
234. O. Foss, *J. Amer. Chem. Soc.*, **69**, 2236 (1947).
235. A. Toshimitsu, S. Uemura and M. Okano, *J. Chem. Soc., Chem. Commun.*, 87 (1982).
236. S. Tomoda, Y. Takeuchi and Y. Nomura, *Chem. Letters*, 1069 (1981).
237. W. S. Cook and R. A. Donia, *J. Amer. Chem. Soc.*, **73**, 2275 (1951).
238. G. Hölzle and W. Jenny, *Helv. Chim. Acta*, **41**, 331 (1958).
239. W. Jenny, *Helv. Chim. Acta*, **35**, 845 (1952).
240. W. Jenny, *Helv. Chim. Acta*, **35**, 1429 (1952).
241. M. A. Umbreit and K. B. Sharpless, *J. Amer. Chem. Soc.*, **99**, 5526 (1977).
242. Y. Nagao, M. Ochiai, K. Kaneko, A. Maeda, K. Watanabe and E. Fujita, *Tetrahedron Letters*, 1345 (1977).
243. J. R. M. Giles, B. P. Roberts, M. J. Perkins and E. S. Turner, *J. Chem. Soc., Chem. Commun.*, 504 (1980).
244. W. Jenny, *Helv. Chim. Acta*, **36**, 1278 (1953).
245. O. Foss and S. R. Svendsen, *Acta Chem. Scand.*, **8**, 1352 (1954).
246. G. Ayrey, D. Barnard and D. T. Woodbridge, *J. Chem. Soc.*, 2089 (1962).
247. G. Hölzle and W. Jenny, *Helv. Chim. Acta*, **41**, 593 (1958).
248. H. J. Emeléus and N. Welcman, *J. Chem. Soc.*, 1268 (1963).
249. N. Welcman and H. Regev, *J. Chem. Soc.*, 7511 (1965).
250. C. Paulmier and P. Lerouge, *Tetrahedron Letters*, **23**, 1557 (1982).
251. M. Jefson and J. Meinwald, *Tetrahedron Letters*, **22**, 3561 (1981).
252. S. V. Ley, B. Lygo, H. Molines and J. A. Morton, *J. Chem. Soc., Chem. Commun.*, 1251 (1982).
253. T. Kobayashi and T. Hiraoka, *Bull. Chem. Soc. Japan*, **52**, 3366 (1979).
254. H. J. Reich and J. M. Renga, *J. Org. Chem.*, **40**, 3313 (1975).
255. C. O. Meese, W. Walter and H.-W. Mueller, *Tetrahedron Letters*, 19 (1977).
256. W. P. Jackson, S. V. Ley and J. A. Morton, *J. Chem. Soc., Chem. Commun.*, 1028 (1980).
257. T. Hori and K. B. Sharpless, *J. Org. Chem.*, **44**, 4204, 4208 (1979).
258. S. V. Ley, N. S. Simpkins and A. J. Whittle, *J. Chem. Soc., Chem. Commun.*, 1001 (1981).
259. R. R. Webb, II and S. Danishefsky, *Tetrahedron Letters*, **24**, 1357 (1983).
260. L. C. F. Blackman and M. J. S. Dewar, *J. Chem. Soc.*, 162 (1957).
261. W. J. Franklin and R. L. Werner, *Tetrahedron Letters*, 3003 (1965).
262. L. Pichat, M. Herbert and M. Thiers, *Tetrahedron*, **12**, 1 (1961).
263. W. H. H. Günther and H. G. Mautner, *J. Amer. Chem. Soc.*, **82**, 2762 (1960).
264. E. Rebane, *Acta Chem. Scand.*, **21**, 652 (1967).
265. L. C. F. Blackman and M. J. S. Dewar, *J. Chem. Soc.*, 165 (1957).
266. A. Fredga, L.-B. Agenäs and E. Jonsson, *Acta Chem. Scand.*, **24**, 3751 (1970).
267. H. Rheinboldt and H. V. de Campos, *J. Amer. Chem. Soc.*, **72**, 2784 (1950).
268. T. van Es and R. L. Whistler, *Tetrahedron*, **23**, 2849 (1967).
269. J. W. Dale, H. J. Emeléus and R. N. Haszeldine, *J. Chem. Soc.*, 2939 (1958).
270. H. Rheinboldt and E. Giesbrecht, *J. Amer. Chem. Soc.*, **72**, 866 (1950).
271. J. Gosselck and E. Wolters, *Chem. Ber.*, **95**, 1237 (1962).
272. E. Mueller, H. B. Stegmann and K. Scheffler, *Justus Liebigs Ann. Chem.*, **657**, 5 (1962).
273. L.-B. Agenäs, *Acta Chem. Scand.*, **17**, 268 (1963).
274. E. E. Aynsley, N. N. Greenwood and M. J. Sprague, *J. Chem. Soc.*, 2395 (1965).
275. L.-B. Agenäs and B. Persson, *Acta Chem. Scand.*, **21**, 837 (1967).
276. F. Wille, A. Ascherl, G. Kaupp and L. Capeller, *Angew. Chem.*, **74**, 753 (1962).
277. G. Hölzle and W. Jenny, *Helv. Chim. Acta*, **41**, 356 (1958).
278. L. Chierici and R. Passerini, *J. Chem. Soc.*, 3249 (1954).
279. E. S. Gould and J. D. McCullough, *J. Amer. Chem. Soc.*, **73**, 1109 (1951).
280. G. Simchen, *Angew. Chem. (Intern. Ed. Engl.)*, **7**, 464 (1968).
281. G. E. Wiseman and E. S. Gould, *J. Amer. Chem. Soc.*, **76**, 1706 (1954).
282. B. Greenberg, E. S. Gould and W. Burlant, *J. Amer. Chem. Soc.*, **78**, 4028 (1956).
283. P. A. Grieco, J. Y. Jaw, D. A. Claremon and K. C. Nicolaou, *J. Org. Chem.*, **46**, 1215 (1981).
284. S. Tomoda, Y. Takeuchi and Y. Nomura, *Chem. Letters*, 1715 (1981).
285. S. Tomoda, Y. Takeuchi and Y. Nomura, *Tetrahedron Letters*, **23**, 1361 (1982).

286. S. Tomoda, Y. Takeuchi and Y. Nomura, *J. Chem. Soc., Chem. Commun.*, 871 (1982).
287. S. Tomoda, Y. Takeuchi and Y. Nomura, *Chem. Letters*, 1733 (1982).
288. D. L. J. Clive, G. Chittatu, N. J. Curtis and S. M. Menchen, *J. Chem. Soc., Chem. Commun.*, 770 (1978).
289. P. A. Grieco, Y. Yokoyama and E. Williams, *J. Org. Chem.*, **43**, 1283 (1978).
290. P. A. Grieco and Y. Yokoyama, *J. Amer. Chem. Soc.*, **99**, 5210 (1977).
291. G. A. Chmutova, I. V. Bautina and A. N. Vereshchagin, *Izv. Akad. Nauk SSSR, Ser. Khim.*, 2469 (1976).
292. A. N. Frolov, E. V. Smirnov, O. V. Kul'bitskaya and A. V. El'tsov, *Zh. Org. Khim.*, **16**, 2302 (1980).
293. G. Simchen and G. Entenmann, *Justus Liebigs Ann. Chem.*, 1249 (1977).
294. H. Suzuki and M. Shinoda, *Synthesis*, 640 (1977).
295. P. D. Woodgate, H. H. Lee, P. S. Rutledge and R. C. Cambie, *Synthesis*, 152 (1978).
296. S. Uemura, N. Watanabe, A. Toshimitsu and M. Okano, *Bull. Chem. Soc. Japan*, **51**, 1818 (1978).
297. A. Arase and Y. Masuda, *Chem. Letters*, 1115 (1976).
298. B. Lindgren, *Acta Chem. Scand.*, **B30**, 941 (1976).
299. A. Arase and Y. Masuda, *Chem. Letters*, 785 (1976).
300. C. Paulmier, *Bull. Soc. Chim. Fr.*, 592 (1979).
301. C. Paulmier, *Bull. Soc. Chim. Fr.*, 237 (1979).
302. A. Toshimitsu, Y. Kozawa, S. Uemura and M. Okano, *J. Chem. Soc., Perkin Trans. 1*, 1273 (1978).
303. T. Kametani, H. Nemoto and K. Fukumoto, *Bioorg. Chem.*, **7**, 215 (1978).
304. M. Sevrin and A. Krief, *J. Chem. Soc., Chem. Commun.*, 656 (1980).
305. M. Takahashi and M. Kurosawa, *Bull. Chem. Soc. Japan*, **53**, 1185 (1980).
306. A. Toshimitsu, T. Aoai, S. Uemura and M. Okano, *J. Org. Chem.*, **45**, 1953 (1980).
307. S. Uemura, A. Toshimitsu, T. Aoai and M. Okano, *Chem. Letters*, 1359 (1979).
308. S. Uemura, A. Toshimitsu, T. Aoai and M. Okano, *J. Chem. Soc., Chem. Commun.*, 610 (1979).
309. H. Suzuki, M. Usuki and T. Hanafusa, *Synthesis*, 705 (1979).
310. T. Otsubo, F. Ogura, H. Yamaguchi, H. Higuchi and S. Misumi, *Synth. Commun.*, **10**, 595 (1980).
311. A. Toshimitsu, S. Uemura and M. Okano, *J. Chem. Soc., Perkin Trans. 1*, 1206 (1979).
312. P. A. Grieco, S. Gilman and M. Nishizawa, *J. Org. Chem.*, **41**, 1485 (1976).
313. D. H. R. Barton, A. G. Brewster, S. V. Ley, C. M. Read and M. N. Rosenfeld, *J. Chem. Soc., Perkin Trans. 1*, 1473 (1981).
314 D. H. R. Barton, A. G. Brewster, R. A. H. F. Hui, D. J. Lester, S. V. Ley and T. G. Back, *J. Chem. Soc., Chem. Commun.*, 952 (1978).
315. D. H. R. Barton, S. V. Ley, P. D. Magnus and M. N. Rosenfeld, *J. Chem. Soc., Perkin Trans. 1*, 567 (1977).
316. D. H. R. Barton, A. G. Brewster, S. V. Ley and M. N. Rosenfeld, *J. Chem. Soc., Chem. Commun.*, 985 (1976).
317. D. H. R. Barton, A. G. Brewster, S. V. Ley and M. N. Rosenfeld, *J. Chem. Soc., Chem. Commun.*, 147 (1977).
318. J. S. E. Holker, E. O'Brien and B. K. Park, *J. Chem. Soc., Perkin Trans. 1*, 1915 (1982).
319. D. H. R. Barton, R. A. H. F. Hui, D. J. Lester and S. V. Ley, *Tetrahedron Letters*, 3331 (1979).
320. D. H. R. Barton, R. A. H. F. Hui and S. V. Ley, *J. Chem. Soc., Perkin Trans. 1*, 2179 (1982).
321. D. H. R. Barton, N. J. Cussans and S. V. Ley, *J. Chem. Soc., Chem. Commun.*, 393 (1978).
322. N. J. Cussans, S. V. Ley and D. H. R. Barton, *J. Chem. Soc., Perkin Trans. 1*, 1650 (1980).
323. K. Yamakawa, T. Satoh, N. Ohba and R. Sakaguchi, *Chem. Letters*, 763 (1979).
324. K. Yamakawa, T. Satoh, N. Ohba, R. Sakaguchi, S. Takita and N. Tamura, *Tetrahedron*, **37**, 473 (1981).
325. D. H.R. Barton, D. J. Lester and S. V. Ley, *J. Chem. Soc., Chem. Commun.*, 130 (1978).
326. M. R. Czarny, *Synth. Commun.*, **6**, 285 (1976).
327. M. R. Czarny, *J. Chem. Soc., Chem. Commun.*, 81 (1976).
328. P. A. Grieco, Y. Yokoyama, S. Gilman and M. Nishizawa, *J. Org. Chem.*, **42**, 2034 (1977).
329. P. A. Grieco, Y. Yokoyama, S. Gilman and Y. Ohfune, *J. Chem. Soc., Chem. Commun.*, 870 (1977).
330. H. J. Reich, F. Chow and S. L. Peake, *Synthesis*, 299 (1978).
331. A. J. Bloodworth, and D. J. Lapham, *J. Chem. Soc., Perkin Trans. 1*, 471 (1983).

332. T. Frejd and K. B. Sharpless, *Tetrahedron Letters*, 2239 (1978).
333. T. G. Back and S. Collins, *Tetrahedron Letters*, **21**, 2213 (1980).
334. T. G. Back, *J. Chem. Soc., Chem. Commun.*, 530 (1981).
335. T. G. Back and R. G. Kerr, *Can. J. Chem.*, **60**, 2711 (1982).
336. T. G. Back, *J. Chem. Soc., Chem. Commun.*, 278 (1978).
337. D. H. R. Barton, X. Lusinchi and P. Milliet, *Tetrahedron Letters*, **23**, 4949 (1982).
338. D. H. R. Barton, R. A. H. F. Hui, S. V. Ley and D. J. Williams, *J. Chem. Soc., Perkin Trans. 1*, 1919 (1982).
339. D. H. R. Barton, D. J. Lester and S. V. Ley, *J. Chem. Soc., Perkin Trans. 1*, 2209 (1980).
340. N. Khôi and J. Polonsky, *Helv. Chim. Acta*, **64**, 1540 (1981).
341. D. H. R. Barton, M. T. Bielska, J. M. Cardoso, N. J. Cussans and S. V. Ley, *J. Chem. Soc., Perkin Trans. 1*, 1840 (1981).
342. D. H. R. Barton, N. J. Cussans and S. V. Ley, *J. Chem. Soc., Chem. Commun.*, 751 (1977).
343. N. J. Cussans, S. V. Ley and D. H. R. Barton, *J. Chem. Soc., Perkin Trans. 1*, 1654 (1980).
344. T. G. Back and N. Ibahim, *Tetrahedron Letters*, 4931 (1979).
345. T. G. Back, *J. Org. Chem.*, **46**, 1442 (1981).
346. D. H. R. Barton, D. J. Lester and S. V. Ley, *J. Chem. Soc., Chem. Commun.*, 445 (1977).
347. D. H. R. Barton, D. J. Lester and S. V. Ley, *J. Chem. Soc., Chem. Commun.*, 276 (1978).
348. D. H. R. Barton, D. J. Lester and S. V. Ley, *J. Chem. Soc., Perkin Trans. 1*, 1212 (1980).
349. J. K. Kice and T. W. S. Lee, *J. Amer. Chem. Soc.*, **100**, 5094 (1978).
350. J. L. Kice, T. W. S. Lee and S. Pan, *J. Amer. Chem. Soc.*, **102**, 4448 (1980).
351. L. G. Faehl and J. L. Kice, *J. Org. Chem.*, **44**, 2357 (1979).
352. R. A. Gancarz and J. L. Kice, *Tetrahedron Letters*, 1697 (1980).
353. A. Cravador and A. Krief, *J. Chem. Soc., Chem. Commun.*, 951 (1980).
354. T. G. Back, N. Ibrahim and D. J. McPhee, *J. Org. Chem.*, **47**, 3283 (1982).
355. R. Paetzold and E. Rönsch, *Z. Anorg. Allgem. Chem.*, **338**, 195 (1965).
356. L.-B. Agenäs, *Acta. Chem. Scand.*, **19**, 764 (1965).
357. H. Rheinboldt and E. Giesbrecht, *Chem. Ber.*, **88**, 666 (1955).
358. A. Fredga and C. Evertsdotter, *Acta Chem. Scand.*, **13**, 1042 (1959).
359. D. De Filippo and F. Momicchioli, *Tetrahedron*, **25**, 5733 (1969).
360. R. Paetzold and D. Lienig, *Z. Anorg. Allgem. Chem.*, **335**, 289 (1965).
361. P. Allen, Jr. and J. W. Brook, *J. Org. Chem.*, **27**, 1019 (1962).
362. R. Paetzold and D. Lienig, *Z. Chem.*, **4**, 186 (1964).
363. M. Schmidt, P. Bornmann and I. Wilhelm, *Angew. Chem. (Intern. Ed. Engl.)*, **2**, 691 (1963).
364. M. Schmidt and I. Wilhelm, *Chem. Ber.*, **97**, 872 (1964).
365. K. Dostál, P. Mošna and Z. Žák, *Z. Chem.*, **6**, 153 (1966).
366. K. Dostál, Z. Žák and M. Černik, *Chem. Ber.*, **104**, 2004 (1971).
367. C. K. Banks and C. S. Hamilton, *J. Amer. Chem. Soc.*, **62**, 1859 (1940).
368. E. Rebane, *Acta Chem. Scand.*, **21**, 657 (1967).
369. E. Rebane, *Acta Chem. Scand.*, **23**, 1817 (1969).
370. V. V. Kozlov and V. M. Malyushina, *Zh. Vses. Khim. O-va.*, **21**, 454 (1976).
371. K. Dostál, M. Černik and M. Sochorcová, *Z. Anorg. Allgen. Chem.*, **417**, 143 (1975).
372. K. Dostál, Z. Žák and A. Haba, *Collect. Czech. Chem. Commun.*, **42**, 1466 (1977).
373. K. J. Irgolic, *The Organic Chemistry of Tellurium*, Gordon and Breach, New York, 1974.
374. R. A. Zingaro and K. J. Irgolic, *Tellurium*, Van Nostrand–Reinhold, New York, 1971, p. 184.
375. K. J. Irgolic, *J. Organometal. Chem.*, **103**, 91 (1975).
376. K. J. Irgolic, *J. Organometal. Chem.*, **130**, 411 (1977).
377. K. J. Irgolic, *J. Organometal. Chem.*, **158**, 235 (1978).
378. K. J. Irgolic, *J. Organometal. Chem.*, **158**, 267 (1978).
379. K. J. Irgolic, *J. Organometal. Chem.*, **189**, 65 (1980).
380. K. J. Irgolic, *J. Organometal. Chem.*, **203**, 367 (1980).
381. A. Baroni, *Atti Accad. Naz. Lincei, Rend., Cl. Sci. Fis. Mat. Nat.*, **27**, 238 (1938).
382. C. W. Sink and A. B. Harvey, *J. Chem. Soc. (D)*, 1023 (1969).
383. C. W. Sink and A. B. Harvey, *J. Chem. Phys.*, **57**, 4434 (1972).
384. K. Hamada and H. Morishita, *Synth. React. Inorg. Met.-Org. Chem.*, **7**, 355 (1977).
385. N. Petragnani, *Chem. Ber.*, **96**, 247 (1963).
386. J. Meinwald, D. Dauplaise and J. Clardy, *J. Amer. Chem. Soc.*, **99**, 7743 (1977).
387. J. Meinwald, D. Dauplaise, F. Wudl and J. J. Hauser, *J. Amer. Chem. Soc.*, **99**, 255 (1977).
388. W. S. Haller and K. J. Irgolic, *J. Organometal. Chem.*, **38**, 97 (1972).

389. Y. Okamoto and T. Yano, *J. Organometal. Chem.*, **29**, 99 (1971).
390. N. Petragnani and M. de Moura Campos, *Chem. Ber.*, **96**, 249 (1963).
391. J. L. Piette and M. Renson, *Bull. Soc. Chim. Belges*, **79**, 353 (1970).
392. J. L. Piette and M. Renson, *Bull. Soc. Chim. Belges*, **79**, 367 (1970).
393. J. L. Piette and M. Renson, *Bull. Soc. Chim. Belges*, **79**, 383 (1970).
394. Yu. A. Boiko, B. S. Kupiu and A. A. Petrov, *Zh. Org. Khim.*, **5**, 1553 (1969).
395. L. Brandsma, H. E. Wijers and J. F. Arens, *Rec. Trav. Chim.*, **81**, 583 (1962).
396. L. Brandsma, H. E. Wijers and C. Jorker, *Rec. Trav. Chim.*, **83**, 208 (1964).
397. D. Seeback and A. K. Beck, *Chem. Ber.*, **108**, 314 (1975).
398. N. Dereu, J. L. Piette, J. Van Coppenolle and M. Renson, *J. Heterocycl. Chem.*, **12**, 423 (1975).
399. K. J. Irgolic, P. J. Busse, R. A. Grigsby and M. R. Smith, *J. Organometal. Chem.*, **88**, 175 (1975).
400. L. M. Klevay, *Pharmacol. Ther., Part A*, **1**, 223 (1976).
401. S. I. Radchenko and A. A. Petrov, *Zh. Org. Khim.*, **13**, 40 (1977).
402. S. I. Radchenko, *Zh. Org. Khim.*, **13**, 2229 (1977).
403. S. R. Buzilova, L. I. Vereshchagin, I. D. Sadekov and V. I. Minkin, *Zh. Obshch. Khim.*, **46**, 932 (1976).
404. S. R. Buzilova, I. D. Sadekov, T. V. Liporich, T. M. Filippova and L. I. Vereshchagin, *Zh. Obshch. Khim.*, **47**, 199 (1977).
405. F. F. Knapp, Jr., K. R. Ambrose and A. P. Callahan, *J. Labelled Compd. Radiopharm.*, **16**, 35 (1979).
406. F. F. Knapp, Jr., K. R. Ambrose and A. P. Callahan, *J. Nucl. Med.*, **21**, 258 (1980).
407. F. F. Knapp, Jr., K. R. Ambrose and A. P. Callahan, *J. Nucl. Med.*, **21**, 251 (1980).
408. G. S. Boyd, M. V. Mervick, R. Monks and I. L. Thomas, *J. Labelled Compd. Radiopharm.*, **16**, 38 (1979).
409. G. P. Basmadjian, G. R. Parker, R. A. Magarian, A. S. Kirschner and R. D. Ice, *J. Labelled Compd. Radiopharm.*, **16**, 33 (1979).
410. G. P. Basmadjian, G. R. Parker, R. A. Magarian, A. S. Kilschner and R. D. Ice, *J. Nucl. Med.*, **19**, 689 (1978).
411. F. F. Knapp, Jr., *J. Org. Chem.*, **44**, 1007 (1979).
412. F. F. Knapp, Jr., K. R. Ambrose, A. P. Callahan, L. A. Ferreu, R. A. Grigsby and K. J. Irgolic, *J. Nucl. Med.*, **22**, 988 (1981).
413. G. Kirsch, M. M. Goodman and F. F. Knapp, Jr., *Organometallics*, **2**, 357 (1983).
414. M. M. Goodman and F. F. Knapp, Jr., *J. Org. Chem.*, **47**, 3004 (1982).
415. M. V. Lakshmikantham, M. P. Cava, M. Albeck, L. Engman, P. Carroll, J. Bergman and F. Wudl, *Tetrahedron Letters*, **22**, 4199 (1981).
416. F. Wudl and E. Aharon-Shalom, *J. Amer. Chem. Soc.*, **104**, 1154 (1982).
417. K. Lerstrup, D. Talham, A. Bloch, T. Poehler and D. Cowan, *J. Chem. Soc., Chem. Commun.*, 336 (1982).
418. L. Engman and M. P. Cava, *J. Org. Chem.*, **47**, 3946 (1982).
419. L. Engman and M. P. Cava, *Organometallics*, **1**, 470 (1982).
420. L. Engman, *Tetrahedron Letters*, **23**, 3601 (1982).
421. S. Uemura and S. Fukuzawa, *Tetrahedron Letters*, **23**, 1181 (1982).
422. S. Uemura, S. Fukuzawa and S. R. Patil, *J. Organometal. Chem.*, **243**, 9 (1983).
423. M. R. Detty, B. J. Murray, D. L. Smith and N. Zumbulyadis, *J. Amer. Chem. Soc.*, **105**, 875 (1983).
424. M. R. Detty and B. J. Murray, *J. Amer. Chem. Soc.*, **105**, 883 (1983).
425. D. L. J. Clive, G. J. Chittattu, V. Farina, W. A. Kiel, S. M. Menchen, C. G. Russell, A. Singh, C. K. Wong and N. J. Curtis, *J. Amer. Chem. Soc.*, **102**, 4438 (1980).
426. O. Foss and S. Hauge, *Acta Chem. Scand.*, **13**, 2155 (1959).
427. O. Foss, S. Husebye and K. Marøy, *Acta Chem. Scand.*, **17**, 1806 (1963).
428. O. Foss and K. Marøy, *Acta Chem. Scand.*, **20**, 123 (1966).
429. O. Foss and K. Husebye, *Acta Chem. Scand.*, **20**, 132 (1966).
430. J.-L. Piette, P. Thibaut and M. Renson, *Tetrahedron*, **34**, 655 (1978).
431. J.-L. Piette, P. Thibaut and M. Renson, *Chem. Scr.*, **8A**, 117 (1975).
432. J. M. Talbot, J.-L. Piette and M. Renson, *Bull. Soc. Chim. Fr.*, 294 (1976).
433. P. Wiriyachitra, S. J. Falcone and M. P. Cava, *J. Org. Chem.*, **44**, 3957 (1979).
434. N. Petragnani and M. de Moura Campos, *Tetrahedron*, **21**, 13 (1965).
435. P. Schulz and G. Klar, *Z. Naturforsch.*, **30b**, 43 (1975).

436. P. Schulz and G. Klar, *Z. Naturforsch.*, **30b**, 40 (1975).
437. T. Kauffmann, *Angew, Chem.* (*Inter. Ed. Engl.*), **21**, 410 (1982).
438. T. Kauffmann and H. Ahlers, *Chem. Ber.*, **116**, 1001 (1983).
439. F. Ogura, H. Yamaguchi, T. Otsubo and K. Chikamatsu, *Synth. Commun.*, **12**, 131 (1982).
440. H. K. Spencer, M. V. Lakshmikantham and M. P. Cava, *J. Amer. Chem. Soc.*, **99**, 1470 (1977).
441. H. K. Spencer and M. P. Cava, *J. Org. Chem.*, **42**, 2937 (1977).
442. S. J. Falcone and M. P. Cava, *J. Org. Chem.*, **45**, 1044 (1980).
443. L. Engman and M. P. Cava, *J. Org. Chem.*, **46**, 4194 (1981).
444. L. Engman, *J. Org. Chem.*, **48**, 2920 (1983).
445. C. H. Fisher and A. Eisner, *J. Org. Chem.*, **6**, 169 (1941).
446. M. de Moura Campos and N. Petragnani, *Tetrahedron Letters*, 11 (1959).
447. M. de Moura Campos and N. Petragnani, *Tetrahedron*, **18**, 527 (1962).
448. J. Bergman and L. Engman, *J. Organometal. Chem.*, **181**, 335 (1979).
449. J.-E. Bäckvall and L. Engman, *Tetrahedron Letters*, **22**, 1919 (1981).
450. J. Bergman and L. Engman, *J. Amer. Chem. Soc.*, **103**, 2715 (1981).
451. S. Uemura, H. Miyoshi and M. Okano, *Chem. Letters*, 1357 (1979).
452. G. T. Morgan and H. D. K. Drew, *J. Chem. Soc.*, 2307 (1925).
453. L. Reichel and K. Ilberg, *Chem. Ber.*, **76**, 1108 (1943).
454. N. Petragnani, *Tetrahedron*, **11**, 15 (1960).
455. G. T. Morgan and R. E. Kellett, *J. Chem. Soc.*, 1080 (1926).
456. G. T. Morgan and O. C. Elvins, *J. Chem. Soc.*, 2625 (1925).
457. G. T. Morgan and H. D. K. Drew, *J. Chem. Soc.*, 731 (1924).
458. M. Ogawa, *Bull. Chem. Soc. Japan*, **41**, 3031 (1968).
459. M. Ogawa and R. Ishioka, *Bull. Chem. Soc. Japan*, **43**, 496 (1970).
460. H. J. Arpe and H. Kuckertz, *Angew. Chem.* (*Intern. Ed. Engl.*), **10**, 73 (1971).
461. S. Uemura and S. Fukuzawa, *J. Chem. Soc., Chem. Commun.*, 1033 (1980).
462. J. Bergman and L. Engman, *Tetrahedron*, **36**, 1275 (1980).
463. J. Bergman and L. Engman, *J. Organomet. Chem.*, **175**, 233 (1979).
464. S. Uemura, M. Wakasugi and M. Okano, *J. Organometal. Chem.*, **175**, 233 (1979).
465. T. Kawamura, K. Kikukawa, M. Takagi and T. Matsuda, *Bull. Chem. Soc. Japan*, **50**, 2021 (1977).
466. W. V. Farrar and J. M. Gulland, *J. Chem. Soc.*, 11 (1945).
467. S. C. Cohen, M. L. N. Reddy and A. G. Massey, *J. Chem. Soc.* (*D*), 451 (1967).
468. G. C. Hayward and P. J. Hendra, *J. Chem. Soc.* (*A*), 1760 (1969).
469. R. H. Vernon, *J. Chem. Soc.*, 86 (1920).
470. R. H. Vernon, *J. Chem. Soc.*, 687 (1921).
471. G. Vicentini, *Chem. Ber.*, **91**, 801 (1958).
472. H. Rheinboldt and G. Vincentini, *Chem. Ber.*, **89**, 624 (1956).
473. N. Petragnani, *Tetrahedron*, **12**, 219 (1961).
474. M. de Moura Campos and N. Petragani, *Tetrahedron*, **18**, 521 (1962).
475. G. T. Morgan and F. H. Burstall, *J. Chem. Soc.*, 2599 (1930).
476. H. D. K. Drew, *J. Chem. Soc.*, 223 (1926).
477. G. T. Morgan and H. Burgess, *J. Chem. Soc.*, 2214 (1929).
478. J. Bergman, R. Carlsson and B. Sjöberg, *Org. Synth.*, **57**, 18 (1977).
479. J. Bergman, *Tetrahedron*, **28**, 3323 (1972).
480. S. Uemura and S. Fukuzawa, *J. Amer. Chem. Soc.*, **105**, 2748 (1983).
481. D. Hellwinkel and G. Fahrbach, *Chem. Ber.*, **101**, 574 (1968).
482. G. Wittig and H. Fritz, *Ann. Chem.*, **577**, 39 (1952).
483. M. de Moura Campos, N. Petragnani and C. Thome, *Tetrahedron Letters*, 5 (1960).
484. K. Lederer, *Ann. Chem.*, **399**, 260 (1913).
485. K. Lederer, *Chem. Ber.*, **49**, 334 (1916).
486. K. Lederer, *Chem. Ber.*, **53**, 1430 (1920).
487. K. Lederer, *Chem. Ber.*, **44**, 2287 (1911).
488. M. P. Balfe, C. A. Chaplin and H. Phillips, *J. Chem. Soc.*, 341 (1938).
489. M. P. Balfe and K. N. Nandi, *J. Chem. Soc.*, 70 (1941).
490. K. Lederer, *Chem. Ber.*, **48**, 1345 (1915).
491. L. Reichel and E. Kirschbaum, *Ann. Chem.*, **523**, 211 (1936).
492. H. Lee and M. P. Cava, *J. Chem. Soc., Chem. Commun.*, 277 (1981).

The Chemistry of Organic
Selenium and Tellurium
Compounds Volume 1
Edited by S. Patai and Z. Rappoport
© 1986 John Wiley & Sons Ltd.

CHAPTER **17**

Stereochemistry and chiroptical properties of organic selenium and tellurium compounds

GÜNTHER SNATZKE

Lehrstuhl für Strukturchemie, Ruhruniversität Bochum, POB 10 21 48, D-4630 Bochum, FRG

I. GENERAL CONSIDERATIONS 667
II. SELENIUM OR TELLURIUM AS PART OF THE CHIRALITY ELEMENT 668
A. Selenonium and Telluronium Compounds 668
B. Selenoxides 668
C. Diselenides and Ditellurides 670
III. SELENIUM OR TELLURIUM NOT PART OF THE CHIRALITY ELEMENT 672
A. Selenides and Tellurides 672
B. Selenoketones. 675
C. Biopolymers Containing Selenium 676
D. Raman Optical Activity 676
E. Selenophosphinic Acid Derivatives 676
IV. ACKNOWLEDGEMENT 676
V. REFERENCES 676

I. GENERAL CONSIDERATIONS

Optical activity, as optical rotation or its dispersion with wavelengths (ORD), circular dichroism (CD) and the development of ellipticity, requires the presence of at least one element of chirality in its molecules[1]. Such may involve the Se or Te atom itself in cases where this is either a centre of chirality (selenonium and telluronium salts, selenoxides), or part of an axis of chirality (diselenides, ditellurides). When this chiral moiety also gives observable Cotton effects, a correlation may be possible between the (local) absolute configuration and the signs of these effects. For the diselenides and ditellurides a 'helicity rule', similar to the one known for disulphides[2], is expected, and for all other cases 'sector rules' may hold.

In order to estimate whether a sector rule may be valid at all, Ruch's[3] general algebraic treatment of chirality has first to be consulted. To this end the number n of ligand places in a transitivity area has to be determined, as well as the chirality order o. The first is the number of ligand places which are equivalent under the symmetry operations of the molecule, the second is the maximal number of alike ligands which can be present in a transitivity area without making this molecule achiral. Only if the difference $(n - o)$ equals 1 is a sector rule possible.

In the onium compounds $n = 3$ and for the selenoxides $n = 2$. As in both cases the chirality order $o = 1$, only for the latter is the difference $(n - o) = 1$ and a general sector rule possible; according to Schellman[4] the simplest one is a hemisphere rule. For the onium compounds, at most a 'local' hemisphere rule may be envisaged if one keeps one ligand constant and thus reduces the C_{3v} to C_s symmetry around the positive ion.

A comprehensive review of the chiroptical properties of Se and Te compounds (at that time mainly optical rotations and ORD data) appeared in 1973[5]; a shorter comparison between UV, CD and MCD (Magneto-CD) data was published more recently[6]. This present review summarizes mainly papers describing Cotton effects (anomalous ORD or CD).

II. SELENIUM OR TELLURIUM AS PART OF THE CHIRALITY ELEMENT

A. Selenonium and Telluronium Compounds

In analogy to the corresponding sulphonium salts a pyramidal arrangement can be assumed, and although crystal structure determinations have shown that selenonium[7] and telluronium[8] compounds contain a five-coordinate central atom, the R_3X moiety (X = Se or Te) has trigonal pyramidal geometry (average bond angles: C—Se—C 101 °, C—Te—C 94 °). NMR measurements have proved that this pyramid is fairly stable[6], so that optical activity can be expected with three different ligands. Indeed compounds **1** and **2** have been

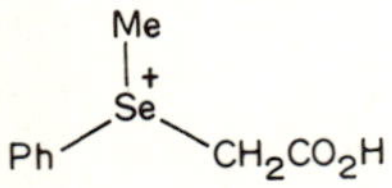

(1)

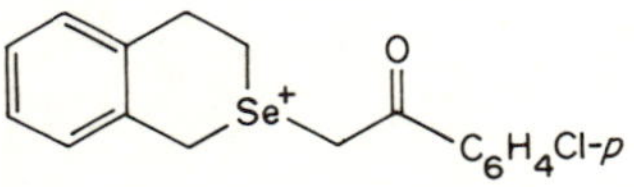

(2)

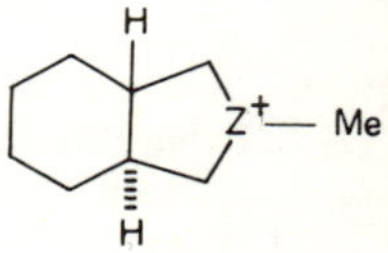

(3)

resolved[9,10] and both enantiomers prepared, although the absolute configuration has not been determined and only the rotation at the Na_D line recorded[9,10]. No Cotton effect in the accessible range has been observed for either type of compound (e.g. for **3**, Z = Se or Te), even with more modern equipment[6], with the exception of a telluronium iodide; the very weak Cotton effect for this compound around 270–290 nm has therefore been associated with some charge transfer between Te and I[6]. The MCD bands come only from the iodine anion[6].

B. Selenoxides

In general selenoxides are rather unstable, but a few optically active ones have been prepared and their chiroptical properties could be studied. Jones and coworkers[11] have succeeded in obtaining a few selenoxides in the steroid series (e.g. **4** and its diastereomer at Se) and their stereochemistry has been determined from their different rates in pyrolyses.

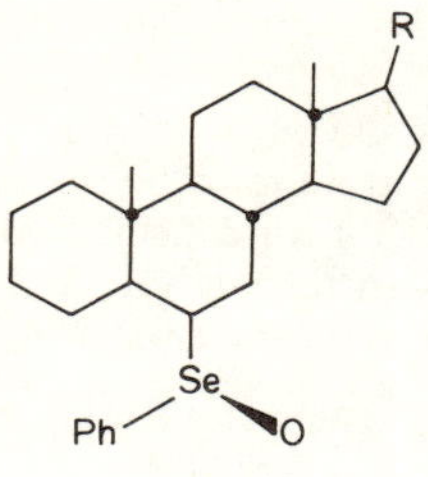

(4)

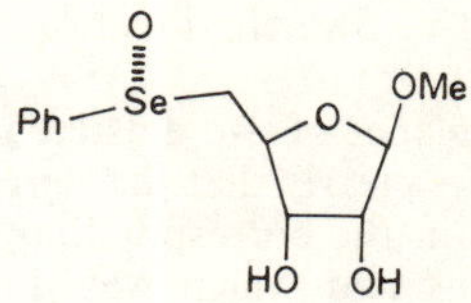

(5)

The (*S*) isomer showed a negative Cotton effect (by ORD) around 260–270 nm and this is in full agreement with the correlation between sign and stereochemistry in the case of the corresponding sulphoxides[11]. Most probably this is not actually a Se transition but one from the chirally perturbed benzene ring (α-band).

Several such chiral selenoxides which are surprisingly thermally stable have been prepared by Zylber and coworkers[12] in the nucleoside series. X-ray analysis has proved the (*S*) configuration of one of the compounds which give a positive CD band around 250 nm. As, however, nucleosides alone give Cotton effects in the same range, the assignment of the CD band is not unambiguous. Hence, for investigation of the chiroptical properties, the simple methyl glycoside **5**, whose chemical and NMR properties suggest the same configuration at the chiral Se atom, was prepared[12]. Its CD spectrum is given in Figure 1. Three Cotton effects around 250, 220 and 200 nm are clearly visible, but the long-wavelength tail of the first CD band is very broad, so that an additional (positive) Cotton effect could still be present around 270 nm. Of course, the older ORD curves did not show such subtleties[11], but the positive sign of the 250 nm CD band agrees with the results found

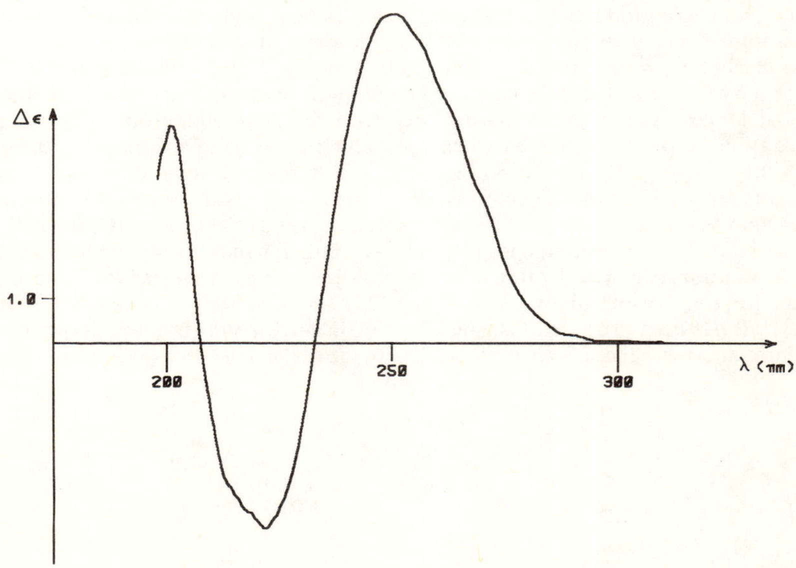

FIGURE 1. CD spectrum of **5** in acetonitrile solution.

for **4**, if indeed **5** and its adenosine analogue have the same absolute configuration at the Se centre.

The existence of two diastereomeric selenoxides which give Cotton effects of opposite signs shows clearly that the stereochemistry around the Se atom must be similar to that around S in the corresponding sulphoxides; X-ray diffraction of **6**, which could have proved this and which was done side-by-side with the corresponding S compound revealed, however, that this compound is present exclusively in the crystal as the ring-closed isomer[13].

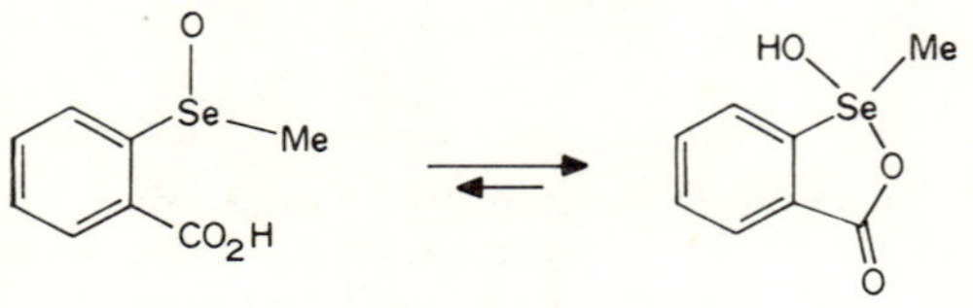

(6)

C. Diselenides and Ditellurides

As in the isologous O and S compounds, the torsional angle (C—)X—X(—C) in diselenides and ditellurides is approximately 90° (e.g., 87.5° for dimethyl diselenide in the gas phase[14] and 87.7° for di-*p*-tolyl telluride in the crystalline phase[15]). The barrier to rotation is, however, smalller than for disulphides[6,16], so that in the absence of chiral ligands the racemates cannot be resolved.

In the presence of a chiral ligand diastereomers are possible, but the deviation from a 1:1 ratio for the P- and M-helical forms is nevertheless expected to be quite small. Any CD measurement for such compounds will thus mirror the chiral perturbation of the diselenide chromophore (and should then follow a sector rule) rather than be due to its inherent helicity. Furthermore, in analogy to the corresponding disulphides, degeneracy of the first two absorption bands is expected for a torsional angle of around 90° and as the two Cotton effects have opposite signs they will approximately cancel each other.

The chiroptical properties of the inherently chiral isologous disulphides have been predicted by theoretical calculations at different levels of sophistication[2,17]; the first two bands at longest wavelengths are associated with the excitations from the n^- and n^+ combination, respectively, of the two energetically highest lying n orbitals on the two S atoms into the σ^* MO of the S—S bond, and a positive torsional angle (ranging approximately from 15° to 75°) leads to a positive first, and negative second, Cotton effect. These same results can also easily be rationalized by using qualitative MO theory[1b]. The same correlation has been found for diselenides[18], but all bands are shifted bathochromically. In addition, two other Cotton effects have been detected at shorter wavelengths, and the data for, e.g., **7** are as follows: 364 (− 3.0), 275 (+ 4.2), 237sh (− 2.0) and 220 nm ($\Delta\varepsilon$ = − 17.1)[6]. It has been proposed that the 237 nm CD corresponds to the $n^- \rightarrow \sigma^*$ (C—Se) transition and the band at shortest wavelengths to the $\sigma \rightarrow \sigma^*$ transition of the Se—Se

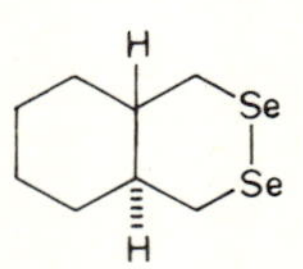

(7)

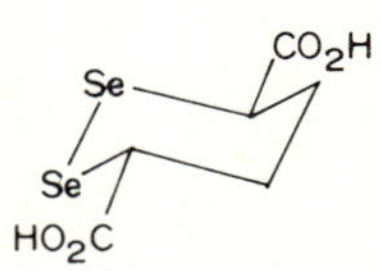

(8)

moiety. Since the rotational strength of the 220 nm band is, however, one order of magnitude larger than that of the 237 nm shoulder, and the excitation from n^- into the energetically lower lying combination with A-symmetry of the two σ^* (C—Se) moieties is associated with the stronger magnetic transition moment, we believe that the assignment of these two bands should be exchanged.

The torsional (C—)Se—Se(—C) angle of the preferred conformation of the diselenane **8** is + 57°, since the two carboxylic groups adopt equatorial position[19]; in agreement with this the first two Cotton effects are positive and negative, respectively: 351 (+ 1.43), 277 (− 2.39) and 249 nm ($\Delta\varepsilon = +4.90$)[18b]. As the band positions and magnitudes of these Cotton effects do not drastically deviate from those of **7** the carboxylic groups in the position α to the Se_2 grouping do not strongly influence the CD.

The chiroptical data of open-chain diselenides are much smaller: for the dialkyl diselenide **9** Cotton effects at 342 (+ 0.10), 286 (− 0.12), 233sh (+ 0.85) and 218 nm ($\Delta\varepsilon =$ + 18) have been reported[6], whereas the isologous ditelluride **10** gives the following

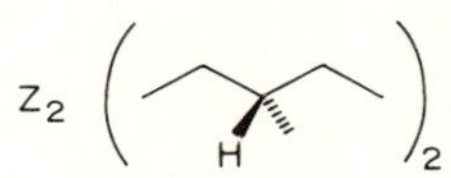

(**9**) Z = Se
(**10**) Z = Te

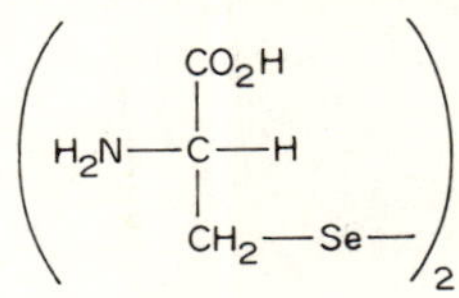

(**11**)

data[6]: 442 (+ 0.032), 350 (− 0.022), 279 (+ 0.35) and 236 nm ($\Delta\varepsilon = +0.74$). Again, strong bathochromicity is found on going from the Se_2 to the Te_2 compound, but all the corresponding Cotton effects have the same sign for these homochirally analogous compounds. When the torsional angle of the (C—)Z—Z(—C) (Z = S, Se or Te) unit is fixed around 60° the UV and CD maxima at long wavelengths coincide approximately. In case of open-chain compounds such as **9** and **10** (and their S isologues) the first UV maximum lies, on the contrary, between two (small) Cotton effects of opposite signs[6], as predicted from theory[20] for an almost perpendicular geometry of the two halves. The position of the first band follows in the Se_2 compounds the easily explained trend[2,17] that the longer is λ_{max} the closer the torsional angle (C—)Se—Se(—C) approaches to 0°.

The CD of the Se isologue **11** of L-cystine and of its enantiomer have been investigated in more detail[21]. In the protonated form (pH = 0.5) the CD bands are at 306 (− 0.24), 241 (− 0.88), 225 (+ 1.10) and 207 nm ($\Delta\varepsilon = -6.1$). Instead of two Cotton effects at longest wavelengths as found for **9**, only one such CD band is observed. Furthermore, the 225 nm Cotton effect is most probably due to the carboxylic acid chromophore, because it is very similar to that found for other L-amino acids. At pH 5.5 the zwitterion is present in solution, and the positions of its Cotton effects resemble more those of the simple reference compound **9**: 320 (− 0.46), 272 (+ 0.32), 232 (− 2.53), 211 (− 3.1) and 200 nm ($\Delta\varepsilon = +3.6$). Preference of M-helicity of the C—Se—Se—C moiety has be inferred from these values[21].

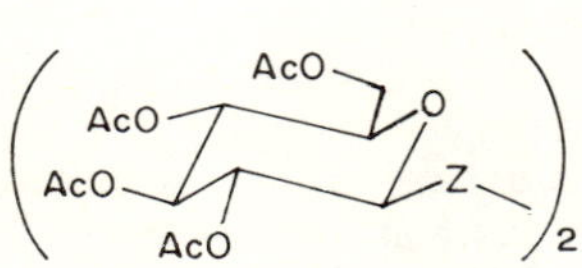

(**12**) Z = Se
(**13**) Z = Te

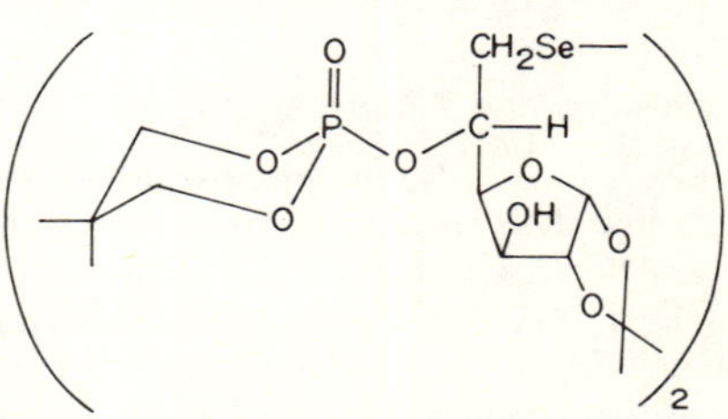

(**14**)

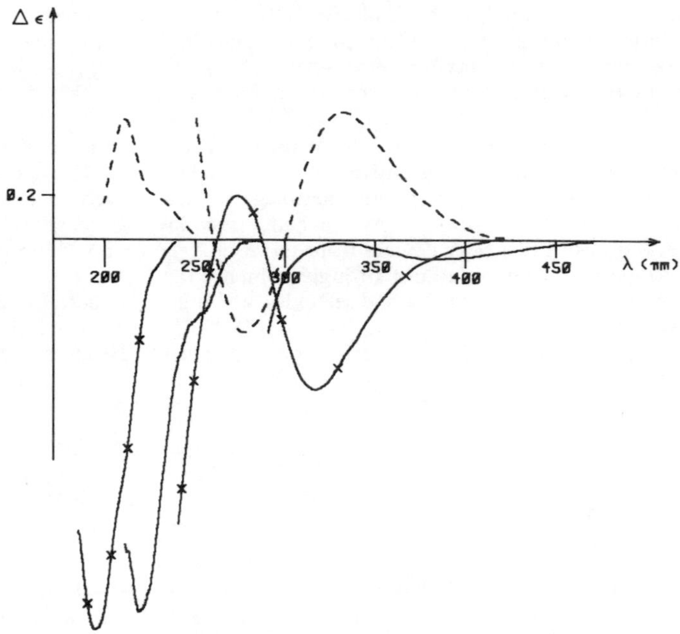

FIGURE 2. CD spectra of **12** (-×-×-), **13** (——) and **14** (----) in acetonitrile solution. Curves at shorter wavelengths were reduced by a factor of 10 for **13** and **14** and by a factor of 25 for **12**.

Many other diselenides have been prepared, but only their rotations at the Na_D line have been reported[22]. We have measured the full CD curve for two diselenides **12** and **14** in the sugar series[23] as well as that of **13**, the Te_2 isologue of **12**[23] (Figure 2). Three Cotton effects have been observed: e.g., **12** gives CD maxima at 314 (−0.66), 273 (+0.22) and 194 nm ($\Delta\varepsilon = -45$). These band positions differ somewhat from those in the CD spectrum of **9**, but resemble more those of **11** in the zwitterion form. The long-wavelength tail of the first Cotton effect extends out to more than 400 nm; it cannot be excluded, therefore, that another weak Cotton effect of same sign as the 315 nm effect is still hidden around 370 to 380 nm although the shape of the CD curve matches perfectly that of a Gaussian curve of half-band width 6170 cm^{-1}. Moreover, for these sugar derivatives the Cotton effects of the Te_2 compound are smaller than those of the Se_2 isologue.

III. SELENIUM OR TELLURIUM NOT PART OF THE CHIRALITY ELEMENT

A. Selenides and Tellurides

For isologous dialkyl sulphides the first three Cotton effects appear around 240, 220 and 200 nm, and theoretical considerations lead to the assignment that the first two correspond to $n \rightarrow \sigma^*$ (C—S) transitions and in the third one a Rydberg transition may be involved[24]. The CD spectrum of the (−)-(*S,S*)-*trans*-selenahydrindane **15** also shows three Cotton effects at 273 (−1.7), 218 (+2.9) and 200 nm ($\Delta\varepsilon = +4.4$)[6], and they even have the same signs as their counterparts in the spectrum of its homochirally analogous S isologue for which, for technical reasons, only first and second Cotton effects have been measured[6].

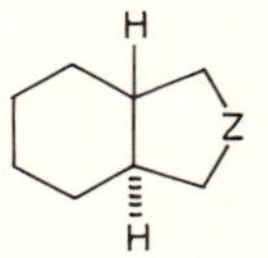

(**15**) Z = Se
(**16**) Z = Te

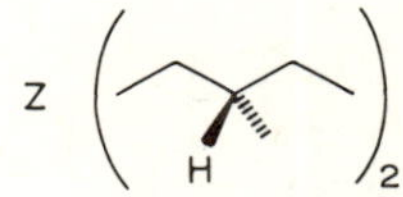

(**17**) Z = Se
(**18**) Z = Te

The corresponding data for the Te compound **16** are 350 (− 1.38), 260 (− 0.35), 240 (+ 4.4) and 200 nm ($\Delta\varepsilon = +2.9$)[6]. The same parentage for the first three transitions as for the sulphides seems therefore probable at least for the selenides but the MCD spectra are somewhat more complex[6]. It is, therefore, tentatively assumed that for the absorptions at longest wavelengths excited triplet states may be involved[6].

Even less clear is the situation with open-chain compounds, because several conformers may be present; the CD spectrum of the selenide **17** contains three bands at 274 (− 0.027), 251 (+ 0.125) and 220 nm ($\Delta\varepsilon = +1.1$) in the accessible range[6], but the corresponding telluride **18** shows five bands: 378 (+ 0.002), 355 (− 0.0055), 312 (+ 0.13), 239 (− 0.48) and 223 nm ($\Delta\varepsilon = -0.28$)[6]. Since at lower temperatures this latter CD spectrum simplifies and the Cotton effects become larger, conformational equilibria may (at least to a great extent) be the reason for the complexity of these CD spectra.

The CD spectra of a few selenoamino acids, such as the Se isologue of lanthionine, have been published[25]; with one exception only a single Cotton effect can be recognized, and all CD maxima are at wavelenths not longer than 220 nm, as for any amino acid. These Cotton effects might, therefore, be mainly due to $n \rightarrow \pi^*$ transitions within the COOR chromophore.

Many optical rotations at the Na_D line but only a very few Cotton effects have been published for sugar and amino acid derivatives (see Ref. 22) in which Se replaces O or S. Selenophenyl esters of amino acids give a Cotton effect around 284 nm, and selenonaphthyl esters at ca. 291 nm[26]. Acetylation at the amino group leads to sign inversion and bathochromicity of this Cotton effect; thus it has been ascribed to an $n \rightarrow \pi^*$ transition mainly within the seleno ester moiety[26]. For the β-phenylselenolacetone **19** small but distinct Cotton effects have been found[27] at 305 (+ 0.19), 278 (+ 0.20), 251 (+ 0.95), 242 (− 0.41) and 223 nm ($\Delta\varepsilon = +3.66$), of which only the last one should come from the lactone chromophore; the others involve both the aromatic π system and the Se atom. It is, therefore, more probable that the 284 nm Cotton effect of the phenylseleno esters also has a large contribution from a $\pi \rightarrow \pi^*$ transition of the benzene ring.

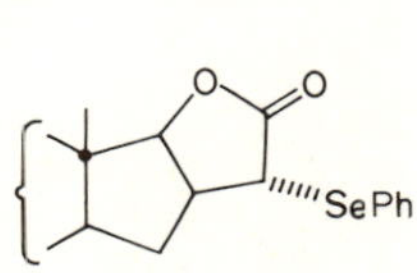

(**19**)

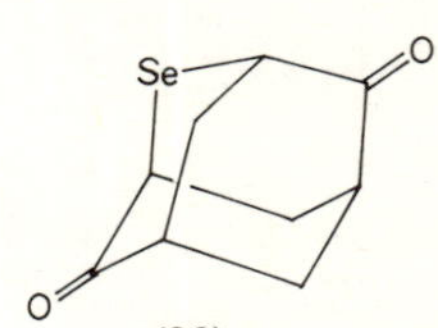

(**20**)

A strong interaction between the Se atom and the two carbonyls of the selenaadamantanedione **20** can be inferred from its CD spectrum[28]: in this case Se acts (similarly to S) in the same way as an axial halogen in a position α to a carbonyl, and therefore the CD is extremely strong for the carbonyl $n \rightarrow \pi^*$ band ($\Delta\varepsilon = +33.20$ at 308 nm); in addition, however, bands are also found at 336 (− 7.70), 322sh (app. + 12), 253 (− 8.00) and 212 nm (+ 8.10). Similar, though slightly smaller, large values are found for the S isologue[28,29].

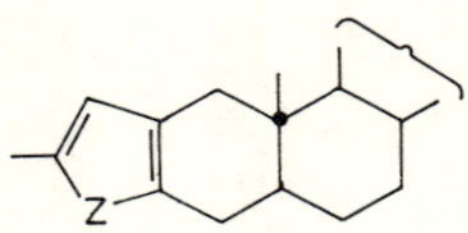

(**21**) Z = Se
(**22**) Z = S

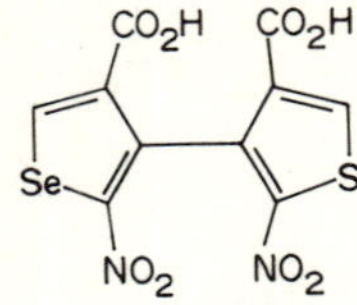

(**23**)

Analogy between the spectra is expected in the series furan, thiophene, selenophene, as overlap between the 2p AOs of the butadiene system and the chalcogen AOs with main quantum number 3 or 4 is not very good. Yet the usual bathochromic shift is found, as seen by comparison of the CD spectra of **21** and its S isologue **22** (Figure 3)[30]. A strong positive Cotton effect (determined by ORD measurements) has been published for the bis-

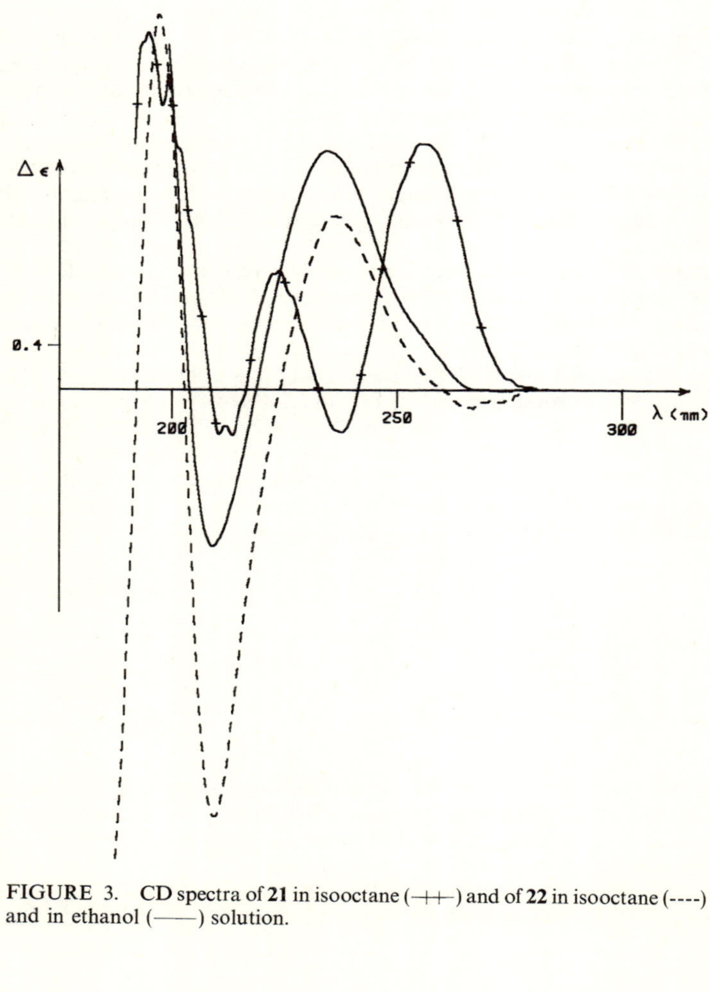

FIGURE 3. CD spectra of **21** in isooctane (-++-) and of **22** in isooctane (----) and in ethanol (——) solution.

selenophene derivative **23** with *R*-axial chirality[31]. Here also the homochirally analogous S isologue gives a very similar Cotton effect[32].

Addition of Br_2 or I_2 to **16** leads to the corresponding dihalogeno adducts. The CD data for the dibromo compound **24** are 285 (−6.3), 265 (−6.2) and 233 nm ($\Delta\varepsilon = +22.4$)[6]. Hence, the Cotton effect at longest wavelengths is strongly blue-shifted as compared to that of its halogen-free parent compound, and all CD values are intensified. The CD spectrum of the iodo product is, however, completely different[6].

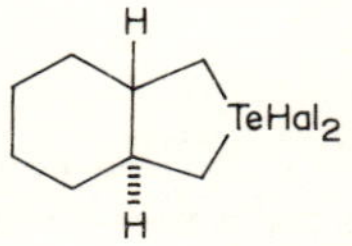

(24)

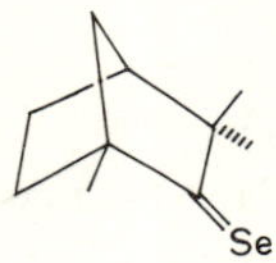

(25)

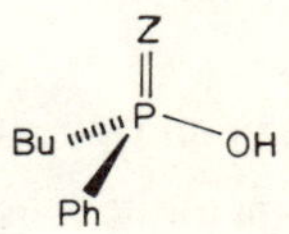

(26) Z = S
(27) Z = Se

B. Selenoketones

The CD of selenofenchone (**25**) derived from (−)-fenchone was recently published[33], together with that of fenchone and its thioketone derivative, but no Cotton effect was given for the absorption band found at 626 nm in the UV/visible spectrum. Obviously this was for instrumental reasons, and remeasurement of a new sample[34] gave the CD spectrum shown in Figure 4. The missing CD band was at 638 nm ($\Delta\varepsilon = +0.033$, cyclohexane solution). In the same stereochemical series the main bands at longest wavelengths have for all 3 compounds the same signs, which, at least for the ketone and its thio derivative, is in agreement with recent calculations[35]. It is possible that the 638 nm Cotton effect comes from an excitation into a triplet state, but it is also not uncommon that $n \rightarrow \pi^*$ CD bands are bisignate for one single electronic transition.

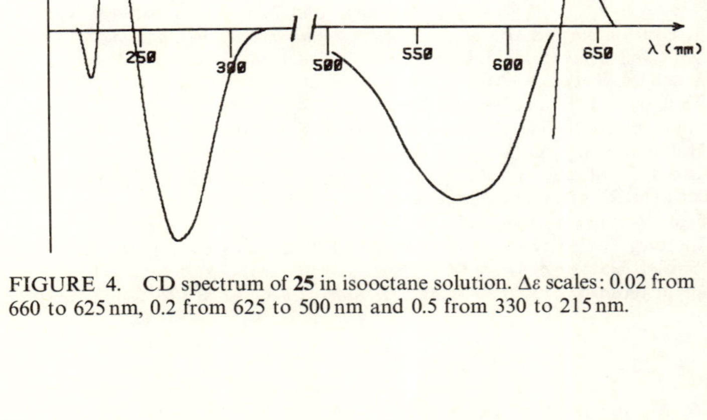

FIGURE 4. CD spectrum of **25** in isooctane solution. $\Delta\varepsilon$ scales: 0.02 from 660 to 625 nm, 0.2 from 625 to 500 nm and 0.5 from 330 to 215 nm.

C. Biopolymers Containing Selenium

In diaminooxytocin the S_2 bridge has been replaced by Se_2 and both possible SeS combinations[36]. The identification of the corresponding CD bands is, however, complicated by the presence of tyrosine in the peptide. The labile S atom of natural parsley ferredoxin has also been replaced by Se (approximately 1 Se per atom Fe)[37], and the CD spectra of the fully oxidized, the 50% reduced and the fully reduced forms have been compared in both series. A slight bathochromic shift is observed by this exchange, but the bands are most probably associated with d–d transitions of the Fe.

D. Raman Optical Activity

Data for Raman optical activity have been published in two spectral ranges for **15**, but no interpretation has been tried[6].

E. Selenophosphinic Acid Derivatives

Alkyl(phenyl)thiophosphinic acids (like **26**) complex *in situ* with dimolybdenum tetraacetate in trifluoracetic acid solution and give CD bands around 480 and 400 nm (and eventually around 540 nm)[38]. The isologous Se compound **27** also forms such a complex and gives (besides the 540 nm CD band) practically the same CD curve as **26** of the same absolute configuration[38].

IV. ACKNOWLEDGEMENT

I thank Mrs. E. Sauerbier and Mr. U. Wagner for most valuable technical assistance, (Mrs.) Prof. Michalska, Prof. Okazaki, Prof. Wicha and Dr. Zylber for providing us with samples, and Fonds der Chemischen Industrie, Deutsche Forschungsgemeinschaft and Hoechst AG for financial support.

V. REFERENCES

1. For recent reviews see (a) M. Legrand and M. J. Rougier, in *Stereochemistry*, Vol. 2 (Ed. H. B. Kagan), Georg Thieme Verlag, Stuttgart, 1977, p. 32; (b) G. Snatzke, *Angew. Chem.*, **91**, 380 (1979), *Intern. Ed.*, **18**, 363 (1979); (c) S. F. Mason, *Molecular Optical Activity and the Chiral Discriminations*, Cambridge University Press, Cambridge, 1982.
2. J. Linderberg and J. Michl, *J. Amer. Chem. Soc.*, **92**, 2619 (1970).
3. E. Ruch and A. Schönhofer, *Theoret. Chim. Acta*, **19**, 225 (1970).
4. J. A. Schellman, *J. Chem. Phys.*, **44**, 55 (1966).
5. G. Snatzke and M. Kajtár, in *Chemistry of Selenium Compounds: their Chemistry and Biology* (Eds. D. L. Klayman and W. H. H. Günther), John Wiley and Sons, New York, 1973, p. 885.
6. P. H. A. Laur, in *Proceedings of the Third International Symposium on Organic Selenium and Tellurium Compounds* (Eds. D. Cagniant and G. Kirsch), Université de Metz, 1981, p. 217.
7. R. V. Mitcham, B. Lee, K. Mertes and R. F. Ziolo, *Inorg. Chem.*, **18**, 3498 (1979).
8. R. F. Ziolo and M. Extine, *Inorg. Chem.*, **19**, 2694 (1980).
9. W. J. Pope and A. Neville, *J. Chem. Soc.*, **81**, 1552 (1902).
10. F. G. Holliman and F. G. Mann, *J. Chem. Soc.*, 37 (1945).
11. D. N. Jones, D. Mundy and R. D. Whitehouse, *J. Chem. Soc. (D)*, 86 (1970).
12. J. Zylber (Thiais), private communication.
13. B. Dahlén, *Acta Cryst.*, **B29**, 595 (1973).
14. P. D'Antonio, C. George, A. H. Lowrey and J. Karle, *J. Chem. Phys.*, **55**, 1071 (1971).
15. M. R. Spirlet, G. v. d. Bossche, O. Dideberg and L. Dupont, *Acta Cryst.*, **B35**, 1727 (1979).
16. V. Renugopalakrishnan and R. Walter, *J. Amer. Chem. Soc.*, **106**, 3413 (1984).

17. (a) M. Carmack and L. A. Neubert, *J. Amer. Chem. Soc.*, **89**, 7134 (1967).
(b) J. P. Casey and R. B. Martin, *J. Amer. Chem. Soc.*, **94**, 6141 (1972).
(c) R. W. Woody, *Tetrahedron*, **29**, 1273 (1973).
(d) R. Nagarajan and R. W. Woody, *J. Amer. Chem. Soc.*, **95**, 7212 (1973).
(e) J. Webb, R. W. Strickland and F. S. Richardson, *J. Amer. Chem. Soc.*, **95**, 4775 (1973).
(f) L. A. Neubert and M. Carmack, *J. Amer. Chem. Soc.*, **96**, 943 (1974).
18. (a) C. Djerassi, A. Fredga and B. Sjöberg, *Acta Chem. Scand.*, **15**, 417 (1961).
(b) C. Djerassi, H. Wolf and E. Bunnenberg, *J. Amer. Chem. Soc.*, **84**, 4552 (1962).
19. O. Foss, K. Johnsen and T. Reistad, *Acta Chem. Scand.*, **18**, 2234 (1964).
20. (a) Th. Bürer, *Helv. Chim. Acta*, **46**, 2388 (1963).
(b) K. M. Wellman, P. H. A. Laur, W. S. Briggs, A. Moscowitz and C. Djerassi, *J. Amer. Chem. Soc.*, **87**, 66 (1965).
21. B. Ringdahl, J. Cymerman Craig, G. Zdansky and A. Fredga, *Acta Chem. Scand.*, **B34**, 735 (1980).
22. For a detailed list of these older papers see Ref. 5.
23. M. Michalska (Łódź), private communication.
24. J. S. Rosenfield and A. Moscowitz, *J. Amer. Chem. Soc.*, **94**, 4797 (1972).
25. J. Cymerman Craig, S.-Y. C. Lee, G. Zdansky and A. Fredga, *J. Amer. Chem. Soc.*, **98**, 6456 (1976).
26. K. Bláha, I. Frič and H.-D. Jakubke, *Collect. Czech. Chem. Commun.*, **32**, 558 (1967).
27. J. Wicha and coworkers (Warszawa), unpublished results.
28. G. Graf, *Dissertation*, Bochum, 1984.
29. G. Snatzke and B. Wolfram, *Tetrahedron*, **28**, 655 (1972).
30. W. Lindner, *Dissertation*, Bochum, 1982.
31. C. dell'Erba, D. Spinelli, G. Carbarino and G. Leandri, *J. Heterocycl. Chem.*, **5**, 45 (1968).
32. S. Gronowitz and P. Gustavson, *Arkiv Kemi*, **20**, 289 (1963).
33. K. K. Andersen, D. M. Gash, J. D. Robertson and F. S. Guziec, Jr., *Tetrahedron Letters*, **23**, 911 (1982).
34. R. Okazaki, A. Ishii and N. Inamoto, *J. Chem. Soc., Chem. Commun.*, 1429 (1983).
35. D. A. Lightner, T. D. Bouman, W. M. D. Wijekoon and A. E. Hansen, *J. Amer. Chem. Soc.*, **106**, 934 (1984).
36. (a) R. Walter and V. du Vigneaud, *J. Amer. Chem. Soc.*, **87**, 4192 (1965); **88**, 1331 (1966).
(b) R. Walter and W. Y. Chan, *J. Amer. Chem. Soc.*, **89**, 3892 (1967).
37. J. A. Fee and G. Palmer, *Biochim. Biophys. Acta*, **245**, 175 (1971).
38. J. Omelańczuk and G. Snatzke, *Angew. Chem.*, **93**, 815 (1981); *Intern. Ed.*, **20**, 786 (1981).

The Chemistry of Organic
Selenium and Tellurium
Compounds Volume 1
Edited by S. Patai and Z. Rappoport
© 1986 John Wiley & Sons Ltd.

CHAPTER **18**

Ligand properties of organic selenium and tellurium compounds

HENRY J. GYSLING

Research Laboratories, Eastman Kodak Company, Rochester, New York 14650, USA

I. INTRODUCTION . . . 680
A. Abbreviations . . . 681
II. ORGANOSELENIUM LIGANDS . . . 683
A. Neutral Monodentate Ligands . . . 683
1. Selenourea . . . 683
2. Diorganoselenides (SeR_2; R = alkyl, aryl) . . . 690
3. Triorganophosphine selenides ($Se{=}PR_3$) . . . 691
4. Selenium heterocycles . . . 699
5. Diorganodiselenides (RSeSeR) and related ligands ($RSeER^1_2$; E = P, As) . . . 709
6. $Se(EMe_3)_2$ (E = Ge, Sn, Pb) . . . 714
7. Selenocarbonyl complexes (M—CSe) . . . 716
8. Diorganoselenoxides (R_2SeO) . . . 721
9. Selenium-functionalized carbene complexes ($M{=}C(SeR)(R^1)$) . . . 722
10. Miscellaneous ligands . . . 725
B. Anionic Monodentate Ligands . . . 730
1. Selenocyanates . . . 730
2. Selenols (RSe^-) . . . 741
3. SeH^- . . . 754
4. $SeER_3^-$ (E = Ge, R = Ph; E = Sn, R = Me) . . . 758
5. Organoseleninates ($RSeO_2^-$) . . . 759
6. Dialkylselenocarbamates ($-SeC(O)NR_2^-$) . . . 762
7. Miscellaneous ligands . . . 765
C. Neutral Bidentate Ligands. . . . 765
1. (Se, C) donors . . . 765
a. CSe_2 complexes . . . 765
b. CSSe complexes . . . 767
c. Selenoformaldehyde complexes . . . 768

2. (Se, N) and (Se, P) donors 770
3. (Se, E) (E = O, S, Se) donors 773
D. Anionic Bidentate Ligands 778
1. $(Se, N)^-$ and $(Se, P)^-$ donors 778
2. $(Se, E)^-$ (E = O, S, Se) donors 782
a. $(Se, O)^-$ donors 785
b. $(Se, S)^-$ donors 786
c. $(Se, Se)^-$ donors. 793
E. Tridentate Ligands 801
F. Tetradentate Ligands 802
G. Complexes Incorporating Se_n. 803
1. μ_2-Se ligand 804
2. η^2-Se_2 ligand 806
3. μ_2-Se_2 and μ_2, η^2-Se_2 ligands 808
4. η^2-Se_4 and η^2-Se_5 ligands 810
5. Se incorporated in metal clusters 811
III. ORGANOTELLURIUM LIGANDS. 815
A. Neutral Monodentate Ligands 815
1. Diorganotellurides 815
a. Dialkyl tellurides 815
b. Diaryl tellurides. 818
2. Diorganoditellurides 818
3. Miscellaneous ligands 819
B. Anionic Monodentate Ligands 824
C. Bidentate Ligands. 828
D. Complexes Incorporating Te_n ($n = 1, 2$) 831
IV. REFERENCES 839

I. INTRODUCTION

Selenium and tellurium ligands have, until recently, been rather neglected in the renaissance of transition-metal coordination and organometallic chemistry[1,2] beginning in the late 1950s. Although selenourea[3] was first described as a ligand in 1886, and dialkyl selenide[4–7] and telluride complexes[6,8] of Pd(II) and Pt(II) were first reported near the beginning of this century, the ligand chemistry of these two elements has been relatively unexplored compared to the lighter Group 5a (N[1,9], P[1,10]) and 6a (O[1,11], S[1,12]) elements. This situation is probably, to a considerable extent, the result of the commonly held assumption that the organometallic derivatives of Se and Te are generally toxic, foul-smelling, air-sensitive materials. Indeed, a review by Murray and Hartley[13] in 1981 of transition-metal cordination complexes of thio, seleno and telluro ethers includes the statement: 'It is difficult to prepare all but the simplest organotellurium ligands, and at the present time only monodentate tellurium ligands are known.'

This general misconception is no doubt based on the fact that the early work in this area did indeed involve rather unstable and foul-smelling Se and Te ligands (e.g. selenourea and low-molecular-weight dialkyl selenides and tellurides). Current methodology, however, allows the synthesis of a wide variety of stable Se[14] and Te[15] ligands.

For example, increasing the chain length of the simple dialkyl derivatives can give air-stable solids (e.g. $Te(C_{16}H_{33}\text{-}n)_2$[16], m.p. 45 °C), and the aromatic derivatives are generally air-stable solids. Ironically, the structural diversity of Se and Te ligands is considerably greater than that of the most widely used ligand class in modern coordination chemistry, triorganophosphines[1,10]. Recent work in Se[17] and Te[15,18] ligand chemistry has centred on the synthesis of new ligand types rather than on the synthesis and detailed studies of the properties of metal complexes with the most common ligands of these elements (e.g. ER_2[13] and ER^-[19], E = Se, Te).

Only a few descriptions of applications-oriented work have appeared. A few examples have been reported of metal selenide electroless plating solutions that contain Se ligands in combination with main-group metal halides (PbSe[20], Sb_2Se_3[21], TlSe[22], CdSe[23]).

A few examples have been reported of the use of Ag(I)[24] and Cu(I)[25] complexes with organotellurium ligands in thermally processed imaging elements. The use of $PtCl_2(SePh_2)_2/SnCl_2$ as a homogeneous catalyst for the hydrogenation of non-aromatic olefins has also been reported[26].

Recently, metal complexes with 1,2-diseleno ligands have been prepared and their conductivities measured[27]. The current interest in 'organic metals'[28] suggests that this chemistry will be the subject of future extensions with both Se ligands and their Te analogues.

A variety of chemical vapour deposition processes for thin-film fabrication (e.g. CdTe, ZnTe, PbTe, SnTe) have been described in which TeR_2 (R = Me, Et)[29] is codeposited on a hot substrate with an appropriate metal alkyl (in corresponding processes for metal selenide deposition, H_2Se is generally used as the chalcogen source[29,30]).

With the recent increased interest in Se[13,14,31] and Te[15g,18,31] ligands, as evidenced by publications describing such complexes, more fundamental studies of such complexes can be expected. Of particular interest are comparative studies of the Se and Te ligand analogues of the many well-established transition-metal phosphine complexes that have applications as homogeneous catalysts[32]. As this ligand chemistry is developed, it can be expected to produce new transition-metal coordination complexes with a variety of applications. The diverse scope of Se and Te ligands should allow for considerable flexibility in fine-tuning the electronic and steric properties of coordination complexes with these types of ligands.

This chapter will summarize the various types of Se and Te ligands that have been reported. Since the emphasis of this volume is on the organic chemistry of these elements, the classification used is based on the number and type of donor atoms incorporated in the ligands. To avoid repetition of structures as well as to define the various bonding modes either established by single-crystal X-ray diffraction or proposed on the basis of spectral data, tables in each section summarize this structural information.

Finally, since comprehensive monographs and reviews have summarized the synthetic methodology associated with many of the Se[14] and Te[15] ligands described here, discussions of this aspect will be limited to recent work.

Table 1 summarizes the notations used to indicate the ligand bonding modes in chelates, dimers and clusters.

A. Abbreviations

acac	= acetylacetonato
bipyr	= 2,2′-bipyridine
COD	= 1,5-cyclooctadiene
Cp	= η^5-cyclopentadienyl
Cp′	= η^5-pentamethylcyclopentadienyl
Cp″	= η^5-methylcyclopentadienyl
detc	= *N,N*-diethyldithiocarbamato; $Et_2NCS_2^-$
DME	= 1,2-dimethoxyethane
DMF	= *N,N*-dimethylformamide
dmgH	= dimethylglyoxime

TABLE 1. Notations for ligand bonding modes

Prefix	Bonding mode
μ_n	Ligand bridges *n* metal centres in the complex
	~M(μ_2-SCN)$_2$M~ = 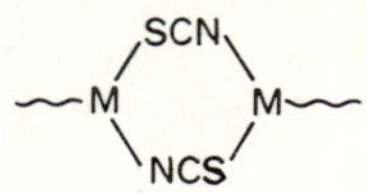
	~M(μ_2-Se$_2$)M~ = 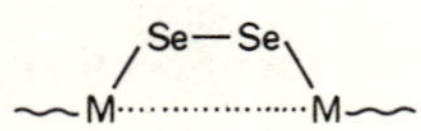
	~M(μ_2-Se)$_2$M~ = 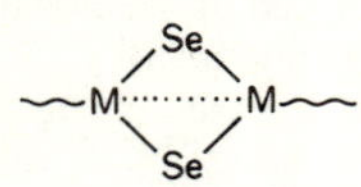
	~M$_3$(μ_3-Se)~ = 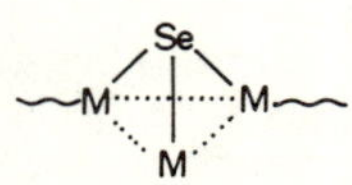
	(M···M; a metal–metal bond may or may not be present)
η^n	Ligand bonds to a single metal atom through *n* sites of the ligand
	(η^1-C$_5$H$_5$)M = (η^1-Cp)M = 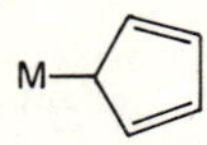
	(η^5-C$_5$H$_5$)M = (η^5-Cp)M = 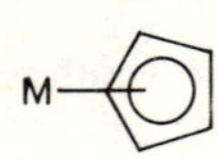
	(η^2-Se$_2$)M~ = 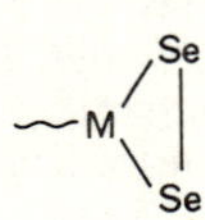
	(η^2-SeCH$_2$)M~ = 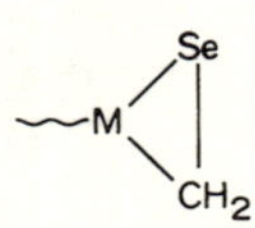
μ_n, η^m	The ligand both bridges *n* metal centres and chelates to a metal centre via *m* sites on the ligand
	(μ_2, η^2-Se$_2$)M$_2$~ = 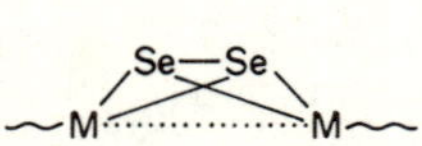

en = ethylenediamine
esu = ethyleneselenourea
F_6acac = hexafluoroacetylacetonato
HMPA = hexamethylphosphoramide
i-Am = isoamyl
dppe = $Ph_2PCH_2CH_2PPh_2$
NBD = norbornadiene
np_3 = $N(CH_2CH_2PPh_2)_3$
phen = 1,10-phenanthroline
PPN = $[(Ph_3P)_2N]^+$
su = selenourea
THF = tetrahydrofuran
triphos = $CH_3C(CH_2PPh_2)_3$
XPS = X-ray photoelectron spectroscopy

II. ORGANOSELENIUM LIGANDS

A. Neutral Monodentate Ligands

1. Selenourea

Although the first reported complexes with a Se ligand involved selenourea (i.e. the poorly characterized complexes $Ag(su)_2Cl$, $Hg(su)_2Cl$ and $Hg(su)Cl)$[3], the oxidative instability of this ligand and its *N*-alkyl-substituted derivatives has limited the study of its coordination chemistry. Indeed, the isolation of these complexes was apparently fortuitous (e.g. the use of acid solutions of su and correct su/metal stoichiometries), since a subsequent paper[33] describing the reaction of AgCl with a concentrated aqueous su solution reported the deposition of elemental Ag and Se and yellow crystals of α,α'-diselenobisformamidinium dichloride via a redox process. The same su oxidation product was obtained by oxidation of a cooled HCl–ethanol solution of su with H_2O_2[33]; it was characterized by a single-crystal X-ray diffraction study (equation 1). In the first detailed study[34] of the ligand properties of su, it was reported to form the yellow $[Bi(su)_9]^{3+}$ and red $[Bi(su)_{12}]^{3+}$ complexes in 0.1M HNO_3 solution. Although the solid complexes were not isolated, the stability constants of the two species were estimated to be 10^{-2} and $10^{-2.5}$, respectively. The stability and intense colour of these complexes suggested their use for qualitative and quantitative detection of this element (e.g. sensitivity limit of 1 μg/ml for the yellow complex and 2.5 μg/ml for the red complex). Qualitative observations of the reactions of su solutions with various metal salts were also reported by these authors[34]; Cu(II), Hg(II), Ag(I), Sn(II), Au(III) and Pd(II) all gave precipitates that dissolved in excess su, and Pb(II) gave a yellow precipitate that was insoluble in excess su.

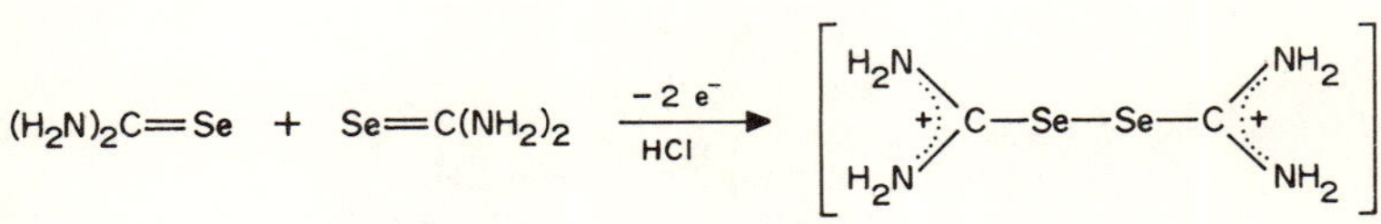

(1)

A spectrophotometric investigation[35] of the aqueous OsO_4/su system established that su initially reduces Os(VIII) to Os(IV) with the subsequent formation of an intense blue-green 8:1 Os(III) complex (equation 2).

$$OsO_4 + 4\,H_2NC(Se)NH_2 \xrightarrow{1\,\text{M HCl}} OsO_2 + 2[\{(H_2N)_2CSe\}_2]^{2+} + 2\,H_2O \quad (2)$$
$$\downarrow 9H_2NC(Se)NH_2$$
$$[Os(Se{=}C(NH_2)_2)_8]^{3+} + \tfrac{1}{2}[\{(H_2N)_2CSe\}_2]^{2+} + 2\,H_2O$$

Thiourea gives a similar five-electron initial reduction but forms a final 6:1 complex, $[Os(thiourea)_6]^{3+}$.

A variety of complexes of selenourea with transition metals and Te(II) have been isolated and characterized (Table 2).

TABLE 2. Data for some typical selenourea complexes

Complex	Reported data[a]	Reference
$[Pd(su)_4]Cl_2$	Wine-red crystals	36
	M.p. 179 °C dec.	
	ν_{Pd-Se}	
	aq. soln., R: 178(P)	
	powder, R: 182s, 176sh	
	IR: 253m, 235s	
$[Pt(su)_4]Br_2$	Yellow crystals	36
	M.p. 160 °C dec.	
	ν_{Pt-Se}	
	powder, R: 202s, 182s	
	IR: 227s, 215w	
$[Co(su)_4](ClO_4)_2$	Air-sensitive, deep olive-green, platelike crystals	
	M.p. 142 °C dec.	
	Strong su IR bands: 1625, 1400, 575, 500, 380	37
	ν_{Co-Se} = 245s, 218m, sh	38
	Visible absorptions	38
$[Co(su)_2Cl_2]$	Air-sensitive, blue-green solid	38, 39
	ν_{Co-Se} = 213s, 189w(sh)	38
	Visible absorptions	38, 39
$[Zn(su)_4](ClO_4)_2$	ν_{Zn-Se} = 208s, 196s	38
$[Zn(su)_2Cl_2]$	ν_{Zn-Se} = 205s, 180m (sh)	38
$[Cd(su)_2Cl_2]$	ν_{Cd-Se} = 188s, 165s(br)	38
	X-Ray powder pattern	40
$[Hg(su)_2Cl_2]$	Moderately air-sensitive, white microcrystalline solid	37, 38
	ν_{Hg-Se} = 176	
	ν_{Hg-Cl} = 204, 200	
$[HgCl_2(su)]_2$	Moderately air-sensitive, white microcrystalline solid	37, 38
	ν_{Hg-Cl} = 276s	37, 38
	$\nu_{Hg-Cl-Hg}$ = 200s	38
	ν_{Hg-Se} = 183s(br)	38
$[Te(su)_4]Cl_2$	Yellow crystals	41
	Single-crystal X-ray diffraction analysis	
	Te—Se bond lengths = 2.814(3) Å, 2.809(3) Å	
	Se—Te—Se angles = 90°	
$[PhTe(su)Cl]$	Orange-red monoclinic prisms	42

[a]Infrared (IR) and Raman (R) data in cm^{-1}; recorded as Nujol mulls unless otherwise indicated; s = strong, m = medium, w = weak, sp = sharp, sh = shoulder, br = broad, P = polarized band.

In both su complexes that have been unequivocally characterized structurally by single-crystal X-ray diffraction, Se bonding was found ($[Te(su)_4]Cl_2$[41], $Te(su)_2(Ph)Cl$[43]). A similar Se bonding mode has been established from the X-ray powder pattern of $CdCl_2(su)_2$, which shows it to be isostructural with the analogous thiourea complex, for which a single-crystal X-ray diffraction study has established a $CdCl_2S_2$ tetrahedral coordination sphere[40]. For the other complexes, infrared spectroscopy[36–39,44] has been most generally used to determine the bonding mode of the potentially ambidentate su ligand, and Se bonding has been assigned in all cases.

In contrast, urea and thiourea coordinate through both N and O or S in their coordination complexes[44]. The tendency of su to coordinate via its Se atom is supported by force-constant calculations[38], which support the increasing importance of canonical forms **B** and **C** in the series X = Se > S > O:

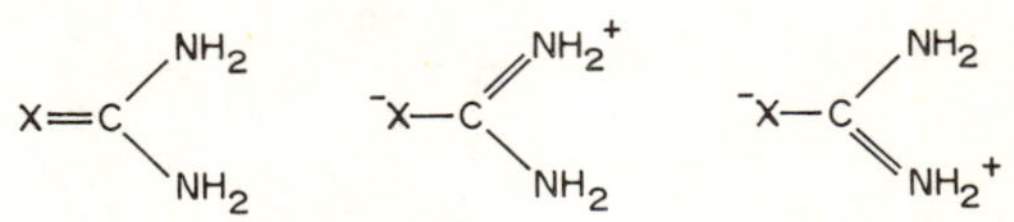

A **B** **C**

Although metal–Se vibrations have been measured for several M–su complexes (M = Pd[36], Pt[36], Co[38], Zn[38], Cd[38], Hg[38]; Table 2), these low-energy bands are often weak and difficult to assign, and careful measurements of infrared shifts of ligand-localized absorptions upon coordination are generally more useful in establishing coordination sites of such potentially ambidentate ligands.

Although there is considerable mixing in the normal modes of su, those bands with a strong CN stretching contribution ($\nu_4 = 1400\,cm^{-1}$, s; $\nu_5 = 1085\,cm^{-1}$, w; $\nu_{16} = 1480\,cm^{-1}$, m) generally increase upon coordination (to 1400–1410, 1090–1120 and 1500–1520 cm^{-1}, respectively), and those with a strong C—Se stretching contribution ($\nu_6 = 640$ and $\nu_7 = 390\,cm^{-1}$) generally decrease 10–20 cm^{-1} upon coordination[38]. The ν_{N-H} vibrations of free su (3320 and 3250 cm^{-1}) also generally sharpen and increase up to 100 cm^{-1} [31] on coordination, owing to the destruction of the weakly hydrogen-bonded structure of the free ligand[45].

The selenourea complexes (Table 2) are readily prepared by reaction of an appropriate metal salt (in aqueous solution) with the stoichiometric amount of selenourea. In aqueous systems a relatively low pH (ca. 3) is used to prevent decomposition of su[34–36]. Non-aqueous solvents include MeOH[42], EtOH[39], *n*-BuOH[37–39] and acetone[37,38]. The syntheses are generally carried out in an inert atmosphere with minimum exposure to light[37]. As with thiourea, reactions of selenourea with reducible metal species (e.g. Te(IV)[41] and Cu(II)[3]) can give complexes in a lower oxidation state (e.g. Te(II) and Cu(I), respectively), the su functioning as both a reducing agent and a stabilizing ligand for the lower oxidation state. With Te, the complex isolated depends on the stoichiometry of su used (equations 3–5).

$$TeO_2 \xrightarrow{HCl} TeCl_6^{2-} \xrightarrow{4\,su} TeCl_2(su)_2\downarrow + [(H_2N)_2C{-}Se]_2^{+2} \quad \text{(Ref. 41)} \qquad (3)$$

$$TeCl_2(su)_2 \xrightarrow{2su} [Te(su)_4]Cl_2$$

$$PhTeTePh + Cl_2 + 2\,su \rightarrow 2\,[PhTeCl(su)] \quad \text{(Ref. 42)} \qquad (4)$$

$$PhTeTePh + Cl_2 + 4\,su \rightarrow 2\,[PhTe(su)_2]Cl \quad \text{(Ref. 42)} \qquad (5)$$

The crystal structure of $[Te(su)_4]Cl_2$[41] has established that the Te(II) is surrounded by four Se atoms in a square planar configuration and the bond lengths and angles in the selenourea ligands are similar to those of free selenourea[45].

The complex $PhTe(su)_2Cl$ contains essentially planar three-coordinate Te(II) (i.e. $[PhTe(su)_2]^+Cl^-$) with a nearly linear three-centre Se—Te—Se system perpendicular to the bonded phenyl C atom[43]. In contrast, the thiourea analogue, $PhTe(tu)_2Cl$, while having a similar structure, has a weak but observable Te···Cl interaction to give a distorted square planar configuration[43]. Such secondary bonding has been observed in a number of Te(II) and Te(IV) complexes[46].

The complex $HgCl_2(su)$ has been formulated as a chloro-bridged dimer (**1**) on the basis of the presence of both terminal Hg—Cl and bridging Hg—Cl—Hg vibrations in the far-infrared region[37,38].

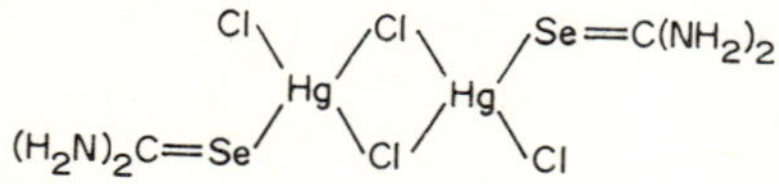

(**1**)

Several electroless plating solutions containing selenourea or *N,N*-dimethylselenourea, which are useful for the deposition of thin films of CdSe[23] and PbSe[20], have been reported. These solutions generally incorporate an additional chelating agent for Cd(II) and are used under alkaline conditions. Such plating processes presumably involve the formation of unstable $M—Se{=}C(NR_2)_2$-type complexes.

In addition to selenourea[3,34–43] itself, several *N,N'*-disubstituted selenoureas[47–49] have been reported to form coordination complexes (Table 3). These substituted derivatives, prepared from the corresponding isoselenocyanates and primary amines[51], form unusual square pyramidal five-coordinate Pd(II) and Pt(II) complexes[47,48] (equation 6). The infrared data support Se bonding in all cases (Table 3). The formulation of the apparently square planar $M(R,R^1\text{-su})_4Cl_2$ complexes (i.e. $[M(R,R^1\text{-su})_4]Cl_2$) as pentacoordinate species (i.e. $[M(R,R^1\text{-su})_4Cl]Cl$)[47,48] is based on their electronic spectra, 1:1 electrolyte behaviour in non-polar organic solvents (Table 3) and the isolation of $[M(R,R^1\text{-su})_4Cl]Y$ ($Y = ClO_4$, BPh_4) complexes on reaction of solutions of the dichloro complexes with excess of the poorly coordinating ClO_4^- or BPh_4^- anions[52] (equation 6). Although they are relatively rare, five-coordinate Pd(II) and Pt(II) complexes have been characterized crystallographically (e.g. $[M(F_6acac)_2(PR_3)]$; M = Pd, R = *o*-Tol; M = Pt, R = *c*-Hex)[53].

$$RNCSe + R^1NH_2 \xrightarrow{C_6H_{12}} RNH\overset{\overset{\displaystyle Se}{\|}}{C}NHR^1 \xrightarrow[\text{acetone}/H_2O]{K_2MCl_4} [M(R,R^1\text{-su})_4Cl]Cl \qquad (6)$$

$R = Ph_2CH$ $\qquad$ $(R,R^1\text{-su})$ $\qquad$ $M = Pd, Pt$

$R^1 = Ph, n\text{-}Bu$

$$[M(R,R^1\text{-su})_4Cl]Cl \xrightarrow[(NaBPh_4)]{HClO_4} [M(R,R^1\text{-su})_4Cl]Y \qquad Y = ClO_4, BPh_4$$

Several Ni(II) complexes with *N,N'*-disubstituted selenoureas have been prepared: tetrahedral $[NiL_2X_2]$, square planar, diamagnetic $[NiL_4](ClO_4)_2$ and the unique tetrahedral $[Ni(N\text{-}Ph_2CH, N'\text{-}Bu\text{-}su)_3Br]Br$[49] (Table 3). The stereochemistries of these complexes have been established by electronic spectroscopy and magnetic susceptibility measurements (Table 3).

The solution and reflectance spectra of the NiL_2X_2 derivatives are identical, indicating that the tetrahedral coordination is retained in solution. The monomeric formulations of these complexes are confirmed by molecular weight measurements, and conductivity measurements show non-electrolyte behaviour.

TABLE 3. Data for some complexes with N,N'-disubstituted selenoureas, RHNC(=Se)NHR[1] (L)

Complex	Reported data[a]	Complex	Reported data[a]
R = *Ph$_2$CH*, *R^1* = *Ph*		*R* = *R^1* = *Ph*	
$[PdL_4Cl]Cl$[47,48]	Red solid M.p. 136–138 °C $\nu_{C-N} = 1565^b$ $\nu_{C=Se} = 788^c$ $\Lambda_M(MeOH) = 105$ (acetone) = 75 $(CH_2Cl_2) = 48$	NiL_2Cl_2[49]	Green solid M.p. 131–134 °C $\mu_{eff} = 3.34$
		$[NiL_4](ClO_4)_2$[49]	Green solid M.p. 126–128 °C Diamagnetic
$[PdL_4Cl]ClO_4$[47,48]	Red–yellow solid M.p. 122–124 °C $\Lambda_M(MeOH) = 125$ $(CH_2Cl_2) = 56$ (acetone) = 170	*R* = *Ph*, *R^1* = *allyl*	
		NiL_2Br_2[49]	Green solid M.p. 118–122 °C $\mu_{eff} = 3.30$
		$[NiL_4](ClO_4)_2$[49]	Green solid M.p. 135–137 °C Diamagnetic
$[PtL_4Cl]Cl$[47,48]	Yellow solid M.p. 147–149 °C $\nu_{C-N} = 1560^b$ $\nu_{C=Se} = 785^c$ $\Lambda_M(MeOH) = 91$ (acetone) = 80 $(CH_2Cl_2) = 59$	*R* = *Ph$_3$C*, *R^1* = *Bu*	
		NiL_2Br_2[49]	Green solid M.p. 160–163 °C $\mu_{eff} = 3.57$
		NiL_2I_2[49]	Brown solid M.p. 128–130 °C $\mu_{eff} = 3.28$
$[RhL_4Cl_2]Cl$[47]	Dark-red solid M.p. 160–163 °C $\nu_{C-N} = 1552^b$ $\nu_{C=Se} = 788^c$ $\Lambda_M(MeOH) = 85$	*R* = *Ph$_3$C*, *R^1* = *Ph$_2$CH*	
		NiL_2Br_2[49]	Green solid M.p. 178–180 °C $\mu_{eff} = 3.33$
R = *Ph$_2$CH*, *R^1* = n-*Bu*			
$[PdL_4Cl]Cl$[48]	Red–brown solid M.p. 120–127 °C	NiL_2I_2[49]	Yellow–brown solid M.p. 134–136 °C $\mu_{eff} = 3.26$
$[PtL_4Cl]Cl$[48]	Light yellow solid M.p. 118–121 °C		
$[NiL_3Br]Br$[49]	Yellow–brown solid M.p. 140–142 °C $\mu_{eff} = 3.40$		

[a] Λ_M = molar conductance ($ohm^{-1}\,cm^2\,mol^{-1}$) of 5×10^{-4} M solution (typical values for 1:1 electrolytes are 80–115 (MeOH)[50], 100–400 (acetone)[50]); μ_{eff} = magnetic moment in Bohr magnetons (B.M.) measured at 293 K.
[b] Corresponding band for free N-Ph$_2$CH,N'-Ph-su is at 1535 cm^{-1}.
[c] Corresponding band for free N-Ph$_2$CH,N'-Ph-su is at 788 cm^{-1}.

Although the $[NiL_4](ClO_4)_2$ complexes (Table 3) have been formulated as square planar complexes in the solid state on the basis of their electronic spectra and magnetic susceptibilities, their instability in solution, even in the presence of excess L, precludes solution studies.

The unique Ni(N-Ph$_2$CH, N'-Bu-su)$_3$Br$_2$ complex has been formulated on the basis of electronic spectroscopy and magnetic susceptibility as the tetrahedral species $[NiL_3Br]^+Br^-$ in the solid state. However, in acetone solution, its conductivity is about zero, and its molecular weight is about half the theoretical value, supporting the dissociation shown in equation (7).

$$[NiL_3Br]Br \rightarrow NiL_2Br_2 + L \tag{7}$$

TABLE 4. Complexes with cyclic selenourea derivatives

Complex	Reported data[a]	Reference
PhTe(esu)$_2$Cl	Orange–red orthorhombic prisms	54
	M.p. 190–191 °C dec.	
	Crystallographic data	
PhTe(esu)Br	Orange–red monoclinic crystals	54
	M.p. 163–164 °C	
	Single-crystal X-ray diffraction study	56
	Te—Se = 2.6160(16) Å	
	Te—Br = 3.0537(16) Å	
	Te—C = 2.118(7) Å	
	∠Se—Te—C = 89.94(19)°	
	∠Br—Te—C = 86.67(19)°	
	∠Se—Te—Br = 175.62(3)°	
	Te···Br (intermolecular) = 3.8490(16) Å	
	$\nu_{Se-Te-Br,sym}$ = 170s,	
	130w (IR)	59
	169m (R)	
	$\nu_{Se-Te-Br,asym}$ < 190m (IR),	
	192w (R)	
PhTe(esu)I	M.p. 137–139 °C	55
	Single-crystal X-ray diffraction study	
	Te—Se = 2.6791(18) Å	
	Te—I = 3.0951(14) Å	
	Te—C = 2.112(7) Å	
	∠Se—Te—C = 89.92(17)°	
	∠I—Te—C = 88.15(17)°	
	∠Se—Te—I = 177.31(2)°	
	$\nu_{Se-Te-I,sym}$ = 150s, 133w (IR)	59
	152vs (R)	
	$\nu_{Se-Te-I,asym}$ = 180m (IR),	
	178w (R)	
Trimethyleneselenourea		
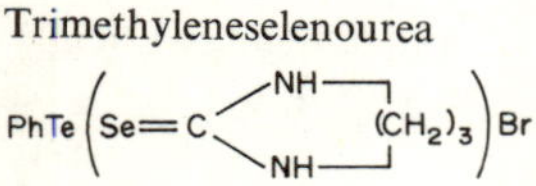	Yellow monoclinic prisms and plates	54
	M.p. 152–153 °C	
	Crystallographic data	

[a]Units as in Table 3.

The only reported complexes with a tetrasubstituted selenourea (tetramethylselenourea, tmsu) were prepared by the route described in equation (4) (i.e. PhTeX (tmsu); X = Cl (m.p. 151–152 °C), X = Br (m.p. 162–163 °C)[54].

Attempted formation of Ni(II)-ethyleneselenourea (esu) complexes resulted in immediate decomposition with deposition of elemental Se. In contrast, ethylenethiourea gave stable NiL_4X_2 complexes[49]. Esu complexes of Te(II), however, have been isolated (PhTe(esu)X; X = Cl[54], Br[54], I[55]; PhTe(esu)$_2$Cl[54]) and characterized by far-infrared spectroscopy and, for X = Br and I, by single-crystal X-ray diffraction (Table 4). The chloro and bromo complexes were prepared[54] by halogen cleavage of Ph_2Te_2 in the presence of esu in methanol solution, as described for the corresponding su complexes (equation 4). The iodo complex was prepared via a metathetical reaction of the bromo analogue and 2 equiv. of NaI in methanol[55]. The formation of these complexes of PhTeX illustrates the stabilizing influence of selenourea ligands, since the parent phenyl tellurenyl halides, although useful as synthetic reagents[57], are unstable[58].

TABLE 5. Ni(II) complexes with polymethylenebis(phenylselenoureas), PhHNC(=Se)NH$(CH_2)_n$NHC(=Se)NHPh (L_n)[60]

Complex	Reported data[a]
$Ni(L_4)_2X_2$	
X = Cl	Green solid, m.p. 161–164 °C
X = Br	Green solid, m.p. 154–156 °C
X = I	Brown–yellow solid, m.p. 132–135 °C
$Ni(L_5)_2X_2$	
X = Cl	Green–yellow solid, m.p. 120–125 °C
X = Br	Green–yellow solid, m.p. 130–135 °C
X = I	Yellow–green solid, m.p. 115–118 °C
$Ni(L_7)_2Cl_2$	Green solid, m.p. 111–114 °C
$Ni(L_7)_2Br_2$	Green solid, m.p. 126–129 °C
$Ni(L_7)_2I_2$	Green solid, m.p. 133–136 °C
$Ni(L_8)_2Cl_2$	Green–yellow solid, m.p. 110–112 °C
$Ni(L_8)_2Br_2$	Green solid, m.p. 115–118 °C
$Ni(L_8)_2I_2$	Green–brown solid, m.p. 125–128 °C

Single-crystal X-ray diffraction studies of PhTe(esu)X (X = Br[56], I[55]) were carried out as part of a study[46a] of the stereochemistry of Te(II) complexes and the *trans* influence of various thio- and seleno-urea ligands (i.e. variation of Te—X_{trans} bond distances in three-coordinate PhTeLX-type complexes). The infrared and Raman spectra of these two complexes have also been recorded[59] in the solid state in the 50–500 cm^{-1} region, and bands due to the asymmetric and symmetric stretching frequencies of the linear Se—Te—X have been assigned (Table 4).

Polymethylenebis(phenylselenourea), PhNHC(Se)NH$(CH_2)_n$NHC(Se)NHPh ($n = 2$–5, 7, 8) (prepared from the corresponding diamines, $H_2N(CH_2)_nNH_2$ and PhNCSe), gives a variety of complexes with Ni(II), the structures being determined by the Ni(II) salt used in the reaction as well as the reaction temperature[60] (Table 5).

With $n = 2$ or 3, and with acetone or ethanol as the reaction solvent, unstable monomeric tetrahedral complexes of the type [$NiLX_2$] containing (Se, Se) chelating selenourea ligands are formed[60] (see Section II.C.3). Use of BuOH as the reaction solvent with L_3 or any of the above solvents with the higher homologues gives complexes of the composition $Ni(L_n)_2X_2$ (Table 5). On the basis of their low solubility, electronic spectra and magnetic susceptibilities, these complexes have been formulated as polymeric, square planar species (**2**) with $NiSe_4$ coordination spheres formed by bridging selenourea ligands with ionic halide.

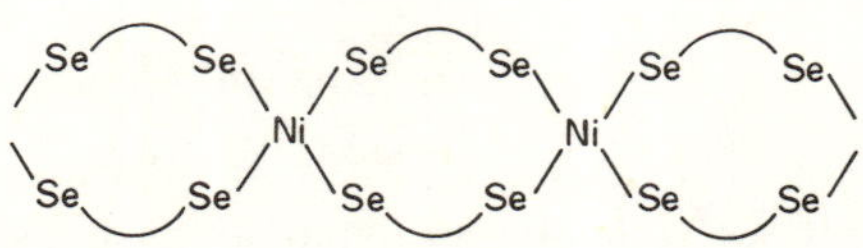

(**2**)

If the reaction temperature is not kept low during the synthesis, the magnetic susceptibility of the products tends to be higher (up to 1.5 B.M.) than the limiting zero value expected for the square-planar diamagnetic products.

Chelating (Se, O) and (Se, S) selenoureas have also been reported (see Section II.D.2).

2. Diorganoselenides (SeR_2; R = alkyl, aryl)

Dialkyl selenides can be synthesized by a variety of standard routes[14]. Recently, Superhydride™, $LiBEt_3H$, has been used as a convenient reagent in such syntheses in non-aqueous solvents (equation 8).

$$Se + LiBEt_3H \xrightarrow{THF} Li_2Se \xrightarrow{2\,RX} SeR_2 \qquad (8)$$

R = n-C_5H_{11}[61], $C_6H_5CH_2$[61], p-$ClC_6H_4CH_2$[61], $PhCH_2CH_2$[61], —$SiMe_3$[62], 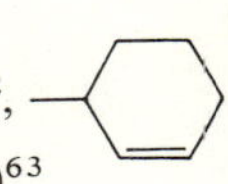 [63],

(analogous reactions of allylic halides with Te^{2-} give coupled 1,5-dienes)[63]

Unsymmetrical diorganoselenides have also been prepared according to equation (9).

$$RSeSeR + 2\,LiBEt_3H \rightarrow 2\,RSeLi + 2\,Et_3B + H_2 \qquad \text{(Ref. 61)} \qquad (9)$$

$$\downarrow 2R^1X$$

$$RSeR^1$$

$$R = Ph,\ R^1 = PhCH_2$$

$$R = PhCH_2,\ R^1 = Me$$

Complexes of several metals with various dialkyl selenides have been reported (Pd[6,7,64–79], Pt[4,5,7,67,68,71,76,77,79–92], Ag[89,93,94], Ga[93], W[95], Re[96,97], Rh[98–100], Ru[101–104], Hg[89,105], Ir[106], Au[107], Cr[108]), and the few reported complexes with diaryl selenides involve $SePh_2$[5,6,69,73,96,101,109–111]. A few examples of metal complexes with alkyl aryl selenides (EtSePh[102,103,105], $(Me_3SiCH_2)SePh$[112]) have also been prepared.

The $MX_2(Se(alkyl)_2)_2$ (M = Pd, Pt; X = Cl, Br, I) complexes have been the subject of the most investigation. The solid-state and solution geometries (*cis*/*trans*) of these square-planar complexes have been studied by infrared[67,68,71,74,78], Raman[67,68,78] and dipole-moment measurements[78,80]. In a number of studies, variable-temperature ^1H-NMR spectroscopy was used to investigate inversion about the Se atom in solutions of such species, e.g. $MX_2(SeEt_2)_2$ (X = Cl, Br, I; M = Pd, Pt[76,77]), $MCl_2(Se(CH_2SiMe_3)_2)_2$ (M = Pd, Pt[79]), $RuX_3(NO)(SeEt_2)_2$ (X = Cl, Br[103]), $RuCl_3(NO)(PhSeEt)_2$[103], $MX_2\{PhSe(CH_2SiMe_3)\}_2$ (M = Pd, X = Cl; M = Pt, X = Br[112]), $PdCl_2(EtSePh)_2$[105]. Such spectra are often complicated by simultaneous intramolecular *cis–trans* isomerization and intermolecular ligand exchange processes[77].

The earliest reports involved the dialkyl selenide complexes of Pt and Pd halides, which readily precipitate upon addition of 2 equiv. of the selenide to an aqueous solution of potassium tetrachlorometallate. The corresponding bromo and iodo complexes are generally prepared by metathetical reactions, although bromo complexes have also been prepared directly (equation 10).

$$K_2MCl_4 + 2\,SeR_2 \xrightarrow{H_2O} [MCl_2(SeR_2)_2]\downarrow \xrightarrow{2\,NaX/acetone} [MX_2(SeR_2)_2] \qquad (10)$$

$$X = Br, I$$

$$M = Pd, Pt$$

$$R = Me, Et, i\text{-}Am$$

The $PtX_2(SeR_2)_2$ complexes generally precipitate as mixtures of the *cis* and *trans* isomers, which can be separated by their solubility differences in $CHCl_3$ (the *cis* isomers are considerably more soluble than the *trans*).

Hieber and coworkers first prepared transition-metal carbonyl complexes with diorganoselenide ligands (equations 11–14). Similar substitution reactions have been used to prepare several other diorganoselenide complexes of transition-metal carbonyls (e.g.

$Mo(CO)_3(phen)(SePh_2)$[110], *trans*-$Rh(CO)(SeEt_2)_2Cl$[98], $Re(CO)_{5-n}(SeEt_2)_nCl$ ($n = 1,2$)[97], $[Ir(CO)_2(SeEt_2)Cl]_n$[106]).

$$Re(CO)_5X + 2\,SeEt_2 \xrightarrow[\Delta]{EtOH} Re(CO)_3X(SeEt_2)_2 \quad \text{(Ref. 96)} \qquad (11)$$

$$X = Br, I$$

$$2\,Re(CO)_5X + 2\,SeR_2 \xrightarrow[\Delta]{EtOH} (OC)_3(R_2Se)Re(\mu\text{-}X)_2Re(CO)_3(SeR_2) \quad \text{(Ref. 96)} \qquad (12)$$

$$X = Cl;\ R = Et$$
$$X = Cl, Br, I;\ R = Ph$$

$$Mn(CO)_5Br + 2\,SePh_2 \xrightarrow[\Delta]{C_6H_6} Mn(CO)_3Br(SePh_2)_2 \quad \text{(Ref. 111)} \qquad (13)$$

$$[Ru(CO)_2I_2]_n + SeR_2 \rightarrow Ru(CO)_2(SeR_2)_2I_2 \quad \text{(Ref. 101)} \qquad (14)$$
$$R = Et, Ph$$

The only metal complex with a diorganoselenide ligand that has been characterized by single-crystal X-ray diffraction is *trans*-$PdCl_2(SeEt_2)_2$[70]. The small decrease in the observed Pd—Se distance (2.424(7) Å) compared to the sum of the covalent radii (2.45 Å) indicates little or no Pd → Se π back-donation.

A novel route to a diorganoselenide ligand involving alkylation of coordinated Se has been reported recently (equation 15).

$$Cr(CO)_6 \xrightarrow[THF]{h\nu} Cr(CO)_5THF \xrightarrow[EtOH]{Li_2Se} [Cr(CO)_5SeH]^- \quad \text{(Ref. 108)} \qquad (15)$$

$$[Cr(CO)_5SeH]^- \xrightarrow{[Et_3O][BF_4]} Cr(CO)_5SeEt_2$$

A more complete discussion of the chemistry and spectroscopic properties of complexes with seleno ethers is given by Murray and Hartley[13].

3. *Triorganophosphine selenides* ($Se{=}PR_3$)

Triorganophosphine selenides are most commonly prepared by the reaction of the phosphine with red Se[113].

The first coordination complexes with a triorganophosphine selenides ligand were reported by Bannister and Cotton in 1960 ($(PdCl_2(SePPh_3)_2$, $SnCl_4(SePPh_3)_2)$[114]. These complexes, which were isolated by reaction of the phosphine selenide with metal chloride complexes (H_2PdCl_4/aqueous ethanol reaction solvent; $SnCl_4(Et_2O)_2$/ether reaction solvent), were not characterized by these authors, but subsequent work has established the utility of infrared spectroscopy as a convenient probe for coordination of such ligands[115]. Upon coordination, the $\nu_{P=Se}$ vibration generally shifts 10–40 cm^{-1} to lower energy vs. the free ligand. Such shifts are smaller than those observed for analogous EPR_3 (E = O, S) complexes ($\Delta\nu_{P=O} = -(38\text{–}70)\,cm^{-1}$; $\Delta\nu_{P=S} = -(40\text{–}50)$), and it has been suggested[116] that these shifts indicate that P—Se π bonding is weaker than in the lighter phosphine chalcogens, although this effect is probably at least partly due to the lower sensitivity of the heavier Se atom to coordination.

A number of complexes with triaryl- and trialkyl-phosphine selenides have been reported, and one such complex, $[HgCl_2(SePPh_3)]_2$ (**3**), has been characterized by single-crystal X-ray diffraction, which established a chloro-bridged dimeric structure with distorted tetrahedral coordination about Hg[117].

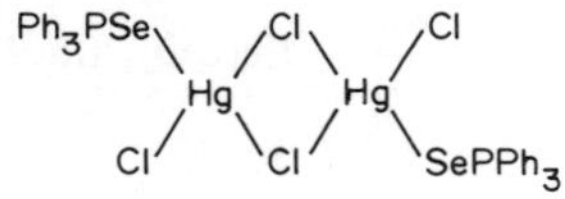

(**3**)

Pd, Pt and Cd complexes of unusual stoichiometries have been reported, the proposed structures (**4**–**7**) assigned primarily on the basis of infrared spectral data.

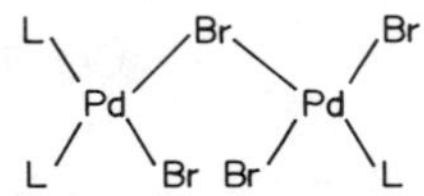

(**4**)[118]

L = $SePPh_3$

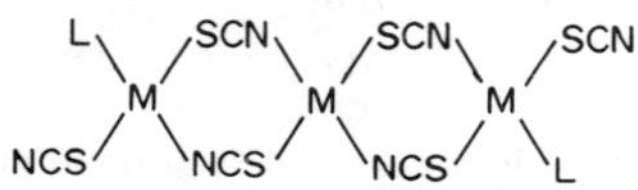

(**5**) M = Pd[119]

(**6**) M = Pt[119]

L = $SePPh_3$

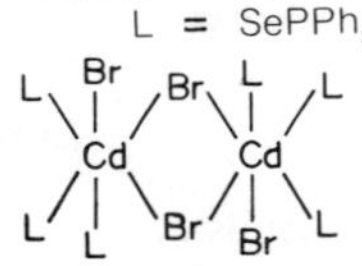

(**7**)[120]

L = $SeP(C_6H_4Me\text{-}p)_3$

TABLE 6. Complexes with triorganophosphine selenides

Complex	Reported data[a]	Reference
$PdCl_2(SePPh_3)_2$	Orange–brown solid	114
	M.p. 206 °C	116
	$\nu_{P=Se} = 543^b$	116
$(PdBr_2)_2(SePPh_3)_3$ (**4**)	Red–brown solid	118
	M.p. 220 °C dec.	
	$\Lambda_M(PhNO_2) = 2.55$	
	$\nu_{P=Se} = 538s$	
$PdCl_2(SeP(C_6H_4Me\text{-}m)_3)_2$	Pale orange solid	119
	$\nu_{P=Se} = 558^c$	
	$\Lambda_M(MeCN) = 3.2$	
$[Pd(SCN)_2]_3(SePPh_3)_2$ (**5**)	$\nu_{P=Se} = 540ms$	119
	$\nu_{C\equiv N} = 2155ms$ (Pd—SCN—Pd bridge)	
	2122mw (terminal Pd—SCN)	

TABLE 6. (*Continued*)

Complex	Reported data[a]	Reference
$[Pd(SCN)_2(SePPh_3)]_2$ (**8**)	$\nu_{P=Se} = 540$ms $\nu_{C\equiv N} = 2154$m, 2122m	121
$PtCl_2(SePPh_3)_2$	Light tan solid $\nu_{P=Se} = 544$	119
	M.p. 201 °C Formulated as *cis* isomer on basis of colour and insolubility	116
$[Pt(SCN)_2]_3(SePPh_3)_2$ (**6**)	$\nu_{C\equiv N} = 2160$m, 2105mw $\nu_{P=Se} = 537$ms	121
$[Ni(SePMe_3)_4](ClO_4)_2$	Red–brown solid $\nu_{P=Se} = 408s^{d}$ $\nu_{Ni-Se} = 250$m $\mu = 3.41$ B.M.	122
$[Cu(SePPh_3)_3]BF_4$	$\nu_{P=Se} = 546$vs $\Lambda_M(MeNO_2) = 69$	123
$[Ag(SePPh_3)_2]ClO_4$	Colourless solid $\Lambda_M(MeCN) = 132$ $\nu_{P=Se} = 551, 542$	119
$[Au(SePPh_3)Cl]$	Colourless solid M.p. 145 °C	116
$[ZnI_2(SePPh_3)_2]$	M.p. 240 °C $\Lambda_M(PhNO_2) = 17.70$ $\nu_{P=Se} = 532$s	120
$[ZnI_2(SeP(C_6H_4Me\text{-}p)_3)_2]$	M.p. 278 °C dec. $\Lambda_M = 15.88$ $\nu_{P=Se} = 520s^{e}$ $\nu_{Se-Zn} = 345$w	120
$[Zn(SePMe_3)_4](ClO_4)_2$	White solid $\nu_{P=Se} = 414$s $\nu_{Zn-Se} = 202$ms	122
$[Zn(SePMe_3)_2Cl_2]$	White solid $\nu_{P=Se} = 419$s $\nu_{Zn-Cl} = 297$s, 285s $\nu_{Zn-Se} = 225$m	122
$CdBr_2(SePBu_3)_2$	M.p. 68–69 °C ^{31}P-NMR[f] (CH_2Cl_2–$CHCl_3$) 33 °C: $\delta = 42.3$ ($^1J_{PSe} = 569$) – 62 °C: ($^1J_{PSe} = 543)^{g}$ – 72 °C: ($^1J_{PSe} = 542$, $J_{PCd} = 40)^{g}$	124
$CdCl_2(SePBu_3)_2$	M.p. 76–77 °C ^{31}P-NMR (CH_2Cl_2–$CHCl_3$) 33 °C: $\delta = 42.5$ ($^1J_{PSe} = 564$) – 59 °C: ($^1J_{PSe} = 544)^{g}$ – 85 °C: ($^1J_{PSe} = 530)^{g}$	124
$CdBr_2(SePPh_3)_2$	^{31}P-NMR (CH_2Cl_2–$CHCl_3$) 33 °C: $\delta = 34.6$ ($^1J_{PSe} = 696)^{h}$	124
$CdI_2(SePPhBu_2)_2$	^{31}P-NMR 33 °C: $\delta = 40.8$ ($^1J_{PSe} = 591)^{i}$	124

TABLE 6. (*Continued*)

Complex	Reported data[a]	Reference
$[CdBr_2(SeP(C_6H_4Me\text{-}p)_3)_3]_2$ (7)	White crystalline solid	120
	M.p. 234 °C	
	$\Lambda_M(PhNO_2) = 6.20$	
	$\nu_{P=Se} = 525s$	
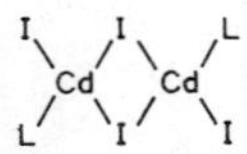	$\Lambda_M(PhNO_2) = 3.85$	120
	$\nu_{P=Se} = 520s$	
	Tetrahedral coordination	
$L = SeP(C_6H_4Me\text{-}p)_3$		
$HgI_2(SePPh_3)$	Light yellow solid	119
	$\nu_{P=Se} = 542$	
	M.p. 221 °C	116
	P 2p binding energy	
	(XPS) = 132.4 ± 0.3 eV	125
	(value for $SePPh_3$ = 133.0 ± 0.2)	
$HgBr_2(SePPh_3)$	Colourless solid	116
	M.p. 260 °C (sublimes)	
	$\nu_{P=Se} = 542$	
$HgCl_2(SePPh_3)$	Colourless solid	116
	M.p. 231 °C	
	$\nu_{P=Se} = 543$	
	Molecular structure **3**	117
$HgCl_2(SePBu_3)_2$	M.p. 91–92 °C	124
	^{31}P-NMR (CH_2Cl_2–$CHCl_3$)[f]	
	33 °C: $\delta = 45.1$	
	($^1J_{PSe} = 551$)	
	– 12 °C: $\delta = 45.4$	
	($^1J_{PSe} = 539$)	
	– 45 °C: $\delta = 45.5$	
	($^1J_{PSe} = 535$)	
	– 55 °C: ($^1J_{PSe} = 526$,	
	$^2J_{PHg} = 157$)	
	– 65 °C: ($^1J_{PSe} = 525$	
	$^2J_{PHg} = 157$)	
	– 96 °C: ($^1J_{PSe} = 524$,	
	$^2J_{PHg} = 159$)	
	$\delta(^{77}Se) = 1535d$ ($J_{SeP} = 518$)	126
	$\delta(^{199}Hg) = 325s$	
$Hg_2Cl_4(SePBu_3)_2$	$\delta(^{31}P) = 45.1$ ($J_{PSe} = 527$)	126
	$\delta(^{77}Se) = 1555d$ ($J_{PSe} = 505$)	
	$\delta(^{199}Hg) = -335s$	
$HgCl_2(SePPhBu_2)_2$	^{31}P-NMR (CH_2Cl_2–$CHCl_3$)	124
	33 °C: $\delta = 43.5$	
	($J_{PSe} = 555$)[j]	
$HgBr_2(SePPhBu_2)_2$	^{31}P-NMR (CH_2Cl_2–$CHCl_3$)	124
	33 °C: $\delta = 42.5$	
	($^1J_{PSe} = 575$)	
	– 29 °C: $\delta = 43.2$	
	($^1J_{PSe} = 563$)	
$Hg_2I_4(SePPh_2Bu)_2)$	$\delta(^{31}P) = 37.4$ ($J_{PSe} = 586$)[k], 33 °C	124
$[HgCl_2(SeP(C_6H_4Me\text{-}p)_3)]_2$	M.p. 205–220 °C dec.	120
	$\Lambda_M(PhNO_2) = 3.56$	
	$\nu_{P=Se} = 520s$	

TABLE 6. (*Continued*)

Complex	Reported data[a]	Reference
$[Co(SePMe_3)_4](ClO_4)_2$	Green solid $\nu_{P=Se} = 409s$ $\nu_{Co-Se} = 273m$ $\mu = 4.40$ B.M.	122
$[Co(SePMe_3)_2Cl_2]$	Blue solid $\nu_{P=Se} = 414s$ $\nu_{Co-Cl} = 321s, 301s$ $\nu_{Co-Se} = 227m$ $\mu = 4.38$ B.M.	122
$[Co(NO)_2(CN)(SePPh_3)]$	M.p. 117 °C $\nu_{C\equiv N} = 2099$ $\nu_{N=O} = 1849, 1791$	127
$[(COD)RhCl(SePPhMe_2)]$	Yellow crystals M.p. 128–130 °C $\nu_{P=Se} = 484^{l}$ $\nu_{Rh-Cl} = 275$	128
$[Fe(NO)_2(CN)(SePPh_3)]$	M.p. 102 °C $\nu_{C\equiv N} = 2081$ $\nu_{N=O} = 1820, 1729$	127
$Ru(CO)_2(SePPh_3)_2I_2$	Brown–yellow solid	98
$Cr(CO)_5(SePPhMe_2)$	Moderately air-stable yellow crystals M.p. 95 °C $\nu_{P=Se} = 483$ $\nu_{C\equiv O} = 2058, 1980, 1941, 1927, 1912$	129
$W(CO)_5(SePPhMe_2)$	Moderately air-stable yellow crystals M.p. 121 °C $\nu_{P=Se} = 482$ $\nu_{C\equiv O} = 2064, 1975, 1924, 1935, 1907$	129
$WCl_5(SePPh_3)$	Air-sensitive yellow–green powder $\nu_{P=Se} = 532$	130
$[W_2Cl_8(SePPh_3)_2]Cl_2$ **(9)**	$\nu_{W-Cl} = 331sh, 329sh, 320vs$ $\Lambda_M(PhNO_2) = 42.2$ $MW(PhNO_2) = 475 \pm 15$ (theor. = 468)	130
$WCl_4(PPh_3)(SePPh_3)$	Yellow solid $\mu = 1.98$ B.M. (W(IV))	131
$SnCl_4(SePPh_3)_2$	Yellow air-sensitive solid M.p. 168 °C $\Lambda_M(DMF) = 5.8$	114 132

[a] NMR chemical shift values in ppm and J values in Hz. ^{31}P: 85% aqueous H_3PO_4 external reference; ^{77}Se: 1 M aqueous selenous acid external reference; ^{199}Hg: 1 M PhHgOAc in DMSO as external reference. Other data are in units given in Table 3. The stretching frequencies for the carbonyl and nitrosyl ligands are given as $\nu_{C\equiv O}$ and $\nu_{N=O}$, respectively, although the actual bond orders in these ligands depend on the specific properties of a given complex (e.g. metal oxidation state, the donor properties of the other ligands present in the coordination sphere etc.; see Ref. 1).
[b] The corresponding value for free $SePPh_3$ is 562 cm^{-1} [117].
[c] The corresponding value for free $SeP(C_6H_4Me\text{-}m)_3$ is 574 cm^{-1}.
[d] The corresponding value for free $SePMe_3$ is 436 cm^{-1} [122].
[e] The corresponding value for free $SeP(C_6H_4Me\text{-}p)_3$ is 535 cm^{-1} [120].
[f] The corresponding values for free $SePBu_3$ are $\delta = 37.0$ ppm, $J_{PSe} = 693$ Hz.
[g] The reference (85% H_3PO_4) freezes, so that accurate chemical shift values were not recorded.
[h] The corresponding value for free $SePPh_3$ is 693 Hz.
[i] The corresponding value for free $SePPhBu_2$ is 693 Hz.
[j] The corresponding value for free $SePPhBu_2$ is 715 Hz.
[k] The corresponding value for free $SePPh_2Bu$ is 730 Hz[124].
[l] The corresponding value for free $SePPhMe_2$ is 495 cm^{-1} [126].

The complex $Pd(SCN)_2(SePPh_3)$ (**8**) has a dimeric structure with both S-bonded terminal and bridging thiocyanate (see infrared data, Table 6).

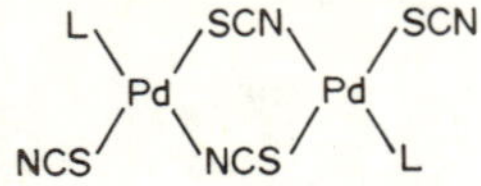

(**8**)[121]

L = Ph_3PSe

These complexes are readily prepared by reaction of the phosphine selenide in a non-aqueous solvent such as ethanol or acetone with an appropriate aqueous or non-aqueous solution of the metal salt. The Cu(I) complex[123] (Table 6), however, was prepared by adding an ethanol solution containing 3 equiv. of the phosphine selenide to an ethanol solution of $Cu(BF_4)_2 \cdot 6\,H_2O$ saturated with SO_2 to effect the Cu(II) → Cu(I) reduction. Triorganophosphines[133] and phosphine sulphides[123,134] are well known to effect this reduction without a supplementary reducing agent. Although the bidentate phosphine selenide, $Ph_2P(Se)CH_2P(Se)Ph_2$, reduces $Cu(ClO_4)_2 \cdot 6\,H_2O$ with the formation of $[Cu(Se,Se)_2]ClO_4$[134], no example of the use of a monodentate R_3PSe derivative as a reductant has been reported.

A trigonal planar three-coordinate Cu(I) complex has been formulated for $[Cu(SePPh_3)_3]BF_4$[123] on the basis of the ionic characterization of the tetrafluoroborate anion by infrared spectroscopy and analogy with the X-ray structure of $[Cu(SPMe_3)_3]ClO_4$, for which this structure has been unequivocally established.

The solution lability of Hg(II) and Cd(II) complexes with $SePR_3$ ligands has been studied by variable-temperature ^{31}P[124], ^{77}Se[126] and ^{199}Hg[126] NMR spectroscopy (see Table 6). In the cases of the $CdX_2(SePBu_3)_2$ complexes, the crystallization of the chloro and iodo complexes at low temperatures prevented freezing out the static tetrahedral structure, but the exchange in the bromo complex was stopped at ca. $-70\,°C$[124]. The complexes $HgX_2(SePBu_3)_n$ ($X = Cl$, Br, I; $n = 1, 2$) have been studied by multinuclear NMR (^{31}P, ^{77}Se, ^{199}Hg)[126]. In all cases the selenide ligand is labile and exchanges rapidly on the NMR time-scale at room temperature, but the exchange can be slowed down at ca. $-100\,°C$, as evidenced by the appearance of $^{31}P–^{199}Hg$ coupling. Halogen exchange is also fast at room temperature in these complexes[126].

The Co(II) complexes (Table 6) were precipitated from concentrated ethanol solutions containing $SePMe_3$ and the appropriate salt by addition of ether (ethyl orthoformate was used for *in situ* chemical dehydration of the metal salts)[122]. Tetrahedral geometries were assigned to these complexes on the basis of their electronic spectra and magnetic susceptibilities. Electronic spectroscopy also showed that the ligand field strength of $EPMe_3$ ($E = O$, S, Se) were comparable in the ionic complexes $[Co(EPMe_3)_4](ClO_4)_2$[122].

The moderately air-stable complex $(COD)RhCl(SePPhMe_2)$ was prepared by a bridge-cleavage reaction (equation 16).

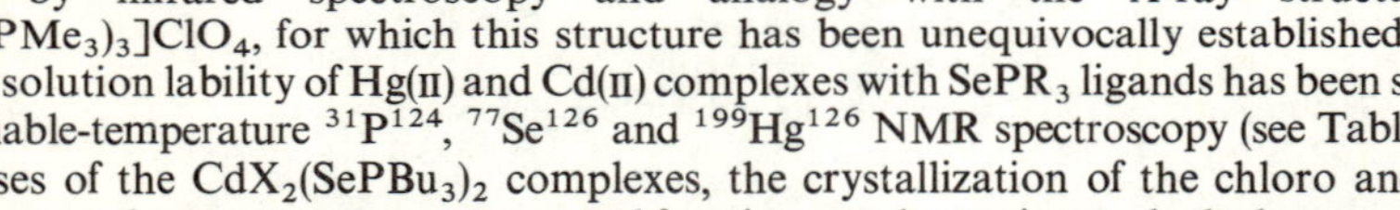

(16)

The only reported Ru complex with a phosphine selenide, $Ru(CO)_2(SePPh_3)_2I_2$[101], undergoes solvent-induced polymerization in DMF (equation 17).

$$Ru(CO)_2(SePPh_3)_2I_2 \xrightarrow{DMF} [Ru_2(CO)(SePPh_3)_2I_4(Me_2NCHO)] \qquad (17)$$

$$\uparrow \; C_6H_6 \; | \; SePPh_3/100\,°C/25\,h$$

$$[Ru(CO)_2I_2]_n$$

$$\nu_{C\equiv O} = 1931\ \text{vs}\ cm^{-1}$$

$$\nu_{C=O}\ (DMF) = 1647\text{s}\ cm^{-1}$$

The decrease in the DMF carbonyl stretching frequency ($-31\ cm^{-1}$ vs. free DMF) is characteristic of the coordination of this solvent[135]. Phosphine selenide complexes of a few transition-metal nitrosyls (Co[127], Fe[127]) and carbonyls (Ru[101], Cr[129]) have also been described briefly. The weak coordination of $SePPhMe_2$ in $Cr(CO)_5(SePPhMe_2)$ is evidenced by its facile replacement by CO or PPh_3 (equation 18). Attempts to prepare $Cr(CO)_5(SePPhMe_2)$ directly by a thermal reaction with $Cr(CO)_6$, rather than the thermally labile THF complex, gave decomposition. This problem of metal-promoted Se extrusion from phosphine selenides has been observed in a number of cases[116,134]. Se transfer between $SePR_3$ and PR_3 in solution is rapid on the NMR time-scale, and a bimolecular process has been proposed[136].

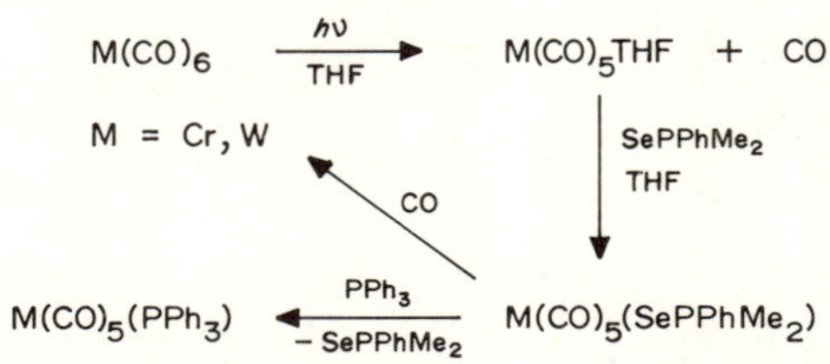

(18)

The problem of Se extrusion from selenocyanate complexes containing organophosphine ligands (i.e. with formation of $SePR_3$ and a cyanide complex) has been discussed in several papers[137–139], and the syntheses of complexes incorporating these two types of ligands must generally be carried out under mild conditions. Indeed, the reaction of KSeCN with PR_3 is a standard preparative route to phosphine selenides[140] (the other route being reaction with red Se[113,122,141]).

The tungsten complex $WCl_5(SePPh_3)$ (**9**) has been formulated, on the bases of infrared spectroscopy, molecular weight and solution conductivity measurements[130] (Table 6), as an ionic dimer with bridging phosphine selenide ligands.

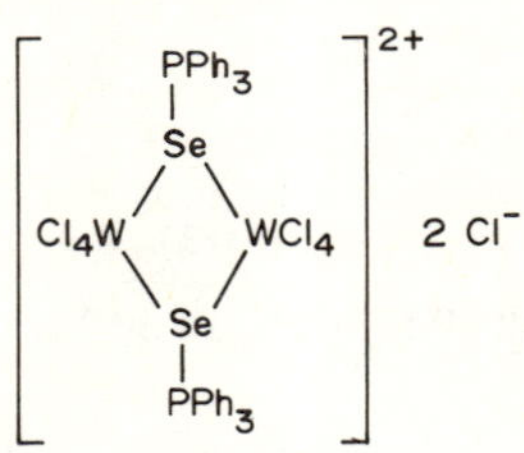

(**9**)

The corresponding $SPPh_3$ complex shows an especially large decrease in its $\nu_{P=S}$ vibration ($\Delta = -96\ cm^{-1}$ vs. ~ -30 for the $SePPh_3$ complex)[130].

The complex $WCl_4(PPh_3)(SePPh_3)$ was formed by an unusual reaction of a coordinated Se atom[131] (equation 19). The evidence for this formulation includes[131]: (*a*) a $\nu_{W=Se}$ vibration was absent in the infrared spectrum of the product, (*b*) hydrolysis of the product gave a 1:1 mixture of PPh_3 and $OPPh_3$, (*c*) the $\nu_{P=S}$ of the coordinated $SPPh_3$ ligand in the S analogue was observed, although the $\nu_{P=Se}$ band was obscured in the infrared spectrum of the $SePPh_3$ complex.

$$WCl_4Se + \text{excess } PPh_3 \xrightarrow[\text{sealed ampoule, 1 week}]{C_6H_6} WCl_4(PPh_3)(SePPh_3) \qquad (19)$$

Although several examples of metal complexes with bidentate bis phosphine selenides ($R_2P(Se)(CH_2)_n(Se)PR_2$, $n = 1$–4) have been reported (see Section II.C.3), in a few cases bridging monodentate coordination modes (**10**–**15**) have been proposed on the basis of spectroscopic evidence.

$PdBr_2(Ph_2P(Se)(CH_2)_4(Se)PPh_2)$[142]

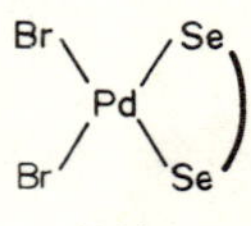

or

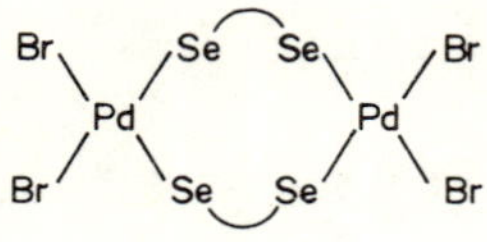

(**10**) (**11**)

$(PdCl_2)_3(Ph_2P(Se)(CH_2)_n(Se)PPh_2)_2$ (n = 2, 6[142])

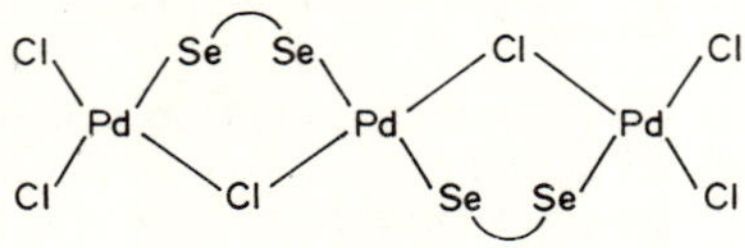

(**12**)

$(PdX_2)_4(Ph_2P(Se)(CH_2)_n(Se)PPh_2)_3$ (X = Cl, n = 4[142]; X = Br, n = 6)

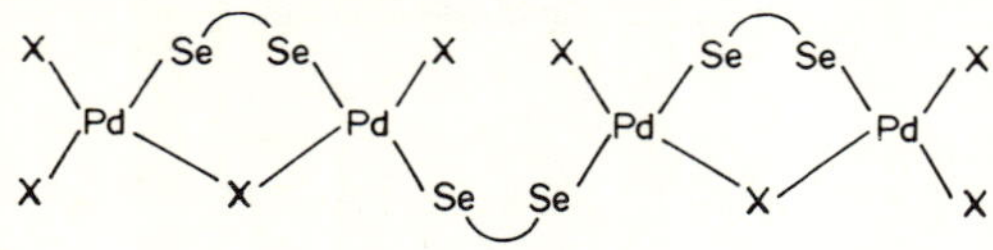

(**13**)

$ZnI_2(Ph_2P(Se)(CH_2)_3(Se)PPh_2)$ (polymeric[120])

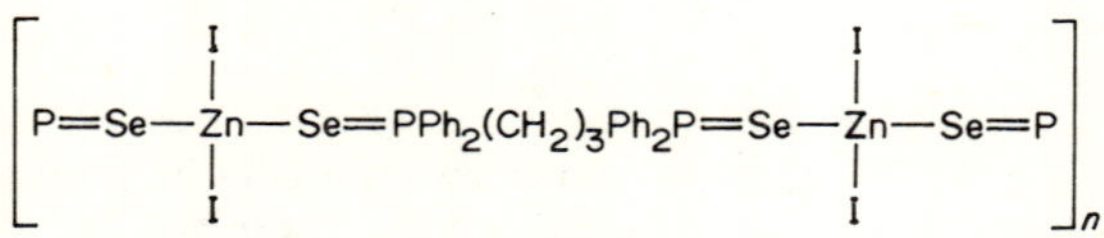

(**14**)

$[PdCl_2(Ph_2P(Se)(CH_2)_3(Se)PPh_2)]_n$[118]

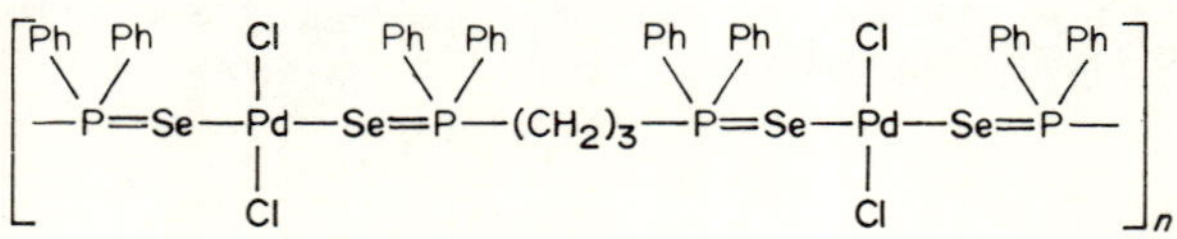

(**15**)

An interesting reaction sequence involving selenation of a coordinated phosphido ligand and thermally or photochemically induced chelation of the phosphido selenide ligand has been described[143] (equation 20).

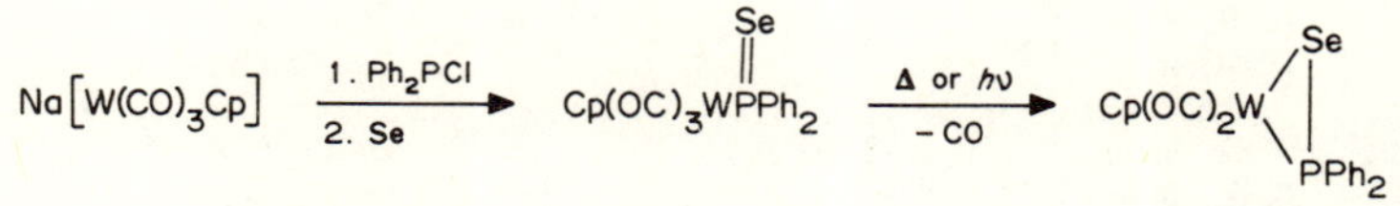

(20)

Bridging phosphido- and arsenido-selenide ligands have also been reported[144] (equation 21). The phosphido selenide complexes **A** are very reactive and readily give the six-membered heterocycles **C**. In the arsenidoselenide systems, both the four-membered ring intermediates (**A**: M = Mn, Re; E = As) and the six-membered ring products (**C**) were isolated.

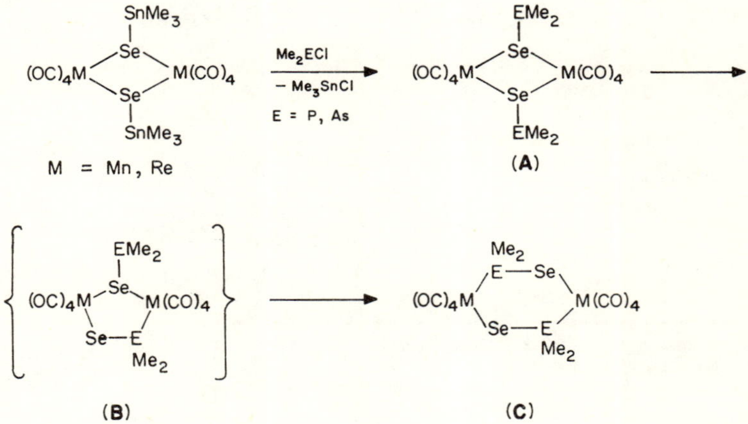

(21)

4. *Selenium heterocycles*

Several Se heterocycles have been shown to function as ligands for transition metals. Metal complexes with 2,1,3-benzoselenadiazole (**16**) and some substituted derivatives (**17**–**19**) have been prepared (Table 7). The 2,1,3-benzoselenadiazole (bsd) complex with palladium, $PdCl_2(bsd)_2$, has been used for the spectrophotometric determination of the metal[145], in which the excess bsd is determined after removal of the insoluble precipitate of the yellow $PdCl_2(bsd)_2$ by centrifugation. This application, which was also the emphasis of much of the early coordination chemistry of various Se ligands, is no longer of interest because of the developments in modern analytical chemistry (e.g. atomic absorption[151] and neutron activation analysis[152]).

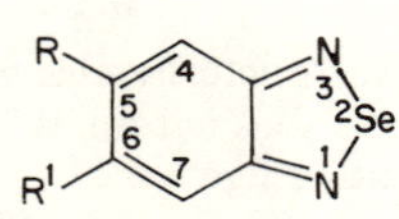

(**16**) $R = R^1 = H$ (bsd)
(**17**) $R = R^1 = Me$ (5,6-Me_2-bsd)
(**18**) $R = Me$; $R^1 = H$ (5-Me-bsd)
(**19**) $R = Cl$; $R^1 = H$ (5-Cl-bsd)

TABLE 7. Metal complexes with 2,1,3-benzoselenadiazoles

Ligand	Metal complexes/properties[a]	Ligand	Metal complexes/properties[a]
bsd (**16**)[b]	$[PdCl_2(bsd)_2]$[145] $[CdCl_2(bsd)]$[148] yellow solid no terminal $\nu_{Cd-Cl} > 250$ $[CdCl_2(bsd)_2]$[148] white solid no terminal $\nu_{Cd-Cl} > 250$ $[HgCl_2(bsd)]_2$[148] yellow solid $\nu_{Hg-Cl} = 305\ cm^{-1}$ (terminal) $[HgCl_2(bsd)_2]$ white solid $\nu_{Hg-Cl} = 304, 326$ (terminal) $CuCl_2(bsd)$[148] brown solid $\nu_{Cu-Cl} = 306$ $CuCl_2(bsd)_2$[148] green solid $\nu_{Cu-Cl} = 316$ $CoCl_2(bsd)$[149] no terminal $\nu_{Co-Cl} > 250$ $\mu = 4.86$ B.M. (O_h)		$NiCl_2(bsd)$[149] no terminal $\nu_{Ni-Cl} > 250$ $\mu = 3.05$ B.M. (O_h) $W(CO)_5(bsd)$[150] purple crystals $\nu_{C\equiv O}$ (hexane): 2078, 1947, 1927 ^{1}H-NMR $\delta(C_7D_8, 27\,°C)$[c] 6.51 (1H, dd), 6.64 (1H, dd), 7.14 (1H, d), 7.24 (1H, d) $^3J(4,5) \approx {}^3J(6,7) = 9.2$ $^3J(5,6) = 6.5$
		5-Me-bsd (**18**)	$CuCl_2(5\text{-Me-bsd})$[148] yellow–brown solid $\nu_{Cu-Cl-Cu} = 318$ $CdCl_2(5\text{-Me-bsd})$[148] $\nu_{Cd-Cl} = 270$
		5-Cl-bsd (**19**)	$CuCl_2(5\text{-Cl-bsd})$[148] yellow–brown solid $\nu_{Cu-Cl-Cu} = 275, 290$ $CdCl_2(5\text{-Cl-bsd})$[148] $\nu_{Cd-Cl} = 310$
		5,6-Me_2-bsd (**17**)	$CdCl_2(5,6\text{-}Me_2\text{-bsd})$[148] $\nu_{Cd-Cl} = 270$

[a] Data are in the units given in Table 3. O_h = octahedral.
[b] This ring is also called 1,2,5-selenadiazole[146,147].
[c] The spectrum at 100 °C had four broad signals at about the same δ values.

A number of Cd(II), Cu(II), Hg(II), Ag(I), Ni(II) and Fe(III) complexes of 2,1,3-benzoselenadiazole (**16**) and some substituted derivatives (**17**–**19**) have been prepared and characterized by elemental analyses and infrared spectroscopy, and their thermal stability has been studied by thermogravimetric and differential thermal analyses under a nitrogen atmosphere[148,149] (Table 7).

These complexes, MCl_2L_n ($n = 1,2$), which are readily precipitated by the addition of an appropriate amount of the selenadiazole to the metal chloride in aqueous ethanol (anhydrous ethanol for Ni and Co), were formulated as dimeric or polymeric species with bridging chloro and/or selenadiazole ligands, except for the monomeric tetrahedral $HgCl_2(bsd)_2$[148,149]. The structural characterization of these complexes is tentative; indeed, the bonding sites of the selenadiazole ligands were not clearly established, although nitrogen bonding was inferred by the trend of the thermal stabilities of the complexes as a function of the substituent in the 5-position of the heterocyclic system. Most likely, the bonding sites of such heterocycles vary according to the nature of the metal ion and other ligands present in the coordination sphere of the complex, as has been well established for other ambidentate ligands[153].

In an ESR study[154] of the complexes $Cu(bsd)_2X_2$ (X = F, Cl, Br) Cu—Se bonding (and Cu—S bonding in the benzothiadiazole (btd) analogues) was proposed. A subsequent investigation[155] of the btd complexes by ESR and infrared spectroscopy and magnetic susceptibility measurements led to the conclusion that these complexes are mononuclear and contain N-bonded btd. Bonding via N(1) has been established by a single-crystal X-ray diffraction analysis[156] of $CdBr_2$ (4-NH_2-btd)$_2$, which has octahedral geometry around the Cd(II) with bridging bromo ligands that form infinite chains and two 4-NH_2-btd ligands coordinated to each Cd in *trans* positions.

This work all illustrates the importance of caution in assigning the structures of complexes containing such ambidentate ligands. Ideally, it is possible to make such assignments on the basis of high-resolution infrared spectroscopy if the appropriate model compounds have been characterized by single-crystal X-ray diffraction. Thiocyanate and selenocyanate complexes, which have the added advantage of having strong $\nu_{C\equiv N}$ bands in a rather isolated infrared spectral region (2000–2200 cm^{-1}), are examples for which this is a fairly safe procedure[153]. More recently, XPS[157] and solid-state NMR[158] have been used to 'fingerprint' bonding sites of ambidentate ligands and should prove more generally useful for such studies.

The observation that the polarographic reduction waves of the 2,1,3-benzo-*X*-diazoles (X = O, S, Se) are shifted only for the S and Se compounds upon addition of Cu(II) and Cd(II) was explained by coordination of only these two ambidentate heterocycles to the metals via the S or Se atoms[159]. In a related study[160], 2,1,3-benzoselenadiazole was shown to function as a catalytically active ligand for the reduction of Ni(II) at a dropping-mercury electrode. A prewave of Ni(II) observed here was assigned to a polarographically active complex formed between the metal ion and the adsorbed ligand.

The radical anion of 2,1,3-benzoselenadiazole (**16**), observed in the two-step polarographic reduction of these heterocycles in DMF[154], has been shown by ESR spectroscopy to form 1:1 and 2:1 complexes with $M(CO)_6$ (M = Cr, Mo, W)[150] (equation 22).

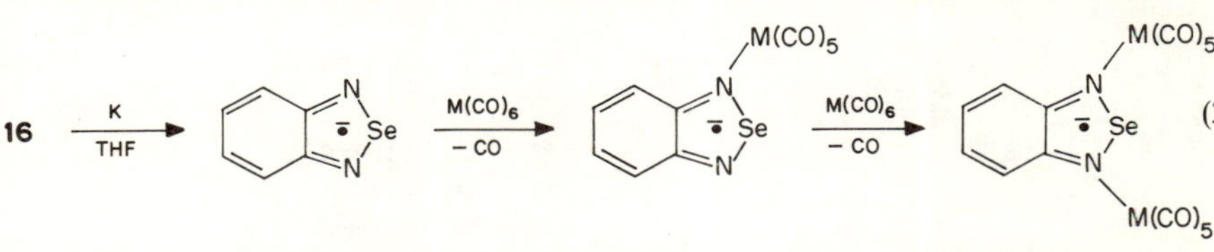

(22)

The formation of the monosubstituted product is rapid, with the formation of the disubstituted product decreasing in the order Cr > Mo > W; $[W(CO)_5]_2(bsd^{\bar{\cdot}})$ forms only after several months at room temperature. These reactions were run in ESR tubes with ca. 5:1 molar ratios of $M(CO)_6$ to bsd. These paramagnetic species are stable in THF for several months at room temperature. The spectra of the Cr and Mo complexes indicated static M—N bonding, whereas the spectrum of $W(CO)_5(bsd^{\bar{\cdot}})$ suggested rapid fluctuation of the $W(CO)_5$ fragment between the two N sites (equation 23). Attempts to freeze out a static structure were hampered by typical anisotropic line-broadening effects below 0 °C.

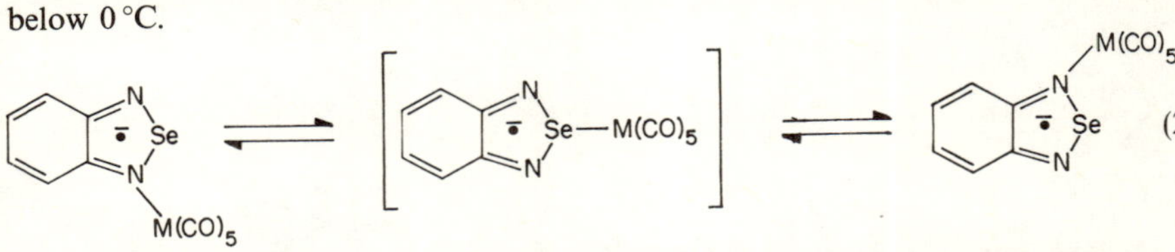

(23)

The ^{1}H-NMR spectrum of the neutral complex $W(CO)_5$(bsd), prepared from $W(CO)_5$(THF) and bsd in THF, gave a resolved spectrum at room temperature (Table 7), and at 120 °C the four resonances were still resolved but slightly broadened with loss of the ^3J coupling.

Complexes with terminal and bridging N-bonded cycloalkenyl-1,2,3-selenadiazoles (**20a–d**) have been obtained with Cr, Mo and W carbonyls[161,162] (Table 8; equations 24–26).

TABLE 8. Metal complexes with cycloalkenyl-1,2,3-selenadiazoles (**20**)

Metal complexes/ properties[a] [162]	Metal complexes/ properties[a] [162]	Metal complexes/ properties[a] [162]
21b: m.p. 111 °C dec. $\nu_{C\equiv O}{}^{b}$ = 2070w, 1989w, 1944s, 1920m	**23b**: m.p. 115 °C dec. $\nu_{C\equiv O}{}^{b}$ = 2070w, 1980w, 1940s, 1920m	**24**: m.p. 90 °C dec. $\nu_{C\equiv O}{}^{d}$ = 2020w, 1907s, 1840s
22b: m.p. 112 °C dec. $\nu_{C\equiv O}{}^{b}$ = 2070w, 1980w, 1935s, 1915m	$\delta(^{1}H)^{c}$: 3.2m and 3.0m (αH); 1.85m (βH)	**25**: m.p. 176 °C dec. $\nu_{C\equiv O}{}^{e}$ = 1905s, 1818s
	$\delta(^{13}C)^{c}$: sp^{2}C, 162.5, 159.5; sp^{3}C, 25.2, 25.1, 22.6, 21.7; *trans*-CO, 202.7; *cis*-CO, 200.4	

[a] Data are in the units given in Table 3. SP = square planar.
[b] In hexane solution.
[c] In $CDCl_3$, internal TMS.
[d] In CH_2Cl_2.
[e] In THF.

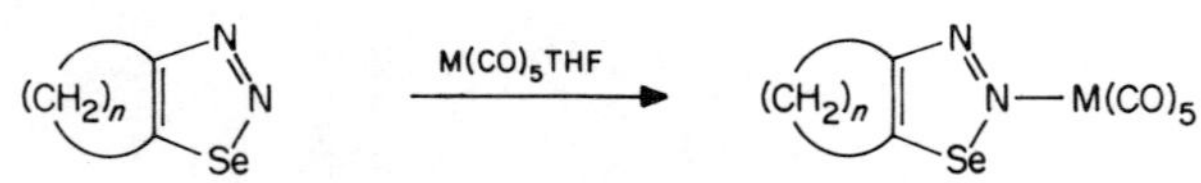

(24)

(**20**)

(**a**) n = 3
(**b**) n = 4
(**c**) n = 5
(**d**) n = 6

M = Cr (**21a–d**)
M = Mo (**22b**)
M = W (**23a–d**)

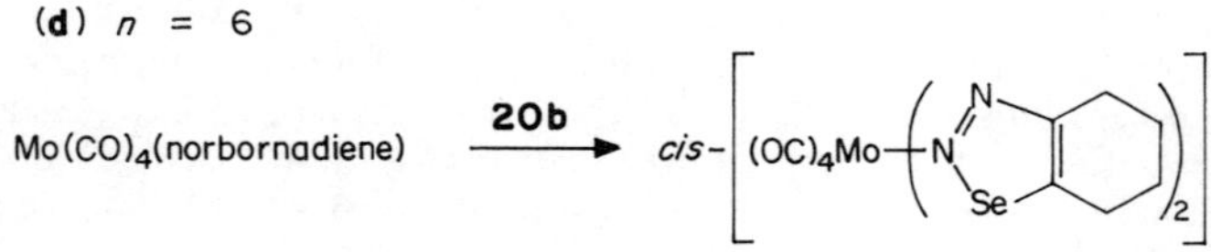

(25)

(**24**)

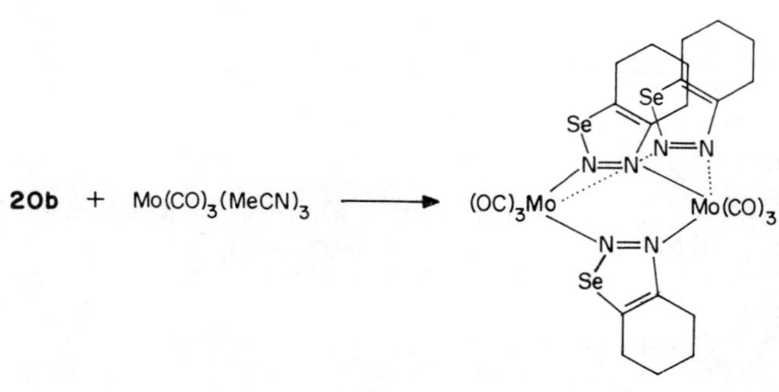

(26)

(**25**)

Reactions of 1,2,3-selenadiazoles with $Fe_2(CO)_9$, however, result in facile extrusion of N_2 from the heterocyclic ring system and formation of iron dimers containing bridging selenoketocarbene ligands[161–164] (equation 27). The isomeric ratios of products obtained from the latter two unsymmetrically substituted derivatives were obtained by ^{1}H-NMR spectroscopy of the crude reaction product obtained after one preparative TLC separation, the major isomer then being isolated by further chromatography and recrystallization[163]. The by-products, $Fe_3(\mu_3\text{-}Se)_2(CO)_9$ and PhC≡C-*p*,-*t*-BuC_6H_4, were obtained only in small amounts in these reactions, although in the reaction with the unsubstituted 1,2,3-selenadiazole, the only iron complex formed was the known cluster $Fe_3(\mu_3\text{-}Se)_2(CO)_9$[163].

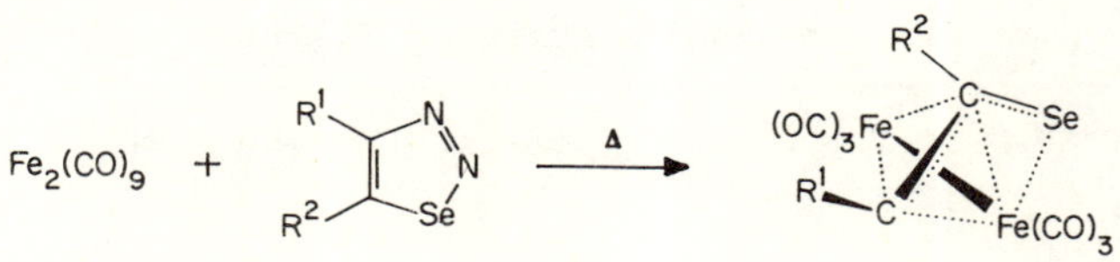

(27)

$R^1 = R^2 = Ph$ — 6%[163]; 21%[164]

$R^1 = Ph, R^2 = p\text{-}t\text{-}BuC_6H_4$ — $R^1 = Ph, R^2 = p\text{-}t\text{-}BuC_6H_4$: 26%[163]; $R^1 = p\text{-}t\text{-}BuC_6H_4, R^2 = Ph$: 8%

$R^1 = p\text{-}t\text{-}BuC_6H_4, R^2 = Ph$ — $R^1 = p\text{-}t\text{-}BuC_6H_4, R^2 = Ph$: 26%[163]; $R^1 = Ph, R^2 = p\text{-}t\text{-}BuC_6H_4$: 8%

In subsequent work[161], such reactions at room temperature gave intermediate selenaferrole complexes that could be isolated by chromatography (equation 28). The selenaferrole complexes (**A**) readily gave the selenoketocarbene complexes (**B**) upon UV irradiation or thermolysis. Examples of both products have been unequivocally characterized by single-crystal X-ray diffraction (**A**, $n = 4$[165]; **B**, $n = 6$[166]).

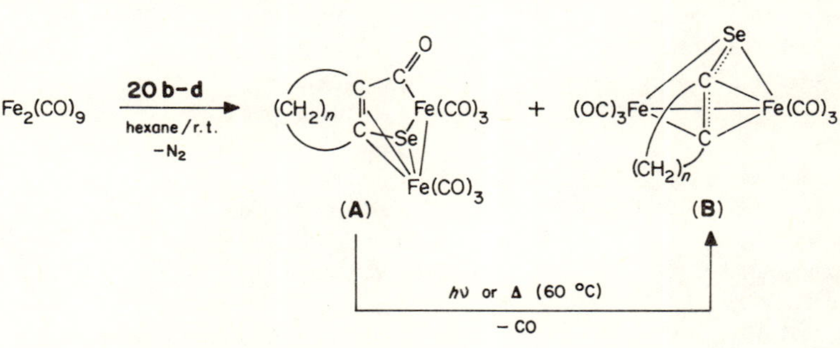

(28)

Beck and coworkers have reported complexes of benzoselenazole (**26a**)[167] and its 2-methyl derivative (**26b**)[168] with Cr, Mo and W carbonyls (Table 9). These complexes were prepared by typical substitution reactions (equations 29 and 30).

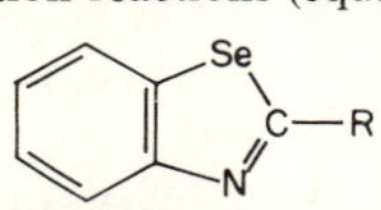

bs, (**26a**) R = H
mbs, (**26b**) R = Me

$$M(CO)_6 + L \xrightarrow[h\nu]{THF} M(CO)_5L \tag{29}$$

$$L = bs^{167}, \qquad mbs^{168}$$

$$Mo(CO)_4(C_7H_8) + bs \xrightarrow[-C_7H_8]{THF} Mo(CO)_4(bs)_2 \tag{30}$$

$$C_7H_8 = \text{cycloheptadiene}$$

Characterization by ^{14}N- and ^{1}H-NMR spectroscopy has indicated that in the bs

TABLE 9. Complexes with benzoselenazoles **26a** (bs) and **26b** (mbs)

Metal complexes/properties[a]	Metal complexes/properties
$Cr(CO)_5(bs)$[167]	m.p. 168 °C
yellow solid	$\nu_{C\equiv O}{}^{b} = 2071, 1930, 1886$
m.p. 153 °C dec.	linkage isomers in acetone solution
$\nu_{C\equiv O}{}^{b} = 2071, 1938, 1895$	W—Se
$\delta(^{14}N)^{b,c}$: + 118 (Δ = + 64)	$\delta(^{14}N)^{b,c}$: + 45 (Δ = − 14)
$\delta(^{1}H)^{d}$: − 10.70 (Δ = − 0.59)	$\delta(^{1}H)^{d}$: N—Me, − 2.83 (Δ = − 0.12)
$Mo(CO)_5(bs)$[167]	W—N
yellow solid	$\delta(^{14}N)^{b,c}$: + 180 (Δ = + 121)
m.p. 145 °C	$\delta(^{1}H)^{d}$: − 3.28 (Δ = − 0.57)
$\nu_{C\equiv O}{}^{b} = 2078, 1944, 1894$	$NiCl_2(mbs)$[169]
$\delta(^{14}N)^{b,c}$: + 117 (Δ = + 63)	red crystals
$\delta(^{1}H)^{d}$: − 10.78 (Δ = − 0.67)	m.p. 216 °C dec.
$W(CO)_5bs$[167]	$Ni(NO_3)_2(mbs)_2$[169]
yellow solid	pale green crystals
m.p. 167 °C	m.p. 222 °C dec.
$\nu_{C\equiv O}{}^{b} = 2078, 1934, 1896$	μ = 3.71 B.M.
$\delta(^{1}H)^{d}$: − 10.94 (Δ = − 0.83)	$NiBr_2(mbs)_2$[169]
$Mo(CO)_4(bs)_2$[167]	pale blue crystals
yellow solid	m.p. 170–171 ° dec.
m.p. 141 °C	$NiI_2(mbs)_2$[169]
$\nu_{C\equiv O}{}^{e} = 2019, 1898, 1859, 1804$	dark green solid
$Cr(CO)_5(mbs)$[168]	m.p. 234–235 °C dec.
green–yellow solid	$Ni(NCS)_2(mbs)_2$[169]
m.p. 110 °C	red solid
$\nu_{C\equiv O}{}^{b} = 2073, 1939, 1887$	m.p. 199 °C (colour change to green at 132 °C)
$\delta(^{14}N)^{b,c}$: + 49 (Δ = − 10)	μ(20–130 °C) = 1.05 B.M.
$\delta(^{1}H)^{d}$: − 2.82 (Δ = − 0.11)	μ(135 °C) = 3.6 B.M.
$W(CO)_5(mbs)$[168]	
yellow crystals	

[a] Data are in the units given in Table 3.
[b] In acetone solution.
[c] Vs. external aqueous $NaNO_3$; Δ = shift vs. free ligand.
[d] Resonance in acetone-d_6 solution (internal TMS) of R in **26**; Δ = shift vs. free ligand.
[e] In Nujol.

complexes the heterocycles coordinate via the nitrogen, whereas $W(CO)_5$mbs contains about equal amounts of the W—N and W—Se linkage isomers, and the Mo analogue, primarily N-bonded, forms a small amount of the Se-bonded isomer on standing in acetone solution (Table 9). The structure of $W(CO)_5$mbs in the solid state has not been established but would be a problem amenable to solid-state Se-NMR[158] or X-ray photoelectron spectroscopy[157].

Several Ni(II) complexes of **26b** have been isolated and characterized by infrared and electronic spectroscopy as well as magnetic susceptibility measurements (Table 9)[169]. Electronic spectroscopy and conductivity measurements indicated significant dissociation of the heterocyclic ligand in methanol. The complex $Ni(NCS)_2(mbs)_2$ exhibited an unusual transition from an equilibrium mixture of paramagnetic tetrahedral–diamagnetic square planar to pure tetrahedral stereochemistry at ~132 °C, a change accompanied by a colour change from red to green. A broad $\nu_{C\equiv N}$ band in the bridging region (2125–2175 cm^{-1}) indicated a dimeric or polymeric formulation for this complex.

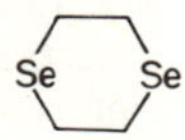

dse, (**27**)

1,4-Diselenan (dse) (**27**), prepared in 10% yield from Al_2Se_3 and $BrCH_2CH_2Br$[170], forms complexes with a variety of transition metals[170,171] (Table 10). This heterocycle can, in principle, function as a monodentate (terminal, **A**, or bridging, **B**) or bidentate ligand

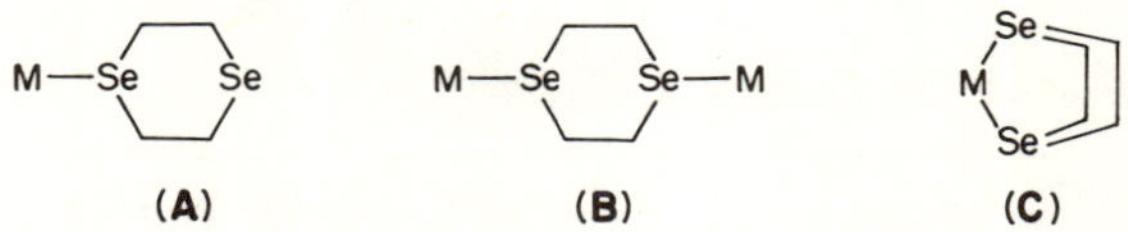

C). Since no definitive single-crystal X-ray diffraction analysis has been reported for such complexes, the proposed bonding modes, in some cases tentative, are based on the stoichiometries as well as infrared[171] and NMR[172] spectroscopy. The complexes of the stoichiometry $(MX_2)_2$(dse) (M = Cd, Hg; X = Cl, Br; Table 10) were formulated as dimers with a bridging dse ligand[171]:

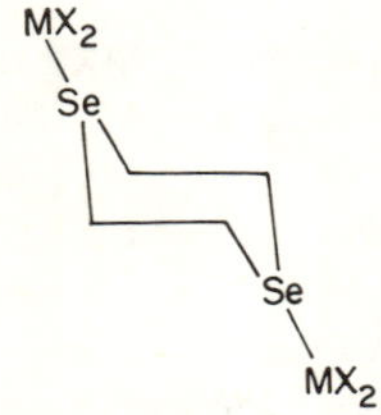

The $\nu_{M—Cl}$ bands in these complexes all occur in the terminal region[171].

A centrosymmetric chair configuration of the bridging ligand has been proposed on the basis of the similarity of its infrared spectrum[171] to that of free dse, for which this conformation has been established by X-ray crystallography[173]. The cuprous chloride complex, CuCl(dse), also gave an infrared spectrum similar to that of the free ligand, but its insolubility suggested a polymeric formulation with dse in a chair conformation. Either bridging chloro or dse ligands (or both) may be present in this complex[171].

TABLE 10. Complexes with 1,4-diselenan (dse) (**27**)

Complex	Properties[a]	Complex	Properties
$[PdCl_2(dse)]$	Bright yellow solid[170]	$(HgBr_2)_2dse$	White solid[171]
	$\nu_{Pd-Cl} = 303, 318$[171]	$(CdCl_2)_2dse$	White solid[171]
$[PdBr_2(dse)]$	Orange yellow solid[171]	$(CuCl_2)dse$	Dark violet crystals[171]
	$\nu_{Pd-Br} = 253, 268$	$(CuBr_2)_2dse$	Brown powder[171]
$[PtCl_2(dse)]$	Pale yellow powder[171]	$AgNO_3(dse)$	White powder[171]
	$\nu_{Pt-Cl} = 314, 324$	CuCl(dse)	White powder[171]
$(HgCl_2)_2dse$	White solid[171]	$AuCl_3(dse)$	Pale yellow solid[171]
	$\nu_{Hg-Cl} = 310$	$AuBr_3(dse)$	Dark brown solid[171]

[a] Data are in the units given in Table 3.

The infrared spectra of the complexes MX_2(dse) (Table 10) are markedly different from the above complexes and have been interpreted in terms of either a monomeric formulation with chelating dse in a boat conformation (**C**) or dimers with two bridging dse ligands in boat conformations. The stoichiometries and infrared spectra of the AuX_3(dse) (X = Cl, Br) complexes suggest monomeric, four-coordinate complexes containing monodentate dse in the chair configuration[171].

The ambidentate 1,4-oxaselenan (ose) (**28**) coordinates through its Se atom as a terminal ligand or via both O and Se atoms as a bridge ligand[174] (Table 11), but no well-established examples of terminal oxygen or chelating (O, Se) coordination have been reported. The essentially unchanged values of the symmetric and unsymmetric C—O—C

TABLE 11. Complexes with 1,4-oxaselenan (ose) (**28**)

Complex	Properties[a]	Complex	Properties[a]
$SnCl_4(ose)_2$[174]	$\nu_{C-O-C,asym} = 1100vs^b$	$NbCl_5(ose)$[174]	$\nu_{C-O-C,asym} = 1100vsb$
	$\nu_{C-O-C,sym} = 818m^c$		$\nu_{C-O-C,sym} = 819s^c$
	δ: $OCH_2 = 4.13$ $(\Delta = 0.08)^d$		δ: $OCH_2 = 4.20$ $(\delta = 0.15)^d$
	$SeCH_2 = 2.91$ $(\Delta = 0.25)^d$		$SeCH_2 = 3.66$ $(\Delta = 1.00)^d$
$SnBr_4(ose)_2$[174]	$\nu_{C-O-C,asym} = 1100vs^b$	$TiCl_4(ose)$[174]	Orange–red solid
	$\nu_{C-O-C,sym} = 891m^c$		$\nu_{C-O-C,asym} = 1043vs^b$
	δ: $OCH_2 = 4.10$ $(\Delta = 0.05)^d$	$MX_2(ose)_2$[175]	$\nu_{C-O-C,sym}$ $790s^c$
	$SeCH_2 = 2.71$ $(\Delta = 0.05)^d$	M = Pd, Pt; X	
$TiCl_4(ose)_2$[174]	Orange solid	= Cl, Br	
	$\nu_{C-O-C,asym} = 1102vs^b$		Variable-temperature ^1H-NMR
	$\nu_{C-O-C,sym} = 812m^c$		
	δ: $OCH_2 = 4.13$ $(\Delta = 0.08)^d$	*trans*-	
	$SeCH_2 = 2.91$ $(\Delta = 0.25)^d$	$PtBr_2(ose)_2$[176]	Molecular structure
			Pt—Se: 2.430 Å
			Pt—Br: 2.442 Å
			<Se—Pt—Br: 93.3°

[a] Data are in the units given in Table 3.
[b] The corresponding value in the free ligand is 1102vs.
[c] The corresponding value in the free ligand is 812s.
[d] Shifts downfield vs. free ose.

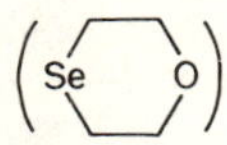

ose, (**28**)

stretching modes of the heterocycle in the complexes $SnX_4(ose)_2$ (X = Cl, Br), $TiCl_4(ose)_2$ and $NbCl_5(ose)$ vs. the free ligand support Se bonding in these complexes[174]. The decrease in the shifts of these bands to lower values for $TiCl_4(ose)$ vs. the free heterocycle (Table 11) and the stoichiometry of this complex support a polymeric formulation with octahedral coordination of the Ti and bridging ose ligands[174].

The weakness of the corresponding CSeC modes precludes direct verification of the proposed Se bonding in these complexes. The larger shift of the methylene protons adjacent to the Se vs. those adjacent to the O in the ^{1}H-NMR of these complexes relative to the free heterocycle (e.g. $\Delta\delta_{SeCH_2} \simeq 1.0$ vs. $\Delta\delta_{OCH_2} \simeq 0.1$) further supports monodentate Se bonding in the above monomeric complexes (Table 11). These results indicate that these halides (Cl, Br) exhibit class B Lewis acid behaviour; the fluorides however, exhibit the class A behaviour expected of such elements in their highest common oxidation states (e.g. $BF_3{\cdot}O(CH_2CH_2)_2S$ exhibits O coordination; the corresponding selenan complex was not reported)[174].

A variable-temperature ^{1}H-NMR study[175] of the complexes $MX_2(ose)_2$ (M = Pt, Pd; X = Cl, Br) showed that the observed coalescence phenomena are caused by site inversion about the Se atom rather than by ring inversion of the heterocyclic ligand.

A single-crystal X-ray diffraction study of the monomeric square-planar complex *trans*-$PtBr_2(ose)_2$ confirmed a monodentate Se bonding mode of the 1,4-oxaselenan ligand with an axial orientation of the Pt—Se bond relative to the ring (the analogous oxathian complex has this bond equatorial to the ring)[176].

Thiazolidine-2-selenone (tzse) (**29**), prepared by the reaction of NaHSe and 1,3-thiazolidine-2-methylthiol, also exhibits ambidentate ligand behaviour[177]. The electronic spectra and magnetic susceptibilities of its Co(II) and Ni(II) complexes (Table 12) indicate tetrahedral geometries in all cases, and infrared spectroscopy supports N bonding in the Co(II) complexes and Se bonding in the Ni(II) complexes. The ν_{N-H} bands for the Co(II) complexes exhibit large negative shifts of $\sim 270\,cm^{-1}$ vs. the free ligand; but the bands associated with the S(ν_{C-S}) and Se ($\nu_{C=Se} + \delta_{N-C-Se}$) are essentially unchanged or show slight positive shifts. More direct evidence for N bonding is found in the far-infrared region, where ν_{Co-N} bands for the three complexes are observed at $\sim 250\,cm^{-1}$; ν_{Co-S} bands would be expected to appear at $\sim 230\,cm^{-1}$. The electronic spectra of the complexes also support CoN_2X_2 (X = Cl, Br, I) tetrahedral coordination spheres.

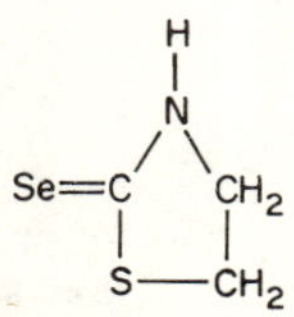

tzse, (**29**)

The ν_{N-H} vibrations of the Ni(II) complexes appear as broad bands in the region 3350–3400 cm^{-1}, near that of the free ligand, and bands in the FIR region, at 170–180 cm^{-1}, are assigned to ν_{Ni-Se} vibrations (ν_{Ni-N} bands would be expected at 220–230 cm^{-1}). The proposed Se bonding mode is further supported by the essentially unchanged position of the NH resonance vs. the free ligand in the ^{1}H-NMR spectra of the Ni(II) complexes, whereas shifts of ~ 1.5 ppm to lower field are observed for the Co(II) complexes.

TABLE 12. Complexes with thiazolidine-2-selenone (tzse) (**29**)[177]

Complex	Properties[a]	Complex	Properties[a]
$[Co(tzse)_2Cl_2]\cdot 2H_2O$	Turquoise solid M.p. > 350 °C $\mu = 5.1$ B.M. $\nu_{N-H} = 3130vs^b$ $\nu_{C-N} = 1530vs^c$ $\nu_{C=Se} + \delta_{N-N-Se}$ $= 1310vs^d$ $\nu_{C-S, asym} = 715^e$ $\nu_{C-S, sym} = 660ms^f$ $\nu_{Co-Cl} = 321m, 305m$ $\nu_{Co-Se} = 252m$	$[NiCl_2(tzse)]$	Olive green solid M.p. > 350 °C $\mu = 3.6$ B.M. $\nu_{N-H} = 3400vs^b$ $\nu_{C-N} = 1560m^c$ $\nu_{C=Se} + \delta_{N-C-Se}$ $= 1310m^d$ $\nu_{C-S, asym} = 710w^e$ $\nu_{Ni-Cl-Ni} = 204m, 154m$ $\nu_{Ni-Cl} = 268m, 240s$ $\nu_{Ni-N} = 180ms$
$[Co(tzse)_2Br_2]\cdot 2H_2O$	Dark green solid M.p. 300 °C dec. $\mu = 4.6$ B.M. $\nu_{N-H} = 3140s^b$ $\nu_{C-N} = 1520vs^c$ $\nu_{C=Se} + \delta_{N-C-Se}$ $= 1300vs^d$ $\nu_{C-S, asym} = 695m^e$ $\nu_{C-S, sym} = 660ms^f$ $\nu_{Co-Br} = 283m, 215m$ $\nu_{Co-Se} = 250m$	$[NiBr_2(tzse)]$	Dark brown solid M.p. > 350 °C $\mu = 3.7$ B.M. $\nu_{N-H} = 3350vs^b$ $\nu_{C-N} = 1510m^c$ $\nu_{C=Se} + \delta_{N-C-Se}$ $= 1295ms^d$ $\nu_{C-S,asym} = 700w^e$ $\nu_{C-S, sym} = 660w^f$ $\nu_{Ni-Br-Ni} = 158m, 148m$ $\nu_{Ni-Br} = 200m$ $\nu_{Ni-N} = 171ms$
$[Co(tzse)_2I_2]\cdot 2H_2O$	Dark green solid M.p. 224 °C dec. $\mu = 4.6$ B.M. $\nu_{N-H} = 3230m^b$ $\nu_{C-N} = 1505vs^c$ $\nu_{C=Se} + \delta_{N-C-Se}$ $= 1290vs^d$ $\nu_{C-S, asym} = 695w^e$ $\nu_{C-S,sym} = 665w^f$ $\nu_{Co-I} = 206ms$ $\nu_{Co-Se} = 255s$	$[NiI_2(tzse)_2]$	Black solid M.p. > 350 °C $\mu = 3.9$ B.M. $\nu_{N-H} = 3130vs^b$ $\nu_{C-N} = 1550m^c$ $\nu_{C-Se} + \delta_{N-C-Se}$ $= 1310w^d$ $\nu_{C-S, asym} = 690w^e$ $\nu_{C-S, sym} = 660^f$ $\nu_{Ni-I} = 210m$ $\nu_{Ni-N} = 176ms$

[a] Data are in the units given in Table 3.
[b] Corresponding band in free tzse = 3100vs (solid) and 3400vs ($CHCl_3$ solution).
[c] 1515vs in free tzse.
[d] 1285vs in free tzse.
[e] 705ms in free tzse.
[f] 655m in free tzse.

Reaction of $Pd(F_6acac)_2$ with 1 or 2 equiv. of phenoselenazine (psz) (**30**) in acetone/toluene gives, on concentration of the reaction solution, blue microcrystals of the composition $Pd(F_6acac)_2(psz)_2$[178] (Table 13). An ionic structure, $[Pd(F_6acac)(psz)_2](F_6acac)$, is supported by the presence of a $\nu_{C=O}$ band at 1670 cm^{-1} (vs.

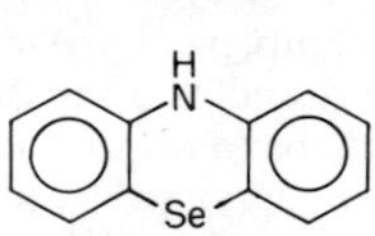

psz , (**30**)

TABLE 13. Complexes with phenoselenazine (psz) (**30**)

Complex	Properties[a]
$[Pd(psz)_2(F_6acac)](F_6acac)$[178]	Deep blue solid λ_{max} (Nujol) = 610, 735 nm
trans-$PtCl_2(psz)_2$[179]	Orange crystals $\nu_{Pt-Cl} = 325$
cis-$[PtCl_2(psz)_2]\cdot MeCN$[179]	Orange needles $\nu_{Pt-Cl} = 310, 285$ $\nu_{C\equiv N} = 2260$ (lattice MeCN) Molecular structure Pt—Se: 2.376(2), 2.400(2) Å
trans-$PdCl_2(psz)_2$[179]	$\nu_{Pd-Cl} = 335$

[a]Data are in the units given in Table 3.

$1603\,cm^{-1}$ in $Pd(F_6acac)_2$). The proposed structure was unequivocally established by a single-crystal X-ray diffraction analysis for the bipy complex ($[Pd(F_6acac)bipy]^+(F_6acac)^-$)[178]. The visible absorption bands of this complex (Table 13) were assigned to ligand→metal charge-transfer processes, since free psz is readily oxidized to a stable radical cation. Dilute acetone solutions of the complex are readily bleached in air[178]. Nucleophilic attack by excess psz at Pd is not observed in this complex although in the cases of L = pyr, Ph_3As and *i*-$PrNH_2$, further substitution to give ultimately $[PdL_4](F_6acac)_2$ has been reported[180].

Reaction of psz with $K[PtCl_3(C_2H_4)]$ in acetonitrile gives an initial orange precipitate of *trans*-$PtCl_2(psz)_2$; after filtration and concentration of the solution, large orange crystals of *cis*-$PtCl_2(psz)_2\cdot MeCN$ are isolated[179]. The *cis* isomer is apparently the thermodynamically stable form, since recrystallization of the *trans* isomer from hot acetonitrile gives the *cis* isomer along with some decomposition products, as evidenced by infrared spectroscopy. Exposure of the *cis* and *trans* complexes to iodine vapour gives the partially oxidized non-crystalline compounds *cis*-$PtCl_2(psz)_2I_{3.9}$ and *trans*-$PtCl_2(psz)_2I_{3.5}$, respectively[179]. These latter complexes have significantly greater electrical conductivity than the insulating precursors. Resonance Raman and XPS indicate that the hole sites are localized on the psz ligand and that the iodine is present predominantly as I_5^-[179].

Tetraphenylselenophene has been obtained in 60% yield by heating a mixture of powered amorphous Se and $Fe_2(CO)_6(C_2Ph_2)_2$ at 200 °C[181]. An ESR study of the interaction of tetrahydroselenophene with the planar Co(II) complex *N, N'*-bis(salicyliden)-*o*-phenylenediaminocobalt(II) (Co(saphen)) has indicated the formation of a 2:1 adduct[182].

5. *Diorganodiselenides (RSeSeR) and related ligands ($RSeER^1_2$; E = P, As)*

Although the most characteristic reaction of these derivatives in transition-metal chemistry involves the cleavage of the Se—Se bond with formal oxidation of the metal centre and formation of terminal or bridging SeR^- ligands (see Section II.B.2), several examples of coordination of intact RSeSeR ligands have been reported (Table 14).

The Pt(IV) dimers were prepared according to equation (31)[183]. The yellow products were isolated by addition of light petroleum after concentration of the reaction solutions. The crystal structural determination[184,185] of the bromo analogue established it as the first example of a complex containing a bridging diorganodiselenide ligand. A similar structure was subsequently established for the iodo analogue[186]. A variable-temperature

TABLE 14. Metal complexes with RSeSeR and $RSeER^1_2$ (E = P, As) ligands

Complex	Reported data	Reference
μ_2-(Alkyl)SeSe(alkyl)		
31a	Yellow solid	183
	M.p. 144–146 °C dec.	
	^{1}H-NMR	183–185
	Molecular structure	184
	Pt—Pt: 3.740(3) Å	
	Pt—Br: 2.629 Å	
	Se—Se: 2.36(1) Å	
31b	Yellow solid	183
	M.p. 150–151 °C dec.	
	MW = 730[a] (calc. 739)	
	^{1}H-NMR	183, 185
31c	Yellow solid	183
	M.p. 120–125 °C dec.	
	MW = 929[a] (calc. 921)	
	^{1}H-NMR	183, 185
	Molecular structure	186
	Pt—Pt: 3.901(2) Å	
	Pt—I: 2.798(2) Å, 2.778(2) Å	
	Se—Se: 2.358(4) Å	
32a	Pale yellow crystals	187
	M.p. 158–159 °C dec.	
	$\nu_{C\equiv O}{}^b$ = 2054, 2038, 1950, 1926	
	Variable-temperature ^{1}H-NMR	
	Molecular structure	
	Re—Re: 3.883 Å	
	Re—Br: 2.644 Å	
	Se—Se: 2.375(4) Å	
32b	Red–orange solid	187
	M.p. 181–182 °C dec.	
	$\nu_{C\equiv O}{}^c$ = 2040, 2027, 1956, 1925	
Terminal (alkyl) SeSe(alkyl)		
$W(CO)_5(R_2Se_2)$	Variable-temperature ^{1}H-NMR	188
$R = CH_2SiMe_3$	E_a (kJ mol^{-1})	
	inversion: 59.8 ± 0.8	
	1,2-shift: 82.6 ± 1.6	
μ_2-*(Aryl)SeSe(aryl)*		
32c	Orange–red solid	189
	$\nu_{C\equiv O}{}^d$ = 2057m, 2041s, 1963s, 1960sh, 1936s	
	Molecular structure	
	Re—Re: 3.899(43) Å	
	Re—Br: 2.656(20) Å	
	Se—Se: 2.411(23) Å	
	Re—Se: 2.604(4) Å	
32d	$\nu_{C\equiv O}{}^d$ = 2051m, 2035s, 2022w, 1961s	190
32e	$\nu_{C\equiv O}{}^d$ = 2053m, 2036s, 2023w, 1977s, 1973sh, 1946s	191
	Molecular structure	
	Mn—Se : 2.478(3) Å	
	Mn—Mn: 3.680(5) Å	
	Se—Se : 2.401(3) Å	
	Mn—Br : 2.542(4), 2.541(3)	
$[Ag_2(Se_2Ph_2)_4](AsF_6)_2$	Molecular structure (Figure 1)	191a

TABLE 14. (*Continued*)

Complex	Reported data	Reference
R_2PSeR^1		
$Cr(CO)_5(Me_2PSeMe)$	Dark yellow air-sensitive crystals $\nu_{C\equiv O}{}^e = 2071m, 1987w, 1960s, 1949vs$ $\delta(^1H) = 1.28\ (\Delta^f = -0.03)$ $1.38\ (\Delta^f = -0.44)$ $\delta(^{31}P) = 38.5\ (\Delta^f = 37.7)$	192, 193
$Cr(CO)_5\{(CF_3)_2PSeMe\}$	Yellow–brown oil $\nu_{C\equiv O}{}^e = 2094m, 2016vw, 1989s, 1984vs$ $\delta(^1H) = 1.80\ (\Delta^f = -0.04)$ $\delta(^{19}F) = -56.6\ (\Delta^f = -2.5)$ $\delta(^{31}P)^g = 99.0\ (\Delta^f = 74.1)$	192
$Mo(CO)_5(Me_2PSeMe)$	$\nu_{C\equiv O} = 2080m, 1994w, 1962s, sh, 1955vs$ $\delta(^1H) = 1.29\ (\Delta^f = 0.04)$ $\delta(^{31}P) = 12.3\ (\Delta^f = 11.5)$	194
$Mo(CO)_5\{(CF_3)_2PSeMe\}$	$\nu_{C\equiv O} = 2095m, 1992s, 1987vs$ $\delta(^1H) = 1.82\ (\Delta^f = 0.06)$ $\delta(^{19}F)^h = -52.0\ (\Delta^f = -2.0)$ $\delta(^{31}P)^g = 72.2\ (\Delta^f = 44.3)$	194
$Mo(CO)_5\{(CF_3)_2PSeCF_3\}$	$\nu_{C\equiv O} = 2096m, 2022w, 1996s, 1986vs$ $\delta(^{19}F)^h = -68.4\ (P(CF_3)_2)$ $(\Delta^f = -14.6)$ $-33.0\ (SeCF_3)$ $(\Delta^f = 1.8)$ $\delta(^{31}P)^g = 129.7\ (\Delta^f = 115.2)$	194
$W(CO)_5(Me_2PSeMe)$	Prepared in quantitative yield via equation (36)	193
R_2AsSeR_1		
$Cr(CO)_5(Me_2AsSeMe)$	Colourless air-sensitive crystals $\nu_{C\equiv O}{}^e = 2073w, 1988vw, 1959s, 1952vs$ $\delta(^1H) = 1.24\ (\Delta^f = 0.04)$ $1.63\ (\Delta^f = -0.27)$	192,193
$Cr(CO)_5\{(CF_3)_2AsSeMe\}$	Brown air-sensitive oil $\nu_{C\equiv O}{}^e = 2077w, 1961s, 1954vs$ $\delta(^1H) = 0.23\ (\Delta^f = 1.52)$ $\Delta(^{19}F)^h = -52.4\ (\Delta^f = -2.4)$	192
$Cr(CO)_5(Me_2AsSeCF_3)$	Red–brown air-sensitive oil $\nu_{C\equiv O}{}^e = 2076m, 1997vw, 1966s, 1956vs$ $\delta^1H) = 1.55\ (\Delta^f = -0.1)$ $\delta(^{19}F)^h = -25.2\ (\Delta^f = 0.6)$	192
$Mo(CO)_5(Me_2AsSeMe)$	$\delta(^1H) = 1.25\ (\Delta^f = 0.02)$, $1.68\ (\Delta^f = -0.22)$	193, 194
$Mo(CO)_5\{CF_3)_2AsSeMe\}$	$\delta(^1H) = 0.23\ (\Delta^f = -1.52)$ $\delta(^{19}F)^h = -52.0\ (\Delta = -2.0)$	194
$Mo(CO)_5(Me_2AsSeCF_3)$	$\delta(^1H) = 1.67\ (\Delta^f = 0.02)$ $\delta(^{19}F)^h = -24.8\ (\Delta^f = 1.0)$	194
$W(CO)_5(Me_2AsSeMe)^i$		193

[a] Determined osmometrically in $CHCl_3$ at 37 °C.
[b] Measured in toluene.
[c] Measured in $CHCl_3$.
[d] Measured in CCl_4.
[e] In cyclohexane solution.
[f] $\Delta = \delta_{complex} - \delta_{free\ ligand}$.
[g] Vs. external 85% H_3PO_4.
[h] Vs. internal CCl_3F.
[i] Prepared in quantitative yield via equation (36).

^{1}H-NMR study[185] gave the barrier energies for pyramidal inversion of Se atoms, Se atom switching between Pt atoms and scrambling of PtMe groups in these three complexes.

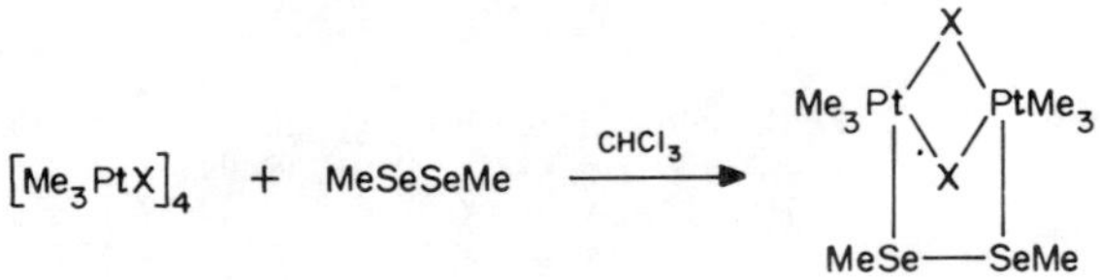

(31)

(31a) X = Br

(31b) X = Cl

(31c) X = I

Re[187,189,190] and Mn[191] carbonyl dimers with similar bridging structures (**32**) have been prepared according to equations (32) and (33). The diselenide bridged dimers **32a**, **32c** and **32d** were also obtained when Se_2R_2 (R = $PhCH_2$, Ph) reacted with $Re_2(CO)_8Br_2$/toluene[187], $Re(CO)_5Br/C_6H_6$[189] or $Re(CO)_5I$/toluene[190] under reflux conditions, which suggests that these dimers are thermodynamically, not kinetically, controlled products.

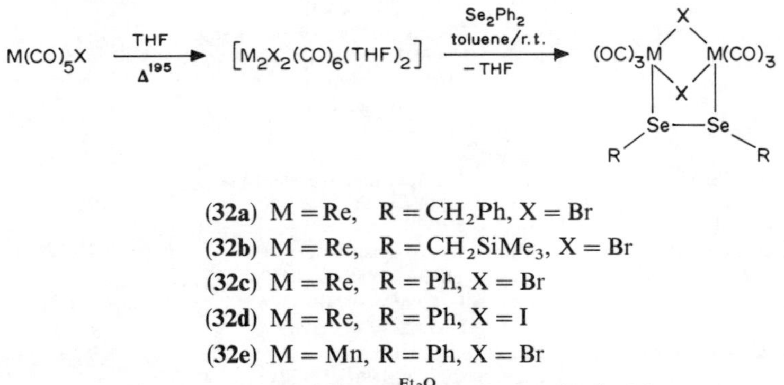

(32)

(32a) M = Re, R = CH_2Ph, X = Br

(32b) M = Re, R = CH_2SiMe_3, X = Br

(32c) M = Re, R = Ph, X = Br

(32d) M = Re, R = Ph, X = I

(32e) M = Mn, R = Ph, X = Br

$$Mn(CO)_5Br + Se_2Ph_2 \xrightarrow[\Delta]{Et_2O} \textbf{32e} \quad \text{(Ref. 191)} \qquad (33)$$

Refluxing a solution of **32e** in diisopropyl ether gave a precipitate of $MnBr_2$ and $(OC)_4Mn(\mu_2\text{-}SePh)_2Mn(CO)_4$ was isolated from the solution[191].

The structures with bridging diorganodiselenide ligands have been unequivocally established by single-crystal X-ray diffraction for **32a**[187], **32c**[189] and **32e**[191] (Table 14).

The reaction of Ph_2Se_2 with $AgAsF_6$ in liquid SO_2 gave a product, $[Ag_2(Se_2Ph_2)_4](AsF_6)_2$, which has been characterized by single-crystal X-ray diffraction[191a]. The structure (Figure 1) contains centrosymmetric 6-membered rings with two Ph_2Se_2 ligands bridging pairs of Ag atoms and three coordination about the Ag atoms being completed by 'intermolecular' Ph_2Se_2 bridges. The Se—Se distances in these bridging ligands are longer than in the free diselenide as generally observed[184,186,187,189,191], but the Se—Se distance in the 6-membered ring (2.344(2) Å) is significantly shorter than that for the other type of bridging ligand (2.360(2) Å).

A single report[188] briefly describes three examples of a terminally bonded diorganodiselenide ligand (equation 34). Variable-temperature ^{1}H-NMR has demonstrated the occurrence of two distinct fluxional processes in these complexes: pyramidal inversion about the coordinated Se atom and, at higher temperature, a novel 1,2-metal shift between

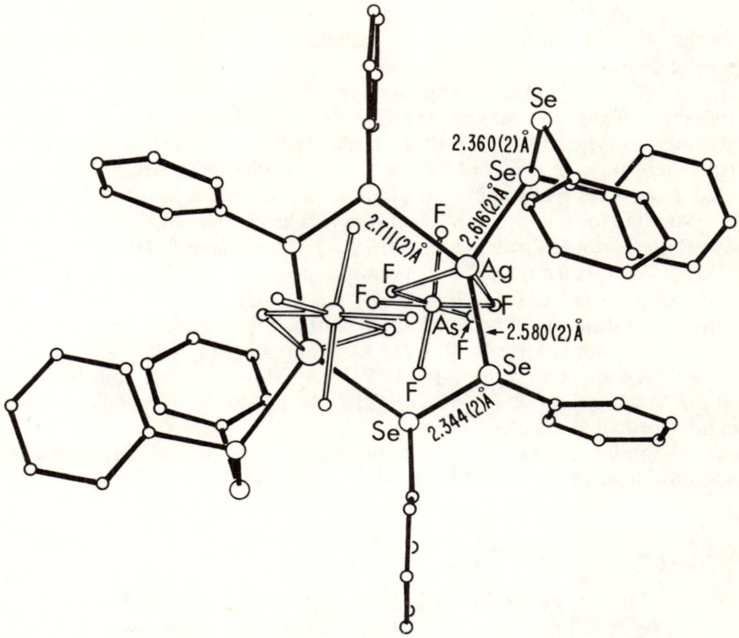

FIGURE 1. Molecular structure of $[Ag_2(Se_2Ph_2)_4](AsF_6)_2$. Reproduced with permission from Ref. 191a

the two chalcogenide atoms of the diselenide[188]. Activation parameters have been reported for the W complex (Table 14).

$$M(CO)_5THF + Me_3SiCH_2SeSeCH_2SiMe_3 \rightarrow (OC)_5M(Me_3SiCH_2SeSeCH_2SiMe_3) \quad M = Cr, Mo, W \qquad (34)$$

Grobe and LeVan[192–194] have studied the coordination chemistry of a wide variety of ambidentate ligands of the type $R_nEE^1R_m$ ($E,E^1 = P$, As; $n = m = 2$. $E = P$, As; $n = 2$; $E^1 = S$, Se, Te; $m = 1$). Included in this work are the synthesis and bonding characterization of some of the R^2ESeR^1 ($E = P$, As) complexes of Cr, Mo and W (Table 14). These complexes were prepared as shown in equations (35) and (36).

$$M(CO)_5THF + R_2ESeR^1 \rightarrow M(CO)_5(ER_2SeR^1) \qquad (35)$$

$M = Cr$ $E = P$; $R = Me, CF_3$; $R^1 = Me$

$E = As$; $R = Me, CF_3$; $R^1 = Me, CF_3$

$M = Mo$ $E = P$; $R = Me, CF_3$; $R^1 = Me, CF_3$

$E = As$; $R = Me, CF_3$; $R^1 = Me$

$R = Me$; $R^1 = CF_3$

$$M(CO)_5(R_2EER_2) + Se_2R^1_2 \rightarrow M(CO)_5(R_2ESeR^1) + R_2ESeR^1 \qquad (36)$$

$M = Cr, Mo, W$; $E = P, As$; $R = R^1 = Me$
$M = Mo$; $E = P, As$; $R = Me$; $R^1 = CF_3$

For the complexes described in equation (35), preparative-scale reactions were run, and products were isolated and characterized spectroscopically (infrared, ^{1}H-, ^{19}F- and ^{31}P-NMR; Table 14). Dismutation reactions (equation 36) were run in NMR tubes, and products were characterized *in situ* by ^{1}H- and/or ^{19}F-NMR.

In the initial work[192], NMR was shown to be useful for assigning the bonding mode of these ambidentate ligands. In the ^{1}H-NMR spectra methyl signals of groups bonded to the metal are shifted to lower fields vs. the free ligand (the absolute values of the PH coupling constants change little (e.g. ca. 12 Hz) but the sign changes). Methyl signals of non-coordinated groups generally shift ~0.4 ppm to higher fields, and the shift is larger for $(CF_3)_2AsSeMe$ than for $(CF_3)_2PSeMe$ complexes. The changes in PH coupling constants for the Me groups of non-coordinated groups vs. the free ligands fall in the range 1.5–5 Hz.

In the ^{19}F-NMR spectra the $\delta(CF_3)$ of coordinated groups shifts ~2–4 ppm to higher field, and the effect is larger for $(CF_3)_2As$— than for $(CF_3)_2P$— ligands. The $\delta(CF_3)$ of non-coordinated groups exhibits only a slight shift to lower field. In general, the coupling constants in both cases change by 2–6 Hz vs. the free ligand.

In the ^{31}P-NMR spectra coordinated R_2P—sites show a strong shift to low field (e.g. ~40 ppm for Me_2P— and ~70 ppm for $(CF_3)_2P$— ligands). Non-coordinated R_2P— groups exhibit only a small shift.

The spectroscopic data (Table 14) indicate that, except for $Cr(CO)_5\{SeMeAs(CF_3)_2\}$, the ambidentate ligands coordinate via the P or As site.

6. $Se(EMe_3)_2$ (E = Ge, Sn, Pb)

The octahedral complexes $M(CO)_5\{Se(EMe_3)_2\}$ (M = Cr, Mo, W; E = Ge, Sn, Pb) have been prepared by reaction (37)[196,197]. The Sn derivatives have also been prepared by direct photochemical substitution reactions (equation 38). All nine of these complexes are yellow solids that are sensitive to atmospheric oxygen and water and decompose in the range 82–105 °C with the thermal stability within each series increasing in the order W > Mo > Cr. They have been characterized by infrared, Raman and ^{1}H-NMR spectroscopy[197]; since they are isostructural, representative data for only one complex, $W(CO)_5\{Se(SnMe_3)_2\}$, are given below:

$$M(CO)_6 \xrightarrow[THF]{h\nu} M(CO)_5THF \xrightarrow{Se(EMe_3)_2} M(CO)_5\{Se(EMe_3)_2\} \qquad \text{(Ref. 197)} \qquad (37)$$

$$M(CO)_6 + Se(SnMe_3)_2 \xrightarrow[h\nu]{THF/r.t.} M(CO)_5\{Se(SnMe_3)_2\} \qquad \text{(Ref. 196)} \qquad (38)$$

$$M = Cr, Mo, W$$

For M = W: M.p. 105 °C

$\nu_{C\equiv O}$ (IR, pentane soln.): 2068, 1973, 1935, 1928, 1917m, 1892 cm^{-1}

Raman, solid (rel. int.): 2067(8), 1974(10), 1940(1), 1924(1), 1918(3), 1883(8) cm^{-1}

$\nu_{Sn_2Se,asym}$ = 228 (IR); 230(1) (Raman) cm^{-1}

$\nu_{Sn_2Se,sym}$ = 220(IR); 224(2) (Raman) cm^{-1}

$\delta(^{1}H) = +409.2$ Hz vs. internal C_6H_6

$J(^{1}HC^{117}Sn)$ = 52.8 Hz

$J(^{1}HC^{119}Sn)$ = 55.0 Hz

$J(^{1}HCSn^{77}Se)$ = 2.4 Hz

This complex has also been structurally characterized by single-crystal X-ray diffraction[198]. Consideration of the W—Se and Se—Sn distances and the Sn—Se—W angles

TABLE 15. Selenocarbonyl complexes

Complex/properties[a]
$RuX_2(CO)(CSe)(PPh_3)_2$ (X = Cl (Figure 2), Br)[214] Colourless air-stable solids $\nu_{C=Se} = 1125$; $\nu_{C\equiv O} = 2030$
$RuI(OH)(CO)(CSe)(PPh_3)_2$[214] $\nu_{C=Se} = 1137$; $\nu_{C\equiv O} = 2070$
$RuCl_2(CSe)(p\text{-}MeC_6H_4CN)(PPh_3)_2$[215] $\nu_{C\equiv N} = 2175$; $\nu_{C=Se} = 1134$
$[RuCl(CO)(CSe)(p\text{-}MeC_6H_4CN)(PPh_3)_2]ClO_4$[215] $\nu_{C\equiv N} = 2201$; $\nu_{C\equiv O} = 2070$; $\nu_{C=Se} = 1135$
$OsCl_2(CSe)(CO)(PPh_3)_2$ (**35**)[216] colourless air-stable solid $\nu_{C=Se} = 1156$vs; $\nu_{C\equiv O} = 2036, 2018$
$IrCl(CSe)(PPh_3)_2$[217] orange needles m.p. 247–249 °C $\nu_{C=Se} = 1198$
$IrHCl_2(CSe)(PPh_3)_2$[217] $\nu_{C=Se} = 1200$; $\nu_{Ir-H} = 2240$
$IrCl_3(PPh_3)_2(CSe)$[217] $\nu_{C=Se} = 1201$
$[Ir(NCMe)(PPh_3)_2(CSe)]ClO_4$[217] $\nu_{C=Se} = 1184$; $\nu_{C\equiv N} = 2320$
$[Ir(CSe)(CO)_2(PPh_3)_2]ClO_4$[217] $\nu_{C=Se} = 1145$; $\nu_{C\equiv O} = 2060, 2000$
$IrCl(PPh_3)_2(CSe)(O_2)$[217] $\nu_{C=Se} = 1165$; $\nu_{O-O} = 848$
Fe(TPP)(CSe)(EtOH)[218] purple crystals $\nu_{C=Se} = 1140$s $\delta(^1H, CDCl_3, TMS)$: 8.88 (s, 8H), 8.15 (m, 8H), 7.75 (m, 12H) TPP = 5, 10, 15, 20-tetraphenylporphinato EtOH: 1.25 (t, $J = 6.5$, 3H), 3.78 (q, $J = 6.5$, 2H) $\delta(^{13}C, CDCl_3, TMS)$ TPP: 145.8, 141.6, 133.6, 132.5, 127.8, 126.9, 122.1 EtOH: 58.4, 18.2 CSe: 320.1
$CpMn(CO)_2(CSe)$[212,219,220] yellow air-stable crystals (moderately light-sensitive) m.p. 67–68 °C $\nu_{C\equiv O}$ (hexane) = 2015s, 1965s $\nu_{C=Se} = 1113$vs δ(Mn—C—Se) = 515s, 508s; $\nu_{Mn-C(Se)} = 363$m (R) $\delta(^{13}C, CS_2, TMS)$ = 85.8 (s, Cp), 205.5 (s, $(CO)_2$), 358.0 (s, CSe) mass spectrum: $[M]^{+\cdot}$, $[M - 2CO]^+$, $[M - 2CO - CSe]^+$[212,219–221]
$Cp''Mn(CO)_2(CSe)$[220] golden oil (b.p. 30–60 °C/0.001 Torr) $\nu_{C\equiv O} = 2003, 1953$; $\nu_{C=Se} = 1106$ mass spectrum: $[M]^{+\cdot}$, $[M - 2CO]^{+\cdot}$, $[MnCSe]^+$[220,221]
$CpRe(CO)_2(CSe)$[220] pale yellow air-stable crystals m.p. 97–100 °C $\nu_{C\equiv O} = 2005, 1946$; $\nu_{C=Se} = 1124$ mass spectrum: $[M]^{+\cdot}$, $[M-2CO]^{+\cdot}$, $[ReCSe]^+$[220,221]
$(\pi\text{-}C_6H_5CO_2Me)Cr(CO)_2(CSe)$[219,220] orange–red air-stable solid m.p. 185 °C dec. $\nu_{C\equiv O} = 1986, 1946$; $\nu_{C=Se} = 1063$
$(\pi\text{-}C_6H_6)Cr(CO)_2(CSe)$[213,220] air-stable orange–yellow crystal (moderately light-sensitive) m.p. 99 °C $\nu_{C\equiv O}(CS_2)$ = 1976s, 1932s; $\nu_{C=Se} = 1061$ $\delta(^{13}C, CH_2Cl_2, TMS)$ = 100.9 (s, C_6H_6), 229.0 (s, $(CO)_2$), 363.7 (s, CSe)
$Cr(CO)_5(CSe)$[213,222] volatile, deep yellow air-stable crystals sublimes before melting moderately light-sensitive $\nu_{C\equiv O}$(hexane) = 2093m, 2031m, 2000 vs; $\nu_{C=Se} = 1077$s $\delta(^{13}C, CH_2Cl_2, TMS)$ = 208.1 (s, *trans*-CO), 211.7 (s, *cis*-CO), 360.7 (s, CSe) $\delta(^{17}O, CH_2Cl_2, {}^{17}OH_2)$ = 373.4 (s, *cis*-CO), 385.3 (s, *trans*-CO) mass spectrum: $[M]^{+\cdot}$, $[M - nCO]^+$ (n = 2–5), Cr^+
$Me_4N[L(OC)_2Mo(CSe)]$[223] L = **37** moderately air-stable solid $\nu_{C\equiv O} = 1913, 1824$vs; $\nu_{C=Se} = 1005$s
$CpCo(CSe)PMe_3$[224] $\nu_{C=Se} = 1128$ $\delta(^1H, C_6D_6)$ = 4.70 (d, Cp) ($J_{PH} = 1.0$); 1.14 (d, Me) ($J_{PH} = 9.9$) m.p. 58 °C dec.
$CpCo(CSe)PPh_3$[224] $\nu_{C=Se} = 1130$ $\delta(^1H, C_6D_6)$ = 4.58 (d, Cp) ($J_{PH} = 0.8$); 7.24 (m, 3H(Ph)); 7.84 (m, 2H(Ph))

[a] Data are in the units given in Table 3.

indicates that the W—Se bonds are single bonds without a significant π-bond contribution.

The ligands, $Se(EMe_3)_2$ (E = Ge, Sn, Pb), used to prepare these complexes were prepared according to equations (39) and (40)

$$(Me_2ESe)_3 + 3\,LiMe \rightarrow 3\,Me_3ESeLi \xrightarrow[-LiCl]{Me_3ECl} (Me_3E)_2Se \quad \text{(Ref. 199)} \quad (39)$$

$$E = Ge, Sn$$

$$2\,Me_3PbCl + Na_2Se \xrightarrow{C_6H_6} (Me_3Pb)_2Se \quad \text{(Ref. 197)} \quad (40)$$

More recently, several derivatives of this type were prepared by simplified routes[200] (Equations 41 and 42).

$$E \xrightarrow[E\,=\,Se,\,Te]{NaBH_4/H_2O/EtOH} Na^+EH^- \xrightarrow[C_6H_6]{Ph_3SnCl} E(SnPh_3)_2 \quad (41)$$

$$E \xrightarrow[THF]{LiBEt_3H} Li_2E \xrightarrow{2\,Me_3SnCl} E(SnMe_3)_2 \quad (42)$$

The derivatives $E(SiR_3)_2$ (E = Se, Te; R = H[201], Me[62], Et[202]) and $EtESiMe_3$[202] have also been prepared, but their ligand properties have not been evaluated.

7. Selenocarbonyl complexes (M—CSe)

Although a vast literature of transition-metal carbonyl complexes[1] exists, considerably less ligand chemistry of thiocarbonyl[203–211] and selenocarbonyl[203,204] has been described. Although the diatomic molecules CO and CS can be readily synthesized, free CSe has never been isolated. It can, however, be stabilized when coordinated in transition-metal complexes[212,213] (Table 15). The instability of free CSe requires that indirect routes

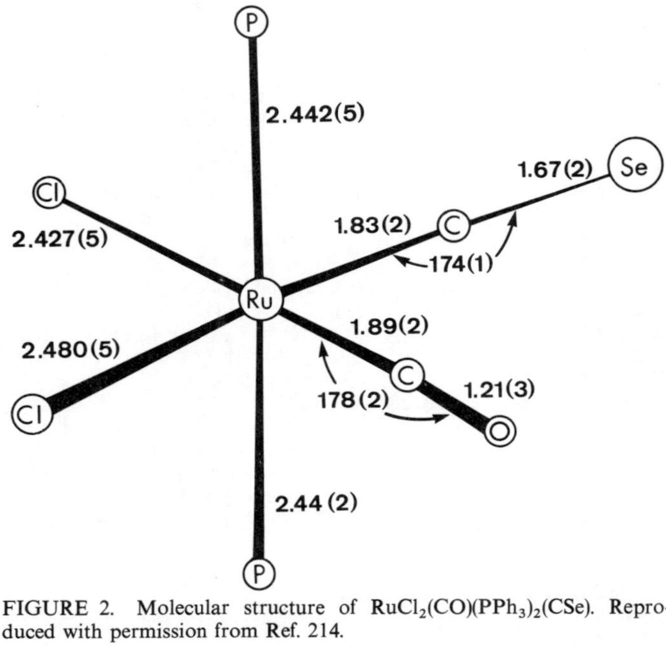

FIGURE 2. Molecular structure of $RuCl_2(CO)(PPh_3)_2(CSe)$. Reproduced with permission from Ref. 214.

be used to prepare selenocarbonyl complexes, two such preparative methods having been used: (1) extrusion of Se from coordinated CSe_2 (η^2-CSe_2, see Section II.C.1.a)[212–214,219,220,224] and (2) substitution reactions of $M{=}CCl_2$[216,219] or $M{\equiv}CCl$[223] complexes.

The first selenocarbonyl complexes were prepared by conversion of a CSe_2 complex by alkylation followed by acidification (equation 43). The crystal structure of the chloro complex (Figure 2) has verified a carbon bonding mode of the potentially ambidentate selenocarbonyl ligand. The rather long Ru—Cl distance *trans* to the CSe ligand (e.g. compared to typical values of 2.29–2.39 Å) indicates a strong *trans* influence of this ligand. The conversion of $RuI_2(CO)(CSe)(PPh_3)_2$ to $RuI(OH)(CO)(CSe)(PPh_3)_2$ on attempted chromatography on alumina is a chemical manifestation of this structural feature[214].

$$Ru(CO)_2(PPh_3)_2 \xrightarrow[CO]{CSe_2} Ru(CO)_2(PPh_3)_2(CSe_2) \xrightarrow[-CO]{MeI} \quad \text{(Ref. 214)} \qquad (43)$$

$$Ru(CO)(PPh_3)_2(CSe_2Me)I \xrightarrow[(-MeSeH)]{HX} RuX_2(CO)(PPh_3)_2(CSe)$$

$$X = Cl, Br$$

A four-step conversion of a coordinated thiocarbonyl ligand to a selenocarbonyl ligand has been described[215] (equation 44).

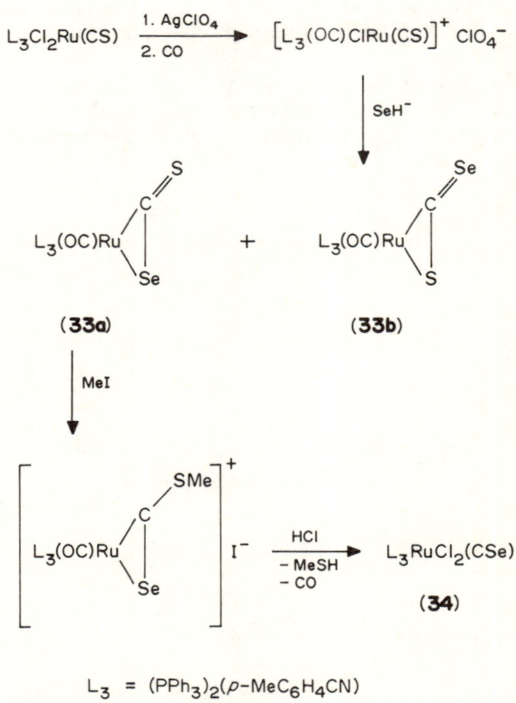

(44)

The linkage-isomeric intermediates **33a** and **33b** were separated by chromatography on silica gel prior to the *S*-methylation and final conversion to the selenocarbonyl complex by treatment with aqueous HCl in refluxing toluene/ethanol solution.

The analogous reaction with $L_3Cl_2Os(CS)$ gave a very slow reaction in the final acidification step with formation of only traces of the selenocarbonyl complex, as evidenced by infrared spectroscopy.

The selenocarbonyl complex (**34**) was converted to the cationic complex $[RuL_3Cl(CO)(CSe)]^+ClO_4^-$ by treatment with $AgClO_4$ followed by CO.

Synthesis of a selenocarbonyl complex via a dichlorocarbene precursor is illustrated by equation (45). The proposed stereochemistry of this complex, confirmed for the CS analogue by single-crystal X-ray diffraction[225] is shown in formula **35**.

$$Cl_2(Ph_3P)_2(OC)OsCCl_2 \xrightarrow{SeH^-} Cl_2(Ph_3P)_2(OC)Os(CSe) \quad \text{(Ref. 216)} \qquad (45)$$

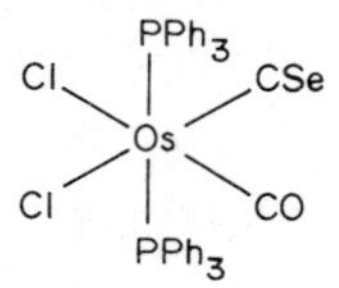

(**35**)

The Ru analogue, $Cl_2(Ph_3P)_2(OC)RuCCl_2$, was reported to react with SeH_2 to give the corresponding $Cl_2(Ph_3P)_2(OC)Ru(CSe)$, although characterization of this product was not reported[226].

A five-step conversion of Vaska's complex, $IrCl(CO)(PPh_3)_2$[227], to its CSe analogue has been described[217] (equation 46). This complex undergoes oxidative addition and chloride substitution reactions typical of the parent Vaska's compound[227] (equation 47).

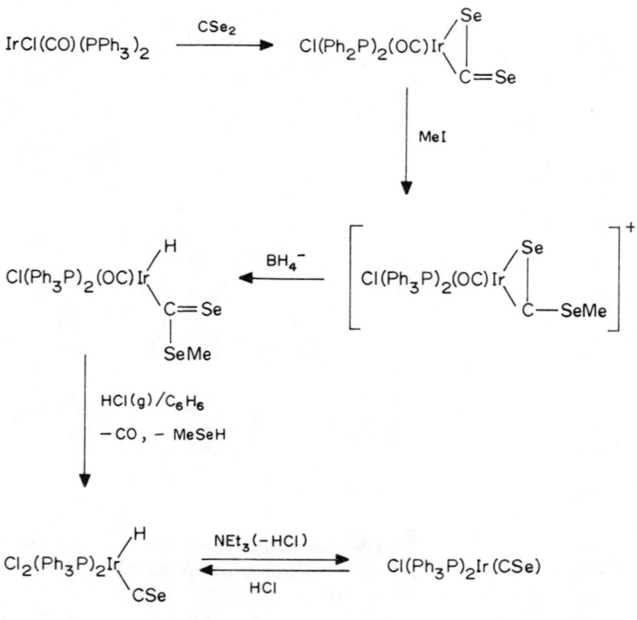

(46)

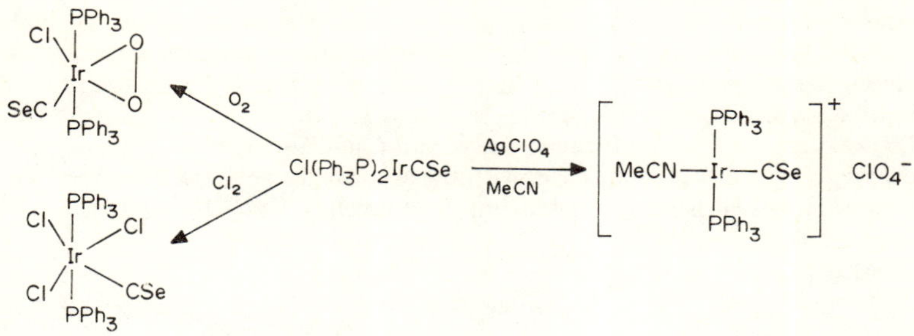

(Ref. 217)

(47)

The $\nu_{C=Se}$ bands in these adducts (Table 15) do not reflect the metal electron density as for the carbonyl analogues[1,227] e.g. in the O_2 adduct $\nu_{C=Se}$ is lowered by 33 cm^{-1} vs. the parent selenocarbonyl complex rather than increased to reflect decreased metal-to-CSe ligand back-donation upon effective metal oxidation. This observation has been attributed to extensive mixing of the $\nu_{C=Se}$ modes with lower energy modes (e.g. $\nu_{Ir-(CSe)}$)[217].

Reaction of the parent selenocarbonyl complex with $NaBH_4$ in the presence of excess PPh_3 gives complete reduction of the CSe ligand rather than the simple hydride formed from the CO and CS analogues (equation 48). Stable complexes containing the proposed selenoformaldehyde reduction intermediate have been isolated recently (see Section II.C.1.c).

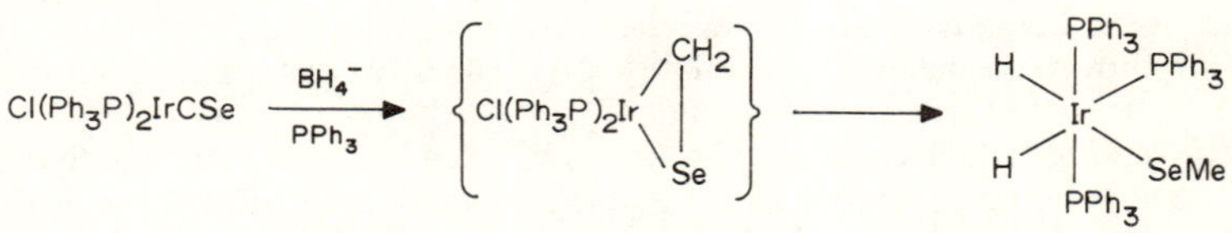

(Ref. 217)

(48)

The complex Fe(TPP)(CSe) (TPP = 5,10,15,20-tetraphenylporphinato) has been prepared from a selenocarbene precursor[218] (equation 49). Addition of a catalytic amount of $FeCl_2$ in acetonitrile to a solution of the carbene complex **36** gave ca. 50% yield of the Fe(TPP)(CSe) product, on the basis of ^{1}H-NMR spectroscopy. The reaction used to isolate this complex, however, was the Fe-powder reduction of $PhCH_2SeCCl_3$ in the presence of Fe(TPP). This reaction produced a product containing ca. equimolar amounts of the carbene complex **36** and the selenocarbonyl complex, which was purified by silica gel thin-layer chromatography and recrystallization from CH_2Cl_2/EtOH to give the octahedral ethanol complex, [Fe(TPP)(CSe)(EtOH)], in 40% yield (Table 15).

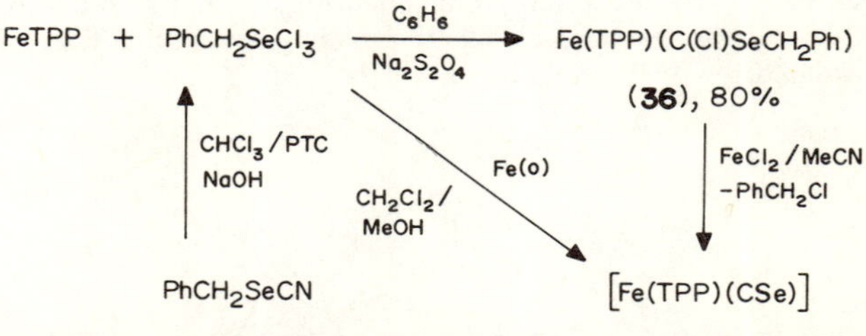

(49)

Electronic spectroscopy showed that the ethanol complex was in equilibrium with the pentacoordinate [Fe(TPP)(CSe)] complex and that more basic ligands (*N*-methylimidazole, pyridine) gave hexacoordinate adducts with enhanced stability[218]. The Fe—CSe bond in these complexes is considerably stronger than in the corresponding

Fe—CO species, as evidenced by the integrity of the former complex in high dilution (2×10^{-8} M) as well as by the fact that the above purification can be done under atmospheric conditions[218]. An EXAFS study of this complex showed linear Fe—C—Se bonding with Se—C = 1.64 Å and Fe—Se = 3.37 Å[228].

Butler and coworkers[212,213,220] have used the reaction of CSe_2 with labile transition-metal complexes (equation 50), in some cases with the addition of PPh_3 as a Se extrusion reagent (equation 51), to synthesize selenocarbonyl complexes.

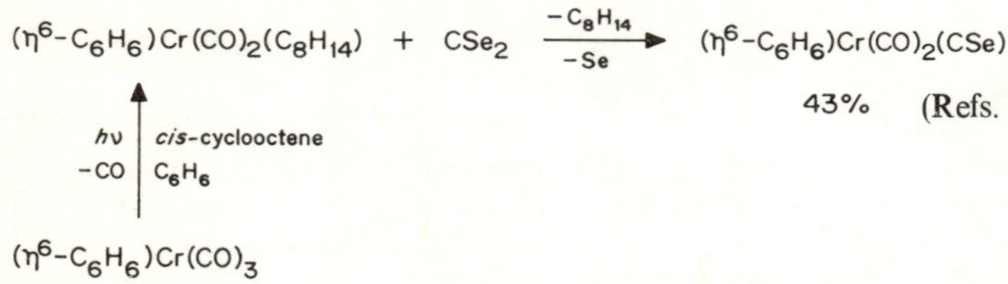

(Refs. 213, 220) (50)

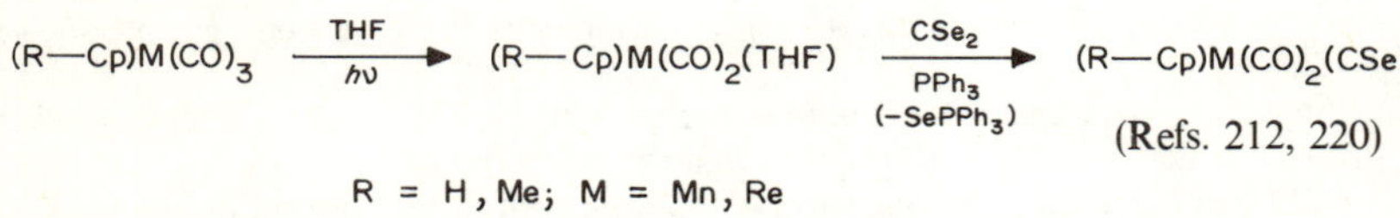

(Refs. 212, 220) (51)

These complexes are air-stable in the solid state, although they darken slowly on exposure to light (rapidly in solution). Their mass spectra[221] are characterized by their tendency to lose CO and organic radicals before the CSe ligand, indicating that the latter is bonded rather strongly to the metal. These complexes have been characterized by infrared and NMR spectroscopy (Table 15).

In a recently reported application of this PPh_3-induced Se extrusion reaction a selenocarbonyl complex incorporating PPh_3 as a ligand was obtained along with the expected product (e.g. $CpCo(\eta^2\text{-}CSe_2)(PMe_3)$ gave a mixture of $CpCo(CSe)(PMe_3)$ and $CpCo(CSe)(PPh_3)$, which was separated by column chromatography)[224].

A unique example of a selenocarbonyl complex containing CO as the only other ligand has been prepared (equation 52).

$$(\eta^6\text{-}C_6H_6)Cr(CO)_2(CSe) + 3\,CO \xrightarrow[20\text{ atm. CO}]{50\,^\circ C} Cr(CO)_5(CSe) + C_6H_6 \quad \text{(Refs. 213, 222)} \quad (52)$$

A detailed study of the vibrational spectra of the complexes $M(CO)_5CE$ (M = Cr, Mo, W; E = S, Se) has also appeared[221].

Nucleophilic displacement by Se^{2-} at the carbon atom of a terminal methylidyne ligand has recently[223] provided a new route (equation 53) to a selenocarbonyl complex (the CS and CTe derivatives were also prepared by this route). This selenocarbonyl complex was isolated in ~60% yield as its Me_4N^+ salt (Table 15).

$$Li_2Se + L(OC)_2Mo{\equiv}CCl \rightarrow L(OC)_2Mo{\equiv}C{-}Se^- \leftrightarrow L(OC)_2Mo^-{-}C{\equiv}Se \quad \text{(Ref. 223)} \quad (53)$$

L =

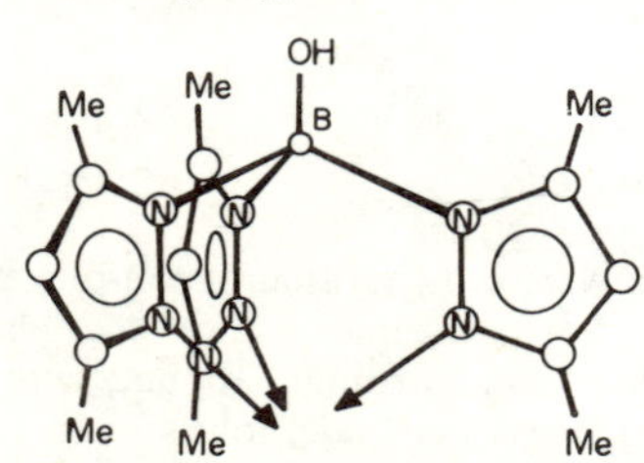

(37)

(hydrotris(3,5-dimethylpyrazol-1-yl)borato)

Although at present only terminally bonded M—CSe complexes are known, the recent development of synthetic methodology for selenocarbonyl complexes should allow further investigations of this ligand chemistry. The carbonyl ligand can coordinate to transition metals in a variety of bonding modes [1,2], and three bonding modes have been found for the thiocarbonyl ligand (**A** and **B** have been characterized by X-ray diffraction)[227]:

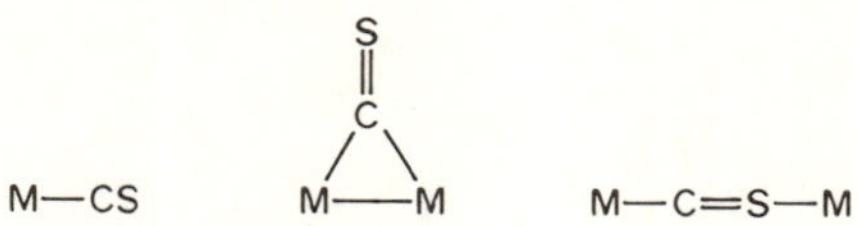

(**A**) (**B**) (**C**)

8. Diorganoselenoxides (R_2SeO)

The coordination chemistry of sulphoxides has received considerable attention[229–235], and the ambidentate character of such ligands has been well established (including the crystal structure determination of both O-bonded (e.g. $[Rh_2(O_2CCF_3)_4(O\text{-}Me_2SO)_2]$[231], $[Th(NO_3)_4(O\text{-}Me_2SO)_3]$[232]) and S-bonded (e.g. $[Rh_2(O_2CEt)_4(S\text{-}Me_2SO)_2]$[231], $[Ru(NH_3)_5(S\text{-}Me_2SO)](PF_6)_2$[233]) complexes, as well as complexes containing both O- and S-bonded Me_2SO (e.g. *cis*-$[RuCl_2(S\text{-}Me_2SO)_3(O\text{-}Me_2SO)]$[234]). In contrast, only a few papers[236–241] have described coordination complexes with diorganoselenoxides, and whenever spectroscopic data were presented, M—O bonding was proposed. The sole exception to this is the first report[236] of metal complexes with diorganoselenoxides, e.g. $HgX_2\cdot(p\text{-}ROC_6H_4)_2SeO$ (R = Me, Et) and HgX_2 (dibenzoselenophene oxide) (X = Cl, Br), where Hg—Se bonding was proposed. These complexes precipitated on cooling hot solutions of ca. equimolar amounts of the reagents in 95% ethanol. Attempts to isolate complexes with $CuCl_2$, $CdCl_2$, $FeCl_3$, $NiCl_2$, $CrCl_3$ and $ZnCl_2$ using water or 95% ethanol gave only the starting materials[236]. Since the ligand field strength of O-bonded selenoxides[242] is less than that of the aquo ligand, this result could be expected.

In a subsequent study using rigorously anhydrous halides of Mn(II), Co(II), Ni(II), Cu(II), Sn(IV) and Zr(IV), diphenyl selenoxide (DPSeO) complexes were isolated[237]. In this work the hydrated halides were thermally dehydrated, although chemical dehydrating agents (e.g. 2,2-dimethoxypropane, ethyl orthoformate) are widely used to form anhydrous metal salts *in situ* in reactions with weak ligands[52,243]. The more basic dimethyl selenoxide (DMSeO) ligand forms complexes with a variety of transition metals, with either the hydrated chloride[238] or perchlorate salt[239]. In the former case the isolated complexes contain dimethyl selenoxide ligands with the chloride still coordinated ($HgCl_2\cdot$(DMSeO); MCl_2(DMSeO)$_2$, M = Cd, Cu, Co; $PdCl_4$(DMSeO)$_2$; $FeCl_3$(DMSeO)$_2$)[238]. In the latter case, however, the weakly coordinating perchlorate anion, as well as the water, is eliminated from the coordination sphere, the complexes formed having the stoichiometry $[M(Me_2SeO)_n]^{m+}(ClO_4^-)_m$ (M = Mn, Co, Ni, Zn, Cd, Mg; $n = 6$, $m = 2$. M = Cr, Fe; $n = 5$, $m = 3$. M = Ag; $n = 2$, $m = 1$. M = Cu; $n = m = 2$).

In all of these cases, the lowering of the Se—O stretching frequency (10–110 cm^{-1}; Table 16) is the basis of the assignment of M—O bonding with these potentially ambidentate ligands, no definitive single-crystal X-ray diffraction study having been reported. The decrease in the ν_{Se-O} vibration vs. the free ligand is a measure of the M—O bond strength, and this value increases, as expected, as the charge on the metal ion increases (e.g. Pd(IV), Sn(IV) and Zr(IV) show the largest decreases). In analogous dimethyl sulphoxide complexes, the $\nu_{S=O}$ decreases on O coordination but increases on S coordination[229–235].

TABLE 16. Infrared data (cm^{-1}) for some metal diorganoselenoxide complexes[a]

Complex[b]	$\nu_{Se=O}$	ν_{C-Se}	ν_{M-O}	Ref.
DMSeO	800vs	585vw		238
$HgCl_2(DMSeO)$	770vs	580vw		238
$CdCl_2(DMSeO)_2$	775vs, 735vs	585w		238
$PdCl_4(DMSeO)_2$	698vs, 690vs	610w	520w, br	238
$CuCl_2(DMSeO)_2$	760vs, 720vs	595w, 580w	515m, 475m	238
$CoCl_2(DMSeO)_2$	760vs, 705s	590w	410s	238
$FeCl_3(DMSeO)_2$	760vs	580w	440s	238
$[Cr(DMSeO)_6](ClO_4)_3$	765vs		440s	239
$[Fe(DMSeO)_6](ClO_4)_3$	772vs	602m, 590m	444s	239
$[Mn(DMSeO)_6](ClO_4)_2$	787vs	593m, 580m	400m	239
$[Co(DMSeO)_6](ClO_4)_2$	787vs, 775vs	595m, 581m	400s	239
$[Ni(DMSeO)_6](ClO_4)_2$	790vs	595s	400s	239
$[Cu(DMSeO)_4](ClO_4)_2$	770vs	610m, 493m	498m, 484m	239
$[Ag(DMSeO)_2]ClO_4$	779vs	593m, 585m	405m	239
$[SnCl_4(DMSeO)_2]$	756vs, 740s	598w, 583w	466s, 445s	241
$[Me_2SnCl_2(DMSeO)_2]$	737vs	594m, 583m	387s	241
$[Et_2SnCl_2(DMSeO)_2]$	744vs	595m, 584m	407s, 200sh	241
DPSeO	831			237
$MnCl_2(DPSeO)_2$	798, 809			237
$CoCl_2(DPSeO)_2$	767, 782			237
$NiCl_2(DPSeO)_4$	795, 800			237
$[Ni(DPSeO)_6]Cl_2$	797, 801			237
$CuCl_2(DPSeO)_2$	790			237
$HgCl_2(DPSeO)$	800			237
$ZrCl_4(DPSeO)_2$	770			237
$SnCl_4(DPSeO)_2$	789			237

[a] Ionic ClO_4^- is confirmed in all cases by strong bands at ~1100 and 625 cm^{-1} [52].
[b] DMSeO = Me_2SeO; DPSeO = Ph_2SeO.

9. *Selenium-functionalized carbene complexes (M=C(SeR)(R¹))*

Since Fischer's original discovery of transition-metal carbene complexes[244] (equation 54), this area has been one of the most actively investigated in organometallic chemistry[245], the main impetus being its mechanistic implications in Fischer–Tropsch and olefin metathesis processes.

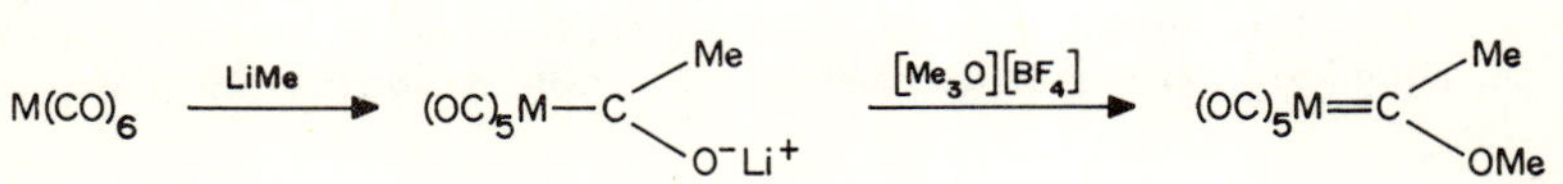

(54)

Included in Fischer's extensive investigations of metal carbene chemistry, and more recently metal carbyne chemistry, have been several papers[246–250] dealing with Se-functionalized derivatives of these two classes of organometallics. The first attempt to prepare selenocarbene complexes by the route successful for thiocarbenes gave products containing M—Se bonds, presumably by nucleophilic replacement of the carbene carbon atom and proton migration (equation 55).

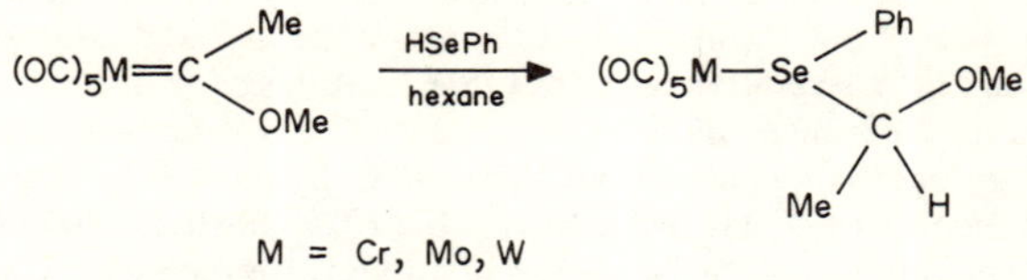

(55)

The three complexes, which had similar properties, were characterized by infrared and ^{1}H-NMR spectroscopy (Table 17). The Se ligand was readily displaced by pyridine, with the formation of M(CO)$_5$pyr.

Successful synthesis of the first selenocarbene complexes involved selenolysis with anhydrous, highly purified methylselenol at low temperature in the dark[247] (equation 56). The triselenoboric acid ester was added to the reaction to scavenge the water and

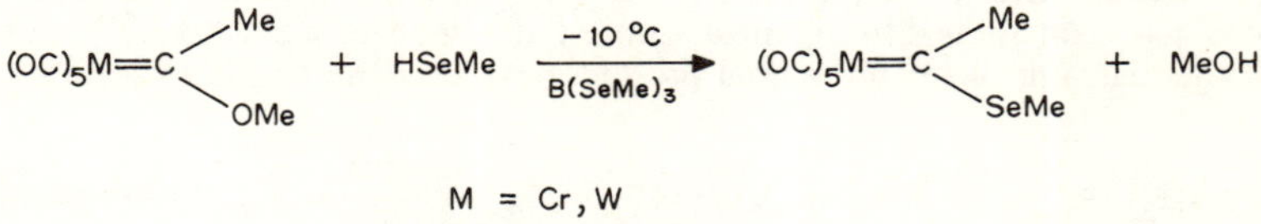

(56)

TABLE 17. Selenium-functionalized carbene complexes

Complex	Complex
(OC)$_5$CrSe(Ph)(CH(Me)(OMe))[246] brown–yellow solid m.p. 74 °C $\nu_{C\equiv O}{}^a$ = 2070, 1990, 1948, 1934 $\delta(^1H)^b$: Me, 1.6d; OMe, 3.55s; CH, 5.20q; Ph, 7.63m	(OC)$_5$WC(SeMe)(PMe$_3$)(Me)[247] yellow crystals $\delta(^1H)^d$: CMe, 2.37d; SeMe, 1.87s; PMe$_3$, 1.82d $\nu_{C\equiv O}{}^f$: 2055w, 1960w, 1903vs, 1878s, sh
(OC)$_5$WC(SeMe)(Me)[247] black crystals m.p. 35 °C $\nu_{C\equiv O}{}^a$ = 2069m, 1956vs $\delta(^1H)^c$: SeMe, 1.66s; Me, 2.80s $\delta(^{13}C)^{d,e}$: W=C, 355.5; *trans*-CO, 205.0; *cis*-CO, 197.86; Se*Me*, 21.79; C*Me*, 54.27	(OC)$_5$CrC(SePh)(NEt$_2$)[248] pale yellow crystals m.p. 74 °C dec. $\nu_{C\equiv O}{}^g$ = 2057, 1977, 1942, 1933 $\nu_{C{\cdots}N}$ = 1518 $\delta(^1H)^h$: NEt$_2$, 1.00t, 1.44t, 3.73q, 4.35q Ph, 7.60m
	(OC)$_5$CrSePhEt[250] yellow needles m.p. 17 °C $\nu_{C\equiv O}{}^a$ = 2075w, 1951vs, 1944s sh, 1938m $\delta(^1H)^i$: Ph, 7.7m; Et, 3.3q, 1.35t

[a] In hexane, measured in cm^{-1}.
[b] In CDCl$_3$, internal TMS.
[c] In toluene-d$_6$.
[d] In acetone-d$_6$.
[e] The corresponding values for the C(Me) (OMe) complex are: 332.9, 203.62, 197.60, 69.69 and 51.51.
[f] In CH$_2$Cl$_2$; cm^{-1}.
[g] In methylcyclohexane; cm^{-1}.
[h] In CCl$_4$, internal TMS.
[i] δ values relative to acetone-d$_6$ = 2.1 ppm.

methanol released in the reaction. The air, light and thermal sensitivity of these complexes required careful purification by low-temperature chromatography (SiO_2/hexane/ $-10\,°C$) followed by sublimation at 0 °C. The two complexes had similar chemical and spectroscopic properties (Table 17).

Hindered rotation about the carbene C—Se bond and inversion at the Se atom were indicated by the reversible broadening of the C—Me and Se—Me NMR singlets as the temperature was decreased. The acidic character of the C—Me protons was reflected in the ^{13}C chemical shift of the Me group as well as its uncatalysed deuteriation above 40 °C by acetone-d_6 (the OMe derivative requires OMe^- catalysis for this reaction). The strong *trans* influence of the carbene ligand was reflected in the ^{13}C chemical shift of the CO *trans* to this ligand (e.g. more deshielded than the *cis*-CO ligands). The low-field shift of the carbene C resonance vs. the OMe analogue (Table 17) indicated enhanced Lewis acidity, and this was reflected in the facile formation of an ylid complex with PMe_3[247] (equation 57).

$$(OC)_5W{=}C(SeMe)(Me) + PMe_3 \xrightarrow{Et_2O/-50\,°C} (OC)_5W{-}C(SeMe)(Me){-}PMe_3 \qquad (57)$$

The synthetically useful transformation carbyne→carbene→carbyne complex in equation (58) has been applied to organoselenium-functionalized derivatives[248], and the kinetics of the rearrangement to the final product have been studied in detail[249].

$$[(OC)_5Cr{\equiv}CNEt_2]^+ BF_4^- \xrightarrow{X^-} (OC)_5Cr{=}C(X)(NEt_2) \xrightarrow[-CO]{\Delta > 30\,°C} (OC)_4XCr{\equiv}CNEt_2 \qquad (58)$$

X = Cl, Br, I, $SnPh_3$, $Se(C_6H_4R\text{-}p)$

R = H, F, Br, Me, OMe, CF_3

The selenocarbene intermediates in this series were isolated by carrying out the initial conversion step and chromatographic purification of the products at low temperature (ca. $-25\,°C$)[248]. These carbenes, which upon warming readily eliminate CO with migration of the SeR^- group from the carbene carbon to the Cr, were characterized spectroscopically (Table 17). An unusual feature of their infrared spectra was the high intensity of the formally infrared-forbidden B_1 bands ($\sim 1977\,cm^{-1}$) in the CO stretching region. This suggested that the ideal C_{4v} symmetry of the $M(CO)_5L$-type complexes was lowered by some distortion. Such a distortion was confirmed for $(OC)_5CrC(SePh)(NEt_2)$ by single-crystal X-ray diffraction[248]. The four equatorial carbonyl ligands are bent back away from

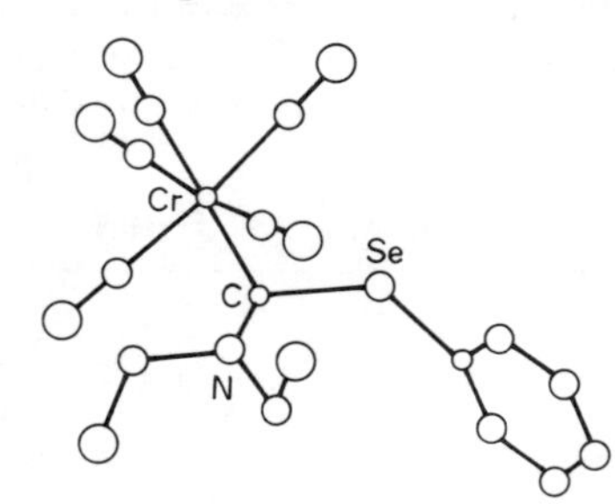

(38)

the carbene ligand with the Cr atom thereby shifted 0.113 Å from the plane of these four ligands (**38**). This complex also has the longest Cr—C (carbene) distance found for such derivatives (2.171 Å).

The double-bond character of the C—N bond, indicated by the value of the $\nu_{C\cdots N}$, is further supported by the appearance of two triplets and quartets for the non-equivalent N—Et groups (Table 17).

Warming CH_2Cl_2 or 1,1,2-trichloroethylene solutions of the selenocarbenes (room temperature to 30 °C) under 10 atm CO gave readily the thermally sensitive carbyne complexes. These were assigned a *trans* configuration on the basis of the observed four $\nu_{C\equiv O}$ bonds[248]. As with the parent selenocarbenes, the intensity ratios of these bands indicated a significant perturbation of the coplanar arrangement of the four carbonyls in solution. Derivatives with considerably enhanced thermal stability can be obtained by carrying out the thermolysis of the carbenes in the presence of PPh_3. The resulting products, $(OC)_3(Ph_3P)(SeAr)CrCNEt_2$, were assigned meridional geometries (i.e. two CO ligands are *trans* and the third is *cis* to the other two) on the basis of the three observed $\nu_{C\equiv O}$ bands[249].

The selenocarbene→carbyne rearrangement followed first-order kinetics and was independent of free-CO concentration and only slightly dependent on solvent polarity and type of substituent R (equation 58). The activation parameters determined are: $\Delta H^{\neq} = 101\text{–}104\,\text{kJ mol}^{-1}$, $\Delta S^{\neq} = 27\text{–}33\,\text{J mol}^{-1}\,\text{K}^{-1}$.

The reactivity of the carbyne complex *trans*-bromotetracarbonyl(phenylcarbyne)chromium varied according to the chalcogen base used, a modest yield of the carbene complex being obtained only with the thiolate[250] (equation 59).

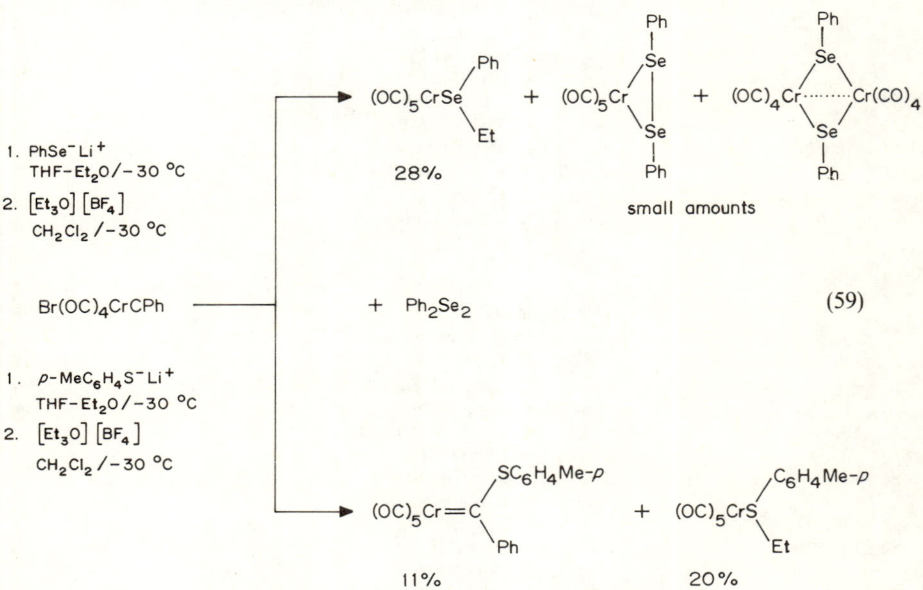

(59)

10. Miscellaneous ligands

A few other metal complexes containing monodentate neutral Se ligands have been reported (Table 18).

TABLE 18. Complexes with miscellaneous monodentate neutral ligands

Complex of	Reported data	Reference
Selenosemicarbazones		
Se ‖ $H_2NCNHN{=}C(H)R$ R-ssc	$[Co(dmg)_2(Ph\text{-}ssc)Cl]\cdot H_2O$	251
	pale-brown microcrystals,	
	sparingly water-soluble	
	$[Co(dmg)_2((C_6H_4OH\text{-}o)\text{-}ssc)_2]Cl$	251
	pale brown solid,	
	sparingly soluble in water	
	$\Lambda_M(10^{-3}M$ aq. sol.$) = 65.3$	
$Se(CN)_2$	$TiCl_4\{Se(CN)_2\}$	252
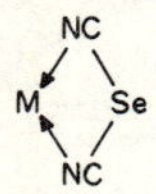	yellow–orange hygroscopic solid	
	m.p. 105 °C dec.	
	$\nu_{C\equiv N}{}^{a,b} = 2270w, 2245s$	
	(R: 2198m, 2145w)	
	$\nu_{C-Se}{}^{c} = 648vs, 638s, sh, 318sh$	
	(R: 657w, 309m, br)	
	$\nu_{Ti-Cl} = 385vs, 370sh$	
	(R: 392vs)	
	$\delta_{Se-C\equiv N} = 345$	
	$\nu_{Ti-N} = 298w, 268w$	
	(R: 272w, br)	
	$ZrCl_4\{Se(CN)_2\}$	252
	bright yellow hygroscopic solid	
	m.p. 125 °C dec.	
	$\nu_{C\equiv N} = 2198m, 2145w$	
	$\nu_{C-Se} = 618w, 570s$	
	$\nu_{Zr-Cl} = 345vs, 315sh$	
	$\nu_{Zr-N} = 290sh, 270sh$	
$M{-}Se(CN)_2$	*trans*-$VCl_4\{Se(CN)_2\}_2$	253
	yellow hygroscopic solid	
	m.p. 196 °C dec.	
	$\nu_{C\equiv N}{}^{a,b} = 2210s, 2160m$	
	$\nu_{C-Se}{}^{c} = 585s, 510m$	
	$\nu_{V-Se} = 302sh$	
	$\delta_{V-SeCN} = 280vw$	
	$MoCl_5\{Se(CN)_2\}$	253
	dark brown hygroscopic solid	
	m.p. 130 °C dec.	
	$\nu_{C\equiv N}{}^{a,b} = 2260s, 2185m$	
	$\nu_{C-Se} = 580w, 510m$	
	$\nu_{Mo-Cl} = 380sh, 350vs, 300w, sh$	
	$\nu_{Mo-Se} = 280vw, sh$	
$Se{=}C(SeEt)_2$	$(OC)_5Cr{-}Se{=}C(SeEt)_2$	254
	red–purple plates	
	m.p. 51 °C	
Se(Ph)(SPh)	$(OC)_5Cr\{SePh(SPh)\}$	255
	molecular structure	
	Cr—Se: 2.531(4) Å	
	Se—S: 2.226(6) Å	
Selenoketones		
$Se{=}CAr_2$	$(OC)_5Cr{-}Se{=}CPh_2$	256
	m.p. 58–59 °C	
	$\nu_{C\equiv O}{}^{d} = 2060, 1991, 1956, 1943$	
	$\delta(^{13}C)^e$: C=Se, 243.1;	
	CO(*trans*), 223.0; CO(*cis*), 214.6	

TABLE 18. (*Continued*)

Complex of	Reported data	Reference
	$(OC)_5W—Se=CPh_2$	256
	black crystals	
	m.p. 45–46 °C	
	$\nu_{C\equiv O}{}^d = 2069, 1987, 1953, 1937$	
	$\delta(^{13}C)^e$: C=Se, 240.0;	
	CO(*trans*), 200.6; CO(*cis*), 196.9	
	$(OC)_5W—Se=C(Ph)C_6H_4CF_3$-*p*	256
	m.p. 70–71 °C	
	$\nu_{C\equiv O}{}^d = 2070, 1991, 1959, 1944$	
	$\delta(^{13}C)^e$: C=Se, 236.7;	
	CO(*trans*), 200.4; CO(*cis*), 196.4	
$RSeCH_2SR$	$Me_3Pt(\mu\text{-}Cl)_2(\mu\text{-}MeSeCH_2SMe)PtMe_3$	257
	white air-stable crystals	
	m.p. 210–212 °C	
	variable-temperature NMR study	
η^1-P_4Se_3	$(np_3)Ni(P_4Se_3)\cdot 2C_6H_6$	258
	air-stable solid	
	insoluble in common organic solvents	
	crystal structure	259
η^3-P_3Se_3	[triphos $Rh(P_3Se_3)$]	260
	orange air-stable crystals	
η^3-P_2Se	$[Co(triphos)(P_2Se)]BF_4\cdot C_6H_6$	261
	orange, moderately air-stable crystals	
	$\Lambda_M(10^{-3}$M, $EtNO_2) = 84$	
	$\delta(^{31}P)^f = 32.70(3)$br; $-145.50(2)$q;	
	$^2J_{PP} = 12$	
	crystal structure	

[a] The infrared spectra were recorded as Nujol mulls.
[b] The corresponding bands in free $Se(CN)_2 = 2183, 2175$.
[c] The corresponding bands in free $Se(CN)_2 = 608, 516$.
[d] In hexane.
[e] In acetone-d_6 at -20 °C; δ values relative to acetone-d_6, $\delta = 205.1$ ppm.
[f] In CD_2Cl_2; chemical shifts relative to 85% H_3PO_4. The spectra are the same at 301 and 203 K.

Selenosemicarbazone complexes of Co(III) have been prepared according to equation (60)[251].

$$trans\text{-}[CoX(dmg)_2(H_2O)] + n\text{-}PhCH{=}NNHC(Se)NH_2 \rightarrow Co(dmg)_2(\text{Ph-ssc})_nX \qquad (60)$$

Ph-ssc

$$X = Cl, Br; \qquad n = 1, 2$$

The Co(III) in these complexes is octahedrally coordinated so that for $n = 1$, the product is a neutral complex, but for $n = 2$, ionic complexes (e.g. $[Co(dmg)_2(\text{Ph-ssc})_2]X$) are obtained. These complexes were prepared by this substitution reaction to avoid the acidic reaction conditions usually used to prepare dmg complexes of this general type [i.e. aerial oxidation of hot aqueous solutions of the corresponding Co(II) salt, dmg-H and the other ligand to be incorporated into the coordination sphere (e.g. $Co(dmg)_2L_nX$, L = thiosemicarbazones[262], thiourea[263])]. Analogous complexes with the selenosemicarbazones of acetone, cyclohexanone, vanillin and salicylaldehyde were also prepared[251].

A remarkable feature of these complexes is the significantly enhanced acid stability of the selenosemicarbazones upon coordination. The complex $[Co(dmg)_2(\text{Ph-ssc})_2]^+$, for example, was reported to be stable in boiling 1:1 aqueous HCl, whereas free selenosemicarbazones readily deposit elemental Se in even weakly acid solutions[251]. This property was cited as evidence for Se bonding of the ambidentate selenosemicarbazone ligands in these complexes. No spectroscopic characterization of the complexes was reported.

Coordination of $Se(CN)_2$, prepared by disproportionation of selenocyanogen[264–266], as an (N,N) bidentate ligand ($MCl_4\{(NC)_2Se\}$; M = Ti, Zr)[252] or a monodentate Se donor (e.g. *trans*-$VCl_4\{Se(CN)_2\}_2$, $MoCl_5\{Se(CN)_2\}$)[253] has been proposed on the basis of infrared and Raman spectroscopy (Table 18). The spectra of $MCl_4\{(NC)_2Se\}$ (M = Ti, Zr) indicate C_{2v} symmetry with bidentate (N,N) coordination of $Se(CN)_2$[252]. For the Ti complex, both infrared and Raman spectra were recorded, but the facile decomposition of the Zr complex in the laser beam precluded the measurement of its Raman spectrum. In both cases, the increases in the $\nu_{C\equiv N}$ and ν_{C-Se} bands vs. the free ligand, the expected number of ν_{M-Cl} bands and the assignments of ν_{M-N} bands all support the $[MCl_4(N,N)]$ C_{2v} structure. Similar reactions with $S(CN)_2$ gave the octahedral complexes $MCl_4\{S(CN)_2\}_2$ containing two S-bonded ligands in *trans* positions[252]. The larger CEC (E = Se, S) angle in $Se(CN)_2$ (99°) vs. $S(CN)_2$ (95.6°) and the larger size of Se vs. S both provide less steric strain for (N,N) bidentate coordination in the former case.

Reactions of $MoCl_5$ and VCl_4 with $Se(CN)_2$, however, gave octahedral complexes with monodentate Se bonding of this ligand ($MoCl_5\{Se(CN)_2\}$, $VCl_4\{Se(CN)_2\}_2$)[253]. The infrared spectra of these complexes show increases in the $\nu_{C\equiv N}$ bands but decreases in the ν_{C-Se} bands vs. the free ligand (Table 18). The *trans* (D_{4h}) symmetry indicated by infrared data for $VCl_4\{Se(CN)_2\}_2$ is supported by the ESR spectrum (V(IV), d^1) in acetone at − 160 °C.

Reactions of $Se(CN)_2$ with some main-group organometallics gave cyano rather than selenocyanato complexes (equations 61 and 62).

$$HgPh_2 + Se(CN)_2 \rightarrow PhHgCN + PhSeCN \quad \text{(Ref. 264)} \qquad (61)$$

$$PbPh_4 + Se(CN)_2 \rightarrow Ph_3PbCN + PhSeCN \quad \text{(Ref. 265)} \qquad (62)$$

The first examples of diselenothiocarbonate and triselenocarbonate ligands were prepared according to equation (63)[254]. Other products were obtained in lower yields, and the mechanism of this complex reaction is not understood.

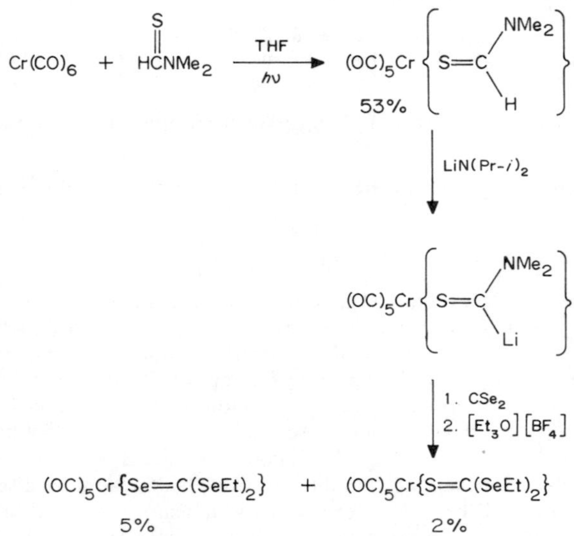

(63)

Coordination of the ambidentate ligand PhSSePh by its Se atom in $(OC)_5Cr(PhSeSPh)$ was established by single-crystal X-ray diffraction[255]. This complex was formed in very low yield by the reaction of the carbyne complex $Br(OC)_4CrCPh$ with $PhSe^-Li^+$ (equation 59) due to the presence (< 1%) of PhSH as an impurity in the PhSeH used to form the lithium reagent. Diffraction-quality crystals of this complex were manually separated from $(OC)_5Cr(Se_2Ph_2)$, with which they crystallized.

Diarylselenoketone complexes of Cr and W carbonyls have been prepared in good yields from the corresponding carbene complexes by use of phenyl isoselenocyanate or potassium selenocyanate as Se insertion reagents[256] (equation 64).

$$(OC)_5M{=}C(Ph)(C_6H_4X\text{-}p) + PhNCSe \xrightarrow[-30\ C]{Et_2O} (OC)_5M{-}Se{=}C(Ph)(C_6H_4X\text{-}p)$$

$$M = Cr;\ R = H$$

$$M = W;\ R = H,\ Br,\ CF_3,\ Me,\ OMe,\ NMe_2 \qquad (64)$$

The dimeric complexes incorporating bridging halo and the hybrid (Se, S) ligand, $Me_3Pt(\mu\text{-}X)_2(\mu\text{-}MeSeCH_2SMe)PtMe_3$ (X = Cl, Br, I), have been prepared[257] by reaction of the (Se, S) ligand with $[Me_3PtX]_4$, as previously described for analogous complexes with bridging RSeSeR ligands (equation 31). Variable-temperature ^1H-NMR studies have demonstrated configurational non-rigidity in these complexes (i.e. ligand ring inversion, pyramidal inversion of both S and Se atoms, ligand switching between Pt atom pairs and scrambling of the Pt methyl environments[257]). The energy barriers for these non-dissociative dynamic processes have been determined.

The P_4E_3 (E = S, Se) cage molecule (Figure 3a) has an unusual coordination chemistry, coordinating as an intact molecule (monodentate coordination through the apical P atom[258,259]; Figure 3b), or undergoing fragmentation to give complexes containing η^3-P_3[267], η^3-P_2E[261,268] (Figure 3c) or η^3-P_3E_3[260] (Figure 3d) ligands.

The type of ligand that is incorporated in a metal complex upon reaction with P_4E_3 depends on the particular transition metal, its oxidation state and the other types of ligands in the coordination sphere. Work to date with the P_4Se_3 system has involved complexes with the tridentate ligands triphos ($MeC(CH_2PPh_2)_3$) and np_3 (equations 65–67).

$$M(np_3) + P_4Se_3 \xrightarrow[35-40\ ^\circ C]{C_6H_6/THF} M(np_3)(\eta^1(P)\text{-}P_4Se_3) \qquad \text{(Ref. 258)} \qquad (65)$$

$$M = Ni,\ Pd$$

$$[RhCl(COD)]_2 + triphos + P_4Se_3 \xrightarrow[BuOH/\Delta]{C_6H_6/THF} [Rh(triphos)(\eta^3(P,P,Se)\text{-}P_3Se_3)]\cdot C_6H_6 \qquad \text{(Ref. 260)} \qquad (66)$$

$$Co(BF_4)_2{\cdot}6\,H_2O + triphos + P_4Se_3 \xrightarrow[\Delta]{C_6H_6/EtOH} [Co(triphos)(\eta^3\text{-}P_2Se)]BF_4\cdot C_6H_6 \qquad \text{(Ref. 261)} \qquad (67)$$

The proposed structures have been confirmed by single-crystal X-ray diffraction for $[Ni(np_3)(\eta^1(P)\text{-}P_4Se_3]\cdot 2\,C_6H_6$[259], $[Rh(triphos)(\eta^3(P,P,S)\text{-}P_3S_3)]\cdot C_6H_6$[260], $[Co(triphos)(\eta^3\text{-}P_2Se)]BF_4\cdot C_6H_6$[261] and $[Co(triphos)(\eta^3\text{-}P_2S)]BF_4\cdot C_6H_6$[268]. Because of the novel structures of these complexes, characterization was limited to elemental analysis, mass spectroscopy and single-crystal X-ray diffraction of typical complexes, and no spectroscopic data have been reported except for the ^{31}P-NMR data for $[Co(triphos)(\eta^3\text{-}P_2Se)]BF_4$[261] (Table 18).

An interesting feature of this spectrum is the resolution of the ^{31}P resonance of the P_2Se

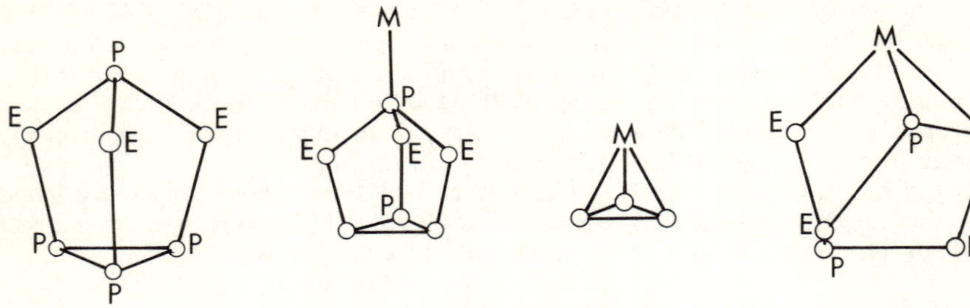

(a) P_4E_3

(b) η^1(P)-P_4E_3

(c) η^3-P_3, η^3-P_2E

(d) η^3(P,P,E)-P_3E_3

FIGURE 3. Bonding modes of P_4E_3 (E = S, Se) and its fragments

ring into a quartet at room temperature, while the triphos resonance is more affected by the cobalt quadrupole moment and remains broad down to − 70 °C. The resonance of the ring system in [Co(triphos)(η^3-P_2S)]BF_4, in contrast, is broad at room temperature but is resolved into a quartet (^2J = 12 Hz) at − 70 °C, while the P_3 resonance of [Co(triphos)(η^3-P_3)] is unresolved even at − 80 °C[261]. The larger quadrupolar effect on the triphos resonance vs. the η^3-P_2E (E = S, Se) resonances has been attributed to the involvement of the former ligand in lower-energy molecular orbitals than the latter cyclic ligands[261].

B. Anionic Monodentate Ligands

1. Selenocyanates

The coordination chemistry of this ligand has been the most studied of any Se or Te ligand, several hundred papers describing its complexes having appeared[269]. Much of the interest in this ligand is related to its ambidentate character[153]. The following bonding modes (*a*)–(*d*) have been unequivocally established by single-crystal X-ray diffraction, and the one-atom bridging mode (*e*) is strongly suggested by the spectroscopic similarity of the complex with the thiocyanate analogue, whose structure has been confirmed by crystallography.

(*a*) M—SeCN, selenocyanato ligand
  NH_4[*trans*-Co(dmg)$_2$(SeCN)$_2$]·3 H_2O[270]
  Me_4N[PhTe(SeCN)$_2$][271]

(*b*) M—NCSe, isoselenocyanato ligand
  [Ni(DMF)$_4$(NCSe)$_2$][272]

(*c*) M—SeCN—M, μ_2-selenocyanato
  HgCo(NCSe)$_4$[273]

M
(*d*) Se—C≡N, μ_2(Se)-selenocyanato
M

[Cu(H_2O)(en)$_2$][Cu_2(CN)$_3$(SeCN)][274]

$[Cu_2L(ECN)_2]$; L = 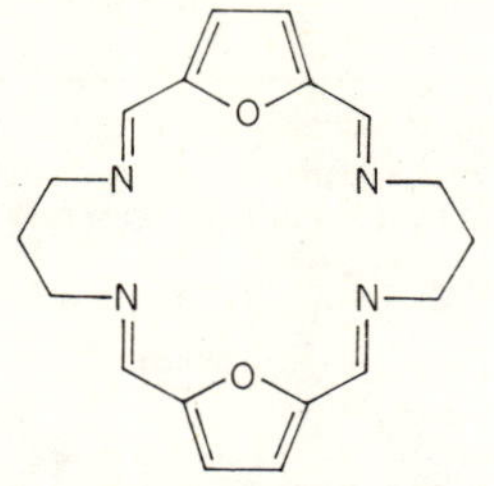 = **(39)**

E = S (crystal structure)[275], Se[276]

(*e*) $\overset{M}{\underset{M}{>}}N—C≡Se$, μ_2(N)-selenocyanate.)

$[Co_2L(OEt)(NCE)_3]$[277]; L = **39**
E = S (crystal structure), Se
(M(CNSe) indicates an undefined bonding mode of selenocyanate.)

Several main-group organometallic complexes with selenocyanate have been prepared[278] (equations 68 and 69). These complexes can also be prepared by metathetical reactions (equations 70 and 71). All of these complexes except Ph_3MNCSe (M = Si, Sn) were formulated as selenocyanato complexes on the basis of infrared spectroscopy (Table 19).

$$Se(SeCN)_2 \begin{cases} \xrightarrow{BiPh_3{}^{279}} Ph_2BiSeCN + PhSeCN + Se \\ \xrightarrow{HgR_2\ (R = Me,\ Ph)^{264}} RHgSeCN + RSeCN + Se \\ \xrightarrow{PbPh_4{}^{265}} Ph_3PbSeCN + PhSeCN + Se \\ \xrightarrow{Pb_2Ph_6{}^{265}} 2\,Ph_3PbNCSe + Se \\ \xrightarrow{TlPh_3{}^{265}} Ph_2TlSeCN + PhSeCN + Se \end{cases} \qquad (68)$$

$$HgR_2 + Hg(SeCN)_2 \rightarrow 2\,RHgSeCN \quad \text{(Ref. 264)} \qquad (69)$$
$$R = Me, Ph$$

$$(p\text{-}MeC_6H_4)_2TlCl + KSeCN \xrightarrow{MeCN/pyr} (p\text{-}MeC_6H_4)_2TlSeCN \quad \text{(Ref. 280)} \qquad (70)$$

$$Me_3SiCl + AgSeCN \xrightarrow{acetone} Me_3SiNCSe + AgCl \quad \text{(Ref. 281)} \qquad (71)$$

Recent studies have examined the bonding mode of the selenocyanate ligand in $Ph_3PbSeCN$ as a function of adduct formation with O and N donors[294] as well as solvent[295]. The solid-state infrared spectrum of the parent compound has been reinterpreted, a polymeric structure with strong Pb—Se and weak Pb—N bonds being proposed[294]. In benzene or CH_2Cl_2 solutions the spectra in the $\nu_{C≡N}$ region indicates monomeric Se-bonded selenocyanate (Table 20). A more detailed study[295] of the infrared spectrum ($\nu_{C≡N}$) of $Ph_3PbSeCN$ in a variety of solvents showed that N-bonded, Se-bonded and ionic pentacoordinate species were formed, depending on the dielectric constant and donor properties of the solvent. These results (Table 20) were interpreted by the following equilibria.

$$Ph_3PbSeCN \underset{}{\overset{L}{\rightleftharpoons}} \begin{matrix} Ph_3PbNCSe{\bullet}L \\ \uparrow\downarrow \\ Ph_3PbSeCN{\bullet}L \end{matrix} \overset{L}{\rightleftharpoons} [Ph_3PbL_2]^+\ SeCN^- \qquad (72)$$

TABLE 19. Selenocyanate complexes (M—SeCN)

Complex	Reported data	Reference
PhHgSeCN	White photosensitive crystals	264
	$\nu_{C\equiv N} = 2129s$	
	$\nu_{C-Se} = 542w$	
	$\delta_{Se-C\equiv N} = 373s, 398m$	
	$\nu_{Hg-Se} = 246s$	
$(p\text{-}MeC_6H_4)_2Tl(SeCN)$	Air-stable white solid	280
	M.p. 212–214 °C dec.	
	$\nu_{C\equiv N} = 2090vs$	
	$\nu_{C-Se} = 571vw$	
	$\delta_{N-C\equiv Se} = 402vw$	
$[Ph_3PbSeCN]_n$	White crystals	265
	M.p. 192 °C	
	$\nu_{C\equiv N} = 2100$	
	Solid state:	294
	$\nu_{C\equiv N} = 2107vs, 2063w$	
	$\nu_{C-Se} = 566m$	
	$2\delta_{N-C\equiv Se} = 759vw, 798vw$	
$(Me_4N)_2[Pt(SeCN)_6]$	$\nu_{C\equiv N} = 2123s$, sp	282
	$\nu_{C-Se} = 528m$	
	$\delta_{Se\equiv C-N} = 398vw, 373m$	
	$\nu_{Pt-Se} = 218w, 210w$	
$[n\text{-}Bu_4N]_2[Pt(SeCN)_4]$	$\nu_{C\equiv N} = 2105s$, sp, 2060	283
	acetone soln.:	
	2117s, sp ($A = 0.59 \times 10^4$)[a]	
	$\nu_{C-Se} = 516w$	
$[Pt(bipy)(SeCN)_2]$	$\nu_{C\equiv N} = 2135s, 2135s$	284
	$\nu_{C-Se} = 532w, 527w$	
$PdL_2(SeCN)_2$	$\nu_{C\equiv N}$ (mull) $\quad \nu_{C\equiv N}$ ($A \times 10^{-4}$)[a,b] $\quad \nu_{C-Se}$	285
L = PPh_3	2127 2122 (0.61) [c]	
4-acetylpyridine	2132 2115 (1.0) 531	
4-methylpyridine	2119 2112 (1.0) [c]	
pyr	2116 2109 (1.0) [c]	
NH_3	2126 2111 (0.93) 528	
L_2 = phen	2123, 2110 [d] 530	
bipyr	2134, 2119 [e] 528	
en	2126, 2108 2112 (0.65) [c]	
$[Ni\{P(OPh)_3\}_2(SeCN)_2]$	red–brown solid	286
	$\nu_{C\equiv N}$(mull) = 2128vs (0.5)[f]	
	(CH_2Cl_2) soln.: 2125	
	$\nu_{C-Se} = 520w$	
	$\delta_{Se-C\equiv N} = 412w$	
	$\nu_{Ni-Se} = 255$	
$K_2[Hg(SeCN)_4]$	$\nu_{C\equiv N} = 2098s$	287
	$\nu_{C-Se} = 543w$	
	$\delta(^{77}Se)$[g]; (D_2O) = −191; −471	288
	(rel. int. = 10:1)	
$NH_4[trans\text{-}Co(dmg)_2(SeCN)_2]\cdot 3H_2O$	$\nu_{C\equiv N} = 2136$	289, 290
	$\nu_{C-Se} = 550$	
	Crystal structure:	270
	Co—Se = 2.4 Å	246

TABLE 19. (*Continued*)

Complex	Reported data	Reference
trans-[Co(dmg)$_2$(PPh$_3$)(SeCN)]	$\nu_{C\equiv N} = 2125$ (2.4)[k] PhNO$_2$ soln: 2120 ($A = 0.68 \times 10^4$)[a]	291
[(Ph$_3$P)AuSeCN]	$\nu_{C\equiv N}$(CH$_2$Cl$_2$) = 2135s ($A = 0.70 \times 10^4$)[a] $\nu_{C-Se} = 529$ $\delta_{N\equiv C-Se} = 372$	292
trans-Pd(PPh$_3$)$_2$(SeCN)$_2$	$\nu_{C\equiv N} = 2134$s, sp CH$_2$Cl$_2$: 2131m, sp ($A = 0.42 \times 10^4$)[a]	293
trans-Pt(PPh$_3$)$_2$(SeCN)$_2$	$\nu_{C\equiv N} = 2130$s, sp CH$_2$Cl$_2$: 2131s, sp (Pt—SeCN) 2100s, br (Pt—NCSe) DMF: 2129s, sp (Pt—SeCN) 2096s, br (Pt—NCSe) 2053w, br (free SeCN$^-$)	293

[a]Solution $\nu_{C\equiv N}$ integrated absorption intensity (see Table 23) in units of $\text{M}^{-1}\,\text{cm}^{-2}$.
[b]In DMF.
[c]Obscured by other vibrations.
[d]Complex insoluble in DMF.
[e]Selenocyanate ligand completely dissociates in DMF.
[f]$\nu_{C\equiv N}$ intensity relative to internal salicyclic acid in the solid state (see Table 23).
[g]Vs. external 1M SeMe$_2$ in CHCl$_3$.
[h]$\nu_{C\equiv N}$ intensity relative to internal 1,4-dicyanobenzene in the solid state (see Table 23).

In non-coordinating non-polar solvents (C_6H_6, CH_2Cl_2) the solution species is exclusively monomeric Ph$_3$PbSeCN. Strong donors (e.g. DMF, OP(OMe)$_3$) and the highly polar but weak donor solvent MeCN cause complete ionization, as reflected in the appearance of only the 'ionic' value of $\nu_{C\equiv N}$ for the selenocyanate. In acetone only partial ionization occurs, with the coordinated selenocyanate being N-bonded, but in THF, a stronger donor but less polar than acetone, an isomeric mixture of N- and Se-bonded monomers is observed. In pyridine partial ionization occurs, and the molecular species is an isomeric mixture.

TABLE 20. Solution infrared spectra in the $\nu_{C\equiv N}$ region of Ph$_3$PbSeCN[295]

			$\nu_{C\equiv N}$ (cm^{-1})			
				Ph$_3$Pb(CNSe)·L)		
Solvent	DN[a]	ε	Ph$_3$PbSeCN	Pb—NCSe	Pb—SeCN	SeCN$^-$
C_6H_6	0.1		2139			
CH_2Cl_2	0.1	9.1	2136			
DMF	26.6	36.1				2066
OP(OMe)$_3$	23.0	20.6				2068
Acetone	17.0	20.7		2056		2066sh
MeCN	14.1	38.0		2066		
THF	20.0	7.6		2036m	2129s	
pyr	33.1	12.3		2117m	2049vs	2064s

[a]Donor number as described by Gutmann[296].

TABLE 21. Infrared spectral data (cm^{-1}) for $Ph_3Pb(CNS)L$ adducts[294]

L	Reported data $\nu_{C\equiv N}$ (solid)	$\nu_{C\equiv N}$ (C_6H_6 soln.)	ν_{C-Se} (solid)
$OPPh_3$	2119s, sp 2055w	2139vs, sp[a] 2125m, sp[b] 2037m, br[c]	[d]
4-picO	2105s, sp 2061vw	2138m, sp[a] 2123vs, sp[b] 2038m, br[c]	547w
pyrO	2106s, sp 2061vw	2138vs, sp[a] 2125m, sp[b] 2031m, br[c]	[d]
HMPA	2053vs, br 2010w	2139s, sp[a] 2123s, sp[b] 2037vs, br[c]	609m

[a] Band of $Ph_3PbSeCN$ formed by dissociation of the neutral ligand.
[b] Pb—SeCN.
[c] Pb—NCSe.
[d] Obscured by other vibrations.

A series of 1:1 adducts, $Ph_3Pb(CNSe)\cdot L$ (L = HMPA, pyridine-*N*-oxide (pyrO), 4-picoline-*N*-oxide (4-picO) and $OPPh_3$), have been prepared and the selenocyanate bonding modes in the solid state and solution (C_6H_6 and CH_2Cl_2) have been determined by infrared spectroscopy[294] (Table 21). In the solid state all the complexes are monomeric with the HMPA adduct containing N-bonded selenocyanate, and the other three oxygen donors give Se-bonded selenocyanate. In solution (C_6H_6 and CH_2Cl_2) all four complexes partially dissociate with the intact soluble adduct being present as both N- and Se-linkage isomers (Table 21).

The derivatives $R_3SiNCSe$ (R = Me, Et, Ph) were formulated as isoselenocyanates (Table 22)[281]. Unlike the heavier main-group organometallic selenocyanates, these compounds readily decompose thermally or hydrolytically, depositing elemental Se[281] (equations 73 and 74). As described for other systems[137–139], Se extrusion can also be effected by PPh_3[281] (equation 75).

$$2\,R_3SiNCSe \xrightarrow{H_2O} (R_3Si)_2O + HCN + Se \tag{73}$$

$$R_3SiNCSe \xrightarrow{\Delta} R_3SiCN + Se \tag{74}$$

$$Me_3SiNCSe + PPh_3 \xrightarrow{-SePPh_3} Me_3SiCN \tag{75}$$

The structure of $Ph_3SnNCSe$, originally proposed to be monomeric with N-bonded selenocyanate[265], has recently been reformulated as polymeric with strong Sn—N bonds and weak Sn—Se bonds (a similar structure has been established by X-ray diffraction for Ph_3SnNCS[301]).

The general approach in structural ambidentate ligand chemistry has been to unequivocally establish the various bonding modes by single-crystal X-ray diffraction and then assign bonding modes in other complexes on the basis of some spectroscopic

TABLE 22. Isoselenocyanate complexes (M—NCSe)

Complex	Reported data	Reference
$R_3SiNCSe$	$\nu_{C\equiv N}$ ν_{C-Se}	281
R = Me	2055vs 625m	
Et	2065vs, 2090sh 600w	
Ph	2070vs 620w	
$Ph_3SnNCSe$	white powder	265
	$\nu_{C\equiv N} = 2108$br	
	$\nu_{C-Se} = 594$m	
$[(n\text{-}Bu)_4N]_3[Y(NCSe)_6]$	$\nu_{C\equiv N} = 2067$s, br; 2030sh	283
	acetone soln.: 2068s, br	
	$(A = 5.1 \times 10^4)^a$	
	$\nu_{C-Se} = 634$m	
	$\delta_{N\equiv C-Se} = 429$m	
$(AsPh_4)_2[Co(NCSe)_4]$	$\nu_{C\equiv N} = 2060$s, br	287, 297
$CpCo(CO)(NCSe)_2$	$\nu_{C\equiv N} = 2136$vs $(1.7)^b$;	298
	2096vs $(2.0)^b$	
	$\nu_{C\equiv O} = 2060$s	
	$\nu_{C-Se} = 575$m	
	$\delta_{N\equiv C-Se} = 540$w	
$Ru(CO)_2(PPh_3)_2(NCSe)_2$	$\nu_{C\equiv N} = 2085$ms (*trans*-$Ru(NCSe)_2$)	299
	$\nu_{C\equiv O} = 2015$vs;	
	2068vs (*cis*-$Ru(CO)_2$)	
	$\nu_{C-Se} = 595$m	
$Cp_2Ti(NCSe)_2$	$\nu_{C\equiv N} = 2050\ (2.4)^b$;	300
	$2015\ (2.3)^b$	
	CH_2Cl_2 soln.: 2050, 2015	
	$\nu_{C-Se} = 600, 593$	
	$\nu_{N\equiv C-Se} = 457, 448$	
$Cp_2V(NCSe)_2$	$\nu_{C\equiv N} = 2083\ (1.8)^b$;	300
	$2065\ (1.9)^b$	
	CH_2Cl_2 soln.: 2080, 2063	
	$\nu_{C-Se} = 620, 596$	
	$\delta_{N\equiv C-Se} = 472, 450$	
$Cp_2Cr(NCSe)(SeCN)$	$\nu_{C\equiv N} = 2120\ (0.4)^b$; Cr—SeCN	300
	$2042\ (1.8)^b$; Cr—NCSe	
	CH_2Cl_2 soln.: 2122, 2038	
	$\nu_{C-Se} = 590$ (Cr—NCSe);	
	540 (Cr—SeCN)	
	$\delta_{N\equiv C-Se} = 435, 425$	
$CpCo(CO)(CN)(NCSe)$	$\nu_{C\equiv N} = 2148$vs $(1.8)^b$; Co—CN	298
	$\nu_{C\equiv N} = 2100$vs $(1.9)^b$; Co—NCSe	
	$\nu_{C\equiv O}$ masked by other bands	
	$\nu_{C-Se} = 575$m	
	CH_2Cl_2 solution:	
	$\nu_{C\equiv N} = 2143$vs (Co—CN)	
	$\nu_{C\equiv N} = 2100$vs (Co—NCSe)	
	$\nu_{C\equiv O} = 2043$vs (Co—CO)	

[a] Solution $\nu_{C\equiv N}$ integrated absorption intensity (see Table 23) in units of $M^{-1}\ cm^{-2}$.
[b] $\nu_{C\equiv N}$ intensity relative to internal salicylic acid in the solid state (see Table 23).

technique, most commonly infrared spectroscopy. This approach has been most developed for thiocyanate complexes. Structural probes include infrared, Raman, electronic, NMR (^1H, ^{13}C, ^{14}N), NQR (^{14}N) and X-ray photoelectron spectroscopy (see Ref. 302 for a review of this work). Recent work in this area includes solid-state quantitative infrared[303] (using an internal reference compound for measuring the relative intensity of the $\nu_{C\equiv N}$), ^{14}N-NQR[304], XPS[157c], and solution[305] and solid-state[158a] ^{13}C-NMR and solution ^{77}Se-NMR[288] spectroscopy. All of these techniques should be applicable to selenocyanate ligand chemistry, although considerably less work has been

TABLE 23. Infrared spectroscopic characterization of the selenocyanate ligand

Bonding mode	$\nu_{C\equiv N}$	$A \times 10^{-4}$ [b]	ISR[c]	ISR[d]	ν_{C-Se} [a]	$\delta_{N\equiv C-Se}$ [a]
M—NCSe	2050–2100 (broad)	5–12	1.5–3	20–30	600–650	410–460
M—SeCN	2070–2130 (sharp)	0.5–1.5	0.3–0.6	2–13	520–550	370–410
M—SeCN—M	2100–2175				550–640	390–410
M_2>Se—C≡N[e]	2105sp				700m	443w, 462w
M_2>N—C≡Se[f]	1995					
Ionic $SeCN^-$ [g]	2068s, sp	1.8			558	416, 424

[a] In cm^{-1}.
[b] Solution $\nu_{C\equiv N}$ integrated absorption intensity in units of $M^{-1} cm^{-1}$. The concentration is based on selenocyanate.
[c] Internal standard ratio: the ratio of the intensity of the $\nu_{C\equiv N}$ vs. salicyclic acid as an internal standard in the solid state[306].
[d] Internal standard ratio using 1,2-dicyanobenzene as an internal reference for $\nu_{C\equiv N}$ in the solid state[303].
[e] The data listed are for the only complex with this bonding mode which has been reported ($Cu_2L(SeCN)_2$; L = **39**)[276].
[f] Only one complex with this bonding mode has been reported: $Co_2L(OEt)(NCSe)_3$ (L = **39**)[277]. The crystal structure of the thiocyanate analogue was reported[277] and showed a structure involving a terminal isothiocyanate ligand on each Co, a bridging ethoxide ligand, and a N-bridged thiocyanate ligand. Two other N-bridging thiocyanate complexes have been structurally characterized by single-crystal X-ray diffraction ($[(n\text{-}Bu)_4N]_3Re_2(NCS)_{10}$[307] and $[Cd_2(NCS)_4(butrz)_3]_n$ (butrz = 4-*t*-butyl-1,2,4-triazole)[308]).
[g] KSeCN in acetone[283].

TABLE 24. NMR spectroscopic characterization of selenocyanate bonding modes

^{13}C-NMR (δ, ppm[a])[304]		
M—SeCN	$SeCN^-$	M—NCSe
105–114	116–121	120–130
^{77}Se-NMR (δ, ppm[b])[288]		
$K_2Hg(SeCN)_4$	KSeCN	$ZnL_2(NCSe)_2$ [c]
−191	−273	−318
		$CdL_2(NCSe)_2$
		−305
		$CdL_2(NCSe)_2$
		−282

[a] Vs. internal TMS. Only three selenocyanate complexes and one isoselenocyanate complex were studied in this work but the $SeCN^-$ resonance tends to shift to high (M—NCSe) or low (M—SeCN) field vs. free $SeCN^-$ upon coordination.
[b] Vs. external 1 M Me_2Se in $CHCl_3$. Only four complexes have been studied, but again there seems to be a high field/low field shift vs. ionic $SeCN^-$ upon coordination.
[c] 4-*t*-Butylpyridine.

done in this area, infrared spectroscopy being the most generally used spectroscopic technique. Infrared and NMR parameters for the assignment of selenocyanate bonding modes are summarized in Tables 23 and 24.

Most of the reported coordination chemistry with this ligand involves transition-metal complexes, and the simple complexes, containing only selenocyanate ligands, follow the bonding pattern expected on the basis of Pearson's hard–soft acid–base theory[309]. This generalization refers to simple complexes containing only selenocyanate ligands (e.g. $[Co(NCSe)_4]^{2-}$ vs. $[Pt(SeCN)_4]^{2-}$), and a variety of factors such as metal oxidation state, steric and electronic effects of other ligands in the coordination sphere, and crystal-packing effects can influence the bonding mode of the ambidentate selenocyanate ligand[153]. Two examples of isolated linkage isomeric pairs with this ligand have been reported: $[Pd(Et_4dien)(SeCN/NCSe)]BPh_4$[310] ($Et_4dien = Et_2NCH_2CH_2NHCH_2CH_2NEt_2$) and $CpFe(CO)(PPh_3)(SeCN/NCSe)$[311]. The limited number of such isomeric pairs compared to the thiocyanate ligand indicates a significantly lower susceptibility of the selenocyanate ligand to the above steric and electronic influences on bonding mode.

Selenocyanate can also form dimeric and polymeric complexes with bridging selenocyanate[312] (Table 25). This bonding mode is most easily detected by an unusually high value of $\nu_{C\equiv N}$ (e.g. 2140–2175 cm^{-1}). The trend from Class A to Class B metal ions is well reflected in the bonding of the selenocyanate ion in the complexes of Group IIb, Zn(II) and Hg(II) giving tetrahedral $Zn(NCSe)_4{}^{2-}$ and $Hg(SeCN)_4{}^{2-}$ complexes with the expected Zn—NCSe (hard acid/hard N site) and Hg—SeCN (soft acid/soft Se site), whereas Cd(II) gives a dimeric complex with N-bonded terminal selenocyanate and bridging selenocyanate[283]:

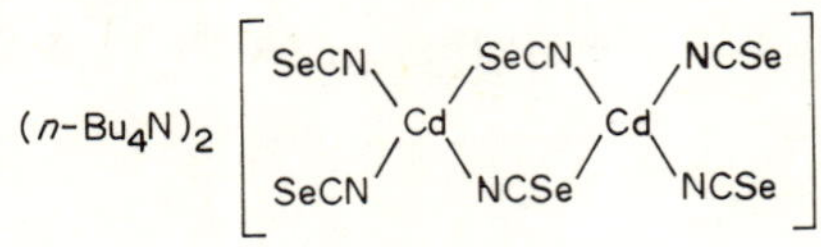

(40)

TABLE 25. Complexes with bridging selenocyanate (M—SeCN—M)

Complex	Reported data	Ref.
$\{AgSeCN\}_n$	$\nu_{C\equiv N} = 2141s$, 2098sh	287
	$\nu_{C-Se} = 580w$	
$HgCo(NCSe)_4$	$\nu_{C\equiv N} = 2146$	
(Co—NCSe—Hg)	$\nu_{C-Se} = 639$	273, 287
$(n\text{-}Bu_4N)_2[Cd_2(SeCN)_6]$	$\nu_{C\equiv N} = 2109s$, br (Cd—NCSe)	283
(40)	2125sh (Cd—NCSe—Cd)	
	acetone soln.:	
	2120s, br ($A = 3.2 \times 10^4$)[a]	
	2076s, br ($A = 2.6 \times 10^4$)[a]	
	$\nu_{C-Se} = 589sh$, 582m	
	$\delta_{N\equiv C-Se} = 417w$, 408m	
$\{Co[Ag(SeCN)_2]_2\}_n$	$\nu_{C\equiv N} = 2125s$, 2150s	
(Co—NCSe—Ag)	$\nu_{C-Se} = 580$, 598sh	313
	$\delta_{N\equiv C-Se} = 395$, 405sh	
	$\nu_{Co-N} = 278s$	
$(SCN)_2Co(NCSeHg(Pr\text{-}n))_2$	$\nu_{C\equiv N} = 2170s$, 2120s	314
	(Co—NCSe—Hg)	
	2090m (Co—NCS)	
	$\nu_{C-S} = 820m$	
	$\nu_{C-Se} = 540s$	

[a]Solution $\nu_{C\equiv C}$ integrated absorption intensity (see Table 23) in units of $M^{-1} cm^{-2}$.

The heterometallic complex $HgCo(NCSe)_4$ is polymeric with bridging selenocyanate, the hard N site bonding to the hard Co(II) Lewis acid and the soft Se site coordinating to the soft Hg(II) acid (Table 25).

The general routes to selenocyanate complexes include metathetical reactions (equations 76–78) and oxidative addition reactions, $(SeCN)_2$ (equations 79–85), $Se(SeCN)_2$ (equations 86–88) and $Se(CN)_2$ (equations 89–91) all having been used as reagents in such reactions.

Metathetical reactions

$$Y(NO_3)_3 \cdot 6\,H_2O + 6\,Bu_4N[SeCN] \xrightarrow{EtOH} (Bu_4N)_3[Y(NCSe)_6] \qquad \text{(Ref. 283)} \quad (76)$$

$$CpCr(NO)_2Cl \xrightarrow[H_2O/MeOH]{AgNO_3} \{CpCr(NO)_2NO_3\} + AgCl \qquad \text{(Ref. 311)} \quad (77)$$

$$\{CpCr(NO)_2NO_3\} \xrightarrow{KSeCN} CpCr(NO)_2NCSe$$

$$trans\text{-}[Co(dmg)_2Cl(H_2O)] + 2\,KSeCN \xrightarrow{H_2O/EtOH} K[trans\text{-}Co(dmg)_2(SeCN)_2] \qquad \text{(Ref. 289)} \quad (78)$$

Oxidative addition reactions
$(SeCN)_2$

$$Pd(PPh_3)_4 \xrightarrow{(SeCN)_2} trans\text{-}Pd(PPh_3)_2(SeCN)_2 \qquad \text{(Ref. 293)} \quad (79)$$

$$(AgSeCN)_x \xrightarrow{I_2/Et_2O} (SeCN)_2$$

$$CpCo(CO)_2 \xrightarrow{(SeCN)_2} CpCo(CO)(NCSe)_2 \qquad \text{(Ref. 298)} \quad (80)$$

$$Ru(CO)_3(PPh_3)_2 \xrightarrow{(SeCN)_2} Ru(CO)_2(PPh_3)_2(NCSe)_2 \qquad \text{(Ref. 299)} \quad (81)$$

$$Cp_2M \xrightarrow{(SeCN)_2} Cp_2V(NCSe)_2 \text{ or } Cp_2Cr(NCSe)(SeCN) \qquad \text{(Ref. 300)} \quad (82)$$

$M = V, Cr$

$$Cp_2Ti(CO)_2 \xrightarrow{(SeCN)_2} Cp_2Ti(NCSe)_2 \qquad \text{(Ref. 300)} \quad (83)$$

$$Ni\{P(OPh)_3\}_4 \xrightarrow{(SeCN)_2} \{P(OPh)_3\}_2Ni(SeCN)_2 \qquad \text{(Ref. 286)} \quad (84)$$

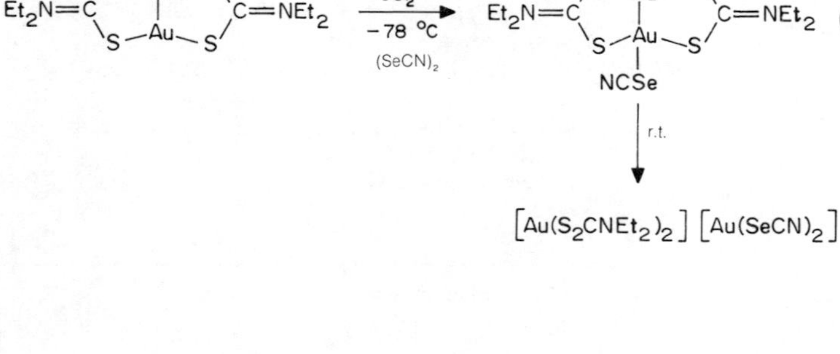

(Ref. 315) (85)

Se(SeCN)₂

$$Cp(OC)(Ph_3P)FeCH_2Ph \xrightarrow{Se(SeCN)_2} Cp(OC)(Ph_3P)FeNCSe \quad \text{(Ref. 137)} \quad (86)$$

$$48\%$$
$$\nu_{C\equiv N}(CHCl_3) = 2120$$
$$A = 5.3 \times 10^4$$
$$\nu_{C-Se} = 663\,\text{m}$$

$$+ Cp(OC)(Ph_3P)FeSeCN + CpFe(CO)(PPh_3)CN + Se + PhCH_2SeCN$$
$$17\% \qquad\qquad 7\%$$
$$\nu_{C\equiv N}(CHCl_3) = 2117$$
$$A = 1.7 \times 10^4$$
$$\nu_{C-Se} = 523\text{w}$$

$$CpFe(CO)_2CH_2Ph \xrightarrow[C_6H_6/\Delta]{Se(SeCN)_2} CpFe(CO)_2(SeCN) \quad \text{(Ref. 311)} \quad (87)$$

$$CpM(CO)_3H \xrightarrow[C_6H_6]{Se(SeCN)_2} CpM(CO)_3SeCN \quad \text{(Ref. 311)} \quad (88)$$
$$M = Mo, W$$

Se(CN)₂

$$CpCo(CO)_2 \xrightarrow{Se(CN)_2} CpCo(CO)(CN)(NCSe) \quad \text{(Ref. 298)} \quad (89)$$
$$\uparrow \text{BrCN/Et}_2\text{O}$$
$$KSeCN$$

$$Cp_2Ti(CO)_2 \xrightarrow[Et_2O]{Se(CN)_2} Cp_2Ti(CN)(NCSe) \quad \text{(Ref. 300)} \quad (90)$$

$$Cp_2M \xrightarrow[Et_2O]{Se(CN)_2} Cp_2M(CN)(NCSe) \quad \text{(Ref. 300)} \quad (91)$$
$$M = V, Cr$$

The complex K[*trans*-$Co(dmg)_2(SeCN)_2$] (equation 78) is an example of significant stabilization of a reagent upon coordination. Addition of concentrated HCl to a hot aqueous solution of this complex causes no deposition of elemental Se. Cooling the resulting solution results in the deposition of crystalline $H[Co(dmg)_2(SeCN)_2]$[289]. A similar coordination stabilization of selenosemicarbazones has been reported for Co(III) complexes of this type[251]. A considerable amount of work has been reported dealing with complexes of the type *trans*-$[Co(dmg)_2L(SeCN)]$ (L = neutral ligands, e.g. PR_3, AsR_3). In such complexes the steric environment of the ambidentate $SeCN^-$ ligand can be held constant while the electronic properties of the ligands in the *trans* position across the $Co(dmg)_2$ plane can be systematically modified to study electronic effects on the $SeCN^-$ bonding[291,316–319].

The first example of isolated linkage isomers of the selenocyanate ligand took advantage of the increased steric requirements of the bent M—SeCN vs. linear M—NCSe linkages, the initially isolated Se-bonded isomer being formed when the substitution reaction was run at low temperature (equation 92).

$$K_2Pd(SeCN)_4 + Et_4dien \xrightarrow{-150\ ^\circ C} [Pd(SeCN)(Et_4dien)]^+ \qquad (92)$$

$$Et_4dien = (Et_2NCH_2CH_2)_2NH$$

$$[Pd(SeCN)(Et_4dien)]^+ \xrightarrow{NaBPh_4} [Pd(SeCN)(Et_4dien)]BPh_4 \xrightarrow{DMF/r.t.} [Pd(NCSe)(Et_4dien)]BPh_4$$

The N-bonded isomer was precipitated from DMF solution by addition of ether[320], but it reisomerized in the solid state at room temperature to the Se-bonded isomer, an unusual example of the influence of a non-coordinated ion on the bonding mode of an ambidentate ligand[310]. The influence of steric effects is demonstrated in this system by the stability of the initially isolated Se-bonded isomer when diethylenetriamine is used in the above reaction in place of the Et_4dien[310]. The linkage isomers $CpFe(CO)(PPh_3)(NCSe/SeCN)$ were prepared with the reagent $Se(SeCN)_2$ (equation 86), and the isomeric products were purified by column chromatography[137]. The two isomers were stable at room temperature, and attempts to cause isomerization by heating resulted in deselenation, this reaction being much faster for the Fe—NCSe isomer (equation 93).

$$CpFe(CO)(PPh_3)(CNSe) \xrightarrow[\Delta]{C_6H_6} CpFe(CO)(PPh_3)CN + Ph_3PSe \qquad (93)$$

The formulation of the linkage isomers involving the isomeric $CpFe(CO)(SePPh_3)CN$ was ruled out by ^{1}H-NMR (the J_{PH} of both isomers and the independently prepared $CpFe(CO)(PPh_3)(CN)$ was 1.3 Hz, indicating the presence of the Fe—PPh_3 moiety)[137].

The infrared data for these two linkage isomers (equation 86) illustrate the ambiguity with bonding assignments based on $\nu_{C\equiv N}$. The values for the two isomers are close, and both could reasonably be assigned to Fe—SeCN bonding. Even the solution-integrated intensities for these bands are at the high and low ends of the ranges for M—SeCN and M—NCSe, respectively. The ν_{C-Se} bands, generally the most unequivocal vibrations for bonding assignment, clearly indicate, however, the presence of linkage isomers. Unfortunately, the spectral region in which the ν_{C-Se} bands are found is often obscured by other ligand vibrations (e.g. Table 19).

Both M—SeCN and M—NCSe products were obtained in the other oxidative addition reactions (equations 79–90), as well as a unique example of a complex containing both bonding modes (equation 82; Table 22). The oxidative addition reaction with the dimeric Au(I) diethyldithiocarbamate complex gave an isolable complex of gold in the unusual +2 oxidation state[291].

The reaction of Mo_2Cl_{10}, a chloro-bridged dimer with a weak Mo—Mo bond, gives a product, $MoCl_5(NCSe)_2$, in which an intact N-bonded $(NCSe)_2$ ligand is proposed on the basis of infrared and ESR spectroscopy[321].

Several metal complexes of organic selenocyanates have also been isolated ($M(CO)_5NCSeR$: M = Cr, R = Me, Et, Ph; M = W, R = Me, Et)[322]. These complexes, which were assigned M—NCSeR bonding modes on the basis of infrared and NMR data, were prepared by substitution reactions (equation 94a) or by alkylation of coordinated isoselenocyanato ligand (equation 94b).

$$Et_4N[M(CO)_5I] \xrightarrow[-AgI/Et_4NBF_4]{AgBF_4/acetone} [M(CO)_5(acetone)] \xrightarrow{RSeCN} [M(CO)_5NCSeR] \qquad (94a)$$

$$(Et_4N)[M(CO)_5NCSe] \xrightarrow[(e.g.\ [Et_3O][BF_4])]{alkylation} [M(CO)_5NCSeR] \qquad (94b)$$

2. Selenols (RSe^-)

Since the first transition-metal complex with an organoselenol ligand, the dimeric doubly selenol-bridged complex $(OC)_3Fe(\mu_2\text{-}SeEt)_2Fe(CO)_3$, was reported in 1960 by Hieber and Beck[323], a number of complexes incorporating terminal (M—SeR) and bridging (—M(μ_2-SeR)$_2$M—) have been described. Crystal structures of two monomeric

TABLE 26. Metal complexes with terminal selenol ligands

Complex[a]	Reported data	Reference
$Cp_2Ti(SePh)_2$	blue–violet crystals	
	M.p. 120–122 °C dec.	324, 325
	153–155 °C	326
	$\delta = 5.99$ (Cp), 7.45 (Ph)	
	MW = 463[b] (calc. 490.2)	
$CpMo(CO)_3SePh$	Red solid	327
	M.p. 92–94 °C	
	$\nu_{C\equiv O}(CS_2) = 2026$vs, sp; 1948vs	
$Ti(SePh)_4$	Black solid	328
$Ti(SeNp)_4$	Black solid	328
$Zr(SePh)_4$	Turquoise solid	328
$Nb(SePh)_3$	Brown microcrystals	328
	MW = 551[b] (calc. 561.1)	
$Nb(SeNp)_3$	Brown–black solid	328
$Ta(SePh)_3$	Brown microcrystals	328
$W(SePh)_4$	Brown solid	328
$(Bu_3P)_2Ni(SePh)_2$	MW = 747[b] (calc. 775.5)	328
$CpFe(CO)_2SePh$	Dark crystals	329
	M.p. 50–52 °C	
	MW = 372[c], 333[d] (calc. 333)	
	$\nu_{C\equiv O}(C_6H_{12}) = 2026$s, 1984s	
	$\delta(CS_2) = 5.21$ (Cp)	
$CpNi(PBu_3)SePh$	Green crystals	330
	M.p. 51–52 °C	
	$\delta(CS_2) = 5.06$ (s, Cp), 7.50–7.41, 6.98–6.71 (m, Ph)	
$Cp_2Ti(SeMe)_2$	Green solid	326
	M.p. 210–211 °C	
	$\delta(CS_2) = 5.97$ (s, Cp), 2.64 (s, Me)	
$Cp_2Zr(SePh)_2$	Yellow solid	326
	M.p. 134–136 °C; 156 °C	326,331
	$\delta(CS_2) = 5.83$ (s, Cp), 7.60–7.46, 7.22–7.07 (m, Ph)	
	MW = 512[b] (calc. 533)	
$Cp_2V(SePh)_2$	Green solid	332
	M.p. 126–130 °C	
$Cp_2Nb(SePh)_2$	Dark green solid	332
	M.p. 101–105 °C	
$Cp_2Mo(SeMe)_2$	Brown crystals	332
	M.p. 197–198 °C	
	$\delta[(CD_3)_2SO] = 5.20$ (s, Cp), 1.55 (s, Me)	
$Cp_2Mo(SePh)_2$	M.p. 230–231 °C	332
	$\delta[(CD_3)_2SO] = 5.22$ (s, Cp), 7.37–6.97 (m, Ph)	

TABLE 26. (*Continued*)

Complex[a]	Reported data	Reference
$Cp_2W(SeMe)_2$	M.p. 207 °C $\delta(CDCl_3) = 5.31$ (s, Cp), 2.53 (s, Me)	332
$Cp_2W(SePh)_2$	M.p. 231–232 °C $\delta(CDCl_3) = 5.36$ (s, Cp), 7.70–6.92 (m, Ph)	332
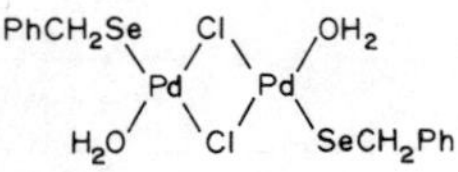	Orange–brown solid M.p. 220 °C dec.	333
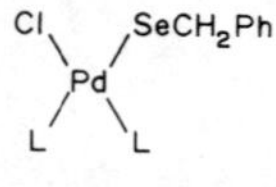 L = p-$H_2NC_6H_4Me$	Yellow crystals M.p. 220 °C $\nu_{Pd-Se} = 300$ $\nu_{Pd-Cl} = 345$ $\nu_{Pd-N} = 475, 510$	333
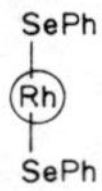 (Rh) = 41	Red crystals Infrared spectrum ^{1}H-NMR	334, 335
	Crystal structure Rh—Se = 2.510(1), 2.544(1)Å	334
$(\eta^7\text{-}C_7H_7)W(CO)_2SePh$	Green crystals M.p. 132 °C $\nu_{C\equiv O} = 1985$vs, 1930vs $\delta(^1H) = 5.11$ (s, C_7H_7), 7.16 (m, Ph)	336
	Crystal structure	337
$Cp_2Hf(SeMe)_2$	M.p. 208 °C $\delta(^1H, CS_2) = 2.35$ (s, Me), 5.95 (s, Cp)	331
$Cp_2Hf(SePh)_2$	M.p. 160 °C $\delta(^1H, CS_2) = 5.70$ (s, Cp), 7.15m, 7.40m	331
$Cp_2Zr(SePh)Me$	$\delta(^1H) = 0.02$ (s, Me), 5.60 (s, Cp), 7.03 (m), 7.60 (m, Ph)	338
$dppe(OC)_2ReSePh$	Yellow crystals $\nu_{C\equiv O}(CHCl_3) = 2020$s, 1945m, 1899m	339
$(OC)_4(PhSe)CrCNEt_2$	Orange solid $\nu_{C\equiv O} = 2078$vw, 2022m, sh; 2007vs, 1976vs $\nu_{C\equiv N} = 1575$	250
$(OC)_3(Ph_3P)(PhSe)CrCNEt_2$	$\nu_{C\equiv O} = 2023$m, 1965s, 1933vs $\nu_{C\equiv N} = 1553$	250
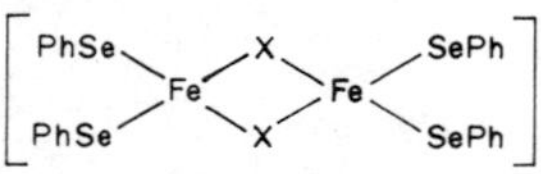 X = S, Se	ESR spectrum	340
$(Ph_4P)[Hg(SePh)_3]$	Yellow solid M.p. 102 °C dec.	341
$(Ph_4P)_2[Hg(SePh)_4]$	Yellow solid	341

[a] Units are as described in Table 3.
[b] Determined cryoscopically in benzene.
[c] Determined by osmometry in $CHCl_3$.
[d] Mass spectroscopy.

TABLE 27. Metal complexes with bridging selenol ligands

Complex	Reported data	Reference
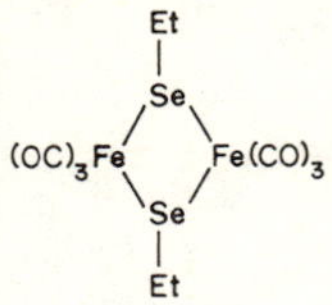	Red crystals M.p. 34 °C MW = 501[a] (calc. 495.8) $\mu = 2.20$ $\nu_{C\equiv O}$ = 2056s, 2027vs, 1985vs	323
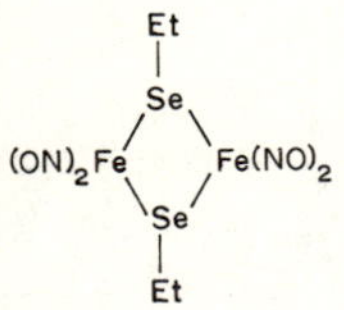	M.p. 97 °C dec. $\mu(C_6H_6) = 0.92$	342
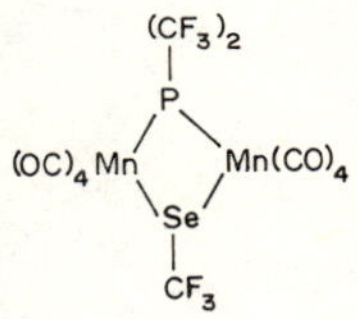	Orange solid $\nu_{C\equiv O}(CCl_4)$ = 2089m, 2082w, 2036s, sh; 2029vs, 2003s $\delta(^{19}F)^b$ = 50.8, 51.3; J_{PF} = 53.8[c] ($-P(CF_3)_2$), 29.3 ($-SeCF_3$)	343
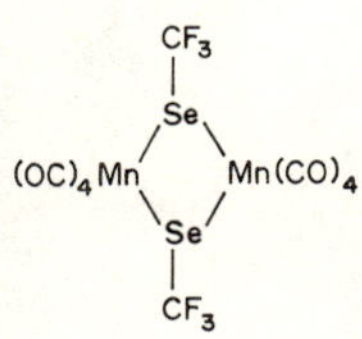	Air-stable orange–brown solid M.p. 96–97 °C $\nu_{C\equiv O}$ = 2100vs, 2060vs, 2020vs, 1955vs	344
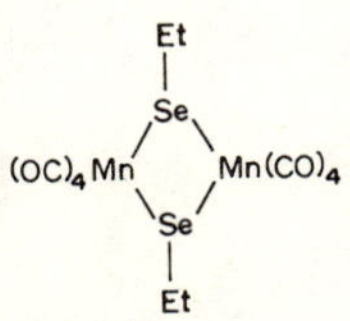	Air-stable red–brown solid M.p. 135 °C dec. $\nu_{C\equiv O}$(KBr) = 2065vs, 2002vs, 1980vs, 1932vs $\delta(CCl_4)$ = 2.82 (q, $J = 7.5$), 1.22 (t, $J = 7.5$)	344
	Orange crystals M.p. 77 °C $\nu_{C\equiv O}(C_6H_{12})$ = 2056m, 2002sh, 1999vs, 1988s, 1957vs	345
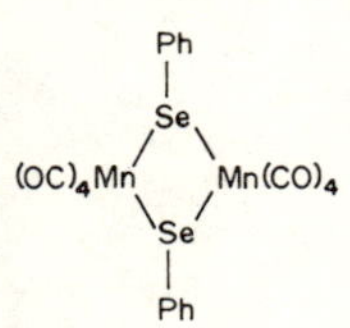	Orange–brown crystals M.p. 135 °C dec. MW = 686 (calc. 646) $\nu_{C\equiv O}(C_6H_{12})$ = 2065s, 2010vs, 1994s, 1962vs	346 345, 347
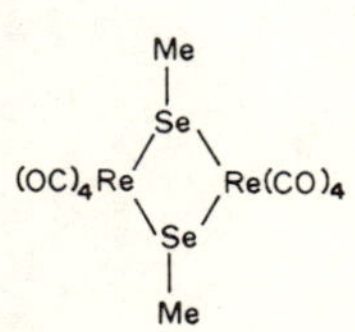	Orange crystals M.p. 172 °C dec. $\nu_{C\equiv O}(C_6H_{12})$ = 2082m, 2002vs, 1990m, 1952vs $\delta(CCl_4)$ = 2.4	345

TABLE 27. (*Continued*)

Complex	Reported data	Reference
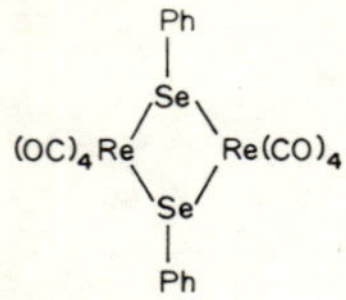	Bright orange crystals M.p. 175 °C dec. $\nu_{C\equiv O}(C_6H_{12}) = 2088m, 2012vs, 1988m, 1952s$	345–347
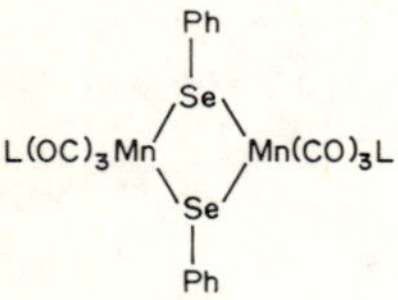 L = PPh_3	$\nu_{C\equiv O}(CHCl_3) = 1988m, 1949s, 1906s$	339
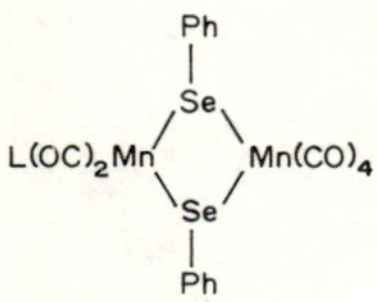 L = dppe	$\nu_{C\equiv O}(CHCl_3) = 2048w, 2022s, 2002s, 1946sh, 1935s, 1908s$	339
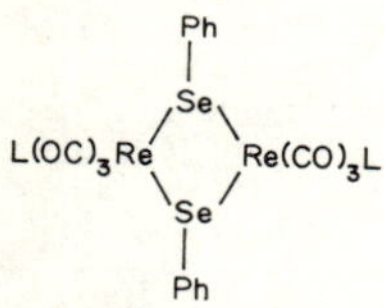 L = PPh_3	$\nu_{C\equiv O}(CHCl_3) = 2028s, 2009s, 1937m, 1917s, 1905s$	339
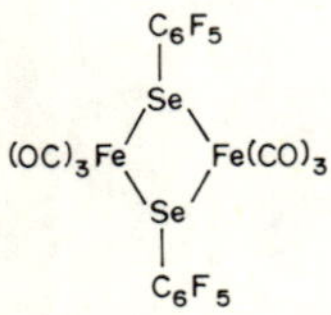	Air-stable solid M.p. 158–159 °C $\nu_{C\equiv O}(CS_2) = 2080, 2051, 2016, 2004$	348
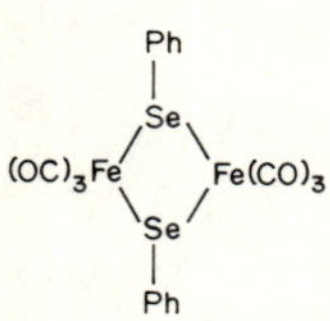	Air-stable red crystals M.p. 94–96 °C MW = 572[d] (calc. 548) $\nu_{C\equiv O}(CS_2) = 2066, 2029, 2000, 1988$ $\nu_{C\equiv O}(C_6H_{12}) = 2061, 2031, 1998, 1991$	348 349
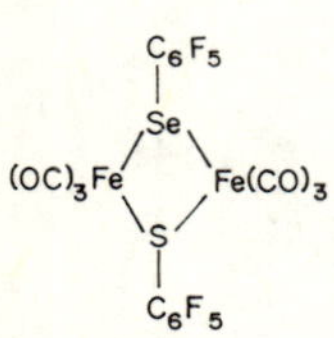	Air-stable solid Mössbauer spectrum[e]: $\delta = 0.36$ $\Delta = 1.35$	348

TABLE 27. (*Continued*)

Complex	Reported data	Reference
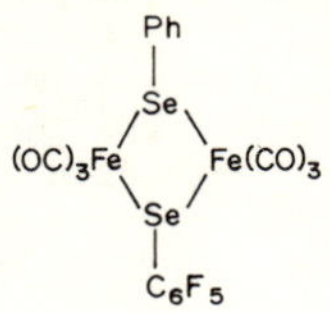	$\nu_{C\equiv O}(CS_2) = 2072, 2038, 2000$	348
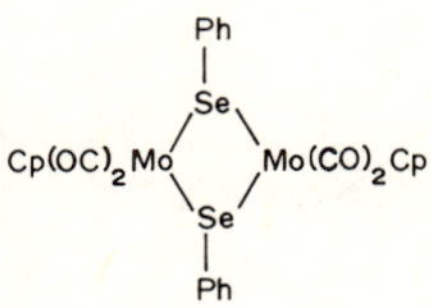	Yellow solid $\nu_{C\equiv O} = 1865, 1874$	327
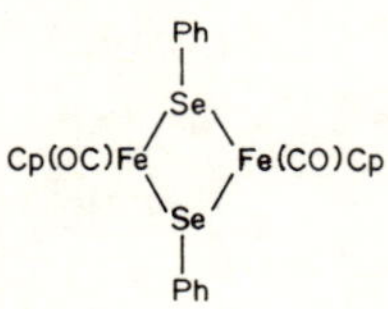	*Isomer A* gold needles m.p. 170–172 °C dec. $\nu_{C\equiv O}(C_6H_{12}) = 1975$s MW = 601[d] (calc. 610)	329
	Isomer B dark brown crystals m.p. 170–172 °C $\nu_{C\equiv O}(C_6H_{12}) = 1947$s, 1931s MW = 589[d]	329
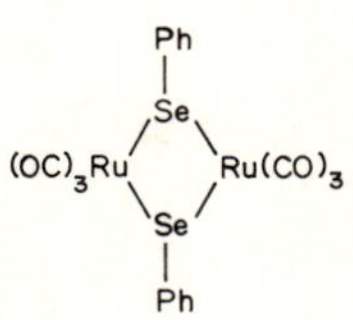	Yellow solid MW = 644[d] (calc. 682) $\nu_{C\equiv O}(CHCl_3) = 2081, 2052, 2006$	349
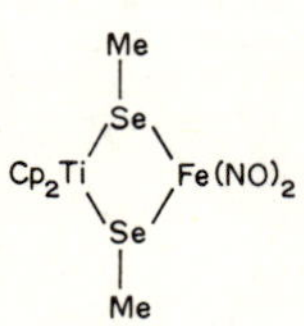	Blue solid M.p. 217–218 °C $\nu_{N=O} = 1670$s, 1715s	350
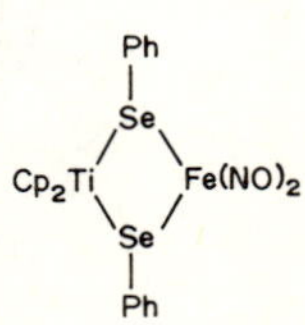	Blue solid M.p. 182–183 °C $\nu_{C\equiv O} = 1688$s, 1706s	350
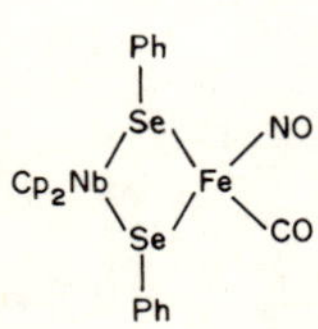	Brown solid M.p. 156–157 °C $\nu_{N=O} = 1610$s $\nu_{C\equiv O} = 1845$s	350

TABLE 27. (*Continued*)

Complex	Reported data	Reference
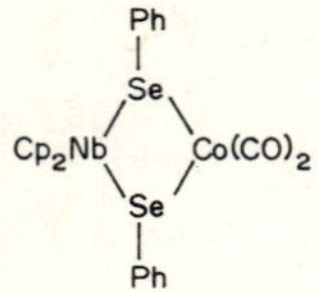	Brown solid M.p. 136–138 °C $\nu_{C\equiv O}$ = 1909s, 1850s	348
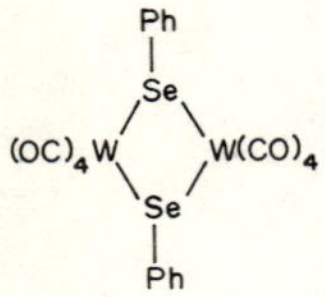	Dark green crystals M.p. 130 °C dec. $\nu_{C\equiv O}$ = 2015s, 1965s	336
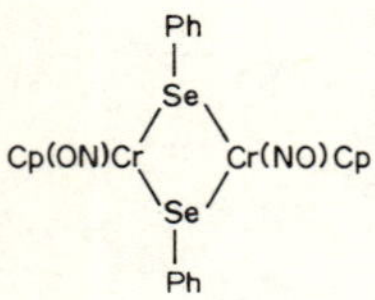	Red–brown solid M.p. 202 °C $\nu_{N=O}$ = 1646 Crystal structure	351
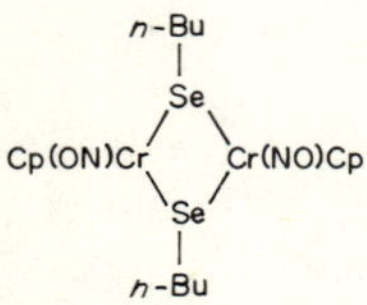	Brown crystals M.p. 134 °C $\nu_{N=O}$ = 1635vs	351
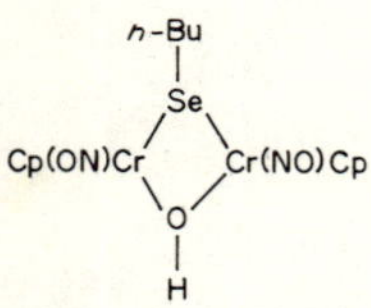	Brown crystals M.p. 125 °C $\nu_{N=O}$ = 1645s, 1612vs ν_{O-H} = 3506m Crystal structure	351
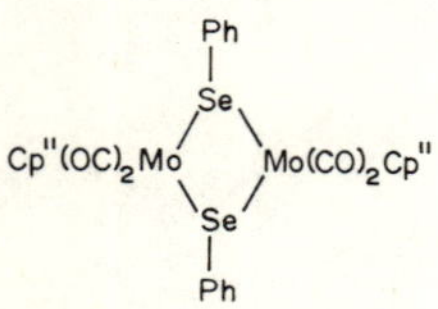	Dark brown solid	352
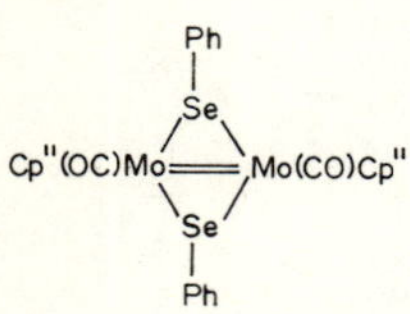	Grey–brown solid M.p. 140 °C dec. $\nu_{C\equiv O}$ = 1878s, 1850m	352

TABLE 27. (*Continued*)

Complex	Reported data	Reference
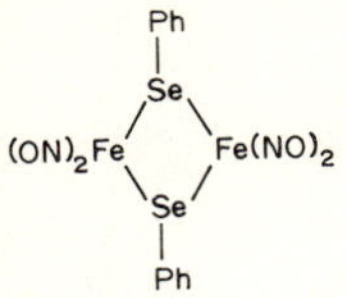	Brown crystals $\nu_{N=O}(CH_2Cl_2) = 1775, 1750$ MW = 546 (EIMS) ^{1}H-NMR	353
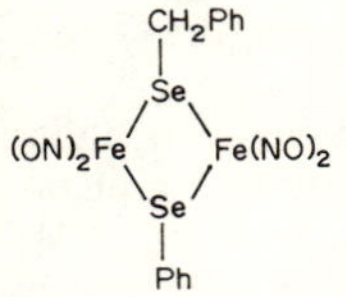	Brown crystals $\nu_{N=O}(CH_2Cl_2) = 1775, 1750$ MW = 574 (EIMS) ^{1}H-NMR	353
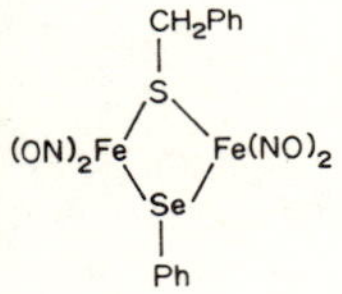	Brown crystals $\nu_{N=O}(CH_2Cl_2) = 1780, 1745$ MW = 482 (EIMS) ^{1}H-NMR	353
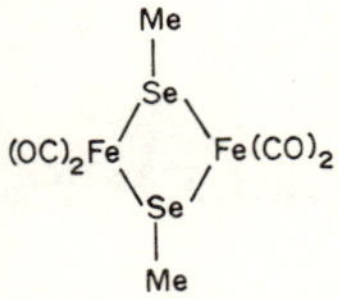	Red air-stable crystals M.p. 134–135 °C $\nu_{C\equiv O}(CHCl_3) = 2066s, 2032vs, 1990vs$ $\delta(^1H, CDCl_3) = 2.10$	354
$[Hg(SeMe)_2]_n$	Yellow crystals Crystal structure	355 356
$[HgClpyr(\mu\text{-}SeEt)]_4$	Crystal structure	356
$Hg(SePh)_2$	Pale yellow solid	357
$Hg(SeC_6F_5)_2$	M.p. 171–173 °C	348
$[Ph_4P]_3[Hg_2(SePh)_7]$	Yellow–green air-stable crystals $\nu_{Hg-Se-Hg} = 132, 140$	341
$Os_3(CO)_{10}(\mu_2\text{-}SePh)_2$ (**45**)	Yellow crystals MW = 1168 (EIMS; ^{192}Os, ^{79}Se) $\nu_{C\equiv O}(C_6H_{12}) = 2104m, 2064s, 2055s, 2050w, 2019vs, 2000w, 1988m, 1980w$	358
$[Mn(CO)_3(\mu_3\text{-}SePh)]_4$ (**42**)	Brown solid $\nu_{C\equiv O}(C_5H_{12}) = 2022vs, 1951s$ MW = 1180 (EIMS)	359

[a] Determined cryoscopically in benzene.
[b] Chemical shift values relative to internal CCl_3F.
[c] The corresponding values for free $(CF_3)_2PSeCF_3$ are: $(CF_3)_2P$—, $\delta = 53.7$ ($J_{PF} = 77.2$).
[d] Determined by osmometry in $CHCl_3$.
[e] Data in mm s^{-1}. Isomer shift relative to sodium nitroprusside.

complexes with terminal M—SeR bonds have been reported (i.e. $(\eta^7\text{-}C_7H_7)W(CO)_2SePh$, *trans*-[(Rh)(SePh)$_2$], (Rh) = **41**; Table 26).

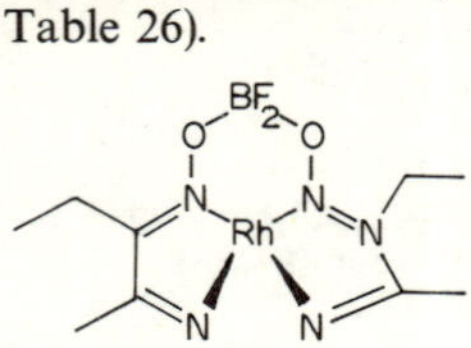

(**41**)

A number of homo- and hetero-nuclear dimeric complexes with bridging selenol ligands have been prepared and characterized by various spectroscopic techniques (Table 27).

A variety of routes have been used to synthesize complexes incorporating these ligands. The most generally used routes involve diorganodiselenides, either directly with a metal substrate in an oxidative addition (equations 95–101) or elimination (equations 102–104) reaction or via an initial reductive cleavage of the Se—Se bond and subsequent metathetical reaction (equations 105 and 106).

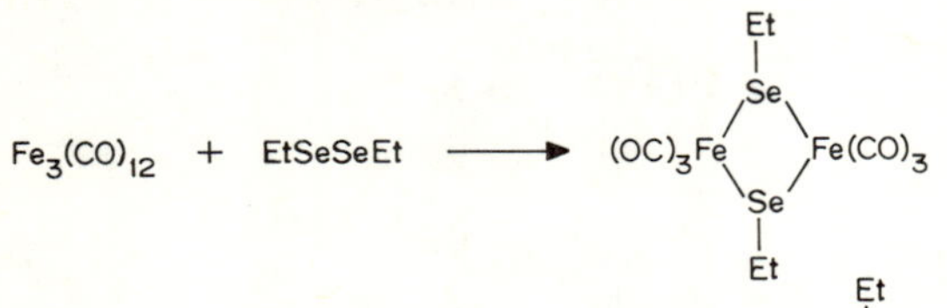

(Ref. 323) (95)

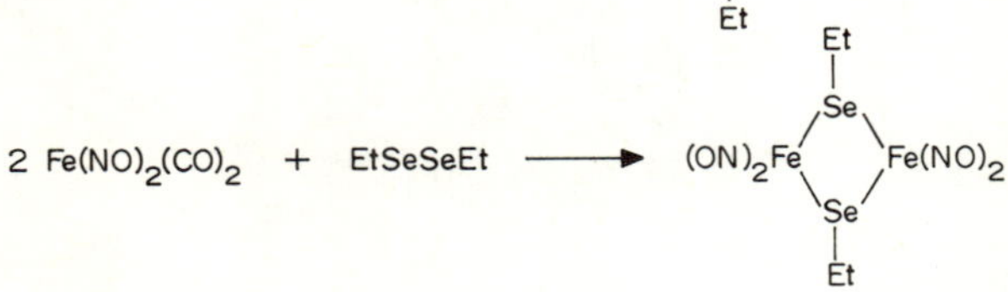

(Ref. 339) (96)

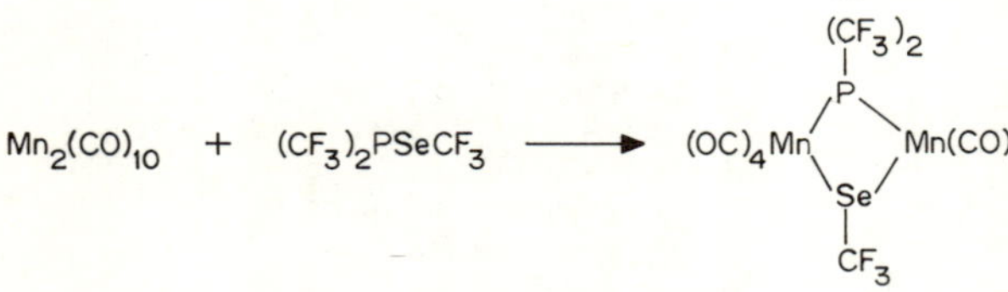

(Ref. 340) (97)

$$Mn_2(CO)_{10} + Se_2(CF_3)_2 \xrightarrow[h\nu]{CH_2Cl_2/C_6H_{14}} (OC)_4Mn(\mu\text{-}SeCF_3)_2Mn(CO)_4$$

(Ref. 360) (98)

$$Fe(CO)_5 + Se_2(CF_3)_2 \xrightarrow{CH_2Cl_2/C_6H_{12}} (OC)_3Fe(\mu\text{-}SeCF_3)_2Fe(CO)_3$$

(Ref. 360) (99)

$$Ru_3(CO)_{12} + Ph_2Se_2 \xrightarrow[60\,^{\circ}C]{C_6H_6} (OC)_3Ru(\mu\text{-}SePh)_2Ru(CO)_3 \quad \text{(Ref. 349)} \quad (100)$$

$$Hg + Ar_2Se_2 \xrightarrow[\text{(b) PhMe/r.t.}]{\text{(a) 180 °C/sealed tube}} Hg(SeAr)_2 \quad (101)$$

(a) $Ar = C_6F_5$[348]

(b) $Ar = Ph$[357]

$$2\,Mn(CO)_5H + R_2Se_2 \xrightarrow[\text{tube}]{\text{sealed}} (OC)_4Mn(\mu\text{-}SeR)_2Mn(CO)_4 + H_2 + 2\,CO \quad \text{(Ref. 344)} \quad (102)$$

$$R = Me, Et, Pr, CF_3, C_2F_5, C_3F_7$$

$$Cp_2ZrMe_2 + PhSeSePh \xrightarrow{\text{heptane}} Cp_2Zr(SePh)Me + PhSeMe \quad \text{(Ref. 338)} \quad (103)$$

$$\downarrow \text{PhSeSePh}$$

$$Cp_2Zr(SePh)_2 + PhSeMe$$

$$Cp_2ZrPh_2 + PhSeSePh \xrightarrow[\text{irradiation}]{\text{heptane/UV}} Cp_2Zr(SePh)_2 + PhPh \quad \text{(Ref. 338)} \quad (104)$$

$$[CpNi(P(Bu\text{-}n)_3)_2]^+Cl^- \xrightarrow{\text{NaSePh}} CpNi(P(Bu\text{-}n)_3)SePh \quad \text{(Ref. 330)} \quad (105)$$

$$Cp_2MCl_2 + 2\,Li^+SeR^- \rightarrow Cp_2M(SeR)_2 + 2\,LiCl \quad \text{(Ref. 331)} \quad (106)$$

$$M = Zr, Hf;\ R = Me, Ph$$

By careful control of reaction conditions, Baddley and coworkers[327,329] demonstrated that monomeric intermediates with terminally bonded selenolate ligands could be isolated (equations 107 and 108). Two isomers of the dimeric iron compound were isolated by column chromatography, differing in the relative orientations of the phenyl groups with respect to the carbonyl ligands (*cis* and *trans*).

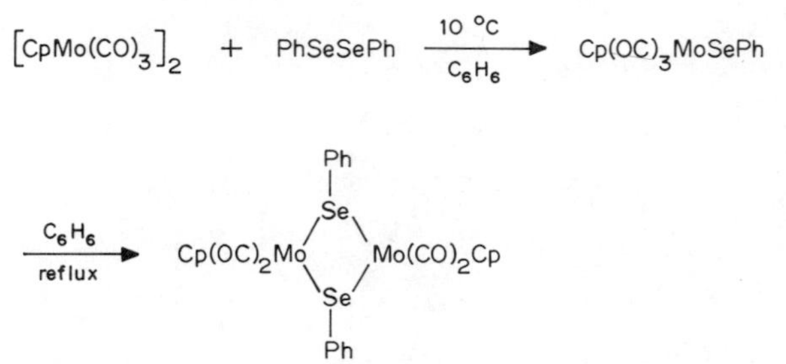

(Ref. 327) (107)

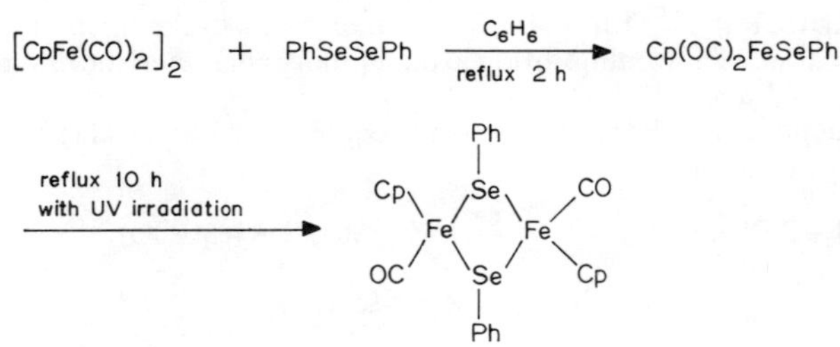

(Ref. 329) (108)

Photolysis of a toluene solution of methylcyclopentadienylmolybdenum tricarbonyl dimer and Ph_2Se_2 gave an isolable dimeric tetracarbonyl, which was further decarbonylated thermally in the solid state to give a formally double-bonded Mo=Mo complex (equation 109).

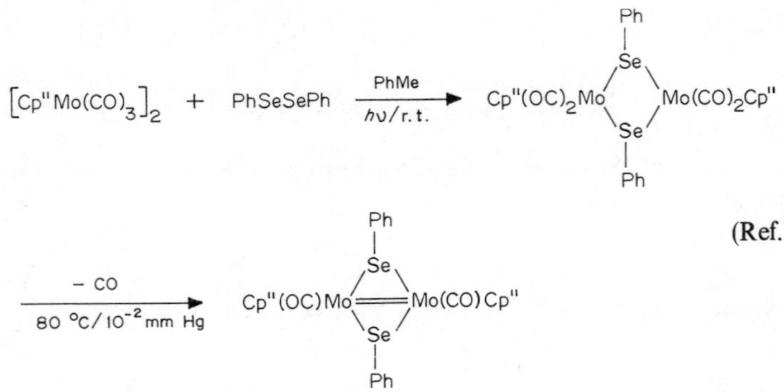

(Ref. 352) (109)

The crystal structure of the analogous complex, $[CpMo(CO)(\mu\text{-}SR)]_2$, showed that the Mo—Mo distance was shorter by ca. 1.3 Å vs. the dicarbonyl precursor, which had no Mo—Mo interaction (Mo—Mo bond distances = 2.616 Å vs. 3.940 Å)[361].

A cubane-type structure (**42**) was proposed for the product obtained by photolysis of a mixture of $Mn_2(CO)_{10}$ and Se_2Ph_2 in pentane[359].

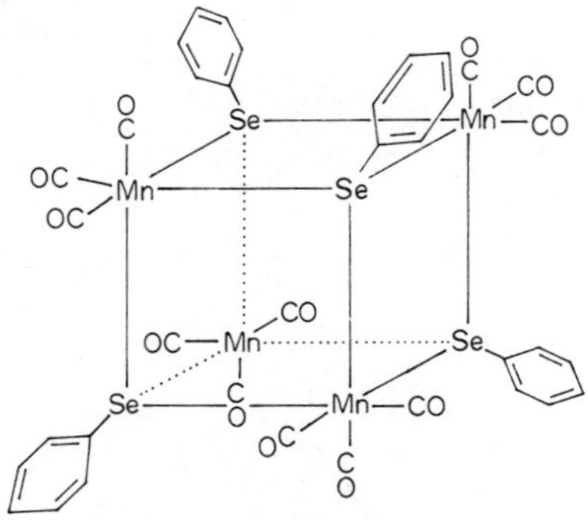

(42)

Oxidative addition of PhSeSePh to the Rh(I) complex **41** gave the octahedral Rh(III) complex *trans*-(Rh)$(SePh)_2$[334] ((Rh) = **41**). The planar macrocyclic ligand in this system precludes the formation of a selenium-bridged dimer more characteristic of such reactions (e.g. equations 95–102 and 107–109).

Elimination reactions with $RSeSnR^1_3$ reagents (equations 110 and 111) have also been used to form M—SeR bonds.

$$2\,M(CO)_5Cl + 2\,RSeSnMe_3 \xrightarrow{-2\,ClSnMe_3} (OC)_4M(\mu\text{-}SeR)_2M(CO)_4 \quad \text{(Ref. 345)} \quad (110)$$

$$M = Mn, Re;\ R = Me, Et, Ph$$

$$2(Ph_3P)_2M(CO)_3Cl + RSeSnMe_3 \xrightarrow[-2\,PPh_3]{-2\,Me_3SnCl} Ph_3P(OC)_3M(\mu\text{-}SeR)_2M(CO)_3PPh_3$$

(Ref. 339) (111)

$$M = Mn, Re;\ R = Me, Ph$$

Metathetical reactions with $Al(SePh)_3$ (equation 112) and $Hg(SeCF_3)_2$ (equation 113) have also been used to form transition-metal selenolate complexes.

$$MCl_n + Al(SePh)_3 \xrightarrow{Et_2O/C_6H_6} M(SePh)_n \qquad \text{(Ref. 328)} \qquad (112)$$

$$n = 3;\ M = Cr$$
$$n = 4;\ M = Ti,\ Zr$$

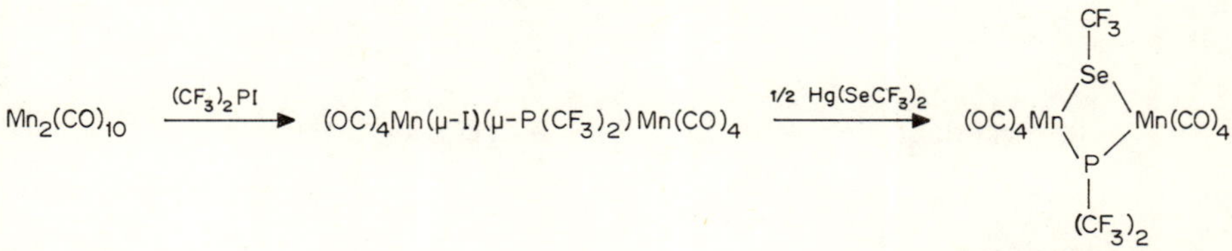

(Ref. 343) (113)

The derivatives prepared by the metathesis route in equation (112) were formulated as monomeric species on the basis of their solubility in organic solvents and molecular weight measurements in certain cases (Table 26). Similar reactions with niobium and tantalum pentachlorides reduced the metals to the +3 states[328] (equation 114).

$$5\,Al(SePh)_3 + 3\,MCl_5 \rightarrow 3\,M(SePh)_5 + 5\,AlCl_3 \qquad (114)$$
$$M = Nb,\ Ta$$
$$(3\,M(SePh)_5 \rightarrow M(SePh)_3 + PhSeSePh)$$

These complexes were also formulated as monomeric, although all of the monomers slowly polymerized in solution[328].

Reaction of metal acetylacetonates with selenophenol also provided a route to the corresponding selenolate complexes (equation 115). These products were formulated as polymeric on the basis of their insolubility in all common organic solvents, but a monomeric Ni(II) complex was formed by reaction with $P(Bu\text{-}n)_3$ (equation 116).

$$M(acac)_2 + 2\,PhSeH \xrightarrow[-2\,Hacac]{C_6H_6} M(SePh)_2 \qquad \text{(Ref. 328)} \qquad (115)$$
$$M = Ni,\ Co$$

$$\{Ni(SePh)_2\}_n \xrightarrow[PhMe/\Delta]{P(Bu\text{-}n)_3} [Ni(SePh)_2(P(Bu\text{-}n)_3)_2] \qquad \text{(Ref. 328)} \qquad (116)$$

Ziegler and coworkers prepared a number of complexes with terminal and bridging SeR^- ligands by metathetical reactions[336,351]. Included in this work is the single-crystal X-ray diffraction characterization of the monomeric complex $(\eta^7\text{-}C_7H_7)W(CO)_2SePh$[337], the dimeric complexes $Cp(ON)Cr(\mu\text{-}SePh)_2Cr(NO)Cp$ and $Cp(ON)Cr(\mu\text{-}SeBu\text{-}n)(\mu\text{-}OH)Cr(NO)Cp$[351] and the unique triply bridged dimer $(\eta^7\text{-}C_7H_7)Mo(\mu_2\text{-}SePh)_3Mo(CO)_3$[337].

The metathetical reactions between a halo complex and a SeR^- reagent generally give a mixture of products, which can be isolated by column chromatography (equation 117).

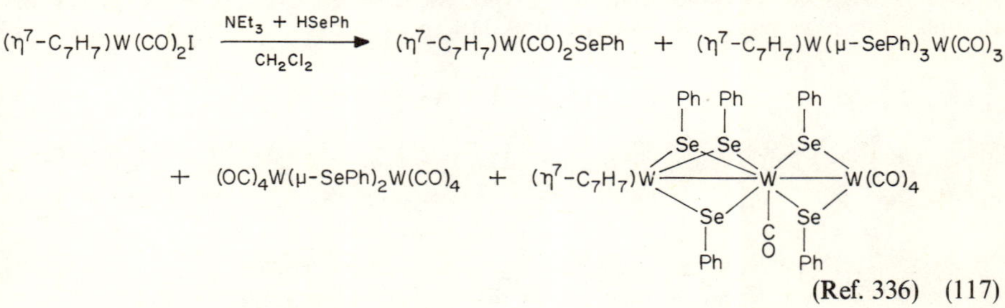

(Ref. 336) (117)

The structure of the sulphur analogue of the non-cyclic trinuclear cluster has been established by X-ray crystallography[362].

The reaction of $CpCr(NO)_2Cl$ with RSeMgBr (R = *n*-Bu, Ph) gave dimeric SeR-bridged products (equations 118 and 119).

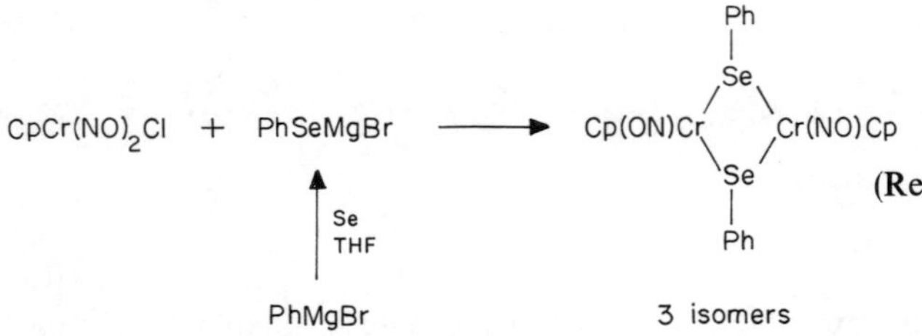

(Ref. 351) (118)

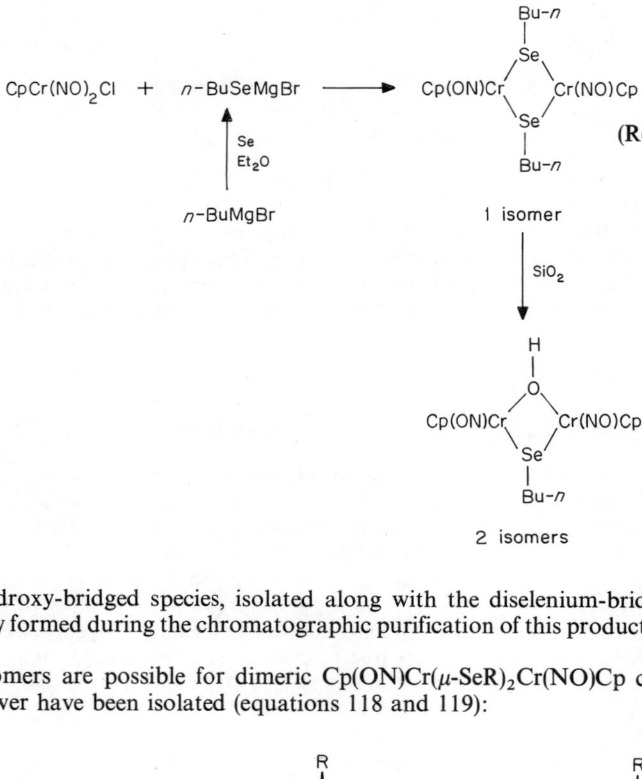

(Ref. 351) (119)

The hydroxy-bridged species, isolated along with the diselenium-bridged complex, apparently formed during the chromatographic purification of this product on a silica gel column.

Five isomers are possible for dimeric $Cp(ON)Cr(\mu\text{-}SeR)_2Cr(NO)Cp$ complexes, although fewer have been isolated (equations 118 and 119):

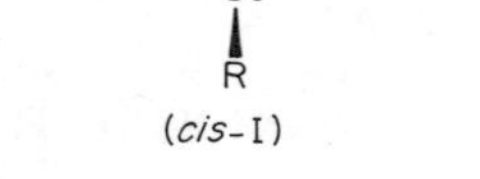

(*cis*-I)

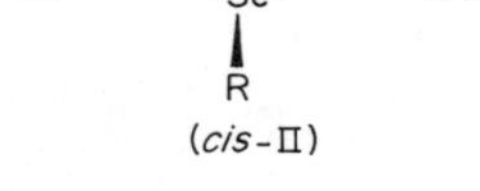

(*cis*-II)

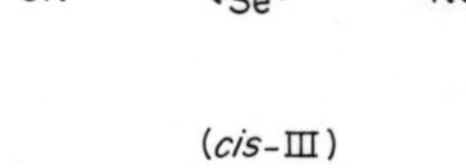

(*cis*-III)

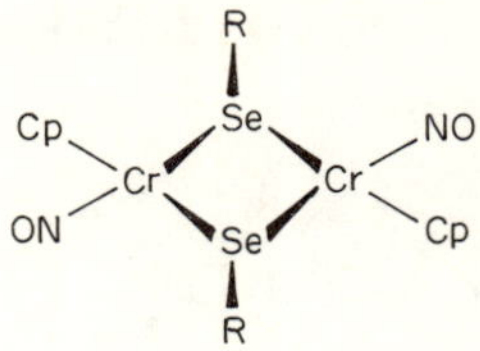

(*trans*-I)

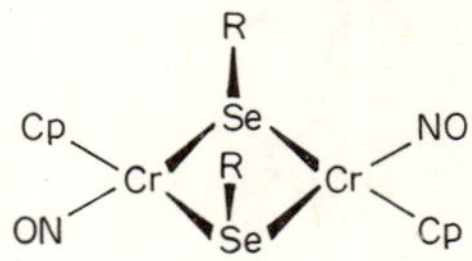

(*trans*-II)

Recently, novel routes involving alkylation of coordinated Se atoms have been described (equations 120–122).

$$(OC)_4Mn(\mu\text{-}SeH)_2Mn(CO)_4 + CH_2N_2 \rightarrow (OC)_4Mn(\mu\text{-}SeMe)_2Mn(CO)_4$$ (Ref. 363) (120)

$$(ON)_2Fe(\mu\text{-}I)_2Fe(NO)_2 + Li_2Se \rightarrow (ON)_2Fe(\mu\text{-}Se)_2Fe(NO)_2{}^{2-}$$ (Ref. 353) (121)

$$Se \xrightarrow{LiBEt_3H,\ THF} Li_2Se; \quad (ON)_2Fe(\mu\text{-}Se)_2Fe(NO)_2{}^{2-} \xrightarrow{PhCH_2Cl} (ON)_2Fe(\mu\text{-}SeCH_2Ph)_2Fe(NO)_2$$

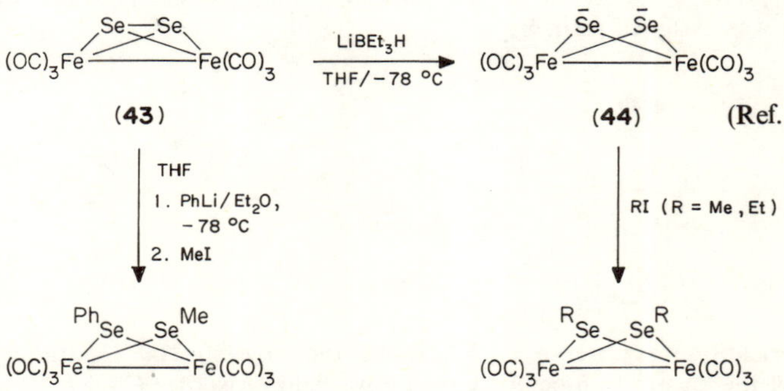

(43) (44) (Ref. 354) (122)

Röll and coworkers[250] reported a reaction involving the displacement of a carbyne ligand with formation of a dimeric complex incorporating $SePh^-$ bridges (equations 123). In this reaction the use of aqueous HCl rather than an alkylating agent enhanced the yields of the first two products (i.e. vs. $(OC)_5Cr(SePhEt)$, equation 59).

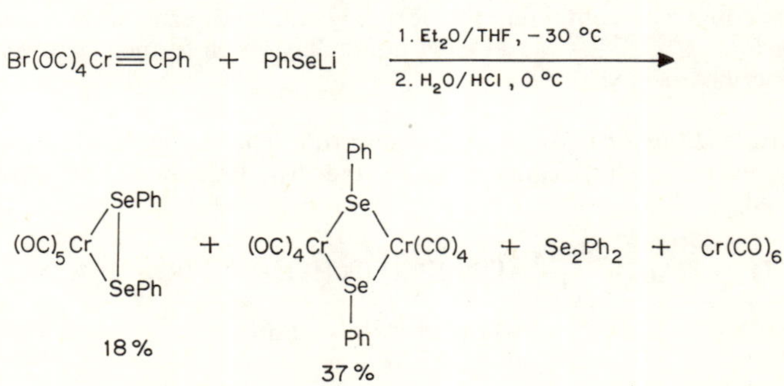

(123)

The only reported transition-metal cluster compound incorporating a bridging selenol ligand was prepared by an oxidative addition reaction[358] (equation 124). The structure **45**

$$Os_3(CO)_{11}(NCMe) \xrightarrow[CH_2Cl_2/r.t.]{Se_2Ph_2} Os_3(CO)_{10}(\mu_2\text{-}SePh)_2 \quad (124)$$

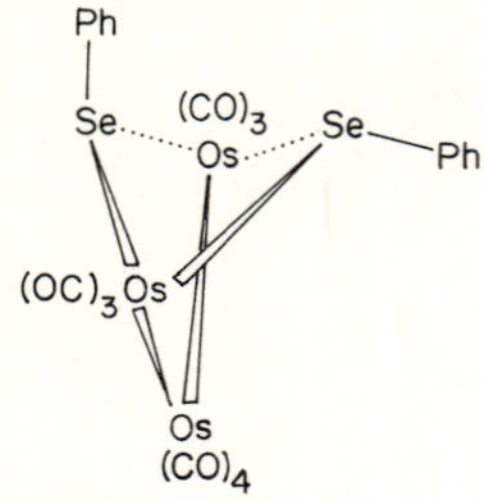

(45)

was proposed on the basis of mass spectroscopy and infrared spectroscopy. The latter was similar to that of $Os_3(CO)_{10}(\mu_2\text{-}X)_2$ (X = Cl, Br, I) which have this general structure. Three isomers, differing in the stereochemistry about the SePh bridges, are possible for this structural type, and, based on the inequivalence of the phenyl groups in its NMR spectrum, it was assigned form **B**:

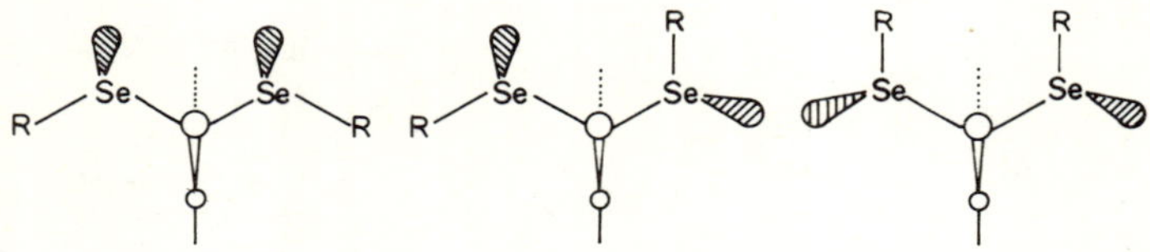

(**A**) (**B**) (**C**)

No isomerization was observed by NMR over the range 30–100 °C, although refluxing a solution of the cluster in *n*-octane gave a low yield of isomer **C** (ca. 3%) as well as a decarbonylation product, $Os_3(CO)_9(\mu_2\text{-}Se)_2$ (ca. 13%; this cluster is discussed in Section II.G.5).

3. SeH^-

Only a few complexes containing the SeH^- ligand, both terminal and bridging, have been reported[108,144,363–368]. Metal–SeH bonds have been formed by several different routes as described below.

a. Oxidative addition of H_2Se to a metal substrate. The oxidative addition of H_2Se to Pt(0) (equation 125), Pt(II) (equation 126) and Ir(I) (equation 127) complexes has been described.

$$[Pt(PPh_3)_2] \xrightarrow[C_6H_6/r.t.]{H_2Se} [Pt(PPh_3)_2(SeH)(H)] \quad \text{(Refs. 364, 365)} \quad (125)$$

white air-stable solid
m.p. 135 °C

$MW(CHCl_3) = 740$ (calc. 798)

$\nu_{Pt-H} = 2140\ cm^{-1}$

$$trans\text{-}PtH(SeH)(PEt_3)_2 \xrightarrow[CH_2Cl_2,\ -85\,^{\circ}C]{H_2Se} Pt(H)_2(SeH)_2(PEt_3)_2 \quad \text{(Ref. 366)} \qquad (126)$$

$\delta(PtH) = -13.1$ ppm
${}^1J_{PtH} = 1004$
${}^2J_{PtH} = -22$
${}^2J_{PH} = -8$
${}^3J_{HH} = 3.5$
$\delta(SeH) = -5.0$ ppm

$$trans\text{-}IrCl(CO)(PPh_3)_2 \xrightarrow[C_6H_6/r.t.]{H_2Se} [IrCl(CO)(PPh_3)_2(SeH)(H)]$$

(Ref. 367) (127)

	air-sensitive solid
	m.p. dec. > 160 °C
$\nu_{C\equiv O} = 1950$vs	2018vs
$\nu_{Ir\ \ Cl} = 325$vs	262s

The ^{1}H-NMR of $[Pt(PPh_3)_2(SeH)(H)]$ (equation 125) in $CDCl_3$ at room temperature has been interpreted in terms of the following equilibrium[365]:

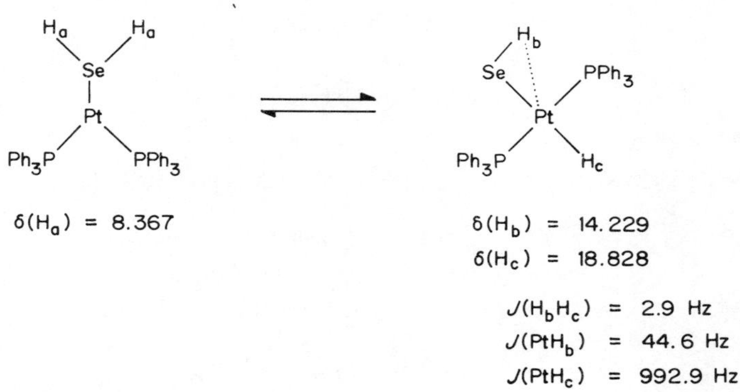

On the basis of detailed analysis of the ^{1}H-NMR spectra and D_2O exchange experiments, the broadening of the resonances as the temperature was decreased was attributed to rapid dissociation of the phosphine ligands rather than exchange of H_b and H_c. With the H_2S analogue, the H_b and H_c resonances split into triplets at -80 °C (J (PH$_b$) = 10 Hz; J(PH$_c$) = 11 Hz). The latter coupling constants also indicate that the two phosphine ligands are mutually *trans*[141].

The reaction described in equation (126) was carried out in an NMR tube at low temperature. On warming the tube to room temperature, H_2 was evolved with the formation of *trans*-$Pt(SeH)_2(PEt_3)_2$. Multinuclear NMR studies (^{1}H, ^{31}P, ^{77}Se, ^{195}Pt) of the Pt(IV) complex indicated a *cis, cis, trans* stereochemistry. Similar reactions between H_2Se and *trans*-$Pt(H)X(PEt_3)_2$ (X = Cl, Br, I) gave the expected Pt(IV) complexes as well as the two redistribution products, on the basis of detailed multinuclear NMR studies (equation 128).

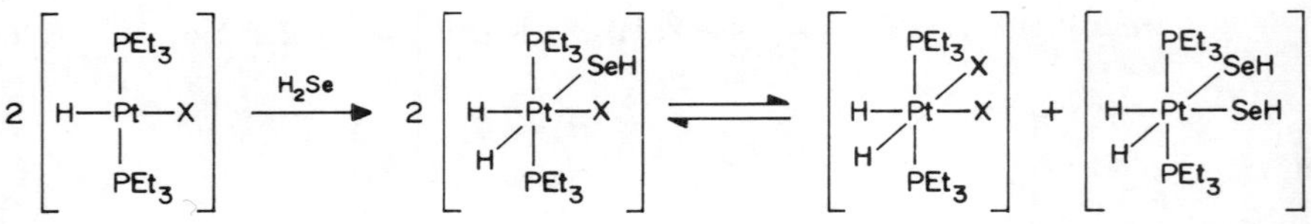

(Ref. 366) (128)

b. Metathetical reactions between a metal halide complex and SeH^-:

$$trans\text{-}PtCl(H)(PEt_3)_2 \xrightarrow[EtOH]{SeH^-} trans\text{-}Pt(H)(SeH)(PEt_3)_2 \quad \text{(Ref. 366)} \quad (129)$$

pale yellow crystals

$\nu_{Pt-H} = 2110$; $\nu_{Se-H} = 2288$w; $\nu_{Pt-Se} = 194$

$\delta(^1H)$:

Pt—H, -10.7 ($^1J_{PtH} = 1108$; $^3J_{HH} = 3$

$^2J_{PH} = -14$)

Se—H, -4.5 ($^2J_{PtH} = -44$; $^3J_{HH} = 3$;

$^3J_{PH} = 13$)

$$trans\text{-}PtCl_2(PEt_3)_2 \xrightarrow[EtOH]{2SeH^-} trans\text{-}Pt(SeH)_2(PEt_3)_2 \quad \text{(Ref. 366)} \quad (130)$$

yellow crystals

$\delta(^1H)$:

SeH, -4.0 ($^2J_{PtH} = -56$;

$^3J_{PH} = 11$)

$$PPN[Cr(CO)_5Cl] \xrightarrow[EtOH]{Na_2Se} PPN[Cr(CO)_5SeH] \quad \text{(Ref. 105)} \quad (131)$$

slightly air-sensitive yellow solid

m.p. 94 °C dec.

$\nu_{C\equiv O}$(THF) = 2035vw, 1904vs, 1850m

$\delta(^1H)$, d_6-DMSO: SeH, -7.47s;

PPN, 7.79m

The complexes $PtH(SeH)(PEt_3)_2$[366] and $Pt(SeH)_2(PEt_3)_2$[366] are very air-sensitive, in contrast to the air-stable $PtH(SeH)(PPh_3)_2$[365]. A *trans* stereochemistry of $PtH(SeH)(PEt_3)_2$ was established by multinuclear NMR spectroscopy. The ^1H spectrum gave a complex pattern characteristic of mutually *trans* PEt_3 groups in square-planar Pt(II) complexes and two triplets each showing coupling to ^{195}Pt (equation 129). In addition, the ^{31}P, $\delta = 19.3$ (vs. external 85% H_3PO_4), ^{195}Pt, $\delta = -205$ (vs. external 0.5M *trans*-$PtCl(H)(PEt_3)_2$ in CH_2Cl_2), $^1J_{PtP} = 2654$, and the ^{77}Se spectra, $\delta = -312$ (vs. 10% $SeMe_2$ in CH_2Cl_2), $^1J_{PtSe} = 322$, $^1J_{SeH}$ not observed, $^2J_{PSe} = 9$, $^2J_{SeH} = 50$, were recorded[366]. Similar data were recorded for *trans*-$Pt(SeH)_2(PEt_3)_2$, but a Se—H coupling constant was measured ($\delta(^{77}Se) = -232$, $^1J_{SeH} = 116$, $^1J_{PtSe} = 166$, $^2J_{PSe} = 12$).

The derivatives $M(CO)_5SeH^-$ (M = Cr, Mo, W) (equation 131) were also obtained by reactions of the $M(CO)_5THF$ derivatives, prepared by photolysis of the hexacarbonyls in THF, with ethanol solutions of Na^+SeH^- followed by precipitation of the anionic complex with PPN^+ or Ph_4As^+. In these reactions addition of ethanol to M_2Se (M = Li, Na, K) results in solvolysis to M^+SeH^-. This latter species can also be formed directly by borohydride reduction of elemental Se in ethanol[369]. These anionic complexes were also formed directly, thermally or photochemically, from solutions of $M(CO)_6$/NaHSe in THF/EtOH[108]. Two reactions of these monomeric complexes were investigated: (*i*) alkylation of the selenol ligand to the neutral diorganoselenide ligand (equation 15) and (*ii*) reaction with the labile $M(CO)_5THF$ species to give selenol-bridged dimers (equation 132).

$$X^+[(OC)_5MSeH]^- + M(CO)_5(THF) \rightarrow X^+[(OC)_5M(\mu\text{-SeH})M(CO)_5] \qquad \text{(Ref. 108)} \quad (132)$$

$$\uparrow h\nu \quad \text{from } M(CO)_6$$

$$M = Cr,\ X = PPN$$
$$M = W,\ X = AsPh_4$$

c. Protonation of coordinated selenium. The reactivity of coordinated chalcogen atoms and rings has been an area of considerable recent interest in organometallic chemistry (see Section II.G), and two examples from this work illustrate the transformation of coordinated Se to the SeH^- ligand. Seyferth and Henderson[354] in their study of the chemistry of $Fe_2(CO)_6(\mu_2,\eta^2\text{-}Se_2)$ (**43**) formed the $Fe_2(CO)_6(\mu_2\text{-}SeH)_2$ dimer in a two-step sequence (equation 133). This product was not isolated but reacted *in situ* with MeI and Et_3N to give $(OC)_3Fe(\mu_2\text{-}SeMe)_2Fe(CO)_3$, a compound which can also be formed by the more direct reaction of **44** with MeI[354].

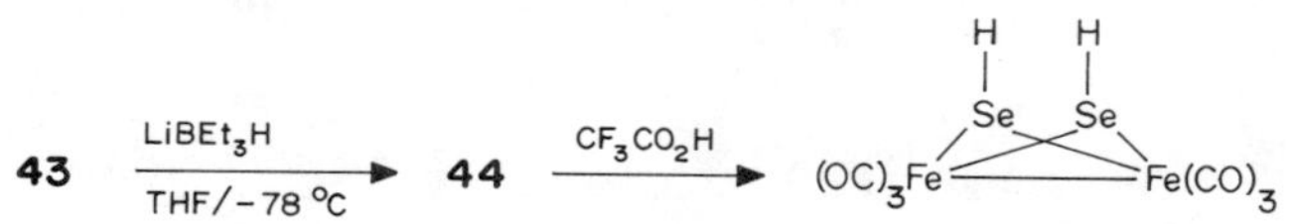

(Ref. 354) (133)

A reaction sequence involving insertion of Se into a metal–metal bond followed by protonation has been used to form a selenol-bridged dimer (equation 134).

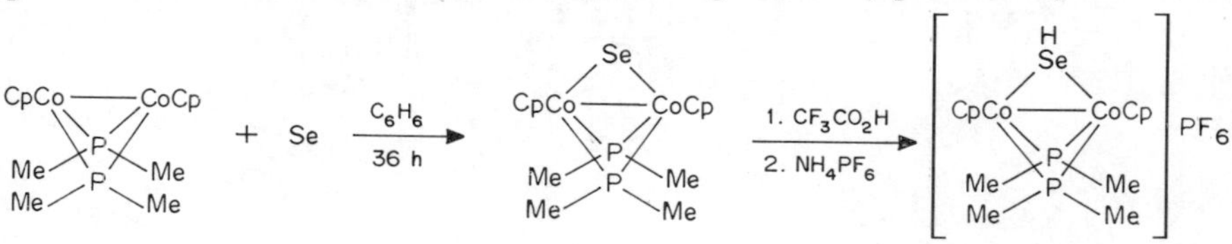

red-brown air-stable salt

(Ref. 368) (134)

d. Elimination reactions. Finally, an elimination reaction involving the bridging —$SeSnMe_3$ ligand has provided a route to the corresponding air-sensitive selenol-bridged dimer (equation 135).

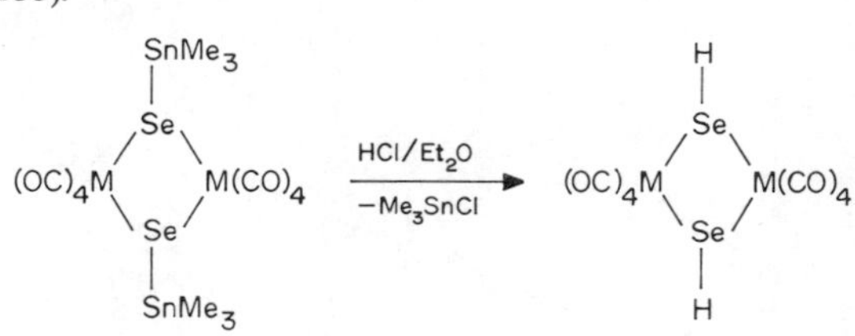

(Ref. 363) (135)

	M = Mn	M = Re
ν_{Se-H} =	2285s, 2263m, 2223s	2280s, 2235s
^{1}H-NMR $\delta(CH_2Cl_2)$:	−6.20 (SeH)	−4.13

These dimers can be *Se*-alkylated to give the corresponding alkylselenol-bridged dimers (equation 120).

4. $SeER_3^-$ (E = Ge, R = Ph; E = Sn, R = Me)

Only a few examples of complexes incorporating ligands of this class, both terminally bonded and bridging, have been reported[144,335,363,370]. Dimeric —$SeSnMe_3$-bridged complexes were prepared by elimination reactions with metal carbonyl halides (equations 136 and 137). These air-sensitive brownish-red crystalline solids were characterized by infrared and ^{1}H-NMR spectroscopy. Reactions for the conversion of the bridging group to —SeH (equation 135) and —$SeEMe_2$ (E = P, As) (equation 21) have been described above. Heating the dimers in refluxing hexane gave high yields of the tetrameric $[Me_3SnSeM(CO)_3]_4$ compounds, which were formulated as cubane-type clusters[363] (e.g. see structure **42**).

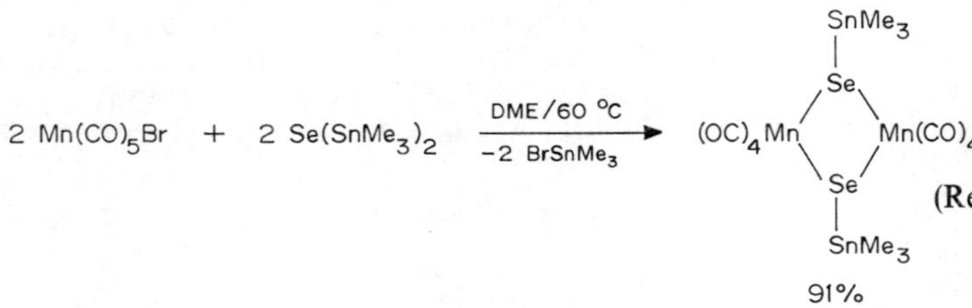

(Ref. 363) (136)

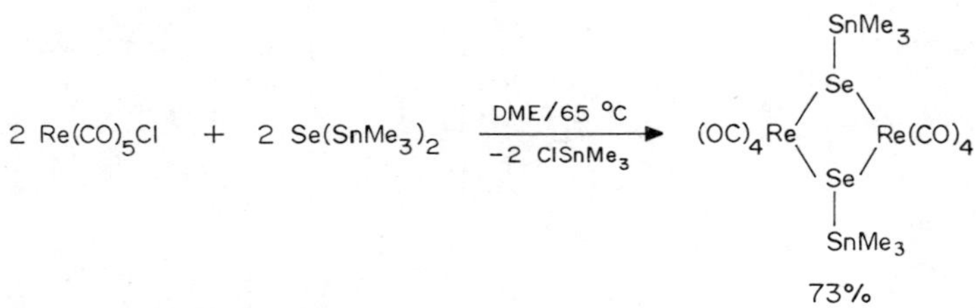

(137)

The reaction of $[(OC)_4ReSeSnMe_3]_2$ with PMe_3 in benzene at room temperature gave $(Me_3P)_2(OC)_3Re\text{-}Se\text{-}Re(CO)_3(PMe_3)_2$, which was characterized by single-crystal X-ray diffraction[370] (Figure 4).

The only reported example of a ligand of this class terminally bonded to a metal was prepared by oxidative addition of a Rh(I) substrate with $Ph_3GeSeSeGePh_3$ or by an equivalent metathetical reaction (equation 138).

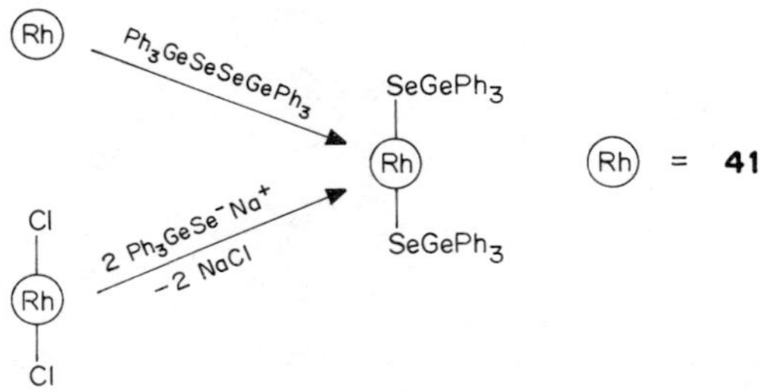

(Ref. 335) (138)

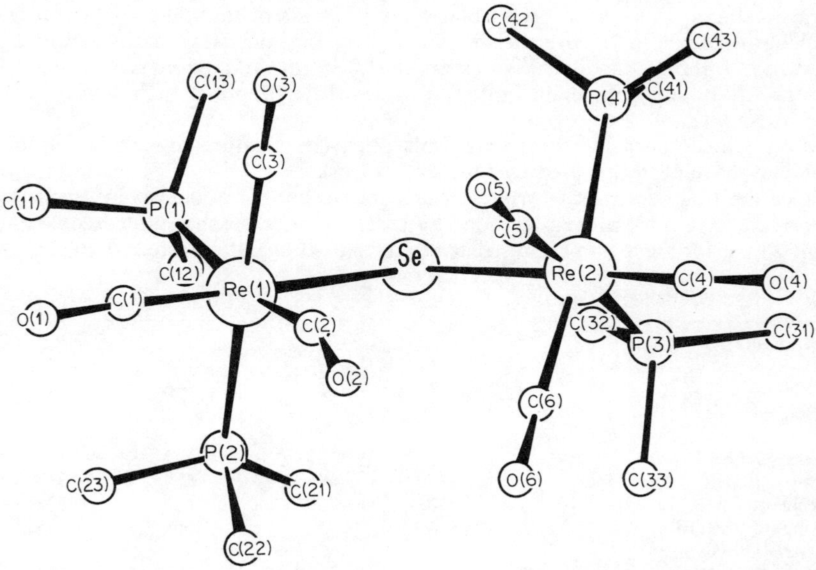

FIGURE 4. Molecular structure of $(Me_3P)_2(OC)_3Re—Se—Re(CO)_3(PMe_3)_2$. Reproduced with permission from Ref. 370

The bis(triphenylgermanium) diselenide was prepared as shown in equation (139)[335].

$$2\,Na_2Se + 2\,Ph_3GeBr \xrightarrow[C_6H_6]{EtOH} 2\,NaSeGePh_3 \xrightarrow[C_6H_6]{MeSO_2Cl} (Ph_3Ge)_2Se_2 \quad (139)$$

5. Organoseleninates ($RSeO_2^-$)

A variety of organoseleninate metal complexes have been prepared by metathetical reactions between a metal chloride salt or complex and a sodium organoseleninate[371–388]. In two cases methylseleninate complexes have been formed by insertion of SeO_2 into a metal–Me bond ($CpFe(CO)_2Me$[389] and η^7-$C_7H_7Mo(CO)_2Me$[390]). No

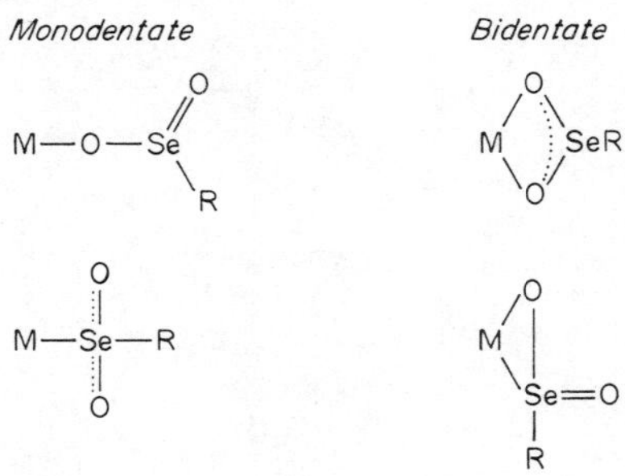

linkage isomeric pairs have been isolated with ligands of this class, as in the case of sulphinato ligands[153e]; nevertheless, they have an interesting ambidentate ligand chemistry. Although no definitive single-crystal X-ray diffraction studies have been reported, the bonding modes illustrated above have been proposed on the basis of infrared spectroscopy (Table 28).

Many transition-metal complexes with phenylseleninato and substituted phenyl analogues have been prepared by Preti and coworkers[371–387].

Since the ring substituents, while causing relatively minor changes in spectroscopic properties, have not changed the bonding mode in any case, the spectroscopic data in Table 28 are, for the most part, restricted to the parent ligand. Infrared spectral changes,

TABLE 28. Infrared spectral data (cm^{-1}) for seleninate complexes[a]

Compound	$\nu_{Se-O,sym}$	$\nu_{Se-O,asym}$	ν_{Se-C}	Ref.
$Na^+SeO_2Ph^-$	768	781	666	371
Monodentate O bonding				
		($\nu_{Se-O-M,asym}$)		
$Cd(OSe(O)Ph)_2$[b]	815	725	675	374
$Co(bipy)_2(OSe(O)Ph)_2 \cdot 3H_2O$[c]	810	698	659	378
$[Ni(phen)_2(OSe(O)Ph)_2]$[d]	850	700	650	378
$Zn(phen)_2(OSe(O)Ph)_2$	866	752	668	386
Monodentate Se bonding				
$[Pt(SeO_2Ph)_2(H_2O)_2]$	730	830	680	371
$Hg(SeO_2—C_6H_4Cl$-*m*$)_2$	830[e]	860[f]	655	374
$Cp(OC)_2FeSeO_2Me$	728	859	570	389
$(\eta^7-C_7H_7)Mo(CO)_2SeO_2Me$	970	887	565	390
Bidentate (O, O) bonding				
$[Cu(O_2SePh)_2]$	725	745	670	372
$[Co(O_2SePh)_2(OH_2)_2]$	690	740	670	371, 377
$[Ni(O_2SePh)_2(OH_2)_2]$	680	750	670	371, 377
$[Rh(O_2SePh)_3] \cdot 2H_2O$	710, 690	740	670	373
$[Ir(O_2SePh)_3] \cdot 2H_2O$	700, 690	740	670	373
$[Zn(O_2SePh)_2(H_2O)_2]$	690	778	670	374
$[Cd(O_2Se(C_6H_4Cl$-*m*$))_2(OH_2)_2]$	730	790	655	374
$[Hg(O_2Se(C_6H_4Cl$-*m*$))_2(OH_2)_2]$	760	790	655	374
$Cr(O_2Se(C_6H_4Br$-*p*$))_3$[g]	726	795, 776	708	381
$Mn(O_2SePh)_2$[h]	735	751	667	381
$Fe(O_2Se(C_6H_4Br$-*p*$))_3$[i]	—	785, 728	708	381
$V(O_2SePh)_2$[j]	714	778	680	388
$Cr(O_2SePh)_3$[k]	726	802	674	388
$Fe(O_2SePh)_3$[l]	723	807	673	388
$Ti(O_2SePh)_4$[m]	738, 724	803, 795	675	388
$Ru(O_2SePh)_3$	729, 700	765	680	383
Bidentate (Se, O) bonding				
$[IrH_2(PPh_3)_2(OSe(O)Ph)]$	865	720	640	375

[a] Measured as KBr discs.
[b] $\nu_{Cd-O} = 450\ cm^{-1}$.
[c] $\nu_{Co-O} = 443\ cm^{-1}$.
[d] $\nu_{Ni-O} = 432\ cm^{-1}$.
[e] The corresponding value for the ionic sodium salt is $795\ cm^{-1}$ [376].
[f] The corresponding value for the ionic sodium salt is $817\ cm^{-1}$.
[g] $\nu_{Cr-O} = 530\ cm^{-1}$.
[h] $\nu_{Mn-O} = 420\ cm^{-1}$.
[i] $\nu_{Fe-O} = 420\ cm^{-1}$.
[j] $\nu_{V-O} = 540\ cm^{-1}$.
[k] $\nu_{Cr-O} = 545\ cm^{-1}$.
[l] $\nu_{Fe-O} = 508\ cm^{-1}$.
[m] $\nu_{Ti-O} = 434\ cm^{-1}$.

TABLE 29. Infrared spectral characterization of organoseleninato ligand bonding modes

Bonding mode	$\nu_{Se-O,sym}$ [a]	$\nu_{Se-O,asym}$ [a]	Δ [b]	ν_{Se-C}
M—Se	Decrease	Increase	80–90	Little change
M—O	Increase	Decrease	80–90	Little change
M—(O, O)	Decrease	Decrease	30–88	Little change
M—(Se, O)	Increase	Decrease	ca. 150	Little change

[a] Shifts vs. the ionic Na salts of the ligands.
[b] $\nu_{Se-O,asym} - \nu_{Se-O,sym}$ of the complexes.

compared to the Na salt of the ligand, have been established as criteria of the bonding mode of the ambidentate organoseleninato ligand (Table 29).

In some cases the proposed bonding mode was further supported by the measurement of a metal–O or metal–Se vibration (Table 28). The insolubility of some of the simple complexes (e.g. Cu[372], Co[371], Ni[371]) suggests polymeric structures involving O,O coordination via bridging ligands rather than intramolecular chelation. For $M(O_2SeAr)_3$ (M = Rh[373], Ir[373], Ru[383], Cr[381], Fe[381]), however, the solubility in alcohols suggested monomeric formulations.

The removal of water from the coordination sphere of the Cd(II) and Hg(II) complexes, $[M(O_2SePh)_2(H_2O)_2]$, changes the bonding mode of the seleninato ligand (to monodentate O bonding in the Cd(II) complex and to monodentate Se in the Hg(II) complex)[374]. With the Zn(II) analogue, however, the O,O bonding mode is retained in the anhydrous compound[374].

The first example of SeO_2 insertion into a metal–alkyl bond involved the facile reaction of freshly sublimed SeO_2 with a benzene solution of a Fe(I) complex[389] (equation 140).

$$CpFe(CO)_2Me + SeO_2 \rightarrow CpFe(CO)_2SeO_2Me + Fe(O_2SeMe)_3 + [CpFe(CO)_2]_2 \quad (140)$$

brown, moderately air-stable solid
MW (CH_2Cl_2): 300
EIMS: $[M]^{+\cdot} = 302, 304$
^{1}H-NMR $\delta(CD_3OD)$: 3.02 (s, Me), 5.05 (s, Cp)

The Me chemical shift, the downfield shift of the Cp resonance vs. the starting Me compound, and the $\nu_{C\equiv O}$ values (2056, 2009; ca. 50 cm^{-1} higher energy than the precursor Me compound) all support the presence of a —SeO_2Me ligand in this product. The infrared spectrum in the 700–900 cm^{-1} region (Table 28) and the independent synthesis of the same complex by a metathetical reaction between $CpFe(CO)_2Cl$ and $NaSeO_2Me$ further confirm the proposed insertion reaction.

A similar SeO_2 insertion reaction, with activated SeO_2 generated in a metal atom reactor, has been reported with a molybdenum alkyl[390] (equation 141).

$$(\eta^7\text{-}C_7H_7)Mo(CO)_2Me + SeO_2 \xrightarrow[\text{8 days}]{\text{THF}, -78\,^\circ\text{C}} (\eta^7\text{-}C_7H_7)Mo(CO)_2SeO_2Me \quad (141)$$

30%
red–brown
$\nu_{C\equiv O}(CH_2Cl_2) = 2000$vs, 1940vs cm^{-1}
ν_{Se-C}(mull) = 565vs cm^{-1}
ν_{Se-O}(mull) = 1070s, 887s, 790s cm^{-1}
^{1}H-NMR $\delta(CDCl_3)$: 5.28 (s, C_7H_7), 2.32 (s, Me)

The Me proton signal in the SeO_2 insertion product is significantly shifted downfield (−0.25 ppm) compared to the precursor Me complex. The product, which was purified by medium-pressure gradient chromatography on SiO_2 at −20 °C, was stable in an inert atmosphere for a few days at −78 °C but decomposed in a few minutes at room temperature. An analogous reaction with $(\eta^7\text{-}C_7H_7)Mo(CO)_2Ph$ failed to give SeO_2 insertion[390].

6. Dialkylselenocarbamates ($—SeC(O)NR_2^-$)

Although the dialkylselenocarbamate hybrid ligand system can bond in a (Se, O) bidentate coordination mode (Section II.D.2a), several examples of monodentate Se coordination have been described[391–400]. Ni(II)[391], Pd(II)[392,393] and Pt(II)[392,393] complexes were prepared by metathetical reactions with selenocarbamate salt generated *in situ* (equation 142).

$$L_2MCl_2 + 2\,R_2NH_2[SeC(O)NR_2] \xrightarrow{C_6H_6} L_2M(SeC(O)NR_2)_2 \qquad (142)$$

$$\begin{array}{lcl} M = Pd,\ Pt & \uparrow \text{CO/THF}^{401} & R = Me,\ Et,\ n\text{-Pr},\ n\text{-Bu};\ L = PPh_3 \\ & Se + R_2NH & R = Me;\ L = PPh_2Me \end{array}$$

As would be expected on the basis of the stoichiometry of the Pd(II) and Pt(II) complexes and the well-established Class B Lewis acid character of these metals[309], the complexes are square-planar with monodentate Se-bonded selenocarbamate. The infrared spectra of these complexes in the solid state show $\nu_{C=O}$ absorptions at ca. 1600 cm^{-1}, characteristic of this bonding mode with a free carbonyl group (Table 30).

TABLE 30. Infrared data (cm^{-1}) for complexes with monodentate Se-bonded dialkylselenocarbamate ligands

Complex	$\nu_{C=O}$	Reference
$(Ph_3P)_2Ni(SeC(O)NEt_2)_2$	1581s (1598m, 1521s)[a]	391
$(Ph_3P)_2Pd(SeC(O)NBu_2)_2$	1600s (1592s, 1532m)[a]	392, 393
$(Ph_3P)_2Pt(SeC(O)NBu_2)_2$	1603s (1595s, 1540sh)[a]	392, 393
$(Ph_2MeP)_2Pd(SeC(O)NMe_2)_2$	1599s (1592s, 1540vw)[a]	393
$(Ph_2MeP)_2Pt(SeC(O)NMe_2)_2$	1603s (1603s)[a]	393
$Me_3SnSeC(O)NMe_2$	1620	394
$(Ph_3P)AuSeC(O)NC_5H_{10}$[b]	1605, 1595[c]	395
$(Ph_3P)AuSeC(O)NMe_2$	1605, 1595[c]	395
$(Ph_3P)_2Rh(CO)SeC(O)NC_5H_{10}$[b]	1585	395
$Mn(CO)_5SeC(O)NMe_2$	1623	396
$[Mn(CO)_4SeC(O)NMe_2]_2$ (**48a**)	1660	397
$Re(CO)_5SeC(O)NMe_2$	1621	398
$[Re(CO)_4SeC(O)NMe_2]_2$ (**48b**)	1665	398
$CpFe(CO)_2SeC(O)NMe_2$	1600	399
$CpMo(CO)_3SeC(O)NMe_2$	1618	400

[a] CH_2Cl_2 solution.
[b] NC_5H_{10} = piperidyl.
[c] Coupled (C⋯O) and (C⋯N) stretching modes.

In CH_2Cl_2 solution, molecular weight, infrared and NMR studies[392,393] have shown that the PPh_3 complexes undergo dissociation of the phosphine ligand and chelation of the selenocarbamate (equation 143). This dissociation reaction occurs to a greater extent for Pd(II) than Pt(II), as expected on the basis of the greater kinetic stability of the latter complexes, whereas the complexes with the more basic phosphine, PPh_2Me, show essentially no dissociation in solution, as evidenced by molecular weight measurements and the presence of only one $\nu_{C=O}$ band due to uncoordinated carbonyl in both the solid state and solution (Table 30).

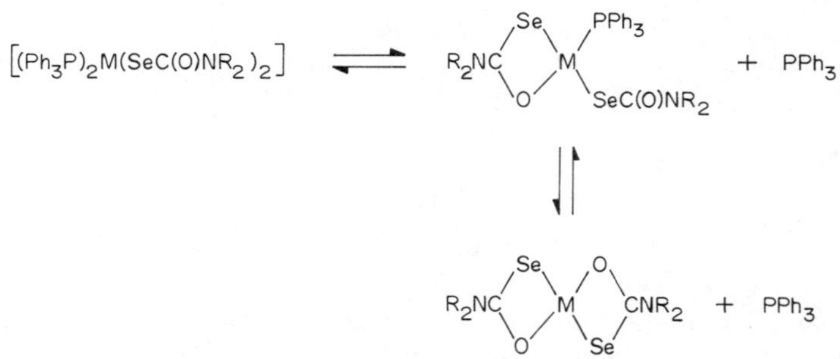

(143)

The exchange process was confirmed by ^{31}P-NMR for $(Ph_3P)_2Pd(SeC(O)NBu_2)_2$, which gave a broad resonance, whereas the other three complexes that were soluble enough for such measurements all gave sharp singlets that did not change with added phosphine [$(Ph_3P)_2Pt(SeC(O)NBu_2)_2$; $(Ph_2MeP)_2M(SeC(O)NMe_2)_2$, M = Pd, Pt]. The ^{31}P–^{195}Pt coupling constants for the Pt compounds indicated a *trans* geometry (2860 and 2713 Hz).

^1H-NMR spectroscopy established free rotation about the C—N bond at room temperature for $(Ph_2MeP)_2M(SeC(O)NMe_2)_2$, but at lower temperature the single broad N—Me resonances were split into two signals (Pd: $\delta = 2.56$ (r.t.), 2.57, 2.46 (−15 °C); Pt: $\delta = 2.57$ (r.t.), 2.56, 2.49 (−15 °C)[393]).

The diamagnetism of the Ni(II) complex, $Ni(PPh_3)_2(SeC(O)NEt_2)_2$, and its infrared spectrum (Table 30) indicated that the complex was square planar with two Se-bonded selenocarbamate ligands[391]. The solution infrared spectrum and low molecular weight in CH_2Cl_2 (439 vs. calculated value of 941) indicated significant dissociation as in the Pd(II) and Pt(II) complexes (equation 143). Treatment of a THF solution of this complex with CO resulted in facile displacement of the selenocarbamate ligand and formation of $Ni(PPh_3)_2(CO)_2$ in ca. 70% yield[391].

Monodentate Se bonding of the selenocarbamate in $Me_3SnSeC(O)NMe_2$[394], prepared by a metathetical reaction from Me_3SnCl, was indicated by its infrared spectrum (Table 30). Related compounds, $Me_2Sn(OSeCNR_2)_2$ and $Me_2ClSn(OSeCNMe_2)$, contain (Se, O) bidentate selenocarbamate (Section II.D.2.a). A tetrahedral structure with monodentate dialkyldithiocarbamate has been established by single-crystal X-ray diffraction of $Me_3SnSC(S)NMe_2$[402].

In contrast to the Pd(II)[393], Pt(II)[393], Au(I)[395] and Rh(I)[395] complexes, restricted rotation about the C—N bond occurs in this Sn complex at room temperature ($\delta(CCl_4)$, NMe_2: 3.28, 3.09)[394].

Manganese[396,397] and rhenium[398] pentacarbonyl complexes containing monodentate Se-bonded selenocarbamate (**46**) undergo successive solid-state thermal decarbonylation reactions to give dimeric complexes with more highly coordinated forms of this ligand (equation 144).

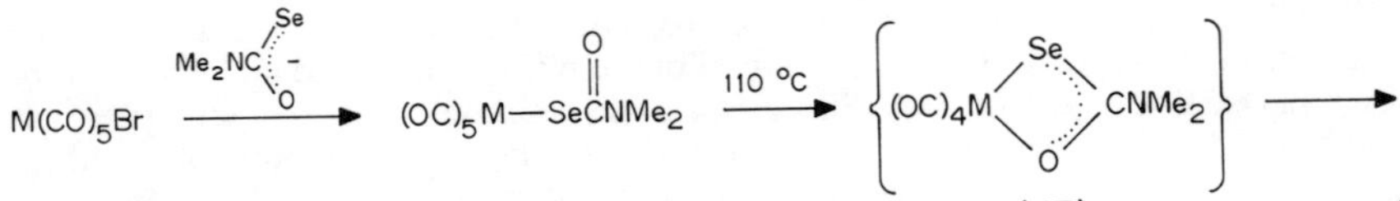

M = Mn, Re (**46**) (**47**) (144)

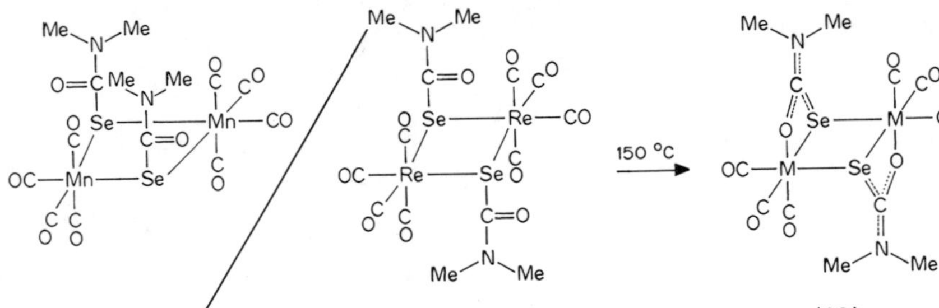

(**48a**) (**48b**) (**49**)

Although the initial report[396] on the thermolysis of the manganese pentacarbonyl complex indicated that the tetracarbonyl complex formed as the initial thermolysis product was monomeric **47** with chelating selenocarbamate ligand, a subsequent report[397] showed that this complex was actually a dimeric species with Se-bridging selenocarbamate (**48**). Thermolysis in cyclohexane did, however, allow the unstable monomeric species **47** to be identified by infrared spectroscopy ($\nu_{C \equiv O} = 1550\,\text{cm}^{-1}$).

Isolation of pure tetracarbonyl dimers required that the thermolysis be carried out under flowing CO[397]. Thermolysis under N_2 always gave contamination with the tricarbonyl dimer **49**. The identification of the tetracarbonyls as dimers with bridging Se-bonded selenocarbamates rather than monomers with chelating selenocarbamates was based on infrared and mass spectroscopy[396–398]. A detailed analysis of the infrared spectra of the dimers in the carbonyl region (M—C≡O, 1900–2100 cm^{-1}) and comparison of the number of observed bands with those expected for the various symmetry types allowed assignments of the detailed stereochemistries (equation 144). The conversion of the manganese tetracarbonyl **48a** to tricarbonyl dimer **49** involves inversion of configuration with respect to the orientation of the selenocarbamate ligand[397], but for the Re dimers this net decarbonylation process is accompanied by retention of configuration[398] (**48b** → **49**).

The Fe(II) and Mo(II) complexes (Table 30) were prepared by substitution reactions (equations 145 and 146).

$$[\text{CpFe(CO)}_2(\text{Me}_2\text{CO})]\text{BF}_4 \xrightarrow{\text{SeC(O)NMe}_2^-} \text{CpFe(CO)}_2\text{SeC(O)NMe}_2 \qquad \text{(Ref. 399)}$$

$$\text{CpFe(CO)}_2\text{Cl} \xrightarrow[\text{Me}_2\text{CO}]{\text{AgBF}_4,\ (-\text{AgCl})} [\text{CpFe(CO)}_2(\text{Me}_2\text{CO})]\text{BF}_4 \qquad (145)$$

$$\text{CpMo(CO)}_3\text{Cl} \xrightarrow[\text{THF}]{\text{SeC(O)NMe}_2^-} \text{CpMo(CO)}_3\text{SeC(O)NMe}_2 \qquad \text{(Ref. 400)} \qquad (146)$$

The electrochemical oxidation of these complexes in MeCN and CH_2Cl_2 has been investigated along with the corresponding analogues with SeSe, SeS, SS and SO donors[399,400].

7. Miscellaneous ligands

A single example of a monodentate Se-bonded dialkyldiselenocarbamate complex has been reported ($CpFe(CO)_2SeC(Se)NMe_2$)[399,403]. This complex has been prepared by a substitution reaction[399] (equation 147a) and a CSe_2 insertion reaction[403] (equation 147b). There is a discrepancy in the infrared data of the two reports. The lower value of ν_{C-N}, $1460\,cm^{-1}$ [399] vs. $1500\,cm^{-1}$ [403], is probably a more reasonable one for monodentate coordination of dialkyldiselenocarbamate, values of ca. $1510–1525\,cm^{-1}$ being generally found for the usual bidentate coordination mode (Section II.D.C).

$$[CpFe(CO)_2(OCMe_2)]BF_4 \xrightarrow{(a)\ [Me_2NH_2][Se_2CNMe_2]} CpFe(CO)_2SeC(Se)NMe_2 \qquad (147)$$

$$CpFe(CO)_2Cl \xrightarrow[-21\,^{\circ}C,\ 2.\ NaPF_6(aq)]{1.\ HNMe_2\ neat} \{CpFe(CO)_2NHMe_2\}PF_6 \xrightarrow{(a)\ CSe/KOBu\text{-}t,\ acetone/-20\,^{\circ}C} CpFe(CO)_2SeC(Se)NMe_2$$

A single example of a complex with monodentate selenothiocarbamate, $CpFe(CO)_2SC(Se)NMe_2$, has also been prepared by the substitution route described in equation (147a). The position of the $\nu_{C\cdots N}$ band, $1460\,cm^{-1}$, supports a monodentate bonding mode, a value of $1515\,cm^{-1}$ being found for the related complex with (S, Se) bidentate coordination ($CpFe(CO)SeSCNMe_2$). This assignment of a Fe—S rather than a Fe—Se bonding mode of this ambidentate ligand was based on the observation of a $[FeS]^+$ fragment peak, but no $[FeSe]^+$ peak in its mass spectrum[399].

The two-step formation of an unusual —SeSeMe ligand has been achieved by alkylation of a coordinated η^2-Se_2 ligand (see equation 215) followed by reductive cleavage of one of the metal—Se bonds[404] (equation 148).

$$Os(\eta^2\text{-}Se_2)(CO)_2(PPh_3)_2 \xrightarrow[2.NaClO_4/EtOH]{1.CF_3SO_3Me/CH_2Cl_2} [Os(\eta^2\text{-}Se_2Me)(CO)_2(PPh_3)_2]ClO_4 \qquad (148)$$

$$\xrightarrow{NaBH_4/EtOH} Os(\eta^1\text{-}Se_2Me)H(CO)_2(PPh_3)_2$$

C. Neutral Bidentate Ligands

1. (Se, C) donors

a. CSe_2 complexes. Because of the instability of CSe_2[405] relatively few complexes[214,217,224,406–408] of this molecule have been isolated compared to the extensive coordination chemistry of CS_2[204,409]. This molecule has also been used in the synthesis of selenocarbonyl complexes (Section II.A.7).

Jensen and Huge-Jensen prepared the first examples of complexes with coordinated CSe_2 by reactions with coordinatively unsaturated low-valent metal species[406] (equations 149–151).

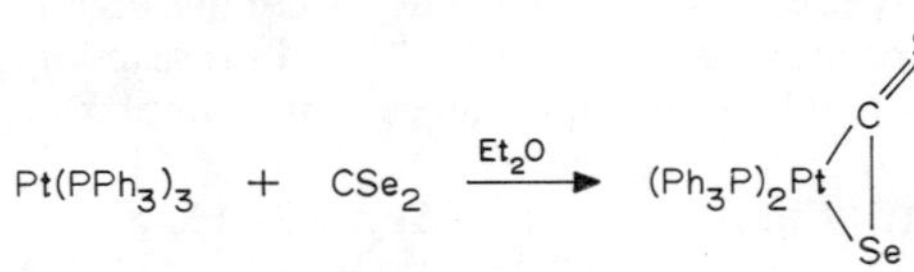

(149)

light green solid

$\nu_{C=Se} = 995\,cm^{-1}$

($1270\,cm^{-1}$ in free CSe_2)

$$\mathrm{Ni(PPh_3)_2(CO)_2 + CSe_2 \rightarrow Ni(PPh_3)_2(CSe_2)_2} \qquad (150)$$
$$\nu_{\mathrm{C{=}Se}} = 870\,\mathrm{cm^{-1}}$$

$$\mathrm{Rh(PPh_3)_3Cl + CSe_2 \rightarrow Rh(PPh_3)_2Cl(CSe_2)_2} \qquad (151)$$
$$\nu_{\mathrm{C{=}Se}} = 865\,\mathrm{cm^{-1}}$$

In the Ni and Rh systems, only complexes with two CSe_2 units could be isolated, and these have been formulated as tetraselenooxalate complexes on the basis of infrared spectroscopy (no $\nu_{\mathrm{C{=}Se}}$ band at $\sim 1000\,\mathrm{cm^{-1}}$ and a strong band at $\sim 870\,\mathrm{cm^{-1}}$ characteristic of the tetraselenooxalate anion).

Although no single-crystal X-ray diffraction structure determination has been reported for a CSe_2 complex, the CS_2 analogue of the product of equation (149) has been shown to have the proposed structure with η^2-CS_2 coordination[410]. Several other complexes with η^2-CS_2 have also been structurally characterized[409].

Similar reactions were subsequently used to prepare Ir[217,407], Pt[407], Pd[408], Ru[214] and Co[224] complexes of CSe_2 (equations 152–156).

$$trans\text{-}\mathrm{IrCl(CO)(PPh_3)_2 \xrightarrow[C_6H_6]{CSe_2} IrCl(CO)(CSe_2)(PPh_3)_2} \qquad \text{(Ref. 407)} \qquad (152)$$

$\nu_{\mathrm{C{\equiv}O}} = 1960\,\mathrm{cm^{-1}}$ $\qquad$ $\nu_{\mathrm{C{=}Se}} = 1008\,\mathrm{cm^{-1}}$

$\nu_{\mathrm{Ir{-}Cl}} = 317\,\mathrm{cm^{-1}}$ $\qquad$ $\nu_{\mathrm{C{\equiv}O}} = 2025\,\mathrm{cm^{-1}}$

$\nu_{\mathrm{Ir{-}Cl}} = 265\,\mathrm{cm^{-1}}$

$$\mathrm{IrCl(CSe)(PPh_3)_2 \xrightarrow{CSe_2} IrCl(CSe)(\eta^2\text{-}CSe_2)(PPh_3)_2} \qquad \text{(Ref. 217)} \qquad (153)$$
$$\nu_{\mathrm{C{=}Se}} = 1175\,\mathrm{cm^{-1}}$$
$$\nu_{\mathrm{C{=}Se(CSe_2)}} = 1010\,\mathrm{cm^{-1}}$$

$$\mathrm{Pt(PPh_2R)_4 \xrightarrow[hexane]{CSe_2} Pt(PPh_2R)_2(CSe_2)} \qquad \text{(Ref. 407)} \qquad (154)$$
$$\mathrm{R = Ph};\ \nu_{\mathrm{C{=}Se}} = 999\,\mathrm{cm^{-1}}$$
$$\mathrm{R = Me};\ \nu_{\mathrm{C{=}Se}} = 985\,\mathrm{cm^{-1}}$$

$$\mathrm{Ru(CO)_2(PPh_3)_3 \xrightarrow{CSe_2} Ru(CSe_2)(CO)_2(PPh_3)_2} \qquad \text{(Ref. 214)} \qquad (155)$$
$$\nu_{\mathrm{C{=}Se}} = 953\,\mathrm{cm^{-1}}$$
$$\nu_{\mathrm{C{\equiv}O}} = 2020,\ 1950\,\mathrm{cm^{-1}}$$

$$\mathrm{Pd(PPh_3)_4 \xrightarrow{CSe_2} Pd(PPh_3)_2(CSe_2)} \qquad \text{(Ref. 408)} \qquad (156)$$
$$\nu_{\mathrm{C{=}Se}} = 990\,\mathrm{cm^{-1}}$$

Coordinated CSe_2 can be readily alkylated, and this reaction has been used to prepare selenocarbonyl complexes (M—CSe) from CSe_2 complexes (equations 43 and 46). Only two other brief reports have described the reactivity of complexes containing η^2-CSe_2: (*i*) phosphine substitution reactions (equation 157)[408] and (*ii*) reaction of the coordinated CSe_2 with a difunctional halide to give a cyclic carbene complex (equation 158)[411].

$$\mathrm{Pd(PPh_3)_2(CSe_2) \xrightarrow[hexane]{L} LPd(CSe_2) + 2\,PPh_3} \qquad (157)$$
$$\mathrm{L = dppe}$$
$$\nu_{\mathrm{C{=}Se}} = 990\,\mathrm{cm^{-1}}$$
$$\mathrm{L} = o\text{-}\mathrm{(Ph_2PCH_2)_2C_6H_4}$$
$$\nu_{\mathrm{C{=}Se}} = 995\,\mathrm{cm^{-1}}$$

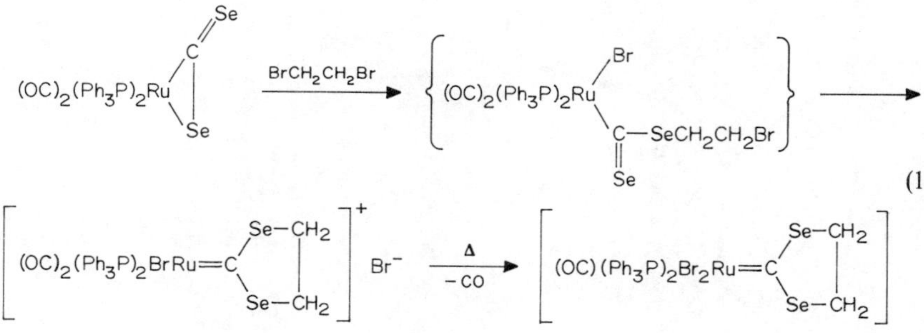

(158)

The initial cationic carbene complex was isolated as its perchlorate salt and characterized by infrared and ^{1}H-NMR spectroscopy. Warming a solution of the cationic complex results in evolution of CO and formation of the neutral dibromo complex.

b. CSSe complexes. Several complexes of CSSe with transition metals (Co[224,412], Rh[224,412], Ru[215], Os[215], Ni[413], Pd[414,415] and Pt[408]) have been prepared by reaction of an appropriate metal precursor and CSSe, and an η^2 bonding mode via C and Se has been found in a recent single-crystal X-ray diffraction characterization of (o-$(Ph_2PCH_2)_2C_6H_4$)Pd(η^2-CSSe)[414] (Figure 5.)

As previously described in connection with the synthesis of selenocarbonyl complexes (Section II.A.7, equation 44) stable linkage isomers of the η^2-CSSe ligand have been isolated (**33a**, $\nu_{C=S} = 1063\ cm^{-1}$; **33b**, $\nu_{C=Se} = 1027\ cm^{-1}$)[215]. The analogous Os linkage isomers have also been isolated (Os(η^2-$\underline{CS}$eS), orange, $\nu_{C=S} = 1066\ cm^{-1}$; Os($\eta^2$-C$\underline{SSe}$), pink, $\nu_{C=Se} = 1015\ cm^{-1}$)[215].

Complexes with η^2-CSSe have been used as precursors to thiocarbonyl complexes via phosphine-induced Se extrusion reactions[412] (equation 159 and 160). A Ni(o) complex with CSSe has been prepared at low temperature by displacement of the labile 1,3-cyclooctadiene (COD) ligand[413] (equation 161).

$$\mathrm{CpCo(PMe_3)_2 \xrightarrow[-20\ ^\circ C/PhMe]{CSeS} CpCo(PMe_3)(\eta^2\text{-}CSeS) + CpCo(PMe_3)(CS)} \quad (159)$$

$$\mathrm{CpCo(PMe_3)(\eta^2\text{-}CSeS) \xrightarrow[-SePPh_3]{PPh_3/C_6H_6,\ 50\ ^\circ C} CpCo(PMe_3)(CS)}$$

$$\mathrm{CpRh(C_2H_4)(PMe_3) \xrightarrow{CSeS} CpRh(\eta^2\text{-}CSeS)(PMe_3) \xrightarrow[-SePPh_3]{PPh_3} CpRh(PMe_3)(CS)} \quad (160)$$

$$\mathrm{Ni(bipy)(COD) + CSeS \xrightarrow{Et_2O/-80\ ^\circ C} Ni(bipy)(CSeS) + COD} \quad (161)$$

$$\nu_{C=S} = 1103\ cm^{-1}$$

The Pd(II)[414] and Pt(II)[408] complexes were prepared by reaction of the appropriate tetrakis(triorganophosphine) complex with CSSe and, as in the CSe_2 analogues[408], the monodentate phosphine ligands can be readily displaced by chelating phosphines (equation 162).

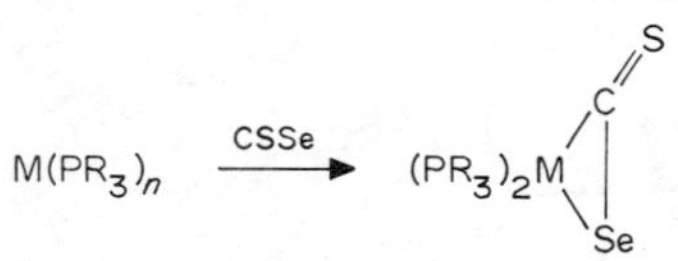

(162)

M = Pd; $n = 2, 3$; R = i-Pr[414]

$n = 4$; R = Ph[414]

M = Pt; $n = 4$; R = Ph[408]

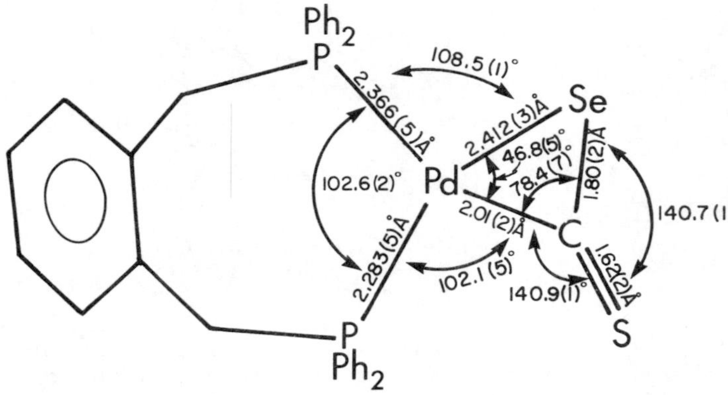

FIGURE 5. Molecular structure of $(o\text{-}(Ph_2PCH_2)_2C_6H_4)Pd(\eta^2\text{-}CSeS)$

Bonding via C and Se was proposed for all of these complexes, based on infrared spectroscopy ($\nu_{C=S} = 1160\text{–}1175\,cm^{-1}$)[414], a conclusion confirmed by single-crystal X-ray diffraction for $(o\text{-}(Ph_2PCH_2)_2C_6H_4)Pd(\eta^2\text{-}CSSe)$[414] (Figure 5).

Noteworthy also is the significant *trans* influence of the C atom on the Pd—$P_{trans(C)}$ distance vs. the Pd—$P_{trans(Se)}$ (Figure 5). A similar *trans* effect has been observed in $Pd(PPh_3)_2(\eta^2\text{-}CS_2)$.

c. Selenoformaldehyde complexes. Although polymeric forms of selenoformaldehyde have been prepared and characterized both in the free state (e.g. the cyclic trimer 1,3,5-triselane[416]) and coordinated to a metal (e.g. $[[(CH_2Se)_3]_2Ag]AsF_6$[417]) monomeric selenoformaldehyde has never been isolated. The latter complex, isolated as yellow crystals in 98% yield, was prepared by the reaction of $AgAsF_6$ and $(CH_2Se)_3$ in liquid SO_2[417]. Single-crystal X-ray diffraction showed the presence of discrete $[Ag\{(CH_2Se)_3\}_2]^+$ cations with all Se atoms irregularly coordinated to the Ag (the next shortest Ag—Se distance is 3.71 Å) (Figure 6).

In 1983 four reports described the first examples of monomeric selenoformaldehyde coordinated as both an η^2-CH_2Se chelate (equations 163[418], 164[419] and 165[420]) and a bridging ligand (equation 166[421]).

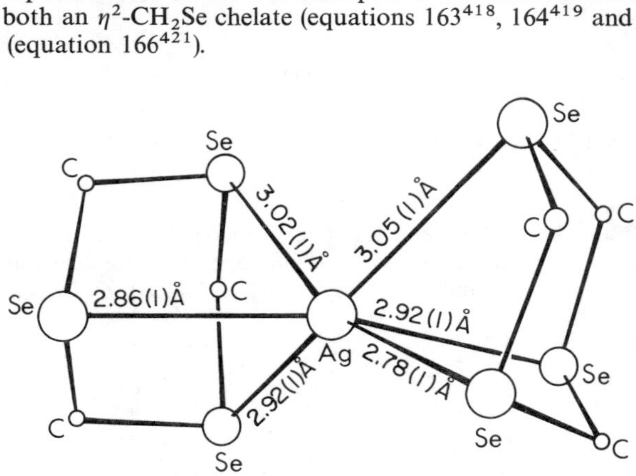

FIGURE 6. Molecular structure of $[Ag\{(CH_2Se)_3\}_2]^+$

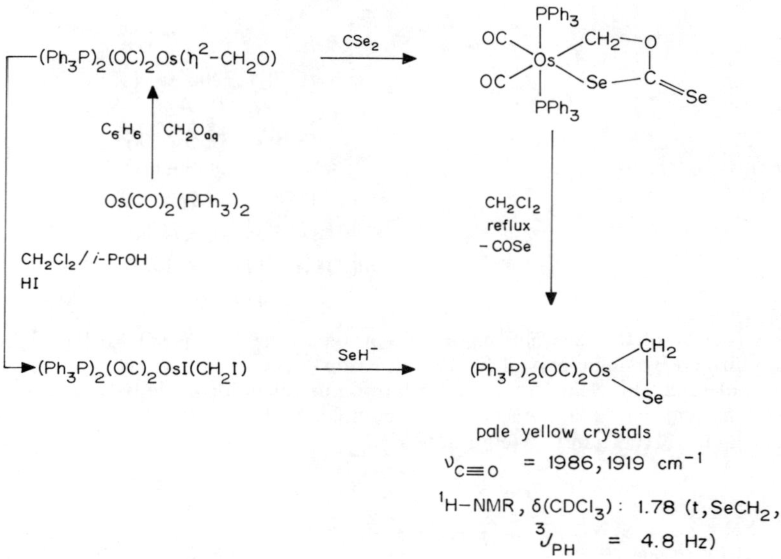

(163)

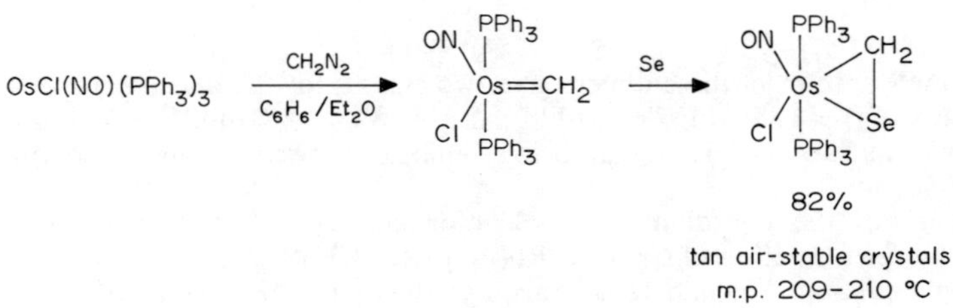

(164)

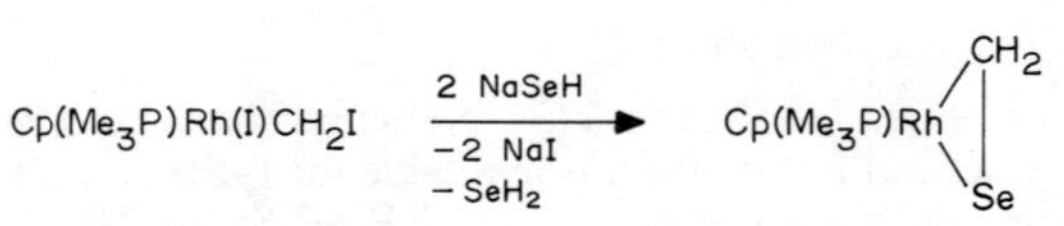

(165)

red–violet slightly air-sensitive solid (72%)
$\delta(CH_2) = 4.26$ (ddd, $J_{HH} = 1.2$ Hz, $J_{RhH} = 0.8$ Hz, $J_{PH} = 8.3$ Hz); 5.31 (ddd, $J_{HH} = 1.2$ Hz, $J_{RhH} = 2.2$ Hz, $J_{PH} = 0.6$ Hz)

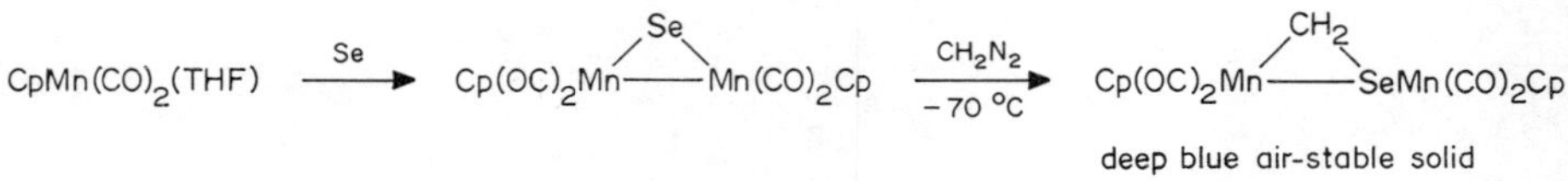

$$\nu_{C\equiv O}(CCl_4) = 2000vs, 1955s, 1934vs, 1882s\,cm^{-1} \qquad (166)$$

$$\nu_{C-H}(KBr) = 3115s, 2965s, 2930vs\,cm^{-1}$$

^{1}H-NMR (270 MHz, C_6D_6)

$\delta(Cp) = 4.34$ (s), 3.71 (s)

$\delta(CH_2) = 4.88$ (d, $^2J_{HH} = 1.5$ Hz)

$= 4.66$ (d, $^2J_{HH} = 1.5$ Hz)

The structure of the selenoformaldehyde-bridged dimer $\{Cp(OC)_2Mn\}_2(CH_2Se)$ has been confirmed by single-crystal X-ray diffraction[421].

Heterometallic binuclear complexes with bridging selenoformaldehyde have also been prepared by acid–base reactions between a complex with η^2-$SeCH_2$ and complexes with the very labile THF ligand[422] (equation 166a).

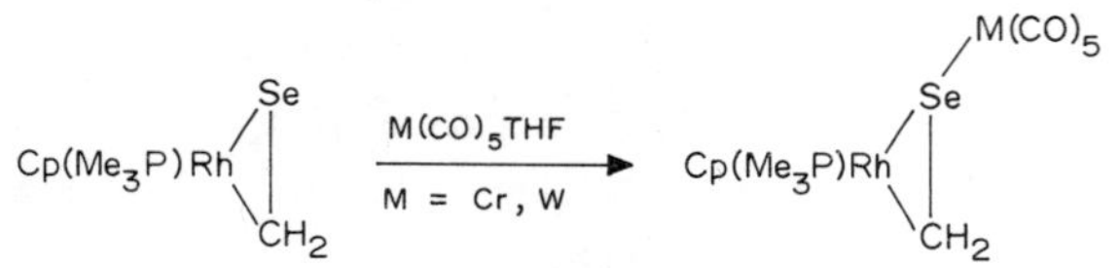

(166a)

The ^{1}H-NMR spectra of these dimers give two signals for the diastereotopic CH_2Se protons (δ(Rh, Cr) = 4.88 (dd), 3.56 (ddd); δ(Rh, W) = 5.18 (ddd), 3.76 (ddd)), the differences in δ values of the CH_2 signals being significantly greater than for the Mn dimer (equation 166).

The basicity of the coordinated η^2-selenoformaldehyde ligand was further demonstrated by the formation of Cp′ (OC)Rh(μ_2,η^2(Rh)-CH_2Se)$M(CO)_5$ by the two-step synthesis (e.g. equations 165 and 166a) starting with Cp′(OC)Rh(I)(CH_2I)[422].

2. (Se, N) and (Se, P) donors

Only a few examples of neutral (Se, N) ligands have been reported (Table 31), and no definitive structural information is available for these complexes.

Complexes of selenosemicarbazide ($RHNC(Se)NHNH_2$; ssc-R, **50**) were readily isolated from a solution of the ligand and the appropriate metal salt. A tetrahedral monomeric structure with (Se, N) bidentate selenosemicarbazide (ssc-H) was assigned to $[Zn(ssc\text{-}H)Cl_2]$[424] on the basis of the similarity of its X-ray powder pattern with that of the analogous thiosemicarbazide complex, the structure of which has been solved by single-crystal X-ray diffraction[435]. A green complex, Cu(ssc-H)Cl_2 was precipitated if an aqueous solution of the ligand, acidified with HCl, was added to cold aqueous $CuCl_2$ in a 1:1 molar ratio. The structure of this complex is unknown, but the similarity of its infrared spectrum with that of the thiosemicarbazide analogue indicates a bidentate (Se, N) coordination mode, and its insolubility suggests a polymeric structure with bridging chloro ligands. The magnetic susceptibility ($\mu_{eff} = 1.78$ B.M.) confirmed the presence of Cu(II).

Reaction of CuX_2 (X = Cl, NO_3) or $CuSO_4$ with ssc-H in a 1:2 molar ratio gave

TABLE 31. Complexes with neutral (Se,N) and (Se,P) donors

Ligand	Complex	Ref.
(Se,N) donor sets		
ssc-R (**50**)	$[Ni(ssc\text{-}Ph)_2]Cl_2$	423
	$[Zn(ssc\text{-}H)Cl_2]$	424
	$[Zn(ssc\text{-}H)_2](NO_3)_2$	424
	$[Cu(ssc\text{-}H)Cl_2]$	424
	$Cu(ssc\text{-}H)_2X_2$; $X = Cl, NO_3, \frac{1}{2}SO_4$	424
	trans-$[Ni(ssc\text{-}H)_2]SO_4 \cdot 3H_2O$ (**51a**)	425
	cis-$[Ni(ssc\text{-}H)_2]SO_4$ (**51b**)	425
	trans-$[Ni(ssc\text{-}H)_2]2X$; $X = Cl, Br, NO_3$	425
	$[Ni(ssc\text{-}H)_3]2X \cdot 2H_2O$; $X = Cl, Br$	425
pbs (**52**)	$[MCl_2(pbs)]$; M = Pd, Pt	426
$MeSeCH_2CH_2NH_2$	$[Co(en)_2(mse)](ClO_4)_3$	427–429
(mse)	$[Co\{N(CH_2CH_2NH_2)_3\}(mse)]Cl_3$	430
(Se,P) donor sets		
spp (**53**)	$[Co(spp)_2Br]ClO_4$	431
	$[Ni(spp)_2Br]ClO_4$	432, 437
	$[Ni(spp)_2(NCS)_2]$	433
	$[Pd(spp)(SCN)_2]$	433, 434
	$[Pd(spp)I_2]$	434
	$[Ru(spp)_2Cl_2]$	434

complexes reported to be $Cu(ssc\text{-}H)_2X_2$ ($X = Cl, NO_3, \frac{1}{2}SO_4$)[424]. The insolubility and red–brown colour of the chloride and nitrate complexes, however, suggest that reduction has occurred with formation of Cu(I) complexes, $[Cu(ssc)]_n$. The magnetic susceptibilities of these complexes were not reported. In contrast, the sulphate complex was reported to be green, characteristic of Cu(II)[424].

Addition of solid ssc-H to a concentrated aqueous solution of $NiSO_4$ gave a precipitate of a diamagnetic red solid, which was assigned a *trans* square-planar structure (**51a**)[425] on the basis of the similarity of its X-ray powder pattern with the structurally characterized thiosemicarbazide analogue[436].

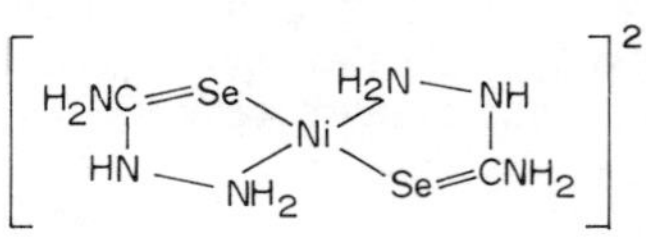

(**51a**)

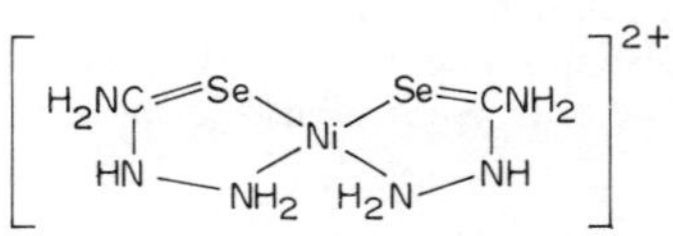

(**51b**)

A pale brown complex, formulated as the *cis* isomer (**51b**), was prepared by addition of a hot ethanol solution of $NiSO_4$ to a hot dilute ethanol solution of ssc-H[425].

The pale brown, diamagnetic square-planar complex $[Ni(ssc\text{-}H)_2]Cl_2$ precipitated on mixing hot ethanol solutions of ssc-H and $NiCl_2 \cdot 6H_2O$ in a 2:1 molar ratio[425]. An aqueous solution of this complex was, however, blue–green and paramagnetic. Concentration of this solution gave green crystals of the octahedral complex $[Ni(ssc\text{-}H)_3]Cl_2 \cdot 2H_2O$ ($\mu_{eff} = 3.14$ B.M.)[425].

Addition of aqueous ammonia to solutions of $[Ni(ssc\text{-}Ph)_2]Cl_2$[423] and $[Ni(ssc\text{-}H)_2]Cl_2$[425] precipitated the corresponding neutral complexes containing the anionic form of the selenosemicarbazide ligand.

A study[426] of the use of 2-(2-pyridyl)benzo[*b*]selenophene (**52**) as a reagent for the

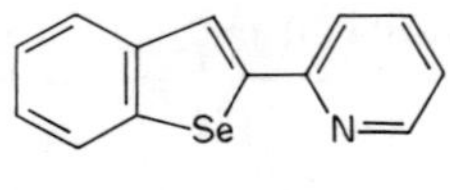

pbs, (**52**)

gravimetric determination of Pd(II) and Pt(II) included the isolation of the yellow complexes $MCl_2(pbs)$. They were formulated as monomeric, square-planar species with (N, Se) bidentate pbs, although no spectroscopic data were presented.

Alkylation of a coordinated Se atom was used to prepare Co(III) complexes with the chelating $RSeCH_2CH_2NH_2$ ligand (equation 167).

$$[Co(en)_2(SeCH_2CH_2NH_2)]^{2+} \xrightarrow{RCl} [Co(en)_2(RSeCH_2CH_2NH_2)]^{3+} \quad (167)$$

R = Me[427,428], Et[428], CH_2Ph[428], $CH_2CH_2C(O)Me$
($MeC(O)CH{=}CH_2$ was used as the alkylating agent[428])

The kinetics of the reaction with MeI was studied, and the relative nucleophilicities of various coordinated S and Se centres were determined: $[(en)_2Co(S(O)CH_2CH_2NH_2)]^{2+}$ < $[(en)_2Co(o\text{-}SC_6H_4NH_2)]^{2+}$ < $[(en)_2Cr(SCH_2CH_2NH_2)]^{2+}$ ~ $[(en)_2Co(SCH_2CO_2)]^{+}$ ≈ $[(en)_2Co(SCH_2CH_2NH_2)]^{2+}$ < $[(en)_2Co(SeCH_2CH_2NH_2]^{2+}$ [427].

Optical isomers of the octahedral complexes $[Co(en)_2(MeSeCH_2CH_2NH_2)]^{2+}$ [429] and $[\{N(CH_2CH_2NH_2)_3\}Co(MeSeCH_2CH_2NH_2)]^{2+}$ [430] have also been isolated and characterized by electronic and circular dichroism spectra. These complexes were synthesized by alkylation of the corresponding selenolato complexes (e.g. equation 167). The complex with the tetradentate amine, $N(CH_2CH_2NH_2)_3$, gave two geometrical isomers differing in the relative orientation of the bidentate (Se, N) ligand, which were separated by chromatography[430].

Although a large number of hybrid bidentate ligands with various (P, E) (E = Group 5a or 6a element) donor sets have been reported[10], only one example with a (Se, P) set is known (**53**). The ligand was prepared by the two-step route shown in equation (168)[431]. This ligand forms complexes with Co(II)[431], Ni(II)[432,433], Ru(II)[434] and Pd(II)[433,434]. The Co(II)[431] and Ni(II)[432] bromo complexes were formulated as square-pyramidal on the basis of solution conductivities (1:1 electrolytes in MeCN and $MeNO_2$), magnetic susceptibilities (Co complex, $\mu_{eff} = 2.54$ B.M.; Ni complex, diamagnetic) and electronic spectroscopy. The thiocyanate complex $[Ni(spp)_2(NCS)_2]$[433] was assigned an octahedral structure with N-bonded thiocyanate, but no spectroscopic data were given in support.

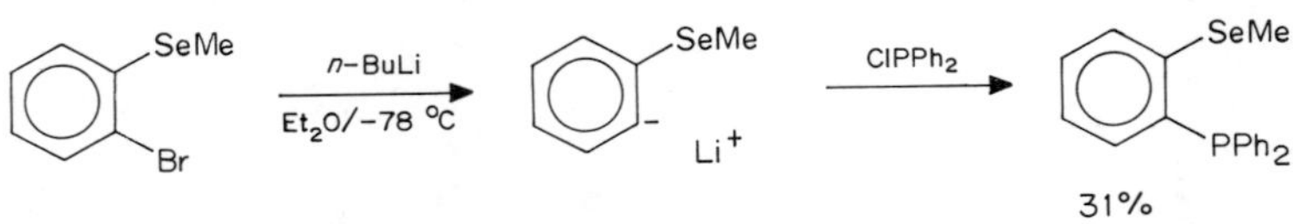

(168)

spp, (**53**)

The Pd(II)[433,434] complexes are square-planar with S-bonded thiocyanate, although no spectroscopic evidence has been presented to support the latter conclusion[434]. These complexes readily undergo thermal *Se*-demethylation in solution to give the corresponding neutral complexes (e.g. $[Ni(ps)_2]$[433], $[(ps)Pd(\mu\text{-SCN})_2Pd(ps)]$[433,434], $[Pd(ps)_2]$[410]; see Section II.D.1). The kinetics of the demethylation of $[Pd(spp)(SCN)_2]$ with SCN^- in CD_3CN/CH_2Cl_2 (3:1) has been studied[434], and under these conditions the reaction goes under milder conditions than in neat solvent (i.e. 55 °C[434] vs. refluxing DMF or *n*-butanol[433]).

3. (Se, E) (E = O, S, Se) donors

Relatively few ligands of this class have been described. The heterocycles 1,4-diselenan and 1,4-oxaselenan, which could coordinate as chelating (Se, Se) and (Se, O) ligands, were discussed in Section II.A.4 (see Tables 10 and 11). No good evidence for such chelation has been published. The Se_n ligands ($n = 2$, 4, 5), which can be formally considered to be neutral (Se, Se) chelates, are discussed in Section II.G.2–4.

The neutral bidentate (Se, E) (E = O, S, Se) ligands and their transition-metal coordination complexes are summarized in Table 32. The complex formation constants of $Se(CH_2CH_2CO_2H)_2$ and its related ligands with various divalent metals were measured in aqueous solution.

Complexes with 1:1 stoichiometries with both neutral and deprotonated forms of the ligands were present in such solutions. Chelation via the Se and O sites of these ligands was suggested, although no spectroscopic data were reported in this work[91].

The hybrid ligand $MeSCH_2SeMe$ functions as a bridging ligand in Pt(IV) complexes (e.g. $Me_3Pt(\mu\text{-}X)_2(\mu\text{-}MeSCH_2SeMe)PtMe_3)$[257]. Abel and coworkers, as a continuation of their detailed NMR studies of pyramidal inversion and intramolecular rearrangements in metal complexes with chalcogen ligands, have shown that increasing the methylene group by one unit allows the resulting ligand to function as a (Se, S) chelate in Pt(IV)[437], Pt(II)[439] and Re(I)[438] complexes (Table 32). This chelating ligand (msmte) was prepared as shown in equation (169)[437]. The metal complexes were prepared from appropriate halo complexes (equations 170–172). Facial geometry of the octahedral Re(I) complexes was established by infrared spectroscopy (e.g. three observed $\nu_{C\equiv O}$ bands of similar intensities).

$$MeSCH_2CH_2Cl + MeSe^-\,Na^+ \rightarrow MeSCH_2CH_2SeMe \qquad (169)$$

$$\uparrow \text{liq. } NH_3 \qquad \text{msmte}$$

$$Me_2Se_2 + Na \qquad \text{pale yellow liquid}$$

b.p. 82 °C (10 mm Hg)
88% yield

$$Me_3PtX \xrightarrow[\text{CHCl}_3\text{ reflux}]{\text{msmte}} Me_3PtX(\text{msmte}) \qquad \text{(Ref. 437)} \qquad (170)$$

$$X = Cl, Br, I$$

$$PtXMe(COD) \xrightarrow[\text{CHCl}_3\text{ reflux}]{\text{msmte}} PtXMe(\text{msmte}) \qquad \text{(Ref. 439)} \qquad (171)$$

$$X = Cl, Br, I$$

$$(OC)_4Re(\mu\text{-}X)_2Re(CO)_4 \xrightarrow[\text{CHCl}_3\text{ reflux}]{\text{msmte}} fac\text{-}[ReX(CO)_3(\text{msmte})]$$

$$\text{(or } [ReX(CO)_3(THF)]) \qquad \text{(Ref. 438)} \qquad (172)$$

$$X = Cl, Br, I$$

The aromatic (Se, S) hybrid ligand msta (**54**) was first prepared by Pierpont[455] (equation 173).

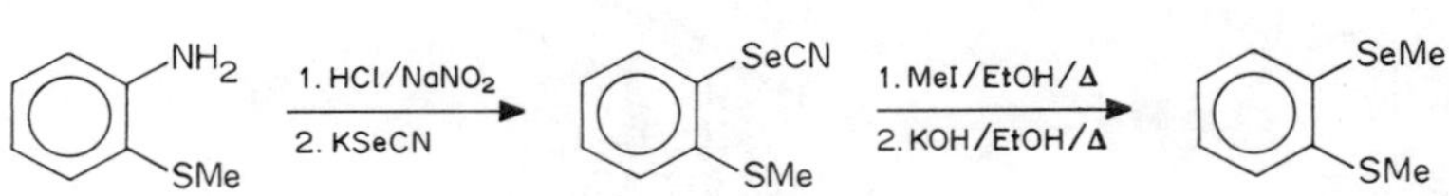

(173)

msta, (**54**)

yellow oil
b.p. 104–105 °C (1 mm Hg)

TABLE 32. Complexes with neutral (Se, E) (E = O, S, Se) ligands

Ligand (L)	Complex	Reference
(Se, O) donors		
$Se(CH_2CH_2CO_2H)_2$	[ML]	91
$Se(CH_2CO_2H)_2$	M = Co(II), Ni(II), Cu(II), Zn(II),	
$Se(CH(Me)CO_2H)_2$	Cd(II), Pb(II), Ag(I)	
(Se, S) donors		
msta (**54**)	$PdCl_2$(msta)	434
	$PtMe_3X$(msta)	437
	X = Cl, Br, I	
	$ReX(CO)_3$(msta)	438
	X = Cl, Br, I	
$MeSeCH_2CH_2SMe$	$PtMe_3X$(msmte)	437
(msmte)	X = Cl, Br, I	
	PtMeX(msmte)	439
	X = Cl, Br, I	
	$ReX(CO)_3$(msmte)	438
	X = Cl, Br, I	
(Se, Se) donors		
$NCSeCH_2CH_2SeCN$	[M(dse)Cl_2] (**56**)	440
(dse)	M = Pd, Pt	
	[Co(dse)$Cl_2]_n$	441
	[M(dse)Cl_3] (**57**)	441
	M = Rh, Ir	
$MeSeCH_2CH_2SeMe$	$MCl_n \cdot L$	442
	$n = 2$; M = Pd, Hg	
	$n = 3$; M = Au, Fe	
	$n = 4$; M = Ti, V, Pt, Sn	
	$M(CO)_3XL$	443
	M = Mn, Re; X = Cl, Br	
	$Cr(CO)_4L$	443
	$[PtXMe_3L]$	444, 445
	X = Cl, Br, I	
	$[ReX(CO)_3L]$	446
i-$PrSeCH_2CH_2SePr$-*i*	$[MX_2L]$	447–450
	M = Pd, Pt; X = Cl, Br	
	$[NiX_2L_2]$	451
	X = Cl, Br	
	$[Ni(NCS)_2L]$	451
	$[M(CO)_4L]$	450, 452
	M = Cr, Mo, W	
$PhSeCH_2CH_2SePh$	$[TiCl_4L]$	453
	$[Ru(CO)_2Br_2L]$	98
$MeSeCH_2CH_2CH_2SeMe$	$MCl_n \cdot L$	442
	$n = 2$; M = Pd, Hg	
	$n = 3$; M = Au	
	$n = 4$; M = Ti, V, Pt, Sn	
	$(FeCl_3)_2 \cdot L$	442
	$TeBr_4 \cdot L$	442
	$[PtXMe_3L]$	444, 445
	X = Cl, Br, I	
	$[ReX(CO)_3L]$	446
	X = Cl, Br, I	
$EtSeCH_2CH_2CH_2SeEt$	$[MCl_2L]$	
	M = Pd	6
	M = Pt	5

TABLE 32. (*Continued*)

Ligand (L)	Complex	Reference
$MeSeCH_2CMe_2CH_2SeMe$	$[M(CO)_4L]$ M = Cr, Mo, W	454
$Ph_2P(Se)CH_2P(Se)Ph_2$	$[CoLX_2]$ X = Cl, Br, I	140d
	$[(COD)RhL]ClO_4$	128
	$[ML_2]ClO_4$ M = Cu, Ag	134
	$[MLX_2]$ M = Pd, Pt; X = Cl; M = Hg; X = Cl, Br	134
$Ph_2P(Se)CH_2CH_2P(Se)Ph_2$	$[AgL_2]ClO_4$	134
	$[HgLCl_2]$	134

A more recent paper[434] described a synthesis via lithiation of *o*-bromoselenoanisole, followed by sulphur insertion and methylation (equation 174). Since the starting *o*-bromoselenoanisole was prepared from *o*-bromoanisole by a route analogous to that described in equation (173), the more direct synthesis via the latter route seems more efficient.

$$o\text{-}BrC_6H_4SeMe \xrightarrow[\text{2. } S_8 \quad \text{3. MeI}]{\text{1. } n\text{-BuLi/hexane/Et}_2\text{O}} \mathbf{54} \qquad (174)$$

The monomeric Pd(II) complex of this hybrid ligand, $[Pd(Se, S)Cl_2]$, shows a typical doublet in the far-infrared region characteristic of a *cis*-$PdCl_2$ group (300, 320 cm^{-1})[434].

Total NMR band-shape fitting methods have been used to determine accurate energy data for inversion barriers at S and Se in the complex $[PtXMe_3(o\text{-}MeSC_6H_4SeMe)]$[437] and *fac*-$[ReX(CO)_3(o\text{-}MeSC_6H_4SeMe)]$[438] (X = Cl, Br, I).

Poorly characterized complexes of Pd(II)[440], Pt(II)[440], Co(II)[441], Rh(III)[441] and Ir(III)[441] with 1,2-diselenocyanatoethane have been reported. This ligand, prepared by reaction of KSeCN and 1,2-dibromoethane in refluxing ethanol, was isolated as white needles after recrystallization from EtOH to remove elemental Se (m.p. 136–137 °C)[440]. The Pt(II) complex[440] gave ν_{Pt-Cl} bands at 320 and 303 cm^{-1}, characteristic of a *cis* terminal $PtCl_2$ group, but no far-infrared data were reported for the Pd complex. Shifts of the $\nu_{C\equiv N}$ and ν_{C-Se} vs. the free ligand suggest that both N and Se sites interact with the metals in these complexes. The insolubility of the complexes precluded molecular weight measurements. The data suggest polymeric structures for these complexes (**56**) rather than monomeric structures with chelating (Se, Se) (**55**).

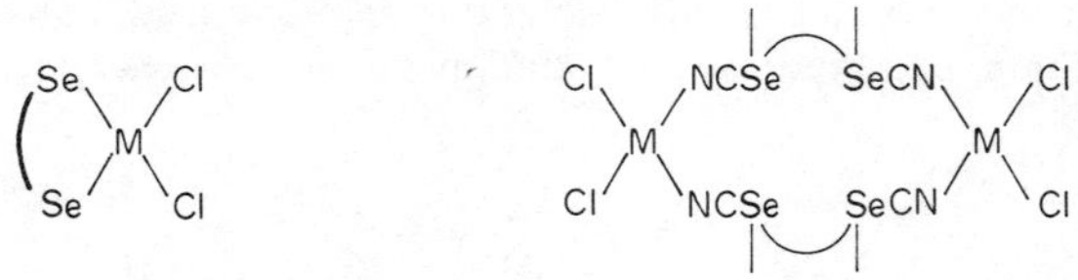

(**55**) (**56**)

Spectroscopic and magnetic data support an octahedral polynuclear formulation for the pink, very hygroscopic complex with $CoCl_2$[441].

The more stable, brown, diamagnetic Rh(III) and Ir(III) complexes, however, were assigned chloro-bridged dimeric structures (**57**) with chelating Se ligands (infrared spectra,

MW in sulpholan solution, non-electrolytes in DMSO). These dimers undergo typical bridge cleavage reactions to give monomeric products[441] (equation 175).

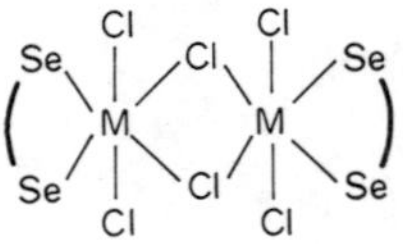

$\nu_{Ir-Cl} = 315, 276\ cm^{-1}$
$\nu_{Rh-Cl} = 339, 278\ cm^{-1}$
$\nu_{Rh-Cl-Rh} = 238\ cm^{-1}$

(**57**)

$$[MCl_3(Se, Se)]_2 \xrightarrow[\text{MeC(O)Et, reflux}]{p\text{-}MeC_6H_4NH_2} Cl_3(p\text{-}MeC_6H_4NH_2)M(Se, Se) \quad (175)$$

$\nu_{Rh-Cl} = 336, 275\ cm^{-1}$
$\nu_{Ir-Cl} = 312, 275\ cm^{-1}$

Complexes of the (Se, Se) chelating ligand, $EtSeCH_2CH_2SeEt$, with Pt(II)[5] and Pd(II)[6] were prepared in the early work of Fritzmann (equations 176 and 177).

$$K_2[PtCl_4] \xrightarrow[\text{(Se, Se)}]{EtSe(CH_2)_3SeEt} cis\text{-}[PtCl_2(Se, Se)] + [Pt(Se, Se)_2][PtCl_4] \quad \text{(Ref. 5)} \quad (176)$$

yellow, m.p. 176.5 °C, soluble in $CHCl_3$. — brown, m.p. 176 °C, insoluble in $CHCl_3$

$$K_2[PdCl_4] \xrightarrow[\text{(Se, Se)}]{EtSe(CH_2)_3SeEt} PdCl_2(Se, Se)] \quad \text{(Ref. 6)} \quad (177)$$

yellow
insoluble in organic solvents
m.p. 181 °C dec.

The pure monomeric *cis*-Pt complex (equation 176) was prepared by using 2 equiv. of the Se ligand and heating the reaction mixture on a steam bath or by heating the reaction mixture of equation (176) to 110 °C, which isomerizes the Magnus-type salt to the monomer[5].

A number of other complexes with $RSe(CH_2)_nSeR$ ($n = 2$, R = Me, *i*-Pr, Ph; $n = 3$, R = Me) have been reported (Table 32). Much of the interest in these complexes centred on variable-temperature ^{1}H-NMR studies of inversion processes in such ring systems[444–446,449,450,452,454].

The complexes $[MX_2(i\text{-}PrSeCH_2CH_2CH_2SePr\text{-}i)]$ (X = Cl, Br; M = Pd, Pt)[447] were isolated from aqueous ethanol reaction solutions as monomers, but on dissolution in $CHCl_3$, dimerization occurred with formation of Se bridges (equation 178). The

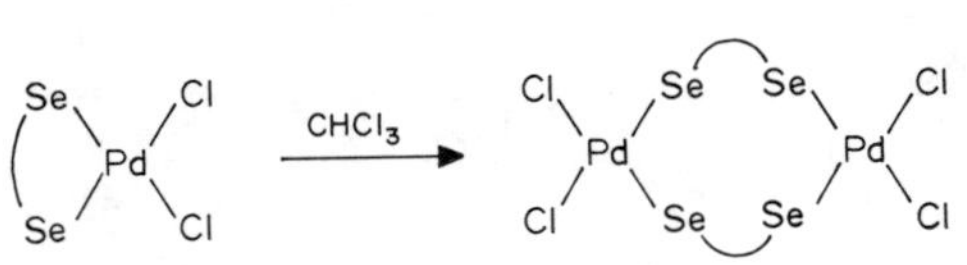

(178)

MW(acetone) = 468 (calc.) = 449.5 — MW = 950

$\nu_{Pd-Cl} = 319, 298\ cm^{-1}$

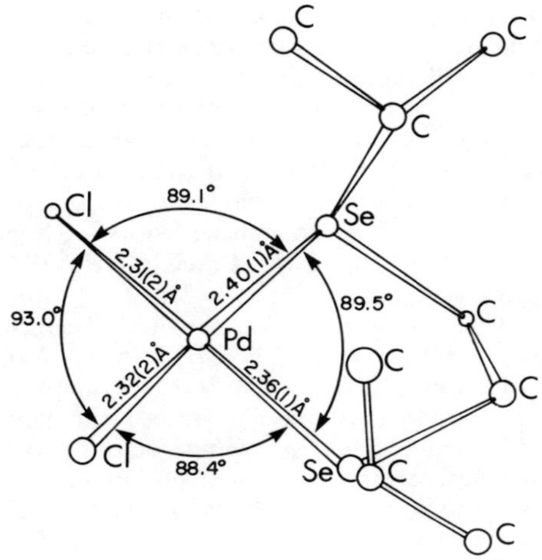

FIGURE 7. Molecular structure of $PdCl_2$(*i*-$PrSeCH_2CH_2SePr$-*i*)

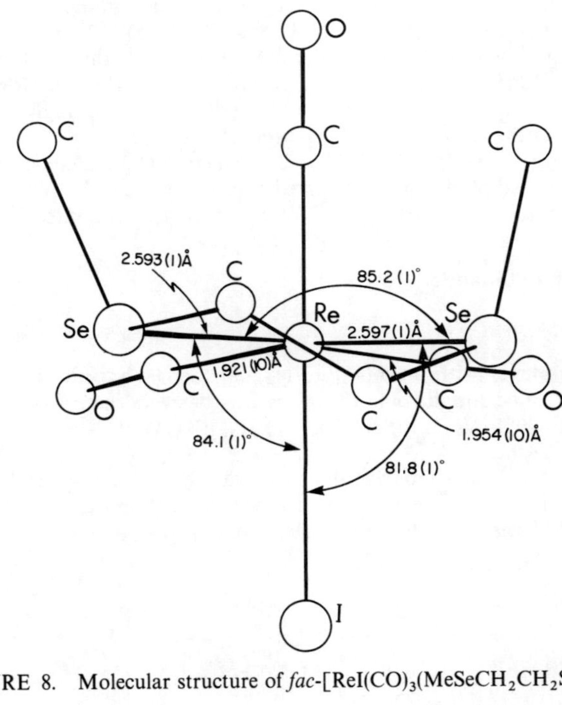

FIGURE 8. Molecular structure of *fac*-[$ReI(CO)_3(MeSeCH_2CH_2SeMe)$]

formulation of the dimers as Se-bridged species rather than the isomeric halo-bridged compounds was based on their failure to give bridge-cleavage monomeric products with *p*-toluidine, a reaction typical of the latter dimers (e.g. equation 175). In addition, the ν_{M-Cl} vibrations are essentially identical in the monomers and dimers, but bridging halo ligands would be expected to have decreased values for such vibrations.

Octahedral monomeric Ni(II) complexes, [NiX_2(*i*-$PrSeCH_2CH_2SePr$-*i*)$_2$], were formulated on the basis of conductometric, magnetic and spectroscopic (infrared and electronic) evidence[451]. Reaction of these complexes with KSCN gave the square-planar, paramagnetic ($\mu_{eff} = 3.40$ B.M.) [$Ni(NCS)_2$(*i*-$PrSeCH_2CH_2SePr$-*i*)], assigned an isothiocyanato bonding mode ($\nu_{C\equiv N} = 2105$, $\nu_{C-S} = 776$, $\delta_{N\equiv C-S} = 467$, $458\,cm^{-1}$).

Two complexes with chelating ligands of this class have been characterized by single-crystal X-ray diffraction: [$PdCl_2$(*i*-$PrSeCH_2CH_2SePr$-*i*)][424] (Figure 7) and *fac*-[$ReI(CO)_3(MeSeCH_2CH_2SeMe)$][422] (Figure 8). Several examples of complexes containing chelating phosphine selenides have been described ($Ph_2P(Se)(CH_2)_nP(Se)Ph_2$, $n =$ 1[128,134], 2[134]; Table 32). Difunctional phosphine selenides with longer methylene chains ($n = 3, 4, 6$) bridge metal centres rather than chelate to a single metal (see Section II.A.3). The difunctional phosphine selenides $Ph_2P(Se)(CH_2)_nP(Se)Ph_2$ ($n = 1, 2$), prepared by reaction of the phosphine with KSeCN in MeCN[140], form (Se, Se) chelate complexes with Cu(I), Ag(I), Hg(II), Pd(II) and Pt(II).

The copper complex [$Cu(Ph_2P(Se)CH_2P(Se)Ph_2)_2$]$ClO_4$[134] precipitated from an $EtOH/CHCl_3$ solution of $Cu(ClO_4)_2 \cdot 6\,H_2O$ and the selenide in a 1:2 molar ratio. In contrast, the analogous phosphine sulphide gave either a Cu(II) or Cu(I) complex, depending on the reaction conditions (e.g. acetone solvent, Cu(II) complex isolated; acetone solvent/hypophosphorous acid or ethanol solvent, Cu(I) complex isolated)[134].

As discussed for monodentate phosphine selenides, a shift of the $\nu_{P=Se}$ band to lower energy is characteristic of Se coordination of this class of ligands. This band was not assigned in [$Co(Ph_2P(Se)CH_2P(Se)Ph_2)X_2$] (X = Cl, Br, I)[140d], but in several other complexes of Rh(I)[128], Cu(I)[134], Ag(I)[134] and Hg(II)[134] with this chelating phosphine selenide, shifts of ca. $10\,cm^{-1}$ to lower energy were observed vs. the value for the free ligand ($Ph_2P(Se)(CH_2)_nP(Se)Ph_2$; $n = 1$: $531\,cm^{-1}$; $n = 2$: $530\,cm^{-1}$). Although no complex with a chelating phosphine selenide ligand has been characterized by single-crystal X-ray diffraction, [$Cu(Ph_2P(S)CH_2P(S)Ph_2)Cl$] has been so characterized and shown to contain bidentate phosphine sulphide in a distorted trigonal-planar Cu(I) complex[134]. The decrease in $\nu_{P=S}$ vs. the free ligand was ca. $60\,cm^{-1}$ for this complex[134].

D. Anionic Bidentate Ligands

1. (Se, N)$^-$ and (Se, P)$^-$ donors

Only a few complexes with anionic (Se, N) and (Se, P) hybrid chelates have been reported (Table 33), but examples of both classes have been characterized by single-crystal X-ray diffraction ([$Co(en)_2(SeCH_2CH_2NH_2)$]$(NO_3)_2$[457] and *trans*-[Ni(*o*-$Ph_2PC_6H_4Se)_2$][464]).

The first reported hybrid ligand of this class was the Se analogue of dithizone, the reagent widely used in trace-metal analysis. This ligand (diphenylselenocarbazone, 3-seleno-1,5-diphenylformazan, 'selenazone', szH, **58**) was prepared in two steps from 3-nitro-1,5-diphenylformazan[456] (equation 179).

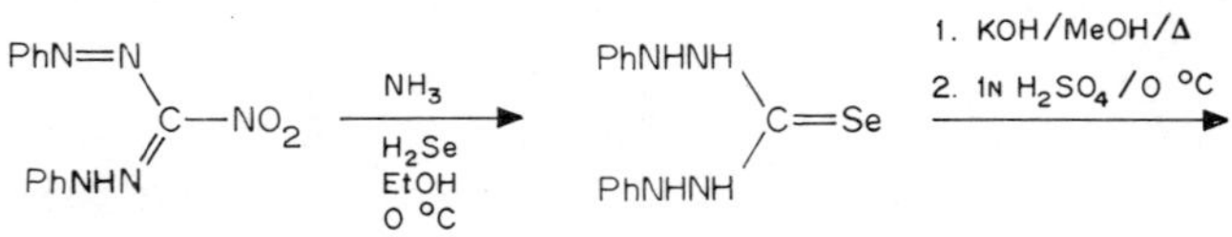

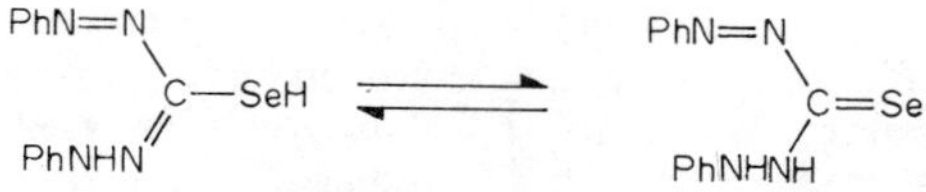

(58a) (58b)

black microcrystals ($CHCl_3$) (179)

m.p. 116 °C

$\nu_{C \cdots Se} = 705\,cm^{-1}$

This reagent forms highly coloured complexes with many metals that can be extracted into organic solvents over a range of pH values, but none of these complexes was isolated and structurally characterized.

A number of dithizone (dzH, i.e. S analogue of **58**) complexes, however, have been characterized, and bonding modes of this ambidentate ligand involving (S, N) chelation ([MeHg(dz)][465], $[Zn(dz)_2]$[466], $[Ni(dz)_2]$[467]) and monodentate S coordination ($[Hg(dz)_2(pyr)_2]$[468]) have been established by single-crystal X-ray diffraction.

An unusual trigonal-bipyramidal structure has been established for $In(dz)_3$[469], one

TABLE 33. $(Se, N)^-$ and $(Se, P)^-$ ligands and their coordination complexes

Ligand	Complex	Reference
(Se, N)⁻ donors		
sz^-, (**58** – H)	$[M(sz)_2]$	456
	M = Mn, Ni, Fe, Co, Cu, Zn, Cd, Hg, Pb	
$H_2NCH_2CH_2Se^-$	$[Co(en)_2(Se,N)]X_2$ (**59**)	
	$X = NO_3$	457[a]
	$X = ClO_4$,	427–429, 458, 459
	$[Co(tren)(Se,N)](ClO_4)_2$	430
	tren = $N(CH_2CH_2NH_2)_3$	
$H_2NCH_2CH(Me)Se^-$	$[Co(en)_2(Se,N)]^{2+}$	460
	$[Co(tren)(Se,N)]^{2+}$	
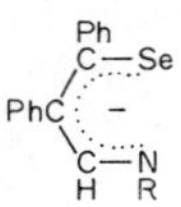	$[Ni(Se,N)_2]$ (**60**)	461
R = Me, Et, *i*-Pr		
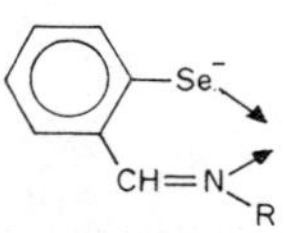	$[M(Se,N)_2]$ (**61**)	
	M = Zn, Ni; R = *p*-Tol, *p*-An	462
	M = Ni; R = Ph, *o*-Tol, *n*-Bu, *i*-Pr, C_6H_{11}	
	M = Co; R = *i*-Pr, C_6H_{11}, *o*-Tol	463
	M = Zn; R = Ph, *n*-Bu	463
(Se, P)⁻ donors		
o-$Ph_2PC_6H_4Se^-$ (ps)	$[Ni(ps)_2]$ (**62**)	433, 464[a]
	$[Pd(ps)SCN]_2$ (**63**)	433
	$[Pd(ps)_2]$ (**64**)	433
R_2P^-(=Se)	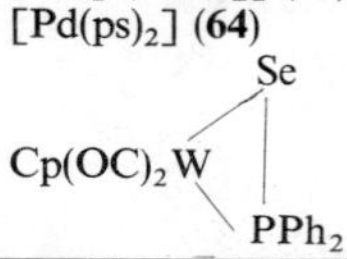	143

[a] Single-crystal X-ray diffraction structure determination.

ligand being coordinated only through a S atom in an equatorial position while the other two dz ligands are bidentate (N, S), spanning axial and equatorial positions.

The tris-chelate Co(III) complex $[Co(en)_2(SeCH_2CH_2NH_2)]^{2+}$ has been prepared by oxidation of a deaerated Co(II)(en) solution with the diselenide[457] (equation 180). The X-ray structural characterization of this complex (**59**) showed a *trans* effect, reflected in a lengthening of the Co—N bond *trans* to the Se donor[457]. This redox route to metastable Co(III) complexes containing reducing ligands has also been applied to the corresponding thiolate complex[457].

$$2\,Co(ClO_4)_2 + 5\,en + (H_3NCH_2CH_2Se)_2SO_4 \xrightarrow{H_2O} 2\,[Co(en)_2(Se,N)](ClO_4)_2 + [enH_2]SO_4 \qquad (180)$$

53%
brown needles

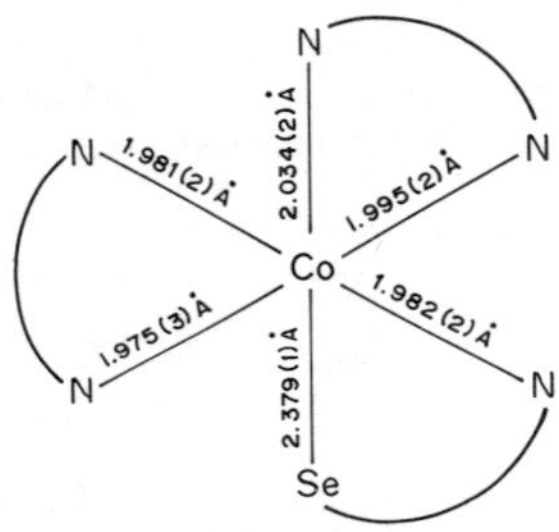

(**59**)

Subsequent studies of this complex have demonstrated the nucleophilicity of the Se atom[427–429] (e.g. alkylation of the coordinated Se; see Section II.C.2). Optical isomers of this chelate system (and the related $[Co\{N(CH_2CH_2NH_2)_3\}SeCH_2CH_2NH_2]^{2+}$ [430] and $[Co\{N(CH_2CH_2NH_2]_3\}SeCH(Me)CH_2NH_2]^{2+}$ [460] chelates) have been isolated and characterized by electronic and circular dichroism spectra. A detailed study of the solution photochemistry of this complex by Adamson and coworkers[458] showed only photoredox decomposition at all wavelengths. The 2-selenolatoethylamine ligand was also oxidized to

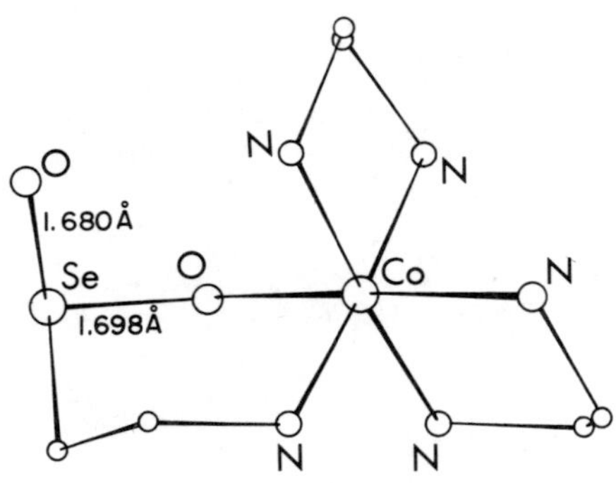

FIGURE 9. Coordination geometry of $[Coen_2(SeO_2CH_2CH_2NH_2)]^{2+}$

the seleninato derivative (—$Se(O)_2CH_2CH_2NH_2$) by treatment of the Co(III) chelate with aqueous H_2O_2. Single-crystal X-ray diffraction of this complex established an (N, O) bonding mode of this ambidentate ligand (Figure 9). The six-membered chelate ring has a chair conformation. The spontaneous resolution of the nitrate salt to give the $(-)^{CD}_{500}$ isomer occurred on cooling an aqueous solution of the racemic salt[470].

Ni(II) aldimine chelates (**60**) have been prepared by reaction of the 1,2-diphenyl-2-formylvinylselenol chelate (equation 181) with primary amines[461].

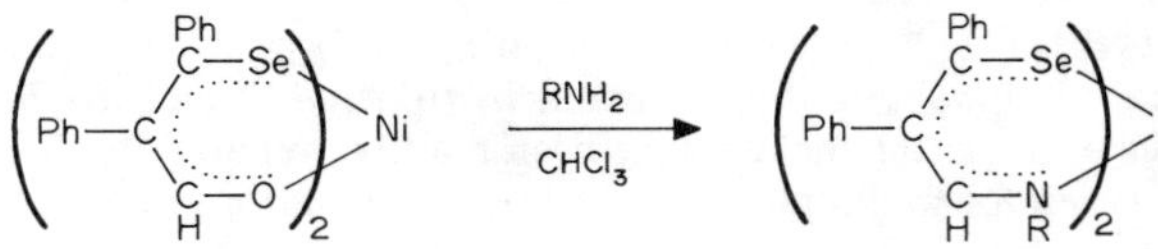

(181)

(**60**)

$\nu_{C\cdots O} = 1560\,cm^{-1}$

R =	Me	Et	*i*-Pr
$\nu_{C\cdots N} =$	1608	1605	$1610\,cm^{-1}$

Selenosalicylideneamine chelates of Ni(II)[462,463] (**61**), Zn(II)[462,463] and Co(II)[463] have been prepared by template-type reactions (equation 182). The air-stable Ni(II) chelates (**61**) can be recrystallized from dioxane, melt without decomposition and are readily soluble in organic solvents. They are diamagnetic in the solid state and solution and have been assigned square-planar geometries. The retention of this stereochemistry in solution contrasts with the square-planar (solid state) → tetrahedral (solution) transition observed in the analogous salicylideneaminato complexes[471,472]. The analogous bis(thiosalicylideneaminato) chelates give tetrahedral paramagnetic complexes at high temperature in DMSO[473]. The order of stability of the square-planar geometry in these complexes, therefore, is (Se, N) > (S, N) > (O, N). The dipole moments of several of these complexes have been determined, the high values (5.8–7.7 D) in C_6H_6 indicating a *cis* configuration.

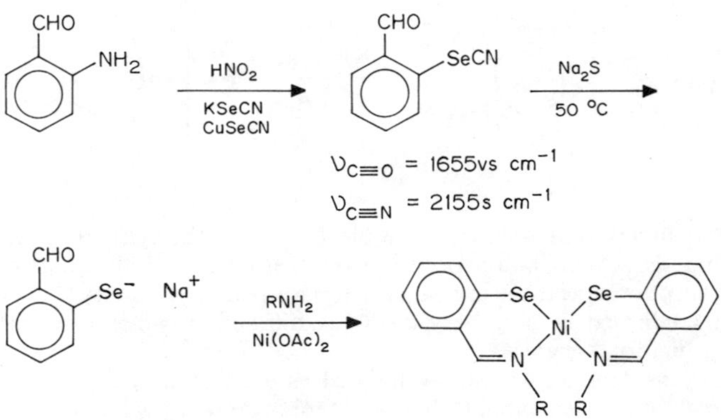

(182)

(**61**)

The Co(II) complexes (Table 33), however, have tetrahedral geometries in the solid state at room temperature ($\mu_{eff} = 4.81$ B.M., $R = C_6H_{11}$; $\mu_{eff} = 4.96$ B.M., R = *i*-Pr; $\mu_{eff} =$ 4.21 B.M., R = *o*-Tol). In the one case where solubility allowed a dipole-moment measurement (R = *c*-HeX) the dipole moment of 4.16 D agreed with the proposed tetrahedral structure.

The Zn(II) chelates have the expected tetrahedral structures.

Axial ligation of $^{-}SeCH_2CH_2NMe_2$ in the bleomycin—Fe(III)—$SeCH_2CH_2NMe_2$ and haemoglobin—Fe(III)—$SeCH_2CH_2NMe_2$ complexes has been confirmed by the rapid-freezing ESR technique[474]. Here the *N,N*-dimethylselenocysteamine functions as a monodentate Se ligand because of the stable planar arrangement of the FeN_4 coordination spheres in these systems.

Other than the tungsten complex with chelating $SePPh_2^{-}$ (equation 20), the only example of an anionic bidentate (Se, P) ligand is diphenyl-*o*-selenatophenylphosphine (ps)[433]. Complexes of the latter chelate have been readily prepared by thermolysis of the diphenyl(*o*-methylselenophenyl)phosphine precursors (equations 183–185).

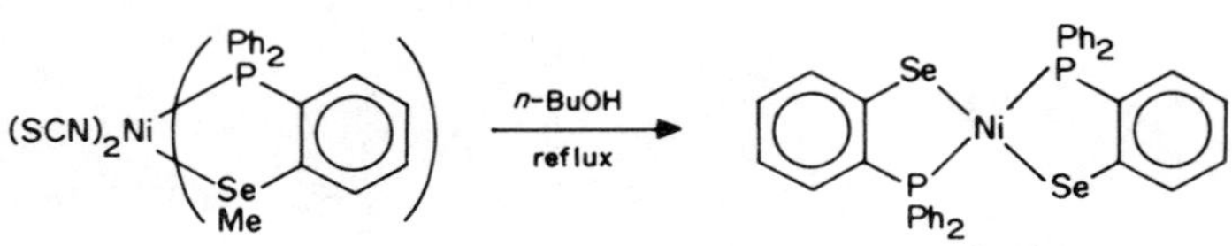

(62) (183)

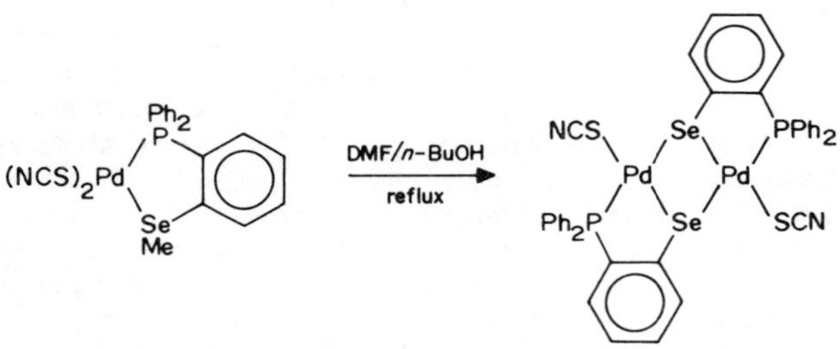

(63) (184)

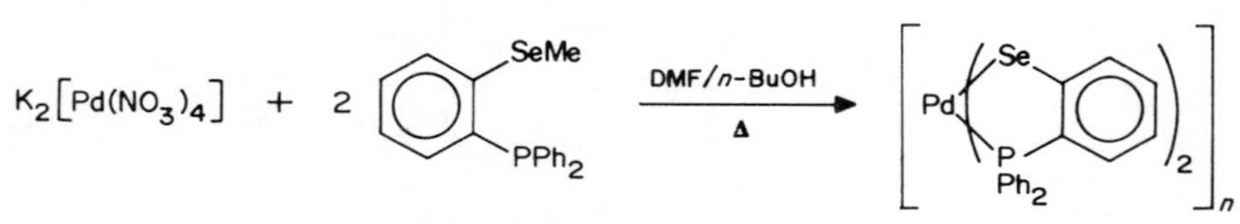

(64) (185)

The Ni(II) complex (**62**) was very soluble in $CHCl_3$, and its ^1H-NMR spectrum confirmed that *Se*-demethylation occurred on thermolysis[433]. Its diamagnetism and electronic absorption spectrum indicated a square-planar geometry[433], a formulation subsequently confirmed by single-crystal X-ray diffraction, which established a *trans*-NiP_2Se_2 coordination sphere[464].

The thermolysis product **63** was formulated as a Se-bridged dimer on the basis of molecular weight measurement in $CHCl_3$ and the single sharp $\nu_{C\equiv N}$ band at 2110 cm^{-1}, characteristic of a terminal S-bonded thiocyanate[433]. The insolubility of the complex **64**, however, indicated a polymeric formulation[433].

2. (Se, E)⁻ (E = O, S, Se) donors

Although not as numerous as the various classes of dithio ligands[12], a number of complexes with anionic (Se, Se) ligands and their (Se, O) and (Se, S) hybrid analogues have

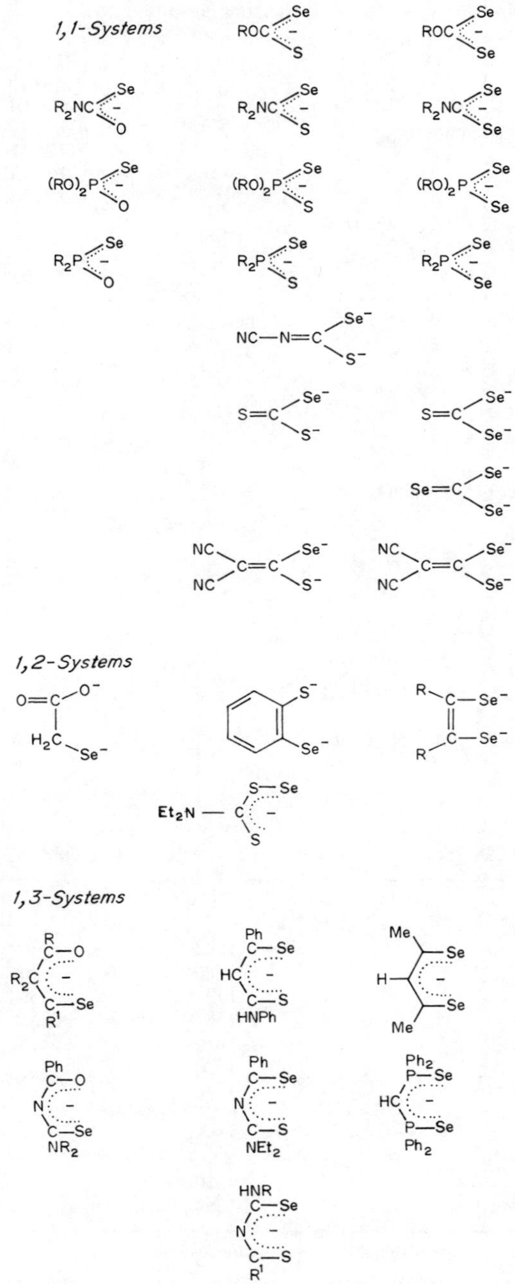

FIGURE 10. Anionic (Se, E)$^{n-}$ (E = O, S, Se; $n = 1, 2$) ligands

TABLE 34. Complexes with anionic (Se,O) chelating ligands

Ligand	Complex	Ligand	Complex
1,1-*(Se,O)*$^-$ *donors*		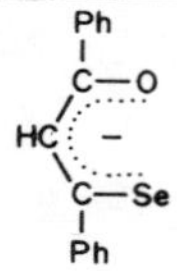 (70)	$[M(Se,O)_n]$[480,481] $n = 1$; M = Tl; $n = 2$; M = Ni, Pd, Hg, Zn, Pb, Cd; $n = 3$; M = Co, In
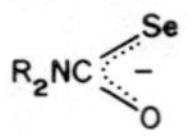 (65)	$Me_2Sn(SeOCNR_2)_2$[394] R = Me, Et		
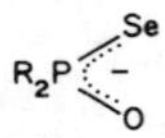 (66)	$Pd(SeOCNEt_2)_2$[393]; $M(SeOPEt_2)_2$[475] M = Zn, Cd, Ni, Pb; $Ph_2Sn(SeOP(OPr\text{-}i)_2)_2$[476]	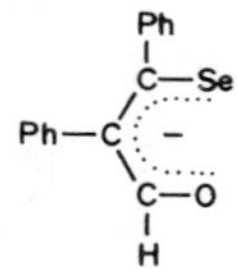 (71)	$[Ni(Se,O)_2]$[461]
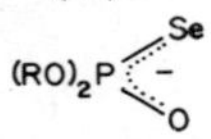 (67)	$M(SeOP(OPr\text{-}i)_2)_n$[477] n = 1; M = Ag; n = 2; M = Zn, Cd, Pb; n = 4; M = Th		
1,2-*(Se, O)*$^-$ *donors*			
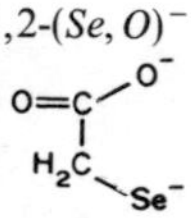 (68)	$[Co(en)_2(Se,O)]ClO_4$[457]	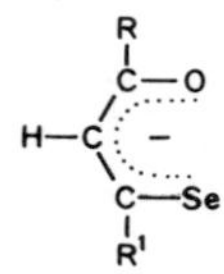 (72)	$[M(Se,O)_2]$[482] M = Fe, Co, Ni, Cu, Zn, Hg, Pb, Cd, Pd; R = CF_3, R^1 = 2-thienyl
1,3-*(Se,O)*$^-$ *donors*			
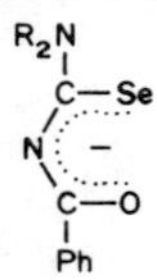 (69)	$[M(Se,O)_n]$ R = Et[478] $n = 2$; M = Ni, Pd; $n = 3$; M = Co; R = *i*-Bu[479] $n = 2$; M = Ni, Cu (73)		

TABLE 35. Infrared data (cm^{-1}) for bidentate selenocarbamate complexes

Complex	$\nu_{C=O}$[a]	Misc. bands	Ref.
$Me_2Sn(SeOCNMe_2)_2$	1592s	337m[b]	394
$Me_2Sn(SeOCNEt_2)_2$	1586s	322m[b]	
$Me_2ClSn(SeOCNMe_2)$	1590s	343m[b]	
$Me_2ClSn(SeOCNEt_2)$	1587s	325m[b]	
$Me_3Sn(SeCONMe_2)$[c]	1620s	328m[b]	
$(SeC(O)NMe_2^-)$	1620		
$(Ph_3P)_2Pd(SeOCN(Pr\text{-}i)_2)_2$	1590s (1589s, 1537m)[d]		393
$(Ph_3P)_2Pt(SeOCN(Pr\text{-}i)_2)_2$	1598s (1599s, 1540sh)[d]		393
$(OC)_4Mn(SeOCNMe_2)$	1550[e]		397
$CpMo(CO)_2(SeOCNMe_2)$	1560		400

[a] Values for $\nu_{C=O}$ for monodentate Se-bonded complexes are given in Table 30.
[b] ν_{Sn-Se}.
[c] Contains monodentate selenocarbamate.
[d] In CH_2Cl_2.
[e] The spectrum of this unstable monomeric chelate was measured in cyclohexane (see equation 144).

been prepared (Figure 10). The following sections summarize the complexes reported for each of these three structural types of ligands along with their characteristic chemical and spectroscopic properties.

a. (Se, O)$^-$ donors. Complexes with monodentate Se-bonded selenocarbamates were discussed in Section II.B.6. Typical complexes containing (Se, O)-chelated selenocarbamates (**65**), selenophosphinates (**66**) and selenophosphates (**67**) along with other anionic (Se, O) ligands are listed in Table 34.

The first reported bidentate selenocarbamate complexes, $Me_2Sn(OSeCNR_2)_2$, $Me_2ClSn(SeOCNR_2)$ (R = Me, Et), are air-stable solids whose infrared spectra give $\nu_{C\cdots O}$ bands (Table 35) at lower energy than that found for the complexes with Se-bonded monodentate selenocarbamate (see Section II.B.6). Weak Sn—O bonds were suggested for these chelates because of the large intensity ratios of $\nu_{sym}(Sn—C)/\nu_{asym}(Sn—C)$ ($Me_2Sn(Se, O)_2$, ca. 0.6–0.95 vs. values of ca. 0.2 for the corresponding $Se_2CNR_2^-$ and $SeSCNR_2^-$ complexes; $Me_2ClSn(Se,E)$ (E = O, S, Se), ca. 0.95).

The ^{1}H-NMR spectra of these chelates show doublets for the N—Me (and doublets of a quartet and a triplet for $N—CH_2CH_3$) due to restricted rotation about the C—N bond. The coalescence temperatures for these proton signals are in the range 50–55 °C ($PhNO_2$ solvent). The latter values, compared to a coalescence temperature of 110 °C for $Me_2ClSn(SSeCNMe_2)$, indicate a rather low double-bond character for these selenocarbamate complexes.

A series of complexes $(Ph_3P)_2M(SeOCNR_2)_2$ (M = Pd, Pt; R = Me, Et, *i*-Pr, *n*-Bu) have been prepared and characterized in the solid state and in solution by infrared, ^{1}H- and ^{31}P-NMR spectroscopy[392,393]. These data, together with molecular weight measurements, established a solution equilibrium (equation 143) involving phosphine dissociation and chelation of the selenocarbamate which is monodentate Se-bonded in the solid state. This reaction is reflected in the appearance of a new $\nu_{C\cdots O}$ band at ca. 1540 cm^{-1} due to the coordinated C=O in the solution infrared spectra of these complexes (the uncoordinated C=O in the solid-state structure has its absorption at ca. 1590 cm^{-1}; Table 35). A similar equilibrium was found for $(Ph_3P)_2Ni(SeC(O)NEt_2)_2$[391] (Table 30).

A brief report of the diethylselenophosphinato (**67**) chelates, $[M(SeOPEt_2)_2]$ (M = Co, Zn, Cd) has appeared[475]. Owing to the instability of these complexes, they were not characterized, and attempts to prepare chelates of Ni(II), Pd(II), Pt(II) and Bi(III) led only to decomposition products[475].

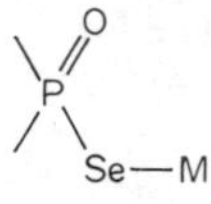

(selenol)

$Me_3SiSeP(O)(OPr\text{-}i)_2$
$Ph_3GeSeP(O)(OPr\text{-}i)_2$
$As(SeP(O)(OPr\text{-}i)_2)_3$

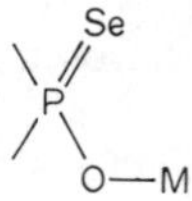

(selenone)

$Me_3SnOP(Se)(OPr\text{-}i)_2$
$P(OP(Se)(OPr\text{-}i)_2)_3$

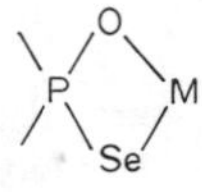

(chelate)

$Ph_2Sn(SeOP(OPr\text{-}i)_2)_2$
$PhSn(SeOP(OPr\text{-}i)_2)_3$

Complexes of diisopropylselenophosphate, $SeOP(OPr\text{-}i)_2^-$, with a variety of main-group elements[476] and transition metals[477] have been isolated (Table 34). The main-group compounds were isolated as air-stable liquids but are very sensitive to heat and light, readily depositing elemental Se[476]. The three bonding modes illustrated above were suggested by the infrared and ^{31}P-NMR spectra of these complexes[476].

The infrared evidence for the bidentate coordination mode was equivocal, but the ^1J values measured for $Me_3Sn(SeOP(OPr\text{-}i)_2)$ (644.1 Hz) and $Ph_2Sn(SeOP(OPr\text{-}i)_2)_2$ (545.9 Hz) supported (Se, O) chelates[476]. Similar spectroscopic evidence suggested (Se, O) chelation in other complexes (Table 34, 850 Hz > $^1J_{PSe}$ > 550 Hz) and monodentate Se coordination ($^1J_{PSe}$ < 550 Hz) in $M(SeP(O)(OPr\text{-}i)_2)_n$ ($n = 2$, M = Hg, Se, Te; $n = 3$, M = As, Sb, Bi)[77]. All of these complexes are unstable.

The tris-chelate $[Co(en)_2SeCH_2CO_2]ClO_4$[457] was prepared by the oxidation of a $Co(ClO_4)_2$/en solution with $(HO_2CCH_2Se)_2$ as previously described (equation 180) for the $^-SeCH_2CH_2NH_2$ analogue. Stable chelates derived from 1,1-dialkyl-3-benzoylselenourea, $R_2NC(Se)NHCOPh$ (**69**) (Table 34) readily precipitated from solutions of the neutral precursor and appropriate metals salts[478,479] (equation 186).

$$KSeCN + PhC(O)Cl + (i\text{-}Bu)_2NH \xrightarrow{\text{acetone}} (i\text{-}Bu)_2NC(Se)NHC(O)Ph \quad (186)$$

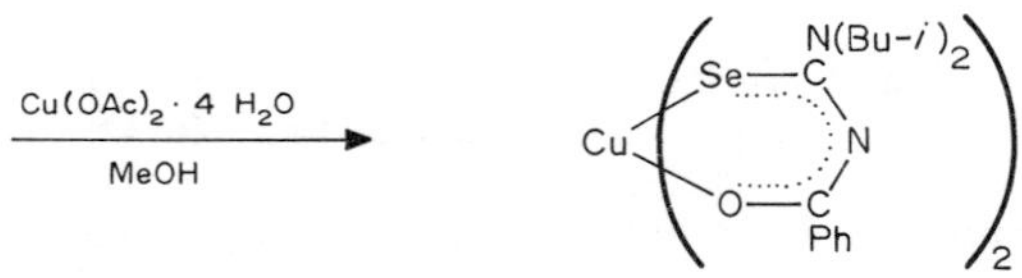

(**73**)

The Ni(II) chelate was also prepared in a one-step procedure in which $Ni(OAc)_2 \cdot 4\,H_2O$ was added to the 1,1-diisobutyl-3-benzoylselenourea formed *in situ*[479].

Hoyer and coworkers, who have extensively studied metal chelates with various anionic (Se, E) (E = O, S, Se) donors, have reported a mass spectroscopic investigation of these complexes[479].

The anion of monoselenodibenzoylmethane (**70**) gives chelates with a variety of metals (Table 34) on reaction of a $CHCl_3$ solution of the ligand with the metal salt in ethanol solution mixed with an aqueous NaOAc buffer[481]. The Ni chelate gave 1:1 adducts with phen and bipyr. Characterization of these complexes was limited to elemental analysis and electronic spectroscopy[481].

The chelate ligand 1,1,1-trifluoro-4-(2-thienyl)-4-seleno-3-butene-2-one (**72**) was obtained as an air-sensitive red oil by the reaction of H_2Se with the corresponding diketone in absolute ethanol in the presence of HCl[482]. This ligand was evaluated as an extractant in $CHCl_3$ solution for various metals salts (Table 34). All of these gave highly coloured solutions, which were characterized by electronic spectroscopy.

b. (Se, S)$^-$ donors. Complexes of several classes of ligands of this general type are known (Table 36) and in two cases molecular structures have been established by single-crystal X-ray diffraction ($[Cu(SeSCNEt_2)_2]$[504], $[(n\text{-}Bu)_4N]_2[Ni(SeSC{=}C(CN)_2)_2]$[485]).

A number of complexes of monoselenoxanthates (**74**) have been prepared (Table 36)[485] (equation 187).

Anionic tris-chelates $[Ni(Se, S)_3]^-$, (Se, S) = **74**, were also isolated by use of the appropriate Ni/selenoxanthate stoichiometry[486]. The magnetic moments and electronic spectra of these complexes indicated a trigonally distorted $NiSe_6$ coordination sphere[486].

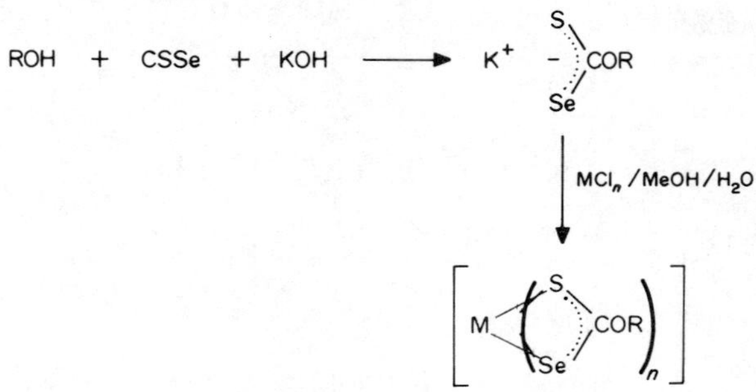
(187)

TABLE 36. Complexes with $(Se, S)^-$ donors

Ligand	Complex	Reference
1, 1-*Selenothio type*		
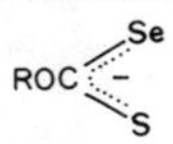	$M(SeSCOEt)_n$	483
	M = Fe, Cu, Pb, Ni; $n = 2$	
	M = Ag; $n = 1$	
(**74**)		
	$Ni(SeSCOEt)_2$	484
	$M(SeSCOCH_2CH_2OMe)_n$	485
	M = Ni, Pd, Pt, Zn; $n = 2$	
	M = Co, Rh, Cr; $n = 3$	
	$[Ni(SeSCOCH_2CH_2OMe)_3]$	486
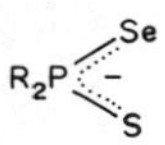	$M(SeSPEt_2)_n$	
	M = Tl; $n = 1$	475, 487
	M = Zn, Cd, Pb, Ni, Pd; $n = 2$	462, 463
(**75**)	M = Bi, Rh; $n = 3$	
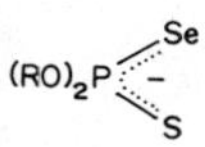	$M(SeSP(OR)_2)_n$	488–490
	M = Pb, Ni, Cu, Zn, Cd, Hg, Pd, Pt; $n = 2$	
	M = Co, Rh, Ir, Cr; $n = 3$	
(**76**)		
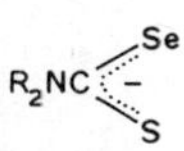	$Me_2Sn(SeSCNMe_2)_2$	491
	$M(SeSCNR_2)_n$	
	$n = 1$	492
(**77**)	M = Tl, Ag; R = Et	
	$n = 2$	
	M = Ni, Pd, Pt; R = Me, Et	492
	M = Zn, Cd, Pb, Ni, Pd, Cu; R = Et	493
	$n = 3$	494
	M = Ga, In, Tl, Fe, Co, Rh, Cr, Mn; R = Et	
	$L_2Au(SSeCNR_2)$	494
	L = Cl, Br, I, Me, Et; R = Me, Et	
	$FeCl(SSeCNR_2)$	495
	R = Me, Et	
	$(OC)_4Mn(SeSCNMe_2)$	396
	$Pd(PR_3)Cl(SeSCNR^1_2)$	496
	$PR_3 = PPh_3, PMePh_2, PMe_2Ph$; R^1 = Me, Et	

TABLE 36. (*Continued*)

Ligand	Complex	Reference
	$M(SeSCNR_2)_2$ M = Ni, Cu; R = C_6H_{11}, Et, CH_2Ph	497
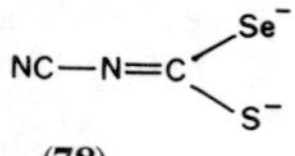 (**78**)	$(Et_4N)_2[M(SeSC{=}NCN)_2]$ M = Ni, Pd	485
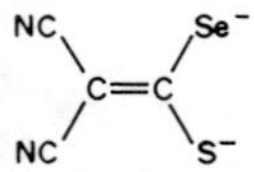 *i*-mnts, (**79**)	$[(n\text{-}Bu)_4N]_a[M(i\text{-mnts})_n]$ M = Au; $a = 1$; $n = 2$ M = Ni, Pd, Pt, Zn, Cd; $a = n = 2$	485
	$(Ph_4As)_n[M(i\text{-mnts})_n]$ M = Se, Te; $n = 2$ M = Cr, Co, Rh, Fe, In; $n = 3$	485
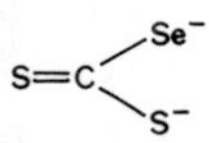 (**80**)	(triphos)CoSSeC=Se	498
1,2-selenothio type		
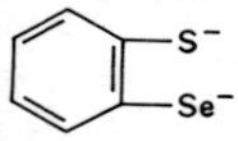 tbs, (**81**)	$Pr_4N[M(tbs)_2]$ M = Ni, Co, Cu	455,499
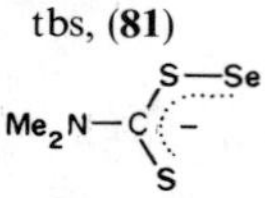 (**82**)	$[Os_2(SeS_2CNMe_2)_2(S_2CNMe_2)_3]PF_6$	500
1,3-Selenothio type		
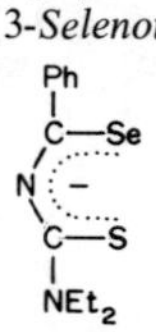 (**83**)	$[Ni(Se,S)_2]$ (**86**)	501
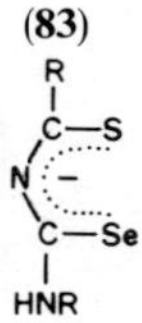 (**84**)	$[Ni(Se,S)_2]$ (**87**) R^1 = Me; R = Ph, *p*-An R^1 = Ph; R = Ph, *p*-Tol, *p*-An	502
	$[M(Se,S)_2]$ (**87**) M = Co, Cu, Zn; R^1 = Ph; R = *p*-An	502
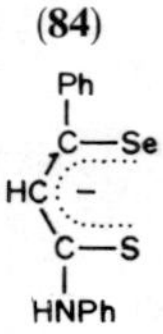	$[M(Se,S)_2]$ (**88**) M = Ni, Cu, Co	502

Because of the poor stability of the selenoxanthate complexes, as well as selenothiophosphates (**76**) and selenothiophosphinates (**75**) little work beyond their isolation has been reported.

The selenothiocarbamate (**77**) complexes are substantially more stable, and spectroscopic data for some of these complexes have been reported. In addition, the complex $[Cu(SeSCNEt_2)_2]$ has been characterized by single-crystal X-ray diffraction[503]. This complex has a *trans*-$[CuS_2Se_2]$ square-planar coordination geometry. The single-crystal ESR spectrum of this complex has also been recorded[504,505].

The rather air-sensitive thioselenocarbamate salts are readily synthesized by reaction of carbon sulphide selenide with a secondary aliphatic amine in a 1:2 molar ratio[492] (equation 188).

$$2\,R_2NH + CSSe \rightarrow [H_2NR_2]^+[SeSCNR_2]^- \qquad (188)$$

Reaction of these salts with appropriate metal salts in aqueous or methanolic solutions gives the metal chelates (Table 36), which are generally air-stable. The Mn(II) and Co(II) salts oxidize during work-up to give the tris-chelates with the trivalent metals[493].

The stereochemistries of these complexes were investigated by the standard methods of coordination chemistry (e.g. electronic spectroscopy and magnetic susceptibility measurements). Because of the nuclear spin of $^{77}Se(I = 1/2$, natural abundance $= 7.58\%)$, ESR spectroscopy can provide useful information on electron density delocalization onto such ligands and therefore has been widely used to study the bonding[504,505] as well as ligand-exchange reactions in these and related Se ligands[480,493,506,507].

Although most of the complexes with selenothiocarbamates were prepared by metathetical reactions, a cleavage reaction involving coordinated Se was used to prepare a Sn(IV) complex[491] (equation 189).

$$3\,Me_2NC(S)Cl + [Me_2SnSe]_3 \xrightarrow{C_6H_6} 3\,Me_2ClSn(SSeCNMe_2) \qquad (189)$$

Repeated extraction of a benzene solution of this product with water gave the bis-chelate[491] (equation 190).

$$2\,Me_2ClSn(SSeCNMe_2) \rightarrow Me_2Sn[SSeCNMe_2]_2 + Me_2SnCl_2 \qquad (190)$$

The strong $\nu_{C\cdots N}$ bands in these complexes above 1500 cm^{-1} (Table 37) and the ν_{Sn-Se} and ν_{Sn-S} bands in the 285–385 cm^{-1} region all support a planar $SnSSeCNC_2$ arrangement:

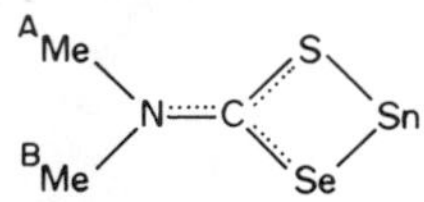

Two Me resonances appear in the ^{1}H-NMR spectra of these complexes (Table 38), indicative of restricted rotation about the C—N bond. The broader lower-field signal is assigned to Me(A) *trans* to the Se atom, since stronger coupling with the ^{77}Se would be expected in this position.

Reaction of $Me_2ClSn(SeSCNR_2)$ (R = Me, Et) with $FeCl_3$ gave the pentacoordinate Fe(III) complexes $[FeCl(SeSCNR_2)_2]$. Characterization by electronic spectroscopy and magnetic susceptibility measurements indicated a square-pyramidal structure for these complexes. In contrast, reaction of $FeCl_3$ with $[Et_2NH_2][SeSCNEt_2]$ gave the tris-chelate[493].

An investigation of the electrochemistry of the chelates $[M(EE^1CNR_2)_2]$ ($E = E^1 = S$, Se; $E = S$, $E^1 = Se$; $R = C_6H_{11}$, Et, CH_2Ph)[497] in acetonitrile showed that the ease of oxidation followed the orders Cu > Ni and $Se_2 > SeS > S_2$.

TABLE 37. Infrared data (cm^{-1}) for some selenothiocarbamate complexes

Complex	$\nu_{C \doteq N}$	ν_{C-S}	ν_{C-Se}	Ref.
$Me_2ClSn(SeSCNMe_2)$	1538s	951s 935s	867w	491
$Me_2Sn(SeSCNMe_2)_2$	1511s	958s	861m	491
$Ni(SeSCNEt_2)_2$	1520s			492
$Pd(SeSCNEt_2)_2$	1515s			492
$Pt(SeSCNEt_2)_2$	1525s			492
$Fe(SeSCNEt_2)_3$	1495			493
$Co(SeSCNEt_2)_3$	1503			493
$Cr(SeSCNEt_2)_3$	1510			493
$Mn(SeSCNEt_2)_3$	1500			493
$Zn(SeSCNEt_2)_2$	1505			493
$Cd(SeSCNEt_2)_2$	1518			493
$Ni(SeSCNEt_2)_2$	1530			493
$Pd(SeSCNEt_2)_2$	1530			493
$Cu(SeSCNEt_2)_2$	1507			493
$Ag(SeSCNEt_2)$	1510			493
$Cl_2Au(SeSCNMe_2)$	1580s			494
$Cl_2Au(SeSCNEt_2)$	1570s			494
$Me_2Au(SeSCNMe_2)$	1545s			494
$Et_2Au(SeSCNMe_2)$	1530s			494
$(OC)_4Mn(SeSCNMe_2)$	1551			396
$(Ph_3P)ClPd(SeSCNMe_2)$	1553			496

TABLE 38. ^{1}H-NMR data for some selenothiocarbamate complexes

Complex	$\delta(-NR_2)$	Ref.
$Me_2Sn(SeSCNMe_2)_2{}^a$	3.40, 3.37	491
$Me_2ClSn(SeSCNMe_2)^a$	3.38, 3.31	491
$Ni(SeSCNEt_2)_2{}^b$	CH_2: 3.65(q), 3.61(q) Me: 1.26(t), 1.23(t)	492
$Pd(SeSCNEt_2)_2{}^b$	CH_2: 3.77(q), 3.72(q) Me: 1.32(t), 1.29(t)	492
$Pt(SeSCNEt_2)_2{}^b$	CH_2: 3.60(q), 3.54(q) Me: 1.32(t), 1.29(t)	492
$Et_2Au(SeSCNMe_2)^b$	3.35, 3.37	494
$Me_2Au(SeSCNMe_2)^b$	3.31, 3.33	494

^{a}In CH_2Cl_2.
^{b}In $CHCl_3$.

Complexes of several metals with the dianionic chelate isomaleonitrile–thioselenolate (*i*-mnts, **79**; Table 36) have been isolated. The ligand was prepared in good yields (80%) from malononitrile and CSSe[485] (equation 191).

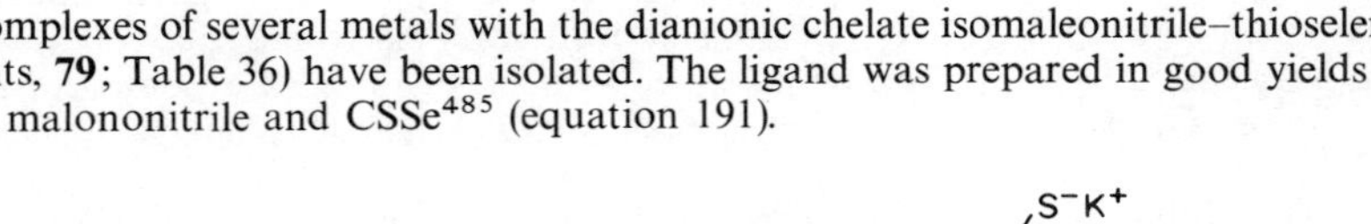

(191)

All of the reported complexes contain the transition metals in their normal oxidation state. In the case of the diseleno analogue **92**, a stable Ni(III) tris-chelate was isolated by oxidation of the Ni(II) precursor (Section II.D.2.c).

A single-crystal X-ray diffraction of $[(n\text{-Bu})_4N]_2[Ni(SeSC{=}C(CN)_2)_2]$ confirmed the proposed *trans* square-planar structure[485]:

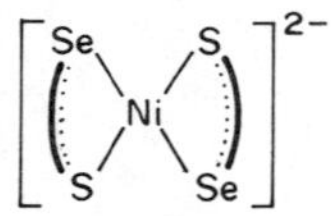

Elemental Se inserts into Co—C bonds of η^2-CS_2 complexes to give new (S, Se) chelates[498] (equations 192 and 193). These facile insertion reactions were cited as evidence for the carbenoid nature of the metal–(η^2-CS_2) and metal–(η^2-CS_2Me) functions[480].

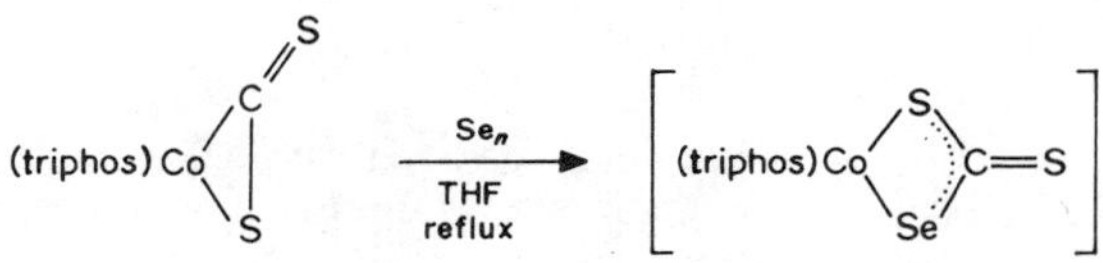

(192)

ν_{C-S} = 790 cm^{-1}

$\nu_{C=S}$ = 1040 cm^{-1}

μ_{eff} (293 K) = 2.03 B.M.

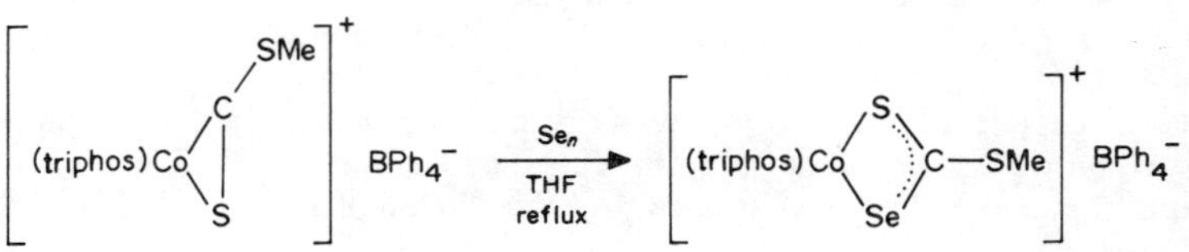

(193)

ν_{C-S} = 960 cm^{-1}

μ_{eff} (293 K) = 2.10 B.M.

The 1,2-selenothio dianion (**81**) (tbs) was prepared by the Na/liquid NH_3 reduction of *o*-methylthioselenoanisole[455,499]. Reaction of this anion with metal halides in ethanol solution gave the $[M(tbs)_2]^{2-}$ (M = Ni, Co, Cu) anions initially, but these oxidized in air to give the monoanionic complexes, which were isolated as their $(n\text{-Pr})_4N^+$ salts. Characterization of these complexes by magnetic susceptibility measurements and electronic spectroscopy indicated square-planar geometries. The polarographic half-wave reduction potentials were also measured as well as the ESR spectrum of the Ni salt. The latter suggested the presence of both *cis* and *trans* isomers at 100 K in DMF–$CHCl_3$ glass.

The first example of a selenodithiocarbamate chelating ligand was prepared recently[500] (equation 194).

$$[Os(S_2CNMe_2)_3] \xrightarrow[\text{2. } NaPF_6/MeOH]{\text{1. } Se_8/DMF \text{ reflux}} [Os_2(\mu\text{-}SeS_2CNMe_2)_2(S_2CNMe_2)_3]PF_6 \qquad (194)$$

10%

$\nu_{C \cdots N} = 1530\ cm^{-1}$

Single-crystal X-ray diffraction established an η^2,μ_2 bonding mode of the selenodithiocarbamate ligands with the dithiocarbamate ligands chelating in the usual manner (**85**)[500].

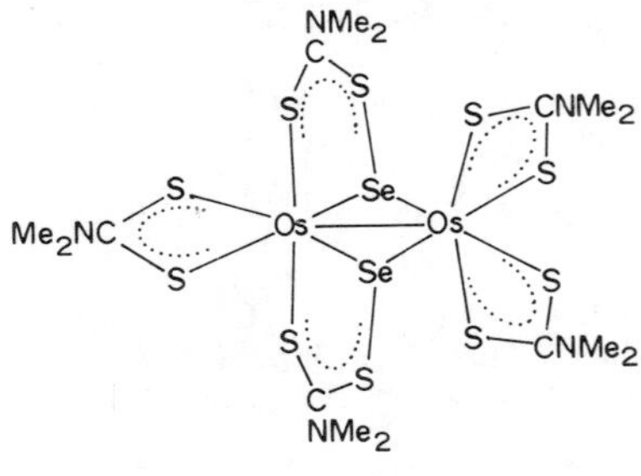

(**85**)

The structure is analogous to the previously characterized trithiocarbamate complex $[Os_2(S_3CNMe_2)_2(S_2CNMe_2)_3]PF_6$[508], although the synthesis of this complex (e.g. equation 194, reaction with elemental S) also gave a product $[Os_2(S_5)(S_3CNMe_2)(S_2CNMe_2)_3]$ with bridging S_5 and S_3CNMe_2 ligands[508].

The location of Se only in the bridging positions of **85** suggests a mechanism involving initial coordination of Se_8 or a lower fragment to $Os(S_2CNMe_2)_3$. A rich and interesting chemistry of complexes with such proposed Se_n ligands is being developed (Section II.G). Brief reports describe the synthesis of the 1,3-selenothio ligands, 1,1-diethyl-3-selenobenzoylthiourea (**83**)[501], 1-thioacyl-3-arylselenoureas (**84**)[502] and 3-selenobenzoylthioacetanilide[502] and some chelates of the deprotonated forms of these ligands (Table 36).

The Ni(II) chelate of 1,1-diethyl-3-selenobenzoylthiourea (**86**)[501] was prepared as shown in equation (195) without isolation of the neutral thiourea[509]. The highest nickel-containing fragment in the mass spectrum of this chelate is m/z 532, $[M - CSSe]^{+\cdot}$.

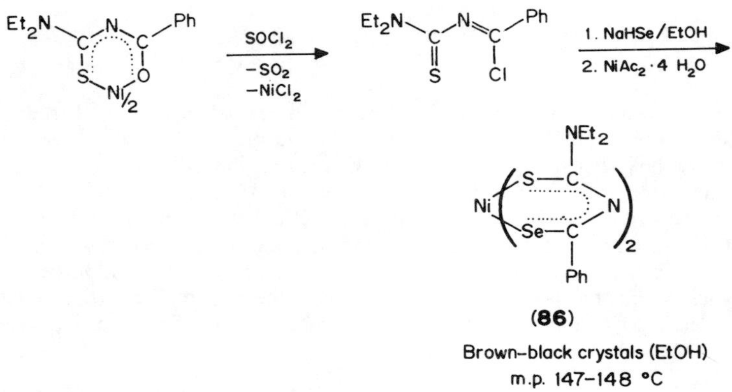

(195)

(**86**)

Brown-black crystals (EtOH)
m.p. 147–148 °C

The 1-thioacyl-3-arylselenourea chelates (**87**)[502] were prepared by addition of an aqueous metal chloride solution to a $CHCl_3$ solution of the freshly prepared selenourea (equation 196). The structurally related 2-selenobenzoylthioacetanilide chelates were prepared as shown in equation (197)[502].

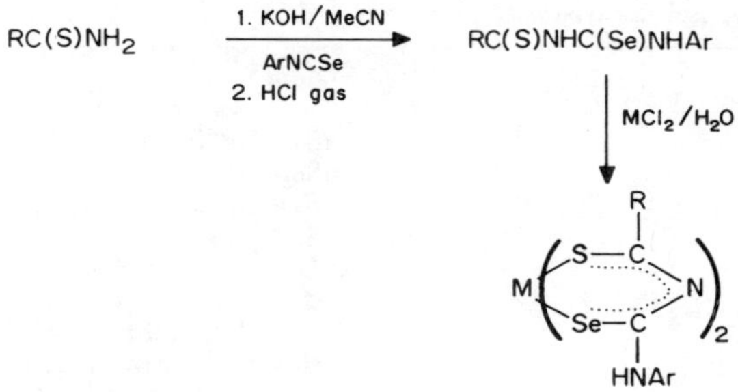

(196)

(**87**)

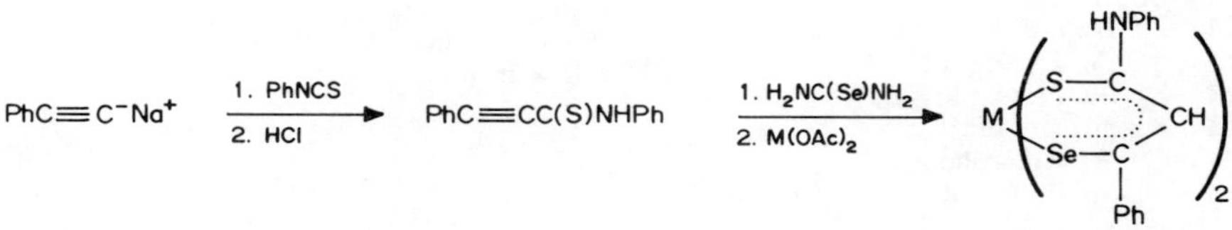

(**88**) (197)

c. (Se, Se)⁻ donors. A variety of complexes incorporating anionic (Se, Se) chelating ligands have been described. Except for the dialkyl diselenocarbamates, the coordination chemistry of these ligands is relatively unexplored, although complexes of several of these types have been structurally characterized by single-crystal X-ray diffraction.

A variety of transition-metal and main-group dialkyldiselenocarbamates have been prepared (Table 39) since the original synthesis of dialkyldiselenocarbamate salts and some transition-metal derivatives by Barnard and Woodbridge[515] (equation 198). In these reactions the dialkyldiselenocarbamate salt solution is used immediately after its preparation.

$$CSe_2 + R_2NH + NaOH \xrightarrow[N_2/-10\,^{\circ}C]{dioxane/H_2O} Na[Se_2CNR_2] \qquad (198)$$

$$R = Me, Et, n\text{-}Bu \qquad \downarrow ZnSO_4/H_2O$$

$$Cu(Se_2CNR_2)_2 \xleftarrow[CuSO_4/H_2O]{CHCl_3} [Zn(Se_2CNR_2)_2]$$

Jensen and coworkers, early investigators in Se ligand chemistry[17b], reported the first detailed study of the synthesis[510] and the infrared[511] and electronic[512] spectra of dialkyldiselenocarbamate complexes. These chelates readily precipitated from solution by the procedure of equation (198), although in this work the $Et_2NH_2[Se_2CNEt_2]$ salt was isolated by carrying out the reaction in ether and then dissolving it in 0.5M NaOH for the subsequent reactions with aqueous solutions of the metal salts[510]. The resulting air-stable chelates can be recrystallized unchanged from organic solvents (e.g. $CHCl_3$)[510] and have d–d transition bands red-shifted vs. the analogous dithiocarbamates[512].

The most characteristic infrared bands of these complexes (Table 40) are the $\nu_{C\cdots N}$ (ca. 1500 cm^{-1}) and the ν_{C-Se} (800–900 cm^{-1}) absorptions, the latter showing more dependence on the nature of the alkyl groups than the former[511]. The location of the $\nu_{C\cdots N}$ band was also dependent on the stereochemistry of the metal complex: square-planar (Ni,

TABLE 39. Dialkyldiselenocarbamate complexes

Complex
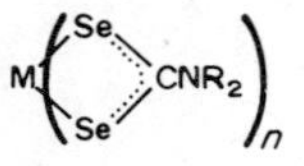
$n = 1$
M = Tl; R = Et[510–512]
M = Ag; R = Et[513]
M = Cu; R = Et[514]
$n = 2$
M = Zn; R = Me, Et, *n*-Bu[515]
M = Cu; R = Et[514,515]
M = Zn, Ni, Pd, Pt; R = Et, *i*-Bu[516–519]
M = Ni, Cu, Zn; R = Et[520a]
M = Ni, Pd, Pt, Cd; R = Et[510]
M = Ni, Pd; R = Et[521]
M = Co; R_2 = morpholino[521]
M = Ni, Pd, Pt, Cu; $R_2 = (CH_2CH_2)_2X$ ($X = CH_2$, O, S)[522]
M = Ni, Cu; R = Et, CH_2Ph, C_6H_{11}[497]
$n = 3$
M = As, Sb, Bi[523]
M = In, Tl, Cr, Rh; R = Et[510]
M = Co, Cr; R = Et[521]
M = Co; R = Et[516]
M = Co, Cr, Fe; $R_2 = (CH_2CH_2)_2X$ ($X = CH_2$, O, S)[524]
$n = 4$
M = Ti, V; R = Et[525]
M = Pt; R = *i*-Bu[516]
$Se(Se_2CN(CH_2CH_2)_2)_2$[526a]
$Pt(Se_2CN(Bu\text{-}i)_2)_2$[516a,518,519,527]
$Pt(Se_2CN(Bu\text{-}i)_2)(PPh_3)Me$[517a,518,519]
$Pt(Se_2CN(Bu\text{-}i)_2)_2(PPh_3)$[518,519,527]
$M(Se_2CNEt_2)(Cl)(PEt_3)$[405,516,519] M = Ni, Pd, Pt
$[M(Se_2CNR_2)_3]BF_4$
M = Ni, Fe, Mn; R = Et[528]
M = Fe; R = Et, CH_2Ph^a; $R_2 = (CH_2CH_2)_2X$ (X = O, NH)[529]
$[Ni(Se_2CN(Bu\text{-}n)_2)_3]Br$[530]
$Cu(Se_2CNEt_2)_2I_3$[514]
$[Pt(Se_2CN(Bu\text{-}i)_2)_2X_2]$[518] X = Br, I

^{a}Characterized by single-crystal X-ray diffraction.

TABLE 40. Infrared data (cm^{-1}) for dialkyldiselenocarbamate complexes

Complex	$\nu_{C\equiv N}$	ν_{C-Se}	Ref.
$Ni(Se_2CNMe_2)_2$	1550vs	890	511
$Co(Se_2CNMe_2)_2$	1528vs	900	511
$Mn(CO)_4(Se_2CNMe_2)$	1537		396
$[Fe(Se_2CNEt_2)_3]BF_4$	1525vs		528
$[Fe(Se_2CN(CH_2Ph)_2)_3]BF_4$	1510vs		529
$Cu(Se_2CNEt_2)$	1489		514
$Cu(Se_2CNEt_2)_2$	1500		514
$Cu(Se_2CNEt_2)_2I_3$	1535		514
$Pt(Se_2CNC_5H_{10})_2{}^a$	1510vs		522
$Pd(Se_2CNC_5H_{10})_2{}^a$	1500vs		522
$Zn(Se_2CNC_5H_{10})_2{}^a$	1480vs		524
$Fe(Se_2CNC_5H_{10})_3{}^a$	1480		524
$Cr(Se_2CNC_5H_{10})_3{}^a$	1488		524

$^a NC_5H_{10}$ = piperidyl.

Cu, Pd, Pt; 1510–1525 cm^{-1}) > tetrahedral (Zn, Cd; 1502–1506 cm^{-1}) > octahedral (Cr, Co, Rh, In, Tl; 1495–1500 cm^{-1}).

The Fe(II) and Co(II) complexes, like the analogous dithiocarbamates, readily react with NO to give $M(Se_2CNEt_2)_2(NO)$, with characteristic strong $\nu_{N=O}$ bands (Co, 1623 cm^{-1}; Fe, 1682 cm^{-1}).

$Ni(Se_2CNEt_2)_2$, again like the dithiocarbamate analogue, reacts with bromide to give the black $Ni(Se_2CNEt_2)_2Br_2$, a formally Ni(IV) complex[510].

The electronic spectra of $CHCl_3$ solutions of these chelates are unchanged on addition of pyridine, indicating the retention of the original coordination spheres in such a solvent system. In contrast, the red $CHCl_3$ solution of the square-planar $Ni(Se_2COEt)_2$ turns yellow on addition of pyridine as a result of formation of the octahedral complex $Ni(Se_2COEt)_2(pyr)_2$, which presumably has the pyridine ligands in the axial positions[510].

The monomeric complex $[Mn(CO)_4(Se_2CNMe_2)]$ undergoes solid-state thermolysis to form the diselenocarbamate-bridged dimer, $(OC)_3Mn(\mu\text{-}Se_2CNMe_2)_2Mn(CO)_3$[396] (e.g. see equation 144). The $\nu_{C\cdots N}$ band of this bridged dimer (1540 cm^{-1}) differs little from that of the monomer (Table 40).

Fackler and coworkers prepared a large number of dialkyldiselenocarbamate complexes[516–519,527] and their characterization of these included ^{77}Se-NMR spectroscopy[518,527] as well as single-crystal X-ray diffraction for $Pt(Se_2CN(Bu\text{-}i)_2)_2$[516], $Pt(Se_2CNEt_2)(PPh_3)(Me)$[517], $Pt(Se_2CNEt_2)(PPh_3)(Cl)$[519] and $Ni(Se_2CNEt_2)(PPh_3)(Cl)$[519].

Diselenocarbamates, like dithiocarbamates, stabilize metal complexes in high formal oxidation states (e.g. Ni(IV)[479,528–530], Cu(III)[479,514], Fe(IV)[528,529]). The structures of the Ni(IV) complex, $[Ni(Se_2CN(Bu\text{-}n)_2)_3]Br$[530], and the Fe(IV) complex, $[Fe(Se_2CN(CH_2Ph)_2)_3]BF_4$[529], were determined by single-crystal X-ray diffraction.

The complex $[Ni(Se_2CN(Bu\text{-}n)_2)_3]$ Br contains an octahedral arrangement of Se atoms about the Ni(IV) centre (Ni—Se = 2.391(5) Å)[530]. The complex $[Fe(Se_2CN(CH_2Ph)_2)_3]BF_4$ has a D_3 macrosymmetry with the $FeSe_6$ coordination core having six Se atoms at the apices of a coordination polyhedron intermediate between the idealized trigonal prismatic and trigonal antiprismatic geometries[529]. Infrared, ESR and Mössbauer spectral data as well as the magnetic susceptibility are all consistent with Fe(IV), but the XPS data (Fe $2P_{3/2}$ binding energy = 708.1 eV) indicate the presence of Fe(II)[529]. The latter problem of *in situ* photoreduction of inorganic compounds involving metal ions in high oxidation states has been observed in a number of other cases.

The other types of 1,1-diseleno ligands and their metal complexes are summarized in Table 41. Only a few reports have described complexes of 1,1-diseleno ligands of the type XSe_2^- (X = ROC, $(RO)_2P$, R_2P) (Table 41).

The complex $Ni(Se_2COEt)_2$ was prepared by the addition of an aqueous solution of $NiCl_2 \cdot 6H_2O$ to a cold solution of freshly prepared KSe_2COEt (from $KOH/CSe_2/EtOH$)[510]. The complex, which immediately precipitated, was isolated by extraction with $CHCl_3$. A dark red $CHCl_3$ solution of this complex instantly became yellow on addition of pyridine[510]. This behaviour was attributed to the conversion of the low-spin square-planar complex to the high-spin octahedral $[Ni(Se_2COEt)_2(pyr)_2]$[510].

Sodium diethyldiselenophosphinate was obtained by the multistep route shown in equation (199).

$$Et_2P(S)P(S)Et_2 \xrightarrow[-PCl_3]{PCl_5/CCl_4/80\,^{\circ}C} Et_2P(S)Cl$$

$$\xrightarrow[-(n\text{-}Bu)_3PS]{(n\text{-}Bu)_3P/150\,^{\circ}C} Et_2PCl \xrightarrow{Se} Et_2P(Se)Cl \qquad (199)$$

$$\xrightarrow[-NaCl]{2\ NaSeH/EtOH} Et_2P(Se)Se^-Na^+$$

TABLE 41. Complexes with miscellaneous 1,1-diseleno ligands

Ligand	Complex	Reference
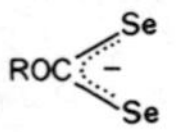 [a]	$M(Se_2COC_{12}H_{25}\text{-}n)_2$	513
	$M = Co, Ni, Zn, Fe, Cu$	
	$Cu(Se_2COEt)_2$	513
(**89**)	$Ni(Se_2COEt)_2$	510
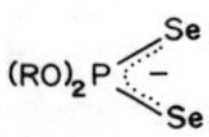 [b]	$M(Se_2P(OEt)_2)_3$	531
	$M = Cr, Rh, Ir$	
	$M(Se_2P(OEt)_2)_n$	532
(**90**)	$n = 1$; $M = Tl$	
	$n = 2$; $M = Pb, Sn$	
	$n = 3$; $M = As, Sb, Bi, In$	
	$n = 3$; $M = Cr$	533
	$Ni(Se_2P(OEt)_2)_2$	534[g]
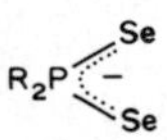 [c]	$M(Se_2PEt_2)_n$	535
	$n = 1$; $M = Tl$	
	$n = 2$; $M = Zn, Cd, Pb, Pd$	
(**91**)	$n = 3$; $M = Bi, In$	
	$Ni(Se_2PPh_2)_2$	536[g]
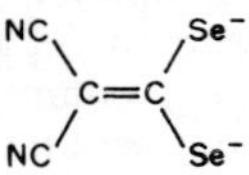 [d]	$(Ph_4P)_n[M(Se_2C{=}C(CN)_2)_2]$	537
	$n = 1$; $M = Au$	
	$n = 2$; $M = Ni, Zn, Cd, Pt, Cu$	
	$n = 3$; $M = Co, Rh, Cr$	
(**92**)	$(Ph_4As)_2[Ni(Se_2C{=}C(CN)_2)_3]$	538, 539[g]
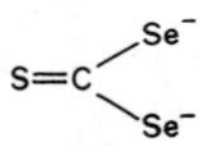 [e]	$Cp(Me_3P)CoSe_2C{=}Se$	224
(**93**)		
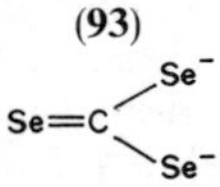 [f, h]	$Cp(Me_3P)CoSe_2C{=}S$	224
	$Cp(Me_2PhP)CoSe_2C{=}S$	
(**94**)		

[a] Diselenoxanthate.
[b] Diorganodiselenophosphate.
[c] Diorganodiselenophosphinate.
[d] Isomaleonitrilediselenolate.
[e] Triselenocarbonate.
[f] Diselenothiocarbonate.
[g] Structure characterized by single-crystal X-ray diffraction.
[h] A complex with (Se, S) coordination of this ambidentate ligand has also been reported (equation 192).

Stable colourless crystals of the hydrate $Et_2P(Se)SeNa \cdot 2H_2O$ were obtained by addition of ligroin to an acetone/water (95:5 v/v) solution of the above product[535]. Complexes (Table 41) readily precipitated from aqueous solutions of the appropriate metal halides by addition of $Et_2P(Se)SeNa \cdot 2H_2O$[535].

Several diethyldiselenophosphate (**91**) complexes (Table 41) were isolated by a similar route[531,532] (equation 200). Because of the instability of the potassium salt, the chelates were generally prepared by addition of the appropriate metal chloride to a freshly prepared solution of the former. The resulting complexes have good stability (e.g. $Cr(Se_2P(OEt)_2)_3$ can be recrystallized from hot EtOH to give air-stable black crystals[533]).

$$4\,EtOH + P_2Se_5 \rightarrow 2\,(EtO)_2P(Se)SeH + H_2Se \xrightarrow{KOH/EtOH} KSe(Se)P(OEt)_2 \xrightarrow[-n\,KCl]{MCl_n} M(Se_2P(OEt)_2)_n \quad (200)$$

A complex of isomaleonitrilediselenolate (**92**) with nickel in the formal +4 oxidation state was prepared by oxidation of the Ni(II) chelate[538] (equation 201). The oxidation product was isolated as the Ph_4As^+ salt in the form of black needles. The other products of this reaction, which involves both oxidation and increase in the coordination number of the Ni, were not characterized[538]. A distorted octahedral coordination of the six Se atoms about the Ni atom has been established for this complex by single-crystal X-ray diffraction (average Ni—Se distance = 2.387(1) Å)[539]. The ligands are planar, and the C=C (1.36 Å) and C≡N (1.15 Å) distances suggest that the stabilization of the Ni(IV) chelate does not result from an extended delocalization of charge on the ligands, but the oxidation should rather be described as loss of electrons from the $NiSe_6$ function exclusively[539].

$$[Ni(Se_2C{=}C(CN)_2)_2]^{2-} \xrightarrow{I_2/MeCN} [Ni(Se_2C{=}C(CN)_2)_3]^{2-} \quad (201)$$

Low yields of diselenothiocarbonate (equations 202 and 203) and triselenocarbonate (equation 204) were obtained by column chromatography of the products obtained in reactions of Co(I) complexes with CSeS and CSe_2, respectively[224].

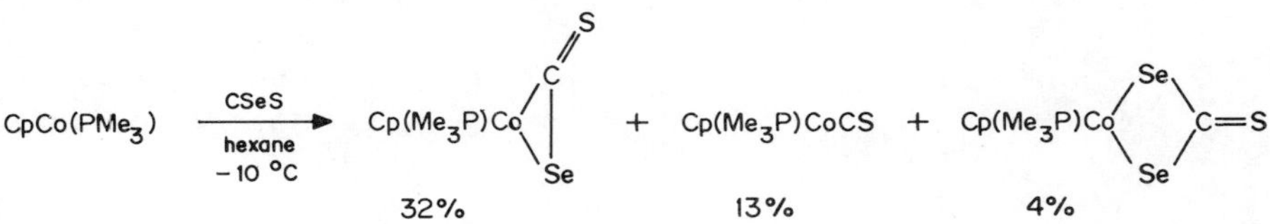

(202)

m.p. 122 °C dec.
$\nu_{C=S}$ (KBr) = 1007 cm^{-1}
ν_{C-Se} (KBr) = 812 cm^{-1}
^{1}H-NMR (δ, acetone-d$_6$): 5.25 (d, Cp), J_{PH} = 0.3 Hz; 1.75 (d, Me), J_{PH} = 10.8 Hz

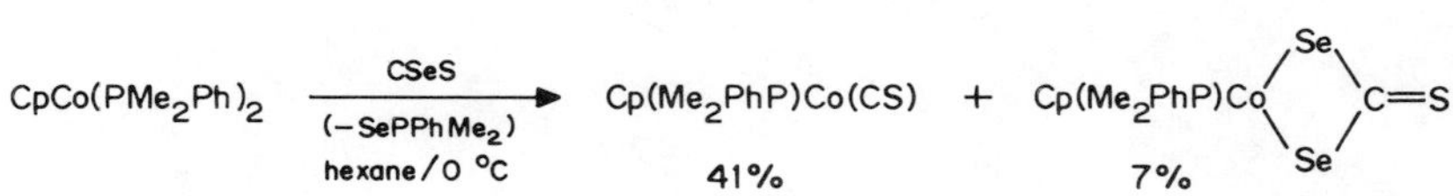

(203)

m.p. 136 °C dec.
$\nu_{C=S}$ (KBr) = 1018 cm^{-1}
ν_{C-Se} (KBr) = 825 cm^{-1}
^{1}H-NMR (δ, acetone-d$_6$): 5.05 (d, Cp), J_{PH} = 0.3 Hz; 1.90 (d, Me), J_{PH} = 10.6 Hz; 7.76 (m, Ph)

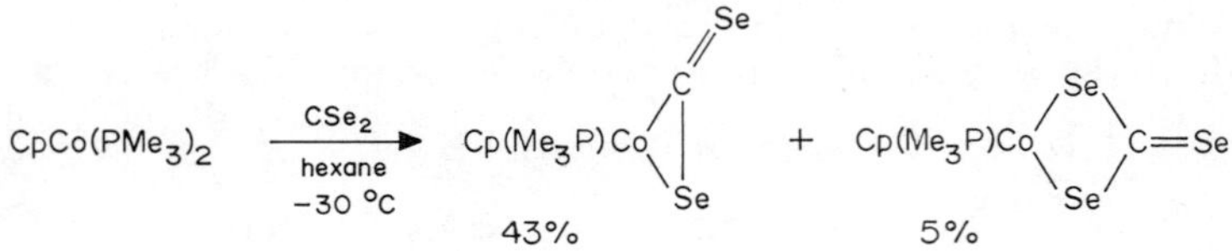

(204)

$\nu_{C=Se}$ (KBr) = 879 cm^{-1}

ν_{C-Se} (KBr) = 810 cm^{-1}

^{1}H-NMR (δ, C_6D_6): 5.30 (d, Cp), J_{PH} = 0.4 Hz; 1.67 (d, Me), J_{PH} = 11.2 Hz

As discussed in Section II.A.7, the η^2-CSeS and η^2-CSe_2 complexes readily undergo PPh_3-induced Se extrusion reactions to give the corresponding Co—CS and Co—CSe complexes under more forcing reaction conditions (e.g. 50 °C)[224].

Only two types of 1,2-diseleno ligands have been reported (Table 42). Davison and Shaw[540] prepared 1,2-diseleno chelates by oxidative addition reactions of metal carbonyls and bis(trifluoromethyl)-1,2-diselentene (equation 205).

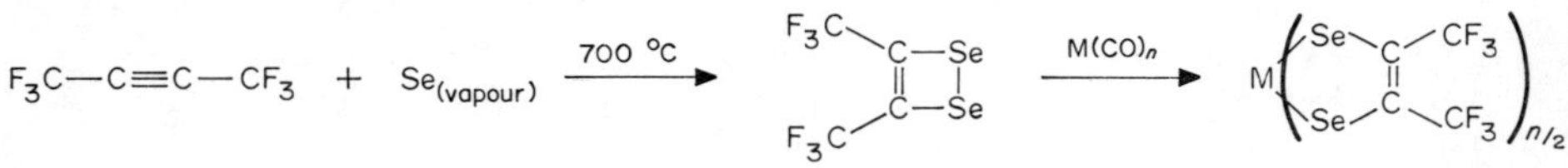

25%

(**96a**) M = Ni, n = 4
(**96b**) M = Mo, n = 6
(**96c**) M = W n = 6

(205)

TABLE 42. Complexes with 1,2- and 1,3-diseleno-type anionic ligands

Ligand	Complex	Ligand	Complex
1,2-*Type*		1,3-*Type*	
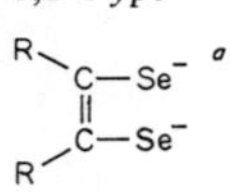 [a] (**95**)	**96a–c**[540] **96b**[541 b] **98**, **99**[542]	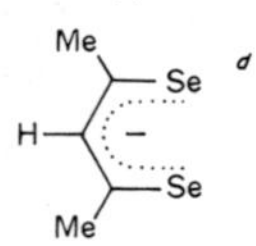 [d] (**102**)	(**103**)[543,544]
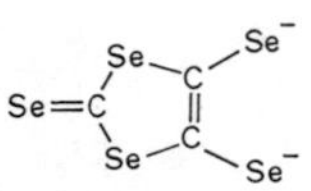 [c] (**100**)	**101**[27]	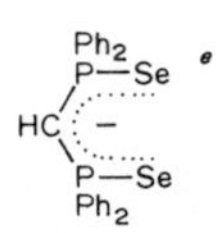 [e] (**104**)	(**105**)[545]

[a] *cis*-1,.2-Di(organo)ethylene-1, 2-diselenato.
[b] Structure characterized by single-crystal X-ray diffraction.
[c] 2-Selenoxo-1, 3-diseleno-4,5-diselenato.
[d] Diselenoacetylacetonato.
[e] Methinotetraphenyldiselenodiphosphino.

The crystal structure of tris(*cis*-1,2-di(trifluoromethyl)ethylene-1,2-diselenato)-molybdenum (**96b**) established a trigonal prismatic arrangement of six Se atoms about the Mo[541]. The relatively short interligand Se····Se distances (3.222(3) Å) compared with the intraligand value of 3.317(5) Å was suggested to indicate a significant interligand Se···· Se interaction, which is important in the stabilization of the relatively unusual trigonal prismatic coordination of this hexacoordinate complex.

Bolinger and Rauchfuss[542] recently described the synthesis of the similar 1,2-diseleno chelates **98** and **99** (equation 206). This route, via the readily prepared Cp_2TiSe_5 (**97**) reagent[542], offers considerable advantage over the reaction of Se vapour at high temperature to give the diselentene (equation 205).

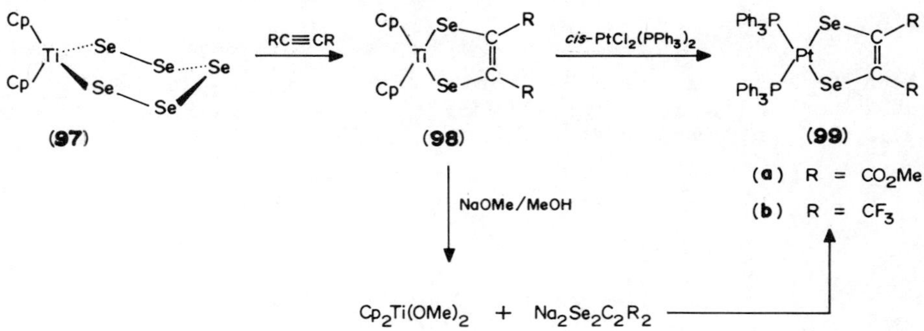

(206)

Chemical or electrochemical reductive coupling of CSe_2 has also provided 1,2-diseleno anions, which react with various metal salts to give the corresponding square-planar bis-chelates[27] (equation 207).

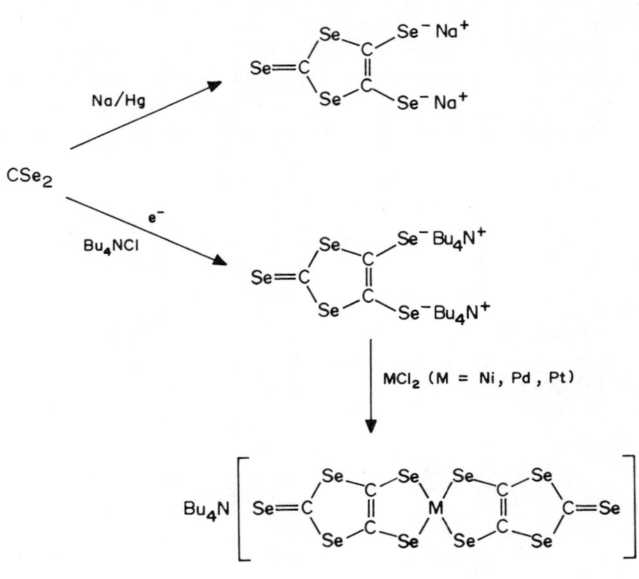

(207)

Ligands of the 1, 3-diseleno type are also restricted to two structural types (**102** and **104**; Table 42).

The diamagnetic square-planar Ni(II) complex (**103**) with the chelating ligand diselenoacetylacetonate (**102**) was prepared by passing H_2Se into an acidified ethanol solution of acetylacetone containing $NiCl_2$[543,544].

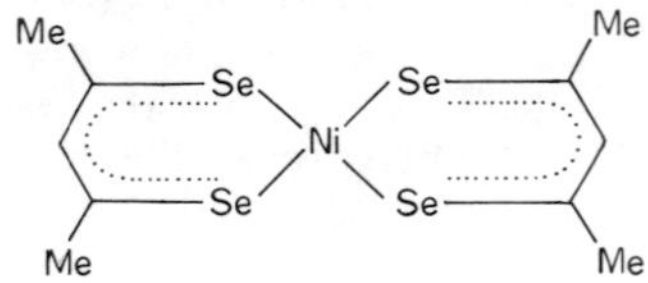

(**103**)

In the absence of the metal salt the reaction presumably gave the dimer of diselenacetylacetone, as has been confirmed by NMR and mass spectroscopy for the reaction of acetylacetone and H_2S in HCl-saturated ethanol (i.e. the product is **106**[544]).

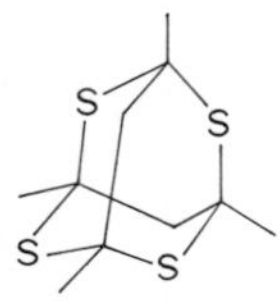

(**106**)

The dark red–brown Ni(II) complex is moderately stable in the solid state at room temperature but slowly decomposes over several weeks. Its thermal instability (decomposes $> 120\,°C$) has precluded the recording of its mass spectrum. It is less stable in solution, but it can be recrystallized from hot $CHCl_3$ if done rapidly to prevent the deposition of elemental Se. The square-planar formulation is supported by its diamagnetism and the similarity of its infrared spectrum with that of the bis-dithioacetylacetonate complex of Co(II), which has been characterized by single-crystal X-ray diffraction[546].

Attempts to prepare the Co(II) complex of **102** by a similar procedure gave the tetrachlorocobaltate(II) salt of the 3,5-dimethyl-1,2-diselenolium cation **107**.

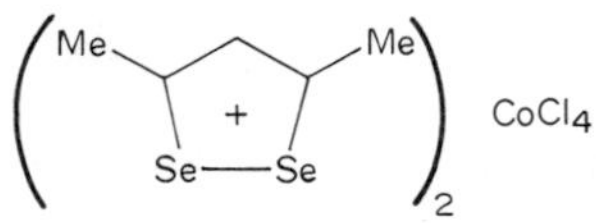

(**107**)

The chelating ligand **104** was prepared by selenation of $(Ph_2P)_2CH_2$ with KSeCN in acetonitrile followed by deprotonation with *n*-BuLi in THF at $-70\,°C$[545]. Reaction of solutions of the lithium salt with $(Et_4N)_2[MBr_2X_2]$ ($M = Fe, Co, X = Cl; M = Ni, X = Br$) gave the neutral bis-chelates **105** after evaporation of the reaction solution and extraction with hot toluene[545].

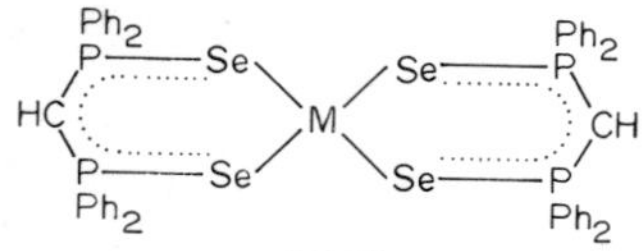

(**105**)

Electronic spectroscopy and magnetic moment measurements (Fe(II), $\mu_{eff} = 5.33$ B.M.; Co(II), $\mu_{eff} = 4.68$ B.M.; Ni(II), diamagnetic) support tetrahedral geometries for the Fe and Co complexes and a square-planar configuration for the Ni complex[545].

E. Tridentate Ligands

The two reported examples of tridentate Se ligands are of the SeN_2 hybrid type (**108**, **109**). Bis(β-(2-pyridyl)ethyl) selenide (bpes, **108**) was prepared according to equation (208)[547]. Several 1:1 complexes of Cu(II) were formed with this ligand in ethanolic or aqueous solution (Cu(bpes)X_2; X = Cl, Br, NO_3, ClO_4)[547].

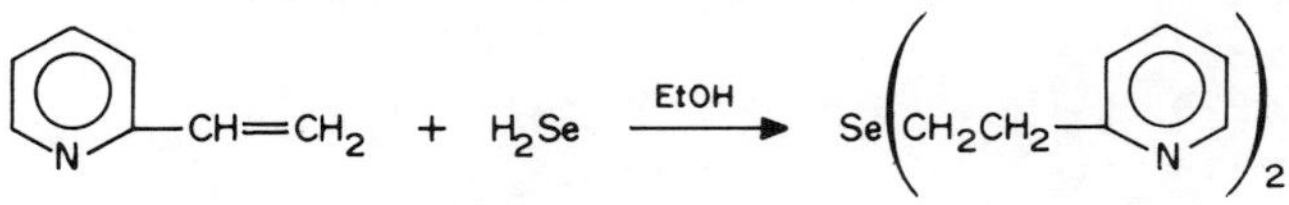

(208)

(**108**)

TABLE 43. Complexes with tetradentate Se ligands

Ligand	Complex	Ref.
Se_4 Donor set		
$MeSe(CH_2)_2Se(CH_2)_3Se(CH_2)_2SeMe$ 1,3-Bis(methylselenoethylseleno) propane (bsep)	[Pd_2(bsep)X_4] X = Cl, Br, I	549
Se_3P Donor set		
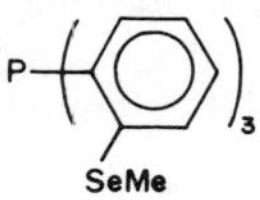	[Ni(TSeP)X]ClO_4 X = Cl, Br, I, NCS	550
Tris(*o*-methylselenophenyl)phosphine (TSeP)	[$Ni(TSeP)_2$]$(ClO_4)_2$	
Se_2N_2 donor sets		
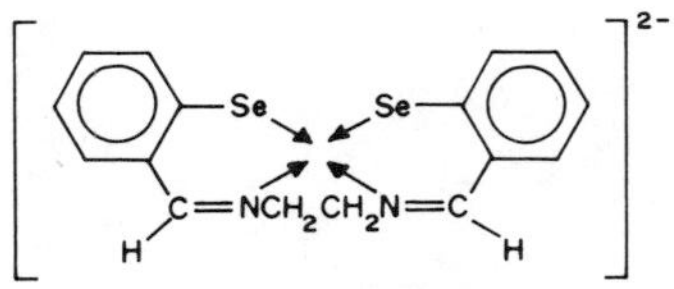	[Ni(Se_2N_2)]	462, 463
α,α'-(Ethylenedinitrilo)di-*o*-tolueneselenolato		
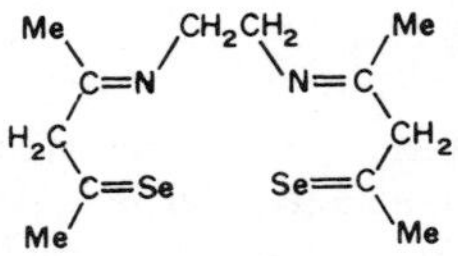	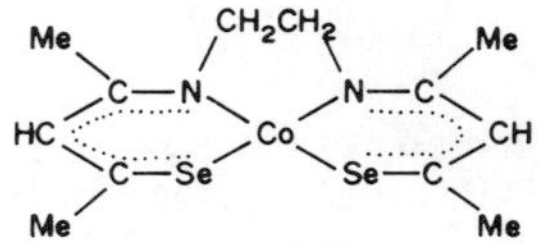	551
N,N'-Ethylenebis(monoseleno-acetylacetonimine) (seacn-H_2)	[$Co(seacn)_2$]	

The other ligand of this type, diacetylselenosemicarbazone oxime (**109**), was prepared by the reaction of diacetyl monoxime and acetone selenosemicarbazone in aqueous ethanol[548]. Complexes with Cu(II) involving coordination of the neutral ligand ($CuX_2(H_2dso)$; X = Cl, Br), the monoanionic form (CuX(Hdso); X = Cl, Br), and the dianionic form ($Cu(dso)\cdot\frac{1}{2}H_2O$), have been isolated[548]. The coordination sites indicated in **109** are based on the similarity of the X-ray powder patterns of these complexes with those of the analogous diacetylthiosemicarbazone oxime complexes, for which these bonding sites have been established. The complexes with the neutral form of the ligand crystallize from methanol solutions of the cupric halide and the ligand, and the ionized forms of the ligand are generated by addition of NaOAc to the reaction solution. These Cu(II) complexes resist reduction to Cu(I) even in boiling solution.

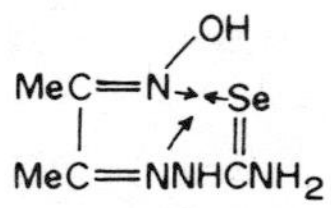

H_2dso, (**109**)

F. Tetradentate Ligands

Tetradentate ligands with Se_4, Se_3P and Se_2N_2 donor sets have been reported (Table 43).

The open-chain tetraseleno ether 1,3-bis(methylselenoethylseleno)propane (bsep) was prepared according to equation (209)[549]. In attempts to prepare the analogous ethylene diselenol intermediate, $HSeCH_2CH_2SeH$, by this route, H_2Se was evolved, presumably by reaction (210). Attempts to distil bsep gave decomposition, and the material was simply heated under vacuum to remove impurities. Characterization of the ligand was limited to ^{1}H-NMR: δ 2.1 (s, Me), 2.7 (m, $(CH_2)_3$), 3.0 (m, —CH_2—).

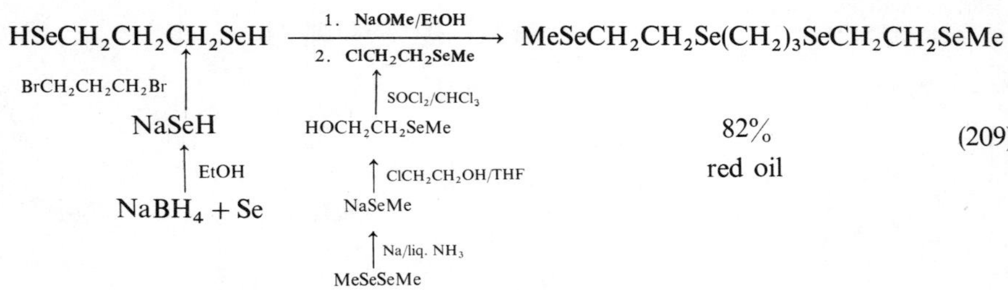

(209)

$$2\,NaSeH + BrCH_2CH_2Br \rightarrow \{HSeCH_2CH_2SeH\} \xrightarrow{-C_2H_4} \{HSeSeH\} \rightarrow H_2Se + Se \quad (210)$$

Reaction of bsep with Na_2PdX_4 (X = Cl, Br, I) in $EtOH/CHCl_3$ gave immediate precipitates of $[Pd_2(bsep)X_4]$ regardless of the molar ratios of the reactants[549]. The chloro complex had ν_{Pd-Cl} bands at 312 and 308 cm^{-1}, characteristic of a *cis* terminal $PdCl_2$ group, and all three complexes gave electronic reflectance spectra characteristic of a $PdSe_2X_2$ square-planar chromophore. Dimeric structures with bridging seleno ether ligands were proposed on the basis of this spectroscopic evidence:

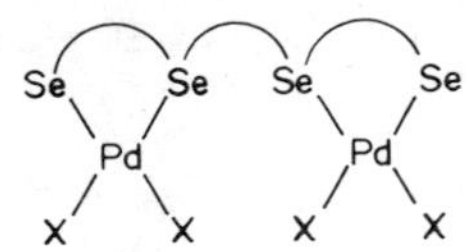

The insolubility, magnetic moment (2.98 B.M.), and electronic reflectance spectrum of $Ni_2(bsep)I_4$ suggest an octahedral polymeric complex with bridging iodo and bsep ligands[549].

The Se_3P hybrid ligand, tris(*o*-methylselenophenyl)phosphine (TSeP), was synthesized in 7% yield as colourless air-stable crystals[550] (equation 211). This ligand forms intensely blue, paramagnetic Ni(II) complexes, $[Ni(TSeP)X]ClO_4$ (X = Cl, Br, I, NCS)[550]. The molar conductances of these complexes in nitromethane solution support their formulation as 1:1 electrolytes ($\Lambda_M = 80–90\,cm^2\,ohm^{-1}\,M^{-1}$), and their electronic spectra are characteristic of trigonal-bipyramidal structures. With the weakly coordinating perchlorate anion a 2:1 electrolyte-type complex, $[Ni(TSeP)_2](ClO_4)_2$, was obtained. The structure of this complex was not determined.

$$o\text{-BrC}_6\text{H}_4\text{NH}_2 \xrightarrow[\text{2. KSeCN}]{\text{1. H}_2\text{SO}_4\text{, NaNO}_2} o\text{-BrC}_6\text{H}_4\text{SeCN} \xrightarrow[\text{2. KOH}]{\text{1. MeI/EtOH}} \qquad (211)$$

$$o\text{-BrC}_6\text{H}_4\text{SeMe} \xrightarrow[\text{Et}_2\text{O, 0}^\circ\text{C}]{n\text{-BuLi}} o\text{-MeSeC}_6\text{H}_4{}^-\ \text{Li}^+ \xrightarrow[\text{Et}_2\text{O}]{\text{PCl}_3} (o\text{-MeSeC}_6\text{H}_4)_3\text{P}$$

(TSeP)

A diamagnetic, square-planar nickel(II) chelate of the Se_2N_2 tetradentate Schiff base derived from selenosalicylaldehyde and ethylenediamine (Table 43) was prepared by the template reaction previously described for related bidentate (Se, N) chelates (equation 182). The dipole moment of this complex (7.28 D in $CHCl_3$) is consistent with the *cis* geometry imposed by the tetradentate structure of the macrocycle[551].

A tetradentate seleno Schiff base derived from acetylacetone has also been reported[551] (equation 212).

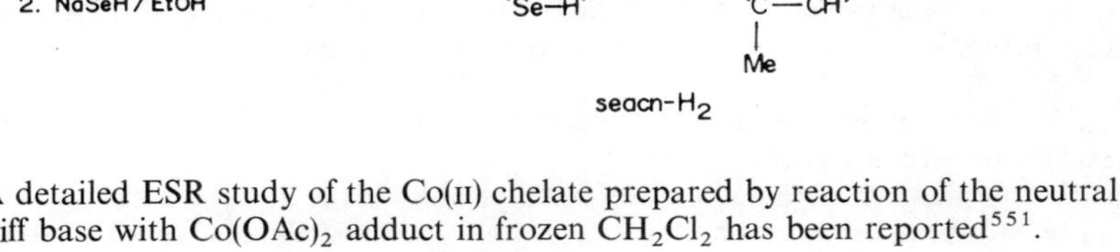

(212)

A detailed ESR study of the Co(II) chelate prepared by reaction of the neutral seleno Schiff base with $Co(OAc)_2$ adduct in frozen CH_2Cl_2 has been reported[551].

G. Complexes Incorporating Se_n

The synthesis of metal complexes with terminal Se^{2-} ligands dates back to the early part of the century (e.g. 1927, $(NH_4)_2[WSe_4]$)[552]. The chemistry of selenometallates has been the subject of several reviews[553]. Within the past few years a number of novel

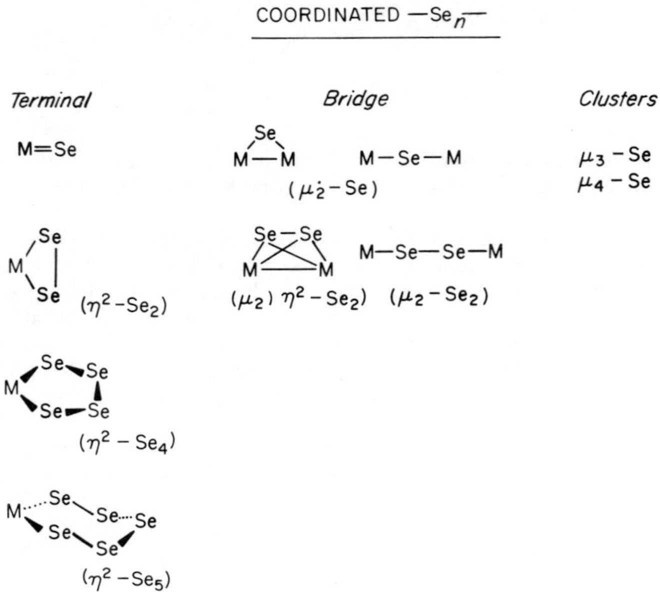

FIGURE 11. Bonding modes of coordinated Se_n

molecular complexes incorporating Se_n (n = 1, 2, 4, 5; Figure 11) have been prepared, and the reactivity of such coordinated Se ligands has been investigated.

In addition to having an interesting chemistry of their own, recent work suggests that such complexes may have considerable utility as precursors of various other organoselenium ligand systems.

1. μ_2-Se ligand

The complex $(Me_3P)_2(OC)_3Re—Se—Re(CO)_3(PMe_3)_2$[370] (Figure 4) was the first structurally characterized complex containing a bridging Se atom. Since this initial report, several dimeric complexes with bridging Se atoms have been prepared by insertion reactions of elemental Se into metal–metal bonds[554–556]. The complex $CpCo(\mu\text{-}PMe_2)_2(\mu\text{-}Se)CoCp$ was prepared in 55% yield by such a route (equation 134). This general reaction has provided a route to related Mo, W and Rh Se-bridged dimers (equations 213–215).

Complexes with labile ligands react with elemental Se (equation 166) or COSe (equation 216) to give Se-bridged dimers.

The μ_2-Se complex was obtained in only small yield from the initial reaction (ca. 10:1 μ_2-Se_2 : μ_2-Se products) but was obtained in high yields by extrusion of a Se atom from the μ_2-Se_2 dimer with triphenylphosphine[557]. The presence of a Mn—Mn bond in the μ_2-Se complex was proposed by analogy with related complexes (e.g. $[Cp''(OC)_2Mn]_2CH_2$ and $[Cp(OC)_2Mn]_2CHR$). The crystal structure of the μ_2-Se complex was reported in the same paper[557].

These two dimers, identical with those isolated by the route of equation (166), were separated by low-temperature (−20 °C) chromatography on SiO_2.

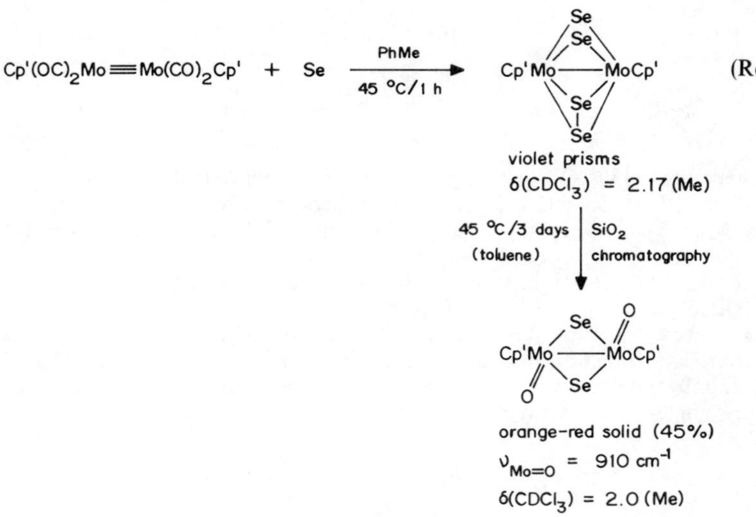

(Ref. 555) (213)

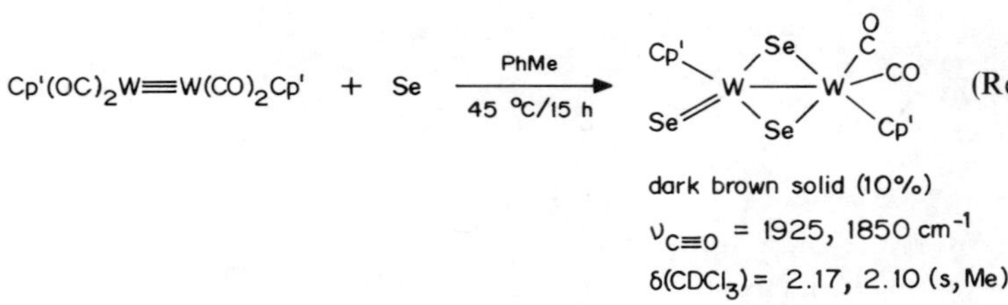

(Ref. 555) (214)

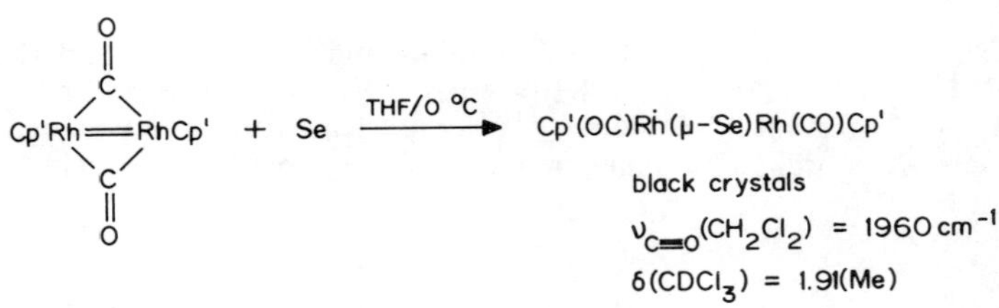

(Ref. 556) (215)

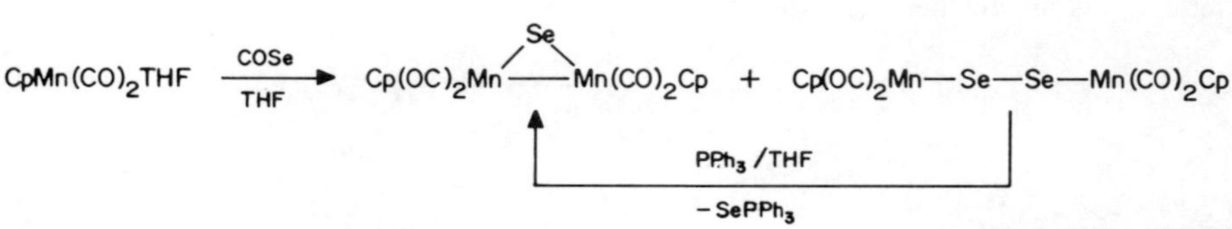

(Ref. 557) (216)

Redox reactions of organometallic anions and sodium selenite have provided another route to Se-bridged dimers (equations 217 and 218).

$$Na^+ [Cp'Cr(CO)_3]^- \xrightarrow[HCl]{Na_2SeO_3} Cp'(OC)_2Cr{\equiv}Se{\equiv}Cr{\equiv}(CO)_2Cp' \qquad \text{(Ref. 558)} \qquad (217)$$

$$Na^+[V(CO)_4(dppe)]^- \xrightarrow[H_3PO_4]{Na_2SeO_3/H_2O} [(OC)_3(dppe)V{\equiv}Se{\equiv}V(dppe)(CO)_3] \quad (218)$$

25% (Ref. 559)

$\nu_{C\equiv O}(CH_2Cl_2)$: 1980, 1930s
1885vs, 1852vs

The crystal structure of $[V(CO)_3(dppe)]_2S$ (e.g. S analogue of product of equation 218) revealed a very short V—S bond (2.172 Å vs. normal single-bond values of ca. 2.30 Å). A bond order between 2 and 3 is suggested for these linear V≡E≡V (E = S, Se, Te) dimers[559].

Reaction of some of these Se-bridged dimers with diazomethane is a convenient route to μ,η^2-selenoformaldehyde complexes[554,558] (e.g. equation 166).

The high oxophilicity of Al(III) has been used to form metal–Se bonds in equation (219)[560]. The dimeric formulation was assigned on the basis of magnetic and infrared spectral data and by analogy with the S analogue, which has been unequivocally characterized by single-crystal X-ray diffraction.

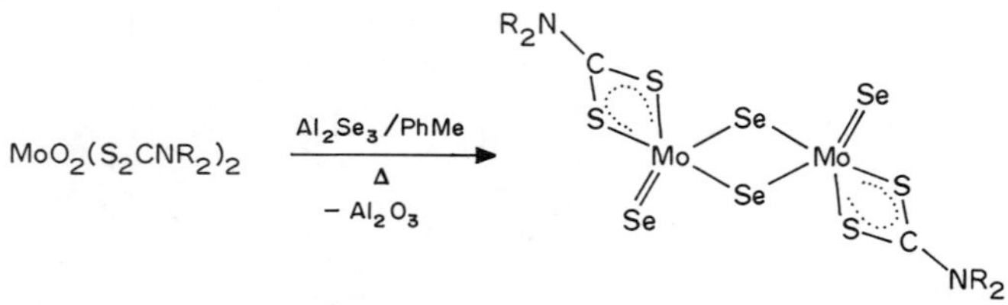

(219)

R = Me, Et
R_2 = pyrrolidinyl, morpholinyl

$\nu_{Mo=Se} = 360\ cm^{-1}$
$\nu_{Mo-Se-Mo} = 250\ cm^{-1}$

Detailed resonance Raman and infrared spectroscopic studies[561] of the dimeric $[(PhE)_2Fe(\mu_2\text{-}Se)_2Fe(EPh)_2]^{2-}$ (E = S, Se) and $[(o\text{-}C_6H_4(CH_2S)_2Fe(\mu_2\text{-}Se)_2Fe(SCH_2)_2C_6H_4\text{-}o)]^{2-}$ and an ESR study[562] of their reduced forms have been reported. These dimers are coordination analogues of the Fe–S proteins, ferredoxins.

2. η^2-Se_2 ligand

The structure of the complex $[Ir(dppe)_2(Se_2)]Cl$, first briefly described in 1971[563] (equation 220), has been reported[564].

$$[Ir(dppe)_2]Cl + Se_8 \xrightarrow[20\ ^\circ C/\ days]{CH_2Cl_2} [Ir(dppe)_2Se_2]Cl \quad (220)$$

green air-stable solid
m.p. 200–201 °C
$\nu_{Se-Se} = 310\ cm^{-1}$

As previously proposed on the basis of its ^{31}P-NMR spectrum[563], single-crystal X-ray diffraction showed the presence of side-on bonded Se_2 (η^2-Se_2; Figure 11).

This complex was prepared more recently with Cp_2TiSe_5 as a Se-transfer agent (equation 221).

$$[Ir(dppe)_2]Cl + Cp_2TiSe_5 \xrightarrow{CH_2Cl_2/THF} [Ir(dppe)_2(Se_2)]Cl \quad \text{(Ref. 565)} \quad (221)$$

A similar bonding mode had previously been found for $Os(\eta^2\text{-}Se_2)(CO)_2(PPh_3)_2$[404] (equation 222).

$$Os(CO)_2(PPh_3)_3 + \text{red Se} \xrightarrow{C_6H_6} Os(\eta^2\text{-}Se_2)(CO)_2(PPh_3)_2 \qquad (222)$$

60%

red–purple crystals ($CHCl_3$/EtOH)

m.p. 192–195 °C

$\nu_{Se—Se} = 310\,cm^{-1}$

The Se—Se bonds in such complexes are reactive to electrophilic alkylating reagents (e.g. equation 148) with retention of the Se—Se bond and to low-valent transition-metal species, which undergo oxidative insertion into the Se—Se bond (equation 223).

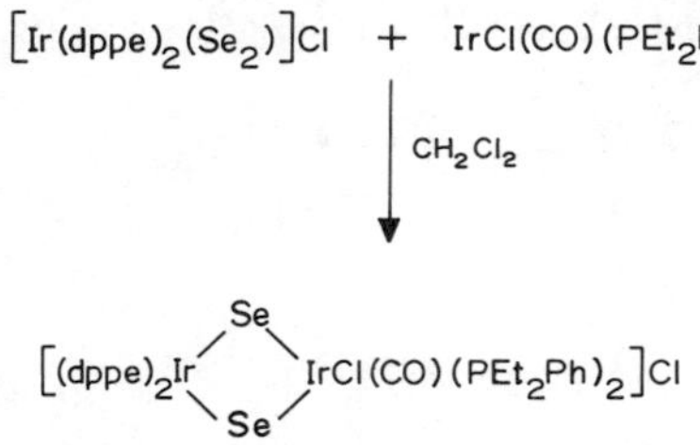

(Ref. 564) (223)

Although the complexes $[(dppe)_2Ir(S_2)]Cl$ and $[(Me_2PCH_2CH_2PMe_2)_2Rh(S_2)]Cl$ undergo oxidative addition reactions with $Pt(PPh_3)_3$ and $Pt(PEtPh_2)_4$ to give the $M(\mu_2\text{-}S)_2Pt$ (M = Rh, Ir) dimers, the reactions with the (η^2-Se_2) complex not only result in addition across the Se—Se bond but also displace the Ir from Se (equations 224 and 225).

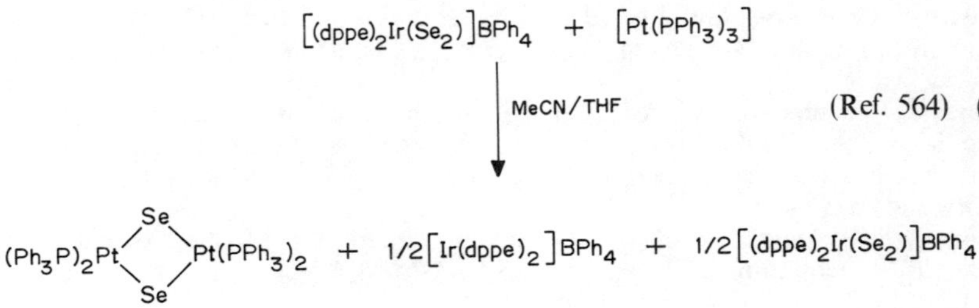

(Ref. 564) (224)

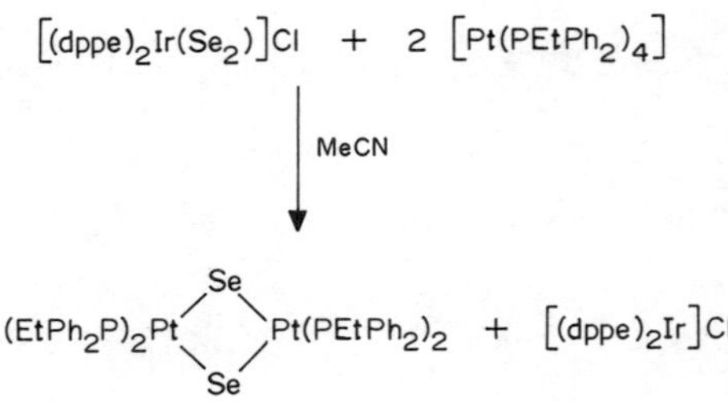

(Ref. 564) (225)

SCF-Xα-SW calculations on the model compounds $[M(X_2)(PH_3)_4]^+$ (M = Rh, Ir; X = S, Se) predict an X—X bond order of about 1 and M—X_2 covalent interaction in the order $RhSe_2 < IrSe_2 < RhS_2 < IrS_2$, both in agreement with the observed properties of the analogues with the bidentate phosphines discussed above[566].

The complex $Cp'Mn(CO)_2(\eta^2\text{-}Se_2)$ has been prepared by reaction of elemental Se with $Cp'Mn(CO)_2THF$[567]. The spectroscopic data for this complex support the 'side-on' coordination of Se_2, an assignment confirmed by preliminary X-ray structure analysis (Se—Se = 2.263(8) Å, Mn—Se = 2.463(4) Å)[567]. These data suggest the Se_2 is best considered as an uncharged two-electron ligand.

3. μ_2-Se_2 and μ_2, η^2-Se_2 ligands

The complex $Cp(OC)_2Mn(SeSe)Mn(CO)_2Cp$ (equation 216) is the only example reported with a bridging M—SeSe—M linkage.

The complex $(OC)_3Fe(\mu_2,\eta^2\text{-}Se_2)Fe(CO)_3$ was first prepared in 1958 by Hieber and Gruber[568]. A detailed description of the synthesis of the dimer, based on their original method, was reported by Seyferth and Henderson[354] (equation 226).

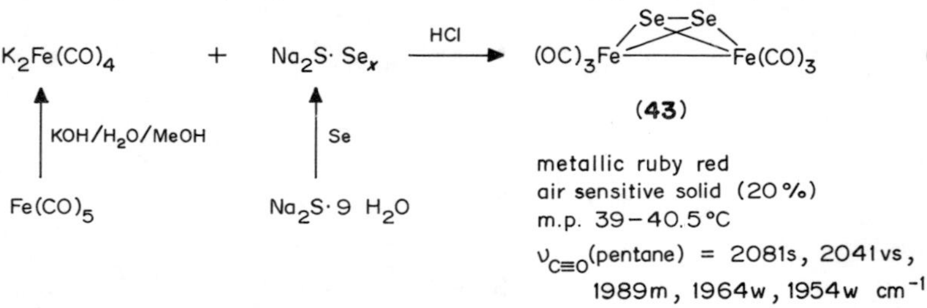

(226)

The proposed structure, involving both Se—Se and Fe—Fe bonds, was confirmed by single-crystal X-ray diffraction[569]. Work by Seyferth and Henderson[354] and Rauchfuss and coworkers[570,571] has established an extensive chemistry of this dimer based on the reactivity of the Se—Se bond (Figure 12). The use of this dimer as a substrate in oxidative addition reactions with low-valent metal (e.g. Pt(O), Co(I), Ni(O)) complexes offers a route to a wide variety of mixed-metal clusters[570,571]. Substitution reactions in which the terminal CO ligands are replaced by phosphines have also been reported[571–573].

The S analogue, $(OC)_3Fe(\mu_2,\eta^2\text{-}S_2)Fe(CO)_3$, exhibits the same general reactivity pattern but has been the subject of more study[570,571,574,575], owing to its enhanced stability and ease of preparation compared to **43**.

The μ_2,η^2-Se_2 ligand also occurs in the dimers $Cp'_2Mo_2Se_4$ (equation 213), $[W_2Se_3Cl_8]^{2-}$ (equation 227) and $[Nb_2Se_4Cl_4(SMe_2)_4]$ (equation 228) as well as the cluster $Nb_4Se_3Br_{10}(MeCN)_4$ (equation 229).

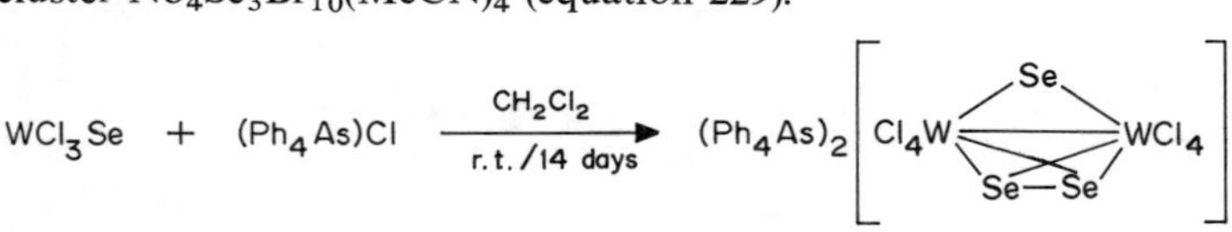

(Ref. 576) (227)

$$NbCl_3Se \xrightarrow{Me_2S} (Me_2S)_2Cl_2Nb(\mu_2,\eta^2\text{-}Se_2)_2NbCl_2(SMe_2)_2$$

(Ref. 577) (228)

$$(3\,NbCl_5 + Sb_2Se_3 \xrightarrow{CSe_2} 3\,NbCl_3Se + 2\,SbCl_3)$$

$$4\,NbBr_4(MeCN)_2 + Sb_2Se_3 \xrightarrow{MeCN} Nb_4(MeCN)_4Br_6(\mu_2\text{-}Br)_4(\mu_2,\eta^2\text{-}Se_2)(\mu_4\text{-}Se)$$

(Ref. 578) (229)

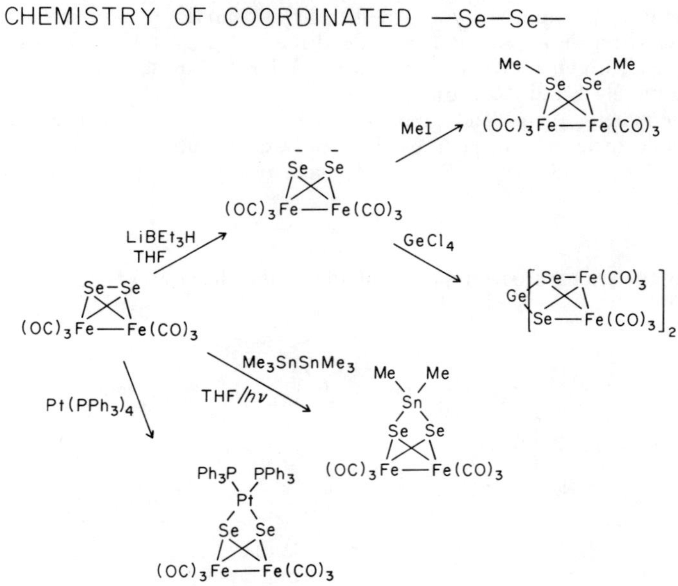

FIGURE 12. The chemistry of $Fe_2Se_2(CO)_6$

The compound $Nb_2Se_2Br_6$, prepared in the form of air-stable, diamagnetic metallic grey crystals from the elements in a sealed quartz tube at 1073 K, has been shown by single-crystal X-ray diffraction to contain the μ_2,η^2-Se_2 ligand[579]. The structure consists of one-dimensional infinite chains of $[Nb_2(Se_2)Br_4]$ units having single side-on bonded Nb_2 and Se_2 dumb-bells forming a quasi-tetrahedral Nb_2Se_2 cluster asymmetrically bridged along the Nb—Nb edge by two bromo ligands (Figure 13).

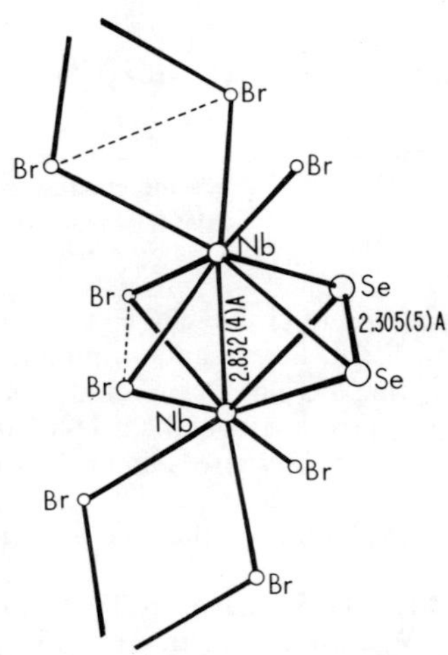

FIGURE 13. Crystal structure of $Nb_2Se_2Br_6$

The interatomic distances show that there is no significant interaction between the chains. The short Nb—Nb and Se—Se distances (Figure 13) support the formal oxidation state assignments Nb(4 +) and Se(1 −). The Se—Se and Nb—Nb bond orders are approximately 1 and 0.62, respectively.

The diamagnetic compounds $NbSe_2X_2$ (X = Cl, Br, I) have been prepared by similar chemical vapour transport reactions and were formulated with similar structural features (i.e. Nb_2^{4+}, $(Se_2)_2^{2-}$, X_4^-) on the basis of crystallographic and XPS data[580].

4. η^2-Se_4 and η^2-Se_5 ligands

Three complexes have been reported involving the chelating η^2-Se_4 ligand (equations 230 and 231).

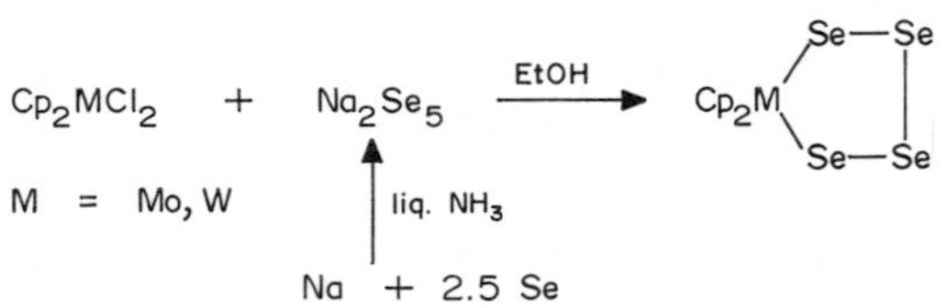

(Ref. 581) (230)

M = Mo

red–black air-stable platelets (36%; recryst. from DMF)

m.p. 215 °C dec.

^{1}H-NMR δ(DMF): 5.17 (s, Cp)

M = *W*

Brownish violet air-stable platelets (86%; recryst. from DMF)

m.p. 225 °C dec.

^{1}H-NMR δ(DMF): 5.22 (s, Cp)

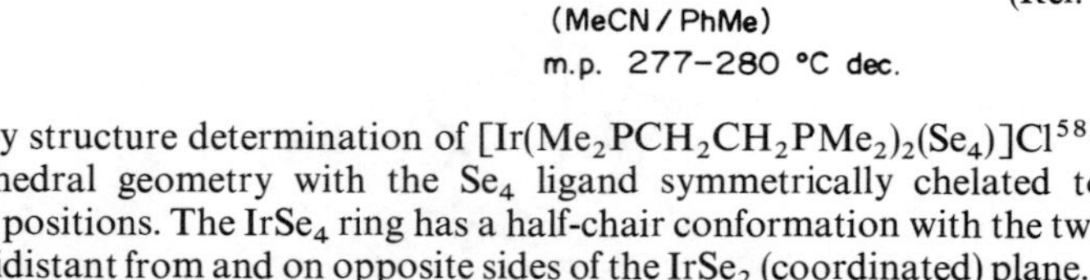

dark red crystals (MeCN/PhMe)

m.p. 277–280 °C dec.

(Ref. 582) (231)

An X-ray structure determination of $[Ir(Me_2PCH_2CH_2PMe_2)_2(Se_4)]Cl$[582] confirmed a *cis*-octahedral geometry with the Se_4 ligand symmetrically chelated to the Ir at equatorial positions. The $IrSe_4$ ring has a half-chair conformation with the two central Se atoms equidistant from and on opposite sides of the $IrSe_2$ (coordinated) plane. The Ir—Se (2.545 Å) and Se—Se (2.307 Å) distances are similar to the corresponding values in $[Ir(dppe)_2(Se_2)]Cl$[564]. As found for the Se_2 ligand[566], SCF-Xα-SW calculations on the model compound $[Ir(PH_3)_4(Se_4)]^+$ indicate that the Se_4 ligand is best described as an excited Se_4 molecule[582].

The reaction of Cp_2TiCl_2 with Na_2Se_5 gave Cp_2TiSe_5 (**97**)[583] (rather than the η^2-Se_4 derivative as in the Mo and W analogues) (equation 230). This dark violet air-stable solid (m.p. 211 °C) gave two signals of equal intensity in the ^{1}H-NMR spectrum; this was the first example of the splitting of the Cp proton signal in a Cp_2Ti^{4+} derivative and indicated that at room temperature and below the $TiSe_5$ ring is in a fixed conformation. The crystal structure of the S_5 analogue has shown that the six-membered TiS_5 ring exists in a cyclohexane-like chair conformation[584]. These two bands broaden with increasing temperature (S_2Me_2 solution) and finally coalesce at ca. 90 °C. This reversible process becomes irreversible above 100 °C, owing to reaction of Cp_2TiSe_5 with the solvent.

The utility of Cp_2TiSe_5 as a convenient reagent for the synthesis of transition-metal ethylene-1,2-diselenate complexes was described previously (equation 206). As with $Fe_2Se_2(CO)_6$, the enhanced stability and ease of preparation of Cp_2TiS_5[584,585] compared to its Se analogue[542] have resulted in more work with the former derivative, although the reactivity patterns for both complexes are similar.

5. *Se incorporated in metal clusters*

The synthesis and structural characterization of transition-metal cluster compounds has been an active area of research for 25 years[1,586]. Several such clusters incorporating Se in the framework have been reported (Table 44). These clusters most commonly incorporate a Se atom bonded to three transition metals in the cluster framework (i.e. a μ_3-Se ligand), although the first example of a μ_4-Se ligand has been structurally characterized[600].

The first example of such a transition-metal cluster compound incorporating Se was reported by Hieber and Gruber in 1958[568] (equation 232). Such cluster compounds generally require a single-crystal X-ray diffraction analysis for an unequivocal structural characterization. Indeed, the crystal structures of most of the known Se-containing clusters have been reported (Table 44). The first such structurally characterized compound was $Fe_3(\mu_3\text{-Se})_2(CO)_9$[587], which consists of an $Se_2Fe_2(CO)_6$ fragment of idealized C_{2v}-2 mm symmetry bonded to an $Fe(CO)_3$ group via two bent Fe—Se bonds and two bent Fe—Fe bonds (Figure 14). The cluster framework can be described as a square pyramid with an Fe at the apex (a seven-coordinated Fe atom) and alternate Se and Fe atoms at the corners of the basal plane.

$$3\,Fe(CO)_4{}^{2-} + 2\,SeO_3{}^{2-} + 10\,H^+ \rightarrow Fe_3(\mu_3\text{-Se})_2(CO)_9 + CO_2 + 2\,CO + 5\,H_2O \qquad \text{(Ref. 568)} \qquad (232)$$

The reaction of $Os_3(CO)_{12}$ with elemental Se in refluxing *n*-octane followed by thin-layer chromatographic separation of the products gave $H_2Os_3Se(CO)_9$, $Os_3Se_2(CO)_9$ (Figure 15), and $H_2Os_4Se_2(CO)_{12}$ (Figure 16)[540]. Carrying out the reaction under CO/H_2 pressure (35 atm, 1:1) gave improved yields of all three clusters, whereas in a CO atmosphere the yields of the hydrido clusters decreased significantly. Initial characterization of these clusters, as for such compounds in general, was by infrared and mass

TABLE 44. Transition-metal cluster compounds containing μ_3-Se ligands

Complex	Ref.	Complex	Ref.
$Fe_3(\mu_3\text{-Se})_2(CO)_9$	568, 587[a]	$Os_3(\mu_3\text{-Se})_2(CO)_9$	593, 594[a]
$Fe_4(\mu_3\text{-Se})_4(CO)_{12}$	588	$H_2Os_4(\mu_3\text{-Se})_2(CO)_{12}$	590[a]
$FeCo_2(\mu_3\text{-Se})(CO)_9$	589[a]	$Co_3(\mu_3\text{-Se})(CO)_9$	589[a], 595
$Fe_2Co(\mu_3\text{-Se})(CO)_9$	589	$Co_6C(\mu_3\text{-Se})_2(CO)_{12}$	595[a]
$Ru_3(\mu_3\text{-Se})_2(CO)_9$	590	$(Me_4N)[Rh_3(\mu_3\text{-Se})_2(CO)_6]$	596[a]
$H_2Ru_3(\mu_3\text{-Se})(CO)_9$	590, 591	$[(Ph_3PAu)_3(\mu_3\text{-Se})](PF_6)$	597[a]
$RuCo_2(\mu_3\text{-Se})(CO)_9$	592	$[(Et_3P)_6Ni_3(\mu_3\text{-Se})_2](BPh_4)_2$	598[a]

[a] Single-crystal X-ray diffraction structure determination.

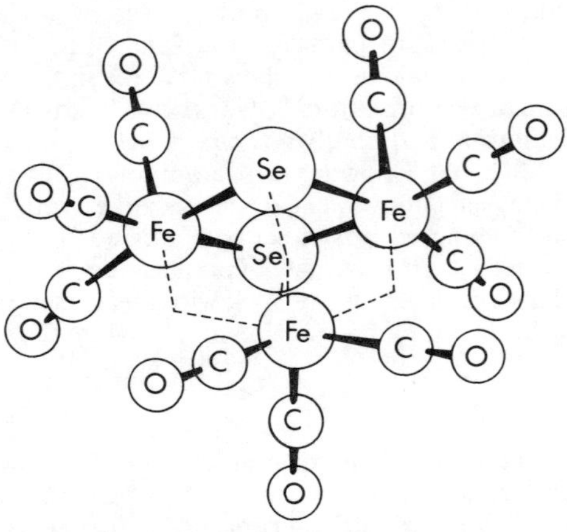

FIGURE 14. Molecular structure of $Fe_3Se_2(CO)_9$. Reproduced with permission from Ref. 587

spectroscopy[590]. The structure of $Os_3Se_2(CO)_9$ (Figure 15) was shown[594] to contain a triangle of Os atoms with one long non-bonding edge (Os—Os = 3.791(1) Å) and Se atoms capping the triangle on both sides to give a trigonal bipyramidal cluster geometry.

The structure of $H_2Os_4Se_2(CO)_{12}$ (Figure 16) has been shown to have a novel trigonal-prismatic Os_4Se_2 core with each Se atom capping a triangle of Os atoms[590]. Two Os—Os distances in each 'Os_3Se' unit are non-bonding (ca. 4 Å). All 12 carbonyl ligands are terminally bonded to the four Os atoms, and the two hydride ligands were formulated as

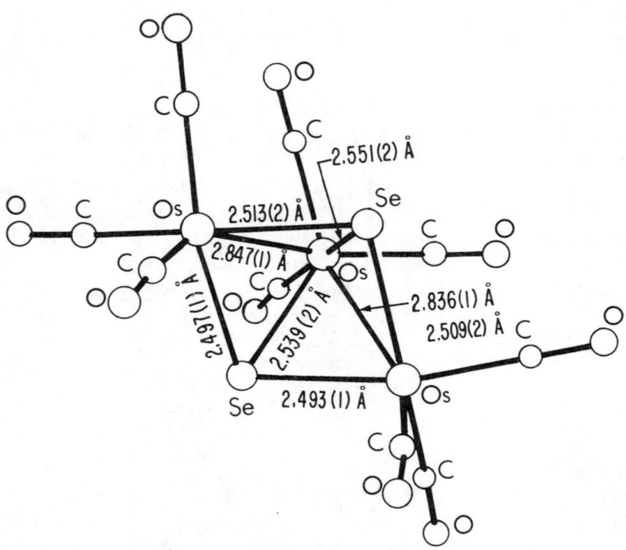

FIGURE 15. Molecular structure of $Os_3Se_2(CO)_9$

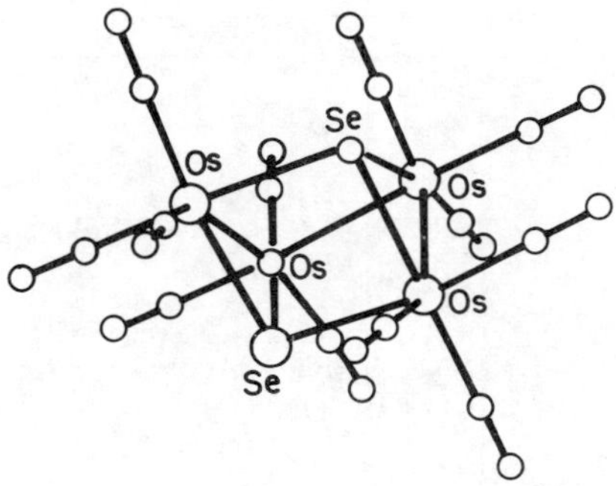

FIGURE 16. Molecular structure of $H_2Os_4Se_2(CO)_{12}$

edge-bridging the two long Os—Os bonds, since the carbonyl ligands bend away from these edges.

Two types of selenide ligands, including a unique μ_4-Se, occur in the cluster $Nb_4(MeCN)_4Br_6(\mu_2\text{-}Br)_4(\mu_2,\eta^2\text{-}Se_2)(\mu_4\text{-}Se)$[578] (Figure 17).

Reaction of H_2Se with $[Ni(OH_2)_6](ClO_4)_2$ and PEt_3 in $H_2O/CHCl_3$ gave the polynuclear complex $[Ni_3(\mu_3\text{-}Se)_2(PEt_3)_6]^{2+}$, which was isolated as its air-stable, red tetraphenylborate salt[598]. The molecular structure[598,599] of this complex contains a triangle of Ni atoms capped on both sides by triply bridging Se atoms forming an almost regular trigonal bipyramid. Each Ni atom is further coordinated to two terminal PEt_3 ligands, the phosphorus atoms lying in the appropriate Ni—Se_2 plane. The average Ni—Ni distance (3.16(2) Å) precludes any significant direct metal–metal interaction, and the cluster geometry can be considered as three d^8 $NiSe_2P_2$ square planes sharing two bridging ligands.

The supertetrahedral clusters, $[(\mu_3\text{-}Se)_4M_{10}(SPh)_{16}]^{4-}$ (M = Zn, Cd) have been prepared in 80–100% yields by reactions of elemental Se with the adamantanoid clusters $[M_4(SPh)_{10}]^{2-}$ and isolated as their tetramethylammonium salts[600].

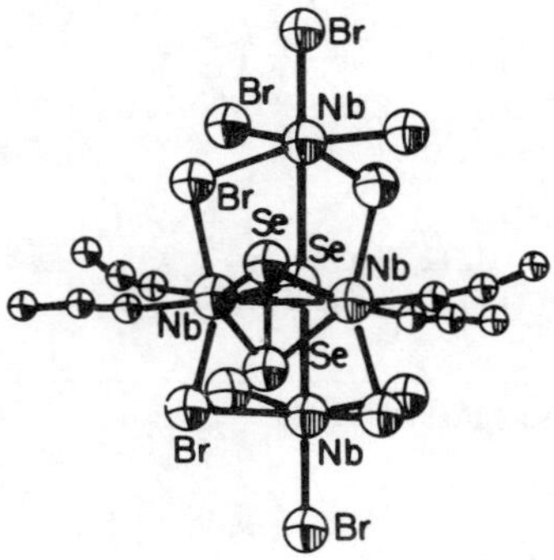

FIGURE 17. Molecular structure of $Nb_4(MeCN)_4Br_6(\mu_2\text{-}Br)_4(\mu_2,\eta^2\text{-}Se_2)(\mu_4\text{-}Se)$. Reproduced with permission from Ref. 578

TABLE 45. Summary of complexes with monodentate tellurium ligands

Ligand	Complex	Ref.
$Te(alkyl)_2$		
terminal	$PtCl_2(Te(CH_2CH_2Ph_2)_2)_2$	605
	$Pd(SCN)_2(Te(CH_2CH_2CH_2SiMe_3)_2)_2$	606 [a]
	$[CpNi(TeMe_2)_2]BF_4$	607
	$(OC)_5Cr(TeEt_2)$	108
bridging	$[n\text{-}Bu_4N]_2[Cl_3Pt(\mu\text{-}TeMe_2)PtCl_3]$	608
	$BrAg(\mu\text{-}Te(Bu\text{-}n)_2)AgBr$	609
	$[CuCl(TeEt_2)]_n$	610 [a]
$Te(aryl)_2$		
terminal	$RhCl(TePh_2)_3$	611
	$HgI_2(TePh_2)]_4$ (**110**)	612, 613 [a]
	$PdCl_2(Te(C_6H_4OEt\text{-}p)_2)_2$	614
	$CuI(Te(C_6H_4Me\text{-}p)_2)_2$	615
bridging	$IAg(\mu\text{-}TePh_2)AgI$	609
Te(alkyl)(aryl)		
terminal	$HgBr_2(TeMePh)$	612
bridging	$BrAg(\mu\text{-}TePhC(O)Ph)AgBr$	609
(alkyl)TeTe(alkyl)		
bridging	$[CuCl(Te_2Et_2)]_n$	616
	$[CpNi(\mu\text{-}Te_2Me_2)NiCp](BF_4)_2$	617
(aryl)TeTe(aryl)		
bridging	$[CuBr(Te_2(C_6H_4OEt\text{-}p)_2)_2]_n$	616
	$[(OC)_3M(\mu\text{-}Br)_2(\mu\text{-}Te_2Ph_2)M(CO)_3$ M = Re, Mn	190 [a], 191
Heterocycles		
L = **111**	$Mn(CO)_3L_2Cl$	618
L = **112**	$LCr(CO)_3$	619
L = **113, 114**	$L_2Rh(CO)Cl$	620
$Te(EMe_3)_2$	$[Cr(CO)_5(Te(EMe_3)_2)]$ E = Ge, Sn, Pb	196
$(CF_3)ETeMe$ E = P, As	$Mo(CO)_5\{(CF_3)_2PTeMe\}$ [b]	194
Tellurourea		
L = **115**	$M(CO)_5L$ M = Cr, Mo, W	621
	$Fe(CO)_4L$	622
CTe [c]	$OsCl_2(CO)(CTe)(PPh_3)_2$	216
	$(Et_4N)[LMo(CO)_2(CTe)]$ L = **37**	223
Telluroketone	$(OC)_5W(Te{=}CPh_2)$	623
Carbenes	$(OC)_5Cr{=}C(NEt_2)(TePh)$	624
Telluroxides	$MeTeCl_3(OTePh_2)_2$ [d]	625
	$(p\text{-}ROC_6H_4)TeCl_3(OTePh_2)_2$ R = Me, Ph	

[a] Structure determined by single-crystal X-ray diffraction.
[b] Mo—P bonding of ambidentate (P, Te) ligand.
[c] M—CTe bonding.
[d] Te—O=TePh₂ bonding.

III. ORGANOTELLURIUM LIGANDS

Because transition-metal complexes with Te ligands have been the subject of recent reviews[15,18] and these complexes are generally prepared by routes described above for the analogous Se complexes, discussion of the former complexes will be limited to a general consideration of the various ligand types and very recent developments.

The extensive ligand chemistry developed for 1,1- and 1,2-dithio[12] and -diseleno (Section II.D.2) anionic ligands has not been extended to the Te analogues. Likewise, no examples of tridentate or tetradentate ligands containing one or more Te donor sites have been reported.

Although several papers have described the syntheses of phosphine tellurides (R_3PTe, $(R_2N)_3PTe$; R = alkyl)[601], no complexes of these donors analogous to those of the lower chalcogens (R_3PX; X = O[602], X = S[603], Se (Section II.A.3)) have been described. Similarly, although several salts of the tellurocyanate anion[604] have been isolated and characterized, no complexes with this ambidentate ligand are known, in contrast to the extensive ligand chemistry of thiocyanate[302] and selenocyanate (Section II.B.1).

Recent advances in the synthetic methodology of organotellurium chemistry as illustrated in other chapters of this volume can be expected to provide impetus for the development of this coordination chemistry. The most diverse class of Te ligands are neutral monodentate (Table 45).

A. Neutral Monodentate Ligands

1. *Diorganotellurides*

a. Dialkyl tellurides. The first reported coordination complex with a Te ligand, *cis*-$PtCl_2(Te(CH_2Ph)_2)_2$, was reported by Fritzmann in 1915[8]. Since then a number of transition-metal and Group IIb complexes with dialkyl and diaryl tellurides have been prepared[18]. Early work in this area generally involved lower molecular weight dialkyl tellurides, and the air sensitivity and foul odour of these ligands no doubt led to the belief among inorganic chemists that these were general characteristics of Te compounds. In fact, the aromatic ligands are air-stable solids that can be readily prepared with a variety of functional groups[15]. The alkyl derivatives can also be made more amenable to convenient handling by increasing the alkyl chain length. The slightly air-sensitive dialkyl tellurides $Te(CH_2CH_2Ph)_2$[605] and $Te((CH_2)_nSiMe_3)_2$[606] ($n = 1, 3$) can be readily prepared in high yields by alkylation of Na_2Te. The complex *trans*-$Pd(SCN)_2(Te(CH_2CH_2CH_2SiMe_3)_2)_2$ has been characterized by single-crystal X-ray diffraction[606]. The complex is square-planar with S-bonded thiocyanate (Figure 18).

The complex *trans*-$[Pd(SCN)_2(Te(CH_2SiMe_2Ph)_2)_2]$ also has S-bonded thiocyanate initially (i.e. $\nu_{C\equiv N} = 2111\ cm^{-1}$), but on aging for several weeks at room temperature, its infrared spectrum indicates a partial Pd—SCN→Pd—NCS isomerization (i.e. $\nu_{C\equiv N} = 2077\ cm^{-1}$) as well as a *trans*→*cis* isomerization of the S-bonded isomer (i.e. $\nu_{C\equiv N} = 2106$, $2098\ cm^{-1}$)[626]. This linkage isomerism is presumably the result of the increased steric bulk introduced around the Pd by the $Te(CH_2SiMe_2Ph)_2$ ligand vs. $Te(CH_2CH_2CH_2SiMe_3)_2$.

^{125}Te-NMR has been used to study the solution structure of the complex $PtCl_2(Te(CH_2CH_2Ph)_2)_2$[605]. This complex has a *cis* configuration in the solid state (a ν_{Pt-Cl} doublet characteristic of C_{2v} symmetry in the far-infrared) (Figure 19a). On dissolution in CH_2Cl_2 the solution initially contains the *cis* isomer (Figure 19b), but *cis–trans* isomerization occurs over several hours to reach an equilibrium mixture

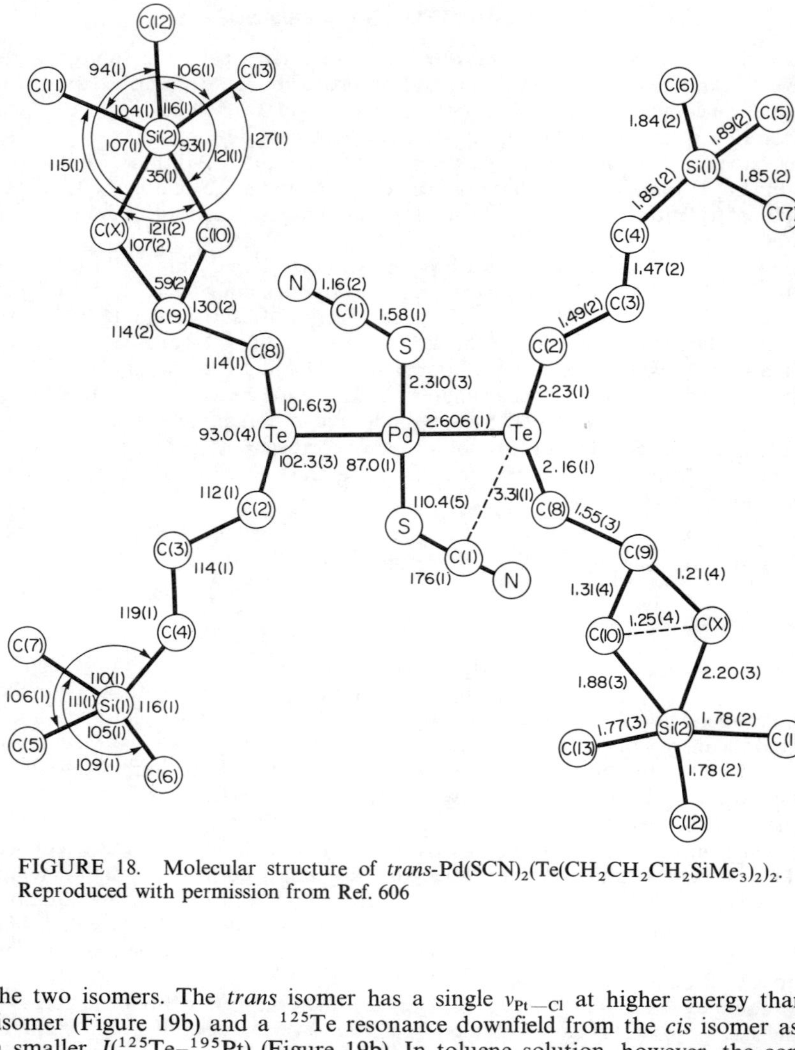

FIGURE 18. Molecular structure of *trans*-$Pd(SCN)_2(Te(CH_2CH_2CH_2SiMe_3)_2)_2$. Reproduced with permission from Ref. 606

of the two isomers. The *trans* isomer has a single ν_{Pt-Cl} at higher energy than the *cis* isomer (Figure 19b) and a ^{125}Te resonance downfield from the *cis* isomer as well as a smaller $J(^{125}Te-^{195}Pt)$ (Figure 19b). In toluene solution, however, the complex exists exclusively as the *trans* isomer ($\nu_{Pt-Cl} = 337\ cm^{-1}$; $\delta(^{125}Te) = -430$ ppm, $J(^{125}Te-^{195}Pt) = 601$ Hz).

The largest number of dialkyl telluride complexes involve Pt(II) and Pd(II). A variety of techniques have been applied to the characterization of their solid-state and solution structures (e.g. infrared, Raman, NMR, dipole-moment measurements)[18]. In such square-planar complexes, $MX_2(TeR_2)_2$ (M = Pd, Pt; X = Cl, Br, I), *cis–trans* isomerization, intramolecular ligand exchange and tellurium inversion processes have all been observed by detailed far-infrared and variable-temperature ^1H-NMR studies[18].

A CH_2Cl_2 solution of $(Bu_4N)[Pt(CNS)_3TeMe_2]$, prepared by equilibration of $(Bu_4N)_2[Pt(SCN)_4]$ with $[Pt(SCN)_2(TeMe_2)_2]$, was shown by $^1H-\{^{195}Pt\}$ INDOR spectroscopy to contain the following four isomers (relative amounts in parentheses)[627]:

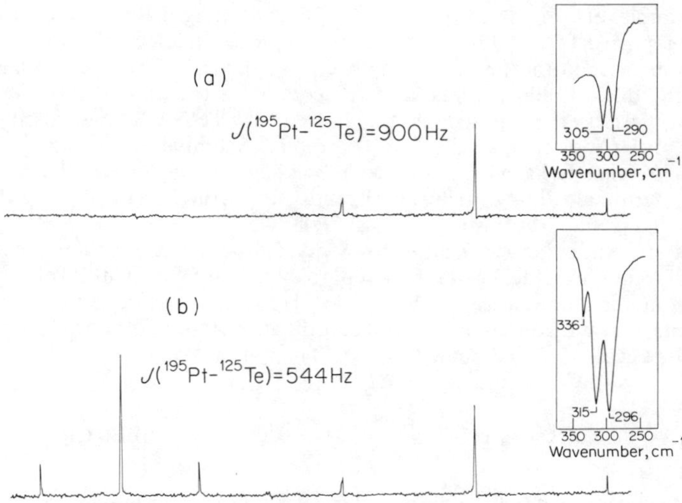

FIGURE 19. ^{125}Te-NMR and far-infrared spectra of $PtCl_2(Te(CH_2CH_2Ph)_2)_2$, (a) in the solid state and (b) in CH_2Cl_2 solution. Reproduced with permission from Ref. 605

		δ(Me)	$^3J(^{195}Pt-^1H)$	$^1J(^{195}Pt-^{14}N)$
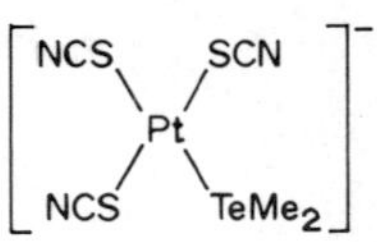	(1)	2.25	33.6 Hz	
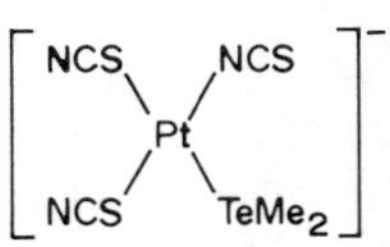	(0.3)	2.24	35.7 Hz	367 ± 5 Hz
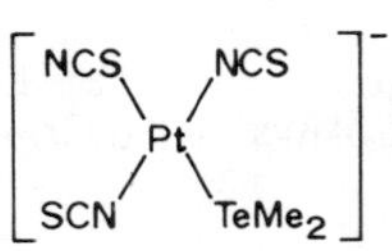	(0.04)	2.20	35 Hz	450 Hz
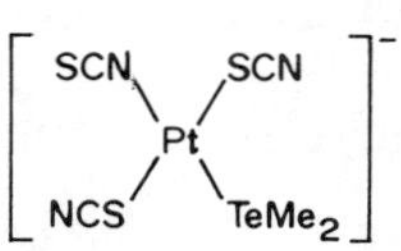	(0.006)	2.14		

The multiplicity of the ^{14}N coupling pattern in the spectrum was used to determine the number of N-bonded thiocyanate ligands in a particular isomer. Specific assignments of the mixed complexes were made by assuming a regular upfield shift of the methyl resonances when S^- is replaced by N-bonded thiocyanate in the position *cis* to the telluride ligand and a somewhat large shift for the *trans* position.

A few examples of bridging diorganotelluride ligands in Ag(I), Pt(II) and Cu(I) complexes have been reported (Table 45). In the Ag(I) complexes ($(AgX)_2TeR_2$; X = Br, I), XPS showed an increase in the Te 3d binding energy of 0.7–0.9 eV vs. the free ligand[609].

A bridging dialkyl telluride ligand in the complexes $(Bu_4N)_2[X_3Pt(\mu\text{-}TeMe_2)PtX_3]$ (X = Cl, Br) has been proposed on the basis of ^{125}Te-NMR spectroscopy[608]. The $J(^{125}Te-^{195}Pt)$ coupling constants in the dimeric complexes, in which both lone pairs of the bridging Te ligand are involved in bonding, are much larger than in complexes containing terminal dialkyl telluride ligands (e.g. $(Bu_4N)_2[Cl_3Pt(\mu\text{-}TeMe_2)PtCl_3]$, 5923 Hz vs. $(Bu_4N)[Cl_3PtTeMe_2]$, − 1553 Hz)[608].

The first structure characterization by X-ray diffraction of a complex containing a bridging diorganotelluride ligand was reported recently for $[CuCl(TeEt_2)]_n$[610]. The structure contains infinite sheets in which two $TeEt_2$ ligands bridge two $(Cu(\mu\text{-}Cl)_2Cu)$ dimeric units to give distorted tetrahedral coordination about both the Cu and Te atoms (Cu—Te distances = 2.535(1) and 2.625(1) Å). The synthesis of this complex involved the *in situ* reduction of Cu(II) (equation 233).

$$2\,CuCl_2 + 3\,TeEt_2 \xrightarrow{EtOH} 2\,[CuCl(TeEt_2)] + TeEt_2Cl_2 \qquad (233)$$

b. Diaryl tellurides. Adducts of diaryl tellurides with mercuric halides were prepared as easily crystallized derivatives in Lederer's early work involving the syntheses of these organotellurides (see Ref. 18 for a discussion of this early work). Few data, other than melting points, were reported for these complexes. More recently some of these complexes have been studied by far-infrared[628] and ^{125}Te Mössbauer spectroscopy[628], and a tetrameric structure (**110**) has been found for $[HgI_2(TePh_2)]$ by X-ray diffraction[613].

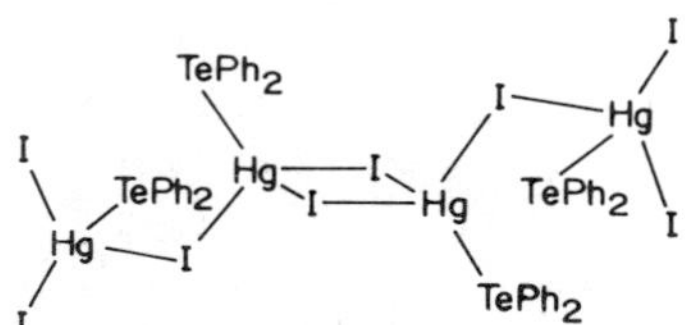

(110)

Relatively few other complexes of transition metals with terminal diaryl telluride ligands (primarily $TePh_2$) have been reported (e.g. Pd(II), Pt(II), Cu(I), Mn(I), Fe(I, II), Ru(II), Rh(I, III), Re(I))[18].

Although diorganotelluride complexes are generally stable, in two cases Te—C bond cleavage has been reported to give complexes with bridging tellurol ligands (equations 234 and 235).

$$Mn_2(CO)_{10} + 2\,TePh_2 \xrightarrow{-Ph_2} (OC)_4Mn(\mu\text{-}TePh)_2Mn(CO)_4 \qquad \text{(Ref. 618)} \quad (234)$$

$$RhCl_3\cdot 3\,H_2O + TePh_2 \xrightarrow[EtOH/reflux]{40\%\ aq.\ CH_2O/KOH} [Rh(TePh)_3]_n \qquad \text{(Ref. 611)} \quad (235)$$

2. Diorganoditellurides

Although ditellurides (and diselenides, Section II.A.5) most commonly undergo cleavage of their E—E (E = Se, Te) bond with formation of terminal or bridging ER^- ligands (e.g. equations 95–104 and 107–109), examples of intact coordinated diorganoditellurides have been described. The crystal structure of the dimeric bridged ditelluride

$[(OC)_3Re(\mu\text{-}Br)_2(\mu\text{-}Te_2Ph_2)Re(CO)_3]$ was reported[184] recently. This complex and the analogous Mn dimer[191] were prepared by the route described for the Se_2R_2 dimers (equation 32, Section II.A.5).

Unique mixed complexes of unsymmetrical diaryl ditellurides were prepared recently by reaction of CuCl with mixtures of the symmetrical ditellurides[629]. The orange–red precipitates formed in these reactions had the composition $PhTeTe(C_6H_4OEt\text{-}p)\cdot 2\,CuX$ (X = Cl, Br), but their insolubility and thermal sensitivity precluded unequivocal characterization. Previous ^{125}Te-NMR studies, however, demonstrated that solutions of two symmetrical ditellurides also contain the unsymmetrical ditelluride[630]. The presence of intact ditelluride was confirmed by the observation of the characteristic ν_{Te-Te} at ca. 175 cm^{-1} in the Raman spectra of these complexes.

Several 1:1 complexes with symmetrical ditellurides have also been prepared ($CuX(Te_2R_2)$: X = Cl, Br; R = Et, *n*-Bu, *n*-C_5H_{11}, Ph, *p*-$EtOC_6H_4$)[616]. The presence of bands assigned to ν_{Te-Te} (170–180 cm^{-1}) and $\nu_{Cu-X-Cu}$ vibrations in the far-infrared spectra of these complexes and their insolubility in organic solvents suggested a polymeric formulation (**116**).

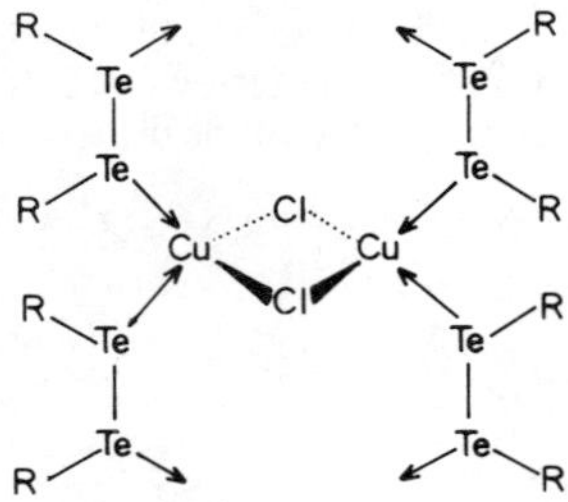

(**116**)

Complexes of Cu(I) and Cu(II) with the potentially chelating ditelluride (*o*-$H_2NC_6H_4$)TeTe($C_6H_4NH_2$-*o*) have also been reported[631], but their structures were not established.

A complex proposed to contain an intact Me_2Te_2 ligand ($[CpNi(Te_2Me_2)]BF_4$) has been prepared by treatment of $[CpNi(C_5H_6)]BF_4$ with the ditelluride in ether[617]. The bonding mode of the Te ligand, however, is not well characterized. Structures involving bridging and chelating MeTeTeMe have been suggested[617].

ESR and infrared spectroscopy were employed in a kinetic and mechanistic study of the oxidative addition of the ditellurides Te_2Ar_2 (Ar = Ph, *p*-MeC_6H_4, *p*-ClC_6H_4, *p*-$EtOC_6H_4$) to *trans*-$(OC)Cl(Ph_3P)_2Ir$[632].

3. Miscellaneous ligands

Only a few examples of transition-metal complexes with Te heterocycles are known. Phenoxtellurine (**111**) forms a complex (Mn—Te bonding) with Mn(I)[618] (equation 236). The reaction of tellurophene (**112**) with Na_2PdCl_4 gave two products, which were separated by their solubility differences in acetone and $CHCl_3$[619] (equation 237).

(236)

(**111**)

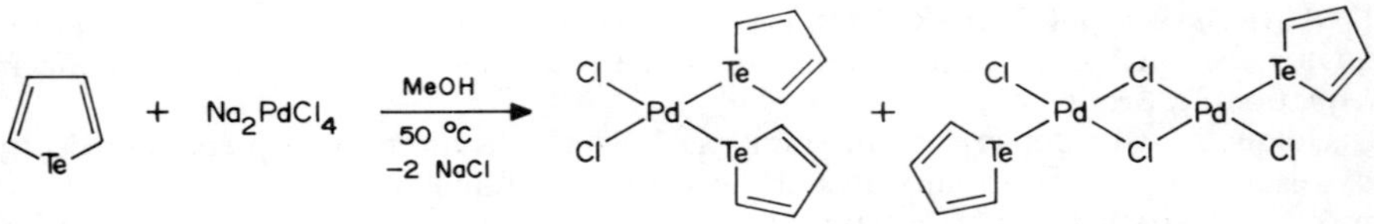

(112) (117) 70% (118) 27% (237)

Treatment of a suspension of **118** in $CHCl_3$ with excess tellurophene gave **117**.

The analogous reaction of Na_2PdCl_4 with tetrachlorotellurophene (tct) gave exclusively *trans*-$PdCl_2(tct)_2$, on the basis of elemental analyses and the observation of only one ν_{Pd-Cl} band in its far-infrared spectrum ($354\,cm^{-1}$)[619].

Reaction of **112** with $Cr(CO)_6$ at high temperature failed to give any substitution product, but the π complex $(\eta^5\text{-}TeC_4H_4)Cr(CO)_3$ was obtained in 80% yield from the reaction of **112** with $Cr(CO)_3(MeCN)_3$ in Bu_2O at 60 °C[619]. The purple–red complex dissolves in benzene as a monomer and can be vacuum sublimed at 65 °C without decomposition.

Monomeric complexes result from the reactions of 9-*H*-telluroxanthene (**113**) and 10-ethyl-3,7-dimethylphenotellurazine (**114**) with the dimeric complex **119**[620] (equation 238).

$$4\,L + (OC)_2Rh(\mu\text{-}Cl)_2Rh(CO)_2 \rightarrow 2\,L_2Rh(CO)Cl \qquad (238)$$

(119)

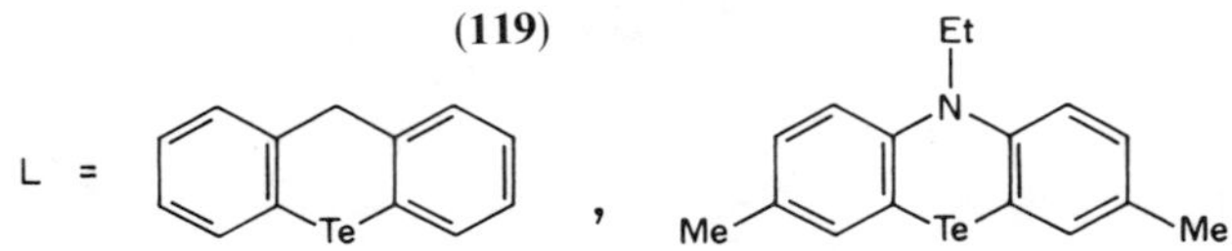

(113) (114)

Chromium complexes with $Te(EMe_3)_2$ (E = Ge, Sn, Pb) were prepared[196] by a two-step route (equation 239) involving the initial photochemical generation of a labile THF adduct rather than by direct substitution (e.g. $Se(SnMe_3)_2$ complexes, equation 38) because of the high photosensitivity of the telluride ligands.

$$Cr(CO)_6 \xrightarrow[THF]{h\nu} \{Cr(CO)_5(THF)\} \xrightarrow[0\,°C]{Te(EMe_3)_2} Cr(CO)_5Te(EMe_3)_2 \qquad (239)$$

$$E = Ge, Sn, Pb$$

These complexes are quite sensitive to air and moisture and slowly decompose in the solid state even at 0 °C. They all exhibited satellites about the singlet Me signal in their ^1H-NMR spectra ($J(^1HC\text{–}^{119}Sn) = 54.5$ Hz, $J(^1HC\text{–}^{117}Sn) = 52.5$ Hz, $J(^1HC\text{–}^{207}Pb) = 57.8$ Hz and a three-bond $J(^1HCGe\text{–}^{125}Te) = 8.5$ Hz).

A similar indirect photochemical substitution route was used to prepare the complexes $(OC)_5Cr\{(CF_3)_2ETeMe\}$ (E = P, As). These monomeric complexes were isolated as air-sensitive red–brown oils in 23% and 14% yields, respectively. The $(CF_3)_2PTeMe$ complex has good thermal and light stability in the solid state and solution, but the complex with the weaker donor, $(CF_3)_2AsTeMe$, decomposes in solution slowly over a few days even in the absence of air and light. Spectroscopic results indicate that the $(CF_3)_2AsTeMe$ complex contains both As- and Te-bonded linkage isomers, whereas the $(CF_3)_2PTeMe$ complex is isomerically pure (Cr—P bonding). The NMR characterization of the bonding mode of these ambidentate ligands was similar to that previously described for the Se analogues (Section II.A.5).

The instability of tellurourea-type compounds has precluded the synthesis of the wide variety of complexes described for selenourea (Section II.A.1). Lappert and coworkers[621,622], however, have reported two examples of complexes with the cyclic tellurourea derivatives (**115**) (Table 45). The Cr, Mo and W complexes were prepared by the substitution route shown in equation (240) using the electron-rich olefin **120**[621].

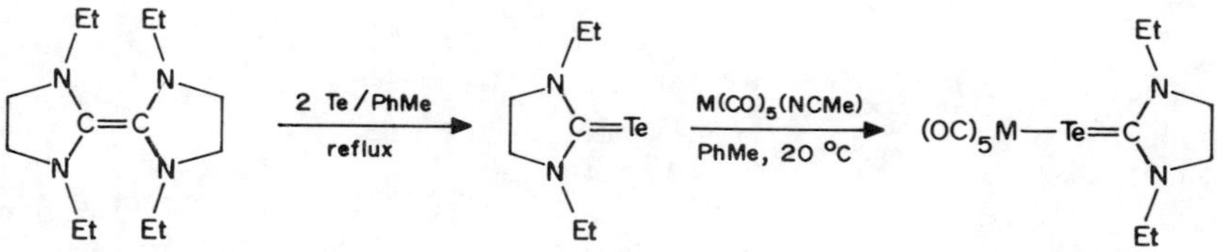

(240)

(**120**) (**115**)

75%, M = Cr
52%, M = Mo
71%, M = W

The Cr complex readily undergoes Te extrusion to give the corresponding carbene complex[621] (equation 241).

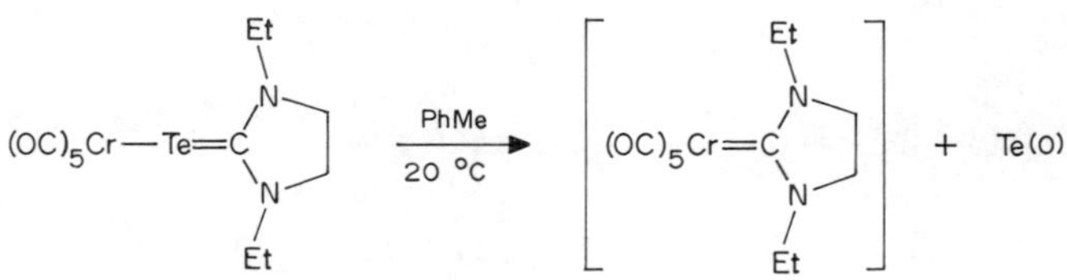

(241)

The Mo and W analogues are somewhat more stable but also undergo the detelluration reaction at high temperatures. The molecular structure of the Cr derivative has been determined by single-crystal X-ray diffraction (Figure 20)[621].

These complexes, an Fe(I) complex with the same ligand and the telluroketone complex $(OC)_5W(Te{=}CPh_2)$, are the only reported complexes with a tellurocarbonyl group ($>$C=Te) (Table 45), although two other examples of organotellurium derivatives of this class were described recently (*o*-alkyltellurocarboxylates, $RC(Te)OR^1$[633] and telluroamides, $RC(Te)NR^1R^2$[634]).

Two complexes of tellurocarbonyl have been prepared as shown in equations (242) and (243).

$$Cl_2(Ph_3P)_2(OC)Os{=}CCl_2 \xrightarrow[C_6H_6,\ EtOH]{Na^+\,TeH^-} Cl_2(Ph_3P)_2(OC)Os{-}C{\equiv}Te \quad \text{(Ref. 216)} \qquad (242)$$

(**121**)
30%
air-stable orange crystals
m.p. 221–223 °C
$\nu_{C\equiv O} = 2040\text{s cm}^{-1}$
$\nu_{C\equiv Te} = 1046\text{s cm}^{-1}$

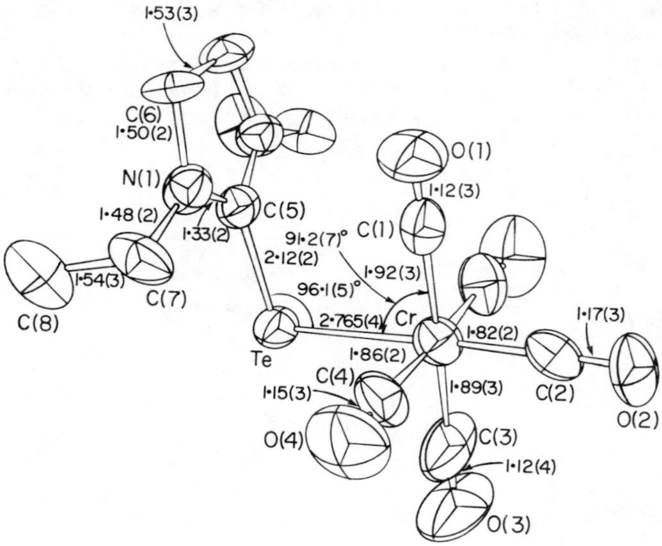

FIGURE 20. Molecular structure of $(OC)_5Cr\{Te{=}\overline{CN(Et)CH_2CH_2N}Et\}$. Reproduced with permission from Ref. 621

$$\begin{array}{lcl} L(OC)_2Mo{\equiv}CCl & \xrightarrow[\text{2. } Et_4NCl]{\text{1. } Li_2Te,\ THF} & Et_4N^+[L(OC)_2Mo{\equiv}C{-}Te^-] \quad \text{(Ref. 223)} \qquad (243) \\ L = \mathbf{37} & & \updownarrow \\ & & Et_4N^+[L(OC)_2Mo^-{-}C{\equiv}Te] \\ & & (\mathbf{122}) \\ & & 60\% \\ & & \nu_{C\equiv O} = 1924,\ 1840\,\text{vs}\,\text{cm}^{-1} \\ & & \nu_{C\equiv Te} = 951\,\text{s}\,\text{cm}^{-1} \end{array}$$

The stereochemistry of the octahedral Os complex **121** was assigned (e.g. **35**, Te in place of Se) on the basis of the similar infrared spectra of all three CX (X = S, Se, Te) derivatives and the characterization of the thiocarbonyl complex by single-crystal X-ray diffraction[225]. The low yield of the tellurocarbonyl complex after silica gel chromatography was attributed to the difficulty of preparing pure TeH^- ($NaBH_4$ reduction of powdered Te in refluxing ethanol followed by cooling to − 20 °C and acidification with HOAc/EtOH[635]).

The Mo complex **122** is less stable than its S and Se congeners, but it was converted to the more stable telluromethylidyne complex[223] (equation 244).

$$\begin{array}{lcl} Et_4N[L(OC)_2Mo{\equiv}C{-}Te] & \xrightarrow[-\,Et_4NI]{MeI} & L(OC)_2Mo{\equiv}C{-}TeMe \qquad (244) \\ L = \mathbf{37} & & 20\% \\ & & \nu_{C\equiv O} = 1992,\ 1911\,\text{s}\,\text{cm}^{-1} \\ & & ^{13}\text{C-NMR}\ (\delta,\ CDCl_3):\ 266.22\ (C{-}Te), \\ & & 226.62\ (CO) \end{array}$$

The first example of a telluroketone complex was prepared recently by insertion of Te from tellurocyanate into a metal–carbene bond[623] (equation 245).

$$(OC)_5W{=}CPh_2 + PPN[TeCN] \xrightarrow[-PPN[CN]]{CH_2Cl_2/-90\,^\circ C} (OC)_5W{-}Te{=}CPh_2 \quad (245)$$

18%

m.p. 35 °C dec.

$\nu_{C{=}O}$ (hexane): 2066m, 1953, 1941 cm^{-1}

^{1}H-NMR δ(acetone-d_6): 7.66 (m)

^{13}C-NMR δ(acetone-d_6): 160.4, 130.4, 128.1, 127.0, 123.8 (Ph); 197.3 (CO_{cis}); 231.9 (Te=C)

The product was isolated as thermally unstable, black crystals by column chromatography (SiO_2/pentane–CH_2Cl_2 10:1) at −50 °C followed by recrystallization from pentane. The formulation of the product as a telluroketone complex was based on spectroscopic evidence, especially the position of the Te=C—^{13}C resonance (equation 245).

Free diaryl telluroketones are not known, although a brief report[636] described the synthesis of a dialkyl telluroketone. This is, therefore, another example of the stabilization of an unstable organic molecule by coordination to a transition-metal centre.

Fischer and coworkers[624] recently reported the first example of a Te-functionalized metal carbene complex and studied its reactivity (equation 246). Rearrangement of the initial Te-functionalized carbene complex **B** occurs spontaneously over a few hours at 30 °C to give quantitatively the carbyne complex **C**, which readily eliminates CO on further stirring in ether at room temperature to give the thermally unstable Te-bridged dimer **D**. Because the latter dimerization is inhibited by free CO, whereas the rearrangement of **B** to **C** is independent of CO concentration, the carbyne **C** can be synthesized in good yield by carrying out the reaction under CO pressure. Heating **B** in an

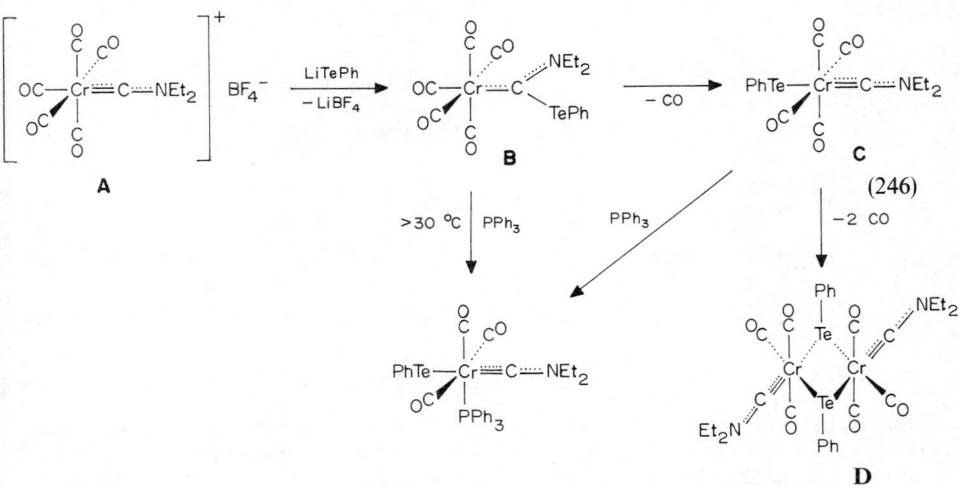

(246)

autoclave at 40 °C under 40 atm of CO for 2 h gave **C**, after recrystallization from 1:1 CH_2Cl_2/pentane, in 77% yield as an orange crystalline solid (m.p. 42 °C dec.). All of the above compounds were characterized by elemental analysis and infrared and ^{1}H-NMR spectroscopy, and the carbene complex **B** was shown to be isostructural with $(OC)_5Cr{=}C(SePh)(NEt_2)$ (**38**), which was previously characterized by single-crystal X-ray diffraction[248]. The rearrangement of **B** to **C** in 1,1,2-trichloroethane follows a first-order rate law ($-d[B]/dt = k[B]$; $\Delta H^{\neq} = 108 \pm 1\ kJ\,mol^{-1}$; $\Delta S^{\neq} = 42 \pm 4\ J\,mol^{-1}\,K^{-1}$).

Diphenyl telluroxide, $Ph_2Te{=}O$, reacts with Sn(IV), Ti(IV) and Sb(V) chlorides to give Ph_2TeCl_2, but 2:1 adducts have been isolated from reactions with organotellurium trichlorides[625] (equation 247). The Te—O=$TePh_2$ bonding mode for these air-stable white complexes was proposed on the basis of infrared spectroscopy ($\nu_{Te=O}$ decreases from 708 cm^{-1} in $Ph_2Te{=}O$ to 658–613 cm^{-1} in the complexes). The presence of two strong $\nu_{Te=O}$ bands in the octahedral complexes suggests a *cis* arrangement of the two $Ph_2Te{=}O$ ligands[625].

$$RTeCl_3 + 2\,Ph_2Te{=}O \xrightarrow{CHCl_3} RCl_3Te[O{=}TePh_2]_2 \qquad (247)$$

$$R = Me,\ C_6H_4OMe\text{-}p,\ C_6H_4OPh\text{-}p$$

B. Anionic Monodentate Ligands

The synthesis of complexes containing terminal and bridging tellurol ligands (Table 46) parallels the routes described above for selenol complexes (Section II.B.2). The most common preparative route involves oxidative addition of a diorganoditelluride to a low-valent metal substrate (equations 248–250).

$$2\,Fe(NO)_2(CO)_2 + ArTeTeAr \xrightarrow[\text{r.t.},\ -4\,CO]{C_6H_6} (ON)_2Fe(\mu\text{-}TeAr)_2Fe(NO)_2 \quad \textbf{(123)} \qquad \text{(Ref. 640)} \quad (248)$$

$$Ar = C_6H_4OMe\text{-}p$$

$$CpFe(\mu\text{-}CO)_2FeCp + PhTePh \xrightarrow[\text{reflux}]{C_6H_6} CpFe(CO)_2TePh \quad \textbf{(124)} \qquad \text{(Ref. 329)} \quad (249)$$

$$\xrightarrow[\text{reflux}/h\nu]{C_6H_6} Cp(OC)Fe(\mu\text{-}TePh)_2Fe(CO)Cp \quad \textbf{(125)}$$

$$Pd(PPh_3)_4 + ArTeTeAr \xrightarrow{C_6H_6}$$

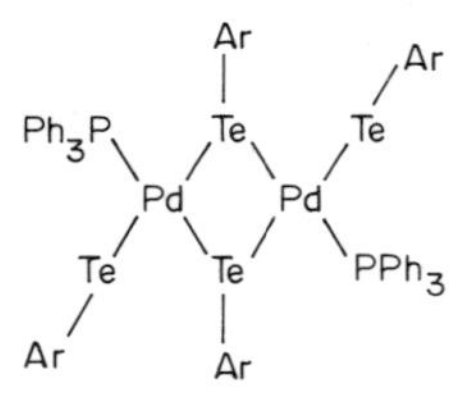

(Ref. 641) (250)

$Ar = C_6H_4OEt\text{-}p$, 2-thienyl

(126)

The diamagnetic dimer **123** is isostructural with Roussin's salt, $[Fe(NO)_2SEt]_2$, the structural determination of which has established tetrahedral coordination about the Fe atoms (two terminal NO ligands and two bridging S atoms; Fe—Fe = 2.72 Å[640]).

TABLE 46. Complexes with anionic tellurium ligands

Ligand	Complex	Ref.
$Te(alkyl)^-$		
terminal	$(OC)_2(Ph_3P)_2(Me)OsTeMe$	418
bridging	$[EtTeCu]_n$	616
	$(ON)_2Fe(\mu\text{-}TeCH_2Ph)_2Fe(NO)_2$	353
	$Cp(ON)_2Cr(\mu\text{-}Te(Bu\text{-}n))_2Cr(NO)_2Cp$	351
$Te(aryl)^-$		
terminal	$CpFe(CO)_2TePh$	305
	$(\eta^7\text{-}C_7H_7)Mo(CO)_2TePh$	337 [a]
	$(\eta^7\text{-}C_7H_7)W(CO)_2TePh$	336
	$[Ph_4P][Hg(TePh)_3]$	637 [a]
	$CpNi(P(Bu\text{-}n)_3)(TeC_6H_4X)$	330
	X = H, p-Cl, p-MeO, p-Me, m-CF_3	
	$Cp_2W(TeAr)_2$	332
	Ar = Ph, C_6H_4Me-p	
	$Cp_2Nb(TePh)_2$	332
bridging	$Cp(OC)Fe(\mu\text{-}TeAr)_2Fe(CO)Cp$	
	Ar = Ph, p-$EtOC_6H_4$	329, 638 [a]
	$(\eta^7\text{-}C_7H_7)W(\mu\text{-}TePh)_3W(CO)_3$	336
	$(OC)_4W(\mu\text{-}TePh)_2W(CO)_4$	336
	$Cp(ON)_2Cr(\mu\text{-}TePh)_2Cr(NO)_2Cp$	351
	$(\eta^7\text{-}C_7H_7)(OC)Mo(\mu\text{-}TePh)_2Mo(CO)(\eta^7\text{-}C_7H_7)$	639
$Te(SnMe_3)^-$		
bridging	$(OC)_4Re(\mu\text{-}TeSnMe_3)_2Re(CO)_4$	363
TeH^-		
terminal	$PPN[Cr(CO)_5TeH]$	108
bridging	$(AsPh_4)[(OC)_5W(\mu\text{-}TeH)W(CO)_5]$	108
$Te(O)_2Me^-$	$(\eta^7\text{-}C_7H_7)Mo(CO)_2Te(O)_2Me$	390

[a] Structure determined by single-crystal X-ray diffraction.

The monomeric complex **124** with terminally bonded $TePh^-$ gave the dimeric complex **125** when the reaction solution was refluxed with infrared irradiation[329]. The dimeric complex was obtained as a mixture of two isomers on the basis of TLC and ^{1}H-NMR evidence, but the pure isomers could not be isolated. Spectroscopic evidence (Table 47) indicated the formulations **127a** and **127b** (Ar = Ph) for the two dimers[329]. Five stereoisomers (**127a**–**e**) of such a dimer are possible, and each of these can exist in two conformational forms for a non-planar Fe_2Te_2 ring[642].

TABLE 47. Spectroscopic data for $Cp(OC)Fe(\mu\text{-}EPh)_2Fe(CO)Cp$ (E = S, Se, Te) dimers (**127a**) and (**127b**)

	127a		**127b**	
E	$\nu_{C\equiv O}$ $(cm^{-1})^a$	δ_{Cp} $(ppm)^b$	$\nu_{C\equiv O}$ $(cm^{-1})^a$	δ_{Cp} $(ppm)^b$
S[c]	1982s	4.43	1953s, 1937s	4.03
Se	1975s	4.46	1947s, 1931s	4.02
Te	1965m	4.48	1937, 1921	4.11

[a] In C_6H_{12} solution.
[b] In CS_2 solution.
[c] The crystal structure of isomer **127a** of this dimer has been reported[643].

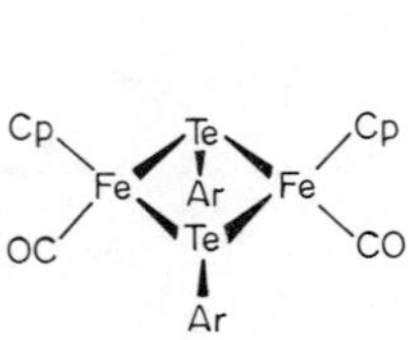

(127a)

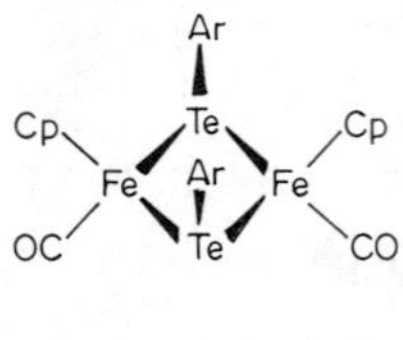

(127b)

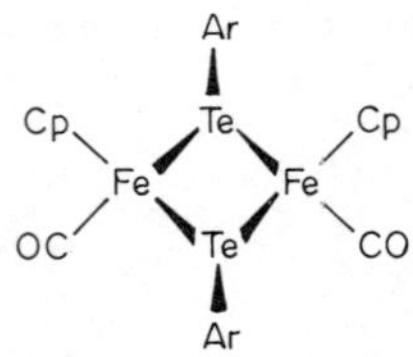

(127c)

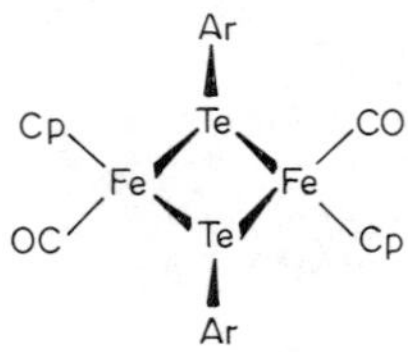

(127d)

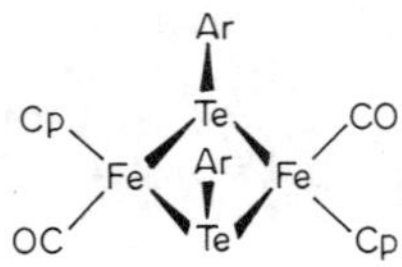

(127e)

The relative stabilities of the isomers **127a** and **127b** vary significantly with the nature of the chalcogen atom. The amounts of isomer **127b** obtained from equation (249) and related reactions for Ph_2E_2 (E = S, Se) are: Te (major product) > Se (~25%) > S (~1%).

The related reaction with $Te_2(C_6H_4OEt\text{-}p)_2$ gave two isomers, which could be mechanically separated after recrystallization from CH_2Cl_2/hexane[638]. Single-crystal X-ray diffraction of the isomer obtained in lower yield (ca. 5%) showed it to be isomer (**127a**) while the other isomer, obtained as 95% of the total yield, has the structure (**127d**)[638].

The unique dimeric complexes **126** contain both terminal and bridging tellurol ligands[641]. These complexes were characterized by elemental analysis and molecular weight measurements in benzene. Although ^{125}Te Mössbauer spectroscopy failed to resolve the chemically inequivalent Te atoms in these dimers, X-ray photoelectron spectroscopy resolved the two types (Te $3d_{5/2} \simeq 572.5, 573.5$ eV)[644].

Elemental Hg also undergoes oxidative addition reactions with diorganoditellurides (equation 251).

$$Hg + ArTeTeAr \rightarrow Hg(TeAr)_2 \qquad (251)$$

$$Ar = Ph^{357}, C_6H_4OEt\text{-}p^{645}$$

A monomeric trigonal planar anionic complex (Figure 21) was obtained from $Hg(TePh)_2$ and isolated as its air-stable Ph_4P^+ salt as shown in equation (252).

$$Hg(TePh)_2 + \underset{\underset{\frac{1}{2}PhTePh}{\Big\uparrow \scriptstyle Na/liq,NH_3}}{NaTePh} \rightarrow Na^+[Hg(TePh)_3]^- \xrightarrow[-NaCl]{Ph_4PCl} Ph_4P[Hg(TePh)_3] \qquad (252)$$

Attempts to form $[Hg(TePh)_4]^{2-}$ by reaction of $[Hg(TePh)_3]^-$ with an equivalent of $[Ph_4P]TePh$ in $CHCl_3$ gave $[Hg_3(TePh)_{11}]^{5-}$, isolated as its orange–red crystalline Ph_4P^+ salt. The structure of this complex is unknown.

Several Cr, Mo and W complexes containing both terminal and bridging $PhTe^-$ have been isolated by metathetical reactions with organometallic halide precursors (Table 46). As discussed for the analogous $SePh^-$ complexes (Section II.B.2; equations 117–119), these reactions generally give mixtures of monomeric and dimeric complexes, which can be separated by column chromatography. The crystal structure of the monomeric

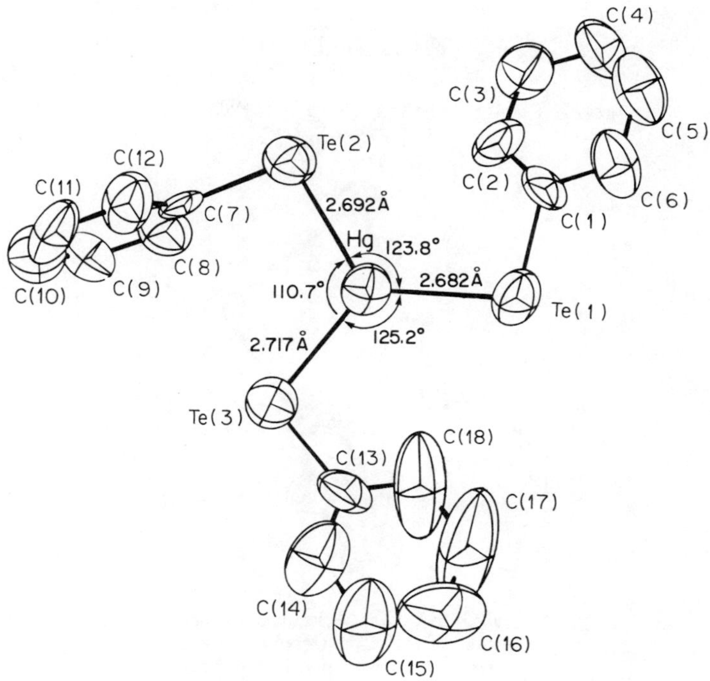

FIGURE 21. Molecular structure of $[Hg(TePh)_3]^-$. Reproduced with permission from Ref. 637

complex $(\eta^7\text{-}C_7H_7)(OC)_2MoTePh$ (Figure 22) has been reported[337]. The Mo—Te bond distance in this complex (2.797 Å) is 0.15 Å shorter than the sum of the covalent radii, indicating a considerable double-bond character.

The monomeric complex $(Ph_3P)_2(OC)_2(Me)OsTeMe$ was prepared by Te methylation (i.e. MeI) followed by borohydride reduction of the telluroformaldehyde complex, $(Ph_3P)_2(OC)_2Os(\eta^2\text{-}CH_2Te)$ (see next section).

The only example of a $—TeO_2R^-$-type ligand was prepared by insertion of TeO_2 into a metal–alkyl bond (equation 253).

$$(\eta^7\text{-}C_7H_7)Mo(CO)_2Me \xrightarrow{TeO_2} (\eta^7\text{-}C_7H_7)Mo(CO)_2Te(O)_2Me \qquad (253)$$

$$\nu_{C\equiv O}\ (CH_2Cl_2) = 1990,\ 1930\,cm^{-1}$$

$$\nu_{Te—C}\ (\text{Nujol mull}) = 470\,cm^{-1}$$

$$\nu_{Te=O}\ (\text{Nujol mull}) = 960,\ 785,\ 750,\ 695\,cm^{-1}$$

$$\delta(CDCl_3) = 2.33\ (Me),\ 5.30\ (C_7H_7)$$

The TeO_2 was an active form of the reagent generated in an ether matrix at $-196\,°C$ in a metal-atom reactor and subsequently reacted with the methyl compound at $-78\,°C$ for 8 weeks. The thermal instability of the red microcrystalline product precluded its X-ray structural characterization, the proposed formulation being based on the similarity of its spectral properties with those of the more stable S and Se analogues.

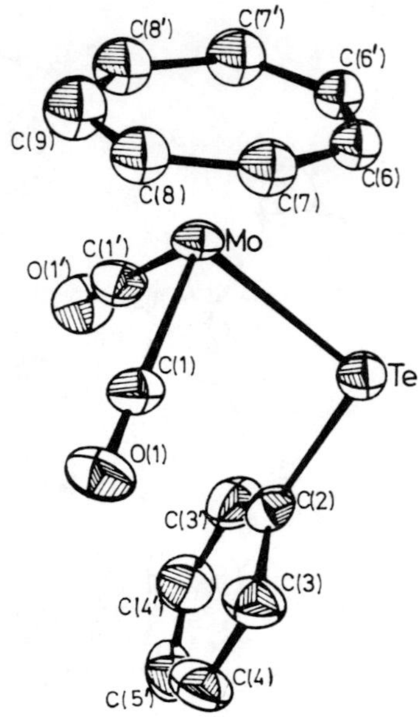

FIGURE 22. Molecular structure of $(\eta^7\text{-}C_7H_7)Mo(CO)_2TePh$. Reproduced with permission from Ref. 337

Complexes containing both terminal (equation 254) and bridging (equation 255) TeH^- have been prepared.

$$M(CO)_5THF + Na_2Te \xrightarrow[-60\,^\circ C]{EtOH} Na[M(CO)_5TeH] \xrightarrow[-NaCl]{X^+Cl^-} X[M(CO)_5TeH] \qquad (254)$$

$$M(CO)_6 \xrightarrow[-40\,^\circ C]{h\nu/THF,\ -CO} M(CO)_5THF$$

X = PPN, M = Cr
X = Ph_4As, M = W

$$X[M(CO)_5TeH] \xrightarrow{M(CO)_5(THF)} X[(CO)_5M(\mu\text{-}TeH)M(CO)_5] \qquad (255)$$

X = PPN, M = Cr
X = Ph_4As, M = W

The terminal TeH^- ligand can be alkylated to the neutral dialkyl telluride analogue (equation 256).

$$PPN[Cr(CO)_5TeH \xrightarrow[EtOH]{[Et_3O][BF_4]} (OC)_5CrTeEt_2 \qquad (256)$$

C. Bidentate Ligands

Only three well-characterized examples of bidentate Te ligands have been reported (Table 48).

TABLE 48. Complexes with bidentate tellurium ligands

Ligand	Complex	Ref.
Telluroformaldehyde		
η^2-CH_2Te	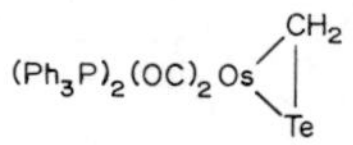	418
	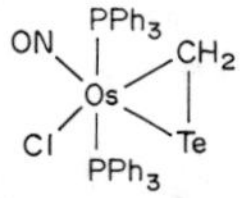	419
	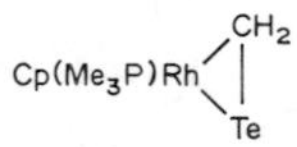	420
η^2,μ-CH_2Te	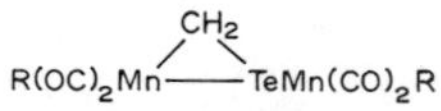	
	R = Cp	421
	R = Cp'	646, 647
η^2,μ-CMe_2Te	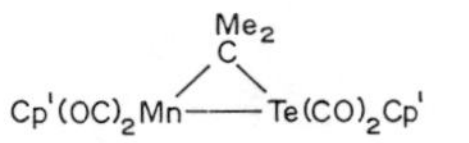	647
Telluroketene	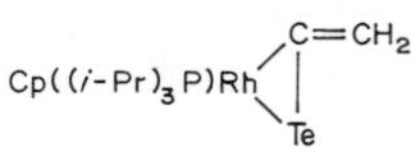	648
(128)	$[Pt(128)_2][Pt(SCN)_4]\cdot 2$ DMF	57[a]

[a] Structure determined by single-crystal X-ray diffraction.

In 1983 the first examples of telluroformaldehyde complexes, involving both chelating η^2-CH_2Te (equations 257–259) and bridging η^2,μ_2-CH_2Te (equations 260 and 261) bonding modes of this ligand, were reported.

These reactions are further examples of coordination stabilization of an unstable organic molecule; monomeric telluroformaldehyde is not known, although a trimer has been reported but not characterized[649]. All four derivatives in the series $(Ph_3P)_2(OC)_2Os(\eta^2\text{-}CH_2E)$ (E = O[650], S[651], Se[418], Te[418]) have been prepared.

The rigid RhCTe three-membered ring in $Cp(Me_3P)Rh(\eta^2\text{-}CH_2Te)$ produces two ^1H-NMR signals for the CH_2Te protons ($\delta = 5.50$ (ddd, $J_{HH} = 0.6$ Hz, $J_{RhH} = 0.6$ Hz, $J_{PH} = 10.2$ Hz), 6.63 (ddd, $J_{HH} = 0.6$ Hz, $J_{RhH} = 2.0$ Hz, $J_{PH} = 0.6$ Hz)[420]). Similarly, the asymmetry of the dimer $Cp(OC)_2Mn(\eta^2,\mu\text{-}TeCH_2)Mn(CO)_2Cp$ results in an AB system for the CH_2 protons in its ^1H-NMR spectrum (equation 260)[421].

The synthesis of μ-alkylidene organotransition-metal complexes (e.g. $M(\eta^2,\mu\text{-}CH_2Te)M$) by carbene transfer from diazoalkane precursors to multiply bonded main-group/transition-metal compounds (e.g. equations 260 and 261) is a recently developed route, which may have general utility for such derivatives[421,554,647].

The previously unknown telluroketene molecule has been stabilized by coordination to Rh(I) in a two-step reaction involving insertion of elemental Te into the metal–carbon bond of an initially formed vinylidene complex[648] (equation 262).

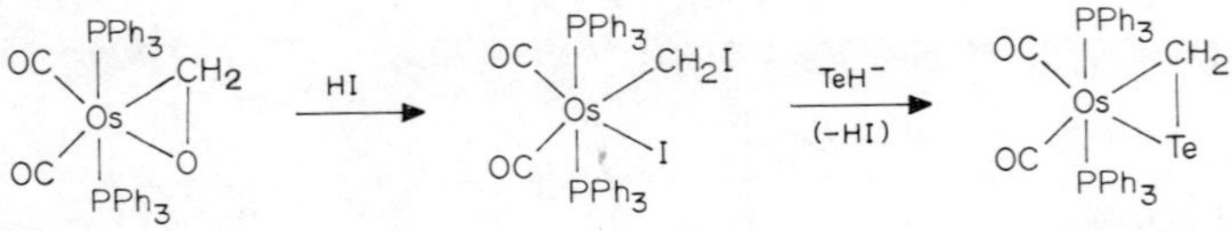

(Ref. 418) (257)

bright yellow air-stable crystals

$\nu_{C\equiv O}$(Nujol mull) = 1991, 1922 cm^{-1}

$\delta(CDCl_3)$: 2.08 (t, CH_2; $^3J(^{31}P-^1H)$ = 6.0 Hz)

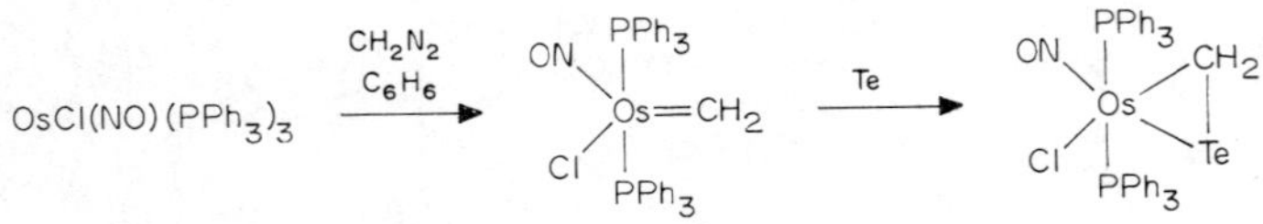

(Ref. 419) (258)

$\nu_{N=O}$ = 1740 cm^{-1}

$\delta(CDCl_3)$: 4.88 (t, CH_2; $J(^{31}P-^1H)$ = 3 Hz)

$$Cp(Me_3P)Rh(C_2H_4) \xrightarrow{CH_2I_2} Cp(Me_3P)Rh(I)(CH_2I) \xrightarrow[\substack{-2\ NaI \\ -TeH_2}]{\substack{2\ NaTeH/THF \\ 25\ ^\circ C}} Cp(Me_3P)Rh(\eta^2\text{-}CH_2Te)$$

24%

dark green, slightly air-sensitive crystals

(Ref. 420) (259)

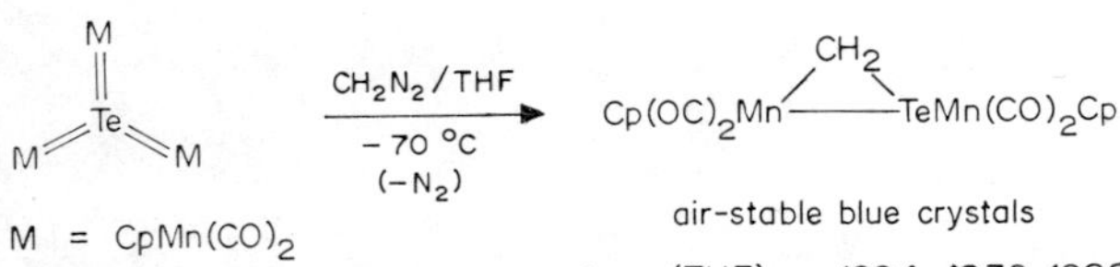

(Ref. 421) (260)

air-stable blue crystals

$\nu_{C\equiv O}$(THF) = 1994, 1939, 1922, 1871 cm^{-1}

$\delta(C_6D_6)$: 4.29 (s, Cp), 3.79 (s, Cp),

5.96 (d, CH_2; $^2J_{HH}$ = 3.7 Hz),

6.02 (d, CH_2; $^2J_{HH}$ = 3.7 Hz)

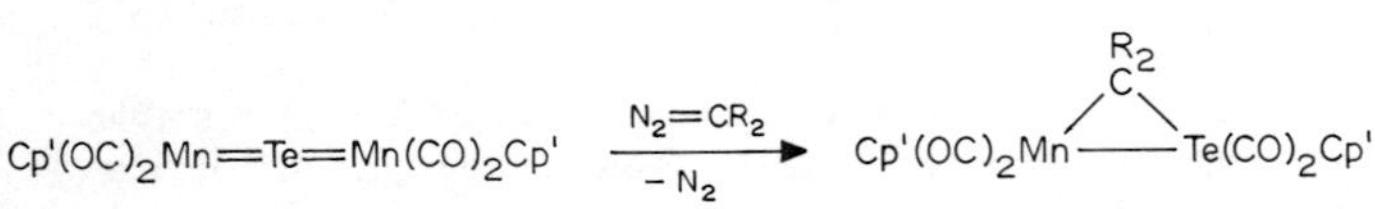

(Ref. 647) (261)

R = H, Me

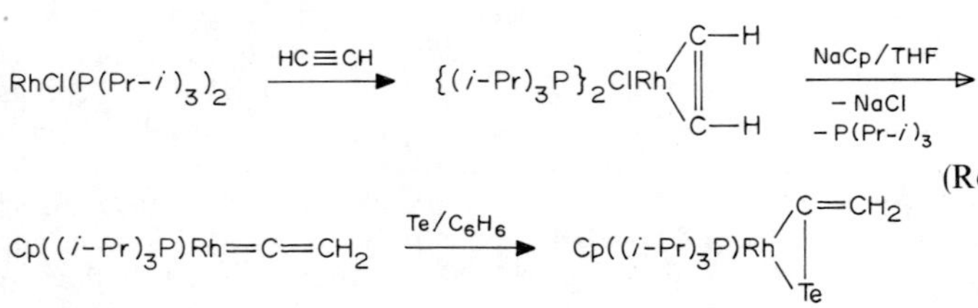

(Ref. 648) (262)

70%

air-sensitive green solid

m.p. 102 °C

$\nu_{C=C}$ = 1572 cm^{-1}

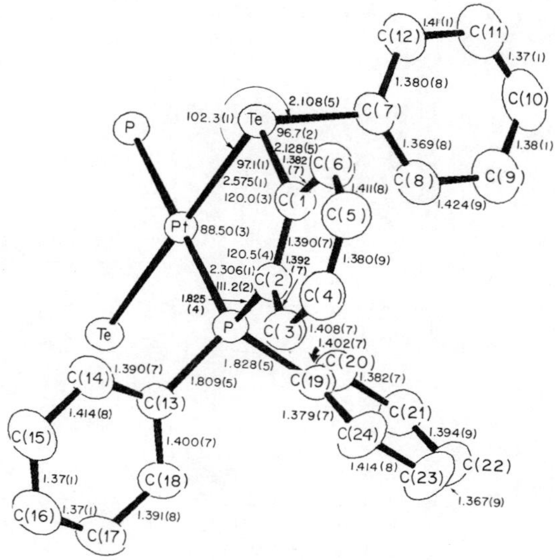

FIGURE 23. Molecular structure of $[Pt(Ph_2PC_6H_4TePh\text{-}o)_2]^{2+}$. Reproduced with permission from Ref. 57

Single-crystal X-ray diffraction has confirmed the proposed coordination mode for the more stable thioketene analogue[648].

The first example of a stable hybrid bidentate ligand containing a Te donor site was prepared recently (equation 263). Reaction of **128** with $K_2Pt(SCN)_4$ gave a complex formulated as the neutral square-planar $[Pt(\mathbf{128})(SCN)_2]$ ($\nu_{C\equiv N} = 2102$, $2118\,cm^{-1}$). Upon recrystallization from hot DMF, however, the complex rearranged to give the Magnus-type salt $[Pt(\mathbf{128})_2][Pt(SCN)_4]\cdot 2\,DMF$[57] ($\nu_{C\equiv N} = 2108\,cm^{-1}$). Single-crystal X-ray diffraction confirmed the proposed bidentate P, Te coordination of the hybrid ligand (Figure 23).

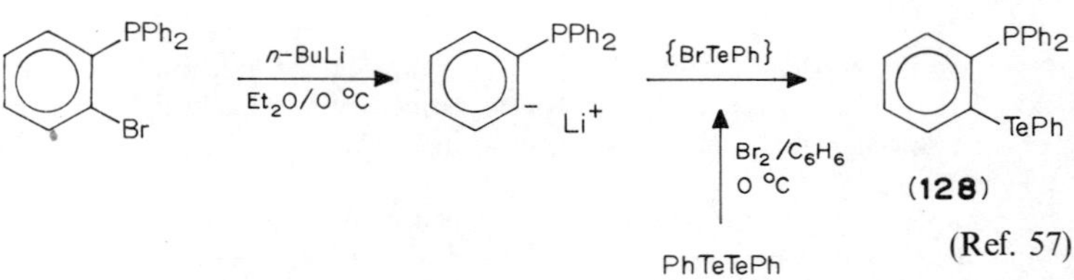

(**128**)

(Ref. 57) (263)

Complexes of Cu(I) and Cu(II) with the potentially chelating ligand di-2-aminophenyl ditelluride (e.g. $[CuCl_n(C_{12}H_{12}N_2Te_2)]_m$; $n = 1, 2$) have been described, but their low solubility, presumably due to polymeric structures, precluded their definitive characterization[631].

D. Complexes Incorporating Te_n ($n = 1, 2$)

Although not as extensive as S_n and Se_n ligand chemistry, several types of transition-metal dimers and cluster compounds incorporating Te atoms are known (Table 49). The

TABLE 49. Complexes with Te_n ligands

Ligand	Complex	Ref.
μ_2-Te	$Li_2[(ON)_2Fe(\mu_2\text{-}Te)_2Fe(NO)_2]$	353
	μ_2-Te[Cp′(OC)Rh]$_2$	556[a]
	μ_2-Te[V(CO)$_3$dppe]$_2$	559
	μ_2-Te[Mn(CO)$_2$Cp′]$_2$	646, 647[a]
	$(OC)_3Fe(\mu_2\text{-}Te\{Pt(PPh_3)_2\})Fe(CO)_3$	570, 571
μ_2, η^2-Te_2	$(OC)_3Fe(\mu_2,\eta^2\text{-}Te_2)Fe(CO)_3$	570, 571, 562
	$Nb_2Te_2X_6$ X = Br, I	579
μ_3-Te	μ_3-Te[Mn(CO)$_2$Cp]$_3$	653
	$Fe_3(\mu_3\text{-}Te)_2(CO)_9$	568, 654–661[a]
	$Fe_3(\mu_3\text{-}Te)_2(CO)_nL_m$ L = P(Bu-n)$_3$, P(OPh)$_3$, AsPh$_3$ $n = 9, 8, m = 1; n = 7, m = 2$	655
	L = PPh$_3$[a], Ph$_2$P(CH$_2$)$_n$PPh$_2$ ($n = 1$–3), CO, t-BuNC	660
	$Fe_3(\mu_3\text{-}E)(\mu_3\text{-}Te)(CO)_9$ E = S, Se	662, 663
	$Fe_3(\mu_3\text{-}E)(\mu_3\text{-}Te)(CO)_nL_m$ E = S, Se $n = 8, m = 1; n = 7, m = 2$ L = AsPh$_3$, P(OPh)$_3$	663
	$Fe_4(\mu_3\text{-}Te)_4(CO)_{12}$	652, 661
	$FeCo_2(\mu_3\text{-}Te)(CO)_9$	589[a], 661
	$CpRhFe_2(\mu_3\text{-}Te)_2(CO)_x$ $x = 6, 7$	664
	$H_2Ru_3(\mu_3\text{-}Te)(CO)_9$	591
	$Ru_3(\mu_3\text{-}Te)_2(CO)_9$	590
	$Os_3(\mu_3\text{-}Te)_2(CO)_9$	590
μ_4-Te	$Co_4(CO)_8(\mu_2\text{-}CO)_2(\mu_2\text{-}Te)_2$	618, 661, 665[a]
	$Co_4Te_2(CO)_{11}$	661

[a] Structures determined by single-crystal X-ray diffraction.

low tendency of Te to form Te_n ($n > 2$) rings and chains comparable to S and Se (e.g. Section II.G.4) is no doubt responsible for this less diverse ligand chemistry of the heaviest member of the chalcogen elements. Te vapour, for example, consists primarily of Te_2 with less than 1% Te_3[666].

Dimeric complexes incorporating bridging Te atoms, **130** and **132**, were prepared recently by insertion of elemental Te into the reactive Rh=Rh double bond of **129** (equation 264) and by reaction of TeH_2 with a substitution labile organomanganese(I) complex (**131**) (equation 265).

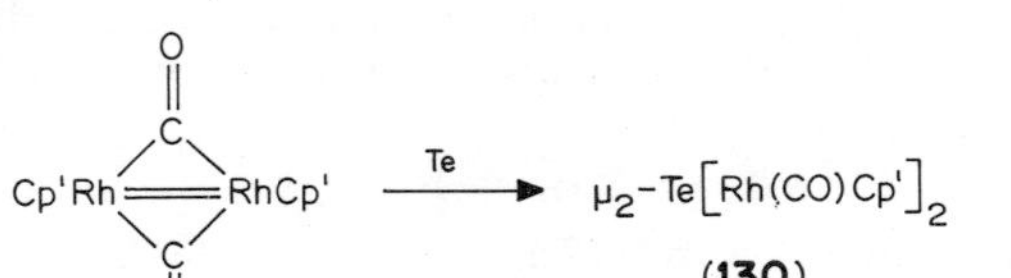

(130) (Ref. 556a) (264)

$\nu_{C\equiv O}$(THF) = 1954 cm^{-1}

(129)

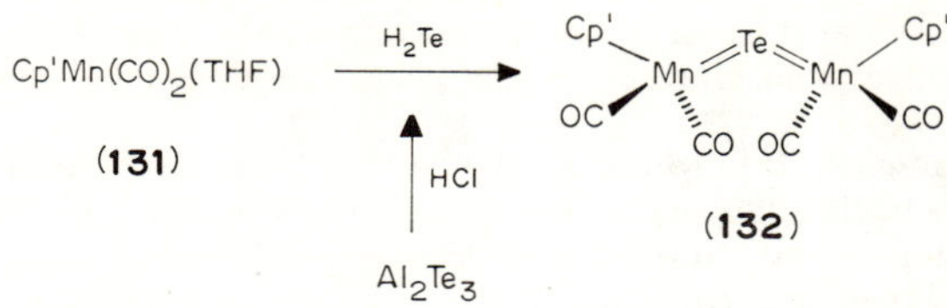

(131) (132) (Ref. 647) (265)

The reactivity of M—M multiple bonds has been shown to be useful for the synthesis of a variety of alkylidene-bridged organometallics[554] as well as S[556], Se (equation 215) and Te (equation 264) insertion products. These reactions generally go under mild conditions in quantitative yields[554,556]. Although elemental S reacts with **129** at -20 °C to give (η^2-S_4)Rh(μ-CO)Rh(η^2-S_4)[556a] with CO elimination, the analogous reactions with Se (equation 215) and Te (equation 264) (-80 to 0 °C) gave products containing only one chalcogen atom. The detailed structures of these latter complexes (three-membered Rh_2E ring systems or Rh=E=Rh geometries) await definitive crystallographic studies.

The structure of **132** has been confirmed by single-crystal X-ray diffraction[647]. The complex contains a bent Mn—Te—Mn framework with multiple bonding (Mn—Te = 2.459(2) Å vs. 2.70 Å for the sum of the covalent radii). The intramolecular Mn—Mn distances (4.209 Å) show that no metal–metal bonding is present. In contrast, the S analogue, [μ-S[Cp'Re(CO)$_2$]$_2$][647], has been shown to contain a Re—Re bond. The dimer **132** readily reacts with diazoalkanes to give alkylidene addition products (equation 261)[647].

Replacement of the Cp′ ligand in **131** by the less sterically demanding unsubstituted Cp ligand in **133** in equation (265) results in the formation of a μ_3-bridged product **134** as well as a small amount of another product (**135**) of uncertain structure (equation 266).

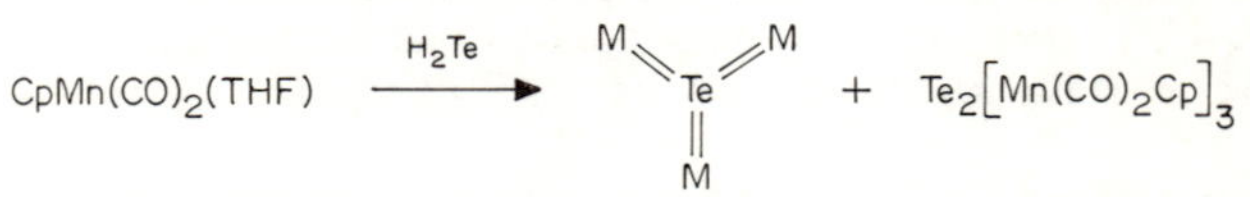

(266)

(133) (134) (135)

M = CpMn(CO)$_2$

	(134)	(135)
	black crystals (47%)	black crystals (0.8%)
$\nu_{C\equiv O}$(Et$_2$O)	1998m, 1992m (sh), 1940s, 1928vs, 1909m, 1886w cm^{-1}	1994s, 1982m (sh), 1935vs, 1919vs, 1901s, 1886m (sh) cm^{-1}
^{1}H-NMR δ((CD$_3$)$_2$CO)	5.23 (s, Cp)	5.23

The propeller-type arrangement of the $MnTe_3$ core of **134** has been confirmed by X-ray diffraction. The central Te atom lies 0.034(1) Å above the centre of the triangular Mn_3 plane, and as in **132**, formal Mn=Te double bonds have been proposed on the basis of bond distances (Mn—Te$_{av.}$ = 2.485 Å vs. 2.54 Å for the sum of the covalent radii) and the inert gas formalism[652].

A redox route was used to prepare the Te-bridged dimer (**136**) (equation 267).

$$\text{Na[V(CO)}_4\text{dppe]} + \text{Na}_2\text{TeO}_3 \xrightarrow{\text{H}_3\text{PO}_4} \text{[V(CO)}_3\text{dppe]}_2\text{Te} \qquad \text{(Ref. 559)} \quad (267)$$

(136)

25%

$\nu_{C\equiv O}$ = 1970m, 1925s, 1880vs cm^{-1}

A linear V$\overset{...}{=}$Te$\overset{...}{=}$V arrangement was proposed for **136** because of its spectral similarity with the S analogue, whose structure has been determined by single-crystal X-ray diffraction.

The dimeric complex $[(ON)_2Fe(\mu\text{-}Te)_2Fe(NO)_2]^{2-}$ was prepared by the previously described metathetical route (e.g. $(ON)_2Fe(\mu\text{-}I)_2Fe(NO)_2 + Li_2Te$; equation 121)[353]. Like the Se analogue, this dimer can undergo μ_2-Te alkylation to give the corresponding neutral dimers (e.g. equation 121)[353].

The complex **137** was first described by Lesch and Rauchfuss[570], who obtained it, contaminated with **138**, in less than 1% yield from fractional sublimation (0.1 mm Hg/45 °C/36 h) of the crude product obtained by Hieber and Gruber's[568] original synthesis of **138** (equation 268).

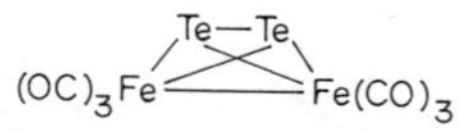

(137)

$$3\,[Fe(CO)_4]^{2-} + 2\,TeO_3^{\,2-} + 10\,H^+ \xrightarrow[-2\,CO]{-5\,H_2O} Fe_3(\mu_3\text{-}Te)_2(CO)_9 + \mathbf{137} \qquad (268)$$

(the $[Fe(CO)_4]^{2-}$ being formed from $3\,Fe(CO)_5$ with NaOH/MeOH; $Fe_3(\mu_3\text{-}Te)_2(CO)_9$ = **(138)**)

$\nu_{C\equiv O}$ (hexane; cm^{-1}):	2045s	2067m
	2025s	2028s
	2004s	1995s

Although the dimer **137** could not be obtained pure, it was characterized by infrared spectroscopy of the sublimate (**137** + **138**), the latter cluster being a well-characterized compound (Table 49). In contrast, the S and Se dimers obtained in reactions analogous to equation (268) were readily separated from $Fe_3(\mu_3\text{-}E)_2(CO)_9$ (E = S, Se) by fractional sublimation.

An effective separation of **137** from **138** was, however, achieved by using the selective reactivity of **137** in an oxidative addition reaction with a Pt(o) substrate[570,571] (equation 269).

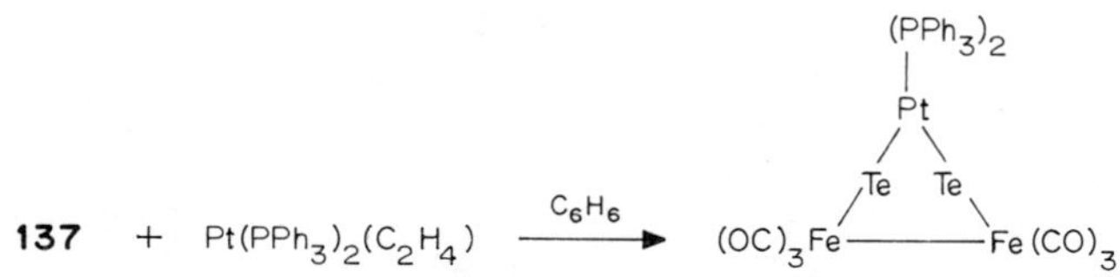

(269)

(139)

red needles (70%)

$\nu_{C\equiv O}$ = 2034s, 1995vs, 1960s cm^{-1}

The cluster **138**, lacking a reactive Te—Te bond, is unreactive to oxidative addition, and the product (**139**) can be readily separated from the former cluster by adsorption chromatography. Surprisingly, Te_2Ph_2 was reported to be unreactive towards $Pt(PPh_3)_2(C_2H_4)$[570]. In contrast, $Pd(PPh_3)_4$ readily undergoes oxidative addition reactions with ditellurides (equation 250)[641], and $Pt(PPh_3)_2(C_2H_4)$ oxidatively adds S_2Ph_2 and Se_2Ph_2[571].

The molecular structure of the Se analogue of **139** has been confirmed by single-crystal X-ray diffraction, and an analogous structure for the Te dimer was inferred from the similar spectral and chemical properties of the two derivatives[571].

In subsequent work[652] non-aqueous gel permeation chromatography with 8% cross-linked polystyrene was an effective method for the separation of the thermally labile $(OC)_3Fe(\mu_2, \eta^2\text{-}Te_2)Fe(CO)_3$ from $Fe_3(\mu_3\text{-}Te)_2(CO)_9$ (products of equation 268). By using this size-exclusion method, yields of ca. 4% of **137**, based on starting $Fe(CO)_5$, were obtained. In an additional modification of the original Hieber and Gruber method (equation 268), the reaction solution was kept at 0 °C to minimize decomposition of the thermally labile **137**. After acidification and CH_2Cl_2 extraction of the black reaction residue, the mixture was applied to a column of Bio-Beads SX-8 resin swelled with CH_2Cl_2 and eluted with the same solvent. A purple band of **138** (58%) eluted first, followed by an orange band of **137** (4.3%). Solutions of **137** were air-stable, but attempts to isolate the solid dimer by evaporation of the solvent under a stream of Ar or CO resulted in decomposition to an insoluble black material, which was tentatively formulated as $Fe_4(\mu_3\text{-}Te)_4(CO)_{12}$. Although this cubane-type cluster was not previously reported, an analogous insoluble and non-volatile cluster, $Co_4(\mu_3\text{-}Sb)_4(CO)_{12}$, has been structurally characterized[667]. Thermolysis of an acetonitrile solution of $Fe_3(\mu_3\text{-}Te)_2(CO)_9$ (80–90 °C) under 2000 psi of CO gave $Fe_4Te_4(CO)_{12}$ in 69% yield (black crystals; $\nu_{C\equiv O}$ (mull): 2043vs, 2035sh, 1994sh, 1988sh, 1982s, 1968w cm^{-1})[661].

The dimer **137** was characterized by mass spectroscopy, which showed peaks for all $Fe_2Te_2(CO)_x$ ($x = 0$–6) fragments, and by infrared and ^{125}Te-NMR spectroscopy[652].

The instability of **137** was rationalized in terms of the considerable strain imposed on the tetrahedral Fe_2Te_2 framework because of the large size of the Te atoms vs. S and Se atoms, which gave stable $Fe_2(\mu_2\text{-}E_2)(CO)_6$ dimers.

Compound (**137**) was later obtained in 40% yield by treatment of methanolic solutions of **138** with NaOMe, followed by acidification with 6M HCl[661].

On the basis of the reactions illustrated in equations (270) and (271), $Fe_2(\mu_2\text{-}Te_2)(CO)_6$ was proposed to be an intermediate in the formation of $Fe_3(\mu_3\text{-}Te)_2(CO)_9$ via $Fe_3(\mu_3\text{-}Te)_2(CO)_{10}$[652].

$$\mathbf{137} + Fe(CO)_5 + 2\,Me_3NO \xrightarrow{CH_2Cl_2} \mathbf{138} + 2\,CO_2 + 2\,Me_3N \qquad (270)$$

$$\mathbf{137} + Fe(CO)_5 \xrightarrow[\text{2. 6M HCl}]{\text{1. MeOH/KOH}} \underset{50\%}{\mathbf{138}} + \underset{(\mathbf{140})\ 50\%}{Fe_3(\mu_3\text{-}Te)_2(CO)_{10}} \qquad (271)$$

The formation of **138** from **137** and $Fe(CO)_5$ in neutral solution occurs only in the presence of the decarbonylation reagent Me_3NO[668] (equation 270). Treatment of **137** with a basic solution of $Fe(CO)_5$ ($HFe(CO)_4^-$ forms under the latter condition) gave about equal amounts of **138** and **140**, the latter presumably forming via oxidative addition of $[HFe(CO)_4]^-$ across the Te—Te bond of **137**, analogous to the reaction in equation (269). The decacarbonyl **140**, which has also been prepared by treatment of **138** with CO[655,660], readily loses CO, thermally (refluxing 0.5 h in hexane) or chemically (CH_2Cl_2 solution of Me_3NO at room temperature), to give **138** quantitatively[652]. The decacarbonyl **140** has been obtained in 70% yield by treatment of a well-stirred mixture of CH_2Cl_2 and aqueous K_2TeO_3 with methanolic $K[HFe(CO)_4]$ at 0 °C, followed by acidification and CH_2Cl_2 extraction[652].

The rather low-yield synthesis of **137** has until recently[661] prevented extensive investigation of its chemistry, but the presence of the reactive Te—Te bond should allow its use in the synthesis of a wide variety of mixed metal clusters (e.g. see Figure 12).

As in the Se_n ligands, the first report of a metal cluster complex incorporating a naked Te atom was Hieber and Gruber's synthesis of $Fe_3(\mu_3\text{-}Te)_2(CO)_9$ (equation 268)[568]. Although this stable black cluster compound was first described in 1958, its X-ray structure determination was not reported until 1982[654]. It was shown to be isostructural

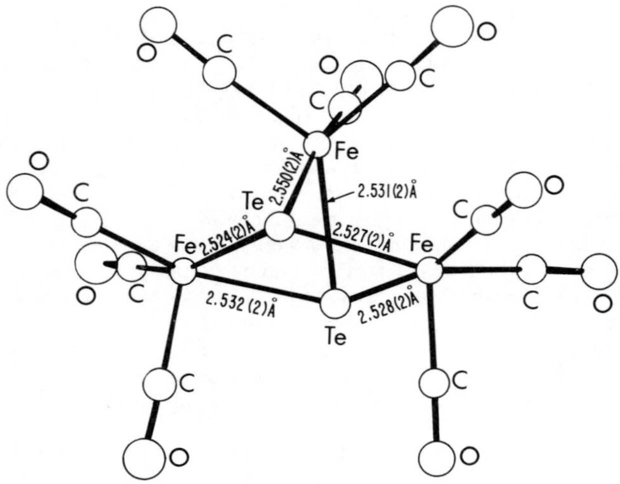

FIGURE 24. Molecular structure of $Fe_3(\mu_3\text{-}Te)_2(CO)_9$

with $Fe_3(\mu_3\text{-}Se)_2(CO)_9$, the Fe_3Te_2 skeleton being square-pyramidal with alternating Fe and Te atoms in the base and an apical Fe atom (Figure 24). These workers[654] obtained the cluster as black needles by cooling a pentane extract of the residue from the reaction of $Fe(CO)_4I_2$ and Na_2Te (equation 272).

$$3\,Fe(CO)_5 + 2\,I_2 + 2\,Na_2Te \xrightarrow[Et_2O]{THF} Fe_3(\mu_3\text{-}Te)_2(CO)_9 + 4\,NaI + 6\,CO \qquad (272)$$

The mass spectrum of **138** shows a molecular peak and nine peaks corresponding to the successive loss of the nine CO ligands, the $[Fe_3Te_2]^+$ peak being the most intense[658].

A variable-temperature ^{13}C-NMR investigation of **138**[659] showed that two discrete CO exchange processes occur in this cluster (Figure 25). At room temperature only two resonances are observed, corresponding to the equivalent carbonyls on the apical iron, Fe(1), and the two basal iron atoms, Fe(2). At $-87\,°C$ the carbonyls on the apical iron, Fe(1), remain equivalent, but two of the three chemically inequivalent types of CO ligands on the basal iron atom, Fe(2), are resolved. Delocalized exchange between CO ligands on the apical and basal iron atoms does not occur.

The ^{125}Te-NMR spectrum of this cluster showed an unexpectedly large downfield shift compared to the $(OC)_3Fe(\mu_2,\eta^2\text{-}Te_2)Fe(CO)_3$ dimer[652] (Table 50). Indeed, ^{125}Te-NMR spectroscopy has been shown to be a useful structural probe for a variety of Te compounds, a range of some 5000 ppm having been observed for the compounds measured to date[669]. In addition, in complexes with metals having NMR active nuclei, ^{125}Te–M coupling constants can provide useful structural information (e.g. Figure 19).

Reactions of **138** with Lewis bases (phosphines, CO, *t*-BuNC) initially give adducts ($Fe_3(\mu_3\text{-}Te)_2(CO)_9L$) followed by substitution reactions under more forcing conditions to give $Fe_3(\mu_3\text{-}Te)_2(CO)_nL_m$ ($n = 8$, $m = 1$ and $n = 7$, $m = 2$; Table 49)[659,660].

The adduct $Fe_3(\mu_3\text{-}Te)_2(CO)_9PPh_3$ has been characterized by single-crystal X-ray diffraction[660]. The structure contains an isosceles triangle of Fe atoms joined by capping μ_3-Te atoms. Each Fe atom contains three terminal CO ligands, and the phosphine coordinates to a unique Fe atom. A variable-temperature ^{13}C-NMR

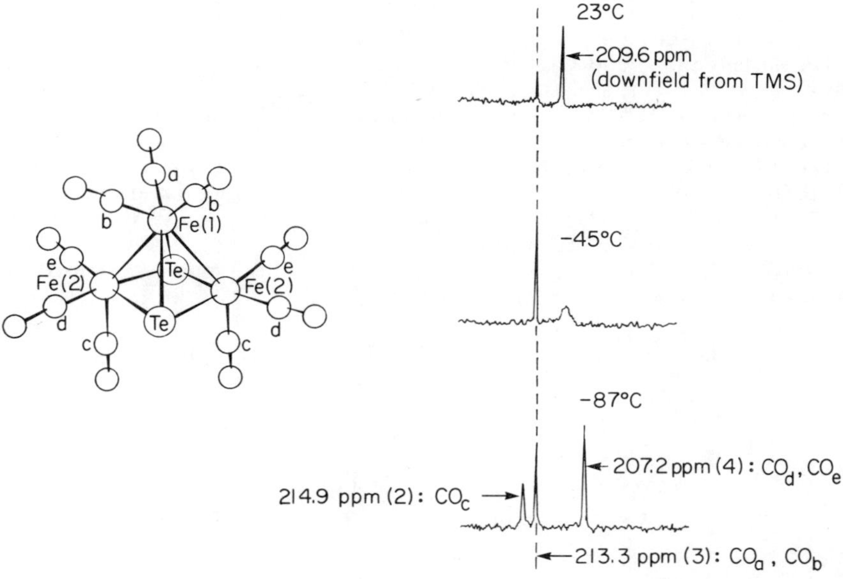

FIGURE 25. Variable-temperature ^{13}C-NMR spectra of $Fe_3(\mu_3\text{-}Te)_2(CO)_9$ ($CDCl_3$ solution). Reproduced with permission from Ref. 659

investigation[659] of the adduct $Fe_3(\mu_3\text{-}Te)_2(CO)_9(P(Bu\text{-}n)_3)$ had been unable to establish the coordination site (Fe or Te) of the phosphine ligand.

Mixed cluster compounds $Fe_3(\mu_3\text{-}Te)(\mu_3\text{-}E)(CO)_9$ (E = S, Se) have been prepared by the route described in equation (268) with appropriate equimolar mixtures of TeO_3^{2-}/EO_3^{2-} [662]. The reactions gave the three possible clusters, $(\mu_3\text{-}E)_2$, $(\mu_3\text{-}Te)_2$, and $(\mu_3\text{-}Te)(\mu_3\text{-}E)$, which were separated by repeated preparative thin-layer chromatography.

TABLE 50. ^{125}Te-NMR data for metal complexes containing Te ligands

Complex	δ (ppm)[a]	$J(^{125}Te\text{–}M)$ (Hz)	Ref.
$Fe_2(\mu_2\text{-}Te_2)(CO)_6$	−733		652
$Fe_3(\mu_3\text{-}Te)_2(CO)_9$	+1123		652
$Fe_2(CO)_6Te_2Pt(PPh_3)_2$	−861	M = ^{195}Pt, 561	571
$CpCoFe_2Te_2(CO)_7$	−825		661
$CpCoFe_2Te_2(CO)_6$	1103, 1087[b]		661
$Fe_3(\mu_3\text{-}Te)_2(CO)_9PPh_3$	−887		664
	−938	M = ^{31}P, 42	
$Fe_3(\mu_3\text{-}Te)_2(CO)_8PPh_3$	1062		664
$CpRhFe_2Te_2(CO)_6PPh_3$	−925	M = ^{103}Rh, 100	664
	−838	M = ^{31}P, 21	
$CpRhFe_2Te_2(CO)_7$	−973	M = ^{103}Rh, 93	664
$CpRhFe_2Te_2(CO)_6$	1081		664
	1109		

[a] Measured in $CDCl_3$ vs. external neat $TeMe_2$; positive chemical shifts are downfield.
[b] The cluster gives two isomers in solution[661].

Like the parent cluster **138**, these mixed derivatives are air-stable and give substitution reactions with $AsPh_3$ and $P(OPh)_3$[663].

The cluster $Fe_3(\mu_3\text{-}Te)_2(CO)_9$ has been shown to be a useful precursor for the synthesis of a variety of mixed metal cluster compounds[661,664]. This reagent is especially attractive since it is one of the relatively few transition-metal carbonyl cluster compounds that can be prepared easily in good yields from inexpensive starting materials[652].

The utility of this cluster has been attributed to its conversion to the reactive $Fe_2(\mu_2,\eta^2\text{-}Te_2)(CO)_6$ in polar solvents (e.g. acetonitrile). Indeed, the mixed metal cluster $Fe_2Te_2Pt(CO)_6(PPh_3)_2$ (**139**), originally prepared from $Fe_2(\mu_2,\eta^2\text{-}Te_2)(CO)_6$ (equation 269), can be isolated in 76% yield from the reaction of $Fe_3(\mu_3\text{-}Te)_2(CO)_9$ (**138**) and $Pt(PPh_3)_2C_2H_4$ in MeCN at room temperature[661]. Other syntheses of mixed metal clusters are illustrated in equations (273)–(276).

$$\mathbf{138} + CpCo(CO)_2 \xrightarrow[\text{reflux}]{\text{MeCN}} \underset{\substack{70\% \\ \text{FDMS: 687 ([M]}^+)}}{CpCoFe_2Te_2(CO)_7} \quad \text{(Ref. 661)} \qquad (273)$$

$$\mathbf{138} + Co_2(CO)_8 \xrightarrow[\substack{180\,^\circ C \\ 1700\ \text{psi CO}}]{\text{hexane}} \underset{76\%}{Co_2FeTe(CO)_9} \quad \text{(Ref. 661)} \qquad (274)$$

$$\mathbf{138} + Co_2(CO)_8 \xrightarrow[\substack{150\,^\circ C \\ 2000\ \text{psi CO}}]{CH_2Cl_2} \underset{64\%}{Co_4Te_2(CO)_{11}} \quad \text{(Ref. 661)} \qquad (275)$$

$$\mathbf{138} + CpRh(CO)_2 \xrightarrow[\text{reflux}]{\text{MeCN}} \underset{\substack{73\% \\ \text{FDMS: 731 ([M]}^{+\cdot})}}{CpRhFe_2Te_2(CO)_7} \quad \text{(Ref. 664)} \qquad (276)$$

The clusters $CpMFe_2Te_2(CO)_7$ (M = Co[661], Rh[664]) can both be decarbonylated by Me_3NO to give $CpMFe_2Te_2(CO)_6$ clusters, which exist in two isomeric forms as evidenced by ^{125}Te-NMR (Table 50), but the isomers cannot be separated by chromatography.

All of these clusters have been characterized by infrared and ^{125}Te-NMR (Table 50) spectroscopy as well as by field-desorption mass spectroscopy (FDMS), a technique especially useful for thermally labile compounds[670].

In two cases (equations 277 and 278), diorganotellurides have been used as precursors to introduce Te into transition-metal cluster frameworks.

$$Co_2(CO)_8 + Fe_3(CO)_{12} + Et_2Te \xrightarrow{\text{hexane}} FeCo_2(\mu_3\text{-}Te)(CO)_9$$

30% (Ref. 589) (277)

air-stable brown crystals

$$2\,Co_2(CO)_8 + 2\,TePh_2 \xrightarrow[\substack{20\,^\circ C \\ -6\ CO}]{C_6H_6} Co_4(\mu_4\text{-}Te)_2(CO)_8(\mu_2\text{-}CO)_2 + PhPh$$

(Ref. 618) (278)

Diphenyl telluride was previously shown to be a precursor for μ_2-$TePh^-$ bridging ligands (equations 234 and 235). The formation of transition-metal cluster compounds containing 'naked' main-group elements by using organometallic compounds of the latter has, however, been demonstrated recently in other systems (e.g. $[Rh_9P(CO)_{21}]^{2-}$ [671], $[Rh_{10}P(CO)_{22}]^{3-}$ [672], $[Rh_{12}Sb(CO)_{27}]^{3-}$ [673] and $[Rh_{10}As(CO)_{22}]^{3-}$ [674] from $Rh(CO)_2$acac and PPh_3, $SbPh_3$ and $AsPh_3$, respectively).

The molecular structure of $FeCo_2(\mu_3\text{-}Te)(CO)_9$ has been described as a tetrahedral $FeCo_3Te$ cluster system formed by the symmetrical coordination of an apical Te atom to a basal $FeCo_2(CO)_9$ fragment containing three $M(CO)_3$ groups at the corners of an equilateral triangle and linked to one another by metal–metal bonds[589].

The cluster $Co_4(\mu_4\text{-}Te)_2(CO)_8(\mu_2\text{-}CO)_2$ (equation 278) is the only reported structure involving a μ_4-Te bridging ligand[665], although other related structures involving a Co_4E framework with an apical main-group element bridging four Co atoms in a basal rectangle are known[665,675].

Attempts to prepare $Ru_3Te_2(CO)_9$ by reaction of $Ru_3(CO)_{12}$ in alkaline tellurite (TeO_2 + aqueous KOH), conditions used for $Fe_3Te_2(CO)_9$ (equation 268), gave instead the hydride cluster $H_2Ru_3Te(CO)_9$ in 0.5% yield after acidification of the reaction solution with 2N H_2SO_4 and CCl_4 extraction of the resulting precipitate[591]. Two structures (**141a** and **b**) were proposed for this cluster. The reaction of $Ru_3(CO)_{12}$ with Te powder in *n*-octane under a pressure of CO/H_2 (35 atm, 1:1) was reported to give a mixture of $H_2Ru_3Te(CO)_9$ and $Ru_3Te_2(CO)_9$[590]. These workers also reported that under a pure CO atmosphere the yield of the hydrido cluster was significantly decreased. These two clusters were separated by thin-layer chromatography, although data (elemental analysis, infrared and mass spectroscopy) are reported only for $H_2Ru_3Te(CO)_9$[590].

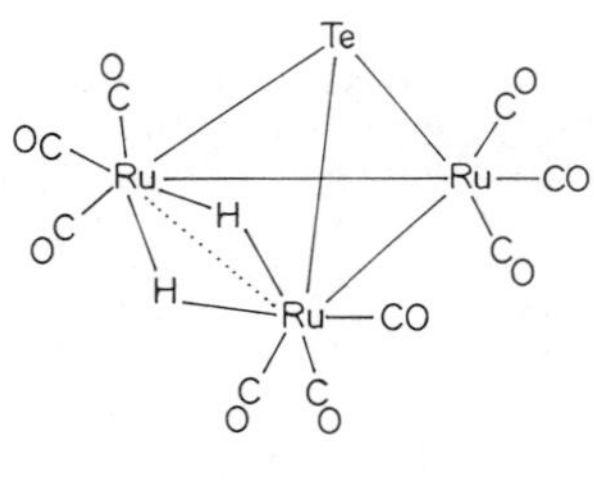

(141a)

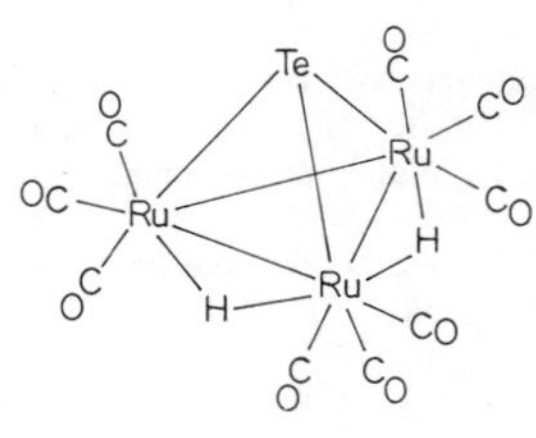

(141b)

A similar reaction of $Os_3(CO)_{12}$ with elemental Te in refluxing *n*-octane gave a mixture of $H_2Os_3Te(CO)_9$, $Os_3Te_2(CO)_9$ and $H_2Os_4Te_2(CO)_{12}$ although again only data for $Os_3Te_2(CO)_9$ were reported[590]. As usual for such complex metal clusters, definitive structural characterization requires single-crystal X-ray diffraction.

IV. REFERENCES

1. F. A. Cotton and G. Wilkinson, *Advanced Inorganic Chemistry*, 4th ed., John Wiley and Sons, New York, 1980.
2. G. Wilkinson, F. G. A. Stone and E. W. Abel (Eds.), *Comprehensive Organometallic Chemistry*, Pergamon Press, Oxford, 1982.
3. M. A. Verneuil, *Ann. Chim. Phys., Ser. 6*, **9**, 292 (1886).
4. J. Petren, *Z. Anorg. Allgem. Chem.*, **20**, 62 (1899).
5. E. Fritzmann, *Z. Anorg. Allgem. Chem.*, **73**, 239 (1912).
6. E. Fritzmann, *Z. Anorg. Allgem. Chem.*, **133**, 119 (1924).
7. E. Fritzmann, *Z. Anorg. Allgem. Chem.*, **133**, 133 (1924).
8. E. Fritzmann, *J. Russ. Phys. Chem. Soc.*, **47**, 588 (1915).
9. (a) M. D. Healy and A. J. Rest, *Adv. Inorg. Chem. Radiochem.*, **21**, 1 (1978).
   (b) S. M. Bloom, M. Green, M. Idelson and M. S. Simon in *The Chemistry of Synthetic Dyes* (Ed. K. Venkataraman), Vol. 8, Academic Press, New York, 1978, p. 331.

(c) M. I. Bruce and B. L. Goodall, in *The Chemistry of the Hydrazo, Azo and Azoxy Groups*, Part 1 (Ed. S. Patai), John Wiley and Sons, London, 1975, p. 259.
(d) G. Booth in *The Chemistry of Synthetic Dyes* (Ed. K. Venkataraman), Vol. 5, Academic Press, New York, 1971, p. 241.
(e) R. Price, in *The Chemistry of Synthetic Dyes* (Ed. K. Venkataraman), Vol. 3, Academic Press, New York, 1970, p. 303.
(f) W. R. McWhinnie and J. D. Miller, *Adv. Inorg. Chem. Radiochem.*, **12**, 135 (1969).

10. (a) C. A. McAuliffe (Ed.), *Transition Metal Complexes of Phosphorus, Arsenic and Antimony Ligands*, John Wiley and Sons, New York, 1973.
(b) C. A. McAuliffe and W. Levason, *Phosphine, Arsine and Stibine Complexes of the Transition Elements*, Elsevier, Amsterdam, 1979.
11. (a) R. M. Izatt and J. J. Christensen (Eds.), *Synthetic Multidentate Macrocyclic Compounds*, Academic Press, New York, 1978.
(b) G. W. Gokel and H. D. Durst, *Aldrichimica*, **9**(1), 3 (1976).
(c) C. Kutal, *J. Chem. Educ.*, **52**, 319 (1975).
12. (a) S. E. Livingstone, *Quart. Rev. (London)*, **19**, 386 (1965).
(b) D. Coucouvanis, *Progr. Inorg. Chem.*, **11**, 233 (1970); **26**, 301 (1979).
(c) R. P. Burns and C. A. McAuliffe, *Adv. Inorg. Chem. Radiochem.*, **22**, 303 (1979).
(d) I. Omae, *Coord. Chem. Rev.*, **28**, 97 (1979).
(e) M. A. Ali and S. E. Livingstone, *Coord. Chem. Rev.*, **13**, 101 (1974).
(f) R. A. Walton, *Coord. Chem. Rev.*, **31**, 79 (1980).
(g) E. Diemann and A. Müller, *Coord. Chem. Rev.*, **10**, 79 (1973).
13. S. G. Murray and F. R. Hartley, *Chem. Rev.*, **81**, 365 (1981).
14. (a) D. L. Klayman and W. H. H. Günther (Eds.), *Organic Selenium Compounds: Their Chemistry and Biology*, John Wiley and Sons, New York, 1973.
(b) K. J. Irgolic and M. V. Kudchadker, 'The Organic Chemistry of Selenium' in *Selenium* (Eds. R. A. Zingaro and W. C. Cooper), Van Nostrand Reinhold, New York, 1974.
15. (a) K. J. Irgolic, *The Organic Chemistry of Tellurium*, Gordon and Breach, London, 1974.
(b) K. J. Irgolic, *J. Organomet. Chem.*, **103**, 91 (1975).
(c) K. J. Irgolic, *J. Organomet. Chem.*, **130**, 411 (1977).
(d) K. J. Irgolic, *J. Organomet. Chem.*, **158**, 235 (1978).
(e) K. J. Irgolic, *J. Organomet. Chem.*, **189**, 65 (1980).
(f) K. J. Irgolic, *J. Organomet. Chem.*, **203**, 367 (1980).
(g) H. J. Gysling, 'The Synthesis of Organotellurium Ligands', *Kodak Lab. Chem. Bull.*, **53**(1) (1982).
16. (a) G. H. Denison and P. C. Condit, California Research Corp., *U.S. Patent*, No. 2,398,414 (1946).
(b) M. R. Detty, *J. Org. Chem.*, **44**, 4528 (1979).
17. (a) V. Krishnan and R. A. Zingaro, in *Selenium* (Eds. R. A. Zingaro and W. C. Cooper), Van Nostrand Reinhold, New York, 1974, pp. 337–407.
(b) K. A. Jensen and C. K. Jørgensen, Ref. 14(a), pp. 1017–1041.
18. H. J. Gysling, *Coord. Chem. Rev.*, **42**, 133 (1982).
19. E. W. Abel and B. C. Crosse, *Organomet. Chem. Rev.*, **2**, 443 (1967).
20. R. A. Zingaro and D. O. Skovlin, *J. Electrochem. Soc.*, **111**, 42 (1964).
21. P. Pramanik and R. N. Bhattacharya, *J. Solid State Chem.*, **44**, 425 (1982).
22. R. N. Bhattacharya and P. Pramanik, *Bull. Mater. Sci.*, **3**, 403 (1981).
23. (a) T. M. Racheva, I. D. Dragieva, D. H. Djoglev and P. P. Dimitrova, *Thin Solid Films*, **17**, 85 (1973).
(b) I. K. Ostrovskaya, L. E. Yatlova, G. A. Kitaev and V. I. Dvoinin, *Zh. Prikl. Khim.*, **49**, 1212 (1976).
24. Asahi Chemical Industry Company, Ltd., *Japanese Patent*, No. 78,65,827 (1978); *Chem. Abstr.*, **89**, 146588 (1978).
25. (a) H. J. Gysling, Eastman Kodak Company, *U.S. Patent*, No. 4,394,318 (1983).
(b) H. J. Gysling, Eastman Kodak Company, *U.S. Patent*, No. 4,287,354 (1981).
26. H. A. Tayim and J. C. Bailar, *J. Amer. Chem. Soc.*, **89**, 4330 (1967).
27. (a) G. C. Papavassiliou, *Mol. Cryst. Liq. Cryst.*, **86**, 159 (1982).
(b) G. C. Papavassiliou, *Z. Naturforsch.*, **37b**, 825 (1982).

28. (a) E. M. Engler, *Proceedings of the 3rd International Symposium on Organic Selenium and Tellurium Compounds*, Metz, France, July 9–12, 1979, p. 357.
(b) J. S. Miller, *Ann. N.Y. Acad. Sci.*, **313**, 25 (1978).
(c) R. N. Lyubovskaya, *Russ. Chem. Rev.*, **52**, 736 (1983).
29. *Organometallics for Electronics*, Thiokol/Ventron Division Alfa Products, 1980.
30. W. Stulius, *Appl. Phys. Letters*, **40**, 246 (1982).
31. H. J. Gysling, *Proceedings of the 4th International Conference on the Organic Chemistry of Selenium and Tellurium*, Birmingham, England, July 25–29, 1983, p. 32.
32. (a) G. W. Parshall, *Homogeneous Catalysis*, Wiley–Interscience, New York, 1980.
(b) 'Catalytic Aspects of Metal Phosphine Complexes', *Advances in Chemistry Series No. 196* (Eds. E. C. Alyea and D. W. Meek), American Chemical Society, Washington, D. C., 1982.
33. A. Chiesi, G. Grossoni, M. Nardelli and M. E. Vidoni, *Chem. Commun.*, 404 (1969).
34. A. T. Pilipenko and I. P. Sereda, *Zh. Anal. Khim.*, **13**, 3 (1958).
35. A. T. Pilipenko and I. P. Sereda, *Russ. J. Inorg. Chem.*, **6**, 209 (1961).
36. P. J. Hendra and Z. Jović, *Spectrochim. Acta*, **24A**, 1713 (1968).
37. G. B. Aitken and G. P. McQuillan, *Inorg. Synth.*, **16**, 83 (1976).
38. G. B. Aitken, J. L. Duncan and G. P. McQuillan, *J. Chem. Soc. (A)*, 2103 (1972).
39. O. Piovesana and C. Furlani, *J. Inorg. Nucl. Chem.*, **30**, 1249 (1968).
40. N. N. Proskina, S. M. Chul'skaya, G. F. Volodina and A. V. Ablov, *Zh. Strukt. Khim.*, **9**, 1095 (1968); *Engl. Ed.*, p. 983.
41. S. Hauge and M. Tysseland, *Acta Chem. Scand.*, **25**, 3072 (1971).
42. S. Hauge and O. Vikane, *Acta Chem. Scand.*, **27**, 3596 (1973).
43. S. Hauge, O. Johannesen and O. Vikane, *Acta Chem. Scand.*, **32A**, 901 (1978).
44. K. Nakamoto, in *Infrared and Raman Spectra of Inorganic and Coordination Compounds*, 3rd ed., Wiley–Interscience, New York, 1978.
45. J. S. Rutherford and C. Calvo, *Z. Kristallogr. Kristallgeo. Kristallphys. Kristallchem.*, **128**, 229 (1969).
46. (a) S. Husebye, *Proceedings of the 4th International Conference on the Organic Chemistry of Selenium and Tellurium*, Birmingham, England, July 25–29, 1983, p. 298.
(b) O. Foss, *Pure Appl. Chem.*, **24**, 31 (1970).
(c) N. W. Alcock, *Adv. Inorg. Chem. Radiochem.*, **15**, 1 (1972).
(d) N. M. Mangion, R. A. Zingaro and E. A. Meyers, *Chem. Scripta*, **8**, 45 (1975).
47. C. Furlani and T. Tarantelli, *Inorg. Nucl. Chem. Letters*, **2**, 391 (1966).
48. T. Tarantelli and C. Furlani, *J. Chem. Soc. (A)*, 1717 (1968).
49. T. Tarantelli and C. Furlani, *J. Inorg. Nucl. Chem.*, **34**, 999 (1972).
50. W. J. Geary, *Coord. Chem. Rev.*, **7**, 81 (1971).
51. T. Tarantelli and D. Leonesi, *Ann. Chim. (Italy)*, **53**, 1113 (1963).
52. M. R. Rosenthal, *J. Chem. Educ.*, **50**, 331 (1973).
53. S. Okeya, T. Milyamoto, S. Ooi, Y. Nakamura and S. Kawaguchi, *Inorg. Chim. Acta*, **45**, L135 (1980).
54. O. Vikane, *Acta Chem. Scand.*, **29A**, 152 (1975).
55. O. Vikane, *Acta Chem. Scand.*, **29A**, 787 (1975).
56. O. Vikane, *Acta Chem. Scand.*, **29A**, 763 (1975).
57. H. J. Gysling and H. R. Luss, *Organometallics*, **3**, 596 (1984).
58. P. Schulz and G. Klar, *Z. Naturforsch.*, **30b**, 40 (1975).
59. P. Klaeboe and O. Vikane, *Acta Chem. Scand.*, **31A**, 120 (1977).
60. T. Tarantelli and B. Chiari, *J. Inorg. Nucl. Chem.*, **39**, 33 (1977).
61. (a) J. A. Gladysz, J. L. Hornby and J. E. Garbe, *J. Org. Chem.*, **43**, 1204 (1978).
(b) J. A. Gladysz, *Aldrichimica Acta*, **12**(1), 13 (1979).
62. M. R. Detty and M. D. Seidler, *J. Org. Chem.*, **47**, 1354 (1982).
63. D. L. J. Clive, P. C. Anderson, N. Moss and A. Singh, *J. Org. Chem.*, **47**, 1641 (1982).
64. J. Chatt and L. M. Venanzi, *J. Chem. Soc.*, 2351 (1957).
65. J. Chatt and L. M. Venanzi, *J. Chem. Soc.*, 2445 (1957).
66. J. Chatt, L. A. Duncanson and L. M. Venanzi, *J. Chem. Soc.*, 3203 (1958).
67. J. R. Allkins and P. J. Hendra, *J. Chem. Soc. (A)*, 1325 (1967).
68. J. R. Allkins and P. J. Hendra, *Spectrochem. Acta*, **24A**, 1305 (1968).
69. L. R. M. Pitombo, *Anal. Chim. Acta*, **46**, 158 (1969).

70. P. E. Skakke and S. E. Rasmussen, *Acta Chem. Scand.*, **24**, 2634 (1970).
71. S. Sergi, F. Faraone, L. Silvestro and R. Pietropaolo, *J. Organomet. Chem.*, **33**, 403 (1971).
72. B. E. Aires, J. E. Fergusson, D. T. Howarth and J. M. Miller, *J. Chem. Soc. (A)*, 1144 (1971).
73. L. R. M. Pitombo, *Anal. Chim. Acta*, **62**, 103 (1972).
74. J. E. Fergusson and K. S. Loh, *Australian J. Chem.*, **26**, 2615 (1973).
75. L. R. M. Pitombo and E. Q. Cartaxo, *Talanta*, **21**, 965 (1974).
76. (a) R. J. Cross, T. H. Green and R. Keat, *Chem. Commun.*, 207 (1974).
(b) R. J. Cross, T. H. Green and R. Keat, *J. Chem. Soc., Dalton Trans.*, 1150 (1976).
77. (a) R. J. Cross, T. H. Green and J. F. Paterson, *Inorg. Nucl. Chem. Letters*, **11**, 145 (1975).
(b) R. J. Cross, T. H. Green, R. Keat and J. F. Paterson, *J. Chem. Soc., Dalton Trans.*, 1486 (1976).
78. R. J. Cross, T. H. Green and R. Keat, *J. Chem. Soc., Dalton Trans.*, 382 (1976).
79. E. W. Abel, G. W. Farrow, K. G. Orrell and V. Šik, *J. Chem. Soc., Dalton Trans.*, 42 (1977).
80. K. A. Jensen, *Z. Anorg. Allgem. Chem.*, **225**, 97 (1935).
81. J. Chatt and L. M. Venanzi, *J. Chem. Soc.*, 2787 (1955).
82. J. Chatt and L. M. Venanzi, *J. Chem. Soc.*, 3858 (1955).
83. J. Chatt, L. A. Duncanson and L. M. Venanzi, *J. Chem. Soc.*, 4461 (1955).
84. J. Chatt, G. A. Gamlen and L. E. Orgel, *J. Chem. Soc.*, 1047 (1959).
85. D. M. Adams, J. Chatt, J. Gerratt and A. D. Westland, *J. Chem. Soc.*, 734 (1964).
86. J. Chatt and A. D. Westland, *J. Chem. Soc. (A)*, 88 (1968).
87. S. Sergi, V. Marsala, R. Pietropaolo and F. Faraone, *J. Organomet. Chem.*, **23**, 281 (1970).
88. F. Faraone, L. Silvestro, S. Sergi and R. Pietropaolo, *J. Organomet. Chem.*, **34**, C55 (1972).
89. S. R. Haddock, *Ph. D. Thesis*, Bristol University, 1975.
90. P. L. Goggin, R. J. Goodfellow and S. R. Haddock, *Chem. Commun.*, 176 (1975).
91. S. J. Anderson, P. L. Goggin and R. J. Goodfellow, *J. Chem. Soc.*, 1959 (1976).
92. S. J. Anderson and R. J. Goodfellow, *J. Chem. Soc., Dalton Trans.*, 1683 (1977).
93. G. E. Coates, *J. Chem. Soc.*, 2003 (1951).
94. D. L. Laing and L. D. Pettit, *J. Chem. Soc., Dalton Trans.*, 2297 (1975).
95. A. M. Noble and J. M. Winfield, *Inorg. Nucl. Chem. Letters*, **4**, 339 (1968).
96. W. Hieber, W. Opavsky and W. Rohm, *Chem. Ber.*, **101**, 2244 (1968).
97. F. Faraone, S. Sergi and R. Pietropaolo, *J. Organomet. Chem.*, **24**, 453 (1970).
98. F. Faraone, R. Pietropaolo and S. Sergi, *J. Organomet. Chem.*, **24**, 797 (1970).
99. F. Faraone, R. Pietropaolo, S. Sergi and P. Pira, *J. Organomet. Chem.*, **24**, 805 (1970).
100. H. C. E. McFarlane, W. McFarlane and R. J. Wood, *Bull. Soc. Chim. Belg.*, **85**, 864 (1976).
101. W. Hieber and P. John, *Chem. Ber.*, **103**, 2161 (1970).
102. J. E. Fergusson and C. T. Page, *Australian J. Chem.*, **29**, 2159 (1976).
103. C. T. Page and J. E. Fergusson, *Australian J. Chem.*, **36**, 855 (1983).
104. C. A. Stein and H. Taube, *Inorg. Chem.*, **18**, 1168 (1979).
105. J. E. Fergusson and K. S. Loh, *Australian J. Chem.*, **26**, 2615 (1973).
106. P. Piraino, F. Faraone and R. Pietropaolo, *Atti Accad. Peloritana Pericolanti*, **51**, 283 (1971); *Chem. Abstr.*, **79**, 13011k (1973).
107. R. Roulet and R. Favez, *Chimia*, **29**, 346 (1975).
108. H. Hausmann, M. Höfler, T. Kruck and H. W. Zimmermann, *Chem. Ber.*, **114**, 975 (1981).
109. K. A. Jensen, *Z. Anorg. Allgem. Chem.*, **231**, 365 (1937).
110. L. W. Houk and G. R. Dobson, *Inorg. Chem.*, **5**, 2119 (1966).
111. W. Hieber and F. Stanner, *Chem. Ber.*, **102**, 2930 (1969).
112. E. W. Abel, A. K. S. Ahmed, G. W. Farrow, K. G. Orrell and V. Šik, *J. Chem. Soc., Dalton Trans.*, 47 (1977).
113. L. Maier, in *Organic Phosphorus Compounds* (Eds. G. M. Kosolapoff and L. M. Maier), Vol. 4, Wiley–Interscience, New York, 1972, Chapt. 7.
114. E. Bannister and F. A. Cotton, *J. Chem. Soc.*, 1959 (1960).
115. T. Singh and S. S. Sandhu, *J. Chem. Sci.*, **4**, 37 (1978).
116. M. G. King and G. P. McQuillan, *J. Chem. Soc.*, 898 (1967).
117. L. S. D. Glasser, L. Ingram, M. G. King and G. P. McQuillan, *J. Chem. Soc. (A)*, 2501 (1969).
118. T. S. Lobana and K. Sharma, *Transition Met. Chem.*, **7**, 333 (1982).
119. P. Nicpon and D. W. Meek, *Chem. Commun.*, 398 (1966).
120. T. S. Lobana, T. R. Gupta and S. S. Sandhu, *Polyhedron*, **1**, 781 (1982).

121. M. G. King, G. P. McQuillan and R. Milne, *J. Inorg. Nucl. Chem.*, **35**, 3039 (1973).
122. A. M. Brodie, G. A. Rodley and C. J. Wilkins, *J. Chem. Soc. (A)*, 2927 (1969).
123. J. A. Tiethof, A. T. Hetey and D. W. Meek, *Inorg. Chem.*, **13**, 2505 (1974).
124. S. O. Grim, E. D. Walton and L. C. Satek, *Can. J. Chem.*, **58**, 1476 (1980).
125. W. E. Morgan, W. J. Stec, R. G. Albridge and J. R. VanWazer, *Inorg. Chem.*, **10**, 926 (1971).
126. R. Colton and D. Dakternieks, *Australian J. Chem.*, **33**, 1463 (1980).
127. W. Hieber and H. Fuhrling, *Z. Anorg. Allgem. Chem.*, **373**, 48 (1970).
128. E. W. Ainscough, A. M. Brodie and E. Mentzer, *J. Chem. Soc., Dalton Trans.*, 2167 (1973).
129. E. W. Ainscough, A. M. Brodie and A. R. Furness, *J. Chem. Soc., Dalton Trans.*, 2360 (1973).
130. P. M. Boorman and K. J. Reimer, *Can. J. Chem.*, **49**, 2926 (1971).
131. D. Britnell, G. W. A. Fowles and D. A. Rice, *J. Chem. Soc., Dalton Trans.*, 213 (1975).
132. K. C. Malhotra, V. P. Mahajan, G. Mehrotra and S. C. Chaudhry, *Chem. Ind. (London)*, 921 (1978).
133. H. J. Gysling, *Inorg. Synth.*, **19**, 92 (1979).
134. E. W. Ainscough, H. A. Bergen, A. M. Brodie and K. A. Brown, *J. Chem. Soc., Dalton Trans.*, 1649 (1976).
135. (a) M. R. Houchin and D. Chapman, *Inorg. Nucl. Chem. Letters*, **16**, 437 (1980).
(b) J. Martin-Gil, F. J. Martin-Gil, A. Perales, J. Fayos and M. Martinez-Ripoll, *Acta Cryst.*, **39C**, 44 (1983).
136. D. H. Brown, R. J. Cross and R. Keat, *J. Chem. Soc., Dalton Trans.*, 871 (1980).
137. M. A. Jennings and A. Wojcicki, *Inorg. Chim. Acta*, **3**, 335 (1969).
138. N. J. DeStefano and J. L. Burmeister, *Inorg. Chem.*, **10**, 998 (1971).
139. J. S. Thayer, *Inorg. Chem.*, **7**, 2599 (1968).
140. (a) P. Nicpon and D. W. Meek, *Inorg. Chem.*, **5**, 1297 (1966).
(b) P. Nicpon and D. W. Meek, *Inorg. Synth.*, **10**, 157 (1967).
(c) A. J. Blake, G. P. McQuillan and I. A. Oxton, *J. Mol. Struct.*, **78**, 265 (1982).
(d) W. E. Slinkard and D. W. Meek, *J. Chem. Soc., Dalton Trans.*, 1024 (1973).
141. (a) R. A. Zingaro and R. E. McGlothlin, *J. Chem. Eng. Data*, **8**, 226 (1963).
(b) P. A. W. Dean, *Can. J. Chem.*, **57**, 754 (1979).
142. S. S. Sandhu and T. Singh, *Transition Met. Chem.*, **1**, 155 (1976).
143. W. Malisch, R. Maisch, I. J. Colquhoun and W. McFarlane, *J. Organomet. Chem.*, **220**, C1 (1981).
144. V. Kullmer and H. Vahrenkamp, *Chem. Ber.*, **110**, 237 (1977).
145. (a) M. Ziegler and O. Glemser, *Z. Anal. Chem.*, **146**, 29 (1955).
(b) T. G. Bunting and C. E. Meloan, *Anal. Chem.*, **40**, 435 (1968).
146. V. G. Pesin, *Russ, Chem. Rev.*, **39**, 923 (1970).
147. J. M. Vernon, M. R. Bryce and T. A. Dransfield, *Tetrahedron*, **39**, 835 (1983).
148. R. H. Hanson and C. E. Meloan, *Inorg. Nucl. Chem. Letters*, **7**, 461 (1971).
149. R. H. Hanson and C. E. Meloan, *Inorg. Nucl. Chem. Letters*, **7**, 467 (1971).
150. W. Kaim, *J. Organomet. Chem.*, **264**, 317 (1984).
151. L. Futekov, R. Parichkova and H. Specker, *Fresenius' Z. Anal. Chem.*, **315**, 342 (1983); *Chem. Abstr.*, **99**, 98243y (1983).
152. (a) K. P. Rudolph, J. Flachowsky, A. Lange and J. Dubnack, *Isotopenpraxis*, **19**, 225 (1983); *Chem. Abstr.*, **99**, 81598v (1983).
(b) S. Grazuliene, V. K. Karandashev and Yu. V. Yakovlev, *Radiochem. Radioanal. Letters*, **57**, 273 (1983); *Chem. Abstr.*, **99**, 132965x (1983).
153. (a) J. L. Burmeister, *Coord. Chem. Rev.*, **1**, 205 (1966).
(b) J. L. Burmeister, *Coord. Chem. Rev.*, **3**, 225 (1968).
(c) A. H. Norbury and A. I. P. Sinha, *Quart. Rev.*, **24**, 69 (1970).
(d) A. H. Norbury, *Adv. Inorg. Chem. Radiochem.*, **17**, 231 (1975).
(e) R. J. Balahura and N. A. Lewis, *Coord. Chem. Rev.*, **20**, 109 (1976).
154. P. M. Solozhenkin, V. Sh. Tsveniashvili, E. V. Semenov and N. S. Khavtasi, *Dokl. Akad. Nauk Tadzh. SSR*, **19**, 38 (1976); *Chem. Abstr.*, **86**, 50 099k (1977).
155. V. V. Davydov, I. N. Marov, V. K. Belyaeva, A. S. Katugin, B. E. Zaitsev and A. K. Molodkin, *Russ. J. Inorg. Chem.*, **26**, 523 (1981).
156. L. G. Kuz'mina, L. P. Grigor'eva, Yu. T. Struchkov, Z. I. Ezhkova, B. E. Zaitsev, V. V. Davydov and A. K. Molodkin, *Russ. J. Inorg. Chem.*, **25**, 1614 (1980).

157. (a) L. J. Gerenser, M. G. Mason and P. J. Trotter, *J. Phys. Chem.*, **80**, 2384 (1976).
(b) P. J. Trotter, M. G. Mason and L. J. Gerenser, *J. Phys. Chem.*, **81**, 1325 (1977).
(c) H. J. Gysling, L. J. Gerenser and M. G. Mason, *J. Coord. Chem.*, **10**, 67 (1980).
158. (a) N. Zumbulyadis and H. J. Gysling, *J. Amer. Chem. Soc.*, **104**, 3246 (1982).
(b) F. DuBois Murphy, W. C. Stevens, T. T. P. Cheung, S. Lacelle, B. C. Gerstein and D. M. Kurtz, *J. Amer. Chem. Soc.*, **103**, 4400 (1981).
159. V. Sh. Tsveniashvili, V. N. Gaprindashvili and N. S. Khavtasi, *Zh. Obshch. Khim.*, **42**, 2049 (1972).
160. V. Sh. Tsveniashvili, M. V. Malsshkhiya, N. S. Khavtasi and V. N. Gaprindashvili, *Zh. Obshch. Khim.*, **53**, 1851 (1983).
161. K. H. Pannell, A. J. Mayr, R. Hoggard and R. C. Pettersen, *Angew. Chem. (Intern. Ed. Engl.)*, **19**, 632 (1980).
162. K. H. Pannell, A. J. Mayr, R. Hoggard, J. S. McKennis and J. C. Dawson, *Chem. Ber.*, **116**, 230 (1983).
163. T. L. Gilchrist, P. G. Mente and C. W. Rees, *J. Chem. Soc., Perkin Trans. 1*, 2165 (1972).
164. G. N. Schrauzer and H. Kisch, *J. Amer. Chem. Soc.*, **95**, 2501 (1973).
165. R. C. Pettersen, K. H. Pannell and A. J. Mayr, *Acta Cryst.*, **36B**, 2434 (1980).
166. R. C. Pettersen, K. H. Pannell and A. J. Mayr, *Cryst. Struct. Commun.*, **9**, 643 (1980).
167. W. Beck, J. C. Weis and J. Wieczorek, *J. Organomet. Chem.*, **30**, 89 (1971).
168. J. C. Weis and W. Beck, *J. Organomet. Chem.*, **44**, 325 (1972).
169. M. V. Artemenko, E. A. Chistyakova, P. A. Suprunenko and G. I. Kal'naya, *Russ. J. Inorg. Chem.*, **17**, 373 (1972).
170. E. S. Gould and W. Burlant, *J. Amer. Chem. Soc.*, **78**, 5825 (1956).
171. P. J. Hendra and N. Sadasivan, *J. Chem. Soc.*, 2063 (1965).
172. G. Hunter and R. C. Massey, *J. Chem. Soc., Dalton Trans.*, 1872 (1974).
173. R. E. Marsh and J. D. McCullough, *J. Amer. Chem. Soc.*, **73**, 1106 (1951).
174. K. L. Baker and G. W. A. Fowles, *J. Chem. Soc. (A)*, 801 (1968).
175. J. C. Barnes, G. Hunter and M. W. Lown, *J. Chem. Soc., Dalton Trans.*, 1227 (1976).
176. J. C. Barnes, G. Hunter and M. W. Lown, *J. Chem. Soc., Dalton Trans.*, 458 (1977).
177. C. Preti, G. Tosi, D. DeFilippo and G. Verani, *Can. J. Chem.*, **52**, 2021 (1974).
178. A. R. Siedle, R. A. Newmark, A. A. Kruger and L. H. Pignolet, *Inorg. Chem.*, **20**, 3399 (1981).
179. A. R. Siedle, M. C. Etter, M. E. Jones, G. Filipovich, H. E. Mishmash and W. Bahmet, *Inorg. Chem.*, **21**, 2624 (1982).
180. A. R. Siedle, R. A. Newmark and L. H. Pignolet, *Inorg. Chem.*, **22**, 2281 (1983).
181. E. H. Braye, W. Hübel and I. Caplier, *J. Amer. Chem. Soc.*, **83**, 4406 (1961).
182. O. Haas and A. Von Zelewsky, *J. Chem. Res.*, 77 (S) (1980); 1201 (M) (1980).
183. E. W. Abel, A. R. Khan, K. Kite, K. G. Orrell and V. Šik, *J. Chem. Soc., Dalton Trans.*, 1169 (1980).
184. E. W. Abel, A. R. Khan, K. Kite, K. G. Orrell, V. Šik, T. S. Cameron and R. Cordes, *Chem. Commun.*, 713 (1979).
185. E. W. Abel, A. R. Khan, K. Kite, K. G. Orrell and V. Šik, *J. Chem. Soc., Dalton Trans.*, 2220 (1980).
186. E. W. Abel, A. R. Khan, K. Kite, M. B. Hursthouse, K. M. A. Malik and M. A. Mazid, *J. Organomet. Chem.*, **235**, 121 (1982).
187. E. W. Abel, S. K. Bhargava, M. M. Bhatti, M. A. Mazid, K. G. Orrell, V. Šik, M. B. Hursthouse and K. M. A. Malik, *J. Organomet. Chem.*, **250**, 373 (1983).
188. E. W. Abel, S. K. Bhargava, P. K. Mittal, K. G. Orrell and V. Šik, *Chem. Commun.*, 535 (1982).
189. J. Korp, I. Bernal, J. L. Atwood, F. Calderazzo and D. Vitali, *J. Chem. Soc., Dalton Trans.*, 1492 (1979).
190. F. Calderazzo, D. Vitali, R. Poli, J. L. Atwood, R. D. Rogers, J. M. Cummings and I. Bernal, *J. Chem. Soc., Dalton Trans.*, 1004 (1981).
191. (a) H. W. Roesky, T. Gries, P. G. Jones, K. L. Weber and G. M. Sheldrick, *J. Chem. Soc., Dalton Trans.*, 1781 (1984).
(b) J. L. Atwood, I. Bernal, F. Calderazzo, L. G. Canada, R. Poli, R. D. Rogers, C. A. Veracini and D. Vitali, *Inorg. Chem.*, **22**, 1797 (1983).
192. J. Grobe and D. LeVan, *Z. Naturforsch.*, **34b**, 1653 (1979).
193. J. Grobe and D. LeVan, *J. Fluorine Chem.*, **19**, 279 (1982).

194. J. Grobe and D. LeVan, *Z. Naturforsch.*, **35b**, 694 (1980).
195. D. Vitali and F. Calderazzo, *Gazz. Chim. Ital.*, **102**, 587 (1972).
196. H. Schumann and R. Weis, *Angew. Chem. (Intern. Ed. Engl.)*, **9**, 246 (1970).
197. H. Schumann, R. Mohtachemi, H.-J. Kroth and U. Frank, *Chem. Ber.*, **106**, 1555 (1973).
198. J. Pickardt, H. Schumann, C. F. Campana and L. F. Dahl, *J. Organomet. Chem.*, **216**, 245 (1981).
199. I. Ruidisch and M. Schmidt, *J. Organomet. Chem.*, **1**, 160 (1963).
200. F. W. B. Einstein, C. H. W. Jones, T. Jones and R. D. Sharma, *Can. J. Chem.*, **61**, 2611 (1983).
201. J. E. Drake, B. M. Glavincevski, R. T. Hemmings and H. E. Henderson, *Inorg. Synth.*, **20**, 171 (1980).
202. A. N. Egorochkin, S. Ya. Khorshev, N. S. Vyazankin, M. N. Bochkarev, O. A. Kruglaya and G. S. Semchikova, *Zh. Obshch. Khim.*, **37**, 2308 (1967).
203. I. S. Butler, *Acc. Chem. Res.*, **10**, 359 (1977).
204. P. V. Yaneff, *Coord. Chem. Rev.*, **23**, 183 (1977).
205. Ref. 1, p. 86.
206. B. D. Dombek and R. J. Angelici, *Inorg. Synth.*, **19**, 183 (1979).
207. I. S. Butler, N. J. Coville, D. Cozak, S. R. Desjardins, A. E. Fenster and K. R. Plowman, *Inorg. Synth.*, **19**, 188 (1979).
208. G. Jaouen and G. Simonneaux, *Inorg. Synth.*, **19**, 197 (1979).
209. M. Kubota and C. O. M. Ho, *Inorg. Synth.*, **19**, 204 (1979).
210. M. Kubota, *Inorg. Synth.*, **19**, 206 (1979).
211. B. W. S. Kolthammer, P. Legzdins and J. T. Malito, *Inorg. Synth.*, **19**, 208 (1979).
212. I. S. Butler, D. Cozak, S. R. Stobart and K. R. Plowman, *Inorg. Synth.*, **19**, 193 (1979).
213. I. S. Butler, A. M. English and K. R. Plowman, *Inorg. Synth.*, **21**, 1 (1981).
214. G. R. Clark, K. R. Grundy, R. O. Harris, S. M. James and W. R. Roper, *J. Organomet. Chem.*, **90**, C37 (1975).
215. P. J. Brothers, C. E. L. Headford and W. R. Roper, *J. Organomet. Chem.*, **195**, C29 (1980).
216. G. R. Clark, K. Marsden, W. R. Roper and L. J. Wright, *J. Amer. Chem. Soc.*, **102**, 1206 (1980).
217. W. R. Roper and K. G. Town, *J. Organomet. Chem.*, **252**, C97 (1983).
218. J. P. Battioni, D. Mansuy and J. C. Chottard, *Inorg. Chem.*, **19**, 791 (1980).
219. I. S. Butler, D. Cozak and S. R. Stobart, *Chem. Commun.*, 103 (1975).
220. I. S. Butler, D. Cozak and S. R. Stobart, *Inorg. Chem.*, **16**, 1779 (1977).
221. D. Cozak and I. S. Butler, *Can. J. Spectry*, **27**, 141 (1982).
222. A. M. English, K. R. Plowman, I. S. Butler, G. Jaouen, P. LeMaux and J. Y. Thépot, *J. Organomet. Chem.*, **132**, C1 (1977).
223. T. Desmond, F. J. Lalor, G. Ferguson and M. Parvez, *Chem. Commun.*, 75 (1984).
224. O. Kolb and H. Werner, *J. Organomet. Chem.*, **268**, 49 (1984).
225. G. R. Clark and K. Marsden, unpublished work cited in Ref. 216.
226. W. R. Roper and A. H. Wright, *J. Organomet. Chem.*, **233**, C59 (1982).
227. J. P. Collman and L. S. Hegedus, *Principles of Organotransition Metal Chemistry*, University Science Books, Mill Valley, California, 1980.
228. J. Goulon, C. Goulon-Ginet, P. Friant, J. L. Poncet, R. Guilard, J. P. Battioni and D. Mansuy, *Proceedings of the 4th International Conference on the Organic Chemistry of Selenium and Tellurium*, Birmingham, England, July 25–29, 1983, p. 379.
229. W. L. Reynolds, *Prog. Inorg. Chem.*, **12**, 1 (1970).
230. J. A. Davies, *Adv. Inorg. Chem. Radiochem.*, **24**, 115 (1981).
231. F. A. Cotton and T. R. Felthouse, *Inorg. Chem.*, **19**, 2347 ((1980).
232. P. J. Alvey, K. W. Bagnall and D. Brown, *J. Chem. Soc., Dalton Trans.*, 2326 (1973).
233. F. C. Marsh and G. Ferguson, *Can. J. Chem.*, **49**, 3590 (1971).
234. A. Mercer and J. Trotter, *J. Chem. Soc., Dalton Trans.*, 2480 (1975).
235. B. R. James and R. H. Morris, *Can. J. Chem.*, **58**, 399 (1980).
236. E. S. Gould and J. D. McCullough, *J. Amer. Chem. Soc.*, **73**, 3196 (1951).
237. R. Paetzold and P. Vordank, *Z. Anorg. Allgem. Chem.*, **347**, 294 (1966).
238. K. A. Jensen and V. Krishnan, *Acta Chem. Scand.*, **21**, 1988 (1967).
239. R. Paetzold and G. Bochmann, *Z. Anorg. Allgem. Chem.*, **368**, 202 (1969).
240. M. Ziegler and H. Schroeder, *Mikrochim. Acta*, 782 (1967).

241. T. Tanaka and T. Kamitani, *Inorg. Chim. Acta*, **2**, 175 (1968).
242. C. K. Jorgensen, *Absorption Spectra and Chemical Bonding in Complexes*, Pergamon Press, Oxford, 1962.
243. A. N. Speca, C. M. Mikulski, F. J. Iaconianni, L. L. Pytlewski and N. M. Karayannis, *Inorg. Chem.*, **19**, 3491 (1980).
244. E. O. Fischer and A. Maasbol, *Chem. Ber.*, **100**, 2445 (1967).
245. (a) E. O. Fischer, *Pure Appl. Chem.*, **24**, 407 (1970).
(b) E. O. Fischer, *Pure Appl. Chem.*, **30**, 353 (1972).
(c) E. O. Fischer, *Adv. Organomet. Chem.*, **14**, 1 (1976).
(d) E. O. Fischer, *Pure Appl. Chem.*, **50**, 857 (1978).
(e) K. H. Dötz, H. Fischer, P. Hofmann, F. R. Kreissl, U. Schubert and K. Weiss, *Transition Metal Carbene Complexes*, Verlag Chemie International, Deerfield Beach, Florida, 1983.
246. E. O. Fischer and V. Kiener, *Angew. Chem. (Intern. Ed. Engl.)*, **6**, 961 (1967).
247. E. O. Fischer, G. Kreis, F. R. Kreissl, C. G. Kreiter and J. Müller, *Chem. Ber.*, **106**, 3910 (1973).
248. E. O. Fischer, D. Himmelreich, R. Cai, H. Fischer, U. Schubert and B. Zimmer-Gasser, *Chem. Ber.*, **114**, 3209 (1981).
249. H. Fischer, E. O. Fischer, D. Himmelreich, R. Cai, U. Schubert and K. Ackermann, *Chem. Ber.*, **114**, 3220 (1981).
250. W. Röll, E. O. Fischer, D. Neugebauer and U. Schubert, *Z. Naturforsch.*, **37b**, 1274 (1982).
251. A. V. Ablov, N. V. Gerbeleu and A. M. Romanov, *Russ. J. Inorg. Chem.*, **13**, 413 (1968).
252. M. S. Delgado and V. Fernandez, *Z. Anorg. Allgem. Chem.*, **455**, 112 (1979).
253. M. S. Delgado, F. Alguacil and V. Fernandez, *Z. Anorg. Allgem. Chem.*, **493**, 187 (1982).
254. H. G. Raubenheimer, G. J. Kruger and A. Lombard, *J. Organomet. Chem.*, **240**, C11 (1982).
255. D. Neugebauer and U. Schubert, *J. Organomet. Chem.*, **256**, 43 (1983).
256. H. Fischer and S. Zeuner, *Z. Naturforsch.*, **38b**, 1365 (1983).
257. E. W. Abel, K. Kite, K. G. Orrell, V. Šik and B. L. Williams, *J. Chem. Soc., Dalton Trans.*, 2439 (1981).
258. M. Di Vaira, M. Peruzzini and P. Stoppioni, *Inorg. Chem.*, **22**, 2196 (1983).
259. M. Di Vaira, M. Peruzzini and P. Stoppioni, *J. Organomet. Chem.*, **258**, 373 (1983).
260. M. Di Vaira, M. Peruzzini and P. Stoppioni, *Chem. Commun.*, 903 (1983).
261. M. Di Vaira, M. Peruzzini and P. Stoppioni, *J. Chem. Soc., Dalton Trans.*, 359 (1984).
262. (a) A. V. Ablov and N. V. Gerbeleu, *Russ. J. Inorg. Chem.*, **6**, 1356 (1961).
(b) A. V. Ablov and N. V. Gerbeleu, *Russ. J. Inorg. Chem.*, **9**, 1260 (1964).
263. A. V. Ablov and N. M. Samus, *Dokl. Akad. Nauk SSSR*, **123**, 457 (1958).
264. E. E. Aynsley, N. N. Greenwood and M. J. Sprague, *J. Chem. Soc.*, 2395 (1965).
265. E. E. Aynsley, N. N. Greenwood, G. Hunter and M. J. Sprague, *J. Chem. Soc. (A)*, 1344 (1966).
266. S. Hauge, *Acta Chem. Scand.*, **25**, 3081 (1971).
267. (a) M. Di Vaira, L. Sacconi and P. Stoppioni, *J. Organomet. Chem.*, **250**, 183 (1983).
(b) M. Di Vaira, M. Peruzzini and P. Stoppioni, *Acta Cryst.*, **39C**, 1210 (1983).
268. M. Di Vaira, M. Peruzzini and P. Stoppioni, *Chem. Commun.*, 894 (1982).
269. (a) A. H. Norbury, *Adv. Inorg. Chem. Radiochem.*, **17**, 231 (1975).
(b) P. P. Singh, *Coord. Chem. Rev.*, **32**, 33 (1980).
(c) V. V. Skopenko, G. V. Tsintsadze and E. I. Ivanova, *Russ. Chem. Rev.*, **51**, 21 (1982).
270. A. V. Ablov and I. D. Samus, *Dokl. Akad. Nauk SSSR*, **146**, 1071 (1962).
271. S. Hauge and O. Vikane, *Acta Chem. Scand.*, **29A**, 755 (1975).
272. G. V. Tsintsadze, M. A. Porai-Koshits and A. S. Antsyshkina, *Zh. Strukt. Khim.*, **8**, 296 (1967).
273. F. A. Cotton, D. M. L. Goodgame, M. Goodgame and T. E. Haas, *Inorg. Chem.*, **1**, 565 (1962).
274. V. Vrábel, J. Lokaj, J. Garaj and F. Pavelcik, *Collect. Czech. Commun.*, **47**, 2623 (1982).
275. S. M. Nelson, F. S. Esho and M. G. B. Drew, *Chem. Commun.*, 388 (1981).
276. J. Devlin and S. M. Nelson, unpublished results cited in Ref. 253.
277. M. G. B. Drew, F. S. Esho and S. M. Nelson, *Inorg. Chim. Acta*, **76**, L269 (1983).
278. J. S. Thayer and R. West, in *Advances in Organometallic Chemistry*, Academic Press, New York, 1967, Vol. 5, p. 169.
279. F. Challenger, A. T. Peters, and J. Halevy, *J. Chem. Soc.*, 1648 (1926).
280. T. N. Srivastava and K. K. Bajpai, *J. Organomet. Chem.*, **31**, 1 (1971).
281. J. S. Thayer, *J. Organomet. Chem.*, **9**, P30 (1967).

282. A. Sabatini and I. Bertini, *Inorg. Chem.*, **4**, 959 (1965).
283. J. L. Burmeister and L. E. Williams, *Inorg. Chem.*, **5**, 1113 (1966).
284. J. L. Burmeister and M. V. Al-Janabi, *Inorg. Chem.*, **4**, 962 (1965).
285. J. L. Burmeister and H. J. Gysling, *Inorg. Chim. Acta*, **1**, 100 (1967).
286. M. Bobadilla and M. Morán, *Z. Naturforsch.*, **38b**, 1069 (1983).
287. A. Turco, C. Pecile and M. Nicolini, *J. Chem. Soc.*, 3008 (1962).
288. W. H. Pan, J. P. Fackler, J. A. Kargol and J. L. Burmeister, *Inorg. Chim. Acta*, **44**, L95 (1980).
289. A. V. Ablov and N. M. Samus, *Dokl. Akad. Nauk SSSR*, **133**, 1327 (1960).
290. A. V. Ablov, N. N. Proskina, O. A. Bologa and N. M. Samus, *Russ. J. Inorg. Chem.*, **15**, 1245 (1970).
291. R. S. Artz and J. L. Burmeister, *Inorg. Chim. Acta*, **40**, 31 (1980).
292. (a) J. L. Burmeister and J. B. Melpolder, *Chem. Commun.*, 613 (1973).
(b) J. B. Melpolder and J. L. Burmeister, *Inorg. Chim. Acta*, **49**, 115 (1981).
293. M. E. Kramer and J. L. Burmeister, *Synth. React. Inorg. Metal.-Org. Chem.*, **7**, 69 (1977).
294. R. Wojtowski, I. Wharf and M. Onyszchuk, *Can. J. Chem.*, **61**, 743 (1983).
295. M. Onyszchuk and I. Wharf, *J. Organomet. Chem.*, **249**, C9 (1983).
296. V. Gutmann and R. Schmid, *Coord. Chem. Rev.*, **12**, 263 (1974).
297. A. Turco, C. Pecile and M. Niccolini, *Proc. Chem. Soc.*, 213 (1961).
298. M. Morán, A. Ibanez and J. R. Masaguer, *Z. Naturforsch.*, **37b**, 292 (1982).
299. F. Faraone and S. Sergi, *J. Organomet. Chem.*, **112**, 201 (1976).
300. M. Moran and V. Fernandez, *J. Organomet. Chem.*, **165**, 215 (1979).
301. A. M. Domingos and G. M. Sheldrick, *J. Organomet. Chem.*, **67**, 257 (1974).
302. J. L. Burmeister, *Chemistry and Biochemistry of Thiocyanic Acid and Its Derivatives* (Ed. H. A. Newman), Academic Press, London, 1975, Chapt. 2, pp. 68–130.
303. W. C. Fultz, J. L. Burmeister, J. J. MacDougall and J. H. Nelson, *Inorg. Chem.*, **19**, 1085 (1980).
304. W. C. Fultz, J. L. Burmeister, C. P. Cheng and T. L. Brown, *Inorg. Chem.*, **20**, 1734 (1981).
305. J. A. Kargol, R. W. Crecely and J. L. Burmeister, *Inorg. Chem.*, **18**, 2532 (1979).
306. R. A. Bailey, T. W. Michelsen and W. N. Mills, *J. Inorg. Nucl. Chem.*, **33**, 3206 (1971).
307. F. A. Cotton, A. Davison, W. H. Ilsley and H. S. Trop, *Inorg. Chem.*, **18**, 2719 (1979).
308. L. R. Groeneveld, G. Vos, G. C. Verschoor and J. Reedijk, *Chem. Commun.*, 620 (1982).
309. R. G. Pearson, *J. Chem. Educ.*, **45**, 581, 643 (1968).
310. (a) J. L. Burmeister, H. J. Gysling and J. C. Lim, *J. Amer. Chem. Soc.*, **91**, 44 (1969).
(b) K. A. Johnson, J. C. Lim and J. L. Burmeister, *Inorg. Chem.*, **12**, 124 (1973).
311. M. A. Jennings and A. Wojcicki, *J. Organomet. Chem.*, **14**, 231 (1968).
312. P. P. Singh, *Coord. Chem. Rev.*, **32**, 33 (1980).
313. P. P. Singh and D. S. Yadav, *J. Mol. Struct.*, **53**, 225 (1979).
314. P. P. Singh and V. P. Singh, *Inorg. Chim. Acta*, **71**, 205 (1983).
315. D. C. Calabro, B. A. Harrison, G. T. Palmer, M. K. Moguel, R. L. Rebbert and J. L. Burmeister, *Inorg. Chem.*, **20**, 4311 (1981).
316. J. A. Kargol, K. D. Lavin, R. W. Crecely and J. L. Burmeister, *Inorg. Chem.*, **19**, 1515 (1980).
317. J. A. Kargol, R. W. Crecely, J. L. Burmeister, P. J. Toscano and L. G. Marzilli, *Inorg. Chim. Acta*, **40**, 79 (1980).
318. E. W. Burkhardt and J. L. Burmeister, *Inorg. Chim. Acta*, **27**, 115 (1978).
319. J. L. Allison and J. L. Burmeister, *Inorg. Chim. Acta*, **16**, 185 (1976).
320. J. L. Burmeister and H. J. Gysling, *Chem. Commun.*, 543 (1967).
321. V. Fernandez, I. Yerga and U. Müller, *Z. Anorg. Allgem. Chem.*, **477**, 205 (1981).
322. M. H. Quick and R. J. Angelici, *Inorg. Chem.*, **15**, 160 (1976).
323. W. Hieber and W. Beck, *Z. Anorg. Allgem. Chem.*, **305**, 265 (1960).
324. H. Köpf, B. Block and M. Schmidt, *Z. Naturforsch.*, **22b**, 1077 (1967).
325. H. Köpf and B. Block, *Z. Naturforsch.*, **23b**, 1536 (1968).
326. W. Sato and T. Yoshida, *J. Organomet. Chem.*, **67**, 395 (1974).
327. E. W. Tillay, E. D. Schermer and W. H. Baddley, *Inorg. Chem.*, **7**, 1925 (1968).
328. K. Andrae, *Z. Anorg. Allgem. Chem.*, **373**, 209 (1970).
329. E. D. Schermer and W. H. Baddley, *J. Organomet. Chem.*, **27**, 83 (1971).
330. M. Sato and T. Yoshida, *J. Organomet. Chem.*, **51**, 231 (1973).
331. B. Gautheron, G. Tainturier and Ph. Meunier, *J. Organomet. Chem.*, **209**, C49 (1981).
332. M. Sato and T. Yoshida, *J. Organomet. Chem.*, **87**, 217 (1975).

333. Z. Hatkevitz and G. Schmuckler, *J. Inorg. Nucl. Chem.*, **38**, 479 (1976).
334. J. P. Collman, R. K. Rothrock, J. P. Sen, T. D. Tullius and K. O. Hodgson, *Inorg. Chem.*, **15**, 2947 (1976).
335. J. P. Collman, R. K. Rothrock and R. A. Stark, *Inorg. Chem.*, **16**, 437 (1977).
336. W. Schulze and M. L. Ziegler, *Z. Anorg. Allgem. Chem.*, **481**, 78 (1981).
337. A. Rettenmeier, K. Weidenhammer and M. L. Ziegler, *Z. Anorg. Allgem. Chem.*, **473**, 91 (1981).
338. S. Pouly, G. Tainturier and B. Gautheron, *J. Organomet. Chem.*, **232**, C65 (1982).
339. E. W. Abel, A. M. Atkins, B. C. Crosse and G. V. Hutson, *J. Chem. Soc. (A)*, 687 (1968).
340. P. Beardwood and J. F. Gibson, *J. Chem. Soc., Dalton Trans.*, 737 (1983).
341. J. Liesk and G. Klar, *Z. Anorg. Allgem. Chem.*, **435**, 103 (1977).
342. W. Hieber and W. Beck, *Z. Anorg. Allgem. Chem.*, **305**, 274 (1966).
343. J. Grobe, *Z. Anorg. Allgem. Chem.*, **331**, 63 (1964).
344. N. Welcman and I. Rot, *J. Chem. Soc.*, 7515 (1965).
345. E. W. Abel, B. C. Crosse and G. V. Hutson, *J. Chem. Soc. (A)*, 2014 (1967).
346. E. W. Abel, B. C. Crosse and G. V. Hutson, *Chem. Ind. (London)*, 238 (1966).
347. E. W. Abel. J. Dalton, I. Paul, P. J. G. Smith and F. G. A. Stone, *J. Chem. Soc. (A)*, 1203 (1968).
348. E. Kostiner, M. L. N. Reddy, D. S. Urch and A. G. Massey, *J. Organomet. Chem.*, **15**, 383 (1968).
349. E. D. Schermer and W. H. Baddley, *J. Organomet. Chem.*, **30**, 67 (1971).
350. M. Sato and T. Yoshida, *J. Organomet. Chem.*, **94**, 403 (1975).
351. J. Rott, E. Guggolz, A. Rettenmeier and M. L. Ziegler, *Z. Naturforsch.*, **37b**, 13 (1982).
352. P. Jaitner, *J. Organomet. Chem.*, **233**, 333 (1982).
353. T. B. Rauchfuss and T. D. Weatherill, *Inorg. Chem.*, **21**, 827 (1982).
354. D. Seyferth and R. S. Henderson, *J. Organomet. Chem.*, **204**, 333 (1981).
355. A. P. Arnold and A. J. Carty, *Inorg. Chim. Acta*, **55**, 181 (1981).
356. A. P. Arnold, A. J. Carty, B. W. Skelton and A. H. White, *J. Chem. Soc., Dalton Trans.*, 607 (1982).
357. Y. Okamoto and T. Yano, *J. Organomet. Chem.*, **29**, 99 (1971).
358. P. V. Broadhurst, B. F. G. Johnson and J. Lewis, *J. Chem. Soc. Dalton Trans.*, 1881 (1982).
359. P. Jaitner, *J. Organomet. Chem.*, **210**, 353 (1981).
360. M. K. Chaudhuri, A. Haas and A. Wensky, *J. Organomet. Chem.*, **116**, 323 (1976).
361. I. B. Benson, S. D. Killops, S. A. R. Knox and A. Welch, *Chem. Commun.*, 1137 (1980).
362. L. R. Krauth-Siegel, W. Schulze and M. L. Ziegler, *Angew. Chem. (Intern. Ed. Engl.)*, **19**, 397 (1980).
363. V. Küllmer and H. Vahrenkamp, *Chem. Ber.*, **110**, 228 (1977).
364. D. Morelli, A. Segre, R. Ugo, G. La Monica, S. Cenini, F. Conti and F. Bonati, *Chem. Commun.*, 524 (1967).
365. R. Ugo, G. La Monica, S. Cenini, A. Segre and F. Conti, *J. Chem. Soc. (A)*, 522 (1971).
366. I. M. Blacklaws, E. A. V. Ebsworth, D. W. H. Rankin and H. E. Robertson, *J. Chem. Soc., Dalton Trans.*, 753 (1978).
367. M. Hsieh, R. A. Zingaro and V. Krishnan, *Int. J. Sulfur Chem.*, **1A**, 197 (1971).
368. W. Hofmann and H. Werner, *Angew. Chem. (Intern. Ed. Engl.)*, **20**, 1014 (1981).
369. P. Shanmugam and T. K. Raja, *Synthesis*, 117 (1977).
370. E. Rottinger, V. Kullmer and H. Vahrenkamp, *J. Organomet. Chem.*, **150**, C6 (1978).
371. C. Preti, G. Tosi, D. De Filippo and G. Verani, *J. Inorg. Nucl. Chem.*, **36**, 2203 (1974).
372. C. Preti, G. Tosi, D. De Filippo and G. Verani, *Inorg. Nucl. Chem. Letters*, **10**, 541 (1974).
373. E. Bertelli, C. Preti and G. Tosi, *J. Inorg. Nucl. Chem.*, **37**, 1421 (1975).
374. C. Preti and G. Tosi, *Spectrochim. Acta*, **31A**, 1139 (1975).
375. C. Preti and G. Tosi, *Z. Anorg. Allgem. Chem.*, **432**, 259 (1977).
376. C. Preti and G. Tosi, *Transition Met. Chem.*, **2**, 1 (1977).
377. C. Preti, G. Razzoli and G. Tosi, *Spectrochim. Acta*, **34A**, 269 (1978).
378. C. Preti and G. Tosi, *Inorg. Chem.*, **16**, 2805 (1977).
379. C. Preti, G. Tosi and P. Zannini, *Transition Met. Chem.*, **5**, 200 (1980).
380. G. Graziosi, C. Preti and G. Tosi, *Transition Met. Chem.*, **5**, 262 (1980).
381. C. Preti and G. Tosi, *Australian J. Chem.*, **33**, 1203 (1980).
382. G. Graziosi, C. Preti and G. Tosi, *Transition Met. Chem.*, **7**, 267 (1982).
383. C. Preti, L. Tassi and G. Tosi, *Spectrochim. Acta*, **39A**, 1 (1983).
384. C. Preti, G. Tosi and P. Zannini, *Transition Met. Chem.*, **2**, 232 (1977).

385. C. Preti, G. Tosi and P. Zannini, *Transition Met. Chem.*, **4**, 123 (1979).
386. C. Preti, G. Tosi and P. Zannini, *J. Coord. Chem.*, **10**, 143 (1980).
387. G. Candrini, W. Malavasi, C. Preti, G. Tosi and P. Zannini, *Spectrochim. Acta*, **39A**, 635 (1983).
388. I. P. Lorenz, *Inorg. Nucl. Chem. Letters*, **15**, 127 (1979).
389. I. P. Lorenz, *Angew. Chem. (Intern. Ed. Engl.)*, **17**, 53 (1978).
390. (a) W. Dell and M. L. Ziegler, *Angew. Chem. (Intern. Ed. Engl.*), **20**, 471 (1981).
(b) W. H. Dell and M. L. Ziegler, *Z. Naturforsch.*, **37b**, 1 (1982).
391. K. Tanaka and T. Tanaka, *Inorg. Nucl. Chem. Letters*, **10**, 605 (1974).
392. K. Tanaka and T. Tanaka, *Inorg. Nucl. Chem. Letters*, **9**, 429 (1973).
393. K. Tanaka and T. Tanaka, *Bull. Chem. Soc. Japan*, **47**, 847 (1974).
394. K. Tanaka, S. Araki and T. Tanaka, *Bull. Chem. Soc. Japan*, **46**, 2136 (1973).
395. F. W. Pijpers, A. H. Dix and J. G. M. Van Der Linden, *Inorg. Chim. Acta*, **11**, 41 (1974).
396. K. Tanaka, Y. Miya-Uchi and T. Tanaka, *Inorg. Chem.*, **14**, 1545 (1975).
397. K. Tanaka, M. Nakamoto and T. Tanaka, *Inorg. Chem.*, **17**, 1380 (1978).
398. M. Nakamoto, K. Tanaka and T. Tanaka, *J. Chem. Soc., Dalton Trans.*, 87 (1979).
399. G. Nagao, K. Tanaka and T. Tanaka, *Inorg. Chim. Acta*, **42**, 43 (1980).
400. K. Tanaka, K. U-Eda and T. Tanaka, *J. Inorg. Nucl. Chem.*, **43**, 2029 (1981).
401. (a) N. Sonoda, T. Yasuhara, K. Kondo, T. Ikeda and S. Tsutsumi, *J. Amer. Chem. Soc.*, **93**, 6344 (1971).
(b) K. Kondo, Y. Sonoda, Y. Yoshida, M. Koishi and S. Tsutsumi, *Chem. Letters*, 401 (1972).
402. G. M. Sheldrick and W. S. Sheldrick, *J. Chem. Soc. (A)*, 490 (1970).
403. E. Román, D. Catheline and D. Astruc, *J. Organomet. Chem.*, **236**, 229 (1982).
404. D. H. Farrar, K. R. Grundy, N. C. Payne, W. R. Roper and A. Walker, *J. Amer. Chem. Soc.*, **101**, 6577 (1979).
405. W.-H. Pan and J. P. Fackler, *Inorg. Synth.*, **21**, 6 (1981).
406. K. A. Jensen and E. Huge-Jensen, *Acta Chem. Scand.*, **27**, 3605 (1973).
407. K. Kawakami, Y. Ozaki and T. Tanaka, *J. Organomet. Chem.*, **69**, 151 (1974).
408. H. Werner and M. Ebner, *J. Organomet. Chem.*, **258**, C52 (1983).
409. C. Bianchini, C. Mealli, A. Meli, A. Orlandini and L. Sacconi, *Inorg. Chem.*, **19**, 2968 (1980).
410. R. Mason and A. I. M. Rae, *J. Chem. Soc. (A)*, 1767 (1970).
411. T. J. Collins, K. R. Grundy, W. R. Roper and S. F. Wong, *J. Organomet. Chem.*, **107**, C37 (1976).
412. H. Werner and O. Kolb, *Angew. Chem. (Intern. Ed. Engl.)*, **18**, 865 (1979).
413. W. Poppitz and E. Uhlig, *Z. Anorg. Allgem. Chem.*, **489**, 67 (1982).
414. H. Werner, M. Ebner, W. Bertleff and U. Schubert, *Organometallics*, **2**, 891 (1983).
415. H. Werner, *Coord. Chem. Rev.*, **43**, 165 (1982).
416. L. Mortillaro, L. Credali, M. Russo and C. DeChecchi, *J. Polym. Sci., Polym. Letters Ed.*, **3**, 581 (1965).
417. H. Hofmann, P. G. Jones, M. Noltemeyer, E. Peymann, W. Pinkert, H. W. Roesky and G. M. Sheldrick, *J. Organomet. Chem.*, **249**, 97 (1983).
418. C. E. L. Headford and W. R. Roper, *J. Organomet. Chem.*, **244**, C53 (1983).
419. A. F. Hill, W. R. Roper, J. M. Waters and A. H. Wright, *J. Amer. Chem. Soc.*, **105**, 5939 (1983).
420. W. Paul and H. Werner, *Angew. Chem. (Intern. Ed. Engl.)*, **22**, 316 (1983).
421. W. A. Herrmann, J. Weichmann, R. Serrano, K. Blechschmitt, H. Pfisterer and M. L. Ziegler, *Angew. Chem. (Intern. Ed. Engl.)*, **22**, 314 (1983).
422. H. Werner and W. Paul, *Angew. Chem. (Intern. Ed. Engl.)*, **23**, 58 (1984).
423. K. A. Jensen and E. Frederiksen, *Z. Anorg. Allgem. Chem.*, **230**, 31 (1936).
424. A. M. Romanov, A. V. Ablov and N. V. Gerbeleu, *Russ. J. Inorg. Chem.*, **14**, 196 (1969).
425. A. V. Ablov, N. V. Gerbeleu and A. M. Romanov, *Russ. J. Inorg. Chem.*, **14**, 512 (1969).
426. L. S. Bark and D. Brandon, *Talanta*, **14**, 759 (1967).
427. M. J. Root and E. Deutsch, *Inorg. Chem.*, **20**, 4376 (1981).
428. L. Roecker, M. H. Dickman, D. L. Nosco, R. J. Doedens and E. Deutsch, *Inorg. Chem.*, **22**, 2022 (1983).
429. T. Konno, K. Okamoto and J. Hidaka, *Chem. Letters*, 535 (1982).
430. K. Nakajima, M. Kojima and J. Fujita, *Chem. Letters*, 925 (1982).
431. G. Dyer and D. W. Meek, *J. Amer. Chem. Soc.*, **89**, 3983 (1967).
432. M. O. Workman, G. Dyer and D. W. Meek, *Inorg. Chem.*, **6**, 1543 (1967).
433. D. W. Meek, *Inorg. Nucl. Chem. Letters*, **5**, 235 (1969).

434. D. M. Roundhill, S. G. N. Roundhill, W. B. Beaulieu and U. Bagchi, *Inorg. Chem.*, **19**, 3365 (1980).
435. L. Cavalca, M. Nardelli and G. Branchi, *Acta Cryst.*, **13**, 688 (1960).
436. R. Gronbaek and S. E. Rasmussen, *Acta Chem. Scand.*, **16**, 2325 (1962).
437. E. W. Abel, S. K. Bhargava, K. Kite, K. G. Orrell, V. Šik and B. L. Williams, *J. Chem. Soc., Dalton Trans.*, 583 (1982).
438. E. W. Abel, S. K. Bhargava, K. Kite, K. G. Orrell, V. Šik and B. L. Williams, *J. Chem. Soc., Dalton Trans.*, 365 (1984).
439. E. W. Abel, S. K. Bhargava, K. Kite, K. G. Orrell, V. Šik and B. L. Williams, *Polyhedron*, **1**, 289 (1982).
440. D. C. Goodall, *J. Inorg. Nucl. Chem.*, **30**, 1269 (1968).
441. D. C. Goodall, *J. Inorg. Nucl. Chem.*, **30**, 2483 (1968).
442. E. E. Aynsley, N. N. Greenwood and J. B. Leach, *Chem. Ind. (London)*, 379 (1966).
443. E. W. Abel and G. V. Hutson, *J. Inorg. Nucl. Chem.*, **31**, 3333 (1969).
444. E. W. Abel, A. R. Khan, K. Kite, K. G. Orrell and V. Šik, *J. Chem. Soc., Dalton Trans.*, 1175 (1980).
445. E. W. Abel, K. G. Orrell and A. W. G. Platt, *J. Chem. Soc., Dalton Trans.*, 2345 (1983).
446. E. W. Abel, S. K. Bhargava, M. M. Bhatti, K. Kite, M. A. Mazid, K. G. Orrell, V. Šik, B. L. Williams, M. B. Hursthous and K. M. A. Malik, *J. Chem. Soc., Dalton Trans.*, 2065 (1982).
447. N. N. Greenwood and G. Hunter, *J. Chem. Soc. (A)*, 1520 (1967).
448. H. J. Whitfield, *J. Chem. Soc. (A)*, 113 (1970).
449. G. Hunter and R. C. Massey, *Chem. Commun.*, 797 (1973).
450. G. Hunter and R. C. Massey, *J. Chem. Soc., Dalton Trans.*, 2007 (1976).
451. N. N. Greenwood and G. Hunter, *J. Chem. Soc. (A)*, 929 (1969).
452. R. Donaldson, G. Hunter and R. C. Massey, *J. Chem. Soc., Dalton Trans.*, 288 (1974).
453. A. D. Westland and L. Westland, *Can. J. Chem.*, **43**, 426 (1965).
454. G. Hunter and R. C. Massey, *J. Chem. Soc., Dalton Trans.*, 209 (1975).
455. C. G. Pierpont, *Ph.D. Thesis*, Brown University, 1972.
456. (a) R. S. Ramakrishna and H. M. N. H. Irving, *Chem. Ind. (London)*, 325 (1969).
(b) R. S. Ramakrishna and H. M. N. H. Irving, *Anal. Chim. Acta*, **48**, 251 (1969).
457. C. A. Stein, P. E. Ellis, R. C. Elder and E. Deutsch, *Inorg. Chem.*, **15**, 1618 (1976).
458. V. H. Houlding, H. Macke and A. W. Adamson, *Inorg. Chem.*, **20**, 4279 (1981).
459. T. Konno, K. Okamoto, H. Einaga and J. Hidaka, *Chem. Letters*, 969 (1983).
460. K. Nakabayashi, K. Doi, M. Kojima and J. Fujita, *Bull. Chem. Soc. Japan*, **57**, 989 (1984).
461. G. Wilke and E. Uhlemann, *Z. Chem.*, **14**, 288 (1974).
462. L. E. Nivorozhkin, V. P. Kurbatov, L. S. Minkina, M. S. Korobov, L. P. Olekhnovich, O. A. Osipov and V. I. Minkin, *Zh. Obshch. Khim.*, **41**, 2726 (1971).
463. V. P. Kurbatov, O. A. Osipov, V. I. Minkin, L. E. Nivorozhkin and L. S. Minkina, *Russ. J. Inorg. Chem.*, **19**, 1200 (1974).
464. R. Curran, J. A. Cunningham and R. Eisenberg, *Inorg. Chem.*, **9**, 2749 (1970).
465. A. T. Hutton, H. M. N. H. Irving, L. R. Nassimbeni and G. Gafner, *Acta Cryst.*, **36B**, 2064 (1980).
466. (a) A. Mawby and H. M. N. H. Irving, *J. Inorg. Nucl. Chem.*, **34**, 109 (1972).
(b) J. McB. Harrowfield, C. Pakawatchai and A. H. White, *Australian J. Chem.*, **36**, 825 (1983).
467. M. Laing and P. A. Alsop, *Talanta*, **17**, 242 (1970).
468. M. M. Harding, *J. Chem. Soc.*, 4136 (1958).
469. J. M. Harrowfield, C. Pakawatchai and A. H. White, *J. Chem. Soc., Dalton Trans.*, 1109 (1983).
470. K. Okamoto, T. Konno, M. Nomoto, H. Einaga and J. Hidaka, *Chem. Letters*, 1941 (1982).
471. R. H. Holm, G. W. Everett and A. Chakravorty, *Progr. Inorg. Chem.*, **7**, 83 (1961).
472. O. A. Osipov, V. I. Minkin and Zh. A. Tumakova, *Zh. Strukt. Khim.*, **5**, 918 (1964).
473. L. E. Nivorozhkin, L. E. Konstantinovsku, M. A. Voronov, V. I. Minkin, O. A. Osipov, V. P. Kurbatov and L. S. Minkina, *Zh. Neorg. Khim.*, **18**, 2456 (1973).
474. Y. Sugiura, J. Kuwahara and T. Suzuki, *Chem. Commun.*, 908 (1982).
475. W. Kuchen and H. Hertel, *Angew. Chem. (Intern. Ed. Engl.)*, **8**, 89 (1969).
476. C. Glidewell and E. J. Leslie, *J. Chem. Soc., Dalton Trans.*, 527 (1977).
477. C. Glidewell, *Inorg. Chim. Acta*, **24**, 255 (1977).
478. L. Beyer, R. Kirmse and E. Hoyer, *Z. Chem.*, **15**, 197 (1975).

479. R. Herzschuh, B. Birner, L. Beyer, F. Dietze and E. Hoyer, *Z. Anorg. Allgem. Chem.*, **464**, 159 (1980).
480. J. Stack, R. Kirmse, A. Heinrich and W. Dietzsch, *Z. Chem.*, **23**, 453 (1983).
481. G. Wilke and E. Uhlemann, *Z. Chem.*, **15**, 66 (1975).
482. T. Honjo, *Chem. Letters*, 481 (1974).
483. A. Stock and E. Willfroth, *Chem. Ber.*, **47**, 152 (1914).
484. W. Dietzsch, J. Sieler and T. Glowiak, *Z. Anorg. Allgem. Chem.*, **435**, 225 (1977).
485. W. Dietzsch, J. Kaiser, R. Richter, L. Golic, J. Siftar and R. Heber, *Z. Anorg. Allgem. Chem.* **477**, 71 (1981).
486. W. Dietzsch, H. Treffurth and E. Hoyer, *Inorg. Chim. Acta*, **27**, L89 (1978).
487. W. Kuchen and B. Knop, *Angew. Chem. (Intern. Ed. Engl.)*, **3**, 507 (1964).
488. L. A. Il'lina, N. I. Zemlyanskii, S. V. Larionov and N. M. Chernaya, *Izv. Akad. Nauk SSSR, Ser. Khim.*, 198 (1969).
489. S. V. Larionov, L. A. Il'lina and T. V. Bulycheva, *Russ. J. Inorg. Chem.*, **17**, 1013 (1972).
490. I. M. Cheremisina, L. A. Il'lina and S. V. Larionov, *Russ. J. Inorg. Chem.*, **18**, 675 (1973).
491. T. Kamitani and T. Tanaka, *Inorg. Nucl. Chem. Letters*, **6**, 91 (1970).
492. T. Tanaka and N. Sonoda, *Inorg. Chem.*, **10**, 2337 (1971).
493. R. Heber, R. Kirmse and E. Hoyer, *Z. Anorg. Allgem. Chem.*, **393**, 159 (1972).
494. N. Sonoda and T. Tanaka, *J. Inorg. Nucl. Chem.*, **35**, 1145 (1975).
495. S. Nakajima and T. Tanaka, *Bull. Chem. Soc. Japan*, **47**, 763 (1974).
496. N. Sonoda and T. Tanaka, *Inorg. Chim. Acta*, **12**, 261 (1975).
497. W. Dietzsch, R. Kirmse and E. Hoyer, *J. Prakt. Chem.*, **322**, 1069 (1980).
498. C. Bianchini and A. Meli, *Chem. Commun.*, 156 (1983).
499. C. G. Pierpont, B. J. Corden and R. Eisenberg, *Chem. Commun.*, 401 (1969).
500. L. J. Maheu, G. L. Miessler, J. Berry, M. Burow and L. H. Pignolet, *Inorg. Chem.*, **22**, 405 (1983).
501. L. Beyer and A. Hantschmann, *Z. Chem.*, **23**, 231 (1983).
502. E. Uhlemann, E. Ludwig, W. Hübner and R. Szargan, *Z. Chem.*, **23**, 32 (1983).
503. J. Kaiser, R. Richter and J. Sieler, *Acta Cryst.*, **31B**, 2600 (1975).
504. (a) J. Stach, R. Kirmse and E. Hoyer, *J. Mol. Struct.*, **46**, 245 (1978).
(b) J. Stach, R. Kirmse, S. Wartewig and R. Böttcher, *Z. Chem.*, **19**, 205 (1979).
505. J. Stach, R. Kirmse, E. Hoyer and S. Wartewig, *J. Inorg. Nucl. Chem.*, **40**, 1529 (1978).
506. W. Dietzsch, J. Lerchner, J. Reinhold, J. Stach, R. Kirmse, G. Steimecke and E. Hoyer, *J. Inorg. Nucl. Chem.*, **42**, 509 (1980).
507. J. Stach, R. Herzschuh, R. Kirmse, W. Dietzsch, A. Heinrich, V. K. Belyaeva and I. N. Marov, *Z. Anorg. Allgem. Chem.*, **508**, 115 (1984).
508. L. J. Maheu and L. H. Pignolet, *J. Amer. Chem. Soc.*, **102**, 6348 (1980).
509. L. Beyer and R. Wildera, *Tetrahedron Letters*, **23**, 1881 (1982).
510. K. A. Jensen and V. Krishnan, *Acta Chem. Scand.*, **21**, 2904 (1967).
511. K. A. Jensen and V. Krishnan, *Acta Chem. Scand.*, **24**, 1088 (1970).
512. K. A. Jensen, V. Krishnan and C. K. Jørgensen, *Acta Chem. Scand.*, **24**, 743 (1970).
513. A. Rosenbaum, H. Kirchberg and E. Leibnitz, *J. Prakt. Chem.*, **19**, 1 (1963).
514. J. G. M. van der Linden and P. J. M. Geurts, *Inorg. Nucl. Chem. Letters*, **8**, 903 (1972).
515. D. Barnard and D. T. Woodbridge, *J. Chem. Soc.*, 2922 (1961).
516. W. H. Pan, J. P. Fackler and H. W. Chen, *Inorg. Chem.*, **20**, 856 (1981).
517. H. W. Chen, J. P. Fackler, A. F. Masters and W. H. Pan, *Inorg. Chim. Acta*, **35**, L333 (1979).
518. W. H. Pan and J. P. Fackler, *J. Amer. Chem. Soc.*, **100**, 5783 (1978).
519. W. H. Pan, *Ph.D. Thesis*, Case Western Reserve University, 1979.
520. M. Bonamico and G. Dessy, *J. Chem. Soc. (A)*, 264 (1971).
521. C. Furlani, E. Cervone and F. D. Camassei, *Inorg. Chem.*, **7**, 265 (1968).
522. D. De Filippo, P. Deplano, F. Devillanova and E. F. Trogu, *Inorg. Chim. Acta*, **17**, 199 (1976).
523. G. E. Manoussakis, C. A. Tsipis and A. G. Christophides, *Inorg. Chem.*, **12**, 3015 (1973).
524. D. De Filippo, P. Deplano, A. Diaz and E. F. Trogu, *Inorg. Chim. Acta*, **17**, 139 (1976).
525. D. Attanasio, C. Bellitto, A. Flamini and G. Pennesi, *Inorg. Chem.*, **21**, 1461 (1982).
526. S. Esperas, S. Husebye and A. Rolandsen, *Acta Chem. Scand.*, **29A**, 608 (1975).
527. J. P. Fackler and W. H. Pan, *J. Amer. Chem. Soc.*, **101**, 1607 (1979).
528. P. Eckstein and E. Hoyer, *Z. Anorg. Allgem. Chem.*, **487**, 33 (1982).

529. P. Deplano, E. F. Trogu, F. Bigoli, E. Leporati, M. A. Pellinghelli, D. L. Perry, R. J. Saxton and L. J. Wilson, *J. Chem. Soc., Dalton Trans.*, 25 (1983).
530. P. T. Beurskens and J. A. Cras, *J. Cryst. Mol. Struct.*, **1**, 63 (1971).
531. C. K. Jørgensen, *Mol. Phys.*, **5**, 485 (1962).
532. N. Krishnan and R. A. Zingaro, *Inorg. Chem.*, **8**, 2337 (1969).
533. M. V. Kudchadker, R. A. Zingaro and K. J. Irgolic, *Can. J. Chem.*, **46**, 1415 (1968).
534. P. J. H. A. M. v. d. Leemput, T. W. Hummelink, J. H. Noordik and P. T. Beurskens, *Cryst. Struct. Commun.*, **4**, 167 (1975).
535. W. Kuchen and B. Knop, *Angew. Chem. (Intern. Ed. Engl.)*, **4**, 244 (1965).
536. L. Gastaldi and P. Porta, *Cryst. Struct. Commun.*, **6**, 175 (1977).
537. K. A. Jensen and V. Krishnan, *Acta Chem. Scand.*, **24**, 1090 (1970).
538. W. Dietzsch, R. Kirmse, E. Hoyer, V. K. Belyaeva and I. N. Marov, *Inorg. Chem.*, **17**, 1665 (1978).
539. J. Kaiser, W. Dietzsch, R. Richter, L. Golic and J. Siftar, *Acta Cryst.*, **36B**, 147 (1980).
540. (a) A. Davison and E. T. Shaw, *Chem. Commun.*, 670 (1967).
(b) A. Davison and E. T. Shaw, *Inorg. Chem.*, **9**, 1820 (1970).
541. C. G. Pierpont and R. Eisenberg, *J. Chem. Soc. (A)*, 2285 (1971).
542. C. M. Bolinger and T. B. Rauchfuss, *Inorg. Chem.*, **21**, 3947 (1982).
543. G. A. Heath, I. M. Stewart and R. L. Martin, *Inorg. Nucl. Chem. Letters*, **5**, 169 (1969).
544. C. G. Barraclough, R. L. Martin and I. M. Stewart, *Australian J. Chem.*, **22**, 891 (1969).
545. A. Davison and D. L. Reger, *Inorg. Chem.*, **10**, 1967 (1971).
546. R. Beckett and B. F. Hoskins, *Chem. Commun.*, 909 (1967).
547. E. Uhlig, B. Borek and H. Glänzer, *Z. Anorg. Allgem. Chem.*, **348**, 189 (1966).
548. A. V. Ablov, N. V. Gerbeleu and N. Ya. Negryatse, *Russ. J. Inorg. Chem.*, **14**, 515 (1969).
549. W. Levason, C. A. McAuliffe and S. G. Murray, *J. Chem. Soc., Dalton Trans.*, 269 (1976).
550. G. Dyer and D. W. Meek, *Inorg. Chem.*, **6**, 149 (1967).
551. G. Labauze and J. B. Raynor, *J. Chem. Soc., Dalton Trans.*, 2388 (1980).
552. V. Lenher and A. G. Fruehan, *J. Amer. Chem. Soc.*, **49**, 3076 (1927).
553. (a) E. Diemann and A. Müller, *Coord. Chem. Rev.*, **10**, 79 (1976).
(b) A. Müller, E. J. Baran and R. O. Carter, *Struct. Bonding (Berlin)*, **26**, 81 (1976).
(c) D. A. Rice, *Coord. Chem. Rev.*, **25**, 199 (1978).
554. W. A. Herrmann, in *Organomet. Compd: Synth. Struct. Theory, Proc. Symp. Ind.-Univ. Coop. Chem. Program, Dept. Chem.* (Ed. B. L. Shapiro), Texas A & M University, 1983, pp. 383–405.
555. H. Brunner, J. Wachter and H. Wintergerst, *J. Organomet. Chem.*, **235**, 77 (1982).
556. (a) W. A. Herrmann, C. Bauer and J. Weichmann, *J. Organomet. Chem.*, **243**, C21 (1983).
(b) W. A. Herrmann, C. Bauer and J. Weichmann, *Chem. Ber.*, **117**, 1271 (1984).
557. M. Heberhold, D. Reiner, B. Zimmer-Gasser and U. Schubert, *Z. Naturforsch.*, **35b**, 1281 (1980).
558. W. A. Herrmann, J. Rohrmann and A. Schäfer, *J. Organomet. Chem.*, **265**, C1 (1984).
559. J. Schiemann, P. Hübener and E. Weiss, *Angew. Chem. (Intern. Ed. Engl.)*, **22**, 980 (1983).
560. K. S. Nagaraja and M. R. Udupa, *Trans. Met. Chem.*, **8**, 191 (1983).
561. P. Beardwood and J. F. Gibson, *J. Chem. Soc., Dalton Trans.*, 1507 (1984).
562. P. Beardwood and J. F. Gibson, *J. Chem. Soc., Dalton Trans.*, 737 (1983).
563. A. P. Ginsberg and W. E. Lindsell, *Chem. Commun.*, 232 (1971).
564. A. P. Ginsberg, W. E. Lindsell, C. R. Sprinkle, K. W. West and R. L. Cohen, *Inorg. Chem.*, **21**, 3666 (1982).
565. (a) C. M. Bolinger, J. E. Hoots and T. B. Rauchfuss, *Organometallics*, **1**, 223 (1982).
(b) J. E. Hoots and T. B. Rauchfuss, *Inorg. Chem.*, **22**, 2806 (1983).
566. A. P. Ginsberg, J. H. Osborne and C. R. Sprinkle, *Inorg. Chem.*, **22**, 254 (1983).
567. M. Herberhold, D. Reiner and V. Théwalt, *Angew. Chem. (Intern. Ed. Engl)*, **22**, 1000 (1983).
568. W. Hieber and J. Gruber, *Z. Anorg. Allgem. Chem.*, **296**, 91 (1958).
569. C. F. Campona, F. Y.-K. Lo and L. F. Dahl, *Inorg. Chem.*, **18**, 3060 (1974).
570. D. A. Lesch and T. B. Rauchfuss, *J. Organomet. Chem.*, **199**, C6 (1980).
571. V. W. Day, D. A. Lesch and T. B. Rauchfuss, *J. Amer. Chem. Soc.*, **104**,1290 (1982).
572. R. Rossetti, G. Gervasio and P. L. Stanghellini, *Inorg. Chim. Acta*, **35**, 73 (1979).
573. S. Aime, G. Gervasio, R. Rossetti and P. L. Stanghellini, *Inorg. Chim. Acta*, **40**, 131 (1980).
574. D. Seyferth, R. S. Henderson and L. C. Song, *Organometallics*, **1**, 125 (1982).
575. L. E. Bogan, D. A. Lesch and T. B. Rauchfuss, *J. Organomet. Chem.*, **250**, 429 (1983).

576. M. G. B. Drew, G. W. A. Fowles, E. M. Page and D. A. Rice, *J. Amer. Chem. Soc.*, **101**, 5827 (1980).
577. M. G. B. Drew, D. A. Rice and D. M. Williams, *J. Chem. Soc., Dalton Trans.*, 1087 (1984).
578. A. J. Benton, M. G. B. Drew and D. A. Rice, *Chem. Commun.*, 1241 (1981).
579. H. F. Franzen, W. Hönle and H. G. V. Schnering, *Z. Anorg. Allgem. Chem.*, **497**, 13 (1983).
580. J. Rijnsdorp, G. J. DeLange and G. A. Wiegers, *J. Solid State Chem.*, **30**, 365 (1979).
581. H. Köpf, W. Kahl and A. Wirl, *Angew. Chem. (Intern. Ed. Engl.)*, **9**, 801 (1970).
582. A. P. Ginsberg, J. H. Osborne and C. R. Sprinkle, *Inorg. Chem.*, **22**, 1781 (1983).
583. H. Köpf, B. Block and M. Schmidt, *Chem. Ber.*, **101**, 272 (1968).
584. E. F. Epstein, I. Bernal and H. Köpf, *J. Organomet. Chem.*, **26**, 229 (1971).
585. (a) R. Steudel and R. Strauss, *J. Chem. Soc., Dalton Trans.*, 1775 (1984).
(b) D. M. Giolando and T. B. Rauchfuss, *Organometallics*, **3**, 487 (1984).
586. D. A. Roberts and G. L. Geoffroy, in *Comprehensive Organometallic Chemistry* (Eds. G. Wilkinson, F. G. A. Stone and E. W. Abel), Pergamon Press, Oxford, 1982, Chapt. 40.
587. L. F. Dahl and P. W. Sutton, *Inorg. Chem.*, **2**, 1067 (1963).
588. L. L. Nelson, F. Y.-K. Lo, A. D. Rae and L. F. Dahl, *J. Organomet. Chem.*, **225**, 309 (1982).
589. C. E. Strouse and L. F. Dahl, *J. Amer. Chem. Soc.*, **93**, 6032 (1971).
590. B. F. G. Johnson, J. Lewis, P. G. Lodge, P. R. Raithby, K. Henrick and M. McPartlin, *Chem. Commun.*, 719 (1979).
591. E. Sappa, O. Gambino and G. Cetini, *J. Organomet. Chem.*, **35**, 375 (1972).
592. E. Roland and H. Vahrenkamp, *Organometallics*, **2**, 1048 (1983).
593. P. V. Broadhurst, B. F. G. Johnson and J. Lewis, *J. Chem. Soc., Dalton Trans.*, 1881 (1982).
594. B. F. G. Johnson, J. Lewis, P. G. Lodge and P. R. Raithby, *Acta Cryst.*, **37B**, 1731 (1981).
595. G. Gervasio, R. Rossetti, P. L. Stanghellini and G. Bor, *Congr. Naz. Chim. Inorg.*, [atti], 16th, 390 (1983); *Chem. Abstr.*, **100**, 28862q (1983).
596. D. Galli, L. Garlaschelli, G. Cioni, A. Fumagalli, S. Martinengo and A. Sironi, *J. Chem. Soc., Dalton Trans.*, 55 (1984).
597. C. Lensch, P. G. Jones and G. M. Sheldrick, *Z. Naturforsch*, **37b**, 944 (1982).
598. F. Cecconi, C. A. Ghilardi and S. Midollini, *Inorg. Chem.*, **22**, 3802 (1983).
599. F. Cecconi, C. A. Ghilardi and S. Midollini, *Cryst. Struct. Commun.*, **11**, 25 (1982).
600. I. G. Dance, A. Choy and M. L. Scudder, *J. Amer. Chem. Soc.*, **106**, 6285 (1984).
601. (a) R. A. Zingaro, *J. Organomet. Chem.*, **1**, 200 (1963).
(b) R. A. Zingaro, B. H. Steeves and K. Irgolic, *J. Organomet. Chem.*, **4**, 320 (1965).
(c) G. N. Chremos and R. A. Zingaro, *J. Organomet. Chem.*, **22**, 637 (1970).
(d) G. N. Chremos and R. A. Zingaro, *J. Organomet. Chem.*, **22**, 647 (1970).
(e) W. W. DuMont and H. J. Kroth, *J. Organomet. Chem.*, **113**, C35 (1976).
(f) N. P. Grechkin, I. A. Nuretdinov and N. A. Buina, *Izv. Akad. Nauk SSSR, Ser. Khim.*, 168 (1969).
(g) A. I. Razumov, B. G. Liorber, M. B. Gazizov and Z. M. Khammatova, *Zh. Obshch. Khim.*, **34**, 1851 (1964).
(h) I. A. Nuretdinov and E. I. Loginova, *Izv. Akad. Nauk SSSR, Ser. Khim.*, 2827 (1973).
(i) E. I. Loginova, I. A. Nuretdinov and Yu. A. Petrov, *Teor. Eksp. Khim.*, **10**, 75 (1974).
(j) R. R. Shagidullin, I. P. Lipatova, I. A. Nuretdinov and S. A. Samartseva, *Dokl. Akad. Nauk SSSR*, **211**, 1363 (1973).
(k) Y. Y. Borovikov, E. V. Ryl'tsev, I. E. Boldeskul, N. G. Feshchenko, Y. P. Makovetskii and Y. P. Egorov, *Zh. Obshch. Khim.*, **40**, 1957 (1970).
(m) P. A. W. Dean, *Can. J. Chem.*, **57**, 754 (1979).
(n) D. H. Brown, R. J. Cross and D. Millington, *J. Organomet. Chem.*, **125**, 219 (1977).
(p) B. Wrackmeyer, *Z. Naturforsch.*, **39b**, 533 (1984).
602. M. Nardelli, C. Pelizzi and G. Pelizzi, *Inorg. Chim. Acta*, **33**, 181 (1979).
603. (a) D. W. Meek, G. Dyer and M. O. Workman, *Inorg. Synth.*, **16**, 168 (1976).
(b) T. S. Lobana and S. S. Sandhu, *J. Chem. Sci.*, **4**, 37 (1978).
604. (a) A. W. Down, *Chem. Commun.*, 1290 (1968).
(b) T. Austad, J. Songstad and K. Åse, *Acta Chem. Scand.*, **25**, 331 (1971).
(c) A. Martinsen and J. Songstad, *Acta Chem. Scand.*, **31A**, 645 (1977).
(d) A. S. Foust, *Chem. Commun.*, 414 (1979).
605. H. J. Gysling, N. Zumbulyadis and J. A. Robertson, *J. Organomet. Chem.*, **209**, C41 (1981).
606. H. J. Gysling, H. R. Luss and D. L. Smith, *Inorg. Chem.*, **18**, 2696 (1979).

607. N. Kuhn and M. Winter, *J. Organomet. Chem.*, **249**, C28 (1983).
608. P. L. Goggin, R. J. Goodfellow and S. R. Haddock, *Chem. Commun.*, 176 (1975).
609. Asahi Chemical Industry Company, Ltd., *Japanese Patent*, No. 53 143216 (1978); *Chem. Abstr.*, **90**, 195607s (1979).
610. R. K. Chadha and J. E. Drake, *J. Organomet. Chem.*, **286**, 121 (1985).
611. S. A. Gardner, *J. Organomet. Chem.*, **190**, 289 (1980).
612. H. H. Glazebrook and T. G. Pearson, *J. Chem. Soc.*, 589 (1939).
613. F. W. B. Einstein, C. H. W. Jones, T. Jones and R. D. Sharma, *Inorg. Chem.*, **22**, 3924 (1983).
614. L. Y. Chia and W. R. McWhinnie, *J. Organomet. Chem.*, **148**, 165 (1978).
615. W. R. McWhinnie and V. Rattanaphani, *Inorg. Chim. Acta*, **9**, 153 (1974).
616. (a) I. Davies and W. R. McWhinnie, *Inorg. Nucl. Chem. Letters*, **12**, 763 (1976).
(b) I. Davies, W. R. McWhinnie, N. S. Dance and C. H. W. Jones, *Inorg. Chim. Acta*, **29**, L217 (1978).
617. N. Kuhn and M. Winter, *J. Organomet. Chem.*, **252**, C86 (1983).
618. W. Hieber and T. Kruck, *Chem. Ber.*, **95**, 2027 (1962).
619. K. Öfele and E. Dotzauer, *J. Organomet. Chem.*, **42**, C87 (1972).
620. A. D. Garnovski, G. M. Abakarov, I. D. Sadekov, V. I. Minkin, T. G. Cherkasova and S. Yu Varshavskii, *Koord. Khim.*, **10**, 234 (1984); *Chem. Abstr.*, **100**, 202373t (1984).
621. M. F. Lappert, T. R. Martin and G. M. McLaughlin, *Chem. Commun.*, 635 (1980).
622. M. F. Lappert, J. J. McQuitty and P. L. Pye, *J. Chem. Soc., Dalton Trans.*, 1583 (1981).
623. H. Fischer and S. Zeuner, *J. Organomet. Chem.*, **252**, C63 (1983).
624. H. Fischer, E. O. Fischer, R. Cai and D. Himmelreich, *Chem. Ber.*, **116**, 1009 (1983).
625. R. C. Paul, R. K. Chadha, B. S. Bhandal and G. Singh, *Inorg. Chim. Acta*, **52**, 125 (1981).
626. H. J. Gysling, *Proceedings of the 3rd International Symposium on Organic Selenium and Tellurium Compounds*, Metz, France, July 9–12, 1979, p. 373.
627. S. J. Anderson and R. J. Goodfellow, *J. Chem. Soc., Dalton Trans.*, 1683 (1977).
628. N. S. Dance and C. H. W. Jones, *J. Organomet. Chem.*, **152**, 175 (1978).
629. A. Z. Al-Rubaie and W. R. McWhinnie, *J. Organomet. Chem.*, **235**, 59 (1982).
630. P. Granger, S. Chapelle, W. R. McWhinnie and A. Al-Rubaie, *J. Organomet. Chem.*, **220**, 149 (1981).
631. R. E. Cobbledick, F. W. B. Einstein, W. R. McWhinnie and F. H. Musa, *J. Chem. Res.*, 1901(M) (1979); 145 (S) (1979).
632. R. T. Mehdi and J. D. Miller, *J. Chem. Soc., Dalton Trans.*, 1065 (1984).
633. A. G. M. Barrett, D. H. R. Barton and R. W. Reed, *Chem. Commun.*, 645 (1979).
634. K. L. Lerstrup and L. Henriksen, *Chem. Commun.*, 1102 (1979).
635. D. H. R. Barton and S. W. McCombie, *J. Chem. Soc., Perkin Trans. 1*, 1574 (1975).
636. R. E. Lyons and E. D. Scudder, *Ber. Dtsch. Chem. Res.*, **64**, 530 (1931).
637. U. Behrens, K. Hoffmann and G. Klar, *Chem. Ber.*, **110**, 3672 (1977).
638. R. E. Cobbledick, N. S. Dance, F. W. B. Einstein, C. H. W. Jones and T. Jones, *Inorg. Chem.*, **20**, 4356 (1981).
639. D. Mohr, H. Wienand and M. L. Ziegler, *J. Organomet. Chem.*, **134**, 281 (1977).
640. W. Hieber and R. Kramolowski, *Z. Anorg. Allgem. Chem.*, **321**, 94 (1963).
641. L. Y. Chia and W. R. McWhinnie, *J. Organomet. Chem.*, **148**, 165 (1978).
642. R. J. Haines, J. A. DeBeer and R. Greatrex, *J. Organomet. Chem.*, **85**, 89 (1975).
643. G. Ferguson, C. Hannaway and K. M. S. Islam, *Chem. Commun.*, 1165 (1968).
644. C. H. W. Jones, N. S. Dance and H. J. Gysling, *Proceedings of the 4th International Conference on the Organic Chemistry of Selenium and Tellurium*, Birmingham, England, July 25–29, 1983, p. 492.
645. N. S. Dance and C. H. W. Jones, *J. Organomet. Chem.*, **152**, 175 (1978).
646. W. A. Herrmann, J. Weichmann, U. Küsthardt, A. Schafer, R. Hörlein, C. Hecht, E. Voss and R. Serrano, *Angew. Chem. (Intern. Ed. Engl.)*, **22**, 979 (1983).
647. W. A. Herrmann, C. Hecht, M. L. Ziegler and B. Balbach, *Chem. Commun.*, 686 (1984).
648. H. Werner, J. Wolf, R. Zolk and U. Schubert, *Angew. Chem. (Intern. Ed. Engl.)*, **22**, 981 (1983).
649. F. D. Williams and F. X. Dunbar, *Chem. Commun.*, 459 (1968).
650. (a) K. L. Brown, G. R. Clark, C. E. L. Headford, K. Marsden and W. R. Roper, *J. Amer. Chem. Soc.*, **101**, 503 (1979).
(b) G. R. Clark, C. E. L. Headford, K. Marsden and W. R. Roper, *J. Organomet. Chem.*, **231**, 335 (1982).

651. T. J. Collins and W. R. Roper, *J. Organomet. Chem.*, **159**, 73 (1978).
652. D. A. Lesch and T. B. Rauchfuss, *Inorg. Chem.*, **20**, 3583 (1981).
653. M. Heberhold, D. Reiner and D. Neugebauer, *Angew. Chem. (Intern. Ed. Engl)*, **22**, 59 (1983).
654. H. Schumann, M. Magerstädt and J. Pickardt, *J. Organomet. Chem.*, **240**, 407 (1982).
655. G. Cetini, P. L. Stanghellini, R. Rossetti and O. Gambino, *J. Organomet. Chem.*, **15**, 373 (1968).
656. G. Cetini, P. L. Stanghellini, R. Rossetti and O. Gambino, *Inorg. Chim. Acta*, **2**, 433 (1968).
657. P. L. Stanghellini, G. Cetini, O. Gambino and R. Rossetti, *Inorg. Chim. Acta*, **3**, 651 (1969).
658. M. K. Chaudhuri, A. Haas, M. Rosenberg, M. Velicescu and N. Welcman, *J. Organomet. Chem.*, **124**, 37 (1977).
659. S. Aime, L. Milone, R. Rossetti and P. L. Stanghellini, *J. Chem. Soc., Dalton Trans.*, 46 (1980).
660. D. A. Lesch and T. B. Rauchfuss, *Organometallics*, **1**, 499 (1982).
661. L. E. Bogan, D. A. Lesch and T. B. Rauchfuss, *J. Organomet. Chem.*, **250**, 429 (1983).
662. R. Rossetti, G. Cetini, O. Gambino and P. L. Stanghellini, *Atti. Accad. Sci. Torino*, **104**, 127 (1969–1970).
663. R. Rossetti, P. L. Stanghellini, O. Gambino and G. Cetini, *Inorg. Chim. Acta*, **6**, 205 (1972).
664. D. A. Lesch and T. B. Rauchfuss, *Inorg. Chem.*, **22**, 1854 (1983).
665. R. C. Ryan and L. F. Dahl, *J. Amer. Chem. Soc.*, **97**, 6905 (1975).
666. (a) J. Berkowitz, *J. Chem. Phys.*, **62**, 4074 (1975).
(b) D. G. Streets and J. Berkowitz, *J. Elec. Spectry Relat. Phenom.*, **9**, 269 (1976).
667. A. S. Faust and L. F. Dahl, *J. Amer. Chem. Soc.*, **92**, 7337 (1970).
668. (a) Y. Shvo and E. Hazum, *Chem. Commun.*, 829 (1975).
(b) J. Elzinga and H. Hogeveen, *Chem. Commun.*, 705 (1977).
(c) R. J. Lawson and J. R. Shapley, *J. Amer. Chem. Soc.*, **98**, 7433 (1976).
669. N. Zumbulyadis and H. J. Gysling, *J. Organomet. Chem.*, **192**, 183 (1980).
670. (a) H. D. Beckey and H. R. Schulten, *Angew. Chem. (Intern. Ed. Engl.)*, **14**, 403 (1975).
(b) D. E. Games, J. L. Gower and L. A. P. Kane-Maguire, *J. Chem. Soc., Dalton Trans.*, 1994 (1981).
(c) L. H. Staal, G. Van Koten, R. H. Fokkens and N. M. M. Nibbering, *Inorg. Chim. Acta*, **50**, 205 (1981).
(d) D. E. Games, L. A. P. Kane-Maguire and D. A. Sweigart, *J. Organomet. Chem.*, **234**, 323 (1982).
671. J. L. Vidal, W. E. Walker, R. L. Pruett and R. C. Schoening, *Inorg. Chem.*, **18**, 129 (1979).
672. J. L. Vidal, W. E. Walker and R. C. Schoening, *Inorg. Chem.*, **20**, 238 (1981).
673. J. L. Vidal and J. M. Troup, *J. Organomet. Chem.*, **213**, 351 (1981).
674. J. L. Vidal, *Inorg. Chem.*, **20**, 243 (1981).
675. R. C. Ryan, C. U. Pittman, J. P. O'Connor and L. F. Dahl, *J. Organomet. Chem.*, **193**, 247 (1980).

Author Index

This author index is designed to enable the reader to locate an author's name and work with the aid of the reference numbers appearing in the text. The page numbers are printed in normal type in ascending numerical order, followed by the reference numbers in parentheses. The numbers in *italics* refer to the pages on which the references are actually listed.

Aarskog, J. 115, 116 (247), *153*
Abakarov, G. 487 (1020), *515*
Abakarov, G. M. 814, 820 (620), *854*
Abdelhamid, A. 467 (814a, 814b), *510*
Abdullaev, G. B. 289, 291 (24), *304*
Abe, H. 623 (48, 49), *657*
Abel, E. W. 114 (245, 246), 115 (246), *153*, 191–193 (16), *239*, 680 (2), 681 (19), 690 (79, 112), 709 (183–186), 710 (183–188), 712 (184–188), 713 (188), 721 (2), 727, 729 (257), 742 (339), 743 (345–347), 744 (339, 345–347), 748 (339), 750 (339, 345), 771 (437), 773 (257, 437–439), 774 (437–439, 443–446), 775 (437, 438), 776 (444–446), 801 (445), 819 (184), *839*, *840*, *842*, *844*, *846*, *848*, *850*
Ablov, A. V. 684–686 (40), 726 (251), 727 (251, 262a, 262b, 263), 728 (251), 730 (270), 732 (270, 289, 290), 738 (289), 739 (251, 289), 770 (424), 771 (424, 425), 778 (424), 802 (548), *841*, *846*, *847*, *849*, *852*
Abragam, A. 193 (20), *239*
Abramenko, P. 437 (420), 438 (436), 439 (443, 445), 472 (875), *502*, *503*, *512*
Abroin, L. 582 (177), *615*
Abronin, I. 431 (378, 379), 434 (396), *502*
Abronin, I. A. 578, 579 (147), 588 (200), 596 (147), *615*, *616*
Acampora, L. A. 334 (18), 340 (38), 341 (18), *341*
Acheson, R. 461 (712), *508*
Ackermann, K. 723–725 (249), *846*
Adams, D. M. 690 (85), *842*
Adams, F. 165, 167–169 (71), *186*
Adams, G. E. 292 (34), 294 (46), 298 (54), 301 (64, 65), *305*
Adams, W. J. 89 (140, 141), *151*
Adamson, A. W. 779, 780 (458), *850*
Addison, A. 483 (983), *514*
Adiwidjaja, G. 456, 458 (640), *507*
Adley, T. J. 261 (64), *284*
Adloff, J.-P. 374 (24), *394*
Adloff, J.P. 371 (13), *394*
Adloff, M. 374 (24), *394*
Adriaenssens, E. 164, 165 (47), *186*
Agenas, L. 493 (1068b), *516*
Agenäs, L.-B. 546 (245), *557*, 641 (266, 273, 275, 276), *661*
Agenäs, L. B. 249 (33), *284*
Ager, I. R. 575 (124), *614*
Agnew, J. E. 392 (132), 393 (158), *397*
Aharon-Shalom, E. 448, 450 (544), 451 (577, 578), *505*, *506*, 651 (416), *664*
Ahern, D. 638 (223), *660*
Ah-Kow, G. 411, 412 (148, 149), 413, 430 (149), 438 (149, 429, 433), *497*, *503*, 581 (181), 586 (194), *615*, *616*
Ahlers, D. 456 (642), *507*
Ahlers, H. 652 (438), *665*
Ahlers, K. 455 (627), 489 (1043), *507*, *515*
Ahmed, A. K. S. 690 (112), *842*
Aime, S. 808 (573), 832, 836, 837 (659), *852*, *855*
Ainscough, E. W. 695 (128, 129), 696 (134), 697 (129, 134), 775, 778 (128, 134), *843*
Aires, B. E. 690 (72), *842*
Aitken, G. B. 684–686 (37, 38), *841*
Ajello, E, 491 (1061a), *516*, 525 (81), *554*
Akahori, Y. 321, 323, 324 (59), *329*
Akasaka, K. 622 (38), *657*
Akhmedov, I.M. 623 (62, 63), 624 (80, 82), *657*, *658*
Akishin, P. A. 76 (80), *149*
Akiyama, H. 383 (72d), *395*
Akulin, Y. 462 (724–726), 463 (727–731), 465 (788b), *509*, *510*
Akulinin, O. 410 (125), *497*
Akulinin, O. B. 89, 90, 128 (143), *151*, 383 (73), *395*
Albanov, A. I. 624 (99), *658*

Albeck, M. 244, 252 (4), *283*, 406 (64, 65), 447 (64), 462 (723a), *495*, *509*, 530 (145), *555*, 651 (415), *664*
Albridge, R. G. 694 (125), *843*
Alcock, N. W. 10 (55), *19*, 537 (172), *556*, 686 (46c), *841*
Aldoshina, M. Z. 95 (171), *151*
Aldrich, J. E. 301 (64), *305*
Aleby, S. 100 (204), *152*
Aleksandrowicz, J. 289 (22, 23), 291 (22), *304*
Aleksanyan, V. T. 383 (69a, 69b, 70), *395*
Alekseev, N. 419 (251, 252), 431 (372), 435 (251, 402), 438 (251, 426), *499*, *502*, *503*
Alexander, J. 165 (48), *186*
Alguacil, F. 726, 728 (253), *846*
Ali, M. A. 680, 782, 815 (12e), *840*
Alicino, J. F. 163, 183 (23), *185*
Aliev, A. Yu. 624 (80, 82), *658*
Aliev, I. A. 563 (37), *612*
Al-Janabi, M. V. 732 (284), *847*
Allaway, W. H. 387 (83), *396*
Allen, D. 462 (723b), *509*
Allen, D. W. 364 (69), *367*
Allen, P. Jr. 643 (361), *663*
Allenmark, S. 421 (280), *500*
Allison, J. L. 739 (319), *847*
Allkins, J. R. 690 (67, 68), *841*
Almenningen, A, 72 (57, 60, 67), 87 (124), 89, 90 (142), *149–151*
Almquist, S. 405, 469 (60b), *495*
Almquist, S. O. 220 (80), *240*
Al'pert, M. L. 563 (36), *612*
Al-Rubaic, A. 221, 222, 226–228, 234 (93), *240*
Al-Rubaie, A. 7 (26), *18*, 819 (630), *854*
Al-Rubaie, A. Z. 819 (629), *854*
Alsop, P. A. 779 (467), *850*
Alstad, J. 303 (72, 73), *305*
Altona, C. 139 (314), *154*
Al-Turaihi, M. A. 165 (55), *186*, 521 (33), 522 (36), *553*
Alvey, P. J. 721 (232), *845*
Ambrose, K. R. 375 (26), 381 (49), 392 (26, 49), *394*, *395*, 651 (405–407, 412), *664*
Amel, R. T. 603 (299), *618*
Amero, R. B. 131 (300), 132, 143 (300, 301), 145 (326), *154*, *155*
Ames, A. 348 (14), *366*
Amis, D. 439, 453 (446), *503*
Amosova, S. 407 (93), 493 (1068c), *496*, *516*
Amundsen, F. 467 (821), *511*
Amundsen, F. A. 117, 118 (263), *153*
Anastaseva, A. 402, 420 (18), *494*
Anastaseva, P. 420 (271), *500*
Andersen, C. 407, 412 (94), *496*
Andersen, J. 244, 269, 283 (3), *283*, 448 (545), 449 (566), *505*
Andersen, J. R. 94 (166), *151*
Andersen, K. 449, 450 (561), *505*
Andersen, K. K. 194, 208, 209 (24), 220 (24, 85), *239*, *240*, 675 (33), *677*
Anderson, J. A. 221 (88), *240*
Anderson, J. W. 349–352 (16), *366*
Anderson, P. C. 690 (63), *841*
Anderson, S.J. 164 (36), *185*, 690 (91, 92), 773, 774 (91), 816 (627), *842*, *854*
Ando, A. 389 (112), *396*
Ando, I. 389 (112), *396*
Ando, R. 10 (53), *19*, 550 (273), *558*, 609 (335), *618*
Andrae, K. 741, 751 (328), *847*
Andren, A.W. 183 (136), *187*
Andressen, J. R. 390 (121), 391 (124), *397*
Andrews, G. C. 604 (321), *618*
Andrews, R. W. 184 (158), *188*
Andrieu, C. 480 (939), *513*
Andrzejewski, S. 458 (667), *507*
Angelelli, J. M. 565 (62), *613*
Angelici, R. J. 716 (206), 740 (322), *845*, *847*
Anger, V. 167 (85), *186*
Anisimova, G. F. 164 (39, 45), 165 (56, 57), *186*
Anokhin, Y. N. 392 (142), *397*
Anonimova, I. V. 106 (227), *153*
Ansorge, U. 9 (42), *18*
Anteunis, M. 404 (52), 480 (52, 943), *495*, *513*
Anthoni, U. 68 (31), *148*, 468 (831), *511*
Antsyshkina, A. S. 730 (272), *846*
Aoai, T. 403 (33), *495*, 632, 633 (187, 189), 636, 638 (217), 639 (187), 641 (306–308), *660*, *662*
Apostolescu, M. 488 (1040a), *515*
Aragi, M. 520, 521 (32), *553*
Araki, S. 356 (38), *366*, 762, 763, 784 (394), *849*
Arase, A. 641 (297, 299), *662*
Arbusow, B. A. 547 (251), *558*
Arbuzov, B. 407 (87), 410 (124), *496*, *497*
Arbuzov, B. A. 106 (226, 227), *153*
Archer, S. 456 (643), *507*
Arcoria, A. 412 (153), *497*, 568, 594 (90), *613*
Ardaillou, N. 393 (151), *397*
Arens, J. F. 651 (395), *664*
Arigoni, D. 639 (124), *658*
Arjunan, P. 469 (844), *511*
Arnau, C. 25 (28, 53), 27 (18), *60*, *61*
Arnold, A. 404 (53b), *495*
Arnold, A. P. 624 (81), *658*, 747 (355, 356), *848*
Arpe, H. J. 527, 528 (112), *555*, 653, 654 (460), *665*

Arshadi, M. R. 158, 159 (12), *160*, 275, 283 (82), *285*
Artemenko, M. V. 704, 705 (169), *844*
Artemov, V. 458 (670), *508*
Artyukhin, V. 457 (659), *507*
Artz, R. S. 733, 739, 740 (291), *847*
Asahi, Y. 589 (203), 592, 593 (203, 236), *616*, *617*
Ascherl, A. 452 (590), *506*, 641 (276), *661*
Åse, K. 13 (73), *19*, 360 (56), *367*, 815 (604b), *853*
Asgarouladi, B. 347 (6), *366*
Asgharian, R. 255, 278, 280 (54), *284*, 448 (536), *505*
Ash, P. 106, 107, 124 (228), *153*
Ashby, J. 493 (1069), *516*
Asinger, F. 485 (998), *514*
Aslam, M. 522 (49), *553*
Asmanova, A. 457 (659), *507*
Assadi, F. 255, 278 (49, 56), 279 (49), 280, 281 (56), *284*, 484 (992), *514*
Astruc, D. 765 (403), *849*
Aswanova, A. B. 624 (78), *657*
Atherton, N. M. 321, 323 (58), *329*
Atkins, A. M. 742, 744, 748, 750 (339), *848*
Atovanyan, L. 448 (539), *505*
Attanasi, O. 426, 459 (316), *501*
Attanasio, D. 794 (525), *851*
Atwood, J. L. 348 (13), *366*, 710, 712 (189, 190, 191b), 814 (190, 191b), 819 (191b), *844*
Aue, W. 183 (135), *187*
Aue, W. A. 167 (89, 92), 170 (89), 171 (93), 177 (89, 92), 183 (89), *187*
Aumann, D. 523, 545 (61), *554*
Austad, T. 13 (73), *19*, 360 (56), *367*, 815 (604b), *853*
Awwad, H. K. 387, 389 (91), *396*
Aygen, S. 403 (46b), *495*
Aynsley, E. E. 641 (274), *661*, 728, 731, 732 (264, 265), 734, 735 (265), 774 (442), *846*, *850*
Ayrey, G. 642, 643 (246), *661*
Azerbaev, I. 457 (659), *507*
Azerbaev, I. N. 624 (77–79), *657*

Babichev, F. 458 (682), 460 (706), *508*
Baccaro, H. R. 308 (4), *328*
Back, T. 467 (812), *510*
Back, T. G. 12 (63), 14 (77), *19*, 318 (39, 40), *329*, 626 (138, 139), 643 (314), 647 (138, 139, 333, 336, 345), 648 (344, 354), 649 (334, 335), *659*, *662*, *663*
Backer, H. 404 (51), 408 (102), 444 (490a), 455 (617), 493 (51), *495*, *496*, *504*, *507*
Backer, H. J. 548 (260, 262), *558*, 600 (279, 280), *617*
Backes, C. 409 (120), *497*
Backvall, J. E. 653, 654 (449), *665*
Bäckvall, J.-E. 403 (24), *494*
Bäckvall, J. E. 527 (113, 114), 529, 533 (114), *555*
Bacq, Z. M. 288, 291 (6), *304*
Baddeley, G. 563 (42), *612*
Baddley, W. H. 741 (327, 329), 744 (349), 745 (327, 329, 349), 748 (349), 749 (327, 329), 824, 825 (329), *847*, *848*
Badiello, R. 234 (120), *241*, 288 (14, 15), 289 (15–20), 290 (29–33), 291 (14–16, 29, 31–33, 35, 37), 292 (31, 35–37, 40), 293 (35, 36, 41–43, 45), 294 (41), 296 (42), 297 (29, 42, 48, 49), 298 (42, 56), 299 (29, 60, 61), 300 (61), 301 (61, 66), 302 (35–37), *304*, *305*
Bagchi, U. 771, 772, 774, 775 (434), *850*
Bagdasarov, K. N. 184 (180, 181), *188*
Bagnall, K. W. 288 (4), *304*, 721 (232), *845*
Bahmet, W. 709 (179), *844*
Baider, L. M. 184 (165), *188*
Bailar, J. C. 681 (26), *840*
Bailey, R. A. 736 (306), *847*
Bailly, P. 455 (615, 616), 456 (616), *507*
Bainbridge, D. R. 381, 393 (61), *395*
Bairov, V. V. 563 (36), *612*
Baiwir, M. 13 (68), *19*, 75–77 (76, 77), 86 (116), 91 (156), *149–151*, 191, 199 (7), 203 (7, 59), 204, 205 (7), 207, 208 (63), 210 (7, 59), 211 (59), 212, 214, 215 (69), 216 (70), 217 (7), *239*, *240*, 454 (604), *506*, 563 (41), 579 (150), 580 (150, 155), *612*, *615*
Bajpai, K. K. 731, 732 (280), *846*
Bak, B. 12 (65), *19*, 65 (18, 19), 89, 90 (146, 147), *148*, *151*, 405 (57), 464 (57, 753), *495*, *509*
Bak, C. 447 (524), *504*
Baker, J. W. 563, 565, 608 (39), *612*
Baker, K. L. 706, 707 (174), *844*
Bakthavachalm, V. 439 (450), *503*
Balaes, G. 165 (63), *186*
Balahura, R. J. 700, 701, 730, 737, 760 (153e), *843*
Balashova, J. 413 (177), *498*
Balbach, B. 829, 830, 832, 833 (647), *854*
Baldini, G. 590, 591 (222), *616*
Baldo, M. 86 (117), *150*
Balfe, M. P. 550 (277), *558*, 655 (488, 489), *665*
Bali, A. 548 (268), *558*
Balli, H. 562 (33), 569, 570 (33, 102), *612*, *614*
Balodis, K. 446 (515), *504*
Balog, I. 418 (238), 481 (238, 950), 494 (1074), *499*, *513*, *516*

Balog, I. M. 530 (143), 531 (147, 148), *555*, *556*
Baltch, A. L. 382 (64), *395*
Balucani, D. 438, 439 (435), 471 (874), *503*, *512*
Balykova, T. 439 (445), *503*
Ban, Y. 622 (38), *657*
Band, A. J. 600 (278), *617*
Bandurski, R. S. 388, 389 (103), *396*
Banerjee, A. 93 (162), *151*
Banks, C. H. 365 (73), *367*
Banks, C. K. 650 (367), *663*
Banks, R. 449 (567), *505*
Bannister, E. 691, 692, 695, 697 (114), *842*
Bannou, T. 627 (149), *659*
Barac, G. 288, 291 (6), *304*
Barak, A. J. 180 (118), *187*, 378 (39), *394*
Baran, E. J. 803 (553b), *852*
Baranov, E. 439 (441), *503*
Baranov, S. 438 (426), 458 (668, 670), *503*, *508*
Barany, H. 458 (677), *508*
Barashenkov, G. 488 (1037), *515*
Barashenkov, G. G. 541 (209–211, 212b), 542 (212b), *557*
Barber, M. 245 (10), *283*
Bard, A. J. 621, 622 (3), *656*
Bardi, R. 118, 119 (262), *153*
Bares, J. E. 565, 604 (61), *613*
Bargon, J. 407 (82b), *496*
Baringer, W. 5 (15), *18*, 457 (649), 485 (994), *507*, *514*
Bark, L. S. 771 (426), *849*
Barker, G. K. 349 (16, 17), 350–352 (16), *366*
Barker, T. 405 (59), *495*
Barker, T. V. 525 (85), *554*
Barmette, W. 403 (46a), *495*
Barnard, D. 642, 643 (246), *661*, 793, 794 (515), *851*
Barnes, D. S. 157–159 (6), *160*
Barnes, J. C. 113 (239), *153*, 706, 707 (175, 176), *844*
Barnes, R. M. 165, 167 (73), *186*
Barnette, W. E. 16 (88), *19*, 627 (150), 636 (220), 638 (220, 224), 639 (230), *659*, *660*
Barni, E. 460 (703), *508*, 573, 574 (119), *614*
Barnum, C. 621, 622 (8), *656*
Baroni, A. 650 (381), *663*
Barr, E. W. 183 (142), *188*
Barraclough, C. G. 798, 800 (544), *852*
Barrett, A. G. M. 13 (69), *19*, 821 (633), *854*
Barrett, G. F. 563, 565, 608 (39), *612*
Barsh, M. K. 577, 601 (139), *614*
Bartell, L. S. 72, 76, 78 (62), 82 (98), 87 (62), 89 (140, 141), *149–151*
Barth, H. 575 (128), *614*
Barth, M. 464 (762), *509*
Barthelat, J. C. 33 (81), *62*
Bartholomaus, R. C. 238 (145), *241*, 382, 392 (66a), *395*
Bartmess, J. E. 565, 604 (61), *613*
Barton, D. 462 (717), 467 (812), *508*, *510*
Barton, D. H. R. 10 (52), 12 (63), 13 (69), 14 (75, 76, 78), 16 (90), *19*, 537 (178, 179), *556*, 643 (314, 315), 644 (313, 315–317), 645 (319, 320), 647 (325, 337–339, 347), 648 (321, 322, 341–343, 346–348), *662*, *663*, 821 (633), 822 (635), *854*
Barton, R. M. 381 (56), *395*
Barton, T. 409 (114, 116), *496*
Barton, T. J. 265 (70), *285*, 378, 386, 387 (38a), *394*
Basch, H. 25 (22, 36), 55 (93), *60–62*
Basmadjian, G. P. 370 (4), 371 (11), 372 (4), 375, 392 (27), 393 (4), *393*, *394*, 651 (409, 410), *664*
Bastani, M. 442 (475), *503*
Bastiansen, O. 72 (67), 87 (124), 139 (312), *149*, *150*, *154*
Battelle, L. 128, 138, 139 (291), *154*
Battioni, J. P. 715, 719 (218), 720 (228), *845*
Battistoni, P. 426, 459 (316), *501*
Baudler, M. 16 (96), *20*, 362 (63), *367*
Baudy, M. 458 (673), *508*
Bauer, C. 804, 805, 832, 833 (556a, 556b), *852*
Bauer, H. 458 (680), 486 (1009), *508*, *514*
Bauer, S. H. 83 (103), 100, 101 (194), *150*, *152*
Baughman, R. H. 332, 334 (3), *341*
Baumane, L. K. 184 (165), *188*
Bausor, H. W. 527 (117), *555*
Bautina, I. V. 642 (291), *662*
Bayet, P. 8 (37), *18*, 603 (302), *618*
Bayly, R. J. 371 (9), 373 (20), 380 (9), 393 (9, 20), *394*
Beagley, B. 72 (58), 83, 84 (104), *149*, *150*
Beardwood, P. 742, 748 (340), 806 (561, 562), 832 (562), *848*, *852*
Beath, O. A. 288 (2), *304*
Beatty, H. A. 6 (25), *18*
Beaulieu, P. 403 (22b), *494*
Beaulieu, P. L. 635 (201, 203, 204, 206), 636, 637 (213), *660*
Beaulieu, W. B. 771, 772, 774, 775 (434), *850*
Bechgaard, K. 17 (102, 103), *20*, 94 (166, 167), *151*, 244 (3), 247 (28), 269, 283 (3), *283*, *284*, 448 (542, 543, 545, 549), 449 (560, 561, 566–568), 450 (542, 543, 549, 561, 569, 570), *505*, *506*, 576 (134), *614*
Beck, A. K. 603 (301), *618*, 638 (227), 651 (397), *660*, *664*

Beck, W. 103 (223), *152*, 703 (167), 704 (167, 168), 741 (323), 743 (323, 342), 748 (323), *844*, *847*, *848*
Becker, E. D. 191 (9), *239*
Becker, W. 464, 465 (765), *509*
Beckett, R. 800 (546), *852*
Beckey, H. D. 838 (670a), *855*
Beecher, J. F. 70 (38), *148*
Beelitz, K. 435 (403–405), 478 (921), *502*, *513*
Beer, R. 467 (820a, 824), *511*
Beer, R. J. S. 118 (265), *153*
Behagel, O. 428 (334), 486 (1001), *501*, *514*
Behaghel, O. 626 (144), 629 (164), 639 (144), *659*
Behan, J. 402 (17), *494*
Behnam, E. 255, 268, 278, 279 (51), *284*, 429 (359), *501*
Behrendt, S. 457 (663), *507*
Behrens, U. 102, 106 (222), *152*, 825, 827 (637), *854*
Behringer, H. 119 (271), *154*
Beijer, B. 455, 461 (631), *507*
Beinert, H. 238 (145), *241*, 382, 392 (66a), *395*
Belchetz, L. 488 (1032), *515*
Belen'kii, G. 405 (62a), *495*
Belen'kii, L. I. 582 (177), *615*
Bell, H. C. 528 (128), *555*
Bell, M. C. 387 (85, 86), *396*
Bellinger, N. 428 (348, 349), 471 (865, 866), 472 (877, 878), 490 (1051), *501*, *512*, *515*
Bellitto, C. 794 (525), *851*
Belyaeva, V. K. 700 (155), 789 (507), 796, 797 (538), *843*, *851*, *852*
Bem, E. W. 163 (28), *185*
Bender, S. 406 (63), 447 (528, 529), 451 (528), *495*, *505*
Bendix, D. 519 (17a), *553*
Benedetti, A. 262 (66, 67), 263 (67), *284*
Benes, J. 165 (76), 167 (76, 86), 168 (86), 169 (76, 86), 170 (76), *186*, 393 (145), *397*
Benfaremo, N. 334, 335 (16), *341*
Bengels, D. 528, 532, 533 (127), *555*
Bennett, F. 455, 457 (626), *507*
Benson, I. B. 750 (361), *848*
Benson, J. V. Jr. 179, 180 (113), *187*
Benson, S. W. 157 (3), 159 (3, 13), *160*
Bentley, M. D. 571 (109), *614*
Bentley, T. W. 248 (30), *284*
Benton, A. J. 808, 813 (578), *853*
Berding, H. 485 (998), *514*
Berg, C. 12 (65), *19*, 405 (57), 449, 450 (561), 464 (57), *495*, *505*
Berg, U. 13 (71), *19*
Bergen, H. A. 696, 697, 775, 778 (134), *843*
Bergenthal, A. 407, 409, 414 (97b), *496*
Bergesen, K. 361 (62), *367*
Bergman, A. 15 (86), *19*
Bergman, J. 135, 143 (304), *154*, 403 (23), 406 (65), 419 (244), 427 (328), 442 (328, 461, 462), 460 (701), 462 (723a), 483 (972), 487 (1021), 492 (701), *494*, *495*, *499*, *501*, *503*, *508*, *509*, *514*, *515*, 519 (18), 520 (21, 22), 522 (18), 525 (95, 218), 527 (114, 115), 528 (136), 529 (114, 137, 141), 530 (136), 531 (149, 151), 532 (157, 158), 533 (114), 534 (157, 158), 543 (151), 544 (115, 219), *553–557*, 585, 590, 596 (187), *616*, 651 (415), 653 (448, 450), 654 (462, 463), 655 (462, 463, 478, 479), *664*, *665*
Bergson, G. 184 (153), *188*, 337 (30), 338 (34), *341*, 444 (488, 489, 491, 493, 494), 480 (941), 484 (987), 493 (1067, 1068a), *504*, *513*, *514*, *516*
Bergstrom, F. W. 520 (67), *554*
Berkowitz, D. S. 25 (40), *61*
Berkowitz, J. 832 (666a, 666b), *855*
Berlin, K. 469 (844, 846), *511*
Berlin, K. D. 253 (44), *284*
Bernal, I. 710, 712 (189, 190, 191b), 811 (584), 814 (190, 191b), 819 (191b), *844*, *853*
Bernier, F. 101 (199), *152*
Berress, L. 575 (128), *614*
Berry, F. 489 (1049), *515*
Berry, F. J. 98 (190), *152*, 225, 226, 228, 229, 233 (103), *241*, 526 (102), *555*
Berry, J. 788, 791, 792 (500), *851*
Bersin, T. 537 (180), *556*
Bertelli, E. 759–761 (373), *848*
Berthier, G. 321, 323 (61), *329*
Berti, M. 163 (34), *185*
Bertin, D. 410 (128, 129), 419 (241), *497*, *499*
Bertini, D. 164 (43), *186*
Bertini, I. 732 (282), *847*
Bertini, V. 454 (607), 466 (797–800), 467 (815), *506*, *510*, 583 (179), 584 (182, 206), 595 (243), 597 (252–254), *615–617*
Bertleff, W. 767, 768 (414), *849*
Bertz, S. 448, 450 (544), *505*
Besnard, M. 297 (52), *305*
Beurskens, P. T. 102 (218), *152*, 794, 795 (530), 796 (534), *852*
Beyer, H. 455 (632, 633), 456 (632–635), *507*
Beyer, L. 784, 786 (478, 479), 788 (501), 792 (501, 509), 795 (479), *850*, *851*
Beyer, W. 162, 165 (13), *185*
Bezoari, M. 415 (210), *498*
Bezzi, S. 445 (503), *504*
Bhandal, B. S. 814, 824 (625), *854*
Bhargava, S. K. 114, 115 (246), *153*, 710, 712 (187, 188), 713 (188), 771 (437), 773 (437–439), 774 (437–439, 446), 775 (437,

438), 776 (446), *844*, *850*
Bhasim, K. K. 522 (45), *553*
Bhattacharya, R. N. 681 (21, 22), *840*
Bhatti, M. M. 114, 115 (246), *153*, 710, 712 (187), 774, 776 (446), *844*, *850*
Bianchini, C. 765, 766 (409), 788, 791 (498), *849*, *851*
Biellmann, J. F. 604 (320), *618*
Bielska, M. T. 648 (341), *663*
Bierschenk, T. R. 355 (31), *366*
Biezais, A. 338 (34), *341*, 444 (493), 493 (1067, 1068a), *504*, *516*
Biezais-Zirnis, A. 476, 492 (907), *512*
Bighi, C. 600 (282), *618*, 621, 622 (5), *656*
Bigoli, F. 794, 795 (529), *852*
Bilderback, J. A. 522 (37), *553*
Bildstein, B. 545 (229), *557*
Bilinski, S. 456 (636–638), 489 (637), *507*
Billmers, J. M. 9 (48), *19*, 624 (148), *659*
Binkley, J. S. 25 (44, 48, 49, 62, 64), 30 (48, 62, 64), 33 (49), 34 (48, 49, 62), 54 (48, 62), 56, 57 (48), *61*
Birchall, T. 196, 199, 205 (33), 217 (74), *239*, *240*
Bird, C. 401 (8a), 465 (792), *494*, *510*
Bird, C. W. 586 (212), *616*
Bird, M. L. 546 (248), *558*, 639, 650 (141), *659*
Bird, P. H. 522 (47), *553*
Bird, R. B. 201 (44), *239*
Birknes, B. 117 (252), *153*
Birner, B. 784, 786, 795 (479), *851*
Bisby, R H. 301 (64, 65), *305*
Bjerre, C. 12 (65), *19*, 405, 464 (57), *495*
Black, D. 403 (47), *495*
Blacklaws, I. M. 754–756 (366), *848*
Blackman, G. L. 90 (150), *151*
Blackman, L. C. F. 641 (260, 265), *661*
Bladon, P. 465 (779), *510*
Blaettner, K. 487, 492 (1029), *515*
Bláha, K. 673 (26), *677*
Blair, L. K. 201 (43), *239*
Blair, W. R. 165, 167, 169 (70), *186*
Blake, A. J. 697, 778 (140c), *843*
Blanpain, P. 320 (52), *329*
Blanquet, G. 64 (10), *148*
Blarzino, C. 181 (121), *187*
Blau, M. 374 (22), 380 (47), 387 (81), 388, 389 (101), *394–396*
Blecher, A. 357 (41, 44), 358 (44, 47, 48), *366*
Blechschmitt, K. 768, 770, 829, 830 (421), *849*
Blinokhvatov, A. 478 (923), *513*
Bloch, A. 17 (104), *20*, 448 (542, 543, 546, 548), 449 (567, 568), 450 (542, 543, 546, 548), 451 (579), *505*, *506*, 651 (417), *664*
Bloch, A. N. 94 (167, 168), *151*, 247 (28), *284*
Block, B. 741 (324, 325), 810 (583), *847*, *853*
Block, E. 11 (57), *19*, 88 (133), *150*
Block, H. D. 344 (2), 347 (11), *365*, *366*, 621, 622 (14), *656*
Blom, R. 72, 90 (65), *149*
Bloodworth, A. J. 646 (331), *662*
Bloom, S. M. 690 (9b), *839*
Blottner, A. 393 (146), *397*
Blount, J. F. 636–638 (210), *660*
Blukis, U. 72, 90 (66), *149*
Blurock, E. S. 25, 29, 30, 33 (71), *61*
Bobadilla, M. 732, 738 (286), *847*
Bobylev, V. 444 (496), *504*
Bobylev, V. A. 624 (85, 91), *658*
Bochacheva, L. 430 (367), *501*
Bochkarev, M. N. 183 (143), *188*, 353 (19), 356 (32, 33), *366*, 716 (202), *845*
Bochmann, G. 546 (247), 548 (247, 264), 549 (247), *558*, 721, 722 (239), *845*
Bock, H. 11 (57), *19*, 78 (90), 86 (118), *150*, 238 (142), *241*, 245 (14, 15), *283*, 403 (46b), *495*
Bodart, F. 163 (35), *185*
Bogan, L. E. 808 (575), 832, 835, 837, 838 (661), *852*, *855*
Bogdanov, V. 430 (366), *501*
Bogert, M. 407, 412 (94), 458 (678, 679), 459 (678), 460 (702), *496*, *508*
Bogert, M. T. 586 (210), *616*
Boggs, J. E. 25 (43, 60), *61*, 126 (287), 139 (312), *154*
Böhm, J. 427 (327), *501*
Böhnke, F. 483 (982), *514*, 521 (34), *553*
Boiko, V. N. 622 (34), *657*
Boiko, Yu. A. 651 (394), *664*
Boldeskul, I. E. 815 (601k), *853*
Bolinger, C. M. 798, 799 (542), 806 (565a), 811 (542), *852*
Bologa, M. 277 (88), *285*
Bologa, O. A. 732 (290), *847*
Boloshchuk, V. G. 565 (53), *612*
Bonamico, M. 102–105 (205), *152*, 794 (520), *851*
Bonati, F. 754 (364), *848*
Bondi, A. 78 (89), *150*
Bonfanti, L. 584 (182), *615*
Bongartz, A. 357, 358 (45), *366*
Bonora, S. 293 (43), 298 (56), *305*
Boord, C. E. 527 (118), *555*
Boorman, P. M. 695, 697 (130), *843*
Booth, G. 680 (9d), *840*
Bor, G. 811 (595), *853*
Borch, G. 68 (31), *148*
Bordeli, R. S. 245 (10), *283*
Bordwell, F. G. 565, 604 (61), *613*

Borek, B. 801 (547), *852*
Bornmann, P. 650 (363), *663*
Bornstein, J. 427 (321), *501*, 592 (230), *616*
Borovikov, Y. Y. 815 (601k), *853*
Borsdorf, R. 457 (663), *507*
Bortnikov, G. N. 183 (143), *188*
Bory, S. 419 (242), *499*
Bos, H. 455 (617), *507*
Bossche, G. v. d. 670 (15), *676*
Bossert, F. 473 (891), *512*
Böttcher, R. 786, 789 (504b), *851*
Bouchier, I. A. D. 392 (132), 393 (158), *397*
Bouman, T. D. 675 (35), *677*
Bourguignon, J. 430 (362), *501*
Bowater, I. C. 106 (224), *152*
Bowden, K. 571 (108), *614*
Bowie, E. J. W. 381, 393 (52), *395*
Bowles, A. 381, 393 (58), *395*
Boyd, G. S. 651 (408), *664*
Bradford Camp, R. Jr. 601, 602 (289), *618*
Bradt, W. E. 520 (68), *554*
Brady, K. R. 179 (112), *187*
Brahler, G. 238 (142), *241*
Bramwell, F. 488 (1042), *515*
Branch, G. E. K. 575 (129), *614*
Branchi, G. 770 (435), *850*
Brandon, D. 771 (426), *849*
Brandsma, L. 355 (27), *366*, 409, 414 (118b), 423 (290b), 447 (527), 481 (953–956, 958), 491 (958), *497*, *500*, *505*, *513*, 622 (41), 651 (395, 396), *657*, *664*
Brandt, C. A. 621, 622 (16), *656*
Braumann, J. I. 201 (43), *239*
Braverman, S. 433 (391), *502*
Bravermann, S. 408 (104), *496*
Braye, E. 407, 409, 414 (97a), *496*
Braye, E. H. 709 (181), *844*
Breccia, A. 234 (120), *241*, 288 (15), 289 (15, 17), 290 (29), 291 (15, 29), 292 (34, 40), 297 (29, 48), 299 (29), *304*, *305*
Bredas, J. L. 332, 334 (3), *341*
Bregadze, V. I. 347 (7), *366*
Bregovec, I. 177 (125), *187*
Bremer, B. 332 (9), 333 (9, 13), 334 (13), *341*
Bremer, J. 380, 388 (44), *395*
Bren, V. A. 522 (44), *553*
Bren', V.A. 574 (120, 121), 577, 601, 603 (138), *614*
Brewster, A. G. 643 (314), 644 (313, 316, 317), *662*
Bridger, H. 487, 488 (1027), *515*
Bridger, H. J. 332 (5), *341*
Briggs, W. S. 671 (20b), *677*
Brinckman, F. E. 162 (15), 165, 167, 169 (70), *185*, *186*
Brintzinger, H. 527, 528, 533 (119), *555*
Briscoe, H. 407 (78, 79), *496*
Britnell, D. 695, 697 (131), *843*
Britten-Kelly, M. 467 (812), *510*
Britten-Kelly, M. R. 12 (63), *19*
Broadhurst, P. V. 747, 753 (358), 811 (593), *848*, *853*
Brodersen, K. 109 (236, 237), 110 (237), *153*
Brodie, A. M. 693 (122), 695 (122, 128, 129), 696 (122, 134), 697 (122, 129, 134), 775, 778 (128, 134), *843*
Brody, K. R. 379 (41), *394*
Bromer, B. 336 (25), *341*
Brook, J. W. 643 (361), *663*
Brooker, L. 421 (277), 455 (611), 480 (940), 481 (952a), *500*, *506*, *513*
Brooker, L. G. S. 315 (24), *328*
Brothers, P. J. 715, 717, 767 (215), *845*
Brown, A. J. 335 (22), *341*
Brown, D. 721 (232), *845*
Brown, D. G. 387, 388 (84), *396*
Brown, D. H. 236 (133), *241*, 697 (136), 815 (601n), *843*, *853*
Brown, D. S. 85 (107), *150*, 164 (36), *185*
Brown, H. C. 561 (17), 563 (34), 565 (54), 593 (239), *612*, *617*
Brown, J. R. 337 (31), *341*
Brown, K. A. 696, 697, 775, 778 (134), *843*
Brown, K. L. 829 (650a), *854*
Brown, N. 465 (779), *510*
Brown, P. 392 (130), *397*
Brown, R. 410 (126), *497*
Brown, R. D. 51, 57, 58 (94), *62*, 89 (144, 145), 90 (145, 150), 106 (224), *151*, *152*, 569, 581 (95), *613*
Brown, T. L. 736 (304), *847*
Brownlee, R. T. 565 (62), *613*
Brownlee, R. T. C. 202 (54b), 204 (56), *240*, 562, 571 (24), *612*
Bruce, M. I. 680 (9c), *840*
Brucker, W. 290, 291 (27, 28), *304*
Brun, P. 565 (64), *613*
Brunner, H. 804, 805 (555), *852*
Bruton, G. 622 (39), *657*
Bryant, H. 465 (771), *510*
Bryant, M. W. R. 86 (109), *150*
Bryce, M. 453 (592–594), 459 (684), 465 (592–594), *506*, *508*
Bryce, M. R. 91 (155), *151*, 700 (147), *843*
Buchanan, A. S. 569, 581 (95), *613*
Buchardt, O. 308 (4), *328*
Buchikhin, E. P. 577, 609 (142), *615*
Buchta, E. 418 (237), *499*
Buck, H. 494 (1071), *516*
Buck, H. M. 624 (101), *658*
Buckler, K. U. 221, 222 (90), *240*
Buckley, D. J. 632 (175), *659*
Bucy, W. E. 70, 73 (42), *149*, 624 (90), *658*

Budzikiewicz, H. 248, 249 (32), 252 (42), *284*, 418 (234), *499*
Bugge, A. 430 (363), *501*
Buhl, H. 464 (762), *509*
Buhlmayer, P. 608 (330), *618*
Buina, N. A. 815 (601f), *853*
Bukharin, B. N. 184 (180, 181), *188*
Bulka, E. 290, 291 (27), *304*, 452 (585, 586), 455 (585, 608, 627, 628, 632, 633), 456 (585, 586, 632–635, 642), 458 (585), 466 (807, 809), 467 (809), 489 (1043), *506*, *507*, *510*, *515*, 583, 586 (204), *616*
Bull, J. R. 623 (58), *657*
Bulycheva, T. V. 786, 787 (489), *851*
Bunnenberg, E. 670, 671 (18b), *677*
Buno, N. Jr. 521 (31), *553*
Bunting, T. G. 699, 700 (145b), *843*
Burden, F. R. 51, 57, 58 (94), *62*, 89 (144), 90 (150), 106 (224), *151*, *152*
Buren, M. S. 165, 177 (74), *186*
Bürer, Th. 671 (20a), *677*
Burger, K. 90 (153), *151*, 463 (732–735, 736a), 484 (733, 736a), *509*
Burgess, H. 468 (830), *511*, 550 (279), *558*, 654 (477), *665*
Burgett, C. A. 183 (144), *188*
Burgiel, M. 487 (1019), *515*
Burk, R. F. 387, 388 (84), 391 (125), *396*, *397*
Burke, S. 629 (171), *659*
Burkett, K. 102, 103 (214), *152*
Burkhardt, E. W. 739 (318), *847*
Burlant, W. 428 (336), 462 (720), *501*, *509*, 642 (282), *661*, 705, 706 (170), *844*
Burlont, W. 521 (35), *553*
Burmeister, J. L. 697 (138), 700, 701, 730 (153a, 153b), 732 (283–285, 288), 733 (291, 292a, 292b, 293), 734 (138), 735 (283), 736 (283, 288, 302–305), 737 (153a, 153b, 283, 310a, 310b), 738 (283, 293, 315), 739 (291, 316–319), 740 (291, 310a, 310b, 320), 815 (302), 825 (305), *843*, *847*
Burns, G. 25 (13), *60*
Burns, R. C. 221, 234 (87), *240*
Burns, R. F. 680, 782, 815 (12c), *840*
Burow, M. 788, 791, 792 (500), *851*
Burstall, F. 404 (50), 418 (230, 233), 444 (50), 468 (828, 832), 480 (230), 489 (832), 491, 492 (828), *495*, *499*, *511*
Burstall, F. H. 333, 335 (14), 337 (28, 29), *341*, 522 (39), *553*, 654 (475), *665*
Burton, A. 319 (47), *329*, 624 (106), *658*
Burton, C. E. 387 (78), *396*
Busetti, V. 96 (176, 177, 181), 118, 119 (262), *151–153*
Bushkov, A. Y. 522, 523 (42), *553*
Bushkov, A. Ya. 518, 520 (4), *553*, 560 (5), 565 (49, 51), 566 (51, 69), 567 (51, 69, 84), 574 (49, 69), *611–613*
Bushkov, A. Yu. 523 (66), *554*
Buso, G. 163 (34), *185*
Buss, B. 567 (79), *613*
Busse, P. J. 249, 251 (40), *284*, 651 (399), *664*
Butler, A. R. 572, 597 (112), *614*
Butler, G. W. 178–180 (104), *187*
Butler, I. S. 715 (212, 213, 219–222), 716 (203, 207, 212, 213), 717 (212, 213, 219, 220), 720 (212, 213, 220–222), 767 (213), *845*
Buu-Hoi, N. P. 255 (47), *284*
Buu-Hoï, N.-P. 413 (167), 420 (264), 421 (278), 426 (264, 313), 428 (345, 346), 438, 439 (435), 465 (770), 479 (931), 486, 487 (1007), *497*, *499–501*, *503*, *510*, *513*, *514*
Buys, H. R. 139 (314), *154*
Buzilova, S. R, 651 (403, 404), *664*
Bycroft, B. M. 183 (129), *187*
Byron, D. J. 573 (114), *614*
Bystrov, V. F. 565 (53), 566 (70), *612*, *613*
Bzesovsky, V. M. 202, 203 (53), *240*
Bzhezovskii, V. M. 563 (36–38), *612*

Caccamese, S. 416 (211), *498*
Cagniant, D. 407 (98), 409 (119, 120), 421 (276), 428 (348, 349), 429 (351), 436 (409), 439 (438, 439), 440 (119), 471 (866), 472 (877, 878), 480 (939), 490 (1051), *496*, *497*, *500–503*, *512*, *513*, *515*, 586 (191), *616*
Cagniant, P. 407 (98–100), 409 (119, 120), 419 (258, 259, 262), 421 (276), 425 (310), 428 (348, 349), 429 (354, 355), 432 (384, 385), 434 (354), 435 (406, 407), 436 (355, 408–410), 439 (438, 439), 440 (119), 471 (100, 865, 866, 868, 870), 472 (868, 877, 878), 490 (870, 1051), *496*, *497*, *499–503*, *512*, *515*, 585 (189), 586 (191), *616*
Cahours, A. 5, 7 (12), *18*
Cai, R. 723–725 (248, 249), 814, 823 (624), *846*, *854*
Caillet, P. 65 (14), *148*
Calabro, D. C. 738 (315), *847*
Calderazzo, F. 710, 712 (189, 190, 191b), 714 (195), 814 (190, 191b), 819 (191b), *844*, *845*
Caldwell, K. A. 171, 179, 181 (115), *187*, 288 (9), *304*
Calhoun, H. P. 546 (242), *557*
Calingaert, G. 6 (25), *18*
Callahan, A. P. 375 (26), 381 (49), 392 (26, 49), *394*, *395*, 651 (405–407, 412), *664*
Callear, A. 402 (12a, 12b), *494*
Callomon, J. H. 51, 57, 58 (94), *62*

Calò, V. 402 (21), *494*
Calusaru, A. 184 (159, 160), *188*
Calvo, C. 67–69 (33), *148*, 685, 686 (45), *841*
Camassei, F. D. 794 (521), *851*
Camasseto, J. V. 6 (19), *18*
Cambie, R. C. 641 (295), *662*
Cambon, A. 249, 251 (39), *284*
Cameron, T. S. 114 (245), 131 (300), 132, 143 (300, 301), 145 (326), *153–155*, 709, 710, 712, 819 (184), *844*
Campana, C. F. 714 (198), *845*
Campbell, C. 453, 465 (592), 483 (981), *506*, *514*
Campbell, I. 482 (968), *514*
Campbell, T. 401 (2), 428 (335), *494*, *501*
Campbell, T. W. 166, 167, 171, 177 (78), *186*, 565 (56), *612*, 629 (159), *659*
Campiglio, A. 164 (41), *186*
Campona, C. F. 808 (569), *852*
Campos, H. V. de 641 (267), *661*
Campsteyn, H. 77 (85), 91 (154), 92, 93 (164), *149*, *151*, 454 (606), *506*
Canada, L. G. 710, 712, 814, 819 (191b), *844*
Candrini, G. 759, 760 (387), *849*
Caneghem, P. van 184 (161), *188*
Cantoni, G. L. 380 (46), 388 (102a), *395*, *396*
Canty, A. 404 (53b), *495*
Capeller, C. 452 (590), *506*
Capeller, L. 641 (276), *661*
Caplier, I. 407, 409, 414 (97a), *496*, 709 (181), *844*
Caplin, G. A. 579 (152), *615*
Capozzi, G. 403 (27, 42, 43), *494*, *495*
Capron, B. 430 (368, 369, 371), 434 (369), *501*, *502*
Carazzolo, G. 96 (176–178), *151*
Carbarino, G. 675 (31), *677*
Cardinale, E. 183 (134), *187*
Cardock, S. 348 (15), *366*
Cardoso, J. M. 648 (341), *663*
Carlowitz, M. von 70, 80 (48), *149*
Carlsen, L. 456, 458 (640), *507*
Carlsson, R. 520 (22), *553*, 655 (478), *665*
Carmack, M. 465 (784), *510*, 670, 671 (17a, 17f), *677*
Carneiro, K. 450 (570), *506*
Caroli, J. 393 (147), *397*
Carpenter, J. H. 67 (27), *148*
Carr, A. 465 (781), *510*
Carr, E. A. Jr. 387 (92), *396*
Carr, S. W. 194 (25), *239*
Carrillo, L. 374 (23), *394*
Carroll, P. 406 (65), 451 (578), 462 (723a), *495*, *506*, *509*, 651 (415), *664*
Carruthers, T. 448, 450 (546), *505*
Cartaxo, E. Q. 690 (75), *842*
Carter, R. O. 803 (553b), *852*
Carty, A. J. 102 (207), *152*, 747 (355, 356), *848*
Caruso, A. J. 623 (67), *657*
Casey, J. P. 670, 671 (17b), *677*
Castellanos, L. T. 542 (215), *557*
Cataliotti, F. 392 (128), *397*
Cataliotti, R. 378, 383 (37a, 37b), *394*
Catheline, D. 765 (403), *849*
Catlin, W. E. 572 (113), *614*
Causley, G. C. 245, 247 (12), *283*
Cauty, A. J. 624 (81), *658*
Cava, M. 334, 335 (16), *341*, 402 (20), 406 (64, 65), 426 (318), 432 (382), 442 (474), 447 (64), 448 (535), 449 (551–553, 556), 450 (551, 553, 556), 451 (578), 458 (671), 462 (553, 723a), 463 (556), 484 (671), *494*, *495*, *501–503*, *505*, *506*, *508*, *509*
Cava, M. P. 6 (20, 22), 10 (50, 54), *18*, *19*, 537 (174, 175), 550, 551 (175, 282), *556*, *558*, 591 (229), *616*, 624 (95), 651 (415, 418, 419), 652 (433, 440–443), 656 (492), *658*, *664*, *665*
Cavalca, L. 770 (435), *850*
Cavallini, D. 181 (120, 122), *187*, 457 (655), *507*, 622 (36), *657*
Cecchetti, E. 289 (19, 20), *304*
Cecconi, F. 811 (598), 813 (598, 599), *853*
Ceder, O. 455, 461 (631), *507*
Cederbaum, L. S. 22 (3), 25 (57), *60*, *61*
Cederlund, B. 246 (20, 22), *283*, *284*, 415 (200, 203), *498*, 599 (268, 269), *617*
Celander, D. R. 381 (50, 51), 393 (50, 51, 160), *395*, *397*
Celander, E. 381 (50, 51), 393 (50, 51, 160), *395*, *397*
Celotti, M. 221–223, 225, 229 (91), *240*
Cenini, S. 754 (364, 365), 755, 756 (365), *848*
Cense, J. M. 580 (156), 604 (318), *615*, *618*
Cernik, M. 547 (257), *558*, 650 (371), *663*
Cerník, M. 650 (366), *663*
Cervone, E. 794 (521), *851*
Cetini, G. 811 (591), 832 (591, 655–657, 662, 663), 835 (655), 837 (662), 838 (663), 839 (591), *853*, *855*
Chabrie, C. 462 (715), *508*
Chacko, E. 427 (321), *501*, 592 (230), *616*
Chadha, R. K. 225, 228, 229, 234 (104), *241*, 522 (45), 541, 542 (214), *553*, *557*, 814 (610, 625), 818 (610), 824 (625), *854*
Chakraborti, C. L. 164 (46), 165 (61), *186*
Chakraborti, D. 162 (13, 14), 165 (13), *185*
Chakravorty, A. 781 (471), *850*
Challenger, F. 546 (248), *558*, 639, 650 (141), *659*, 731 (279), *846*
Chambers, J. 449, 450 (554), *505*
Chambers, J. C. 165 (52), *186*
Chambers, V. E. M. 371 (9), 373 (19, 20),

380 (9), 393 (9, 19, 20), *394*
Chan, C. 145 (326), *155*
Chan, L. 519 (14), *553*
Chan, S. I. 87 (127), *150*
Chan, T. 402 (15), *494*
Chan, W. Y. 676 (36b), *677*
Chan, Y. K. 165, 167–169 (65, 66), *186*
Chance, R. R. 332, 334 (3), *341*
Chandler, C. 465 (793), *510*
Chandler, G. S. 25 (58), *61*
Chang, M.-M. Y. 107–109, 124 (234), *153*
Chang, M. Y. 261 (62), *284*
Chapelle, S. 7 (26), *18*, 206, 210, 211 (61), 221 (61, 92, 93), 222 (92, 93), 226–228, 234 (93), *240*, 819 (630), *854*
Chaplin, C. A. 550 (277), *558*, 655 (488), *665*
Chapman, D. 697 (135a), *843*
Chapman, N. B. 561, 563, 565 (10, 11), 568 (11), 571, 579 (10, 11), 597 (11), *611*
Charov, A. I. 353 (19), *366*
Charton, M. 202 (55), *240*, 562 (30), 565 (57, 58), 568, 569 (30), *612*, *613*
Chastain, B. 465 (771), *510*
Chatrousse, A. 411 (142–145), *497*
Chatrousse, A. P. 590 (223, 224), 591 (224, 225), *616*
Chatt, J. 690 (64–66, 81–86), *841*, *842*
Chaudhry, S. C. 695 (132), *843*
Chaudhuri, M. K. 281 (90–92), *285*, 748 (360), 832, 836 (658), *848*, *855*
Chauvin, P. 413 (188, 189), 438 (189), 455 (188, 612, 613), 460 (613), *498*, *506*, 584 (180), *615*
Cheeseman, G. 401 (8a), 465 (792), 487 (1015), *494*, *510*, *514*
Cheeseman, G. W. H. 586 (212), *616*
Chellappa, K. L. 316 (31), *329*
Chen, C.-C. 392 (131), *397*
Chen, C.-S. 391 (127), *397*
Chen, G. C. 365 (73), *367*
Chen, H.-W. 102 (220), *152*
Chen, H. W. 794, 795 (516, 517), *851*
Chen, Y. 458, 459 (678), *508*
Chen, Y.-Y. 638 (226), *660*
Chen-Fu Hsu 588 (202), 589 (203), 592, 593 (203, 235, 236), *616*, *617*
Cheng, C. P. 736 (304), *847*
Cheng, F. H. F. 393 (148), *397*
Cheng, K. L. 163 (31), *185*
Cheng, P.-T. 127 (288), *154*
Cheremisina, I. M. 787 (490), *851*
Cherkasova, T. G. 814, 820 (620), *854*
Cherkinskaya, M. L. 574 (120–122), 575 (122), *614*
Chernaya, N. M. 787 (488), *851*
Chernish, Yu. E. 574, 575 (122), *614*
Cheshko, R. S. 575 (125), *614*
Cheung, T. T. P. 701, 705 (158b), *844*
Chi, P. B. 364 (70), *367*
Chia, L. Y. 814 (614), 824, 826, 834 (641), *854*
Chianelli, D. 621–623 (6), *656*
Chiang, J. F. 87 (131), *150*
Chiang, L. 449 (558b), *505*
Chiang, L.-Y. 446 (523), 448, 450 (548), *504*, *505*
Chiari, B. 689 (60), *841*
Chierici, L. 411 (139, 141), 414 (141), 428 (341), *497*, *501*, 589, 591 (216), 606 (326), *616*, *618*, 642 (278), *661*
Chiesi, A. 295 (47), *305*, 683 (33), *841*
Chiesi Villa, A. 84, 85 (105), *150*
Chikamatsu, K. 652 (439), *665*
Chin, M. F. 321, 323, 324 (56), *329*
Chirazi, A. 461 (711), *508*
Chittattu, G. 636 (207, 215, 221, 222), 637 (207, 215), 638 (221), 642 (288), *660*, *662*
Chittattu, G. J. 651 (425), *664*
Chiu, M. F. 237 (141), *241*, 321, 323, 324 (60), *329*
Chiuguk, V. 467 (813), *510*
Chmeclewski, J. 456 (636, 637), 489 (637), *507*
Chmutova, G. 402 (18), 420 (18, 269, 271), 471 (269), 472 (879), *494*, *500*, *512*
Chmutova, G. A. 70, 74–76 (46), 78 (46, 90), 86 (118), *149*, *150*, 245 (15), *283*, 563 (38), 565 (65), 574, 576 (133), *612–614*, 642 (291), *662*
Cholpankulova, S. 457 (659), *507*
Chottard, J. C. 715, 719 (218), *845*
Chovnikova, N. G. 602 (292, 293), 603 (293), *618*
Chow, F. 603, 604, 609, 611 (304), *618*, 624, 625, 627 (117), 645 (330), *658*, *662*
Choy, A. 811, 813 (600), *853*
Chremos, G. N. 815 (601c, 601d), *853*
Christe, K. O. 129 (298), *154*, 545 (290), *558*
Christen, D. 66, 67 (26), *148*
Christensen, D. 89, 90 (146, 147), *151*
Christensen, G. C. 303 (72, 73), *305*
Christensen, J. J. 680 (11a), *840*
Christensen, M. J. 387, 388 (98), *396*
Christiaens, L. 17 (101), *20*, 86 (116), *150*, 191, 199 (7), 203 (7, 59), 204, 205 (7), 207, 208 (63), 210 (7, 59), 211 (59), 212, 214, 215 (69), 216 (70), 217 (7), *239*, *240*, 244 (2), 265 (71), 266 (2), 267 (72), 268 (74, 75), *283*, *285*, 315 (27), *328*, 402 (11), 419 (261), 420 (261, 263, 265, 270), 421 (279), 422 (285–285), 423 (285, 292, 298–300), 424 (292, 299, 300, 302–307), 425 (285, 299, 307, 308, 309a, 310–312), 426 (298, 312, 314), 427 (11, 325), 429 (350, 351),

431 (374), 432 (383, 386), 433 (325, 387), 434 (374, 383), 439 (444), 442 (473), 449, 451 (559), 454 (601), 458, 461 (683), 471 (868, 869), 472 (868, 869, 882, 886, 887), 475 (283, 897), 476 (473, 905, 908, 910), 477, 478 (914), *494*, *499–503*, *505*, *506*, *508*, *512*, 563 (40, 41), 579 (150), 580 (150, 155), 584 (183, 205), 585 (183, 188–190), 592 (233, 234), 596, 597 (251), *612*, *615–617*, 621, 622 (27), 623 (70), *656*, *657*
Christian, G. D. 184 (177), *188*
Christie, R. 271 (78), *285*, 467 (826a), *511*
Christofferson, G. D. 567 (74, 75), *613*
Christokletov, V. N. 609 (332), *618*
Christophides, A. G. 794 (523), *851*
Christyakova, E. A. 704, 705 (169), *844*
Chu, J. Y. C. 236 (132), *241*, 520 (19), 524 (80), *553*, *554*
Chu, S.-H. 427 (326), 457 (650), *501*, *507*
Chu, S. H. 288 (13), *304*
Chul'skaya, S. M. 684–686 (40), *841*
Cífka, J. 371 (10a), *394*
Cini, C. 181 (121), *187*
Cioni, G. 811 (596), *853*
Ciurdaru, Gh. 277 (87, 88), *285*
Claeson, G. 444 (488), *504*
Clardy, J. 446 (522), *504*, 650 (386), *663*
Claremon, D. A. 16 (88, 89), *19*, 639 (230), 640 (283), *660*, *661*
Clark, A. H. 96 (182), *152*
Clark, E. R. 165 (55), *186*, 521 (33), 522 (36), *553*
Clark, G. R. 715 (214, 216), 716 (214), 717 (214, 216), 718 (225), 765, 766 (214), 814, 821 (216), 829 (650a, 650b), *845*, *854*
Clark, L. 458 (674a, 681), 459 (674a), *508*
Clark, M. C. 604 (319), *618*, 622 (42), *657*
Clark, P. S. 629 (162), *659*
Claxton, T. A. 25 (55), *61*
Clayton, J. 466 (795), *510*
Clayton, P. 408 (103), *496*
Clegg, D. E. 183 (129), *187*
Clementi, E. 25 (12, 41, 51), 26 (12), 27 (17, 19), *60*, *61*
Clementi, L. 25 (59), *61*
Clementi, S. 177 (128), *187*, 410 (132), 416 (213), 438 (432), *497*, *498*, *503*, 560 (7–9), 571 (7, 9, 105), 572 (105), 573, 577, 578 (117), 581 (165, 166), 582 (165), 586 (195), 589, 591 (165), 593, 594 (117), *611*, *614–616*
Clifton, E. C. 601, 602 (289), *618*
Clive, D. 402 (14), 403 (22a, 22b), *494*
Clive, D. L. J. 6 (24), 8 (36), 10 (24), 16 (99), *18*, *20*, 347 (8–10), *366*, 536 (165), *556*, 560, 599 (6), *611*, 624 (112), 630 (166), 632, 633 (185), 636 (207, 209, 215, 218, 219, 221, 222), 637 (207, 209, 215), 638 (218, 219, 221), 639 (166, 185), 640 (166), 642 (288), 651 (425), *658–660*, *662*, *664*, 690 (63), *841*
Close, R. 409, 440 (119), 480 (939), *497*, *513*
Clouston, A. 545 (231), *557*
Coates, G. E. 348 (12), *366*, 690 (93), 842
Cobbledick, R. 467 (816), *510*
Cobbledick, R. E. 819 (631), 825, 826 (638), 831 (631), *854*
Coccia, R. 181 (121, 122), *187*, 457 (655), *507*
Cody, R. 465, 468 (772), *510*
Cohen, G. N. 387, 388 (99), *396*
Cohen, M. L. 600 (277), 604 (322), *617*, *618*, 621–623 (2), 629 (162), *656*, *659*
Cohen, R. L. 806, 807, 810 (564), *852*
Cohen, S. 484 (985), *514*
Cohen, S. C. 269 (77), *285*, 654 (467), *665*
Cohen, V. 455 (614), 460 (705), 464 (766), 466 (806), 484 (992), *506*, *508–510*, *514*
Cohen, V. I. 255, 278 (49, 53), 279 (49), 280 (53), *284*
Cohen, Y. 297 (52), *305*
Colantti, P. 163 (34), *185*
Collienne, R. 427 (323), *501*
Collin, J. E. 22 (7), *60*
Collins, S. 14 (77), *19*, 626 (138, 139), 647 (138, 139, 333), *659*, *663*
Collins, T. J. 766 (411), 829 (651), *849*, *855*
Collman, J. P. 102 (219), *152*, 718, 719, 721 (227), 742 (334, 335), 750 (334), 758, 759 (335), *845*, *848*
Colombetti, G. 238 (143), *241*, 299 (57, 58), *305*
Colonna, F. 483 (974), *514*
Colonna, F. P. 246 (18), 247 (24, 25), *283*, *284*
Colquhoun, I. J. 699, 779 (143), *843*
Colter, M. A. 606 (328), *618*
Colton, R. 194 (25), *239*, 694–696 (126), *843*
Comasseto, J. V. 542 (216), *557*, 621, 622 (16), 624 (75), *656*, *657*
Combrink, W. 464 (745, 746), *509*
Comrie, A. 457 (662), 458 (664), *507*
Conde, A. 70 (36), 101 (198, 199), *148*, *152*
Condell, R. A. 390 (115b), *396*
Condit, P. C. 680 (16a), *840*
Cone, J. E. 390 (119), *397*
Connor, J. 403 (25), *494*
Consiglio, G. 591 (226, 227), *616*
Conti, F. 754 (364, 365), 755, 756 (365), *848*
Cook, D. B. 26 (15), *60*
Cook, W. S. 639, 640 (237), *661*
Cooke, M. 453, 465 (592), *506*
Cooper, F. W. 94 (170), *151*
Cooper, J. 449 (568), *505*

Cooper, W. 401 (4), *494*
Cooper, W. C. 2 (2), *17*, 163 (22, 30), 165 (30), 183 (22), *185*, 546 (246), *558*
Cope, F. F. 527 (118), *555*
Coppenolle, J. van 269 (76), *285*
Coppens, P. 94 (170), *151*
Coppinger, G. 401 (2), *494*
Coppola, J. 462 (723b), *509*
Coppola, J. C. 364 (69), *367*
Cordella, G. 487 (1012), *514*
Corden, B. J. 788, 791 (499), *851*
Cordes, A. W. 64 (1b), *148*
Cordes, R. 114 (245), *153*, 709, 710, 712, 819 (184), *844*
Cordes, R. E. 131 (300), 132, 143 (300, 301), 145 (326), *154*, *155*
Corey, E. J. 638 (225), *660*
Corey, E. R. 11 (57), *19*
Cornelius, W. 486, 487 (1005), *514*
Cort, L. 493 (1069), *516*
Cosgrove, D. O. 392 (135), *397*
Cossu, P. 166 (84), 171 (117), 178 (108), 180 (117), 181 (120), *186*, *187*
Costello, S. 461 (707), *508*
Cotton, F. A. 680, 681 (1), 691, 692, 695, 697 (114), 716 (1, 205), 719 (1), 721 (1, 231), 730 (273), 736 (307), 737 (273), 811 (1), *839*, *842*, *845–847*
Cottrall, M. F. 392 (134), *397*
Cotul, S. 393 (156), *397*
Coucouvanis, D. 680, 782, 815 (12b), *840*
Coulson, C. A. 604 (314), *618*
Courtot, C. 428 (333), 442 (475), *501*, *503*
Courtoy, C. P. 64 (6, 10), *148*
Couture, A. 473 (889), *512*
Coville, N. J. 716 (207), *845*
Cowan, D. 17 (104), *20*, 448 (542, 543, 546, 548), 449 (557, 558b, 565, 567, 568), 450 (542, 543, 546, 548, 572a), 451 (579), *505*, *506*, 651 (417), *664*
Cowan, D. O. 93 (165), 94 (167, 168), *151*, 247 (28), *284*, 576 (134), *614*
Cowie, D. B. 387, 388 (99), *396*
Cowley, A. H. 247 (26), *284*
Cox, J. D. 157–159 (2), *160*
Cox, S. 451 (578), *506*
Cozak, D. 715 (212, 219–221), 716 (207, 212), 717 (212, 219, 220), 720 (212, 220, 221), *845*
Cradock, S. 66 (22), 72 (58), *148*, *149*, 245 (13), *283*
Craig, P. 487 (1010), *514*
Craik, D. J. 204 (56), *240*
Cras, J. A. 102 (218), *152*, 794, 795 (530), *852*
Cravador, A. 319 (47), *329*, 623 (55), 624, (106), 649 (353), *657*, *658*, *663*
Craven, R. 449 (566), *505*
Crecely, R. W. 736 (305), 739 (316, 317), 825 (305), *847*
Credali, I. 487 (1024), *515*
Credali, L. 332 (6, 7), 333 (6, 11), 334 (12), 340 (11), *341*, 487 (1022, 1028), 488 (1022), 492 (1028), *515*, 768 (416), *849*
Criddle, R. S. 184 (157), *188*, 370 (2a), *393*, 600 (283), *618*
Crofts, J. 410 (126), *497*
Crofts, J. G. 89, 90 (145), *151*
Croise, A. 584 (205), *616*
Croisy, A. 244 (2), 265 (71), 266 (2), 267 (72), 268 (74), 276 (86), *283*, *285*, 422 (282, 283), 423, 426 (298), 471 (874), 475 (283), 479 (931), *500*, *512*, *513*
Crompton, T. R. 162 (12), 163 (26), *185*
Cross, R. J. 236 (133, 134), *241*, 690 (76a, 76b, 77a, 77b, 78), 697 (136), 786 (77a, 77b), 815 (601n), *842*, *843*, *853*
Crosse, B. C. 681 (19), 742 (339), 743 (345, 346), 744 (339, 345, 346), 748 (339), 750 (339, 345), *840*, *848*
Crossley, S. L. 217 (74), *240*
Cruickshank, F. R. 157, 159 (3), *160*
Csizmadia, I. G. 24 (10), 25 (14, 37–39, 50, 63), 30 (63, 73–75), 31 (10), *60*, *61*
Cullen, E. R. 194, 208, 209 (24). 220 (24, 85), *239*, *240*
Cullinane, N. 428, 442, 483 (337), 484 (986), *501*, *514*
Cummings, J. M. 710, 712, 814 (190), *844*
Cummins, L. M. 179, 180 (110), *187*
Cundall, R. B. 301 (64, 65), *305*
Cunningham, J. A. 102 (216), *152*, 778, 779, 782 (464), *850*
Curl, R. F. 64 (11), *148*
Curran, R. 102 (216), *152*, 778, 779, 782 (464), *850*
Current, S. 636, 637 (214), *660*
Curtis, N. J. 636 (222), 642 (288), 651 (425), *660*, *662*, *664*
Curtis, R. 407 (89), *496*
Curtiss, C. F. 201 (44), *239*
Cusachs, L. C. 565, 604 (46), *612*
Cushley, R. J. 276 (85), *285*
Cussans, N. J. 648 (321, 322, 341–343), *662*, *663*
Cuthbertson, E. 442 (470), 494 (1072, 1073), *503*, *516*
Cutter, G. A. 165, 167–169 (67). *186*
Cymerman Craig, J. 671 (21), 673 (25), *677*
Cyvin, B. N. 87 (128), *150*
Cyvin, S. J. 66 (23), 71, 72 (54), 87 (128), 140 (316), *148–150*, *154*
Czarny, M. R. 647 (326, 327), *662*
Czauderna, M. 298 (55), *305*, 539 (188), *556*

Czyzewski, J. 467 (827b), *511*

Dahl, L. F. 714 (198), 808 (569), 811 (587–589), 812 (587), 832 (589, 665), 835 (667), 838 (589), 839 (589, 665, 675), *845*, *852*, *853*, *855*
Dahlén, B. 127 (289), 128, 129 (292, 293), *154*, 670 (13), *676*
Dakternieks, D. 694–696 (126), *843*
Dale, J. W. 629 (161), 641, 643 (269), *659*, *661*
Dallacker, F. 420 (266), *499*
Dal Monte, D. 590 (219), 597 (261, 262), 598 (262), *616*, *617*
Dalton, J. 743, 744 (347), *848*
Dan, C. 371, 392 (12a), *394*
Dance, I. G. 811, 813 (600), *853*
Dance, N. 492 (1063), *516*
Dance, N. S. 254 (46), *284*, 358 (46), *366*, 520 (26, 29), *553*, 814 (616b), 818 (628), 819 (616b), 825 (616b, 638), 826 (638, 644, 645), *854*
Danchura, W. 624 (86), *658*
Daniel, J. R. 365 (72), *367*
Danielson, R. 184 (147), *188*
Danishefsky, S. 634 (195), 642 (259), *660*, *661*
D'Antonio, P. 82, 83 (101a), *150*, 670 (14), *676*
Danze, J.-M. 471, 472 (864), *512*
Dao, T. V. 465 (786), *510*
Darmadi, A. 11 (61), 16 (87), *19*, 66, 67 (26), *148*, 199 (52), *240*
Darwent, B. de B. 159 (15), *160*
Dasgupta, H. S. 281 (92), *285*
Daskalova, K. 165 (53), 171 (102), *186*, *187*
Daub, G. 490 (1056), *515*
Daudel, R. 25 (50, 63), 30 (63, 73–75), *61*
Daugherty, C. M. 603 (307), *618*
Dauplaise, D. 238 (142), *241*, 446 (521, 522), 447 (521), *504*, 650 (386, 387), *663*
Davalian, D. 377, 386 (35), *394*, 447, 464 (532), *505*
Davaliani, D. 581, 595 (162), *615*
Davidoff, E. F. 198, 211 (39), *239*
Davidon, W. C. 32 (79), *62*
Davies, I. 814, 819, 825 (616a, 616b), *854*
Davies, J. A. 721 (230), *845*
Davies, W. 455 (630), *507*
Davis, F. A. 9 (48), *19*, 624 (148), *659*
Davis, M. 407 (91, 92), 413 (182b), 414 (192b), 425 (309b), 433 (182b), 455 (620), 461 (707), 464 (750, 752a), *496*, *498*, *500*, *507–509*, 592 (231), *616*
Davis, M. A. 378, 392 (36), *394*
Davis, P. 466 (796), *510*
Davis, R. W. 75 (79), *149*
Davison, A. 405 (56), *495*, 736 (307), 798 (540a, 540b, 545), 800 (545), 811 (540a, 540b), *847*, *852*
Davydov, V. V. 700 (155), 701 (156), *843*
Dawson, J. C. 701–703 (162), *844*
Dawson, W. H. 191 (10, 11), 192, 193, 195, 197, 199, 205, 209 (11), *239*
Day, V. W. 808, 832, 834, 837 (571), *852*
Dayal, S. K. - 575 (123), *614*
Deady, L. 459 (699), *508*
Deady, L. W. 592 (231, 232), *616*, *617*
Dean, J. 5 (11), *18*
Dean, P. A. W. 194, 199 (27), *239*, 697, 755 (141b), 815 (601m), *843*, *853*
De Battisti, A. 600 (282), *618*, 621, 622 (5), *656*
DeBeer, J. A. 825 (642), *854*
De Checchi, C. 332 (6, 7), 333 (6), 334 (12), *341*, 487 (1022, 1024, 1028), 488 (1022), 492 (1028), *515*
DeChecchi, C. 768 (416), *849*
Dechter, J. J. 191, 192 (14), *239*
De Cicco, P. 621, 622 (15), *656*
Deckart, H. 393 (146), *397*
Deconninck, G. 163 (35), *185*
Decroix, B. 413 (176, 190, 191, 192a), 414 (193), 415 (176, 209), 438 (434), 461 (192a), *498*, *503*
Dedeyne, R. 480 (943), *513*
Dedov, V. I. 392 (136–141, 143), *397*
Deelstra, H. 164, 165 (47), *186*
Deeva, V. S. 387 (89), *396*
De Filippo, D. 166 (82), *186*, 580 (159, 160), 599, 600 (159), *615*, 759–761 (371, 372), 794 (522, 524), *848*, *851*
DeFilippo, D. 643 (359), *663*, 707, 708 (177), *844*
Defrees, D. J. 25 (49, 64, 71), 29 (71), 30 (64, 71), 33 (49, 71), 34 (49), *61*
Degani, I. 423 (290a), 469 (838, 839), 470 (838, 839, 856), 471 (867), 472 (867, 881, 885), 473 (881, 888), 476, 477 (867), 478 (885, 888, 930), 479 (885), *500*, *511–513*, 592 (238), 598 (264, 266, 267), 599 (267), *617*
Degering, E. F. 522 (37), *553*
Deguchi, Y. 444 (492), *504*
De Jong, R. 409, 414 (118b), *497*
De Ketelaere, R. F. 565 (67), *613*
Del'Acqua, A. 163 (17), *185*
DeLange, G. J. 810 (580), *853*
Delgado, M. S. 726, 728 (252, 253), *846*
Dell, W. 9 (43), *18*, 759–762, 825 (390a, 390b), *849*
Della Fiorentina, H. 488, 489, 492 (1039b), *515*
Dell'Erba, C. 411 (136, 138–141), 412 (159),

414 (141), 416 (159), *497*, 567–569 (87), 572 (87, 111), 578 (145, 146), 582 (87), 589 (146, 216–218), 591 (145, 216, 217), *613–616*
Del Pra, A. 96 (178–180), *151*, *152*
De Lucchi, O. 403 (27, 42), *494*, *495*
Delwicke, J. 22 (7), *60*
De Marco, C. 171 (116, 117), 178 (108), 179 (116), 180 (117), 181 (120–122), *187*, 622 (32, 36), *657*
DeMarco, C. 166 (84), *186*
De Mayo, P. 621, 622 (29), *657*
Demerseman, P. 413 (167), *497*
De Munno, A. 466 (799), *510*
Demuynck, J. 25, 28, 30, 33, 34 (67), *61*, 604 (313), *618*, 624 (87), *658*
Denbigh, D. 13 (68), *19*
Denis, J.-N. 403 (32), *495*
Denis, J. N. 624 (107), 632, 633 (191), *658*, *660*
Denison, G. H. 680 (16a), *840*
Denney, D. B. 191, 192, 221, 222, 232 (15), *239*, 537 (181), 538 (181, 184), 539 (181), *556*
Denney, D. Z. 191, 192, 221, 222, 232 (15), *239*, 537 (181), 538 (181, 184), 539 (181), *556*
Dennis, R. 465, 468 (772), *510*
Denniston, M. L. 249 (35), *284*
Denoel, J. 191, 199, 203–205, 210, 217 (7), *239*, 563 (41), *612*
De Nunno, A. 584 (182, 206), 597 (252), *615–617*
Denyer, C. 402 (14), *494*
Deplano, P. 794 (522, 524, 529), 795 (529), *851*, *852*
Dereu, N. 203, 221 (57), 222 (57, 105), 223–225 (57), 226, 227 (105), *240*, *241*, 438 (424), 443 (483, 484), 452 (424), 471 (871), 472 (880), 473 (484), 474 (896), 475 (896–898), 484 (989, 990), *502*, *504*, *512*, *514*, 563 (40), *612*, 651 (398), *664*
Dereu, N. L. M. 77 (84), 101 (200), *149*, *152*, 227 (106), 231 (112), 232–234 (106), *241*
Derkach, N. 488 (1037), *515*
Derkach, N. Y. 16 (93), *20*
Derkach, N. Ya. 541 (203, 209–211, 212a, 212b, 213), 542 (212a, 212b, 213), 548 (270, 271), *557*, *558*
Dernini, S. 166 (84), 171 (116, 117), 179 (116), 180 (117), 181 (120), *186*, *187*, 622 (32, 36), *657*
De Roos, K. 465 (780), *510*
De Saint-Laurent, J. 393 (147), *397*
Desauvage, S. 320 (50, 52), *329*
De Shazo, M. 601, 602 (288), *618*
Desjardins, C. D. 253 (45), *284*, 523 (64, 65), 537 (65), 545 (64, 65, 240), *554*, *557*
Desjardins, S. R. 716 (207), *845*
Desmond, T. 715, 717, 720, 822 (223), *845*
Dessy, G. 102–105 (205), *152*, 794 (520), *851*
Dessy, M. R. 171, 179 (116), *187*, 622 (32), *657*
DeStefano, N. J. 697, 734 (138), *843*
Destro, R. 403 (44), *495*
Detty, M. 410 (123), 445 (501, 506b), 447 (528, 529), 451 (528), 469 (848), 470 (857, 861), 473 (894), 475 (899, 900), *497*, *504*, *505*, *511*, *512*
Detty, M. R. 10 (49), *19*, 354 (20, 24), 355 (25), *366*, 539 (185), 544 (220), *556*, *557*, 621, 622 (22), 624 (76, 109), 651 (423, 424), *656–658*, *664*, 680 (16b), 690, 716 (62), *840*, *841*
Deuten, K. von 101 (197), 147 (330), *152*, *155*, 540 (193–196), *556*, *557*
Deutsch, E. 102, 103 (213, 214), *152*, 771, 772 (427, 428), 778, 779 (457), 780 (427, 428, 457), 784, 786 (457), *849*, *850*
Devaud, M. 469, 470 (849), *511*
Develotte, J. 458, 459 (676), 486 (1002), *508*, *514*
Devilanova, F. 794 (522), *851*
Devillanova, F. 457 (657), *507*
Devlin, J. 731, 736 (276), *846*
Dewan, J. 469 (854a, 854b), *511*
Dewan, J. C. 97 (183–185, 187, 188), 98 (183–185), 132 (183), *152*, 525 (89–93), *554*
Dewar, D. 405, 469 (60a), *495*
Dewar, D. H. 525 (86), *554*
Dewar, M. J. S. 562 (27), 571 (107, 109), *612*, *614*, 641 (260, 265), *661*
Dexter, D. D. 78 (92), *150*
Dhawan, S. 464 (752b), *509*
Dhillon, D. S. 548 (269), *558*
Diaz, A. 794 (524), *851*
Dickman, M. H. 771, 772, 780 (428), *849*
Dickson, R. C. 288 (10), *304*
Dideberg, O. 75, 76 (76, 77), 77 (76, 77, 83), 86 (113, 114), 91 (156), 95 (175), *149–151*, 454 (604), *506*, 670 (15), *676*
Diemann, E. 680, 782 (12g), 803 (553a), 815 (12g), *840*, *852*
Dienstbier, Z. 289 (25), 290 (26), 291 (25), *304*, 392 (144), *397*
Diercksen, G. H. F. 25 (57), *61*
Dietrich, C. O. 16 (94), *20*, 541 (208), *557*
Dietz, R. 321, 323 (58), *329*
Dietz, W. 456 (635), *507*
Dietze, F. 784, 786, 795 (479), *851*
Dietzsch, W. 102 (217), *152*, 624 (92), *658*, 784 (480), 786 (485, 486), 787 (484–486), 788 (485, 497), 789 (480, 497, 506, 507),

790 (485), 791 (480, 485), 794 (497), 796, 797 (538, 539), *851*, *852*
Dilger, W. 184 (166), *188*, 427 (330), *501*
Dilli, S. 183 (141), *188*
Dilthey, W. 547 (252), *558*
DiLuzio, S. 621, 622 (15), *656*
Dilworth, G. L. 370, 379, 388 (5), 391 (122), *393*, *397*
Di Maggio, D. 290, 291 (33), *305*
Di Marco, C. 457 (655), *507*
Dimitrova, P. P. 681, 686 (23a), *840*
Di Modica, G. 573, 574 (119), *614*
Dingwall, D. 457 (662), 458 (664), *507*
Dingwall, J. 445 (507a), 467 (826b), *504*, *511*
Dinilenko, V. A. 602 (296), *618*
Dippy, F. J. 575 (130, 132), *614*
Di Somma, A. A. 178 (105), *187*
Distefano, G. 246 (16–18), 247 (24, 25), *283*, *284*, 416 (212), 483 (974), *498*, *514*
Distetano, G. 44, 54 (86), *62*
Ditchfield, R. 25 (29, 32, 35), *60*, *61*
Dittmer, D. C. 403 (26b), *494*
Dittrich, G. 465 (768), *509*
Di Vaira, M. 727 (258–261), 729 (258–261, 267a, 267b, 268), 730 (261), *846*
Divnich, T. F. 624 (89), *658*
Dix, A. H. 762, 763 (395), *849*
Dixon, D. A. 26, 27 (16), *60*
Dixon, K. 493 (1066), *516*
Dixon, W. B. 89, 90 (146), *151*
Djerassi, C. 252 (42), *284*, 418 (234), *499*, 670 (18a, 18b), 671 (18b, 20b), *677*
Djoglev, D. H. 681, 686 (23a), *840*
Djumanazarova, A. 431 (378, 379), 435 (400), *502*
Djumanazarova, A. Z. 578, 579, 596 (147), *615*
Dobrowolski, J. 289 (23), *304*
Dobryakova, G. V. 389 (110), *396*
Dobson, G. R. 690, 691 (110), *842*
Doddi, G. 590 (220–222), 591 (222), *616*
Doedens, R. J. 771, 772, 780 (428), *849*
Doi, K. 779, 780 (460), *850*
Dolenko, G. N. 197 (38), *239*
Dombek, B. D. 716 (206), *845*
Domcke, W. 25 (57), *61*
Domenicano, A. 121 (281), *154*
Domingos, A. M. 734 (301), *847*
Donaldson, R. 774, 776 (452), *850*
Donia, R. A. 639, 640 (237), *661*
Do Phuoc Hien 426, 465 (315), *501*
Dorofeenko, G. 480 (937), *513*
Dorofeeva, O. V. 87 (126, 128), *150*
Dosseto, M. 381, 393 (59), *395*
Dostàl, K. 547 (256, 257), *558*, 650 (365, 366, 371), (372), *663*
Dötz, K. H. 722 (245e), *846*
Dotzauer, E. 814, 819, 820 (619), *854*
Down, A. W. 815 (604a), *853*
Drager, M. 244, 263 (6), *283*, 357 (41), *366*
Dräger, M. 357 (44), 358 (44, 47), *366*
Dragieva, I. D. 681, 686 (23a), *840*
Draguet, C. 457 (652–654, 656), 475 (902), 484 (993), 488, 489, 492 (1039b), *507*, *512*, *514*, *515*
Drake, J. E. 349 (16, 17), 350–352 (16), 353 (18), *366*, 621, 622 (20), *656*, 716 (201), 814, 818 (610), *845*, *854*
Drakenberg, T. 230, 232 (109, 110), *241*, 579, 580 (150), *615*
Dransfield, T. 453, 465 (594), *506*
Dransfield, T. A. 700 (147), *843*
Dreeskamp, H. 233 (116), *241*
Dreizler, H. 51, 57, 58 (94), *62*, 70, 72 (37), 73 (70, 71), 90 (37), *148*, *149*
Drevko, B. 469 (840a), *511*
Drew, H. 405 (59), 469 (851), 482 (960, 961, 969, 970), 483 (961), *495*, *511*, *513*, *514*
Drew, H. D. K. 519 (12), 525 (84, 85), 526 (100), 543 (12), 550 (280), *553–555*, *558*, 653 (452, 457), 654 (452, 476), *665*
Drew, M. G. B. 731 (275, 277), 736 (277), 808 (576–578), 813 (578), *846*, *853*
Driscoll, J. 478 (927), *513*
Drory, A. 427 (322), *501*
Drucker, G. E. 565, 604 (61), *613*
Druyan, M. E. 86 (120), *150*
Duar, I. 408 (104), *496*
Dubnack, J. 699 (152a), *843*
DuBois Murphy, F. 701, 705 (158b), *844*
Dubus, P. 413 (190, 192a), 415 (209), 461 (192a), *498*
Ducep, J. B. 604 (320), *618*
Duffield, A. 418 (234), *499*
Duffield, A. M. 252 (42), *284*
Dufour, M. 426 (313), *500*
Dufour, R. 424 (303), *500*, 584, 585 (183), *615*
Dugger, D. L. 334, 341 (18), *341*
Dulenko, V. 419 (251, 252), 435 (251, 402), 438 (251, 426), *499*, *502*, *503*
Dumont, W. 8 (37), *18*, 319 (47), *329*, 526 (99), *555*, 603 (302), *618*, 623 (60), 624 (104, 106, 110, 111), *657*, *658*
DuMont, W. W. 815 (601e), *853*
duMont, W.-W. 222, 223 (97), *240*
duMont, W. W. 8, 16 (41), *18*
Dunbar, F. 488, 491 (1034), *515*
Dunbar, F. X. 829 (649), *854*
Duncan, J. L. 71 (53), *149*, 684–686 (38), *841*
Duncanson, L. A. 690 (66, 83), *841*, *842*
Dunkin, I. 466 (801), *510*
Dunlap, J. S. 387 (95), *396*
Dunlap, R. B. 194, 195 (23, 29), 196 (29),

206 (60, 62), 207 (62, 64), 208 (64), 210, 211, 228 (60), *239*, *240*
Dunn, A. 467 (826b), *511*
Dunning, T. H. Jr. 25 (30, 33, 54, 65), 30 (65), *60*, *61*
Dupont, L. 75, 76 (76, 77), 77 (76, 77, 83, 85), 86 (113, 114), 91 (154, 156), 92, 93 (164), *149–151*, 454 (604, 606), *506*, 670 (15), *676*
Durand, P. 33 (81), *62*
Durant, F. 99 (191), *152*
Durell, J. 380 (46), *395*
Durig, J. R. 70, 73 (42), *149*, 624 (90), *658*
Durst, H. D. 680 (11b), *840*
Duus, F. 11, 12 (56), *19*
Du Vigneaud, V. 182 (124), *187*
Dvoinin, V. I. 681, 686 (23b), *840*
Dyer, G. 771, 772 (431, 432), 801, 803 (550), 815 (603a), *849*, *852*, *853*
Dzafarov, A. I. 289, 291 (24), *304*
Dzhumaev, I. 438 (437), 449 (558a), 464 (755), *503*, *505*, *509*
Dzhumanazarova, A. 434 (396), 435 (400), *502*
Dzurilla, M. 485 (996), *514*

Ebner, M. 765, 766 (408), 767 (408, 414), 768 (414), *849*
Ebsworth, E. A. V. 348 (15), *366*, 754–756 (366), *848*
Echter, T. 462 (722), *509*
Eckerson, B. A. 577 (140, 141), 601 (140), *614*
Eckstein, P. 794, 795 (528), *851*
Edmonds, J. W. 94 (170), *151*
Edwardly, J. S. 387 (96), *396*
Edwards, T. H. 72 (56, 59, 64), *149*
Efremenko, V. 458 (665), *507*
Efremov, E. A. 167 (90), 169 (90, 91), 171, 172 (91), *187*
Efros, L. 429, 430 (356), 437 (415), 462 (724–726), 463 (727–731), 465 (787, 788a, 788b), *501*, *502*, *509*, *510*
Efros, L. S. 586 (211), 587 (197), *616*
Eggert, H. 221 (89), *240*
Egorochkin, A. N. 716 (202), *845*
Egorov, Y. P. 16 (93), *20*, 815 (601k), *853*
Egsgaard, H. 244, 269, 283 (3), *283*
Ehlers, D. 466 (807, 809), 467 (809), *510*
Ehlig, C. F. 387 (83), *396*
Ehrenson, S. 202 (54a, 54b), *240*, 561 (18, 22), 562 (22, 24), 571 (24), *612*
Einaga, H. 779 (459), 781 (470), *850*
Einstein, F. 519 (13, 14), 543 (13), *553*
Einstein, F. B. W. 15 (85), *19*
Einstein, F. W. B. 716 (200), 814, 818 (613), 819 (631), 825, 826 (638), 831 (631), *845*, *854*
Eisenberg, R. 102 (212, 216), *152*, 778, 779, 782 (464), 788, 791 (499), 798, 799 (541), *850–852*
Eisner, A. 653, 654 (445), *665*
Eisner, U. 493 (1069), *516*
Eklund, N. 520 (21), *553*
Elder, R. C. 102, 103 (213, 214), *152*, 778–780, 784, 786 (457), *850*
Elgad, U. 545 (224), *557*
Ellermann, J. 363 (68), *367*
Ellis, P. D. 191–193, 195, 197, 199, 205, 209 (11), 218 (76), *239*, *240*
Ellis, P. E. 102, 103 (213), *152*, 778–780, 784, 786 (457), *850*
Elmaleh, D. 267 (73), *285*, 434 (399), *502*, 527 (107, 108), *555*
Elokhina, N. 413 (186), *498*
Elsenbaumer, R. L. 332, 334 (3), *341*
El Sheraki, A. S. 387, 389 (91), *396*
El Shetaiwi, S. 387, 389 (91), *396*
El'tsov, A. 429 (357), 439 (441), 472 (876), *501*, *503*, *512*
El'tsov, A. V. 587 (196), *616*, 641 (292), *662*
Elvidge, J. 493 (1069), *516*
Elvidge, J. A. 586 (192), *616*
Elvins, O. C. 523, 524 (70), *554*, 653, 654 (456), *665*
Elzinga, J. 835 (668b), *855*
Emeleus, H. J. 520 (69), *554*
Emeléus, H. J. 362 (65), *367*, 629 (161), 640 (248), 641, 643 (269), *659*, *661*
Emel'yanova, N. N. 602, 603 (293), *618*
Emge, T. J. 93 (165), 94 (168), *151*
Emma, T. 334, 341 (18), *341*
Emrick, D. 490 (1055), *515*
Engler, E. 244, 269, 283 (3), *283*, 448 (538), 449 (554, 555, 562, 563b, 566), 450 (538, 554, 571, 572b, 574, 576), 451 (574), 464 (563b), *505*, *506*
Engler, E. M. 310 (16), 311 (16, 17), *328*, 681 (28a), *841*
Engler, R. 381, 393 (57), *395*
English, A. M. 715 (213, 222), 716, 717 (213), 720 (213, 222), 767 (213), *845*
Engman, L. 6 (20–22), 10 (54), 15 (86), *18*, *19*, 135, 143 (304), *154*, 403 (23, 24), 406 (64, 65), 419 (244), 427 (328), 442 (328, 461, 462, 474), 447 (64), 462 (723a), 487 (1021), *494*, *495*, *499*, *501*, *503*, *509*, *515*, 525 (95), 527 (113–115), 528 (136), 529 (114, 137, 141), 530 (136), 532 (157, 158), 533 (114), 534 (157, 158), 537 (174), 544 (115), 550, 551 (282), *554–556*, *558*, 585, 590, 596 (187), *616*, 651 (415, 418–420), 652 (443, 444), 653 (448–450), 654 (449,

462, 463), 655 (462, 463), *664*, *665*
Enstein, F. 467 (816), *510*
Entenmann, G. 484 (991b), *514*, 642 (293), *662*
Entwistle, I. D. 639 (142), *659*
Enzmann, P. J. 382 (63), *395*
Epifanov, V. S. 171 (103), *187*
Epstein, A. 470 (859, 860), *511*
Epstein, E. F. 811 (584), *853*
Erba, C. dell' 675 (31), *677*
Eremeeva, G. 465 (787, 788a, 788b), *510*
Eriksen, R. 77 (81, 82), *149*
Ermakov, A. N. 163–165, 183 (19), *185*
Ernst, R. 191 (8), *239*
Ernstbrunner, E. E. 602 (294), *618*
Ershova, I. I. 533 (159), *556*
Es, T. van 641 (268), *661*
Esaki, N. 621, 622 (18), *656*
Esho, F. S. 731 (275, 277), 736 (277), *846*
Esperas, S. 794 (526), *851*
Esperås, S. 12 (67), *19*, 102, 103 (209), 145 (329), 147 (329, 332), *152*, *155*
Es-Seddiki, S. 469, 470 (849), *511*
Estes, S. A. 165, 167 (73), *186*
Etter, M. C. 709 (179), *844*
Evans, C. S. 166–170 (80), *186*
Evans, D. A. 604 (321), *618*
Evans, T. C. 393 (148), *397*
Evatt, B. 393 (150), *397*
Evdomikov, A. 462 (726), *509*
Everett, G. W. 781 (471), *850*
Evers, M. 268 (74), *285*, 422 (283), 458, 461 (683), 475 (283), *500*, *508*, 623 (70), *657*
Evertsdotter, C. 643, 650 (358), *663*
Evrard, G. 320 (52), *329*
Ewert, H. 455 (627), *507*
Ewing, V. 72 (67), *149*
Exner, O. 319 (46), *329*, 562 (23), 565 (48), *612*
Extine, M. 107 (232), *153*, 668 (8), *676*
Extine, M. W. 114 (244), *153*
Ezhkova, Z. I. 701 (156), *843*

Fabre, C. 470 (858), *511*
Fackler, J. P. 102 (220), *152*, 191–193 (12), *239*, 732, 736 (288), 765 (405), 794 (405, 516–518, 527), 795 (516–518, 527), *847*, *849*, *851*
Fadkina, S. 429 (357), *501*
Faehl, L. G. 648, 650 (351), *663*
Fagioli, F. 600 (282), *618*, 621, 622 (5), *656*
Fahel, L. G. 599 (275), *617*
Fahrbach, G. 428 (338), 443 (477, 478, 480), *501*, *504*, 535, 536, 544 (163a–c), *556*, 655 (481), *665*
Falcone, S. J. 652 (433, 442), *664*, *665*
Falconi, G. 299 (57), *305*
Faller, P. 420 (267), 423 (299, 300), 424 (299, 300, 304), 425 (299), 429 (350–353), 435 (267), *499–501*, 585 (188), *616*
Fanghaenel, E. 448 (540), *505*
Fanghängel, E. 450 (573), *506*
Faraone, F. 690 (71, 87, 88, 97–99, 106), 691 (97, 98, 106), 695 (98), 735, 738 (299), 774 (98), *842*, *847*
Farina, V. 636, 638 (219), 651 (425), *660*, *664*
Farnier, M. 436 (411, 412), 460 (412), *502*
Farrar, D. H. 765, 807 (404), *849*
Farrar, T. C. 191 (9), *239*
Farrar, W. 418 (232), 440 (455), 468, 481 (232), *499*, *503*
Farrar, W. V. 654 (466), *665*
Farrow, G. W. 690 (79, 112), *842*
Fauconnier, A. 438, 452, 453 (425), *502*
Faust, A. S. 835 (667), *855*
Fava, G. 426, 459 (316), *501*
Favez, R. 690 (107), *842*
Favo, G. 383 (72c), *395*
Fayos, J. 697 (135b), *843*
Fazakerley, G. V. 221–223, 225, 229 (91), *240*
Fazio, R. 638 (223), *660*
Featherman, S. 468 (836), *511*
Federici, G. 621, 622 (15), *656*
Federsel, H. 460, 492 (701), *508*
Fedorov, V. A. 167 (90), 169 (90, 91), 171, 172 (91), *187*
Fedorova, N. A. 632 (176), *659*
Fedotov, M. A. 197, 204, 205 (37), *239*
Fee, J. A. 238 (147), *241*, 382, 392 (67), *395*, 676 (37), *677*
Fefe, I. 585 (190), *616*
Fehlhammer, W. P. 103 (223), *152*
Feigl, F. 163 (16), *185*
Feiring, A. E. 632, 635 (183, 184), *660*
Fellous, R. 562 (29), *612*
Felthouse, T. R. 721 (231), *845*
Felton, D. 467 (817), *510*
Fenster, A. E. 716 (207), *845*
Ferguson, G. 715, 717, 720 (223), 721 (233), 822 (223), 825 (643), *845*, *854*
Fergusson, J. 405, 469 (60a), *495*
Fergusson, J. E. 525 (86), *554*, 690 (72, 74, 102, 103, 105), 756 (105), *842*
Fernandez, F. J. 165 (69), *186*
Fernandez, J. 87 (127), *150*
Fernandez, V. 726, 728 (252, 253), 735, 738, 739 (300), 740 (321), *846*, *847*
Fernholt, L. 72 (60), 137 (308), *149*, *154*
Ferraris, J. 449 (565), 483 (975), *505*, *514*
Ferraris, J. P. 247 (28), *284*
Ferreira, J. T. B. 6 (19), *18*, 621, 622 (16), 624 (75), *656*, *657*
Ferreu, L. A. 651 (412), *664*

Ferrier, W. G. 99 (192a), *152*
Feshchenko, N. G. 815 (601k), *853*
Fialkov, Yu. A. 565 (43), *612*
Fichtner, M. 447 (529), *505*
Fielden, E. M. 290 (30), 293, 294 (41), 299–301 (61), *304*, *305*
Fijal, D. 387 (87), *396*
Filer, C. N. 638 (223), *660*
Filipovich, G. 709 (179), *844*
Filippov, E. P. 167 (90), 169 (90, 91), 171, 172 (91), *187*
Filippova, T. M. 651 (404), *664*
Findlay, R. H. 55 (92), *62*, 308 (13), *328*
Finkenbine, J. 402 (15), *494*
Firl, J. 463 (734, 736a), 484 (736a), *509*
Fischer, E. O. 722 (244, 245a–e), 723 (246–250), 724 (247–249), 725 (248–250), 732 (246), 742, 753 (250), 814, 823 (624), *846*, *854*
Fischer, G. 167 (85), *186*
Fischer, H. 723–725 (248, 249), 726, 727, 729 (256), 814, 823 (623, 624), *846*, *854*
Fisher, A. 470 (859), *511*, 575 (131), *614*
Fisher, C. H. 653, 654 (445), *665*
Fissi, A. 466 (799), *510*, 597 (252), *617*
Fitjer, L. 275 (84), *285*, 418 (236), *499*
Flachowsky, J. 699 (152a), *843*
Flamini, A. 794 (525), *851*
Flato, J. 423 (291), *500*, 528 (127), 532 (127, 152), 533 (127), 534 (152), *555*, *556*
Fletcher, R. 32 (78), *61*
Flinn, C. G. 167 (89, 92), 170 (89), 171 (93), 177 (89, 92), 183 (89), *187*
Flohé, L. 389 (114), *396*
Floor, J. 623 (58), *657*
Florey, J. B. 565, 604 (46), *612*
Flumignan, E. 622 (31), *657*
Fochi, R. 423 (290a), 469 (838), 470 (838, 856), 471 (867), 472 (867, 881, 885), 473 (881, 888), 476, 477 (867), 478 (885, 888, 930), 479 (885), *500*, *511–513*, 592 (238), 598 (264, 266, 267), 599 (267), *617*
Fogarasi, G. 25 (43, 60), *61*
Föhr, M. 547 (255), *558*
Fokkens, R. H. 838 (670c), *855*
Forchioni, A. 86 (117), *150*
Ford, J. A. 315 (24), *328*
Ford, S. 421 (277), *500*
Forstrom, J. W. 390 (115a, 115b), *396*
Foss, O. 14 (80), 16 (98), *19*, *20*, 84, 85 (106), 86 (111, 112), 119 (275, 276), *150*, *154*, 166 (79), *186*, 639 (234), 640, 642 (245), 651 (426–429), *661*, *664*, 671 (19), *677*, 686 (46b), *841*
Foster, S. J. 244 (7), *283*
Fourrey, J.-L. 622 (30), *657*
Foust, A. S. 815 (604d), *853*
Fowles, G. W. A. 695, 697 (131), 706, 707 (174), 808 (576), *843*, *844*, *853*
Foxman, B. 446 (516), *504*
Foxton, M. 461 (712), *508*
Fradkina, S. P. 587 (196), *616*
Fraga, S. 22, 25, 28 (8), *60*, 202 (51), *239*
Francis, M. 404 (54), *495*
Francl, M. M. 25, 30 (64), *61*
François, F. 478 (922), *513*
Frank, A. 90 (153), *151*, 463 (735), *509*
Frank, U. 359 (49), *366*, 714, 716 (197), *845*
Franklin, W. J. 641 (261), *661*
Franzen, H. F. 809, 832 (579), *853*
Franzi, R. 236 (135), 237 (135, 137, 139), *241*
Frapart, P. 297 (52), *305*
Fraser, A. 66 (24a), *148*
Fraser, G. W. 545 (223, 231–237), *557*
Fraser, R. R. 571 (110), *614*
Frederiksen, E. 771 (423), *849*
Fredga, A. 191, 196, 198–200, 205, 210 (5), 217 (71), 218 (72), *238*, *240*, 338 (33), *341*, 418 (235), 468 (833), 476 (907), 480 (942), 492 (907), *499*, *511–513*, 622 (35), 641 (266), 643, 650 (358), *657*, *661*, *663*, 670 (18a), 671 (21), 673 (25), *677*
Freeman, B. H. 602 (295), *618*
Freeman, H. C. 308 (11), *328*
Frejd, F. 595 (244, 248, 250, 256), 596 (248–250), *617*
Frejd, T. 407 (82a), 413 (182b), 414 (196), 416 (217, 218, 220, 222), 417 (223–227, 229), 418 (222), 419 (248), 433 (182b, 248), *496*, *498*, *499*, 621, 622 (28), 647 (332), *657*, *662*
Fremdling, H. 547 (255), *558*
Fresco, J. 308 (7), *328*
Freund, M. 433 (391), *502*
Friant, P. 720 (228), *845*
Frič, I. 673 (26), *677*
Frick, C. E. 528 (121), *555*
Friedheim, E. A. H. 539 (187), *556*
Friedlander, A. 434 (397), *502*
Friend, N. A. C. 528 (122), *555*
Fringuelli, F. 44, 54 (86, 87), *62*, 221, 222 (114), 230, 232 (109), 233 (114, 117), *241*, 246 (17, 18), 264 (69), *283*, *284*, 378, 383 (37b), 385 (75), *394*, *395*, 406 (75), 409 (112, 113), 410 (127–129, 131, 132), 411 (134, 135, 137), 414 (75, 112, 113), 416 (113, 211–213, 215, 216), 419 (241), 442 (465, 466), *496–499*, *503*, 560 (4, 7–9), 563 (4), 567, 568 (85, 88, 89), 569 (91, 92), 571 (7, 9, 105), 572 (4, 91, 105), 573 (117), 577 (117, 143), 578 (117), 579, 580 (153, 154), 581 (143, 165, 174), 582, 589, 591 (165, 176), 593, 594 (117), 597 (91, 260), *611*, *613–615*, *617*

Fritsch, K. 459 (692), *508*
Fritz, H. 534 (162), *556*, 655 (482), *665*
Fritzmann, E. 680 (5–8), 690 (5–7), 774, 776, (5, 6), 815 (8), *839*
Frolov, A. N. 641 (292), *662*
Frolow, F. 409 (117, 118a), *496*, *497*
Fruehan, A. G. 803 (552), *852*
Fruhling, J. 392 (129), *397*
Fueno, T. 320 (51), *329*
Fugier, C. 413 (191), *498*
Fugnitto, R. 470 (858), *511*
Fuhrling, H. 695, 697 (127), *843*
Fujimori, K. 370 (3a, 3b), *393*, 623 (52, 53), *657*
Fujita, E. 317 (37), *329*, 642 (242), *661*
Fujita, J. 771, 772 (430), 779, 780 (430, 460), *849*, *850*
Fujiwara, I. 303 (70, 71), *305*
Fukae, S. 393 (155), *397*
Fukumoto, K. 642 (303), *662*
Fukumura, M. 165 (54), *186*, 404, 446 (511), *504*, 541 (206), *557*
Fukuzawa, S. 10 (51), *19*, 537 (176), 551 (283, 284), *556*, *558*, 650, 651 (421, 422), 653 (461), 655 (480), *664*, *665*
Fukuzawa, S.-I. 530, 533 (142), *555*
Füllgrabe, H.-J. 356 (34), *366*
Fultz, W. C. 736 (303, 304), *847*
Fumagalli, A. 811 (596), *853*
Funk, H. 489 (1050), *515*, 522 (38), 528 (38, 123, 124), 530, 532 (123, 124), 533 (123), *553*, *555*
Furin, G. G. 197 (37, 38), 204, 205 (37), *239*
Furlani, C. 684, 685 (39), 686 (39, 47–49), 688 (49), 794 (521), *841*, *851*
Furmanova, N. G. 16 (92), *19*, 541 (200), *557*
Furness, A. R. 695, 697 (129), *843*
Fürstenberg, G. 16 (96), *20*, 362 (63), *367*
Furukawa, N. 541 (206), *557*
Fusco, R. 466, 467 (810), *510*
Futekov, L. 261 (60), *284*, 525 (83), *554*, 699 (151), *843*

Gacnarz, R. A. 626 (130), *659*
Gadzhiev, G. G. 624 (82), *658*
Gadzhiev, S. N. 158 (9), *160*
Gafner, G. 779 (465), *850*
Gagne, P. 340 (38), *341*
Gaidis, J. 428 (339), *501*
Gais, H.-J. 622 (43, 44), *657*
Galambos, G. 316 (32), *329*
Galambos, J. T. 387 (80), *396*
Galasso, V, 221, 222 (114), 233 (114, 117), *241*, 247 (24, 25), *284*, 483 (974), *514*
Galateanu, J. 297–299 (51), *305*
Gal'bershtam, M. 407 (86), 412 (86, 164, 165), 413 (86, 170, 179, 181), *496–498*
Gal'bershtam, M. A. 573 (118), 578 (144), 581 (144, 170), 593 (118), *614*, *615*
Galina, V. M. 494 (1074), *516*
Gall, J. 494 (1073), *516*
Galla, S. 423 (297), *500*
Galla, S. V. 527 (105), *555*
Galla-Bobic, S. 423 (293), 482 (959), *500*, *513*
Galla-Bobik, S. V. 528 (125, 126, 133), 532 (153), 533 (159), 534 (153), *555*, *556*
Gallaher, K. L. 100, 101 (194), *152*
Galli, D. 811 (596), *853*
Gallorini, M. 165 (75), *186*
Gambino, O. 811 (591), 832 (591, 655–657, 662, 663), 835 (655), 837 (662), 838 (663), 839 (591, *853*, *855*
Games, D. E. 838 (670b, 670d), *855*
Gamlen, G. A. 690 (84), *842*
Gancanz, R. A. 648 (352), *663*
Gandel'sman, L. Z. 565 (43), *612*
Gansow, O. A. 191, 192 (14), *239*
Ganther, H. E. 162 (11), *185*, 244 (7), *283*, 389 (113), *396*
Gaprindashvili, V. N. 184 (167, 169, 175), *188*, 701 (159, 160), *844*
Garaj, J. 730 (274), *846*
Garazzolo, G. 332 (8), *341*
Garbarino, G. 411 (138), 412, 416 (159), *497*, 567–569, 572, 582 (87), 589, 591 (217), *613*, *616*
Garbe, J. E. 690 (61a), *841*
Garbuglio, C. 445 (503, 504), *504*
Gardner, S. 524 (79), *554*
Gardner, S. A. 145, 146 (328), *155*, 356 (36), *366*, 814, 818 (611), *854*
Garito, A. 449 (552, 556), 450, 463 (556), *505*
Garkin, V. P. 540 (197), 541 (199), *557*
Garlaschelli, L. 811 (596), *853*
Garnovski, A. D. 814, 820 (620), *854*
Garnovskii, A. 487 (1020), *515*
Garratt, D. 403 (29–31, 45), *495*, 528, 533 (130), *555*
Garratt, D. G. 127 (288), *154*, 529 (138–140), 533 (138, 139, 160), *555*, *556*, 632 (177, 180, 181), 633 (180, 181), 634 (177), 635 (201, 203, 204, 206), 636 (213, 229), 637 (213), *659*, *660*
Gasanov, B. R. 184 (145), *188*
Gasanov, F. G. 623 (62, 63), 624 (80), *657*, *658*
Gasanov, G. G. 289, 291 (24), *304*
Gasco, A. 573, 574 (119), *614*
Gash, D. M. 675 (33), *677*
Gasic, G. P. 638 (224), *660*
Gastaldi, L. 796 (536), *852*
Gasteiger, J. 456 (639), *507*

Gastmans, D. F. 163 (17), *185*
Gatien, L. 393 (160), *397*
Gattavecchia, E. 289, 291 (16), *304*
Gattermayer, R. 357, 358 (40, 43), *366*
Gattow, G. 158 (8), *160*, 244, 263 (6), *283*
Gautheron, B. 741 (331), 742, 749 (331, 338), *847*, *848*
Gavar, R. A. 184 (165), *188*
Gaydon, A. G. 159 (14), *160*
Gazieva, G. B. 632 (176), *659*
Gazizov, I. G. 245 (15), *283*
Gazizov, M. B. 815 (601g), *853*
Geary, W. J. 687 (50), *841*
Gebhardt, O. 71, 72 (54), *149*
Geens, A. 404, 480 (52), *495*
Geiger, C. 419 (250), *499*
Geise, H. J. 89 (140, 141), *151*
Gelius, U. 117 (268, 269), *153*
Gel'mont, M. 463 (728, 729, 731), *509*
Genard, J.-C. 440 (456), *503*
Genard, J. C. 584 (184), *615*
Geoffroy, M. 236 (135), 237 (135–139), *241*, 301 (62, 63), *305*
Geofroy, G. L. 811 (586), *853*
George, C. 82, 83 (101a), *150*, 670 (14), *676*
Georges, B. 320 (52), *329*
Georgiou, K. 11 (59), *19*, 65 (21), *148*
Georgoulis, C. 604 (317), *618*
Gerbeleu, N. V. 726 (251), 727 (251, 262a, 262b), 728, 739 (251), 770 (424), 771 (424, 425), 778 (424), 802 (548), *846*, *849*, *852*
Gerdil, R. 321, 323 (55), *329*
Gerding, H. 383 (72b), *395*
Gerenser, L. J. 701, 705 (157a–c), 736 (157c), *844*
Gerhold, J. 565, 604 (61), *613*
Gerlach, M. L. 178–181 (106), *187*, 387 (90), *396*
German, L. 405 (62a), *495*
Gerratt, J. 690 (85), *842*
Gerry, M. C. L. 65 (16), 75 (79), *148*, *149*
Gerstein, B. C. 701, 705 (158b), *844*
Gervasio, G. 808 (572, 573), 811 (595), *852*, *853*
Geserich, H. 465 (786), *510*
Geserich, H.-P. 90 (152), *151*
Geurts, P. J. M. 794, 795 (514), *851*
Gewald, K. 448 (537a), *505*
Gfeller, J.-C. 638 (227), *660*
Ghadendehari, M. H. 581, 595 (162), *615*
Ghanbarpour, A. 469 (842), *511*
Ghandehari, M. 447, 464 (532), *505*
Ghandehari, M. H. 377, 386 (35), *394*
Ghapgaran, F. 469 (842), *511*
Ghilardi, C. A. 811 (598), 813 (598, 599), *853*
Gianolio, L. 25 (41, 59), *61*
Gibson, C. 480 (944, 946), *513*
Gibson, C. S. 527 (117), 528 (128), *555*
Gibson, J. F. 742, 748 (340), 806 (561, 562), 832 (562), *848*, *852*
Gibson, J. S. 86, 87 (122), *150*
Gidali, M. 393 (156), *397*
Giddings, P. J. 624 (97), *658*
Gieren, A. 90 (152), *151*, 465 (786), *510*
Giesbrecht, E. 420 (273), *500*, 548 (266, 267), *558*, 567 (80), *613*, 626 (132–134), 629 (134, 156, 158), 641 (134, 270), 643 (357), 652, 653, 655 (158), *659*, *661*, *663*
Giese, B. 326 (70), 327 (72), *329*
Giese, R. 461 (707), *508*
Giese, R. W. 272 (80), *285*
Gil, R. 425 (309b), *500*
Gilbert, B. C. 234 (124), 235 (127), 237 (141), *241*, 321, 323, 324 (56, 60), *329*
Gilbert, F. L. 7 (32), *18*, 550 (278), *558*
Gilchrist, T. 403 (39), 488 (1040b), *495*, *515*
Gilchrist, T. L. 703 (163), *844*
Giles, J. R. H. 325 (65), *329*
Giles, J. R. M. 642 (243), *661*
Gillespie, J. R. 196, 199, 205 (33), *239*
Gillespie, R. 518 (8), *553*
Gillespie, R. J. 67, 71, 125, 129 (28), *148*, 217 (74), *240*
Gillissen, H. 494 (1071), *516*
Gillissen, H. M. J. 624 (101), *658*
Gillman, G. P. 337 (31), *341*
Gillson, J. 450 (575), *506*
Gilman, S. 642 (312), 645 (328, 329), *662*
Gilow, H. M. 601, 602 (287–289), *618*
Gilraldi, G. 383 (72c), *395*
Ginak, A. 458 (669), *508*
Ginet, L. 237 (137), *241*
Ginsberg, A. P. 806 (563, 564), 807 (564), 808 (566), 810 (564, 566, 582), *852*, *853*
Ginsburg, S. R. 519 (16), *553*
Gioaba, A. 482 (963–967), 483 (966, 973), *514*
Giolando, D. M. 811 (585b), *853*
Giorgini, M. G. 378, 383 (37b), *394*
Girardin, F. 429 (350, 351), *501*
Giudicelli, J.-F. 457 (660, 661), *507*
Gladyshev, E. N. 356 (35), *366*
Gladysz, J. A. 690 (61a, 61b), *841*
Glänzer, H. 801 (547), *852*
Glasebrook, A. 488 (1031), *515*
Glasser, L. S. D. 692, 694, 695 (117), *842*
Glavincevski, B. M. 716 (201), *845*
Glazebrook, H. H. 814 (612), *854*
Gleiter, R. 119 (271, 272), *154*, 247 (28), *284*
Glemser, O. 699, 700 (145a), *843*
Gliborskii, D. 407 (83), *496*
Glick, R. E. 561 (18), *612*
Glidewell, C. 66 (25), 72 (58, 61, 63), 76

(86), 78 (86, 87), *148–150*, 784, 786 (476, 477), *850*
Glover, J. S. 373, 393 (19), *394*
Glover, S. A. 536 (164), *556*
Glowiak, T. 787 (484), *851*
Glustrakov, V. 467 (813), *510*
Godfrey, P. D. 51, 57, 58 (94), *62*, 89 (144), *151*
Godovikov, N. N. 347 (7), *366*
Goerdeler, J. 465 (767), *509*
Goethals, E. 426 (317), *501*
Goggin, P. L. 690 (90, 91), 773, 774 (91), 814, 818 (608), *842*, *854*
Gokel, G. W. 680 (11b), *840*
Gold, L. 466 (805), *510*
Gold, L. P. 90 (149), *151*
Gold, P. M. 603 (308), *618*, 639 (119), *658*
Goldberg, I. 433 (391), *502*
Golden, D. M. 157, 159 (3), *160*
Gol'dfarb, Ya. 430 (364–367), 431 (373, 377), 434 (394, 396), *501*, *502*
Goldfarb, Ya. L. 597 (257), *617*
Goldsmith, D. 636, 637 (212), *660*
Golgolab, H. 464 (744, 747), *509*
Golic, L. 786–788, 790, 791 (485), 796, 797 (539), *851*, *852*
Golič, L. 102 (217), *152*, 624 (92), *658*
Golmohammadi, R. 622 (35), *657*
Golubinskii, A. V. 70, 74, 75 (46), 76 (46, 78b), 78 (46), *149*
Gombler, W. 71, 72 (55), *149*, 197, 199, 201, 205, 207 (41), 208, 209 (65), 210, 211, 218 (41), 219 (41, 77, 78), 220 (65), 224, 226, 229 (41), *239*, *240*
Gomper, R. 451 (583), *506*
Gonda, J. 486 (1000b), *514*
Gonzalez, C. L. 166 (83), *186*
Goodall, B. L. 680 (9c), *840*
Goodall, D. C. 774, 775 (440, 441), 776 (441), *850*
Goodfellow, R. J. 690 (90–92), 773, 774 (91), 814 (608), 816 (627), 818 (608), *842*, *854*
Goodgame, D. M. L. 730, 737 (273), *846*
Goodgame, M. 730, 737 (273), *846*
Goodman, M. M. 522 (52), *554*, 651 (413, 414), *664*
Gordienko, V. G. 573 (115), *614*
Gordon, G. 393 (160), *397*
Gordon, K. M. 317 (36), *329*
Gordon, M. S. 25, 30 (62, 64), 34, 54 (62), *61*
Gorgon, O. 459 (689), *508*
Goridis, C. 381, 393 (59), *395*
Gorissen-Hervens, F. 326 (71), *329*
Gorokhovskaya, V. I. 184 (162, 163), *188*
Gorton, J. E. 576 (135), *614*
Gosselck, J. 420 (274), 469 (841), 471 (862, 872), *500*, *511*, *512*, 574 (127), 575 (127, 128), *614*, 629 (165), 641 (271), *659*, *661*
Goth, H. 463, 484 (736a), *509*
Gould, E. 428 (335, 336, 344, 347), 462 (720, 721), 480, 481 (947), *501*, *509*, *513*
Gould, E. S. 521 (35), *553*, 580 (157, 158), 600 (157), *615*, 642 (279, 281, 282), 643 (279), 649 (281), *661*, 705, 706 (170), 721 (236), *844*, *845*
Gould, R. O. 102 (208), *152*
Goulden, P. D. 165, 167–169 (65), *186*
Goulon, J. 720 (228), *845*
Goulon-Ginet, C. 720 (228), *845*
Gower, J. L. 838 (670b), *855*
Grabley, F. F. 12 (66), *19*
Grachev, S. A. 172 (97), 173, 174 (97, 98), 175 (98), 176 (97, 98), *187*
Gracheva, L. N. 173, 175, 176 (99), *187*
Graf, G. 673 (28), *677*
Graham, W. A. G. 546 (242), *557*
Grandjean, D. 65 (13, 14), *148*
Grandolini, G. 421 (278), *500*, 582, 589, 591 (176), *615*
Granger, P. 7 (26), *18*, 206, 210, 211 (61), 221 (61, 87, 92, 93), 222 (92, 93), 226–228 (93), 234 (87, 93), *240*, 819 (630), *854*
Grant, D. M. 191 (17), *239*
Grant, H. 489 (1046), *515*
Grant, J. L. 310–312 (18), *328*
Grant, P. 404 (54), *495*
Grant, Z. A. 172–174, 176 (97), *187*
Granzhan, V. A. 567 (82), *613*
Gray, G. W. 573 (114), *614*
Gray, M. R. 184 (179), *188*
Graziosi, G. 759, 760 (380, 382), *848*
Grazuliene, S. 699 (152b), *843*
Greatrex, R. 825 (642), *854*
Grechkin, N. P. 815 (601f), *853*
Green, D. 449, 450 (554), 451 (580), *505*, *506*
Green, J. F. 520 (68), *554*
Green, M. 680 (9b), *839*
Green, T. E. 163–165, 183 (18), *185*
Green, T. H. 690 (76a, 76b, 77a, 77b, 78), 786 (77a, 77b), *842*
Green, W. H. 88 (138), *151*, 371, 382 (6), *393*
Greenberg, B. 428 (336), *501*, 521 (35), *553*, 642 (282), *661*
Greenhouse, J. A. 88 (138), *151*
Greenwood, N. N. 641 (274), *661*, 728, 731, 732 (264, 265), 734, 735 (265), 774 (442, 447, 451), 776 (447), 778 (451), *846*, *850*
Gregory, D. C. 106, 128 (225), *152*
Gregory, P. E. 391 (125), *397*
Greiciute, D. 403 (26a), *494*
Greiner, J. 249, 251 (39), *284*
Greiner, K. 418 (237), *499*

Grekova, E. 412–416 (157), *497*
Grekova, E. A. 595 (247), *617*
Grenan, M. M. 291 (39), *305*
Grewe, F. 462 (718), *508*
Grey, G. 403 (25), *494*
Grid, A. V. 601, 602 (290), *618*
Grieco, P. A. 16 (89), *19*, 604 (312), *618*, 629 (171), 640 (283), 642 (289, 290, 312), 645 (328, 329), *659*, *661*, *662*
Grier, R. A. 184 (158), *188*
Gries, T. 710, 712, 713, 814, 819 (191a), *844*
Griffin, D. A. 184 (178), *188*
Grigor'eva, L. P. 701 (156), *843*
Grigsby, R. A. 222, 226, 227 (105), *241*, 249, 251 (40), *284*, 375, 392 (26), *394*, 651 (399, 412), *664*
Grim, S. O. 198, 211 (39), *239*, 693–696 (124), *843*
Grimm, H. G. 335 (20), *341*
Grisdale, P. J. 562 (27), *612*
Grobe, J. 362, 363 (64), *367*, 711, 713 (192–194), 714 (192), 743, 751 (343), 814 (194), *844*, *845*, *848*
Groeneveld, L. R. 736 (308), *847*
Gronbaek, R. 771 (436), *850*
Gronowitz, S. 17 (100), *20*, 44, 54 (87), *62*, 191, 196 (3–6), 198–200 (5), 204 (6), 205, 210 (5), 213 (4), 217 (71), 218 (72), 230, 232 (109, 110), *238–241*, 385 (74), *395*, 407 (82a, 95), 411 (151b), 412 (151b, 166), 413 (182a, 182b), 414 (196), 415 (197, 206, 207, 210), 416 (215, 217–220, 222), 417 (223–225, 227, 229), 418 (222), 419 (248), 430 (370), 431 (375, 376), 433 (182a, 182b, 248, 389, 390, 392), 434 (393), 435 (403, 404), 436 (413), 437 (414, 417), 438 (430, 432), 478 (921), *496–499*, *501–503*, *513*, 579 (148–151, 153, 154), 580 (150, 151, 153, 154), 586 (195), 587 (198, 199), 588 (198–201), 591 (228), 595 (244–246, 248, 250, 256), 596 (248–250), 597 (199), *615–617*, 621, 622 (28), *657*, 675 (32), *677*
Gropen, O. 25, 30 (72), *61*
Groshev, V. 458 (665, 666), 490 (666), *507*
Gross, D. 465 (767), *509*
Gross, M. L. 252, 261 (43), *284*
Gross, R. 459 (697), *508*
Grossman, S. J. 519 (16), *553*
Grossoni, G. 295 (47), *305*, 683 (33), *841*
Groves, J. 487 (1018), *515*
Groves, L. G. 51 (91), *62*
Gruber, J. 808, 811, 834, 835 (568), *852*
Grummon, G. D. 371, 392 (12b), *394*
Gründel, D. 457 (663), *507*
Grundy, K. R. 715–717 (214), 765 (214, 404), 766 (214, 411), 807 (404), *845*, *849*
Grütjen, W. 519 (17a), *553*
Guanti, G. 411 (136, 138), *497*, 572 (111), 589, 591 (217), *614*, *616*
Guareschi, A. 411 (139, 140), *497*, 578 (145), 589 (216), 591 (145, 216), *615*, *616*
Guest, A. 408 (103), *496*
Guggolz, E. 746, 751, 752, 825 (351), *848*
Guha, P. 455 (618), *507*
Guilard, R. 720 (228), *845*
Guillaume, M. 268 (75), *285*, 442 (473), 476 (473, 908), *503*, *512*
Guimon, C. 458 (673), *508*
Gulland, J. 418, 468, 481 (232), *499*
Gulland, J. M. 654 (466), *665*
Gunning, H. 402 (13), *494*
Gunsalus, I. C. 238 (144, 145), *241*, 382, 392 (66a, 66b), *395*
Gunter, W. H. H. 560, 563 (2), *611*
Günther, H. 442 (467), *503*
Günther, W. 401 (3), 427 (324), 442 (471), 492, 493 (1061b), *494*, *501*, *503*, *516*
Günther, W. H. H. 2 (1), 5 (6), *17*, *18*, 162 (2), 166, 168, 177 (77), 179 (112), 182 (77), *185–187*, 236 (132), *241*, 288 (3, 12), 291 (3), *304*, 307 (2), *328*, 331 (1), 339, 340 (35), *341*, 379 (41), *394*, 518 (2), 519 (15), 520 (19), *552*, *553*, 620 (1), 621, 622 (1, 13), 624 (92), 629 (1), 641 (263), 650 (1), *656*, *658*, *661*, 680, 681, 690 (14a), *840*
Günzler, W. A. 389 (114), *396*
Gupta, T. R. 693–695, 698 (120), *842*
Guseinov, M. M. 623 (62, 63), 624 (82), *657*, *658*
Gustavson, P. 675 (32), *677*
Gutman, G. A. 381, 393 (58), *395*
Gutmann, R. 466 (803), *510*
Gutmann, V. 733 (296), *847*
Guziec, F. 467 (812), *510*
Guziec, F. S. 194, 196, 209, 220 (26), *239*
Guziec, F. S. Jr. 12 (63), *19*, 194, 208, 209 (24), 220 (24, 85), *239*, *240*, 675 (33), *677*
Gvozdeva, H. A. 604 (316), *618*
Gwinn, W. D. 87 (127), 126 (283), *150*, *154*
Gysling, A. J. 7 (27), *18*
Gysling, H. J. 64 (2b), 145, 146 (328), *148*, *155*, 222, 223 (98, 99), 231 (98), *240*, 344 (1), 356 (36), *365*, *366*, 524 (78, 79), *554*, 680 (15g), 681 (15g, 18, 25a, 25b, 31), 685 (31), 688 (57), 696 (133), 701, 705 (157c, 158a), 732 (285), 736 (157c, 158a), 737 (310a), 740 (310a, 320), 814 (605, 606), 815 (15g, 18, 605, 606, 626), 816 (18, 606), 817 (605), 818 (18), 826 (644), 829, 831 (57), 836 (669), *840*, *841*, *843*, *844*, *847*, *853–855*

Haaland, A. 72, 90 (65), *149*
Haas, A. 11 (61), 16 (87), *19*, 66, 67 (26), 87

(129), *148*, *150*, 199 (52), *240*, 281 (90, 91), *285*, 405 (61), *495*, 748 (360), 832, 836 (658), *848*, *855*
Haas, O. 709 (182), *844*
Haas, T. E. .730, 737 (273), *846*
Haase, W. 455 (627), *507*
Haba, A. (372), *663*
Hacker, N. 466 (802), *510*
Hadchouel, P. 393 (147), *397*
Haddad, P. R. 183 (141), *188*
Haddock, S. R. 690 (89, 90), 814, 818 (608), *842*, *854*
Haddon, R. 488 (1042), *515*
Hafeman, D. G. 389 (113), *396*
Hagelberg, L. 337 (27), *341*, 444, 492 (487), *504*
Hagen, G. 139, 140 (315), *154*
Haginiwa, J. 455 (610, 625), 456 (625, 641), 458 (674b, 675), 459 (674b, 686), *506–508*, 586 (209), 590 (209, 215), *616*
Haines, R. J. 825 (642), *854*
Haissinsky, M. 302 (67–69), *305*
Hajdu, R. 403 (40), 464 (764), *495*, *509*
Hajdu, R. A. 624 (93), *658*
Halazy, S. 321 (54), *329*, 624 (102), 635 (205), *658*, *660*
Halevy, J. 731 (279), *846*
Haley, N. 406 (63), 447 (528, 529), 451 (528), *495*, *505*
Hallberg, A. 412 (158), 417 (227, 229), *497*, *499*, 595 (256), *617*
Haller, R. 469 (843, 845a, 845b), *511*
Haller, W. S. 244, 253, 254 (5), *283*, 651 (388), *663*
Hamada, K. 71 (52), *149*, 650 (384), *663*
Hambrecht, J. 427 (331, 332), *501*
Hamburger, G. 518 (7), *553*
Hamer, F. 459 (695), *508*
Hamilton, C. S. 650 (367), *663*
Hamm, D. J. 387 (83), *396*
Hammar, G. 455, 457 (626), *507*
Hammer, M. 382 (64), *395*
Hammett, L. P. 561, 570 (13), *611*
Hammond, Ph. J. 537–539 (181), *556*
Hammond, P. J. 191, 192, 221, 222, 232 (15), *239*
Hanafusa, T. 642 (309), *662*
Hancock, J. R. 167, 177 (92), *187*
Hannaway, C. 825 (643), *854*
Hanocq, M. 164 (38), *186*, 581 (163), 598 (163, 263), *615*, *617*
Hansch, C. 419 (250), *499*, 563, 565 (35), 570 (35, 103), *612*, *614*
Hänsel, W. 469 (845a, 845b), *511*
Hansen, A. E. 675 (35), *677*
Hansen, L. 467 (821), *511*
Hansen, L. K. 116 (251), 117 (251, 253, 263, 270), 118 (263), 119 (251, 270), *153*, *154*
Hansen, M. R. 606 (328), *618*
Hansen, P. 321, 323, 324 (60), *329*
Hansen, P. E. 79 (94), *150*
Hansen, R. E. 238 (145), *241*, 382, 392 (66a), *395*
Hansen, R. S. 217, 218 (75), *240*
Hansen-Nygaard, L. 89, 90 (146, 147), *151*
Hanson, P. 91 (155), *151*, 237 (141), *241*, 453 (592, 593), 459 (684), 465 (592, 593), *506*, *508*
Hanson, R. 414 (192b), 455 (620), 461 (707), 464 (750, 752a), *498*, *507–509*
Hanson, R. H. 700 (148, 149), *843*
Hanson, R. N. 272 (80), *285*, 378, 392 (36), *394*
Hantschmann, A. 463 (736b), 469 (840b), 484 (736b), *509*, *511*, 788, 792 (501), *851*
Harding, M. M. 779 (468), *850*
Hardy, W. A. 64 (5), *148*
Hargittai, I. 64 (3, 4), 66 (4), 75 (78a), 76 (3, 78a), 81, 82 (3, 100), 83, 84 (3), 86 (4), 87 (3, 126, 132), 106 (225), 113 (240), 116 (4), 125 (282), 126 (285), 128 (225), 129 (282), 139, 140 (315), *148–150*, *152–154*
Haris, M. 492 (1062), *516*
Harkema, S. 320 (52), *329*
Harper, P. V. 387 (79), *396*
Harris, A. W. 381, 393 (58), *395*
Harris, D. O. 86, 87 (122), *150*
Harris, J. C. 600 (276), *617*
Harris, R. K. 190, 191 (1), 220 (79), 223 (101), 233 (113), *238*, *240*, *241*
Harris, R. O. 715–717, 765, 766 (214), *845*
Harris, W. C. 88 (133), *150*
Harrison, B. A. 738 (315), *847*
Harrison, W. D. 10 (55), *19*, 537 (172), *556*
Harrit, H. 466 (801), *510*
Harrowfield, J. M. 779 (469), *850*
Harrowfield, J. McB. 779 (466b), *850*
Hartl, H. 545 (239), *557*
Hartley, F. R. 680, 681, 691 (13), *840*
Hartmann, H. 408 (101), 453 (596), 455 (629), 463 (736b), 469, 470 (850), 484 (736b), 485 (997), *496*, *506*, *507*, *509*, *511*, *514*
Hartough, H. 406 (71), *495*
Harts, G. 465 (780), *510*
Harvey, A. B. 88 (138), *151*, 371 (6, 14, 15), 382 (6, 15), *393*, *394*, 650 (382, 383), *663*
Hasan, G. 459, 460 (685), *508*
Hase, T. A. 632 (174), *659*
Hasnain, S. 407 (89), *496*
Hassaneen, H. 466 (811), 467 (811, 814a, 814b), *510*
Hassel, O. 139 (313), *154*
Hastings, C. R. 183 (135), *187*

Haszeldine, R. N. 629 (161), 641, 643 (269), *659*, *661*
Hatkevitz, Z. 742 (333), *848*
Hatton, J. 467 (820a), *511*
Hatton, J. R. 118 (265), *153*
Hauge, S. 70 (35), 77 (81), 85 (108), 86 (121), 102 (210, 211), 115 (35, 247), 116 (35, 247–250), 119 (210, 280), 120 (280), 124 (210, 280), *148–150*, *152–154*, 651 (426), *664*, 684 (41, 42), 685, 686 (41–43), 728 (266), 730 (271), *841*, *846*
Haugen, G. R. 157, 159 (3), *160*
Hauser, J. 446, 447 (521), *504*
Hauser, J. J. 650 (387), *663*
Hausmann, H. 690, 691, 754, 757, 814, 825 (108), *842*
Hautschmann, A. 457 (663), *507*
Havezov, I. 165 (64), *186*
Havinga, E. 139 (314), *154*
Hayama, T. 632, 633 (190), *660*
Hayashi, M. 70 (41, 50), 71, 72 (50), 73 (41, 69), 81 (41), *149*
Hayashi, Y. 631 (172), *659*
Hayes, F. 490 (1056), *515*
Hayes, N. V. 575 (129), *614*
Hayon, E. 297 (53), *305*
Hayward, G. C. 654 (468), *665*
Haywood, B. 487 (1018), *515*
Hazell, A. C. 144 (324), *155*
Hazum, E. 835 (668a), *855*
Headford, C. E. L. 715, 717, 767 (215), 768, 825 (418), 829 (418, 650a, 650b), 830 (418), *845*, *849*, *854*
Healy, M. D. 680 (9a), *839*
Heath, G. 444 (498, 499), *504*
Heath, G. A. 308 (5, 6), *328*, 798, 800 (543), *852*
Heber, R. 624 (92), *658*, 786 (485), 787 (485, 493), 788 (485), 789 (493), 790 (485, 493), 791 (485), *851*
Heberhold, M. 804, 805 (557), 832 (653), *852*, *855*
Hebtchek, P. R. 525 (86), *554*
Hecht, C. 829 (646, 647), 830 (647), 832 (646, 647), 833 (647), *854*
Hedberg, K. 72 (57, 67), 117 (266), 126 (286), *149*, *153*, *154*
Hedberg, L. 126 (286), *154*
Hefferman, M. L. 383 (72e), *395*
Hegedus, L. S. 718, 719, 721 (227), *845*
Hegsted, D. M. 387 (77a), *395*
Hehre, W. J. 25 (27, 29, 32, 35, 48, 56, 62, 64, 68, 71), 29 (27, 68, 71), 30 (48, 62, 64, 68, 71), 33 (27, 68, 71), 34, 54 (48, 62), 56 (48, 68), 57 (48), *60*, *61*
Heibl, C. 12 (64), *19*
Heinrich, A. 784 (480), 789 (480, 507), 791 (480), *851*
Heinrich, P. 357, 358 (45), *366*
Heintz, G. 462 (714), *508*
Hellrung, B. 562 (33), 569, 570 (33, 102), *612*, *614*
Hellwinkel, D. 8 (35), *18*, 428 (338), 443 (477, 478, 480), *501*, *504*, 535, 536, 544 (163a–d), *556*, 655 (481), *665*
Helmi, S. 387, 389 (91), *396*
Hemmerich, P. 308 (11), *328*
Hemmings, R. T. 349 (17), 353 (18), *366*, 621, 622 (20), *656*, 716 (201), *845*
Hende, J. H. van den 117 (258), *153*
Henderson, H. E. 716 (201), *845*
Henderson, R. S. 747, 753, 757 (354), 808 (354, 574), *848*, *852*
Hendra, P. J. 654 (468), *665*, 684–686 (36), 690 (67, 68), 705, 706 (171), *841*, *844*
Hengge, E. 355 (29), *366*
Henglein, A. 293 (44), *305*
Henrick, K. 811, 812, 832, 839 (590), *853*
Henriksen, L. 9 (46), 12 (67), 13 (70, 71), *18*, *19*, 405 (58), 406 (70), 447 (524), 448 (533, 534, 543), 450 (543), 468 (827c), 480 (533, 534), *495*, *504*, *505*, *511*, 821 (634), *854*
Henriksen, U. 308 (3), *328*, 444 (497), *504*
Henrio, G. 415 (202, 208), 435 (208, 401), *498*, *502*
Henry, G. 622 (30), *657*
Hentschel, P. 405, 469 (60a), *495*
Herberg, S. 523, 545 (61–63), *554*
Herberhold, M. 808 (567), *852*
Herbert, M. 641, 650 (262), *661*
Herd, M. 469 (844), *511*
Hermann, L. 113 (240), *153*
Herrmann, D. 363 (67), *367*
Herrmann, W. A. 768, 770 (421), 804 (554, 556a, 556b), 805 (556a, 556b, 558), 806 (554, 558), 829 (421, 554, 646, 647), 830 (421, 467), 832 (556a, 556b, 646, 647), 833 (554, 556a, 556b, 647), *849*, *852*, *854*
Hertel, H. 784, 785, 787 (475), *850*
Herzberg, G. 22 (2), 51 (94), 54 (2), 57, 58 (94), *60*, *62*
Herzig, C. 456 (639), *507*
Herzschuh, R. 784, 786 (479), 789 (507), 795 (479), *851*
Hetey, A. T. 693, 696 (123), *843*
Hetzel, K. R. 370, 372, 393 (4). *393*
Heuvel, C. van den 248 (31), *284*
Hevesi, L. 319 (42, 43, 47, 48), 320 (50, 52), 321 (53, 54), *329*, 526 (99), *555*, 605 (324), *618*, 623 (55, 57), 624 (106), 626 (137), 635 (205), *657–660*
Hewitt, T. G. 96 (182), *152*

Hidaka, J. 771, 772 (429), 779 (459), 780 (429), 781 (470), *849*, *850*
Hieber, W. 690, 691 (96, 101, 111), 695 (127), 696 (101), 697 (101, 127), 741 (323), 743 (323, 342), 748 (323), 808, 811 (568), 814, 818, 819 (618), 824 (640), 832 (618), 834, 835 (568), 838 (618), *842*, *843*, *847*, *848*, *852*, *854*
Higgins, P. J. 381 (56), *395*
Higuchi, H. 492 (1064b), *516*, 642 (310), *662*
Hill, A. F. 768, 829, 830 (419), *849*
Hill, R. A. 72 (59), *149*
Hillers, S. 184 (164), *188*
Hillman, D. C. 162 (14), *185*
Hilti, B. 446 (520), *504*
Himmelreich, D. 723–725 (248, 249), 814, 823 (624), *846*, *854*
Hinrichs, W. 101 (197), *152*, 483 (984), *514*
Hinsberg, O. 5 (17), *18*, 465 (769), *510*
Hinsel, W. 469 (843), *511*
Hinsken, H. 562, 569, 570 (33), *612*
Hirabayashi, T. 11 (57), *19*
Hiraoka, K. 202 (50), *239*
Hiraoka, T. 642 (253), *661*
Hirooka, T. 387 (80), *396*
Hirose, C. 64 (11), *148*
Hirota, E. 51, 57, 58 (94), *62*
Hirschfelder, J. D. 201 (44), *239*
Hisada, K. 389 (112), *396*
Hladík, V. 371, 382 (10b), *394*
Hlavacek, R. 455 (622), *507*
Ho, C. O. M. 716 (209), *845*
Ho, Tse-Lok 370 (2b), *393*
Hoan, N. G. 428 (345, 346), *501*
Hocking, W. H. 65 (16), *148*
Hodgeman, D. K. C. 235 (127), *241*
Hodgson, K. O. 102 (219), *152*, 742, 750 (334), *848*
Hodkinson, B. A. 393 (150), *397*
Hoeger, C. 444 (490b), *504*
Hoeger, C. A. 625 (118), *658*
Hoffman, J. L. 391 (126a, 126b), *397*
Hoffmann, K. 102, 106 (222), *152*, 825, 827 (637), *854*
Hoffmann, R. 119 (272), *154*
Höfler, M. 690, 691, 754, 757, 814, 825 (108), *842*
Hofmann, G. 5 (16), *18*, 455 (609), *506*
Hofmann, H. 768 (417), *849*
Hofmann, K. 428 (334), *501*
Hofmann, P. 722 (245e), *846*
Hofmann, W. 9 (45), *18*, 754, 757 (368), *848*
Hogeveen, H. 835 (668b), *855*
Hoggard, R. 701 (161, 162), 702 (162), 703 (161, 162), *844*
Hogue, D. E. 387 (83), *396*
Hohlneicher, G. 22 (3), *60*
Hojo, Y. 600 (284), *618*
Holker, J. S. E. 644 (318), *662*
Holliman, F. G. 550 (281), *558*, 668 (10), *676*
Holliman, R. 476 (906), *512*
Hollo, Z. M. 289, 291 (21), *304*
Holm, A. 12 (65), *19*, 65 (18, 19), *148*, 405 (57), 464 (57, 753), 468 (827c), *495*, *509*, *511*
Holm, B. 415 (206, 207), *498*, 595 (245, 246), *617*
Holm, R. H. 781 (471), *850*
Holmes, E. 469 (852), *511*
Holmes, T. 470 (859, 860), *511*
Holt, D. 449 (558b), *505*
Holtz, D. 202 (48, 49), *239*
Hölzle, G. 629 (157), 639 (238), 642 (247, 277), *659*, *661*
Homfeld, E. 592 (231), *616*
Hommes, H. 409, 414 (118b), 423 (290b), *497*, *500*
Homsany, R. 316 (30, 31), *329*, 333–336, 339, 340 (15), *341*
Honjo, T. 784, 786 (482), *851*
Hönle, W. 809, 832 (579), *853*
Hoots, J. E. 806 (565a, 565b), *852*
Hooz, J. 635 (202), *660*
Hope, H. 92, 93 (160), 138, 139, 143 (310, 311), *151*, *154*
Hopf, G. 548 (265), *558*
Hopkins, H. 460 (702), *508*
Hopkins, H. P. Jr. 600 (281), *617*, 621, 622 (4), *656*
Hordvik, A. 116 (251), 117 (251–257, 263, 270), 118 (259, 261, 263), 119 (251, 270, 273, 274), *153*, *154*, 308 (8–10, 12), *328*, 467 (821), *511*
Hori, M. 478 (924, 928, 929), *513*, 588 (202), 589 (203), 592, 593 (203, 235, 236), *616*, *617*
Hori, T. 16 (94), *20*, 541 (208), *557*, 624, 625, 627 (131), 640, 641 (257), *659*, *661*
Hörlein, R. 829, 832 (646), *854*
Hornback, C. J. 25 (26), *60*
Hornby, J. L. 690 (61a), *841*
Hornfeldt, A. B. 385 (74), *395*
Hörnfeldt, A.-B. 191, 196 (3–6), 198–200 (5), 204 (6), 205, 210 (5), 213 (4), 217 (71), 218 (72), 230, 232 (109, 110), *238–241*, 246 (20–22), *283*, *284*, 411, 412 (151b), 415 (197, 199, 200, 203, 210), 416 (215), 419 (248), 433 (248, 389), 434 (393), 436 (413), 437 (414, 417), 438 (430, 432), *497–499*, *502*, *503*
Hörnfeldt, A. B. 560, 563 (3), 579 (148, 149, 151, 153, 154), 580 (151, 153, 154), 586

(195), 588 (201), 595 (3), 599 (268–271), *611*, *615–617*
Hoskins, B. F. 800 (546), *852*
Houchin, M. R. 697 (135a), *843*
Houk, L. W. 690, 691 (110), *842*
Houlding, V. H. 779, 780 (458), *850*
Hout, R. F. Jr. 25, 29, 30, 33 (71), *61*
Howard, A. G. 184 (179), *188*
Howarth, D. T. 690 (72), *842*
Howell, H. E. 165, 177 (74), *186*
Hoye, T. R. 623 (67), *657*
Hoyer, E. 784 (478, 479), 786 (478, 479, 486, 504a), 787 (486, 493), 788 (497), 789 (493, 497, 504a, 505, 506), 790 (493), 794 (497, 528), 795 (479, 528), 796, 797 (538), *850–852*
Hsieh, M. 754, 755 (367), *848*
Hsu, C. F. 478 (924), *513*
Hsu, Y. F. 191, 192, 221, 222, 232 (15), *239*, 537 (181), 538 (181, 184), 539 (181), *556*
Huang, C.-K. 203, 221–225 (57), *240*
Huang, F. 458 (672), *508*
Huang, J. J. 238 (146), *241*, 382, 392 (68), *395*
Huang, W. S. 334, 335 (16), *341*
Hubel, W. 407, 409, 414 (97a), *496*
Hübel, W. 709 (181), *844*
Hübener, P. 806, 832, 833 (559), *852*
Huber, R. E. 184 (157), *188*, 370 (2a), *393*, 600 (283), *618*
Hübner, W. 788, 792 (502), *851*
Hubrecht, G. 426 (317), *501*
Huestis, L. 465 (793), *510*
Huge-Jensen, E. 765 (406), *849*
Hughes, G. 428, 443, 484 (340), *501*
Huheey, J. E. 129 (297), *154*
Hui, R. A. H. F. 643 (314), 645 (319, 320), 647 (338), *662*, *663*
Hulfnagel, J. 449 (564), *505*
Humffray, A. 383 (72e), *395*
Humffray, A. A. 569, 581 (95), *613*
Hummel, G. J. van 320 (52), *329*
Hummelink, T. W. 796 (534), *852*
Hundt, R. 357, 358 (40, 43), *366*
Hunger, B. 406, 447 (67), *495*
Hünig, S. 459 (692), *508*
Hunter, G. 113 (239), *153*, 705 (172), 706, 707 (175, 176), 728, 731, 732, 734, 735 (265), 774 (447, 449–452), 775 (454), 776 (447, 449, 450, 452, 454), 778 (451), *844*, *846*, *850*
Hunter, R. 459, 460 (685), *508*
Huppmann, P. 545 (239), *557*
Hursthouse, M. B. 114, 115 (246), *153*, 709 (186), 710, 712 (186, 187), 774, 776 (446), *844*, *850*
Huseby, S. 540 (190–192), *556*
Husebye, K. 651 (429), *664*
Husebye, S. 12 (67), *19*, 102, 103 (209), 119 (275), 145 (329), 147 (329, 331–333), *152*, *154*, *155*, 651 (427), *664*, 686, 689 (46a), 794 (526), *841*, *851*
Hutchins, S. R. 183 (141), *188*
Hutchinson, M. 11 (60), *19*, 67, 69 (29), *148*
Hutson, G. V. 742 (339), 743 (345, 346), 744 (339, 345, 346), 748 (339), 750 (339, 345), 774 (443), *848*, *850*
Hutton, A. T. 779 (465), *850*
Huygens, A. V. 522 (53, 55), *554*
Huzinaga, S. 25 (20, 24, 28, 45, 47, 53, 61, 66, 69, 70), 27 (18), 28 (66), 30 (47, 61, 66), 34 (47, 61, 69), 54 (61, 69), 60 (69), *60*, *61*
Hyde, W. 183 (137), *187*

Iaconianni, F. J. 721 (243), *846*
Ibahim, N. 648 (344), *663*
Iball, J. 99 (192a), *152*
Ibanez, A. 735, 738, 739 (298), *847*
Ibragimov, A. A. 76 (32), *394*
Ibragimova, A. 413 (180), *498*
Ibragimova, A. B. 581 (171), *615*
Ibrahim, N. 318 (39, 40), *329*, 648 (354), *663*
Ice, R. D. 370 (4), 371 (11), 372 (4), 375, 392 (27), 393 (4), *393*, *394*, 651 (409, 410), *664*
Ichihara, R. 446 (512b), *504*
Idelson, M. 680 (9b), *839*
Ignatenko, E. G. 184 (180, 181), *188*
Iijima, T. 70, 72 (51), 89 (148), 90 (51), *149*, *151*
Ikeda, T. 762 (401a), *849*
Ikeda, Y. 203, 204, 218, 220 (58), *240*, 446 (512c, 512d), *504*, 519 (17b), *553*
Iksanova, S. V. 537 (183), *556*
Ilberg, K. 653 (453), *665*
Il'chenko, A. Ya. 566 (71), *613*
Il'lina, L. A. 786 (489), 787 (488–490), *851*
Illuminati, G. 590 (220–222), 591 (222), *616*
Ilsley, W. H. 736 (307), *847*
Il'yasov, A. V. 565, 566 (50), *612*
Imhoff, D. 391 (124), *397*
Inamoto, N. 675 (34), *677*
Inamura, H. 446 (512c, 512d), *504*
Ingold, K. U. 234 (121), 235 (121, 130, 131), 236 (121), *241*, 325 (66, 67), 326 (68), *329*
Ingram, L. 692, 694, 695 (117), *842*
Innorta, G. 44, 54 (86), *62*, 246 (17), *283*, 416 (212), *498*
Inoue, M. 453, 486, 487 (595), *506*
Ionin, B. I. 624 (91), *658*
Iordanov, N. 165 (53), *186*
Irgolic, K. 16 (95), *20*, 401 (5, 6), 406 (6), 483 (974), *494*, *514*, 815 (601b), *853*
Irgolic, K. J. 2 (3), *17*, 86 (117), *150*, 162 (1,

4–6, 13, 14), 163 (1, 4, 6, 29), 165 (13), *185*, 203, 221 (57), 222 (57, 105), 223–225 (57), 226, 227 (105), *240*, *241*, 244 (5), 247 (24), 249, 251 (40), 253, 254 (5), *283*, *284*, 362 (66), 365 (73), *367*, 375, 392 (26), *394*, 518, 520, 540 (3), 550 (275), *553*, *558*, 651 (388, 399, 412), 656 (373–380), *663*, *664*, 680, 681 (14b, 15a–f), 690 (14b), 796, 797 (533), 815 (15a–f), *840*, *852*
Irving, H. M. N. H. 778 (456a, 456b), 779 (456a, 456b, 465, 466a), *850*
Irving, J. G. 571 (108), *614*
Isaeva, L. S. 601, 602 (290), *618*
Isbell, A. F. 360 (52), *366*
Ishida, T. 453, 486, 487 (595), *506*
Ishii, A. 675 (34), *677*
Ishii, H. 354 (21, 22), 355 (26), *366*, 621, 622, 624 (19), *656*
Ishii, Y. 520 (20, 32), 521 (32), *553*
Ishioka, R. 527, 528 (111), *555*, 653 (459), *665*
Islam, K. M. S. 825 (643), *854*
Israel, M. 624 (108), *658*
Issaq, H. J. 183 (142), *188*
Itagaki, K. 634 (199), *660*
Iteke, E. 267 (72), *285*, 432 (383), 433 (387), 434 (383), *502*
Iteke, F. 431 (374), 432 (386), 434 (374), *502*
Iteke, F. B. 596, 597 (251), *617*
Itoh, K. 634 (198, 199), *660*
Ivanova, E. I. 730 (269c), *846*
Iversen, A. J. 360 (58), *367*
Ives, D. J. G. 335 (21), *341*
Iwama, Y. 520 (20, 32), 521 (32), *553*
Iwamura, H. 203, 204, 218, 220 (58), *240*, 519 (17b), *553*
Izatt, R. M. 680 (11a), *840*

Jackson, W. P. 642 (256), *661*
Jacob, R. 462 (715), *508*
Jacobs, A. L. 178 (105), *187*
Jacobs, P. 407 (91, 92), *496*
Jacobsen, C. 449 (560, 561), 450 (561, 570), *505*, *506*
Jacobsen, C. S. 94 (166), *151*
Jacobsen, H.-J. 357, 358 (44), *366*
Jacobsen, M. 458 (827c), *511*
Jacobus, D. P. 291 (39), *305*
Jacoby, B. 381, 393 (60), *395*
Jacquignon, P. 244 (2), 255 (47), 265 (71), 266 (2), 267 (72), 268 (74), 276 (86), *283–285*, 422 (282, 283), 423 (298), 426 (298, 313–315), 465 (315), 475 (283), 479 (931), *500*, *501*, *513*
Jacquo, M. Jr. 393 (160), *397*
Jafari-Namin, R. 469 (842), *511*
Jaffari, G. 464 (757), *509*
Jaffé, H. H. 561, 565 (15), 569 (99), *611*, *614*
Jaitner, P. 746 (352), 747 (359), 750 (352, 359), *848*
Jakobsen, H. 416 (221), *499*
Jakobsen, H. J. 196, 208 (36), 217 (75), 218 (75, 76), 222, 232 (36), *239*, *240*
Jakubke, H.-D. 673 (26), *677*
James, B. R. 721 (235), *845*
James, F. G. 624 (100), *658*
James, S. M. 715–717, 765, 766 (214), *845*
Janghorbani, M. 387, 388 (98), *396*
Janickis, V. 86 (111, 112), *150*
Janne, D. 473 (892), *512*
Janušek, Z. 326 (69, 71), *329*
Jaouen, G. 715 (222), 716 (208), 720 (222), *845*
Jaquignon, P. 584 (205), *616*
Jaw, J. Y. 16 (89), *19*, 640 (283), *661*
Jayle, M. F. 381, 393 (57), *395*
Jefson, M. 642 (251), *661*
Jen, K.-Y. 624 (95), *658*
Jen, K. Y. 334, 335 (16), *341*
Jenkins, K. J. 387 (97), *396*
Jennings, K. R. 245 (9), *283*
Jennings, M. A. 697, 734 (137), 737, 738 (311), 739 (137, 311), 740 (137), *843*, *847*
Jennings, W. 469 (854a), *511*
Jenny, W. 626 (145, 151), 629 (151, 157), 639 (151, 238–240), 642 (244, 247, 277), *659*, *661*
Jensen, H. P. 639 (123, 124), *658*
Jensen, K. 405 (58), 406 (70), 444 (497), 448, 480 (534), *495*, *504*, *505*
Jensen, K. A. 5 (7), 7 (27), 9 (46), *18*, 68 (32), *148*, 308 (3, 4), *328*, 335 (23), *341*, 681 (17b), 690 (80, 109), 721, 722 (238), 765 (406), 771 (423), 793 (17b, 510–512), 794 (510–512), 795 (510), 796 (510, 537), *840*, *842*, *845*, *849*, *851*, *852*
Jensen, L. S. 387 (95), *396*
Jerome, D. 449 (568), 450 (569), *505*, *506*
Jérome, D. 17 (102), *20*
Jessep, H. F. 348 (15), *366*
Jesson, J. P. 546 (243), *557*
Jewett, K. J. 162 (15), *185*
Jham, G. N. 272 (80), *285*
Jiang, S. 165, 167–169 (71), *186*
Jilek, J. 490 (1053), *515*
Jiménez-Garay, R. 101 (198), *152*
Joanny, M. 488 (1036), *515*
Johannesen, O. 685, 686 (43), *841*
Johannesen, Ø. 102, 119, 124 (210), *152*
Johannsen, I. 221 (89), *240*, 448 (549), 449 (560), 450 (549), *505*
Johansen, R. 82 (99), *150*
Johansson, B. L. 184 (147), *188*
Johansson, E. 184 (151), *188*

John, D. I. 624 (97), *658*
John, P. 690, 691, 696, 697 (101), *842*
Johnsen, J. 84, 85 (106), *150*
Johnsen, K. 671 (19), *677*
Johnson, A. W. 603 (299), *618*
Johnson, B. F. G. 747, 753 (358), 811 (590, 593, 594), 812 (590, 594), 832, 839 (590), *848*, *853*
Johnson, C. D. 561 (14, 16), *611*
Johnson, C. M. 166–170 (80), *186*
Johnson, D. 468 (834, 835), *511*
Johnson, D. H. 376 (29, 30), 384 (30), 385 (29), *394*, 546 (249), *558*
Johnson, D. R. 64 (7), *148*
Johnson, H. R. 51 (90), *62*
Johnson, I. 191, 196 (3, 4), 213 (4), 230, 232 (109), *238*, *241*, 385 (74), *395*, 416 (215), *498*, 579 (148, 149, 153, 154), 580 (153, 154), *615*
Johnson, J. 480 (944, 946), *513*
Johnson, K. A. 737, 740 (310b), *847*
Johnson, R. A. 163 (31), *185*
Johnson, R. E. 387 (81), *396*
Johnston, J. 622, 624 (21), *656*
Johnstone, R. 402 (17), *494*
Johnstone, R. A. W. 248 (30), *284*
Jones, C. 489 (1049), *515*, 526 (102), *555*
Jones, C. H. W. 358 (46), *366*, 716 (200), 814 (613, 616b), 818 (613, 628), 819 (616b), 825 (616b, 638), 826 (638, 644, 645), *845*, *854*
Jones, C. L. 102 (208), *152*
Jones, D. N. 668, 669 (11), *676*
Jones, G. 414 (192b), *498*
Jones, H. L. 569 (99), *614*
Jones, J. B. 391 (122), *397*
Jones, J. R. 586 (192), *616*
Jones, M. E. 709 (179), *844*
Jones, P. G. 710, 712, 713 (191a), 768 (417), 811 (597), 814, 819 (191a), *844*, *849*, *853*
Jones, R. N. 523 (73), *554*
Jones, T. 15 (85), *19*, 716 (200), 814, 818 (613), 825, 826 (638), *845*, *854*
Jong, R. de 481, 491 (958), *513*
Jonghe, W. de 165, 167–169 (71), *186*
Jonsson, E. 641 (266), *661*
Jordanov, N. 165 (64), 171 (102), *186*, *187*
Jordanovska, V. 486 (1008), 487 (1008, 1013), *514*
Jorgensen, C. K. 721 (242), *846*
Jørgensen, C. K. 7 (27), *18*, 681 (17b), 793 (17b, 512), 794 (512), 796 (531), *840*, *851*, *852*
Jorker, C. 651 (396), *664*
Jouille, M. 403 (46a), *495*
Jouin, P. 622 (30), *657*
Jović, Z. 684–686 (36), *841*
Joy, C. A. 5 (11), *18*
Juhlke, T. J. 355 (31), *366*
Julshamn, K. 117 (255), 118 (259), *153*
Junk, G. A. 247 (29), *284*
Jynge, K. 118 (260), *153*

Kabo, A. 632, 633 (180), 636 (229), *660*
Kada, R. 459 (693), *508*
Kadis, S. 381 (54), *395*
Kaeriyama, K. 334, 335 (19), *341*
Kagan, J. 464 (752b), *509*
Kahl, W. 810 (581), *853*
Kahn, L. R. 25, 33, 34 (49), *61*
Kahn, N. 165 (69), *186*
Kahudo, M. 404, 446 (511), *504*
Kaim, W. 700, 701 (150), *843*
Kaiser, E. 420 (266), *499*
Kaiser, J. 102 (206, 217), *152*, 624 (92), *658*, 786–788 (485), 789 (503), 790, 791 (485), 796, 797 (539), *851*, *852*
Kajtár, M. 668 (5), 672 (22), *676*, *677*
Kalabin, G. 407 (93), 493 (1068c), *496*, *516*
Kalabin, G. A. 194, 195 (28), 202, 203 (53), 210 (28, 67), 219 (73), 220 (84), 222, 223, 225, 230, 233 (100), *239*, *240*, 563 (36–38), 565–567, 603 (52), *612*
Kalasinsky, V. F. 88 (133), *150*
Kal'naya, G. I. 704, 705 (169), *844*
Kamada, T. 165 (49, 51), *186*
Kambara, H. 87 (125), *150*
Kametani, T. 623 (61), 642 (303), *657*, *662*
Kaminskii, V. F. 95 (171–174), *151*
Kaminskii, Y. L. 569 (97), *613*
Kaminskii, Yu. L. 562, 569 (31), *612*
Kamitani, T. 721, 722 (241), 787, 789, 790 (491), *846*, *851*
Kamiya, M. 321, 323, 324 (59), *329*
Kamlet, M. J. 194 (30), *239*
Kampel, V. Ts. 347 (7), *366*
Kanada, Y. 446 (512d), *504*
Kanahara, S. 459 (698), *508*
Kandror, L. L. 578, 581 (144), *615*
Kaneko, K. 317 (37), *329*, 642 (242), *661*
Kane-Maguire, L. A. P. 838 (670b, 670d), *855*
Kanetsura, H. 334, 335 (19), *341*
Kanevskaya, Z. M. 567 (82), *613*
Kapkan, L. M. 565 (43), *612*
Kaplan, L. 419, 435, 438 (251), *499*
Kaplan, M. 488 (1042), *515*
Kapoor, K. 522 (48), *553*
Kapovits, I. 75, 76 (78a), *149*
Kappe, T. 461 (711), *508*
Karakida, K. 87 (123), *150*
Karandashev, V. K. 699 (152b), *843*
Karayannis, N. M. 721 (243), *846*

Karchenko, V. 469 (840a), 478 (923), *511*, *513*
Kargin, Yu. M. 565 (50, 65, 68), 566 (50), *612*, *613*
Kargol, J. A. 732 (288), 736 (288, 305), 739 (316, 317), 825 (305), *847*
Kari, R. 30 (73, 75), *61*
Kari, R. E. 25 (14, 37, 38), 60, *61*
Karle, I. L. 64 (1a), *148*
Karle, J. 64 (1a), 82, 83 (101a, 101b), *148*, *150*, 670 (14), *676*
Karle, J. A. 390 (116), *396*
Karlsson, O. 417 (224), *499*
Karmann, W. 293 (44), *305*
Karnaukhova, R. 413 (186), *498*
Karrer, P. 486 (1006), 487 (1016), *514*, *515*
Karup Nielsen, I. 7 (26), *18*, 519 (17a), *553*
Karwowski, J. 22, 25, 28 (8), *60*
Kasai, P. H. 72, 90 (66), *149*
Kasai, T. 444, 484 (495), *504*
Kaschani-Motlach, M. 16 (87), *19*
Kaschau, A. 483 (977), *514*
Kataev, E. 402 (18), 407 (87), 419 (247), 420 (18, 269, 271), 471 (269), 481 (952b), *494*, *496*, *499*, *500*, *513*
Kataev, E. G. 635 (200), *660*
Kataeva, L. 419 (247), 481 (952b), *499*, *513*
Kataeva, L. M. 565 (68), *613*
Kataoka, T. 478 (924, 928, 929), *513*, 588 (202), 589 (203), 592, 593 (203, 235, 236), *616*, *617*
Katar, E. 410 (124), *497*
Kato, M. 334, 335 (19), *341*
Kato, Y. 6 (23), *18*
Katritzky, A. R. 565 (62), *613*
Katsura, S. 165 (54), *186*
Katugin, A. S. 700 (155), *843*
Kauffmann, T. 652 (437, 438), *665*
Kaunziger, E. 537, 540 (171), *556*
Kaunzinger, E. 488 (1035), *515*
Kaupp, G. 452 (590), *506*, 641 (276), *661*
Kawaguchi, S. 686 (53), *841*
Kawakami, K. 765, 766 (407), *849*
Kawamura, T. 654 (465), *665*
Kawazsuo, Y. 542 (216), *557*
Keat, R. 690 (76a, 76b, 77b, 78), 697 (136), 786 (77b), *842*, *843*
Kebarle, P. 202 (50), *239*, 402 (13), *494*
Keck, G. E. 638 (225), *660*
Keimatsu, S. 339 (36), *341*, 463 (737, 738), 483 (737, 738, 978), 487 (1017), *509*, *514*, *515*
Keller, P. 460 (700), *508*
Kellett, R. E. 526 (101), *555*, 653, 654 (455), *665*
Kendall, R. 466 (804), *510*
Kennard, O. 78 (91), *150*, 364 (69), *367*, 462 (723b), *509*
Kenny, N. C. 94 (170), *151*
Kent, J. E. 90 (150), *151*
Kerber, J. D. 165 (69), *186*
Kerr, R. G. 14 (77), *19*, 626, 647 (138), 649 (335), *659*, *663*
Keske, R. 468 (835), *511*
Keske, R. G. 376 (28, 29), 377, 384 (28), 385 (29), *394*
Kewley, R. 65 (17), *148*
Keyes, G. 455 (611), *506*
Khait, S. E. 387 (88, 89), *396*
Khaletskii, A. 465 (782), *510*
Khaletskii, M. 465 (776), *510*
Khammatova, Z. M. 815 (601g), *853*
Khan, A. R. 114 (245), *153*, 709, 710 (183–186), 712 (184–186), 774, 776 (444), 819 (184), *844*, *850*
Khan, M. M. 252, 261 (43), *284*
Khangazheev, S. Kh. 624 (99), *658*
Kharasch, N. 629, 642 (153), *659*
Kharchenko, V. 408 (106), *496*
Khathing, D. T. 281 (92), *285*
Khattack, R. 458 (672), *508*
Khaverzov, I. 165 (53), *186*
Khavtasi, N. S. 184 (169), *188*, 700 (154), 701 (154, 159, 160), *843*, *844*
Khidekel', M. L. 95 (171, 173), *151*
Khilya, V. 486 (1003), *514*
Khmel'nitskii, L. 407 (84), *496*
Khôi, N. 644 (340), *663*
Khorami, J. 464 (751), *509*
Khorshev, S. Ya. 716 (202), *845*
Khotsyanova, T. L. 108 (235), *153*, 567 (81), *613*
Khozeeva, R. 439 (441), *503*
Khripak, S. 482 (959), *513*
Khripak, S. M. 528 (125), *555*
Khristova, N. 487 (1011), *514*
Khysalnov, M. M. 565 (68), *613*
Kiaeay, G. 255, 278, 279 (57), *284*
Kiaevay, G. 461 (713), *508*
Kice, J. 599 (275), *617*
Kice, J. L. 626 (130, 136), 648 (349–352), 650 (351), *659*, *663*
Kiel, W. A. 636 (218, 219, 222), 638 (218, 219), 651 (425), *660*, *664*
Kiener, V. 723, 732 (246), *846*
Kihwan Kim 570 (103), *614*
Kikukawa, K. 654 (465), *665*
Killops, S. D. 750 (361), *848*
Kilschner, A. S. 651 (410), *664*
Kimel'fel'd, Ya. M. 88 (137), *151*
Kimel'fel'd, Y. M. 383 (69a, 69b, 70), *395*
Kimura, M. 70, 72 (51), 89 (148), 90 (51), *149*, *151*

Kimura, T. 238 (146), *241*, 382, 392 (68), *395*
King, C. 483 (974), *514*
King, C. E. 247 (24), *284*
King, F. 467 (817), *510*
King, H. F. 25 (21), *60*
King, L. 455 (622), *507*
King, M. G. 691 (116), 692 (116, 117), 693 (116, 121), 694 (116, 117), 695 (117), *842*, *843*
Kinker, J. 183 (137), *187*
Kiprianov, A. 486 (1003), *514*
Kirchberg, H. 794, 796 (513), *851*
Kirilash, A. 431 (372), *502*
Kirkien, A. M. 264 (68), *284*
Kirmse, R. 784 (478, 480), 786 (478, 504a, 504b), 787 (493), 788 (497), 789 (480, 493, 497, 504a, 504b, 505–507), 790 (493), 791 (480), 794 (497), 796, 797 (538), *850–852*
Kirsch, G. 407 (98–100), 409 (119, 120), 419 (258, 259, 262), 421 (276), 425 (310), 429 (354, 355), 432 (384, 385), 434 (354), 435 (407), 436 (355, 408–410), 439 (438, 439), 440 (119), 471 (100, 870), 490 (870), *496*, *497*, *499–503*, *512*, 522 (52), *554*, 651 (413), *664*
Kirschbaum, E. 482 (962), *513*
Kirschner, A. S. 651 (409), *664*
Kirsh, G 585 (189), 586 (191), *616*
Kirshbaum, E. 655 (491), *665*
Kisch, H. 703 (164), *844*
Kishore, R. 382 (64), *395*
Kiss, I. 419 (255), *499*
Kistenmacher, T. 450 (572a), *506*
Kistenmacher, Th. 449 (557), *505*
Kistenmacher, T. J. 93 (165), 94 (167, 168), *151*
Kitaev, G. A. 177 (127), *187*, 681, 686 (23b), *840*
Kite, K. 114 (245, 246), 115 (246), *153*, 709, 710 (183–186), 712 (184–186), 727, 729 (257), 771 (437), 773 (257, 437–439), 774 (437–439, 444, 446), 775 (437, 438), 776 (444, 446), 819 (184), *844*, *846*, *850*
Kivekaes, R. 447 (526), *505*
Kivekäs, R. 79 (95, 96), 92 (157–159), *150*, *151*, 525 (94, 98), *554*
Kjøge, H. M. 308 (8), *328*
Klaeboe, P. 68 (31), *148*, 688, 689 (59), *841*
Klages, C.-P. 237 (140), *241*
Klar, G. 15 (82), *19*, 101 (197), 102, 106 (222), 147 (330), *152*, *155*, 483 (982, 984), *514*, 521 (34), 539 (186), 540 (193–196), *553*, *556*, *557*, 652 (435, 436), *664*, *665*, 688 (58), 742, 747 (341), 825, 827 (637), *841*, *848*, *854*
Klayman, D. 401 (3), 457 (651), 461 (708, 709), *494*, *507*, *508*
Klayman, D. L. 2 (1), *17*, 162 (2), *185*, 288 (3, 11), 291 (3, 39), *304*, *305*, 307 (2), *328*, 518 (2), *552*, 560, 563 (2), 599 (274), *611*, *617*, 620–622 (1), 626 (152), 629, 650 (1), *656*, *659*, 680, 681, 690 (14a), *840*
Kleczek, E. 458 (667), *507*
Kleinpeter, E. 457 (663), *507*
Klevay, L. M. 651 (400), *664*
Klimova, L. I. 184 (162, 163), *188*
Klimova, V. A. 164 (45), 165 (56, 57), *186*
Klinga, M. 79 (96), 92 (157, 158), *150*, *151*, 525 (98), *554*
Klingen, M. 447 (526), *505*
Klingsberg, E. 117 (258), *153*, 445 (507b), 446 (509), *504*
Klinke, H. 465 (767), *509*
Kloeboe, P. 15 (83), *19*
Klofutar, C. 537 (173), *556*
Klyne, W. 88 (136), *151*
Klyshevich, R. P. 577, 609 (142), *615*
Knapp, F. F. Jr. 203, 221 (57), 222 (57, 105), 223–225 (57), 226, 227 (105), *240*, *241*, 374 (25), 375 (26), 381 (49), 392 (26, 49), 393 (25), *394*, *395*, 522 (52), *554*, 651 (405–407, 411–414), *664*
Knapp, G. 163 (33), *185*
Knapp, K. 460 (700), *508*
Knauer, K. H. 308 (11), *328*
Knaust, D. 336 (26), *341*
Knerr, G. D. 126 (284), *154*
Kneubuhler-Hof, A. 569, 570 (102), *614*
Kniezo, L. 457 (657), *507*
Knizhnik, A. 458 (669), *508*
Knobler, C. 92, 93 (160), 128 (291), 135, 136 (305, 306), 138 (291, 305, 306, 309–311), 139 (291, 309–311), 140 (317), 143 (305, 306, 309–311, 317), *151*, *154*
Knoblock, E. C. 184 (177), *188*
Knol, J. 487 (1018), *515*
Knop, B. 360 (51), *366*, 787 (487), 796 (535), *851*, *852*
Knoppova, V. 459 (693), *508*
Knott, E. 455, 460 (623), *507*
Knox, S. A. R. 750 (361), *848*
Knunyants, I. 403 (26a), 405 (62a), *494*, *495*
Kobayashi, M. 247 (28), *284*
Kobayashi, S. 338 (32), *341*
Kobayashi, T. 483 (978), *514*, 642 (253), *661*
Kobelt, D. 129, 131, 143 (299), 145, 146 (327), *154*, *155*, 528, 533 (134), *555*, 567 (76, 78), *613*
Kober, F. 364 (70), *367*
Koch, B. 87 (129), *150*
Koch, W. 191–193 (13), *239*
Kochendorfer, P. 487, 492 (1029), *515*
Koerber, J. 472 (884), *512*
Kohne, B. 196, 208, 222, 232 (36), *239*, 446

(512a), *504*
Kohri, Y. 70, 73, 81 (44), *149*
Koishi, M. 762 (401b), *849*
Kojima, M. 771, 772 (430), 779, 780 (430, 460), *849*, *850*
Kojima, T. 72 (68), *149*
Kokhimidis, M. 465 (786), *510*
Kokkinidis, M. 90 (152), *151*
Kolar, F. L. 249, 251 (40), *284*
Kolb, O. 715, 717, 720, 765 (224), 767 (224, 412), 796–798 (224), *845*, *849*
Kolesnikova, N. M. 724 (78), *657*
Kollmann, G. 381 (54), *395*
Kolodina, N. S. 569 (97), *613*
Kolonits, M. 106, 128 (225), *152*
Koloskova, N. 411 (133), 497
Koloskova, N. M. 602 (291), *618*
Koloskova, N. .N. 582 (178), *615*
Kolshorn, H. 196 (35), *239*
Kolthammer, B. W. S. 716 (211), *845*
Komarov, V. Ya. 624 (91), *658*
Komatsu, K. 458 (675), *508*, 590 (215), *616*
Kometani, T. 623 (69), *657*
Komin, A. 465 (784), *510*
Komina, T. V. 245 (15), *283*
Komppa, G. 419 (254), *499*
Konar, A. 191, 196, 204 (6), *239*, 407 (95), 413 (182a), 431 (375, 376, 378, 379), 432 (381), 433 (182a, 389, 390, 392), 434 (393, 395), 435 (400), *496*, *498*, *502*, 578 (147), 579 (147, 151), 580 (151), 587 (198, 199), 588 (198–200), 596 (147), 597 (199), *615*, *616*
Kondo, A. 393 (155), *397*
Kondo, K. 354 (21, 23), 355 (26), *366*, 403 (35), *495*, 621, 622 (7, 19), 624 (19), *656*, 762 (401a, 401b), *849*
Kondranina, V. Z. 565 (88), *613*
Kondratenko, N. B. 566 (71), *613*
Kondroi, I. 412 (165), *497*
Konjaeva, I. 430 (366), *501*
Konjajeva, I. P. 597 (257), *617*
Konno, T. 771, 772 (429), 779 (459), 780 (429), 781 (470), *849*, *850*
Kononenko, V. 458 (668, 670), *508*
Konstantinov, P. 411 (133), *497*
Konstantinov, P. A. 582 (178), *615*
Konstantinovsku, L. E. 781 (473), *850*
Konwer, D. 548 (269), *558*
Konyaeva, I. 431 (373, 377), 434 (394), *502*
Konyakhina, L. V. 171 (103), *187*
Koolpe, G. A. 605 (323), *618*
Koopman, T. 35 (84), *62*
Kooyman, E. C. 522 (53, 56), *554*
Kopaevich, Y. 405 (62a), *495*
Köpf, H. 741 (324, 325), 810 (581, 583), 811 (584), *847*, *853*
Kopoldova, J. 297, 299 (50), *305*
Kopylov, S. D. 164 (39), *186*
Korobov, M. S. 577, 601, 603 (138), *614*, 779, 781, 787, 801 (462), *850*
Korolev, G. K. 389 (104a, 104b), *396*
Korolov, M. S. 522 (44), *553*
Korp, J. 710, 712 (189), *844*
Korzhenevskaya, N. G. 565 (43), *612*
Koshelev, Y. 439 (440), *503*
Koshelev, Yu. 437 (415), *502*
Koshelev, Yu. N. 587 (197), *616*
Koshhy, K. M. 319 (45), *329*
Kosina, A. P. 356 (35), *366*
Kossmehl, G. 431 (380), *502*
Kostiner, E. 744–747 (348), *848*
Kotia, N. K. 371 (17), *394*
Kotov, A. I. 95 (173), *151*
Kovac, J. 459 (693), *508*
Kovacic, P. 415 (210), *498*
Kow, R. 603 (306), *618*
Kowald, J. A. 163, 183 (23), *185*
Koyama, E. 165 (54), *186*
Kozak, I. 289 (25), 290 (26), 291 (25), *304*, 392 (144), *397*
Kozawa, Y. 641 (302), *662*
Kozikowski, A. P. 348 (14), *366*, 632, 633 (188), *660*
Kozikowsky, A. P. 638 (226), *660*
Kozlov, V. V. 649, 650 (370), *663*
Koz'min, A. 404 (55), 418 (238), 423 (297), 481 (238, 950), 494 (1075), *495*, *499*, *500*, *513*, *516*
Koz'min, A. S. 530 (144, 146), 531 (147, 148), *555*, *556*
Kozmin, A. S. 527 (105), *555*
Krackov, M. 427 (326), *501*
Krafft, F. 5 (13), *18*, 483 (977), *514*, 547 (254), *558*
Kramer, M. E. 733, 738 (293), *847*
Kramolowski, R. 824 (640), *854*
Krantz, A. 403 (40, 41), 464 (764), *495*, *509*, 624 (93), *658*
Krasilnikova, E. V. 356 (35), *366*
Krasnov, V. P. 16 (91, 92), *19*, 541 (200), *557*
Krasochka, O. 448 (539), *505*
Krasovec, F. 537 (173), *556*
Kraus, R. J. 244 (7), *283*
Krauth-Siegel, L. R. 752 (362), *848*
Krebs, B. 357, 358 (44), *366*, 567 (79), *613*
Kreis, G. 723, 724 (247), *846*
Kreissl, F. R. 722 (245e), 723, 724 (247), *846*
Kreiter, C. G. 723, 724 (247), *846*
Krief, A. 8 (37), *18*, 307 (2), 319 (47, 49), *328*, *329*, 402 (19), 403 (32), *494*, *495*, 526 (99), *555*, 599 (272), 603 (302), *617*, *618*, 623 (55, 60), 624 (73, 102–107, 110, 111), 626 (137), 632, 633 (191), 642 (304), 649

(353), *657–660*, *662*, *663*
Krilanovich, N. J. 166, 167, 171, 177 (78), *186*
Krishnan, R. 25 (44, 49), 33, 34 (49), *61*, 796 (532), *852*
Krishnan, V. 7 (27), *18*, 681 (17a), 721, 722 (238), 754, 755 (367), 793, 794 (510–512), 795 (510), 796 (510, 537), *840*, *845*, *848*, *851*, *852*
Krispan, C. 481 (952c), *513*
Kristian, P. 485 (996), 486 (1000b), 489 (1044), *514*, *515*, 580 (161), *615*
Kristiansen, N. A. 65 (19), *148*
Krivun, S. 470 (855), *511*
Kronenbitter, J. 221, 222 (90), *240*
Kronrad, L. 289 (25), 290 (26), 291 (25), *304*, 392 (144), *397*
Kronrád, L. 371 (10a, 10b), 382 (10b), *394*
Krosec, R. 340 (37), *341*
Kroth, H.-J. 222, 223 (97), *240*, 359 (49), 361 (60), *366*, *367*, 714, 716 (197), *845*
Kroth, H. J. 8, 16 (41), *18*, 815 (601e), *853*
Kroto, H. W. 11 (58–60), *19*, 65 (21), 67 (29, 30), 69 (29), *148*
Kruck, T. 690, 691, 754, 757 (108), 814 (108, 618), 818, 819 (618), 825 (108), 832, 838 (618), *842*, *854*
Kruger, A. A. 708, 709 (178), *844*
Kruger, G. J. 726, 728 (254), *846*
Krugh, N. 466 (805), *510*
Krugh, W. D. 90 (149), *151*
Kruglaya, O. A. 716 (202), *845*
Krupoder, S. A. 197 (38), *239*
Kruse, F. H. 86 (115), *150*, 165 (58), *186*
Ku, A. T. 7 (28), *18*
Kubota, M. 716 (209, 210), *845*
Kuchen, W. 360 (51), *366*, 784, 785 (475), 787 (475, 487), 796 (535), *850–852*
Kuchitsu, K. 51, 57, 58 (94), *62*, 87 (123, 125), *150*
Kuckertz, H. 527, 528 (112), *555*, 653, 654 (460), *665*
Kucsman, Á. 75, 76 (78a), *149*
Kudchadker, M. 162 (5), *185*
Kudchadker, M. V. 680, 681, 690 (14b), 796, 797 (533), *840*, *852*
Kudinova, M. 469 (847), 470 (855), 472 (883), *511*, *512*
Kudryashova, M. 407 (83), *496*
Kuebler, N. A. 25 (22), *60*
Kuhn, N. 814 (607, 617), 819 (617), *854*
Kukkola, P. 632 (174), *659*
Kul'bitskaya, O. V. 641 (292), *662*
Kuliev, F. A. 184 (145), *188*
Kuliev, M. 563 (37), *612*
Kulik, W. 409, 414 (118b), *497*
Kulkarni, P. S. 252, 261 (43), *284*
Kulkowit, S. 632 (175), *659*
Kull, F. J. 390 (116), *396*
Kullmer, V. 669, 754 (144), 758 (144, 370), 759, 804 (370), *843*, *848*
Küllmer, V. 753, 754, 757, 758, 825 (363), *848*
Kumar, M. 393 (153), *397*
Kumar, R. 442 (472), *503*, 522 (50), *553*
Kumar, R. C. 126 (284), *154*
Kumar, V. 522 (47), *553*
Kumari, S. 465 (785), *510*
Kunckell, F. 523 (71, 72), 525 (72), *554*
Kung, H. 374 (22), *394*
Kunitskaya, G. P. 540, 541 (198), *557*
Kupiu, B. S. 651 (394), *664*
Kurbanov, S. 407, 468 (85), *496*
Kurbanov, S. B. 623 (62, 63), *657*
Kurbatov, V. P. 779 (462, 463), 781 (462, 463, 473), 787, 801 (462, 463), *850*
Kurbatova, E. B. 164 (39), *186*
Kurobe, H. 623 (61), *657*
Kurosawa, M. 642 (305), *662*
Kurtz, D. M. 701, 705 (158b), *844*
Kurz, J. L. 600 (276), *617*
Kusar, M. 537 (173), *556*
Kuscik, D. 486 (1000b), *514*
Kushnarev, D. F. 194, 195 (28), 202, 203 (53), 210 (28, 67), 219 (73), 220 (84), 222, 223, 225, 230, 233 (100), *239*, *240*, 565–567, 603 (52), *612*
Küsthardt, U. 829, 832 (646), *854*
Kusuyama, Y. 519 (17b), *553*
Kutal, C. 680 (11c), *840*
Kutulya, L. A. 569 (94), 573 (115, 116), *613*, *614*
Kuwada, K. 73 (69), *149*
Kuwahara, J. 238 (148), *241*, 782 (474), *850*
Kuwajima, I. 10 (53), *19*, 549 (272), 550 (273), *558*, 609 (333–335), *618*, 625, 628 (125–129), 642 (147), *659*
Kuz'mina, L. G. 701 (156), *843*
Kuznetsov, S. I. 108 (235), *153*, 567 (81), *613*
Kvitko, I. 429 (356, 357), 430 (356), 437 (415), 439 (440–442), 472 (876), *501–503*, *512*
Kvitko, I. Ya. 587 (196, 197), *616*
Kyandzhetsian, R. A. 108 (235), *153*, 261 (63), *284*, 565, 566 (52), 567 (52, 81), 602 (293, 296), 603 (52, 293), *612*, *613*, *618*
Kzakov, M. I. 166, 171, 172, 174, 176 (81), *186*

Labar, D. 626 (137), *659*
Labauze, G. 801, 803 (551), *852*
Lablache-Combier, A. 473 (889), *512*
Laboureur, J. L. 319 (49), *329*
Laceile, S. 701, 705 (158b), *844*

Laconi, A. 392 (128), *397*
Ladatko, A. 487 (925), 479 (935, 936), 480 (936–938), *513*
Lafferty, J. 487 (1010), *514*
Lafferty, W. J. 51, 57, 58 (94), *62*
Lagow, R. J. 355 (31), *366*
Laing, D. L. 690 (94), *842*
Laing, M. 779 (467), *850*
Laishev, V. 406 (66, 69), 410 (69, 121a, 122), 447 (66), 448 (537b), *495*, *497*, *505*
Laishev, V. Z. 621, 622 (26), 624 (84), *656*, *658*
Laitalainen, T. 79 (95, 96), 92 (157–159), *150*, *151*, 447 (526), *505*, 525 (94, 96–98, 218), *554*, *557*
Laitem, L. 191, 199, 203–205, 210, 217 (7), *239*, 420 (270), 421 (279), 424 (307), 425 (307, 308, 309a, 311), *500*, 563 (41), *612*
Laitinen, R. 86 (119), *150*
Lakhsmikantham, M. 462 (723a), *509*
Lakomy, J. 562 (23), *612*
Lakshmikantham, M. 402 (20), 406 (64, 65), 447 (64), 448 (535), 449 (551–553, 556), 450 (551, 553, 556), 451 (578), 462 (553), 463 (556), *494*, *495*, *505*, *506*
Lakshmikantham, M. V. 652 (440), *665*
Lakshmikanthan, M. V. 334, 335 (16), *341*
Lakshmikkanthem, M. V. 651 (415), *664*
Lalezari, I. 255, 278 (48, 55), 279 (48), *284*, 406 (68), 409 (107–109, 121b), 447 (68, 531), 449 (563a), 452 (587), 455 (619), 461 (619, 710), 463 (739–741), 464 (107, 108, 531, 563a, 740, 741, 744, 747–749, 751, 754, 760), 466, 467 (808), 469 (842), *495–497*, *505–511*, 621, 622 (23, 24), *656*
Lalor, F. J. 715, 717, 720, 822 (223), *845*
Lam, D. H. 606 (239), *618*
Lambert, C. 449, 451 (559), *505*
Lambert, J. 468 (834–836), *511*
Lambert, J. B. 376 (28–30), 377 (28), 384 (28, 30), 385 (29), *394*, 546 (249), *558*
Lamm, V. 90 (152), *151*
La Monica, G. 754 (364, 365), 755, 756 (365), *848*
Lamotte, B. 321, 323 (61), *329*
Lamotte, J. 77 (83, 85), 92, 93 (164), *149*, *151*, 454 (604), *506*
Lamotte-Brasseur, J. 91 (154), *151*, 454 (606), *506*
Landolt-Börnstein 65, 81, 89, 128 (20), *148*
Landsberg, B. M. 11 (58, 59), *19*, 65 (15, 21), 67 (30), 70, 73, 81 (43), *148*, *149*
Lange, A. 699 (152a), *843*
Langer, E. 427 (329), *501*
Langmyhr, F. J. 165 (62), *186*
Lanovaya, E. 491 (1059), *516*
Lansinger, J. 427 (320), *501*
Lansinger, J. M. 318 (38), *329*
Lanthier, G. F. 269 (77), *285*
Lantz, R. 246 (21, 22), *284*, 415 (199, 203), *498*, 599 (270, 271), *617*
Lapham, D. J. 646 (331), *662*
Lappert, M. F. 13 (72), *19*, 102, 160 (221), *152*, 814, 821 (621, 622), 822 (621), *854*
Lardon, M. 195 (32), 196 (32, 34), 199 (34), 205 (32), *239*
Larionov, S. V. 786 (489), 787 (488–490), *851*
Larsen, C. 468 (831), *511*
Larsen, E. 244, 269, 283 (3), *283*
Larsen, R. D. 537 (168, 169), *556*
Lassigne, C. R. 234 (118), *241*
Lathan, W. A. 25 (56), *61*
Lathrop, K. A. 387 (79, 81), *396*
Lattman, M. 247 (26), *284*
Latyaeva, V. N. 356 (35), *366*
Latypova, P. V. 565, 566 (50), *612*
Lau, C. 523, 545 (64), *554*
Laue, H. A. H. 234 (124), *241*
Lauer, R. F. 624 (114), 629, 630 (114, 154), 632, 633 (114), 639 (114, 120), *658*, *659*
Laur, P. 7 (26, 33), 9 (33), *18*, 519 (17a), *553*
Laur, P. H. 231 (111), *241*
Laur, P. H. A. 668, 670 (6), 671 (6, 20b), 672, 673, 675, 676 (6), *676*, *677*
Laureni, J. 403 (40, 41), 464 (764), *495*, *509*, 624 (93), *658*
Laurenson, G. S. 66 (22, 24a), *148*
Lautenschlager, F. 403 (28), *495*, 527, 530 (120), *555*
Lavin, K. D. 739 (316), *847*
Lavrushin, V. F. 569 (94, 100), 573 (115), *613*, *614*
Lawitz, Z. 417 (224), *499*
Lawson, D. D. 629, 642 (153), *659*
Lawson, R. J. 835 (668c), *855*
Layer, M. 464 (746), *509*
Leach, J. B. 774 (442), *850*
Leandri, G. 412, 416 (159), *497*, 675 (31), *677*
Lebedeva, G. 472 (876), *512*
Le Beuze, A. 65 (14), *148*
Lebez, D. 381 (55), *395*
Le Coustumer, G. 469, 470 (849), 480 (939), *511*, *513*
Le Coustumer, J. 475 (898), *512*
Leddicotte, G. W. 379 (40), *394*
Lederer, C. M. 370 (1), *393*
Lederer, K. 655 (484–487, 490), *665*
Lee, B. 668 (7), *676*
Lee, C. 636, 637 (212), *660*
Lee, H. 10 (50), *19*, 537, 550, 551 (175), *556*, 656 (492), *665*
Lee, H. H. 641 (295), *662*

Lee, J. 470 (860), *511*
Lee, J. D. 85 (107), 86 (109), *150*
Lee, J.-S. 106 (228), 107 (228–231), 124 (228, 229), 132, 134, 143 (303), *153*, *154*
Lee, M. 449 (557), 450 (572a), *505*, *506*
Lee, S.-Y. C. 673 (25), *677*
Lee, T. W. S. 648 (349, 350), *663*
Lee, V. 449 (555), 450 (572b), *505*, *506*
Leemput, P. J. H. A. M. v. d. 796 (534), *852*
Lefohn, A. 376 (31), *394*, 418 (231), 468 (829), 480 (231), *499*, *511*
Lefur, D. 449 (568), *505*
Legendre, P. 419 (262), *499*
Legrand, M. 667 (1a), *676*
Legzdins, P. 716 (211), *845*
Lehmann, G. 621, 622 (17), *656*
Lehn, J.-M. 25, 28, 30, 33, 34 (67), *61*, 488, 489, 492 (1039a), *515*, 624 (87), *658*
Lehn, J. M. 604 (313), *618*
Leibnitz, E. 794, 796 (513), *851*
Leicester, H. M. 520 (67), *554*
Lemaire, C. 476 (908), *512*
Le Marechal, A. 447 (525), *504*
LeMaux, P. 715, 720 (222), *845*
Lendel, V. 404 (55), 418 (238), 423 (293, 297), 481 (238, 950), 494 (1074, 1075), *495*, *499*, *500*, *513*, *516*
Lendel, V. G. 528 (133), 530 (143, 144, 146), 531 (147, 148), 532, 534 (153), *555*, *556*
Lendel, V. L. 527 (105), *555*
Lenher, V. 803 (552), *852*
Lensch, C. 811 (597), *853*
Lentz, D. 545 (239), *557*
Lentzener, H. L. 576 (135), *614*
Leo, A. 570 (103), *614*
Leo, A. J. 563, 565, 570 (35), *612*
Leon, S. A. 381 (54), *395*
Leonesi, D. 686 (51), *841*
Lepage, L. 407 (90), *496*
Lepage, Y. 407 (90), *496*
LePeintre, C. 184 (176), *188*
Leporati, E. 794, 795 (529), *852*
Lepper, J. P. J. de 522 (53), *554*
Lepri, S. 438 (432), *503*, 586 (195), *616*
Leray, N. 237 (137–139), *241*, 301 (63), *305*
Lerchner, J. 789 (506), *851*
Lerouge, P. 642 (250), *661*
Leroux, Y. 377 (34), *394*
Lerstrup, K. 17 (104), *20*, 449 (557), 450 (572a), 451 (579), *505*, *506*, 651 (417), *664*
Lerstrup, K. A. 13 (70, 71), *19*
Lerstrup, K. L. 821 (634), *854*
Lesbre, M. 488 (1036), *515*
Lesch, D. A. 808 (570, 571, 575), 832 (570, 571, 652, 660, 661, 664), 833 (652), 834 (570, 571), 835 (652, 660, 661), 836 (652, 660), 837 (571, 652, 661, 664), 838 (652, 661, 664), *852*, *855*
Leslie, E. J. 784, 786 (476), *850*
Lesser, R. 419, 421 (253), 446 (508), 454 (599, 603), 477 (912, 915), *499*, *504*, *506*, *512*, *513*, 547 (253), *558*
Lester, D. J. 14 (76), *19*, 643 (314), 645 (319), 647 (325, 339, 347), 648 (346–348), *662*, *663*
Leung, M. 319 (41), *329*, 565, 574, 575, 606 (47), *612*
Le Van, D. 362, 363 (64), *367*
LeVan D. 711, 713 (192–194), 714 (192), 814 (194), *844*, *845*
Levander, O. A. 387 (94), *396*
Levason, W. 622 (33), *657*, 680, 681, 772 (10b), 801–803 (549), *840*, *852*
Levchenko, A. V. 173 (98–100), 174 (98), 175, 176 (98–100), *187*
Levchenko, E. S. 541 (210, 212a, 212b), 542 (212a, 212b), 548 (270), *557*, *558*
Levi, B. A. 25, 29, 30, 33, 56 (68), *61*
Levin, G. E. 393 (158), *397*
Levin, R. D. 245 (11), *283*
Levine, D. 466 (805), *510*
Levine, D. M. 90 (149), *151*
Levkoev, I. 437 (421), *502*
Levshin, I. 458 (669), *508*
Levshin, I. B. 164 (39), *186*
Levy, G. C. 191 (18), *239*
Lewicki, J. W. 524 (80), *554*
Lewis, I. C. 561 (18, 21), 571 (21), *612*
Lewis, J. 747, 753 (358), 811 (590, 593, 594), 812 (590, 594), 832, 839 (590), *848*, *853*
Lewis, N. A. 700, 701, 730, 737, 760 (153e), *843*
Ley, S. 462 (717), *508*
Ley, S. V. 10 (52), 14 (75, 76, 78), *19*, 537 (178, 179), *556*, 642 (252, 256, 258), 643 (314, 315), 644 (313, 315–317), 645 (319, 320), 647 (325, 338, 339, 347), 648 (321, 322, 341–343, 346–348), *661–663*
Li, T. 483 (983), *514*
Lias, S. G. 245 (11), *283*
Lichwa, J. 165 (69), *186*
Lide, D. R. 51 (88), *62*
Lie, G. L. 25 (51), *61*
Liebscher, J. 408 (101), 453 (596), 455 (629), 463 (736b), 469, 470 (850), 484 (736b), 485 (997), *496*, *506*, *507*, *509*, *511*, *514*
Lied, T. 622 (44), *657*
Liegeois, C. 475 (897), *512*
Liegeois, Ch. 563 (40), *612*
Liehr, G. 109 (236, 237), 110 (237), *153*
Lien, E. J. 570 (103), *614*
Lienig, D. 462, 492 (719), *509*, 643 (360), 650 (362), *663*
Liesk, J. 742, 747 (341), *848*

Lietz, M. 363 (68), *367*
Lightner, D. A. 675 (35), *677*
Likhitskaya, V. 458 (682), 460 (706), *508*
Liljefors, S. 412 (158), *497*
Lillicrap, S. C. 299–301 (61), *305*
Lim, J. C. 737, 740 (310a, 310b), *847*
Linda, P. 246 (23), *284*, 410 (132), 416 (213), 442 (466), *497*, *498*, *503*, 560, 571 (7), 573, 577, 578 (117), 581 (164, 165), 582, 589, 591 (165), 593, 594 (117), *611*, *614*, *615*
Linden, J. G. M. van der 794, 795 (514), *851*
Lindenauer, S. 487 (1018), *515*
Linderberg, J. 668, 670, 671 (2), *676*
Lindgren, B. 128, 129 (292, 293), *154*, 403 (37), 404 (53a), 493 (1068b), *495*, *516*, 548 (261b), *558*, 641 (298), *662*
Lindner, E. 9 (42), *18*
Lindner, U. 520 (27), *553*
Lindner, W. 674 (30), *677*
Lindøy, S. 70 (49), *149*
Lindsell, W. E. 806 (563, 564), 807, 810 (564), *852*
Linyova, A. N. 356 (35), *366*
Liorber, B. G. 815 (601g), *853*
Liotta, C. L. 600 (281), *617*, 621, 622 (4), *656*
Liotta, D. 6, 10 (24), *18*, 621 (8, 10), 622 (8, 10, 21), 623 (10, 64, 66), 624 (21), 632 (173, 182), 633 (182), 634 (192, 193), 636, 637 (212), *656*, *657*, *659*, *660*
Lipatova, I. P. 815 (601j), *853*
Lipnitskii, V. 456 (644), *507*
Liporich, T. V. 651 (404), *664*
Lipshan, Y. 448 (539), *505*
Lipsky, S. R. 276 (85), *285*
Lisiewicz, J. 289 (23), *304*
Litvinenko, L. M. 565 (59), 575 (59, 125, 126), *613*, *614*
Litvinov, V. 430 (364–367), 431 (373, 375–379), 432 (381), 434 (394, 396), 435 (400), 438 (437), 449 (558a), 464 (755), *501–503*, *505*, *509*
Litvinov, V. P. 578, 579 (147), 587 (198, 199), 588 (198–200), 596 (147), 597 (199, 257), *615–617*
Livdane, A. 446 (515), *504*
Livingston, H. K. 340 (37), *341*
Livingstone, S. E. 680, 782, 815 (12a, 12e), *840*
Ljungdahl, L. G. 390 (121), *397*
Llabador, Y. 303 (74, 75), *305*, 371 (13), *394*
Llabres, G. 75–77 (76, 77), 86 (113, 116), *149*, *150*, 191, 199 (7), 203 (7, 59), 204, 205 (7), 207, 208 (63), 210 (7, 59), 211 (59), 212, 214, 215 (69), 216 (70), 217 (7), *239*, *240*, 425 (311), *500*, 563 (41), 579 (150), 580 (150, 155), *612*, *615*
Llabres, G. L. 13 (68), *19*
Llaguno, E. 467 (820a, 820b), *511*
Llaguno, E. C. 118 (265), *153*
Lloyd, D. L. 602 (294, 295), *618*
Lo, B. W. N. 202 (51), *239*
Lo, F. Y.-K. 808 (569), 811 (588), *852*, *853*
Lobana, T. S. 692 (118), 693–695 (120), 698 (118, 120), 815 (603b), *842*, *853*
Lodge, P. G. 811, 812 (590, 594), 832, 839 (590), *853*
Loevenich, J. 547 (255), *558*
Loew, G. H. 25 (40), *61*
Loginova, E. I. 815 (601h, 601i), *853*
Loh, K. S. 690 (74, 105), 756 (105), *842*
Lohner, W. 196, 208 (36), 222 (36, 95, 96), 231 (95, 96), 232 (36), *239*, *240*, 409 (115), 416 (214), 421, 440 (275), 441 (459), 479 (933, 934), 480 (934), *496*, *498*, *500*, *503*, *513*
Lokaj, J. 730 (274), *846*
Lombard, A. 726, 728 (254), *846*
Lombart, C. 381, 393 (57), *395*
Long, R. E. 94 (169), *151*
Loon, J. C. van 165 (69), *186*
Loon, J. van 165, 167–169 (68), *186*
Lopez, L. 402 (21), *494*
Lopez-Alonso, H. 166 (83), *186*
López-Castro, A. 70 (36), *148*
Lorck, M. 462 (722), *509*
Lorenz, I. P. 759, 760 (388, 389), 761 (389), *849*
Loth-Compere, M. 268 (75), *285*
Loth-Compère, M. 442, 476 (473), *503*
Lott, P. F. 183 (139), *188*
Lou, C. 545 (240), *557*
Lovas, F. J. 64 (9), *148*
Low, B. W. 78 (93), *150*
Low, K. 404 (54), *495*
Löwig, C. J. 5 (9), *18*
Lown, J. W. 626 (152), *659*
Lown, M. W. 113 (239), *153*, 706, 707 (175, 176), *844*
Lowrey, A. H. 82, 83 (101a), *150*, 670 (14), *676*
Lowry, T. M. 7 (32), *18*, 550 (278), *558*
Lu, K. C. 87 (131), *150*
Lucchesini, F. 454 (607), 466 (800), 467 (815), *506*, *510*, 583 (179), 584 (206), 597 (253, 254), *615–617*
Lucchetti, J. 319 (49), *329*, 624 (73, 102), *657*, *658*
Lucchini, V. 403 (27, 42, 43), *494*, *495*
Lucken, E. A. S. 321, 323 (55), *329*
Lüdesdorf, R. 477 (917), *513*
Ludwig, E. 788, 792 (502), *851*
Luettke, W. 418 (236), *499*
Luft, R. 562 (29), *612*
Lugovkin, B. 459 (694), *508*

Luknitskii, F. 485 (995), 488 (1038), *514*, *515*
Lulek, R. 452 (588), *506*
Lumbroso, H. 410 (128, 129), 419 (241), 475 (897), *497*, *499*, *512*, 563 (40), *612*
Lumme, P. O. 597 (259), *617*
Lunde, G. 302 (67), *305*
Lungu, V. V. 297–299 (51), *305*
Lun'kova, M. 403 (26a), *494*
Luppold, E. 414 (195), 417 (228), 442 (468, 469), 443 (481, 482), *498*, *499*, *503*, *504*, 595 (255), *617*
Lupton, E. C. 562, 571, 579 (25), *612*
Lusinchi, X. 647 (337), *663*
Luss, H. 406 (63), 445 (506b), *495*, *504*
Luss, H. R. 145, 146 (328), *155*, 524 (79), *554*, 688 (57), 814–816 (606), 829, 831 (57), *841*, *853*
Luthra, N. P. 194, 195 (23, 29), 196 (29), 198, 199, 205 (40), 206 (60, 62), 207 (62, 64), 208 (64), 210, 211 (60, 68), 221 (68), 228 (60), *239*, *240*
Lutskii, A. E. 565 (44), 567 (82, 83), *612*, *613*, 624 (88), *658*
Luttke, W. 275 (84), *285*
Lutz, O. 191–193 (13), 221, 222 (90), *239*, *240*
Luxen, A. 268 (75), *285*, 442 (473), 458, 461 (683), 474 (895), 475 (897), 476 (473, 905, 908, 910), *503*, *508*, *512*, 563 (40), *612*
Lyapina, T. 488 (1037), *515*
Lyapina, T. V. 16 (93), *20*, 541 (203, 212a, 213), 542 (212a, 213), 548 (271), *557*, *558*
Lyerla, J. R. 191 (17), *239*
Lygo, B. 642 (252), *661*
Lyons, R. E. 547 (254), *558*, 823 (636), *854*
Lysenko, Z. 403 (46a), *495*, 636–638 (211), *660*
Lysy, R. 14 (79), *19*
Lyubimova, E. 413 (178), *498*
Lyubovskaya, P. N. 95 (171), *151*
Lyubovskaya, R. 448 (539), *505*
Lyubovskaya, R. N. 681 (28c), *841*

Ma, T. S. 165 (59), *186*
Maartmann-Moe, K. 13 (74), *19*, 531, 543 (151), 544 (219), *556*, *557*
Maasbol, A. 722 (244), *846*
Maatman, R. W. 565 (56), *612*
Maccarone, E. 412 (153), *497*, 568 (90), 569 (93), 594 (90, 240), *613*, *617*
Macda, A. 317 (37), *329*
MacDiamid, A. G. 334, 335 (16), *341*
MacDougall, J. J. 736 (303), *847*
Maciel, G. E. 201 (45a), *239*
Mack, W. 409 (111), 412 (155), 414 (111, 194), *496–498*
Macke, H. 779, 780 (458), *850*
Mackrodt, W. C. 22 (9), *60*
MacMullin, E. C. 356 (37), *366*
MacNicol, D. 494 (1072, 1073), *516*
MacNicol, D. D. 624 (96), *658*
Maeda, A. 642 (242), *661*
Magarian, R. A. 651 (409, 410), *664*
Magdesieva, N. 406 (72, 73), 407 (88), 410 (125), 412 (88), 413 (168, 172), 419 (256, 257), 421 (256), 424 (256, 301), 427 (319), *496–501*
Magdesieva, N. N. 88 (135, 137), 89, 90 (143), 108 (235), 128 (143), *151*, *153*, 261 (63), *284*, 376 (32), 377 (33a, 33b), 383 (69a, 69b, 70, 71, 73), 385, 386 (33b, 71), *394*, *395*, 565, 566 (52), 567 (52, 81), 573 (115, 116), 581, 591, 594, 595 (167), 602 (291–293, 296), 603 (52, 293), 604 (167), *612–615*, *618*
Magerstadt, M. 228, 229 (107), *241*
Magerstädt, M. 832, 835, 836 (654), *855*
Magnus, P. D. 14 (81), *19*, 643, 644 (315), *662*
Magolda, R. 403 (46a), *495*
Magolda, R. L. 634 (197), 636, 638 (220), *660*
Mahajan, V. P. 695 (132), *843*
Maheu, L. J. 788, 791 (500), 792 (500, 508), *851*
Maier, H. 447 (530), *505*
Maier, L. 691, 697 (113), *842*
Maior, O. 482 (966, 967), 483 (966, 973), *514*
Maiorova, L. P. 356 (33), *366*
Maisch, R. 699, 779 (143), *843*
Makarenko, V. G. 167 (90), 169 (90, 91), 171, 172 (91), *187*
Maki, A. G. 51, 57, 58 (94), *62*, 64 (7, 8, 12), *148*
Makovetskii, Y. P. 815 (601k), *853*
Maksimenko, A. A. 552 (41), 523 (75), 537 (177), *553*, *554*, *556*
Malavasi, W. 759, 760 (387), *849*
Malek-Yazdi, F. 464 (761), *509*, 621, 622 (25), *656*
Maletras, C. 414 (193), *498*
Malhotra, K. C. 548 (268), *558*, 695 (132), *843*
Malik, K. M. A. 114, 115 (246), *153*, 709 (186), 710, 712 (186, 187), 774, 776 (446), *844*, *850*
Malisch, W. 699, 779 (143), *843*
Malito, J. T. 716 (211), *845*
Malkinson, F. D. 387 (79), *396*
Mallaki, J. 520 (28, 29), 523 (28), *553*
Mallet, J. W. 5 (11), *18*
Malmquist, P. Å. 117 (268, 269), *153*
Malmstrom, M. C. 99 (192a), *152*
Malone, S. F. 102 (207), *152*

Malsshkhiya, M. V. 701 (160), *844*
Maltesson, B. 430 (370), *501*
Malyushina, V. M. 649, 650 (370), *663*
Mamedov, E. 407, 468 (85), *496*
Mamedov, E. G. 624 (80), *658*
Mamleev, A. H. 88 (135), *151*
Mammi, M. 96 (176–179, 181), 100, 102 (203), *151*, *152*, 332 (8), *341*, 445 (503, 504), 487, 488 (1022), *504*, *515*
Mammo, A. 568 (90), 569 (93), 594 (90), *613*
Mamo, A. 412 (153), *497*, 594 (240), *617*
Manaresi, P. 333 (10), *341*
Manassen, J. 409 (117, 118a), *496*, *497*
Mancinelli, C. S. 132, 133, 143 (302), *154*
Mangane, M. 255 (47), *284*
Mangion, M. M. 64 (2a), 101 (202), 141 (202, 320–322), 142, 143 (320–322), *148*, *152*, *154*, *155*, 518 (11), *553*
Mangion, N. M. 686 (46d), *841*
Mann, B. E. 190, 191 (1), 220 (79), 223 (101), 233 (113), *238*, *240*, *241*
Mann, B. R. 575 (131), *614*
Mann, F. 462 (723b), 476 (906), *509*, *512*
Mann, F. G. 364 (69), *367*, 550 (281), *558*, 668 (10), *676*
Mannafov, T. G. 70 (47), 74 (72), 75 (47), 76 (78b), *149*, 219 (73), *240*, 565 (65), *613*, 635 (200), *660*
Manoussakis, G. E. 794 (523), *851*
Mansuy, D. 715, 719 (218), 720 (228), *845*
Mantovani, F. 420 (267), 423 (299), 424 (299, 304), 425 (299), 435 (267), *499*, *500*, 585 (188), *616*
Mao, M. 608 (330), *618*
Marcevicziojewski, B. 456 (638), *507*
March, J. 244 (1), *283*
Marchenko, T. N. 624 (88), *658*
Marechal, G. 422 (282), *500*, 584 (205), *616*
Maréchal, G. 420 (264), 423 (298), 426 (264, 298, 313–315), 465 (315), *499–501*
Mareev, Y. 420, 471 (269), *500*
Marino, G. 44, 54 (86), *62*, 246 (17, 18, 23), *283*, *284*, 385 (75), *395*, 406 (75, 76), 410 (76, 127, 130–132), 411 (134, 135), 414 (75), 416 (212, 213), 442 (466), *496–498*, *503*, 560, 563 (4), 567, 568 (88, 89), 569 (91, 92, 96), 572 (4, 91), 573 (117), 577 (117, 143), 578 (117), 581 (96, 143, 164–166), 582 (96, 165, 176), 584 (96), 589, 591 (165, 176), 593 (96, 117), 594 (117), 597 (91), *611*, *613–615*
Marino, J. H. 537 (169), *556*
Marino, J.-P. 462 (716), *508*
Markiewicz, W. 621, 622 (10), 623 (10, 64), *656*, *657*
Markovskii, L. N. 523 (66), 537 (183), 539 (189), 540 (189, 198), 541 (198), *554*, *556*, *557*
Markovtseva, O. 478 (923), *513*
Markowska, A. 16 (97), *20*
Marks, C. 465 (777), *510*
Markushina, I. 408 (106), *496*
Marøey, K. 651 (427, 428), *664*
Marolt, P. 314 (22, 23), *328*
Marov, I. N. 700 (155), 789 (507), 796, 797 (538), *843*, *851*, *852*
Marøy, K. 119 (275), *154*
Marquardt, A. 534 (161), *556*
Marquet, A. 419 (242, 243), *499*
Márquez, R. 70 (36), 101 (198, 199), *148*, *152*
Marsala, V. 690 (87), *842*
Marschalk, C. 446 (519), *504*
Marsden, C. J. 70 (39, 45), 71, 72 (39), 74 (45), 81 (39), 82 (98, 102), 83 (102, 104), 84 (104), 111, 112 (238), 144 (323), *149*, *150*, *153*, *155*
Marsden, K. 715, 717 (216), 718 (225), 814, 821 (216), 829 (650a, 650b), *845*, *854*
Marsh, D. G. 236 (132), *241*, 524 (80), *554*
Marsh, F. C. 721 (233), *845*
Marsh, R. E. 86 (110, 115), *150*, 705 (173), *844*
Marshall, J. 488 (1042), *515*
Martani, A. 438, 439 (435), *503*
Martens, J. 257, 258 (58), *284*, 477 (916, 917), *513*, 624 (94), *658*
Mårtensson, N. 117 (268, 269), *153*
Martin, D. R. 249 (35), *284*
Martin, G. J. 377 (34), *394*
Martin, J. C. 518 (5), 545 (5, 241), 546 (241), *553*, *557*
Martin, J. L. 162 (11), 178 (106), 179, 180 (106, 110), 181 (106), *185*, *187*, 387 (90), *396*
Martin, M. L. 221, 222 (114), 233 (114, 117), *241*
Martin, R. 427 (320), 444 (498, 499), *501*, *504*
Martin, R. B. 670, 671 (17b), *677*
Martin, R. L. 308 (5, 6), 318 (38), *328*, *329*, 798, 800 (543, 544), *852*
Martin, T. R. 13 (72), *19*, 102, 106 (221), *152*, 814, 821, 822 (621), *854*
Martin, W. G. 387, 388 (93), *396*
Martinengo, S. 811 (596), *853*
Martinez, J. 455, 460 (613), *506*
Martinez-Ripoll, M. 697 (135b), *843*
Martin-Gil, F. J. 697 (135b), *843*
Martin-Gil, J. 697 (135b), *843*
Martinsen, A. 382 (65), *395*, 815 (604c), *853*
Maruyama, K. 321, 323 (62), *329*
Maryanoff, B. 477, 478 (911), *512*

Maryanoff, B. E. 278 (89), *285*, 593 (237), *617*
Marynick, D. S. 26, 27 (16), *60*
Maryott, A. A. 51 (88), *62*
Marzilli, L. G. 739 (317), *847*
Masaguer, J. R. 735, 738, 739 (298), *847*
Mason, M. G. 701, 705 (157a–c), 736 (157c), *844*
Mason, R. 766, 772 (410), *849*
Mason, R. S. 245 (9), *283*
Mason, S. F. 667 (1c), *676*
Massey, A. 428, 443 (340), 484 (340, 985), *501*, *514*
Massey, A. G. 85 (107), *150*, 269 (77), *285*, 654 (467), *665*, 744–747 (348), *848*
Massey, R. C. 705 (172), 774 (449, 450, 452), 775 (454), 776 (449, 450, 452, 454), *844*, *850*
Masson, M. R. 163 (27), 164 (27, 42), 165 (42), *185*, *186*
Masters, A. F. 102 (220), *152*, 794, 795 (517), *851*
Mastryukov, V. S. 87 (126, 128), *150*
Masuda, Y. 641 (297, 299), *662*
Matamedi, A. 428 (333), *501*
Mathey, F. 402 (16), *494*
Mathiasch, B. 357 (39, 41, 42), 358 (48), *366*
Matsuda, T. 654 (465), *665*
Matsui, K. 520, 521 (32), *553*
Matsumoto, S. 446 (512c), *504*
Matsunaga, Y. 487 (1014), *514*
Matsuura, Y. 404, 446 (511), *504*
Matthews, W. S. 565, 604 (61), *613*
Mattii, M. 234 (120), *241*, 288 (14, 15), 289 (15–20), 291 (14–16), *304*
Matyushcheva, G. I. 567, 568 (86), *613*
Mautner, H. 427 (326), 457 (650), *501*, *507*
Mautner, H. G. 166, 168, 177 (77), 179 (112), 182 (77), *186*, *187*, 276 (85), *285*, 288 (12, 13), *304*, 379 (40, 41), *394*, 641 (263), *661*
Mawby, A. 779 (466a), *850*
Maxwell, W. 542 (215), *557*
Mayer, C. 446 (520), *504*
Mayer, R. 406, 447 (67), 448 (537a, 541), *495*, *505*
Mayr, A. J. 113 (241), 114 (242), *153*, 701 (161, 162), 702 (162), 703 (161, 162, 165, 166), *844*
Mazalov, L. N. 197 (38), *239*
Mazaud, A. 450 (569), *506*
Mazerolles, P. 488 (1036), *515*
Mazid, M. A. 114, 115 (246), *153*, 709 (186), 710, 712 (186, 187), 774, 776 (446), *844*, *850*
Mazloumi, A. 255, 278, 280 (52, 53), *284*, 439 (447), 460 (705), *503*, *508*
Mazza, F. 407, 420 (77), 440 (454), *496*, *503*
M'Buyi, M. 454 (598), 458 (683), 459 (598), 461 (683), 486 (598), *506*, *508*
McAfee, F. 626 (136), *659*
McAuliffe, C. A. 622 (33), *657*, 680 (10a, 10b, 12c), 681, 772 (10a, 10b), 782 (12c), 801–803 (549), 815 (12c), *840*, *852*
McCarthy, D. M. 392 (130), *397*
McCausland, J. H. 548 (263), *558*
McClellan, B. E. 165 (52), *186*
McClelland, R. A. 319 (41), *329*, 565, 574, 575 (47), 605 (325), 606 (47, 325), *612*, *618*
McCollum, G. J. 565, 604 (61), *613*
McCombie, S. W. 822 (635), *854*
McConnell, K. P. 179, 180 (111), *187*, 379 (40), 381 (56), 391 (126a, 126b), *394*, *395*, *397*
McConnell, W. B. 171, 182 (123), *187*
McCready, V. R. 392 (135), *397*
McCullough, I. 468 (829), *511*
McCullough, J. 418 (231), 428 (335), 443 (479), 480 (231, 945, 947), 481 (947–949), *499*, *501*, *504*, *513*
McCullough, J. D. 86 (115), 92, 93 (160, 161), 128 (291), 135, 136 (305–307), 137 (307), 138 (291, 305–307, 309–311), 139 (291, 309–311), 140 (318), 141 (318, 319), 142 (319), 143 (305–307, 309–311, 318, 319), *150*, *151*, *154*, 166, 167 (78), 171 (78, 101), 177 (78), *186*, *187*, 518 (7), *553*, 567 (74, 75), 577 (139–141), 580 (157, 158), 600 (157), 601 (139, 140), *613–615*, 629 (159), 642, 643 (279), *659*, *661*, 705 (173), 721 (236), *844*, *845*
McCullough, L. D. 376 (31), *394*
McCurry, P. M. Jr. 634 (195), *660*
McDaniel, D. H. 565 (54), *612*
McDonald, W. S. 80 (97), *150*
McElhill, E. A. 202 (46), *239*
McFarlane, H. C. E. 221–226, 228, 229, 232, 233 (94), *240*, 690 (100), *842*
McFarlane, W. 194–197 (22), 198 (22, 39), 199, 203, 205, 207 (22), 211 (22, 39), 218, 219 (22), 220 (81, 83), 221 (86, 94), 222–224 (94), 225, 226, 228, 229 (94, 103), 232 (94), 233 (94, 103, 115), *239–241*, 690 (100), 699, 779 (143), *842*, *843*
McGarry, R. 456 (643), *507*
McGlothin, R. E. 360 (53), *366*
McGlothlin, R. E. 697, 755 (141a), *843*
McIntyre, P. 423 (291), *500*, 532, 534 (152), *556*
McIntyre, P. A. 393 (150), *397*
McKean, D. C. 71 (53), *149*
McKennis, J. S. 701–703 (162), *844*
McKenzie, S. 445 (507a), *504*

McKervey, A. 632 (175), *659*
McKinney, J. D. 99, 101 (192b), *152*
McKinnon, B. J. 621, 622 (29), *657*
McLafferty, F. W. 245 (8), *283*
McLaren, J. 455 (630), *507*
McLaughlin, G. M. 13 (72), *19*, 102, 106 (221), *152*, 814, 821, 822 (621), *854*
McLean, A. D. 25 (40, 58), *61*
McLellan, A. L. 51 (89), *62*
McMahon, F. 409 (110), *496*
McManus, S. P. 606 (329), *618*
McMurray, W. J. 276 (85), *285*
McNicol, D. 442 (470), *503*
McPartlin, M. 811, 812, 832, 839 (590), *853*
McPhee, D. J. 318 (40), *329*
Mcphee, D. J. 648 (354), *663*
McQuillan, G. P. 684–686 (37, 38), 691 (116), 692 (116, 117), 693 (116, 121), 694 (116, 117), 695 (117), 697, 778 (140c), *841–843*
McQuitty, J. J. 814, 821 (622), *854*
McWhinnie, W. 467 (816), 492 (1063), *510*, *516*
McWhinnie, W. R. 7 (26), *18*, 221, 222, 226–228, 234 (93), *240*, 254 (46), *284*, 358 (46), *366*, 518 (9), 520 (23, 24, 26, 28, 29), 522 (24), 523 (28), 536 (9), 542 (23), *553*, 567 (77), *613*, 680 (9f), 814 (614, 615, 616a, 616b), 819 (616a, 616b, 629–631), 824 (641), 825 (616a, 616b), 826 (641), 831 (631), 834 (641), *840*, *854*
Meakin, P. 546 (243), *557*
Mealli, C. 765, 766 (409), *849*
Mechin, B. 377 (34), *394*
Medne, R. 446 (515), *504*
Medvedev, S. Y. 164 (40), *186*
Medynya, B. A. 184 (165), *188*
Meek, D. W. 360 (54), *367*, 692 (119), 693 (119, 123), 694 (119), 696 (123), 697 (140a, 140b, 140d), 771, 772 (431–433), 775 (140d), 778 (140a, 140b, 140d), 779, 782 (433), 801, 803 (550), 815 (603a), *842*, *843*, *849*, *852*, *853*
Meerholz, C. 462 (717), *508*
Meerholz, C. A. 10 (52), *19*, 537 (178, 179), *556*
Meerwein, H. 537 (180), *556*
Meese, C. O. 622 (45), 642 (255), *657*, *661*
Meessen, J. 423, 473, 476 (286), *500*
Mehdi, R. T. 819 (632), *854*
Mehler, E. L. 25 (42), *61*
Mehrotra, G. 695 (132), *843*
Meier, F. 455 (627), *507*
Meier, H. 309 (14), 311, 312 (15), *328*, 451 (584), 462 (722), 464 (742, 743, 745, 746, 756, 758, 759, 762, 763), *506*, *509*
Meijer, J. 447 (527), 481 (953–957), *505*, *513*
Meikle, G. D. 545 (223, 233–236), *557*
Meinwald, J. 238 (142), *241*, 446 (521–523), 447 (521), *504*, 642 (251), 650 (386, 387), *661*, *663*
Meir, H. 196 (35), *239*
Meixner, J. 326 (70), *329*
Mekthiev, M. A. 289, 291 (24), *304*
Mel'chekova, Z. E. 177 (127), *187*
Melchiona, L. 440 (454), *503*
Melenteva, T. 419 (249), *499*
Melfi, F. 392 (128), *397*
Meli, A. 765, 766 (409), 788, 791 (498), *849*, *851*
Meller, A. 356 (34), *366*
Mellini, M. 90 (151), *151*
Mellon, F. A. 248 (30), *284*
Melmed, R. N. 392 (132), *397*
Meloan, C. E. 699 (145b), 700 (145b, 148, 149), *843*
Melpolder, J. B. 733 (292a, 292b), *847*
Menchen, S. 403 (22a), *494*
Menchen, S. M. 347 (8–10), *366*, 636, 638 (218, 219), 642 (288), 651 (425), *660*, *662*, *664*
Mencken, S. M. 16 (99), *20*
Menconi, A. 466 (799), *510*, 597 (252), *617*
Menin, J. 457 (660, 661), *507*
Menkin, V. L. 16 (91), *19*
Mente, P. 403 (38, 39), *495*
Mente, P. G. 703 (163), *844*
Mentzer, E. 695, 775, 778 (128), *843*
Menzel, I. 309 (14), *328*, 447 (530), 464 (742, 763), *505*, *509*
Meranger, J. C. 165 (50), *186*
Mercer, A. 721 (234), *845*
Merényi, R. 326 (69), *329*
Merkle, U. 462 (722), *509*
Merlino, S. 90 (151), *151*
Merten, H. 157, 158 (5), *160*
Mertes, K. B. 106, 107, 124 (228), *153*, 668 (7), *676*
Mervick, M. V. 651 (408), *664*
Meth-Cohn, O. 401 (8b, 10), *494*
Metlushenco, V. V. 567 (84), *613*
Metlushenko, V. P. 574, 575 (122), *614*
Metys, J. 477, 478 (913), 490 (1053), *512*, *515*
Metysova, J. 477, 478 (913), 490 (1052–1054), *512*, *515*
Metzger, H. 335 (20), *341*
Metzger, J. 455 (615, 616), 456 (616), *507*
Meunier, J. C. 377 (34), *394*
Meunier, Ph. 741, 742, 749 (331), *847*
Meyer, J. 464, 465 (765), *509*
Meyer, N. 603 (301), *618*
Meyerheim, G. 465 (783), *510*
Meyers, C. Y. 565 (63), *613*
Meyers, E. 484 (990, 991a), *514*

Meyers, E. A. 9 (44), *18*, 64 (2a), 69 (34), 77 (84), 100 (195, 201), 101 (193, 195, 196, 200–202), 141 (202, 320–322), 142, 143 (320–322), 144, 145 (325), *148*, *149*, *152*, *154*, *155*, 227, 232–234 (106), *241*, 518 (11), *553*, 686 (46d), *841*
Mezentsova, N. 412 (154), 413 (169, 177), *497*, *498*
Mezey, P. G. 25 (37–39, 63), 30 (63, 74), *61*
Michaelis, A. 523 (71), 543 (161), *554*, *556*
Michalak, R. S. 518 (5), 545 (5, 241), 546 (241), *553*, *557*
Michalska, M. 672 (23), *677*
Michalski, J. 16 (97), *20*
Michel, A. 99 (191), *152*, 320 (52), *329*
Michels, W. 456 (645), *507*
Michelsen, T. W. 736 (306), *847*
Michl, J. 668, 670, 671 (2), *676*
Micklem, H. S. 381, 393 (62), *395*
Midollini, S. 811 (598), 813 (598, 599), *853*
Miessler, G. L. 788, 791, 792 (500), *851*
Migalina, Y. 418 (238), 423 (293, 297), 481 (238, 950), 482 (959), 494 (1075), *499*, *500*, *513*, *516*
Migalina, Yu. 404 (55), *495*
Migalina, Yu. V. 527 (105), 528 (125, 126, 133), 530 (143, 144, 146), 531 (147, 148), 532 (153), 533 (159), 534 (153), *555*, *556*
Mikawa, H. 446 (517, 518), *504*
Mikulski, C. M. 721 (243), *846*
Milazzo, G. 383 (72a, 72b), *395*
Milburn, G. H. W. 308 (11), *328*
Milhaud, G. 393 (147), *397*
Millar, J. B. 545 (232), *557*
Millar, K. R. 171, 181 (119), *187*
Miller, E. 595 (255), *617*
Miller, G. 461 (712), *508*
Miller, J. D. 680 (9f), 819 (632), *840*, *854*
Miller, J. M. 225, 228, 229, 234 (104), *241*, 269 (77), *285*, 541, 542 (214), *557*, 690 (72), *842*
Miller, J. S. 681 (28b), *841*
Miller, S. I. 571, 572 (106), *614*
Milliet, P. 647 (337), *663*
Millington, D. 236 (133, 134), *241*, 815 (601n), *853*
Mills, S. L. 375, 392 (27), *394*
Mills, W. N. 736 (306), *847*
Milne, G. 461 (708), *508*
Milne, R. 693 (121), *843*
Milone, L. 832, 836, 837 (659), *855*
Milyamoto, T. 686 (53), *841*
Mincuzzi, A. 402 (21), *494*
Mingaev, R. M. 522 (43), *553*
Minh, T.-Q. 473 (889), *512*
Minh, T. Q. 419, 420 (261), 423 (292, 299, 300), 424 (292, 299, 300, 306), 425 (299, 312), 426 (312), *499*, *500*
Minkin, V. 442 (463), 479 (935, 936), 480 (936–938), 487 (1020), *503*, *513*, *515*
Minkin, V. I. 518, 520 (4), 522 (42–44), 523 (42, 66), 532, 534 (155), 540 (197), *553*, *554*, *556*, *557*, 560 (5), 563 (36), 565 (51), 566 (51, 69, 72, 73), 567 (51, 69, 73), 574 (69, 120–122), 575 (122), 577, 601, 603 (138), *611–614*, 651 (403), *664*, 779 (462, 463), 781 (462, 463, 472, 473), 787, 801 (462, 463), 814, 820 (620), *850*, *854*
Minkina, L. S. 779 (462, 463), 781 (462, 463, 473), 787, 801 (462, 463), *850*
Min'tyau, N. 569 (94), *613*
Minyaev, R. M. 566 (72, 73), 567 (73), *613*
Mironyuk, L. V. 16 (93), *20*
Mirrashed, M. 464 (748), *509*
Mirri, A.M. 73 (71), *149*
Mirskov, R. G. 624 (99), *658*
Mirzai, H. 622 (45), *657*
Mishiev, R. 407, 468 (85), *496*
Mishmash, H. E. 709 (179), *844*
Mishulina, R. A. 76 (80), *149*
Misiuna, D. 456 (636, 637), 489 (637), *507*
Mislow, K. 278 (89), *285*, 477, 478 (911), *512*, 593 (237), *617*
Misumi, S. 492 (1064b), *516*, 642 (310), *662*
Mitcham, R. V. 668 (7), *676*
Mitchell, R. 492 (1064a), 493 (1065, 1066), *516*
Mitichkin, A. I. 624 (88), *658*
Mitra, R. 420 (268), *499*
Mitran, M. 482, 483 (966), *514*
Mitsui, A. 74 (73), *149*
Mittal, P. K. 710, 712, 713 (188), *844*
Mixan, C. 468 (834, 835), *511*
Mixan, C. E. 376 (29, 30), 384 (30), 385 (29), *394*, 546 (249), *558*
Miyashita, M. 624 (74), *657*
Miya-Uchi, Y. 762–764, 787, 790, 794, 795 (396), *849*
Miyazaki, H. 428 (342), *501*
Miyoshi, H. 532, 534 (156), *556*, 653 (451), *665*
Miyoshi, N. 317 (33), *329*, 354 (21–23), 355 (26), *366*, 403 (35), *495*, 621, 622, 624 (19), 640 (140, 143), *656*, *659*
Mizuhara, Y. 456 (648), *507*
Mizukami, T. 393 (155), *397*
Mizuta, E. 589 (203), 592, 593 (203, 236), *616*, *617*
Moberg-Ogard, A. 407 (82a), *496*
Mock, W. L. 548 (263), *558*
Modena, G. 403 (27, 42, 43), *494*, *495*
Modica, G. di 460 (703), *508*
Moguel, M. K. 738 (315), *847*
Mohammed, J. 334, 341 (18), *341*

Mohmand, S. 11 (57), *19*, 407 (82b), *496*
Mohr, D. 825 (639), *854*
Mohtachemi, R. 359 (49), *366*, 714, 716 (197), *845*
Moldovan, Z. 277 (87, 88), *285*
Molines, H. 642 (252), *661*
Molle, L. 164 (38), *186*, 581 (163), 598 (163, 263), *615*, *617*
Moller, J. 271 (78), 274, 275 (81), *285*
Mollier, Y. 469, 470 (849), 475 (898), 480 (939), *511–513*
Molodkin, A. K. 700 (155), 701 (156), *843*
Molz, T. 462 (722), *509*
Momicchioli, F. 580, 599, 600 (159), *615*, 643 (359), *663*
Monakhova, A. 412 (154), 413 (172), *497*, *498*
Monako, G. 541 (202), *557*
Moncur, N. K. 72 (56, 64), *149*
Monks, R. 371 (7, 9), 372 (7, 18), 373 (20), 380 (9), 393 (7, 9, 18, 20), *394*, 651 (408), *664*
Monsef-Mirzai, Z. 520 (29), *553*
Mont, W.-W. du 360 (59), 361 (60, 61), *367*
Montag, R. A. 247 (26), *284*
Montanari, F. 606 (326), *618*
Montanucci, M. 621–623 (6), *656*
Montaudo, S. 416 (211), *498*
Monti, S. 238 (143), *241*, 299 (58), *305*
Moore, C. E. 22 (1), *60*
Moradpour, A. 448, 450 (549), *505*
Moran, M. 735, 738, 739 (300), *847*
Morán, M. 732 (286), 735 (298), 738 (286, 298), 739 (298), *847*
Morel, J. 412 (160, 161), 413 (173–176, 188–191, 192a), 414 (193), 415 (175, 176, 198, 201, 202, 204, 205, 208, 209), 430 (361, 362), 435 (205, 208, 401), 438 (189, 427, 428, 434), 455 (188, 612, 613), 460 (613), 461 (192a), *497*, *498*, *501–503*, *506*, 584 (180), *615*
Morelli, D. 754 (364), *848*
Moretti, E. 580 (160), *615*
Moretti, S. 288 (14), 289 (17, 18), 291 (14), *304*
Morgan, G. 404 (50), 405 (59), 418 (230, 233), 444 (50), 468 (828, 830, 832), 469 (851–853), 480(230), 489 (832), 491, 492 (828), *495*, *499*, *511*
Morgan, G. T. 333, 335 (14), 337 (28, 29), *341*, 522 (39), 523, 524 (70), 525 (84, 85), 526 (100, 101), 550 (279), *553–555*, *558*, 653 (452, 455–457), 654 (452, 455, 456, 475, 477), *665*
Morgan, W. E. 694 (125), *843*
Mori, I. 420 (273), *500*
Mori, Y. 8 (39), *18*
Morishita, H. 71 (52), *149*, 650 (384), *663*
Morisset, V. M. 635 (201, 203, 206), *660*
Morita, S. 393 (155), *397*
Moriyama, H. 303 (70, 71), *305*
Mörner, M. 455, 456 (633), *507*
Morris, R. H. 721 (235), *845*
Mortensen, K. 449 (560), *505*
Mortillaro, L. 331 (2), 332 (6, 7), 333 (6, 10), 334 (12), *341*, 487 (1022, 1024, 1028), 488 (1022), 492 (1028), *515*, 768 (416), *849*
Mortimer, C. T. 157 (6), 158 (6, 11), 159 (6), *160*
Mortimer, R. D. 635 (202), *660*
Mortola, A. P. 25 (36), *61*
Morton, J. A. 642 (252, 256), *661*
Moschini, G. 163 (34), *185*
Moscowitz, A. 671 (20b), 672 (24), *677*
Moses, P. 416 (219), *499*
Moskalev, Y. I. 389 (105, 106), *396*
Moskowitz, J. W. 25 (36), *61*
Mosna, P. 547 (256), *558*
Mošna, P. 650 (365), *663*
Moss, N. 690 (63), *841*
Motherwell, W. 462 (723b), *509*
Motherwell, W. D. S. 364 (69), *367*
Moura Campos, M. de 527 (109, 110), 528 (110), 529 (109, 110), 532 (154), 533 (110), *555*, *556*, 636 (208), 651 (390), 652 (208, 434), 653 (446, 447), 654 (446, 474), 655 (483), *660*, *664*, *665*
Mousseron, M. 565 (64), *613*
Moustakis, C. A. 194, 196, 209, 220 (26), *239*
Moza, A. K. 393 (153), *397*
Mrotzek, H. 622 (45), *657*
Mudd, S. H. 388 (102a), *396*
Mueller, E. 184 (166), *188*, 641 (272), *661*
Mueller, H.-W. 642 (255), *661*
Mueller, P. 623 (50), *657*
Mueller, R. G. 184 (155), *188*
Muetterties, E. L. 546 (243), *557*
Mukai, K. 238 (146), *241*, 382, 392 (68), *395*
Mullen, G. P. 207, 208 (64), *240*
Muller, A. 406, 447 (67), *495*
Müller, A. 680, 782 (12g), 803 (553a, 553b), 815 (12g), *840*, *852*
Müller, E. 414 (195), 417 (228), 427 (329–332), 430 (360), 442 (468, 469), 443 (360, 482), *498*, *499*, *501*, *503*, *504*
Muller, G. 402 (16), *494*
Müller, J. 723, 724 (247), *846*
Muller, J.-F. 246 (19), *283*
Müller, K. 448 (541), *505*
Muller, M. 469 (840b), *511*
Müller, P. 486, 487 (1007), *514*
Müller, U. 740 (321), *847*
Müller, W. 626, 639 (144), *659*
Mulliken, R. S. 35 (85), *62*

Mulvey, D. 466 (796), *510*
Mundy, D. 668, 669 (11), *676*
Munno, A. de 466 (800), 467 (815), *510*
Murahashi, S. 623 (51), *657*
Murai, S. 317 (33), *329*, 354 (21–23), 355 (26), *366*, 403 (35), *495*, 621, 622 (7, 19), 624 (19), 634 (194), 640 (140, 143), *656*, *659*, *660*
Murakami, K. 321, 323 (62), *329*
Murata, H. 74 (73), *149*
Murata, S. 446 (512d), *504*
Muravnik, R. S. 586 (213), *616*
Murdoch, J. D. 72 (61), *149*
Murphy, C. J. 194, 208, 209 (24), 220 (24, 85), *239*, *240*
Murphy, J. C. 390 (120), *397*
Murray, B. 410 (123), 445 (501), 469 (848), 470 (857, 861), 475 (899, 900), *497*, *504*, *511*, *512*
Murray, B. J. 539 (185), *556*, 651 (423, 424), *664*
Murray, S. G. 622 (33), *657*, 680, 681, 691 (13), 801–803 (549), *840*, *852*
Murray-Rust, P. 121 (281), *154*
Musa, F. H. 819, 831 (631), *854*
Musa, F. M. 520, 522 (24), *553*
Musa, P. 467 (816), *510*
Musante, C. 466, 467 (810), *510*
Musher, J. I. 107–109, 124 (234), *153*, 261 (62), *284*, 517 (1), *552*, 565, 604 (45), *612*
Mushkako, L. 486 (1004), *514*
Mushkalo, L. 491 (1058–1060), *516*
Musina, A. 402 (18), 420 (18, 271), *494*, *500*
Musorin, G. 407 (93), 493 (1068c), *496*, *516*
Muszkat, K. 434 (398), *502*
Muszkat, K. A. 527 (106), *555*
Muth, B. 419 (255), 478 (926), *499*, *513*
Muth, O. H. 162 (9), *185*
Myers, R. J. 72, 90 (66), *149*
Mysov, E. 405 (62a), *495*

Naaktgeboren, A. 481 (956), *513*
Nadaka, V. I. 16 (91, 92), *19*
Naddaka, V. I. 540 (197), 541 (199, 200), *557*
Nagao, G. 762, 764, 765 (399), *849*
Nagao, Y. 317 (37), *329*, 642 (242), *661*
Nagaraja, K. S. 806 (560), *852*
Nagarajan, R. 670, 671 (17d), *677*
Nagarayan, S. 439 (449), *503*
Nagashima, K. 165 (54), *186*
Nagel, A. 94 (170), *151*
Náhlovská, Z. 88 (134, 139), 89 (139), 90 (134, 139), *150*, *151*
Náhlovský, B. 88 (134, 139), 89 (139), 90 (134, 139), *150*, *151*
Nain Java, K. 436 (411), *502*
Najean, Y. 393 (151), *397*
Najer, H. 457 (660, 661), *507*
Nakabayashi, K. 779, 780 (460), *850*
Nakagawa, J. 70 (41), 73 (41, 69), 81 (41), *149*
Nakajima, K. 771, 772, 779, 780 (430), *849*
Nakajima, S. 787 (495), *851*
Nakamoto, K. 685 (44), *841*
Nakamoto, M. 762–764 (397, 398), 784 (397), *849*
Nakamura, N. 600 (285), *618*
Nakamura, Y. 686 (53), *841*
Nakanishi, W. 203, 204, 218, 220 (58), *240*, 446 (512c, 512d), *504*, 519 (17b), *553*
Nakashima, S. 183 (130), *187*
Nakatsuji, H. K. 25 (46), *61*
Nakayama, S. 393 (155), *397*
Nakhdjavan, B. 539 (186), *556*
Nakhmanovich, A. 413 (186), *498*
Nalewajek, D. 13 (68), *19*, 448, 450 (547), *505*
Namtuedt, M. J. 238 (144), *241*
Namtvedt, M. J. 382, 392 (66b), *395*
Nandi, K. N. 655 (489), *665*
Nanjappan, P. 253 (44), *284*, 469 (844, 846), *511*
Nardelli, M. 84, 85 (105), *150*, 295 (47), *305*, 383 (72c), *395*, 683 (33), 770 (435), 815 (602), *841*, *850*, *853*
Narimatsu, S. 634 (198), *660*
Nasonova, L. I. 197 (38), *239*
Nassiff, S. J. 374 (23), *394*
Nassimbeni, L. R. 779 (465), *850*
Natalis, P. 22 (7), *60*
Natori, Y. 379 (43), 380 (44), 388 (43, 44), *395*
Naumann, D. 523, 545 (62, 63), *554*
Naumov, V. A. 106 (226, 227), *153*
Nazarenko, I. I. 163–165, 183 (19), *185*
Nazareth, L. 32 (79), *62*
Nealey, R. 478 (927), *513*
Nefedov, V. D. 166 (81), 167, 170 (87, 88), 171 (81, 87, 88, 95), 172 (81, 87, 88, 95–97), 173 (87, 96–100), 174 (81, 87, 88, 95, 97, 98), 175 (87, 98–100), 176 (81, 88, 95, 97–100), *186*, *187*
Negryatse, N. Ya. 802 (548), *852*
Neidlein, R. 90 (152), *151*, 465 (786), 472 (884), *510*, *512*
Neiland, O. 446 (515), *504*
Nelson, D. 234 (125), *241*
Nelson, J. H. 736 (303), *847*
Nelson, L. L. 811 (588), *853*
Nelson, R. D. 51 (88), *62*
Nelson, R. E. 522 (37), 523 (73), *553*, *554*
Nelson, S. M. 731 (275–277), 736 (276, 277), *846*
Nemcova, L. 297, 299 (50), *305*

Nemoto, H. 623 (61), 642 (303), *657*, *662*
Neonilina, V. I. 574, 576 (133), *614*
Nepywoda, J. 520 (19), *553*
Nercesian, D. 464 (748), *509*
Nesmeyanov, N. N. 601, 602 (290), *618*
Nesterova, A. G. 586 (214), *616*
Neta, P. 297 (53), *305*
Neubert, L. A. 670, 671 (17a, 17f), *677*
Neugebauer, D. 723, 725 (250), 726, 729 (255), 742, 753 (250), 832 (653), *846*, *855*
Neuhaus, L. 547 (252), *558*
Neve, J. 164 (38), *186*, 581 (163), 598 (163, 263), *615*, *617*
Neville, A. 7 (31), *18*, 668 (9), *676*
Newmark, R. A. 708 (178), 709 (178, 180), *844*
Nezentsova, N. 419 (249), *499*
Nguyen Ngi, Le. 602 (291), *618*
Nguyen-Thi, M. P. 623 (50), *657*
Niazi, M. 469 (842), *511*
Nibbering, N. 248 (31), *284*
Nibbering, N. M. M. 838 (670c), *855*
Niccolini, M. 735 (297), *847*
Nicholson, D. 183 (137), *187*
Nicolaou, K. 403 (46a), *495*
Nicolaou, K. C. 16 (88, 89), *19*, 627 (150), 634 (197), 636 (210, 211, 216, 220), 637 (210, 211), 638 (210, 211, 216, 220, 224), 639 (216, 230), 640 (216, 283), *659–661*
Nicolini, M. 732, 735, 737 (287), *847*
Nicpon, P. 360 (54), *367*, 692–694 (119), 697, 778 (140a, 140b), *842*, *843*
Nielsen, C. J. 15 (83), *19*
Nielsen, O. 464 (753), *509*
Nielsen, O. J. 12 (65), *19*, 65 (18), *148*
Nielsen, P. 468 (831), *511*
Nielsen, P. H. 68 (31, 32), *148*, 308 (4), *328*, 335 (23), *341*
Niemczura, W. 624 (86), *658*
Niessen, W. von 22 (3), *60*
Nigam, S. N. 171, 182 (123), *187*
Nikaitani, D. 570 (103), *614*
Nimmesgern, H. 456, 458 (640), *507*
Nishi, T. 303 (70, 71), *305*
Nishikawa, T. 72 (68), *149*
Nishikida, K. 235 (126), *241*, 325 (64), *329*
Nishiyama, H. 634 (198, 199), *660*
Nishizawa, M. 642 (312), 645 (328), *662*
Nitsche, W. 483 (982), *514*, 521 (34), *553*
Nivorozhkin, L. E. 779 (462, 463), 781 (462, 463, 473), 787, 801 (462, 463), *850*
Niwa, I. 634 (194), *660*
Noble, A. M. 690 (95), *842*
Nockolds, C. E. 308 (11), *328*
Nogami, T. 446 (517, 518), *504*
Noggle, J. H. 191 (19), *239*
Nölken, E. 475 (901), *512*
Nolle, A. 191–193 (13), 221, 222 (90), *239*, *240*
Noltemeyer, M. 356 (34), *366*, 768 (417), *849*
Nomoto, M. 781 (470), *850*
Nomura, Y. 632, 633 (190), 639 (236), 641 (284–287), *660–662*
Noordik, J. H. 102, 103 (215), *152*, 796 (534), *852*
Norbury, A. H. 164 (36), *185*, 360 (56), *367*, 700, 701 (153c, 153d), 730 (153c, 513d, 269a), 737 (153c, 153d), *843*, *846*
Norcross, B. 427 (320), *501*
Norcross, B. E. 318 (38), *329*
Nordhoff, E. 361 (61), *367*
Norets, T. A. 387 (88, 89), 392 (136–139, 141–143), *396*, *397*
Norman, R. O. C. 234 (124), 235 (127), *241*
Norton, H. 407 (91), *496*
Norton, J. R. 565 (55), *612*
Nosco, D. L. 771, 772, 780 (428), *849*
Nötke, D. 545 (227), *557*
Noto, R. 591 (227), *616*
Nowlan, V. J. 319 (44), *329*
Noyes, W. 467 (820b), *511*
Nsunda, K. M. 319 (48), *329*
Nuhn, P. 454 (600), *506*
Numata, T. 316 (29), *329*
Nunn, A. 464 (757), *509*
Nuretdinov, I. A. 815 (601f, 601h–j), *853*
Nuti, V. 164 (43), *186*
Nuzhdina, Y. A. 16 (93), *20*
Nyburg, S. C. 127 (288), *154*
Nygard, B. 184 (146–154, 156), *188*
Nygärd, B. 299 (59), *305*
Nyholm, R. 518 (8), *553*
Nyman, G. 419 (254), *499*

Oae, S. 316 (29), *329*, 370 (3a, 3b), *393*, 404 (48, 49, 511), 418 (240), 446 (48, 49, 511, 512b), *495*, *499*, *504*, 541 (205, 206), *557*, 623 (52, 53), *657*
Oberhammer, H. 66, 67 (26), 71, 72 (55), 87 (129), 126 (284, 287), 129 (298), *148–150*, *154*
O'Brien, D. H. 203, 221 (57), 222 (57, 102, 105), 223 (57, 102), 224 (57), 225 (57, 102), 226 (105), 227 (102, 105), 229, 230, 233 (102), *240*, *241*, 524 (77), *554*
O'Brien, E. 644 (318), *662*
Obukhova, E. M. 565 (44), 567 (82, 83), *612*, *613*
O'Callaghan, W. 402 (13), *494*
Ochiai, E. 458 (674b, 675), 459 (674b), *508*, 586 (209), 590 (209, 215), *616*
Ochiai, M. 317 (37), *329*, 642 (242), *661*
Ockwell, J. N. 321, 323 (58), *329*
O'Connor, J. P. 839 (675), *855*

Odom, J. D. 191 (10, 11), 192, 193 (11), 194 (23, 29), 195 (11, 23, 29), 196 (29), 197 (11), 198 (40), 199, 205 (11, 40), 206 (60, 62), 207 (62, 64), 208 (64), 209 (11), 210, 211 (60, 68), 218 (76), 221 (68, 88), 228 (60), *239*, *240*
Öfele, K. 814, 819, 820 (619), *854*
Offermanns, H. 485 (998), *514*
Ogawa, M. 520 (20, 32), 521 (32), 527 (111), 528 (111, 135), *553*, *555*, 653 (458, 459), 654 (458), *665*
Ogloblin, A. N. 537 (183), *556*
Ogura, F. 536 (167), 541 (201), *556*, *557*, 642 (310), 652 (439), *662*, *665*
Ohashi, O. 70, 73, 81 (44), *149*
O'Haver, T. C. 165, 167, 170 (72), *186*
Ohba, N. 644 (323, 324), *662*
Ohfune, Y. 645 (329), *662*
Ohmasa, N. 623 (49), *657*
Ohnishi, S. 446 (517), *504*
Ohno, K. 74 (73), *149*
Ohno, S. 403 (35), *495*
Ohta, K. 25 (46), *61*
Ohtake, H. 393 (155), *397*
Okamato, Y. 451 (581, 582), *506*
Okamoto, K. 771, 772 (429), 779 (459), 780 (429), 781 (470), *849*, *850*
Okamoto, Y. 316 (30, 31), *329*, 331 (1), 333–335 (15), 336 (15, 24), 339, 340 (15), *341*, 561 (17), *612*, 651 (389), *664*, 747 (357), *848*
Okano, M. 403 (33), 481, 489, 491, 493 (951), *495*, *513*, 528 (131, 132), 532, 534 (156), *555*, *556*, 632, 633 (187, 189), 636, 638 (217), 639 (187, 235), 641 (296, 302, 306–308, 311), 653 (451), 654 (464), *660–662*, *665*
Okazaki, R. 675 (34), *677*
Okeya, S. 686 (53), *841*
Okhlobystin, O. 480 (938), *513*
Okutani, H. 70, 73, 81 (41), *149*
Okuyama, T. 320 (51), *329*
Olah, G. A. 7 (28), *18*, 310–312 (18), *328*
Oldendorf, W. H. 393 (157), *397*
Oldfield, J. E. 162 (9), *185*
Olekhnovich, L. P. 779, 781, 787, 801 (462), *850*
Oliveira, A. R. M. de 6 (19), *18*
Olmsted, M. 487 (1010), *514*
Olofson, R. A. 603 (307), *618*
Olofsson, J. 184 (151, 153), *188*
Olofsson, R. 466 (804), *510*
Olsen, M. 450 (570), *506*
Olson, E. C. 164 (37), *185*
Olson, O. E. 163, 183 (21, 25), *185*
Olson, R. E. 603 (308), *618*, 639 (119), *658*
Olsson, K. 220 (80), *240*, 405, 496 (60b), *495*
Omae, I. 680, 782, 815 (12d), *840*
Omelańczuk, J. 676 (38), *677*
O'Neal, H. E. 157, 159 (3), *160*
Onkelinx, C. 288, 291 (6), *304*
Ono, M. 627 (149), *659*
Onoe, A. 528 (131, 132), *555*
Onyamboko, N. V. 437 (423), 438 (424, 425), 452 (423–425), 453 (423, 425), 454 (423), *502*
Onyszchuk, M. 731 (294, 295), 732 (294), 733 (295), 734 (294), *847*
Ooi, S. 686 (53), *841*
Ool, P. J. J. M. van 624 (101), *658*
Opavsky, W. 690, 691 (96), *842*
Opedal, D. 115, 116 (247), *153*
Oppermann, R. 455 (628), *507*
Oprean, I. 277 (87), *285*
Orgel, L. E. 690 (84), *842*
Orlandini, A. 765, 766 (409), *849*
Orlov, V. Y. 167, 169 (90), *187*
Orme-Johnson, W. H. 238 (145), *241*, 382, 392 (66a), *395*
Orrell, K. G. 114 (245, 246), 115 (246), *153*, 191–193 (16), *239*, 690 (79, 112), 709 (183–185), 710 (183–185, 187, 188), 712 (184, 185, 187, 188), 713 (188), 727, 729 (257), 771 (437), 773 (257, 437–439), 774 (437–439, 444–446), 775 (437, 438), 776 (444–446), 801 (445), 819 (184), *842*, *844*, *846*, *850*
Orvini, E. 165 (75), *186*
Osaki, K. 99, 101 (192b), *152*
Osborne, J. H. 808 (566), 810 (566, 582), *852*, *853*
Osipov, O. A. 779 (462, 463), 781 (462, 463, 472, 473), 787, 801 (462, 463). *850*
Ospici, A. 344 (3), *366*
Ostman, B. 581 (173), *615*
Ostrovskaya, I. K. 681, 686 (23b), *840*
Osuka, A. 8 (38, 39), *18*, 623 (48, 49), *657*
Otsu, T. 338 (32), *341*
Otsubo, T. 536 (167), 541 (201), *556*, *557*, 642 (310), 652 (439), *662*, *665*
Ott, D. 490 (1056), *515*
Ottlinger, R. 90 (153), *151*, 463 (732–735, 736a), 484 (733, 736a), *509*
Otto, P. P. H. L. 373 (21), 379 (42), 392 (21), *394*, *395*
Outurquin, F. 411, 412 (148), 437 (422), 438 (427–429, 431), *497*, *502*, *503*, 586 (193, 194), *616*
Owada, H. 403 (33), *495*, 632, 633 (187, 189), 639 (187), *660*
Owen, C. A. Jr. 381, 393 (52), *395*
Oxton, I. A. 697, 778 (140c), *843*
Ozaki, Y. 765, 766 (407), *849*
Ozolin, S. 430 (364), *501*

Paal, C. 5 (18), *18*, 407 (96), *496*
Packer, K. J. 362 (65), *367*
Paetzold, R. 336 (26), *341*, 462, 492 (719), *509*, 520 (27), 537 (182), 546 (247), 548 (247, 264, 265), 549 (247), *553*, *556*, *558*, 629 (155), 643 (355, 360), 650 (362), *659*, *663*, 721, 722 (237, 239), *845*
Page, C. T. 690 (102, 103), *842*
Page, E. M. 808 (576), *853*
Page, G. 16 (90), *19*
Page, H. L. 249 (36), *284*
Page, J. E. 575 (132), *614*
Paige, H. L. 523, 537, 545 (65), *554*
Painter, E. 401 (1), *494*
Painter, E. P. 371 (16), *394*
Pakawatchai, C. 779 (466b, 469), *850*
Pakzad, B. 477 (917), 478 (919), *513*
Paliani, G. 378, 383 (37a, 37b), *394*
Palibroda, N. 277 (88), *285*
Palmer, G. 238 (147), *241*, 382, 392 (67), *395*, 676 (37), *677*
Palmer, G. T. 738 (315), *847*
Palmer, I. S. 163, 183 (25), *185*
Palmer, M. H. 308 (13), *328*
Palyulin, V. 481 (950), *513*
Palyulin, V. A. 531 (147), *555*
Pan, F. 379 (43), 388 (43, 102b), *395*, *396*
Pan, S. 648 (350), *663*
Pan, W.-H. 102 (220), *152*, 191–193 (12), *239*, 765, 794 (405), *849*
Pan, W. H. 732, 736 (288), 794, 795 (516–519, 527), *847*, *851*
Pandey, G. K. 51, 57, 58 (94), *62*, 70, 72, 90 (37), *148*
Pang, F. 25 (43), *61*
Pankiewicz, J. 413 (191), *498*
Pannell, K. H. 113 (241), 114 (242), *153*, 701 (161, 162), 702 (162), 703 (161, 162, 165, 166), *844*
Panov, V. 480 (938), *513*
Pant, B. C. 261 (64), *284*, 520 (26), 522 (46, 47), *553*
Papavassiliou, G. C. 681, 798, 799 (27a, 27b), *840*
Papenroth, W. 489 (1050), *515*, 522 (38), 528 (38, 124), 530, 532 (124), *553*, *555*
Pappalardo, G. 483 (974), *514*
Pappalardo, G. C. 86 (117), *150*, 247 (24), *284*, 362 (66), *367*
Paquer, D. 467 (818), *510*
Paramonov, V. 444 (496), *504*
Paramonov, V. A. 624 (85), *658*
Parichkova, R. 699 (151), *843*
Parizek, J. 167–169 (86), *186*, 393 (145), *397*
Park, B. K. 644 (318), *662*
Parker, G. R. 651 (409, 410), *664*
Parkin, S. 449 (555), 450 (572b), *505*, *506*
Parks, G. L. 97 (189), *152*, 525 (88), *554*
Parlova, V. L. 566, 567, 574 (69), *613*
Parr, W. J. E. 624 (86), *658*
Parris, G. E. 165, 167, 169 (70), *186*
Parshall, G. W. 681 (32a), *841*
Partovi, M. 447, 464 (532), *505*
Partovi, M. H. 377, 386 (35), *394*
Partovi, M. N. 581, 595 (162), *615*
Parvez, M. 715, 717, 720, 822 (223), *845*
Pasmurtseva, N. A. 16 (93), *20*, 548 (270, 271), *558*
Passerini, R. 428 (341), 442 (476), *501*, *503*, 585 (207), *616*, 642 (278), *661*
Passmore, J. 249 (36), 253 (45), *284*, 523 (64, 65), 537 (65), 545 (64, 65, 240), *554*, *557*
Pastour, P. 411 (142, 149), 412 (149, 160, 161, 163), 413 (149, 163, 171, 173–176, 188–191, 192a), 415 (163, 175, 176, 198, 201, 202, 204, 205, 208, 209), 430 (149, 361, 362, 368, 369), 434 (369), 435 (205, 208), 438 (149, 189, 427, 428, 433, 434), 455 (188, 612, 613), 460 (613), 461 (192a), *497*, *498*, *501*, *503*, *506*, 581 (181), 584 (180), 586 (193), 590 (223), *615*, *616*
Patai, S. 267 (73), *285*, 434 (397–399), *502*, 527 (106–108), *555*
Patel, P. 492 (1062), *516*
Patel, V. 448 (538), 449 (562), 450 (538, 571, 574, 576), 451 (574), *505*, *506*
Patel, V. V. 310 (16), 311 (16, 17), *328*
Paterson, J. F. 690, 786 (77a, 77b), *842*
Patigny, P. 302 (67), *305*
Patil, S. R. 650, 651 (422), *664*
Patrick, H. 387, 388 (93), *396*
Patrick, R. 160 (16), *160*
Patsaev, A. K. 624 (77–79), *657*
Pattenden, G. 401 (7), *494*
Patterson, J. A. 179, 180 (113), *187*
Patwardahom, B. H. 403 (26b), *494*
Paty, P. B. 622, 624 (21), *656*
Patzwaldt, H.-G. 455 (632), 456 (632, 635), *507*
Paul, I. 467 (820a, 820b), *511*, 743, 744 (347), *848*
Paul, I. C. 64, 66, 86 (4), 107–109 (234), 116 (4), 118 (265), 124 (234), *148*, *153*, 261 (62), *284*
Paul, R. C. 522 (45), 548 (269), *553*, *558*, 814, 824 (625), *854*
Paul, W. 768 (420, 421), 770 (421), 829, 830 (420, 421), *849*
Pauling, L. 78 (88), *150*
Paulmier, C. 206, 210, 211, 221 (61), *240*, 377 (34), *394*, 411 (142–145, 148, 149), 412 (148, 149, 160, 161, 163), 413 (149, 163, 171, 173–175, 183–185, 189), 414 (193), 415 (163, 175, 198, 201, 204, 205),

430 (149, 361, 362, 368, 369, 371), 434 (369), 435 (205), 436 (411, 412), 437 (416, 418, 419, 422), 438 (149, 189, 424, 427–429, 431, 433, 434), 452 (424), 460 (412, 419), *497*, *498*, *501–503*, 581 (181), 586 (193, 194), 590 (223, 224), 591 (224, 225), *615*, *616*, 641 (300, 301), 642 (250), *661*, *662*
Paulus, E. F. 129, 131, 143 (299), 145, 146 (327), *154*, *155*, 528, 533 (134), *555*, 567 (76, 78), *613*
Pavani, R. 25 (41), *61*
Pavelcik, F. 730 (274), *846*
Pavloff, A. 487 (1010), *514*
Pavlova, V. L. 522, 523 (42), *553*, 574 (120, 121), *614*
Payne, N. C. 621, 622 (29), *657*, 765, 807 (404), *849*
Peach, M. E. 356 (37), *366*
Peachey, S. J. 7 (30), *18*
Peacock, R. D. 545 (231, 237), *557*
Peake, S. L. 645 (330), *662*
Pearson, P. S. 229 (108), *241*
Pearson, R. 64 (8), *148*
Pearson, R. G. 737, 762 (309), *847*
Pearson, T. 409 (110), 487 (1026), 488 (1033), *496*, *515*
Pearson, T. G. 814 (612), *854*
Pecile, C. 732 (287), 735 (287, 297), 737 (287), *847*
Pedaja, P. 412 (158), 417 (224), *497*, *499*
Pedersen, B. 361 (62), *367*
Pedersen, C. 271 (78), 275 (83), *285*, 465 (789–791), 466 (802), *510*
Pedersen, C. L. 274, 275 (81), *285*
Pedersen, C. T. 117, 119 (267), *153*
Pedersen, E. B. 595 (258), *617*
Pedersen, H. 450 (570), *506*
Pedersen, N. D. 391 (123), *397*
Pedersen, S. 465 (793), *510*
Pedersen, T. 117, 119 (267), *153*
Peel, J. 407 (78, 79), *496*
Pegg, W. J. 624 (108), *658*
Peleties, N. 603, 604 (300, 309), *618*
Pelizzi, C. 815 (602), *853*
Pelizzi, G. 815 (602), *853*
Pellinghelli, M. A. 794, 795 (529), *852*
Pellini, G. 544 (221), *557*
Peñéñory, A. B. 621, 622 (12), *656*
Penn, R. E. 11 (57), *19*
Penner, J. A. 393 (149), *397*
Pennesi, G. 794 (525), *851*
Peper, F.-K. 455, 456 (632), *507*
Perales, A. 697 (135b), *843*
Perdue, E. M. 600 (281), *617*, 621, 622 (4), *656*
Perin, F. 421 (278), *500*
Perin, P. 407 (98–100), 429 (354, 355), 434 (354), 436 (355), 471 (100), *496*, *501*
Peringar, P. 545 (226), *557*
Perkins, M. J. 623 (54, 56), 624 (98, 100), 642 (243), *657*, *658*, *661*
Perlstein, J. 449 (565), 470 (861), *505*, *511*
Perrier, M. 271 (79), *285*, 467 (818, 819, 822, 823, 825), *510*, *511*, 639 (231, 233), *661*
Perrin, L. 432 (385), *502*
Perry, D. A. 11 (62), *19*
Perry, D. L. 794, 795 (529), *852*
Persson, B. 184 (147), *188*, 641 (275), *661*
Peruzzini, M. 727 (258–261), 729 (258–261, 267b, 268), 730 (261), *846*
Perveev, F. F. 407 (83), *496*
Pesce, G. 402 (21), *494*
Pesch, R. 248, 249 (32), *284*
Pesin, V. 465 (776, 782), *510*
Pesin, V. G. 586 (213, 214), *616*, 700 (146), *843*
Peters, A. T. 731 (279), *846*
Peters, P. 465 (793), *510*
Petersen, C. 466 (801), *510*
Petersen, H. 311, 312 (15), *328*, 464 (743, 759), *509*
Petersen, M. 451 (584), *506*
Peterson, M. R. 32 (76), *61*
Peterson, R. E. 393 (148), *397*
Peterson, R. J. 178–180 (104), *187*
Petit, A. 440 (457), *503*
Petit, M. G. 86, 87 (122), *150*
Petragnani, N. 483 (976), *514*, 522 (51), 523 (76), 527 (109, 110), 528 (110), 529 (109, 110), 532 (154), 533 (110), 542 (215–217), *554–557*, 621, 622 (16), 624 (75), 636 (208), 650 (385), 651 (390), 652 (208, 434), 653 (446, 447, 454), 654 (446, 473, 474), 655 (483), *656*, *657*, *660*, *663–665*
Petren, J. 680, 690 (4), *839*
Petrov, A. 406 (66, 69), 410 (69, 121a, 122), 444 (496), 447 (66), 448 (537b), *495*, *497*, *504*, *505*
Petrov, A. A. 609 (331, 332), *618*, 621 (9, 26), 622 (9, 26, 40), 624 (83–85), 651 (40, 394, 401), *656–658*, *664*
Petrov, M. 406 (66, 69), 410 (69, 121a, 122), 444 (496), 447 (66), 448 (537b), *495*, *497*, *504*, *505*
Petrov, M. L. 621, 622 (26), 624 (83–85, 91), *656*, *658*
Petrov, Yu. A. 815 (601i), *853*
Petrova, L. 407, 412 (88), *496*
Petrova, L. Y. 377 (33a), *394*
Petrzilka, M. 634 (196), *660*
Pettersen, R. C. 113 (241), 114 (242), *153*,

701 (161), 703 (161, 165, 166), *844*
Pettersson, K. 415 (197), 417 (224), *498*, *499*
Pettit, L. D. 80 (97), *150*, 565, 574, 575, 608 (60), *613*, 690 (94), *842*
Peymann, E. 768 (417), *849*
Peyrussan, V. 448, 450 (549), *505*
Pfander, W. H. 387 (77b), *396*
Pfannstiel, K. 527, 528, 533 (119), *555*
Pfisterer, G. 233 (116), *241*
Pfisterer, H. 768, 770, 829, 830 (421), *849*
Phillips, H. 550 (277), *558*, 655 (488), *665*
Phillips, L. 575 (124), *614*
Pianka, M. 458 (677), *508*
Picci, M. 584 (182), *615*
Pichat, L. 641, 650 (262), *661*
Pickardt, J. 714 (198), 832, 835, 836 (654), *845*, *855*
Pierce, L. 70–72 (50), *149*
Pierini, A. B. 622 (46), *657*
Pierpont, C. G. 102 (212), *152*, 773 (455), 788, 791 (455, 499), 798, 799 (541), *850–852*
Pierres, M. 381, 393 (59), *395*
Pietra, S. 333 (10), *341*, 445 (504), *504*
Pietro, W. J. 25 (62, 64, 68, 71), 29 (68, 71), 30 (62, 64, 68, 71), 33 (68, 71), 34, 54 (62), 56 (68), *61*
Pietropaolo, R. 690 (71, 87, 88, 97–99, 106), 691 (97, 98, 106), 695, 774 (98), *842*
Piette, J.-L. 86 (116), *150*, 191, 199 (7), 203 (7, 59), 204, 205 (7), 207, 208 (63), 210 (7, 59), 211 (59), 212, 214, 215 (69), 216 (70), 217 (7), 230, 232 (110), *239–241*, 416 (213), 439 (453), 440 (453, 456–458), 441 (458, 460), 442 (460, 466), 443 (483), 444 (458), 454, 458 (605), 471 (871, 873), 476 (905), 479 (932), *498*, *503*, *504*, *506*, *512*, *513*
Piette, J. L. 13 (68), 14 (79), 17 (105), *19*, *20*, 75, 76 (76, 77), 77 (76, 77, 83), *149*, 255 (47), *284*, 563 (41), 579 (150), 580 (150, 155), 584 (184, 186), 585 (186), *612*, *615*, *616*, 629 (163), 651 (391–393, 398), 652 (430–432), 655 (391, 392), *659*, *664*
Piettre, S. 326 (71), 327 (73), *329*
Pignataro, S. 44, 54 (86), *62*, 246 (16–18, 23), *283*, *284*, 416 (212), *498*
Pignolet, L. H. 708 (178), 709 (178, 180), 788, 791 (500), 792 (500, 508), *844*, *851*
Pihl, A. 288 (1), *304*
Pijpers, F. W. 762, 763 (395), *849*
Pilati, T. 403 (44), *495*
Pilcher, G. 157–159 (2), *160*
Pilipenko, A. T. 683, 685, 686 (34, 35), *841*
Pilz, W. 455 (627), *507*
Pinel, R. 271 (79), *285*, 467 (822), *511*
Pinkert, W. 768 (417), *849*
Piovesana, O. 684–686 (39), *841*
Piquard, J. L. 319 (42, 43), *329*, 605 (324), *618*
Pira, P. 690 (99), *842*
Piraino, P. 690, 691 (106), *842*
Pirson, P. 244, 266 (2), *283*, 315 (26), *328*, 439 (444), 476, 477 (909), *503*, *512*, 598, (265), *617*
Pitombo, L. R. M. 622 (31), 629 (158), 639, 640 (232), 652, 653, 655 (158), *657*, *659*, *661*, 690 (69, 73, 75), *841*, *842*
Pittman, C. U. 839 (675), *855*
Pittman, R. 487, 488 (1027), *515*
Pittman, R. W. 332 (5), 335 (21), *341*
Pivovarevich, L. P. 573 (115, 116), *614*
Platt, A. W. G. 191–193 (16), *239*, 774, 776, 801 (445), *850*
Plé, G. 435 (401), *502*
Plimmer, J. R. 264 (68), *284*
Plowman, K. R. 715 (212, 213, 222), 716 (207, 212, 213), 717 (212, 213), 720 (212, 213, 222), 767 (213), *845*
Plücken, U. 311, 312 (15), *328*, 451 (584), *506*
Pluim, H. 624 (72), *657*
Plummer, C. 428, 442, 483 (337), *501*
Podkovyrina, T. A. 565 (65), *613*
Podolesov, B. 486 (1008), 487 (1008, 1013), *514*
Poehler, T. 17 (104), *20*, 449 (567), *505*, 651 (417), *664*
Poehler, Th. 448, 450 (548), 451 (579), *505*, *506*
Pogonowski, C. S. 629 (171), *659*
Pogorelyi, V. K. 624 (89), *658*
Poirier, R. A. 25 (14, 50, 63), 30 (63, 73–75), 32 (76), 33 (83), *60–62*
Poleschner, H. 448 (540), *505*
Poletti, A. 378, 383 (37b), *394*
Poli, R. 710, 712, 814 (190, 191b), 819 (191b), *844*
Poliakoff, M. 466 (801), *510*
Polidoro, G. 621, 622 (15), *656*
Poloni, L. 383 (72a), *395*
Polonsky, J. 644 (340), *663*
Pomeroy, R. K. 546 (242), *557*
Pomykacek, J. 490 (1053), *515*
Poncet, J. L. 720 (228), *845*
Ponet, M. J. 580 (156), *615*
Ponikarova, T. M. 389 (109), *396*
Ponomareva, O. B. 347 (7), *366*
Ponomareva, T. 439 (445), *503*
Poole, A. 467 (824), *511*
Pope, A. L. 389 (113), *396*
Pope, W. J. 7 (30, 31), *18*, 527 (117), *555*,

668 (9), *676*
Popik, M. V. 70 (47), 75 (47, 75), 76 (75), *149*
Pople, J. A. 25 (27, 29, 32, 35, 44, 48, 49, 62, 64), 29 (27), 30 (48, 62, 64), 33 (27, 49), 34 (48, 49, 62), 54 (48, 62), 56, 57 (48), *60*, *61*
Popov, A. F. 575 (126), *614*
Popov, D. K. 389 (109), *396*
Popova, L. L. 16 (91), *19*
Popova, R. S. 575 (126), *614*
Poppe, E. 460 (704), *508*
Poppitz, W. 767 (413), *849*
Porai-Koshits, M. A. 730 (272), *846*
Porta, O. 624 (100), *658*
Porta, P. 796 (536), *852*
Portanova, J. P. 178 (109), *187*
Porten, J. A. 118 (261), *153*
Posner, G. H. 623 (59), *657*
Post, B. 462 (721), *509*
Potchen, E. J. 393 (159), *397*
Pote, C. S. 51, 57, 58 (94), *62*
Potts, A. W. 22 (5, 6), *60*
Potts, K. 458 (672), 465, 468 (772), *508*, *510*
Pouly, S. 742, 749 (338), *848*
Pourcelot, G. 604 (317, 318), *618*
Pourveur, A. 473 (893), 475 (903), *512*
Powell, M. J. D. 32 (78, 80), *61*, *62*
Powers, D. E. 88 (133), *150*
Pozdeev, N. 410 (125), *497*
Pozdeev, N. M. 88 (135), 89, 90, 128 (143), *151*, 383 (73), *395*
Pràefcke, K. 196, 208 (36), 222 (36, 95, 96), 231 (95, 96), 232 (36), *239*, *240*, 257, 258 (58), *284*, 355 (30), *366*, 409 (115), 416 (214), 421 (275), 435 (403–405), 440 (275), 441 (459), 446 (512a), 447 (524), 477 (916–918), 478 (919–921), 479 (933, 934), 480 (934), *496*, *498*, *500*, *502–504*, *513*, 624 (94), *658*
Pramanik, P. 681 (21, 22), *840*
Prelicz, D. 458 (667), *507*
Prelog, V. 88 (136), *151*
Preston, R. L. 387 (77b), *396*
Preti, C. 166 (82), *186*, 262 (66, 67), 263 (67), *284*, 580 (160), *615*, 707, 708 (177), 759, 760 (371–387), 761 (371–374, 381, 383), *844*, *848*, *849*
Price, M. J. 571 (108), *614*
Price, R. 680 (9e), *840*
Price, W. C. 22 (4–6), *60*
Prince, M. 332 (9), 333 (9, 13), 334 (13), 336 (25), *341*
Pritzkow, H. 129–131, 143 (296), *154*
Prochazkova, V. 165, 167, 169, 170 (76), *186*
Prokof'eva, A. F. 573, 593 (118), *614*
Proksch, A. 463 (734), *509*
Pronicheva, L. D. 106 (226), *153*
Prophet, M. 157 (1), *160*
Proskina, N. N. 684–686 (40), 732 (290), *841*, *847*
Protiva, M. 477, 478 (913), 490 (1052–1054), *512*, *515*
Prousa, R. 406, 447 (67), *495*
Pruett, R. L. 839 (671), *855*
Puchkova, S. M. 389 (111), *396*
Puckett, J. 523 (58), *554*
Puff, H. 357, 358 (40, 43), *366*
Pulay, P. 25 (43), *61*
Pulidori, F. 600 (282), *618*, 621, 622 (5), *656*
Puls, M. 469 (840b), *511*
Pummerer, R. 316 (28), *329*
Purcell, R. 487 (1026), 488 (1033), *515*
Purdy, W. C. 184 (177), *188*
Purello, G. 442 (476), *503*, 585 (207), *616*
Puri, J. K. 548 (269), *558*
Pye, P. L. 814, 821 (622), *854*
Pyle, R. 449 (567), *505*
Pyles, R. A. 362 (66), *367*
Pytlewski, L. L. 721 (243), *846*

Quick, M. H. 740 (322), *847*
Quintiliani, M. 290, 291 (33), *305*

Raasch, M. 405 (62b), *495*
Rabet, F. 409, 464 (108), *496*
Rabindran, K. 420 (268), *499*
Rabinovich, D. 409 (117, 118a), *496*, *497*
Racheva, T. M. 681, 686 (23a), *840*
Radchenko, S. I. 609 (331, 332), *618*, 621 (9), 622 (9, 40), 651 (40, 401, 402), *656*, *657*, *664*
Radlick, P. 480 (945), *513*
Radziuk, B. 165 (68, 69), 167–169 (68), *186*
Rae, A. D. 811 (588), *853*
Rae, A. I. M. 766, 772 (410), *849*
Raimondi, D. L. 25, 26 (12), *60*
Rainer, R. 455 (624), *507*
Rainville, D. 100 (195), 101 (193, 195, 196), 144, 145 (325), *152*, *155*, 483 (975), 484 (991a), *514*
Rainville, D. P. 100, 101 (201), *152*
Raithby, P. R. 811, 812 (590, 594), 832, 839 (590), *853*
Raja, T. 439 (449–452), 446 (513, 514), 469 (845c), *503*, *504*, *511*
Raja, T. K. 757 (369), *848*
Rakhlin, V. I. 624 (99), *658*
Raksha, M. A. 629, 632 (160), *659*
Ralph, J. 464 (757), 465 (777), *509*, *510*
Ramage, R. 408 (103), *496*
Ramakrishna, R. S. 778, 779 (456a, 456b), *850*

Ramalingam, K. 253 (44), *284*, 469 (844, 846), *511*
Ramasamy, K. 443 (485, 486), *504*
Ramsey, N. F. 194 (21), *239*
Rankin, D. W. H. 66 (22, 23, 24a, 24b), 72 (58, 61), *148*, *149*, 754–756 (366), *848*
Ranneva, Y. I. 377, 385, 386 (33b), *394*
Rappoport, Z. 267 (73), *285*, 434 (399), *502*, 527 (107, 108), *555*
Raptis, S. E. 163 (33), *185*
Raseo, P. 584 (182), *615*
Rashidbaigi, A. 409 (109), *496*
Rasmussen, F. 450 (570), *506*
Rasmussen, S. E. 690, 691 (70), 661 (436), *842*, *850*
Rasteikiene, L. 403 (26a), *494*
Raston, C. L. 97, 98, 132, 143 (186), *152*, 525 (87), *554*
Rastrup-Andersen, J. 89, 90 (146, 147), *151*
Rathke, B. 5 (10), *18*
Rathke, M. W. 603 (306), *618*
Rattanaphani, V. 814 (615), *854*
Ratts, R. W. 603 (297), *618*
Ratushnaya, S. K. 183 (143), *188*
Raubenheimer, H. G. 726, 728 (254), *846*
Raucher, S. 403 (34), *495*, 605 (323), 606 (327, 328), *618*, 632, 633 (178, 179), *659*
Rauchfuss, T. B. 747, 753 (353), 798, 799 (542), 806 (565a, 565b), 808 (570, 571, 575), 811 (542, 585b), 825 (353), 832 (353, 570, 571, 652, 660, 661, 664), 833 (652), 834 (353, 570, 571), 835 (652, 660, 661), 836 (652, 660), 837 (571, 652, 661, 664), 838 (652, 661, 664), *848*, *852*, *853*, *855*
Rauer, R. F. 621, 622, 624, 625 (11), *656*
Ray, D. 319 (45), *329*
Ray, F. 428 (343), 465 (781), *501*, *510*
Raymond, A. H. F. H. 14 (78), *19*
Raynor, J. B. 801, 803 (551), *852*
Razumov, A. I. 815 (601g), *853*
Razuvaev, G. A. 356 (33, 35), *366*
Razzoli, G. 759, 760 (377), *848*
Read, C. M. 644 (313), *662*
Read, R. W. 13 (69), *19*
Reamer, D. C. 165, 167, 170 (72), *186*, 387 (94), *396*
Rebane, E. 249 (34, 37, 38), 252 (34, 41), 255–257 (38), 258 (59), 260, 261 (37), 262, 263 (65), *284*, 547, 548 (250), *558*, 641, 643 (264), 650 (368, 369), *661*, *663*
Rebbert, R. L. 738 (315), *847*
Recca, A. 416 (211), *498*
Rechowicz, K. 393 (152), *397*
Reddy, M. 484 (985), *514*
Reddy, M. L. N. 654 (467), *665*, 744–747 (348), *848*
Redpath, J. L. 301 (64, 65), *305*
Reed, R. W. 821 (633), *854*
Reedijk, J. 522 (56), *554*, 736 (308), *847*
Rees, A. 428, 442, 483 (337), *501*
Rees, C. 403 (38, 39), 453, 465 (592), 483 (981), 488 (1040b), *495*, *506*, *514*, *515*
Rees, C. W. 703 (163), *844*
Reger, D. L. 798, 800 (545), *852*
Regev, H. 640 (249), *661*
Reich, H. 403 (36), 420 (272), *495*, *500*
Reich, H. J. 6, 10 (24), *18*, 220 (82), *240*, 317 (34, 35), *329*, 518 (6), 536 (166), 547 (258), 548 (261a), 549 (258), *553*, *556*, *558*, 599 (273), 600 (277), 603 (303–305, 308, 310, 311), 604 (304, 310, 315, 319, 322), 605 (310), 609, 611 (304), *617*, *618*, 621 (2), 622 (2, 42), 623 (2), 624 (113, 115, 117), 625 (116–118), 626 (115, 135), 627 (117), 629 (115, 162, 168), 630 (115, 167, 168), 631 (113, 115, 167), 632, 633 (186), 639 (119, 121, 122, 186), 640 (186, 254), 642 (146), 645 (330), *656–662*
Reich, I. 317 (34), *329*
Reich, I. L. 624, 626 (115), 629 (115, 168), 630 (115, 167, 168), 631 (115, 167), 642 (146), *658*, *659*
Reichel, L. 482 (962), *513*, 653 (453), 655 (491), *665*
Reichenbacher, M. 537 (182), *556*
Reid, D. 271 (78), *285*, 401, 406 (9), 444 (500), 445 (505, 506a, 507a), 467 (826a, 826b, 827a, 827b), 475 (500), *494*, *504*, *511*
Reid, D. H. 325 (66), *329*
Reif, L. 594 (242), *617*
Reimer, K. J. 695, 697 (130), *843*
Reiner, D. 804, 805 (557), 808 (567), 832 (653), *852*, *855*
Reinhold, J. 789 (506), *851*
Reis, E. 547 (252), *558*
Reisch, H. 444 (490b), *504*
Reisch, J. 407, 409, 414 (97b), *496*
Reistad, T. 671 (19), *677*
Rémion, J. 623 (60), *657*
Remizov, A. B. 74 (72), *149*
Renard, M. 321 (53), *329*
Renaud, R. N. 571 (110), *614*
Renga, J. M. 317 (34), *329*, 624, 626 (115), 629 (115, 168), 630 (115, 167, 168), 631 (115, 167), 640 (254), *658*, *659*, *661*
Renken, T. L. 11 (57), *19*
Renson, M. 9 (44), 14 (79), 17 (105), *18–20*, 255 (47), 267 (72), 268 (74, 75), 269 (76), 276 (86), *284*, *285*, 315 (26, 27), *328*, 419 (245, 260, 261), 420 (260, 261, 263–265), 421 (281), 422 (281–285), 423 (285, 286, 288, 289, 292, 298–300), 424 (292, 299, 300, 302, 303, 305, 306), 425 (285, 299, 312), 426 (260, 264, 298, 312–315), 427

(323, 325), 431 (374), 432 (383, 386), 433 (325, 387), 434 (374, 383), 435 (406), 438 (424, 425), 439 (453), 440 (453, 456–458), 441 (458, 460), 442 (460, 473), 443 (245, 484), 444 (458), 446 (597), 452 (424, 425, 591), 453 (425, 591, 597), 454 (597, 602, 605), 457 (652–654, 656), 458 (605), 465 (315), 471 (863, 864, 868, 869, 871, 873), 472 (863, 864, 868, 869, 877, 882, 886, 887), 473 (286, 484, 863, 890, 892, 893), 474 (245, 896), 475 (283, 896, 898, 902, 903), 476 (286, 473, 904, 905, 909, 910), 477 (909, 914), 478 (914), 479 (931, 932), 481 (289), 484 (993), 488, 489 (1039b), 490 (863, 1051), 491 (602), 492 (1039b), *499–504*, *506*, *507*, *512–515*, 584 (183–186, 205), 585 (183, 185, 186, 190), 586, 590 (208), 592 (233, 234), 595 (185), 596 (185, 251), 597 (208, 251), 598 (265), *615–617*, 621, 622 (27), 629 (163), 651 (391–393, 398), 652 (430–432), 655 (391, 392), *656*, *659*, *664*
Renugopalakrishnan, V. 86 (120), *150*, 670 (16), *676*
Rest, A. J. 680 (9a), *839*
Reston, C. 469 (854c), *511*
Rettenmeier, A. 742 (337), 746 (351), 751 (337, 351), 752 (351), 825 (337, 351), 827, 828 (337), *848*
Reynolds, C. 453, 465 (593), *506*
Reynolds, C. D. 91 (155), *151*
Reynolds, G. 459 (691), *508*
Reynolds, W. L. 721 (229), *845*
Reznichenko, A. 437 (415), *502*
Reznichenko, A. V. 587 (197), *616*
Rezvukhin, A. I. 197, 204, 205 (37), *239*
Rheinboldt, H. 162 (3), *185*, 307 (1), *328*, 525 (82), 546 (244), 548 (266, 267), 550 (274), *554*, *557*, *558*, 567 (80), *613*, 626 (132–134), 629 (134, 156), 639 (231, 233), 641 (134, 267, 270), 643 (357), 654 (472), *659*, *661*, *663*, *665*
Ribaud, M. 450 (569), *506*
Ricci, A. 21 (278), 438, 439 (435), 471 (874), *500*, *503*, *512*
Rice, D. A. 695, 697 (131), 803 (553c), 808 (576–578), 813 (578), *843*, *852*, *853*
Rice, F. 488 (1031), *515*
Rice, J. 451 (582), *506*
Richardson, F. S. 670, 671 (17e), *677*
Richter, R. 102 (206, 217), *152*, 624 (92), *658*, 786–788 (485), 789 (503), 790, 791 (485), 796, 797 (539), *851*, *852*
Riddel, F. 488, 489, 492 (1039a), *515*
Rideal, E. 488 (1032), *515*
Ried, W. 423 (294), 459 (697), *500*, *508*
Riegel, F. 345 (4), 346 (5), *366*
Righetti, B. 299 (60), *305*
Rihs, G. 446 (520), *504*
Rijnsdorp, J. 810 (580), *853*
Riley, A. L. M. 371, 372, 393 (7), *394*
Riley, F. 528, 532, 533 (127), *555*
Riley, H. L. 528 (122), *555*
Riley, R. 423 (291), *500*
Riley, R. F. 532, 534 (152), *556*
Rimala, T. S. 117 (256, 257), *153*
Rimondi, C. 292 (34), *305*
Rinaldi, A. 166 (84), 171 (116, 117), 178 (108), 179 (116), 180 (117), 181 (120, 122), *186*, *187*, 457 (655), *507*, 622 (32, 36), *657*
Rindorf, G. 94 (166), *151*, 450 (570), *506*
Ringdahl, B. 671 (21), *677*
Rio, R. M. d. 390 (119), *397*
Rips, R. 486, 487 (1007), *514*
Rishman, G. 487 (1015), *514*
Ritchie, C. D. 25 (21), *60*, 562 (26), *612*
Rivkin, B. B, 523 (75), 537 (170, 177), *554*, *556*
Rizov, N. 171 (102), *187*
Robert, A. 447 (525), 458 (673), *504*, *508*
Robert, R. V. D. 165 (63), *186*
Roberts, B. P. 325 (65), *329*, 642 (243), *661*
Roberts, D. A. 811 (586), *853*
Roberts, J. D. 202 (46), *239*
Robertson, H. E. 754–756 (366), *848*
Robertson, J. A. 222, 223 (99), *240*, 814, 815, 817 (605), *853*
Robertson, J. D. 675 (33), *677*
Robiette, A. G. 66 (25), 72 (58), *148*, *149*
Robin, M. B. 25 (22), *60*
Robinson, M. F. 387 (78), *396*
Robinson, P. 407 (79), 409 (110), *496*
Rød, T. 360 (56), *367*
Rodekirch, G. 456 (634), *507*
Rodger, M. 349–352 (16), *366*
Rodgers, A. S. 157, 159 (3), *160*
Rodionova, L. S. 624 (83), *658*
Rodley, G. A. 693, 695–697 (122), *843*
Roecker, L. 771, 772, 780 (428), *849*
Roesky, H. W. 710, 712, 713 (191a), 768 (417), 814, 819 (191a), *844*, *849*
Roetti, C. 27 (19), *60*
Rogers, D. Z. 623 (59), *657*
Rogers, M. T. 565 (56), *612*
Rogers, R. D. 710, 712, 814 (190, 191b), 819 (191b), *844*
Rogozina, S. 494 (1070), *516*
Rohde, H. G. 290, 291 (28), *304*
Rohm, W. 690, 691 (96), *842*
Rohner, C. 381, 393 (59), *395*
Rohrmann, J. 805, 806 (558), *852*
Roland, E. 811 (592), *853*
Rolandsen, A. 794 (526), *851*

Rolandsen, Å. 12 (67), *19*, 102, 103 (209), *152*
Röll, W. 723, 725, 742, 753 (250), *846*
Rollmann, M. 486 (1001), *514*
Román, E. 765 (403), *849*
Romanov, A. M. 726–728, 739 (251), 770 (424), 771 (424, 425), 778 (424), *846*, *849*
Romers, C. 139 (314), *154*
Rømming, C. 79 (94), *150*, 360 (58), *367*
Rönsch, E. 643 (355), *663*
Roos, B. 25 (31), *60*
Root, M. J. 771, 772, 780 (427), *849*
Roper, W. R. 715 (214–217), 716 (214), 717 (214–216), 718 (217, 226), 719 (217), 765 (214, 217, 404), 766 (214, 217, 411), 767 (215), 768 (418, 419), 807 (404), 814, 821 (216), 825 (418), 829 (418, 419, 650a, 650b, 651), 830 (418, 419), *845*, *849*, *854*, *855*
Ropez, R. L. 387 (77b), *396*
Rose, M. L. 381, 393 (62), *395*
Rosenbaum, A. 794, 796 (513), *851*
Rosenberg, M. 281 (90), *285*, 832, 836 (658), *855*
Rosenfeld, I. 288 (2), *304*
Rosenfeld, M. N. 643 (315), 644 (313, 315–317), *662*
Rosenfield, J. S. 672 (24), *677*
Rosenkilde, S. 65 (19), *148*
Rosenthal, M. 109 (236, 237), 110 (237), *153*
Rosenthal, M. R. 686, 721, 722 (52), *841*
Ross, B. 483 (979), *514*
Rossetti, R. 808 (572, 573), 811 (595), 832 (655–657, 659, 662, 663), 835 (655), 836 (659), 837 (659, 662), 838 (663), *852*, *853*, *855*
Rossi, C. 471 (874), *512*
Rossi, R. A. 621 (12), 622 (12, 46, 47), *656*, *657*
Rossmark, H. H. 383 (72b), *395*
Rot, I. 743, 749 (344), *848*
Roth, R. 409 (114, 116), *496*
Roth, R. W. 265 (70), *285*, 378, 386, 387 (38a), *394*
Rothchill, E. O. 387 (81), *396*
Rothrock, R. K. 102 (219), *152*, 742 (334, 335), 750 (334), 758, 759 (335), *848*
Rotruck, J. T. 389 (113), *396*
Rott, J. 746, 751, 752, 825 (351), *848*
Rottinger, E. 758, 759, 804 (370), *848*
Rougier, M. J. 667 (1a), *676*
Roulet, R. 690 (107), *842*
Roundhill, D. M. 771, 772, 774, 775 (434), *850*
Roundhill, S. G. N. 771, 772, 774, 775 (434), *850*
Rouvier, E. 249, 251 (39), *284*
Roy, A. 455 (618), *507*
Roy, N. 281 (92), *285*
Royer, R. 413 (167), *497*
Royston, A. 565, 574, 575, 608 (60), *613*
Rozantsev, G. 411 (147), *497*
Rozantsev, G. G. 581 (172), *615*
Rozsondai, B. 81, 82 (100), *150*
Ruban, G. 431 (380), *502*
Rubner, M. 334, 341 (18), *341*
Rubnet, M. 334, 335 (17), *341*
Ruch, E. 668 (3), *676*
Rudko, A. D. 78 (93), *150*
Rudolph, H. D. 73 (70), *149*
Rudolph, K. P. 699 (152a), *843*
Ruf, H. 487, 492 (1029), *515*
Ruge, B. 621, 622 (29), *657*
Ruidisch, I. 716 (199), *845*
Rund, J. 460 (700), *508*
Ruppel, W. 90 (152), *151*, 465 (786), *510*
Ruppert, I. 15 (84), *19*, 523, 545 (60), *554*
Russell, B. R. 245, 247 (12), *283*
Russell, C. G. 636, 637 (209), 651 (425), *660*, *664*
Russell, G. A. 321, 323 (57), *329*
Russo, M. 331 (2), 332 (6, 7), 333 (6, 10, 11), 334 (12), 340 (11), *341*, 487 (1022, 1024, 1028), 488 (1022), 492 (1028), *515*, 768 (416), *849*
Rutherford, J. S. 67–69 (33), *148*, 685, 686 (45), *841*
Rutledge, P. S. 641 (295), *662*
Ruwet, A. 420 (264), 421, 422 (281), 423 (286), 426 (264, 313, 315), 465 (315), 473 (286, 890, 892, 893), 475 (902, 903), 476 (286, 904), 479 (931), *499–501*, *512*, *513*
Ryan, M. D. 635 (204), 636 (213, 229), 637 (213), *660*
Ryan, R. C. 832 (665), 839 (665, 675), *855*
Rycroft, D. S. 220 (81, 83), *240*
Rye, R. T. 261 (64), *284*
Ryl'tsev, E. V. 815 (601k), *853*
Ryu, I. 634 (194), *660*

Saatsazov, V. V. 108 (235), *153*, 567 (81), *613*
Sabatini, A. 732 (282), *847*
Sacconi, L. 729 (267a), 765, 766 (409), *846*, *849*
Sachs, F. 465 (783), 488 (1041), *510*, *515*
Sadasivan, N. 705, 706 (171), *844*
Sadeghi-Milani, S. 464 (749, 754), *509*
Sadeh, T. 413 (182b), 425 (309b), 433 (182b), *498*, *500*
Sadek, S. A. 371 (11), *394*
Sadekov, I. 442 (463), 478 (925), 479 (935, 936), 480 (936–938), 487 (1020), *503*, *513*, *515*

Sadekov, I. D. 184 (181), *188*, 518, 520 (4), 522 (41–44), 523 (42, 66, 75), 532, 534 (155), 537 (170, 177), 540 (197), *553*, *554*, *556*, *557*, 560 (5), 563 (36), 565 (49, 51), 556 (51, 69, 73), 567 (51, 69, 73, 84), 574 (49, 69, 120–122), 575 (122), 577, 601, 603 (138), *611–614*, 651 (403, 404), *664*, 814, 820 (620), *854*
Sadekova, E. 478 (925), 480 (937), *513*
Sadovaya, N. 407 (86), 412 (86, 152), 413 (86, 170), 414 (152), *496–498*
Sadovaya, N. K. 581 (168–171), 595 (169), *615*
Sadovaya, N. N. 595 (247), *617*
Sadoyava, N. 412 (157, 162, 164), 413 (157, 178–181), 414 (157, 162), 416(157), *497*, *498*
Sadykhov, S. T. 289, 291 (24), *304*
Saeed, K. 165 (48), *186*
Saelinger, D. A. 391 (126a), *397*
Saethre, L. J. 117 (256, 257, 268–270), 119 (270, 274), *153*, *154*
Saftner, R. W. 165 (58), *186*
Sager, W. F. 562 (26), *612*
Said, A. K. 387 (77a), *395*
Saigh, G. 487 (1026), 488 (1033), *515*
Saillard, J,-Y. 65 (13, 14), *148*
Saindane, M. 621, 622 (8), 634 (192), *656*, *660*
Saint-Ruf, G. 438, 439 (435), *503*
Sakaguchi, R. 644 (323, 324), *662*
Sakai, Y. 25 (24, 61, 69), 30 (61), 34, 54 (61, 69), 60 (69), *60*, *61*
Sakaizumi, T. 70, 73, 81 (44), *149*
Sakaki, K. 404 (48, 49, 511), 418 (240), 446 (48, 49, 511), *495*, *499*, *504*, 541 (204, 205), *557*
Sakakibara, J. 453, 486, 487 (595), *506*
Sakeh, T. 292 (38), *305*
Sakurai, H. 195 (31), *239*
Sakuta, K. 634 (199), *660*
Salemink, C. 465 (780), *510*
Salez, C. 25 (23), *60*
Salih, R. 586 (192), *616*
Salmond, W. G. 622 (37), *657*
Salzman, M. 492, 493 (1061b), *516*
Salzman, M. N. 339, 340 (35), *341*
Samartseva, E. 439 (440), *503*
Samartseva, S. A. 815 (601j), *853*
Samitov, Yu. Yu. 635 (200), *660*
Samochocka, K. 298 (55), *305*, 539 (188), *556*
Sams, R. L. 64 (8, 12), *148*
Samuelson, L. 334 (17, 18), 335 (17), 341 (18), *341*
Samus, I. D. 730, 732 (270), *846*
Samus, N. M. 727 (263), 732 (289, 290), 738, 739 (289), *846*, *847*
Sanderud, K. A. 13 (74), *19*
Sandhu, S. S. 691 (115), 693–695 (120), 698 (120, 142), 815 (603b), *842*, *843*, *853*
Sandman, D. 470 (859, 860), *511*
Sandman, D. J. 334 (17, 18), 335 (17), 340 (38), 341 (18), *341*
Sandman, S. 446 (516), *504*
Sandri, E. 590 (219), 597 (261, 262), 598 (262), *616*, *617*
Sandrolini, F. 333 (10), *341*
Sandström, J. 13 (71), *19*
Sanina, L. P. 356 (32), *366*
Sanner, T. 288 (1), 304
Santiesteban, H. 621, 622 (10), 623 (10, 64, 66), *656*, *657*
Sapora, O. 290, 291 (32, 33), *305*
Sappa, E. 811, 832, 839 (591), *853*
Sapru, R. P. 393 (153), *397*
Sardella, D. 427 (321), *501*
Sardella, D. J. 592 (230), *616*
Saris, I. 458, 484 (671), *508*
Saris, L. 426 (318), 432 (382), *501*, *502*
Saris, L. E. 591 (229), *616*
Sarsfield, A. 465 (792), *510*
Sarsfield, A. A. 586 (212), *616*
Sasse, K. 462 (718), *508*
Sastry, K. V. L. N. 65 (17), *148*
Satek, L. C. 693–696 (124), *843*
Sato, M. 334, 335 (19), *341*, 574, 576 (136), 577, 601 (137), *614*, 741 (330, 332) 742 (332), 745 (350), 749 (330), 825 (330, 332), *847*, *848*
Sato, W. 741 (326), *847*
Satoda, I. 339 (36), *341*, 483 (978), *514*
Satoh, T. 644 (323, 324), *662*
Sattari, S. 429, 439 (358), *501*
Satyamurthy, N. 469 (846), *511*
Savabi, F. 255, 278, 279 (48), *284*, 409 (121b), 449, 464 (563a), *497*, *505*, 621, 622 (23, 24), *656*
Savage, W. J. 102 (208), *152*
Savel'eva, O. P. 393 (154), *397*
Savelli, G. 177 (128), *187*, 410 (131, 132), 416 (213), 442 (466), *497*, *498*, *503*, 560 (7, 8), 571 (7, 105), 572 (105), 573 (117), 577 (117, 143), 578 (117), 581 (143, 165), 582, 589, 591 (165), 593, 594 (117), *611*, *614*, *615*
Savich, I. G. 609 (331), *618*
Sawicki, E. 428 (343), 465 (771, 773–775, 778, 781), *501*, *510*
Sawicki, R. A. 636 (228), *660*
Saxena, K. M. 202 (51), *239*
Saxena, K. M. S. 22, 25, 28 (8), *60*
Saxton, R. J. 794, 795 (529), *852*
Sayapina, L. M. 566, 567 (73), *613*

Scaiano, J. C. 234 (121), 235 (121, 130, 131), 236 (121), *241*, 325 (67), 326 (68), *329*
Scala, J. 179, 181 (114), *187*
Scarborough, R. M. Jr. 623 (65, 68, 71), 627 (150), *657*, *659*
Scarlata, G. 594 (240), *617*
Schaal, R. 411 (142, 144), *497*
Schaefer, M. 429 (353), *501*
Schaefer, T. 624 (86), *658*
Schaefer, W. 44, 54 (87), *62*
Schaeffer, J. 466 (796), *510*
Schafer, A. 829, 832 (646), *854*
Schäfer, A. 805, 806 (558), *852*
Schaffrin, R. M. 93 (163), *151*
Schaumann, E. 12 (66), *19*, 456, 458 (640), *507*
Scheffel, U. 393 (150), *397*
Scheffler, K. 641 (272), *661*
Scheiter, M. 7 (26), *18*
Scheller, A. 430, 443 (360), *501*
Schellman, J. A. 668 (4), *676*
Schermer, E. D. 741 (327, 329), 744 (349), 745 (327, 329, 349), 748 (349), 749 (327, 329), 824, 825 (329), *847*, *848*
Schiemann, J. 806, 832, 833 (559), *852*
Schilling, F. 488 (1042), *515*
Schings, U. 16 (96), *20*, 362 (63), *367*
Schinner, A. 332 (4), *341*, 487 (1025), *515*
Schipper, P. 494 (1071), *516*, 624 (101), *658*
Schirmer, R. E. 191 (19), *239*
Schlegel, H. B. 25 (49), 32 (77), 33, 34 (49), *61*
Schleser, H. 73 (70), *149*
Schlesinger, D. H. 178–181 (107), *187*
Schlierf, C. 314 (21), *328*
Schloter, K. 314 (21), *328*
Schlüter, H. 157, 158 (5), *160*
Schmid, G. 403 (29–31, 45), *495*
Schmid, G. H. 127 (288), *154*, 528 (130), 529 (138–140), 533 (130, 138, 139, 160), *555*, *556*, 632, 633 (181), *660*
Schmid, R. 733 (296), *847*
Schmidt, D. 119 (271), *154*, 477 (918), *513*
Schmidt, M. 288 (4), *304*, 344 (2), 347 (11), *365*, *366*, 483 (980), 484 (988), 487 (1029), 489 (1047), 492 (1029), *514*, *515*, 621, 622 (14), 650 (363, 364), *656*, *663*, 716 (199), 741 (324), 810 (583), *845*, *847*, *853*
Schmidt, U. 446 (510), *504*
Schmiesing, R. J. 632, 633 (188), *660*
Schmitz-Du Mont, O. 483 (979), *514*
Schmuckler, G. 742 (333), *848*
Schnabel, W. 147 (330), *155*, 540 (193–196), *556*, *557*
Schnering, H. G. V. 809, 832 (579), *853*
Schniepp, S. 464 (745, 746), *509*
Schnöckel, H. 11 (61), *19*
Schoeller, A. 419, 421 (253), *499*
Schoening, R. C. 839 (671, 672), *855*
Schoental, R. 387 (82), *396*
Schommer, W. 547 (252), *558*
Schöneshöfer, M. 293 (44), *305*
Schönhofer, A. 668 (3), *676*
Schottländer, M. 89, 90 (146), *151*
Schoufs, M. 447 (527), 481 (955), *505*, *513*
Schrauzer, G. N. 703 (164), *844*
Schrobilgen, G. J. 221, 234 (87), *240*
Schroeder, H. 721 (240), *845*
Schubert, U. 90 (153), *151*, 463 (735), *509*, 722 (245e), 723 (248–250), 724 (248, 249), 725 (248–250), 726, 729 (255), 742, 753 (250), 767, 768 (414), 804, 805 (557), 829–831 (648), *846*, *849*, *852*, *854*
Schuh, W. 357, 358 (45), *366*
Schukat, G. 450 (573), *506*
Schulte, K. 407, 409, 414 (97b), *496*
Schulte, R. 403 (46c), *495*
Schulten, H. R. 838 (670a), *855*
Schultz, G. 75, 76 (78a), 113 (240), 139, 140 (315), *149*, *153*, *154*
Schultz, P. 15 (82), *19*
Schulz, J. 487 (1018), *515*
Schulz, P. 652 (435, 436), *664*, *665*, 688 (58), *841*
Schulz, R. 247 (27), *284*
Schulze, U. 257, 258 (58), *284*, 478 (920), *513*
Schulze, W. 742, 746, 751 (336), 752 (362), 825 (336), *848*
Schumaker, R. 449 (555, 563b), 450 (572b), 464 (563b), *505*, *506*
Schumann, H. 228, 229 (107), *241*, 359 (49), *366*, 483 (980), 484 (988), *514*, 714 (196–198), 716 (197), 814, 820 (196), 832, 835, 836 (654), *845*, *855*
Schuster, G. 355 (29), *366*
Schüzendübel, J. 450 (573), *506*
Schwartz, A. 462 (716), *508*
Schwartz, I. L. 178–181 (107), *187*
Schwartz, J. 631 (172), *659*
Schwarz, H. 257, 258 (58), *284*
Schwarzmann, G. 489 (1048), *515*
Schwedt, G. 183 (138, 140), *187*, *188*
Schweig, A. 44, 54 (87), *62*, 247 (27), *284*, 403 (46c), *495*
Schwetlick, K. 385 (76), *395*
Schwetlick, Von K. 569, 581 (98), *614*
Sciotto, D. 569 (93), *613*
Scott, B. 449 (566), *505*
Scott, J. 450 (570), *506*
Scott, J. D. 245, 247 (12), *283*
Screttas, C. 360 (52), *366*
Scrimin, P. 403 (43), *495*
Scudder, E, D. 823 (636), *854*

Scudder, M. L. 811, 813 (600), *853*
Seale, S. K. 348 (13), *366*
Sealy, R. C. 234 (124), *241*
Sebastiani, G. 438 (432), *503*
Sebastiani, G. V. 586 (195), *616*
Secomb, R. 469 (854c), *511*
Secomb, R. J. 97, 98, 132, 143 (186), *152*, 525 (87), *554*
Sedgwick, R. D. 245 (10), *283*
Sedova, L. N. 565 (43), *612*
Seebach, D. 603 (300, 301, 309), 604 (300, 309), *618*, 638 (227), 651 (397), *660*, *664*
Seeger, R. 25 (44, 49), 33, 34 (49), *61*
Segre, A. 754 (364, 365), 755, 756 (365), *848*
Seibert, H. 629 (164), *659*
Seidler, M. 410 (123), 469 (848), *497*, *511*
Seidler, M. D. 690, 716 (62), *841*
Seip, H. M. 70 (49), 72 (60), 82 (99), 88 (134, 139), 89 (139, 142), 90 (134, 139, 142), 139 (312), *149–151*, *154*
Seip, R. 72 (57, 65), 87 (130), 90 (65), *149*, *150*
Seitz, S. P. 16 (88), *19*, 636–638 (210), 639 (230), *660*
Sekido, E. 600 (285), *618*
Selig, H. 545 (224), *557*
Sell, G. 423 (294), *500*
Semard, D. 415 (201, 205), 430 (361), 435 (205), *498*, *501*
Semchikova, G. S. 716 (202), *845*
Semenov, E. V. 700, 701 (154), *843*
Semilia, M. 392 (128), *397*
Sen, J. P. 102 (219), *152*, 742, 750 (334), *848*
Senkler, G. 477, 478 (911), *512*
Senkler, G. H. Jr. 278 (89), *285*, 593 (237), *617*
Senma, M. 99, 101 (192b), *152*
Seppelt, K. 545 (222, 227, 238, 239), *557*
Serafini, A. 33 (81), *62*
Serebryanskaya, A. I. 377, 385, 286 (33b), *394*
Sereda, I. P. 683, 685, 686 (34, 35), *841*
Serena, B. 411 (137), 416 (216), *497*, *499*, 567, 568 (85), *613*
Sergeev, V. A. 184 (167), *188*, 586 (214), *616*
Sergi, S. 690 (71, 87, 88, 97–99), 691 (97, 98), 695 (98), 735, 738 (299), 774 (98), *842*, *847*
Sernard, D. 415 (198), *498*
Serrano, R. 768, 770 (421), 829 (421, 646), 830 (421), 832 (646), *849*, *854*
Seshadri, R. 624 (108), *658*
Sevrin, M. 526 (99), *555*, 624 (103, 107, 110), 642 (304), *658*, *662*
Seybold, G. 12 (64), *19*
Seyferth, D. 747, 753, 757 (354), 808 (354, 574), *848*, *852*
Seyhan, M. 459 (696), *508*
Sgustov, I. 431 (380), *502*
Shaal, R. 590 (223, 224), 591 (224), *616*
Shabang, M. 158, 159 (12), *160*
Shack, C. J. 545 (290), *558*
Shacklette, L. W. 332, 334 (3), *341*
Shafiee, A. 255 (48–57), 268 (51), 278 (48–57), 279 (48–51, 57), 280 (52–54, 56), 281 (56), *284*, 406 (68), 409 (107–109, 121b), 429 (358, 359), 439 (358, 447, 448), 447 (68, 531), 448 (536), 449 (563a), 452 (587), 455 (619), 460 (705), 461 (619, 710, 713), 463 (739–741), 464 (107, 108, 531, 563a, 740, 741, 744, 748, 749, 751, 760), 466, 467 (808), 484 (992), *495–497*, *501*, *503*, *505–510*, *514*, 621, 622 (23, 24), *656*
Shagidullin, R. R. 815 (601j), *853*
Shah, S. K. 317 (35), *329*, 603 (303, 304, 308, 311), 604, 609, 611 (304), *618*, 639 (119, 122), *658*
Shaik, S. 244, 252 (4), *283*
Shakhgel'diev, M. A. 184 (145), *188*, 563 (37), *612*
Shakhtakhtinskii, T. 407, 468 (85), *496*
Shakir, M. 522 (49), *553*
Shamberger, R. J. 162 (8), 163–165, 183 (24), *185*
Shandalo, M. 586 (192), *616*
Shanmugam, P. 439 (449–452), 443 (485, 486), 446 (513). *503*, *504*, 757 (369), *848*
Shapiro, I. O. 377, 385, 386 (33b), *394*
Shapkin, A. 410 (125), *497*
Shapkin, A. A. 89, 90, 128 (143), *151*, 383 (73), *395*
Shapley, J. R. 835 (668c), *855*
Sharma, K. 465 (785), *510*, 692, 698 (118), *842*
Sharma, R. D. 716 (200), 814, 818 (613), *845*, *854*
Sharpless, K. B. 16 (94), *20*, 317 (36), *329*, 541 (207, 208). *557*, 621, 622 (11), 624 (11, 114, 131), 625 (11, 131), 627 (131), 629 (114, 154), 630 (114, 154, 169), 632, 633 (114), 636, 637 (214), 639 (114, 120, 123, 124), 640, 641 (257), 642 (241), 647 (332), *656*, *658–662*
Sharts, C. M. 202 (47), *239*
Shatenshtein, A. I. 377, 385, 386 (33b), *394*, 594 (241), 604 (316), *617*, *618*
Shaw, E. T. 798, 811 (540a, 540b), *852*
Shaw, R. 157, 159 (3), *160*
Shawali, A. 466 (811), 467 (811, 814a, 814b), *510*
Shawl, E. 405 (56), *495*
Shchepina, N. E. 167, 170–172, 174, 176 (88), *186*
Shealy, Y. 466 (795), *510*

Shefter, E. 78 (91), *150*
Sheiko, D. 467 (813), 486 (1004), 491 (1058–1060), *510*, *514*, *516*
Sheldrick, G. M. 66 (25), 70 (39, 45), 71 (39), 72 (39, 58), 74 (45), 81 (39), 82, 83 (102), 111, 112 (238), 144 (323), *148–150*, *153*, *155*, 710, 712, 713 (191a), 734 (301), 763 (402), 768 (417), 811 (597), 814, 819 (191a), *844*, *847*, *849*, *853*
Sheldrick, W. S. 763 (402), *849*
Shell, J. W. 164 (37), *185*
Shelton, E. J. 638 (223), *660*
Shemanina, V. N. 632 (176), *659*
Shen, Q. 117 (266), *153*
Sheppard, W. A. 202 (47), *239*
Sheremeteva, G. I 624 (88), *658*
Sherrington, C. 565, 574, 575, 608 (60), *613*
Sherwin, P. F. 11 (57), *19*
Shetta, A. 466 (811), 467 (811, 814a), *510*
Shibaeva, R. P. 95 (171–174), *151*
Shibata, Y. 453, 486, 487 (595), *506*
Shimazu, F. 234 (119), *241*, 288, 291 (7, 8), *304*
Shimizu, H. 8 (39), *18*, 478 (924, 928, 929), *513*, 592, 593 (236), *617*
Shimizu, M. 10 (53), *19*, 549 (272), 550 (273), *558*, 609 (333–335), *618*, 625, 628 (125–129), 642 (147), *659*
Shimoishi, Y. 164, 165 (44), 183 (131–133), *186*, *187*
Shine, R. J. 264 (68), *284*
Shinoda, M. 641 (294), *662*
Shiraishi, T. 165 (51), *186*
Shirley, V. S. 370 (1), *393*
Shlaider, I. 478 (923), *513*
Shorter, J. 561 (10–12), 563, 565 (10, 11), 568 (11), 571, 579 (10, 11), 597 (11), *611*
Shreeve, J. M. 126 (284), *154*
Shrift, A. 178 (109), *187*, 390 (116), *396*
Shtacher, G. 292 (38), *305*
Shu, P. 93 (165), *151*, 448 (546), 449 (558b), 450 (546), *505*, 576 (134), *614*
Shulezhko, A. A. 567, 568 (86), *613*
Shulezhko, L. 472 (883), *512*
Shulezhko, L. M. 567, 568 (86), *613*
Shum, A. C. 390 (120), *397*
Shupik, R. 411 (133), *497*
Shupik, R. I. 582 (178), *615*
Shvaika, O. 456 (644), 458 (670), *507*, *508*
Shvo, Y. 835 (668a), *855*
Siaglo, H. 458 (667), *507*
Siddiqi, Z. A. 522 (49), *553*
Sidén, J. 531, 543 (151), 544 (219), *556*, *557*
Siebert, W. 70 (49), 82 (99), *149*, *150*, 288 (4), *304*, 344 (3), 345 (4), 346 (5), 347 (6), *366*
Siedle, A. R. 708 (178), 709 (178–180), *844*
Siegbahn, B. 25 (31), *60*
Siegbahn, K. 117 (268, 269), *153*
Sieler, J. 102 (206), *152*, 787 (484), 789 (503), *851*
Siemens, C. 5 (11), *18*
Siftar, J. 102 (217), *152*, 624 (92), *658*, 786–788, 790, 791 (485), 796, 797 (539), *851*, *852*
Siimann, O. 308 (7), *328*
Šik, V. 114 (245, 246), 115 (246), *153*, 690 (79, 112), 709 (183–185), 710 (183–185, 187, 188), 712 (184, 185, 187, 188), 713 (188), 727, 729 (257), 771 (437), 773 (257, 437–439), 774 (437–439, 444, 446), 775 (437, 438), 776 (444, 446), 819 (184), *842*, *844*, *846*, *850*
Silver, J. 97 (183–185, 187, 188), 98 (183–185, 190), 132 (183), *152*, 469 (854a, 854b), *511*, 525 (89–93), *554*
Silverman, R. B. 315 (25), *328*
Silverstein, A. 393 (157), *397*
Silvestro, L. 690 (71, 88), *842*
Silvey, G. 64 (5), *148*
Simchen, G. 484 (991b), 485 (999, 1000a), *514*, 642 (280, 293), *661*, *662*
Simek, M. 459 (689), *508*
Simic, M. 297 (53), *305*
Simmonnin, M. P. 580 (156), *615*
Simon, A. 462 (714), *508*
Simon, H. 257, 258 (58), *284*, 477 (916), 478 (919), *513*, 624 (94), *658*
Simon, M. S. 680 (9b), *839*
Simone, G. 290, 291 (32), *305*
Simonen, T. 79 (96), 92 (157–159), *150*, *151*, 447 (526), *505*, 525 (98), *554*
Simonetta, M. 403 (44), *495*
Simonidesz, V. 316 (32), *329*
Simonneaux, G. 716 (208), *845*
Simonnin, M. 411 (143), *497*
Simons, G. 33 (82), *62*
Simov, D. 487 (1011), *514*
Simpkins, N. S. 642 (258), *661*
Sindelar, K. 477, 478 (913), 490 (1052–1054), *512*, *515*
Singer, M. I. C. 602 (295), *618*
Singer, S. P. 541 (207), *557*
Singh, A. 439, 453 (446), *503*, 636 (209, 219), 637 (209), 638 (219), 651 (425), *660*, *664*, 690 (63), *841*
Singh, G. 814, 824 (625), *854*
Singh, H. 418 (239), 469 (837), *499*, *511*
Singh, M. 418 (239), *499*, 520 (25), 522 (25, 40), *553*
Singh, P. 99, 101 (192b), *152*
Singh, P. P. 730 (269b), 737 (312–314), *846*, *847*
Singh, R. 465 (785), *510*

Singh, T. 691 (115), 698 (142), *842*, *843*
Singh, V. P. 737 (314), *847*
Sinha, A. I. P. 700, 701, 730, 737 (153c), *843*
Sink, C. W. 650 (382, 383), *663*
Sinomova, E. N. 167, 170–175 (87), *186*
Sinotova, E. N. 166, 171 (81), 172 (81, 94, 96), 173 (96), 174, 176 (81), *186*, *187*
Sipio, W. 403 (46a), *495*
Sipio, W. J. 634 (197), 636–638 (210), *660*
Sironi, A. 811 (596), *853*
Sisler, H. H. 371 (17), *394*
Sisson, W. B. 393 (157), *397*
Sivin, B. N. 87 (128), *150*
Sivin, S. I. 87 (128), *150*
Sjöberg, B. 520 (22), *553*, 655 (478), *665*, 670 (18a), *677*
Skaarup, S. V. 117, 119 (267), *153*
Skakke, P. E. 690, 691 (70), *842*
Skancke, P. N. 25 (60), *61*, 87 (124), *150*
Skelton, B. W. 747 (356), *848*
Skinner, H. A. 157, 158 (4), *160*
Skopenko, V. V. 730 (269c), *846*
Skotnicki, A. 289 (23), *304*
Skovlin, D. O. 681, 686 (20), *840*
Sladky, F. 545 (225, 226, 228–230), *557*
Slater, J. C. 25 (11), *60*
Slebocka-Tilk, H. 626 (136), *659*
Sletten, E. 119 (273), *154*, 308 (10), *328*
Sletten, J. 116 (249), 119 (273), *153*, *154*
Slinkard, W. E. 697, 775, 778 (140d), *843*
Sliwa, W. 466 (794), *510*
Slouka, V. 389 (108), *396*
Slyusarenko, E. I. 541 (211), *557*
Smedslund, T. 528 (129), *555*
Smiles, S. 7 (29), *18*
Smirnov, E. V. 641 (292), *662*
Smirnov-Zamkov, I. 423 (295, 296), *500*
Smith, A. B. III 623 (65, 68, 71), 627 (150), *657*, *659*
Smith, A. C. 521 (31), *553*
Smith, B. 489 (1049), *515*
Smith, B. C. 225, 226, 228, 229, 233 (103), *241*, 526 (102), *555*
Smith, B. V. 623 (54, 56), 624 (100), *657*, *658*
Smith, C. S. 132, 134, 143 (303), *154*, 521 (31), *553*
Smith, D. 445 (501), *504*
Smith, D. L. 651 (423), *664*, 814–816 (606), *853*
Smith, G. F. W. 381, 393 (60), *395*
Smith, G. P. 160 (16), *160*
Smith, K. V. 261 (61), *284*
Smith, M. R. 101 (202), 141 (202, 321, 322), 142, 143 (321, 322), *152*, *155*, 651 (399), *664*
Smith, N. A. 25 (55), *61*
Smith, P. J. G. 743, 744 (347), *848*
Smith, R. P. 382 (64), *395*
Smith, Z. 87 (130), *150*
Smits, J. M. M. 102, 103 (215), *152*
Smyk, B. 289 (23), *304*
Smyth, C. P. 519 (16), *553*
Smyth, W. 439, 453 (446), *503*
Snagoshenko, L. P. 575 (126), *614*
Snatzke, G. 667 (1b), 668 (5), 670 (1b), 672 (22), 673 (29), 676 (38), *676*, *677*
Snyder, L. C. 55 (93), *62*
Snyder, L. E. 72 (56), *149*
Sochorcová, M. 650 (371), *663*
Soda, K. 621, 622 (18), *656*
Soja, P. 630, 631 (170), *659*
Sokolov, S. V. 76 (78b), *149*
Sokolova, N. 429 (356, 357), 430 (356), 439 (442), *501*, *503*
Sokolova, N. B. 587 (196), *616*
Sokolovsky, M. 434 (397, 398), *502*, 527 (106), *555*
Sokov, L. A. 389 (107, 111), *396*
Solazzo, L. 407, 420 (77), *496*
Solenghi, M. D. 290–292 (31), *304*
Soling, H. 94 (166), *151*
Solouki, B. 11 (57), *19*, 403 (46b), *495*
Solozhenkin, P. M. 700, 701 (154), *843*
Soltani, A. 464 (751), *509*
Song, L. C. 808 (574), *852*
Songstad, J. 13 (73, 74), *19*, 360 (56, 58), 361 (62), *367*, 382 (65), *395*, 815 (604b, 604c), *853*
Sonoda, N. 317 (33), *329*, 354 (21–23), 355 (26), *366*, 403 (35), *495*, 621, 622 (7, 19), 624 (19), 634 (194), 640 (140, 143), *656*, *659*, *660*, 762 (401a), 787 (492, 494, 496), 789 (492), 790 (492, 494, 496), *849*, *851*
Sonoda, Y. 762 (401b), *849*
Sorgi, K. L. 632, 633 (188), *660*
Sorokina, L. A. 565 (68), *613*
Soth, S. 436 (411, 412), 460 (412), *502*
Sotoda, I. 463, 483 (737, 738), *509*
Soundararayan, N. 439 (450), *503*
Spallholz, J. E. 162 (11), *185*
Spanget-Larsen, J. 247 (28), *284*
Sparatore, F. 487 (1012), *514*
Sparks, R. A. 94 (169), *151*, 567 (74), *613*
Speca, A. N. 721 (243), *846*
Specker, H. 261 (60), *284*, 525 (83), *554*, 699 (151), *843*
Spencer, H. 449, 450, 463 (556), *505*
Spencer, H. K. 652 (440, 441), *665*
Spencer, R. P. 179 (112), *187*, 379 (41), 380 (48), 388, 389 (101), 392 (48), *394–396*
Spierenburg, J. 522 (54, 56), *554*
Spies, H. 393 (146), *397*, 448 (537a), *505*
Spinelli, D. 411 (136, 139–141), 412 (159), 414 (141), 416 (159), *497*, 541 (202), *557*,

567–569 (87), 572 (87, 111), 578 (145, 146), 582 (87), 589 (146, 216, 218), 591 (145, 216, 226, 227), *613–616*, 675 (31), *677*
Spiridonov, V. P. 76 (80), *149*
Spirlet, M. R. 86 (114), *150*, 670 (15), *676*
Sprague, M. J. 641 (274), *661*, 728, 731, 732 (264, 265), 734, 735 (265), *846*
Sprague, R. 481 (952a), *513*
Sprinkle, C. R. 806, 807 (564), 808 (566), 810 (564, 566, 582), *852*, *853*
Spunta, G. 472 (885), 478 (885, 930), 479 (885), *512*, *513*, 592 (238), 598 (264, 267), 599 (267), *617*
Srivastava, P. 442 (472), *503*
Srivastava, R. 418 (239), 469 (837), *499*, *511*
Srivastava, R. C. 520 (25), 522 (25, 40, 48, 50), *553*
Srivastava, T. 418 (239), 442 (472), 469 (837), *499*, *503*, *511*
Srivastava, T. N. 520 (25), 522 (25, 40, 48, 50), *553*, 731, 732 (280), *846*
Staal, L. H. 838 (670c), *855*
Stach, J. 786 (504a, 504b), 789 (504a, 504b, 505–507), *851*
Stack, J. 784, 789, 791 (480), *851*
Stackhouse, J. 278 (89), *285*, 477, 478 (911), *512*, 593 (237), *617*
Stadman, T. C. 288 (5), *304*
Stadtman, T. C. 390 (117–119), 391 (117, 122, 127), *396*, *397*
Stahr, H. M. 183 (137), *187*
Stall, D. R. 157 (1), *160*
Stanghellini, P. L. 808 (572, 573), 811 (595), 832 (655–657, 659, 662, 663), 835 (655), 836 (659), 837 (659, 662), 838 (663), *852*, *853*, *855*
Staninets, V. 423 (293, 296), 481 (950), 482 (959), 494 (1074, 1075), *500*, *513*, *516*
Staninets, V. I. 528 (125, 126, 133), 530 (143, 144), 531 (147), 532 (153), 533 (159), 534 (153), *555*, *556*
Stanner, F. 690, 691 (111), *842*
Stark, J. 446 (516), *504*
Stark, J. C. 340 (38), *341*
Stark, R. A. 742, 758, 759 (335), *848*
Stec, W. J. 694 (125), *843*
Steeves, B. H. 16 (95), *20*, 815 (601b), *853*
Stefanac, Z. 177 (125, 126), *187*
Stefko, R. 489 (1045), *515*
Stegel, F. 590 (220–222), 591 (222), *616*
Stegmann, H. B. 641 (272), *661*
Steimecke, G. 789 (506), *851*
Stein, C. A. 102, 103 (213), *152*, 690 (104), 778–780, 784, 786 (457), *842*, *850*
Steinke, F. H. 387, 388 (98), *396*
Stejskal, R. 371 (10a), *394*
Stella, L. 326 (69), *329*
Stenlake, J. 457 (662), 458 (664), *507*
Stephenson, T. A. 102 (208), *152*
Steudel, R. 811 (585a), *853*
Stevens, E. 118 (264), *153*
Stevens, R. A. J. 381, 393 (60), *395*
Stevens, W. 408 (102), *496*
Stevens, W. C. 701, 705 (158b), *844*
Stewart, I. 444 (498, 499), *504*
Stewart, I. M. 308 (5, 6), *328*, 798, 800 (543, 544), *852*
Stewart, R. F. 25 (25, 27, 68, 71), 29 (27, 68, 71), 30 (68, 71), 33 (27, 68, 71), 56 (68), *60*, *61*
Stievano, B. M. 163 (34), *185*
Stijve, T. 183 (134), *187*
Stirling, I. 622 (39), *657*
Stobart, S. R. 715 (212, 219, 220), 716 (212), 717 (212, 219, 220), 720 (212, 220), *845*
Stock, A. 787 (483), *851*
Stock, L. M. 563 (34), 593 (239), *612*, *617*
Stockton, R. A. 162, 165 (13), *185*
Stojanov, S. 261 (60), *284*, 525 (83), *554*
Stoklosa, H. J. 359 (50), *366*
Stollé, R. 466 (803), *510*
Stone, A. J. 324 (63), *329*
Stone, F. G. A. 680, 712 (2), 743, 744 (347), *839*, *848*
Stoppioni, P. 727 (258–261), 729 (258–261, 267a, 267b, 268), 730 (261), *846*
Storm, H. 466 (807), *510*
Stothers, J. B. 201 (45b), 210 (66), *239*, *240*
Stradins, J. 184 (164), *188*
Stradyn', Y. P. 184 (165), *188*
Strandberg, M. W. P. 51 (90), *62*
Strating, J. 548 (262), *558*
Straub, H. 427 (331), *501*
Strauss, R. 811 (585a), *853*
Strausz, O. 402 (13), 403 (25), *494*
Streets, D. G. 832 (666b), *855*
Streitweiser, A. 594 (242), *617*
Streitwieser, A. 562 (32), *612*
Strelets, B. 462 (724–726), 463 (727–731), 465 (787, 788a, 788b), *509*, *510*
Strickland, R. W. 670, 671 (17e), *677*
Stricks, W. 184 (155), *188*
Stridsberg, B. 421 (280), *500*
Stringer, O. D. 9 (48), *19*, 624 (148), *659*
Strom, E. T. 321, 323 (57), *329*, 565 (55), *612*
Stromberg, A. 25, 30 (72), *61*
Strouse, C. E. 811, 832, 838, 839 (589), *853*
Struchkov, Yu. T. 701 (156), *843*
Strzelecka, H. 470 (858), *511*
Stukalo, E. A. 537 (183), 539 (189), 540 (189, 198), 541 (198), *556*, *557*
Stulius, W. 681 (30), *841*
Stull, A. 458 (679), *508*

Sturgeon, G. O. 252, 261 (43), *284*
Styrman, K. 468 (833), *511*
Subramanian, K. S. 165 (50), *186*
Suchar, G. 489 (1044, 1045), *515*, 580 (161), *615*
Suchi, R. 15 (83), *19*
Suchomel, H. 16 (96), *20*, 362 (63), *367*
Suda, Y. 334, 335 (19), *341*
Suemitsu, R. 6 (23), *18*
Suffolk, R. J. 11 (58), *19*
Suganara, T. 446 (512c, 512d), *504*
Sugden, S. 51 (91), *62*
Suginome, H. 407, 412 (80), *496*
Sugiura, Y. 238 (148), *241*, 600 (284), *618*, 782 (474), *850*
Suiyama, M. 520, 521 (32), *553*
Sukach, Yu. A. 573 (116), *614*
Sukhai, R. 447 (527), 481 (957, 958), 491 (958), *505*, *513*
Sukhai, R. S. 355 (27), *366*
Sukhov, S. G. 167, 170–172, 174, 176 (88), *186*
Sukiasjan, A. 430 (366, 367), *501*
Sukkai, R. S. 622 (41), *657*
Sukurai, H. 600 (286), *618*
Summers, L. 489 (1046), *515*
Sunay, U. 623 (64), *657*
Sundberg, M. 79 (96), 92 (157), *150*, *151*
Suprunenko, P. A. 704, 705 (169), *844*
Surov, Yu. N. 569 (100), 573 (116), *614*
Suschitzky, H. 401 (8b, 10), *494*
Sutphen, N. T. 382 (64), *395*
Suttle, J. F. 165 (58), *186*
Sutton, P. W. 811, 812 (587), *853*
Suzuki, H. 8 (38, 39), *18*, 428 (342), *501*, 623 (48, 49), 641 (294), 642 (309), *657*, *662*
Suzuki, K. 623 (61), *657*
Suzuki, T. 238 (148), *241*, 621, 622 (18), *656*, 782 (474), *850*
Svaeren, S. E. 147 (331), *155*, 540 (191), *556*
Svanholm, U. 190, 220 (2), *238*
Svanholt, H. 12 (65), *19*, 65 (18, 19), *148*, 405 (57), 464 (57, 753), *495*, *509*
Svatek, E. 477, 478 (913), 490 (1054), *512*, *515*
Svec, H. J. 247 (29), *284*
Svendsen, S. R. 166 (79), *186*, 640, 642 (245), *661*
Svensson, S. 117 (268, 269), *153*
Svoboda, J. J. 7 (28), *18*
Swain, C. G. 562, 571, 579 (25), *612*
Swanberg, S. C. 180 (118), *187*, 378 (39), *394*
Swanson, A. B. 389 (113), *396*
Swanson, C. A. 387 (94), *396*
Sweigart, D. A. 838 (670d), *855*
Symons, M. C. R. 234 (122, 123, 125), *241*
Syrova, G. P. 565 (53), 566 (70), *612*, *613*
Syskova, V. P. 171 (103), *187*
Szantai, I. 393 (156), *397*
Szargan, R. 788, 792 (502), *851*
Székely, I. 638 (225), *660*
Szmant, H. H. 9 (47), *18*
Szperl, L. 427 (327), *501*

Taber, R. 490 (1056), *515*
Tadino, A. 315 (27), *328*, 471 (868, 869), 472 (868, 869, 882, 886, 887), *512*, 592 (233, 234), *617*
Taft, R. W. 194 (30), 202 (54a, 54b), *239*, *240*, 562 (24, 28), 568 (28), 571 (24), 575 (123), *612*, *614*
Taft, R. W. Jr. 561 (18, 20–22), 562 (22), 571 (21), *612*
Taga, T. 99, 101 (192b), *152*
Tainturier, G. 741 (331), 742, 749 (331, 338), *847*, *848*
Taira, Z. 99, 101 (192b), *152*
Takabayashi, F. 87 (125), *150*
Takagi, M. 654 (465), *665*
Takahashi, M. 444, 484 (495), *504*, 642 (305), *662*
Takai, Y. 640 (140), *659*
Takarada, M. 621, 622 (7), *656*
Takeda, R. 625, 628 (125–127), *659*
Takeuchi, Y. 623 (69), 632, 633 (190), 639 (236), 641 (284–287), *657*, *660–662*
Takeuky, Y. 571 (107), *614*
Takita, S. 644 (324), *662*
Talaty, E. R. 33 (82), *62*
Talbot, J.-M. 230, 232 (110), *241*, 440 (456, 458), 441 (458, 460), 442 (460, 464), 443 (464), 444 (458), *503*
Talbot, J. M. 579, 580 (150), 584 (184, 186), 585 (186), *615*, *616*, 652 (432), *664*
Talham, D. 17 (104), *20*, 451 (579), *506*, 651 (417), *664*
Talmi, Y. 183 (136), *187*
Tamagaki, S. 404 (48, 49, 511), 418 (240), 446 (48, 49, 511, 512b), *495*, *499*, *504*, 541 (204, 205), *557*
Tamagawa, K. 89 (148), *151*
Tamai, Y. 600 (284), *618*
Tamamushi, B. 383 (72d), *395*
Tamary, T. 530 (145), *555*
Tamba, M. 289 (16), 290 (31, 32), 291 (16, 31, 32, 35, 37), 292 (31, 35–37), 293 (35, 36, 42, 43, 45), 296 (42), 297 (42, 49), 298 (42, 56), 299 (60), 301 (66), 302 (35–37), *304*, *305*
Tamelen, E. E. van 378, 386, 387 (38b), *394*
Tamura, N. 644 (324), *662*
Tanaka, H. 195 (31), *239*, 536 (167), *556*, 600 (284, 286), *618*, 621, 622 (18), *656*
Tanaka, K. 312 (19), *328*, 356 (38), *366*, 762

(391–394, 396–400), 763 (391–394, 396–398), 764 (396–400), 765 (399), 784 (393, 394, 397, 400), 785 (391–393), 787, 790, 794, 795 (396), *849*
Tanaka, S. 334, 335 (19), *341*
Tanaka, T. 312 (19), *328*, 356 (38), *366*, 487 (1014), *514*, 721, 722 (241), 762 (391–394, 396–400), 763 (391–394, 396–398), 764 (396–400), 765 (399, 407), 766 (407), 784 (393, 394, 397, 400), 785 (391–393), 787 (396, 491, 492, 494–496), 789 (491, 492), 790 (396, 491, 492, 494, 496), 794, 795 (396), *846*, *849*, *851*
Tappel, A. I. 171, 179, 181 (115), *187*
Tappel, A. L. 234 (119), *241*, 288 (7–10), 291 (7, 8), *304*, 390 (115a, 115b), *396*
Tarantelli, T. 686 (47–49, 51), 688 (49), 689 (60), *841*
Tarver, H. 379 (43), 388 (43, 102b), *395*, *396*
Tasmurtseva, N. A. 541 (203), *557*
Tassi, L. 262, 263 (67), *284*, 759–761 (383), *848*
Tatewaki, H. 25 (47, 61, 69), 30 (47, 61), 34 (47, 61, 69), 54 (61, 69), 60 (69), *61*
Taticchi, A. 44, 54 (86, 87), *62*, 221, 222 (114), 230, 232 (109), 233 (114, 117), *241*, 246 (17, 18), 264 (69), *283*, *284*, 378, 383 (37b), 385 (75), *394*, *395*, 406 (75), 409 (112, 113), 410 (127–129, 131, 132), 411 (134, 135, 137), 414 (75, 112, 113), 416 (113, 211–213, 215, 216), 419 (241), 442 (465, 466), *496–499*, *503*, 560, 563 (4), 567, 568 (85, 88, 89), 569 (91, 92), 572 (4, 91), 573 (117), 577 (117, 143), 578 (117), 579, 580 (153, 154), 581 (143, 165, 174), 582, 589, 591 (165, 176), 593, 594 (117), 597 (91, 260), *611*, *613–615*, *617*
Taube, B. 488 (1038), *515*
Taube, D. 485 (995), *514*
Taube, H. 690 (104), *842*
Tavaarwerk, G. J. M. 381, 393 (53), *395*
Tavouktsoglou, A. N. 25 (45), *61*
Tayim, H. A. 681 (26), *840*
Taylor, A. 408 (103), 469 (853), *496*, *511*
Taylor, D. M. 392 (134, 135), *397*
Taylor, J. 407 (89), *496*
Taylor, N. J. 102 (207), *152*
Taylor, P. 523, 537, 545 (65), *554*
Taylor, R. 144 (323), *155*
Taylor, S. E. 586 (192), *616*
Tebbe, F. N. 546 (243), *557*
Tebbe, K. 199 (52), *240*
Teranishi, A. Y. 629, 630 (154), *659*
Terem, B. 623 (54), *657*
Terent'ev, P. B. 261 (63), *284*
Terrier, F. 411 (142–145), *497*, 590 (223, 224), 591 (224, 225), *616*
Testaferri, L. 621–623 (6), *656*
Tezuka, T. 428 (342), *501*
Thavornyutikarn, P. 165 (60), *186*, 520, 542 (23), *553*, 567 (77), *613*
Thayer, J. S. 261 (61), *284*, 697 (139), 731 (278, 281), 734 (139, 281), 735 (281), *843*, *846*
Thépot, J. Y. 715, 720 (222), *845*
Théwalt, V. 808 (567), *852*
Thibaut, P. 479 (932), *513*, 652 (430, 431), *664*
Thibaut, Ph. 268 (74, 75), *285*, 422 (283, 285), 423 (285), 424 (306), 425 (285, 309a), 442 (473), 475 (283), 476 (473), *500*, *503*
Thiele, G. 109, 110 (237), *153*
Thiers, M. 641, 650 (262), *661*
Thomas, A. 466 (794), *510*
Thomas, C. H. 70, 72 (40), *149*
Thomas, E. J. 624 (97), *658*
Thomas, I. L. 372, 393 (18), *394*, 651 (408), *664*
Thomason, R. 482, 483 (961), *513*
Thomassen, Y. 165 (48, 62), *186*
Thome, C. 655 (483), *665*
Thompson, M. 483 (971), *514*
Thomson, C. D. 387 (78), *396*
Thornton, C. J. 521 (31), *553*
Thorstad, O. 246 (20–22), *283*, *284*, 415 (203), *498*
Thorstand, O. 599 (268, 271), *617*
Thorup, N. 94 (166), *151*
Tideswall, N. 481 (949), *513*
Tidwell, T. T. 319 (44, 45), *329*
Tiecco, M. 621–623 (6), *656*
Tiemann, E. 64 (9), *148*
Tiethof, J. A. 693, 696 (123), *843*
Tihange, G. 458, 461 (683), *508*
Tilak, B. 420 (268), *499*
Tillay, E. W. 741, 745, 749 (327), *847*
Timm, U. 311, 312 (15), *328*, 451 (584), *506*
Timofeev, S. A. 166, 171 (81), 172 (81, 94), 174, 176 (81), *186*, *187*
Timofeeva, T. 465 (788b), *510*
Tingoli, M. 621–623 (6), *656*
Titov, E. V. 565 (43), *612*
Titova, Z. 419 (247), 481 (952b), *499*, *513*
Titus, D. 470 (860), *511*
Titus, D. D. 8 (34), *18*, 106 (228), 107 (228–231, 233), 124 (228, 229), 132 (302, 303), 133 (302), 134 (303), 143 (302, 303), *153*, *154*, 520 (30), 521 (31), *553*
Tobiason, F. 465 (793), *510*
Toder, B. H. 623 (68), *657*
Todres, Z. V. 184 (170–174), *188*
Todres-Selektor, Z. V. 586 (211), *616*

Toei, K. 164, 165 (44), 183 (130, 131, 133), *186*, *187*
Toelg, G. 163 (33), *185*
Tolles, W. M. 126 (283), *154*
Tolley, M. 469 (854a), *511*
Tolmachev, A. 469 (847), 470 (855), 472 (883), *511*, *512*
Tolmachev, A. I. 567, 568 (86), *613*
Tolstaya, T. P. 601, 602 (290), *618*
Toma, Š. 569, 592 (101), *614*
Tomasik, P. 561 (16), *611*
Tomaskovic, M. 177 (125, 126), *187*
Tomimatsu, K. 478 (928, 929), *513*
Tomkiewicz, Y. 449 (566), *505*
Tomoda, S. 632, 633 (190), 639 (236), 641 (284–287), *660–662*
Toncheva, V. 165 (53), *186*
Topiol, S. 25, 33, 34 (49), *61*
Topsom, R. D. 565 (62), *613*
Torii, S. 627 (149), *659*
Toropova, M. A. 167, 170 (88), 171, 172 (88, 95), 173 (98–100), 174 (88, 95, 98), 175 (98–100), 176 (88, 95, 98–100), *186*, *187*
Torre, M. 569 (93), 594 (240), *613*, *617*
Torres, L. 542 (217), *557*
Toscano, P. J. 739 (317), *847*
Toshimitsu, A. 403 (33), 481, 489, 491, 493 (951), *495*, *513*, 530, 533 (142), *555*, 632, 633 (187, 189), 636, 638 (217), 639 (187, 235), 641 (296, 302, 306–308, 311), *660–662*
Tosi, G. 262 (66, 67), 263 (67), *284*, 707, 708 (177), 759, 760 (371–387), 761 (371–374, 381, 383), *844*, *848*, *849*
Tötsch, W. 545 (225, 226, 228–230), *557*
Toussaint, J. 95 (175), *151*
Town, K. G. 715, 718, 719, 765, 766 (217), *845*
Traetteberg, M. 72 (67), *149*
Tran, Q. M. 621, 622 (27), *656*
Tran-Viet, D. 90 (152), *151*
Traverso, G. 445 (502, 503), *504*
Treffurth, H. 786, 787 (486), *851*
Trege, L. 407 (82a), *496*
Tregnoghi, C. 163 (34), *185*
Treibs, W. 460 (704), *508*
Trend, J. 403 (36), 420 (272), *495*, *500*
Trend, J. E. 220 (82), *240*, 624 (117), 625 (116, 117), 627 (117), *658*
Trenta, A. 234 (120), *241*, 288 (14, 15), 289 (15, 17, 18), 291 (14, 15), *304*
Treppendahl, S. 459 (687), *508*
Trierweiler, M. 233 (117), *241*
Trierweiller, M. 221, 222, 233 (114), *241*
Tripathy, S. 334, 341 (18), *341*
Trofimov, B. 407 (93), 493 (1068c), *496*, *516*
Trofimov, B. A. 563 (36–38), *612*
Trogu, E. F. 794 (522, 524, 529), 795 (529), *851*, *852*
Trop, H. S. 736 (307), *847*
Troshenko, E. 458 (665), *507*
Trost, B. M. 603 (298), 608 (330), *618*
Trotter, J. 93 (163), *151*, 519, 543 (13), *553*, 721 (234), *845*
Trotter, P. J. 356 (36), *366*, 701, 705 (157a, 157b), *844*
Troup, J. M. 114 (243, 244), *153*, 518 (10), *553*, 839 (673), *855*
Truce, W. 490 (1055), *515*
Trueblood, K. N. 94 (169), *151*
Truesdale, L. K. 16 (94), *20*, 541 (208), *557*
Trushule, M. A. 184 (165), *188*
Tschmutova, G. A. 202, 203 (53), 210 (67), *240*
Tschmutowa, G. 78 (90), 86 (118), *150*, 245 (14), *283*
Tseimakh, I. D. 522 (44), *553*, 577, 601, 603 (138), *614*
Tsibris, J. C. M. 238 (144, 145), *241*, 382, 392 (66a, 66b), *395*
Tsintsadze, G. V. 730 (296c, 272), *846*
Tsipis, C. A. 794 (523), *851*
Tskalobadze, L. A. 184 (167, 175), *188*
Tsoi, L. 457 (659), *507*
Tsoi, L. A. 624 (77–79), *657*
Tsuchiya, S. 70, 72, 90 (51), *149*
Tsuda, K. 338 (32), *341*
Tsukerman, S. V. 569 (94, 100), 573 (115, 116), *613*, *614*
Tsuno, Y. 565, 571 (104), *614*
Tsurkan, A. 455 (621), 458 (665, 666, 669), 490 (666), *507*, *508*
Tsurkan, A. A. 164 (39), *186*
Tsurkan, T. S. 164 (39), *186*
Tsutsumi, S. 762 (401a, 401b), *849*
Tsveniashvili, V. S. 184 (167–175), *188*
Tsveniashvili, V. Sh. 700 (154), 701 (154, 159, 160), *843*, *844*
Tudor, R. 373, 393 (19), *394*
Tullius, T. D. 102 (219), *152*, 742, 750 (334), *848*
Tully, C. 409 (116), *496*
Tully, C. R. 265 (70), *285*, 378, 386, 387 (38a), *394*
Tumakova, Zh. A. 781 (472), *850*
Tupistin, I. F. 569 (97), *613*
Tupitsyn, I. F. 562, 569 (31), *612*
Turco, A. 732 (287), 735 (287, 297), 737 (287), *847*
Turley, M. 163–165, 183 (18), *185*
Turnbo, R. G. 523 (74), *554*
Turner, C. J. 220 (83), *240*
Turner, D. C. 390 (118), *397*
Turner, E. 482 (968), 483 (971), *514*

Turner, E. S. 623 (54, 56), 624 (98), 642 (243), *657*, *658*, *661*
Tuve, T. W. 387, 388 (100), *396*
Tvedten, O. 84, 85 (106), *150*
Tweed, W. T. 563, 565, 608 (39), *612*
Tyerman, W. 402 (12a, 12b, 13), *494*
Tyler, A. N. 245 (10), *283*
Typke, V. 128 (290), *154*
Tysseland, M. 102 (211), *152*, 684–686 (41), *841*

Uamaguchi, H. 563 (167), *556*
Uchida, H. 164, 165 (44), *186*
Uddrich, P. 420 (266), *499*
Uden, P. C. 165, 167 (73), *186*
Udupa, M. R. 806 (560), *852*
U-Eda, K. 762, 764, 784 (400), *849*
Ueda, T. 453, 486, 487 (595), *506*
Uemura, S. 10 (51), *19*, 403 (33), 481, 489, 491, 493 (951), *495*, *513*, 528 (131, 132), 530 (142), 532 (156), 533 (142), 534 (156), 537 (176), 551 (283–285), *555*, *556*, *558*, 621, 622 (18), 632, 633 (187, 189), 636, 638 (217), 639 (187, 235), 641 (296, 302, 306–308, 311), 650, 651 (421, 422), 653 (451, 461), 654 (464), 655 (480), *656*, *660–662*, *664*, *665*
Uggla, R. 79 (96), 92 (157), *150*, *151*
Ugo, R. 754 (364, 365), 755, 756 (365), *848*
Uhlemann, E. 779, 781 (461), 784 (461, 481), 786 (481), 788, 792 (502), *850*, *851*
Uhlig, E. 767 (413), 801 (547), *849*, *852*
Ujjainwalla, M. 528, 533 (130), *555*, 635 (203), *660*
Ukai, T. 459 (698), *508*
Ulman, A. 409 (117, 118a), *496*, *497*
Umbreit, M. A. 642 (241), *661*
Umemura, Y. 622 (31), *657*
Umezaki, N. 393 (155), *397*
Umezawa, S. 407, 412 (80, 81), 413 (81), 419 (246), 433 (388), *496*, *499*, *502*
Umizaqa, S. 383 (72d), *395*
Undheim, K. 246 (20–22), *283*, *284*, 415 (203), *498*, 599 (268, 271), *617*
Unerstenhofer, G. 462 (718), *508*
Uneyama, K. 627 (149), *659*
Unger, S. H. 570 (103), *614*
Unverferth, K. 385 (76), *395*, 569, 581 (98), *614*
Uppstrom, B. 591 (228), *616*
Uppström, P. 412 (166), *497*
Urabe, H. 625, 628 (128), *659*
Urch, D. S. 744–747 (348), *848*
Usachev, A. I. 522 (44), *553*, 577, 601, 603 (138), *614*
Usiatinsky, A. Ya. 347 (7), *366*
Usmanov, A. 88 (137), *151*
Ustingov, A. N. 565, 566 (50), *612*
Usuki, M. 642 (309), *662*

Vafai, M. 419, 420, 426 (260), *499*
Vafina, A. A. 565, 566 (50), *612*
Vahrenkamp, H. 699 (144), 753 (363), 754 (144, 363), 757 (363), 758 (144, 363, 370), 759, 804 (370), 811 (592), 825 (363), *843*, *848*, *853*
Val, J. A. F. 6 (19), *18*
Valeev, R. B. 194, 195, 210 (28), 222, 223, 225, 230, 233 (100), *239*, *240*
Valle, G. 96 (176–181), 100, 102 (203), 118, 119 (262), *151–153*, 487, 488 (1022), *515*
Van Allan, J. 459 (691), *508*
Van Bekkum, H. 561, 575 (19), *612*
Vancea, L. 546 (242), *557*
Van Cleave, W. C. 601, 602 (288), *618*
Van Coppenolle, J. 423 (287–289), 471 (871), 473, 476 (287), 481 (289), *500*, *512*, 651 (398), *664*
Van Dam, W. 600 (279, 280), *617*
Van den Bossche, G. 86 (114), *150*
Van den Hende, J. 445 (507b), *504*
Van Der Kelen, G. P. 565 (66, 67), *613*
Van Der Linden, J. G. M. 762, 763 (395), *849*
Van der Puy, M. 565, 604 (61), *613*
Van DeVondel, D. F. 565 (66), *613*
Van Dormael, A. 456 (647), *507*
Van Duijneveldt, F. B. 25 (34), *60*
Van Ende, D. 402 (19), *494*, 624 (104, 105), *658*
Vanhare, E. 421 (277), *500*
Vanier, N. R. 565, 604 (61), *613*
Vanino, J. 332 (4), *341*
Vanino, L. 487 (1025), *515*
Van Koten, G. 838 (670c), *855*
Van Lare, E. J. 315 (24), *328*
Van Sickle, D. E. 594 (242), *617*
VanWazer, J. R. 694 (125), *843*
Varshavskii, S. Yu. 814, 820 (620), *854*
Vasella, A. 639 (124), *658*
Vasilev, V. V. 624 (91), *658*
Vasiliu, G. 482 (963–965), *514*
Vaskowskii, V. 413 (169), *498*
Vaughan, D. 488 (1040b), *515*
Vaughan, J. 575 (131), *614*
Vdovin, V. 419 (256, 257), 421 (256), 424 (256, 301), 427 (319), *499–501*
Vdovin, V. A. 88 (137), *151*
Vedejs, E. 11 (62), *19*
Veillard, A. 25 (23, 52), *60*, *61*
Veillon, C. 387 (94), *396*
Vekris, S. L. 196, 199, 205 (33), *239*
Velicescu, M. 281 (90), *285*, 832, 836 (658), *855*
Venanzi, L. M. 690 (64–66, 81–83), *841*, *842*

Veracini, C. A. 710, 712, 814, 819 (191b), *844*
Verani, G. 457 (658), *507*, 707, 708 (177), 759–761 (371, 372), *844*, *848*
Verboom, W. 447 (527), 481 (955, 957), *505*, *513*
Vereshchagin, A. N. 642 (291), *662*
Vereshchagin, L. I. 651 (403, 404), *664*
Vergoni, M. 177 (128), *187*
Verkade, J. G. 247 (26), *284*
Verkade, P. E. 561, 575 (19), *612*
Verkruijsse, H. 409, 414 (118b), 481 (954, 955, 958), 491 (958), *497*, *513*
Verkruysse, N. 423 (290b), *500*
Verlinden, M. 164, 165 (47), *186*
Vermeer, P. 481 (953, 954, 956), *513*
Vermeire, M. 77 (85), 91 (154), 92, 93 (164), *149*, *151*, 454 (606), *506*
Verneuil, A. 5 (14), *18*
Verneuil, M. A. 680, 683, 686 (3), *839*
Vernon, J. 453, 465 (592–594), *506*
Vernon, J. M. 91 (155), *151*, 700 (147), *843*
Vernon, M. 459 (684), *508*
Vernon, R. H. 550 (276), *558*, 654 (469, 470), *665*
Vernon, W. D. 191, 192 (14), *239*
Verschoor, G. C. 736 (308), *847*
Verschuere, C. 426 (317), *501*
Verweij, J. 522 (57), *554*
Vialle, J. 271 (79), *285*, 467 (818, 819, 822, 823, 825), *510*, *511*
Vicens, J. 403 (32), *495*, 632, 633 (191), *660*
Vicentini, G. 629, 652, 653 (158), 654 (471, 472), 655 (158), *659*, *665*
Vicenzi, C. 423 (290a), *500*
Vickrey, T. M. 165, 177 (74), *186*
Vidal, J. L. 839 (671–674), *855*
Vidoni, N. E. 295 (47), *305*, 683 (33), *841*
Vidoni Tani, M. E. 84, 85 (105), *150*
Viehe, H. G. 326 (69, 71), 327 (73), *329*
Viervoll, H. 139 (313), *154*
Vigneaud, V. du 676 (36a), *677*
Vikane, O. 15 (83), *19*, 102 (210), 119 (210, 277–280), 120 (277–280), 121–123 (278), 124 (210, 277–280), *152*, *154*, 684 (42), 685, 686 (42, 43), 688 (54–56, 59), 689 (55, 56, 59), 730 (271), *841*, *846*
Vilkov, L. V. 70 (46, 47), 74 (46), 75 (46, 47, 75), 76 (46, 75, 78b), 78 (46), 87 (126, 128), *149*, *150*
Villadsen, B. 196, 208, 222, 232 (36), *239*
Villares, P. 101 (198), *152*
Vincenzi, C. 469 (838, 839), 470 (838, 839, 856), 472, 473 (881), *511*, *512*, 598 (266), *617*
Viorel, D. 297–299 (51), *305*
Vitali, D. 710, 712 (189, 190, 191b), 714 (195), 814 (190, 191b), 819 (191b), *844*, *845*
Vlasakova, V. 167–169 (86), *186*
Vobetsky, M. 167, 170, 171 (87), 172, 173 (87, 96), 174, 175 (87), *186*, *187*
Vodden, A. 11 (58), *19*
Vogel, H. 527, 528, 533 (119), *555*
Voicu, V. 184 (159, 160), *188*
Voigt, E. 464 (758), *509*
Volashchuk, V. G. 565 (44), *612*
Volchenok, S. A. 567 (82), *613*
Volkov, M. 411 (133), *497*
Volkov, M. N. 582 (178), *615*
Volodina, G. F. 684–686 (40), *841*
Volodina, M. A. 164 (40), *186*
Voloshchuk, V. G. 547, 548 (259), *558*, 566 (70), 567 (82), *613*, 622 (34), *657*
Von Niessen, W. 25 (57), *61*
Von Zelewsky, A. 709 (182), *844*
Vordank, P. 721, 722 (237), *845*
Voronin, S. 408 (106), *496*
Voronkov, M. G. 624 (99), *658*
Voronov, M. A. 781 (473), *850*
Vorster, W. 5 (13), *18*
Vos, D. de 522 (53), *554*
Vos, G. 736 (308), *847*
Vosooghi, M. 255, 278 (54, 55, 57), 279 (57), 280 (54), *284*, 448 (536), 461 (713), *505*, *508*
Voss, E. 829, 832 (646), *854*
Voss, J. 237 (140), *241*
Vouros, P. 272 (80), *285*
Vovsi, B. 485 (995), 488 (1038), *514*, *515*
Vrábel, V. 730 (274), *846*
Vreshova, E. 419 (249), *499*
Vtyurina, N. N. 245 (15), *283*
Vyazankin, N. S. 183 (143), *188*, 353 (19), *366*, 716 (202), *845*
V'yunov, K. 458 (669), *508*
Vzazankin, N. S. 356 (33), *366*

Wabnitz, C. H. 179, 180 (111), *187*
Wachter, J. 804, 805 (555), *852*
Wade, K. 467 (826b), *511*
Wadsworth, D. 473 (894), *512*
Wadsworth, D. H. 624 (76), *657*
Wagner, G. 454 (600), *506*, 621, 622 (17), *656*
Wahlgren, U. 25, 30 (72), *61*
Wakamatsu, T. 622 (38), *657*
Wakasugi, M. 654 (464), *665*
Walatka, V. 449 (565), *505*
Walker, A. 765, 807 (404), *849*
Walker, B. J. 387 (92), *396*
Walker, G. L. 601, 602 (287), *618*
Walker, H. 401 (2), *494*
Walker, R. 467 (827a), *511*

Walker, W. E. 839 (671, 672), *855*
Wallmark, I. 427 (326), *501*
Walrand, J. 64 (10), *148*
Walsh, R. 157, 159 (3), *160*
Walter, E. D. 387 (95), *396*
Walter, R. 86 (120), *150*, 178–181 (107), 182 (124), *187*, 670 (16), 676 (36a, 36b), *676*, *677*
Walter, W. 622 (45), 642 (255), *657*, *661*
Waltman, R. 407 (82b), *496*
Walton, E. D. 693–696 (124), *843*
Walton, J. C. 325 (66), *329*
Walton, R. A. 680, 782, 815 (12f), *840*
Wang, J. T. 235 (128, 129), *241*
Wang, K.-C. 392 (131), *397*
Wanzelburger, J. 485 (1000a), *514*
Wardlaw, W. 335 (21), *341*
Warner, N. L. 381, 393 (58), *395*
Wartewig, S. 786 (504b), 789 (504b, 505), *851*
Wasson, J. R. 359 (50), *366*
Watanabe, K. 317 (37), *329*, 642 (242), *661*
Watanabe, N. 641 (296), *662*
Watanabe, S. 444, 484 (495), *504*
Waterhouse, J. 158 (11), *160*
Waters, A. J. 184 (179), *188*
Waters, J. M. 768, 829, 830 (419), *849*
Watkins, P. M. 545 (237), *557*
Watkinson, J. H. 163, 183 (20), *185*
Watson, D. 462 (723b), *509*
Watson, D. G. 364 (69), *367*
Watson, K. 403 (47), *495*
Watts, H. G. 393 (159), *397*
Watts, W. E. 576 (135), *614*
Wautier, H. 319 (43), 320 (50), *329*, 605 (324), *618*
Wazewsda-Czyzewska, M. 289, 291 (22), *304*
Weary, D. K. 376, 377, 384 (28), *394*
Weast, R. C. 201 (42), *239*
Weatherill, T. D. 747, 753, 825, 832, 834 (353), *848*
Weaver, A. 380, 392 (48), *395*
Weaver, J. L. 524 (80), *554*
Weavers, R. 404 (54), *495*
Webb, J. 670, 671 (17e), *677*
Webb, R. L. 202 (46), *239*
Webb, R. R. II 642 (259), *661*
Weber, J. 429 (352, 353), *501*
Weber, J.-V. 439 (438, 439), *503*
Weber, J. V. 586 (191), *616*
Weber, K. L. 710, 712, 713, 814, 819 (191a), *844*
Weber, R. 17 (105), *20*, 268 (74), 276 (86), *285*, 422 (283), 438 (424, 425), 446 (597), 452 (424, 425, 591), 453 (425, 591, 597), 454 (597, 602, 604, 605), 458 (605), 475 (283), 491 (602), *500*, *502*, *506*, 586, 590, 597 (208), *616*
Webert, J. M. 439 (438, 439), *503*, 586 (191), *616*
Webster, D. 467 (827a), *511*
Webster, R. 444 (500), 445 (506a), 475 (500), *504*
Weger, M. 449 (568), *505*
Wegler, R. 462 (718), *508*
Wegscheider, W. 163 (33), *185*
Wehrung, T. 87 (129), *150*
Wei, K. H. 261 (62), *284*
Wei, K.-T. H. 107–109, 124 (234), *153*
Weibust, G. 165 (62), *186*
Weichmann, J. 768, 770 (421), 804, 805 (556a, 556b), 829 (421, 646), 830 (421), 832 (556a, 556b, 646), 833 (556a, 556b), *849*, *852*, *854*
Weichsel, C. 355 (30), *366*, 446 (512a), *504*
Weidenhammer, K. 742, 751, 825, 827, 828 (337), *848*
Weiler, T. 483 (983), *514*
Weinstock, L. 466 (796), *510*
Weis, J. C. 703 (167), 704 (167, 168), *844*
Weis, R. 714, 814, 820 (196), *845*
Weisenfels, M. 469 (840b), *511*
Weiss, E. 806, 832, 833 (559), *852*
Weiss, K. 722 (245e), *846*
Weiss, R. 314 (21–23), *328*, 446 (508), 451 (583), 454 (599, 603), 477 (912, 915), *504*, *506*, *512*, *513*, 547 (253), *558*
Weiss, T. 483 (982), *514*, 521 (34), *553*
Weiss, U. 455 (624), *507*
Weiss, W. 528, 530, 532, 533 (123), *555*
Weissflog, E. 487 (1023, 1030), 492 (1023), *515*
Welch, A. 750 (361), *848*
Welcman, N. 87 (129), *150*, 281 (90), *285*, 362 (65), *367*, 640 (248, 249), *661*, 743, 749 (344), 832, 836 (658), *848*, *855*
Wellman, K. M. 671 (20b), *677*
Wells, E. J. 234 (118), *241*
Wells, P. R. 561, 562 (22), *612*
Wells, R. R. 202 (54a), *240*
Welter, A. 425 (308), 454 (601), *500*, *506*
Wendelborn, D. F. 624, 625, 627 (117), *658*
Wensky, A. 281 (91), *285*, 748 (360), *848*
Wepster, B. M. 561, 575 (19), *612*
Werchan, H. 465 (768), *509*
Werner, H. 9 (45), *18*, 715, 717, 720 (224), 754, 757 (368), 765 (224, 408), 766 (408), 767 (224, 408, 412, 414, 415), 768 (414, 420), 770, 778 (422), 796–798 (224), 829, 830 (420, 648), 831 (648), *845*, *848*, *849*, *854*
Werner, R. L. 641 (261), *661*
West, K. W. 806, 807, 810 (564), *852*

West, R. 8 (40), *18*, 355 (28), *366*, 731 (278), *846*
Westerlund, C. 411, 412 (151b), 436 (413), 437 (414, 417), 438 (430, 432), *497*, *502*, *503*, 586 (195), 588 (201), *616*
Westland, A. D. 690 (85, 86), 774 (453), *842*, *850*
Westland, L. 774 (453), *850*
Weswig, P. H. 162 (9), *185*, 391 (123), *397*
Whalley, E. 335 (22), *341*
Whanger, P. D. 391 (123), *397*
Wharf, I. 731 (294, 295), 732 (294), 733 (295), 734 (294), *847*
Wheeler, G. L. 183 (139), *188*
Wheeler, O. H. 166 (83), *186*
Wheland, R. 450 (575), *506*
Whewell, R. J. 565, 574, 575, 608 (60), *613*
Whistler, R. L. 641 (268), *661*
White, A. 469 (854c), *511*
White, A. H. 97, 98, 132, 143 (186), *152*, 525 (87), *554*, 747 (356), 779 (466b, 469), *848*, *850*
White, F. 455 (611), *506*
Whiteford, R. A. 245 (13), *283*
Whitehead, E. I. 163, 183 (25), *185*
Whitehouse, R. D. 668, 669 (11), *676*
Whiteside, R. A. 25, 33, 34 (49), *61*
Whitesides, T. H. 378, 386, 387 (38b), *394*
Whitfield, H. J. 774 (448), *850*
Whitman, D. R. 25 (26), *60*
Whittle, A. J. 642 (258), *661*
Wicha, J. 673 (27), *677*
Widdowson, D. A. 16 (90), *19*
Wieber, M. 488 (1035), 489 (1047, 1048), *515*, 537, 540 (171), *556*
Wieczorek, J. 703, 704 (167), *844*
Wiegers, G. A. 810 (580), *853*
Wiegert, P. E. 371, 392 (12b), *394*
Wienand, H. 825 (639), *854*
Wijekoon, W. M. D. 675 (35), *677*
Wijers, H. E. 651 (395, 396), *664*
Wilckens, R. 90 (152), *151*, 465 (786), *510*
Wildera, R. 792 (509), *851*
Wiley, G. R. 571, 572 (106), *614*
Wilhelm, I. 650 (363, 364), *663*
Wilke, G. 779, 781 (461), 784 (461, 481), 786 (481), *850*, *851*
Wilkins, C. 405, 469 (60a), *495*
Wilkins, C. J. 525 (86), *554*, 693, 695–697 (122), *843*
Wilkinson, G. 680 (1, 2), 681 (1), 716 (1, 205), 719 (1), 721 (1, 2), 811 (1), *839*, *845*
Wilkinson, L. 455 (630), *507*
Willadsen, T. 89, 90 (142), *151*
Wille, F. 452 (590), *506*, 641 (276), *661*
Willfroth, E. 787 (483), *851*
Williams, B. L. 114, 115 (246), *153*, 727, 729 (257), 771 (437), 773 (257, 437–439), 774 (437–439, 446), 775 (437, 438), 776 (446), *846*, *850*
Williams, D. J. 542 (217), *557*, 624 (97), 647 (338), *658*, *663*
Williams, D. M. 808 (577), *853*
Williams, E. 642 (289), *662*
Williams, F. 235 (126, 128, 129), *241*, 325 (64), *329*, 488, 491 (1034), *515*
Williams, F. D. 829 (649), *854*
Williams, H. H. 179, 181 (114), *187*, 387, 388 (100), *396*
Williams, L. E. 732, 735–738 (283), *847*
Williams, P. 495, 479 (60a), *495*
Williams, P. P. 525 (86), *554*
Williams, R. F. 575 (130), *614*
Willis, W. 444 (490b), *504*
Willis, W. W. Jr. 599 (273), 604 (315, 319), *617*, *618*, 622 (42), 625 (118), 626 (135), *657–659*
Williston, C. 519, 543 (13), *553*
Willner, H. 11 (61), *19*, 71, 72 (55), *149*
Willson, P. D. 72 (64), *149*
Willson, R. L. 301 (64, 65), *305*
Wilson, D. R. 624 (96), *658*
Wilson, L. G. 388, 389 (103), *396*
Wilson, L. J. 794, 795 (529), *852*
Wilson, M. K. 371 (14, 15), 382 (15), *394*
Wilson, R. C. 573 (114), *614*
Wilson, S. R. 518, 545 (5), *553*, 636 (228), *660*
Wilzbach, K. E. 380 (45), *395*
Winfield, J. M. 690 (95), *842*
Winnewisser, G. 65 (16), *148*
Winnewisser, M. 65 (17), *148*
Winter, H. 404 (51), 444 (490a), 493 (51), *495*, *504*
Winter, H. J. 548 (260), *558*
Winter, M. 814 (607, 617), 819 (617), *854*
Winter, W. 414 (195), 430 (360), 442 (468, 469), 443 (360, 481, 482), *498*, *501*, *503*, *504*, 595 (255), *617*
Wintergerst, H. 804, 805 (555), *852*
Wipff, G. 25, 28, 30, 33, 34 (67), *61*, 604 (313), *618*, 624 (87), *658*
Wiriyachitra, P. 652 (433), *664*
Wirl, A. 810 (581), *853*
Wirtz, P. 454 (601), *506*
Wiseman, G. 428 (344, 347), *501*
Wiseman, G. E. 642, 649 (281), *661*
Wisløff Nilssen, E. 82 (99), *150*
Wittig, G. 534 (162), *556*, 655 (482), *665*
Wiygul, F. 449 (557), 450 (572a), *505*, *506*
Wobschall, D. 449 (564), *505*
Wöhler, F. 5 (8, 11), *18*
Wojcicki, A. 697, 734 (137), 737, 738 (311), 739 (137, 311), 740 (137), *843*, *847*

Wojciechowski, P. 451 (581, 582), *506*
Wojciechowski, P. S. 336 (24), *341*
Wojnowska, M. 8 (40), *18*, 355 (28), 356 (34), *366*
Wojnowski, W. 8 (40), *18*, 355 (28), *366*
Wojtowski, R. 731, 732, 734 (294), *847*
Wolf, H. 670, 671 (18b), *677*
Wolf, J. 829–831 (648), *854*
Wolff, M. 491 (1057), *516*
Wolfram, B. 673 (29), *677*
Wolfram, E. 629 (155), *659*
Wollowitz, S. 599 (273), *617*, 624, 625 (117), 626 (135), 627 (117), 642 (146), *658*, *659*
Woltermann, G. M. 359 (50), *366*
Wolters, E. 469 (841), 471 (872), *511*, *512*, 641 (271), *661*
Wolters, J. 522 (53, 56), *554*
Wong, C. K. 636 (218, 219, 221, 222), 638 (218, 219, 221), 651 (425), *660*, *664*
Wong, P. T. S. 165, 167–169 (65, 66), *186*
Wong, S. F. 766 (411), *849*
Wong, S. H. 380, 392 (48), *395*
Wong, T. C. 194 (24, 26), 196 (26), 208 (24), 209 (24, 26), 220 (24, 26, 85), *239*, *240*
Wood, R. J. 194–199, 203, 205, 207, 211, 218, 219 (22), *239*, 690 (100), *842*
Woodard, C. 428, 443, 484 (340), *501*
Woodard, C. M. 85 (107), *150*
Woodbridge, D. T. 642, 643 (246), *661*, 793, 794 (515), *851*
Woodgate, P. D. 641 (295), *662*
Woods, T. 461 (709), *508*
Woody, R. W. 670, 671 (17c, 17d), *677*
Workman, M. O. 771, 772 (432), 815 (603a), *849*, *853*
Wrackmeyer, B. 815 (601p), *853*
Wright, A. H. 718 (226), 768, 829, 830 (419), *845*, *849*
Wright, L. J. 715, 717, 814, 821 (216), *845*
Wright, M. 402 (17), *494*
Wright, P. L. 387 (85, 86), *396*
Wright, W. B. 69 (34), *148*
Wudl, F. 13 (68), *19*, 94 (170), 128 (294, 295), 130 (294), *151*, *154*, 406 (65), 446, 447 (521), 448 (544, 547, 550), 449 (564), 450 (544, 547), 451 (577, 578), 462 (723a), *495*, *504–506*, *509*, 526 (103, 104), *555*, 650 (387), 651 (415, 416), *663*, *664*
Wuyts, L. F. 565 (66), *613*
Wynberg, H. 624 (72), *657*
Wynne, K. J. 229 (108), *241*, 523 (58, 59), 542 (215, 217), *554*, *557*
Wyssbrod, H. R. 86 (120), *150*

Xyazankin, N. S. 356 (32), *366*

Yadav, D. S. 737 (313), *847*
Yagubskii, É. B. 95 (173), *151*
Yagupol'skii, L. M. 565 (43, 44, 53), 566 (70, 71), 567 (82, 83, 86), 568 (86), *612*, *613*, 622 (34), *657*
Yagupolskii, L. M. 547, 548 (259), *558*
Yakhoontov, L. 565 (62), *613*
Yakobson, G. G. 197 (37, 38), 204, 205 (27), *239*
Yakovlev, Yu. V. 699 (152b), *843*
Yakovleva, O. G. 565, 566 (50), *612*
Yalpani, M. 377, 386 (35), *394*, 406 (68), 409 (107, 108), 447 (68, 532), 463 (739–741), 464 (107, 108, 532, 740, 741, 760, 761), *495*, *496*, *505*, *509*, 581, 595 (162), *615*, 621, 622 (25), *656*
Yamagata, S. 392 (133), *397*
Yamaguchi, H. 541 (201), *557*, 642 (310), 652 (439), *662*, *665*
Yamaguchi, I. 70, 73, 81 (44), *149*
Yamahira, A. 446 (518), *504*
Yamakawa, K. 644 (323, 324), *662*
Yamamoto, Y. 165 (49, 51), *186*, 519 (17b), *553*
Yamashita, M. 6 (23), *18*
Yamidagni, R. 202 (50), *239*
Yan, C. F. 634 (195), *660*
Yaneff, P. V. 716, 765 (204), *845*
Yanilkin, V. V. 565 (68), *613*
Yano, K. 393 (155), *397*
Yano, T. 316 (30), *329*, 333–336, 339, 340 (15), *341*, 651 (389), *664*, 747 (357), *848*
Yao, A. N. 603 (297), *618*
Yao, H. 371, 392 (12a), *394*
Yarilin, A. A. 392 (142), *397*
Yarno, T. 623 (51), *657*
Yashin, Y. I. 183 (143), *188*
Yasuhara, T. 762 (401a), *849*
Yatlova, L. E. 681, 686 (23b), *840*
Yazdany, S. 447, 464 (531), *505*
Yerga, I. 740 (321), *847*
Yokota, K. 497 (1017), *515*
Yokoyama, A. 195 (31), *239*, 600 (286), *618*
Yokoyama, Y. 604 (312), *618*, 642 (289, 290), 645 (328, 329), *662*
Yokozeki, A. 83 (103), *150*
Yonezawa, T. 25 (46), *61*
Yorovenko, N. N. 629 (160), 632 (160, 176), *659*
Yoshida, M. 321, 323 (62), *329*
Yoshida, T. 574, 576 (136), 577, 601 (137), *614*, 741 (326, 330, 332), 742 (332), 745 (350), 749 (330), 825 (330, 332), *847*, *848*
Yoshida, Y. 762 (401b), *849*
Yoshida, Z. 314 (20), *328*
Yoshii, E. 623 (69), *657*
Yoshikoshi, A. 624 (74), *657*
Yoshimoto, H. 623 (52, 53), *657*

Yoshimoto, Y. 370 (3a, 3b), *393*
Young, M. W. 630 (169), *659*
Young, V. R. 387, 388 (98), *396*
Youngs, G. R. 393 (158), *397*
Yu-Gwan Chen 586 (210), *616*
Yukawa, Y. 565, 571 (104), *614*
Yu'rev, Y. K. 184 (164), *188*
Yur'ev, Y. K. 377 (33a), 383 (69a, 69b), *394*, *395*
Yur'ev, Yu. 407 (86, 88), 408 (105), 411 (146, 147, 150, 151a), 412 (86, 88, 152, 154, 157, 162, 164, 165), 413 (86, 157, 168–170, 172, 177–181, 187), 414 (150, 152, 157, 162), 416 (157), 418 (105), 419 (249), *496–499*
Yur'ev, Yu. K. 407 (84), *496*, 573 (118), 578 (144), 581 (144, 168–172, 175), 593 (118), 595 (169, 247), *614*, *615*, *617*
Yureva, E. M. 539, 540 (189), *556*
Yur'eva, V. S. 565 (51), 566, 567 (51, 69), 574 (69), *612*, *613*
Yureva, Y. L. 522, 523 (42), *553*

Zaidi, S. A. A. 522 (49), *553*
Zaitsev, B. E. 700 (155), 701 (156), *843*
Zaitseva, E. 411 (146, 147, 150, 151a), 414 (150), *497*
Zaitseva, E. L. 581 (172, 175), *615*
Žák, Z. 547 (256, 257), *558*, 650 (365, 366), (372), *663*
Zakowski, J. J. 390 (115a, 115b), *396*
Zanati, G. 491 (1057), *516*
Zankowska-Jasinska, W. 487 (1019), *515*
Zannini, P. 262 (66), *284*, 759, 760 (379, 384–387), *848*, *849*
Zanotti, G. 96 (180), 100, 102 (203), *152*
Zaripov, N. M. 70 (46, 47), 74 (46, 74), 75 (46, 47, 74, 75), 76 (46, 75, 78b), 78 (46), 106 (226), *149*, *153*
Zatsepina, N. N. 569 (97), *613*
Zborovskii, I. 423 (295, 296), *500*
Zdansky, G. 671 (21), 673 (25), *677*
Zefirov, N. 404 (55), 406 (73), 418 (238), 423 (297), 481 (238, 950), 494 (1070, 1075), *495*, *496*, *499*, *500*, *513*, *516*
Zefirov, N. S. 527 (105), 530 (144, 146), 531 (147, 148), *555*, *556*
Zeil, W. 66, 67 (26), *148*
Zellers, E. T. 128 (294, 295), 130 (294), *154*
Zemlyanskii, N. I. 787 (488), *851*
Zeuner, S. 726, 727, 729 (256), 814, 823 (623), *846*, *854*
Zhdanov, S. I. 184 (168, 170–174), *188*
Zhetar, B. 458 (668), *508*
Zhidomirov, G. 434 (396), *502*
Zhiryakov, V. 437 (420, 421), 438 (436), 472 (875), *502*, *503*, *512*
Zhizhin, G. N. 88 (137), *151*
Zhuravlev, V. E. 171, 172 (95), 173 (98–100), 174 (95, 98), 175 (98–100), 176 (94, 98–100), *187*
Ziegler, E. 461 (711), 475 (901), *508*, *512*
Ziegler, M. 699, 700 (145a), 721 (240), *843*, *845*
Ziegler, M. L. 9 (43), *18*, 742 (336, 337), 746 (336, 351), 751 (336, 337, 351), 752 (351, 362), 759–762 (390a, 390b), 768, 770 (421), 825 (336, 337, 351, 390a, 390b, 639), 827, 828 (337), 829, 830 (421, 647), 832, 833 (647), *848*, *849*, *854*
Zilverschoon, A. 481 (953), *513*
Zima, G. 621 (8), 622 (8, 21), 624 (21), 632 (173, 182,), 633 (182), 634 (192, 193), 636, 637 (212), *656*, *659*, *660*
Zimmer, O. 464 (756), *509*
Zimmer, R. 357, 358 (40, 43, 45), *366*
Zimmer-Gasser, B. 723–725 (248), 804, 805 (557), *846*, *852*
Zimmermann, H. W. 690, 691, 754, 757, 814, 825 (108), *842*
Zimmermann, R. 523, 525 (72), *554*
Zingaro, R. 455, 457 (626), 483 (975), 484 (989, 990, 991a), *507*, *514*
Zingaro, R. A. 2 (2), 7 (27), 9 (44), 16 (95), *17*, *18*, *20*, 64 (2a), 77 (84), 100 (195, 201), 101 (193, 195, 196, 200–202), 141 (202), 144, 145 (325), *148*, *149*, *152*, *155*, 162 (4, 14), 163 (4), *185*, 227 (106), 231 (112), 232–234 (106), *241*, 360 (53, 55, 57), 365 (71–73), *366*, *367*, 518 (11), 546 (246), *553*, *558*, 656 (374), *663*, 681 (17a, 20), 686 (20, 46d), 697 (141a), 754 (367), 755 (141a, 367), 796 (532, 533), 797 (533), 815 (601a–d), *840*, *841*, *843*, *848*, *852*, *853*
Zinn, J. 87 (127), *150*
Ziolo, R. 442 (471), *503*
Ziolo, R. F. 8 (34), *18*, 106 (228), 107 (228, 230–233), 114 (243, 244), 124 (228), 132 (302, 303), 133 (302), 134 (303), 140 (317), 143 (302, 303, 317), *153*, *154*, 518 (10), 519 (15), 520 (30), 521 (31), *553*, 668 (7, 8), *676*
Ziriakus, W. 469 (843), *511*
Zirkle, C. 487 (1010), *514*
Zlatarov, S. 289, 291 (21), *304*
Zobel, D. 431 (380), *502*
Zoellner, W. G. 165 (59), *186*
Zolk, R. 829–831 (648), *854*
Zoller, U. 425 (309b), *500*
Zoller, W. H. 165, 167, 170 (72), *186*
Zolotova-Zolotukhina, L. 465 (776, 782), *510*

Zoretic, P. A. 630, 631 (170), *659*
Zountsas, J. 464 (759), *509*
Zozulin, A. J. 218 (76), 221 (88), *240*
Zumbulyadis, N. 222, 223 (98, 99), 231 (98), *240*, 445 (501), *504*, 651 (423), *664*, 701, 705, 736 (158a), 814, 815, 817 (605), 836 (669), *844*, *853*, *855*
Zylber, J. 668, 669 (12), *676*

Subject Index

Ab initio computational methods 30–33
Acetylacetone complexes, with $SeCl_4$ or $TeCl_4$ 97
Acyclic diselenides,
geometry of 82–86
isotopically labelled 371, 372
Acyclic disulphides, geometry of 83–85
Acyclic ditellurides, geometry of 86
Acyclic selenides,
geometry of 70–82
isotopically labelled 371–374, 392
Acyclic sulphides, geometry of 72
Acyclic tellurides,
geometry of 72, 76–78
isotopically labelled 373–376
Addition reactions, directing and activating effects in 591–593
Aldimine chelates 781
Alkyl aryl selenides,
mass spectra of 249–251
^{77}Se-NMR chemical shifts for 202–204
Alkyl aryl tellurides, mass spectra of 249–251
Alkylideneseleniranium ions 321, 322
Alkyltellurium trihalides 403
Allyl alcohols, synthesis of 627
Allyl selenides 634
cyclization of 402, 420
Aluminium methylselenolates 348
Ambidentate ligands 700, 701, 713
dithizone 779
1,4-oxaselenan 706
PhSSePh 729
$R^1_2ESeR^2$ (E = As,P) 711, 714
$R^1_2ETeR^2$ (E= As,P) 820
$Se(CN)_2$ 726,728
seleninate 760, 761
selenocyanate 730, 734
selenosemicarbazone 726, 728
selenoxide 721
$^{-}Se(O)_2CH_2CH_2NH_2$ 781
tellurocyanate 815
thiazolidine-2-selenone 707
Aminoselenazoles, radiation chemistry of 290, 291
Aminoselenoazolines, radiation chemistry of 289–291
Appearance energies 248
Arsine selenides 364
Arsinous acid esters 365

Basis sets 23–30
3-21G/LWD 34, 36, 41–45, 49–60
MINI-1 34, 36, 39, 40, 43–45, 47, 48, 51–60
STO-3G 29, 30, 33–35, 37, 38, 43–47, 51–60
Beckmann rearrangement 458
Benzene ring deformation 107, 109, 121, 136, 145
Benzeneselenenyl derivatives, geometry of 75–77
Benzenesulphenyl derivatives, geometry of 75, 76
Benzenetellurenyl derivatives, geometry of 77
Benzoanthracene, Se analogues of 429
Benzodichalcogenazolium salts 462, 463
Benzodioxatelluranes 488
Benzoisoselenazoles 458
directing and activating effects in 586, 590
Benzoisoselenazolinones 453, 454
Benzoisoselenazolinthiones 454
Benzoselenadiazoles 426, 453, 463, 465, 483
complexes of 699–701
directing and activating effects in 592, 597, 598
Benzoselenadiazolium salts 465
Benzoselenazaboroles 460
Benzoselenazepines 454
Benzoselenazines 459, 485, 486
Benzoselenazoles 458–460, 486
chemical shifts in 573
complexes of 703–705

directing and activating effects in 586, 590, 592
Benzoselenazolines 460
Benzoselenazolium ions 459, 460
Benzoselenazolopyridines 461
Benzoselenepins 489, 490
Benzoseleninoindoles 479
Benzoseleninones 477, 478
Benzoseleninopyridinones 478
Benzoselenolobenzothiophenes 432
Benzoselenoloindoles 438, 439, 453
Benzoselenolooxazoles 438
Benzoselenolopyridines 439
Benzoselenolopyrroles 438
Benzoselenoloselenadiazoles 438, 464
Benzoselenolothiazoles 438
Benzoselenolothiepins 436
Benzoselenolothiins 436
Benzoselenolothiophenes 431
Benzoselenophenes 532
as reaction products 454, 472, 473
complexes of 771, 772
detection and determination of 177, 183
directing and activating effects in 584, 585, 591, 592, 597
geometry of 92, 93
mass spectra of 265–269
metalated 425
reactions of 424–427
^{77}Se-NMR chemical shifts for 212, 214, 215
synthesis of 419–423, 426
Benzoselenophenones, mass spectra of 268
Benzotelluroloindoles 444
Benzotellurophenes 439–442, 532
directing and activating effects in 584, 585, 590, 597
Benzotellurophenones, mass spectra of 268
Benzothiaselenins 481
Benzothiaselenolthiones 446
Benzothiepinoselenazoles 461
Benzoxepinoselenazoles 461
Bernthsen synthesis 426
Beyer–Combes synthesis 426
Biopolymers, Se-containing, chiroptical properties of 676
2-Biphenylyltellurium trihalides, geometry of 135–137
Bischalcogenopyranylidenes 470
Bis(dichalcogeno)tetracenes 446
Bis(diseleno)anthracenes 446
Bis(diselenocarbamato)selenium(II) compounds 406
Bis(diseleno)naphthalenes 446
1,3-Bis(methylselenoethylseleno)propane complexes 801, 802
Bis-phosphine selenides, complexes of 698
Bis(β-(2-pyridyl)ethyl) selenide complexes 801
Bis(seleno)carbenium ions 320, 321
Bis-selenophenes, chiroptical properties of 674, 675
Bis(selenourea) cation 84
Bond/bond interactions 134
Bond configuration,
of Se 77, 102
of Te 77, 102
Bond dissociation energies 158–160
Borolanes 345, 346
Bradsher reaction 428
Brain-imaging agents 392

Carbene complexes,
Se-functionalized 722–725
Te-functionalized 814, 823
Carbenium ions, Se-stabilized 308–321
Carbonate analogues, mass spectra of 263, 264
Carbon dioxide analogues 11
geometry of 64
Carbon monoxide analogues, geometry of 64
Carbonyl compounds — *see also* Aldehydes, Ketones
α,β-unsaturated — *see* α,β-Unsaturated carbonyl compounds
Carbonyl difluoride analogues, geometry of 66, 67
Carbyne complexes 723, 823
Catalysts, for the hydrogenation of non-aromatic olefins 681
Chalcogen–arsenic compounds 16
Chalcogen–chalcogen bond lengths 117, 118
Chalcogen–halogen compounds 15, 16
Chalcogen–nitrogen compounds 16
Chalcogen–phosphorus compounds 16
Chalcogen sulphides 539
Charge distribution 35, 45–50, 55, 56
Charge localization theory 247
Charge-transfer salts 95
Chemical shift anisotropy mechanism 193
Chemical shift referencing 196
Chemical shifts,
for Se compounds 193–220
isotope effects on 217–220
medium effects on 194–196
for Te compounds 221–233
medium effects on 222, 223, 228, 229
Chemical vapour deposition 681
Chromium complexes, geometry of 65, 105
Circular dichroism, general considerations of 667, 668
Claisen rearrangement 421
α-Cleavage, in mass spectral fragmentation 249, 251

Cluster compounds,
  of Au 811
  of Cd 813
  of Co 811, 832, 837, 838
  of Fe 811, 812, 832, 834–837
  of Mn 747, 750
  of Nb 808, 809, 813, 832
  of Ni 811, 813
  of Os 747, 754, 811–813, 832
  of Pt 834
  of Rh 811, 837, 838
  of Ru 811, 832, 839
  of Zn 813
Cobalt complexes, geometry of 69, 104
Comproportionation 625
Concentration effects, on ^{77}Se-NMR chemical shifts 195
Conformational stabilities 34, 45, 55
Copolymerization 426
Copper complexes, geometry of 103–105
Correlation energy 22
Cotton effects, general considerations of 667, 668
Coupling constants,
  for Se compounds 220, 221
  for Te compounds 233, 234
CSe_2 complexes 717, 765–767
CSSe complexs 767, 768
Cyanine dyes 460
Cycloalkenylselenodiazoles, complexes of 702
Cyclodehydration 428
Cyclodehydrohalogenation 428
Cyclofunctionalization 636–638
Cyclohexan-1,3-diones 525, 526
Cyclopentan-1,3-diones 525
Cycloreversion, in mass spectral fragmentations 275

Decarbonylation reagents 835
Se-Demethylation 772, 782
Determination,
  of Se 163–165
  of Te 165
Dialkoxytellurides 491
Dialkyl selenides,
  complexes of 690, 691
  mass spectra of 249
  reactions of 403
  ^{77}Se chemical shifts for 198–202
Dialkyl tellurides,
  complexes of 814–818
  mass spectra of 249
  radiation chemistry of 303
  reactions of 403
Diaryl dichalcogenides, conformational properties of 86
Diaryl selenides
  mass spectra of 252
  ^{77}Se-NMR chemical shifts for 204, 205
Diaryl tellurides,
  complexes of 814–818
  mass spectra of 252
  radiation chemistry of 303
Dibenzoselenapyrilium ions, equilibrium constants for 598
Dibenzoselenazocines 492
Dibenzoselenepins 490
Dibenzoselenophenes 428, 429
Dibenzoselenophenium halides 428
Dibenzotellurophenes 442, 443, 484
  directing and activating effects in 585
Dibromotellurides, chiroptical properties of 675
Dichalcogen dihalides 527
1,3-Dichalcogenolylium ions 309–312
Dieckmann condensation 420
Dihalides 15
  detection and determination of 171, 174, 175
  mass spectra of 252, 260, 261
  preparation and properties of 519–528, 530, 542, 654, 655
  ^{125}Te-NMR chemical shifts for 228, 229
Dihaloaryltellurates 543
Diimides 540, 541
Dimethyltellurium compounds, geometry of 129–131
Diorganotellurium dicarboxylates, mass spectra of 261, 262
Dioxaselenanes 487
Dioxaselenolan oxides 462
Dioxides,
  radiation chemistry of 291–293, 302
  reactions of 529, 532
Diphenyl-*o*-selenatophenyl phosphine complexes 782
Dipole–dipole relaxation mechanism 191
Dipole moments 45, 51, 55–57
Directional effects 129
Diselenadiazocanes 488
Diselenadiazocines 492
Diselenafulvenes 447, 448, 464
Diselenanes 480, 481
  chiroptical properties of 671
Diselenaneselones 480
Diselenans, complexes of 705, 706
Diselenates, complexes of 798
Diselenatetrastannanes 357
Diselenatristannolanes 357
Diselenazines 488, 492
Diselenazolium ions 463
Diselenepanes 491
Diselenetanes 405, 469, 525

Diselenetenes 798
Diselenides,
acyclic — *see* Acyclic diselenides
chiroptical properties of 670–672
complexes of 709–714
detection and determination of 166–168, 184
mass spectra of 253–255
organometallic 759
reactions of 402, 405, 406, 444, 625, 626, 748–750, 753
^{77}Se-NMR chemical shifts for 204–206
Diselenins 464, 481
Diselenoacetylacetonates, complexes of 798, 800
α,α′-Diselenobisformamidinium ions 295
Diselenocarbamates, complexes of 765, 793, 794
Diselenocyanatoalkanes, complexes of 775
Diselenodiphosphines, complexes of 798
Diselenolanes 444
1,3-Diselenolan-2-ylium ions 312, 314
Diselenolates, complexes of 796, 797
Diselenoles 445–447
Diselenoleselones 464
Diselenolethiones 464
Diselenolium ions 800
Diselenolones 444
Diselenoloquinolines 446
Diselenolothiopyranones 435
Diselenolylium ions 308–312, 444, 448, 451
Diselenophosphates, complexes of 796
Diselenophosphinates, complexes of 796
Diselenothiocarbonates, complexes of 728, 796, 797
Diselenoxanthates, complexes of 796
Dispersion effect 225
Disproportionation reactions 627
Dissociative electron capture 275
Disulphides,
acyclic — *see* Acyclic disulphides
chiroptical properties of 670
Ditellurafulvenes 406, 447
Ditelluretanes 406
Ditellurides,
acyclic — *see* Acyclic ditellurides
chiroptical properties of 670–672
complexes of 814, 818, 819, 831
detection and determination of 171
mass spectra of 253–255
reactions of 818, 819, 824, 826
^{125}Te-NMR chemical shifts for 224–227
Ditelluroles 446, 447
Ditellurolylium ions 451
Ditellurones 552
Dithiaselenoloquinoxaline oxides 472
Dithiocarbamates, detection and determination of 183
1,2-Dithiolylium ions 308, 309
Dithizone complexes 779
Doebner–Miller reaction 460
Dual Substituent Parameter equations 202

Electroless plating solutions 681, 686
Electron density distribution 114
Electron diffraction techniques, in structure determinations 66, 67, 74–76, 80, 82, 83, 87, 88, 96, 106, 113–119
Electronegativity,
effect on chemical shifts 198, 211
effect on ionization potentials 245
Electronic delocalization 427
Electronic energies 22
Electronic transmission,
through chalcogen atoms 574–577
through heterocyclic systems 572–574
Electron-pair repulsions 64
Electrophilic aromatic substitutions, directing and activating effects in 581–589
Electrophilic cyclization 521
Enzymatic synthesis 379
Episelenides 402
Episelenuranes 403
Equilibrium reactions, directing and activating effects in 593
ESR spectroscopy, of Se compounds 234–238, 292, 301, 321, 323–327
N,N′-Ethylenebis(monoselenoacetylacetonimine) complexes 801
α,α′-(Ethylenedinitrilo)di-*o*-tolueneselenolate complexes 801
Exchange reactions, in ditellurides 226
Exponential-type functions (ETF) 24

Fatty acid analogues, isotopically labelled 392
Fermi contact term 220, 233, 234
Fiesselmann–Hauptmann reaction 429
Fiesselmann reaction 408, 409
Fischer indole synthesis 426, 439, 444
Fischer reaction 438
Flash photolysis 402, 403
Folate derivatives, ^{75}Se-containing 393
Free radicals, Se-stabilized 321–328
Friedel–Crafts reactions 411, 413, 428
Furans,
chiroptical properties of 674
reactions of 409

Gamma effect 201, 225
Gaussian-type orbitals (GTO) 24, 25, 28–30, 33
Gel permeation chromatography 835

Geometries,
calculated 37–42, 59
experimental 56–59
Glycerine triethers, isotopically labelled 373, 392

Hantzsch reaction 455, 457, 458
Hartree–Fock energy 22
Hartree–Fock limit 25, 31
Heats of formation 157, 158
Hemetsberger cyclization 436
Heterocyclic compounds 17
chemical shifts for 212–217, 230–232
complexes of 699–709, 814, 819, 820
isotopically labelled 376–378, 383–387, 392
mass spectra of 252, 253, 264–281
radiation chemistry of 288–292, 299–301
structural chemistry of 86–102
Hexamethylcyclotristannatelluranes 358
Hinsberg ring-closure reaction 408
HMO calculations 324
Homochromans 490
Homochromenes 490
Homoselenochromanols 490
Homoselenochromanones 489
Hot-atom chemistry 287, 303
Hybrid ligands 773, 774, 831
Hydrogenation, energies of 35, 42, 52–54
Hydrogen bonding,
effect on chemical shifts 198
in selenides 79
in selenosemicarbazones 69, 70
in selenourea 68
β-Hydroxyselenides 624–627
Hyperfine coupling 235
Hyperfine coupling constants 321, 323–325, 327

Imaging elements, thermally processed 681
Imidazoloselenatriazoles 468
Immunological investigations 393
Insertion compounds 8, 9
Insertion into a metal–alkyl bond,
of SeO_2 759, 761
of TeO_2 827
Insertion into a metal–carbon bond, of elemental Te 829
Interactions,
non-bonded 64, 72, 76, 78, 87, 111, 112, 143, 144
Se···C 106
Se···Cl 143
Se···N 106
Se···O 76
Te···C 136, 137
Te···I 136, 140, 142
Te···N 91
Te···O 76
Te···S 132
Te···Te 97, 98
Ionization energies 245–248
Ionization potentials 22, 35, 44, 54, 55
for dimethyl chalcogenides 245
Iron carbonyl compounds, mass spectra of 281
Iron complexes, geometry of 113
IR spectroscopy,
of 1,2-dichalcogenolylium ions 308, 309
of isotopically labelled compounds 382, 383
Isodesmic reactions 34, 35, 43
Isoindolobenzoselenazoles 460
Isoselenazoles — *see also* Benzoisoselenazoles 452, 453, 458
geometry of 91
mass spectra of 276, 277
Isoselenazolium ions 453, 454
Isoselenazolopyrimidines 453
Isoselenochromans 476
Isoselenochromylium ions 477
Isoselenocoumarins 476
Isoselenocyanates,
complexes of 730, 734, 735
geometry of 66
Isoselenocyanic acid, geometry of 65
Isoselenothiadiazoles, mass spectra of 279
Isoselenouronium salts, radiation chemistry of 289, 291
Isotellurazoles 454
geometry of 91
Isotellurazolium salts 454
Isotellurochromans 476
Isotellurocoumarins 476
Isothiocyanates, geometry of 66
Isothiocyanic acid, geometry of 65
Isotope effects, on ^{77}Se-NMR chemical shifts 217–220
Isotopic exchange 385, 386

Kaufmann reaction 414
Ketene selenoacetals 320
Koopman's theorem 35, 44, 54

Librational motion 64, 112
Ligand electronegativity 71, 83, 124, 133, 134, 147
Linkage isomers,
of benzoselenazole 705
of $(CF_3)_2AsTeMe$ 820
of CSSe 717, 767
of selenocyanates 737, 739, 740
of thiocyanates 815, 817
Lone electron pairs 70, 86, 89, 97, 98, 114, 125, 126, 133–135, 140–145, 147

Lone-pair/bonding-pair interactions 126, 134

Manganese carbonyl compounds,
geometry of 111, 112
mass spectra of 281
Markownikoff and *anti*-Markownikoff adducts, of selenenyl halides to olefins 633
Mass spectrometry 243–283
McLachlan calculations 324
Meisenheimer complexes 411, 590, 591
Mercury complexes 104, 105
Metalation,
of selenolothiophenes 431
of selenophenes and tellurophenes 415, 416, 425, 428
Metalation reactions, directing and activating effects in 595–597
Metal tellurides, reactions of 409
Methyl group, tilt of 70, 72, 73
Microwave spectroscopy,
in structure determinations 64–67, 70–73, 88–90, 92, 119, 128
of isotopically labelled selenophenes 383
Molecular orbital theory 24
Molecular vibrations 64, 66, 71
Molybdenum complexes 104
Monohalides 15
addition to alkenes 403
^{125}Te-NMR chemical shifts for 228, 229
Monoselenocarbenium ions 315–319
Monoselenodibenzoylmethane complexes 786
Mössbauer isomer shift values 234
Mössbauer spectroscopy 795, 818, 826
Mulliken Population Analysis 35, 45–50, 55, 56
Myocardial imaging agents 392

Naphthoselenazoles 460
Naphthoselenophenes 420, 428
Naphthotellurophenes 442
Neutron activation analysis 378
Nickel complexes, geometry of 103, 104
NMR spectroscopy,
of isotopically labelled compounds 383–385
of Se compounds 191–221, 310–313, 320, 625, 736, 789
of Te compounds 221–234, 815, 817–820, 835–837
Nomenclature of chalcogen compounds 3–5
Non-haemeiron proteins 382, 391, 392
Non-Relativistic Limit (NRL) 31
NQR spectroscopy 736
Nuclear Overhauser Enhancement 191
Nucleophilic aromatic substitutions, directing and activating effects in 589, 590
Nucleosides, chiroptical properties of 669
Nucleotides, ^{75}Se-labelled 372, 393

Octahedral configuration 138, 140, 141, 143, 144
Optical activity, general considerations of 667, 668
Organometallic complexes, structural chemistry of 103–106
Organometallic compounds,
containing elements of Groups IIIa, IVa and Va 343–365
containing transition metals 281
Ortho effects 265–267
Oxadiselenanes 462, 487
Oxaselenanes 480
complexes of 113
Oxaselenans, complexes of 706, 707
Oxaselenazines 488
Oxaselenepanes 491
Oxaselenetanes 404
Oxaselenoles 447
Oxaselenolones 446
Oxaselenolylium ions 444, 445
Oxatellurolylium ions 444, 475
Oxathiane complexes 113
Oxidative addition reactions,
of H_2Se 754, 755
of Se complexes 807–809
of Se ligands 738–740, 748–750, 753, 758, 798
of Te complexes 834, 835
of Te ligands 818, 819, 824, 826
Oxygen atom transfer reactions 257, 258
Oxyhalides, reactions of 527, 528
Oxyselenation 626, 641
Oxyselenenylation 640

Paal–Knorr synthesis 407, 433
Pancreas-imaging agents 392
Pentagonal bipyramidal configuration 145
Pentahalides, reactions of 405
Pentalenes — *see also* Selenapentalenes, Tellurapentalenes
geometry of 116–119
Pentaselenecanes 492
Peptides, Se analogues of 78
Perselenanes 517, 518
Pertelluranes 517, 518, 544–546
pH effects, on ^{77}Se-NMR chemical shifts 195
Phenanthroselenadiazoles 464
Phenetole rearrangement 250
Phenoselenazines 486, 487
Phenotellurazines 487
Phenothiatellurins 483
Phenoxaselenins 482, 483

Phenoxatellurins 482, 483
  geometry of 141, 142
Phenyl migration, in mass spectral fragmentations 257
α-Phenylselenenyl carbonyl compounds 629, 632
Phenylselenium compounds, detection and determination of 167
Phenyltellurium compounds,
  detection and determination of 167
  geometry of 119–123, 132, 133, 144, 145, 147
Phosphidoselenides, complexes of 699
Phosphines, geometry of 66
Phosphine selenides 349, 360
  reactions of 402
Phosphine tellurides 360
Phosphorotellurates 402
Phosphorus/selenium ligands 727
Phosphoselenothioates 360
Photoelectron spectroscopy 245
Phthalide analogues, mass spectra of 276
Piazselenols 164, 177
  detection and determination of 183
Platinum complexes, geometry of 105, 113, 114
Polarizability 201
Polarography, in determination of Se and Te compounds 163, 164
Poly(diselenoalkylene)s 336–338
Polymers,
  O- and Se-containing 333, 334
  Se-containing 332–340
  Te-containing 340, 341
Polyselenoalkylenes 332, 333
Poly-Wurtz reaction 416
Pomeranz–Fritsch cyclization 438
Proton affinities 43, 45, 52
Protophilic reactions, directing and activating effects in 594, 595
Pschorr reaction 428, 439
Pulse radiolysis 292, 302
  of selenoamino acids 298, 299
  of selenourea 294
Pummerer rearrangement 316–318, 427
Pyrroloquinoxalines 426
Pyrroloselenazoles 460

Quadruple average angle 126, 129, 140
Quasiequilibrium theory 248
Quinoxalines 426

Radiation chemistry,
  of Se compounds 287–302
  of Te compounds 302, 303
Radiation damage 287
Radical anions, ESR spectroscopy of 237, 321, 323, 324
Radical cations, ESR spectroscopy of 235, 237, 238, 321, 323–325
Radical scavenging 536
Radioactive equilibrium 374
Radioprotectors 287
Radiosensitizers 288
Raman optical activity 676
Raman spectroscopy, of isotopically labelled chalcogen compounds 382, 383
Reductive elimination, in mass spectral fragmentations 251, 260, 268
Relativistic energy 22
Relaxation times,
  for Se compounds 191–193, 234
  for Te compounds 221
Retro-Diels–Alder reactions 268
Rieche formylation 424
Ring-closure reactions 402–410, 419–494
Ring-contraction, of benzochalcogen-azines 459
Ring-opening,
  by selenols 623, 624
  of isoselenazoles 453
  of selenadiazoles 464, 465, 467
  of selenetanes 404, 406
  of selenolothiophenes 431
  of selenophenes 416–418, 425
  of tellurophenes 442, 443
Ring-opening–ring-closure reactions 420
Ring-puckering 86, 87, 96, 138
t-RNA, Se-containing 391
Robinson–Stork annelation 429
Rotational barriers,
  in acyclic selenides 72, 73
  in bis(seleno)carbenium ions 320
$RSeCH_2SR$ ligands 727
Ruthenium complexes 105

Sandmeyer reaction 459
Sandmeyer–Schiemann reaction 456
Scintiscanning 392, 393
$Se(CN)_2$ complexes 726, 728
$Se{=}C(SeEt)_2$ complexes 726
$Se(EMe_3)_2$ (E = Ge,Sn,Pb) ligands 714, 716
$SeER_3^-$ (E = Ge, R = Ph; E = Sn, R = Me) ligands 758, 759
SeH^- ligand 754–758
Selenaadamantanediones 673
Selenaazapentalenes 467, 468
Selenabenzenes 477
  mass spectra of 278
Selena-bridged heterocycles 493, 494
Selenabutanolides, geometry of 128
Selenacyclohexanes 363, 468, 518, 546, 547
  geometry of 96, 128, 138, 139

isotopically labelled 376, 384, 385
mass spectra of 252
polarographic reduction of 184
Selenacyclopentanes 363
isotopically labelled 376
mass spectra of 252, 255
polarographic reduction of 184
Selenadiazepines 491
Selenadiazines 488, 489
Selenadiazoles — *see also* Benzoselenadiazoles, Benzoselenoloselenadiazoles, Phenanthroselenadiazoles, Selenoloselenadiazoles
directing and activating effects in 581, 584, 595
geometry of 90
isotopically labelled 377, 378, 386, 392
mass spectra of 274, 275
polarographic reduction of 184
reactions of 403, 405, 409, 447, 449, 462, 464–467, 481
synthesis of 463–466
Selenadiazolidinediones 465
Selenadiazolines 467
Selenadiazolopyrimidinones 466, 467
Selenadiborolenes 347
Selenadigermanes 488, 489
Selenadiphosphoranes 362
Selenadisilanes 489
Selenaferrole complexes 703
Selenaferrole ring 114
Selenafulvalenes 449, 450
coupling constants for 221
geometry of 93, 94
mass spectra of 269, 270
Selenahydrindanes, chiroptical properties of 672, 673
Selenals 11, 12
geometry of 67
Selenametacyclophanes 492, 493
Selenanes (conventional nomenclature) — *see* Selenacyclohexanes
Selenanes (Musher's nomenclature) — *see also* Selenuranes 517–519, 537, 538
Selenanols 469
Selenanones 469
mass spectra of 253
Selenanthrenes 483, 484
geometry of 99–101
Selenaoxazines 542
Selenapentalenes 445, 449
geometry of 116–119
mass spectra of 271, 272
radicals produced from 325
^{77}Se-NMR chemical shifts for 216
Selenaphthalides, geometry of 128
Selenapurines 465
Selenapyrylium ions 523
Selenates, radiation chemistry of 289, 291
Selenathietans 530
Selenation 68
of a coordinated phosphido ligand 699
Selenatriazoles 468
Selenazaphospholines, geometry of 90
Selenazepines 454, 491
Selenazine carboxylates, mass spectra of 278, 279
Selenazines 459, 484–488, 492
complexes of 708, 709
Selenazinium dyes 315
Selenazinium ions 485
Selenaziridines 404
Selenazoles — *see also* Aminoselenazoles, Benzoselenazoles, Benzothiepinoselenazoles, Benzoxepinoselenazoles, Naphthoselenazoles, Pyrroloselenazoles, Thienoselenazoles 413, 452, 455, 456, 458–461, 486
directing and activating effects in 583
mass spectra of 272–277, 279, 280
radiation chemistry of 290, 291
Selenazolidines 456–458
Selenazolines 456, 457, 460
Selenazoloazepines 461
Selenazolopyrimidines 461
Selenazolotriazines 461
Selenenamides 640
Selenenic acid anhydrides,
generation of 624–626
reactions of 626–628
Selenenic acids 14, 599
generation of 624–626
electrochemical 627
reactions of 626–628
Selenenic esters 639, 640
Selenenoyl radicals 237
Selenenyl compounds, detection and determination of 166
Selenenyl halides,
geometry of 70, 75–77
reactions of 403, 421, 428, 626, 629–639
with alkenes 632–635
with alkynes 635
with allenes 635
with carbonyl compounds 629–632
with nucleophiles 639
^{77}Se-NMR chemical shifts for 211, 212
synthesis of 629
Selenenyl sulphides, ^{77}Se-NMR chemical shifts for 206, 207
Selenepanes 489, 491
Selenepins 489–491
Selenetanes 404, 405, 444, 469, 525
geometry of 86, 87

Selenetanols 404
Selenic acid 13
Selenides 6, 7
acyclic — *see* Acyclic selenides
alkyl aryl — *see* Alkyl aryl selenides
allyl — *see* Allyl selenides
arsine — *see* Arsine selenides
bis-phosphine — *see* Bis-phosphine selenides
chiroptical properties of 672–675
detection and determination of 167, 169, 170, 177, 183
dialkyl — *see* Dialkyl selenides
diaryl — *see* Diaryl selenides
directing and activating effects in 603–609
phosphine — *see* Phosphine selenides
triorganophosphine — *see* Triorganophosphine selenides
vinyl — *see* Vinyl selenides
Selenienyl compounds 414, 415
Selenienylethyl acetates 572, 593, 594
Selenienylpropenoic acids 430
Selenimides 541
Seleninates 548
complexes of 759–762
Seleninic acid anhydrides,
as oxidizing agents 643–648
deprotection by 648, 649
epoxidation of olefins using 645, 646
α-hydroxylation of ketones by 644, 645
synthesis of 643
Seleninic acids 14
as radiolytic products 296
detection and determination of 166, 184
dissociation constants of 599–601
mass spectra of 262, 263
properties of 642
reactions of 626, 643–649
substituent effects in 580
synthesis of 643
Seleninic esters 548
Seleninium ions 470
Seleninopyrazoles 472
Seleninopyridines 472
Selenins 464, 469, 476, 478, 481–483
Seleninyl halides, geometry of 106, 128
Seleninyl radicals 237
Seleniranes 402, 404
Seleniranium ions 403, 632, 633
Seleniren ions 403
Selenirenium ions 403, 424
Selenites 183
cyclic organic 462
geometry of 106
isotopically labelled 387–391, 393
radiation chemistry of 289, 291
Selenium,
colloidal, radiation chemistry of 288, 290–292, 296
coordinated,
alkylation of 691
protonation of 757
elemental,
as radiolytic product 296, 297
radioactive 370, 371
reactions of 406, 407, 419, 428
five-coordinated 143
four-coordinated 125–130, 139, 140, 517–542
metabolism of 387–389
one-coordinated 64–70
red form of 292
six-coordinated 546–550
three-coordinated 106–121, 124, 125
two-coordinated 70–106
Selenium bond angles 70, 72–77, 79, 85, 86, 88–96, 99–102, 107, 108, 113–116, 127–129
Selenium–carbon bond lengths 64–67, 69, 70, 75–83, 85–96, 99–103, 107, 108, 111, 115–118, 121, 124, 128, 139, 320, 321
Selenium–carbon double bonds, ^{77}Se-NMR chemical shifts for compounds containing 208, 209
Selenium extrusion, in mass spectral fragmentations 249
Selenium ligands,
bidentate,
anionic 778–801
neutral 765–778
incorporating Se_n 803–814
monodentate,
anionic 730–765
neutral 683–730
tetradentate 802, 803
tridentate 801, 802
Selenium–oxygen compounds, ^{77}Se-NMR chemical shifts for 212, 217, 218
Selenium–selenium bond lengths 83–86, 95, 103, 104, 118
Selenoacetals 346
ketene — *see* Ketene selenoacetals
Seleno alcohols 348
Selenoalkanoic acids, ^{77}Se-NMR chemical shifts for 198, 200
Selenoalkyl radicals 235, 236
Selenoallylcarbenium ions 321, 322
Selenoallyl cations 321, 322
Selenoaluminates 349
Selenoamides 13, 355
Selenoamines, geometry of 80
Selenoamino acids 457
chiroptical properties of 671, 673
detection and determination of 171, 177–182, 184

ESR spectroscopy of 238
isotopically labelled 378–382, 388–390, 392, 393
radiation chemistry of 288–291, 293, 296–299
Selenoanisoles,
geometry of 70, 74, 75
^{77}Se-NMR chemical shifts for 202, 203
Selenoarsines 362
Selenoaryl esters of amino acids, chiroptical properties of 673
Selenoaurones 421, 429, 440, 474
Selenoazolidines, radiation chemistry of 292
Selenoazolineselones 456
Selenobenzamides 454
Selenobenzofurans, mass spectra of 268
Selenobenzoic acids 477
Selenobenzoyl halides 454
Selenobenzyl halides 477
Selenobenzyl ketones, cyclodehydration of 422
Selenoboranes 344, 345
geometry of 70, 81, 82
Selenocarbamates 356, 448
complexes of 762–764, 784, 785
structure of 103–105
Selenocarbazides, radiation chemistry of 290, 291
Selenocarbazones, complexes of 778, 779
Selenocarbenes 719
Selenocarbenium ions,
aliphatic 315–321
cyclic 308–315
Selenocarbonyl complexes 715–721
Selenocarbonyl compounds, coupling constants for 220
Selenocarboxylates 639, 640
Selenocarboxylic acids 12
Selenocarboxylic esters 12
Selenochromanols 471
Selenochromanones 470, 471, 479
Selenochromans 472
Selenochromenes 423, 470–473
Selenochromones 423, 473–475
radiation chemistry of 288, 291
Selenochromylium ions 315, 423, 472
directing and activating effects in 592
equilibrium constants for 598
Selenocinnamic acids 419, 474
Selenocoumarines 423, 473
Selenocoumarins 475, 476
Selenocyanate complexes 103, 116, 697, 730–734, 736–740
Selenocyanates,
detection and determination of 166, 167
geometry of 70, 73, 74, 80, 86, 120
mass spectra of 249, 262, 263
radiation chemistry of 291–293, 302
reactions of 402, 404
^{77}Se-NMR chemical shifts for 209–211
Selenocyanides 641, 642
Selenocyanine dyes 315
Selenocyclopropenylium salts 314, 315
Selenocysteamines 782
Selenodialkanoic acids, mass spectra of 249
Selenodihydroarsanthrenes 364
Selenodithiocarbamates, complexes of 791, 792
Selenodithiocarbonate anion, reactions of 402
Seleno esters 622
^{77}Se-NMR chemical shifts for 207, 208
Seleno ethers,
complexes of 680, 691
reactions of 444
Selenoflavones 472–474
Selenoformaldehyde complexes 768–770
α-Seleno free radicals 321–328
addition to olefins 326–328
conformational effects in 326
Selenogermanes 349–351, 354, 356
Selenoindigoid dyes 421
Selenoindoxyls 421, 474, 475
Selenoisocyanates, ^{77}Se-NMR chemical shifts for 210, 219
Selenoisothiocyanates, ^{77}Se-NMR chemical shifts for 210, 219
Selenoketenes 12, 403, 405, 464
geometry of 65
Selenoketocarbene complexes 703
Selenoketocarbene moiety 113
Selenoketones,
chiroptical properties of 675
complexes of 726, 729
reactions of 408
synthesis of 628, 634
Selenolactones 433
chiroptical properties of 673
Selenolactonization 422
Selenolanes 444
Selenolates 348, 409, 410, 621, 623
^{77}Se-NMR chemical shifts for 196–198
Selenolatoethylamine complexes 780
Selenol complexes,
bridging bonding in 743–748, 750–754
structure of 103–105
terminal bonding in 741, 742, 749–751
Selenol esters 348
Selenolobenzofurans, mass spectra of 279
Selenolobenzothiophenes 431, 432
Selenolofurans 429, 430
Selenoloindoles 438, 439, 453
Selenoloisoselenazoles 437, 438
Selenoloisothiazoles 437

Selenoloisoxazoles 437
Selenolooxazoles 438
Selenolopyrans 435
Selenolopyrazoles 437, 439
Selenolopyridazines 438
Selenolopyridines,
  mass spectra of 266, 267
  synthesis of 438, 439
Selenolopyrimidines 438
Selenolopyrroles 436, 438, 439
Selenolopyrylium ions 420, 435, 438
Selenoloquinolines 439
Selenoloselenadiazoles 438, 464
Selenoloselenophenes 433–435
Selenolothiadiazoles, mass spectra of 278
Selenolothiazinones 437
Selenolothiazoles,
  mass spectra of 280
  synthesis of 437, 438
Selenolothiophenes 430–433
Selenolothiopyranones 435
Selenols 5, 6, 355, 356
  detection and determination of 165, 166, 184
  dissociation constants of 600
  geometry of 70, 73, 74
  isotopically labelled 371, 372, 382
  properties of 620, 621
  reactions of 622–625
  ^{77}Se-NMR chemical shifts for 196–198
  synthesis of 621, 622
Selenomalonates 475
Selenomethionines, mass spectra of 249
Selenomorpholines 485
Selenones 9, 10, 183, 546–550
  complexes of 707, 708
  directing and activating effects in 609–611
  mass spectra of 258–260
  radiation chemistry of 301
Selenone–seleninate rearrangement 259, 283
Selenonic acids 14, 547
  detection and determination of 166
  directing and activating effects in 599
  reactions of 650
  synthesis of 649, 650
Selenonium ions 7, 8
  chiroptical properties of 668
  detection and determination of 167, 171, 176
  directing and activating effects in 601, 602
  geometry of 106, 107
  radiation chemistry of 301
  ^{77}Se-NMR chemical shifts for 217, 219
Selenonium ylides 418, 602, 603
Selenophenes — *see also* Benzoselenophenes, Bis-selenophenes, Dibenzoselenophenes, Naphthoselenophenes, Selenoloselenophenes, Tellurolobenzoselenophenes, Telluroloselenophenes 17
  chiroptical properties of 674, 675
  detection and determination of 177, 183, 184
  directing and activating effects in 581, 582, 587–589, 591, 594–597, 599
  geometry of 88, 89, 92, 93, 128–130
  ground-state aromaticity of 410
  hydrogenated 418, 419
  isotopically labelled 377, 383, 385
  mass spectra of 264–269, 279
  metalated 414–418
  photolysis of 265, 386
  reactions of 410–414
  ^{77}Se-NMR chemical shifts for 212–215
  substituent electronic effects in 572–574, 578, 579
  synthesis of 407–409, 464
  tautomerism of 246
Selenophenols 419, 475
  radiation chemistry of 288, 291
Selenophenopyridines 586
Selenophenothiophenes 587, 588, 597
Selenophosphates 359
  complexes of 785, 786
Selenophosphinates, complexes of 785
Selenophosphinic acids 676
Selenophosphoranes 463
Selenophthalic anhydrides 427
Selenophthalides 427
Selenopropionic acids 471
Selenoproteins 381, 382, 389, 390
Selenopyranones 471
Selenopyrans 469–471
Selenopyrones 469
Selenopyrylium ions 266, 315, 445, 470
Selenosalicylideneamine chelates 781
Selenosemicarbazides 13
  complexes of 770
Selenosemicarbazones 455
  complexes of 726, 727
  radiation chemistry of 290, 291
Selenosilanes 349–352, 354, 355
Selenostannanes 349, 350, 356–358
Selenosultones 482
Selenothiocarbamates,
  complexes of 765, 789, 790
  mass spectra of 264
Selenothio complexes 786–793
Selenothiophosphates, complexes of 789
Selenothiophosphinates, complexes of 789
Selenoureas 13, 448
  complexes of 683–689, 786, 792
    structure of 69, 70, 103, 104, 115
  detection and determination of 107, 177, 184

geometry of 67–70, 84, 85, 119, 120
isotopically labelled 382
mass spectra of 264
radiation chemistry of 288–295, 297
Selenous acids,
as radiolytic products 297
isotopically labelled 371, 374
Selenoxanthates, complexes of 786, 787
Selenoxanthenes 478, 489
radiation chemistry of 288, 291, 299–301
Selenoxanthenols 478
Selenoxanthones 477, 478
radiation chemistry of 288, 291
Selenoxanthylium dyes 315
Selenoxanthylium salts 478
directing and activating effects in 588, 589
equilibrium constants for 598
Selenoxide elimination 631
Selenoxides 9, 10, 536, 537
chiroptical properties of 668–670
complexes of 721, 722
detection and determination of 177
directing and activating effects in 609, 610
geometry of 127
mass spectra of 255–257
reactions of 421, 427, 428, 624, 625
synthesis of 419
Selenoxide–selenate rearrangement 256, 257, 262, 282, 639
Selenoxobenzothiazoles 402
2-Selenoxo-1,3-diseleno-4,5-diselenates, complexes of 798
Selenoxodiselenoles 448
Selenoxodithioles, mass spectra of 280
Selenoxoselenazolidinones 456
Selenuranes — *see also* Selenanes (Musher's nomenclature) 403, 566, 567
geometry of 107, 108
Selenuranyl-type radicals 325
Selones 11, 12, 403
Se(Ph)(SPh) complexes 726
$Se(SR)_4$ 539
Side-chain reactions, directing and activating effects in 593, 594
Sigma constants 319
Silanes, geometry of 66
Simonis synthesis 473
Skraup synthesis 426
Slater determinant 31
Slater's rules 25
Slater-type orbitals (STO) 24–28
Solvent effects, on ^{77}Se-NMR chemical shifts 194, 195
Spin densities 321
Spin-orbit coupling constants 324
Spin-orbit line broadening mechanisms 326
Spin-rotation mechanism 326
Spin-rotation relaxation 193, 238
Spiroselenium heterocycles 493
Square pyramidal configuration 132
Steroids, isotopically labelled 372–375
Structural chemistry,
of five- and higher-coordinated Se and Te 143–147
of four-coordinated Se and Te 125–143
of one-coordinated Se and Te 64–70
of three-coordinated Se and Te 106–125
of two-coordinated Se and Te 70–106
Substituent Chemical Shift values 202
Substituent constants 560–563
for chalcogenazidinium salts 569, 570
for heterocyclic groups 567, 568
for Se(II)- and Te(II)-containing groups 563–566
for Se(IV)- and Te(IV)-containing groups 566, 567
hydrophobic 570
replacement 568, 569
Substituent effects,
by NMR spectroscopy 579, 580
in aromatic electrophilic and nucleophilic reactions 577
Sugar diselenides, chiroptical properties of 672
Sulphenyl halides, geometry of 75, 76
Sulphides 6, 7
Sulphonium salts 7, 8
Sulphoxides 9
Sulphur bond angles 72, 76, 83, 85, 89–93, 95, 96, 101, 113, 126
Sulphur–carbon bond lengths 64–67, 76, 78, 81–83, 85, 89, 90, 92–96, 101, 103, 117
Sulphur–sulphur bond lengths 83–85, 95, 117
Sulphur tetrafluorides, geometry of 126

$Te(ArS)_2$ 540
$Te(EMe_3)_2$ (E = Ge,Sn,Pb) ligands 814–820
Tellones 11, 12
Tellura-bridged heterocycles 494
Telluracycloheptanes 491
Telluracyclohexanes 468, 469, 526
geometry of 97, 98, 132, 138, 139
isotopically labelled 377, 384
Telluracyclopentanes 519
geometry of 140
Telluradiazoles 467
directing and activating effects in 584
Tellurafulvalenes 450, 451
Tellurals 11, 12
Telluranes (conventional nomenclature) — *see* Telluracyclohexanes
Telluranes (Musher's nomenclature) 517–519, 537, 539, 540, 566, 567
Telluranthrenes 484

geometry of 99–101
Tellurapentalenes 445
mass spectra of 272
Tellurasilanes 480
Tellurathiaporphyrines 409
Tellurazines 487
complexes of 820
Tellurazoles 461
directing and activating effects in 583
Tellurenic acid derivatives 14
Tellurenyl halides 440, 651, 652
Tellurenyl tellurinyl selenides, ^{125}Te-NMR chemical shifts for 232, 233
Tellurepins 491
Telluric acid 14
Tellurides 6, 7
acyclic — *see* Acyclic tellurides
alkyl aryl — *see* Alkyl aryl tellurides
chiroptical properties of 672–675
detection and determination of 167, 171–173, 183
dialkyl — *see* Dialkyl tellurides
diaryl — *see* Diaryl tellurides
directing and activating effects in 609
metal — *see* Metal tellurides
phosphine — *see* Phosphine tellurides
^{125}Te-NMR chemical shifts for 222–225, 233
Tellurienylethyl acetates 573, 593, 594
Tellurimides 540, 541
Tellurinic acids 14, 655
Tellurins 475, 482, 483, 489
Tellurinyl trihalides 651, 653, 654
Telluriranes 402, 403
Tellurium,
elemental,
as radiolytic product 303
radioactive 371
five-coordinated 107, 144–146, 534, 535, 542–544
four-coordinated 125, 129–143, 517–542
metabolism of 389
one-coordinated 64
seven-coordinated 145, 147
six-coordinated 107, 145, 146, 517, 518, 540, 544–546, 550–552
three-coordinated 106, 107, 114, 119–124
two-coordinated 72, 77, 78, 86, 89–93, 95, 97–101, 105, 106
Tellurium bond angles 72, 77, 89–91, 93, 95, 97, 101, 107, 114, 131–135, 138, 142, 143
Tellurium–carbon bond lengths 64, 72, 77, 78, 86, 89–91, 93, 95, 97, 101, 106, 107, 114, 120, 121, 124, 131, 139, 142–145, 147
Tellurium ligands,
bidentate 828–831
incorporating Te_n 831–839
monodentate,
anionic 824–828
neutral 814–824
Tellurium–tellurium bond lengths 86, 95
Tellurium tetrahalides 403
Telluroalkanoic acids, isotopically labelled 375
Telluroamides 13
Telluroamino acids, isotopically labelled 381, 392
Telluroarsines 362
Telluroaurones 440
Tellurocarbonyl complexes 814, 821, 822
Tellurocarboxylic acids 13
Tellurocarboxylic esters 13
Tellurochromanols 471
Tellurochromanones 470, 471, 474
Tellurochromans 472
Tellurochromenes 470, 471, 473
Tellurochromones 474, 475
Tellurocoumarins 476
Tellurocyanates 13
reactions of 823
Tellurocyanides 652
Tellurocyclopropenylium salts 315
Tellurodesulphonation 442
Tellurodihydroarsanthrenes 364
Telluro ethers, complexes of 680
Telluroflavones 475
Telluroformaldehyde complexes 827, 829, 830
Tellurogermanes 353, 356
Tellurohydrazides 13
Telluroindigoid dyes 441
Telluroindoxyls 440, 443, 444
Telluroketenes 12, 406, 462
complexes of 829, 830
Telluroketones, complexes of 814, 823
Tellurolates 353, 650, 651
reactions of 409
Tellurol ligands 824–828
Tellurolobenzofurans 443
Tellurolobenzoselenophenes 443
Tellurolofurans 443
Tellurolopyridazines 443
Telluroloquinolines 443
Telluroloselenophenes 443
Tellurolothiepins 443
Tellurolothiophenes 443
Tellurols 5, 6
detection and determination of 165, 166
properties of 650
Telluromethylidyne complexes 822
Tellurones 9, 10, 550–552
Telluronium ions 7, 8
chiroptical properties of 668

detection and determination of 167, 171, 176
geometry of 107, 114
Telluronium ylides 602, 603
Telluronyl compounds 656
Tellurophenes — *see also* Benzotellurophenes, Dibenzotellurophenes, Naphthotellurophenes 17, 528
complexes of 820
detection and determination of 177
directing and activating effects in 581, 582, 593, 595–597
geometry of 89, 92, 93
ground-state aromaticity of 410
hydrogenated 418, 419
isotopically labelled 378, 383, 385
mass spectra of 264, 265
metalated 414–418
photolysis of 386, 387
reactions of 410–414
substituent electronic effects in 572–574, 580
synthesis of 409, 469, 484
^{125}Te-NMR chemical shifts for 230, 232
Tellurophosphinous anhydrides 361
Tellurophthalic anhydrides 442
Telluropyrones 469
Telluropyrylium ions 470, 480
Tellurosilanes 353–355
Tellurostannanes 353
Telluroureas 13
complexes of 814, 821
structure of 106
Telluroxanthenes 479, 480
complexes of 820
Telluroxanthones 479
Telluroxides 9, 10, 419, 537
complexes of 814, 824
Temperature effects, on NMR spectra 195, 320
Template reactions 781
$Te(SR)_4$ 539
Tetracenes, geometry of 95
Tetrahalides,
geometry of 129, 130
radiation chemistry of 303
reactions of 405, 409, 423, 519, 520, 522, 533–535, 540, 544, 653
with alkenes 526–531
with alkynes 532
with carbonyl compounds 523–526
Tetrahaloorganyltellurates 542–544
Tetraorganyltellurium compounds 534–536
Tetraselenocanes 492
Tetraselenooxalates 766
Tetratellurins 489
Tetratellurofulvalenes 651
Thiadiazoles 278, 279, 426
Thials 11, 12
geometry of 67
Thianes,
geometry of 96, 138, 139
mass spectra of 252
NMR spectra of 384, 385
Thianones, mass spectra of 253
Thianthrenes, geometry of 99–101
Thiaselenafulvenes 448
Thiaselenanes 480
Thiaselenazines 488
Thiaselenetanes 404
Thiaselenins 481
1,3-Thiaselenolan-2-ylium ions 312, 314
Thiaselenoles 444, 445, 453
mass spectra of 278, 280, 281
Thiaselenolethiones 448, 449
Thiaselenolylium ions 451
Thiatelluranes 481
Thienoselenazoles 452, 460
Thienoselenoxoselenoles 449
Thietanes, geometry of 87, 88
Thiiranes, desulphurization of 402
Thiiranium salts 403
Thiirenium salts 403
Thioamines, geometry of 82
Thioanisoles, geometry of 74, 75
Thiocyanates,
geometry of 86, 120
rotational barriers in 73
Thioketenes 12
geometry of 65
Thiols 5, 6
Thiones 11, 12
Thiophenes,
chiroptical properties of 674
geometry of 88, 89, 92, 93
synthesis of 408, 409
Thioureas, geometry of 119–123
Topomerization 518
Trans effect 103
Triarylselenium halides 520, 521
Triaryltellurium halides 520, 521
Triaryltellurium pseudohalides 520
1,1,1-Trifluoro-3-(2-thienyl)-4-seleno-3-butene-2-one complexes 786
Trigonal bipyramidal configuration 125, 129, 132, 135, 142, 147
Trihalides 15
detection and determination of 166, 171
geometry of 143, 144
reactions of 403, 523, 526, 527, 529, 530, 533, 654
synthesis of 521, 522, 653
^{125}Te-NMR chemical shifts for 228
Triorganophosphine selenides, complexes

of 891–899
Triorganyltellurium compounds 655
Triselenadiborolanes 345
Triselenanes 487
Triselenocarbonates, complexes of 728, 796, 797
Triselenocyanate anion 116
Triselenolanes 462
Triselenoles 462
Tris(*o*-methylselenophenyl)phosphine complexes 801, 803
Tris(selenourea) cation 115
Tritelluranes 488
Tritelluroles 462

Ullmann–Fetiadjan cyclization 426
Ullmann reaction 414
Unimolecular dissociation 247
α,β-Unsaturated carbonyl compounds,
reactions of 632
synthesis of 629–631
UV spectroscopy, of 1,2-dichalcogenolylium ions 308, 309

Valence-shell electron-pair repulsion (VSEPR) model 63, 67, 71, 129, 134, 147
Van der Waals' radii 78
Vaska's complex 718
Vibrational amplitudes 140
Vilsmeier–Haack formylation 410, 411, 413, 424, 431, 434, 437
Vinyl selenides 319, 635

XPS spectroscopy 736, 795
X-ray diffraction crystallography, in structure determinations 67–70, 75–82, 84–86, 90–98, 107–124, 127, 129–147, 320

Ylide resonance structures 107
Ylides — *see also* Selenonium ylides, Telluronium ylides 8
mass spectra of 261

Zinc complexes 103, 105
Zwitterions 544

I C [redacted]

24 0119723 3

Lyon Playfair Library

IMPERIAL COLLEGE, LONDON SW7 2AZ

To be returned by last date stamped.
The loan period will be reduced to THREE WEEKS if the book is reserved by another reader.

24 JUN 1988	19 OCT 1992	26 APR 1995
-7 OCT 1988	-8 OCT 1993	30 SEP 1998
11 JAN 1989	14 JAN 1994	-5 OCT 2000
28 APR 1989	27 APR 1994	11/11/2001
30 JUN 1989	30 SEP 1994	11 JAN 2002
12 JAN 1990	26 JUN 1998	21 JUN 2002
27 APR 1990		-7 OCT 2005
26 APR 1991		
4 OCT 1991		

BNA
14.03.88
JM/4

The chemistry of

organic selenium and tellurium compounds

Volume 2

THE CHEMISTRY OF FUNCTIONAL GROUPS

A series of advanced treatises under the general editorship of
Professor Saul Patai

The chemistry of alkenes (2 volumes)
The chemistry of the carbonyl group (2 volumes)
The chemistry of the ether linkage
The chemistry of the amino group
The chemistry of the nitro and nitroso groups (2 parts)
The chemistry of carboxylic acids and esters
The chemistry of the carbon–nitrogen double bond
The chemistry of amides
The chemistry of the cyano group
The chemistry of the hydroxyl group (2 parts)
The chemistry of the azido group
The chemistry of acyl halides
The chemistry of the carbon–halogen bond (2 parts)
The chemistry of the quinonoid compounds (2 parts)
The chemistry of the thiol group (2 parts)
The chemistry of the hydrazo, azo and azoxy groups (2 parts)
The chemistry of amidines and imidates
The chemistry of cyanates and their thio derivatives (2 parts)
The chemistry of diazonium and diazo groups (2 parts)
The chemistry of the carbon–carbon triple bond (2 parts)
The chemistry of ketenes, allenes and related compounds (2 parts)
The chemistry of the sulphonium group (2 parts)
Supplement A: The chemistry of double-bonded functional groups (2 parts)
Supplement B: The chemistry of acid derivatives (2 parts)
Supplement C: The chemistry of triple-bonded functional groups (2 parts)
Supplement D: The chemistry of halides, pseudo-halides and azides (2 parts)
Supplement E: The chemistry of ethers, crown ethers, hydroxyl groups and their sulphur analogues (2 parts)
Supplement F: The chemistry of amino, nitroso and nitro compounds and their derivatives (2 parts)
The chemistry of the metal–carbon bond (4 volumes)
The chemistry of peroxides
The chemistry of organic Se and Te compounds Volume 1
The chemistry of the cyclopropyl group (2 parts)

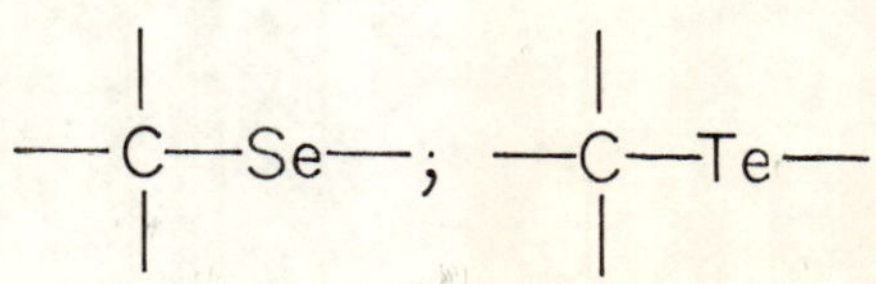

The chemistry of
organic selenium and tellurium compounds
Volume 2

Edited by

SAUL PATAI

The Hebrew University, Jerusalem

1987

JOHN WILEY & SONS

CHICHESTER – NEW YORK – BRISBANE – TORONTO – SINGAPORE

An Interscience ® Publication

547.269 CHE

Copyright © 1987 by John Wiley & Sons Ltd.

All rights reserved.

No part of this book may be reproduced by any means, or transmitted, or translated into a machine language without the written permission of the publisher.

Library of Congress Cataloging-in-Publication Data:

(Revised for vol. 2)

The Chemistry of organic selenium and tellurium compounds.

(The Chemistry of functional groups)
'An Interscience publication.'
Includes indexes.
1. Organoselenium compounds. 2. Organotellurium compounds. I. Patai, Saul. II. Rappoport, Zvi. III. Series.
QD412.S5C53 1986 547'.0572 85-17868

ISBN 0 471 90425 2

British Library Cataloguing in Publication Data:

The Chemistry of organic selenium and tellurium compounds.—(The Chemistry of functional groups)
Vol. 2
1. Organoselenium compounds
2. Organotellurium compounds
I. Patai, Saul II. Rappoport, Zvi.
III. Series
547'.05724 QD412.S5

ISBN 0 471 91020 1

Printed and bound in Great Britain
at The Bath Press, Avon

Contributing Authors

F. J. Berry	Department of Chemistry, University of Birmingham, P.O. Box 363, Birmingham B1S 2TT, UK
T. G. Back	Faculty of Science, Department of Chemistry, University of Calgary, 2500 University Drive N.W., Calgary, Alberta, Canada T2N 1N4
C. Carletti	Universita' degli Studi di Roma, 'La Sapienza', Dipartimento di Chimica, Piazzale Aldo Moro 5, 00185 Roma, Italy
L. Castle	Department of Chemistry, Royal Holloway and Bedford New College, University of London, Egham TW20 OEX, UK
L. Christiaens	Laboratoire de Chimie Organique, University of Metz, Faculte des Sciences, Ile du Saulcy 57045 – Metz Cedex, France
L. Christiaens	Department of Organic Chemistry, State University of Liège, Place du 20-Aõut 7, 4000-Liège, Belgium
D. O. Cowan	Department of Chemistry, The Johns Hopkins University, Remsen Hall – Dunning Hall, Baltimore, Maryland 21218, USA
G. Distefano	C.N.R. Laboratorio dei composti del Carbonio contenenti Eteroatomi, Via Tolara di Sotto 89, 400064 Ozzano Emilia (BO), Italy
N. Esaki	Institute for Chemical Research, Kyoto University, Uji, Kyoto-Fu 611, Japan
Z. Goldschmidt	Department of Chemistry, Bar Ilan University, Ramat Gan 52100, Israel
F. Guziec	Department of Chemistry, New Mexico State University, Box 3C/Las Cruces, New Mexico 88003, USA
L. Henriksen	Department of General and Organic Chemistry, The H. C. Ørsted Institute, DK-2100 Copenhagen, Denmark
R. Hensel	Fachbereich Chemie, Universität Oldenburg, Carl-von-Ossietzky Str. 9–11, D-2900 Oldenburg, FRG
A. Kini	Department of Chemistry, The Johns Hopkins University, Remsen Hall – Dunning Hall, Baltimore, Maryland 21218, USA
G. Kirsch	Laboratoire de Carbochimie et de Synthèse Organique, Faculté des Sciences, Université de Metz, Ile du Saulcy, 57045-Metz, France
A. Krief	Department of Chemistry, University of Namur, Rue de Bruxelles 61, B-5000 Namur, Belgium

S. Kubiniok	Fachbereich Chemie, Universität Oldenburg, Carl-von-Ossietzky Str. 9–11, D-2900 Oldenburg, FRG
L. Lange	Fachbereich Chemie, Universität Oldenburg, Carl-von-Ossietzky Str. 9–11, D-2900 Oldenburg, FRG
T. Masukawa	Department of Pharmacology, Faculty of Pharmacy, Setsunan University, Nagaotoge-cho, Hirakata, Osaka-Fu 573-01, Japan
W. R. McWhinnie	Department of Chemistry, University of Aston in Birmingham, Gosta Green, Birmingham B4 7ET, UK
W.-W. du Mont	Fachbereich Chemie, Anorganische Chemie, Universität Oldenburg, Ammerlander Heerstrasse 67–99, D-2900 Oldenburg, Federal Republic of Germany
M. J. Perkins	Royal Holloway and Bedford Colleges, University of London, Egham Hill, Egham TW20 0EX, UK
T. B. Rauchfuss	Department of Chemistry, University of Illinois at Urbana-Champaign, 505 South Mathews Avenue, Urbana, Illinois 61801, USA
T. Sadeh	Department of Radiopharmacology, Israel Atomic Energy Commission, Soreq Nuclear Research Centre, Yavne, Israel
K. Soda	Institute for Chemical Research, Kyoto University, Uji, Kyoto-Fu 611, Japan
H. Tanaka	Institute for Chemical Research, Kyoto University, Uji, Kyoto-Fu 611, Japan
A. Toshimitsu	Institute for Chemical Research, Kyoto University, Uji, Kyoto-Fu 611, Japan
S. Uemura	Institute for Chemical Research, Kyoto University, Uji, Kyoto-Fu 611, Japan
Z. Witczak	Whistler Centre for Carbohydrate Research, Purdue University, West Lafayette, Indiana 47907, USA

Foreword

The first volume of '*The Chemistry of Organic Selenium and Tellurium Compounds*' was published in 1986. For various reasons, several chapters which were planned for that volume, had to be omitted. All these chapters are included in the present volume, with the exception of a chapter on UV, visible, IR and Raman spectroscopy, which did not materialize.

Thus I believe the coverage of the organic chemistry of the derivatives of selenium and tellurium is reasonably complete in these two volumes. The coverage of the literature in the present second volume is, in most chapters, up to the end of 1985 or even to the middle of 1986.

I would be very grateful to readers who would let me know about omissions or mistakes in this volume as well as in other volumes of the Functional Groups series.

Jerusalem
July 1987

SAUL PATAI

The Chemistry of Functional Groups
Preface to the Series

The series 'The Chemistry of Functional Groups' is planned to cover in each volume all aspects of the chemistry of one of the important functional groups in organic chemistry. The emphasis is laid on the functional group treated and on the effects which it exerts on the chemical and physical properties, primarily in the immediate vicinity of the group in question, and secondarily on the behaviour of the whole molecule. For instance, the volume *The Chemistry of the Ether Linkage* deals with reactions in which the C—O—C group is involved, as well as with the effects of the C—O—C group on the reactions of alkyl or aryl groups connected to the ether oxygen. It is the purpose of the volume to give a complete coverage of all properties and reactions of ethers in as far as these depend on the presence of the ether group but the primary subject matter is not the whole molecule, but the C—O—C functional group.

A further restriction in the treatment of the various functional groups in these volumes is that material included in easily and generally available secondary or tertiary sources, such as Chemical Reviews, Quarterly Reviews, Organic Reactions, various 'Advances' and 'Progress' series as well as textbooks (i.e. in books which are usually found in the chemical libraries of universities and research institutes) should not, as a rule, be repeated in detail, unless it is necessary for the balanced treatment of the subject. Therefore each of the authors is asked *not* to give an encyclopaedic coverage of his subject, but to concentrate on the most important recent developments and mainly on material that has not been adequately covered by reviews or other secondary sources by the time of writing of the chapter, and to address himself to a reader who is assumed to be at a fairly advanced post-graduate level.

With these restrictions, it is realized that no plan can be devised for a volume that would give a *complete* coverage of the subject with *no* overlap between chapters, while at the same time preserving the readability of the text. The Editor set himself the goal of attaining *reasonable* coverage with *moderate* overlap, with a minimum of cross-references between the chapters of each volume. In this manner, sufficient freedom is given to each author to produce readable quasi-monographic chapters.

The general plan of each volume includes the following main sections:

(a) An introductory chapter dealing with the general and theoretical aspects of the group.

(b) One or more chapters dealing with the formation of the functional group in question, either from groups present in the molecule, or by introducing the new group directly or indirectly.

(c) Chapters describing the characterization and characteristics of the functional groups, i.e. a chapter dealing with qualitative and quantitative methods of determination including chemical and physical methods, ultraviolet, infrared, nuclear magnetic resonance and mass spectra: a chapter dealing with activating and directive effects exerted by the group and/or a chapter on the basicity, acidity or complex-forming ability of the group (if applicable).

(d) Chapters on the reactions, transformations and rearrangements which the functional group can undergo, either alone or in conjunction with other reagents.

(e) Special topics which do not fit any of the above sections, such as photochemistry, radiation chemistry, biochemical formations and reactions. Depending on the nature of each functional group treated, these special topics may include short monographs on related functional groups on which no separate volume is planned (e.g. a chapter on 'Thioketones' is included in the volume *The Chemistry of the Carbonyl Group*, and a chapter on 'Ketenes' is included in the volume *The Chemistry of Alkenes*). In other cases certain compounds, though containing only the functional group of the title, may have special features so as to be best treated in a separate chapter, as e.g. 'Polyethers' in *The Chemistry of the Ether Linkage*, or 'Tetraaminoethylenes' in *The Chemistry of the Amino Group.*

This plan entails that the breadth, depth and thought-provoking nature of each chapter will differ with the views and inclinations of the author and the presentation will necessarily be somewhat uneven, Moreover, a serious problem is caused by authors who deliver their manuscript late or not at all. In order to overcome this problem at least to some extent, it was decided to publish certain volumes in several parts, without giving consideration to the originally planned logical order of the chapters. If after the appearance of the originally planned parts of a volume it is found that either owing to non-delivery of chapters, or to new developments in the subject, sufficient material has accumulated for publication of a supplementary volume, containing material on related functional groups, this will be done as soon as possible.

The overall plan of the volumes in the series 'The Chemistry of Functional Groups' includes the titles listed below:

The Chemistry of Alkenes (two volumes)
The Chemistry of the Carbonyl Group (two volumes)
The Chemistry of the Ether Linkage
The Chemistry of the Amino Group
The Chemistry of the Nitro and Nitroso Groups (two parts)
The Chemistry of Carboxylic Acids and Esters
The Chemistry of the Carbon–Nitrogen Double Bond
The Chemistry of the Cyano Group
The Chemistry of Amides
The Chemistry of the Hydroxyl Group (two parts)
The Chemistry of the Azido Group
The Chemistry of Acyl Halides
The Chemistry of the Carbon–Halogen Bond (two parts)
The Chemistry of the Quinonoid Compounds (two parts)
The Chemistry of the Thiol Group (two parts)
The Chemistry of Amidines and Imidates
The Chemistry of the Hydrazo, Azo and Azoxy Groups (two parts)
The Chemistry of Cyanates and their Thio Derivatives (two parts)
The Chemistry of Diazonium and Diazo Groups (two parts)
The Chemistry of the Carbon–Carbon Triple Bond (two parts)

Supplement A: The Chemistry of Double-bonded Functional Groups (*two parts*)
The Chemistry of Ketenes, Allenes and Related Compounds (*two parts*)
Supplement B: The Chemistry of Acid Derivatives (*two parts*)
Supplement C: The Chemistry of Triple-Bonded Functional Groups (*two parts*)
Supplement D: The Chemistry of Halides, Pseudo-halides and Azides (*two parts*)
Supplement E: The Chemistry of Ethers, Crown Ethers, Hydroxyl Groups and their Sulphur Analogues (*two parts*)
The Chemistry of the Sulphonium Group (*two parts*)
Supplement F: The Chemistry of Amino, Nitroso and Nitro Groups and their Derivatives (*two parts*)
The Chemistry of the Metal–Carbon Bond (*four volumes*)
The Chemistry of Peroxides
The Chemistry of Organic Se and Te Compounds Vol. 1
The Chemistry of the Cyclopropyl Group (*two parts*)
The Chemistry of Se and Te Compounds Vol. 2

Titles in press:

The Chemistry of the Quinonoid Compounds Vol. 2
The Chemistry of Sulphones and Sulphonides Vol. 2
The Chemistry of Organo-silicon Compounds

Advice of criticism regarding the plan and execution of this series will be welcomed by the Editor.

The publication of this series would never have started, let alone continued, without the support of many persons. First and foremost among these is Dr Arnold Weissberger, whose reassurance and trust encouraged me to tackle this task. The efficient and patient cooperation of several staff-members of the Publisher also rendered me invaluable aid (but unfortunately their code of ethics does not allow me to thank them by name). Many of my friends and colleagues in Israel and overseas helped me in the solution of various major and minor matters, and my thanks are due to all of them, especially to Professor Z. Rappoport. Carrying out such a long-range project would be quite impossible without the non-professional but none the less essential participation and partnership of my wife.

The Hebrew University
Jerusalem, Israel

SAUL PATAI

Contents

1. Photoelectron spectroscopy of organic derivatives containing selenium and tellurium — 1
   C. Cauletti and G. Distefano
2. The application of Mössbauer spectroscopy to the study of organotellurium compounds — 51
   F. J. Berry
3. Preparative uses of organoselenium and organotellurium compounds — 91
   T. G. Back
4. Seleno- and telluro-carbonyl compounds — 215
   F. S. Guziec, Jr.
5. Photochemistry of organic compounds of selenium and tellurium — 275
   Z. Goldschmidt
6. Acidity, hydrogen bonding and self-association in organic and organometallic compounds of selenium and tellurium — 339
   T. B. Rauchfuss
7. Biochemistry of physiologically active selenium compounds — 349
   K. Soda, H. Tanaka and N. Esaki
8. Biological and biochemical aspects of tellurium derivatives — 367
   T. Sadeh
9. Pharmacological and toxicological aspects of inorganic and organic selenium compounds — 377
   T. Masukawa
10. Insertion and extrusion reactions — 393
    L. Henriksen
11. Compounds with Se—N and Te—N bonds — 421
    G. Kirsch and L. Christiaens
12. Synthesis of organic conductors containing selenium and tellurium — 463
    D. Cowan and A. Kini
13. Organoselenium and organotellurium analogues of ethers and peroxides — 495
    W. R. McWhinnie

14. Organic selenocyanates and tellurocyanates and related compounds 541
A. Toshimitsu and S. Uemura

15. Organic compounds containing bonds between Se or Te with P, As, Sb and Bi 591
W.-W. du Mont, R. Hensel, S. Kubiniok and L. Lange

16. Free radical reactions of organoselenium and organotellurium compounds 657
L. Castle and M. J. Perkins

17. Synthesis of selenium and tellurium ylides and carbanions: application to organic synthesis 675
A. Krief

18. Selenium and tellurium derivatives of carbohydrates and nucleoxide analogs 765
Z. Witczak

Author Index 795

Subject Index 553

The Chemistry of Organic Selenium and Tellurium Compounds Volume 2
Edited by S. Patai
© 1987 John Wiley & Sons Ltd.

CHAPTER **1**

Photoelectron spectroscopy of organic derivatives containing selenium and tellurium

CARLA CAULETTI

Dipartimento di Chimica, Università La Sapienzà di Roma, Piazzale Aldo Moro 5,00185 Rome, Italy

GIUSEPPE DISTEFANO

Istituto dei Composti del Carbonio Contenenti Eteroatomi e loro Applicazioni, Ozzano Emilia, Via Tolara di Sotto 89, 40064 Ozzano Emilia, Italy

I. INTRODUCTION . . . 2
A. Preliminaries and Scope . . . 2
B. Ionization Energies, Ionic States and Molecular Orbitals . . . 3
C. Analysis of a UP Spectrum . . . 3
II. SMALL MOLECULES . . . 6
A. Diatomic Species . . . 6
1. Homonuclear molecules . . . 6
2. Heteronuclear molecules . . . 8
B. Triatomic Species . . . 10
1. Linear molecules . . . 10
2. Bent XA_2 molecules . . . 13
C. Pseudohalide Derivatives . . . 17
III. CHALCOGEN HETEROCYCLOPENTADIENES AND THEIR DERIVATIVES . . . 21
A. Monocyclic Compounds . . . 21
1. Experimental data . . . 21
2. Comparison with theoretical computations . . . 24
3. Electron affinity values . . . 24
B. Bicyclic Derivatives . . . 26
1. Benzo derivatives . . . 26
2. Thienothiophenes and selenolo analogues . . . 26

IV. SELENIDES AND TELLURIDES 27
A. Alkyl Chalcogenides 27
B. Open-chain and Cyclic Ethylene Derivatives. 30
C. Alkyl Phenyl Chalcogenides 32
D. Cyclic Phenyl Chalcogenides 34
V. DISELENIDES AND DITELLURIDES 35
A. Methyl and Phenyl Dichalcogenides 35
B. *Peri* Dichalcogenides 36
VI. SELENOKETENES, SELENOCARBONYLS AND SELENOFULVALENES 37
A. Selenoketenes 37
B. Selenocarbonyl Derivatives 39
1. Selenoaldehydes 39
2. Selenoketones 40
C. Selenofulvalenes. 41
VII. GROUP 5A DERIVATIVES 42
A. Phosphoroselenenic Acid *O,O,O*-Trimethyl Ester 42
B. Phosphine Derivatives 42
C. Me_2YXMe 44
VIII. APPENDIX: ELECTRON TRANSMISSION SPECTROSCOPY 45
IX. REFERENCES 46

I. INTRODUCTION

A. Preliminaries and Scope

Photoelectron spectroscopy is now a well established technique which permits the direct measurement of the energies required to remove electrons from atoms and molecules. When a photon of sufficient energy impinges on a neutral molecule, M, it is possible to eject an electron, leaving behind a radical cation, M^+, in one of several possible electronic states. If $h\nu$ is the energy of the incident photons, IE the ionization energy involved in a given ionization event and KE the excess kinetic energy of the emitted photoelectron, the equation:

$$h\nu = IE + KE \qquad (1)$$

provides a means of finding IE by measuring KE. When the ionizing source is of medium or high energy, such as those produced in a inert gas discharge (for instance 21.22 eV in He I $2p \rightarrow 1s$ or 40.81 eV in He II $2p \rightarrow 1s$) and the target is in the gas phase, the information obtained concerns the ionization energies of valence shells of free molecules and the technique is named UV photoelectron spectroscopy (UPS). When $h\nu$ is in the fields of the soft X-rays (up to a few thousand electronvolts) the primary information concerns core electrons and the technique is named X-ray photoelectron spectroscopy (XPS).

A deeper discussion of the general aspects and the chemical applications of UPS may be found in some reviews and books [1–11].

This present chapter reports UPS studies on selenium- and tellurium-containing organic molecules. The number of investigations on such compounds by this technique is limited, probably owing to the instability of most of them. Nevertheless, recent developments in the analysis and interpretation of the experimental data allowed valuable information to be obtained on the electronic structure, i.e. the energy and composition of molecular orbitals (MOs), even of transient molecules.

In the presentation and discussion of the results, we shall often refer to the oxygen and sulphur analogues of the examined compounds for the sake of completeness and comparison.

B. Ionization Energies, Ionic States and Molecular Orbitals

A photoelectron spectrum consists of a plot of the number of electrons with a given kinetic energy, KE (which may be converted into ionization energy, IE, through equation 1), versus that energy. Hence the primary information obtained by UPS is the energy of the molecular ion states relative to the ground-state molecule and the IEs are strictly energy differences between states of M^+ and the ground state of M. There is therefore a correlation between the energy sequence of ionic levels, which is directly drawn out by a UP spectrum, and the sequence of the MOs of the corresponding molecular ground state, but they are not necessarily coincident. However, the approximation of Koopmans' theorem[12] states that the ionization energies of a molecule are equal to the negative of the self-consistent field (SCF) orbital energies, ε_i:

$$IE_i = -\varepsilon_i \qquad (2)$$

Equation 2 obviously does not apply to open-shell molecules.

Following this theorem, a photoelectron spectrum should be a reflection of the molecular orbital diagram, the sequence of UPS bands matching the sequence of the negative of the energies of the occupied molecular orbitals. Koopmans' theorem would be fully valid if the electronic configuration of the remainder of the molecule remained unchanged or 'frozen' during the ionization process, neglecting relaxation phenomena, which may be important, e.g. when ionization of metal-centred orbitals is involved. However, in most organic molecules, relaxation effects are nearly constant in the ionization of all valence orbitals, so that Koopmans' approximation allows correct qualitative correlations, in addition to semiquantitative comparisons.

The following may give more bands in a UP spectrum than there are valence orbitals in a molecule:

(i) ionization of one electron with simultaneous excitation of a second electron to an unoccupied excited orbital, i.e. a two-electron process, referred to as shake-up or configuration interaction (CI) processes;
(ii) removal of degeneracy in the molecular ion by mechanisms such as spin–orbit coupling and the Jahn–Teller effect;
(iii) the presence of one or more open shells, the ionization of which may give rise to a multiplicity of ionic states.

C. Analysis of a UP Spectrum

The analysis of a UP spectrum consists in assigning each spectral band to an electronic state of the molecular ion and identifying the orbital from which electrons are ejected. In this interpretative work some approaches may be of help, as follows.

a. *Spectral fine structure.* The observation of fine structure in spectral bands, such as that arising from spin–orbit coupling and the Jahn–Teller effect, or vibrational progressions can lead to positive identification of the nature of the ionized MO. In particular, analysis of the vibrational structure, whenever present, of a photoelectron band may give precise information on the non-bonding, bonding or antibonding character of the orbital. In fact, removal of a non-bonding electron does not cause changes in the geometry of the molecular ion with respect to the neutral molecule, so that the corresponding UPS band is sharp and without any vibrational progression. In contrast ionization of a bonding or antibonding orbital decreases or increases the bond-lengths, this being reflected in a change in the vibrational spacing (in this case a vibrational structure does exist) on passing from the molecule to the ion. For instance, the first band of the UP spectrum of the hydrides H_2X[13,14] (X = O, S, Se and Te) (See Figure 1) is very sharp, indicating its origin from the ionization of the lone pair orbital of X. The second ($2a_1^{-1}$) and third ($1b_2^{-1}$) bands show progression in the symmetrical bending (ν_2) and stretching (ν_1) vibrations, respectively. Their frequencies are reduced from the corresponding values in the neutral state indicating a strong H—X bonding character for these orbitals.

b. *Intensity arguments.* The areas of photoelectron bands in a photoelectron spectrum are approximately proportional to the relative probabilities of ionization to the different ionic states. These probabilities are called relative partial photoionization cross-sections (for details, see, e.g., Refs. 7 and 10). The most important factors that determine the cross-sections and consequently the UPS band intensities are the following.

(i) The partial cross-section for ionization from a given orbital is proportional to the number of equivalent electrons that are available to be ionized (in the case of closed-shell molecules) or to the statistical weight of the ionic state produced (in the case of atoms or open-shell molecules).

(ii) The photoionization cross-section (σ) depends on the character of the ionized orbital, i.e. its size, number of nodes, location in the molecule, etc. Further, it varies with the wavelength of the ionizing radiation, the variation following different trends for the various orbitals. This point is most meaningful when intensity

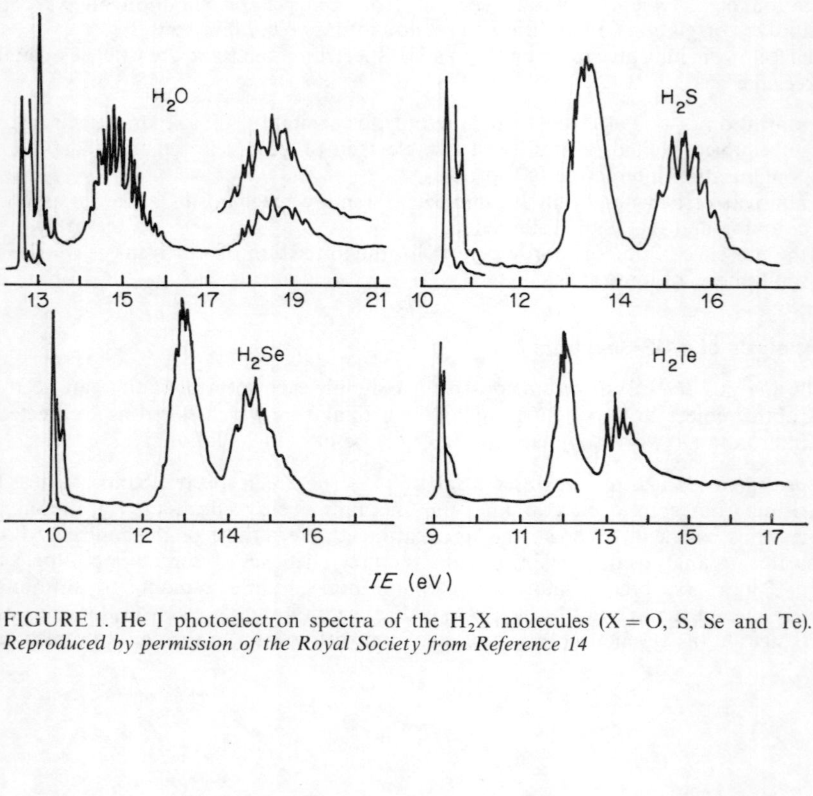

FIGURE 1. He I photoelectron spectra of the H_2X molecules (X = O, S, Se and Te). *Reproduced by permission of the Royal Society from Reference 14*

arguments are used as interpretative tools. The large number of UPS experiments performed with two or more ionizing sources (mainly He I and He II) allow some empirical rules to be drawn. Firstly, in photoionization by He I, the cross-section of the valence atomic orbital increases on going down a Group of the Periodic Table. This trend is a least partially reversed when light of shorter wavelength, particularly the He II line, is used for ionization. Thus, the cross-sections of p orbitals of chlorine, bromine and sulphur decrease significantly on passing from He I to He II, whereas that of carbon 2p orbitals remains roughly unchanged and those of nitrogen 2p and oxygen 2p orbitals show an increase. Not many data are available on the behaviour of Se 4p and Te 5p orbitals, whose photoionization cross-sections are, however, expected to decrease relative, for instance, to that of C 2p on passing from 21.22 to 40.81 eV ionizing radiation. Theoretical calculations suggest a $\sigma_{\mathrm{He\,I}}/\sigma_{\mathrm{He\,II}}$ ratio of ca. 17 for Se 4p orbitals[15] and ca. 14 for Te 4p orbitals[16], whereas there is experimental evidence that this ratio is roughly 1 for C 2p orbitals. As for molecular orbitals, the Gelius model[16a] allows σ to be evaluated starting from that of the contributing atomic orbitals (AOs):

$$\sigma_j = \sum_{A,i} P_{j,i_A} \sigma_i^{A} \tag{3}$$

where σ_j is the one-electron photoionization cross-section for the jth MO and the summation extends over atomic orbitals, ϕ_i^{A}, on the different atomic centres A. The σ_i^{A} are one-electron atomic cross-sections and P_{ji_A} are factors describing the effective occupancy of the MO.

The main application of equation 3 in UPS has been in identifying the predominant AO character in the MO giving rise to a particular band. Hence bands corresponding to halogen lone pairs are easily recognizable by their strong decrease in intensity in He II relative to bands corresponding to C 2p-based orbitals. The same occurs for sulphur-, selenium- and tellurium-centred orbitals.

As an example of use of intensity considerations in the analysis of the nature of valence orbitals in molecules, we can quote UPS studies on the isomeric compounds MeNCS[17] and MeSCN[18]. Of the first two bands of the spectrum of the former compound, assigned to π non-bonding and π bonding orbitals, respectively, that at lower *IE* does not change in intensity on switching from He I to He II, whereas the other shows an interesting increase. Since for nitrogen and sulphur p orbitals the expected intensity changes are in opposite directions, the experimental findings suggest that the π non-bonding orbital has contributions from both atoms, whereas the π bonding MO has a large charge density on the nitrogen atom. This behaviour contrasts with that for methyl thiocyanate, where the first band shows a marked intensity decrease in the He II spectrum (see Figure 10), which implies that the corresponding orbital is largely localized on the sulphur atom. The behaviour is the same in the spectrum of MeSeCN, indicating that the highest occupied MO (HOMO) is essentially a lone-pair orbital of selenium.

c. *UP spectroscopic comparison of chemically related compounds.* This is often the most valuable tool in the assigment. The observation of trends, variations, etc., depending on various factors such as electronegativities, size and steric hindrance of substituents or the presence or absence of conjugative interactions, is usually of great help in the understanding of the valence-shell structure even of large molecules. Many examples of such an analysis can be found in the following sections of this review.

d. *Theoretical calculations.* Sometimes molecular orbital calculations are the only means of achieving at least a tentative assignment of the UP spectra. Calculation of the molecular ground-state energies and derivation of ionization energies through the

Koopmans theorem is by far the most commonly used approach. It is usual to distinguish four types of molecular orbital calculations:

(i) Exact Hartree–Fock (HF) calculations. In these, made by the self-consistent-field (SCF) method, no terms are neglected and no further improvement in the total energy can be gained either by increasing the number of iterations or by expanding the basis set. These calculations are unfortunately limited to relatively small molecules because of their cost in terms of computer time.

(ii) *Ab initio* SCF calculations. These are made possible by mathematical simplification through contraction of the basis set, but do not involve any arbitrary parameters. The general theory of *ab initio* calculations for closed-shell molecules was derived by Roothaan[19] and by Hall[20] in 1951 and are becoming increasingly common for polyatomic molecules[21,22].

(iii) Semi-empirical calculations. These calculations are simplified further by the neglect of certain integrals and the empirical rather than the theoretical evaluation of others. The different methods that have been extensively used are generally known by their acronyms, CNDO[23–26], INDO[27], MINDO[28–32] and SPINDO[33], where NDO = neglect of differential overlap, C = complete, I = intermediate, M = modified and SP = spectroscopic potentials adjusted. Because of the simplifications achieved, these calculations can be applied to large and complex molecules.

(iv) Empirical calculations. In empirical calculations no attempt is made to mimic the HF method, the aim being rather to evolve the simplest possible theory, using only a small number of arbitrary parameters. The Hückel molecular orbital (HMO) method[34] concerns only π electron systems of conjugated and aromatic molecules. The extended HMO (EHMO) method as developed by Hoffmann[35] includes all valence electrons and is of considerable use in assigning spectra.

The simplest Hückel-type method is the linear combination of bond orbitals (LCBO)[36], which starts not from atomic orbitals but from the occupied group orbitals of the building blocks of a molecule, thus reducing the number of terms of the secular determinant. The matrix elements are not mathematically evaluated but are empirical parameters which can compensate for the neglected integrals and the lack of empty orbitals. In particular, the numerical values of the self energies (α) and of the interaction parameters (β) can be derived from the IEs of reference systems.

II. SMALL MOLECULES

In this section we shall describe some examples of UPS studies of small molecules, often discussing non-organic molecules which will be useful as comparisons in the following sections of this chapter.

A. Diatomic Species

1. Homonuclear molecules

The photoelectron spectra of the open-shell system X_2 (X = O, S, Se and Te) are characterized by a large number of bands because each ionization process gives rise to a multiplicity of ionic states. Further, spin–orbit coupling is expected to produce significant splittings.

The electronic configuration of O_2^+ has been extensively investigated both theoretically and experimentally by UPS[37,38].

In the heavier Group VI homonuclear diatomics, spin–orbit splitting is expected to play

TABLE 1. Vertical ionization energies (eV) for S_2, Se_2 and Te_2[a]

State	S_2[b]	Se_2[c]	Te_2[c]
$^2\Pi_{g,1/2}$	9.41	8.93	8.23
$^2\Pi_{g,3/2}$		9.10	8.73
$^4\Pi_u$	11.82	10.65	9.38
$^2\Pi_u(1)$	12.33	11.19	10.07
$^4\Sigma_g$	13.20	12.25	11.01
$^2\Pi_u(2)$	[d]	(13)	11.53
$^2\Sigma_g$	14.62	13.33	11.84
$^2\Pi_u(3)$	15.58	14.00	12.34
CI	—	—	13.43[e]
$^4\Sigma_u$	17.70	(17.2)[e]	16.36[e]
$^2\Sigma_u$	18.10[f]	18.44[e]	18.16[e]
	18.66	(23.68)[e]	(20.56)[e]
	23.33		(23.63)[e]
	25.99		

[a]Parentheses indicate uncertainty in measurement.
[b]From Reference 40.
[c]From Reference 42.
[d]Predicted to be very weak in S_2.
[e]Feature observed in He II spectrum only.
[f]Assignment of this band is uncertain.

an increasingly important role as the atomic weight of the chalcogen increases. Despite the severe difficulties in avoiding contamination of the diatomic species in the vapour phase, UP spectra were obtained for S_2[39–42], Se_2 and Te_2[39,41,42]. Table 1 reports the observed ionization energies.

The valence molecular orbital configuration of X_2 can be represented, in the ground state, by $\ldots(\sigma_g ns)^2(\sigma_u ns)^2(\sigma_g np)^2(\pi_u np)^4(\pi_g np)^2$.

Ionization of the π_g orbital leads to a $^2\Pi_g$ state, which can be split into the two spin–orbit components $^2\Pi_{g,1/2}$ and $^2\Pi_{g,3/2}$. This splitting is actually not appreciable in S_2^+, whilst being evident in Se_2^+ (0.17 eV[42]) and Te_2^+ (0.50 eV[42]). Of the five possible X_2^+ states following ionization of the π_u orbital ($^4\Pi_u$, $^2\Phi_2$ and three $^2\Pi_u$ states), four may be reached by a direct one-electron transition. The corresponding UPS bands are observed in all the members of the series. The multiplet splittings in the Π_u ion states of these species are reported in Table 2. It can be seen that the $^4\Pi_u - {}^2\Pi_u(1)$ gap initially decreases with

TABLE 2. Multiplet splitting data (eV) for $^{4,2}\Pi_u$ $(\pi_u np)^{-1}$ states of O_2, S_2, Se_2 and Te_2

	O_2[a]	S_2[b]	Se_2[c]	Te_2[c]
$^4\Pi_u - {}^2\Pi_u(1)$	1.03	0.51	0.54	0.69
$^4\Pi_u - {}^2\Pi_u(2)$	2.6	—	2.35	2.15
$^4\Pi_u - {}^2\Pi_u(3)$	7.3	3.76	3.35	2.96

[a] From Reference 38.
[b] From Reference 40.
[c] From Reference 41.

TABLE 3. He II/He I photoionization cross-section ratios for Se_2 and Te_2

Molecule	Orbital	$\sigma_{HeII}/\sigma_{HeI}$	Theoretical ratio for the valence p shells of atomic Se and Te[b]
Se_2[a]	π_g	0.046	0.06
	π_u	0.066	
	σ_g	0.105	
Te_2[a]	π_g	0.074	0.07
	π_u	0.103	
	σ_g	0.21	

[a] From Reference 42
[b] From References 15 and 16

increasing atomic weight until the effects of large spin–orbit coupling cause the gap to increase. The same decrease is observed for the $^4\Pi_u - {}^2\Pi_u(3)$ separation.

The analysis of the ionization phenomena accompanying the expulsion of an electron from the $\sigma_u$ns and $\sigma_g$ns orbitals is definitely more complex and the occurrence of two-electron processes, suggested by Streets and Berkowitz[41], should be further confirmed by more refined experiments, for instance by using synchrotron radiation as the ionizing source.

The He II spectra of Se_2 and Te_2, studied by Potts and Novak[42], gave interesting information on the relative photoionization cross-sections of the various orbitals. Table 3 reports the measured He II/He I cross-section ratios compared with theoretical values for atomic Se[15] and Te[16]. It is of interest that the He II Se 4p photoionization cross-section is markedly lower than the He I cross-section, whereas the He II Se 4s cross-section is predicted to be markedly higher than the He I cross-section.

2. Heteronuclear molecules

The electronic configuration of TeO, which is the predominant species in the vapour produced by heating solid TeO_2 at 773 K, is analogous to that of the above molecules and

TABLE 4. Ionization energies for TeO[a]

State	*IE* (eV)
$^2\Pi_{1/2}$	8.72
$^2\Pi_{3/2}$	9.32
$^4\Pi$	10.80
$^2\Pi(1) + {}^4\Sigma$	11.17
$^2\Sigma$	12.00
$^2\Pi(2)$	12.7
$^2\Pi(3)$	13.49

[a] From Reference 43.

the assignment of its photoelectron spectrum[43] follows the same lines (see Table 4). In this case the highest π orbital is largely localized on tellurium and is antibonding in character.

The Group IV–VI diatomics GeX (X = O[44], S[44,45], Se[44,45], Te[44]), SnX (X = S[44,45], Se[44], Te[44,45]) and PbTe[45] have been extensively studied by UPS. One $^2\Pi_{1/2,3/2}$ and three $^2\Sigma^+_{1/2}$ ionic states correspond to the electronic configuration.

$$(1\sigma)^2(2\sigma)^2(3\sigma)^2(1\pi)^4$$

of these molecules. Tables 5 and 6 report the experimental ionization energies for these molecules together with the results of CNDO calculations. An interesting aspect is the appearance in the spectra of features arising from many-electrons effects (CI) involving the $(2\sigma)^{-1}$ primary hole state. For both GeX and SnX series, an energy stabilization of the 3σ orbital relative to the 1π orbital develops with increasing molecular weight. Wu and Fehlner[45] suggested that this is due to a sharp decrease in the π-type interaction between the atoms whereas White and coworkers[44] support this hypothesis on the basis of the CNDO overlap population and charge distribution. The 1π orbital is localized principally

TABLE 5. Experimental and theoretical (CNDO) ionization energies for GeX (X = O, S, Se, Te) molecules[a]

Molecule	MO	Ionic state	Experimental IE (eV)[b]	Theoretical IE (eV)[c]
GeO	3σ	$^2\Sigma^+_{1/2}$	11.25(1)	11.61
	1π	$^2\Pi_{1/2,3/2}$	11.40(5)	12.16
	2σ	$^2\Sigma^+_{1/2}$	15.16(2)	15.21
	1σ	$^2\Sigma^+_{1/2}$	—	30.97
GeS	1π	$^2\Pi_{1/2,3/2}$	10.36(5)	10.17
	3σ	$^2\Sigma^+_{1/2}$	10.43(2)	10.24
	2σ	$^2\Sigma^+_{1/2}$	14.00(4)	14.40
	2σ	CI	14.6(1)	—
	2σ	CI	15.28(5)	—
	2σ	CI	18.86(6)	—
	1σ	$^2\Sigma^+_{1/2}$	—	22.22
GeSe	1π	$^2\Pi_{3/2}$	9.8(1)	10.14
	1π	$^2\Pi_{1/2}$	9.95(5)	
	3σ	$^2\Sigma^+_{1/2}$	10.16(1)	10.37
	2σ	$^2\Sigma^+_{1/2}$	13.56(5)	13.92
	2σ	CI	14.0(1)	—
	2σ	CI	14.88(5)	—
	2σ	CI	16.31(5)	—
	1σ	$^2\Sigma^+_{1/2}$	—	22.50
GeTe	1π	$^2\Pi_{3/2}$	9.1(1)	8.67
	1π	$^2\Pi_{1/2}$	9.32(2)	
	3σ	$^2\Sigma^+_{1/2}$	9.76(1)	9.55
	2σ	$^2\Sigma^+_{1/2}$(?)	13.31(1)	13.56
	2σ	CI	13.88(6)	—
	2σ	CI	15.20(5)	—
	1σ	$^2\Sigma^+_{1/2}$	—	20.63

[a] From Reference 44.

[b] Vertical IEs. Uncertainties in parentheses.

[c] Values empirically reduced by 0.8 to reflect final ionic state relaxation.

TABLE 6. Experimental and theoretical ionization energies for SnX (X = S, Se, Te) and PbTe molecules

Molecule	MO	Ionic state	Experimental IE (eV)[a]	Theoretical IE (eV)[b]
SnS[c]	1π	$^2\Pi_{1/2,3/2}$	9.42(5)	9.33
	3σ	$^2\Sigma^+_{1/2}$	9.73(3)	9.53
	2σ	CI	11.93(5)	—
	2σ	$^2\Sigma^+_{1/2}$	13.09(5)	13.08
	2σ	CI	14.03(5)	—
	2σ	CI	15.46(6)	—
	2σ	CI	18.12(6)	—
	1σ	$^2\Sigma^+_{1/2}$	—	20.80
SnSe[c]	1π	$^2\Pi_{3/2}$	9.0(1)	9.30
	1π	$^2\Pi_{1/2}$	9.26(3)	
	3σ	$^2\Sigma^+_{1/2}$	9.56(1)	9.77
	2σ	$^2\Sigma^+_{1/2}$	12.7(1)	12.89
	2σ	CI	13.65(5)	—
	2σ	CI	14.99(5)	—
	1σ	$^2\Sigma^+_{1/2}$	—	20.85
SnTe[c]	1π	$^2\Pi_{3/2}$	8.61(4)	8.58
	1π	$^2\Pi_{1/2}$	8.91(2)	
	3σ	$^2\Sigma^+_{1/2}$	9.30(1)	9.57
	2σ	CI	10.55(5)	—
	2σ	CI	11.02(5)	—
	2σ	$^2\Sigma^+_{1/2}$ (?)	11.91(1)	12.48
	2σ	CI	12.8(1)	—
	2σ	CI	14.33(5)	—
	1σ	$^2\Sigma^+_{1/2}$	—	18.89
PbTe[d]	1π	$^2\Pi_{3/2}$	8.04	—
	1π	$^2\Pi_{1/2}$	8.34	—
	3σ	$^2\Sigma^+_{1/2}$	9.01	—

[a] Vertical IEs. Uncertainties in parenthese.
[b] Value empirically reduced by 0.85 to reflect final ionic state relaxation.
[c] From Reference 44.
[d] From Reference 45.

on the chalogenide whereas the 3σ orbital is more equally distributed between both atoms. Hence, as one proceeds down the GeX or SnX series, the IEs of the $(3\sigma)^{-1}$ and $(1\pi)^{-1}$ states decrease, following the decrease in IE of the chalogenide ns and np atomic orbitals, the $(1\pi)^{-1}$ IE decreasing faster owing to its higher atomic character.

B. Triatomic Species

1. Linear molecules

The photoelectron spectra of the linear molecules X=C=Y (X = O, Y = Se[46,47]; X = S, Y = Se[46,47]; X = Y = Se[46,48]) have been investigated and compared with those of the analogous well known species CO_2, COS and CS_2, already studied by UPS[49,50]. Figure 2 shows the He I UP spectra of these molecules.

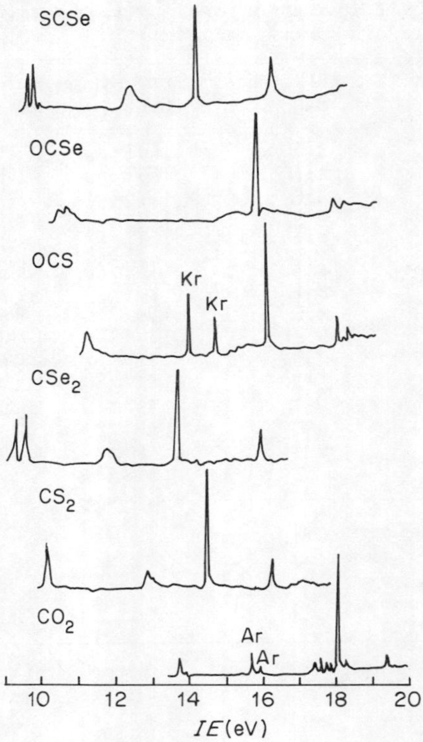

FIGURE 2. He I photoelectron spectra of some X=C=Y molecules. *Reproduced by permission of the American Institute of Physics from Reference 46*

The assignment of the spectra is the same for all members of the series, as expected in view of their chemical similarity. The first band corresponds to the $^1\pi_g$ level, essentially non-bonding, consistent with its sharpness and poor vibrational structure. It shows a spin–orbit splitting increasing in the order $CO_2 < OCS < CS_2 < SCSe < OCSe < CSe_2$. The second band is broad with a vibrational progression indicating a strongly bonding character of the ionised level, $^1\pi_u$. The third and fourth bands correspond again to essentially non-bonding orbitals, $^2\sigma_u$ and $^2\sigma_g$ respectively.

Frost and coworkers[46] measured the vibrational Franck–Condon factors for each ionic state, of which they estimated the geometry, using the method of Coon and collaborators[51]. A selection of the results obtained are reported in Table 7.

Figure 3 is a correlation diagram of the ionic states which shows several interesting trends. The energy of each ionic state is lowered if one end atom is replaced by a heavier atom. The effect of replacement of S by Se is much smaller than that of O by S or Se. The latter replacements result in a decrease in energy of more than 1 eV, whereas the former produces a change of far less than 1 eV. This reflects the well established fact that the resemblance between two successive elements in a Periodic Group becomes closer as one proceeds down the Group.

TABLE 7. Bond lengths of the ionic states of some X=C=Y estimated from Franck–Condon factors[a]

Molecule XCY	Electronic state	r(C—X) (Å)	r(C—Y) (Å)
SeCSe	$X^1\Sigma_g^+$	1.711[b]	1.711[b]
$SeCSe^+$	$X^2\Pi_g$	1.711	1.711
	$A^2\Pi_u$	1.799	1.799
	$B^2\Sigma_u^+$	1.731	1.731
	$C^2\Sigma_g^+$	1.711	1.711
OCSe	$X^1\Sigma^+$	1.159[a]	1.1709[a]
$OCSe^+$	$X^2\Pi$		
	$A^2\Pi$	1.90	1.890
	$B^2\Sigma^+$	1.145	1.718
	$C^2\Sigma^+$	1.111	1.776
SCSe	$X^1\Sigma^+$	1.557[a]	1.709
$SCSe^+$	$X^2\Pi$	1.537	1.750
	$A^2\Pi$	1.620	1.813
	$B^2\Sigma^+$	1.550	1.738
	$C^2\Sigma^+$	1.540	1.759

[a] From Reference 46.
[b] From Reference 52.

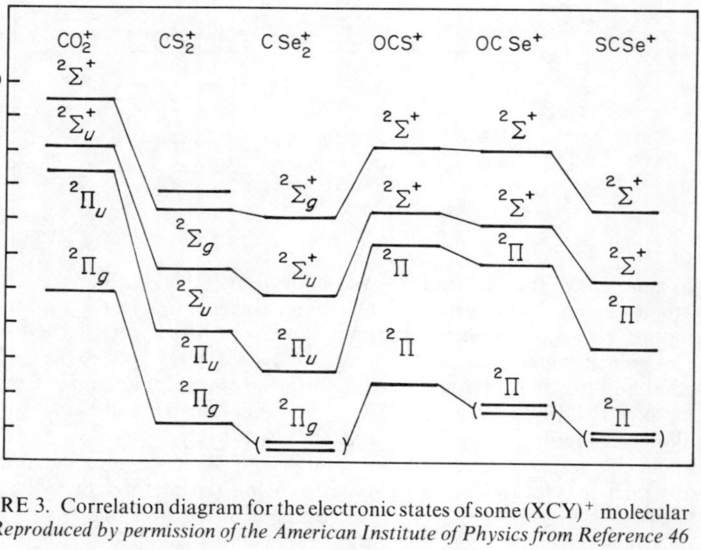

FIGURE 3. Correlation diagram for the electronic states of some $(XCY)^+$ molecular ions. *Reproduced by permission of the American Institute of Physics from Reference 46*

Cradock and Duncan found in the UP spectra of OCSe[47], SCSe[47] and CSe_2[48] some weak extra bands, not observed by Frost *et al.*[46]. They attributed these features to shake-up processes, leading to $^2\Pi_u$ states derived from the $(1\pi_g)^2\,(2\pi_u)^1$ configuration of the ion.

2. Bent XA_2 molecules

The valence shell configuration of the XO_2 dioxides (X = S, Se, Te) is

$$(1a_1)^2(1b_2)^2(2a_1)^2(1b_1)^2(3a_1)^2(2b_2)^2(3b_2)^2(1a_2)^2(4a_1)^2$$

The photoelectron spectrum of SeO_2[53] strongly resembles that of SO_2, which has been widely discussed[50,54–56].

Figure 4 shows the He I and He II spectra of both molecules and Table 8 reports the observed *IE*s for SEO_2 with the proposed assignment. The first band in SeO_2, at 11.76 eV, strongly decreases in intensity compared with the second band on passing from He I to

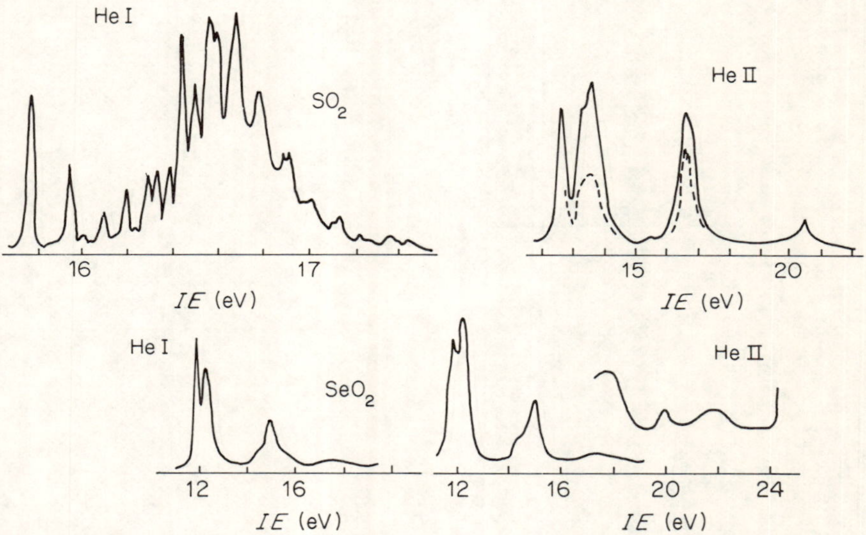

FIGURE 4. Photoelectron spectra of SO_2 (*reproduced by permission of Taylor and Francis from Reference 55*) and SeO_2 (*reproduced by permission of Elsevier Science Publishers from Reference 53*)

TABLE 8. Ionization energies for SeO_2[a]

IE[b] (eV)	Assignment
11.76	$4a_1(n_{Se})$
12.18	$1a_2(\pi_{Se-O}) + 3b_2(\pi_{O-O})$
14.56	$2b_2(\sigma_{O-Se-O})$
14.95	$3a_1(\pi_{O-Se-O}) + 1b_1(\pi_{O-Se-O})$
17.61	CI
19.90	$2a_1(4s_{Se})$
21.80	CI

[a] From Reference 53.
[b] Values refer to band maxima.

He II ionizing radiation. This supports the assignment to the ionization of a substantially non-bonding orbital ($4a_1$) with the main contribution from the Se 4p atomic orbital. The second band, at 12.18 eV, accounts for two ionizations, $(1a_2)^{-1}$ and $(3b_2)^{-1}$, of a π Se—O bonding orbital and an antibonding O—O orbital. The third, broad band, with a peak at

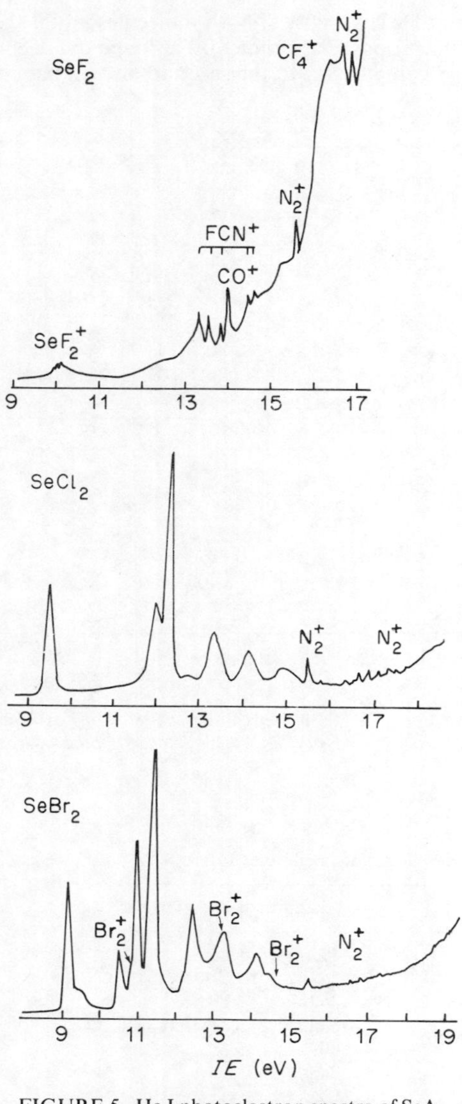

FIGURE 5. He I photoelectron spectra of SeA_2 molecules (A = F, Cl and Br). *Reproduced by permission of North-Holland Physics Publishing from Reference 61*

14.95 eV and a shoulder at 14.56 eV, arises from ionization of the $2b_2$, of σ nature, $3a_1$ and $1b_1$, both bonding between oxygen and selenium.

The weak band at 19.9 eV observed in the He II spectrum corresponds to the non-bonding $2a_1$ orbital. The features at 17.61 and 21.80 eV derive from configuration interaction (CI) processes.

In the spectrum of TeO_2[53], the vapour in equilibrium with this substance was shown to consist largely of TeO (see Section II.A.2). However, analysis showed that bands at 11.15 eV, presumably due to $(3b_2)^{-1} + (1a_2)^{-1} + (4a_1)^{-1}$ ionization, and 12.7 eV, due to $(2b_2)^{-1} + (3a_1)^{-1} + (1b_1)^{-1}$ ionization, could probably be attributed to the TeO_2 molecule.

In recent years significant progress has been made in the study of transient species in the gas phase by means of UPS[57]. Because of the importance of short-lived species in various branches of gas-phase chemistry, and since physical information is often scarce, the study of transients by means of spectroscopic techniques does not lack incentive.

Related small molecules studied extensively by UPS are the XA_2 (X = O, S, Se, Te; A = F, Cl, Br) compounds. Only the three stable members of this series were studied with conventional experimental methods, viz. OF_2[58], OCl_2[58] and SCl_2[59,60]. Application of UPS to the remaining members of the series has been met with varying degrees of experimental difficulty. He I photoelectron spetra of SeF_2[61], $SeCl_2$[61,62] and $SeBr_2$[61,63] were described and interpreted on the basis of comparison with the spectra of the related sulphur and oxygen dihalogenides and with the results of theoretical calculations. In the UP spectrum of these molecules seven bands are expected from 14 p electrons. In fact, each halogen atom contributes two lone pairs and one unpaired electron, while the selenium atom contributes one lone pair and two unpaired electrons. The combined lone-pair orbitals on both halogen atoms transform as a_1, a_2, b_1, b_2 in C_{2v} symmetry, while the selenium lone-pair orbital transforms as b_2. The out-of-plane b_2 orbitals interact strongly to form the widely separated antibonding $2b_2$ and bonding $1b_2$ combinations. The remaining a_1, a_2 and b_1 halogen orbitals are relatively unshifted with respect to their positions in the atomic halogens and lie between the two b_2 MOs. The remaining two unpaired selenium electrons combine with those of the halogen atoms to form the strongly bonding $3a_1$ and $2b_1$ MOs, which will lie below $1b_2$. The UP spectra, shown in Figure 5, are consistent with this description. Unfortunately, only the first ionization of $SeFe_2$ could be observed. In the spectrum of this molecule bands due to CF_4, N_2, CO and FCN were also detected.

The LCBO model, applied to the series $SeCl_2$, SCl_2 and OCl_2, gives good results[61] but fails for $SeBr_2$ because insufficient UPS data are available for OBr_2 and SBr_2.

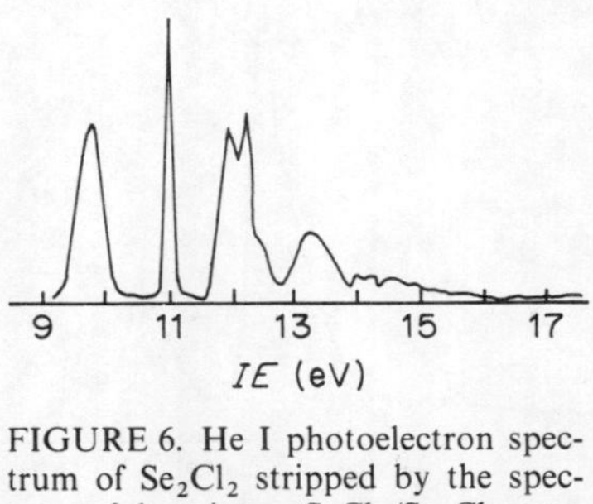

FIGURE 6. He I photoelectron spectrum of Se_2Cl_2 stripped by the spectrum of the mixture $SeCl_2/Se_2Cl_2$, product of the reaction $Se + Cl_2$. *Reproduced by permission of the Royal Society of Chemistry from Reference 62*

TABLE 9. Ionization energies (eV) for S_2Cl_2[a] and Se_2Cl_2[b]

S_2Cl_2			Se_2Cl_2		
Assignment[c]	Experimental	Calculated (CNDO)	Assignment[c]	Experimental	Calculated (VEOMP)
n_S^-(17*a*)	(10.1)[d]	11.02	n_{Se}^-(26*a*)	9.81	10.00
n_S^+(16*b*)	(10.3)	11.30	n_{Se}^+(25*b*)	9.81	10.30
n_{Cl} (16*a*)	11.43	12.32	n_{Cl} (25*a*)	11.05	11.01
n_{Cl} (15*b*)	12.20	13.59	n_{Cl} (24*b*)	12.03	12.52
n_{Cl} (15*a*)	12.52	14.05	n_{Cl} (23*b*)	12.28	12.98
n_{Cl} (14*b*)	(12.6)	14.13	n_{Cl} (24*a*)	12.50	13.00
σ_{SS}(14*a*)	14.07	15.88	σ_{Se-Se}(23*a*)	13.41	13.61
σ_{SCl}^-(13*b*)	15.65	16.05	σ_{SeCl}^-(22*b*)	13.65	13.66
σ_{SCl}^+(13*a*)	(17.02)	16.72	σ_{SeCl}^+(22*a*)	14.66	14.34
$3s_S$(12*b*)	(19.3)	23.29			

[a] From Reference 64.
[b] From Reference 62.
[c] The figures and letters in parentheses refer to the numbering and symmetry of MOs.
[d] Values in parentheses are ill-defined maxima.

Nagy-Felsobuki and Peel[62] detected, in the He I spectrum of the gaseous products of the reaction of chlorine with powdered selenium at room temperature, bands arising from Se_2Cl_2 species. Figure 6 shows the stripped spectrum of Se_2Cl_2. It is very similar to that of S_2Cl_2, reported by Solouki and Bock[64], and the assignment follows the same lines [see Table 9, where the results of valence-electron-only-model-potential (VEOMP) calcu-

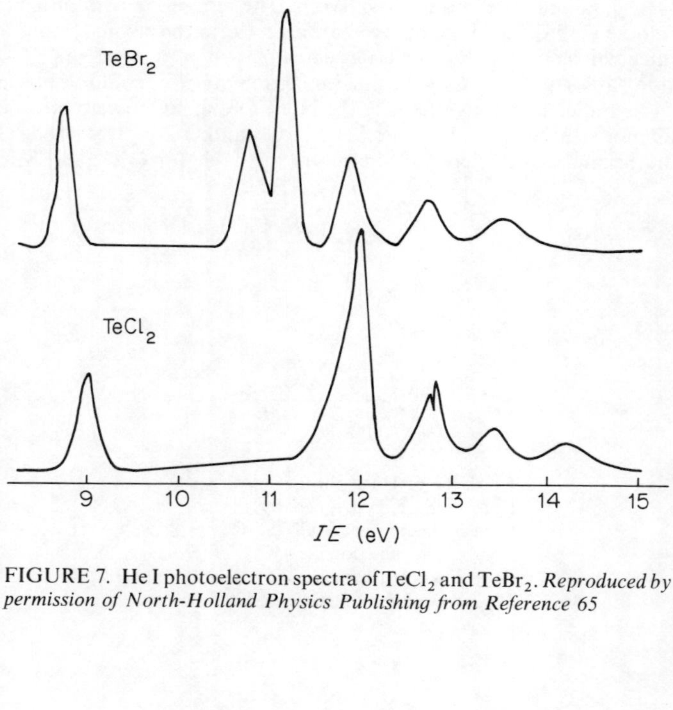

FIGURE 7. He I photoelectron spectra of $TeCl_2$ and $TeBr_2$. *Reproduced by permission of North-Holland Physics Publishing from Reference 65*

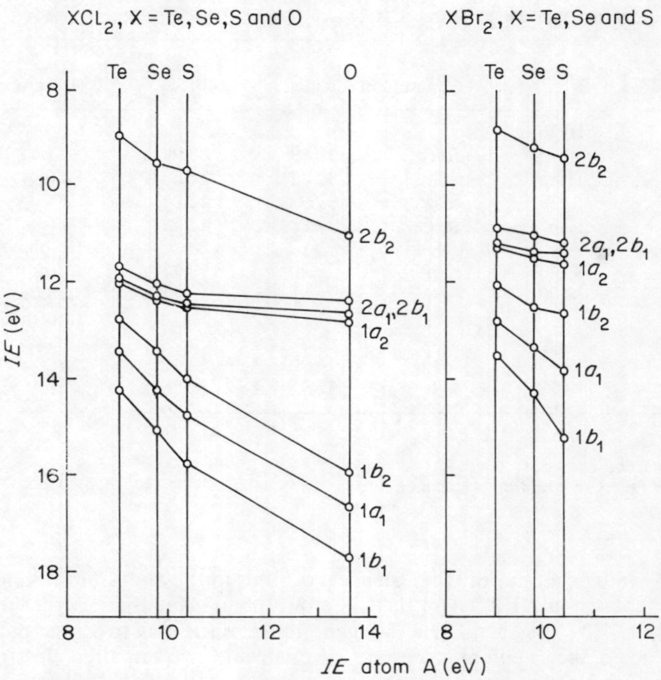

FIGURE 8. Correlation diagram of the experimental ionization energies of XCl_2 (X = Te, Se, S and O) and XBr_2 (X = Te, Se and S) molecules versus the first atomic ionization energies. *Reproduced by permission of North-Holland Physics Publishing from Reference 65*

lations are also reported]. The agreement between the experimental and theoretical data confirms the validity of the geometry of Se_2Cl_2 optimized by the VEOMP calculations: C_2 symmetry with Se—Se and Se—Cl bond lengths of 2.40 and 2.20 Å, respectively, a Cl—Se—Se bond angle of 98° and a dihedral angle of 90°.

The He I spectra of $TeCl_2$ and $TeBr_2$, measured by Jonkers and coworkers[65], are shown in Figure 7. The assignment, supported by both non-relativistic and relativistic Hartree–Fock–Slater (HFS) calculations, is analogous to that for $SeCl_2$ and $SeBr_2$. In Figure 8 the trends found in the He I spectra of XCl_2 and XBr_2 molecules are presented.

C. Pseudohalide Derivatives

The transient molecules ASeCN (A = Cl, Br) were prepared and detected by UPS for the first time by Jonkers and coworkers[66], by passing the corresponding molecular halogens over freshly prepared silver selenocyanate (AgSeCN) deposited in the tip of the spectrometer inlet system. The photoelectron spectrum was assigned on the ground of comparison with the already known UPS data for the ASCN (A = Cl, Br) molecules[67] and with the results of HFS transition-state calculations. From a simple MO approach with the molecules constructed from three fragments, i.e. two separate atoms and one cyanide

TABLE 10. Ionization energies (eV) for AXCN (A = Cl, Br; X = S, Se)

Band label	ClSCN[a]	BrSCN[a]	Assignment	ClSeCN[b]	BrSeCN[b]	Assignment
	(10.45)	(10.26)				
1	10.52	10.32	$3a''(\pi_{AS})$	10.19	9.99	$3a''(\pi_{ASe})$
2	12.67	11.89	$8a'(n_A)$	12.43	11.69	$8a'(n_A)$
3	13.05	12.39	$7a'(\pi_A)$	~12.9	12.24	$2a''(\pi_A)$
4	13.28	12.65	$2a''[n_A, \pi_{C\equiv N}$ (in-plane)]	12.82	12.56	$7a'[\pi_{C\equiv N}$ (in-plane)]
5	13.66	13.43	$6a'(p\sigma_{SCN})$	13.49	13.25	$6a'(n_N)$
6	14.47	14.00	$1a'(\pi_b)$[c]	13.85	13.54	$1a''[\pi_{C\equiv N}$ (out-of-plane)]
7	15.44	14.93	$5a'(\pi_b)$[c]	14.90	14.41	$5a'(\pi_b)$[c]
8	17.55	17.18	$4a'(\sigma_{S-C})$	16.95	16.76	$4a'(\sigma_{Se-C})$

[a] From Reference 67.
[b] From Reference 66.
[c] Bonding orbital delocalized on the whole molecule.

group, taking into account only the valence p orbitals for Se and X atoms and the 2σ, 3σ and 1π fragment orbitals (FROs) for the CN group, eight orbitals are expected to give rise to bands in the He I spectrum. The halogen atom contributes two lone pairs and one unpaired electron, the selenium atom one lone pair and two unpaired electrons and the cyanide fragment one lone pair ($2\sigma = n_N$), two filled π and one half-filled 3σ FRO. The selenium atom shares one of its unpaired electrons with the halogen atom and the other with the CN fragment (3σ) giving rise to two MO ionization phenomena, while six more or less lone-pair or π-type ionizations are expected. Table 10 reports the observed *IE*s and the proposed assignment for both ASeCN and ASCN. The overall UP spectra and even various features of the corresponding bands in both series show detailed agreement. Spectral comparison between ClXCN and BrXCN (X = S or Se) indicates that bands 1, 6, 7 and 8 show shifts due to significant central atom character, in agreement with the proposed assignments.

The same authors[66,68] measured the He I and He II spectra of the stable compound $Se(CN)_2$, which are shown in Figure 9 and whose assignment parallels that for SeA_2 (A = Cl, Br)[65], with the obvious difference that the halogen lone pairs are replaced by π_{CN} FROs and two nitrogen lone-pair ionizations (n_N) appear.

At variance with the ASeCN molecules, methyl selenocyanate (MeSeCN) is a stable compound. Its photoelectron spectrum was studied[18] together with that of the thio analogue, (MeSCN), investigated by UPS also by other workers[69]. The spectra of the two compounds look very similar (see Figure 10), the only appreciable difference being a parallel shift by about 0.3 eV of all bands to lower ionization energy in the selenium derivative, consistent with the lower electronegativity of Se compared with S. The assignment, supported by CNDO and *ab initio* calculations and comparison with related simpler molecules, therefore follows the same general lines. The first band, significantly decreasing in intensity with respect to the following bands in He II, arises from the ionization of an essentially non-bonding orbital, i.e. the halogen lone pair (π_X; X = S, Se). The calculations suggest a certain degree of conjugation with a $\pi_{C\equiv N}$ component. The second band corresponds to a p-type orbital of X, which is close to bisecting the CXC angle. The shape and intensity of the third band (a very intense adiabatic transition,

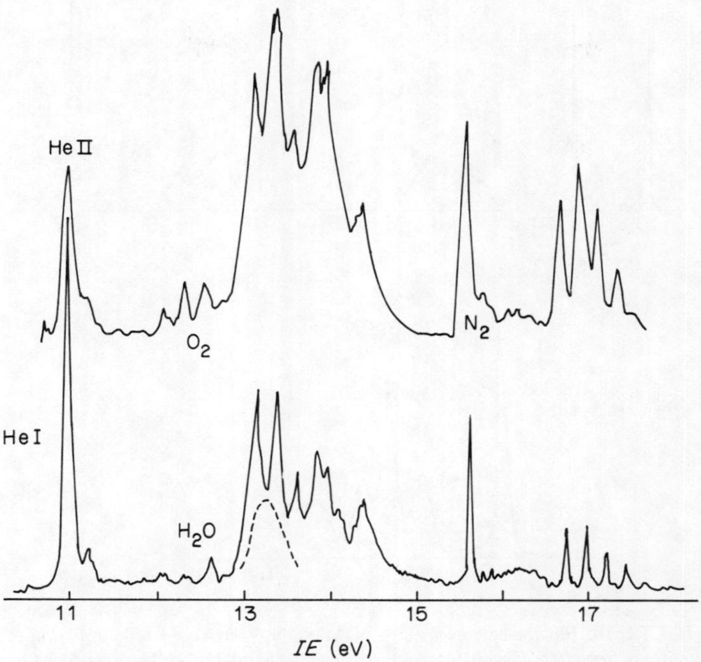

FIGURE 9. Photoelectron spectra of $Se(CN)_2$. *Reproduced by permission of Taylor and Francis from Reference 66*

followed by a broad structure with an irregular vibrational progression) suggest that it should be attributed to two ionization processes, namely of the two π orbitals of the XCN group, mainly formed by C and N atomic contribution. The fourth band is due to ionization of a σ-bonding orbital of the system C—X—C—N.

It is interesting to study the effect on the MOs of changing A in the series AXCN (A = Cl, Br, CH_3; X = S, Se). Figure 11 shows a correlation diagram between ClSeCN, BrSeCN and MeSeCN. Apart from the obvious appearance of the halogen lone-pair ionizations in the first two compounds, the following points deserve attention:

(i) the HOMO does not shift much along the series (10.19 eV for A = Cl, 9.99 eV for A = Br, 9.68 eV for A = Me), despite the opposite inductive effect of Cl and Br versus Me. This is probably due to the destabilizing interaction of the selenium lone pair with the halogen lone pairs, which partly counterbalances the electron-withdrawing effect of the halogen.

(ii) the same does not occur for the second MO (the fourth of the halogen derivatives) which ionizes at 12.82 eV in ClSeCN and 12.56 eV in BrSeCN, moving to 11.76 eV in MeSeCN. This is in agreement with the nature of this orbital, lying in the plane of molecule.

(iii) the most affected MO is the fifth one of MeSeCN (seventh of ClSeCN and BrSeCN, *IE* 14.90 eV in ClSeCN, 14.41 eV in BrSeCN and 13.37 eV in MeSeCN), consistent with the significant contribution of AOs of all the three groups.

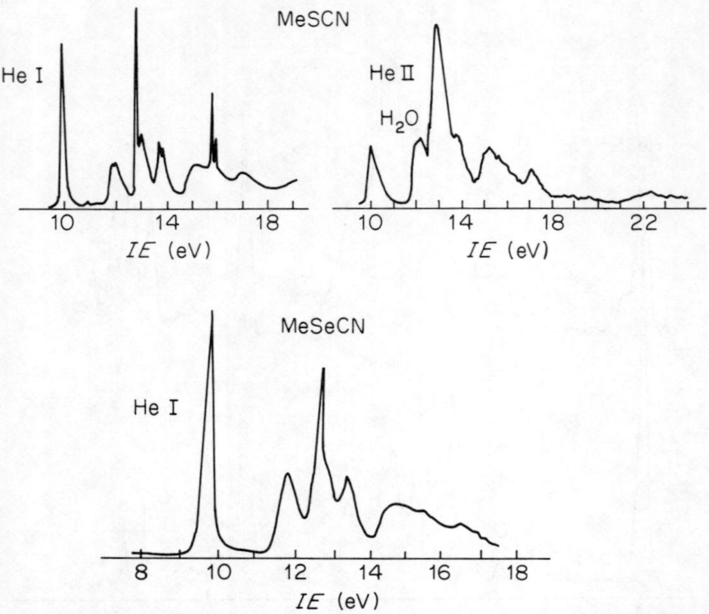

FIGURE 10. Photoelectron spectra of MeSCN (*reproduced by permission of the Royal Society of Chemistry from Reference 18*) and MeSeCN (*reproduced by permission of Elsevier Science Publishers from Reference 69*)

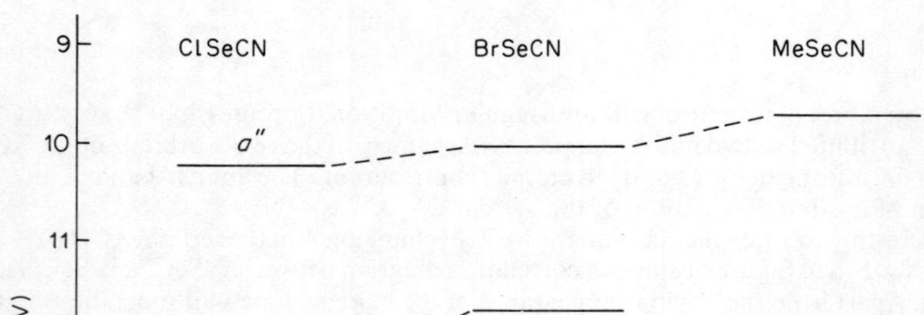

FIGURE 11. Correlation diagram of the ionization energies of the ASeCN (A = Cl, Br and Me) molecules. ClSeCN and BrSeCN from Reference 66; MeSeCN from Reference 18

TABLE 11. Vertical ionization energies (eV) for PF_2X (X = —NCS, —NCSe)

X = —NCS[a]	X = —NCSe[b]	Assignment
10.2	9.8	π_{NCX}
11.9	11.7	n_P
13.9	13.6	π_{NCX}
15.5	15.2	$\sigma_{P—N}$
16.4	16.4	π_F
18.0		$\sigma_{P—F}$

[a] From Reference 71.
[b] From Reference 70.

Andreocci and collaborators[18] also measured the UP spectra of phenyl-substituted selenocyanates and isoselenocyanates, together with those of the corresponding thio analogues. Unfortunately, the dominant features of the spectra, corresponding to ionizations of phenyl-based orbitals, completely masked the ionizations of the —XCN and —NCX groups, preventing any detailed assignment.

The moderately stable (in the vapour phase) difluoro(isoselenocyanato)phosphine, $PF_2(NCSe)$, was studied by UPS, IR spectroscopy and NMR[70]. The photoelectron spectrum was assigned by analogy with those of $PF_2(NCO)$ and $PF_2(NCS)$[71] (see Table 11). The first and third bands, peaked at 9.8 and 13.6 eV, respectively, arise from $(\pi)^{-1}$ ionizations of the NCSe moiety, the second band, at 11.7 eV, from ionization of the phosphorus lone pair, the fourth band, at 13.6 eV, is related to the P—N bond and the fifth band, at 15.2 eV, is related to a fluorine lone pair.

III. CHALCOGEN HETEROCYCLOPENTADIENES AND THEIR DERIVATIVES

A. Monocyclic Compounds

1. Experimental data

The filled $1a_2$, $2b_1$ and $1b_1$ and empty $3b_1$ and $2a_2$ π MOs of the pentaatomic heteroaromatic rings furan (**1**), thiophen (**2**), selenophen (**3**) and tellurophen (**4**) can be considered to derive from the interaction between those of *cis*-butadiene and the p_π AO of the heteroatom X. By symmetry, the latter can participate only in the b_1 MOs (see Figure 13). Therefore, the influence of X is expected to be much larger on these than on the a_2 MOs.

A comparison of the UPS data for the first two bands of the four compounds shows[72,73] that the energy of one is constant, whereas that of the other changes with the heteroatom. They have therefore been assigned to the $1a_2$ and $2b_1$ MOs, respectively (see Table 12). This assumption, which was also based on the non-bonding shape of the first band in the spectrum of tellurophen, and on the variation of the *IE* values with the heteroatom electronegativity (or *IE* value)[72,73], leads to a reversal of the energy sequence of the π $1a_2$ and π $2b_1$ MOs in tellurophen with respect to the other congeners where the HOMO is the $1a_2$ MO (see Figure 12).

This assignment has been confirmed by the data of the α-monosubstituted derivatives[74]. In fact, methyl or halogen substitution at $C_{(2)}$ (see Table 12) causes an increase in the

TABLE 12. Experimental π_3 and π_2 ionization energy values and their difference (Δ) for furan, thiophen, selenophen and tellurophen and for their 2-methyl derivatives (eV)[a]

Compound	π_3 ($1a_2$)	π_2 ($2b_1$)	$\Delta(\pi_2 - \pi_3)$
Furan	8.88	10.31	1.43
2-Methylfuran	8.37	10.13	1.76
Thiophen	8.87	9.52	0.65
2-Methylthiophen	8.43	9.23	0.80
Selenophen	8.88	9.14	0.26
2-Methylselenophen	8.40	8.96	0.56
Tellurophen	8.88	8.40	0.48
2-Methyltellurophen	8.43	8.20	0.23

[a] The ionization energies were taken from References 72–77.

separation of the first two bands in **1**–**3** and a decrease in **4**, in agreement with the expectation that electron-releasing substituents in the α-position exert a more pronounced destabilizing effect on the energy of the $1a_2$ MO (because of its larger wavefunction coefficient) than on the $2b_1$ MO.

More conclusive support for the reversed sequence in **4** has been obtained[74] from an examination of the correlation between the corresponding IE values for the various congeners. In fact, slopes close to one, very high correlation coefficients and low values of standard errors were obtained when the IE values of $1a_2$ and $2b_1$ bands of a series of substituted furans (selenophens or tellurophens) were plotted against the corresponding values for substituted thiophens. The very poor correlations obtained by plotting the IE_1 ($2b_1$) of tellurophens against IE_1 ($1a_2$) of thiophens and the IE_2 ($1a_2$) of tellurophens versus IE_2 ($2b_1$) of thiophens confirmed the change of ordering[74].

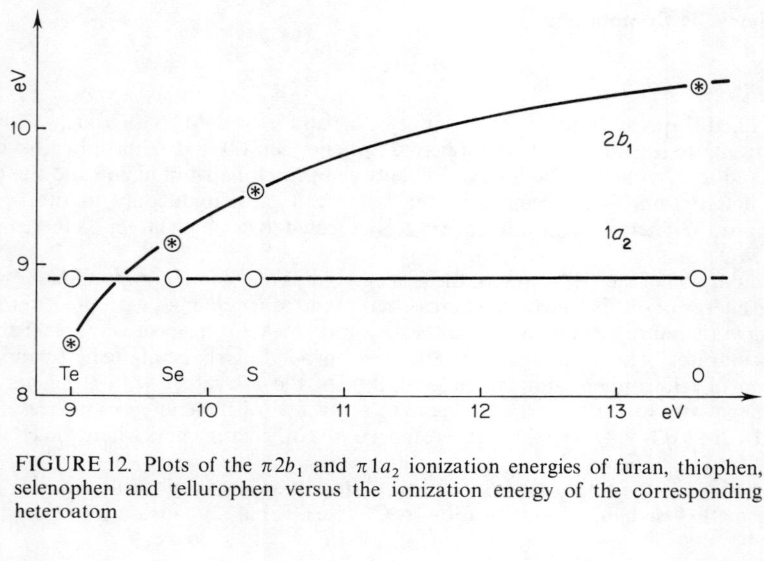

FIGURE 12. Plots of the $\pi 2b_1$ and $\pi 1a_2$ ionization energies of furan, thiophen, selenophen and tellurophen versus the ionization energy of the corresponding heteroatom

The first two bands in the UP spectrum of **3** partly overlap, indicating that the corresponding MOs are close in energy. The vibrational structures are only partly resolved so that there were some doubts[72,73] about the assignment of IE_1 and IE_2 and their precise vertical values.

A contribution toward the solution of this problem has been obtained[74] by plotting the IE_1 and IE_2 values for α-substituted selenophens against the corresponding values for the other congeners. Average interpolated values (for α-substituent = H) of 8.88 (π $1a_2$) and 9.14 (π $2b_1$) eV were obtained, which are in excellent agreement with the values and the assignment reported[72].

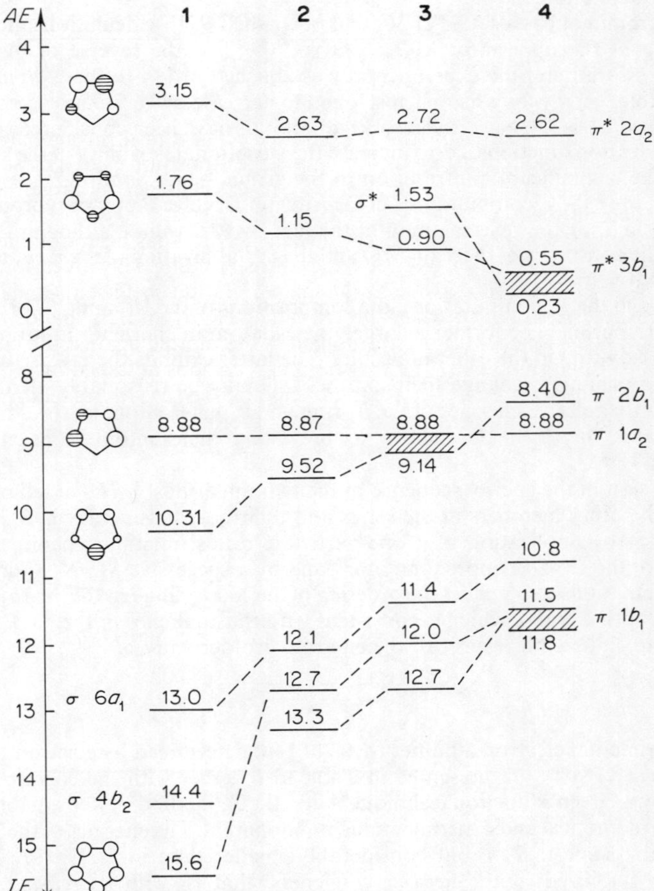

FIGURE 13. Ionization energy (eV) and attachment energy correlation diagram for furan (**1**), thiophen (**2**), selenophen (**3**) and tellurophen (**4**) *Reproduced by permission of North-Holland Physics Publishing from Reference 81*

The assignments of the next bands of **3** and **4** have been attempted[74] by comparing their UP spectra with those of the corresponding tetrahydro derivatives[80]. On this basis, it has been suggested that ionization from the $1b_1$ π MO contributes to the fourth band in both spectra. The third and the remaining component of the fourth band in **4** and the fifth band in **3** have been ascribed to σ_{X-C} ionizations.

According to this assignment, the *IE* value of the innermost π MO of the five-membered congener compounds decreases with increasing heteroatom electronegativity (see Figure 13).

2. *Comparison with theoretical computations*

Several attempts have been made to confirm the empirical assignments on the basis of theoretical calculations.

Findlay's minimal basis set STO-3G and open-shell RHF calculations indicate[82] that the ordering of the uppermost MOs of **3** is $2b_1 < 1a_2$, the reverse of that found for thiophen and furan and the experimentally established order. In these *ab initio* computations, the total energy for selenophen is found to decrease by 76.5 kJ mol^{-1} on going from an sp to an spd basis set. Nevertheless, the 4d orbitals have been considered to assume the role of polarization functions, i.e. to increase the variational flexibility of the system rather than to make a significant contribution to the ground-state bonding[82].

Similarly, the MS X_α computations of De Alti and Decleva[83] do not reproduce the near degeneracy, in 3, or the reversal, in **4**, of the first two *IE* values, although they correctly reproduce the constancy of the π $1a_2$MO and the destabilization of the π $1b_1$ and the σ $6a_1$ and σ $4b_2$ MOs from **1** to **4**.

According to the X_α computations, the compositions of the $1b_1$ and $2b_1$ orbitals change with the heteroatom. The former is largely heteroatom in character in furan, whereas it becomes mostly ring in the other molecules. The latter exhibits the reverse behaviour, so that the two orbitals exchange their nature. This may be reasonably attributed to the difference in the relative energies of the b_1 ring and p_π heteroatom levels along the series. The $6a_1$ and $4b_2$ σ MOs show a prevailing heteroatom character deduced experimentally[74].

The inversion in the energy sequence of the two uppermost π *IE*s of tellurophen, with respect to the other members in the series and their near degeneracy in **3**, are correctly accounted for by application of a one-particle Green's function technique within the framework of the CNDO approximation[84] and by a successive MS X_α approach[81]. The former computations also predict the ordering of the succeeding *IE*s ($6a_1 < 1b_1 \approx 4b_2$) to be the same in both **3** and **4**, which is consistent with the analogies in their UP spectra[72,74] and the band shifts with respect to their tetrahydro derivatives[74,80].

3. *Electron affinity values*

The experimental electron affinities (*EA*s) of **1**–**4**, determined by electron transmission spectroscopy (ETS)[81†], are presented in Table 13 together with the corresponding data computed by a Green's function technique[84] and the X_α method[81]. Reasonable agreement between the theoretical and experimental *EA*s is apparent. The energies of the $2a_2$ (π^*) MO are nearly the same in **2**–**4**, but considerably smaller than in **1**. The $3b_1$ (π^*) MO is increasingly stabilized in the heavier congeners, that is, with *decreasing* heteroatom

† For a brief description of the ETS technique, see Section VIII.

TABLE 13. Electron affinities (eV) of heterocyclopentadienes

Molecule	MO	EA[a] (Green's function)	EA[b] (MS X_α)	EA[b] (exptl.)
Furan	$2a_2$	−3.06	−3.19	−3.15
	$3b_1$	−1.95	−1.82	−1.76
Tiophen	$2a_2$	−2.84	−2.81	−2.63
	$3b_1$	−1.47	−1.57	−1.15
Selenophen	$2a_2$	−2.69	−2.76	−2.72
	$3b_1$	−1.19	−1.53	−0.90
Tellurophen	$2a_2$	−2.59	−2.67	−2.62
	$3b_1$	−0.95	−1.02	−0.55

[a] From Reference 84.
[b] From Reference 81.

electronegativity (see Figure 13)[†]. Both the $3b_1$ and $2a_2$ (π^*) MOs can mix with heteroatom d orbitals of proper symmetry and be stabilized. This interaction is absent in furan where d orbitals at low energy are not available.
A correct evaluation of the effective d–π^* mixing has been obtained by X_α computations. This method, in fact, solves numerically the Schrödinger equation in partial waves. In the LCAO methods the contributions from d orbital basis functions can arise through lack of convergence of the lower angular momentum functions[86–88].

According to the results of the calculations, the d–π^* mixing is negligible in furan, while the two Π anion states of **2**–**4** have some heteroatom d orbital character, in agreement with the observed stabilization. It is to be noted, however, that the largest d orbital character is calculated for the $^2A_2$ anion state of thiophen (ca. 16% of the atomic sphere charge density), suggesting that geometric parameters affect the observed attachment energies (*AE*s, the negative of the electron affinities). The large increase in the chalcogen—carbon bond length (O—C, 1.362 Å[89]; S—C, 1.714 Å[90]; Se—C, 1.885 Å[91]; Te—C, 2.055 Å[92]) can reduce the antibonding interaction between the heteroatom lone pair orbital and the adjacent carbon atoms, and also that between $C_{(2)}$ and $C_{(5)}$ (see the orbital sketches in Figure 13), resulting in a stabilization of the $3b_1$ and $2a_2$ MOs, respectively[81]. This effect is apparent only when vacant orbitals are considered probably because of their more diffuse nature with respect to filled orbitals.

The X_α calculations predict in **2–4** a Σ^2B_2 anion state with a very large heteroatom d character close in energy to the Π^2B_1 anion state. The corresponding resonance is observed between the two σ resonances in the spectrum of **3** and at the lowest energy in the spectrum of **4**. It is not observed, however, in the spectrum of thiophen[85], probably because it overlaps with the π resonance. In fact, the spectra of thiazole and isothiazole display an additional resonance centred at about 1.6 eV[85].

[†] The *IE* values have been taken from References 72–79, 83 and 84 and the *EA* values from References 81 and 85. In the energy region shown in the figure there are some other σ *IE*s deriving from MOs mainly localized at the ring and therefore almost insensitive to the change in heteroatom. They have been omitted for clarity.

B. Bicyclic Derivatives

1. Benzo derivatives

The assignment of the UP spectra of benzo[*b*]selenophen (**5**) and benzo[*b*]tellurophen has been made[93] on the basis of semiempirical PPP calculations and the analysis of relative band intensities observed by changing the ionizing radiation from He I to He II. The first three bands have been found to correspond to the uppermost π orbitals by analogy with the assignment of the spectra of the corresponding oxygen and sulphur derivatives[94–97]. The variation of the ionization energy values ($IE_1 - IE_3$) with the electronegativity of the ionization energy of the heteroatom is similar, but smaller, than that observed for the simple pentaatomic compounds. This smaller sensitivity has been ascribed to the larger number of carbon atoms present per heteroatom and to the lack of symmetry[93].

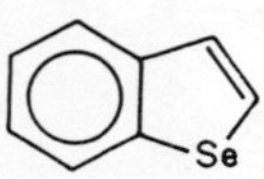

(5)

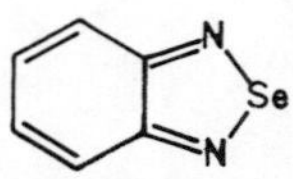

(6)

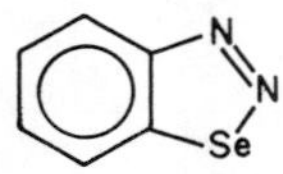

(7)

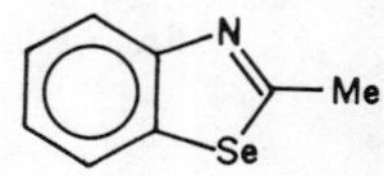

(8)

The photoelectron spectrum of 2, 1, 3-benzoselenadiazole (**6**) is remarkably similar to that of the corresponding thiadiazole, both in overal shape and in the position of the maxima. The same situation occurred with selenophen and thiophen, and has been related to the similar electronegativity of the heteroatoms and to the similarity of their orbital energies[98]. Analogously, the spectra of 1, 2, 3-selenadiazole[99,100], 1, 2, 3-benzoselenadiazole (**7**)[100] and 2-methylbenzoselenazole (**8**)[101] very closely resemble those of the corresponding sulphur compounds[102,103]. The assignment of the first few bands has been reported[99,101].

2. Thienothiophenes and selenolo analogues

A more detailed analysis of the electronic effects of the heteroatom on the orbital energies of five-membered heterocycles has been reported by Gleiter and coworkers[104]. They compared the UP spectra of thieno[2, 3-*b*]thiophen (**9**) and thieno[3, 2-*b*]thiophen (**12**) with the selenolo analogues **10–14**.

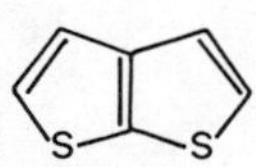

(9)

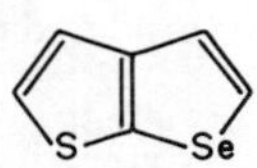

(10)

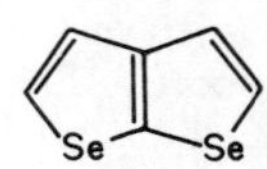

(11)

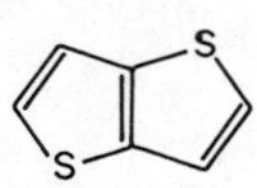

(12)

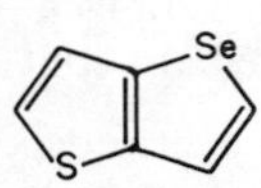

(13)

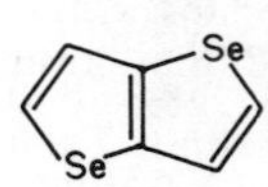

(14)

The UP spectra of the series **9–10–11** all appear very similar, as do the UP spectra of the series **12–13–14**. HMO calculations reproduce the significant trends observed (see Table 14).

TABLE 14. Measured vertical ionization energies (*IE*s) and Hückel MO theory orbital energies (ε) (eV) of **9**–**14**[a]

Compound	Assignment	ε	*IE*
9	$4b_1$	-8.48	8.32
	$3a_2$	-8.60	8.41
10	$8a''$	-8.32	8.28
	$7a''$	-8.49	8.28
11	$4a_2$	-8.16	8.16
	$5b_1$	-8.45	8.16
12	$4a_u$	-8.25	8.10
	$3a_u$	-8.94	8.61
13	$8a''$	-8.19	8.08
	$7a''$	-8.61	8.39
14	$5a_u$	-8.09	8.05
	$4a_u$	-8.50	8.20

[a] *Reproduced by permission of the American Chemical Society from Reference 104. Copyright (1977) American Chemical Society.*

Replacement of sulphur by selenium has been considered as a minor perturbation. Its effects have in fact been described by simple perturbation theory applied to the MOs of **9** and **12** computed according to the Hückel method. The HOMO of **9** has a small localization at the sulphur atoms, while the next MO has a large amplitude in these positions. The two orbitals are therefore destabilized to different extents by substitution of selenium for sulphur. This gives rise to a reversal of the ordering in the series **9**–**10**–**11**. Similar shifts are shown by the two highest occupied orbitals in **12**. They, however, have the same symmetry and interact considerably, but the crossing (along the series **12**–**14**) is not allowed. Their shape corresponds to the initial stages of an 'avoid crossing'[104].

IV. SELENIDES AND TELLURIDES

A. Alkyl Chalcogenides

The Group IV hydrides have the valence shell structure $(1a_1)^2$ $(1b_2)^2$ $(2a_1)^2$ $(1b_1)^2$. Ionization from the $1b_1$ orbital results[13,14] in a sharp and intense peak, characteristic of non-bonding electrons, followed by a short vibrational progression. The second ($2a_1$) and third ($1b_2$) band systems show long progressions, in agreement with the strong H—X bonding character for these orbitals (see Figure 1).

The $(1a_1)^{-1}$ photoelectron bands are all extremely weak because they are related to orbitals which have mainly s AO character. For these orbitals the photoionization cross-section near the ionization threshold is very much less than that of p AOs.

The *IE* values (see Table 15) exhibit a clear dependence on the atomic number of the heteroatom. In particular, IE_1 and IE_4 are fairly close to the first and the second *IE* values[105] of the corresponding heteroatom.

There is a sudden loss of vibrational structure around the maximum of the second band of H_2S, H_2Se and H_2Te. It may be due, at least in part, to predissociation into $X^+ + H_2$, which also is found at these energies[106]. In contrast, the $^2A_1$ state of H_2O^+ does not predissociate. The formation of O^+ ions from H_2O occurs at much higher energy than that

TABLE 15. X lone pair and X—C vertical IE values (eV) of alkyl chalogenides, R_2X (X = O, S, Se, Te)[a]

R	$X_{lp}(b_1)$	$\sigma_{(X-C)}(a_1)$	$\sigma_{(X-C)}(b_2)$
Oxygen derivatives			
H	12.62	14.74	18.51
Me	10.04	11.91	13.43
SiH_3	11.17	11.2	14.5
GeH_3	10.40	10.9	13.5
$CH_2—CH_2—$ | $CH_2—CH_2—$	9.65	11.4	13.0
Sulphur derivatives			
H	10.47	13.33	15.47
Me	8.71	11.28	12.68
SiH_3	9.70	11.15	11.71
GeH_3	9.25	10.66	11.30
$CH_2—CH_2—$ | $CH_2—CH_2—$	8.42	10.9	$\geqslant 11.9$
Selenium derivatives			
H	9.88	12.93	14.62
Me	8.40	11.0	12.0
SiH_3	9.18	10.85	11.29
GeH_3	8.84	10.44	10.88
$CH_2—CH_2—$ | $CH_2—CH_2—$	8.14	10.5	$\geqslant 11.4$
Tellurium derivatives			
H	9.14	12.00	13.25
Me	7.89	10.35	11.32
SiH_3	8.63	10.23	10.83
GeH_3	8.34	9.93	10.56
$CH_2—CH_2—$ | $CH_2—CH_2—$	7.73	10.0	10.7

[a] R = H, from Reference 14; R = MH_3, from Reference 107; R = C_4H_8, from Reference 108.

of the $^2A_1$ state and the $^2A_1$ band exhibits vibrational structure extending over the whole band[13,14].

Analogously, the sharp, intense band at lowest IE in each of the $(MH_3)_2X$ molecules (M = C, Si, Ge; X = O, S, Se, Te) has been assigned to the $1b_1$ MO, the X_{np} lone pair orbital[107]. The trend of IE_1 parallels that observed[14] for the corresponding hydrides (see Table 15).

The second and the third bands correspond to the $3a_1$ and $2b_2$ bonding orbitals, respectively. They shift to lower values with increasing atomic number of X and M. All are smaller than the corresponding hydride values, indicating an electron-releasing capability of the MH_3 groups.

The variation of the first IE with M is C < Ge < Si in each series, rather than C > Si $\geqslant$ Ge, as might have been expected on the basis of electronegativity changes. This effect has been ascribed to π-donation from the X lone pair towards the nd AOs of M.

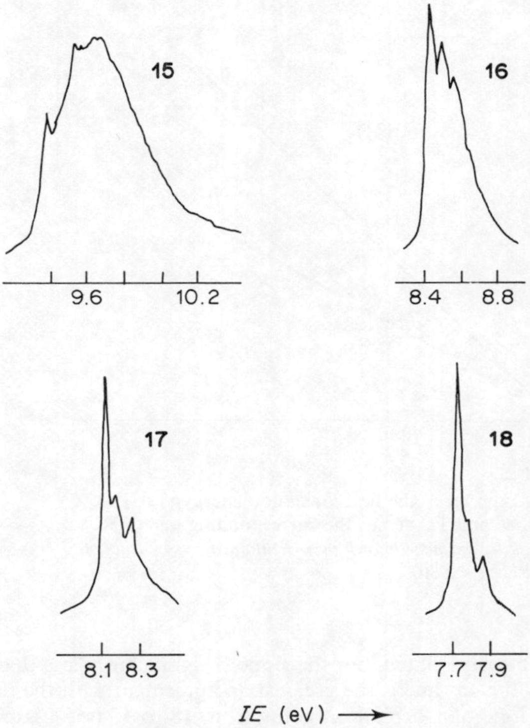

FIGURE 14. First band of the He I photoelectron spectra of $(CH_2)_4X$ (X = O, S, Se and Te). *Reproduced by permission of North-Holland Physics Publishing from Reference 80*

The X_{lp} and the two X—C bonding orbitals of the $(CH_2)_4X$ derivatives (**15**–**18**) are pushed to even higher energy with respect to the corresponding MOs of the dimethyl

(15) **(16)** **(17)** **(18)**

derivatives. However, their relative ordering in each compound and by changing X are mantained[108] (see Table 15).

The most peculiar aspect of the spectra of these cyclic compounds is the shape of the first band. The characteristic non-bonding aspect observed in the previous series is still present in **18** and, to a lesser extent, in **17**, while the band assumes an increasingly bonding structure on going towards tetrahydrofuran (see Figure 14).

The X_{lp} orbital can interact by symmetry with the in-phase combination of the pseudo π orbitals of the CH_2 groups. For overlap and energy reasons this interaction gradually increases on going from **18** to **15**, causing a gradual broadening of the first band. This is confirmed by the plot of the IE_1 values against those of the corresponding hydrides[14].

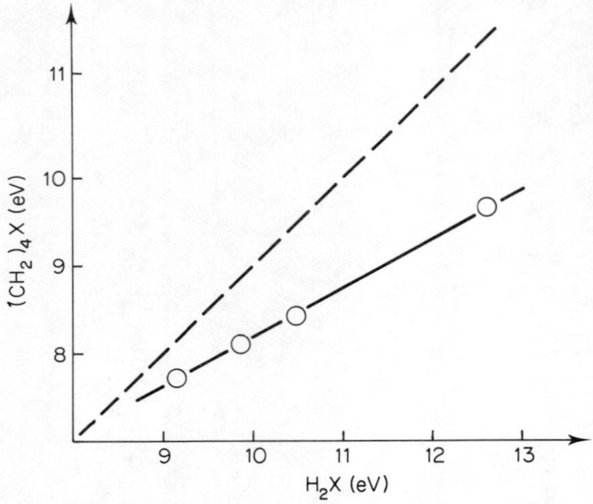

FIGURE 15. Plot of the first ionization energy of $(CH_2)_4X$ (X = O, S, Se and Te) versus the corresponding one of H_2X. *Reproduced by permission of North-Holland Physics Publishing from Reference 80*

The two sets of data are linearly related, but the slope is significantly less than one. The distance of $(CH_2)_4O$ from the 45° line is the greatest, in agreement with the increasingly bonding character of the lone pair orbital on going from **18** to **15** (see Figure 15).

B. Open-chain and Cyclic Ethylene Derivatives

1, 4-Dithiin (**19**) and the related compounds 1, 4-thiaselenin (**20**) and 1, 4-thiatellurin (**21**) posses four π MOs whose energy patterns are shown in Figure 16[110].

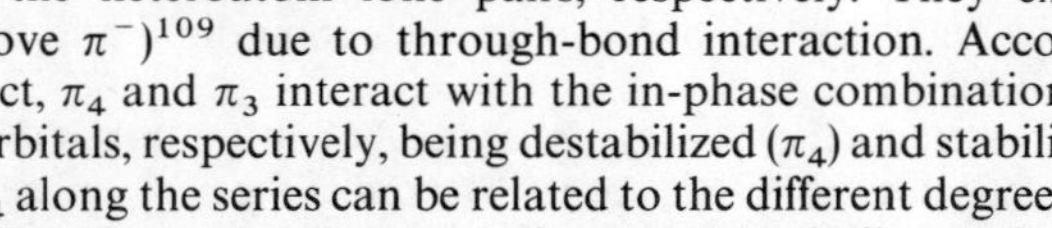

(19) **(20)** **(21)**

The first two bands, π_4 and π_3, are related to the in-phase (π^+) and out-of-phase (π^-) combinations of the heteroatom lone pairs, respectively. They exhibit the 'reverse ordering' (π^+ above π^-)[109] due to through-bond interaction. According to *ab initio* calculations, in fact, π_4 and π_3 interact with the in-phase combinations of the filled and empty ethylenic orbitals, respectively, being destabilized (π_4) and stabilized (π_3). The small perturbation of π_4 along the series can be related to the different degree of mixing with the ethylenic bonds. The energy and the trend of π_3 are very similar to the first *IE* of the free heavier chalcogen atom[105].

π_2 is related to the out-of-phase combination of the π C=C orbitals. The decrease in its ionization energy in the series **19** > **20** > **21** is in agreement with the decrease in the inductive effect of the chalcogen atom due to the change in the order of electronegativities: 3.5 (O), 2.5 (S), 2.4 (Se), 2.1 (Te)[11].

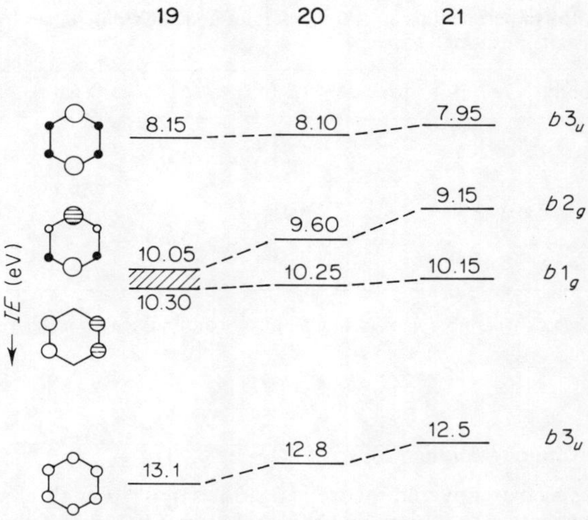

FIGURE 16. Partial energy level diagrams (eV) showing the evolution of the π MOs in 1, 4-dithiin (**19**), 1, 4-thiaselenin (**20**) and 1, 4-thiatellurin (**21**). Three σ levels at energies intermediate between the π_1 and π_2 levels have been omitted for clarity

The UP spectra of some 1-acceptor-2-donor-substituted ethylenes have been analysed to study the 'push–pull' interaction and its effect on the electronic structure[112].

The first three bands of **22**–**25** have been ascribed in order of increasing *IE* value, to ionization from the HOMO, a π MO delocalized over all the molecule, from the out-of-phase combination of the chalcogen lone pair (n^-), and from a carbon—chalcogen σ MO.

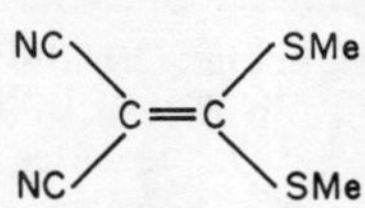

(22)

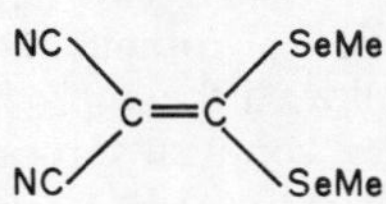

(23)

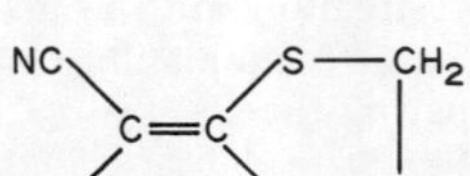

(24)

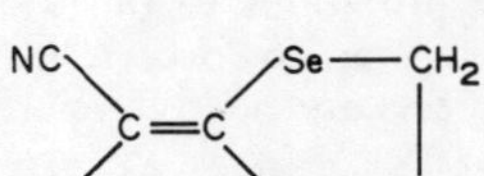

(25)

The spectra of the seleno derivatives are similar to those of the corresponding thio analogues, but the considered *IE* values are generally smaller, in agreement with the difference in the *IE*s of the sulphur and selenium atoms (see Table 16). The shifts are larger (0.4–0.5 eV) for IE_2 and IE_3, whose corresponding MOs have prevailing chalcogen character[112], than for the HOMO (0.3 eV from **22** to **25**). The close similarity ($\Delta = 0.06$ eV) of the IE_1 of **24** and **25** is probably related to the different sizes of the sulphur and selenium atoms, which cause a grater deviation from planarity of the pentaatomic ring in **24** than in **25**. This allows additional mixing of the sulphur outer shell p_z orbitals with other orbitals.

TABLE 16. Experimental ionization energies and their differences (eV) for some push–pull ethylenes[a]

Compound	$IE_1(\pi)$	ΔIE_1	$IE_2(n^-)$	ΔIE_2	$IE_3(\sigma)$	ΔIE_3
22	9.16		9.89		11.20	
		0.31		0.50		0.47
23	8.85		9.39		10.73	
24	8.88		9.94		11.55	
		0.06		0.40		0.40
25	8.82		9.54		11.15	

[a] *Reporduced by permission* of Acta Chemica Scandinavica *from Reference 112.*

C. Alkyl Phenyl Chalcogenides

Photoelectron spectroscopy and theoretical studies[113–115] on alkyl phenyl ethers and thio ethers have suggested that in phenyl chalcogenides a balance of electronic and steric factors may exist which leads to the prevalence of one of the two limiting forms designed as planar (I) and perpendicular (II). In the former, the conjugation between the ring π orbitals and the X_{lp} orbital is more extensive than in the latter, producing a larger splitting of relevant MOs.

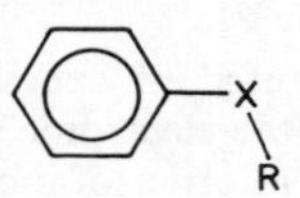

(I)

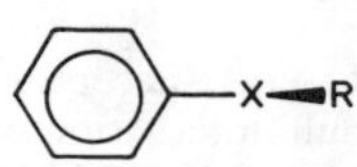

(II)

Schweig and Thon[114] have shown that by increasing the source temperature from 20 to 500 °C the population of rotamer II in thioanisole increases significantly. Mellor and collaborators[115] followed the approach of changing the size of R. They found that the importance of the less conjugated rotamers of phenyl alkyl sulphides and ethers increases in the order H < Me < Et < *i*-Pr < *t*-Bu[115]. In both cases, however, the bands of the two rotamers are not clearly discernible in the spectra, so that arguments must reside on the appearance of shoulders.

Much more propitious to this kind of investigation are the alkyl phenyl selenides studied by Baker and coworkers[116]. The following example illustrates their approach.

The steric effects are clearly apparent comparing the spectra of **26**–**28**.

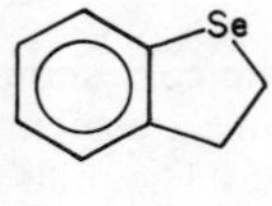

(26)

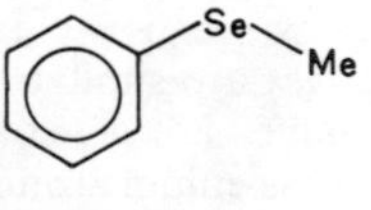

(27)

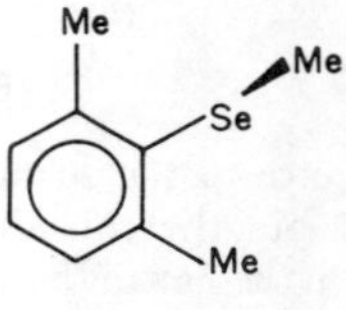

(28)

In each compound the first band corresponds to the HOMO which has prevailing Se_{4p} character. It shows a doublet structure (intensity ratio 1:1) in the spectrum of **27**. The lowest *IE* component (8.0 eV) corresponds to the HOMO of the planar rotamer I and the second (8.3 eV) to the rotamer with reduced conjugation. The single band observed for **26** must correspond to form I, the only rotamer allowed by the constrained structure. Its low *IE* value (7.6 eV) is in fact compatible only with a conjugated system.

The *ortho* disubstitution in **28** makes II the only form allowed. The IE_1 value, 8.7 eV, clearly shows the absence of p-π conjugation.

On going through the series PhXMe (X = O, S, Se, Te) the ratio between the planar and perpendicular conformers gradually decreases. In fact, at room temperature no perpendicular conformer can be seen in the spectrum of anisole[115], the I:II ratio is about 1.5:1 in thioanisole[114] and 1:1 and < 1:2 when X = Se[116] and Te[117], respectively.

This trend is probably related to the balance of several geometric and electronic factors. In fact, the C—X—C angle is much larger when X = O than for the heavier chalcogens favouring the planar rotamer, while the increasing bond length acts in the opposite direction. The n–π overlap decreases when the size of the heteroatom increases, while the n-π energy matching is the best for X = S (benzene, 9.24 eV; the X_{1p} *IE* values of the dimethyl derivatives are listed in Table 15). Finally, the $\pi \rightarrow \sigma^*_{X-C}$ charge-transfer interaction (which favours rotamer II) is energetically favoured in the heavier chalcogenides, where the empty orbital is stabilized by mixing with the heteroatom d orbitals (see Section III.A.3).

Other studies have been carried out on phenyl chalcogenides. The IE_1 of *p*-$NO_2C_6H_4XMe$ (X = O, S, Se) have been correlated with structure and MO characteristics. The ν_{max} value in the UV–visible spectra has been found to increase with increasing value of the first *IE*[118,119]. This finding is in agreement with the fact that the HOMO is mainly localized on the ring-chalcogen part of the molecule and therefore it is sensitive to the nature of X, whereas the lowest unoccupied MO (LUMO) is localized at the NO_2 group in all cases, and therefore its energy is nearly constant. MO calculations on *o*-, *m*- and *p*-$MeSeC_6H_4Y$ (Y = OMe and NH_2) indicate that the substituent with the lowest *IE* of the lone pair (SeMe) makes the largest contribution to the HOMO, and that a substituent with high *IE* contributes more to the second HOMO than to the HOMO[120].

Five bands, two arising from the benzene-like π MOs, two from the two (in-phase and out-of-phase) CO lone pair n^+ and n^- and one from the π MO localized mainly at the heteroatom (π_X) are present in the low *IE* region of the UP spectra of phthalic anhydride and its chalcogen analogues (**29**–**32**)[121].

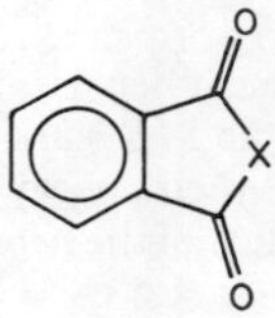

(29) X = O
(30) X = S
(31) X = Se
(32) X = Te

The benzene-like π MOs are nearly insensitive to X. Their *IE* values range from 10.25 and 10.63 eV in **29** to 10.02 and 10.28 eV in **32** indicating that the CO—X—CO groups have a similar electronic effect on the benzene ring. In contrast, the π_X MOs change in a parallel way with the $2b_1$ MO of the related pentaatomic heterocycles going from 11.73 eV in **29** to 8.59 eV in **32**[74,121].

The two lone-pair MOs n^+ and n^- are similarly destabilized because of their effective mixing with the σ_{X-C} orbitals and the changing inductive effect. The electronic distribution at the CO groups also depends on the mesomeric interaction with the p_π AO of X. Other observables, such as the $\nu_{C=O}$ stretching frequencies and the ^{13}C NMR chemical shifts, depend on the electronic distribution in the CO groups. It has been found that these three properties (IE_n, ν and δ) vary in a parallel manner[121].

D. Cyclic Phenyl Chalcogenides

Replacement of an oxygen atom with a heavier chalcogen causes changes in the electronic and geometric structures, the effects of which are visible in the UP spectra. In the previous section we have seen that in phenyl alkyl chalcogenides steric hindrance can cause rotation about the ring—X bond. In cyclic derivatives, relief of strain about the chalcogen atoms can occur by a different conformational change, as shown by the following examples.

From the value of appropriate *IE*s, Pfister-Guillouzo and collaborators[122] obtained information on the preferred conformation of some organometallic zirconium compounds. They suggested that in $(RC_5H_4)_2Zr(Me)SePh$ the benzene ring plane forms a large angle (ca. 90°) with the plane defined by the Zr, Se and $C_{(Ph)}$ atoms, while in $(RC_5h_4)_2Zr(SePh)_2$ the corresponding angles are small (ca. 0°) and therefore a significant n_{Se}–π_{ring} conjugation is present.

In the cyclic derivative **33**, the question arises of whether the heterocycle is planar or not.

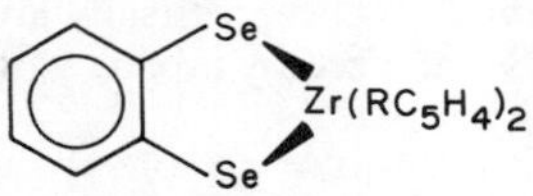

(33)

A qualitative interaction diagram between the selenium unshared pairs and the two uppermost degenerate π benzene orbitals indicated[122] that only in a conformation folded along the Se—Se line the two uppermost π MOs can be nearly degenerate as found experimentally. This result is in agreement with an EHT calculation and the solid-state structure[123].

The UP spectra of the homonuclear and heteronuclear chalcanthrenes (**34**) have been recently analyzed[124–126]. The six bands present in the low-energy region of the spectra

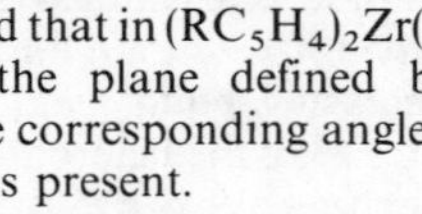

X, Y = O, S, Se, Te

(34)

derive from the interaction of two π MOs of each benzene ring and the heteroatom lone-pair orbitals. Two of the resulting MOs (IE_3 and IE_4) are localized only on the benzene ring (in a perfectly planar conformation), and their energy is only slightly modified from the benzene value (9.24 eV) by the inductive effect of the heteroatoms. The *IE*s obtained, in fact, range from 9.5 and 9.7 eV in dibenzo-*p*-dioxin (X = Y = O) to 9.12 and 9.30 eV in telluranthrene (X = Y = Te)[126] (see Table 17).

The two outermost MOs are the antibonding combinations between $\pi^+_{X,Y}$ and $\pi^-_{X,Y}$ orbitals and appropriate combinations of the benzene π orbitals. If the chalcanthrenes where planar or had the same angle of fold Φ along the line connecting the two heteroatoms, these MOs would become progressively destabilized with increasing size of the heteroatoms (as is well established, see Section III). The HOMO, on the other hand, is only slightly destabilized in phenoxachalcogenins (**34**, X = O, Y = O, S, Se, Te) [7.78 (O), 7.72 (S), 7.74 (Se), 7.61 (Te)eV[124]] and, in the homonuclear chalcanthrenes (**34**, X = Y), the HOMO energy does not decrease monotonically along the series [7.78 (O), 7.94 (S), 7.93 (Se), 7.52 (Te)eV[127]] (see Table 17).

Dipole moment analyses showed that the folding is small and nearly constant in phenoxachalcogenins, whereas it can not be neglected and increases with the mass of the

TABLE 17. Experimental ionization energies (eV) and angles of fold (°) for chalcanthrenes (**34**)[a]

Heteroatoms X	Y	IE_1 ($\pi^+_{X,Y}$)	IE_2 ($\pi^-_{X,Y}$)	IE_3 (π_{ring})	IE_4 (π_{ring})	$\Delta IE_{1,2}$	Φ
O	O	7.78	8.76	9.5	9.7	0.98	163.8
O	S	7.72	8.71	9.4	9.6	0.99	163.4
O	Se	7.74	8.67	9.33	9.5	0.93	162.6
O	Te	7.61	8.66	9.24	9.4	1.05	172.2
S	S	7.94	8.43	9.30	9.45	0.49	142.4
Se	Se	7.93	8.18	9.26	9.4	0.25	139.0
Te	Te	7.52	7.72	9.12	9.30	0.20	124.6
S	Se	7.93	8.32	9.27	9.45	0.39	135.0
S	Te	7.70	8.25	9.23	9.4	0.55	133.3
Se	Te	7.67	8.07	9.21	9.5	0.40	134.0

[a] From References 124–126.

heteroatoms in homonuclear chalcanthrenes[124–126]. The folding causes a reduction of the $n_{X,Y}$–π_{ring} interaction, decreasing the energy of the HOMO.

In a non-planar conformation, the separation of MOs into π and σ is no longer strictly valid and the $\pi^-_{X,Y}$ MOs can also mix with formally σ orbitals. This is reflected, for example, in the energy separation (Δ) between IE_1 and IE_2. Δ is ca. 1 eV in phenoxachalcogenins, whereas it decreases from 1 to 0.2 eV in the homonuclear chalcanthrenes[125].

The conformation of diphenylchalcogenides (PhXPh) and some chalcanthrenes has been investigated by means of UPS and theoretical computations also by Traven and coworkers[127,128]. In the latter compounds, the Φ values are different from those determined in solution[124–126], even though the general trend is maintained.

V. DISELENIDES AND DITELLURIDES

A. Methyl and Phenyl Dichalcogenides

Dichalcogenides represent an ideal series for studying the interactions between p orbitals of adjacent atoms as the geometry and general characteristics of the system are changed[129].

For aliphatic dichalcogenides the first two bands in the UP spectra correspond to the symmetric and antisymmetric combinations of the X_{1p} orbitals n^+ and n^-. When the dihedral angle C—X—X—C (ω) deviates from 90°, the energy separation between the n^+ and n^- orbitals (ΔIE) will increase. In fact, ΔIE has been found in disulphides to be about 0.2–0.3 eV when R = Me, Et and Pr and 0.65 eV in di-*t*-butyl derivatives[130–132]. The value of ω has been computed to be close to 90° and to 98°[129] or 110°[131,132], respectively.

Dimethyl diselenide assumes a stable skew conformation with a dihedral angle ω of 87.5°[133]. Accordingly, the ΔIE value is small (0.23 eV)[134]. EHMO calculations indicate that this Δ value derives in part from through-space splitting and spin–orbit coupling (ca. 0.07 and 0.06 eV, respectively) and in part (ca. 0.1 eV) from interaction of the n^+ and n^- orbitals with the low-lying σ^-_{Se-C} and σ_{Se-Se} orbitals, respectively[134]. According to calculations, these through-bond mixings are larger than the corresponding interactions in dimethyl disulphide. The n^+/n^- splitting does not change in diisopropyl diselenide (as inferred from Figures 1 and 2 in Ref. 135).

In agreement with the shift to lower IE values observed for π, non-bonding and σ_{X-C}

orbitals when selenium is substituted for sulphur, the σ_{X-X} orbital is equally destabilized. It goes from 11.28 eV in dimethyl disulphide to 10.67 eV in dimethyl diselenide[134].

In this class of compounds, the conformation with a dihedral angle of 90° between the two electron lone pairs on the chalcogen atoms minimize lone pair–lone pair repulsion. Alternatively, the n–n repulsion is reduced if the lone electron pairs are delocalized on the substituents as in diphenyl dichalcogenides, Ph—X—X—Ph. Their splitting in fact is about 0.4–0.5 eV in diphenyl diselenide and ditelluride (see Table 18), which corresponds to an ω value of about 70°[134,136].

In the diphenyl dichalcogenides the $n^{\pm}$ orbitals mix with the $\pi_s^{\pm}$ combinations of appropriate symmetry, giving rise to the $IE_{1,2}$ and $IE_{5,6}$ bands[136]. The modest splittings $\Delta IE_{1,2}$ and $\Delta IE_{5,6}$ (unresolved) and the similarity of their average values $\overline{IE}_{1,2}$ and $\overline{IE}_{5,6}$ to the corresponding IEs in PhXMe[114–117] indicate that very little interaction takes place between the two phenyl groups through the dichalcogen bridge.

In addition, the splittings between the π_{ring} (asymmetric or non-interacting) MOs 3 and 4 and the π MOs 5 and 6 can be taken as a measure of the X–ring interaction, which, in turn, depends on conformation. Their absolute energy values and splittings are very similar to those reported for the planar conformer of the PhXMe systems[114–117] (see Table 18). One can therefore infer that the combination of the two PhX moieties to give Ph_2X_2 will not substantially influence the interaction between the chalcogen $\pi_{lone\ pair}$ and the π_{ring} orbitals and hence will not change the conformation of each PhX—fragment.

B. *Peri* Dichalcogenides

The naphthalene-1,8-dichalcogenides (**35a–f**) have received renewed interest as potential electron donors in one-dimensional organic charge-transfer conductors[137,138].

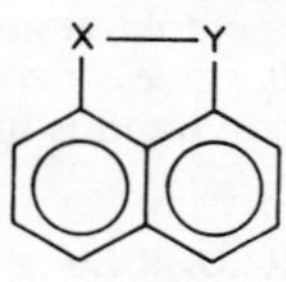

X,Y	X,Y
(**a**) S,S	(**d**) Se,Se
(**b**) S,Se	(**e**) Se,Te
(**c**) S,Te	(**f**) Te,Te

(35)

TABLE 18. IE values, average values and their splitting for some diphenyl dichalcogenides (eV) (from Reference 136). The values in parentheses refer to ionization from the corresponding MOs of the planar conformation of the PhXMe derivatives

	Ph—S—S—Ph			Ph—Se—Se—Ph			Ph—Te—Te—Ph		
Band	IE	$\overline{IE}$	ΔIE	IE	$\overline{IE}$	ΔIE	IE	$\overline{IE}$	ΔIE
1 (n)	8.21			8.06			7.68		
	(8.07)[a]	8.33	0.24	(8.0)[b]	8.30	0.48	(7.6)[c]	7.88	0.40
2 (n)	8.45			8.54			8.08		
3 (π_a)									
	9.36			9.31			9.22		
4 (π_a)	(9.3)[a]			(9.3)[b]			(9.2)[c]		
			0.80			0.52			0.28
5 (π_s)			(0.85)			(0.6)			(0.3)
	10.16			9.83			9.50		
6 (π_s)	(10.15)[a]			(9.9)[b]			(9.5)[c]		

[a] From Reference 115.
[b] From Reference 116.
[c] Estimated from Figure 2 in Reference 117.

Compounds **35a** and **35b** have been found to be planar by Raman studies[139] and X-ray diffraction data[137], respectively, so that coplanar chalcogen–chalcogen lone pair (through-space) interaction is expected to be large.

The first four photoelectron bands of the UP spectrum have been assigned to π MOs on the basis of perturbational arguments and semiempirical calculations[137,138]. The spectra are very similar for all the derivatives, showing in the low *IE* region one isolated band at 7.03–7.14 eV followed by three partially overlapped bands at 8.7–9.3 eV. The n^- orbital is destabilized by 1 eV[138] or 2 eV[137] with respect to n^+ by through-space interaction. Its coupling with the naphthalene HOMO gives rise to the first band and to one component of the second structure.

The HOMO of **35a–f** is destabilized by more than 1 eV with respect to the naphthalene HOMO. Therefore, the destabilizing effect of the *peri* dichalcogenide substitution is similar to that of two amino groups[137].

The constancy of the first *IE* values and the similarity of the spectra indicate that the electron-releasing inductive effect of the heavier chalcogens is counteracted by the reduced second-order perturbation, and that all compounds **35** are planar or nearly planar.

VI. SELENOKETENES, SELENOCARBONYLS AND SELENOFULVALENES

Some of the compounds presented in this section are short-lived or unstable species which have been generated *in situ* from appropriate precursors, at high temperature, inside or close to the ionization chamber of the spectrometer. Most of these studies, which have often been accompanied by high-level theoretical calculations, have been performed by the research teams of Bock and of Schweig.

A. Selenoketenes

Gas-phase thermal decomposition of 1,2,3-selenadiazole afforded the short-lived selenoketene **36a** (yield 95%, $T = 800\,K$[99,140,141]) which has been identified by mass spectrometry and by comparing its UP spectrum with those of thioketene **36b**[142,143] and ketene **36c**[144] and with the results of *ab initio* SCF calculations.

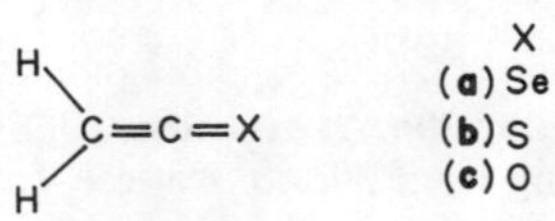

(36)

In particular, the in-phase and out-of-phase combination of the selenium $4p_\pi$ and $\pi_{C=C}$ orbitals give rise to IE_3 (11.7 eV) and IE_1 (8.7 eV), respectively. IE_2 (10.7 eV) corresponds to the in-plane $\pi_{C=Se}$ MO. The succeeding three bands have been assigned to the σ_{C-C-Se} (14.1), σ_{CH_2} (15.3) and σ_{C-CH_2} (17.0 eV) MOs (see Table 19).

The MO ordering is equal to that found for thioketene[142,143], while the relative ordering of IE_4 and IE_5 is reversed with respect to ketene[140,144]. Substitution of selenium for sulphur destabilizes corresponding MOs by 0.2–0.6 eV. Shifts as large as 3.5 eV have been reported for the comparison between the selenium and oxygen derivatives. The largest shifts are associated with MOs whose localization at the heteroatom greatly depends on the nature of X.

Analogously, the products of the gas-phase thermal decompositions of cyclohexene-1,2,3-selenodiazole and cyclooctene-1,2,3-selenodiazole have been identified as cyclopentylideneselenoketene (**37**) and cycloheptylideneselenoketene (**38**), respectively[145]; **37** is formed in high yield (ca. 100%), while cyclooctyne is the main product of the second reaction.

TABLE 19. Experimental ionization energies (eV) of selenoketene (**36a**), thioketene (**36b**) and ketene (**36c**) with assignments

Assignment		**36a**[a]	**36b**[b]	**36c**[c]
$\pi_{X/C=C}$	(b_1)	8.72	8.89	9.8
$\pi_{C=X}$	(b_2)	10.75	11.32	14.2
$\pi_{X/C=C}$	(b_1)	11.6	12.14	15.0
σ_{CCX}	(a_1)	14.15	14.55	16.8
σ_{CH_2}	(b_2)	(15.3)	(15.5)	16.3
σ_{C-CH_2}	(a_1)	(17.0)	(17.5)	18.2

[a] From References 99, 140 and 141.
[b] From References 142 and 143.
[c] From Reference 144.

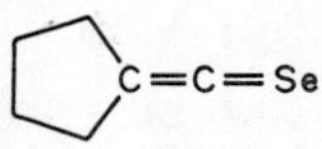

(37)

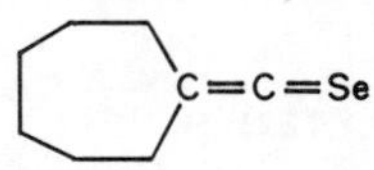

(38)

The spectrum of **37** has been assigned by analogy with those of the corresponding thio and oxa derivatives. The sequence of the three uppermost MOs is the same as that of **36a**. The cycloalkyl substituent destabilizes the MOs by 0.8–1.0 eV. Only the first *IE* value has been reported for **38**.

6-Fulveneketene (**39c**), 6-fulvenethione (**39b**)[146,147] and 6-fulveneselone (**39a**)[99,147] have been generated *in situ* at high temperature from appropriate precursors by means of 'variable-temperature photoelectron spectroscopy'[148]. The UP spectra are very similar to

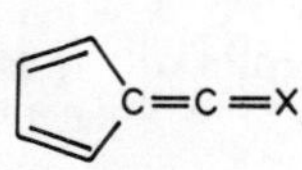

X
(a) Se
(b) S
(c) O

(39)

each other, mainly those of **39b** and **39a**. As expected, a constant shift to lower values of the *IE*s has been observed on going from **39c** to **39a** (see Table 20).

TABLE 20. Experimental ionization energies (eV) of 6-fulveneketene (**39c**), 6-fulvenethione (**39b**) and 6-fulveneselone (**39a**)[a]

Ionization energy		**39a**	**39b**	**39c**
IE_1	(b_1)	8.36	8.37	8.56
IE_2	(a_2)	8.46	8.57	9.06
IE_3	(b_1)	10.78	11.07	12.12
IE_4	(b_2)	10.78	11.07	12.70
IE_5	(a_1)	12.50	12.60	13.00
IE_6	(b_1)	13.00	13.15	

[a] *Reproduced by permission of Verlag der* Zeitschrift fur Naturforschung *from Reference 147.*

The π MOs of **39** can be considered to derive from the symmetry-allowed interaction between the orbitals of *cis*-butadiene and corresponding ketenes. In particular, IE_1, IE_3 and IE_6 in **39b** and **39a** derive from the mixing of the in-phase combination of the π orbitals of the ring with IE_1 and IE_3 of thioketene or selenoketene. IE_2 corresponds to the out-of-phase combination of the ethylenic orbitals, while IE_4 derives from the in-plane $\pi_{C=X}$ MO[147]. This assignment has been obtained experimentally, with the help of semiempirical calculations[99,147].

The thermal decomposition of 1, 2, 3-benzoselenadiazole at intermediate temperature (below about 600 °C) and low pressure produces a short-lived intermediate which, from its UP spectrum, has been identified as benzo [1, 2*b*]selenirene (**40**)[147], an isomer of the stable final product **39a**.

The assignment of the spectrum has been made on the basis of that of benzocyclopropane (**41**)[149,150] and semiempirical calculations on the latter and on (the unknown) benzothioirene.

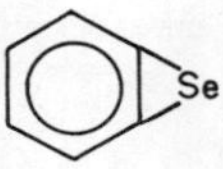

(40)

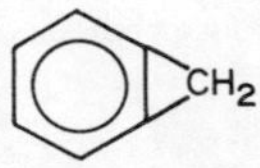

(41)

It is interesting that selenium exerts on the benzene π orbitals of **40** a σ electron-withdrawing effect (0.2 eV) larger than that exerted by the CH_2 groups in **41**, and a π donor effect.

B. Selenocarbonyl Derivatives

1. Selenoaldehydes

Selenoformaldehyde, $H_2C{=}Se$ (**42**), has been prepared (together with other decomposition products) by pyrolysis of its trimer *s*-triselenane (1100 K, 10^{-4} mbar) in a very short pipe close to the target chamber of the spectrometer[151].

The first three bands in its spectrum have been assigned to the corresponding MOs (n_{Se}, $\pi_{C=Se}$ and σ_{Se-C} in order of increasing IE values) on the basis of ΔE^{SCF} restricted Hartree–Fock calculations, taking into account the different electron correlations in the various states. The 'suggested values' thus obtained[151] are compared with the experimental values in Table 21.

TABLE 21. Experimental IE values (eV) for selenoformaldehyde compared with the corresponding data for thioformaldehyde and formaldehyde and with the results of theoretical calculations

		$H_2C{=}Se^a$		$H_2C{=}S^b$		$H_2C{=}O^c$	
Assignment		Exptl.	Theor.	Exptl.	Theor.	Exptl.	Theor.
n_X	(b_2)	8.95	8.8	9.34	9.08	10.9	10.81
$\pi_{C=X}$	(b_1)	11.10	11.1	11.78	11.49	14.5	14.62
σ_{X-C}	(a_1)	13.25	12.9	13.90	13.75	16.2	16.20
σ_{CH_2}	(b_2)	15.1		15.15	15.78	~ 17.0	17.36

[a] From Reference 151.
[b] From Reference 154.
[c] From References 156 and 157.

TABLE 22. Ionization energies (eV) and CNDO/S orbital assignment of tetramethylthiourea (**43**) compared with ionization energies of tetramethyl selenourea (**44**)[a]

43		**44**
IE (eV)	Assignment and electron density (%)	*IE* (eV)
7.80	$n_S(92)$	7.35
8.15	$\pi_S(77)$	7.70
8.60	$\pi_N-(61)$	8.60
11.25	$\sigma_{S-C}(C=22, S=48)$	11.05
12.10	$\pi_{N_2CS}(N=30, C=24, S=7)$	11.95

[a] *Reproduced by permission of* Acta Chemica Scandinavica *from Reference 160.*

The formation of **42** and the assignment of its UP spectrum have been confirmed, producing selenoacetaldehyde, MeC(H)=Se, and selenocarbonyl difluoride, $F_2C{=}Se$, by thermal monomerization of $[MeC(H)Se]_3$ and $(F_2C{=}Se)_2$ at 1000 and 1140 K, respectively[151].

As expected, methyl substitution destabilizes the $\pi_{C=Se}$ MO more than the HOMO, mainly deriving from the Se_{4p} AO (0.8 and 0.45 eV, respectively). Fluorine substitution, instead, stabilizes the n MO (by 0.85 eV) but does not influence the energy of the $\pi_{C=Se}$ orbital, in line with expectation based on the 'perfluoro effect'[152].

Further support for the interpretation of the UP spectrum of the main product obtained from the thermal decomposition of *s*-triselenane has been obtained by the comparison[151] of the bands of its spectrum with the corresponding bands in the spectra of the chemically equivalent compounds thioformaldehyde[153,155] and formaldehyde[156,157]. Table 21 compares the uppermost *IE* values by changing the heteroatom. The lowering of the *IE* values with increasing effective nuclear charge (Se < S ≪ O) is self-evident.

2. Selenoketones

The assignment of the spectrum of tetramethylthiourea (**43**), initially proposed by Pfister-Guillouzo and collaborators[158,159] on the basis of experimental evidence, was subsequently confirmed with the help of semiempirical calculations[160] to be n_S, π_S, π_{N^-}, σ_{S-C} and π_{N_2CS} in order of increasing *IE* values (see Table 22).

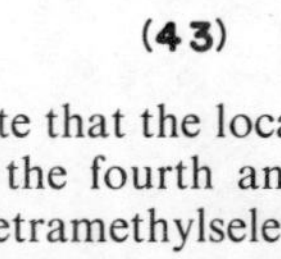

(**43**)

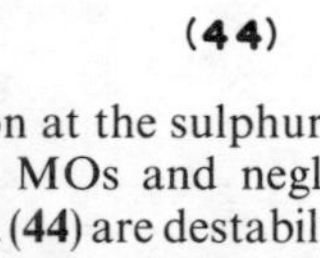

(**44**)

The calculations indicate that the localization at the sulphur atom is large for the first two MOs, moderate for the fourth and fifth MOs and negligible for the third MO. Accordingly, the MOs of tetramethylselenourea (**44**) are destabilized by about 0.45, 0.2 and 0.0 eV, in that order, with respect to the corresponding MOs of **43**. Therefore, both the n_{Se}

and π_{Se} ionizations in the amino-substituted selenocarbonyl group are small (7.35 and 7.70 eV, respectively[160]).

As observed for **43** and **44**, even the n_X and π_X ionizations of the C=X group of compounds **45** are lowered by 0.4–0.5 eV on going from the thio- to the corresponding selenoketones[161].

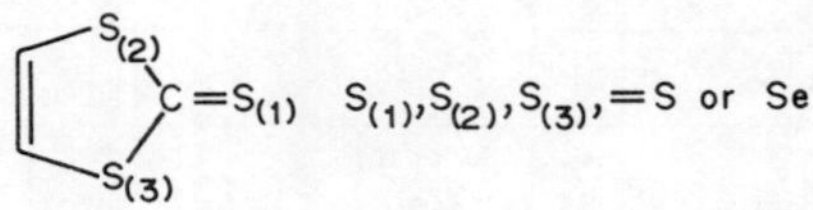

(45)

The absolute *IE* values of the selenocarbonyl group in compounds **45** are 0.35–0.50 eV higher than the corresponding values of **44** because of the smaller electron-releasing capability of the chalcogen substituents with respect to the dimethylamino groups. The first six ionization events from these compounds have been assigned on the basis of semiempirical calculations, perturbational arguments and the comparison of the energy trends of corresponding bands along the series[161].

C. Selenofulvalenes

Schweig and coworkers[162] have found that in the series **46a–c** the ionizations related to MOs largely localized on the heteroatoms are distinctly more affected by selenium for sulphur substitution than the rest of them.

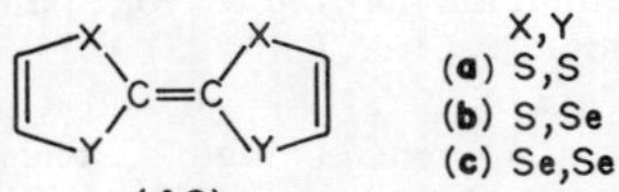

(46)

Surprisingly, the HOMO is stabilized by the selenium atoms (**46a** = 6.70, **46b** = 6.75, **46c** = 6.90 eV). We think that a possible explanation could be based on the shape of the HOMO[162] shown below.

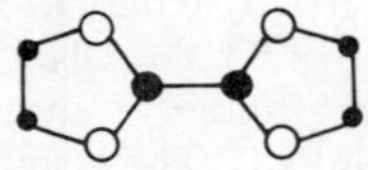

(47)

The two antibonding interactions between each heteroatom and the adjacent carbon atoms could be relieved by the increased X—C bond lengths. The resulting stabilization should overwhelm the destabilization connected with the smaller electronegativity of selenium.

Similarly, the stabilization of the HOMO observed on going from tetramethyltetrathiafulvalene (6.40 eV) to tetramethyltetraselenafulvalene (6.58 eV) has been ascribed to a reduction of the X—C resonance integral, $\beta_{X\ C}$, from − 1.8 to − 1.5 eV[163]. The spectra of both compounds have been interpreted on the basis of molecular orbital calculations based on a zero differential overlap (ZDO) model and semiempirical calculations. The assignment of the first four ionizations agrees with those of **46a** and **46c**[162], whilst the fifth and sixth ionizations are reversed.

TABLE 23. Ionization energies (eV) for $(MeO)_3P$ and $(MeO)_3PX$ (X = O, S, Se)[a]

Compound	$n_P(a)$	$n_O(a)$	$n_O(e)$	$\sigma/n_O(e)$	$\sigma/n_O(a)$
$(MeO)_3P$	9.22	10.54	11.11, 11.3	12.3	13.0
	$n_X(e)$	$n_O(a)$	$n_O(e)$	$\sigma_{PX}(a)$	$\sigma/n_O(a+e)$
$(MeO)_3PO$	10.82	11.36	11.9, 12.1	12.4	12.9
$(MeO)_3PS$	9.16	11.15	11.56, 11.8	12.0	12.7
$(MeO)_3PSe$	8.67	10.93	11.5	—[b]	—[b]

[a] From Reference 164.
[b] Overlapping peaks.

VII. GROUP 5A DERIVATIVES

A. Phosphoroselenoic Acid *O,O,O*-Trimethyl Ester

$(MeO)_3PX$ (X = O, S, Se) have been investigated by UPS[164] in the framework of an extensive study on some phosphite esters[164,165]. On passing from $(MeO)_3P$ to the adducts with oxygen, sulphur and selenium the phosphorus lone pair is stabilized by ca. 3 eV as it becomes the dative $P \rightarrow X$ σ bond, σ_{PX} (see Table 23). Each of the acceptors X features a doubly degenerate lone-pair MO, $n_X(e)$, at low *IE* and a singly degenerate MO, $n_X(a)$, comprising largely s-character, at appreciably high *IE*. The acceptor $n_X(e)$ lone-pair MO replaces the phosphorus lone-pair MO as the lowest *IE* peak when coordination to $(MeO)_3P$ occurs.

B. Phosphine Derivatives

Despite the low molecular symmetry (C_s), the UP spectra of the compounds $R_2P(X)Y$ and $RP(X)Y_2$ (R = Me or F; X = O, S or Se; Y = Cl or Br)[166] were assigned empirically, by comparison with related molecules. Figure 17 shows the He I spectra of the species $MeP(X)Cl_2$, together with that of $MePCl_2$. The assignment and the correlations between the various MOs are also indicated. Table 24 reports the *IE*s for some series of sixteen-electron compounds. In Figure 17 and Table 24 the symbols R, V and T indicate orbitals with radial (R), horizontal (T, tangential) and vertical (V) orientations relative to the bond axes (see Figure 18). The first band of $R_nP(X)Y_{3-n}$ compounds is assigned to an MO exhibiting predominant lone-pair character on X (n_X) in each case. On coordination to X all original R_2PY[69,167,168] and RPY_2[167,168] energies are appreciably stabilized by the strong electron-withdrawing effect of the acceptor X and partially by hyperconjugation. In the spectra of the compounds $MeP(X)Y_2$ and $Me_2P(X)Y$, the energy range beyond the n_X bands may be further subdivided into n_Y (sharper intense band, 11–13 eV) and Me regions (broad featureless band, ca. 14–16 eV) joined by bands due to orbitals possessing large P—R, P—Y and P—X σ-bonding contributions.

A marked low-energy shift of the first band only takes place for the transition $X = O \rightarrow S$ (second to third period); n_{Se} is less shifted relative to n_S and thus reflects the trend in p valence ionization energy.

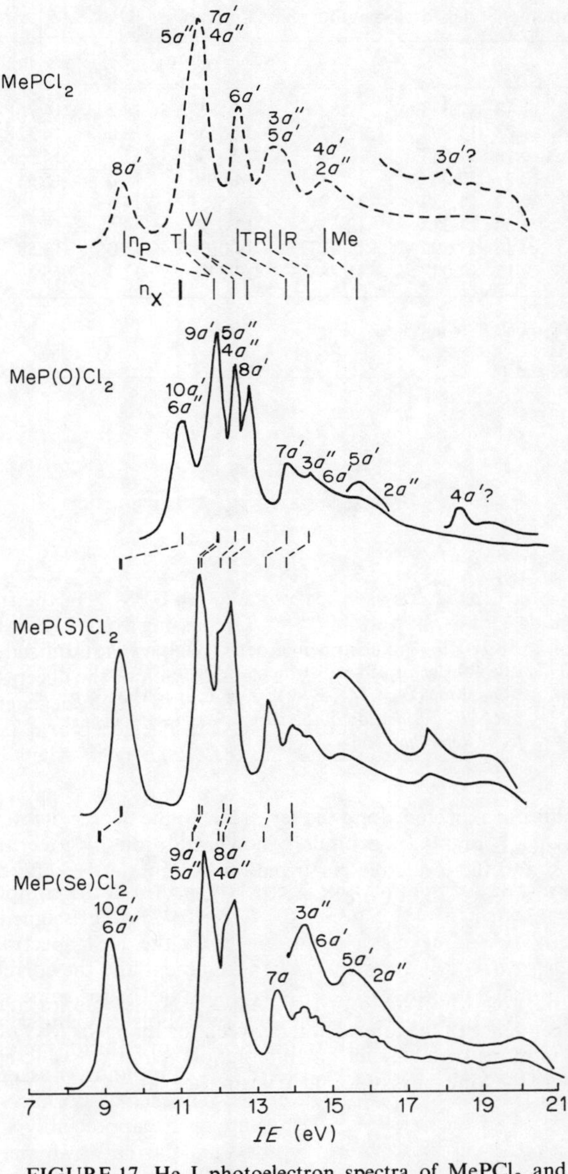

FIGURE 17. He I photoelectron spectra of $MePCl_2$ and $MeP(X)Cl_2$ (X = O, S and Se). The figures and letters on the bands refer to the numbering and symmetry of MOS. *Reproduced by permission of the Royal Society of Chemistry from Reference 166*

TABLE 24. Ionization energies for some phosphine derivatives[a]

Compound	*IE* (eV)							
$MeP(O)Cl_2$[b]	11.43	12.33T	12.82V	13.17V	14.23T	14.75R	15.23?R	16.15
$MeP(S)Cl_2$[b]	9.73	11.89T	12.47V	12.65V	13.66T	14.32R	15.6	17.9
$MeP(Se)Cl_2$[b]	9.16	11.64T		12.47V	13.6T	14.25R	15.62	19.94
$Me_2P(O)Cl$[b]	10.77	12.0T	13.28V	14.12R	15.0	15.53	18.12	
$Me_2P(S)Cl$[b]	9.12	11.53T	12.69V	13.5R	14.55	17.39	19.2	
$Me_2P(Se)Cl$[b]	8.64	11.31T	12.57V	13.67R	14.08	19.6		
$MePCl_2$[c]	9.86	11.89TVV	12.91T	14.0R	15.06	18.58		
Me_2PCl[c]	9.15	11.0T	11.74V	12.72R	13.9	15.3	16.98	

[a] For the meaning of symbols T, V and R, see text.
[b] From Reference 166.
[c] From References 167 and 168.

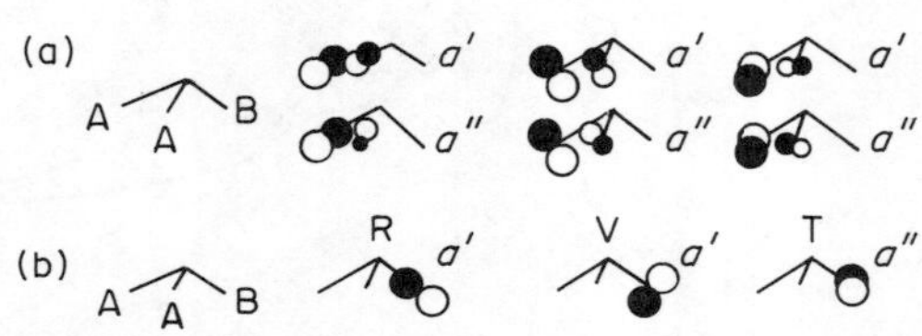

FIGURE 18. Substituent group orbitals for two (a) or one (b) decoupled substituents (A_2 and B) in A_2PB phosphines (A = Me, B = Y; A = Y, B = Me). *Reproduced by permission of the Royal Society of Chemistry from Reference 166*

Differing n_Y shifts are expected along the series due to the degree of interaction between n_Y and the $\sigma_{P\ X}$, $\sigma_{P\ Me}$, and $\sigma_{P\ Y}$ orbitals, which is additionally governed by the energy separation $\alpha_Y \leftrightarrow \alpha_X$ and the inductive perturbation. Again, the *IE* difference is larger on going from X = O to X = S than from X = S to X = Se.

C. Me_2YXMe

In the UPS study on Me_2YXMe (Y = P, As; X = S, Se), Böhm and coworkers[169] emphasize the dependence of the interaction between the lone pair MO of Y and X on the molecular conformation. In fact, such an interaction is not possible either in *A* or *C* conformation where the lone pairs of X and Y lay in the plane XYC_1 and perpendicular to

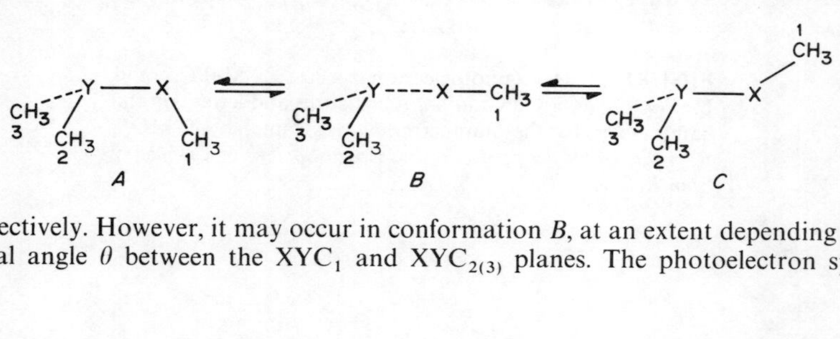

it, respectively. However, it may occur in conformation *B*, at an extent depending on the dihedral angle *θ* between the XYC_1 and $XYC_{2(3)}$ planes. The photoelectron spectra,

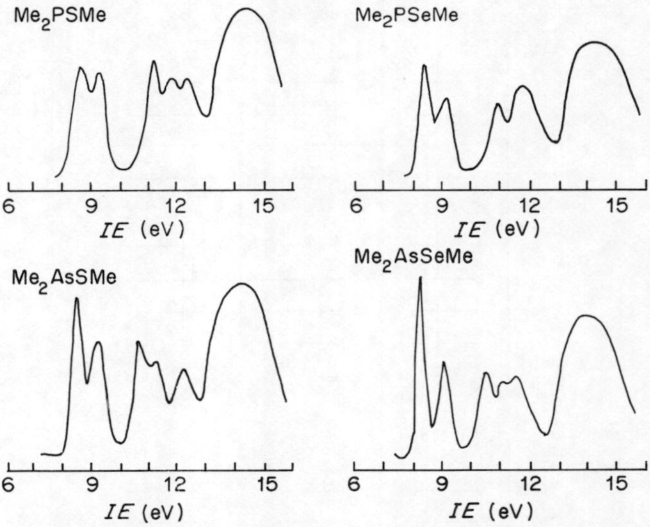

FIGURE 19. He I photoelectron spectra of the Me_2YXMe (Y = P, As; X = S, Se) molecules. *Reproduced by permission of VCH Verlagsgesellschaft from Reference 169*

showing two bands between 8 and 9.5 eV (see Figure 19), which account for the ionization of the out-of-phase and in-phase combination of the two lone pairs, indicate for these molecules a *B*-type conformation. INDO calculations suggest that the conjugation is maximum for θ between 80 and 90°.

VIII. APPENDIX: ELECTRON TRANSMISSION SPECTROSCOPY

From a theoretical and chemical point of view, the electron affinities (*EA*s), associated with electron capture into the normally unoccupied MOs, are as important as the ionization energies. Information on the anion states, however, is scarce. In part this derives from experimental difficulties connected with measuring the electron affinities of molecules which possess stable anions.

One of the most powerful tools of *EA* measurements is electron transmission spectroscopy (ETS)[170–172]. In this technique, an electron beam, selected in energy by a trochoidal monochromator and aligned by a magnetic field, is passed through a gas-filled collision chamber. Electrons of appropriate energy and angular momentum can be temporarily trapped in unoccupied MOs. A retarding voltage is responsible for the rejection of those scattering electrons which have lost a given value of axial velocity. ETS, therefore, makes it possible to determine the energies at which an electron is temporarily trapped in normally unoccupied MOs, from the sharp variations in the electron scattering cross-section.

The most important difference with respect to UPS is that in UPS all the cation states whose energy difference from the neutral ground state is smaller than the energy of the radiation used can be detected, whereas in ETS there is an energy region which cannot be examined. In fact, anion states more stable than the neutral ground state cannot be detected. (see Scheme 1).

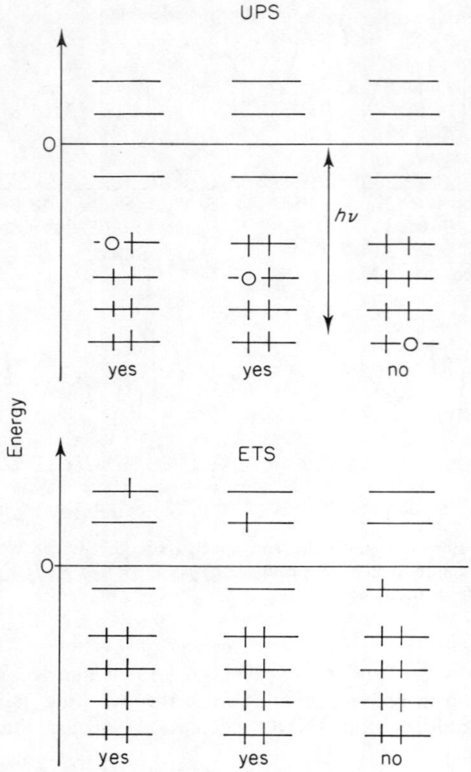

SCHEME 1

IX. REFERENCES

1. A. D. Baker, *Acc. Chem. Res.*, **3**, 17 (1970).
2. A. Hammett and A. F. Orchard, in *Electronic Structure and Magnetism of Inorganic Compounds*, Vol. 1, Chemical Society, London, 1972, p. 1.
3. A. D. Baker, C. R. Brundle and M. Thompson, *Chem. Soc. Rev.*, **3**, 355 (1972).
4. S. Evans and A. F. Orchard, in *Electronic Structure and Magnetism of Inorganic Compounds*, Vol. 2, Chemical Society, London, 1973, p. 1.
5. A. Hammett and A. F. Orchard, in *Electronic Structure and Magnetism of Inorganic Compounds*, Vol. 3, Chemical Society, London, 1974, p. 218.
6. A. F. Orchard, in *Electronic States of Inorganic Compounds: New Experimental Techniques* (Ed. P. Day), Reidel, Dordrecht, 1975, p. 267.
7. R. G. Egdell and A. W. Potts, in *Electronic Structure and Magnetism of Inorganic Compounds*, Vol. 6, Chemical Society, London, 1980, p. 1.
8. D. W. Turner, C. Baker, A. D. Baker and C. R. Brundle, *Molecular Photoelectron Spectroscopy*, Wiley–Interscience, London, 1970.
9. J. H. D. Eland, *Photoelectron Spectroscopy*, 2nd ed., Butterworths, London, 1974.
10. J. W. Rabalais, *Principles of Ultraviolet Photoelectron Spectroscopy*, Wiley–Interscience, New York, 1977.

11. R. E. Ballard, *Photoelectron Spectroscopy and Molecular Orbital Theory*, Adam Hilger, Bristol, 1978.
12. T. Koopmans, *Physica*, **1**, 104 (1934).
13. J. Delwiche, P. Natalis and J. E. Collin, *Int. J. Mass Spectrom. Ion Phys.*, **5**, 443 (1970).
14. A. W. Potts and W. C. Price *Proc. R. Soc. London, Ser. A*, **326**, 181 (1972).
15. S. M. Goldberg, C. S. Fadley and S. Kono, *J. Electron Spectrosc. Relat. Phenom.*, **21**, 285 (1981).
16. S. T. Manson, A. Msezane, A. F. Starace and S. Shahabi, *Phys. Rev. A*, **20**, 1005 (1979).
16a. U. Gelius and K. Siegbahn, *Faraday Discuss. Chem. Soc.*, **54**, 257 (1972).
17. P. H. Connington and N. S. Ham, *J. Electron Spectrosc. Relat. Phenom.*, **31**, 175 (1983).
18. M. V. Andreocci, M. Bossa, C. Furlani, M. N. Piancastelli, C. Cauletti and T. Tarantelli, *J. Chem. Soc., Faraday Trans. 2*, 105 (1979).
19. C. C. J. Roothaan, *Rev. Mod. Phys.*, **23**, 69 (1951).
20. G. G. Hall, *Proc. Roy. Soc. London, Ser. A*, **205**, 541 (1951); **213**, 102, 113 (1952).
21. A. Golebiewski and H. S. Taylor, *Annu. Rev. Phys. Chem.*, **18**, 353 (1952).
22. E. Clementi, *Chem. Rev.*, **68**, 341 (1968).
23. J. A. Pople and D. L. Beveridge, *Approximate Molecular Orbital Theory*, McGraw-Hill, New York, 1970.
24. J. A. Pople, D. P. Santry and G. A. Segal, *J. Chem. Phys.*, **43**, S129 (1965).
25. J. A. Pople and G. A. Segal, *J. Chem. Phys.*, **43**, S136 (1965); **44**, 3289 (1966).
26. D. P. Santry and G. A. Segal, *J. Chem. Phys.*, **47**, 158 (1967).
27. J. A. Pople, D. L. Beveridge and P. A. Dobosh, *J. Chem. Phys.*, **47**, 2026 (1967).
28. N. C. Baird and M. J. S. Dewar, *J. Chem. Phys.*, **50**, 1262 (1969).
29. N. C. Baird, M. J. S. Dewar and R. Sustmann, *J. Chem. Phys.*, **50**, 1275 (1969).
30. M. J. S. Dewar and E. Haselbach, *J. Am. Chem. Soc.*, **92**, 590 (1970).
31. N. Bodor, M. J. S. Dewar, A. Harget and E. Haselbach, *J. Am. Chem. Soc.*, **92**, 3854 (1970).
32. M. J. S. Dewar and G. Klopman, *J. Am. Chem. Soc.*, **89**, 3089 (1967).
33. C. Fridh, L. Åsbrink and E. Lindholm, *Chem. Phys. Lett.*, **15**, 408 (1972).
34. A. Streitwiesser, Jr., *Molecular Orbital Theory for Organic Chemistry*, Wiley, New York, 1961.
35. R. Hoffmann, *J. Chem. Phys.*, **39**, 1397 (1963).
36. A. Modelli and G. Distefano, *Z. Naturforsch., Teil A*, **36**, 1344 (1981), and references cited therein.
37. R. N. Dixon and S. E. Hull, *Chem. Phys. Lett.*, **3**, 367 (1969).
38. O. Edqvist, E. Lindholm, L. E. Selin and L. Åsbrink, *Phys. Scr.* **1**, 25 (1970).
39. J. Berkowitz, *J. Chem. Phys.*, **62**, 4074 (1975).
40. J. M. Dyke, L. Golob, N. Jonathan and A. Morris, *J. Chem. Soc., Faraday Trans. 2*, 1026 (1975).
41. D. G. Streets and J. Berkowitz, *J. Electron Spectrosc. Relat. Phenom.*, **9**, 269 (1976).
42. A. W. Potts and I. Novak, *J. Electron Spectrosc. Relat. Phenom.*, **28**, 267 (1983).
43. A. W. Potts and T. A. Williams, *Chem. Phys. Lett.* **42** 550 (1976).
44. M. G. White, R. A. Rosenberg, S. T. Lee and D. A. Shirley, *J. Electron Spectrosc. Relat. Phenom.*, **17**, 323 (1979).
45. M. Wu and T. P. Fehlner, *J. Am. Chem. Soc.*, **98**, 7578 (1976).
46. D. C. Frost, S. T. Lee and C. A. McDowell, *J. Chem. Phys.*, **59**, 5484 (1973).
47. S. Cradock and W. Duncan, *J. Chem. Soc., Faraday Trans.* **2**, 1262 (1975).
48. S. Cradock and W. Duncan, *Mol. Phys.*, **27**, 837 (1974).
49. Reference 8, p. 66.
50. J. H. D. Eland and C. J. Danby, *Int. J. Mass. Spectrom. Ion Phys.*, **1**, 111 (1968).
51. J. B. Coon, R. E. De Wames and C. M. Lloyd, *J. Mol. Spectrosc.*, **8**, 285 (1962).
52. T. Wentink, *J. Chem. Phys.*, **30**, 109 (1959).
53. A. W. Potts, *J. Electron Spectrosc. Relat. Phenom.*, **11**, 157 (1977).
54. Reference 8, p. 84.
55. D. R. Lloyd and P. J. Roberts, *Mol. Phys.*, **26**, 225 (1973).
56. I. H. Hillier and V. R. Saunders, *Mol. Phys.*, **22**, 193 (1971).
57. For a review, see J. M. Dyke, N. Jonathan and A. Morris, in *Electron Spectroscopy, Theory, Technique and Applications*, Vol. 3, Academic Press, New York, 1979, p. 189.
58. A. B. Cornford, D. C. Frost, F. G. Herring and C. A. McDowell, *J. Chem. Phys.*, **55**, 2820 (1971).
59. B. Solouki, P. Rosmus and H. Bock, *Chem. Phys. Lett.*, **26**, 20 (1974).
60. R. J. Colton and J. W. Rabalais, *J. Electron Spectrosc. Relat. Phenom.*, **3**, 345 (1974).
61. D. M. Leeuw, R. Mooyman and C. A. de Lange, *Chem. Phys.*, **38**, 21 (1979).

62. E. Nagy-Felsobuki and J. B. Peel, *J. Chem. Soc., Faraday Trans. 2*, 148 (1980).
63. E. Nagy-Felsobuki and J. B. Peel, *Chem. Phys.*, **45**, 189 (1980).
64. B. Solouki and H. Bock, *Inorg. Chem.*, **16**, 665 (1977).
65. G. Jonkers, R. Mooyman and C. A. de Lange, *Chem. Phys.*, **50**, 11 (1980).
66. G. Jonkers, R. Mooyman and C. A. de Lange, *Mol. Phys.*, **43**, 655 (1981).
67. D. C. Frost, C. B. MacDonald, C. A. McDowell and N. P. C. Westwood, *J. Am. Chem. Soc.*, **103**, 4423 (1981).
68. G. Jonkers, C. A. de Lange, L. Noodleman and E. J. Baerends, *Mol. Phys.*, **46**, 609 (1982).
69. B. J. M. Neijzen and C. A. de Lange, *J. Electron Spectrosc. Relat. Phenom.*, **18**, 179 (1980).
70. S. Cradock, E. A. V. Ebsworth, M. L. McConnell, D. W. H. Rankin and M. R. Todd, *J. Chem. Soc., Dalton Trans.*, 1925 (1977).
71. S. Cradock and D. W. H. Rankin, *J. Chem. Soc., Faraday Trans. 2*, 940 (1972).
72. G. Distefano, S. Pignataro, G. Innorta, F. Fringuelli, G. Marino and A. Taticchi, *Chem. Phys. Lett.* **22**, 132 (1973).
73. W. Schäffer, A. Schweig, S. Gronowitz, A. Taticchi and F. Fringuelli, *J. Chem. Soc., Chem. Commun.*, 541 (1973).
74. F. Fringuelli, G. Marino, A. Taticchi, G. Distefano, F. P. Colonna and S. Pignataro, *J. Chem. Soc., Perkin Trans.* **2**, 276 (1976).
75. P. J. Derrick, L. Åsbrink, O. Edqvist, B.-Ö. Jonsson and E. Lindholm, *Int. J. Mass Spectrom. Ion Phys.*, **6**, 161 (1971).
76. P. J. Derrick, L. Åsbrink, O. Edqvist, B.-Ö. Jonsson and E. Lindholm, *Int. J. Mass Spectrom. Ion Phys.*, **6**, 177 (1981).
77. J. A. Sell and K. Kuppermann, *Chem. Phys. Lett.*, **61**, 355 (1979).
78. W. von Niessen, W. P. Kraemer and L. S. Cederbaum, *J. Electron Spectrosc. Relat. Phenom.*, **8**, 179 (1976).
79. T. Munakata, K. Kuchitsu and Y. Harada, *J. Electron Spectrosc. Relat. Phenom.*, **20**, 235 (1980).
80. S. Pignataro and G. Distefano, *Chem. Phys. Lett.*, **26**, 356 (1974).
81. A. Modelli, M. Guerra, D. Jones, G. Distefano, K. J. Irgolic, K. French and G. C. Pappalardo, *Chem. Phys.*, **88**, 455 (1984).
82. R. H. Findlay, *J. Chem. Soc., Faraday Trans. 2*, 1397 (1975).
83. G. De Alti and P. Decleva, *Chem. Phys. Lett.*, **77**, 413 (1981).
84. V. Galasso, *J. Mol. Struct. Theochem.*, **3**, 231 (1982).
85. A. Modelli, G. Distefano and P. D. Burrow, unpublished results.
86. K. H. Johnson, *Adv. Quantum Chem.*, **7**, 143 (1973).
87. J. C. Slater, *Adv. Quantum Chem.*, **6**, 1 (1972).
88. R. Roberge and D. R. Salahub, *J. Chem. Phys.*, **70**, 1117 (1979).
89. B. Bak. D. Christensen, W. B. Dixon, L. Hansen-Nygaard, J. Rastrup-Andersen and M. Scottlander, *J. Mol. Spectrosc.*, **9**, 124 (1962).
90. B. Bak, D. Christensen, L. Hansen–Nygaard and J. Rastrup–Andersen, *J. Mol. Spectrosc.*, **7**, 58 (1961).
91. N. M. Pozeev, O. B. Akulinin, A. A. Schapkin and N. N. Magdesieva, *Dokl. Akad. Nauk SSSR, Phys. Chem. Sect.*, **185**, 384 (1969).
92. R. D. Brown and J. C. Crofts, *Chem. Phys.*, **1**, 217 (1973).
93. J.-F. Muller, *Helv. Chim. Acta*, **58**, 2646 (1975).
94. J. H. Eland, *Int. J. Mass. Spectrom. Ion Phys.*, **2**, 417 (1969).
95. P. A. Clark, R. Gleiter and E. Heilbronner, *Tetrahedron*, **29**, 3085 (1973).
96. J. P. Maier and D. W. Turner, *J. Chem. Soc., Faraday Trans. 2*, 196 (1973).
97. M. H. Palmer and M. S. F. Kennedy, *J. Chem. Soc., Perkin Trans.* **2**, 1893 (1974).
98. M. H. Palmer and M. S. F. Kennedy, *J. Mol. Struct.*, **43**, 203 (1978).
99. R. Schulz and A. Schweig, *Angew. Chem., Int. Ed. Engl.*, **19**, 69 (1980).
100. H. Bock, S. Aygen, P. Rosmus and B. Solouki, *Chem. Ber.*, **113**, 3187 (1980).
101. C. Guimon, G. Pfister–Guillouzo, G. Salmona and E. J. Vincent, *J. Chim. Phys. Phys.-Chim. Biol.*, **75**, 859 (1978).
102. M. H. Palmer, R. H. Findlay, J. Neil, A. Ridyard, A. Barrie and P. Swift, *J. Mol. Struct.*, **39**, 189 (1977).
103. R. A. W. Johnston and F. A. Mellon, *J. Chem. Soc., Faraday Trans. 2*, 1155 (1973); M. H. Palmer and M. S. F. Kennedy, *J. Mol. Struct.*, **43**, 203 (1978).

104. R. Gleiter, M. Kobayashi, J. Spanget–Larsen, S. Gronowitz, A. Konar and M. Farnier, *J. Org. Chem.*, **42**, 2230 (1977).
105. R. C. Weast (Ed.), *Handbook of Chemistry and Physics*, Chemical Rubber Co., Cleveland, OH, 1971.
106. V. H. Diebeler and S. R. Liston, *J. Chem. Phys.*, **49**, 482 (1968).
107. S. Cradock and R. A. Whiteford, *J. Chem. Soc., Faraday Trans. 2*, 281 (1972).
108. S. Pignataro and G. Distefano, *Chem. Phys. Lett.*, **26**, 356 (1974).
109. E. Heilbronner, *Isr. J. Chem.*, **10**, 143 (1972); E. Heilbronner and A. Schmelzer, *Helv. Chim. Acta*, **58**, 936 (1975).
110. F. P. Colonna, G. Distefano and V. Galasso, *J. Electron Spectrosc. Relat. Phenom.*, **18**, 75 (1980).
111. V. Shomaker and D. P. Stevenson, *J. Am. Chem. Soc.*, **63**, 37 (1941)
112. D. Betteridge, L. Henriksen, J. Sandstroem, I. Wennerbeck and M. A. Williams, *Acta Chem. Scand., Ser. A*, **31**, 14 (1977).
113. N. L. Allinger, J. J. Maul and M. J. Hickey, *J. Org. Chem.*, **36**, 2747 (1971).
114. A. Schweig and N. Thon, *Chem. Phys. Lett.*, **38**, 482 (1976).
115. P. S. Dewar, E. Ernstbrunner, J. R. Gilmor, M. Godfrey and J. M. Mellor, *Tetrahedron*, **30**, 2455 (1974).
116. A. D. Baker, G. Horozoglu Armen, Y. Guang–di, D. Liotta, N. Flannagan, C. Barnum, M. Saindane, G. C. Zima and J. Grossman, *J. Org. Chem.*, **46**, 4127 (1981).
117. G. Tschmutova and H. Bock, *Z. Naturforsch., Teil B*, **31**, 1611 (1976).
118. C. A. Tschmutova, N. N. Vtyurina and H. Bock, *Dokl. Akad. Nauk SSSR*, **244**, 1138 (1979).
119. F. Bernardi, G. Distefano, A. Mangini, S. Pignataro and G. Spunta, *J. Electron Spectrosc. Relat. Phenom.*, **7**, 457 (1975).
120. C. A. Tschmutova, N. N. Vtyurina, T. V. Komina, I. G. Gazizov and H. Bock, *Zh. Obshch. Khim.*, **49**, 192 (1979).
121. A. Bigotto, V. Galasso, G. Pellizer, G. Distefano, G. C. Pappalardo, J. Bergman and L. Engman, *Spectrochim. Acta. Part A*, **38**, 185 (1982).
122. C. Guimon, G. Pfister–Guillouzo, Ph. Meunier, B. Gautheron, G. Tainturier and S. Pouly, personal communication.
123. B. Gautheron, G. Tainturier, S. Pouly, F. Theobald, H. Vivier and A. Laarif, *Organometallics*, **3**, 1495 (1984).
124. F. P. Colonna, G. Distefano, V. Galasso, K. J. Irgolic, C. E. King and G. C. Pappalardo, *J. Organomet. Chem.*, **146**, 235 (1978).
125. G. Distefano, V. Galasso, K. J. Irgolic and G. C. Pappalardo, *J. Chem. Soc., Perkin Trans. 2*, 1109 (1983).
126. G. Distefano, V. Galasso, K. J. Irgolic and G. C. Pappalardo, unpublished results.
127. O. G. Rodin, V. F. Traven, V. V. Redchenko, M. Yu. Eismont and B. I. Stepanov, *Zh. Obshch. Khim.*, **53**, 2537 (1983).
128. V. F. Traven, O. G. Rodin, M. Yu. Eismont and B. I. Stepanov, in *Proceedings of the European Workshop on UV Molecular Photoelectron Spectroscopy, Rimini, Italy, 1983*, p. 138.
129. A. D. Baker, M. Brisk and M. Gellender, *J. Electron Spectrosc. Relat. Phenom.*, **3**, 227 (1974).
130. H. Bock and G. Wagner, *Angew. Chem., Int. Ed. Engl.*, **11**, 150 (1972).
131. G. Wagner and H. Bock, *Chem. Ber.*, **107**, 68 (1974).
132. M. F. Guimon, C. Guimon and G. Pfister–Guillouzo, *Tetrahedron Lett.*, 441 (1975).
133. P. D'Antonio, C. George, A. H. Lowerey and J. Karle, *J. Chem. Phys.*, **55**, 1071 (1971).
134. G. Tschmutowa and H. Bock, *Z. Naturforsch., Teil B*, **31**, 1616 (1976).
135. T. Hirabayashi, S. Mohamand and H. Bock, *Chem. Ber.*, **115**, 483 (1982).
136. F. P. Colona, G. Distefano, V. Galasso, K. J. Irgolic, G. C. Pappalardo and L. Pope, *J. Chem. Soc., Perkin Trans. 2*, 281 (1981).
137. D. J. Sandman, G. P. Ceasar, P. Nielsen, A. J. Epstein and T. J. Holmes, *J. Am. Chem. Soc.*, **100**, 202 (1978), and references cited therein.
138. H. Bock, G. Brähler, D. Dauplase and J. Meinwald, *Chem. Ber.*, **114**, 2622 (1981).
139. H. E. van Wart and H. A. Sheraga, *J. Phys. Chem.*, **80**, 1823 (1980).
140. H. Bock, S. Aygen, P. Rosmus and B. Solouki, *Chem. Ber.*, **113**, 3187 (1980).
141. H. Bock, B. Solouki, S. Aygen, T. Hirabayashi, S. Mohamand, P. Rosmus and J. Wittmann, *J. Mol. Struct.*, **60**, 31 (1980).
142. H. Bock, B. Solouki, G. Bert and P. Rosmus, *J. Am. Chem. Soc.*, **99**, 1663 (1977).

143. P. Rosmus, B. Solouki and H. Bock, *Chem. Phys.*, **22**, 453 (1977).
144. D. Hall, J. P. Maier and P. Rosmus, *Chem. Phys.*, **24**, 373 (1977).
145. R. Shulz and A. Schweig, *Z. Naturforsch., Teil B*, **39**, 1536 (1984).
146. R. Shulz and A. Schweig, *Tetrahedron Lett.*, 59 (1979).
147. H. Bock, S. Aygen and B. Solouki, *Z. Naturforsch., Teil B*, **38**, 611 (1983).
148. V. Eck, C. Müller, R. Shulz, A. Schweig and H. Verner, *J. Electron Spectrosc., Relat. Phenom.*, **17**, 67 (1969).
149. G. Bieri, F. Burger, E. Heilbronner and J. P. Maier, *Helv. Chim. Acta*, **60**, 2213 (1977).
150. F. Brogli, E. Giovannini, E. Heilbronner and R. Schuster, *Chem. Ber.*, **106**, 961 (1973).
151. H. Bock, S. Aygen, P. Rosmus, B. Solouki and E. Weissflog, *Chem. Ber.*, **117**, 187 (1984).
152. C. R. Brundle, N. A. Kuebler, M. B. Robin and H. Basch, *J. Am. Chem. Soc.*, **94**, 1451 (1972).
153. B. Solouki, P. Rosmus and H. Bock, *J. Am. Chem. Soc.*, **98**, 6054 (1976).
154. H. Bock, T. Hirabayashi and S. Mohamand, *Chem. Ber.*, **115**, 492 (1982).
155. E. Block, E. R. Corey, R. E. Penn, T. L. Renken, P. F. Sherwin, H. Bock, T. Hirabayashi, S. Mohamand, and B. Solouki, *J. Am. Chem. Soc.*, **104**, 3119 (1982).
156. W. Domcke and L. S. Cederbaum, *J. Chem. Phys.*, **64**, 612 (1976).
157. K. Kimura, S. Katsumata, Y. Achiba, T. Yamazaki and S. Iwata, *Handbook of He I Photoelectron Spectra of Fundamental Organic Molecules*, Japan Scientific Society Press, Tokyo, 1981, p. 144.
158. C. Guimon, G. Pfister–Guillouzo, M. Arbelot and M. Chanon, *Tetrahedron*, **30**, 3831 (1974).
159. C. Guimon, D. Gonbeau, G. Pfister–Guillouzo, L. Åsbrink and J. Sandstroem, *J. Electron Spectrosc. Relat. Phenom.*, **4**, 49 (1974).
160. G. Borch, P. Klaeboe, P. H. Nielsen and L. M. Pedersen, *Acta Chem. Scand., Ser. A*, **32**, 259 (1978).
161. J. Spanget–Larsen, R. Gleiter, M. Kobayashi, E. M. Engler, P. Sju and D. O. Cowan, *J. Am. Chem. Soc.*, **99**, 2855 (1977).
162. A. Schweig, N. Thon and E. M. Engler, *J. Electron Spectrosc. Relat. Phenom.*, **12**, 335 (1977).
163. R. Gleiter, M. Kobayashi, J. Spanget–Larsen, J. P. Ferraris, A. N. Bloch, K. Bechgaard and D. O. Cowan, *Ber. Bunsenges. Phys. Chem.*, **79**, 1218 (1975).
164. A. H. Cowley, M. Lattman, R. A. Montag and J. G. Verkade, *Inorg. Chim. Acta*, **25**, L151 (1977).
165. A. H. Cowley, M. Lattman, R. A. Montag and J. G. Verkade, *Inorg. Chem.*, **23**, 3378 (1984).
166. S. Elbel and H. tom Dieck, *J. Chem. Soc., Dalton Trans.*, 1762 (1976).
167. S. Elbel and H. tom Dieck, *Z. Naturforsch.*, **316**, 178 (1976).
168. G. K. Barker, M. F. Lappert, J. B. Pedley, G. J. Sharp and N. P. C. Westwood, *J. Chem. Soc., Dalton* Trans., 1765 (1975).
169. M. C. Böhm, M. Eckert–Maksić, R. Gleiter, J. Grober and D. Le Van, *Chem. Ber.*, **114**, 2300 (1981).
170. L. Sanche and G. J. Schulz, *Phys. Rev. A*, **5**, 1672 (1972).
171. G. J. Schulz, *Rev. Mod. Phys.*, **45**, 378, 423 (1973).
172. K. D. Jordan, and P. D. Burrow, *Acc. Chem. Res.*, **11**, 341 (1978).

The Chemistry of Organic Selenium and Tellurium Compounds Volume 2
Edited by S. Patai
© 1987 John Wiley & Sons Ltd.

CHAPTER **2**

The application of Mössbauer spectroscopy to the study of organotellurium compounds

FRANK J. BERRY

Department of Chemistry, University of Birmingham, Birmingham B15 2TT, UK

I. INTRODUCTION 52
II. MÖSSBAUER SPECTROSCOPY 52
A. The Mössbauer Effect. 52
B. Tellurium-125 Mössbauer Spectroscopy 53
1. General principles 53
a. Sources. 53
b. Instrumentation 54
2. Tellurium-125 Mössbauer parameters 58
a. Chemical isomer shifts 58
b. Quadrupole splittings. 59
c. Magnetic splitting 61
3. Other features of tellurium-125 Mössbauer spectra 61
a. Time-dependent effects 61
b. Absorbers. 62
c. Reference standards 62
d. Limitations 62
III. APPLICATION TO THE STUDY OF ORGANOTELLURIUM COMPOUNDS 62
A. Organo-tellurides(II) and -ditellurides(II) 62
1. Dialkyl and diaryl tellurides and ditellurides. 62
2. Bis(organyltelluro)methanes 66
3. Cyclic derivatives 66
B. Organotellurium(IV) Halides 67
1. Diorganotellurium(IV) dihalides 67
2. Organotellurium(IV) trihalides 71
3. Alkylaryltellurium(IV) halides 72
4. Bis(organyltelluro)methane halides 73
5. Bis(trifluoromethyl)- and bis(pentafluorophenyl)-tellurium(IV) dihalides 74

6. Tetrahaloaryltellurates 75
7. Diaryltellurium(IV) dicarboxylates 76
8. Additivity model for the quadrupole splittings 77
9. Colour in organotellurium(IV) halides 80
C. Heterocyclic Tellurium Compounds. 80
1. Cyclic tellurides. 80
2. Heterocyclic tellurium dichlorides 82
D. Organotellurium Ligands 83
E. Compounds with Tellurium to Metal Bonds. 84
F. Tellurium-containing Charge-transfer Complexes 85
G. Iodine-125 Emission Mössbauer Studies 85
H. Tellurium Complexes with Thiourea and Related Compounds . . . 86
IV. CONCLUSION 87
V. REFERENCES 87

I. INTRODUCTION

The discovery in 1958 by Rudolf Mössbauer of recoil-free nuclear resonance fluorescence[1] gave rise to the technique which is now known as Mössbauer spectroscopy. Since the late 1950s, Mössbauer spectroscopy has developed from an elegant experiment in nuclear physics to a technique which has made important contributions in many areas of science. As an experimental method of scientific investigation it successfully complements other techniques such as visible-light spectroscopy, infrared spectroscopy, nuclear magnetic resonance and the various diffraction methods but, because it has several advantages which give it a special power in a number of important situations and applications, Mössbauer spectroscopy has developed as a particularly useful technique in solid-state and structural chemistry. This chapter initially outlines the basic theory of Mössbauer spectroscopy and the type of instrumentation which is involved, and then considers the application of the technique to the study of various types of organotellurium compounds.

II. MÖSSBAUER SPECTROSCOPY

Several texts[2–8] which deal in detail with the theory and practice of Mössbauer spectroscopy, and also a shorter article[9], are recommended to the reader requiring further information on this technique.

A. Mössbauer Effect

Atomic resonant fluorescence was predicted and discovered shortly after the turn of the century. The process may be envisaged in simple terms as involving the decay of an atom in an excited electronic state to the ground state by the emission of a photon which can then be absorbed by another atom during electronic excitation. The subsequent de-excitation and re-emission of the photon in random directions gives rise to scattering or resonant fluorescence.

Nuclear resonant absorption might be considered in analogous terms. The decay of many radioactive nuclei in an excited state occurs by the emission of gamma-rays and the possibility that these might excite another stable nucleus of the same isotope and give rise to nuclear resonant absorption and fluorescence was recognized early in the 1920s. Although the initial attempts to detect these resonant processes were unsuccessful, the inhibiting role of nuclear recoil and Doppler broadening effects were accurately identified.

Subsequent experiments in the quest were inspired by the awareness that the emitted gamma-rays would be an unusually good source of monochromatic radiation but the degradation of the gamma-ray energies by nuclear recoil and thermal energy constraints persisted as insuperable problems.

Mössbauer successfully eliminated these destructive effects by considering the behaviour of the recoiling nucleus when it was no longer isolated but was fixed in a crystal lattice. When, under these circumstances, the recoil energy is less than the lowest quantized lattice vibrational energy the gamma-ray is emitted without loss of energy due to the recoil of the nucleus. Such recoilless emission is optimized for low-energy gamma-rays from nuclei strongly bound in a crystal lattice at low temperatures and, if incident on another identical matrix containing the same isotope in the ground state, will be resonantly absorbed and subsequently re-emitted in a random direction, i.e. by resonant fluorescence. It is also relevant to note that the energy distribution, or line width, of the gamma-ray depends on the lifetime of the excited nuclear state such that the ease with which the Mössbauer effect can be observed for a particular isotope, and the likelihood of recording chemically significant data, is strongly related to the gamma-ray line width. It will be appreciated from the foregoing that whether or not an isotope of an element exhibits the Mössbauer effect depends on inherent properties of the nucleus which cannot be changed. Selenium, for example, has no Mössbauer active isotopes whereas the tellurium-125 isotope is Mössbauer sensitive with the resonance process being first observed[10,11] in 1962.

B. Tellurium-125 Mössbauer Spectroscopy

1. General principles

a. Sources. The energy levels of an atomic nucleus are modified by the electronic environment of that nucleus. Mössbauer spectroscopy is a means by which these energy levels may be examined and the results interpreted in terms of the local environment of the nucleus. Hence, to probe the energy states of the tellurium-125 nucleus by the Mössbauer effect it is necessary to have a source containing the tellurium-125 nucleus in an excited state. The three convenient nuclear decay schemes which populate the $I = +3/2$ excited state of ^{125}Te emanate from ^{125}Sb, ^{125m}Te and ^{125}I and are depicted in Figure 1. The radioactive excited ^{125}Te nucleus emits a gamma-ray of energy 35.48 keV when decaying from the first excited nuclear state, $I = +3/2$, to the ground state, $I = 1/2$, and it is this 35.48 keV gamma radiation which is used in ^{125}Te Mössbauer spectroscopy. Hence the preparation of a good source is an essential prerequisite of ^{125}Te Mössbauer spectroscopy. It will be seen from Figure 1 that the decay of ^{125}Sb by β-emission with a half-life of 2.7 years involves a complex decay scheme which includes the metastable ^{125m}Te isotope. The complexity is a disadvantage but the method has been used successfully when the ^{125}Sb has been diffused into a copper or rhodium[13] matrix to minimize the nuclear recoil. The ^{125m}Te, state, with a half-life of 58 days, can also be populated by neutron irradiation and single line sources have been obtained from matrices of PbTe[14], ZnTe[15], electrodeposited Te/Pt[16] and TeO_3[17,18], but there are recorded instances of radiation damage affecting the emission line. ^{125}I decays with a half-life of 60 days by electron capture directly to the 35.48 keV level and is therefore the most efficient precursor with a very satisfactory source being conveniently made by the diffusion of ^{125}I into copper foil[19].

The present author has been involved in tellurium-125 Mössbauer studies using an ^{125}I/Cu source, a Pb^{125m}Te source prepared by neutron irradiation of a Pb^{125}Te sample in a thermal neutron flux and both ^{125}Sb/Cu and ^{125}Sb/Rh sources. The relatively short half-lives of the ^{125}I and ^{125m}Te isotopes are a considerable disadvantage in the use of ^{125}I/Cu and Pb^{125m}Te, sources and, given the narrow line widths and relatively large

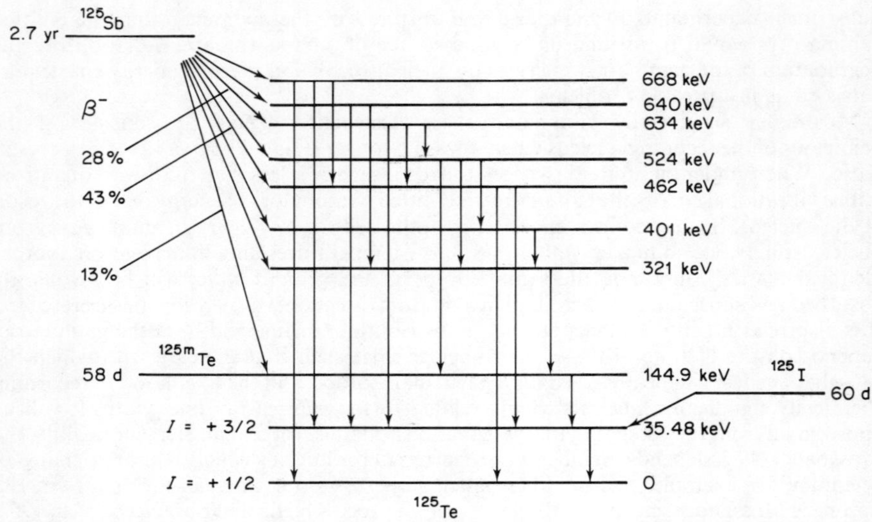

FIGURE 1. Nuclear decay schemes for ^{125}Sb, 125^{m}Te and ^{125}I which populate the 35.48 keV level of tellurium-125

recoil free fractions obtained from ^{125}Sb/Cu and ^{125}Sb/Rh, the sources produced from ^{125}Sb seem to be the most favourable. It is relevant to note, however, that a source[16] prepared from ^{125m}Te which had been radiochemically milked from the parent ^{125}Sb and deposited on platinum gave large recoil fractions and narrow line widths and may have been the best source yet used in ^{125}Te Mössbauer spectroscopy. Such a preparation avoids the high-level background from the ^{125}Sb but retains the advantageously long half-life of the parent. It must be appreciated, however, that the radiochemical manipulations which are required in preparing such a source are a significant disadvantage.

The 35.48 keV excited state of ^{125}Te has a half-life[20] of $1.535 + 10^{-9}$ s, the transition is highly converted $\alpha_T = 12.7$, and the gamma-rays are sometimes difficult to resolve from the intense Kα (27.4 keV) and Kβ (31.00 keV) X-rays, although copper filters can be used to reduce the background radiation, as will be discussed later.

b. Instrumentation. The investigation of tellurium-containing materials by Mössbauer spectroscopy involves the exposure of the samples, called the absorber, to the radiation from the source. An absorber containing tellurium atoms in the same chemical environment as the source will, as a consequence of the presence of ^{125}Te nuclei with 6.99% natural abundance, absorb this 35.48 keV radiation and become raised from the nuclear ground state to the nuclear first excited state. However, the energy states of nuclei depend on the interaction between the nuclei and their electronic environments. Hence, when the environments of the tellurium nuclei in the source and absorber are different, as will occur for example when the source and absorber are different chemical compounds, the nuclear energy levels will no longer coincide and absorption can only occur when the energy of the gamma-ray emitted by the source is modulated by the application of a Doppler velocity (Figure 2). This is achieved by oscillating the source backwards and forwards with respect

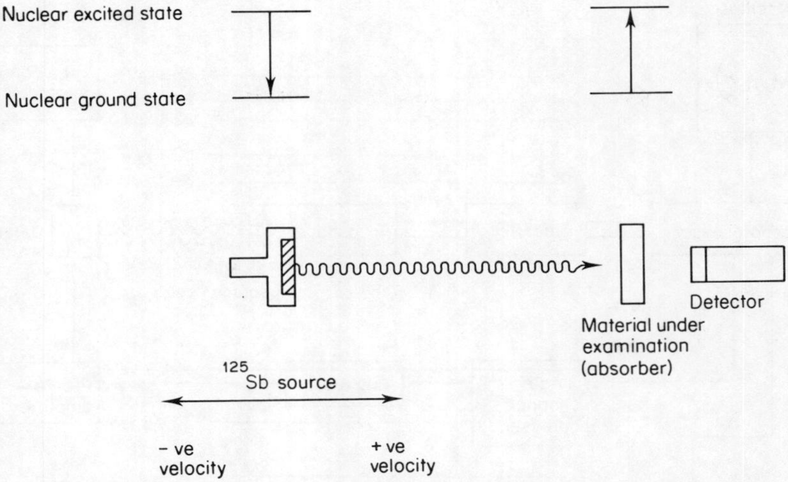

FIGURE 2. Energy of the gamma ray emitted by the source modified by oscillating the source backwards and forwards with respect to the stationary absorber. Reproduced with permission from *Phys. Bull.*, **34**, 517 (1984)

to the stationary absorber and resonant absorption occurs when the energy of the incident gamma-ray just matches the nuclear transition energy of the ^{125}Te nuclei in the absorber. A range of velocities are scanned until maximum absorption occurs and a Mössbauer spectrum is composed of a plot of gamma-ray counts against the velocity of the source with respect to the absorber measured in millimetres per second (Figure 3).

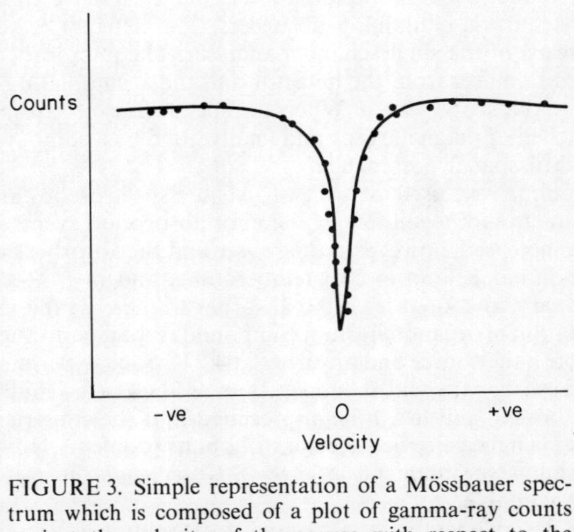

FIGURE 3. Simple representation of a Mössbauer spectrum which is composed of a plot of gamma-ray counts against the velocity of the source with respect to the absorber measured in millimetres per second

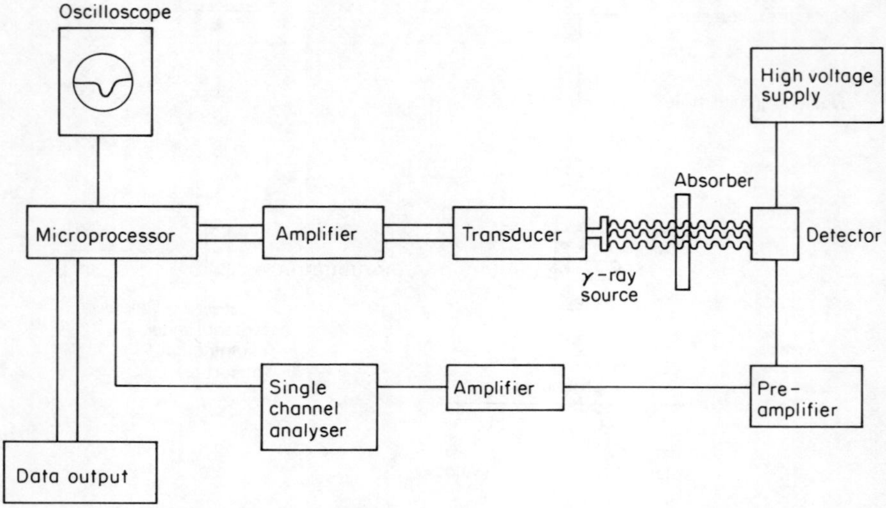

FIGURE 4. Schematic representation of a microprocessor controlled Mössbauer spectrometer

The Doppler velocities of ca. $10 mm s^{-1}$ can be generated by electromechanical transducers, electronic drives and loudspeaker devices and, in the modern Mössbauer spectrometer (Figure 4), are controlled by a microprocessor which also collects the data. The detection of the 35.48 keV gamma-ray can be achieved indirectly through the escape peak using a thin NaI/Th scintillation detector or a gas-filled proportional counter, or directly using a high-resolution germanium or lithium-drifted germanium detector. The persent author has found that a Xe/CO_2 proportional counter, coupled with a copper critical absorber to reduce background radiation under the 6 keV escape peak, gives good results with most tellurium-containing absorbers. The detector is set to monitor the gamma-rays by means of the single-channel analyser. The microprocessor receives and stores the amplified impulses from the detector and the accumulating spectrum may be monitored on the oscilloscope screen. When the spectrum is of satisfactory quality it is read out on to punched or magnetic tape and finally fitted by a computer which produces the spectrum and Mössbauer parameters.

The 35.48 keV gamma-rays used in ^{125}Te Mössbauer spectroscopy are sufficiently high in energy that the fraction of recoil-free emission or absorption events is low. The recoil-free fraction can be increased, however, if the source and the absorber are cooled to liquid nitrogen (77 K) or liquid helium (4.2 K) temperatures and in ^{125}Te Mössbauer spectroscopy it is necessary to cool at least the absorber to increase the recoilless fractions. Indeed, the examination of organotellurium compounds is best performed at liquid helium temperature with both the source and absorber at 4.2 K. Such experiments in which both the source and the absorber are held at temperatures as low as a few millikelvin still require the source to be vibrated, and it will be appreciated that the construction of cryogenic facilities to perform such measurements is not without its problems. However, a number of cryostats for performing measurements at 77 or 4.2 K, such as is illustrated in Figure 5, are now commercially available.

Cryostats with a variable temperature control are used in many studies, such as those involving the investigation of phase transitions, and, for investigations where it is desirable

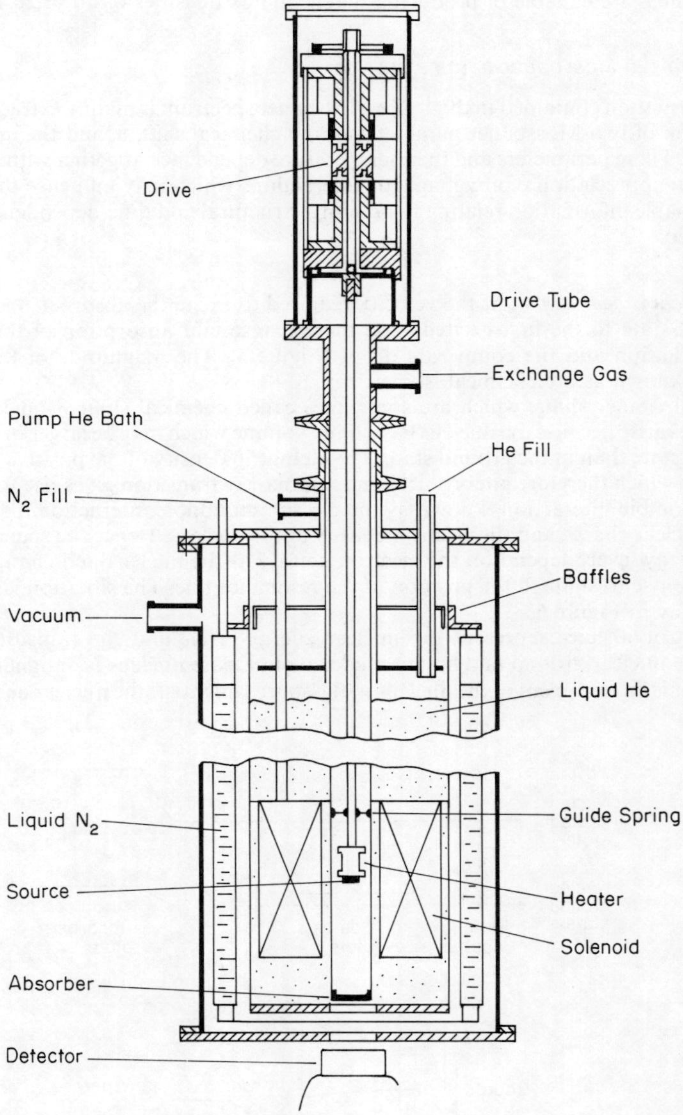

FIGURE 5. Helium cryostat with vertical beam geometry. Both source and absorber can be cooled. The system is equipped with a superconducting solenoid. If the magnet coil is not in use, the drive tube is extended to bring the source closer to the absorber in order to obtain a larger solid angle. Reproduced by permission of North-Holland Publishing Company, Amsterdam, from Shenoy and Wagner (Eds.), *Mössbauer Isomer Shifts*, 1978

to apply a large external magnetic field to the absorber, superconducting magnets are required which are capable of producing magnetic flux densities of up to 10 T.

2. Tellurium-125 Mössbauer parameters

The information contained in the ^{125}Te Mössbauer spectrum is mainly extracted by the computation of two Mössbauer parameters—the chemical shift, δ, and the quadrupole splitting, Δ. These parameters and their temperature dependence, together with line width data and an appreciation of any magnetic interactions which may influence the spectral patterns, enable information relating to bonding, structural and time-dependent effects to be examined.

a. Chemical isomer shifts. At the velocity required to excite the absorber nucleus from the ground state to the first excited state there is resonant absorption of the incident gamma radiation and the count rate drops (Figure 3). The magnitude of the applied velocity is known as the chemical isomer shift, δ.

Chemical isomer shifts, which are sometimes called chemical shifts, isomer shifts or centre shifts, arise because the nucleus has a finite volume which may be larger or smaller in the excited state than in the ground state. The change in density of the positive charge on the nucleus which therefore alters during the gamma-ray transition gives rise to a change in the coulombic interaction, known as the electric monopole interaction, between the positive nuclear charge and the electron density at the nucleus. Hence the spacings of the nuclear energy levels depend on the electron density at the nucleus and changes in this density give rise to shifts in the position of the resonance line. The situation is described schematically in Figure 6.

The horizontal lines represent the nuclear energy levels and the transition energy between the nuclear ground and first excited state in a bare nucleus is designated by E_n. The electronic environment of the nucleus in the source might lift the nuclear energy levels

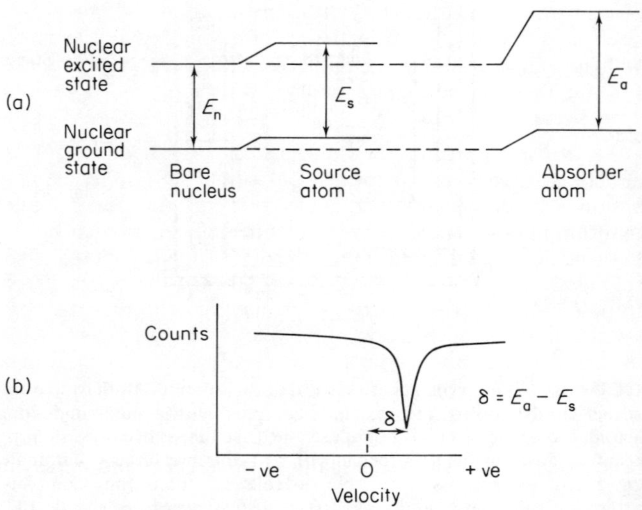

FIGURE 6. Schematic representation of chemical isomer shifts in Mössbauer spectroscopy

such that the transition energy for the source is E_s. When the absorber nucleus is in a different chemical environment the displacement of the energy levels will differ from those in the source such that the transition energy E_a will also differ. The application of a Doppler velocity to the source to attain resonance therefore results in a shift of the position of the resonance line in the spectrum. The applied velocity is known as the chemical isomer shift, δ.

To a first approximation, the electron density about the nucleus is appreciably large only for s-electrons. By making certain assumptions[3] it can be shown that the chemical isomer shift δ can be described by the expression

$$\delta = \text{constant} \times \frac{\Delta R}{R}\left(|\Psi_s(0)|_a^2 - |\Psi_s(0)|_s^2 \right)$$

where ΔR is the change in the nuclear radius during its transition from the excited to the ground state and R is the radius of the ground state. The value of $\Delta R/R$ is characteristic of each Mössbauer transition and may be of either sign, such that a positive $\Delta R/R$ indicates that the nucleus shrinks on de-excitation. After some controversy it is now accepted that $\Delta R/R$ is positive for the ^{125}Te 35.48 keV γ-ray transition[21].

The terms $|\Psi_s(0)|_a^2$ and $|\Psi_s(0)|_s^2$ refer to the s electron densities at the nuclei in the absorber and source, or reference, respectively. Since $\Delta R/R$ is positive for the ^{125}Te transition the chemical isomer shift increases with increasing s electron density at the absorber tellurium nucleus. Although the term $|\Psi_s(0)|^2$ includes contributions from all the occupied s electron orbitals in the atom, it is also sensitive to p and d orbital electron density as a result of shielding and penetration effects. Mössbauer spectroscopy therefore provides a means of monitoring s-electron density at the nucleus which is dependent on p, d and f electron disposition.

It will now be appreciated that the chemical isomer shift, δ, is an important means by which atomic oxidation states, which have sometimes in the past been difficult to determine, may now be directly investigated. Similarly, covalency effects and the shielding of one set of electrons by another which also influences the electronic environment of the nucleus may also be reflected in changes in δ. The chemical isomer shift data can sometimes be used to assess quantitatively the electron-withdrawing power of substituent electronegative groups in addition to the degree of π-bonding and back-donation from metal atoms to ligands in coordination complexes. The interpretation of the isomer shift data in terms of the Townes and Dailey theory[22] will be discussed later.

b. Quadrupole splittings. The principles outlined during the discussion of the electric monopole interaction which gives rise to chemical isomer shifts assumed that the nuclear charge distribution is spherical. However, nuclei in energy states with a nuclear angular momentum quantum number $I > 1/2$ have non-spherical charge distributions which are characterized by a nuclear quadrupole moment, Q, i.e. the nuclear charge distribution may be elongated along the intrinsic axis of symmetry labelled the z-axis, in which case the nuclear quadrupole moment Q is positive, or it may be compressed along this axis, in which case Q is negative. The interaction of the nuclear charge density with asymmetric extranuclear electric fields, i.e. non-symmetric arrays of electronic charge, ligands on ions, which are characterized by a tensor quantity called the electric field gradient (EFG), is called the electric quadrupole interaction. The axes of the electric field gradient are labelled such that $V_{zz} > V_{xx} > V_{yy}$ and the EFG is normally expressed in terms of the principal component V_{zz}, which is usually written as eq, and an asymmetry parameter η, which is described by

$$\eta = \frac{V_{xx} - V_{yy}}{V_{zz}}$$

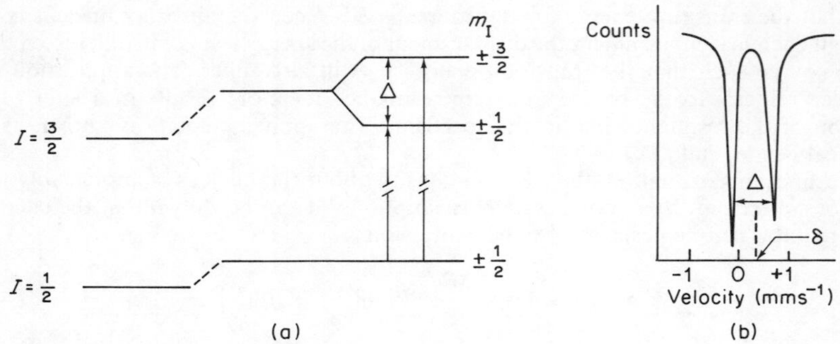

FIGURE 7. The Mössbauer quadrupole splitting in tellurium-125: (a) the excited state, $I = 3/2$, splits into two; (b) the resulting Mössbauer spectrum

The electric quadrupole interaction involves the nuclear quadrupole moment aligning itself either with or across the electric field gradient, i.e. the coupling of eQ with eq, which is expressed as the quadrupole coupling constant e^2qQ, and gives rise to a splitting of the nuclear energy levels. Hence, in ^{125}Te the excited state has $I = 3/2$ and in the presence of an asymmetric electric field splits into two substates characterized by $M_I \pm 1/2$ and $M_I = \pm 3/2$, as is shown schematically in Figure 7. The transitions from the degenerate ground state ($I = 1/2$) to the excited state produce a two-line spectrum in which the separation between the two lines, measured in millimetres per second, is a measure of the quadrupole splitting Δ. The centroid of the two peaks represents the chemical isomer shift δ.

Hence the quadrupole splitting obtained from the Mössbauer spectrum involves both a nuclear quantity, the quadrupole moment, and an electronic quantity, the electric field gradient, and, for Mössbauer transitions between nuclear states with spin 1/2 and 3/2, may be expressed in terms of the quadrupole coupling constant for the $I = 3/2$ state:

$$\Delta = \tfrac{1}{2}e^2qQ\left(1 + \frac{\eta^2}{3}\right)^{1/2}$$

The electric field gradient contains contributions from several different components. One of the major contributions may be described as that due to the electronic environment about the nucleus and is called the valence term (q_{val}). This may be envisaged as arising from the valence electrons of the Mössbauer atom itself and originates from asymmetry in the electronic structure which may derive from the unfilled or partly filled electronic shells occupied by the valence electrons. Another contribution to the electric field gradient originates from surrounding charged entities and is called the lattice contribution (q_{lat}). This arises from the asymmetry in the arrangement of atoms around the Mössbauer nuclei and is most important in ionic systems. Other contributions to the electric field gradient include the effects of molecular orbitals and any polarisation of the core electrons of the Mössbauer atom.

The quadrupole splitting therefore reflects the symmetry of the bonding environment and the local structure in the vicinity of the Mössbauer atom. There have been several attempts to formulate semi quantitative treatments of the electric field gradient which have recently been well summarized[8]. One of these, based on the theory of Townes and Dailey[22], is particularly well suited for the treatment of tellurium-containing compounds. The method assumes that the electric field gradient arises entirely within the valence shell of the tellurium atom, that the bonding involves only the p orbitals or has a constant s

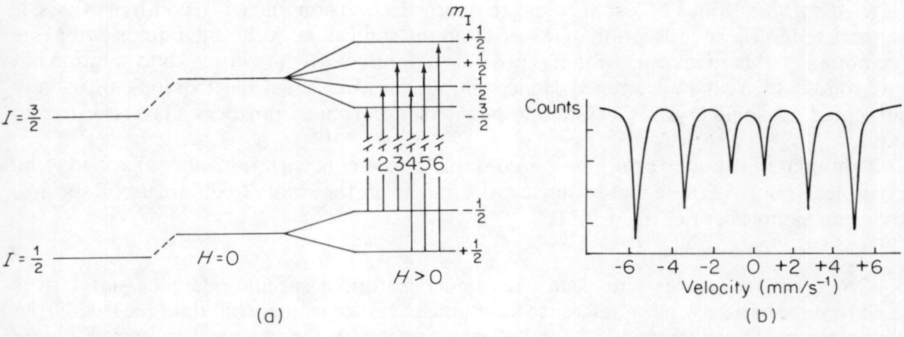

FIGURE 8. Effect of magnetic splitting on the nuclear energy levels of iron-57: (a) splitting of the nuclear energy levels in a magnetic field; (b) the resulting Mössbauer spectrum

character, and that π-bonding interactions can be ignored. From the Mössbauer quadrupole splitting and quadrupole coupling constants the p orbital imbalance, designated as U_p, can be derived which gives information about the electron population of the p orbitals. The method also allows the calculation of a quantity, h_p, which is pertinent to isomer shift calculations and which describes the number of holes in the shell.

c. Magnetic splitting. When a nuclear state with spin quantum number I is placed in a magnetic field its energy levels split into $2I + 1$ levels. Hence metallic iron gives a six-line Mössbauer spectrum (Figure 8). The spacing of the levels is directly proportional to the magnetic field at the nucleus and Mössbauer spectroscopy is therefore a simple method by which the magnitude and direction of the field may be measured.

Although the gamma-ray transition in ^{125}Te is also between nuclear states with spin 1/2 and 3/2 the magnetic splitting of the ^{125}Te resonance is less helpful than in, for example, ^{57}Fe or ^{119}Sn Mössbauer spectroscopy because of the poorer resolution which is obtained. The magnetic splitting has not been used in the interpretation of ^{125}Te Mössbauer spectra recorded from organotellurium compounds.

3. Other features of tellurium-125 Mössbauer spectra

a. Time-dependent effects. The hyperfine interactions observed in Mössbauer spectra occur on characteristic time scales and the spectrum observed from a specific system depends on whether the nuclear environment or the position of the nucleus is changing relative to these times. These time-dependent effects can influence both the spectral line shapes and the observed Mössbauer parameters. Time-dependent changes in the nuclear environment are often called relaxation processes and may relate to time-dependent structural changes or to changes involving the electronic configuration. It is also relevant to note that any dynamic properties of Mössbauer nuclei arising from the lattice dynamics of the solid in which the nucleus is situated, or from the motion of a localized part of the system such as molecular motion or the motion of the whole system within its environment, may also be reflected in changes in the Mössbauer spectra. Studies of these types of phenomena usually require the monitoring of the spectral changes as a function of temperature.

b. Absorbers. Since Mössbauer spectroscopy depends on nuclear recoil-free events, it is restricted to the investigation of materials in the solid state. Although liquids and gases are not amenable to investigation it is possible to freeze both pure liquids and solutions to give solids which can be examined. Hence samples for Mössbauer spectroscopy are usually presented as single crystals, crystalline powders, amorphous powders, glasses or frozen liquids.

If the material under examination contains a low concentration of ^{125}Te it may be possible to improve the signal-to-noise ratio, and hence the ability to obtain useful spectra, by using samples enriched in ^{125}Te.

c. Reference standards. For each Mössbauer isotope a specific reference standard is used to serve as a zero-point reference for all the chemical isomer shift data recorded from that element. In tellurium-125 Mössbauer spectroscopy the chemical isomer shifts are usually referred to ^{125}I/Cu, mainly because this source was used in many early Mössbauer studies and because most chemical isomer shift values have normally been referred to this standard. However, an alternative standard[16] is tetragonal tellurium dioxide, which has some advantages since it is one of the most readily obtainable pure well characterized compounds of tellurium and gives an excellent quadrupole split absorption.

d. Limitations. The chemical isomer shifts recorded in ^{125}Te Mössbauer spectra span a rather narrow range of about 3 mm s^{-1}, which is less than the natural line width[20] of 5.02 mm s^{-1}. However, good quality spectra have been obtained by careful experimental methods and the use of a good source[13,16] such that differences of only 0.1 mm s^{-1} may be shown to be significant. The low resolution of the spectra frequently precludes the interpretation of data in terms of a heterogeneity of tellurium sites and the detection of different tellurium-containing materials in a multi-phased sample is usually difficult.

III. APPLICATION TO THE STUDY OF ORGANOTELLURIUM COMPOUNDS

The initial ^{125}Te Mössbauer investigations of organotellurium compounds were performed in the mid-1970s and involved the examination[23–27] of a variety of alkyl and aryl derivatives of tellurium-(II) and -(IV). The early Mössbauer studies were summarized[28] in 1976 and the most recent developments were reviewed[29] in 1983.

A. Organo-tellurides(II) and -ditellurides (II)

1. Dialkyl and diaryl tellurides and ditellurides

The diorgano-tellurides and -ditellurides give ^{125}Te Mössbauer spectra which are characterized by small positive chemical isomer shifts and large quadrupole splittings (Figure 9, Table 1).

Investigations of the structural properties of simple aryltellurium(II) compounds have shown them to be covalently bonded molecules with bond angles within the range expected for sp^3 or p^3 hybridized tellurium. For example, the C—Te—C bond angle in[31] $(p\text{-Tol})_2Te$ is 101.0 ± 2.7° whereas the C—Te—Te bond angle in $(p\text{-ClC}_6H_4)_2Te_2$ is[32] 94.4° and in Ph_2Te_2 is[33] 99° (Figure 10).

Since the nuclear radius term is positive for the ^{125}Te transition[21], the chemical isomer shift data become more positive as the s electron density, $|\psi_s(0)|^2$, at the tellurium nucleus increases. Such an increase can occur by the removal of 5p electron density from the tellurium atom, which results in a deshielding of the 5s electrons from the nucleus. On the

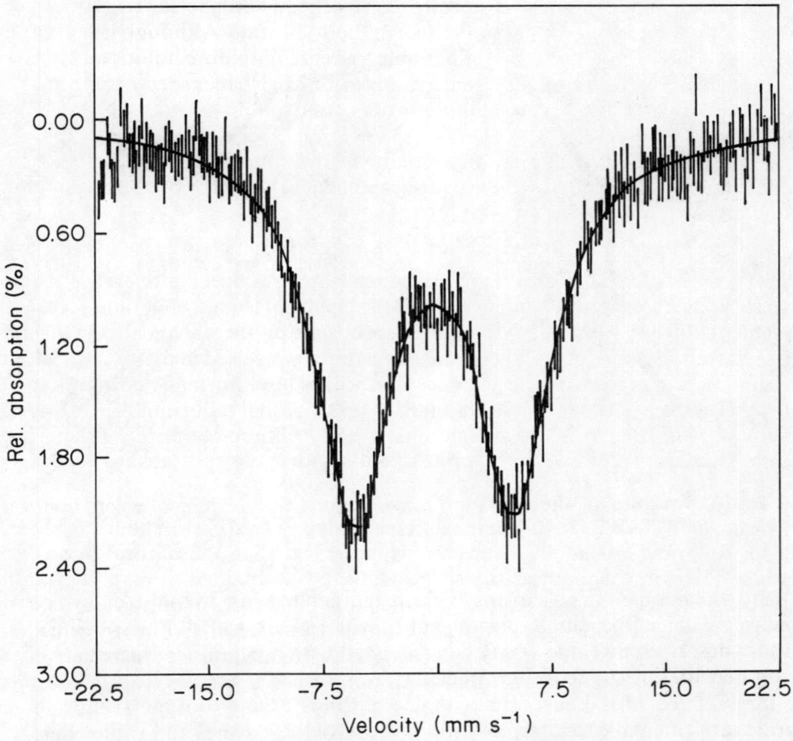

FIGURE 9. ^{125}Te Mössbauer spectrum of *i*-Pr_2Te_2 measured at 4.2 K. Reproduced with permission from *J. Organomet. Chem.*, **255**, 61 (1983)

TABLE 1. ^{125}Te Mössbauer parameters recorded from some organo-tellurides and -ditellurides at 4.2 K

Compound	$\delta^a \pm 0.08$ (mm s^{-1})	$\Delta \pm 0.1$ (mm s^{-1})	Ref.
Me_2Te	0.06	10.5	25
t-Bu_2Te	0.19	10.5	30
Ph_2Te	0.18	10.5	25
p-Tol_2Te	0.7	10.1	23
p-An_2Te	0.3	11.3	23
Me_2Te_2	0.26	12.3	30
i-Pr_2Te_2	0.37	11.0	30
i-Bu_2Te_2	0.32	11.0	30
p-$TolTe_2$	0.6	9.9	23
Ph_2Te_2	0.37	10.7	25
p-$AnTe_2$	0.3	10.3	23
(*p*-$EtOC_6H_4)_2Te_2$	0.28	10.6	25
(*p*-$C_6H_5OC_6H_4)_2Te_2$	0.3	10.3	23
(*neo*-Hex)$_2Te_2$	0.38	10.5	30

[a] δ with respect to I/Cu.

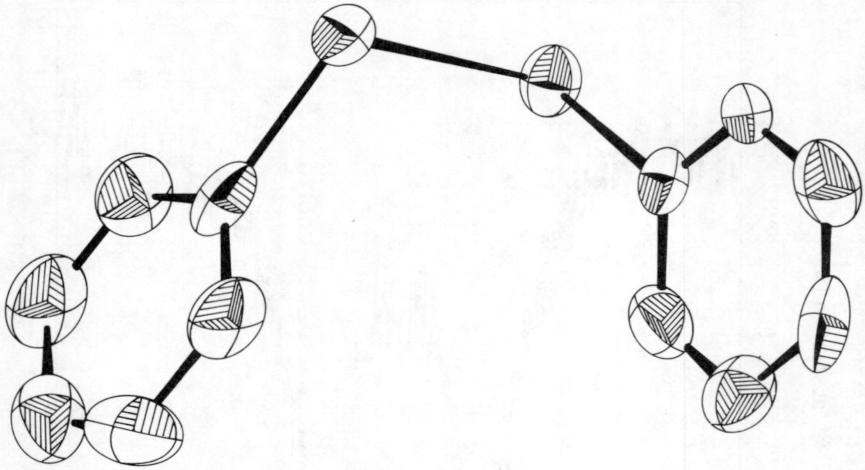

FIGURE 10. Structure of diphenyl ditelluride. Reproduced with permission from *Acta Crystallogr., Sect. B*, **28**, 2438 (1972)

other hand, the removal of s electrons leads to a direct decrease in the s electron density at the nucleus. Hence the small positive chemical isomer shifts of the tellurides and ditellurides described in Table 1 may be associated with compounds with relatively small s electron densities at the tellurium nuclei.

The large ^{125}Te Mössbauer quadrupole splittings of the diorgano-tellurides and -ditellurides are amenable to interpretation in terms of the Townes and Dailey theory[22,25], for which the basic tenets, as it is generally applied, are that the electric field gradient derives predominantly from any imbalance in the tellurium 5p orbital populations and that the lattice terms, the d orbital contributions, and Sternheimer shielding and anti-shielding factors[12] can all, to a first approximation, be conveniently ignored. The quadrupole coupling constants, e^2qQ, are then linearly related to the p orbital imbalance, U_p, through the expression

$$\frac{e^2qQ}{e^2q_0Q} = -U_p$$

$$\Delta = \tfrac{1}{2}e^2qQ\left(1 + \frac{\eta^2}{3}\right)^{1/2}$$

where η is the asymmetry parameter and e^2q_0Q is the quadrupole coupling constant for one 5p electron on tellurium, which is assumed to remain constant with changes in oxidation state and has[34] a value of ca. $+24$ mm s^{-1}. The p orbital imbalance is defined as

$$U_p = -U_z + \frac{U_x + U_y}{2}$$

The p-orbital populations U_x, U_y, and U_z can be related to the bonding orbital populations through a consideration of the hybrid orbitals which describe the bonding to the tellurium atom and, by using this approach[25], the large quadrupole splittings of the diorgano-tellurides and -ditellurides may be considered as reflecting a considerable imbalance in the tellurium 5p orbital population. The method[25] accommodates the structural properties[31–33] of the compounds and the Mössbauer chemical isomer shift data[23,25,30] which indicate that the tellurium 5s electrons participate in the bonding such that the bonding may be viewed as intermediate between pure p^3 and sp^3 to generate bond angles θ in the range $90° < \theta < 109.5°$ and concludes[25] that

$$\frac{e^2qQ}{e^2q_0Q}(1-\cos\theta) = 2 - U_c$$

$$\eta = 3\cos\theta$$

where θ is the C—Te—C bond angle and U_c is the Te—C bond orbital population. By assuming that the bonding is essentially p in character, i.e. $\theta = 90°$ with only limited s character in the Te—C bonds, the lone pairs on the tellurium atom may be visualized as occupying the 5s and the $5p_z$ orbitals, the latter being directed out of the bonding plane. Since this would correspond to an excess of p electron density along the z-axis, V_{zz}, the principal component of the electric field gradient tensor, may be assumed to be negative. Under such circumstances, e^2qQ would be positive since Q, the quadrupole moment of the $I = 3/2$ excited state, is negative. For the tellurides the quadrupole splitting Δ is the same, within the errors, for the different alkyl and aryl groups and the average value for the quadrupole splitting of 10.6 mm s^{-1} corresponds to an orbital population of ca. 1.1 for the Te—C bond. The ditellurides have very similar splittings to the tellurides, indicating a similar covalency of the Te—C and Te—Te bonds and, since the bond orbital population for the latter is presumably 1.0, this is consistent with a value of 1.1 for the Te—C bond.

Given that the large ^{125}Te Mössbauer quadrupole splittings recorded from the diorgano-tellurides and -ditellurides arise from the p-orbital imbalance between the Te—C bonds and the lone pairs of electrons on the tellurium atom, it would be reasonable to except that increasing electronegativity in the organic groups would result in 5p electron density being removed from the tellurium and to be reflected in increases in quadrupole splitting. It might also be expected that, depending on the relative s–p character of the Te—C bond, the Mössbauer chemical isomer shift might also be influenced. It is interesting, therefore, that the data compiled in Table 1 show that the Mössbauer chemical isomer shifts and quadrupole splittings are essentially similar for all the diorgano-tellurides and -ditellurides. However, it is notable that dimethyl ditelluride has a larger quadrupole splitting, which is consistent with enhanced electron withdrawal by the methyl group and an increased removal of 5p electron density from tellurium. This molecule would also presumably have a larger asymmetry parameter, which would further enhance the magnitude of the quadrupole splitting.

In these respects it is interesting to note the Mössbauer parameters more recently recorded[35] from bis(trifluoromethyl) and bis(pentafluorophenyl) tellurides, which contain significantly more electronegative organic ligands (Table 2). The ^{125}Te Mössbauer chemical isomer shifts are not too dissimilar from those of the dialkyl and diaryl tellurides (Table 1) and, given the small range of isomer shifts for ^{125}Te, the similarity in the chemical isomer shift data is not too surprising. On the other hand, the quadrupole splitting data recorded from $(CF_3)_2Te$ and $(C_6F_5)_2Te$ are significantly larger than those of the dialkyl and diaryl tellurides. These larger quadrupole splittings are indicative of a considerable p orbital imbalance on the tellurium atom and reflect the high electronegativity of the CF_3 and C_6F_5 ligands.

It is interesting that there appear[30] to be no simple relationships between the ^{125}Te Mössbauer parameters recorded from dialkyl tellurides and ditellurides in the solid state, or as frozen liquids, and the ^{125}Te NMR data recorded from the species in solution. Although the absence of any such correlation may reflect the different environments about tellurium in the solid state and in solution, it must be acknowledged that there is no

TABLE 2. ^{125}Te Mössbauer parameters recorded[35] from bis(trifluoromethyl) and bis(pentafluorophenyl) tellurides at 4.2 K

Compound	$\delta^a \pm 0.08$ ($mm\,s^{-1}$)	$\Delta \pm 0.08$ ($mm\,s^{-1}$)
$(CF_3)_2Te$	−0.14	14.02
$(C_6F_5)_2Te$	+0.14	13.40

^{a} δ relative to I/Cu.

evidence from the crystal structure data of any intermolecular attractions in the solid state in either the tellurides or the ditellurides which might lead to such a difference.

Hence the large quadrupole splittings of the tellurides and ditellurides correspond to a considerable imbalance in the 5p orbital population in the Te—C bonds and to the presence of lone pairs of electrons on the tellurium atoms. The data are indicative of considerable covalent character in the Te—C and Te—Te bonds and suggest that the tellurium atoms have a significant number of holes in the 5p shells. Such a situation would be expected to give rise to increased s electron densities at the tellurium nuclei and consequently to large and positive chemical isomer shifts. Hence the small chemical isomer shifts, when considered in conjunction with the quadrupole splitting data, imply that the Te—C and Te—Te bonds possess some s character and that the bonding in these compounds is intermediate between p^3 and sp^3. It is also pertinent to record that the Mössbauer data are consistent with the C—Te—C and C—Te—Te bond angles, which are indicative of the presence of stereochemically active lone pairs in these compounds.

2. *Bis(organyltelluro)methanes*

Another group of compounds closely related to the diorganotellurides are the bis(organyltelluro)methanes of composition $(RTe)_2CH_2$. In a study of bis(methyltelluro)- and bis(phenyltelluro)-methanes, the ^{125}Te Mössbauer quadrupole splittings were shown[36] to be large, Δ 10.4 and 11.1 $mm\,s^{-1}$, respectively, and the chemical isomer shifts small and positive. The data are similar to those recorded from the simple diorganotellurides and give no evidence for the second tellurium atom in the Te—CH_2—Te linkage having any significant effect on the electronic environment at tellurium. The Mössbauer parameters appear to be dominated by the bonds of each tellurium to two carbon atoms, indeed the data recorded[36] from the ethyl and isopropyl derivatives show no measurable differences.

3. *Cyclic derivatives*

It is also relevant to comment on the compound of formula CH_2Te_2, which was previously described[37] as a three-membered cyclic species called ditelluromethane (Figure 11a) and which has more recently[38] been shown to be the six-membered cyclic compound 1, 2, 4, 5-tetratelluracyclohexane (Figure 11b). The ^{125}Te Mössbauer parameters[38], δ 0.42 $mm\,s^{-1}$, Δ 9.3 $mm\,s^{-1}$, indicate the presence of only one tellurium site in the compound and are similar to those of the diorganoditellurides (Table 1).

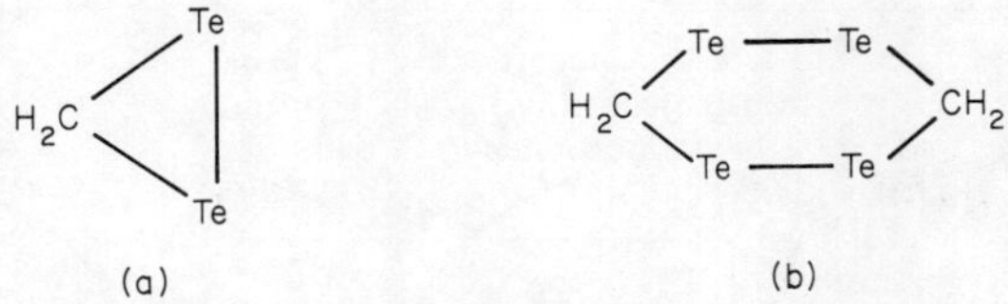

FIGURE 11. Schematic representation of (a) ditelluromethane and (b) 1, 2, 4, 5-tetratelluracyclohexane

B. Organotellurium(IV) Halides

1. Diorganotellurium(IV) dihalides

The diorganotellurium(IV) dihalides were among the first organotellurium(IV) compounds to be examined by ^{125}Te Mössbauer spectroscopy. The Mössbauer parameters recorded from selected series of diorganotellurium(IV) dihalides are collected in Table 3 and a representative spectrum is illustrated in Figure 12.

Several structural studies of the crystalline diorganotellurium(IV) dihalides Ph_2TeBr_2[39], Me_2TeCl_2[40], $(p\text{-}ClC_6H_4)_2TeI_2$[41], α-$Me_2TeI_2$[42], and Ph_2TeF_2[43] have shown the tellurium atoms to occupy distorted trigonal bipyramidal coordination (Figure 13). In these structures the halogen atoms occupy *trans* axial positions and the organic groups are situated in the equatorial plane in which the third position is occupied by a lone pair of electrons which may sometimes be involved in intermolecular bonding through bridging

TABLE 3. ^{125}Te Mössbauer parameters recorded from some diorganotellurium(IV) dihalides at 4.2 K

Compound	$\delta^a \pm 0.08$ ($mm\,s^{-1}$)	$\Delta \pm 0.1$ ($mm\,s^{-1}$)	Ref.
Me_2TeCl_2	0.58	9.4	25
Me_2TeBr_2	0.65	8.5	25
Me_2TeI_2	0.52	7.6	25
Ph_2TeF_2	0.52	10.4	26
$p\text{-}An_2TeF_2$	0.54	10.0	26
$(p\text{-}EtOC_6H_4)_2TeF_2$	0.36	10.1	26
Ph_2TeCl_2	0.5	9.2	23
$p\text{-}An_2TeCl_2$	0.68	9.1	26
$(p\text{-}EtOC_6H_4)_2TeCl_2$	0.7	9.1	23
Ph_2TeBr_2	0.5	8.1	23
$p\text{-}An_2TeBr_2$	0.72	7.8	26
$(p\text{-}EtOC_6H_4)_2TeBr_2$	0.52	7.6	25
$(CH_2)_4TeBr_2$	0.54	7.1	25
$(C_6F_5)_2TeBr_2$	0.73	5.3	25
$p\text{-}An_2TeI_2$	0.51	6.1	26
$p\text{-}Tol_2TeI_2$	0.6	6.3	23

[a] δ with respect to I/Cu.

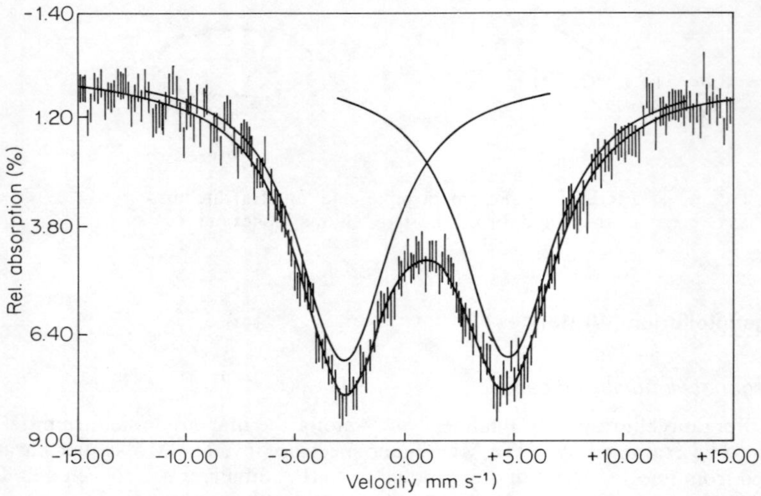

FIGURE 12. ^{125}Te Mössbauer spectrum recorded from $(p\text{-}EtOC_6H_4)_2TeCl_2$ at 4.2 K. Reproduced with permission from *Can. J. Chem.*, **54**, 3234 (1976)

halogens. Indeed the structures of all the diorganotellurium(IV) dihalides examined so far are essentially similar despite their various descriptions as distorted trigonal bipyramidal, distorted tetrahedral and distorted octahedral, depending on the significance which has been attributed to intermolecular associations through bridging halogens and the mixing of 5s with p electrons. Low-frequency infrared and Raman spectra of many diaryltellurium(IV) dihalides are also consistent with Ψ-trigonal bipyramidal structures[44].

The chemical isomer shifts of the diorganotellurium(IV) dihalides are more positive than those of the diorganotellurides and are consistent with the halogen ligands removing predominantly p electron density from the tellurium and the conversion of tellurium(II) to tellurium(IV). However, within a given series of dihalides the chemical isomer shift does not change significantly with increasing electronegativity of the halogen. For example, the compounds $p\text{-}An_2TeX_2$ (X = F, Cl, Br, I; *p*-An = $p\text{-}MeOC_6H_4$) all have the same chemical isomer shifts within the limits of experimental accuracy. On first inspection this may seem unusual, since it would be reasonable to expect that, within a series of dihalides, the chemical isomer shift would change as the electronegativity of the halogens increases from iodine through bromine and chlorine to fluorine such that increasing amounts of 5p electron density are removed from the vicinity of the tellurium nucleus. However, the relatively small value of $\Delta R/R$ for the ^{125}Te Mössbauer transition gives rise to a small range of chemical isomer shifts and it must be appreciated that δ increases[34] by only *ca.* $+0.4\,mm\,s^{-1}$ on the removal of one tellurium 5p electron. This, together with the broad natural line width of the ^{125}Te Mössbauer transition and the consequently relatively large errors in δ, means that the chemical isomer shift is not a particularly sensitive probe to small changes in the tellurium 5p orbital populations. It is also possible that the s–p hybrid character of the bonds changes as the electronegativity of the halogens varies and that this effect is also reflected in the consistency of the chemical isomer shift data.

The quadrupole splittings of the dihalides lie in the order F > Cl > Br > I, and the trend is well illustrated by a consideration of the data for the *p*-methoxyphenyl derivatives. The order is consistent with an increasing imbalance in the p orbital electron population about

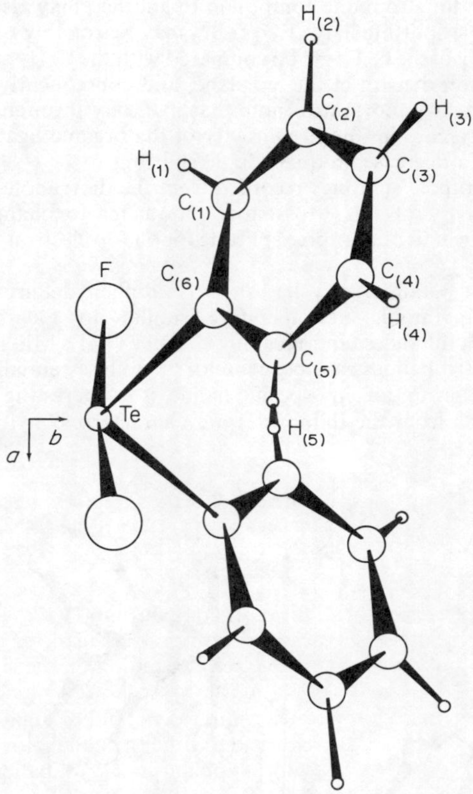

FIGURE 13. Distorted trigonal bipyramidal structure of diphenyltellurium(IV) difluoride. Reproduced with permission from *J. Chem. Soc., Dalton Trans.*, 2306 (1980)

tellurium as the electronegativity of the halogen increases. The trend may be rationalized if the bonding in these compounds is envisaged as primarily involving the p-orbitals such that V_{zz}, the principal component of the electric field gradient tensor, lies along the X—Te—X bond axis and is positive. Thus in the difluorides the fluorine ligand may be viewed as removing considerable electron density from the tellurium $5p_z$ orbital leading to an electron deficit along that axis compared with the *xy* equatorial plane. As the electronegativity of the halogen ligand decreases the occupation of the tellurium $5p_z$ orbital would be expected to increase and hence give rise to a decrease in the quadrupole splitting.

Although the data recorded for most halides show the Mössbauer parameters to be independent of the nature of the organic group, it is interesting to note the order of quadrupole splittings in the series of dibromides: $Me_2TeBr_2 > Ph_2TeBr_2 >$ *p*-$An_2TeBr_2 >$ (*p*-$EtOC_6H_4)_2TeBr_2 > (CH_2)_4TeBr_2 > (C_6F_5)_2TeBr_2$. These differences may arise from variations in the σ donor character of the organic ligands although contributions arising from changes in the s–p hybrid character of the Te—C bonds, or variations in the degree of intermolecular bonding from one compound to another, may also be relevant. Hence the smaller quadrupole splitting in $(C_6F_5)_2TeBr_2$ may reasonably be associated with the lower σ donor capacity of the C_6F_5 group compared with the C_6H_5 entity which results in a decreased 5p electron density in the *xy* plane, and consequently produces a smaller *p*-orbital imbalance. It is important to note that if V_{zz} lay through the lone pair in the equatorial plane a decrease in donor character of the organic ligand would lead to an increase, rather than a decrease, in quadrupole splitting.

The range of quadrupole splittings recorded from the diorganotellurium(IV) dihalides illustrates the sensitivity of this Mössbauer parameter to changes in the p orbital population with Δ_0, the unit quadrupole splitting for one 5p electron, being estimated[25] to be ca. 12 mm s^{-1}.

Given that X-ray crystallography has shown[43] diphenyltellurium(IV) difluoride to be a Ψ-trigonal bipyramid, and therefore similar to the structures of other diorganotellurium(IV) dihalides, and recalling that the ^{125}Te Mössbauer parameters of the diaryltellurium(IV) dihalides have been found[25,26] to be essentially independent of the nature of the organic group for any specific halide, it is interesting that the Mössbauer parameters recorded[26] from the difluorides are characterized by low chemical isomer

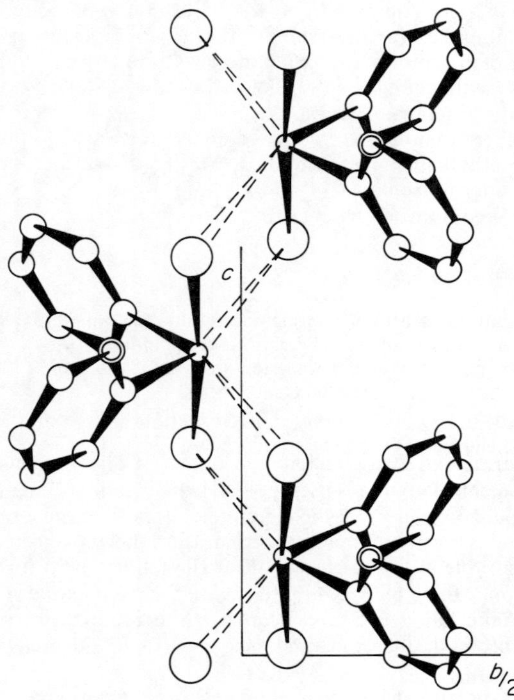

FIGURE 14. Projection of a part of the structure of diphenyltellurium(IV) difluoride down [100] showing the weak intermolecular interactions (hydrogen atoms are not shown for clarity). Reproduced with permission from *J. Chem. Soc., Dalton Trans.*, 2306 (1980)

shifts and large quadrupole splittings. The results are indicative of compounds which, although adopting essentially similar structures to those of the heavier dihalides, have subtle differences in bonding and structure, e.g. the low chemical isomer shifts may indicate a greater stereochemical activity of the 5s electrons in the difluorides. Although the larger quadrupole splittings of the difluorides may easily be associated with the dominating effect of electronic, rather than ligand, asymmetry in these compounds, it is also likely that the larger value of Δ reflects the lower degree of intermolecular association in the difluorides which is achieved by, for example in Ph_2TeF_2, two long bridging contacts from tellurium to fluorine of 3.208 Å (Figure 14)[43]. The ^{125}Te Mössbauer parameters of these compounds may therefore also be considered in terms of the extent of intermolecular association as measured by the ratio of the crystallographically determined tellurium—halogen intermolecular distance to that of the intramolecular distance, i.e. ratios of 1.60:1 for Ph_2TeF_2, 1.39:1 for Me_2TeCl_2, 1.47:1 for Ph_2TeBr_2, 1.41:1 for $(p\text{-}ClC_6H_4)_2TeI_2$ and 1.34:1 for Me_2TeI_2. The difluoride is clearly the dihalide with least intermolecular association. It is also clear that intermolecular association is more significant in alkyl than in the corresponding aryl compounds and this is presumably a reflection of the closer packing in the former species due to the smaller size of the alkyl group. It is also interesting to note that the chemical isomer shifts for the tellurium tetrahalides TeX_4 (X = Cl, Br, or I)[45] and the antimony halides SbX_3 (X = F, Cl, Br or I)[46] are, like those of the diaryltellurium(IV) dihalides, essentially similar within each series of compounds. This relative constancy of s electron density at the tellurium, and antimony, nuclei within a given series of compounds is further evidence that the stereochemical activity of the lone pair and the degree of intermolecular association in the diaryltellurium(IV) dihalides vary in a complex way with changes in the electronegativity of the halogen and in the coordination about the tellurium atom.

Finally, it is pertinent to note that ^{125}Te Mössbauer data have been used[47] to show that the compound dimethyltellurium tetraiodide, Me_2TeI_4, is an adduct of Me_2TeI_2 with iodine linked by intermolecular I—I bonds and weak Te—I bonds and that the compound does not contain tellurium(VI).

2. *Organotellurium(IV) trihalides*

The ^{125}Te Mössbauer parameters recorded from some aryltellurium(IV) trihalides are collected in Table 4.

TABLE 4. ^{125}Te Mössbauer parameters recorded from some aryltellurium(IV) trihalides at 4.2 K

Compound	$\delta^a \pm 0.08$ (mm s^{-1})	$\Delta \pm 0.1$ (mm s^{-1})	Ref.
p-AnTeF$_3$	0.56	8.6	26
$(p\text{-}EtOC_6H_4)TeF_3$	0.6	8.6	26
p-AnTeCl$_3$	0.9	9.2	23
$(p\text{-}EtOC_6H_4)TeCl_3$	0.91	9.4	25
$PhTeBr_3$	0.91	7.8	25
$(p\text{-}EtOC_6H_4)TeBr_3$	1.0	8.0	23
$PhTeI_3$	0.9	3.9	23
$(p\text{-}EtOC_6H_4)TeI_3$	1.0	5.2	23

^{a} δ with respect to I/Cu.

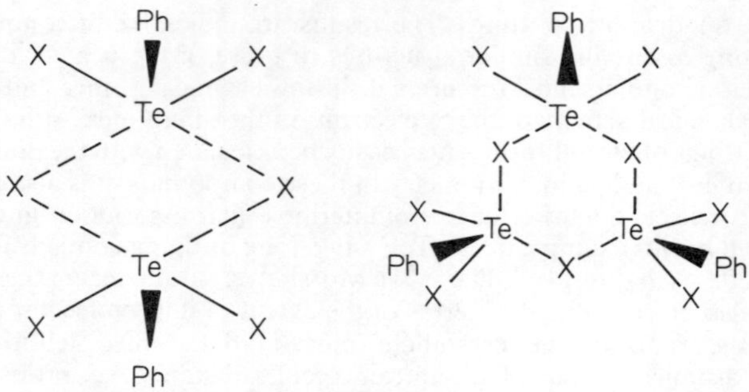

FIGURE 15. Illustration of typical structures found in aryltellurium(IV) trihalides. Reproduced with permission from *J. Chem. Soc., Dalton Trans.*, 551 (1972)

The aryltellurium(IV) trihalides have been described[48], on the basis of infrared and Raman spectroscopic data, as associated structures as shown in Figure 15.

The trihalides $RTeX_3$ (X = Cl, Br, I) have more positive chemical isomer shifts than those of the corresponding dihalides which are consistent with the removal of predominantly p electrons by the bonded halogens. However, the chemical isomer shifts of the difluorides and trifluorides are similar and, in this respect, the data suggest that the trifluorides are different from the other trihalides.

The quadrupole splittings recorded from the trihalides when X = Cl and Br are the same, within experimental error, as those from the corresponding dihalides. The data are consistent with the essentially similar arrangement of atoms about tellurium in these halides in which the bonding is predominantly concerned with the involvement of p orbitals. For the iodides the agreement is not as close, although the two measurements on the triiodides are not in good agreement. For the fluorides the data are consistent within themselves and show that the quadrupole splittings for the trifluorides are significantly smaller than those for the difluorides. The data recorded from the trifluorides are consistent with a more symmetrical environment about the tellurium and, given the smaller chemical isomer shifts which are indicative of greater stereochemical activity in the lone pairs, suggest that the coordination about the tellurium atom in the trifluorides must be significantly different from that in the other trihalides.

3. Alkylaryltellurium(IV) halides

The ^{125}Te Mössbauer parameters recorded[26] from some alkylaryltellurium(IV) iodides are collected in Table 5. The compound $PhMeTeI_2$ has essentially similar Mössbauer parameters as p-An_2TeI_2, although a smaller quadrupole splitting than α-Me_2TeI_2 (Table 3). Hence the presence of both alkyl and aryl substituents on the tellurium atom fails to produce any exceptional changes in the Mössbauer parameters and the results indicate that the alkyl and aryl ligands have very similar properties.

The compounds $PhMe_2TeI$ and Ph_2MeTeI probably contain the $PhMe_2Te^+$ and Ph_2MeTe^+ cations, since X-ray crystallography has shown[49] that Ph_3SeCl contains the Ph_3Se^+ cation and infrared spectroscopy has shown[50] Me_3SeCl to contain Me_3Se^+. Further, conductance measurements on the compounds R_3TeX (R = Me, Ph; X = Cl, Br, I) in a variety of solvents, including the compound Ph_2MeTeI, also support the formulation of these compounds as $R_3Te^+X^-$ in solution[51–53].

TABLE 5. ^{125}Te Mössbauer parameters recorded[26] from some alkylaryl-tellurium(IV) iodides at 4.2 K

Compound	$\delta^a \pm 0.09$ (mm s^{-1})	$\Delta \pm 0.15$ (mm s^{-1})
Ph_2MeTeI	0.38	5.6
$PhMe_2TeI$	0.32	5.4
$PhMeTeI_2$	0.56	6.4

^{a} δ with respect to I/Cu.

The compounds $PhMe_2TeI$ and Ph_2MeTeI have, as might be expected, identical Mössbauer parameters, within the limits of experimental accuracy. The small chemical isomer shifts recorded from these compounds are consistent with the presence of R_3Te^+ cations since the coordination of iodine to the tellurium atom would be expected to lead to a more positive value of δ than those recorded from the diorganotellurides, R_2Te, for which $\delta \approx 0.2$ mm s^{-1} (Table 1). It therefore seems that the covalent Te—C bond has significant s character but the removal of s and p electron density is such as to lead to a small net increase in $|\Psi_s(0)|^2$ through deshielding of the remaining 5s electrons at the nucleus. The small quadrupole splittings of these compounds are also consistent with s character in the Te—C bonds since the data imply some p character in the lone pair of electrons on tellurium. Although other possible contributions to the electric field gradient at the tellurium nucleus could arise from inequalities in the Te—C bonds in the Ph_2MeTe^+ and $PhMe_2Te^+$ cations or from the lattice terms, the main contributor to the electric field gradient is probably the lone pair of electrons on the tellurium atom.

It is interesting that the quadrupole splittings of *ca.* 5 mm s^{-1} for Ph_2MeTeI and $PhMe_2TeI$ correspond to a coupling constant of 350 MHz, assuming $\eta = 0$, whilst the coupling constant for Ph_3Sb is 530 MHz[54]. Taking the ratio of the principal components of the field gradient tensor per 5p electron for antimony and tellurium to be[55] 0.87, the ratio of the coupling constants corresponds to a ratio of Q_{gnd} (^{121}Sb)/Q_{ex} (^{125}Te) of 1.7, which is not inconsistent with previous estimates of the magnitude of the quadrupole moments of these two states.[54,56]. Thus the magnitudes of the quadrupole splittings for Ph_2MeTeI and $PhMe_2TeI$ are consistent with the compounds containing pyramidal Ph_2MeTe^+ and $PhMe_2Te^+$ cations which are isoelectronic and isostructural with Ph_3Sb.

4. *Bis(organyltelluro)methane halides*

The ^{125}Te Mössbauer chemical isomer shifts recorded[36] from the halide derivatives of the bis(organyltelluro)methanes have been found to be more positive than those in the parent telluromethanes (Section III.A.2) corresponding to an increase in the s electron density at the tellurium nucleus, $|\Psi_s(0)|^2$, and consistent with the tellurium—halogen bonds being predominantly p in character such that the removal of 5p electron density from tellurium leads to a deshielding of the 5s electrons from the nucleus and an increase in $|\Psi_s(0)|^2$.

The quadrupole splittings for the halide derivatives reflect the p orbital imbalance resulting from the tellurium—carbon bonds on the one hand and the tellurium—halogen bonds on the other. As the electronegativity of the halogen decreases the covalency of the tellurium—halogen bond increases and the quadrupole splitting decreases. The quadrupole splittings for the $CH_2[Te(X)_2Me]_2$ compounds[36] are very similar to those of the corresponding R_2TeX_2 halides[23–26]. However, the quadrupole splittings for the phenyl compounds, $CH_2[Te(X)_2Ph]_2$, are[36] smaller, at least for the chloride and bromide. The origin of this difference may lie in the stereochemical requirements of accommodating the bulky phenyl group and the halogen ligands about the two tellurium atoms in these molecules.

For $CH_2(TeCl_3)_2$ the problem of packing in the solid state may be even more marked. The organotellurium(IV) trichlorides[23,25,26] generally have chemical isomer shifts of ca. 0.9 mm s^{-1} and quadrupole splittings of ca. 9.0 mm s^{-1}. However, $CH_2(TeCl_3)_2$ has[36] Mössbauer parameters δ 0.9 mm s^{-1}, Δ 7.6 mm s^{-1}. It would seem that the polymeric nature of the $RTeCl_3$ compounds involving bridging Cl—Te bonds and tellurium bonded to four chlorines roughly in planar coordination and the Te—R bond directed out of the plane may be difficult to achieve in $CH_2(TeCl_3)_2$ where the two tellurium atoms are separated only by a methylene group.

5. Bis(trifluoromethyl)- and bis(pentafluorophenyl)-tellurium(IV) dihalides

The ^{125}Te Mössbauer parameters recorded from[35] the bis(trifluoromethyl)- and bis(pentafluorophenyl)-tellurium(IV) dihalides are collected in Table 6. The dihalides of $(CF_3)_2Te$ and $(C_6F_5)_2Te$ exhibit quadrupole splittings which are systematically smaller than those generally observed for dialkyl- and diaryl-tellurium dihalides. For example, $(C_6F_5)_2TeF_2$ has a quadrupole splitting of 8.8 mm s^{-1} in comparison with 10.4 mm s^{-1} recorded for Ph_2TeF_2. Similarly, the quadrupole splittings of $(C_6F_5)_2TeCl_2$ and $(CF_3)_2TeCl_2$, Δ 6.9 mm s^{-1}, are significantly smaller than the corresponding parameter for Ph_2TeCl_2 and Me_2TeCl_2, Δ 9.25 mm s^{-1}. Within each series of dihalides the quadrupole splittings follow a simple trend $F > Cl > Br$. These compounds, like the diaryltellurium(IV) dihalides previously discussed, can readily be interpreted in terms of the Ψ-trigonal bipyramidal structures where the quadrupole splittings reflect the p orbital imbalance between the X—Te—X axial linkage and the Te—C bonds, and the lone pair in the equatorial plane. As the electronegativity of the organic ligand increases this p orbital imbalance undergoes a consequent decrease consistent with the smaller quadrupole splitting in $(C_6F_5)_2TeF_2$ as compared with that recorded from $(C_6H_5)_2TeF_2$ and consistent with a greater removal of electron density from the tellurium in the equatorial plane by the C_6F_5 ligand.

TABLE 6. ^{125}Te Mössbauer parameters recorded[35] from bis(trifluoromethyl)- and bis(pentafluorophenyl)-tellurium(IV) dihalides at 4.2 K

Compound	$\delta^a \pm 0.08$ (mm s^{-1})	$\Delta \pm 0.08$ (mm s^{-1})
$(CF_3)_2Te$	−0.14	14.02
$(CF_3)_2TeF_2$	+0.59	7.36
$(CF_3)_2TeCl_2$	+0.58	6.89
$(CF_3)_2TeBr_2$	+0.64	5.60
$(C_6F_5)_2Te$	+0.14	13.40
$(C_6F_5)_2TeFe_2$	+0.45	8.78
$(C_6F_5)_2TeCl_2$	+0.59	6.91
$(C_6F_5)_2TeBr_2$	+0.59	5.38

$^a\delta$ relative to I/Cu.

The chemical isomer shifts recorded from the $(CF_3)_2TeX_2$ and $(C_6F_5)_2TeX_2$ compounds are all similar and more positive than those of the parent tellurides and, as observed in the other diaryltellurium(IV)dihalides, are consistent with the removal of predominantly p electron density from tellurium in the X—Te—X linkage leading to a deshielding of the 5s electrons from the nucleus.

6. *Tetrahaloaryltellurates*

The tetrahaloaryltellurate anions and a number of mixed tetrahaloaryltellurate anions have recently been examined[57]. The parent tetrahalotellurate anions, $RTeX_4^-$ (R = *p*-$C_2H_5OC_6H_4^-$; X = Cl, Br or I), all give quadrupole split Mössbauer spectra. The electric field gradient in these ions, which are square pyramidal with the organic ligand occupying the axial position[58] with bonding to the halogen ligands being viewed as three-centre, four-electron bonds involving only tellurium 5p orbitals, may be considered as deriving from p orbital imbalance between the *z*-axis, as defined by the Te—C bond, and the *x—y* plane in which the four halogen ligands lie. This p orbital imbalance arises because of the difference in the ionicity of the Te—C and Te—halogen bonds and also because of the significant tellurium s character associated with the Te—C bond, which confers significant p character on the lone pair. The stereochemically active lone pair is then viewed as occupying the sixth coordinate position *trans* to the aryl group. The interpretation is supported by the observation that the quadrupole splittings decrease in the order $RTeCl_4^- > RTeBr_4^- > RTeI_4^-$. As the electronegativity of the halogen ligands decreases, the removal of p electron density from the tellurium in the *x–y* plane decreases and the quadrupole splittings likewise decrease. The isomer shift of $RTeCl_4^-$ is significantly

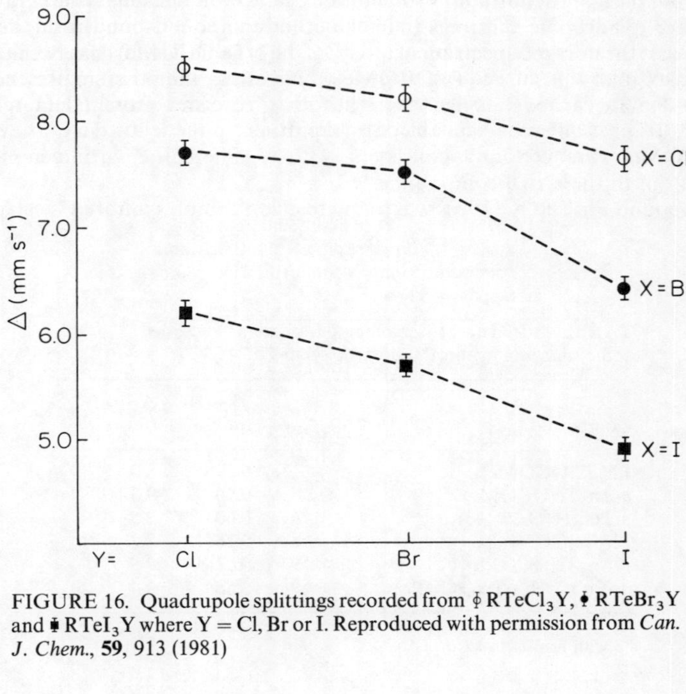

FIGURE 16. Quadrupole splittings recorded from ϕ $RTeCl_3Y$, ● $RTeBr_3Y$ and ■ $RTeI_3Y$ where Y = Cl, Br or I. Reproduced with permission from *Can. J. Chem.*, **59**, 913 (1981)

smaller than that of $TeCl_6^{2-}$ or $TeCl_5^-$, and this reflects the marked decrease in nuclear s electron density as a result of removal of tellurium 5s electrons into the Te—C bond.

In the mixed tetrahaloaryltellurates $RTeX_3Y$, where X = Cl and Y = Cl, Br or I, Δ decreases in the series Y = Cl > Br > I. For the bromide series, i.e. X = Br, Y = Cl, Br or I, the Δ values again decrease in the series Y = Cl > Br > I and a similar trend is again observed for the iodides, i.e. $RTeI_3Cl^- > RTeI_3Br^- > RTeI_4^-$ (Figure 16). These trends again reflect a decrease in the removal of p orbital electron density from the tellurium in the x–y plane as the electronegativity of the halogen ligand, Y, decreases. It would appear that in these compounds Δ is a simple additive property of the ligands attached to tellurium. To a first approximation, and ignoring any asymmetry in the $RTeX_3Y$ anions, the difference in quadrupole splitting between $RTeCl_4^-$ and $RTeBr_4^-$ should be twice that between $RTeCl_3Br^-$ and $RTeClBr_3^-$. This relationship has been confirmed[57].

The findings are consistent with the application of an additivity model to rationalise the values of the quadrupole splittings in the organotellurides and organotellurium dihalides and trihalides, as will be discussed later (Section 8).

7. Diaryltellurium(IV) dicarboxylates

The ^{125}Te Mössbauer parameters recorded[26] from some diaryltellurium(IV) diacetates and dibenzoates are collected in Table 7 and a representative spectrum is illustrated in Figure 17.

The infrared spectra recorded[59] from the diaryltellurium(IV) dicarboxylates have been interpreted in terms of Ψ-trigonal bipyramidal structures in which the carboxylate ligands occupy the *trans* axial positions. The ^{125}Te Mössbauer parameters recorded from these compounds are generally consistent with such a description with the chemical isomer shifts and quadrupole splittings being essentially similar to the Mössbauer parameters recorded from the diaryltellurium(IV) dichlorides. However, the phenyl derivatives exhibit much smaller quadrupole splittings than the other compounds and in this respect it is interesting that the infrared spectrum of $(p\text{-EtOC}_6H_4)_2Te(OCOMe)_2$ has been found to be more complex than that of $Ph_2Te(OCOMe)_2$, suggesting the possible presence of both uni- and bi-dentate carboxylate ligands in the former case and only unidentate ligands in the latter. The different quadrupole splittings of $p\text{-An}_2Te(OCOMe)_2$ and $Ph_2Te(OCOMe)_2$ are certainly consistent with a different environment about the tellurium atom in these two compounds.

The compound $Ph_2Te(OCOPh)_2$ was found to give a complex infrared spectrum similar

TABLE 7. ^{125}Te Mössbauer parameters recorded[26] from some diaryltellurium(IV) dicarboxylates at 4.2 K

Compound	$\delta^a \pm 0.07$ (mm s^{-1})	$\Delta \pm 0.1$ (mm s^{-1})
$Ph_2Te(OCOMe)_2$	0.59	8.0
$p\text{-An}_2Te(OCOMe)_2$	0.50	9.3
$p\text{-Tol}_2Te(OCOMe)_2$	0.60	9.5
$Ph_2Te(OCOPh)_2$	0.76	8.2
$p\text{-An}_2Te(OCOPh)_2$	0.52	9.5
$(p\text{-EtOC}_6H_4)_2Te(OCOPh)_2$	0.56	9.4

^{a} δ with respect to I/Cu.

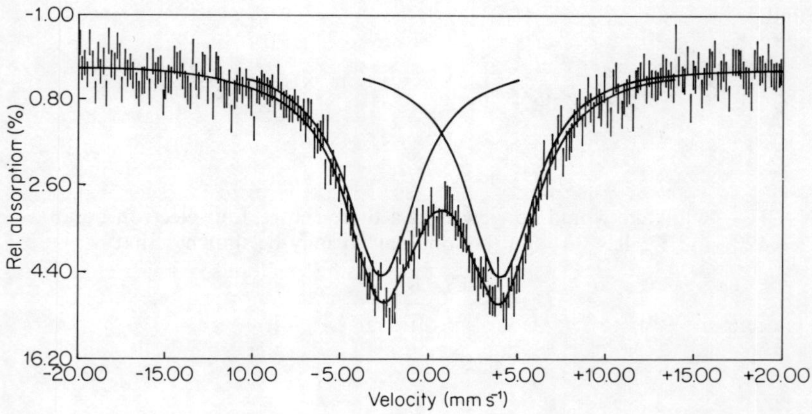

FIGURE 17. ^{125}Te Mössbauer spectrum recorded from $Ph_2Te(OCOMe)_2$. Reproduced with permission from *Can. J. Chem.*, **54**, 3737 (1976)

to that of $(p\text{-}EtOC_6H_4)_2Te(OCOMe)_2$ rather than that of the diphenyltellurium diacetate. However, the Mössbauer data suggest that the tellurium environments in the diphenyltellurium diacetate and dibenzoate are very similar. Thus the infrared and Mössbauer data are not wholly consistent but both suggest that the structures of the diacetates and dibenzoates may show subtle differences with changes in the nature of the aryl ligands. A fuller interpretation of the Mössbauer data from these compounds must await structural studies on the carboxylates.

8. Additivity model for the quadrupole splittings

The ^{125}Te Mössbauer quadrupole splittings recorded from the diorganotellurium(IV) dihalides and organotellurium(IV) trihalides have been found[25] to be amenable to consideration within the theory of Townes and Dailey. By considering the dihalides to be trigonal bipyramidal structures with nominally sp^2 hybrids in the equatorial plane and axial pd hybrids defining the z-axis,

$$\frac{e^2qQ}{e^2q_0Q}\left(1-\frac{\eta}{3}\right)=U_X-U_C$$

and

$$\frac{e^2qQ}{e^2q_0Q}\eta=\tfrac{3}{2}(U_C-U_2)(1-\cot^2{}_\gamma)$$

where U_X, U_C and U_2 are the Te—X, Te—C and non-bonding orbital populations, respectively, η is the asymmetry parameter and 2_γ is the C—Te—C bond angle. This equation is applicable only if $(1-\cot^2{}_\gamma)$ is small, i.e. if the lone pair is predominantly 5s in character, which appears to be valid[25] for example in Me_2TeCl_2 ($2_\gamma = 98°$), Ph_2TeBr_2 ($2_\gamma = 94°$) and α-Me_2TeI_2 ($2_\gamma = 91-97°$). For large values of 2_γ, approaching 120°, and for the U_X and U_C values observed in these compounds, V_{zz} would lie through the lone pair. By assuming[25] that the bonding occurs only through tellurium 5p orbitals ($2_\gamma = 90°$), then

$\eta = 0$ and

$$\frac{e^2qQ}{e^2q_0Q} = -U_p = U_{X\text{—Te—}X} - U_c$$

i.e.

$$U_p = U_C - U_{X\text{—Te—}X}$$

The X—Te—X linkage would be viewed as a three-centre, four-electron bond.

If $2_\gamma = 120°$ and V_{zz} lies through the lone pair, it may be shown[25] that

$$U_p = -\frac{4}{3} + \frac{U_C}{6} + \frac{U_{X\text{—Te—}X}}{2}$$

$$\eta = \frac{3}{2}\frac{(U_X - U_C)}{U_p}$$

In the trihalides, assuming the structure in Figure 15 and the presence of sp^3d^2 hybrids, the application[25] of the Townes and Dailey theory predicts

$$\frac{e^2qQ}{e^2q_0Q} = -U_X + \left(1 + \frac{U_C}{2}\right)$$

$$\eta = 0$$

If the bonding is assumed to be pure p in character with the $5s^2$ electrons constituting the lone pair,

$$U_p = U_{X\text{—Te—}X} - U_C$$

Thus, if the Te—C and X—Te—X bonds each have predominantly p character and constant covalent character in the dihalides and the trihalides, then Δ should have the same magnitude but opposite signs in the two cases. Since it would be expected that $U_C > U_{X\text{—Te—}X}$, then Δ should be negative in the dihalides and positive in the trihalides. The experimental results (Tables 3 and 4) show that for the chlorides, bromides and iodides the quadrupole splittings for the dihalides are of similar magnitude to those of the trihalides but the results give no indication of the sign of Δ.

The application of the additivity model as described[25] suggests that, where the model is applicable, the X—Te—X bond has the same covalent character whether X is terminal or bridging. The analysis also suggests that, in those instances where the dihalides and trihalides have the same magnitude for Δ, intermolecular bonding in the dihalides is not a major factor in determining Δ. The observation[25] that (p-EtOC_6H_4)TeCl_3 and $(pyH)^+$ (p-EtOC_6H_4)TeCl_4^- have the same Δ, within the errors, supports the assumption that the field gradient derives from the valence shell orbital populations and that the lattice terms are not important. Thus, for the compounds R_2Te, R_2TeX_2 and $RTeX_3$ the quadrupole splittings appear to be an additive property of the ligands.

The ^{125}Te Mössbauer parameters recorded[25] from the adducts of p-AnTeCl_3 with pyridine and tetramethylthiourea (tmtu), together with those from tellurium tetrachloride with pyridine, provide a further test of the additivity model for the ^{125}Te Mössbauer quadrupole splittings. The starting point in such an analysis is the Te—C bond orbital population obtained from the tellurides. From the quadrupole splittings recorded from the dichlorides it is then possible to estimate $U_{Cl\text{—Te—}Cl}$ and, from Δ recorded from $TeCl_4$(py), U_{py}. These data, together with the orbital population[34] of the Te—S bond, $U_S = 0.73$, give orbital populations as summarized in Table 8 and leads to the conclusion that $U_C = 1.11$, $U_{Cl\text{—Te—}Cl} = 0.35$ and $U_{py} = 0.73$.

TABLE 8. Orbital populations derived[25,34] using the Townes and Dailey theory

Compound	Δ^a (mm s^{-1})	Orbital populations			
		$U_p{}^b$	$U_x{}^c$	$U_y{}^c$	$U_z{}^c$
Ar_2Te	(+)10.6	−0.89	1.11	1.11	[2.0]
p-An_2TeCl_2	(−) 9.1	+0.76	[1.11]	[1.11]	0.35
p-$AnTeCl_3$	(+) 9.2	−0.76	0.35	0.35	[1.11]
$TeCl_4$.py	(+) 4.5	−0.38	[0.35]	[0.35]	0.73
Te(thiourea)$_2^{2+}$	(+)15.6	−1.30	0.70	0.70	[2.0]

^{a}The signs of Δ assumed are shown in parentheses.
$^b U_p = -e^2Q/e^2q_0Q$ and $e^2qQ = +24$ mm s^{-1}.
^{c}The values shown in brackets were assumed in any one case.

TABLE 9. Predicted and observed quadrupole splittings[25]

Compound	Δ(mm s^{-1})	
	Observed	Calculated
p-$AnTeCl_3$.py	7.6	(+)7.3
p-$AnTeCl_3$.tmtu	7.9	(+)7.3

TABLE 10 Bound orbital populations, U_L, for different ligands[25,26]

Estimator	Parameter	U_L
Ph_2TeCl_2	$U_{Cl—Te—Cl}$	0.4
Ph_2TeBr_2	$U_{Br—Te—Br}$	0.5
p-Tol_2TeI_2	$U_{I—Te—I}$	0.6
$Ph_2Te(OCOMe)_2$	$U_{O\ Te\ O}$	0.4
p-$An_2Te(OCOMe)_2$	$U_{O—Te—O}$	0.3
$Ph_2Te(OCOPh)_2$	$U_{O—Te—O}$	0.4
p-$An_2Te(OCOPh)_2$	$U_{O—Te—O}$	0.3
$TeCl_4$(py)	U_{py}	0.7
Te(thiourea)$_2^{2+}$	U_S	0.7
$(C_6F_5)_2TeBr_2$	$U_{C_6F_5}$	0.8
Tellurides	U_C	1.1

The values of U were used[25] to estimate values of Δ for p-$AnTeCl_3$(py) and p-$AnTeCl_3$(tmtu) and the results found to compare well with the experimentally determined values of Δ as summarized in Table 9.

The orbital populations for a number of ligands derived by these methods are summarized in Table 10. Given the many assumptions implicit in the analysis of the Δ values and the fact that e^2q_0Q is not accurately known, the U values have relative rather than absolute significance.

9. *Colour in organotellurium(IV) halides*

Many organotellurium compounds are highly coloured and, on the basis of structural data, the occurrence of colour has been associated with the overlap of orbitals of intermolecularly associated heavy atoms[60]. Using these concepts the ^{125}Te Mössbauer parameters recorded from a variety of organotellurium compounds have been used[27] to examine the occurrence of colour in terms of the population of low energy conductance bands by non-bonding electrons.

C. Heterocyclic Tellurium Compounds

Two ^{125}Te Mössbauer studies of organotellurium heterocycles have appeared in recent years[61,62]. In these compounds the bond lengths and bond angles are largely determined by the constraint of accommodating the tellurium atom in the ring.

1. Cyclic tellurides

The structural formulae of the cyclic tellurides (a) dibenzotellurophene ($C_{12}H_8Te$), (b) phenoxtellurine ($C_{12}H_8OTe$), (c) tellurium acetylacetonate ($C_5H_6O_2Te$) and (d) some substituted tellurium(II) acetylacetonates are illustrated in Figure 18. The ^{125}Te Mössbauer parameters are recorded in Table 11 and some crystal structure data, which are compared with that of $(p\text{-Tol})_2Te$, are collected in Table 12. In the relatively simple molecules (a)–(c) the C—Te—C bond angles vary slightly and some variation is observed in the bond lengths. The ^{125}Te Mössbauer quadrupole splitting data, and possibly the chemical isomer shift data, show small differences from one compound to another but, given the variation in the C—Te bond lengths and hence the tellurium bond orbital populations, a simple dependence of Δ on the C—Te—C bond angle θ is not to be expected. It is interesting, however, that both $C_{12}H_8OTe$ and $C_5H_6O_2Te$, for which $\theta \approx 90°$, exhibit larger values of δ and Δ than the simple diorganotellurides (Table 1). The results are consistent with a greater p character in the Te—C bonds and a greater lone pair character for the tellurium 5s electrons. The inductive effect of the oxygen atom in $C_{12}H_8OTe$ and of the carbonyl group in $C_5H_6O_2Te$ may be important and influential factors in influencing Δ for these compounds.

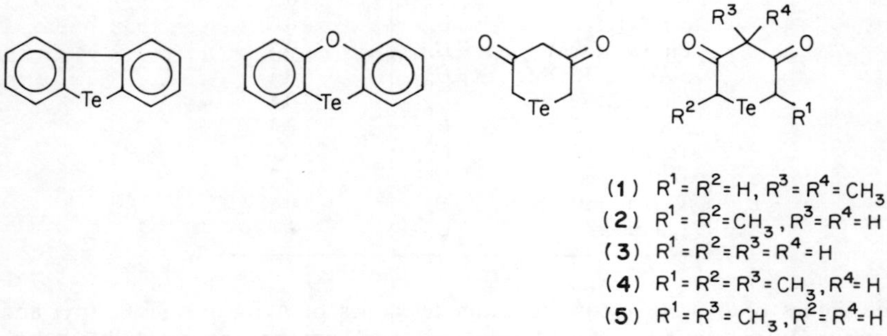

FIGURE 18. Structural formulae of the cyclic tellurides (a) dibenzotellurophene ($C_{12}H_8Te$), (b) phenoxtellurine ($C_{12}H_8OTe$) (c) tellurium acetylacetonate ($C_5H_6O_2Te$) and (d) substituted tellurium(II) acetylacetonates

TABLE 11. ^{125}Te Mössbauer parameters recorded from some cyclic tellurides at 4.2 K

Compound (see Figure 18)			$\delta^a \pm 0.08$ ($mm\,s^{-1}$)	$\Delta \pm 0.2$ ($mm\,s^{-1}$)	Ref.
(a)		$C_{12}H_8Te$	0.14	9.3	61
(b)		$C_{12}H_8OTe$	0.24	11.2	61
(c)		$C_5H_6O_2Te$	0.39	11.7	61
(d)	**(1)**	$C_7H_{10}O_2Te$	0.70	12.77	62
	(2)	$C_7H_{10}O_2Te$	0.39	10.99	62
	(3)	$C_5H_6O_2Te$	0.42	9.81	62
	(4)	$C_8H_{12}O_2Te$	0.47	10.58	62
	(5)	$C_7H_{10}O_2Te$	0.41	11.25	62

^{a} δ with respect to I/Cu.

TABLE 12. Crystallographic data for some tellurides

Compound	C—Te—C bond angle (°)	C—Te bond length (Å)	Ref.
$(p\text{-Tol})_2Te$	101	2.05 (5)	31
(a)^a $C_{12}H_8Te$	81.7 (2)	2.087 (5)	63
(b)^a $C_{12}H_8OTe$	89.4 (3)	2.10 (9)	64
(c)^a $C_5H_6O_2Te$	89.5 (4)	2.17 (1)	65

^{a} See Figure 18.

The C—Te—C bond angle of 81.7° in $C_{12}H_8Te$ implies that a simple combination of s and p orbitals cannot be used to rationalize the bonding about tellurium. The quadrupole splitting for $C_{12}H_8Te$ is smaller than that of other cyclic tellurides and may indicate significant π delocalization around the essentially planar five-membered ring.

The ^{125}Te Mössbauer parameters recorded for all the tellurium(II) acetylacetonates except **1** (Figure 18) fall within a narrow range and reflect the similar structures[65–69] of compounds **2**–**5** where the C—Te—C bond angles of ca. 90° suggest that the bonding occurs primarily through the p orbitals. The main structural difference between these compounds is the smaller[70] C—Te—C bond angle of 86.4° in **1** which cannot be simply rationalised in terms of the tellurium s and p orbitals. On first inspection, the small bond angle may reasonably be associated with enhanced repulsion by the lone pair of electrons on the tellurium atom. Both Mössbauer parameters for this compound are larger than those recorded from compounds **2**–**5** and are indicative of a relatively higher s electron density at the tellurium nucleus and a larger p orbital imbalance around the tellurium atom. Clearly it is difficult to correlate the crystallographically determined small bond angle and the proposed high stereochemical activity of the tellurium lone pair with the Mössbauer parameters which imply that the lone pair is 5s in character and contributes significantly to a high electron density at the tellurium nucleus. Indeed, it would be reasonable to expect that any significant stereochemical activity of the 5s lone pair would be reflected in a smaller chemical isomer shift in **1** compared with **2**–**5**. However, it is interesting that the intermolecular packing of the tellurium(II) acetylacetonates[66–69] involves the yellow materials **2**–**5** having intermolecular Te···Te associations of ca. 4 Å which give rise to polymeric chains of tellurium atoms. The methyl groups in these compounds are equatorial to the heterocyclic ring. The colour of **1** is less intense and, although the molecular structure is essentially the same, there is no evidence[70] of short intermolecular Te···Te associations as in **2–5**. It would appear that one of the two methyl groups which occupy the positions designated by R^3 and R^4 in **1** is axial to the heterocyclic ring and sterically prevents the formation of short intermolecular Te···Te associations.

The significance of such close intermolecular distances in coloured tellurium compounds has been noted during structural investigations in the past[60] and it has also been suggested[71–76] that colour in other p block elements, particularly those of tin and antimony, which also give lower than expected Mössbauer chemical isomer shifts, may be explained by the direct population of low-energy conduction bands by non-bonding valence electrons. It therefore seems that the low ^{125}Te Mössbauer chemical isomer shifts in the yellow associated tellurium(II) acetylacetonates **2–5** may be indicative of the population by $5s^2$ non-bonding tellurium electrons of conductance bands formed by overlap of orbitals on intermolecularly associated tellurium atoms. The donation and movement of these non-bonding valence electrons in the conductance bands may give rise to the colour, and the consequent reduction of the 5s electron density at the tellurium nucleus is reflected in the chemical isomer shift which is smaller than expected. The smaller quadrupole splittings in the tellurium(II) acetylacetonate compounds **2–5**, compared with unassociated compound **1**, presumably reflects the higher degree of coordination and symmetry of the tellurium atom in the associated species.

Hence the delocalization of electrons into a π-bonded aromatic ring or into a low-energy conductance band formed by overlap of orbitals on intermolecularly associated heavy atoms may be as important in explaining the ^{125}Te Mössbauer parameters of heterocyclic tellurium compounds as is a consideration of the bond angle alone.

2. *Heterocyclic tellurium dichlorides*

The heterocyclic tellurium dichlorides, with structural formulae as illustrated in Figure 19, (a) $C_{12}H_8TeCl_2$, (b) $C_{12}H_8OTeCl_2$ and (c) $C_5H_6O_2TeCl_2$, give the ^{125}Te Mössbauer parameters summarized in Table 13. The crystal structure[65] of $C_5H_6O_2TeCl_2$ shows the tellurium to be in a Ψ-trigonal bipyramidal geometry with the halogen ligands in *trans* axial positions. The dihalides exhibit more positive chemical isomer shifts than the parent tellurides, the halogen ligands removing predominantly p electron density from the tellurium, and the values of δ are comparable to those recorded from the diaryltellurium(IV) dichlorides (Table 3). However, the quadrupole splittings recorded from the cyclic dichlorides are all significantly smaller which, although partly arising from differences in the asymmetry parameter, η, may be better understood in terms of the additivity model. According to these principles, and the appropriate discussion in Sections

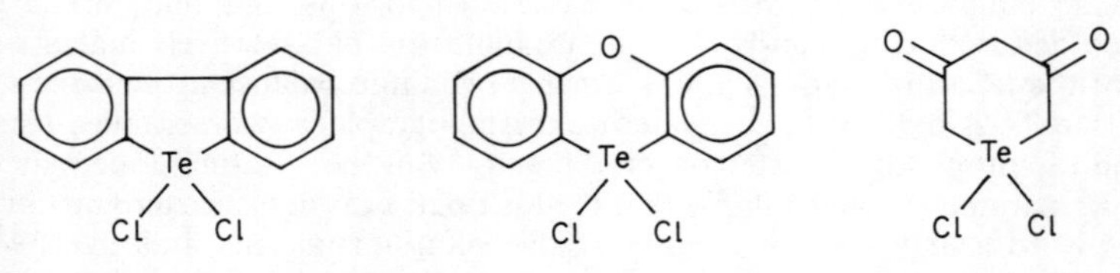

FIGURE 19. Structural formulae of (a) $C_{12}H_8TeCl_2$, (b) $C_{12}H_8OTeCl_2$ and (c) $C_5H_6O_2TeCl_2$

TABLE 13. ^{125}Te Mössbauer parameters recorded[61] from some heterocyclic tellurium dichlorides at 4.2 K

Compound (see Figure 19)		$\delta^a \pm 0.08$ ($mm\,s^{-1}$)	$\Delta \pm 0.2$ ($mm\,s^{-1}$)
(a)	$C_{12}H_8TeCl_2$	0.75	8.1
(b)	$C_{12}H_8OTeCl_2$	0.70	8.0
(c)	$C_5H_6O_2TeCl_2$	0.71	8.2

[a] δ relative to I/Cu.

III.A and III.B, it is reasonable to expect the cyclic dihalides to have quadrupole splittings which are smaller than those recorded from the parent tellurides.

It is also interesting that the cyclic dinitrate related to Figure 19b of composition $C_{12}H_8OTe(NO_3)_2$ gave[61] ^{125}Te Mössbauer parameters of δ 0.56 $mm\,s^{-1}$ and Δ 9.3 $mm\,s^{-1}$, in which the large quadrupole splitting is comparable to that recorded from the dichloride and is consistent with the Ψ-trigonal bipyramidal array about tellurium in which the nitrate groups adopt the *trans* axial positions[64].

D. Organotellurium Ligands

The use of Mössbauer spectroscopy to examine the ligand properties of various species including tellurium has recently been reviewed[77]. A survey of the chemical isomer shift and quadrupole splitting data recorded from several metal complexes of (p-$EtOC_6H_4)_2Te$ showed that the $R_2Te \rightarrow M$ donation involves primarily the tellurium 5p orbital and that changes in chemical isomer shift are due to deshielding.

It is also interesting that whereas carbonyl groups on a metal are generally able to enhance the Lewis acidity of the metal centre allowing a greater degree of ligand-to-metal donation, the trend does not appear to hold for R_2Te ligands from which carbonyl systems apparently remove ca. 0.33 electrons whilst non-carbonyl acceptors take ca. 0.4–0.6 electrons. It is possible that the presence of two lone pairs on tellurium, only one of which is involved in coordination, is significant such that R_2Te can be regarded as a potential π-donor ligand.

It is relevant to record the examination by ^{125}Te Mössbauer spectroscopy of complexes of di-(ethoxyphenyl) telluride, (p-$EtOC_6H_4)_2Te$, with various metal halides[78]. By assuming that the quadrupole interaction is dominated by an imbalance in the tellurium p-orbital population the quadrupole splitting data were interpreted in terms of the donor–acceptor interaction of the telluride with the metal halides and used to show that the order of Lewis acidity towards (p-$EtOC_6H_4)_2Te$ is Hg(II) > Pt(II) > Pd(II) > Cu(I). In a subsequent report[79] some diorganotelluride–mercury(II) complexes of the type $R_2Te.HgX_2$ (R = Ph, $EtOC_6H_4$, X = Cl, Br, I) were examined by ^{125}Te Mössbauer, infrared and Raman spectroscopy. The results indicated that the tellurium atom in these complexes adopts a similar environment to that of tellurium in a triorganotellurium salt and the complexes were described by the formulation $[R_2Te^+—HgX]X^-$. The structure of $(Ph_2Te).HgI_2$ was subsequently investigated[80] by X-ray crystallography and found to exhibit a novel tetrameric structure involving two different types of iodine bridges.

In another report[81], a new series of copper(I) derivatives of organotellurols of composition RTeCu, where R is an alkyl or aryl group, were described and the ^{125}Te Mössbauer parameters interpreted in terms of only limited electron transfer from tellurium to copper. A series of diorganoditelluride complexes of copper(I) halides of composition $R_2Te_2.CuX$, where R = alkyl or aryl and X = Cl, Br or I, gave[81] Mössbauer data which indicated the tellurium atoms to be only weakly coordinated to copper with the halide ligands involved in intermolecular bridging. The complexes of some diorganoditellurides with mercury(II) halides have also been investigated[79] and the compounds were found to correspond to two distinct classes of complex. The majority, of stoichiometry $RTeHgX_2$ or $R_2Te_2HgX_2$, gave ^{125}Te Mössbauer parameters which indicated the presence of a single trigonally coordinated tellurium site. The second group gave Mössbauer parameters indicative of two-coordinate tellurium in complexes of the type $R_2Te_2.HgX_2$ where both the Te—Te and the Hg—X bonds remain essentially similar to those present in the reactants.

Some other organotellurium–mercury(II) complexes of the type α-$Me_2TeI_2.HgBr_2$, $Ph_2Hg.(Me_2TeI_2)_2$, (p-$EtOC_6H_4Te)_2.Hg$ and p-$EtOC_6H_4Te.HgCl$ have also been examined[79] and the Mössbauer data interpreted in terms of the possible structures of the complexes.

E. Compounds with Tellurium to Metal Bonds

Several Mössbauer studies of compounds containing tellurium–tin bonds have appeared and are particularly important because they allow the examination of the compounds by both ^{125}Te and ^{119}Sn Mössbauer spectroscopy.

Some triphenylstannyl(aryl) tellurides prepared by the reaction of triphenyltin hydride with various diaryl ditellurides were amongst the first compounds to be studied by both ^{125}Te and ^{119}Sn Mössbauer spectroscopy[82]. The ^{119}Sn chemical isomer shifts for Ph_3SnXPh, X = O, S, Se or Te, were found to follow the order $Ph_4Sn < Ph_3SnOPh < Ph_3SnSPh \approx Ph_3SnSePh \approx Ph_3SnTePh$, thereby demonstrating a greater p character in the Sn—XPh bonds than in the Sn—Ph bonds. However, the Sn—XPh bonds must retain significant s character and the relative chemical isomer shifts of Ph_3Sn—OPh and Ph_3Sn—TePh suggested the presence of greater s electron density and enhanced covalency of the S—Te bond relative to the Sn—O bond. The ^{125}Te Mössbauer chemical isomer shifts recorded from the compounds Ph_3SnTeR, R = Ph, p-An, p-$EtOC_6H_4$, were less positive than those of the diaryl ditellurides, R_2Te_2, suggesting a greater occupancy of the tellurium 5p orbital in the Sn—Te bond than in the Te—Te bond. The smaller ^{125}Te quadrupole splittings for the triphenylstannyl compounds were consistent with this model. Hence both the ^{119}Sn and ^{125}Te Mössbauer data are consistent with the donation of electron density from the tin to the tellurium and give no evidence for any significant $5p\pi$ Te $\rightarrow 5d\pi$ Sn overlap such that only an s–p bonding model gives a self-consistent explanation of the combined ^{119}Sn and ^{125}Te Mössbauer data.

A ^{119}Sn and ^{125}Te Mössbauer and NMR study of some Group IV organotellurides is another example of the successful combined use of two Mössbauer isotopes in chemical investigations[83]. In these compounds of composition$(R_3X)_2Te$, R = Me, X = C, Si, Ge and Sn; R = Ph, X = Ge and Sn; R_3MTePh, R = Me, X = Si, Ge and Sn; R = Ph, X = Ge, Sn, Pb; $R_2Sn(TePh)_2$, R = Me and t-Bu; and the cyclic compounds $(Me_2SnTe)_3$ $(Me_2Sn)_3Te_2$ and $(t$-$Bu_2SnTe)_2$, the ^{119}Sn and ^{125}Te Mössbauer and NMR data have provided evidence that there is little transmission of bonding effects through the tin—tellurium bond as the chemical environments about the tin or the tellurium atoms are changed.

In a recent study[84] of some tellurium- and transition metal-containing compounds the ^{125}Te Mössbauer spectra were recorded from $(\mu$-Te)[V(CO)$_3$diphos]$_2$, where diphos is ethylenebis(diphenylphosphane), $(\mu_3$-Te)[Mn(CO)$_2(\eta^5$-$C_5H_5)]_3$, $(\mu$-Te)[Mn(CO)$_2(\eta^5$-$C_5Me_5]_2$, $(\mu,\eta^2$-$TeCH_2)$[Mn(CO)$_2(\eta^5$-$C_5Me_5)]_2$ and $(\mu$-Te)[Cr(CO)$_3(\eta^5$-$C_5H_5)]_2$. The chemical isomer shifts, δ, were found to be essentially similar and independent of whether the tellurium is considered as acting as a six-, four- or two-electron donor. The first three compounds gave very small quadrupole splittings, which were interpreted in terms of multiple bond character in the tellurium—metal bonds.

F. Tellurium-containing Charge-transfer Complexes

In a study[85] of the reaction of NaTeR, R = *p*-EtOC$_6$H$_4$ or Ph, with organic dihalides, $(CH_2)_nX_2$, $n = 1$–4, which gives tellurium salts when $n = 3$ or 4 and X = Cl or Br, some charge-transfer complexes of stoichiometry $(RTe)_2CH_2.CH_2X_2$ were formed when $n = 1$ and X = Br or I. The Mössbauer spectrum recorded from the complex $[(p\text{-}EtOC_6H_4)Te]_2CH_2.CH_2Br_2$ was dissimilar to those recorded from the telluronium salts. The more positive chemical isomer shift was associated with a greater s electron density at tellurium as a result of the withdrawal of 5p electron density. The quadrupole splitting was smaller than that recorded from simple tellurides (Table 1), but larger than that recorded from telluronium salts. The description of the material, which also gave a broad ESR signal centred on $g = 2.18$, as a charge-transfer complex is consistent with the Mössbauer data since the transfer of electron density from the spare-pair p orbital on tellurium would decrease the p orbital occupation imbalance and give rise to a smaller quadrupole splitting and more positive chemical isomer shift.

Similar effects were subsequently observed[86] in the Mössbauer spectra recorded from organotellurium complexes with 7, 7, 8, 8-tetracyanoquinodimethane (tcnq). In these investigations a range of organotellurium(II) compounds were used as donors to form donor–acceptor complexes with tcnq. Although most reactions gave 1:1 complexes, the ditelluride gave complexes of the type $R_2Te_2.2$tcnq, where R is Ph or (*p*-EtOC$_6$H$_4$). Although the complexes were insulators the conductivities did vary over several orders of magnitude and the two ditelluride complexes showed evidence of semiconducting properties. Infrared and ESR data indicated the degree of charge transfer to be small in most cases. However, the tcnq complexes of 1, 3-dihydro-2-telluraindene and diphenyl ditelluride gave ^{125}Te Mössbauer spectra which were interpreted in terms of significant charge transfer and the results have led to the development of a method whereby the Mössbauer parameters can be used to assess the degree of charge transfer for donor–acceptor complexes containing organotellurium compounds. Given the likely growth in this area of chemistry, it is important that the potential use of ^{125}Te Mössbauer spectroscopy as an investigative technique for these types of compounds be explored more fully. Some evidence for this development is reflected in a more recent study[13] of some 2-phenylazophenyl-*C*, *N'*-tellurium(II) dithiocarbamates of composition $(C_{12}H_9N_2)$Te(II)(dtc) and the corresponding tellurium(IV) tris compounds $(C_{12}H_9N_2)$Te(dtc)$_3$, where dtc is dimethyl, diethyl or dibenzyl dithiocarbamate. The ^{125}Te Mössbauer parameters recorded from these complexes were used to formulate the tris compounds as loose charge-transfer complexes of the type $(C_{12}H_9N_2)$Te(II)(dtc).$[R_2NC(S)SS(S)CNR_2]$.

G. Iodine-125 Emission Mössbauer Studies

In Mössbauer emission spectroscopy the sample to be studied is prepared with a radioactive isotope and is used as a source. The radiation emitted from the sample is absorbed by a standard absorber that contains the corresponding Mössbauer isotope. Since an iodine-125 nucleus undergoes radioactive decay to populate the 35.48 keV excited state of tellurium-125, compounds labelled with iodine-125 can be used as sources in Mössbauer emission spectroscopy. It must be appreciated, however, that the decay processes may create unstable but long-living changes in the surroundings of the Mössbauer nucleus and that such source experiments are often complicated and the

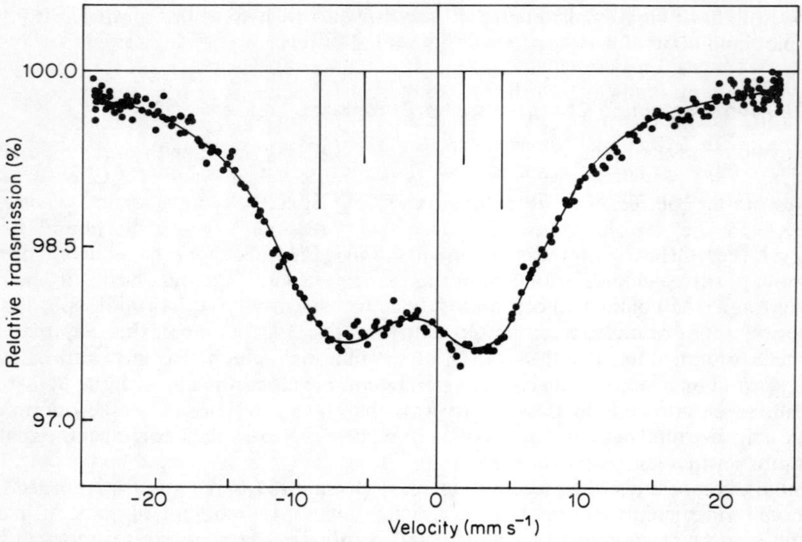

FIGURE 20. Mössbauer emission spectrum recorded from [^{125}I]iodobenzene in hexane at 4.2 K. Reproduced with permission from *Chem. Phys. Lett.*, **94**, 227 (1983)

consequences of nuclear transformations which involve both oxidation and reduction are not fully understood.

A Mössbauer emission spectroscopy study of ^{125}I-labelled iodobenzene, PhI, benzyl iodide, $PhCH_2I$, and iodine in benzene reported[87] similar spectra from all samples showing quadrupole split absorptions, $\Delta \approx 11.3$ mm s^{-1}, which were associated with the formation of Te—C bonds. In a subsequent study[88] by ^{125}I Mössbauer emission techniques and ^{127}NQR spectroscopy a series of *o*-, *m*- and *p*-substituted iodobenzenes were found to undergo at least partial decomposition to diorganotellurides. The most recent Mössbauer emission investigation of ^{125}I-labelled iodobenzene, methyl iodide and their dilute solutions in benzene and hexane[89] gave spectra (Figure 20) which, by careful computer analysis, were interpreted in terms of the presence of two species. One of the species was tentatively described as a diaryl, arylalkyl or dialkyl telluride whilst the other was described as a tellurium atom attached to a single aryl or alkyl moiety with a positive charge, e.g. $[PhTe]^+$, $[IC_6H_4Te]^+$ or $[C_6H_{13}Te]^+$. The observations were rationalized in terms of the ^{125}I in MeI or PhI undergoing decay by electron capture with the rupture of the Te—C bond and the excited and highly reactive tellurium ion undergoing reaction with neighbouring molecules of iodobenzene or methyl iodide.

H. Tellurium Complexes with Thiourea and Related Compounds

It is relevant to mention here several significant studies of thiourea and related compounds. ^{125}Te Mössbauer data recorded from $TeTu_4X_2$, where Tu is thiourea and X is halide, were used in the initial derivations of relationships between the ^{125}Te Mössbauer chemical isomer shifts and holes in the tellurium 5p shell in tellurium compounds which are p bonded[90]. In a subsequent study[91], asymmetry in the Mössbauer spectra of tellurium(II) thiourea complexes was shown to arise predominantly from single crystal orientation effects and the asymmetry in the spectrum recorded from $TeTu_4Cl_2.2H_2O$ was shown to be consistent with a negative sign for the nuclear quadrupole moment of the 3/2 excited state in ^{125}Te. In other studies[34,56,92] of tellurium(II) and tellurium(IV) complexes with sulphur-containing ligands the Mössbauer parameters have been interpreted in terms of bonding and the structural properties of the complexes. Finally, it is pertinent to record that the ^{125}Te Mössbauer emission spectra recorded from a number of tellurium(II) thiourea complexes have given evidence[93] of extensive molecular fragmentation during the isomeric transition which precedes the Mössbauer transition.

IV. CONCLUSION

Tellurium-125 Mössbauer spectroscopy is a technique which is well suited for the investigation of organotellurium compounds. The different types of compounds which have been examined to date illustrate the variety of materials which are amenable to study by the technique. The results which have been interpreted in terms of the bonding and structural properties of the compounds exemplify the power of the technique to give unique information and also demonstrate the capacity of Mössbauer spectroscopy to complement other methods of investigation. The currently developing interst in tellurium-containing compounds with potentially technologically important properties will undoubtedly lead to a further growth in the application of Mössbauer spectroscopy, in both conventional and unconventional modes of operation, to the study of organotellurium compounds.

V. REFERENCES

1. R. L. Mössbauer, *Z. Phys.*, **151**, 124 (1958).
2. G. K. Wertheim, *Mössbauer Effect: Principles and Applications*, Academic Press, New York, 1964.
3. N. N. Greenwood and T. C. Gibb, *Mössbauer Spectroscopy*, Chapman and Hall, London, 1971.
4. L. May (Ed.), *An Introduction to Mössbauer Spectroscopy*, Plenum Press, New York, 1971.
5. G. M. Bancroft, *Mössbauer Spectroscopy, An Introduction for Inorganic Chemists and Geochemists*, McGraw-Hill, Maidenhead, 1973.
6. U. Gonser (Ed.), *Topics in Applied Physics, Vol. V: Mössbauer Spectroscopy*, Springer-Verlag, Berlin, 1975.
7. T. C. Gibb, *Principles of Mössbauer Spectroscopy*, Chapman and Hall, London, 1976.
8. F. J. Berry and D. P. E. Dickson (Eds.), *Mössbauer Spectroscopy*, Cambridge University Press, Cambridge, 1986.
9. F. J. Berry, in *Physical Methods of Chemistry* (Eds. B. W. Rossiter and J. F. Hamilton), 5th ed., Wiley, New York, in press.
10. Pham Zuy Hein, V. G. Shapiro and V. S. Shpinel, *Sov. Phys. JETP*, **15**, 489 (1962).
11. N. Shikazono, T. Shorji, H. Takekoshi and P. Tseng, *J. Phys. Soc. Jpn.*, **17**, 1205 (1962).
12. G. M. Bancroft, *Mössbauer Spectroscopy, An Introduction for Inorganic Chemists and Geochemists*, McGraw-Hill, Maidenhead, 1973, p. 27.
13. M. A. K. Ahmed, A. E. McCarthy, W. R. McWhinnie and F. J. Berry, *J. Chem. Soc., Dalton Trans.*, 771 (1986).
14. E. P. Stepanov, K. P. Aleshine, R. A. Manapov, B. N. Samoilau, V. V. Sklyarevskii and V. G. Stankevich, *Phys. Lett.*, **6**, 155 (1963).
15. M. Pasternak and A. L. Spijkervat, *Phys. Rev.*, **181**, 574 (1969).
16. N. E. Erickson and A. G. Maddock, *J. Chem. Soc. A*, 1665·(1970).
17. R. N. Kuzmin, A. A. Opalenko, V. S. Shpinel and I. A. Avenarius, *Sov. Phys. JETP*, **29**, 94 (1969).
18. V. A. Lebedev, R. A. Lebedev, A. M. Babeshkim and A. N. Nesmeyanov, *Vestn. Mosk. Univ. Khim.*, **24**, 128 (1969).
19. C. E. Violet, R. Booth and F. Wooten, *Phys. Lett.*, **5**, 230 (1963).
20. H. Voorthius, W. Beens and H. Verheul, *Physica*, **33**, 695 (1967).

21. B. Martin and R. Schule, *Phys. Lett. B*, **46**, 367 (1973).
22. B. P. Dailey and C. H. Townes, *J. Chem. Phys.*, **23**, 118 (1955).
23. F. J. Berry, E. H. Kustan and B. C. Smith, *J. Chem. Soc., Dalton Trans.*, 1323 (1975).
24. K. V. Smith, J. S. Thayer and B. J. Zambransky, *Inorg. Nucl. Chem. Lett.*, **11**, 441 (1975).
25. C. H. W. Jones, R. Schultz, W. R. McWhinnie and N. S. Dance, *Can. J. Chem.*, **54**, 3234 (1976).
26. F. J. Berry and C. H. W. Jones, *Can. J. Chem.*, **54**, 3737 (1976).
27. F. J. Berry and J. Silver, *J. Organomet. Chem.*, **129**, 437 (1977).
28. F. J. Berry, C. H. W. Jones and W. R. McWhinnie, in *Mössbauer Effect Methodology* (Eds. I. J. Gruverman and C. W. Seidel), Vol. 10, Plenum Press, New York, 1976, p. 227.
29. F. J. Berry, C. H. W. Jones and W. R. McWhinnie, in *Proceedings of the Fourth International Conference on the Organic Chemistry of Selenium and Tellurium* (eds. F. J. Berry and W. R. McWhinnie), University of Aston Press, Birmingham, 1983, p. 492.
30. C. H. W. Jones and R. D. Sharma, *J. Organomet. Chem.*, **255**, 61 (1983).
31. W. R. Blackmore and S. C. Abrahams, *Acta Crystallogr.*, **8**, 317 (1955).
32. F. H. Kruse, R. E. Marsh and J. D. McCullough, *Acta Crystallogr.*, **10**, 201 (1957).
33. G. Llabres, O. Dideberg and L. Dupont, *Acta Crystallogr., Sect. B*, **28**, 2438 (1972).
34. R. M. Cheyne, C. H. W. Jones and S. Husebye, *Can. J. Chem.*, **53**, 1855 (1975).
35. C. H. W. Jones, R. D. Sharma and D. Naumann, *Can. J. Chem.*, **64**, 987 (1986).
36. C. H. W. Jones and R. D. Sharma, *Organometallics*, **5**, 805 (1986).
37. G. T. Morgan and H. D. K. Drew, *J. Chem. Soc.*, 531 (1925).
38. F. J. Berry, B. C. Smith and C. H. W. Jones, *J. Organomet. Chem.*, **110**, 201 (1976).
39. G. D. Christofferson and J. D. McCullough, *Acta Crystallogr.*, **11**, 249 (1958).
40. G. D. Christofferson, R. A. Sparks and J. D. McCullough, *Acta Crystallorg.*, **11**, 782 (1958).
41. G. Y. Chao and J. D. McCullough, *Acta Crystallogr.*, **15**, 887 (1962).
42. L. Y. Y. Chan and F. W. B. Einstein, *J. Chem. Soc., Dalton Trans.*, 316 (1972).
43. F. J. Berry and A. J. Edwards, *J. Chem. Soc., Dalton Trans.*, 2306 (1980).
44. W. R. McWhinnie and M. G. Patel, *J. Chem. Soc., Dalton Trans.*, 199 (1972).
45. J. J. Johnstone, C. H. W. Jones and P. Vasudev, *Can J. Chem.*, **50**, 3037 (1972).
46. L. H. Bowen, J. G. Stevens and G. G. Long, *J. Chem. Phys.*, **51**, 2010 (1969).
47. H. Pritzkow, *Inorg. Chem.*, **18**, 311 (1979).
48. W. R. McWhinnie and P. Thavornyutikarn, *J. Chem. Soc., Dalton Trans.*, 551 (1972).
49. J. D. McCullough and P. F. Marsh, *J. Am. Chem. Soc.*, **72**, 4556 (1950).
50. K. J. Wynne and J. W. George, *J. Am. Chem. Soc.*, **91**, 1649 (1969).
51. D. A. Couch, P. S. Elmes, J. E. Ferguson, M. C. Greenfield and C. J. Wilkens, *J. Chem. Soc. A*, 1813 (1967).
52. M. T. Chen and J. W. George, *J. Am. Chem. Soc.*, **90**, 4580 (1968).
53. K. J. Wynne and P. S. Pearson, *Inorg. Chem.*, **9**, 106 (1970).
54. L. H. Bowen, in *Mössbauer Effect Data Index for 1972* (Eds. J. G. Stevens and V. E. Stevens), Plenum Press, New York, 1973, p. 71.
55. R. G. Barnes and W. V. Smith, *Phys. Rev.*, **93**, 9 (1954).
56. R. M. Cheyne, C. H. W. Jones and P. Vasudev, *Can. J. Chem.*, **50**, 3677 (1972).
57. N. S. Dance, P. Dobud and C. H. W. Jones, *Can. J. Chem.*, **59**, 913 (1981).
58. N. S. Dance and C. H. W. Jones, *Can. J. Chem.*, **56**, 1745 (1978).
59. B. C. Pant, W. R. McWhinnie and N. S. Dance, *J. Organomet. Chem.*, **63**, 305 (1973).
60. J. D. McCullough, *Inorg. Chem.*, **12**, 2669 (1973).
61. N. S. Dance and C. H. W. Jones, *Can. J. Chem.*, **56**, 1747 (1978).
62. F. J. Berry and J. Silver, *J. Organomet. Chem.*, **188**, 255 (1980).
63. J. D. McCullough, *Inorg. Chem.*, **14**, 2639 (1975).
64. M. R. Smith, M. M. Mangion, R. A. Zingaro and E. A. Mengers, *J. Heterocycl. Chem.*, **10**, 527 (1973).
65, C. L. Raston, R. J. Secomb and A. H. White, *J. Chem. Soc., Dalton Trans.*, 2307 (1976).
66. J. C. Dewan and J. Silver, *J. Organomet. Chem.*, **125**, 125 (1977).
67. J. C. Dewan and J. Silver, *Aust. J. Chem.*, **30**, 487 (1977).
68. J. C. Dewan and J. Silver, *Acta Crystollogr., Sect. B*, **33**, 1469 (1977).
69. J. C. Dewan and J. Silver, *Acta Crystallogr., Sect. B*, **33**, 261 (1977).
70. J. C. Dewan and J. Silver, *J. Chem. Soc., Dalton Trans.*, 644 (1977).
71. J. Barrett, S. R. A. Bird, J. D. Donaldson and J. Silver, *J. Chem. Soc. A*, 3105 (1971).
72. J. D. Donaldson, D. Laughlin, S. D. Ross and J. Silver, *J. Chem. Soc., Dalton Trans.*, 1985 (1971).

73. J. D. Donaldson and J. Silver, *Inorg. Nucl. Chem. Lett.*, **10**, 537 (1974).
74. J. D. Donaldson, J. Silver, S. Hadjiminolis and S. D. Ross, *J. Chem. Soc., Dalton Trans.*, 1500 (1975).
75. J. D. Donaldson and J. Silver, *J. Solid State Chem.*, **18**, 117 (1976).
76. J. D. Donaldson, D. R. Laughlin and M. J. K. Thomas, *Radiochem. Radioanal. Lett.*, **28** 91 (1977).
77. R. V. Parish, *Coord. Chem. Rev.*, **42**, 1 (1982).
78. I. Davies, W. R. McWhinnie, N. S. Dance and C. H. W. Jones, *Inorg. Chim. Acta Lett.*, **29**, L203 (1978).
79. N. S. Dance and C. H. W. Jones, *J. Organomet. Chem.*, **152**, 175 (1975).
80. F. W. B. Einstein, C. H. W. Jones, T. Jones and R. D. Sharma, *Inorg. Chem.*, **22**, 3924 (1983).
81. I. Davies, W. R. McWhinnie, N. S. Dance and C. H. W. Jones, *Inorg. Chim. Acta Lett.*, **29**, L217 (1978).
82. N. S. Dance, W. R. McWhinnie and C. H. W. Jones, *J. Organomet. Chem.*, **125**, 291 (1977).
83. C. H. W. Jones, R. D. Sharma and S. P. Taneja, *Can. J. Chem.*, **64**, 980 (1986).
84. C. H. W. Jones and R. D. Sharma, *Organometallics*, **5**, 1194 (1986).
85. K. G. K. De Silva, Z. Monsef–Mirzai and W. R. McWhinnie, *J. Chem. Soc., Dalton Trans.*, 2143 (1983).
86. H. B. Singh, W. R. McWhinnie, R. F. Ziolo and C. H. W. Jones, *J. Chem. Soc., Dalton Trans.*, 1267 (1984).
87. C. H. W. Jones and A. F. Skinner, in *Proceedings of the International Conference on Mössbauer Spectroscopy* (Eds. D. Barb and D. Tarina), Vol. 1, Bucharest 1977, p. 283.
88. G. L. Shtyrkov, N. G. Iroshnikova, V. V. Khrapov, T. A. Babushikina and V. I. Stanko, *Zh. Fiz. Khim.*, 5322 (1979).
89. Ch. Sauer, M. Rieche, A. Halpern and A. Nath, *Chem. Phys. Lett.*, **94**, 227 (1983).
90. R. M. Cheyne, J. J. Johnstone and C. H. W. Jones, *Chem. Phys. Lett.*, **14**, 545 (1972).
91. R. M. Cheyne and C. H. W. Jones, *Can. J. Chem.*, **52**, 564 (1974).
92. P. W. C. Barnarc, J. D. Donaldson, R. M. A. Grimsey, G. Dennes, U. Russo and S. Calogero, *Inorg. Chim. Acta*, **51**, 217 (1981).
93. C. H. W. Jones and M. Dombsky, *J. Phys.* (*Paris*), *Colloq. C2*, **40**, 417 (1979).

The Chemistry of Organic Selenium and Tellurium Compounds Volume 2
Edited by S. Patai
© 1987 John Wiley & Sons Ltd.

CHAPTER **3**

Preparative uses of organoselenium and organotellurium compounds

THOMAS G. BACK

Department of Chemistry, University of Calgary, Calgary, Alberta, Canada, T2N 1N4

I. INTRODUCTION . . . 94
II. OLEFIN PREPARATIONS BY ELIMINATION . . . 95
A. Selenoxide *syn*-Elimination . . . 95
1. General comments . . . 95
2. Stereochemistry . . . 95
3. Regiochemistry . . . 96
4. Techniques and side reactions . . . 98
5. Acetylenes and allenes from vinyl selenoxides . . . 100
B. Telluroxide Elimination . . . 100
C. Elimination of Selenonium Salts . . . 100
III. DEHYDROGENATION OF CARBONYL COMPOUNDS . . . 101
A. Via α-Selenenylated Derivatives . . . 101
1. Aldehydes and ketones . . . 101
2. Esters and lactones . . . 103
3. Amides and lactams . . . 105
4. β-Dicarbonyl compounds . . . 106
5. Enol acetates, enol silyl ethers, enamines and related compounds . . . 107
B. With Selenium Dioxide, Benzeneseleninic Anhydride and Benzeneseleninyl Chloride . . . 110
1. Selenium dioxide . . . 110
2. Benzeneseleninic anhydride . . . 111
3. Benzeneseleninyl chloride . . . 112
IV. DEHYDROGENATION OF OTHER FUNCTIONAL GROUPS . . . 113
A. Nitrogen Compounds . . . 113
B. Nitriles . . . 113
C. Phosphorus Compounds . . . 114
D. Sulfur Compounds . . . 114

V. OXYGENATION REACTIONS 115
A. Oxyselenenylation and Oxytellurenylation 115
1. Olefins 115
2. Enol ethers 117
3. Acetylenes. 118
B. Allylic Oxidation 118
C. Benzylic Oxidation. 120
D. Acetoxymethylation 120
E. Oxidation of Phenols 121
F. Oxidation of Ketones 121
1. α-Oxygenation 121
2. Baeyer–Villiger reactions 122
G. Epoxidation, Hydroxylation and Acetoxylation of Olefins 123
1. Epoxidation 123
2. Hydroxylation 124
3. Acetoxylation 124
H. Oxidation of Selenides 124
1. Pummerer reactions 124
2. Via selenones and tellurones 125
3. In 1,2-carbonyl transpositions. 125
I. Hydrolysis of Vinyl Selenides 126
VI. AMINATION AND AMIDATION REACTIONS. 126
A. Of Olefins 126
1. 1,2-Additions of Selenenic electrophiles 126
2. Allylic amidation 127
B. Of Phenols and Catechols 128
C. Of Michael Acceptors. 128
VII. HALOGENATION 129
A. By Electrophilic Addition of Selenium or Tellurium Halides to Olefins and Acetylenes. 129
1. Olefins 129
2. Acetylenes. 131
B. Allylic Halogenation 131
C. Halogenolysis of Selenides and Tellurides 132
D. Other Halogenation Methods 133
VIII. INTRODUCTION OF OTHER FUNCTIONAL GROUPS INTO OLEFINS AND ACETYLENES 134
A. Vinyl Selenides 134
B. Electrophilic Additions of Selenenyl Pseudohalides 136
C. Free-radical Selenosulfonation. 137
1. Olefins 137
2. Allenes 138
3. Acetylenes. 138
IX. CYCLIZATION. 139
A. Unsaturated Alcohols. 139
B. Unsaturated Carboxylic Acids. 141
C. Unsaturated Nitrogen Compounds 142
D. Unsaturated Thiols. 142
E. Allenic Phosphonates 143
F. Unsaturated β-Dicarbonyl Compounds 143
G. Dienes 144
H. Free-radical Cyclizations. 145
I. Hydroxyl-substituted Vinyl Selenones 146

3. Preparative uses of organoselenium and organotellurium compounds 93

X. OXIDATIONS OF FUNCTIONAL GROUPS 146
A. Alcohols, Hydroquinones and Catechols 146
B. Nitrogen Compounds 147
1. Amines 147
2. Indolines and indoles 148
3. Hydrazines 149
4. Hydrazones, oximes and semicarbazones 150
5. Nitro compounds 151
6. Azasteroid lactams 151
C. Phosphorus and Sulfur Compounds 152
1. Thiols 152
2. Sulfides and thioketals 152
3. Thiocarbonyl compounds 153
4. Sulfinic acids 153
5. Phosphorus compounds 153
XI. CARBON—CARBON BOND-FORMING REACTIONS 154
A. Connective Reactions with Selenium- and Tellurium-stabilized Anions 154
1. Anions from selenides, selenoacetals and their tellurium analogues 154
2. Anions from allylic and propargylic selenides 157
3. Anions from vinyl selenides and tellurides 159
4. Anions from selenoxides 160
5. Dianions from α-phenylselenocarboxylic acids 160
6. Anions from α-seleno nitriles, esters and lactones 161
7. Selenium and tellurium ylides 161
B. Connective Reactions from Substitution of Selenium and Tellurium Groups 162
1. Vinyl and allyl selenides and tellurides 162
2. Selenoesters 163
3. Selenones 163
C. Tellurium-mediated Aryl and Allyl Coupling Reactions 164
D. Selenium and Tellurium Extrusion Reactions 165
1. C—C bond formation 165
2. C=C bond formation 166
E. Other Connective Reactions 167
1. With selenoacetals activated by Lewis acids 167
2. Free-radical allylation 168
3. Phenolic coupling 168
4. Arylation of olefins 168
5. Wittig reactions of α-selenophosphoranes 169
XII. CLEAVAGE REACTIONS WITH SELENIUM AND TELLURIUM NUCLEOPHILES 169
A. Cleavage of Esters and Lactones 169
B. Cleavage of 2-Oxazolines and 2-Oxazines 171
C. Cleavage of Epoxides 171
D. Cleavage of Ethers and Amines 172
E. Cleavage of Cyclopropanes 172
XIII. TRANSFORMATIONS OF ALCOHOLS VIA SELENIDES . . . 173
XIV. REDUCTIVE DESELENIZATION AND DETELLURIZATION . 174
A. Reagents 174
B. Deoxygenation of Alcohols 174
C. Deoxygenation of Aldehydes and Ketones 174
D. Reduction of Selenoesters and Telluroesters 174

1. *Se*-Phenyl selenoesters and selenocarbonates 174
2. *O*-Alkyl seleno- and telluro-esters 175
XV. REDUCTIONS WITH SELENIUM AND TELLURIUM COMPOUNDS 175
A. Reduction of Halides 175
B. Reduction of Epoxides 176
C. Reduction of Carbonyl Compounds 177
D. Reduction of Nitrogen Compounds 177
E. Reduction of Sulfoxides 177
XVI. REARRANGEMENTS 178
A. [2,3] Sigmatropic Rearrangements of Allylic Selenoxides 178
B. Other Sigmatropic Reactions 182
C. Ring Expansions and Contractions 182
1. Ring expansions 182
2. Ring contractions 183
D. Other Rearrangements 184
XVII. TRANSFORMATIONS OF ALDEHYDES, KETONES AND ENONES 184
A. Via α-Selenenylated Derivatives 184
B. Reactions of Enones 187
XVIII. TRANSFORMATIONS OF DIAZO COMPOUNDS 189
A. Diazoalkanes 189
B. α-Diazo Ketones and α-Diazo Esters 189
C. Diazopenicillinates 191
D. Reaction with Selenoesters 191
XIX. CARBONYLATIONS 191
XX. MISCELLANEOUS SYNTHETIC APPLICATIONS 191
A. Decarboxylations 191
B. Alkylations with Trimethylselenonium Hydroxide 193
C. Telluroxides as Aldol Catalysts 193
D. Protecting Group for Alcohols 193
XXI. REFERENCES 194

I. INTRODUCTION

Although organoselenium chemistry had been widely studied and exhaustively reviewed[1,2] by the early 1970s, only a few selenium-based synthetic methods existed in the literature of that time. The discovery and subsequent recognition of the selenoxide *syn*-elimination as an exceptionally efficient olefin-forming reaction did much to enhance the appreciation of selenium compounds by synthetic organic chemists. Increasing interest in the field soon led to other discoveries of widespread synthetic utility based on the unique properties of this element. The resulting proliferation of the literature on organoselenium chemistry prompted several excellent reviews in the late 1970s[3–8] as well as a number of later ones[9–11]. Very recently, two new books (in addition to Volume 1 of the present series) have appeared on the subject which are based on reaction types[12] and applications in the synthesis of natural products[13]. A number of more specialized reviews have been written recently and these will be cited at appropriate points in this chapter to direct the reader to additional information.

The history of organotellurium chemistry is also lengthy, but again contributions of broad synthetic potential have only begun to appear in the past few years. The general subject has been reviewed periodically[8,14–22] and appears to be generating increasing interest. However, broad acceptance of tellurium-based synthetic methodology lags behind that of selenium at present.

The aim of this chapter is to provide coverage of recent preparative uses of selenium and tellurium compounds. In keeping with the general theme of this series, and in order to provide a complementary treatment to existing reviews, this chapter will be organized principally according to the types of functional group and skeletal transformations which can be achieved. Considerations of mechanisms and of the strategic role played by selenium and tellurium reagents in specific syntheses of natural products will be kept to a minimum. No attempt is made to provide exhaustive coverage of all known examples of the more widely employed procedures, although illustrative examples will be shown as required to demonstrate their scope and limitations. Coverage of the literature up to early 1985 is provided.

II. OLEFIN PREPARATIONS BY ELIMINATION

A. Selenoxide *syn*-Elimination

1. General comments

The selenoxide *syn*-elimination (equation 1) constitutes one of the most expedient methods for introducing a carbon—carbon double bond into a saturated substrate. It is formally akin to older methods such as ester pyrolyses and the Chugaev, Cope and sulfoxide eliminations, but has the advantage of proceeding under far milder conditions, often at or even below room temperature. Further, the required selenide starting materials are readily available from a variety of sources (see the chapter by McWhinnie) and are efficiently oxidized to their selenoxides by many common reagents. Over-oxidation to selenones is seldom a complication, in contrast to sulfur analogues where competing sulfone formation is more difficult to avoid.

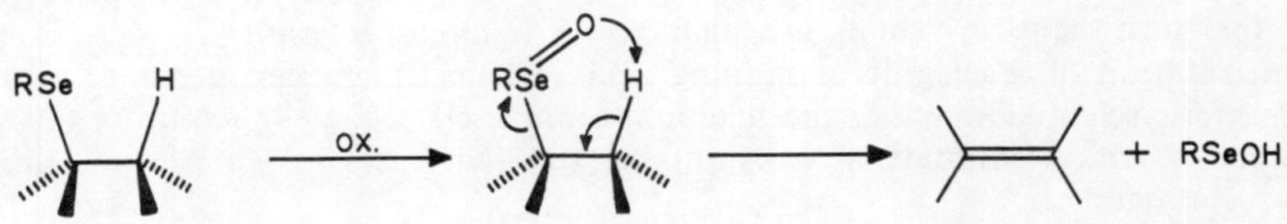

(1)

Despite several earlier reports of successful but isolated selenoxide eliminations[23–25], the synthetic potential of the method remained largely unrecognized until 1973. In that year Sharpless and coworkers[26–28], Reich and coworkers[29] and Clive[30] independently demonstrated its merits. Subsequent applications have been rapid and now number many hundreds. The subject has been reviewed recently[31].

2. Stereochemistry

The selenoxide *syn*-elimination requires the coplanarity of all five participating centers and so is highly stereospecific. This is illustrated by the examples in equations 2 and 3. Thus, in the first example, the *erythro*- and *threo*-selenides **1** and **2** produced the corresponding (*E*)- and (*Z*)-olefins, respectively, on oxidation[28]. In the second example, the *cis*-selenide **3**, where the PhSe group is *cis* to the ring-fusion hydrogen, eliminated to afford mainly the more stable *endo* olefin, while the *trans*-selenide **4** produced only the *exo* isomer[32]. The latter result is attributed to the inability of the system to achieve the required planar conformation for *endo* elimination.

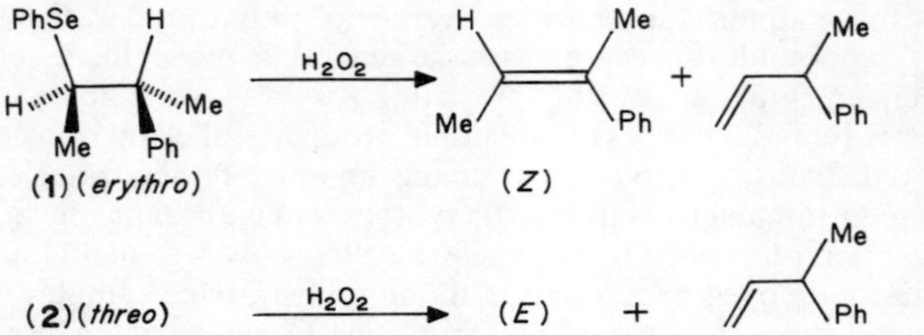

(2)

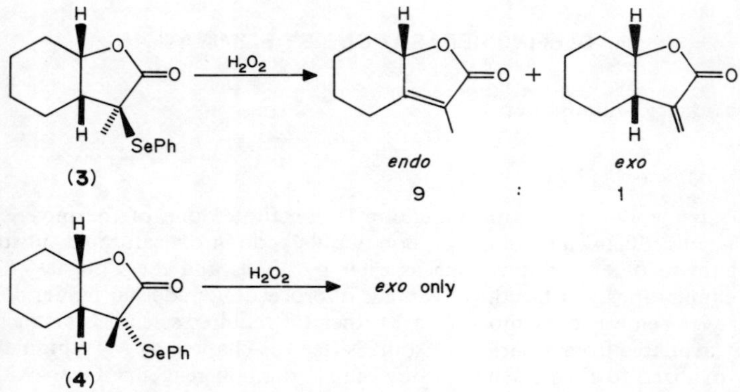

(3)

When the product of elimination is a 1,2-disubstituted olefin, it is generally formed preferentially with the *E* configuration[26–28,33–35]. The stereoselectivity is usually high unless the substituents are small, in which case *E*, *Z* mixtures result[34].

The oxidation of a selenide containing a chiral substituent can generate a pair of diasteromeric selenoxides, since the selenium atom itself becomes a chiral center. Such diastereomers may eliminate at substantially different rates[24,36], or follow different pathways altogether[37].

3. Regiochemistry

In the case of branched selenoxides, elimination towards the less substituted position is generally favoured[28,34] (e.g. equation 4[28]).

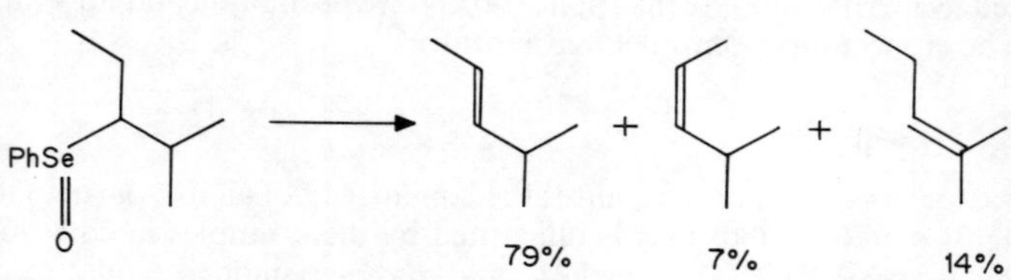

(4)

In selenoxides of general structure **5**, α-alkyl substituents tend to accelerate elimination whereas β-[38,39] and even γ-alkyl[39] groups retard it. Phenyl substituents in either the α- or β-position enhance the rate[38].

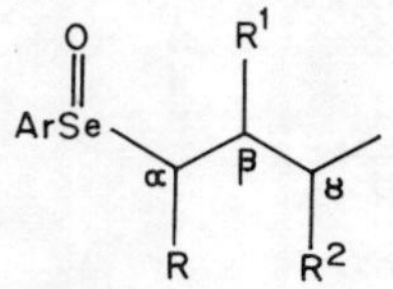

(5)

The regiochemistry of elimination can be strongly influenced by the presence of a heteroatom in the β-position of the selenoxide. Oxygen substituents in the form of hydroxyl, ether or acetoxy groups strongly direct elimination toward the allylic site and numerous examples demonstrating this effect are known (e.g. equation 5[40]). When an allylic hydrogen is unavailable elimination towards the vinylic position can occur, although more forcing conditions are often required. Enol ethers[36,41,42], enol acetates[43], ketones[43], ketene acetals[44–47] and an enol phosphate[48] have been prepared in this manner (equation 6). In a few exceptional cases secondary selenoxides have been observed to eliminate preferentially toward β-acetoxy[49,50] or acetal[51] functions to give the vinylic instead of allylic products. Presumably conformational or other factors suppress reaction at the allylic site.

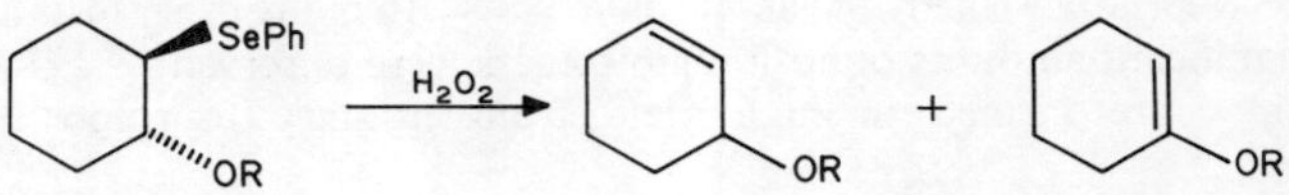

(5)

R=H, Ac or alkyl $\geqslant 97$: $\leqslant 3$

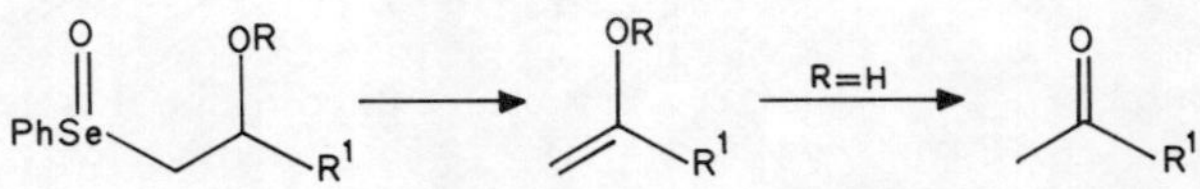

(6)

R=alkyl, Ac, $P(O)(OMe)_2$

β-Nitrogen substituents also exert a directing effect, although its direction and magnitude depend on the nature of the functional group. β-Amido selenoxides produce allylic amides exclusively[52,53] whenever an allylic hydrogen atom is available. Otherwise, enamides[54,55] or dehydroamino acids[25] are formed. Allylic elimination is also preferred in the case of cyclic[56,57] and acyclic[58] amines, although the regiospecificity is lower than in the case of oxygen substituents. Azides[59] and isothiocyanates[60] favour allylic and vinylic elimination, respectively. Examples are shown in equations 7 and 8[58]. When intramolecular hydrogen bonding is possible between the selenoxide and a β-hydroxyl[61] or β-amido function[52,53] (e.g. **6**), then the resulting stabilization necessitates the use of elevated temperatures to effect elimination.

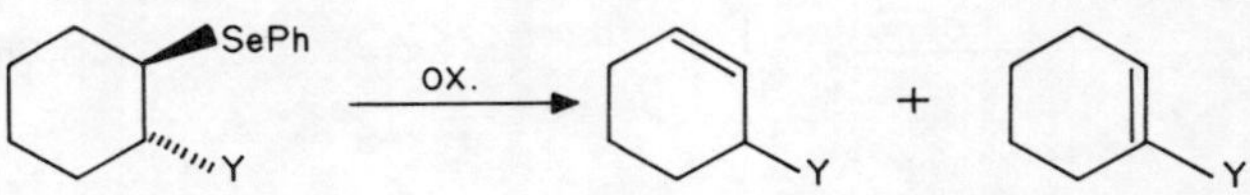

(7)

Y=NHAc:	100	:	0	(Refs. 52,53)
Y=N_3:	60	:	40	(Ref. 59)
Y=NCS:	17	:	83	(Ref. 60)

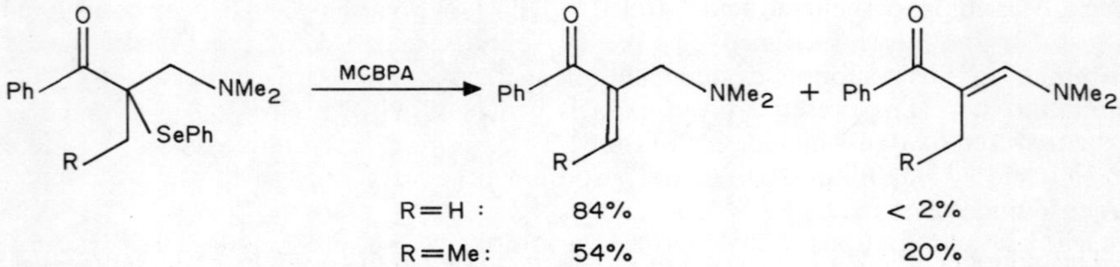

(8)

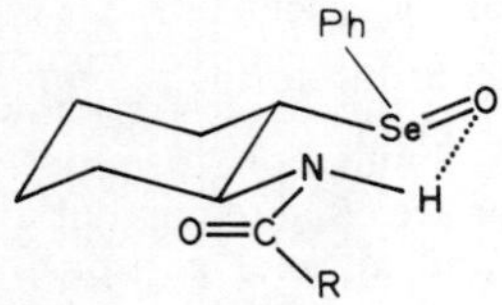

(6)

The effect of halogens on the regiochemistry is less predictable. It has been reported that a chloro substituent in either the α- or β-position enhances elimination[38]. However, there are numerous examples where elimination occurs preferentially away from a chlorine or bromine atom to produce chiefly the allylic halide[62–66]. In many other instances mixtures containing significant amounts of both regioisomers were observed[40,63,65] (equation 9). Vinyl bromides are formed in high yield from primary β-bromo selenoxides[62] (equation 10).

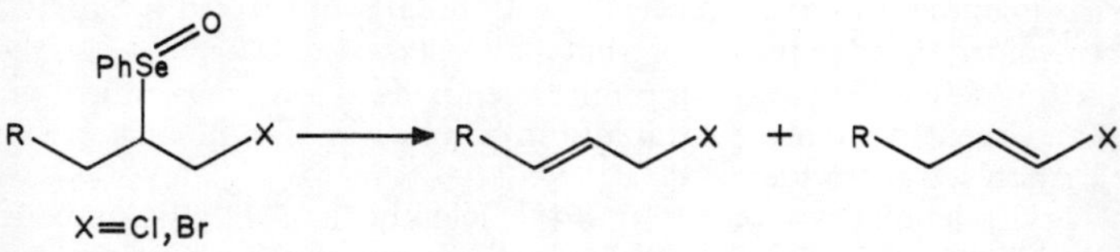

(9)

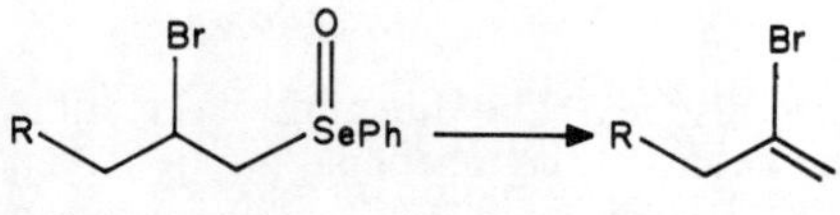

(10)

γ-Stannyl substituents are reported to have a mild directing effect towards the homoallylic rather than allylic site for selenoxide elimination[67].

Groups which have an acidifying effect on an adjacent hydrogen atom (e.g. carbonyl[28,68,69], sulfone[70–78], sulfoxide[77,78], nitro[79,80] and cyano[81] groups) strongly accelerate elimination towards that site, often to the complete exclusion of other regioisomers (equation 11).

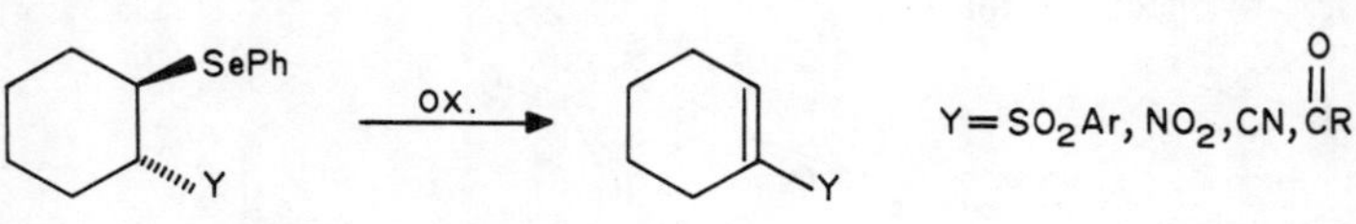

(11)

4. Techniques and side reactions

The oxidation of selenides to selenoxides can be achieved with diverse reagents. Those most commonly employed are hydrogen peroxide, ozone, sodium periodate and peracids, especially *m*-chloroperbenzoic acid (MCPBA). Their relative merits have been considered in various reviews (see Section I). The use of *t*-butyl hydroperoxide has been especially recommended[82] as it suppresses side reactions such as epoxidation (see below) during the elimination step. The presence of alumina is beneficial with this reagent[83]. Other less frequently used oxidants include Chloramine-T[82], *N*-chlorosuccinimide (NCS)[84], *t*-butyl hypochlorite[84,85], thallium(III) nitrate[86], sodium perborate[87], oxaziridines[88,89], singlet oxygen[90] and Jones reagent[91,92].

Selenoxide eliminations are sometimes accompanied by undesirable side reactions. Detailed studies by Reich and coworkers[38] and Sharpless and coworkers[28,82] have resulted in effective countermeasures. The selenenic acid (RSeOH) or one of its disproportionation products which forms during the elimination (equation 1) can add to the double bond of the desired olefin[38,82,93], particularly under neutral or acidic conditions[38]. This can be suppressed by the presence of an amine which removes the selenenic acid as it forms[38,93]. Other scavengers which have been employed for this purpose are reactive olefins such as enol ethers[51,94,95], enol acetates[95] and norbornadiene[41].

Competing decomposition of the selenoxide via other routes can occur when the elimination is slow, and so conditions should be chosen to enhance the elimination rate to the fullest extent. Water deactivates selenoxides through hydrate formation and so anhydrous conditions are advantageous. The addition of anhydrous $MgSO_4$ has been recommended for this purpose[28].

Other protic solvents also suppress elimination and should be avoided in troublesome cases[38]. When hydrogen peroxide is employed in the oxidation step, its catalytic decomposition by the selenoxide may require a large excess of the oxidant[38]. Epoxidation of the olefin is sometimes observed, especially with highly substituted products[38,82], but can usually be avoided through the use of *t*-butyl hydroperoxide[82]. In particularly difficult cases, it is sometimes expedient to pre-form the selenoxide at low temperature with ozone or a peracid and then to pyrolyse it rapidly in a refluxing non-polar solvent such as hexane[38] or carbon tetrachloride[38,60,96].

In general, aryl selenoxides eliminate more efficiently than alkyl selenoxides[83] and electron-withdrawing substituents on the aryl group further enhance the process[38,39,97–99]. The 2-pyridylseleno group has also been reported to give particularly efficacious results[100–102].

Other side reactions include the reduction of the selenoxide back to the selenide[25,103] by divalent selenium byproducts, isomerization of a previously existing double bond (*cis* to *trans*) in the substrate[104] and the formation of alcohols or other oxygenated products[105–107], particularly when elimination is rendered difficult by strain or other factors such as the formation of stabilized carbonium ions[107] from C—Se cleavage (e.g. equation 12).

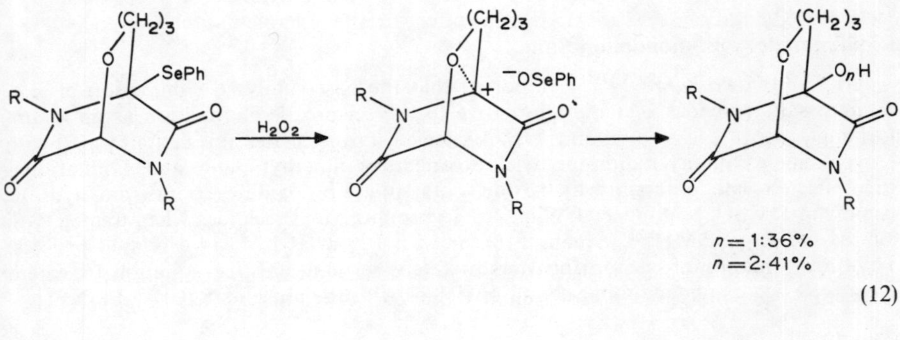

$n = 1$: 36%
$n = 2$: 41%

(12)

5. Acetylenes and allenes from vinyl selenoxides

The *syn*-eliminations of vinyl selenoxides were studied by Reich and Willis[108] and provide a route to acetylenes and allenes. The former compounds are produced preferentially, provided that a *cis*-hydrogen atom is available for elimination (equation 13). The selenoxides derived from β-(phenylseleno)vinyl sulfones can be employed in the preparation of acetylenic[109–111] or allenic[112] sulfones (see Sections VIII.C.2 and 3).

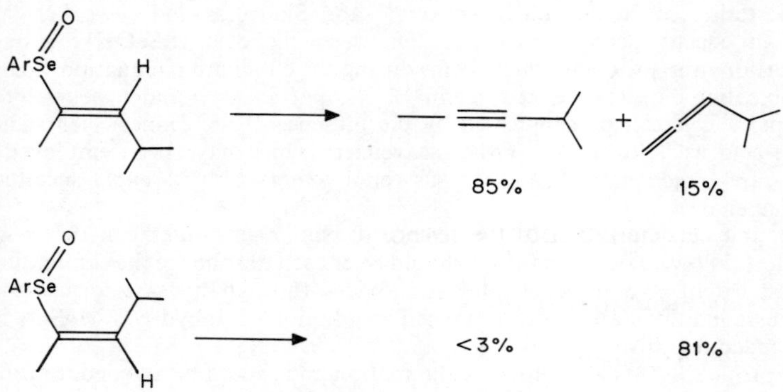

(13)

B. Telluroxide Elimination

Only a few examples of olefin preparations via telluroxide elimination have been reported[113–120]. Telluroxides form stable hydrates which may require pyrolysis at temperatures of ca. 200 °C[116,118]. They are also readily reduced by the byproduct tellurenic acid (RTeOH) back to the original tellurides[113], and overoxidized to tellurones[113], resulting in the formation of oxygenated byproducts[113,114,120]. The greater difficulty in controlling the desired oxidation state of telluroxides makes them less attractive than selenoxides for olefin preparations in most cases.

Telluroxides, like selenoxides, tend to eliminate towards the less substituted position[118,120] and to the allylic position with respect to an oxygen function[116–119]. A series of monosubstituted olefins were prepared in high yields from primary tellurides, using Chloramine-T as the oxidant[115] (equation 14).

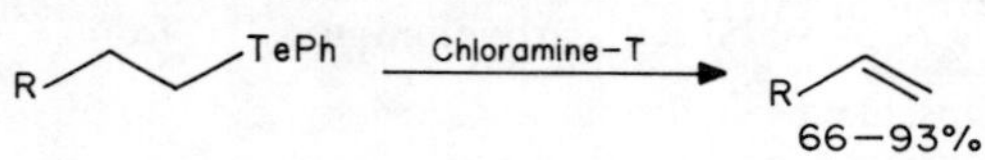

(14)

C. Elimination of Selenonium Salts

Krief and coworkers[91,121–124] reported that the base-catalysed elimination of selenonium salts provides a viable alternative to the selenoxide elimination as an olefin-forming procedure (e.g. equation 15[121]). A selenide starting material is first alkylated with reagents such as methyl iodide, methyl fluorosulfate or dimethyl sulfate, and elimination is then effected with potassium hydroxide, potassium *t*-butoxide or potassium hydride, usually in DMSO. Selenoacetals eliminate to produce vinyl selenides when treated with methyl iodide in DMF[125] (equation 16), or with PI_3 or P_2I_4[126] in a related process. The transformation of selenoorthoesters to ketene selenoacetals (i.e. equation 16, where $R^3 = SeR$) occurs under similar conditions[126] or with the reagents $SnCl_4$–i-Pr_2NEt[127].

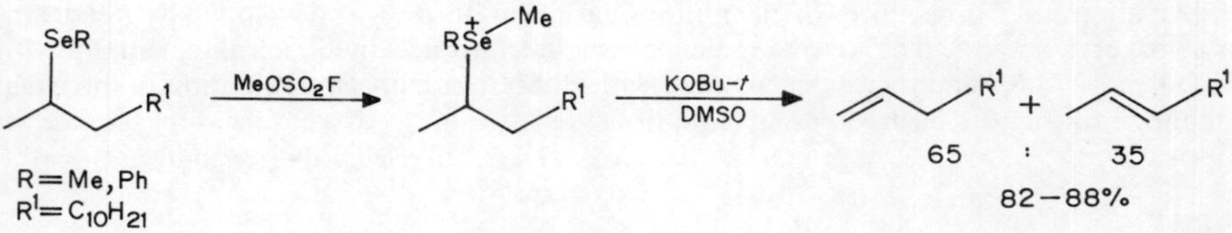

(15)

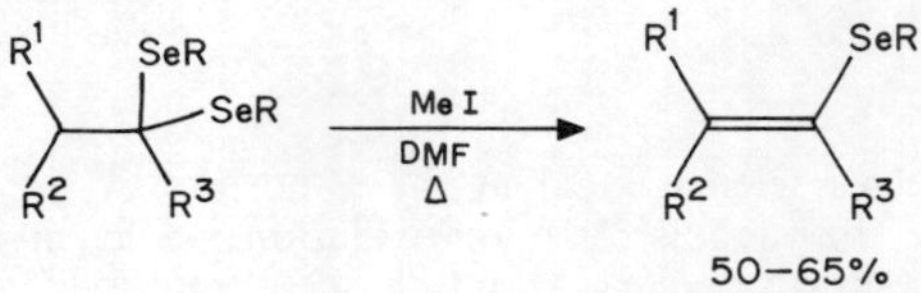

(16)

III. DEHYDROGENATION OF CARBONYL COMPOUNDS

A. Via α-Selenenylated Derivatives

1. *Aldehydes and ketones*

The dehydrogenation of aldehydes and ketones to their α,β-unsaturated derivatives can be effected by α-selenenylation followed by oxidation and selenoxide elimination. Several different protocols have been devised for the selenenylation step. Sharpless and coworkers[27] discovered that the treatment of aldehydes or ketones with benzeneselenenyl chloride (PhSeCl) in ethyl acetate provides the α-selenides in high yield (equation 17).

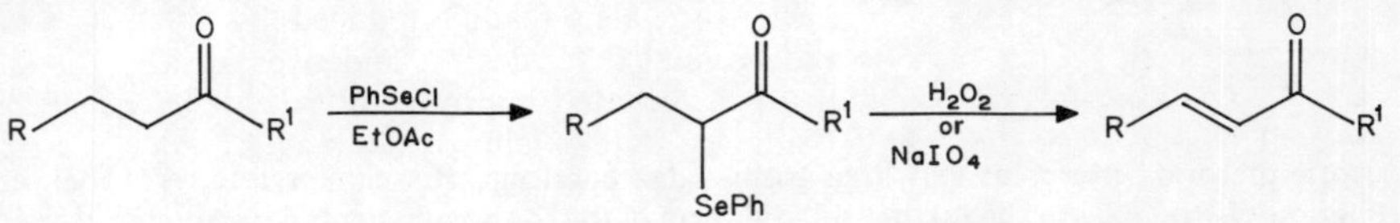

(17)

The reaction proceeds via the enol form of the carbonyl compound and is sometimes aided by catalysts such as HCl[27], $BF_3 \cdot OEt_2$[128,129] or Dowex 50-X8 (H^+)[130]. The direction of enolization dictates the regiochemistry of selenenylation in unsymmetrical ketones and the procedure is compatible with a large variety of functionalities including other carbonyl groups, olefins, hydroxyl and carboxyl groups, ethers and epoxides. Basic amino groups, however, can create complications, as demonstrated by equation 18, where the codeinone derivative **8** failed to undergo selenenylation whereas, the urethane analog **7** reacted smoothly[131,132].

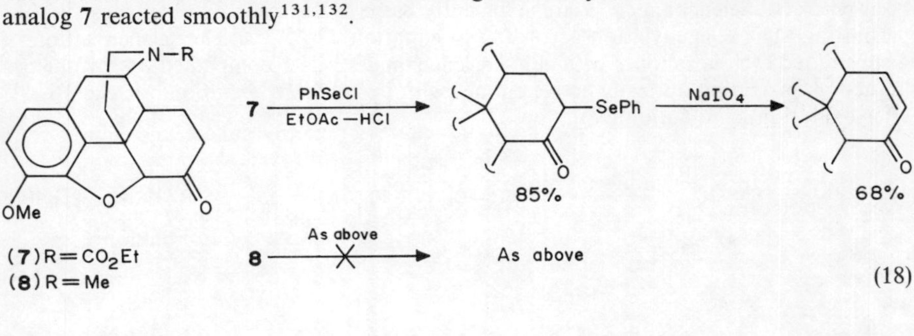

(18)

An alternative procedure which employs basic conditions was developed by Reich and coworkers[29,133,134]. The ketone is first converted into its kinetic enolate, usually with LDA at $-78\,°C$, and then selenenylated with PhSeCl or PhSeBr. Oxidation in the usual manner affords the desired enone (equation 19).

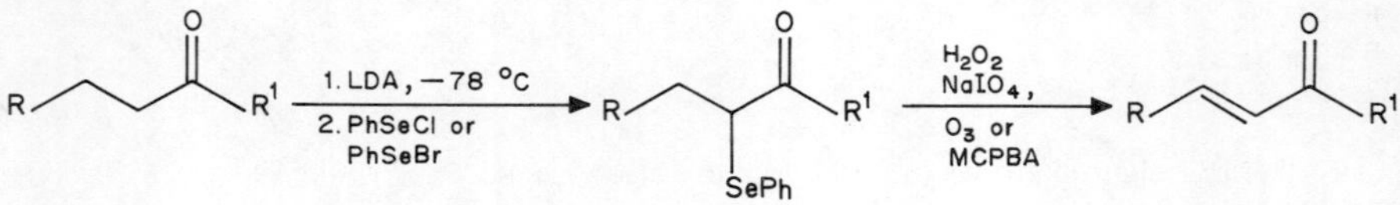

(19)

Enones are themselves selenenylated at the α′-position[135–137] and produce cross-conjugated dienones[135] or phenols[136,137] on oxidation. γ-Selenenylation occurs[138,139] if the α′-position is fully substituted (e.g. equations 20[135] and 21[138]).

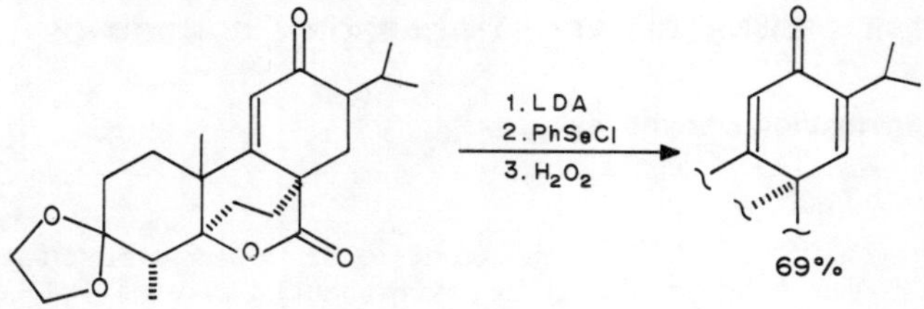

(20)

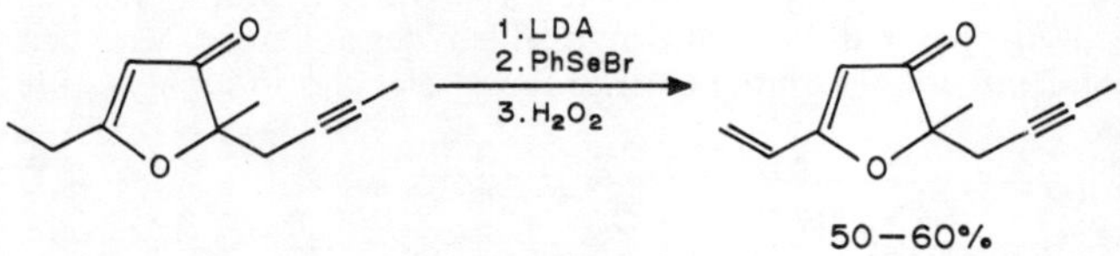

(21)

Both methods are relatively free from side reactions. Bisselenenylated ketones are sometimes formed from the further selenylation of the monosubstituted products[29,140,141] but seldom in significant amounts. The oxidation step may be accompanied by Pummerer reactions of the selenoxide intermediates[133,134], which can be suppressed by the addition of an amine. Baeyer–Villiger reactions of the product enones[114,142,143] or of remote ketone groups[102] sometimes occur under these conditions. In some cases these processes are promoted deliberately if the oxygenated byproducts are desired instead of the enone (see Sections V.F.2 and V.H.1).

Although the above procedures have proved the most versatile and widely employed, several other useful variations for the selenenylation of aldehydes and ketones have also been reported. Selenenamides **9** are sufficiently basic to deprotonate aldehydes[144], thus permitting their selenenylation according to equation 22[144–147]. The selenenylation of 4-substituted cyclohexanones with chiral selenenamides was accompanied by asymmetric induction, providing optically active enones with up to 26.2% enantiomeric excesses[148] after oxidation–elimination (equation 23).

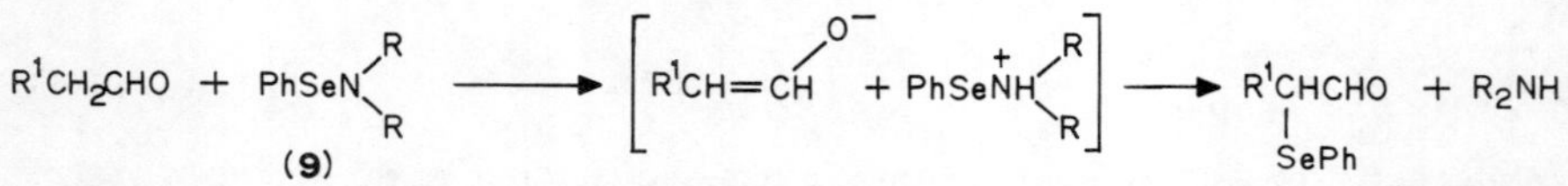

(22)

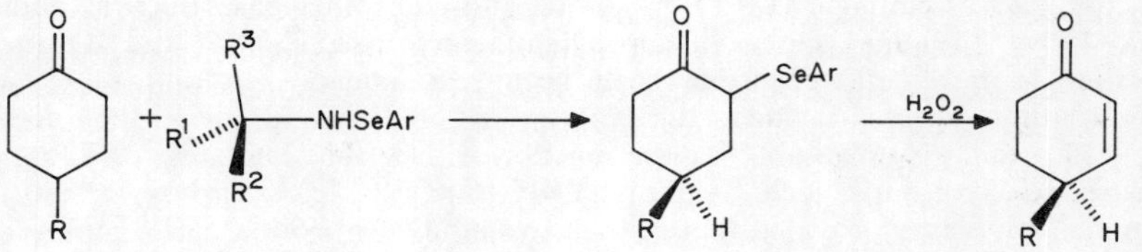

(23)

α-(2-Pyridylseleno) ketones or aldehydes are easily prepared with the reagents **10**, **11** or **12** (equation 24) and permit particularly efficient elimination to enones[96,149], even in the case of cycloheptenones and cyclooctenones which are difficult to prepare from their 2-phenylseleno derivatives[134]. A polymer-supported areneselenenyl chloride was also reported to function as an effective selenenylating reagent[150]. A number of other selenenylation procedures employing elemental selenium[151], selenophosphoranes[152], PhSeSePh–selenium dioxide[153] and selenenic electrophiles generated from the anodic oxidation of PhSeSePh[154] are shown in equation 25.

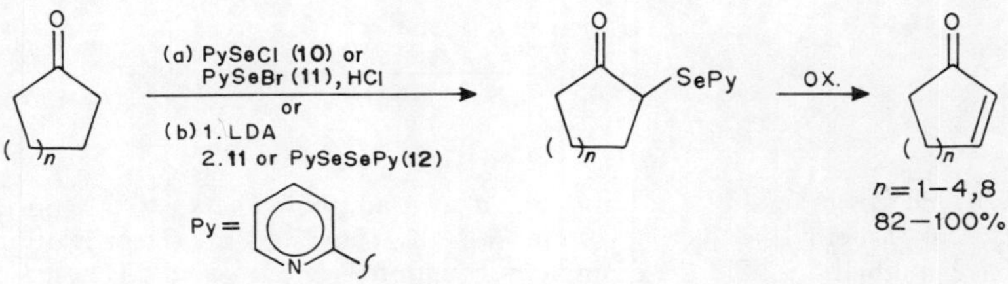

(24)

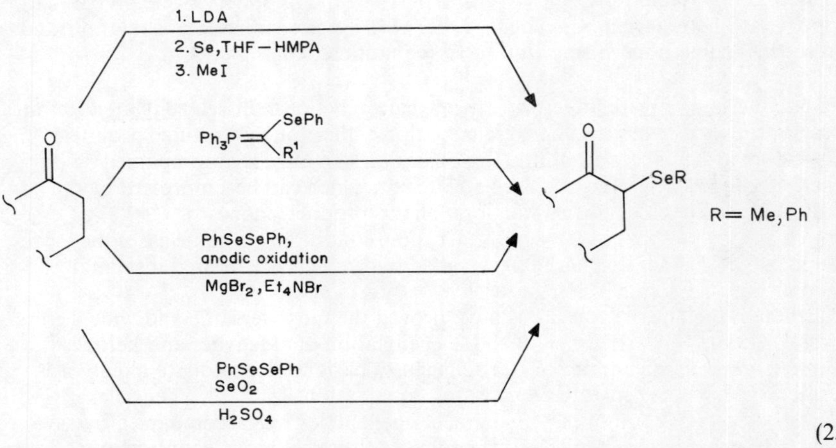

(25)

The further transformation of enones to other products with selenium- or tellurium-based methods is discussed separately in Section XVII.

2. Esters and lactones

In contrast to aldehydes and ketones, esters and lactones fail to undergo the direct selenenylation shown in equation 17, and prior generation of the enolate is required[27].

Typically LDA is employed for this purpose, although other bases such as potassium hydride[155–157], lithium cyclohexylisopropylamide[158–160], LiTMP[161] and lithium[162] or potassium[163] hexamethyldisilazide have been recommended in difficult situations. Selenenylating agents include PhSeCl and PhSeBr, and also the diselenide PhSeSePh, which is unsuitable for the reaction with ketone enolates[134]. The method tolerates acidic groups such as the hydroxyl[155–157,164,165], carboxyl[166] and sulfonamide[160] moieties provided that at least two equivalents of base are employed. The dehydrogenations of esters and lactones containing a variety of α-heteroatoms including oxygen[167], chlorine[167], nitrogen[168], sulfone sulfur[169] and phosphinate phosphorus[170,171] proceed normally (equation 26).

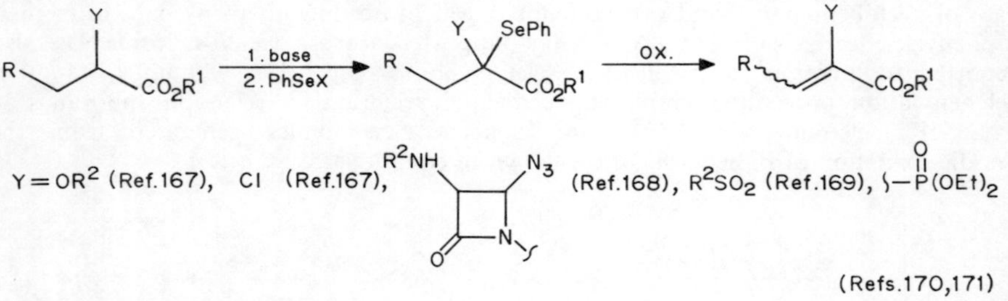

(26)

Grieco and coworkers[32,172,173] and Yamakawa and coworkers[174,175] reported the synthesis of α-methylene lactones from α-methyl lactones by selenenylation and selenoxide elimination, as in the example in equation 3. Unsaturated γ-lactones can be further transformed into furans by reduction with diisobutylaluminum hydride (DIBAL)[176,177] (equation 27[176]). Furans, pyrroles and thiophenes were prepared from the appropriate heterocyclic vinylidene esters (**13**) by selenenylation, oxidation and acid-catalysed double-bond isomerization[178] (equation 28).

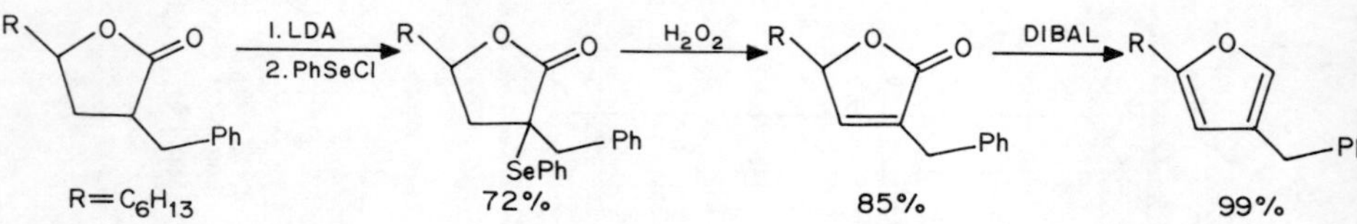

(27)

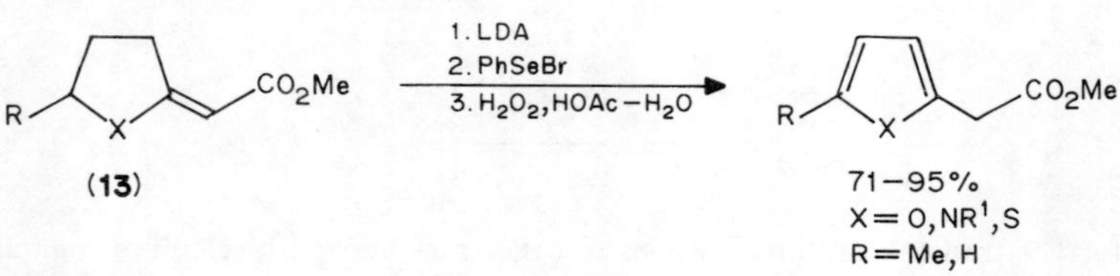

(28)

An interesting isomerization of the double bond of the enamine ester **14** was accomplished by selenenylation of the corresponding aluminum enolate generated with DIBAL, followed by selenoxide elimination[179] (equation 29). The simultaneous selenenylation and decarboxylation of ester **15** was performed without affecting the keto group[180] (equation 30).

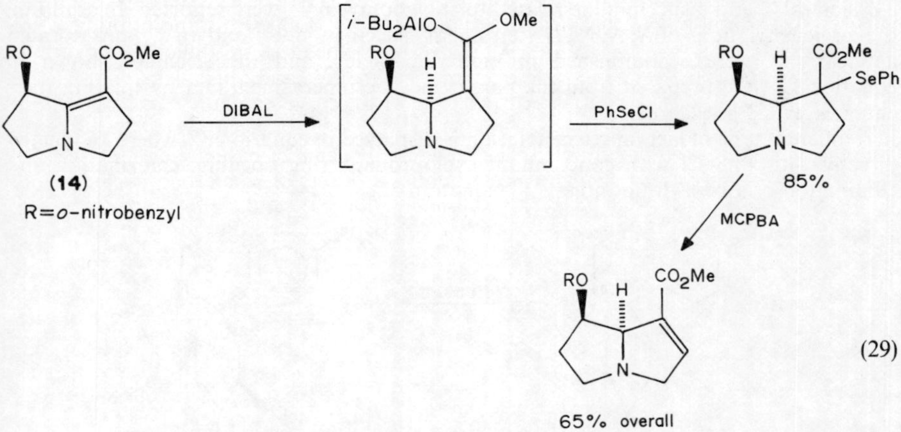

(29)

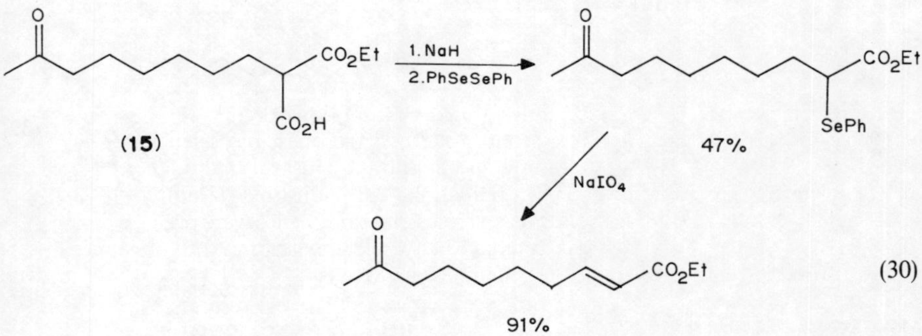

(30)

3. Amides and lactams

Relatively few examples of the dehydrogenation of amides and lactams are known. Zoretic and Soja[181,182] found that excess of base is required for the monoselenenylation of *N*-methylpyrrolidin-2-one. With only one equivalent of base, bisselenenylation dominates, as shown in equation 31. The dehydrogenations of several other γ-lactams[57,183,184],

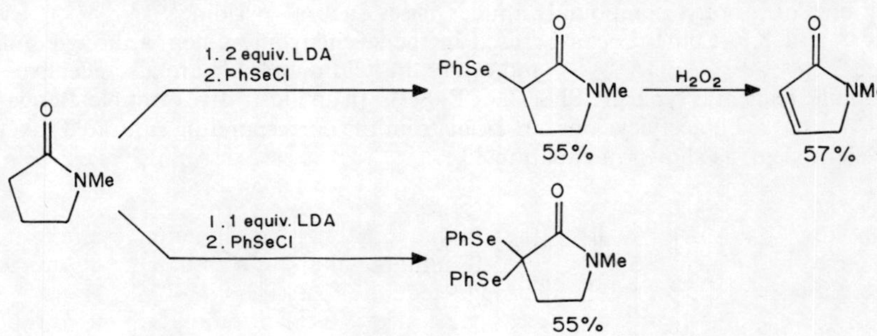

(31)

a δ-lactam[185] and a succinimide related to showdomycin[186] were reported. In addition, the α-selenenylations of N, N, N′, N′-tetramethylsuccinamide[187] and an α, β-unsaturated amide[188,189] were accomplished in moderate yields, and the selenides shown in equation 12 were prepared from the parent bicyclic piperazinediones by treating their enolates with PhSeCl[107].

A different type of lactam selenenylation is displayed in equation 32, where **16** reacted directly with PhSeCl with concomitant cyclopropane ring opening, chlorination and elimination of a benzylthio moiety[190] (equation 32).

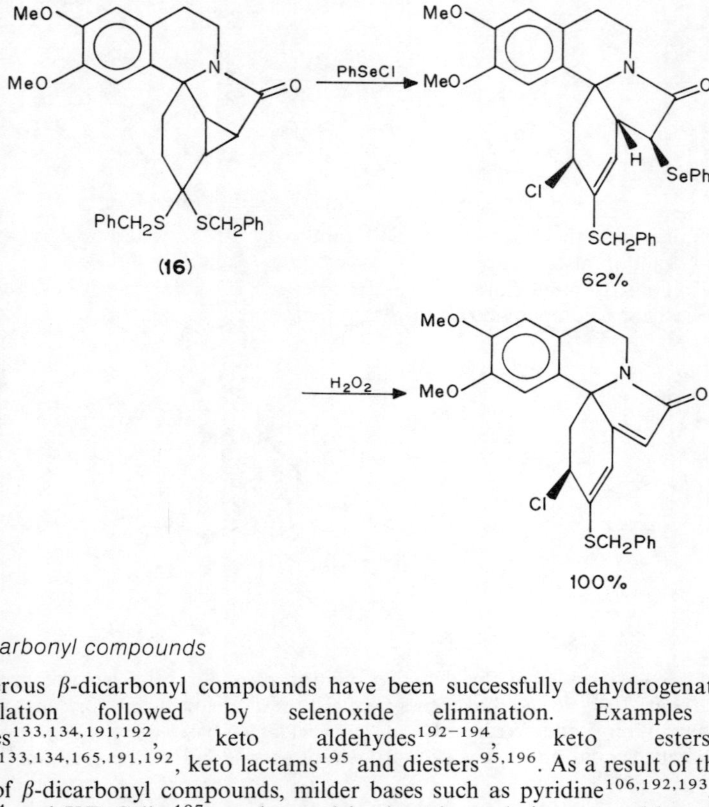

(32)

4. β-Dicarbonyl compounds

Numerous β-dicarbonyl compounds have been successfully dehydrogenated by α-selenenylation followed by selenoxide elimination. Examples include diketones[133,134,191,192], keto aldehydes[192–194], keto esters and lactones[133,134,165,191,192], keto lactams[195] and diesters[95,196]. As a result of the greater acidity of β-dicarbonyl compounds, milder bases such as pyridine[106,192,193], triethylamine[194] and KF–Celite[197] can be used in the selenenylation step, although sodium hydride[133,134,191,195], LDA[165,198] and even n-butyllithium[95,196] are also effective. The electrophilic reagent is typically PhSeCl or PhSeBr. In an illustrative example, Renga and Reich[191] prepared 2-acetylcyclohexen-2-one from the corresponding saturated diketone in excellent yield, as shown in equation 33.

(33)

Other selenenylation procedures applicable to β-dicarbonyl compounds include the use of selenenamides[58], elemental selenium followed by methyl iodide[199] and PhSeSePh either electrolytically[154] or in the presence of selenium dioxide[153]. These processes are analogous to those depicted in equations 22 and 25.

The elimination step can be effected by the usual oxidative techniques via the selenoxide, or by a non-oxidative procedure in which the starting material is converted directly into the unsaturated product with excess PhSeCl in pyridine[192] (e.g. equation 34). One example of a similar non-oxidative elimination has also been reported in the case of an isolated aldehyde[200].

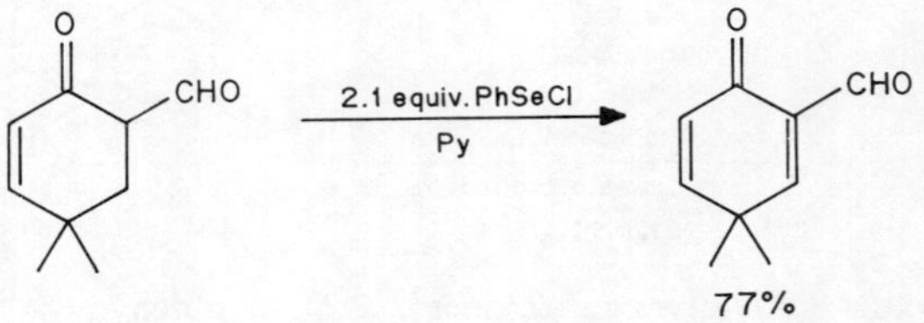

(34)

The dianion from the β-keto lactone **17** underwent chiefly γ-selenenylation, although significant reaction also occurred at the α-position[201,202] (equation 35).

An anomalous, although potentially useful, selenenylation was observed with the diketo lactone **18**, where accompanying fragmentation produced the α-phenylseleno enone **19**[203] (equation 36).

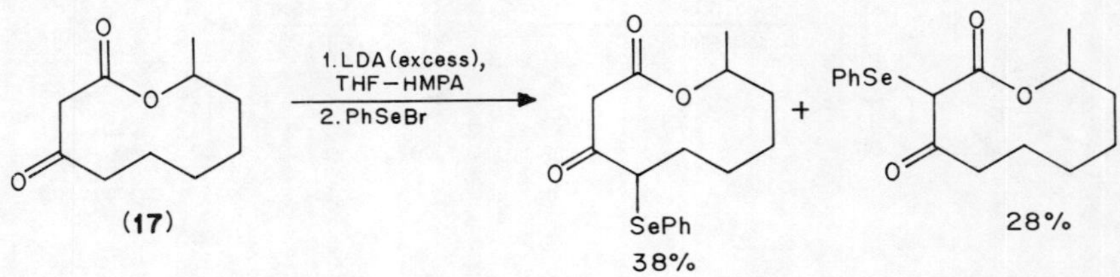

(35)

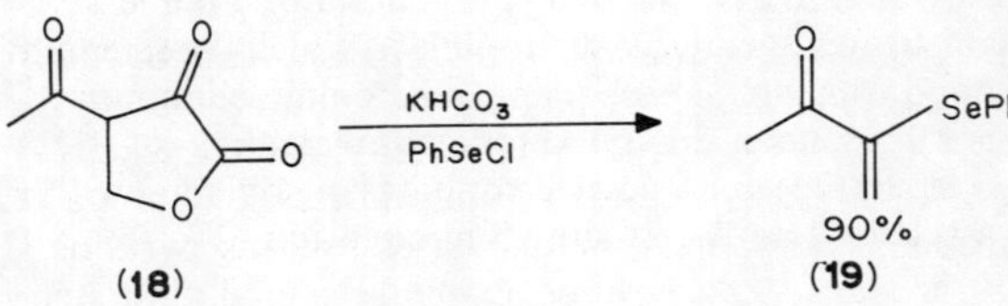

(36)

5. *Enol acetates, enol silyl ethers, enamines and related compounds*

As shown in equation 19, unsymmetrical ketones can be selenenylated on their less substituted flank via their kinetic enolates. On the other hand, enol acetates and enol silyl ethers can be prepared from ketones on the thermodynamically favored more substituted side. Their selenenylation with an appropriate electrophile therefore permits the synthesis of α-seleno ketones (and hence enones) which are regioisomers of those obtained via equation 19.

Enol acetates can be selenenylated in this manner by their sequential treatment with methyllithium and PhSeBr[133], by their direct reaction with benzeneselenenyl acetate

(PhSeOAc)[49] or trifluoroacetate ($PhSeOCOCF_3$)[30,133,134,204] and electrochemically with PhSeSePh[205,206] (equation 37). Methyl ketones afford α-selenomethyl derivatives via the reaction of their enol acetates with PhSeCl[207,208].

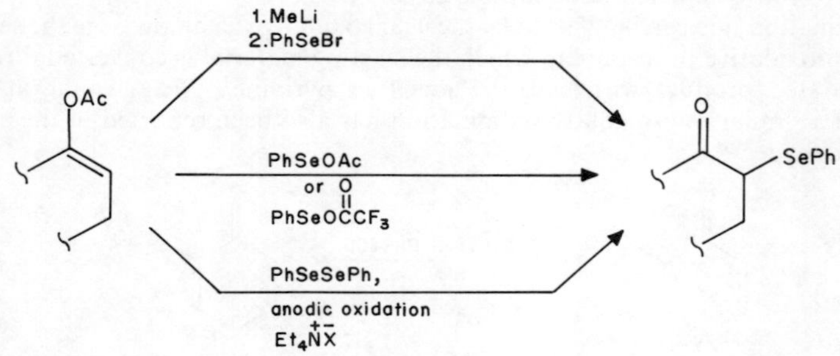

(37)

The enol silyl ether derivatives of ketones[209–213] or aldehydes[209,214] are typically selenenylated with PhSeCl or PhSeBr, and also with PySeBr (**11**)[96]. An example of the preparation of the more substituted enone from an unsymmetrical ketone in this manner is given in equation 38[213].

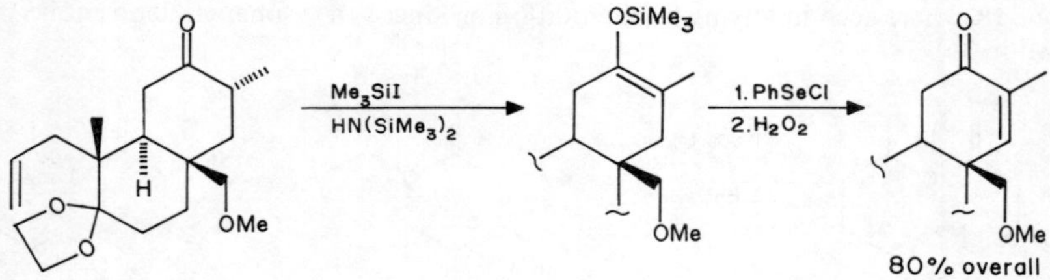

(38)

Danishefsky and coworkers demonstrated that the selenenylated products of dienol silyl ethers are useful in Diels–Alder reactions[215–217]. The α-seleno ketone **20** in equation 39 was resilylated to give the highly functionalized diene **21**[216,217], whereas the similar ketone **22** constitutes a valuable dienophile in which the selenium moiety provides a latent double bond through subsequent selenoxide elimination[215] (equation 40). Compound **22** thus represents a divinyl ketone equivalent.

Cyclopropyl silyl ethers (**23**) undergo electrophilic ring opening with PhSeCl and $TiCl_4$ catalyst to afford β-seleno ketones, as shown in equation 41[218].

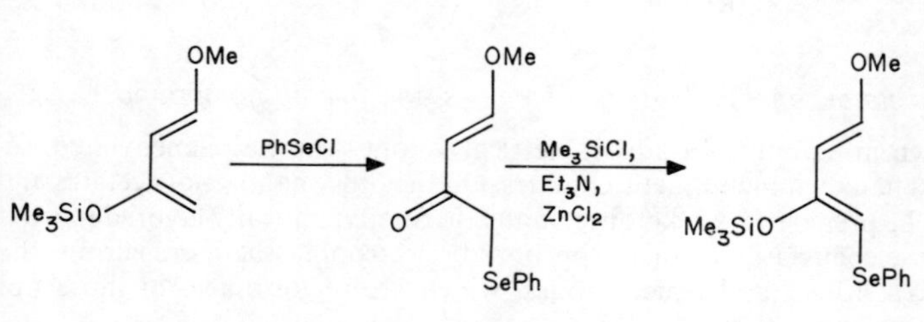

82% (**20**) 80% (**21**)

(39)

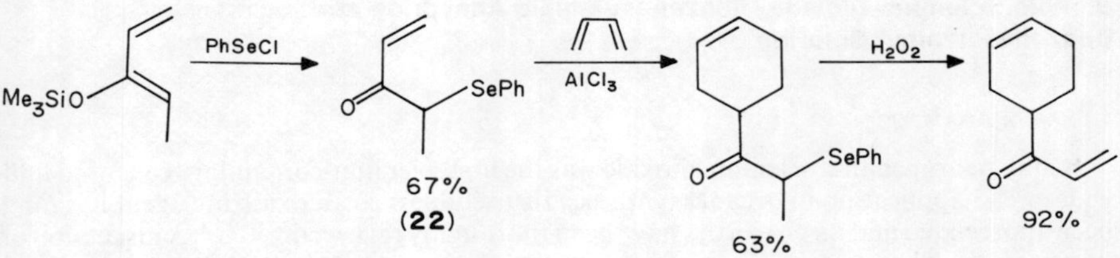

(40)

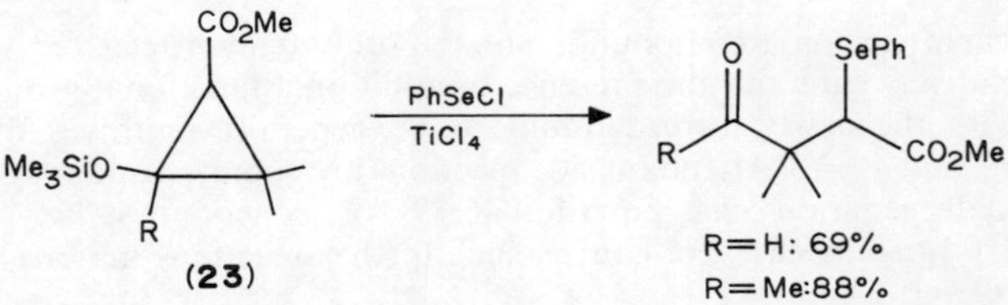

(41)

Enamines[219–222], enamides[223], enol boranes[224], enol ethers[183,225,226] and ketals[225,227] all react with electrophiles such as PhSeCl to provide the corresponding selenenylated products. Several aldehydes were thus dehydrogenated by subsequent selenoxide elimination[183,219] and an example is given in equation 42[219]. The enol ether **24** was converted into the naphthol **26** via its α-seleno ketone **25** (equation 43)[226].

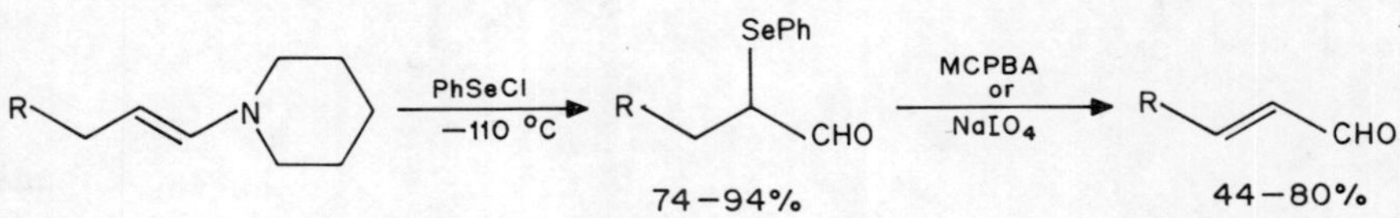

(42)

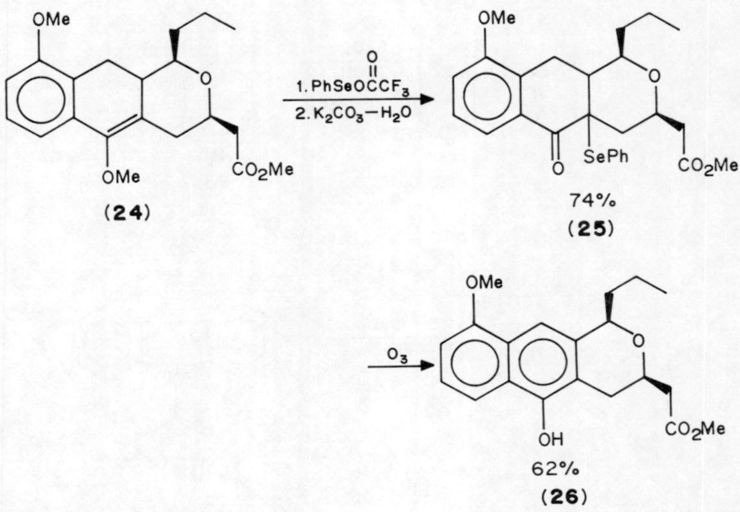

(43)

B. With Selenium Dioxide, Benzeneseleninic Anhydride and Benzeneseleninyl Chloride

1. Selenium dioxide

It may be argued that selenium dioxide was the first selenium-containing reagent to find widespread application in organic synthesis. Its usefulness as an oxidizing agent has long been appreciated and its properties have been thoroughly reviewed[228–233]. A discussion of its use as a dehydrogenating agent for carbonyl compounds, and other applications described in other sections, will therefore be brief and confined to recent illustrative examples.

Instances where carbonyl compounds are smoothly dehydrogenated with selenium dioxide are relatively rare and the presence of additional functional groups capable of conjugating with the newly formed double bond generally improves the yields (e.g. equations 44[234] and 45[235]). Attendant side reactions are common and may include more extensive dehydrogenation (e.g. equation 46[236]) or concomitant oxygenation (e.g. equation 47[237]). In some cases, of course, such further reactions may be desirable (see Sections V.B, C and F).

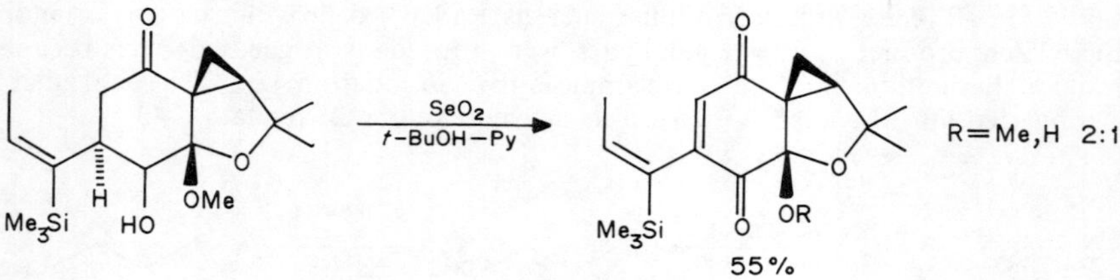

(44)

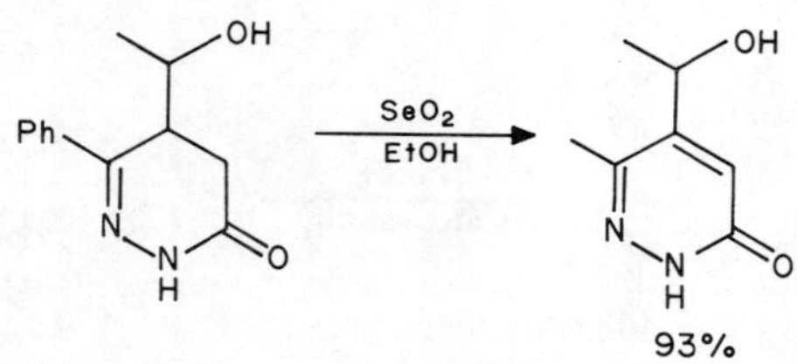

(45)

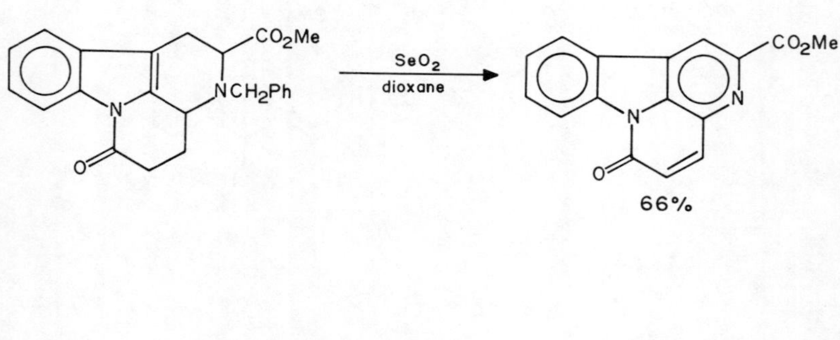

(46)

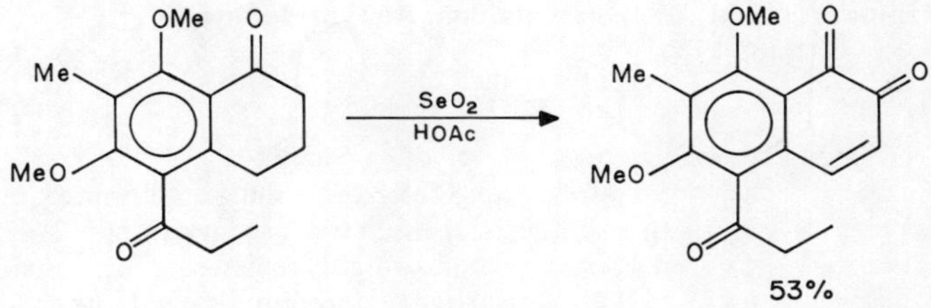

(47)

2. Benzeneseleninic anhydride

The use of benzeneseleninic anhydride (BSA) (**27**) as a reagent for the dehydrogenation of carbonyl compounds was first reported by Barton. For instance, 3-cholestanone and its Δ^1 and Δ^4 derivatives all produced the corresponding dienone **28** in high yield when heated with BSA in chlorobenzene[238,239] (equation 48). Since BSA also oxidizes alcohols to ketones (see Section X.A), it is possible to convert cholestan-3-ol and related alcohols directly into the dienones in one step[240]. In general, BSA is a cleaner reagent than selenium dioxide, although under forcing conditions the formation of side products can occur. For example, A-nordiketones were produced along with the expected enone product in the case of 4,4-dimethyl steroidal ketones such as α- or β-amyrone.

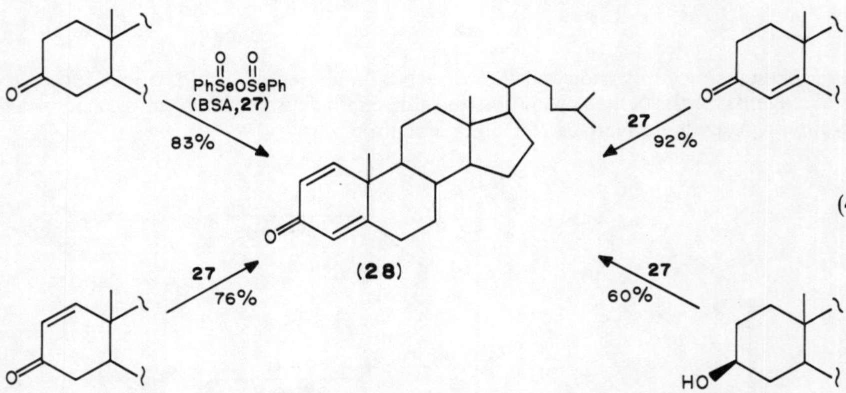

(48)

An attractive feature of this method is that BSA can be employed in catalytic amounts in the presence of a co-oxidant such as iodoxybenzene or *m*-iodoxybenzoic acid[241,242]. These reagents regenerate the anhydride from reduced selenium byproducts as they are formed, and PhSeSePh may also be used as the catalyst as it too is oxidized to BSA under these conditions.

Several other types of ketones[242-244], and also lactones[245-248] and lactams[249-250], were similarly dehydrogenated with BSA and representative examples are shown in equations 49[244], 50[248] and 51[250].

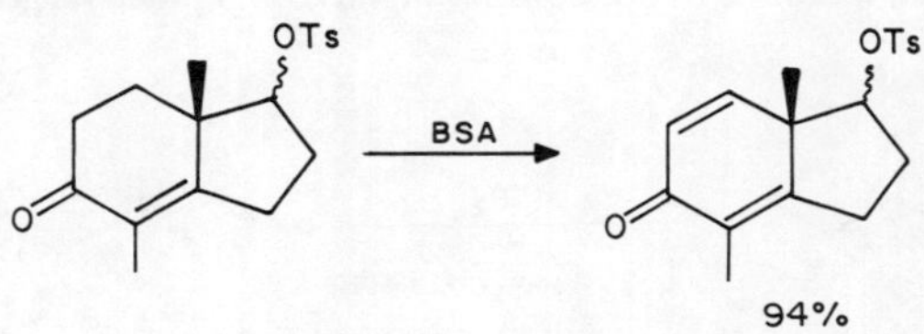

(49)

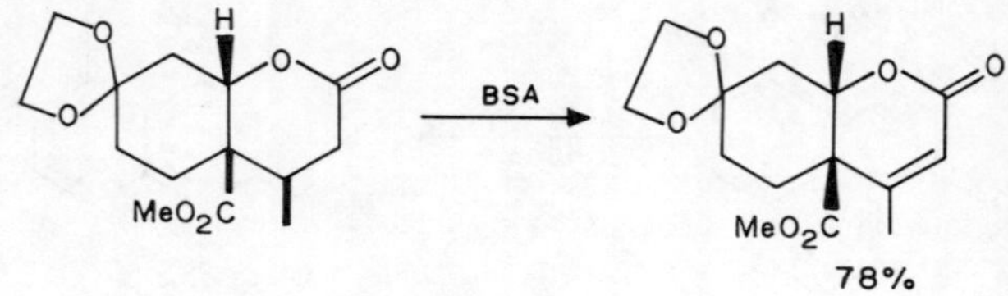

(50)

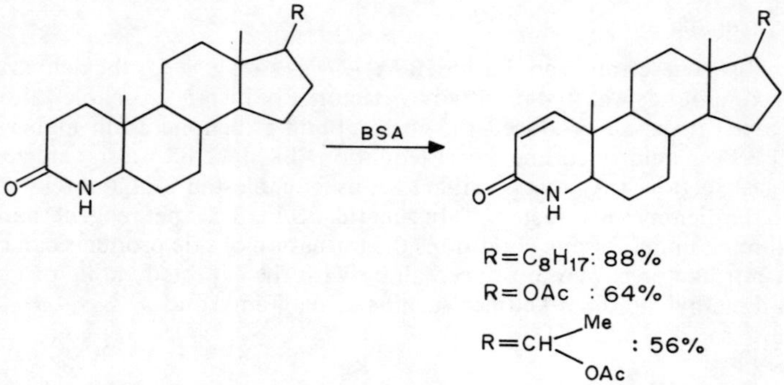

(51)

Benzeneseleninic anhydride and the corresponding seleninic acid ($PhSeO_2H$) serve as useful reagents for the dehydrogenation and side-chain degradation of cholic acids via the 4, 5-dihydrooxazole derivatives **29**[251] (e.g. equation 52).

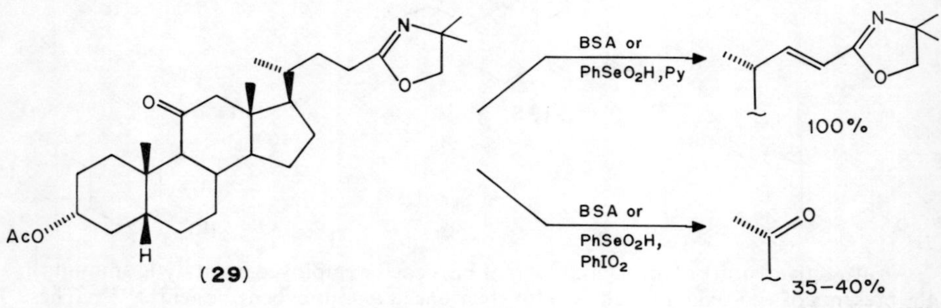

(52)

3. Benzeneseleninyl chloride

Benzeneseleninyl chloride (**30**) was briefly investigated as a possible reagent for the dehydrogenation of ketones and esters[134] according to equation 53. However, difficulties in the preparation and handling of this compound have precluded its general acceptance in this regard.

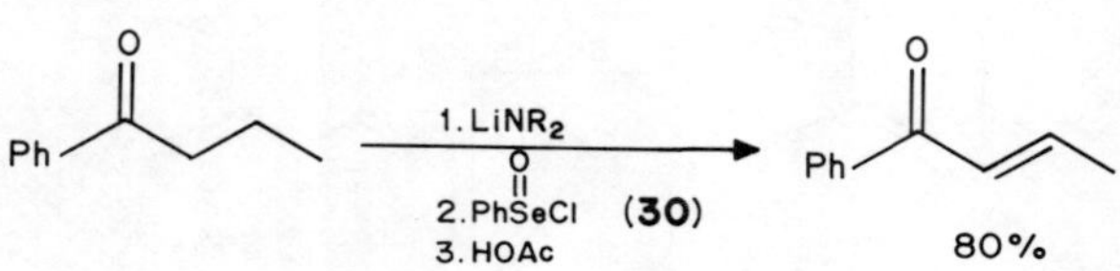

(53)

IV. DEHYDROGENATION OF OTHER FUNCTIONAL GROUPS

A. Nitrogen Compounds

Apart from carbonyl groups, other functionalities which stabilize an adjacent carbanion can be selenenylated and dehydrogenated by means of selenoxide eliminations. Thus, aliphatic nitro compounds were converted into their α-seleno derivatives[252–255] and subsequently to nitroolefins by oxidation of the selenide moiety[253,254]. Aldol condensations of the selenides with aldehydes can be performed prior to oxidation[254], thereby providing hydroxylated nitroolefins as shown in equation 54.

Similarly, carbanions derived from *N*-nitroso compounds[256,257] and amidines[258] were selenenylated and dehydrogenated as indicated in equations 55[257] and 56[258].

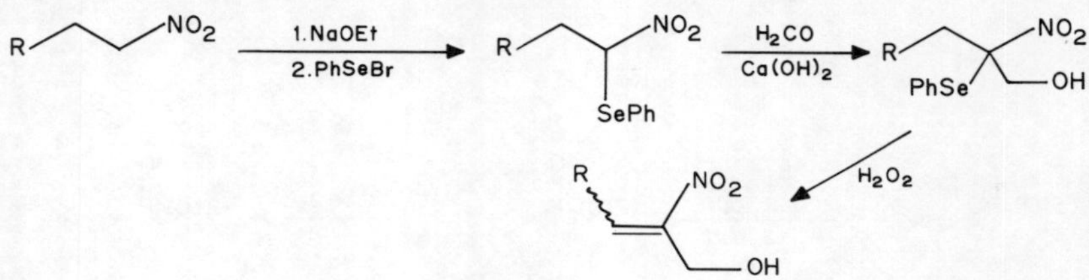

(54)

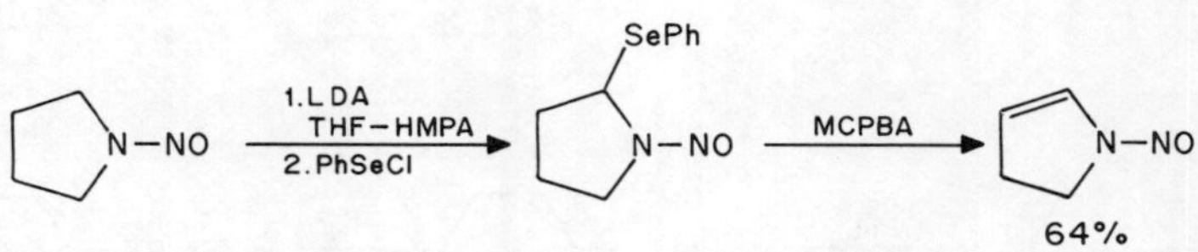

(55)

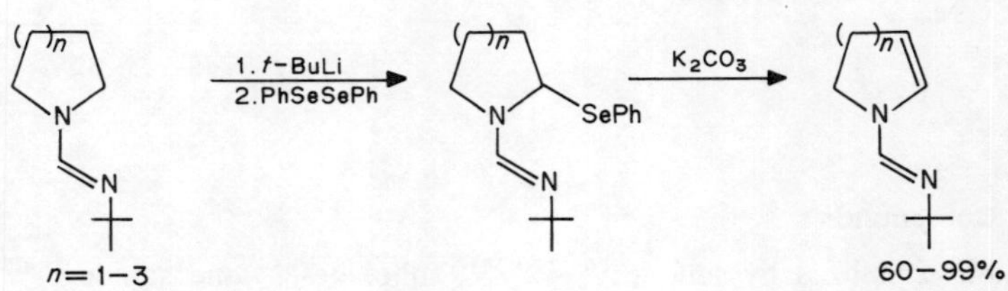

(56)

B. Nitriles

Nitriles, like lactams (see Section III.A.3), require the presence of an excess of base[259,260]. Monoselenenylation followed by selenoxide elimination provides a convenient route to α,β-unsaturated nitriles[259–261]. The latter compounds can in turn be dehydrogenated to diene nitriles in high yield (equation 57[262]).

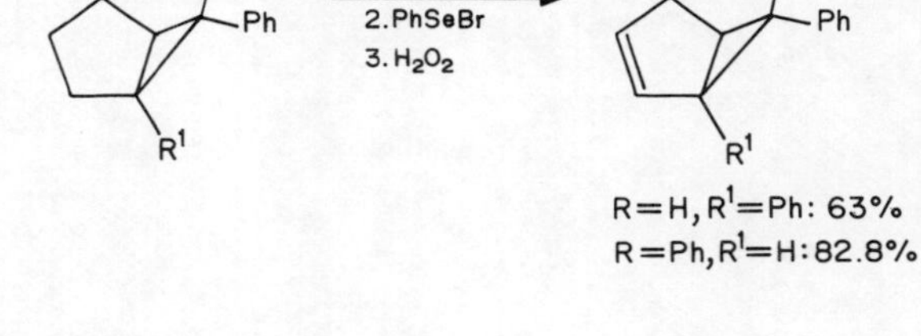

(57)

C. Phosphorus Compounds

Phosphoranes[263–267], phosphonates[170,171,268] and phosphine oxides[265,269] have been selenenylated with selenenyl chlorides and bromides, and with elemental selenium and methyl iodide[268]. In several instances the corresponding unsaturated compounds were prepared by selenoxide elimination[170,171,265,266,268]. Examples are given in equations 58[266], 59[268] and 60[265].

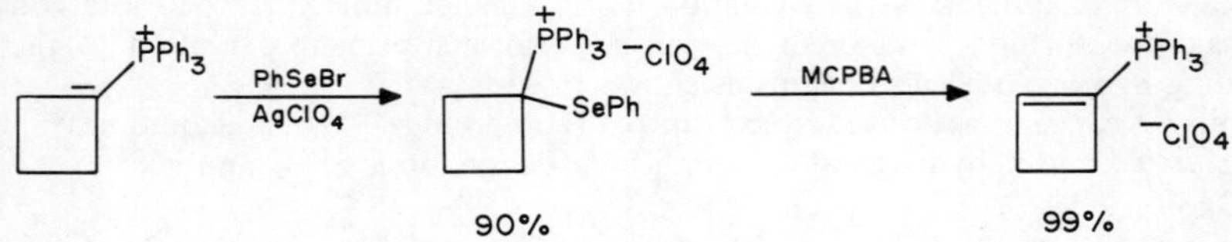

(58)

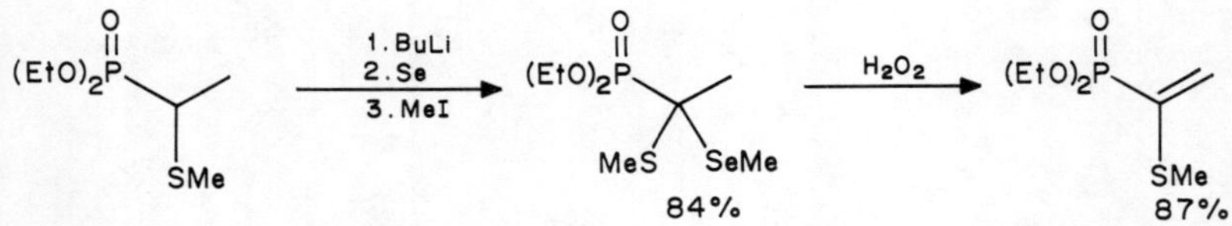

(59)

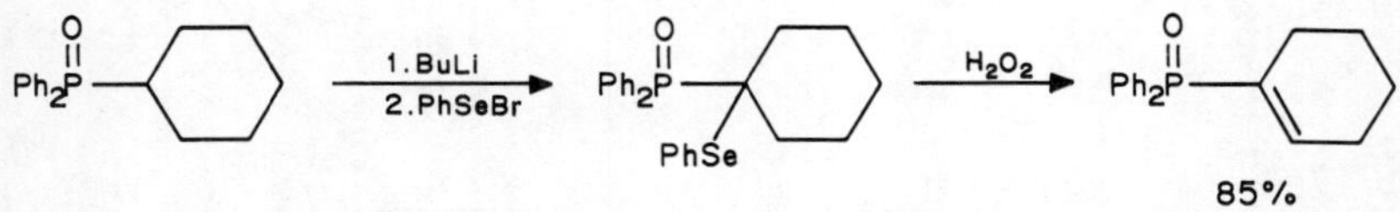

(60)

D. Sulfur Compounds

Carbanions stabilized by sulfones[169,270,271], sulfoxides[134] and sulfides[268,272] were selenenylated in much the same way and dehydrogenated via the usual oxidative procedures[134,169,268]. Oxidation of the α-seleno-α-silyl sulfone **31** afforded the carboxylic acid **32** instead of the elimination product[270] (equation 61).

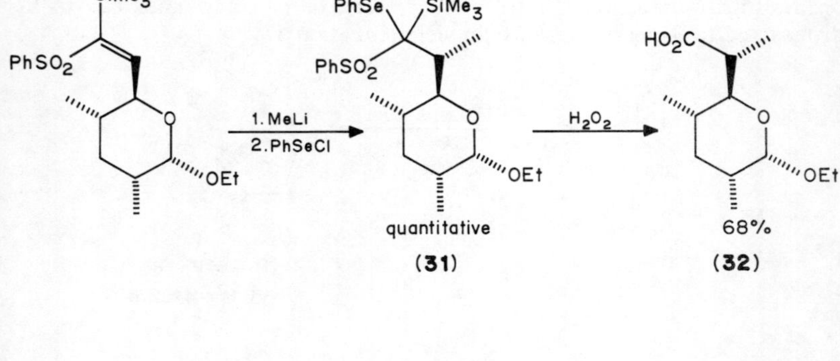

(31) **(32)**

(61)

V. OXYGENATION REACTIONS

A. Oxyselenenylation and Oxytellurenylation

1. Olefins

Olefins react with various selenium electrophiles to produce 1, 2-adducts via seleniranium ion intermediates (**33**) (equation 62). This may be exploited in the introduction of a new oxygen function into the olefin in either of two ways. First, if the electrophile is a selenenic* acid[38,82,273,274] or a derivative such as the selenenyl acetate PhSeOAc[40] or the trifluoroacetate $PhSeOCOCF_3$[275,276], then direct incorporation of the oxygen function occurs.

Alternatively, the olefin may be treated with other selenenic electrophiles in the presence of a nucleophilic oxygen-containing solvent such as water or an alcohol. The solvent then intercepts the bridged intermediate **33** and so affords the β-oxygenated selenide (equation 62). Similarly, electrophilic tellurium species produce β-oxygenated organotellurium products from olefins. The numerous types of electrophiles which have been used in this manner are listed in Table 1.

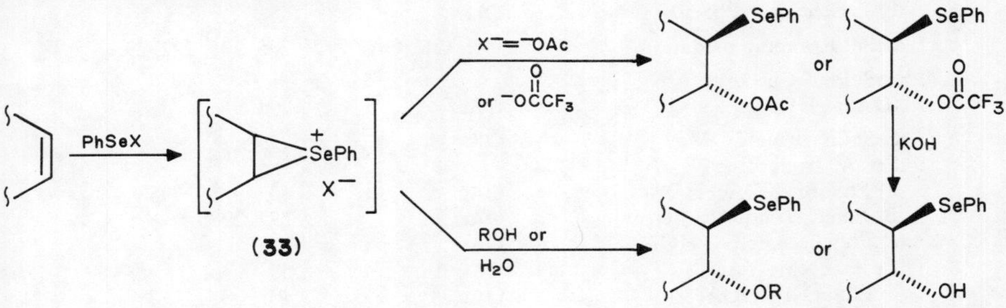

(62)

In general, oxyselenenylation proceeds stereospecifically via *anti* addition. Markovnikov adducts usually predominate, but their regioisomers can form in substantial amounts. Since the adducts undergo selenoxide elimination away from the oxygen function (cf. equation 5), the method provides an excellent synthesis of allylic alcohols, ethers and acetates (equation 63).

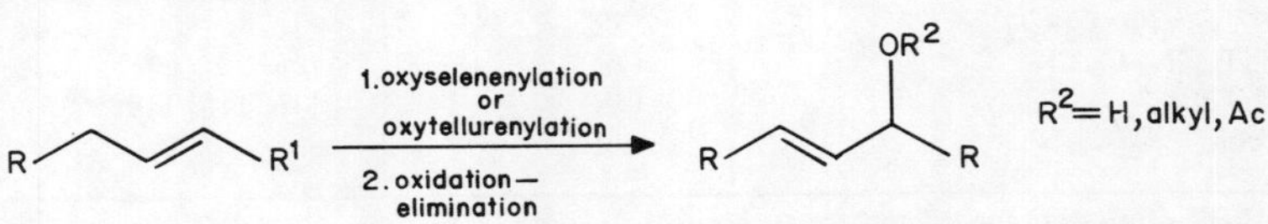

(63)

Under some circumstances the oxygen function in the initial 1, 2-adduct may be further oxidized to a carbonyl group. Kuwajima and coworkers found that the putative electrophile PhSeOSePh[306–309] [generated *in situ* from the oxidation of PhSeSePh with

*Selenenic acids RSeOH, and also the selenenyl acetate and trifluoroacetate, are unstable and must be generated *in situ*. The acids disproportionate readily according to the equation[38,82] $3RSeOH \rightleftharpoons RSeO_2 + RSeSeR + H_2O$ and dehydrate spontaneously to the anhydrides RSeOSeR[277,278]. The precise nature of the active electrophile in a given reaction is therefore equivocal and it may be present in only minute amounts[279]. However, such species are often designated as the parent selenenic acids for the sake of convenience.

TABLE 1. Reagents for oxyselenenylation and oxytellurenylation

Electrophilic reagent	Type of oxygen function	Ref.
PhSeCl	OH	208
	OR	40, 280–283
PhSeBr	OH	59
	OR	40, 284
MeSeBr	OH	59
PySeBr (**11**)	OR	101
PhSeN 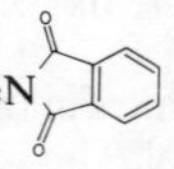 (N-PSP) (**34**)	OH	285–288
	OR	286, 289
$PhSeCN–CuCl_2$	OH, OR, OAc	290, 291
RSeOH or RSeOSeR from:		
(a) $RSeO_2H + H_3PO_2$	OH	273, 274
(b) $PhSeSePh + H_2O_2$	OH	82
(c) $PhSeSePh + PhSeO_2H$	OH	38
PhSeSePh, anodic oxidation	OH, OR, OAc	205, 206, 292, 293
$PhSeSePh + Cu(OAc)_2$ or $Pb(OAc)_2$, HOAc	OAc	294
PhSeSePh, BrNHCOMe	OR	295
PhSeOAc from:		
(a) PhSeBr, KOAc, HOAc	OAc	40, 49, 50, 296
(b) PhSeCl, NaOAc, HOAc	OAc	297, 298
(c) PhSeCl, $LiClO_4$, HOAc	OAc	299
(d) $PhSeNMe_2$, Ac_2O	OAc	58
(e) $PhSeO_2H$, HOAc	OAc	40, 300
MeSeOAc from:		
(a) $Me_2Se{=}O$, HOAc	OAc	301, 302
(b) $MeSeCH_2OAc$, H_2O_2, HOAc	OAc	303
$PhSeOCOCF_3$ from PhSeCl or PhSeBr + $AgOCOCF_3$	$OCOCF_3$, OH[a]	84, 275, 276, 296, 304
$PhTeTePh–Br_2$	OR	116, 118, 119
$PhTeTePh–CuCl_2$	OR	116
$PhTeCN–CuCl_2$	OR	116
$PhTeBr_3$[b]	OR	116, 118, 119
$TeCl_4$	OR	305

[a]After hydrolysis of the trifluoroacetate group.
[b]The 1,2-adduct $RO{-}\overset{|}{\underset{|}{C}}{-}\overset{|}{\underset{|}{C}}{-}TeBr_2Ph$ can be reduced to the telluride with N_2H_4, $Na_2S_2O_3$, Na_2S or $NaHSO_3$, or hydrolysed to the telluroxide with NaOH.

BSA (**27**) or *t*-butyl hydroperoxide] and $PhSeOSnBu_3$[308,310] [from PhSeSePh, Br_2 and $(Bu_3Sn)_2O$] afford α-seleno ketones and aldehydes from olefins (equation 64). The former products dominate in DMSO[306], whereas the latter are preferred in the presence of an already present allylic oxygen function[307,309].

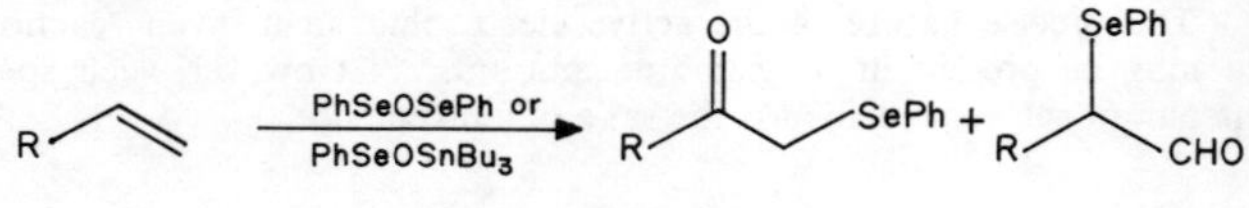

(64)

The conversion of terminal olefins into 1-phenylselenoalkan-2-ones was also accomplished through the sodium periodate oxidation of their Markovnikov alkoxyselenenylation products[311] and by the DMSO-mediated oxidation of their PhSeBr adducts[312].

Oxyselenenylation of olefins may also be performed with nucleophilic selenenium species. Olefins react with copper(II) halides and potassium selenocyanate (KSeCN) in alcoholic media to produce β-alkoxy selenocyanates[313] (equation 65). The corresponding β-halo derivatives are formed in the absence of alcohols. A related procedure involves thallation of terminal olefins followed by treatment with KSeCN[314].

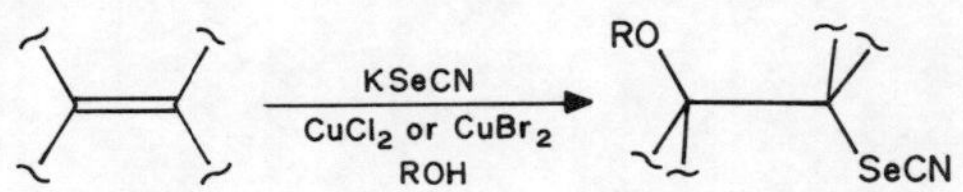

(65)

2. Enol ethers

The oxyselenenylation of enol ethers with alcohols followed by selenoxide elimination provides a convenient route to dihydropyran acetals[315–319] (e.g. equation 66[318]). The method constitutes an effective glycosylation procedure when the alcohol is an appropriate carbohydrate[51,320]. Ketene acetals are formed similarly. When they are derived from allylic alcohols and used in conjunction with a Claisen rearrangement[44–47], they provide access to γ,δ-unsaturated esters or acids (e.g equation 67[44]). Hydrolysis instead of

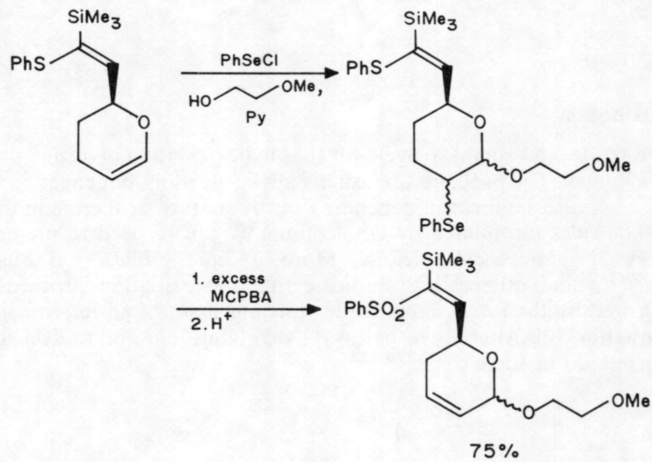

(66)

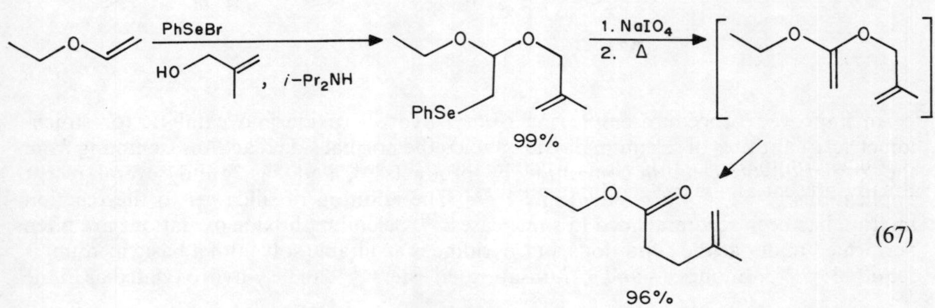

(67)

oxidation–elimination of the acetal **35** affords synthetically useful phenylselenoacetaldehyde (**36**)[321] (see Section XVII.A) via equation 68.

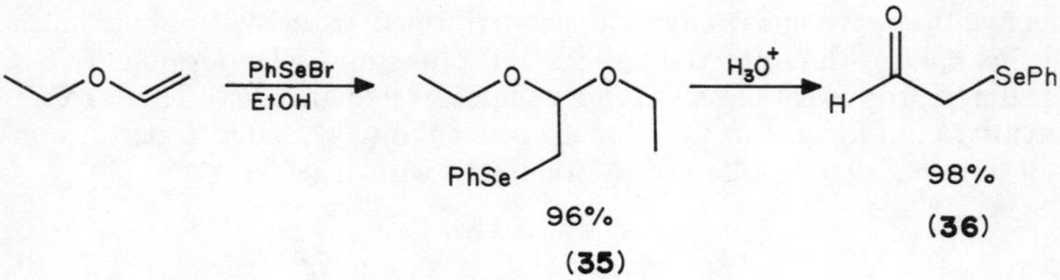

(68)

3. *Acetylenes*

Acetylenes produce α-seleno ketones (and hence enones) when treated with $PhSeOCOCF_3$ followed by alkaline hydrolysis[134,276] (equation 69).

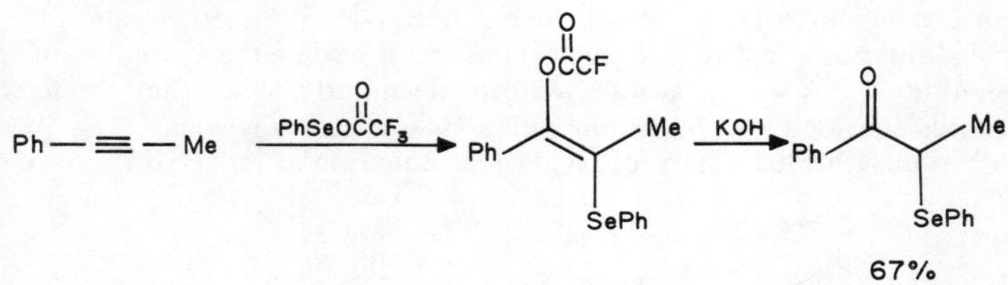

(69)

B. Allylic Oxidation

Selenium dioxide is a valuable reagent for the allylic oxidation of olefins, and numerous examples are known. The products are usually allylic alcohols, but enals[322,323], esters[324] or enones[325] can also be formed, depending on the nature of the olefin and the exact conditions. The rules formulated by Guillemonat[326] can be used to predict the site of oxidation in unsymmetrical olefins. More recent studies by Sharpless and coworkers[327–330] and others[331,332] indicate that the oxidation proceeds by an ene reaction followed by the [2, 3] sigmatropic rearrangement of an intermediate seleninic acid (**37**) (equation 70). Alternative pathways involving ionic or radical intermediates have been proposed in some cases[330,331].

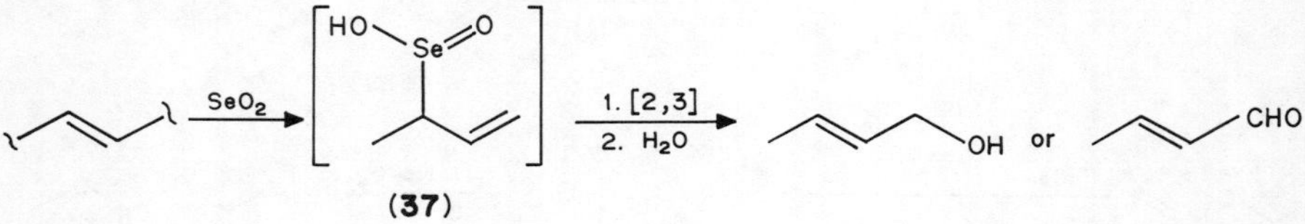

(70)

An improved procedure employing *t*-butyl hydroperoxide and catalytic (or stoichiometric) quantitites of selenium dioxide avoids the normal side reactions stemming from the formation of selenium-containing by products[333], and has found several recent applications[236,334–338] (e.g. equation 71[338]). The addition of silica gel to the reaction mixture has been recommended in some cases[339]. Selenium dioxide oxidations are often performed under acidic conditions, but pyridine is a suitable solvent if a basic medium is required[325,340]. Enones[341] and α, β-unsaturated esters[342] can be γ-hydroxylated as in the example in equation 72[342]. Dehydrogenation and aromatization (see Section III.B.1) sometimes occur simultaneously[236,334,343] (equation 73[236]) with allylic oxidation. Acetylenes furnish propargylic alcohols and ketones with selenium dioxide and *t*-butyl hydroperoxide[344] (equation 74).

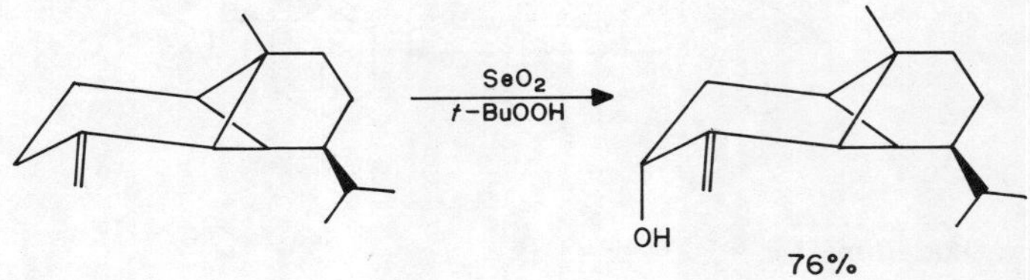

(71)

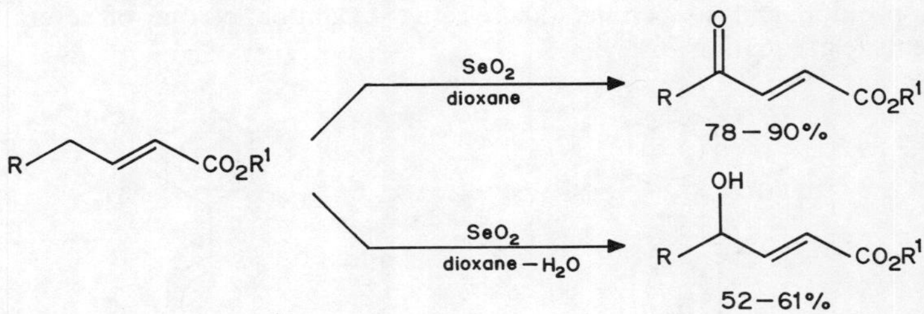

(72)

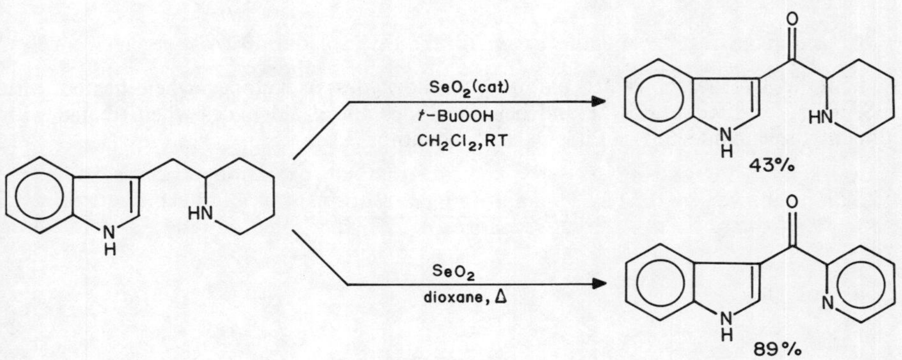

(73)

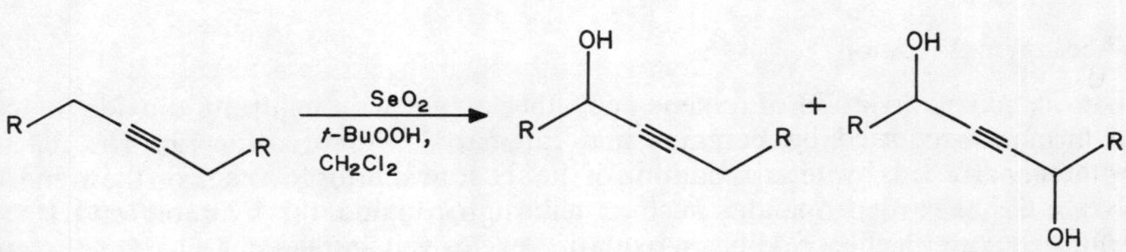

(74)

An alternative reagent to selenium dioxide was recently reported for the allylic oxidation of olefins. This consists of catalytic amounts of PySeSePy (**12**) in the presence of iodoxyarenes as indicated in the example in equation 75[345].

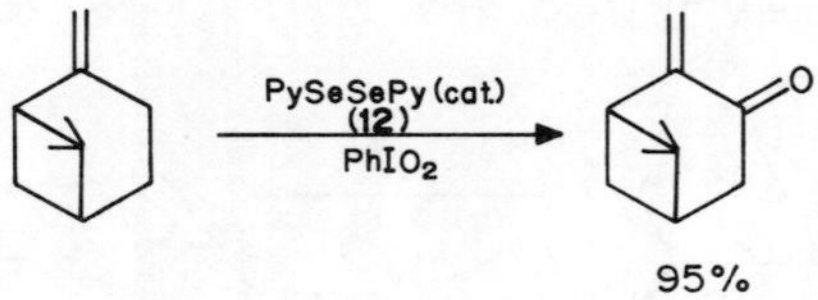

(75)

C. Benzylic Oxidation

Selenium dioxide oxidizes heterocyclic aryl methyl groups to aldehydes[346–349]. The presence of an electron-withdrawing group increases the reactivity of a methyl group in the *para*-position and so permits the selective oxidation of one of several methyl substituents (e.g. equation 76[348]).

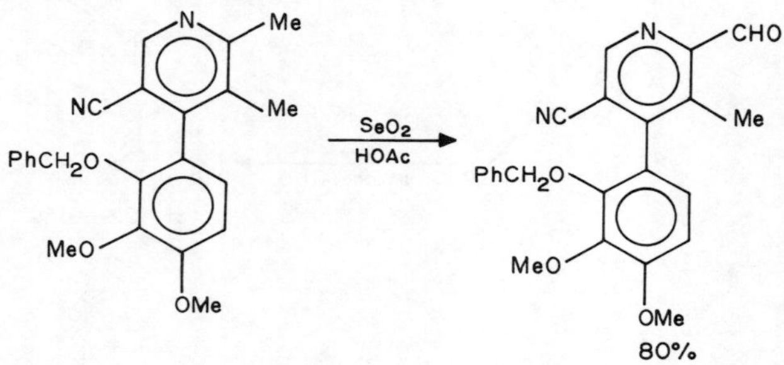

(76)

Benzylic hydrocarbons are oxidized to aldehydes or ketones when heated with BSA (**27**)[245,350] (equation 77), and benzyl halides afford aldehydes when treated with dimethyl selenoxide (**38**) or $PhSeO_2H$[351] (equation 78).

$$ArCH_2R \xrightarrow[\Delta]{BSA} ArC(=O)R \quad (77)$$

$$\underset{X=Cl,Br}{ArCH_2X} \xrightarrow[\text{or } PhSeO_2K,\ K_2HPO_4]{MeSe(=O)Me\ (\mathbf{38}),\ K_2HPO_4} ArCHO \quad (78)$$

D. Acetoxymethylation

The acetoxymethylation of toluene and other arenes with tellurium dioxide in acetic acid has been reported by Bergman and Engman[352,353] (equation 79). The reaction presumably proceeds by initial oxidation of the acetic acid prior to attack on the arene. It is interesting that related oxidants such as tellurium trioxide, the hydrate $Te(OH)_6$ and selenium dioxide all effect side-chain oxidation exclusively instead of acetoxymethylation under these conditions[353].

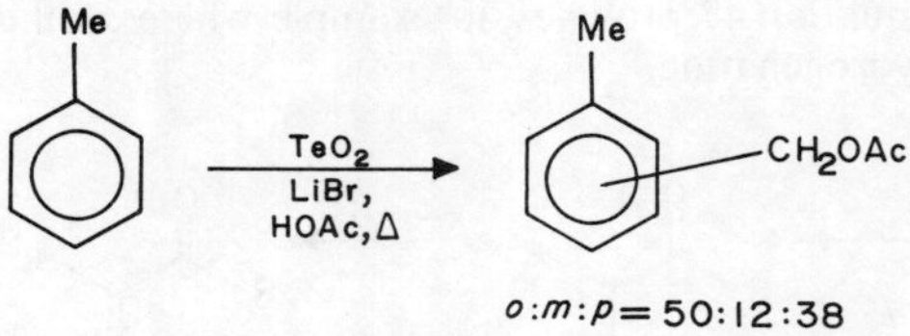

(79)

E. Oxidation of Phenols

Barton and coworkers reported that phenols are oxidized to orthoquinones[354,355] or hydroxydienones[356,357] with BSA (**27**) (e.g. equations 80[355] and 81[357]). In some instances, phenolic coupling products were also observed (see Section XI.E.3). Several polyaromatic hydrocarbon quinones were prepared from phenolic precursors[358,359] with this procedure, and the naphthol **39** yielded the corresponding hydroxyenone **40** with high regioselectivity[360] (equation 82). A polymer-supported seleninic acid was recently employed as an alternative to BSA for the similar oxidation of phenols[361].

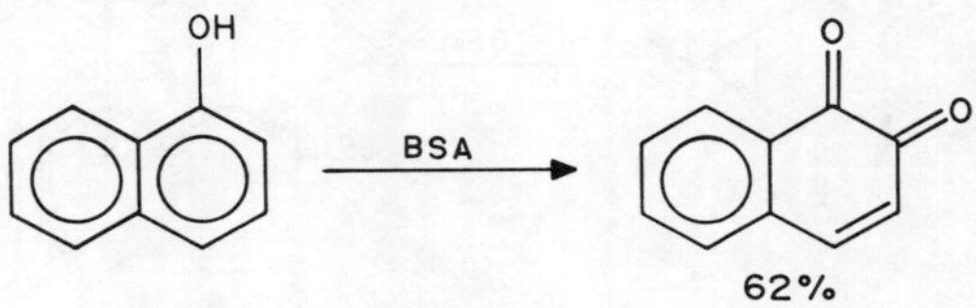

(80)

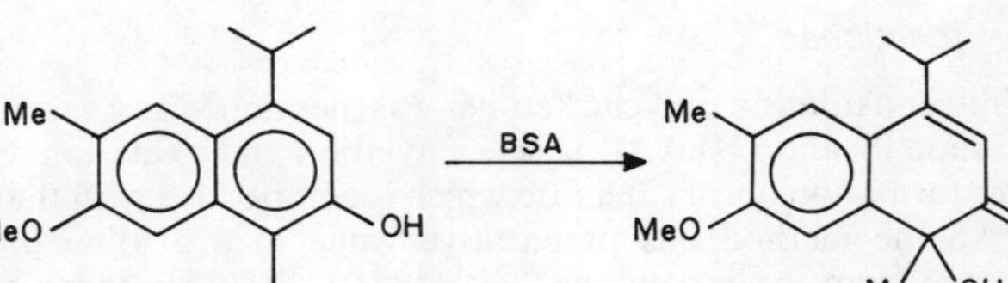

(81)

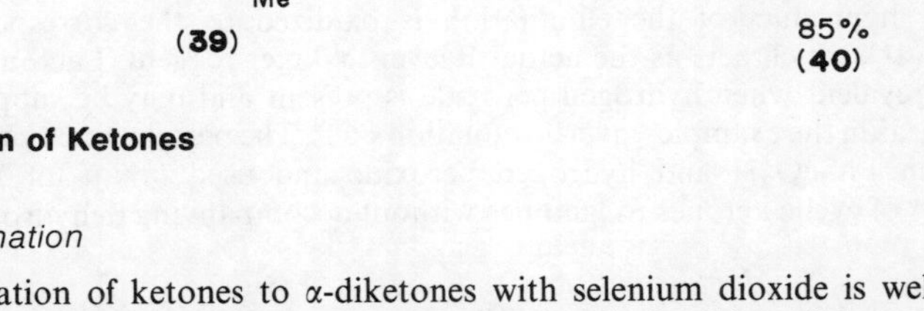

(82)

F. Oxidation of Ketones

1. α-Oxygenation

The oxidation of ketones to α-diketones with selenium dioxide is well known and literature examples abound. Although several rationales for this useful reaction have been proposed, the most recent study provided evidence for the pathway in equation 83[362]. In addition, several methyl ketones were converted into the corresponding keto aldehydes[363] in this fashion and equation 47 provides an example where α-diketone formation was accompanied by dehydrogenation.

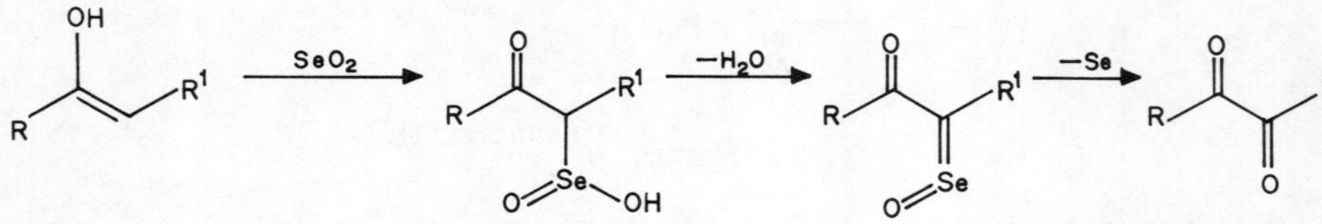

(83)

Yamakawa and coworkers[243,364,365] employed BSA for the angular α-hydroxylation of ketones as in equation 84[364], while others have noted γ-hydroxylation during the reactions of BSA with an enone[247], and an α, β-unsaturated lactone which was itself formed by the dehydrogenation of the parent compound[246].

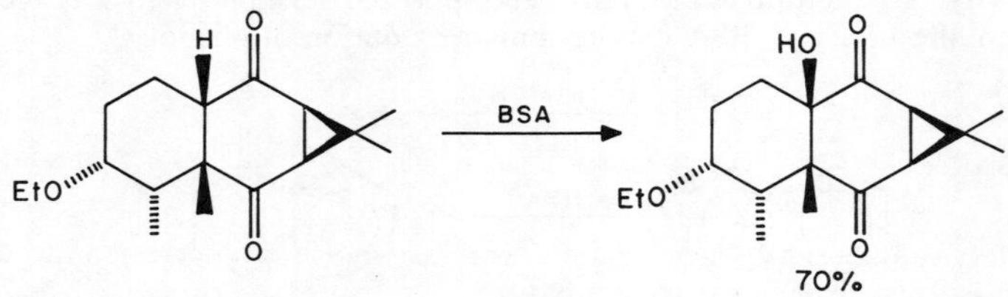

(84)

Carbonyl compounds are α-acetoxylated with tellurium dioxide or trioxide in acetic acid in the presence of LiBr[353] (e.g. equation 85).

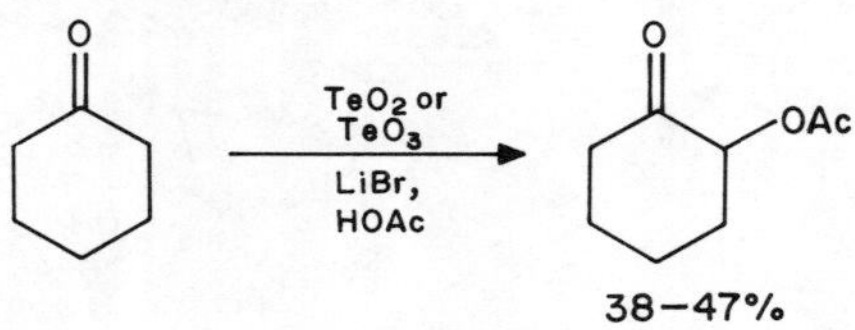

(85)

2. Baeyer–Villiger reactions

The Baeyer–Villiger oxidation of cyclic ketones was mentioned as a side reaction during their dehydrogenation (Section III.A.1) via selenenylation and oxidation–elimination. In some cases lactone formation occurs at a rate which is competitive with that of selenoxide elimination and so the method has preparative value in providing direct access to unsaturated lactones from cyclic ketones[69,94,114,142,143]. Under these conditions, the selenenic acid byproduct of the elimination is oxidized to the corresponding perseleninic acid (**41**), which acts as the actual Baeyer–Villiger reagent. Lactone formation is especially prevalent when hydrogen peroxide is present and may be suppressed with other oxidants, as in the examples given in equation 86[69]. The perseleninic acid can also be generated from $PhSeO_2H$ and hydrogen peroxide and used *in situ* for the efficient transformation of cyclic ketones to lactones without accompanying dehydrogenation[366] (equation 87).

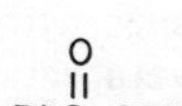

(41)

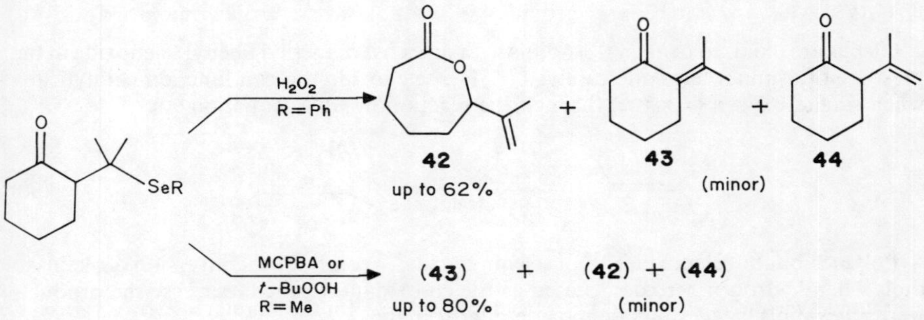

(86)

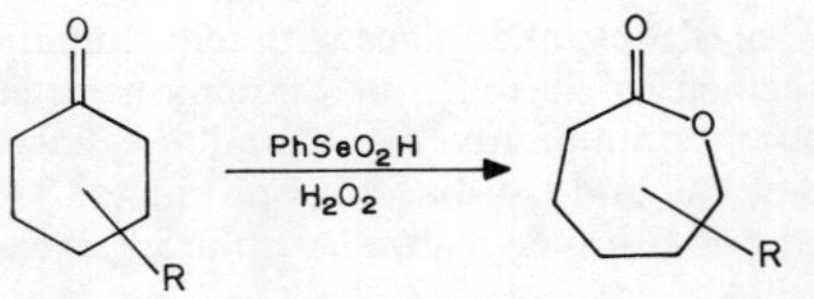

(87)

Other effective Baeyer–Villiger reagents are a polymer-bound perseleninic acid[361] and selenium dioxide–hydrogen peroxide mixtures[367–371] in which the perseleninic acid **45** is a probable intermediate (e.g. equation 88[371]). The use of the latter reagent often results in relatively complex product mixtures.

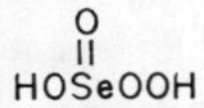

(45)

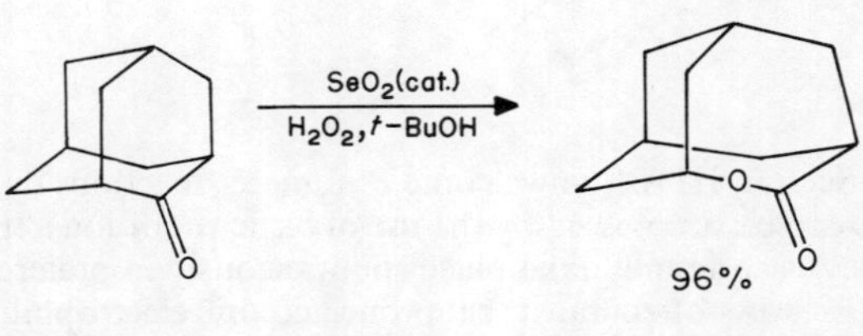

(88)

G. Epoxidation, Hydroxylation and Acetoxylation of Olefins

1. Epoxidation

Perseleninic acids smoothly epoxidize olefins[38,82,372–375] (equation 89) even in catalytic amounts. They can be generated *in situ* for this purpose in the same way as when they are required for Baeyer–Villiger reactions. As with other peracid-mediated epoxidations, more highly substituted olefins react more efficiently than less substituted ones[372].

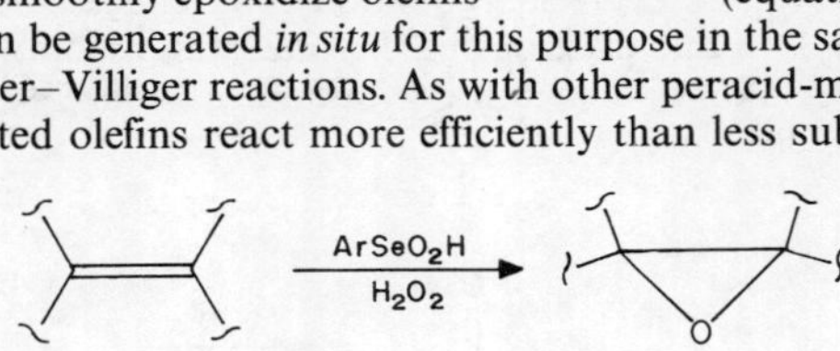

(89)

2. Hydroxylation

Olefins are oxidized to vicinal *cis*-diols with diphenyl or methyl phenyl selenoxide in the presence of osmium tetroxide catalyst[376]. The selenoxide too can function catalytically when singlet oxygen is passed through the reaction mixture (equation 90).

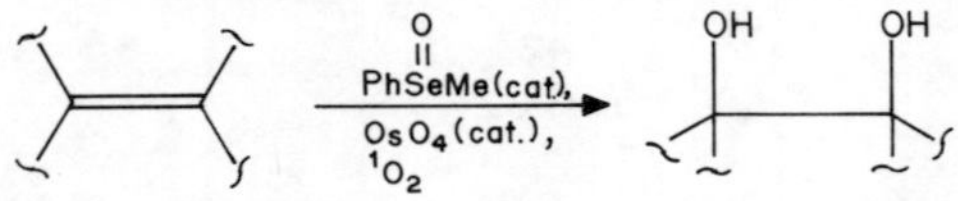

(90)

Polymer-bound selenoxides[150] or seleninic acids[361] catalyse the conversion of olefins to diols when hydrogen peroxide is used as the co-oxidant. In the latter case the products were *trans*-diols, suggesting an epoxide intermediate.

3. Acetoxylation

Vicinal diacetoxylation of olefins takes place with tellurium dioxide and LiBr in acetic acid[305]. The *syn*-stereospecificity of the reaction was demonstrated by the fact that *cis*- and *trans*-but-2-ene produced predominantly *meso*- and *d,l*-diacetates, respectively. 1,3-Dienes afforded mixtures of 1,2- and 1,4-diacetoxy products[377] (equation 91), with ratios of up to 9:1 in favor of the 1,4-isomer realized with a 5:1 excess of LiBr over TeO_2.

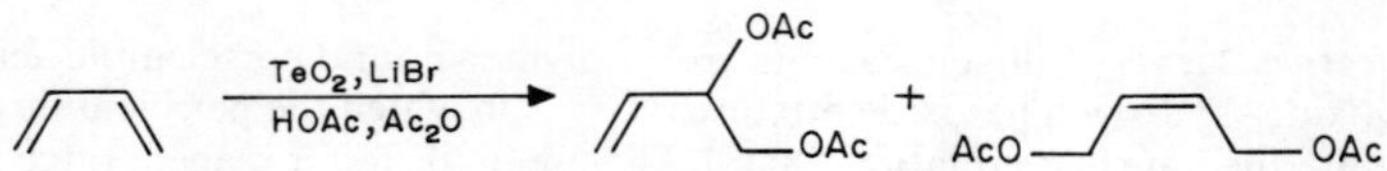

(91)

H. Oxidation of Selenides

Although the oxidation of selenides usually leads to selenoxide elimination, under some circumstances oxygenated products result instead. Several synthetically useful adaptations of the latter type of process are described below. The preparation of allylic alcohols from the oxidation and [2, 3] sigmatropic rearrangement of allylic selenides is discussed in Section XVI.A.

1. Pummerer reactions

As mentioned in Section III.A.1, unwelcome Pummerer reactions during the dehydrogenation of ketones can be suppressed by the use of basic conditions. In situations where they are desired, they occasionally take place spontaneously in preference to selenoxide eliminations[378], but more often must be promoted by electrophiles such as acetic anhydride[167,379–381] or BSA[223,382]. Pummerer reactions have proved useful for introducing new keto[223,381,382] or acetoxy groups[167,379,380] into the substrate, as shown in equation 92. A specific example of the preparation of an α-diketone from an α-seleno ketone by this method is presented in equation 93[381].

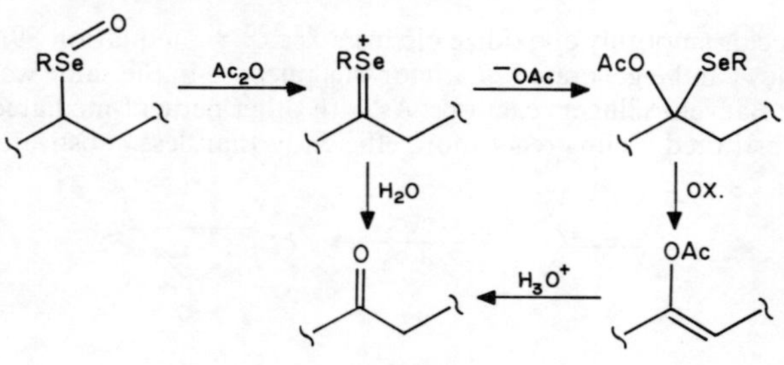

(92)

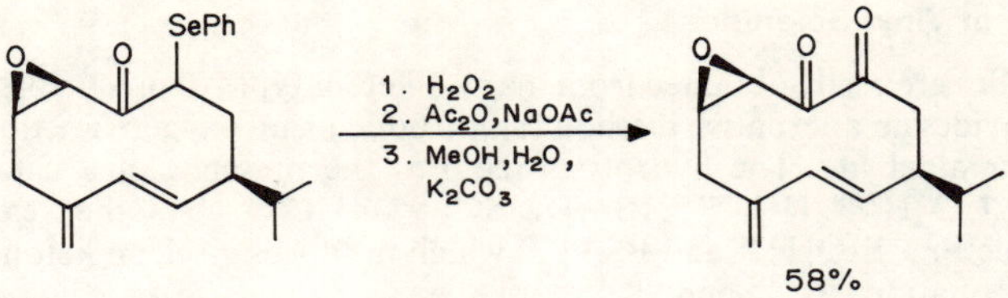

(93)

α-Silyl selenoxides undergo Pummerer rearrangements and provide access to silyl ketones[383], ketones[384] and aldehydes[385,386] (equation 94).

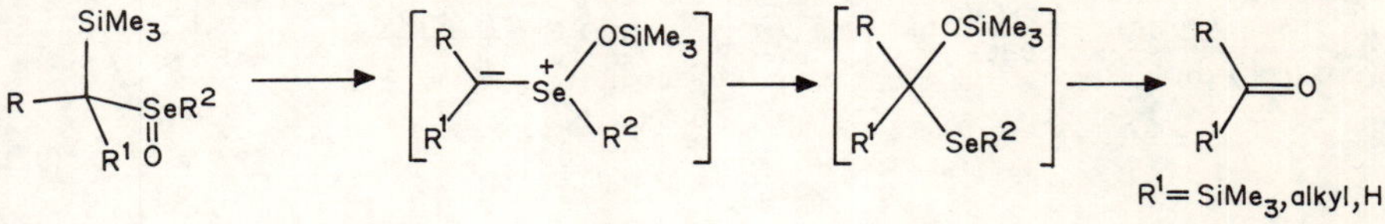

(94)

2. Via selenones and tellurones

Selenones and tellurones react readily with nucleophiles in substitution reactions because the anions $RSeO_2^-$ and $RTeO_2^-$ are excellent leaving groups. These compounds are accessible from the oxidation of selenides and tellurides with excess of strong oxidants such as MCPBA. When primary or secondary alkyl phenyl selenides[283,387] or tellurides[117,283] are thus oxidized in methanol solvent, solvolysis occurs *in situ* to afford methyl ethers in high yields (equation 95). Rearranged products result when β-substituents with high migratory aptitudes (e.g. phenyl) are present, and cyclic compounds undergo ring contractions (see Section XVI.C.2).

$$\begin{array}{c}\text{RSePh}\\ \text{or}\\ \text{RTePh}\end{array} \xrightarrow[\text{MeOH}]{\text{excess MCPBA}} \text{ROMe} \qquad (95)$$

3. In 1,2-carbonyl transpositions

α-Seleno ketones produce α-ketals when oxidized with MCPBA[129] or when treated with mercury(II) perchlorate[128] in methanol. This α-oxygenation procedure can be used in conjunction will deoxygenation of the original carbonyl group to effect an overall 1,2-carbonyl transposition[129] (equation 96).

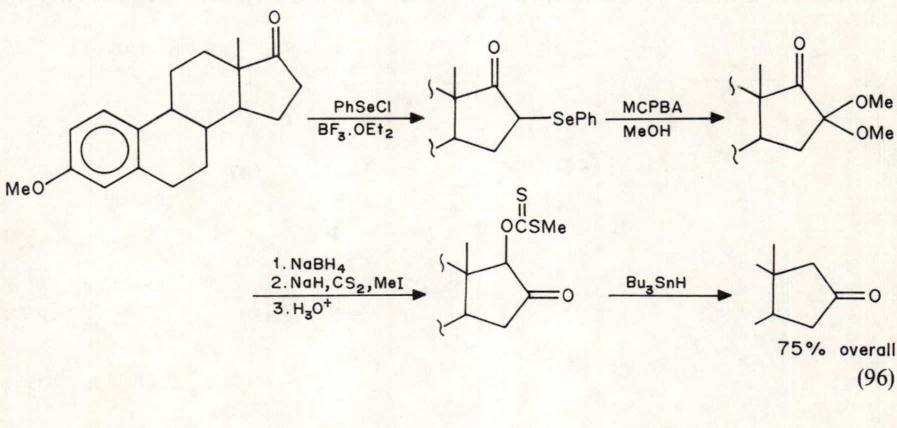

(96)

1. Hydrolysis of Vinyl Selenides

Vinyl selenides are easily obtained from many different types of precursors[388] and their hydrolysis provides an alternative method for introducing an oxygen function in the form of a ketone or aldehyde. The hydrolysis requires the presence of a catalyst such as $HgCl_2$[269], CF_3CO_2H[269], HCl[389], HBr–DMSO[390] or $HClO_4$[111]. Other reagents for this purpose include Br_2–EtOH[391] and BSA[392], which produces α-seleno ketones. Examples are shown in equations 97[389] and 98[392].

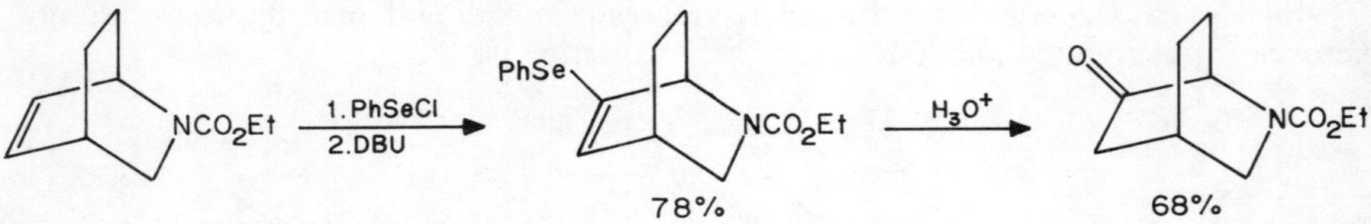

(97)

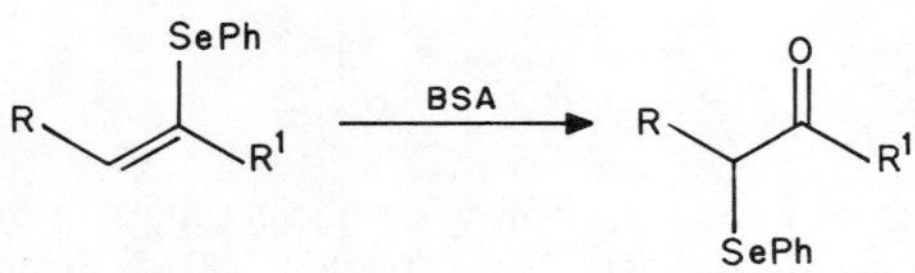

(98)

VI. AMINATION AND AMIDATION REACTIONS

A. Of Olefins

1. 1,2-Additions of selenenic electrophiles

In a manner reminiscent of oxyselenenylation, the reaction of olefins with selenium electrophiles in the presence of nitrogen nucleophiles results in the incorporation of both species into the olefin. Uemura and coworkers[53,393] treated a series of alkenes and cycloalkenes with PhSeCl in the presence of nitriles to generate the presumed intermediate **46**, followed by hydrolysis to afford β-amido selenides (**47**) (equation 99). The latter products were further transformed into allylic or saturated amides by oxidation[52,53] or reduction[53], respectively. The selenium-induced amidation of dienes is accompanied by cyclization (see Section IX.G). An electrochemical variation of this method has been reported[394] and consists in the anodic oxidation of PhSeSePh is acetonitrile in the presence of the olefin. Products **47** (R = Me) were obtained in generally high yields.

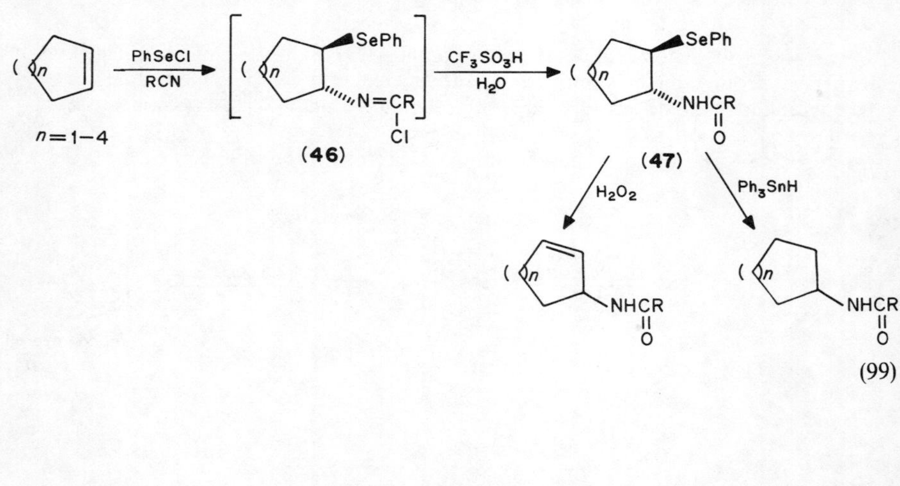

(99)

The sequential reaction of cyclohexene with RSeBr, sodium azide and lithium aluminum hydride provides an example of a potentially useful amination method[59] (equation 100). Oxidation instead of reduction of the β-seleno azide **48** also provides access to unsaturated azides (see equation 7). A different approach, shown in equation 101, affords tosylamides from olefins with PhSeSePh and Chloramine-T[395].

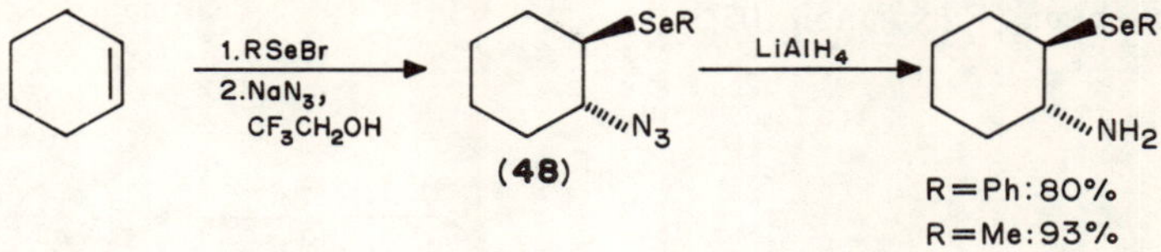

(100)

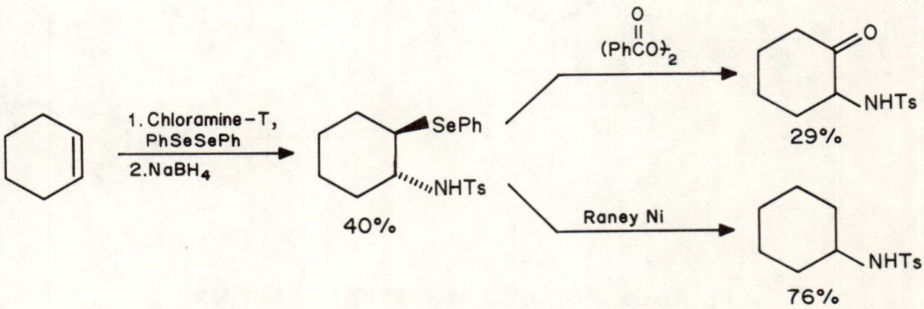

(101)

2. Allylic amidation

Sharpless and coworkers[160,396] reported that the selenium diimide reagent **49** (and also its sulfur analog) smoothly effects the allylic amidation of olefins and acetylenes. The diimide resembles selenium dioxide in its mode of action, which involves an ene reaction followed by a [2, 3] sigmatropic rearrangement (cf. equation 70). A typical example is provided in equation 102[396]. The reaction of 1, 3-dienes with **49** furnished 1, 2-diamidation products in modest yields[397].

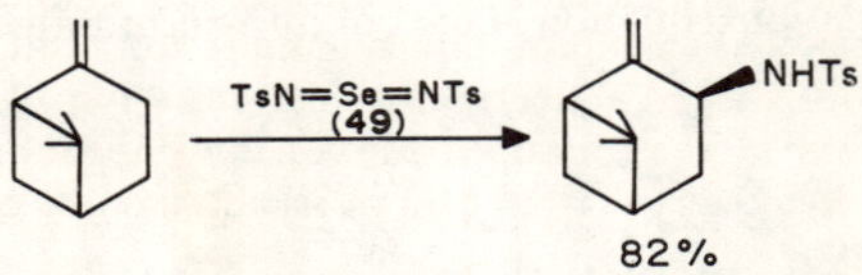

(102)

Rearranged allylic tosylamides[398] or carbamates[399–402] were obtained by Hopkins and coworkers from allylic selenides via their reaction with Chloramine-T or *N*-chlorocarbamates. The [2, 3] shift of a selenium imide (**50**) was implicated in this process[399] (equation 103) and applications include the preparation of β,γ-unsaturated amino acids[401] and optically active amino acids from chiral allylic selenides[402].

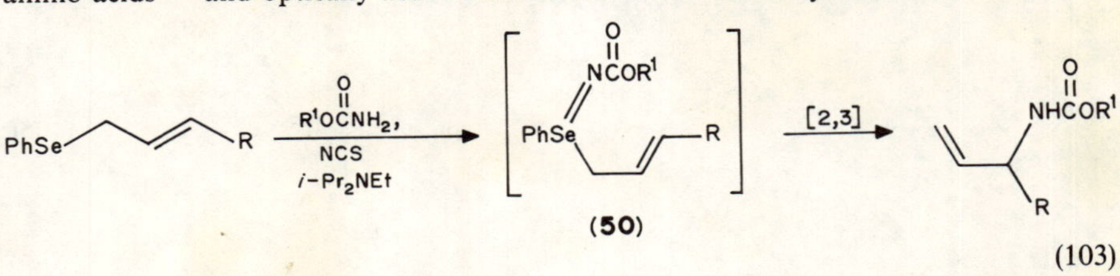

(103)

B. Of Phenols and Catechols

When the Barton procedure for the oxidation of phenols with BSA is carried out in the presence of hexamethyldisilazane, selenoimides such as **51** are produced[403,404]. These can be reduced to the corresponding anilines or anilides as indicated in equation 104. Catechols undergo amination when oxidized with diphenyl selenoxide (**52**) in the presence of aniline[405] (equation 105).

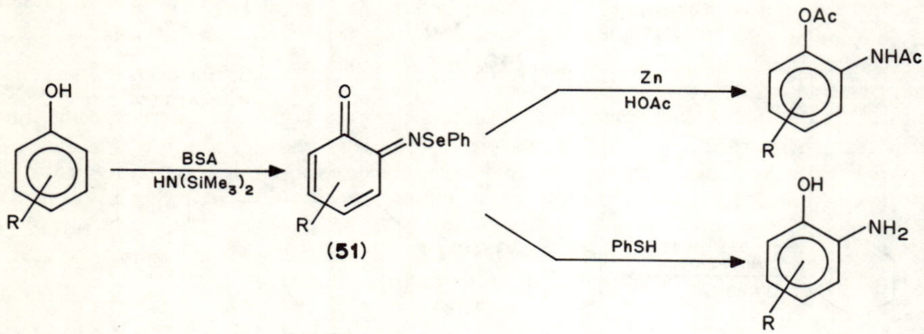

(104)

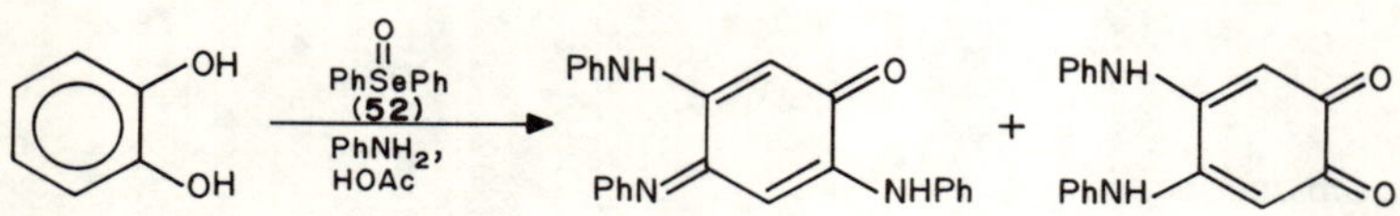

(105)

C. Of Michael Acceptors

Selenenamides are sufficiently nucleophilic to attack Michael acceptors such as enones[58], enals[406] or dimethyl acetylenedicarboxylate[407]. After intramolecular selenenylation of the α-position, 1, 2-adducts are obtained. These can be further converted into α-seleno enones by deamination (equation 106[58]), or into unsaturated amines by selenoxide elimination (see equation 8).

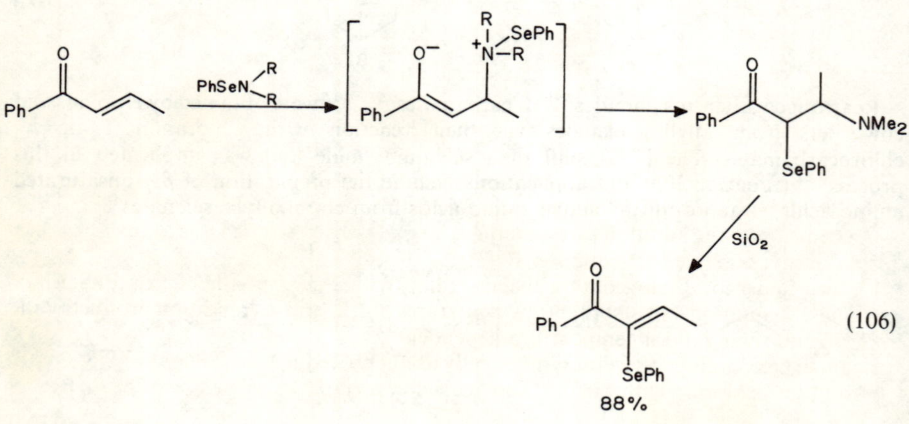

(106)

VII. HALOGENATION

A. By Electrophilic Addition of Selenium or Tellurium Halides to Olefins and Acetylenes

1. Olefins

In the absence of other nucleophiles, selenenyl halides add to olefins to furnish β-halo selenides (equation 107). The direct halogenation of the olefin is thus achieved and the adjacent selenium residue provides the means for further transformations. The mechanism, stereochemistry and regiochemistry of these processes have been extensively studied[62,63,408,409] and reviewed[410]. In general they are highly *anti*-stereospecific, but their regiochemistry depends on the nature of the olefin and the conditions employed. Complementary regioisomers are sometimes available in a high state of purity by choosing conditions which favor either the product of kinetic or that of theromodynamic control. Stereospecific reactions are also observed with aryl- or alkyl-selenium trichlorides[411], whereas tellurium electrophiles (e.g., $TeCl_4$, $RTeCl_3$) display more complex behavior[22,412], often producing both *syn*- and *anti*-adducts.

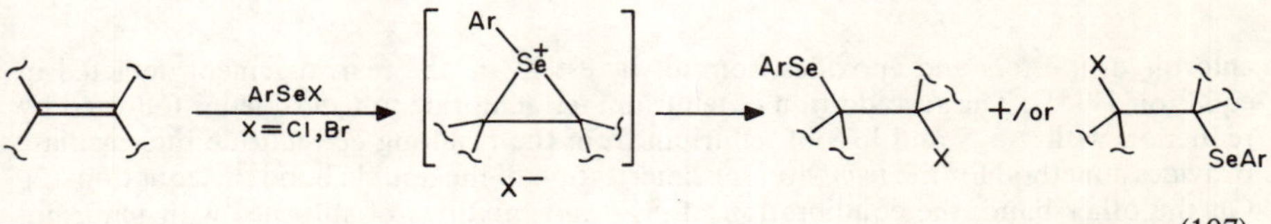

(107)

Numerous examples have been reported with selenenyl chlorides and bromides, and recently the first instance of an olefin fluorination was described with N-PSP (**34**) and pyridinium fluoride[286] (equation 108). The 1, 2-addition of a selenenyl iodide to an olefin remains undocumented, although the electrophile PhSeI adds to hex-1-yne (see equation 113) and has found use in the cyclization of dienes (see equation 161).

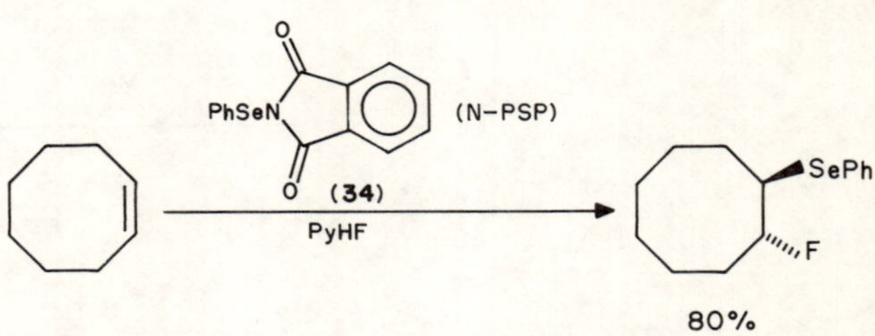

(108)

When the β-chloro- or bromo-selenides thus obtained are subjected to selenoxide elimination, vinyl or allyl halides are formed[62–66] (see Section II.A.3). Since the regiochemistry of the addition can sometimes be precisely controlled (see above), a choice of products is possible (e.g. equation 109[62]).

Enones undergo α-halogenation when treated with excess of PhSeCl or PhSeBr in pyridine[413] (equation 110). Isoprene gives mixtures of 1, 2- and 1, 4-addition products with PhSeCl, and their ratio is temperature dependent[414].

Engman has developed several synthetically useful procedures based on the reactions of tellurium tetrachloride with olefins. This reagent was used to prepare stereospecifically

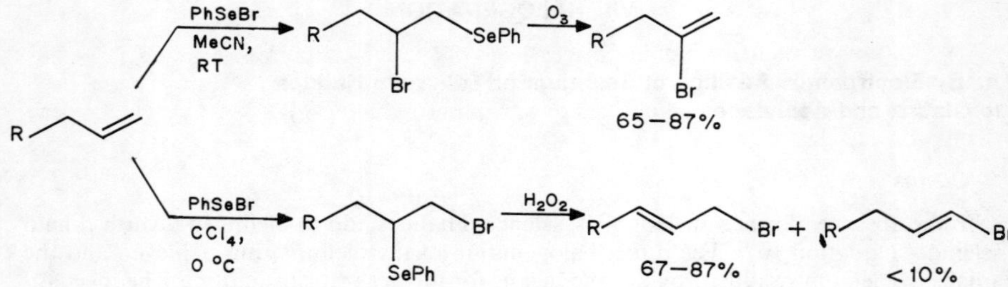

(109)

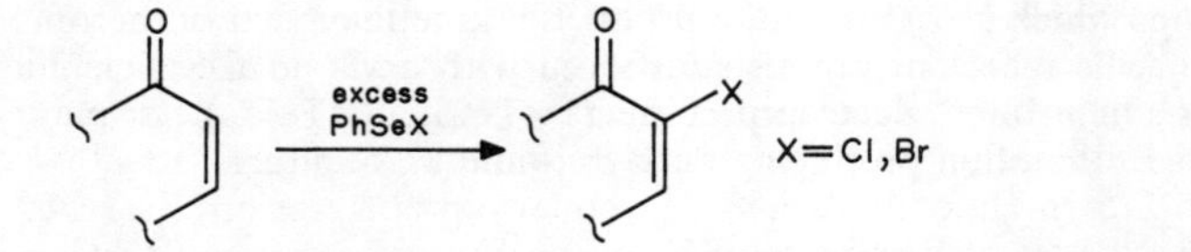

(110)

chlorohydrin esters and epoxides from allylic esters via the rearrangement depicted in equation 111[415]. The *syn*-addition of tellurium tetrachloride to *trans*-olefins followed by reduction with Na_2S and loss of tellurium from the resulting epitelluride intermediate provides a method for the *trans*-to-*cis* isomerization of the double bond[416] (equation 12). On the other hand, the equilibration of *cis–trans* mixtures of stilbenes with tellurium tetrachloride affords the pure *trans* isomers[417].

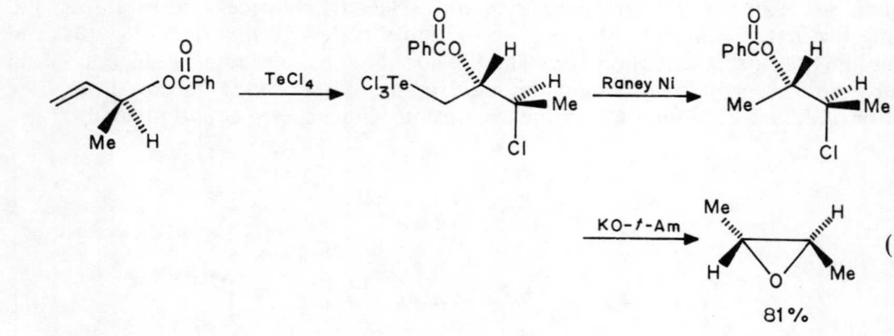

(111)

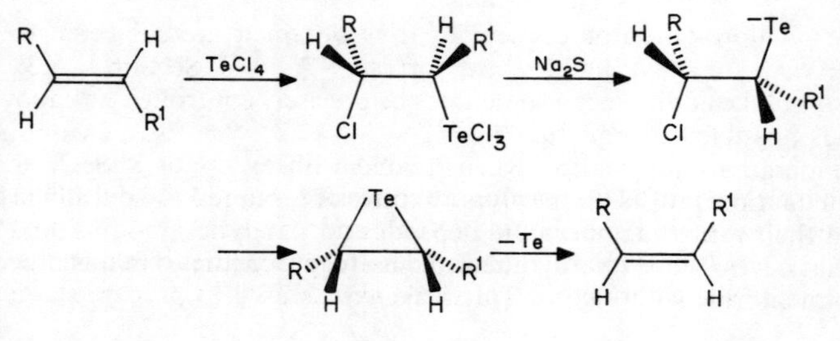

(112)

2. Acetylenes

The additions of selenenyl halides to acetylenes produce β-halovinyl selenides[418] which can then be further transformed into other useful products. Most such additions have been restricted to chlorides and bromides, but one example involving a selenenyl iodide is given in equation 113[147].

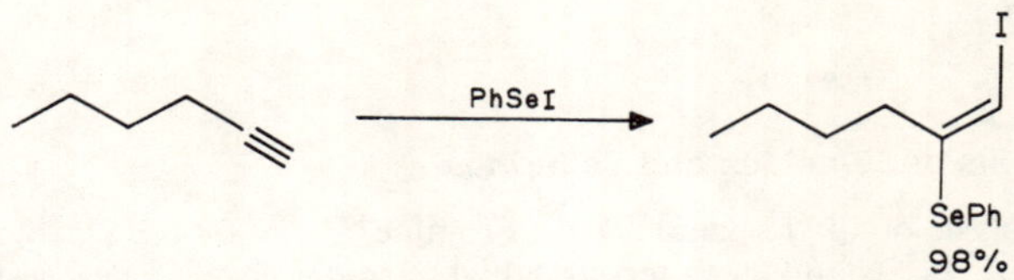

(113)

1,4-Dichlorobut-2-yne afforded highly functionalized 1,3-dienes when treated with PhSeCl or PhSeBr[419–421] (equation 114). The products proved of value in Diels–Alder reactions[420,421]. Similarly, the additions of PhSeCl to alkynols[422,423] and to the stannylynamine **53**[424] furnished adducts of potential synthetic utility. In the latter example, elimination of the stannyl moiety occurred together with addition (equation 115). β-Halovinyl selenides were also prepared from the reaction of acetylenes with phenyl selenocyanate (PhSeCN) in the presence of $CuCl_2$ or $CuBr_2$ and triethylamine[425].

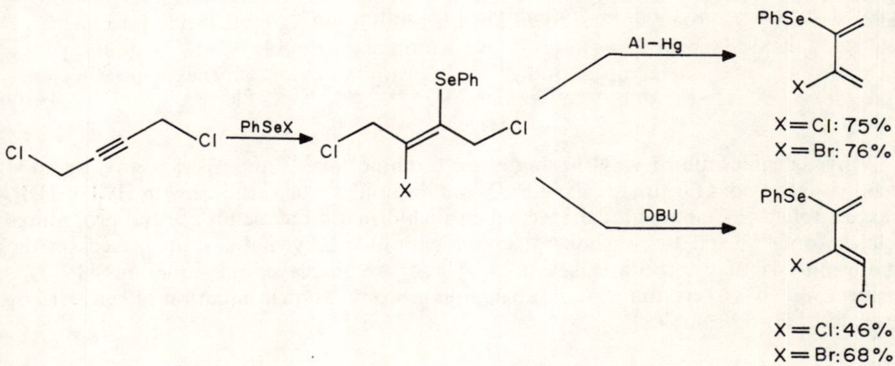

(114)

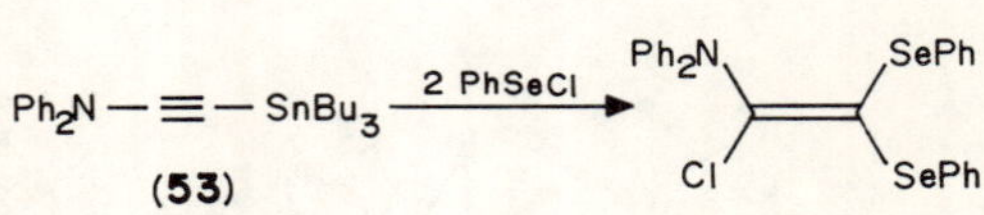

(115)

B. Allylic Halogenation

Hori and Sharpless reported that olefins are converted into rearranged allylic halides by their reaction with *N*-chlorosuccinimide (NCS) and catalytic amounts of PhSeCl or PhSeSePh[426] (e.g. equation 116). Allylic selenides are intermediates in this process and so they too produce allylic chlorides when treated with NCS[427].

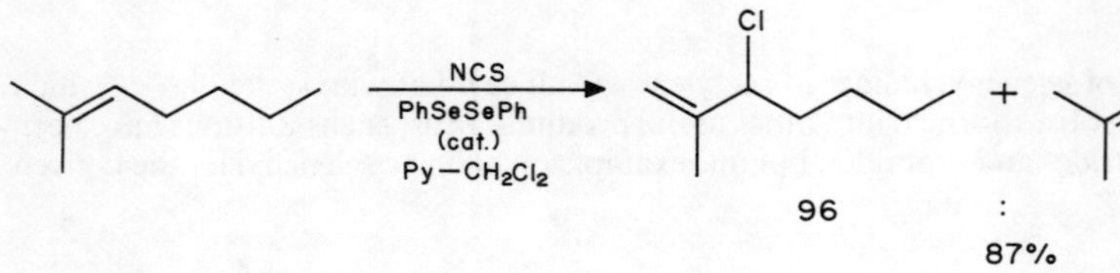

(116)

C. Halogenolysis of Selenides and Tellurides

The brominolysis of alkyl selenides with bromine[428–430] or NBS[428] was reported by Krief and coworkers. Secondary or tertiary alkyl selenides gave the best results[428] and the corresponding alkyl bromides were formed with inversion of configuration[430] (equation 117).

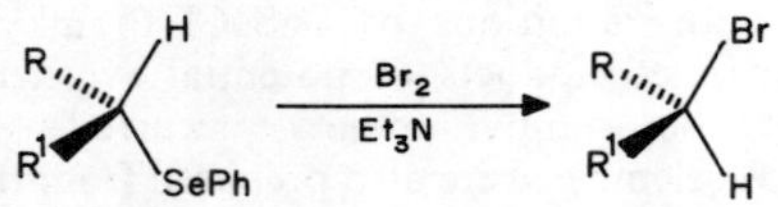

(117)

An alternative procedure consists of the alkylation of the selenide with methyl iodide, followed by displacement of the resulting selenonium salt with iodide ion[102,428,431] to provide primary alkyl iodides in high yield (equation 118[431]).

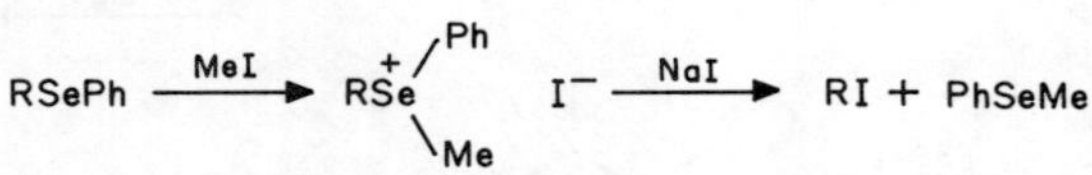

(118)

Vinyl selenides afford vinyl bromides on brominolysis[391] and alkyl phenyl or alkyl methyl selenoxides furnish alkyl chlorides and bromides when reacted with HCl or HBr, respectively[432]. A method for the vicinal *cis*-dichlorination of olefins[433], and procedures for the one-[428] and two-carbon[434] homologation of alkyl halides, are based on the halogenolysis of appropriate selenides. These are displayed in equations 119–121, respectively. It is interesting that the analogous dibromination in equation 119 gave *trans*- instead of *cis*-dibromides[433].

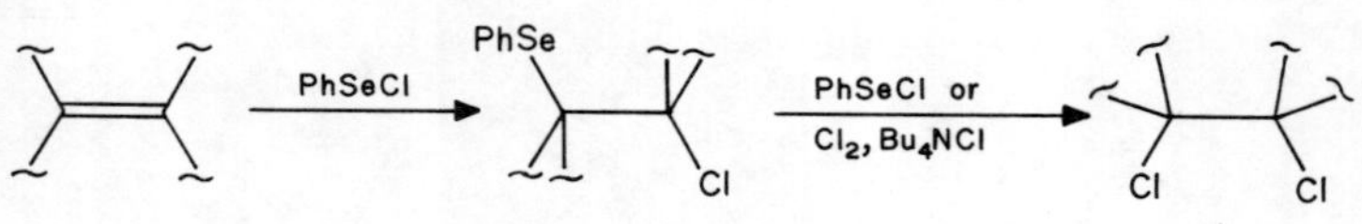

(119)

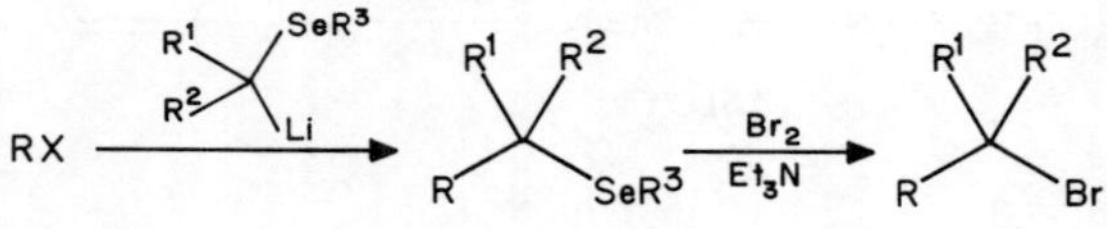

(120)

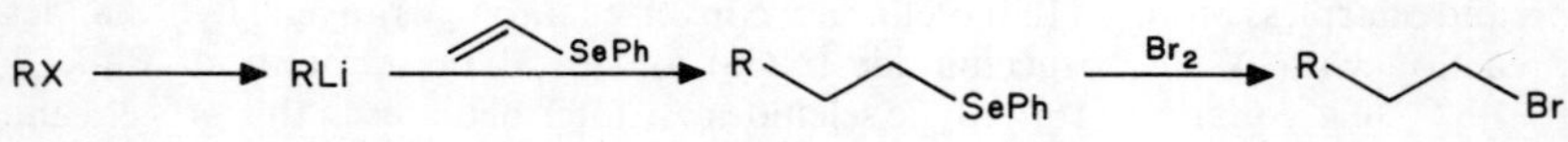

(121)

Alkyl halides are accessible from alkyl phenyl tellurides by halogenolysis with sulfuryl chloride, bromine and iodine, or via their telluronium salts, in a manner akin to equation 118[435]. Studies by Uemura and coworkers indicate that aryl-, vinyl- and alkyl-tellurium trihalides produce the corresponding organohalides with retention of configuration when they are photolysed in benzene[436,437] or when oxidized with *t*-butyl hydroperoxide[437,438]. The required vinyl- and alkyl-tellurium trihalide precursors were in turn obtained by the addition of $TeCl_4$ to acetylenes and olefins, respectively. Aryl-[439] and vinyl-tellurium[440] trichlorides were also converted into aryl and vinyl iodides and bromides by halogenolysis with iodine, bromine or NBS. In some cases, diaryltellurium dihalides can be used in place of the monoaryl trihalides[436,437,439].

Several illustrative examples are given in equations 122–124.

$$ArTeX_3 \xrightarrow[h\nu,\ C_6H_6]{t\text{-BuOOH} \text{ or}} ArX \qquad X = Cl, Br, I \tag{122}$$

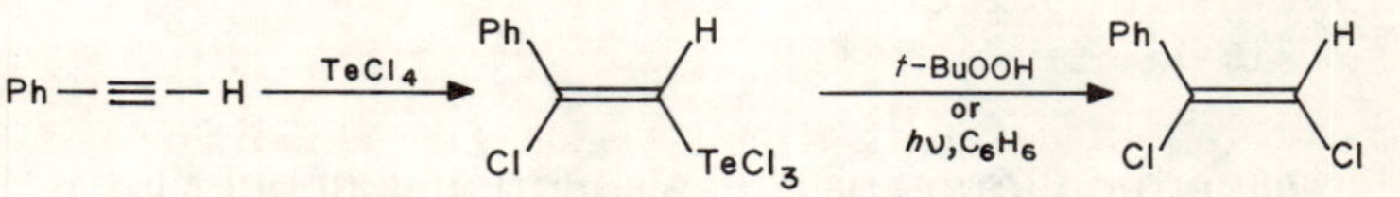

(123)

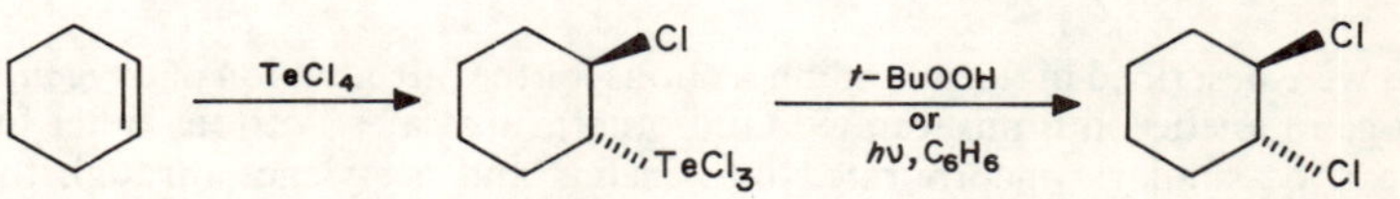

(124)

D. Other Halogenation Methods

Selenium and tellurium halides are capable of halogenating arenes and certain other hydrocarbons. Electron-rich arenes are preferentially chlorinated at the *para*-position by PhSeCl[441] (equation 125) while anthracene and other aromatics produce mixtures of mono- and di-halo products with tellurium(IV) halides[442]. Also, substituted cycloheptatrienes furnish the corresponding benzyl chlorides with $TeCl_4$[443].

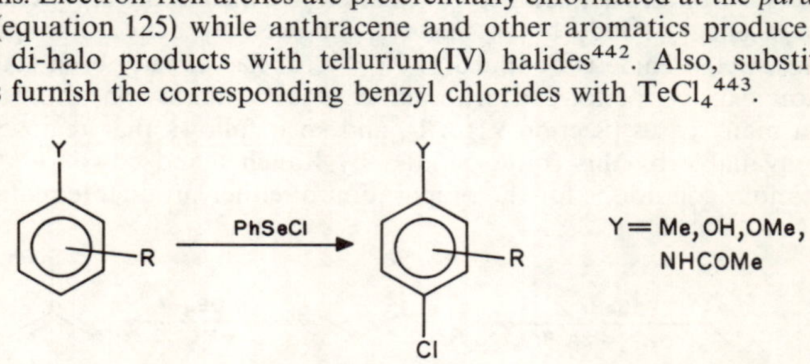

(125)

An interesting fluorination technique employing SeF_4 or its pyridine complex was reported by Olah and coworkers[444]. This reagent converts ketones into *gem*-difluoroalkanes, carboxylic acids or anhydrides into acyl fluorides and alcohols into fluoroalkanes (equation 126).

Several heteroaromatic hydrazines have been oxidized with selenium dioxide in the presence of HCl and Cu_2Cl_2 to produce poor to moderate yields of aryl chlorides[445] (equation 127).

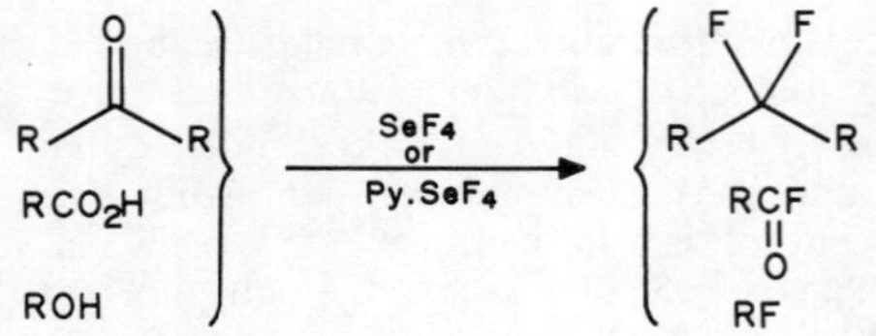

(126)

$$ArNHNH_2 \xrightarrow[HCl,\ Cu_2Cl_2]{SeO_2} ArCl \qquad (127)$$

Finally, a halogenolysis of a different type was reported by Detty[446] and Detty and Seidler[447], who noted that phenyl trimethylsilyl selenide (**54**) reacts with iodine (and other halogens) to generate trimethylsilyl iodide according to equation 128. The latter reagent can then be employed *in situ* for such purposes as the silylation of alcohols or the elimination of epoxides.

$$\underset{(\mathbf{54})}{2\ PhSeSiMe_3} + I_2 \longrightarrow 2\ Me_3SiI + PhSeSePh \qquad (128)$$

VIII. INTRODUCTION OF OTHER FUNCTIONAL GROUPS INTO OLEFINS AND ACETYLENES

Methods were described in the preceding sections for the introduction of oxygen, nitrogen and halogen moieties into unsaturated (and other) substrates. Various other functional groups can be similarly incorporated into olefins and acetylenes through the use of organo-selenium and -tellurium reagents. These methods are discussed below.

A. Vinyl Selenides

Vinyl selenides are key intermediates in the preparations of many types of compounds[388]. They are in turn available from selenium-free olefins and acetylenes. When the addition of a selenenyl halide to an olefin is followed by dehydrohalogenation, the corresponding vinyl selenide is obtained[408,448–450]. It has already been stated that the initial addition can be performed with either Markovnikov or anti-Markovnikov orientation in many cases (Section VII.A.1), and so it follows that regiosomeric vinyl selenides are available by this route. Studies by Raucher and coworkers[408,448] have indicated optimum conditions for the preparation of either product (equation 129).

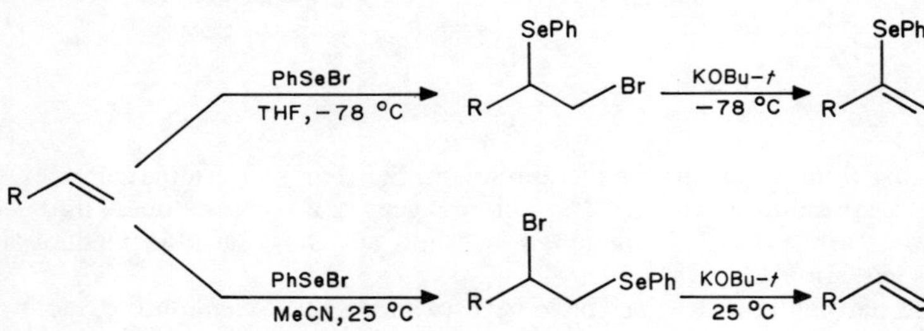

(129)

Fluorinated olefins furnish vinyl selenides in a similar manner[451,452], and in the case of 3,3,3-trifluoropropene[452] give the product **55**, which has interesting properties as a Michael acceptor (equation 130). Deactivated olefins such as enones[453,454] and also α,β-unsaturated esters[455], sulfones[454], nitriles[453] and nitro compounds[454] undergo additions with selenenyl halides, although in some cases very slowly and in poor yield[455]. Dehydrohalogenation of the adducts provides α-seleno enones[453,454] (e.g. equation 131) or other captodative olefins of current interest. An alternative route to α-seleno enones involves the electrochemical selenenylation and rearrangement of propargyl alcohols[456] (equation 132). These products served as useful dienophiles in cycloadditions. Alkynyl borates (**56**) react with PhSeCl, accompanied by the boron to carbon alkyl migration[457,458] shown in equation 133. The adducts **57** furnish α-seleno ketones (and therefore indirectly enones via selenoxide elimination) when oxidized with trimethylamine oxide[458]. This procedure thereby provides a useful synthesis of enones from acetylenes. The reactions of silylallenes such as **58** with PhSeCl afforded the silyl ketones **59** and the functionalized butadiene **60** as indicated in equation 134[459,460].

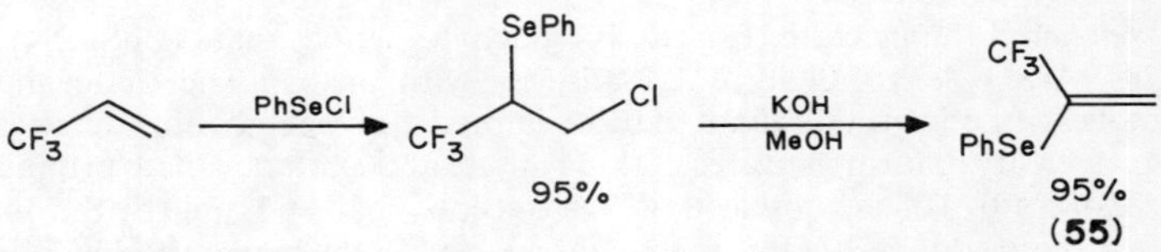

(130)

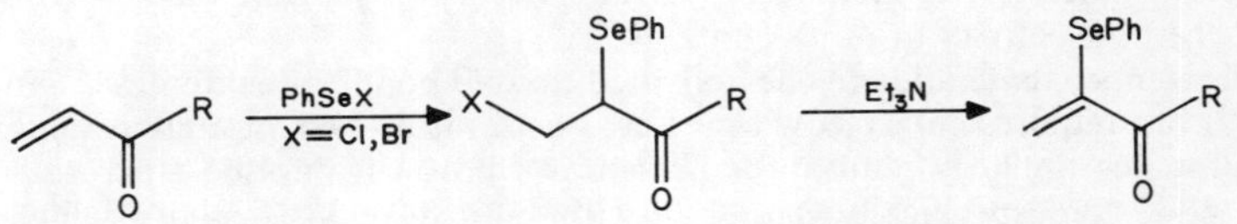

(131)

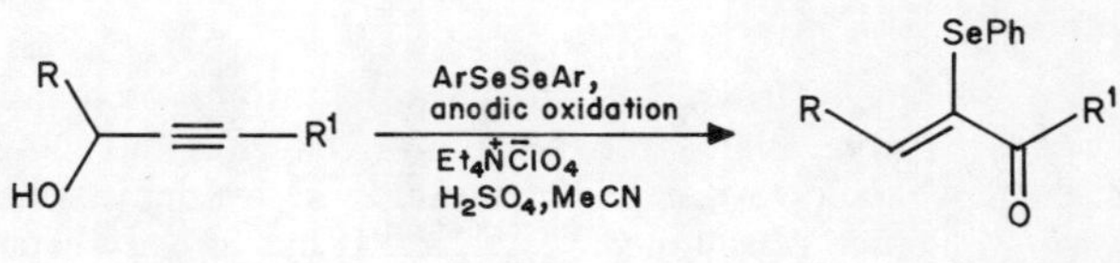

(132)

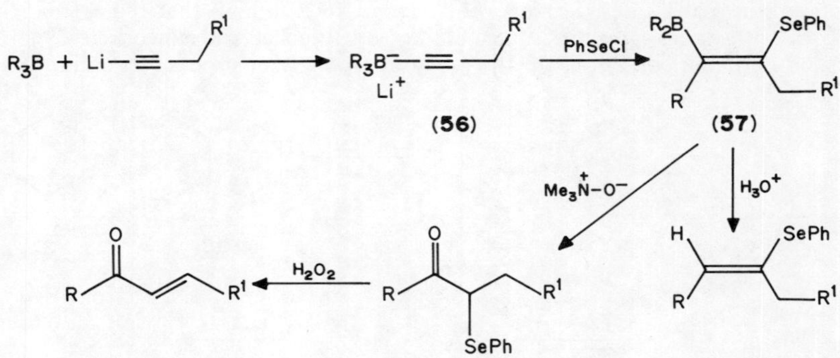

(133)

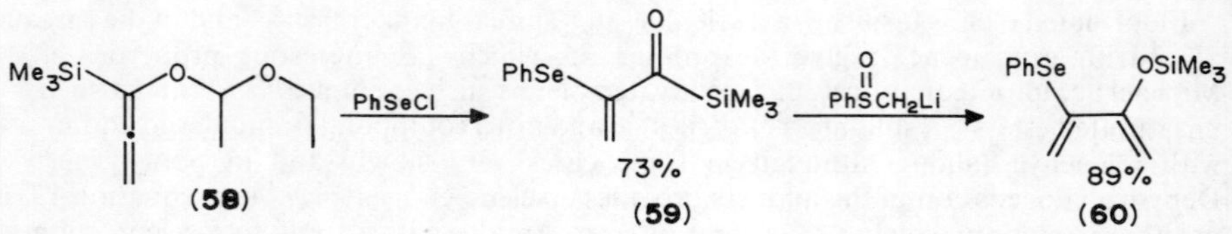

(58) (59) (60)

(134)

B. Electrophilic Additions of Selenenyl Pseudohalides

Many divalent selenium compounds of general structure RSeX, where X is a non-halide leaving group, add to olefins or acetylenes electrophilically, as in the case of the selenenyl halides themselves. This provides a convenient method for introducing the functionality X, along with the accompanying selenium residue, into the unsaturated substrate.

Benzeneselenenyl thiocyanate (PhSeSCN), phenyl selenocyanate (PhSeCN) and selenosulfonates ($ArSO_2SePh$) all form 1,2-adducts with olefins. The thiocyanate reacts rapidly, but its synthetic utility is limited by its unpredictable regio- and stereo-chemistry, and by its propensity to form mixtures of thiocyanate and isothiocyanate products[461,462]. Some degree of control of chemoselectivity was recently made possible by the observation that isothiocyanate adducts are formed in high yield with the reagent PhSeCl–$Hg(SCN)_2$ and long reaction times, while the thiocyanates are favoured with PhSeCl–NaSCN and brief reaction times[60]. The method of preparation of this reagent was also observed to affect its chemoselectivity in other contexts[463].

Phenyl selenocyanate adds to the activated double bonds of enamines[464] and ketene acetals[465], but requires catalysis with a Lewis acid in the case of unactivated olefins[81]. Selenosulfonates similarly require the presence of boron trifluoride etherate[70,71]. Both reagents add stereospecifically *anti*, but with limited regioselectivity which favours

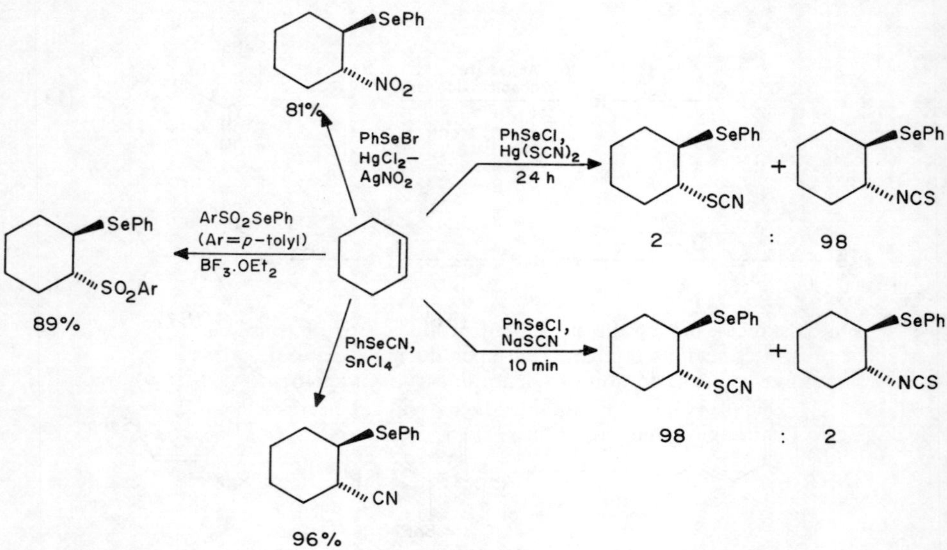

(135)

Markovnikov orientation. The free-radical additions of $ArSO_2SePh$ to olefins produce regioisomeric anti-Markovnikov adducts exclusively (see equation 136 and Section VIII.C).

Although the species $PhSeNO_2$ has not been isolated, its formal addition to olefins can be achieved with PhSeBr or PhSeCl and $AgNO_2$[79,80,466]. Further, the presence of $HgCl_2$ suppresses the competing formation of oxyselenenylated products[80,466]. *Anti* addition and Markovnikov orientation again prevail, and in all four types of additions subsequent selenoxide eliminations yield the corresponding vinyl isothiocyanates, nitriles, sulfones and nitro compounds, respectively (see equations 7 and 11). The results of the additions of all four selenenyl pseudohalides to cyclohexene are shown in equation 135.

C. Free-radical Selenosulfonation

1. Olefins

Gancarz and Kice[72,73] and Back and Collins[70,71] independently reported that selenosulfonates ($ArSO_2SePh$) add to olefins by a free-radical chain mechanism. The process may be initiated photochemically[72,73], or by pyrolysis in refluxing benzene or chloroform[70,71], in which case the addition of a radical initiator such as AIBN is beneficial[71]. Monosubstituted olefins react efficiently whereas more highly substituted olefins give lower yields of addition products. As expected in a free-radical process, the addition is non-stereospecific but highly regioselective in the antiMarkovnikov sense. The product β-phenylseleno sulfones are readily converted into vinyl sulfones in virtually quantitative yield by selenoxide elimination[70–73] (equation 136). The latter products are complementary regioisomers of those produced by electrophilic selenosulfonation and elimination (see preceding section). Several vinyl sulfones prepared from olefins by free-radical selenosulfonation have found applications as dienophiles in Diels–Alder reactions[74–76].

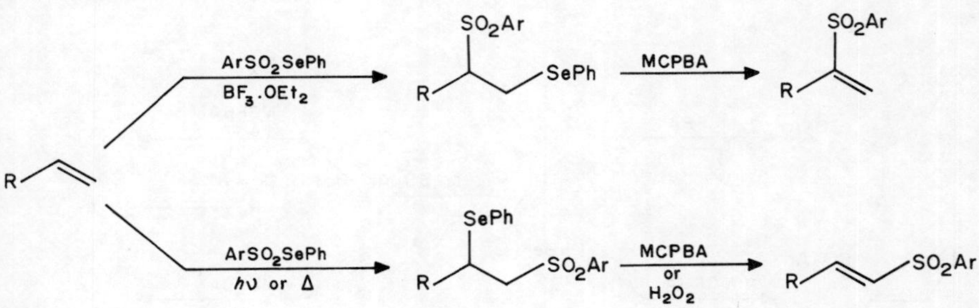

(136)

In some instances, dienes afford cyclized products (see Section IX.H)[73,467], whereas strained substrates such as β-pinene ring-open during selenosulfonation[467].

A related free-radical addition of selenothiocarboxylates (**61**) to olefins was recently reported[468]. The process serves to introduce vicinal phenylseleno and thiobenzoate groups into the double bond (equation 137).

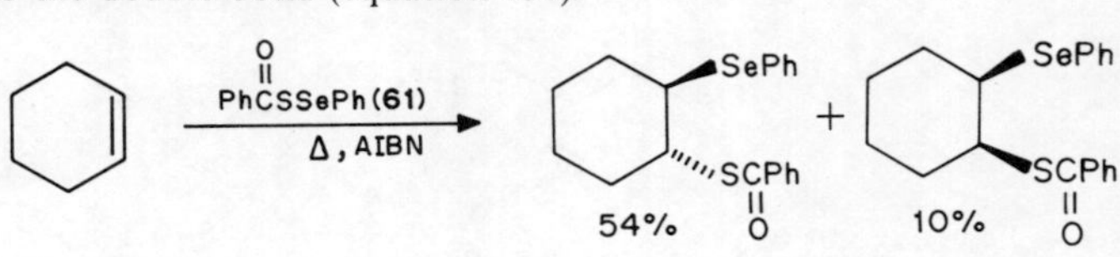

(137)

2. Allenes

Allenes undergo addition to the less substituted double bond, with attack by the sulfonyl radical occurring at the central sp-hybridized carbon atom[469,470]. The corresponding selenoxides react via a [2, 3] sigmatropic shift which provides high yields of β-sulfonyl allylic alcohols (equation 138).

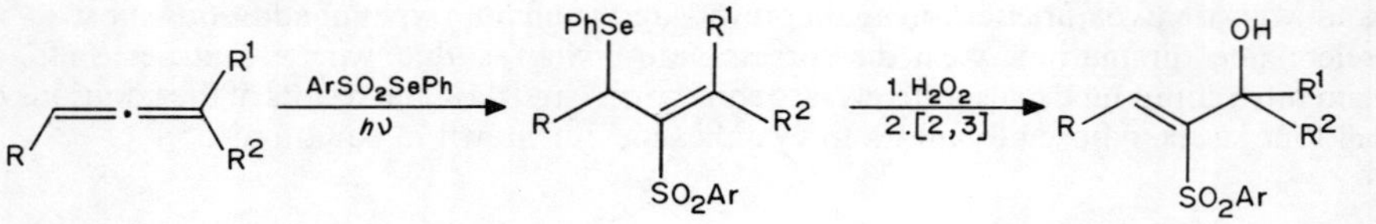

(138)

3. Acetylenes

The thermal selenosulfonation of terminal or disubstituted acetylenes[109–112] proceeds efficiently and affords 1, 2-adducts both regio- and, surprisingly, stereo-specifically, as shown in equation 139. The addition products can be converted into a variety of useful compounds which include acetylenic sulfones[109–111], allenic sulfones[112], β-keto sulfones or their ketal derivatives[111] and enamine sulfones[222] as depicted in equation 140. The adducts also undergo substitution of the PhSe group by organocuprates, as described in Section XI.B.1.

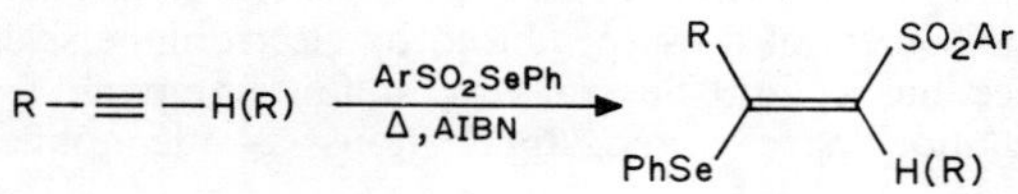

(139)

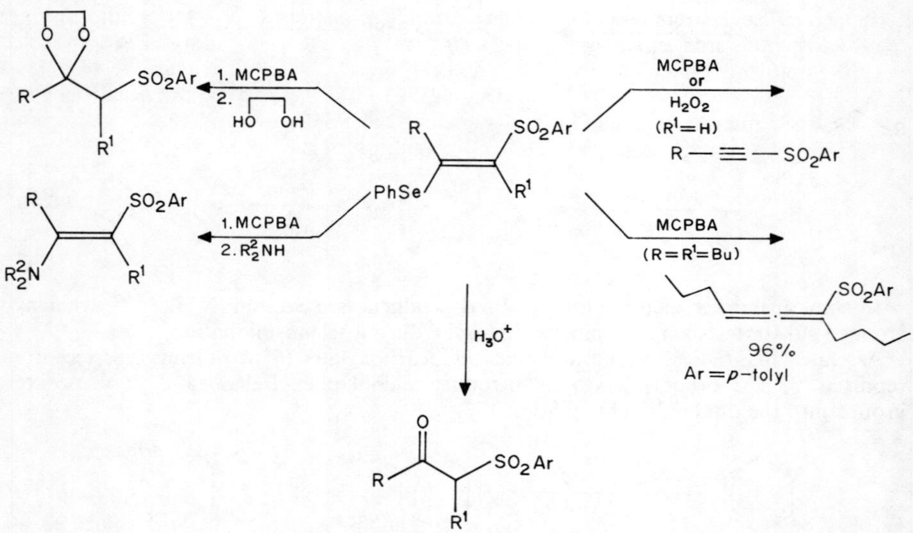

(140)

IX. CYCLIZATION

As seen in previous sections, the reactions of olefins with selenium electrophiles in the presence of external nucleophiles result in the addition of both the selenium residue and the nucleophile. However, when the nucleophile is part of the unsaturated substrate, the process becomes intramolecular and results in cyclization (equation 141). In most cases the cyclization step is followed by selenoxide elimination or reductive deselenization to afford unsaturated or saturated products, respectively. The term 'cyclofunctionalization' was introduced by Clive and coworkers[471,472] to describe such processes in general. The subject has recently been reviewed[473].

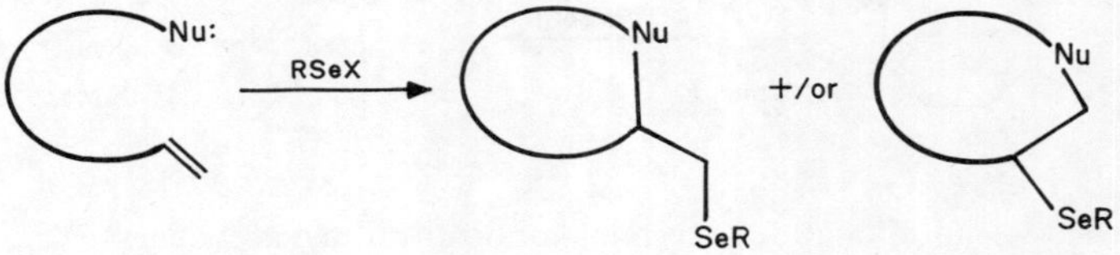

(141)

Certain selenium-induced cyclizations can be reversed with sodium in liquid ammonia[78,474] or with Me_3SiCl–NaI[475]. Cyclization can thus provide a method for the protection of the original olefin.

Although most cyclization procedures are mediated by electrophilic selenium species, several free-radical procedures have also appeared in recent years, as described in Section IX.H.

A. Unsaturated Alcohols

The cyclization of unsaturated alcohols was first reported independently by Corey and coworkers[476], Clive and coworkers[477] and Nicolaou and coworkers[478,479], and consists of an intramolecular version of oxyselenenylation. The method affords tetrahydrofurans, tetrahydropyrans and other cyclic ethers efficiently. Numerous examples are known, mostly employing PhSeCl as the electrophilic species[477–481]. Other effective reagents include PhSeBr[476,482], N-PSP (**34**)[286], TeO_2–LiCl in acetic acid[483], electrochemically generated selenium electrophiles from diselenides[484] and SeO_2 in an oxidative cyclization procedure[485]. A typical example is shown in equation 142[478].

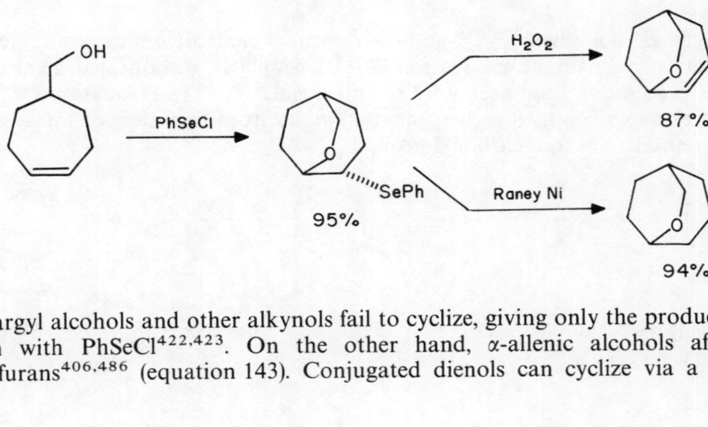

(142)

Propargyl alcohols and other alkynols fail to cyclize, giving only the products of 1,2-addition with PhSeCl[422,423]. On the other hand, α-allenic alcohols afford 2,5-dihydrofurans[406,486] (equation 143). Conjugated dienols can cyclize via a conjugate addition mechanism[177] (equation 144), and *o*-alkenylphenols produce benzofurans or benzopyrans when treated with PhSeCl[471] (equation 145).

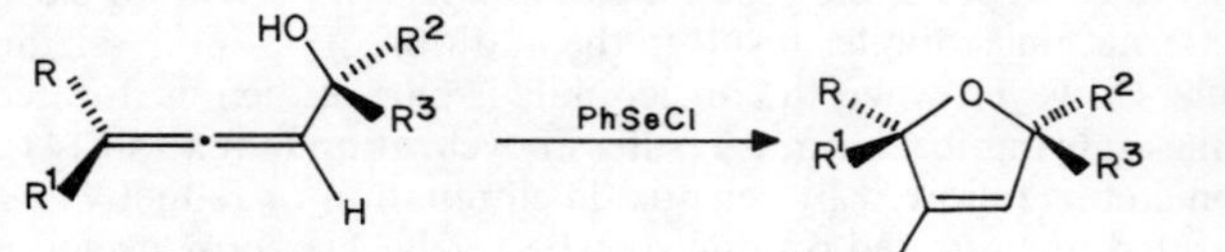

(143)

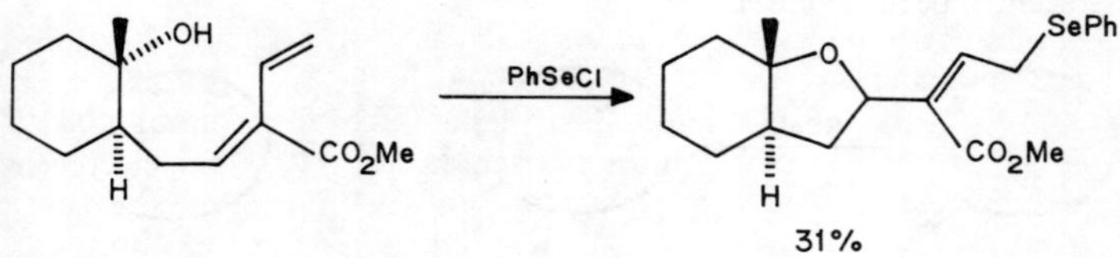

(144)

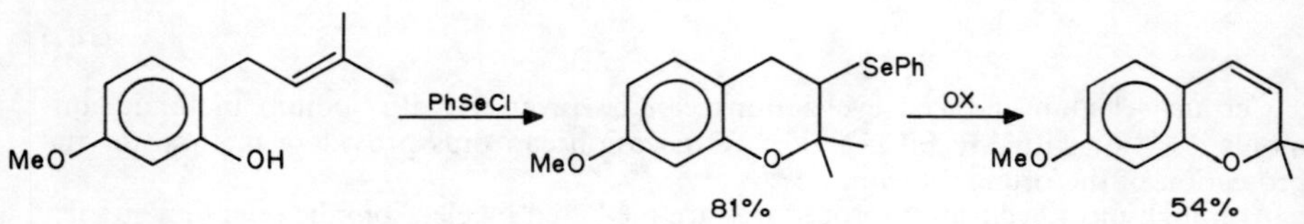

(145)

When the cyclization is followed by cleavage of the original C—O bond, the net effect is the transposition of an oxygen function to the initial olefinic site[487] (equation 146). The method thus provides an alternative to oxyselenenylation of the double bond.

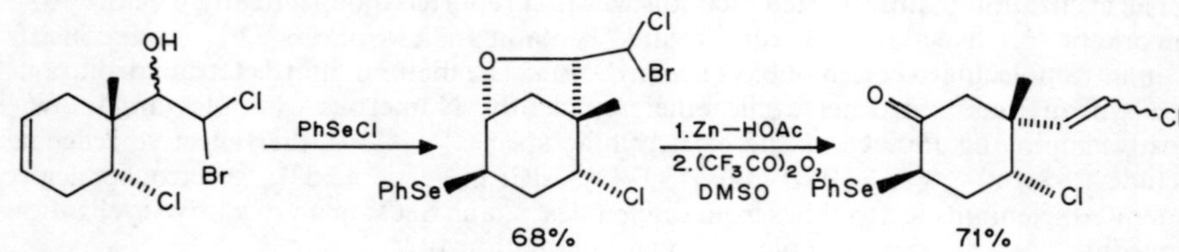

(146)

Hydroxyalkyl enol ethers[378,488] and hydroxyalkyl enamides[379] cyclize to ketals and amino ethers respectively (e.g. equation 147[378]). Similarly, unsaturated hemiacetals or hemiketals produce cyclized acetals[489] or spiroketals[490,491] (e.g. equation 148[491]). The required precursors can in turn be generated *in situ* from the inter- or intra-molecular addition of an alcohol to a carbonyl group.

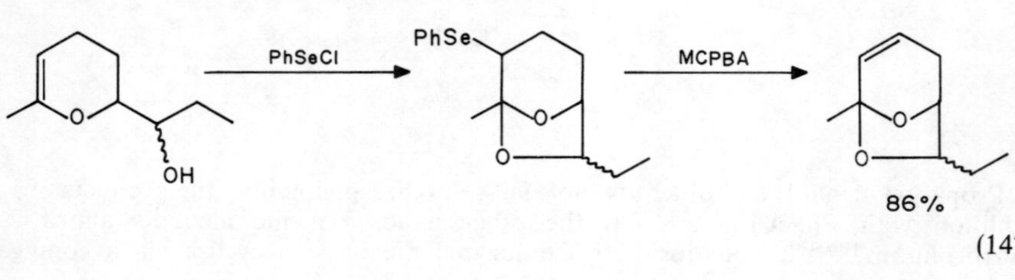

(147)

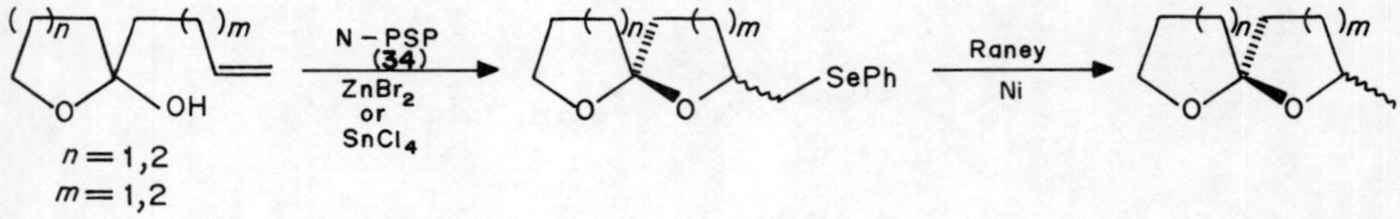

(148)

B. Unsaturated Carboxylic Acids

The selenium-induced cyclization of unsaturated carboxylic acids to lactones was first reported by de Moura Campos and Petragnani[492]. As in the preparation of cyclic ethers, PhSeCl again serves as the most popular electrophile[479,493–497], although N-PSP (**34**) is also convenient[285,286,496,498]. Aryltellurium trichlorides[492,499] and electrochemically generated selenium electrophiles[206] were employed in several examples. Extensive work by Clive and coworkers[472] and Nicolaou and coworkers[474] has demonstrated that γ-lactones are formed preferentially to δ-lactones, and that the latter are preferred to larger ring sizes. An illustrative example is provided in equation 149[472]. Larger rings are accessible, however, and several examples of macrolide ring closures with N-PSP or its succinimide analog were reported[285].

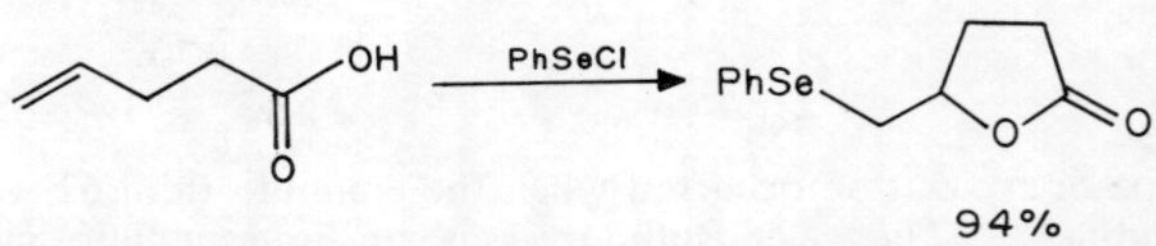

(149)

Both allenic[500] and acetylenic[498] carboxylic acids afford γ-lactones (equations 150 and 151), the latter in contrast to the corresponding alcohols, which failed to cyclize. Silyl esters can be used instead of free carboxylic acids in cyclizations promoted by PhSeCl[501]. An oxygen transposition procedure related to equation 146 is based on lactonization followed by lactone hydrolysis[104]. Anomalous reactions were observed with β,γ-unsaturated carboxylic acids containing β- but not γ-substituents. These compounds undergo decarboxylation instead of lactonization[495] when treated with PhSeCl (equation 152).

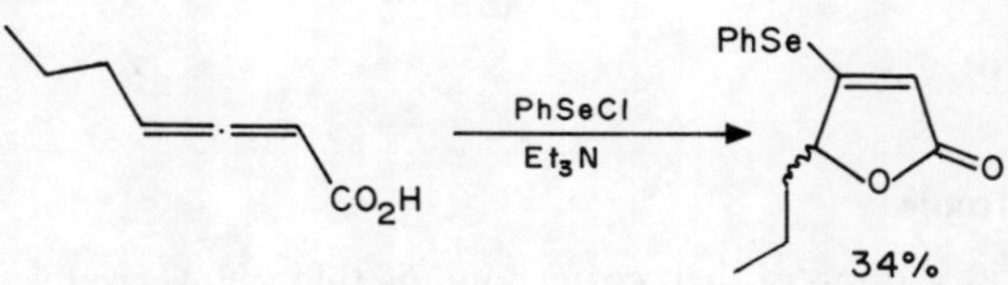

(150)

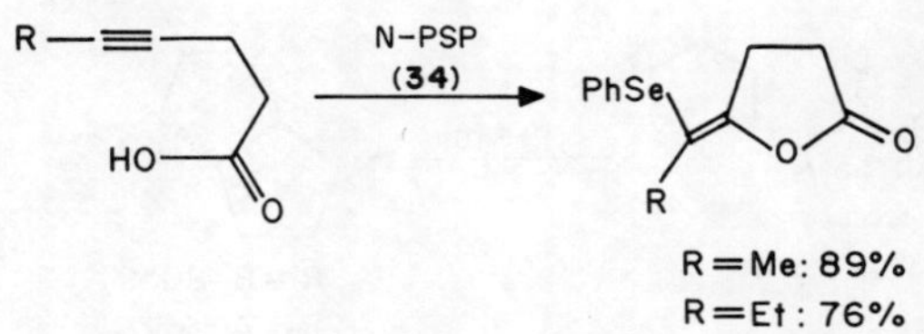

(151)

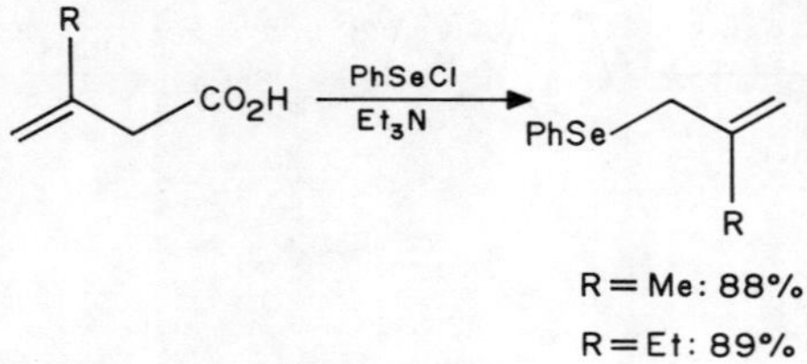

(152)

C. Unsaturated Nitrogen Compounds

Several unsaturated urethanes[502–504] and amides[505] were cyclized with PhSeCl[502,503,505] or N-PSP (**34**)[504] (e.g. equation 153[503]). Difficulties have been encountered with free amines[503], although isolated examples of the cyclization of a secondary amine[506] and an aniline[56] are known.

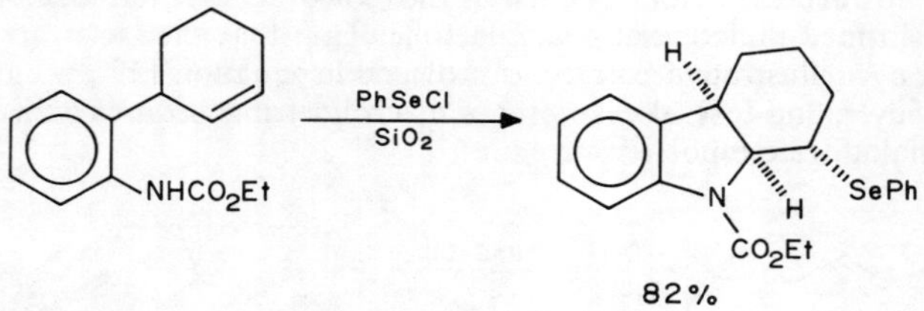

(153)

A different type of cyclization occurred when the enamido thiol **62** was treated with PhSeBr in pyridine[507]. The selenosulfide **63** is an intermediate in this reaction (equation 154).

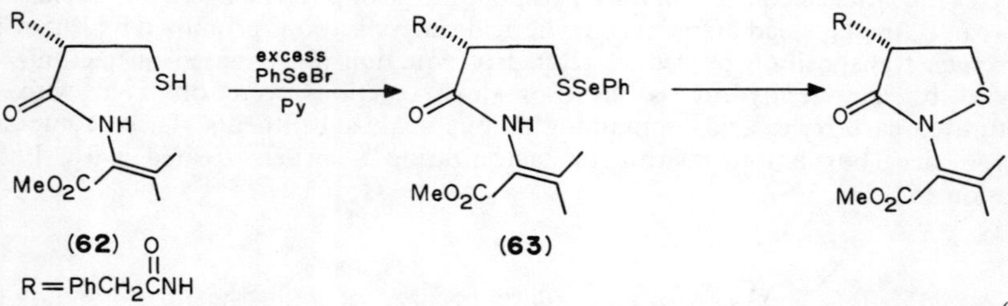

(154)

D. Unsaturated Thiols

Several unsaturated thiols or their corresponding thioacetates were cyclized[77,78,508] as in equation 155[508].

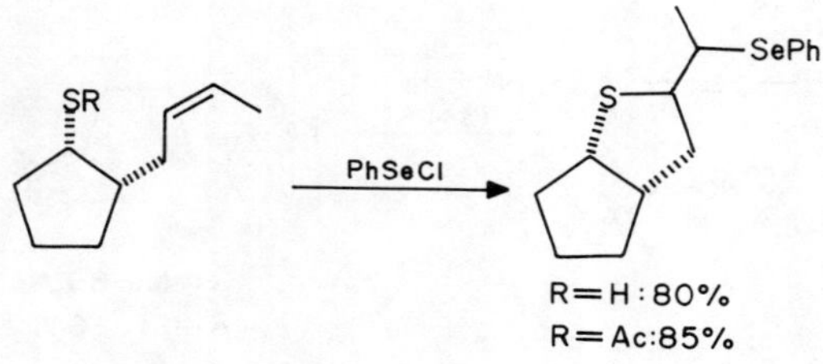

(155)

E. Allenic Phosphonates

Allenic phosphonates afford cyclic phosphonates according to equation 156[509,510].

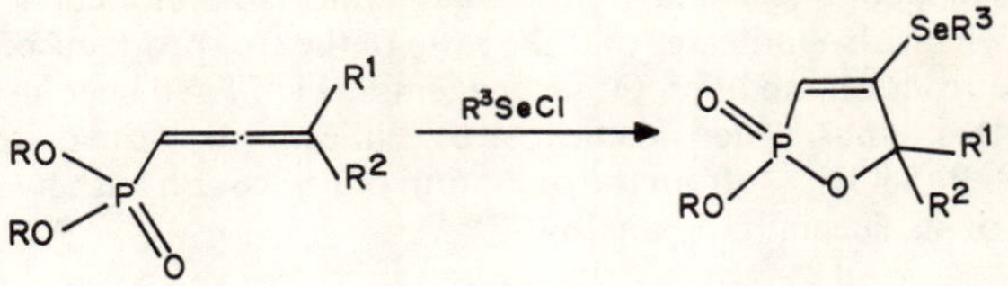

(156)

F. Unsaturated β-Dicarbonyl Compounds

Carbocyclic ring closures of unsaturated β-keto esters[511–516] with N-PSP (**34**) and catalysts such as $SnCl_4$[512–516], p-toluenesulfonic acid[511,512], iodine[511] and ZnI_2[511] were developed by Ley and coworkers for a number of synthetic applications. Cyclization occurs either through the ketone oxygen or the enolic carbon. The former products are favoured under the conditions of kinetic control, but are converted into the corresponding carbocycles on further equilibration. An example is provided in equation 157[515]. Benzeneselenenyl chloride (PhSeCl) with $AlCl_3$ catalyst also serve as an effective reagent combination for this type of transformation[517].

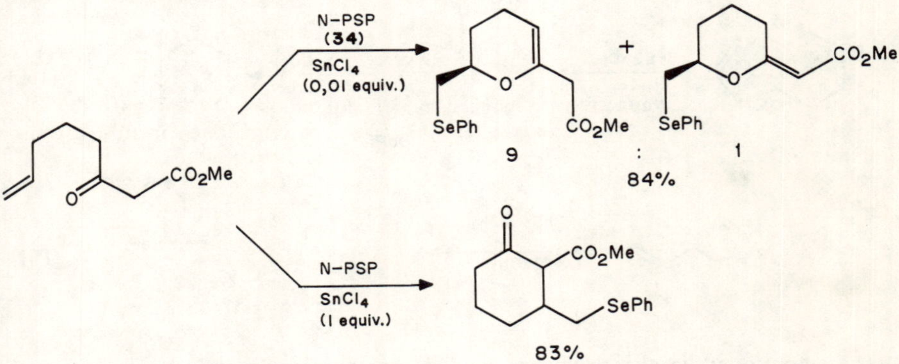

(157)

Unsaturated β-diketones can be similarly cyclized to either the corresponding cyclic enol ethers or carbocycles with N-PSP[511] or benzeneselenenyl hexafluorophosphate ($PhSe^+ PF_6^-$)[518]. Equilibration was promoted with $SnCl_4$[518].

The acid-catalysed addition of the β-keto ester moiety in **64** to the vinyl selenide provides the basis for a different cyclization as shown in equation 158[519].

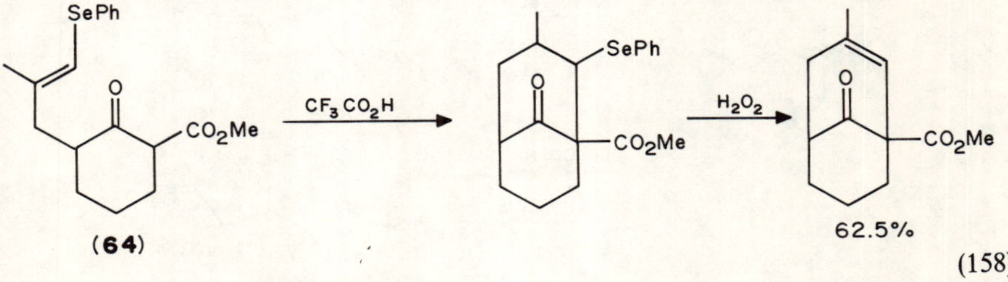

(158)

G. Dienes

Dienes can be converted into cyclic ethers with PhSeCN or KSeCN in the presence of $CuCl_2$ in aqueous or alcoholic media[520–522]. The example of cycloocta-1,4-diene is shown in equation 159, which also indicates that the ratio of the two products **65** and **66** is highly dependent on the reaction medium. Other reagents which have been used in the similar cyclization of this and other dienes are PhSeCl in aqueous acetonitrile[523], $PhSeOCOCF_3$[84], PhSeOH[524] (from the oxidation of PhSePh with hydrogen peroxide) and N-PSP (**34**) or its succinimide analog[285,286].

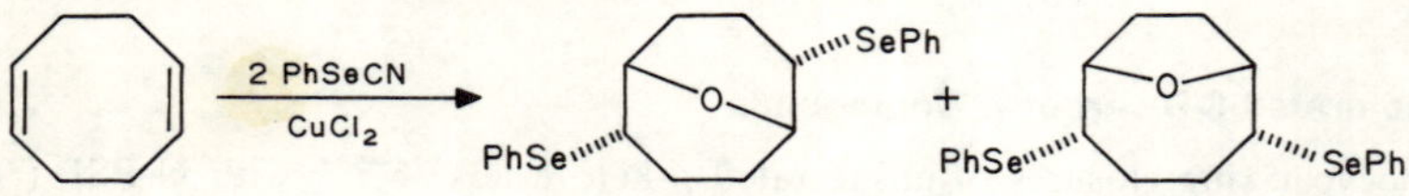

(159)

(**65**) (**66**)

95 : 5 (in MeOH)

0 : 100 (in THF–H_2O)

Carbocyclic products containing acetoxy and acetamido substituents were prepared from dienes using PhSeCl in acetic acid[299,525], and PhSeCl[526] or PhSeI[527] in acetonitrile, respectively (e.g. equations 160[525] and 161[527]).

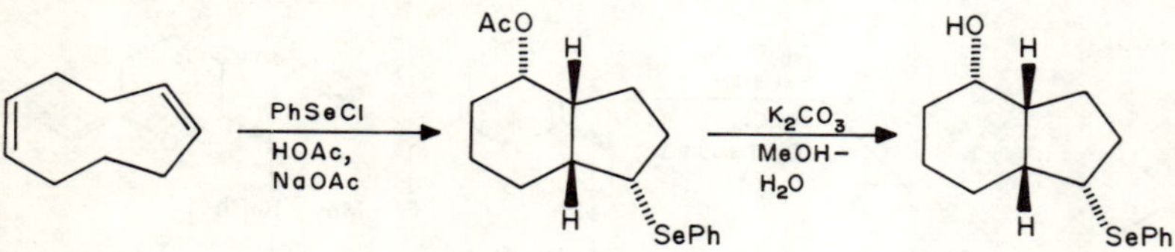

(160)

68%

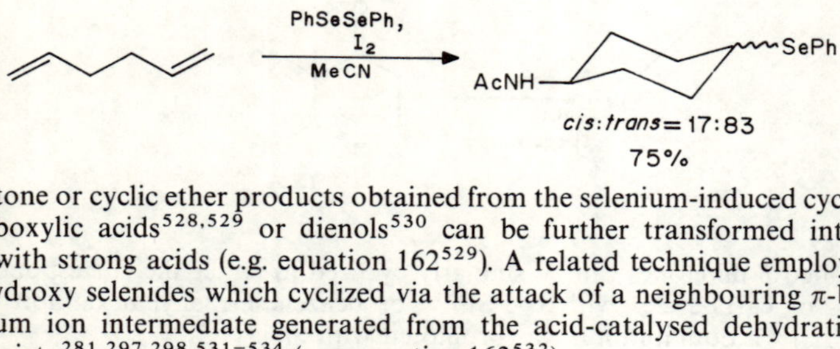

(161)

cis:*trans* = 17:83

75%

The lactone or cyclic ether products obtained from the selenium-induced cyclization of diene carboxylic acids[528,529] or dienols[530] can be further transformed into bicyclic products with strong acids (e.g. equation 162[529]). A related technique employs unsaturated β-hydroxy selenides which cyclized via the attack of a neighbouring π-bond on a seleniranium ion intermediate generated from the acid-catalysed dehydration of the alcohol moiety[281,297,298,531–534] (e.g. equation 163[532]).

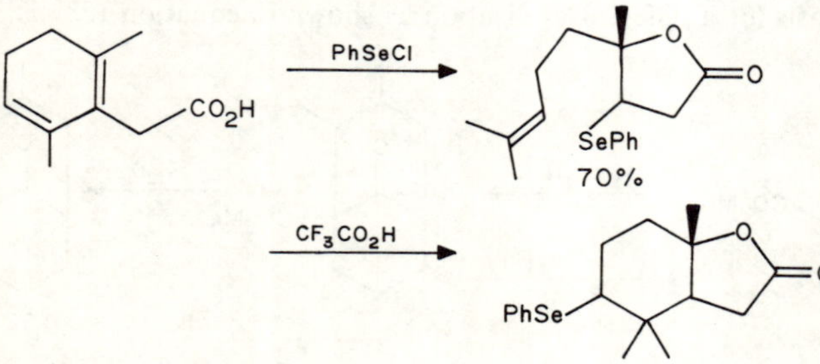

(162)

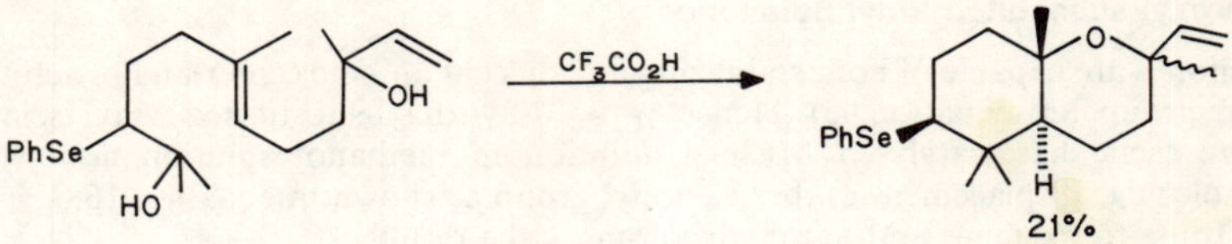

(163)

H. Free-radical Cyclizations

Carbon-centered radicals formed by the free-radical deselenization of alkyl phenyl selenides with tin hydrides (see Section XIV) undergo inter-[535] or intra-molecular capture by olefins[196,536–538], allenes[539] and acetylenes[536,540,541]. Some recent examples are illustrated in equations 164[536], 165[539] and 166[540].

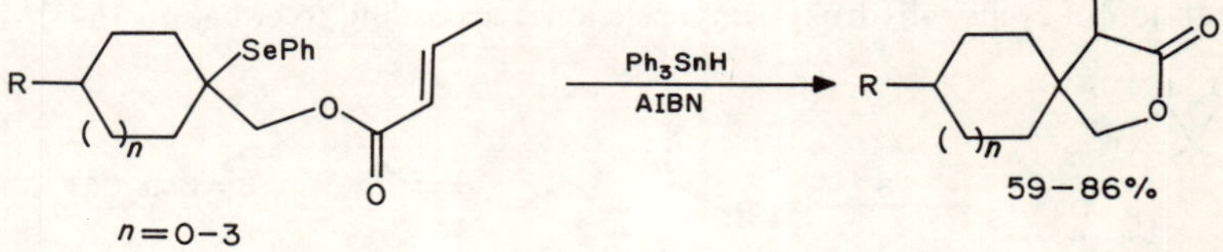

(164)

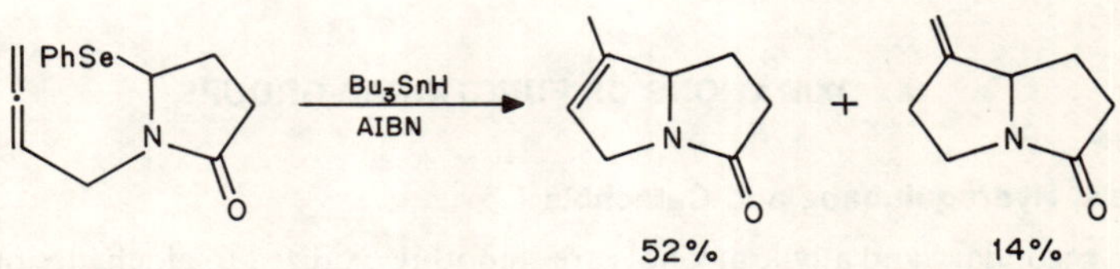

(165)

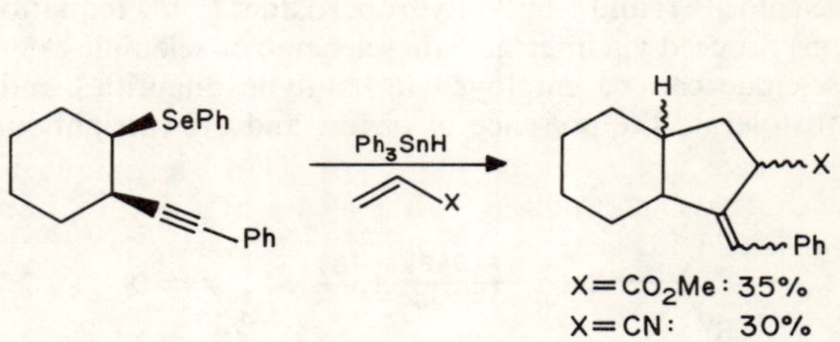

(166)

The free-radical selenosulfonation of dienes affords carbocycles[73,467] together with mono- and di-1, 2-addition products. Cyclization is enhanced by increasing the dilution (equation 167[467]).

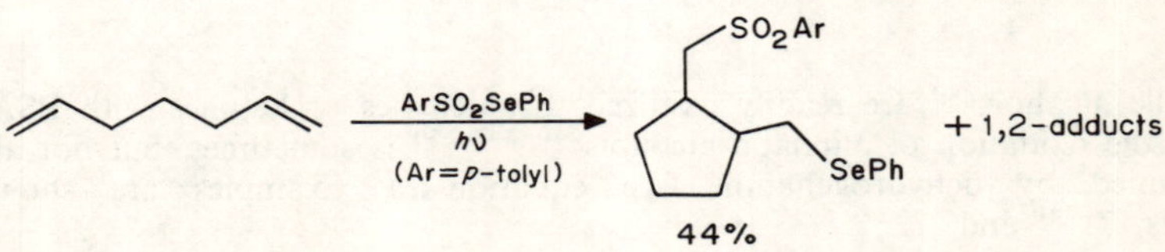

(167)

I. Hydroxyl-substituted Vinyl Selenones

Selenones are capable of both stabilizing an adjacent anionic center and of acting as a leaving group (cf. equation 95). Hence γ- or δ-hydroxyl-substituted vinyl selenones undergo facile base-catalysed Michael addition in methanol solution, followed by intramolecular displacement of the selenonyl group as shown in equation 168[542,543]. A useful route to oxetanes and tetrahydrofurans is the result.

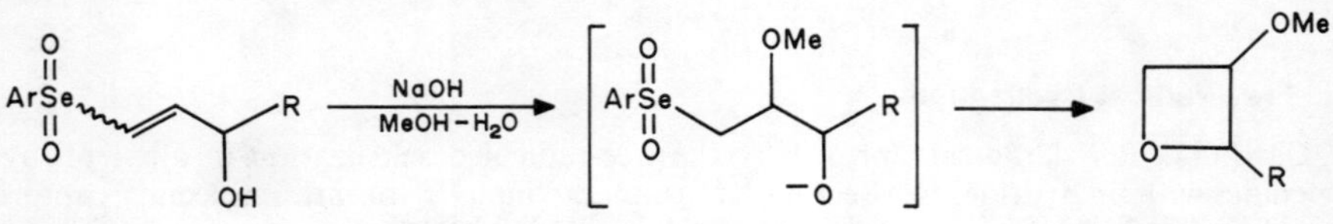

(168)

It is appropriate to mention that the above properties of selenones also permit the fragmentation of cyclic γ-hydroxyl vinyl selenones according to equation 169[543,544].

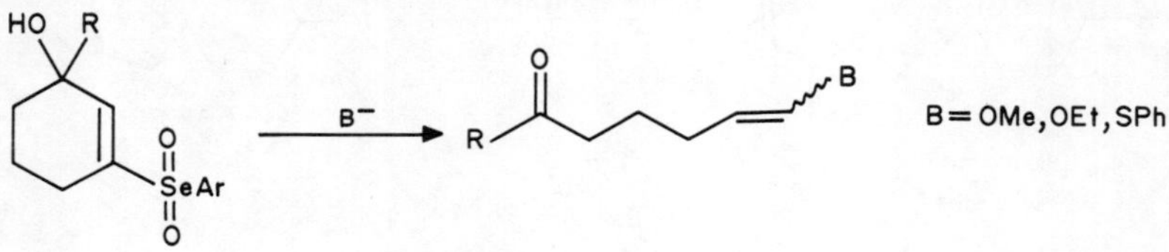

(169)

X. OXIDATIONS OF FUNCTIONAL GROUPS

A. Alcohols, Hydroquinones and Catechols

Primary, secondary and allylic alcohols are smoothly oxidized to aldehydes or ketones with dimesityl diselenide (**67**) and *t*-butyl hydroperoxide[545–547] (equation 170). Presumably, these reactions proceed via intermediate selenenic or seleninic esters of the starting alcohols. The diselenide can be employed in catalytic quantities and the method is sufficiently mild to tolerate the presence of olefins and even neighbouring sulfide and selenide groups.

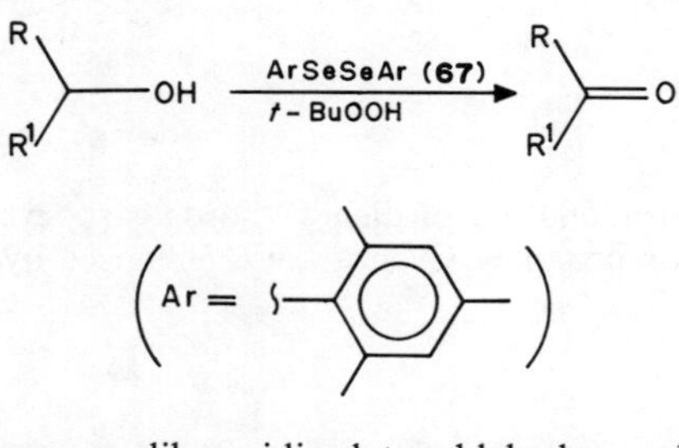

(170)

Benzylic alcohols[240] are readily oxidized to aldehydes or ketones with BSA (**27**), whereas the oxidation of aliphatic alcohols[240–242,250] is sometimes, but not always, accompanied by dehydrogenation (see equation 48). Examples are shown in equations 171[240] and 172[250].

$$PhCH_2OH \xrightarrow{BSA} \underset{99.5\%}{PhCHO} \qquad (171)$$

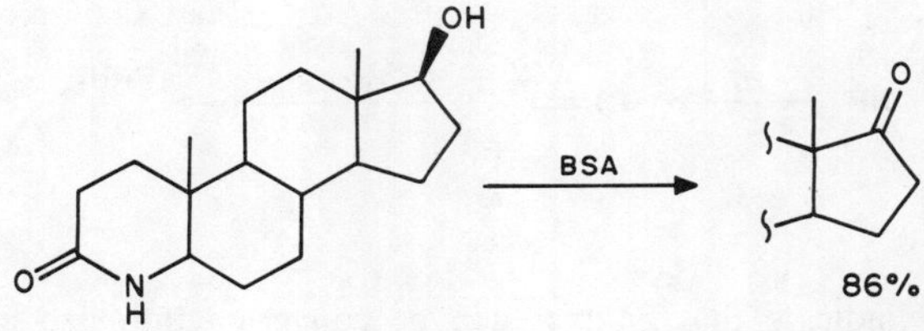

(172)

Other reagents which are reported to oxidize alcohols to carbonyl compounds are dimethyl selenoxide (**38**)[351], di(*p*-methoxyphenyl) selenoxide (**68**) in the presence of selenium dioxide[548], di(*p*-methoxyphenyl) tellurone (**69**)[549], the species **71** which are generated *in situ* from selenides and NCS[550] and a polymer-bound seleninic acid[361].

Hydroquinones and catechols afford *p*- or *o*-quinones, respectively, without accompanying oxygenation (see Section V.E) on oxidation with the telluroxide **70**[551] or with selenoxides **68**[552] and **52**[553]. The last reagent has also been employed in the conversion of adrenaline to adrenochrome[554] and ascorbic acid to its dehydro derivative[555].

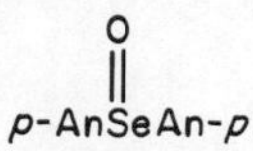

(**68**)

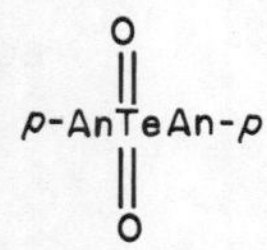

(**69**)

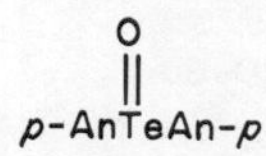

(**70**)

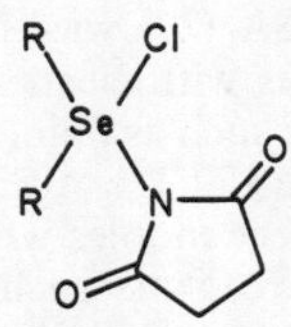

(**71**)

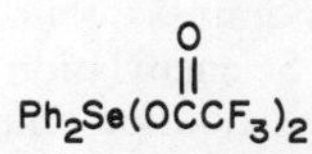

(**72**)

B. Nitrogen Compounds

1. Amines

Amines can be oxidized to imines with a variety of selenium oxidants. These include BSA (**27**)[556–558], diphenylselenium bistrifluoroacetate (**72**)[559] and benzeneseleninyl chloride (**30**)[560]. The imines so obtained can be used in the preparation of other compounds. Thus, hydrolysis affords ketones[556,560] (equation 173), further oxidation provides nitriles[556,560] (equation 174) and the addition of NaCN or Me_3SiCN furnishes α-cyanoamines[557] (equation 175).

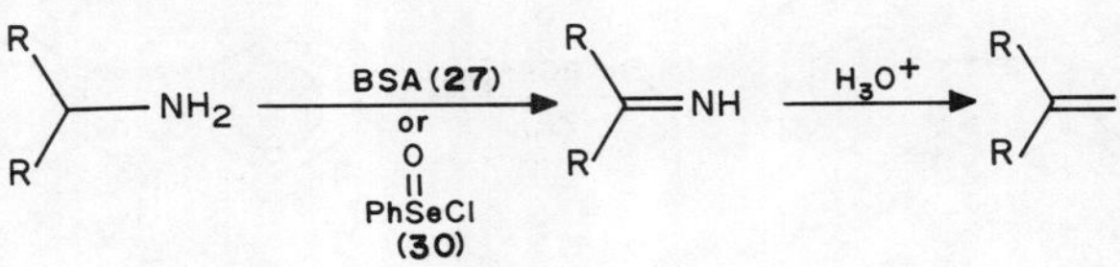

(173)

$$PhCH_2NH_2 \xrightarrow[\mathbf{27}]{\mathbf{30}\text{ or}} PhCN \quad 85\text{–}96\% \tag{174}$$

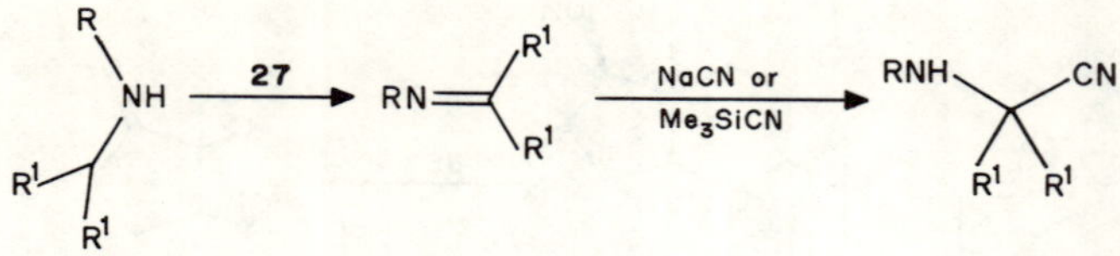

(175)

Other studies indicate that anilines can be converted into azo compounds with Ph_2SeCl_2, or with diphenyl selenoxide (**52**) and $ZnCl_2$[561]. Tertiary amines afford *N*-oxides when treated with selenoxides[562]. In a different approach, the cephalosporin selenenamide **73** was used to prepare the methoxy derivative **74** by oxidation with MnO_2 and methanolysis[563] (equation 176).

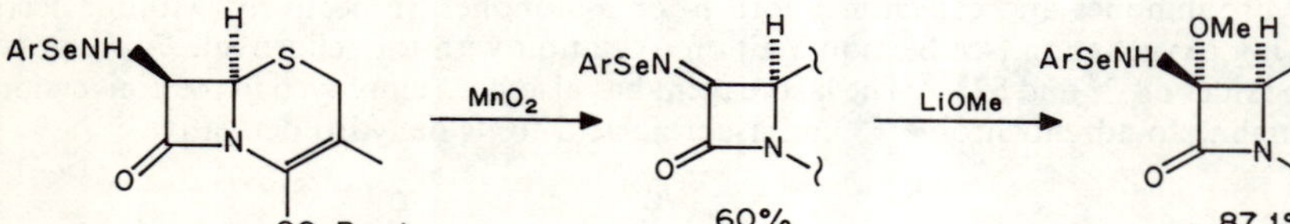

(73) 60% 87.1% (74)

(176)

2. Indolines and indoles

The oxidation of indolines to indoles with BSA (**27**) was studied by Barton and coworkers[558,564–566]. Selenenylation of the indoles with electrophilic selenium byproducts accompanied the oxidation unless scavengers such as indole or dihydropyran were added to the reaction mixture[565,566] (e.g. equation 177[565]). Alternatively, the selenenylated byproducts were reductively deselenized to free indoles with nickel boride[564,566]. Various indoles undergo further oxidation with reagents such as **72**[559] or selenium dioxide[236,334,343], as in the example shown in equation 178[559].

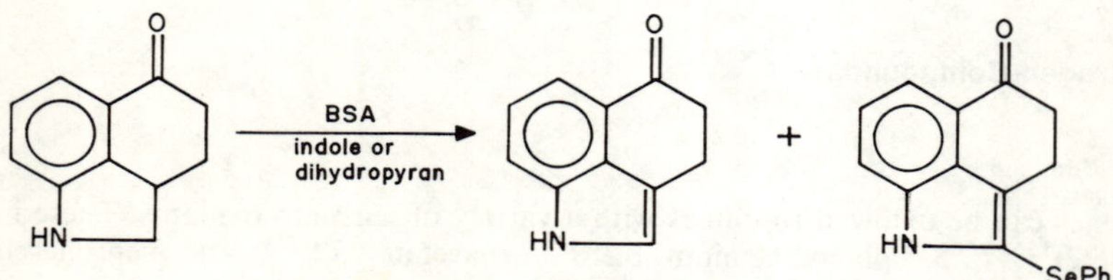

(177)

(178)

3. Hydrazines

Hydrazines are oxidized to a wide array of products with selenium and tellurium oxidants, often in a preparatively useful manner. Hydrazine hydrate generates the hydrogenating agent diimide when treated with $PhSeO_2H$[249,567] or elemental selenium in the presence of oxygen[568]. Monoarylhydrazines produce mixtures of arenes and aryl phenyl selenides with BSA or $PhSeO_2H$[249,567,569,570] (equation 179). Arenes and tellurides are formed similarly when the telluroxide **70**[571] or the reagent TeO_2–LiCl–HOAc[572] is employed as the oxidant.

$$ArNHNH_2 \xrightarrow[\text{or } PhSeO_2H]{BSA} ArH + ArSePh \qquad (179)$$

Hydrazides are smoothly oxidized to 1,2-diacylhydrazines with diphenyl selenoxide (**52**)[573], the telluroxide **70**[571] or $PhSeO_2H$[567,574] if the seleninic acid is introduced by slow addition. On the other hand, if the hydrazide and triphenylphosphine are added slowly to the seleninic acid, then diacylhydrazine formation is suppressed and selenoesters are obtained in good yield[567,574] (equation 180). Similarly, sulfonhydrazides produced selenosulfonates in excellent yield without the need for triphenylphosphine[575] (equation 181). Hydrazides bearing γ- or δ-hydroxy substituents underwent nearly quantitative cyclization to lactones when oxidized with $PhSeO_2H$[472,567] (equation 182), and phthalhydrazide formed the dimer **75** when treated with the selenoxide **52** in acetic acid[576] (equation 183).

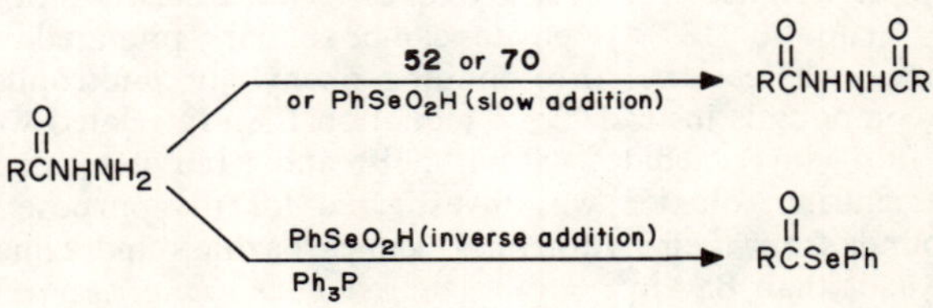

(180)

$$RSO_2NHNH_2 \xrightarrow{PhSeO_2H} \underset{65-100\%}{RSO_2SePh} \qquad (181)$$

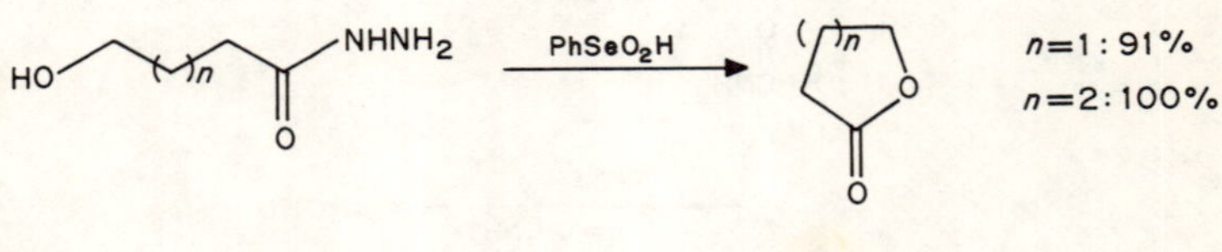

(182)

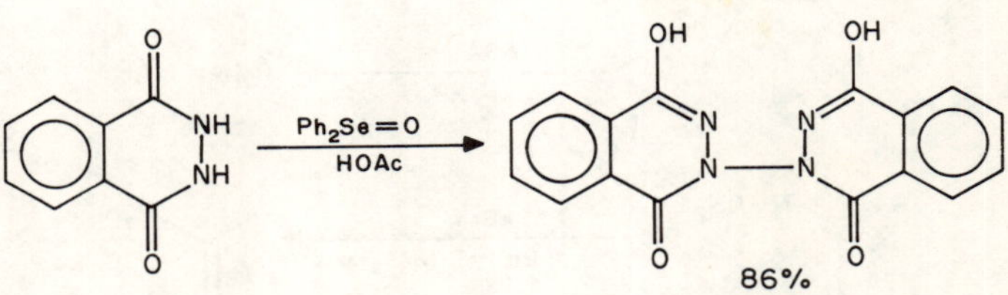

(183)

86%
(**75**)

1,1-Disubstituted hydrazines dimerize to tetrazenes when treated with $PhSeO_2H$[577] or SeO_2[577] (equation 184). The products were obtained in generally high yield unless an *N*-aryl or *N*-sulfonyl substituent was present. 1,2-Disubstituted hydrazines and hydroxylamines produced azo compounds[249,567,569,570], including Cookson's reagent[578], and nitroso compounds[551,569,570,571], respectively, on oxidation with BSA, $PhSeO_2H$ or the telluroxide **70** (equations 185, 186).

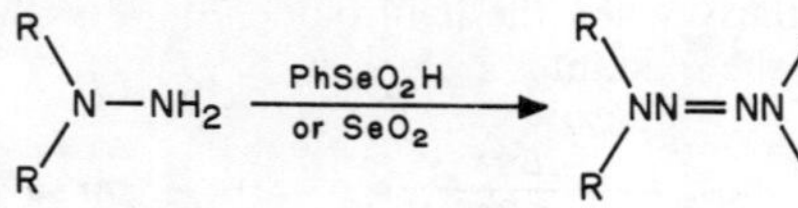
(184)

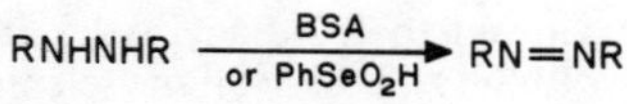
(185)

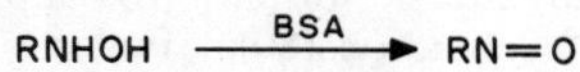
(186)

4. Hydrazones, oximes and semicarbazones

Barton and coworkers reported that BSA (**27**) is an exceptionally effective reagent for the regeneration of ketones from their oximes, semicarbazones or aryl- or tosyl-hydrazones[570,579] (equation 187). Aldehydes can be similarly prepared from their oximes or tosylhydrazones, whereas the corresponding phenyl- or *p*-nitrophenyl-hydrazones afford keto azo compounds instead[569,570] (equation 188). In related work, hydrazones were converted into vinyl selenides with PhSeBr and *t*-butyltetramethylguanidine[580] (equation 189). Tellurium dioxide was investigated for the purpose of regenerating carbonyl compounds from their hydrazones, semicarbazones and azines, but generally provides lower yields than BSA[572].

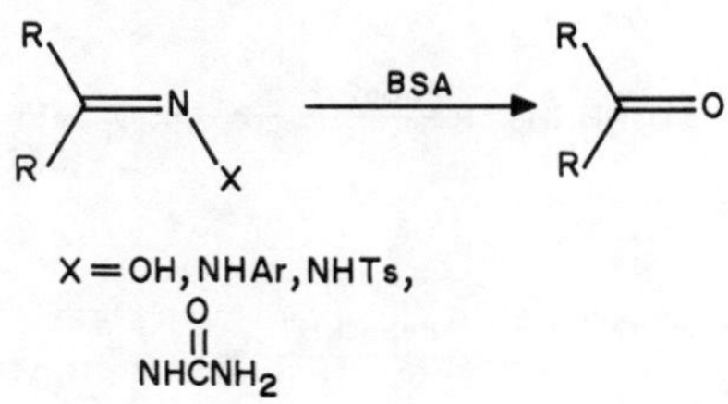
(187)

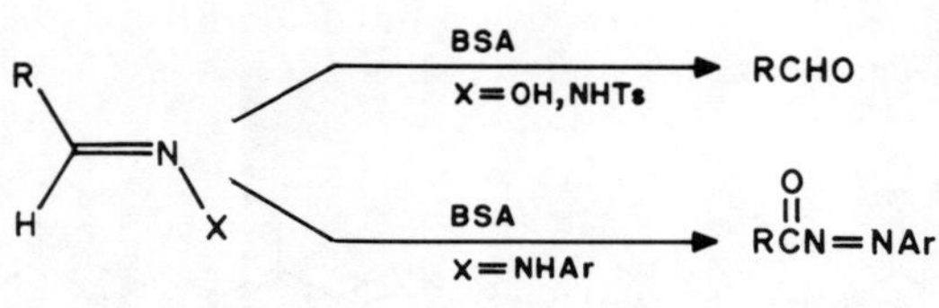
(188)

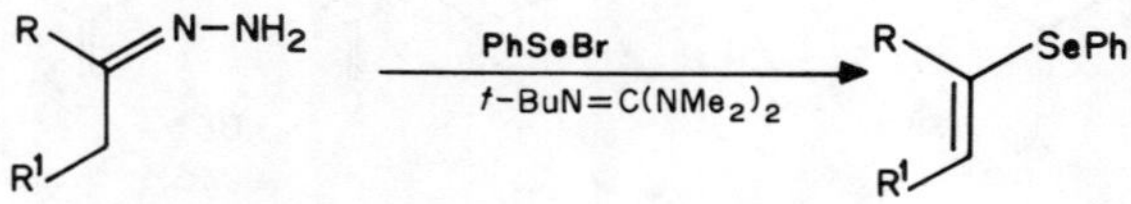
(189)

The oxidation of semicarbazones with selenium dioxide constitutes an important route to acetylenes via the photolytic or pyrolytic fragmentation of selenadiazole intermediates (equation 190). Lalezari and coworkers[581–589] and Meier and coworkers[590–599] have studied this process extensively. They and others[600,601] prepared diverse acetylenes including diynes[585], sulfonylacetylenes[587], acetylenic sulfides[588], selenides[589] and ethers[588], radioactively labelled acetylenes[601] and a large number of highly strained cyclic acetylenes[590,591,593,595–599].

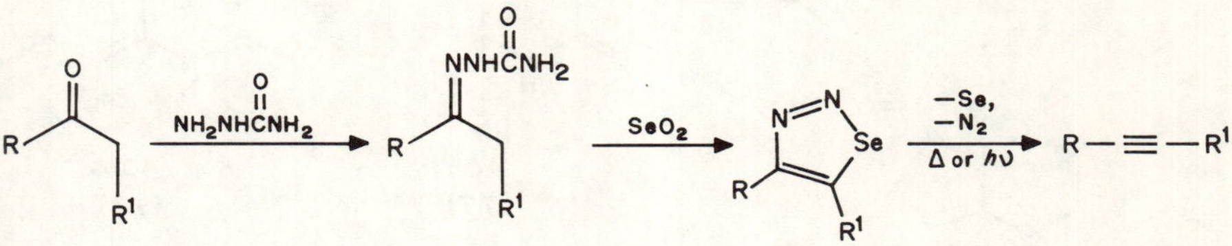

(190)

The dehydration of aldoximes with selenium dioxide[602–604], Se_2Cl_2[605], PhSeCl[606] and the seleninyl chloride **30**[606] provides a useful preparative route to nitriles (equation 191).

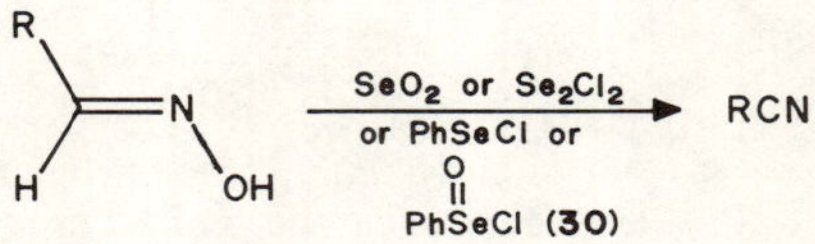

(191)

5. Nitro compounds

Primary aliphatic nitro compounds undergo oxidation at the α-carbon and simultaneous reduction of the nitro group when treated with selenium dioxide and triethylamine[607]. This reaction permits the isomerization of such nitro compounds to hydroxamic acids (equation 192).

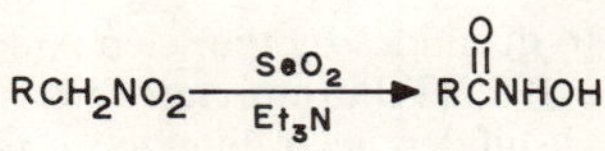

(192)

6. Azasteroid lactams

Azasteroid lactams display several modes of reaction when oxidized with BSA (**27**). When conformational considerations permit selenoxide elimination to take place across the C—N bond, *N*-acylimines are first formed and then oxidized further to imides[250] (e.g. equation 193). If imine formation is precluded by conformational effects, then dehydrogenation occurs at higher temperatures according to equation 51.

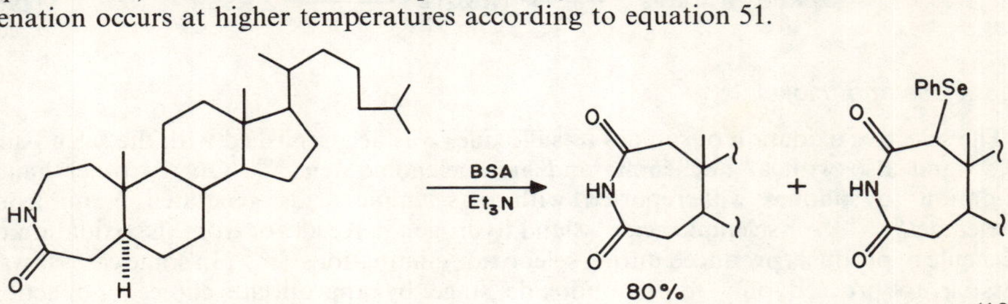

(193)

Azasteroid enamides react in a more complex fashion. Keto carbinolamides such as **77** and hydroxylated products **78**, were formed from azasteroids such as **76** (equation 194)[223,382]. The carbinolamides were obtained in improved yield by selenenylation of the 6-position, with PhSeCl, followed by oxidation with excess MCPBA[223].

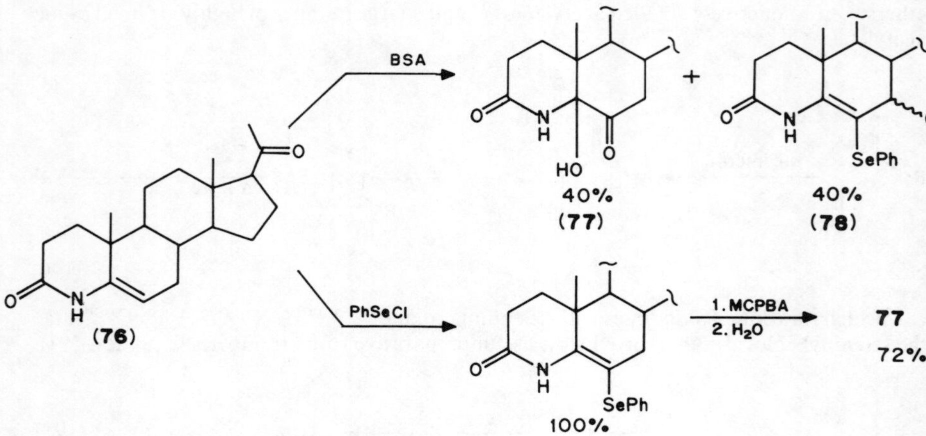

(194)

C. Phosphorus and Sulfur Compounds

Mild and selective oxidants of sulfur compounds are of importance, as mixtures of products with different oxidation states are often formed indiscriminately. A number of selenium and tellurium reagents have proved valuable in such reactions.

1. *Thiols*

Thiols are cleanly oxidized to disulfides by the selenoxide **68**[552], telluroxide **70**[551,571] and tellurone **69**[549] (equation 195). Tellurium tetrachloride forms tetrathiotellurium species which decompose to disulfides according to equation 196[608], and which are effective in converting dithiols to bisdisulfides.

$$2\ RSH \xrightarrow{\textbf{68,69}\text{ or }\textbf{70}} RSSR \qquad (195)$$

$$4\ RSH + TeCl_4 \longrightarrow (RS)_4Te \xrightarrow[\Delta]{} 2\ RSSR \qquad (196)$$

2. *Sulfides and thioketals*

The selective oxidation of sulfides to sulfoxides was accomplished with the selenoxide **68**[552] and also with areneseleninic and areneselenonic acids[609]. On the other hand, oxidation to sulfones was reported with perseleninic acids generated *in situ* from diselenides[610–614] or seleninic acids[373] and hydrogen peroxide, or from the oxidation of selenenic byproducts produced during selenoxide eliminations[78,508]. In some cases it was possible to proceed only to the sulfoxide stage by appropriate choice of reaction conditions[373,508]. Although perseleninic acids are known to epoxidize olefins (see

Section V.G.1), the selective oxidation of sulfides in the presence of double bonds is possible, as demonstrated by the example in equation 197[614]. Sulfides can also be cleanly converted into either sulfoxides[615] or sulfones[616] with selenium dioxide and hydrogen peroxide.

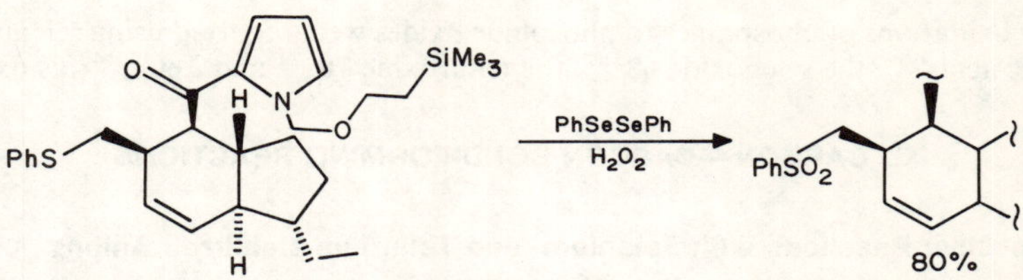

(197)

Free ketones and aldehydes are unmasked from their dithioketal or dithioacetal protecting groups with BSA (**27**)[617–620] (equation 198). The reaction is applicable to hindered substrates and also to diselenoketals[621,622].

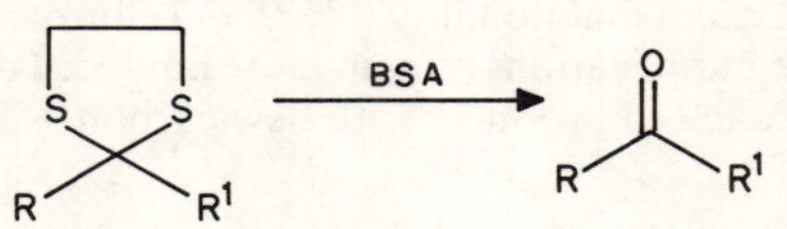

(198)

3. Thiocarbonyl compounds

Many types of thiocarbonyl compounds cleanly afford their carbonyl analogues on oxidation with BSA (**27**)[623,624], the telluroxide **70**[551,571] or various selenoxides[552,625,626] (equation 199). A particularly useful variation of this process involves the chlorination of a catalytic amount of a telluride with 1, 2-dibromo-1, 1, 2, 2-tetrachloroethane, followed by *in situ* hydrolysis of the resulting diaryltellurium dichloride to the corresponding telluroxide, which then functions as the actual oxidant of the thiocarbonyl compound[571,627]. The conversion of telluroesters to esters was similarly effected with BSA[628].

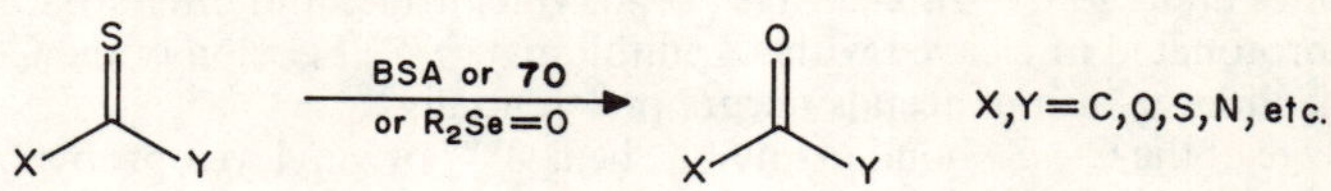

(199)

4. Sulfinic acids

The oxidation of sulfinic acids with $PhSeO_2H$ provides a useful preparation of selenosulfonates[73,629] (equation 200).

$$RSO_2H \xrightarrow{PhSeO_2H} RSO_2SePh \qquad (200)$$

5. Phosphorus compounds

Various trivalent phosphorus compounds, and also their sulfides and selenides, produce the corresponding oxides when treated with dimethyl selenoxide (**38**)[626] (equation 201). Interestingly, acyclic systems undergo inversion of configuration at phosphorus whereas cyclic systems react with retention.

$$\left.\begin{array}{l} R_3P \\ R_3P{=}S \\ R_3P{=}Se \end{array}\right\} \xrightarrow{\mathbf{38}} \begin{array}{l} R_3P{=}O \\ R{=}Ph,OEt,NEt_2 \end{array} \qquad (201)$$

Other oxidations of phosphines to phosphine oxides were reported using seleninic and selenonic acids[609], the selenoxide **68**[552], the telluroxide **70**[571] and $TeCl_4$[442] as oxidants.

XI. CARBON—CARBON BOND-FORMING REACTIONS

A. Connective Reactions With Selenium- and Tellurium-stabilized Anions

The stabilization of carbanions by adjacent selenium residues facilitates their formation while permitting the retention of a high degree of nucleophilic character. The synthetic applications of these species have been extensively studied by Krief, Seebach, Reich and others, and are generally based on their reactions with electrophiles, followed by further transformations of the selenium functionality[630,631]. α-Telluro carbanions have also been investigated, but so far their applications remain more limited. These procedures provide a powerful connective approach to products with new carbon—carbon single and double bonds.

1. Anions from selenides, selenoacetals and their tellurium analogues

Selenium-stabilized alkyl carbanions can be conveniently generated by the deprotonation of alkyl selenides, diselenoacetals or triselenoorthoformates with hindered amide bases. A different approach to these anions is based on the facile C—Se cleavage reactions of diselenoketals and triselenoorthoesters with alkyllithiums. These processes are summarized in Scheme 1.

When the α-carbon atom is substituted with another heteroatom besides selenium, alkyllithium-induced C—Se cleavage occurs preferentially to C—Si or C—S scission in α-selenosilanes[384,386,647,650] or α-selenosulfides[662], respectively, whereas α-bromoselenides undergo C—Br cleavage[663]. Alkyl tellurides and ditelluroacetals can be similarly deprotonated or cleaved with alkyllithiums[115,664]. Scission of the C—Te bond in mixed selenium–tellurium acetals occurs preferentially[665].

The cleavage of the C—Se bond in allyl[659], benzyl[660] or vinylcyclopropyl[661] selenides provides a convenient method for the *in situ* generation of allyl-, benzyl- or vinylcyclopropyl-lithiums, which are difficult to make otherwise.

The anions thus generated react with a wide array of electrophiles. They can be alkylated with alkyl halides[35,115,121,122,124,385,386,435,621,633,637,645,656–658,661,663–665], epoxides[91,122,429,621,637,648] and oxetanes[429], acylated with DMF[644,652,666], carbon dioxide[34,644,661] or other acylating reagents[644], silylated with trimethylsilyl chloride[386,635,642,647,650,654] and sulfenylated with sulfenyl halides[639,653] or disulfides[642]. They condense readily with aldehydes and ketones to afford β-hydroxy selenides[34,122,384,386,536,621,632,633,635–641,643,648,649,651,653–656,661,663,664,666–671], even with hindered or easily enolized substrates. Enones react via either 1,2-or 1,4-addition and good regioselectivity for either mode can be achieved by the appropriate choice of conditions[634,646,672–677]. Intramolecular alkylation of γ-chloro or γ-tosyl diselenoacetal anions affords the corresponding cyclopropanes[648].

Further transformations provide access to many useful types of compounds. The alkylated diselenoketals can be converted into ketones either hydrolytically with $HgCl_2$[622] or $CuCl_2$–CuO[621,622,637,673,676] or oxidatively with BSA[621,622] or

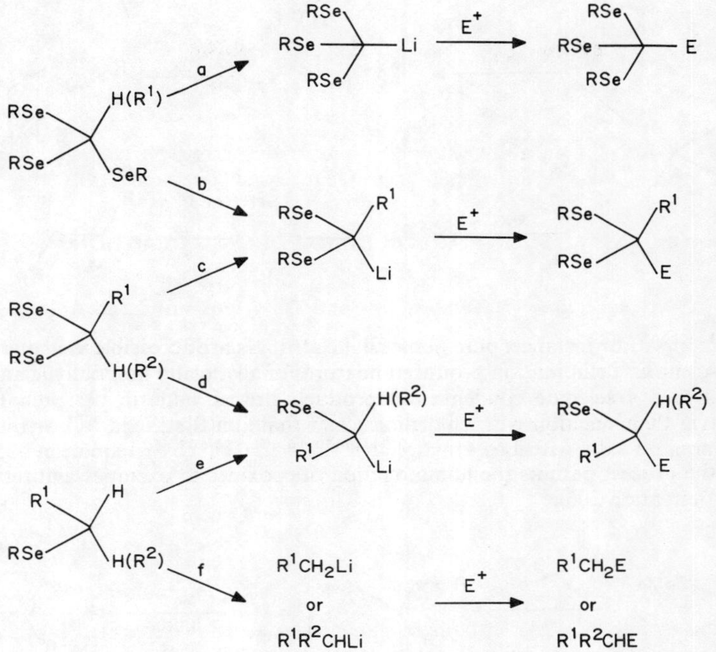

a. $LiN(i\text{-}Bu)_2$[632,633]; LDA[386,621]
b. BuLi[386,621,633,634]
c. $LiN(i\text{-}Bu)_2$[632,633]; LDA[621,635,636]; LiTMP[621]; KDA[637]
d. BuLi[91,122,124,384,536,633,639–655]; *t*-BuLi[639]; MeMgBr[638]
e. LDA[35,385,656,657]; LiTMP[656,659]; $LiN(SiMe_3)_2$[658]; BuLi–TMEDA[633]; *sec*-BuLi–TMEDA[33]
f. BuLi[647,650,654,659–661]; *t*-BuLi[660]

SCHEME 1

H_2O_2[622]. Similarly, ketones were obtained from alkylated ditelluroacetals with I_2–NaI[435] and aldehydes from α-silyl selenides with $NaIO_4$[386] or H_2O_2[385,386]. Since the starting seleno- or telluro-acetals or themselves prepared from aldehydes, the overall procedure permits the alkylation of an acyl anion equivalent (equation 202).

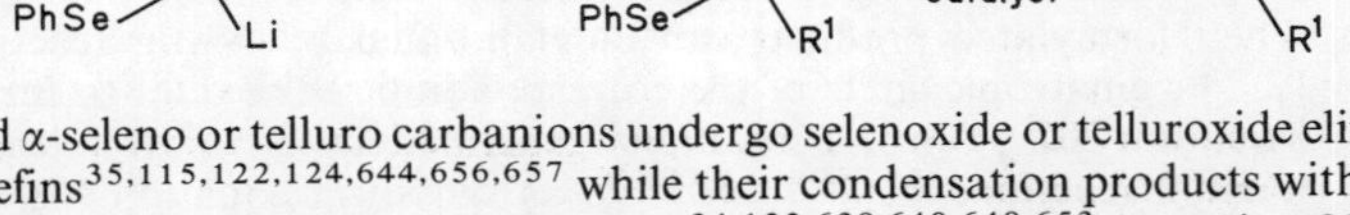

(202)

Alkylated α-seleno or telluro carbanions undergo selenoxide or telluroxide eliminations to afford olefins[35,115,122,124,644,656,657] while their condensation products with carbonyl compounds similarly provide allylic alcohols[34,122,639,640,649,653] (equation 203).

In some cases, eliminations of the corresponding selenonium salts gave results which were as effective as or even superior to those obtained with selenoxide elimin-

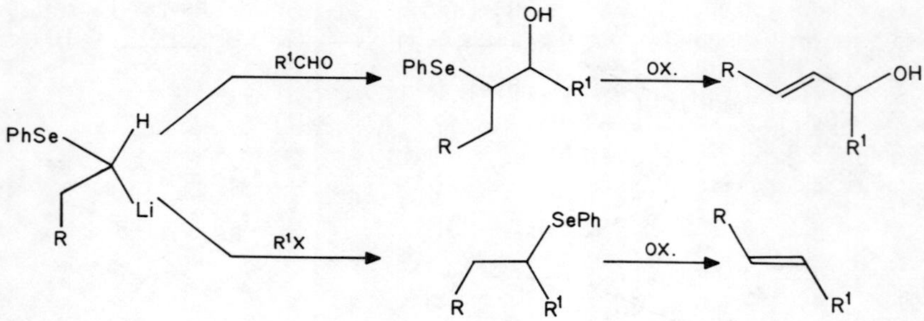

(203)

ations[91,121,122]. For instance, pure homoallylic alcohols are accessible via equation 204, whereas selenoxide elimination produced mixtures of allylic and homoallylic alcohols[91].

The β-hydroxy selenide condensation products are of value in the preparation of epoxides via their reaction with chloroform and thallium(I) ethoxide[678] or by treating their selenonium salts with bases[122,639,641,643,646,651,653,662,671] (equation 205). Repetition of the process permits the homologation of epoxides to oxetanes and tetrahydrofurans[429] (equation 206).

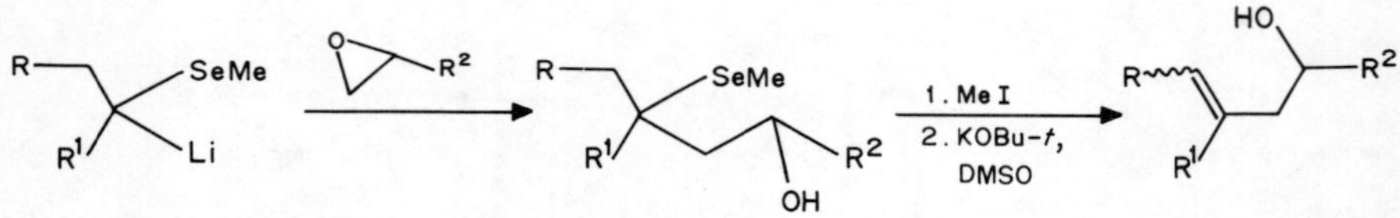

(204)

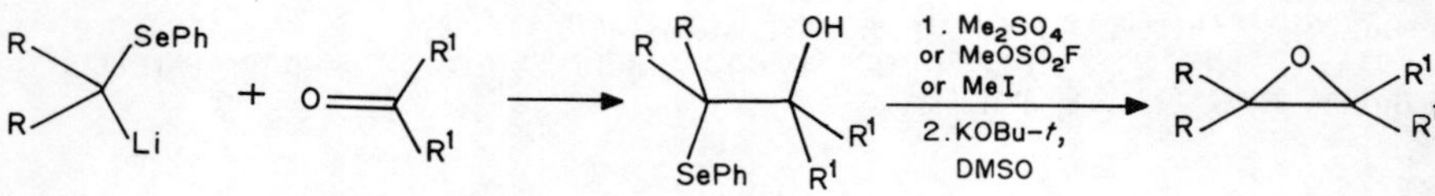

(205)

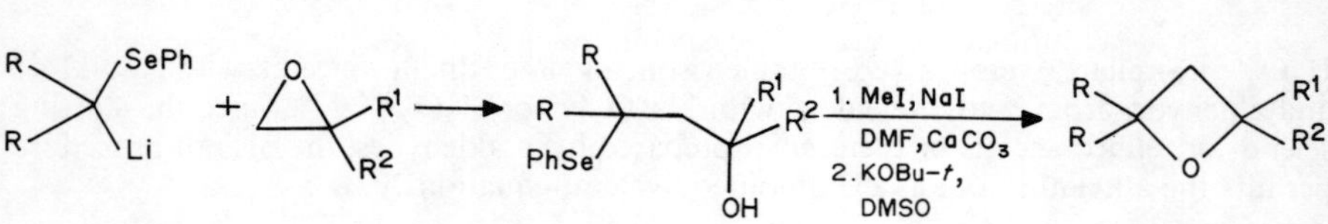

(206)

Selenium-stabilized carbanions provide a connective route to olefins through several approaches. Their formylated products can be elaborated by Wittig reactions[652,666], followed by [2, 3] sigmatropic shifts of the corresponding selenoxides to furnish allylic alcohols[652] (equation 207). α-Silyl carbanions generated from α-silyl selenides afford olefins via Peterson reactions[384,386,635,647,650], as shown in equation 208. Numerous examples have been reported where β-hydroxy selenides were prepared as in equation 209 and converted into olefins by reductive elimination. This procedure constitutes an alternative to the Wittig reaction and is less susceptible to steric effects. Further, reductive elimination often takes place *anti* stereospecifically. Effective reagents for this transformation are: methanesulfonyl chloride (MsCl)–Et_3N[34,636,655,656,679,680], TsOH[681,682] $HClO_4$[681], $(CF_3CO)_2O$–Et_3N[681], $POCl_3$–Et_3N[384,672,683], $POCl_3$–$SnCl_2$[34], $SOCl_2$–Et_3N[214,384,683–685], PI_3–Et_3N[122,650,651,666,668,685], P_2I_4–Et_3N[391,666,685,686], 1,2-phenylenephosphorochloridite[653,684], Me_3SiCl–NaI[475] and *N,N'*-carbonylbisimidazole[122,668]. Finally, symmetrical olefins are accessible through the direct coupling of selenium-stabilized carbanions with CuI·SMe_2[124] or CuI[687] (equation 210).

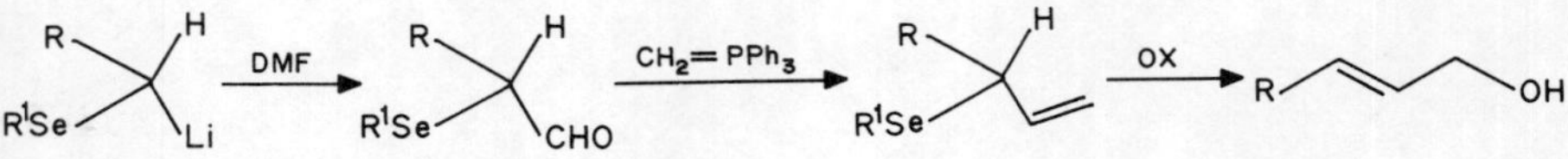

(207)

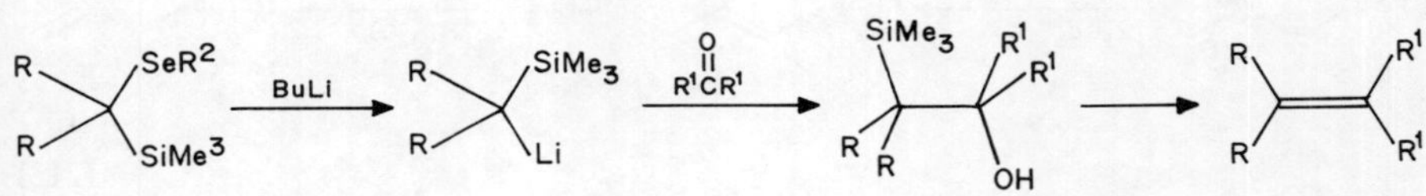

(208)

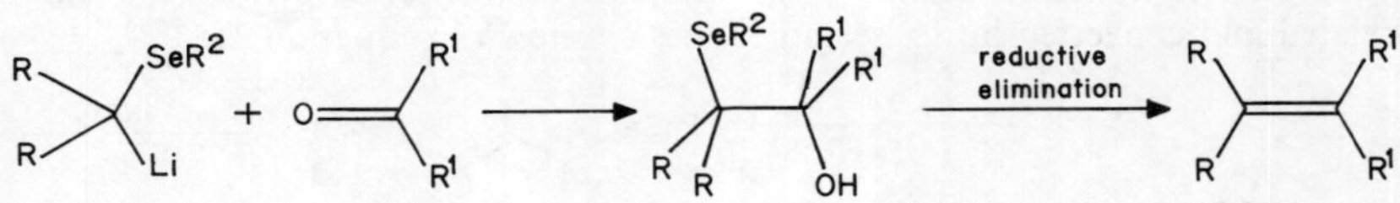

(209)

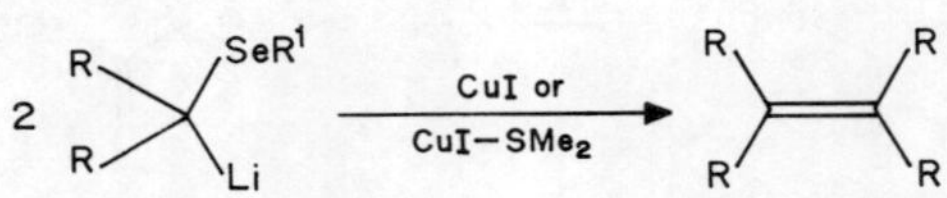

(210)

Metalated triselenoorthoformates function as carbene equivalents in the cyclopropanation of ketene dithioacetals according to equation 211[632,633].

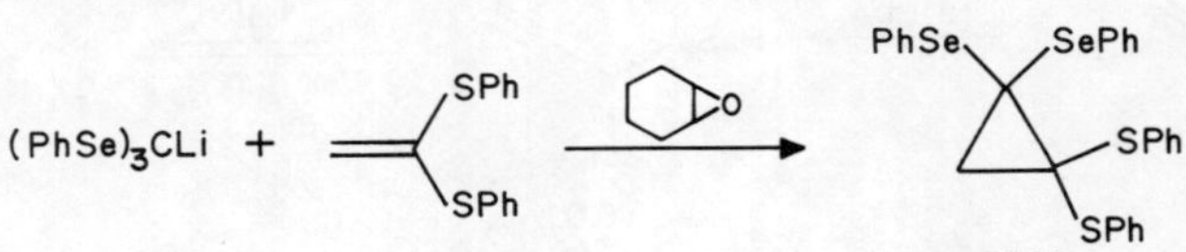

(211)

Ring expansions of β-hydroxy selenides prepared via equations 203, 205 and 209 are discussed in Section XVI.C.1.

2. Anions from allylic and propargylic selenides

Allylic selenides are deprotonated with hindered amide bases such as LDA[336,680,682,688–692], $LiNEt_2$[680,691], LiTMP[659,690] and KDA[659]. The resulting anions react chiefly at their α-position with electrophiles which include alkyl halides[336,680,691], epoxides[680,691], silyl chlorides[680,689,691] and boranes[688]. The alkylated products are useful in the preparation of rearranged allylic alcohols after oxidation and [2,3] sigmatropic shifts[691,692] (equation 212). Allylic selenides were employed in a variation of the Mannich reaction as shown in equation 213[692].

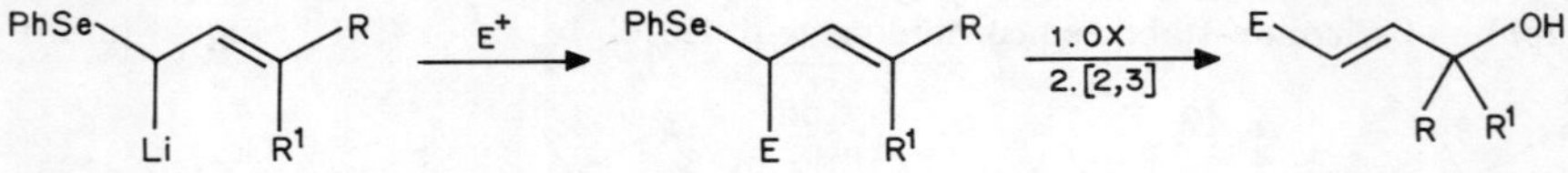

(212)

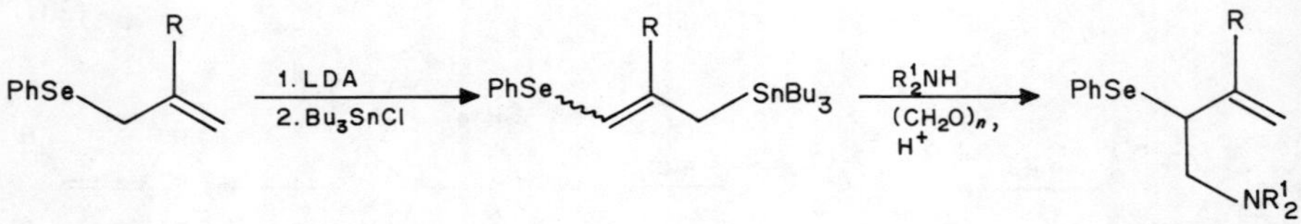

(213)

1,3-Bis(phenylseleno)propene (**79**) can be alkylated at both the 1- and 3-positions and affords α,β-unsaturated silyl ketones[680] according to equation 214 (or gives enals if the silylation step is omitted). Similarly, the dianion of the propargyl selenide **80** was dialkylated and converted into α-seleno enones as shown in equation 215[693–695].

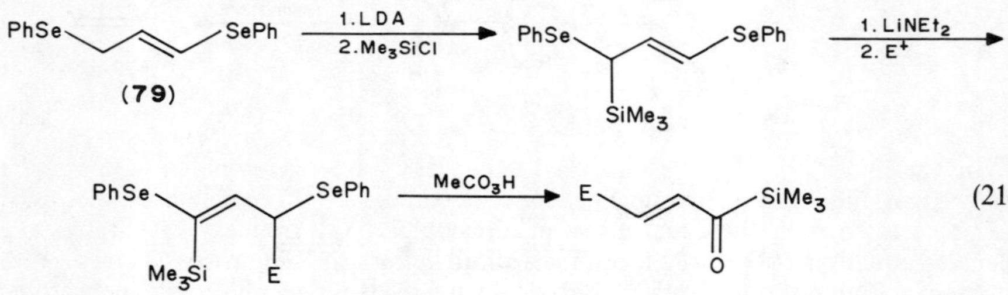

(214)

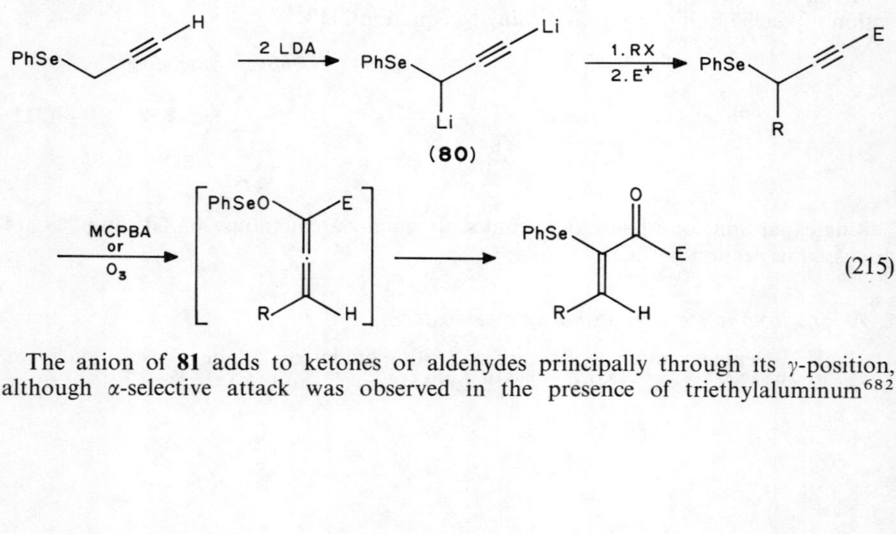

(215)

The anion of **81** adds to ketones or aldehydes principally through its γ-position, although α-selective attack was observed in the presence of triethylaluminum[682] (equation 216). The same anion reacted with cyclopentenone mainly via 1,2-addition in THF at −78 °C, but exclusively via 1,4-addition in the presence of HMPA[696]. In both cases, the α-position of the anion was the favoured site of attack. In contrast to equation 216, the anion of **82** reacted with acetone through its α-position and furnished the allylic alcohol **83** after [1,5] migration of the selenide moiety[697] (equation 217).

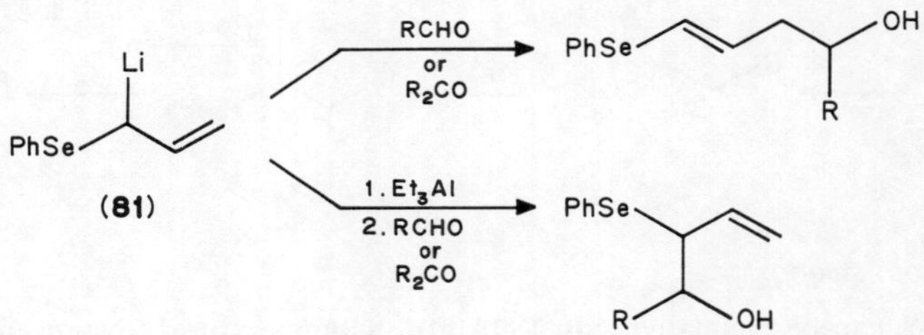

(81)

(216)

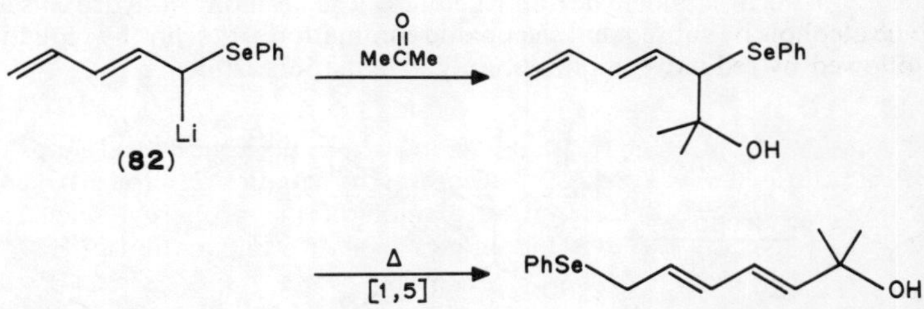

(82)

(83)

(217)

3. Anions from vinyl selenides and tellurides

Vinyl selenides[431,636,637,690] and vinyl tellurides[698] are deprotonated with hindered amide bases, and KDA is recommended for reluctant cases[637]. The metalated species can then be treated with the usual array of electrophiles to provide α-substituted vinyl selenides (equation 218). Since the products afford ketones after hydrolysis, vinyl selenides represent acyl anion equivalents[637]. 2-Pyridyl vinyl selenides are reported to deprotonate more easily than others and the subsequent alkylation takes place stereospecifically with retention of configuration[101]. Ketene diselenoacetals provide another source of selenium-stabilized vinyl anions through cleavage of one C—Se bond with *n*-butyllithium[391,699] (equation 218).

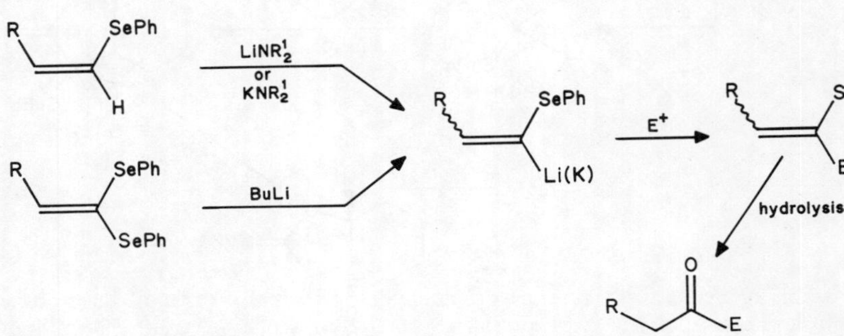

(218)

The Michael additions of alkyllithiums to phenyl vinyl selenide produce α-metalated selenides, which react with various electrophiles as shown in equation 219[431,700]. If the latter step is followed by selenoxide elimination to the corresponding olefin, the selenide becomes the equivalent of the synthon $^{+}CH{=}CH^{-}$.

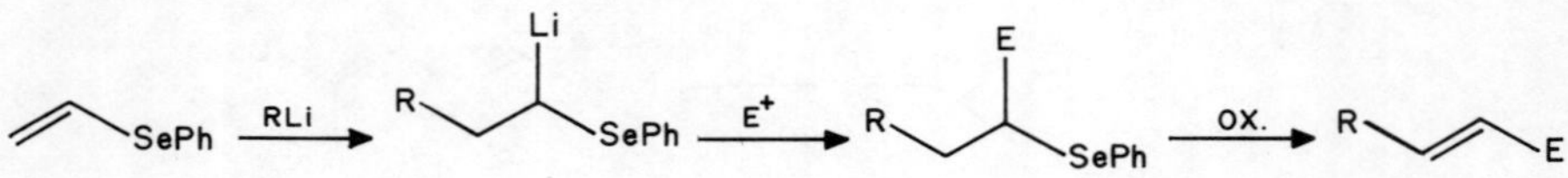

(219)

4. Anions from selenoxides

Selenoxides can be metalated with LDA at low temperatures (to prevent premature selenoxide elimination), and the resulting anions react with various electrophiles[33,34,679,701] such as aldehydes and ketones. The resulting β-hydroxy selenoxides afford allylic alcohols by subsequent selenoxide elimination, or olefins by reduction to the selenide followed by reductive elimination[34,679] (equation 220).

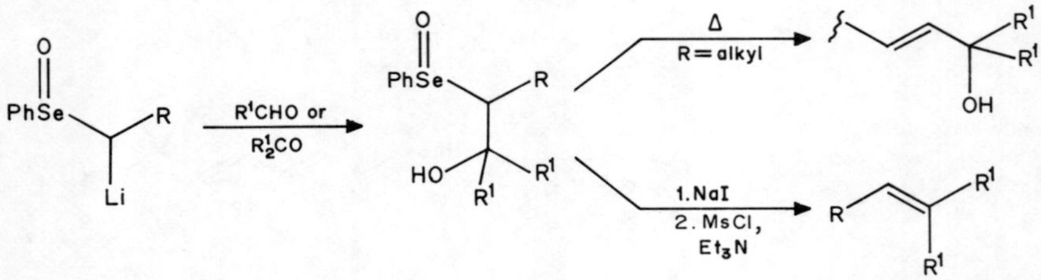

(220)

5. Dianions from α-phenylselenocarboxylic acids

The dianion **84** of 2-phenylselenopropanoic acid was exploited by Petragnani and coworkers[702,703] in several approaches to α-methylene lactones. One example is shown in equation 221[702]. The condensation of the dianion of α-phenylselenoacetic acid with benzaldehyde was also reported[656].

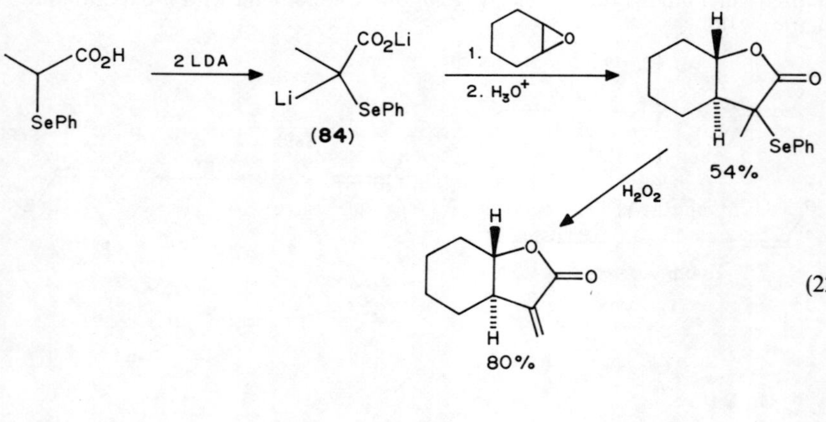

(221)

6. Anions from α-seleno nitriles, esters and lactones

Selenides bearing additional stabilizing groups can also be alkylated via their anions. These include α-seleno nitriles[261,704], esters[497,683] and lactones[683,705]. These reactions are often used in conjunction with selenoxide or reductive eliminations to yield unsaturated products. Two interesting examples are shown in equations 222[261] and 223[705]. Anions derived from α-seleno ketones are considered in Section XVII.A.

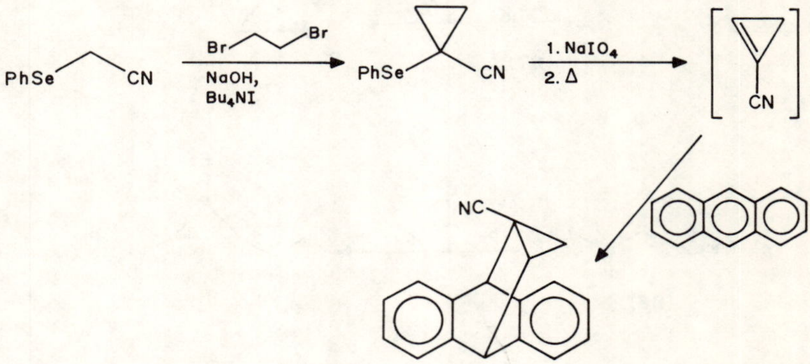

(222)

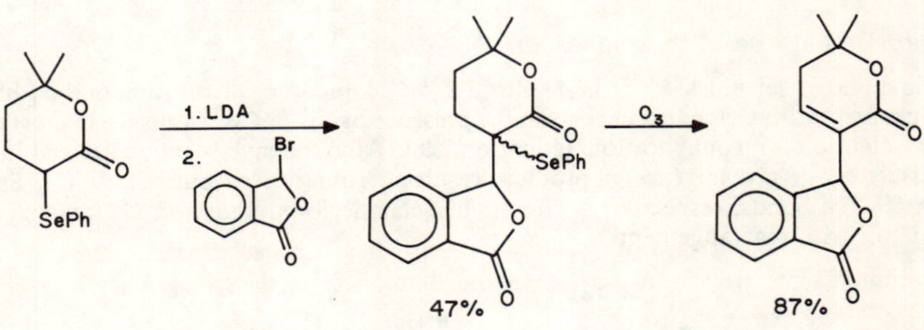

(223)

7. Selenium and tellurium ylides

Although selenium and tellurium ylides have been little investigated with respect to synthetic utility, several recent examples suggest potential value. Thus, the tellurium ylides **85** and **86** reacted with carbonyl compounds to afford α,β-unsaturated esters[706] and epoxides[707], respectively. The selenium ylides **87** also gave epoxides with non-enolizable aldehydes and ketones[708], including chalcone and cinnamaldehyde, whereas the ylide **88** cyclopropanated chalcone instead[709] (equations 224–227).

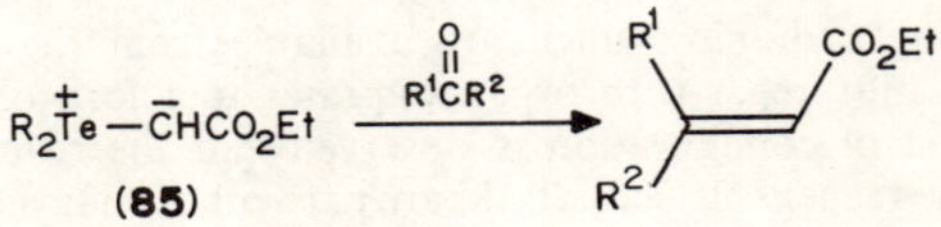

(224)

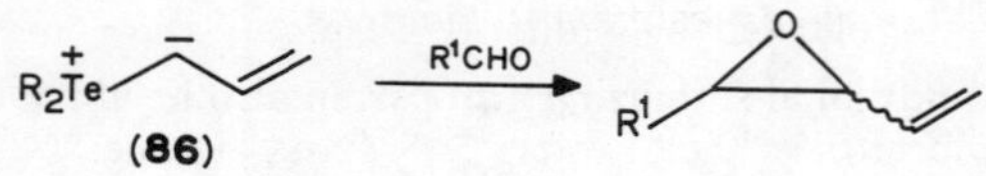

(225)

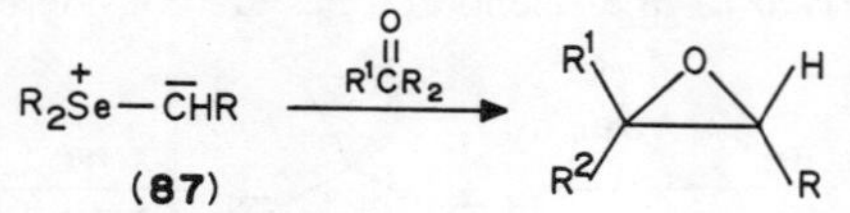

(226)

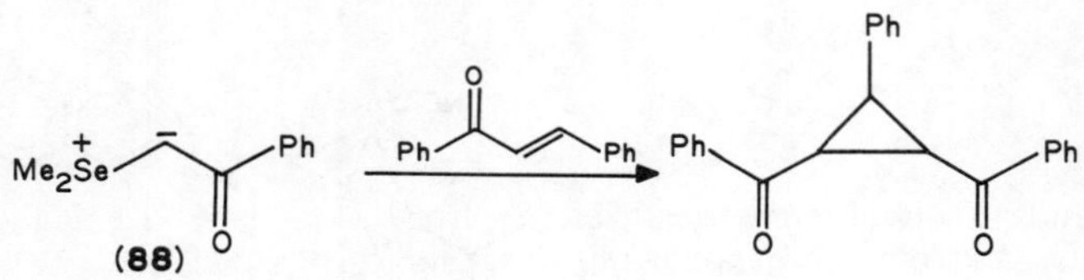

(227)

B. Connective Reactions from Substitution of Selenium and Tellurium Groups

Another approach to carbon—carbon bond formation is based on the substitution of ArSe or ArTe groups with organometallic reagents, or on the direct displacement of nucleofugal selenonyl groups with various nucleophiles.

1. Vinyl and allyl selenides and tellurides

Vinyl phenyl selenides[710,711] and tellurides[712,713] undergo substitution of the PhSe or PhTe moiety with Grignard reagents in the presence of Ni^{II} or Co^{II} catalysts, in most cases with retention of configuration (equation 228). Allyl phenyl selenide[711] and diaryl tellurides[712,713] give mixtures of products resulting from the cleavage of both C—Se and both C—Te bonds, respectively. The allylic selenide **89** in equation 229 reacted with Me_2CuLi with rearrangement[691].

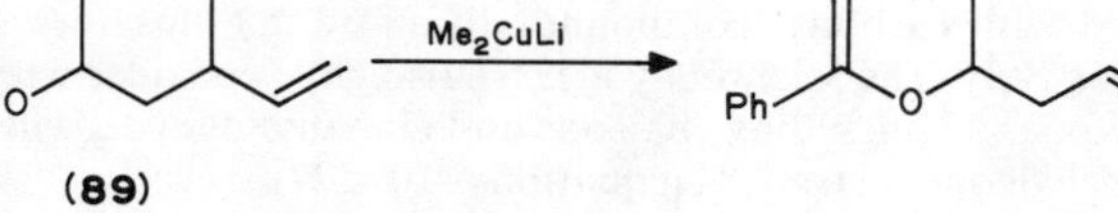

(228)

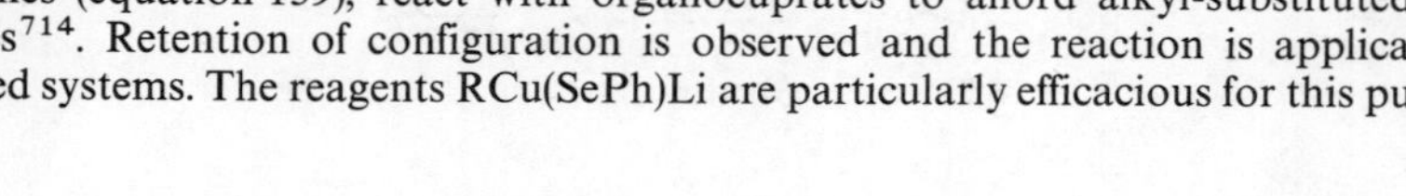

(229)

(89)

β-(Phenylseleno)vinyl sulfones, which are available from the selenosulfonation of acetylenes (equation 139), react with organocuprates to afford alkyl-substituted vinyl sulfones[714]. Retention of configuration is observed and the reaction is applicable to hindered systems. The reagents RCu(SePh)Li are particularly efficacious for this purpose.

When this procedure is followed by reductive desulfonylation, it permits the overall 2-alkylation of acetylenes as shown in equation 230.

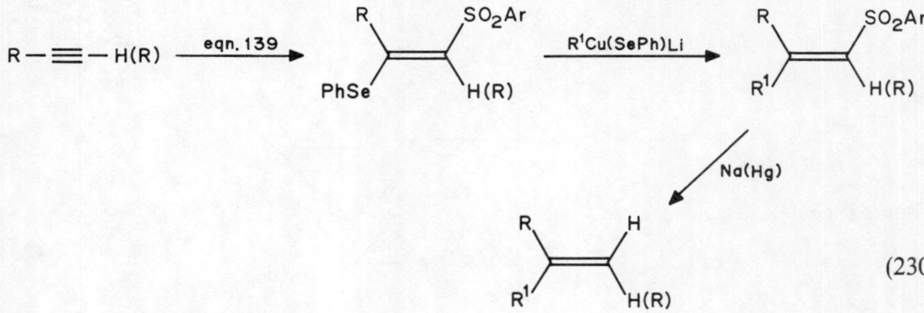

(230)

2. Selenoesters

The preparation of unsymmetrical ketones from selenoesters was reported with organocuprates and other organometallic reagents[715,716] (equation 231). Even vinyl cuprates are effective, permitting the synthesis of enones with little further Michael addition to the products[716].

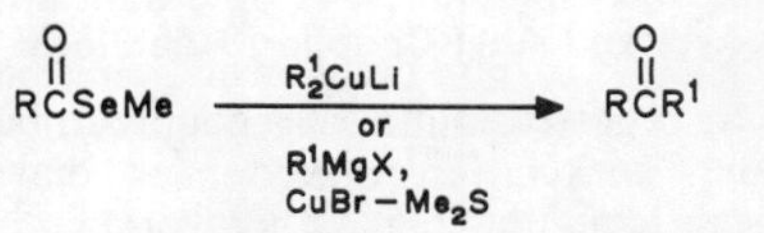

(231)

Selenoesters function as acylating agents when activated by 'selenophilic' metal catalysts such as Cu^{I}, Cu^{II} and Hg^{II} species. They perform Friedel–Crafts acylations of electron-rich aromatic compounds[717,718] and produce cycloadducts **90** with isonitriles[717,718] (equation 232). Under similar conditions, selenoesters also acylate alcohols and amines to afford esters and amides, respectively[718,719].

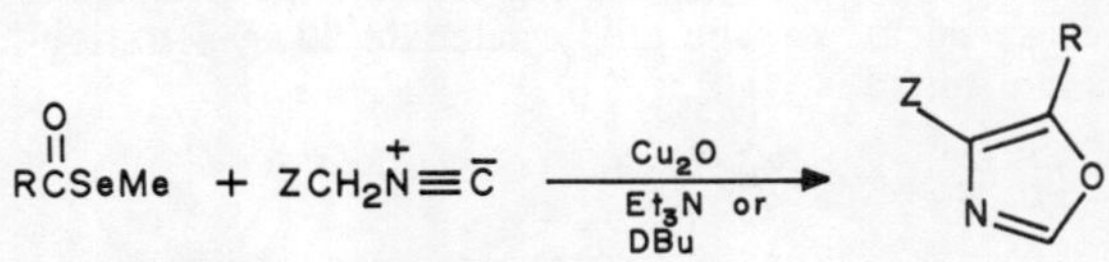

(232)

(Z = electron-withdrawing group)

(90)

3. Selenones

As mentioned earlier (Sections V.H.2 and IX.I), selenones have strong nucleofugal character, as well as the ability to stabilize an adjacent carbanionic center. This combination of properties was exploited by Kuwajima and coworkers[720–722] in a cyclopropane synthesis based on the Michael addition of enolates or active methylene compounds to vinyl selenones, followed by the nucleophilic displacement of the selenonyl leaving group (equation 233). Several examples of the related cyclopropanation of vinyl selenoxides were also reported[722,723]. Equation 234 depicts an epoxide-forming reaction based on the condensation of a selenone with benzaldehyde[724] in a manner reminiscent of the Darzen condensation. Selenones also undergo displacement with other nucleophiles such as halides[102,724], CN^-[724], H_2O[724], MeONa[724] and NaSPh[724].

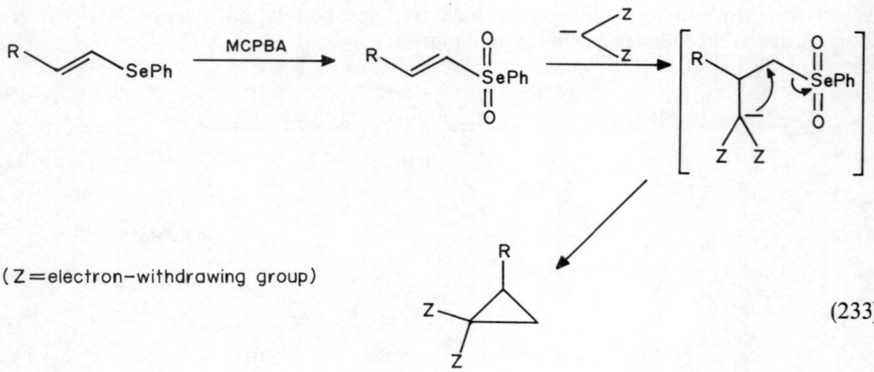

(Z=electron-withdrawing group)

(233)

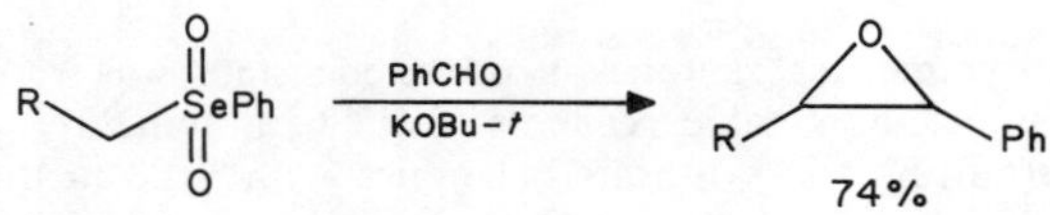

R=nonyl (234)

C. Tellurium-mediated Aryl and Allyl Coupling Reactions

Bergman and coworkers reported that arenes couple to biphenyls when treated with $TeCl_4$ followed by Raney nickel[725–727]. Aryl- and diaryl-tellurium chlorides are intermediates in this process (equation 235). Other workers[728] have found that the latter compounds are similarly converted into biphenyls in the presence of Pd^{II} catalysts. The intramolecular coupling of diaryl ethers with $TeCl_4$ affords dibenzofurans[729], as shown in equation 236. This reagent also catalysed the cationic polymerization of olefins such as stilbene[730]. Biphenyls were produced together with diaryl tellurides when tetraaryltellurides (**91**) were pyrolysed[731,732] (equation 237).

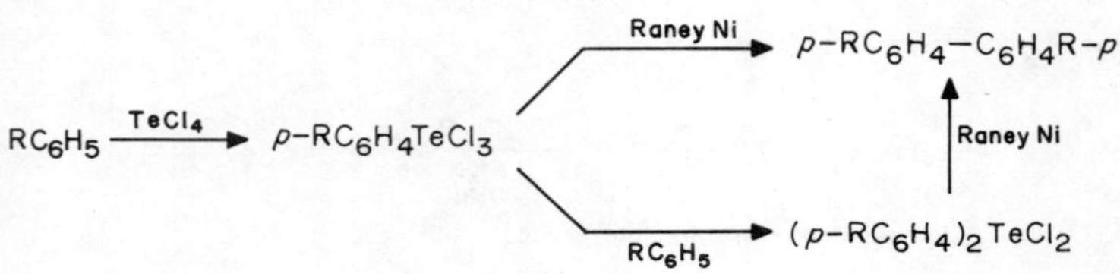

(235)

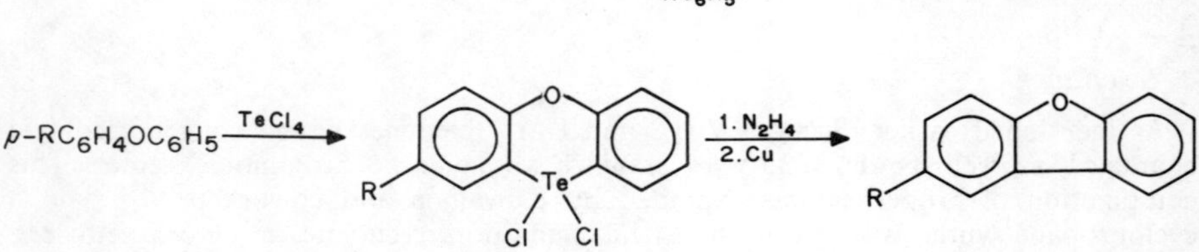

(236)

$$TeCl_4 + 4\,ArLi \longrightarrow \underset{(\mathbf{91})}{Ar_4Te} \xrightarrow{\Delta} ArAr + ArTeAr \qquad (237)$$

Allyl halides were dimerized to 1,5-dienes by treatment with the telluride dianion (Te^{2-}), probably via radical intermediates[733] (e.g. equation 238). In the case of unsymmetrical starting materials, varying amounts of all three coupled products (head-to-head, head-to-tail and tail-to-tail) were sometimes observed. Several sulfone-stabilized anions were dimerized to olefins with elemental tellurium[417] (equation 239).

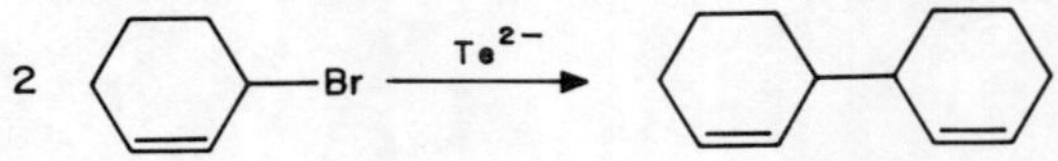

86%

(238)

$$ArCH_2SO_2Ph \xrightarrow[\text{2. Te(cat.)}]{\text{1. BuLi}} ArCH{=}CHAr \qquad (239)$$

D. Selenium and Tellurium Extrusion Reactions

1. C—C bond formation

The flash vacuum pyrolysis or photolysis of selenides, diselenides and tellurides results in extrusion of the selenium or tellurium atom and concomitant coupling of the two α-carbon atoms. Applications have been found in the synthesis of bibenzyls[734] (equation 240) and various strained hydrocarbons. These include benzo- and naphthocyclobutane[735], cyclophanes[736–739] such as **93**, biplanene (**95**)[740], lepidopterene (**96**)[740], and norcaradiene[741,742], the last compound reversibly. Illustrative examples are given in equations 241[735], 242[738] and 243[740]. Compounds **95** and **96** were produced from the dimerization of anthrylmethyl radicals generated by the extrusion of selenium from **94** followed by photochemical [4 + 4] or thermal [4 + 2] cycloadditions, respectively[740].

$$ArCH_2SeCH_2Ar \xrightarrow[(600\ ^\circ C)]{\Delta} ArCH_2CH_2Ar + Se \qquad (240)$$

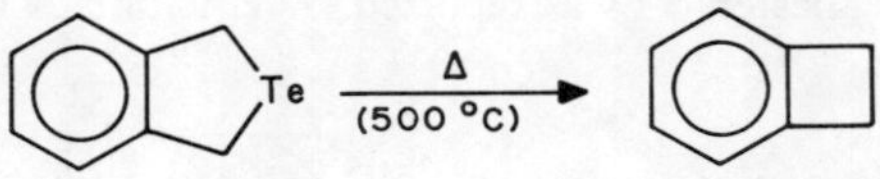

74%
(**92**)

(241)

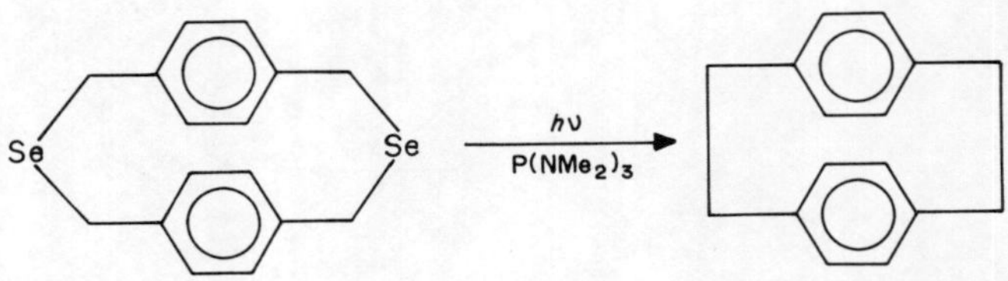

93%
(**93**)

(242)

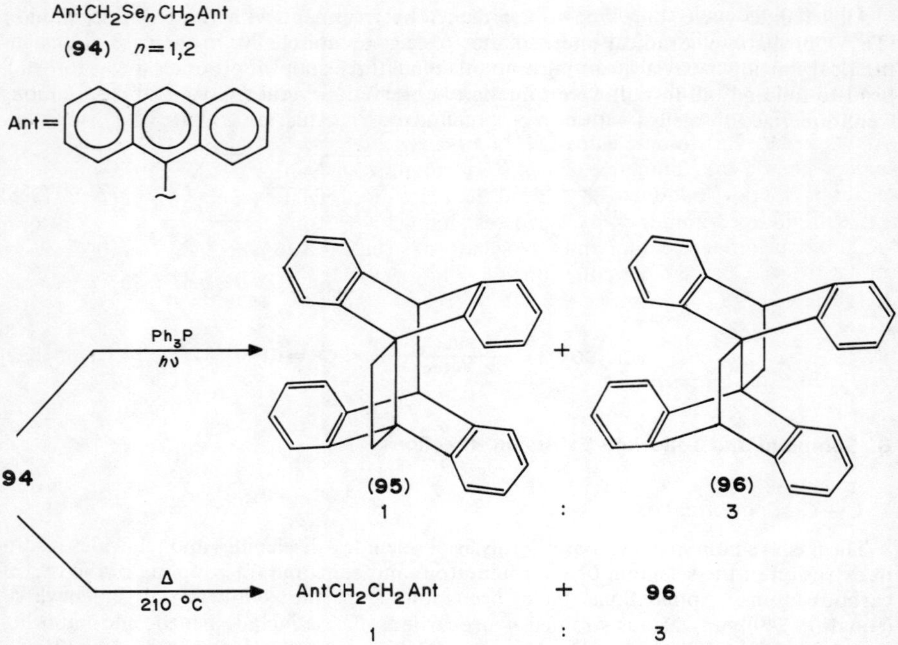

(243)

A related approach to bibenzyls and hydrocarbons such as **92**, **93**, **95** and **96** employs the disproportionation of aryl benzyl selenides induced by flash vacuum pyrolysis or photolysis, as shown in general form in equation 244[743,744].

$$2\ ArCH_2SePh \xrightarrow[h\nu]{\Delta\ or} ArCH_2CH_2Ar + PhSeSePh \tag{244}$$

The preparation of β-diketones or β-keto esters from the base-catalysed extrusion of selenium from the selenoesters **97** was reported[745] (equation 245).

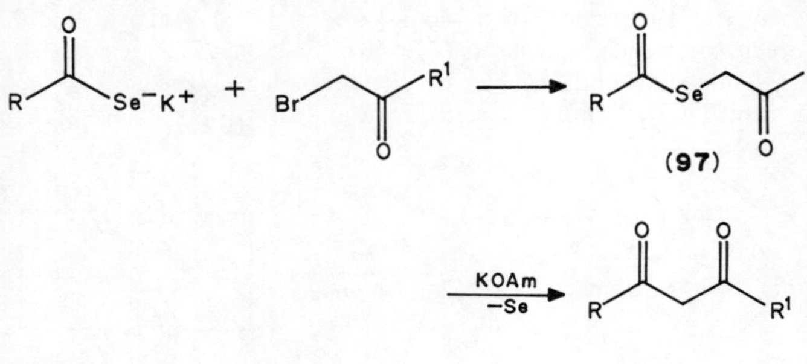

(245)

2. C=C bond formation

A particularly efficacious variation of the Barton two-fold extrusion technique for the synthesis of hindered olefins is based on the cycloadditions of diazo compounds (or their phosphoranylidenehydrazones) with selenoketones[746,747]. The resulting selenadiazolines (**98**) fragment readily by loss of nitrogen to furnish episelenide intermediates, which collapse to the desired olefins by the spontaneous extrusion of selenium (equation 246). Extensive further studies by Guziec and coworkers[748–752] have resulted in the preparation of a number of exceptionally hindered olefins such as **99** and **100**, but tetra *t*-butylethylene has to date eluded all preparative attempts by this or other methods. In some related work, hindered azines were obtained from the photolytic extrusion of selenium from cycloadducts **98**[751], and the reaction of phenyl azide with selenoketones afforded hindered imines[753] by a process analogous to equation 246.

The extrusion of selenium and nitrogen from selenadiazoles shown in equation 190 is similar to the process in equation 246 and provides a nonconnective synthesis of acetylenes.

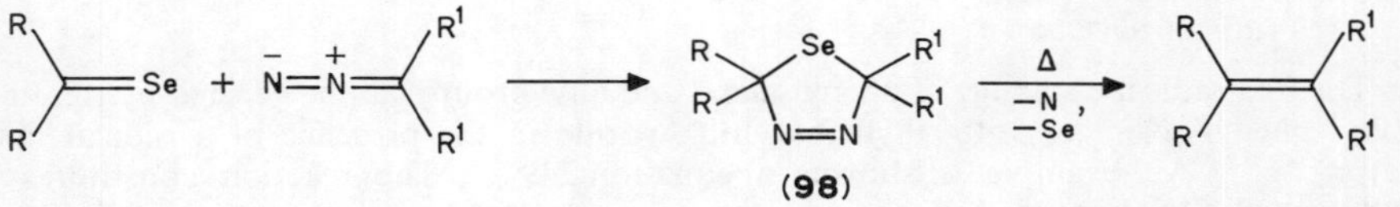

(246)

(99) (100)

E. Other Connective Reactions

1. *With selenoacetals activated by Lewis acids*

Several connective methods employing the reactions of nucleophilic carbon centers with selenium-stabilized cationic species have been reported. The intermediates are normally formed *in situ* by the abstraction of a second selenium moiety or a halide from an appropriate precursor with a Lewis acid. Enol silyl ethers thus produce *C*-alkylated products with selenoacetals and $SnCl_4$[69], triselenoorthoformates and $TiCl_4$[754], and with α-halo selenides and $TiCl_4$[754] or β-halo selenides and $ZnBr_2$[755]. An example is provided in equation 247[69].

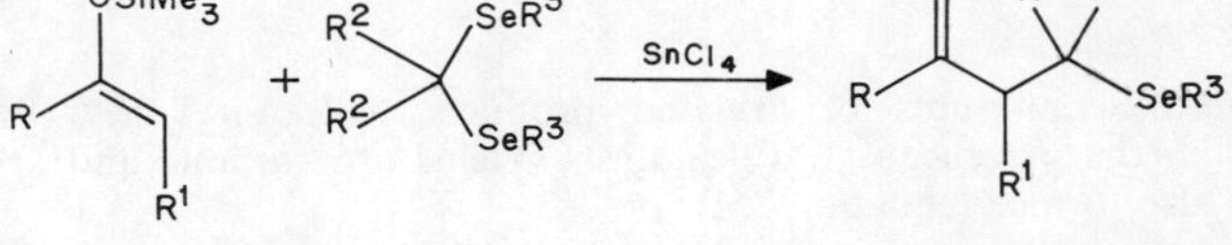

(247)

Pyrrole, furan, thiophene and 1, 3, 5-trimethoxybenzene undergo Friedel–Crafts alkylations with 1-seleno-[756] or 2-seleno-substituted[757] allyl cations generated according to equation 248.

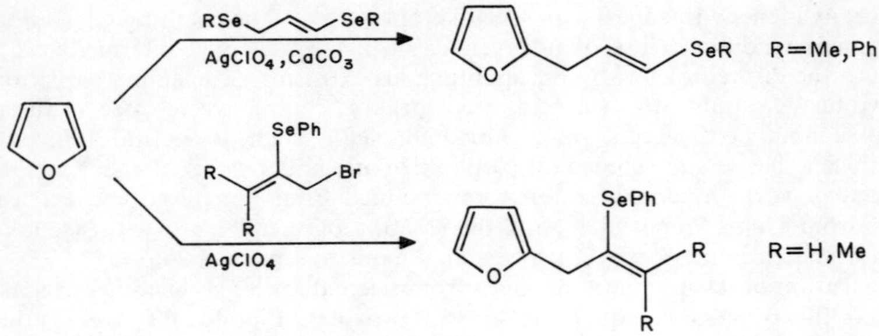

(248)

2. *Free-radical allylation*

The free-radical exchange of phenylseleno and allyl groups can be effected by treating alkyl phenyl selenides with allyltributyltin hydride in the presence of a radical initiator[504,758]. An example is shown in equation 249[758]. The reaction constitutes a convenient allylation procedure.

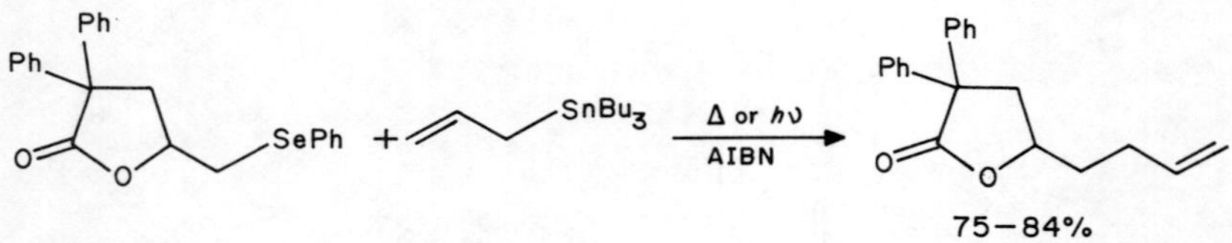

(249)

3. *Phenolic coupling*

Phenolic coupling reactions are sometimes observed during the oxidations of phenols and catechols with BSA (**27**)[356] or diphenyl selenoxide (**52**)[553] (see Section V.E.). In some cases the yields of coupled products are high enough to be of preparative value (e.g. equation 250[356]).

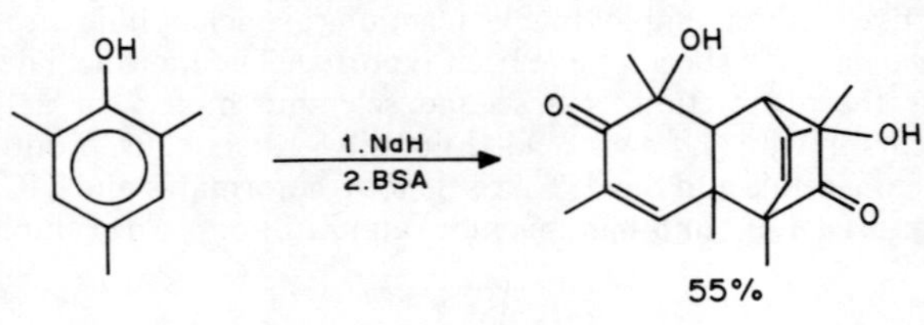

(250)

4. *Arylation of olefins*

Olefins produce mixtures of arylated products with Ph_2Te[759], Ph_2Se[759] or Ph_2TeCl_2[728] in the presence of Pd^{II} catalysts. Yields are variable and aryl coupling products are also formed (equation 251).

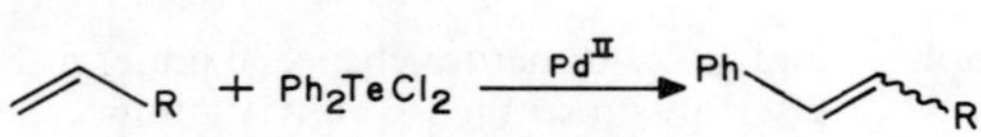

(251)

5. *Wittig reactions of α-selenophosphoranes*

Petragnani, Comasseto and coworkers reported a connective approach to vinyl selenides via the Wittig reaction of α-selenophosphoranes with aldehydes[152,263,760], or the Horner–Wittig reaction of α-selenophosphonates with aldehydes or ketones[269,760]. The further transformation of the products to ketones by hydrolysis was carried out in several cases[269] (equations 252 and 253).

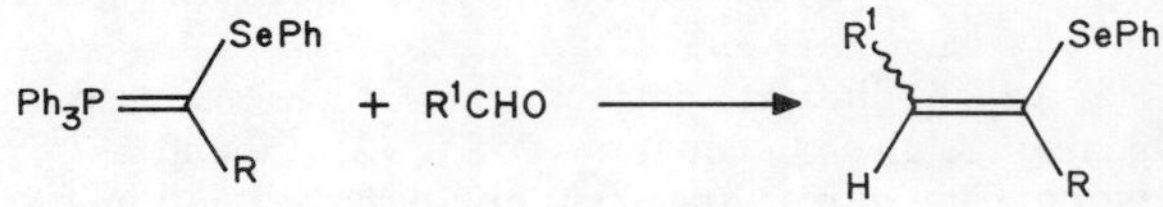

(252)

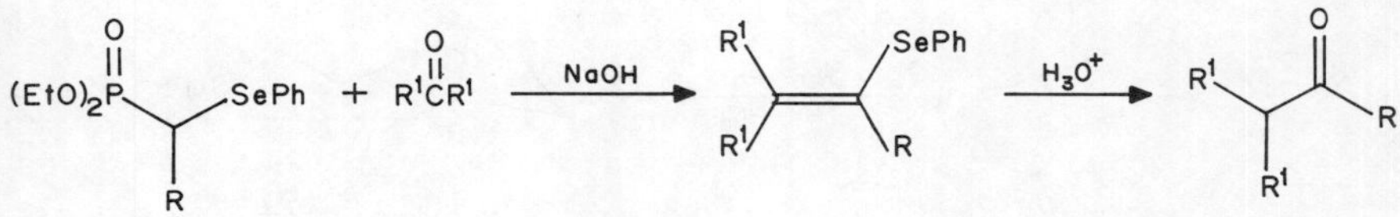

(253)

XII. CLEAVAGE REACTIONS WITH SELENIUM AND TELLURIUM NUCLEOPHILES

Selenolates (RSe^-) are weak bases but exceptionally powerful, soft nucleophiles. These properties make them valuable and sometimes unique reagents in a number of synthetic operations. Their ability to displace even relatively poor leaving groups such as carboxylates, alkoxides and amines from carbon enable them to carry out the cleavage of O— or N—alkyl bonds in esters, lactones, epoxides, ethers and amines. Liotta and coworkers[761,762] demonstrated that the reactivity of selenolates is highly dependent on their degree of solvation, the nature of the counter ion and the method of preparation, which must therefore be chosen with care. (For instance, the popular procedure for preparing NaSePh from PhSeSePh and $NaBH_4$ in ethanol probably affords a borane–selenolate complex with considerably suppressed nucleophilic strength[762]. This method should therefore be avoided if a more potent selenolate nucleophile is required.)

Several other reagents which either liberate selenolates *in situ* or act as their equivalents have been introduced recently. These include the silyl selenide $PhSeSiMe_3$ (**54**), its tellurium analog $PhTeSiMe_3$ (**101**), the aluminum selenolates $Me_2AlSeMe$ (**102**) and *i*-$Bu_2AlSeMe$ (**103**), and the selenoboranes $B(SeMe)_3$ (**104**) and $B(SePh)_3$ (**105**).

A. Cleavage of Esters and Lactones

Selenolates attack esters and lactones at the softer alkyl rather than the harder acyl carbon atom, resulting in the displacement of the carboxylate anion (equation 254). Reagents include LiSeMe in DMF[763], NaSePh in DMF[204,764,765] or THF[766], usually with added HMPA[177,761,762,767–771], as well as the more reactive potassium salt KSePh[769,770]. The silyl selenide **54**[654,772,773] or telluride **101**[774] can be used in a similar capacity in the presence of KF and 18-crown-6[772], ZnI_2[773,774] or $TiCl_4$[654]. The reaction is sensitive to steric hindrance and so methyl esters can be cleaved in the presence of others[762]. In contrast to the above reagents, the aluminum selenolate **102** attacks esters and lactones at the acyl carbon atom to afford high yields of selenoesters[718,719] (equation 254).

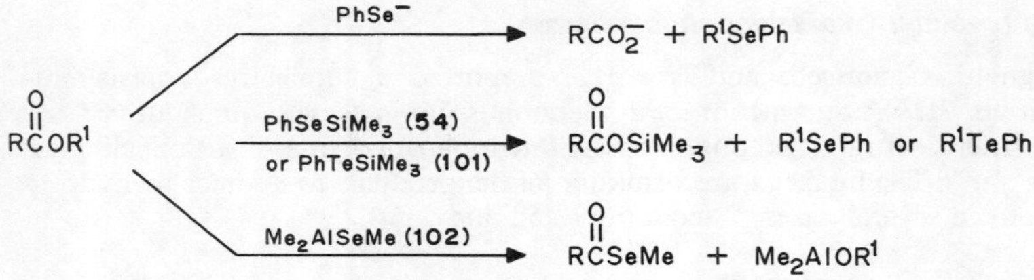

(254)

Several examples which illustrate further transformations used in concert with selenolate ring opening are given in equations 255[771], 256[765] and 257[768]. These include selenoxide elimination, Claisen rearrangement and decarboxylation, respectively.

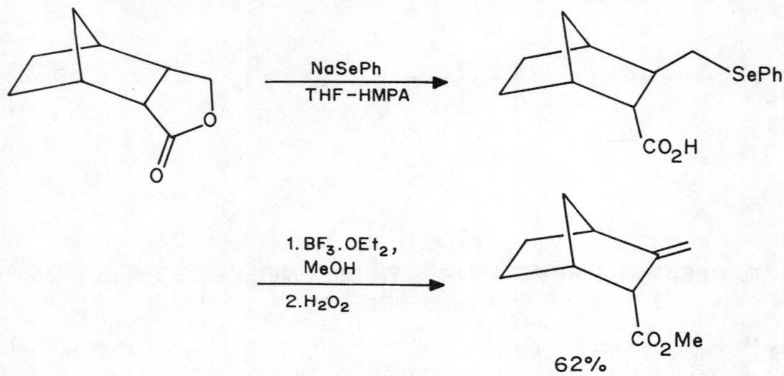

(255)

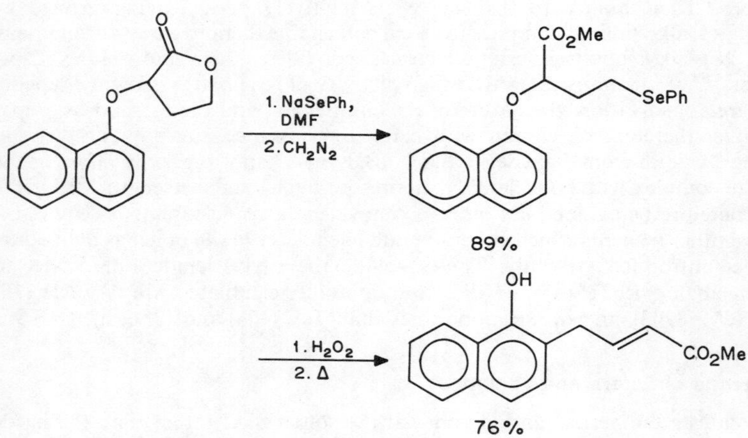

(256)

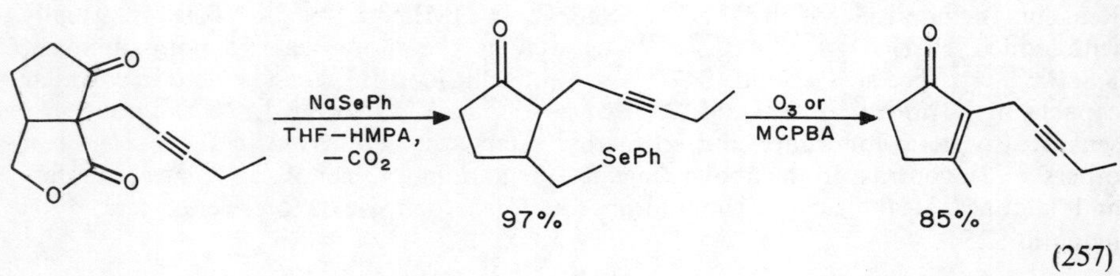

(257)

B. Cleavage of 2-Oxazolines and 2-Oxazines

2-Methyl-2-oxazolines and 2-methyl-2-oxazines[775,776] undergo ring opening with NaSePh in DMF or with the silyl selenide **54**. A synthetic route to secondary carboxamides was reported on the basis of prior *N*-alkylation of the oxazoline **106**, ring opening and degradation of the resulting β-selenoethyl substituent[776] (equation 258).

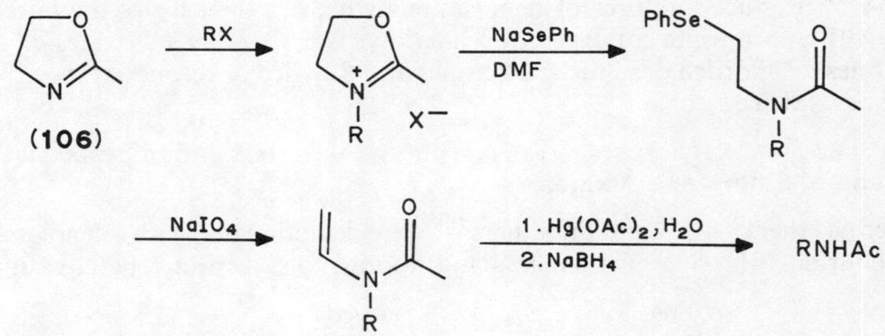

(258)

C. Cleavage of Epoxides

The cleavage of epoxides with a selenolate was first reported by Sharpless and Lauer[26] and provides an excellent preparative route to allylic alcohols via equation 259. When the β-hydroxy selenides thus formed are subjected to reductive deselenization (see equation 209), the method constitutes an efficient procedure for the stereospecific deoxygenation of epoxides[655,681,684].

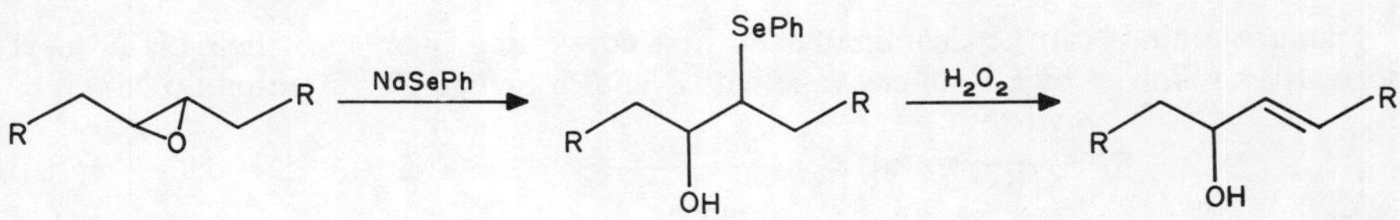

(259)

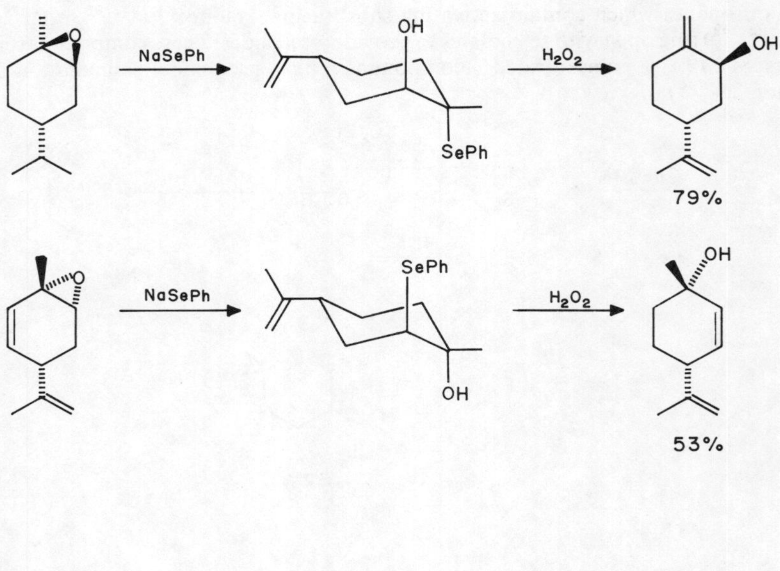

(260)

Effective reagents for epoxide opening include selenolates[49,61,204,531–534,655,777–779], NaTePh[780], the reagents **54**[772,781], **101**[774], **102**[718,719], **103**[48,782], **104**[783] and **105**[783] and benzeneselenol (PhSeH) in the presence of alumina[784]. Attack by the selenolate generally occurs at the less hindered epoxide carbon and is *anti* stereospecific. In cyclic systems, a strong preference for approach from the axial direction may outweigh other factors, as in the example in equation 260[61]. The selenoborane reagents **104** and **105**[783] and the silyl selenide **54**[781] produce mixtures of regioisomers which in the case of the latter can be regulated with appropriate catalysts. In related work, α,β-epoxy sulfoxides[779] and α,β-epoxy ketones[785] afforded desulfurized ketones and β-hydroxy ketones, respectively, with NaSePh and NaHTe.

D. Cleavage of Ethers and Amines

Aryl methyl ethers[786–789] and thioethers[788] were demethylated with selenolates or with the silyl telluride **101**[774] to give phenols and thiophenols, respectively (equation 261).

$$\begin{matrix}\mathrm{ArOMe}\\ \mathrm{or}\\ \mathrm{ArSMe}\end{matrix} \xrightarrow{\mathrm{NaSeR}} \begin{matrix}\mathrm{ArOH}\\ \mathrm{or}\\ \mathrm{ArSH}\end{matrix} + \mathrm{RSeMe} \qquad (261)$$

Quaternary ammonium salts were demethylated with NaSePh[790] while amines were similarly dealkylated in the presence of ruthenium catalyst[791]. Selenols are also effective reagents for amine dealkylations as they are sufficiently acidic to protonate the amine, thereby activating it toward C—N cleavage[792] (equation 262).

$$\mathrm{R_3N + PhSeH \longrightarrow R_3\overset{+}{N}H\ {}^{-}SePh \longrightarrow R_2NH + RSePh} \qquad (262)$$

Primary amines can be deaminated by first converting them into their bis-*N*-tosyl derivatives followed by C—N cleavage with NaSePh or NaSPh[793] (equation 263).

$$\mathrm{RNH_2 \longrightarrow RNTs_2 \xrightarrow[\mathrm{NaSPh}]{\mathrm{NaSePh\ or}} RSePh\ or\ RSPh} \qquad (263)$$

E. Cleavage of Cyclopropanes

Cyclopropanes which contain activating substituents (ketone[764,794,795], ester[794,796], cyano[764,794]) ring open with selenolates to provide γ-selenides. These compounds serve as precursors of β-γ-unsaturated compounds by selenoxide elimination (e.g. equation 264[796]).

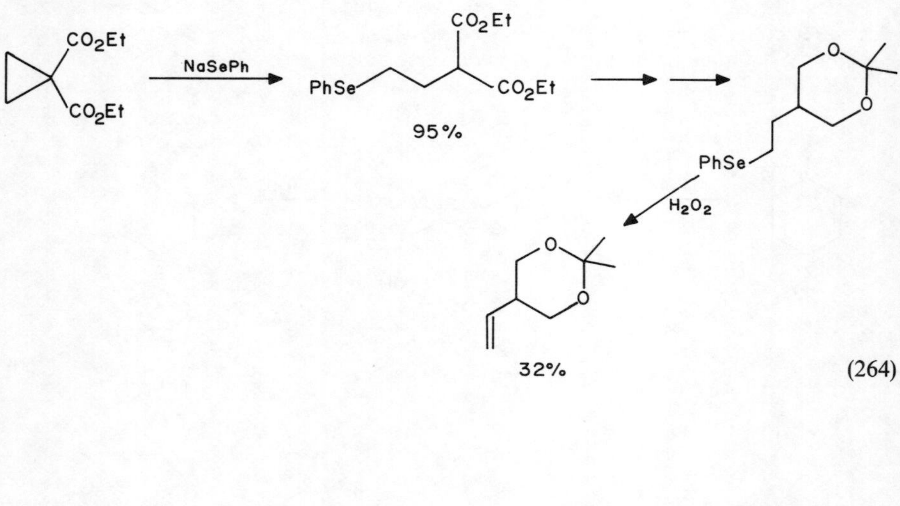

(264)

XIII. TRANSFORMATIONS OF ALCOHOLS VIA SELENIDES

Alcohols are readily converted into selenides by the displacement of their tosylates or mesylates with selenolates, or by their direct reaction with selenols in the presence of H_2SO_4 or $ZnCl_2$[797]. Subsequent oxidation and elimination produce the corresponding olefins, thus achieving the overall dehydration of the alcohol. This procedure generally avoids the rearrangements and other side reactions which frequently accompany classical dehydration procedures. Similarly, alkyl halides can be dehydrohalogenated via their selenides. Numerous examples of both processes have appeared in the literature since the early work of Grieco and coworkers[98,99,798–800] and others demonstrated their synthetic utility (equation 265).

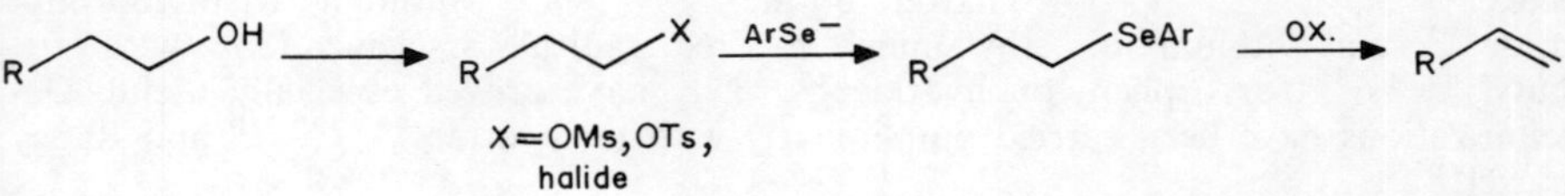

(265)

An alternative procedure for the direct transformation of primary or secondary alcohols to selenides was reported by Grieco and coworkers[39,173,801–803] and it obviates the need for intermediate tosylates, mesylates or halides. The alcohol is treated with an aryl selenocyanate (ArSeCN; Ar is usually *o*-nitrophenyl) and tributylphosphine in solvents such as pyridine or THF. Selenoxide elimination then completes the dehydration process as usual (equation 266). More recently, N-PSP (**34**) has been used in place of aryl selenocyanates. This reagent works effectively for primary alcohols[286,804–807], but in the case of a secondary alcohol the formation of the corresponding alkylphthalimide occurred instead of the expected selenide[808]. Alcohols can also be converted into tellurides with PhTeCN and tributylphosphine and then subjected to telluroxide elimination[809].

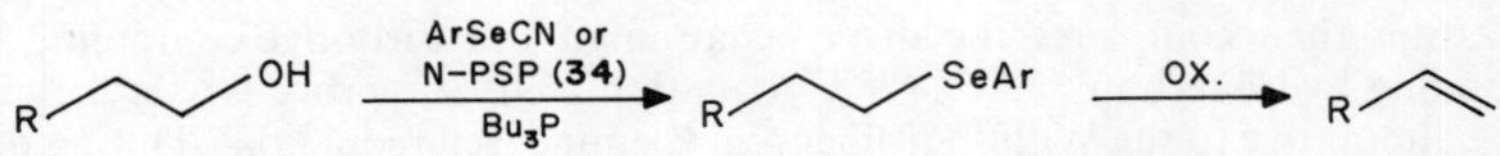

(266)

Since selenide formation in equation 266 occurs with inversion of configuration at the α-carbon, subsequent brominolysis of the product via equation 117 (where inversion also takes place) provides an overall method for the conversion of chiral alcohols into alkyl bromides with overall retention of configuration[430].

In a closely related variation of this work, it was found that the treatment of carboxylic acids with PhSeCN[810] or N-PSP[286,804] and tri-butylphosphine affords selenoesters (equation 267). Other selenium electrophiles which have been employed for this purpose are PhSeCl[811], PhSeSePh[811], $Bu_3P^+SeMe^-OSO_2R$[812] and PhSeNMeAc[147]. When N-PSP was used in the presence of primary or secondary amines, the corresponding carboxamides were formed in high yield[804].

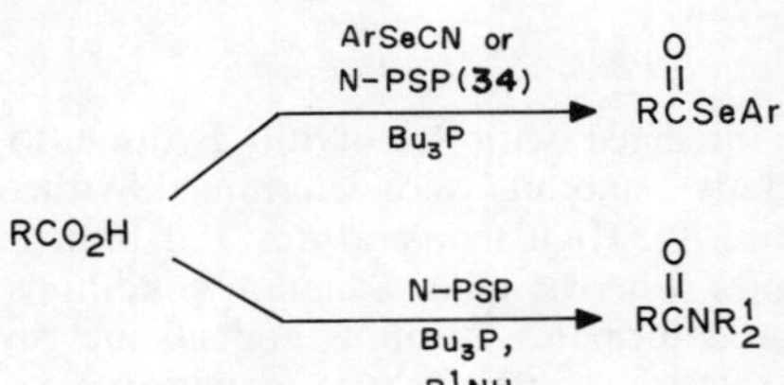

(267)

The deoxygenation of alcohols and carboxylic acids via their selenides and selenoesters is discussed in the following section.

XIV. REDUCTIVE DESELENIZATION AND DETELLURIZATION

A. Reagents

Many selenium- or tellurium-based transformations require the reductive removal of the selenium or tellurium residue after it has served its purpose. A number of procedures have been developed for this reduction. Deselenizing reagents include Raney nickel[78,395,477,493,513,645,813], nickel boride[112,395,566,814], lithium triethylborohydride[815] and lithium in ethylamine[645]. Free radical deselenizations with tributyl[285,480,816] or triphenyltin hydride[525,780,817] have proved especially useful. Detellurizations have been carried out similarly with tin hydrides[483,499,780] and Raney nickel[415].

B. Deoxygenation of Alcohols

Since alcohols are efficiently converted to selenides via equations 265 and 266, subsequent reductive deselenization[645,780,817] results in their overall deoxygenation to alkanes (equation 268).

$$\text{ROH} \longrightarrow \text{RSeAr} \xrightarrow{\text{reductive deselenization}} \text{RH} \qquad (268)$$

C. Deoxygenation of Aldehydes and Ketones

Selenoacetals are readily available from their parent aldehydes or ketones[631]. Reductive deselenization then completes the deoxygenation of the carbonyl compound to the corresponding hydrocarbon[634,645,780,817] (equation 269). The procedure is therefore an attractive alternative to the Wolff–Kishner and Clemmensen reductions. During detailed studies of the utility of triphenyltin hydride in this respect, Clive and coworkers[780,817] also demonstrated that the corresponding tin deuteride made possible the preparation of deuterium-labelled hydrocarbons.

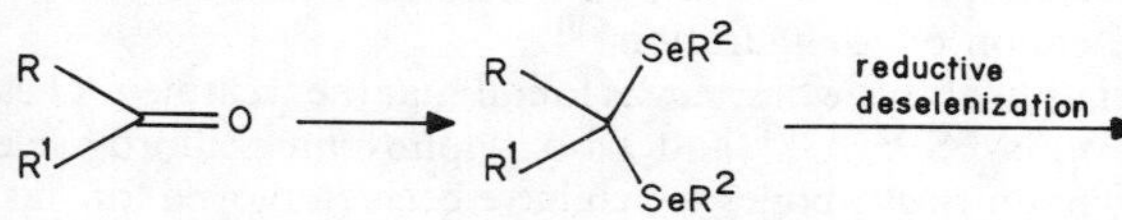

(269)

D. Reduction of Selenoesters and Telluroesters

1. *Se-Phenyl selenoesters and selenocarbonates*

Se-Phenyl selenoesters reacted with tributyltin hydride to afford aldehydes and alkanes[818–820] whose relative amounts were determined by the nature of the substrates and conditions (equation 270). High temperatures and branched selenoesters favored decarbonylation to alkanes, whereas mild reduction conditions favored the aldehydes. Selenocarbonates produced formates, alcohols and alkanes by deselenization, decarbonylation and decarboxylation, respectively[818] (equation 271).

$$RC(=O)SePh \xrightarrow[\text{AIBN, } \Delta \text{ or } h\nu]{Bu_3SnH} RCHO + RH \quad (270)$$

$$ROC(=O)SePh \xrightarrow[\text{AIBN, } \Delta \text{ or } h\nu]{Bu_3SnH} ROC(=O)H + ROH + RH \quad (271)$$

2. O-Alkyl seleno- and telluro-esters

The selenoester **107** afforded cholestene as the principal product after tin hydride reduction[821] (equation 272), and *O*-alkyl telluroesters produced monomeric and dimeric ethers[628,822] with NaHTe (equation 273).

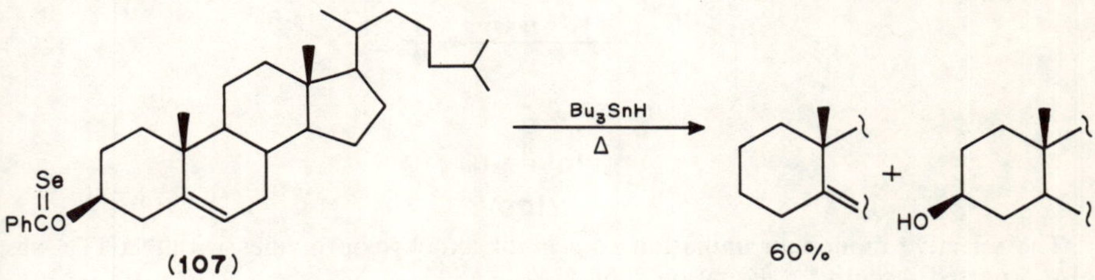

(272)

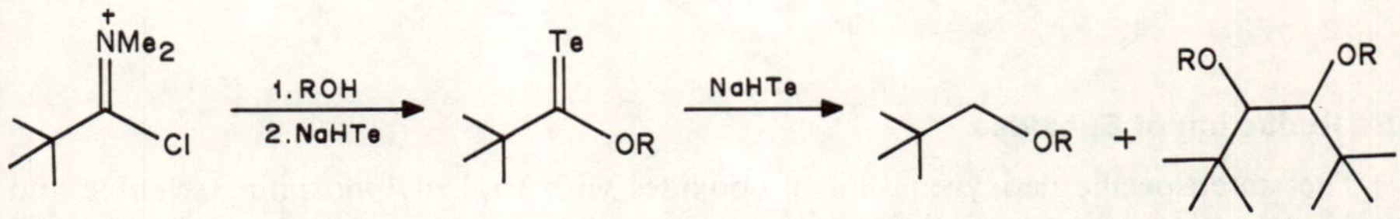

(273)

XV. REDUCTIONS WITH SELENIUM AND TELLURIUM COMPOUNDS

Selenolates and tellurolates, and also inorganic species such as NaHSe and NaHTe, function as efficient and sometimes highly selective reducing agents for many organic functionalities. In many cases the reducing species can be formed *in situ* in catalytic amounts in the presence of a less expensive co-reductant.

A. Reduction of Halides

Vicinal dihalides and also β-chloro- and β-bromo-selenides undergo reductive elimination to olefins with NaHSe[823], NaSePh[824], NaSeMe[824], Na_2Te[825], NaHTe[826,827], KTeAr[828] and the tellurolate **108**[829] (equation 274). The products are often formed stereospecifically by *anti*-elimination. Conjugated dienes were regenerated from their 1,4-dibromo adducts with reagent **108**[830]. The sequence of bromination–debromination can therefore be used in the protection of olefins and dienes. Further reduction of the olefin product was observed with excess of NaHTe in the case of styrene derivatives[827].

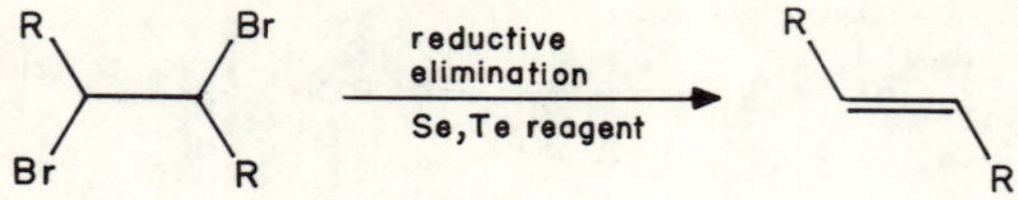

(274)

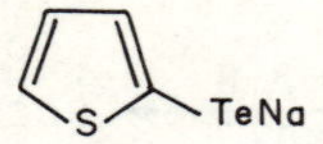

(108)

α-Halo ketones[831–834], esters[833], amides[832,833] and carboxylic acids[832] were dehalogenated with PhSeH[834], NaHTe[833], the tellurolate **108**[832] or the tellurophosphate **109**[831] (equation 275). In some cases α-acetoxy, α-mesyloxy or α-phenylthio substituents were similarly reduced[832].

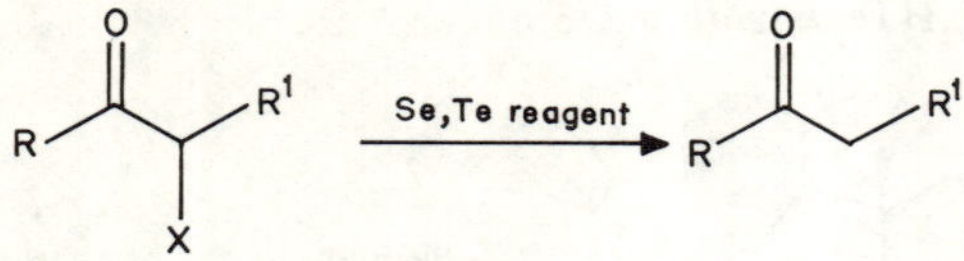

(275)

$(EtO)_2P(O)TeNa$

(109)

The selective monodebromination of *gem*-dibromocyclopropanes with NaHTe was also reported recently[835] (equation 276).

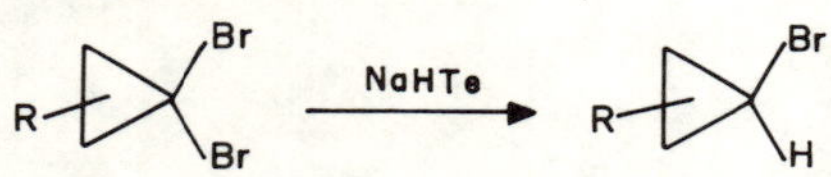

(276)

B. Reduction of Epoxides

The stereospecific deoxygenation of epoxides with triphenylphosphine selenide and trifluoroacetic acid provides olefins with retention of configuration[836,837] via the pathway shown in equation 277. Other reagents which were later employed in a similar manner include the more reactive phosphine selenides **110** and **111**[838], the selenoamides **112**[839]

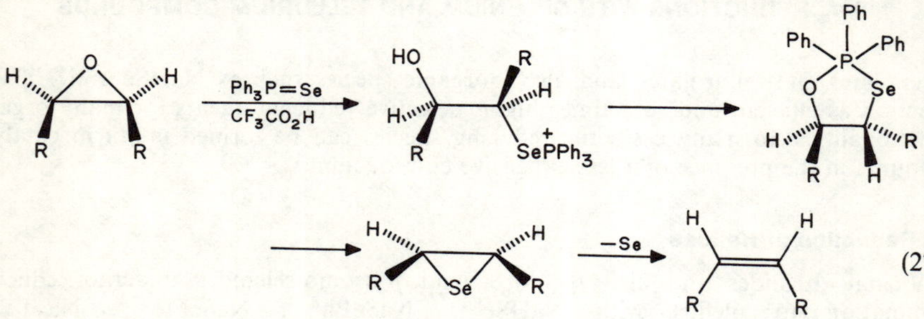

(277)

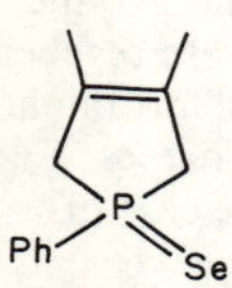

(110)

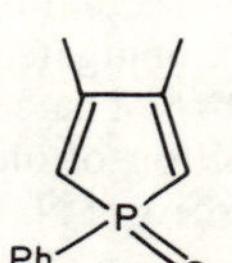

(111)

$PhC(=Se)NH_2$

(112)

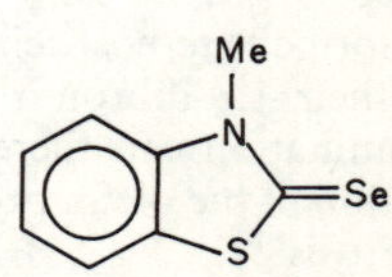

(113)

and **113**[840], the tellurophosphate **109**[841] and KSeCN[842]. Alternatively, the deoxygenation of epoxides can be carried out by ring opening with a selenolate followed by reductive elimination (see Section XII.C).

Potassium selenocyanate (KSeCN) was reported to effect the $E \rightleftharpoons Z$ isomerization of olefins via their bromohydrins in a process resembling equation 277[843]. This provides an alternative to the $TeCl_4$ -mediated isomerization shown in equation 112.

C. Reduction of Carbonyl Compounds

Aldehydes and ketones are efficiently reduced to alcohols with hydrogen telluride produced *in situ* from the hydrolysis of Al_2Te_3[844] or by photolysis with hydrogen selenide[844] (equation 278). Non-enolizable aldehydes afford primary alcohols with PhSeH[845] or with the magnesium selenolate PhSeMgBr[846]. Enones and other α,β-unsaturated carbonyl compounds undergo preferential reduction of the carbon—carbon double with NaHTe[847], H_2Te[844] or photochemically with PhSeH[848] (equation 279).

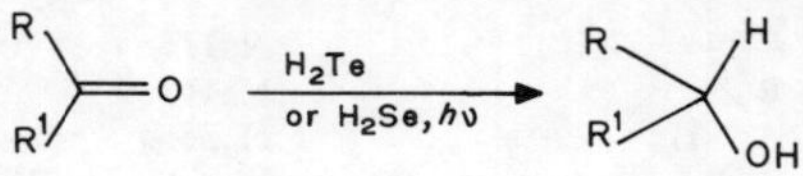

(278)

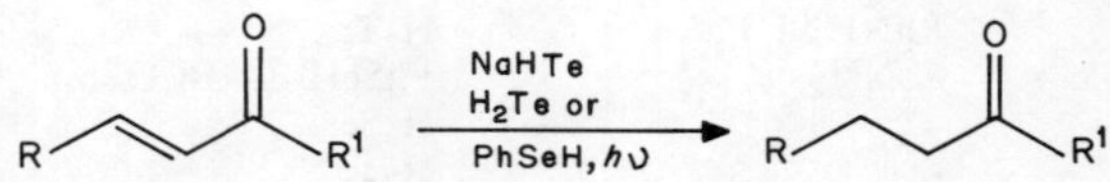

(279)

D. Reduction of Nitrogen Compounds

Selenium and tellurium reducing agents perform several types of synthetically useful transformations of nitrogen compounds. These generally involve reduction of the N—O, N—N, N=N, $-\overset{+}{N}\equiv N$ or C=N bonds. Examples are summarized in Table 2.

E. Reduction of Sulfoxides

Sulfoxides were efficiently deoxygenated to sulfides with selenols[861], the selenoboranes **104** and **105**[862,863], the selenophosphate **114**[864] and the silyl selenides **54**[865] and **115**–**117**[866]. The last reagents were also used to deoxygenate selenoxides and telluroxides (equation 280).

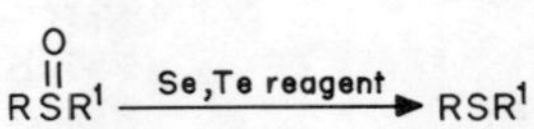

(280)

$(EtO)_2P(O)SeNa$ (**114**)

$Me_3SiXSiMe_3$
(**115**) X=S
(**116**) X=Se
(**117**) X=Te

XVI. REARRANGEMENTS

A. [2, 3] Sigmatropic Rearrangements of Allylic Selenoxides

The facile [2, 3] shifts of allylic selenoxides provide an efficient method for introducing new allylic oxygen functions into an organic substrate. Several examples have already been noted in previous sections in connection with other overall transformations (see equations 70, 138, 207, 214 and 215). In general, [2, 3] sigmatropic rearrangements of

TABLE 2. Reduction of nitrogen compounds

Nitrogen compound	Product	Reagent	References
$ArNO_2$	$ArNH_2$	H_2Se	849, 850
"	"	H_2Te	851
"	"	NaHTe	845, 852
"	"	PhTeH	845, 853
$ArNO_2$	$ArN{=}\overset{+}{N}(O^-)Ar$	NaHTe	852
RNO_2	RNH_2	PhSeH–DABCO	854
RNO_2	$(RNO)_2$	NaHTe	852
$R\text{-}\overset{+}{N}(O^-){=}CHR'$	$R\text{-}NHCH_2R'$	NaHTe (pH 6)	855
$R\text{-}\overset{+}{N}(O^-){=}CHR'$	$R\text{-}N{=}CHR'$	NaHTe (pH 10–11)	855
$R_3\overset{+}{N}\text{-}\overset{-}{O}$	R_3N	NaHTe	855
$ArN{=}O$	$ArNH_2$	H_2Se	850
"	"	PhSeH–DABCO	854
$PhN{=}O$	PhNHNHPh	H_2Te	851
ArNHOH	$ArNH_2$	PhSeH–DABCO	854
"	"	H_2Se	850
"	"	H_2Te	851
$ArN{=}\overset{+}{N}(O^-)Ar$	ArNHNHAr	H_2Te	851
"	"	NaHTe	845
$ArN{=}\overset{+}{N}(O^-)Ar$	$ArNH_2$	PhSeH–DABCO	854
$ArN{=}NAr$	ArNHNHAr	H_2Te	851
$ArN{=}NAr$	$ArNH_2$	PhSeH–DABCO	854
RN_3	RNH_2	NaHTe	856
$ArN_2{}^+BF_4{}^-$	$ArNHNH_2$	$PhSeH\text{–}CH_2Cl_2$	857
$ArN_2{}^+BF_4{}^-$	ArSeH + ArH	PhSeH–acetone	857
ArNHNHAr	$ArNH_2$	PhSeH–DABCO	854
$R_2C{=}N\text{-}R'$	R_2CHNHR'	PhSeH	848, 858
"	"	NaHTe	845, 859
"	"	PhTeH	845
$R_2C{=}N\text{–}X$ ($X = OH$ or NR^1_2)	R_2CHNHX	PhSeH	848
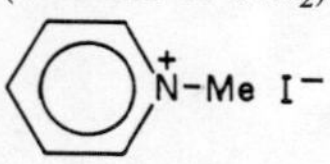	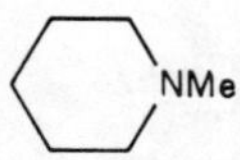	NaHTe (pH 6)	859
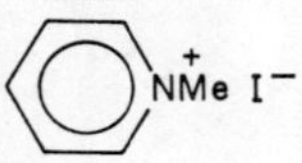	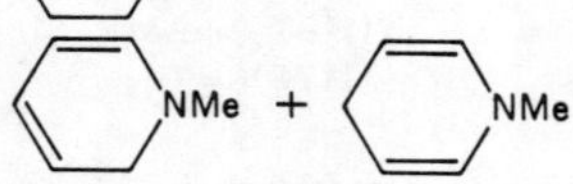 2 : 1	NaHTe (pH 10–11)	859
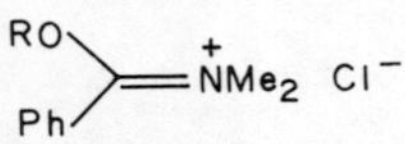	$PhCH_2OR$	NaHTe	821, 860

selenoxides are stereospecific and more rapid than competing selenoxide eliminations. However, if the rearrangement is impeded by steric or conformational effects, then diene products resulting from elimination can become dominant[37,138].

Kametani and coworkers[374,375] and Clive and coworkers[867] independently developed an allylic alcohol transposition process based on conversion of the initial alcohol to a selenide with Grieco and coworkers' ArSeCN–Bu_3P reagent, followed by oxidation and rearrangement (equation 281). Several recent applications are depicted in equations 282[808], 283[868] and 284[123].

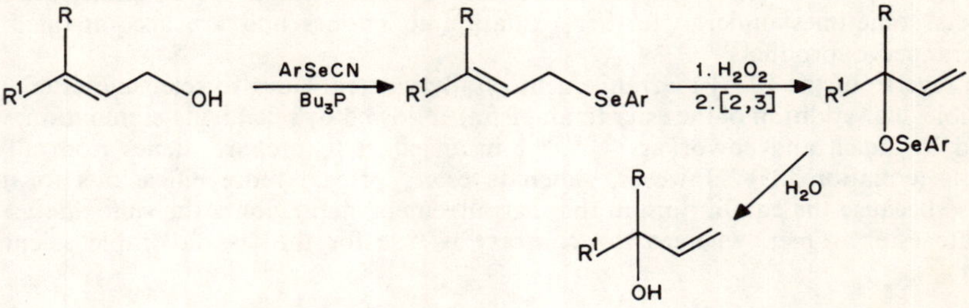

(281)

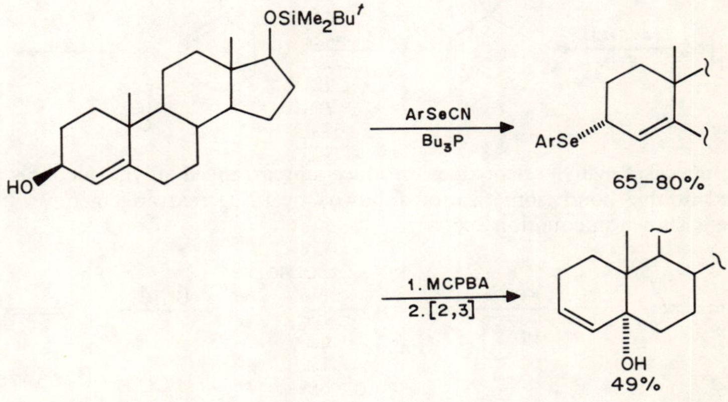

(282)

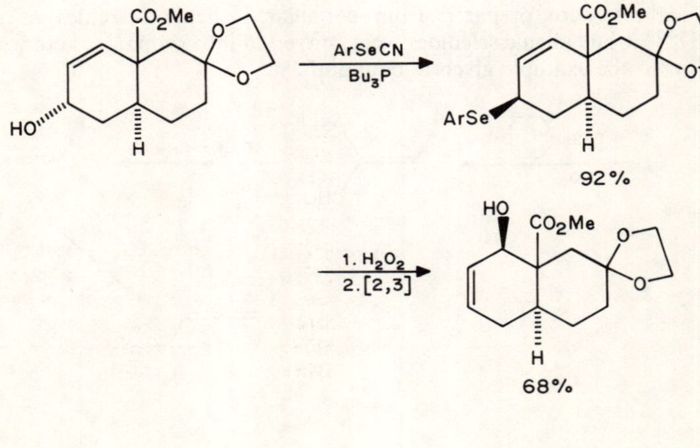

(283)

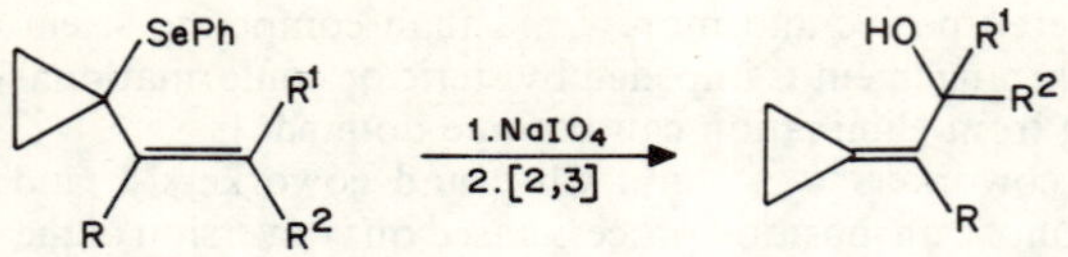

(284)

Oxaziridines are particularly effective reagents for the oxidation step[88] and chiral oxaziridines resulted in chirality transfer to the product allylic alcohol with enantiomeric excesses of up to 12.8%[89]. Allylic telluroxides undergo similar [2, 3] shifts, but the products sometimes undergo further oxidation to enones and are accompanied by unrearranged byproducts[869].

The reverse of the [2, 3] rearrangement of allylic selenoxides to selenenic esters (i.e. intramolecular addition of the ester to an olefin), followed by selenoxide elimination, was studied by Reich and coworkers[66,870,871] in an effort to prepare dienes from allylic alcohols (equation 285). However, sulfenate esters proved more efficacious for this purpose because the equilibrium in the rearrangement step favours the sulfoxide vs. its sulfenate ester isomer, whereas the contrary is true for the less favorable selenium system[872].

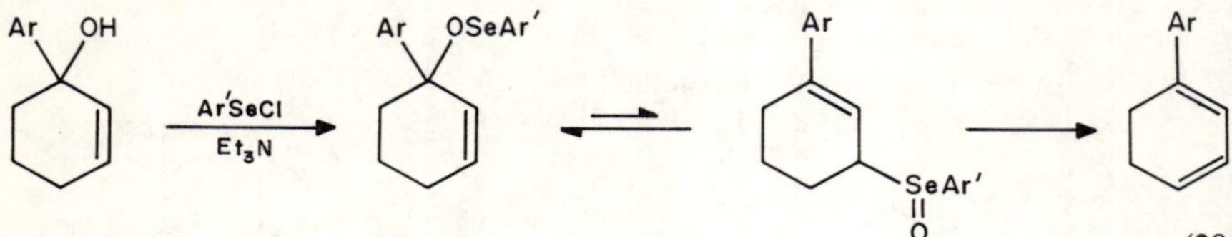

(285)

In some cases, vinyl selenoxides produce unrearranged allylic alcohols by base-catalysed double bond isomerization followed by [2, 3] rearrangement[112,222,693]. An example is shown in equation 286[112,222].

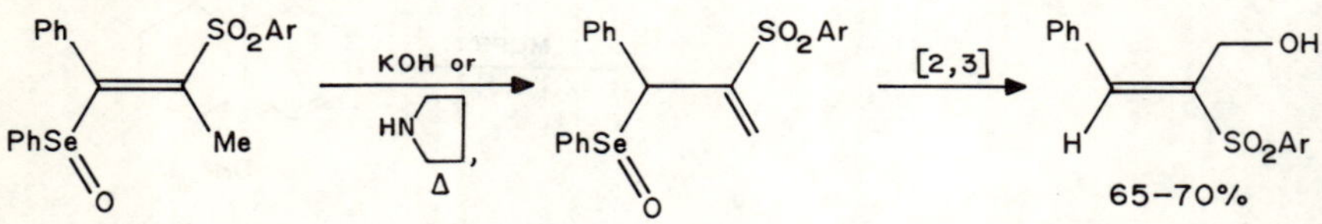

(286)

Allenic alcohols were prepared from conjugated dienyl selenides as shown in equation 287[406], while allenic selenides were converted into propargyl ketones[460,873] or alcohols[874] as in the example given in equation 288[873].

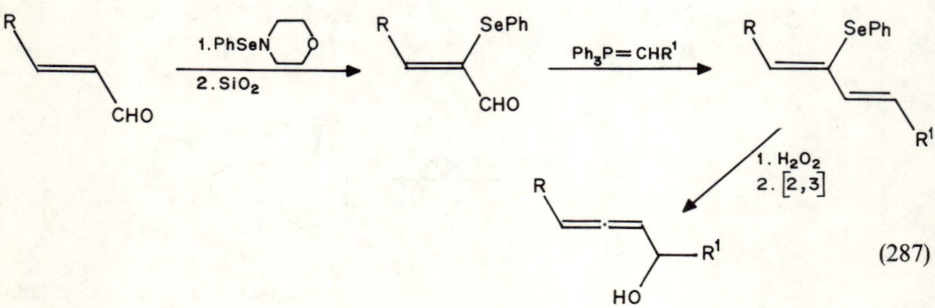

(287)

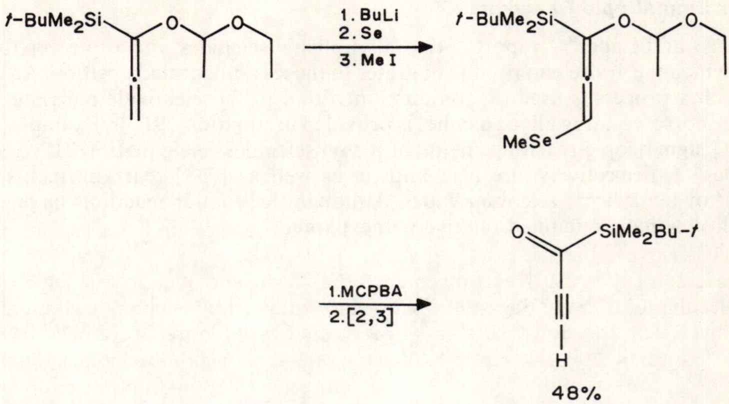

(288)

Allyl silanes serve as convenient precursors of rearranged alcohols[875–877]. For example, silanes **118** were treated with PhSeCl and $SnCl_2$ to afford the selenides **119**, which in turn produced rearranged alcohols on oxidation and [2,3] shift, as shown in equation 289[877]. An added feature of the method is that sulfenyl chlorides can be similarly employed but give the isomeric unrearranged allylic alcohols via sulfides **120**.

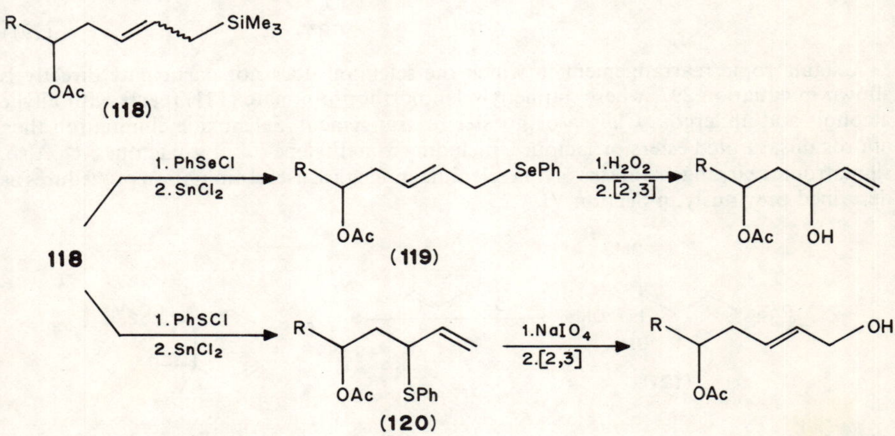

(289)

A different reaction in which an allyl silane was transformed into an unrearranged alcohol is illustrated in equation 290[878].

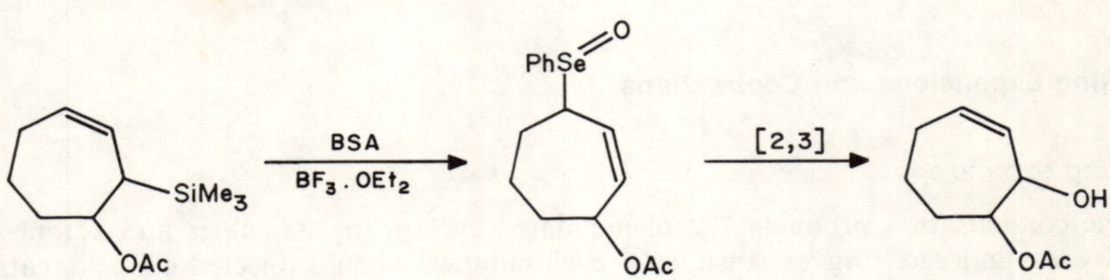

(290)

B. Other Sigmatropic Reactions

Sharpless and Lauer[879] reported that allyl phenyl selenides undergo reversible [1, 3] shifts in which the arylseleno residue migrates to the less substituted position. An example where such a process is used on conjunction with a [2, 3] selenoxide rearrangement to obtain the corresponding allylic alcohol is provided in equation 291[692]. Examples of [1, 5] and [2, 3] sigmatropic rearrangements of dienyl selenides (see equation 217) and allylic diselenides[879], respectively, are also known, as well as [2, 3] rearrangements of allylic[880–882] or benzylic[883] selenium ylides. Although these latter reactions have synthetic potential, they have remained relatively unexploited.

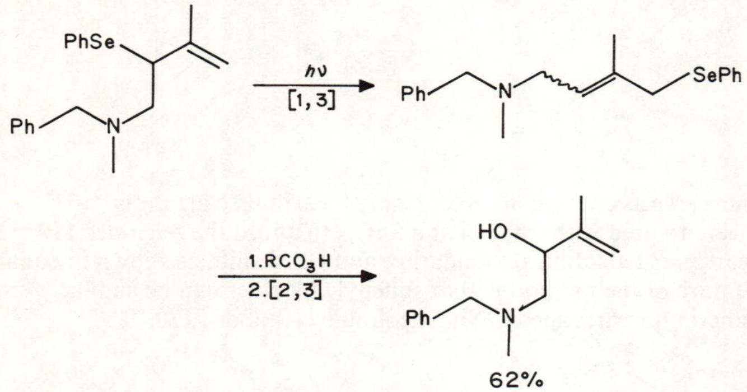

(291)

A sigmatropic rearrangement in which the selenium does not participate directly is shown in equation 292, where 3-(phenylseleno)orthopropionate (**121**) reacts with allylic alcohols and undergoes Claisen orthoester rearrangement. Selenoxide elimination then affords unsaturated esters or lactones, including α-methylene-γ-butyrolactones[884]. Also, sigmatropic rearrangements involving selenium imides are useful amination procedures as described previously in Section VI.A.2.

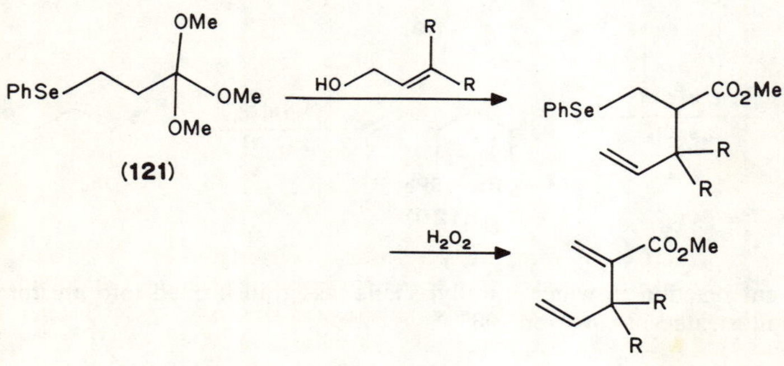

(292)

C. Ring Expansions and Contractions

1. Ring expansions

The condensation products **122** of metalated cyclopropyl selenides and ketones or aldehydes undergo ring expansion to cyclobutanes when subjected to acid catalysis[668,670,671] (equation 293). The cyclopropyl selenides **123** behave similarly[661]. The procedure can be employed in an iterative manner as the product cyclobutanones[670,671] as well as cyclic ketones of larger ring sizes[667,669], are themselves easily converted into β-hydroxy selenides capable of ring expansion. Several related selenium-mediated ring expansions of cyclopropyl[94,777] and cyclobutylcarbinols[701] were also reported and one example is given in equation 294[94]. A different type of ring expansion of cyclic dithioketals is shown in equation 295[885]. A ring expansion of cyclic enones is described in Section XVII.A (see equation 300).

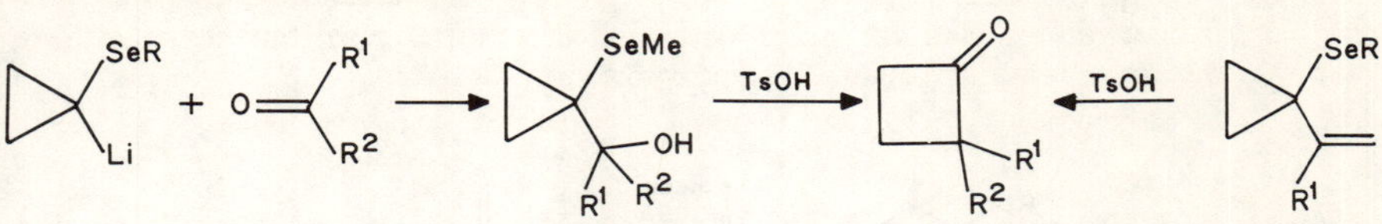

(122) (123)

(293)

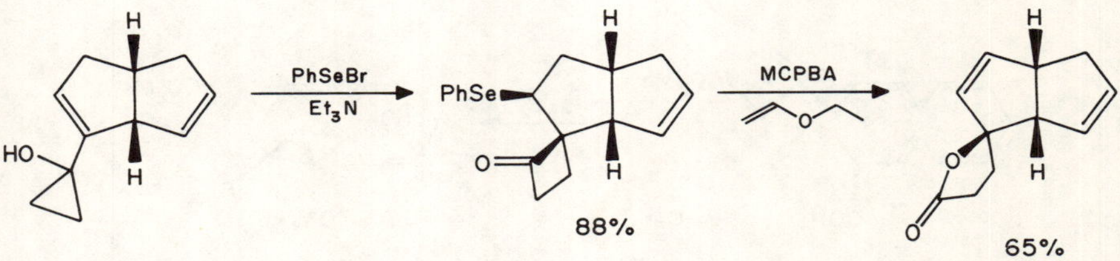

(294)

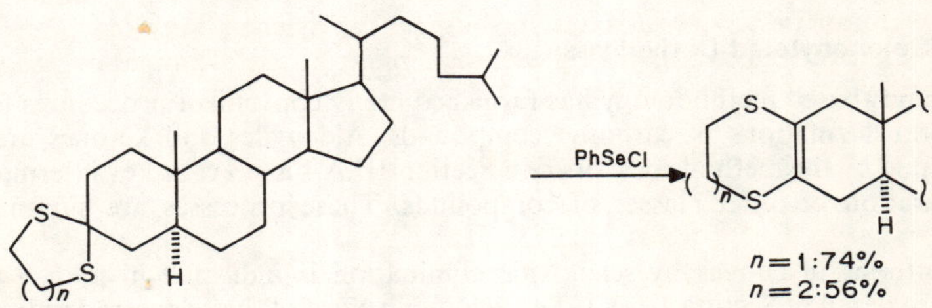

(295)

2. Ring contractions

The ring contraction of β-methoxycycloalkyl selenides **124**[283,387] and tellurides **125**[117,283] is promoted by oxidation to the corresponding selenones and tellurones with MCPBA (equation 296). Selenoxide elimination was observed instead when hydrogen peroxide was used as the oxidant, presumably because oxidation beyond the selenoxide or telluroxide level proceeds more slowly than elimination.

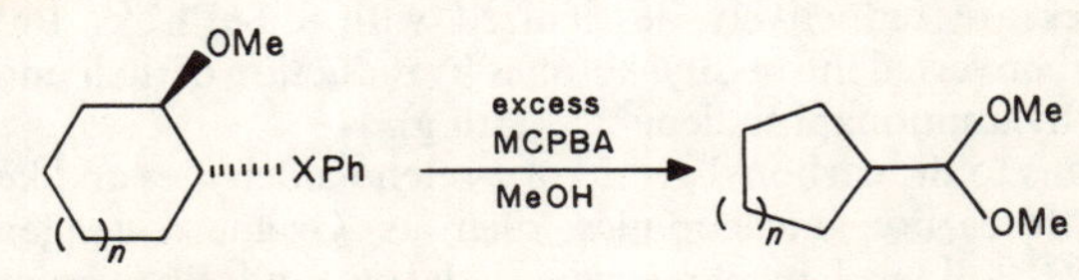

(296)

(**124**) X=Se
(**125**) X=Te

D. Other Rearrangements

Several examples of isolated cases of selenium-induced rearrangements with synthetic potential are provided in equations 297[886], 298[887] and 299[888].

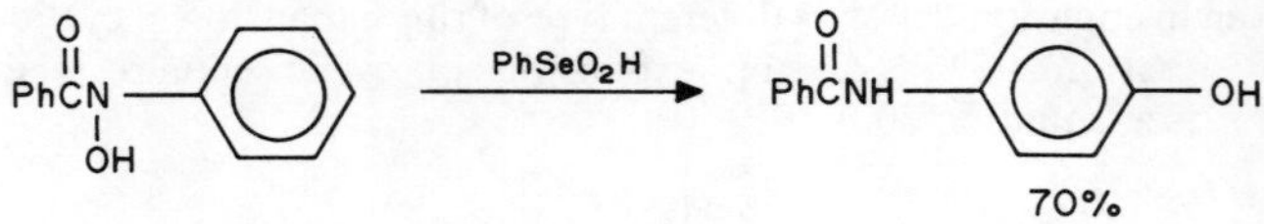

(297)

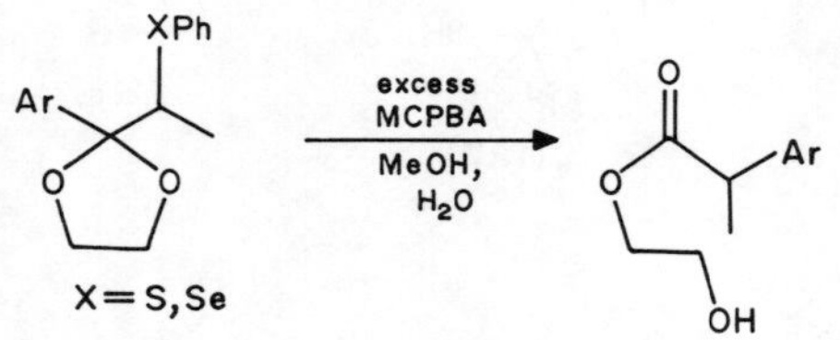

(298)

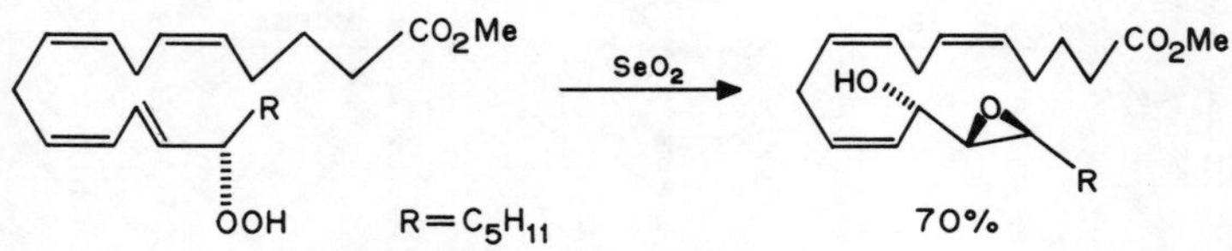

(299)

XVII. TRANSFORMATIONS OF ALDEHYDES, KETONES AND ENONES

A. Via α-Selenenylated Derivatives

Selenenium-based methodology has furnished many convenient procedures for effecting key transformations of carbonyl compounds. Aldehydes and ketones are readily selenenylated by the methods described in Section III.A.1 and act as key intermediates in the preparation of other classes of compounds. These processes are summarized in Scheme 2.

The synthesis of enones by selenoxide elimination is indicated in path a and was discussed in detail in Section III.A.1. The acidifying effect of the selenium residue permits the regioselective alkylation of α-seleno ketones via their enolates[212,311,799,802,803,880,889] (path b). The alkylated products can be further converted to exo-[889] or endo-cyclic enones[799,802,803] (path c).

The treatment of α-seleno ketones with 0.5 equivalent of LDA in THF containing HMPA results in the transselenenylation of the selenium moiety from the α to the α'-position, if it is less substituted than the α-position[889,890] (path d). Similar transselenenylation were noted in α-seleno-β-ketoesters and their enamines[891]. Selenoxide elimination of the rearranged products affords α',β'-unsaturated ketones via path e[212,890].

α-Seleno ketones are reductively deselenized with LiSePh[892], PhSH[311,694,695] or HBr[893] (path f), or converted into α-silyl ketones by reduction of their enol silyl ethers with lithium and dimethylaminonaphthalene[894] (path g).

Addition reactions to the carbonyl group of α-seleno aldehydes and ketones occur with $LiAlH_4$[895] or with carbon nucleophiles such as Grignard reagents[214,221,321,895], Me_3SiCH_2Li[896,897] (followed by dehydroxysilylation and PhSe migration) and enolates[898–901] to furnish β-hydroxy selenides (path h). The diastereoselectivity of such

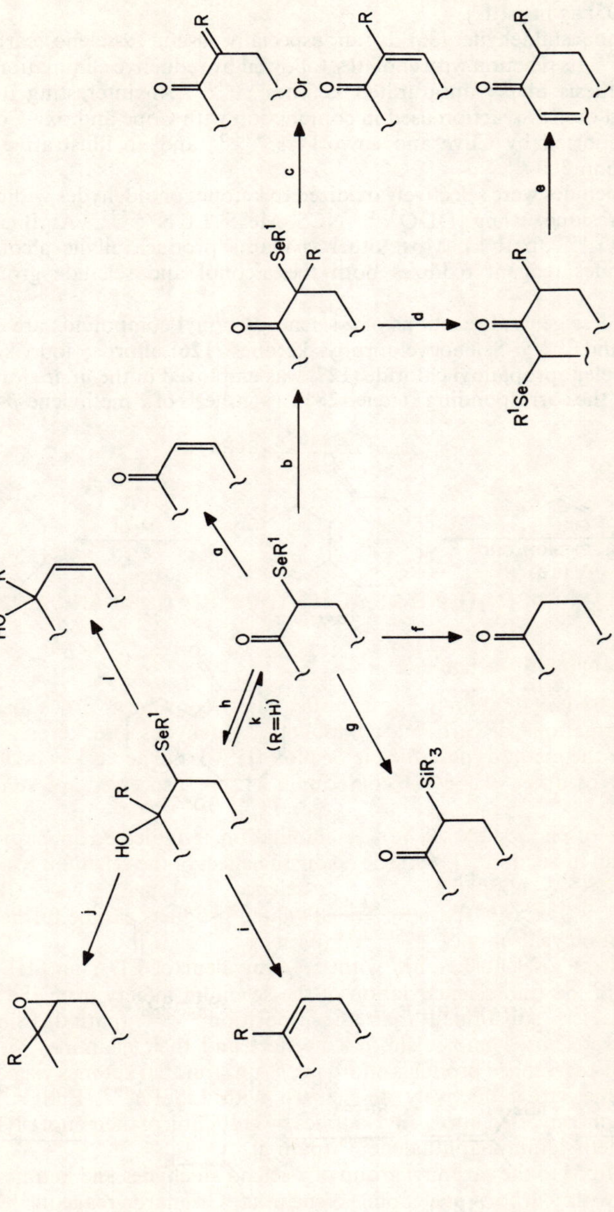

SCHEME 2

additions favors *threo* products with $LiAlH_4$ (i.e. R = H in path h) and *erythro* isomers with Grignard reagents[895]. The products in turn afford olefins by reductive elimination[321,895] (path i) or epoxides via the corresponding selenonium salts[214,221,321,895] (see equation 205) as in path j.

α-Phenylselenoacetaldehyde (**36**) is an especially useful α-seleno carbonyl compound[321,898–901]. Its reaction with enolates, followed by reductive elimination, provides a convenient synthesis of β,γ-unsaturated ketones[898–901]. An interesting iterative ring expansion based on this reaction used in conjunction with Cope and oxy-Cope rearrangements was reported by Clive and coworkers[899,901] and an illustrative example is shown in equation 300[901].

β-Hydroxy selenides were selectively oxidized to ketones or aldehydes without affecting the selenium residue using DDQ[321], $NCS–Me_2S–Et_3N$[92,321], Ar_3BiCO_3[902] and $CCl_3CHO–Al_2O_3$[903] (path k). Most other oxidants produce allylic alcohols instead (path l), and Jones reagent oxidizes both the alcohol and selenide groups to give enones[91,92].

Several other less general reactions of α-seleno carbonyl compounds are displayed in equations 301 and 302. α-Selenocyclopropyl ketones (**126**) afford γ-iodo ketones with P_2I_4[904], and α-selenopropanoyl chloride (**127**) was employed in the *in situ* formation and cycloaddition of the corresponding ketene **128** in a synthesis of α-methylene-β-lactams[905].

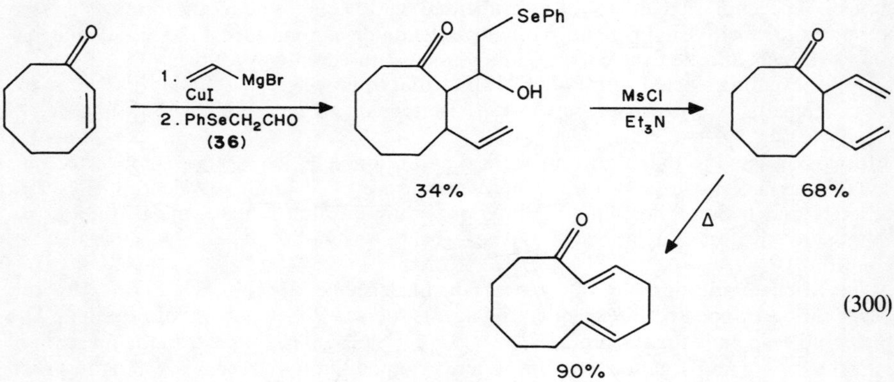

(300)

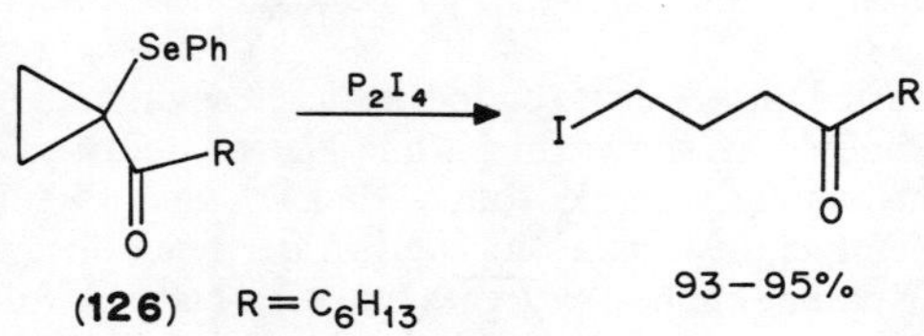

(301)

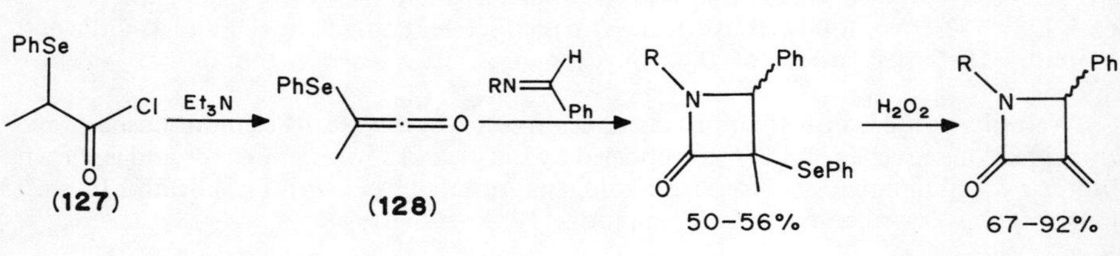

(302)

B. Reactions of Enones

Synthetically useful selenium-mediated transformations of enones are described in Scheme 3. Enones[91,906–908] (and also α, β-unsaturated lactones[906,909–913], propargylic esters[914,915] and nitroolefins[916]) act as Michael acceptors of selenols and selenolates (path a). Since the reaction can be easily reversed by selenoxide elimination (path b) the procedure serves as a protective method for sensitive functionalities such as α-methylene lactones[909] and α-methylene cyclic ketones[908]. Selenols add to cyclohexenones in the presence of (−)-cinchonidine to afford optically active β-seleno ketones with an enantiomeric excess of 11–43%[917]. The β-seleno ketones obtained by path a react with hydride-reducing agents[907] or with Grignard reagents[91] to afford the corresponding alcohols (or lactols from β-seleno lactones[910]), as shown in path c.

Michael additions to enones were also studied with the silyl selenide **54**[68,446,772,918,919], the aluminum selenolate **102**[718,719,920,921] and a selenoborane[922]. In some cases, reagents **54** and **102** were used to deliver the equivalent of the selenolate anion RSe^- to the enone to give β-seleno ketones in much the same manner as path a[718,719,772,919]. The silyl selenide **54** also afforded isolable β-seleno enol silyl ethers from enones in the presence of Ph_3P[918] or iodine[446] (path d; $X = SiMe_3$). In other instances, the corresponding enol aluminates[920,921] (path d; $X = AlMe_2$) and enol boronates[922] (path d; $X = BR^2_2$) were employed in subsequent aldol condensations (path e). The enol silyl ethers also condensed with ketals and orthoformates in the presence of trimethylsilyl triflate[68]. Finally, selenoxide elimination of the aldol condensation products afforded the corresponding α-substituted enones (path e).

Cyclohexenone reacted with PhSeCN and tributylphosphine in a similar manner to the above reagents to furnish the γ-seleno-α, β-unsaturated nitrile product shown in path f in 50% yield[704].

In work related to that shown in paths d and f, simple aldehydes were converted into silylated hemiselenoacetals with the silyl selenide **54**[446,918] and to α-seleno nitriles with ArSeCN and tributylphosphine[704], thus providing an alternative route to these compounds to the selenenylation of nitriles. These reactions are shown separately in equation 303.

The Michael additions of various carbon nucleophiles to enones generate enolates which can be trapped with selenenylating agents such as PhSeCl or PhSeBr (path g). The nucleophiles include organocuprates[133,134,923–926], Grignard reagents in the presence of Cu^I salts[927–930] and organozirconium[931] and organoaluminum reagents[931]. In most such examples, the selenenylation step was followed by selenoxide elimination, resulting in the overall β-alkylation of the original enone, or by reductive deselenization to the saturated ketone (path h).

Enones can be converted into α-seleno enones with selenenamides (see equation 106) or with PhSeCl in pyridine[932] as shown in path i. The related α-selenenylation of α, β-unsaturated esters was similarly carried out with LDA and PhSeBr[933]. The α-seleno enones are in turn useful as dienophiles in Diels–Alder reactions[934]. They can also be reconverted into their parent enones by treatment with NaSePh followed by oxidation with H_2O_2[695] (path j), or subjected to Michael addition with organocuprates via path k[694,695,889,892] followed by oxidative or reductive removal of selenium as indicated in path h. A brief review of the transformations of α-seleno enones has recently appeared[934].

Two other useful transformations of enones are worthy of note. A selenium-based enone transposition sequence (path l) was reported by Liotta and coworkers[935,936] and is shown in more detail in equation 304. The α-halogenation of enones with PhSeBr and PhSeCl (path m) was described earlier (see equation 110).

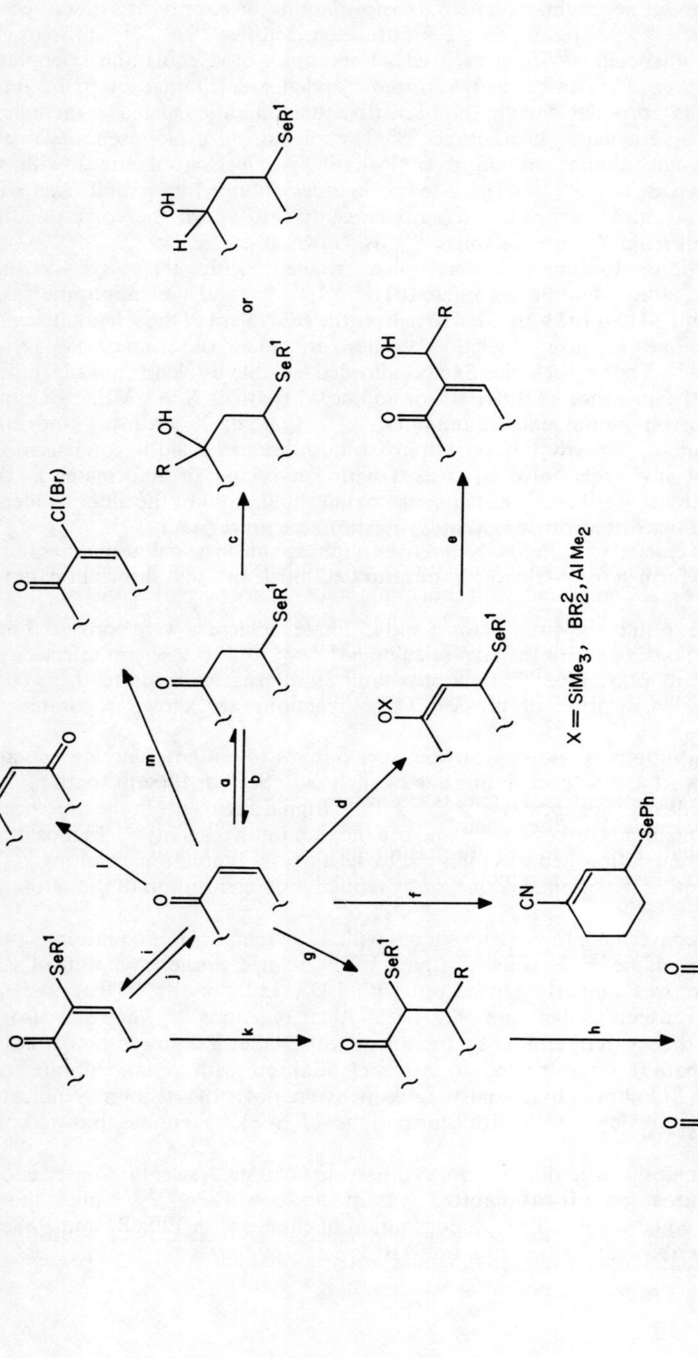

SCHEME 3

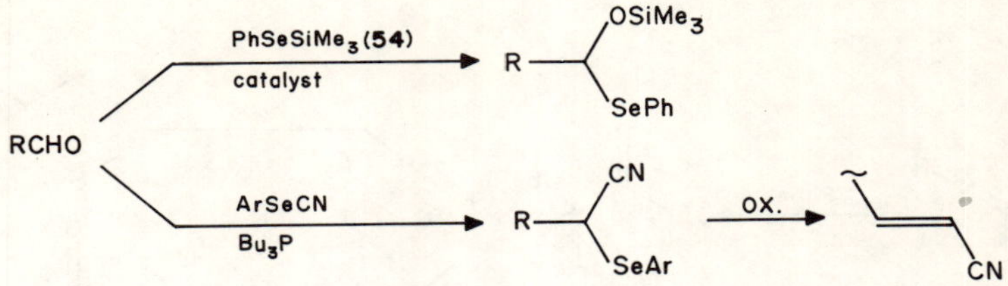

(303)

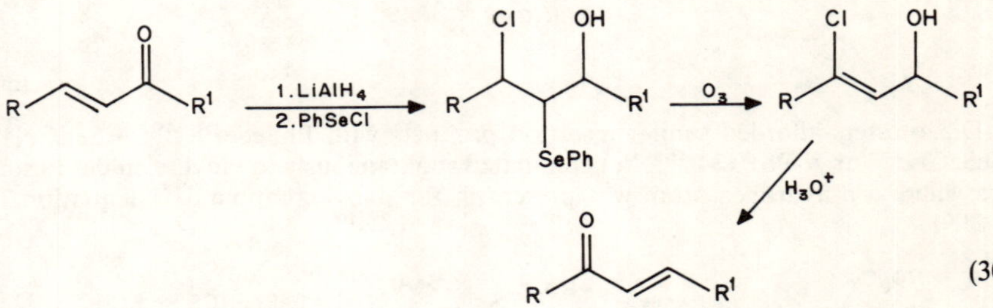

(304)

XVIII. TRANSFORMATIONS OF DIAZO COMPOUNDS

Several types of selenium and tellurium compounds undergo insertion reactions by the carbon atom of a diazo compound, with concomitant loss of nitrogen. It is thus possible to introduce a selenium or tellurium moiety, together with an additional functional group, to the diazo carbon atom. Although formally related, such reactions proceed via a number of different pathways, including ionic, carbene or carbenoid, and free-radical mechanisms, depending on the nature of the reactants and the conditions.

A. Diazoalkanes

Diazomethane and other diazoalkanes insert into the Se—Se, Te—Te, Se—Br or Se—X linkage of diselenides[937], ditellurides[937,664], selenenyl bromides[263,665] and various selenenyl pseudohalides[463] as shown in equation 305, while a selenosulfonate afforded an anomalous double-insertion product with diazomethane under photochemical conditions[271] (equation 306).

$$CH_2N_2 + ArSeX \longrightarrow ArSeCH_2X \quad X = ArSe, Br, NCS, NCSe, \text{phthalimido} \tag{305}$$

$$\underset{\text{(excess)}}{CH_2N_2} + p\text{-TolSO}_2\text{SePh} \xrightarrow{h\nu} \underset{60\%}{p\text{-TolSO}_2CH_2CH_2\text{SePh}} \tag{306}$$

B. α-Diazo Ketones and α-Diazo Esters

Cyclic and acyclic α-diazo ketones were conveniently converted into α-substituted enones by their reaction with selenenyl halides or pseudohalides, followed by selenoxide elimination or dehydrohalogenation[938,939] (equation 307).

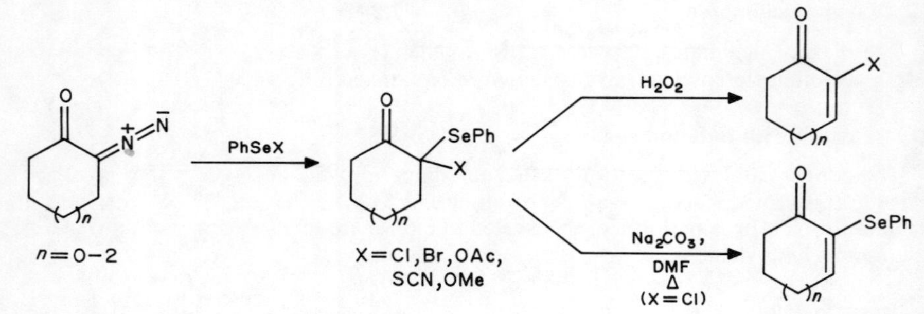

(307)

α-Diazo esters afforded similar insertion products with PhSeSePh[940], ArSeSCN[463], ArSeSeCN[463] or N-PSP (**34**)[463], but eliminated spontaneously to vinyl selenides in some cases where a β-hydrogen atom was present in the diazo compound[463] (equation 308 and 309).

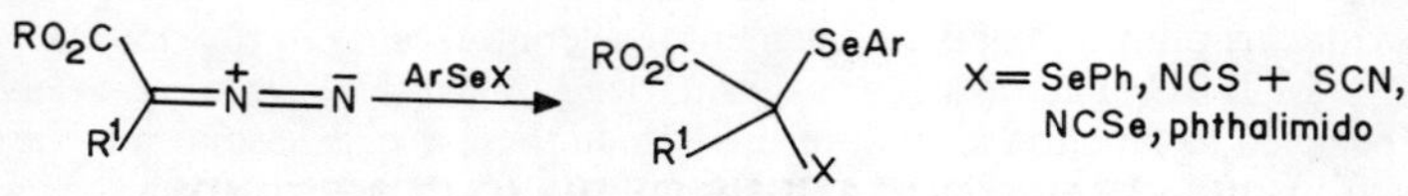

(308)

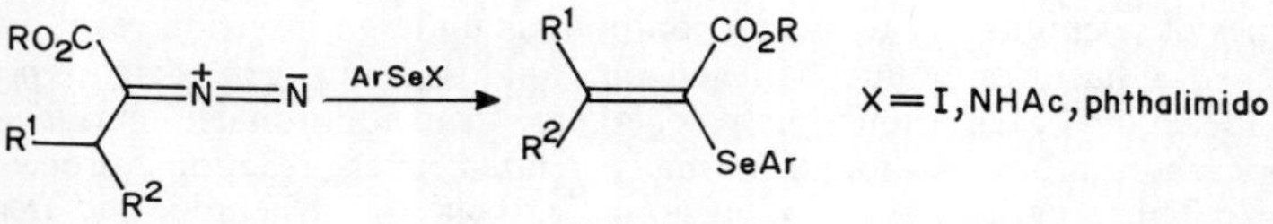

(309)

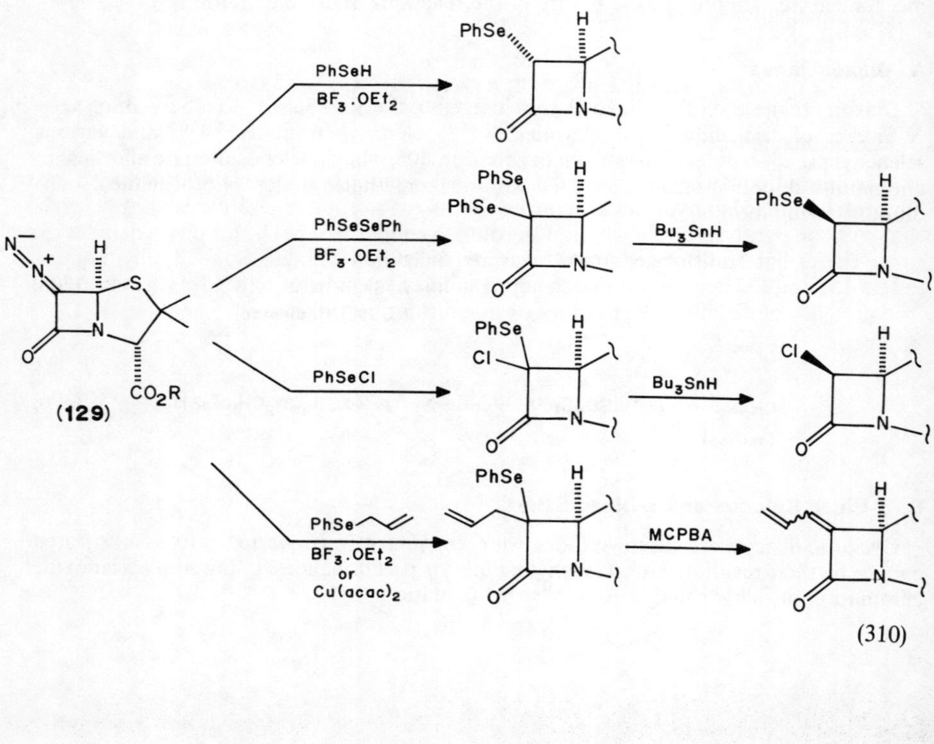

(310)

C. Diazopenicillinates

The diazopenicillinates **129** were transformed into a variety of selenium-containing products, often stereoselectively, as shown in equation 310[881,882,941].

D. Reaction with Selenoesters

The copper-catalysed insertion of diazomethane into the acyl—selenium linkage of selenoesters provides access to α-seleno ketones[893,942] via equation 311. When a workup with aqueous HBr is performed, the α-seleno ketones are further converted into methyl ketones in high yield[893].

$$RC(=O)SeR^1 \xrightarrow[\text{Cu or } Cu^{I}]{CH_2N_2} RC(=O)CH_2SeR^1 \xrightarrow{HBr} RC(=O)Me \qquad (311)$$

XIX. CARBONYLATIONS

A method for the carbonylation of nucleophilic compounds such as amines, alcohols and thiols was developed by Sonoda and coworkers and involves treating the organic substrate with carbon monoxide and elemental selenium, often in the presence of a base such as Et_3N or DBU. The results are summarized in Table 3. Carbonyl selenide (Se=C=O) is believed to function as an intermediate in these processes and may be prepared separately from carbon monoxide and selenium[961]. The selenium can be employed catalytically in carbonylations in the presence of oxygen, which reoxidizes the byproduct hydrogen selenide back to the fee element. The carbonylation of amines was also reported with tellurium instead of selenium[962,963]. Formamide byproducts were produced unless nitrobenzene was added to suppress their formation[963].

The carbonylations of aryltellurium chlorides and of vinyl and alkyl tellurides were effected with nickel carbonyl[964] or with carbon monoxide and $PdCl_2$–LiCl catalyst[965] to afford principally carboxylic acid products.

XX. MISCELLANEOUS SYNTHETIC APPLICATIONS

A. Decarboxylations

Barton and coworkers[102] reported a general decarboxylation method in which acid chlorides or mixed anhydrides are converted into the selenohydroxamic esters **130**, which fragment on pyrolysis or photolysis according to equation 312. The product selenides can then be further transformed oxidatively or otherwise. Similarly, the thiohydroxamic esters **131** and **132** decarboxylated to afford sulfides, selenides or tellurides when heated in the presence of disulfides, diselenides or ditellurides, respectively[966].

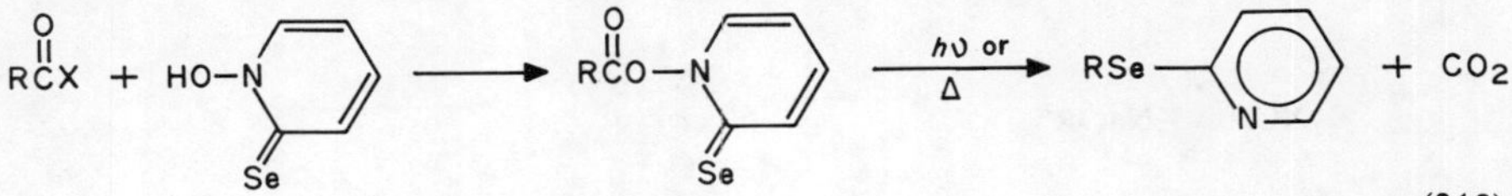

(130)

(312)

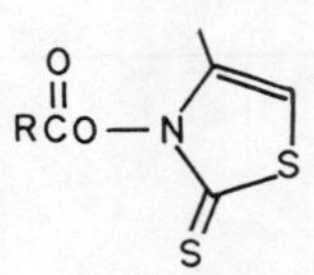

(131)

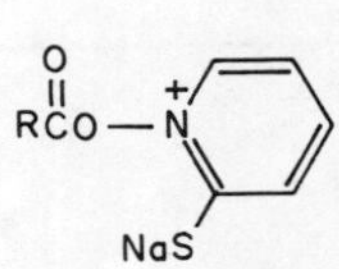

(132)

TABLE 3. Carbonylations with carbon monoxide and selenium

Starting materials	Carbonylation products	Ref.
RNH_2(or $ArNH_2$)	$RNHC(=O)NHR$	943–946
R_2NH	$R_2NC(=O)NR_2$	947
$ArNH_2$	$ArNHCHO$	948
$RNH_2 + R'SSR'$	$RNHC(=O)SR'$	949
$R_2NH + R'X$	$R_2NC(=O)SeR'$	950
$RNH_2 + R'OH$	$RNHC(=O)OR'$	951
Me_2NNH_2	$Me_2NNHC(=O)NHNMe_2$	952
RONa	$ROC(=O)OR$	953, 954
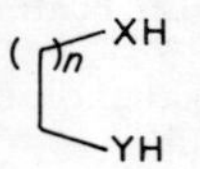 ($n = 1-3$)	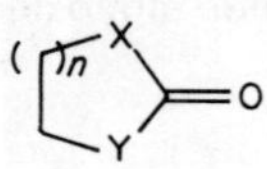 (X,Y = NH,O,S)	945, 955, 956
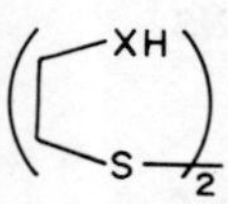	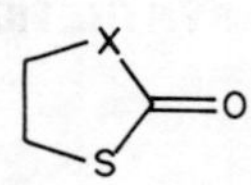 (X = NH,O)	955
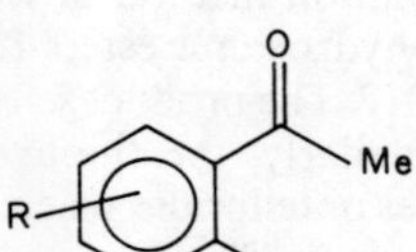	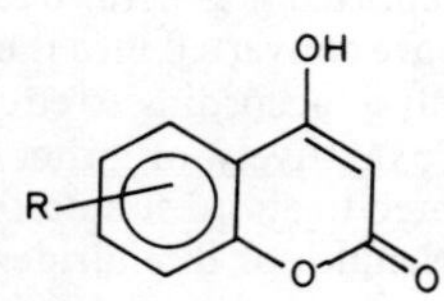	957, 958
$R_2NCHO + NaOR'$	$R_2NC(=O)OR'$	959
$ROCHO + NaOR'$	$ROC(=O)OR'$	960

B. Alkylations with Trimethylselenonium Hydroxide

The selenonium hydroxide **133** is a strong methylating agent which converts carboxylic acids into methyl esters, alcohols and phenols to methyl ethers, thiols to methyl sulfides, and amines and azoles to their *N*-methyl derivatives[967,968] (equation 313).

$$\left.\begin{array}{l} RCO_2H \\ ROH \\ RSH \\ RNH_2 \end{array}\right\} + \underset{(\mathbf{133})}{Me_3\overset{+}{Se}\,\overset{-}{O}H} \longrightarrow \left\{\begin{array}{l} RCO_2Me \\ ROMe \\ RSMe \\ RNHMe \end{array}\right. \qquad (313)$$

C. Telluroxides as Aldol Catalysts

Telluroxides are effective catalysts for aldol condensations[969,970], as illustrated by the example in equation 314[969].

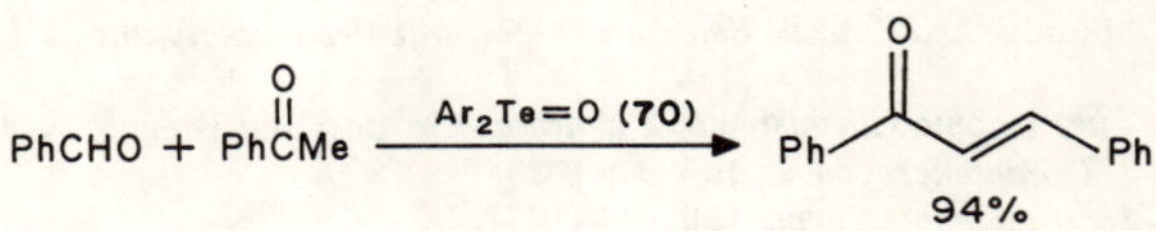

(314)

D. Protecting Group for Alcohols

The β-phenylselenoethyl group was recently employed as an alcohol protective group[971]. It can be introduced via the β-bromoethyl selenide **134** and removed by oxidation and hydrolysis (equation 315). The protected alcohol withstands ^-OH, $NaBH_4$, $LiAlH_4$ and Grignard reagents. The similarly protected dienol ether **135** has recently found application in Diels–Alder reactions[972].

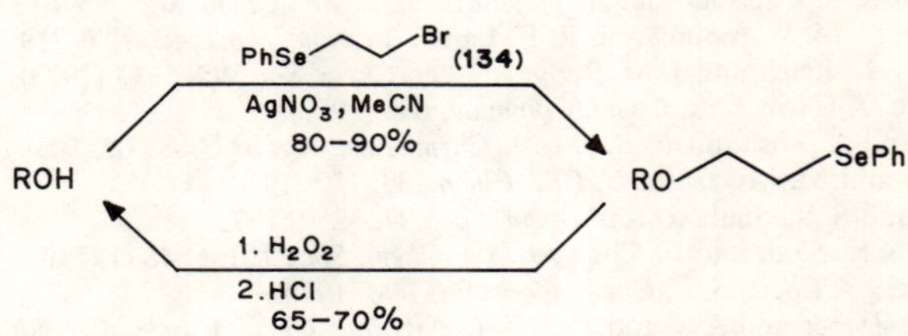

(315)

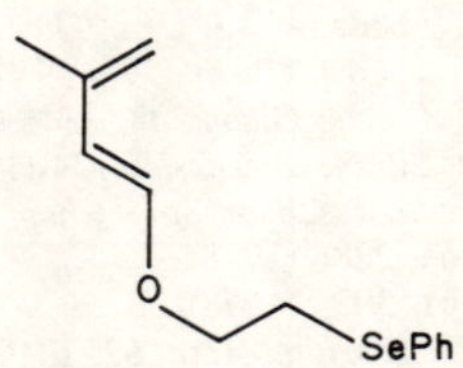

(135)

XXI. REFERENCES

1. D. L. Klayman and W. H. H. Gunther, (Eds.), *Organic Selenium Compounds: Their Chemistry and Biology*,. Wiley–Interscience, New York, 1973.
2. K. J. Irgolic and M. V. Kudchadker, in *Selenium* (Eds. R. A. Zingaro and W. C. Cooper), Van Nostrand Reinhold, New York, 1974, Ch. 8.
3. K. B. Sharpless, K. M. Gordon, R. F. Lauer, D.W. Patrick, S. P. Singer and M. W. Young, *Chem. Scr.*, **8A**, 9 (1975).
4. H. J. Reich, in *Oxidation in Organic Chemistry, Part C* (Ed. W. S. Trahanovsky), Academic Press, New York, 1978, Ch. 1.
5. H. J. Reich, *Acc. Chem. Res.*, **12**, 22 (1979).
6. D. L. J. Clive, *Tetrahedron*, **34**, 1049 (1978).
7. D. L. J. Clive, *Aldrichim. Acta*, **11**, 43 (1978).
8. P. D. Magnus, in *Comprehensive Organic Chemistry* (Eds. D. H. R. Barton and W. D. Ollis), Vol. 3, Part 12, Pergamon Press, Oxford, 1979.
9. S. V. Ley, *Annu. Rep. Chem. Soc., Sect. B*, **77**, 233 (1980).
10. D. Liotta, *Acc. Chem. Res.*, **17**, 28 (1984).
11. D. Liotta, (Ed.), *Recent Aspects of Organoselenium Chemistry*, Tetrahedron Symposium-in-Print, *Tetrahedron*, **41**, 4727–4889 (1985).
12. D. Liotta (Ed.), Organoselenium Chemistry, Wiley, New York, in press.
13. K. C. Nicolaou and N. A. Petasis, *Selenium in Natural Products Synthesis*, CIS, Philadelphia, 1984.
14. K. J. Irgolic, *The Organic Chemistry of Tellurium*, Gordon and Breach, New York, 1974.
15. K. J. Irgolic, *J. Organomet. Chem.*, **103**, 91 (1973).
16. K. J. Irgolic, *J. Organomet. Chem.*, **130**, 411 (1977).
17. K. J. Irgolic, *J. Organomet. Chem.*, **158**, 235 (1978).
18. K. J. Irgolic, *J. Organomet. Chem.*, **158**, 267 (1978).
19. K. J. Irgolic, *J. Organomet. Chem.*, **189**, 65 (1980).
20. K. J. Irgolic, *J. Organomet. Chem.*, **203**, 367 (1980).
21. S. Uemura, *Kagaku* (*Kyoto*), **36**, 381 (1981).
22. L. Engman, *Acc. Chem. Res.*, **18**, 274 (1985).
23. J. L. Huguet, *Adv. Chem. Ser.*, No. 76, 345 (1967).
24. D. N. Jones, D. Mundy and R. D. Whitehouse, *J. Chem. Soc., Chem. Commun.*, 86 (1970).
25. R. Walter and J. Roy, *J. Org. Chem.*, **36**, 2561 (1971).
26. K. B. Sharpless and R. F. Lauer, *J. Am. Chem. Soc.*, **95**, 2697 (1973).
27. K. B. Sharpless, R. F. Lauer and A. Y. Teranishi, *J. Am. Chem. Soc.*, **95**, 6137 (1973).
28. K. B. Sharpless, M. W. Young and R. F. Lauer, *Tetrahedron Lett.*, 1979 (1973).
29. H. J. Reich, I. L. Reich and J. M. Renga, *J. Am. Chem. Soc.*, **95**, 5813 (1973).
30. D. L. J. Clive, *J. Chem. Soc., Chem. Commun.*, 695 (1973).
31. S. Uemura and A, Toshimitsu, *Bull. Inst. Chem. Res., Kyoto Univ.*, **62**, 105 (1984).
32. P. A. Grieco and M. Miyashita, *J. Org. Chem.*, **39**, 120 (1974).
33. H. J. Reich and S. K. Shah, *J. Am. Chem. Soc.*, **97**, 3250 (1975).
34. H. J. Reich, S. K. Shah and F. Chow, *J. Am. Chem. Soc.*, **101**, 6648 (1979).
35. R. H. Mitchell, *J. Chem. Soc., Chem. Commun.*, 990 (1974).
36. N. Zylber, J. Zylber and A. Gaudemer, *J. Chem. Soc., Chem. Commun.*, 1084 (1978).
37. W. G. Salmond, M. A. Barta, A. M. Cain and M. C. Sobala, *Tetrahedron Lett.*, 1683 (1977).
38. H. J. Reich, S. Wollowitz, J. E. Trend, F. Chow and D. F. Wendelborn, *J. Org. Chem.*, **43**, 1697 (1978).
39. P. A. Grieco, S. Gilman and M. Nishizawa, *J. Org. Chem.*, **41**, 1485 (1976).
40. K. B. Sharpless and R. F. Lauer, *J. Org. Chem.*, **39**, 429 (1974).
41. B. M. Trost and D. M. T. Chan, *J. Org. Chem.*, **48**, 3346 (1983).
42. P. Rollin, V. V. Bencomo and P. Sinaÿ, *Synthesis.*, 134 (1984).
43. K. Furuichi, S. Yogai and T. Miwa, *J. Chem. Soc., Chem. Commun.*, 66 (1980).
44. M. Petrzilka, *Helv. Chim. Acta*, **61**, 2286 (1978).
45. M. Petrzilka, *Helv. Chim. Acta*, **61**, 3075 (1978).
46. R. Pitteloud and M. Petrzilka, *Helv. Chim. Acta*, **62**, 1319 (1979).
47. P. Metz and H. -J. Schäfer, *Tetrahedron Lett.*, **23**, 4067 (1982).

48. P. A. Bartlett and P. M. Chouinard, *J. Org. Chem.*, **48**, 3854 (1983).
49. T. Arunachalam and E. Caspi, *J. Org. Chem.*, **46**, 3415 (1981).
50. B. Umezawa, O. Hoshino, S. Sawaki, H. Sashida, K. Mori, Y. Hamada, K. Kotera and Y. Iitaka, *Tetrahedron*, **40**, 1783 (1984).
51. G. Jaurand, J. -M. Beau and P. Sinaÿ, *J. Chem. Soc., Chem. Commun.*, 701 (1982).
52. A. Toshimitsu, H. Owada, T. Aoai, S. Uemura and M. Okano, *J. Chem. Soc., Chem. Commun.*, 546 (1981).
53. A. Toshimitsu, T. Aoai, H. Owada, S. Uemura and M. Okano, *J. Org. Chem.*, **46**, 4727 (1981).
54. U. Schmidt, A. Lieberknecht, H. Bökens and H. Griesser, *J. Org. Chem.*, **48**, 2680 (1983).
55. J. V. Heck and B. G. Christensen, *Tetrahedron Lett.*, **22**, 5027 (1981).
56. S. Danishefsky, E. M. Berman, M. Ciufolini, S. J. Etheredge and B. E. Segmuller, *J. Am. Chem. Soc.*, **107**, 3891 (1985).
57. C. G. Chavdarian, *J. Org. Chem.*, **48**, 1529 (1983).
58. H. J. Reich and J. M. Renga, *J. Org. Chem.*, **40**, 3313 (1975).
59. J. N. Denis, J. Vicens and A. Krief, *Tetrahedron Lett.*, 2697 (1979).
60. A. Toshimitsu, S. Uemura, M. Okano and N. Watanabe, *J. Org. Chem.*, **48**, 5246 (1983).
61. R. W. Rickards and W. P. Watson, *Aust. J. Chem.*, **33**, 451 (1980).
62. S. Raucher, *Tetrahedron Lett.*, 3909 (1977).
63. P. -T. Ho and R. J. Kolt, *Can. J. Chem.*, **60**, 663 (1982).
64. J. -M. Beau, S. Aburaki, J. -R. Pougny and P. Sinaÿ, *J. Am. Chem. Soc.*, **105**, 621 (1983).
65. M. Miyano, *J. Org. Chem.*, **46**, 1846 (1981).
66. I. L. Reich and H. J. Reich, *J. Org. Chem.*, **46**, 3721 (1981).
67. V. J. Jephcote and E. J. Thomas, *Tetrahedron Lett.*, **26**, 5327 (1985).
68. M. Suzuki, T. Kawagishi and R. Noyori, *Tetrahedron Lett.*, **22**, 1809 (1981).
69. K. M. Nsunda and L. Hevesi, *Tetrahedron Lett.*, **25**, 4441 (1984).
70. T. G. Back and S. Collins, *Tetrahedron Lett.*, **21**, 2215 (1980).
71. T. G. Back and S. Collins, *J. Org. Chem.*, **46**, 3249 (1981).
72. R. A. Gancarz and J. L. Kice, *Tetrahedron Lett.*, **21**, 4155 (1980).
73. R. A. Gancarz and J. L. Kice, *J. Org. Chem.*, **46**, 4899 (1981).
74. L. A. Paquette and W. A. Kinney, *Tetrahedron Lett.*, **23**, 5127 (1982).
75. L. A. Paquette and G. D. Crouse, *J. Org. Chem.*, **48**, 141 (1983).
76. W. A. Kinney, G. D. Crouse and L. A. Paquette, *J. Org. Chem.*, **48**, 4986 (1983).
77. K. C. Nicolaou, W. E. Barnette and R. L. Magolda, *J. Am. Chem. Soc.*, **103**, 3486 (1981).
78. K. C. Nicolaou, R. L. Magolda, W. J. Sipio, W. E. Barnette, Z. Lysenko and M. M. Joullie, *J. Am. Chem. Soc.*, **102**, 3784 (1980).
79. T. Hayama, S. Tomoda, Y. Takeuchi and Y. Nomura, *Chem. Lett.*, 1109 (1982).
80. T. Hayama, S. Tomoda, Y. Takeuchi and Y. Nomura, *Tetrahedron Lett.*, **23**, 4733 (1982).
81. S. Tomoda, Y. Takeuchi and Y. Nomura, *J. Chem. Soc., Chem. Commun.*, 871 (1982).
82. T. Hori and K. B. Sharpless, *J. Org. Chem.*, **43**, 1689 (1978).
83. D. Labar, L. Hevesi, W. Dumont and A. Krief, *Tetrahedron Lett.*, 1141 (1978).
84. M. R. Detty, *J. Org. Chem.*, **45**, 274 (1980).
85. E. Vedejs, M. J. Mullins, J. M. Renga and S. P. Singer, *Tetrahedron Lett.*, 519 (1978).
86. Y. Nagao, M. Ochiai, K. Kaneko, A. Maeda, K. Watanabe and E. Fujita, *Tetrahedron Lett.*, 1345 (1977).
87. G. Lesma, G. Palmisano and S. Tollari, *J. Chem. Soc., Perkin Trans.* **1**, 1593 (1984).
88. F. A. Davis, O. D. Stringer and J. M. Billmers, *Tetrahedron Lett.*, **24**, 1213 (1983).
89. F. A. Davis, O. D. Stringer and J. P. McCauley Jr., *Tetrahedron*, **41**, 4747 (1985).
90. L. Hevesi and A. Krief, *Angew. Chem., Int. Ed. Engl*, **15**, 381 (1976).
91. M. Sevrin and A. Krief, *Tetrahedron Lett.*, 187 (1978).
92. J. Lucchetti and A. Krief, *C. R. Acad. Sci., Ser. C*, **288**, 537 (1979).
93. H. J. Reich and J. E. Trend, *J. Org. Chem.*, **41**, 2503 (1976).
94. B. M. Trost, J. M. Balkovec and M. K. -T. Mao, *J. Am. Chem. Soc.*, **105**, 6755 (1983).
95. P. E. Eaton, G. D. Andrews, E. -P. Krebs and A. Kunai, *J. Org. Chem.*, **44**, 2824 (1979).
96. A. Toshimitsu, H. Owada, K. Terao, S. Uemura and M. Okano, *J. Org. Chem.*, **49**, 3796 (1984).
97. K. B. Sharpless and M. W. Young, *J. Org. Chem.*, **40**, 947 (1975).
98. P. A. Grieco, K. Hiroi, J. J. Reap and J. A. Noguez, *J. Org. Chem.*, **40**, 1450 (1975).
99. P. A. Grieco, Y. Masaki and D. Boxler, *J. Am. Chem. Soc.*, **97**, 1597 (1975)..

100. A. Toshimitsu, H. Owada, S. Uemura and M. Okano, *Tetrahedron Lett.*, **21**, 5037 (1980).
101. A. Toshimitsu, H. Owada, K. Terao, S. Uemura and M. Okano, *J. Chem. Soc., Perkin Trans. 1*, 373 (1985).
102. D. H. R. Barton, D. Crich, Y. Hervé, P. Potier and J. Thierry, *Tetrahedron*, **41**, 4347 (1985).
103. R. D. Clark and C. H. Heathcock, *J. Org. Chem.*, **41**, 1396 (1976).
104. J. E. Baldwin, N. V. Reed and E. J. Thomas, *Tetrahedron*, **37**, 263 (1981).
105. H. O. House, W. A. Kleschick and E. J. Zaiko, *J. Org. Chem.*, **43**, 3653 (1978).
106. P. E. Eaton, W. H. Bunnelle and P. Engel, *Can. J. Chem.*, **62**, 2612 (1984).
107. R. M. Williams and J. -S. Dung, *Tetrahedron Lett.*, **26**, 37 (1985).
108. H. J. Reich and W. W. Willis, Jr., *J. Am. Chem. Soc.*, **102**, 5967 (1980).
109. T. Miura and M. Kobayashi, *J. Chem. Soc., Chem. Commun.*, 438 (1982).
110. T. G. Back and S. Collins, *Tetrahedron Lett.*, **22**, 5111 (1981).
111. T. G. Back, S. Collins and R. G. Kerr, *J. Org. Chem.*, **48**, 3077 (1983).
112. T. G. Back, S. Collins, U. Gokhale and K. -W. Law, *J. Org. Chem.*, **48**, 4776 (1983).
113. H. Lee and M. P. Cava, *J. Chem. Soc., Chem. Commun.*, 277 (1981).
114. K. B. Sharpless, K. M. Gordon, R. F. Lauer, D. W. Patrick, S. P. Singer and M. W. Young, *Chem. Scr.*, **8A**, 9 (1975).
115. T. Otsubo, F. Ogura, H. Yamaguchi, H. Higuchi, Y. Sakata and S. Misumi, *Chem. Lett.*, 447 (1981).
116. S. Uemura, S. Fukuzawa, A. Toshimitsu and M. Okano, *Tetrahedron Lett.*, **23**, 1177 (1982).
117. S. Uemura and S. Fukuzawa, *Tetrahedron Lett.*; **24**, 4347 (1983).
118. S. Uemura and S. Fukuzawa, *J. Am. Chem. Soc.*, **105**, 2748 (1983).
119. S. Uemura, S. Fukuzawa and A. Toshimitsu, *J. Organomet. Chem.*, **250**, 203 (1983).
120. S. Uemura, K. Ohe and S. Fukuzawa, *Tetrahedron Lett.*, **26**, 895 (1985).
121. S. Halazy and A. Krief, *Tetrahedron Lett.*, 4233 (1979).
122. S. Halazy and A. Krief, *Tetrahedron Lett.*, **21**, 1997 (1980).
123. S. Halazy and A. Krief, *Tetrahedron Lett.*, **22**, 2135 (1981).
124. M. Clarembeau, J. L. Bertrand and A. Krief, *Isr. J. Chem.*, **24**, 125 (1984).
125. M. Sevrin, W. Dumont and A. Krief, *Tetrahedron Lett.*, 3835 (1977).
126. J. N. Denis and A. Krief, *Tetrahedron Lett.*, **23**, 3407 (1982).
127. K. M. Nsunda and L. Hevesi, *J. Chem. Soc., Chem. Commun.*, 1000 (1985).
128. Y. Tsuda, S. Hosoi, A. Nakai, T. Ohshima, Y. Sakai and F. Kiuchi, *J. Chem. Soc., Chem. Commun.*, 1216 (1984).
129. Y. Tsuda and S. Hosoi, *Chem. Pharm. Bull.*, **33**, 1745 (1985).
130. P. A. Grieco, T. Oguri, S. Burke, E. Rodriguez, G. T. DeTitta and S. Fortier, *J. Org. Chem.*, **43**, 4552 (1978).
131. I. Iijima, K. C. Rice and J. V. Silverton, *Heterocycles*, **6**, 1157 (1977).
132. D. D. Weller and H. Rapoport, *J. Med. Chem.*, **19**, 1171 (1976).
133. H. J. Reich, J. M. Renga and I. L. Reich, *J. Org. Chem.*, **39**, 2133 (1974).
134. H. J. Reich, J. M. Renga and I. L. Reich, *J. Am. Chem. Soc.*, **97**, 5434 (1975).
135. D. L. Snitman, R. J. Himmelsbach and D. S. Watt, *J. Org. Chem.*, **43**, 4758 (1978).
136. J. Blumbach, D. A. Hammond and D. A. Whiting, *Tetrahedron Lett.*, **23**, 3949 (1982).
137. L. F. Tietze, G. V. Kiedrowski and B. Berger, *Tetrahedron Lett.*, **23**, 51 (1982).
138. A. B. Smith, III, P. A. Levenberg, P. J. Jerris, R. M. Scarborough, Jr., and P. M. Wovkulich, *J. Am. Chem. Soc.*, **103**, 1501 (1981).
139. G. Quinkert, G. Dürner, E. Kleiner, F. Adam, E. Haupt and D. Leibfritz, *Chem. Ber.*, **113**, 2227 (1980).
140. C. Romming and P. E. Hansen, *Acta Chem. Scand., Ser. A*, **33**, 265 (1979).
141. C. H. Heathcock, C. M. Tice and T. C. Gemroth, *J. Am. Chem. Soc.*, **104**, 6081 (1982).
142. J. R. Williams and J. D. Leber, *Synthesis*, 425 (1977).
143. V. Balogh, J. -C. Beloeil and M. Fetizon, *Tetrahedron*, **33**, 1321 (1977).
144. M. Jefson and J. Meinwald, *Tetrahedron Lett.*, **22**, 3561 (1981).
145. C. Paulmier and P. Lerouge, *Tetrahedron Lett.*, **23**, 1557 (1982).
146. P. Lerouge and C. Paulmier, *Tetrahedron Lett.*, **25**, 1983 (1984).
147. S. Tomoda, Y. Takeuchi and Y. Nomura, *Synthesis*, 212 (1985).
148. K. Hiroi and S. Sato, *Synthesis*, 635 (1985).
149. A. Toshimitsu, H. Owada, S. Uemura and M. Okano, *Tetrahedron Lett.*, **23**, 2105 (1982).

150. R. Michels, M. Kato and W. Heitz, *Makromol. Chem.*, **177**, 2311 (1976).
151. D. Liotta, G. Zima, C. Barnum and M. Saindane, **21**, 3643 (1980).
152. N. Petragnani, J. V. Comasseto, R. Rodrigues and T. J. Brocksom, *J. Organomet. Chem.*, **124**, 1 (1977).
153. N. Miyoshi, T. Yamamoto, N. Kambe, S. Murai and N. Sonoda, *Tetrahedron Lett.*, **23**, 4813 (1982).
154. S. Torii, K. Uneyama and K. Handa, *Tetrahedron Lett.*, **21**, 1863 (1980).
155. A. L. Cossey, L. Lombardo and L. N. Mander, *Tetrahedron Lett.*, **21**, 4383 (1980).
156. L. Lombardo, L. N. Mander and J. V. Turner, *J. Am. Chem. Soc.*, **102**, 6626 (1980).
157. L. Lombardo and L. N. Mander, *J. Org. Chem.*, **48**, 2298 (1983).
158. G. Ohloff, W. Giersch, K. H. Schulte-Elte and C. Vial, *Helv. Chim. Acta*, **59**, 1140 (1976).
159. J. Tsuji, K. Masaoka, T. Takahashi, A. Suzuki and N. Miyaura, *Bull. Chem. Soc. Jpn.*, **50**, 2507 (1977).
160. S. P. Singer and K. B. Sharpless, *J. Org. Chem.*, **43**, 1448 (1978).
161. W. Oppolzer and K. Thirring, *J. Am. Chem. Soc.*, **104**, 4978 (1982).
162. E. Pfaff and H. Plieninger, *Chem. Ber.*, **115**, 1967 (1982).
163. A. B. Smith, III, and R. E. Richmond, *J. Am. Chem. Soc.*, **105**, 575 (1983).
164. M. F. Semmelhack and S. Tomoda, *J. Am. Chem. Soc.*, **103**, 2427 (1981).
165. M. F. Semmelhack, S. Tomoda, H. Nagaoka, S. D. Boettger and K. M. Hurst, *J. Am. Chem. Soc.*, **104**, 747 (1982).
166. A. R. Battersby, A. L. Gutman, C. J. R. Fookes, H. Günther and H. Simon, *J. Chem. Soc., Chem. Commun.*, 645 (1981).
167. N. Ikota and B. Ganem, *J. Org. Chem.*, **43**, 1607 (1978).
168. M. J. Pearson, *J. Chem. Soc., Perkin Trans. 1*, 2544 (1981).
169. E. Gipstein, C. G. Willson and H. S. Sachdev, *J. Org. Chem.*, **45**, 1486 (1980).
170. W. A. Kleschick and C. H. Heathcock, *J. Org. Chem.*, **43**, 1256 (1978).
171. B. B. Snider and G. B. Phillips, *J. Org. Chem.*, **48**, 3685 (1983).
172. P. A. Grieco and M. Nishizawa, *J. Chem. Soc., Chem. Commun.*, 582 (1976).
173. P. A. Grieco and M. Nishizawa, *J. Org. Chem.*, **42**, 1717 (1977).
174. K. Yamakawa, K. Nishitani and T. Tominaga, *Tetrahedron Lett.*, 2829 (1975).
175. K. Yamakawa, T. Tominaga and K. Nishitani, *Tetrahedron Lett.*, 4137 (1975).
176. P. A. Grieco, C. S. Pogonowski and S. Burke, *J. Org. Chem.*, **40**, 542 (1975).
177. T. R. Hoye and A. J. Caruso, *J. Org. Chem.*, **46**, 1198 (1981).
178. C. A. Wilson, II, and T. A. Bryson, *J. Org. Chem.*, **40**, 800 (1975).
179. E. Vedejs, S. Larsen and F. G. West, *J. Org. Chem.*, **50**, 2170 (1985).
180. J. Tsuji, K. Masaoka and T. Takahashi, *Tetrahedron Lett.*, 2267 (1977).
181. P. A. Zoretic and P. Soja, *J. Org. Chem.*, **41**, 3587 (1976).
182. P. A. Zoretic and P. Soja, *J. Heterocycl. Chem.*, **14**, 681 (1977).
183. Y. Ohfune and M. Tomita, *J. Am. Chem. Soc.*, **104**, 3511 (1982).
184. Y. Tsuda, A. Nakai, K. Ito, F. Suzuki and M. Haruna, *Heterocycles*, **22**, 1817 (1984).
185. J. Ficini, A. Guingant and J. d'Angelo, *J. Am. Chem. Soc.*, **101**, 1318 (1979).
186. A. P. Kozikowski and A. Ames, *J. Am. Chem. Soc.*, **103**, 3923 (1981).
187. K. K. Mahalanabis, M. Mumtaz and V. Snieckus, *Tetrahedron Lett.*, **23**, 3971 (1982).
188. J. A. Oakleaf, M. T. Thomas, A. Wu and V. Snieckus, *Tetrahedron Lett.*, 1645 (1978).
189. M. Majewski, G. B. Mpango, M. T. Thomas, A. Wu and V. Snieckus, *J. Org. Chem.*, **46**, 2029 (1981).
190. K. Ito, F. Suzuki and M. Haruna, *J. Chem. Soc., Chem. Commun.*, 733 (1978).
191. J. M. Renga and H. J. Reich, *Org. Synth.*, **59**, 58 (1980).
192. D. Liotta, C. Barnum, R. Puleo, G. Zima, C. Bayer and H. S. Kezar, III, *J. Org. Chem.*, **46**, 2920 (1981).
193. D. J. Goldsmith and H. S. Kezar, III, *Tetrahedron Lett.*, **21**, 3543 (1980).
194. S. C. Welch, C. P. Hagan, D. H. White, W. P. Fleming and J. W. Trotter, *J. Am. Chem. Soc.*, **99**, 549 (1977).
195. M. L. Durrant and E. J. Thomas, *J. Chem. Soc., Perkin Trans. 1*, 901 (1984).
196. R. L. Sobczak, M. E. Osborn and L. A. Paquette, *J. Org. Chem.*, **44**, 4886 (1979).
197. R. Bloch and P. Orvane, *Synth. Commun.*, **11**, 913 (1981).
198. R. F. C. Brown, K. J. Coulston, F. W. Eastwood, B. M. Gatehouse, L. W. Guddatt, M.

Pfenninger and I. Rainbow, *Aust. J. Chem.*, **37**, 2509 (1984).
199. D. Liotta, M. Saindane, C. Barnum, H. Ensley and P. Balakrishnan, *Tetrahedron Lett.*, **22**, 3043 (1981).
200. M. R. Roberts and R. H. Schlessinger, *J. Am. Chem. Soc.*, **103**, 724 (1981).
201. T. Ishida and K. Wada, *J. Chem. Soc., Chem. Commun.*, 337 (1977).
202. T. Ishida and K. Wada, *J. Chem. Soc., Perkin Trans. 1*, 323 (1979).
203. G. M. Ksander, J. E. McMurry and M. Johnson, *J. Org. Chem.*, **42**, 1180 (1977).
204. M. Arita, K. Adachi, Y. Ito, H. Sawai and M. Ohno, *J. Am. Chem. Soc.*, **105**, 4049 (1983).
205. S. Torii, K. Uneyama and M. Ono, *Tetrahedron Lett.*, **21**, 2741 (1980).
206. S. Torii, K. Uneyama, M. Ono and T. Bannou, *J. Am. Chem. Soc.*, **103**, 4606 (1981).
207. F. G. Bordwell, J. E. Bares, J. E. Bartmess, G. E. Drucker, J. Gerhold, G. J. McCollum, M. Van Der Puy, N. R. Vanier and W. S. Mathews, *J. Org. Chem.*, **42**, 326 (1977).
208. A. Toshimitsu, T. Aoai, H. Owada, S. Uemura and M. Okano, *J. Chem. Soc., Chem. Commun.*, 412 (1980).
209. I. Ryu, S. Murai, I. Niwa and N. Sonoda, *Synthesis*, 874 (1977).
210. S. Danishefsky, R. Zamboni, M. Kahn and S. J. Etheredge, *J. Am. Chem. Soc.*, **102**, 2097 (1980).
211. S. Danishefsky, R. Zamboni, M. Kahn and S. J. Etheredge, *J. Am. Chem. Soc.*, **103**, 3460 (1981).
212. J. Olliver and J. Salaün, *J. Chem. Soc., Chem. Commun.*, 1269 (1985).
213. P. A. Grieco, H. L. Sham, J. Inanaga, H. Kim and P. A. Tuthill, *J. Chem. Soc., Chem. Commun.*, 1345 (1984).
214. G. D. Crouse and L. A. Paquette, *J. Org. Chem.*, **46**, 4272 (1981).
215. S. Danishefsky and C. F. Yan, *Synth. Commun.*, **8**, 211 (1978).
216. S. Danishefsky, C. F. Yan and P. M. McCurry, Jr., *J. Org. Chem.*, **42**, 1819 (1977).
217. S. Danishefsky, C. F. Yan, R. K. Singh, R. B. Gammill, P. M. McCurry, Jr., N. Fritsch and J. Clardy, *J. Am. Chem. Soc.*, **101**, 7001 (1979).
218. H. -U. Reissig and I. Reichelt, *Tetrahedron Lett.*, **25**, 5879 (1984).
219. D. R. Williams and K. Nishitani, *Tetrahedron Lett.*, **21**, 4417 (1980).
220. D. R. Williams, B. A. Barner, K. Nishitani and J. G. Phillips, *J. Am. Chem. Soc.*, **104**, 4708 (1982).
221. L. A. Paquette, M. J. Coghlan and P. C. Hayes, *J. Org. Chem.*, **49**, 4516 (1984).
222. T. G. Back, S. Collins and K. -W. Law, *Can. J. Chem.*, **63**, 2313 (1985).
223. T. G. Back, N. Ibrahim and D. J. McPhee, *J. Org. Chem.*, **47**, 3283 (1982).
224. J. Hooz and J. Oudenes, *Synth. Commun.*, **10**, 667 (1980).
225. K. C. Nicolaou, R. L. Magolda and W. J. Sipio, *Synthesis*, 982 (1979).
226. M. F. Semmelhack and A. Zask, *J. Am. Chem. Soc.*, **105**, 2034 (1983).
227. A. G. González, C. Betancor, C. G. Francisco, R. Hernández, J. A. Salazar and E. Suárez, *Tetrahedron Lett.*, 2959 (1977).
228. G. R. Waitkins and C. W. Clark, *Chem. Rev.*, **36**, 235 (1945).
229. N. Rabjohn, *Org. React.*, **5**, 331 (1949).
230. N. Rabjohn, *Org. React.*, **24**, 261 (1976).
231. E. N. Trachtenberg, in *Oxidation* (Ed. R. L. Augustine), Vol. 1, Marcel Dekker, New York, 1969, Ch. 3.
232. R. A. Jerussi, in *Selective Organic Transformations* (Ed. B. S. Thyagarajan), Vol. 1, Wiley, New York, 1970, p. 301.
233. T. Laitalainen, *Ann. Acad. Sci. Fenn., Ser. A2*, 195 (1982).
234. E. R. Koft and A. B. Smith, III, *J. Am. Chem. Soc.*, **104**, 2659 (1982).
235. N. A. Shams, *J. Prakt. Chem.*, **327**, 536 (1985).
236. M. Cain, O. Campos, F. Guzman and J. M. Cook, *J. Am. Chem. Soc.*, **105**, 907 (1983).
237. H. Nagaoka, G. Schmid, H. Iio and Y. Kishi, *Tetrahedron Lett.*, **22**, 899 (1981).
238. D. H. R. Barton, D. J. Lester and S. V. Ley, *J. Chem. Soc., Chem. Commun.*, 130 (1978).
239. D. H. R. Barton, D. J. Lester and S. V. Ley, *J. Chem. Soc., Perkin Trans. 1*, 2209 (1980).
240. D. H. R. Barton, A. G. Brewster, R. A. H. F. Hui, D. J. Lester, S. V. Ley and T. G. Back, *J. Chem. Soc., Chem. Commun.*, 952 (1978).
241. D. H. R. Barton, J. W. Morzycki, W. B. Motherwell and S. V. Ley, *J. Chem. Soc., Chem. Commun.*, 1044 (1981).
242. D. H. R. Barton C. R. A. Godfrey, J. W. Morzycki, W. B. Motherwell and S. V. Ley, *J. Chem. Soc., Perkin Trans. 1*, 1947 (1982).
243. K. Yamakawa, T. Satoh, N. Ohba, R. Sakaguchi, S. Takita and N. Tamura, *Tetrahedron*, **37**, 473 (1981).

244. S. J. Foster, C. W. Rees and D. J. Williams, *J. Chem. Soc., Perkin Trans. 1*, 711 (1985).
245. D. H. R. Barton, R. A. H. F. Hui, D. J. Lester and S. V. Ley, *Tetrahedron Lett.*, 3331 (1979).
246. D. H. R. Barton, R. A. H. F. Hui, S. V. Ley and D. J. Williams, *J. Chem. Soc., Perkin Trans. 1*, 1919 (1982).
247. N. Khoi and J. Polonsky, *Helv. Chim. Acta*, **64**, 1540 (1981).
248. T. Matsui, S. Tahara and M. Nakayama, *Bull. Chem. Soc. Jpn.*, **57**, 3355 (1984).
249. T. G. Back, *J. Chem. Soc., Chem. Commun.*, 278 (1978).
250. T. G. Back, *J. Org. Chem.*, **46**, 1442 (1981).
251. D. H. R. Barton, W. B. Motherwell, J. Wozniak and S. Z. Zard, *J. Chem. Soc., Perkin Trans. 1*, 1865 (1985).
252. V. I. Erashko, O. M. Sazonova, A. A. Tishaninova and A. A. Fainzil'berg, *Izv. Akad. Nauk SSSR, Ser. Khim.*, 161 (1982); *Chem. Abstr.*, **96**, 142370b (1982).
253. T. Sakakibara, I. Takai, E. Ohara and R. Sudoh, *J. Chem. Soc., Chem. Commun.*, 261 (1981).
254. T. Sakakibara, S. Ikuta and R. Sudoh, *Synthesis*, 261 (1982).
255. T. Sakakibara, M. D. Manandhar and Y. Ishido, *Synthesis*, 920 (1983).
256. D. Seebach and D. Enders, *J. Med. Chem.*, **17**, 1225 (1974).
257. R. Kupper and C. J. Michejda, *J. Org. Chem.*, **44**, 2326 (1979).
258. A. I. Meyers, P. D. Edwards, T. R. Bailey and G. E. Jagdmann, Jr., *J. Org. Chem.*, **50**, 1019 (1985).
259. D. N. Brattesani and C. H. Heathcock, *Tetrahedron Lett.*, 2279 (1974).
260. D. N. Barattesani and C. H. Heathcock, *J. Org. Chem.*, **40**, 2165 (1975).
261. Y. Masuyama, Y. Ueno and M. Okawara, *Chem. Lett.*, 835 (1977).
262. H. E. Zimmerman and D. R. Diehl, *J. Am. Chem. Soc.*, **101**, 1841 (1979).
263. N. Petragnani, R. Rodrigues and J. V. Comasseto, *J. Organomet. Chem.*, **114**, 281 (1976).
264. A. L. Braga, J. V. Comasseto and N. Petragnani, *Synthesis*, 240 (1984).
265. G. Saleh, T. Minami, Y. Ohshiro and T. Agawa, *Chem. Ber.*, **112**, 355 (1979).
266. T. Minami, H. Sako, T. Ikehira, T. Hanamoto and I. Hirao, *J. Org. Chem.*, **48**, 2569 (1983).
267. H. Schmidbaur, C. Zybill, C. Krüger and H. -J. Kraus, *Chem. Ber.*, **116**, 1955 (1983).
268. M. Mikolajczyk, S. Grzejszczak and K. Korbacz, *Tetrahedron Lett.*, **22**, 3097 (1981).
269. J. V. Comasseto and N. Petragnani, *J. Organomet. Chem.*, **152**, 295 (1978).
270. M. Isobe, Y. Ichikawa and T. Goto, *Tetrahedron Lett.*, **22**, 4287 (1981).
271. T. G. Back, *J. Org. Chem.*, **46**, 5443 (1981).
272. B. Harirchian and P. Magnus, *J. Chem. Soc., Chem. Commun.*, 522 (1977).
273. D. Labar, A. Krief and L. Hevesi, *Tetrahedron Lett.*, 3967 (1978).
274. D. Labar and A. Krief, *Bull. Soc. Chim. Belg.*, **93**, 1005 (1984).
275. D. L. J. Clive, *J. Chem. Soc., Chem. Commun.*, 100 (1974).
276. H. J. Reich, *J. Org. Chem.*, **39**, 428 (1974).
277. H. J. Reich, W. W. Willis, Jr., and S. Wollowitz, *Tetrahedron Lett.*, **23**, 3319 (1982).
278. J. L. Kice, F. McAfee and H. Slebocka-Tilk, *Tetrahedron Lett.*, **23**, 3323 (1982).
279. R. A. Gancarz and J. L. Kice, *Tetrahedron Lett.*, **22**, 1661 (1981).
280. D. G. Garratt and A. Kabo, *Can. J. Chem.*, **58**, 1030 (1980).
281. M. Perrier and R. Rouessac, *C. R. Acad. Sci., Ser. II*, **295**, 729 (1982).
282. A. Toshimitsu, S. Uemura and M. Okano, *J. Chem. Soc., Chem. Commun.*, 965 (1982).
283. S. Uemura and S. Fukuzawa, *J. Chem. Soc., Perkin Trans. 1*, 471 (1985).
284. R. K. Boeckman, Jr., and S. S. Ko, *J. Am. Chem. Soc.*, **102**, 7146 (1980).
285. K. C. Nicolaou, D. A. Claremon, W. E. Barnette and S. P. Seitz, *J. Am. Chem. Soc.*, **101**, 3704 (1979).
286. K. C. Nicolaou, N. A. Petasis and D: A. Claremon, *Tetrahedron*, **41**, 4835 (1985).
287. P. S. Liu, V. E. Marquez, J. A. Kelley and J. S. Driscoll, *J. Org. Chem.*, **45**, 5225 (1980).
288. K. Katsuura and V. Snieckus, *Tetrahedron Lett.*, **26**, 9 (1985).
289. A. B. Smith, III, B. H. Toder, R. E. Richmond and S. J. Branca, *J. Am. Chem. Soc.*, **106**, 4001 (1984).
290. A. Toshimitsu, S. Uemura and M. Okano, *J. Chem. Soc., Chem. Commun.*, 166 (1977).
291. A. Toshimitsu, T. Aoai, S. Uemura and M. Okano, *J. Org. Chem.*, **45**, 1953 (1980).
292. S. Torii, K. Uneyama and M. Ono, *Tetrahedron Lett.*, **21**, 2653 (1980).
293. K. Uneyama, M. Ono and S. Torii, *Phosphorus Sulfur*, **16**, 35 (1983).
294. N. Miyoshi, Y. Ohno, K. Kondo, S. Murai and N. Sonoda, *Chem. Lett.*, 1309 (1979).
295. K. Isobe, J. Taga and Y. Tsuda, *Tetrahedron Lett.*, 2331 (1976).
296. L. A. Paquette, T. J. Nitz, R. J. Ross and J. P. Springer, *J. Am. Chem. Soc.*, **106**, 1446 (1984).

297. T. Kametani, K. Suzuki, H. Kurobe and H. Nemoto, *J. Chem. Soc., Chem. Commun.*, 1128 (1979).
298. T. Kametani, K. Suzuki, H. Kurobe and H. Nemoto, *Chem. Pharm. Bull.*, **29**, 105 (1981).
299. D. G. Garratt, M. D. Ryan and A. Kabo, *Can. J. Chem.*, **58**, 2329 (1980).
300. N. Miyoshi, Y. Takai, S. Murai and N. Sonoda, *Bull. Chem. Soc. Jpn.*, **51**, 1265 (1978).
301. N. Miyoshi, S. Furui, S. Murai and N. Sonoda, *J. Chem. Soc., Chem. Commun.*, 293 (1975).
302. N. Sonoda, N. Miyoshi and S. Murai, *Chem. Scr.*, **8A**, 109 (1975).
303. N. Miyoshi, S. Murai and N. Sonoda, *Tetrahedron Lett.*, 851 (1977).
304. D. Seebach, G. Calderari and P. Knochel, *Tetrahedron*, **41**, 4861 (1985).
305. J. Bergman and L. Engman, *J. Organomet. Chem.*, **181**, 335 (1979).
306. M. Shimizu, R. Takeda and I. Kuwajima, *Tetrahedron Lett.*, 419 (1979).
307. M. Shimizu, R. Takeda and I. Kuwajima, *Tetrahedron Lett.*, 3461 (1979).
308. M. Shimizu, and I. Kuwajima, *Bull. Chem. Soc. Jpn.*, **54**, 3100 (1981).
309. M. Shimizu, R. Takeda and I. Kuwajima, *Bull. Chem. Soc. Jpn.*, **54**, 3510 (1981).
310. I. Kuwajima and M. Shimizu, *Tetrahedron Lett.*, 1277 (1978).
311. T. Takahashi, H. Nagashima and J. Tsuji, *Tetrahedron Lett.*, 799 (1978).
312. S. Raucher, *Tetrahedron Lett.*, 2261 (1978).
313. A. Toshimitsu, Y. Kozawa, S. Uemura and M. Okano, *J. Chem. Soc., Perkin Trans. 1*, 1273 (1978).
314. S. Uemura, A. Toshimitsu, M. Okano and K. Ichikawa, *Bull. Chem. Soc. Jpn.*, **48**, 1925 (1975).
315. A. P. Kozikowski, K. L. Sorgi and R. J. Schmiesing, *J. Chem. Soc., Chem. Commun.*, 477 (1980).
316. A. P. Kozikowski, R. J. Schmiesing and K. L. Sorgi, *J. Am. Chem. Soc.*, **102**, 6580 (1980).
317. M. Isobe, M. Kitamura and T. Goto, *Tetrahedron Lett.*, **22**, 239 (1981).
318. M. Kitamura, M. Isobe, Y. Ichikawa and T. Goto, *J. Org. Chem.*, **49**, 3517 (1984).
319. G. Bérubé, E. Luce and K. Jankowski, *Bull. Soc. Chim. Fr., Ser. 2*, 109 (1983).
320. G. Jaurand, J. -M. Beau and P. Sinaÿ, *J. Chem. Soc., Chem. Commun.*, 572 (1981).
321. R. Baudat and M. Petrzilka, *Helv. Chim. Acta*, **62**, 1406 (1979).
322. H. J. Bestmann and P. Ermann, *Justus Liebigs Ann. Chem.*, 1740 (1984).
323. N. Satyanarayana and U. R. Nayak, *Synth. Commun.*, **15**, 331 (1985).
324. S. F. Martin, S. R. Desai, G. W. Phillips and A. C. Miller, *J. Am. Chem. Soc.*, **102**, 3294 (1980).
325. H. Ogawa, H. Morita, S. Suemura, T. Imoto, Y. Nogami, T. Koga, Y. Sakurgai, I. Miyamoto, H. Kato and Y. Taniguchi, *Tetrahedron Lett.*, **26**, 5567 (1985).
326. A. Guillemonat, *Ann. Chim. (Paris)*, **11**, 143 (1939).
327. K. B. Sharpless and R. F. Lauer, *J. Am. Chem. Soc.*, **94**, 7154 (1972).
328. D. Arigoni, A. Vasella, K. B. Sharpless and H. B. Jensen, *J. Am. Chem. Soc.*, **95**, 7917 (1973).
329. H. P. Jensen and K. B. Sharpless, *J. Org. Chem.*, **40**, 264 (1975).
330. M. A. Warpehoski, B. Chabaud and K. B. Sharpless, *J. Org. Chem.*, **47**, 2897 (1982).
331. L. M. Stephenson and D. R. Speth, *J. Org. Chem.*, **44**, 4683 (1979).
332. W. -D. Woggon, F. Ruther and H. Egli, *J. Chem. Soc., Chem. Commun.*, 706 (1980).
333. M. A. Umbreit and K. B. Sharpless, *J. Am. Chem. Soc.*, **99**, 5526 (1977).
334. O. Campos and J. M. Cook, *Tetrahedron Lett.*, 1025 (1979).
335. M. Haruna and K. Ito, *J. Chem. Soc., Chem. Commun.*, 483 (1981).
336. W. C. Still and D. Mobilio, *J. Org. Chem.*, **48**, 4785 (1983).
337. D. R. Williams, M. L. Bremmer, D. L. Brown and J. D'Antuono, *J. Org. Chem.*, **50**, 2807 (1985).
338. B. B. Snider and Y. S. Kulkarni, *Tetrahedron Lett.*, **26**, 5675 (1985).
339. B. R. Chhabra, K. Hayano, T. Ohtsuka, H. Shirahama and T. Matsumoto, *Chem. Lett.*, 1703 (1981).
340. F. Camps, J. Coll and A. Parente, *Synthesis*, 215 (1978).
341. Y. -W. Lee, E. Lee and K. Nakanishi, *Tetrahedron Lett.*, **21**, 4323 (1980).
342. H. -J. Bestmann and R. Schobert, *Angew. Chem., Int. Ed. Engl.*, **24**, 791 (1985).
343. O. Campos, M. DiPierro, M. Cain, R. Mantei, A. Gawish and J. M. Cook, *Heterocycles*, **14**, 975 (1980).
344. B. Chabaud and K. B. Sharpless, *J. Org. Chem.*, **44**, 4202 (1979).
345. D. H. R. Barton and D. Crich, *Tetrahedron*, **41**, 4359 (1985).
346. A. S. Kende, D. P. Lorah and R. J. Boatman, *J. Am. Chem. Soc.*, **103**, 1271 (1981).
347. T. Sakamoto, T. Sakasai and H. Yamanaka, *Chem. Pharm. Bull.*, **29**, 2485 (1981).
348. T. K. Liao, P. J. Wittek and C. C. Cheng, *J. Heterocycl. Chem.*, **13**, 1283 (1976).
349. P. J. Wittek, T. K. Liao and C. C. Cheng, *J. Org. Chem.*, **44**, 870 (1979).
350. D. H. R. Barton, R. A. H. F. Hui and S. V. Ley, *J. Chem. Soc., Perkin Trans. 1*, 2179 (1982).

351. L. Syper and J. Mlochowski, *Synthesis*, 747 (1984).
352. J. Bergman and L. Engman, *Tetrahedron Lett.*, 3279 (1978).
353. J. Bergman and L. Engman, *J. Org. Chem.*, **47**, 5191 (1982).
354. D. H. R. Barton, A. G. Brewster, S. V. Ley and M. N. Rosenfeld, *J. Chem. Soc., Chem. Commun.*, 985 (1976).
355. D. H. R. Barton, A. G. Brewster, S. V. Ley, C. M. Read and M. N. Rosenfeld, *J. Chem. Soc., Perkin Trans. 1*, 1473 (1981).
356. D. H. R. Barton, P. D. Magnus and M. N. Rosenfeld, *J. Chem. Soc., Chem. Commun.*, 301 (1975).
357. D. H. R. Barton, S. V. Ley, P. D. Magnus and M. N. Rosenfeld, *J. Chem. Soc., Perkin Trans. 1*, 567 (1977).
358. K. B. Sukumaran and R. G. Harvey, *J. Am. Chem. Soc.*, **101**, 1353 (1979).
359. K. B. Sukumaran and R. G. Harvey, *J. Org. Chem.*, **45**, 4407 (1980).
360. P. W. Jeffs and D. G. Lynn, *Tetrahedron Lett.*, 1617 (1978).
361. R. T. Taylor and L. A. Flood, *J. Org. Chem.*, **48**, 5160 (1983).
362. K. B. Sharpless and K. M. Gordon, *J. Am. Chem. Soc.*, **98**, 300 (1976).
363. I. Simiti, A. Mureson and M. Coman, *Arch. Pharm.*, **314**, 744 (1981).
364. K. Yamakawa, T. Satoh, N. Ohba and R. Sakaguchi, *Chem. Lett.*, 763 (1979).
365. K. Yamakawa and T. Satoh, *Heterocycles*, **15**, 337 (1981).
366. P. A. Grieco, Y. Yokoyama, S. Gilman and Y. Ohfune, *J. Chem. Soc., Chem. Commun.*, 870 (1977).
367. S. R. Dutta and B. P. Pradhan, *Indian J. Chem., Sect. B*, **21**, 575 (1982).
368. S. R. Dutta and B. P. Pradhan, *Indian J. Chem., Sect. B*, **22**, 680 (1983).
369. B. P. Pradhan and S. R. Dutta, *Indian J. Chem., Sect. B*, **23**, 565 (1984).
370. V. Anjaneyulu and G. S. Rao, *Indian J. Chem., Sect. B*, **23**, 663 (1984).
371. E. E. Gilbert, *Synth. Commun.*, **15**, 53 (1985).
372. P. A. Grieco, Y. Yokoyama, S. Gilman and M. Nishizawa, *J. Org. Chem.*, **42**, 2034 (1977).
373. H. J. Reich, F. Chow and S. L. Peake, *Synthesis*, 299 (1978).
374. T. Kametani, H. Nemoto and K. Fukumoto, *Heterocycles*, **6**, 1365 (1977).
375. T. Kametani, H. Nemoto and K. Fukumoto, *Biorg. Chem.*, **7**, 215 (1978).
376. A. G. Abatjoglou and D. R. Bryant, *Tetrahedron Lett.*, **22**, 2051 (1981).
377. a) S. Uemura, S. Fukuzawa and M. Okano, *Tetrahedron Lett.*, **22**, 5331 (1981). b) S. Uemura, S. Fukuzawa, S. R. Patil and M. Okano, *J. Chem. Soc., Perkin Trans. 1*, 499 (1985).
378. B. P. Mundy and W. G. Bornmann, *J. Org. Chem.*, **49**, 5264 (1984).
379. T. Fukuyama, B. D. Robins and R. A. Sachleben, *Tetrahedron Lett.*, **22**, 4155 (1981).
380. J. A. Marshall and R. D. Royce, Jr., *J. Org. Chem.*, **47**, 693 (1982).
381. S. L. Schreiber and C. Santini, *J. Am. Chem. Soc.*, **106**, 4038 (1984).
382. T. G. Back and N. Ibrahim, *Tetrahedron Lett.*, 4931 (1979).
383. H. J. Reich and S. K. Shah, *J. Org. Chem.*, **42**, 1773 (1977).
384. W. Dumont, D. Van Ende and A. Krief, *Tetrahedron Lett.*, 485 (1979).
385. K. Sachdev and H. S. Sachdev, *Tetrahedron Lett.*, 4223 (1976).
386. D. Van Ende, W. Dumont and A. Krief, *J. Organomet. Chem.*, **149**, C10 (1978).
387. S. Uemura, S. Fukuzawa and A. Toshimitsu, *J. Chem. Soc., Chem. Commun.*, 1501 (1983).
388. J. V. Comasseto, *J. Organomet. Chem.*, **253**, 131 (1983).
389. G. R. Krow and D. A. Shaw, *Synth. Commun.*, **12**, 313 (1982).
390. W. Dumont, M. Sevrin and A. Krief, *Tetrahedron Lett.*, 183 (1978).
391. J. N. Denis and A. Krief, *Tetrahedron Lett.*, **23**, 3411 (1982).
392. A. Cravador and A. Krief, *J. Chem. Soc., Chem. Commun.*, 951 (1980).
393. A. Toshimitsu, T. Aoai, S. Uemura and M. Okano, *J. Chem. Soc., Chem. Commun.*, 1041 (1980).
394. A. Bewick, D. E. Coe, G. B. Fuller and J. M. Mellor, *Tetrahedron Lett.*, **21**, 3827 (1980).
395. D. H. R. Barton, M. R. Britten-Kelly and D. Ferreira, *J. Chem. Soc., Perkin Trans. 1*, 1682 (1978).
396. K. B. Sharpless, T. Hori, L. K. Truesdale and C. O. Dietrich, *J. Am. Chem. Soc.*, **98**, 269 (1976).
397. K. B. Sharpless and S. P. Singer, *J. Org. Chem.*, **41**, 2504 (1976).
398. J. E. Fankhauser, R. M. Peevey and P. B. Hopkins, *Tetrahedron Lett.*, **25**, 15 (1984).
399. J. N. Fitzner, R. G. Shea, J. E. Fankhauser and P. B. Hopkins, *Synth. Commun.*, **14**, 605 (1984).
400. R. G. Shea, J. N. Fitzner, J. E. Fankhauser and P. B. Hopkins, *J. Org. Chem.*, **49**, 3647 (1984).
401. J. N. Fitzner, D. V. Pratt and P. B. Hopkins, *Tetrahedron Lett.*, **26**, 1959 (1985).
402. J. N. Fitzner, R. G. Shea, J. E. Fankhauser and P. B. Hopkins, *J. Org. Chem.*, **50**, 417 (1985).

403. D. H. R. Barton, A. G. Brewster, S. V. Ley and M. N. Rosenfeld, *J. Chem. Soc., Chem. Commun.*, 147 (1977).
404. J. S. E. Holker, E. O'Brien and B. K. Park, *J. Chem. Soc., Perkin Trans. 1*, 1915 (1982).
405. I. Perina, N. Bregant and K. Balenovic, *Bull. Sci. Cons. Acad. Sci. Arts RSF Yougosl., Sect. A*, **18**, 3 (1973); *Chem. Abstr.*, **79**, 5096b (1973).
406. P. Lerouge and C. Paulmier, *Tetrahedron Lett.*, **25**, 1987 (1984).
407. H. J. Reich, J. M. Renga and J. E. Trend, *Tetrahedron Lett.*, 2217 (1976).
408. S. Raucher, *J. Org. Chem.*, **42**, 2950 (1977).
409. D. Liotta and G. Zima, *Tetrahedron Lett.*, 4977 (1978).
410. G. H. Schmid and D. G. Garratt, in *The Chemistry of Double-Bonded Functional Groups. Supplement A, Part 2* (Ed. S. Patai), Wiley, Chichester, 1977, Ch. 9.
411. D. G. Garratt and G. H. Schmid, *J. Org. Chem.*, **42**, 1776 (1977).
412. J. -E. Bäckvall, J. Bergman and L. Engman, *J. Org. Chem.*, **48**, 3918 (1983).
413. S. V. Ley and A. J. Whittle, *Tetrahedron Lett.*, **22**, 3301 (1981).
414. R. S. Brown, S. C. Eyley and P. J. Parsons, *J. Chem. Soc., Chem. Commun.*, 438 (1984).
415. L. Engman, *J. Am. Chem. Soc.*, **106**, 3977 (1984).
416. J. E. Bäckvall and L. Engman, *Tetrahedron Lett.*, **22**, 1919 (1981).
417. L. Engman, *J. Org. Chem.*, **49**, 3559 (1984).
418. G. H. Schmid, in *The Chemistry of the Carbon—Carbon Triple Bond* (Ed. S. Patai), Wiley, Chichester, 1978, Ch. 8.
419. A. J. Bridges and J. W. Fischer, *Tetrahedron Lett.*, **24**, 445 (1983).
420. A. J. Bridges and J. W. Fischer, *Tetrahedron Lett.*, **24**, 447 (1983).
421. A. J. Bridges and J. W. Fischer, *J. Org. Chem.*, **49**, 2954 (1984).
422. C. N. Filer, D. Ahern, R. Fazio and E. J. Shelton, *J. Org. Chem.*, **45**, 1313 (1980).
423. D. G. Garratt, P. L. Beaulieu and V. M. Morisset, *Can. J. Chem.*, **59**, 927 (1981).
424. G. Himbert and M. Feustel, *J. Chem. Res. (S)*, 240 (1984).
425. S. Tomoda, Y. Takeuchi and Y. Nomura, *Chem. Lett.*, 1715 (1981).
426. T. Hori and K. B. Sharpless, *J. Org. Chem.*, **44**, 4204 (1979).
427. T. Hori and K. B. Sharpless, *J. Org. Chem.*, **44**, 4208 (1979).
428. M. Sevrin, W. Dumont, L. Hevesi and A. Krief, *Tetrahedron Lett.*, 2647 (1976).
429. M. Sevrin and A. Krief, *Tetrahedron Lett.*, **21**, 585 (1980).
430. M. Sevrin and A. Krief, *J. Chem. Soc., Chem. Commun.*, 656 (1980).
431. M. Sevrin, J. N. Denis and A. Krief, *Angew. Chem., Int. Ed. Engl.*, **17**, 526 (1978).
432. L. Hevesi, M. Sevrin and A. Krief, *Tetrahedron Lett.*, 2651 (1976).
433. A. M. Morella and A. D. Ward, *Tetrahedron Lett.*, **25**, 1197 (1984).
434. T. Kauffmann, H. Ahlers, H. -J. Tilhard and A. Woltermann, *Angew. Chem., Int. Ed. Engl.*, **16**, 710 (1977).
435. K. Chikamatsu, T. Otsubo, F. Ogura and H. Yamaguchi, *Chem. Lett.*, 1081 (1982).
436. S. Uemura and S. Fukuzawa, *Chem. Lett.*, 943 (1980).
437. S. Uemura and S. Fukuzawa, *J. Organomet. Chem.*, **268**, 223 (1984).
438. S. Uemura and S. Fukuzawa, *J. Chem. Soc., Chem. Commun.*, 1033 (1980).
439. S. Uemura, S. Fukuzawa, M. Wakasugi and M. Okano, *J. Organomet. Chem.*, **214**, 319 (1981).
440. S. Uemura, H. Miyoshi and M. Okano, *Chem. Lett.*, 1357 (1979).
441. F. O. Ayorinde, *Tetrahedron Lett.*, **24**, 2077 (1983).
442. M. Albeck and S. Shaik, *J. Chem. Soc., Perkin Trans. 1*, 1223 (1975).
443. M. Albeck, T. Tamari and M. Sprecher, *J. Org. Chem.*, **48**, 2276 (1983).
444. G. A. Olah, M. Nojima and I. Kerekes, *J. Am. Chem. Soc.*, **96**, 925 (1974).
445. Y. A. Sedov, *Khim. Geterotsikl. Soedin.*, 265 (1983).
446. M. R. Detty, *Tetrahedron Lett.*, 4189 (1979).
447. M. R. Detty and M. D. Seidler, *J. Org. Chem.*, **46**, 1283 (1981).
448. S. Raucher, M. R. Hansen and M. A. Colter, *J. Org. Chem.*, **43**, 4885 (1978).
449. E. G. Kataev, T. G. Mannafov, A. B. Remizov and O. A. Komarovskaya, *Zh. Org. Khim.*, **11**, 2322 (1975).
450. E. G. Kataev, T. G. Mannafov, E. A. Berdnikov and O. A. Komarovskaya, *Zh. Org. Khim.*, **9**, 1983 (1973).
451. A. E. Feiring, *J. Org. Chem.*, **45**, 1958 (1980).
452. A. E. Feiring, *J. Org. Chem.*, **45**, 1962 (1980).

453. Z. Janousek, S. Piettre, F. Gorissen-Hervens and H. G. Viehe, *J. Organomet. Chem.*, **250**, 197 (1983).
454. S. Piettre, Z. Janousek, R. Merenyi and H. G. Viehe, *Tetrahedron*, **41**, 2527 (1985).
455. M. J. Crimmin, P. J. O'Hanlon and N. H. Rogers, *J. Chem. Soc., Perkin Trans. 1*, 541 (1985).
456. K. Uneyama, K. Takano and S. Torii, *Tetrahedron Lett.*, **23**, 1161 (1982).
457. J. Hooz and R. Mortimer, *Tetrahedron Lett.*, 805 (1976).
458. J. Hooz and R. D. Mortimer, *Can. J. Chem.*, **56**, 2786 (1978).
459. H. J. Reich, J. J. Rusek and R. E. Olson, *J. Am. Chem. Soc.*, **101**, 2225 (1979).
460. H. J. Reich, M. J. Kelly, R. E. Olson and R. C. Holtan, *Tetrahedron*, **39**, 949 (1983).
461. D. G. Garratt, M. D. Ryan and M. Ujjainwalla, *Can. J. Chem.*, **57**, 2145 (1979).
462. D. G. Garratt, *Can. J. Chem.*, **57**, 2180 (1979).
463. T. G. Back and R. G. Kerr, *J. Organomet. Chem.*, **286**, 171 (1985).
464. S. Tomoda, Y. Takeuchi and Y. Nomura, *Tetrahedron Lett.*, **23**, 1361 (1982).
465. S. Tomoda, Y. Takeuchi and Y. Nomura, *Chem. Lett.*, 1733 (1982).
466. T. Hayama, S. Tomoda, Y. Takeuchi and Y. Nomura, *J. Org. Chem.*, **49**, 3235 (1984).
467. Y. -H. Kang and J. L. Kice, *J. Org. Chem.*, **49**, 1507 (1984).
468. T. Toru, T. Seko and E. Maekawa, *Tetrahedron Lett.*, **26**, 3263 (1985).
469. Y. -H. Kang and J. L. Kice, *Tetrahedron Lett.*, **23**, 5373 (1982).
470. J. L. Kice and Y. -H. Kang, *Tetrahedron*, **41**, 4739 (1985).
471. D. L. J. Clive, G. Chittattu, N. J. Curtis, W. A. Kiel and C. K. Wong, *J. Chem. Soc., Chem. Commun.*, 725 (1977).
472. D. L. J. Clive, C. G. Russell, G. Chittattu and A. Singh, *Tetrahedron*, **36**, 1399 (1980).
473. K. C. Nicolaou, *Tetrahedron*, **37**, 4097 (1981).
474. (a) K. C. Nicolaou, S. P. Seitz, W. J. Sipio and J. F. Blount, *J. Am. Chem. Soc.*, **101**, 3884 (1979); (b) K. C. Nicolaou, W. J. Sipio, R. L. Magolda and D. A. Claremon, *J. Chem. Soc., Chem. Commun.*, 83 (1979).
475. D. L. J. Clive and V. N. Kalè, *J. Org. Chem.*, **46**, 231 (1981).
476. E. J. Corey, G. E. Keck and I. Székely, *J. Am. Chem. Soc.*, **99**, 2006 (1977).
477. D. L. J. Clive, G. Chittattu and C. K. Wong, *Can. J. Chem.*, **55**, 3894 (1977).
478. K. C. Nicolaou and Z. Lysenko, *Tetrahedron Lett.*, 1257 (1977).
479. K. C. Nicolaou and W. E. Barnette, *J. Chem. Soc., Chem. Commun.*, 331 (1977).
480. K. C. Nicolaou, W. E. Barnette, R. L. Magolda, P. A. Grieco, W. Owens, C. -L. J. Wang, J. B. Smith, M. Ogletree and A. M. Lefer, *Prostaglandins*, **16**, 789 (1978).
481. K. C. Nicolaou, W. E. Barnette and R. L. Magolda, *J. Am. Chem. Soc.*, **103**, 3480 (1981).
482. Y. Naruta, H. Uno and K. Maruyama, *J. Chem. Soc., Chem. Commun.*, 1277 (1981).
483. J. Bergman and L. Engman, *J. Am. Chem. Soc.*, **103**, 5196 (1981).
484. K. Uneyama, S. Fujibayashi and S. Torii, *Tetrahedron Lett.*, **26**, 4637 (1985).
485. F. M. Hauser and R. P. Rhee, *J. Am. Chem. Soc.*, **101**, 1628 (1979).
486. P. L. Beaulieu, V. M. Morisset and D. G. Garratt, *Tetrahedron Lett.*, **21**, 129 (1980).
487. P. G. Williard and S. E. de Laszlo, *J. Org. Chem.*, **50**, 3738 (1985).
488. K. Suzuki and T. Mukaiyama, *Chem. Lett.*, 683 (1982).
489. S. Current and K. B. Sharpless, *Tetrahedron Lett.*, 5075 (1978).
490. S. V. Ley and B. Lygo, *Tetrahedron Lett.*, **23**, 4625 (1982).
491. A. M. Doherty, S. V. Ley, B. Lygo and D. J. Williams, *J. Chem. Soc., Perkin Trans. 1*, 1371 (1984).
492. M. de Moura Campos and N. Petragnani, *Chem. Ber.*, **93**, 317 (1960).
493. K. C. Nicolaou and Z. Lysenko, *J. Am. Chem. Soc.*, **99**, 3185 (1977).
494. D. L. J. Clive and G. Chittattu, *J. Chem. Soc., Chem. Commun.*, 484 (1977).
495. D. Goldsmith, D. Liotta, C. Lee and G. Zima, *Tetrahedron Lett.*, 4801 (1979).
496. A. Rouessac, F. Rouessac and H. Zamarlik, *Tetrahedron Lett.*, **22**, 2641 (1981).
497. S. W. Rollinson, R. A. Amos and J. A. Katzenellenbogen, *J. Am. Chem. Soc.*, **103**, 4114 (1981).
498. T. Toru, S. Fujita and E. Maekawa, *J. Chem. Soc., Chem. Commun.*, 1082 (1985).
499. J. V. Comasseto and N. Petragnani, *Synth. Commun.*, **13**, 889 (1983).
500. G. B. Gill and M. S. H. Idris, *Tetrahedron Lett.*, **26**, 4811 (1985).
501. D. P. Curran and D. M. Rakiewicz, *J. Am. Chem. Soc.*, **107**, 1448 (1985)
502. D. L. J. Clive, C. K. Wong, W. A. Kiel and S. M. Menchen, *J. Chem. Soc., Chem. Commun.*, 379 (1978).

503. D. L. J. Clive, V. Farina, A. Singh, C. K. Wong, W. A. Kiel and S. M. Menchen, *J. Org. Chem.*, **45**, 2120 (1980).
504. R. P. Webb, II and S. Danishefsky, *Tetrahedron Lett.*, **24**, 1357 (1983).
505. A. Toshimitsu, K. Terao and S. Uemura, *Tetrahedron Lett.*, **25**, 5917 (1984).
506. S. R. Wilson and R. A. Sawicki, *J. Org. Chem.*, **44**, 287 (1979).
507. J. E. Baldwin, S. B. Haber and J. Kitchin, *J. Chem. Soc., Chem. Commun.*, 790 (1973).
508. K. C. Nicolaou, W. E. Barnette and R. L. Magolda, *J.. Am. Chem. Soc.*, **100**, 2567 (1978).
509. C. M. Angelov, *Heterocycles*, **20**, 791 (1983).
510. K. M. Angelov and C. Tancheva, *Zh. Obshch. Khim.*, **55**, 53 (1985).
511. W. P. Jackson, S. V. Ley and J. A. Morton, *J. Chem. Soc., Chem. Commun.*, 1028 (1980).
512. W. P. Jackson, S. V. Ley and J. A. Morton, *Tetrahedron Lett.*, **22**, 2601 (1981).
513. S. V. Ley and P. J. Murray, *J. Chem. Soc., Chem. Commun.*, 1252 (1982).
514. S. V. Ley, B. Lygo, H. Molines and J. A. Morton, *J. Chem. Soc., Chem. Commun.*, 1251 (1982).
515. S. V. Ley, B. Lygo and H. Molines, *J. Chem. Soc., Perkin Trans. 1*, 2403 (1984).
516. S. V. Ley, P. J. Murray and B. D. Palmer, *Tetrahedron*, **41**, 4765 (1985).
517. M. Alderdice and L. Weiler, *Can. J. Chem.*, **59**, 2239 (1981).
518. W. P. Jackson, S. V. Ley and A. J. Whittle, *J. Chem. Soc., Chem. Commun.*, 1173 (1980).
519. D. Gravel, R. Déziel and L. Bordeleau, *Tetrahedron Lett.*, **24**, 699 (1983).
520. S. Uemura, A. Toshimitsu, T. Aoai and M. Okano, *Chem. Lett.*, 1359 (1979).
521. (a) S. Uemura, A. Toshimitsu, T. Aoai and M. Okano, *J. Chem. Soc., Chem. Commun.*, 610 (1979); (b) A. Toshimitsu, T. Aoai, S. Uemura and M. Okano, *J. Org. Chem.*, **46**, 3021 (1981).
522. A. Toshimitsu, S. Uemura and M. Okano, *J. Chem. Soc., Perkin Trans. 1*, 1206 (1979).
523. S. Uemura, A. Toshimitsu, T. Aoai and M. Okano, *Tetrahedron Lett.*, **21**, 1533 (1980).
524. R. M. Scarborough, Jr., A. B. Smith, III, W. E. Barnette and K. C. Nicolaou, *J. Org. Chem.*, **44**, 1742 (1979).
525. D. L. J. Clive, G. Chittattu and C. K. Wong, *J. Chem. Soc., Chem. Commun.* 441 (1978).
526. R. V. Stevens and K. F. Albizati, *J. Org. Chem.*, **50**, 632 (1985).
527. A. Toshimitsu, S. Uemura and M. Okano, *J. Chem. Soc., Chem. Commun.*, 87 (1982).
528. F. Rouessac and H. Zamarlik, *Tetrahedron Lett.*, **22**, 2643 (1981).
529. H. Zamarlik, N. Gnonlonfoun and F. Rouessac, *Can. J. Chem.*, **62**, 2326 (1984).
530. A. Rouessac and F. Rouessac, *Tetrahedron*, **37**, 4165 (1981).
531. T. Kametani, H. Kurobe and H. Nemoto, *J. Chem. Soc., Chem. Commun.*, 762 (1980).
532. T. Kametani, K. Fukumoto, H. Kurobe and H. Nemoto, *Tetrahedron Lett.*, **22**, 3653 (1981).
533. T. Kametani, H. Kurobe and H. Nemoto, *J. Chem. Soc., Perkin Trans. 1*, 756 (1981).
534. T. Kametani, H. Kurobe, H. Nemoto and K. Fukumoto, *J. Chem. Soc., Perkin Trans. 1*, 1085 (1982).
535. S. D. Burke, W. F. Fobare and D. M. Armistead, *J. Org. Chem.*, **47**, 3348 (1982).
536. L. Set, D. R. Cheshire and D. L. J. Clive, *J. Chem. Soc., Chem. Commun.*, 1205 (1985).
537. D. L. J. Clive and P. L. Beaulieu, *J. Chem. Soc., Chem. Commun.*, 307 (1983).
538. (a) M. D. Bachi and C. Hoornaert, *Tetrahedron Lett.*, **22**, 2693 (1981); (b) M. D. Bachi, F. Frolow and C. Hoornaert, *J. Org. Chem.*, **48**, 1841 (1983).
539. D. A. Burnett, J. -K. Choi, D. J. Hart and Y. -M. Tsai, *J. Am. Chem. Soc.*, **106**, 8201 (1984).
540. A. G. Angoh and D. L. J. Clive, *J. Chem. Soc., Chem. Commun.*, 980 (1985).
541. A. G. Angoh and D. L. J. Clive, *J. Chem. Soc., Chem. Commun.*, 941 (1985).
542. M. Shimizu and I. Kuwajima, *J. Org. Chem.*, **45**, 4063 (1980).
543. M. Shimizu, R. Ando and I. Kuwajima, *J. Org. Chem.*, **49**, 1230 (1984).
544. M. Shimizu, R. Ando and I. Kuwajima, *J. Org. Chem.*, **46**, 5246 (1981).
545. M. Shimizu and I. Kuwajima, *Tetrahedron Lett.*, 2801 (1979).
546. M. Shimizu, H. Urabe and I. Kuwajima, *Tetrahedron Lett.*, **22**, 2183 (1981).
547. I. Kuwajima, M. Shimizu and H. Urabe, *J. Org. Chem.*, **47**, 837 (1982).
548. F. Ogura, T. Otsubo, K. Ariyoshi and H. Yamaguchi, *Chem. Lett.*, 1833 (1983).
549. L. Engman and M. P. Cava, *J. Chem. Soc., Chem. Commun.*, 164 (1982).
550. K. Takaki, M. Yasumura and K. Negoro, *J. Org. Chem.*, **48**, 54 (1983).
551. D. H. R. Barton, S. V. Ley and C. A. Meerholz, *J. Chem. Soc., Chem. Commun.*, 755 (1979).
552. F. Ogura, H. Yamaguchi, T. Otsubo and H. Tanaka, *Bull. Chem. Soc. Jpn.*, **55**, 641 (1982).
553. J. P. Marino and A. Schwartz, *Tetrahedron Lett.*, 3253 (1979).
554. K. Balenovic, N. Bregant and I. Perina, *Synthesis*, 172 (1973).

555. I. Perina, N. Bregant and K. Balenovic, *Bull. Sci. Cons. Acad. Sci. Arts RSF Yougosl., Sect. A*, **18**, 4 (1973); *Chem. Abstr.*, **78**, 160007w (1973).
556. M. R. Czarny, *J. Chem. Soc., Chem. Commun.*, 81 (1976).
557. D. H. R. Barton, A. Billion and J. Boivin, *Tetrahedron Lett.*, **26**, 1229 (1985).
558. D. H. R. Barton, X. Lusinchi and P. Milliet, *Tetrahedron*, **41**, 4727 (1985).
559. J. P. Marino and R. D. Larsen, Jr., *J. Am. Chem. Soc.*, **103**, 4642 (1981).
560. M. R. Czarny, *Synth. Commun.*, **6**, 285 (1976).
561. V. I. Naddaka, V. P. Garkin and V. I. Minkin, *Zh. Org. Khim.*, **12**, 2481 (1976).
562. M. Poje and K. Balenovic, *Bull. Sci. Cons. Acad. Sci. Arts RSF Yougosl., Sect. A*, **20**, 1 (1975); *Chem. Abstr.*, **83**, 43558u (1975).
563. T. Kobayashi and T. Hiraoka, *Bull. Chem. Soc. Jpn.*, **52**, 3366 (1979).
564. D. H. R. Barton, X. Lusinchi and P. Milliet, *Tetrahedron Lett.*, **23**, 4949 (1982).
565. I. Ninomiya, T. Kiguchi, C. Hashimoto, D. H. R. Barton, X. Lusinchi and P. Milliet, *Tetrahedron Lett.*, **26**, 4183 (1985).
566. I. Ninomiya, C. Hashimoto, T. Kiguchi, D. H. R. Barton, X. Lusinchi and P. Milliet, *Tetrahedron Lett.*, **26**, 4187 (1985).
567. T. G. Back, S. Collins and R. G. Kerr, *J. Org. Chem.*, **46**, 1564 (1981).
568. K. Kondo, S. Murai and N. Sonoda, *Tetrahedron Lett.*, 3727 (1977).
569. D. H. R. Barton, D. J. Lester and S. V. Ley, *J. Chem. Soc., Chem. Commun.*, 276 (1978).
570. D. H. R. Barton, D. J. Lester and S. V. Ley, *J. Chem. Soc., Perkin Trans. 1*, 1212 (1980).
571. S. V. Ley, C. A. Meerholz and D. H. R. Barton, *Tetrahedron*, **37**, Suppl. 1, 213 (1981).
572. J. Bergman and L. Engman, *Z. Naturforsch., Teil B*, **35**, 882 (1980).
573. K. Balenovic, R. Lazic, V. Polak and P. Stern, *Bull. Sci. Cons. Acad. Sci. Arts RSF Yougosl., Sect. A*, **17**, 147 (1972); *Chem. Abstr.*, **77**, 139499c (1972).
574. T. G. Back and S. Collins, *Tetrahedron Lett.*, 2661 (1979).
575. T. G. Back and S. Collins, *Tetrahedron Lett.*, **21**, 2213 (1980).
576. N. Bregant, I. Perina and K. Balenovic, *Bull. Sci. Cons. Acad. Sci. Arts RSF Yougosl., Sect. A*, **17**, 148 (1972); *Chem. Abstr.*, **77**, 152092j (1972).
577. (a) T. G. Back, *J. Chem. Soc., Chem. Commun.*, 530 (1981); (b) T. G. Back and R. G. Kerr, *Can. J. Chem.*, **60**, 2711 (1982).
578. D. H. R. Barton, X. Lusinchi and J. S. Ramirez, *Tetrahedron Lett.*, **24**, 2995 (1983).
579. D. H. R. Barton, D. J. Lester and S. V. Ley, *J. Chem Soc., Chem. Commun.*, 445 (1977).
580. D. H. R. Barton, G. Bashiardes and J. -L. Fourrey, *Tetrahedron Lett.*, **25**, 1287 (1984).
581. I. Lalezari, A. Shafiee and M. Yalpani, *Angew. Chem., Int. Ed. Engl.*, **9**, 464 (1970).
582. I. Lalezari, A. Shafiee and M. Yalpani, *J. Org. Chem.*, **36**, 2836 (1971).
583. I. Lalezari, A. Shafiee and M. Yalpani, *J. Heterocycl. Chem.*, **9**, 1411 (1972).
584. I. Lalezari, A. Shafiee, F. Rabet and M. Yalpani, *J. Heterocycl. Chem.*, **10**, 953 (1973).
585. I. Lalezari, A. Shafiee and H. Golgolab, *J. Heterocycl. Chem.*, **10**, 655 (1973).
586. H. Golgolab and I. Lalezari, *J. Heterocycl. Chem.*, **12**, 801 (1975).
587. I. Lalezari, A. Shafiee, J. Khorrami and A. Soltani, *J. Pharm. Sci.*, **67**, 1336 (1978).
588. A. Shafiee, S. Toghraie, F. Aria and G. Mortezaei-Zandjani, *J. Heterocycl. Chem.*, **19**, 1305 (1982).
589. I. Lalezari, *Synthesis*, 660 (1984).
590. H. Meier and I. Menzel, *J. Chem. Soc., Chem. Commun.*, 1059 (1971).
591. H. Meier and E. Voigt, *Tetrahedron*, **28**, 187 (1972).
592. H. Meier and I. Menzel, *Tetrahedron Lett.*, 445 (1972).
593. H. Meier, M. Layer and A. Zetzsche, *Chem. Ztg.*, **98**, 460 (1974).
594. H. Meier, S. Schniepp and W. Combrink, *Chem. Ztg.*, **99**, 461 (1975).
595. H. Meier and H. Gugel, *Synthesis*, 338 (1976).
596. H. Meier, M. Layer, W. Combrink and S. Schniepp, *Chem. Ber.*, **109**, 1650 (1976).
597. H. Meier, T. Echter and H. Petersen, *Angew. Chem., Int. Ed. Engl.*, **17**, 942 (1978).
598. J. Zountsas and H. Meier, *Justus Liebigs Ann. Chem.*, 1366 (1982).
599. N. Hanold and H. Meier, *Chem. Ber.*, **118**, 198 (1985).
600. B. A. Keay and R. Rodrigo, *J. Am. Chem. Soc.*, **104**, 4725 (1982).
601. S. N. Dhawan and J. Kagan, *J. Lebelled Compd. Radiopharm.*, **19**, 331 (1982).
602. G. Sosnovsky and J. A. Krogh, *Synthesis*, 703 (1978).
603. G. Sosnovsky, J. A. Krogh and S. G. Umhoefer, *Synthesis*, 722 (1979).
604. A. G. H. Wee, A. Y. L. Shu, E. Bunnenberg and C. Djerassi, *J. Org. Chem.*, **49**, 3327 (1984).

605. C. L. Pedersen, *Acta Chem. Scand., Ser. B*, **30**, 675 (1976).
606. G. Sosnovsky and J. A. Krogh, *Z. Naturforsch., Teil B*, **34**, 511 (1979).
607. G. Sosnovsky and J. A. Krogh, *Synthesis*, 654 (1980).
608. B. Nakhdjavan and G. Klar, *Justus Liebigs Ann. Chem.*, 1683 (1977).
609. L. G. Faehl and J. L. Kice, *J. Org. Chem.*, **44**, 2357 (1979).
610. K. C. Nicolaou, R. L. Magolda and W. E. Barnette, *J. Chem. Soc., Chem. Commun.*, 375 (1978).
611. K. C. Nicolaou, D. A. Claremon, D. P. Papahatjis and R. L. Magolda, *J. Am. Chem. Soc.*, **103**, 6969 (1981).
612. K. C. Nicolaou, D. P. Papahatjis, D. A. Claremon, R. L. Magolda and R. E. Dolle, *J. Org. Chem.*, **50**, 1440 (1985).
613. M. P. Edwards, S. V. Ley, S. G. Lister and B. D. Palmer, *J. Chem. Soc., Chem. Commun.*, 630 (1983).
614. M. P. Edwards, S. V. Ley, S. G. Lister, B. D. Palmer and D. J. Williams, *J. Org. Chem.*, **49**, 3503 (1984).
615. J. Drabowicz and M. Mikolajczyk, *Synthesis*, 758 (1978).
616. J. Drabowicz, P. Lyzwa and M. Mikolajczyk, *Phosphorus Sulfur*, **17**, 169 (1983).
617. D. H. R. Barton, N. J. Cussans and S. V. Ley, *J. Chem. Soc., Chem. Commun.*, 751 (1977).
618. N. J. Cussans, S. V. Ley and D. H. R. Barton, *J. Chem. Soc., Perkin Trans. 1*, 1654 (1980).
619. D. H. R. Barton, M. T. Bielska, J. M. Cardoso, N. J. Cussans and S. V. Ley, *J. Chem. Soc., Perkin Trans. 1*, 1840 (1981).
620. J. -C. Barrière, A. Chiaroni, J. Cléophax, S. D. Géro, C. Riche and M. Vuilhorgne, *Helv. Chim. Acta*, **64**, 1140 (1981).
621. D. VanEnde, A. Cravador and A. Krief, *J. Organomet. Chem.*, **177**, 1 (1979).
622. A. Burton, L. Hevesi, W. Dumont, A. Cravador and A. Krief, *Synthesis*, 877 (1979).
623. D. H. R. Barton, N. J. Cussans and S. V. Ley, *J. Chem. Soc., Chem. Commun.*, 393 (1978).
624. N. J. Cussans, S. V. Ley and D. H. R. Barton, *J. Chem. Soc., Perkin Trans. 1*, 1650 (1980).
625. S. Tamagaki, I. Hatanaka and S. Kozuka, *Bull. Chem. Soc. Jpn.*, **50**, 3421 (1977).
626. M. Mikolajczyk and J. Luczak, *J. Org. Chem.*, **43**, 2132 (1978).
627. S. V. Ley, C. A. Meerholz and D. H. R. Barton *Tetrahedron Lett.*, **21**, 1785 (1980).
628. A. G. M. Barrett, D. H. R. Barton and R. W. Read, *J. Chem. Soc., Chem. Commun.*, 645 (1979).
629. R. A. Gancarz and J. L. Kice, *Tetrahedron Lett.*, **21**, 1697 (1980).
630. A. Krief, *Tetrahedron*, **36**, 2531 (1980).
631. M. Clarembeau, A. Cravador, W. Dumont, L. Hevesi, A. Krief, J. Lucchetti and D. Van Ende, *Tetrahedron*, **41**, 4793 (1985).
632. D. Seebach and N. Peleties, *Angew. Chem., Int. Ed. Engl.*, **8**, 450 (1969).
633. D. Seebach and N. Peleties, *Chem. Ber.*, **105**, 511 (1972).
634. J. Lucchetti and A. Krief, *Tetrahedron Lett.*, **22**, 1623 (1981).
635. B. -T. Gröbel and D. Seebach, *Chem. Ber.*, **110**, 852 (1977).
636. H. J. Reich, W. W. Willis, Jr. and P. D. Clark, *J. Org. Chem.*, **46**, 2775 (1981).
637. S. Raucher and G. A. Koolpe, *J. Org. Chem.*, **43**, 3794 (1978).
638. K. Hirai, Y. Iwano and K. Fujimoto, *Tetrahedron Lett.*, **23**, 4021 (1982).
639. D. Seebach and A. K. Beck, *Angew. Chem., Int. Ed. Engl.*, **13**, 806 (1974).
640. W. Dumont, P. Bayet and A. Krief, *Angew. Chem., Int. Ed. Engl.*, **13**, 804 (1974).
641. W. Dumont and A. Krief, *Angew. Chem., Int. Ed. Engl.*, **14**, 350 (1975).
642. A. Anciaux, A. Eman, W. Dumont, D. Van Ende and A. Krief, *Tetrahedron Lett.*, 1613 (1975).
643. D. Van Ende, W. Dumont and A. Krief, *Angew. Chem., Int. Ed. Engl.*, **14**, 700 (1975).
644. J. N. Denis, W. Dumont and A. Krief, *Tetrahedron Lett.*, 453 (1976).
645. M. Sevrin, D. Van Ende and A. Krief, *Tetrahedron Lett.*, 2643 (1976).
646. D. Van Ende and A. Krief, *Tetrahedron Lett.*, 457 (1976).
647. W. Dumont and A. Krief, *Angew. Chem., Int. Ed. Engl.*, **15**, 161 (1976).
648. S. Halazy, J. Lucchetti and A. Krief, *Tetrahedron Lett.*, 3971 (1978).
649. D. Labar, W. Dumont, L. Hevesi and A. Krief, *Tetrahedron Lett.*, 1145 (1978).
650. S. Halazy, W. Dumont and A. Krief, *Tetrahedron Lett.*, **22**, 4737 (1981).
651. D. Labar and A. Krief, *J. Chem. Soc., Chem. Commun.*, 564 (1982).
652. T. DiGiamberardino, S. Halazy, W. Dumont and A. Krief, *Tetrahedron Lett.*, **24**, 3413 (1983).
653. D. Seebach, N. Meyer and A. K. Beck, *Justus Liebigs Ann. Chem.*, 846 (1977).
654. I. Kuwajima, S. Hoshino, T. Tanaka and M. Shimizu, *Tetrahedron Lett.*, **21**, 3209 (1980).

655. L. A. Paquette, T. -H. Yan and G. J. Wells, *J. Org. Chem.*, **49**, 3610 (1984).
656. H. J. Reich, F. Chow and S. K. Shah, *J. Am. Chem. Soc.*, **101**, 6638 (1979).
657. R. H. Mitchell, *Can. J. Chem.*, **58**, 1398 (1980).
658. D. J. Goldsmith, D. C. Liotta, M. Volmer, W. Hoekstra and L. Waykole, *Tetrahedron*, **41**, 4873 (1985).
659. M. Clarembeau and A. Krief, *Tetrahedron Lett.*, **25**, 3629 (1984).
660. M. Clarembeau and A. Krief, *Tetrahedron Lett.*, **26**, 1093 (1985).
661. S. Halazy and A. Krief, *Tetrahedron Lett.*, **22**, 4341 (1981).
662. A. Anciaux, A. Eman, W. Dumont and A. Krief, *Tetrahedron Lett.*, 1617 (1975).
663. W. Dumont, M. Sevrin and A. Krief, *Angew. Chem., Int. Ed. Engl.*, **16**, 541 (1977).
664. D. Seebach and A. K. Beck, *Chem. Ber.*, **108**, 314 (1975).
665. C. A. Brandt, J. V. Comasseto, W. Nakamura and N. Petragnani, *J. Chem. Res.* (*S*), 156 (1983).
666. S. Halazy and A. Krief, *Tetrahedron Lett.*, **22**, 1833 (1981).
667. J. L. Laboureur and A. Krief, *Tetrahedron Lett.*, **25**, 2713 (1984).
668. S. Halazy and A. Krief, *J. Chem. Soc., Chem. Commun.*, 1136 (1979).
669. D. Labar, J. L. Laboureur and A. Krief, *Tetrahedron Lett.*, **23**, 983 (1982).
670. S. Halazy, F. Zutterman and A. Krief, *Tetrahedron Lett.*, **23**, 4385 (1982).
671. S. Halazy and A. Krief, *J. Chem. Soc., Chem. Commun.*, 1200 (1982).
672. J. Lucchetti and A. Krief, *Tetrahedron Lett.*, 2697 (1978).
673. J. Lucchetti, W. Dumont and A. Krief, *Tetrahedron Lett.*, 2695 (1979).
674. J. Lucchetti and A. Krief, *J. Organomet. Chem.*, **194**, C49 (1980).
675. J. Lucchetti and A. Krief, *J. Chem. Soc., Chem. Commun.*, 127 (1982).
676. J. Lucchetti and A. Krief, *Synth. Commun.*, **13**, 1153 (1983).
677. W. Dumont, J. Lucchetti and A. Krief, *J. Chem. Soc., Chem. Commun.*, 66 (1983).
678. J. L. Laboureur, W. Dumont and A. Krief, *Tetrahedron Lett.*, **25**, 4569 (1984).
679. H. J. Reich and F. Chow, *J. Chem. Soc., Chem. Commun.*, 790 (1975).
680. H. J. Reich, M. C. Clark and W. W. Willis, Jr., *J. Org. Chem.*, **47**, 1618 (1982).
681. J. Rémion, W. Dumont and A. Krief, *Tetrahedron Lett.*, 1385 (1976).
682. Y. Yamamoto, Y. Saito and K. Maruyama, *Tetrahedron Lett.*, **23**, 4597 (1982).
683. J. Lucchetti and A. Krief, *Tetrahedron Lett.*, 2693 (1978).
684. J. Rémion and A. Krief, *Tetrahedron Lett.*, 3743 (1976).
685. J. N. Denis, S. Desauvage, L. Hevesi and A. Krief, *Tetrahedron Lett.*, **22**, 4009 (1981).
686. J. N. Denis, W. Dumont and A. Krief, *Tetrahedron Lett.*, 4111 (1979).
687. J. Lucchetti, J. Rémion and A. Krief, *C. R. Acad. Sci., Ser. C*, **288**, 553 (1979).
688. Y. Yamamoto, Y. Saito and K. Maruyama, *J. Org. Chem.*, **48**, 5408 (1983).
689. H. Wetter, *Helv. Chim. Acta*, **61**, 3072 (1978).
690. H. J. Reich and W. W. Willis, Jr., *J. Org. Chem.*, **45**, 5227 (1980).
691. H. J. Reich, *J. Org. Chem.*, **40**, 2570 (1975).
692. H. J. Reich, M. C. Schroeder and I. L. Reich, *Isr. J. Chem.*, **24**, 157 (1984).
693. H. J. Reich and S. K. Shah, *J. Am. Chem. Soc.*, **99**, 263 (1977).
694. H. J. Reich, P. M. Gold and F. Chow, *Tetrahedron Lett.*, 4433 (1979).
695. H. J. Reich, S. K. Shah, P. M. Gold and R. E. Olson, *J. Am. Chem. Soc.*, **103**, 3112 (1981).
696. M. R. Binns and R. K. Haynes, *J. Org. Chem.*, **46**, 3790 (1981).
697. T. Kauffmann and K. -R. Gaydoul, *Tetrahedron Lett.*, **26**, 4071 (1985).
698. T. Kauffmann and H. Ahlers, *Chem. Ber.*, **116**, 1001 (1983).
699. B. -T. Gröbel and D. Seebach, *Chem. Ber.*, **110**, 867 (1977).
700. S. Raucher and G. A. Koolpe, *J. Org. Chem.*, **43**, 4252 (1978).
701. R. C. Gadwood, *J. Org. Chem.*, **48**, 2098 (1983).
702. N. Petragnani and H. M. C. Ferraz, *Synthesis*, 476 (1978).
703. N. Petragnani, H. M. C. Ferraz and M. Yonashiro, *Synthesis*, 27 (1985).
704. P. A. Grieco and Y. Yokoyama, *J. Am. Chem. Soc.*, **99**, 5210 (1977).
705. J. N. Marx and P. J. Dobrowolski, *Tetrahedron Lett.*, **23**, 4457 (1982).
706. A. Osuka, Y. Mori, H. Shimizu and H. Suzuki, *Tetrahedron Lett.*, **24**, 2599 (1983).
707. A. Osuka and H. Suzuki, *Tetrahedron Lett.*, **24**, 5109 (1983).
708. W. Dumont, P. Bayet and A. Krief, *Angew. Chem., Int. Ed. Engl.*, **13**, 274 (1974).
709. W. W. Lotz and J. Gosselck, *Tetrahedron*, **29**, 917 (1973).
710. E. Wenkert, T. W. Ferreira and E. L. Michelotti, *J. Chem. Soc., Chem. Commun.*, 637 (1979).

711. H. Okamura, M. Miura, K. Kosugi and H. Takei, *Tetrahedron Lett.*, **21**, 87 (1980).
712. S. Uemura and S. Fukuzawa, *Tetrahedron Lett.*, **23**, 1181 (1982).
713. S. Uemura, S. Fukuzawa and S. R. Patil, *J. Organomet. Chem.*, **243**, 9 (1983).
714. T. G. Back, S. Collins and K. -W. Law, *Tetrahedron Lett.*, **25**, 1689 (1984).
715. A. F. Sviridov, M. S. Ermolenko, D. V. Yashunsky and N. K. Kochetkov, *Tetrahedron Lett.*, **24**, 4355 (1983).
716. A. F. Sviridov, M. S. Ermolenko, D. V. Yashunsky and N. K. Kochetkov, *Tetrahedron Lett.*, **24**, 4359 (1983).
717. A. P. Kozikowski and A. Ames, *J. Am. Chem. Soc.*, **102**, 860 (1980).
718. A. P. Kozikowski and A. Ames, *Tetrahedron*, **41**, 4821·(1985).
719. A. P. Kozikowski and A. Ames, *J. Org. Chem.*, **43**, 2735 (1978).
720. R. Ando, T. Sugawara and I. Kuwajima, *J. Chem. Soc., Chem. Commun.*, 1514 (1983).
721. I. Kuwajima, R. Ando and T. Sugawara, *Tetrahedron Lett.*, **24**, 4429 (1983).
722. R. Ando, T. Sugawara, M. Shimizu and I. Kuwajima, *Bull. Chem. Soc. Jpn.*, **57**, 2897 (1984).
723. M. Shimizu and I. Kuwajima, *J. Org. Chem.*, **45**, 2921 (1980).
724. A. Krief, W. Dumont and J. N. Denis, *J. Chem. Soc., Chem. Commun.*, 571 (1985).
725. J. Bergman, *Tetrahedron*, **28**, 3323 (1972).
726. J. Bergman, R. Carlsson and B. Sjöberg, *Org. Synth.*, **57**, 18 (1977).
727. J. Bergman and L. Engman, *Tetrahedron*, **36**, 1275 (1980).
728. S. Uemura, M. Wakasugi and M. Okano, *J. Organomet. Chem.*, **194**, 277 (1980).
729. J. Bergman, *Chem. Scr.*, **8A**, 116 (1975).
730. M. Albeck and T. Tamari, *J. Organomet. Chem.*, **238**, 357 (1982).
731. D. H. R. Barton, S. A. Glover and S. V. Ley, *J. Chem. Soc., Chem. Commun.*, 266 (1977).
732. S. A. Glover, *J. Chem. Soc., Perkin Trans. 1*, 1338 (1980).
733. D. L. J. Clive, P. C. Anderson, N. Moss and A. Singh, *J. Org. Chem.*, **47**, 1641 (1982).
734. T. Otsubo, F. Ogura, H. Yamaguchi, H. Higuchi and S. Misumi, *Synth. Commun.*, **10**, 595 (1980).
735. E. Cuthbertson and D. D. MacNicol, *Tetrahedron Lett.*, 1893 (1975).
736. R. H. Mitchell, *Can. J. Chem.*, **54**, 238 (1976).
737. H. Higuchi and S. Misumi, *Tetrahedron Lett.*, **23**, 5571 (1982).
738. H. Higuchi, M. Kugimiya, T. Otsubo, Y. Sakata and S. Misumi, *Tetrahedron Lett.*, **24**, 2593 (1983).
739. T. Kaneda, S. Umeda, H. Tanigawa, S. Misumi, Y. Kai, H. Morii, K. Miki and N. Kasai, *J. Am. Chem. Soc.*, **107**, 4802 (1985).
740. A. Couture, A. Lablache-Combier, R. Lapouyade and Félix, *J. Chem. Res. (S)*, 258 (1979).
741. E. Cuthbertson and D. D. MacNicol, *J. Chem. Soc., Chem. Commun.*, 498 (1974).
742. E. Cuthbertson, J. H. Gall and D. D. MacNicol, *Tetrahedron Lett.*, 3203 (1977).
743. H. Higuchi, Y. Sakata, S. Misumi, T. Otsubo, F. Ogura and H. Yamaguchi, *Chem. Lett.*, 627 (1981).
744. H. Higuchi, T. Otsubo, F. Ogura, H. Yamaguchi, Y. Sakata and S. Misumi, *Bull. Chem. Soc. Jpn.*, **55**, 182 (1982).
745. H. Ishihara and Y. Hirabayashi, *Chem. Lett.*, 1007 (1978).
746. T. G. Back, D. H. R. Barton, M. R. Britten-Kelly and F. S. Guziec, Jr., *J. Chem. Soc., Chem. Commun.*, 539 (1975).
747. T. G. Back, D. H. R. Barton, M. R. Britten-Kelly and F. S. Guziec, Jr., *J. Chem. Soc., Perkin Trans. 1*, 2079 (1976).
748. F. S. Guziec, Jr., and C. J. Murphy, *J. Org. Chem.*, **45**, 2890 (1980).
749. E. R. Cullen, F. S. Guziec, Jr., M. I. Hollander and C. J. Murphy, *Tetrahedron Lett.*, **22**, 4563 (1981).
750. E. R. Cullen, F. S. Guziec, Jr., and C. J. Murphy, *J. Org. Chem.*, **47**, 3563 (1982).
751. F. S. Guziec, Jr., C. J. Murphy and E. R. Cullen, *J. Chem. Soc., Perkin Trans. 1*, 107 (1985).
752. F. S. Guziec, Jr., L. J. San Filippo, C. J. Murphy, C. A. Moustakis and E. R. Cullen, *Tetrahedron*, **41**, 4843 (1985).
753. F. S. Guziec, Jr., and C. A. Moustakis, *J. Chem. Soc., Chem. Commun.*, 63 (1984).
754. C. C. Silveira, J. V. Comasseto and V. Catani, *Synth. Commun.*, **15**, 931 (1985).
755. R. P. Alexander and I. Paterson, *Tetrahedron Lett.*, **24**, 5911 (1983).
756. M. Renard and L. Hevesi, *Tetrahedron Lett.*, **24**, 3911 (1983).
757. S. Halazy and L. Hevesi, *J. Org. Chem.*, **48**, 5242 (1983).
758. G. E. Keck and J. B. Yates, *J. Am. Chem. Soc.*, **104**, 5829 (1982).

759. T. Kawamura, K. Kikukawa, M. Takagi and T. Matsuda, *Bull. Chem. Soc. Jpn.*, **50**, 2021 (1977).
760. J. V. Comasseto and C. A. Brandt, *J. Chem. Res. (S)*, 56 (1982).
761. D. Liotta, W. Markiewicz and H. Santiesteban, *Tetrahedron Lett.*, 4365 (1977).
762. D. Liotta, U. Sunay, H. Santiesteban and W. Markiewicz, *J. Org. Chem.*, **46**, 2605 (1981).
763. M. Tiecco, L. Testaferri, M. Tingoli, D. Chianelli, and M. Montanucci, *Synth. Commun.*, **13**, 617 (1983).
764. (a) R. M. Scarborough, Jr., and A. B. Smith, III, *Tetrahedron Lett.*, 4361 (1977); (b) R. M. Scarborough, Jr., B. H. Toder and A. B. Smith, III, *J. Am. Chem. Soc.*, **102**, 3904 (1980).
765. T. Kometani, Y. Takeuchi and E. Yoshii, *J. Org. Chem.*, **48**, 2630 (1983).
766. S. Takano, N. Tamura and K. Ogasawara, *J. Chem. Soc., Chem. Commun.*, 1155 (1981).
767. D. Liotta and H. Santiesteban, *Tetrahedron Lett.*, 4369 (1977).
768. D. J. Goldsmith and J. K. Thottathil, *Tetrahedron Lett.*, **22**, 2447 (1981).
769. P. Dowd and P. Kennedy, *Synth. Commun.*, **11**, 935 (1981).
770. T. R. Hoye and A. J. Caruso, *Tetrahedron Lett.*, 4611 (1978).
771. K. P. Lok, I. J. Jakovac and J. B. Jones, *J. Am. Chem. Soc.*, **107**, 2521 (1985).
772. M. R. Detty, *Tetrahedron Lett.*, 5087 (1978).
773. N. Miyoshi, H. Ishii, S. Murai and N. Sonoda, *Chem. Lett.*, 873 (1979).
774. K. Sasaki, Y. Aso, T. Otsubo and F. Ogura, *Tetrahedron Lett.*, **26**, 453 (1985).
775. S. Saito, H. Tamai, Y. Usui, M. Inaba and T. Moriwake, *Chem. Lett.*, 1243 (1984).
776. M. Inaba, T. Moriwake and S. Saito, *Tetrahedron Lett.*, **26**, 3235 (1985).
777. B. M. Trost and P. H. Scudder, *J. Am. Chem. Soc.*, **99**, 7601 (1977).
778. H. M. Sirat, E. J. Thomas and J. D. Wallis, *J. Chem. Soc., Perkin Trans. 1*, 2885 (1982).
779. T. Satoh, Y. Kaneko, T. Kumagawa, T. Izawa, K. Sakata and K. Yamakawa, *Chem. Lett.*, 1957 (1984).
780. D. L. J. Clive, G. J. Chittattu, V. Farina, W. A. Kiel, S. M. Menchen, C. G. Russell, A. Singh, C. K. Wong and N. J. Curtis, *J. Am. Chem. Soc.*, **102**, 4438 (1980).
781. N. Miyoshi, K. Kondo, S. Murai and N. Sonoda, *Chem. Lett.*, 909 (1979).
782. P. A. Bartlett and L. A. McQuaid, *J. Am. Chem. Soc.*, **106**, 7854 (1984).
783. A. Cravador and A. Krief, *Tetrahedron Lett.*, **22**, 2491 (1981).
784. G. H. Posner and D. Z. Rogers, *J. Am. Chem. Soc.*, **99**, 8208 (1977).
785. A. Osuka, K. Taka-Oka and H. Suzuki, *Chem. Lett.*, 271 (1984).
786. R. Ahmad, J. M. Saá and M. P. Cava, *J. Org. Chem.*, **42**, 1228 (1977).
787. Z. Ahmed and M. P. Cava, *J. Am. Chem. Soc.*, **105**, 682 (1983).
788. M. Evers and L. Christiaens, *Tetrahedron Lett.*, **24**, 377 (1983).
789. D. L. Boger and J. S. Panek, *J. Am. Chem. Soc.*, **107**, 5745 (1985).
790. V. Šimánek and A. Klásek, *Tetrahedron Lett.*, 3039 (1969).
791. S. Murahashi and T. Yano, *J. Am. Chem. Soc.*, **102**, 2456 (1980).
792. H. J. Reich and M. L. Cohen, *J. Org. Chem.*, **44**, 3148 (1979).
793. P. Müller and M. P. N. Thi, *Helv. Chim. Acta*, **63**, 2168 (1980).
794. A. B. Smith, III, and R. M. Scarborough, Jr., *Tetrahedron Lett.*, 1649 (1978).
795. S. Masamune, T. Kaiho and D. S. Garvey, *J. Am. Chem. Soc.*, **104**, 5521 (1982).
796. P. Kocienski and C. Yeates, *J. Chem. Soc., Perkin Trans. 1*, 1879 (1985).
797. M. Clarembeau and A. Krief, *Tetrahedron Lett.*, **25**, 3625 (1984).
798. P. A. Grieco, J. A. Noguez and Y. Masaki, *Tetrahedron Lett.*, 4213 (1975).
799. P. A. Grieco, M. Nishizawa, S. D. Burke and N. Marinovic, *J. Am. Chem. Soc.*, **98**, 1612 (1976).
800. P. A. Grieco, J. A. Noguez and Y. Masaki, *J. Org. Chem.*, **42**, 495 (1977).
801. G. Majetich, P. A. Grieco and M. Nishizawa, *J. Org. Chem.*, **42**, 2327 (1977).
802. P. A. Grieco, M. Nishizawa, T. Oguri, S. D. Burke and N. Marinovic, *J. Am. Chem. Soc.*, **99**, 5773 (1977).
803. M. Nishizawa, P. A. Grieco, S. D. Burke and W. Metz, *J. Chem. Soc., Chem. Commun.*, 76 (1978).
804. P. A. Grieco, J. Y. Jaw, D. A. Claremon and K. C. Nicolaou, *J. Org. Chem.*, **46**, 1215 (1981).
805. S. V. Ley, N. S. Simpkins and A. J. Whittle, *J. Chem. Soc., Chem. Commun.*, 1001 (1981).
806. S. V. Ley, D. Neuhaus, N. S. Simpkins and A. J. Whittle, *J. Chem. Soc., Perkin Trans. 1*, 2157 (1982).
807. S. V. Ley, N. S. Simpkins and A. J. Whittle, *J. Chem. Soc., Chem. Commun.*, 503 (1983).
808. T. G. Back and D. J. McPhee, *J. Org. Chem.*, **49**, 3842 (1984).
809. F. Ogura, H. Yamaguchi, T. Otsubo and K. Chikamatsu, *Synth. Commun.*, **12**, 131 (1982).
810. P. A. Grieco, Y. Yokoyama and E. Williams, *J. Org. Chem.*, **43**, 1283 (1978).

811. S. Masamune, Y. Hayase, W. Schilling, W. K. Chan and G. S. Bates, *J. Am. Chem. Soc.*, **99**, 6756 (1977).
812. R. K. Haynes and C. Indorato, *Aust. J. Chem.*, **37**, 1183 (1984).
813. B. Danieli, G. Lesma, G. Palmisano and S. Tollari, *J. Chem. Soc., Perkin Trans. 1*, 1237 (1984).
814. T. G. Back, *J. Chem. Soc., Chem. Commun.*, 1417 (1984).
815. R. O. Hutchins and K. Learn, *J. Org. Chem.*, **47**, 4380 (1982).
816. E. J. Corey, H. L. Pearce, I. Székely and M. Ishiguro, *Tetrahedron Lett.*, 1023 (1978).
817. D. L. J. Clive, G. Chittattu and C. K. Wong, *J. Chem. Soc., Chem. Commun.*, 41 (1978).
818. J. Pfenninger, C. Heuberger and W. Graf, *Helv. Chim. Acta*, **63**, 2328 (1980).
819. J. Pfenninger and W. Graf, *Helv. Chim. Acta*, **63**, 1562 (1980).
820. R. E. Ireland, D. W. Norbeck, G. S. Mandel and N. S. Mandel, *J. Am. Chem. Soc.*, **107**, 3285 (1985).
821. D. H. R. Barton and S. W. McCombie, *J. Chem. Soc., Perkin Trans. 1*, 1574 (1975).
822. A. G. M. Barrett, R. W. Read and D. H. R. Barton, *J. Chem. Soc., Perkin Trans. 1*, 2191 (1980).
823. T. K. Raja, *Indian J. Chem.*, **19B**, 812 (1980).
824. M. Sevrin, J. N. Denis and A. Krief, *Tetrahedron Lett.*, **21**, 1877 (1980).
825. H. Suzuki and M. Inouye, *Chem. Lett.*, 225 (1985).
826. K. Ramasamy, S. K. Kalyanasundaram and P. Shanmugam, *Synthesis*, 311 (1978).
827. K. Ramasamy, S. K. Kalyanasundaram and P. Shanmugam, *Synthesis*, 545 (1978).
828. H. Suzuki, A. Kondo and A. Osuka, *Bull. Chem. Soc. Jpn.*, **58**, 1335 (1985).
829. L. Engman, *Tetrahedron Lett.*, **23**, 3601 (1982).
830. L. Engman and S. E. Byström, *J. Org. Chem.*, **50**, 3170 (1985).
831. D. L. J. Clive and P. L. Beaulieu, *J. Org. Chem.*, **47**, 1124 (1982).
832. L. Engman and M. P. Cava, *J. Org. Chem.*, **47**, 3946 (1982).
833. A. Osuka and H. Suzuki, *Chem. Lett.*, 119 (1983).
834. R. Seshadri, W. J. Pegg and M. Israel, *J. Org. Chem.*, **46**, 2596 (1981).
835. A. Osuka, K. Takechi and H. Suzuki, *Bull. Chem. Soc. Jpn.*, **57**, 303 (1984).
836. D. L. J. Clive and C. A. Denyer, *J. Chem. Soc., Chem. Commun.*, 253 (1973).
837. T. H. Chan and J. R. Finkenbine, *Tetrahedron Lett.*, 2091 (1974).
838. F. Mathey and G. Muller, *C. R. Acad. Sci., Ser. C*, **281**, 881 (1975).
839. A. Ogawa, J. Miyake, S. Murai and N. Sonoda, *Tetrahedron Lett.*, **26**, 669 (1985).
840. V. Calò, L. Lopez, A. Mincuzzi and G. Pesce, *Synthesis*, 200 (1976).
841. (a) D. L. J. Clive and S. M. Menchen, *J. Chem. Soc., Chem. Commun.*, 658 (1977); (b) D. L. J. Clive and S. M. Menchen, *J. Org. Chem.*, **45**, 2347 (1980).
842. J. M. Behan, R. A. W. Johnstone and M. J. Wright, *J. Chem. Soc., Perkin Trans. 1*, 1216 (1975).
843. D. Van Ende and A. Krief, *Tetrahedron Lett.*, 2709 (1975).
844. (a) N. Kambe, K. Kondo, S. Morita, S. Murai and N. Sonoda, *Angew. Chem., Int. Ed. Engl.*, **19**, 1009 (1980); (b) N. Kambe, K. Kondo, S. Murai and N. Sonoda, *Angew. Chem., Int. Ed. Engl.*, **19**, 1008 (1980).
845. M. Akiba and M. P. Cava, *Synth. Commun.*, **14**, 1119 (1984).
846. T. Kametani, M. Aizawa, H. Kurobe, T. Matsuura, H. Nemoto and K. Fukumoto, *Chem. Pharm. Bull.*, **30**, 1493 (1982).
847. M. Yamashita, Y. Kato and R. Suemitsu, *Chem. Lett.*, 847 (1980).
848. M. J. Perkins, B. V. Smith and E. S. Turner, *J. Chem. Soc., Chem. Commun.*, 977 (1980).
849. T. Miyata, K. Kondo, S. Murai, T. Hirashima and N. Sonoda, *Angew. Chem., Int. Ed. Engl.*, **19**, 1008 (1980).
850. K. Kondo, N. Sonoda and H. Sakurai, *J. Chem. Soc., Chem. Commun.*, 42 (1975).
851. N. Kambe, K. Kondo and N. Sonoda, *Angew. Chem., Int. Ed. Engl.*, **19**, 1009 (1980).
852. A. Osuka, H. Shimizu and H. Suzuki, *Chem. Lett.*, 1373 (1983).
853. N. Ohira, Y. Aso, T. Otsubo and F. Ogura, *Chem. Lett.*, 853 (1984).
854. K. Fujimori, H. Yoshimoto and S. Oae, *Tetrahedron Lett.*, 4397 (1979).
855. D. H. R. Barton, A. Fekih and X. Lusinchi, *Tetrahedron Lett.*, **26**, 4603 (1985).
856. H. Suzuki and K. Takaoka, *Chem. Lett.*, 1733 (1984).
857. F. G. James, M. J. Perkins, O. Porta and B. V. Smith, *J. Chem. Soc., Chem. Commun.*, 131 (1977).
858. K. Fujimori, H. Yoshimoto and S. Oae, *Tetrahedron Lett.*, **21**, 3385 (1980).
859. D. H. R. Barton, A. Fekih and X. Lusinchi, *Tetrahedron Lett.*, **26**, 3693 (1985).
860. A. G. M. Barrett, R. W. Read and D. H. R. Barton, *J. Chem. Soc., Perkin Trans. 1*, 2184 (1980).

861. M. J. Perkins, B. V. Smith, B. Terem and E. S. Turner, *J. Chem. Res. (S)*, 341 (1979).
862. D. L. J. Clive and S. M. Menchen, *J. Chem. Soc., Chem. Commun.*, 168 (1979).
863. D. L. J. Clive and S. M. Menchen, *J. Org. Chem.*, **44**, 4279 (1979).
864. D. L. J. Clive, W. A. Kiel, S. M. Menchen and C. K. Wong, *J. Chem. Soc., Chem. Commun.*, 657 (1977).
865. M. R. Detty, *J. Org. Chem.*, **44**, 4528 (1979).
866. M. R. Detty and M. D. Seidler, *J. Org. Chem.*, **47**, 1354 (1982).
867. D. L. J. Clive, G. Chittattu, N. J. Curtis and S. M. Menchen, *J. Chem. Soc., Chem. Commun.*, 770 (1978).
868. P. A. Zoretic, R. J. Chambers, G. D. Marbury and A. A. Riebiro, *J. Org. Chem.*, **50**, 2981 (1985).
869. S. Uemura, S. Fukuzawa and K. Ohe, *Tetrahedron Lett.*, **26**, 921 (1985).
870. H. J. Reich, I. L. Reich and S. Wollowitz, *J. Am. Chem. Soc.*, **100**, 5981 (1978).
871. H. J. Reich and S. Wollowitz, *J. Am. Chem. Soc.*, **104**, 7051 (1982).
872. H. J. Reich, K. E. Yelm and S. Wollowitz, *J. Am. Chem. Soc.*, **105**, 2503 (1983).
873. H. J. Reich and M. J. Kelly, *J. Am. Chem. Soc.*, **104**, 1119 (1982).
874. A. Haces, E. M. G. A. van Kruchten and W. H. Okamura, *Tetrahedron Lett.*, **23**, 2707 (1982).
875. S. I. Pennanen, *Synth. Commun.*, **10**, 373 (1980).
876. H. Nishiyama, K. Itagaki, K. Sakuta and K. Itoh, *Tetrahedron Lett.*, **22**, 5285 (1981).
877. H. Nishiyama, S. Narimatsu and K. Itoh, *Tetrahedron Lett.*, **22**, 5289 (1981).
878. P. Magnus, F. Cooke and T. Sarkar, *Organometallics*, **1**, 562 (1982).
879. K. B. Sharpless and R. F. Lauer, *J. Org. Chem.*, **37**, 3973 (1972).
880. H. J. Reich and M. L. Cohen, *J. Am. Chem. Soc.*, **101**, 1307 (1979).
881. P. J. Giddings, D. I. John and E. J. Thomas, *Tetrahedron Lett.*, **21**, 395 (1980).
882. P. J. Giddings, D. I. John, E. J. Thomas and D. J. Williams, *J. Chem. Soc., Perkin Trans. 1*, 2757 (1982).
883. P. G. Gassman, T. Miura and A. Mossman, *J. Chem. Soc., Chem. Commun.*, 558 (1980).
884. S. Raucher, K. -J. Hwang and J. E. Macdonald, *Tetrahedron Lett.*, 3057 (1979).
885. C. G. Francisco, R. Freire, R. Hernández, J. A. Salazar and E. Suárez, *Tetrahedron Lett.*, **25**, 1621 (1984).
886. T. Frejd and K. B. Sharpless, *Tetrahedron Lett.*, 2239 (1978).
887. S. Uemura, S. Fukuzawa, T. Yamauchi, K. Hattori, S. Mizutaki and K. Tamaki, *J. Chem. Soc., Chem. Commun.*, 426 (1984).
888. E. J. Corey, W. Su and M. M. Mehrotra, *Tetrahedron Lett.*, **25**, 5123 (1984).
889. D. Liotta, C. S. Barnum and M. Saindane, *J. Org. Chem.*, **46**, 4301 (1981).
890. D. Liotta, M. Saindane and D. Brothers, *J. Org. Chem.*, **47**, 1598 (1982).
891. S. J. Falcone and M. E. Munk, *Synth. Commun.*, **9**, 719 (1979).
892. G. Zima, C. Barnum and D. Liotta, *J. Org. Chem.*, **45**, 2736 (1980).
893. T. G. Back and R. G. Kerr, *Tetrahedron*, **41**, 4759 (1985).
894. I. Kuwajima and R. Takeda, *Tetrahedron Lett.*, **22**, 2381 (1981).
895. A. M. Léonard-Coppens and A. Krief, *Tetrahedron Lett.*, 3227 (1976).
896. H. Nishiyama, K. Itagaki, N. Osaka and K. Itoh, *Tetrahedron Lett.*, **23**, 4103 (1982).
897. H. Nishiyama, T. Kitajima, A. Yamamoto and K. Itoh, *J. Chem. Soc., Chem. Commun.*, 1232 (1982).
898. C. J. Kowalski and J. -S. Dung, *J. Am. Chem. Soc.*, **102**, 7950 (1980).
899. D. L. J. Clive, A. G. Angoh, S. C. Suri, S. N. Rao and C. G. Russell, *J. Chem. Soc., Chem. Commun.*, 828 (1982).
900. D. L. J. Clive and C. G. Russell, *J. Chem. Soc., Chem. Commun.*, 434 (1981).
901. D. L. J. Clive, C. G. Russell and S. C. Suri, *J. Org. Chem.*, **47**, 1632 (1982).
902. D. H. R. Barton, D. J. Lester, W. B. Motherwell and M. T. B. Papoula, *J. Chem. Soc., Chem. Commun.*, 246 (1980).
903. G. H. Posner and M. J. Chapdelaine, *Tetrahedron Lett.*, 3227 (1977).
904. J. N. Denis and A. Krief, *J. Chem. Soc., Chem. Commun.*, 229 (1983).
905. T. Agawa, M. Ishida and Y. Ohshiro, *Synthesis*, 933 (1980).
906. M. Miyashita and A. Yoshikoshi, *Synthesis*, 664 (1980).
907. M. Miyashita, T. Suzuki and A. Yoshikoshi, *J. Org. Chem.*, **50**, 3377 (1985).
908. A. Nickon, A. D. Rodriguez, R. Ganguly and V. Shirhatti, *J. Org. Chem.*, **50**, 2767 (1985).
909. P. A. Grieco and M. Miyashita, *Tetrahedron Lett.*, 1869 (1974).

910. K. Mori, *Tetrahedron*, **32**, 1101 (1976).
911. J. C. Sih and D. R. Graber, *J. Org. Chem.*, **43**, 3798 (1978).
912. R. H. Wollenberg, *Tetrahedron Lett.*, **21**, 3139 (1980).
913. R. C. Nickolson and H. Vorbrüggen, *Tetrahedron Lett.*, **24**, 47 (1983).
914. D. H. Wadsworth and M. R. Detty, *J. Org. Chem.*, **45**, 4611 (1980).
915. L. A. Tsoi, A. K. Patsaev, V. Z. Ushanov and L. V. Vyaznikovtsev, *Zh. Org. Khim.*, **20**, 2081 (1984).
916. N. Shinmon, M. P. Cava and R. F. C. Brown, *J. Chem. Soc., Chem. Commun.*, 1020 (1980).
917. H. Pluim and H. Wynberg, *Tetrahedron Lett.*, 1251 (1979).
918. D. Liotta, P. B. Paty, J. Johnston and G. Zima, *Tetrahedron Lett.*, 5091 (1978).
919. N. Miyoshi, H. Ishii, K. Kondo, S. Muraì and N. Sonoda, *Synthesis*, 300 (1979).
920. A. Itoh, S. Ozawa, K. Oshima and H. Nozaki, *Tetrahedron Lett.*, **21**, 361 (1980).
921. A. Itoh, S. Ozawa, K. Oshima and H. Nozaki, *Bull. Chem. Soc. Jpn.*, **54**, 274 (1981).
922. W. R. Leonard and T. Livinghouse, *J. Org. Chem.*, **50**, 730 (1985).
923. A. Pelter, R. S. Ward, D. Ohlendorf and D. H. J. Ashdown, *Tetrahedron*, **35**, 531 (1979).
924. Y. Hayashi, T. Matsumoto, T. Hyono, N. Nishikawa, M. Uemura, M. Nishizawa, M. Togami and T. Sakan, *Tetrahedron Lett.*, 3311 (1979).
925. K. Ishii, P. Mathies, T. Nishio, H. R. Wolf, B. Frei and O. Jeger, *Helv. Chim. Acta*, **67**, 1175 (1984).
926. K. Mori and S. Takechi, *Tetrahedron*, **41**, 3049 (1985).
927. H. E. Zimmerman and M. C. Hovey, *J. Org. Chem.*, **44**, 2331 (1979).
928. H. E. Zimmerman and S. A. Fleming, *J. Org. Chem.*, **50**, 2539 (1985).
929. W. Oppolzer and K. Bättig, *Helv. Chim. Acta*, **64**, 2489 (1981).
930. R. K. Hill and L. A. Renbaum, *Tetrahedron*, **38**, 1959 (1982).
931. J. Schwartz and Y. Hayasi, *Tetrahedron Lett.*, **21**, 1497 (1980).
932. G. Zima and D. Liotta, *Synth. Commun.*, **9**, 697 (1979).
933. T. A. Hase and P. Kukkola, *Synth. Commun.*, **10**, 451 (1980).
934. D. Liotta, M. Saindane, C. Barnum and G. Zima, *Tetrahedron*, **41**, 4881 (1985).
935. D. Liotta and G. Zima, *J. Org. Chem.*, **45**, 2551 (1980).
936. D. Liotta, G. Zima and M. Saindane, *J. Org. Chem.*, **47**, 1258 (1982).
937. N. Petragnani and G. Schill, *Chem. Ber.*, **103**, 2271 (1970).
938. D. J. Buckley, S. Kulkowit and A. McKervey, *J. Chem. Soc., Chem. Commun.*, 506 (1980).
939. D. J. Buckley and M. A. McKervey, *J. Chem. Soc., Perkin Trans. 1*, 2193 (1985).
940. R. Pellicciari, M. Curini, P. Ceccherelli and R. Fringuelli, *J. Chem. Soc., Chem. Commun.*, 440 (1979).
941. P. J. Giddings, D. I. John and E. J. Thomas, *Tetrahedron Lett.*, **21**, 399 (1980).
942. T. G. Back and R. G. Kerr, *Tetrahedron Lett.*, **23**, 3241 (1982).
943. N. Sonoda, T. Yasuhara, K. Kondo, T. Ikeda and S. Tsutsumi, *J. Am. Chem. Soc.*, **93**, 6344 (1971).
944. K. Kondo, N. Sonoda and S. Tsutsumi, *J. Chem. Soc., Chem. Commun.*, 307 (1972).
945. K. Kondo, S. Yokoyama, N. Miyoshi, S. Murai and N. Sonoda, *Angew. Chem. Int. Ed. Engl.*, **18**, 692 (1979).
946. K. Kondo, K. Murata, N. Miyoshi, S. Murai and N. Sonoda, *Synthesis*, 735 (1979).
947. K. Kondo, N. Sonoda, K. Yoshida, M. Koishi and S. Tsutsumi, *Chem. Lett.*, 401 (1972).
948. K. Kondo, N. Sonoda and H. Sakurai, *J. Chem. Soc., Chem. Commun.*, 853 (1973).
949. P. Koch, *Tetrahedron Lett.*, 2087 (1975).
950. K. Kondo, M. Takarada, S. Murai and N. Sonoda, *Synthesis*, 597 (1979).
951. K. Kondo, N. Sonoda and S. Tsutsumi, *Chem. Lett.*, 373 (1972).
952. K. Kondo, N. Sonoda and H. Sakurai, *Chem. Lett.*, 1429 (1974).
953. K. Kondo, N. Sonoda and S. Tsutsumi, *Tetrahedron Lett.*, 4885 (1971).
954. K. Kondo, N. Sonoda and H. Sakurai, *Bull. Chem. Soc. Jpn.*, **48**, 108 (1975).
955. P. Koch and E. Perrotti, *Tetrahedron Lett.*, 2899 (1974).
956. N. Sonoda, G. Yamamoto, K. Natsukawa, K. Kondo and S. Murai, *Tetrahedron Lett.*, 1969 (1975).
957. A. Ogawa, K. Kondo, S. Murai and N. Sonoda, *J. Chem. Soc., Chem. Commun.*, 1283 (1982).
958. A. Ogawa, N. Kambe, S. Murai and N. Sonoda, *Tetrahedron*, **41**, 4813 (1985).
959. K. Kondo, N. Sonoda and H. Sakurai, *J. Chem. Soc., Chem. Commun.*, 160 (1974).
960. K. Kondo, N. Sonoda and H. Sakurai, *Tetrahedron Lett.*, 803 (1974).
961. K. Kondo, S. Yokoyama, N. Miyoshi, S. Murai and N. Sonoda, *Angew. Chem., Int. Ed. Engl.*, **18**, 691 (1979).

962. N. Kambe, K. Kondo, H. Ishii, S. Murai and N. Sonoda, *Angew. Chem., Int. Ed. Engl.*, **18**, 547 (1979).
963. N. Kambe, K. Kondo, H. Ishii and N. Sonoda, *Bull. Chem. Soc. Jpn.*, **54**, 1460 (1981).
964. J. Bergman and L. Engman, *J. Organomet. Chem.*, **175**, 233 (1979).
965. S. Uemura, K. Ohe, J. -R. Kim, K. Kudo and N. Sugita, *J. Chem. Soc., Chem. Commun.*, 271 (1985).
966. D. H. R. Barton, D. Bridon and S. Z. Zard, *Tetrahedron Lett.*, **25**, 5777 (1984).
967. K. Yamauchi, K. Nakamura and M. Kinoshita, *Tetrahedron Lett.*, 1787 (1979).
968. K. Yamauchi, K. Hattori and M. Kinoshita, *J. Chem. Soc., Perkin Trans. 1*, 1327 (1985).
969. L. Engman and M. P. Cava, *Tetrahedron Lett.*, **22**, 5251 (1981).
970. M. Akiba, M. V. Lakshmikantham, K. Y. Jen and M. P. Cava, *J. Org. Chem.*, **49**, 4819 (1984).
971. T. -L. Ho and T. W. Hall, *Synth. Commun.*, **5**, 367 (1975).
972. T. Mandai, K. Osaka, M. Kawagishi, M. Kawada and J. Otera, *Synth. Commun.*, **14**, 797 (1984).

The Chemistry of Organic Selenium and Tellurium Compounds Volume 2
Edited by S. Patai
© 1987 John Wiley & Sons Ltd.

CHAPTER **4**

Seleno- and telluro-carbonyl compounds

FRANK S. GUZIEC, JR.

Department of Chemistry, New Mexico State University, Las Cruces, New Mexico 88003, USA

I. INTRODUCTION . . . 216
A. Overview . . . 216
B. Nomenclature of Se/Te Carbonyl Compounds . . . 217
C. Carbon—Chalcogen Multiple Bonds . . . 218
D. Reagents for the Synthesis of Se/Te Carbonyl Compounds . . . 220
II. Se/Te KETONES AND ALDEHYDES . . . 221
A. Early Reports . . . 221
B. Selenoketones (Selones) . . . 222
1. Preparation . . . 222
2. Reactions . . . 224
C. Telluroketones . . . 227
D. Selenoaldehydes—Preparation and Detection . . . 227
E. Telluroaldehydes—Preparation and Reactions . . . 228
III. Se/Te ESTERS . . . 229
A. Selenoesters . . . 229
1. Preparation . . . 229
2. Reactions . . . 231
B. Telluroesters—Preparation and Reactions . . . 233
IV. Se/Te AMIDES . . . 234
A. Selenoamides . . . 234
1. Preparation . . . 234
2. Reactions . . . 238
B. Telluroamides—Preparation and Reactions . . . 240
V. Se/Te CARBOXYLIC ACIDS AND SALTS . . . 240
VI. Se/Te KETENES . . . 242
A. Selenoketenes—Preparation and Reactions . . . 242
B. Telluroketenes—Preparation and Reactions . . . 245
VII. ISOSELENOCYANATES . . . 245
A. Preparation . . . 245
B. Reactions . . . 246

VIII. SELENOSEMICARBAZIDES 247
A. Preparation. 247
B. Reactions 247
IX. SELENOCARBONATES 248
A. Preparation. 248
B. Reactions 251
X. Se/Te CARBAMATES AND HYDRAZIDES 253
A. Selenocarbamates and Selenohydrazides—Preparation and Reactions 253
B. Tellurohydrazides and Tellurocarbazates 254
XI. Se/Te UREAS 255
A. Selenoureas. 255
1. Preparation 255
2. Reactions 257
B. Telluroureas—Preparation and Reactions 258
XII. METAL AND RESONANCE STABILIZED Se/Te DERIVATIVES. . 259
XIII. SELENOCARBONYL COMPOUNDS OF BIOLOGICAL INTEREST 262
XIV. SPECTROSCOPIC AND ELECTROCHEMICAL STUDIES OF Se/Te CARBONYL COMPOUNDS 263
A. NMR Spectra 263
B. Infrared Spectra 264
C. Ultraviolet and Visible Spectra 264
D. Mass Spectra 265
E. Dipole Moment Studies 265
F. Chiroptical Properties 265
G. Electrochemical Studies 266
H. Photoelectron Spectra 266
I. X-Ray and Microwave Studies 266
XV. ACKNOWLEDGEMENTS 267
XVI. REFERENCES 267

I. INTRODUCTION

A. Overview

The chemistry of the carbonyl group is well known as one of the cornerstones of organic chemistry, and that of the thiocarbonyl group has recently become much more familiar to the practising organic chemist. The corresponding selenium and tellurium compounds, however, are generally considered to be exotic and perhaps out of the mainstream of organic chemistry. While it would be easy to blame these views on pervasive 'seleno- and telluro-phobia', it is clear that these analogues are much less well known than the corresponding oxygen and sulfur compounds.

In many cases, early preparations of seleno- or telluro-carbonyl compounds involved the use of toxic and foul-smelling reagents. In addition, often the selenium and tellurium reagents used in attempted preparations of these compounds behave differently than their oxygen or sulfur analogs, leading to errors and confusion in the early literature.

Recent advances in spectroscopy, the ability to do manipulations routinely and carry out observations at low temperatures and in inert atmospheres and the use of more convenient and less hazardous procedures have opened up new areas of investigation, and have made seleno- and telluro-carbonyl compounds much more available for study.

Books by Klayman and Gunther[1] and Zingaro and Cooper[2] provide useful chapters[3–7] on the chemistry of selenocarbonyl compounds up to about 1972. Sections on tellurocarbonyl compounds appear in the books by Irgolic[8] in 1974 and that of Cooper[9] in 1971. More recent coverage of these areas can be found in the *Specialist Periodical Reports* of the Royal Society of Chemistry (London)[10] and in the Barton and Ollis compendium[11]. The use of seleno- and telluro-carbonyl compounds in heterocyclic synthesis has recently been described by Renson[12].

B. Nomenclature of Se/Te Carbonyl Compounds

A major difficulty in finding seleno- and telluro-carbonyl compounds in the literature occurs as a result of the sometimes ambiguous nomenclature associated with these compounds. For example, 'selenoketone' may refer to selenium substitution on a normal ketone in addition to the selenium analog of a ketone. Both uses for this name have appeared in recent titles and abstracts. Another major difficulty results from attempts by authors, especially in the older literature, to apply sulfur nomenclature directly to selenium compounds. For example, the selenium analogue of a thione is *not* a selenone.

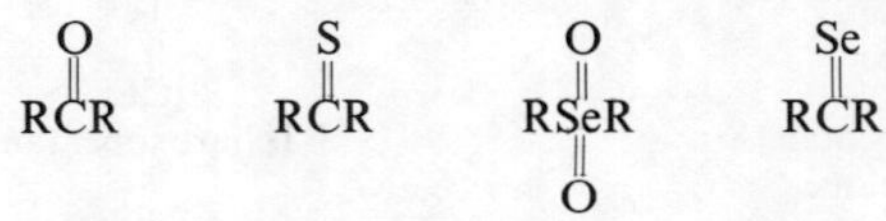

Ketone Thione Selenone Selone

The most recent IUPAC *Nomenclature of Organic Chemistry* lists rules for naming organic selenium and tellurium compounds[13]. These rules state that 'organic compounds of

TABLE 1. Suggested functional group nomenclature of selenocarbonyl and tellurocarbonyl compounds

Functional group	Prefix	Suffix or functional group name
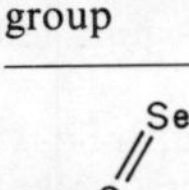	Selenoformyl-	-selenal -carboselenaldehyde
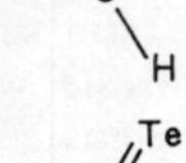	Telluroformyl-	-tellural -carbotelluraldehyde
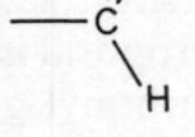	Selenoxo-	-selone
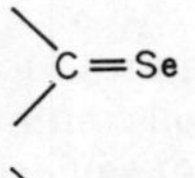	Telluroxo-	—
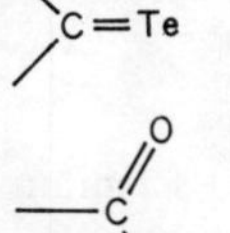	—	-carboselenoic acid

TABLE 2. Examples of selenocarbonyl and tellurocarbonyl nomenclature

$Me(CH_2)_3CHSe$ (a) Pentaneselenal (b) 1-Butanecarboselenaldehyde	$Me(CH_2)_3CHTe$ (a) Pentanetellural (b) 1-Butanecarbotelluraldehyde
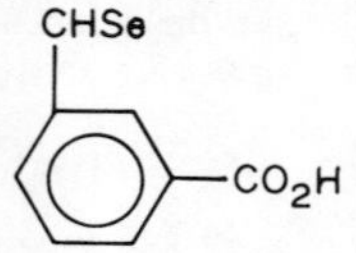 *m*-Selenoformylbenzoic acid	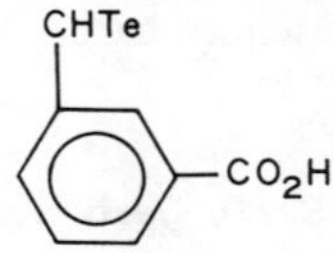 *m*-Telluroformylbenzoic acid
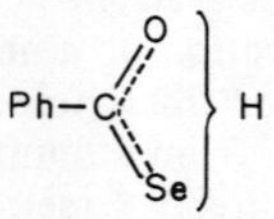 Selenobenzoic acid $Se{=}C(NH_2)_2$ Selenourea	$Te{=}C(NH_2)_2$ Tellurourea
$EtC(=Se)Et$ (a) 3-Selenoxopentane (b) 3-Pentaneselone	$EtC(=Te)Et$ 3-Telluroxopentane

selenium are named as far as possible analogously to the corresponding sulfur compounds' and 'organic compounds of tellurium are named in the same way as those of selenium with "tellur-" in place of "selen-"'. Application of these rules to seleno- and telluro-carbonyl groups are summarized in Table 1. An additional rule states that compounds which cannot be named using the above scheme are named by placing 'seleno-' or 'telluro-' before the name of the corresponding oxygen compound. Examples of seleno- and telluro-carbonyl compounds named using these rules are listed in Table 2.

One complication arises because of the inability of this system to distinguish Se- verses O-protonated isomers of selenobenzoic or, more importantly, the Se- versus O-alkylated ester derivatives. To avoid such ambiguity, it may be worthwhile to use the term 'selone' wherever a selenocarbonyl moiety exists. For example, the O-alkylated seleno ester would be designated as a selone ester and the Se-alkylated compound would be designated as a selenol ester. Accordingly seloneamide, selonecarbonate, selonecarbamate, etc., would denote selenocarbonyl-containing compounds. One difficulty with such a proposal is that no nomenclature analogue of 'selone' has been adopted for the tellurocarbonyl group ('tellone').

Until these nomenclature problems are resolved, and consistent nomenclatures adopted in the current literature, finding seleno- and telluro-carbonyl compounds will continue to be difficult. The naming of compounds in this chapter will follow the IUPAC guidelines.

C. Carbon—Chalcogen Multiple Bonds

Recent successes in the preparations of compounds with carbon—selenium and carbon—tellurium double bonds may, at first glance, appear surprising. Poor overlap in the 2p–3p π bond of thiones has been used to explain the lowered stabilities of these compounds relative to ketones[14]. An extension of this line of thought made the existence of selones, with even poorer 2p–4pπ overlap, doubtful. It appeared that significant resonance interaction or metal coordination, lowering the bond order of carbon—selenium or carbon—tellurium multiple bonds, would be necessary for stability in seleno- or telluro-carbonyl compounds.

The preparation of selones with a 'pure' carbon—selenium double bond showed such stabilization was not necessary[15]. Subsequent preparation of monomeric telluroesters shows that compounds with formal 2p–5pπ bonding can exist with only moderate resonance stabilization[16].

Comparison of the ^{17}O and ^{77}Se shifts of carbonyl and selenocarbonyl compounds indicate that the bond orders of the carbonyl and selenocarbonyl groups are similar[17]. The absorption spectra of selenocarbonyl[18] and tellurocarbonyl[16] compounds parallel those of their carbonyl and thiocarbonyl analogs. Both of these observations suggest that there is true 'double bond character' in these functional groups.

Data dealing with bonding in carbon—chalcogen double bonds are summarized in Table 3. As expected, the carbon—chalcogen double bond length increases significantly in the series O, S, Se, Te. With this increase in size comes greater polarizability. Because of this polarizability, seleno- and telluro-carbonyl compounds are weaker bases than their sulfur and oxygen analogs, but show increased nucleophilicity at the selenium and tellurium atoms. This increased polarizability can also be used to explain the increased carbon electrophilicity of the seleno- and telluro-carbonyl groups.

The polarity of the seleno- and telluro-carbonyl groups deserves special comment. Whereas the carbonyl group is highly polar, the thiocarbonyl group is much less so. Thiophilic addition—nucleophilic addition at the sulfur center of a thiocarbonyl group—was originally explained by a 'reversed polarity' of the thiocarbonyl moiety with carbon being δ^- relative to sulfur. This 'reversed polarity' should be even more pronounced in the seleno- and telluro-carbonyl groups. A comparison of orbital electronegativities suggests that this is not the case[20]. Both the σ and π components of the seleno- and telluro-carbonyl groups should show 'normal' polarity with the chalcogen being δ^- relative to carbon. A dipole moment study of 'pure' carbon—chalcogen double bonds in ketone analogues suggests 'normal polarity' for these functional groups (Section XIV.E)[21].

While many seleno- and telluro-carbonyl compounds are reasonably stable, their chemistry is best characterized by reactions which readily convert the carbon—selenium and carbon—tellurium double bonds into more stable selenium- or tellurium-containing single-bonded species, or which bring about the replacement of selenium or tellurium by oxygen, sulfur or nitrogen. These reactions often make the preparation and characteriz-

TABLE 3. Steric and electronic factors involved in carbon—chalcogen double bonds

	Covalent radii (Å)[19]					
	C	O	S	Se	Te	
Single-bonded	0.77	0.74	1.04	1.17	1.37	
Double-bonded	0.67	0.62	0.94	1.07	1.27	
	Orbital electronegativities[20] (Pauling scale)					
		C	O	S	Se	Te
σ	sp^2	2.75	5.54	3.46	3.29	3.17
π	p	1.68	3.19	2.40	2.31	2.31

ation of seleno- and telluro-carbonyl compounds difficult. A useful tactic in increasing the stabilities of seleno- and telluro-carbonyl compounds, aside from resonance stabilization, has been to shield these group sterically, thereby eliminating some possible decomposition pathways.

D. Reagents for the Synthesis of Se/Te Carbonyl Compounds

As was previously mentioned, two of the major difficulties associated with the investigations of seleno- and telluro-carbonyl compounds are the toxicity and unpleasant odors of reagents used in their preparation.

Hydrogen selenide is a colorless gas, b.p. $-41\,°C$, with an unpleasant odor resembling that of radishes. It is extremely toxic and irritating and should be treated with great care[22]. Hydrogen selenide is a moderately strong acid[23], pK_1 3.88, $pK_2 \approx 11$. It is readily oxidized by atmospheric oxygen affording finely divided red selenium, and should be used in an inert atmosphere. Hydrogen selenide is commercially available in cylinders, or it can be generated in small quantities by hydrolysis of commercially available aluminum selenide[24]. Alternatively, hydrogen selenide can be prepared by reduction of selenium by carbon monoxide in water in the presence of triethylamine[25]. These conditions are mild enough to allow *in situ* formation of hydrogen selenide.

Although hydrogen selenide is used directly in many procedures, the generation of ethanolic solutions of sodium hydrogen selenide using the procedure of Klayman and Griffin[26] is generally far more convenient (equation 1). This procedure utilizes sodium borohydride and elemental selenium in ethanol. The triethylborate formed, in general, does not interfere with further reactions. Aqueous solutions of sodium hydrogen selenide can be prepared similarly.

$$Se + NaBH_4 \xrightarrow{EtOH} NaHSe + B(OEt)_3 + H_2 \qquad (1)$$

Sodium and lithium selenides can be prepared by metal reduction of selenium in liquid ammonia (equation 2)[27,28]. The diselenides can be prepared similarly with a 1:1 molar ratio of reagents (equation 3).

$$Se + 2Na \xrightarrow{NH_3} Na_2Se \qquad (2)$$

$$2Se + 2Na \xrightarrow{NH_3} Na_2Se_2 \qquad (3)$$

Suspensions of lithium selenide and lithium diselenide in tetrahydrofuran can be easily prepared by reaction of selenium with lithium triethylborohydride (equations 4 and 5)[29].

$$Se + 2LiEt_3BH \xrightarrow{THF} Li_2Se + 2Et_3B + H_2 \qquad (4)$$

$$2Se + 2LiEt_3BH \xrightarrow{THF} Li_2Se_2 + 2Et_3B + H_2 \qquad (5)$$

Carbon diselenide is a substance with a remarkably unpleasant odor[30]. It is commercially available and can be most conveniently prepared from carbon tetrachloride or dichloromethane and selenium at elevated temperature. An improved technique for the preparation of carbon diselenide from dichloromethane affords this reagent in ca. 85% yield (equation 6)[31,32]. While the reactions of carbon diselenide resemble those of carbon disulfide, the former is more likely to polymerize in the presence of nucleophiles[30].

$$CH_2Cl_2 + 2Se \xrightarrow{565\,°C} CSe_2 + 2HCl \qquad (6)$$

Phosphorus pentaselenide has not proved to be particularly useful in the preparation of selenocarbonyl compounds. Although phosphorus pentaselenide has not been well characterized[33], it appears to be much less soluble than its sulfur analogue, perhaps leading to lowered reactivity. Attempts to prepare the selenium analog of the well known thionating reagent *p*-methoxyphenylthioxophosphine sulfide dimer (**1**) using phosphorus pentaselenide were unsuccessful[34].

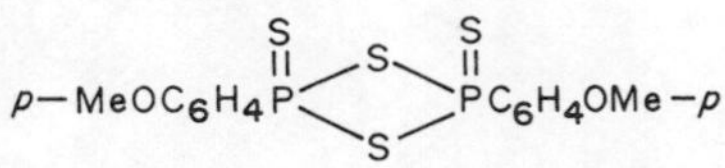

(**1**)

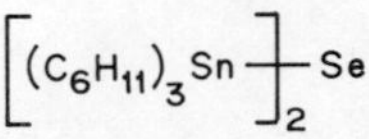

(**2**)

Selenium(I) bromide can be prepared by reaction of selenium dioxide, selenium, concentrated hydrobromic acid and concentrated sulfuric acid[34].

Boron triselenide can be prepared *in situ* in a non-aggregated form suitable for selenation by reaction of bis(tricyclohexyltin) selenide (**2**) with boron trichloride[35]. Compound **2** can be prepared from tricyclohexyltin chloride and sodium selenide.

Hydrogen telluride is an unstable toxic gas most easily prepared by hydrolysis of aluminum telluride[36]. It is more acidic than hydrogen selenide (pK_1 2.64, $pK_2 \approx 10.8$) and deposits elemental tellurium on exposure to air[23]. Lithium telluride and sodium hydrogen telluride can be prepared analogously to the preparations of the corresponding selenium compounds[37,38]. Carbon ditelluride has not been reported.

In general, both organo-selenium and -tellurium compounds should be considered toxic and treated with extreme care. This author finds that treatment of glassware and apparatus with permanganate solution is useful in removing persistent odors. Concentrated nitric acid or potassium hydroxide in isopropanol are useful for cleaning red selenium residues from glassware. Traps containing sodium hydroxide–hydrogen peroxide are useful in carbon diselenide work[31].

II. Se/Te KETONES AND ALDEHYDES

A. Early Reports

Although selenium and tellurium analogues of aldehydes and ketones have been described since the 1870s, many of these early reports should be viewed with skepticism since adequate analytical techniques were not available for the characterization of these compounds. Many early claims for the preparation of seleno- and telluro-aldehydes and -ketones involved the addition of hydrogen selenide or hydrogen telluride to aldehydes or ketones. Monomeric, dimeric and trimeric seleno- and telluro-carbonyl compounds were claimed to be products of these reactions[39–44].

Only the selenium reactions have been reinvestigated in detail. Margolis and Pittman[45] showed that, at least in the case of ketones, the products were red selenium and bis(dialkylmethyl) diselenides, presumably derived from an intermediate selone. They suggested that the reaction of hydrogen selenide with a carbonyl compound initially involves an addition, affording an intermediate hydroselenohydrin (**3a** or **3b**). In the case of aldehydes, trimerization of this intermediate is rapid and no selenium is formed, the product being the trimeric cyclic triselenide **4** (equation 7)[46]. X-ray analysis subsequently confirmed the nature of trimeric 'selenoformaldehyde'[47] and 'selenoacetaldehyde'[48]; linear polymeric forms of 'selenoformaldehyde'[49,50] and trimeric 'selenobenzaldehyde'[51] have also been described.

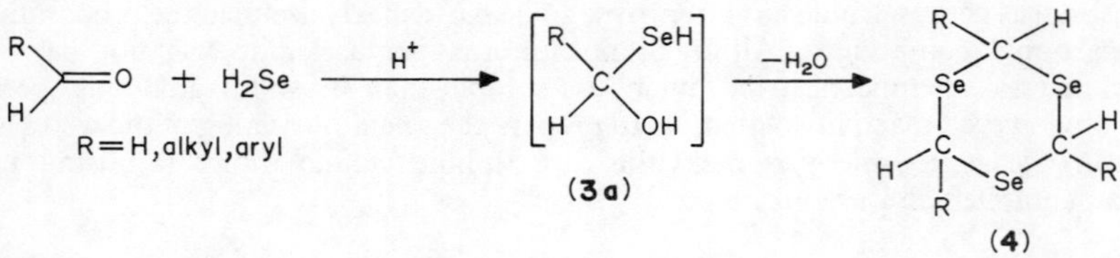

(7)

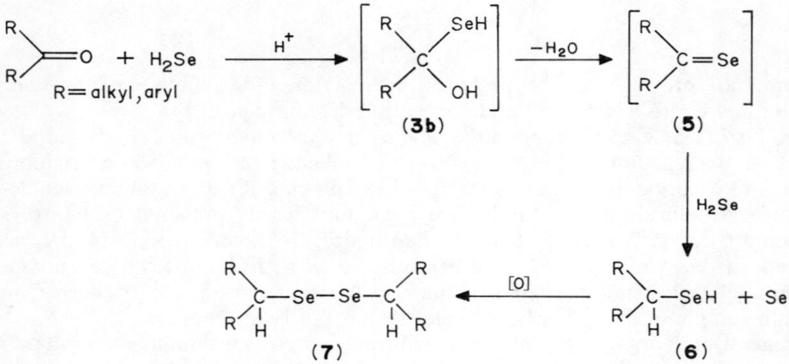

(8)

In the case of the ketones, dehydration of **3b** occurs, affording an intermediate selone which is rapidly reduced by hydrogen selenide (an excellent reducing agent) to the selenol **6**. Oxidation of **6** affords the diselenide **7** (equation 8)[45]. Although no reinvestigation of the hydrogen telluride reaction has been reported, it is unlike that monomeric telluroketones were in fact isolated under similar conditions.

A detailed discussion of the early reports on the preparation and reactions of selenocarbonyl analogues of aldehydes and ketones (and the problems associated with these reports) has appeared previously[3].

B. Selenoketones (Selones)

1. Preparation

The first true monomeric selones isolated, not stabilized by resonance interactions or metal ligands, were di-*tert*-butylselone and selenofenchone. These were prepared by pyrolysis of a phosphoranylidine hydrazone (**8**) (prepared from the corresponding hydrazone) in the presence of selenium powder. The selone **9**, nitrogen and triphenylphosphine selenide were obtained (equation 9)[15,52]. A variety of other sterically hindered selones have been prepared using this method[53–55]. The method was subsequently applied to the preparation of sterically hindered thiones, and it was found that sulfur was more reactive than selenium in these reactions[56,57].

While the mechanism of this reaction has not been investigated in detail, it has been suggested that the first step involves cleavage of **8** to a diazo compound (**10**) and triphenylphosphine (equation 10)[21]. No selone-forming reaction is observed at temperatures below the thermal decomposition point of the phosphoranylidene hydrazone. The resulting diazo compound may react with powered selenium, or more likely, with an

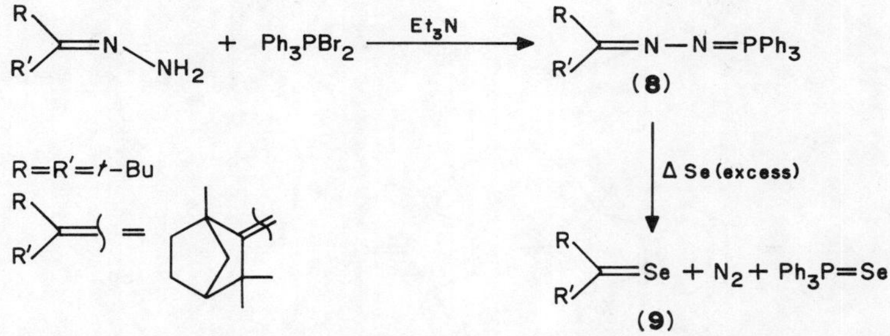

(9)

activated form of selenium generated on reaction of selenium with the liberated phosphine. Since there is no evidence of selenium reacting directly with carbenes or carbenoid species, a likely intermediate in the reaction is the *N*-selenonitrimine **11**, which loses nitrogen to afford the selone.

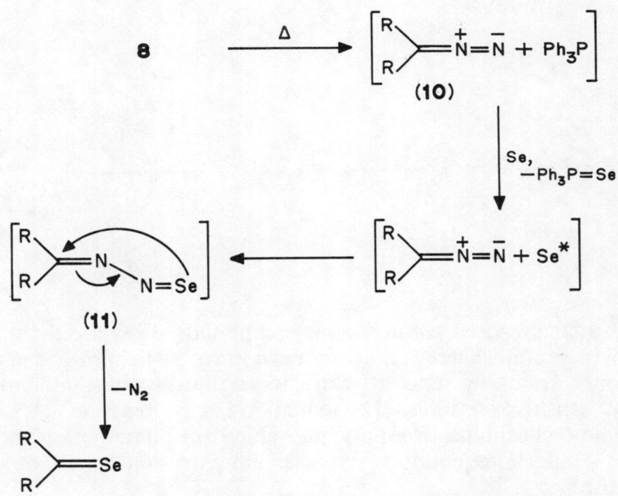

(10)

Among the by-products of this reaction are diazo compound decomposition products and symmetrical olefins formed by an *in situ* 'two-fold extrusion' reaction (see below). Unless the reaction conditions are carefully controlled, these by-products may predominate. These compounds are also the only isolated products from the attempted synthesis of moderately sterically hindered selones such as selenobenzophenone and selenocamphor[52].

A more convenient alternative to the phosphoranylidene hydrazone method involves the reaction of the hydrazone directly with selenium(I) bromide in the presence of a tertiary amine[34] (equation 11) or the reaction of a hydrazone dimagnesium salt with selenium(I) chloride[58]. An *N*-selenonitrosimine may again be an intermediate in this reaction. Again, only sterically hindered selones could be prepared using the selenium bromide method; no selones were obtained from camphor, benzophenone or fluorenone hydrazones.

A third method for the preparation of selones utilizes bis(tricyclohexyltin) selenide (**2**)–boron trichloride as an *in situ* source of non-aggregated boron selenide (equation 12)[35]. Only selenofenchone has been prepared using these conditions. A comparison of preparations of selones using these methods has recently been reported[21].

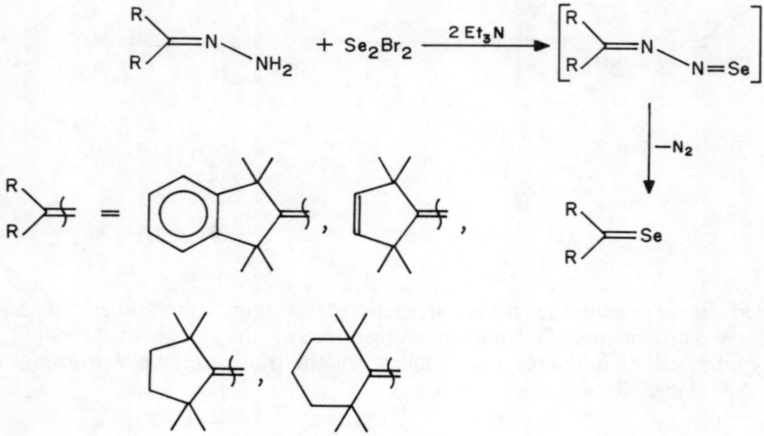

(11)

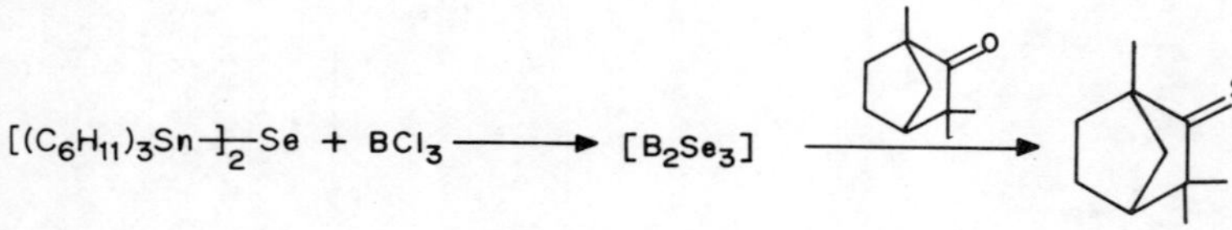

(12)

2. Reactions

All selones so far prepared are deep blue compounds. They are thermally stable up to 150 °C in an inert atmosphere and do not react in visible light provided that oxygen is excluded. Selones are easily reduced with sodium borohydride, sodium amalgam or sodium to air-sensitive selenols (**12**) (equation 13)[52]. Reaction of a selone with tris(dimethylamino)phosphine, tri-*n*-butyl phosphine or sodium potassium alloy leads to reduction to the alkane (equation 14)[52]. Selenium extrusion occurs on reaction with pentacarbonyliron.

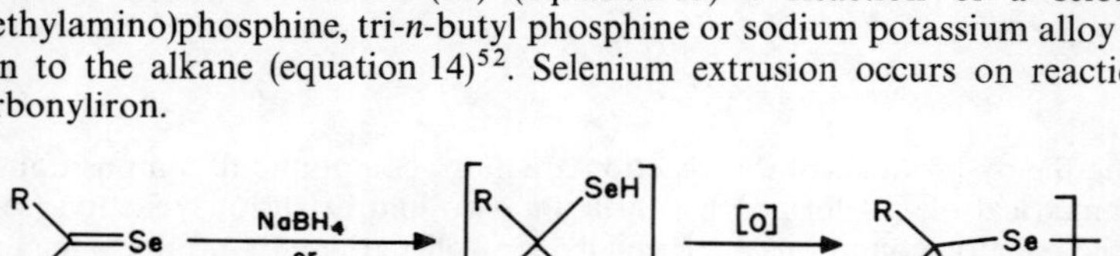

(13)

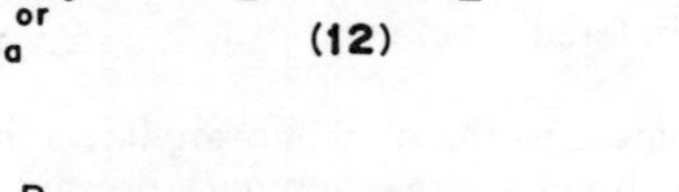

(14)

Oxidation of a selone with peracids leads to the corresponding ketone, presumably via a selenine intermediate (**13**) (equation 15). When a selone is heated with excess of sulfur, exchange of sulfur and selenium takes place, affording the corresponding thione[59]. This method provides a convenient route to *S*-labelled thiones (**14**), and can also be carried out on selone esters and selenoamides (equation 16).

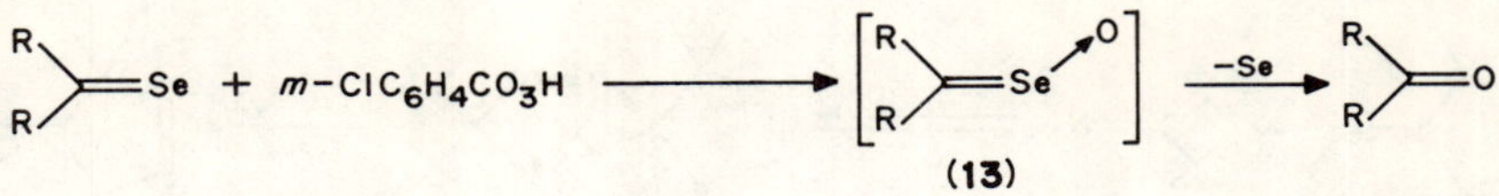

(15)

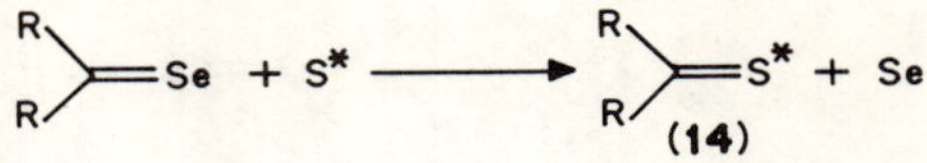

(16)

Selones are less basic than thiones. Under conditions where thiones are protonated in superacid media, selones lead only to polymeric materials[60].

Selones react with sterically hindered diazo compounds in a two-fold extrusion reaction to afford extremely sterically hindered olefins. The cycloaddition of a selone with a diazo compound is much more facile than the corresponding reaction of a thione (equation 17)[52].

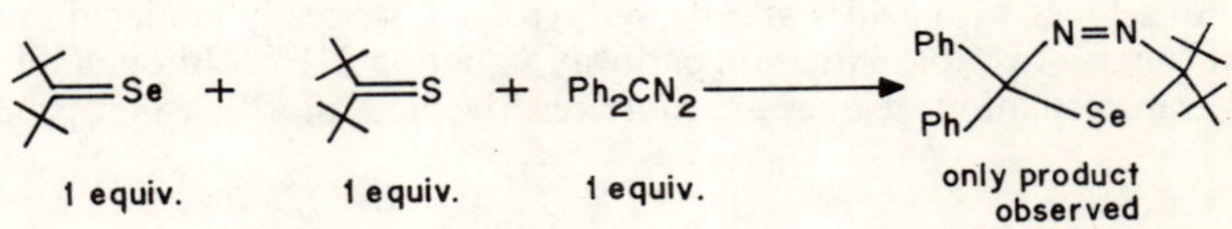

(17)

The intermediate heterocyclic selenadiazoline **15** thermally extrudes nitrogen, affording a thermally unstable episelenide, which then extrudes selenium to give the olefin (equation 18)[52].

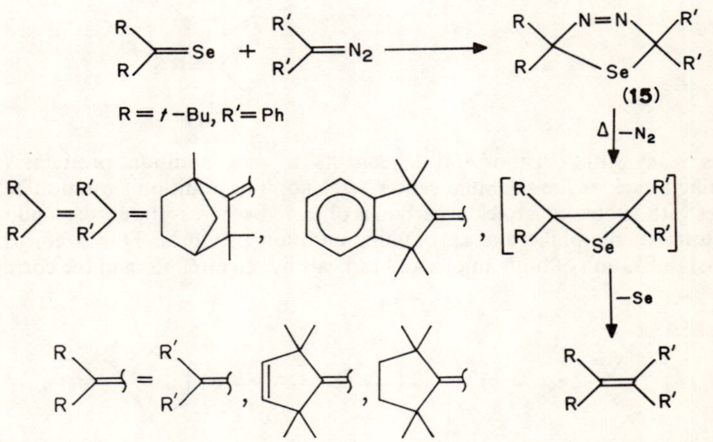

(18)

Although a variety of sterically hindered olefins have been prepared using this method, in some extreme cases retrocyclization of the selenadiazoline can compete with the desired extrusion, affording complex mixtures of products (equation 19). Olefin-forming two-fold extrusion reactions using selone intermediates have been recently summarized[21].

Photolysis of the isolated selenadiazoline intermediates affords symmetrical and unsymmetrical sterically hindered azines **16** (equation 20)[61].

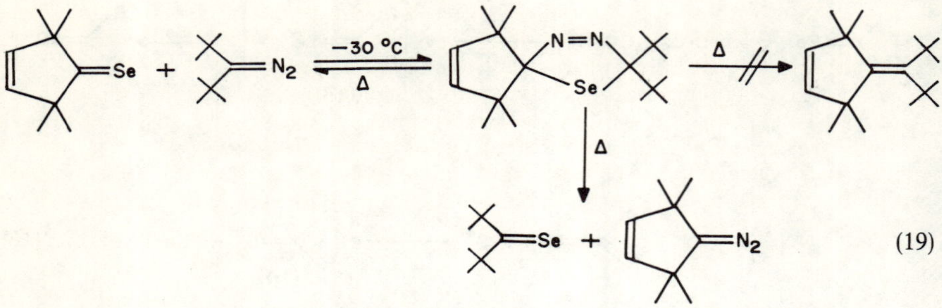

(19)

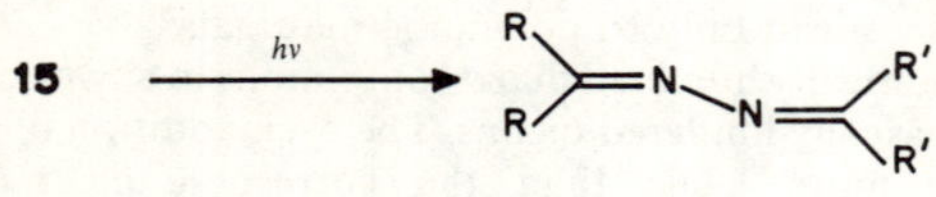

(20)

(16)

Treatment of selones with azides affords, on pyrolysis, sterically hindered imines (**17**), presumably through a two-fold extrusion pathway (equation 21)[62]. Although thiones also undergo this transformation, selones are more reactive and afford cleaner products.

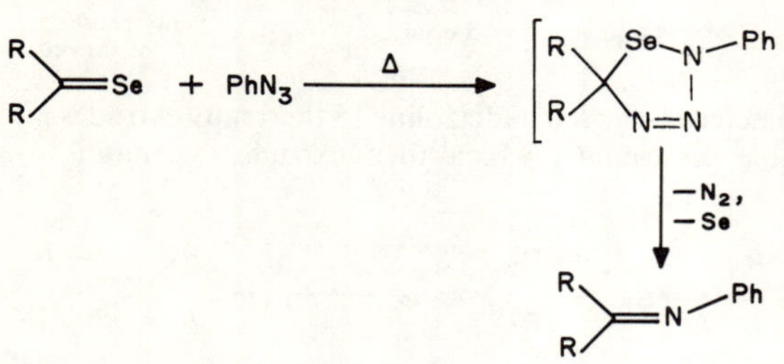

(21)

(17)

Selones react with organometallic reagents to give addition products (**18**) from nucleophilic attack at the selenium center—selenophilic addition (equation 22)[63–65]. In most cases with selones no attack at carbon is observed, whereas the corresponding thiones give mixtures of thiophilic and carbophilic addition products. This is consistent with greater α-stabilization of both anions and radicals by selenium than in the corresponding sulfur cases.

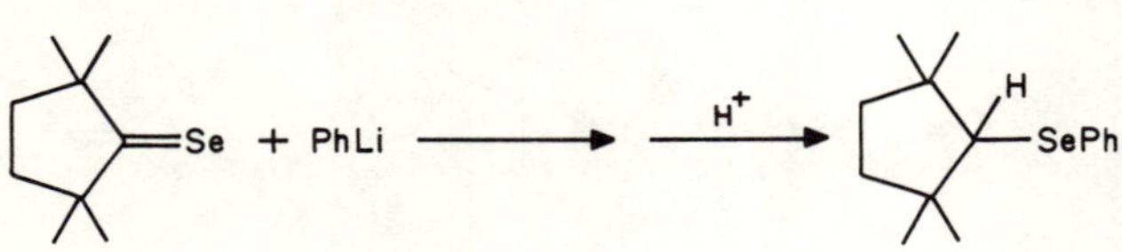

(22)

(18)

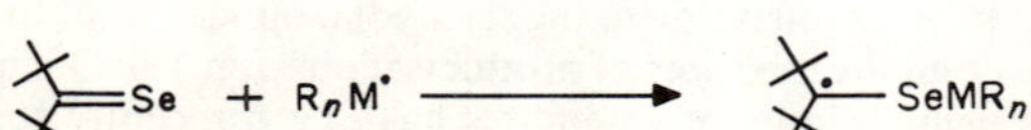

(23)

(19)

M=C,Se,O,Si,S,Sn,Ge

Selones react with a variety of radical species to afford persistent selenoalkyl radicals (**19**) (equation 23)[66–68]. Low spin density was observed on selenium in these radicals. In these reactions selones are better radical traps than the corresponding thiones[69].

The detailed photochemistry of di-*tert*-butyl selone and selenofenchone has been investigated and compared with that of the corresponding thiones[70,71]. Irradiation of di-*tert*-butyl selone in a variety of solvents leads to the diselenide (equation 24). An intermediate radical (**20**) has been detected on low-temperature irradiation. This radical was also subsequently detected in the low-temperature radical-induced reaction of this selone (equation 25)[67].

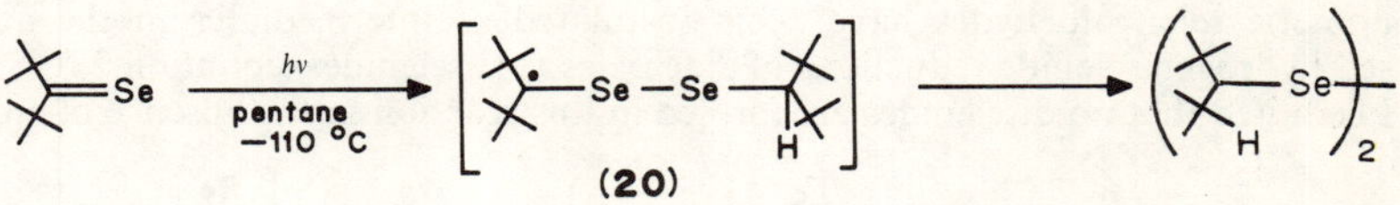

(24)

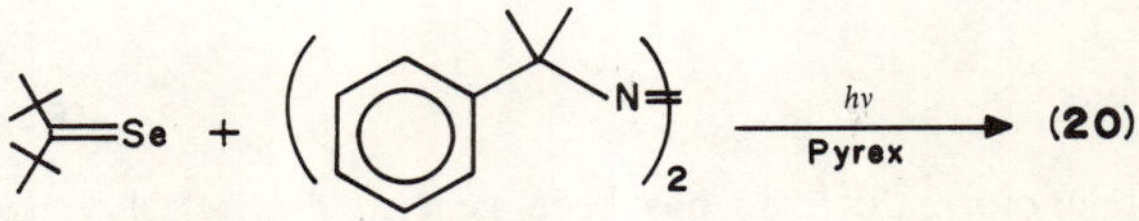

(25)

Irradiation of selenofenchone leads to a mixture of diselenides, including some resulting from intermolecular hydrogen transfer (equation 26). No such reaction was noted on irradiation of thiofenchone. No cycloadducts were noted when irradiations of sterically hindered selones were carried out in the presence of olefins.

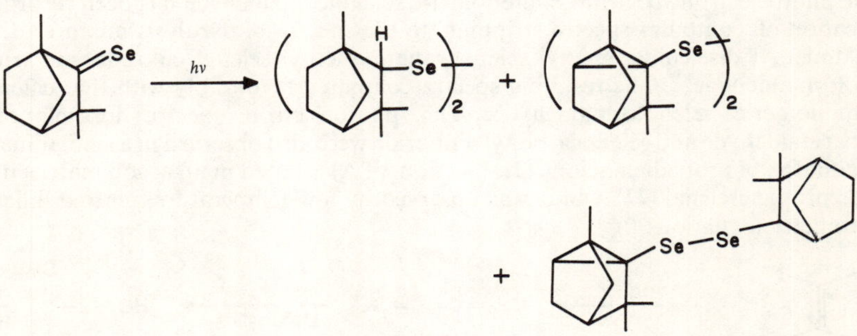

(26)

C. Telluroketones

Apart from the previously described early report[43], no well characterized telluroketone has appeared in the literature. No telluroketone is formed on pyrolysis of a phosphoranylidine hydrazone in the presence of tellurium[72].

D. Selenoaldehydes—Preparation and Detection

No monomeric selenoaldehydes unstabilized by metal complexation or resonance interactions had been isolated and characterized up to 1985.

Apart from the previously mentioned attempts at the acid-catalysed preparations of selenoformaldehyde (Section II.A), reactions of methylene with selenium mirrors afforded a volatile product presumed to be selenoformaldehyde which polymerized on cooling[73,74]. It was subsequently shown that methylene generated by diazomethane thermolysis reacted with a selenium mirror to give, on cooling, the trimeric 'selenoformaldehyde' (equation 27)[75].

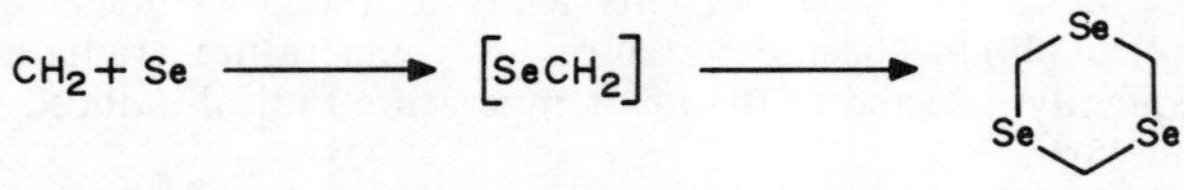

(27)

Monomeric selenoaldehydes have been postulated as intermediates in the amine-catalysed hydrogen selenide reductions of aldehydes to diselenides (equation 28)[76–80]. It should be noted that no diselenides are formed in this reaction in the absence of amines.

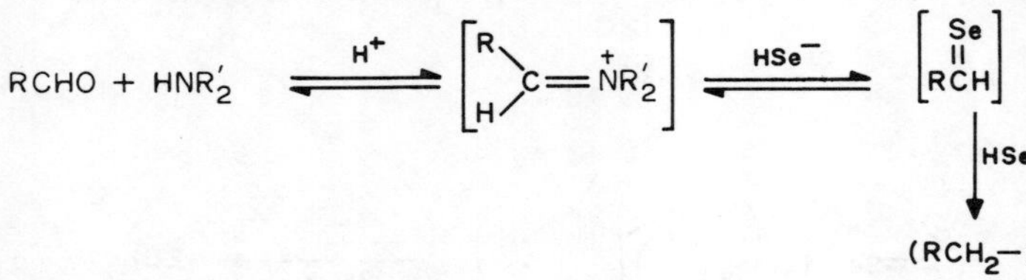

(28)

Monomeric selenoformaldehyde has been generated by flash pyrolysis of dimethylselenide at 700 °C (equation 29). Its spectrum in the near-infrared region has been recorded and compared with that of monomeric thioformaldehyde and formaldehyde[79].

$$Me_2Se \xrightarrow{700\ ^\circ C} [H_2C=Se] + CH_4 \tag{29}$$

The photoelectron spectrum of monomeric selenoformaldehyde has been recorded by application of 'computer spectra stripping' to the spectra of pyrolysis mixtures derived from dimethyl diselenide, methyl selenocyanate, methyl selenyl chloride and trimeric selenoformaldehyde[80]. The resulting spectra compared favourably with that calculated for monomeric selenoformaldehyde. The photoelectron spectra for monomeric selenoacetaldehyde and selenocarbonyl difluoride were also observed in a similar manner.

Irradiation of propadieneselone (**21**) (Section VI.A) isolated in an argon matrix at 12K affords propyneselenal (**22**), which was observed by low-temperature infrared difference spectrometry (equation 30)[81].

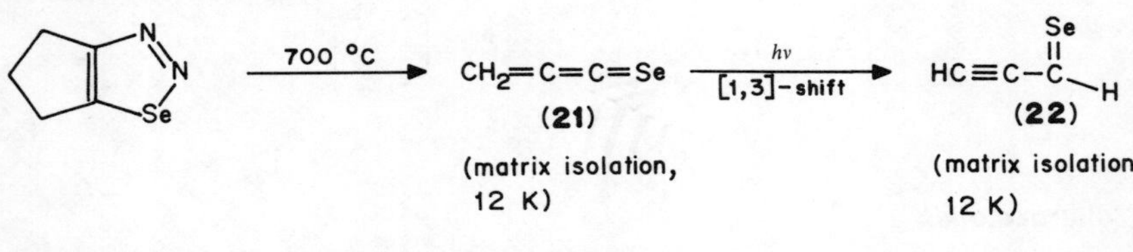

(30)

Metal-stabilized selenoformaldehyde and selenobenzaldehyde are discussed later (Section XII).

(See Note, p. 273).

E. Telluroaldehydes—Preparation and Reactions

The reaction of a tellurium mirror with methylene generated by thermal decomposition of diazomethane or methane, or by photolysis of ketene, was claimed to afford telluroformaldehyde as a monomeric gas[73,74]. This material trimerized to tritellurofor-

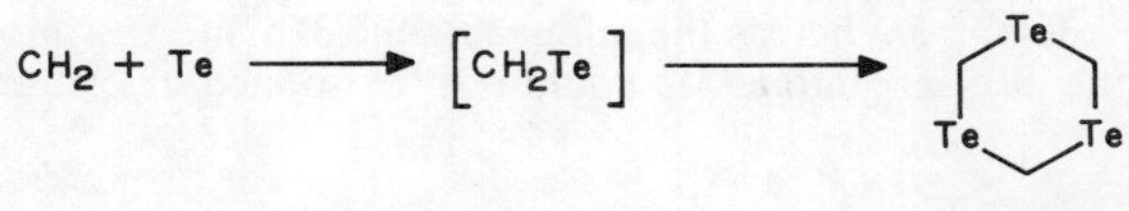

(31)

(23)

maldehyde (**23**) (equation 31)[75]. Gaseous telluroformaldehyde (or the trimer) reacted with bromine and iodine to afford the corresponding dihalomethanes[74].

III. Se/Te ESTERS

A. Selenoesters

1. Preparation

Selenoesters may be conveniently prepared by the reactions of *N,N*-dialkyliminium esters with sodium hydrogen selenide[38] or by the reactions of imidate esters with hydrogen selenide[82,83]. The first procedure is probably the most convenient[38]. The iminium esters can be easily prepared by the reaction of a dialkylamide with phosgene. The chloroiminium salt **24** is then treated with an alcohol, and the resulting iminium ester **25** treated with ethanolic sodium hydrogen selenide to afford the selenoester **26** (equation 32). A great variety of *O*-alkyl and *O*-aryl selenoesters were obtained in good yields using this procedure. One limitation of this route was noted; in the case of the hindered *O*-aryl iminium ester (R = *i*-Bu, R″ = Ph) only selenoamide was obtained.

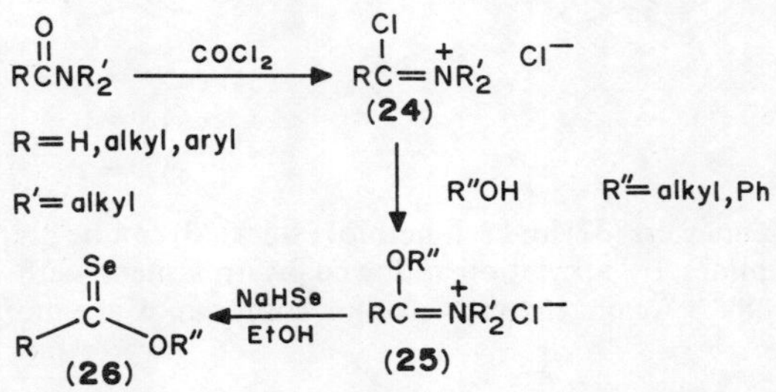

(32)

A second route starts with the conversion of a nitrile to the imidate ester **27**, followed by reaction with hydrogen selenide (equation 33). This method is limited to selenoesters of low molecular weight alcohols, and uses a large excess of gaseous hydrogen selenide[82,83].

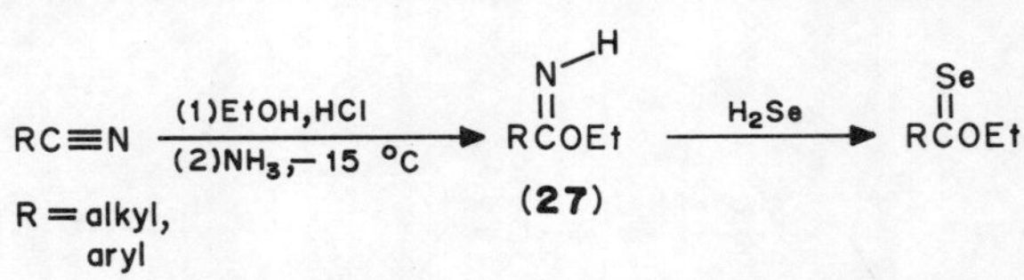

(33)

Selenoesters unsubstituted at the 2-position can be prepared by reaction of a selenoketene intermediate (**29**) with an alcohol (equation 34). The selenoketene (Section VI.A) can be prepared by pyrolysis or base-catalysed fragmentation of 4-substituted-1, 2, 3-selenadiazolines (**28**)[84].

Selenoesters can also be prepared in low yield by the reaction of pentacarbonylchromium(0) arylalkoxycarbenes (**30**) with selenium (equation 35)[85].

Selenocarbonyl phthalides and derivatives (**31**) can be prepared from the corresponding imidates (equation 36)[86,87]. Whereas the selone lactone **31a** and the thiol selone lactone **31b** were stable, the diselenolactone **31c** could not be isolated or characterized*.

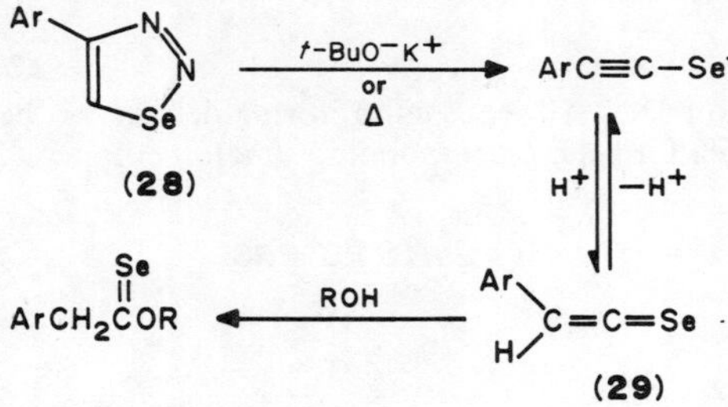

(34)

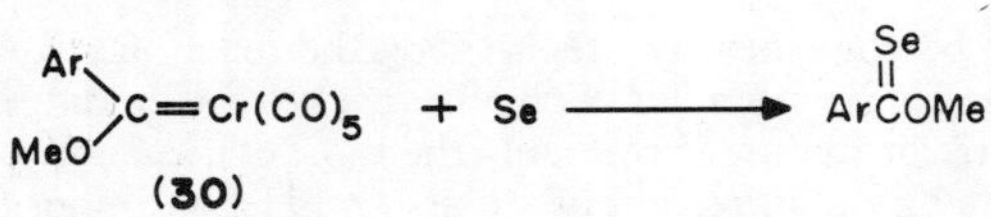

(35)

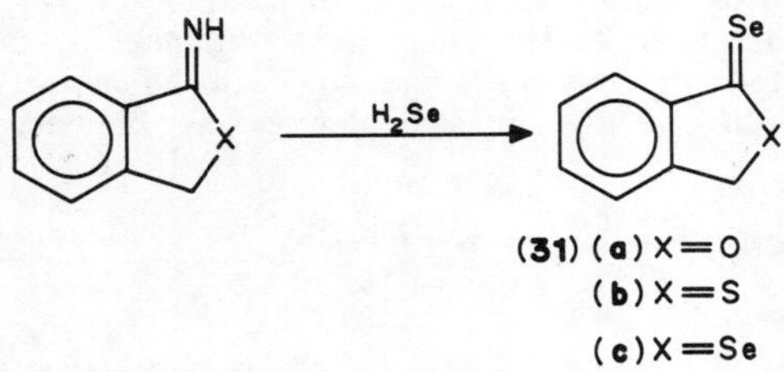

(36)

The related diselenoesters (**32**) and selonethiol esters (**33**) can be prepared from tertiary seleno- and thio-amides by alkylation followed by treatment with hydrogen selenide (equations 37 and 38)[88]. Again, the mono-sulfur compounds are more stable than their diselenium analogs.

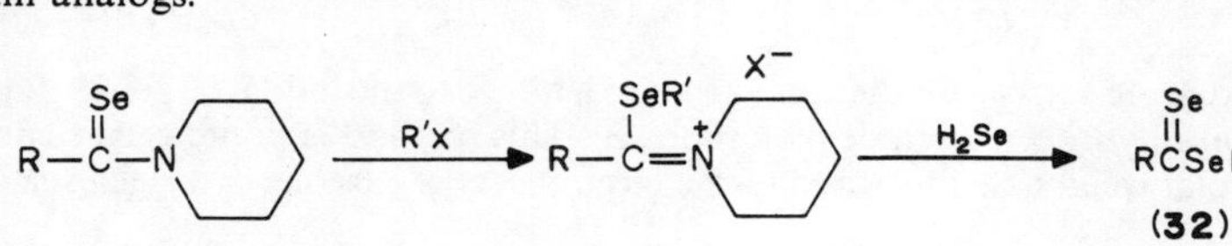

(37)

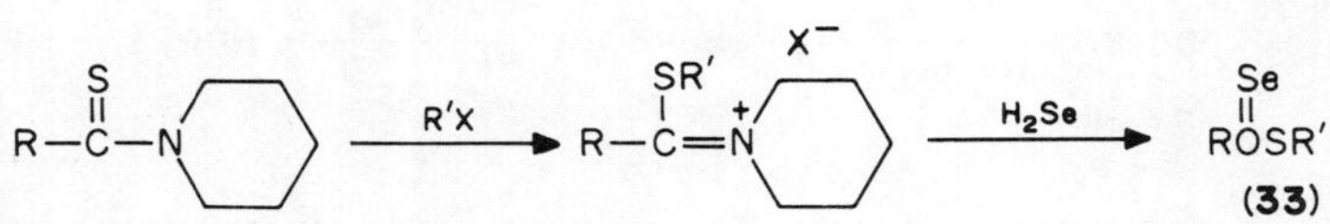

(38)

*To avoid ambiguity, in accordance with the proposal on p. 253, these and related selenocarbonyl compounds (Section X.A) are designated as selone derivatives. This nomenclature has not been adopted by IUPAC.

A diselenoformate intermediate (**35**) generated by reduction of a triselenocarbonate (**34**) is reported to be unstable, oligomerizing on formation (equation 39)[89].

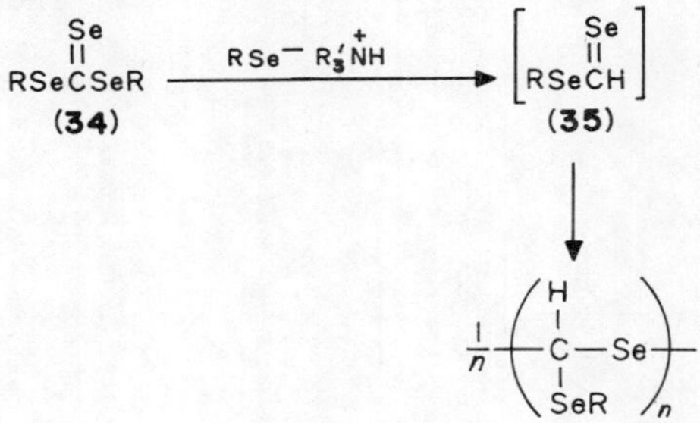

(39)

2. *Reactions*

The reactions of selenoesters have been extensively studied by Barton and coworkers. Selenoesters can be reduced by Raney nickel to the corresponding ethers (equation 40)[90]. The method appears to be useful for the selective formation of ethers in steroids and aminoglycosides.

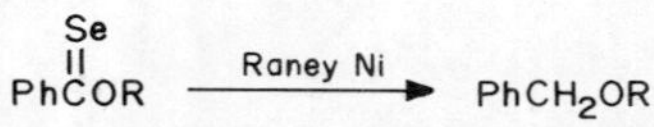

(40)

Sodium borohydride reduction of selenoesters under an inert atmosphere led to a remarkably stable tetrahedral intermediate (**36**), which oxidatively dimerized on workup (equation 41). Treatment of **36** with methyl iodide afforded the alkoxymethyl alkylselenide **37**. Treatment of the selone ester with borohydride followed by triethylphosphine addition led to formation of the corresponding ethyl ether (equation 42)[90].

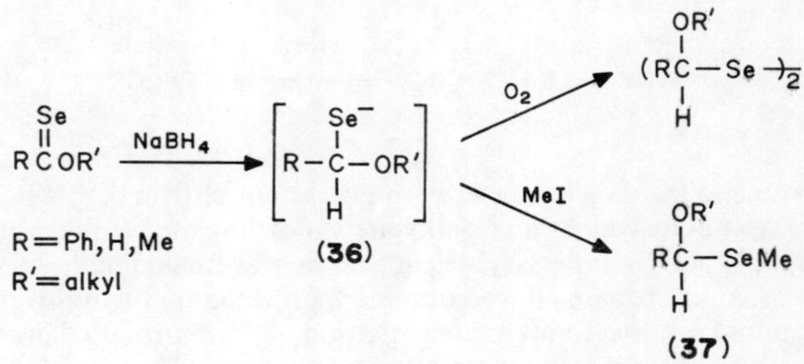

(41)

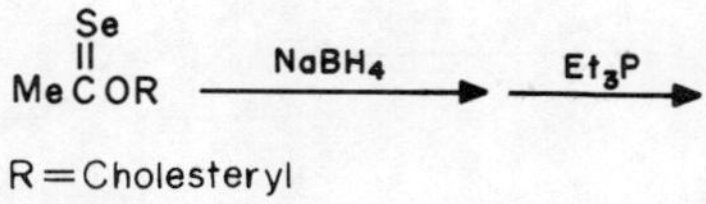

(42)

The latter reaction failed in the case of the selenoformate and the selenobenzoate, presumably because of a rapid reaction of these compounds directly with triethylphosphine. Treatment of a selenobenzoate with a trialkylphosphine affords an interesting moderately stable compound (**38**), which can undergo a variety of reactions, presumably via ylid and carbene intermediates (equation 43)[91].

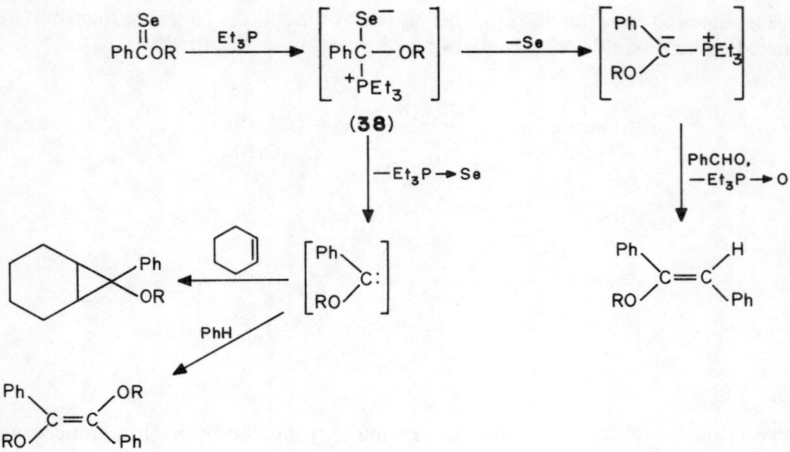

(43)

Although selenobenzoates can be used as substrates in the tributylstannane-mediated deoxygenation procedure developed by Barton and McCombie (equation 44), xanthates and thione benzoates appear to be more useful intermediates in this reaction[38].

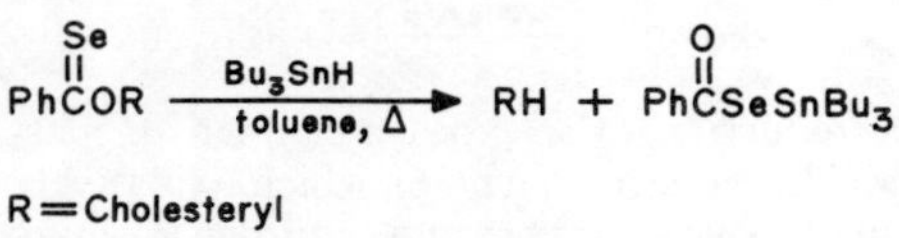

(44)

Treatment of selenobenzoates with methylenetriphenylphosphorane affords enol ethers (equation 45)[90].

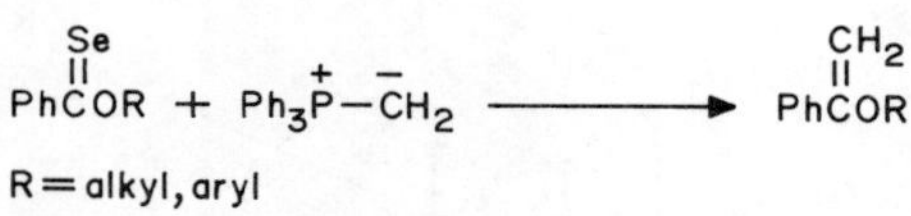

(45)

Treatment of an enolizable selenoester with potassium bis(trimethylsilyl)amide affords the intermediate selenolate which can be alkylated affording the ketone monoselenoacetal **39** (equation 46)[90]. When *O*-cholesteryl selenoacetate was treated with the same hindered base, the alkoxyselenocrotonate **40** was obtained, presumably owing to more favorable explusion of selenium relative to alkoxide (equation 47)[90]. Normal Claisen products are typically obtained from *O*-alkyl thione esters.

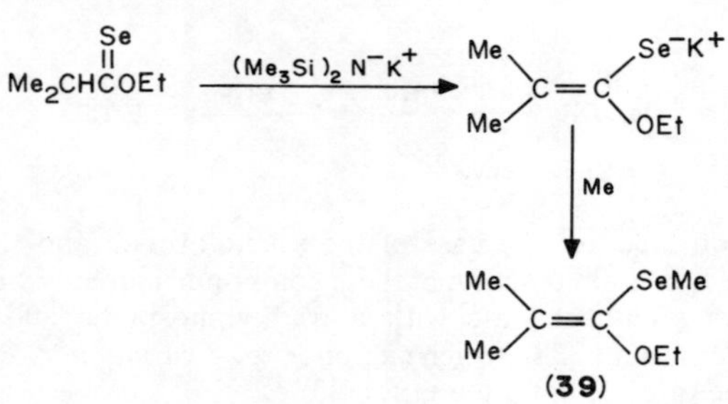

(46)

Selenoesters can be used as reagents in the preparation of selenoamides[83]. They also can be converted into thione esters by a thermal reaction with sulfur (equation 48)[59]. The uses of selenoesters in heterocyclic synthesis have also been reported[83].

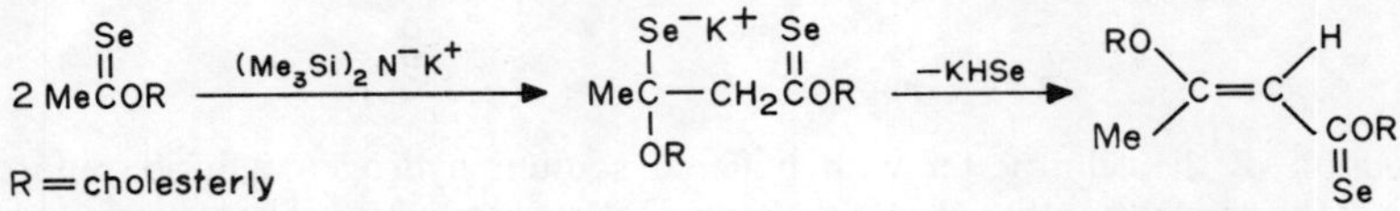

(40)

(47)

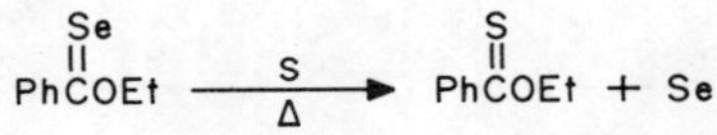

(48)

B. Telluroesters—Preparation and Reactions

When the phenyl dimethyliminium salt of cholesterol was treated with excess sodium hydrogen telluride at $-20\,^{\circ}C$ under conditions analogous to those used to prepare the selenoester, reduction occurred affording the benzyl ether in good yield[38]. The authors suggested that this reaction proceeded through a tellurobenzoate intermediate (**41**), which was further reduced by hydrogen telluride to the ether and tellurium (equation 49). The method could be used for the preparation of ether derivatives of complex aminoglycosides[92].

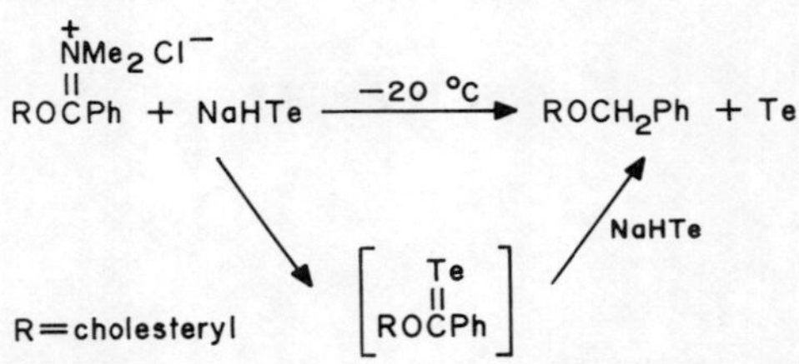

(41)

(49)

When the analogous reaction was carried out on the unsubstituted iminium compound, a presumed unstable telluroformate and *N,N*-dimethyltelluroformamide were observed, via partitioning of the tetrahedral intermediate[38].

The use of the more hindered pivaloyl derivative **42** in these reactions led to the first stable characterizable telluroesters (**43**) (equation 50)[16,93].

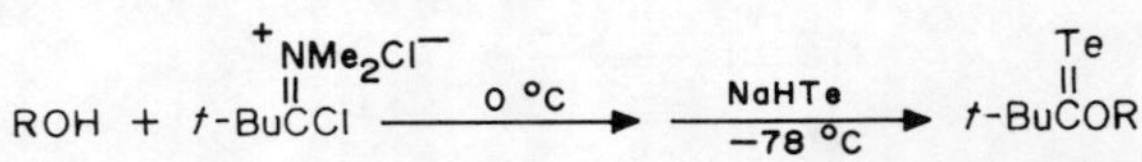

(42) (43)

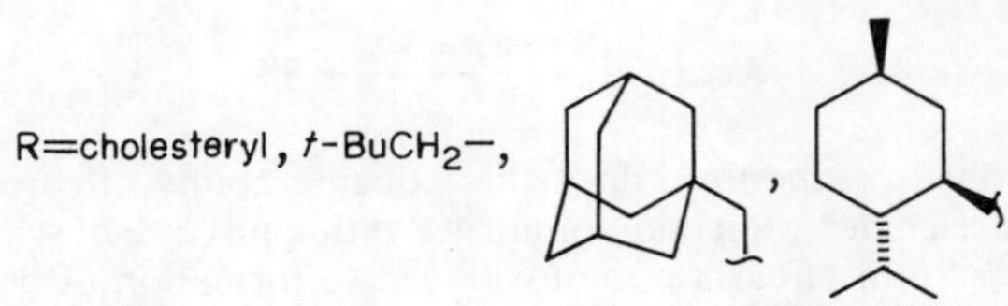

(50)

When pure, the telluroesters were stable to air in the dark or on irradiation (>500 nm) when oxygen was excluded. Oxidation of the telluroester with diphenylseleninic anhydride afforded the pivalate in excellent yield (equation 51)[93].

$$\mathbf{43} + (\mathrm{Ph}\overset{\mathrm{O}}{\overset{\|}{\mathrm{Se}}})_2\mathrm{O} \longrightarrow t\text{-}\mathrm{Bu}\overset{\mathrm{O}}{\overset{\|}{\mathrm{C}}}\mathrm{OR} \qquad (51)$$

Reduction of the telluroester with buffered sodium hydrogen telluride afforded a mixture of the neopentyl ether **46** and the pinacol ether product **47**. The radicals **44** and **45** have been suggested as intermediates in these hydrogen telluride reductions (equation 52)[93].

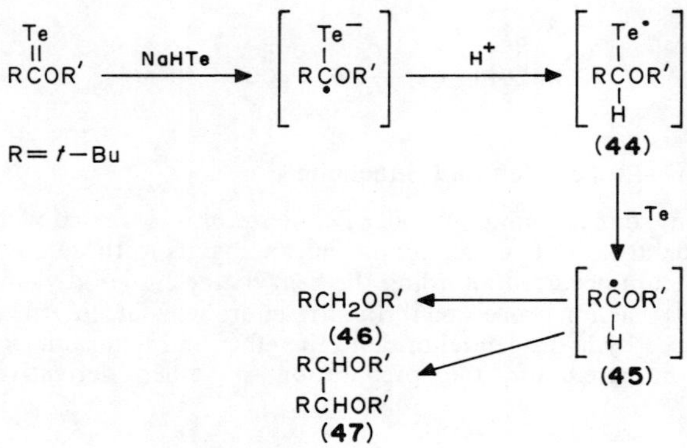

(52)

IV. Se/Te AMIDES

A. Selenoamides

1. *Preparation*

Primary selenoamides (**48**) can be most conveniently prepared from nitriles. While the earliest procedures involved direct addition of gaseous hydrogen selenide to nitriles (equation 53)[94,95], two more convenient procedures involve *in situ* formation of hydrogen selenide or its salts. The first utilizes aluminum selenide in a mixture of aqueous pyridine and triethylamine, affording a variety of aromatic and heterocyclic aromatic primary selenoamides in good yields (equation 54)[96].

$$\mathrm{ArC{\equiv}N} \xrightarrow{\mathrm{H_2Se}} \underset{(\mathbf{48})}{\mathrm{Ar}\overset{\mathrm{Se}}{\overset{\|}{\mathrm{C}}}\mathrm{NH_2}} \qquad (53)$$

$$\mathrm{ArC{\equiv}N} \xrightarrow[\mathrm{H_2O,\ Et_3N}]{\mathrm{Al_2Se_3,\ pyridine,}} \mathbf{48} \qquad (54)$$

A recently described procedure utilizes the reducing ability of carbon monoxide to generate hydrogen selenide[97]. Reaction of nitriles with a mixture of selenium, water and triethylamine under 5 atm of carbon monoxide led to formation of both aliphatic and aromatic selenoamides (equation 55). The aromatic compounds were generally obtained in good yield, with sterically hindered compounds giving lower yields. The aliphatic derivatives were obtained in approximately 35% yield, presumably owing to a decreased stability of these compounds relative to their aromatic analogues.

$$RC{\equiv}N + CO + Se + H_2O \xrightarrow[Et_3N]{100\,^{\circ}C} R\overset{\overset{\displaystyle Se}{\|}}{C}NH_2 \qquad (55)$$

R = alkyl or aryl

Secondary and tertiary selenoamides can generally be prepared in moderate to good yield by aminolysis of the corresponding selenoesters[77]. While secondary amines react slowly with selenoesters to form tertiary selenoamides without difficulty, the magnesium halide salts of primary amines must be used to avoid imido ester formation (equations 56 and 57). These products were observed when primary amines were used directly in these aminolysis reactions.

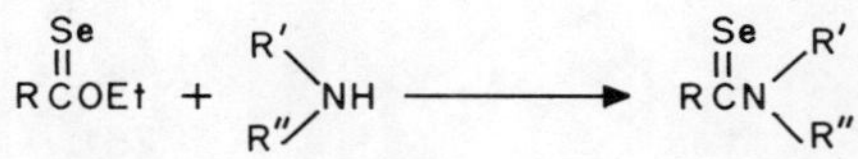

(56)

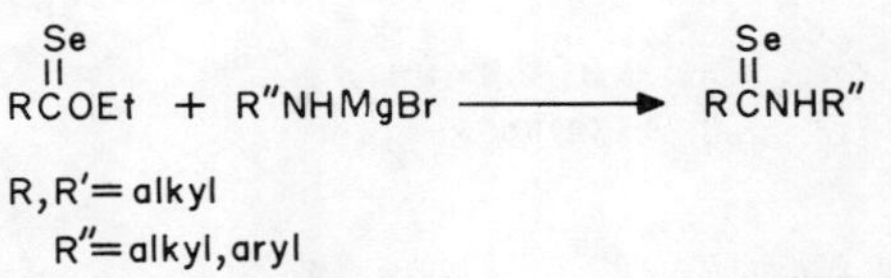

(57)

Secondary and tertiary selenoamides unsubstituted at the α-position can be prepared via acetylenic selenol–selenoketene intermediates (**49** and **50**) analogous to procedures used for the preparation of selenoesters (equations 58 and 59)[98,99] (cf. Section III.A.1).

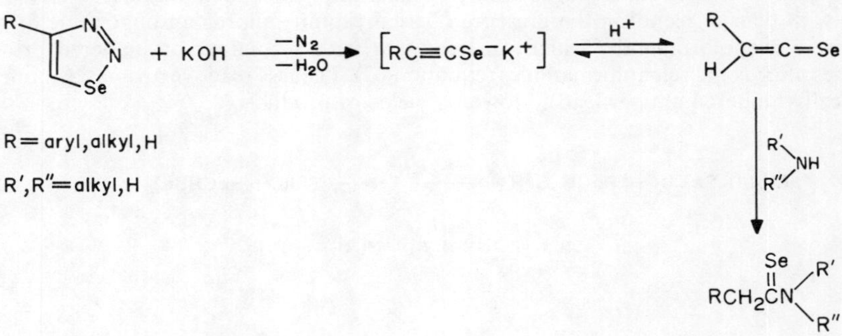

(58)

A number of heterocyclic selenoformamides (**51**) have been prepared from the corresponding aminoheterocycles by treatment with dimethylformamide acetals followed by hydrogen selenide treatment (equation 60)[100]. Heterocyclic vinylogous selenoformamides (**52**) have been prepared by Vilsmeier–Haack formylation and treatment with hydrogen selenide (equation 61)[101].

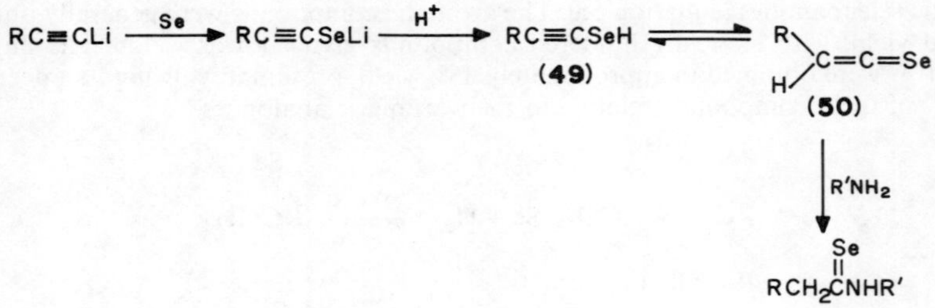

(59)

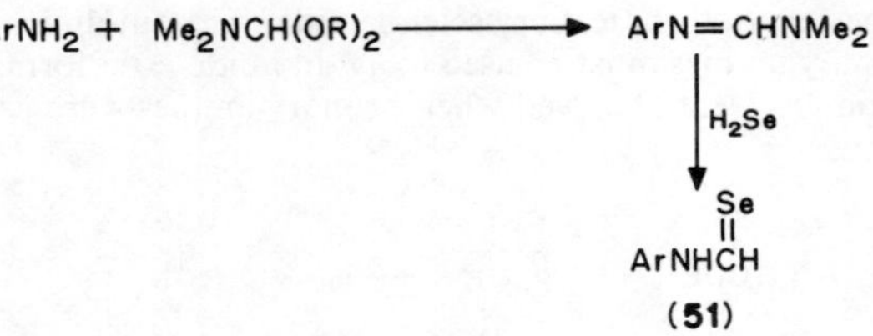

(60)

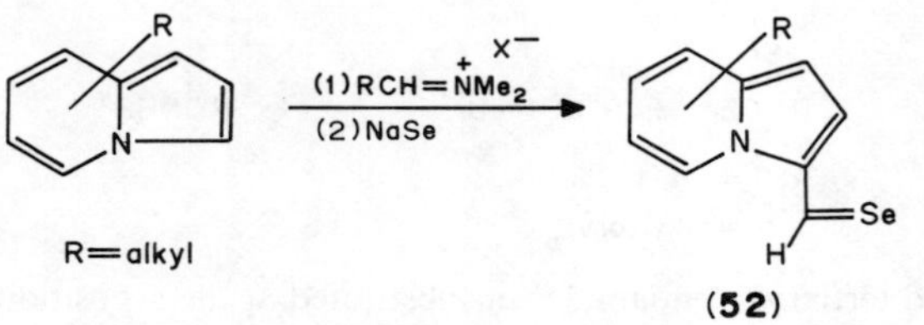

(61)

Another interesting procedure for the preparation of selenoformamides utilizes dibenzyltriselenocarbonate (**53**) as a starting material[89]. Treatment of the triselenocarbonate with benzyl selenol in the presence of triethylamine affords an oligomeric form of benzyl diselenoformamide, which on treatment with secondary or hindered primary amines affords the selenoformamides (equation 62). The less reactive aromatic amines or sterically hindered amines lead to lowered yields of products.

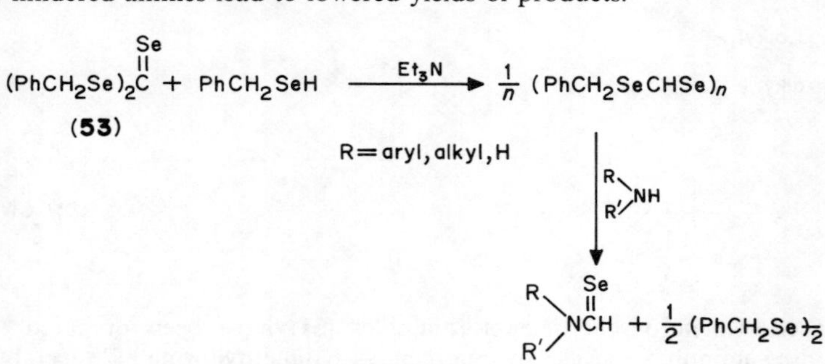

(62)

The reactions of chlorimidates (**54**)[102], *S*-alkylated thioamides (**55**)[103] and iminium salts (**56**)[104] with sodium hydrogen selenide have also been used to prepare selenoamides and vinylogous selenoamides (equations 63–65).

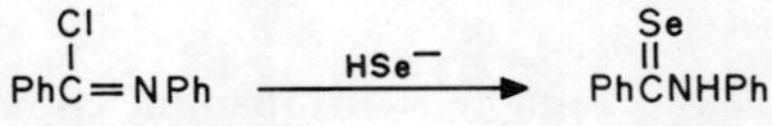
(54) (63)

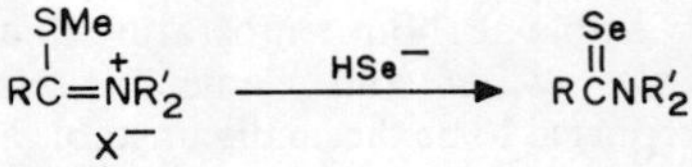
(55) (64)

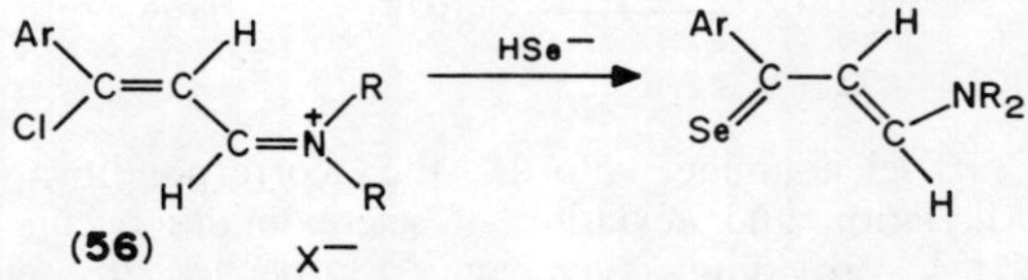
(56) (65)

A number of *N*-substituted selenopyridinones (**58**) have been prepared by reaction of the potassium salt of glucondialdehyde (**57**) with isoselenocyanates (equation 66)[105].

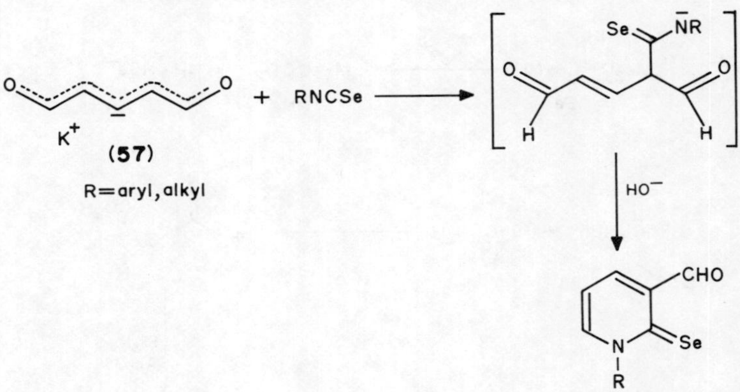
(57) R=aryl, alkyl (58) (66)

Dimethylselenoformamide has been prepared as a by-product in the synthesis of 1, 3, 4-selenadiazole from dimethylformamide azine (equation 67)[106].

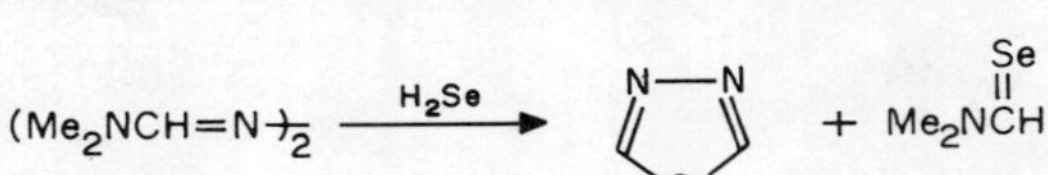
(67)

Selenoamides have also been prepared in low yields using phosphorus pentaselenide[107] or by halogen displacement using hydrogen selenide or hydroselenide ion (equation 68)[108].

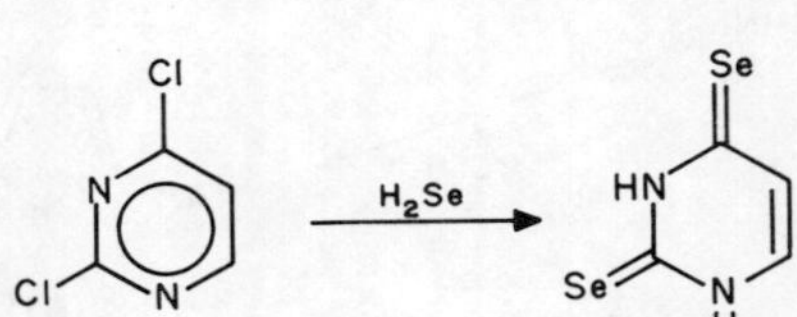
(68)

2. Reactions

Selenoamides are reported to be more stable than the corresponding selenoesters[77]. Primary selenoamides are less stable than secondary or tertiary selenoamides owing to slow loss of hydrogen selenide to afford nitriles (equation 69)[97]. The aromatic primary selenoamides are moderately stable at room temperature under nitrogen; in air they are slowly converted into nitriles, water and elemental selenium. Aliphatic primary selenoamides, however, are reported to be thermally unstable and highly sensitive to air[97].

$$RC(=Se)NH_2 \rightleftharpoons RC{\equiv}N + H_2Se \qquad (69)$$

R=aryl, alkyl

Hydrolysis of a selenoamide affords the corresponding carboxylic acid. (equation 70)[109]. Alkylation and acylation of selenoamides occurs in high yield at selenium[102,104,110–112], providing convenient approaches to selenium-containing heterocycles (**59–61**) (equations 71 and 72)[102,104,112], as do oxidations of selenoamides with iodine (e.g. **62** and **63**, equations 73 and 74)[95,113]. Reactions of selenoamides with hydrazine derivatives also afford interesting heterocyclic systems[83].

$$Me_2C(NH_2)-C(=Se)-NH_2 \xrightarrow{H_3O^+} Me_2C(NH_3^+)CO_2^- + NH_4^+HSe^- \qquad (70)$$

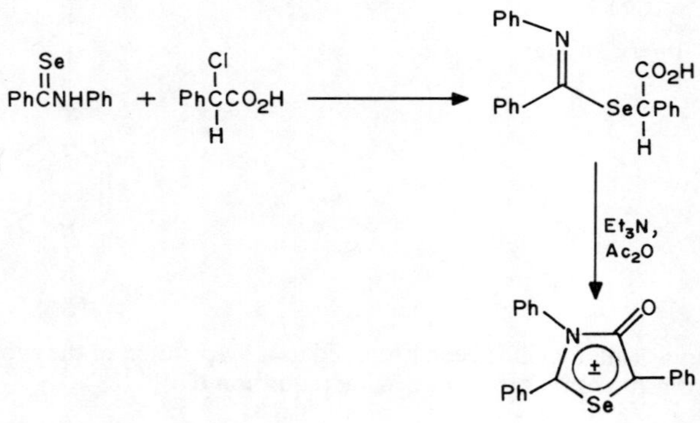

(59)

(71)

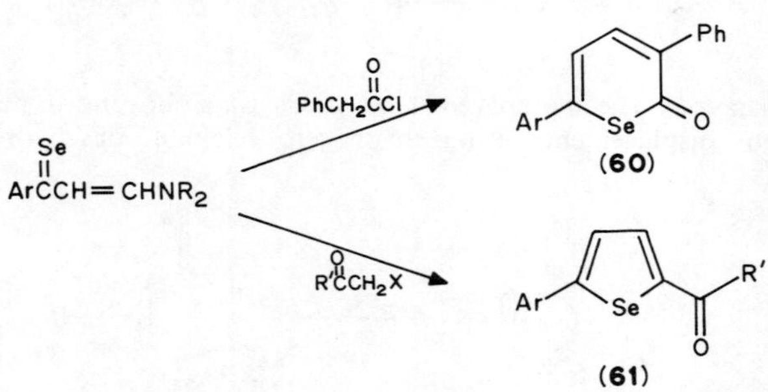

(60)

(61)

(72)

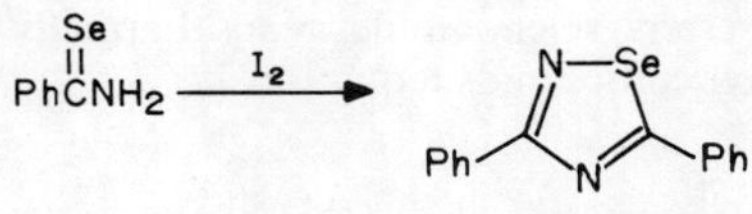

(62)

(73)

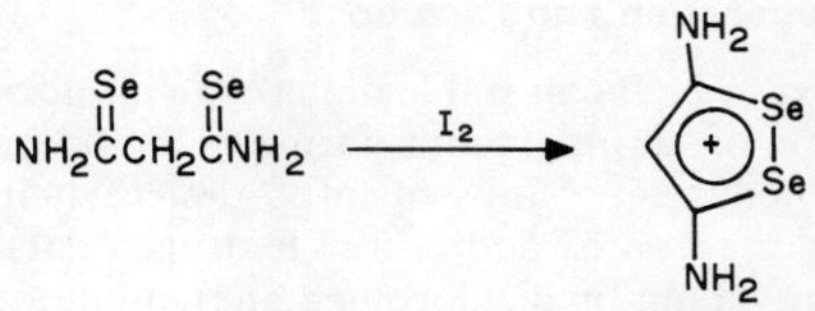

(63)

(74)

Tertiary selenoamides with α-hydrogens may be conveniently trimethylsilylated at selenium in good yield to afford **64** (equation 75)[114]. Tertiary selenoamides are also used in the preparation of diselenoesters (see Section III.A.1, equation 38)[88].

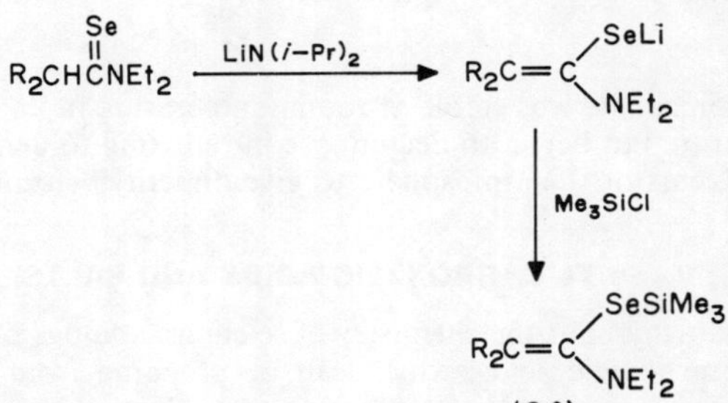

(64)

(75)

Selenoamides have recently been used as reagents in the stereospecific deoxygenation of epoxides[115]. This reaction is acid catalysed and proceeds with retention, presumably via the cyclic intermediate **65** (equation 76).

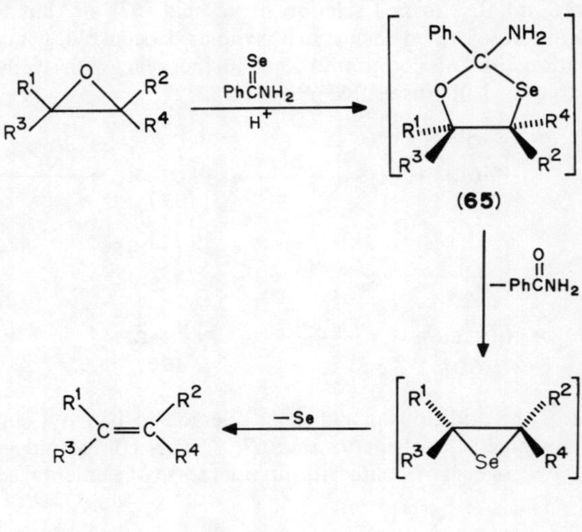

(65)

(76)

Like selones and selone esters, selenoamides react thermally with sulfur, affording the corresponding thiocarbonyl compounds (equation 77)[59].

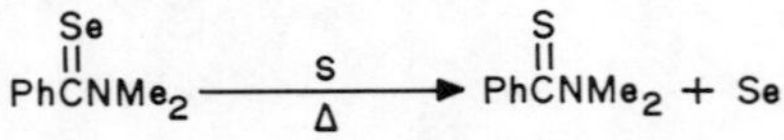

(77)

B. Telluroamides—Preparation and Reactions

A compound presumed to be the unstable dimethyltelluroformamide was observed in the attempted preparation of steroidal telluroformates, by treatment of the iminium salt with sodium hydrogen telluride[38]. Subsequently, dimethyltellurobenzamide (**66**) was prepared at $-40\,^{\circ}C$ under argon by addition of hydrogen telluride to a solution of the iminium species and triethylamine in dichloromethane (equation 78)[116]. These conditions must be more rigorously controlled than those required for the preparation of selenoamides.

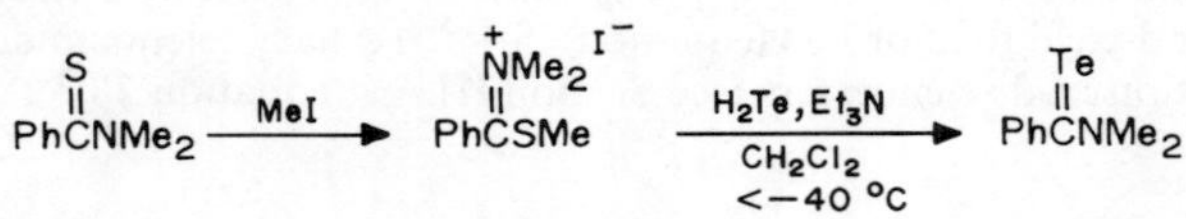

(78)

Dimethyltellurobenzamide was stable at room temperature in an inert atmosphere in the absence of moisture but began to decompose in refluxing toluene[116]. It was rapidly attacked by wet solvents or the atmosphere to give dimethylbenzamide and tellurium.

V. Se/Te CARBOXYLIC ACIDS AND SALTS

Relatively little is known about the chemistry of selenium analogs of carboxylic acids[88]. In contrast to thiocarboxylic acids, which can be prepared from acid chlorides and hydrogen sulfide or its salts, attempts to prepare selenocarboxylic acids using an analogous procedure led, in general, to diacyldiselenides. Early descriptions of the preparation and properties of selenobenzoic acid were in error[117,118], the product described being in fact dibenzoyl diselenide[119].

Jensen and coworkers[119] suggested that this product formed either by oxidative dimerization of the initially formed selenobenzoic acid (**67**), or that loss of hydrogen selenide from the initially formed selenocarboxylic acid occurred, generating dibenzoyl selenide (**68**) (Equation 79). This compound could further react with selenobenzoic acid via a redox reaction to afford the diselenide **69**[119].

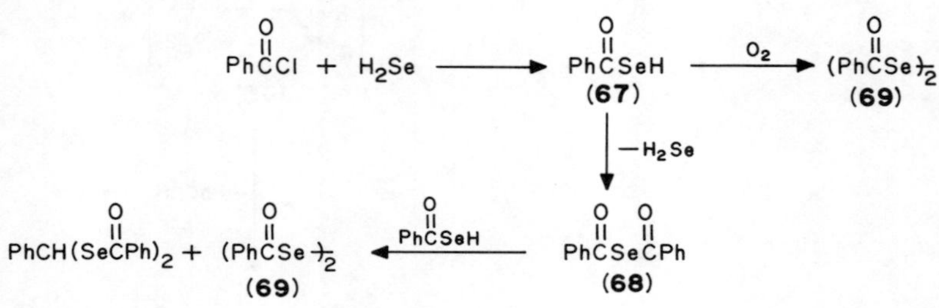

(79)

Careful control of the addition of hydrogen selenide to benzoyl chloride led to the formation of the unstable selenobenzoic acid **67**[119]. This compound is readily soluble in aqueous sodium hydrogen carbonate without liberation of elemental selenium, and can be precipitated by addition of sulfuric acid (equation 80). As expected, in the presence of air, dibenzoyl diselenide was formed from the selenoacid. The selenoacid rapidly lost hydrogen selenide with formation of dibenzoyl selenide. The selenoacid reacted with *m*-nitrobenzaldehyde to give the acylselenoacetal **70**. Sodium selenobenzoate could be alkylated to the *Se*-alkyl ester **71**[119].

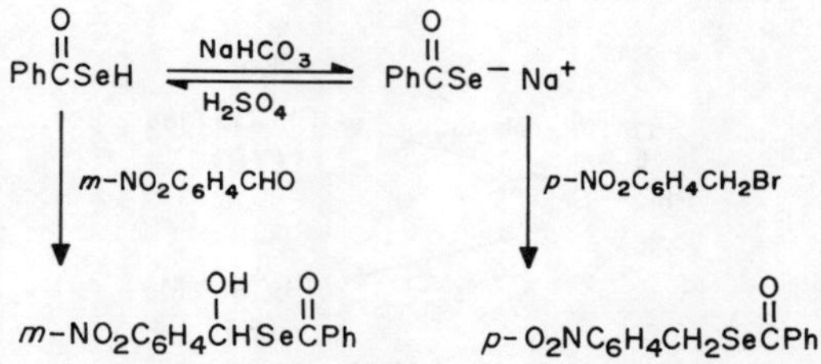

(70) (71)

(80)

Attempts to prepare aliphatic selenoacids similarly were unsuccessful, leading to low yields of diacyl selenides[119]. Dipropionyl selenide (**72**) reacts with two equivalents of aniline affording two moles of the anilide and hydrogen selenide (equation 81), indicating that selenoacids, like thioacids, can act as acylating agents.

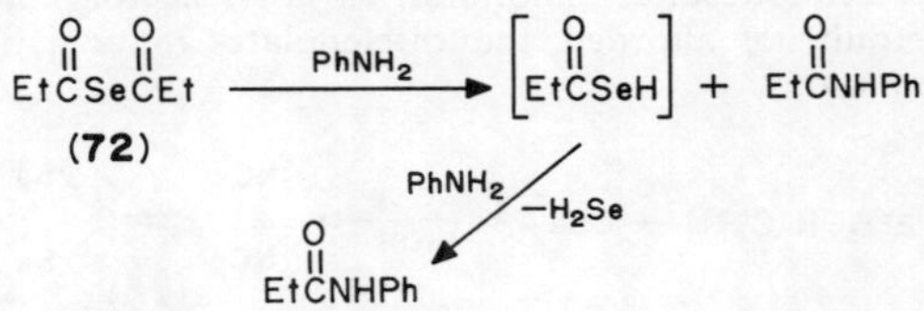

(81)

Salts of selenocarboxylic acids may be prepared by treatment of the relatively unstable diacyl selenides (**73**) with potassium hydroxide or piperidine (equation 82)[120,121]. A more convenient procedure utilizes aminolysis of the more stable diacyl diselenides (**74**) (equation 83)[122]. The diacyl selenides and diacyl diselenides can be prepared from the corresponding acid chlorides[120–122].

$$\underset{(\mathbf{73})}{RC(=O)SeC(=O)R} + KOH \longrightarrow RCOSe^-K^+ + RCO_2{}^-K^+ \qquad (82)$$

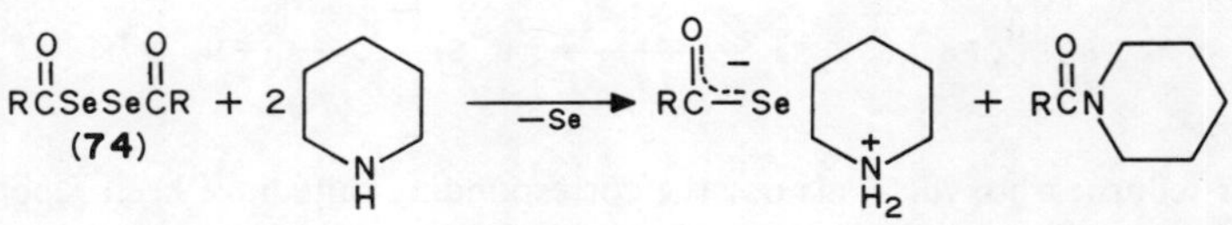

(83)

Selenocarboxylate salts can be alkylated on selenium or oxidized to diacyl diselenides. Two equivalents of selenostearate react with dichloromethane to afford bis(selenostearoyl)methane (**75**) (equation 84)[120].

$$2\ Me(CH_2)_{16}C(=O)Se^-K^+ + CH_2Cl_2 \longrightarrow \underset{(\mathbf{75})}{[Me(CH_2)_{16}C(=O)Se]_2CH_2} \qquad (84)$$

Potassium selenobenzoate reacts with trimethylsilyl chloride at the oxygen center to afford the very water-sensitive *O*-trimethylsilyl ester **76** (equation 85)[123]. Presumably the strength of the oxygen—silicon bond overrides the more nucleophilic nature of selenium. The corresponding reactions with trimethyltin chloride and trimethylgermanium chloride afford the Se-metallated derivatives **77**. This behavior of the selenocarboxylate parallels that of the corresponding thiocarboxylate.

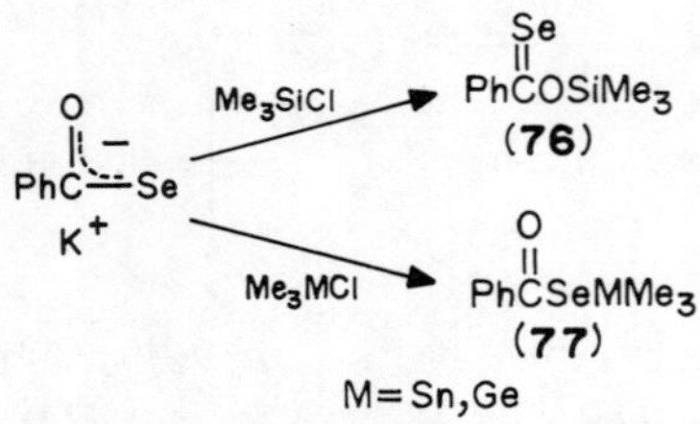

(85)

Free diselenocarboxylic acids have not been described. In attempts to prepare these compounds by the reaction of organometallic compounds with carbon diselenide, only the dialkylzinc reagents gave diselenocarboxylate salts[124]. It is likely that these insoluble zinc salts are polymeric[125].

Reactions of carbon diselenide with less basic nucleophilic carbon species with pK_a 8–20 (malonic ester, ethyl cyanoacetate, malononitrile, ethyl acetoacetate, fluorene, acetophenone, phenylacetonitrile) afforded the diselenolates **78**, which could be readily dialkylated (equation 86)[126].

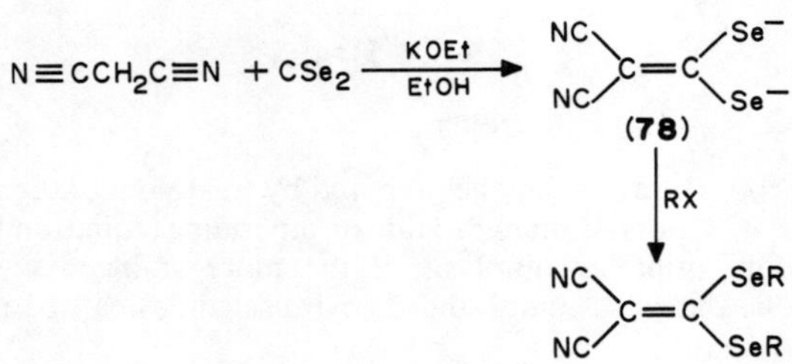

(86)

Dipotassium diselenooxalate (**79**) has been prepared by the reaction of potassium selenide with diphenyl oxalate (equation 87), and its structure has been determined by X-ray diffraction[127].

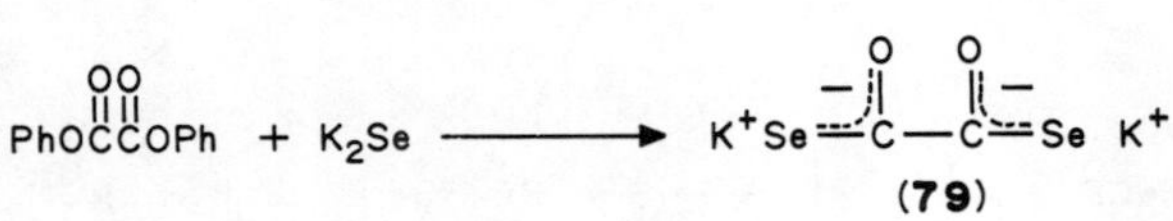

(87)

Neither tellurocarboxylic acids nor the corresponding salts have been reported.

VI. Se/Te KETENES

A. Selenoketenes—Preparation and Reactions

Selenoketenes have been postulated as intermediates in the photolysis and base-catalysed fragmentation of 4-substituted-1, 2, 3-selenadiazoles. They have proved to be valuable intermediates in the preparation of α-unsubstituted selenoesters and selenoamides (Section III.A.1, equation 34; Section IV.A.1, equations 58 and 59).

Vapor-phase pyrolysis of selenadiazoles (**80**) at 500–600 °C and trapping of the intermediates at − 190 °C afforded selenoketenes (**81**) (equation 88)[128–130]. The parent unsubstituted selenoketene polymerized at − 80°C, but could be converted into the selenoamide by reaction at low temperature with dimethylamine vapor. The substituted compounds dimerized thermally to the diselenetanes (**82**)[130].

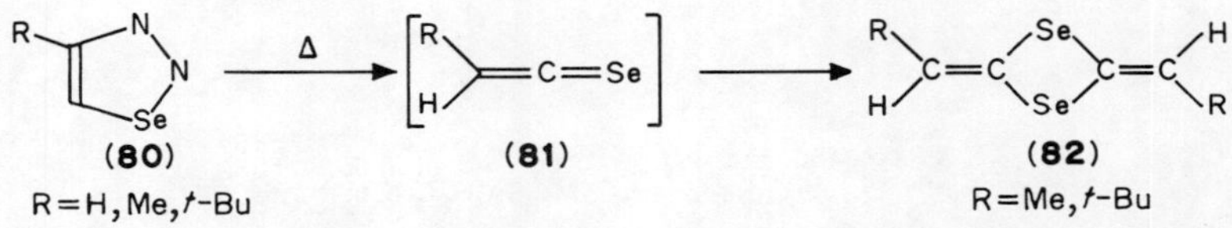

(**80**) R=H, Me, *t*-Bu (**81**) (**82**) R=Me, *t*-Bu (88)

A similar dimer (**83**) has been obtained by the base-catalysed reaction of dimethyl malonate with carbon diselenide, presumably through an intermediate selenoketene (equation 89)[126].

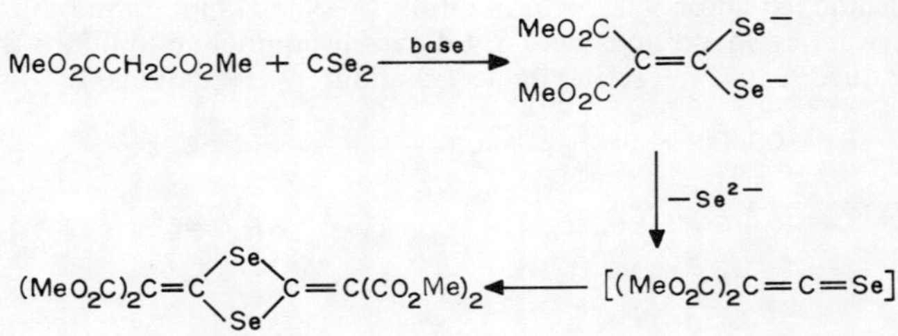

(**83**) (89)

Alkaline or photochemical generation of selenoketenes in concentrated solution affords isomeric mixtures of dimeric 2, 6-disubstituted-1, 4-diselenadihydrofulvenes (**84**) (equation 90)[131–134].

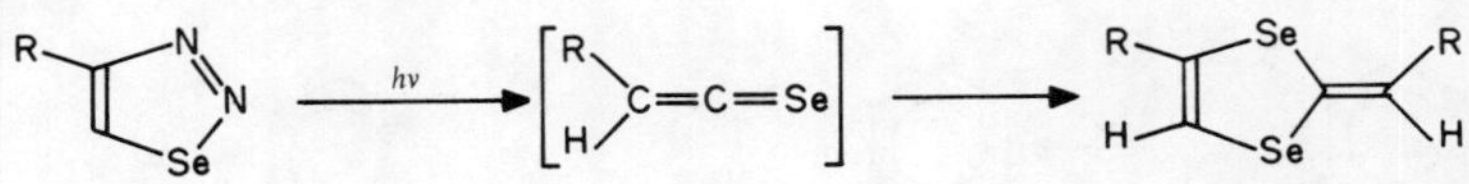

(**84**) (90)

Treatment of lithium acetylides with selenium affords acetylenic selenolates (**85**), which on protonation isomerize to selenoketenes (equation 91)[99]. This provides an alternative to selenadiazole pyrolyses in the preparation of selenoamides (see Section IV.A, equation 59).

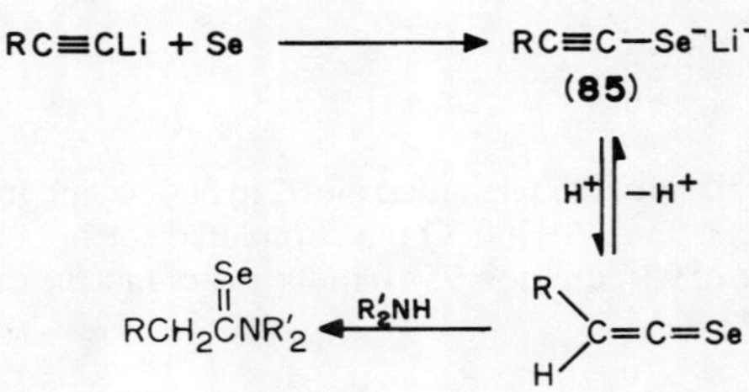

(**85**) (91)

Bistrimethylsilyl selenoketene (**86**) is remarkably stable, and could be distilled at 90 °C[135]. It could be readily prepared by trimethylsilylation of the acetylenic selenolate (equation 92). Reaction of this selenoketene with diethylamine led to selenoamides **87** and **88**.

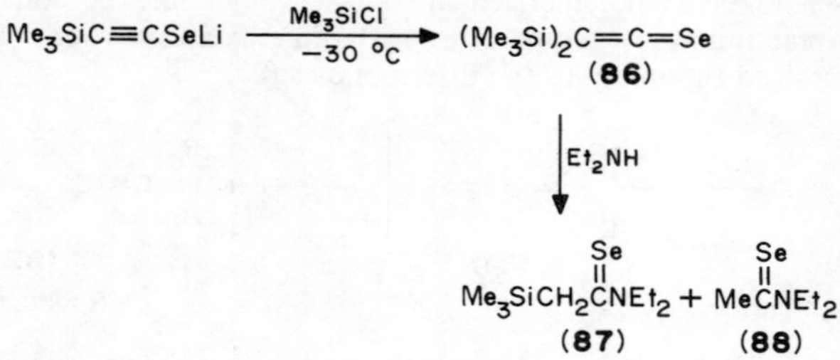

(92)

Other thermally stable, sterically hindered selenoketenes (**89**) could also be prepared starting from the acetylenic selenolate through a selena-Cope reaction of the allylic acetylenic selenide (equation 93)[136]. In addition to selenoamide formation, these compounds underwent cycloadditions with 3,4-dihydroisoquinoline to afford the seleno-β-lactam 1:1 adduct **90** and the selenoamide 1:2 adduct **91** (equation 94).

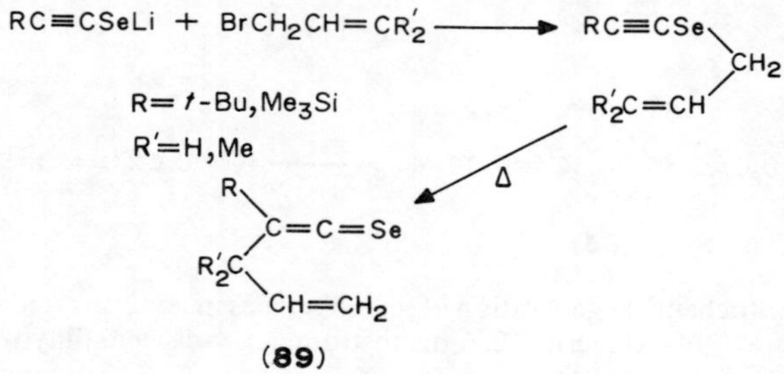

(93)

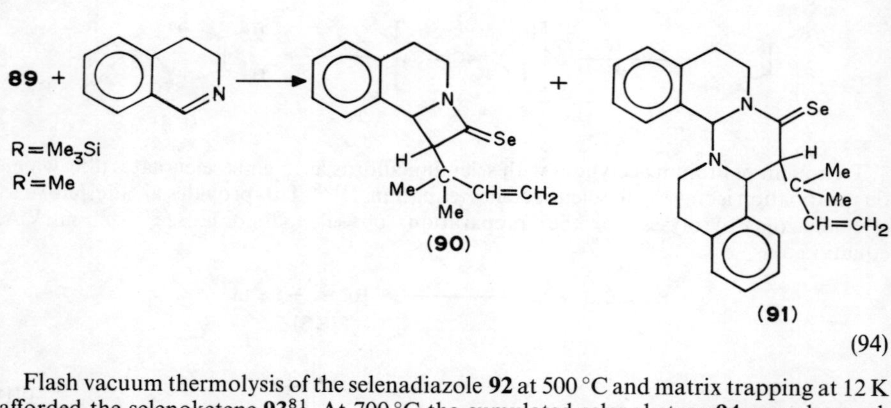

(94)

Flash vacuum thermolysis of the selenadiazole **92** at 500 °C and matrix trapping at 12 K afforded the selenoketene **93**[81]. At 700 °C the cumulated selenoketene **94** was observed, presumably by cleavage of **93** (equation 95). Irradiation of matrix-isolated **94** at 12 K gave the acetylenic selenal **95**.

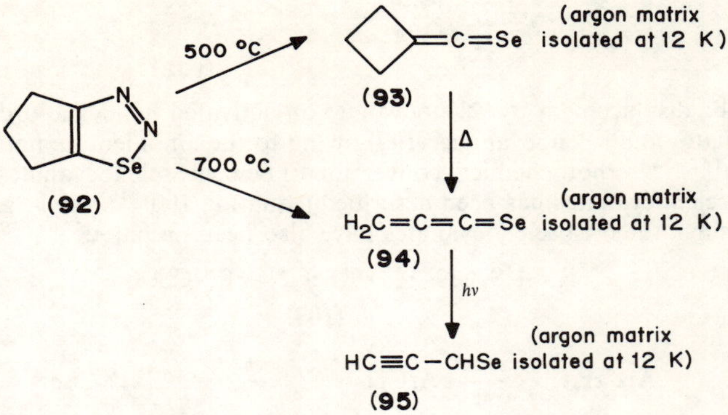

(95)

B. Telluroketenes—Preparation and Reactions

The reaction of the sodium salt of phenylacetylene with tellurium, followed by acidification, leads to a mixture of products in which the amounts vary, depending on the reaction conditions[137–140]. The reactions are presumed to go through acetylenic tellurolate and telluroketene intermediates **96** and **97** (equation 96). The telluroketene **97** can dimerize directly to **98** or ditellurodihydrofulvene (**99**), or react with ditellurate to afford **100**. The most recent reference describes this complex reaction most clearly[140].

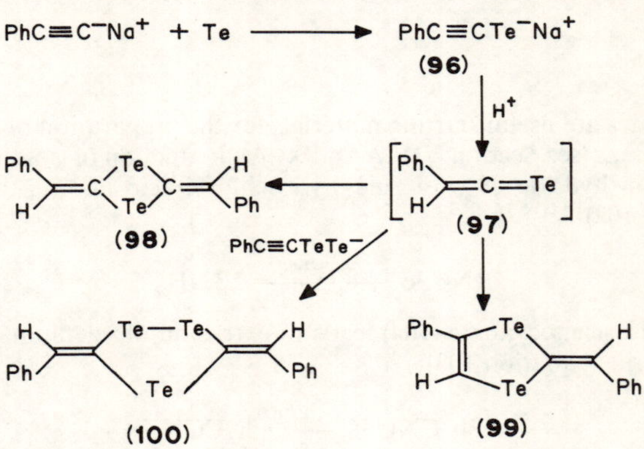

(96)

VII. ISOSELENOCYANATES

A. Preparation

The isoselenocyanates **101** may be prepared by reaction of an isonitrile with selenium (equation 97)[141,142]. The sterically hindered isoselenocyanate **102** was best prepared by reaction of the isonitrile anion with selenium at low temperature (equation 98)[52].

$$R\overset{+}{N}\equiv\overset{-}{C} \xrightarrow{Se} RN{=}C{=}Se \quad (97)$$

(101)

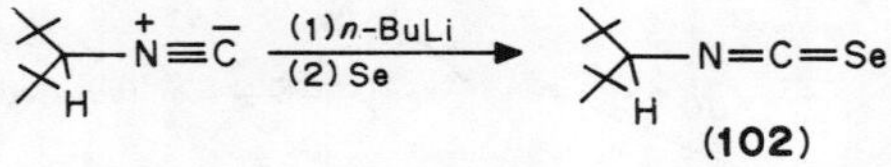

(98)

(102)

Nucleophilic displacement by selenocyanate on activated halides affords mixtures of isoselenocyanates and selenocyanates (**103**) owing to the ambidentate nature of the ion (equation 99)[142–144]. Photochemical conversion of benzylic selenocyanates to the corresponding isoselenocyanates has been described (equation 100)[145]. Silyl-, germanyl- and phosphorus-substituted isoselenocyanates have also been prepared[146–149].

$$RX + SeCN^- \longrightarrow RSeCN + RNCSe \qquad (99)$$

(103)

$$ArCH_2Cl \xrightarrow[\text{DMF}]{\text{KSeCN}} ArCH_2SeCN \xrightarrow{h\nu} ArCH_2NCSe \qquad (100)$$

$$RNH_2 + CSe_2 + 2Et_3N + HgCl_2 \longrightarrow RNCSe + HgSe + 2Et_3N{\cdot}HCl \qquad (101)$$

Treatment of a primary amine with carbon diselenide in the presence of a tertiary amine and mercury(II) acetate affords isoselenocyanates in moderate to good yields (equation 101)[150]. Isoselenocyanates have been postulated as intermediates in the reaction of carbon diselenide or triselenocarbonates with primary amines to form selenoureas (see Section XI, equation 152).

Other methods of preparation of isoselenocyanates include the reaction of an isocyanate with phosphorus pentaselenide and the reaction of dichloroisocyanate with sodium selenide[151].

B. Reactions

Isoselenocyanates are useful starting materials for the preparation of selenosemicarbazides and selenoureas (see Section VIII.A and XI.A). Reduction of isoselenocyanates with lithium aluminium hydride or with zinc–hydrochloric acid affords the corresponding amines (equation 102)[152,153].

$$RNCSe \xrightarrow{\text{reduction}} RNH_2 \qquad (102)$$

Heating trityl isoselenocyanate (**104**) leads to extrusion of selenium and formation of triphenylacetonitrile (equation 103)[152].

$$Ph_3CNCSe \xrightarrow[-\,\text{Se}]{\Delta} Ph_3CCN \qquad (103)$$

(104)

Isoselenocyanates are useful synthetic intermediates in a recently described deamination procedure[154]. Treatment of the steroidal isoselenocyanate **105** with tri-*n*-butylstannane affords the product **106** in moderate yield (equation 104). Isothiocyanates and isocyanides can also act as substrates in this reduction.

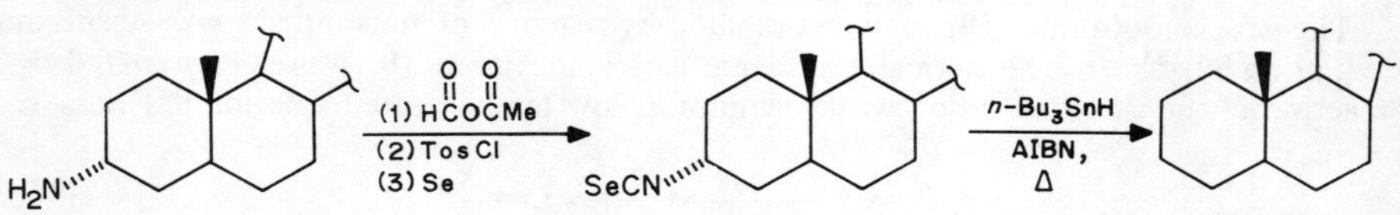

(105) **(106)**

(104)

A cycloaddition of phenyl isoselenocyanate to the enamine **107** in the presence of sulfur affords the cyclic selone thiocarbamate **108** (equation 105)[155].

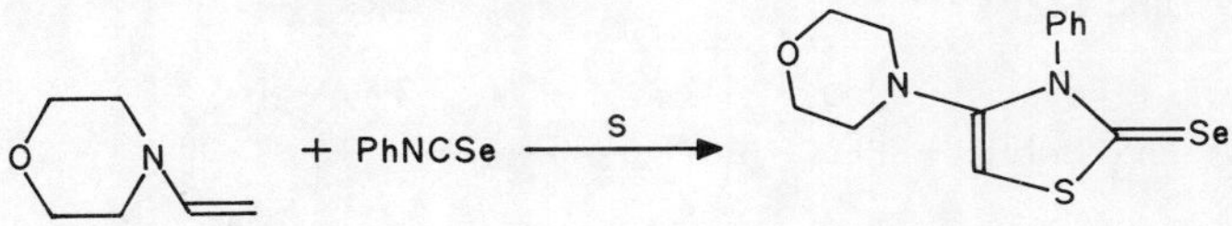

(105)

(107) **(108)**

VIII. SELENOSEMICARBAZIDES

A. Preparation

Selenosemicarbazides (**109**) can be prepared by the reaction of a isoselenocyanate with a hydrazine (equation 106)[156–162]. They can also be prepared from isonitriles and hydrazines in the presence of selenium (equation 107)[163]. The highly substituted selenosemicarbazide **111** could be prepared by reaction of bis(N,N,N'-trimethylselenocarbazoyl) diselenide (**110**) (see Section X.A) with dimethylamine (equation 108)[164].

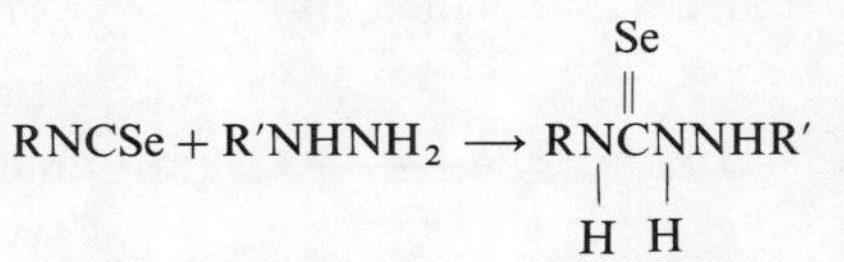

(106)

(109)

$$RNC + Se + R'NHNH_2 \longrightarrow \mathbf{109} \qquad (107)$$

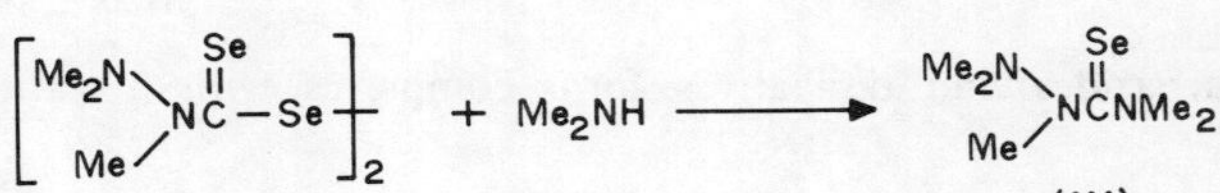

(108)

(110) **(111)**

Selenosemicarbazides can also be prepared by addition of hydrogen selenide to cyanohydrazines or aromatic diazonium cyanides[158,165].

While selenosemicarbazones (**112**) could be isolated from the reaction of hydrazine selenocyanate with aldehydes or ketones (equation 109) the parent selenosemicarbazide could not be directly isolated by rearrangement of hydrazine selenocyanate[157].

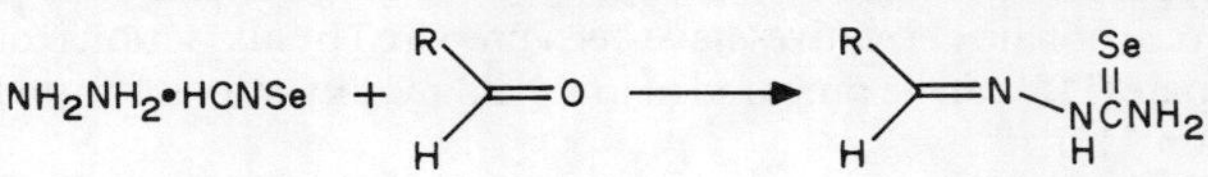

(109)

(112)

B. Reactions

Selenosemicarbazones can be prepared from aldehydes or ketones and selenosemicarbazides under acid catalysis (equation 110)[158,159,162,165]. Aldehyde derivatives can be prepared by exchange with acetone selenosemicarbazone[157].

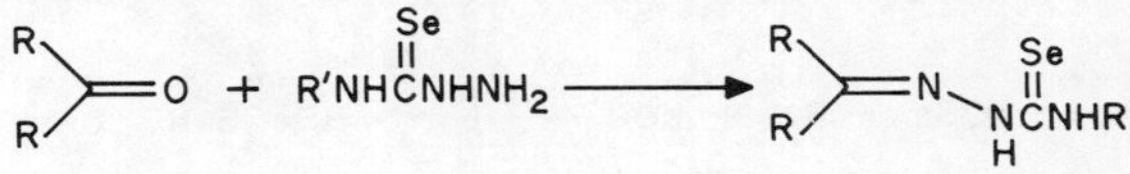

(110)

Selenosemicarbazides and selenosemicarbazones are also widely used in the preparation of a variety of complex selenium-containing heterocycles[5,12,166] (e.g. **113** and **114**, equations 111 and 112)[159,167].

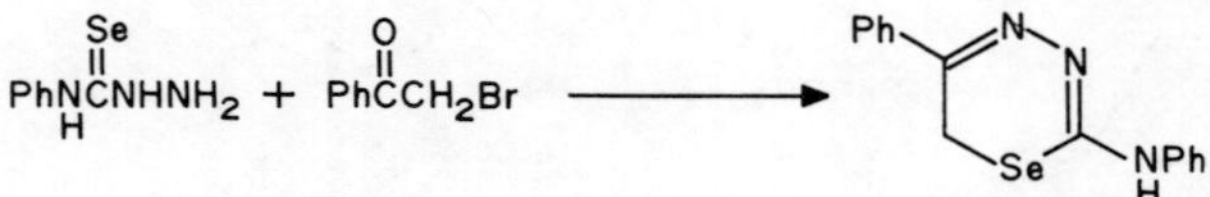

(113) (111)

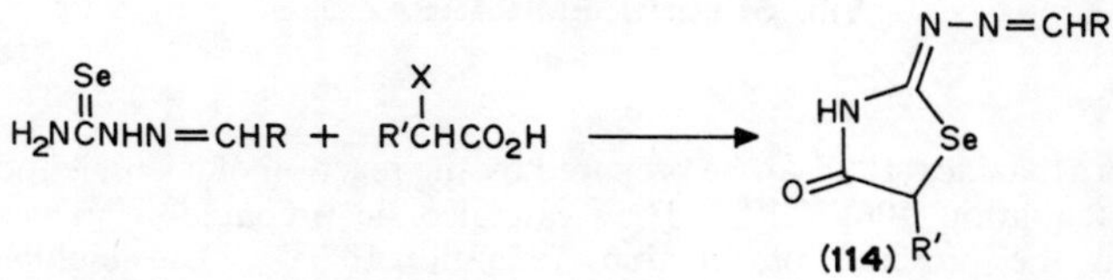

(114) (112)

Alkylation of a selenosemicarbazide or a selenosemicarbazone occurs at selenium (equations 111–113)[158,159,167]. Reaction of a selenocyanate with a selenosemicarbazide leads to the *N*-acylated product **115** (equation 114)[162].

$$C_4H_9NHC(=Se)N(Me)NMe_2 + (MeO)_2SO_2 \longrightarrow C_4H_9\overset{+}{N}H{=}C(SeMe)N(Me)NMe_2\text{—}MeOSO_3^- \quad (113)$$

$$p\text{-Tol-NHC(=Se)NHNH}_2 + p\text{-Tol-N=C=Se} \longrightarrow (p\text{-Tol-NHC(=Se)NH})_2 \quad (114)$$

(115)

Selenosemicarbazides and derivatives form complexes with a variety of metal ions[168–170].

IX. SELENOCARBONATES

A. Preparation

Selenium derivatives of carbonic acids are not well characterized and decompose with liberation of hydrogen selenide. The chemistry of these acids has been reviewed[169]. A variety of selenocarbonate derivatives have been prepared by alkylation reactions. Diethyl triselenocarbonate (**116**) can be prepared in low yield by alkylation of barium triselenocarbonate (equation 115)[170].

Other dialkyl triselenocarbonates can be prepared by reaction of an alkyl halide, carbon diselenide and base in DMSO (equation 116)[171].

$$BaCSe_3 + 2\,EtI \longrightarrow EtSeC(=Se)SeEt \quad (115)$$

(116)

$$CSe_2 + RX + KOH \xrightarrow[H_2O]{DMSO} RSeC(=Se)SeR \quad (116)$$

While the reaction of a selenolate with carbon diselenide followed by alkylation does not afford triselenocarbonates[171], the *O*- and *S*-alkyl derivatives can be satisfactorily prepared starting from the alkoxide or thiolate (equations 117–119)[172,173].

$$RSe^- + CSe_2 \not\longrightarrow RSeC(=Se)Se^- \quad (117)$$

$$RS^- + CSe_2 \longrightarrow RSC(=Se)Se^- \xrightarrow{ClCH_2CO_2^-} RSC(=Se)SeCH_2CO_2^- \quad (118)$$

$$RO^- + CSe_2 \longrightarrow ROC(=Se)Se^- \xrightarrow{ClCH_2CO_2^-} ROC(=Se)SeCH_2CO_2^- \quad (119)$$

Cyclic selenocarbonates have been extensively studied as starting materials in the preparation of organic semiconductors[174]. Alkylation of dithio- or diseleno-carbamate anions (see Section X.A) followed by acid-catalysed cyclization and hydrogen selenide treatment provides a convenient route to disubstituted triselenocarbonates[175] and selonedithiocarbonates (**117**) (equation 120)[176].

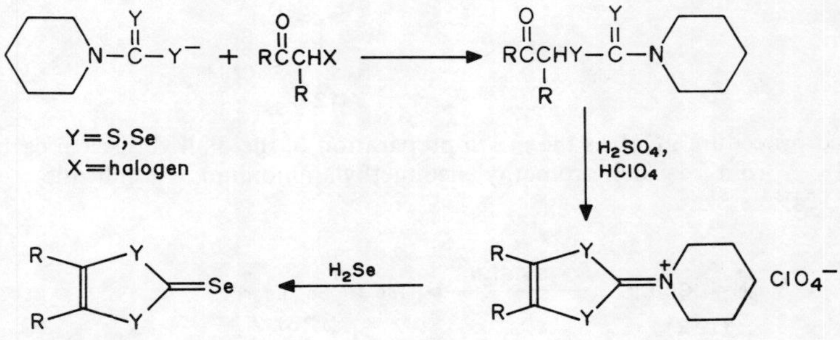

(**117**)

(120)

The unsubstituted triselenocarbonate **118** can be prepared by reaction of sodium acetylide with carbon diselenide and selenium, followed by acidification (equation 121)[177].

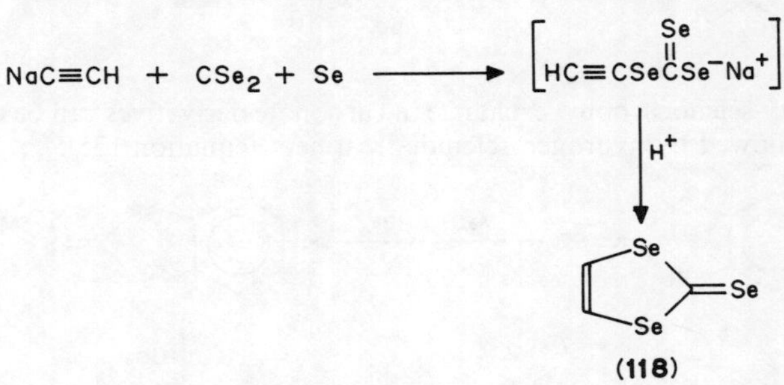

(**118**)

(121)

This compound can also be prepared via the reaction of bis(*N*,*N*-dialkylselenocarbamoyl) selenide (**119**) (see Section X.A) with the diselenide **120** (equation 122)[178].

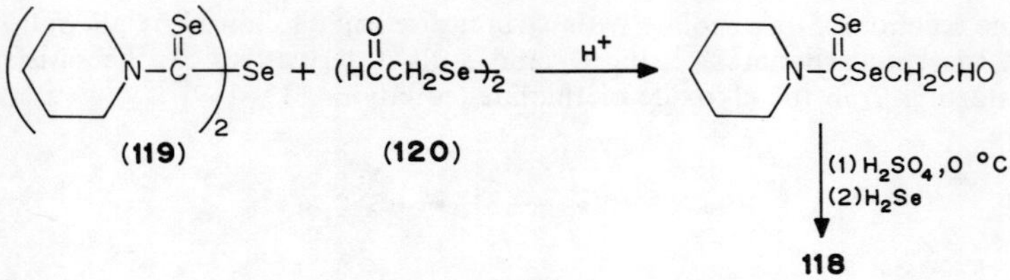

(122)

Two more recent routes to triselenocarbonates avoid the use of the noxious carbon diselenide as a selenium source. The first utilizes alkylation of tetramethylselenourea followed by hydrogen selenide treatment to afford the diselenourethane intermediate **121** (equation 123)[179].

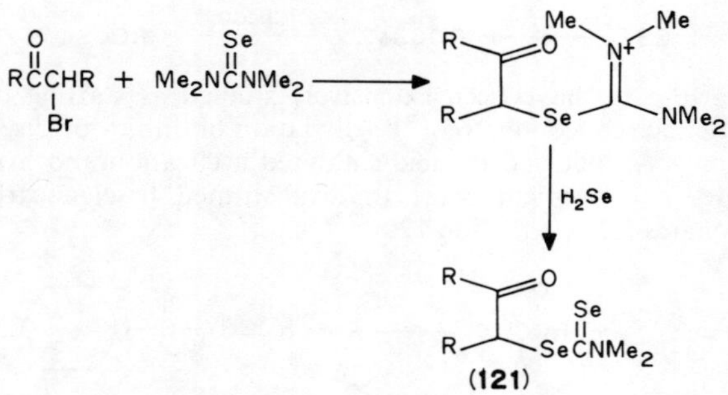

(123)

A related procedure involves the *in situ* preparation of the dialkyl diselenocarbamate anion **123** from dichloromethylenedimethylammonium chloride (**122**) (equation 124)[180,181].

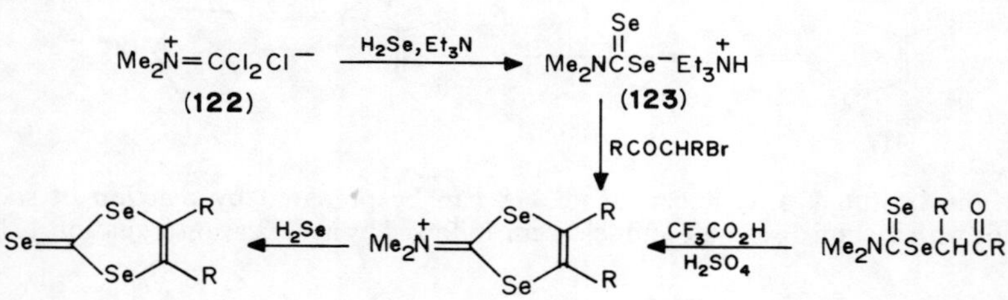

(124)

Thiocarbonyl–selenocarbonyl exchange in carbonate derivatives can be carried out by *S*-alkylation followed by hydrogen selenide treatment (equation 125)[182].

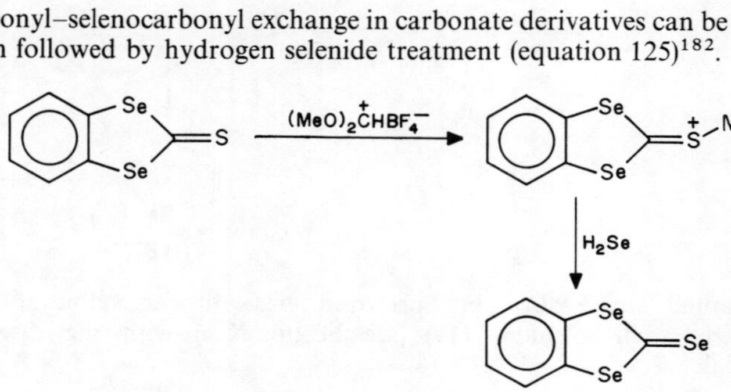

(125)

Thermolysis of 1, 2, 3-benzoselenadiazole (**124**) in the presence of excess of carbon diselenide provides an alternative route to benzotriselenocarbonate (**125**) (equation 126)[183].

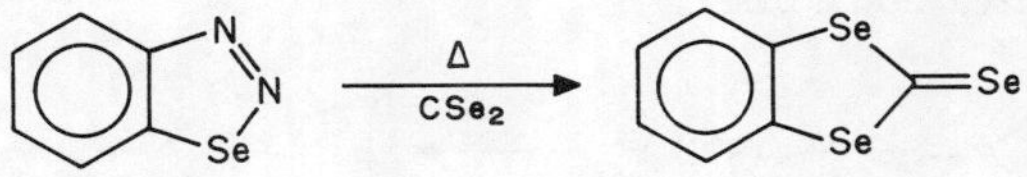

(126)

(124) **(125)**

Electrochemical reduction of carbon diselenide affords the diselenolate **126**, which can be alkylated to the cyclic triselenocarbonate **127** (equation 127)[184].

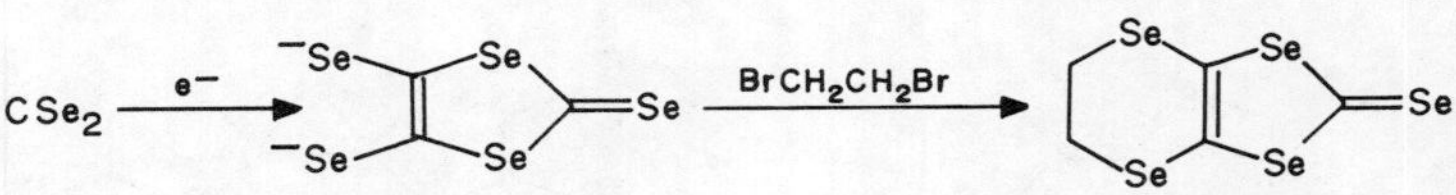

(127)

(126) **(127)**

The selenocarbonate derivative **128** can be prepared by selenation of the corresponding carbonyl compound using bistricyclohexyltin selenide–boron trichloride (equation 128)[176].

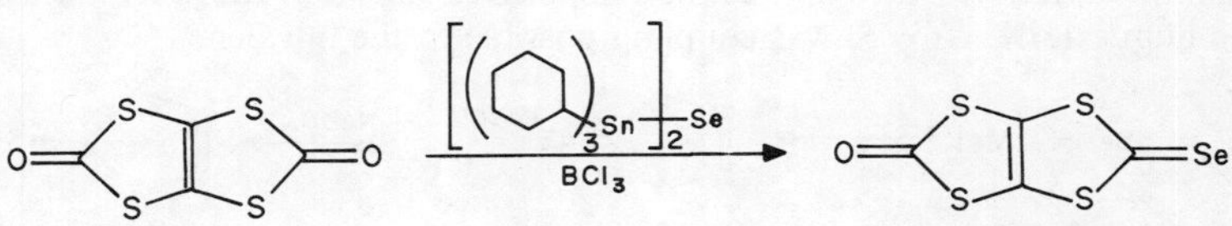

(128)

(128)

Selenopyrones **130** and **132** can be prepared from the corresponding carbonyl and thiocarbonyl compounds **129** and **131** (equations 129 and 130)[185,186].

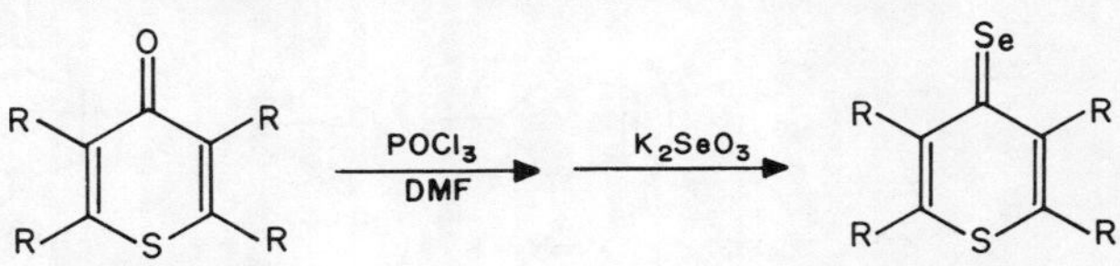

(129)

(129) **(130)**

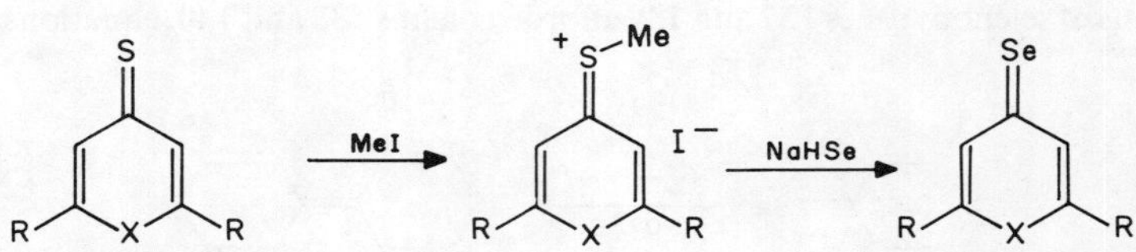

(130)

(131) **(132)**

X=O,S

B. Reactions

Treatment of a selenocarbonate with a tertiary phosphine or phosphite affords the dihydrofulvalene derivatives **134**[177,182,183]. This coupling reaction is typically much more successful for selenocarbonyl derivatives than the corresponding thiocarbonyl compounds, although an exception has been noted[184]. It is likely that the desired coupling occurs through a 1, 3-dipolar intermediate **133** formed by nucleophilic attack on the thio- or seleno-carbonyl derivative (equation 131)[176,187]. Such 'heterophilic attacks' would be more facile in the case of the selenocarbonyl compounds (see Section II.B.2). No reaction was observed in the carbonyl case.

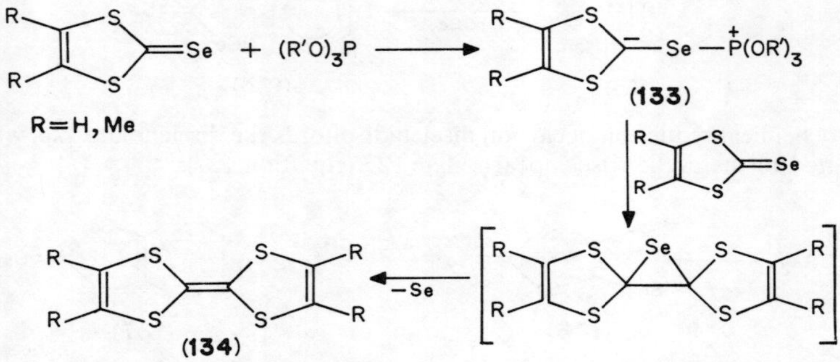

(131)

The selenocarbonyl group of selenocarbonates can be readily alkylated and the alkylated product reduced with borohydride (equation 132)[188,189]. Treatment of the resulting dithiole **135** with acid affords the stabilized carbocation **136**; only in the case of the dithiolium salt (X = Y = S) was coupling possible to the fulvalene.

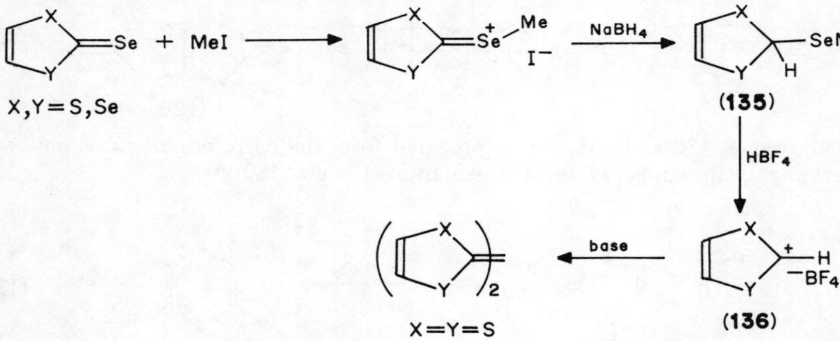

(132)

Pyrolysis of selenopyrones **137** and **139** afforded olefins **138** and **140** (equations 133 and 134)[186,190].

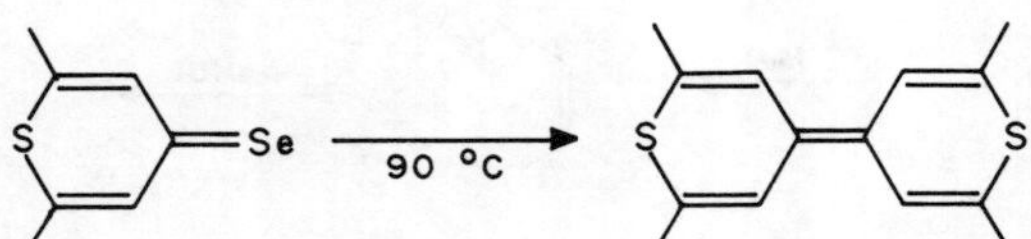

(133)

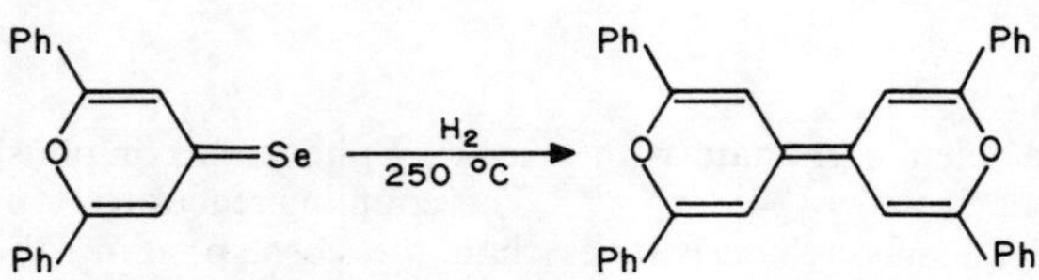

(134)

Dialkyl triselenocarbonates react with amines in a variety of ways. Reaction of dialkyl triselenocarbonates with secondary amines afford diselenocarbamates (see Section X.A). Reaction of the triselenocarbonate with primary amines presumably affords a selenocarbamate intermediate. This intermediate further reacts generating the selenoisocyanate and ultimately the selenourea. As previously described, dialkyl triselenocarbonates react with amines in the presence of a selenolate to afford selenoformamides (see Section IV.A.1, equation 62)[88]. These differences in reactivity have been explained by differences in the electrophilic nature of the triselenocarbonate depending on whether the nucleophile is 'hard' (reaction at the selenocarbonyl carbon) or 'soft' (reaction at the selenium of the Se alkyl group) (equation 135)[191].

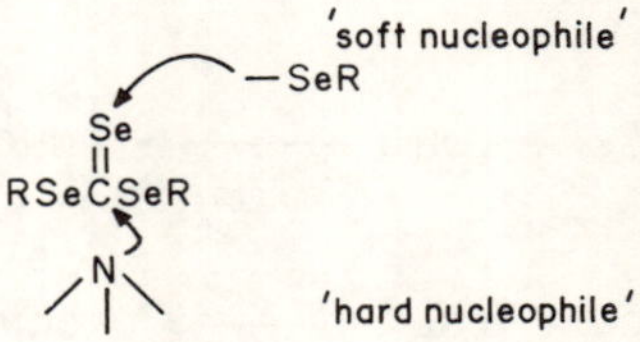

(135)

X. Se/Te CARBAMATES AND HYDRAZIDES

A. Selenocarbamates and Selenohydrazides—Preparation and Reactions

The reactions of secondary amines[192–196] or hydrazines[197,198] with carbon diselenide afford salts of diselenocarbamic acid (**141**, **142**) (equations 136 and 137). These can be readily alkylated to selenocarbamates, which have been previously described as important intermediates in the preparation of cyclic selenocarbonates (Section IX.A, equation 120).

$$2R_2NH + CSe_2 \longrightarrow \underset{\textbf{(141)}}{R_2NCSe_2^-\ R_2NH_2^+} \qquad (136)$$

$$R_2NNH_2 + CSe_2 \longrightarrow \underset{\textbf{(142)}}{R_2\overset{+}{N}NHCSe_2^-} \qquad (137)$$

On mild oxidation, dialkyl selenocarbamate salts afford equimolar mixtures of the bisselenocarbamoyl selenide and the triselenide[192,198,199]. The reaction has been shown to go through a diselenide intermediate **143**, which is in equilibrium with the selenide **144** and triselenide **145** (equation 138)[199]. Treatment of **143** or **145** with triethyl phosphite affords **144**; treatment of **144** with selenium affords **145** (equation 139).

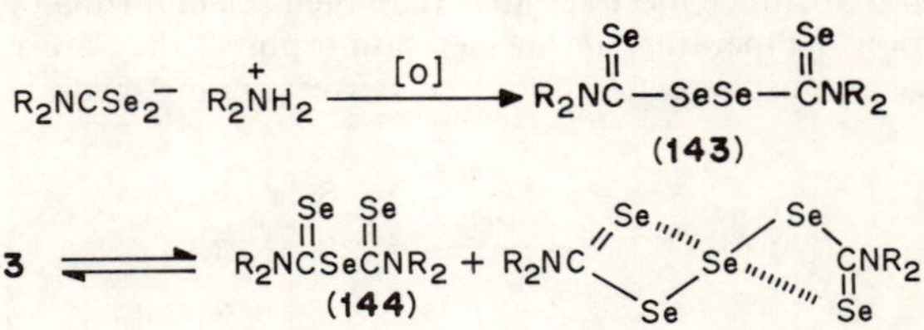

(138)

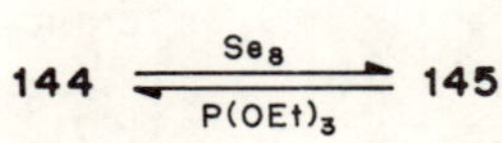

(139)

Selenourethane derivatives can be prepared by ammonolysis or hydrazinolysis of selenocarbonates (equation 140)[191,200,201] or by reaction of an alcohol with a selenoisocyanate (equation 141)[202]. Unsubstituted monoselenourethanes can be obtained by addition of hydrogen selenide to alkyl cyanates (**146**) (equation 142)[203].

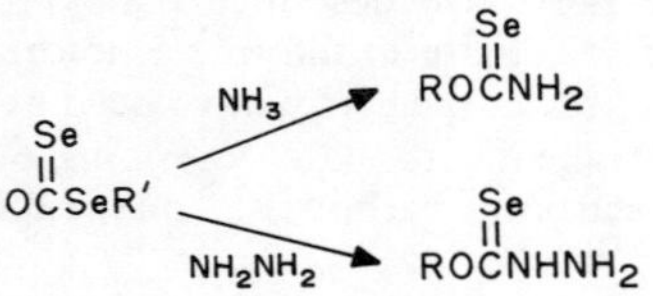

(140)

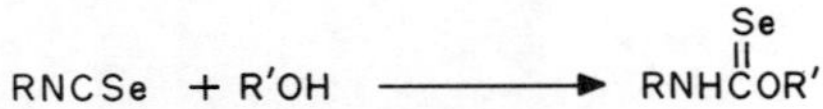

(141)

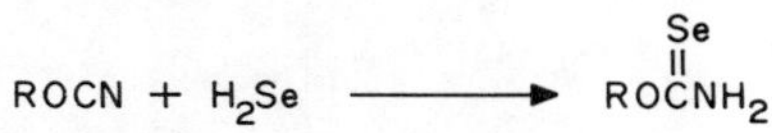

(142)

(**146**)

Selenothiocarbamates can be prepared from thioureas by alkylation, followed by reaction with hydrogen selenide at pH 5–6 (equation 143)[204] (cf. reaction at alkaline pH, Section XI.A.1, equation 147).

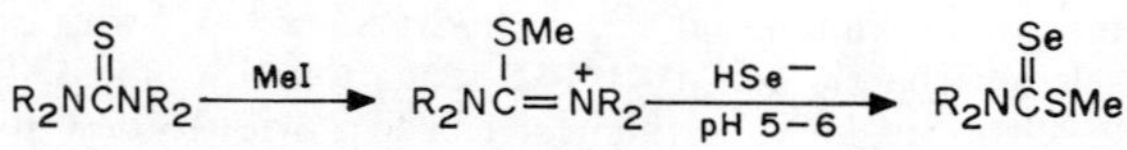

(143)

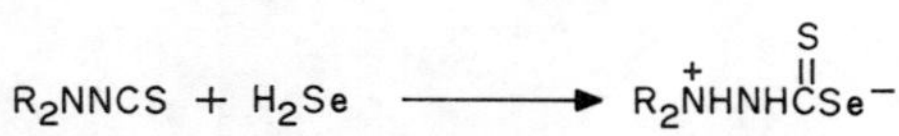

(144)

Addition of hydrogen selenide to an *N*-isothiocyanatoamine affords the selenothiocarbazate salt (equation 144)[205].

B. Tellurohydrazides and Tellurocarbazates

Tellurohydrazides (**147**) have been prepared from the corresponding thiohydrazides by alkylation and careful treatment of the *S*-alkylated product with triethylamine–hydrogen telluride in dichloromethane at $-40\,^{\circ}C$ under an inert atmosphere (equation 145)[116]. These compounds are much more sensitive than their selenium analogues. Although they are stable at room temperature in an inert atmosphere, they are rapidly attacked by moisture or the atmosphere to afford the corresponding hydrazides and elemental tellurium.

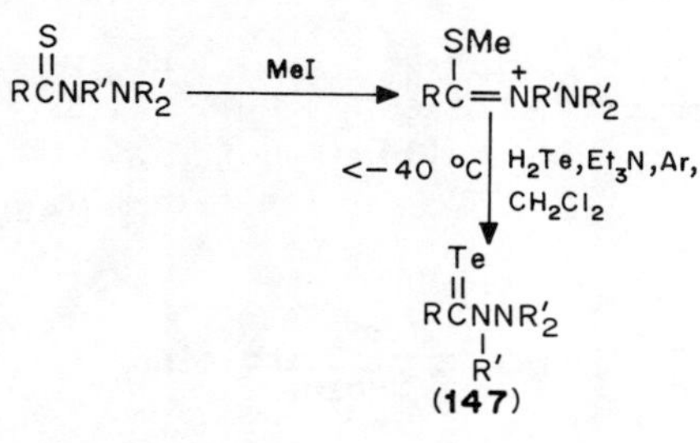

(145)

(**147**)

Treatment of the *N*-isothiocyanatoamine **148** with hydrogen telluride afforded the ditelluride **150**, presumably through an unstable tellurothiocarbazic acid (**149**) (equation 146)[205].

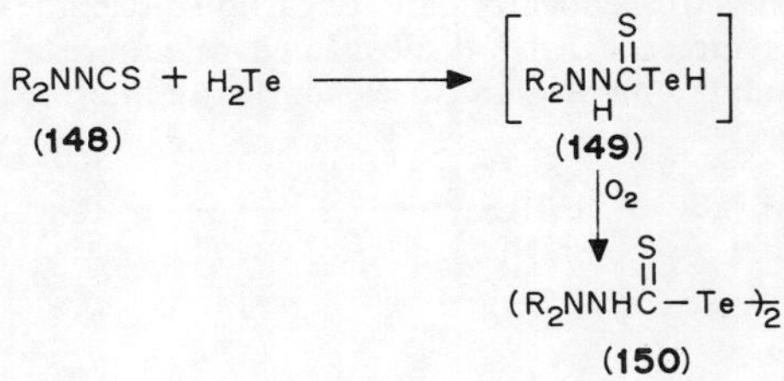

(146)

XI. Se/Te UREAS

A. Selenoureas

1. *Preparation*

A variety of methods are available for the preparation of selenoureas (**151**). They may be prepared from thioureas via alkylation followed by treatment with hydroselenide ion at pH 8–9 (equation 147)[204]. Reaction of the salt at more acidic pH affords the selenocarbamate (see Section X.A.1, equation 143). This method provides a convenient route for the preparation of tetrasubstituted selenoureas. Trisubstituted selenoureas could be prepared via a similar procedure[206].

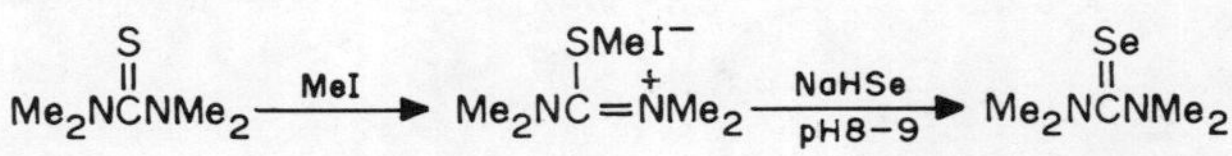

(147)

(151)

Mono-, di- and tri-substituted selenoureas can be prepared by the reaction of an isoselenocyanate with ammonia, primary or secondary amines (equation 148)[156,159,162,207]. Reaction of selenocyanate ion with amines also yields selenoureas (equation 149)[208].

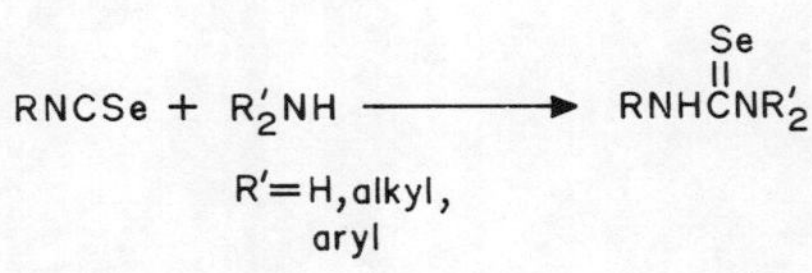

(148)

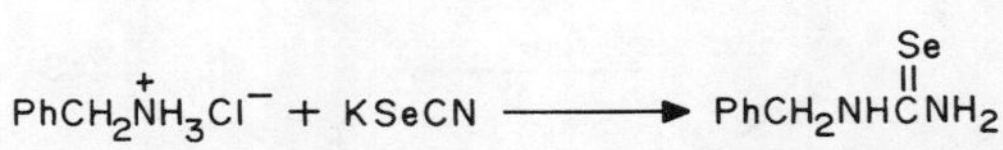

(149)

Reaction of cyanamide with hydrogen selenide provides a convenient route to the unsubstituted selenourea (equation 150)[209]. Substituted cyanamides afford mono- or 1,1-di-substituted selenoureas[210–212]. Carbodiimides also react with hydrogen selenide to afford 1,3-disubstituted selenoureas (equation 151)[210].

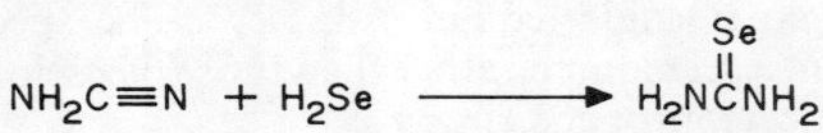

(150)

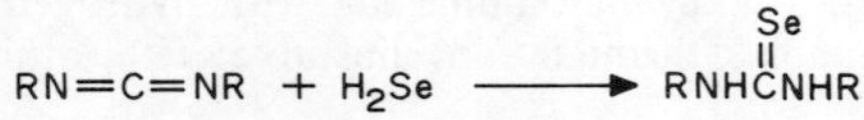

(151)

The reactions of dialkyl triselenocarbonates or carbon diselenide with excess of primary amines both afford symmetrical 1,3-disubstituted selenoureas (equation 152)[191,213]. These reactions presumably involve an isoselenocyanate intermediate.

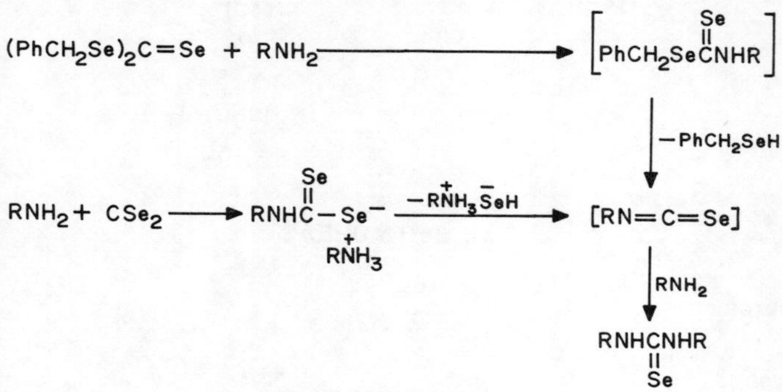

(152)

Phosphorus pentaselenide converts urea into selenourea in poor yield[214]. Selenobiurets (**152**) could be prepared from thiobiurets by the previously described alkylation–sodium hydrogen selenide procedure (equation 153)[204,215].

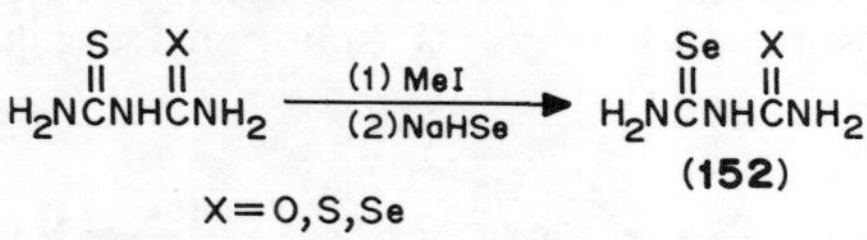

(153)

Sterically hindered tetrasubstituted selenoureas can be prepared using diselenocarbamoyl selenides or triselenides (**153**) (equation 154)[199]. This selenide mixture may also be generated *in situ*.

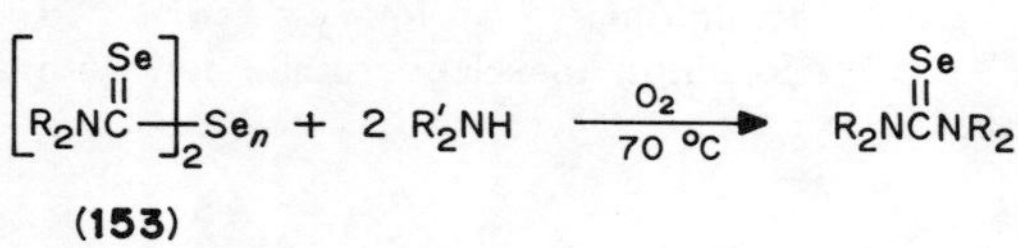

(154)

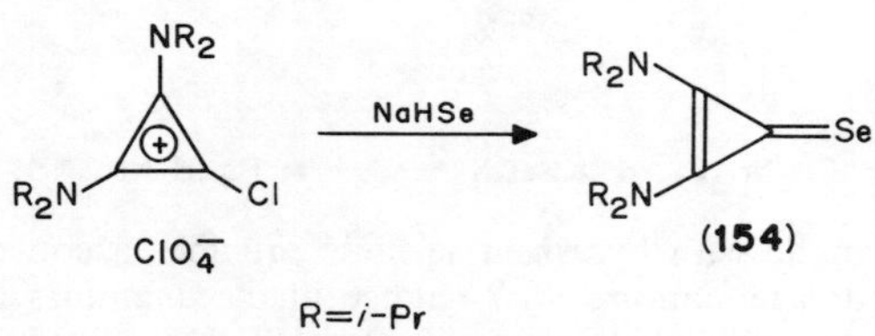

(155)

The 'quasi-selenourea' **154** was prepared in quantitative yield from the cyclopropenium salt and sodium hydrogen selenide (equation 155)[126]. The spectral properties of this molecule resemble those of a selenourea rather than those of a selone because of the strong conjugation through the cyclopropene ring.

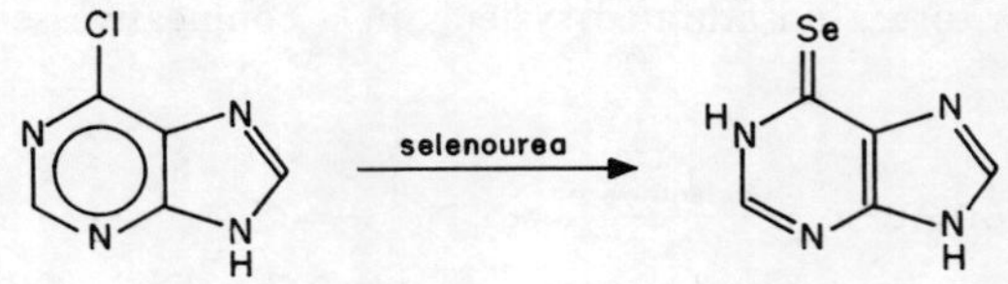

(156)

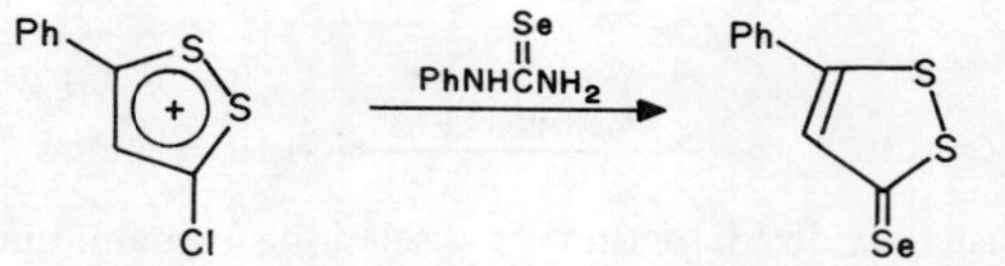

(157)

2. Reactions

Selenoureas are widely used for introducing selenium into organic molecules (equation 156 and 157). As expected, the selenium center is especially nucleophilic. Selenoureas are very useful as reagents in the synthesis of selenium-containing heterocycles[5,12,166] (e.g. **155**–**158**, equations 158–161).

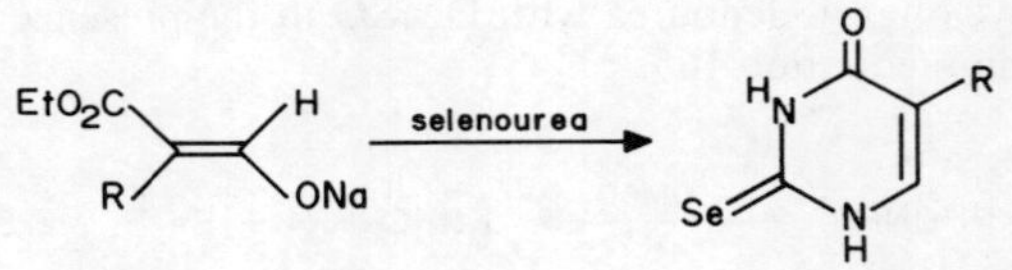

(158)

(155)

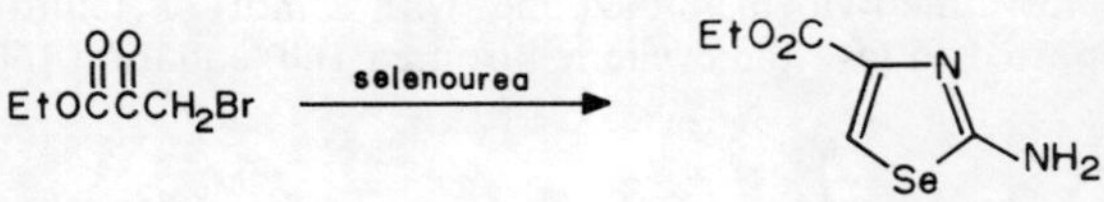

(159)

(156)

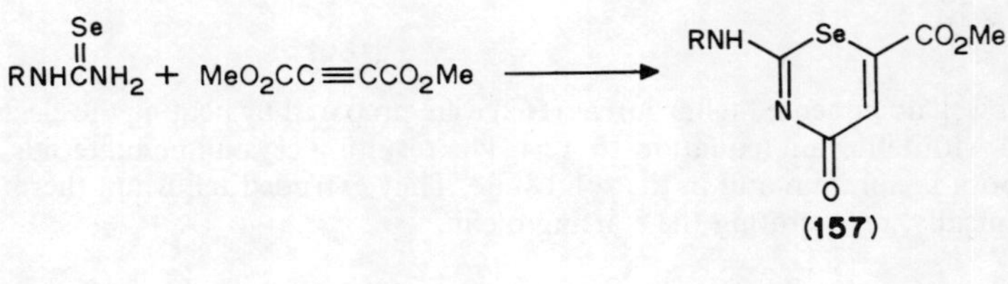

(160)

(157)

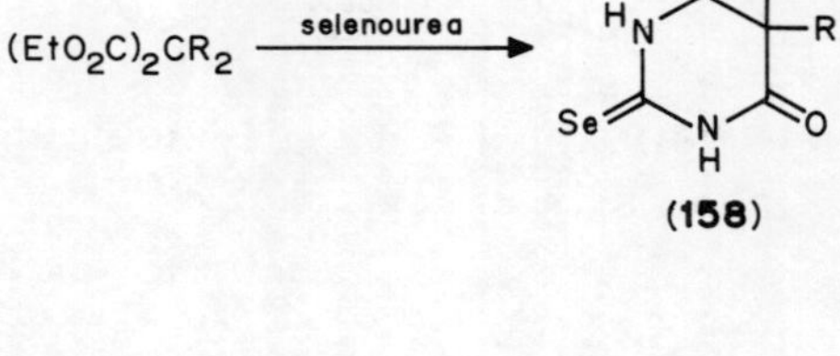

(161)

(158)

Selenourea adds to α,β-unsaturated systems in a conjugative sense (equations 162 and 163)[221,222].

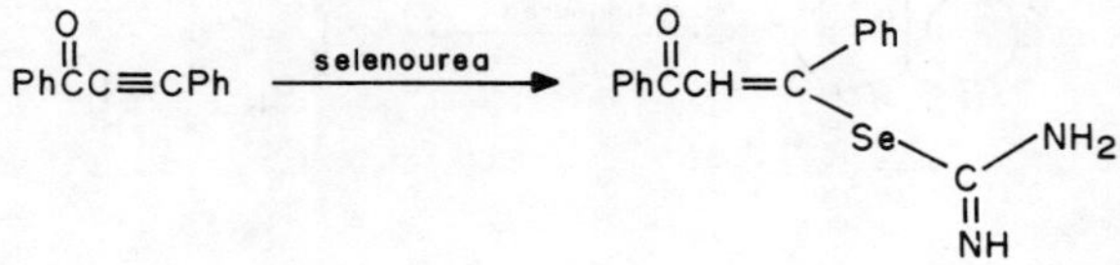

(162)

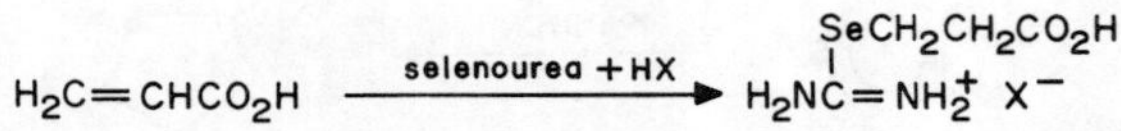

(163)

Selenourea is easily oxidized to the α,α-diselenobisformamidium cation **159** using hydrogen peroxide, hexacyanoferrate(III), *p*-benzoquinone or by electrochemical oxidation (equation 164)[223,224].

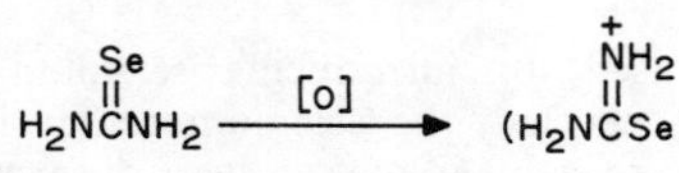

(164)

(159)

Treatment of dicyclohexylselenourea with DMSO in the presence of acid leads to an oxidative deselenation (equation 165)[225].

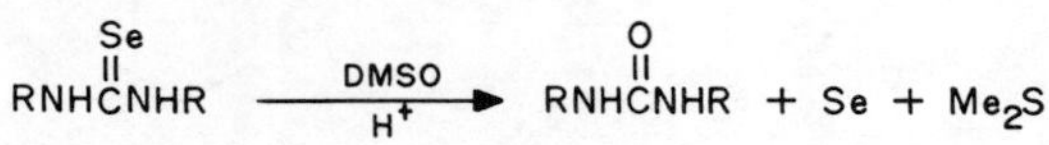

(165)

B. Tellurourea—Preparation and Reactions

Treatment of 1,3-dimethylbenzimidazoline with 2 mol of tellurium at elevated temperature is reported to give the cyclic tellurourea **160** (equation 166)[226].

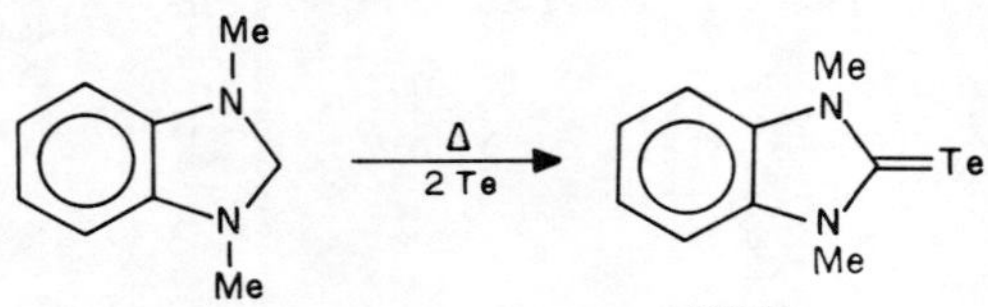

(166)

(160)

The first well documented tellurourea (**162**) were prepared by heating the electron-rich olefins **161** with tellurium (equation 167)[227]. The resulting crystalline materials were air sensitive both in solution and in the solid state. They extruded tellurium thermally and photochemically, regenerating the starting olefin.

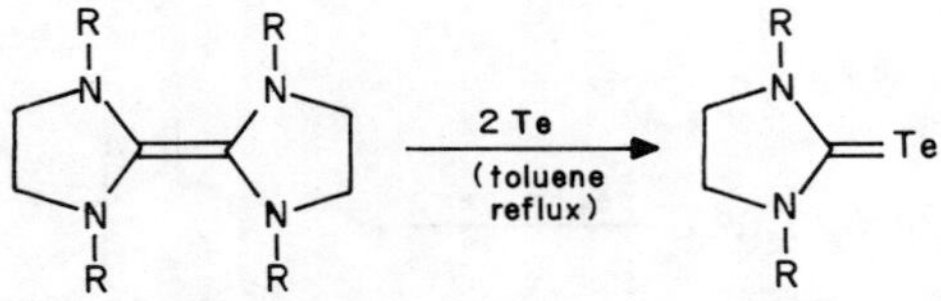

(167)

(161) **(162)**

R=Me, Et

Metal complexes of these telluroureas have been prepared (Section XII, equation 175)[227]. The authors suggest that the facile detelluration of these telluroureas and their metal complexes may lead to their use in synthesis as 'masked' nucleophilic carbenes, amines and acylcarbanions.

XII. METAL AND RESONANCE STABILIZED Se/Te DERIVATIVES

Many otherwise unstable seleno- and telluro-carbonyl compounds can be isolated as metal complexes. Complexes of both seleno- and telluro-formaldehyde have recently been isolated using a variety of methods. These include reactions of sodium hydrogen selenide or telluride with diiodo-rhodium and -osmium complexes (equation 168)[228,229], and carbon diselenide addition to the osmium–formaldehyde complex, followed by extrusion of carbonyl selenide (equations 169 and 170)[230].

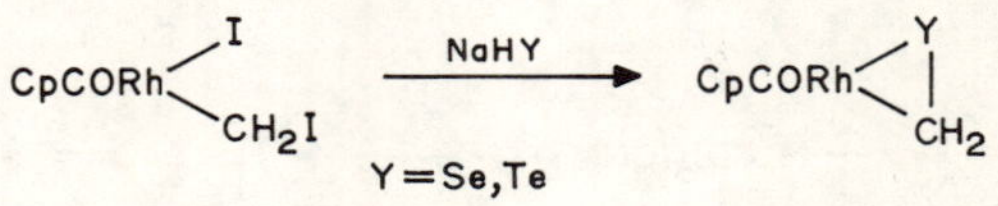

(168)

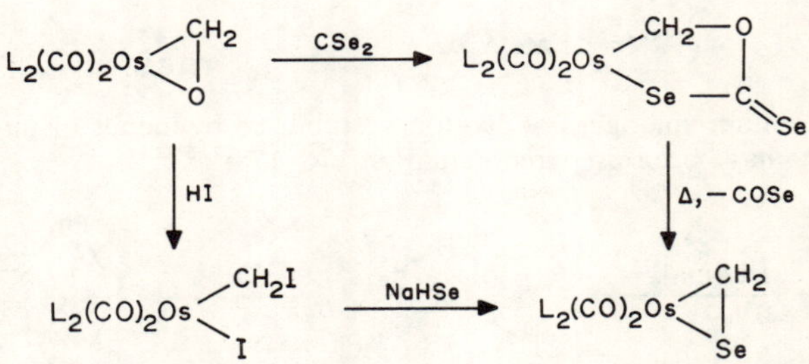

(169)

(170)

Alternatively, addition of diazomethane to organometallic manganese complexes of selenium and tellurium affords the seleno- and telluro-formaldehyde derivatives (equations 171 and 172)[231].

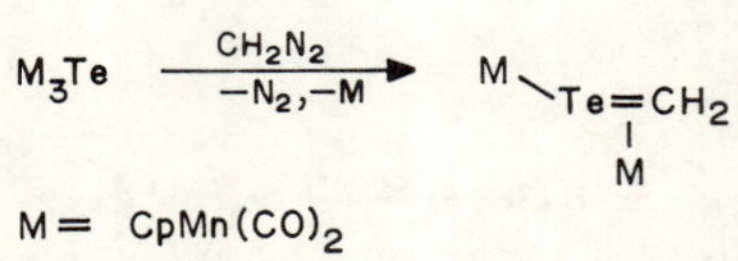

(171)

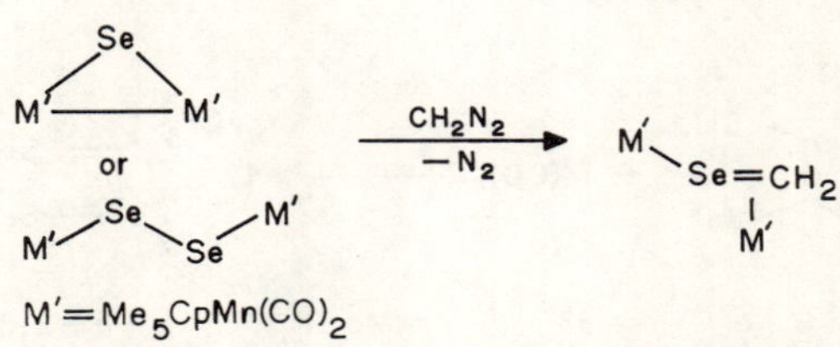

(172)

Chromium and tungsten complexes of selenobenzaldehyde[232], selenobenzophenone[233] and tellurobenzophenone[234] have been prepared by carbenoid abstraction of selenium or tellurium from seleno- and telluro-cyanates (equations 173 and 174). The complex bonding in the selenobenzaldehyde case has been studied[232]. A number of stable metal complexes of the telluroureas **163** have been prepared (equation 175)[227].

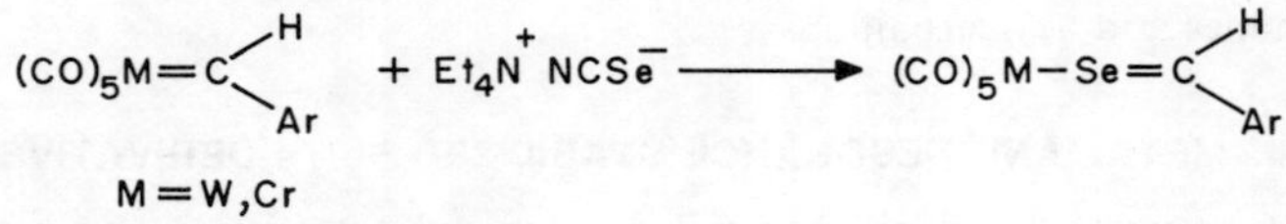

M = W, Cr (173)

$$(CO)_5M{=}CPh_2 + Y{=}C{=}N^- \longrightarrow (CO)_5M{-}Y{=}CPh_2 \qquad Y = Se, Te \tag{174}$$

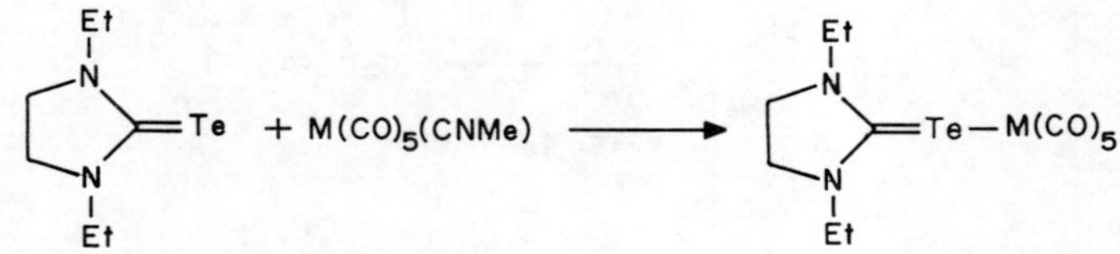

M = Cr, Mo, W **(163)** (175)

A number of selenium analogues of diketones, stabilized by metals or intramolecular hydrogen bonding, have been prepared (equation 176–178)[235–237].

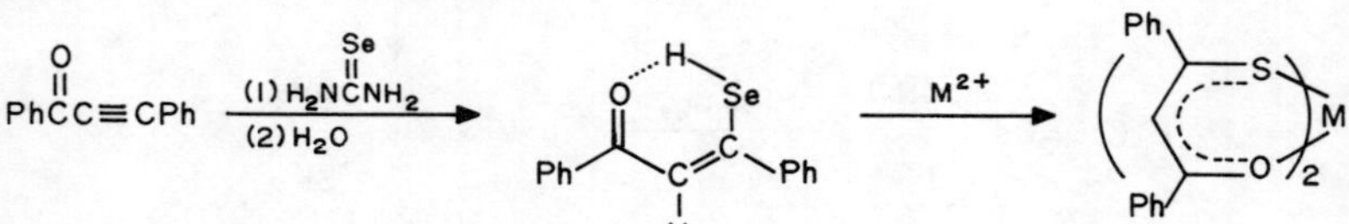

M = Zn, Cd, Hg (176)

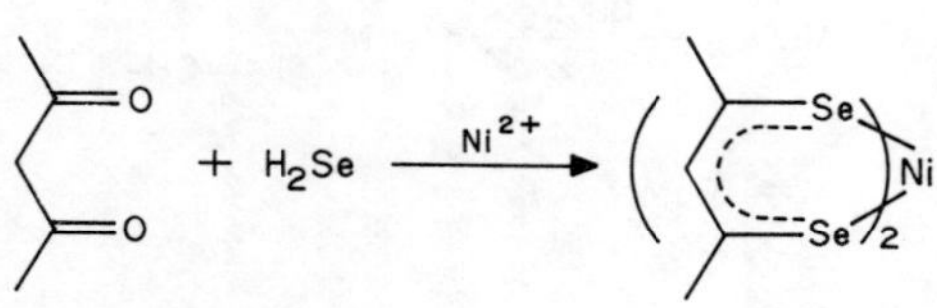

(177)

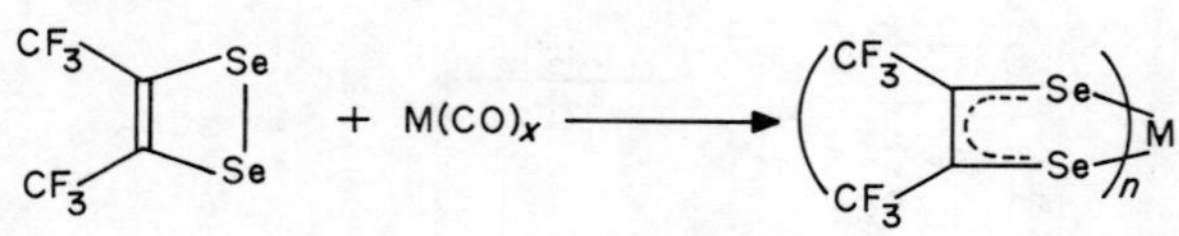

M = Mo, Ni, W $n = 2, 3$ (178)

The especially well stabilized selenolate **164** can be prepared by hydrogen selenide addition to a bromocyclobutenedione[238]. Alkylation and protonation occur at selenium while aniline addition occurs at a carbonyl center (equation 179).

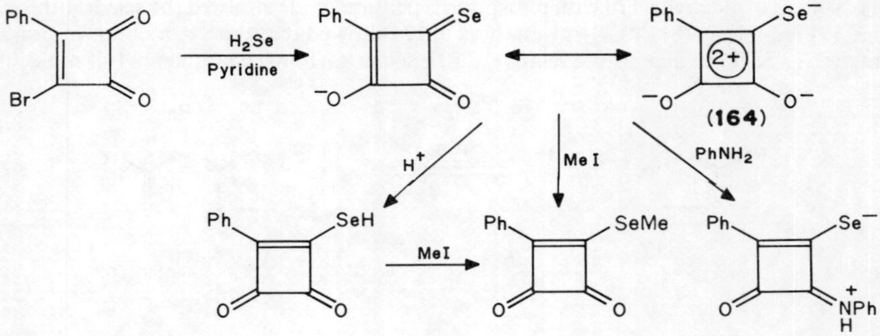

(179)

A number of selenium compounds have been obtained which formally contain selenocarbonyl groups stabilized by 'no-bond resonance'. Treatment of the selenopyrone **165** with sodium selenide affords a mixture of the dieneselenol **166** and its oxidation product **167**. Treatment of this material with phosphorus pentasulfide afforded a mixture of selenothiophthenes **168** and **169** (equation 180)[239–241].

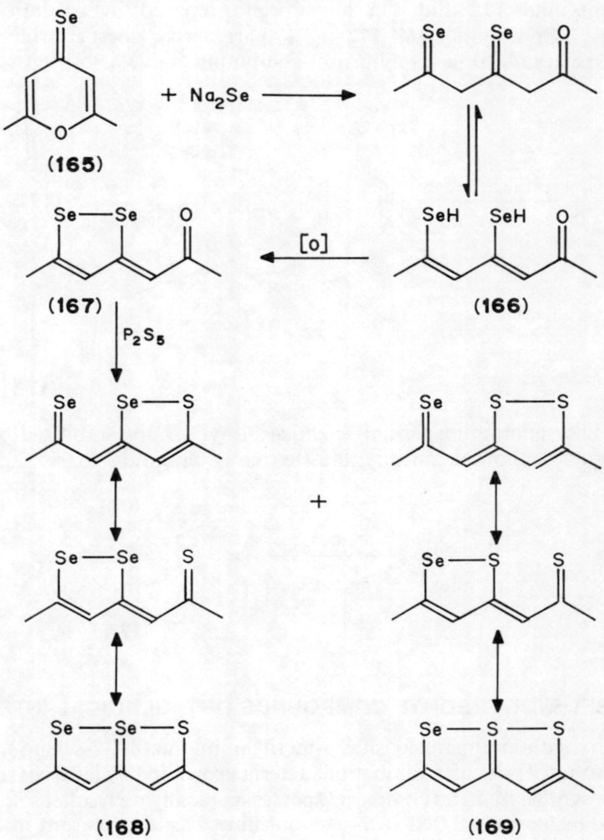

(180)

Treatment of aldehyde **170** with phosphorus pentaselenide afforded the selenodithiophthene **171** (equation 181)[242]. X-ray analysis of **171** showed that the S—S bond was longer than the S—Se bond, suggesting relatively little selenocarbonyl character in this molecule.

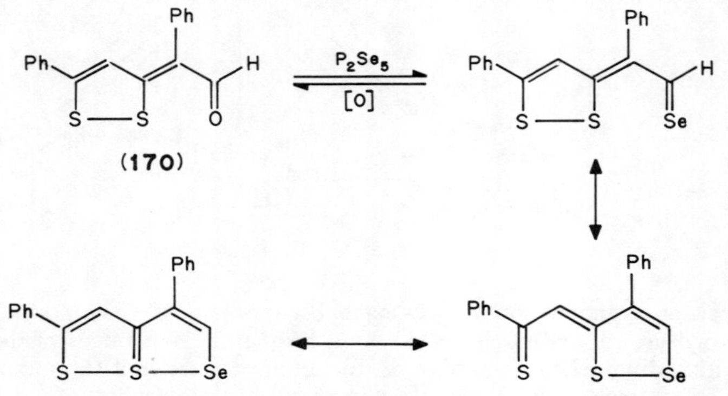

(170) (171) (181)

Related compounds **172** and **173** have been prepared by oxidation of dianions (equation 182)[185]. Compounds **168**, **172** and **173** are perhaps best considered as hypervalent selenium species. Analogous tellurium compounds have also been reported[243].

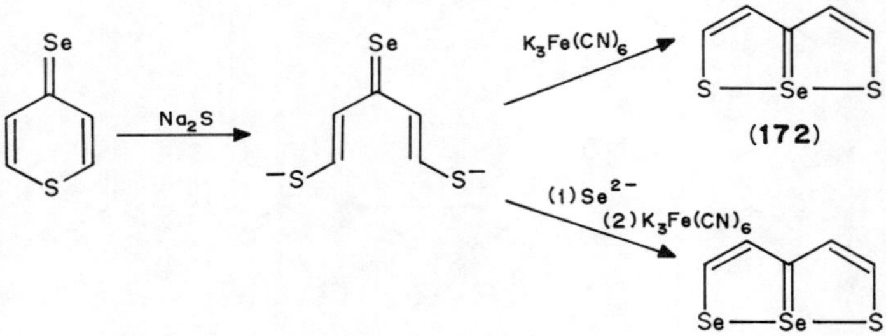

(172) (173) (182)

Other molecules containing formal selenocarbonyl groups stabilized by metals (**174**) have been produced photochemically and thermally (equation 183)[244–246].

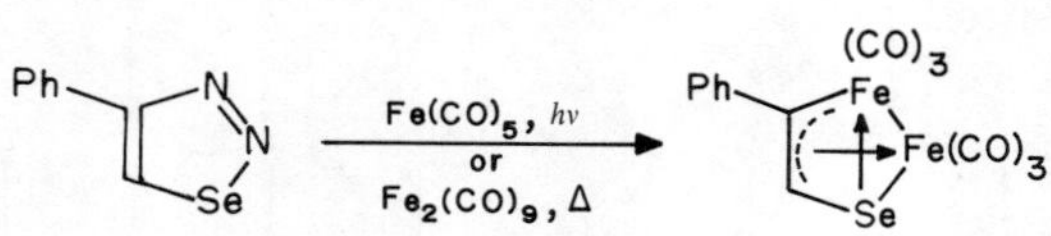

(174) (183)

XIII. SELENOCARBONYL COMPOUNDS OF BIOLOGICAL INTEREST

A selenocarbonyl-containing nucleoside, 5-methylaminomethyl-2-selenouridine (**175**) has been isolated from t-RNAs of a number of bacteria grown in low levels of selenite[247]. This compound was synthesized and compared spectroscopically with other selenonucleosides. Another selenonucleoside related to 2-selenouridine was also present in these bacterial t-RNAs but could not be identified. There is also some evidence for naturally occurring 4-selenouridine[248,249].

Other selenocarbonyl analogues of compounds of biochemical and medicinal interest have also been prepared. A number of selone nucleoside derivatives have been prepared and evaluated as antineoplastic agents. Treatment of 2-aminoadenosine with hydrogen selenide in aqueous pyridine led to displacement of a heterocyclic amine group affording 6-selenoguanosine (**176**)[250–252]. Other 6-selenopurine nucleoside derivatives could be similarly prepared. A number of other selenocarbonyl-containing purine and pyridine derivatives, including selenocytosine, 5-methylselenocytosine, diselenothymine, 6-selenopurine, 2-selenouracil and 2-selenothymine, have also been prepared and evaluated as purine and pyrimidine antagonists[108,253–256]. Selenosemicarbazones have also been evaluated as antineoplastic agents[257]. A number of selenocarbonyl derivatives have been evaluated as antifungal and antimicrobial agents[255,258,259]. Selenobarbituric acid derivatives have also been prepared[220].

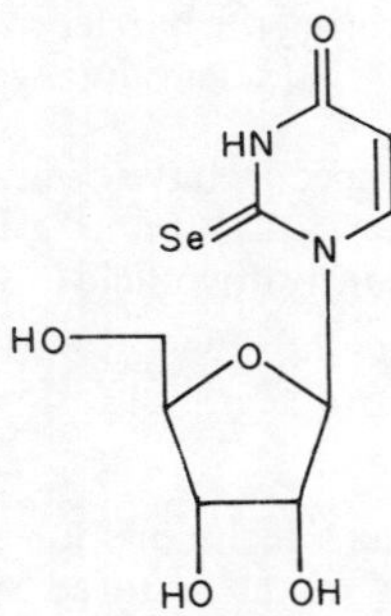

(175)

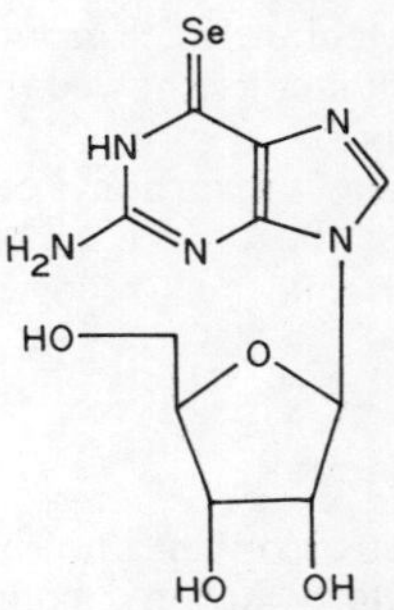

(176)

XIV. SPECTROSCOPIC AND ELECTROCHEMICAL STUDIES OF Se/Te CARBONYL COMPOUNDS

A. NMR Spectra

^{13}C NMR studies of selenocarbonyl compounds show that the selenocarbonyl carbon is shifted significantly downfield relative to the corresponding thiocarbonyl and carbonyl compounds[260]. The selenocarbonyl carbons of selones are reported to be the most deshielded carbons observed in neutral molecules (287–295 ppm)[59,260]. ^{77}Se NMR studies of selenocarbonyl compounds show that the δ ^{77}Se shifts of the selenocarbonyl group are also shifted significantly downfield (ca. 100 ppm relative to dimethyl selenide)[18]. The ^{77}Se shift closely parallels the λ_{max} of the $n \rightarrow \pi^*$ transition of the selenocarbonyl group, and is very sensitive to changes in electronic structure.

The ^{77}Se–^{13}C coupling constants of selenocarbonyl compounds have been determined and found to be much larger than previously observed (209–221 Hz)[18,261]. The ^{13}C isotope effect on ^{77}Se shielding has been determined and correlated with C—Se bond distances for a number of selenocarbonyl compounds[262]. A comparison of the ^{77}Se and ^{77}O chemical shifts of selenocarbonyl and carbonyl compounds has also been made[17]. Deshielding of $\delta(^{77}\text{Se})$ is always accompanied by shielding of $\delta(^{13}\text{C=Se})$, paralleling the change of $\delta(^{77}\text{O})$ versus $\delta(^{13}\text{C=O})$. Both ^{77}Se and ^{17}O shifts are dominated by the local paramagnetic screening term. Based on these observations, the bond order term of the C=Se bond of selones was shown to closely resemble the bond order term of the

C=O bond of ketones, indicating the true double bond nature of the selenocarbonyl group[17].

^1H NMR studies of amide analogues show an increased barrier to rotation about the C—N bond on going from amide to thioamide to selenoamide[263,264]. This indicates an increased contribution of the dipolar resonance form **177b** in the order O < S < Se. This result is consistent with ^{14}N and ^{13}C NMR studies of these compounds[260,265]. Similar results were observed in the case of selenosemicarbazides[264].

(177a) (177b)

X=O,S,Se

Studies of tetramethylselenourea indicate that there is free rotation about the C—N bond in this compound even at −120 °C[263]. This has been explained by the increased steric effect of selenium in the ground state effectively lowering the barrier to rotation. ^{14}N decoupled NMR studies showed that the rotational barrier of selenourea was the same as that of thiourea[266].

In the case of tellurocarbonyl compounds, ^{13}C NMR spectra have been reported only for telluroesters, the C=Te carbon resonance being at ca. 229 ppm[16]. The *N*-methyl groups of telluroamides are reported to be shifted significantly downfield (δ 3.6–3.75)[116] in the ^1H NMR.

B. Infrared Spectra

Infrared spectroscopy has not been especially useful as a technique for characterizing seleno- and telluro-carbonyl compounds. Comparison of the infrared spectra of telluroamides and tellurohydrazides with those of the corresponding selenium compounds showed them to be almost superimposable[16]. Relatively little change had previously been reported in a comparison of the infrared spectra of thiolactams and selenolactams. Reasons for the difficulties in determining the C=Se (and hence C=Te) frequencies have been discussed in detail[267]. Suffice it to say that the C=Se and C=Te absorption bands would be expected to be much weaker than C=O bands. Owing to significant resonance interactions in the compounds studied (amides, hydrazides and semicarbazide derivatives), the main absorptions due to the seleno- and telluro-carbonyl groups would be in the C—Se and C—Te single-bonded regions. In addition, coupling of these bonds with other vibrations would be expected, and would complicate detailed interpretations.

C. Ultraviolet and Visible Spectra

The ultraviolet and visible spectra of a number of selenocarbonyl and tellurocarbonyl compounds have been reported. Selones are typically deep blue compounds which exhibit an $n \rightarrow \pi^*$ transition in the visible region ($\lambda_{max} \approx 600$–$700$ nm, $\varepsilon \approx 20$–$40\ l\ mol^{-1}\ cm^{-1}$)[18,268]. An intense $\pi \rightarrow \pi^*$ transition ($\lambda_{max} \approx 270$ nm, $\varepsilon \approx 10^4\ l\ mol^{-1}\ cm^{-1}$) as well as a presumed $n \rightarrow \sigma^*$ transition ($\lambda_{max} \approx 230$ nm, $\varepsilon \approx 10^3\ l\ mol^{-1}\ cm^{-1}$) occur in the ultraviolet region. The long-wavelength band is dominated by a singlet-triplet component[269]. The UV–visible data for selenofenchone, thiofenchone and fenchone have been compared[268,269].

The ultraviolet spectra of methyl phenylacetate and its thione and selone analogs have been compared, with a red shift observed on going from the ester ($\lambda_{max} = 214$ nm, $\varepsilon = 6.0 \times 10^3\ l\ mol^{-1}\ cm^{-1}$) to the thione ester $\lambda_{max} = 240$ nm, $\varepsilon = 7.1 \times 10^3\ l\ mol^{-1}\ cm^{-1}$) to the selone ester ($\lambda_{max} = 275$ nm, $\varepsilon = 6.7 \times 10^3\ l\ mol^{-1}\ cm^{-1}$)[84]. Similar shifts have been observed for the amide, semicarbazide and urea derivatives[270]. This trend has been explained by a decrease in the energy difference between ground and excited states, due to an enhanced contribution of the dipolar resonance species **177b**[108,165,220] (cf. NMR Spectra, Section XIV.A, and Dipole Moment Studies, Section XIV.E).

Telluro-amides and -hydrazides, although much less well studied than their selenium analogs, also show red shifts of both the $\pi \rightarrow \pi^*$ and $n \rightarrow \pi^*$ bonds relative to the selenium analogues [For X = Te ($\lambda_{max} = 375$ nm, $\varepsilon = 10^4$ mol^{-1} cm^{-1}; $\lambda_{max} = 540$ nm, $\varepsilon = 10^3$ l mol^{-1} cm^{-1}), X = Se ($\lambda_{max} = 316$ nm, $\varepsilon = 10^4$ l mol^{-1} cm^{-1}; $\lambda_{max} = 442$ nm, $\varepsilon = 4 \times 10^2$ l mol^{-1} cm^{-1})][116]. Telluroesters exhibit maxima at ca. 240 nm ($\varepsilon \approx 4 \times 10^3$ l mol^{-1} cm^{-1}), ca. 265 nm(sh) ($\varepsilon = 10^3$ l mol^{-1} cm^{-1}), 345 nm ($\varepsilon \approx 7.5 \times 10^3$ l mol^{-1} cm^{-1}) and ca. 590 nm ($\varepsilon \approx 3 \times 10^2$ l mol^{-1} cm^{-1}).

D. Mass Spectra

Comparisons of the mass spectral behavior of selenoureas[271] and selones[272] with their sulfur analogs have been made. Data allowing a similar comparison for selenoamides and thioamides have also been published[96]. The molecular ion abundance was less for selenocarbonyl compounds than for their thiocarbonyl analogs, yet in general, thiocarbonyl compounds give more intense molecular ions than their oxygen analogs. Fragments containing multiply bonded selenium were absent, or of much lower intensity than the corresponding sulfur fragments. It is likely that the greater ability of selenium to stabilize a positive charge is more than counterbalanced by the lowered stability of multiply bonded selenium in cationic species, accounting for the differences observed in the oxygen, sulfur and selenium spectra[272]. Similar conclusions resulted from mass spectral studies of furan, thiophene, selenophene and tellurophene[273].

E. Dipole Moment Studies

A comparison of dipole moments of fenchone derivatives (**178**) has been carried out[21]. The carbonyl moieties of these molecules are not stabilized by resonance and may be considered as 'pure' carbon—chalcogen double bonds. The results are consistent with a decreased, but not reversed, polarity of the chalcogen—carbon bond in the order O > S > Se. In contrast, in the resonance-stabilized pyridone and phthalide series, dipole moments increased in the order O < S < Se, presumably owing to increased dipolar resonance contributions in the thione and selone cases[274].

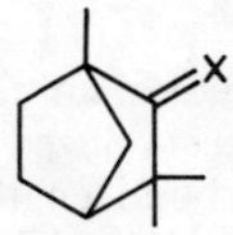

(178)

X=O,S,Se

F. Chiroptical Properties

The CD and ORD spectra of chiral selenofenchone have been reported[268,269]. The chiroptical properties of (−)-fenchone and the thione and selone derived from this ketone are parallel, showing long-wavelength negative Cotton effects. The long-wavelength CD and UV bonds in selenofenchone are well separated. The magnetic circular dichroism spectrum of racemic selenofenchone has also been reported and discussed in terms of a dominant singlet–triplet component in the $n \rightarrow \pi^*$ transition[269].

G. Electrochemical Studies

In a comparison of the reductions of *O*-methyl selenobenzoate and the corresponding thione benzoate using zinc and hydrochloric acid, the selenium compound was more easily reduced[275]. Polarographic studies of the reduction of these esters and the corresponding imidate indicate that the order of ease of reduction is C=Se > C=NH > C=S > C=O. A comparison of the reductions of selenobenzamide and thiobenzamide was consistent with this series.

The electrochemical reductions of di-*t*-butyl selone and 1, 1, 3, 3-tetramethylindane-2-selone to the corresponding radical anions have been carried out, and the half-wave potentials for these reactions have been determined[276,277]. The reduction to the radical anion of the indaneselone is electrochemically reversible. The reduction product was isolated as the corresponding diselenide.

H. Photoelectron Spectra

A comparison of the photoelectron spectra of pyridone derivatives (**179**) has been reported[278]. Replacement of oxygen by sulfur or selenium affects both the HOMO energy and electron distribution. In the latter cases the HOMO is concentrated on the chalcogen atom, paralleling the increased nucleophilicity of the chalcogen. These results were consistent with ^{13}C NMR studies of these compounds, and also with calculations describing similar compounds[279,280].

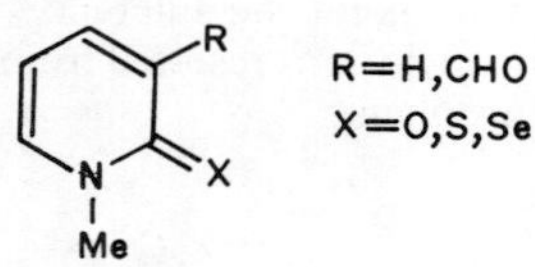

(179)

The photoelectron spectra of the unstable selenoformaldehyde, selenoactaldehyde and selenocarbonyl difluoride have been obtained by 'computer spectra stripping'[80]. The photoelectron spectrum of selenoformaldehyde compared favorably with its calculated spectrum.

I. X-Ray and Microwave Studies

A limited number of structural studies on selenocarbonyl compounds have been reported[281,282]. As yet, no data on an unstabilized selenocarbonyl moiety (e.g. in a selone) have appeared. The selenium—carbon bond length in resonance-stabilized selenoureas is typically in the range 1.82–1.89 Å[283–285]. The selenocarboxylate of dipotassium diselenooxalate has a C—Se bond distance of 1.87 Å[127]. These bonds are probably intermediate between carbon—selenium single and double bonds. In diselenouracil (**180**) two widely differing C—Se bond lengths were observed[286]. In addition to the usual selenourea-like C—Se bond at the 2-position of uracil there is a very long C—Se bond at the 4-position. This lengthening has been explained by an enhanced single bond character due to resonance.

Hydrogen bonding in selenocarbonyl compounds has been reported for selenourea (Se···HN bond distance of 3.51 Å)[285], 1-benzoyl-3-phenyl-2-selenourea (3.83 Å)[283] and diselenouracil (3.47 and 3.75 Å)[286]. No hydrogen bonds were observed in 1-acetyl-3-phenyl-2-selenourea[284].

Structural studies of metal complexes of selenocarbonyl compounds have also

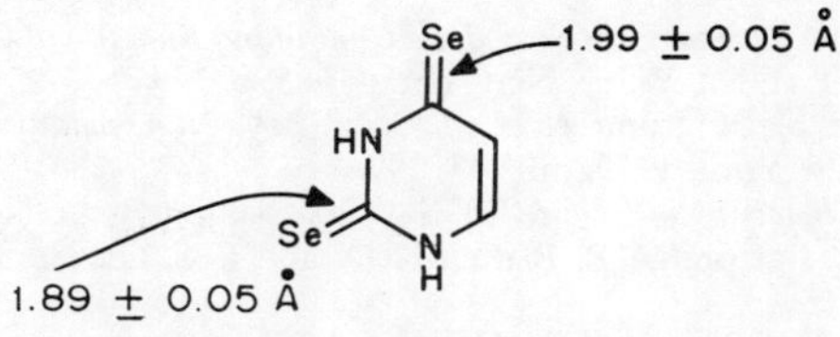

(180)

appeared. Nickel bisdiethylaminodiselenocarbonate exhibits two unequal carbon—selenium bond lengths (1.84 and 1.97 Å)[287].

A bisisoselenocyanato nickel complex (**181**) has a carbon—selenium bond length of 1.71 Å[288], in line with that expected for a C—Se double bond ca. 1.74 Å, see Table 3). Microwave studies of selenoketene indicate a carbon—selenium double bond length of about 1.70 Å[128].

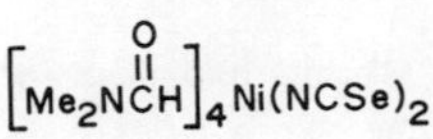

(181)

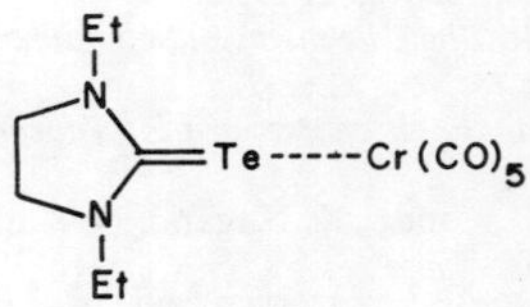

(182)

An X-ray structure of a tellurourea derivative (**182**) with the tellurocarbonyl tellurium complexed to chromium exhibits a carbon—tellurium bond length, 2.12 Å, typical of a carbon—tellurium single bond[227]. This is consistent with stabilization of the tellurocarbonyl moiety due to a lowering of the carbon—tellurium bond order by resonance and interaction with the metal.

XV. ACKNOWLEDGEMENTS

I gratefully acknowledge many talented coworkers in my research on organoselenium and organotellurium chemistry. I especially thank Lynn J. SanFilippo and Loide M. Wasmund for helpful comments and suggestions in the preparation of the manuscript.

XVI. REFERENCES

1. D. L. Klayman and W. H. H. Gunther (Eds.), *Organic Selenium Compounds: Their Chemistry and Biology*, Wiley–Interscience, New York, 1973.
2. R. A. Zingaro and W. C. Cooper (Eds.), *Selenium*, Van Nostrand Reinhold, New York, 1974.
3. R. B. Silverman, in Ref., 1 p. 245.
4. K. A. Jensen, in Ref. 1, p. 263.
5. R. J. Shine, in Ref.1, p. 273.
6. H. G. Mautner, in Ref. 1, p. 497.
7. K. J. Irgolic and M. V. Kuchadker, in Ref. 2, p. 408.
8. K. J. Irgolic, *The Organic Chemistry of Tellurium*, Gordon and Breach, New York, 1974, pp. 52–55, 242–243.
9. W. Cooper, *Tellurium*, Van Nostrand, New York, 1971, p. 255.
10. D. Reid, *Organic Chemistry of Sulphur, Selenium and Tellurium*, Vols. 1–6, Royal Society of Chemistry, London, 1970–81.
11. P. D. Magnus, in *Comprehensive Organic Chemistry* (Eds. D. H. R. Barton and W. D. Ollis), Vol. 3 (Ed. D. N. Jones), Pergamon Press, Oxford, 1979, p. 489.

12. M. Renson, in *Organic Chemistry of Se and Te Containing Functional Groups* (Eds. S. Patai and Z. Rappaport), Vol. 1, (Eds.), Wiley, New York, 1985, p. 399.
13. International Union of Pure and Applied Chemistry, *Nomenclature of Organic Chemistry*, Pergamon Press, New York, 1979, pp. 247, 299.
14. A. Ohno, *Organic Chemistry of Sulfur* (Ed. S. Oae), Plenum Press, New York, 1977, p. 189.
15. T. G. Back, D. H. R. Barton, M. R. Britten-Kelly and F. S. Guziec, Jr., *J. Chem. Soc., Chem. Commun.*, 539 (1975).
16. A. G. Barrett, D. H. R. Barton and R. W. Read, *J. Chem. Soc., Chem. Commun.*, 645 (1979).
17. T. C. Wong, F. S. Guziec, Jr., and C. A. Moustakis, *J. Chem. Soc., Perkin Trans. 2*, 1471 (1983).
18. E. R. Cullen, F. S. Guziec, Jr., C. J. Murphy, T. C. Wong and K. K. Andersen, *J. Am. Chem. Soc.*, **103**, 7055 (1981).
19. L. Pauling, *The Nature of the Chemical Bond*, 3rd ed., Cornell University Press, Ithaca, N.Y., 1960, p. 224.
20. J. Hinze and H. H. Jaffe, *J. Am. Chem. Soc.*, **84**, 540 (1962); *J. Phys. Chem.*, **67**, 1501 (1963).
21. F. S. Guziec, Jr., L. J. SanFilippo, C. J. Murphy, C. A. Moustakis and E. R. Cullen, *Tetrahedron*, **41**, 4843 (1985).
22. W. H. Günther, in Ref. 1, p. 35.
23. K. W. Bagnall, *The Chemistry of Selenium, Tellurium and Polonium*, Elsevier, Amsterdam, 1966, pp. 10, 45.
24. F. Fehler, *Handbook of Preparative Inorganic Chemistry*, Vol. I, Academic Press, New York, 1963, p. 418.
25. N. Sonoda, K. Kondo, K. Nagano, N. Kambe and F. Morimoto, *Angew. Chem., Int. Ed. Engl.*, **19**, 308 (1980).
26. D. L. Klayman and T. S. Griffin, *J. Am. Chem. Soc.*, **95**, 197 (1973).
27. F. Fehler, in Ref. 24, p. 421.
28. J. E. Drake and B. M. Glavincevski, *Inorg. Synth.*, **20**, 171 (1980).
29. J. A. Gladysz, J. L. Hornby and J. E. Garbe, *J. Org. Chem.*, **43**, 1205 (1978).
30. J. S. Warner, *J. Org. Chem.*, **28**, 1642 (1963).
31. L. Henriksen and E. S. Kristiansen, *Int. J. Sulfur Chem. Part A*, **2**, 133 (1972).
32. W. -H. Pan, J. P. Fackler and H. W. Chen, *Inorg. Chem.*, **20**, 856 (1981).
33. W. H. Gunther, in Ref. 1, p. 44.
34. F. S. Guziec, Jr., and C. A. Moustakis, *J. Org. Chem.*, **49**, 189 (1984).
35. K. Steliou and M. Mrani, *J. Am. Chem. Soc.*, **104**, 3104 (1982).
36. N. Kambe, K. Kondo, S. Morita, S. Murai and N. Sonoda, *Angew. Chem., Int. Ed. Engl.*, **19**, 1009 (1980).
37. M. R. Detty and M. D. Seidler, *J. Org. Chem.*, **47**, 1354 (1982).
38. D. H. R. Barton and S. W. McCombie, *J. Chem. Soc., Perkin Trans. 1*, 1574 (1975).
39. A. Schröder, *Chem. Ber.*, **4**, 400 (1871).
40. G. Cole, *Chem. Ber.*, **8**, 1165 (1875).
41. L. Vanino and A. Schinner, *J. Prakt. Chem.*, **91**, 116 (1915).
42. R. E. Lyons and W. E. Brandt, *Chem. Ber.*, **60**, 824 (1927).
43. R. E. Lyons and E. D. Scudder, *Chem. Ber.*, **64**, 530 (1931).
44. R. Kuhn, *J. Chem. Soc.*, 605 (1938).
45. D. S. Margolis and R. W. Pittman, *J. Chem. Soc.*, 799 (1957).
46. H. J. Bridger and R. W. Pittman, *J. Chem. Soc.*, 1371 (1950).
47. L. Mortillaro, L. Credali, M. Mammi, and G. Valle, *J. Chem. Soc.*, 807 (1965).
48. L. Credali, M. Russo, L. Mortillaro, C. De Checchi, G. Valle and M. Mammi, *J. Chem. Soc. B*, 117 (1967).
49. L. Mortillaro, L. Credali, M. Russo and C. De Checchi, *J. Polym. Sci., Part B*, **3**, 581 (1965).
50. M. Russo, L. Mortillaro, L. Credali and C. De Checchi, *J. Polym. Sci., Part A1*, **4**, 248 (1966).
51. L. Szperl and W. Wiorogorski, *Rocz. Chem.*, **12**, 270 (1932).
52. T. G. Back, D. H. R. Barton, M. R. Britten-Kelly and F. S. Guziec, Jr., *J. Chem. Soc., Perkin Trans. 1*, 2079 (1976).
53. F. S. Guziec, Jr., and C. J. Murphy, *J. Org. Chem.*, **45**, 2890 (1980).
54. E. R. Cullen, F. S. Guziec, Jr., M. I. Hollander and C. J. Murphy, *Tetrahedron Lett.*, 4563 (1981).
55. E. R. Cullen, F. S. Guziec, Jr., and C. J. Murphy, *J. Org. Chem.*, **47**, 3563 (1982).
56. P. de Mayo, G. L. R. Petrasiunas and A. C. Weedon, *Tetrahedron Lett.*, 4621 (1978).
57. N. Y. M. Fung, *MSc Thesis*, University of Western Ontario, 1979.

58. R. Okazaki, A. Ishii, N. Fukuda, H. Oyama, and N. Inamoto, *J. Chem. Soc., Chem. Commun.*, 1187 (1982).
59. C.-P. Klages and J. Voss, *Angew. Chem., Int. Ed. Engl.*, **16**, 725 (1977).
60. G. A. Olah, T. Nakajima and G. K. Surya Prakash, *Angew. Chem., Int. Ed. Engl.*, **19**, 811 (1980).
61. F. S. Guziec, Jr., C. J. Murphy and E. R. Cullen, *J. Chem. Soc., Perkin Trans. 1*, 107 (1985).
62. F. S. Guziec, Jr., and C. A. Moustakis, *J. Chem. Soc., Chem. Commun.*, 63 (1984).
63. C. A. Moustakis and F. S. Guziec, Jr., American Chemical Society 187th National Meeting, St. Louis, Mo, April 1984, Abstract 135.
64. R. Okazaki, A. Ishii and N. Inamoto, *Tetrahedron Lett.*, **25**, 5147 (1984).
65. F. S. Guziec, Jr., C. A. Moustakis and L. J. SanFilippo, submitted for publication.
66. J. C. Scaiano and K. U. Ingold, *J. Chem. Soc., Chem. Commun.*, 205 (1976).
67. J. C. Scaiano and K. U. Ingold, *J. Phys. Chem.*, **80**, 1901 (1976).
68. J. C. Scaiano and K. U. Ingold, *J. Am. Chem. Soc.*, **99**, 2079 (1977).
69. J. C. Scaiano, *J. Am. Chem. Soc.*, **99**, 1494 (1977).
70. B. J. McKinnon, P. de Mayo, N. C. Payne and B. Ruge, *Nouv. J. Chim.*, **2**, 91 (1978).
71. N. Y. M. Fung, P. de Mayo, B. Ruge, A. C. Weedon and S. K. Wong, *Can, J. Chem.*, **58**, 6 (1980).
72. D. H. R. Barton and F. S. Guziec, Jr., unpublished work.
73. F. O. Rice and A. L. Glasebrook, *J. Am. Chem. Soc.*, **56**, 2381 (1934).
74. T. G. Pearson, R. H. Purcell and G. S. Saigh, *J. Chem. Soc.*, 409 (1938).
75. F. D. Williams and F. X. Dunbar, *J. Chem. Soc., Chem. Commun.*, 459 (1968).
76. W. H. Gunther, *J. Org. Chem.*, **32**, 3929 (1967).
77. V. I. Cohen, *J. Org. Chem.*, **42**, 2510 (1977).
78. J. W. Lewicki, W. H. Gunther and J. Y. Chu, *J. Org. Chem.*, **43**, 2672 (1978).
79. R. H. Judge and D. C. Moule, *J. Am. Chem. Soc.*, **106**, 5406 (1984).
80. H. Bock, S. Aygen, P. Rosmus, B. Solouki and E. Weissflog, *Chem. Ber.*, **117**, 187 (1984).
81. W. W. Sander and O. L. Chapman, *J. Org. Chem.*, **50**, 543 (1985).
82. C. Collard-Charon and M. Renson, *Bull. Soc. Chim. Belg.*, **71**, 563 (1962).
83. V. I. Cohen, *J. Org. Chem.*, **42**, 2645 (1977); *J. Heterocycl. Chem.*, **15**, 237 (1978); **15**, 1113 (1978); **16**, 365 (1979).
84. F. Malek-Yazdi and M. Yalpani, *J. Org. Chem.*, **41**, 729 (1976).
85. E. O. Fisher and S. Riedmüller, *Chem. Ber.*, **107**, 915 (1974).
86. M. Renson and R. Collienne, *Bull. Soc. Chim. Belg.*, **73**, 491 (1964).
87. I. Wallmark, M. H. Krackov, S.-H. Chu and H. G. Mautner, *J. Am. Chem. Soc.*, **92**, 4447 (1970).
88. K. A. Jensen, *Q. Rep. Sulfur Chem.*, **5**, 45 (1970).
89. L. Henriksen, *Synthesis*, 501 (1974).
90. D. H. R. Barton, P. E. Hansen and K. Picker, *J. Chem. Soc., Perkin Trans. 1*, 1723 (1977).
91. P.-E. Hansen, *J. Chem. Soc., Perkin Trans. 1*, 1627 (1980).
92. A. G. Barrett, R. W. Read and D. H. R. Barton, *J. Chem. Soc., Perkin Trans. 1*, 2184 (1980).
93. A. G. Barrett, R. W. Read and D. H. R. Barton, *J. Chem. Soc., Perkin Trans. 1*, 2191 (1980).
94. F. von Dechend, *Chem. Ber.*, **7**, 1273 (1874).
95. W. Becker and J. Meyer, *Chem. Ber.*, **37**, 2550 (1904); V. I. Cohen, *Synthesis*, 768 (1978).
96. V. I. Cohen, *Synthesis*, 668 (1978).
97. A. Ogawa, J.-I. Miyake, Y. Karasaki, S. Murai and N. Sonada, *J. Org. Chem.*, **50**, 384 (1985).
98. F. Malek-Yazdi and M. Yalpani, *Synthesis*, 328 (1977).
99. R. S. Sukhai, R. DeJong and L. Brandsma, *Synthesis*, 888 (1977).
100. M. Tisler, B. S. Tanovnik, Z. Zrimsek and C. Stropnik, *Synthesis*, 299 (1981).
101. D. H. Reid, R. G. Webster and S. McKenzie, *J. Chem. Soc., Perkin Trans. 1*, 2334 (1979).
102. M. P. Cava and L. E. Saris, *J. Chem. Soc., Chem. Commun.*, 617 (1975).
103. H. Hartmann, *Z. Chem.*, **11**, 60 (1971).
104. J. Liebscher and H. Hartmann, *Synthesis*, 521 (1976).
105. J. Becher, E. G. Frandsen, C. Dreier and L. Henriksen, *Acta Chem. Scand., Ser. B*, **31**, 843 (1977).
106. R. V. Kendall and R. A. Olofson, *J. Org. Chem.*, **35**, 806 (1970).
107. *Ger. Pat.*, 910, 199 (1954); *Chem. Abstr.*, **53**, 936h (1959).
108. H. G. Mautner, *J. Am. Chem. Soc.*, **78**, 5293 (1956).
109. F. Asinger, H. Berding and H. Offermanns, *Monatsh. Chem.*, **99**, 2072 (1968).
110. K. A. Jensen and P. H. Nielsen, *Acta Chem. Scand.*, **20**, 597 (1966).
111. C. O. Maese, W. Walter, H. Mrotzek and H. Mirzai, *Chem. Ber.*, **109**, 956 (1976).
112. J. Liebscher and H. Hartmann, *Tetrahedron*, **33**, 731 (1977).

113. K. A. Jensen and L. Henriksen, *Acta Chem. Scand.*, **21**, 1991 (1967).
114. R. S. Sukhai and L. Brandsma, *Synthesis*, 455 (1979).
115. A. Ogawa, J. -I. Miyake, S. Murai and N. Sonada, *Tetrahedron Lett.*, **26**, 669 (1985).
116. K. A. Lerstrup and L. Henriksen, *J. Chem. Soc., Chem. Commun.*, 1102 (1979).
117. Q. Mingoia, *Gazz. Chim. Ital.*, **56**, 835 (1926).
118. D. T. Lewis, *J. Chem. Soc.*, 831 (1940).
119. K. A. Jensen, L. Boje and L. Henriksen, *Acta Chem. Scand.*, **26**, 1465 (1972).
120. H. Ishihara and Y. Hirabayashi, *Chem. Lett.*, 203 (1976).
121. H. Ishihara, S. Sato and Y. Hirabayashi, *Bull. Chem. Soc. Jpn.*, **50**, 3007 (1977).
122. H. Ishihara, S. Muto and S. Kato, *Synthesis*, 128 (1986).
123. H. Ishihara and S. Kato, *Tetrahedron Lett.*, 3751 (1972).
124. K. A. Jensen, J. B. Carlsen, A. Holm and P. H. Nielsen, *Acta Chem. Scand.*, **17**, 550 (1963).
125. K. A. Jensen, in Ref. 1, p. 268.
126. K. A. Jensen and L. Henriksen, *Acta Chem. Scand.*, **24**, 3213 (1970).
127. C. Matz and R. Mattes, *Z. Naturforsch., Teil B*, **33**, 461 (1978).
128. B. Bak, O. J. Nielsen, H. Svanholt and A. Holm, *Chem. Phys. Lett.*, **55**, 36 (1978).
129. B. Bak, O. J. Nielsen, H. Svanholt and A. Holm, *Chem. Phys. Lett.*, **53**, 374 (1978).
130. A. Holm, C. Berg, C. Bjerre, B. Bak and H. Svanholt, *J. Chem. Soc., Chem. Commun.*, 99 (1979).
131. I. Lalezari, A. Shafiee and M. Yalpani, *Tetrahedron Lett.*, 5105 (1969).
132. H. Meier and I. Menzel, *Tetrahedron Lett.*, 445 (1972).
133. I. Lalezari, A. Shafiee and M. Yalpani, *J. Org. Chem.*, **38**, 338 (1973).
134. M. H. Ghandehari, D. Davalion, M. Yalpani and M. H. Partovi, *J. Org. Chem.*, **39**, 3906 (1974).
135. R. S. Sukhai and L. Brandsma, *Recl. Trav. Chim. Pays-Bas*, **98**, 55 (1979).
136. E. Schaumann and F. F. Grabley, *Tetrahedron Lett.*, **21**, 4251 (1980).
137. M. L. Petrov, V. Z. Laishev and A. A. Petrov, *Zh. Org. Khim.*, **15**, 2596 (1979).
138. S. L. Bender, N. F. Haley and H. R. Luss, *Tetrahedron Lett.*, **22**, 1495 (1981).
139. M. V. Lakshmikanthan, M. P. Cava, M. Albeck, L. Engman, F. Wudl and E. Ahran-Shalom, *J. Chem. Soc., Chem. Commun.*, 829 (1981).
140. M. V. Lakshmikanthan, M. P. Cava, M. Albeck, L. Engman, P. Carroll, J. Bergman and F. Wudl, *Tetrahedron Lett.*, **22**, 4199 (1981).
141. E. Bulka and K. -D. Ahlers, *Z. Chem.*, **3**, 348 (1963).
142. C. Collard-Charon and M. Renson, *Bull. Soc. Chim. Belg.*, **71**, 531 (1962).
143. I. B. Douglass, *J. Am. Chem. Soc.*, **59**, 740 (1937).
144. C. T. Pedersen, *Acta Chem. Scand.*, **17**, 1459 (1963).
145. H. Suzuki, M. Usuki and T. Hanafusa, *Synthesis*, 705 (1979).
146. E. A. Ebsworth and M. J. Mays, *J. Chem. Soc.*, 3893 (1963).
147. J. S. Thayer, *J. Organomet. Chem.*, **9**, 30 (1967).
148. H. Burger and U. Goetze, *J. Organomet. Chem.*, **10**, 380 (1967).
149. T. Gabrio and G. Barnikov, *Z. Chem.*, **9**, 183 (1969).
150. L. Hendriksen and J. Ehrbor, *Synthesis*, 519 (1976).
151. C. Collard-Charon and M. Renson, *Bull. Soc. Chim. Belg.*, **71**, 531 (1962).
152. T. Tarantelli and C. Pecile, *Ann. Chim. (Rome)*, **52**, 75 (1962).
153. W. J. Franklin and R. L. Werner, *Tetrahedron Lett.*, 3003 (1965).
154. D. H. R. Barton, G. Bringmann, G. Lamotte, W. B. Motherwell, R. S. Motherwell and A. E. Porter, *J. Chem. Soc., Perkin Trans. 1*, 2657 (1980).
155. H. Spies, K. Gewald and R. Mayer, *J. Prakt. Chem.*, **314**, 646 (1972).
156. K. A. Jensen and E. Fredericksen, *Z. Anorg. Allg. Chem.*, **230**, 31 (1936).
157. R. Huls and M. R. Renson, *Bull. Soc. Chim. Belg.*, **65**, 684 (1956); **66**, 265 (1957).
158. C. Collard-Charon, R. Huls and M. Renson, *Bull. Soc. Chim. Belg.*, **71**, 541 (1962).
159. E. Bulka and K. -D. Ahlers, *Z. Chem.*, **3**, 348 (1963).
160. C. T. Pedersen, *Acta Chem. Scand.*, **17**, 1459 (1963).
161. K. A. Jensen, G. Felbert, C. T. Pedersen and U. Svanholm, *Acta Chem. Scand*, **20**, 278 (1966).
162. E. Bulka, K. -D. Ahlers and E. Tucek, *Chem. Ber.*, **100**, 1373 (1967).
163. M. Lipp, F. Dallacker, I. Meir zu Köchen, *Monatsh. Chem.*, **90**, 41 (1959).
164. L. Henriksen, *J. Chem. Soc., Chem. Commun.*, 773 (1982).
165. H. G. Mautner and W. D. Kumler, *J. Am. Chem. Soc.*, **78**, 97 (1956).
166. E. Bulka, in Ref. 1, p. 459.
167. A. M. Comrie, D. Dingwall and J. B. Stenlake, *J. Pharm. Pharmacol.*, **16**, 268 (1964).

168. K. A. Jensen and C. K. Jorgensen, in Ref. 1, p. 1017; V. Krishnan and R. A. Zingaro, in Ref. 2, p. 337; H. J. Gysling, in *The Chemistry of Organic Se/Te Compounds* (Eds. S. Patai and Z. Rappaport).
169. M. Dräger and G. Gattow, *Angew. Chem., Int. Ed. Engl.*, **11**, 868 (1968).
170. M. Dräger and G. Gattow, *Chem. Ber.*, **104**, 1429 (1971).
171. L. Henriksen, *Acta Chem. Scand.*, **21**, 1981 (1967).
172. K. A. Jensen and U. Anthoni, *Acta Chem. Scand.*, **24**, 2055 (1970).
173. K. A. Jensen, P. A. Fredericksen and L. Henriksen, *Acta. Chem. Scand.*, **24**, 2061 (1970).
174. M. Narita and C. U. Pittman, Jr., *Synthesis*, 489 (1976).
175. K. Bechgaard, D. O. Cowan, A. N. Bloch and L. Henriksen, *J. Org. Chem.*, **40**, 746 (1975).
176. F. S. Guziec, Jr., G. C. Long and J. M. Russo, unpublished work.
177. E. M. Engler, B. A. Scott, S. Etemad, T. Penney and V. V. Patel, *J. Am. Chem. Soc.*, **99**, 5909 (1977).
178. I. Johannsen, K. Lerstrup, L. Henriksen and K. Bechgaard, *J. Chem. Soc., Chem. Commun.*, 89 (1984).
179. P. Shu, A. N. Block, T. F. Carruthers and D. O. Cowan, *J. Chem. Soc., Chem. Commun.*, 505 (1977).
180. F. Wudl and D. Nalewajek, *J. Chem. Soc., Chem. Commun.*, 866 (1980).
181. L. -Y. Chiang, T. O. Poehler, A. N. Block and D. O. Cowan, *J. Chem. Soc., Chem. Commun.*, 866 (1980).
182. K. Lerstrup, M. Lee, F. M. Wiygul, T. J. Kistenmacher and D. O. Cowan, *J. Chem. Soc., Chem. Commun.*, 294 (1983).
183. I. Johannsen, K. Bechgaard, K. Mortensen and C. Jacobsen, *J. Chem. Soc., Chem. Commun.*, 295 (1983).
184. V. Y. Lee, E. M. Engler, R. R. Schumaker and S. S. Parkin, *J. Chem. Soc., Chem. Commun.*, 235 (1983).
185. D. H. Reid, *J. Chem. Soc. C*, 3187 (1971).
186. D. J. Sandman, A. J. Epstein, T. J. Holmes and A. P. Fisher, III, *J. Chem. Soc., Chem. Commun.*, 177 (1977).
187. G. Scherowsky and J. Weiland, *Chem. Ber.*, 107, 3155 (1974).
188. E. M. Engler and V. V. Patel, *J. Chem. Soc., Chem. Commun.*, 671 (1975).
189. E. M. Engler and V. V. Patel, *Tetrahedron Lett.*, 1259 (1975).
190. G. Traverso, *Ann. Chim. (Rome)*, **47**, 1244 (1957).
191. L. Henriksen and E. S. Kristiansen, *Ann. N. Y. Acad. Sci.*, 101 (1972).
192. D. Barnard and D. T. Woodbridge, *J. Chem. Soc.*, 2922 (1961).
193. G. M. Larin, P. M. Solozhenkin, N. I. Kopitsya and H. Kirspuu, *Dokl. Akad. Nauk. Tadzh. SSR*, **12**, 32 (1969); *Chem. Abstr.*, **72**, 84787j (1970).
194. K. A. Jensen and V. Krishnan, *Acta Chem. Scand.*, **21**, 2904 (1967).
195. B. Lorenz and E. Hoyer, *Z. Chem.*, **8**, 230 (1968).
196. M. L. Shankaranarayana, *Acta Chem. Scand.*, **24**, 251 (1970).
197. U. Anthoni, *Acta Chem. Scand.*, **20**, 2742 (1966).
198. U. Anthoni, B. M. Dahl, C. Larsen and P. H. Nielsen, *Acta Chem. Scand.*, **24**, 959 (1970).
199. L. Henriksen, *Synthesis*, 771 (1982).
200. K. A. Jensen and A. Holm, *Acta Chem. Scand.*, **18**, 2417 (1964).
201. K. A. Jensen, P. A. Frederiksen and L. Henriksen, *Acta Chem. Scand.*, **24**, 2061 (1970).
202. N. Sonoda, G. Yamamoto and S. Tsutsumi, *Bull. Soc. Chem. Jpn.*, **45**, 2937 (1972).
203. K. A. Jensen, M. Due, A. Holm and C. Wentrup, *Acta Chem. Scand.*, **20**, 2091 (1966).
204. D. L. Klayman and R. J. Shine, *J. Org. Chem.*, **34**, 3549 (1969).
205. U. Anthoni, C. Larsen and P. H. Nielsen, *Acta Chem. Scand.*, **21**, 2571 (1967).
206. C. Collard-Charon and M. Renson, *Bull. Soc. Chim. Belg.*, **72**, 149 (1963).
207. V. I. Cohen, *Synthesis*, 60 (1980).
208. M. Giua and R. Bianco, *Gazz. Chim. Ital.*, **89**, 693 (1959).
209. H. Hope, *Acta Chem. Scand.*, **18**, 1800 (1964).
210. R. A. Zingaro, F. C. Bennett, Jr., and G. W. Hammar, *J. Org. Chem.*, **18**, 292 (1953).
211. F. C. Bennett, Jr., and R. A. Zingaro, *Org. Synth.*, Coll. Vol. **4**, 359 (1963).
212. L. C. King and R. J. Hlavacek, *J. Am. Chem. Soc.*, **73**, 1864 (1951).
213. H. G. Grimm and H. Metzger, *Chem. Ber.*, **69**, 1356 (1936).
214. K. A. Jensen, G. Felbert and B. Kägi, *Acta Chem. Scand.*, **20**, 281 (1966).

215. T. S. Griffin and D. L. Klayman, *J. Org. Chem.*, **39**, 3161 (1974).
216. Z. I. Yoshida, H. Konishi and H. Ogoshi, *J. Chem. Soc., Chem. Commun.*, 359 (1975).
217. H. Spies, K. Gewald and R. Mayer, *J. Prakt. Chem.*, **313**, 804 (1971).
218. A. Shafiee and I. Lalezari, *J. Heterocycl. Chem.*, **12**, 675 (1975).
219. A. Shafiee, F. Assadi and V. I. Cohen, *J. Heterocycl. Chem.*, **15**, 39 (1978).
220. H. Mautner and E. M. Clayton, *J. Am. Chem. Soc.*, **81**, 6270 (1959).
221. G. Wilke and E. Uhlemann, *Z. Chem.*, **14**, 288 (1974); **15**, 66 (1975).
222. Ref. 11, p. 523.
223. P. W. Preisler and T. N. Scortia, *J. Am. Chem. Soc.*, **80**, 2309 (1958).
224. A. Chiesi Villa, G. Grossoni, M. Nardelli and M. E. Vidoni Toni, *J. Chem. Soc., Chem. Commun.*, 404 (1969).
225. M. Mikolajczyk and J. Luczak, *Chem. Ind.* (*London*), 76 (1972).
226. M. Z. Girshovich and V. El'tsov, *Zh. Obshch. Khim.*, **39**, 941 (1969).
227. M. F. Lappert, T. R. Martin and G. M. McLaughlin, *J. Chem. Soc., Chem. Commun.*, 635 (1980).
228. W. Paul and H. Werner, *Angew Chem., Int. Ed. Engl.*, **22**, 316 (1983).
229. H. Werner and W. Paul, *Angew. Chem., Int. Ed. Engl.*, **23**, 58 (1984).
230. C. E. Headford and W. A. Roper, *J. Organomet. Chem.*, **244**, C53 (1983).
231. W. A. Hermann, J. Weichmann, R. Serrano, K. Blechschmitt, H. Pfisterer and M. L. Ziegler, *Angew. Chem., Int. Ed. Engl.*, **22**, 314 (1983).
232. H. Fischer, S. Zeuner and J. Riede, *Angew. Chem., Int. Ed. Engl.*, **23**, 726 (1984).
233. H. Fischer and S. Zeuner, *Z. Naturforsch., Teil B*, **38**, 1365 (1983).
234. H. Fischer and S. Zeuner, *J. Organomet. Chem.*, **252**, C63 (1983).
235. C. G. Barraclough, R. L. Martin and I. M. Stewart, *Aust. J. Chem.*, **22**, 891 (1969); A. Davison and E. T. Shawl, *J. Chem. Soc., Chem. Commun.*, 670 (1967); *Inorg. Chem.*, **9**, 1820 (1970).
236. V. A. Alekseevskii, O. A. Ramazanova and L. A. Deryabina, *Zh. Neorg Khim.*, **23**, 545 (1978); *Chem. Abstr.*, **88**, 142408 (1978).
237. E. Uhlemann, *Int. Symp. Specific Interact. Mol. Ions, Proc., 3rd*, **2**, 539 (1976); *Chem. Abstr.*, **88**, 79947 (1978).
238. A. H. Schmidt, W. Ried and P. Pustoslemsek, *Angew. Chem., Int. Ed. Engl.*, **15**, 704 (1976).
239. G. Traverso, *Ann. Chim.* (*Rome*), **47**, 3 (1957).
240. S. Bezzi, *Gazz Chim. Ital.*, **92**, 859 (1962).
241. S. Pietra, C. Garbuglio and M. Mammi, *Gazz. Chim. Ital.*, **94**, 48 (1964).
242. J. H. van den Hende and E. Klingberg, *J. Am. Chem. Soc.*, **88**, 5045 (1966).
243. M. R. Detty and H. R. Luss, *J. Org. Chem.*, **48**, 5149 (1983).
244. T. L. Gilchrist, P. G. Mente and C. W. Rees, *J. Chem. Soc., Perkin Trans. 1*, 2165 (1972).
245. G. N. Schrauzer and H. J. Kisch, *J. Am. Chem. Soc.*, **95**, 2501 (1973).
246. K. H. Pannell, A. J. Mayr, R. Hoggard and R. C. Pettersen, *Angew. Chem., Int. Ed. Engl.*, **19**, 632 (1980).
247. A. J. Wittwer, L. Tsai, W. -M. Ching and T. Stadtman, *Biochemistry*, **23**, 4650 (1984).
248. J. W. Hoffman and K. P. McConnell, *Biochim. Biophys. Acta*, **366**, 109 (1974).
249. Y. S. Prasado Rao and J. D. Cherayil, *Life Sci.*, **14**, 2051 (1974).
250. C. -Y. Shiue and S. -H. Chu, *J. Chem. Soc., Chem. Commun.*, 319 (1975).
251. L. B. Townsend and G. H. Milne, *J. Heterocycl. Chem.*, **7**, 753 (1970).
252. S. -W. Chu, *J. Med. Chem.*, **14**, 254 (1971); **15**, 1088 (1972).
253. F. Bergmann and M. Rashi, *Isr. J. Chem.*, **7**, 63 (1969).
254. H. G. Mautner and J. J. Jaffe, *Cancer Res.*, **18**, 294 (1958).
255. H. G. Mautner, *Biochem. Pharmacol.*, **1**, 169 (1959).
256. S. -H. Chu, C. -Y. Shiue and M. Y. Chu., *J. Med. Chem.*, **18**, 559 (1975); *J. Pharm. Sci.*, **64**, 1343 (1975).
257. K. C. Agrawal, B. A. Booth, R. L. Michand, E. C. Moore and A. C. Santorelli, *Biochem. Pharmacol.*, **23**, 2421 (1976).
258. H. G. Mautner, W. D. Kumler, Y. Okono and R. Pratt, *Antibiot. Chemother.*, **6**, 51 (1956).
259. K. Bednarz, *Diss. Pharm.*, **9**, 249 (1957); *Chem. Abstr.*, **52**, 8083f (1958).
260. E. R. Cullen, F. S. Guziec, Jr., C. J. Murphy and T. C. Wong, *J. Chem. Soc., Perkin Trans. 1*, 473 (1982).
261. W. Gombler, *Z. Naturforsch., Teil B*, 1561 (1981).
262. W. Gombler, *J. Am. Chem. Soc.*, **104**, 6616 (1982).
263. K. A. Jensen and J. Sandstrom, *Acta Chem. Scand.*, **23**, 1911 (1969).

264. U. Svanholm, *Ann. N. Y. Acad. Sci.*, **192**, 124 (1972).
265. P. Hampson and A. Mathias, *Mol. Phys.*, **13**, 361 (1967).
266. W. Walter, E. Schaumann and H. Rose, *Tetrahedron*, **28**, 3233 (1972).
267. K. A. Jensen, L. Henriksen and P. H. Nielsen, in Ref. 1, pp. 843–851.
268. K. K. Andersen, D. M. Gash, J. D. Robertson and F. S. Guziec, Jr., *Tetrahedron Lett.*, **23**, 911 (1982).
269. W. M. Wijekoon, E. Bunnenberg, R. Records and D. A. Lightner, *J. Phys. Chem.*, **87**, 3034 (1983).
270. J. E. Kuder, in Ref. 1, pp. 872–874.
271. A. M. Kirkien, R. J. Shine and J. R. Plimmer, *Org. Mass Spectrom.* **7**, 233 (1973).
272. F. S. Guziec, Jr., and C. A. Moustakis, *Phosphorus Sulfur*, **21**, 189 (1984).
273. F. Fringuelli and A. Taticchi, *J. Heterocycl. Chem.*, **15**, 137 (1978).
274. Ref. 6, p. 500.
275. R. Mayer, S. Scheithauer and D. Kunz, *Chem. Ber.*, **99**, 1393 (1966).
276. C. -P. Klages and J. Voss, *Chem. Ber.*, **113**, 2255 (1980).
277. W. E. Britton, E. Alaydros, R. Kashyap, J. Blair and F. S. Guziec, Jr., unpublished work.
278. C. Dreier, J. Becher, E. G. Frandsen and L. Henriksen, *Tetrahedron*, **37**, 2663 (1981).
279. H. G. Mautner and G. Bergson, *Acta Chem. Scand.*, **17**, 1694 (1963).
280. A. Azman, M. Drofenik, D. Hadzi and B. Lukman, *J. Mol. Struct.*, **1**, 181 (1968).
281. I. J. Karle and J. Karle, in Ref. 1, p. 989.
282. A. W. Cordes, in Ref. 2, p. 298.
283. H. Hope, *Acta Crystallogr.*, **18**, 259 (1965).
284. M. Perez-Rodriguez and A. Lopez-Castro, *Acta Crystallogr., Sect. B*, **25**, 532 (1969).
285. J. S. Rutherford and C. Calvo, *Z. Kristallogr.*, **128**, 229 (1969).
286. E. Shefter, M. N. James and H. G. Mautner, *J. Pharm. Sci.*, **55**, 643 (1966).
287. M. Bonamico and G. Dessy, *J. Chem. Soc., Chem. Commun.*, 1114 (1967).
288. G. U. Tsintsadze, M. A. Porai-Koshits and A. S. Antsyshkina, *J. Struct. Chem.*, **8**, 253 (1967).

NOTE ADDED IN PROOF

Recently three papers have appeared dealing with the preparation and trapping of intermediate monomeric selenoaldehydes:

R. Okazaki, A. Ishii and N. Inamoto, *J. Chem. Soc., Chem. Commun.*, 71 (1986).
G. W. Kirby and A. N. Trethewey, *J. Chem. Soc., Chem. Commun.*, 1152 (1986).
G. A. Krafft and P. T. Meinke, *J. Am. Chem. Soc.*, **108**, 1314 (1986).

The Chemistry of Organic Selenium and Tellurium Compounds Volume 2
Edited by S. Patai
© 1987 John Wiley & Sons Ltd.

CHAPTER **5**

Photochemistry of organic compounds of selenium and tellurium

ZEEV GOLDSCHMIDT

Department of Chemistry, Bar Ilan University, Ramat-Gen, Israel

I. INTRODUCTION 276
II. PHOTODISSOCIATIVE PROCESSES 276
A. Photofragmentation 276
1. Cleavage of Se—C and Te—C bonds 277
2. Cleavage of Se—Se and Te—Te bonds 279
3. Cleavage of bonds between Se and other atoms 280
a. Selenium—sulfur 280
b. Selenium—germanium 281
B. Photoeliminations 281
1. Deselenation and detelluration 281
2. Nitrogen eliminations 288
3. Cycloreversions 290
4. Cheletropic reactions 292
III. PHOTOASSOCIATIVE PROCESSES 293
A. Photosubstitutions 294
1. Nucleophilic substitution 294
a. Aromatic substitutions 294
b. Aliphatic substitution 297
c. Intramolecular substitution 299
2. Organometallic substitutions 302
a. Free-radical displacements 302
b. Ligand-exchange reactions 306
B. Photoaddition reactions 309
C. Photocycloadditions 314
D. Insertion Reactions 317
IV. PHOTOREARRANGEMENTS AND PHOTOISOMERIZATIONS. 318
A. Rearrangements 318
B. *Cis–Trans* Isomerization 326
V. PHOTOOXIDATIONS. 327

VI. PHOTOREDUCTIONS 329
A. Hydrogen Abstraction 329
B. Oxygen Transfer 333
VII. ACKNOWLEDGEMENTS 334
VIII. REFERENCES. 334

I. INTRODUCTION

The photochemistry of organic selenium and tellurium compounds has attracted much less attention than the light-induced reactions of the early organochalcogen derivatives of oxygen[1,2] and sulfur[3,4]. Although the sensitivity to sunlight of dibenzyl diselenide (**1**), one of the most common selenium compounds, has been known since 1875[5], the mechanistic details of this photofragmentation lay dormant for nearly a century[6]. It seems that the frequently observed deposition of elemental selenium and tellurium during photolyses, although a very useful process in photography, deterred chemists from exploring the fate of the organic residue. Accordingly, it is perhaps not surprising that photoreactions of the heavier organochalcogens are hardly mentioned in comprehensive treatises on the chemistry of these compounds[7,8], and that the single photochemical review appeared only in 1980[9].

$$PhCh_2SeSeCH_2Ph$$

(1)

The last decade has witnessed substantial progress in the understanding of the physical and chemical consequences of the electronic excitation of organoselenium and tellurium compounds. In general, it is found that whereas in organic oxygen compounds carbonyl excited states play the major role in the photochemical transformations, selenium and tellurium derivatives, like their sulfur counterparts, react mainly via homolytic cleavage of sigma bonds.

This chapter intends to provide a comprehensive review of the synthetic and mechanistic aspects of the photochemistry of the heavier organochalcogen compounds. It is organized according to reaction types, that is, dissociative and associative processes, rearrangements and isomerizations, oxidations and reductions. Because of the relatively few photoreactions of tellurium compounds, and their resemblance to those of selenium derivatives, both groups will be discussed together. It is hoped that this chapter will provide the reader with a framework for identifying unexploited areas of the organic photochemistry of selenium and tellurium compounds, and help stimulate further progress in this growing field of research.

II. PHOTODISSOCIATIVE PROCESSES

A. Photofragmentation

The homolytic cleavage of a sigma bond between a chalcogen and another atom, to form radicals, is the most common primary process in the photochemistry of organo-selenium and tellurium compounds. In the absence of ground-state molecules to react with these radicals, they may recombine to form the starting material, or dispropor-tionate by a variety of routes. The chemistry of these disproportionation reactions will be examined in this section.

1. Cleavage of Se—C and Te—C bonds

A unique cleavage of an aliphatic selenide bond is observed when selenide **2** is irradiated in benzene solutions in the presence of *p*-toluenesulfonic acid, yielding acetylacetone (**3**) and PhSeOTs (**4**) (equation 1)[10].

$$\underset{(\mathbf{2})}{\text{MeCOCH(SePh)COMe}} \xrightarrow[\text{PhH}, p\text{-TsOH}]{h\nu} \underset{(\mathbf{3})}{\text{MeCOCH}_2\text{COMe}} + \underset{(\mathbf{4})}{\text{PhSeOTs}} \quad (1)$$

The photochemistry of aromatic selenol esters has been studied extensively. These esters usually undergo photo-Fries type rearrangements (Section IV.A). However, in some cases the main products are formed only by cleavage. Irradiation of the diselenol ester **5** gave high yields of anthraquinone (**6**) and bis(*p*-tolyl selenide) (**7**) (equation 2)[11].

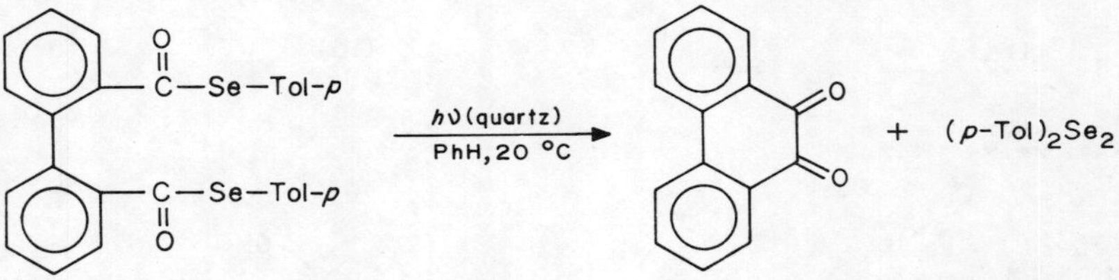

(2)

(**5**) (**6**) (**7**)

When the selenoester **8** was photolysed, the major products observed were methyl sulfinyl acid (**9**) (41% yield) and **7** (49%) (equation 3)[12]. Photolysis of selenoester **10** gave, among other substitution products (Section III.A.1.c), a 22% yield of **11** in a presumably photo-Friedel–Crafts reaction[13] (equation 4). No cyclization products were obtained when aliphatic sulfides were irradiated. In a similar photo-Friedel–Crafts reaction of the nicotinic ester **12**, selenoxanthone (**13**) was obtained in 15% yield[14] (equation 5).

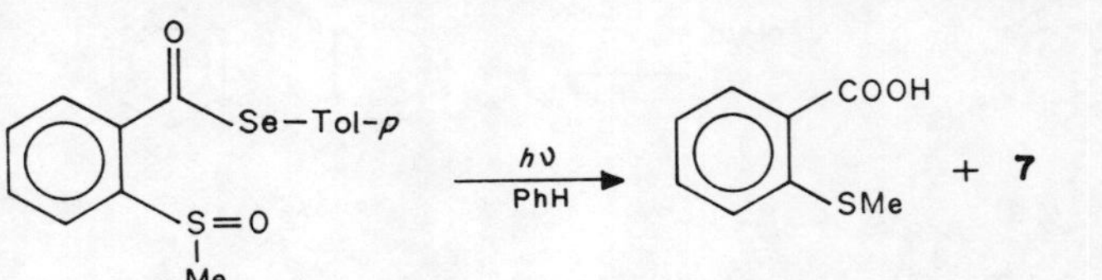

(3)

(**8**) (**9**)

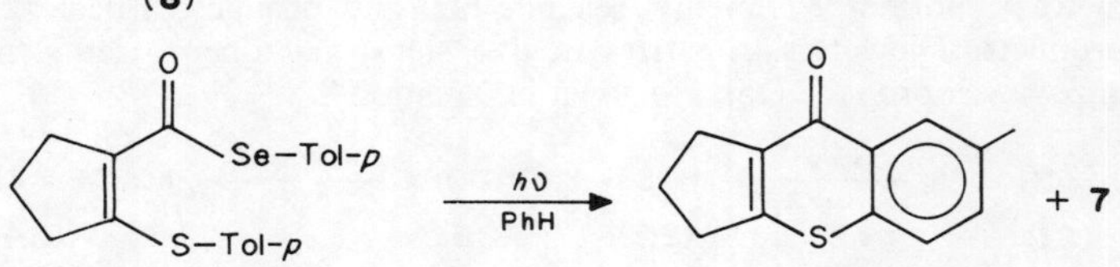

(4)

(**10**) (**11**)

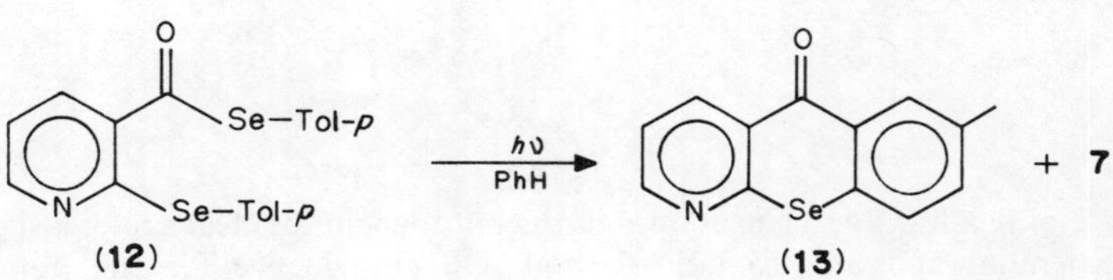

(5)

(**12**) (**13**)

Telluroester (**14**) undergo homolysis on irradiation in dry benzene solutions[15] to give aldehydes (**17**) and bis(*p*-tolyl telluride) (**18**) (equation 6). The mechanism is believed to involve α-cleavage to radicals **15** and **16**, followed by hydrogen abstraction by **15** and dimerization of **16**, to give the observed products. Although the origin of the hydrogen atom abstracted was not determined, it may be derived from the solvent. As with the selenium analogs, small amounts of photo-Fries rearrangement products are obtained (Section IV.A) together with some elemental tellurium. In the case of the photolysis of **19**[16], an intramolecular substitution occurred giving **20** (Section III.A.1) (equation 7).

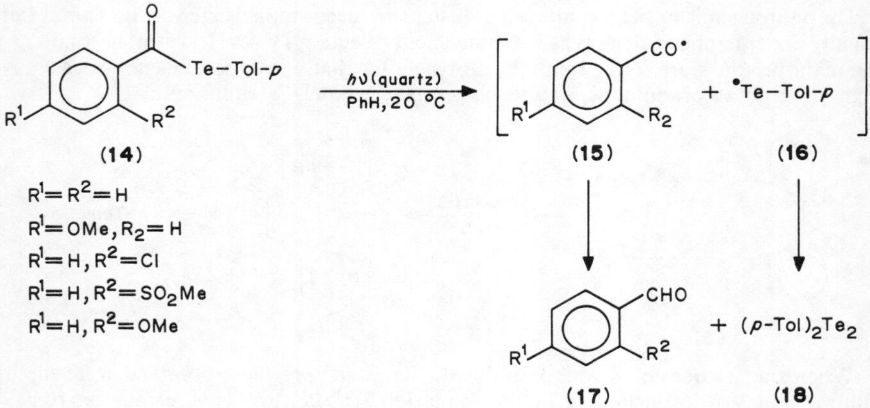

(6)

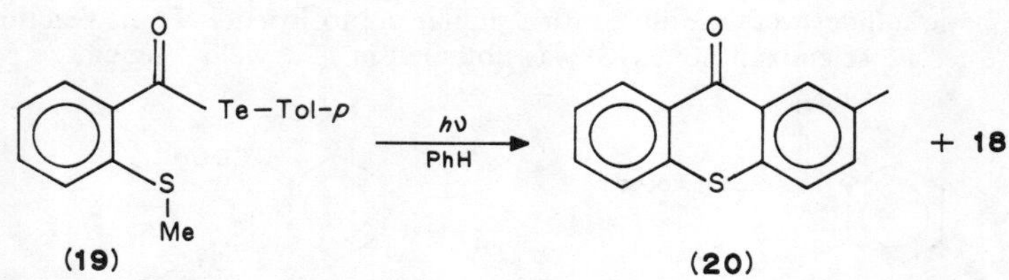

(7)

An unusual ylid selenium—carbon bond rupture on irradiation of **21** at 238 nm in chloroform was reported[17]. Dimethyl selenide (**22**) and phenacylcarbene (**23**) are the primary products. The latter trimerizes in two steps, via dibenzoylethylene (**24**), to *trans*-tribenzoylcyclopropane (**25**), as shown in equation 8.

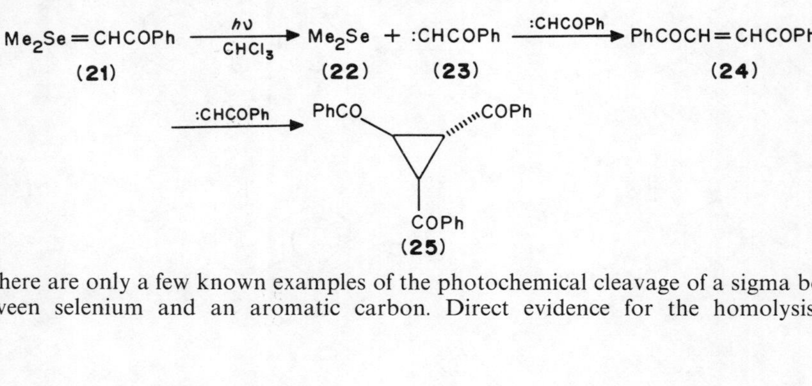

(8)

There are only a few known examples of the photochemical cleavage of a sigma bond between selenium and an aromatic carbon. Direct evidence for the homolysis of bis(*p*-methoxyphenyl) selenide (**26**) to the selenide radical **27** was obtained from pulse radiolysis studies[18], in which the transition absorption spectrum of **27** (λ_{max}535 nm) was observed in neutral or acidic methanolic solutions (equation 9). Similarly, pulse irradiated solutions of the diselenide **28** give the transient absorption spectrum of the diselenide radical **29** (λ_{max} 600 nm) (equation 10). Here the spectrum of radical **27** also appeared as a result of the homolytic cleavage of the Se—Se bond. The decay of the 535 nm absorption in irradiated solutions of **26** followed pseudo-first-order kinetics with a rate constant $k = 8 \times 10^5 \, l\,mol^{-1}\,s^{-1}$. Pulse radiolysis of methanolic solutions of bis(*p*-methoxyphenyl) telluride (**30**) similarly gave telluride radicals **31** (λ_{max} 508 nm), which decay faster ($k = 1.5 \times 10^{10} \, l\,mol^{-1}\,s^{-1}$) (equation 11)[18].

$$\underset{(\mathbf{26})}{Ar_2Se} \xrightarrow[MeOH]{e_s^-} \underset{(\mathbf{27})}{ArSe^{\bullet}} + ArH \qquad (9)$$

$$\underset{(\mathbf{28})}{Ar_2Se_2} \xrightarrow[MeOH]{e_s^-} \underset{(\mathbf{29})}{ArSeSe^{\bullet}} + ArH$$

$$\underset{(\mathbf{28})}{Ar_2Se_2} \xrightarrow[MeOH]{e_s^-} \underset{(\mathbf{27})}{ArSe^{\bullet}} + ArSeH \qquad (10)$$

$$\underset{(\mathbf{30})}{Ar_2Te} \xrightarrow[MeOH]{e_s^-} \underset{(\mathbf{31})}{ArTe^{\bullet}} + ArH \qquad Ar = p\text{-methoxyphenyl} \qquad (11)$$

Cleavage of the Se—Ph bond was also observed in the sunlight-induced transfer reaction between triphenylselenonium fluoroborate (**32**) and triphenylphosphine (**33**) to form diphenyl selenide (**34**) and tetraphenylphosphonium fluoroborate (**35**)[19] (equation 12). The homolytic character of this reaction is evident from the isolation of small amounts of benzene, which apparently result from hydrogen abstraction from the solvent by phenyl radicals.

$$\underset{(\mathbf{32})}{Ph_3SeBF_4} + \underset{(\mathbf{33})}{Ph_3P} \xrightarrow[acetone]{Sunlight} \underset{(\mathbf{34})}{Ph_2Se} + \underset{(\mathbf{35})}{Ph_4PBF_4} + PhH \qquad (12)$$

2. Cleavage of Se—Se and Te—Te bonds

Shortwave UV irradiation of diphenyl diselenide (**36**) in high vacuum gave phenylselenyl radicals (**37**), which were detected by ESR measurements (equation 13)[20]. These radicals are thermally more stable than the corresponding sulfide radicals, and recombine back to **36** only at temperatures above 77 K.

$$\underset{(\mathbf{36})}{PhSeSePh} \xrightarrow{h\nu} \underset{(\mathbf{37})}{PhSe^{\bullet}} \qquad (13)$$

As has been shown above (equation 10), pulse radiolysis of bis(*p*-methoxyphenyl selenide) (**28**) affords selenide (**27**) and diselenide (**29**) radicals. The corresponding ditellurides (**38**) on radiolysis gave only telluride radicals (**31**), which absorb at λ_{max} 508 nm (equation 14)[18].

$$\text{ArTeTeAr} \xrightarrow[\text{MeOH}]{e_s^-} \text{ArTe}^{\bullet} + \text{ArTeH} \tag{14}$$

(**38**) (**31**)

Ar = *p*-methoxyphenyl

A kinetic study of the photochemical degradation of bis(*p*-ethoxyphenyl telluride) (**38**, Ar = *p*-ethoxyphenyl) in toluene solutions containing ethanol revealed[21] that in the rigorous exclusion of oxygen no reaction occurred. However in the presence of even catalytic amounts of oxygen the ditelluride decomposes after a short induction period, with a rate half order in ditelluride (equation 15). The reaction is also first order in ethanol, but independent of temperature.

$$[\text{Ar}_2\text{Te}_2]^{1/2} = \text{constant} - k_{\text{obs}}t \tag{15}$$

ESR findings indicate that a homolytic Te—Te bond cleavage took place, yielding ArTe radicals. Since the overall reaction is not independent of ditelluride concentration, two ditelluride molecules must be involved in the radical production. Molecular oxygen is obviously also involved at an early stage of the reaction. Consequently, a reaction sequence is proposed which includes the essential steps shown in equations 16–18. Both reaction products, **39** and **40**, are unstable and oxidize immediately on exposure to oxygen, to give **41**.

$$\text{Ar}_2\text{Te}^* + \text{O}_2 \longrightarrow \text{Ar}_2\text{Te}_2\text{O}_2 \tag{16}$$

$$\text{Ar}_2\text{Te}_2\text{O}_2 + \text{Ar}_2\text{Te}_2 \longrightarrow 2\text{ArTe}^{\bullet} + \text{Ar}_2\text{Te}_2 + \text{O}_2 \tag{17}$$

$$[2\text{ArTe}^{\bullet} + \text{EtOH} \longrightarrow \text{ArTeCH(Me)OH} + \text{ArTeH} \xrightarrow{[\text{O}]} (\text{ArTe})_2\text{O}] \tag{18}$$

(**39**) (**40**) (**41**)

Ar = *p*-ethoxyphenyl

3. Cleavage of bonds between Se and other atoms

a. Selenium–sulfur. Phenyl areneselenosulfonates (**42**) are unusually photosensitive. Irradiation of **42** through Pyrex, in degassed carbon tetrachloride, led to its complete decomposition. The identifiable products of decomposition were diphenyl diselenide (**36**) and arenesulfonic anhydride (**43**) (32% yield) (equation 19)[22]. These products are believed to arise as a result of the reaction sequence shown in equations 20–23, initiated by the homolytic photodissociation of **42** into PhSe and $ArSO_2$ radicals (equation 20). It should be noted that at some stage oxidation must occur, presumably of the intermediate dimer of $ArSO_2$ (equation 22).

$$\text{PhSeSO}_2\text{Ar} \xrightarrow[\text{CCl}_4]{h\nu(\text{Pyrex})} \text{PhSeSePh} + \text{ArSO}_2\text{OSO}_2\text{Ar} \tag{19}$$

(**42**) (**36**) (**43**)

Ar = Ph, *p*-tolyl

$$PhSeSO_2Ar \longrightarrow PhSe^\bullet + ArSO_2^\bullet \tag{20}$$

$$PhSe^\bullet + PhSeSO_2Ar \longrightarrow PhSeSePh + ArSO_2^\bullet \tag{21}$$

$$2ArSO_2^\bullet \longrightarrow [ArSOOSO_2Ar] \xrightarrow{[O]} ArSO_2OSO_2Ar \tag{22}$$

$$2PhSe^\bullet \longrightarrow PhSeSePh \tag{23}$$

b. Selenium–germanium. The irradiation of triethylgermylselenol (**44**) in the presence of olefins usually gives addition products[23] (Section III.B). However, when **44** was irradiated in the presence of equimolar amounts of acrylonitrile, a 76.7% yield of hexaethyldigermselenane (**45**) was obtained, clearly indicating cleavage of the Se—Ge bond (equation 24).

$$\underset{(\mathbf{44})}{Et_3GeSeH} \xrightarrow{h\nu(\text{Pyrex})} \underset{\substack{77\% \\ (\mathbf{45})}}{(Et_3Ge)_2Se} \tag{24}$$

B. Photoeliminations

1. Deselenation and detelluration

The loss of a selenium or tellurium atom during the irradiation of organochalcogen compounds is a very common reaction. Early gas-phase flash photolysis studies of CSe_2[24,25] (equation 25) and COSe[24,26] (equation 26) revealed the presence of excited selenium atoms in the photolysis mixture. These may react with a variety of hydrocarbons and olefins to give insertion products, (Section III.D).

$$CSe_2 \xrightarrow{\text{flash } h\nu} CSe(X'\Sigma) + Se(4^3P_y) \tag{25}$$

$$COSe \xrightarrow{\text{flash } h\nu} CO + Se(4^1D_2) \tag{26}$$

Flash photolysis of dimethyl telluride (**46**) is a useful source of triplet excited tellurium atoms (equation 27)[27]. Thus, inspection of flashed mixtures of **46** vapor (10^{-3}–10^{-5} Torr with CO_2 diluent), using kinetic absorption spectroscopy, showed intense atomic absorptions at λ214.3 and 225.9 nm, corresponding to transitions of $Te(^3P_2)$, and at 238.6 and 238.3 nm, corresponding to $Te(^3P_1)$ and $(^3P_0)$, respectively. In addition, Σ–Σ type transitions of the Te_2 molecule appeared between 360 and 430 nm. At low initial pressures of **46**, the 216 nm absorption of the methyl radical was detected. From flash energy variations, the carrier of a series of four bands between 224 and 243 nm appears to be the primary photoproduct, the MeTe radical.

$$\underset{(\mathbf{46})}{Me_2Te} \xrightarrow[CO_2]{\text{flash } h\nu} Te(^3P_{2,1,0}) \tag{27}$$

As an alternative source of Te atoms, the flash photolysis of H_2Te was examined. Ground-state and spin–orbit excited Te atoms were observed along with Te_2. A transient species absorbing at 215.5 nm was detected and was assigned as the HTe radical[27]. This represents a unique case of a photohomolytic cleavage of a Te—H bond (equation 28).

$$H_2Te \xrightarrow{h\nu} HTe^{\bullet} + H^{\bullet} \longrightarrow Te \tag{28}$$

Dibenzyl diselenide (**1**) was the first organoselenium compound whose photochemistry was studied in solution[28]. Irradiation in benzene at 350 nm, in the absence of atmospheric oxygen, gave a 60% yield of isolable monoselenide (**47**) (equation 29).

$$\underset{(\mathbf{1})}{PhCH_2SeSeCH_2Ph} \xrightarrow[PhH]{h\nu\ (350\,nm)} \underset{(\mathbf{47})}{PhCH_2SeCH_2Ph} + Se \tag{29}$$

Detailed kinetic and mechanistic studies of the deselenation of **1** were carried out in acetonitrile using light of $\lambda > 280$ nm (Pyrex filter)[6]. First, it was established that **47** and a stoichiometric amount of red amorphous selenium are the sole products of the reaction. The observed quantum yield for the disappearance of **1** (at 313 nm) was $\phi = 0.16 \pm 0.02$ at low conversions, and the initial rates of the photodecomposition were found to be dependent on the light intensity, but not on the concentration (equation 30).

$$d(\mathbf{1})/dt = kI_a \tag{30}$$

where I_a = intensity of light absorption by **1**;

$$\underset{(\mathbf{1})}{PhCH_2SeSeCH_2Ph} \overset{h\nu}{\rightleftharpoons} \underset{(\mathbf{48})}{2PhCH_2Se^{\bullet}} \tag{31}$$

$$\underset{(\mathbf{1})}{PhCH_2SeSeCH_2Ph} \overset{h\nu}{\rightleftharpoons} \underset{(\mathbf{49})}{PhCH_2^{\bullet}} + \underset{(\mathbf{50})}{PhCH_2SeSe^{\bullet}} \tag{32}$$

Since under the irradiation conditions (λ310 nm $\approx$ 92 kcal) both the Se—Se bond (44 kcal) and the C—Se bond (57 kcal) may cleave, it is reasonable to suppose that the two primary processes involve homolytic cleavage of these bonds in **1** (equations 31 and 32). The presence of benzyl selenide radicals (**48**) in the reaction mixture was confirmed by trapping **48** with carbon tetrachloride, to give benzyl chloride. However, since the disappearance of **1** is dependent on the light intensity but not on the concentration, only the homolytic cleavage of the C—Se bond to produce benzyl (**49**) and benzyl diselenyl (**50**) radicals leads to products, whereas benzyl selenide (**48**) radicals are inactive except for recombination to **1**. This is consistent with the non-chain radical mechanism shown in equations 33–40, for which the rate equation has the form depicted in equation 30, where $k = k_3/(k_1 + k_2 + k_3)$.

$$RSeSeR \xrightarrow{h\nu} RSeSeR^* \tag{33}$$

$$R = PhCH_2$$

$$RSeSeR^* \xrightarrow{k_1} RSeSeR \tag{34}$$

$$RSeSeR^* \xrightarrow{k_2} 2RSe^{\bullet} \tag{35}$$

$$RSeSeR^* \xrightarrow{k_3} R^{\bullet} + RSeSe^{\bullet} \tag{36}$$

$$RSeSe^{\cdot} \xrightarrow{k_4} RSe^{\cdot} + Se \qquad (37)$$

$$R^{\cdot} + RSeSeR \xrightarrow{k_5} RSe^{\cdot} + RSeR \qquad (38)$$

$$R^{\cdot} + RSe^{\cdot} \xrightarrow{k_6} RSeR \qquad (39)$$

$$2RSe^{\cdot} \xrightarrow{k_7} RSeSeR \qquad (40)$$

An important conclusion that can be drawn from the above scheme is that benzylselenyl radicals (**48**) do not dissociate to Se and benzyl radicals under normal photolytic conditions. However, in the presence of Ph_3P, the diselenation reaction takes a different course[29]. In addition to the expected dibenzyl selenide (**47**) and triphenylphosphine selenide (**52**), bibenzyl (**51**) is formed (equation 41). Further, the quantum yields for disappearance of **1** are dependent on the concentration of Ph_3P, and rise to $\phi > 1.0$. This suggests a radical chain mechanism, shown in equations 42–46, in which benzylselenyl radicals (**38**), obtained from the homolytic dissociation of **1**, play a crucial part in the reaction.

$$\underset{(\mathbf{1})}{PhCH_2SeSeCH_2Ph} + Ph_3P \xrightarrow[MeCN]{h\nu} \underset{(\mathbf{47})}{PhCH_2SeCH_2Ph} + \underset{(\mathbf{51})}{PhCH_2CH_2Ph} + \underset{(\mathbf{52})}{Ph_3PSe} \qquad (41)$$

$$RSeSeR \longrightarrow 2RSe^{\cdot} \qquad (42)$$

$$RSe^{\cdot} + Ph_3P \longrightarrow Ph_3PSeR \qquad (43)$$

$$Ph_3PSeR \longrightarrow R^{\cdot} + Ph_3PSe \qquad (44)$$

$$R^{\cdot} + RSeSeR \longrightarrow RSeR + RSe^{\cdot} \qquad (45)$$

$$2R^{\cdot} \longrightarrow RR \qquad (46)$$

Dibenzyl ditelluride (**53**) and diethyl ditelluride (**54**), like their selenium counterparts, react quantitatively under UV irradiation to give tellurium and the corresponding tellurides (**55** and **56**)[30,31] (equation 47). The benzyl derivatives photolyse more rapidly than the corresponding ethyl compounds[31]. Dibenzyl telluride (**55**) itself is both thermal and light sensitive, decomposing readily to Te and bibenzyl (**51**). Cleavage of the ditellurides, like that of selenides, is accelerated in the presence of tertiary phosphines (**57**), affording the corresponding phosphine tellurides (**58**). The latter decompose back to **57** and Te, forming an equilibrium mixture shown in equation 48. However, unlike selenides, the increase in reaction rates is not marked, suggesting that C—Te bond cleavage in alkyltellurium compounds predominates over the rupture of Te—Te bonds. In general, it appears that the difference in reactivity between the selenium and tellurium compounds lies in the fact that Se—Se bond cleavage occurs more readily than that of a C—Se bond, whereas the opposite is true for the tellurium analogs.

$$RTeTeR \xrightarrow[CDCl_3]{h\nu} RTeR + Te \qquad (47)$$

(**53**) (**55**) $R = PhCH_2$

(**54**) (**56**) $R = Et$

$$RTeTeR + R^1R^2R^3P \xrightarrow{h\nu} RTeR + R^1R^2R^3PTe \rightleftharpoons R^1R^2R^3P + Te \quad (48)$$

(57) (58)

$R = Et, PhCH_2; R^1 = R^2 = Ph, R^3 = Me; R^1 = R^2 = Ph, R^3 = CH_2PPh_2$

An interesting photoextrusion of selenium from bis(9-anthrylmethyl selenide) (**59**) in the presence of Ph_3P has been reported to give two isomeric dimers, hydrocarbons **60** and **61**, in a ratio of 3:1 (equation 49)[32,33]. The proposed mechanism involves the initial deselenation of **59** to the anthrylmethyl radical **62**, which first dimerizes to a mixture of hydrocarbons **63** and **64**, followed by a thermal intramolecular 4 + 2 cycloaddition (Diels–Alder reaction) of **63** to give **60**, and a photochemical 4 + 4 cycloaddition of **64**[34] to dimer **61**.

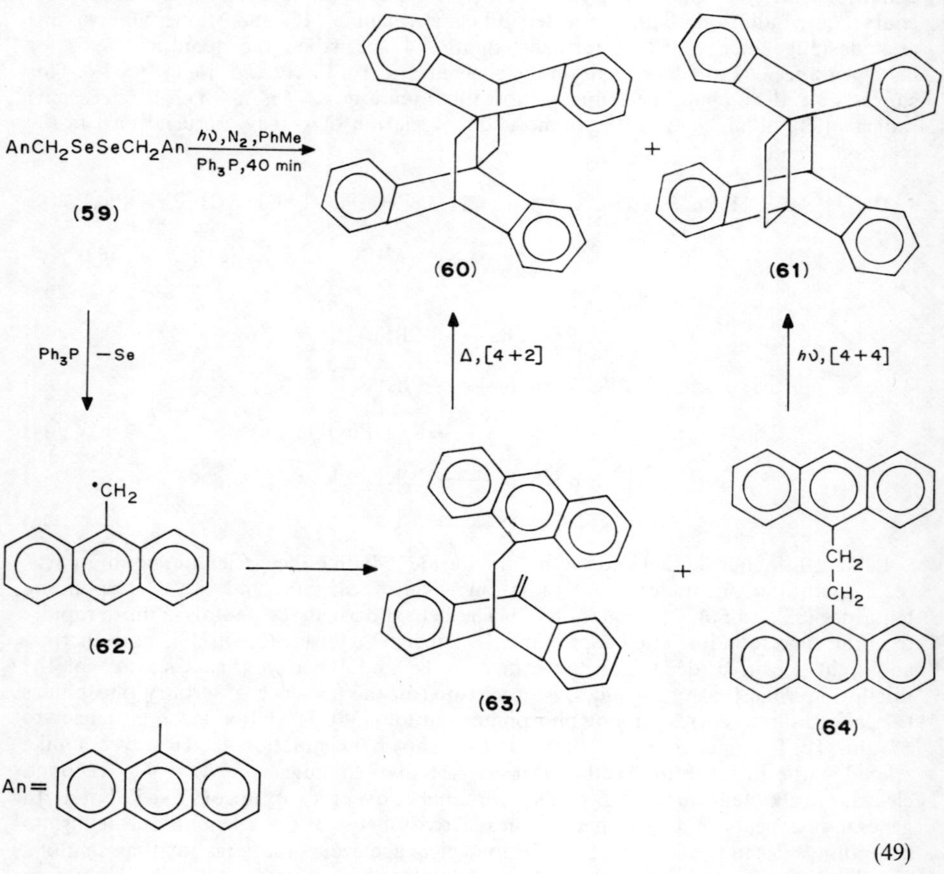

(49)

The photolysis of bis(2-cyanoethyl) selenide (**65**) and bis[2-(methoxycarbonyl)ethyl] selenide (**69**) gave a complex mixture of products derived from the homolytic cleavage of the Se—C bonds[35]. The composition of the mixture was strongly solvent dependent. In methanol, **65** decomposed mainly to elemental Se and traces of H_2Se and acrylonitrile (**66**) (equation 50). In THF solutions, the major product was bis(2-cyanoethyl selenide) (**67**) together with small amounts of adduct **68**, H_2Se and **66** (equation 51). Photolysis of **69** in methanol solutions afforded mainly Se together with small amounts of methyl adipate (**70**) and methyl acrylate (**71**). In THF solutions, moderate amounts of **70** were obtained with essentially quantitative formation of elemental Se (equation 52).

$$\underset{(\mathbf{65})}{Se(CH_2CH_2CN)_2} \xrightarrow[\text{MeOH}]{h\nu} \underset{66\%}{Se} + \underset{\text{traces}}{H_2Se} + \underset{\underset{\text{traces}}{(\mathbf{66})}}{CH_2=CHCN} \qquad (50)$$

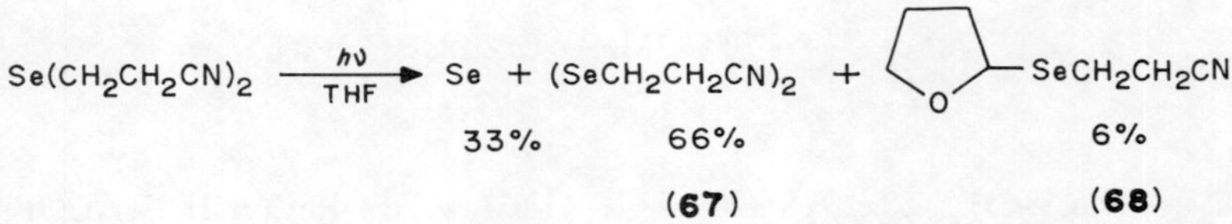

(51)

$$Se(CH_2CH_2COOMe)_2 \xrightarrow{h\nu} Se + (CH_2CH_2COOMe)_2 + CH_2=CHCOOMe \qquad (52)$$

(**69**)		(**70**)	(**71**)
MeOH:	54%	1%	1%
THF:	100%	10%	traces

Excitation of bis(benzoylmethyl)tellurium dichloride (**72**) with 313 nm light in degassed solutions of benzene, perfluorobenzene or acetonitrile yielded chloroacetophenone (**73**) and tellurium as the principal photoproducts[36]. In addition, smaller amounts of acetophenone (**74**) and dibenzoylethane (**75**) were obtained (equation 53). When the photolysis was performed in hydrogen-donating solvents such as tetrahydrofuran, acetophenone (**74**) and tellurium were the two major products formed. Trace amounts of **73** and 2-phenylacetyltetrahydrofuran (**76**) were also identified in this reaction.

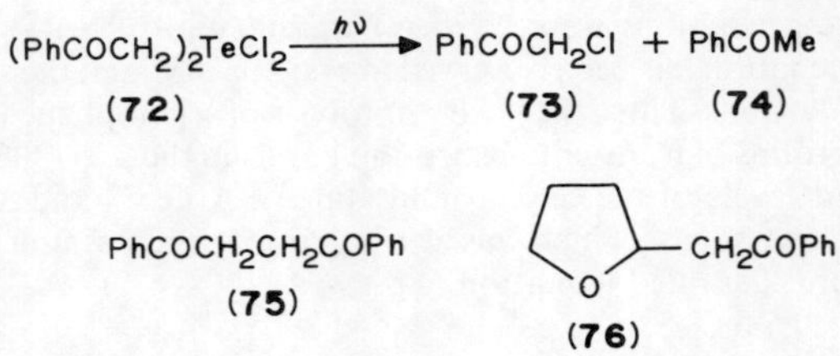

(53)

Quantum yields (in acetonitrile) for acetophenone formation are less than those for chloroacetophenone. The extrapolated minimum quantum yields for **73** and **74** are 0.01 and 0.1, respectively. These values correspond to non-free-radical concerted processes that are independent of the concentration of **72**. In the presence of the efficient radical trap CCl_4, the quantum yield of **73** increases, whereas that of **74** decreases to the limiting value of 0.01 ($[CCl_4] > 5$ M).

No room temperature emission was observed for **72**, but at 77 K in methyltetrahydrofuran glass, phosphorescence is detected with quantum yield $\phi_p = 0.18$ and lifetime $\tau_p = 24$ms. The spectroscopic triplet is assigned as originating principally from the $^3\pi,\pi^*$

state with appreciable $^3n,\pi^*$ character. The triplet state of **72** ($E_T = 72$ kcal) cannot be quenched by molecules with lower energy triplet states such as naphthalene ($E_T = 61$ kcal) and fluorene ($E_T = 68$ kcal), but the reaction can be sensitized by the higher energy 9-xanthone ($E_T = 74$ kcal).

These results are consistent with a mechanism in which the decomposition of the triplet π,π^* state of **72** may lead to products **73** and **74** by a concerted type II γ-hydrogen abstraction and by β-cleavage reactions. They may also be derived by reactions of phenacyl radicals **77** (produced from β-cleavage) which abstract hydrogen from the solvent or a chlorine atom from **72**. Recombination of **77** with solvent radicals affords **76**. In inert solvents **77** recombines to form dibenzoylethane (**75**). Tellurium metal formation may be accounted for by disproportionation of $TeCl_2$ (equations 54–57).

$$\mathbf{72}^* \xrightarrow[\text{type II}]{} \mathrm{PhCOMe} + [\mathrm{Cl_2TeCHCOPh}] \quad (54)$$

(**74**)

$$\mathbf{72}^* \longrightarrow \mathrm{PhCOCH_2^{\bullet}} + \mathrm{TeCl_2} \xrightarrow[\beta\text{-cleavage}]{} \mathrm{TeCl_4} + \mathrm{Te} + \mathrm{PhCOCH_2CH_2COPh}$$

(**77**) (**75**)

(55)

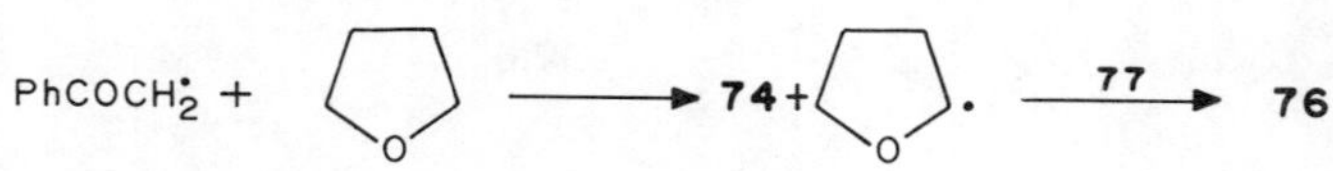

(56)

$$\mathrm{PhCOCH_2^{\bullet}} + \mathbf{72} \longrightarrow \mathbf{73} \quad (57)$$

Kinetic expressions for the quantum yields of acetophenone and chloroacetophenone formation were derived using steady-state approximations for excited states and radicals[36]. The quantum yield of **74** in good hydrogen-donating solvents was predicted to be proportional to the concentration of **72**. In poor hydrogen-donating solvents the quantum yields of **73** (ϕ_{ca}) and **74** (ϕ_a) were expected to be linearly dependent on the initial concentration of **72**, and on the inverse square root of the absorbed light (ϕ_a, $\phi_{ac} \propto [\mathbf{72}]/I_a^{1/2}$). These predictions were verified experimentally[36].

Photochemical deselenation has been observed in systems in which the selenium atom is flanked by two vinylic carbons. Thus, the flash photolysis of selenophene (**77**) diluted with argon resulted in two groups of transient absorption bands in the 350–390 nm region, due to vibrationally excited levels of the Se_2 ground state[37]. A new band system was also observed in the 397–418 nm region which was attributed to the remaining C_4H_4 fragment, whose structure was not identified (equation 58).

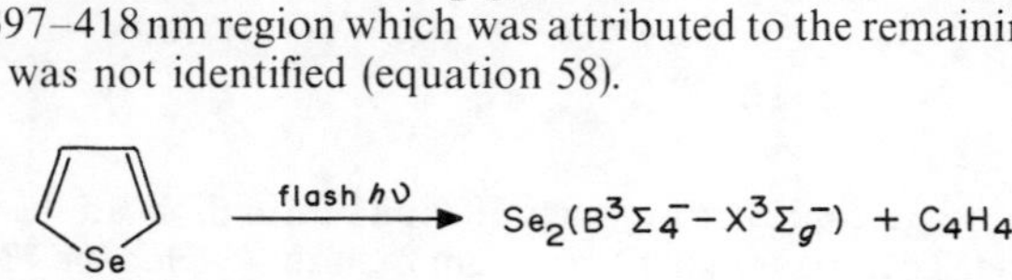

(**77**)

(58)

The solution photochemistry of 2-phenylselenophene (**78**) resulted in the formation of a mixture containing selenium and phenylvinylacetylene (**79**) as the deselenation products, together with the isomeric 3-phenylselenophene (**80**) (Section IV.A) (equation 59)[38].

Irradiation of 2-phenyltellurophene (**81**) gave only the fragmentation products, tellurium and **79** (equation 60)[38].

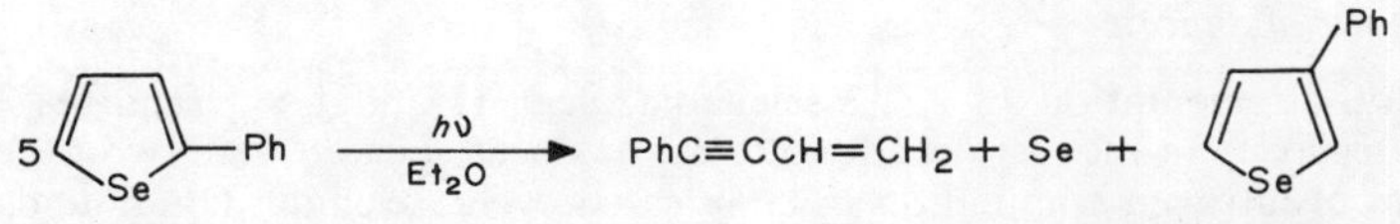

(78) (79) (80)

(59)

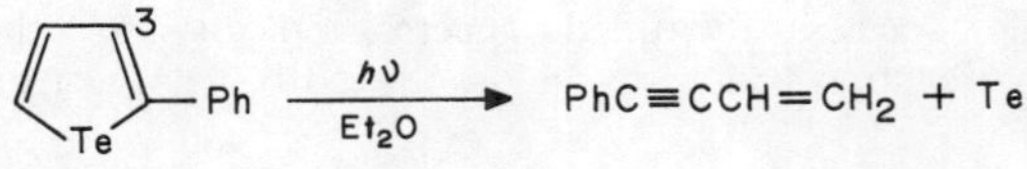

(81)

(60)

Deuterium labeling experiments with 5-d-**78** and 3, 4, 5-d_3-**81** revealed that the formation of **79** did not arise from intramolecular hydrogen shift in the intermediate radical but rather via hydrogen exchange with the solvent. The mechanistic scheme proposed for the deselenation (equation 61) involves the initial cleavage of the Se—C bond to form the diradical **82**, which loses a hydrogen atom to either an excited starting material or a solvent radical, forming the vinyl selenide radical **83**. This monoradical then extrudes selenium to give **84**, which in turn abstracts a hydrogen from the solvent, to give **79**.

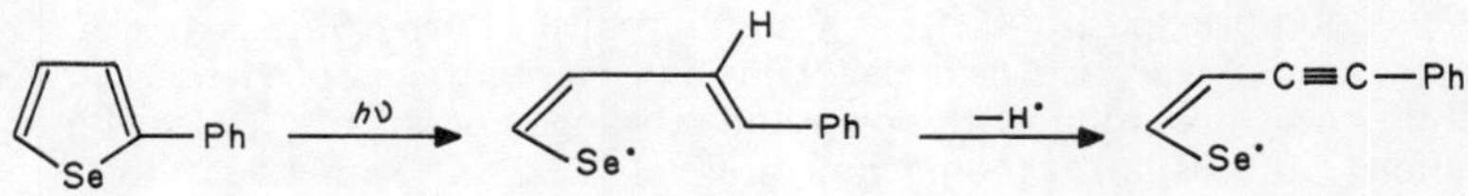

(78) (82) (83)

(84) (79)

(61)

Two interesting variations of selenium extrusion in related divinyl derivatives are worth mentioning. First, the irradiation of the 2, 1, 3-benzoselenadiazole 1-oxide (**85**) in methylene chloride solutions, which gave benzofurazan (**87**) in 94% yield (equation 62)[49], has been shown to occur in two steps via cyclization of the intermediate **86** (Section II.B.3). The second deselenation is that of 3, 3′-bis(1-methylindolyl) selenide (**88**), which on irradiation in benzene afforded the biindolyl **89** and 1-methylindole (**90**)[40] (equation 63). In this case the first step is a homolytic cleavage.

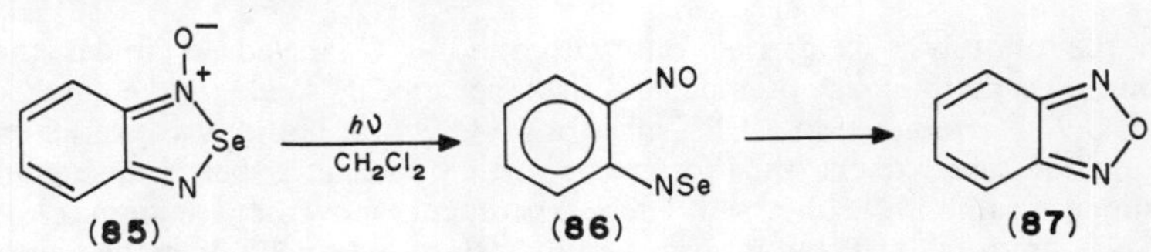

(85) (86) (87)

(62)

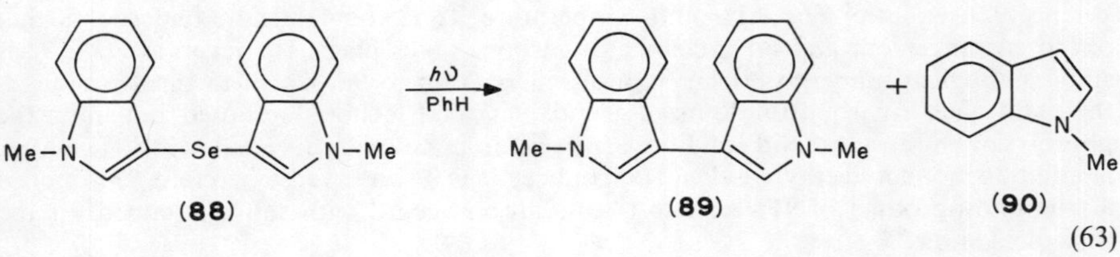

(88) (89) (90)

(63)

2. Nitrogen eliminations

The photofragmentation of 1,2,3-selenadiazole (**91**) and its derivatives has been studied extensively in the last decade. Interest has been focused mainly on the detailed mechanism of nitrogen elimination in conjugated cyclic selenium compounds *vis-à-vis* analogous heterocyclic systems. Irradiation (235–280 nm) of argon or nitrogen matrix-isolated 1,2,3-selenadiazole (**91**) at 8 K produced selenoketene (**92**), ethynylselenol (**93**) and a photolabile species, identified spectroscopically as the selenirene (**95**) (equation 64)[41]. The reaction progress was monitored by IR spectroscopy.

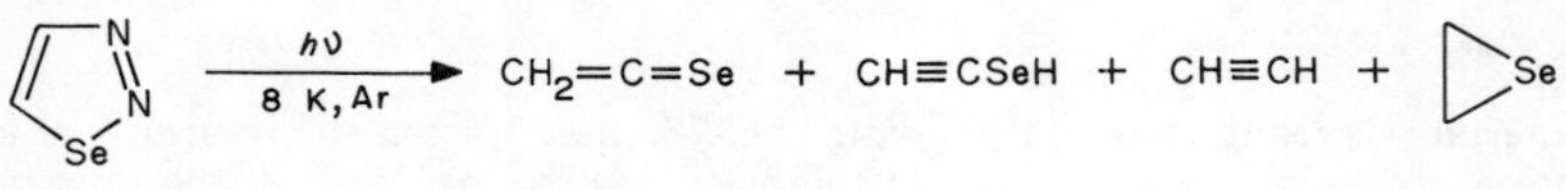

(64)

(**91**) (**92**) (**93**) (**94**) (**95**)

Irradiation of the mixture of products at 275–325 nm led to the conversion of **91** into selenoketene and acetylene. If, however, irradiation was continued, using light of 235–280 nm, the intensity of the IR bands characteristic of **92** and **94** was reduced, whereas the bands assigned to **95** reappeared and those belonging to **93** were enhanced. This suggests that **93** is not a primary product but a photoproduct of **95**.

The question of whether selenirene (**95**) is a primary product of **91** was clarified only recently by a combination of trapping and labeling experiments[42]. It was found that irradiation (290 ± 10 nm) at room temperature of **91** (λ_{max} 285 nm, ε 905 $l mol^{-1} cm^{-1}$) at low concentrations (ca. 5×10^{-4}M) in cyclohexane or diethyl ether–isopentane–ethanol (5:5:1) (EPA), and in the presence of 1% diethylamine afforded only diethylselenoacetamide (**96**) in 90–100% yield. The quantum yield of disappearance of **91** is independent of the diethylamine concentration and is not quenched by oxygen. This implies that **91** undergoes a unimolecular fragmentation from the singlet excited state to form selenoketene (**92**) (either directly or via short-lived intermediates), which subsequently traps diethylamine to give **96** (equation 65).

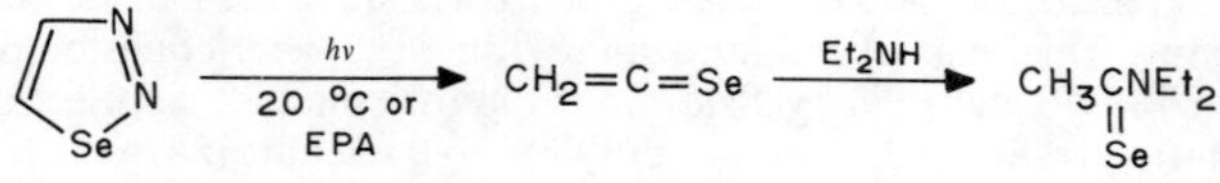

(65)

(**91**) (**92**) (**96**)

When the photolysis is carried out with the 5-^{13}C-labeled **91** under the same conditions, no scrambling is observed, and the label appears solely on the amide carbon. However, if **91** is irradiated in a EPA glass or PVC film at liquid nitrogen temperature (80 K), and the selenoketene thus formed is further irradiated before quenching with diethylamine, scrambling is observed. These experiments show that selenirene (**95**) is not a primary photoproduct of **91** but is formed as an unstable intermediate in the photolysis of ketene (**92**) (equation 66). At room temperature **92** is too labile to undergo further photorearrangement and no exchange of carbons takes place. However, at 80 K **92** is stable enough to undergo further photoisomerization to **95**, which in turn rearranges thermally back to carbon-randomized selenoketene (**92**). It should be noted that unlike the photolysis at 8 K under solid conditions, no selenirene could be detected with 80 K matrix irradiation, nor could ethynylselenol (**93**) or acetylene by observed. Therefore, **93** is indeed a true photoproduct of **91**[41]. These results are in accord with those obtained for the thiazole analogs[43].

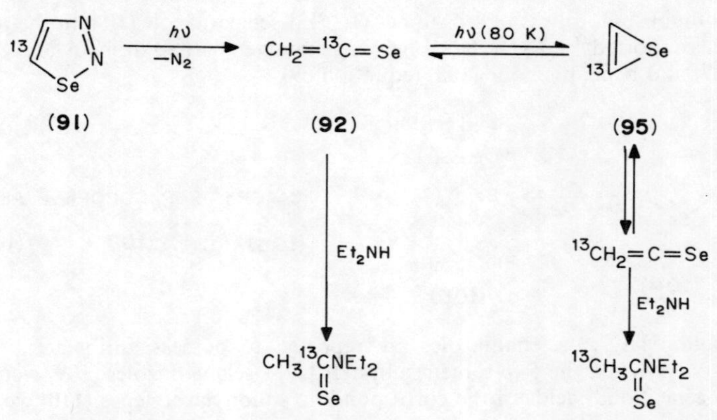

(66)

In a related study[44], matrix- isolated cyclopenteno-1, 2, 3-selenadiazole (**97**) in argon at 12 K was subjected to UV irradiation ($\lambda < 300$ nm), giving trimethyleneselenoketene (**98**) (equation 67) as the sole product.

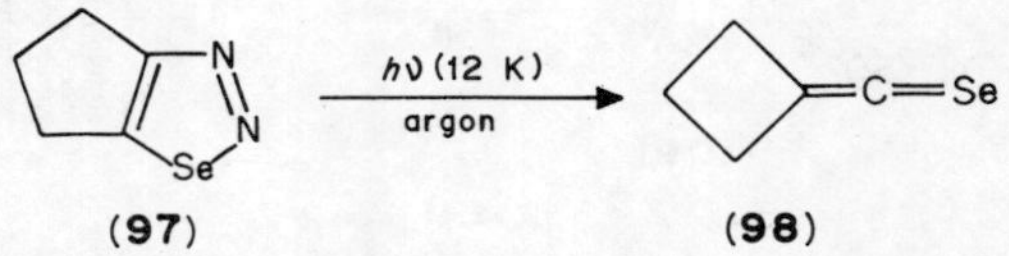

(67)

Substitution of the selenadiazole ring by electron-withdrawing groups such as ethoxycarbonyl (**99**), arylsulfonyl (**109**) or phenyl (**104**), while not affecting the primary photoextrusion of nitrogen, results in the isolation of products different from those derived from the parent compound. Thus, irradiation of **99** in benzene gave a 1:1 mixture of the two isomeric selenafulvenes **100** and **101**, in 18% (R = H) and 12% (R = Me) yields[45] (equation 68). The mechanism of this reaction is assumed to involve extrusion of nitrogen to produce diradical **102**, which undergoes a Wolff rearrangement to the ketene **103** (the same steps which presumably occur in the parent compound **91**). The two intermediates **102** and **103** combine in a formal 3 + 2 cycloaddition to give the products shown in equation 68.

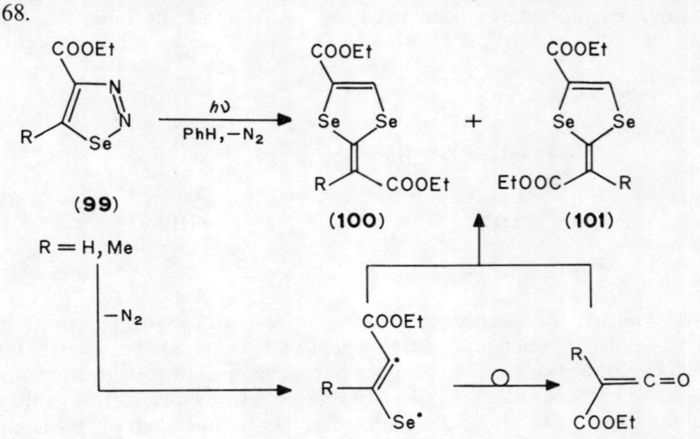

(68)

The analogous photolysis of 4, 4-diphenyl-1, 2, 3-selenadiazole (**104**) in the presence of oxygen was reported[46] to give tetraphenylselenafulvene (**105**), diphenylacetylene (**106**), benzil (**107**) and benzophenone (**108**) (equation 69).

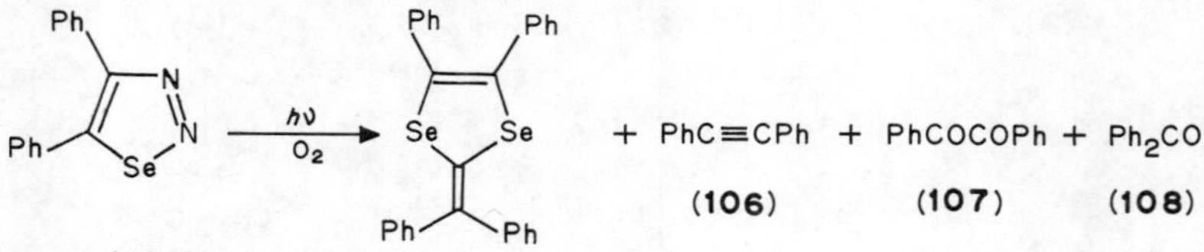

(69)

(104) (105)

5-Arylsulfonyl-1, 2, 3-selenadiazoles are reported to possess antibacterial and antifungal activity. When the 5-(*p*-toluenesulfonyl)-1, 2, 3-selenadiazoles (**109**) were photolysed in benzene, good yields of the corresponding sulfonylacetylenes (**110**) were isolated (equation 70)[48].

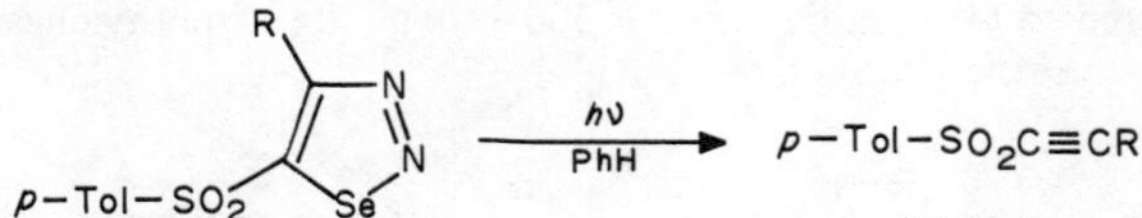

(70)

(109) (110)

R = Me 42%

R = Ph 64%

Although diazoketones seem to be good candidates for photochemical extrusion of nitrogen to form carbenes or ketenes (by subsequent rearrangement), there is only a single case in which such a reaction was utilized with the heavier organochalcogen compounds. Thus, the irradiation of the *ortho*-substituted diazoacetophenone **111** in dry benzene afforded telluro-3-coumaranone (**112**), albeit in poor yield (4%)[49] (equation 71). It should be noted that when the reaction was carried out under thermal, CuO-catalysed conditions, substantially higher yields of **112** (and also of the analogous selenocoumaranone) were obtained[49]. The mechanism of this intramolecular carbene reaction is not clear, although attempts have been made to understand the details in the oxygen analogs[50].

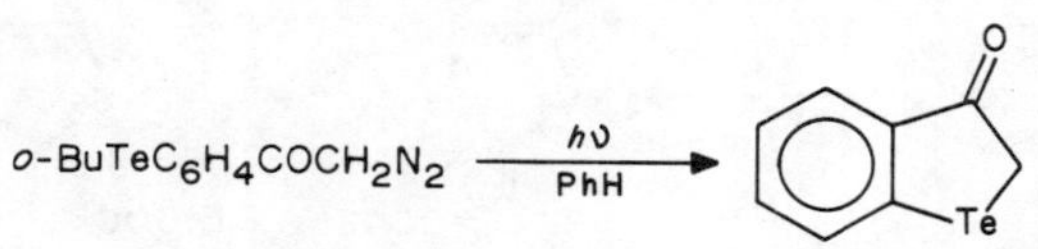

(71)

(111) (112)

3. Cycloreversions

The most extensive photochemical studies on cycloreversion reactions have been performed on isomers of selenadiazoles in which there is at least one Se—N bond in the molecule. Except for a few cases, the primary photochemical process in selenadiazoles is a formal 3 + 2 cycloreversion. In 1, 2, 3-selenadiazoles (previous section) this results in nitrogen elimination. The 1, 2, 5-selenadiazoles (**113**) and their 1, 2, 4-isomers (**114**) eliminate the corresponding nitriles.

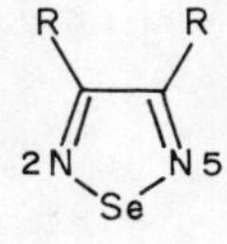

(113)

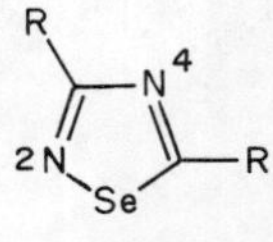

(114)

Irradiation at room temperature of 1, 2, 5-selenadiazoles (**115–117**) in 95% ethanol at 300 ± 20 nm produced elemental selenium and nitriles **118–120**, respectively, in high chemical and quantum yields, as shown in equation 72[51–53].

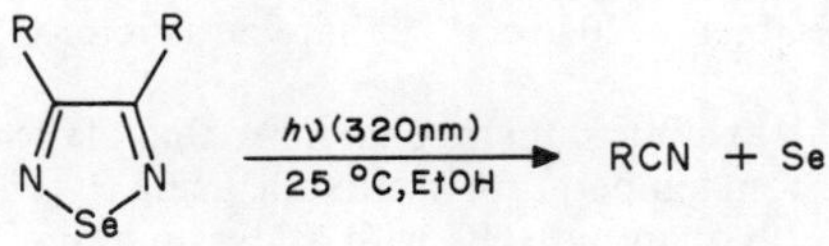

(72)

(**115**)	(**118**)	R = H	85%;	$\Phi_r 0.71$	(285 nm)
(**116**)	(**119**)	R = Me	100%;	$\Phi_r 0.80$	(290 nm)
(**117**)	(**120**)	R = Ph	100%;	$\Phi_r 0.96$	(317 nm)

The multiplicity of the excited states involved in the reactions was determined from quenching and sensitization experiments. The rate of photolysis of the aliphatic selenadiazoles **115** (in heptane) and **116** (in benzene) was found to be insensitive to 1 atm oxygen ($E_S = 23$ kcal) or piperylene ($E_T = 57$–59 kcal) at concentrations up to 0.3M. However, the reactions could be triplet-sensitized by xanthone ($E_T = 74$ kcal). Therefore, although the reactions may occur from either a singlet or triplet excited state, the triplet state is apparently not populated upon direct excitation at room temperature, and the photoactive state in these reactions is the singlet.

The photolysis of diphenylselenadiazole (**117**) in ethanol was partially quenched by oxygen ($\phi_r = 0.33$). The degree of quenching of **117** (in benzene) was studied as a function of added piperylene. The Stern–Volmer plot indicated that two photoactive states (S_1 and T_1) are involved in the unsensitized reaction, and only the T_1 state is quenched by piperylene. The ratio of triplet to singlet reaction, $T_1/S_1 = 4$, was deduced from the kinetic expression[51], and the lifetime (τ) of T_1 in benzene at room temperature was calculated to be in the range $1.6 \times 10^{-7}\,\text{s} < \tau < 1.6 \times 10^{-6}\,\text{s}$.

Both selenadiazoles **116** and **117** phosphoresce in EPA (at 77 K), whereas the parent diazole (**115**) does not. Compound **116** emitted at λ_{max} 455 nm (excited at 280–290 nm) with a quantum yield $\phi_p = 0.14$. The calculated lifetime of the triplet state was $\tau_p = 2.5 \times 10^{-3}$ s. For **117**, the emission was measured at λ_{max} 538 nm (excited at 310–320 nm) with a quantum yield $\phi_p = 0.33$ and a lifetime of $\tau_p = 1.7 \times 10^{-3}$ s. These results suggest that the lowest triplet state of the selenadiazoles is π, π^* in character.

Low-temperature experiments provided an insight into the transient species formed in the photolysis. Irradiation of **117** was performed in PVC or in EPA (diethyl ether – isopentane–ethanol, 5:5:2) at 85K and in solid nitrogen at 20K, and the reaction was followed by UV and IR spectroscopy. In EPA with irradiation at 300 nm, new absorptions appeared at 254, 325 (sh), 357 and 384 nm. Simultaneously, characteristic absorptions of benzonitrile (**120**) appeared. On prolonged photolysis (or heating above 240 K) the former absorptions disappeared with the concomitant enhancement of the intensity of benzonitrile absorption bands. Similar results were obtained with irradiation in PVC and in nitrogen. The IR spectrum consists of two bands in the 2300–1900 cm^{-1} region, benzonitrile bands at 2235 cm^{-1} and a band at 2200 cm^{-1} assigned to the transient benzonitrile selenide (**121**, R = Ph), which disappeared on continuous irradiation or heating.

$$\mathrm{RC{\equiv}\overset{+}{N}{-}\overset{-}{Se}}$$

(**121**)

R = H, Me, Ph

Similar low-temperature experiments were carried out in PVC, EPA and argon, to detect the transient nitrile selenides obtained from the aliphatic selenadiazoles **115** and **116**. The absorption bands assigned to HCNSe appear at 237 and 255 (sh) nm in PVC, and those assigned to MeCNSe appear to 233 and 252 nm. Interestingly, the IR spectra in both experiments (in argon) display a doublet at 2046 cm^{-1}, characteristic of CN radicals of unknown origin[51].

Diphenyl-1, 2, 4-selenadiazole (**122**) exhibits similar behavior to that of the 1, 2, 5-isomer (**117**)[51,52]. The photolysis of **122** at room temperature in ethanol, using Pyrex-filtered light (λ 300 nm), produced benzonitrile (**120**) quantitatively, with a low quantum yield of $\phi_r =$ 0.085 (at 310nm) (equation 73). Attempts to trap the intermediate benzonitriles selenide (**121**, R = Ph) [k(decay) = $10\,960 \pm 140\,\mathrm{s}^{-1}$ at 25 °C)[53] with dimethyl acetylenedicarboxylate failed. The photolysis could not be quenched with piperylene, and could not be sensitized by xanthone. Therefore, the photoactive state is the excited singlet state[52]. However, absorption spectra taken during the photolysis of **122** at 85 K in PVC film showed absorptions due to PhCNSe, identical with those exhibited in the irradiation of **117** under the same conditions. A similar spectrum was also obtained in frozen EPA at 85 K. IR spectra taken in a solid nitrogen matrix at 20 K again proved the presence of the transient PhCNSe, by showing the characteristic 2200 cm^{-1} absorption band.

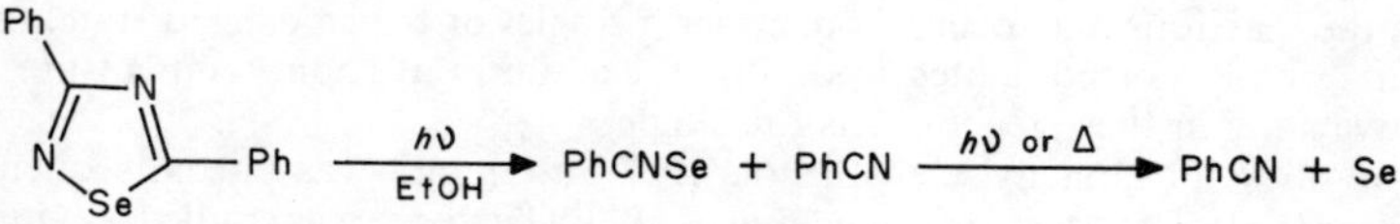

(**122**) (73)

4. Cheletropic reactions

2, 1, 3-Benzoselenadiazole (**123**) does not undergo the usual ring cleavage observed in 1, 2, 5-selenadiazoles (previous section). Instead, irradiation of **123** in 95% ethanol through a Pyrex filter at room temperature led to slow extrusion of selenium and the formation of the photoequilibrium mixture of the three isomeric hexadienedinitriles **124**–**126**[51], in a formal cheletropic reaction[54] (equation 74).

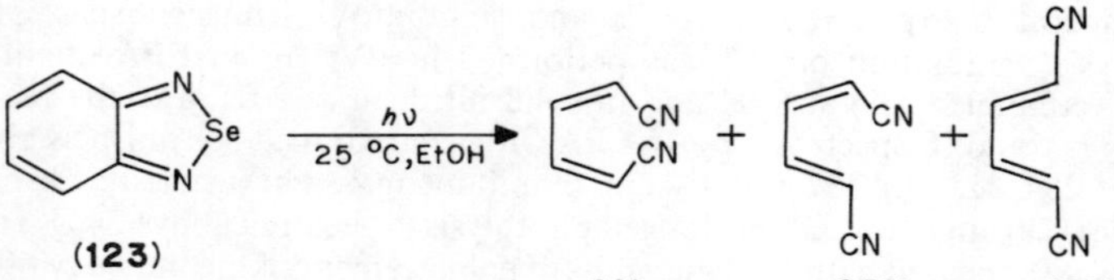

(**123**) (**124**) (**125**) (**126**) (74)

Highly substituted 2, 5-dihydro-1, 3, 4-selenadiazoles (**127**) are the only members of this heterocyclic ring system which are known to be photoreactive[55]. On irradiation at 300 nm in pentane solutions, these selenadiazoles undergo a cheletropic deselenation to give tetrasubstituted azines (**128**) in good yields (equation 75). It is interesting that under thermal conditions a two-fold extrusion reaction occurred with loss of both nitrogen and selenium, affording the corresponding highly crowded tetrasubstituted olefins.

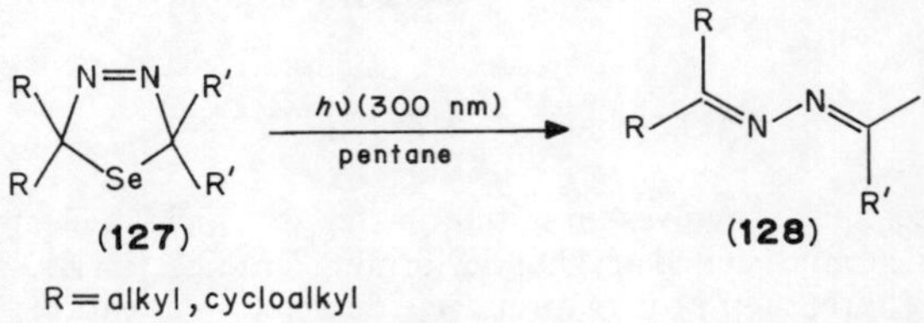

(75)

An unusual variation of a photochemical cheletropic reaction in which a phosphorus bridge is eliminated has recently been described[56]. Irradiation in methanol of the adducts **129** and **130**, obtained by thermal Diels – Alder cycloaddition of phenyl selenoxophosphole with maleic anhydride and phenylmaleimide, respectively, gave the corresponding dihydrophthalic anhydride **131** and phthalimide (**132**) together with dimethyl phenylselenophosphonate (**135**) (equation 76). The unstable phenylselenoxophosphan **133** was shown to be the primary photoproduct of the elimination. It first adds methanol to form **134**, which in turn reacts photochemically with methanol to give **135**.

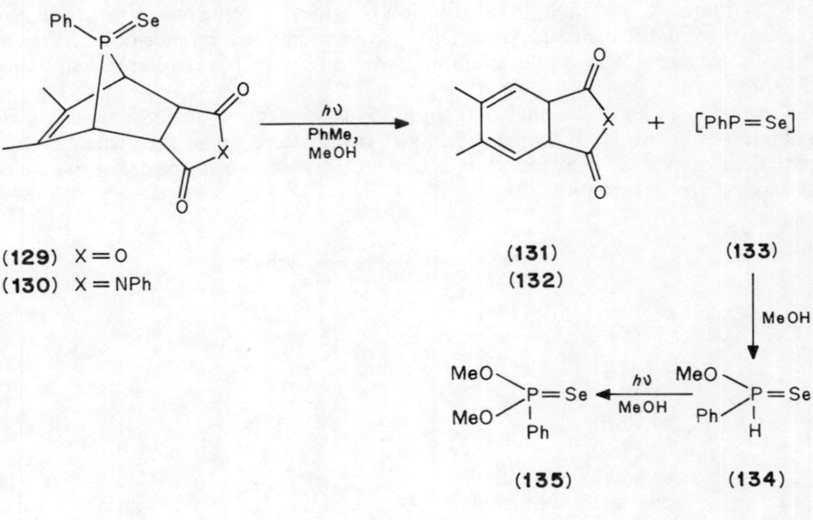

(76)

III. PHOTOASSOCIATIVE PROCESSES

In many reactions described in this section the primary photoproducts result from homolytic cleavage of organochalcogen σ bonds. Nevertheless, such reactions will be covered here if the chalcogenide radicals thus formed react with ground-state molecules to form new chalcogen—carbon σ bonds.

A. Photosubstitution

1. Nucleophilic substitution

a. Aromatic substitutions. Photolysis of aryl halides in the presence of nucleophiles leads to substitution by a radical chain mechanism commonly known as the aromatic $S_{RN}1$ reaction (equation 77)[57,58].

$$ArX + Nu^- \xrightarrow{h\nu} ArNu + Y^- \qquad (77)$$

The reaction mechanism involves an initiation step in which an electron is transfered from the nucleophile to the excited aryl halide, forming a radical ion $ArX^{-\cdot}$ (equation 78), which enters into a three step propagation cycle depicted in equations 79–81.

$$ArX + Nu^- \longrightarrow ArX^{-\cdot} + Nu^{\cdot} \qquad (78)$$

$$ArX^{-\cdot} \longrightarrow Ar^{\cdot} + X^- \qquad (79)$$

$$Ar^{\cdot} + Nu^- \longrightarrow ArNu^{-\cdot} \qquad (80)$$

$$ArNu^{-\cdot} + ArX \longrightarrow ArNu + ArX^{-\cdot} \qquad (81)$$

The use of phenyl selenide (**137**)[59,60] and phenyl telluride (**142**)[59,61] as nucleophiles in the photostimulated nucleophilic substitution provides a simple and direct route for the synthesis of symmetrical and unsymmetrical diaryl selenides and tellurides. The reactions are usually carried out in liquid ammonia solutions, with Pyrex-filtered irradiation at $\lambda < 350$ nm.

The reactivity order of the nucleofugal group in the reaction of $PhSe^-$ with a series of aryl halides, as determined from the chemical yields, decreases in the order I > Br > Cl. Naphthalene derivatives were found to be more reactive than benzene derivatives with the same leaving group (equation 82).

$$\underset{(\mathbf{136})}{ArX} + \underset{(\mathbf{137})}{PhSe^-} \xrightarrow[\text{liq NH}_3\text{ (Pyrex)}]{h\nu(\lambda < 350\text{ nm})} \underset{(\mathbf{138})}{ArSePh^-} + X^- \qquad (82)$$

ArX	Yields	
(a) PhCl	–	1%
(b) PhBr	1%	4%
(c) PhI	55%	66%
(d) 1-chloronaphthalene	73%	88%
(e) 4-chlorobiphenyl	52%	59%
(f) 9-bromophenanthrene	72%	85%
(g) 2-chloroquinoline	40%	54%
(h) 4-chlorobenzophenone	–	5%

The above results conform with the general photo-$S_{RN}1$ mechanism outlined above and parallel the reactions observed with the sulfur analogs[62–64]. However, there appears to be one exceptional case, 4-chlorobenzophenone, which with selenide gave very poor yields compared with sulfide, probably because of competing nucleophilic addition of $PhSe^-$ to the carbonyl group[59] (equation 83).

$$4\text{-ClC}_6\text{H}_4\text{COPh} + \text{PhSe}^- \xrightarrow{\Delta} 4\text{-ClC}_6\text{H}_4\text{C}(\text{O}^-)(\text{Ph})\text{SePh} \quad (83)$$

The photostimulated reaction of $PhSe^-$ with dihaloarenes[59] gave the disubstituted products. Use of *p*-bromoiodobenzene (**139**) as substrate afforded 70% yield of *p*-bis(phenylselenyl)benzene (**140**) and > 80% of iodide and bromide ions. Only traces of the monoadduct **141** were observed, although some of the starting material (12%) was recovered (equation 84). This suggests a propagation mechanism (equations 85–88) in which the radical anion of the monosubstituted product (equation 86) decomposes to the *p*-(phenylselenyl)phenyl radical and Br^- (equation 87) much faster than electron transfer to the substrate.

$$\underset{(\mathbf{139})}{4\text{-IC}_6\text{H}_4\text{Br}} + \text{PhSe}^- \xrightarrow{h\nu} \underset{(\mathbf{140})}{1,4\text{-(PhSe)}_2\text{C}_6\text{H}_4} + \text{Br}^- + \text{I}^- + \underset{(\mathbf{141})}{4\text{-BrC}_6\text{H}_4\text{SePh}} \quad (84)$$

$$4\text{-BrC}_6\text{H}_4\text{I}^{-\bullet} \rightarrow 4\text{-BrC}_6\text{H}_4^{\bullet} + \text{I}^- \quad (85)$$

$$4\text{-BrC}_6\text{H}_4^{\bullet} + \text{PhSe}^- \rightarrow 4\text{-BrC}_6\text{H}_4\text{SePh}^{-\bullet} \quad (86)$$

$$4\text{-PhSeC}_6\text{H}_4\text{Br}^{-\bullet} \rightarrow 4\text{-PhSeC}_6\text{H}_4^{\bullet} + \text{Br}^- \quad (87)$$

$$4\text{-PhSeC}_6\text{H}_4^{\bullet} + \text{PhSe}^- \rightarrow 4\text{-PhSeC}_6\text{H}_4\text{SePh}^{-\bullet} \quad (88)$$

The photonucleophilic substitution reaction of aryl halides with phenyl telluride (**142**) is in general more efficient than the analogous reaction with $PhSe^-$[59]. The reaction with bromobenzene gave a 20% yield of diphenyl telluride (**143**), and a 90% yield of **143** was obtained starting from iodobenzene. However, with more nucleofugal aryl halides such as *p*-iodoanisole and with 1-chloro- or 1-bromo-naphthalene, two unexpected products, **145** and **143**, were observed, indicating the existence of mechanistic pathways which enable the scrambling of the aryl groups (equation 89).

$$\text{ArX} + \underset{(\mathbf{142})}{\text{PhTe}^-} \xrightarrow[\text{liq. NH}_3]{h\nu} \underset{(\mathbf{143})}{\text{Ph}_2\text{Te}} + \underset{(\mathbf{144})}{\text{ArTePh}} + \underset{(\mathbf{145})}{\text{Ar}_2\text{Te}} \quad (89)$$

ArX	Yields		
	(143)	(144)	(145)
(a) bromobenzene	20%	—	—
(b) iodobenzene	90%	—	—
(c) *p*-iodoanisole	15%	73%	11%
(d) 1-chloronaphthalene	9%	41%	6%
(e) 1-bromonaphthalene	16%	53%	16%

The remarkable difference in behavior between the two nucleophiles was ascribed to the reversibility of coupling in the case of $PhTe^-$ (equation 90)[59]. The radical anion formed from coupling of the initially formed aryl radical with $PhTe^-$ may undergo three competing reactions: reversion to starting material (k_f), fragmentation to form aryltelluride ion and phenyl radical (k'_f) and electron transfer to form aryl phenyl telluride. The two new species formed, $ArTe^-$ and $Ph^{\bullet}$ may recombine (k'_c) to the aryl phenyl telluride radical anion or, alternatively, react with $Ar^{\bullet}$ and $PhTe^-$, respectively, to give the symmetrical tellurides. The reason why the phenylselenyl anion does not undergo nucleophilic scrambling reactions is that the fragmentation of the initially formed mixed radical anion is much slower than electron transfer, $k'_f \ll k_t[\mathrm{ArX}]$, and thus only the straightforward substitution product is formed.

$$\mathrm{Ar}^{\bullet} + {}^{-}\mathrm{TePh} \underset{}{\overset{k_f}{\rightleftharpoons}} (\mathrm{ArTePh})^{-\bullet} \underset{k'_c}{\overset{k'_f}{\rightleftharpoons}} \mathrm{ArTe}^{-} + \mathrm{Ph}^{\bullet} \qquad (90)$$

$$(\mathrm{ArTePh})^{-\bullet} \xrightarrow{k_t[\mathrm{ArX}]} \mathrm{ArTePh}$$

It is important to note from the above mechanistic interpretation that by decreasing the bond dissociation energy of the C—E bond (E = Se, Te), the chances of competing scrambling reactions increases. On the other hand, the reduction potentials of the aromatic compounds will be of importance in the electron-transfer processes. The reduction potentials of aromatic compounds are related to their π^* MO energy levels. The more negative the reduction potential, the higher is the π^* in energy. The bond dissociation energy parallels the σ^* MO energy levels and decreases on going down the Periodic Table. Hence, in reactions of a particular aryl radical with PhE^- nucleophiles, the difference in energy between σ^* and π^* MOs decreases from Se to Te, and the competition of bond fragmentation becomes more important[57]. Indeed, when the photoreaction of $PhSe^-$ was carried out with 4-iodoanisole and 2-bromopyridine, scrambling mixtures were obtained (equation 91)[65]. Moreover, use of iodobenzene and 1-adamantyl selenide (1-$AdSe^-$) (**146**) as nucleophile gave an equimolar mixture of the three corresponding selenides **147**–**149** (equation 92)[66]. The same reaction with 1-adamantyl telluride (1-$AdTe^-$) gave only **143** (80%) and **150** (16%) (equation 93).

In order to quantify further the mechanistic interpretation, competition experiments were carried out under conditions of irreversible coupling, using the reaction between various nucleophiles PhE^- (E = S, Se, Te) with 2-quinolyl radicals. The results indicate increasing reactivity on going down Group 6A in the order PhO (0.0) < PhS (1.0) < PhSe (5.8) < PhTe (28)[65].

$$\mathrm{ArX} + \mathrm{PhSe}^- \xrightarrow[\text{liq. NH}_3,\ -33\,^\circ\mathrm{C}]{h\nu} \mathrm{Ph_2Se} + \mathrm{ArSePh} + \mathrm{Ar_2Se} \qquad (91)$$

	Ph_2Se	ArSePh	Ar_2Se
4-iodoanisole	20%	25%	19%
2-bromopyridine	5%	72%	2%

$$\mathrm{PhI} + 1\text{-AdSe}^- \xrightarrow[\mathrm{NH_3}]{h\nu} \mathrm{Ph_2Se} + 1\text{-AdSePh} + (1\text{-Ad})_2\mathrm{Se} \qquad (92)$$

(**146**) (**147**) (**148**) (**149**)

$$\mathrm{PhI} + 1\text{-AdTe}^- \xrightarrow[\mathrm{NH_3}]{h\nu} \mathrm{PhTe} + 1\text{-AdTePh} \qquad (93)$$

1-Ad = 1-adamantyl (**143**) (**150**)

Nucleophilic substitution of aryl halides by selenocyanate ion (**151**) may also be induced by UV light in aqueous *tert*-butanol. Irradiation of **151** with 2-amino-5-chlorobenzonitrile and 1-chloro-2-naphthylamine gave the corresponding selenocyanates (**152**), whereas the photolysis with 4-chloroanisole and 5-bromo-2-(dimethylamino)pyrimidine afforded the corresponding diaryl diselenides (**153**) (equation 94)[67]. The mechanism of these reactions proceeds probably via aromatic $S_{RN}1$ substitution. The diaryl diselenides may be derived from recombination of ArSe$^{\bullet}$ formed by cleavage of the intermediate aryl selenocyanate radical anion (equation 95).

$$ArX + SeCN^- \xrightarrow[t\text{-BuOH}]{h\nu} ArSeCN\ (\mathbf{152}) + ArSeSeAr\ (\mathbf{153}) \qquad (94)$$

(**151**)

$$Ar^{\bullet} + SeCN^- \rightarrow ArSeCN^{-\bullet} \rightarrow ArSe^{\bullet} + CN^- \rightarrow ArSeSeAr \qquad (95)$$

b. Aliphatic substitution. Photochemical reactions of alkyl halides in the gas phase[68,69] and in solution[70,71] usually involve homolytic cleavage of carbon—halogen bonds. For example, direct chemical evidence for the intermediacy of trifluoromethyl radicals has been provided in the photolysis of CF_3I with halobenzenes[72]. When perfluoroalkyl iodides (**154**) are photolysed with dimethyl telluride (**155**) at 310 nm, good conversions of the iodides into perfluoroalkyl substituted tellurides (**156** and **157**) was observed (equation 96)[73]. When equimolar ratios of telluride to iodide were used, tellurium metal separated from the reaction mixture. However, when excess of iodide was utilized (3:1 molar ratio), iodine was formed.

$$R_FI + Me_2Te \xrightarrow{h\nu(310\,nm)} R_FTeMe + (R_F)_2Te \qquad (96)$$

(**154**) (**155**) (**156**) (**157**)

$R_F = CF_3, C_2F_5$

Trifluoromethyl radicals generated by photolysis of CF_3I in liquid ammonia may also be trapped efficiently (34–83%) by selenophenols (**158**) to give the corresponding trifluoromethyl aryl selenides (**159**) (equation 97)[74]. Similarly, irradiation of $PhSe^-$ (produced in liquid ammonia from Ph_2Se_2 and Na metal) with perfluoroalkyl iodides (**154**) gave the corresponding perfluoroalkyl selenides (**160**) (equation 98)[74].

$$CF_3I + 4\text{-}RC_6H_4SeH \xrightarrow[liq.NH_3]{h\nu} 4\text{-}RC_6H_4SeCF_3 \qquad (97)$$

(**158**) (**159**)

$R = H, Br, Me, CF_3$

$$R_FI + PhSe^- \xrightarrow[liq.NH_3]{h\nu} PhSeR_F \qquad (98)$$

(**154**) (**137**) (**160**)

$R_F = CF_3, C_3F_7$

It is reasonable to suppose that the reactions in liquid ammonia proceed by an $S_{RN}1$ type chain mechanism whereby the initially formed alkyl radicals react with aryl selenide anion to give the corresponding radical anion, which decays to the product by electron transfer (ET) (equation 99).

$$R^{\bullet} + ArSe^{-} \rightarrow RSeAr^{-\bullet} \xrightarrow{ET} RSeAr \qquad (99)$$

It is well known that bridgehead-substituted halides are very unreactive towards nucleophilic substitution reactions[75]. It has recently been reported, however, that 1-haloadamantanes and 9-halotriptycene react with nucleophiles by the photochemical $S_{RN}1$ mechanism[76,77]. 1-Iodoadamantane (1-AdI) (**161**) reacts with disodium selenide and telluride under photolytic conditions in ammonia to give, after air oxidation, bis(1-adamantyl selenide) and bis(1-adamantyl telluride) (**162**) in good yields (equation 100). In both cases the reaction mixture also contained moderate yields of the hydrocarbon adamantane (**163**). 1-Adamantyl selenide and telluride anions (**164**) are believed to be the primary photosubstitution products, which dimerize oxidatively to the observed products (**162**) (equation 101). Adamantane was probably obtained from the intermediate adamantyl radicals by reaction with the solvent.

$$\text{1-AdI} + E^{2-} \xrightarrow[NH_3]{h\nu} \text{1-AdEEAd-1} + \text{AdH} + I^{-} \qquad (100)$$

(**161**)	(**162**)	(**163**)	
(a) E = Se	46%	23%	70%
(b) E = Te	47%	20%	69%

$$\text{1-AdI} + E^{2-} \xrightarrow[S_{RN}1]{h\nu} \text{1-AdE}^{-} + I^{-} \xrightarrow{[O]} \text{1-AdEEAd-1} \qquad (101)$$

(**161**) (**164**) (**162**)

The photostimulated reaction of 1-iodoadamantane with PhS^{-} gave three products (**147**–**149**), the same scrambling mixture that was obtained when iodobenzene reacted with 1-$AdSe^{-}$ (equation 92), but in a different ratio (equation 102)[66]. Similarly, the photolysis of 1-AdI with $PhTe^{-}$ gave a different ratio of the same two products (**143** and **145**) which were obtained (equation 93) in the reaction of PhI with 1-$AdTe^{-}$ (equation 103). With 1-naphthyl selenide ion (**165**), only the unsymmetrical selenide **166** was observed (equation 104), and no selenides were formed in the photolysis of 1-AdI with 1-$AdTe^{-}$.

$$\text{1-AdI} + PhSe^{-} \xrightarrow[\text{liq. } NH_3]{h\nu} Ph_2Se + \text{1-AdSePh} + (\text{1-Ad})_2Se \qquad (102)$$

(**161**)	(**147**)	(**148**)	(**149**)
relative yields:	10%	74%	16%

$$\text{1-AdI} + PhTe^{-} \xrightarrow{h\nu} Ph_2Te + \text{1-AdTePh} \qquad (103)$$

	(**143**)	(**158**)
relative yields:	31%	39%

$$\text{1-AdI} + \text{1-NaphSe}^{-} \xrightarrow{h\nu} (\text{1-Naph})Se(\text{1-Ad}) \qquad (104)$$

(**165**) (**166**)

These results confirm the mechanistic scheme outlined in equation 105, in which an equilibrium is generally established between the mixed aralkyl radical ion **167** and the two ion – radical pairs which are obtained from the alternative photostimulated reactions.

Fragmentations of **167** on both sides permit the formation of the symmetrical scrambling products, and compete with electron transfer to form the mixed product. Since the rates of electron transfer from **167** to the substrate halides should be similar, and probably diffusion controlled, the relative amount of scrambling products will be determined by the different fragmentation rates of the R—E [k_f(RE)] and Ar—E [k_f(ArE)] bonds in **167**[66]. The ratio of fragmentation rates for the adamantyl – phenyl pair, k_f(AdE)/k_f(PhE), was found to be 3.7 (E = S), 9.5 (E = Se) and 13 (E = Te). This increase in the ratio of fragmentation rates on going from sulfur to tellurium suggests that in the photostimulated reactions the products obtained depend on the energy levels of the σ^* and π^* of the C—E bonds of the radical anion intermediate[66,77].

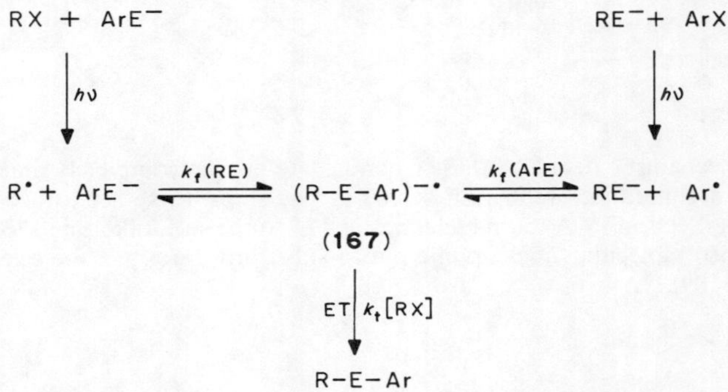

(105)

c. *Intramolecular substitution.* Excitation of *Se*-aryl esters of aromatic selenoacids **168**, which are *ortho*-substituted by appropriate leaving groups, often results in cyclization by intramolecular substitution. This reaction provides an important synthetic entry to novel heterocyclic seleno compounds.

ArCOSeAr′
(**168**)

In the first typical example, selenoxanthone (**170**) was obtained in 19% yield by irradiating through quartz a benzene solution of *Se*-phenyl 2-chloroselenobenzoate (**169**) with a high-pressure mercury lamp (equation 106)[11]. Similarly, irradiation of the 2-chloroselenonicotinate ester (**171**) leads via loss of HCl to a mixture of selenoxanthone (**13**), (25%) and bis(*p*-tolyl selenide) (**7**) (57%) (equation 107)[14,78]. The selenophene analog **173** was obtained in low yields (4%) together with **7** (66%) irradiation of the *Se*-*p*-tolyl ester of 3-bromoselenophene-2-selenocarboxylic acid (**172**) (equation 108)[79].

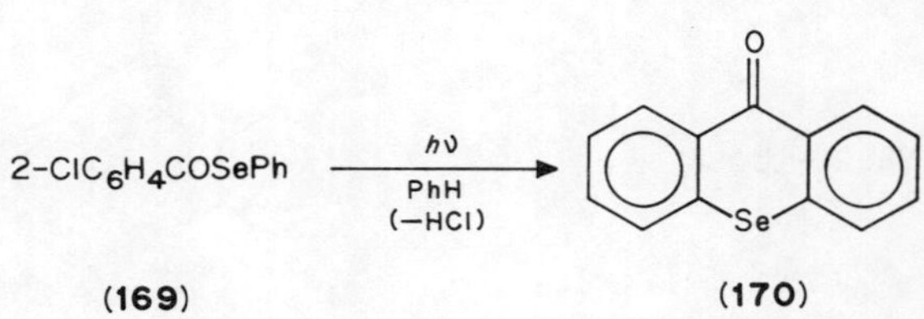

(106)

(**169**) (**170**)

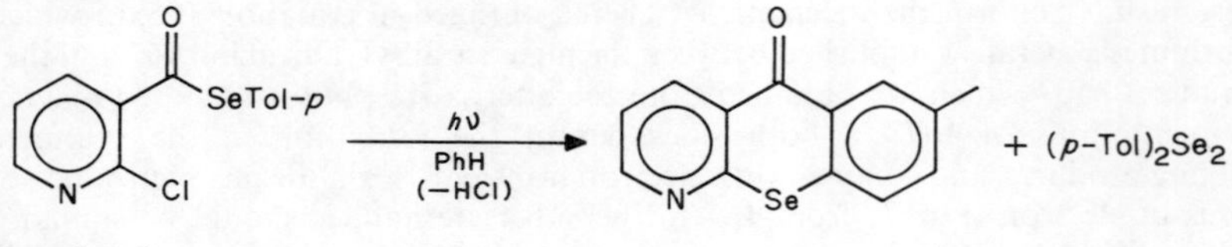

(171) (13) (7)

(107)

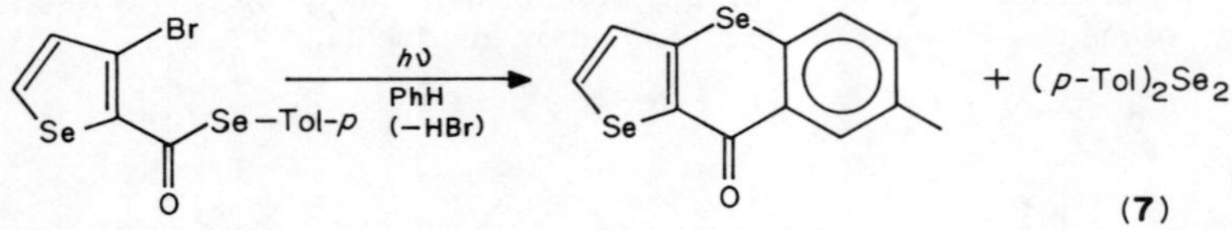

(172) (173) (7)

(108)

While the mechanistic details of these reactions are not yet completely understood, it is reasonable to assume that cyclization occurs in two steps. First, a photochemical Fries rearrangement (Section IV.A) of the selenoester **174** to the selenolketone **175** takes place, followed by intramolecular nucleophilic photo-substitution to give the selenoxanthone **176** (equation 109).

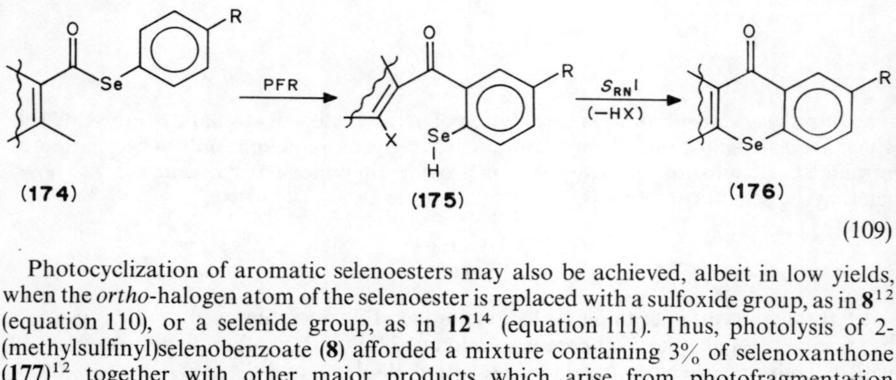

(174) (175) (176)

(109)

Photocyclization of aromatic selenoesters may also be achieved, albeit in low yields, when the *ortho*-halogen atom of the selenoester is replaced with a sulfoxide group, as in **8**[12] (equation 110), or a selenide group, as in **12**[14] (equation 111). Thus, photolysis of 2-(methylsulfinyl)selenobenzoate (**8**) afforded a mixture containing 3% of selenoxanthone (**177**)[12] together with other major products which arise from photofragmentation (Section II.A.1). Similarly, irradiation of selenonicotinic ester **12** gave 15% yield of **13**[14]. Here, however, an alternative mechanistic pathway has been proposed[14], which involves a photoinduced homolytic fragmentation of **12** to radicals **178** and **179**, followed by an intramolecular Friedel – Crafts cyclization[80] of **178** to **13** and dimerization of **179** to the diselenide **7** (equation 111).

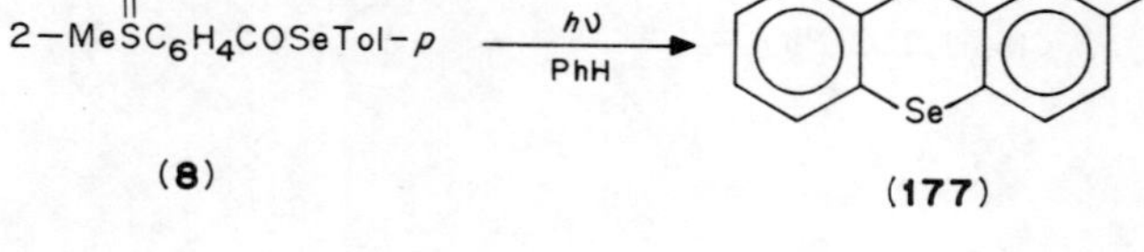

(8) (177)

(110)

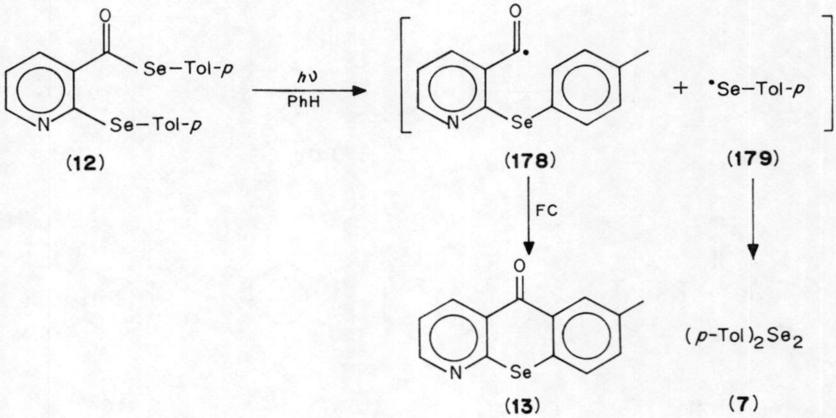

(111)

An interesting example in which competition between a photo-Fries rearrangement and Friedel–Crafts acylation apparently occurred has been reported for the photoinduced reaction of **10** (equation 112)[13]. On irradiation, two isoelectronic heterocyclic systems were formed in a 2:3 ratio; the selenopyrone **180** via the photo-Fries pathway and the thiopyrone **11** by the Friedel–Crafts mechanism. These products were accompanied by the corresponding disulfide and the diselenide. Further experimental evidence for the occurrence of both mechanistic pathways in this reaction was deduced from the observation that when the aromatic sulfide moiety in **10a** was replaced by the inferior methyl sulfide leaving group of **10b**, no cyclization products were detected and only the diselenide **7** could be isolated from the reaction mixture.

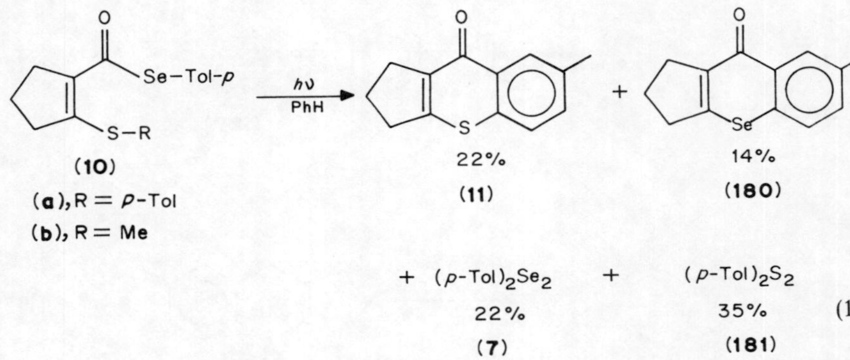

(112)

Telluroesters do not usually undergo photochemical cyclization reactions, although both fragmentation (Section II.A.1) and photo-Fries rearrangements (Section IV.A.1) have often been observed in these compounds. However, the methylthio-substituted telluroester **19** undergoes photocyclization to give, among other products, 22% of the unexpected thioxanthone **20** (equation 113)[16]. Here, unlike with the selenoester **181**, the rearrangement is followed by displacement of the tellurium by sulfur. The rearrangement products, **182** and **183**, also readily undergo photocyclization to thioxanthone **20**, demonstrating again the ability of excited sulfides to displace tellurides[16].

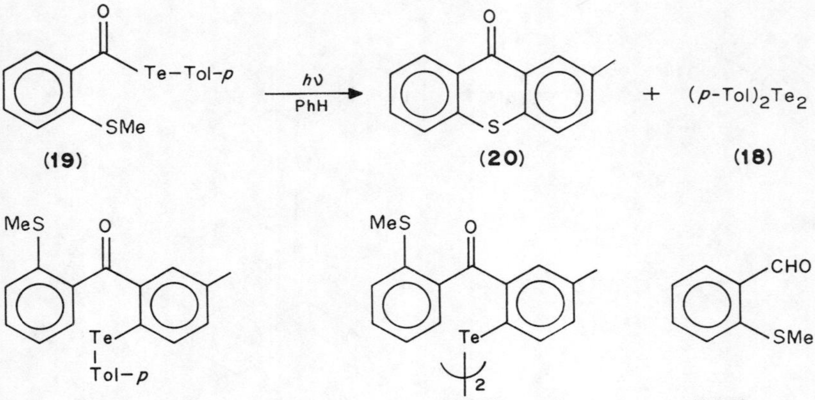

(19) (20) (18)

(182) (183) (184)

(113)

A unique example of a photocyclization at an unsubstituted vinylic carbon of a selenophene ring has been reported[79]. Irradiation of the selenoester **185** in benzene (Solidexglas filter) afforded the novel 6-methyl-4*H*-selenolo[2, 3-*b*] [1]benzoselenin-4-one (**186**) in 16% yield, together with diselenide **7** (34%) and selenophene-3-carboxaldehyde (**187**) (2%) (equation 114). A similar reaction occurred when the analogous selenophene thioester **188** photocyclized to thieninone (**189**) (equation 115)[81]. In both cases a formal oxidation by loss of a hydrogen molecule occurred, but this could not be established experimentally.

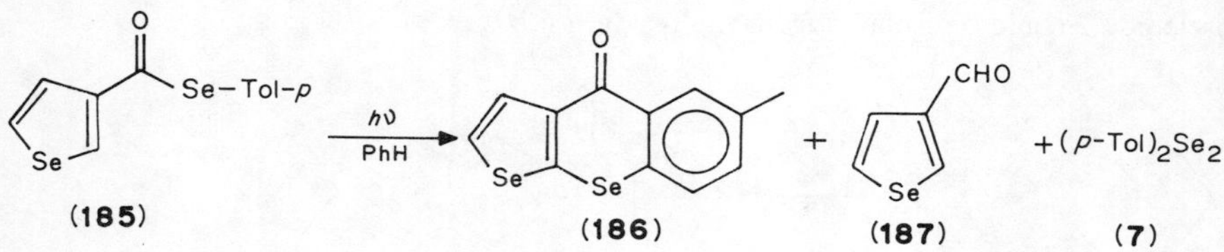

(185) (186) (187) (7)

(114)

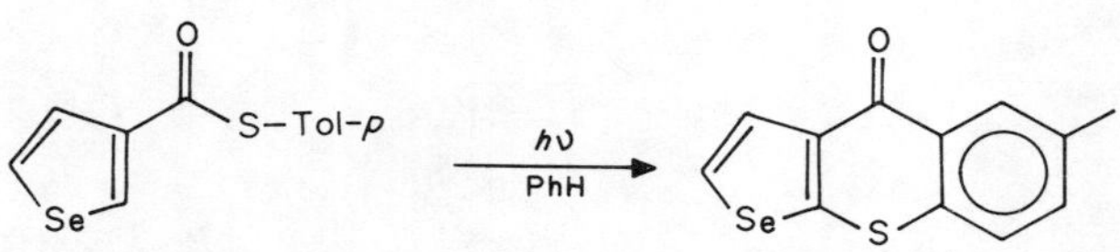

(188) (189)

(115)

2. Organometallic substitutions

a. Free-radical displacements. Free-radical chain reactions involving substitution at a metal center may in general be separated into bimolecular homolytic substitutions (S_H2) and addition–elimination reactions, represented by equations 116 and 117, respectively. Such radical chain reactions are formally analogous to the well known S_N2 and S_N1

heterolytic reactions, and have ample precedence in the free-radical chemistry of organometallic compounds[82].

$$LMR + X^{\bullet} \xrightarrow{S_H2} LMX + R^{\bullet} \quad (116)$$

$$LMR + X^{\bullet} \xrightarrow{\text{addition}} [LMRX] \xrightarrow{\text{elimination}} RX + LM^{\bullet} \quad (117)$$

Vinylmercury halides (**190**) readily undergo photostimulated free-radical chain substitution reactions in the presence of bis(phenyl selenide) and bis(phenyl telluride) to give the corresponding vinyl phenyl selenides and tellurides (**191**) and mercury(II) salts (**192**) in high yields (equation 118)[83]. The reactions were carried out in benzene solutions using a 275-W sunlamp at 34–45 °C, and were completely inhibited by radical quenchers such as di-*tert*-butyl nitroxide and galvinoxyl.

$$\underset{(\mathbf{190})}{RCH=CHHgCl} + PhEEPh \xrightarrow[\text{PhH}]{h\nu} \underset{(\mathbf{191})}{RCH=CHEPh} + \underset{(\mathbf{192})}{PhEHgCl} \quad (118)$$

		yield
R = H,	E = Se	91%
R = *t*-Bu,	E = Se	95%
R = Ph,	E = Se	90%
R = *t*-Bu,	E = Te	89%

Since these reactions fail to occur with benzyl- or phenyl-mercury halides, the initial formation of a carbon-centered radical intermediate to propagate an S_H2 reaction was excluded. Instead, a free-radical addition–elimination[82,84] mechanism was suggested, with initiation by the photohomolytic cleavage of the chalcogenide (equation 119), and the subsequent propagation sequence shown in equations 120–121.

$$PhEEPh \xrightarrow{h\nu} PhE^{\bullet} \quad (119)$$

$$(E = Se, Te)$$

$$RCH=CHHgX + PhE^{\bullet} \longrightarrow RCHCH(EPh)HgX \longrightarrow RCH=CHEPh + HgX^{\bullet} \quad (120)$$

$$HgX^{\bullet} + PhEEPh \longrightarrow PEHgX + PhE^{\bullet} \quad (121)$$

Irradiation of allyltributylstannane (**193**) with PhSeSePh in benzene (Pyrex filter) afforded allyl phenyl selenide (**194**) and the selenide **195** (equation 122)[85]. The reaction did not occur in the dark, but was initiated by azobis(isobutyronitrile) (AIBN) at 70 °C. The photostimulated reaction was completely inhibited by nitroxide or galvinoxyl. When crotyltributylstannane (**196**) was irradiated with bis(phenyl sulfide), the allylic rearrangement product **197** was obtained (equation 123).

$$\underset{(\mathbf{193})}{CH_2=CHCH_2SnBu_3} + \underset{(\mathbf{36})}{PhSeSePh} \xrightarrow[\text{PhH}]{h\nu\ (\text{Pyrex})} \underset{(\mathbf{194})}{PhSeCH_2CH=CH_2} + \underset{(\mathbf{195})}{Bu_3SnSePh} \quad (122)$$

$$MeCH=CHCH_2SnBu_3 + PhSSPh \xrightarrow{h\nu} PhSCH(Me)CH=CH_2 + Bu_3SnSePh$$

(196) **(197)**

(123)

These results suggest that unlike the vinyl mercurials **190**, the stannanes operative via an S_H2 free-radical chain mechanism induced by PhSe radicals. Propagation is achieved by allylic displacement of a Bu_3Sn radical, followed by reaction of the latter with the diselenide to regenerate PhSe radicals (equations 124 and 125).

$$PhSe^{\bullet} + CH_2=CHCH_2SnBu_3 \longrightarrow PhSeCH_2CH=CH_2 + Bu_3Sn^{\bullet} \quad (124)$$

$$Bu_3Sn^{\bullet} + PhSeSePh \longrightarrow Bu_3SePh + PhSe^{\bullet} \quad (125)$$

A closer look at the intermediate radicals obtained during the photochemical metal displacement with selenide was provided by studies of the photoinduced reaction of bis(phenyl selenide) (**36**) with alkyl, allyl and allenyl derivatives of bis(dimethylglyoximato)pyridinecobalt(III) complexes (**198**) (cobaloximes). When equimolar amounts of the optically active (*R*)-*s*-octyl-cobaloxime (**198a**) and **36** in dichloromethane were irradiated through pyrex with tungsten light, the racemic *s*-octyl phenyl selenide **199a** was obtained in 62% yield, together with phenylselenyl-cobaloxime (**200**) (equation 126)[86]. Good yields of selenides were also obtained from benzylic and allylic cobaloximes **198b**–**d**.

$$RCo(dmgH)_2py + PhSeSePh \xrightarrow[CH_2Cl_2]{h\nu} PhSeR + PhSeCo(dmgH)_2py \quad (126)$$

(198) **(36)** **(199)** **(200)**

(**a**) R = (*R*)-*s*-octyl R = racemic *s*-octyl
(**b**) R = PhCHMe
(**c**) $R = Me_2C=CHCH_2$
(**d**) $R = PhCH=CHCH_2$

In all the above reactions only a single isomeric selenide (**199**) was observed. However, when the but-2-enyl-cobaloxime **201** was irradiated, two isomers, **202** and **203**, were isolated (equation 127). The allenyl-cobaloxime **204** likewise gave two isomers, propadienyl selenide (**205**) and propynyl selenide (**206**) (equation 128).

$$MeCH=CHCH_2Co(dmgH)_2py + PhSeSePh \xrightarrow{h\nu} MeCH=CHCH_2SePh$$

(201) **(202)**

$$+ CH_2=CHCH(Me)SePh + \mathbf{200}$$

(203)

(127)

$$Me_2C=C=CHCo(dmgH)_2py + PSeSePh \xrightarrow{h\nu} Me_2C=C=CHSePh +$$

(204) **(205)**

$$HC\equiv CC(Me)_2SePh + \mathbf{200}$$

(206)

(128)

Based on the fact that cobaloximes readily form radicals under photolytic conditions[87], a non-chain free-radical mechanism is suggested (equations 129–131)[86], in which an initial photoinduced homolysis of the cobaloximes takes place to give the corresponding alkyl radicals, which attack diphenyl diselenide to form alkyl phenyl selenide and phenyl selenide radicals. The latter recombine with the metal fragment to form phenyl selenide cobaloxime. The formation of isomeric mixtures of selenides in the reaction of **201** and **204** supports this mechanism, although a combination of the S_H2[82,85] and addition–elimination[83,84] pathways cannot be excluded. However, racemization of the (*R*)-*s*-octyl group during the photolysis of cobaloxime **198a** seems to exclude these pathways and to favor the non-chain mechanism suggested.

$$RCo^{III}(dmgH)_2py \longrightarrow R^{\bullet} + Co^{II}(dmgH)_2py \qquad (129)$$

$$R^{\bullet} + PhSeSePh \longrightarrow RSePh + PhSe^{\bullet} \qquad (130)$$

$$Co^{III}(dmgH)_2py + PhSe^{\bullet} \longrightarrow PhSeCo(dmgH)_2py \qquad (131)$$

Aromatic ligands are photolytically displaced by selenides in zirconocene derivatives[88]. Irradiation of heptane solutions of diphenylzirconocene (**207**) with bis(phenyl selenide) (**36**) afforded bis(phenylselenyl)zirconocene (**208**) and biphenyl (equation 132). No reaction occurred in the dark. With dimethylzirconocene (**209**), monosubstitution by a PhSe group took place thermally in the dark, forming the thermally stable monoselenide **210** (equation 133), which on irradiation with **36** gave the expected **207**. These reactions are assumed to proceed via the bimolecular homolytic substitution (S_H2) mechanism.

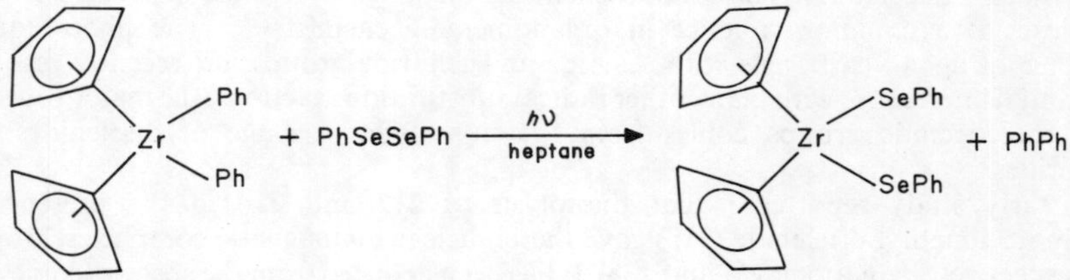

(207) **(36)** **(208)**

(132)

$$(\eta^5\text{-}RC_5H_4)ZrMe_2 + PhSeSePh \xrightarrow{\Delta} (\eta^5\text{-}RC_5H_4)_2Zr(SePh)Me + PhSeMe \qquad (133)$$

(209) **(210)**

$$(\eta^5\text{-}RC_5H_4)_2Zr(SePh)Me + PhSeSePh \xrightarrow{h\nu} (\eta^5\text{-}RC_5H_4)_2Zr(SePh)_2 \qquad (134)$$

(210) **(208)**

R = H, *t*-Bu

The same group[89] succeeded in achieving photosubstitution of phenyl and methyl groups in zirconocene and hafnocene derivatives using elemental (grey) selenium. Thus, when pentane solutions of **211** (M = Zr, Hf) were irradiated with UV light in the presence of grey selenium, a dinuclear selenium-bridged complex (**212**) was isolated, together with an equimolar amount of biphenyl. The same products were obtained when the dimethyl analogs $(\eta^5\text{-}t\text{-}BuC_5H_4)_2M(CH_3)_2$ (M = Zr, Hf) were irradiated in the presence of Se, and when **208** was irradiated with **211** (equations 135 and 136). The yields were lower, however, and the by-products were not characterized.

$$(\eta^5\text{-}t\text{-BuCp})_2\text{MPh}_2 + \text{Se (grey)} \xrightarrow{h\nu} [(\pi^5\text{-}t\text{-BuCp})_2\text{M}(\mu\text{-Se})]_2 + \text{PhPh} \quad (135)$$

(211) **(212)**

$$(\eta^5\text{-}t\text{-BuCp})_2\text{MPh}_2 + (\eta^5\text{-}t\text{-BuCp})\text{M(SePh)}_2 \xrightarrow{h\nu} [(\eta^5\text{-}t\text{-BuCp})_2\text{M}(\mu\text{-Se})]_2 \quad (136)$$

(211) **(208)** **(212)**

M = Zr, Hf

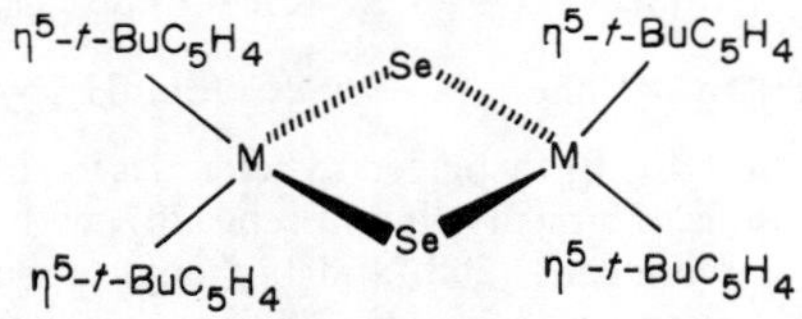

(212)
M=Zr, Hf

b. Ligand-exchange reactions. Photochemical exchange of ligands in metal carbonyl complexes is a common practice in organometallic chemistry[90]. The photoinduced exchange of ligand carbonyls with selenide and telluride groups has recently regained substantial interest. As with many other radical substitution reactions, the major source of these chalcogenide groups comes from the homolytic cleavage of diselenides and ditellurides.

An early study reported[91] that photolysis of **213** and **214** in the presence of hexafluorodimethyl diselenide (**215**) gave the binuclear diamagnetic complexes **216** and **217**, respectively (equations 137 and 138). It has been deduced from the spectroscopic data that these complexes contain bridging CF_3Se groups and terminal CO groups. A tentative free-radical mechanism is suggested in which $SeCF_3$ radicals combine with $M(CO)_n$ fragment (M = Fe, $n = 4$; M = Mn, $n = 5$) to give a labile metal selenide $[M(CO)_n(SeCF_3)]$ which dimerizes to the products with the expulsion of CO.

$$\text{Fe(CO)}_5 + \text{CF}_3\text{SeSeCF}_3 \xrightarrow[\text{CH}_2\text{Cl}_2,\ \text{hexane}]{h\nu} [\text{Fe(CO)}_3(\mu\text{-SeCF}_3)]_2 \quad (137)$$

(213) **(215)** **(216)**

$$\text{Mn}_2\text{(CO)}_{10} + \text{CF}_3\text{SeSeCF}_3 \xrightarrow{h\nu} [\text{Mn(CO)}_4(\mu\text{-SeCF}_3)]_2 \quad (138)$$

(214) **(217)**

More recently, a cubane-type tetranuclear cluster **218** was isolated[92] in the photolysis of **214** in the presence of PhSeSePh (equation 139). In addition, small amounts of the dinuclear complex **219** were isolated. Attempts to prepare mixed metal compounds having both sulfur and selenium bridges in the same molecule by irradiating **214** in the presence of both diphenyl diselenide and disulfide failed. Only the selenium derivatives **218** and **219** were obtained[93].

$$\underset{(\mathbf{214})}{Mn_2(CO)_{10}} + PhSeSePh \xrightarrow[\text{pentane}]{h\nu} \underset{(\mathbf{218})}{[Mn(CO)_3(\mu\text{-}SePh)]_4} + \underset{(\mathbf{219})}{[Mn(CO)_4(\mu\text{-}SePh)]_2} \qquad (139)$$

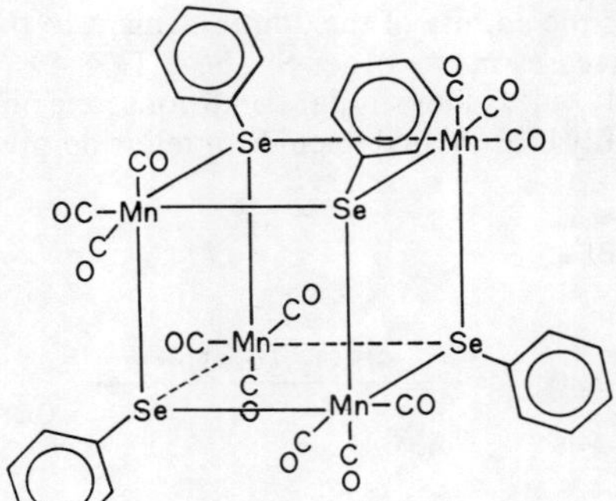

(**218**)
Structure suggested for
$[Mn(CO)_3SeC_6H_5]_4$

With Group VIB carbonyls, the dimolybdenum complexes **221** and **222** were isolated when the corresponding dimolybdenum hexacarbonyl complex **220** was irradiated with PhEEPh (E = Se, Te) (equation 140)[94]. The reaction proceeded in two steps, photolysis of **220** giving **221** as the primary product, followed by decarbonylation of the latter in vacuum under mild thermolysis conditions. The structure of **222** was deduced from the IR and mass spectra.

$$\underset{(\mathbf{220})}{[(\eta^5\text{-}MeCp)Mo(CO)_3]_2} \xrightarrow[\text{toluene}]{h\nu,\ +(PhE)_2} \underset{(\mathbf{221})}{[(\eta^5\text{-}MeCp)Mo(CO)_2(\mu\text{-}EPh)]_2} \xrightarrow{-CO} \underset{(\mathbf{222})}{[(\eta^5\text{-}MeCp)Mo(CO)(\mu\text{-}EPh)]_2} \qquad (140)$$

E = Se, Te

MeCp = methylcyclopentadienyl

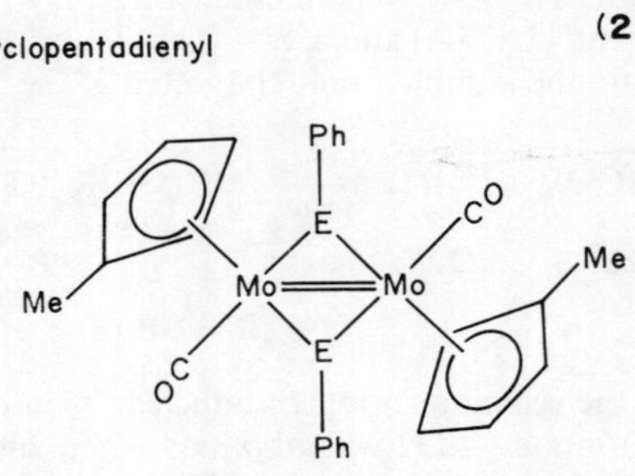

(**222**)
E = Se, Te
Structure suggested for
$[(\eta^5\text{-}MeCp)Mo(CO)(\mu\text{-}EPh)]_2$

Substitution of a carbonyl group by dimethyl selenide and telluride (and also sulfide) to give cyclopentadienyliron fluoroborate complexes (**224**) was accomplished by irradiation at room temperature of dichloromethane solutions of complexes **223** and the corresponding chalcogenides[95] (equation 141). The half-life of the exchange reaction in going from sulfur to tellurium compounds increases in the order $\tau_{1/2} = 1.7\,h\,(S) < 2.3\,h\,(Se) < 11.6\,h\,(Te)$. This reflects the affinity of the donor element to the coordination center of the complexes, which decreases in the order S < Se < Te.

In a closely related study, and under similar conditions, one phosphite group of **225** was substituted photochemically by dimethyl selenide or telluride, giving **226** (equation 142)[96].

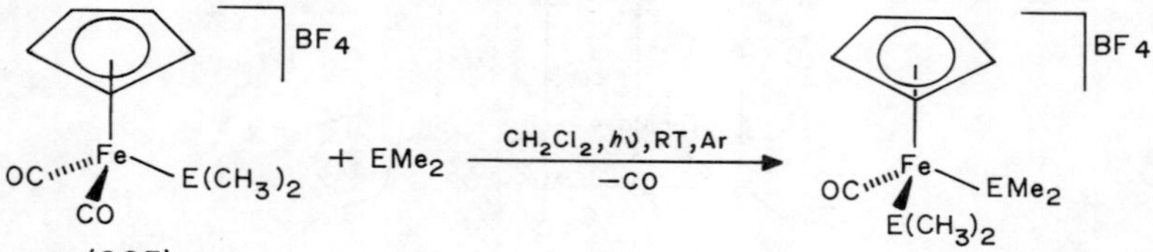

(141)

(223)

E=S,Se,Te

(224)

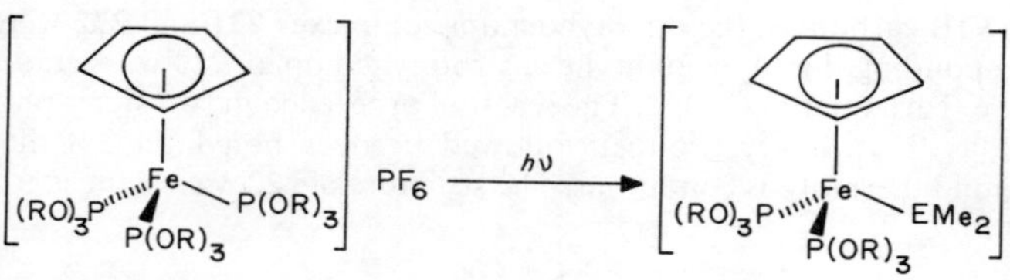

(142)

(225)

E=S,Se,Te; R=Me,Ph

(226)

The first stable tellurophosphosphorane complex (**229**) were prepared in almost quantitative yields by irradiation of Group VIB transition metal hexacarbonyls (**227**) with tri-*tert*-butyltellurophosphorane (**228**) in THF (equation 143)[97]. The complex products (**229**) are fairly stable in air. The X-ray crystal structure of $W(CO)_5(R_3PTe)$ (R = *t*-Bu) exhibited a short W–Te bond (2.875 Å) and a W—Te—P angle close to 120°, suggesting an sp^2-type tellurium atom and double bond character of the W—Te bond[97].

$$M(CO)_6 + R_3PTe \xrightarrow[THF]{h\nu} M(CO)_5(R_3PTe) \qquad (143)$$

(**227**) (**228**) (**229**)

M = Cr, Mo, W; R = *t*-Bu

The X-ray structure of the selenium-bridged dinuclear rhenium complex **232** (E = Se) has been established[98]. Compound **232** was obtained when **230** was irradiated in THF solution. The key intermediate in the photolysis of **230** is the labile rhenium complex **231**, in which one of the original carbonyls is replaced by a THF molecule. This complex has been shown to react thermally with elemental selenium or tellurium to give the bridged products **232**. This may represent a general mechanism of photoinduced carbonyl exchange reactions (equation 144).

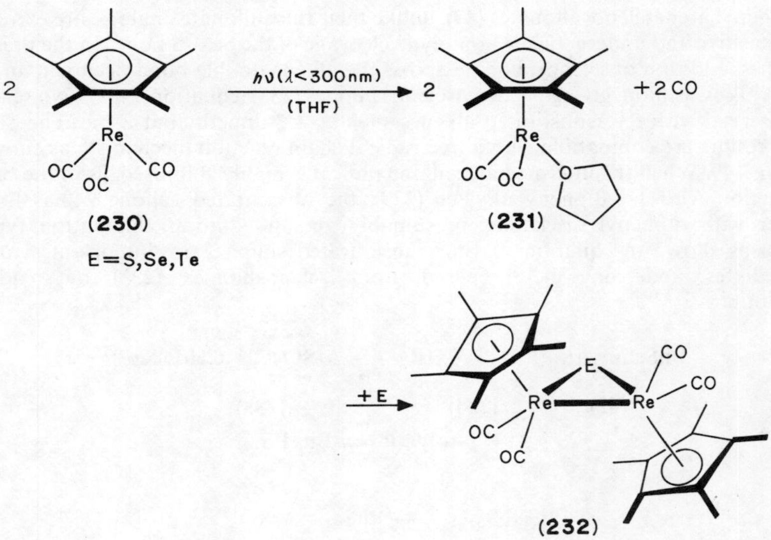

(144)

B. Photoaddition Reactions

Photolysis of silicon- and germanium-bonded selenols with olefins affords anti-Markovnikov addition products which result from homolytic cleavage of the Se—H bond[23]. Triethylsilaneselenol (**233**) reacts with styrene (**234**, R = Ph) and with hexene (**234**, R = Bu), giving the 1:1 adducts **235** in 90% and 72% yield, respectively (equation 145). Similarly, triethylgermaneselenol (**236**) reacts with styrene to give triethyl(phenethylselenyl)germane (**237**, R = Ph). However, the reaction of **236** with ethyl acrylate (**234**, R = COOEt) gave both the expected 3-[(triethylgermyl)selenoyl]propionate (**237**, R = COOEt) and **45** (equation 146). Irradiation of **236** with acrylonitrile (**66**) resulted in the isolation of only the symmetrical product 3, 3′-selenodipropionitrile (**67**) (equation 147). The presence of these products in the reaction mixture clearly indicates that cleavage of the Ge–Se bond competes with that of the Se—H bond (Section II.A.3b).

$$Et_3SiSeH + CH_2=CHR \xrightarrow{h\nu\ (Pyrex)} Et_3SiSeCH_2CH_2R \qquad (145)$$

(**233**) (**234**) (**235**)

R = *t*-Bu, Ph

$$Et_3GeSeH + CH_2=CHR \xrightarrow{h\nu} Et_3GeSeCH_2CH_2R + (Et_3Ge)_2Se \qquad (146)$$

(**236**) (**234**) (**237**) (**45**)

R = Ph, COOEt

$$Et_3GeSeH + CH_2=CHCN \xrightarrow{h\nu} Se(CH_2CH_2CN)_2 + \mathbf{45} \qquad (147)$$

(**66**) (**67**)

Se-Phenyl areneselenosulfonates (**42**), unlike their thiosulfonate analogs, are extremely photosensitive and undergo facile homolytic cleavage of the Se—S bond. In the presence of alkenes, addition of selenosulfonate across the alkene double bond occurs in an anti-Markovnikov fashion, giving β-phenylseleno sulfones (**238**) (equation 148)[22]. No reaction was observed with tetrasubstituted alkenes such as 2, 3-dimethylbut-2-ene. The experimental results are compatible with a free-radical chain reaction mechanism as shown in equation 149. When the intermediate sulfone radical is highly stabilized, as in the case of the reaction with 1, 1-diphenylethylene (**239**), the unsaturated sulfone **240** is formed together with diphenyl diselenide, presumably via an addition–elimination type of reaction as shown in equation 50. Such unsaturated sulfones are important synthetic intermediates and may be prepared from selenosulfones (**238**) by oxidative elimination[22,99,100].

$$\underset{(\mathbf{42})}{PhSeSO_2Ar} + \underset{(\mathbf{234})}{CH_2{=}CHR} \xrightarrow[CCl_4]{h\nu} \underset{(\mathbf{238})}{ArSO_2CH_2CH(R)SePh} \tag{148}$$

Ar = *p*-tolyl; R = *n*-Bu, Ph

$$\begin{aligned} PhSeSO_2Ar &\xrightarrow{h\nu} PhSe^{\bullet} + ArSO_2^{\bullet} \\ ArSO_2^{\bullet} + CH_2{=}CHR &\longrightarrow ArSO_2CH_2\dot{C}HR \\ ArSO_2CH_2\dot{C}HR + PhSeSO_2Ar &\longrightarrow ArSO_2CH_2CH(R)SePh + ArSO_2^{\bullet} \end{aligned} \tag{149}$$

$$ArSO_2^{\bullet} + \underset{(\mathbf{239})}{CH_2{=}CPh_2} \longrightarrow ArSO_2CH_2\dot{C}Ph_2 \xrightarrow{-H^{\bullet}} \underset{(\mathbf{240})}{ArSO_2CH{=}CPh_2} \tag{150}$$

When a large excess of olefin over selenosulfonate was used, polymerization occurred. Thus, irradiation of **42** in 10-fold excess of hex-1-ene gave polymers of general structure **241**.

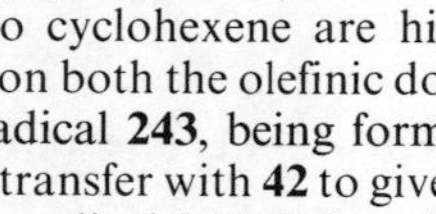

$n > 1$

(**241**)

The photoaddition of **42** to cyclohexene was found to be highly stereospecific, giving rise to 80% of the *trans*-adduct **242** (equation 151) with no significant amounts of the *cis*-adduct. Free-radical additions to cyclohexene are highly stereoselective, exhibiting a strong preference for axial attack on both the olefinic double bond and the derived radical intermediate[101,102]. Therefore, radical **243**, being formed by axial attack of $ArSO_2^{\bullet}$ on cyclohexene, must undergo chain transfer with **42** to give **242** more rapidly than ring flip to the conformationally more stable radical **244**. Otherwise, the *cis*-isomer **245** would have been formed (equation 152).

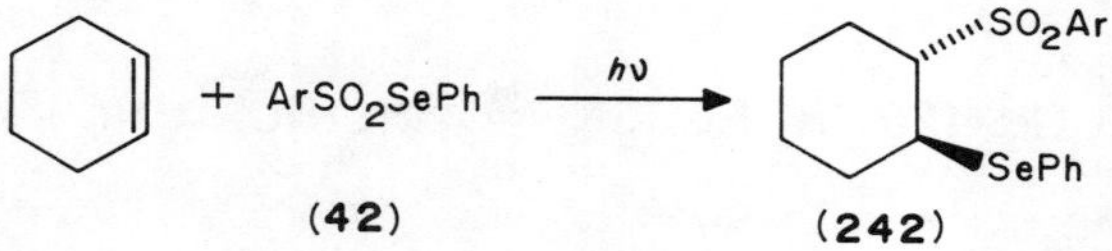

(**42**) (**242**) (151)

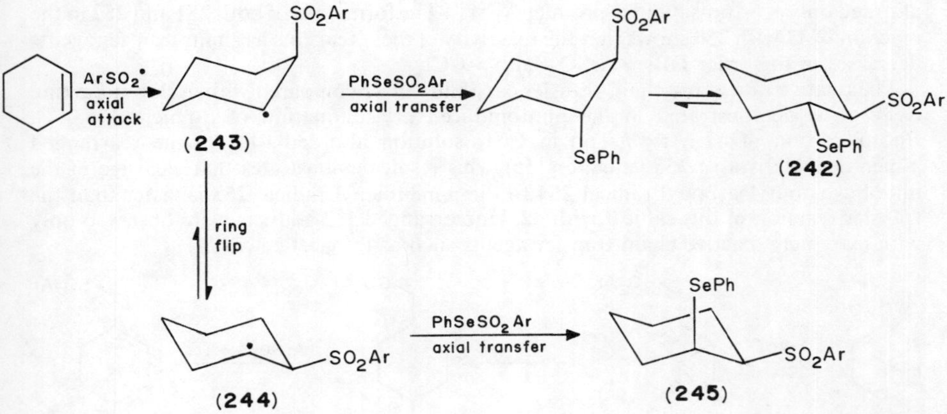

(152)

Further information regarding the reactivity of the intermediate arylsulfonyl radical in the chain transfer process is provided from studies of the photoinduced selenosulfonation of non-conjugated dienes[22,103]. Photoaddition of **42** to hepta-1, 6-diene (**246**) in CCl_4 (equation 153) afforded a mixture of three products in a ratio depending on the initial concentration of reactants. At 1 M the open-chain products **247** and **248** predominate over the cyclic product **249** (9:1 ratio), while in more dilute solutions (0.1 M) equal amounts of open-chain to cyclic products are observed[103]. This indicates that **42** is reactive enough as a chain-transfer agent to compete with the transannular reaction to give **249**.

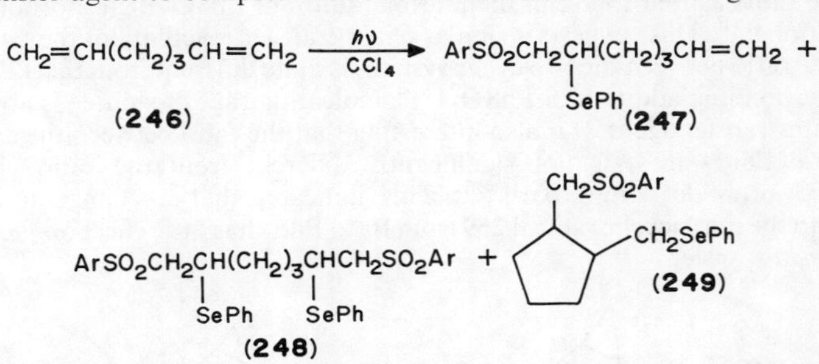

(153)

Likewise, free-radical addition of **42** to cycloocta-1, 5-diene (**250**) gives a mixture of approximately equal amounts of the 1, 2-adduct **251** and the 1, 5-transannular product **252** (equation 154)[22].

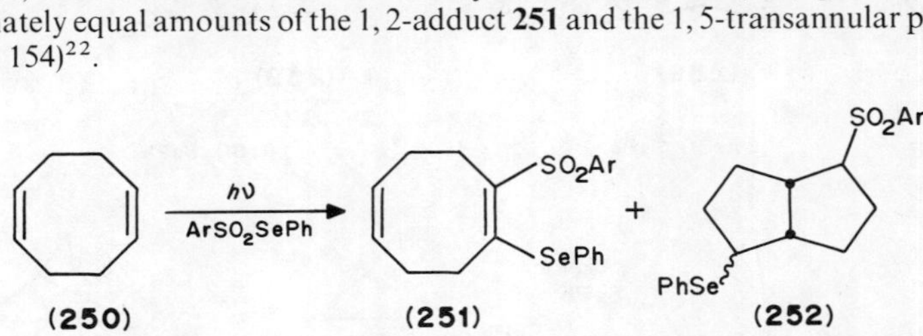

(154)

It is instructive to compare the reactivity of selenosulfonates (**42**) with other chain-transfer agents such as HBr and CCl_4. Photoaddition of HBr to cycloocta-1, 5-diene (**250**) is reported to give only 1, 2-adducts[104], whereas free-radical addition of CCl_4 to **250** afforded only 1, 5-transannular products[105,106]. The formation of both **251** and **252** in the reaction of **42** with **250** shows that the reactivity of these reagents as chain-transfer agents decreases in the order HBr > $ArSO_2SePh$ > CCl_4.

The relative rate of the chain-transfer capability of **42** compared with cyclobutane ring cleavage is demonstrated in the photoinduced arylsulfonation of β-pinene (**253**)[103]. Photoaddition of **42** (Ar = *p*-Tol) in CCl_4 solution afforded 91% of the rearranged *p*-menthene derivative **256** (equation 155). This result demonstrates that cleavage of the initial cyclobutylcarbonyl radical **254** to the *p*-menthenyl radical **255** is faster than the transfer reaction of this radical with **42**. Unrearranged 1, 2-adducts were observed only with extremely reactive chain-transfer agents such as thioacetic acid[107].

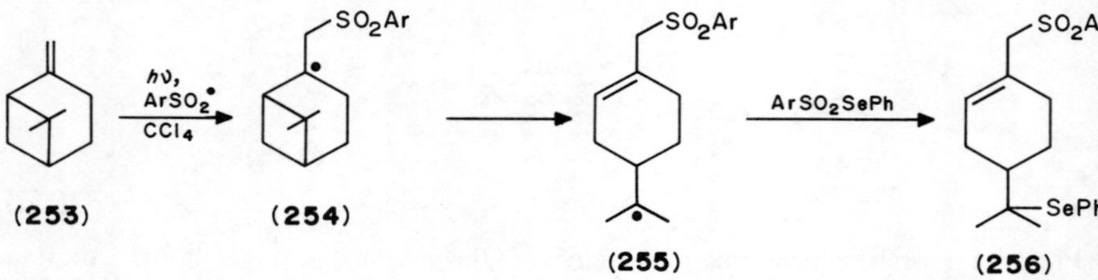

(253) (254) (255) (256)

(155)

The chain-transfer reactivity of selenosulfonate **42** (Ar = *p*-Tol) was studied in the norbornane – nortricyclane system. The interconversion between radicals **258** and **259** is known to be very rapid, with the nortricyclenyl radical being favored at equilibrium[108,109]. Photoaddition of **42** (Ar = *p*-Tol) to norbornadiene (**257**) gives a 75% yield of a mixture of adducts consisting of ca. 90% of **261** and **262** and only 10% of **260**. This ratio is virtually the same as that found in the photoaddition of $PhSO_2Br$ to norbornadiene at similar dilutions[110]. This suggests a similar chain-transfer reactivity of the two reagents. However, the ratio between the two cyclic systems is quite different from that (27:73) found for the corresponding addition of $PhSO_2I$[110], indicating that the iodide is a much more reactive chain-transfer agent. It is also interesting that the ratio between the *exo*-**261** and *endo*-**262** tricyclanes (6:4) is not significantly different from the ratio (4:3) of the arenesulfonyl bromide counterparts[111]. This indicates that a change in the group transferred to the nortricyclic radical **259** from Br to PhSe has little effect on the preference of *exo* vs. *endo* transfer.

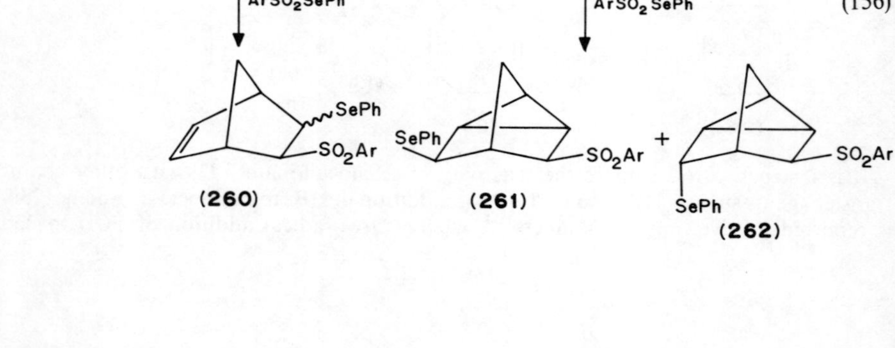

(257) (258) (259) (260) (261) (262)

(156)

Selenosulfonation of olefins may also be effected by alkaneselenosulfonates such as *Se*-phenyl-1-dodecaneselenosulfonate (**263**), which adds to cyclohexene in the usual 1,2-manner in 71% yield[103]. On the other hand, the photolysis of the corresponding benzylsulfonate **264** in the presence of cyclohexene did not result in an addition product. Instead, benzyl phenyl selenide (**265**) was isolated in high yields. Apparently, the benzylsulfonyl radical initially formed undergoes rapid desulfonylation to the more stable benzyl radical, which undergoes a transfer reaction with another molecule of selenosulfonate in preference to addition to an alkene molecule (equation 157).

$$\underset{\textbf{(263)}}{C_{12}H_{25}SO_2SePh} \qquad \underset{\textbf{(264)}}{PhCH_2SO_2SePh}$$

$$PhCH_2SO_2^{\bullet} \rightarrow PhCH_2^{\bullet} + SO_2$$

$$PhCH_2^{\bullet} + PhCH_2SO_2SePh \rightarrow \underset{\textbf{(265)}}{PhCH_2SePh} + PhCH_2SO_2^{\bullet} \qquad (157)$$

Another variation of the selenosulfonation reaction was observed in the photolysis of bis(*p*-tolylsulfonyl) selenide (**266**)[103] with cyclohexene (equation 158). In addition to the expected 1,2-adduct **267**, the diselenide **268** and some elemental selenium precipitate were obtained. The diselenide **268** is believed to be formed from the primary selenosulfonate product **267** in a subsequent photodissociation of the Se—S bond, followed by dimerization. Interestingly, the photoinduced reaction of **266** with styrene gave only high yields of the selenide **269** and none of the expected primary 1,2-addition product (equation 159).

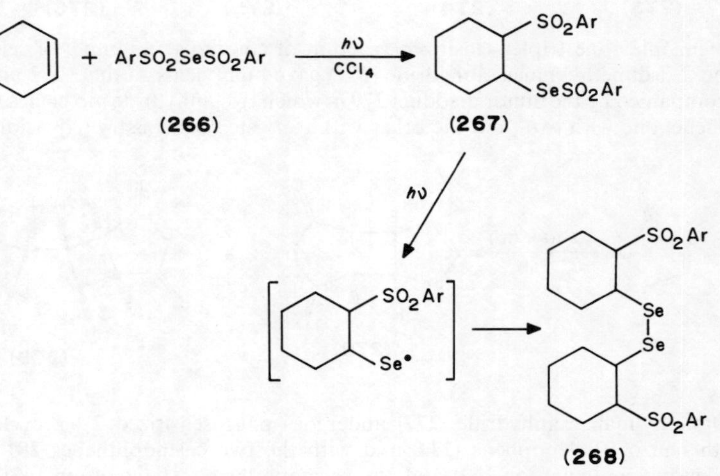

(158)

$$PhCH{=}CH_2 \xrightarrow[266]{h\nu} \underset{\textbf{(269)}}{(ArSO_2CH_2CHPh)_2Se} \qquad (159)$$

We conclude this section with the unusual photoaddition of ArSe radicals **271** (Ar = Ph, *o*-$NO_2C_6H_4$) to nitrosodurene (**270**)[112], which gave the corresponding spin adducts **272** (equation 160). Experiments were performed in a quartz cell at $-50\,^{\circ}C$. The ESR spectra of these selenylnitroxide radicals show hyperfine coupling with ^{77}Se ($\alpha = 7.8$–8.7 G), indicating a low spin density on the selenium. Similar results were obtained when diphenyl selenide was photolysed in the presence of **270**. This reconfirmed previous evidence of the homolytic nature of the cleavage of Se—Se and Se—C bonds in aromatic diselenides.

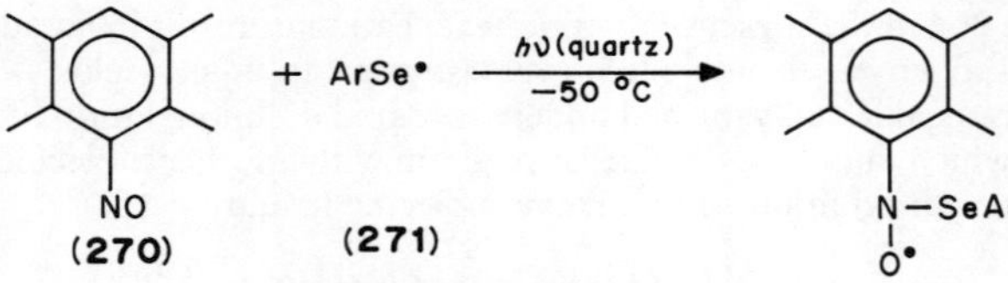

(270) (271) (272)

Ar=Ph,o-$O_2NC_6H_4$

(160)

C. Photocycloadditions

All photocycloadditions of organoselenium compounds involve the 2 + 2 cycloaddition of olefins and ketones to selenophene (**273**) and benzo[*b*]selenophene (**274**) derivatives. Olefins yield cyclobutanes[113] and ketones yield oxetanes (Paterno–Buchi reaction)[114]. Light-induced cycloaddition reactions of tellurophene derivatives are as yet unknown. Both tellurophene (**275**) and 2, 5-diphenyltellurophene (**276**) were stable under irradiation conditions in which selenophene derivatives react[115,116].

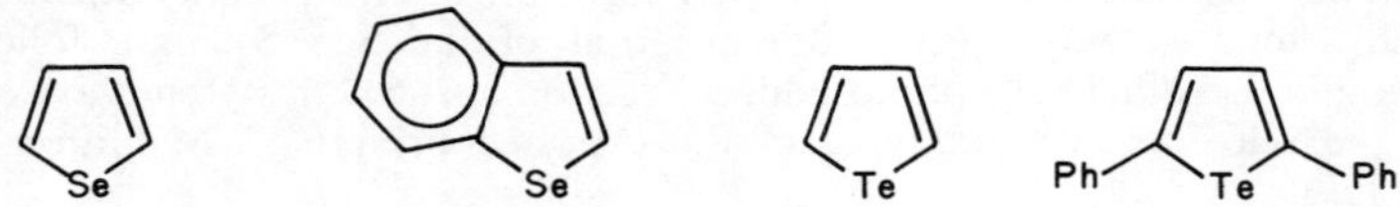

(273) (274) (275) (276)

The benzophenone triplet-sensitized reaction of equimolar amounts of selenophene (**273**) and 2, 3-dimethylmaleic anhydride (**277**) gave high yields of the 2 + 2 anti-adduct **278**, accompanied by the minor diadduct **279** in which two anhydride molecules add to the selenophene, one with *exo-* and the other with *endo*-stereochemistry (equation 161)[117].

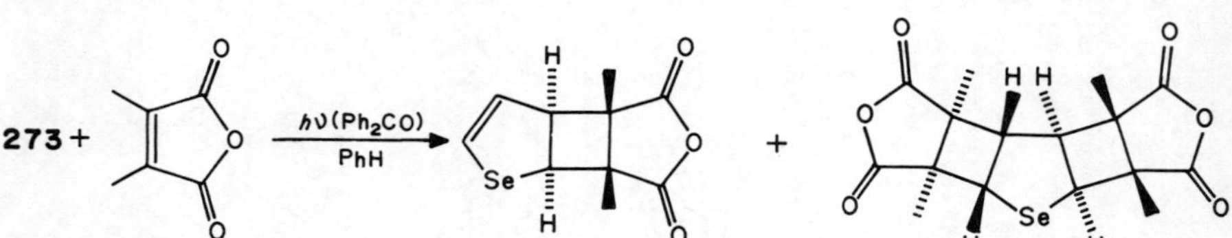

(277) (278) (279)

(161)

2, 3-Dimethylmaleic anhydride (**277**) undergoes photosensitized 2 + 2 cycloaddition also with benzo[b]selenophene (**274**) and with the two selenophthenes **281** and **283**. Benzoselenophene (equation 162) and the selenophthene **281** (equation 163) gave the corresponding monoadducts **280** and **282**, whereas the selenophthene **283** (equation 164) gave a mixture of the monoadduct **284** and diadduct **285**[118].

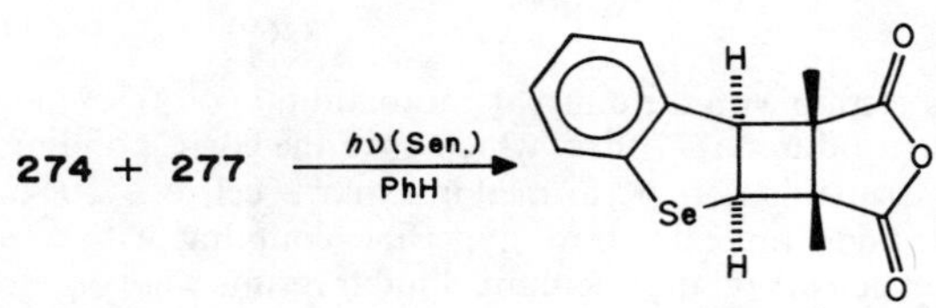

(162)

(Sen. = photosensitizer) (280)

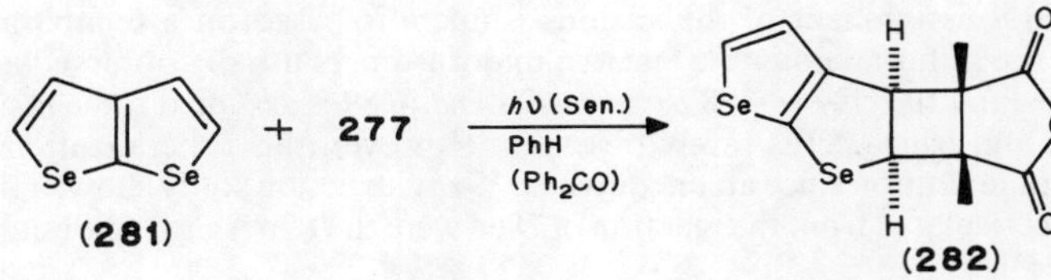

(163)

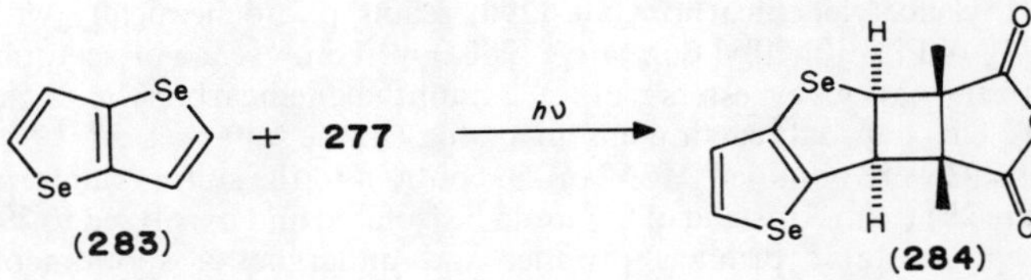

(164)

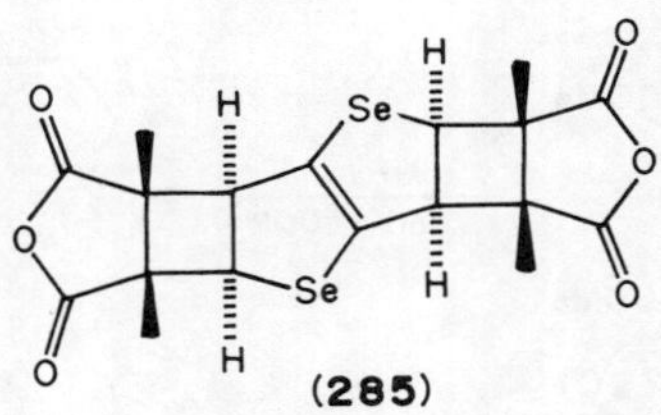

Acetophenone-sensitized photocycloaddition of the benzoselenophene derivatives **286** (triplet energy $E_T = 69$ kcal) with 1, 2-dichloroethylene (**287**) (*cis* and *trans* isomers) gave a mixture of cyclobutanes **288** and **289** (equation 165)[119]. As with the sulfur analogs[120], four such isomeric 2 + 2 adducts may be obtained, a pair of *trans-* and *cis-exo* isomers **288** and a pair of *trans-* and *cis-endo* isomers **289**. A mixture of all four isomeric products was obtained only in the photoreaction of **286c**. Three isomers were identified in the reaction of the monomethyl derivatives **286a** and **286b**, and two isomers were detected with 3-acetoxybenzoselenophene (**286d**). A single isomeric 2 + 2 adduct was observed in the photoexcitation of **286e**.

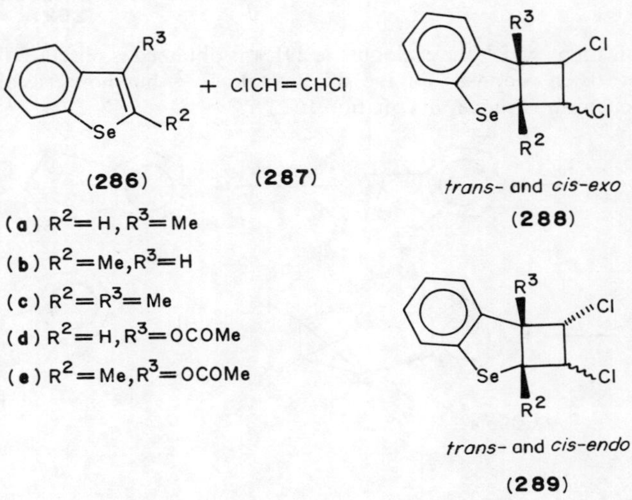

(165)

The structural assignment of the various isomers is based on a comparative NMR analysis with the sulfur analogs[120]. The two major isomers usually observed are believed to be the *trans*- and the *cis-exo*-**288** isomers. *Cis-endo*-**289** is assumed to be formed in low yields (1–3%) and *trans*-**289** is rarely observed. However, this generalization should be treated with some caution since an unequivocal X-ray structure study showed that the two acetoxy adducts isolated from the reaction of **286d** were the two *trans*-cyclobutane isomers **288d** and **289d**[119].

Unlike 1,2-dichloroethylene and 2,3-dimethylmaleic anhydride, which give isolable 2 + 2 adducts, dimethyl acetylenedicarboxylate (**290**) reacts photochemically with benzoselenophene (**274**) and its 3-methyl derivative (**286a**) in the presence of acetophenone as a triplet sensitizer to give esters of 1,2-naphthalenedicarboxylic acid (**292**)[119] (equation 166). This unusual reaction has an analogy in the sulfur series[121,122], where the same products **292** were obtained. However, in contrast to the sulfur counterparts, where the intermediate **291** (with S instead of Se) could be isolated and pyrolysed to **292, 291** itself is apparently an unstable primary product and undergoes a spontaneous thermal deselenation to **292**.

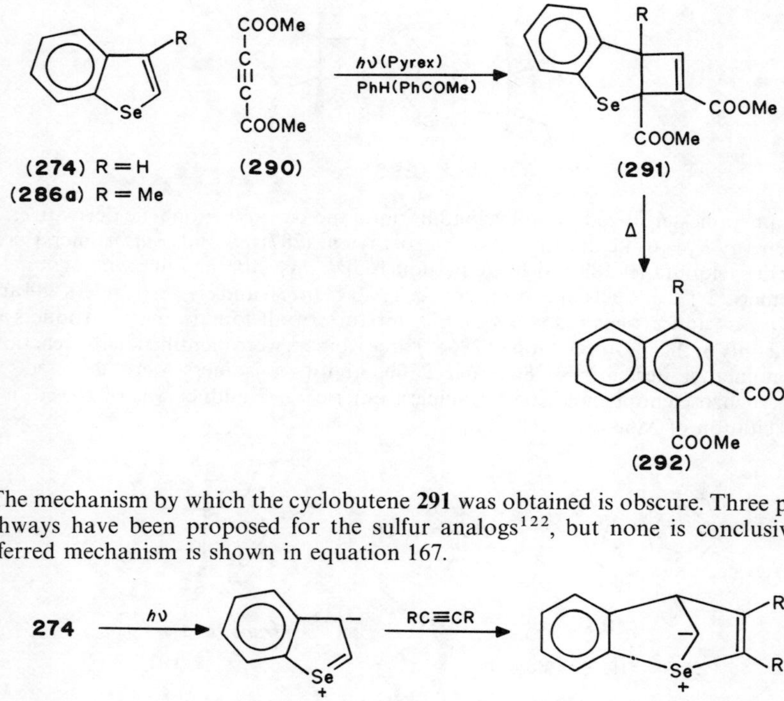

(166)

The mechanism by which the cyclobutene **291** was obtained is obscure. Three possible pathways have been proposed for the sulfur analogs[122], but none is conclusive. The preferred mechanism is shown in equation 167.

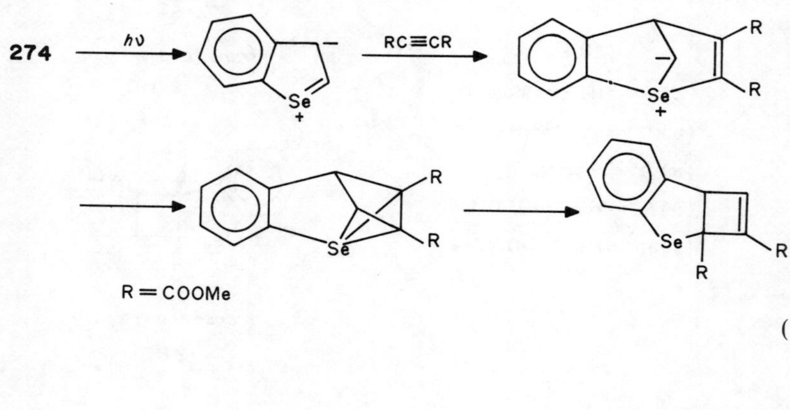

(167)

The reluctance of unsubstituted selenophene, thiophene and tellurophene to photodimerize or give oxetanes was first attributed to their aromatic character[116,123]. Later[115,124,125], however, it was shown that the lack of reactivity results from the ability of these compounds to act as quenchers of the ketones used. In order to test the quenching ability of selenophene on the excited state of benzophenone, a kinetic study of the photoreduction of benzophenone by isopropanol in the presence of selenophene was carried out[124]. Stern–Volmer plots of ϕ_o/ϕ_Q versus [Q] (where ϕ_o and ϕ_Q are the quantum yields for the photoreduction of benzophenone in the absence and presence of the quencher, respectively, and [Q] is the concentration of selenophene as quencher) indicate that selenophene and also thiophene, pyrrole and imidazole derivatives are good quenchers of the triplet state of benzophenone. It is perhaps important to mention that unlike imidazole and the thiophene derivatives, which give a normal linear Stern–Volmer plot, selenophene and pyrrole derivatives gave quadratic plots. This is consistent with a mechanism in which an exciplex (Q–B)* is initially formed between triplet benzophenone (^{3}B) and ground-state selenophene (Q), and subsequently quenched by another quencher molecule. The Stern–Volmer equation will then obtain the quadratic form shown in equation 168[126].

$$^3B + Q \longrightarrow (Q\text{–}B)^* \xrightarrow{Q} B + Q^* + Q \qquad (168)$$

$$\phi_o/\phi = (1 + k_q\tau[Q])\,(1 + k'_q\tau'[Q])$$

Although unsubstituted selenophene, like thiophene, is inert towards 2 + 2 cycloadditions with benzophenone, 2-methylselenophene (**293**) is active. Irradiation of **293** in benzene using benzophenone as sensitizer gave 34% of the oxetane **294** (equation 169)[127]. Thiophene has been shown to require two methyl groups in order to participate in the Paterno–Buchi oxetane cycloaddition[128].

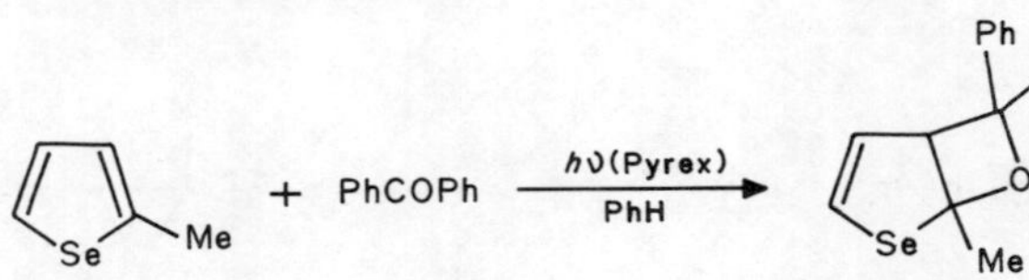

(**293**) (**294**) (169)

D. Insertion Reactions

Diazomethane (**295**) reacts with dibenzyl diselenide (**1**) in sunlight to give quantitatively the insertion product **296** (equation 170)[129]. The corresponding ditelluride **53** undergoes the same insertion reaction in the dark.

$$\underset{(\mathbf{1})}{PhCH_2SeSeCH_2Ph} + \underset{(\mathbf{295})}{CH_2N_2} \xrightarrow[Et_2O,\ 0^\circ C]{\text{sunlight}} \underset{(\mathbf{296})}{(PhCH_2Se)_2CH_2} \qquad (170)$$

A plausible reaction pathway involves light-induced homolytic cleavage of the Se—Se bond of **1** (Section II.A.2), forming benzyl selenide radicals, which react with diazomethane to give a labile diazo radical. Loss of nitrogen and recombination with a benzylselenyl radical affords the observed product (equation 171). Other mechanistic pathways, such as the initial formation of carbene cannot be entirely ruled out.

$$(PhCH_2Se)_2 \xrightarrow{h\nu} PhCH_2Se^{\bullet} \xrightarrow{CH_2N_2^{\bullet}} PhCH_2SeCH_2N_2^{\bullet}$$

$$\xrightarrow{-N_2} PhCH_2SeCH_2^{\bullet} \xrightarrow{PhCH_2Se^{\bullet}} (PhCH_2Se)_2CH_2 \qquad (171)$$

In the gas phase, atomic selenium was monitored in flashed CSe_2 and COSe (Section II.B.1) by kinetic absorption spectroscopy, and its rate of reaction with alkanes and alkenes was measured in the temperature range 302–412 K[24–26]. The excited Se(4^1D_2) atoms insert into the C—H bond of saturated hydrocarbons such as propane, cyclopropane, cyclobutane, ethane and isobutane, and also possibly into the Si—H bonds of methylsilane[26]. When mixtures of COSe and propane were flashed, no hexane was formed, although small amounts of HSe were detected. This indicates that selenomercaptan formation in these reactions is a single-stage process (equation 172). Had it been a two-step hydrogen abstraction followed by combination of HSe and C_3H_7, hexane formation would have been expected.

$$RH + COSe \xrightarrow{h\nu} RSeH + CO + Se_2 \tag{172}$$

Generation of excited selenium atoms in the presence of olefins resulted in the formation of adducts which are probably the corresponding episelenides (**297**) (equation 173)[24,25]. The activation energies were shown to correlate with the olefin ionization potentials. All the episelenides were found to be unstable at room temperature but could be trapped at 77 K in a fast flow system. They decay at room temperature with half-lives between 30 ms and a few seconds. Decomposition of the episelenides, if it occurs in the gas phase, is probably bimolecular, giving Se_2 and two olefin molecules.

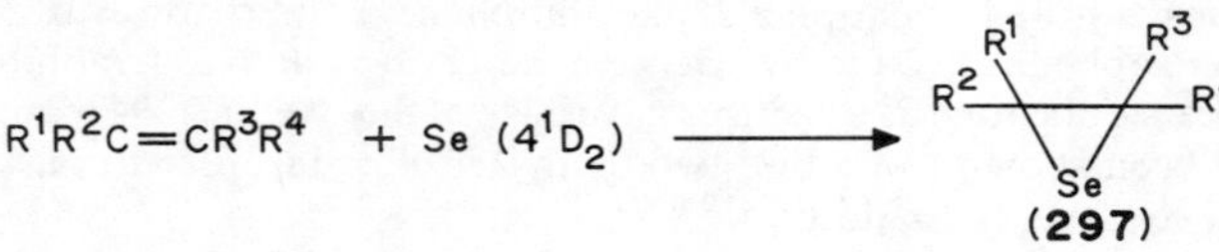

(173)

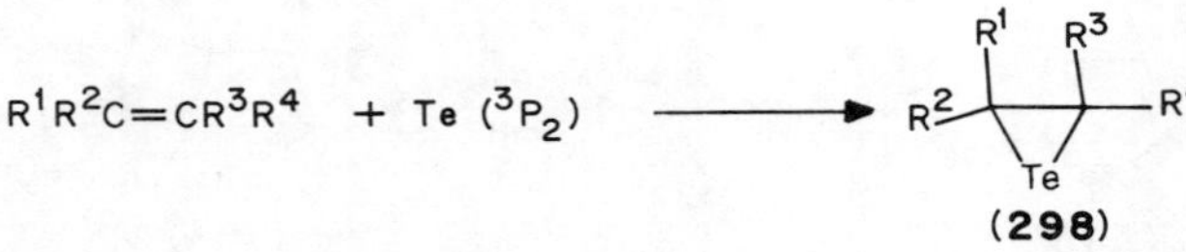

(174)

Similarly, flashing dimethyl telluride produced excited Te($^3P_2$) atoms (Section II.B.1), which in the presence of olefins give unstable adducts. These adducts have lifetimes of several milliseconds, and have been assigned as the corresponding epitellurides **298** (equation 174)[26,27]. The formation of the transient epitelluride molecules has been confirmed by kinetic mass spectrometry, and has been shown to be slower and more selective than the addition of O, S or Se atoms to olefins. It is interesting that the rate constants for the addition of Te($^3P_2$) to tetramethylethylene $3.9 \times 10^9\,l\,mol^{-1}\,s^{-1}$ (25 °C) and $2.6 \times 10^9\,l\,mol^{-1}\,s^{-1}$ (80 °C) correspond to an activation energy $E_a = -1.6 \pm 1.4$ kcal mol^{-1}, which is the first reported case of a negative temperature coefficient for an addition reaction in its second-order region[27].

IV. PHOTOREARRANGEMENTS AND PHOTOISOMERIZATIONS

A. Rearrangements

There are in principle two groups of rearrangement reactions, those in which a bond between a chalcogen and another atom is cleaved during the reaction, and others in which the bonds to selenium or tellurium remain intact and rearrangement occurs at a different molecular site. It is often observed that primary rearrangement products in which an Se—C bond was cleaved are air sensitive and can be isolated only as oxidation products. In other cases further photochemical (or thermal) transformations occur.

Photolysis of 2-phenylselenophene (**78**) in dilute ethereal solutions gave a 15% yield of the isomeric 3-phenylselenophene (**80**) together with deselenation products (equation 175)[38]. The corresponding 2-phenyltellurophene, however, gave only fragmentation products (Section II.B.1). 3-Phenylselenophene (**80**) proved to be stable and could not be rearranged back to **78** under the reaction conditions. It is assumed that the rearrangement occurs in a manner analogous to that suggested for thiophene rearrangements (equation 176)[130]. This involves cleavage of the Se—C bond to form the intermediate cyclopropenyl selone **299**, which rearranges by a formal 1, 3-sigmatropic shift to **80**[38].

$$\mathbf{78} \xrightarrow[Et_2O]{h\nu} \mathbf{80} \qquad (175)$$

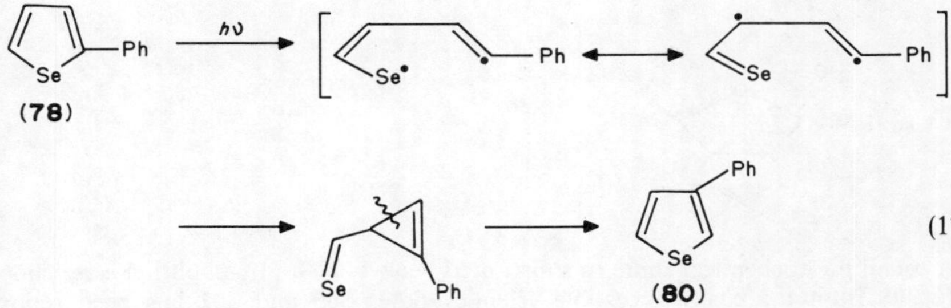

(78) (299) (80)

(176)

The photoinduced electrocyclic ring opening of selenochromene **300** was examined as a potential photochromic system in relation to the oxygen and sulfur analogs (equation 177)[131]. Irradiation of **300** (R = H) ($c = 10^{-4}$ mol l^{-1}) in glass solutions (isopentene–isopropanol, 5:2) at 313 nm resulted in color formation. The UV spectrum displayed characteristic features (λ_{max}355 and 660 nm) of the conjugated system **301** observed for the lower chalcogen chromenes. However, unlike these chromenes no fluorescence of **300** was detected and the reaction could neither be sensitized nor quenched by triplet donors or acceptors. Further, unlike the lower analogs the characteristic bleaching of the solutions by thermal recyclization of **301** could only be effected at temperatures above 150 K. It was concluded that the selenochromenes **300** differ in their photochromic properties, and perhaps also in their photocoloring mechanism, from the corresponding chromenes and thiochromenes.

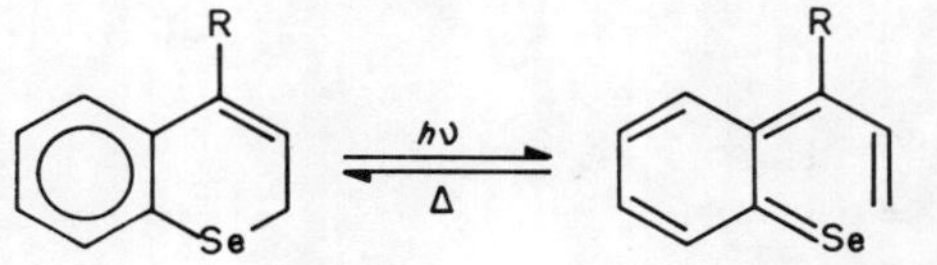

(177)

(300) (301)

The photorearrangement of alkyl phenyl selenides **302** led via a formal 1, 3-alkyl shift to *ortho*-substituted phenyl selenols **303**. These selenols oxidize during workup, and were isolated as the corresponding diselenides **304** (equation 178)[132]. By application of this reaction to phenyl ribosyl selenide derivatives (**302b** and **c**), pseudonucleosides (C-nucleosides) of potential biological interest were obtained as mixtures of α- and β-anomers.

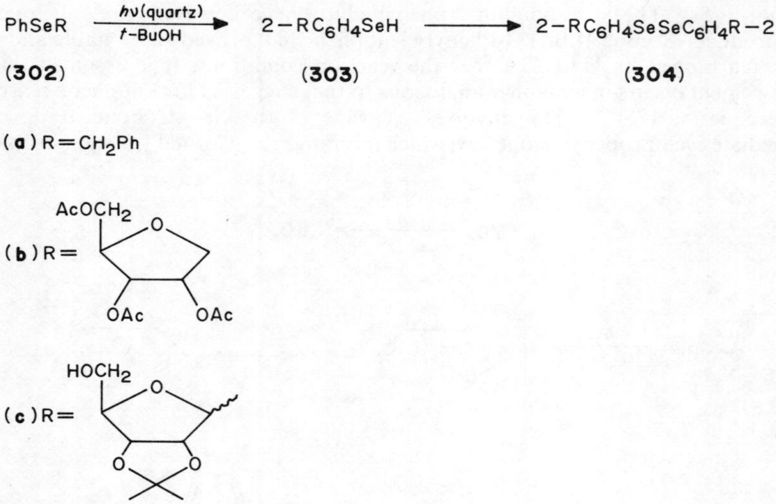

(178)

A useful photochemical route to substituted benzyl **306** and 1-naphthyl isoselenocyanate **308** from the easily accessible selenocyanates **305** and **307** has been reported (equations 179 and 180)[133]. The photochemical transformation is carried out by irradiation through quartz of 0.1–1.0 mM solutions of the selenocyanates in dry acetic acid with a high-pressure mercury lamp. A photoequilibrium is reached in which the isoselenocyanates predominate. This formal sigmatropic 1,3-benzyl shift may occur by a concerted pathway or via homolytic Se—C bond cleavage to benzyl and selenocyanate radicals, which then recombine at either carbon or nitrogen.

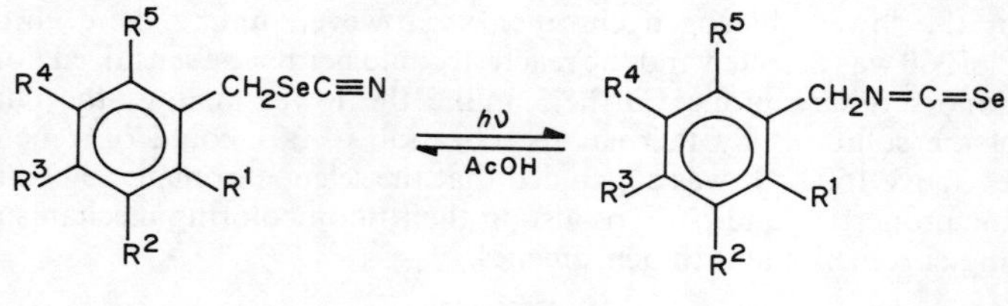

(305) (306)

$R^{1,2,3,4,5} = H$
$R^{1,2,4,5} = H, R^3 = Cl$
$R^{1,2,4,5} = Me, R^3 = Br$
$R^{1,2,3,5} = Me, R^4 = Cl$
$R^{1,3,5} = Me, R^{2,4} = H$
$R^{1,2,3,4,5} = Me$

(179)

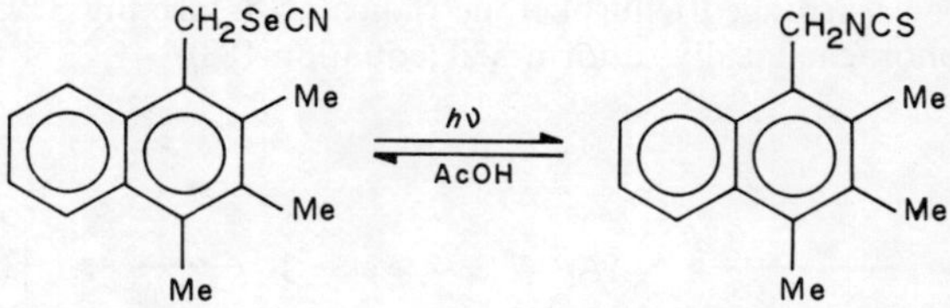

(307) (308) (180)

Aromatic selenoesters undergo an efficient photo-Fries rearrangement[134] to ketones. In the first published example of a seleno-photo-Fries reaction[11], *Se-p*-tolyl selenobenzoate (**309**, R = H) was irradiated in benzene at room temperature, affording a complex mixture containing the substituted benzophenone **310** (R = H) together with *p*-selenocresol (**311**, R = H), benzaldehyde (**312**, R = H) and the two diselenides **313** and **7** obtained by oxidative dimerization of the corresponding selenols (equation 181). Similarly, the selenophthalic ester **309** (R = COOMe) rearranges photochemically to the benzophenone **310** (R = COOMe), which was isolated as the diselenide **313** (R = COOMe)[135]. When the *ortho*-position of the selenoester is substituted by appropriate leaving groups, such as halogens, the reaction could not be stopped at the rearrangement step but instead selenoxanthones derived from further intramolecular substitution were obtained (see Section III.A.1.c). If the *ortho*-position was substituted by an aromatic selenide or telluride group, photo-Friedel–Crafts reaction occurred without rearrangement (see Section II.A.1).

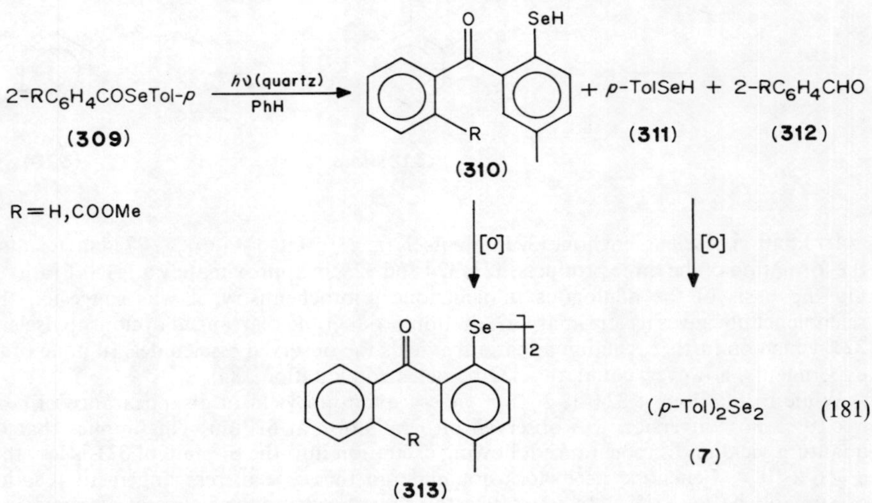

(309) (310) (311) (312) (313) (7) (181)

These results suggest that the primary photoreaction involves a homolytic cleavage of the CO—Se bond of the ester **314** to form the acyl (**315**) and selenyl (**316**) radicals, which recombine to give the benzophenone **317** (equation 182). Hydrogen abstraction from the solvent by the selenyl radical **316** gives **311**, and that by the acyl radical **315** gives **312**.

Photo-Fries rearrangement is also observed with the telluroesters **14**, albeit in low yields compared with the competitive fragmentation reactions (see Section II.A.1)[15,16]. These reactions follow essentially the same route as those of selenoesters **309** with perhaps one exception, namely the isolation of small amounts of the mixed monotelluride **320** from the reaction mixture. This is presumably obtained by combination of the telluride fragment radical **16** with the rearranged telluroketone radical **318**, forming **319**, which loses one tellurium atom photochemically to form **320** (equation 183).

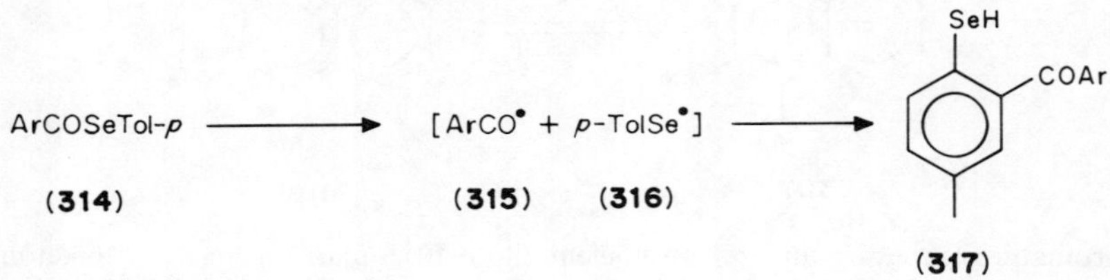

(182)

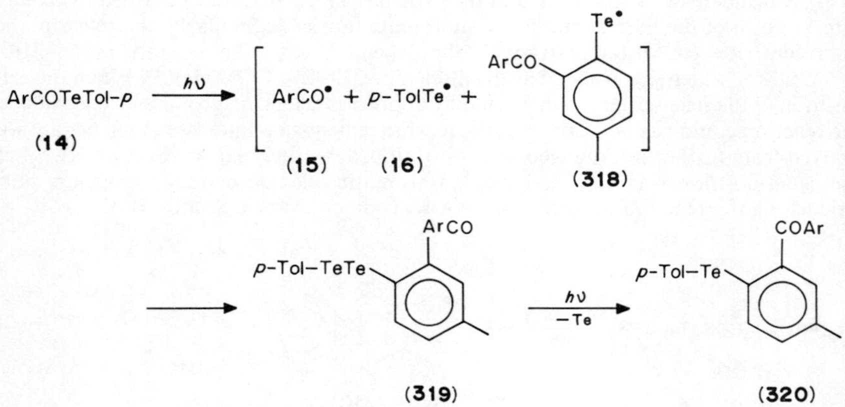

(183)

Irradiation of selenofenchone (**321**) into its S_2 (π, π^*) excited state (λ_{max} 272 nm) results in the formation of the three products, **323**, **324** and **325**, in approximately a 1:1:0.1 ratio[136]. On the basis of the analogous thiofenchone photochemistry, it was suggested that selenofenchone gives as a primary irradiation product the rearranged cyclopropylselenol **322**, which on further reaction apparently yields the observed diselenides. In none of the experiments, however, could any **322** be detected (equation 184).

While irradiation of **321** at 268 nm gave a quantum yield of disappearance of $\phi_2 = 5 \times 10^{-3}$, no conversion was observed on irradiation at 620 nm. This implies that the quantum yield of disappearance following excitation into the S_1 state of **321** is less than $\phi = 6 \times 10^{-4}$. Fenchone itself does not undergo this type of rearrangement reaction, presumably because the S_2 (π, π^*) state is too short-lived and the S_1 (n, π^*) state possesses the wrong geometry for β-hydrogen abstraction to occur[136]. In selenofenchone (and thiofenchone[137]), if the S_2 state is π, π^* in nature, the partially filled, and hence electrophilic, π orbital is of the correct geometry to insert into the βC—H bond.

A new method of preparing alkenes from alkanoic acids has been described recently[138], utilizing a Hunsdiecker- type photodecarboxylative rearrangement of carboxylic esters of 2-selenopyridine *N*-oxide (**326**) to 2-pyridylselenides (**327**) (equation 185). This decarboxylative rearrangement was shown to proceed efficiently with simple saturated aliphatic acid derivatives (e.g. **328** $\longrightarrow$ **329**) (equation 186), steroidal acid derivatives (e.g. **330** $\longrightarrow$ **331**) (equation 187) and amino acid derivatives (e.g. **332** $\longrightarrow$ **333**) (equation 188), and

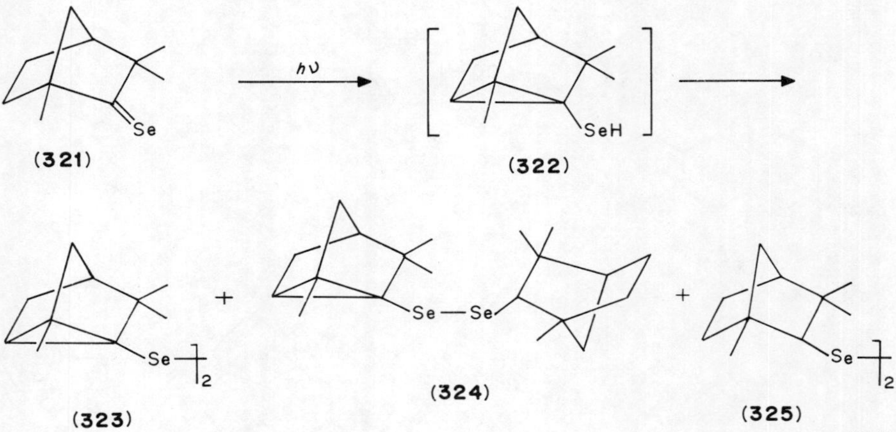

(184)

is of particular importance for chiral molecules since the configuration of the allylic carbon of the olefinic product is retained. Since decarboxylation can also be effected thermally in refluxing benzene, the mechanism is assumed to involve a free-radical pathway. However, unlike the radicals obtained from esters of the thio derivatives[139], the radicals derived from the selenoesters could not be intercepted.

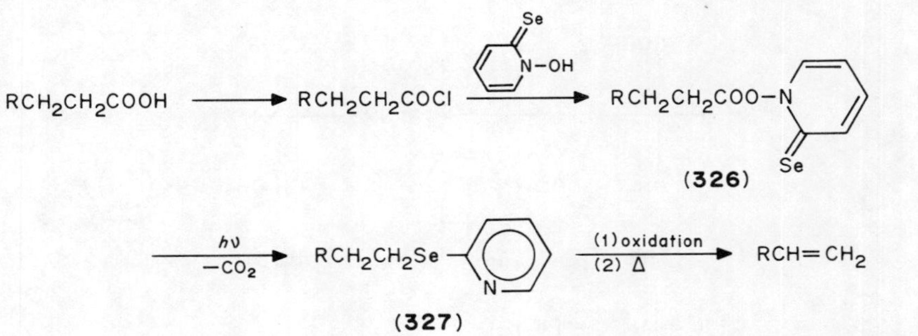

(185)

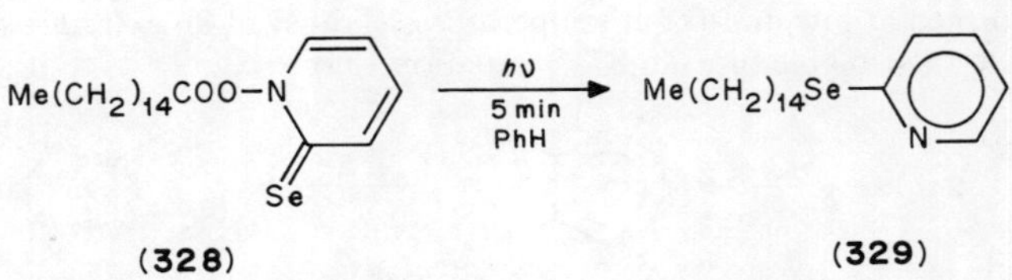

(186)

Irradiation ($\lambda > 390$ nm) of 2,1,3-benzoselenadiazole 1-oxide (**85**) produced benzofurazan (**87**) (see Section II.B.1)[39,140]. Flash photolysis of **85** in both aerated and degassed solutions (96% ethanol and cyclohexane) using Pyrex-filtered light ($\lambda > 300$ nm) gave rise to absorptions due to the transient **86** obtained by 6-electron cycloreversion, which decayed in a first-order reaction. From the UV spectra it was concluded that the same

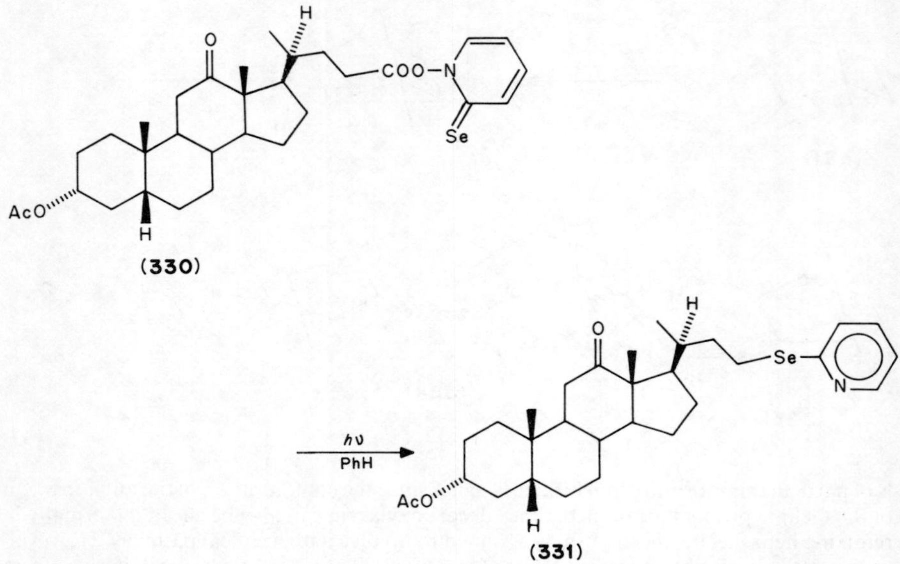

(330) (331)

(187)

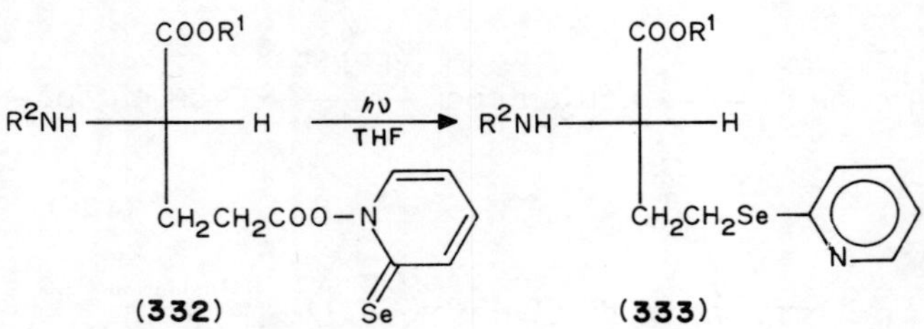

(332) (333)

(188)

$R^1 = Me, R^2 = PhCH_2O$
$R^1 = CH_2Ph, R^2 = t\text{-}Bu$

transient is formed at low and room temperatures. Transient **86** extrudes selenium and cyclizes to give the isolable furazan **87**[53] (equation 189).

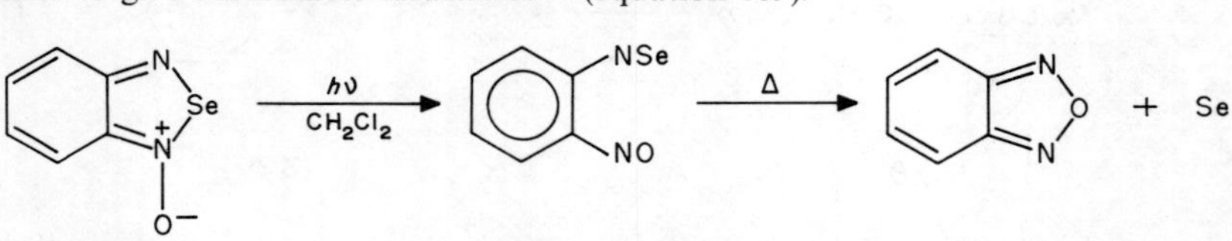

(85) (86) (87)

(189)

Selenophene dicarboxaldehydes **334** and **337** undergo an intramolecular photo-Cannizaro disproportionation, analogous to the photochemical transformation of phthalaldehyde[141], to the corresponding γ-lactones **335**–**338** as shown in equation 190 and 191[142].

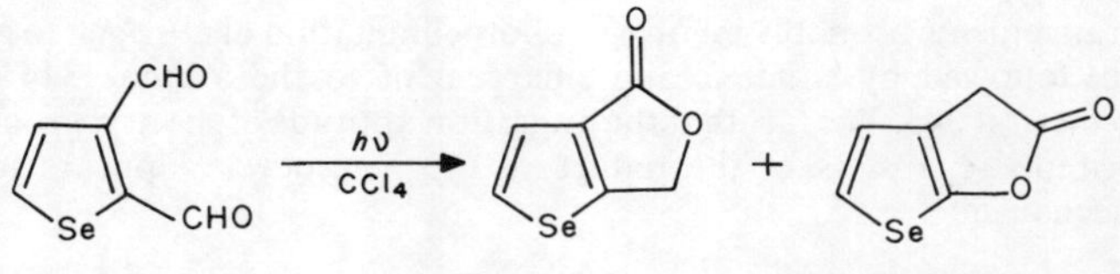

(334) (335) (336)

(190)

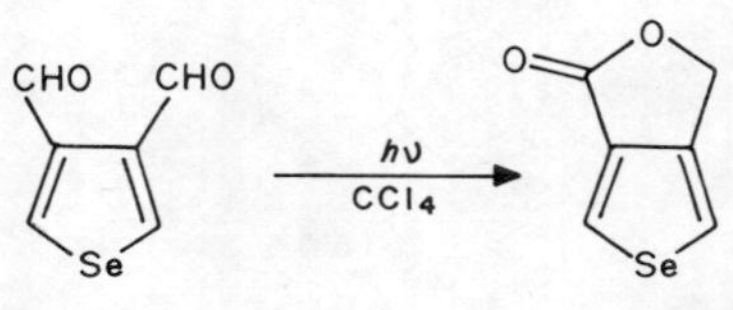

(337) (338)

(191)

The phenyl selenide substituted acetylacetone **2** undergoes an acid-catalysed photocyclization to the dihydro benzoselenophene **341**, which dehydrates under the reaction conditions to give 2-acetyl-3-methylbenzo[*b*]selenophene (**342**) in 60% yield[10]. The mechanism for this transformation presumably involves a photoinduced electrocyclic reaction of the enol **339** to the selenocarbonyl ylid **340**, which undergoes a further 1,4-hydrogen shift to the dihydroselenophene **341** (equation 192). Analogous cyclizations with divinyl systems having heteroatoms other than selenium, such as oxygen[143], sulfur[144] and nitrogen[145], have also been reported.

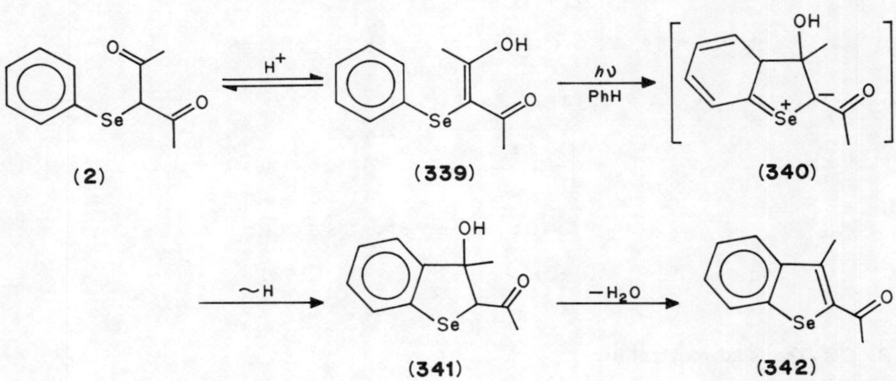

(2) (339) (340)

(341) (342)

(192)

In a search for dibenzoheteroazepines having pharmacodynamic activity, a series of 9-azido-9-arylselenoxanthenes (**343**) and their oxygen and sulfur analogs were photolysed[146,147] (equation 193). Irradiation of 9-azido-9-phenylselenoxanthene (**343**, X = Se, Ar = Ph) gave a 70% yield of a mixture containing 60% of the dibenzoselenoazepine **344** (X = Se, Ar = Ph) and 40% of the isomeric anil **345**[146]. When 9-azido-9-(4-pyridyl)selenoxanthene (**343**, X = Se, Ar = 4-pyridyl) was irradiated, 32% of the selenoazepine **344** (X = Se, Ar = 4-pyridyl) was obtained together with 25% of the isomeric anil **345**[147]. The best results were obtained when 1-methylnaphthalene was added to the irradiated reaction mixture. It is considered to serve as both a singlet sensitizer and triplet quencher, and as such it serves as a protector for the reaction products[147]. The mechanism of these rearrangements probably involves a photoelimination of nitrogen to give a nitrene (or nitrenoid) followed by either a ring enlargement to the azepine **344** or aryl 1,2-migration to the anil **345**. The fact that the migration aptitude of the aryl group (4-pyridyl) remains constant, regardless of the nature of the heterocycle, appears to support a concerted mechanism[147,148].

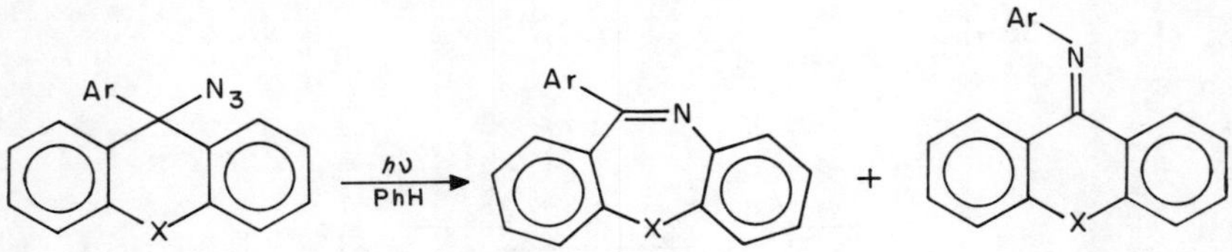

(343) **(344)** **(345)**

X = O, S, Se

Ar = Ph, 4-pyridyl

(193)

Irradiation ($\lambda > 220$ nm) of propadieneselone (**346**) matrix isolated in argon at 12K gives propyneselenal (**347**) (selenal C=Se stretching band at 1058 cm^{-1})[44]. The formation of **347** may involve either a 1,3-hydrogen shift, or loss of selenium giving allenylidene (**348**), which rearranges to propargylene (**349**) and recombines with selenium (equation 194). The starting material (**346**) was obtained by flash vacuum thermolysis of cyclopenteno-1,2,3-selenadiazole (Section II.B.2).

$$\begin{array}{ccc} \underset{(\mathbf{346})}{CH_2=C=C=Se} & \xrightarrow[\sim 1,3\text{-H}]{h\nu\ (12\ K)} & \underset{(\mathbf{347})}{HC\equiv CCH=Se} \\ \downarrow -Se & & \uparrow +Se \\ \underset{(\mathbf{348})}{CH_2=C=C\colon} & \xrightarrow{\sim H} & \underset{(\mathbf{349})}{CH\equiv CCH\colon} \end{array} \qquad (194)$$

B. *Cis–Trans* Isomerization

The photoisomerization of selenoindigo (**350**) was studied as a potential photochromic indigoid[149]. Typical of the majority of photoisomerizable indigoids, *trans*-selenoindigo shows a red fluorescence at about 620 nm on irradiation at its principal visible absorption band. By irradiation of **350** in toluene at 562 nm a photostationary state (**350**$\rightleftharpoons$**351**) was achieved (equation 195). The solution contained 57% of the *trans* isomer **350**, compared with 41% of the *trans* thio-analog at photoequilibrium (542 nm). The photoequilibrium mixture is solvent dependent, containing more of the *cis* isomer in chloroform than in toluene. This has been attributed to the lowering of the quantum yield for *cis–trans* photoisomerization on changing the solvent to chloroform.

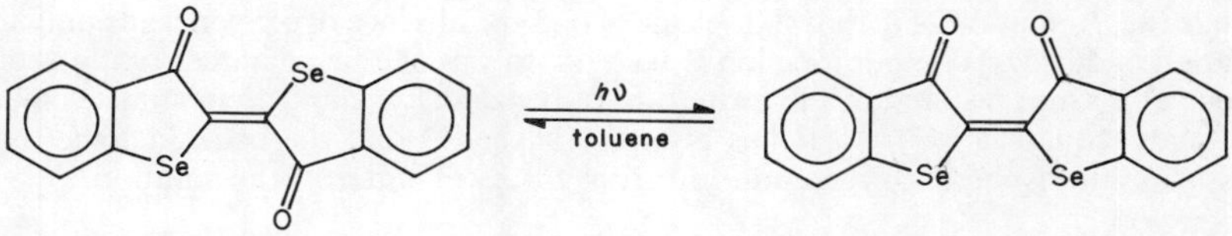

(350) (351) (195)

The quantum yields for the *trans–cis* ($\phi_t = 0.025$ at 562 nm) and *cis–trans* ($\phi_c = 0.275$ at 485 nm) photoisomerizations, respectively, coincide with the values obtained for ϕ_t and ϕ_c of the thio analogs. However they differ considerably from the quantum yield of isomerization of the oxoindigoid ($\phi_t = 0.63$ at 413 nm, $\phi_c = 0.35$ at 396 nm). The quantum yields for *cis–trans* photoisomerizations over spectral regions which include long-wavelength absorption remained constant.

The photoelectron spectra of several aryl alkyl selenides have been measured[150]. Some show evidence of the existence of two predominant rotamers, one with a maximum p–π overlap and one with minimal overlap. It has been shown that steric factors affect the rotamer population while the detection of rotamers by the photoelectron technique is dependent on electronic factors.

The photoelectron spectra of simple selenium-substituted olefins has been studied[151] and compared with other push–pull olefins containing electron-donor (e.g. thio and amino) and -acceptor (e.g. cyano, carbonyl and carboxyl) groups. In olefins with electron-donating groups ionization potentials (IPs) decrease with the IPs of the corresponding atoms, namely in the order S < Se < N. Similarly, the IPs and electron affinities of chalcogen cyanates and isocyanates[152] and also those of non-isomerizable chalcogen heterocyclic olefins[153,154] have been determined by UV photoelectron spectroscopy.

A useful photoinduced diphenyl diselenide-catalysed isomerization of the Δ^6-double bond of ergocalciferol (**352**) to 5,6-*trans*-ergocalciferol (**353**)[155] was reported (equation 196). This reaction probably proceeds via the addition of phenylselenyl radical to $C_{(19)}$, isomerization of the derived allylic radical and loss of the phenylselenyl radical. An analogous double-bond isomerization of (−)-caryophyllene to (−)-isocaryophyllene was reported using diphenyl disulfide[156].

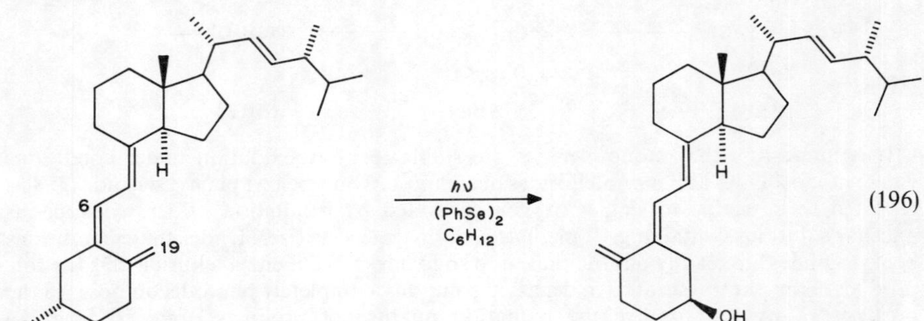

(352) (353) (196)

V. PHOTOOXIDATIONS

Although oxidative decompositions of organoselenium compounds by UV light were reported as early as 1974[28], only a few quantitative studies have been made since then which contain a mechanistic analysis of the results.

While the irradiation of dibenzyl diselenide (**1**) in the absence of oxygen leads only to the formation of dibenzyl selenide (Section II.B.1), photolysis of dilute aerated solutions of **1** in benzene at 350 nm resulted in the formation of benzaldehyde and elemental selenium in high yields (equation 197)[6,28]. It has been established[28] that the benzaldehyde oxygen arises from atmospheric oxygen and not from traces of water in the solution.

$$\underset{\textbf{(1)}}{PhCH_2Se_2CH_2Ph} \xrightarrow[\text{PhH}]{h\nu\ (350\,\text{nm}),\ O_2} PhCHO + Se \qquad (197)$$

When the diselenide **1** or the ditelluride **53** was photolysed in $CDCl_3$ under limited amounts of oxygen, complex mixtures of products were obtained containing benzyl alcohol, toluene and 1, 2-diphenylethylene in addition to benzaldehyde (equation 198)[30]. These were all presumably derived from incomplete oxidation of intermediate benzyl radicals formed by homolytic deselenation.

$$PhCH_2EECH_2Ph \xrightarrow[\text{CHCl}_3]{h\nu} PhCHO + PhCH_2OH + PhMe + PhCH_2CH_2Ph \qquad (198)$$

$E = Se, Te$

In the absence of oxygen, benzyl tellurocyanate (**354**) is photochemically stable. However, irradiation of **354** under atmospheric oxygen gave a black precipitate of elemental tellurium and the solution contained a mixture of 60% benzaldehyde and 40% benzyl alcohol (equation 199)[157]. The speculative mechanism proposed included a singlet oxygen attack on tellurium to form the 1, 3-dipolar peroxide **355** followed by rearrangement to the unstable benzylperoxytellurocyanate **356**. Peroxide **356** in turn decomposes to benzyloxy radicals **357**, which disproportionate to the observed products.

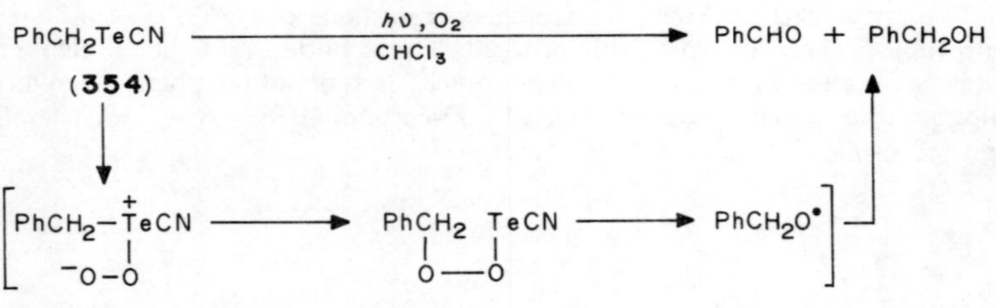

(199)

It is interesting that monoselenides are stable to photooxidation under conditions where monosulfides and monotellurides decompose. Thus, benzyl phenyl selenide (**358**) is reported to be stable to singlet oxygen generated by irradiation of aerated aqueous micellar solutions containing 10-methylphenothiazine sensitizer. Under these conditions benzyl sulfides are readily photooxidized to sulfoxides[158]. Dibenzyl telluride (**55**), the sole product of the photolysis of **53** in degassed solutions, completely photodecomposes in the presence of oxygen to give the oxidation mixture of products obtained from **53** (equation 198)[30].

$$\underset{\textbf{(358)}}{PhCH_2SePh} \qquad \underset{\textbf{(55)}}{PhCH_2TeCH_2Ph}$$

The tellurophenopyridazine **359** is photooxidized in chloroform solutions to give a 40% yield of 4, 5-dibenzoylpyridazine (**361**). The peroxide **360** arising from a Diels–Alder reaction of singlet oxygen[159] with the tellurophene ring is suggested as the probable intermediate, which readily decomposes thermally by loss of elemental tellurium to **361** (equation 200)[160]. Analogously, 4, 5-dibenzoyl-2, 7-diphenyltropone (**363**) was formed in 43% yield by visible light photooxidative detelluration of the tellurophene **362** (equation 201). When a high-pressure UV mercury lamp was used, only polymers and unidentified decomposition products were observed[160].

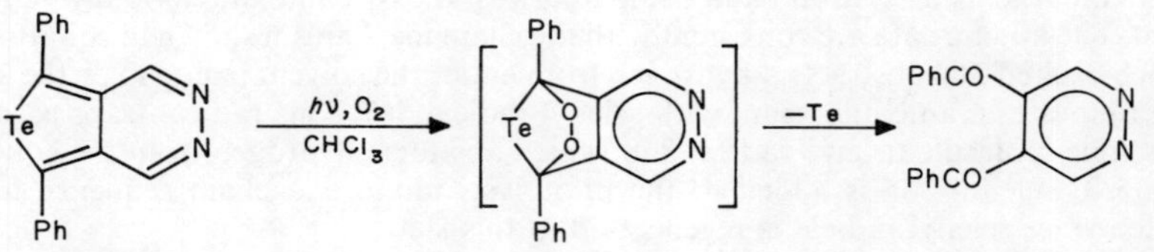

(359) (360) (361)

(200)

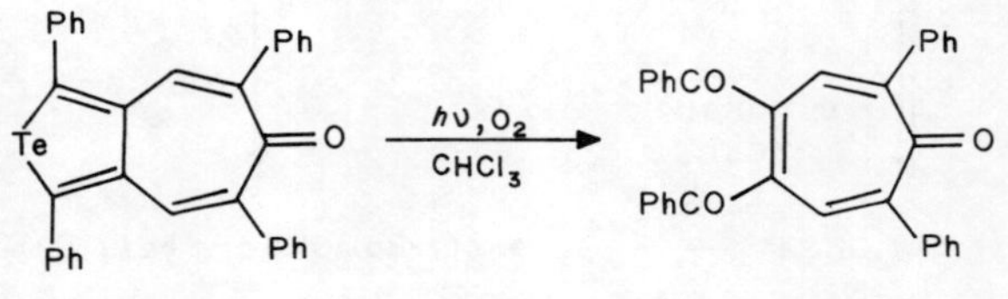

(201)

(362) (363)

VI. PHOTOREDUCTIONS

A. Hydrogen Abstraction

In contrast to the rich literature on the photoreduction of carbonyl compounds[1,2] and the well documented photochemistry of thioketones (thiones)[4], reports on the photochemical behavior of selones are scarce[136,161]. Di-*tert*-butyl selone (**364**), first prepared in 1976[161], was found to be stable when irradiated with visible light under nitrogen. However, under UV irradiation in a variety of hydrogen-donating solvents such as hexane, dioxane and benzene, **364** was smoothly reduced to the diselenide **365** (equation 202). In solvents other than benzene, substantial amounts of unidentified compounds together with selenium metal deposition were observed[136]. The presence of a better hydrogen donor such as the selenol **366** improved the yields of **365** up to 85%.

$$(t\text{-Bu})_2C{=}Se \xrightarrow[RH]{h\nu} (t\text{-Bu})_2CHSeSeCH(t\text{-Bu})_2 \qquad (202)$$

(**364**) (**365**)

The quantum yield of disappearance of **364** in hexane increases from $\phi = 5.5 \times 10^{-3}$ in the absence of the selenol **366** to $\phi = 8 \times 10^{-2}$ in the presence of 9×10^{-2} M **366**. The photoreduction is dependent on the wavelength. Excitation into the S_1 state ($\lambda = 689$ nm) is extremely inefficient, $\phi_1 = 1.2 \times 10^{-3}$ compared with $\phi_2 = 3.2 \times 10^{-2}$ from the higher S_2 ($\lambda = 266$ nm) excited state under the same conditions. The 266 nm band tails to wavelengths beyond 300 nm and consequently irradiation of **364** in Pyrex using a medium-pressure lamp allows excitation into both S_1 and S_2 states in the same experiment. This suggests that, like the photoreduction of thiones[4], the reaction occurred from the relatively longlived π, π^* transition state and unlike ketones for which the lower n, π^* state is reactive in hydrogen abstraction reactions[162].

When **364** was irradiated in pentane at $-110\,^{\circ}C$ under conditions such that the diselenide **365** is formed, ESR signals are observed[136] which correspond to the presence of radical **368** in the solution. The intensity of the ESR signals increases 60-fold when **364** is irradiated in the S_2 excitation band compared with the S_1 band and 20-fold when **366** is present. The kinetic data are consistent with the chain mechanism shown in equation 203, in which excited **364** abstracts a hydrogen from either the solvent (route a) or the selenol (route b) to attain an equilibrium with selenyl radical **367**. This radical traps a ground-state selone molecule to give radical **368**, which abstracts hydrogen from the solvent to give **365**. If selenol **366** is added to the photolysis mixture, a chain sequence may be expected since selenyl radical is regenerated in this step.

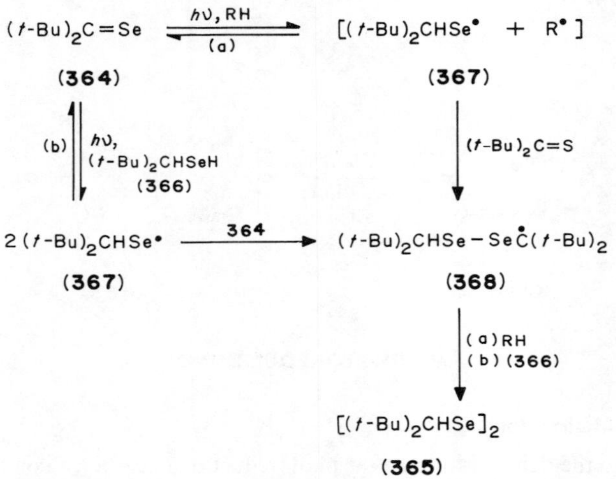

(203)

Quenching experiments with biacetyl gave Stern–Volmer plots for the quantum yields of formation of **365** and disappearance of the selone **364** as a function of the quencher concentration. Assuming $k_q = 10^{10}\,l\,mol^{-1}\,s^{-1}$ (hexane), the lifetime of the S_2 excited state is of the order of 10^{-10} s, which is close to the lifetime of the reactive S_2 state of thioketones[4].

Irradiation of **364** into both the S_1 and S_2 states in the presence of olefins such as acrylonitrile, ethyl vinyl ether, fumaronitrile and dimethyl maleate failed to produce detectable amounts of adducts. This is in contrast to the analogous di-*tert*-butylthione[163] and other thioketones[164]. The inertness of **364** has been attributed to steric hindrance[136].

Unlike di-*tert*-butyl selone (**364**), the photochemistry of the bicyclic selenofenchone (**321**) did not lead to reduction by intermolecular hydrogen abstraction, even in reasonably good hydrogen atom-donating solvents such as pentane. Instead, excitation of **321** into the $S_2\,(\pi, \pi^*)$ state gave a mixture of diselenides (see Section IV.A).

Esters of steroidal carboselenoic acids (**369–371**) undergo photoinduced reduction with tributyltin hydride to the corresponding aldehydes (**372–374**) and alkanes (**375**, **376**) which are shorter by one side-chain carbon (equations 204–206)[165]. Since these reactions may also be initiated thermally with azobis(isobutyronitrile), a radical-chain mechanism is suggested, involving either hydrogen atom abstraction from the stannane by the acyl radical (**377**) intermediate to give aldehyde, or alternatively, decarbonylation of the acyl radical followed by hydrogen abstraction to give the alkane (equation 207).

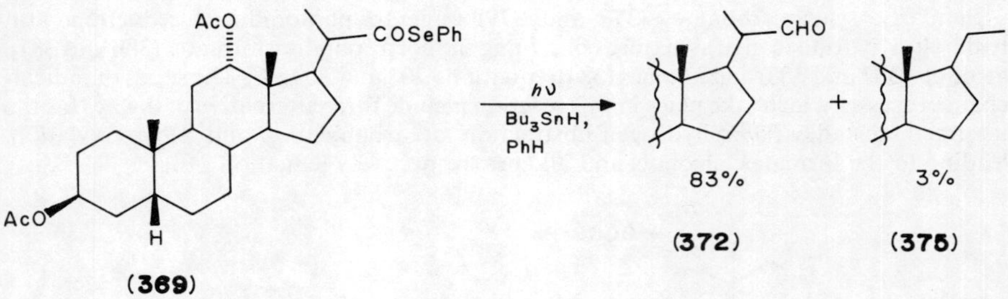

(204)

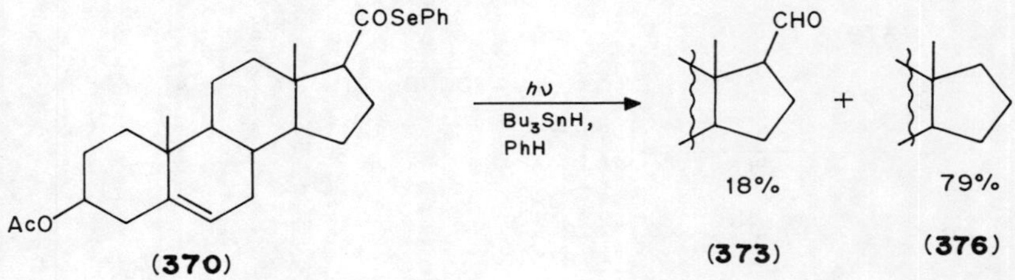

(205)

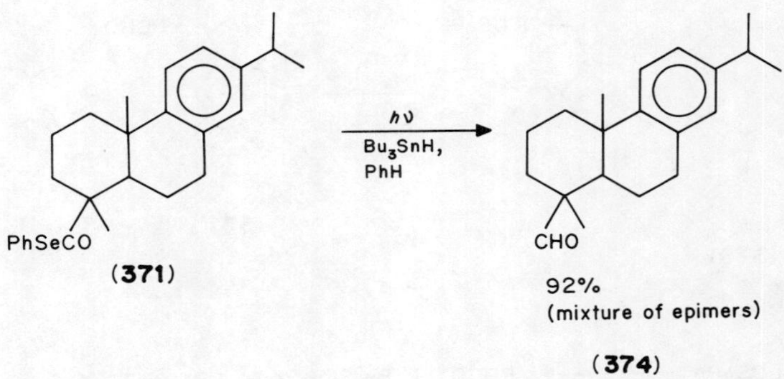

(206)

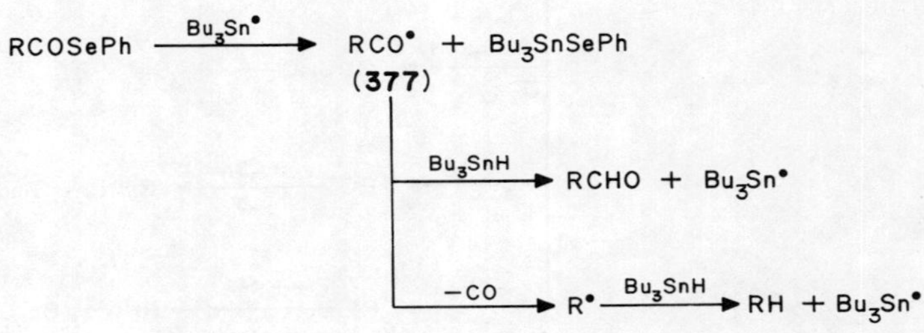

(207)

Similarly, selenocarbonates (**378** and **379**) undergo photoinduced reduction with tributyltin hydride to give a mixture containing the corresponding formates (**380** and **381**), alcohols (**382** and **383**) and alkanes (**384**) (equations 208 and 209)[165]. The series of radical-chain processes which take place in this reaction include three alternative pathways for the intermediate radical **385**: hydrogen abstraction, decarbonylation and decarboxylation, leading to the formates, alcohols and alkanes, respectively (equation 210).

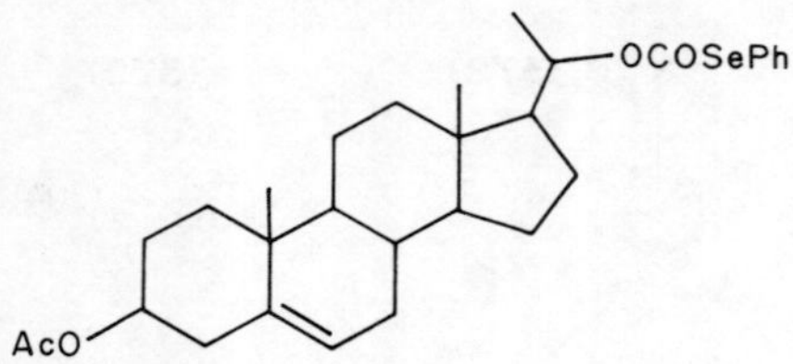

(378)

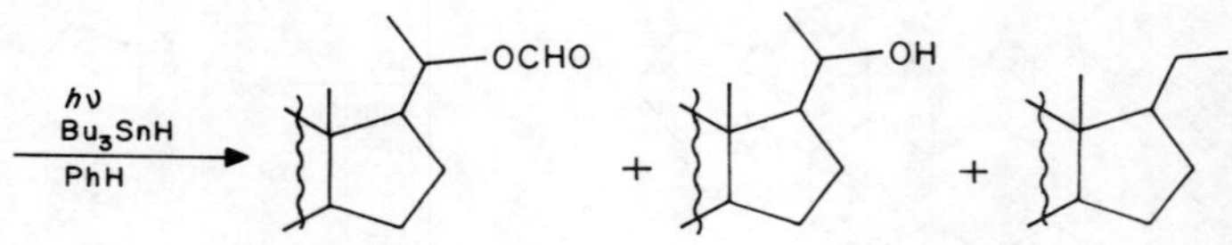

87% (**380**) 1% (**382**) 9% (**384**)

(208)

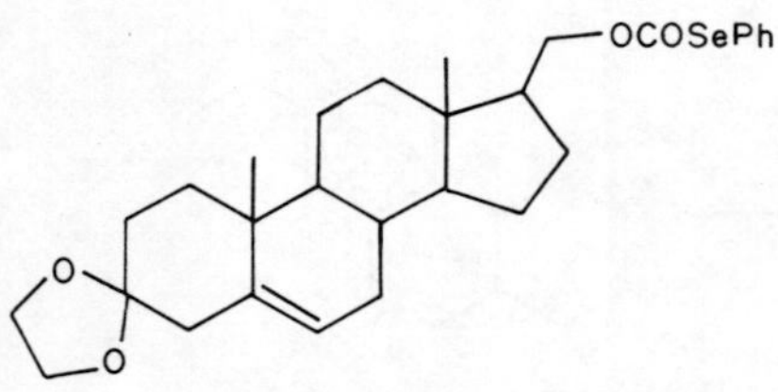

(379)

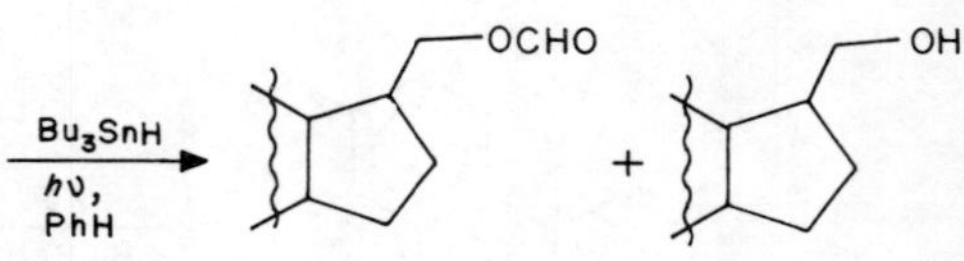

81% (**381**) 19% (**383**)

(209)

$$ROCOSePh \xrightarrow{Bu_3Sn^\bullet} \underset{(\mathbf{385})}{ROCO^\bullet} + Bu_3SnSePh$$

$$ROCO^\bullet \xrightarrow{Bu_3SnH} ROCHO + Bu_3Sn^\bullet$$

$$ROCO^\bullet \xrightarrow{-CO} RO^\bullet \xrightarrow{Bu_3SnH} ROH + Bu_3Sn^\bullet \qquad (210)$$

$$ROCO^\bullet \xrightarrow{-CO_2} R^\bullet \xrightarrow{Bu_3SnH} RH + Bu_3Sn^\bullet$$

B. Oxygen Transfer

Diaryl selenoxides are photochemically reduced by sulfides to selenides and the corresponding sulfoxides. Thus, when a 3:1 mixture of dibenzoselenophene oxide (**386**) and the sulfide **387** in methanol (or dichloromethane) was irradiated under argon through a Pyrex filter with a high-pressure mercury lamp, dibenzoselenophene (**388**) was obtained in high yields together with the sulfoxide **389** (equation 211)[166]. Similarly, diphenyl selenoxide (**390**) is reduced by phenyl methyl sulfide to diphenyl selenide (**36**) in moderate yields. The mechanistic pathway for the photochemical transfer of oxygen in these reactions has been suggested to involve a bimolecular intermediate between the excited selenoxide and sulfide, which collapses to the products. Other possibilities, such as the involvement of atomic oxygen, were excluded.

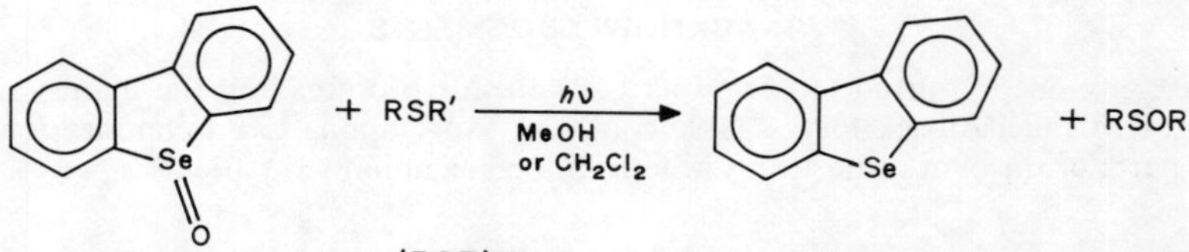

(**386**) (**387**) (**388**) (**389**)

R = Ph, R′ = Me
R = R′ = *n*-Bu
R = R′ = $(CH_2)_5$

(211)

$$Ph_2Se=O + PhSMe \xrightarrow[CH_2Cl_2]{h\nu} Ph_2Se + PhSOMe \qquad (212)$$

(**390**) (**36**)

Deoxygenation has also been noted[167] when 4-methyl-2′-nitrodiphenyl selenoxide (**391**) was irradiated in benzene solution through quartz to give the selenide **392** (equation 213). The oxygen acceptor in this reaction was not indicated. In contrast, the analogous nitrosulfoxide **393**, on irradiation into the singlet excited state, is reported[168] to undergo an intramolecular oxygen transfer from nitrogen to sulfur, forming the nitrososulfone **394** (equation 214). However, in the presence of triphenylamine as a triplet sensitizer, **393** is reduced to the corresponding sulfide in low yields (1–4%).

$$2\text{-}NO_2C_6H_4\overset{O}{\overset{\|}{Se}}\text{-Tol-}p \xrightarrow[PhH]{h\nu\ (quartz)} 2\text{-}NO_2C_6H_4Se\text{-Tol-}p \qquad (213)$$

(**391**) (**392**)

$$2\text{-}NO_2C_6H_4\overset{O}{\overset{\|}{S}}\text{-Tol-}p \xrightarrow{h\nu} 2\text{-}NOC_6H_4\underset{\underset{O}{\|}}{\overset{O}{\overset{\|}{S}}}\text{-Tol-}p \qquad (214)$$

(**393**) (**394**)

Photochemical deoxygenation was observed when 2, 1, 3-benzoselenadiazole 1-oxide (**85**) was irradiated ($\lambda > 390$ nm) in isopropanol to give a 5% isolated yield of 2, 1, 3-benzoselenadiazole (**395**)[53]. The major product in the photolysis was benzofurazan (**87**) (equation 215) (Section II.B.1). Irradiation (410 nm) of **85** at 100 K in methanol–ethanol glass (2:5) showed the development of both **395** and **87** in a 40:30 ratio. Interestingly, the thermal reaction gives higher yields of **395**[140]. The oxygen is probably transferred to the solvent by an analogous mechanism to that suggested for the photo-deoxygenation of amine *N*-oxides[169].

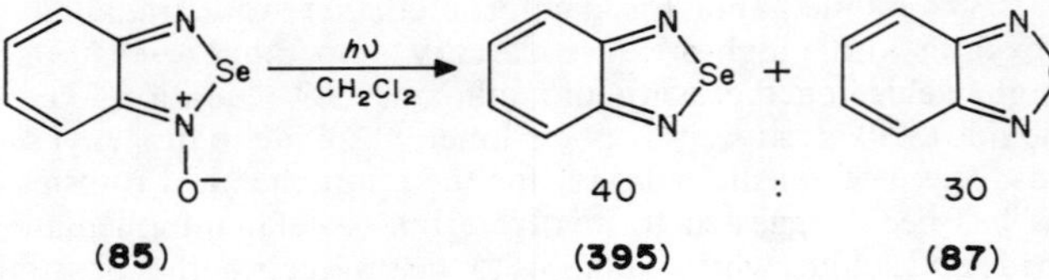 (215)

(**85**) (**395**) (**87**)

VII. ACKNOWLEDGEMENTS

I express my deep gratitude to Dr. Joel L. Wolk for proofreading the manuscript, to Professor Michael Albeck for valuable comments, to Daphna Goldschmidt for typing major parts of this work and to my wife Riki for her unlimited patience.

VIII. REFERENCES

1. P. J. Wagner, in *Triplet States III* (*Topics in Current Chemistry*, 66), Springer Verlag, Berlin, 1976, pp. 1–52.
2. A. Schönberger, *Preparative Organic Photochemistry*, Springer Verlag, Berlin, 1968.
3. H. Dürr, in *Houben-Weyl, Methoden der Organischen Chemie* (ed. E. Müller), Teil IV/5b, Georg Thieme Verlag, Stuttgart, 1975, pp. 1008–1076.
4. P. de Mayo, *Acc. Chem. Res.*, **9**, 52 (1976).
5. C. L. Jackson, *Justus Liebigs Ann. Chem.*, **179**, 1 (1875).
6. J. Y. C. Chu, D. G. Marsh and W. H. H. Gunther, *J. Am. Chem. Soc.*, **97**, 4905 (1975).
7. D. L. Klayman and W. H. H. Gunther, *Organic Selenium Compounds: Their Chemistry and Biology*, Wiley, New York, 1973.
8. K. J. Irgolic, *The Organic Chemistry of Tellurium*, Gordon and Breach, New York, 1974.
9. J. Martens and K. Praefcke, *J. Organomet. Chem.*, **198**, 321 (1980).
10. A. G. Schultz, *J. Org. Chem.*, **40**, 3466 (1975).
11. J. Martens, K. Praefcke and H. Simon, *Z. Naturforsch., Teil B*, **31**, 1717 (1976).
12. R. Ludersdorf, J. Martens, P. Pakzad and K. Praefcke, *Justus Liebigs Ann. Chem.*, 1992 (1977).
13. K. Praefcke and D. Schmidt, *J. Heterocycl. Chem.*, **16**, 47 (1979).
14. K. Praefcke and U. Schulze, *J. Organomet. Chem.*, **184**, 189 (1980).
15. W. Lohner, J. Martens, K. Praefcke and H. Simon, *J. Organomet. Chem.*, **154**, 263 (1978).
16. G. Hohne, W. Lohner, K. Praefcke, U. Schulz and H. Simon, *Tetrahedron Lett.*, 613 (1978).
17. W. W. Lotz and J. Gosselck, *Tetrahedron*, **29**, 917 (1973).
18. J. Bergman, N. Eklund, T. E. Eriksen and J. Lind, *Acta Chem. Scand., Ser. A*, **32**, 455 (1978).
19. O. A. Ptitsina, M. E. Gurskii and O. A. Reutov, *Zh. Org. Khim.*, **10**, 2246 (1974); *J. Org. Chem. USSR*, **10**, 2262 (1974).
20. U. Schmidt, A. Muller and K. Markau, *Chem. Ber.*, **97**, 405 (1964).
21. R. T. Mehdi, J. D. Miller and T. A. Tahir, *Inorg. Chim. Acta*, **90**, 85 (1984).
22. R. A. Gancarz and J. L. Kice, *J. Org. Chem.*, **46**, 4899 (1981).
23. N. S. Vyazankin, M. N. Bochkarev and L. P. Maiorova (Sanina), *Zh. Obshch. Khim.*, **39**, 468 (1969); *J. Gen. Chem. USSR*, **39**, 444 (1969).
24. A. B. Callear and W. J. R. Tyerman, *Trans. Faraday Soc.*, **61**, 2395 (1965).
25. A. B. Callear and W. J. R. Tyerman, *Trans. Faraday Soc.*, **62**, 371 (1966).
26. W. J. R. Tyerman, W. B. O'Callaghan, P. Kebarle, O. P. Strausz and H. E. Gunning, *J. Am. Chem. Soc.*, **88**, 4277 (1966).
27. J. Connor, G. Grieg and O. P. Strausz, *J. Am. Chem. Soc.*, **91**, 5695 (1969).
28. W. Stanley, M. R. VanDeMark and P. L. Kumler, *J. Chem. Soc., Chem. Commun.*, 700 (1974).
29. J. Y. C. Chu and D. G. Marsh, *J. Org. Chem.*, **41**, 3204 (1976).
30. H. K. Spencer and M. P. Cava, *J. Org. Chem.*, **42**, 2937 (1977).

31. D. H. Brown, R. J. Cross and D. Millington, *J. Organomet. Chem.*, **125**, 219 (1977).
32. A. Couture, A. Lablache-Combier, R. Laypouyade and G. Felix, *J. Chem. Res. (S)*, 258 (1979).
33. G. Vermeersch, N. Fabvey-Garot, S. Caplain, A. Couture and A. Lablache-Combier, *Tetrahedron Lett.*, 609 (1979).
34. R. Livingston and K. S. Wei, *J. Am. Chem. Soc.*, **89**, 3098 (1967).
35. G. Ahlgren, B. Akermark and M. Nilsson, *J. Organomet. Chem.*, **30**, 303 (1971).
36. D. G. Marsh, J. Y. C. Chu, J. W. Lewicki and J. L. Weaver, *J. Am. Chem. Soc.*, **98**, 8432 (1976).
37. S. L. N. G. Krishnamachari and T. V. Vinkitachalam, *Chem. Phys. Lett.*, **67**, 69 (1979).
38. T. A. Barton, C. R. Tully and R. W. Roth, *J. Organomet. Chem.*, **108**, 183 (1976).
39. C. L. Pedersen, *Acta Chem. Scand., Ser. B*, **30**, 675 (1976).
40. J. Bergman, *Acta Chem. Scand.*, **22**, 1883 (1968).
41. A. Krantz and J. Laureni, *J. Am. Chem. Soc.*, **99**, 4842 (1977).
42. N. Harrit, S. Rosenkilde, B. D. Larsen and A. Holm, *J. Chem. Soc., Perkin Trans. 1*, 907 (1985).
43. A. Krantz and J. Laureni, *J. Am. Chem. Soc.*, **103**, 480 (1981).
44. W. W. Sander and O. L. Chapmann, *J. Org. Chem.*, **50**, 543 (1985).
45. H. Meier and I. Menzel, *Tetrahedron Lett.*, 445 (1972).
46. F. Malek-Yazdi and M. Yalpani, unpublished results (1976), quoted in Ref. 47.
47. I. Lalezari, A. Shafiee and M. Yalpani, *Adv. Heterocycl. Chem.*, **24**, 109 (1979).
48. I. Lalezari, A. Shafiee, J. Khorrami and A. Soltani, *J. Pharm. Sci.*, **67**, 1336 (1978).
49. W. Lohner and K. Praefcke, *J. Organomet. Chem.*, **194**, 173 (1980).
50. T. Ibata, K. Ueda and M. Takebayashi, *Bull. Chem. Soc. Jpn.*, **46**, 2897 (1973).
51. C. L. Pedersen, N. Harrit, M. Poliakoff and I. Dunkin, *Acta Chem. Scand., Ser. B*, **31**, 848 (1977).
52. C. L. Pedersen and N. Hacker, *Tetrahedron Lett.*, 3981 (1977).
53. C. L. Pedersen, *Tetrahedron Lett.*, 745 (1979).
54. R. B. Woodward and R. Hoffmann, *The Conservation of Orbital Symmetry*, Verlag Chemie, Weinheim, 1970.
55. F. S. Guziec, Jr., C. J. Murphy and E. R. Cullen, *J. Chem. Soc., Perkin Trans. 1*, 107 (1985).
56. R. Hussong, H. Heydt and M. Regitz, *Phosphorus Sulfur*, **25**, 201 (1985).
57. J. F. Bunnet, *Acc. Chem. Res.*, **11**, 413 (1978).
58. R. A. Rossi, *Acc. Chem. Res.*, **15**, 164 (1982).
59. A. B. Pierini and R. A. Rossi, *J. Org. Chem.*, **44**, 4667 (1979).
60. A. B. Pierini and R. A. Rossi, *J. Organomet. Chem.*, **144**, C12 (1978).
61. A. B. Pierini and R. A. Rossi, *J. Organomet. Chem.*, **168**, 163 (1979).
62. J. F. Bunnett and X. Creary, *J. Org. Chem.*, **39**, 3173 (1974).
63. J. F. Bunnett and X. Creary, *J. Org. Chem.*, **40**, 3740 (1975).
64. R. A. Rossi, R. H. de Rossi and A. F. Lopez, *J. Am. Chem. Soc.*, **98**, 1252 (1976).
65. A. B. Pierini, A. B. Penenory and R. A. Rossi, *J. Org. Chem.*, **49**, 486 (1984).
66. S. M. Palacios, R. A. Alonso and R. A. Rossi, *Tetrahedron*, **41**, 4147 (1985).
67. A. N. Frolov, A. V. El'tsov, E. V. Smirnov and O. V. Kul'bitzkaya, *Zh. Org. Khim.*, **13**, 2007 (1977); *Chem. Abstr.*, **88**, 6482 (1978).
68. H. Okabe, *Photochemistry of Small Molecules*, Wiley–Interscience, New York, 1978.
69. L. Andrews, *Annu. Rev. Phys. Chem.*, **30**, 79 (1979).
70. G. Lodder, in *The Chemistry of Functional Groups, Supplement D* (Eds. S. Patai and Z. Rappoport), Wiley, New York, 1983, Chapt. 29, pp. 1605–1683.
71. P. G. Sammes, in *The Chemistry of the Carbon–Halogen Bond* (Ed. S. Patai), Wiley, New York, 1973, Chapt. 11.
72. J. M. Birchall, G. P. Irvin and R. A. Boyson, *J. Chem. Soc., Perkin Trans. 2*, 435 (1975).
73. M. L. Denniston and D. R. Martin, *J. Inorg. Nucl. Chem.*, **37**, 1871 (1975).
74. V. G. Voloshchuk, V. N. Bioko and L. M. Yagupol'skii, *Zh. Org. Khim.*, **13**, 2008 (1977); *Chem. Abstr.*, **88**, 6483 (1978).
75. R. C. Fort and P. v. R. Schleyer, *Adv. Alicycl. Chem.*, **1**, 283 (1966).
76. R. A. Rossi, S. M. Palacios and A. N. Santiago, *J. Org. Chem.*, **47**, 4654 (1982).
77. S. M. Palacios, A. N. Santiago and R. A. Rossi, *J. Org. Chem.*, **49**, 4609 (1984).
78. B. Pakzad, K. Praefcke and H. Simon, *Angew. Chem., Int. Ed. Engl.*, **16**, 319 (1977).
79. K. Beelitz, K. Praefcke and S. Gronowitz, *J. Organomet. Chem.*, **194**, 167 (1980).
80. J. Martens, K. Praefcke and U. Schulze, *Synthesis*, 532 (1976).
81. K. Beelitz, K. Praefcke and S. Gronowitz, *Justus Liebigs Ann. Chem.*, 1597 (1980).
82. M. F. Lappert and P. W. Lendor *Adv. Organomet. Chem.*, **14**, 345 (1976).

83. G. A. Russell and J. Hershberger, *J. Am. Chem. Soc.*, **102**, 7604 (1980).
84. J. Hershberger and G. A. Russell, *Synthesis*, 475 (1980).
85. G. A. Russell and L. L. Herold, *J. Org. Chem.*, **50**, 1037 (1985).
86. J. Deniau, K. N. V. Duong, A. Gaudemer, P. Bougeard and M. D. Johnson, *J. Chem. Soc., Perkin Trans. 2*, 393 (1981).
87. P. Mailbard, J. C. Massot and C. Giannotti, *J. Organomet. Chem.*, **159**, 219 (1978).
88. S. Pouly, G. Tainturier and B. Gautheron, *J. Organomet. Chem.*, **232**, C56 (1982).
89. B. Gautheron, G. Tainturier and S. Pouly, *J. Organomet. Chem.*, **268**, C56 (1984).
90. V. Balzani and V. Carassiti, *Photochemistry of Coordination Compounds*, Academic Press, London, 1970.
91. M. K. Chandhuri, A. Haas and A. Wensky, *J. Organomet. Chem.*, **116**, 326 (1976).
92. P. Jaitner, *J. Organomet. Chem.*, **210**, 353 (1981).
93. P. Jaitner, *J. Organomet. Chem.*, **231**, 225 (1982).
94. P. Jaitner, *J. Organomet. Chem.*, **233**, 333 (1982).
95. N. Kuhn and H. Schumann, *J. Organomet. Chem.*, **276**, 55 (1984).
96. H. Schumann, *J. Organomet. Chem.*, **293**, 75 (1985).
97. N. Kuhn, H. Schumann and G. Wolmershauser, *J. Chem. Soc., Chem. Commun.*, 1595 (1985).
98. M. Herberhold, B. Schmidkonz, U. Thewalts, A. Razavi, H. Schollhorn, W. A. Herrmann and C. Hecht, *J. Organomet. Chem.*, **299**, 213 (1986).
99. H. J. Reich, *Acc. Chem. Res.*, **12**, 22 (1979).
100. D. L. J. Clive, *Tetrahedron*, **34**, 1049 (1978).
101. B. A. Bohm and P. I. Abell, *Chem. Rev.*, **62**, 599 (1962).
102. J. L. Traynham, A. G. Lane and N. Bhacca, *J. Org. Chem.*, **34**, 1302 (1969).
103. Y.-H. Kang and J. L. Kice, *J. Org. Chem.*, **49**, 1507 (1984).
104. L. H. Gale, *J. Org. Chem.*, **33**, 3643 (1968).
105. R. Dowbenko, *Tetrahedron*, **20**, 1843 (1964); *J. Am. Chem. Soc.*, **86**, 946 (1964).
106. L. Friedman, *J. Am. Chem. Soc.*, **86**, 1885 (1964).
107. F. G. Bordwell and W. A. Hewett, *J. Am. Chem. Soc.*, **79**, 3493 (1957).
108. M. S. Alnajar and H. G. Kuivila, *J. Org. Chem.*, **46**, 1053 (1981).
109. S. J. Cristol and R. W. Gleason, *J. Org. Chem.*, **34**, 1762 (1969).
110. S. J. Cristol and D. I. Davies, *J. Org. Chem.*, **29**, 1282 (1964).
111. S. J. Cristol, J. K. Harrington and H. S. Singer, *J. Am. Chem. Soc.*, **88**, 1529 (1964).
112. R. Franzi and M. Geoffroy, *J. Organomet. Chem.*, **218**, 321 (1981).
113. O. L. Chapman, *Org. Photochem.*, **1**, 33 (1967).
114. D. R. Arnold, *Adv. Photochem.*, **6**, 301 (1968).
115. R. A. Bolivar and C. Rivas, *J. Photochem.*, **19**, 95 (1982).
116. T. J. Barton and A. J. Nelson, *Tetrahedron Lett.*, 5037 (1969).
117. C. Rivas, D. Pacheco and F. Vargas, *J. Heterocycl. Chem.*, **17**, 1151 (1980).
118. D. Pacheco, C. Rivas and F. Vargas, *J. Heterocycl. Chem.*, **20**, 1465 (1983).
119. T. Q. Minh, L. Christiaens, P. Grandclaudon and A. Lablache-Combier, *Tetrahedron*, **33**, 2225 (1977).
120. D. C. Neckers, J. H. Dopper and H. Wynberg, *J. Org. Chem.*, **35**, 1582 (1970).
121. D. C. Neckers, J. H. Dopper and H. Wynberg, *Tetrahedron Lett.*, 2913 (1969).
122. J. H. Dopper and D. C. Neckers, *J. Org. Chem.*, **36**, 3756 (1971).
123. G. R. Evanega and E. Payo, *J. Org. Chem.*, **32**, 2918 (1967).
124. R. A. Bolivar, R. Machado, L. Montero, F. Vargas and C. Rivas, *J. Photochem.*, **22**, 91 (1983).
125. R. A. Bolivar and C. Rivas, *J. Photochem.*, **17**, 91 (1981).
126. J. A. Barltrop and J. D. Coyle, *Excited States in Organic Chemistry*, Wiley, New York, 1975, p. 101.
127. C. Rivas, D. Pacheco and F. Vargas, *Acta Sud Am. Quim.*, **2**, 1 (1982); *Chem. Abstr.*, **99**, 38394 (1983).
128. C. Rivas and R. A. Bolivar, *J. Heterocycl. Chem.*, **10**, 967 (1973).
129. N. Petragnani and G. Schill, *Chem. Ber.*, **103**, 2271 (1970).
130. R. M. Kellogg, *Tetrahedron Lett.*, 1429 (1972).
131. B. S. Lukjanow, M. I. Knjazschanski, J. W. Rewinski, L. E. Niworozschkin and W. I. Minkin, *Tetrahedron Lett.*, 2007 (1973).
132. J.-L. Fourrey, G. Henry and P. Jouin, *Tetrahedron Lett.*, 455 (1980).
133. H. Suzuki, M. Usuki and T. Hanafusa, *Synthesis*, 705 (1979).

134. D. Bellus, *Adv. Photochem.*, **8**, 109 (1971).
135. J. Martens, K. Praefcke and H. Simon, *Chem. -Ztg.*, **102**, 108 (1978).
136. N. Y. M. Fung, P. de Mayo, B. Ruge, A. C. Weedon and S. K. Wong, *Can. J. Chem.*, **58**, 6 (1980).
137. D. S. L. Blackwell, K. H. Lee, P. de Mayo, G. L. R. Petrasiunas and G. Reverdy, *Nouv. J. Chim.*, **3**, 123 (1979).
138. D. H. R. Barton, D. Crich, Y. Herve, P. Potier and J. Thierry, *Tetrahedron*, **41**, 4347 (1985).
139. D. H R. Barton, D. Crich and G. Kretzschmar, *Tetrahedron Lett.*, **25**, 1055 (1984).
140. C. L. Pedersen, *J. Chem. Soc., Chem. Commun.*, 704 (1974).
141. J. Kagan, *Tetrahedron Lett.*, 6097 (1966).
142. C. Paulmier, J. Bourguignon, J. Morel and P. Pastour, *C. R. Acad. Sci., Ser. C*, **270**, 494 (1974).
143. A. G. Schultz and R. P. Lucci, *J. Org. Chem.*, **40**, 1371 (1975).
144. A. G. Schultz, *J. Org. Chem.*, **39**, 3185 (1974).
145. O. L. Chapman and G. L. Eian, *J. Am. Chem. Soc.*, **90**, 5329 (1968).
146. J.-P. Le Roux, P. L. Desbene and M. Seguin, *Tetrahedron Lett.*, 3141 (1976).
147. P. L. Desbene and N. Jehanno, *J. Heterocycl. Chem.*, **21**, 1321 (1984).
148. R. A. Abramovitch and E. P. Kyba, *The Chemistry of the Azido Group* (Ed. S. Patai), Interscience, New York, 1971, p. 306.
149. D. L. Ross, J. Blanc and F. J. Maticoli, *J. Am. Chem. Soc.*, **92**, 5750 (1970).
150. A. D. Baker, G. H. Armen, Y. Guang-di, D. Liotta, N. Flannagan, C. Barnum, M. Saindane, G. C. Zima and J. Grossman, *J. Org. Chem.*, **46**, 4127 (1981).
151. D. Betteridge, L. Hendriksen, J. Sandstrom, I. Wennerbeck and M. A. Williams, *Acta Chem. Scand., Ser. A*, **31**, 14 (1977).
152. M. V. Andreocci, M. Bossa, C. Furlani, M. N. Piancastelli, C. Cauletti and T. Tarantelli, *J. Chem. Soc., Faraday Trans. 2*, **75**, 105 (1979).
153. V. Galasso, *J. Mol. Struct.*, **86**, 231 (1982).
154. F. P. Colona and G. Distefano, *J. Electron Spectrosc.*, **18**, 75 (1980).
155. A. G. M. Barrett, D. H. R. Barton, G. Johnson and S. Nagubandi, *Synthesis*, 741 (1978).
156. K. H. Schulte and G. Ohloff, *Helv. Chim. Acta*, **51**, 548 (1968).
157. H. K. Spencer, M. V. Lakshmikantham and M. C. Cava, *J. Am. Chem. Soc.*, **99**, 1470 (1977).
158. M. C. Hovey, *J. Am. Chem. Soc.*, **104**, 4196 (1982).
159. A. A. Frimer, *Chem. Rev.*, **79**, 359 (1979).
160. E. Luppold, W. Winter and E. Muller, *Chem. Ber.*, **109**, 3886 (1976).
161. T. G. Back, D. H. R. Barton, M. R. Britten-Kelly and F. S. Guziec, *J. Chem. Soc., Perkin Trans. 1*, 2079 (1976).
162. N. J. Turro, *Modern Molecular Photochemistry*, Benjamin-Cummings, Menlo Park, CA, 1978, Chapt. 10, pp. 362–395.
163. A. Ohno, M. Uohama, K. Nakamura and S. Oka, *Tetrahedron Lett.*, 1905 (1977).
164. C. C. Liao and P. de Mayo, *J. Chem. Soc., Chem. Commun.*, 1525 (1971).
165. J. Pfenninger, C. Henberger and W. Graf, *Helv. Chim. Acta*, **63**, 2328 (1980).
166. T. Tezuka, H. Suzuki and H. Miyazaki, *Tetrahedron Lett.*, 4885 (1978).
167. J. Martens, K. Praefcke, U. Schulze, H. Schwarz and H. Simon, *Tetrahedron*, **32**, 2467 (1976).
168. R. Tanikaga and A. Kaji, *Bull. Chem. Soc. Jpn.*, **46**, 3814 (1973).
169. G. G. Spence, E. C. Taylor and O. Buchardt, *Chem. Rev.*, **70**, 231 (1970).

The Chemistry of Organic Selenium and Tellurium Compounds Volume 2
Edited by S. Patai
© 1987 John Wiley & Sons Ltd.

CHAPTER **6**

Acidity, hydrogen bonding and self-association in organic and organometallic compounds of selenium and tellurium

THOMAS B. RAUCHFUSS

School of Chemical Sciences, University of Illinois, Urbana, Illinois 61801, USA

I. INTRODUCTION 339
II. ACIDITY AND HYDROGEN BONDING IN SELENOLS AND TELLUROLS 339
III. SELF-ASSOCIATION OF DIORGANO-SELENIUM AND -TELLURIUM COMPOUNDS 341
IV. COMPLEX FORMATION. 342
A. Selenium(II) and Tellurium(II) Compounds 342
B. Selenium(IV) and Tellurium(IV) Compounds. 343
V. Se···Se AND Te···Te INTERACTIONS IN ORGANOTRANSITION METAL CHEMISTRY 345
VI. REFERENCES 346

I. INTRODUCTION

This chapter is primarily concerned with association phenomena in the organic and organometallic chemistry of selenium and tellurium. Since the classical example of self-association involves hydrogen bonding, the review begins with this topic. However, more important examples of self-association occur in more delocalized systems and in the coordination chemistry of organotellurium compounds.

II. ACIDITY AND HYDROGEN BONDING IN SELENOLS AND TELLUROLS

The chemistry of compounds containing SeH and particularly TeH functionalities is so limited in itself that there is only meager information about their physical and structural properties. It is useful to refer to the data on the inorganic hydrides; the first acid dissociation constants of H_2S, H_2Se and H_2Te[1] are 1×10^{-7}, 2×10^{-4} and 2.3×10^{-3}. These values indicate that selenols and tellurols should be substantially stronger acids than the corresponding thiols.

Few pK_a measurements have been made on selenols and no such measurements have been made for tellurols[2]. The simplest alkyltellurols[3] and the organometallic anions $M(CO)_5(TeH)^-$ (M = Cr, Mo, W) and $[M(CO)_5]_2(\mu\text{-TeH})^-$[4], are the only TeH-containing compounds known to be stable at room temperature. It is generally accepted that selenols are more acidic than the corresponding thiols by 2–3 pK_a units. Unfortunately, direct pK_a measurements of simple aryl- or alkyl-selenols have not been described although ^{13}C NMR measurements show that tertiary amines such as $(i\text{-Pr})_2EtN$ are >90% protonated by one equivalent of PhSeH in $CDCl_3$ solution[5]. The pK_a of $(i\text{-Pr})_2EtNH^+$ is estimated to be 10.5.

The solution behaviour of 8-quinolineselenol indicates a pK_a of 4.94 for the non-zwitterionic form; the corresponding pK_a values for 8-quinolinethiol and 8-quinolinol are 7.68 and 9.72, respectively[6]. An equally revealing effect is also seen in the tautomeric constants shown in equation 1.

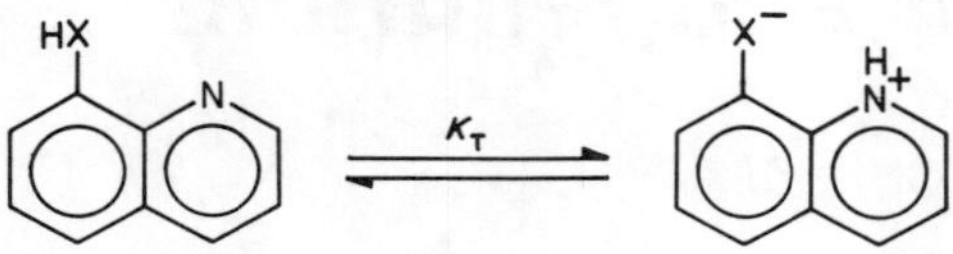

(1)

X	K_T
O	0.04
S	3.8
Se	1740

These and related measurements on selenols are hampered by their facile aerial oxidation to diselenides[7].

2-Aminoethylselenol has been studied in solution in part because of its potential as an antioxidant in nuclear medicine. Like 8-quinolineselenol, the selenium analog of cysteamine exists almost exclusively as a zwitterion at neutral pH[8]. The pK_a of selenacysteamine is 5.01 and that of cysteamine is 8.27. The pK_a of $HSeCH_2CH_2NH_2$ has been confirmed by ^{77}Se NMR spectroscopy, where the ^{77}Se chemical shift moves almost 200 ppm upfield as the pH of its aqueous solution changes from 3.5 to 6[9]. The pK_a values of MeO_2CCH_2SeH and MeO_2CCH_2SH have been determined to be 4.7 and 8.1, respectively[10].

There is little information available concerning hydrogen bonding in selenols and tellurols. Given that the strength of the RSH···SR(H) hydrogen bond[11] is only 1.5–2.0 kcal mol^{-1}, the corresponding selenol–selenol interaction is expected to be smaller since the strength of the hydrogen bond decreases as one descends a column in the Periodic Table. An early infrared measurement showed that ν_{SeH} in simple alkylselenols occurs in the range 2300–2380 cm^{-1} and exhibits no solvent or concentration dependence[12]. On the other hand, the poor correlation of $\delta(^{77}Se)$ vs. Hammett σ constants for arylselenols has been attributed to hydrogen bonding effects[13]. Hydrogen bonding involving Se···H···N is well documented through crystallographic studies on selenourea, 1-benzoyl-3-phenyl-2-selenourea and diselenouracil. These materials are all expected to contain very polarized C=Se bonds which should stabilize hydrogen bonding[14].

III. SELF-ASSOCIATION OF DIORGANO-SELENIUM AND -TELLURIUM COMPOUNDS

There is considerable crystallographic evidence for intermolecular Se···Se and Te···Te interactions in organo-Se and -Te compounds. This topic has assumed great importance because of its bearing on the electrical properties of the Se- and Te-based organic metals[15]. In general it appears that Te···Te interactions are structurally more important than Se···Se interactions as judged by the overlap of van der Waals radii. Most examples of this structural effect contain the chalcogen in the divalent state.

Bis(2-naphthyl) ditelluride has been obtained in two crystalline modifications which show substantial structural differences. Polymorph A consists of a transoid Ar_2Te_2 moiety whose closest Te···Te contact is > 4.1 Å. The Ar_2Te_2 moieties are cisoid in the lattice of polymorph B and the intermolecular Te···Te contacts are 3.71 and 4.00 Å[16]. Similarly, short Te···Te contacts of 3.70 and 3.63 Å are observed in monoclinic 5,6:11,12-tetratelluro(tetracene)[16] and *trans*-2,4-dibenzylidene-1,3-ditelluretane, $(PhCHC)_2Te_2$[17], respectively. The latter is a poorly soluble compound formed in low yield from the reaction of sodium phenylacetylide and elemental Te. Although not discussed in the literature, a striking structural feature of this material is the intramolecular Te···Te contact of 3.23 Å.

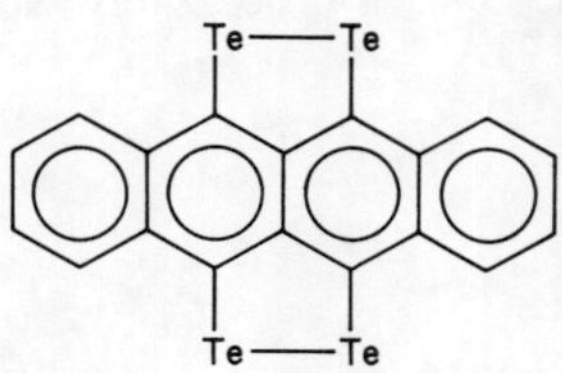

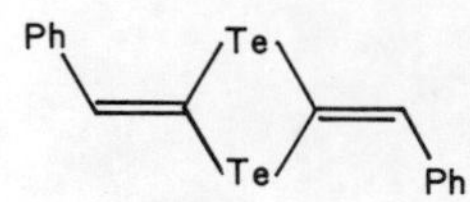

The shortest intermolecular Te···Te contact is 3.58 Å, observed in hexamethylenetetratellurafulvalene[18]. This distance is comparable to the two intramolecular Te···Te contacts of 3.65 and 3.52 Å calculated on the basis of the published data[18a]. In other words, the solid-state structure of this tetratellurafulvalene may be viewed as a Te superlattice into which are fitted organic substituents. The nearly square array of the Te atoms in this compound is reminiscent of Te_4^{2+} (see Section V).

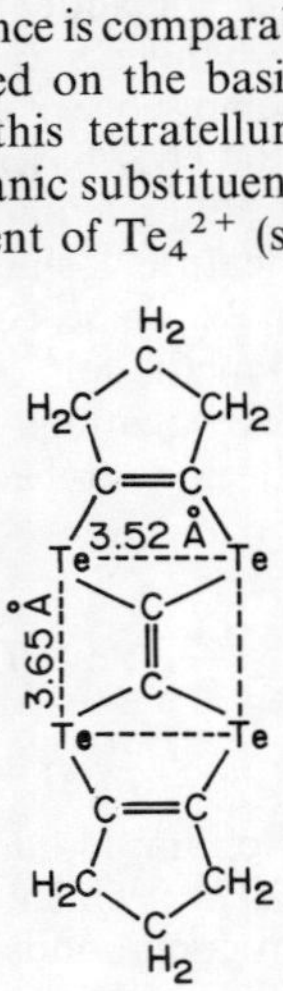

Bis(dimethylthieno)tetratellurafulvalene features Te···Te contacts of 3.66 and 3.76 Å[18b]. These intermolecular interactions may be responsible for the striking non-planarity of this molecule in the lattice where the two $Me_2SC_4Te_2$ planes are bent by 47

and 16° from the plane defined by the central C_2Te_4 core.

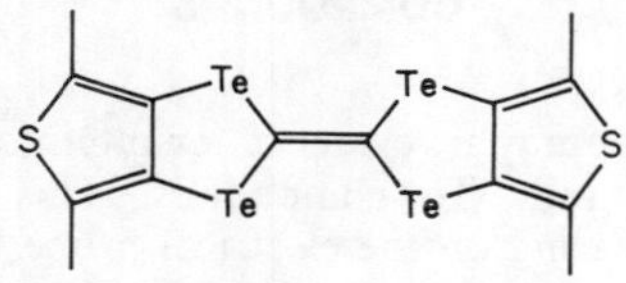

Stereochemically significant intermolecular Te···Te contacts have also been observed in 1,3-ditellurole. The intramolecular Te···Te interaction has been discussed in light of $J(^{125}Te, ^{125}Te)$ values[19].

IV. COMPLEX FORMATION

A. Selenium(II) and Tellurium(II) Compounds

The Lewis acidity of organotellurium(II) halides has been known for many years and has been briefly reviewed by Gysling[20]. Whereas many simple aryltellurium(II) halides are reported to be thermally unstable[21], derivatives containing donor substituents in the *ortho* position are robust. Examples of this type of compound are shown below [X = Cl, Br; D = CHO[22], NO_2[23], N_2Ph[24], $CONH_2$, C(O)R[25]]. The X-ray structure of 2-C_6H_4(CHO)TeBr shows a Te···O distance of 2.31 Å[26].

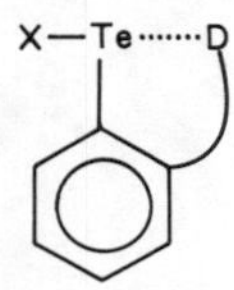

In the same way that *ortho* donor substituents stabilize aryltellurenyl halides, this class of compound has been extended to include adducts with soft donor ligands. In fact, the instability of simple organotellurenyl halides may be due to associative decomposition pathways which are suppressed by the presence of donor ligands[27]. Treatment of Ph_2Te_2 with halogens in the presence of soft Lewis bases gives the compounds PhTeLX and $[PhTeL_2]X$. In most studies thioureas and selenoureas have been employed as the Lewis bases. The structures of the adducts feature T-shaped complexes with a ligand vacancy *trans* to the phenyl group (equation 2). These compounds are also formed from the reduction of $ArTeCl_3$ by thioureas in methanol[28]. This class of compound has been the subject of several crystallographic investigations but there are few surprises and few corroborative solution studies have been described.

$$Ph_2Te_2 + Br_2 \xrightarrow{L} Ph{-}\overset{\displaystyle X}{\underset{\displaystyle L}{Te}} + \left[Ph{-}\overset{\displaystyle L}{\underset{\displaystyle L}{Te}}\right]^+ \tag{2}$$

The addition of tetraorganophosphonium iodides to solutions prepared from Ar_2Te_2 and I_2 affords crystals of $(R_4P)ArTeI_2$[29], which also has the T-geometry about Te. Catenation of simple organotellurium compounds does not seem to have been studied although $Ph_3Te_3^-$ should be stable. Ph_3Te^- has not been observed although it would be an electronic relative of $PhTeI_2^-$ and $Ph_3Te_3^-$ (Evidence has been presented for Ph_2I^-[30]). Related salts of $PhSeBr_2^-$ have been isolated from the reaction of thioureas with $PhSeBr_3$[31].

B. Selenium(IV) and Tellurium(IV) Compounds

Monoorganotellurium(IV) halides are generally prepared via one of the routes shown below[32–34]:

$R_2Te_2 + X_2 \longrightarrow$

$ArHgX + TeCl_4 \longrightarrow$

$ArOR + TeCl_4 \longrightarrow$

$RCH{=}CH_2 + TeCl_4 \longrightarrow$

These compounds probably aggregate in solution and exist as dimers and polymers in the solid state. The majority of the most definitive work has involved the aryl derivatives.

In the solid state, the compounds $RTeCl_3$ (R = 2-$ClCH_2CH_2$[34], Ph[35], 4-EtOPh[33]) are polymers involving square-pyramidal Te centers interconnected by single chloride bridges. As in the aryltellurenyl systems, the coordination site *trans* to the organic substituent is vacant. The compounds $ArTeBr_3$ and $ArTeI_3$ form molecular dimers in the solid state and two isomers of $PhTeI_3$ have been characterized crystallographically[36], although it is not known if such species have significant stability in solution.

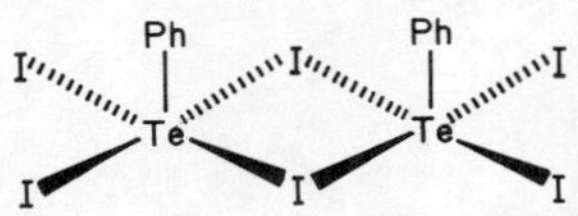

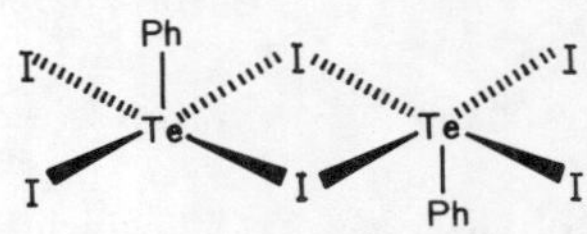

The structural differences between the trichlorides vs. the heavier trihalides have been attributed to the large size of tellurium which precludes the formation of $Te_2(\mu\text{-}Cl)_2$ rings. $PhTeCl_3$ is ionic in solution presumably due to the equilibrium shown in equation 3[30].

$$ArTeCl_3 \longrightarrow ArTeCl_2{}^+ + Cl^- \qquad (3)$$

In a related way, the organotransition metal ditelluride $[(C_5H_5)Mo(CO)_2][Fe(CO)_3](Te_2)SbF_6$ is a good Lewis acid and forms adducts with pseudohalides[37] (Scheme 1). Most of the reactivity of this cation is Te-localized. The Lewis base adducts feature 4-coordinate, hypervalent Te centers.

$PhTeCl_3$ forms stable pentacoordinate adducts with thioureas[28], but reduction of the tellurium(IV) occurs when this reaction is conducted in methanolic solution[27]. Attempted complexation of $(Me_2N)_2CS$ with $PhSeBr_3$ gave $[\{(Me_2N)_2CS\}_2][PhSeBr_2]_2$[28].

The thermochromic compound $PhTe(S_2CNR_2)_3$[38] can be prepared via two routes (equation 4). In the lattice, $PhTe(S_2CNR_2)_3$ adopts a pentagonal pyramidal structure.

$$PhTeCl_3 + 3NaS_2CNR_2 \longrightarrow PhTe(S_2CNR_2)_3 \longleftarrow Ph_2Te_2 + 3[R_2NCCS_2]_2 \qquad (4)$$

Many diorganotellurium halides have been prepared via the routes shown below[35,36].

$$Ar_2Te_2 + RCHBrCHBrR \longrightarrow Ar_2TeBr_2 + RCH{=}CHR + Te$$
$$ArTeCl_3 + Ar_2Te_2 \longrightarrow Ar_2TeCl_2 + Te$$
$$TeCl_4 + Ar_2Hg \longrightarrow Ar_2TeCl_2 + Hg$$
$$R_2Te + X_2 \longrightarrow R_2TeX_2$$
$$Ar_2TeO + 2HI \longrightarrow Ar_2TeI_2$$

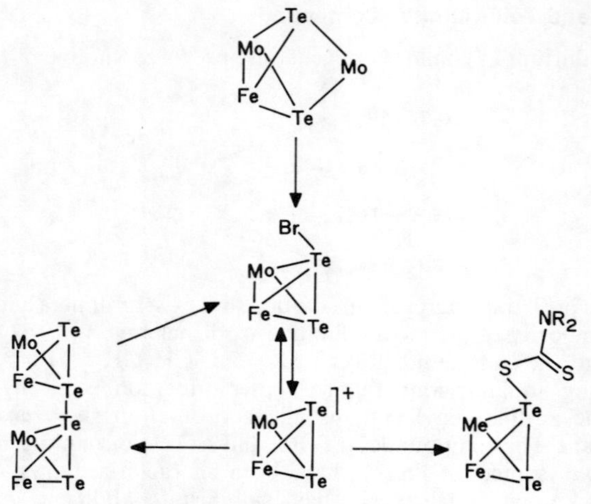

Mo=$(RC_5H_4)Mo(CO)_2$; Fe=$Fe(CO)_3$

Simplest of all is Me_2TeCl_2, which, in the lattice, contains *cis*-Me groups, *trans*-Cl groups and long intermolecular Te···Cl interactions. The crystallographic results for Ph_2TeCl_2 show that a Cl ligand from another molecule approaches the equatorial plane but this secondary Te···Cl distance is 3.67 Å[35].

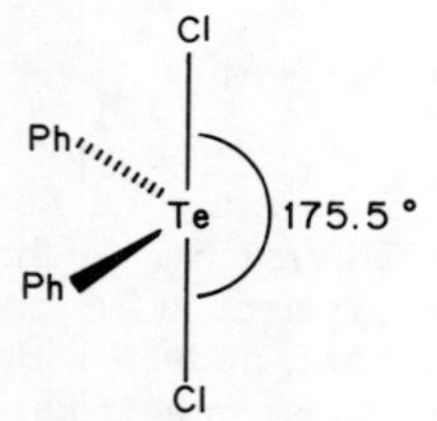

It should be noted that while the aforementioned Me_2TeCl_2 structure is general for other tellurium halides as well as Me_2SeBr_2, Me_2SeI_2 adopts a different geometry[39].

Treatment of $TeCl_4$ with acetophenones[40] and 1,3-diketones[41] gives functionalized dialkyltellurium(IV) dihalides (equations 5 and 6).

$$2ArC(O)Me + TeCl_4 \longrightarrow [ArC(O)CH_2]_2TeCl_2 + 2HCl \qquad (5)$$

$$MeC(O)CH_2C(O)Me + TeCl_4 \longrightarrow Cl_2\underline{TeCH_2C(O)CH_2C(O)C}H_2 + 2HCl \qquad (6)$$

In contrast, certain functionalized ketones when treated with $TeCl_4$ give monoalkyl derivatives[42] (equation 7).

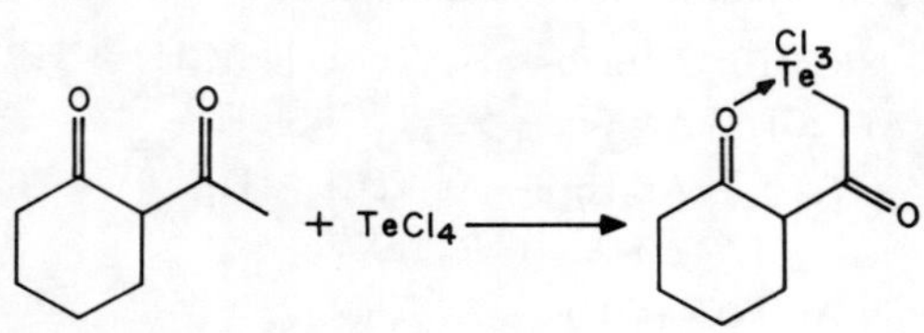

(7)

The remarkable $[Ph_3Te][BPh_4]$ has been prepared via two routes (equation 8). In the lattice, this salt consists of well isolated Ph_3Te^+ ions of approximate C_{3v} symmetry[43].

$$Ph_4Te + BPh_3 \rightarrow Ph_3Te^+BPh_4^- \leftarrow Ph_3TeCl + BPh_4^- \qquad (8)$$

V. Se···Se AND Te···Te INTERACTIONS IN ORGANOTRANSITION METAL CHEMISTRY

The presence of short intramolecular Te–Te contacts has been demonstrated crystallographically for a number of organometallic derivatives of tellurium. Such interactions may be classified as association since, on the basis of the Te—Te distance (r_{TeTe}) they are weaker than simple two-center two-electron bonds of the type found in diphenylditelluride ($r_{TeTe} = 2.72$ Å)[44]. The variability of r_{TeTe} in related compounds is highlighted by the structural results on a series of iron–molybdenum clusters, shown below[37,45,46].

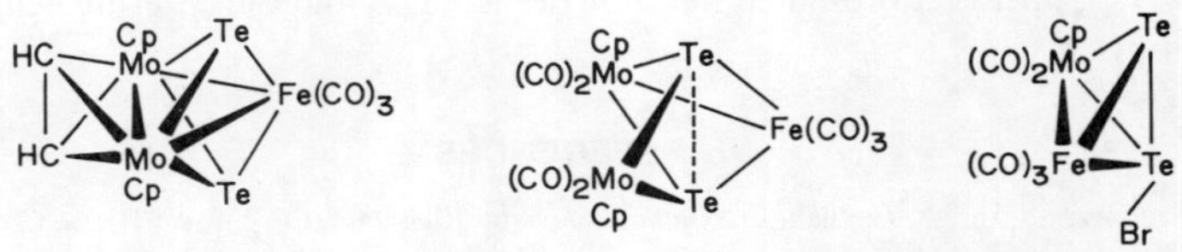

$r_{TeTe} = 3.80$ Å $\quad r_{TeTe} = 3.14$ Å $\quad r_{TeTe} = 2.81$ Å

In such cases, the concept of oxidation state becomes useless and the materials must be viewed as highly delocalized systems[45]. Recall that intermolecular TeTe distances of 3.8 Å are considered short in simpler organotellurium compounds. The results for the organotransition metal tellurides lead one to examine the corresponding intramolecular Te–Te contacts in organotellurium compounds. As mentioned in Section III, hexamethylenetetratellurafulvalene[18] is an interesting standard in this regard; the four Te atoms form a rectangle with edges of 3.52 and 3.65 Å. $[Te_4]^{2+}$ is a square ion with Te···Te distances of 2.66 Å[47].

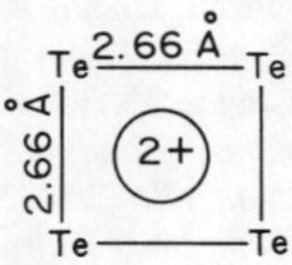

Short Te–Te contacts may be the basis of two effects in organotransition metal–Te cluster chemistry. The easily prepared compound $Fe_3Te_2(CO)_9$ forms stable adducts with a variety of Lewis bases such as phosphines, amines, isocyanides and carbon monoxide[48]. In contrast, $Fe_3S_2(CO)_9$ and $Fe_3Se_2(CO)_9$, which are nearly isostructural with the ditelluride, show no Lewis acidity although they do undergo CO substitution[49]. It has been hypothesized that the facility of the adduct forming reaction of $Fe_3Te_2(CO)_9$ derives in part from an attractive Te···Te interaction[45] (equation 9).

Even more striking is the conversion of octahedral $Co_4Te_2(CO)_{10}$ ($r_{TeTe} = 3.3$ Å) into $Co_4Te_2(CO)_{11}$ ($r_{TeTe} = 3.06$ Å)[50]; in the latter the r_{TeTe} is elongated by only 10% relative to elemental tellurium[51].

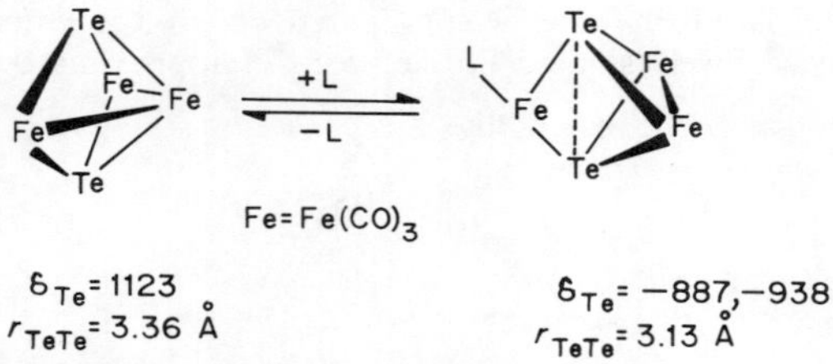

The ^{125}Te chemical shifts for clusters of the type $M_3Te_2L_n$ (L = ligand) vary in a highly systematic way over a 2000 ppm range[52]. Those clusters with short Te···Te distances (3.15–2.8 Å) exhibit δ_{Te} 1000 ± 100 ppm upfield of neat Me_2Te, whereas those clusters with Te···Te distances in the range 3.3–3.8 Å have δ_{Te} of 1000 ± 100 ppm downfield of Me_2Te. The large upfield shifts apparently arise because of magnetic anisotropy due to the short Te—Te distances.

Associative phenomena are prevalent in the chemistry of the polychalcogen cations[47]. Because of their very high reactivity[53], this chemistry has only recently been extended to the organometallic realm. The ion $[W_2(CO)_{10}Se_4]^{2+}$, prepared from $W(CO)_5$THF and $[Se_4]^{2+}$, may be considered as a dimer of two $[W(CO)_5Se_2]^+$ radical cations[54]. In support of this view, r_{SeSe} for the coordinated Se_2 moieties is 2.21 Å and the 'intermolecular' r_{SeSe} is 3.02 Å.

VI. REFERENCES

1. N. N. Greenwood and A. Earnshaw, *Chemistry of the Elements*, Pergamon Press, Oxford, 1984, p. 900.
2. (a) K. Hamada and M. Morishita, *Synth. React. Inorg. Met.-Org. Chem.*, **7**, 355 (1970); (b) C. W. Sink and A. B. Harvey, *Chem. Commun.*, 1023 (1969); (c) J. E. Drake and R. T. Hemmings, *Inorg. Chem.*, **19**, 1879 (1980).
3. K. J. Irgolic, *The Organic Chemistry of Tellurium*, Gordon and Breach, New York, 1974.
4. H. Hausmann, M. Höfler, T. Kruck and H. W. Zimmermann, *Chem. Ber.*, **114**, 975 (1983); V. Küllmer, E. Röttinger and H. Vahrenkamp, *Z. Naturforsch., Teil B*, **34**, 217 (1979). For a more recent example, see W. A. Herrmann, J. Rohrmann, M. L. Ziegler and T. Zahn, *J. Organomet. Chem.*, **273**, 221 (1984).
5. H. J. Reich and M. L. Cohen, *J. Org. Chem.*, **44**, 3149 (1979).
6. N. Nakamura and E. Sekido, *Talanta*, **17**, 515 (1970).
7. D. G. Foster, *Org. Synth.*, Coll. Vol. **III**, 771 (1951).
8. H. Tanaka, H. Sakurai and A. Yokohama, *Chem. Pharm. Bull.*, **18**, 1015 (1970).
9. J. D. Odom, W. H. Dawson and P. D. Ellis, *J. Am. Chem. Soc.*, **101**, 5815 (1979).
10. J. H. Kurz and J. C. Harris, *J. Org. Chem.*, **35**, 3086 (1970).
11. M. D. Joesten and L. J. Schaad, *Hydrogen Bonding*, Marcel Dekker, New York, 1974.
12. N. Sharghi and I. Lalezari, *Spectrochim. Acta*, **20**, 237 (1964).
13. W. McFarlane and R. J. Wood, *J. Chem. Soc., Dalton Trans.*, 1397 (1972).
14. I. L. Karle and J. Karle, in *Organic Selenium Compounds: Their Chemistry and Biology* (Eds. D. L. Klayman and W. H. H. Günther), Wiley, New York 1973, p. 67.
15. M.-H. Whangbo, J. M. Williams, M. A. Beno and J. R. Dorfman, *J. Am. Chem. Soc.*, **105**, 645 (1983).
16. D. J. Sandman, J. C. Stark, M. Rubner, L. A. Acampora and L. A. Samuelson *Mol. Cryst. Liq. Cryst.*, **93**, 293 (1983); D. J. Sandman, J. C. Stark and B. M. Foxman, *Organometallics*, **1**, 739 (1982).
17. S. L. Bender, N. F. Haley and H. R. Loss, *Tetrahedron Lett.*, **22**, 1495 (1981); M. V. Lakshmikantham, M. P. Cava, M. Albeck, L. Engman, P. Carroll, J. Bergman and F. Wudl, *Tetrahedron Lett.*, **22**, 4199 (1981); for related work see S. L. Bender, M. R. Detty, M. W. Fichtner and N. F. Haley, *Tetrahedron Lett.*, **24**, 237 (1983); see also S. L. Bender, M. R. Detty and N. F. Haley, *Tetrahedron Lett.*, **23**, 1531 (1982).

18a. P. J. Carroll, M. V. Lakshmikantham, M. P. Cava, F. Wudl, E. Aharon-Shalom and S. D. Cox, *J. Chem. Soc., Chem. Commun.*, 1316 (1982).
18b. K. Lerstrup, D. O. Cowan and T. J. Kistenmacher, *J. Am. Chem. Soc.*, **106**, 8303 (1984).
19. M. R. Detty, P. M. Henrichs and J. A. Whitefield, *Organometallics*, **5**, 1544 (1986); M. R. Detty, N. F. Haley, R. S. Eachus, J. W. Hassett, H. R. Luss, J. M. McKelvey and A. A. Wernberg, *J. Am. Chem. Soc.*, **107**, 6298 (1985).
20. H. J. Gysling, *Coord. Chem. Rev.*, **42**, 133 (1982).
21. N. Petragnani and M. de Moura Campos, *Chem. Ber.*, **94**, 1759 (1961).
22. J. L. Piette, R. Lysy and M. Renson, *Bull. Soc. Chim. Fr.*, 3559 (1972).
23. P. Wiriyachitra, S. J. Falcone and M. P. Cava, *J. Org. Chem.*, **44**, 3957 (1979); J. L. Piette, P. Thibaut and M. Renson, *Tetrahedron Lett.*, 655 (1978).
24. R. E. Cobbledick, F. W. B. Einstein, W. R. McWhinnie and F. H. Musa, *J. Chem. Res. (S)*, 145 (1979).
25. J. L. Piette and M. Renson, *Bull. Soc. Chim. Belg.*, **80**, 669 (1971).
26. M. Baiwir, G. Llabres, O. Dideberg, L. DuPont and J. L. Piette, *Acta Crystallogr., Sect. B*, **30**, 139 (1974).
27. (a) O. Foss, S. Husebye and K. Maroy, *Acta Chem. Scand.*, **17**, 1806 (1963); (b) O. Foss and K. Maroy, *Acta Chem. Scand.*, **20**, 123 (1966). (c) O. Foss and S. Husebye, *Acta Chem. Scand.*, **20**, 132 (1966). (d) S. Hauge and O. Vikane, *Acta Chem. Scand.*, **27**, 3596 (1973). (e) O. Vikane, *Acta Chem. Scand., Ser. A*, **29**, 738, 763, 787 (1975). (f) S. Hauge, O. Johannesen and O. Vikane, *Acta Chem. Scand., Ser. A*, **32**, 901 (1978). (g) O. Foss and S. Hauge, *Acta Chem. Scand.*, **13**, 2155 (1959). (h) P. Klaeboe and O. Vikane, *Acta Chem. Scand., Ser. A*, **31**, 120 (1977).
28. K. J. Wynne and P. S. Pearson, *Inorg. Chem.*, **10**, 2735 (1971).
29. (a) P. Klaeboe, C. J. Nielson, R. Suchi and O. Vikane, *Acta Chem. Scand., Ser. A*, **32**, 565 (1978); (b) S. Hauge and O. Vikane, *Acta Chem. Scand., Ser. A*, **29**, 755 (1975). (c) K. Maartmann-Moe and J. Songstad, *Acta Chem. Scand., Ser. A*, **36**, 829 (1982).
30. H. J. Reich, N. H. Phillips and I. L. Reich, *J. Am. Chem. Soc.*, **107**, 4104 (1985).
31. K. J. Wynne and P. S. Pearson, *Inorg. Chem.*, **11**, 1192, 1197 (1972).
32. N. Petragnani, *Tetrahedron*, **11**, 15 (1960).
33. P. H. Bird, V. Kumar and B. C. Paul, *Inorg. Chem.*, **19**, 2487 (1980).
34. D. Kobelt and E. F. Paulus, *Angew. Chem., Int. Ed. Engl.*, **10**, 73, 74 (1971).
35. N. W. Alcock and W. D. Harrison, *J. Chem. Soc., Dalton Trans.*, 251 (1982).
36. (a) N. W. Alcock and W. D. Harrison, *J. Chem. Soc., Dalton Trans.*, 869 (1984); (b) A. C. Hazell *Acta Chem. Scand.*, **26**, 1510 (1972).
37. L. E. Bogan, Jr., T. B. Rauchfuss and A. L. Rheingold, *Inorg. Chem.*, **24**, 3720 (1985).
38. S. Esperas and S. Husebye, *Acta Chem. Scand.*, **26**, 3293 (1972).
39. (a) G. D. Cristofferson, R. A. Sparks and J. D. McCullough, *Acta Crystallogr.*, **11**, 782, 249 (1958); (b) G. C. Hayward and P. J. Hendra, *J. Chem. Soc. A*, 1760 (1969).
40. D. G. Marsh, J. Y. C. Chu, J. W. Lewicki and J. L. Weaver, *J. Am. Chem. Soc.*, **98**, 8432 (1976).
41. C. L. Raston, R. J. Secomb and A. H. White, *J. Chem. Soc., Dalton Trans.*, 2307 (1976).
42. H. J. Gysling, H. R. Luss and S. A. Gardner, *J. Organomet. Chem.*, **184**, 417 (1980).
43. R. F. Ziolo and J. Troup, *Inorg. Chem.*, **18**, 2271 (1979).
44. G. Llabres, O. Dideberg, and L. Dupont, *Acta Crystallogr., Sect. B*, **28**, 2438 (1972).
45. L. E. Bogan, Jr., T. B. Rauchfuss and A. L. Rheingold, *J. Am. Chem. Soc.*, **106**, 3843 (1985).
46. L. E. Bogan, Jr., G. R. Clark and T. B. Rauchfuss, *Inorg. Chem.*, **25**, 4050 (1986).
47. R. J. Gillespie, *Chem. Soc. Rev.*, **8**, 315 (1979).
48. D. A. Lesch and T. B. Rauchfuss, *Organometallics*, **1**, 494 (1982).
49. S. Aime, L. Milone, R. Rossetti and P. L. Stanghellini, *J. Chem. Soc., Dalton Trans.* 500 (1980).
50. R. C. Ryan and L. F. Dahl, *J. Am. Chem. Soc.*, **97**, 6904 (1975).
51. J. Donohue, *The Structures of the Elements*, Wiley, New York, 1971.
52. D. A. Lesch and T. B. Rauchfuss, *Inorg. Chem.*, **22**, 1854 (1983).
53. A. M. Rosan, *J. Chem. Soc., Chem. Commun.*, 377 (1985).
54. C. Belin, T. Makani and J. Roziere, *J. Chem. Soc., Chem. Commun.*, 118 (1985).

The Chemistry of Organic Selenium and Tellurium Compounds Volume 2
Edited by S. Patai
© 1987 John Wiley & Sons Ltd.

CHAPTER **7**

Biochemistry of physiologically active selenium compounds

KENJI SODA, HIDEHIKO TANAKA AND NOBUYOSHI ESAKI

Institute for Chemical Research, Kyoto University, Uji, Kyoto-Fu 611, Japan

I. SELENIUM AMINO ACIDS AND SELENOENZYMES 349
A. Introduction 349
B. Selenium Enzymes 350
C. Enzymological Aspects of Selenium Amino Acid Metabolism . . . 351
1. Synthesis of selenocysteine 351
2. Occurrence of selenocysteine and its enzymological properties . . 352
3. Reaction mechanism of selenocysteine β-lyase 355
D. Enzymatic Synthesis of Selenium Amino Acids 356
1. Methionine γ-lyase. 356
2. Tryptophan synthase 358
3. *O*-Acetylhomoserine sulphydrylase 359
4. Comparison between pyridoxal-P enzymes catalysing sulphur and selenium amino acid synthesis. 360
II. SELENIUM NUCLEIC ACIDS 360
A. Introduction 360
B. Natural Occurrence of Selenium-containing tRNA 360
C. Structure of Selenium Nucleosides 361
D. Biological Function of Seleno-tRNA 362
III. REFERENCES 363

I. SELENIUM AMINO ACIDS AND SELENOENZYMES

A. Introduction

Selenium is recognized as an essential micronutrient for mammals, birds, fish and several bacteria. The physiological functions of selenium in mammals and birds can be attributed mostly to the action of glutathione peroxidase, which contains the essential selenocysteine residue in the polypeptide chain. Several microbial enzymes have also been demonstrated to contain in their polypeptide chains selenocysteine residues which play an integral role in catalysis. Various other natural selenium amino acids occur in the free state and have been reported to give protection to mammals from radiation injury and carcinogenesis, as described in Masukawa's chapter in this volume.

In this section we describe enzymological aspects of selenium-containing enzymes and the metabolism of selenium amino acids. The mechanism of the enzymatic synthesis of selenium amino acids is also described here.

B. Selenium Enzymes

Enzymes that contain selenium as an essential component are called selenoenzymes. Table 1 summarizes the selenoenzymes so far studied.

Glutathione peroxidase (E.C. 1.11.1.9) catalyses the decomposition of hydrogen peroxide or organic peroxides by reduction with glutathione as shown in the equation

$$2GSH + H_2O_2 \text{ (or ROOH)} \rightarrow GSSG + H_2O \text{ (or ROH} + H_2O) \quad (1)$$

The enzyme is distributed widely in mammalian tissues, and is responsible for the protection of erythrocyte membrane and other tissues from oxidation damage. The crystalline enzyme has been prepared from bovine erythrocytes to study its three-dimensional structure by X-ray analysis at 2.0 Å resolution[1]. The partial amino acid sequence of rat liver enzyme determined by Edman degradation coincides with the corresponding part of the bovine erythrocyte enzyme deduced by X-ray analysis[1].

Several molecular weights (76000–92000) have been reported for the native enzyme from various sources. The enzyme is composed of four identical subunits (M_r 19000–23000), each of which contains a selenocysteine residue. The selenol group of selenocysteine has lower redox potential and higher nucleophilicity than the thiol group of cysteine. The high reactivity of selenol substantially contributes to the specific catalytic function of glutathione peroxidase. The selenium-containing moiety of xanthine dehydrogenase and nicotinic acid hydroxylase has not been elucidated. Recently, Sliwkowski and

TABLE 1. Selenium enzymes

Enzyme	Chemical form of selenium	Other components	Molecular weight	Source
Glutathione peroxidase	Selenocysteine	–	76000–92000	Rat liver Bovine erythrocytes
Glycine reductase	Selenocysteine	Fe	462000	*Clostridium sticklandii*
Formate dehydrogenase	Selenocysteine	Mo, Fe/S	500000	*Methanococcus vannielii*
	Selenocysteine	W, Mo, Fe/S	300000	*Clostridium thermoaceticum*
	N.d.[a]	Mo, Fe/S, haeme	590000	*E. coli*
Hydrogenase	Selenocysteine	Ni	340000	*Methanococcus vannielii*
Nicotinate hydroxylase	N.d.[a]	Fe/S	300000	*Clostridium barkeri*
Xanthine dehydrogenase	N.d.[a]	Fe, Mo	N.d.[a]	*Clostridium cylindrosporum*

[a]Not determined

Stadtman[2] showed that selenomethionine residues are distributed throughout the primary structure of thiolase; methionine is replaced by selenomethionine non-specifically. Therefore, they suggested that no specific function of selenomethionine is required for the enzyme. All the other selenoenzymes contain a selenocysteine residue. The properties and catalytic functions of these selenoenzymes have been reviewed in detail by Stadtman[3].

C. Enzymological Aspects of Selenium Amino Acid Metabolism

Selenium amino acids are thought to be synthesized through an analogous pathway to the sulphur counterparts. Various enzymes acting on sulphur amino acids (e.g. mammalian cystathionine γ-lyase and bacterial methionine γ-lyase) work on the selenium analogues, although enzymes that act specifically on selenium compounds have been considered[4]. The indiscriminate catalytic action of enzymes on sulphur and selenium compounds is probably involved at least partly in the toxicity of selenium compounds. Provided that the selenium compounds are present at high concentrations, indiscriminate substitution of selenium for sulphur in proteins, nucleic acids and complex carbohydrates could occur and cause various toxic effects on organisms. The metabolism of selenium amino acids was discussed in detail in earlier reviews by Stadtman[4] and Shrift[5], and is not treated here. We shall discuss mainly the metabolism of selenocysteine recently elucidated in our laboratory.

1. Synthesis of selenocysteine

Selenomethionine has been shown to occur in wheat and some other grains, but there are few reports of the occurrence of selenocysteine and selenocystine[6]. In mammalian tissues, selenocysteine synthesis was assumed to be analogous to that of cysteine. Selenomethionine serves as a better substrate than methionine for ATP:L-methionine *S*-adenosyltransferase of rabbit liver and other sources[7,8]. *Se*-Adenosylselenomethionine was shown to be an efficient methyl donor in various methylation systems[9]. Evidence for the enzymatic synthesis of selenocysteine has been reported[10]. Cystathionine β-synthase of rat liver catalyses the formation of selenocystathionine (equation 2) in addition to the cystathionine synthesis (equation 3), as summarized in Table 2.

$$\text{L-serine} + \text{L-selenohomocysteine} \rightarrow \text{L-selenocystathionine} + H_2O \quad (2)$$

$$\text{L-serine} + \text{L-homocysteine} \rightarrow \text{L-cystathionine} + H_2O \quad (3)$$

Selenohomocysteine is as susceptible as homocysteine; the relative V_{max} value is about 0.7. L-Homocysteine inhibits the selenocystathionine synthesis in the presence of 5 mM L-serine as a competitive inhibitor for L-selenohomocysteine, and L-selenohomocysteine also inhibits the cystathionine synthesis in the same manner. The K_i values are substantially consistent with the K_m values (Table 2), indicating that both reactions 2 and 3 are carried out at the same active site.

Cystathionine γ-lyase can catalyse α,β-elimination of L-cystine in addition to α,γ-elimination of L-cystathionine, as described by Cavallini *et al.*[11]. Esaki *et al.*[10] have found that the α,β-elimination of L-cystathionine (equation 4) proceeds much more slowly ($>3\%$) than the α,γ-elimination (equation 5).

$$\text{L-cystathionine} + H_2O \rightarrow \text{pyruvate} + \text{L-homocysteine} + NH_3 \quad (4)$$

$$\text{L-cystathionine} + H_2O \rightarrow \alpha\text{-ketobutyrate} + \text{L-cysteine} + NH_3 \quad (5)$$

$$\text{L-selenocystathionine} + H_2O \rightarrow \text{pyruvate} + \text{L-selenohomocysteine} + NH_3 \quad (6)$$

$$\text{L-selenocystathionine} + H_2O \rightarrow \alpha\text{-ketobutyrate} + \text{L-selenocysteine} + NH_3 \quad (7)$$

TABLE 2. Kinetic parameters for cystathionine β-synthase

	V_{max} (μmol mg^{-1} min^{-1})	K_m (mM)	K_i (mM)
Selenocystathionine synthesis			
(reaction 2)	0.11		
L-Serine		0.3	
L-Selenohomocysteine		2.5	3.4
Cystathionine synthesis			
(reaction 3)	0.16		
L-Serine		1.2	
L-Homocysteine		1.2	0.85

TABLE 3. Kinetic parameters for cystathionine γ-lyase

	Type of reaction	K_m (mM)	V_{max}(μ mol mg^{-1} min^{-1})
L-Selenocystathionine	α,γ-Elimination (reaction 7)	1.7	19
	α,β-Elimination (reaction 6)		9.5
L-Cystathionine	α,γ-Elimination (reaction 5)	1.3	6.2
	α,β-Elimination (reaction 4)		0.18
L-Selenohomocysteine	α,γ-elimination	1.4	0.62
L-Homocysteine	α,γ-Elimination	2.9	0.25
DL-Selenocysteine	α,β-Elimination	1.2	0.51
L-Cysteine	α,β-Elimination	1.1	0.17

However, the α,β-elimination of selenocystathionine proceeds at a comparable rate to the α,γ-elimination of the same substrate (Table 3). Cystathionine γ-lyase can eliminate further both the amino acids formed from selenocystathionine and cystathionine by elimination reactions, although slowly. All the selenium amino acids are decomposed 2.5–3 times more rapidly than the corresponding sulphur analogues (Table 3). Braunstein *et al.*[12] reported that cystathionine β-synthase of chicken liver catalyses synthesis of cysteine from serine and H_2S (equation 8). The rat liver enzyme also catalyses the reaction at a rate of 12% of cystathionine synthesis (equation 3), but selenocysteine cannot be synthesized directly from L-serine and H_2Se (equation 9)[10]. This is probably due to the low reactivity of selenide as a substituting agent in the replacement reaction.

$$\text{L-serine} + H_2S \rightarrow \text{L-cysteine} + H_2O \quad (8)$$

$$\text{L-serine} + H_2Se \rightarrow \text{L-selenocysteine} + H_2O \quad (9)$$

2. Occurrence of selenocysteine β-lyase and its enzymological properties

Selenocysteine is synthesized by the coupled reactions with cystathionine β-synthase (E.C. 4.2.1.22) and cystathionine γ-lyase (E.C. 4.4.1.1) purified from rat liver, and also by reaction with a rat liver homogenate[10], with which the synthesis proceeds far less efficiently. This is due to the presence of a novel enzyme in the homogenate that decomposes selenocysteine specifically into alanine and H_2Se[13]. The enzyme catalyses the removal of elemental selenium from L-selenocysteine; the formation of H_2Se is due to reduction of the elemental selenium by the unreacted selenocysteine. Thus, the enzyme has been termed selenocysteine β-lyase, or systematically selenocysteine selenium-lyase (alanine-forming)[13].

Selenocysteine β-lyase is distributed widely in mammalian tissues (Table 4)[13]. The enzyme activities of livers and kidneys are higher than those of other tissues in several animals. Significant activity is found in pancreas and adrenal, but none in blood and fat. The enzyme has also been demonstrated in various bacterial strains such as *Citrobacter freundii, Alcaligenes viscolactis* and *Pseudomonas alkanolytica.* However, no significant activity was found in yeasts and fungi (Table 5)[14].

Selenocysteine β-lyase has been purified to homogeneity from pig liver and *Citrobacter freundii* and characterized[13,15]. The bacterial enzyme is remarkably different from the mammalian enzyme in its physicochemical properties (Table 6) and amino acid composition (Table 7). In contrast, both the enzymes are very similar in their enzymological properties: both contain pyridoxal 5′-phosphate (pyridoxal-P) as a coenzyme, exhibit strict specificity for L-selenocysteine and have similar K_m values for the substrates. L-Cysteine behaves as a competitive inhibitor against L-selenocysteine for both the enzymes. Based on the K_m value for L-selenocysteine and the K_i value for L-cysteine, the enzyme probably acts on selenocysteine very slowly *in vivo*, because the concentration of selenocysteine in the tissues is lower than the K_m value. However, the total activity of enzyme is probably sufficient to metabolize a small amount of selenocysteine in the tissues (Table 4). The localization and compartmentation of the enzyme, the substrate and the inhibitors probably affect the enzyme activity *in vivo.*

Selenomethionine and selenocysteine are toxic to animals, and H_2Se is the most toxic selenium compound studied so far[16]. Selenocysteine can be synthesized from selenomethionine derived from a diet[6] as described above. H_2Se is produced from selenomethionine through selenocysteine by catalysis of selenocysteine β-lyase. The lack of specificity of the enzymes acting on the biosynthetic pathway of cysteine from methionine (e.g. cysta-

TABLE 4. Distribution of selenocysteine β-lyase in tissues

Tissue	Specific activity ($\times 10^3$)								
	Rat	Dog	Mouse	Guinea pig	Pig	Cat	Rabbit	Bovine	Monkey
Liver	5.5	10	9.7	15	8.2	1.6	17	3.5	9.9
Kidney	4.9	4.5	8.9	14	3.6	1.9	17		
Pancreas	8.0	3.0	0.21	0.34					
Adrenal	3.5								
Heart	0.90	0.81							
Lung	2.0	1.2							
Testis	0.83								
Brain	0.93	0.96							
Thymus	1.7	6.0							
Spleen	1.4	1.8							
Muscle	0.57								
Fat	0								
Blood	0								

TABLE 5. Distribution of selenocysteine β-lyase in microorganisms

Microorganism	Activity (nmol min^{-1} mg^{-1}) determined by		
	H_2Se formed	Alanine formed	Selenocysteine consumed
Alcaligenes viscolactis	32.0	28.5	33.9
Citrobacter freundii	17.5	20.5	31.3
Corynebacterium pseudodiphtheriticum	15.0	19.5	12.8
Pseudomonas alkanolytica	16.7	17.9	17.2
Brevibacterium leucinophagum	22.5	28.3	29.2
Escherichia coli	1.90	1.31	
Erwinia carotovora	2.80	2.10	
Serratia marcescens	2.70	1.54	
Alcaligenes bookeri	5.40	5.24	
Aspergillus ficuum	3.80	1.33	
Aspergillus sojae	0.70	0.00	
Absidia corymbifera	1.40	0.94	
Neurospora crassa	0.56	0.26	
Penicillium expansum	1.00	0.48	
Saccharomyces cerevisiae	2.60	1.30	
Kluyveromyces fragilis	5.60	2.52	
Candida albicans	8.10	1.13	
Hansenula beckii	0.70	0.55	
Schwanniomyces occidentalis	2.90	1.62	

TABLE 6. Properties of selenocysteine β-lyase

Property	Pig liver enzyme	*Citrobacter freundii* enzyme
Molecular weight	93000	63000
Subunit structure	Dimer	Monomer
Isoelectric point	5.9	6.6
Absorption maxima	280, 420	280, 420
Pyridoxal-P content (mol per mol of enzyme)	1	1
Optimum pH	8.4	7.0
Substrate (K_m)	L-Selenocysteine (0.8 mM)	L-Selenocysteine (0.95 mM)
Competitive inhibitor (K_i)	L-Cysteine (1.0 mM)	L-Cysteine (0.65 mM)
Inactivation by abortive transamination	Yes	Yes
Irreversible inactivation by β-chloroalanine	Yes	Yes

TABLE 7. Amino acid composition of selenocysteine β-lyase

Amino acid	*Citrobacter freundii* enzyme		Pig liver enzyme	
	mol-%	mol per mol of subunit	mol-%	mol per mol of subunit
Aspartic acid	14.12	82	8.15	36
Threonine	7.53	43	5.55	25
Serine	12.44	72	5.45	24
Glutamic acid	6.30	36	11.84	52
Proline	3.82	22	5.68	25
Glycine	9.99	58	9.06	40
Alanine	5.93	34	10.66	47
Cysteine	1.05	6	0.84	4
Valine	2.47	14	7.58	33
Methionine	1.63	9	2.26	10
Isoleucine	2.03	12	3.30	15
Leucine	4.05	23	9.48	42
Tyrosine	4.53	26	1.72	8
Phenylalanine	2.69	16	2.82	13
Lysine	7.30	42	3.20	14
Histidine	4.51	26	2.95	13
Arginine	7.89	46	7.19	32
Tryptophan	1.73	10	2.98	7

thionine γ-lyase and cystathionine β-synthase) and the presence of selenocysteine β-lyase may contribute in part to the selenium toxicity.

The mechanism of incorporation of selenium into the enzymes that contain selenocysteine residues has been studied. Sunde and Hoekstra[17] suggested that serine or cysteine residues in polypeptide chains are converted, for example, into dehydroalanine residues, and then H_2Se is added to the α, β-unsaturated bond to give the selenocysteine residues. In contrast, Hawkes *et al.*[18] proposed a direct incorporation mechanism, according to which a tRNA and a codon specific for selenocysteine are present, and selenocysteine is directly incorporated into the protein in the stage of translation. According to the post-translational selenium incorporation hypothesis[17], selenocysteine β-lyase can function to give the direct selenium precursor that is incorporated into the proenzyme to form the selenocysteine residue. Alternatively, according to the direct incorporation hypothesis, selenocysteine β-lyase functions to decompose an excess of selenocysteine to give H_2Se, which can be detoxified through methylation or some other reactions[18].

3. Reaction mechanism of selenocysteine β-lyase

The selenocysteine β-lyase reaction is exceptional among those of the pyridoxal-P enzymes studied so far. The enzyme resembles bacterial aspartate β-decarboxylase (E.C. 4.1.1.12)[19] and kynureninase (E.C. 3.7.1.3)[20] in the reaction mechanism where a moiety binding to $C_{(3)}$ of the substrate is cleaved to produce alanine. Esaki *et al.*[21] and Chocat *et al.*[15] have proposed mechanisms for the reactions catalysed by selenocysteine β-lyase. The selenohydryl group of selenocysteine is substantially in an anionic form under physiological conditions because its pK_a is 5.28[22]. Cysteine is not a substrate of selenocysteine β-lyase, but it inhibits the enzyme reaction competitively with selenocysteine between pH 7.0 and 9.0. In this pH range, a thiol group of cysteine is dissociated at least partially since its pK_a is about 8.8. Thus, the difference in the enzymatic reactivities of selenol and thiol is not derived from differences in their dissociation states. The deuterium isotope effect at the α position determined by Esaki *et al.*[21] and Chocat *et al.*[15] indicates that an α-hydrogen release occurs in the enzyme reaction and is rate limiting. The α-hydrogen of selenocysteine is abstracted by a base at the enzyme active site, and then selenium is removed in an elemental form. L-Cysteine can bind the enzyme active site, but elemental sulphur is not removed from a cysteine–pyridoxal-P aldimine complex owing to the strong bond between the β-carbon and the sulphur of cysteine, which is stronger than that between the β-carbon and selenium of selenocysteine (the dissociation energies are 272 kJ mol^{-1} for C—S and 243 kJ mol^{-1} for C—Se)[23]. The situation is similar in various organic reactions, such as the reactions of episulphides and episelenides[24].

The selenocysteine β-lyase reactions have been studied in deuterium oxide by Esaki *et al.*[21] and Chocat *et al.*[15] to show deuterium incorporation into alanine. The ^1H and ^{13}C NMR spectra of the alanine produced indicate the formation of [β-$^1H_1$]- and [β-$^1H_2$]-alanines. Therefore, in addition to the incorporation of one deuterium atom into the β-position of alanine after removal of elemental selenium, one of the two β-hydrogen atoms of selenocysteine is exchanged with a solvent deuterium atom at a frequency of 0.5. The enzyme catalyses no hydrogen exchange at the α- and β-positions of alanine with a solvent deuterium atom, and the α-hydrogen of selenocysteine is fully retained at the α-position of alanine. Thus, a two-base mechanism has been proposed for the enzyme reaction: the α-protonation and deprotonation are performed by one base, and the other base mediates the β-protonation. A similar mechanism has been proposed by Chang *et al.*[25] for the reactions catalysed by aspartate β-decarboxylase.

Selenocysteine β-lyase is inactivated through transamination between selenocysteine and the bound pyridoxal-P to produce pyridoxamine 5′-phosphate (pyridoxamine-P) and a keto analogue of selenocysteine when the enzyme is incubated with L-selenocysteine in the absence of added pyridoxal-P, as reported by Esaki *et al.*[21] and Chocat *et al.*[15]. Analogous transaminations catalysed by pyridoxal-P enzymes have been reported involving serine hydroxymethyltransferase[26], arginine racemase[27], tryptophan synthase[28], aspartate β-decarboxylase[25] and kynureninase[20]. All three enzymes catalysing β-elimination (kynureninase, aspartate β-decarboxylase and selenocysteine β-lyase) also catalyse the transamination, although very slowly.

Chocat *et al.*[15] reported that selenocysteine β-lyase catalyses the α, β-elimination of β-chloro-L-alanine to form NH_3, pyruvate and Cl^-, and is irreversibly inactivated during the reaction in a suicide fashion. The relatively low partition ratio (825) of the α, β-elimination to the inactivation is similar to that reported for alanine racemase[29] and amino acid racemase with low substrate specificity[30], and indicates highly efficient inactivation.

D. Enzymatic Synthesis of Selenium Amino Acids

The biological role of optically active selenium amino acids has received considerable attention, but studies of their metabolism have been hampered by difficulties in their synthesis. Recently, facile synthetic procedures giving optically active *Se*-substituted selenocysteines and *Se*-substituted selenohomocysteines were developed by Soda and coworkers[31–33,39–41] by means of the microbial pyridoxal-P enzymes methionine γ-lyase, tryptophan synthase and *O*-acetylhomoserine sulphydrylase.

1. *Methionine γ-lyase*

L-Methionine γ-lyase (E.C.4.4.1.11) is a pyridoxal-P enzyme catalysing the conversion of L-methionine into α-ketobutyrate, methanethiol and ammonia, and plays an important role in the bacterial metabolism of methionine. The enzyme is widely distributed in psuedomonads. The enzyme is inducibly produced by addition of L-methionine to the medium, and was purified to homogeneity from the crude extract of *Pseudomonas putida* ICR 3460, the best producer of the enzyme[31]. Recently, *Aeromonas* sp. isolated from a lake was also found to produce L-methionine γ-lyase abundantly, and the *Aeromonas* enzyme was purified to homogeneity by a similar procedure to that for the *Pseudomonas* enzyme[32]. Table 8 summarizes the physicochemical properties of the enzyme purified from both strains. The enzyme has multiple catalytic functions: it catalyses α,γ-elimination and γ-replacement reactions of L-methionine and its analogues and α,β-elimination and β-replacement reactions of L-cysteine and its analogues. The enzyme also catalyses the α,γ-elimination of selenomethionine to yield α-ketobutyrate, ammonia and methaneselenol, and also γ-replacement reactions with various thiols to produce *S*-substituted homocysteines[33]. Selenomethionine is a better substrate than methionine for α,γ-elimination based on the V_{max} and K_m values, but is less effective for γ-replacement. In addition, L-methionine and its derivatives, which are substrates for the α,γ-elimination, react with selenols to form the corresponding *Se*-substituted selenohomocysteines, although selenols are less efficient substituent donors than thiols. The enzymatic β-replacement reaction also occurs between *S*-substituted cysteines or *O*-substituted serines and selenols.

Davis and Metzler[34] have proposed that a ketimine intermediate of pyridoxal-P and vinylglycine (2-amino-3-butenoate) is the key intermediate of α,γ-elimination and γ-replacement reactions catalysed by pyridoxal-P enzymes. Vinylglycine has been reported to inactivate several transaminases as a suicide substrate[35]. This is not the case with methionine γ-lyase. It catalyzes the deamination reaction of vinylglycine to produce α-ketobutyrate and ammonia (equation 10), but is not inactivated by vinylglycine[36]. The enzyme also catalyses the γ-addition reaction of various thiols or selenols to yield the corresponding *S*- or *Se*-substituted homocysteines (equation 11). The relative activities of the enzyme for a variety of thiols and selenols in the γ-addition reaction of vinylglycine are close to those in the γ-replacement reaction of methionine. Incubation of the enzyme with vinylglycine results in the appearance of a new absorption band at 480 nm, which is also observed with substrates such as methionine, *O*-acetylhomoserine and selenomethionine

TABLE 8. Properties of methionine γ-lyase

Property	Source: *Pseudomonas putida*	Source: *Aeromonas* sp.
$S^0_{20,w}$	8.3S	–
Molecular weight:		
Sedimentation equilibrium	165000	–
Gel permeation chromatography	–	149000
Low-angle light scattering	174000	159000
Absorption maxima	278 nm ($\varepsilon = 134000$) 420 nm ($\varepsilon = 38900$)	278 nm ($\varepsilon = 159000$) 423 nm ($\varepsilon = 31300$)
Number of subunits	4	4
Molecular weight of subunits	43000	41000
Pyridoxal-P content (mol per mol of enzyme)	4	4

for α,γ-elimination, but not with α,β-elimination substrates. These findings support the mechanism through a vinylglycine–pyridoxal-P quinonoid intermediate proposed by Davis and Metzler[34].

$$CH_2=CHCH(NH_2)COOH + H_2O \longrightarrow EtCOCOOH + NH_3 \quad (10)$$

$$CH_2=CHCH(NH_2)COOH + RXH \longrightarrow RXCH_2CH_2CH(NH_2)COOH \quad (11)$$

$$(X = S \text{ or } Se)$$

2. Tryptophan synthase

Tryptophan synthase (E.C. 4.2.1.20) is a pyridoxal-P enzyme with a variety of catalytic functions (Table 9, reactions 12–17), among which reaction 12, the synthesis of tryptophan from indole glycerol phosphate, is of physiological importance. Reactions 13 and 14 are regarded as partial reactions of reaction 12 (Table 9). Tryptophan synthase is found widely in various bacteria, yeasts, moulds and plants. The enzyme of *E. coli* is composed of two kinds of proteins, α (M_r29000) and β (M_r44200). Pyridoxal-P is bound to the β-subunit through a Schiff base. Two α-subunits combine with one β_2-dimer to form an $\alpha_2\beta_2$-complex (M_r147000) that catalyses the physiological reaction. Each of the subunits also catalyses its own specific reaction (Table 9). The crystalline $\alpha_2\beta_2$-complex is obtained after a six-fold purification from *E. coli* trp R$^-$ ΔtrpED102/F′ΔtrpED102. About 16% of the intracellular soluble protein of this mutant is the tryptophan synthase complex. Goldberg and Baldwin[37] and Miles *et al.*[38] have shown that methanethiol and β-mercaptoethanol serve as *S*-substituent donors to serine to yield *S*-methyl-L-cysteine and *S*-(β-hydroxyethyl)-L-cysteine, respectively, by the $\alpha_2\beta_2$- and β_2-complexes (Table 9, reaction 16).

Esaki *et al.*[39] have studied the enzymatic synthesis of various *S*-substituted-L-cysteines from L-serine and its derivatives (e.g. β-chloro-L-alanine and *O*-methyl-L-serine) with the $\alpha_2\beta_2$-complex. Thiols such as α-toluenethiol, 1-propanethiol and 1-butanethiol are efficient *S*-substituent donors. When L-threonine and L-vinylglycine are used as *S*-substituent acceptors of thiols, the corresponding *S*-substituted β-methyl-L-cysteines are synthesized[39]. The enzyme also catalyses the β-replacement reactions of L-serine with selenols to produce the corresponding *Se*-substituted L-selenocysteines[40]. *Se*-Benzyl-L-selenocysteine and *Se*-methyl-L-selenocysteine are synthesized from L-serine and α-tolueneselenol and methaneselenol in a similar way with yields of 44 and 16%, respectively,

TABLE 9. Reactions catalysed by tryptophan synthase

No.	Reaction	Catalysed by
12	Indole-3-glycerol phosphate + L-serine → L-tryptophan + D-glyceraldehyde 3-phosphate + H_2O	$\alpha_2\beta_2$
13	Indole-3-glycerol phosphate → indole + D-glyceraldehyde 3-phosphate	α; $\alpha_2\beta_2$
14	Indole + L-serine → L-tryptophan + H_2O	β_2; $\alpha_2\beta_2$
15	L-Serine → pyruvate + ammonia	β_2
16	β-Mercaptoethanol + L-serine → *S*-(β-hydroxyethyl-L-cysteine + H_2O	β_2; $\alpha_2\beta_2$
17	β-Mercaptoethanol + L-serine + pyridoxal-P → *S*-pyruvylmercaptoethanol + pyridoxamine-P + H_2O	β_2

based on L-serine. The relative activity of indole to methanethiol (100) is approximately 150. Those of methaneselenol and α-tolueneselenol are 24 and 30%, respectively. The production of *Se*-methyl- and *Se*-benzyl-L-selenocysteines proceeds much more rapidly than tryptophan synthesis, the inherent reaction of the enzyme. L-Serine can be replaced with a variety of β-substituted L-alanines such as β-chloroalanine and *O*-acetylserine in the reaction system containing these selenols.

According to the general mechanism for the β-replacement reaction catalysed by pyridoxal-P enzymes[34], nucleophilic addition of selenols occurs in an intermediate derived from the substrate. Although selenols are more nucleophilic than thiols, selenols are less reactive substituent donors than thiols in the enzymatic β-replacement reactions catalysed by tryptophan synthase. This is compatible with the reactivities of α-tolueneselenol and α-toluenethiol, although methaneselenol is a slightly more efficient substituent donor than methanethiol. Some physicochemical properties of methaneselenol and methanethiol, such as volatility and solubility, may affect their reactivity in the enzyme reaction.

3. *O-Acetylhomoserine sulphydrylase*

O-Acetylhomoserine sulphydrylase [*O*-acetylhomoserine (thiol)-lyase, E.C. 4.2.99.10] is a pyridoxal-P enzyme that catalyses the synthesis of cysteine and homocysteine from H_2S with *O*-acetyl-L-serine (OAS) and *O*-acetyl-L-homoserine (OAH), respectively. The enzyme of baker's yeast has been purified and characterized. It is involved in the synthesis *in vivo* of cysteine.

Chocat *et al.*[41] have shown that OAH sulphydrylase catalyses the β- and γ-replacement reactions between the *O*-acetyl groups of OAS and OAH and Na_2Se_2. Serine *O*-sulphate also serves as a substrate of the β-replacement reaction, although its reactivity is lower than that of OAS (Table 10). The selenium amino acids produced have been isolated and identified as L-selenocystine and L-selenohomocystine, but this does not necessarily mean that the primary enzymatic product is the diselenide. The initial products of the enzyme reaction are probably *Se*-(selenohydryl) derivatives, i.e. $^-SeSeCH_2CH(NH_2)COOH$ and $^-SeSeCH_2CH_2CH(NH_2)COOH$, which are non-enzymatically oxidized to the corresponding diselenides.

TABLE 10. Kinetic parameters of reactions catalysed by *O*-acetylhomoserine sulphydrylase

Substituent acceptor	Substituent donor	Relative V_{max}	K_m (mM) substituent	
			Acceptor	Donor
O-Acetylhomoserine	NaHS	100	4.1	0.52
O-Acetylhomoserine	Na_2Se_2	17	5.3	8.9
O-Acetylserine	NaHS	14	2.5	0.70
O-Acetylserine	Na_2Se_2	8.4	5.0	N.d.[a]
Serine *O*-sulphate	NaHS	5.1	4.0	0.70
Serine *O*-sulphate	Na_2Se_2	1.4	4.0	10
Serine *O*-sulphate	NaHSe	1.3	N.d.[a]	1.2

[a]Not determined

4. Comparison between pyridoxal-P enzymes catalysing sulphur and selenium amino acid synthesis

Methionine γ-lyase is useful for the production of various sulphur and selenium amino acids. However, the enzyme catalyses not only replacements but also eliminations; the yield depends mainly on the concentrations of the substrates, thiols or selenols. The amino acids produced may be decomposed by the elimination reactions, which constitutes a weak point of the enzyme from the point of view of amino acid production. Therefore, an excess of thiols or selenols have to be added to prevent the elimination reaction.

Tryptophan synthase surpasses methionine γ-lyase with regard to reaction specificity: it does not catalyse the elimination reaction that leads to a decrease in a yield of sulphur and selenium amino acids. However, tryptophan synthase is inferior to methionine γ-lyase in substrate specificity: it does not catalyse the synthesis of *S*-substituted homocysteines and *Se*-substituted selenohomocysteines by the γ-replacement reactions.

O-Acetylhomoserine (thiol)-lyase catalyses both the β- and γ-replacement reactions between *O*-acetylhomoserine or *O*-acetylserine and various thiols or selenols, but does not catalyse the elimination reactions. Therefore, the enzyme is also useful for the synthesis of optically active sulphur and selenium amino acids. However, *O*-acetylhomoserine (thiol)-lyase is poorly produced by yeasts, and shows a high substrate specificity with respect to substituent acceptors. Hence we have to use expensive substrates such as *O*-acetylhomoserine and *O*-acetylserine for the desired amino acid syntheses.

II. SELENIUM NUCLEIC ACIDS

A. Introduction

Amino acid transfer ribonucleic acids (tRNAs) undergo a variety of modifications after transcription from deoxyribonucleic acid. More than 50 modified nucleosides including sulphur-containing derivatives of uridine, cytidine and adenosine have been identified in tRNA to date. Recently, it was discovered that certain tRNAs also contain selenium nucleosides as a specific constituent. The selenium is incorporated through a highly specific process. Selenium tRNAs are the second groups of biologically active selenium-containing macromolecules, selenium enzymes described in Section I being the first. The selenium nucleic acids have been studied mainly by Stadtman and coworkers. In the following we describe the occurrence, structure and function of the selenium tRNAs.

B. Natural Occurrence of Selenium-containing tRNA

In 1972, Saelinger *et al.*[42] reported that growth of *Escherichia coli* in a medium containing [^{75}Se]selenite results in incorporation of ^{75}Se into tRNA bases. One of the nucleosides isolated by enzymatic digestion of ^{75}Se-labelled tRNAs was identified as 4-selenouridine. The authors assumed that selenium incorporation results from the non-specific substitution through the pathway of sulphur transfer to a uracil residue of *E. coli* tRNA. However, incorporation of selenium into the tRNA of *E. coli* B grown in the presence of 0.08 μM selenite was not affected by the addition of 1 mM sulphate[43]. Chen and Stadtman found that in *Clostridium sticklandii*, selenium is incorporated into tRNAs and also into selenoprotein A, one of the components of glycine reductase, in the presence of a large excess of sulphate[44]. Selenium incorporation was shown to be specific for selenium; no dilution effects of varying ratios of the concentrations of sulphur to that of selenium in the medium were observed. Selenium occurs in at least four different tRNA species. The lability of the incorporated selenium in these species towards CNBr, KBH_4 and iodoacetate and its stability under conditions in which aminoacylated tRNA is de-esterified indicate that the selenium is located in the polynucleotide portion of the molecules, but not in esterified amino acids. The most prominent seleno-tRNA corresponding to 80% of the total seleno-tRNAs is the major glutamate-accepting species, which contains one selenium atom per tRNA molecule[45]. The seleno-$tRNA^{Glu}$ has been purified in its acylated form by repeated reversed-phase chromatography from an enriched sample prepared with an organomercurial affinity column[46]. A proline isoacceptor[44] and a valine isoacceptor[47] are also seleno-tRNAs.

Synthesis of seleno-tRNA by a highly specific process, which is distinct from the mechanism of sulphur incorporation, was also found in bacteria such as *E. coli*[48] and *Methanococcus vannielii*[49], in cultured mammalian cells[50] and in plants[51]. In *E. coli*, the amount of selenium incorporated is unchanged by 10–20-fold variations in selenium or sulphate concentrations or by the addition of excess of cysteine, sulphide and sulphite. Further *E. coli* mutants that do not synthesize an abundant sulphur-modified base, 4-thiouracil, produce normal levels of selenium-modified tRNAs. Two major *E. coli* tRNA species modified with selenium (over 50% of the total seleno-tRNAs) have been identified as a lysyl-tRNA and a glutamyl-tRNA, but they are minor isoacceptors for lysine and glutamate[48]. In *M. vannielii*, 13–20% of the total tRNA population is modified with selenium[49]. The amount of seleno-tRNA in *M. vannielli*, a strict anaerobe, is higher than those in *C. sticklandii* (5–8%), a less strictly anaerobic bacterium and in *E. coli* (up to 6%), a facultative anaerobe[49]. No seleno-tRNA is detected in *Bacillus subtilis*, a strict aerobe[52]. Hence the biosynthesis of seleno-tRNA may be associated with the concentration of oxygen. The amino acid-accepting activities of *M. vannielii* seleno-tRNA have not been determined exactly, although the possibility of a glutamyl-tRNA is presumed.

The tRNA of mouse leukaemia cells has a low concentration of selenium, and the chemical properties of the seleno-tRNA are different from those of the bacterial species described above[50]. In plants, the occurrence of seleno-tRNA in *Astragalus bisulcatus*, a selenium accumulator plant[43], and wild carrot cells, *Daucus carota* L.[51], has been demonstrated.

C. Structure of Selenium Nucleosides

The most abundant selenium nucleoside found in cells of three bacterial species has been identified as 5-[(methylamino)methyl]-2-selenouridine (mnm^5se^2U), and was compared with the authentic compound (Fig. 1). The latter was chemically synthesized from selenourea and characterized by its UV and ^1H NMR spectral properties[52]. The authentic

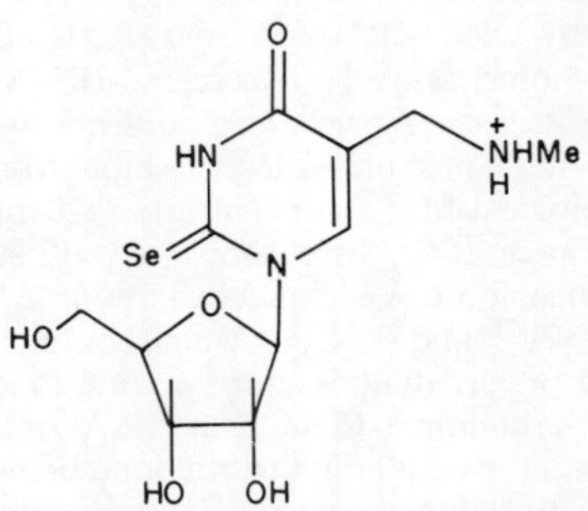

FIGURE 1. Structure of 5-methylamino-methyl-2-selenouridine (mnm^5se^2U).

and natural compounds show identical UV spectra and chromatographic behaviour, and are decomposed by anaerobic treatment with HCl to the same products[52]. All the ^{75}Se label in the bulk tRNA in *E. coli* and about half of the ^{75}Se found in [^{75}Se]tRNA from *C. sticklandii* and *M. vannielii* are present as mnm^{5}se^{2}U[49,52]. Each of the latter two bacterial species contains an additional uncharacterized 2-selenouridine derivative. 4-Selenouridine, which was initially reported to be a constituent of *E. coli* tRNAs[43,53,54], has not been detected in the tRNA populations of *E. coli, C. sticklandii* or *M. vannielii*, which produce about the same amount of 4-thiouridine[52]. 4-Selenouridine may be synthesized only when ratio of the concentration of selenium to that of sulphur in the growth medium is abnormally high.

The structure of the selenonucleoside of the major seleno-tRNA found in cultured mouse leukaemia cells has not been determined. It exhibits strong hydrophobic character, and resembles chromatographically highly modified thioadenosine derivatives such as 2-methyl-N^6-isopentyladenosine, 2-methylthioribosylzeatin and *N*-[(9-β-D-ribofuranosyl-2-methylthiopurine-6-yl)carbamoyl]threonine[50]. Therefore, selenium analogues of these or related thioadenosines are probably found in mammalian cells.

The mechanism of the incorporation of selenium into tRNA and the biosynthesis of the 5-methylaminomethyl side-chain of mnm^{5}se^{2}U residue on seleno-tRNA^{Glu} described above are unknown. Presumably, the nucleoside of the precursor tRNA is modified post-transcriptionally by enzymes.

D. Biological Function of seleno-tRNA

Modified tRNA bases have been shown to play an important role in codon recognition, and some show a regulatory function. Removal of selenium from seleno-tRNA^{Glu} of *C. sticklandii* by exposure to alkali[45] or by treatment with CNBr[46] results in a loss of glutamate-accepting activity. The glutamate-accepting activity of a partially purified *M. vannielii* tRNA preparation also disappears on release of the selenium[49]. These results suggest that the presence of selenium in tRNA^{Glu} species is essential for aminoacylation by its cognate glutamyl-tRNA synthetase. Similar findings have been reported for the removal of sulphur from *E. coli* sulphur modified tRNA$_2{}^{Glu}$, which contains 5-methylaminomethyl-2-thiouridine (mnm^{5}s^{2}U) in the first position of the anticodon[55]. Nucleoside sequence analysis of the seleno-tRNA^{Glu} from *C. sticklandii* indicates that the mnm^{5}se^{2}U residue is also located at this site (the 'wobble position')[56]. The sequence containing the anticodon (25 bases long) is homologous with that of *E. coli* sulphur-modified tRNA$_2{}^{Glu}$ from residues 27 to 50, except that mnm^{5}se^{2}U replaces mnm^{5}s^{2}U in the first position of the anticodon (Fig. 2)[56]. Similar results have been shown for the relationship between seleno-tRNA^{Lys} and non-seleno-tRNA^{Lys}[57].

The codon recognition by seleno-tRNA^{Glu} from *C. sticklandii* was studied with the standard trinucleotide-ribosome assay[46]. The seleno-tRNA^{Glu} recognizes both glutamate codons (GAA and GAG) equally well, and does not interact with the termination codons (UAA and UAG) at a non-physiological high Mg^{2+} concentration (20 mM). However, at a near-physiological Mg^{2+} concentration (10 mM), the GAA codon is only slightly favored over GAG (ca. 15%). In contrast, *E. coli* sulphur-modified tRNA$_2{}^{Glu}$ shows a four-fold preference for GAA[46]. Several reports have shown that sulphur-modified tRNAs such as tRNA^{Glu} and tRNA^{Lys}, which contain 5-alkyl-2-thiouridine at the wobble position, recognize preferentially codons ending in A (i.e. glutamate GAA and lysine AAA) rather than those ending in G (GAG and AAG)[58]. This codon preference has been explained on the basis of the weaker hydrogen bond between the 2-thio group and G compared with the standard U–A base pair; in U–A pairing, the $C_{(4)}$ oxygen and $N_{(3)}$ nitrogen of U participate, whereas, according to the wobble hypothesis, the $C_{(2)}$ oxygen and $N_{(3)}$ nitrogen of U participate in U–G pairing[58]. This explanation, however, conflicts

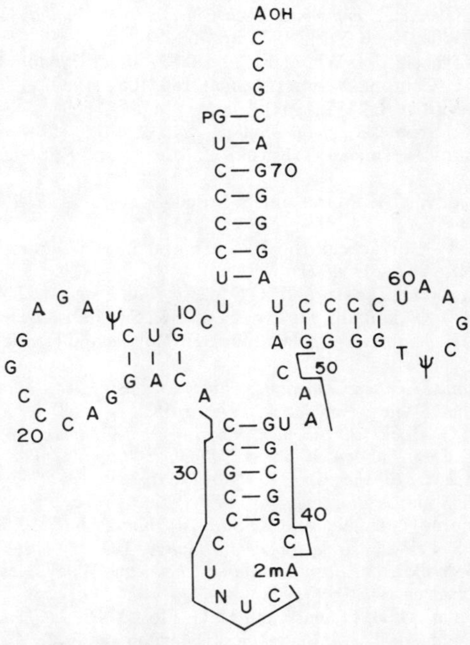

FIGURE 2. Comparison of *E. coli* $tRNA_2^{Glu}$ sequence with the sequence of seleno-$tRNA^{Glu}$ from *C. sticklandii*. Total sequence of *E. coli* $tRNA_2^{Glu}$ is shown in a cloverleaf structure. Bold line indicates region of homology between the two bacterial tRNAs. N represents mnm^5s^2U in *E. coli* $tRNA_2^{Glu}$ and mnm^5se^2U in *C. sticklandii* seleno-$trRNA^{Glu}$[56].

with the results of the nearly equal recognition of GAA and GAG by seleno-$tRNA^{Glu}$ described above, because the hydrogen bond involving a seleno group is expected to be even weaker than that with a thio group (i.e. mnm^5se^2U should show a stronger preference of GAA over GAG). Ching[46] have shown that the mnm^5se^2U residue is significantly ionized at neutral pH, whereas the sulphur analogue is not ionized under the same conditions[46]. However, it has also been found that the ionization of the mnm^5se^2U residue is not relevant to the difference in codon preference between the seleno-$tRNA^{Glu}$ and sulphur-containing $tRNA^{Glu}$[46]. Some other factors such as the interaction of the 5-methylaminomethyl side-chain with the ribosome site and the different atomic radii of selenium and sulphur may cause the difference[46].

[^{3}H]Glutamylseleno-$tRNA^{Glu}$ from *C. sticklandii* is effectively utilized for protein synthesis in a wheat germ extract *in vitro* translation system[46].

III. REFERENCES

1. D. Epp, R. Ladenstein and A. Wendel, *Eur. J. Biochem.*, **133**, 51 (1983).
2. M. Sliwkowski & T. C. Stadtman, *J. Biol. Chem.*, **260**, 3140 (1985).
3. T. C. Stadtman, *Annu. Rev. Biochem.*, **49**, 93 (1980).
4. T. C. Stadtman, *Adv. Enzymol.*, **48**, 1 (1979).

5. A. Shrift, in *Organic Selenium Compounds: Their Chemistry and Biology* (Eds. D. L. Klayman and W. H. H. Gunther), Wiley, New York, 1973 pp. 763–814.
6. O. E. Olson, E. J. Novacek, E. I. Whitehead and I. S. Palmer, *Phytochemistry*, **9**, 1181 (1970).
7. S. H. Mudd and G. L. Cantoni, *Nature* (*London*), **180**, 1052 (1957).
8. R. C. Greene, *Biochemistry*, **8**, 2255 (1969).
9. E. H. Coch and R. C. Greene, *Biochim. Biophys. Acta*, **230**, 223 (1971).
10. N. Esaki, T. Nakamura, H. Tanaka, T. Suzuki, Y. Morino and K. Soda, *Biochemistry*, **20**, 4492 (1981).
11. D. Cavallini, B. Mondovi, C. DeMarco and S. Sciscia-Santoro, *Arch. Biochem. Biophys.*, **96**, 456 (1962).
12. A. E. Braunstein, E. V. Goryachenkova, E. A. Tolosa, I. H. Willhardt and L. L. Yefremova, *Biochim. Biophys. Acta*, **242**, 247 (1971).
13. N. Esaki, T. Nakamura, H. Tanaka and K. Soda, *J. Biol. Chem.*, **257**, 4386 (1982).
14. P. Chocat, N. Esaki, T. Nakamura, H. Tanaka and K. Soda, *J. Bacteriol.*, **156**, 455 (1983).
15. P. Chocat, N. Esaki, K. Tanizawa, K. Nakamura, H. Tanaka and K. Soda, *J. Bacteriol.*, **163**, 669 (1985).
16. J. L. Martin, in *Organic Selenium Compounds: Their Chemistry and Biology* (Eds. D. L. Klayman and W. H. H. Gunther), Wiley, New York, 1973, p. 663.
17. R. A. Sunde and W. G. Hoekstra, *Biochem. Biophys. Res. Commun.*, **93**, 1181 (1980).
18. W. C. Hawkes, D. E. Lyons and A. L. Tappel, *Biochim. Biophys. Acta*, **699**, 183 (1982).
19. S. S. Tate and A. Meister, *Biochemistry*, **8**, 1660 (1969).
20. K. Tanizawa and K. Soda, *J. Biochem.*, **86**, 1199 (1979).
21. N. Esaki, T. Nakamura, H. Tanaka and K. Soda, *J. Biol. Chem.*, **257**, 4386 (1982).
22. R. E. Huber and R. S. Criddle, *Arch. Biochem. Biophys.*, **122**, 164 (1967).
23. F. A. Cotton and G. Wilkinson, *Basic Inorganic Chemistry*, Wiley, New York, 1976, p. 10.
24. D. L. J. Clive, *Tetrahedron*, **34**, 1049 (1978).
25. C. C. Chang, A. Laghai, M. H. O'Leary and H. G. Floss, *J. Biol. Chem.*, **257**, 3564 (1982).
26. J. G. Voet, D. M. Hindenlag, T. J. J. Blanck, R. J. Ulevitch and R. G. Kallen, *J. Biol. Chem.*, **248**, 841 (1973).
27. T. Yorifuji and K. Soda, *J. Biol. Chem.*, **246**, 5093 (1971).
28. H. C. Dunathan and J. G. Voet, *Proc. Natl. Acad. Sci. USA*, **71**, 3888 (1974).
29. B. Badet, D. Roise and C. T. Walsh, *Biochemistry*, **23**, 5188 (1984).
30. D. Roise, K. Soda, T. Yagi and C. T. Walsh, *Biochemistry*, **23**, 5195 (1984).
31. T. Nakayama, N. Esaki, K. Sugie, T. T. Berezov, H. Tanaka and K. Soda, *Anal. Biochem.*, **138**, 421 (1984).
32. T. Nakayama, N. Esaki, W.-J. Lee, I. Tanaka, H. Tanaka and K. Soda, *Agric. Biol. Chem.*, **48**, 1991 (1984).
33. N. Esaki, H. Tanaka, S. Uemura, T. Suzuki and K. Soda, *Biochemistry*, **18**, 407 (1979).
34. L. Davis and D. E. Metzler, *Enzymes*, 3rd ed., Vol. 7, Academic Press, New York, 1972, p. 33.
35. C. Walsh, *Tetrahedron*, **38**, 871 (1982).
36. N. Esaki, T. Suzuki, H. Tanaka, K. Soda and R. R. Rando, *FEBS Lett.*, **84**, 309 (1977).
37. M. E. Goldberg and R. L. Baldwin, *Biochemistry*, **6**, 2113 (1967).
38. E. W. Miles, M. Hatanaka and I. P. Crawford, *Biochemistry*, **7**, 2742 (1968).
39. N. Esaki, H. Tanaka, E. W. Miles and K. Soda, *Agric. Biol. Chem.*, **47**, 2861 (1983).
40. N. Esaki, H. Tanaka, E. W. Miles and K. Soda, *FEBS Lett.*, **161**, 207 (1983).
41. P. Chocat, N. Esaki, H. Tanaka and K. Soda, *Agric. Biol. Chem.*, **49**, 1143 (1985).
42. D. A. Saelinger, J. L. Hoffman and K. P. McConnell, *J. Mol. Biol.*, **69**, 9 (1972).
43. P. A. Young and I. I. Kaiser, *Plant Physiol.*, **63**, 511 (1979).
44. C.-S. Chen and T. C. Stadtman, *Proc. Natl. Acad. Sci. USA*, **77**, 1403 (1980).
45. W.-M. Ching and T. C. Stadtman, *Proc. Natl. Acad. Sci. USA*, **79**, 347 (1982).
46. W.-M. Ching, *Arch. Biochem. Biophys.*, **244**, 137 (1986).
47. C.-S. Chen, T. N. Wen and H. M. Tuan, *Biochim. Biophys. Acta*, **699**, 92 (1982).
48. A. J. Wittwer, J. Biol. Chem., **258**, 8637 (1983).
49. W.-M. Ching, A. J. Wittwer, L. Tsai and T. C. Stadtman, *Proc. Natl. Acad. Sci. USA*, **81**, 57 (1984).
50. W.-M. Ching, *Proc. Natl. Acad. Sci. USA*, **81**, 3010 (1984).
51. C.-S. Chen, N. W. Tuan and J.-H. Chang, *Curr. Top. Cell. Regul.*, **27**, 509 (1985).
52. A. J. Wittwer, L. Tsai, W.-M. Ching and T. C. Stadtman, *Biochemistry*, **23**, 4650 (1984).

53. J. L. Hoffman and K. P. McConnell, *Biochim. Biophys. Acta*, **366**, 109 (1974).
54. Y. S. Prasada Rao and J. D. Cherayil, *Life Sci.*, **14**, 2051 (1974).
55. P. F. Agris, D. Soll and T. Sens, *Biochemistry*, **12**, 4331 (1973).
56. W.-M. Ching, B. Alzner-DeWeerd and T. C. Stadtman, *Proc. Natl. Acad. Sci. USA*, **82**, 347 (1985).
57. W.-M. Ching, L. Tsai and A. J. Wittwer, *Curr. Top. Cell. Regul.*, **27**, 497 (1985).
58. S. Nishimura, in *Transfer-RNA: Structure, Properties and Recognition* (eds. P. R. Schimmel, I. Soll and J. N. A. Belson), Cold Spring Harbor Laboratory, Cold Spring Harbor, NY 1979, p. 59.

The Chemistry of Organic Selenium and Tellurium Compounds Volume 2
Edited by S. Patai
© 1987 John Wiley & Sons Ltd.

CHAPTER **8**

Biological and biochemical aspects of tellurium derivatives

TSVI SADEH*

Institute of Chemical Research, University of Kyoto, Uji, Kyoto-Fu 611, Japan

I. INTRODUCTION 367
II. Te COMPOUNDS IN BIOLOGY AND BIOCHEMISTRY 368
A. Toxicology. 368
B. Biological Interactions 369
C. Organotellurium Compounds in Nature and Their Synthetic Analogues 371
1. Carbohydrate derivatives 371
2. Fatty acid analogues 371
3. Tellura-steroids 372
4. Telluro-amino acids 372
5. Porphyrin analogues 372
6. Analogues of drugs and antibiotics 373
III. A LOOK AT THE FUTURE 373
A. Teratogenicity 373
B. Tellurium and the Nervous System 373
C. Tellurium Bio-organic Chemistry 373
D. Application of Organotellurium Compounds in Biology and Medicine 374
E. Tellurium as an Essential Trace Element 374
IV. ACKNOWLEDGEMENT 374
V. REFERENCES 374

I. INTRODUCTION

Within the context of biologically oriented research on Group VI elements, tellurium can be regarded as the biological Cinderella of the group. The position of oxygen and sulphur is, of course, incontestable with respect to their biological importance, and the last few years have seen the emergence of selenium as a major factor in biochemistry, playing a

*On Sabbatical leave from the Soreq Nuclear Research Center, 70600 Yavneh, Israel.

specific role as a key component in certain vitally important enzymes and being regarded, therefore, as an essential trace element. In comparison, no similar functions have yet been discovered in the case of tellurium. However, the available facts point to a so far unexplored territory in tellurium research, involving its biological interactions with livings systems, which may require a different approach to its problems compared with sulphur and selenium. The underlying reasons can be sought in the specificity of its chemical interactions, which differ in many ways from those of sulphur and selenium. The causes can be found in the more metallic properties of the tellurium, which are expressed by a distinct weakening of its covalent bonds with hydrogen and carbon in comparison with sulphur and selenium. Also, the larger covalent, electrovalent and coordination sphere radii of tellurium must be considered in its chemical interactions, and will inevitably influence its biochemical behaviour.

Tellurium has an extensive organic chemistry with distinctive reaction patterns[1], but its biochemistry lags behind that of selenium and can be regarded as having reached a state of development comparable to that of selenium 30 years ago. However, the toxicology of tellurium has been the subject of investigations which might lend themselves as a suitable starting point for more fundamental biochemical research. The very fact that tellurium compounds are toxic points to their biological interactions with metabolic processes of major importance to cell life and indicates the involvement of tellurium-containing molecular species with vital cell components, possibly enzymes, in a so far unknown manner. Whereas selenium can be found incorporated into numerous kinds of biomolecules, the list of such tellurium compounds is very short. This does not necessarily prove that none are formed in living systems; very possibly the methods applied for their isolation were not adequate and did not take sufficient account of their great sensitivity to air and light.

II. TE COMPOUNDS IN BIOLOGY AND BIOCHEMISTRY

We shall divide the discussion of tellurium in biological and biochemical research into three sections: (a) toxicity of tellurium; (b) biological interactions of tellurium, including those with other elements encountered in the environment; and (c) the occurrence of tellurium and its compounds in nature and tellurium-containing analogues of naturally occurring compounds.

A. Toxicology

The toxicology of tellurium has received less attention than that of selenium, partly because contact between tellurium compounds and humans or animals in industry or agriculture is less frequent, leading to fewer accidental poisonings. Possibly the toxicity of tellurium in general may be lower than that of selenium. One of the reasons for the smaller potential hazards from tellurium compounds may be their lower solubility at physiological pH. For instance, elemental tellurium is less oxidizable than elemental selenium, and the tellurium dioxide formed is almost insoluble in water at physiological pH. Also, hydrogen telluride, a potential industrial pollutant, is even more easily decomposed than hydrogen selenide by light and oxygen, yielding relatively harmless elemental tellurium, before the H_2Te can do any damage. The toxicology of tellurium has been reviewed[2], experimental toxicity studies have been conducted[2–6] and accidental tellurium poisoning has been described[7,8], but it may safely be stated that the danger from occupational or environmental tellurium poisoning does not constitute one of the major hazards of human existence[9].

B. Biological Interactions

Clearly, tellurium and its compounds are not biologically inert; they interact with living systems in specific ways. Toxicity is caused only if the detoxification mechanism available to living organisms is overwhelmed by excessive amounts of tellurium compounds. Living organisms have various methods of ridding themselves of unrequired tellurium compounds. Bacteria, for instance, deal with them by a reductive pathway leading to elemental black tellurium, which is biologically inert[10], and which forms a black precipitate within the bacterial cell. The capability for tellurite reduction varies with different kinds of microorganisms and can be employed for their identification[11], such as for *Vibrio* species[12]. Fungi and animals can carry the reduction of tellurites and tellurates further. The tellurium is reduced to the telluride stage, Te^{2-}, and subsequently methylated[19,13–15]. The dimethyl telluride produced is volatile and is thus removed from the organisms. The production of methylated compounds from Group V and V_I elements and also from certain metals in other groups has been well documented[14,15]. Not only fungi, but also mammals, including humans, are able to methylate tellurium compounds and to produce dimethyl telluride[16], which imparts a characteristic smell to persons who come into contact with inorganic tellurium compounds[9]. The biological interaction of microorganisms with tellurium is not confined to biological methylation.

Certain microorganisms are able to decompose metallic ores oxidatively and liberate anionic Group VI elements, including tellurium, as water-soluble salts[17–19]. It seems that sulphur bacteria are implicated in these oxidative reactions, leading to mineral breakdown and solubilization. However, it was found that the sulphur bacterium *Thiobacillus thiooxidans* is sensitive to relatively moderate tellurium concentrations[20]. In its interactions with higher animals, it was found that tellurium acts as an antagonist to selenium, and is able to cause the symptoms of selenium deficiency[21]. The animals undergoing these tests were fed tellurium compounds and subsequently showed the characteristic lesions of the internal organs associated with selenium deficiency. In the case of pigs fed tellurium, a marked decrease in glutathione peroxidase activity of the blood was noted[22]. This can be taken as an indication that tellurium interferes in an unknown manner with the selenium-incorporating mechanism of mammalian protein synthesis.

In living systems, tellurium tends to become attached to proteins in a manner yet to be established[23], and has a special affinity for nervous tissue. Prolonged treatment with tellurium will cause degeneration of the nervous system and demyelinization of nerve strands[24].

An effect typical of tellurium was found in rats. In animals fed tellurium for prolonged periods, lipofuscin pigment developed in their brains in the form of granules, interspersed with elemental tellurium particles[25]. Tellurium has been found to affect adversely the learning ability of rats[26]. If administered to pregnant animals, it acts as a teratogen[27] and in particular may cause hydrocephalus of foetuses. It acts in the same manner towards embryos in birds' eggs[28].

Additional biological effects of tellurium have been noted. Tellurium salts administered to rats influence the development of experimental caries, which in turn can be manipulated by administration of additional chemicals, such as glutathione[29], cysteine, ascorbic acid, thioctic acid[30] and dimercaptosuccinic acid[31]. Also, a change in the oral streptococcal flora was observed on administration of tellurite[32].

Continuing with the evaluation of the interaction of tellurium salts with rat tissues, or rat tissue components, the influence of tellurite and related chemicals on the respiration and oxidative phosphorylation of rat liver mitochondria was investigated[33]. At relatively high concentrations, tellurite was found to be an inhibitor of substrate oxidation, whereas at lower concentrations it proved to be an uncoupler for oxidative phosphorylation. Rats exposed to drinking water contaminated with tellurium tetrachloride showed that tellurium is a cumulative contaminant, but that typical tellurium neurochemical effect was revealed only after a threshold concentration in the brain was exceeded[34]. By employing radioactive ^{127m}Te, Duckett[35] was able to demonstrate a physiological difference in the penetration rate of blood vessels by tellurium in rats which had been exposed previously to tellurium, in comparison with rats not previously exposed to tellurium[35]. Blood vessels of rats to whom tellurium had been administered previously permitted a more rapid penetration of radiotellurium into the sciatic nerve than in rats without prior exposure.

The specific damage to the nervous tissue of rats fed tellurium was also demonstrated[36]. Tellurium tetrachloride fed to ducklings caused the symptoms of selenium–vitamin E deficiency, which became evident as myocardial ultrastructural alterations. However, some kind of repair mechanism for the damaged heart muscle seems to come into play also[37]. Bacteria have also been the subject of studies concerning tellurium. Resistance to tellurium poisoning in *Salmonella*[38] and certain strains of *Escherichia coli*[39] is carried by IncH plasmids, which also confer resistance to other toxic agents. In other bacterial strains, resistance to tellurite is carried in IncP plasmids[40]. Reistance to heavy metals and drugs, colicin production and the biochemical characteristics of selected bovine and porcine strains of *E. coli* all seem to depend on belonging to certain O serogroups, with corresponding resistance to tellurium toxicity[41]. Autotrophic, photosynthetic cyanobacteria species also show patterns of resistance to groups of toxic substances, including tellurium[42]. This also seems to indicate genetic control of these properties.

Aspects of environmental pollution seemed to be the reason for investigating the binding of tellurium and heavy metals by the proteins contained in isolated gill preparations of certain mussels[43]. Likewise, the determination of tellurium in cereal crops[44] could conceivably have a bearing on the problem of Keshan desease[45], since tellurium is known to act as an antagonist to selenium, and can cause the symptoms of selenium deficiency[22].

Tellurium can show certain heavy metal binding properties resembling those of selenium if administered to various animal species. For instance, tellurium administered to mice will cause enhanced accumulation of inhaled mercury, particularly in living tissue[46].

Some physiological effects of tellurium have been discovered that point to potentially significant functions in relation to the chemistry of cell membranes. Tellurite and also selenite caused the haemolysis of sheep erythrocytes[47]. This haemolysis seems to depend on an interaction with reduced glutathione, since sheep erythrocytes depleted of reduced glutathione showed increased resistance to haemolysis. This effect was independent of biochemical lesions responsible for reduced glutathione deficiency. Another approach to the problem of the causes of the interaction of tellurium compounds with erythrocyte cell membranes was made by De Meio and Doughty[48], who also found a dependence of erythrocyte haemolysis on reduced glutathione concentration. This reduced glutathione-induced haemolysis was inhibited by disodium 4-acetamido-4″-isothiocyanatostilbene 2,2′-disulphonate. The inhibition could be abolished by excess of reduced glutathione. Anaerobic incubation of tellurite with reduced glutathione produced a haemolytic agent, but air prevented its formation. Mercury-containing haemolytic agents, such as *p*-hydroxymercuribenzoate or *p*-hydroxymercuriphenyl sulphonate, caused haemolysis which did not involve reduced glutathione.

An effect of considerable significance involving tellurite was discovered recently. Tellurite prevents sickling of erythrocytes, and it was found to be a potent membrane-acting agent *in vitro*[49]. Tellurite is effective in low doses, and the anti-sickling effect depends on the incubation time. Tellurite causes swelling of erythrocytes and the anti-sickling effect can be attributed to a decreased mean cell haemoglobin concentration[49]. The problem which remains to be solved is the mechanism of this interesting effect, and to establish what can be learned from it in relation to blood diseases such as sickle-cell anaemia.

An *in vitro* effect of tellurite which closely resembles that of selenite is represented by its catalytic action on the reduction of methaemoglobin by glutathione (GSH). The catalytic action is explained by the possible formation of selenol groups attached to sulphur if excess of GSH acts on selenite. It must therefore be postulated, if indeed the explanation is correct, that excess of GSH would form tellurol groups attached to sulphur if acting on tellurite[50,51]. Support for this suggestion can be inferred from the fact that heavy metals and other sulphydryl inhibitors are able to block this catalytic reaction by both selenite and tellurite.

C. Organotellurium Compounds in Nature and Their Synthetic Analogues

Unlike organosulphur and organoselenium compounds, no organotellurium derivatives more complicated than dimethyl telluride seem to have been isolated so far from natural sources. This does not necessarily indicate that they are not produced by various organisms that have to deal with tellurium absorbed from the environment, but rather that suitable isolation methods have not been employed so far. Every chemist who has worked with organotellurium compounds, particularly of the aliphatic kind, has experienced their sensitivity towards light and air. Isolation procedures for organotellurium species in materials derived from living matter must take account of these factors and be adapted accordingly. Analogously, it took many years for techniques to be developed sufficiently to enable organoselenium compounds to be isolated from natural sources. However, synthetic tellurium analogues of naturally occurring compounds have been prepared, and are described below.

1. Carbohydrate derivatives

Sugar complexes with hexavalent telluric acid have been described[52]. It could well be that the primary reaction of hexavalent tellurium freshly absorbed by plants, microorganisms or animals could be with carbohydrate derivatives, such as glucose. These complexes might well be an intermediate stage of tellurium metabolism, before the element undergoes further biochemical reactions and incorporation into compounds with carbon—tellurium bonds. True telluro-carbohydrates have also been synthesized recently[53]. The compounds concerned are essentially 1-β-D-telluroglucosides made from acetobromoglucose.

The synthesis of the telluroglucoside described involves a novel method of introducing the tellurium atom by means of 2-tellurido-2-oxo-1, 3, 2-dioxaphosphorinane. This method might be applied with advantage also to other syntheses with tellurium not involving carbohydrate chemistry. The sensitivity of the telluroglucosides described towards oxygen and moisture, and presumably towards light, show again the necessity to take this into account when attempting to isolate tellurium-containing natural products from living systems fed tellurium salts. It should be noted in passing that naturally occurring carbohydrates containing selenium have been found and identified[54].

2. Fatty acid analogues

A group of organotellurium compounds which have found use in biomedical research are tellura-fatty acids. These are long-chain fatty acid analogues in which a methylene group has been replaced with a bivalent tellurium atom, which thus forms a telluro-ether linkage between two carbon chains constituting the molecule[55].

Tellura-fatty acids have been synthesized with γ-ray-emitting tellurium isotopes for organ-imaging purposes in nuclear medicine, and have been designed specifically for the investigation of cardiac disorders[55–69].

It was also found that a tellurium atom enhances retention of the fatty acid analogue in heart muscle tissue. The reason for this is assumed to be connected with oxidation of the tellurium atom *in vivo* to a hydrated telluroxide moiety, which interacts with the carboxy group of the fatty acid molecule to form a cyclic hemiketal ester-like entity, which probably prevents β-oxidative breakdown of the fatty acid analogue molecule. It could be demonstrated that the analogous selena-fatty acid analogues do not form a similarly hydrated selenoxide moiety, and evidently a cyclic hemiketal ester structure cannot be expected to be formed[62]. It seems, therefore, that tellurium as a labelling atom has potential advantages, because of its specific chemical reactivity, over selenium, and future work along these lines might well reveal additional features of interest to biomedicine.

The utility of double labelling has been demonstrated[58], the tellurium atom employed being non-radioactive natural tellurium, with the radioactivity being carried by one of the γ-ray-emitting radioisotopes of bromine[67] or iodine[61]. The radiohalogens are attached preferentially to an aromatic[60] or vinylic[64] moiety forming the end of the carbon chain of the fatty acid analogue. The reason for this is the superior stability of a halogen atom bound to an aromatic or vinylic structure to loss by hydrolysis or exchange compared with a halogen bound to an aliphatic moiety.

3. *Tellura-steroids*

In addition to fatty acid analogues containing tellurium, other lipids have also been synthesized with an included tellurium atom[55,70–74]. Whereas Wolff and Zanati[70] prepared a telluro-steroid derivative as a potential androgen analogue, Knapp and coworkers[55,71–75] synthesized their compounds for intended use as adrenal imaging agents by employing the γ-ray-emitting radioisotope ^{123m}Te.

4. *Telluro-amino acids*

Telluro-α-amino acids including telluro-methionine have been described[75]. In the case of telluro-methionine, the difficulty of working with functionalized aliphatic tellurium derivatives was again demonstrated. Although the method used to synthesize telluro-methionine was essentially identical with one of the established methods for preparing ordinary 'sulphur' methionine, the compound failed to crystallize during the final isolation step, despite various approaches. No such difficulties were encountered during the crystallization of the analogous selenomethionine[76] from its mother liquor. However, various analytical methods (e.g. NMR, TLC and mass spectometry) confirmed the presence of telluromethionine. Telluro-α-amino acids with a telluro-ether link between the aliphatic moiety of the functional part of the amino acid and an aromatic ring were found to be more stable and could be isolated[77]. For some of them, anticarcinogenic and antileukaemic properties have been claimed[78]. Heterocyclic ring-derived α-amino acids have also been prepared. Thus a benzo[b]tellurophene derivative of glycine, with the α-carbon of the amino acid linked to one of the carbon atoms of the heterocyclic part of the benzo[b]tellurophene moiety, has recently been prepared[79].

5. *Porphyrin analogues*

Macro-ring nitrogen-, sulphur- and tellurium-containing heterocycles as analogues of porphyrins have also been prepared. These compounds show strong interactions between the internal heteroatoms, in particular between the heterocyclic tellurium and sulphur atoms situated opposite each other at the apexes of the five-membered rings which form part of the porphyrin system[80].

6. Analogues of drugs and antibiotics

Tellurium-containing analogues of drugs and antibiotics have also been prepared. Examples are tellurium-substituted barbiturates[81] and a tellurium-containing analogue of the antibiotic chloromycetin[82].

Various organotellurium compounds synthesized in the past have been tested for antimicrobial activity and have been listed by Irgolic[83]. More recent work seems to indicate a renewal of interest in the antibacterial properties of tellurium compounds[84,85].

Tellurium-containing chelates of organic metal complexes have also been investigated as potential antimicrobial derivatives. They include tellurium sulphonamide Schiff base complexes[86], aromatic imine complexes of selenium and tellurium[87], tellurium complexes with substituted chalcones[88] and bimetallic complexes of 2, 4-diketonates[89], in addition to selenium and tellurium complexes with 2-substituted benzimidazoles[90] and thiopicolinamide complexes of selenium and tellurium[91].

III. A LOOK AT THE FUTURE

In the field of bioscience-oriented tellurium work, the following research topics might well be rewarding.

A. Teratogenicity

Nothing is really known about the mechanism by which simple inorganic tellurium compounds cause teratogenicity. The reason is the absence of knowledge concerning the molecular basis of the interaction of tellurium compounds with biological systems. For instance, the fate of simple inorganic tellurium compounds inside various organisms has not been adequately investigated. It is not known whether organisms are able to synthesize organotellurium compounds more complicated than dimethyl telluride, nor has the exact manner in which tellurium binds to biopolymers such as proteins, nucleic acids and polysaccharides been determined.

B. Tellurium and the Nervous System

Tellurium is known to have an affinity for the nervous system and cerebral tissue and has been found to cause hydrocephalus in foetuses. Here, too, no research in depth has been carried out to determine the underlying metabolic interactions which might throw light on the biochemical processes responsible for the influence of tellurium on the nervous system.

C. Tellurium Bio-organic Chemistry

Whereas with selenium an area of life science-oriented bio-organic chemistry has evolved, selenium-specific and independent of that of sulphur, little has been achieved yet in the parallel field of organotellurium chemistry. Much needs to be done in the identification and synthesis of tellurium containing amino acids, peptides and carbohydrates. Whereas selenium compounds in these categories have been found in nature and have also been synthesized, no such organotellurium analogues have been mentioned in the literature, with the exception of the few tellura-fatty acids, tellura-steroids, telluro-amino acids and a lone telluro-carbohydrate, covered in this review. Whereas with organoselenium compounds numerous selenoanalogues of sulphur derivatives have been prepared and tested for biological activity, no such efforts seem to have been made to prepare the corresponding tellurium analogues.

D. Application of Organotellurium Compounds in Biology and Medicine

Certain specific properties of the tellurium atom of potential usefulness to biomedicine have already been discovered[62]. Further efforts in this direction may lead to new developments which may lend themselves to possible applications to pharmacology. As examples, recent developments in selenium pharmacology may be cited, such as that of ebselen[92–95] and the anti-viral drug selenazofurin[96–99].

E. Tellurium as an Essential Trace Element

Selenium has been proved to be an essential trace element, but the possible role played by tellurium in biological systems has not been evaluated sufficiently. One possible reason for this might be the difficulty of removing completely the tellurium content from the nutrients supplied to the living organisms to be tested, because only if nutrients free from traces of tellurium are available will it be possible to assess its potential role as an essential trace element.

IV. ACKNOWLEDGEMENT

This contribution cannot be regarded as complete without due expression of sincere thanks to Professor Kenji Soda, who coordinated the various section of this review, and whose creation of a suitable atmosphere of collaboration was essential in encouraging the contributors to give their best efforts.

V. REFERENCES

1. K. J. Irgolic, *The Organic Chemistry of Tellurium*, Gordon and Breach, New York, 1974.
2. L. Fishbein in *Advances in Modern Toxicology, Vol. 2, Toxicology of Trace Elements, Selenium and Tellurium* (Eds. R. A. Goyer and M. A. Mehlman), Wiley, New York, 1977, pp. 191–240.
3. R. H. De Meio, *J. Ind. Hyg. Toxicol.*, **29**, 393 (1946).
4. R. H. De Meio and F. C. Henriques, Jr., *J. Biol. Chem.*, **169**, 609 (1947).
5. R. H. De Meio and W. W. Jetter, *J. Ind. Hyg. Toxicol.*, **30**, 53 (1948).
6. M. L. Amdur, *Arch. Ind. Health*, **17**, 665 (1958).
7. J. H. H. Keall, N. H. Martin and R. E. Tunbridge, *Br. J. Ind. Med.*, **3**, 175 (1946).
8. E. S. Blackadder and W. G. Manderson, *Br. J. Ind. Med.*, **32**, 5961 (1975).
9. H. J. Einbrodt and S. Michels, in *Metalle in der Umwelt* (Ed. E. Merian), Verlag Chemie. Weinheim, 1984, Chapt. II.21, pp. 561, 564, 566, 567.
10. W. A. Konetzka, in *Microbiol. Series, 1977, 3, Review: Microbiology of Metal Transformations, Selenium and Tellurium*, Chapt. 8, pp. 323, 338, 339.
11. F. J. Baker, in *Handbook of Bacteriological Techniques*, Butterworth, London, 1967, p. 98.
12. M. O'Brien and R. Colwell, *J. Clin. Microbiol.*, **22**, 1011 (1985).
13. F. Challenger, in *Organometals and Organometalloids* (Eds. F. E. Brinckman and J. M. Bellama), ACS Symp. Ser. No. 82, American Chemical Society, Washington, DC, 1978, p. 1.
14. Y. K. Chau, and P. T. S. Wong, 1978, in *Organometals and Organometalloids* (Eds. F. E. Brinckman and J. M. Bellama), ACS Symp. Ser. No. 82, American Chemical Society, Washington, DC, 1978, p. 39.
15. N. R. Bottino, E. R. Cox, K. J. Irgolic, S. Maeda, W. J. McShane, R. A. Stockton and R. A. Zingaro, in *Organometals and Organometalloids* (Eds. F. E. Brinckman and J. M. Bellama), ACS Symp. Ser. No. 82, American Chemical Society, Washington, DC 1978, p. 116.
16. R. W. Fleming and M. Alexander *Appl. Microbiol.*, **24**, 424 (1972).
17. N. N. Lyalikova, M. F. Kulikova and N. Z. Kurbanova, *Geokhim. Deyat. Mikroorg. Vodooemakh. Mestorozhz. Polezn. Inskop.*, 293 (1973).
18. M. Miskey and T. Lakatos, *Banyasz. Kohasz. Lapok Kohasz.*, **108**, 35 (1975); *Chem. Abstr.*, **83**, 118974n (1975).

19. J. F. Spisak, in *Biotechnology and Minerals: Legitimate Challenge or Myth*? (Eds. J. F. Spisak and G. Jergensen, II), AIME, New York, 1985, pp. 3–12.
20. O. H. Tuovinen, S. I. Niemala and H. G. Gyllenberg, *Antonie van Leeuwenhoek J. Microbiol Serol.*, **37**, 489 (1971).
21. R. J. Shamberger, in *Biochemistry of Selenium, Biochemistry of the Elements Series* (Ed. E. Frieden) Vol. 2, Plenum, New York, 1983, Chapt. 5, pp. 154–155.
22. J. F. Van Vleet, G. D. Boon and V. J. Ferrans, *Am. J. Vet. Res.*, **42**, 789 (1981).
23. W. F. Agnew and J. T. Cheng, *Toxicol. Appl. Pharmacol.*, **20**, 346 (1971).
24. G. Said and S. Duckett, *Muscle Nerve*, **4**, 319 (1981).
25. S. Duckett and R. White, *Brain Res.*, **73**, 205 (1974).
26. D. Dru, W. F. Agnew and E. Greene, *Psychopharmacologia*, **24**, 508 (1972).
27. S. Duckett, *Ann. N. Y. Acad. Sci.*, **192**, 220 (1972).
28. W. W. Carlton and W. A. Kelly *Toxicol. Appl. Pharmacol.*, **2**, 203 (1967).
29. Y. Matsumoto, T. Saruta, T. Higa, N. Fujimoto and S. Tamura, *Shika Gakuho*, **81**, 797 (1981).
30. T. Saruta, T. Higa, N. Fujimoto and Y. Matsumoto, *Shika Gakuho*, **82**, 447 (1982).
31. S. Tamura, Y. Matsumoto, T. Saruta, E. Nakayasu, T. Kano and N. Haseda, *Bull. Tokyo Dent. Coll.*, **26**, 95 (1985).
32. H. Baba and Y. Matsumoto, *Bull. Tokyo Dent. Coll.*, **26**, 103 (1985).
33. T. Watanabe, Y. Osawa and S. Horie, *Kitasato Igaku*, **15**, 35 (1985).
34. S. Valkonen and H. Savolainen, *Bull. Environ. Contam. Toxicol.*, **34**, 170 (1985).
35. S. Duckett, *Neurotoxicology*, **3**, 63 (1982).
36. G. Said and S. Duckett, *Muscle Nerve*, **4**, 319 (1981).
37. J. F. Van Vleet and V. J. Ferrans, *Am. J. Vet. Res.*, **43**, 2000 (1983).
38. P. Pohl, P. Lintermans, B. Corbio, J. Gledel, L. Le Minor, M.-L. Chasseur and G. Ghysels *Ann. Microbiol. (Paris)*, **132B**, 399 (1981).
39. E. Chaslus-Dancla and J.-P. Lafont *Appl. Environ. Microbiol.*, **49**, 1016 (1985).
40. D. E. Bradley, *J. Gen. Microbiol.*, **131**, 3135 (1985).
41. N. M. Harnett and G. L. Gyles, *Appl. Environ. Microbiol.*, **48**, 930 (1984).
42. M. G. Lorenz and W. E. Krumbein, *Appl. Microbiol. Biotechnol.*, **20**, 422 (1984).
43. C. Nolan, E. Duke, G. Lorenzon, E. Sabbioni and E. Marafante, *Sci. Total. Environ.*, **40**, 83 (1984).
44. S. Tang, X. Chen and J. Li, *Huanjing Huaxne*, **2**, 35 (1983).
45. X. Chen, G. Yang, J. Chen, X. Chen, Z. Wen and K. Ge, *Biol. Trace Elem. Res.*, **2**, 91 (1980); *Chem. Abstr.*, 93, 237390k (1980).
46. A. I. Y. Khayat, *Acta Pharm. Suec.*, **22**, 301 (1985).
47. J. D. Young, C. Crowley and E. M. Tucker, *Biochem. Pharamacol.*, **30** 2527 (1981).
48. R. H. De Meio and P. F. Doughty, *Biol. Trace Elem. Res.*, **1**, 131 (1979).
49. T. Asakura, Y. Shibutani, M. P. Reilly and R. H. De Meio, *Blood*, **64**, 305 (1984).
50. T. Masukawa and H. Iwata, *Life Sci.*, **21**, 695 (1977).
51. H. Iwata, T. Masukawa. S. Kasamatsu, K. Inoue and H. Okamoto, *Experientia*, **33**, 678 (1977).
52. M. Jolocam, *Carbohydr. Res.*, **140**, 151 (1985).
53. J. Czyzewska-Chlebny and M. Michalska, *J. Chem. Soc., Chem. Commun.*, 693 (1985).
54. R. A. Zingaro, J. Ekelund-Price and C. R. Benedict, *J. Carbohydr. Nucleos. Nucleot.*, **4**, 271 (1977).
55. F. F. Knapp, Jr., in *Radiopharmaceuticals, Structure–Activity Relationships* (Ed. R. P. Spencer), Grune and Stratton, New York, 1981, Chapt. 16, pp. 345–391.
56. J. A. Bianco, L. A. Pape, J. S. Alpert, M. Zheng, D. Hnatowich, M. M. Goodman and F. F. Knapp, Jr., *J. Am. Coll. Cardiol.*, **4**, 80 (1984).
57. D. R. Elmaleh, F. F. Knapp, Jr., T. Yasuda, S. Kopiwoda, K. A. McKusick and H. W. Strauss, *J. Nucl. Med., Proc. 27th Annu. Meet.*, **21**, 58 (1980).
58. M. M. Goodman, F. F. Knapp, Jr., A. P. Callahan and L. A. Ferren, *J. Med. Chem.*, **25**, 613 (1982).
59. M. M. Goodman F. F. Knapp, Jr., A. P. Callaher and L. A. Ferren, *J. Nucl. Med.*, **23**, 904 (1982).
60. M. M. Goodman and F. F. Knapp, Jr., *J. Org. Chem.*, **47**, 3004 (1982).
61. M. M. Goodman, A. P. Callahan, F. F. Knapp, Jr., M. D. Strauss, D. Elmaleh, P. Richards and L. F. Mausner, in *Proceedings of Symposium on Single-Photon, Ultra-Short-Lived Radionuclides*, DOE Symp. Ser., 1985, 56; *Nucl. Med. Pract.*, 488; *Chem. Abstr.*, **103**, 156525f (1985).
62. G. Kirsch, M. M. Goodman and F. F. Knapp, Jr., *Organometallics*, **2**, 357 (1983).
63. F. F. Knapp, Jr., K. R. Ambrose, A. P. Callahan, L. A. Ferren R. A. Grigsby and K. J. Irgolic, *J. Nucl. Med.*, **22**, 988 (1981).

64. F. F. Knapp, Jr., M. M. Goodman, A. P. Callahan, L. A. Ferren, G. W. Kabalka and K. A. R. Sastry, *J. Med. Chem.*, **26**, 1293 (1983).
65. F. F. Knapp, Jr., P. C. Srivastava, A. P. Callahan, E. B. Cunningham, G. W. Kabalka and K. A. R. Sastry, *J. Med. Chem.*, **27**, 57 (1984).
66. R. D. Okada, F. F. Knapp, Jr., D. R. Elmaleh, D. Yasuda, C. A. Boucher and H. W. Strauss, *Circulation*, **65**, 305 (1982).
67. P. C. Srivastava, F. F. Knapp, Jr., A. P. Callahan, B. A. Owen, G. W. Kabalka and K. A. R. Sastry, *J. Med. Chem.*, **28**, 408 (1985).
68. T. Yasuda, F. F. Knapp, Jr., D. R. Elmaleh, S. Kopiwoda, K. A. McKusick and H. W. Strauss, *J. Nucl. Med., Proc. 27th Annu. Meet.*, **21**, 58 (1980).
69. Y. Yonekura, P. Som, S. C. Srivastava, K. Yamamoto, G. E. Meineken, A. B. Brill, D. R. Elmaleh, H. W. Strauss, M. M. Goodman and F. F. Knapp, Jr., *J. Nucl. Med., Proc. 30th Annu. Meet.*, **24**, 125 (1983).
70. M. E. Wolff and G. Zanati, *Experientia*, **26**, 1115 (1970).
71. F. F. Knapp, Jr., *J. Labelled Compd. Radiopharm.*, **17**, 81 (1979).
72. F. F. Knapp, Jr., K. R. Ambrose and A. P. Callahan, *J. Nucl. Med.*, **21**, 251 (1980).
73. F. F. Knapp, Jr., K. R. Ambrose and A. P. Callahan, *J. Nucl. Med.*, **21**, 258 (1980).
74. D. V. Woo F. F. Knapp, Jr., K. R. Ambrose, A.-P. Callahan and J. L. Coffey, *J. Nucl. Med.*, **21**, 454 (1980).
75. F. F. Knapp, Jr., *J. Org. Chem.*, **44**, 1007 (1979).
76. G. Zdansky, in *Organic Selenium Compounds. Their Chemistry and Biology* (Eds. D. L. Klayman and W. H. H. Günther), Wiley, New York, 1973, Chapt. XII, p. 593.
77. F. F. Knapp, Jr., K. R. Ambrose and A. P. Callahan, *J. Med. Chem.*, **24**, 794 (1981).
78. Ajinomoto Co. Inc., *Japan Kokai Tokkyo Koho*, JP 59216868 (84216868), 1984; *Chem. Abstr.*, **102**, 185497h (1985).
79. T. Sadeh, L. Christiaens, J. L. Piette, M. Guillaume and K. Soda, *Abstr. Jpn. Agric. Biochemi. Soc. Meet., April 2, 1986*, 1986, p. 80, 2G-10.
80. A. Ulman, J. Menassen, F. Frolow and D. Rabinovich, *Tetrahedron Lett.*, 1885 (1978).
81. R. A. Grigsby, F. F. Knapp, Jr., A. P. Callahan, L. A. Ferren and K. J. Irgolic, *J. Nucl. Med., Proc. 28th Annu. Meet.*, **22**, 12 (1981).
82. F. Rogoz, *Diss. Pharm.*, **16**, 157 (1962).
83. K. J. Irgolic, *The Organic Chemistry of Tellurium*, Gordon and Breach, New York, 1974, Chapt. XVI, pp. 369–379.
84. T. N. Srivastava, P. C. Srivastava and R. Kumar, *Bokin Bobai*, **10**, 523 (1982).
85. Y. D. Kulkarni and S. Srivastava, *Indian J. Chem., Sect. A*, **24**, 429 (1985).
86. W. E. Rudzinski, T. M. Aminabhavi, N. S, Biradar and C. S. Patil, *Inorg. Chim. Acta*, **67**, 177 (1982).
87. W. E. Rudzinski, T. M. Aminabhavi, N. S. Biradar and C. S. Patil, *Inorg. Chim. Acta*, **69**, 83 (1983).
88. W. E. Rudzinski, T. M. Aminabhavi, N. S. Biradar and C. S. Patil, *Inorg. Chim. Acta*, **70**, 175 (1983).
89. N. S. Biradar, M. C. Divakar and G. V. Karajagi, *J. Karnatak Univ. Sci.*, **28**, 89 (1983).
90. N. S. Biradar, T. M. Aminabhavi, C. S. Patil and W. E. Rudzinski, *Inorg. Chim. Acta*, **78**, 47 (1983).
91. T. M. Aminabhavi, W. E. Rudzinski, N. S. Biradar and C. S. Patil, *Inorg. Chim. Acta*, **78**, 51 (1983).
92. A. Mueller, E. Cadenas, P. Graf and H. Sies, *Biochem. Pharmacol.*, **33**, 3235 (1984).
93. A. Wendel, M. Fausel, H. Safayhi, G. Tiegs and R. Otter, *Biochem. Pharmacol.*, **33**, 3241 (1984).
94. M. J. Parnham and S. Kindt, *Biochem. Pharmacol.*, **33**, 3247 (1984).
95. A. Mueller, H. Gabriel and H. Sies, *Biochem. Pharmacol.*, **34**, 1185 (1985).
96. J. J. Kirsi, J. A. North, P. A. McKernan, B. K. Murray, P. G. Canonico, J. W. Huggins, P. C. Srivastava and R. K. Robins, *Antimicrob. Agents Chemother.*, **24**, 353 (1983).
97. J. L. Huey, K. Pawlak, B. T. Nguyen, R. K. Robins and W. Sadee, *Cancer Res.*, **45**, 5512 (1985).
98. R. W. Sidwell, J. H. Huffman, E. W. Call, H. Alaghamandan, P. D. Cook and R. K. Robins, *Antimicrob. Agents Chemother.*, **28**, 375 (1985).
99. R. K. Robins, *Chem. Eng. News*, **64**, No. 4, 29 (1986).

The Chemistry of Organic Selenium and Tellurium Compounds Volume 2
Edited by S. Patai
© 1987 John Wiley & Sons Ltd.

CHAPTER **9**

Pharmacological and toxicological aspects of inorganic and organic selenium compounds

TOHRU MASUKAWA

Department of Pharmacology, Faculty of Pharmaceutical Sciences, Setsunan University, 45-1, Nagaotoge-cho, Hirakata, Osaka 573-01, Japan

I.	INTRODUCTION	377
II.	BIOLOGICAL EFFECTS ON ORGANS	378
	A. Cardiovascular System	378
	B. Pulmonary System	380
	C. Liver	380
	D. Gastrointestinal System	382
	E. Haematological System	382
	F. Others	383
III.	RELATION TO INFLAMMATION AND IMMUNITY	383
IV.	RELATION TO CANCER	384
V.	INTERACTION WITH HEAVY METALS	387
VI.	MEDICAL APPLICATIONS	387
VII.	REFERENCES	388

I. INTRODUCTION

Since selenium intoxication such as alkali disease and blind staggers and its deficiency syndromes such as white muscle disease and cardiac myopathy have long been known in animals, the toxicological and nutritional aspects of selenium in farm animals have become major subjects of selenium research. During the early period of these investigations, carcinogenic properties of selenium were reported. Later, the opposite effect was observed and selenium has been demonstrated to exhibit anticarcinogenic activity in experiments with animals and in epidemiological studies with humans. Even so, confusion and controversy over the relationship between selenium and cancer has continued for a long time. On the other hand, selenium deficiency syndrome was recognized in animals and humans, and selenium was found to be an important element in human nutrition.

Most of the investigations on the biological activity of selenium have employed nutritional methods using physiological doses of selenium. One of the possible reasons for this is the narrow safety margin between an effective and a toxic dose and a lingering doubt as to the carcinogenicity of selenium. The pharmacological effects of selenium remain obscure. Recently, considering that selenium exhibits various beneficial effects as described below, attempts have been made to synthesize organic selenium compounds with lower toxicity than inorganic selenium compounds and to apply them in human medication.

This review is mainly concerned with the physiological and pharmacological aspects of selenium.

II. BIOLOGICAL EFFECTS ON ORGANS

A. Cardiovascular System

The cardiac effects of selenium vary with the amount of selenium administered or ingested. Heinrich and MacCanon[1] demonstrated that a decrease in systemic blood pressure and a rise in pulmonary artery pressure were produced by a toxic dose of selenite ($2.0\,mg\,kg^{-1}$ i.v.) in dogs, accompanied by increased heart rate, arrhythmic bradycardia and occasionally tachycardia and fibrillation. Lower doses produced moderate elevations of both systemic and pulmonary pressure. However, according to Aviado *et al.*[2], selenite (0.04–$20\,mg\,kg^{-1}$ i.v.) did not elicit any signs of cardiac depression in dogs. Selenite caused a positive inotropic effect in isolated dog ventricular strip perfused with either blood or a krebs–hydrogen carbonate solution that was deficient in oxygen. They thought that this effect may relate to the protective action of selenite against the lethal effects of oxygen deficiency, ouabain and 2,4-dinitrophenol in mice. However, in an isolated and perfused guinea-pig heart, selenite showed a negative inotropic effect, accompanied by mitochondrial alterations[3].

In animals fed on a low selenium diet, it has been demonstrated that abnormal electrocardiograms (ECGs) accompanied by blood pressure changes appear. Histological changes in the cardiac muscle of rats reveal that the abnormal ECG pattern is related to a change in the properties of the muscle fibres rather than to electrolyte disturbances or other changes in plasma constituents[4]. In lambs, ECG showed the progressive development of a characteristic abnormality. Just before death, the ECG pattern became grossly abnormal in some cases, a rise in the T-wave giving way to an elevated S–T segment, similar to that seen frequently in myocardial infarction in humans. From a comparison of the histopathological pictures, entirely different syndromes appear to develop in the rat and the lamb, but the ECG changes in both species are similar[5]. These abnormalities can be prevented by the administration of adequate amounts of selenium. Similar myocardial disease is frequently seen in other species, e.g. pigs, sheep, calves and ducklings, with selenium–vitamin E deficiency[6].

In humans, associations between selenium intake and cardiovascular diseases have been suggested from numerous epidemiological studies[6]. According to Shamberger[6], both the male and female mortalities due to cardiovascular, renal, cerebrovascular, coronary and hypertensive heart diseases were significantly lower in a high than in a low selenium area. In another study on the relationship between trace element intakes and cardiovascular diseases in 25 countries, a significant negative correlation was seen between selenium and ischemic heart disease, and a positive correlation was observed with cadmium and ischaemic heart disease. An even more significant correlation existed between the ratio of cadmium/selenium intake and ischemic heart disease[6]. An association between mode rate exposure to cadmium and hypertension in experimental animals has also been reported[7], but cadmium-induced hypertension in rats could be alleviated by feeding selenium. The preventive effect of selenium can be explained by its antagonistic action towards cadmium[8]. Further, prophylactic application of selenite (1.0 mg kg^{-1} s.c. per day) was shown to lower the blood pressure of rats with renal hypertension and to suppress a rise in blood pressure following infusion of angiotensin II[9].

Another epidemiological study showed that low serum selenium resulting from low dietary selenium intake is associated with an excess risk of death from acute coronary heart disease and myocardial infarction. Salonen *et al.*[10] investigated case-control pairs coming from a population of 11 000 persons in eastern Finland. The subjects were aged 35–59 years and had either died of acute coronary heart disease or other atherosclerotic cardiovascular diseases or had a non-fatal myocardial infarction during a 7-year follow-up. Controls were matched for sex, age, tobacco consumption, serum cholesterol, diastolic blood pressure and history of angina pectoris. The mean serum selenium levels in all cases and all controls were 51.8 and 55.3 $\mu g\,l^{-1}$, respectively. Since there was no significant relationship between serum selenium and the risk of death of the diseases studied at Se levels above 45 $\mu g\,l^{-1}$, this level was used as a reference. The relative risk of death markedly increased with the decrease of serum selenium. Regarding the relationship between the development of coronary heart diseases and selenium intake, the following mechanisms of action were proposed: Firstly, the decrease in glutathione peroxidase activity due to selenium deficiency may lead to an accumulation of lipid peroxides with subsequent injury to various tissues including arterial vessels. Secondly, thrombus formation in the coronary artery may be involved in the development of the diseases. In fact, platelet aggregatability was enhanced in selenium-deficient animals, as evidenced by the aggravation of arachidonate-induced respiratory distress in mice and enhanced platelet aggregation by arachidonate, ADP and collagen *in vitro*[11]. Selenium-deficient platelets have been observed to be associated with a marked alteration in the lipoxygenase metabolism of arachidonate[12]. The platelets of patients with acute myocardial infarction had significantly lower activity of glutathione peroxidase[13]. Additionally, the formation of PGI_2 with a potent antiaggregatory and vasodilating activity was suppressed in the aorta from selenium-deficient rats[11]. These findings suggest that dietary selenium may function in vascular haemostasis and thrombosis by maintaining the metabolism of arachidonate.

It was recently reported that Keshan disease with cardiomyopathy developed in some areas of China is a selenium deficiency syndrome[14]. Keshan disease could be prevented by selenium supplementation[15], but its mechanism remains obscure. Other cases of cardiomyopathy similar to Keshan disease have been observed in patients given long-term parenteral alimentation without selenium supplementation[16].

Experimental cardiomyopathy related to dietary selenium has also been investigated. Long-term administration of adriamycin, an antineoplastic agent, resulted in the development of cardiomyopathy in various animals, including humans. The toxic effect may be the result of membrane lipid peroxidation. Revis and Marusic[17] demonstrated that a marked decrease in glutathione peroxidase activity and selenium content occurred in the hearts of adriamycin-treated rabbits. They suggested that the increase in lipid peroxides in the heart may be the result of a decrease in glutathione peroxidase activity, in turn caused by an alteration in the selenium flux in the myocardial cells. The protective effects of vitamin E and selenium supplementation on cardiomyopathy were reported to be found in rabbits[18] but not in dogs[19] and rats[20]. Cadmium also caused cardiomyopathy in rats. In cadmium-treated rats fed a low selenium diet, an increase in heart weight with histopathological changes was produced, accompanied by a decrease in glutathione peroxidase activity and a rise in the lipid peroxide level in the heart. Increased dietary selenium markedly decreased the level of lipid peroxide. Since the cardiotoxic effect of cadmium was observed without concomitant peroxidative damage to the kidney and liver, the heart is the most susceptible organ to cadmium-induced damage[21].

B. Pulmonary System

The injection of selenite (0.1–2.0 mg kg^{-1} i.v.) to anaesthetized dogs produced respiratory stimulation, increased minute volume and decreased tidal volume, which appear to result from reflexes originating in the thorax. The early elevation of oxygen consumption after selenite may be due to the release of epinephrine. From the results, Heinrich and MacCanon[22] thought that the primary cause of death in acute selenium poisoning is not respiratory failure.

The lung is directly exposed to oxidizing pollutants in the atmosphere. Inhalation of oxygen, ozone and nitrogen dioxide causes pulmonary tissue injury in animals, including humans. These oxidants readily attack unsaturated lipids in cell membranes, and the resulting lipid peroxidation is an important process in lung injury. After 3 days of exposure to 80% O_2, 35% of selenium-deficient rats died in respiratory distress, but no respiratory damages were observed in selenium-supplemented rats[23]. This indicates that dietary selenium reduces the susceptibility of pulmonary tissue to oxidant-induced damage as a nutritional factor. Glutathione peroxidase activity in the lung after exposure to various oxidant gases was shown to increase in selenium-supplemented[23–25] but not in selenium-deficient animals[23,24]. According to Elsayed *et al.*[24], since the selenium content in the lung increased with exposure to ozone, selenium might be mobilized to the lung from other body sites under oxidant stress. However, considering that other antioxidative defence factors such as superoxide dismutase, catalase, non-selenium glutathione peroxidase(glutathione S-transferases) and vitamin E also play an important role in the protective effect of oxidant-induced injury, the effect of dietary selenium could not be explained merely by the alteration of glutathione peroxidase activity alone.

Compounds that undergo redox cycling have been shown to produce the superoxide anion ($O_2^{-\bullet}$) in the process. Injury by paraquat, a herbicide, is thought to be due to the peroxidative damage of membrane lipids by $O_2^{-\bullet}$ formed during the redox cycling of the parent compound. Lung injury developed because paraquat was selectively retained in the lungs. Both dietary selenium and vitamin E provided protection against the toxicity of paraquat[26,27]. The acute toxicity of paraquat in the chick was especially highly responsive to selenium status but not to vitamin E status[28]. A similar redox cycling mechanism underlies the pulmonary injury caused by nitrofuranotoin, a urinary tract antibiotic. In protecting against the acute toxicity of nitrofurantoin in the chick, selenium status was shown to be more important than vitamin E status[29]. These findings suggest that the toxicity of redox cycling compounds may be enhanced by selenium deficiency.

C. Liver

The main effects of chronic excess selenium poisoning are depressed growth, decreased survival and damage to the liver and other organs. Although the detailed mechanism still remains unknown, perturbations of hepatic glutathione status or resulting impairment of the redox state of the cells may be involved, since the liver is the main organ involved in the reduction of excess selenium[30].

The relationship between liver necrosis and lipid peroxidation has been demonstrated under various experimental conditions. Dietary deficiency of selenium and vitamin E in growing rats caused massive liver necrosis[31]. Rats with nutritional liver necrosis exhaled large amounts of ethane as an index of *in vivo* lipid peroxidation, suggesting that liver necrosis is the result of lipid peroxidation[32]. Concerning liver necrosis as a result of drug toxicity, Gallagher[33] first reported that the injection of selenite and also vitamin E and other antioxidants protects mice from the lethal toxicity of carbon tetrachloride. According to Hafeman and Hoekstra[34], dietary selenium inhibited carbon tetrachloride-induced evolution of ethane from rats. This is also true of paraquat- and diquat-induced liver necrosis in selenium-deficient animals[35,36]. The protective effect of selenium against liver necrosis is considered to be due to the decomposition of lipid peroxides via glutathione peroxidase. However, in the case of diquat toxicity, selenium injection provided significant protection against lipid peroxidation and mortality within 10 h, even though this treatment did not result in a rise in tissue glutathione peroxidase activity. Thus, a selenium-dependent factor in addition to glutathione peroxidase is thought to exist that protects against lipid peroxidation[36].

Lipid peroxidation does not always correlate with liver necrosis. Both iodipamide and acetaminophen were shown to cause liver necrosis but only minor ethane production. The hepatic toxicity of these drugs was diminished by selenium deficiency[37]. Since the detoxification of these drugs was carried out either by binding to glutathione S-transferases or by conjugation with glutathione catalysed by the enzymes, an increased activity of hepatic glutathione S-transferases in selenium deficiency may result in enhanced detoxification of these drugs, leading to reduced toxic effects. The hepatotoxicity of aflatoxin B_1 was also markedly attenuated in a selenium-deficient status[38]. Since aflatoxin B_1 is detoxified by glutathione conjugation, a similar mechanism may be involved. Bromobenzene-induced hepatic damage was prevented by the acute injection of selenite (12.5, 30 $\mu g\,kg^{-1}$ i.p.), but the mechanism remains obscure[39].

Selenium is known to influence microsomal drug metabolizing systems via alterations of the haeme metabolism. Pharmacological or toxicological doses of selenite (10–100 $\mu mol\,kg^{-1}$ s.c.) in rats with a normal selenium status caused the induction of δ-aminolevulinate synthase and haeme oxygenase in the liver. The effect of selenium was rapid; the cellular content of haeme was significantly increased 30 min after injection and subsequently returned to normal levels. The alteration of the haeme metabolism with excess of selenium resulted in the marked inhibition of the microsomal drug metabolism[40]. Nutritional doses of selenium play an important role in the maintenance of drug metabolism. The effect of selenium deficiency on the microsomal drug metabolism was investigated by Burk and Masters[41]. Selenium deficiency impaired the induction of cytochrome P-450 by phenobarbital, but had no effect on basal levels of the haemoprotein.

Ethylmorphine demethylase activity was affected by selenium deficiency, but NADPH–cytochrome c reductase activity and biphenyl-4-hydroxylase activity were not. When 3-methylcholanthrene was used in place of phenobarbital, selenium deficiency had no effect on the system. In addition, aminopyrine *N*-demethylation, monocrotane metabolism and aniline hydroxylation were depressed by selenium deficiency. The depression was greater in second generation rats with severe selenium deficiency[42].

From a comparison of hepatic haeme metabolisms in selenium-deficient rats and in control rats, Burk and Correia[43] showed that phenobarbital increased the hepatic microsomal haeme oxygenase activity in selenium-deficient rats. After phenobarbital administration, the haeme synthesis and catabolism increased strikingly in selenium-deficient liver, whereas haeme utilization in the formation of cytochrome P-450 was impaired in selenium deficiency. The resulting abnormal excess of haeme induced microsomal haeme oxygenase. These findings may be related to the finding that the induction of cytochrome P-450 system in rats and chicks was accompanied by an increased requirement for selenium[44,45]. Thus, selenium may play a key role in the homeostasis of microsomal haeme through normalization of haeme synthesis and/or its utilization. Concerning the mode of action of selenium involved in haeme metabolism, injection of selenite corrected the abnormality in haeme metabolism within 12 h, although there was not detectable recovery of glutathione peroxidase activity in this period. Therefore, the improvement of the abnormality in haeme metabolism is considered to be due to an unrecognized function of selenium other than glutathione peroxidase[43].

D. Gastrointestinal System

Selenoamino acids are thought to be absorbed from the gastrointestinal tract in the same way as other amino acids[46], and inorganic selenium such as selenite and selenate to be likewise absorbed from the intestine as are selenoamino acids[47]. In the ligated intestinal segments of rats, the absorption of selenite and selenomethionine occurred from the duodenum rather than from the jejunum or ileum[47]. As a mechanism of absorption, McConnell and Cho[48] demonstrated that selenite is absorbed by diffusion, but not by active transport. According to Wolffram *et al.*[49], selenate was absorbed markedly faster than selenite and its absorption showed a saturable process, indicating that the absorption of selenate may occur by a carrier-mediated mechanism.

Hadjimarkos[50] has reported that dental caries is more prevalent in seleniferous areas than non-seleniferous areas.

Prevention of the formation and detoxification of lipid peroxides may be especially important in the gastrointestinal tract because the tissue is frequently exposed to such substances. Oral administration of hydrogen peroxide in the chick could have a detrimental effect on the intestinal mucosa, resulting in a decreased rate of selenium absorption and reduced glutathione peroxidase activity[51]. On the other hand, Vilas *et al.*[52], observing that dietary peroxides increased glutathione peroxidase activity in the gastric mucosa, suggested that glutathione peroxidase may be involved in protecting the gastric mucosa from damage caused by dietary peroxides. Further, according to Negishi *et al.*[53], when mice were orally given autoxidized methyl linoleate with a low vitamin E diet, an increase in glutathione peroxidase activity in the gastrointestinal tract occurred in proportion to the peroxide value of the oil, but the activity in the liver remained unchanged. With intraperitoneal injection of the oil, an increase in enzyme activity was observed in the liver, but not in the gastrointestinal tract. From the results, they suggested that most of the orally administered oil is reduced in the mucosa of the gastrointestinal tract.

Dietary selenium deficiency produced an increase in glutathione S-transferase activity in the duodenal mucosa in addition to the liver and kidney[54]. The enhanced activity was restored to the control value 48 h after injection of selenite (1.0 $\mu g\,kg^{-1}$ s.c.), but the total glutathione peroxidase activity, including non-selenium glutathione peroxidase (glutathione S-transferases), remained unchanged owing to an increase in selenium-dependent glutathione peroxidase activity. This suggests that glutathione S-transferases with non-selenium glutathione peroxidase activity may function as a substitute for selenium-dependent glutathione peroxidase in the duodenal mucosa in addition to the liver and kidney of selenium-deficient rats. Further, selenium deficiency caused a decrease in the cytochrome P-450 level in the small intestinal mucosa of rats, resulting in profound decreases in aryl hydrocarbon hydroxylase and ethoxyresorufin *O*-deethylase activities[55]. These findings indicate that selenium is necessary for the metabolism in the gastrointestinal tract of ingested xenobiotics.

E. Haematological System

Anaemia is known to occur in animals as a result of selenium intoxication and also selenium deficiency. Rats fed a diet containing a high concentration of selenium developed anaemia, which was attributed to haemolysis rather than to a defect in erythrocyte synthesis[56]. In contrast, in experiments with erythrocytes from selenium-deficient rats, dietary selenium was shown to reduce ascorbic acid-induced haemolysis, oxidation of haemoglobin and the proportion of cells with Heinz bodies in the presence of glucose[57]. The effects of dietary selenium were explained by a defect of the antioxidative system due to a deficiency of glutathione peroxidase[58]. In contrast to the above *in vitro* findings, Hu *et al.*[59] reported that haemolytic anaemia or oxidation of haemoglobin does not occur in rats fed a selenium-deficient diet for 7 months. They thought that the above effect of dietary selenium may not have an important physiological significance *in vivo* under normal conditions. However, an anaemia associated with the presence of Heinz bodies and selenium deficiency was recently reported to develop in cattle grazing in the Florida Everglades[60].

On the other hand, according to Iwata *et al.*[61], when injected at high levels with aniline or phenylhydrazine, pharmacological doses of selenite (0.5, 2.0 mg kg^{-1} s.c.) was shown to suppress drug-induced methaemoglobinaemia in rats. The effect of selenite was not attributed to the activity of glutathione peroxidase but to the selenite-induced catalytic reduction of methaemoglobin by glutathione[62,63]. The catalytic action was highly specific to selenium.

F. Others

A deficient detoxification of peroxides may be associated with the pathogenic mechanism of neuronal ceroid lipofuscinosis (NCL), characterized by visual failure and progressive cerebral injury. As a result, ceroid and lipofuscin pigments as an endo-product of lipid peroxidation were accumulated in the nerve cells and other cells. A decreased plasma selenium content and erythrocyte glutathione peroxidase activity in NCL patients in Finland were corrected by prolonged administration of selenite[64,65]. The symptoms of the disease were also improved by this treatment. In other neurological degenerative diseases associated with increased oxidative damage, e.g. Down's syndrome and geriatric disease, antioxidant therapy with selenium and vitamin E seems to be effective[66].

The skeletal muscle injury in patients with myotonic dystrophy, Duchenne muscular dystrophy and long-term parenteral alimentation therapy was also associated with selenium deficiency[66]. The symptoms due to this injury were shown to be alleviated by selenium supplementation.

Kaschin–Beck disease in China, characterized by a disorder of cartilage development, was demonstrated to be associated with selenium deficiency and to be prevented by oral selenite supplementation[67,68].

Selenium is recognized as a constituent of sperm and to be essential for spermatogenesis. The effect on spermatogenesis was specific to selenium and could not be replaced by vitamin E[69].

III. RELATION TO INFLAMMATION AND IMMUNITY

In 1963, an inorganic substance with anti-inflammatory effects was isolated and reported to be selenium[70]. The effectiveness of various organic selenium compounds tested varied with the organic group attached to the selenium atom. Selenite was also shown to exhibit a stabilizing action against heat-induced protein denaturation[71] and lysosomal labilization[72], the action of which was correlated with anti-inflammatory activity.

Macrophages and neutrophils involved in the inflammatory process are both capable of releasing large amounts of reactive oxygen species at sites of inflammation. According to Parnham *et al.*[73], reduced glutathione peroxidase activity in macrophages from selenium-deficient mice was associated with enhanced macrophage hydrogen peroxide release on zymosan stimulation. Hydrogen peroxide-mediated cell injury may account for the reduction in lymphocyte mitogenesis and enhancement of adjuvant arthritis in selenium-deficient animals. Thus, enhancement of glutathione peroxidase activity is thought to be beneficial in improving inflammatory and immune diseases. Recently, an organic selenium compound, 2-phenyl-1,2-benzoisoselenazole-3(2*H*)-one (ebselen), with very low toxicity was reported to have anti-inflammatory activity in various inflammatory models, e.g. cobra venom factor- or carrageenin-induced paw oedema (ID_{50} 60–100 mg kg^{-1}, oral), cotton pellet granuloma (ID_{30} 4 mg kg^{-1}, oral) and adjuvant arthritis (ID_{30} 2 mg kg^{-1}, oral)[74]. Ebselen itself exhibited glutathione peroxidase-like activity *in vitro* and antioxidant activity independent of exogenous glutathione[75,76]. Additionally, both the generation of chemiluminescence by macrophages (an index of reactive oxygen species) and the formation of leukotriene B_4 as a potent mediator of chemotaxis and aggregation of neutrophils were inhibited dose dependently by ebselen[77]. Parnham and Kindt[77] thought that scavenging of peroxides at inflamed sites by oral ebselen is a possible new approach to anti-inflammatory therapy. Further, they suggested the possibility of the therapeutic usefulness of this type of selenium compounds in treating various diseases associated with overproduction of hydrogen peroxide or lipid peroxides. Rudzinski and coworkers[78–80] synthesized various chelate complexes of selenium with sulphonamide Schiff bases, 2-substituted benzimidazoles and thiopicolinamide, which exhibited a variety of biological effects, and examined their pharmacological activities. Some of these complexes exhibited mild anti-inflammatory activity in addition to antibacterial and hypoglycaemic activities.

Selenium acted as immunoadjuvant when it was fed to animals in amounts in nutritional excess[81] or when administered by injection[82]. Dietary selenium at levels above that generally accepted as nutritionally adequate (0.1 ppm) enhanced the primary immune response in mice as measured by the plaque-forming cell test and by haemaggulutination[81]. The enhancement of the primary immune response by selenite administration (3, 5 μg of Se) was greatest when selenium was administered prior to or simultaneously with sheep red blood cells[82]. Mice fed selenium (1–3 ppm) supplemented diets showed a markedly increased formation of IgG and IgM antibody to the sheep red blood cell antigen[83].

Selenium may possibly be involved in both cellular and humoral immunity. The involvement of selenium in cellular immunity was supported by impaired microbicidal activity, increased dinitrochlorobenzene (DNCB) hypersensitivity and the insensitivity of lymphocytes to selenium depletion. Serfass and Ganther[84] reported that phargocytic cells of selenium-deficient rats, although capable of ingestion of yeast cells *in vitro*, are unable to kill them. The finding was supported in other animal species deficient in selenium[85]. Guinea-pigs fed diets with 1–3 ppm selenium supplements appeared more sensitive than controls to DNCB, showing that selenium enhances the delayed type hypersensitivity[86]. The mitogenesis of lymphocytes to various mitogens was suppressed in selenium-deficient animals[73,87]. Recently, it was demonstrated that the supplementation of selenium (0.5, 2.0 ppm) in drinking water enhances the cytotoxic response of rat splenic natural killer cells, which are considered to play a part in immunosurveillance against tumours[88].

IV. RELATION TO CANCER

In 1943, Nelson *et al.*[89] reported the development of hepatic cell adenoma and low-grade carcinoma in rat liver, beginning 18 months after the rats had been fed on a seleniferous diet. Later, several reports supporting the carcinogenicity of selenium appeared[90,91]. In contrast, in 1949, Clayton and Baumann[92] presented the first evidence that selenium may have an anticarcinogenic effect; dietary selenium reduced liver tumours caused by 3-methyl-4-dimethylaminoazobenzene. Owing to the apparently conflicting reports, confusion and controversy about the relationship between selenium and cancer continued for a long time.

In the numerous epidemiological studies in humans that have subsequently been reported, selenium has been demonstrated to exhibit an anticarcinogenic effect. There was an inverse relationship between human cancer incidence and the selenium content of plants in the local area and between blood selenium levels and cancer deaths, as demonstrated by the studies of Shamberger and coworkers[93,94]. Further, Schrauzer *et al.*[95] showed significant inverse correlations between selenium intake, estimated from food consumption data in 27 countries, and the incidence of cancers of the large intestine, rectum, prostate, breast, ovary and lung. Similar results were obtained in the relationship between cancer and selenium[96]. However, since there is no higher incidence of cancer in low-selenium districts such as Finland, New Zealand and China, other factors must be considered[97,98]. In this respect, Schrauzer[99] thought that selenium deficiency does not cause cancer but merely increases the susceptibility to cancer induction.

In recent years, evidence has accumulated that selenium can prevent or retard the growth of chemically induced, viral-induced and transplantable tumours in experimental animals[98,100,101]. Table 1 summarizes the anticarcinogenic effects of selenium in various cancer models. Selenium is effective against carcinogenesis by a number of chemical carcinogens. Most of these studies involved the use of inorganic selenium (mainly selenite) supplemented either in the drinking water or in the diet. Selenium intake was for the entire duration of the experiments and the doses were subtoxic, ranging from 0.5 to 6 ppm. To establish the time at which selenium is most effective against chemically induced

TABLE 1. Anticarcinogenic effects of selenium in various cancer models

Carcinogenesis	Species	Tissue of tumours	Ref.
Chemically induced:			
3-Methyl-4-dimethylaminoazobenzene	Rat	Liver	92
2-Acetylaminofluorene	Rat	Liver	104
Aflatoxin B_1	Rat	Liver	128
Dimethylnitrosamine	Rat	Liver	129
3-Methylcholanthrene	Mouse	Skin	130
Benzopyrene	Mouse	Skin	130
7, 12-Dimethylbenz[*a*]anthracene	Mouse	Skin	130
	Mouse	Mammary gland	131, 132
	Rat	Mammary gland	102, 133, 134
1-Methyl-1-nitrosourea	Rat	Mammary gland	135
Estrone, progesterone	Mouse	Mammary gland	136
1, 2-Dimethylhydrazine	Rat	Colon	103, 105
Methylazoxymethanol	Rat	Colon	137
Bis(2-oxopropyl)nitrosamine	Rat	Colon	138
	Rat	Lung	138
Azoxymethane	Rat	Intestine	139
Virus-induced:			
	Mouse (C_3H/St)	Mammary gland	106
	Mouse (BALB/cf. C_3H)	Mammary gland	107
Transplantable:			
Ehrlich ascites cells	Mouse		108
L1210 leukaemic cells	Mouse		109

carcinogenesis, Ip[102] designed an experiment in which rats were given selenium either before, during or after various combinations of exposure to 7,12-dimethyl[a]anthracene. The results indicated that selenium can inhibit both the initiation and promotion phases of carcinogenesis. The inhibitory effect of selenium in the early promotion phase was probably reversible, but a continuous intake of selenium was necessary in order to achieve maximal effect. Similar results were found for the effects of selenium against 1,2-dimethylhydrazine-induced colon tumour[103]. Since selenium is effective in inhibiting tumours induced by a variety of carcinogens, the primary action of selenium is probably not through interference with the carcinogen metabolism. However, in the case of 2-acetylaminofluorene, increased ring hydroxylation and decreased *N*-hydroxylation of the carcinogen may be involved in the protection by selenium against the carcinogen-induced tumour[104]. With 1,2-dimethylhydrazine-induced tumours, the protection by selenium may be partly attributed to enhanced detoxification due to an increase in glutathione S-transferase activity[105].

Selenium exhibits an anticarcinogenic effect against spontaneous or virally induced tumours[98,101]. Subtoxic amounts of selenium (1–15 ppm) were shown to prevent the genesis of spontaneous mammary tumours in C_3H/St mice[106]. Additionally, mammary carcinogenesis in BALB/cfC_3H mice, containing the highly oncogenic exogenic murine mammary tumour virus, was also prevented by selenium (2, 6 ppm) in the drinking water[107]. Since the tumour models may be related to human breast cancer development, these results are of considerable interest.

With regard to transplantable tumours, the injection of selenium was reported to prevent the development of tumours in Ehrlich ascites tumour cell-inoculated mice[108]. Of various selenium compounds used (0.25–2.0 mg kg^{-1} i.p.), selenite was the most effective. Similar results were also obtained with L1210 leukaemic cell-inoculated mice[109]. These results clearly indicate that selenium may have both preventive and therapeutic benefits in the etiology of cancer.

When comparing the activities of various selenium compounds, selenomethionine or selenocystine as organic selenium compounds are exclusively used, whereas their anticarcinogenic effect was less effective[108–110]. In general, the degree to which the biological activity of drugs can be altered by replacing sulphur or other atoms with selenium has received considerable attention. In developmental research on antitumour agents, a number of organic selnium compounds have been synthesized with the purpose of enhancing or improving antitumour activity by modifications to parent compounds. For example, various selenium analogues of nucleoside- and nucleotide-related compounds[111–117], aromatic seleno lactones[118], retinyl phenyl selenoether[119] and the analogues of amino acids and steroids bearing the —$SeAsMe_2$ group[120] have been shown to possess antitumour activity. In particular, 2-β-D-ribofuranosylselenazole-4-carboxamide (selenazofurin), derived from the corresponding 2-β-D-ribofuranosylthiazole-4-carboxamide (tiazofurin) synthesized as a novel potential antitumour and antiviral agent[121], exhibited remarkable effects in doses of 6–24 mg kg^{-1} i.p. against Lewis lung carcinoma in mice and was about 10 times more cytotoxic than tiazofurin towards L-1210 and P-388 cells in culture[111,122]. Selenazofurin was metabolized into ribonucleoside monophosphate and subsequently into an analogue of NAD in which the nicotinamide portion of the molecule was replaced with selenazofurin. The resulting NAD analogue stopped the proliferation of tumour cells by depressing guanosine nucleotide synthesis as a result of the inhibition of IMP dehydrogenase, similarly to tiazofurin[123,124]. Further, selenazofurin also possessed a broad-spectrum antiviral activity against DNA and RNA viruses[125]. Since both activities of selenazofurin were noticeably better than those of tiazofurin, selenazofurin may be expected to be useful as a new antitumour and antiviral agent. However, in these studies, the activity of the organic selenium compounds was exclusively compared with that of the corresponding parent compounds, while their mechanism of action may be different from that of inorganic selenium.

Little information is available on the mode of action of selenium, but there is some support for its anticarcinogenic properties: exposure to a high concentration of selenium inhibited DNA synthesis[126], and modulation of mitochondria function by selenium was involved in one of the early effects of growth inhibition[127]. Further, the doses of selenium required for anticarcinogenic activity agree with the doses for potentiating the immune response[88].

V. INTERACTION WITH HEAVY METALS

There have been many reports on the biological interaction between selenium and a number of heavy metals. The first report of such an interaction was presented in 1938, arsenic counteracting the toxic effects of seleniferous grain[140,141]. So far, the metals which alleviate selenium toxicity are known to be arsenic[140,141], silver[142], tin[143], copper[142], lead[143], mercury[144], cadmium[144], thallium[143] and tungsten[141]. Conversely, the toxicities of mercury[145–147], cadmium[148], silver[149], lead[150], tin[151], thallium[152] and platinum[153] were suppressed by selenium. The detoxification of heavy metals by selenium is thought to be one of its important roles.

Inorganic selenium compounds such as selenite may be reduced in the body to selenide (Se^{2-}), which may subsequently react with heavy metals. Examples of the direct interaction are the formation of inactive complexes or compounds such as HgSe and CdSe detected after coadministration of selenium with these metals[154–156]. The direct interactions may result in changes in the distribution of the metals in various organs, which may be relevant to the mechanisms of the metal detoxification. Heavy metals administered alone may react with endogenous Se^{2-} to cause the status of selenium deficiency, including a decrease in glutathione peroxidase activity, resulting in damage of physiological functions dependent on selenium[149,157]. In the case of methylmercury intoxication, bismethylmercury(II) selenide(BMS) was formed temporarily soon after injection of selenite[158,159]. Since BMS is a non-ionic, lipid-soluble substance, it may function as a diffusible form in the process of selenite-induced redistribution of methylmercury[160]. However, it remains obscure whether this phenomenon may be responsible for the protective action of selenium against methylmercury toxicity. The mechanisms underlying the interaction of selenium with heavy metals are unknown, although suggestions for the mechanisms have been provided by many investigators.

The side effects of drugs undergoing redox cycling have already been described to be suppressed by selenium supplementation, and the same is true for the platinum-containing drug *cis*-diamminedichloroplatinum (cisplatin). Renal toxicity as a side effect of cisplatin, a widely used antitumour agent, was markedly prevented by injection of selenite, without masking its antitumour activity[153]. Since the reduction of nephrotoxicity by selenium may improve the therapeutic value of cisplatin, this action is of considerable interest.

VI. MEDICAL APPLICATIONS

Selenium compounds for human use may be classified into two groups according to their purpose, one being as antidotes and the other as diagnostic drugs for diseases. At present, the single drug used as an antidote is selenium disulphide for treating seborrheic dermatitis and common dandruff. The use of selenium compounds is strictly controlled because of their high toxicity. If the deleterious effect of selenium could be controlled or alleviated, selenium compounds might be employed as possible antidotes for various human diseases. Several attempts at the chemical modification of biologically active compounds by replacing sulphur or other atoms with selenium have been made in the past to induce antagonism or augment the action of the parent compounds[161–163]. Recently, good results have been achieved in such investigations; the excellent character of ebselen and selenazofurin as possible candidates in human medicine has already been mentioned.

Concerning the toxicity of organic selenium compounds, if selenium atoms or selenium-containing fragments are released by the metabolic degradation of the organic compounds, even though the latter seem to be of low toxicity, the effective toxicity of these selenium species may be expected to become enhanced. In particular, chronic toxicity due to accumulation in the organs of released selenium may be a major problem in long-term administration. Thus, in order to introduce organic selenium compounds as human medicines, extensive metabolic investigations including absorption, metabolism, accumulation and elimination, and also toxicity studies, are especially important. Further, a comprehensive evaluation of their effective dose:toxicity ratio is necessary.

Another application of selenium compounds is as radiopharmaceuticals for imaging organs and tumours. Although selenium-75 may not be an ideal radionuclide for diagnostic use, it offers several advantages, e.g. long half-life (118.5 d), ease of incorporation of selenium into organic molecules, increased stability *in vivo* of organic selenium compounds compared with the corresponding halogenated analogues and the possibility of preliminary feasibility studies by preparing selenium-73 labelled compounds ($t_{1/2} = 7.1$ h). Thus, selenium-75 labelled radiopharmaceuticals are of considerable value in diagnostic nuclear medicine. [^{75}Se]Selenomethionine has been used for pancreatic imaging for many years. Additionally, 6-(methyl[^{75}Se]selenomethyl)-19-norcholest-5(10)-en-3β-ol (scintadren) is also used clinically as an adrenal imaging agent[164]. For the same purpose, selenium-75 labelled compounds such as 19-selenocholesterol[165] and 24-(isopropylseleno)chol-5-en-3β-ol[166] as cholesterol analogues and 2[3,4-dimethoxyphenyl)ethyl]dimethylselenonium iodide[167] as a dopamine analogue have been synthesized. As a breast tumour-imaging agent, among the seleno derivatives of estrogen and estradiol examined, 16α-(methyl[^{75}Se]seleno)-17β-estradiol[168] is considered to be a good candidate. Selenium-75 labelled tertiary diamines, bis(β-morpholinoethyl)selenide and bis(β-piperidinoethyl)selenide, with high brain uptakes have been investigated as brain-imaging agents[169]. Further, 23-[^{75}Se]selena-25-homotaurocholate has been evaluated as a radiopharmaceutical for investigating the enterohepatic circulation[170].

VII. REFERENCES

1. M. A. Heinrich, Jr., and D. M. MacCanon, *Toxicol. Appl. Pharmacol.*, **2**, 33 (1960).
2. D. M. Aviado, J. Drimal, T. Watanabe and P. M. Lish, *Cardiology*, **60**, 113 (1975).
3. F. Fanconi, N. Manghi, A. Giotti, F. Martini and G. Dini, *Acta Pharmacol. Toxicol.*, **46**, 98 (1980).
4. K. O. Godwin, *Q. J. Exp. Physiol.*, **50**, 282 (1965).
5. K. O. Godwin and F. J. Fraser, *Q. J. Exp. Physiol.*, **51**, 94 (1966).
6. R. J. Shamberger, in *Selenium in Biology and Medicine* (Eds. J. E. Spallholz, J. L. Martin and H. E. Ganther), Avi, Westport, 1981, p. 391.
7. H. M. Perry, Jr., *J. Am. Diet. Assoc.*, **62**, 631 (1973).
8. H. M. Perry, Jr., E. F. Perry and M. W. Erlanger, in *Trace Substances in Environmental Health—VIII* (Ed. D. D. Hemphill), University of Missouri, Columbia, 1974.
9. H. Hilse, P. Oehme, W. Krause and K. Hecht, *Acta Physiol. Pharmacol. Bulg.*, **5**, 47 (1979).
10. J. T. Salonen, G. Alfthan, J. K. Huttunen, J. Pikkarainen and P. Puska, *Lancet*, **2**, 175 (1982).
11. T. Masukawa, J. Goto and H. Iwata, *Experientia*, **39**, 405 (1983).
12. R. W. Bryant, T. C. Simon and J. M. Bailey, *Biochem. Biophys. Res. Commun.*, **117**, 183 (1983).
13. Y. X. Wong, K. Bocker, H. Reuter, J. Kiem, G. V. Iyenger, F. Loogen, R. Gross and L. E. Feinendegen, *Klin. Wochenschr.*, **59**, 817 (1981).
14. Keshan Disease Research Group, *Chin. Med. J.*, **92**, 477 (1979).

15. Keshan Disease Research Group, *Chin. Med. J.*, **92**, 471 (1979).
16. R. A. Johnson, S. S. Baker, J. T. Fallon, E. P. Maynard, J. N. Ruskin, Z. Wen, K. Ge and H. J. Cohen, *N. Engl. J. Med.*, **304**, 1210 (1981).
17. N. W. Revis and N. Marusic, *J. Mol. Cell Cardiol.*, **10**, 945 (1978).
18. J. F. Van Bleet, L. Greenwood V. J. Ferrans and A. H. Rebar, *Am. J. Vet. Res.*, **39**, 997 (1978).
19. J. F. Van Bleet, V. J. Ferrans and W. E. Weirich, *Am. J. Pathol.*, **99**, 13 (1980).
20. T. Facchnetti, F. Delaini, M. Salmona, M. B. Donati, S. Feuerstein and A. Wendel, *Toxicol. Lett.*, **15**, 301 (1983).
21. I. S. Jamall and J. C. Smith, *Toxicol. Appl. Pharmacol.*, **80**, 33 (1985).
22. M. A. Heinrich, Jr., and D. M. MacCanon, *Toxicol. Appl. Pharmacol.*, **3**, 174 (1961).
23. C. E. Cross, G. Hasegawa, K. A. Reddy and S. T. Omaye, *Res. Commun. Chem. Pathol. Pharmacol.*, **16**, 695 (1977).
24. N. M. Elsayed, A. D. Hacker, K. Kuehn, M. G. Mustafa and G. N. Schrauzer, *Toxicol. Appl. Pharmacol.*, **71**, 398 (1983).
25. T. Ichinoe and M. Sagai, *Toxicol. Appl. Pharmacol.*, **66**, 1 (1982).
26. J. S. Bus, S. D. Aust and J. E. Gibson, *Res. Commun. Chem. Pathol. Pharmacol.*, **11**, 31 (1975).
27. S. T. Omaye, K. A. Reddy and C. E. Cross, *Toxicol. Appl. Pharmacol.*, **43**, 237 (1978).
28. G. F. Combs and F. J. Peterson, *J. Nutr.*, **113**, 538 (1983).
29. F. J. Peterson, G. F. Combs, J. L. Holtzman and R. P. Mason, *J. Nutr.*, **112**, 1741 (1982).
30. R. A. LeBoeuf and W. G. Hoekstra, *J. Nutr.*, **113**, 845 (1983).
31. K. Schwartz and C. M. Foltz, *J. Am. Chem. Soc.*, **79**, 3292 (1957).
32. D. F. Hafeman and W. G. Hoekstra, *J. Nutr.*, **107**, 666 (1977).
33. C. H. Gallagher, *Nature* (*London*), **192**, 881 (1961).
34. D. G. Hafeman and W. G. Hoekstra, *J. Nutr.*, **107**, 656 (1977).
35. S. Z. Cagen and J. E. Gibson, *Toxicol. Appl. Pharmacol.*, **40**, 193 (1977).
36. R. F. Burk, R. A. Lawrence and J. M. Lane, *J. Clin. Invest.*, **65**, 1024 (1980).
37. R. F. Burk and J. M. Lane, *Toxicol. Appl. Pharmacol.*, **50**, 467 (1979).
38. R. F. Burk and J. M. Lane, *Fundam. Appl. Toxicol.*, **3**, 218 (1983).
39. B. A. Merrick, M. H. Davies, K. L. Johnson and R. C. Schnell, *Toxicol. Appl. Pharmacol.*, **72**, 102 (1982).
40. M. D. Maines and A. Kappas, *Proc. Natl. Acad. Sci. USA*, **73**, 4428 (1976).
41. R. F. Burk and B. S. S. Masters, *Arch. Biochem. Biophys.*, **170**, 124 (1975).
42. L. R. Shull, G. W. Buckmaster and P. R. Cheeke, *J. Environ. Pathol. Toxicol.*, **2**, 1127 (1979).
43. R. F. Burk and M. A. Correia, in *Selenium in Biology and Medicine* (Eds. J. E. Spallholz, J. L. Martin and H. E. Ganther), Avi, Westport, 1981, p. 86.
44. G. Siami, A. R. Schulert and R. A. Neal, *J. Nutr.*, **102**, 857 (1972).
45. G. F. Combs and M. L. Scott, *Poult. Sci.*, **54**, 1152 (1975).
46. K. P. MacConnell and G. L. Cho, *Am. J. Physiol.*, **213**, 150 (1967).
47. P. D. Whanger, N. D. Pedersen, J. Harfield and P. H. Weswig, *Proc. Soc. Exp. Biol. Med.*, **153**, 295 (1976).
48. K. P. McConnell and G. J. Cho, *Am. J. Physiol.*, **208**, 1191 (1965).
49. S. Wolffram, F. Arduser and E. Scharrer, *J. Nutr.*, **115**, 454 (1985).
50. D. M. Hadjimarkos, *Arch. Environ. Health*, **10**, 893 (1965).
51. J. W. Fidler, E. C. Naber and J. D. Latshaw, *Poult. Sci.*, **59**, 141 (1980).
52. N. N. Vilas, R. R. Bell and H. H. Draper, *J. Nutr.*, **106**, 589 (1976).
53. H. Negishi, K. Fujimoto and T. Kaneda, *J. Nutr. Sci. Vitaminol.*, **26**, 309 (1980).
54. T. Masukawa, T. Nishimura and H. Iwata, *Biochem. Pharmacol.*, **33**, 2635 (1984).
55. G. A. Pascoe, J. Sakai-Wong, E. Soliven and M. A. Correia, *Biochem. Pharmacol.*, **32**, 3027 (1983).
56. A. W. Halverson, D-T, Tsay, K. C. Triebwasser and E. I. Whitehead, *Toxicol. Appl. Pharmacol.*, **17**, 151 (1970).
57. J. T. Rotruck, A. L. Pope, H. E. Ganther and W. A. Hoekstra, *J. Nutr.*, **102**, 689 (1972).
58. J. T. Rotruck, A. L. Pope, H. E. Ganther, A. B. Swanson, D. G. Hafeman and W. G. Hoekstra, *Science*, **179**, 588 (1973).
59. M.-L. Hu, C. Chung and J. E. Spallholz, *J. Inorg. Biochem.*, **22**, 165 (1984).
60. J. G. Morris, W. S. Cripe, H. L. Chapman, Jr., D. F. Walker, J. B. Armstrong, J. D. Alexander, Jr., R. Miranda, A. Sanchez, J. R. Blair-West and D. A. Denton, *Science*, **223**, 491 (1984).
61. H. Iwata, T. Masukawa and S. Nakaya, *Biochem. Pharmacol.*, **28**, 2209 (1979).

62. H. Iwata, T. Masukawa, S. Kasamatsu, K. Inoue and H. Okamoto, *Experientia*, **33**, 678 (1977).
63. T. Masukawa and H. Iwata, *Life Sci.*, **21**, 695 (1977).
64. T. Westermark and M. Sandholm, *Acta Pharmacol. Toxicol.*, **40**, 70 (1977).
65. T. Westermark, *Acta Pharmacol. Toxicol.*, **41**, 121 (1977).
66. J. Veve, F. Vertongen and L. Molle, *Clin. Endocrinol. Metab.*, **14**, 629 (1985).
67. S. Hau and Z. Zhu, *Huanjing Kexue*, **3**, 18 (1982); *Chem. Abstr.*, **97**, 37802s (1982).
68. J. Li, S. Ren and D. Chen, *Huanjing Kexue Xuebao*, **2**, 91 (1982); *Chem. Abstr.*, **97**, 71210g (1982).
69. S. H. Wu, J. E. Oldfield, P. D. Whanger and P. H. Weswig, *Biol. Reprod.*, **8**, 625 (1973).
70. M. E. Roberts, *Toxicol. Appl. Pharmacol.*, **5**, 500 (1963).
71. J. L. Taylor and J. H. Brown, *Proc. Soc. Exp. Biol. Med.*, **145**, 32 (1974).
72. J. H. Brown and S. H. Pollock, *J. Nutr.*, **102**, 1413 (1972).
73. M. J. Parnham, J. Winkelmann and S. Leyck, *Int. J. Immunopharmacol.*, **5**, 455 (1983).
74. M. J. Parnham, S. Leyck, N. Dereu, J. Winkelmann and E. Graf, *Adv. Inflam. Res.*, **10**, 397 (1986).
75. A. Muller, E. Cadenas, P. Graf and H. Sies, *Biochem. Pharmacol.*, **33**, 3235 (1984).
76. A. Wendel, M. Fausel, H. Safayhi, G. Tiegs and R. Otter, *Biochem. Pharmacol.*, **33**, 3241(1984).
77. M. J. Parnham and S. Kindt, *Biochem. Pharmacol.*, **33**, 3247 (1984).
78. W. E. Rudzinski, T. M. Aminabhavi, N. S. Biradar and C. S. Patil, *Inorg. Chim. Acta*, **67**, 177 (1982).
79. N. S. Biradar, T. M. Aminabhavi, C. S. Patil and W. E. Rudzinski, *Inorg. Chim. Acta*, **78**, 47 (1983).
80. T. M. Aminabhavi, W. E. Rudzinski, N. S. Biradar and C. S. Patil, *Inorg. Chim. Acta*, **78**, 51 (1983).
81. J. E. Spallholz, J. L. Martin, M. L. Gerlach and R. H. Heinzerling, *Proc. Soc. Exp. Biol. Med.*, **143**, 685 (1973).
82. J. E. Spallholz, J. E. Martin, M. L. Gerlach and R. H. Heinzerling, *Proc. Soc. Exp. Biol. Med.*, **148**, 37 (1975).
83. J. E. Spallholz, J. E. Martin, M. L. Gerlach and R. H. Heinzerling, *Infect. Immun.*, **8**, 841 (1973).
84. R. E. Serfass and H. E. Ganther, *Nature* (*London*), **255**, 640 (1975).
85. R. Boyne and J. R. Arthur, *J. Comp. Pathol.*, **89**, 151 (1979).
86. J. E. Spallholz, *Adv. Exp. Med. Biol.*, **135**, 43 (1981).
87. B. E. Sheffy and R. D. Schultz, *Fed. Proc. Fed. Am. Soc. Exp. Biol.*, **38**, 2139 (1979).
88. P. A. Talcott, J. H. Exon and L. D. Koller, *Cancer Lett.*, **23**, 313 (1983).
89. A. A. Nelson, O. G. Fitzhugh and H. O. Calvery, *Cancer Res.*, **3**, 230 (1943).
90. M. N. Volgarev and C. A. Tscherkes, in *Selenium in Biomedicine* (Ed. O. H. Muth), Avi, Westport, 1967, p. 179.
91. H. A. Schroeder, D. V. Frost and J. J. Balassa, *J. Chron. Dis.*, **23**, 227 (1970).
92. C. C. Clayton and C. A. Baumann, *Cancer Res.*, **9**, 575 (1949).
93. R. J. Shamberger and D. V. Frost, *Can. Med. Assoc. J.*, **100**, 682 (1969).
94. R. J. Shamberger and C. E. Willis, *CRC Crit. Rev. Clin. Lab. Sci.*, **2**, 211 (1971).
95. G. N. Schrauzer, D. A. White and C. J. Schneider, *Bioinorg. Chem.*, **7**, 23 (1977).
96. L. C. Clark, *Fed. Proc. Fed. Am. Soc. Exp. Biol.*, **44**, 2584 (1985).
97. C. D. Thomson and M. F. Robinson, *Am. J. Clin. Nutr.*, **33**, 303 (1980).
98. P. D. Whanger, *Fundam. Appl. Toxicol.*, **3**, 424 (1983).
99. G. N. Schrauzer, *Am. J. Clin. Nutr.*, **33**, 1892 (1980).
100. C. Ip, *Fed. Proc. Fed. Am. Soc. Exp. Biol.*, **44**, 2573 (1985).
101. J. A. Milner, *Fed. Proc. Fed. Am. Soc. Exp. Biol.*, **44**, 2568 (1985).
102. C. Ip, *Cancer Res.*, **41**, 4386 (1981).
103. M. M. Jacobs, C. F. Frost and F. A. Beams, *Cancer Res.*, **41**, 4458 (1981).
104. M. V. Marshall, M. S. Arnott, M. M. Jacobs and A. C. Griffin, *Cancer Lett.*, **7**, 331 (1979).
105. M. M. Jacobs, *Cancer Res.*, **43**, 1646 (1983).
106. G. N. Schrauzer and D. Ishmael, *Ann. Clin. Lab. Sci.*, **4**, 441 (1974).
107. D. Medina and F. Shepherd, *Cancer Lett.*, **8**, 241 (1980).
108. G. A. Greeder and J. A. Milner, *Science*, **209**, 825 (1980).
109. J. A. Milner and C. Y. Hsu, *Cancer Res.*, **41**, 1652 (1981).
110. H. J. Thompson, L. D. Meeker and S. Kokoska, *Cancer Res.*, **44**, 2803 (1984).
111. P. C. Srivastava and R. K. Robins, *J. Med. Chem.*, **26**, 445 (1983).
112. H. G. Mautner, S.-H., Chu, J. J. Jaffe and A. C. Sartorelli, *J. Med. Chem.*, **6**, 36 (1963).
113. S-H, Chu, *J. Med. Chem.*, **14**, 254 (1971).

114. S-H, Chu and D. D. Davidson, *J. Med. Chem.*, **15**, 1088 (1972).
115. S.-H, Chu, C.-Y. Shiue and M.-Y. Chu, *J. Med. Chem.*, **17**, 406 (1974).
116. S.-H. Chu, C.-Y. Shiue and M.-Y. Chu, *J. Med. Chem.*, **18**, 559 (1975).
117. M. Maeda, N. Abiko and T. Sasaki, *J. Med. Chem.*, **24**, 167 (1981).
118. W. C. Groutas, M. C. Theodorakis, W. A. F. Tomkins, G. Herro and T. Gaynor, *J. Med. Chem.*, **27**, 548 (1984).
119. S. C. Welch and J. M. Gruber, *J. Med. Chem.*, **22**, 1532 (1979).
120. C. H. Banks, J. R. Daniel and R. A. Zingaro, *J. Med. Chem.*, **22**, 572 (1979).
121. R. K. Robins, P. C. Srivastava, V. L. Narayanan, J. Plowman and K. D. Paull, *J. Med. Chem.*, **25**, 107 (1982).
122. D. G. Streeter and R. K. Robins, *Biochem. Biophys. Res. Commun.*, **115**, 544 (1983).
123. H. N. Jayaram, G. S. Ahluwalia, R. L. Dion, G. Gebeyehu, V. E. Marquez, J. A. Kelley, R. K. Robins, D. A. Cooney and D. G. Johns, *Biochem. Pharmacol.*, **32**, 2633 (1983).
124. T. J. Boritzki, D. A. Berry, J. A. Besserer, P. D. Cook, D. W. Fry, W. R. Leopold and R. C. Jackson, *Biochem. Pharmacol.*, **34**, 1109 (1985).
125. J. J. Kirsi, J. A. North, P. A. McKernan, B. K. Murray, P. G. Ganonico, J. W. Huggins, P. C. Srivastava and R. K. Robins, *Antimicrob. Agents Chemother.*, **24**, 353 (1983).
126. D. Medina and C. J. Oborn, *Cancer Lett.*, **13**, 333 (1981).
127. D. Medina, H. W. Lane and C. M. Tracey, *Cancer Res.*, **43**, 2460s (1983).
128. K. E. Grant, M. W. Conner and P. M. Newberne, *Toxicol. Appl. Pharmacol.*, **41**, 166 (1977).
129. R. A. LeBoeuf, B. A. Laishes and W. G. Hoekstra, *Fed. Proc. Fed. Am. Soc. Exp. Biol.*, **42**, 669 (1983).
130. R. J. Shamberger, *J. Natl. Cancer Inst.*, **44**, 931 (1970).
131. D. Medina, H. W. Lane and F. Shepherd, *Carcinogenesis*, **4**, 1159 (1983).
132. D. Medina and F. Shepherd, *Carcinogenesis*, **2**, 451 (1981).
133. C. Ip, *Cancer Res.*, **41**, 2683 (1981).
134. H. J. Thompson, L. D. Meeker, P. J. Becci and S. Kokoska, *Cancer Res.*, **42**, 4954 (1982).
135. H. J. Thompson and P. J. Becci, *J. Natl. Cancer Inst.*, **65**, 1299 (1980).
136. C. W. Welsch, M. Goodrich-Smith, C. K. Brown, H. D. Greene and E. J. Hamel, *Carcinogenesis*, **2**, 519 (1981).
137. M. M. Jacobs, B. Jansson and A. C. Griffin, *Cancer Lett.*, **2**, 133 (1977).
138. D. F. Birt, T. A. Lawson, A. D. Julius, C. E. Runice and S. Salmasi, *Cancer Res.*, **42**, 4455 (1982).
139. B. K. Soullier, P. S. Wilson and N. D. Nigro, *Cancer Lett.*, **12**, 343 (1981).
140. A. L. Moxon, *Science*, **88**, 81 (1938).
141. A. L. Moxon and K. P. DuBois, *J. Nutr.*, **18**, 447 (1939).
142. L. S. Jensen, *J. Nutr.*, **105**, 769 (1975).
143. G. O. Howell and C. H. Hill, *Environ. Health Perspect.*, **25**, 147 (1978).
144. C. H. Hill, *J. Nutr.*, **104**, 593 (1974).
145. J. Parizek and I. Ostadalova, *Experientia*, **23**, 142 (1967).
146. H. E. Ganther, C. Goudie, M. L. Sunde, M. J. Kopecky, P. Wagner, S. H. Oh and W. H. Hoekstra, *Science*, **175**, 1122 (1972).
147. H. Iwata, H. Okamoto and Y. Ohsawa, *Res. Commun. Chem. Pathol. Pharmacol.*, **5**, 673 (1973).
148. A. B. Kar, R. P. Das and B. Mukerji, *Proc. Natl. Inst. Sci., India, Part B*, **26**, 40 (1960).
149. P. A. Wagner, W. G. Hoekstra, and H. E. Ganther, *Proc. Soc. Exp. Biol. Med.*, **148**, 1106 (1975).
150. S. C. Rastogi, J. Clausen and K. C. Srivastava, *Toxicology*, **6**, 377 (1976).
151. M. Chiba, N. Fujimoto and M. Kikuchi, *Toxicol. Lett.*, **24**, 235 (1985).
152. Z. M. Hollo and S. Slatarov, *Naturwissenschaften*, **47**, 87 (1960).
153. A. Naganuma, M. Satoh, M. Yokoyama and N. Imura, *Res. Commun. Chem. Pathol. Pharmacol.*, **42**, 127 (1983).
154. A. Naganuma and N. Imura, *Pharmacol. Biochem. Behav.*, **15**, 449 (1981).
155. T. A. Gasiewicz and J. C. Smith, *Biochim. Biophys. Acta*, **428**, 113 (1976).
156. C. H. Hill, *Fed. Proc. Fed. Am. Soc. Exp. Biol.*, **34**, 2096 (1975).
157. P. D. Whanger and P. H. Weswig, *Nutr. Rep. Int.*, **18**, 421 (1978).
158. A. Naganuma and N. Imura, *Res. Commun. Chem. Pathol. Pharmacol.*, **27**, 163 (1980).
159. H. Iwata, T. Masukawa, H. Kito and M. Hayashi, *Biochem. Pharmacol.*, **30**, 3159 (1981).
160. T. Masukawa, H. Kito, M. Hayashi and H. Iwata, *Biochem. Pharmacol.*, **31**, 75 (1982).
161. K. A. Scott and H. G. Mautner, *Biochem. Pharmacol.*, **13**, 907 (1964).
162. S.-H. Chu, G. R. Hillman and H. G. Mautner, *J. Med. Chem.*, **15**, 760 (1972).

163. R. N. Hauson, R. W. Giese, M. A. Davis and S. M. Costello, *J. Med. Chem.*, **21**, 496 (1978).
164. J. F. Chatal, B. Charbonnel and D. Guihard, *Clin. Nucl. Med.*, **3**, 71 (1978).
165. S. D. Sarkar, R. D. Ice, W. H. Beierwaltes, S. P. Gill, S. Balachandran and G. P. Basmadjian, *J. Nucl. Med.*, **17**, 212 (1976).
166. F. F. Knapp, Jr., T. A. Butler, L. A. Ferren, A. P. Callahan, C. E. Guyer and J. L. Coffey, *J. Med. Chem.*, **26**, 1538 (1983).
167. S. A. Sadek, G. P. Basmadjian, P. M. Huu and J. A. Rieger, *J. Med. Chem.*, **26**, 946 (1983).
168. S. A. Sadek, W. V. Kessler, S. M. Shaw, J. N. Anderson and G. C. Wolf, *J. Med. Chem.*, **25**, 1488 (1982).
169. H. F. Kung and M. Blau, *J. Nucl. Med.*, **21**, 147 (1980).
170. M. V. Merrick, M. A. Eastwood, J. R. Anderson and H. M. Ross, *J. Nucl. Med.*, **23**, 126 (1982).

The Chemistry of Organic Selenium and Tellurium Compounds Volume 2
Edited by S. Patai
© 1987 John Wiley & Sons Ltd.

CHAPTER **10**

Insertion and extrusion reactions

LARS HENRIKSEN

Department of General and Organic Chemistry, The H. C. Ørsted Institute, University of Copenhagen, DK-2100 Copenhagen, Denmark

I. INTRODUCTION 394
A. Scope and Limitations 394
B. Theoretical Background 394
II. HOMOGENEOUS BONDS 395
A. Carbon—Carbon Bonds 395
1. Insertions 395
2. Extrusion reactions. 396
B. Other Group IV Elements 398
1. Si—Si bonds 398
2. Ge—Ge and Sn—Sn bonds 399
C. Group V Elements 399
1. N—N bonds 399
2. P—P bonds 400
3. As—As and Sb—Sb bonds 400
D. Group VI Elements 401
1. General remarks 401
2. Selenium and interchalcogen compounds 402
3. Dialkyl(aryl) polyselanes and thiaselanes 403
4. Selenium(II) and tellurium(II) dichalcogenocarbamates 404
E. Bonds Between Group VI and VIII Metals 405
III. HETEROGENEOUS BONDS INVOLVING CARBON 406
A. Bonds to Hydrogen or Group IA or IIA Elements 406
1. General remarks 406
2. Alkynes 406
3. Arylmetal compounds. 407
4. Alkyl- and alkenyl-metals 409
5. Insertions not involving carbanions 409
B. Covalent Carbon—Metal Bonds (Group III—VIII Metals) 410
1. Aluminium 410
2. Group IV metals 410
3. Group VI metals 410
4. Group VIII metals 411

C. Bonds to Group VB—VIIB Elements 411
1. Nitrogen 411
2. Group VIB elements 412
3. Group VIIB elements 414
IV. HETEROGENEOUS BONDS INVOLVING ELEMENTS OTHER THAN CARBON 414
A. Bonds to Hydrogen and to Alkali Metals 414
1. Elements outside Group VI 414
2. Group VI elements 415
B. Miscellaneous 416
V. REFERENCES 416

I. INTRODUCTION

A. Scope and Limitations

The title 'selenium and tellurium insertion and extrusion reactions' strictly refers to the equation

$$\text{A—B} + \text{E} \rightleftharpoons \text{A—E—B} \quad (1)$$

$$(\text{E} = \text{Se, Te; A, B} = \text{any elements})$$

In accord with this definition, only reactions involving the free elements, E, are reviewed extensively. Reactions which establish or disrupt an A—E—B bridge connecting two independent A—Y/B—X species, or involve selenium and tellurium reagents other than the free elements, are included only when relevant to the main title.

The fields of organoselenium[1] and organotellurium[2,3] chemistry have been reviewed thoroughly up to 1971 and 1973, respectively, and the references cited should be consulted for early data. This review is concerned mainly with developments from these dates up to the present.

As a means of systematization, the reactions are classified according to the types of bond broken in the insertion or established in the extrusion reaction.

B. Theoretical Background

As discussed in the chapters on theoretical aspects and thermochemistry in the first volume of this book, quantum mechanical and thermodynamic data on organic selenium and tellurium compounds are rather meagre and certainly have not developed to an extent which allows generalized predictions with respect to kinetics or equilibria for organo-chalcogen compounds. In a qualitative sense, however, selenium and tellurium can be characterized as highly polarizable elements. The polarizability, α, can be expressed as[4]

$$\alpha = \frac{2}{3}\sum_{n} \frac{(m_x)_{n0}{}^2 + (m_y)_{n0}{}^2 + (m_z)_{n0}{}^2}{E_n - E_0}$$

where m = transient moment, 0 designates the ground state and n the excited states.

Going from second row to higher row elements in the Periodic Table, the energy gap between the potential excited states and the ground state diminishes, leading to a progressive increase in α. With respect to chemical reactivity, the polarizability serves as an indicator of the ability of a given species to tolerate disturbances of the electronic system, i.e. a high polarizability corresponds to low activation barriers for chemical reactions. Since selenium and tellurium are located roughly in the middle of the electronegativity scale, their high polarizabilities allow facile reaction routes involving radical as well as cationic and anionic intermediates, all of which have been encountered. In addition, the ability of these elements to form tetracovalent (selenurane and tellururane) compounds makes pericyclic cycloaddition–cycloreversion reactions a fourth possible reaction path (see Section II.E).

II. HOMOGENEOUS BONDS

A. Carbon—Carbon Bonds

1. *Insertions*

The reaction normally proceeds towards the right in both thermal and photochemical processes and examples on insertions of the chalcogens into C—C bonds are only found in very special cases.

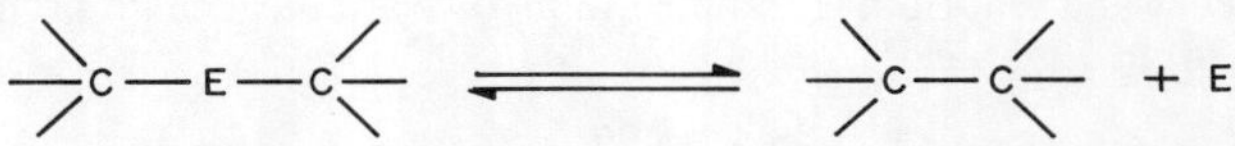

The benzocyclobutadiene **1** reacts with selenium in a thermal process to give the benzoselenophene **2**. Under similar conditions sulphur gives the isobenzothiophene **3** whereas tellurium is unreactive[5].

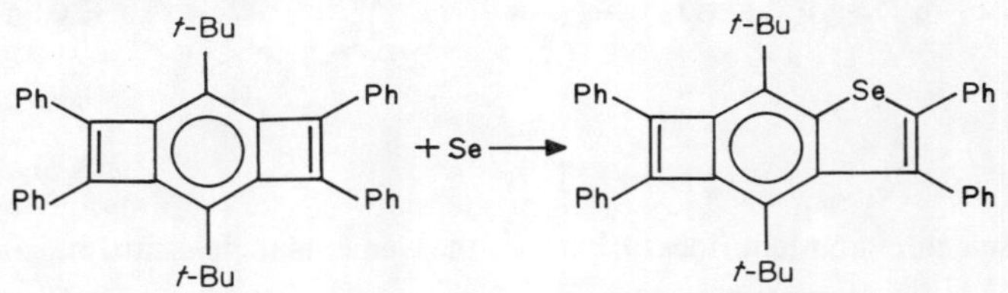

(**1**) (**2**)

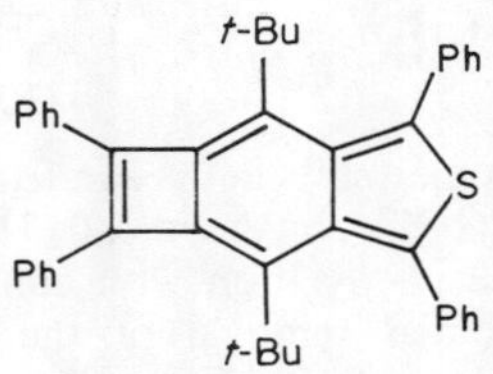

(**3**)

Bicyclo[5.1.0]octa-2, 5-diene (**4**) reacts with selenium (205 °C, 5h) to give 9-selenabicyclo[3.3.1]nona-2, 6-diene (**5a**) in a suggested biradical process. Prolonged heating leads to selenium extrusion with the formation of ethylbenzene and *o*-xylene[6]. In contrast, the tellura analogue **5b** prepared by an independent route extrudes tellurium at 175 °C with clean formation of **4**[7].

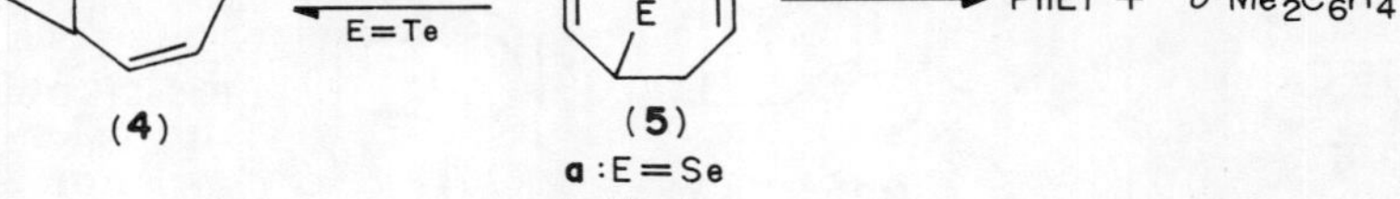

(**4**) (**5**)

a : E = Se

b : E = Te

1, 1′, 3, 3′-Tetrasubstituted biimidazolidines (**6a** and **b**) on heating insert tellurium in the double bond with reversible formation of cyclic telluroureas (**7a** and **b**)[8]. Selenium insertion is reported for **6c**. The reverse process in this case requires the presence of copper powder[9]. Although not proved, the process might involve intermediary chalcogen insertion in the C—C double bond.

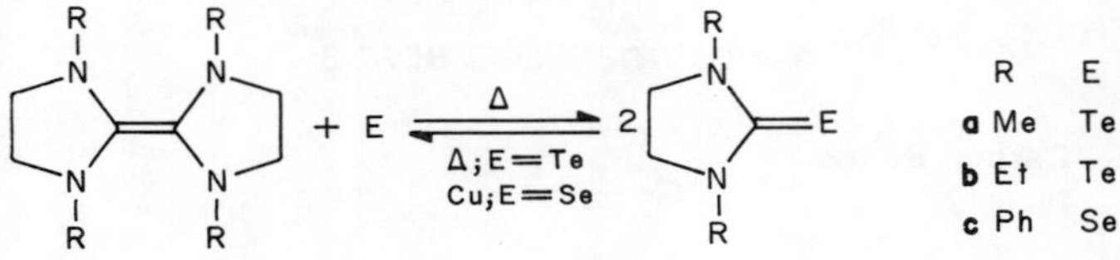

(**6**) (**7**)

Selenium and tellurium have been inserted in aryl—aryl bonds in an aided reaction, by heating the elements with diaryl sulphones[10]. A similar reaction between hexameric *o*-phenylenemercury and tellurium is assumed to involve a radical chain process and gives telluranthrene (**8**) in 72% yield[11].

$$ArSO_2Ar + E \xrightarrow{\sim 300\ °C} ArEAr + SO_2$$

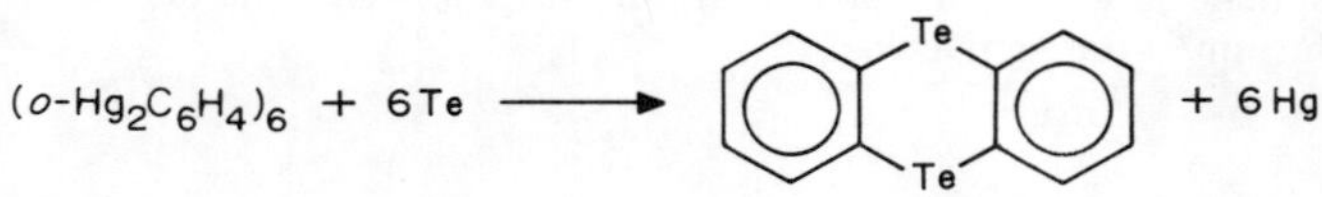

(**8**)

2. Extrusion reactions

Benzylic selenides and tellurides (**9**) extrude the elements to give, fairly cleanly, the 1, 2-diarylethanes (**10**).

$$\underset{(\mathbf{9})}{ArCH_2\text{-}E\text{-}CH_2Ar} \longrightarrow \underset{(\mathbf{10})}{ArCH_2CH_2Ar} + E$$

The reaction has been carried out thermally in the condensed phase at ca. 200 °C[12–14], by flash vacuum thermolysis at 600 °C[15–17], by the so-called benzyne hydrogenolysis method (an aided extrusion involving Se phenylation with concurrent Sterens rearrangement followed by Rany nickel mediated removal of the phenylseleno group) and by photolysis[14,18,19]. The ease of formation of benzylic chalcogenides (e.g. from alkali metal chalcogenides and the reactive benzylic halides) and the cleanness of the extrusion reaction make it an excellent route to *m*- (**11**) and *p*-2, 2-cyclophanes (**12**). Benzyl phenyl selenides also give diarylethanes but in this case selenium is retained as diphenyl diselenide. *o*-Bis(phenylselenomethyl)benzene gives benzocyclobutene (**13**) instead of an *o*-cyclophane[16,20]. Arenocyclobutenes are also formed on vacuum pyrolysis of the corresponding areno [c]-2, 5-dihydrotellurophenes[21].

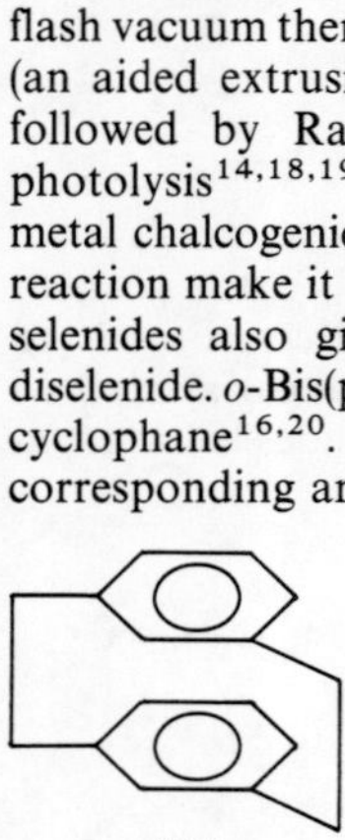

(**11**)

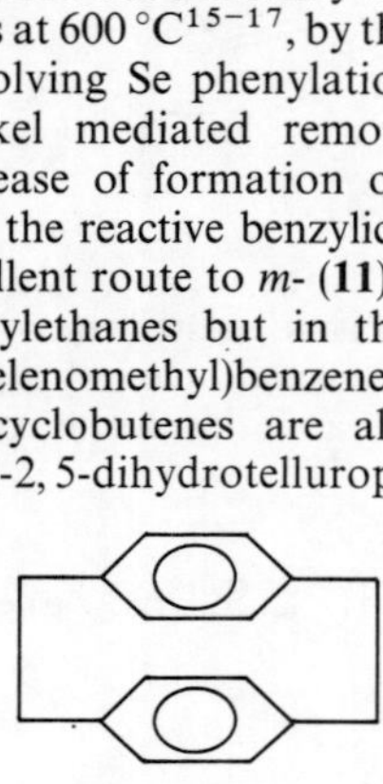

(**12**)

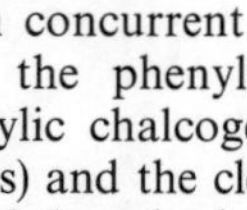

(**13**)

$PhSeCH_2CH{=}CH_2$
(**14**)

$PhSeCH_2C{\equiv}N$
(**15**)

The reactions are undoubtedly of radical character. Scrambled products are observed from non-symmetric benzylic selenides and other compounds which may give rise to stabilized carbon radicals, e.g. diallyl tellurides[22] and also **14** and **15** display the C—C coupling reaction while diaryl and di-*n*-alkyl selenides do not[16].

The primary thermal and photochemical reactions of diselenides and ditellurides are the loss of one chalcogen atom to give the corresponding selenides and tellurides (see Section III.C). On sufficiently vigorous treatment the former compounds therefore give rise to the same products as the latter.

While diaryl selenides and tellurides do not extrude the chalcogen atom directly this atom is removed fairly cleanly with the formation of biaryls on treatment with degassed Raney nickel[23]. This reaction can also be carried out in a single operation from diaryltellurium dichlorides[23,24]. These and related methods for C—C bond formation from tellurides have recently been reviewed[25].

$$\left.\begin{array}{l} \mathrm{ArEAr}\ (\mathrm{E = Se, Te}) \\ \\ \mathrm{Ar_2TeCl_2} \end{array}\right\} \xrightarrow{\text{Raney Ni}} \mathrm{ArAr}$$

Recently the reactions of diaryl, arylalkynyl and arylalkyl tellurides with organomagnesium bromides in the presence of a Ni^{II} or Co^{II} phosphine catalyst have been described[26]. The reactions lead to C—C coupling with extrusion of tellurium. Products of coupling to the Grignard reagent are also observed. An intermediate (**16**) corresponding to metal insertion in a C—Te bond is suggested to account for the cross-coupled products.

$$\mathrm{Ar_2Te + RMgBr} \xrightarrow[\mathrm{Ni^{II}\ or\ Co^{II}}]{\mathrm{THF}} \mathrm{ArR + ArAr + RR}$$

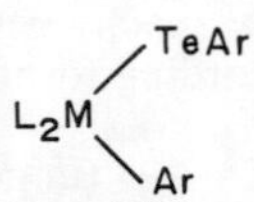

(**16**)

L = R_3P, M = Ni, Co

A series of α-(acylseleno)ketones (**17**) was found to undergo selenium extrusion with the formation of *β*-diketones on treatment with strong base (*t*-BuOK)[27]. The mechanism is suggested to involve enolate ion formation followed by internal displacement on the acylseleno carbon and subsequent loss of selenium with formation of the stabilized anion of the *β*-diketone (**18**)

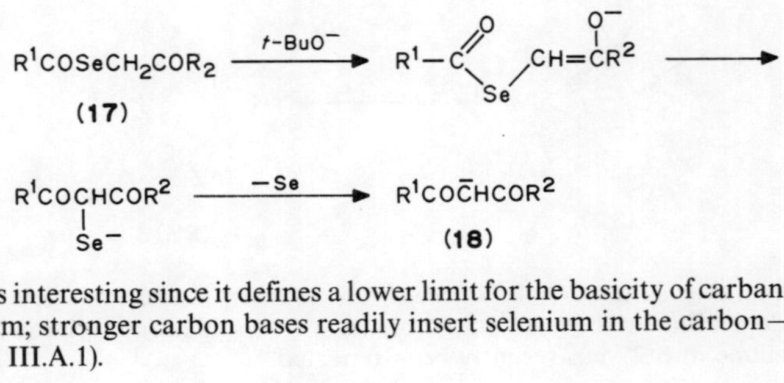

This result is interesting since it defines a lower limit for the basicity of carbanions reacting with selenium; stronger carbon bases readily insert selenium in the carbon—metal bond (see Section III.A.1).

Tetraalkylselenadiazolines (**19**) (available through a cycloaddition reaction between a selenoketone and a diazoalkane) undergo thermolysis with extrusion both of selenium and of nitrogen and formation of a C=C bond[28–31]. In contrast to the sulphur analogues where the thermal reaction stops with formation of a thiirane and a phosphine is needed for the final extrusion of sulphur, the assumed selenirane intermediate (**20**) directly extrudes selenium with alkene formation. This reaction is noted to be superior to the sulphur route for the formation of highly crowded ethenes[27]. The photolysis of **19**, in contrast to the thermal reaction, gives selenium extrusion alone with formation of azines (**21**)[31].

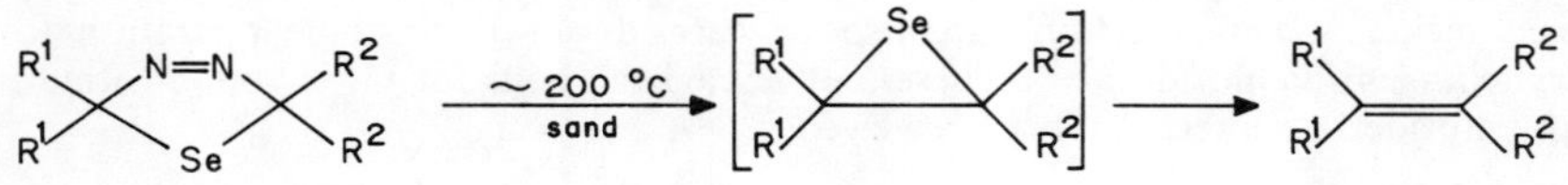

(**19**) (**20**)

$$\mathbf{19} \xrightarrow{h\nu} R^1_2C=NN=CR^2_2$$

(**21**)

Selenium extrusion from **20** at or below room temperature is also reported in the conversion of oxiranes to alkenes with triphenylphosphine selenide[32], selenocyanate ion[33] and selenoamides[34].

B. Other Group IV Elements

Insertion reactions have been reported for Si—Si, Ge—Ge and Sn—Sn bonds. Kinetic comparisons indicate that the order of reactivities is S > Se > Te[35]. This order also appears in an exchange experiment according to the equation[36]

$$(Et_3M)_2E^1 + E^2 \longrightarrow (Et_3M)_2E^2 + E^1 \qquad (1)$$

$$M = Si, Ge, Sn; (E^1, E^2) = (Te, Se); (Te, S); (Se, S)$$

1. Si—Si bonds

In 1980 Wojnowska *et al.*[37] reported that one selenium atom was inserted in decamethylcyclopentasilane (**22**) to give the cycloselenapentasilane **23**, whereas the dodecamethylcyclohexasilane was unreactive.

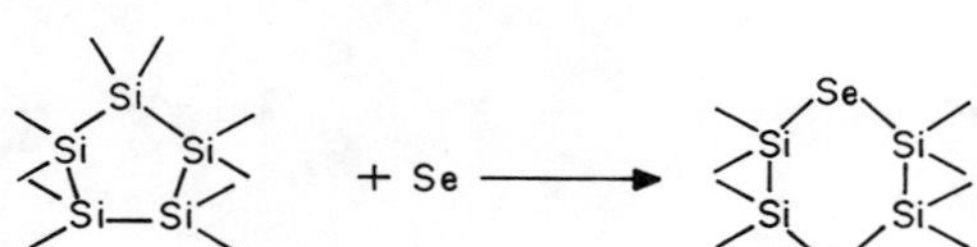

(**22**) (**23**)

Later, octamethylcyclotetrasilane (**24**) was found to insert one or two selenium atoms to give **25** or **26** depending on the reaction temperature[38,39], and the conversion of hexa-*tert*-butylcyclotrisilane into **27** has recently been reported[40].

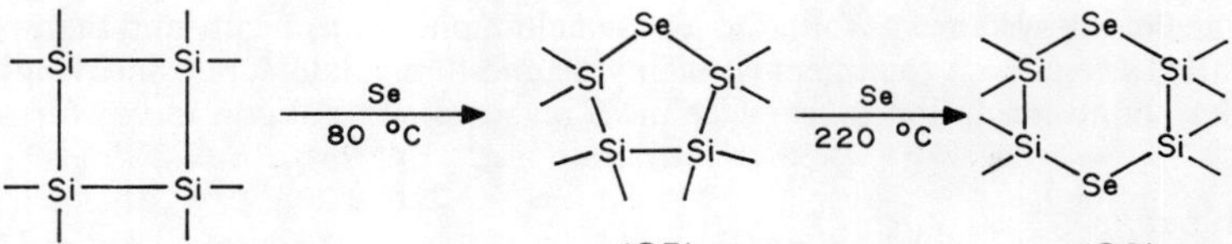

(24) (25) (26)

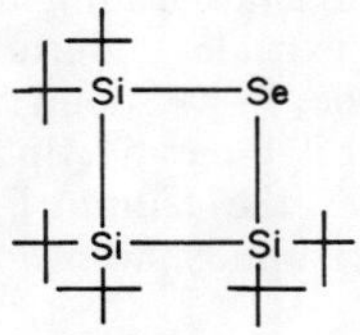

(27)

The tellura analogues of **23** and **25**–**27** have not been reported, although bis(trialkylsilyl)tellurides are known[36] and tellurium insertion in a Si—Si bond by an indirect route involving a silylphosphine has been reported (see Section II.C).

2. Ge—Ge and Sn—Sn bonds

Bochkarev and coworkers[35,36,41,42] have studied the reaction of hexa-substituted digermanes and observed the insertion of selenium while tellurium was unreactive[35].

$$R^1_3GeGeR^2_3 + Se \xrightarrow{100\ ^{\circ}C} R^1_3GeSeGeR^2_3$$

$$R^1, R^2 = C_6F_5, \text{alkyl}$$

More recently the insertion of both selenium and tellurium into the Ge—Ge bond of 1, 1, 2, 2-tetra-*tert*-butyldigermane to give the four-membered rings (**28**) (M = Ge, E = Se, Te) has been reported[43].

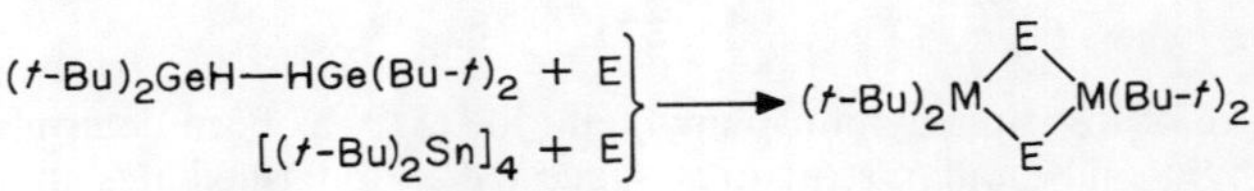

(30) (28)

E=Se,Te; M=Ge,Sn

Hexaethyldistannane reacts with selenium insertion to give bis(triethylstannyl)selenide (**29**)[44], while octa-*tert*-butylcyclotetrastannane (**30**) inserts both selenium and tellurium with concomitant ring scission to give **28** (M = Sn, E = Se, Te)[45].

$$Et_3SnSeSnEt_3$$
(29)

C. Group V Elements

1. N—N bonds

There appear to be no reports of the insertion of selenium or tellurium into an N—N bond or of the generation of such a bond by the extrusion of the elements. The 1, 2, 5-

selenadiazole ring system was found to lose selenium photochemically and thermally[46,47] or, as dialkyl selenide, on treatment with alkylmagnesium halide[48], but since ring opening rather than bond formation takes place these reactions are not considered further here.

2. P—P bonds

In the reaction of trimethylsilylphosphines (**30**) with tellurium[49,50], the primary insertion to give **31** is followed by an exchange reaction leading to diphosphinyl (**32**) and disilyl telluride (**33**). Compound **32** (R = Ph) extrudes tellurium to give **34** (R = Ph) whereas **34** (R = *t*-Bu) inserts tellurium in the P—P bond with formation of the corresponding **32**. The reaction of bis(trimethylsilyl)-*tert*-butylphosphine with tellurium through a similar series of steps gives either the telluratriphosphetane (**35**) (low-temperature product) or the unstable telluradiphosphirane (**36**) (high-temperature product).

$$\underset{(\mathbf{30})}{R_2PSiMe_3} \xrightarrow{Te} \underset{(\mathbf{31})}{R_2PTeSiMe_3} \rightleftharpoons \underset{(\mathbf{32})}{(R_2P)_2Te} + \underset{(\mathbf{33})}{(Me_3Si)_2Te}$$

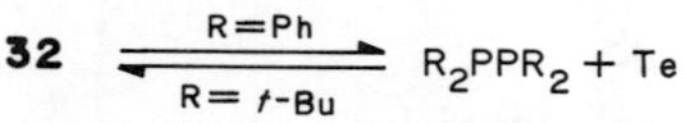

(**34**)

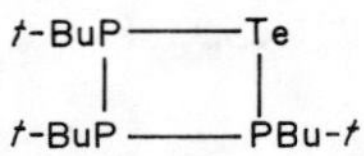

(**35**)

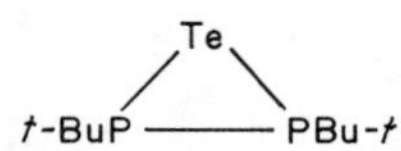

(**36**)

$Ph_3PTePPh_3$

(**37**)

The structure of bis(triphenylphosphinyl)telluride (**37**) has been determined by X-ray crystallography[51] and similar structures suggested as intermediates in the tellurium exchange between phosphines and phosphinetellurides[52]. These are, however, not true insertion products since they are formed from and revert to two moles of phosphine.

The insertion of selenium in the P=P double bond of a hindered diphosphene (**38**) to give a selenadiphosphirane (**39**) has been described[53]. The selenium atom is extruded with regeneration of **38** on treatment with a phosphorus triamide.

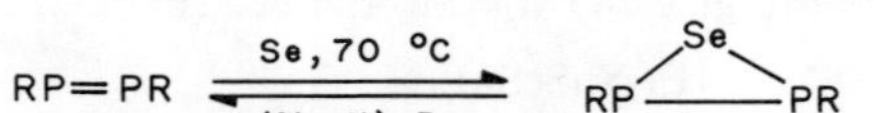

(**38**) (**39**)

$-(Me_2N)_3PSe$ $R = 2,4,6-(t\text{-Bu})_3C_6H_2$

3. As—As and Sb—Sb bonds

Hexaphenylhexarsane (**40**) and pentamethylpentarsolane (**41**) insert selenium thermally in the As—As bonds with concomitant ring formation and the formation of 2, 5-diphenyl-

1, 3, 4-triselena-2, 5-diarsolane (**42**) and 2, 4, 6-trimethyl-1, 3, 5-triselena-2, 4, 6-triarsane (**43**), respectively[54].

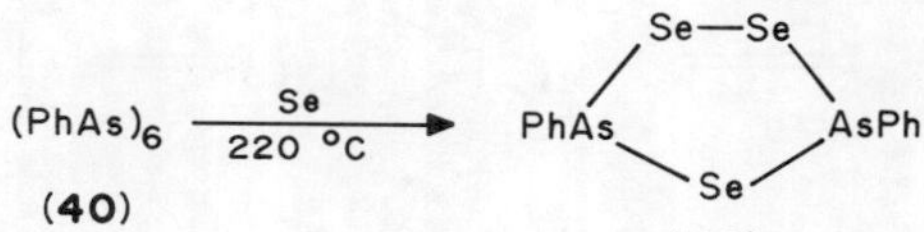

(**40**) (**42**)

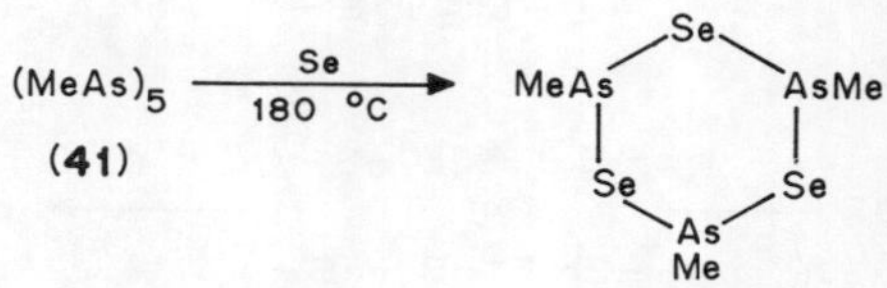

(**41**) (**43**)

Selenium inserts in the As—As bonds of **44** to give a diselenatriarsolane ring (**45**)[55], the structure of which has been determined by X-ray crystallography[56]. Curiously, selenium was inserted faster than sulphur while attempted tellurium insertion was unsuccessful.

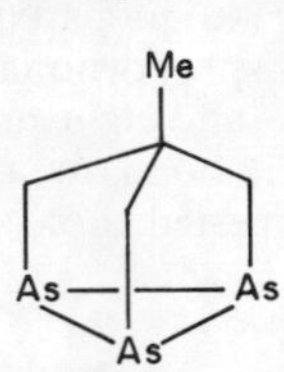

(**44**)

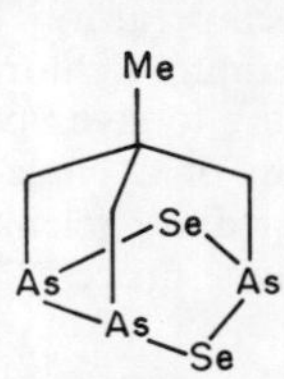

(**45**)

The insertion of selenium and tellurium into Sb—Sb bonds with the formation of distibinyl selenides and tellurides, respectively, has recently been reported[57].

$R_2SbESbR_2$	R	E
	Me	Se, Te
	Et	Se, Te
	Ph	Se

D. Group VI Elements

1. General remarks

Compounds containing an E^2—E^1—E^2 moiety ($E^1 = Se$, $E^2 = S, Se$) as selenium rings and chains, as mixed sulphur—selenium rings or as diorgano polyselanes or polythiaselanes freely undergo exchange reactions under mild conditions which leave bonds to second-row elements untouched. These reactions often give rise to complex equilibrium mixtures and neat selenium insertion or extrusion may result. The high polarizability of selenium opens up four alternative reaction routes, all of which are experimentally supported. The exchange reactions may proceed uncatalysed via a (thermally or photochemically induced) free-radical pathway (equation 2), or in a concerted process (equation 3). Alternatively, generalized acids or bases may catalyse the exchange through cationic (equation 4) or anionic (equation 5) intermediates, respectively.

$$—E^1—E^2—E^1— \rightleftharpoons —E^1—E^{2\bullet} + {}^{\bullet}E^1— \longrightarrow \text{reaction chain} \quad (2)$$

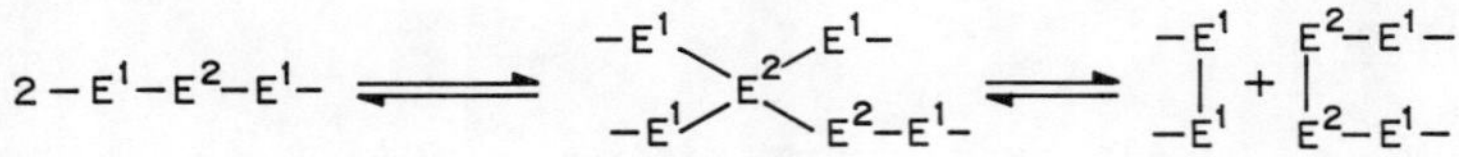
(3)

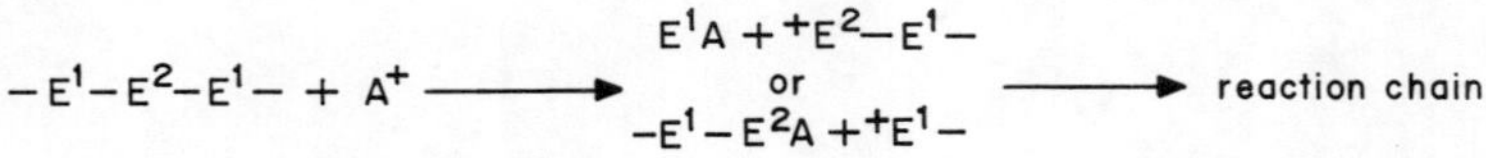
(4)

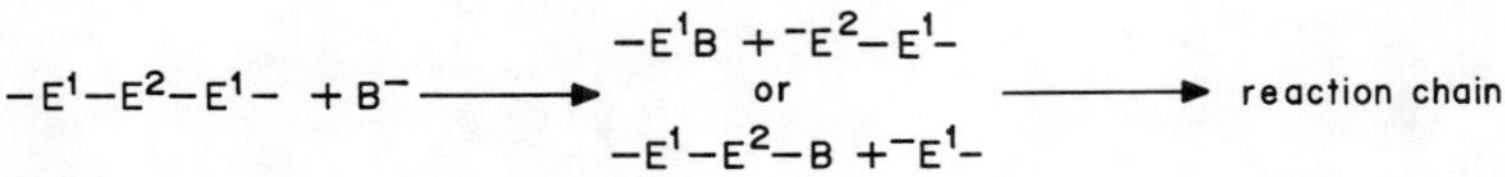
(5)

2. Selenium and interchalcogen compounds

On dissolution in CS_2, pure Se_8 has been found to equilibrate with Se_7 and Se_6 rings within 5 min at room temperature[58]. The Se_2S_5 ring (**46**) prepared from diethyltitanium pentasulphide and selenium dichloride undergoes disproportionation with $t_{1/2} \approx 1$ h in CS_2 at room temperature to give, specifically, the six- and eight-membered rings, **47** and **48**. Concerted addition of a chalcogen—chalcogen bond to a selenium atom with formation of an intermediary selenurane (**49**) is suggested to account for the ease and specificity of the exchange process[59].

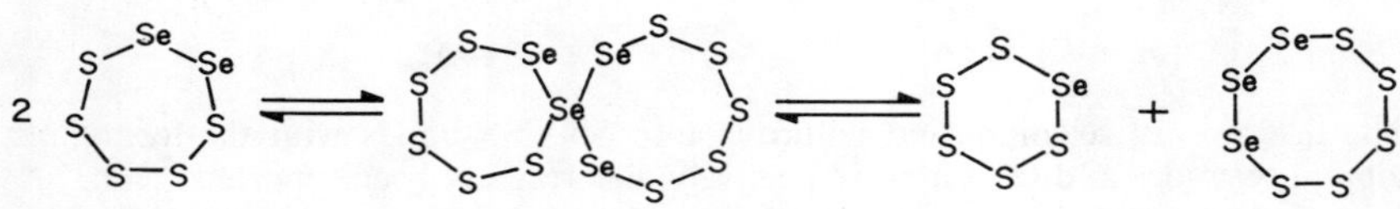
(46) (49) (47) (48)

A similar type of mechanism is suggested by mechanistic studies of the decomposition of tetraorganotellururanes (**50**) to give diorganotellurides and C—C coupled products[60] and may also be involved in the extrusion of tellurium from tetrakis(alkylthio)tellururanes (**51**) with formation of disulphides[61].

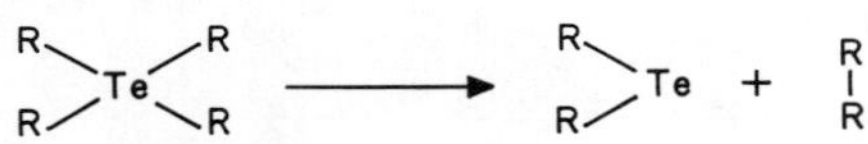
(50)

$$(RS)_4Te \xrightarrow{\sim 100\,^{\circ}C} 2RSSR + Te$$

(51)

A rapid base-catalysed exchange process in selenium chains has been observed by ^{77}Se NMR[62]. A solution of polyselenide ion in DMF with an average chain length of 4 shows a single broad signal at ca. 50°C but distinct signals from tetra- and penta-selenide ion at − 60 °C. Obviously these anionic exchange processes are even faster than those observed in neutral rings.

3. Dialkyl(aryl) polyselanes and thiaselanes

The reactions of 1, 3-diorgano 1, 3-dithia-2-selanes (**52**) [bis(alkylthio) selenides] with nucleophiles have been studied. With hydroxides the reaction proceeds according to equation 6[63], indicating attack on selenium with subsequent disproportionation of the Se^{II} species formed. In contrast, sulphur nucleophiles attack on S with formation of a disulphide and selenium (equation 7). This latter reaction is suggested to be involved in the handling of selenium by biological systems[63–65].

$$\underset{(\mathbf{52})}{2\,RSSeR} + 6\,OH^- \longrightarrow 4\,RS^- + Se + SeO_3{}^{2-} \tag{6}$$

$$\mathbf{52} + R^1S^- \longrightarrow RSSR^1 + RSSe^- \longrightarrow RS^- + Se \tag{7}$$

A study[66] of the selenium extrusion from the two isomeric purinyl-1, 3-dithia-2-selanes **53** and **54** in an aqueous medium showed a curious difference between the two isomers: **53** was stable at all pH values, whereas the rate of selenium extrusion from **54** increased from low pH to reach a maximum at pH 6.7 followed by a decrease to a minimum at pH 8.9 and a steady rise at higher pH values

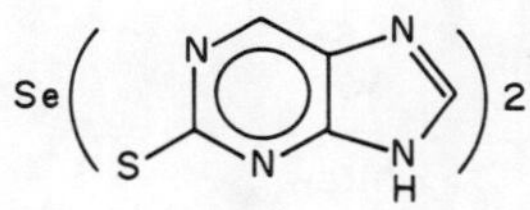

(**53**)

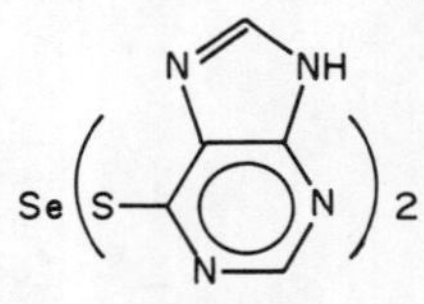

(**54**)

The photochemical formation of mixed dimethylthiaselanes from dimethyl disulphide and selenium has been mentioned but no experimental details were given[67].

It is a long established fact that diorgano polyselanes (**55**) are formed in thermal reactions between selenides or diselenides and elemental selenium[1] and that the pyrolysis of benzylic diselenides at temperatures below those leading to total selenium extrusion and C–C coupling (see Section III.C) gives a mixture of monoselenide and **55**[12,13].

$$RSe_2R + Se \longrightarrow \underset{\substack{(\mathbf{55})\\(n \geqslant 3)}}{RSe_nR} \xleftarrow[-Se]{RX} \underset{(\mathbf{56})}{Se_m{}^{2-}}$$

The alkylation of polyselenide ions (**56**; $n \geqslant 3$) in DMF or DMSO solution[68] is accompanied by precipitation of selenium and the average selenium content in **55** is ca. 2.4 atoms per molecule regardless of the chain length in **56**. ^{77}Se NMR has revealed the presence of all the polyselanes with chain lengths 2–6 and the coupling pattern in the ^{77}Se–^{77}Se satellite spectrum has proved the chain structure. An identical mixture of **55** has been observed on treatment of pure diselenide with selenium in the presence of an acid catalyst, indicating that an equilibrium mixture is formed. Thus the Se–Se exchange reaction catalysed by selenolate and/or polyselenide ion (equation 4, Section II.D.1) has a rate comparable to or higher than that of the alkylation of these ions.

Attempts to separate the individual compounds (**55**) by chromatography normally result in selenium extrusion and concentration of the diselenide. Undoubtedly column materials such as silica gel and alumina act as acid catalysts for the Se–Se exchange processes. Instability of **55** to TLC has been reported previously[13].

A few benzylic triselanes, notably with halogen substituents (**57**), are able to pass through a silica gel column unchanged. Some of these have been isolated from the alkylation of polyselenide ion[68,69] and one structure (**57**, Ar = 2,4-$Cl_2C_6H_3$) determined by X-ray crystallography[69]. However, even these compounds revert to the equilibrium mixture on treatment with strong acid. The bis(*o*-bromo- and -iodo-benzyl)tetraselanes **58** also pass unchanged through silica gel as shown by ^{77}Se NMR, but have not been separated from the corresponding triselanes.

$$\underset{(\mathbf{57})}{ArCH_2SeSeSeCH_2Ar} \quad \underset{(\mathbf{58})}{o\text{-}XC_6H_4CH_2Se_4CH_2C_6H_4X\text{-}o}$$

$$X = Br, I$$

4. Selenium(II) and tellurium(II) dichalcogenocarbamates

Dialkyldiselenocarbamate ions (**59**), in contrast to their thio analogues, on mild oxidation form a mixture of a bis(dialkylselenocarbamoyl)selenide (**60**) and a bis(dialkyldiselenocarbamato)selenium(II) (**61**)[70,71]. The latter structure has been confirmed by X-ray crystallography[72,73]. *N*-Alkyldiselenocarbamates (**59**; R^1 = H) give the unstable **61** (R^1 = H)[74], while trimethyldiselenocarbazate is oxidized to the diselenide (**62**; R^1 = Me, R^2 = NMe_2)[71].

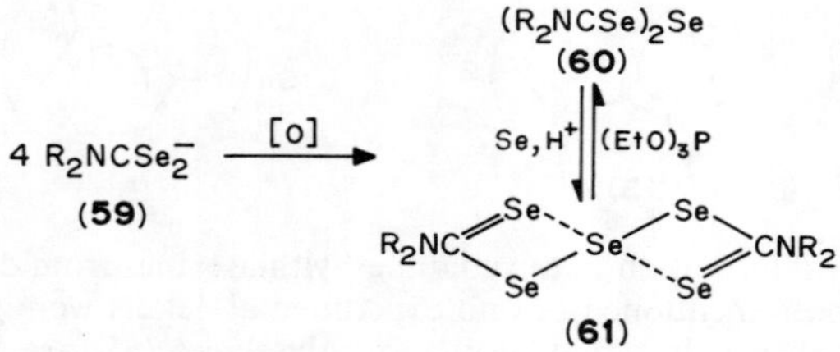

$$(R^1R^2NCSe_2)_2$$
(**62**)

Compounds **60** and **61** are readily interconverted; **60** is equilibrated with selenium by acid catalysis to give **61** almost quantitatively; the deselenation **61** ⟶ **60** requires that selenium is removed, e.g. by a phosphite. The equilibrium 2 **62** ⇌ **60** + **61** was observed for R^1 = Ph, R^2 = Me. In this case **61** spontaneously extrudes selenium to give **62** on attempted isolation[71].

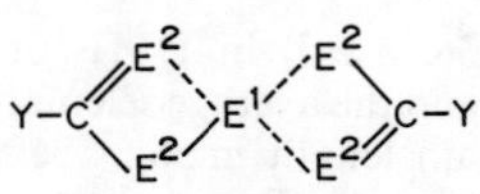

(**63**)

Y = RO, R_2N
E^1 = Se, Te
E^2 = S, Se

A number of compounds of the general structure **63** are known (see Ref. 75 for a recent review). These are often prepared from the YCE_2^- ions and an Se^{IV} or Te^{IV} species. In an alternative preparation[76] the anions react with selenium or tellurium pentathionate, a reaction which may be considered as an analogue of the anionic exchange processes in selenium chains (Section II.D.2).

$$^-O_3SSE^1SSO_3^- + 2YCE_2^- \longrightarrow \mathbf{63} + 2S_2O_3^{2-}$$
$$E^1 = Se, Te$$

In addition to **61**, one compound of structure **63** [$Y = (CH_2)_5N, E^2 = S, E^1 = Se$] has been prepared by acid-catalysed selenium insertion in the corresponding disulphide and there is reason to believe that this insertion is general as far as the resulting **63** is a stable structure[77].

E. Bonds Between Group VI and VIII Metals

In recent years it has been demonstrated that selenium and tellurium can be inserted in metal—metal bonds of the Group VI–VIII metals. The insertion of one, two and four selenium atoms in the Cr—Cr bond of dicyclopentadienylhexacarbonyldichromium has recently been reported[78]. Brunner *et al.*[79] found that four selenium atoms are inserted in the Mo—Mo triple bond of **64** to give **65**, whereas a different type of compound (**66**) is obtained by the insertion in the W—W triple bond of the tungsten analogue of **64**.

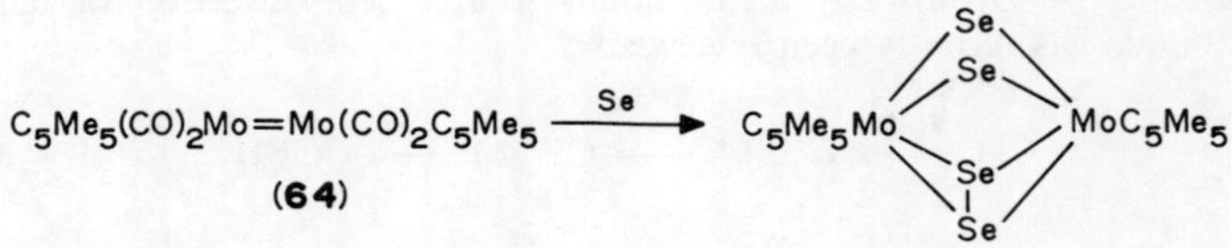

(64) **(65)**

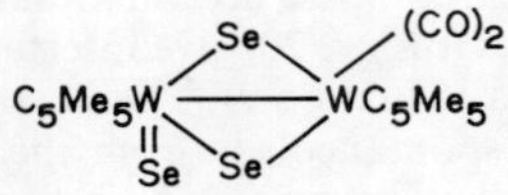

(66)

Hofmann and Werner[80] investigated the insertion of chalcogens in a Co—Co single bond, viz. the reaction **67** ⟶ **68**, and found the order of reactivities S > Se > Te.

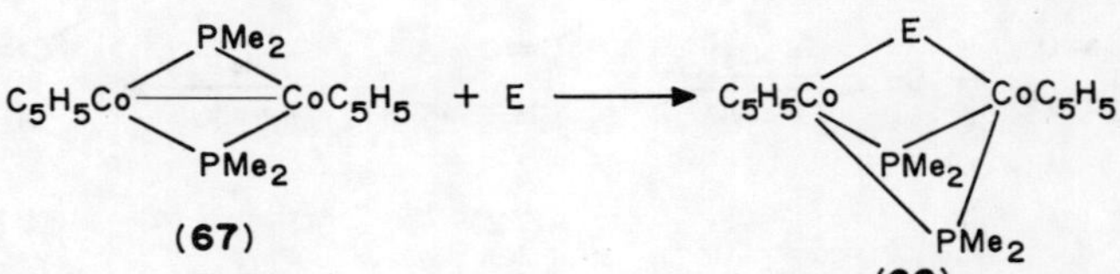

(67) **(68)**

E = S, Se, Te

The facile insertion of selenium in Co—Co and Rh—Rh double bonds has been reported by Herrmann *et al.*[81], who observed the conversion **69** ⟶ **70**.

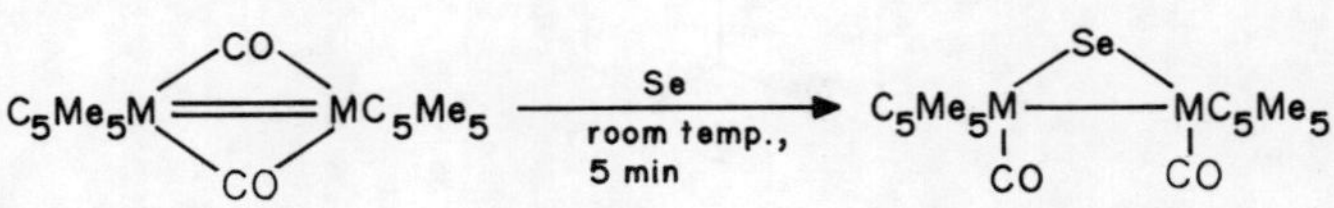

(69) **(70)**

M = Co, Rh

The insertion of both selenium and tellurium in osmium and ruthenium cluster compounds has also been reported[82].

III. HETEROGENEOUS BONDS INVOLVING CARBON

A. Bonds to Hydrogen or Group IA or IIA Elements

1. General remarks

The most common and reliable method for the generation of selenols and tellurols (**71**) involves the insertion of the free elements in the (partially ionic) bonds between carbon and the Group IA and IIA metals. The reactive carbanionic intermediates may be generated by direct hydrogen–metal exchange or via halogenation and halogen–metal exchange (equation 8). Moreover, the air-sensitive selenols and tellurols are usually not isolated but derivatized *in situ*, directly from the metal salts or with preceding protonation. Therefore, it is natural to treat the insertion of selenium and tellurium in the C—H and the C—Group IA and C—Group IIA metal bonds jointly. The insertion of tellurium in C—metal bonds has recently been reviewed[83].

$$\mathrm{CH} \xrightarrow[\text{or (i)}X_2\text{ (ii)M}]{B^-H^+} \mathrm{C^-M^+} \xrightarrow{E} \mathrm{CE^-M^+} \xrightarrow{H^+} \underset{(\mathbf{71})}{\mathrm{CEH}} \qquad (8)$$

An indication of the strength of a carbon base needed for insertion of selenium is found in the observation that metal enolates ($pK_A \approx 20$) give insertion products (**72**) isolated as methyl selenides (**73**) in high yields (equation 9)[84] while the anions of β-dicarbonyl compounds ($pK_A \approx 10$) are formed spontaneously from the corresponding selenolates (Section II.A.2). On the other hand, alkali cyanides ($pK_A \approx 9$) readily insert both selenium and tellurium with formation of the corresponding chalcogenocyanates[1,3]. This shows that not only the strength but also the softness (cf. Pearson[85]) of the anion influence the insertion reaction.

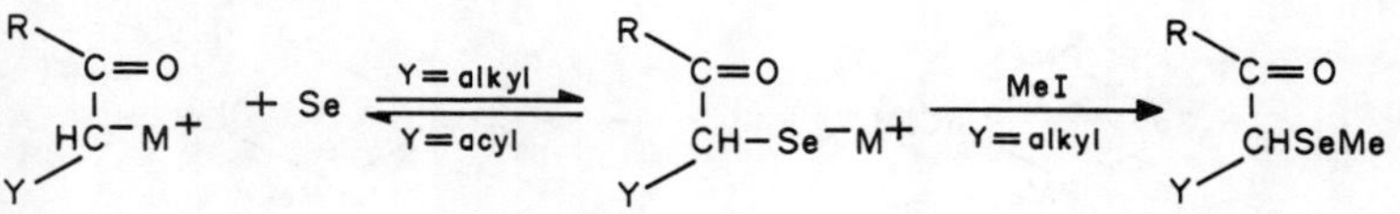

(9)

(**72**) (**73**)

The insertion depicted in equation 9 probably also constitutes the initial reaction in the reported[86] formation of **74** from pentanone, selenium and aziridines.

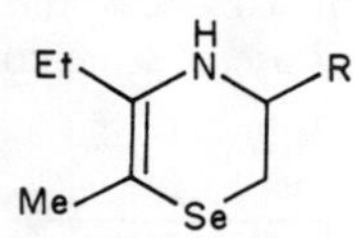

(**74**)

R=H,Me

2. Alkynes

The insertion of selenium and tellurium in metal—alkyne bonds (equation 10) is well documented[1,3].

$$RC{\equiv}C^{-}M^{+} + E \longrightarrow RC{\equiv}CE^{-}M^{+}$$
$$E = Se, Te;\ M = Li, Na, K, MgX \qquad (10)$$

A number of recent examples on the use of these reactions are collected in Table 1.

Trofimov and coworkers[100,101] investigated the reaction of selenium and tellurium with acetylene in the presence of potassium hydroxide in dipolar aprotic solvents (DMSO and HMPA) at about 100 °C. These conditions do not lead to insertion. Instead, a formal addition of H_2E giving divinyl chalcogenides (**77**) predominates. The mechanism involved is still obscure. Pre-formed selenide ion does not give the addition product[100]. A mechanism involving nucleophilic addition of a $ROSe^-$ species was proposed[101] but seems improbable in view of the rapid disproportionation of Se^{II} to give selenite and polyselenide ion[102] in the media employed. The latter ions might, however, add to the acetylene and a rapid Se–Se exchange (Section II.D.2) will then produce the assumed ethynylselenolate intermediate.

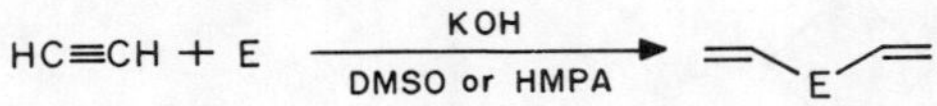

(**77**) E=Se,Te

3. *Arylmetal compounds*

The insertion of selenium[1] and tellurium[3] in the C—metal bonds of arylmetals is by far the most widespread and reliable route to seleno- and telluro-phenols and their derivatives. Recently, diphenyl diselenide and ditelluride have received renewed interest as the stable precursors of several important reagents for organic synthesis, e.g. selenols, tellurols, selenyl halides and seleninic acids. A number of authors have reinvestigated their formation from arylmetals. The results are compared in Table 2. The data show that phenyllithium in THF offers the best choice of conditions and that the advantage is particular striking for the tellurium compound.

Another recent contribution[108] introduces the purification of areneselenolates via *Se*-arylthioselenocarbamates (**78**). The aryllithium path has also been used for the introduction of selenium[109,110] and tellurium[111] into five-membered heterocycles. In thiophenes the α-lithium reagent can be prepared by H–Li exchange while the β-reagent requires halogen–Li exchange[109,112]. Two chalcogen atoms have been introduced by a two-step insertion procedure as in **79**[113,114], **80**[112] and **81**[112,115,116], while tellurium atoms were introduced in **82** in a single step[117]. The ferrocene derivative **83** was obtained by selenium insertion in 1, 1′-ferrocenyldilithium[118].

$ArSeO(S)NMe_2$
(**78**)

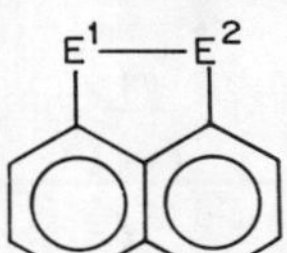

(**79**)

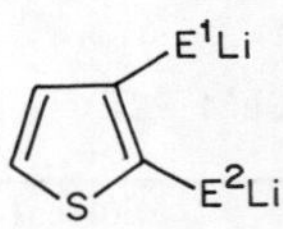

(**80**)

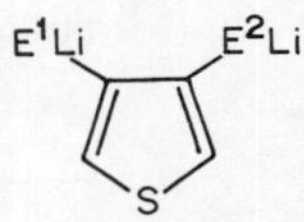

(**81**)

E^1,E^2 = S and/or Se and/or Te

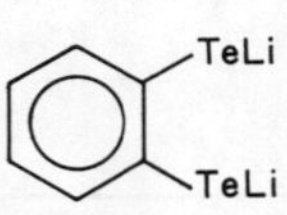

(**82**)

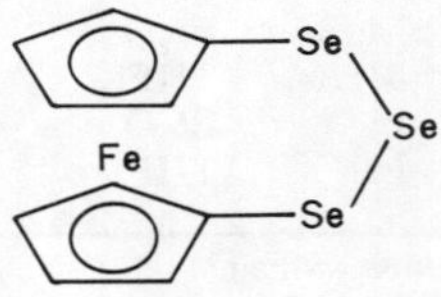

(**83**)

TABLE 1. Products from 1-alkynylchalcogenates ($RC{\equiv}CE^-M^+$) formed according to equation 10

Reagent	Product				Ref.
		R	E	Y	
CY_2	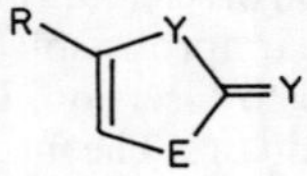	Ph	Se	Se	87
		H	Se	Se	88
		H	Se	S/Se	89
		Ph	Se	Se	90
H^+	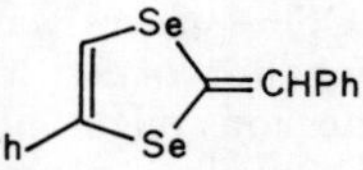 (**75**)				91
	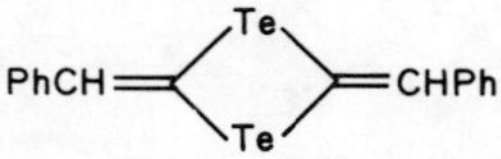 (**76**)[a]				92, 93
(i) ICH_2Cl (ii) Te^{2-}	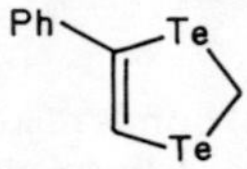				94
(i) BuLi (ii) BuLi/*t*-BuOK (iii) Se	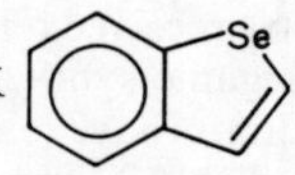				95
R^2X	$RC{\equiv}CER^2$	R	E	R^2	
		t-Bu	Se	$CH_2CH{=}CH_2$	96, 97
		Me_3Si	Se	$CH_2CH{=}CH_2$	96
		Me, *t*-Bu	Se	$SiMe_3$	97
		Me_3Si	Se	$CH_2C{\equiv}CH$	97
		H	Se	Me	98
		H	Te	Me	98

[a] Compound **76** was originally assumed to have the ditellurolan structure analogous to **75**[99]. Structure **76** has subsequently been determined by X-ray diffraction[93].

TABLE 2. Preparation of diphenyl diselenide and diphenyl ditelluride

$$2PhM + 2E \longrightarrow 2PhE^-M^+ \xrightarrow{[O]} PhEEPh$$

E	M	Solvent	Yield (%)	Ref.
Se	MgBr	Et_2O	55–71[a]	103
Se	MgBr	Et_2O	64–70[b]	104
Se	Li	THF	80	105
Te	MgBr	Et_2O	48	106
Te	MgBr	THF	55–85	107
Te	Li	Et_2O	51	106
Te	Li	THF	100	105

[a]Oxidation with air.
[b]Oxidation with bromine.

4. Alkyl- and alkenyl-metals

The reaction of benzylmagnesium chloride with selenium is reported to give up to 71% of products derived from benzylselenomagnesium chloride[119] and vinylmagnesium bromide reacts with tellurium to give vinyltelluromagnesium bromide, from which 72% of Te-vinyl thiotellurocarbamate (**84**) was isolated[120].

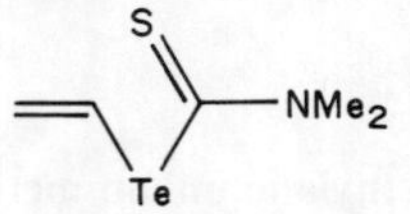

(**84**)

The reaction of methyl- and butyl-lithium in diethyl ether is reported to give the selenolates pure and in almost quantitative yields (room temperature, 15 min)[121]. The reaction of alkyllithiums with tellurium proceeds readily in THF. Yields of 89 and 84% of dibutyl and di-*tert*-butyl ditelluride, respectively, are reported by this route[122]. Sodium and lithium carboranes insert selenium in the C—metal bond to give after oxidation the dicarboranyl diselenides[123].

5. Insertions not involving carbanions

The reaction of benzylic halides with sodium polyselenide in DMSO may lead to selenium insertion in the benzylic C—H bonds with formation of a diselenobenzoate (**85**)[102]. The mechanism probably involves a series of 1,2-eliminations in benzylpolyselenide anions (equation 11) since it is only observed with acidic α-protons and it requires the presence of excess of base.

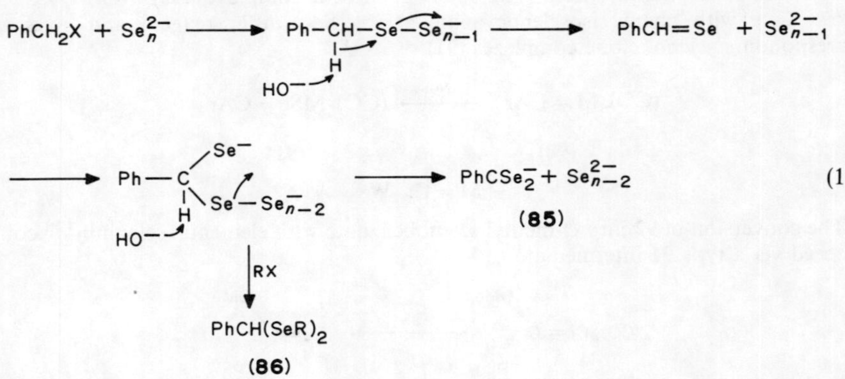

(11)

The first elimination step seems documented since the reaction can be arrested and diselenoacetals (**86**) trapped. Whether the last step forming **85** is another elimination as depicted or alternatively a hydride transfer to selenium is unknown.

The insertion of selenium in the C—H bond of dimethylformamide in the presence of sodium hydroxide has been described[124] (equation 12). The authors discount a carbanionic intermediate since no products from the condensation of such species were found. On the other hand, an addition-elimination sequence such as equation 11 is highly improbable. Polyselenide ion is formed under the reaction conditions[102] but the formyl hydrogen atom has hydride rather than acidic character. The observation of sodium selenide which is not formed by selenium disproportionation makes it plausible that polyselenide addition in this case is followed by hydride transfer to selenium.

$$Me_2NCHO + Se \xrightarrow{HO^-} Me_2NC(O)Se^- + Se^{2-} \quad (12)$$

B. Covalent Carbon—Metal Bonds (Group III–VIII Metals)

1. Aluminium

An efficient preparation of dimethylaluminium methyl selenide (**87**) has appeared[125].

$$Me_3Al + Se \xrightarrow[\text{reflux, 3 h}]{PhMe} Me_2AlSeMe$$

(87)

2. Group IV metals

The reported[126] conversion **88** ⟶ **89** may involve preliminary Se insertion in the C—Sn bonds.

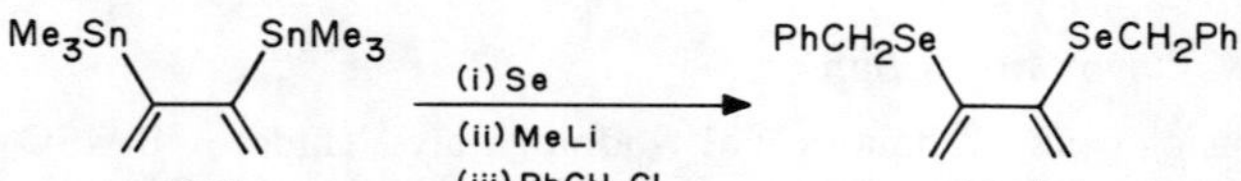

(88) **(89)**

3. Group VI metals

Insertion of selenium in the C—M bonds of carbene complexes (**90**) with M = Cr, W takes place with phenyl isoselenocyanate as the Se-transfer reagent and gives the corresponding selenoketone complexes (**91**)[127].

$$(CO)_5M = CAr_2 \xrightarrow{PhNCSe} (CO)_5MSe = CAr_2$$

(90) **(91)**

M = Cr, W

The conversion of **92** into *O*-methyl selenobenzoate with elemental selenium[128] could proceed via a type **91** intermediate.

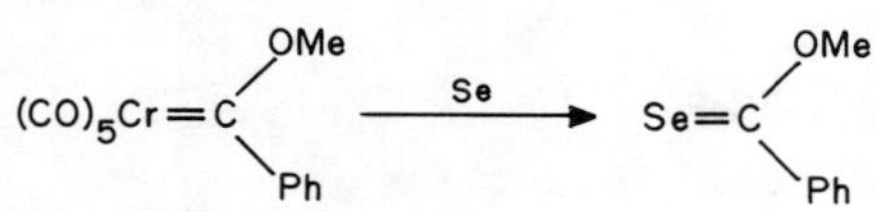

(92) **(93)**

Metal carbyne complexes (**94**) readily insert two selenium atoms in the C—M bond with formation of diselenocarboxylato complexes (**95**)[129].

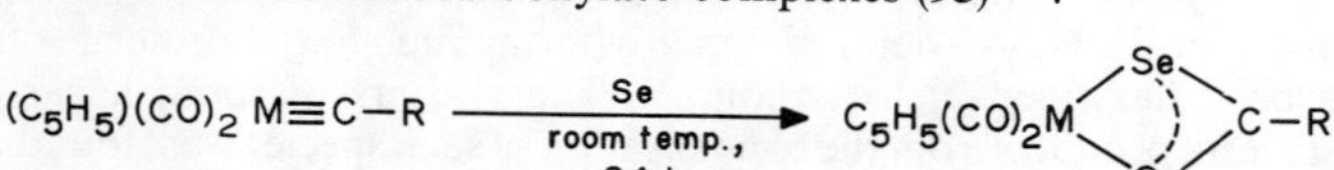

(94) **(95)**

M=Mo; R=CH₂Bu-*t*

M= W ; R=C₆H₄Me-*p*

4. Group VIII metals

The cobalt complex **96** inserts selenium directly in the C—Co bond with formation of **97**, while the trinuclear bis(carbyne) cobalt clusters **98** react with selenium insertion in C—Co bonds and simultaneous carbon–carbon coupling to give **99**[130].

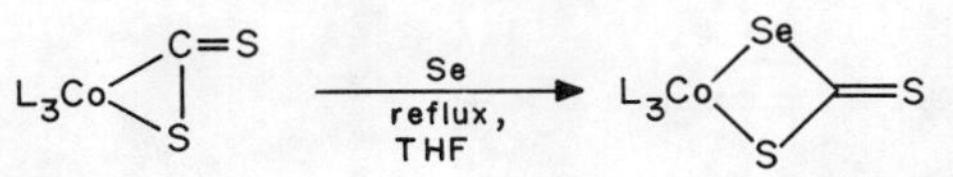

(96) **(97)** L_3=MeC(CH_2PPh_3)$_3$

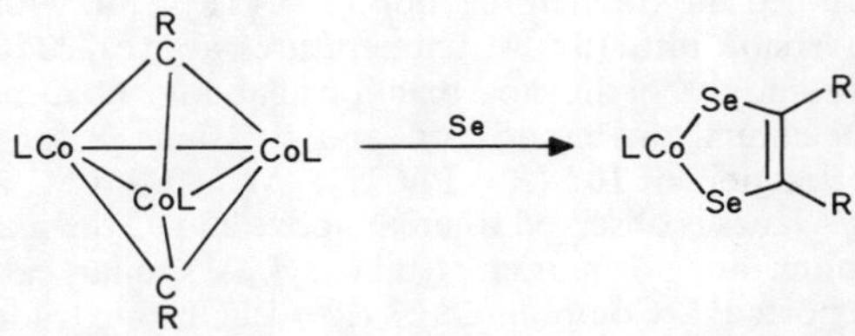

L=C_5H_5

(98) **(99)**

Reaction of the rhodium complex **100** with sulphur, selenium and tellurium results in ultimate replacement of the metal with a chalcogen atom to give **101**. The order of reactivities is S > Se > Te[131]. A reaction path involving initial insertion in the C—Rh bond could be envisaged.

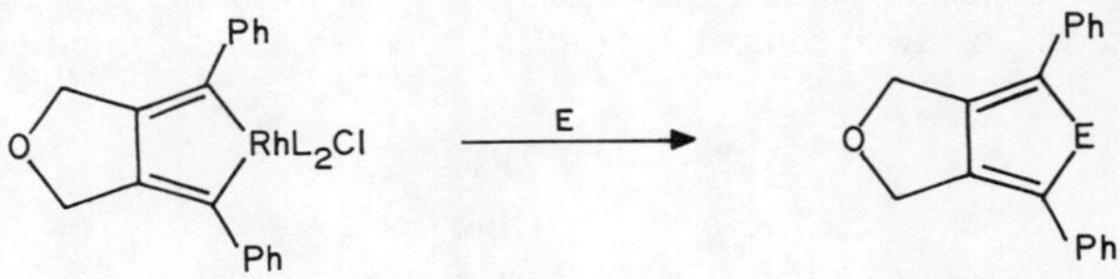

(100) **(101)**

C. Bonds to Group VB–VIIB Elements

1. Nitrogen

The occurrence of the C—E—N moiety is mainly restricted to heterocycles of the selenazole type. Such compounds readily extrude selenium in thermal and photochemical processes. The chemistry of the 1,2,3- and 1,2,4-selenadiazole and the 1,2,3,4-selenatriazole rings has recently been reviewed[132]. The selenium extrusion is part of a fragmentation pattern initiated by the loss of a Y≡N unit (equation 13) and will not be treated in further detail here.

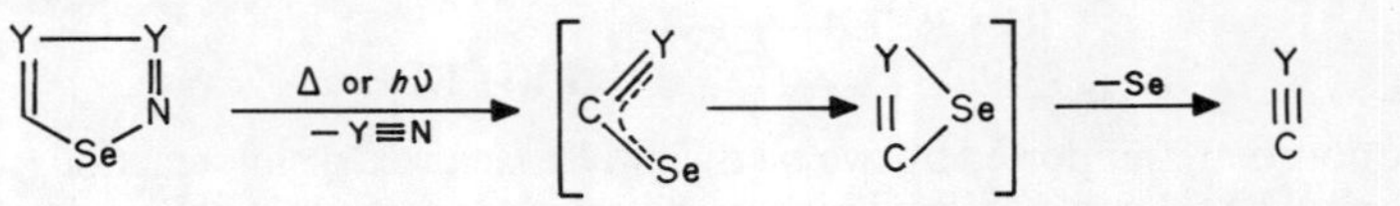

(13)

(Y,Y=RC,N or N,N)

2. Group VIB elements

The thermally or photochemically induced extrusion of selenium and tellurium from diorgano diselenides (**102**) and ditellurides (**103**) to give, respectively, diorgano-selenides (**104**) and -tellurides (**105**) and the converse insertions (equation 14) have received much attention.

$$\text{REER} \underset{}{\overset{\Delta \text{ or } h\nu}{\rightleftharpoons}} \text{RER} + \text{E} \qquad (14)$$

$$(\mathbf{102})\,\text{E}=\text{Se} \qquad (\mathbf{104})\,\text{E}=\text{Se}$$

$$(\mathbf{103})\,\text{E}=\text{Te} \qquad (\mathbf{105})\,\text{E}=\text{Te}$$

Krafft and Lyons[133] studied the thermal reaction of **104** (R = Ph) with selenium and obtained about 33% conversion into **102**. At temperatures above 220 °C the converse reaction took place. A later study[134] of the same reaction at about 290 °C indicated that an equilibrium mixture of selenium and mono-, di- and tri-selane is formed. Lardon[12] investigated the thermal reaction of **102** (R = $PhCH_2$). At 150–170 °C an equilibrium between **102**, **104** and polyselanes is observed whereas above 200 °C the reaction proceeds with further loss of selenium and formation of bibenzyl. A similar behaviour of **102** (R = Ph_2CH) has been reported[13]. Compound **103** (R = $PhCH_2$) extrudes tellurium to give **105** at a lower temperature (ca. 120 °C) than its selenium analogue[18]. Flash vacuum thermolysis of **102** (R = Me) at 550 °C gives rise to a mixture of the corresponding **104**, methane and methaneselenol[135]. On UV irradiation in the absence of oxygen **102** and **103** (R = $PhCH_2$) cleanly extrude one chalcogen atom with formation of **104**[136,137] and **105**[18,138], respectively. It has been noted that **103** reacts faster than **102** and that the rate for **103** (R = $PhCH_2$) is about 10 times that of **103** (R = Et) in the photochemical process.

The thermal and photochemical chalcogen extrusion from **102** and **103** is generally discussed in terms of two competing radical type scissions (equations 15 and 16, respectively).

$$\text{REER} \begin{cases} \longrightarrow 2\text{RE}^{\bullet} \; (\mathbf{106}) & (15) \\ \longrightarrow \text{REE}^{\bullet} + \text{R}^{\bullet} & (16) \end{cases}$$

Equation 15 has the lower activation energy but produces only the low-energy chalcogen-based radicals (**106**). The fate of these apart from recombination can only be the attack on a chalcogen atom of the starting material giving a trivial reaction when the two R groups are identical but gives scrambling of different R groups (equation 17). Equation 16 with the higher activation energy also produces the more reactive carbon-based radicals which give rise to extrusion products **104**, **105** and hydrocarbons (equation 18).

$$\text{R}^1\text{E}^{\bullet} + \text{R}^1\text{EER}^2 \longrightarrow \begin{cases} \text{R}^1\text{EER}^1 + \text{R}^2\text{E}^{\bullet} \\ \text{R}^1\text{EER}^2 + \text{R}^1\text{E}^{\bullet} \end{cases} \qquad (17)$$

$$\text{R}^{\bullet} + \text{REER} \longrightarrow \begin{cases} \text{RER} + \text{RE}^{\bullet} \\ \text{RR} + \text{REE}^{\bullet} \end{cases} \qquad (18)$$

These concepts are supported by two facts: firstly, a clean extrusion reaction from **102** is observed only when $R^{\bullet}$ is a stabilized carbon radical; and secondly, a pronounced increase in the rate of the conversion **102** to **104** is observed on introduction of a phosphorus(III) species. The use of a tris(dialkylamino)phosphite in the thermal extrusion of selenium from

102 (R = Me) has been reported[139]. The quantum yield in the photochemical conversion of **102** ($R = PhCH_2$) rises from 0.16 in the absence[137] to 4.7 in the presence of triphenylphosphine[140]. The rates of the photochemical selenium extrusion from **102** ($R = PhCH_2$, Et, Me, Ph) is raised by several orders of magnitude on addition of methyldiphenylphosphine. The relative half-lives of the four compounds are given as 1:6:24:240, indicating that C—Se bond breaking is still involved in or before the rate-determining step[141,142]. Similar but less pronounced trends are reported for **103** ($R = PhCH_2$, Et)[138]. In accord with these observations, the role of the phosphine is assumed to be the trapping of the organoselenyl radical **106** as **107** in competition with its recombination. Further, the formation of a phosphine chalcogenide facilitates the C—E bond breaking and a radical chain is set up through equations 18 and 19.

$$\mathbf{106} + PR_3 \longrightarrow \underset{\mathbf{(107)}}{R^1EPR_3} \longrightarrow R^{\bullet} + EPR_3 \qquad (19)$$

The acid-catalysed insertion of selenium in a C—Se bond in bis(selenocarbamoyl)selenides (**60**) has been discussed in another connection (Section II.D). A mechanistically related reaction with ultimate selenium extrusion takes place between dialkyl(aryl) diselenides and **60** (equation 20)[102,143].

$$\underset{\mathbf{(60)}}{(R^1_2NCSe)_2Se} + 2\,\mathbf{102} \rightleftharpoons 2R^1_2NCSeSeR + Se \qquad (20)$$

The reaction is initiated by electrophilic activation of **60**. It is reversible and various dialkyl and alkylacyl polyselanes are observed in the reaction mixture[102]. The removal of selenium by coupling to the conversion **60** to **61** displaces the equilibrium to the right and makes the reaction a useful synthetic route to diselenocarbamates **108** inaccessible by nucleophilic substitution reactions (equation 21)[143]. The equilibrium **20** has also been observed with diselenides **109**[102] and **110**[102] and appears to be general for acyl-type selenides and diselenides.

$$2\underset{\mathbf{(60)}}{(R^1_2NCSe)_2Se} + 3\,\underset{\mathbf{(102)}}{R^2SeSeR^2} \longrightarrow 4\,\underset{\mathbf{(108)}}{R^1_2NCSe\text{-}SeR^2} + \underset{\mathbf{(61)}}{(R^1_2NCSe_2)_2Se} \qquad (21)$$

$$\underset{\mathbf{(109)}}{(EtOCSe)_2} \qquad \underset{\mathbf{(110)}}{(PhCOSe)_2}$$

The reactions shown in equation 22 have been reported. These reactions may involve selenium insertion in a C—S bond in the thiazet **111** and its subsequent extrusion during different reactions[144].

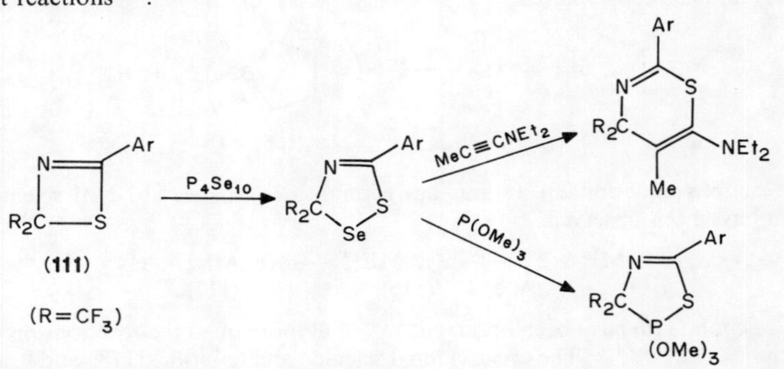

(22)

(111)

(R=CF$_3$)

In the presence of bromine selenium is inserted in one double bond of carbon diselenide with simultaneous bromination of the other selenium atom with formation of 1,2,3-tri-selenetan-4-(Se,Se-dibromoselone) (**112**), the structure of which was determined by X-ray crystallography[145]. The reaction is specifically catalysed by bromide ion and involves (1) the formation of selenium bromide, (2) the reversible generation of bromodiselenide (**113**) and tribromide ions and (3) attack by **113** on carbon diselenide followed by cyclization and bromine addition[102,145].

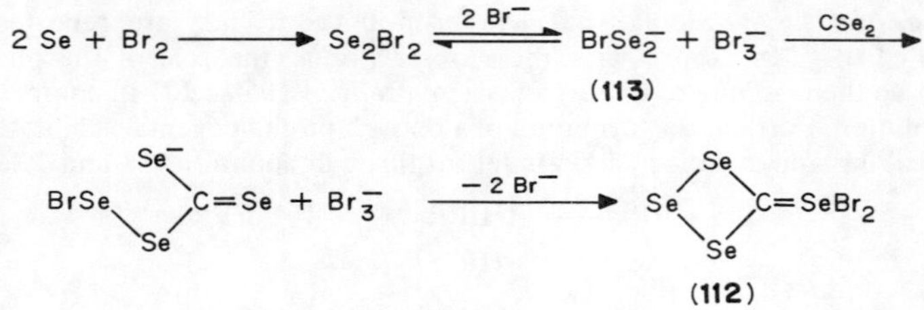

3. *Group VIIB elements*

The insertion of tellurium in C—I bonds is reported as a synthetic route to diorganotellurium diiodides (**114**)[3,83].

$$2RI + Te \longrightarrow \underset{(\mathbf{114})}{R_2TeI_2}$$

IV. HETEROGENEOUS BONDS INVOLVING ELEMENTS OTHER THAN CARBON

A. Bonds to Hydrogen and to Alkali Metals

1. *Elements outside Group VI*

By analogy with the finding for carbon (Section III.A), the base-induced insertion of selenium and tellurium in A—H bonds appears to be a general reaction provided that A^- is sufficiently nucleophilic. However, insertions under neutral conditions are also recorded in particular for third-row and higher row elements.

The insertion of selenium into a boron—hydrogen bond with formation of diboranyldiselenide (**115**) has been reported[146].

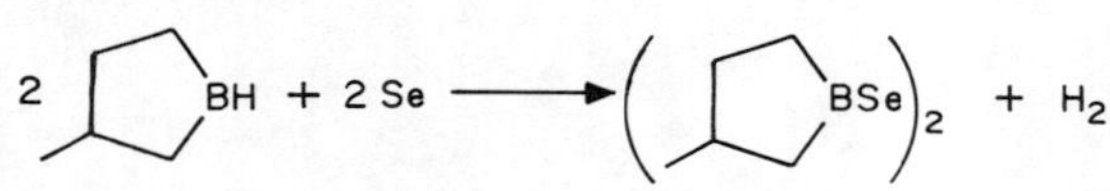

(115)

Similar reactions (equation 23) of stannanes, germanes and silanes with both selenium and tellurium have been observed.

$$2\,R_3MH + 2\,E \longrightarrow \underset{(\mathbf{116})}{2\,R_3MEH} \longrightarrow \underset{(\mathbf{117})}{(R_3M)_2E} + H_2 \qquad (23)$$

Some selenols (**116**) have been observed[44,147], but more often the reactions proceed to give selenides **117**[36,44,147]. The cyclic stannyl selenide and tellurides (**118a** and **b**) and **119** were obtained from the elements and dimethylstannane[148,149].

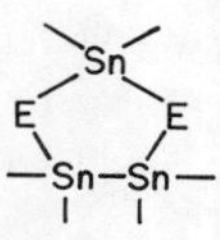

(118)

(a) E=Se
(b) E=Te

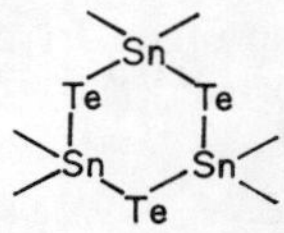

(119)

Selenium dissolves in boiling piperidine and morpholine in the presence of certain lead salts, notably Pb_3O_4. Diaminopolyselanes (**120**) have been isolated from the solutions[150,151]. The reaction could involve insertion in the N—H bond followed by oxidation.

$$R_2N\text{-}Se_n\text{-}NR_2$$

(**120**)

$$R_2 = (CH_3)_5 \ (n = 4)$$
$$R_2 = (CH_2)_2O(CH_2)_2 \ (n = 2, 3, 4)$$

2. Group VI elements

Selenium readily inserts in the E—metal bonds of thiolate and selenolate species. The formation of sodium selenosulphate from selenium and sodium sulphite and the dissolution of selenium in alkali metal selenides with formation of alkali metal polyselenides are well known[1] examples on this type of reaction.

In spite of the vast number of papers claiming the diselenide ion (**121**) as an intermediate in the preparation of dialkyl diselenides (**122**) (equation 24), the occurrence of this ion is dubious at least in some solvents. Sharp and Koehler[152] studied the reaction of selenium and sodium selenide in liquid ammonia and observed by UV spectroscopy the tri-, tetra- and hexa-selenide ions. Their results further indicate that the equilibrium constant for disproportionation of **121** into mono- and tri-selenide ion is of the order of 10^7. A reinvestigation[62] of the formation of **122** from selenium, hydrazine, base and alkyl halides in DMF[153] (equation 25) showed that the reduction of selenium stopped at the triselenide ion stage and further reduction to **122** only took place after the introduction of the alkyl halide.

$$Se^{2-} + Se \longrightarrow \underset{(\mathbf{121})}{Se_2{}^{2-}} \xrightarrow{2\,RX} \underset{(\mathbf{122})}{RSeSeR} \qquad (24)$$

$$4RO^- + 4Se + N_2H_4 + 4RX \longrightarrow 2\,\mathbf{122} + N_2 + 4ROH + 4X^- \qquad (25)$$

Preliminary low-temperature ^{77}Se NMR results in DMF[62] indicated the presence in this solvent of the tetra-, penta- and hexa-selenide ion whereas no sign of **121** was observed.

This type of insertion is particularly rapid in dipolar aprotic solvents. A solution of sodium phenylmethaneselenolate in DMF was found to dissolve two atoms of selenium[62].

$$PhCH_2SeNa \xrightarrow[DMF]{Se} PhCH_2Se_3Na$$

The insertion reactions are reversible and selenium is rapidly extruded on protonation of the salts.

Recently, the reaction of 1, 1-ferrocene-dithiolate and -diselenolate with selenium to give **123** has been reported[118,154].

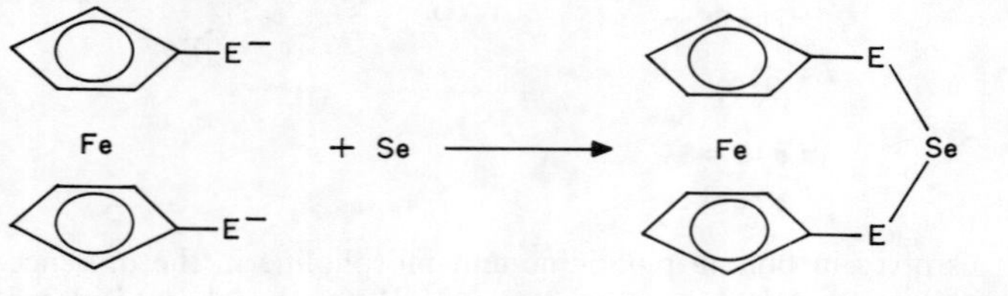

(**123**) E=S,Se

Tellurium shows less tendency than selenium to undergo these reactions. However, the strongly nucleophilic sodium telluride reacts with tellurium to give sodium di- and tetratelluride[3,83].

$$Na_2Te \xrightarrow{Te} Na_2Te_2 \xrightarrow{Te} Na_2Te_4$$

B. Miscellaneous

The insertions of selenium in boron—mercury[155,156] and boron—iodine bonds[146] have been reported. The latter report also contains an example on selenium extrusion from a diboranyl diselenide (equation 26).

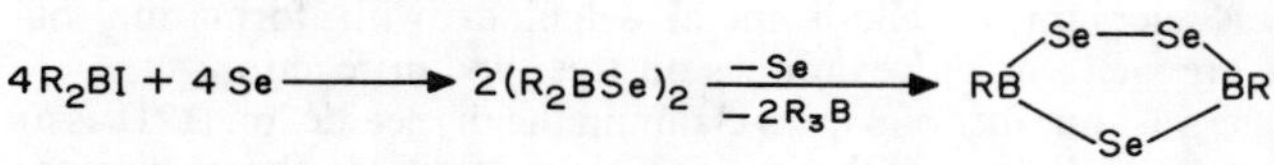

(26)

The insertion of tellurium in a Si—P bond was mentioned earlier (Section II.C).

A ready insertion of selenium in the nitrogen—chlorine bonds of two *N,N*-dichloroamides, i.e. the conversions **124** to **125** and **126** to **127**, have been described[157].

$$\underset{(\mathbf{124})}{C_6F_5SO_2NCl_2} \xrightarrow{Se} \underset{(\mathbf{125})}{C_6F_5SO_2N=SeCl_2}$$

$$\underset{(\mathbf{126})}{C_6F_5CONCl_2} \xrightarrow{Se} \underset{(\mathbf{127})}{C_6F_5CON=SeCl_2}$$

Dimanganyl diselenide (**128**) on treatment with a phosphine extrudes one selenium atom with simultaneous metal—metal coupling to give **129**[158].

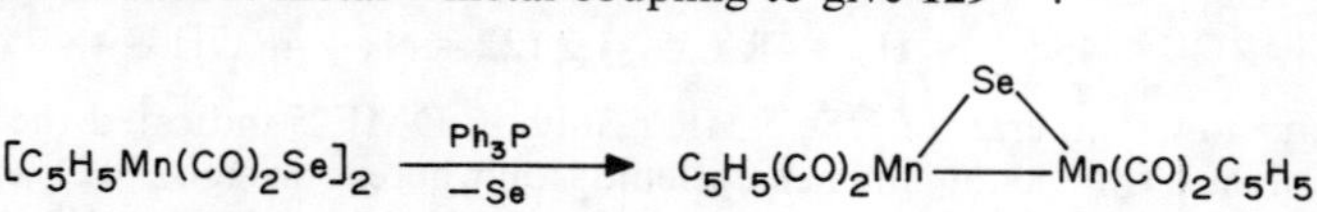

(**128**) (**129**)

V. REFERENCES

1. D. L. Klayman and W. H. H. Günther (Eds.), *Organic Selenium Compounds*, Wiley , New York, 1973.
2. W. C. Cooper (Ed.), *Tellurium*, Van Nostrand Reinhold, New York, 1971.

3. K. J. Irgolic, *The Organic Chemistry of Tellurium*, Gordon and Breach, New York, 1974.
4. K. B. Wiberg, *Physical Organic Chemistry*, Wiley, New York, 1963, p. 138.
5. F. Toda and K. Tanaka, *Chem. Lett.*, 1451 (1979).
6. E. Culbertson, J. H. Gall and D. D. MacNicol, *Tetrahedron Lett.*, **36**, 3207 (1977).
7. E. Culbertson and D. D. MacNicol, *J. Chem. Soc., Chem. Commun.*, 498 (1974).
8. M. F. Lappert, T. R. Martin and G. M. McLaughlin, *J. Chem. Soc., Chem. Commun.*, 635 (1980).
9. A. Schönberg, E. Singer and W. Stephan, *Chem. Ber.*, **116**, 2068 (1983).
10. Ref. 1, p. 31.
11. N. L. M. Dereu and R. A. Zingaro, *J. Organomet. Chem.*, **212**, 141 (1981).
12. M. A. Lardon, *Ann. N. Y. Acad. Sci.*, **192**, 132 (1972).
13. J. Y.C. Chu and J. W. Lewicki, *J. Org. Chem.*, **42**, 2491 (1977).
14. A. Couture, A. Lablanche-Combier, R. Lapouyade and G. Felix, *J. Chem. Res. (S)*, 258 (1979).
15. T. Otsubo, F. Ogara, H. Yamaguchi, H. Higuchi and S. Misumi, *Synth. Commun.*, **10**, 595 (1980).
16. H. Higuchi, T. Otsubo, F. Ogara, H. Yamaguchi, Y. Sakata and S. Misumi, *Bull. Chem. Soc. Jpn.*, **55**, 182 (1982).
17. H. Higuchi and S. Misumi, *Tetrahedron Lett.*, **23**, 5571 (1982).
18. H. K. Spencer and M. P. Cava, *J. Org. Chem.*, **42**, 2937 (1977).
19. H. Higuchi, M. Kugimiya, T. Otsubo, Y. Sakata and S. Misumi, *Tetrahedron Lett.*, **24**, 2593 (1983).
20. H. Higuchi, Y. Sakata, S. Misumi, T. Otsubo, F. Ogura and H. Yamaguchi, *Chem. Lett.*, 627 (1981).
21. E. Cuthbertson and D. D. MacNicol, *Tetrahedron Lett.*, 1893 (1975).
22. D. L. J. Clive, P. C. Anderson, N. Moss and A. Singh, *J. Org. Chem.*, **47**, 1641 (1982).
23. J. Bergman, *Tetrahedron*, **28**, 3323 (1972).
24. J. Bergman, R. Carlson and B. Sjøberg, *Org. Synth.*, **57**, 18 (1977).
25. J. Bergman, in *Proceedings of the Fourth International Conference on the Organic Chemistry of Selenium and Tellurium* (Eds. F. J. Berry and W. R. McWhinnie), University of Aston in Birmingham, Birmingham, 1983, p. 215.
26. S. Uemura, S. I. Fukuzawa and S. R. Patil, *J. Organomet. Chem.*, **243**, 9 (1983).
27. H. Ishihara and Y. Hirabayashi, *Chem. Lett.*, 1007 (1978).
28. T. G. Back, D. H. R. Barton, M. R. Britten-Kelly and F. S. Guziec, Jr., *J. Chem. Soc., Perkin Trans. 1*, 2079 (1976).
29. E. R. Cullen, F. S. Guziec, Jr., and C. T. Murphy, *J. Org. Chem.*, **47**, 3563 (1982).
30. A. Krebs, W. Rüger, B. Ziegenhagen, M. Hebold, I. Hardte, R. Müller, M. Schütz, M. Wietzke and M. Wilke, *Chem. Ber.*, **117**, 277 (1984).
31. F. S. Guziec, Jr., C. J. Murphy and E. R. Cullen, *J. Chem. Soc., Perkin Trans. 1*, 107 (1985).
32. D. L. Clive and C. V. Denyer, *J. Chem. Soc., Chem. Commun.*, 253 (1973).
33. J. M. Behan, R. A. W. Johnstone and M. J. Wright, *J. Chem. Soc., Perkin Trans. 1*, 1216 (1975).
34. A. Ogawa, J. Miyake, S. Murai and N. Sonoda, *Tetrahedron Lett.*, **26**, 669 (1985).
35. M. N. Bochkarev, N. S. Vyazankin, L. N. Bochkarev and G. A. Razuvaev, *J. Organomet. Chem.*, **110**, 149 (1976).
36. N. S. Vyazankin, M. N. Bochkarev and L. P. Sanina, *Zh. Obshch. Khim.*, **38**, 414 (1968).
37. M. Wojnowska, W. Wojnowski and R. West, *J. Organomet. Chem.*, **199**, C1 (1980).
38. E. Hengge and H. G. Schuster, *J. Organomet. Chem.*, **231**, C17 (1982).
39. E. Hengge and H. G. Schuster, *J. Organomet. Chem.*, **240**, C65 (1982).
40. H. Weidenbruch and A. Schafer, *J. Organomet. Chem.*, **269**, 231 (1984).
41. M. N. Bochkarev, L. P. Mairova, N. S. Vyazankin and G. A. Razuvaev, *J. Organomet. Chem.*, **82**, 65 (1974).
42. M. N. Bochkarev, G. A. Razuvayev and N. S. Vyazankin, *Izv. Akad. Nauk SSSR. Ser. Khim.*, 1820 (1975).
43. M. Wojnowska, M. Noltemeyer, H. -J. Füllgrabe and A. Meller, *J. Organomet. Chem.*, **228**, 229 (1982).
44. N. S. Vyazankin, M. N. Bochkarev and L. P. Sanina, *Zh. Obshch. Khim.*, **36**, 1961 (1966).
45. H. Puff, R. Gattermayer, R. Hundt and R. Zimmer, *Angew. Chem.*, **89**, 556 (1977).
46. C. L. Pedersen and N. Hacker, *Tetrahedron Lett.* 3981 (1977).
47. C. L. Pedersen, *J. Chem. Soc., Chem. Commun.*, 704 (1974).
48. V. Bertini, A. DeMunno, A. Menconi and A. Fissi, *J. Org. Chem.*, **39**, 2294 (1974).
49. W. -W. duMont, R. Hensel and T. Severengis, *Phosphorus Sulfur*, **18**, 73 (1983).

50. W. -W. duMont, T. Severengis and B. Meyer, *Angew. Chem.*, **95**, 1025 (1983); *Angew. Chem., Int. Ed. Engl.*, **22**, 983 (1983).
51. T. Austad, T. Rød, K. Åsa, J. Songstad and A. H. Norburg, *Acta Chem. Scand.*, **27**, 1939 (1973).
52. W. -W. duMont and H. -J. Kroth, *J. Organomet. Chem.*, **113**, C35 (1976).
53. M. Yoshifiyi, K. Shibayama and N. Inamoto, *Chem. Lett.*, 603 (1984).
54. J. Ellermann and L. Brehm, *Chem. Ber.*, **117**, 2675 (1984).
55. D. Herrmann, *Z. Anorg. Allg. Chem.*, **416**, 50 (1975).
56. G. Thiele, H. W. Rotter, M. Lietz and J. Ellermann, *Z. Naturforsch., Teil B*, **39**, 1344 (1984).
57. H. J. Breunig and H. Jawad, *J. Organomet. Chem.*, **277**, 257 (1984).
58. R. Steudel and E. -M. Strauss, *Z. Naturforsch., Teil B*, **36**, 1085 (1981).
59. R. Steudel and E. -M. Strauss, *Angew. Chem.*, **96**, 356 (1984); *Angew. Chem., Int. Ed. Engl.*, **23**, 362 (1984).
60. S. A. Glover, *J. Chem. Soc., Perkin Trans. 1*, 1338 (1980), and references cited therein.
61. B. Nakhdjavan and G. Klar, *Justus Liebigs Ann. Chem.*, 1683 (1977).
62. L. Henriksen, O. Nielsen and H. Eggert, in preparation.
63. K. L. Jenkins, *Can. J. Biochem.*, **46**, 1417 (1968).
64. H. E. Ganther, *Biochemistry*, **10**, 4089 (1971).
65. J. L. Kice and H. Slebocka-Tilk, *J. Am. Chem. Soc.*, **104**, 7123 (1984).
66. M. Czauderna and K. Samochoka, *J. Chem. Soc., Perkin Trans. 2*, 765 (1983).
67. L. W. Hillen and R. L. Werner, *J. Chromatogr.*, **79**, 319 (1973).
68. H. Eggert, O. Nielsen and L. Henriksen, *J. Am. Chem. Soc.*, **108**, 1725 (1986).
69. S. Larsen and L. Henriksen, in preparation.
70. D. Barnard and D. T. Woodbridge, *J. Chem. Soc.*, 2922 (1961).
71. L. Henriksen, *Synthesis*, 771 (1982).
72. S. Esperås, S. Husebye and Å. Rolandson, *Acta Chem. Scand., Ser. A*, **29**, 608 (1975).
73. R. O. Gould, *J. Chem. Soc., Dalton Trans.*, 908 (1976).
74. L. Henriksen, *Int. J. Sulfur Chem.*, **8**, 389 (1973).
75. S. Husebye, in Ref. 25, p. 298.
76. O. Foss, *Inorg. Synth.*, **4**, 91 (1953).
77. L. Henriksen, unpublished results.
78. L. Y. Goh, C. Wei and E. Sinn, *J. Chem. Soc., Chem. Commun.*, 462 (1985).
79. H. Brunner, J. Wachter and H. Wintergerst, *J. Organomet. Chem.*, **235**, 77 (1982).
80. W. Hofmann and H. Werner, *Angew. Chem.*, **93**, 1088 (1981); *Angew. Chem., Int. Ed. Engl.*, **20**, 1014 (1981).
81. W. A. Herrmann, C. Bauer and J. Weichmann, *J. Organomet. Chem.*, **243**, C21 (1983); *Chem. Ber.*, **117**, 1271 (1984).
82. B. F. G. Johnson, J. Lewis, P. G. Lodge, P. R. Raithby, *J. Chem. Soc., Chem. Commun.*, 719 (1979).
83. N. Petragnani and J. V. Comasseto in Ref. 25, p. 98.
84. D. Liotta, G. Zima, C. Barnum and M. Saindane, *Tetrahedron Lett.*, **21**, 3643 (1980).
85. R. G. Pearson (Ed.), *Hard and Soft Acids and Bases*, Dowden, Hutchinson and Ross, Strondsburg, PA, 1973.
86. F. Asinger and M. K. Schmitz, *Monatsh. Chem.*, **113**, 1191 (1982).
87. R. Mayer and A. K. Müller, *Z. Chem.*, **4**, 384 (1964).
88. E. M. Engler and V. V. Patel, *J. Am. Chem. Soc.*, **96**, 7376 (1974).
89. E. M. Engler and V. V. Patel, *J. Org. Chem.*, **40**, 387 (1975).
90. H. Poleschner and E. Fanghängel, *J. Prakt. Chem.*, **324**, 691 (1982).
91. R. Myer, B. Hunger, R. Prousa and A. K. Müller, *J. Prakt. Chem.*, **35**, 294 (1967).
92. M. V. Lakshmikanthan, M. P. Cava, M. Albeck, L. Engman and P. Carrol, *Tetrahedron Lett.*, **22**, 4199 (1981).
93. S. L. Bender, N. F. Haley and H. R. Luss, *Tetrahedron Lett.*, **22**, 1495 (1981).
94. S. L. Bender, M. R. Detty and N. F. Haley, *Tetrahedron Lett.*, **23**, 1531 (1982).
95. H. Hommes, H. D. Verkruijsse and L. Brandsma, *J. Chem. Soc., Chem. Commun.*, 366 (1981).
96. E. Schaumann and F. -F. Grabley, *Tetrahedron Lett.*, **21**, 4251 (1980).
97. R. S. Sukhai and L. Brandsma, Recl. Trav. Chim. Pays-Bas, **98**, 55 (1979).
98. S. I. Radchenko, *Zh. Org. Khim.*, **13**, 2229 (1977).
99. M. L. Petrov, V. J. Laishev and A. A. Petrov, *Zh. Org. Khim.*, **15**, 2596 (1979).

100. B. A. Trofimov, S. V. Amosova, N. K. Gusarova and G. K. Musorin, *Tetrahedron*, **38**, 713 (1982), and references cited therein.
101. B. A. Trofimov, S. V. Amosova, N. K. Gusarova, V. A. Potapor and A. A. Tatarinova, *Sulfur Lett.*, **1**, 151 (1983).
102. L. Henriksen, in Ref. 25, p. 242.
103. D. C. Foster, *Org. Synth.*, Coll. Vol. III, 771 (1955).
104. H. J. Reich, M. L. Cohen and P. S. Clark, *Org. Synth.*, **59**, 141 (1980).
105. D. Seebach and A. K. Beck, *Chem. Ber.*, **108**, 314 (1975).
106. N. Petragnani and M. de Moura Compos, *Chem. Ber.*, **96**, 249 (1963).
107. W. S. Haller and K. J. Irgolic, *J. Organomet. Chem.*, **38**, 97 (1972).
108. K. -Y. Jen and M. P. Cava, *J. Org. Chem.*, **48**, 1449 (1983).
109. S. Gronowitz, A. Konor and A. -B. Hörnfeldt, *Chem. Scr.*, **10**, 159 (1976).
110. K. -Y. Jen, N. Benforemo, M. P. Cava, W. -S. Huang and A. G. MacDiarmid, *J. Chem. Soc., Chem. Commun.*, 633 (1983).
111. L. Engman and M. P. Cava, *Organometallics*, **1**, 470 (1982).
112. V. P. Litvinov and I. A. Dzhunaev, *Izv. Akad. Nauk SSSR, Ser. Khim.*, 777 (1982).
113. D. Dauplaise, J. Meinwald, J. C. Scott, H. Temkin and J. Clardy, *Ann. H. Y. Acad. Sci.*, **313**, 382 (1978).
114. J. Meinwald, D. Dauplaise and J. Clardy, *J. Am. Chem. Soc.*, **99**, 7743 (1977).
115. L. -Y. Chiang, P. Shu, D. Holt and D. Cowan, *J. Org. Chem.*, **48**, 4713 (1983).
116. K. Lerstrup, D. O. Cowan and T. J. Kistenmacher, *J. Am. Chem. Soc.*, **106**, 8303 (1984).
117. K. Lerstrup, D. Talham, A. Bloch, T. Poehler and D. Cowan, *J. Chem. Soc., Chem. Commun.*, 336 (1982).
118. R. E. Hollands, A. G. Osborne and I. Townsend, *Inorg. Chim. Acta*, **371**, L541 (1979).
119. V. P. Krasnov, V. I. Naddaka, V. P. Gar'kin and V. I. Minkin, *Zh. Org. Khim.*, **14**, 2620 (1978).
120. F. Wudl and D. Nalewajek, *J. Organomet. Chem.*, **217**, 329 (1981).
121. M. Tiecco, L. Testaferri, M. Tingoli, D. Chianelli and M. Montanucci, *Synth. Commun.*, **13**, 617 (1983).
122. L. Engman and M. P. Cava, *Synth. Commun.*, **12**, 163 (1982).
123. L. I. Zakharin, N. Yu. Krainova, G. G. Zhigarev and I. V. Pisareva, *Izv. Akad. Nauk SSSR, Ser. Khim.*, 1650 (1982).
124. K. Kondo, N. Sonoda and H. Sakurai, *J. Chem. Soc., Chem. Commun.*, 160 (1974).
125. A. P. Kozikowski and A. Ames, *J. Org. Chem.*, **43**, 2735 (1978).
126. H. J. Reich, in Ref. 25, p. 258.
127. H. Fischer and S. Zeuner, *Z. Naturforsch., Teil B*, **38**, 1365 (1983).
128. E. O. Fischer and S. Riedmüller, *Chem. Ber.*, **107**, 915 (1974).
129. D. S. Gill, M. Green, K. Marsden, I. Moore, A. Guy Orpen, F. Gordon, A. Stone and P. Woodward, *J. Chem. Soc., Dalton Trans.*, 1343 (1984).
130. (a) C. Bianchini and A. Meli, *J. Chem. Soc., Chem. Commun.*, 156 (1983); (b) K. Vollhardt, C. Peter and E. C. Walborsky, *J. Am. Chem. Soc.*, **105**, 5507 (1983).
131. A. Scheller, W. Winther and E. Müller, *Justus Liebigs Ann. Chem.*, 1448 (1976).
132. A. R. Katrtzky and C. W. Rees (Eds.), *Comprehensive Heterocyclic Chemistry* (Eds. A. R. Katritzky and C. W. Rees), Vol. 6, Pergamon Press, Oxford, 1984, pp. 349, 354 and 587.
133. F. Krafft and R. E. Lyons, *Chem. Ber.*, **27**, 1761 (1894).
134. A. Silverwood and M. Orchin, *J. Org. Chem.*, **27**, 3401 (1962).
135. T. Hirabayashi, S. Mohmand and H. Bock, *Chem. Ber.*, **115**, 483 (1982).
136. W. Stanley, M. R. van De Mark and P. L. Kumler, *J. Chem. Soc., Chem. Commun.*, 700 (1974).
137. J. Y. C. Chu, D. G. Marsh and W. H. H. Günther, *J. Am. Chem. Soc.*, **97**, 4905 (1975).
138. D. H. Brown, R. J. Cross and D. Millington, *J. Organomet. Chem.*, **125**, 219 (1977).
139. I. A. Nuretdinov, E. V. Bayandina and G. M. Vinokurova, *Zh. Obshch. Khim.*, **44**, 2588 (1974).
140. J. Y. C. Chu and D. G. Marsh, *J. Org. Chem.*, **41**, 3204 (1976).
141. R. J. Cross and D. Millington, *J. Chem. Soc., Chem. Commun.*, 455 (1975).
142. D. H. Brown, R. J. Cross and D. Millington, *J. Chem. Soc., Dalton Trans.*, 159 (1977).
143. I. Johannsen, K. Lerstrup, L. Henriksen and K. Bechgaard, *J. Chem. Soc., Chem. Commun.*, 89 (1984).
144. K. Burger and R. Ottlinger, *Synthesis*, 44 (1978).
145. S. Larsen and L. Henriksen, *Acta Chem. Scand., Ser. A*, **38**, 289 (1984).
146. F. Riegel and W. Siebert, *Z. Naturforsch., Teil B*, **29**, 719 (1974).

147. N. S. Vyazankin and M. N. Bochkarev, *Zh. Obshch. Khim.*, **37**, 1037 (1967).
148. B. Mathiasch, *Z. Anorg. Allg. Chem.*, **432**, 269 (1977).
149. A. Blecher and B. Mathiasch, *Z. Naturforsch., Teil B*, **33**, 246 (1978).
150. O. Foss and V. Janickis, *J. Chem. Soc., Chem. Commun.*, 833 (1977).
151. O. Foss and V. Janickis, *J. Chem. Soc., Dalton Trans.*, 620 (1980).
152. K. W. Sharp and W. H. Koehler, *Inorg. Chem.*, **16**, 2258 (1977).
153. L. Syper and J. Mlochowski, *Synthesis*, 439 (1984).
154. A. G. Osborne, R. E. Hollands, J. A. K. Howard and R. F. Bryan, *J. Organomet. Chem.*, **205**, 395 (1981).
155. V. I. Bregadze, V. Ts. Kampel, A. Ya. Usiatinsky and O. B. Ponomareva, *Izv. Akad. Nauk SSSR, Ser. Khim.*, 1434 (1982).
156. V. I. Bregadze, V. Ts. Kampel, A. Ya Usiatinsky, O. B. Ponomareva and N. N. Godovikov, *J. Organomet. Chem.*, **233C**, 33 (1982).
157. A. V. Zibarev, G. N. Dolenko, S. A. Krupoder, L. N. Malazow, O. K. Poelshcuk, G. G. Furin and G. G. Yakobson, *Zh. Org. Khim.*, **16**, 390 (1980).
158. M. Herberhold, D. Reiner, B. Zimmer-Gasser and U. Schubert, *Z. Naturforsch., Teil B*, **35**, 1281 (1980).

The Chemistry of Organic Selenium and Tellurium Compounds, Volume 2
Edited by S. Patai
© 1987 John Wiley & Sons Ltd.

CHAPTER **11**

Compounds with Se—N and Te—N bonds

G. KIRSCH

Laboratoire de Carbochimie et Synthèse Organique, Université de Metz, Faculté des Sciences, Ile du Saulcy, 57045-Metz Cédex, France

L. CHRISTIAENS

Institut de Chimie Organique, Université de Liège, Sart-Tilman, 4000-Liège, Belgium

---

Part One: Non-Cyclic Compounds with Se—N and Te—N Bonds

I. INTRODUCTION 423
II. NITROGEN DERIVATIVES OF SELENENIC, SELENINIC AND SELENONIC ACIDS 424
A. Selenenamides $RSeNR^1R^2$ 424
1. Preparation 424
2. Chemical properties 425
3. Selenenylation reactions 426
4. Addition to carbonyl compounds 427
5. Chemical properties of *N*-phenylselenophthalimide (N-PSP) and *N*-phenylselenosuccinimide 427
B. Seleninamides 428
C. Selenonamides 428
III. SELENENIMINES AND SELENINIMINES 429
A. Preparation 429
B. Chemical Properties 430
IV. SELENIMIDES AND DERIVATIVES 430
A. Selenimides 430
1. Preparation 430
a. Preparation from selenides (Se^{II}) (method A) 430
b. Preparations from Se^{IV} species 431
i. Method B 431
ii. Method C 431
iii. Method D 431

2. Chemical properties . . . 432
a. Reduction . . . 432
b. Base and acid treatments . . . 433
c. Reaction with halogenating agents . . . 433
d. Action of oxidizing agents . . . 433
e. Thermal behaviour . . . 434
f. Physicochemical properties . . . 435
B. *Se*-Chloroselenimides . . . 435
1. Preparation . . . 435
2. Chemical properties . . . 435
3. *Se*-dichloro-*N*-arylseleninyl *Se*-chloroselenimides $ArSeCl_2N=SeClAr$ . . . 437
C. Selenoximides . . . 437
1. Preparation . . . 437
2. Chemical reactivity . . . 437
V. NITROGEN DERIVATIVES OF SELENIUM DIOXIDE . . . 438
A. *N*-Selenium Diimides RN=Se=NR . . . 438
1. Preparation . . . 438
2. Chemical properties . . . 438
B. *Se*-Dichloroseleniumimide $RN=SeCl_2$ . . . 439
1. Preparation . . . 439
2. Chemical properties . . . 439
C. *Se*-Diamido-*Se*-dichloroselenium $(RCONH)_2SeCl_2$ . . . 440
D. Miscellaneous . . . 440
1. Amidoselenious esters . . . 440
2. Aminoselanes . . . 441
VI. COMPOUNDS WITH Te—N BONDS . . . 441
A. Tellurimides $R_2^1TeNR^2$ . . . 441
1. Preparation . . . 441
2. Chemical properties . . . 443
a. Reduction . . . 443
b. Reaction with electrophilic reagents . . . 443
c. Reaction with active methylene compounds . . . 443
d. Reaction with aldehydes . . . 443
B. *Te*-Chloro-*N*-tosyltellurimide RTe(Ce)=NTos . . . 443
C. Nitrogen Derivatives of Tellurium Dioxide . . . 444

Part Two: Heterocyclic Compounds with Se—N and Te—N Bonds

VII. INTRODUCTION . . . 445
VIII. THREE- AND FOUR-MEMBERED RINGS . . . 445
IX. FIVE-MEMBERED RINGS . . . 446
A. One Se—N Bond . . . 446
1. 1,2-Selenazoles, (*d*)-condensed systems and derivatives . . . 446
a. Synthesis . . . 446
i. 1,2-Selenazoles and (*d*)-condensed systems . . . 446
ii. Isoselenazolium salts and (*d*)-condensed systems . . . 446
iii. Benzisoselenazolinones . . . 446
b. Properties . . . 448
i. 1,2-Selenazoles and (*d*)-condensed systems . . . 448
ii. Benzisoselenazolinones . . . 448
2. (*c*)-Condensed 1,2-selenazoles . . . 448

B. One Se—N Bond with One Other Heteroatom in the Ring 449
1. Thiaselenazoles, diselenazoles and derivatives 449
2. 1,2,3-Selenadiazoles and derivatives[79] 449
a. Synthesis 449
b. Properties 450
3. 1,2,4-Selenadiazoles 451
C. One Se—N Bond with Two Other Heteroatoms in the Ring . . . 451
1. 1,2,3,4-Selenatriazole 451
D. Two Se—N Bonds 452
1. 1,2,5-Selenadiazoles and derivatives 452
a. Synthesis 452
b. Properties 452
E. Two Se—N Bonds with Two Other Heteroatoms in the Ring . . . 453
1. 1,2,3,5-Selenatriazole 453
F. One Te—N Bond 454
1. 1,2-Tellurazole (isotellurazole) and derivatives 454
G. Two Te—N Bonds 454
1. 1,2,5-Telluradiazole and derivatives 454
X. SIX-MEMBERED RINGS 455
A. One Se—N Bond with One Oxygen in the Ring. 455
B. One Se—N Bond with One Sulphur in the Ring 455
C. One Se—N Bond with One Other Nitrogen in the Ring 455
D. Two Se—N Bonds 456
XI. LARGER RINGS. 456
XII. MISCELLANEOUS: AZASELENAPENTALENES 456
XIII. REFERENCES 456

This chapter deals with organic compounds with selenium—nitrogen and tellurium—nitrogen bonds, and is divided into two parts. Part One relates to compounds with non-cyclic Se—N and Te—N bonds and Part Two treats heterocyclic derivatives containing Se—N or Te—N bonds.

The literature for the two parts, gathered together at the end of the chapter in Section XIII, has been reviewed since 1972 up to the end of Volume 102 for *Chemical Abstracts*. Several books and reviews discussing this subject have appeared[1–14], including an overview of the chemistry of selenium and tellurium heterocycles[15]. Some unreported results are also mentioned, together with comparisons with the sulphur-containing analogues.

PART ONE. NON-CYCLIC COMPOUNDS WITH Se—N AND Te—N BONDS

I. INTRODUCTION

Among compounds with Se—N and Te—N bonds, the former have been the most studied. Sometimes in the literature different names have been chosen for designating the same derivatives, especially in the Se—N series. As a rule, there is no ambiguity with regard to selenenamides, seleninamides and selenonamides, but the compounds $RSeN=CR^1R^2$ have been named selenoximes, selenylimines, selenoimines and selenenamides. The same problem arises with $R_2Se=NR^1$ which have been named selenimides, selenilimines and selenium imides. In this chapter we have used the following nomenclature for the main structures:

$RSeNR^1R^2$	Selenenamides
$RSeONR^1R^2$	Seleninamides
$RSeO_2NR^1R^2$	Selenonamides
(Amides from organotellurium acids are not known)	
$RSeN=CR^1R^2$	Selenenimines
$RSe(=O)N=CR^1R^2$	Seleninimines
(Tellureninimes and tellurinimes are not known)	
$R^1R^2Se=NR^3$	Selenimides and derivatives
$R^1R^2Te=NR^3$	Tellurimides
$RN=Se=NR$	Seleniumdiimides
$RN=Te=NR$	Telluriumdiimides

II. NITROGEN DERIVATIVES OF SELENENIC, SELENINIC AND SELENONIC ACIDS

The amidic nitrogen compounds derived from these acids are named selenenamides, seleninamides and selenonamides, respectively.

A. Selenenamides $RSeNR^1R^2$

1. Preparation

Selenenamides (**1**) the amides of selenenic acids, are obtained from the corresponding selenenic acid, RSeOH, or from the acid chloride or bromide. Most of the preparations start from an aryl or alkyl selenenyl chloride or bromide (**2**)[1,4,16,17] (equation 1).

$$\underset{(\mathbf{2})}{RSeX} + R^1R^2R^3N \longrightarrow \underset{(\mathbf{1})}{RSeNR^1R^2} \quad (1)$$

$$X = Cl, Br;\ R = \text{aryl, alkyl};\ R^1, R^2, R^3 = H, CH_3, \text{etc.}$$

Instead of selenenyl chloride, selenocyanate can be used[4]. Phenylselenenic acid has also been used to prepare selenenamides. In this case, the acid is generated *in situ* from a selenoxide elimination reaction[16] (equation 2).

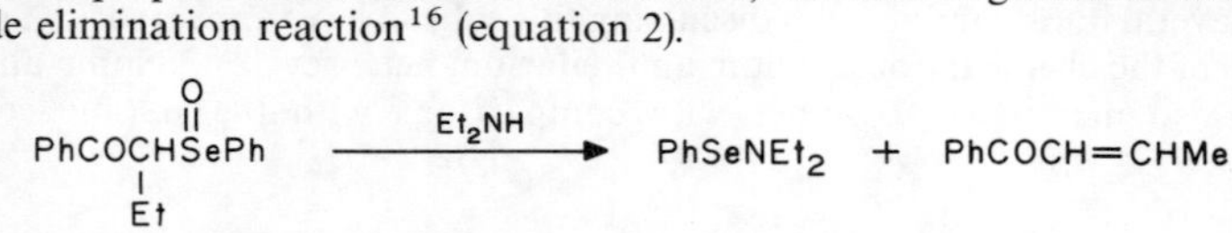

(2)

Table 1 summarizes the selenenamides prepared by these methods.

N-Phenylselenophthalimide (**3**) and *N*-phenylselenosuccinimide (**4**) can be considered as particular selenenamides. They are prepared by the reaction of phenylseleninyl chloride with potassium phthalimide (equation 3) and of phenyl allyl selenide (or diphenyl selenide) with *N*-chlorosuccinimide (equation 4), respectively[17].

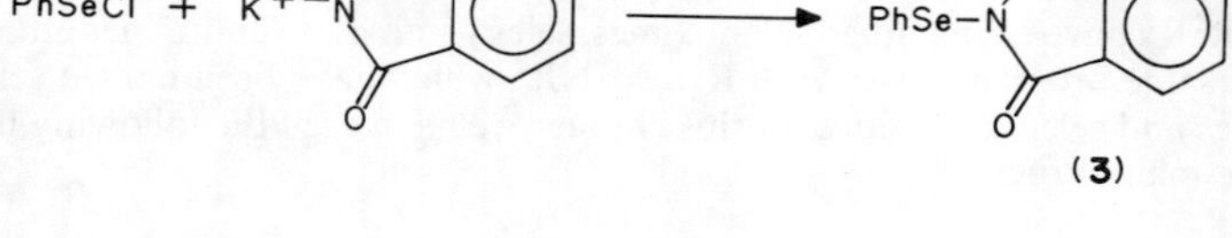

(3)

(**3**)

TABLE 1. Selenenamides prepared by reaction 1:
$RSeX + R^1R^2R^3N \longrightarrow RSeR^1R^2$

R	X	R^1, R^2, R^3	Ref.
DNP[a]	Br	Ph, H, H	1
DNP[a]	BR	H, H, H	1
DNP[a]	OMe	H, H, H	1
Pr	Cl	Me, Me, H	4
Et	Cl	Me, H, H[b]	4
CF_3	Cl	Ph, H, H[c]	4
1-Naph	Cl	2-Naph, H, H	4
Ph	Cl	Me, Me, H	
Ph	Br	Et, Et, H	16a, b
Ph	OH	*i*-Pr, *i*-Pr, H	
DNP[a]	Cl	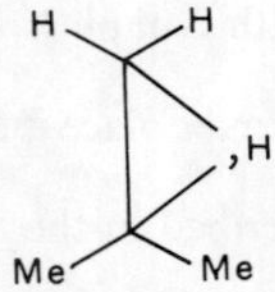	21
Ph	Cl	Et, Et, $SiMe_3$	16c
Ph	Cl	Me, COMe, SiMe	16c

[a]DNP = 2, 4-dinitrophenyl.
[b]Mono- or di-substitution at Se occurs depending on the ratio of the reactants.
[c]The corresponding selenenamide is obtained as an intermediate in the reaction but the rearrangement product is isolated (see Section II.A.2).

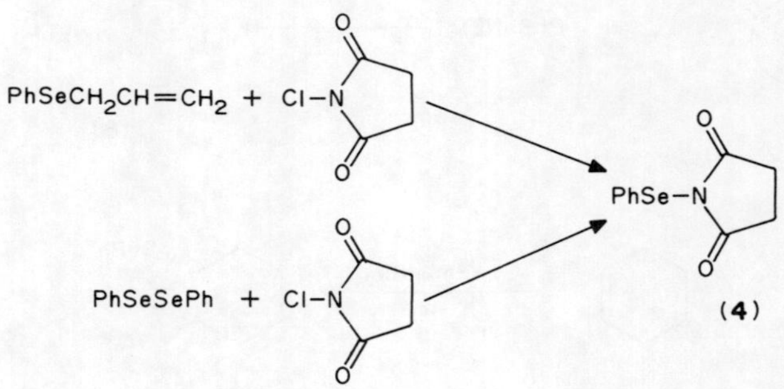

(4)

$$\tag{4}$$

2. *Chemical properties*

Selenenamides present interesting chemical properties themselves and in their synthetic applications. Hydrolysis of selenenamides gave back the selenenic acids[16]. In certain cases selenenamides rearrange in a Hofmann–Martins-type reaction. For example, reaction of 2, 4-dinitrophenylselenenyl bromide (**5**) with 1-naphthylamine does not lead to the expected selenenamide **6** but to 4-(1-aminonaphthyl)-2′, 4′-dinitrophenyl selenide (**7**) (equation 5).

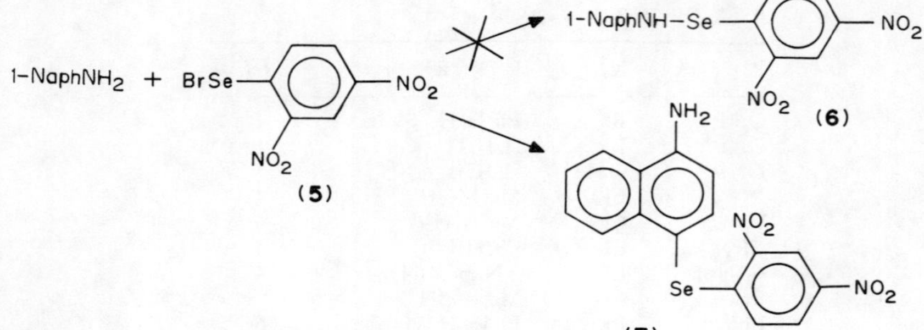

(5) (6) (7) (5)

This reaction takes place also with 2-naphthylamine; then the rearrangement leads to the 1-(2-aminonaphthyl) selenide.

In general, when rearrangement takes place it affects the *para* position if this position is free.

Various other examples are described in the literature[1,18].

3. Selenenylation reactions

N, *N*-Disubstituted phenylselenenenamides are useful selenenylation reagents. They react with β-dicarbonyl compounds to give dicarbonyl selenides (**8**)[16] (equation 6). With alkenes in the presence of acetic anhydride they give benzeneselenenyl acetate (**9**)[16] (equation 7). With some enones they give α-(phenylseleno-β-dialkylamino) ketones (**10**)[16] (equation 8), and with aldehydes they give α-phenylselenoaldehydes (**11**)[19a,b] (equation 9).

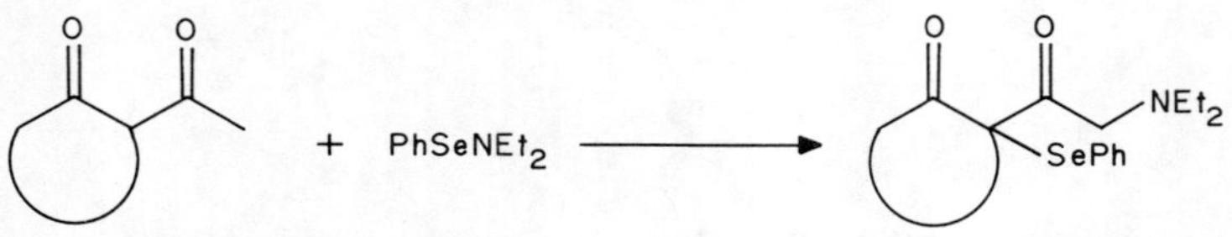

(6)

(**8**)

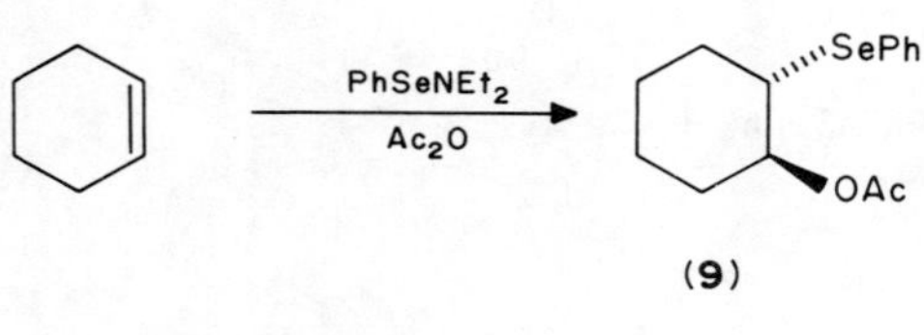

(7)

(**9**)

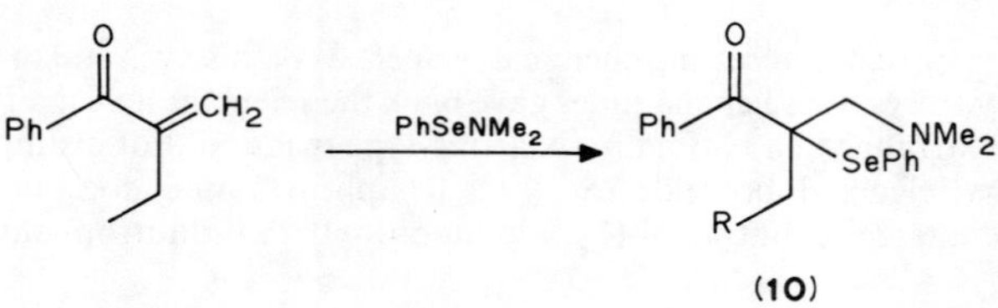

(8)

(**10**)

$$Ph(CH_2)_3CHO \xrightarrow{PhSeNEt_2} \underset{\displaystyle SePh}{Ph(CH_2)_2\underset{|}{C}HCHO} \tag{9}$$

(11)

An extensive study of the addition of *N*, *N*-dimethylbenzeneselenenamide to dimethylacetylene dicarboxylate has also been made[20] (equation 10). The product of the reaction was a mixture of the maleate and fumarate esters **12** and **13**. However, a kinetic study in chloroform showed that the addition was a second-order reaction and that in the early stage only one product, the maleate ester **12**, is formed. The fumarate ester **13** appears later and slowly approaches the equilibrium concentration. A simple thermal rotation around the double bond seems to be the mechanism involved in the equilibration of the two isomers.

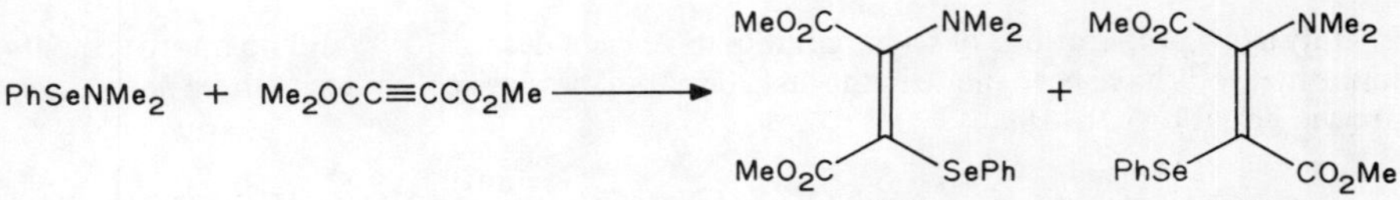

(12) **(13)**

(10)

4. Addition to carbonyl compounds

Like sulphenamides, selenenamides condense with carbonyl compounds to give the selenenimine derivatives[22]. For example, 2,4-dinitrophenylselenenylamide (**14**) condenses with acetone to give the corresponding selenenimine (**15**) (equation 11).

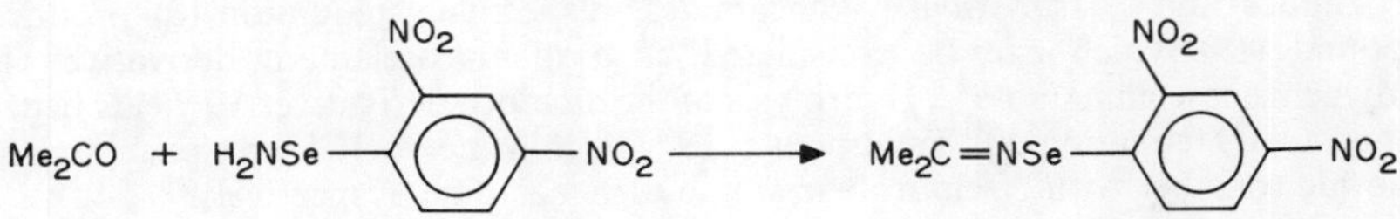

(14) **(15)**

(11)

5. Chemical properties of N-phenylselenophthalimide (N-PSP) and N-phenylselenosuccinimide

These reagents, in presence of tributylphosphine, have been used for the conversion of alcohols into the corresponding phenyl selenides[23] (equation 12). However, when the reaction was applied to an allylic 3-sterol, the 3-phthalimido derivative was formed instead of the expected selenide[24].

$$ROH + N\text{-}PSP \rightarrow RSePh \tag{12}$$

In the presence of water and under catalysis, N-PSP adds to alkenes to form a hydroxyselenide (**16**) (equation 13) or phenylselenolactone (**17**)[25a,b,c] (equation 14).

$$PhCH=CH_2 + N\text{-}PSP \xrightarrow[H^+]{H_2O} \underset{\displaystyle OH}{Ph\underset{|}{C}HCH_2SePh} \tag{13}$$

(16)

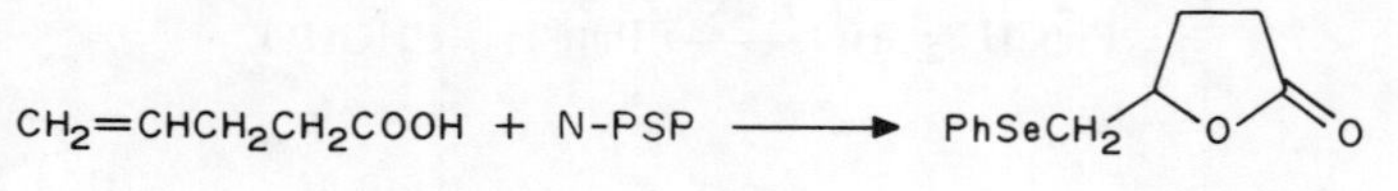 (14)

(17)

ω-Ethylenic amines have been cyclized by use of N-PSP[25d]. Transformation of carboxylic acids into their phenyl selenoesters can also be achieved by use of N-PSP[26]. N-PSP was recently used for the preparation of *S*-acylphenylselenenyl sulphides (**18**) from thiocarboxylic acids[27] (equation 15).

$$RC(=O)SH + N\text{-}PSP \longrightarrow \underset{\mathbf{(18)}}{RC(=O)SSePh} \tag{15}$$

B. Seleninamides

Only a few preparations of seleninamides have been described[1,4], and no improvements in the methods have been made in the last 20 years. The classical routes to these derivatives are via equations 16–18.

$$p\text{-}NO_2C_6H_4SeO_2H \xrightarrow{SOCl_2} p\text{-}NO_2C_6H_4SeOCl \xrightarrow{\text{conc. }NH_3} p\text{-}NO_2C_6H_4SeONH_2 \tag{16}$$

$$MeCONHC_6H_4SeCN \xrightarrow[\text{2. conc. }NH_3]{\text{1. }HNO_3,\ 0\,^{\circ}C} MeCONHC_6H_4SeONH_2 \tag{17}$$

$$(p\text{-}NO_2C_6H_4Se)_2 \xrightarrow[\text{2. conc. }NH_3]{\text{1. }HNO_3,\ -5\,^{\circ}C} p\text{-}NO_2C_6H_4SeONH_2 \tag{18}$$

The reaction of phenylselenenyl chloride or phenylselenium trichloride with *N*-chloroamides and *N*-trimethylsilylbenzamides, respectively (equation 19) produces a compound (**19**) which can be considered as a dichloroselenium derivative of *N*-benzoylbenzeneseleninamide[28]. Hydrolysis of the dichloro derivatives **19** yields benzeneseleninic acid[28] (equation 20). Treatment of **19** with chlorine or HCl gives phenylselenium trichloride together with *N*-chlorobenzamide or benzamide, respectively[28].

$$PhSeCl + PhCONHCl \longrightarrow \underset{\mathbf{(19)}}{PhSeCl_2NHCOPh} \longleftarrow PhSeCl_3 + PhCONHSiMe_3 \tag{19}$$

$$PhSeCl_2NHCOPh + H_2O \rightarrow PhSeO_2H \tag{20}$$

C. Selenonamides

Very few studies of selenonamides have been made. Preparations of the amides by reaction of the corresponding methyl selenoesters with ammonia or amines have been described[29,30] (equation 21).

$$PhSeO_3Me + R^1R^2NH \longrightarrow PhSO_2NR^1R^2 \quad (21)$$

$$R^1 = R^2 = H;\ R^1,\ R^2 = H,\ Me;\ R^1 = R^2 = Me$$

Studies of the chemical properties of selenonamides have not yet been published.

III. SELENENIMINES AND SELENINIMINES

Among the simple structures containing the Se—N moiety, selenenimines and seleninimines have been very little studied. The first examples of these compounds were described in 1976[31].

A. Preparation

The first selenenimines (**20**) and seleninimines (**21**) were prepared by reaction of the corresponding chlorides with the appropriate imine[31] (equation 22).

$$RC_6H_4Se(O)_nCl + NH = C(p\text{-Tol})_2 \longrightarrow RC_6H_4Se(O)_nN = C(p\text{-Tol})_2 \quad (22)$$

$$R = H, NO_2 \qquad n = 0\ (\mathbf{20}) \qquad n = 1\ (\mathbf{21})$$

Compounds **20** were obtained easily but the reaction failed for the preparation of **21** with R = H. Preparation of **21** from **20** by oxidation with *m*-chloroperbenzoic acid also failed. The imines **20** and **21** were prepared for the study of the planar inversion at the nitrogen atom.

At the same time another study was published on the same subject[21], in which the preparation was effected by reaction of ketones with phenylselenenamides[22] (equation 23). Table 2 summarizes the compounds prepared by this method.

$$R^1R^2CO + NH_2SeR^3 \longrightarrow R^1R^2C = NSeR^3 \quad (23)$$

(22)

The same condensation applied to orthoesters yields some particular selenenimines (**23**) (equation 24). The different imines prepared by this method are reported in Table 3.

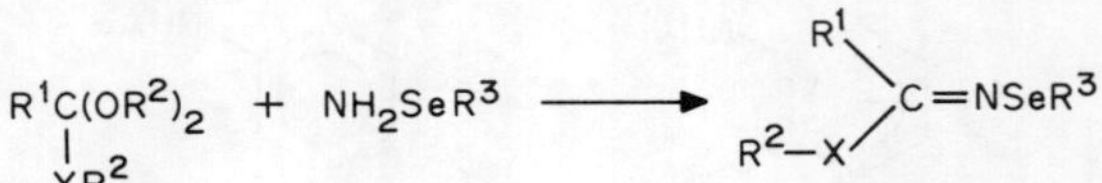

(24)

(23)

Oxidation of phenols by phenylseleninic anhydride (Barton's oxidation) in the presence of hexamethyldisilazane gives another preparation of phenylselenenimine (**24**)[32] (equation 25).

TABLE 2. Selenenimines (**22**) prepared by condensation of ketones with selenenamides (equation 23)

R^1	R^2	R^3
t-Bu	*p*Tol	*o*-$NO_2C_6H_4$
t-Bu	*p*-Tol	2,4-di-$NO_2C_6H_3$
Me	Me	*o*-$NO_2C_6H_4$

TABLE 3. Selenenimines (**23**) prepared by condensation of selenenamides with orthoesters[a] (equation 24)

R^1	R^2	R^3	X
H	Et	o-$NO_2C_6H_4$	O
Ph	Me	o-$NO_2C_6H_4$	O
Ph	Et	o-$NO_2C_6H_4$	O
OEt	Et	o-$NO_2C_6H_4$	O
$oHOC_6H_4$	$PhCH_2$	o-$NO_2C_6H_4$	S
Me	Ph	o-$NO_2C_6H_4$	S
Ph	$PhCH_2$	o-$NO_2C_6H_4$	S
2-Tol	$PhCH_2$	o-$NO_2C_6H_4$	S

[a]In the starting orthoesters R is Me or Et.

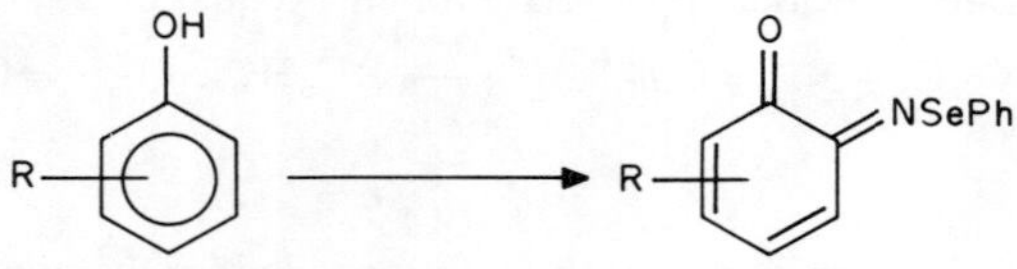

(24)

(25)

B. Chemical Properties

The chemical properties of the imines have been less studied, but some physicochemical studies have been carried out[21,31]. However, some chemical properties of **24** have been described[32].

Treatment of **24** with thiophenol at room temperature yields the corresponding aminophenol (**25**), whereas treatment with zinc in acetic anhydride gives the acetamidoacetate (**26**).

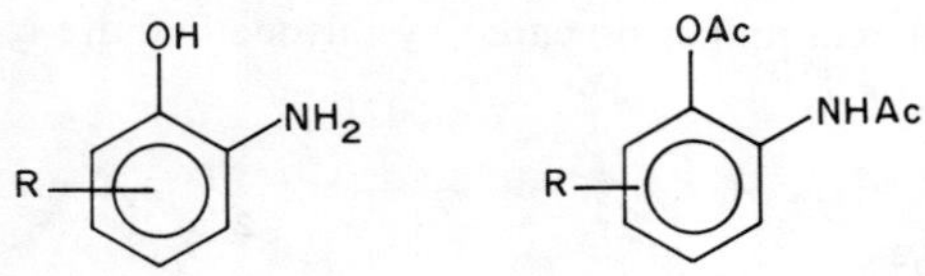

(25) (26)

IV. SELENIMIDES AND DERIVATIVES

A. Selenimides

1. Preparation

N-Substituted selenimides have been prepared by different methods depending on the oxidation state of the selenium in the starting material.

a. Preparation from selenides (Se^{II}) *(method A)*. The selenimides **27** and **28** are obtained by reaction of selenides with chloramine-T or -B[33–37] (equations 26 and 27).

$$R^1SeR^2 + \underset{\text{Chloramine-T}}{NaNClSO_2Tol\text{-}p} \longrightarrow \underset{(\mathbf{27})}{R^1R^2Se=NSO_2Tol\text{-}p} \qquad (26)$$

$$R^1SeR^2 + \underset{\text{Chloramine-B}}{NaNClSO_2Ph} \longrightarrow \underset{(\mathbf{28})}{R^1R^2Se=NSO_2Ph} \qquad (27)$$

b. Preparations from Se^{IV} species. Selenimides can be prepared in different ways starting from Se^{IV} species.

(i) Method B. The addition of *t*-BuOCl to a selenide gives an Se^{IV} addition product, which on treatment with an *N*-sodioarylsulphonamide or benzamide yields the corresponding selenimide[35,38] (equation 28). This synthesis can be realized in a one-pot reaction.

$$R^1SeR^2 \xrightarrow{t\text{-BuOCl}} R^1R^2Se(OBu\text{-}t)Cl \xrightarrow[(NaNHCOAr)]{NaNHSO_2Ar} R^1R^2Se=NSO_2Ar$$
$$(R^1R^2Se=NCOAr) \qquad (28)$$

However, when the reaction is attempted with diphenyl selenide, the expected selenimide is not obtained but another Se^{IV} compound, **29**, is isolated[35].

$$\underset{(\mathbf{29})}{Ph_2Se(OH)NHSO_2Tol\text{-}p}$$

(ii) Method C. Treatment of a dichloroselenide with sodium L(−)-menthylate followed by reaction with *N*-sodiophenylsulphonamide gives the selenimide **30**[29] (equation 29). This method allows the preparation of the selenimide **30** with one of the enantiomers in slight excess.

$$R^1R^2SeCl_2 \xrightarrow{Na\,L(-)\text{-menthylate}} R^1R^2Se(Cl)\text{-}L(-)\text{-menthylate}$$
$$\xrightarrow{PhSO_2NHNa} \underset{(\mathbf{30})}{R^1R^2Se=NSO_2Ph} \qquad (29)$$

An attempt to prepare the diphenylselenimide **31** directly by the action of ammonia on diphenylselenium dichloride failed and gave instead the quaternary ammonium salt **32**[40].

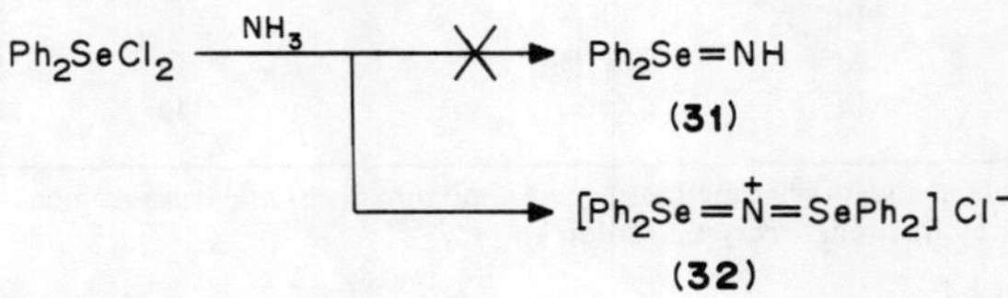

(iii) Method D. Condensations of selenoxides with amides give the corresponding selenimides[35,37] in good yields (equation 30).

$$R^1R^2Se{=}O + \begin{matrix} NH_2SO_2R \\ \text{or} \\ NH_2COR \end{matrix} \longrightarrow \begin{matrix} {R^1 \atop R^2}{>}Se=NSO_2R \\ \text{or} \\ {R^1 \atop R^2}{>}Se=NCOR \end{matrix} \qquad (30)$$

TABLE 4. Selenimides $R^1R^2Se = NR^3$ prepared by different methods

R^1	R^2	R^3	Method	Refs.
Ph	Ph	SO_2Ph	A	34, 35
Ph	Ph	SO_2Tol-*p*	A	34, 35
Ph	Ph	$SO_2C_6H_4Cl$-*p*	A	34, 35
Ph	Ph	$SO_2C_6H_4NO_2$-*p*	A	34, 35
Ph	Ph	$SO_2C_6H_4NO_2$-*o*	A	34, 35
p-Tol	*p*-Tol	SO_2Ph	A	34, 35
p-Tol	*p*-Tol	SO_2Tol-*p*	A	34, 35
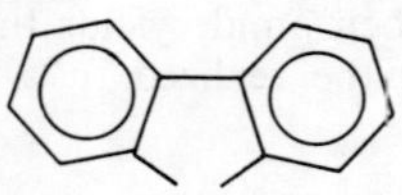		SO_2Tol-*p*	A	33
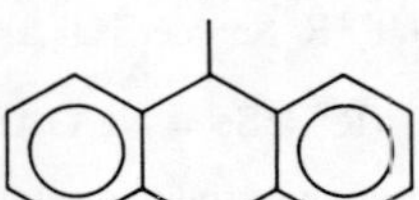 [a]		SO_2Tol-*p*	A	36
$—(CH_2)_4—$		SO_2Tol-*p*	A	35, 37
$—(CH_2)_5—$		SO_2Tol-*p*	A	37
$PhCH_2$	$PhCH_2$	SO_2Tol-*p*	A	37
Ph	Ph	SO_2Ph	B	38
Ph	Ph	COPh	B	38
$PhCH_2$	$PhCH_2$	SO_2Tol-*p*	B	35
Me	Ph	SO_2Tol-*p*	B	35
$—(CH_2)_4—$		SO_2Tol-*p*	B	35
Me	Ph	SO_2Tol-*p*	C	39
Me	*p*-BrC_6H_5	SO_2Tol-*p*	C	39
$PhCH_2$	$PhCH_2$	SO_2Tol-*p*	D	35
$—(CH_2)_4—$		SO_2Tol-*p*	D	35
Ph	Ph	Ph	D	37
Ph	Ph	*p*-ClC_6H_4	D	37
Ph	Ph	CCl_3	D	37
Ph	*p*-An	Ph	D	37
Ph	*p*-Tol	Ph	D	37
$PhCH_2$	$PhCH_2$	CCl_3	D	37
p-BrC_6H_4	Me	SO_2Ph	D	41
p-BrC_6H_4	Me	$COCF_3$	D	41

[a]In the case of selenoxanthene, method A gave a mixture of *cis*- and *trans*-isomers which could be separated by fractional recrystallization.

The reactions are run in chloroform with the sulphonamides and in acetic anhydride with the carboxamides. These condensations work better when an electron-donating group is present in the selenoxide and an electron-withdrawing group in the amide.

Table 4 summarizes the selenimides prepared by the various methods.

2. Chemical properties

a. Reduction. Reduction of *N*-tosyl selenimides to yield the selenide can be achieved quantitatively by refluxing the imide with triphenylphosphine in chloroform, ethanol or a mixture of benzene and acetic acid[36] (equation 31).

$$\mathrm{Ph(Me)Se{=}NTos} \xrightarrow[\text{solvent}]{\mathrm{Ph_3P}} \mathrm{PhSeMe} \tag{31}$$

9-Phenylselenoxanthene *N*-tosylimide has been reduced by $NaBH_4$ to the 9-phenylselenoxanthene[37].

b. Base and acid treatments. Basic hydrolysis of *N*-tosyl and *N*-acyl selenimides in methanol gives access to the corresponding selenoxides[34,37] (equation 32).

$$\begin{matrix} \mathrm{R^1R^2Se{=}NTos} \\ \text{or} \\ \mathrm{R^1R^2Se{=}NAcyl} \end{matrix} \xrightarrow[\mathrm{MeOH}]{\mathrm{KOH(NaOH)}} \mathrm{R^1R^2Se{=}O} + \begin{matrix} \mathrm{H_2NTos} \\ \text{or} \\ \mathrm{H_2NAcyl} \end{matrix} \tag{32}$$

Treatment of *N*-tosyl selenimide with water in the presence of silica gel also gives the selenoxide[36].

In one particular case, *N*-tosyl pentamethyleneselenimide (**33**) was treated with concentrated sulphuric acid at $-50\,°C$, but instead of the expected *N*-unsubstituted selenimides a crystalline product was isolated, for which spectral and elemental analyses indicated the selenurane structure **34**[37] (equation 33). The reaction took place by a Se—N bond cleavage rather than by the expected $N—SO_2$ cleavage.

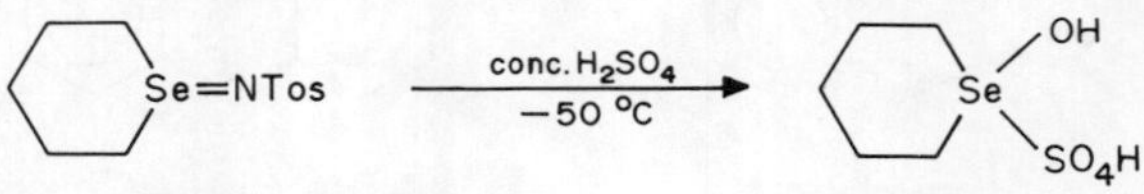

(**33**) (**34**) (33)

c. Reaction with halogenating agents. Treatment of selenimides with hydrogen chloride or chlorine produces the corresponding disubstituted selenium dichloride in quantitative yield[34]. Disubstituted dihalogenoselenium derivatives can also be obtained by treatment of *N*-tosyl or *N*-acyl selenimides with acetyl chloride, triphenylphosphine dibromide or chlorotrimethylsilane[75] (equation 34).

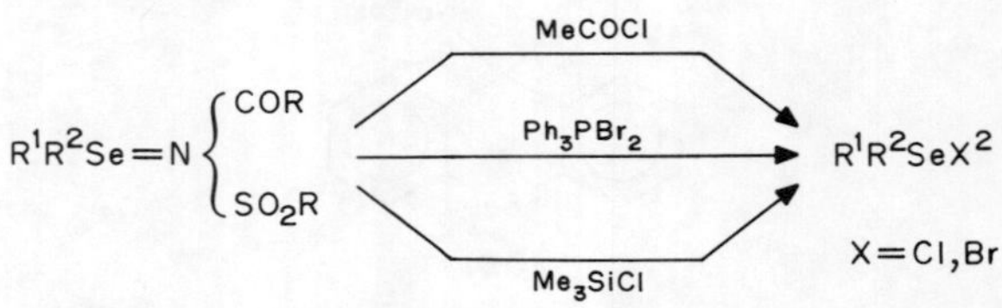

(34)

d. Action of oxidizing agents. Some oxidations of selenimides have been tried with potassium permanganate[37], hydrogen peroxide[34,37] or nitrogen oxide[34]. While different workers found that oxidation by potassium permanganate in water yields the corresponding selenones, they do not agree about the oxidation with 30% hydrogen peroxide. Some workers[34] found that this oxidation gave the selenones, whereas others[37] obtained the selenoxide or the selenimide hydrate (**35**) (equation 35).

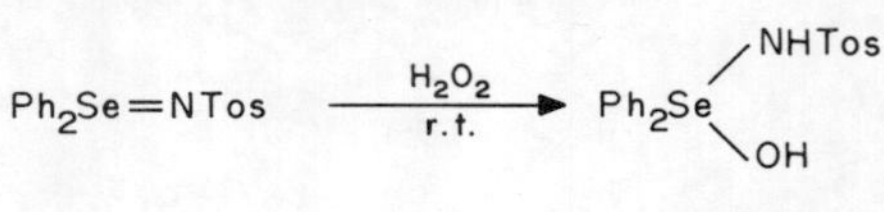

(**35**) (35)

However, when diphenyl *N*-phenylsulphonylselenimide is oxidized under the same conditions, the diphenyl selenoxide is obtained.

e. Thermal behaviour. The thermal behaviour of the different *N*-substituted selenimides has been studied. When *N*-tosyl selenimides are heated in a solvent in sealed tubes, reduction occurs and the corresponding selenides and sulphonamides are obtained[37] (equation 36).

$$R^1R^2Se{=}NTos \longrightarrow R^1R^2Se + TosNH_2 \qquad (36)$$

Heating *N*-acyl selenimides at 140–160 °C in DMF produces the selenide together with the isocyanate[37]. This last product is isolated as its phenylcarbamate, as shown in equation 37. Only in the case of *N*-trichloroacetyl diphenylselenimides is the corresponding imide obtained instead the carbamate.

$$R^1R^2Se = NCOR^3 \xrightarrow[(2)\ PhNH_2]{(1)\ 160\,^{\circ}C/DMF} R^1R^2Se + PhNHCONR^3 \qquad (37)$$

9-Phenylselenoxanthenetosylimine has been isolated as the *cis*- and *trans*-isomers **36** and **37**.[36]

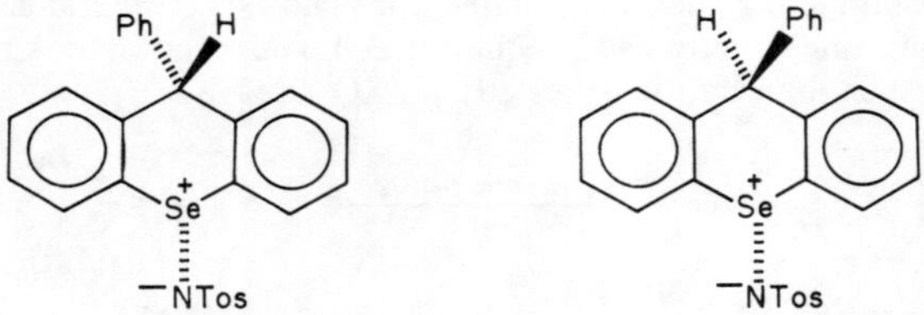

(**36**) (**37**)

Refluxing the *trans*-isomer **37** in toluene yields the rearrangement product **38**. Under the same conditions the *cis*-isomer **36** does not rearrange or isomerize to **37**. When the reaction is catalysed by DABCO (1, 4-diazabicyclo[2.2.2]octane), rearrangement takes place with either isomer even in benzene at room temperature.

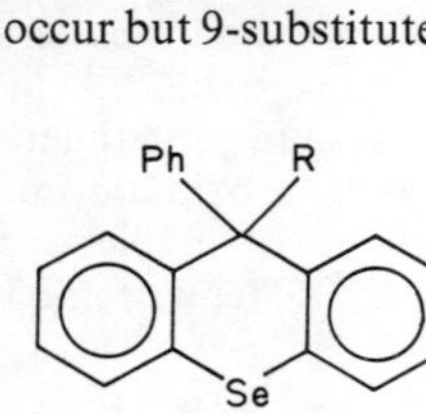

(**38**)

When other bases are used, such as sodium methoxide or carbanions from Grignard reagents, rearrangement does not occur but 9-substituted 9-phenylselenoxanthene (**39**) are obtained.

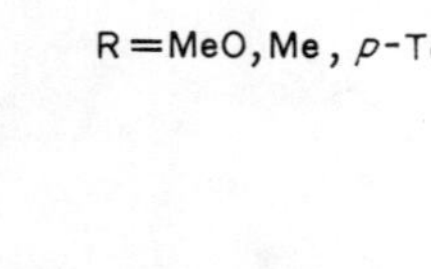

(**39**)

R=MeO, Me, *p*-Tol

f. Physicochemical properties. The pK_a values of different *N*-tosyl selenimides (**40**) have been measured in acetonitrile[42]. Their values range from 6.63 to 10.03. A $\rho\sigma$ analysis indicated that the pK_a values were determined by the inductive and mesomeric effects of R.

$$\begin{array}{c} RC_6H_4SeCH_2Ph \\ \| \\ NTos \end{array}$$

(**40**)

R = *p*-Me, *m*-Me, *p*-MeO, *m*-MeO, *p*-NMe$_2$, *p*-Br, *m*-Br, *p*-Cl

B. Se-Chloroselenimides

1. Preparation

The first examples of these compounds were prepared by reaction of *N,N*-dichlorophenylsulphonamide with arylselenotrimethylsilane[43,44] (equation 38). This synthesis has been extended to *N,N*-dichloro-*N*-acyl derivatives[44].

$$ArSe\,SiMe_3 + Ar'SO_2NCl_2 \longrightarrow \underset{\displaystyle Cl}{\overset{|}{ArSe}} = NSO_2Ar' \qquad (38)$$

Several other methods have also been used for the preparation of these compounds. Equations 39 show the different syntheses used and Table 5 gives the compounds prepared:

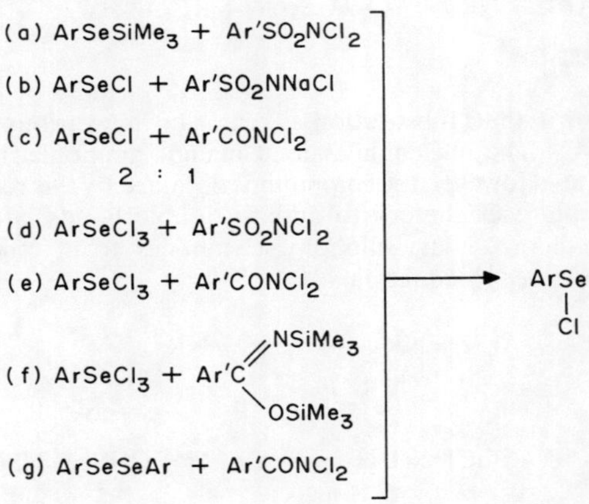

(39)

2. Chemical properties

Some of the chemical properties of *Se*-chloroarylselenimides have been studied. Hydrolysis by water with or without the presence of silver oxide yields the corresponding seleninic acids[43,44,45] (equation 40).

$$\underset{\displaystyle Cl}{\overset{|}{ArSe}} = NR \xrightarrow[(Ag_2O)]{H_2O} ArSeO_2H + RCONH_2 \qquad (40)$$

TABLE 5. *Se*-Chloroarylselenimides synthesized.

$$\underset{\displaystyle\mathrm{Cl}}{\mathrm{ArSe}}=\mathrm{NAr'}$$

Ar	Ar′	Method in equation 39	Ref.
Ph	$PhSO_2$	a, b, c, e	43, 44, 45
Ph	p-$TolSO_2$	a	44
Ph	p-$ClC_6H_4SO_2$	a	44
Ph	p-$NO_2C_6H_4SO_2$	a	44
p-Tol	$PhSO_2$	a	44
p-BrC_6H_4	$PhSO_2$	a	44
p-NO_2	$PhSO_2$	a	44
Ph	PhCO	c, e, f, g	45
Ph	p-TolCO	g	45
Ph	p-BrC_6H_4CO	g	45
Ph	p-$NO_2C_6H_4CO$	g	45
Ph	3, 5-$(NO_2)_2C_6H_3CO$	g	45
p-Tol	p-BrC_6H_4CO	g	45
p-ClC_6H_4	p-ClC_6H_4CO	g	45

Reaction of hydrogen chloride or chlorine leads to the formation of the corresponding arylselenium trichloride (equation 41).

$$\underset{\displaystyle\mathrm{Cl}}{\mathrm{ArSe}}=\mathrm{NR}\xrightarrow[\text{or } Cl_2]{\text{HCl}}\mathrm{ArSeCl_3}+\mathrm{RCONH_2} \tag{41}$$

It has also been shown that substitution of chlorine by various nucleophiles is possible[31,32] (equation 42). The nucleophiles used include ammonia, imines, alcohols, alcoholates and phenolates. However, the compounds obtained by this route are unstable.

The property of *Se*-chloroselenimides to add to double bonds is interesting. The addition of various *Se*-chloro *N*-arylsulphonylselenimides to styrene to yield new selenimides has been described[44] (equation 43).

$$\underset{\displaystyle\mathrm{Cl}}{\mathrm{ArSe}}=\mathrm{NR}\xrightarrow{\text{Nu}}\underset{\displaystyle\mathrm{Nu}}{\mathrm{ArSe}}=\mathrm{NR} \tag{42}$$

$$\underset{\displaystyle\mathrm{Cl}}{\mathrm{ArSe}}=\mathrm{NSO_2Ar'}+\mathrm{PhCH}=\mathrm{CH_2}\longrightarrow \mathrm{PhCH}=\mathrm{CH}\underset{\displaystyle\mathrm{Ar}}{\mathrm{Se}}=\mathrm{NSO_2Ar'} \tag{43}$$

When the reaction is carried out with *Se*-chloro *N*-acylselenimides, 1, 4-cycloaddition products are isolated[45] (equation 44).

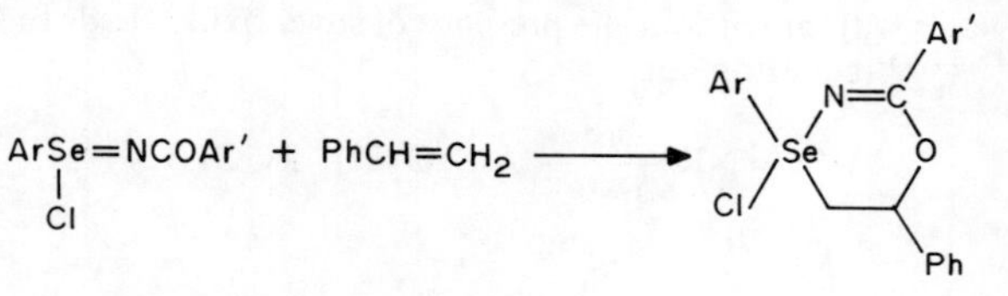

(44)

3. *Se-dichloro-N-arylseleninyl Se-chloroselenimides* $ArSeCl_2N=SeClAr$

These compounds are obtained by reaction of *Se*-dichloro diarylselenides with bis(trimethylsilyl)amines[46] or arylseleninyl trichlorides with *N*-chloro-bis(trimethylsilyl)amines[47] (equation 45).

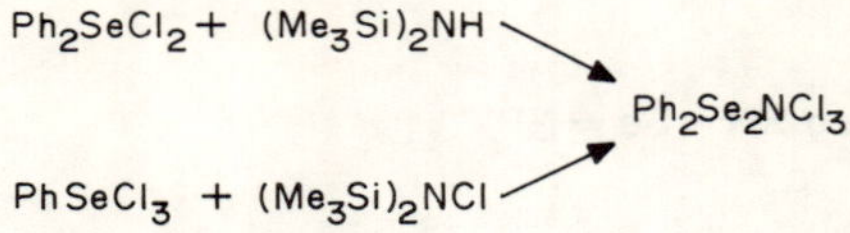
(45)

It is not yet known whether the structure of the product is covalent (**41**) or ionic (**42**).

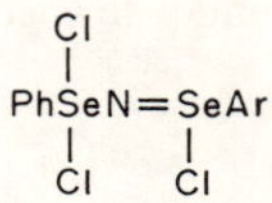
(**41**)

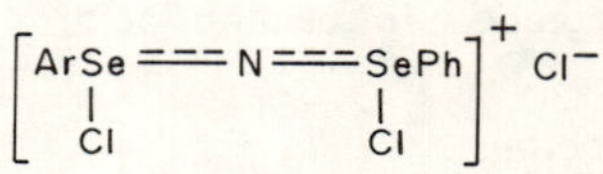
(**42**)

Hydrolysis of the phenyl derivative leads to phenylseleninic acid. Treatment with hydrochloric acid or with chlorine gives phenylselenium trichloride[47]. The product decomposes on heating giving a mixture of phenylselenium trichloride and phenylselenenyl chloride (equation 46).

$$Ph_2Se_2NCl_3 \xrightarrow{\Delta} PhSeCl + PhSeCl_3 + N_2 \qquad (46)$$

C. Selenoximides

1. *Preparation*

The only synthesis of this type of compound which has been published is the condensation of diarylselenones with *N*-sulphurylarenesulphonamides[48,49] (equation 47). All attempts to prepare these compounds by condensations of diarylselenones with *N*-sulphinylarylamines, *N*-sulphinyltrimethylsylilamines or *N*-sulphinylarylcarboxamides failed[47].

$$Ar_2SeO_2 + Ar'SO_2NSO \longrightarrow SO_2 + Ar_2\underset{\underset{O}{\|}}{Se} = NSO_2Ar' \qquad (47)$$

$$Ar = Ph,\ p\text{-Tol},\ Ar' = Ph,\ p\text{-Tol},\ p\text{-}NO_2C_6H_4$$

The products prepared were solids and were characterized by IR spectroscopy.

2. *Chemical reactivity*

Only the action of bases and acids on selenoximides has been studied[49]. Basic hydrolysis of the oximides yields the corresponding selenones, whereas treatment with hydrochloric acid (4 moles per mole) gave the *Se*-dichloro diarylselenide, arylsulphonamide and chlorine.

V. NITROGEN DERIVATIVES OF SELENIUM DIOXIDE

Among these compounds *N*-substituted selenium diimides have been the most studied derivatives. Only one example of an *N*-substituted selenium oximide (**43**) is known[50].

$$PhNH_2 + SeOCl_2 \longrightarrow PhN=Se=O$$
(**43**)

A. *N*-Selenium Diimides RN = Se = NR

1. Preparation

Selenium diimides are prepared by condensation of amines, amides and their derivatives with selenium, selenium tetrahalides or oxyselenohalides[51–53]. Table 6 summarizes the different methods available for the preparation. The diimides formed are sensitive to moisture and decompose to selenious acid[52].

2. Chemical properties

Selenium diimides have been reported to be good aminating agents. Thus, olefins can be aminated in the allylic position by *N*-substituted selenium diimides (equation 48) and many examples of such reactions have been described[51].

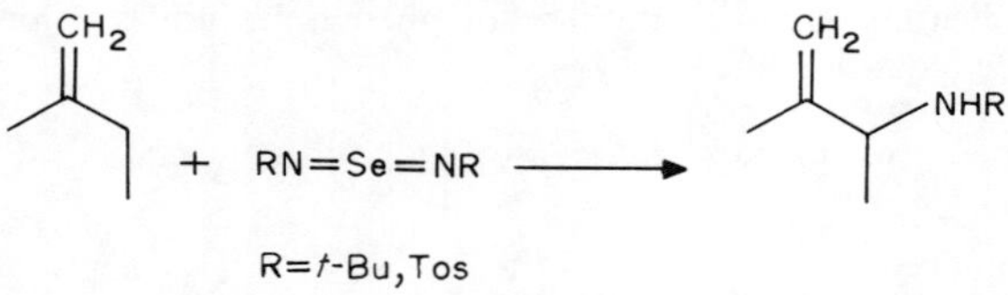

(48)

Treatment of a diene with *N*-tosyl selenium diimide yields the 1, 2-diamination product (equation 49).

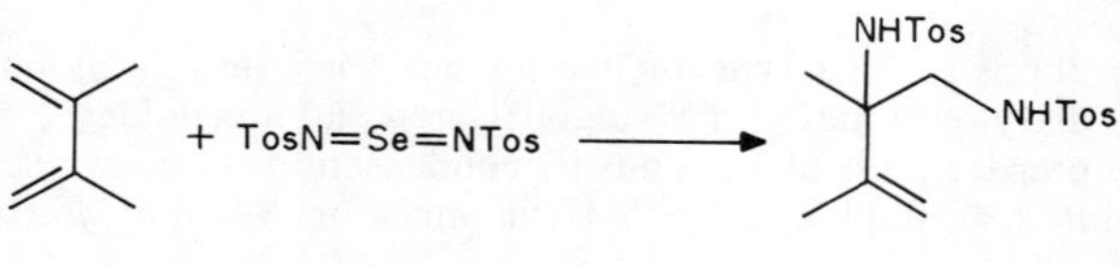

(49)

TABLE 6. Preparation of *N*-substituted selenium diimides RN = Se = NR

Starting material	Selenium vector	Product	Ref.
Chloramine-T	Se	$(TosN)_2Se$	51
t-$BuNH_2$	$SeCl_4$	$(t\text{-}BuN)_2Se$	51
$TosNH_2$	$SeCl_4$	$(TosN)_2Se$	51
$RN(SiMe_3)_2$ ($R = PhSO_2$, PhCO, MeCO)	SeF_4	$(RN)_2Se$	52
$RSO_2N(SiMe_3)_2$ R = Me, Ph)	SeF_4	$(RSO_2N)_2Se$ $+ RSO_2N=SeF_2$	53
$(Me_3Si)_2O$	$(PhSO_2N)SeF_2$	$(PhSO_2N)_2Se$	53

The reaction with the selenium diimides certainly goes through a [4 + 2] cycloaddition compound (**44**), and the same reaction with sulphur diimide actually stops at the cycloaddition stage[54].

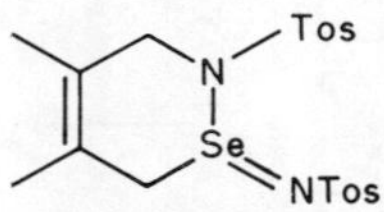

(**44**)

Diphenylsulphonyl selenium diimide (**45**) reacts with compounds having a double-bonded oxygen moiety to give the corresponding *N*-sulphonyl derivatives (**46**)[55] (equation 50).

$$PhSO_2N=Se=NSO_2Ph + E=O \longrightarrow 2PhSO_2N=E + SeO_2 \qquad (50)$$

(**45**) (**46**)

E = RCH with R = C_6H_5, *p*-BrC_6H_4, *p*-$NO_2C_6H_4$; Me_2NCH; Me_2S; Ph_3P; Ph_2Se; Ph_2SeO, PhN; *p*-$Et_2NC_6H_4N$

Diacyl selenium diimides are less reactive in this condensation. Dibenzoyl selenium diimide reacts only with dimethyl sulphoxide. Diacetyl derivatives react rapidly with aldehydes. The compounds **47** obtained by treatment of triphenylphosphine oxide with diacyl selenium diimides yield the nitrile **48** when treated with triphenylphosphine[42] (equation 51).

$$RCON=PPh_3 \xrightarrow[Ph_3P]{} RCN \qquad (51)$$

(**47**) (**48**)

Ditosyl and diacyl selenium diimides have also been used for the synthesis of selenadiazole and oxaselenazine (see Part Two, Section III.A).

B. *Se*-Dichloroseleniumimide RN = SeCl$_2$

1. Preparation

The first preparation of these compounds was described in 1967[56] by the reaction *N*, *N*-dichloroamines or *N*, *N*-dichloroamides with selenium or with selenium tetrachloride (equation 52).

$$RNCl_2 + Se\ (SeCl_4) \rightarrow RNSeCl_2\ (+Cl_2) \qquad (52)$$

The method has been extended to *N*, *N*-dichloroarylsulphonamides (R = $ArSO_2$), *N*, *N*-dichlorophosphonamides [R = $(Et_2O)PO$] and *N*, *N*-dichlorocarbonates (R = CO_2Et).

2. Chemical properties

Some chemical properties of the derivatives have been studied[57]; they were found to be very reactive. They hydrolyse to the amines or amides and selenious acid. On acidolysis with formic acid, *Se*-dichloroarylsulphonyl selenium imide gave the arylsulphonamide and selenium oxychloride (equation 53).

$$ArSO_2NSeCl_2 + HCOOH \longrightarrow ArSO_2NH_2 + SeOCl_2 \qquad (53)$$

The chlorine atom can be substituted by various nucleophiles such as amines, alcoholates, phenolates, glycol and ethylene oxide. For example, *Se*-dichlorophenylsulphonyl selenium imide (**49**) reacts with *N*-trimethylsilylmorpholine to give the *Se*-dimorpholinophenylsulphonyl selenium imide (**50**) (equation 54).

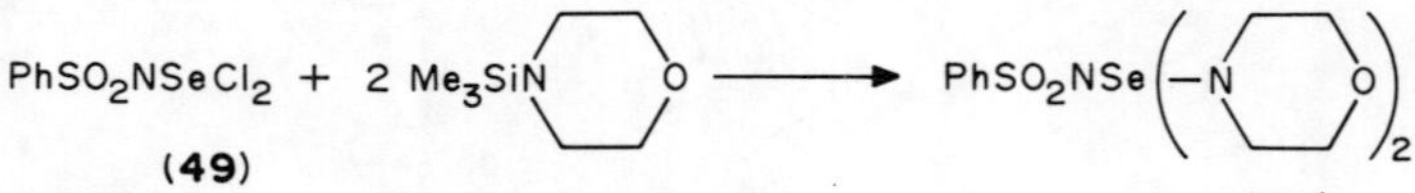

(54)

(49) **(50)**

An addition product (**51**) with styrene has also been prepared (equation 55).

$$PhSO_2NSeCl_2 + 2PhCH = CH_2 \longrightarrow \underset{\textbf{(53)}}{PhSO_2NSe(CH_2CHClPh)_2} \quad (55)$$

C. *Se*-Diamido-*Se*-dichloroselenium $(RCONH)_2SeCl_2$

Aromatic and aliphatic chloroamides react with selenium in a 2:1 ratio to give *Se*-diamido-*Se*-dichloroselenium derivatives[58,59] (equation 56).

$$2RCONHCl + Se \longrightarrow (RCONH)_2SeCl_2 \quad (56)$$

$$R = Me, Ph, p\text{-}Tol, p\text{-}BrC_6H_4, p\text{-}ClC_6H_4, 2,4\text{-}Cl_2C_6H_3$$

Another preparation is possible by reaction of *N*-trimethylsilylamides with selenium tetrachloride (equation 57). The products are thermally unstable and are cleaved by hydrochloric acid or chlorine to the corresponding amides or *N*-chloroamides, together with selenium tetrachloride[59].

$$2\,ArCONHSiMe_3 + SeCl_4 \longrightarrow (PhCONH)_2\,SeCl_2 \quad (57)$$

D. Miscellaneous

1. *Amidoselenious esters*

Diamidoselenious acid derivatives are prepared by reaction of selenious esters with a primary amine[60] (equation 58).

$$(MeO)_2SeO + 2MeNH_2 \rightleftharpoons (MeNH)_2SeO \quad (58)$$

Amidoselenious acid esters are obtained by reaction of selenium oxide with *N*-trimethylsilylamines[61] (equation 59).

$$R_2NSiMe_3 + SeO_2 \longrightarrow R_2N\underset{\underset{O}{\|}}{Se}OSiMe_3 \quad (59)$$

$$R_2N = \text{morpholine, piperidine, diethylamine}$$

Treatment of latter compounds with water yields the corresponding pyroselenites (**52**) (equation 60).

$$R_2N\underset{\underset{O}{\|}}{Se}OSiMe_3 \xrightarrow{H_2O} \underset{\textbf{(52)}}{[R\overset{+}{N}H_2]_2\,Se_2O_5{}^{2-}} \quad (60)$$

2. Aminoselanes

An example of preparation of an aminotetraselane is given in equation 61[62].

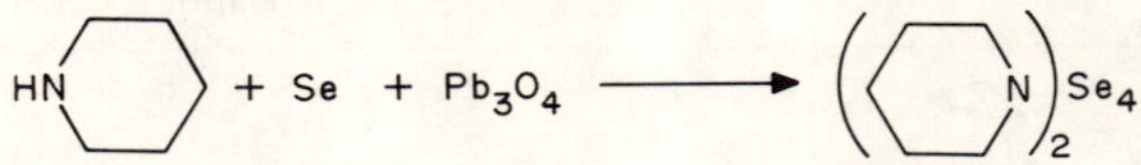

(61)

The same treatment with morpholine gives a mixture of tetra- and tri-aminoselanes (selanes are structures with three or more consecutive Se atoms).

VI. COMPOUNDS WITH Te—N BONDS

The only type of derivatives prepared in this series are tellurimides and tellurium imides.

A. Tellurimides $R^1_2TeNR^2$

1. Preparation

The first tellurimide was obtained by reaction of chloramine-T with 2, 2′-biphenylene telluride (**53**)[63] (equation 62).

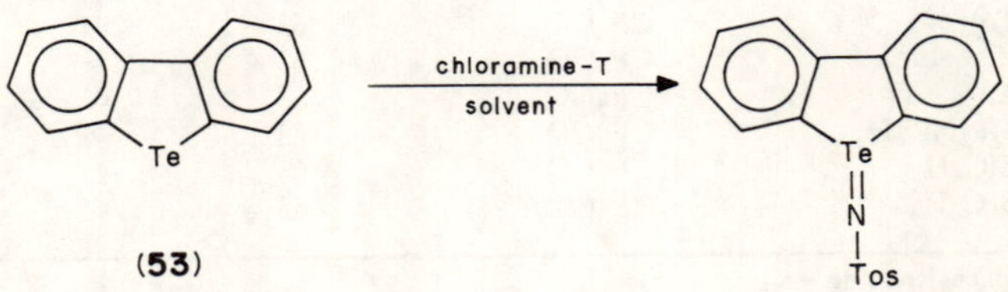

(**53**) (62)

Like the selenimides, tellurimides have also been prepared by reaction of telluroxide with *N*-substituted sulphonamides[41,64] or amides[41,65,66] (equation 63) and by reaction of *Te*-chloro-*Te*-butoxytelluride with *N*-sodiosulphonamides[65] (equation 64). A recent method allows also the synthesis directly from tellurides[67] (equation 65).

$$R_2{}^1TeO \xrightarrow{R^2NH_2} R_2{}^1TeNR^2 \qquad (63)$$

$$(R^2NH_2 = \text{amide or sulphonamide})$$

$$R_2{}^1Te + Me_3COCl \longrightarrow R_2{}^1Te\begin{matrix}Cl \\ OCMe_3\end{matrix} \longrightarrow R_2{}^1Te{=}NSO_2R^2 \qquad (64)$$

$$R_2{}^1Te \xrightarrow{TosN_3} R_2{}^1TeNTos \qquad (65)$$

Condensation with amides (equation 63) takes place when the telluroxide substituents are sufficiently electron donating and the amide susbtituents sufficiently electron withdrawing.

Table 7 summarizes the tellurimides prepared by the different methods.

TABLE 7. Tellurimides $R_2{}^1TeNR^2$ prepared

R^1	R^2	Method (equation)	Ref.
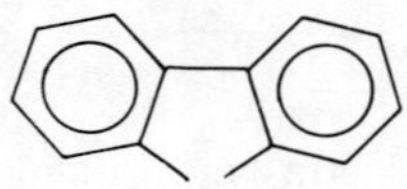 [a]	Tos	62	63
p-An	Tos	63–65	64, 41, 66
p-An	$PhSO_2$	63–65	64, 4, 646
p-An	$PhCH_2SO_2$	63–65	64, 41, 66
Ph	Tos	64, 65	41, 65
Ph	$PhSO_2$	64, 65	41, 65
Ph	$PhCH_2SO_2$	64, 65	41, 65
p-Tol	$PhCH_2SO_2$	63	65
p-ClC_6H_4	$PhCH_2S_2O_2$	63	65
p-BrC_6H_4	$PhCH_2SO_2$	63	65
p-An	CCl_3CO	63	41
p-An	*m*- or *p*-$NO_2C_6H_4CO$	63	41
p-$NMe_2C_6H_4$	CCl_3CO	63	41
p-$NMe_2C_6H_4$	*m*- or *p*-$NO_2C_6H_4CO$	63	41
Ph	CF_3CO	63	41
p-Tol	CF_3CO	63	41
m-Tol	CF_3CO	63	41
p-An	CF_3CO	63	41
p-$Me_2NC_6H_4$	CF_3CO	63	41
p-ClC_6H_4	CF_3CO	63	41
p-BrC_6H_4	CF_3CO	63	41

[a]Biphenylene ring = $R_2{}^1$

Spectroscopic and X-ray diffraction studies of these compounds showed that they are highly bipolar with a considerable positive charge on tellurium. This fact allows one to write the structures in the form **54** or **55** when necessary.

$$R_2{}^1Te=NSO_2R \longleftrightarrow R_2{}^1Te^+{-}N^-SO_2R$$

(54)

$$R_2{}^1Te=NCOR \longleftrightarrow R_2{}^1Te^+{-}N^-COR \longleftrightarrow R_2{}^1Te^+{-}N{=}\overset{O^-}{\overset{|}{C}}R$$

(55)

As a considerable double bond character is possible in structure **55**, the authors[66] claimed that the separation of the *syn* (**56**) and *anti* (**57**) isomers was possible by recrystallization. The *syn* isomer is stabilized by charge interaction.

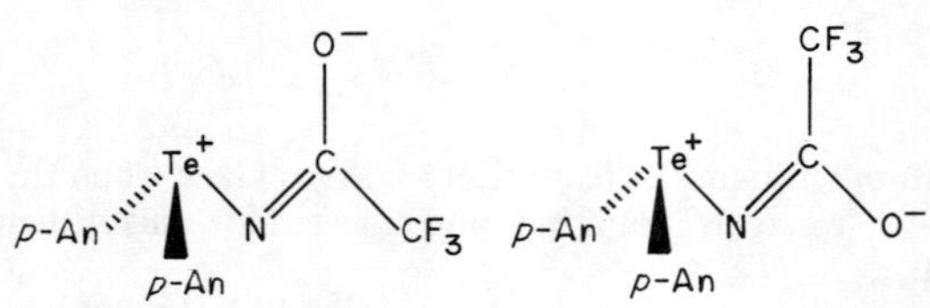

(56) **(57)**

2. Chemical properties

a. Reduction. Tellurimides are reduced to the corresponding tellurides with triphenylphosphine[68] (equation 66).

$$R_2{}^1Te = NR^2 \xrightarrow{Ph_3P} (R^1)_2Te \tag{66}$$

$$R^1 = Ph,\ p\text{-An};\ R^2 = p\text{-Tos},\ CF_3CO$$

On reduction of *N*-tosylimides, $Ph_3P^+N^-Tos$ is obtained as a by-product, whereas the reduction of *N*-trifluoroacetylimide leads to triphenylphosphine oxide via the hygroscopically unstable $Ph_3\overset{+}{P}\overset{-}{N}COCF_3$.

b. Reaction with electrophilic reagents. Tellurimides are cleaved at the Te—N bond by electrophilic reagents[69] (equation 67).

$$(p\text{-}RC_6H_4)_2Te = NR^1 + X^-Y^+ \longrightarrow (p\text{-}RC_6H_4)_2TeX_2 + R^1NY_2 \tag{67}$$
$$R = H,\ OMe;\ R^1 = PhSO_2,\ CF_3CO;$$
$$XY = HCl,\ Cl_2,\ AcCl,\ Ac_2O,\ AcOH$$

c. Reaction with active methylene compounds. Tellurimides react with dimedone at room temperature to yield the telluronium ylide[70] (equation 68).

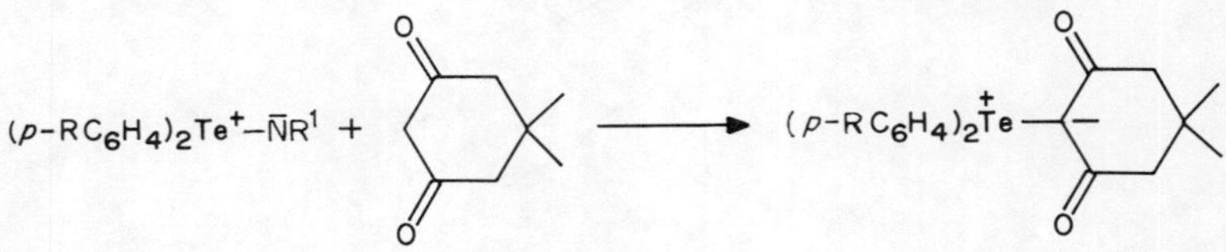

(68)

$R = H, OMe;\ R^1 = Tos, PhSO_2, CF_3SO_2$

d. Reaction with aldehydes. Reaction of aldehydes and *N*-tosyltellurimides gave the *N*-tosylaldimines and telluroxides[57,71] (equation 69).

$$R_2{}^1Te^+\text{—}NTos + RCHO \longrightarrow p\text{-}RC_6H_4CH = NTos + (p\text{-}RC_6H_4)_2TeO \tag{69}$$
$$R^1 = Ph,\ p\text{-An};\ R = Me_2NC_6H_4,\ p\text{-}NO_2C_6H_4$$

However, there are some limitations to the reaction. Thus trifluoroacetyltellurimides do not react[71], and although the reaction occurs well with aromatic aldehydes and with aldehydes having no α-hydrogen, in other cases it leads to unidentified mixtures[67].

B. *Te*-Chloro-*N*-tosyltellurimide RTe(Cl) = NTos

These compounds have been prepared by three different routes[72] (equation 70).

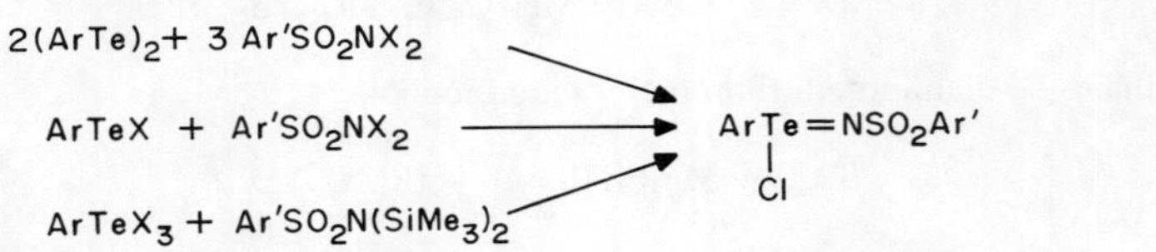

(70)

Ar=Ph, *p*-An, *p*-EtOC$_6$H$_4$, *p*-ClC$_6$H$_4$; Ar'=*p*-Tol; X=Cl,Br

Different chemical properties were investigated[72]. The chlorine atom in the molecule can undergo substitution by activated amines, e.g. *N*-trimethylsilylmorpholine. Treatment with hydrochloric acid or water gives the aryltellurium trichloride and the aryltellurinyl halide, respectively, together with the corresponding sulphonamide.

The *Te*-chlorotellurimide **56** reacts with dimedone to give the condensation product **57**, which in turn can react with another dimedone molecule in the presence of a base to give **58** (equation 71).

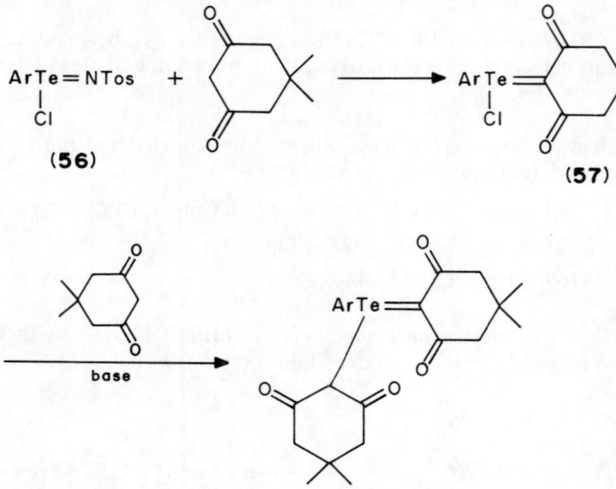

(56) (57) (58) (71)

C. Nitrogen Derivatives of Tellurium Dioxide

Four types of derivatives have been prepared, starting from tellurium tetrahalides: (a) Dihalotellurium imides (**59**)[73] (equation 72).

$$TeX_4 + (Me_3Si)_2NR \longrightarrow \underset{(\mathbf{59})}{RN = TeX_2} \quad (72)$$

$X = F, Cl; R = MeCO, PhSO_2, Tos$

(b) Tellurium diimides (**60**)[73] (equation 73).

$$TeX_4 + 2(Me_3Si)_2NR \longrightarrow \underset{(\mathbf{60})}{RN = Te = NR} \quad (73)$$

$X = F, Cl; R = MeCO, PhSO_2, Tos$

(c) Dihalo-*Te*-diaminotellurium (**61**)[74] (equation 74).

$$TeX_4 + Me_3SiNR_2 \longrightarrow \underset{(\mathbf{61})}{(R_2N)_2TeX_2} \quad (74)$$

$X = Cl, F; R = Et; R_2N = morpholine, piperidine$

(d) Diamino-*Te*-tellurium imides (**62**)[74] (equation 75).

$$(R_2N)_2TeX_2 + (Me_3Si)_2NR^1 \longrightarrow \underset{(\mathbf{62})}{(R_2N)_2Te=NR} \tag{75}$$

$$X = F;\ R_2N = \text{morpholine};\ R^1 = PhSO_2,\ \text{Tos}$$

Derivatives **59** and **60** behave in the same manner on hydrolysis or on treatment with acids. In both cases, tellurium oxide is obtained on hydrolysis together with the amide, and tellurium tetrachloride with the amide on treatment with hydrochloric acid. When treated with *N*-trimethylsilylmorpholine, **59** yields **62**, and so does **61** when treated with *N*,*N*-trimethylsilylarylsulphonamides.

PART TWO. HETEROCYCLIC COMPOUNDS WITH Se—N AND Te—N BONDS

VII. INTRODUCTION

This second part deals only with heterocyclic systems bearing at least one Se—N or Te—N bond. Other heterocycles containing Se (or Te) and nitrogen but not bound to each other were recently reviewed by Renson[15] in Volume 1 of this book and are not covered in this chapter. Thus this part completes Renson's chapter. We refer frequently to Renson's chapter; in order to present a complete and readable review here, some repetition and overlap was necessary, but this has been kept to the minimum.

VIII. THREE- AND FOUR-MEMBERED RINGS

The chemistry of this type of molecule is poorly documented. Selenaziridines cannot be obtained by treatment of oxaziridines with potassium selenoxanthate[76]. Decomposition of 1,2-benzodithiole-3-selenoximide (**63**) leads to the intermediate **64**, which rearranges with extrusion of sulphur[77] (equation 76).

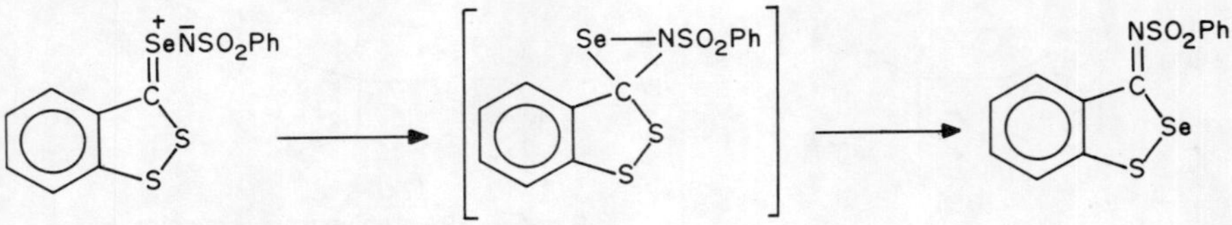

(**63**) (**64**) (76)

An interesting spiro heterocycle (**66**) has been synthesized[78] through the reaction of $SeOCl_2$ with **65** (equation 77).

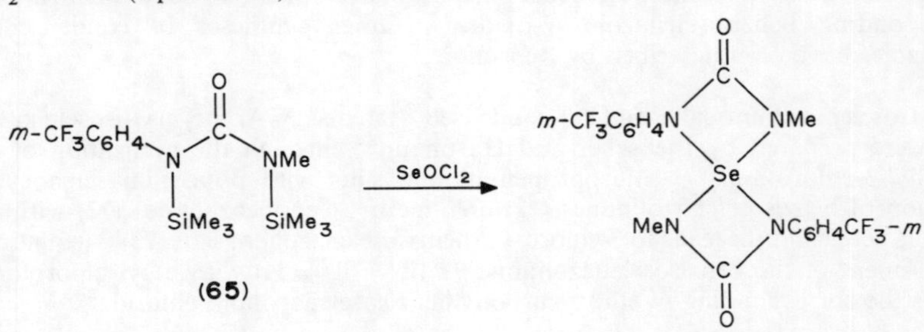

(**65**) (**66**) (77)

IX. FIVE-MEMBERED RINGS

A. One Se—N bond

In addition to the work of Lalezari *et al.*[79], Renson[15] has reviewed heterocycles containing N and either Se or Te.

1. 1,2-Selenazoles, (d)-condensed systems and derivatives

a. Synthesis. (i) 1,2-Selenazoles and (*d*)-condensed systems. The parent heterocycle (1,2-selenazole or isoselenazole) and its 3-methyl derivative were first mentioned in 1962[80]. They were obtained by reaction of selenocyanic acid with acrolein or but-3-yn-2-one. Using similar methodology, a general synthesis of isoselenazoles and the corresponding (*d*)-condensed systems was described in 1975[15]. More recently[81], preparations of 3-, 5- or 3,5-substituted isoselenazoles (**68**) were reported. The ring formation step consists in an intramolecular nucleophilic substitution in hydroxylamine-*O*-sulphonic acids of α,β-acetylenic carbonyl compounds (**67**) (equation 78).

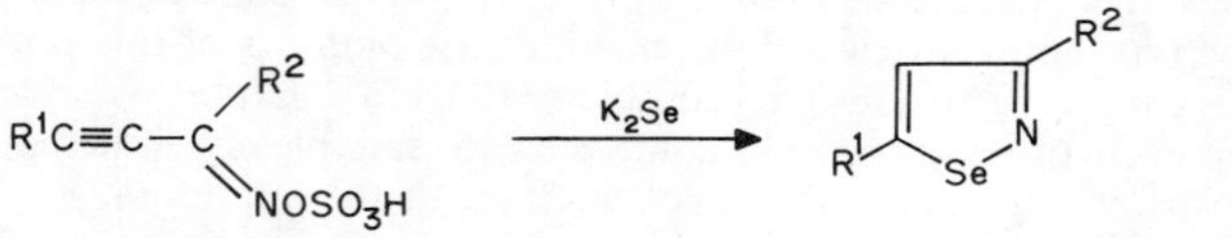

(78)

(67) **(68)**

Benzeneselenenyl halides containing *ortho*-carbonyl groups can be involved in similar cyclizations, affording 1,2-benzisoselenazoles. This reaction is fairly general and several condensed systems have been described[15]. In some cases electrophilic cyclization of aldoximes (**69**) also affords (*d*)-condensed 1,2-selenazoles (**70**)[82] (equation 79).

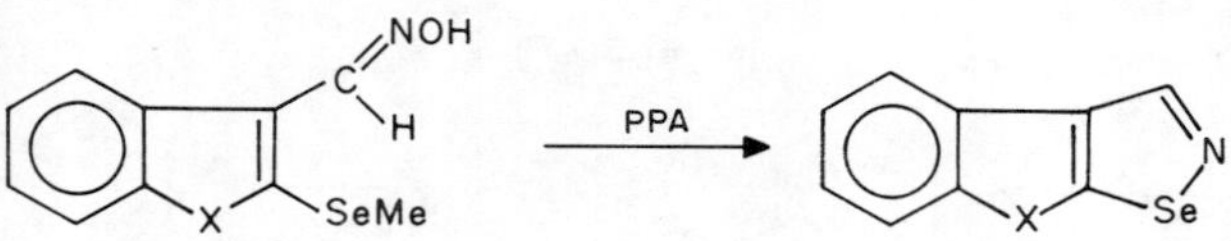

(79)

(69) **(70)**

X=S or Se; PPA= polyphosphoric acid

The polyphosphoric acid treatment of the corresponding ketoximes leads to condensed 1,3-selenazole systems through an initial Beckman rearrangement[99].

The oximation of 2-selenocyanobenzophenone or acetophenone was found to afford the corresponding benzisoselenazole *N*-oxides[83]. Other syntheses of condensed 1,2-selenazoles have been described by Renson[15].

(ii) Isoselenazolium salts and (*d*)-condensed systems. *N*-Aryl-5-aryl-isoselenazolium salts were prepared by Liebscher and Hartmann[84] through the cyclization of *N,N*-dimethyl-3-chloro-3-aryl-2-chloropropeniminium salts with potassium selenocyanate. Reaction of benzisoselenazolinones (**71**) or *o*-methylselenobenzamides (**72**) with phosphorous pentachloride leads to 3-chloro-1,2-benzisoselenazolium salts (**73**)[85] (equation 80).

Treatment of the benzisoselenazolinone **71** ($R^1 = R^2 = H$) with ethyl chloroformate affords the corresponding *N*-ethoxycarbonylbenzisoselenazolium chloride[86].

(iii) Benzisoselenazolinones. In 1924, Lesser and Weiss[87] were the first to synthesize

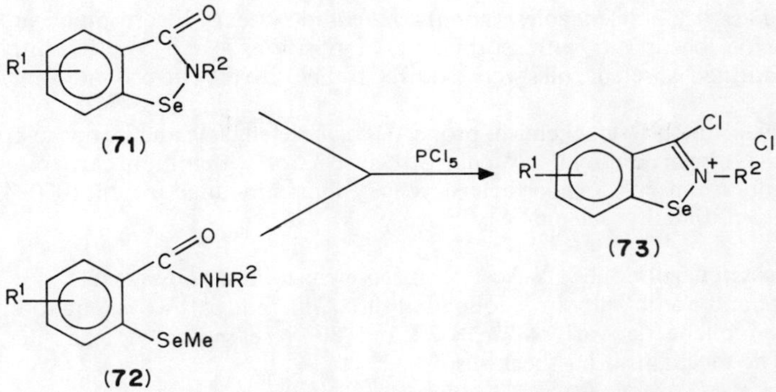

(80)

benzisoselenazolinone derivatives (**75**) by condensation of dichlorides (**74**) with ammonia or amines (equation 81).

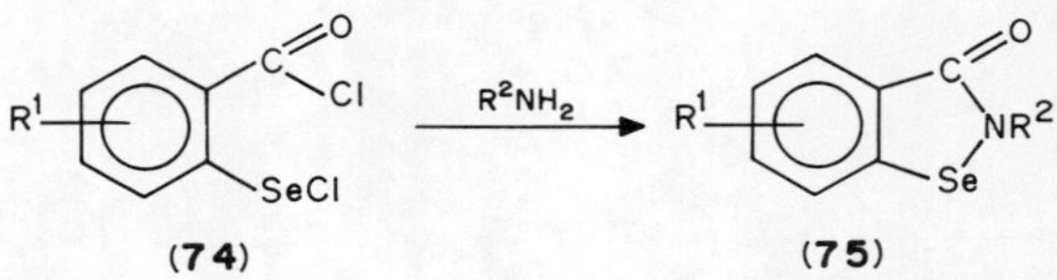

(81)

$R^1 = H; R^2 = H, Me, Et, Ph$

More recently, it has been shown that the hydrolysis of 3-chlorobenzisoselenazolium salts (**73**) or the reaction of *o*-methylselenobenzamides (**72**) with bromine in basic medium also leads to various benzisoselenazolinone derivatives. During the last five years, and owing to interest in the anti-inflammatory and glutathion peroxidase-like activities[88–92] of this class of compounds, their chemistry has been considerably investigated and patented[93–95]. It is not noteworthy that the sulphur analogues of **75** do not exhibit such pronounced activities. Important improvements in the synthesis of benzisoselenazolinones substituted in the benzene nucleus involve *ortho*-lithiation reactions[14,96], which afford faster chemical pathways[97,98] to some starting materials as shown in equation 82.

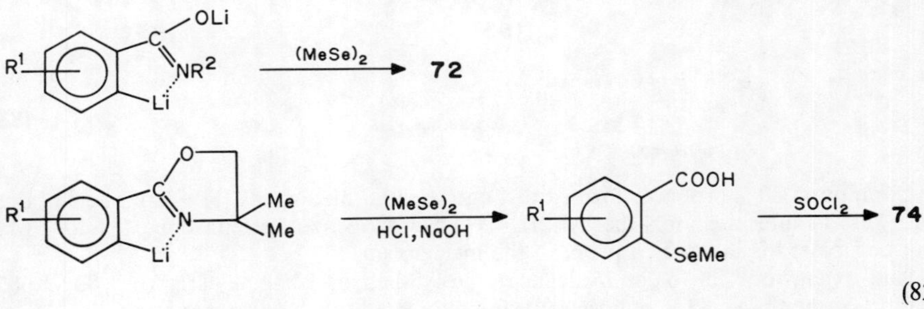

(82)

Various substituents were introduced into the benzene ring or on the nitrogen atom of **75** to try to improve the activity of the so-called PZ 51 or Ebselen **75** ($R^1 = H$; $R^2 = Ph$). The variations of the nature of R^2 in **75** included alkyl, aryl, heteroaryl and amino acid groups[98]. Other (*d*)-condensed isoselenazolinone systems (naphtho, thieno, furo and pyrido) were also obtained[99].

b. Properties. (i) 1, 2-Selenazoles and (*d*)-condensed systems. Electrophilic substitution of isoselenazole occurs, as with isothiazole, at positions 4 and 5[15]. Recently, several methyl-substituted isoselenazoles were oxidized by SeO_2 to the corresponding acids in low yields[100].

The chemical and physicochemical properties of isoselenazole and its benzo-condensed derivatives have been described particularly their reaction with lithium carbanions[15]. ^{77}Se chemical shifts are specific for such heterocyclic systems and range around 1000–1050 ppm downfield from dimethyl selenide[101,102].

(ii) Benzisoselenazolinones. *N*-Acetylbenzisoselenazolinone (**76a**) reacts[103] with potassium acetate and acetic anhydride following three different pathways (equation 83). The benzo(*b*)selenophene derivatives **77**, **78**, **79** and **80** were separated from the reaction products. The mechanism has been discussed[103].

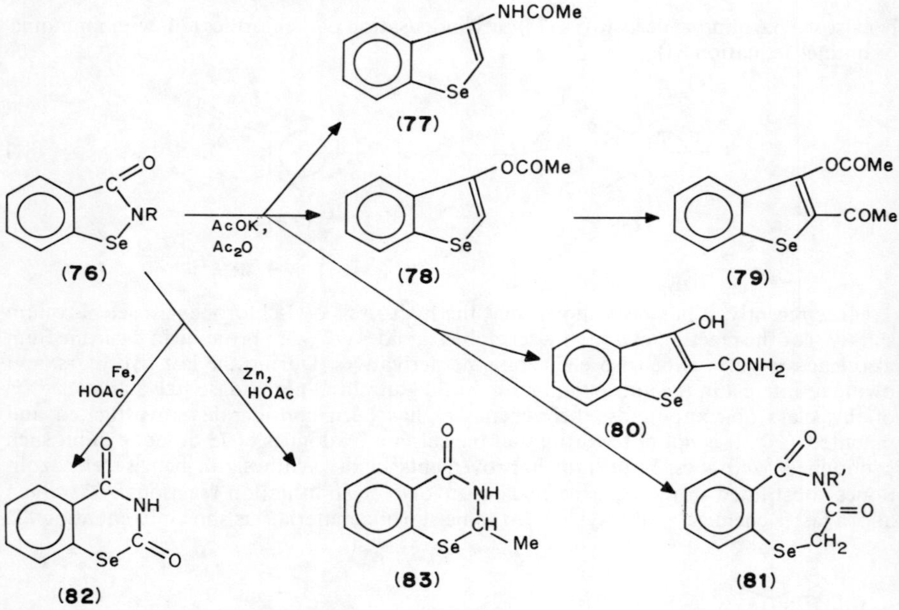

(**a**) R = COMe

(**b**) R = Me (83)

(**c**) R = COOEt

Compound **80** is probably formed through the intermediate **81** (R′ = H). The fact that with **76b** the intermediate 4-methyl-2*H*, 4*H*-3, 5-dioxobenzo(*b*)selenazepine (**81**, R′ = Me) can be isolated is taken as a proof of the mechanism.

Ring expansion[86] of **76a** or **76c** leads to the 1, 3-benzoselenazine derivatives **83** and **82**. With diazomethane, **82** is regioselectively methylated on the nitrogen atom.

2. (*c*)-Condensed 1, 2-selenazoles

SeO_2 oxidation of 5-amino-6-methyl-3-phenyl-4-(2*H*)-pyrimidone (**84**) (equation 84) leads to isoselenazolo[4, 3-*d*]pyrimidine (**85**), the only known (*c*)-condensed 1, 2-

selenazole[104,105a,b]. Unlike the analogous benzo[*c*]isothiazoles, several attempts to synthesize benzo[*c*]isoselenazoles were unsuccessful[99].

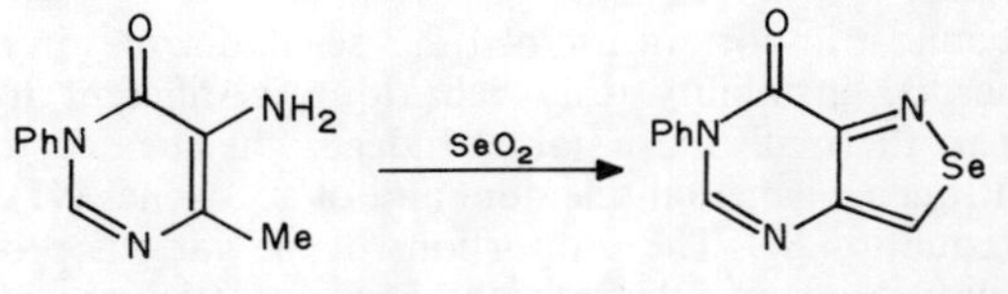

(84)

(84) **(85)**

Attempted ring closures of *o*-toluidine with electrophilic selenium reagents ($SeOCl_2$, $SeCl_4$, $SeBr_4$, SeO_2, Se_2Cl_2) and the oxidation of *o*-aminoselenobenzamide or selenol benzoate did not lead to the desired products.

B. One Se—N Bond with One Other Heteroatom in the Ring

So far now no derivative of oxaselenazole has been described.

1. *Thiaselenazoles, diselenazoles and derivatives*

Benzo-condensed systems (1, 2, 3-benzodichalcogenazolium salts are the only known derivatives. They were reviewed by Renson (Ref. 15, pp. 462–463).

2. *1, 2, 3-Selenadiazoles and derivatives*[79]

a. Synthesis. The chemistry of 1, 2, 3-selenadiazoles has been extensively studied since 1969, when Lalezari and coworkers proved that these heterocycles are easily obtained by SeO_2 oxidation of semicarbazones[106] (equation 85), and later showed their usefulness as synthons for access to alkynes[79].

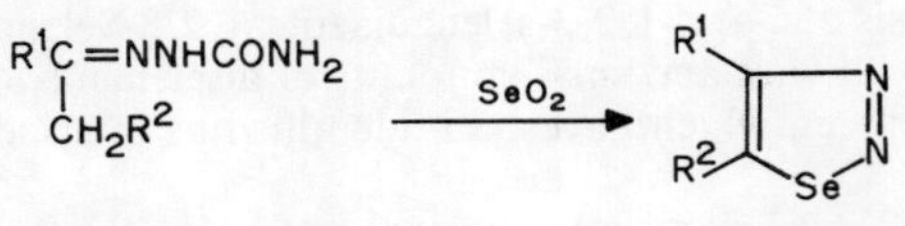

(85)

(86) **(87)**

The mechanism of the SeO_2 oxidation of the semicarbazones **86** to yield 1, 2, 3-selenadiazoles (**87**)[106] has been proved[79] and the regioselectivity of the oxidative ring closure has been studied[107].

$^{75}SeO_2$ oxidation[108,109] has also been performed, affording ^{75}Se-labelled 1, 2, 3-selenadiazoles. ^{14}C-Labelled[110] and deuteriated[112] derivatives are known. Treatment of 7-hydroxycycloocteno-1, 2, 3-selenadiazole affords a condensed 1, 2, 3-selenadiazoline as rearrangement product[113].

Aromatic condensed 1, 2, 3-selenadiazoles cannot be obtained by the reaction pathway of equation 85. It has been shown[15,114] that the diazotization of *o*-aminoselenophenols (**88**) leads to 1, 2, 3-benzo[*d*]selenadiazoles (**89**) (equation 86).

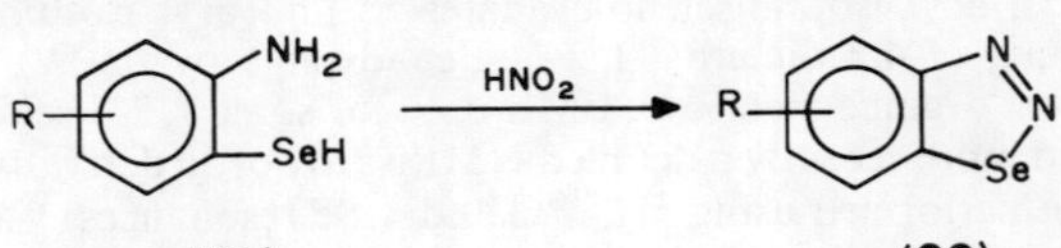

(86)

(88) **(89)**

In our experience, diazotization of *o*-aminophenyl tellurides gives immediate evolution of nitrogen, and no derivative of 1, 2, 3-benzo[*d*]telluradiazole can be isolated[115].

b. Properties. The most important use of 1, 2, 3-selenadiazoles in organic syntheses is based on their thermal instability (this behaviour is different from that of 1, 2, 3-thiadiazoles, which are thermally more stable). Indeed, when heated, they decompose with the formation of nitrogen, elemental selenium and of acetylenes (**91**), diselenins (**92**) and selenophenes (**93**) (equation 87). The proportions of the various products are generally dependent on the nature of the substrate, the concentration, the temperature and the medium[15].

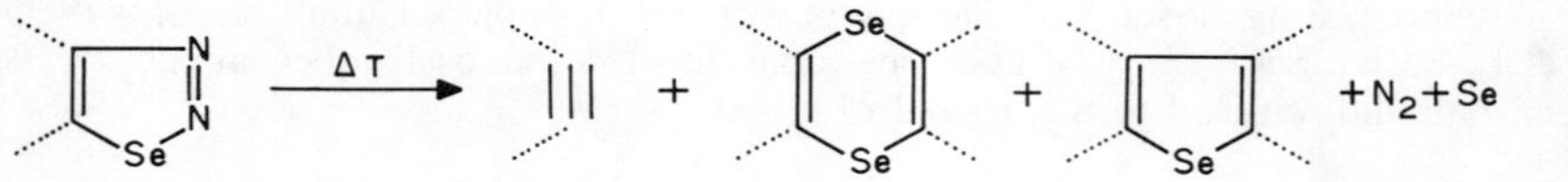

(**90**) (**91**) (**92**) (**93**) (87)

The mechanism of this reaction has been studied[116] and alkyneselenol, selenoketone and diselenetane[15,117] were recognized as intermediate or products of the reaction. 1, 2, 3-Benzoselenadiazole undergoes a thermal ring cleavage leading to 6-fulvene selenone[118]. In vacuum flash thermolysis, cyclopenta[*d*]-1, 2, 3-selenadiazole decomposes through propadiene selone to a selenopropargyl aldehyde derivative[119], whereas higher rings give tetrahydrofulvene selone derivatives[120]. For eight and higher membered cycloalkeno-1, 2, 3-selenadiazoles, the reaction may be considered as the method of choice for the transformation of a ketonic to an acetylenic function[121].

Cycloalkeno-1, 2, 3-selenadiazoles with smaller than six-membered rings give unstable acetylenes which can be trapped with tetraphenylcyclopentadienone[122,123] or with α-pyrone derivatives[124], in the latter case affording a new general method for benzoannellation. An analogous synthesis has been developed with 1, 4-quinone complexes[125].

In other cases, dimerization occurs, leading to **92**, and subsequent loss of a selenium atom leads finally to **93**. In this way, a mixture of 2, 4- and 2, 5-diarylselenophenes is obtained by thermolysis of 4-aryl-1, 2, 3-selenadiazole. 1, 2, 3-Selenadiazoles also undergo photolysis, leading to alkynes and small amounts of diselenafulvalenes through alkyne-selenols and selenoketenes, which have been identified as intermediates in the reaction (ref. 15, p. 463).

The behaviour of 1, 2, 3-selenadiazoles towards bases also illustrates the synthetic utility of these heterocycles. It has been shown that the abstraction of the 5-proton in 4-substituted 1, 2, 3-selenadiazole is followed by ring opening and leads to an alkyne selenolate (**94**) (equation 88).

Alkyne selenolates (**94**) are versatile synthons and allow the synthesis of a wide variety of molecules[15]. In acidic media, **94** can dimerize through nucleophilic addition and give diselenafulvene derivatives (**96**). Intermediates **94** also undergo carbon disulphide or diselenide cycloaddition, affording key molecules for the preparation of 'organic metals' (Ref. 15, p. 448). Alkyne selenolates react in dilute solutions with amines or alcohols, leading to selenoamides (**97**) or selenoesters (**98**)[15,126]. Cycloaddition with acetylene derivatives yields substituted selenophenes (**99**)[15].

The nucleophilic character of the selenolates **94** has been used in other condensations with ω-halo acids[127], acyl isothio(seleno)cyanates[128] and aryl isothiocyanates[129,130]. Recently, ring opening of 4, 5-dimethyl-1, 2, 3-selenadiazole by tin(IV) derivatives also afforded precursors of 'organic metal'[132]. PMR data for some 1, 2, 3-selenadiazoles have been reported[131] and more recently a detailed NMR study of 36 derivatives of the parent heterocycle has been performed using ^{1}H, ^{13}C and ^{77}Se resonances[130].

Photoelectronic and mass spectrometric spectra were also analysed for simple compounds[133]. Complexes with pentacarbonyl metals which are stable in air are available

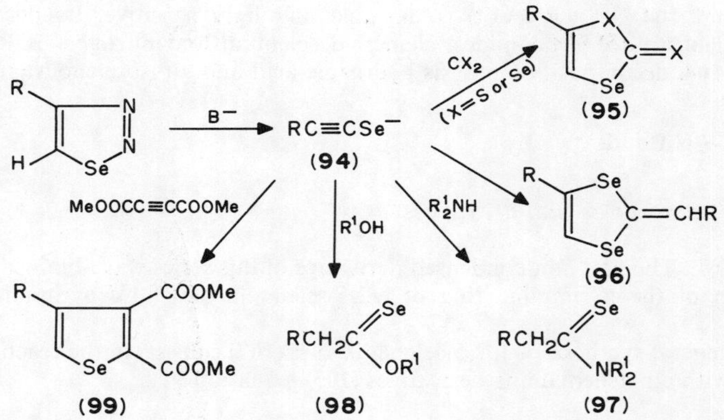

(88)

with $Cr(CO)_5$[134], $W(CO)_5$[134] or $Fe(CO)_3$[135,136,137]. Some derivatives of 1, 2, 3-selenadiazoles exhibit antibacterial and antifungal activity[15,138].

3. 1, 2, 4-Selenadiazoles

The first synthesis of 1, 2, 4-selenadiazole derivatives was published in 1904[139a] and developed later by Cohen[139b].

Iodine oxidation of the arylselenoamide **100** affords an unstable intermediate which immediately cyclizes, with loss of H_2Se, to give 3, 5-diaryl-1, 2, 4-selenadiazoles (**101**) (equation 89).

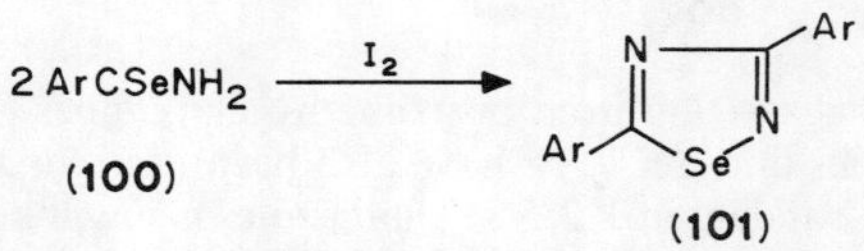

(89)

Analogously to the synthesis of 1, 2, 4-thiadiazoles, potassium selenocyanate reacts with *N*-haloamidines and gives 5-amino-3-substituted-1, 2, 4-selenadiazoles[139c].

C. One Se—N Bond with Two Other Heteroatoms in the Ring

1. 1, 2, 3, 4-Selenatriazole

1, 2, 3, 4-Selenatriazole is the only five-membered ring system, (relevant to this chapter) known with four heteroatoms. Some of its derivatives (**104**) have recently been obtained[140] through condensation of 4, 4-dialkylselenosemicarbazides (**102**) with nitrous acid or with an aza transfer reagent (equation 90). Reaction of the diselenides **103** with azide anion affords a similar result.

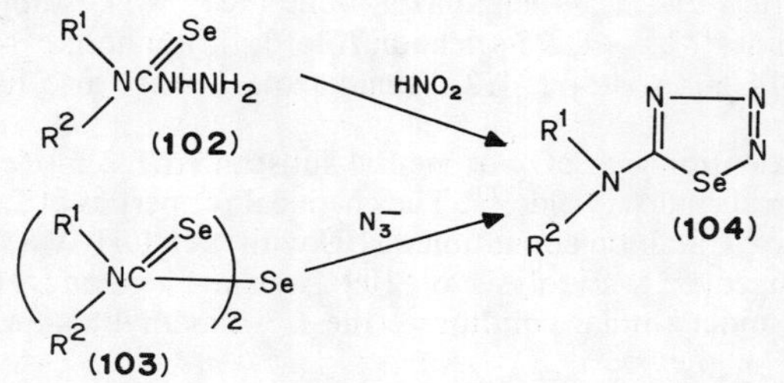

(90)

Compound **104** ($R^1 = R^2 = Et$) is unstable and light sensitive. Its decomposition furnishes disubstituted cyanamides, elemental selenium and nitrogen. If $R^1 = H$ and $R^2 = Et$ in **104**, decomposition affords hydrazoic acid and an isoselenocyanate.

D. Two Se—N Bonds

1. 1,2,5-Selenadiazoles and derivatives

a. Synthesis. The first non-condensed derivative of this series was obtained in 1967 by degradation of the pyrimidine ring of 1,2,5-selenadiazolo[3,4-*d*]pyrimidine-7(6*H*)-6-one[141].

A more general synthesis of 1,2,5-selenadiazoles (**107**) consists in the reaction of SeO_2 or Se_2Cl_2 with ethylenediamine derivatives (**105**) (equation 91)[15].

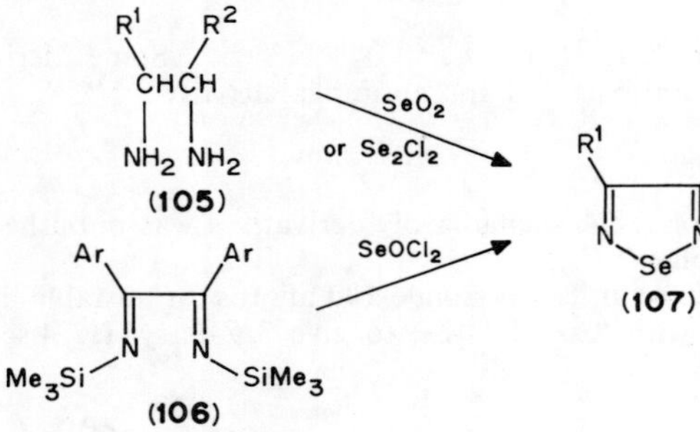

(91)

Two similar reactions starting from bis(trimethylsilyl)imines (**106**) and $SeOCl_2$[142] or from unprotected imines and $SeCl_2$[143] have also been used for the synthesis of **107**.

The chemistry of (*c*)-condensed 1,2,5-selenadiazoles is much better documented and a recent review[144] has been devoted to them. 2,1,3-Benzoselenadiazole (**109**) (piaselenole or piazselenole) is one of the oldest known organoselenium compounds. It was synthesized in 1889[145] by reaction of *o*-phenylenediamine (**108**) with SeO_2 (equation 92). Owing to the high reproducibility and the quantitative yields furnished by this reaction, it has been extensively used (ref. 15, p. 465).

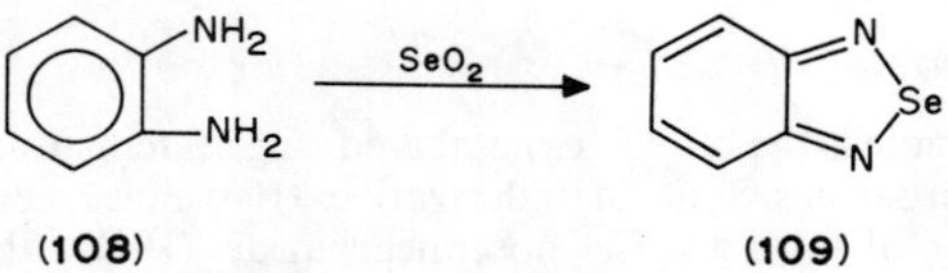

(92)

1,2,5-Selenadiazolohalogeno[3,4-*b*]- and -[3,4-*c*]-pyridine[146], 9-tosylhydrazononaphtho[2,3-*c*]-1,2,5-selenadiazol-4-one[147], 4,7-diphenyl-1,2,5-selenadiazolo[3,4-*c*]pyridines[148], 1,2,5-selenadiazolo[3,4-*c*]quinoline[149] and 2,1,3-benzoselenadiazolo[4,5-*c*:6,7-*c'*]bis-1,2,5-thiadiazole[150] have also been described.

b. Properties. Oxidation by SeO_2 of methyl-substituted 1,2,5-selenadiazoles affords the corresponding acids in low yields[100]. The chemical properties of 1,2,5-selenadiazoles have been reviewed[15]. Catalytic amination of **109** with $NH_2OH–V_2O_5$ affords 4-amino-2,1,3-benzoselenadiazole[151]. Pyrolysis of **109** *N*-oxide generates **109** and benzofuroxan[152], whereas, under similar conditions, the 1,3-dimethyl bis-quaternary salt gives benzimidazole and its 1-methyl derivative[153]. Photolysis of **107** derivatives leads to nitrile selenide as the initial product[154,155], which is further decomposed to nitrile and elemental selenium. Owing to the easy access to (*c*)-condensed 1, 2, 5-selenadiazoles, these molecules have been exhaustively examined by physical and analytical methods. Theoretical calculations (MO, LCAO and PPP) have been carried out on the parent compound and its benzo derivatives[156]. Polarized electronic and electric field spectra of **107** have also been reported[157]. The parent heterocycle **107** ($R^1 = R^2 = H$) has been studied by photoelectron spectroscopy[158]. Several ^{1}H NMR studies have been reported. The effect of substituents on the coupling constants in the benzene ring of **109** has been described[159].

It has been shown that the electron-donating ability follows the order NMe > O > S or Se in *N*-quaternized 2, 1, 3-benzoselenadiazoles and their analogues[160]. A conformational study of 3-phenyl-1, 2, 5-selenadiazole in the nematic phase at 100 MHz was reported[161]. The structures of the parent compound[162] **107** and of its benzo derivative **109**[163] were determined under similar conditions on the basis of ^{1}H, ^{13}C and ^{77}Se NMR resonances. ^{13}C chemical shifts and ^{13}C–H coupling constants in 2, 1, 3-benzoselenadiazole are also available[164]. Weak solvent effects, detected in the ^{13}C NMR[164] and ^{15}N NMR spectra, were described[165] for **109** and its other chalcogeno analogues. Magnetic circular dichroism and electronic spectra of piazselenole have been reported[166]. UV spectra[167] and pK_a determinations by UV–visible techniques were described[168]. Selenium-containing radicals, derived from 1, 2, 5-selenadiazoles, were investigated by ESR techniques[169]. The behaviour of theses heterocycles under electronic impact showed an important loss of the corresponding nitrile and chalcogen extrusion ($Se > S \gg O$)[170,171]. Meisenheimer complexes[172] and metal carbonyl derivatives[173] were investigated. Shermann and coworkers discussed the redox behaviour of 2, 1, 3-benzoselenadiazole[174] and the polarographic reduction of (*c*)-condensed 1, 2, 3-selenadiazoles was studied[175,176]. The investigation of **109** by electrical dipole moment measurements allowed the determination of the importance of its mesomeric forms[177] and the increase in mesomeric charge transfer from oxygen to selenium has been confirmed recently[178]. Complexes with tetracyanoquinodimethane (TCNQ) have been studied[179].

Kinetic data for the formation of (*c*)-condensed selenadiazoles have been reported[180,181] and allow the use of this reaction for the determination of Se^{IV} derivatives (even in trace amounts) or of *o*-diamines[182–184]. Quantitative removal of chlorine or bromine from air streams is possible with piazselenole[185], and the complex so obtained can be thermally regenerated[186a]. The pH-dependent stability of Pd^{2+} complexes with 2, 1, 3-benzoselenadiazolyl-α-amino acids has been studied[186b].

E. Two Se—N Bonds with Two Other Heteroatoms in the Ring

1. 1, 2, 3, 5-Selenatriazole

Reaction of 1, 2-diamino-5-phenylimidazoles with SeO_2[187a] affords 5-phenylimidazolo[1, 2-*c*]-1, 2, 3, 5-selenatriazole derivatives (equation 93). Benzimidazolo[1, 2-*c*]-1, 2, 3, 5-selenatriazole is obtained similarly.

(110) (111)

(93)

F. One Te—N Bond

1. 1,2-Tellurazole (isotellurazole) and derivatives

3-Substituted isotellurazoles were obtained recently in 5–10% yields by reaction of α, β-acetylenic ketones with hydroxylamine-θ-sulphonic acid derivatives and K_2Te[187b]. This method is analogous to that one used in the Se series[81] but fails for the preparation of the unsubstituted heterocycle.

Benzisotellurazole (**114**) was first obtained in 1978[188] by the reaction of *o*-bromotellurobenzaldehyde (**112**) with ammonia or the polyphosphoric acid (PPA) ring closure of *o*-butyltellurobenzaldoxime (**113**) (equation 94).

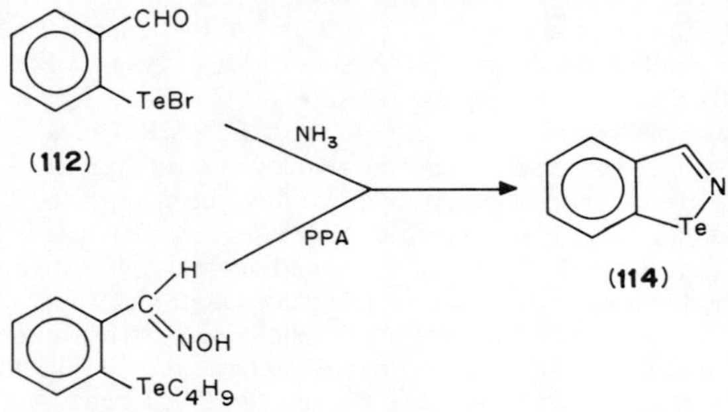

(113)

(94)

Thienoisotellurazoles could not be obtained by this method, the corresponding ditellurides being formed. Benzisotellurazole reacts with MeI to give *N*-methylbenzisotellurazolium iodide[188]. X-ray analysis[189] of the structure of 114 shows very short Te $\cdots$ N intermolecular contacts (2.4 Å), explaining its higher melting point and its lower solubility compared with the selenium analogue.

G. Two Te—N Bonds

1. 1,2,5-Telluradiazole and derivatives

In 1982, Bertini and coworkers[190,191] described the synthesis of the first examples of 1, 2, 5-telluradiazoles (**116**) (equation 95). The ring-opening reaction of 1, 2, 5-thiadiazoles (or better their selenium analogue) affords an intermediate dimetalloaldimine (**115**), which can be trapped with $TeCl_4$ in the presence of Et_3N to give **116**.

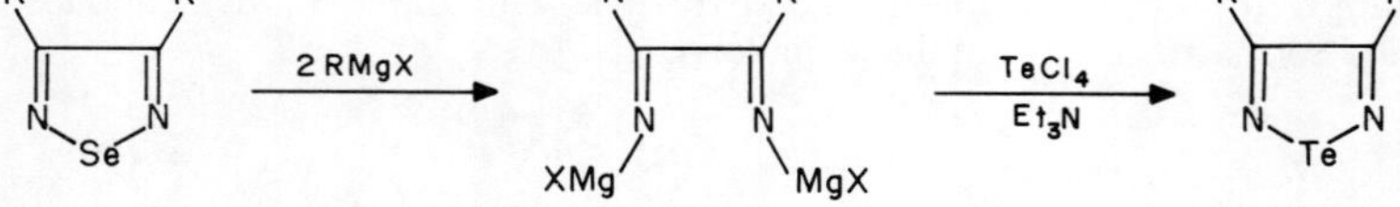

(95)

(115) + R_2Se (116)

As for benzisotellurazole (**114**), X-ray analysis of the structure of **116** also confirms the presence of close intermolecular quasi-polymeric contacts[192], explaining the large difference between the melting points of **116** ($R^1 = R^2 = H$; 185–188 °C) and its Se analogue (21 °C). Compound **116** and its derivatives are stable in air and water and are weak bases. They are unstable to light and are hydrolysed by acids to α-diketones, NH_4^+, Te and H_2TeO_3. Elemental tellurium reacts with *o*-chloromercuridiazobenzene to give a compound reducible to di(*o*-aminophenyl) ditelluride, the X-ray data for which are in agreement with a quasi 1-chloro-1, 2, 5-telluradiazole ring system[193].

X. SIX-MEMBERED RINGS

A. One Se—N Bond with One Oxygen in the Ring

Only derivatives of 5, 6-dihydro-1, 4, 3-oxaselenazine (**117**) are available[194–197] through 1, 4-dipolar cycloaddition of styrene (equation 96).

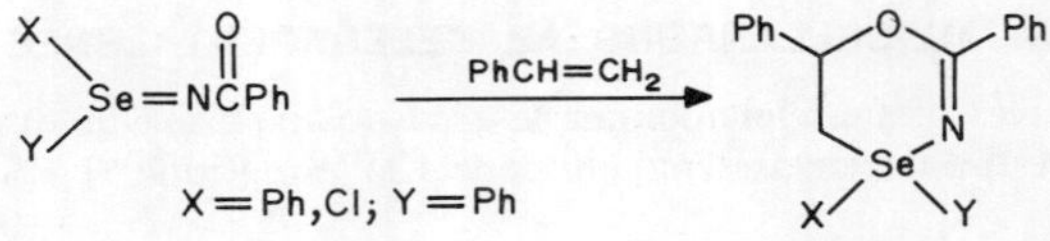

(96)

B. One Se—N Bond with One Sulphur in the Ring

Recently, ring expansion of 2-azido-2-ethylthio-1, 3-thiaselenols (**118**) afforded 2-ethylthio-1, 4, 3-thiaselenazine (**119**) derivatives with good yields[198] (equation 97).

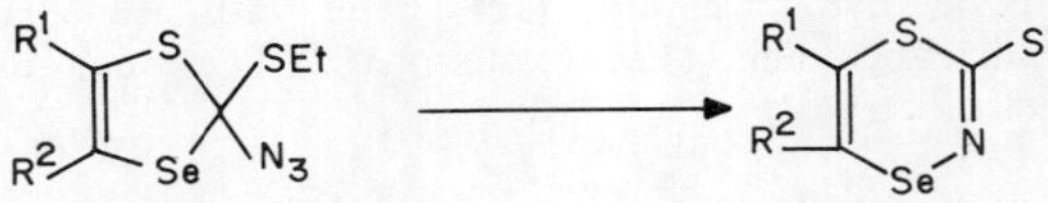

(97)

These new heterocyclic systems were well characterized by the usual techniques. They are thermally unstable and react with HI to give 2-amino-1, 3-thiaselenolium iodides.

C. One Se—N Bond with One Other Nitrogen in the Ring

3-Thioxo-6-chloromethyl-1, 2, 4-perhydroselenadiazine is produced by the reaction of allylthiourea with Se_2Cl_2[199]. Condensation[200] of benzeneselenenyl chloride with **120** leads to an ylid (**121**), which can be thermolysed to the benzoselenadiazine derivative **122** (equation 98).

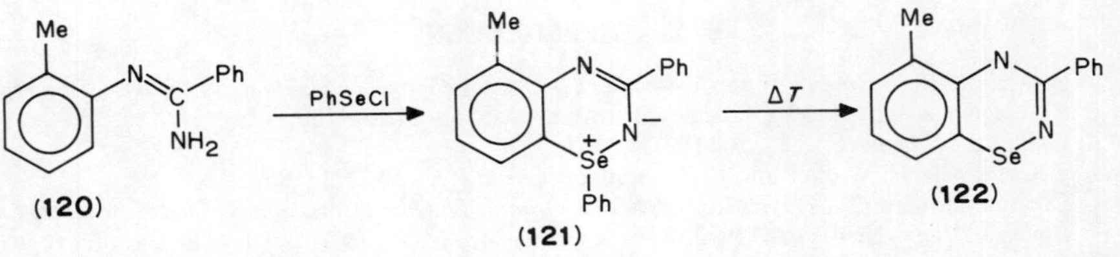

(98)

D. Two Se—N Bonds

As with *o*-phenylenediamine, the reaction of 1,8-diaminonaphthalene with SeO_2 affords 1*H*,3*H*-naphtho[1,8-*c*,*d*]-1,2,6-selenadiazine[201,202]. This reaction was irreproducible but, by reaction with $SeOCl_2$, the corresponding selenoxide could be isolated in 51% yield. The same methodology was applied to 1,4,5,8-tetraaminonaphthalene and afforded naphtho[1,8-*c*,*d*:4,5-*c*′,*d*′]bis-(1,2,6-selenadiazine)[15]. Its electronic structure has been studied by ESCA and theoretical calculations[203].

XI. LARGER RINGS

Condensation of $SeOCl_2$ (or Se_2Cl_2) with lithium trisilylamide affords eight- and fifteen-membered rings. They have been isolated and characterized by physicochemical methods and by X-ray diffraction[204].

XII. MISCELLANEOUS: AZASELENAPENTALENES

Only a few derivatives of azaselenapentalene are known. Bisarylhydrazones (**123**) react with SeO_2 and give the tetraazaselenapentalene **124** (equation 99)[205].

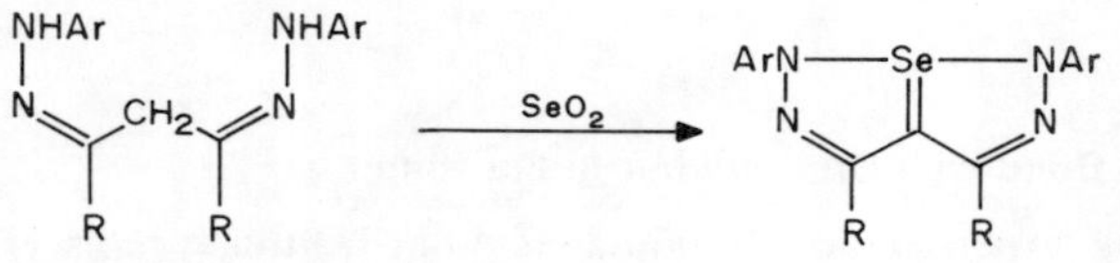

(123) (124)

(99)

3-Methylene-1,2-diselenolylium cations (**125**) react with arenediazonium salts[206] and lead to diselenadiazapentalene (**126**) (equation100). These compounds have been studied by mass spectrometry[111].

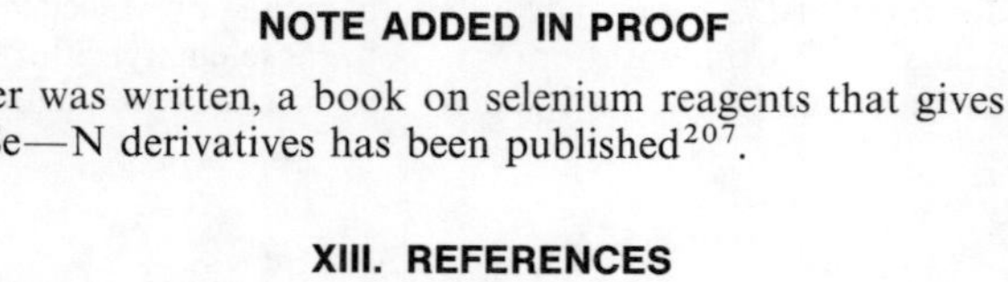

(125) (126)

(100)

NOTE ADDED IN PROOF

Since this chapter was written, a book on selenium reagents that gives some synthetic applications of Se—N derivatives has been published[207].

XIII. REFERENCES

1. H. Rheinholt, in *Houben Weyl, Methoden der Organischen Chemie, Schwefel-, Selen-, Tellurverbindungen*, Vol. IX, Georg Thieme, Stuttgart, 1955, pp. 1124–1180.
2. E. Painter, *Chem. Rev.*, **28**, 200 (1941).
3. T. Campbell, H. Walker and G. Coppinger, *Chem. Rev.*, **50**, 279 (1952).
4. D. Klaymann and W. H. H. Gunther (Eds.), *Organic Selenium Compounds: Their Chemistry and Biology*, Wiley, New York, 1973.
5. R. A. Zingaro and W. Cooper, *Selenium*, Van Nostrand Reinhold, New York, 1974.

6. K. Irgolic, *The Organic Chemistry of Tellurium*, Gordon and Breach, New York, 1974.
7. W. Cooper, *Tellurium*, Van Nostrand Reinhold, New York, 1977.
8. K. Irgolic, *J. Organomet. Chem.*, **103**, 91 (1975); **130**, 411 (1977); **158**, 269 (1978); **189**, 65 (1980); **203**, 367 (1980).
9. G. Pattenden, *Saturated Heterocyclic Chemistry*, Chemical Society, London, 1973–78.
10. C. Bird and G. Cheeseman, *Aromatic and Heteroaromatic Chemistry*, Vols. 1–5, Chemical Society, London, 1973–77.
11. H. Suschitzky and O. Meth-Cohn, *Aromatic and Hetercoaromatic Chemistry*, Vols. 6 and 7, Chemical Society, London, 1978–79.
12. D. Reid, *Organic Compounds of Sulphur, Selenium and Tellurium*, Vols. 1–6, Chemical Society, London, 1970–81.
13. H. Suschitzky and O. Meth-Cohn, *Heterocyclic Chemistry*, Chemical Society, London, 1980.
14. L. Christiaens, in *Proceedings of the Third International Symposium on Organic Selenium and Tellurium Compounds* (eds. D. Cagniant and G. Kirsch), University of Metz, Metz, 1979, pp. 69–118.
15. M. Renson, in *The Chemistry of Organic Selenium and Tellurium Compounds* (Eds. S. Patai and Z. Rappoport), Vol. 1, Wiley, Chichester, 1986, pp. 399–516.
16. (a) H. J. Reich and J. M. Renga, *J. Org. Chem.*, **40**, 3313 (1975); (b) H. J. Reich, *Acc. Chem. Res.*, **12**, 22 (1979); (c) S. Tomoda, Y. Takeuchi and Y. Nomura, *Synthesis*, 212 (1985).
17. T. Hori and K. B. Sharpless, *J. Org. Chem.*, **44**, 4208 (1979).
18. P. T. Southwellkelly, F. L. Johnstone and E. R. Cole, *Phosphorus Sulfur*, **1**, 261 (1976).
19. (a) M. Jefson and J. Meinwald, *Tetrahedron Lett.*, **22**, 3561 (1981); (b) C. Paulmier and P. Lerouge, *Tetrahedron Lett.*, **23**, 1557 (1982).
20. H. J. Reich, J. M. Renga and J. E. Trend, *Tetrahedron Lett.*, 2217 (1976).
21. C. O. Meese, W. Walter and H. Schmidt, *Tetrahedron Lett.*, 3133 (1976).
22. C. O. Meese, *Dissertation*, University of Hamburg, 1978.
23. P. A. Grieco, Y. Yokuyama and E. Williams, *J. Org. Chem.*, **43**, 1283 (1978).
24. T. G. Back and D. J. McPhee *J. Org. Chem.*, **49**, 3842 (1984).
25. (a) K. C. Nicolaou, D. A. Claremon, W. E. Barnette and S. P. Seitz, *J. Am. Chem. Soc.*, **101**, 3704 (1979); (b) S. V. Ley and B. Lygo, *Tetrahedron Lett.*, 4625 (1982); (c) T. Toru, S. Fujita and E. Maekawa, *J. Chem. Soc., Chem. Commun.*, 1082 (1985); (d) R. R. Webb, II, and S. Danishefsky, *Tetrahedron Lett.*, 1357 (1983).
26. P. A. Grieco, D. Y. Jaw, D. A. Claremon and K. C. Nicolaou, *J. Org. Chem.*, **46**, 1215 (1981).
27. T. Toru, M. Nischigaki, T. Seko, T. Kanefusa and E. Maekawa, *Synthesis*, 878 (1985).
28. N. Ya. Derkach, T. V. Lyapina and E. S. Levchenko, *Zh. Org. Khim.*, **14**, 280 (1978); *J. Org. Chem. USSR*, **14**, 256 (1978).
29. K. Dostal and M. Sochorkowa, *Z. Chem.*, **12**, 46 (1972).
30. K. Dostal, Z. Zak and A. Haba, *Collect. Czech. Chem. Commun.*, **42**, 1466 (1977).
31. F. A. Davis and E. W. Kluger, *J. Am. Chem. Soc.*, **98**, 302 (1976).
32. D. H. R. Barton, A. G. Brewster, S. V. Ley and M. N. Rosenfeld, *J. Chem. Soc., Chem. Commun.*, 147 (1977).
33. D. Hellwinckel and G. Fahrbach, *Justus Liebigs Ann. Chem.*, **715**, 68 (1962).
34. N. Ya. Derkach, T. V. Lyapina and N. A. Pasmurtseva, *Zh. Org. Khim.*, **10**, 807 (1974); *J. Org. Chem. USSR*, **10**, 810 (1974).
35. S. Tamagaki, S. Oae and K. Sakaki, *Tetrahedron Lett.*, 649 (1975).
36. M. Hori, T. Kataoka, H. Shimizu and K. Tomimatsu, *Tetrahedron Lett.*, 901 (1982); *Phosphorus Sulfur*, **16**, 97 (1983).
37. S. Oae, M. Fukumura and N. Furukawa, *Phosphorus Sulfur*, **10**, 153 (1981).
38. N. Ya Derkach, N. P. Tischenko, and V. G. Volonchuk, *Zh. Org. Khim.*, **14**, 958 (1978); *J. Org. Chem. USSR*, 14, 896 (1978).
39. V. P. Krasnov, V. I. Naddaka and V. I. Minkin, *Zh. Org. Khim.*, **17**, 445 (1981); *Chem. Abstr.*, **95**, 42551x.
40. R. Appel and G. Buchler, *Z. Anorg. Allg. Chem.*, **348**, 175 (1966).
41. V. I. Naddaka, V. P. Gar'kin, I. D. Sadekov, V. P. Krasnov and V. I. Minkin, *Zh. Org. Khim.*, **15**, 896 (1979); *J. Org. Chem. USSR*, **15**, 804 (1979).
42. V. I. Naddaka, L. L. Popova, and V. I. Minkin, *Zh. Org. Khim.*, **18**, 259 (1982); *J. Org. Chem. USSR*, **18**, 226 (1982).

43. N. Ya. Derkach, N. A. Pasmurtseva and E. S. Levchenko, *Zh. Org. Khim.*, **7**, 1543 (1971); *J. Org. Chem. USSR*, **7**, 1600 (1971).
44. N. Ya. Derkach, N. A. Pasmurtseva, T. V. Lyapina and E. S. Levchenko *Zh. Org. Khim.*, **10**, 1873 (1974); *J. Org. Chem. USSR*, **10**, 1884 (1974).
45. N. Ya. Derkach, T. V. Lyapina and E. S. Levchenko, *Zh. Org. Khim.*, **16**, 33 (1980); *J. Org. Chem. USSR*, **16**, 31 (1980).
46. V. Horn and R. Paetzold, *Z. Anorg. Allg. Chem.*, **404**, 213 (1974).
47. N. Ya. Derkach, T. V. Lyapina and E.S. Levchenko, *Zh. Org. Khim.*, **17**, 622 (1981); *J. Org. Chem. USSR*, **17**, 536 (1981).
48. N. Ya. Derkach, N. A. Pasmurtseva and E. S. Levchenko, *Zh. Obsch. Khim.*, **42**, 1652 (1972); *J. Gen. Chem. USSR*, **42**, 1646 (1972).
49. N. Ya. Derkach, T. V. Lyapina and N. A. Pasmurtseva, *Zh. Org. Khim.*, **10**, 807 (1974); *J. Org. Chem. USSR*, **10**, 810 (1974).
50. A. Baroni, *Atti R. Accad. Naz. Lincei Mem. Cl. Sci. Fis. Mat. Nat.*, **6**, 255 (1932); *Chem. Abstr.*, **27**, 1870 (1933).
51. K. B. Sharpless, T. Hori, L. K. Truesdale and C. O. Dietrich, *J. Am. Chem. Soc.*, **98**, 269 (1976).
52. N. Ya. Derkach and G. G. Barashenkov, *Zh. Org. Khim.*, **12**, 2484 (1976); *J. Org. Chem. USSR*, **12**, 2408 (1976).
53. N. Ya. Derkach, G. G. Barashenkov and E. S. Levchenko, *Zh. Org. Khim.*, **18**, 964 (1982); *J. Org. Chem. USSR*, **18**, 837 (1982).
54. G. Kresze and W. Wucherpfennig, *Angew. Chem., Int. Ed. Engl.*, **6**, 149 (1967).
55. N. Ya. Derkach, G. G. Barashenkov and E. I. Slyusarenko, *Zh. Org. Khim.*, **18**, 70 (1982); *J. Org. Chem. USSR*, **18**, 60 (1982).
56. Ya. G. Balton and E. S. Levchenko, *Zh. Org. Khim.*, **3**, 772 (1967); *J. Org. Chem. USSR*, **3**, 742 (1967).
57. N. Ya. Derkach and N. A. Pasmurtseva, *Zh. Org. Khim.*, **9**, 1414 (1973); *J. Org. Chem. USSR*, **9**, 1443 (1973).
58. N. Ya. Derkach, T. V. Lyapina, and E. S. Levchenko, *Zh. Org. Khim.*, **10**, 139 (1974); *J. Org. Chem. USSR*, **10**, 145 (1974).
59. N. Ya. Derkach, T. V. Lyapina, and E. S. Levchenko, *Zh. Org. Khim.*, **14**, 280 (1978); *J. Org. Chem. USSR*, **14**, 256 (1978).
60. G. Hopf and R. Paetzold, *Z. Anorg. Allg. Chem.*, **401**, 179 (1973).
61. G. G. Barashenkov and N. Ya. Derkach, *Z. Obshch. Khim.*, **48**, 1110 (1978); *J. Gen. Chem. USSR*, **48**, 1012 (1978).
62. O. Foss and V. Janickis, *J. Chem. Soc., Dalton Trans.*, 620 (1980).
63. D. Hellwinckel and G. Fahrbach, *Justus Liebigs Ann. Chem.*, **712**, 1 (1968).
64. V. I. Naddaka, V. P. Gar'kin, I. D. Sadekov and V. I. Minkin, *Zh. Org. Khim.*, **13**, 220 (1977); *J. Org. Chem. USSR*, **13**, 202 (1977).
65. V. I. Naddaka, V. P. Gar'kin, A. A. Maksimenko and I. D. Sadekov, *Zh. Org. Khim.*, **16**, 2581 (1980); *J. Org. Chem. USSR*, **16**, 2210 (1980).
66. V. I. Naddaka and V. P. Gar'kin, *Zh. Org. Khim.*, **17**, 401 (1981); *J. Org. Chem. USSR*, **17**, 341 (1981).
67. H. Suzuki, S. Takeda and Y. Hanazaki, *Chem. Lett.*, 679 (1985).
68. V. I. Naddaka, V. P. Gar'kin, K. V. Obayan and V. I. Minkin, *Zh. Org. Khim.*, **17**, 669 (1981); *J. Org. Chem. USSR*, **17**, 579 (1981).
69. V. I. Naddaka, V. P. Gar'kin and V. I. Minkin, *Zh. Org. Khim.*, **16**, 2619 (1980); *Chem. Abstr.*, **94**, 208474f.
70. V. I. Naddaka, V. P. Gar'kin, K. V. Obayan and V. I. Minkin, *Zh. Org. Khim.*, **17**, 883 (1981); *J. Org. Chem. USSR*, **17**, 777 (1981).
71. V. I. Naddaka, V. P. Gar'kin and K. V. Obayan, *Zh. Org. Khim.*, **17**, 2020 (1981); *J. Org. Chem. USSR*, **17**, 1803 (1981).
72. E. A. Stukalo, A. A. Kisilenko and L. N. Markowski, *Zh. Org. Khim.*, **16**, 2554 (1980); *J. Org. Chem. USSR*, **16**, 2184 (1980).
73. L. N. Markovski, E. A. Stukalo and G. P. Kunitskaya, *Zh. Org. Khim.*, **13**, 2055 (1977); *J. Org. Chem. USSR*, **13**, 1911 (1977).
74. L. N. Markovski, E. A. Stukalo and G. P. Kunitskaya, *Zh. Org. Khim.*, **13**, 2514 (1977); *J. Org. Chem. USSR*, **13**, 2360 (1977).

75. V. I. Naddaka, V. P. Krasnov and V. I. Minkin, *Zh. Org. Khim.*, **19**, 2632 (1983); *J. Org. Chem. USSR*, **19**, 2314 (1983).
76. D. Black and K. Watson, *Aust. J. Chem.*, **26**, 2167 (1973).
77. S. Tamagaki, K. Sakaki and S. Oae, *Heterocycles*, **2**, 39 (1974).
78. H. Roesky and K. Ambrosius, *Z. Naturforsch., Teil B*, **33**, 759 (1978).
79. I. Lalezari, A. Shafiee and M. Yalpani, *Adv. Heterocycl. Chem.* (Ed. A. R. Katritzky), **24**, 109–150, Academic Press, Orlando (1979).
80. F. Wille, A. Ascherl, G. Kaupp and L. Capeller, *Angew. Chem.*, **74**, 753 (1962).
81. F. Lucchesini, V. Bertini and A. De Munno, *Tetrahedron*, **40**, 931 (1984).
82. N. Onyamboko, R. Weber, A. Fauconnier and M. Renson, *Bull. Soc. Chim. Belg.*, **92**, 53 (1983).
83. D. Ames, A. Singh and W. Smyth, *Tetrahedron*, **39**, 831 (1983).
84. J. Liebscher and H. Hartmann, *Synthesis*, 273 (1976).
85. R. Weber and M. Renson, *Bull. Soc. Chim. Fr.*, 1124 (1976).
86. M. M'Buyi, *Doctoral Thesis*, University of Liège, 1983.
87. R. Lesser and R. Weiss, *Chem. Ber.*, **57**, 1079 (1924).
88. A. Muller, E. Cadenas, P. Graf and H. Sies, *Biochem. Pharmacol.*, **33**, 3235 (1984).
89. R. Reiter and A. Wendel, *Biochem. Pharmacol.*, **33**, 1923 (1984).
90. M. Parnham and S. Kindt, *Biochem. Pharmacol.*, **33**, 3247 (1984).
91. A. Muller, H. Gabriel and H. Sies, *Biochem. Pharmacol.*, **34**, 1185 (1985).
92. P. Kuhl, H. Borbe, A. Romer, H. Fischer and M. Parnham, Paper presented at the 7th European Workshop on Inflammation, Capri, Italy, 1985.
93. A. Welter, L. Christiaens and P. Wirtz, *Eur. Pat. Appl.*, EP 44453; *Chem. Abstr.*, **96**, 199699v (1982).
94. A. Welter, S. Leyck and E. Etschenberg, *Ger. Offen.*, DE 3226286; *Chem. Abstr.*, **100**, 191883f (1984).
95. A. Welter, S. Leyck and E. Etschenberg, *Ger. Offen.*, DE 3226284; *Chem. Abstr.*, **101**, 38454t (1984).
96. H. Gschwend and H. Rodriguez, *Org. React.*, **26**, 1–360, Wiley, New York (1979).
97. L. Christiaens, A. Luxen, M. Evers, Ph. Thibaut, M. M'Buyi and A. Welter, *Chem. Scr.*, **24**, 178 (1984).
98. C. Lambert, University of Liège, unpublished results, 1986.
99. R. Weber, *Doctoral Thesis*, University of Liège, 1975.
100. F. Lucchesini, V. Bertini and A. De Munno, *Heterocycles*, **23**, 127 (1985).
101. L. Christiaens, J.-L. Piette, L. Laitem, M. Baiwir, J. Denoel and G. Llabres, *Org. Magn. Reson.*, **8**, 354 (1976).
102. N. Onyamboko, M. Renson, S. Chapelle and P. Granger, *Org. Magn. Reason.*, **19**, 74 (1982).
103 R. Weber and M. Renson, *Bull. Soc. R. Sci. Liège*, **48**, 146 (1979).
104. T. Ueda, Y. Shibata, J. Sakakibara, M. Indue and T. Ishida, *Chem. Pharm. Bull.*, **30**, 3424 (1982).
105. (a) H. Ito, J. Sakakibara and T. Veda, *Cancer Lett.*, **28**, 61 (1985); (b) T. Veda, Y. Shibata and J. Sakakibara, *Chem. Pharm. Bull.*, **33**, 3065 (1985).
106. I. Lalezari, A. Shafiee and M. Yalpani, *Tetrahedron Lett.*, **10**, 5105 (1969).
107. O. Zimmer and H. Meier, *Chem. Ber.*, **114**, 2938 (1981).
108. R. Hanson and M. Davis, *J. Labelled Compd. Radiopharm.*, **16**, 31 (1979).
109. R. Hanson and M. Davis, *J. Pharm. Sci.*, **70**, 91 (1981).
110. S. Dhawan and J. Kagan, *J. Labelled Compd. Radiopharm.*, **19**, 331 (1982).
111. J. Moller, R. Christie, C. Pedersen and D. Reid, *Acta Chem. Scand., Ser. B*, **30**, 600 (1976).
112. B. Bak, O. Nielsen, H. Svanholt and A. Holm, *Chem. Phys. Lett.*, **55**, 36 (1978).
113. H. Petersen and H. Meier, *Chem. Ber.*, **111**, 3423 (1978).
114. H. Spencer, M. Lakshmilkantham, M. Cava and A. Garito, *J. Chem. Soc., Chem. Commun.*, 867 (1975).
115. L. Christiaens, unpublished results.
116. M. Ghandehari, D. Davalian, M. Yalpani and M. Partovi, *J. Org. Chem.*, **39**, 3906 (1974).
117. A. Holm, C. Berg, C. Bjerre, B. Bak and H. Svanholt, *J. Chem. Soc., Chem. Commun.*, 99 (1979).
118. H. Bock, S. Aygen and B. Solouki, *Z. Naturforsch., Teil B*, **38**, 611 (1983).
119. W. Sander and O. Chapman, *J. Org. Chem.*, **50**, 543 (1985).
120. (a) R. Schulz and A. Schweig, *Z. Naturforsch., Teil B*, **39**, 1536 (1984); (b) R. Schulz and A. Schweig, *Tetrahedron Lett.*, **25**, 2337 (1984).

121. A. Shafiee, S. Toghraie, F. Aria and G. Mortezaei-Zandjani, *J. Heterocycl. Chem.*, **19**, 1305 (1982).
122. H. Meier, M. Layer and A. Zetzeke, *Chem. Ztg.*, **98**, 460 (1975).
123. H. Meier, N. Hanold and H. Kolshorn, *Angew. Chem., Int. Ed. Engl.*, **21**, 66 (1982).
124. H. Meier, T. Molz, U. Mekle, T. Echter and M. Loreh, *Justus Liebigs Ann. Chem.*, 914 (1982).
125. E. Muller and G. Odenigbo, *Justus Liebigs Ann. Chem.*, 1435 (1975).
126. F. Malek-Yazdi and M. Yalpani, *J. Org. Chem.*, **41**, 729 (1976).
127. I. Lalezari, *Synthesis*, 660 (1984).
128. A. Shafiee and G. Fanaii, *Synthesis*, 512 (1984).
129. V. Laishev, M. Petrov and A. Petrov, *Zh. Org. Khim.*, **18**, 514 (1982).
130. H. Meier, J. Zountsas and O. Zimmer, *Z. Naturforsch., Teil B.*, **36**, 1017 (1981),
131. A. Caplin, *J. Chem. Soc., Perkin Trans. 1*, 30 (1974).
132. R. Arad-Yellin and F. Wudl, *J. Organomet. Chem.*, **280**, 197 (1985).
133. H. Boek, S. Aygen, P. Rosmus and B. Solouki, *Chem. Ber.*, **113**, 3187 (1980).
134. N. Batzel and R. Boese, *Z. Naturforsch., Teil B*, **36**, 172 (1981).
135. P. Mente and C. Rees, *J. Chem. Soc., Chem. Commun.*, 418 (1972).
136. T. Gilchrist, P. Mente and C. Rees, *J. Chem. Soc., Perkin Trans. 1*, 2165 (1972).
137. G. Schrauzer and H. Kisch, *J. Am. Chem. Soc.*, **95**, 2501 (1973).
138. A. Eid and A. Salama, *Egypt. J. Pharm. Sci.*, **20**, 41 (1979); *Chem. Abstr.*, **98**, 143332u (1983).
139. (a) W. Becker and J. Meyer, *Chem. Ber.*, **37**, 2550 (1904); (b) V. Cohen, *Synthesis*, 768 (1978); (c) J. Goerdeler, D. Gross and H. Klinke, *Chem. Ber.*, **96**, 1289 (1963).
140. M. Jacobsen, L. Henriksen and A. Holm, *Acta Chem. Scand., Ser. B*, **37**, 585 (1983).
141. Y. Shealy and J. Clayton, *J. Heterocycl. Chem.*, **4**, 96 (1967).
142. H. Buchwald and K. Ruhlmann, *J. Organomet. Chem.*, **166**, 25 (1979).
143. H. Roesky and H. Hofmann, *Z. Naturforsch., Teil B*, **39**, 1315 (1984).
144. W. Sliwa and A. Thomas, *Heterocycles*, **20**, 71 (1983).
145. O. Hinsberg, *Chem. Ber.*, **22**, 862, 2895 (1889).
146. A. Thomas and W. Sliwa, *Heterocycles*, **20**, 1043 (1983).
147. R. Neidlein, D. Droste-Tran-Viet, A. Gieren, M. Kokkinidis, R. Wilckens, H. Geserich and W. Ruppel, *Helv. Chim. Acta*, **67**, 574 (1984).
148. S. Mataka, K. Takahashi, T. Imura and M. Tashiro, *J. Heterocycl. Chem.*, **19**, 1481 (1982).
149. K. Sharma, R. Singh and S. Kumari, *Synthesis*, 581 (1983).
150. A. Komin and M. Carmack, *J. Heterocycl. Chem.*, **12**, 829 (1975).
151. V. Sergeev, V. Pesin and N. Notikova, *Khim. Geterotsikl. Soedin.*, 328 (1972).
152. C. Pedersen, *Tetrahedron Lett.*, **20**, 745 (1979).
153. G. Eremeeva, Y. Akulin, T. Timofeeva, B. Strelets and L. Efros, *Khim. Geterotsikl. Soedin.*, 1129 (1982).
159. C. Pedersen, H. Harrit, M. Poliakoff and I. Dunkin, *Acta Chem. Scand., Ser B*, **31**, 848 (1977).
155. C. Pedersen and N. Hacker, *Tetrahedron Lett.*, **17**, 3981 (1977).
156. A. Gyul'maliev, I. Stenkevich and Z. Todres, *Khim. Geterotsikl. Soedin.*, 1055 (1975).
157. T. Lin and J. Braun, *Chem. Phys.*, **26**, 403 (1977).
158. H. Bock, B. Solouki, S. Aygen, T. Hirabayashi, S. Mohmada, P. Rosmus and J. Wittmann, *J. Mol. Struct.*, **60**, 31 (1980).
159. A. Katritzky and Y. Takeuchi, *J. Chem. Soc., Perkin Trans. 2*, 1682 (1972).
160. M. Davis, L. Deady and E. Homfeld, *J. Heterocycl. Chem.*, **11**, 1011 (1974).
161. C. Veracini, A. De Munno, V. Bertini, M. Congeri and G. Chidichimo, *J. Chem. Soc., Perkin Trans. 2*, 561 (1977).
162. G. Chidichimo, F. Lelj, P. Barili and C. Veracini, *Chem. Phys. Lett.*, **55**, 519 (1978).
163. C. Khetrapal and A. Kunwar, *J. Mol. Struct.*, **74**, 343 (1981).
164. G. Cheeseman and C. Turner, *Org. Magn. Reson.*, **6**, 430 (1974).
165. I. Yavari, R. Botto and J. Roberts, *J. Org. Chem.*, **43**, 2542 (1978).
166. H. Yamaguchi, A. Uchida, F. Yoneda and H. Baumann, *J. Chem. Soc., Faraday Trans. 2*, **77**, 947 (1981).
167. R. D. Gordon, *Spectrochim. Acta, Part A*, **36**, 1089 (1980).
168. J. Neve, M. Hanocq and L. Molle, *Talanta*, **26**, 15 (1979).
169. M. Chiu and B. Gilbert, *J. Chem. Soc., Perkin Trans. 2*, 258 (1973).
170. C. Pedersen and J. Moeller, *Acta Chem. Scand., Ser. B*, **29**, 483 (1975).
171. M. Arshadi, *Org. Mass. Spectrom.*, **13**, 379 (1978).

172. C. Deicha and F. Terrier, *J. Chem. Res.* (*S*), 312 (1981).
173. K. Pannell, A. Mayr, R. Hoggard, J. McKennis and J. Dawson, *Chem. Ber.*, **116**, 230 (1983).
174. E. Shermann, S. Lambert and K. Pilgram, *J. Heterocycl. Chem.*, **11**, 763 (1974).
175. V. Tsveniashvili, L. Tskalobadze and V. Gaprindashvili, *Soobsheh, Akad. Nauk Gruz. SSSR*, **71**, 625 (1973); *Chem. Abstr.*, **80**, 70049 (1974).
176. V. Tsveniashvili, M. Malashkhiya, N. Khavtasi and V. Gaprindashvili, *Zh. Obshch. Khim.*, **53**, 1851 (1983).
177. F. Tobiason, L. Huestis, C. Chandler, S. Pedersen and P. Peters, *J. Heterocycl. Chem.*, **10**, 773 (1973).
178. A. Gieren, H. Betz, T. Hubner, V. Lamm, R. Neillein and D. Droste, *Z. Naturforsch., Teil B*, **39**, 485 (1984).
179. H. Singh, W. McWhinnie, R. Ziolo and C. Jones, *J. Chem. Soc., Dalton Trans.*, 1267 (1984).
180. J. Neve, M. Hanocq and L. Molle, *Talanta*, **26**, 1173 (1979).
181. J. Neve, M. Hanocq and L. Molle, *Mikrochim. Acta*, 41 (1980).
182. G. Schwedt, *Fresenius. Anal. Chem.*, **288**, 50 (1977).
183. A. Montiel, *Analusis*, **9**, 102 (1981).
184. Y. Shibata, M. Morita and K. Fuwa, *Anal. Chem.*, **56**, 1527 (1984).
185. J. Blaha, W. Brittain and C. Meloan, *Anal. Lett.*, 1391 (1980).
186. (a) J. Blaha, W. Brittain, C. Meloan and W. Fateley, *Appl. Spectrosc.*, **34**, 636 (1980); (b) S. D'yachenko, M. Bureneva, M. Papirnik, V. Pesin, N. Ostashkova and A. Stetsenko, *Z. Obshch. Khim.*, **51**, 1912 (1981).
187. (a) K. Potts, R. Cody and R. Dennis, *J. Org. Chem.*, **46**, 4065 (1981); (b) F. Lucchesini and V. Bertini, Synthesis, 824 (1983).
188. R. Weber, J.-L. Piette and M. Renson, *J. Heterocycl. Chem.*, **15**, 865 (1978).
189. H. Campsteyn, L. Dupont, J. Lamotte-Brasseur and M. Vermeire, *J. Heterocycl. Chem.*, **15**, 745 (1978).
190. V. Bertini, F. Lucchesini and A. De Munno, *Synthesis*, 681 (1982).
191. M. Muniz-Miranda, G. Sbrana, V. Bertini, F. Lucchesini, E. Benedetti and A. De Munno, *Spectrochim. Acta, Part A*, **40**, 847 (1984).
192. V. Bertini, P. Dapporto, F. Lucchesini, A. Sega and A. De Munno, *Acta Crystallog., Sect. C*, **40**, 653 (1984).
193. R. Cobbledick, F. Enstein, W. McWhinnie and F. Musa, *J. Chem. Res.* (*S*), 145 (1979).
194. N. Derkach, T. Lyapina and G. Barashenkov, *Zh. Org. Khim.*, **15**, 788 (1979).
195. N. Derkach, T. Lyapina and G. Barshenkov, *Zh. Org. Khim.*, **16**, 33 (1980).
196. N. Derkach, G. Barashenkov and E. Levchenko, *Zh. Org. Khim.*, **19**, 1622 (1983).
197. N. Derkach, T. Krasnyanskaya, E. Levchenko and T. Zabolotnaya, *Zh. Org. Khim.*, **20**, 2377 (1984).
198. H. Poleschner, J. Bottger and F. Franghanel, *Synthesis*, 667 (1984).
199. M. Apostolescu, *Bull. Inst. Politech, Iasi*, **20**, 9 (1974); *Chem. Abstr.*, **82**, 112049x (1975).
200. T. Gilchrist, C. Rees and D. Vaughan, *J. Chem. Soc., Perkin Trans. 1*, 55 (1983).
201. F. Sachs, *Justus Liebigs Ann. Chem.*, **365**, 142 (1909).
202. M. Kaplan, R. Haddon, F. Schilling, J. Marshall and F. Bramwell, *J. Am. Chem. Soc.*, **101**, 3306 (1979).
203. J.-P. Boutique, J. Riga, J. Verbist, J. Delhalle, J. Fripiat, J.-M. Andre, R. Haddon and M. Kaplan, *J. Am. Chem. Soc.*, **104**, 2691 (1982).
204. H. Roesky, K. Weber and J. Bats, *Chem. Ber.*, **117**, 2686 (1984).
205. M. Perrier and J. Vialle, *Bull. Soc. Chim. Fr.*, **II**, 205 (1979).
206. M. Renson, in *The Chemistry of Organic Selenuim and Tellurium Compounds* (Eds. S. Patai and Z. Rappoport), Vol. 1, Wiley, Chichester, 1986, pp. 467–468.
207. C. Paulmier, *Selenium Reagents and Intermediates in Organic Synthesis*, Pergamon Press, Oxford, 1986.

The Chemistry of Organic Selenium and Tellurium Compounds Volume 2
Edited by S. Patai
© 1987 John Wiley & Sons Ltd.

CHAPTER **12**

Synthesis of organic conductors containing selenium and tellurium

D. COWAN and A. KINI*

The Johns Hopkins University, Baltimore, Maryland, USA

I. INTRODUCTION 463
A. What Are Organic Metals? 463
B. Sulfur-, Selenium- and Tellurium--Based Organic Metals 466
II. SYNTHESIS 467
A. Heterofulvalene Donors 467
1. Selenium donors 469
a. Engler's tetraselenafulvalene synthesis 472
b. Selenocarbamate route 473
c. Addition of CSe_2 to acetylenes 476
d. Triselenocarbonate–acetylene route 477
e. Selenodiazole route 477
f. Selenium insertion route 478
g. Electrochemical reduction of CSe_2 483
2. Tellurium donors 485
B. Other selenium/tellurium donors 485
1. Tetraseleno- and tetratelluro-polyacenes 485
2. Dichalcogenopyranylidenes 485
3. Miscellaneous Se/Te donors 489
III. ABBREVIATIONS USED 490
IV. ACKNOWLEDGEMENTS 491
V. REFERENCES 491

I. INTRODUCTION

A. What Are Organic Metals?

The electrical conductivity (σ) of solids is a physical property with a range of values greater than 20 orders of magnitude. It is usually determined by a four-probe measurement

*Present address: Argonne National Laboratory, Argonne, Illinois 60439, USA.

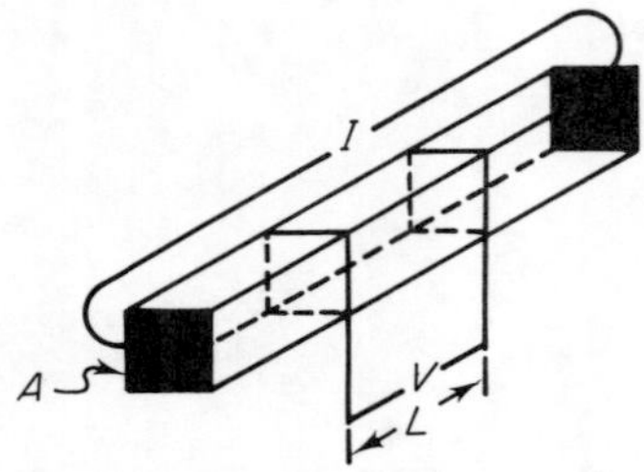

FIGURE 1. Arrangement of four probes for electrical conductivity measurements on single crystals

depicted in Figure 1. This type of measurement minimizes any error due to contact resistance since little current flows in the voltage measuring circuit. The electrical conductivity is then calculated from the known or measured current (I), the voltage drop (V) and the sample dimensions (A and L) (equation 1).

$$\sigma = (1/R)(L/A) = (I/V)(L/A)\Omega^{-1}\text{cm}^{-1} \tag{1}$$

The best insulators such as PTFE have room temperature conductivities of about 10^{-16} to $10^{-20}\Omega^{-1}\text{cm}^{-1}$, whereas semiconductors such as silicon have conductivities in the range 10^{-2}–$10^{-8}\,\Omega^{-1}\text{cm}^{-1}$ and metals such as copper and silver have conductivities that approach $10^6\Omega^{-1}\text{cm}^{-1}$. Materials such as mercury that become superconducting at low temperatures have no resistance below the critical temperature (T_c) and therefore have infinite conductivity.

Even more characteristic than the magnitude of the conductivity is the temperature dependence of the conductivity. As the temperature is increased, the conductivity increases for insulators and semiconductors (see equation 2, where σ_0, E_g, k and Δ are constants)

$$\sigma = \sigma_0 e^{-Eg/kT} = \sigma_0 e^{-\Delta/T} \tag{2}$$

while the conductivity of metals decreases as the temperature increases (see equation 3, where ρ is the resistivity and ρ_0, B and γ are constants).

$$\rho = 1/\sigma = \rho_0 + BT^{\gamma} \tag{3}$$

Since the electrical conductivity of a solid is determined by its electronic structure, an elementary band description is presented. This model will be shown to be valid for many of the organic conductors later in the chapter. On bringing together two atomic p orbitals on adjacent atoms or molecular fragments, a π bonding and a π^* anti-bonding set of molecular orbitals are formed. If there is one electron per basis function, the π bonding molecular orbital is filled and the π^* anti-bonding molecular orbital is empty. Now, if we bring together or stack two of these molecules, the π bonding level is split into two new molecular orbitals and the π^* anti-bonding level is also split into two new orbitals. Since there are two electrons per bonding π orbital, both of the new π levels are filled and the π^* unfilled. If n of the molecules are stacked as in a crystal, then we will form one band from the π orbital (HOMO) and one from the π^* orbital (LUMO). Each band will contain n states or levels. Since there are $2n$ electrons in the band formed from the HOMO, this band will be completely filled while the band from the LUMO will be completely empty—there were no electrons associated with the π^* basis functions. The width of the band will be determined by the extent to which the orbitals overlap (see Figure 2). This band picture describes most insulators and semiconductors. As the temperature is increased, electrons

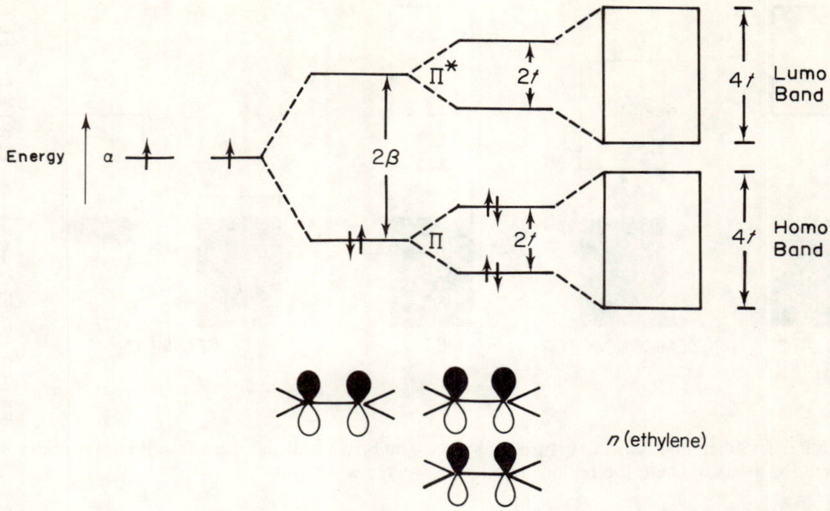

FIGURE 2. Formation of electronic bands from a stack of ethylene molecules. The band width is given by the value of $4t$ and is approximately twice the value of the dimer splitting

are thermally excited from the filled to empty band. Both the electrons in the LUMO and holes in the HOMO band can contribute to the electrical conductivity. Intrinsic conductivity is then largely controlled by the band gap (E_g, see equation 2).

If it is possible to change the electronic structure by adding electrons to the band formed from the LUMO or remove electrons from the band formed from the HOMO, then a metal should result. The metallic structure is a consequence of the fact that it is possible in an incompletely filled band to shift some of the electrons to infinitesimally higher energy states within the band, giving rise to a net velocity or conductivity. This conductivity in metals (and also in semiconductors) is determined by the concentration and the mobility of the charge carriers. Unlike semiconductors, there is no energy or band gap in metals, and therefore the temperature dependence of the conductivity is dominated by the mobility term, which depends on the scattering of the electrons caused by interactions with the lattice vibrations. As the temperature is reduced, the amplitude of the lattice vibrations is diminished and the mobility and conductivity increase. Figure 3 provides a schematic drawing for insulators, semiconductors, metals and semimetals. The semimetals result from band overlaps which produce incompletely filled bands. It is obvious that to form either a semimetal or metal we need incompletely filled bands, and this translates into the need for stable organic radicals. The other obvious feature is that these radicals must be planar molecules with delocalized π molecular orbitals to allow stacking. The production of a stable radical by steric protection will not allow effective interaction of the radical species in the solid state. These stacks of planar radical species are quasi-one-dimensional in character since the strong interaction is along the stack direction. This results in a structure that is subject to phase transitions which convert metals into semiconductors. In order to increase the interchain interaction (more 2D character) and suppress these transitions, heteroatoms (S, Se, Te) have been added to the ring. The heteroatoms have three other beneficial effects: 1, because of their polarizability they stabilize radical (cation)

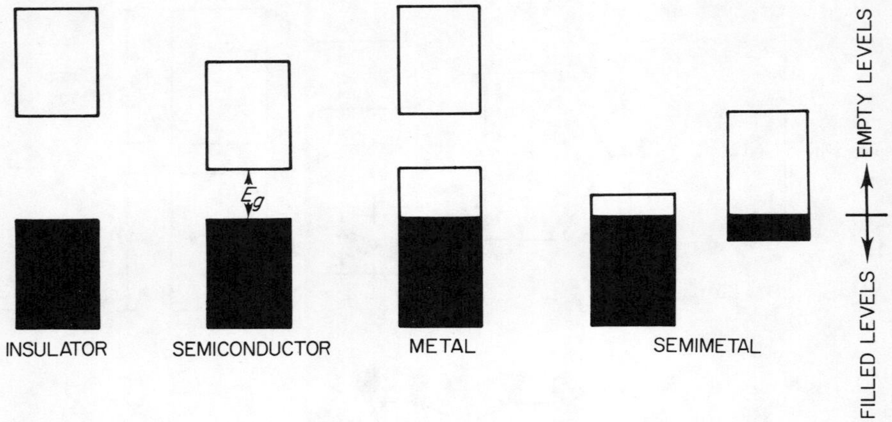

FIGURE 3. Schematic band structures of solids. The boxes indicate the allowed energy states and the shaded areas indicate the regions filled with electrons

species; 2, again because of their polarizability they help to stabilize a stack of like-charged species; and 3, because their p orbitals extend further in space than those centered on carbon there is greater orbital overlap in the stacking direction producing wider bands (a solid which is more metallic in nature). In 1976, we proposed a set of design constraints for the preparation of good organic conductors[1,2]. These constraints with a few modifications which we made in 1982 are as follows[3]:

1. Stable open-shell (free radical) species in order to form a partially filled band.
2. Planar molecules with delocalized π-molecular orbitals so that effective overlap of HOMO and/or LUMO can occur.
3. Inhomogeneous charge and spin distribution to reduce the repulsion when like charged molecules are stacked.
4. Segregated stacks of radical species—regardless of charge transfer, a mixed stack will always have a completely filled band.
5. No periodic distortion which opens a gap at the Fermi level (uniform stacks).
6. Little or no disorder (symmetric radicals and/or radical anions and cations)—disorder tends to produce a potential which localizes the wave function.
7. Molecular components of similar size.
8. Fractional charge (mixed-valence)—by adjusting the charge, it is possible to minimize the on-site Coulomb repulsion (U_{eff}), which should be small compared with the band width ($4t$).
9. Relatively strong interchain coupling to suppress phase transitions.
10. Cation and/or anion nominally divalent—unless the molecular components can support doubly charged species, only a correlated type of conductivity is possible.
11. Polarizable species.

More detailed descriptions of the solid-state chemistry and physics can be found in a number of recent reviews[4–6].

B. Sulfur-, Selenium- and Tellurium-based Organic Metals

Table 1 gives a list of some of the most conducting organic compounds and Figure 4 shows their structures. One of the striking features of this list is that so many are formed

TABLE I. Organic metals

Compound[a]	z^b	σ_{RT} ($\Omega^{-1}cm^{-1}$)	σ_{max} ($\Omega^{-1}cm^{-1}$)	$T_{max}{}^c$(K)	Ref.
TTF–TCNQ	0.59	500	2×10^4	59	7–13
TMTTF–TCNQ	0.65	350	5×10^3	60	1, 2, 14, 15
HMTTF–TCNQ	0.72	500	2×10^3	75	16
TSF–TCNQ	0.63	800	1×10^4	40	17–20
TMTSF–TCNQ	0.57	1200	7×10^3	61	1, 2, 21
HMTSF–TCNQ	0.74	2000	7×10^3	(32)	1, 2, 22, 23
HMTTeF–TCNQ	—	550	9×10^2	(73)	24, 25
HMTTeF–DMTCNQ	—	460	1×10^3	(83)	24, 25
HMTSF–TNAP	—	2900	2×10^4	50	1, 2, 26, 27
$(TTT)_2I_3$	0.5	1000	3×10^3	(60)	28, 29
$(TSeT)_2I$	0.5	1500	1×10^4	(35)	30, 31
$(TSeT)_2Cl$	0.5	2100	2×10^4	(26)	32–36
$(TMTSF)_2PF_6$	0.5	550	7×10^4	(18)	37–44
$(TMTSF)_2ClO_4$	0.5	650	5×10^5	(6)	45, 47
$(BEDT–TTF)_2I_3$	0.5	30	1.5×10^4	(4.2)	48–51
$(Perylene)_2(PF_6)_{1.1}$	0.55	900	1×10^3	200	52–54

[a]For structures see Figure 4.
[b]z≡fractional charge per cation; see Ref. 2 and references cited therein.
[c]Parentheses indicate a broad conductivity maximum.

from sulfur and selenium heterocycles. We expect that this list will grow to include more tellurium compounds as more work is done on these materials. As a rule, the selenium compounds tend to be more conducting than the sulfur compounds and stay metallic to lower temperatures.

From the magnitude of the electrical conductivity and the crystal structure, it is possible to calculate the effective mean free path of the carriers. For TTF–TCNQ, the mean free path at room temperature is almost one lattice spacing and at 59 K it is about ten lattice spacings (40 Å). For comparison, the mean free path in Pb at 60 K is about 28 lattice spacings (100 Å). While a band picture is consistent with all of the experimental evidence on organic metals, it is clear that a hopping model is not adequate to explain the conductivity of these materials at low temperatures. To date, the only superconducting organic compounds have been salts prepared from TMTSF [for example, $(TMTSF)_2ClO_4$ where $T_c = 1.2$ K] and BEDT-TTF [for example, $(BEDT\text{-}TTF)_2I_3$ where $T_c = 1.4$K].

II. SYNTHESIS

A. Heterofulvalene Donors

Heterofulvalene donors containing four chalcogen atoms (**1**) are the most widely studied class of compounds in the realm of organic conductors. Tetrathiafulvalene (TTF, **1a**) has provided much of the impetus for the later developments in organic synthesis and in physics with the discovery that its charge-transfer salt with TCNQ displays metal-like electrical behaviour[7–13]. In addition, TTF represents a prototype structure, ripe for structural and elemental modifications, thus providing a fertile ground for synthetic organic chemists to explore new methodologies. The selenium and tellurium analogs, TSF (**1b**) and TTeF (**1c**) respectively, are widely sought-after compounds since the charge-transfer solids derived from them are expected to be even better electrical conductors. So

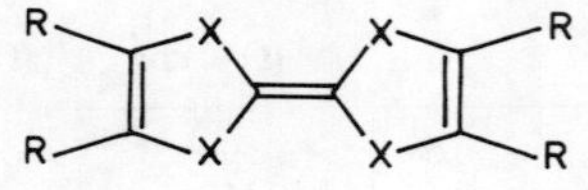

R	X	Compound
H	S	TTF
H	Se	TSF
CH_3	S	TMTTF
CH_3	Se	TMTSF
$—CH_2CH_2CH_2—$	S	HMTTF
$—CH_2CH_2CH_2—$	Se	HMTSF
$—CH_2CH_2CH_2—$	Te	HMTTeF
$—SCH_2CH_2S—$	S	BEDT−TTF

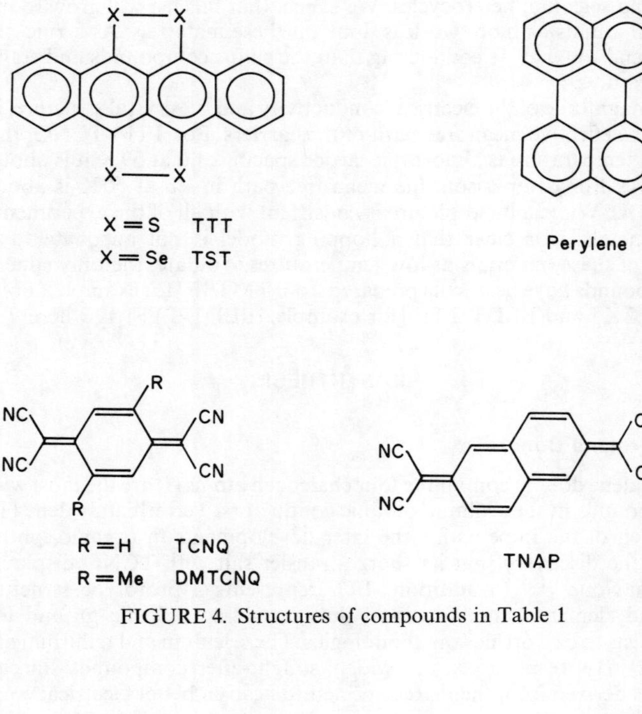

FIGURE 4. Structures of compounds in Table 1

far, TSF, its derivatives and some derivatives of TTeF have been prepared. The parent tellurium compound, TTeF, has remained elusive.

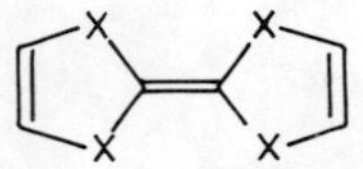

(**1a**) X = S TTF
(**1b**) X = Se TSF
(**1c**) X = Te TTeF

To a large measure, the synthetic activity has focused on various methods for the synthesis for TMTSF and HMTSF. This is primarily due to the superconducting property[37–47] of salts $(TMTSF)_2X$ where $X = PF_6$, ClO_4 and the absence of a metal-to-insulator transition in HMTSF–TCNQ as the temperature approaches zero[22,23].

1. Selenium donors

Tetraselenafulvalene and its derivatives, owing to their symmetry (D_{2h}), are conveniently constructed by coupling appropriate 1,3-diselenole units wherein the tetraselenoalkene bond is formed in the last step. A widely employed strategy consists in coupling 1,3-diselenole-2-one, -thione or -selone with a trivalent phosphorus reagent, Ph_3P or $(MeO)_3P$ (Scheme 1).

While the exact mechanistic details of the coupling reaction are yet to be unravelled, a general pathway can be written as in Scheme 2. Some of the intermediates shown have been either isolated[55,56] or trapped[57] from the coupling reaction. In spite of the lack of understanding of the mechanism, some empirical guidelines for successful coupling with phosphorus reagents have emerged:

1. Selones, in general, can be coupled under milder conditions than and with superior yields to their thio or oxo counterparts. Exceptions seem to be the compounds with electron-withdrawing substituents (e.g. tetracyano-TTF[58]) and BEDS-TSF[110]; in these cases the oxo precursors couple more efficiently.
2. Selones usually do not yield products with chalcogen scrambling. For example, certain 1,3-diselenole-2-thiones were found to give triselenathiafulvalenes as a by-product (R = Me) and as the only product ($R = CF_3$, COOMe) (Scheme 3)[59]. Such scrambling was not observed for 1,3-dithiole-2-selones.
3. Triarylphosphines are superior to trialkylphosphites since the latter can undergo an Arbuzov-type reaction with thiones[55].

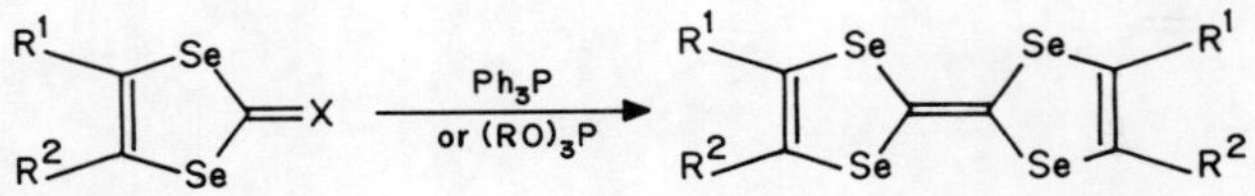

X = O,S,Se

(**1b**) $R^1 = R^2 = H$ TSF
(**2**) $R^1 = R^2 = Me$ TMTSF
(**3**) $R^1 = R^2 = -(CH_2)_3-$ HMTSF
(**4**) $R^1 + R^2 =$ benzo DBTSF
(**5**) $R^1 = R^2 = COOMe$

SCHEME 1

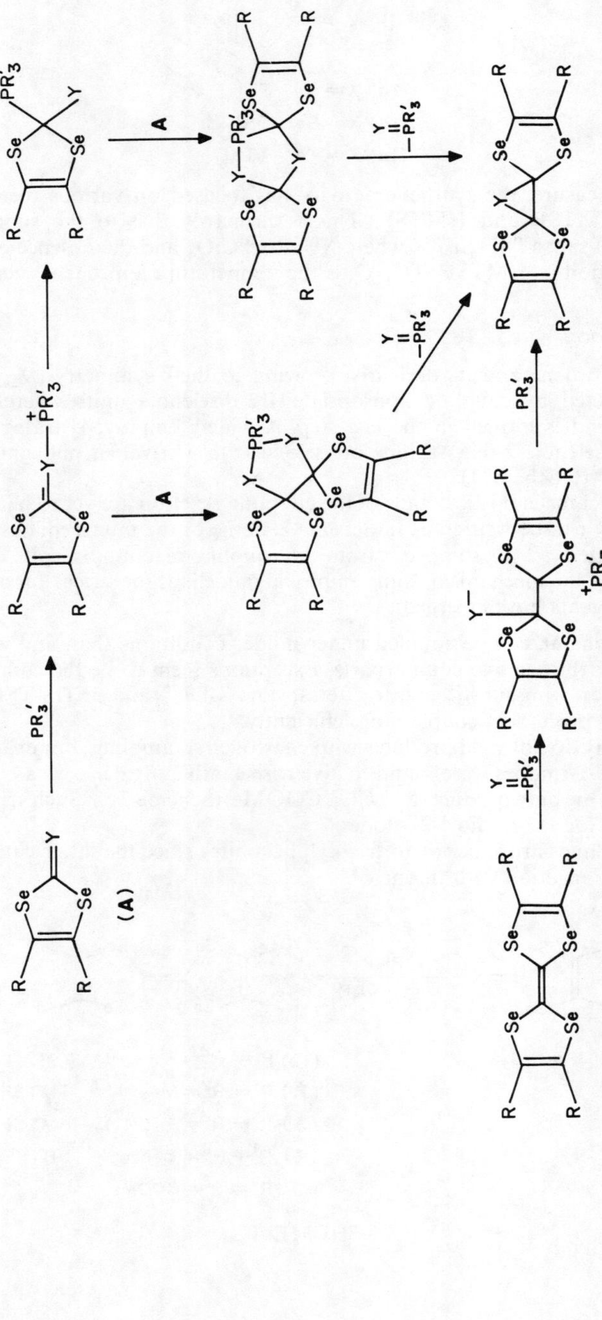

SCHEME 2

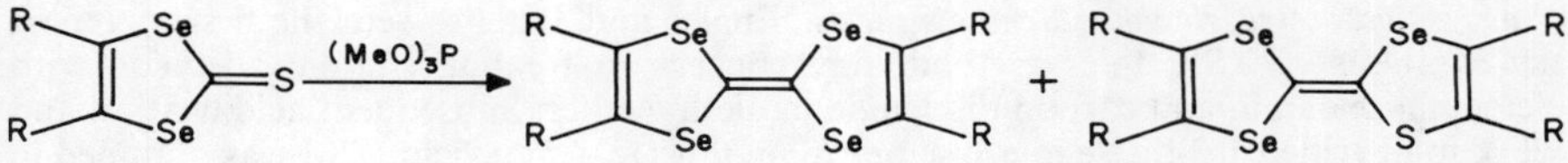

R = Me	85%	15%
R = CF_3	—	100%
R = COOMe	—	100%

SCHEME 3

In practice, however, experimentation with different phosphorus reagents, solvents and substrates (oxo, thioxo, selenoxo) is often necessary to determine the optimum combination that yields the desired heterofulvalene.

Deprotonation of the 1,3-dithiolium ion **7**, generated by peracid treatment of 1,3-dithiole-2-thione (**6**), has been known to give the corresponding TTF derivative via a carbene intermediate (Scheme 4)[60–62]. An analogous approach in the case of 1,3-diselenole-2-thione, however, was unsuccessful (Scheme 5)[63,66].

A recently described[93] non-coupling synthesis of HMTSF has the potential of being applicable to other TSF derivatives. Here, the 1,2-diselenol dianion is reacted with tetrachloroethylene to give HMTSF (Scheme 6).

The next section will describe various reported methods of synthesizing the 1,3-diselenole-2-thiones and the 1,3-diselenole-2-selones.

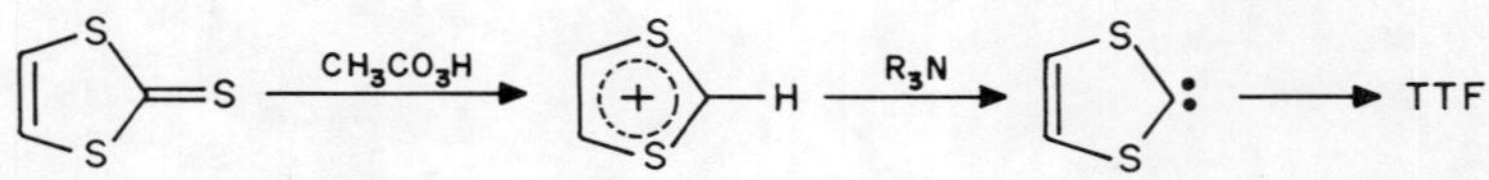

(6) (7)

SCHEME 4

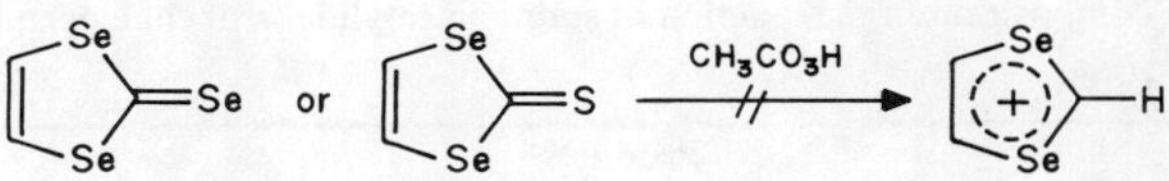

(8) (9)

SCHEME 5

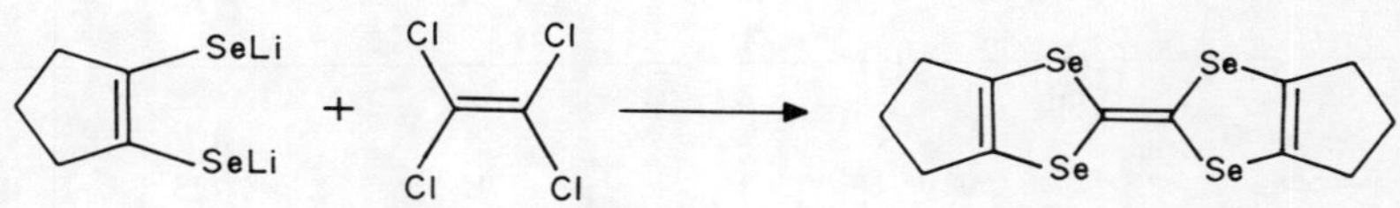

SCHEME 6

a. Engler's tetraselenafulvalene synthesis. Engler and Patel[63] were the first to report the synthesis of TSF; their methodology consists in treating sodium acetylide with elemental selenium and carbon diselenide in diethyl ether. Subsequent acidification and workup provided 1, 3-diselenole-2-selone (**8**) in low (15–25%) yield. TSF was obtained in 70–80% yield by triphenylphosphine coupling[63]. As the mechanism (Scheme 7) suggests, the method is limited to the synthesis of unsubstituted or monosubstituted TSF[63,64]. The reaction is general inasmuch as any combination of sulfur/selenium and carbon disulfide/carbon diselenide can be employed to obtain the corresponding heterocycles. However, sulfur/carbon diselenide and selenium/carbon disulfide combinations result in extensive scrambling (see Table 2)[65].

Coupling of 1, 3-thiaselenole-2-selone (**12**) with trimethyl phosphite yielded a mixture of *cis*- and *trans*-dithiadiselenafulvalene (**13**). The lower symmetry and the presence of cis/trans isomers in DSDTF (**13**) causes a structural disorder in the TCNQ salt. It was expected that its conductivity would be much lower than that of either TTF–TCNQ or TSF–TCNQ. However, neither the room temperature conductivities nor the

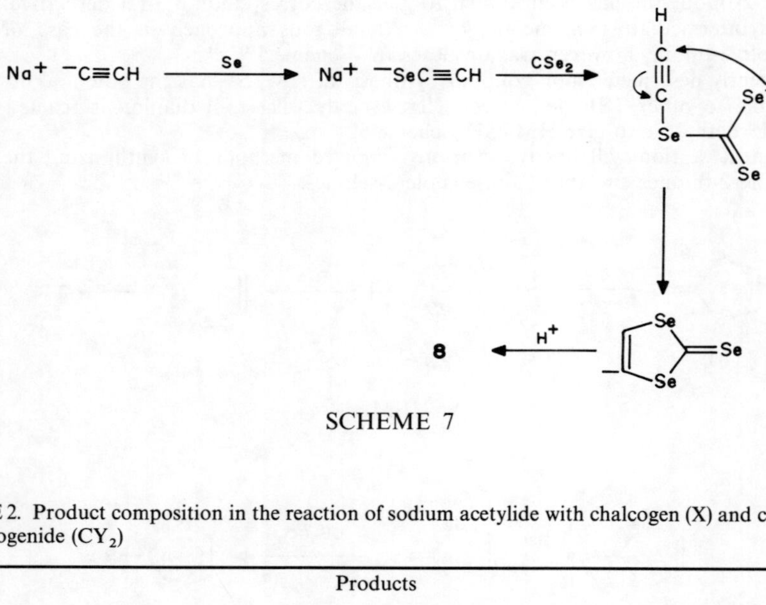

SCHEME 7

TABLE 2. Product composition in the reaction of sodium acetylide with chalcogen (X) and carbon dichalcogenide (CY_2)

	Products					
NaC≡CH $\xrightarrow{X,\ CY_2}$	**6**	**8**	**9**	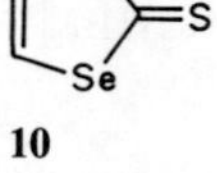 **10**	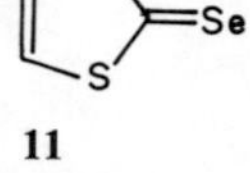 **11**	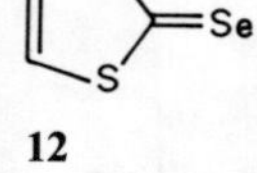 **12**
X = Y = S	Only	—	—	—	—	—
X = Y = Se	—	Only	—	—	—	—
X = S; Y = Se	1	50	8	1	15	25
X = Se; Y = S	22	0	1	70	5	2

conductivity–temperature profiles were dramatically different for the three salts. Thus, it was concluded that intramolecular structural disorder plays an important but not dominant role in the solid-state electrical properties[67].

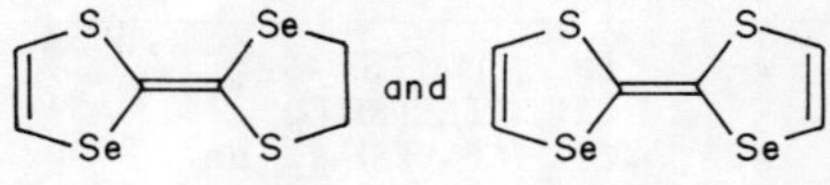

(13) DSDTF

b. Selenocarbamate route. This has been one of the most widely used synthetic strategies for the construction of TSF derivatives[66]. As described in Scheme 8, it involves the intermediacy of 2-oxoalkyl -*N*, *N*-dialkyldiselenocarbamate (**15**), which is generated by the nucleophilic attack of *N*, *N*-dialkyldiselenocarbamate ion (**14**)[68] on an appropriate α-haloketone. H_2SO_4-catalysed cyclization of **15** was followed by anion exchange with $HClO_4$ or HBF_4 (for stability reasons) to provide the 2-imminium-1, 3-diselenole salt **16**. Conversion of **16** to 1, 3-diselenole-2-selone (**17**) or -thione (**18**) was achieved by bubbling H_2Se or H_2S into a methanolic solution of **16**. The selone or the thione was then coupled in the usual manner [Ph_3P or $(MeO)_3P$] to obtain the corresponding TSF derivative[66].

A large number of di- and tetra-substituted TSF derivatives have been prepared by this method (Table 3).

Various synthetic approaches to the *N*, *N*-dialkyldiselenocarbamate ion (**14**) have been investigated owing to the experimental difficulties associated with the use of carbon diselenide, which is toxic, very expensive, extremely fetid and ill-ventable.

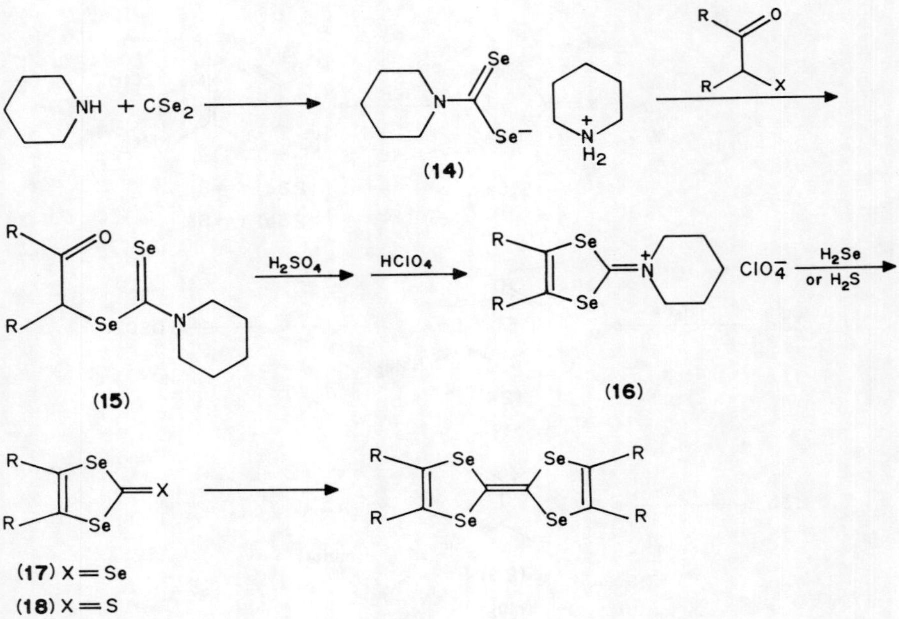

SCHEME 8

TABLE 3. Various TSF derivatives synthesized by the diselenocarbamate route

R_1	R_2	Product	Ref.
CH_3	H	DMTSF(**19**)	66
CH_3	CH_3	TMTSF (**2**)	66, 74
CD_3	CD_3	TMTSF-d_{12}(**20**)	69
Ph	H	DPTSF(**21**)	66
Ph	Ph	TPTSF(**22**)	66
—$(CH_2)_3$—		HMTSF(**3**)	66, 74
—$(CH_2)_4$—		OMTSF	Unpublished
—CH_2CH_2S—		α-DTTSeF	77

Shu *et al.*[70] reported the use of thio- and seleno-urea for the synthesis of diselenadithiafulvalene (DSDTF) and TSF derivatives (Scheme 9). Here, the α-haloketone was reacted with thio- or seleno-urea in perchloric acid to obtain the corresponding thio- or selenouronium salt (**23**). Treatment of thiouronium salt **23a** with H_2Se gave 2-oxoalkyl-*N*, *N*-dimethylthioselenocarbamate (**24**) in 75% yield. Further elaboration, as described in Scheme 7, furnished DSDTF in ca. 40% overall yield. The selenouronium salt **23b**, however, yielded 2-oxoalkyl-*N*, *N*-dimethyldiselenocarbamate (**15**) as a minor product. The major reaction pathway was the elimination of 2-oxoalkylselenide ion (**25**).

A further refinement for the generation of the diselenocarbamate ion, avoiding the use of CSe_2, came independently and simultaneously from the groups of Wudl[71] and Cowan[72]. As shown in Scheme 10, hydrogen selenide was reacted with phosgene imminium chloride

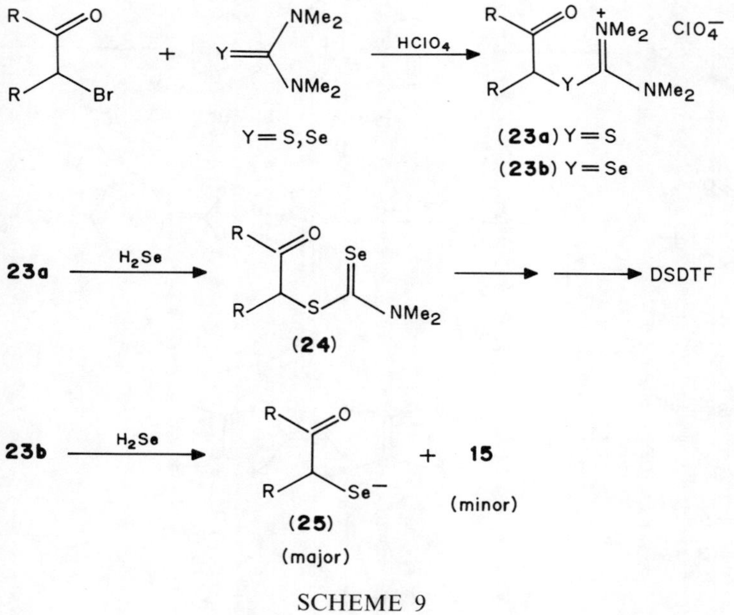

SCHEME 9

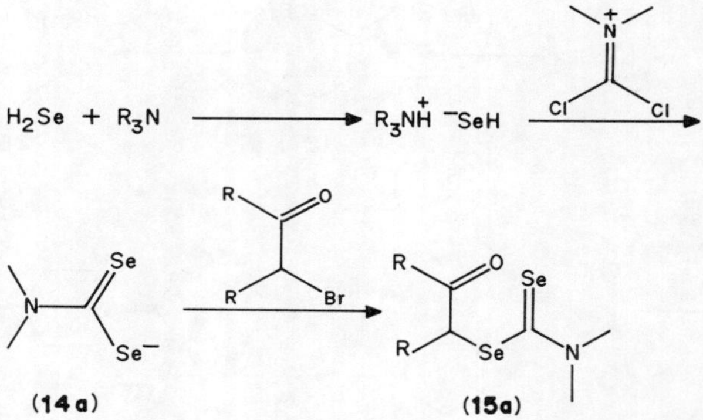

(14a) **(15a)**

SCHEME 10

in the presence of a di- or tri-alkylamine to give the dialkyldiselenocarbamate ion **14a**; the latter was then treated with an α-haloketone to form the key intermediate 2-oxoalkyl-*N*, *N*-dialkyldiselenocarbamate (**15a**).

Moradpour *et al.*[73] reported a slight modification to the Wudl–Cowan route which avoids the use of gaseous selenium reagents, thus making the synthesis of TSF derivatives much less hazardous. In this method, sodium selenide (generated *in situ* from elemental selenium and $NaBH_4$ in DMF) was partially acidified with *t*-BuOH and the resulting NaSeH was allowed to react with phosgene imminium chloride. The product, *N*, *N*-dimethyldiselenocarbamate ion (**14a**), was converted to the 2-imminium-1, 3-diselenole salt **16a** in the usual manner and treated with NaSeH in EtOH (instead of bubbling in H_2Se) to give 1, 3-diselenole-2-selone (**17a**) in good yield (Scheme 11).

Notably absent among the selenafulvalene donors synthesized by the selenocarbamate route is TSF. There are several reasons for this. First, α-haloacetaldehyde is extremely unstable. Second, neither α-haloacetaldehyde nor its acetal (which is stable and commercially available) have been shown to react with the diselenocarbamate ion **14** to form the key intermediate, diselenocarbamate ester.

Johannsen *et al.*[75] reported a novel acid-catalysed exchange reaction between diselenides and bis(*N*, *N*-dialkylselenocarbamoyl)selenide, which furnished the key carbamate ester in near quantitative yield (Scheme 12). Chloroacetaldehyde diethylacetal (**26**),

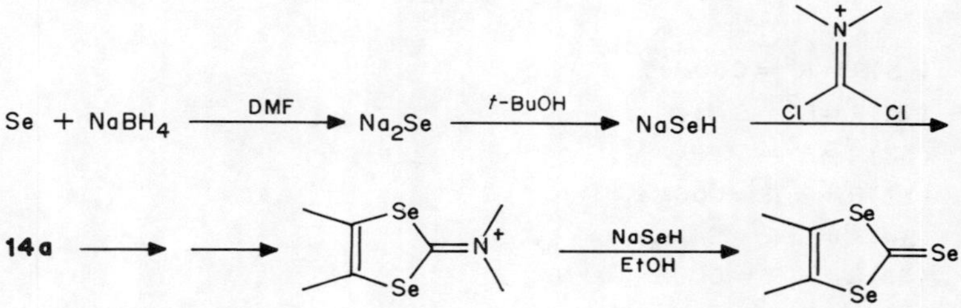

(17a)

SCHEME 11

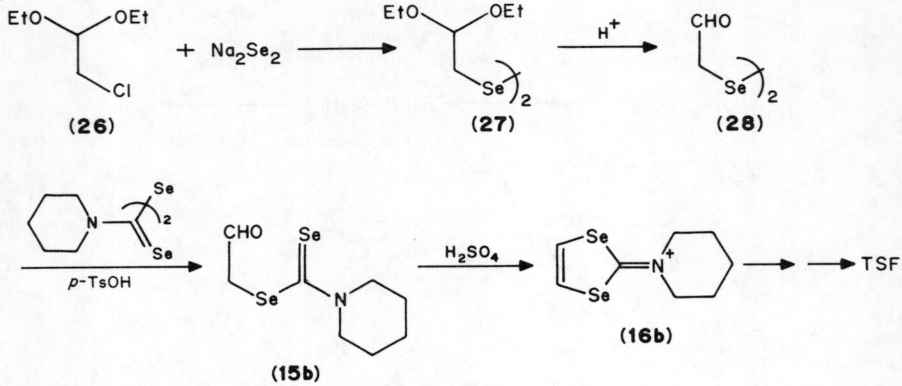

SCHEME 12

while inert to diselenocarbamate ion **14**, reacts smoothly with sodium diselenide, which is significantly more nucleophilic. The resultant bis(diethoxyethyl)diselenide (**27**) was hydrolysed to bis(2-oxoethyl)diselenide (**28**) in 85% yield. Then, following the method of Henriksen[76], **28** was reacted with bis(*N*, *N*-dialkylselenocarbamoyl)selenide in a 2:3 molar ratio (with acid catalysis) to obtain the desired diselenocarbamate ester **15b**. Cyclization of **15b** within strong acid furnished the 2-imminium-1, 3-diselenole salt **16b** in 65% yield. This salt could be transformed to TSF via known procedures. This method represents the most efficient synthetic route to TSF.

c. Addition of CSe_2 to acetylenes. Certain substituted tetraselenafulvalenes and diselenadithiafulvalenes can be synthesized in a one-step procedure. Okamoto and his group found that CSe_2 adds to acetylenes bearing at least one electron-withdrawing group under high pressure (ca. 5–10 kbar)[78]. The addition is believed to result in a carbene (via a dipolar intermediate **29**), which then dimerizes to the corresponding TSF derivative (Scheme 13)[78–80].

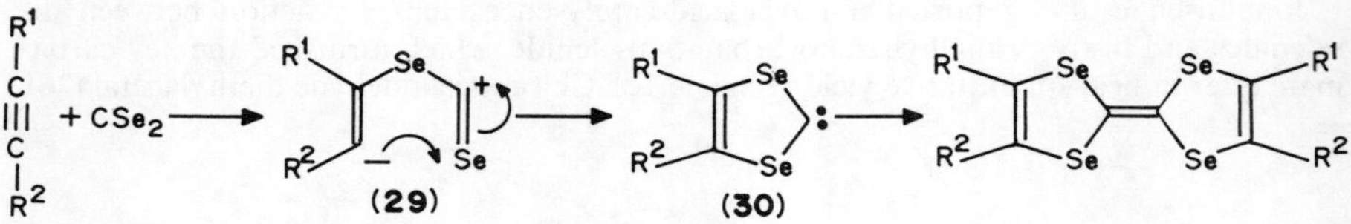

(**5**) $R^1 = R^2 =$ COOMe
(**31**) $R^1 = R^2 =$ $SiMe_3$
(**32**) $R^1 = R^2 =$ $CONH_2$
(**33**) $R^1 =$ H; $R^2 =$ COOMe
(**34**) $R^1 =$ H; $R^2 =$ $SiMe_3$
(**35**) $R^1 =$ H; $R^2 =$ COOH
(**36**) $R^1 =$ H; $R^2 =$ $CONH_2$

SCHEME 13

Failure of unsubstituted acetylene to react in this manner (to give TSF) supports the formation of the dipolar intermediate **29**. The reaction is general in scope since DSDTF and TTF derivatives can similarly be obtained from CSSe[79] and CS_2[78–81], respectively. TSF could be indirectly obtained by this method since Lakshmikantam and Cava[85] converted tetrakis(carbomethoxy)TSF (**5**) to TSF by the action LiBr in HMPA. However, the yield in this latter reaction is more subject to the whims of nature than are most reactions.

d. Triselenocarbonate–acetylene route. Easton and Leaver[82] reported in 1965 that when ethylene trithiocarbonate and dimethyl acetylenedicarboxylate were refluxed in toluene, an exchange of the trithiocarbonate unit giving ethylene and 4,5-dicarbomethoxy-1,3-dithiole-2-thione (**37**) had occurred (Scheme 14). The reaction appears to involve a dipolar addition and fails in the case of mixed sulfur–oxygen analogs of ethylene trithiocarbonate[83].

A successful extension of the above reaction to the selenium analog was subsequently reported by Cava's group[84,85]. While 1,3-diselenolane-2-thione (**38**) gave the 1,3-thiaselenole-2-selone derivative, a precursor to DSDTF derivative[84], the corresponding 1,3-diselenolane-2-selone **39** gave the 1,3-diselenole-2-selone derivative (Scheme 15)[84,85]. Again, the presence of at least one electron-withdrawing group on acetylene is essential.

e. Selenadiazole route. Thermolysis of 1,2,3-selenadiazoles in the presence of either CS_2 or CSe_2 is a direct route to 1,3-thiaselenole-2-thione or 1,3-diselenole-2-selone. Cava

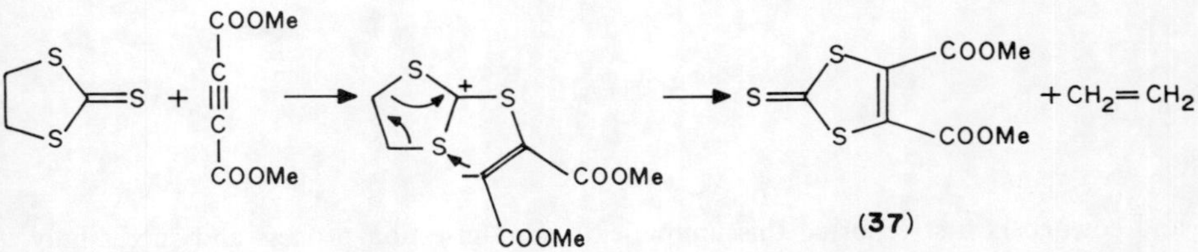

SCHEME 14

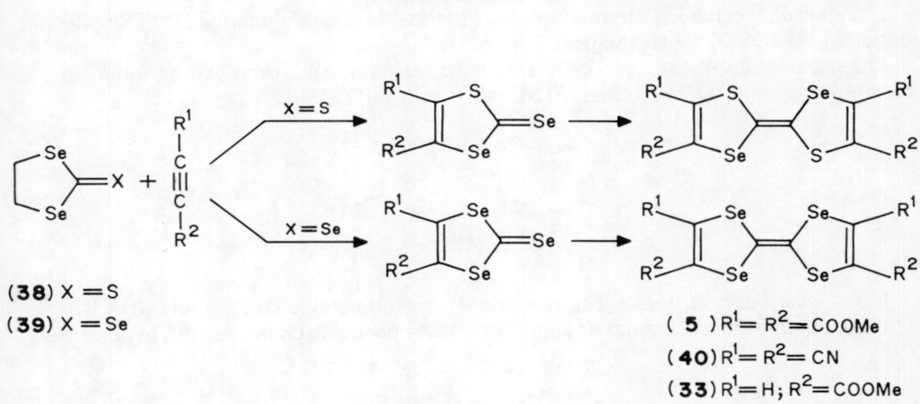

SCHEME 15

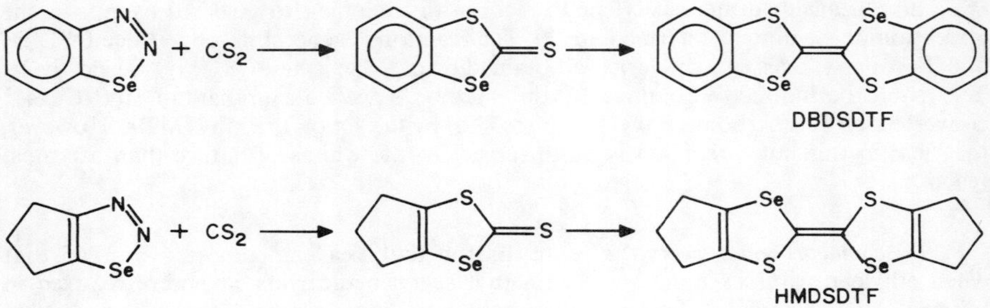

SCHEME 16

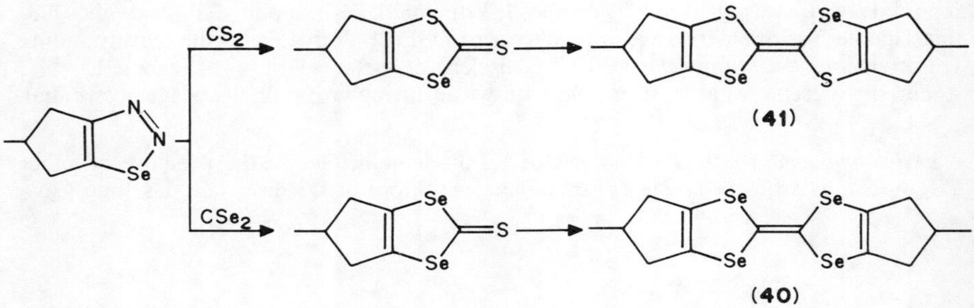

SCHEME 17

and coworkers first reported this unique extrusion–insertion process and successfully applied it to the syntheses of dibenzodiselenadithiafulvalene (DBDSDTF) and hexamethylenediselenadithiafulvalene (HMDSDTF) (Scheme 16)[86].

Engler *et al.*[77] extended this methodology to the synthesis of dimethyl HMTSF (**40**) and dimethyl HMDSDTF (**41**) (Scheme 17).

Other examples of the successful use of the selenadiazole route can be found in the formation of di(2, 5-dihydrothieno)TSF (**42**)[87] and DBTSF (**4**)[88].

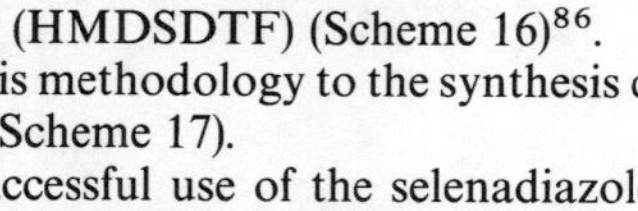

(42)

f. Selenium insertion route. The reaction of organometallic compounds (RLi, RMgX) with elemental sulfur, selenium or tellurium usually produces chalcogenates in good yield.

$$R^-M^+ + X \longrightarrow RX^-M^+$$
$$M = Na, Li, K, Mg, Etc.$$
$$X = S, Se, Te$$

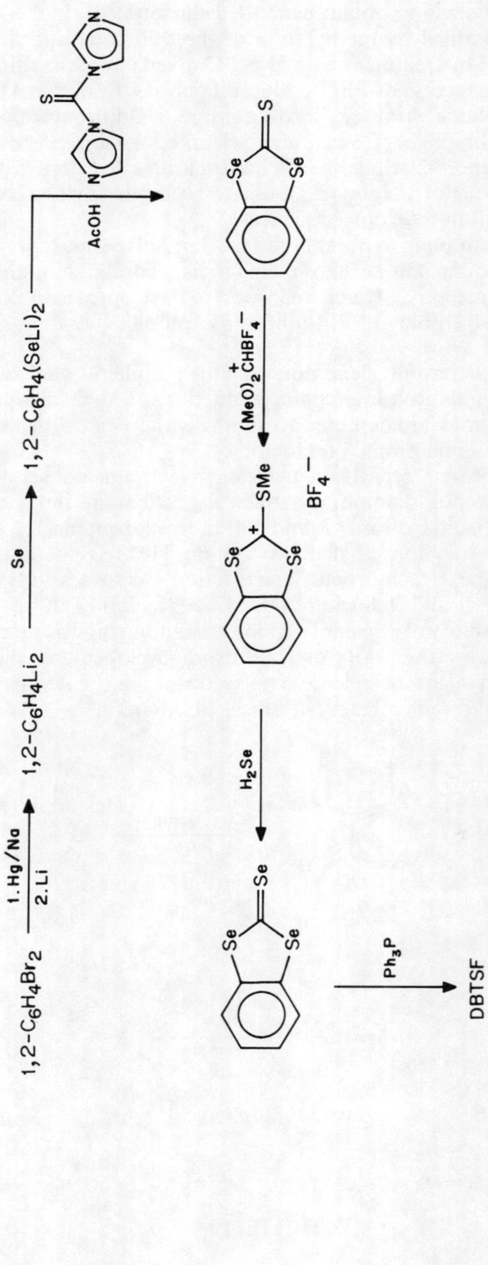
1,2-$C_6H_4Br_2$
1. Hg/Na
2. Li
1,2-$C_6H_4Li_2$
Se
1,2-$C_6H_4(SeLi)_2$
AcOH
S
N
Se
$(MeO)_2\overset{+}{C}HBF_4^-$
SMe
BF_4^-
H_2Se
Ph_3P
DBTSF

SCHEME 18

It has been used with great success in the syntheses of several heterocycles and TSF precursors. Synthesis of DBTSF (**4**) is a representative example (Scheme 18)[89]. *o*-Dilithiobenzene, generated by the method of Wittig via the mercury compound, was treated with elemental Se and then the 1, 2-diselenol dianion was protonated and treated with thiocarbonyldiimidazole to obtain benzo-1, 3-diselenole-2-thione. The corresponding selone was then obtained by methylation of the thione with dimethoxycarbenium tetrafluoborate followed by treatment with H_2Se. Conversion of the thione to selone was found necessary for the success of Ph_3P-induced coupling to obtain DBTSF.

With a similar synthetic strategy, starting from 3, 4-dibromothiophene and 2, 5-dimethyl-3- 4-dibromothiophene, Cowan and coworkers have reported the synthesis of thieno[3, 4-*d*]-1, 3-diselenole-2-selone (**43a**)[90] and its dimethyl derivative **43b**[91], respectively. Generation of diselenol dianion was found to be feasible by two consecutive metal–halogen exchange/selenium insertion sequences.

Triphenylphosphine coupling to obtain the TSF derivative was, however, possible only with the dimethyl compound **43b**, perhaps owing to its favorable solubility characteristics (Scheme 19)[91]. More recently, Ketcham *et al.*[92] have provided the synthesis of bis(selenopheno)TTF (**44**) employing 3, 4-dibromoselenophene as the starting compound (Scheme 20).

The interest in these heterofulvalene donors with peripheral chalcogen atoms stems from the possibility of enhanced interchain interactions in their charge-transfer solids. Such interchain interactions are expected to suppress the Peierls' instability commonly associated with the low-dimensional conductors.

Two reports have appeared recently on methods that circumvent several steps involved in converting the diselenol dianion (obtained by selenium insertion) to the TSF derivatives. McCullough and Cowan[93] found that the cyclopentene-1, 2-diselenol dianion can be treated with tetrachloroethylene to obtain HMTSF in a one-pot reaction (Scheme 21). Although the tetrachloroethylene reaction was successfully employed earlier for sulfur (DBTTF)[94,95] and tellurium (HMTTeF[96], DBTTeF[97], BDMT-TTeF[98]) donors, the apparent inability of diselenol dianions to follow suit was puzzling. The critical factors for the success were the purity of 1, 2-dibromocyclopentene, the temperature at which the second equivalent of selenium is added and the temperature at which tetrachloroethylene is allowed to react with diselenol dianion[93].

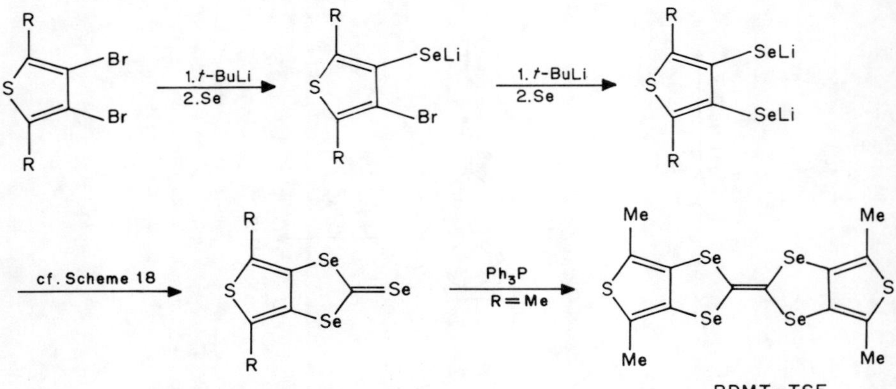

(**43a**) R = H

(**43b**) R = Me

SCHEME 19

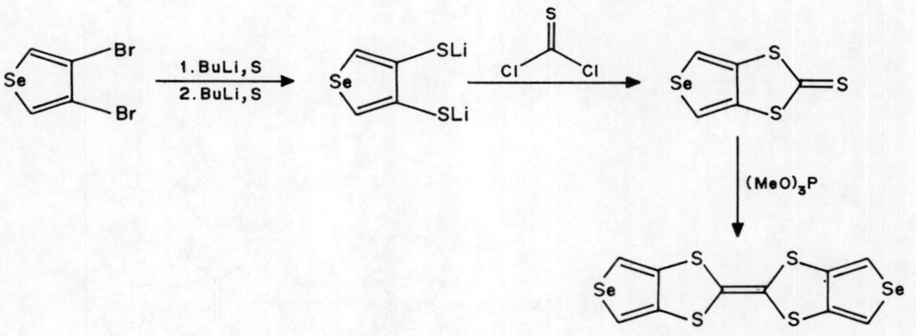

(44)

SCHEME 20

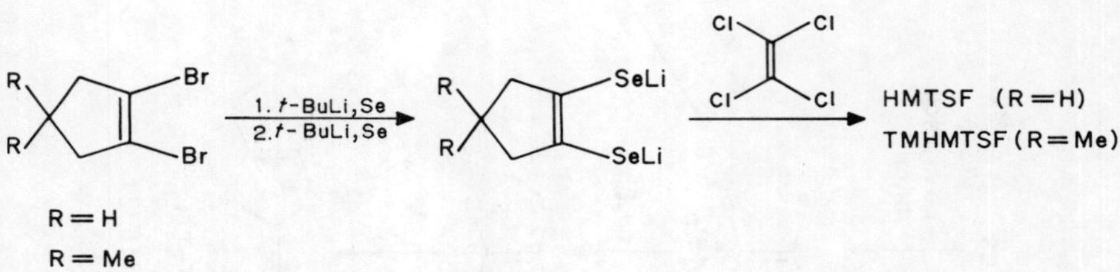

SCHEME 21

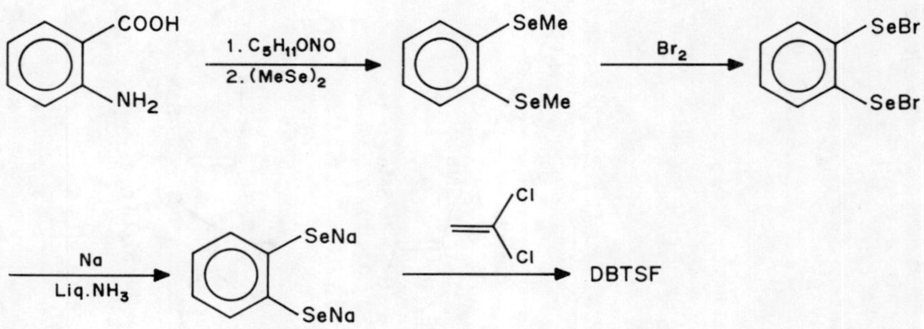

SCHEME 22

In a close structural analogy to the above reaction, Lambert and Christiaens[99] have developed a synthesis of DBTSF. As shown in Scheme 22, it involved the addition of dimethyl diselenide to benzyne to obtain bis(methylseleno)benzene, which was converted to diselenol dianion via bromination and treatment with sodium–liquid NH_3. The disodium salt of benzene-1, 2-diselenol thus generated was reacted with vinylidene

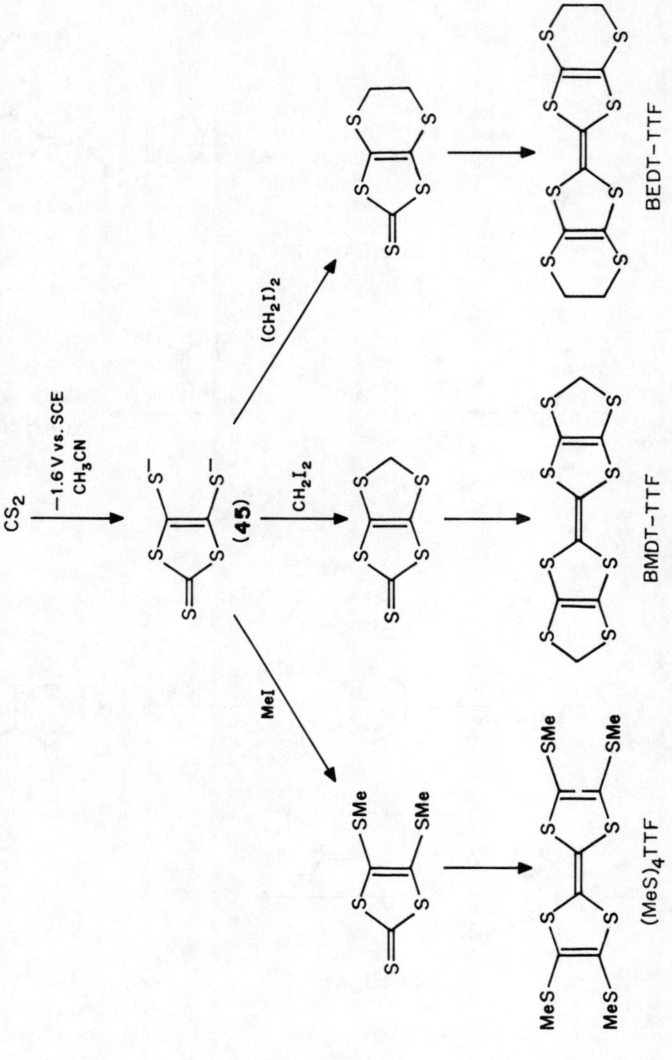

CS_2
−1.6 V vs. SCE
CH_3CN
(45)
MeI
CH_2I_2
$(CH_2I)_2$
(MeS)$_4$TTF
BMDT–TTF
BEDT–TTF

SCHEME 23

dichloride to obtain DBTSF. An unspecified acetylenic intermediate was postulated by these authors[99].

g. Electrochemical reduction of CSe_2. Carbon disulfide has been known to undergo electrochemical reduction (CH_3CN, − 1.6 V vs SCE) to give the dithiol dianion **45**[100,101]. With this common intermediate, various sulfur-rich TTF derivatives, tetrakis(thiomethoxy)TTF[102], BMDT-TTF[1,103] and BEDT-TTF[104,105] were prepared as shown in Scheme 23. Interest in these TTF derivatives, BEDT-TTF in particular, has mounted recently owing to the observation of high electrical conductivity in two dimensions and the consequent suppression of the metal–insulator transition in (BEDT-TTF)$_2$ClO$_4$[106]. Furthermore, superconductivity has been discovered at 7 Kbar in (BEDT-TTF)$_2$ReO$_4$ at 1.5 K[107]. More recently, ambient pressure superconductivity was observed in (BEDT-TTF)$_2$I$_3$ at 1.4 K[48–51]. Changing the anion from I_3^- to AuI_2^- increased T_c to 5 K[108]. These developments, naturally, have focused attention on the selenium analog, BEDS-TSF.

In fact, electrochemical reduction of CSe_2 (DMF, − 1.35 V vs. SCE, 40 °C)[109,110] followed a close parallel to CS_2 reduction and 1,3-diselenole-2-selone-4,5-diselenol dianion (**46**) was indeed formed (Scheme 24). The reduction could also be effected chemically with potassium naphthalide in THF[110]. The intermediate **46** was then alkylated with 1,2-dibromoethane to 4,5-ethylenediseleno-1,3-diselenole-2-selone (**47**). Since the direct coupling of **47** to BEDS-TSF was unsuccessful, it had first to be transformed to the oxo analog **48**, which was smoothly coupled to the desired selenium donor[110].

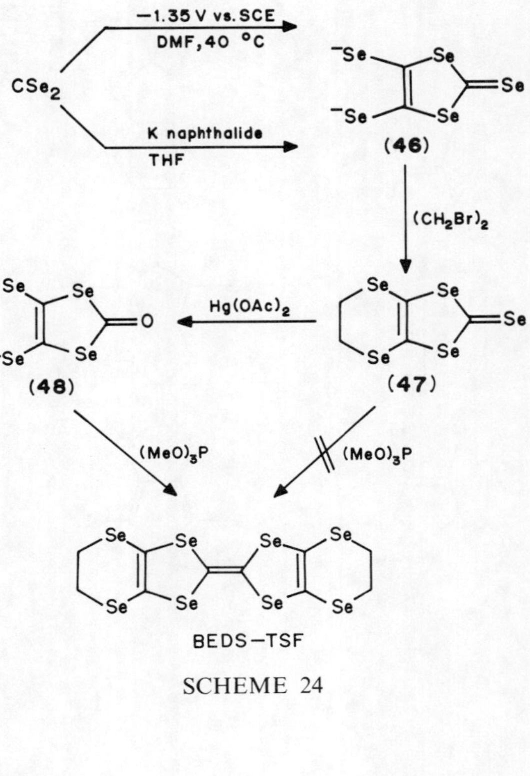

SCHEME 24

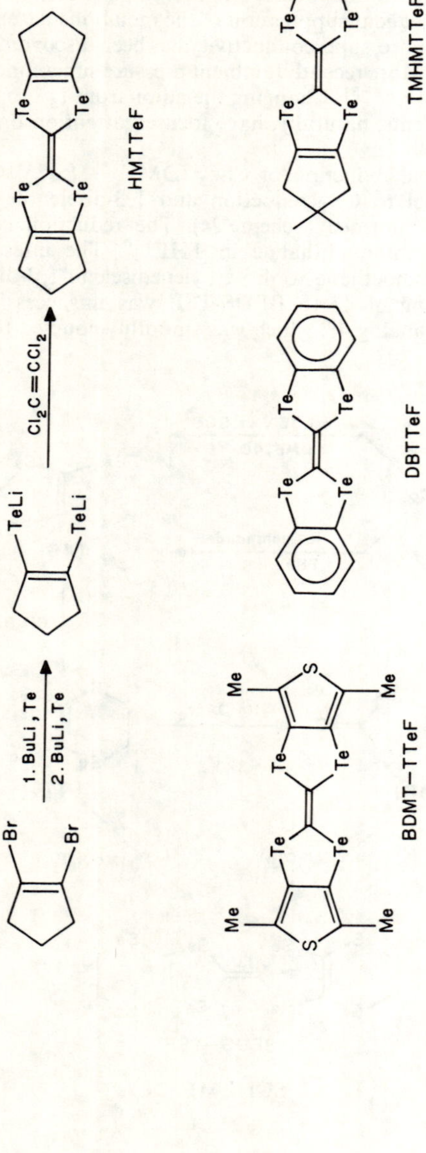
Br
Br
1. BuLi, Te
2. BuLi, Te
TeLi
TeLi
$Cl_2C{=}CCl_2$
Te
Te
HMTTeF
TMHMTTeF
BDMT–TTeF
DBTTeF
Me
S

SCHEME 25

2. Tellurium donors

Unlike their selenium counterparts, very few tetratellurafulvalenes are known and the parent compound TTeF has not been prepared. The only known tellurafulvalenes to date are HMTTeF[96], DBTTeF[97], BDMT-TTeF[98] and TMHMTTeF[111].

All four compounds share a common synthetic procedure—tellurium insertion into a carbon—lithium bond and subsequent reaction of ditellurolate with tetrachloroethylene. Synthesis of HMTTeF is shown in Scheme 25 as a representative example.

B. Other Selenium/Tellurium Donors

Electrical conductivity in organic solids is not limited to those comprising heterofulvalene donors discussed in Section II.A. Other sulfur, selenium and tellurium heterocyclic donors and polycyclic aromatic hydrocarbons have also been found to be excellent molecular components of good electrical conductors. Their charge-transfer salts and cation-radical salts have been widely investigated. In this section, the synthetic methods employed in their preparation are presented.

1. Tetraseleno- and tetratelluro-polyacenes

Reports of high metal-like conductivity in salts derived from TTT (**51a**)[28,29] and TTN (**49a**)[112] stimulated synthetic efforts on the selenium and tellurium counterparts. In addition, tetrachalcogenoanthracenes (**50a**–**c**) were envisaged by Endres *et al.*[113] to be perylene-like and were expected to yield solids with enhanced conductivity properties compared with those derived from perylene itself[52–54,114,130]. The presence of peripheral chalcogen atoms bound to each other, a distinctly different structural feature from the heterofulvalenes, was expected to result in higher dimensionality and stabilization of the metallic state.

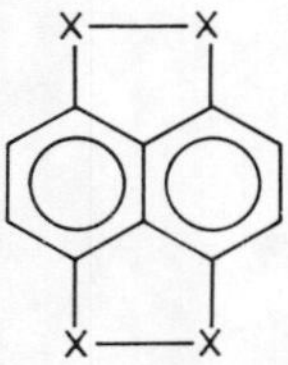

(**49a**) X=S TTN
(**49b**) X=Se TSeN
(**49c**) X=Te TTeN

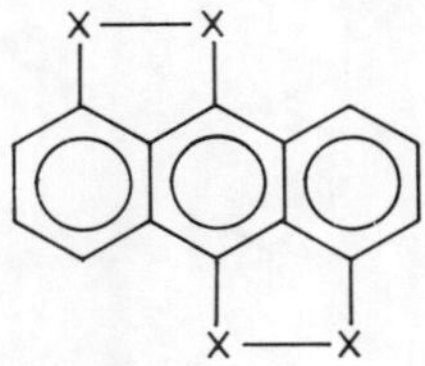

(**50a**) TTA
(**50b**) TSeA
(**50c**) TTeA

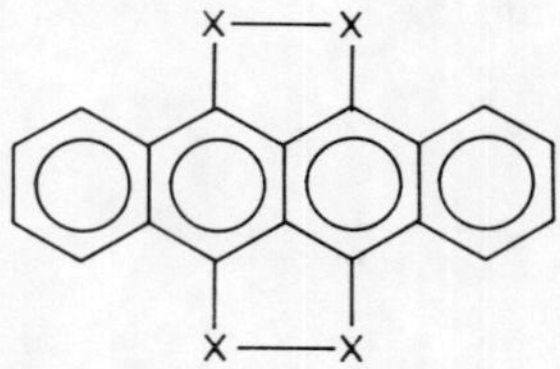

(**51a**) TTT
(**51b**) TSeT
(**51c**) TTeT

Tetrachalcogenopolyacenes are prepared by a general method which involves reacting the corresponding tetrachloro aromatic compound with an alkali metal dichalcogenide in a dipolar aprotic solvent (Scheme 26). To date, TSeN[115,116], TSeA[113], TSeT[117] and TTeT[118] have been prepared by the above method.

Delhaes *et al.*[30] have also reported the synthesis of TSeT by the reaction of selenium with 5,11-dichlorotetracene in boiling Dowtherm (a eutectic of biphenyl and diphenyl ether, b.p. ca. 260 °C) (Scheme 27).

2. Dichalcogenopyranylidenes

Dichalcogenopyranylidenes are isoelectronic with tetrachalcogenafulvalenes and consequently form stable cation radical salts. Once again, the discovery of electrical

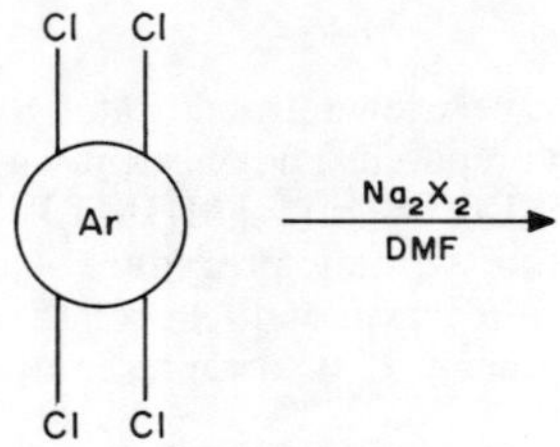

SCHEME 26

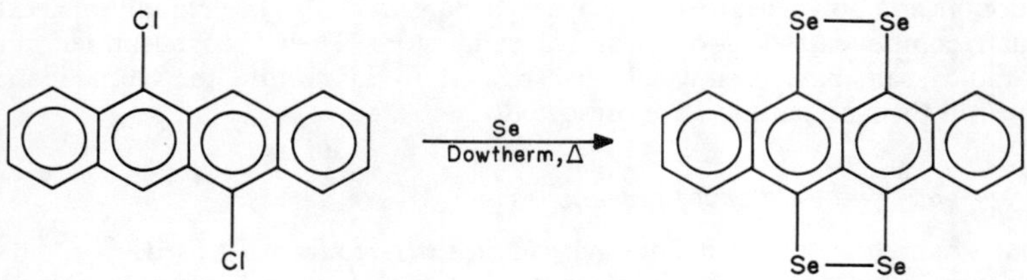

SCHEME 27

conductivity in charge-transfer salts derived from 4,4′-bithiopyranylidene (**52a**)[119–121] and its derivatives[120,122] spurred the interest in the synthesis of the selenium/tellurium analogs.

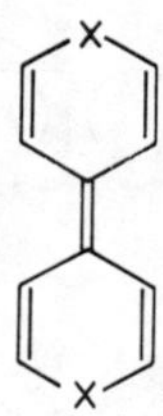

(**52a**) X=S BTP
(**52b**) X=Se BSeP
(**52c**) X=Te BTeP

Unlike BTP, the unsubstituted compounds BSeP and BTeP are at present unknown; however, several of their derivatives have been synthesized recently.

Tetraphenyl and tetramethyl derivatives of BSeP were first prepared by Mollier and co-workers according to Scheme 28[123,124]. An appropriately substituted acetylene was first converted to a penta-1,4-diyne-3-one derivative (**53**). Addition of H_2Se in a double Michael fashion gave the selenopyranone **54**, which was converted to the thione **55** by boron sulfide. Coupling of **55** with copper powder in xylene afforded the BSeP derivatives **56** and **57** in good yield.

An alternative method, reported by the same authors, consisted of coupling selenopyry-

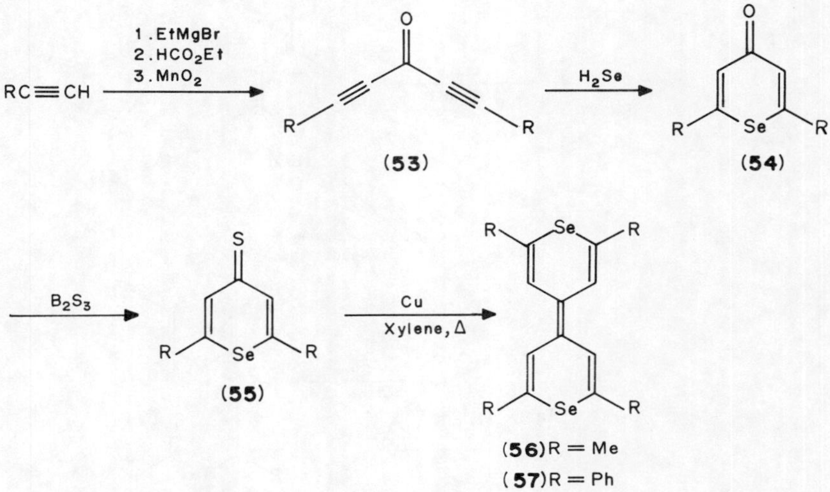

SCHEME 28

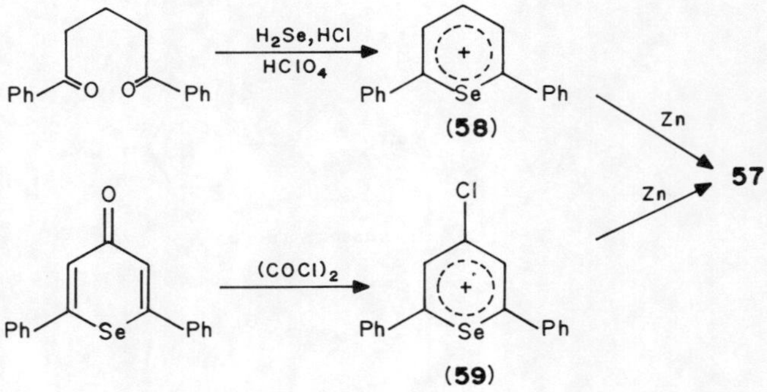

SCHEME 29

lium salts (**58** or **59**) with Zn or $TiCl_3$ (Scheme 29)[124]. The selenopyrylium salt **58** was derived from 1, 5-diphenylpentane-1, 5-dione and H_2Se under acidic conditions, while treatment of the selenopyranone **54** with oxalyl chloride provided the chloro derivative **59**.

Tetraphenyl-BSeP (**57**) forms a 1:1 complex with TCNQ that has a conductivity of $0.5\,\Omega^{-1}\,cm^{-1}$ (compacted powder), which is comparable to that of TTF–TCNQ[124,125].

Synthesis of tellurium compounds in this series was first reported in 1982 with a synthetic sequence that closely parallels the route to selenium compounds (Scheme 30)[126,127]. The only modifications were the use of lithium telluride (generated from tellurium and lithium triethylborohydride) instead of hydrogen chalcogenide and carrying out the reaction under basic conditions (with added NaOEt). These modifications were found to be essential to suppress the formation of the so-called 'anti-Michael' adduct **61** and to facilitate the isolation of desired telluropyranone **60** in high yield[128]. Thionation

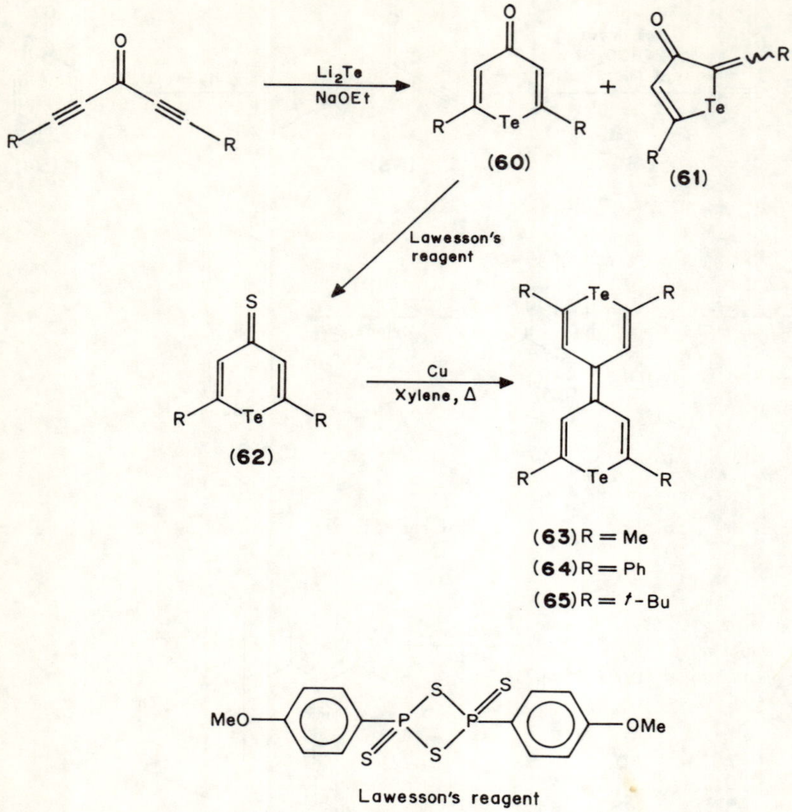

SCHEME 30

with Lawesson's reagent [2,4-bis(4-methoxyphenyl)-1,3-dithia-2,4-diphosphetane-2,4-disulfide] converted **60** to the telluropyran-4-thione **62**, which when refluxed with copper powder in xylene culminated in BTeP derivatives. The dibenzo BTeP derivative **66** was also prepared[127].

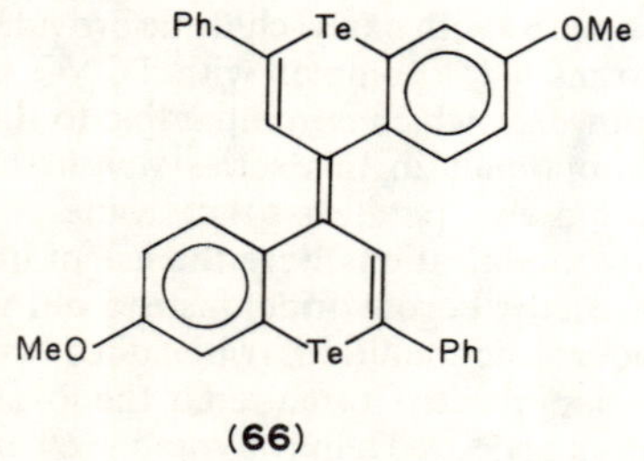

(66)

3. Miscellaneous selenium/tellurium donors

In this section, syntheses of donors that do not fall into the previous categories are discussed. The major thrust of the search for new, novel donors is the striking discovery that aromatic hydrocarbons (perylene, pyrene, etc.) yield cation radical salts, similar to the Bechgaard salts [$(TMTSF)_2X$], and these salts can be highly conducting[52–54,113,114]. In fact, these salts were discovered much earlier[129,130] than the Bechgaard salts. They are typically represented as $(D^+X^-)_aD_b$ where D and X are a donor and an inorganic diamagnetic ion, respectively, and $a = b = 1$. Thus, the importance of studying diverse classes of compounds has been clearly recognized and has prompted investigations towards that end.

Incorporation of chalcogen atoms in aromatic carbon frameworks offers an attractive avenue of research. Trichalcogenatriphenylenes (**67**) and dichalcogenapyrenes (**68**) represent two sets of donors examined in that context.

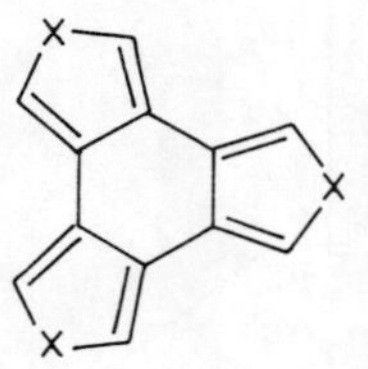

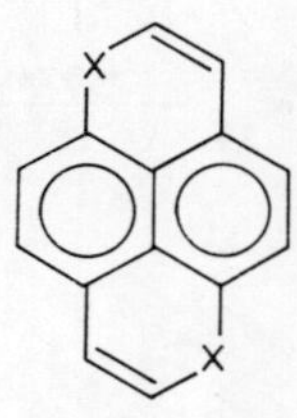

(**67a**) X=S BTT
(**67b**) X=Se BTSe
(**67c**) X=Te BTTe

(**68a**) DTP
(**68b**) DSeP
(**68c**) DTeP

Benzotrithiophene (BTT, **67a**) was first reported by Hart and Sasaoka[131] as a model sulfur-bridged hexaradialene. The donor properties of BTT were evident from its ability to form charge-transfer complexes with electron acceptors (TCNQ, DDQ) and radical cation salts with Lewis acids ($SbCl_5$, $FeCl_3$). Subsequently the 1:1 complexes BTT–TCNQ and BTT–$TCNQF_4$ were prepared and both were found to be insulators[3]. Since the nitrile stretching frequencies in these complexes indicated partial charge transfer[132], a requirement for high electrical conductivity[133], the crystals of the complexes were inferred to be of mixed-stack type.

The selenium analog BTSe (**67b**) was synthesized as shown in Scheme 31[3]. However, the reaction of lithium telluride with hexakis(bromomethyl)benzene to obtain hexahydro-BTTe (**70**) proved to be difficult. Precipitation of elemental tellurium on mixing the reactants was the typical outcome, although in a few instances small amounts of **70** were isolated. Its propensity to form hexaradialene was evident in the mass spectral fragmentation pattern.

Bechgaard recently reported an improved synthesis of DTP (**68a**) and the electrical properties of DTP–TCNQ[134]. The salt is very one-dimensional and stays metallic down to 1.2 K. This result is interesting because it supports the prediction that a 'purely' one-dimensional system is not subject to Peierls' distortion[135]. (Peierls' distortion is a lattice distortion commonly encountered in quasi-one-dimensional solids resulting in a band gap. Its chemical analog is the well known Jahn–Teller distortion).

The selenium and tellurium analogs DSeP (**68b**) and DTeP (**68c**) are at present unknown.

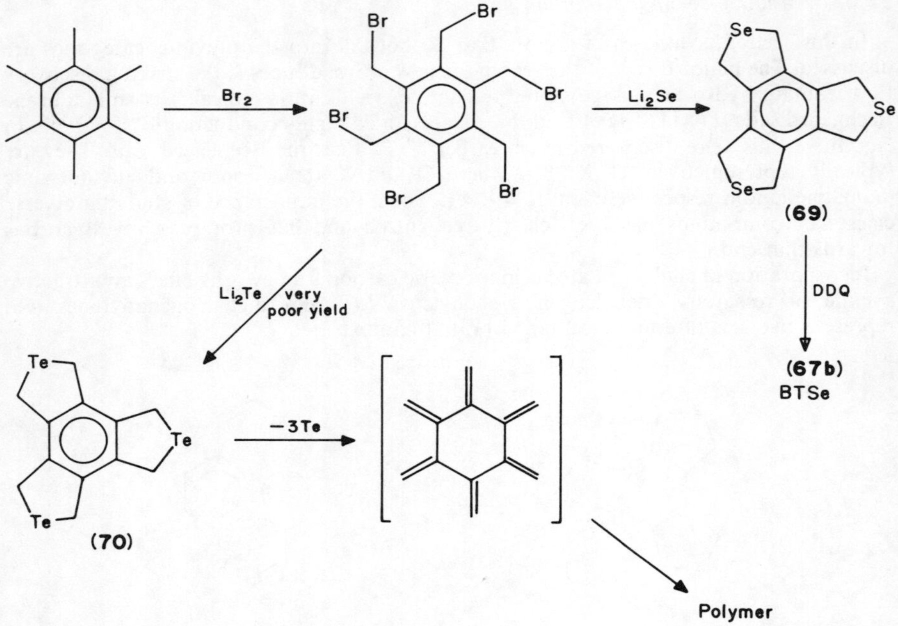

SCHEME 31

III. ABBREVIATIONS USED

BDMT-TSF	Bis(2, 5-dimethylthieno)tetraselenafulvalene
BDMT-TTF	Bis(2, 5-dimethylthieno)tetrathiafulvalene
BDMT-TTeF	Bis(2, 5-dimethylthieno)tetratellurafulvalene
BEDS-TSF	Bis(ethylenediseleno)tetraselenafulvalene
BEDT-TTF	Bis(ethylenedithio)tetrathiafulvalene
BMDT-TTF	Bis(methylenedithio)tetrathiafulvalene
BSeP	4, 4′-Biselenopyranylidene
BTP	4, 4′-Bithiopyranylidene
BTSe	Benzotriselenophene
BTT	Benzotrithiophene
BTTe	Benzotritellurophene
BTeP	4, 4′-Bitelluropyranylidene
DBDSDTF	Dibenzodiselenadithiafulvalene
DBTSF	Dibenzotetraselenafulvalene
DBTTF	Dibenzotetrathiafulvalene
DBTTeF	Dibenzotetratellurafulvalene
DDQ	2, 3-Dichloro-5, 6-dicyanoquinone
DMTCNQ	2, 5-Dimethyltetracyanoquinodimethane
DMTSF	Dimethyltetraselenafulvalene
DPTSF	Diphenyltetraselenafulvalene
DSDTF	Diselenadithiafulvalene

DSeP	1,6-Diselenapyrene
DTP	1,6-Dithiapyrene
DTeP	1,6-Ditellurapyrene
HMDSDTF	Hexamethylenediselenadithiafulvalene
HMTSF	Hexamethylenetetraselenafulvalene
HMTTF	Hexamethylenetetrathiafulvalene
HMTTeF	Hexamethylenetetratellurafulvalene
OMTSF	Octamethylenetetraselenafulvalene
TCNQ	Tetracyanoquinodimethane
$TCNQF_4$	Tetrafluorotetracyanoquinodimethane
TMHMTTeF	Tetramethylhexamethylenetetratellurafulvalene
TMTSF	Tetramethyltetraselenafulvalene
TMTTF	Tetramethyltetrathiafulvalene
TNAP	Tetracyano-2,6-naphthoquinodimethane
TPTSF	Tetraphenyltetraselenafulvalene
TSF	Tetraselenafulvalene
TSeA	Tetraselenoanthracene
TSeN	Tetraselenonaphthalene
TSeT	Tetraselenotetracene
TTA	Tetrathioanthracene
TTF	Tetrathiafulvalene
TTN	Tetrathionaphthalene
TTT	Tetrathiotetracene
TTeA	Tetratelluroanthracene
TTeF	Tetratellurafulvalene
TTeN	Tetratelluronaphthalene
TTeT	Tetratellurotetracene

IV. ACKNOWLEDGEMENTS

This review has benefited greatly by support from the National Science Foundation (DMR-8615305). We thank Richard McCullough and Knud Lerstrup for their helpful comments on the manuscript.

V. REFERENCES

1. D. Cowan, P. Shu, C. Hu, W. Krug, T. Carruthers, T. Poehler and A. Bloch, *Chemistry and Physics of One-Dimensional Metals* (Ed. H. J. Keller), Plenum Press, New York, 1977, pp. 25–46.
2. A. N. Bloch, T. F. Carruthers, T. O. Poehler and D. O. Cowan, *Chemistry and Physics of One-Dimensional Metals* (Ed. H. J. Keller), Plenum Press, New York, 1977, pp. 47–85.
3. D. O. Cowan, A. Kini, L.-Y. Chiang, K. Lerstrup, D. Talham, T. Poehler and A. Bloch, *Mol. Cryst. Liq. Cryst.*, **86**, 1 (1982); D. O. Cowan and F. M. Wiygul, *Chem. Eng. News*, **64**(29), 28 (1986).
4. D. Jerome and H. J. Schulz, *Adv. Phys.*, **31**, 299 (1982).
5. A. J. Berlinsky, *Contemp. Phys.*, **17**, 331 (1976).
6. R. L. Greene and P. M. Chaikin, *Physica*, **B126**, 431 (1984).
7. J. H. Perlstein, J. P. Ferraris, V. W. Walatka, D. O. Cowan and G. A. Candela, *AIP Conf. Proc.* (*Magnetism and Magnetic Materials*), **10**, 1494 (1972).
8. J. Ferraris, D. O. Cowan, V. Walatka and J. Perlstein, *J. Am. Chem. Soc.*, **95**, 948 (1973).
9. A. N. Bloch, J. P. Ferraris, D. O. Cowan and T. O. Poehler, *Solid State Commun.*, **13**, 753 (1973).
10. L. B. Coleman, J. M. Cohen, D. J. Sandman, F. G. Yamagishi, A. F. Garito and A. J. Heeger, *Solid State Commun.*, **12**, 1125 (1973).
11. D. E. Schafter, F. Wudl, G. A. Thomas, J. P. Ferraris and D. O. Cowan, *Solid State Commun.*, **14**, 347 (1974).

12. G. A. Thomas *et al.*, *Phys. Rev. B*, **13**, 5105 (1976).
13. R. V. Gemmer, D. O. Cowan, T. O. Poehler, A. N. Bloch, R. E. Pyle and R. H. Banks, *J. Org. Chem.*, **40**, 3544 (1975).
14. J. P. Ferraris, T. O. Poehler, A. N. Bloch and D. O. Cowan, *Tetrahedron Lett.*, 2553 (1973).
15. T. Ishiguro, H. Sumi, S. Kagoshima, K. Kajimura and H. Anzai, *J. Phys. Soc. Jpn.*, **48**, 456 (1980).
16. R. L. Greene, J. J. Mayerle, R. R. Schumaker, G. Castro, P. M. Chaikin, S. Etemad and S. J. LaPlaca, *Solid State Commun.*, **20**, 943 (1976).
17. E. M. Engler and V. V. Patel, *J. Am. Chem. Soc.*, **96**, 7376 (1974).
18. S. Etemad, T. Penney, E. M. Engler, B. A. Scott and P. E. Seiden, *Phys. Rev. Lett.*, **34**, 741 (1975).
19. S. Etemad, *Phys. Rev. B*, **13**, 2254 (1976).
20. E. M. Engler, B. A. Scott, S. Etemad, T. Penney and V. V. Patel, *J. Am. Chem. Soc.*, **99**, 5909 (1977).
21. K. Bechgaard, D. O. Cowan and A. N. Bloch, *Chem. Commun.*, 937 (1974).
22. A. N. Bloch, D. O. Cowan, K. Bechgaard, R. E. Pyle, R. H. Banks and T. O. Poehler, *Phys. Rev. Lett.*, **34**, 1561 (1975).
23. J. R. Cooper, M. Weger, D. Jerome, D. Lefur, K. Bechgaard, A. N. Bloch and D. O. Cowan, *Solid State Commun.*, **19**, 749 (1976).
24. D. O. Cowan, M. Mays, M. Lee, R. McCullough, A. Bailey, K. Lerstrup, F. Wiygul, T. Kistenmacher, T. Poehler and L. -Y. Chiang, *Mol. Cryst. Liq. Cryst.*, **125**, 191 (1985).
25. G. Saito, H. Kumogai, J. Tanaka, T. Enoki and H. Inokuchi, *Mol. Cryst. Liq. Cryst.*, **120**, 337 (1985).
26. K. Bechgaard, C. S. Jacobsen and N. Hessel Andersen, *Solid State Commun.*, **25**, 875 (1978).
27. B. Mogensen, R. H. Friend, D. Jerome, K. Bechgaard and K. Carneiro, *Solid State Commun.*, **26**, 919 (1978).
28. G. Mihaly, A. Janossy and G. Gruner, *Solid State Commun.*, **22**, 771 (1977).
29. L. C. Isett and E. A. Perez-Albuerne, *Solid State Commun.*, **21**, 433 (1977).
30. B. Hilti, C. W. Mayer and G. Rihs, *Helv. Chim. Acta*, **61**, 1462 (1978).
31. P. Delhaes, C. Coulon, S. Flandrois, H. Hilti, C. W. Mayer and G. Rihs, *J. Chem. Phys.*, **73**, 1452 (1980).
32. S. P. Zolotukhin, V. F. Kaminskii, A. I. Kotov, R. B. Lyubovskii, M. L. Khidekel, R. P. Shibaeva, I. F. Shchegolev and E. B. Yagubskii, *Piśma Zh. Eksp. Teor. Fiz.*, **25**, 480 (1977); *JETP Lett.*, **25**, 451 (1977).
33. V. N. Laukhin, A. I. Katov, M. L. Khidekel, I. F. Shchegolev, and E. B. Yagubskii, *Piśma Zh. Eksp. Teor. Fiz.*, **28**, 284 (1978).
34. S. P. Zolotukhin, Y. S. Karimov and I. F. Shchegolev, *Sov. Phys. JETP*, **49**, 192 (1979).
35. V. B. Ginodman, L. N. Zherikhina and E. E. Yurchakevich, *JETP Lett.*, **30**, 268 (1980).
36. V. N. Laukhin and I. F. Shchegolev, *Sov. Phys. JETP*, **51**, 1170 (1980).
37. D. Jerome, A. Mazaud, M. Ribault and K. Bechgaard, *J. Phys. Lett.*, **41**, L95 (1980).
38. M. Ribault, G. Benedek, D. Jerome and K. Bechgaard, *J. Phys. Lett.*, **41**, 397 (1980).
39. K. Andres, F. Wudl, D. B. McWhan, G. A. Thomas, D. Nalewajek and A. L. Stevens, *Phys. Rev. Lett.*, **45**, 1449 (1980).
40. R. L. Greene and E. M. Engler, *Phys. Rev. Lett.*, **45**, 1587 (1980).
41. H. J. Schulz, D. Jerome, M. Ribault, A. Mazaud and K. Bechgaard, *J. Phys. Lett.*, **42**, 51 (1981).
42. D. Jerome, *Chem. Scr.*, **17**, 13 (1981).
43. F. Wudl, K. Andres, D. B. McWhan, G. A. Thomas, D. Nalewajek, W. M. Walsh, Jr., L. W. Rupp, Jr., F. J. DiSalvo, J. V. Wazczak and A. I. Stevens, *Chem. Scr.*, **17**, 19 (1981).
44. C. S. Jacobsen, K. Mortensen, N. Thorup, D. B. Tanner, M. Weger and K. Bechgaard, *Chem. Scr.*, **17**, 103 (1981).
45. K. Bechgaard, K. Carneiro, M. Olsen, F. Rasmussen and C. S. Jacobsen, *Phys. Rev. Lett.*, **46**, 852 (1981).
46. K. Bechgaard, K. Carneiro, F. Rasmussen, M. Olsen, G. Rindorf, C. S. Jacobsen, H. J. Pedersen and J. C. Scott, *J. Am. Chem. Soc.*, **103**, 2440 (1981).
47. D. U. Gubser, W. W. Fuller, T. O. Poehler, D. O. Cowan, M. Lee, R. S. Potember, L. -Y. Chiang and A. N. Bloch, *Phys. Rev. B*, **24**, 478 (1981).
48. E. B. Yagubskii, I. F. Shchegolev, V. N. Laukhin, P. A. Kononovich, M. V. Karatsovnik, A. V. Zvarykina and L. I. Buravov, *JETP Lett.*, **39**, 12 (1984).
49. V. F. Kaminskii, T. G. Prokhorova, R. P. Shibaeva and E. B. Yagubskii, *JETP Lett.*, **39**, 17 (1984).

50. G. W. Crabtree, K. D. Carlson, L. N. Hall, T. Copps, H. H. Wang, T. J. Emge, M. A. Beno and J. Williams, *Phys. Rev. B*, **30**, 2958 (1984).
51. T. Emge, P. Leung, M. Beno, A. Schultz, A. Wang, L. Sowa and J. Williams, *Phys. Rev. B*, **30**, 6780 (1984).
52. H. J. Keller, D. Nothe, H. Pritzkow, D. Wehne, M. Werner, P. Koch and D. Schweitzer, *Mol. Cryst. Liq. Cryst.*, **62**, 181 (1980).
53. H. J. Keller, D. Nothe, H. Pritzkow, D. Dehe, M. Werner, R. H. Harms, P. Koch and D. Schweitzer, *Chem. Scr.*, **17**, 101 (1981).
54. L. Alcacer, D. Chasseau and J. Gaultier, *Solid State Commun.*, **35**, 945 (1980).
55. S. Yoneda, T. Kawase, M. Inaba and Z. Yoshida, *J. Org. Chem.*, **43**, 595 (1978).
56. Z. Yoshida, T. Kawase and S. Yondea, *Tetrahedron Lett.*, 331 (1975).
57. G. Scherowsky and J. Weiland, *Chem. Ber.*, **107**, 3155 (1974).
58. M. G. Miles, J. D. Wilson, D. J. Dahn and J. H. Wagenknecht, *J. Chem. Soc., Chem. Commun.*, 751 (1974).
59. E. M. Engler, V. V. Patel and R. R. Schumaker, *J. Chem. Soc., Chem. Commun.*, 835 (1977).
60. E. Klingbserg, *J. Am. Chem. Soc.*, **84**, 3410 (1962).
61. D. Leaver, W. A. H. Robertson and D. M. McKinnon, *J. Chem. Soc.*, 5104 (1962).
62. H. Prinzbach, H. Berger and A. Luttinghaus, *Angew. Chem.*, **77**, 453 (1965).
63. E. M. Engler and V. V. Patel, *J. Am. Chem. Soc.*, **96**, 7376 (1974).
64. H. Poleschner and E. Fanghänel, *J. Prakt. Chem.*, **324**, 691 (1982).
65. E. M. Engler and V. V. Patel, *J. Org. Chem.*, **40**, 487 (1975).
66. K. Bechgaard, D. O. Cowan, A. N. Bloch and L. Henriksen, *J. Org. Chem.*, **40**, 746 (1975).
67. E. M. Engler and V. V. Patel, *J. Chem. Soc., Chem. Commun.*, 672 (1975).
68. L. Henriksen and E. S. Kristiansen, *Int. J. Sulfur Chem., Part A*, **2**, 13 (1972).
69. F. Wudl, E. Aharon-Shalom and S. H. Bertz, *J. Org. Chem.*, **46**, 4612 (1981).
70. P. Shu, A. N. Bloch, T. F. Carruthers and D. O. Cowan, *J. Chem. Soc., Chem. Commun.*, 505 (1977).
71. F. Wudl and D. Nalewajek, *J. Chem. Soc., Chem. Commun.*, 866 (1980).
72. L. -Y. Chiang, T. O. Poehler, A. N. Bloch and D. O. Cowan, *J. Chem. Soc., Chem. Commun.*, 866 (1980).
73. A. Moradpour, V. Peyrussan, I. Johannsen and K. Bechgaard, *J. Org. Chem.*, **48**, 388 (1983).
74. L. -Y. Chiang, D. O. Cowan, T. O. Poehler and A. N. Bloch, *Mol. Cryst. Liq. Cryst.*, **86**, 27 (1982).
75. I. Johannsen, K. Lerstrup, L. Henriksen and K. Bechgaard, *J. Chem. Soc., Chem. Commun.*, 89 (1984).
76. L. Henriksen, *Synthesis*, 771 (1982).
77. E. M. Engler, V. V. Patel, J. R. Andersen, R. R. Schumaker and A. A. Fukushima, *J. Am. Chem. Soc.*, **100**, 3769 (1978).
78. Y. Okamoto and P. S. Wojciechowski, *J. Chem. Soc., Chem. Commun.*, 669 (1981).
79. J. E. Rice, P. Wojciechowski and Y. Okamoto, *Heterocycles*, **18**, 191 (1982).
80. Y. Okamoto, H. S. Lee and S. T. Attarwala, *J. Org. Chem.*, **50**, 2788 (1985).
81. H. D. Hartzler, *J. Am. Chem. Soc.*, **95**, 4374 (1973).
82. D. B. J. Easton and D. Leaver, *J. Chem. Soc., Chem. Commun.*, 585 (1965).
83. B. R. O. Connor and F. N. Jones, *J. Org. Chem.*, **35**, 2002 (1970).
84. M. V. Lakshmikantam, M. P. Cava and A. F. Garito, *J. Chem. Soc., Chem. Commun.*, 383 (1975).
85. M. V. Lakshmikantam and M. P. Cava, *J. Org. Chem.*, **41**, 882 (1976).
86. H. K. Spencer, M. V. Lakshmikantam, M. P. Cava and A. F. Garito, *J. Chem. Soc., Chem. Commun.*, 867 (1975).
87. C. Berg, K. Bechgaard, J. R. Andersen and C. S. Jacobsen, *Tetrahedron Lett.*, 1719 (1976).
88. I. Johannsen, K. Bechgaard, K. Mortensen and C. Jacobsen, *J. Chem. Soc., Chem. Commun.*, 295 (1983).
89. K. Lerstrup, M. Lee, F. M. Wiygul, T. J. Kistenmacher and D. O. Cowan, *J. Chem. Soc., Chem. Commun.*, 294 (1983).
90. L. -Y. Chiang, P. Shu, D. Holt and D. O. Cowan, *J. Org. Chem.*, **48**, 4713 (1983).
91. K. Lerstrup and D. O. Cowan, *J. Phys.* (*Paris*) *Colloq. C3*, 1247 (1983).
92. R. Ketcham, A. B. Hornfeldt and S. Gronowitz, *J. Org. Chem.*, **49**, 1117 (1984).
93. R. D. McCullough and D. O. Cowan, *J. Org. Chem.*, **50**, 4646 (1985).
94. W. R. H. Hurtley and S. Smiles, *J. Chem. Soc.*, 2263 (1926).

95. G. S. Bajwa, K. D. Berlin and H. A. Pohl, *J. Org. Chem.*, **41**, 145 (1976).
96. F. Wudl and E. Aharon-Shalom, *J. Am. Chem. Soc.*, **104**, 1154 (1982).
97. K. Lerstrup, D. Talham, A. Bloch, T. Poehler and D. Cowan, *J. Chem. Soc., Chem. Commun.*, 336 (1982).
98. K. Lerstrup, D. O. Cowan and T. J. Kistenmacher, *J. Am. Chem. Soc.*, **106**, 8303 (1984).
99. C. Lambert and L. Christiaens, *Tetrahedron Lett.*, 833 (1984).
100. S. Wawzonek and S. Heilmann, *J. Org. Chem.*, **39**, 511 (1974).
101. G. Bontempelli, F. Nago, G. Mazzochin and R. Seeber, *J. Electroanal. Chem. Interfacial Electrochem.*, **63**, 231 (1974).
102. P. R. Moses and J. Q. Chambers, *J. Am. Chem. Soc.*, **96**, 945 (1974).
103. W. P. Krug, A. N. Bloch, T. O. Poehler and D. O. Cowan, *Ann. N. Y. Acad. Sci.*, **313**, 366 (1978).
104. M. Mizuno, A. F. Garito and M. P. Cava, *J. Chem. Soc., Chem. Commun.*, 18 (1978).
105. W. Krug, *PhD Dissertation*, The Johns Hopkins University, Baltimore, 1977.
106. G. Saito, T. Enoki, K. Toriumi and H. Inokuchi, *Solid State Commun.*, **42**, 557 (1982).
107. S. S. P. Parkin, E. M. Engler, R. R. Schumaker, R. Lagier, V. Y. Lee, J. C. Scott and R. L. Greene, *Phys. Rev. Lett.*, **50**, 270 (1983).
108. H. H. Wang, M. A. Beno, U. Geiser, M. A. Firestone, K. S. Webb, L. Nunez, G. W. Crabtree, K. D. Carlson, J. M. Williams, L. J. Azvedo, J. F. Kwak and J. E. Schirber, *Inorg. Chem.*, **24**, 2465 (1985).
109. E. M. Engler, D. C. Green and J. Q. Chambers, *J. Chem. Soc., Chem. Commun.*, 148 (1976).
110. V. Y. Lee, E. M. Engler, R. R. Schumaker and S. S. P. Parkin, *J. Chem. Soc., Chem. Commun.*, 235 (1983).
111. D. Cowan, A. Bailey and K. Lerstrup, unpublished results.
112. F. Wudl, D. E. Shafer and B. Miller, *J. Am. Chem. Soc.*, **98**, 252 (1976).
113. H. Endres, H. J. Keller, J. Quekborner, J. Veigel and D. Schweitzer, *Mol. Cryst. Liq. Cryst.*, **86**, 1851 (1982).
114. C. Krohnke, V. Enkelman and G. Wegner, *Angew. Chem., Int. Ed. Engl.*, **19**, 912 (1980).
115. A. Yamahira, T. Nohami and H. Mikawa, *J. Chem. Soc., Chem. Commun.*, 904 (1983).
116. J. C. Stark, R. Reed, L. A. Acampora, D. J. Sandman, S. Jensen, M. T. Jones and B. M. Foxman, *Organometallics*, **3**, 732 (1984).
117. K. A. Balodis, A. D. Livdane, R. S. Medne and O. V. Neiland, *J. Org. Chem. USSR* (*Engl. Transl.*), **15**, 343 (1979).
118. D. J. Sandman, J. C. Stark and B. M. Foxman, *Organometallics*, **1**, 739 (1982).
119. S. Hunig, B. J. Garner, G. Ruider and W. Schenk, *Justus Liebigs Ann. Chem.*, 1036 (1973).
120. J. H. Perlstein, J. A. Van Allen, L. C. Iset and G. A. Reynolds, *Ann. N. Y. Acad. Sci.*, **313**, 61 (1977).
121. D. J. Sandman, A. J. Epstein, T. J. Holmes, J.-S. Lee and D. D. Titus, *J. Chem. Soc., Perkin Trans. 2*, 1578 (1980).
122. H. Strezelecka and J. Rivory, *Mater. Res. Bull.*, **15**, 899 (1980).
123. S. Es-Seddiki, G. Le Coustumer, Y. Mollier and M. Davoud, *Tetrahedron Lett.*, 2771 (1981).
124. S. Es-Seddiki, G. Le Costumer, Y. Mollier and M. Davoud, *Mol. Cryst. Liq. Cryst.*, **86**, 1811 (1982).
125. J. Amzil, M. Brutus, G. LeCostumer, M. Hakiki, J. F. Hemidy, Y. Mollier, J. P. Sauve and M. Stavaux, *J. Phys.* (*Paris*) *Colloq. C*, 1249 (1983).
126. M. R. Detty, B. J. Murray and J. H. Perlstein, *Tetrahedron Lett.*, 539 (1983).
127. M. R. Detty and B. J. Murray, *J. Org. Chem.*, **47**, 1146 (1982).
128. M. R. Detty, B. J. Murray and M. D. Seidler, *J. Org. Chem.*, **47**, 1968 (1982).
129. T. C. Chiang, A. H. Reddock and D. F. Williams, *J. Chem. Phys.*, **54**, 2051 (1971).
130. H. Akamatu, H. Inokucki and Y. Matsunaga, *Nature* (*London*), **173**, 168 (1954).
131. H. Hart and M. Sasaoka, *J. Am. Chem. Soc.*, **100**, 4326 (1978).
132. J. S. Chappell, A. N. Bloch, W. A. Bryden, M. Maxfield, T. O. Poehler and D. O. Cowan, *J. Am. Chem. Soc.*, **103**, 2442 (1981).
133. J. B. Torrance, *Acc. Chem. Res.*, **12**, 79 (1979).
134. K. Bechgaard, in *Proceedings of Symposium* '*Horizons in the Chemistry and Properties of Low Dimensional Solids*', *Honolulu, Hawaii, December 16–21, 1984; Mol. Cryst. Liq. Cryst.*, **125**, 81 (1985).
135. R. E. Peierls, *Quantum Theory of Solids*, Oxford University Press, London, 1955, p. 108.

The Chemistry of Organic Selenium and Tellurium Compounds Volume 2
Edited by S. Patai
© 1987 John Wiley & Sons Ltd.

CHAPTER **13**

Organoselenium and organotellurium analogues of ethers and peroxides

WILLIAM R. McWHINNIE

Department of Molecular Sciences, Aston University, Aston Triangle, Birmingham B4 7ET, UK

I. INTRODUCTION 496
II. SYNTHESIS 497
A. Symmetrical and Unsymmetrical Selenides 497
1. Elemental selenium 497
2. Selenium(IV) compounds. 497
3. Selenium(II) compounds 498
4. Organyl selenides 500
5. Selenides from diselenides 500
6. Miscellaneous 500
B. Symmetrical, Unsymmetrical and Cyclic Diselenides; Triselenides . 501
1. Symmetrical diselenides 501
2. Unsymmetrical diselenides 503
3. Cyclic diselenides 503
4. Triselenides and tetraselenides 504
C. Symmetrical and Unsymmetrical Tellurides 504
1. Elemental tellurium 505
2. Organyl tellurides 506
3. Tellurium(II) compounds. 507
4. Tellurium(IV) compounds 507
5. Miscellaneous 509
D. Diorganyl Ditellurides and Related Compounds 510
1. Syntheses based on tellurium 510
2. Syntheses based on tellurium(II) compounds 511
3. Syntheses based on tellurium(IV) compounds 511
4. Polytellurides. 512
III. STRUCTURE. 512
A. Selenides 513
B. Di- and Poly-selenides. 513

C. Tellurides . . . 515
D. Ditellurides . . . 516
IV. REACTIONS . . . 516
A. Diorganyl Selenides and Tellurides . . . 516
B. Diorganyl Diselenides and Ditellurides . . . 518
C. Applications in Organic Synthesis . . . 520
1. Coupling reactions . . . 520
2. Lithium reagents . . . 521
3. Removal of —ER (E = Se, Te) . . . 522
4. Carbon to oxygen bond formation . . . 523
D. Photochemical Reactions . . . 523
1. Selenides and tellurides . . . 523
2. Diselenides and ditellurides . . . 525
E. Lewis Base Reactions . . . 527
1. Selenides and tellurides as ligands . . . 527
2. Diselenides and ditellurides as ligands . . . 528
3. Bi- and poly-dentate ligands . . . 528
4. Scales of Lewis acidity . . . 529
5. Conclusion . . . 529
F. Charge-transfer Complexes . . . 529
V. SPECTROSCOPY . . . 530
A. Nuclear Magnetic Resonance Spectroscopy . . . 530
B. Vibrational Spectra . . . 533
C. ^{125}Te Mössbauer Spectra . . . 533
VI. CONCLUSION . . . 533
VII. REFERENCES . . . 533

I. INTRODUCTION

Diorganyl selenides and diorganyl tellurides, the selenium and tellurium analogues of the ethers, probably constitute the largest single class of known organic derivatives of the heavier chalcogens. The scope of this chapter could, therefore, by very great. To avoid a totally superficial coverage of a large bulk of material, some constraints will be introduced; thus most heterocyclic systems which could, strictly, be classified as examples of R_2E (E = Se, Te) will be omitted. This is sensible, particularly when, as with tellurophene for example, it is wise to consider reactions of the total ring system.

Diorganyl diselenides and diorganyl ditellurides present less classification problems since, although some dichalcogenide linkages appear as constituents of a heterocyclic ring, e.g. **1**, the reactivity of such compounds is often dominated by the reactivity of the dichalcogenide linkage.

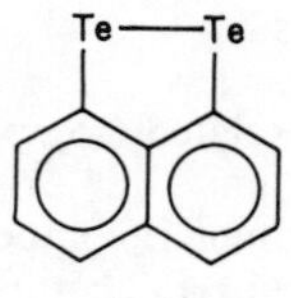

(1)

This chapter will cover the general methods of synthesis of R_2E and R_2E_2 (E = Se, Te), the structures of the compounds, the reactions they undergo and their spectroscopic properties. The objective under each heading will be to bring out the salient features rather than to attempt a comprehensive coverage. The fact that interest in the compounds has been steadily growing over the past decade reflects amongst other things their growing importance in organic synthesis, their role in formulating novel materials of high electrical conductivity and their amenity to study by techniques such as X-ray crystallography and multinuclear NMR spectroscopy. Not surprisingly, therefore, a number of books[1-3], review articles and annual surveys[4-6] and the proceedings of three relevant international conferences[7-9] do, between them, provide a very comprehensive entrée to the literature. Further, the present series of texts will also provide specialist chapters relevant to topics covered here; consequently, in dealing with spectroscopic properties, for example, this chapter will confine attention to those data which provide extra chemical insight to the molecules R_2E and R_2E_2 and will not be concerned to elucidate the detailed theory of the techniques. Although, inevitably, some overlap with other chapters will occur, it is hoped that this contribution can stand alone as a helpful introduction to the chemistry of the title compounds.

II. SYNTHESIS

A. Symmetrical and Unsymmetrical Selenides

A variety of inorganic and organic selenium compounds may be employed as starting materials for the synthesis of selenides, R_2Se and RR′Se. Examples of synthetic methods are conveniently presented as a function of these various materials.

1. Elemental selenium

Selenium is normally converted to selenide. This may be achieved with a variety of reducing agents in aqueous or non-aqueous media. Typical examples are

$$\mathrm{Se} \xrightarrow{\mathrm{NaOH/H_2O}} \mathrm{Na_2Se} \xrightarrow{\mathrm{RX}} \mathrm{R_2Se} \qquad \text{(Ref. 10)}$$

$$\text{e.g. } \mathrm{R} = \text{alkyl}$$

and

$$\mathrm{Se} \xrightarrow[\mathrm{NH_3(liq.)}]{\mathrm{Na}} \mathrm{Na_2Se} \xrightarrow{\mathrm{RX}} \mathrm{R_2Se} \qquad \text{(Ref. 11)}$$

$$\text{e.g. } \mathrm{R} = \mathrm{Me, Et}$$

A recently developed procedure[12] uses DMF as a non-aqueous solvent for the reaction of sodium and elemental selenium. It is always advisable to work with solutions of selenide (and telluride) under a strictly inert atmosphere.

An interesting example of the use of the Se^{2-}/DMF reagent is provided by the synthesis of poly(*p*-phenylene selenide)[13,14]:

$$p\text{-}\mathrm{BrC_6H_4Br} + \mathrm{Na_2Se} \xrightarrow[120-140\,^{\circ}\mathrm{C}]{\mathrm{DMF}} (\mathrm{C_6H_4Se})_n$$

The experimental conditions were milder than those required for poly(*p*-phenylene sulphide) and it is believed that an $S_{RN}1$ mechanism is operative[14].

2. Selenium(IV) compounds

Selenium tetrachloride is readily available and reacts with a variety of organometallic compounds to give, under appropriate reaction conditions, diorganylselenium dichlorides, which may then be reduced to the selenide. In some cases reduction is not necessary since the selenide results directly. Appropriate organometallic compounds for the transmetallation reactions include Grignard reagents, organolithiums, organothalliums (e.g. R_2TlBr), organotins and organomercurides. Some examples are as follows:

$$SeCl_4 + RLi \longrightarrow RSeCl_3 \xrightarrow{RLi} R_2SeCl_2 \qquad \text{(Ref. 15)}$$

$$SeCl_4 + 4\,RMgBr \longrightarrow R_2Se + RR + 2\,MgCl_2 \rightarrow 2\,MgBr_2 \qquad \text{(Ref. 16)}$$

$$SeCl_4 + (F_5C_6)_2TlBr \longrightarrow (F_5C_6)_2SeCl_2 + TlCl_2Br \qquad \text{(Ref. 17)}$$

$$SeCl_4 + Ph_4Sn \longrightarrow Ph_2SeCl_2 + Ph_2SnCl_2 \qquad \text{(Ref. 18)}$$

$$SeBr_4 + 3\,Ph_2Hg \longrightarrow Ph_2Se + 3\,PhHgBr + PhBr \qquad \text{(Ref. 19)}$$

An alternative selenium(IV) compound is the dioxide; a neat example of its use involves a triorganyl hydroborate reagent[20]:

$$SeO_2\,\text{(suspension)} \xrightarrow[\text{THF}]{R_3BH^-} R_2Se$$

$$R = \text{allyl}$$

Selenium tetrachloride may also react directly with some organic compounds. Arylselenium compounds result from the direct reaction of $SeCl_4$ with those aromatic compounds which are activated to electrophilic attack, e.g.[21]

$$2\,EtOC_6H_5 + SeCl_4 \xrightarrow{CHCl_3} (p\text{-}EtOC_6H_4)_2SeCl_2 + 2\,HCl$$

Carbon—carbon double bonds are also susceptible to attack; selenium tetrachloride adds across the bond to give a 2-chloroalkylselenium trichloride, which may then react with further alkene to give bis(2-chloroalkyl)selenium dichloride[22].

In many of the reactions illustrated above the product is a diorganylselenium dihalide which must be reduced to the selenide. A variety of reducing agents may be used but aqueous $Na_2S_2O_5$ and hydrazine are currently among the most popular, e.g.

$$R_2SeCl_2 \xrightarrow[H_2O]{Na_2S_2O_5} R_2Se$$

3. *Selenium(II) compounds*

The availability of organyl selenium compounds of the type RSeX (the tellurium analogues are much less convenient) make them all an excellent choice for the synthesis of unsymmetrical selenides. These compounds also play a prominent role in the application or organoselenium chemistry to organic synthesis[3,5,23]. The reactions of RSeX are illustrated for a variety of X groups; examples are taken from the more recent literature, earlier examples being well documented[1].

X = Cl, Br

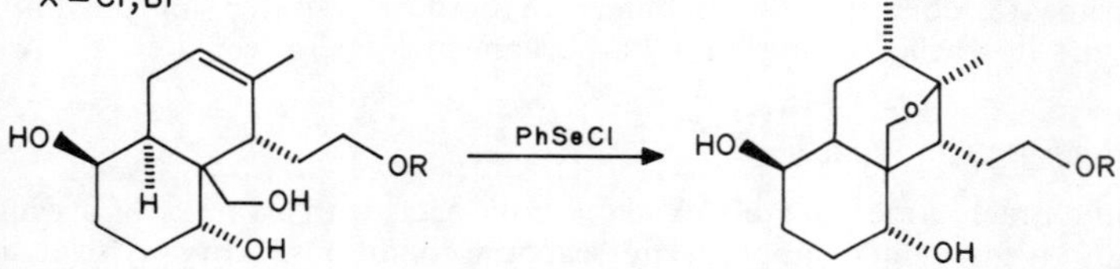

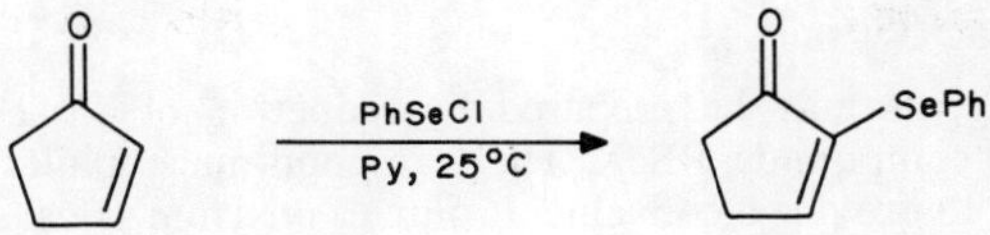

$$F_3CCH=CH_2 + PhSeCl \xrightarrow{-HCl} F_3CCH=CHSePh \xrightarrow{BuLi} PhSeBu \text{ (Ref. 26)}$$

$$RCH=CH_2 + PhSeBr \longrightarrow PhSeCHRCH_2Br \quad \text{(Ref. 27)}$$

(kinetic control)

($R = C_1$–C_4 alkyl) $RCHBrCH_2SePh$

(thermodynamic control)

X = $SiMe_3$:

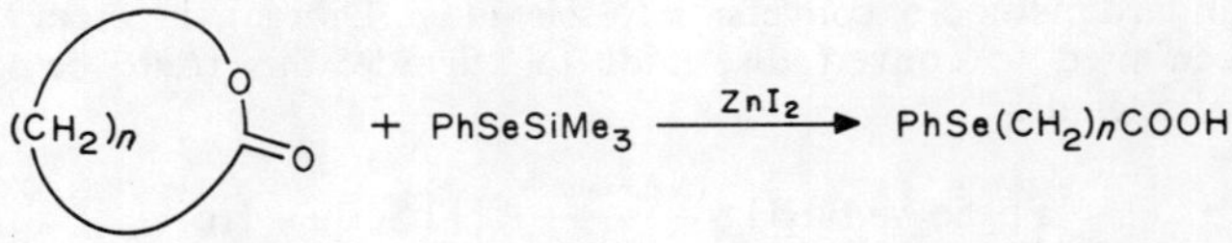

X = H:

$$2\,RSeH + Cu_2O \longrightarrow 2\,RSeCu + H_2O$$

$$\downarrow 2\,R^1I$$

$$2\,RR^1Se + Cu_2I_2$$

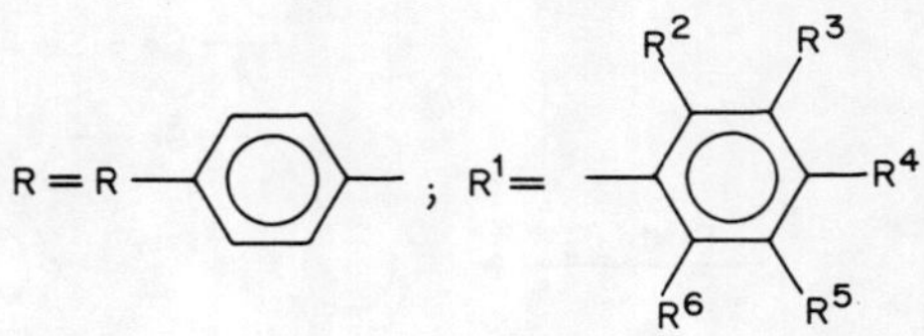

X = CN:

$$RCH_2OH + ArSeCN \xrightarrow[Bu_3P]{\text{THF or Py}} (Ar)(RCH_2)Se$$

e.g. 1-dodecanol reacts with o-$NO_2C_6H_4SeCN$ to give 94% selenide[30].

Organylselenium(II) halides will react with unsaturated organic molecules, e.g. Kataev and coworkers[31,32] employed this reaction for the synthesis of vinyl selenides:

$$p\text{-}RC_6H_4SeCl + C_2H_4 \longrightarrow p\text{-}RC_6H_4SeCH_2CH_2Cl$$

$$\xrightarrow[(Me_3COK)]{-HCl} p\text{-}RC_6H_4Se(CH=CH_2)$$

$$p\text{-}RC_6H_4SeCl + C_2H_2 \longrightarrow p\text{-}RC_6H_4SeCH=CHCl$$

R = MeO, Me, Cl, Br

4. Organyl selenides, RSe^-

These useful reagents may be prepared by reduction of diselenides, R_2Se_2, or of organylselenium(II) compounds, RSeX. They are good nucleophiles and will react with organic halides R′X to provide a convenient route to unsymmetrical selenides, RR′Se. An example of this reaction type which affords a potentially terdentate selenium ligand is[33]

$$(BrCH_2CH_2)_2Se + 2PhSe^- \longrightarrow (PhSeCH_2CH_2)_2Se$$

An interesting synthesis of unsymmetrical selenides from primary, secondary or tertiary amines with $PhSe^-$ over a ruthenium catalyst ($RuCl_3$/K) was recently reported[34], e.g.

$$R^1R^2R^3N + PhSeNa \xrightarrow[2.\ H_2O]{1.\ Ru,\ diglyme,\ Ar} R^1PhSe + R^2R^3NH$$

5. Selenides from diselenides

Some diselenides are photochemically unstable with respect to elimination of selenium, but these reactions will be dealt with in Section IV.D. Heating diselenides with Raney nickel is a standard method of conversion to selenide[35]. Diorganyl mercury compounds have also been used to convert diselenide to selenide, this route being useful for unsymmetrical selenides[36]:

$$Ph_2Se_2 + Bu_2Hg \xrightarrow[reflux]{dioxane} 2PhSeBu + Hg$$

Mitchell[37] has prepared 2, 3, 12, 13-tetraselena(4, 4)metacyclophane in 8% yield using the following reaction and has also demonstrated the conversion to a bis-selenide on treatment with tris(diethylamino)phosphine:

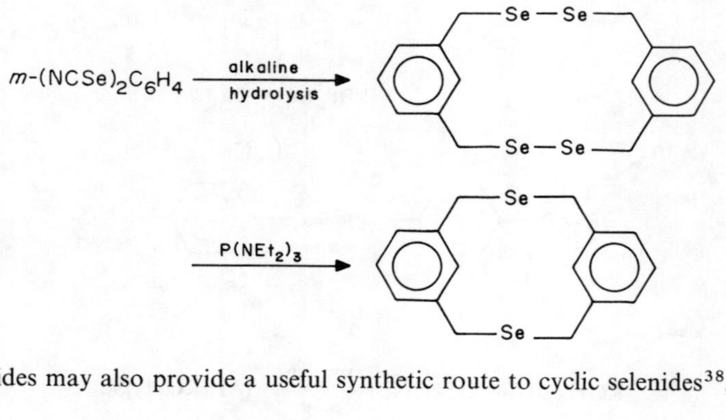

Diselenides may also provide a useful synthetic route to cyclic selenides[38], e.g.

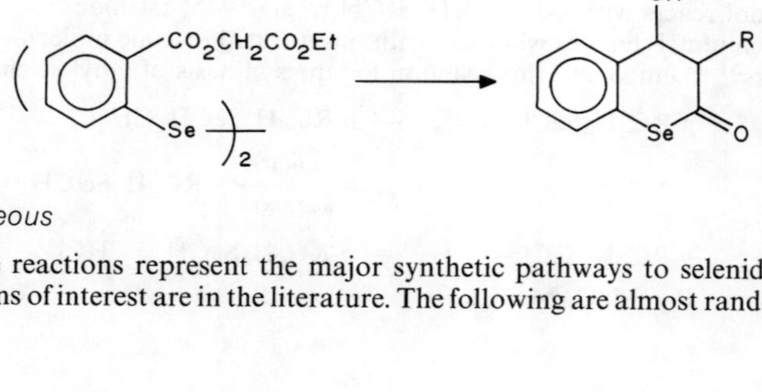

6. Miscellaneous

The above reactions represent the major synthetic pathways to selenides, but many other reactions of interest are in the literature. The following are almost random examples.

An attempted Wittig reaction of selenophosphoranes with ketones resulted in a general method for the synthesis of unsymmetrical selenides[39], e.g.

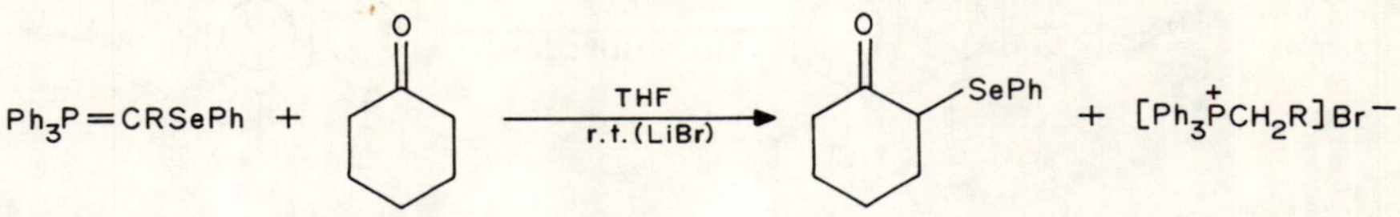

The mechanism of the reaction is believed to be as follows:

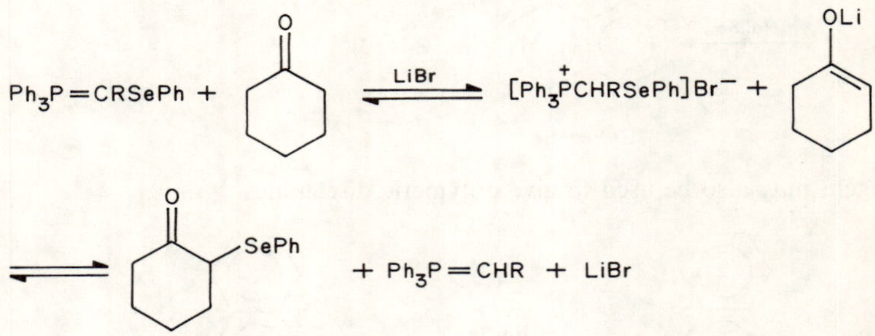

Polyselenides will react with halogens such as bromine in a complex reaction to give monomeric selenides[40], e.g.

$$(CH_2CHSe)_n + Br_2 \longrightarrow (BrCH_2)_2Se \xrightarrow[\text{excess}]{Br_2} (BrCH_2)_2SeBr_2$$

Selenium, $SeBr_4$ and dibromomethane are also formed in the reaction.

B. Symmetrical , Unsymmetrical and Cyclic Diselenides; Triselenides

Diselenides are the selenium analogues of the organic peroxides. They are yellow to orange–red in colour and, unless constrained in a small ring, are structurally analogous to hydrogen peroxide (see Section III). They are interesting compounds in their own right and are, as illustrated in Section II.A.5 above, useful synthetic intermediates in the synthesis of other organoselenium compounds.

1. Symmetrical diselenides

Standard methods of synthesis may be found in Ref. 1. Elemental selenium is a useful starting material since reduction with either sodium or rongalite ($HOCH_2OSO_2Na$) can give Se_2^{2-}, which can then react with various organic reagents to give the diselenide. Specific examples follow:

$$Na_2Se_2 + 2MeX \longrightarrow Me_2Se_2 + 2NaX \qquad \text{(Ref. 41)}$$

The diselenide may be formed in liquid ammonia solution using sodium as reductant or it may be prepared in aqueous solution using rongalite. The recent introduction of DMF as a convenient medium for the reaction of sodium and selenium[12] should greatly increase the utility of this reaction. Although it is generally true that aromatic halides will react with Se_2^{2-} only when activated by a nitro-group[42], there are exceptions when the Se^{2-}/DMF reagent is used as the following sequence illustrates[43].

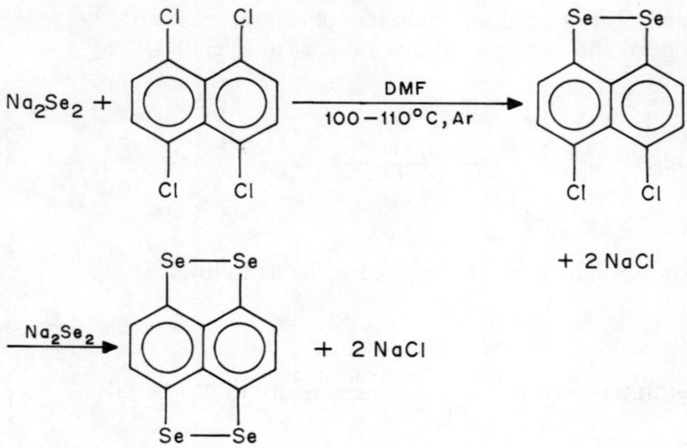

The reagent may also be used to give polymeric diselenides[44], thus:

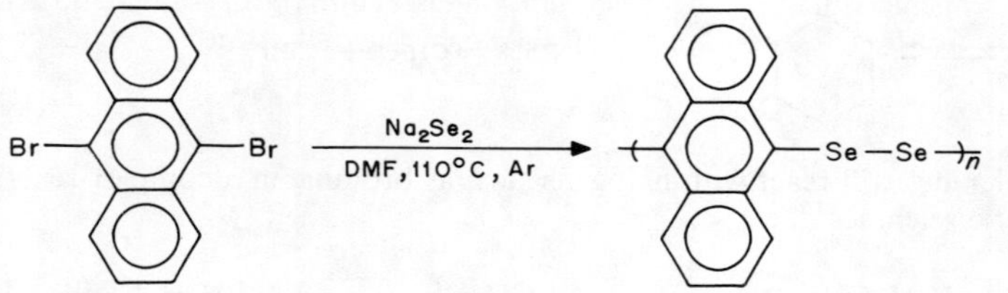

However, diazonium salts may be considered more standard reagents for the synthesis of diaryl diselenides, e.g.[45]

$$2R\overset{+}{N}_2Cl^- + Na_2Se_2 \longrightarrow R_2Se_2 + N_2 + 2NaCl$$

$$R = 2\text{- and }4\text{-carboxyphenyl}$$

Two other inorganic selenium reagents that may be used for the synthesis of diselenides are Se_2Cl_2 and H_2Se. Diselenium dichloride will react with unsaturated organic compounds, e.g. alkynes[46]:

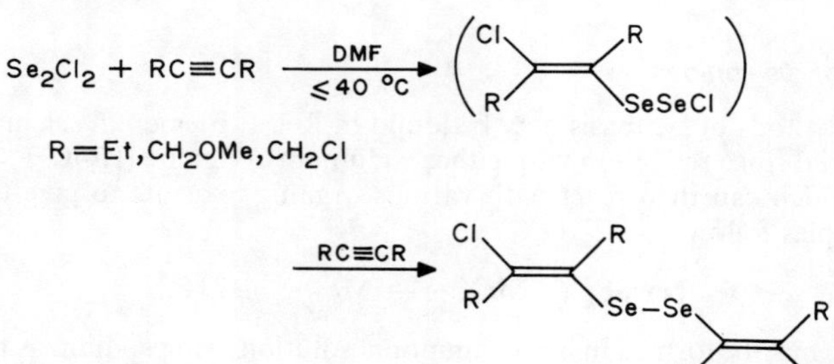

Complications may arise if phenylacetylenes, PhC≡CR, are used in that anything up to a 41% yield of a benzoselenophene may be obtained, viz.:

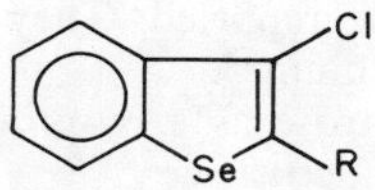

Aldehydes will react with H_2Se under reducing conditions to give diselenides[47], e.g.

$$ArCHO + H_2Se \xrightarrow[2.\ NaBH_4]{1.\ base} (ArOH_2\text{-}Se)_2$$

92%

Ar = 1-naphthyl

Organic selenium compounds may also be used as precursors for diselenides. For example, the hydrolysis of organyl selenocyanates gives diorganyl diselenides[48] (see also Section II.A.5):

$$2RSeCN \xrightarrow[H_2O\ or\ EtOH]{NaOH} R_2Se_2$$

Grignard reagents are useful since selenium will insert into the C—Mg bond; hydrolysis of the reagent followed by oxidation with dioxygen then gives the diselenide, e.g.[49]

$$p\text{-}ClC_6H_4MgBr + Se \longrightarrow p\text{-}ClC_6H_4SeMgBr \xrightarrow[2.\ O_2]{1.\ H_2O} (p\text{-}CLC_6H_4)_2Se_2$$

A variation on this theme, but involving organolithium compounds, has recently been claimed to give yields and purity for diselenides which are superior to other methods[50]:

$$RLi + Se \longrightarrow RSeLi \xrightarrow{Me_2NC(S)Cl} Me_2NC(S)SeR \xrightarrow[Fe(CN)_6^{3-}]{OH^-} R_2Se_2$$

R = aryl or heteroaryl

The intermediate is readily crystallized and purified. Other workers[51] have also used an intermediate with a selenium—sulphur bond, e.g.

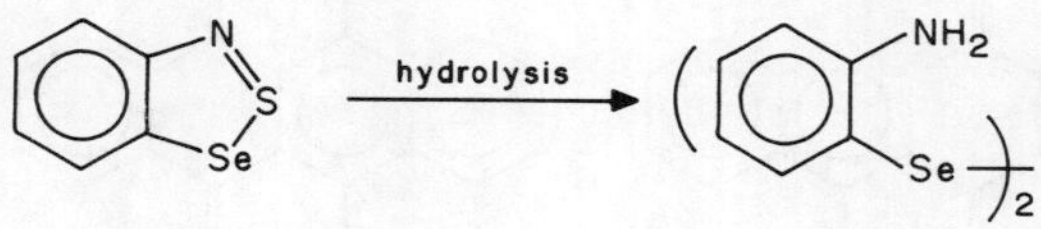

2. *Unsymmetrical diselenides*

Few compounds have been isolated, but the greater stability of RSeX and RSeH than their tellurium counterparts do provide synthetic pathways to unsymmetrical diselenides which are not available for ditellurides. The reaction generally used is[42,52]

$$RSeX + R^1SeH \longrightarrow RSeSeR^1 + HX$$

X = halide, CN

3. *Cyclic diselenides*

Naphtho[1,8-*cd*:4,5-c^1d^1]bis-1,2-diselenole has been considered in Section II.B.1, but there has been growing interest recently in cyclic diselenides and this has led to the publication of several papers describing their synthesis. *Peri*-bridged naphthalenes were the first of their class to be reported. They were prepared by the reaction of 1,8-dilithionaphthalene with selenium[53].

An interesting cyclic diselenide arose as a by-product from a reaction designed to produce a new tetraselenafulvalene derivative[54]:

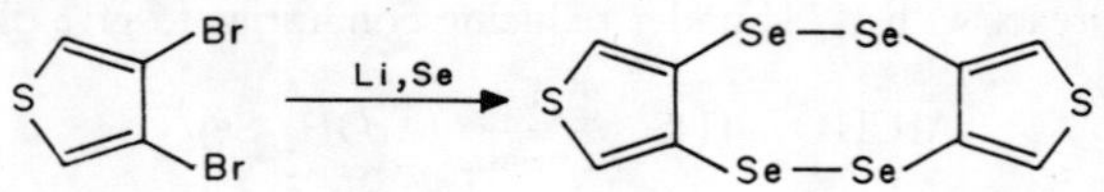

4. Triselenides and tetraselenides

Diethyl triselenide has been reported[42] and a number of aromatic triselenides are known[55]. A general method of preparation is

$$2RSeSeCN + 2EtSH \longrightarrow 2HCN + Se + Et_2S_2 + R_2Se_3$$

$$R = p\text{-}XC_6H_4, \text{ where } X = H, Me, Cl, Br, MeO, NO_2$$

Other earlier examples will be found in Ref. 1. Foss and Janickis[56] heated piperidine or morpholine with selenium powder in the presence or red lead. The morpholine reaction gave di-, tri- and tetra-selenium dimorpholine whereas the piperidine gave the tetra-selenium compound.

A search for novel materials with high electrical conductivity led to an unexpected triselenide isolated as a complex with benzonitrile[57]:

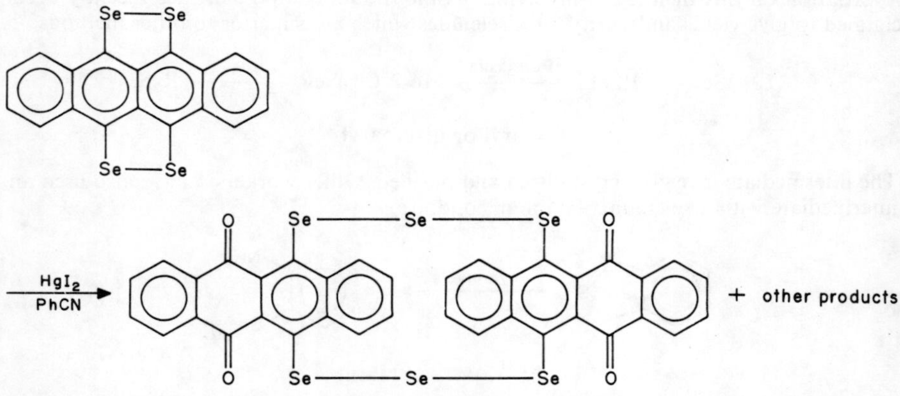

Another novel example of a triselenide based on ferrocene has been reported[58], namely

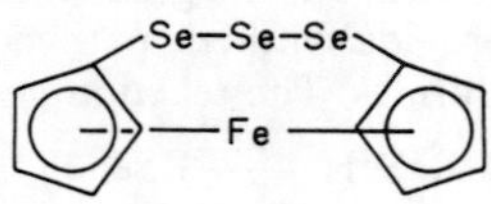

C. Symmetrical and Unsymmetrical Tellurides

Methods broadly similar to those used for diorganyl selenides may be used to synthesis diorganyl tellurides, but the sections are separated in this chapter for greater ease of reference. Some options available in selenium chemistry may be less convenient in tellurium chemistry. For example, organyl tellurenyl compounds, RTeX, are less convenient than their Se analogues. Some may be prepared by careful addition of 1 mol of halogen to 1 mol of ditelluride (e.g. PhTeI, p-PhC_6H_4TeBr[59]), but the reaction is not general since RTeX tends to disproportionate to Te and R_2TeX_2. The isolation of stable, well defined examples of RTeX usually requires the presence of a coordinating group in the molecule, e.g. **2**[60].

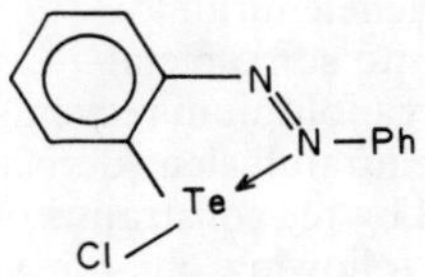

(**2**)

Thus, although the earlier literature contains several references to the use of 'TeX_2' ($X = Cl$, Br) as starting materials for the synthesis of R_2Te, it is now unusual for tellurium(II) compounds to be selected as the precursors of choice.

1. Elemental tellurium

Elemental tellurium is now a common starting material in organotellurium chemistry, particularly for the synthesis of more sophisticated tellurides. It may be used in one of two reaction types: prior reduction to telluride or insertion into a carbon—metal or carbon—halogen bond (see Section II.C).

The first organyl tellurium compound to be reported, Et_2Te, was prepared by reaction of potassium telluride with ethyl sulphate[61]; the telluride was prepared by an inconvenient route (Te + potassium hydrogen D-tartrate at red heat), but as Irgolic[2] suggests, another early use of a binary metal telluride by Natta[62] could repay further investigation and exploitation:

$$3Te + 2Al \longrightarrow Al_2Te_3 \xrightarrow[ROH \text{ or } R_2O]{250-300\,^\circ C} R_2Te$$

Usually the telluride is prepared *in situ* by reduction of elemental tellurium in aqueous on non-aqueous media. An example is[63]

$$\underset{\text{('rongalite')}}{Te + HOCH_2OSO_2Na} \xrightarrow[\text{air-free, } 70-80\,^\circ C]{10\%\ NaOH} Na_2Te \xrightarrow{PhCH_2Cl} (PhCH_2)_2Te$$

A recent variation[64,65] of this reaction has employed $SnCl_2$ as reducing agent: (the method will also work for selenium; thus divinylselenide may be prepared at ambient pressure[66]).

$$2\,CH_2 = CH_2 + Te \xrightarrow[90-120\,^\circ C/10-15\,atm]{SnCl_2, OH^-/H_2O} (CH_2 = CH)_2Te$$

The reduction of tellurium in liquid ammonia is conveniently carried out with metallic sodium. The Na_2Te formed will react readily with aliphatic halides, e.g.[67]

$$2\,EtBr + Na_2Te \xrightarrow{NH_3} \underset{80\%}{Et_2Te}$$

The recent exploitation of dipolar aprotic solvents such as DMF as a medium for the synthesis of sodium selenide is no less successful for the synthesis of sodium telluride from the elements[12], e.g.

$$p\text{-}IC_6H_4I + Na_2Te \xrightarrow[110-120\,^{\circ}C]{DMF} (\ C_6H_4Te\ )_n$$

(the product contains some residual iodine, the quantity suggesting a molecular weight of around 8000 for the poly-*p*-phenylenetelluride[14]).

Elemental tellurium will insert into some metal—carbon bonds. The most important examples involve Grignard and organolithium reagents. Recent examples of importance include the first synthesis of tetratellurafulvalene derivatives[68,69]. Although, in principle, those examples should be excluded by the constraints of the Introduction to this chapter, an exception will be made since the following 'one-pot' synthesis[69] elegantly illustrates the use of the RTeLi reagent:

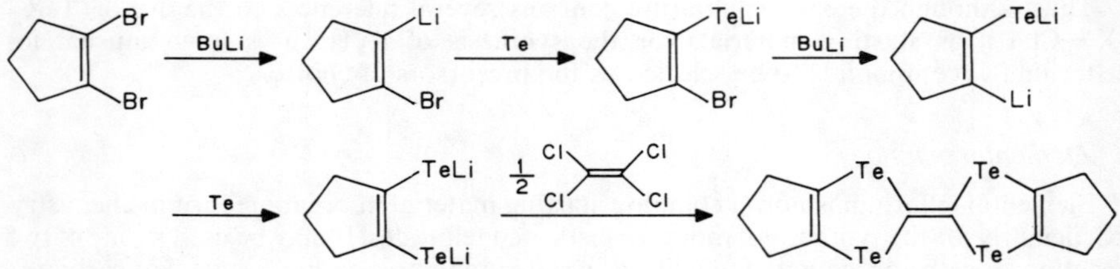

The lithium reagent is also valuable for the synthesis of unsymmetrical tellurides, e.g.[70]

$$PhLi \xrightarrow{Te} PhTeLi \xrightarrow{RX} PhTeR + LiX$$

$$PhTeLi \xrightarrow{CH_2I_2} PhTeCH_2TePh$$

also[71]

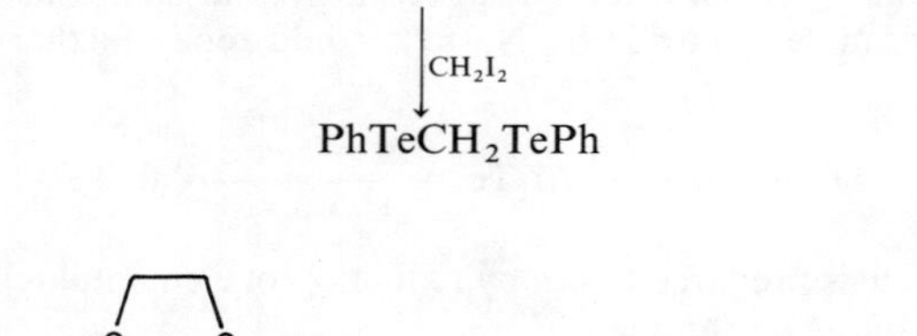

2. Organyl tellurides, RTe⁻

Clearly, the reactions of organolithium reagents given above proceed via the *in situ* synthesis of RTe^-, hence Sections II.c.1 and II.c.2 overlap. A frequently used alternative synthesis of RTe^- is by the alkaline sodium borohydride reduction of diorganylditellurides, a route that is particularly attractive for aryl compounds. The following is a recent example of the use of the method to form a series of bis-tellurides[72,73]:

$$R_2Te_2 \xrightarrow[EtOH/C_6H_6]{NaBH_4/OH^-} NaTeR \xrightarrow{(CH_2)_nBr_2} RTe(CH_2)_nTeR$$

$$R = p\text{-}EtOC_6H_4 \qquad\qquad n = 1, 5, 6, 7, 9, 10$$

For $n = 1$, if the organic dihalide (CH_2Br_2 or CH_2I_2) is in excess, a charge-transfer complex is isolated, i.e. $(RTe)_2CH_2{\cdot}X_2CH_2$. When $n = 2$, the bis-telluride is unstable with respect to elimination of ditelluride; however, this does provide a general reaction for the conversion of organic 1, 2-dihalides to alkenes. When $n = 3$ or 4, internal quaternarization giving a telluronium salt is more rapid that nucleophilic attack by RTe^- on the second C—Br bond, e.g.

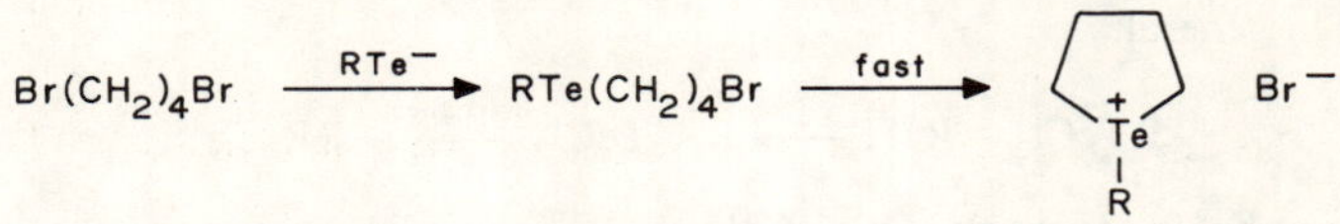

For the single case of $n = 5$, experimental conditions may be manipulated to prepare either the bis-telluride or the telluronium salt.

The reactions of RTe^- are not restricted to organic halogen substrates. Thus, for example, NaTeR (R = Ph, p-MeC_6H_4) will react with a range of carboxyl-substituted alkynes to give the expected products arising from nucleophilic attack at the β-carbon atom[74], e.g.

$$RTeNa + PhC{\equiv}CCOPh \xrightarrow{EtOH} RTeC(Ph) = CHCOPH$$

$$R = Ph,\ p\text{-}MeC_6H_4$$

Although sodium borohydride is the commonly used reducing agent to prepare RTe^- from a diorganyl ditelluride, others such as lithium aluminium hydride may be used. An example is[75]

$$(RC_6H_4)_2Te_2 \xrightarrow[Et_2O,\ Ar]{LiAlH_4} LiTeC_6H_4R \xrightarrow[UV]{C_3F_7I} RC_6H_4TeC_3F_7$$

$$(R = H,\ p\text{-}F,\ p\text{-}Br,\ m\text{-}F)$$

3. Tellurium(II) compounds

Following the reported synthesis of PhTeI[59], the compound has been used in conjunction with vinyl and alkynyl Grignard reagents to prepare unsymmetrical vinyl and alkynylaryl tellurides[76]:

$$CH_2 = CHMgCl + PhTeI \xrightarrow[-10\,^{\circ}C]{THF} CH_2 = CHTePh$$

$$CH{\equiv}CMgBr + PhTeI \xrightarrow[r.t.]{THF} CH{\equiv}CTePh$$

4. Tellurium(IV) compounds

Tellurium tetrachloride is the most common reagent. It may be prepared from the elements[77] or obtained commercially as such. It may be prepared *in situ* by reacting chlorine with a suspension of tellurium in refluxing tetrachloroethane[78]. The compound has electrophilic character and will react with aromatic compounds activated to electrophilic attack[79], e.g.

$$p\text{-}EtOC_6H_5 \xrightarrow[CHCl_3]{TeCl_4} (p\text{-}EtOC_6H_4)TeCl_3 \xrightarrow[neat]{p\text{-}EtOC_6H_5} (p\text{-}EtOC_6H_4)_2TeCl_2$$

$$\xrightarrow[e.g.\ Na_2S.9H_2O]{Reduce} (p\text{-}EtOC_6H_4)_2Te$$

The reaction is more sluggish than in the selenium case and more forcing conditions must be used to obtain the diorganyl tellurium dihalide. However, the fact that the reaction may be stopped at the trihalide stage does mean that it has utility in the synthesis of unsymmetrical tellurides either by changing the organic substrate in the second stage or, alternatively, reacting another trihalide with an appropriately activated aromatic compound, in general:

$$RTeCl_3 + R^1H \longrightarrow RR^1TeCl_2 \xrightarrow{\text{reduce}} RR^1Te$$

where R^1 may be, for example, p-$EtOC_6H_4$

A wide range of reducing agents will convert diorganyltellurium dihalides to diorganyl tellurides. Very frequently used are $Na_2S{\cdot}9H_2O$, $Na_2S_2O_5$ and hydrazine hydrate. Also used from time to time have been $LiAlH_4$, $SnCl_2$, Zn and $Na_2S_2O_3$.

In some cases condensation of $TeCl_4$ with aromatic compounds gives cyclic products. A well known example for which no mechanistic studies appear to have been made is[80]

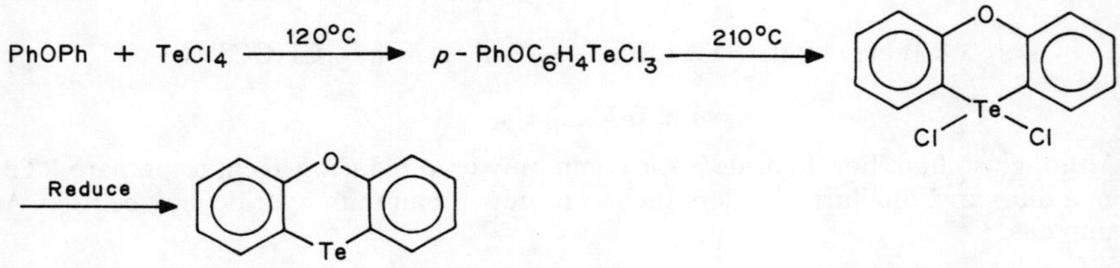

The generality of the reaction of $TeCl_4$ with aromatic compounds has been extended by using Lewis acid promoters such as aluminium trichloride[81]:

$$2C_6H_6 + TeCl_4 \xrightarrow{AlCl_3(3\text{ parts})} Ph_2TeCl_2 \xrightarrow{\text{reduce}} Ph_2Te$$

Methylene groups adjacent to carbonyl functions are also susceptible to direct attack, e.g.[82]

$$PyH^+PhTeCl_4^- + Me_2CO \longrightarrow Ph(MeCOCH_2)TeCl_2 + PyH^+Cl^- + HCl$$

The dihalide may be reduced to an unsymmetrical telluride.

Transmetallation reactions with tellurium tetrachloride are commonly used. The classic examples involve the use of Grignard, organolithium or organomercury reagents. Tellurium tetrachloride will, for example, react with phenylmagnesium bromide as follows[83,84]:

$$TeCl_4 + 4PhMgBr \longrightarrow Ph_2Te + Ph_2 + 4MgBr$$

The reaction is convenient but is inefficient in terms of moles of reagent consumed. There has been speculation that it proceeds via a tetraorganyl tellurium compound. It is certainly known that methylmagnesium bromide will reduce diaryltellurium dichlorides to diaryl tellurides with the evolution of ethane, which presumably is formed by the decomposition of Ar_2TeMe_2[85].

Organylmercury(II) halides are valuable for the synthesis of unsymmetrical (usually aryl) tellurides; the following is typical[86]:

$$RTeCl_3 + R^1HgCl \xrightarrow[\text{reflux}]{\text{dioxane}} RR^1TeCl_2 \xrightarrow[90-95\,^{\circ}C]{Na_2S{\cdot}9H_2O} RR^1Te$$

Organotin[18] and organolead[87] compounds may also be employed to transfer organic groups to $TeCl_4$, in general

$$R_4M + TeCl_4 \longrightarrow R_2MCl_2 + R_2TeCl_2 \xrightarrow{\text{reduce}} R_2Te$$

(M = Sn, Pb)

Other processes exist in which tellurium(IV) compounds occur as isolated intermediates, although the actual starting material is elemental tellurium. The method is valuable for the synthesis of cyclic tellurides[88]:

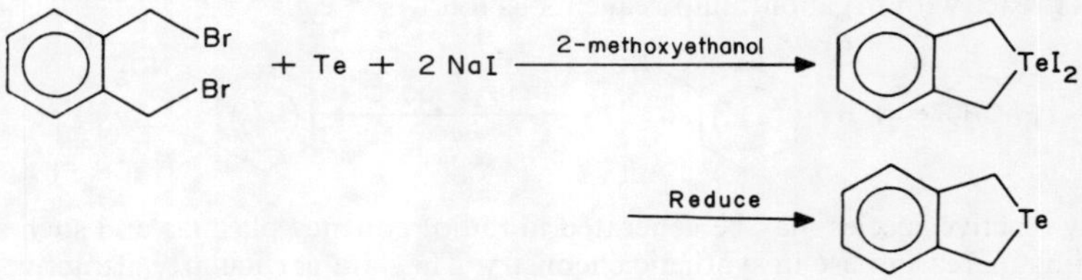

However, in some cases the method can succeed with monofunctional organic halides, for example with cetyl chloride ($C_{16}H_{33}Cl$)[89]:

$$2C_{16}H_{33}Cl + Te + 2NaI \longrightarrow (C_{16}H_{33})_2TeI_2 \xrightarrow{Na_2S\cdot 9H_2O} (C_{16}H_{33})_2Te$$

5. Miscellaneous

A very large number of reactions can yield tellurides. Those listed above are the most generally useful but in this section some examples are quoted of more recent reports using novel reagents or new conditions.

Aliphatic diazo compounds may be used both for initial synthesis and for the modification of an existing telluride. The following examples illustrate these points:

$$Ph_2Te_2 + CH_2N_2 \longrightarrow N_2 + (PhTe)_2CH_2 \qquad \text{(Ref. 70)}$$

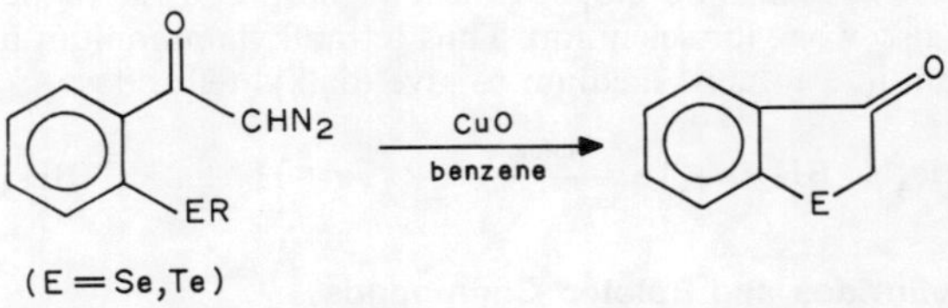

(Ref. 90)

Diazonium salts are useful in the aryl series. For example, reaction of aryldiazonium tetrafluoroborates with ditellurides using 18-crown-6 as a phase-transfer catalyst provides a convenient route to aryl alkyl tellurides[91], e.g.

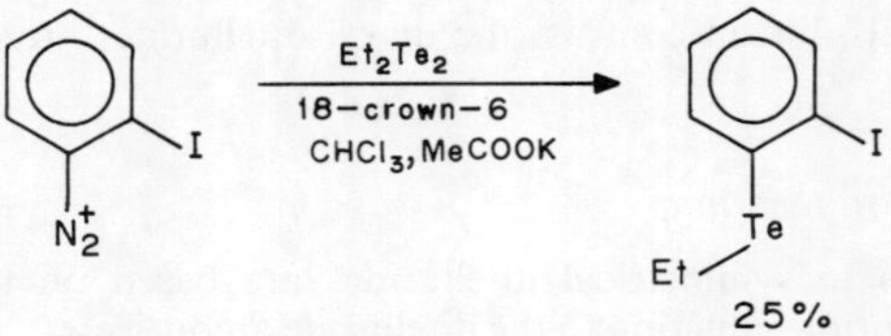

The same diazonium salts react with potassium tellurocyanate at room temperature to give symmetric tellurides[92].:

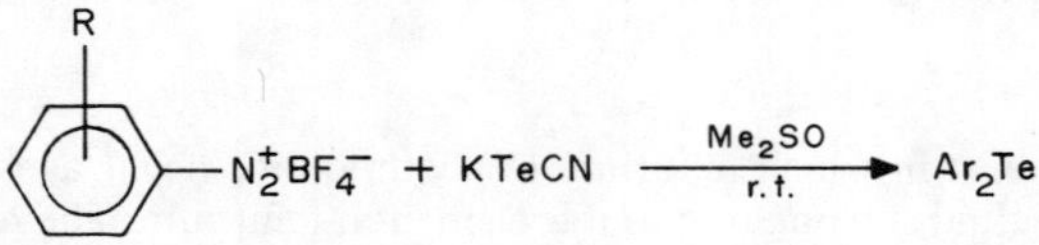

Organic tellurocyanates [which could also be seen as organyltellurium(II) cyanides, RTeCN] react with organolithium reagents as follows[93], e.g.:

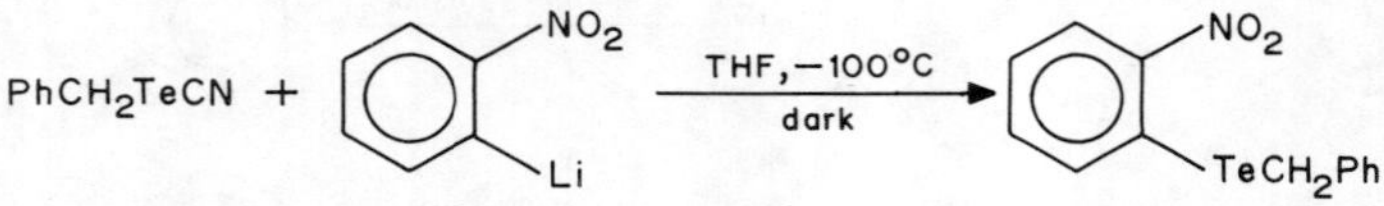

Many reactive species may be generated in radiofrequency plasmas and such systems are finding increasing use in synthetic chemistry. They are particularly attractive for the synthesis of perfluoro compounds such as $(CF_3)_2Te$[94], but the method is general for the synthesis of metal alkyls (including tellurium). The reactive radicals are co-condensed with metal vapour on a cold finger at − 196 °C[95]. The products often include ditellurides.

Tellurium dioxide has been little used as a source of tellurium in the synthesis of tellurides. Bergman and Engman[96] have used their valuable TeO_2–glacial acetic acid – lithium halide reagent to produce aryl alkyl tellurides by the following sequence of reactions:

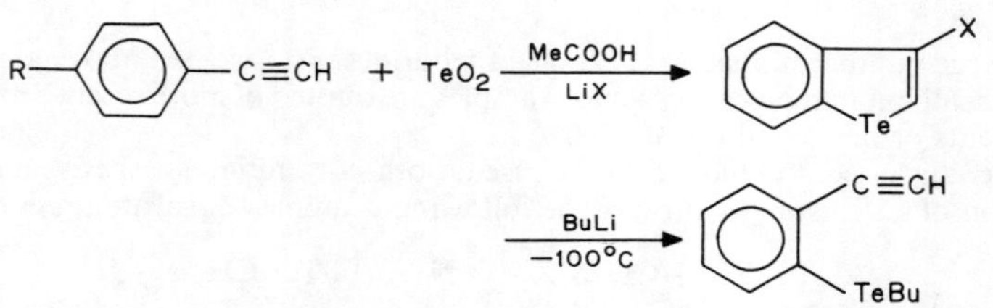

The same workers[97] have also developed a neat variation on the reduction of tellurium (the reaction will also work for selenium). Thus tetraalkylammonium borohydrides will react with tellurium in a toluene medium to give dialkyl tellurides:

$$2R_4N^+BH_4^- + Te \xrightarrow{\text{toluene}} R_2Te + H_2 + 2R_3NBH_3$$

D. Diorganyl Ditellurides and Related Compounds

Like diselenides, the diorganylditellurides are strongly coloured, usually orange–red or deep red. To date no pure unsymmetrical ditelluride has been isolated although their existence in solution is now well documented (see Section V.A). The lower molecular weight dialkyl ditellurides (and for that matter the corresponding dialkyl tellurides) are foul-smelling materials, but in contrast the diaryl ditellurides are odourless crystalline solids.

1. Syntheses based on tellurium

Two major routes to symmetrical ditellurides are based on tellurium as starting material. In the first, prior reduction to the ditelluride anion, Te_2^{2-}, is required. This may be achieved in aqueous alkaline medium by treating tellurium with rongalite[98] followed by an organic halide. Alternatively, the reaction could be carried out in liquid ammonia using sodium as reducing agent, e.g.[99]

$$2Te + 2Na + 2MeI \xrightarrow{NH_3(l)} Me_2Te_2$$

However, the recently developed method using solvents such as DMF is the preferred route, e.g.[44]

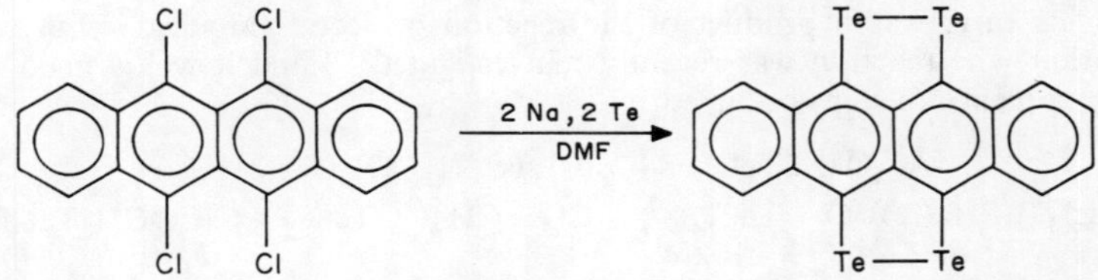

The other major route based on elemental tellurium involves insertion of a Te atom into a C—Mg or C—Li, followed by oxidation with dioxygen. In the case of the Grignard route, the method is applicable only in the aromatic series[100], e.g.

$$PhMgBr + Te \longrightarrow PhTeMgBr \xrightarrow{O_2} Ph_2Te_2 \quad (80\%)$$

The organolithium route is identical and a recent application offered a simple preparation of the sterically crowded di[2, 4, 6-tri(*tert*-butyl)phenyl] telluride[101]:

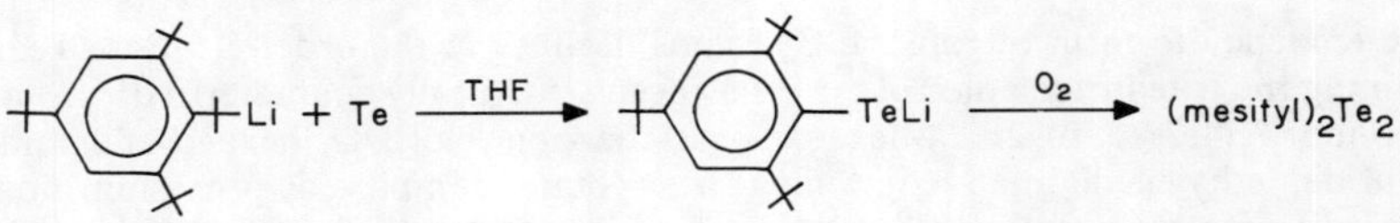

2. Syntheses based on tellurium(II) compounds

Although tellurium(II) compounds are almost certainly intermediates in the reduction of organyl tellurium(IV) trihalides (see Section II.D.3), they are of little practical importance for the synthesis of ditellurides. One of the few examples in the literature involves the preparation of di(2-naphthyl) ditelluride by the alkaline hydrolysis of 2-naphthyltellurium(II) iodide or alternatively by reacting the iodide with triethylamine or with triisopropyl phosphite[102].

3. Syntheses based on tellurium(IV) compounds

The reaction of an organyl tellurium(IV) trihalide with a wide variety of reducing agents will give good yields of ditelluride (e.g. $Na_2S{\cdot}9H_2O$, $Na_2S_2O_5$ and N_2H_4 are commonly used. With a suitable choice of organic groups, intermediate stages in the reduction may be identified, e.g. the reduction of (phenylazophenyl-C, N[1])tellurium(IV) trichloride[60,103]:

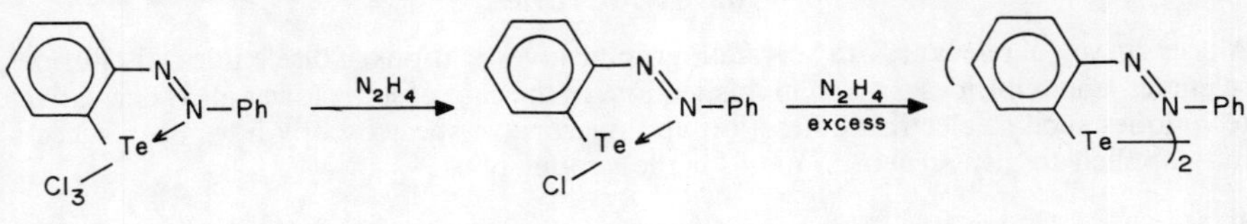

If sodium borohydride is selected rather than hydrazine as the reducing agent, the azo linkage is reduced and di(*o*-aminophenyl) ditelluride results[60], illustrating the possibility of modifying the organic group without destroying the Te—Te linkage.

The two variations on this route involve selection of reducing agent and choice of synthesis for the organyl tellurium(IV) trihalide. Organomercury halides react with tellurium tetrahalides to give the $RTeX_3$ compounds, a route favoured in at least one early classic paper on diorganyl ditelluride chemistry[104].

Morgan and Drew[105] reported the synthesis of CH_2Te_2 in poor yield. It was obtained by the potassium disulphite reduction of bis(trichlorotelluro)methane, $(Cl_3Te)_2CH_2$, which, in its turn, was a product of the reaction of acetic anhydride with tellurium tetrachloride. The reaction was recently reinvestigated[106] and it was argued that the following sequence of steps occurred:

$$Ac_2O + 2TeCl_4 \longrightarrow (Cl_3TeCH_2CO)_2O + 2HCl$$

$$(Cl_3TeCH_2CO)_2O + TeCl_4 \longrightarrow Cl_3TeCH_2COTeCl_3 + ClCOCH_2TeCl_3$$

$$Cl_3TeCH_2COTeCl_3 \xrightarrow{-CO_2} Cl_3TeCH_2TeCl_3$$

Hence, by use of the stoichiometric ratio of $Ac_2O + 3TeCl_4$, a 50% yield of the bis(trichlorotelluro)methane was obtained that could be reduced to CH_2Te_2, which is a red powder when initially prepared but which converts to a grey solid.

4. *Polytellurides*

The tendency to form oligomeric E_n chains declines in the order $S > Se \gg Te$. A salt containing the tritelluride anion, Te_3^{2-}, has been structurally characterized[107]. Thus the compound (4, 7, 13, 16, 21, 24-hexaoxa-1, 10-diazobicyclo[8.8.8]hexacosane)potassium tritelluride, ethylenediamine containing a cryptate complexed potassium ion and ethylenediamine of crystallization contains Te_3^{2-} anions with Te—Te 2.692 and 2.720 Å with $Te\hat{T}eTe = 113.1°$, but no organyl tritellurides appear to have been reported in the literature. Very recently, in the author's laboratory[108], in the reduction of (phenyl-2-pyridyl-C, N^1)tellurium(IV) tribromide, a reaction in which some elemental tellurium was released, a product analysing as $2\text{-}pyC_6H_4TeTeTeC_6H_4py\text{-}2$ was isolated. The material has been crystallographically characterized*.

Mixed selenium and tellurium trichalcogenides are known; a recent example of such a compound is[109]

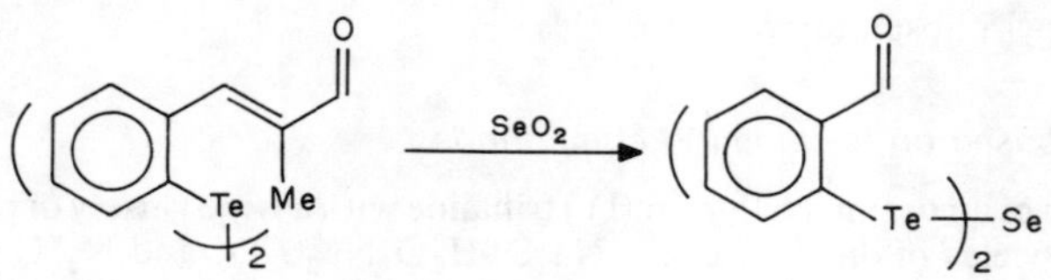

III. STRUCTURE

A brief survey of relevant X-ray crystallographic investigations of diselenides, ditellurides, selenides, and tellurides is given in this section. In the case of some lower alkyl derivatives, techniques such as electron diffraction and microwave spectroscopy have been successfully applied to give structural data for the vapour phase.

*$[(Me_3Si)_3C]_2Te_3$ has now been reported[108a].

A. Selenides

Simple VSEPR[110] considerations for Se^{II} (and Te^{II}) predict non-linear structures for selenides (and tellurides) and, indeed, this is confirmed by the available structural data. Estimates of the Se—C single bond length vary slightly depending on the choice of covalent radii. Thus the commonly quoted Pauling values[111] suggest 1.93 Å, whereas estimates due to Van Vechten and Phillips[112] predict 1.999 Å for bonding to sp^3 carbon (strictly, the selenium radius is 'tetrahedral', but with two spare pairs this may not be an unreasonable estimate). Small differences (0.03 Å) may be expected between E—alkyl and E—aryl (E = Se, Te). Some representative data for selenides are given in Table 1. It is noted that Se—C varies from 1.916 to 1.97 Å (but phase differences are involved, so no great significance can be deduced from the differences). Bond angles, C$\hat{Se}$C, vary more and may be determined by repulsive forces [cf. 96.2° for Me_2Se and 104.4° for $(CF_3)_2Se$]. The observed angles are between the extremes expected for bonding involving pure selenium p orbitals (90°) and sp^3 hybridization with strongly directed lone pairs. It must be concluded that significant s character is included in the selenium bonding orbitals. When selenium is constrained in a ring structure, C_3H_7NSeBr (Table 1) for example, understandably the C$\hat{Se}$C angle is much reduced, to 86° in this case; yet in the complex of diiodine with selenacyclopentane the C$\hat{Se}$C angle is 94°, which is comparable to values in non-cyclic selenides.

B. Di- and Poly-selenides

Diselenides (and ditellurides), when the Se—Se linkage is not constrained within a ring structure, have the 'hydrogen peroxide' structure, the dihedral angle (θ, Figure 1) varying between 75° and 82°. Attempts have been made to calculate rotational barriers in Me_2Se_2[118]. It is generally accepted that torsional barriers are greater for the *cis* ($\theta = 0°$) configuration than for the *trans* ($\theta = 180°$) configuration, and indeed calculations indicated differences of the order of 28 kJ mol^{-1}.

Estimates of the Se—Se bond length based on Pauling covalent radii[111] suggest values around 2.32 Å, a figure in reasonable agreement with data in Table 2. It is possible that the Se—Se linkage is part of a ring structure. If the ring is sufficiently large and flexible, e.g. 13, 13-dimethyl-8, 13-dihydro-5H-dibenzo[*d*, *g*]-1, 2-diselena-6-silonine (**3**)[119], this

TABLE 1. Some structural data for selenides, R_2Se

Compound	Se—C (Å)	C$\hat{Se}$C (°)	Phase	Ref.
Me_2Se	1.943	96.2	g	113
$(CF_3)_2Se$	1.96	104.4	g	114
p-Tol_2Se	1.92, 1.93	106.2	s	115
$C_3H_7N_2SeBr$: 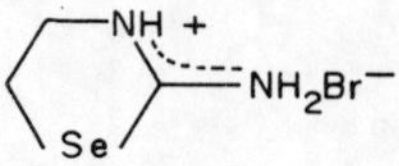	1.88, 1.99	86	s	116
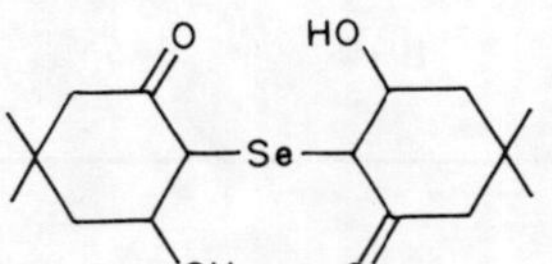	1.916	100.9	s	117

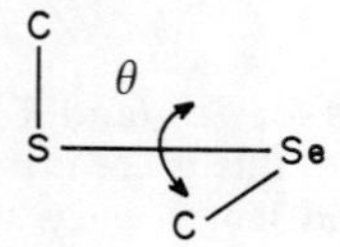

makes little difference to the geometry around selenium. However, if the system is much more constrained, e.g. 1, 8:4, 5-bis(diseleno)naphthalene[43], θ in Figure 1 is effectively zero and the observed Se—Se bond length is marginally longer at 2.364 Å than the sum of the Pauling covalent radii.

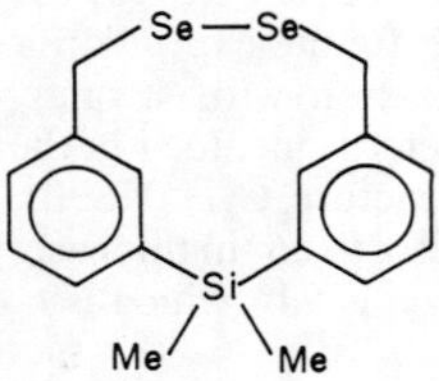

(3)

The structures of some tri- and tetra-selenides have been determined[56]. Thus pure tetraselenium dipiperidide and tri- and tetra-selenium dimorpholide have been structurally characterized. N—Se—Se—Se—Se—N and N—Se—Se—Se—N chains occur in extended helix rotational–isomeric forms; variations of Se—Se bond lengths occur along the chains. In triselenium dimorpholide an intermolecular contact of 3.404 Å occurs. Some data for di- and poly-selenides are given in Table 2.

TABLE 2. Some structural data for di- and poly-selenides, R_2Se_n

Compound	Se—C(Å)	Se—Se (Å)	$C\hat{Se}C$(°)	θ (°)[a]	Ref.
$(PhCH_2)_2Se_2$	1.97	2.285	100.1	82	120
Ph_2Se_2	1.93	2.29	105	82	121
$(p\text{-}ClC_6H_4)_2Se_2$	1.9	2.33	100	74.5	122
$(CF_3)_2Se_2$	1.93	2.34	103.5	—	114
3 (see text)	2.01	2.288	104.4	73.7	119
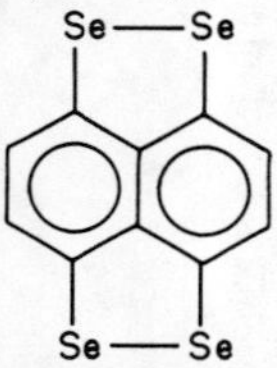	1.905 1.915	2.364	91.8 91.3	0	43
$(C_5H_{10}N)_2Se_4$ dipiperidide)	1.832	2.328 2.346	107.2	88.6 76.3	56
$(C_4H_8ON)_2Se_4$ (dimorpholide)	1.836[b]	2.336 2.356	106.8	90.2 75	56
$(C_4H_8ON)_2Se_3$ (dimorpholide)	1.845[b]	2.346	109.0	94.6	56

[a]θ, Dihedral angle (see Figure 1).
[b]Se—N.

C. Tellurides

Tellurides closely resemble selenides in the structural sense and some relevant information is given in Table 3. The sum of the Pauling[111] covalent radii for Te and C is 2.12 Å and most observed bond lengths are close to this estimate. Both inter- and intra-molecular interactions can be important, e.g. for 3,4-quinoxalino-1-telluracyclopentane (**4**)[123], Te···Te contacts of 3.998 Å are less than the sum of the Pauling estimates of Van der Waals radii (4.40 Å[111]); in contrast, the Te···N distances of 4.087 and 4.090 Å are too long to suggest any interaction.

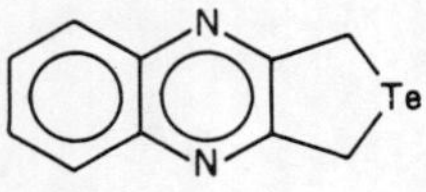

(**4**)

TABLE 3. Structural data for tellurides, R_2Te

Compound	Te—C (Å)	$C\hat{Te}C$(°)	Ref.
p-Tol_2Te	2.05	101	126
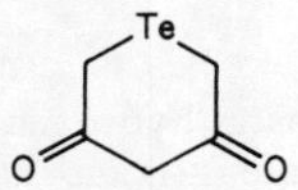	2.16	89.5	127, 128
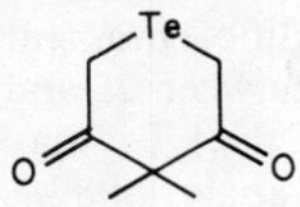	2.168	86.4	129
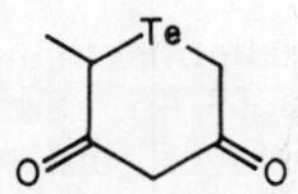	2.202 2.157 (Te···Te = 4.042 Å)	88.4	130
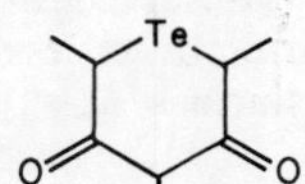	2.184 (Te···Te = 4.068 Å)	89.7	131
4 (see text)	2.145 2.123 (Te···Te = 3.998 Å)	80.7	123
5 (see text)	2.05	94	124
6 (see text)	2.08		125
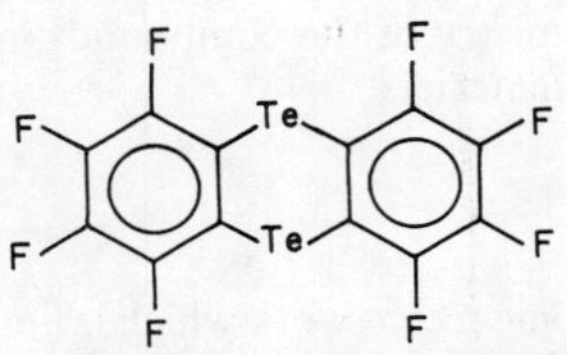	2.117 2.111	92.9	132

For 5[124], the unit cell contains two independent molecules with Te···O contacts of 2.577 and 2.574 Å (Van der Waals distance 3.6 Å), giving a distorted three-fold coordination about tellurium. This situation is also observed in 6[125], with Te···N contacts of 2.773 Å, well within the Van der Waals distance of 3.7 Å[111]. The tellurium atom in **5** and **6** therefore shows weak Lewis acidity.

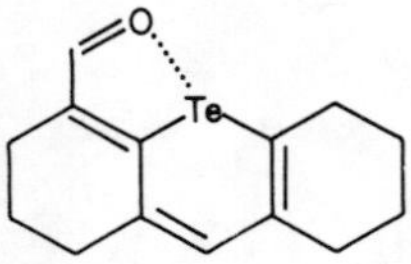

(5)

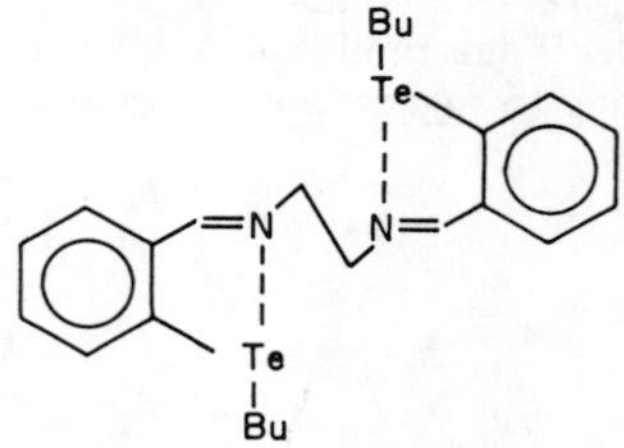

(6)

Another way in which tellurium in a telluride may extend its coordination number is by acting as a Lewis base. This is dealt with in another Section IV.E.

D. Ditellurides

Similar general comments apply to ditellurides as to diselenides. Thus a 'hydrogen peroxide' structure is the expected configuration unless the ditelluride linkage is constrained in a small ring. Dihedral angles often approach 90°, although a dipole moment study[133] suggests that significant variation of the angle may occur. Thus between 25 and 45 °C free rotation about the Te—Te bond in $(p\text{-}BrC_6H_4)_2Te_2$ and $(p\text{-}CH_3C_6H_4)_2Te_2$ may occur, but for Ph_2Te_2, $(p\text{-}FC_6H_4)_2Te_2$ and $(m\text{-}FC_6H_4)_2Te_2$ rigid dihedral angles of 89.7, 47.7 and 89.7°, respectively are calculated.

Intermolecular Te···Te contacts are more important and frequent for ditellurides than for diselenides and sometimes they can be remarkably short. Thus contacts of 3.701 Å are seen for tetratelluratetracene[134] and for one of the polymorphs of bis(2-naphthyl)ditelluride, namely the 'cisoid' form, a Te···Te contact of 3.707 Å is observed. In contrast, the other polymorph ('transoid') has no Te···Te intermolecular contract less than 4.128 Å; the transoid structure is currently unique amongst the ditellurides.

Some representative structural data are given in Table 4.

IV. REACTIONS

The general reactions of the heavier diorganyl chalcogenides and dichalcogenides have been very adequately summarized in previous reviews[4–6] and books[1–3]; there seems little point in duplicating that material in this chapter. Consequently, attention will be devoted to updating the surveys already available by concentrating on more recent contributions. This section is arranged to reflect areas of major current effort; thus much reaction chemistry of selenides and tellurides relates to their use in organic synthesis. Also, there are an increasing number of papers reporting on the photochemistry of the compounds; a further area of rapid growth is the ligand chemistry of the materials.

A. Diorganyl Selenides and Tellurides

This sub-section (and the following one) gathers together some recent work which is not readily classifiable under the major growth areas mentioned above.

TABLE 4. Structural data for ditellurides, R_2Te_2

Compound	Te—C (Å)	Te—Te (Å)	$C\hat{Te}C$ (°)	θ (°)	Ref.
Ph_2Te_2	2.15 2.08	2.712	97.4 100.3	88	135
$(p\text{-}ClC_6H_4)_2Te_2$	2.16	2.702			122
$p\text{-}Tol_2Te_2$	2.13	2.697		85.7	136
$p\text{-}An_2Te_2$	2.14	2.72	{ 99.8 100.7	81.2	137
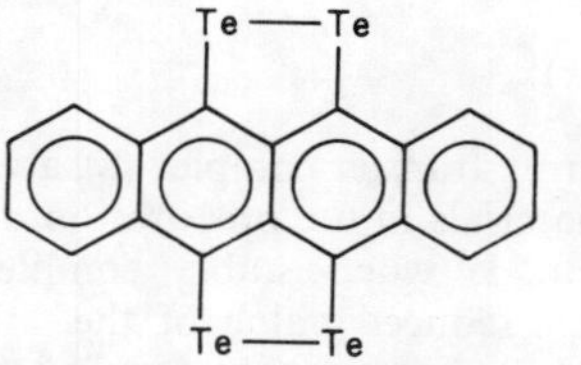	2.114 2.111	2.673	87.5	0	134, 138

One theme detectable in recent publications is a growing interest in the mechanisms of the reactions of selenides and tellurides. Systematic kinetic studies remain scarce but will surely increase in number in the near future. Russian workers[139] have investigated the thermal decomposition of Me_2Te at 350 °C under static conditions. The products are CH_4, C and Te. It is considered that $(CH_2Te)_n$ is an intermediate which arises from the intramolecular disproportionation of methyl groups. At 500 °C, radical intermediates become more prominent. Tributyltin radicals will react with $(alkyl)_2E$ (E = Se, Te) via an S_H2 reaction[140].

The acid-catalysed hydrolysis of vinyl selenides has been investigated by two groups. The rate-determining step for the hydrolysis of $RSeCH{=}CH_2$ (R = aryl) is the protonation of the double bond to give a selenium stabilized carbonium ion. When R = Ph, the order of reactivity is RO > RS > RSe (relative rates 42:7:1)[141]. For the specific case of the following reaction[142]:

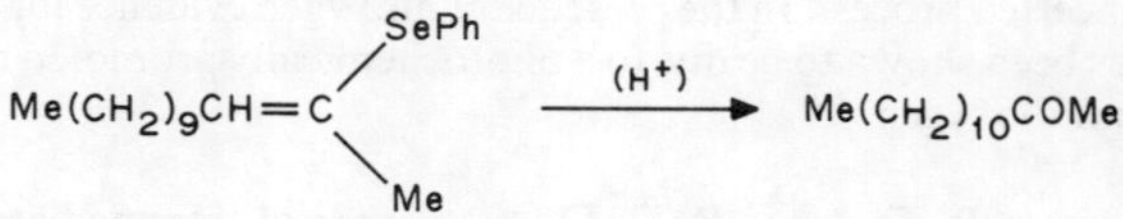

(7)

a kinetic solvent deuterium isotope effect is observed, suggesting that the protonation is partially reversible. Also, hydrolysis of a mixture of the *E* and *Z* isomers of (7) in the presence of D_2O gave 25% incorporation of D into the unreacted vinyl selenides.

Oxidation of selenides and tellurides is a commonly observed reaction. In some cases this might be achieved electrochemically, e.g. 1, 3-dihydro-2-telluraindene may be oxidized in MeCN solution in the presence of tetraalkylammonium salts to C_8H_8TeX (X = ClO_4, PF_6)[143]. More conventional means of oxidation have also been employed, e.g. $(CF_3)_2Te$ will react with Cl_2, Br_2, O_2 and $ClONO_2$ to give $(CF_3)_2TeX_2$ (X = Cl, Br, NO_3) and $CF_3)_2TeO$. Ozone and $(CF_3)_2Te$ are reported to give an unstable compound[144]. A further variation on this theme involves the oxidative addition of interhalogens to unsymmetrical diaryl tellurides; no disproportionation to symmetrical compounds was observed[145].

The well known reaction of an alkyl halide with a telluride has been the subject of further investigation[146]. Experiments in the presence of a spin trap [PhCH = N(O)Bu^t] have detected radical intermediates and when the organic radical enjoys some resonance stabilization, e.g. allyl, the tellurium product is often R_2TeX_2 (X = halogen) rather than the telluronium salt; thus,

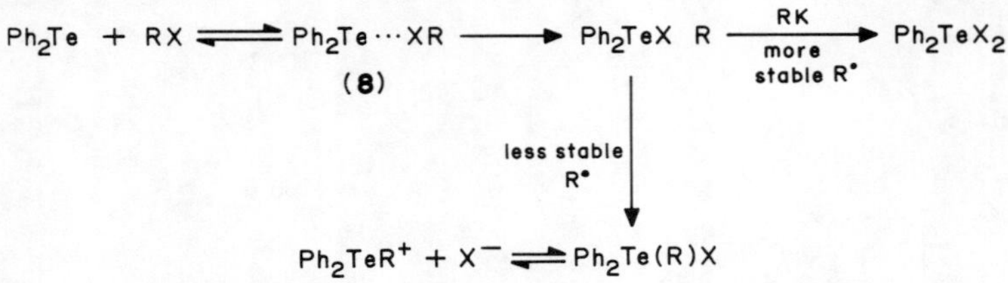

The initial step is considered to be the formation of a charge-transfer complex (**8**), a view that gained recent support from the isolation of such materials in the preparation of *p*-AnTe(CH_2)TeAn-*p* from $ArTe^-$ and CH_2X_2 (X = Br, I) when stable complexes ArTe(CH_2)TeAr·CH_2X_2 were isolated. The significantly reduced value of the ^{125}Te Mössbauer quadrupole splitting (7.58 mm s^{-1} for the CH_2Br_2 complex), compared with values of ca. 10 mm s^{-1} expected for a telluride, indicates a relatively strong interaction between donor and acceptor[72].

B. Diorganyl Diselenides and Ditellurides

In contrast to selenium chemistry where solid RSeSeR^1 compounds are known (Section II.B.2), no unsymmetrical ditellurides have been isolated as solids, but they are well known in solution where they are conveniently identified by ^{125}Te NMR spectroscopy[147] (Section V.A). In organic solvents such as chloroform a slow exchange reaction occurs:

$$R_2Te_2 + R^1_2Te_2 \rightleftharpoons 2RTeTeR^1$$

with an equilibrium constant close to 4 and independent of temperature, implying that the process is under entropy control. It is speculated[148] that the exchange occurs via a non-radical mechanism and involves a square $R_2Te_2{\cdot}R^1_2Te_2$ intermediate which affords RTeTeR^1 in a concerted process. In the presence of dioxygen evidence for radical species is found, but this has been shown to be due to a photochemically promoted reaction between the ditelluride and dioxygen[149]:

$$R_2Te_2 \xrightarrow{h\nu} R_2Te_2{}^* \xrightarrow{O_2} R_2Te_2O_2 \longrightarrow \text{radical intermediates}$$

This reaction may also involve a square intermediate and is strongly catalysed by ethanol.

When diselenides and ditellurides react with aqueous alkali under phase transfer conditions (PTC) some oxidized products are observed[150]:

$$2ArSeSeAr + 4KOH \longrightarrow 3ArSeK + ArSeO_2K + 2H_2O$$

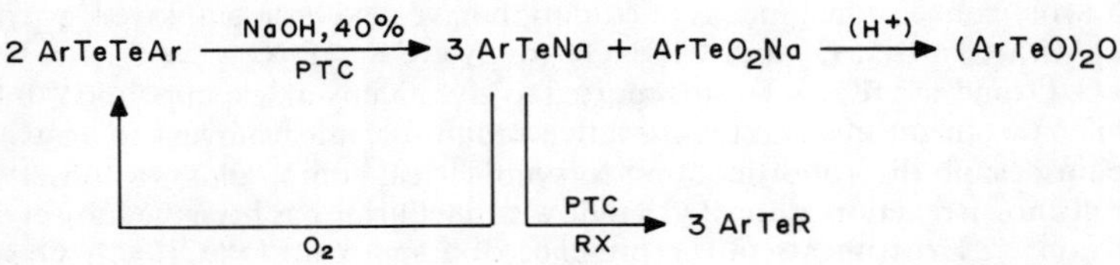

The oxidation of ditellurides by halogen is a well documented means of synthesizing $RTeX_3$ (X = halogen)[2]. Recently the mechanism of the reaction between $(p\text{-}EtOC_6H_4)_2Te_2$ and diiodine was investigated[151] and kinetic data were consistent with the following mechanism:

$$R_2Te_2 + I_2 \xrightarrow{k_1} 2RTeI \qquad (k_1 = 56 \pm 2\,dm^3\,mol^{-1}\,s^{-1} \text{ at } 7\,°C)$$

$$RTeI + I_2 \overset{k_2}{\rightleftharpoons} RTe^{II}(I_3) \qquad (k_2 = 1570\,dm^3\,mol^{-1} \text{ at } 25\,°C)$$

$$RTe^{II}(I_3) \xrightarrow{k_3} RTe^{IV(I)}{}_3 \qquad (k_3 = 6.2 \times 10^{-4}\,s^{-1} \text{ at } 25\,°C)$$

Again, a square intermediate is suggested to produce RTeI via a concerted mechanism. RTeI then forms a 1:1 complex with a molecule of I_2, which is probably the aryltellurenyl triiodide which can undergo unimolecular rearrangement to the aryltellurium(IV) triiodide. Ditellurides may also react with organyltellurium(IV) trihalides, e.g.[152]

$$2R_2Te_2 + 2RTeX_3 \longrightarrow 6RTeX \longrightarrow 3Te + 3R_2TeX_2$$

The stability of the organyltellurenyl halide is for R $Ph > 4\text{-}MeOC_6H_4 \approx 3,4\text{-}(MeO)_2C_6H_3$ and for X $I > Br > Cl$.

A reaction of a triselenide with diiodine was recently reported. Thus $Se_3[C(Se)R]_2$ (R = morpholino) reacted with excess of I_2 in CH_2Cl_2 to give a stable 1:1 charge-transfer complex which reacted further in solution with a rate constant of $10.1 \times 10^{-6}\,s^{-1}$ at 25 °C to give $[RC(Se)_2]SeI$[153]. The oxidation of $(C_2F_5)_2Se_2$ by SbF_5 or by $O_2{}^+$ produces evidence of cationic species such as $(C_2F_5)_3Se^+$ and $\{[Se(C_2F_5)]_{4n}\}^{2n+}$. The value of n is likely to be 1, hence again a square intermediate may be involved, the oxidation of which gives the tetraselena cation[154].

Thus it would seem that, contrary to earlier belief[104], ditellurides (and diselenides) do not readily give rise to radical species. However, two recent reports[155,156] define conditions under which relatively stable radicals may be produced. If $(p\text{-}XC_6H_4)_2Se_2$ (X = H, F, Cl, Br) is reacted in solvents degassed at 77K and under an argon atmosphere with $AlBr_3$, ESR evidence is obtained for species such as $(p\text{-}XC_6H_4Se^{\bullet})AlBr_3$ and $(p\text{-}XC_6H_4SeSe^{\bullet})AlBr_3$, which are said to be stable for several months in the absence of dioxygen and moisture. Gallium halides are also effective.

Another reaction which has attracted attention is the insertion of an atom or group into the Se—Se or Te—Te bond. Depending on the group, this may result in the formal oxidation or reduction of the chalcogen. The reaction with diazomethane[70] has already been mentioned (Section II.C.5) and a variation on the reaction was recently reported[157].

$$N_2CHCO_2Et + Ph_2Se_2 \xrightarrow[\text{xylene, 100 °C, 5 h}]{\text{Cu bronze}} (PhSe)_2CHCO_2Et$$

45%

The products of such reactions are formally Se^{II} compounds. However, when, say, a ditelluride reacts with a low-valent transition metal complex such as $Pd(PPh_3)_4$, complexes $(Ph_3P)Pd(TeR)_2$ ($R = p\text{-}EtOC_6H_4$ or $2\text{-}C_4H_3S$) which are probably dimeric are formed[158] and the formal oxidation state of tellurium is now −1. The reaction of $(p\text{-}EtOC_6H_4)_2Te_2$ with Vaska's compound. $Ir(PPh_3)_2(CO)Cl$, has been the subject of a detailed mechanistic study[159]. It is considered that the first step is an addition to Ir^I followed by homolytic fission of the Te—Te bond to give radical intermediates which ultimately combine to give *trans*-$[IrCl(TeR)_2(CO)(PPh_3)_2]$:

$$[IrCl(CO)(PPh_3)_2] + R_2Te_2 \overset{k_1}{\rightleftharpoons} [IrCl(CO)(PPh_3)_2]{\cdot}R_2Te_2$$

$$[IrCl(CO)(PPh_3)_2]{\cdot}R_2Te_2 \underset{k_{-2}}{\overset{k_2}{\rightleftharpoons}} [IrCl(CO)(TeR)(PPh_3)_2] + RTe^{\bullet}$$

$$[IrCl(CO)(RTe)(PPh_3)_2] + RTe^{\bullet} \xrightarrow{k_3} trans\text{-}[IrCl(CO)(RTe)_2(PPh_3)_2]$$

If the radical $RTe^{\bullet}$ is confined to a solvent cage and the first equilibrium lies in favour of the reagents, then second-order kinetics are expected, and observed. The scheme is also in accord with spectroscopic and preparative data, although ^{125}Te NMR data do suggest that under some conditions the reverse reductive elimination of ditelluride might occur.

C. Applications in Organic Synthesis

A major growth area in selenium (and to a lesser extent, tellurium) chemistry is the development of synthetic methodologies based on organoselenium compounds. Many of the reagents are of the type RSeX and therefore not strictly within the scope of this chapter. However, tellurides or selenides are often intermediates and some new chemistry has been developed to convert these intermediates through to the final product. This section gives examples of such reactions as well as selecting other procedures which do not require RSeX as the starting reagent. It is worth pointing out that several workers who have made distinguished contributions in this field are also authors of recent review articles, e.g. Clive[3,5], Reich[23,160], Krief *et al.*[161] and Liotta[25,162].

1. Coupling reactions

Degassed Raney nickel is a powerful reagent for the coupling of aryl groups in compounds such as Ar_2TeCl_2[163]. The coupling probably involves Ar_2Te as an intermediate. If RR^1TeCl_2 (RR^1 = aryl) is used, Raney nickel produces non-specific coupling giving relatively high yields of the symmetric products RR and R^1R^1; however, if a stoichiometric reaction of RR^1TeCl_2 and $Pd(PPh_3)_4$ is carried out, excellent (ca. 90%) yields of the unsymmetric coupling product are obtained[164], e.g.

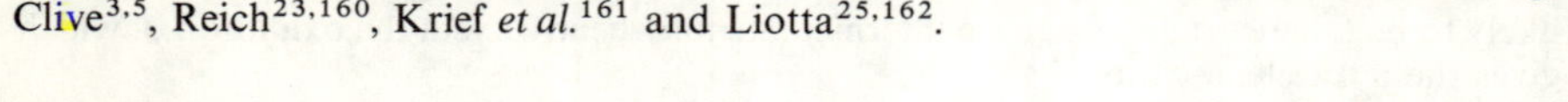

Bergman[164] attributes the difference to the differing interactions of RR^1Te with the polyatomic nickel surface and the monomeric Pd(0) complex, i.e.

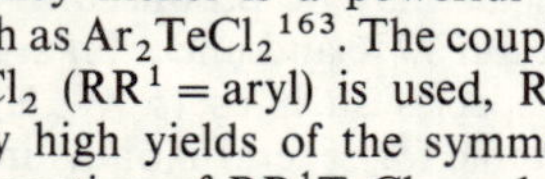

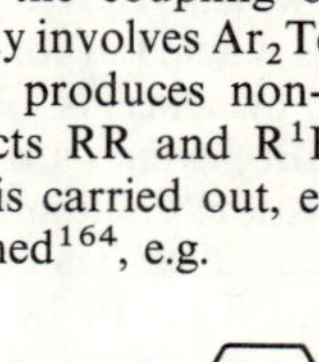

Benzylic and allylic couplings occur readily via tellurides which can be prepared by nucleophilic attack of Te^{2-} on an appropriate organic halide[165], e.g.

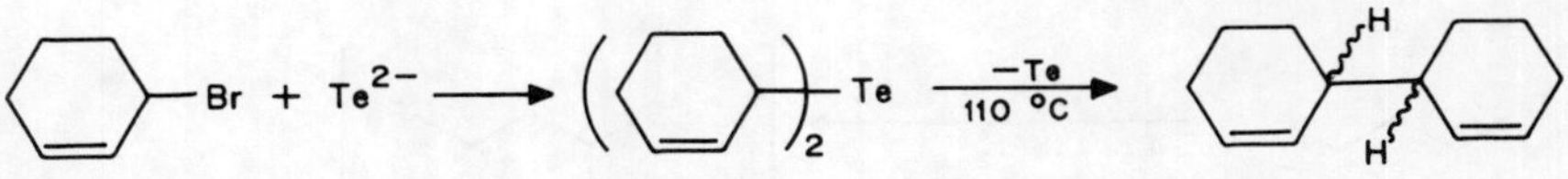

However, in contrast, the corresponding di(2-cyclohexenyl) selenide is stable at 110 °C, showing only 2% decomposition over 3h. Although many of these coupling reactions may be promoted thermally, some may be induced photochemically, e.g.

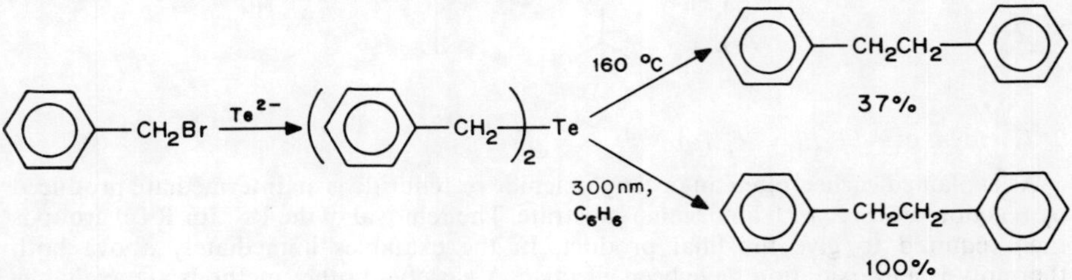

Allylic couplings may also be induced from ditellurides formed by nucleophilic attack of Te_2^{2-} on an allylic halide. These results have been the subject of an excellent review[164].

2. Lithium reagents

The α-protons of selenides are mildly acidic and this fact has been exploited for synthetic purposes via lithiation of the α-carbon atom; for example, it opens up a new route to alkenes[166]:

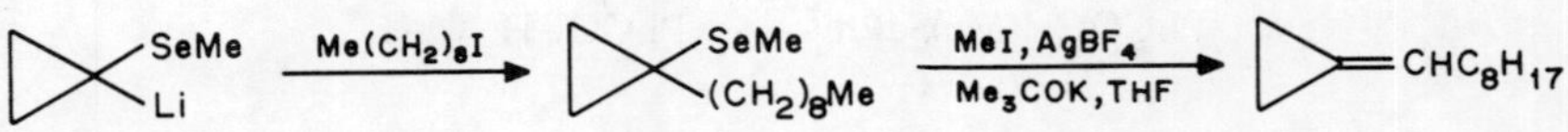

A variation on the theme is the addition of RLi to a vinylic selenide[167]:

$$PhSeCH=CH_2 \xrightarrow{RLi} PhSeCHLiCH_2R \xrightarrow[(R)']{\text{electrophile}}$$

$$PhSeCHR'CH_2R \xrightarrow{-PhSeH} RCH=CHR'$$

An earlier work describes a route to allylic alcohols[168], e.g.

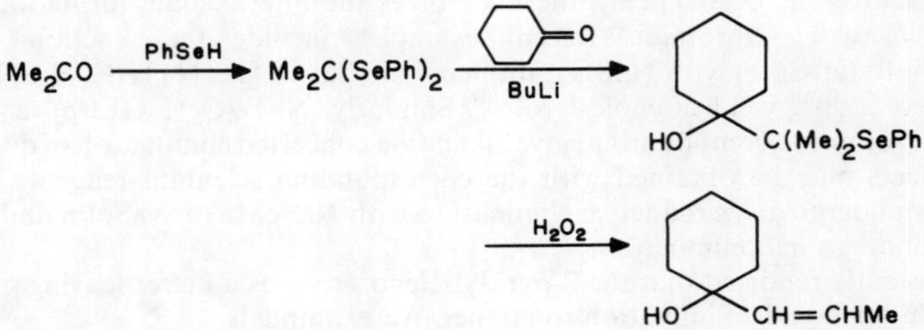

Reich and coworkers[23,169,170] point out that the α-protons of selenoxides are more acidic than those of selenides, hence there is some advantage in oxidizing the selenide; an example of the methodology is[169]

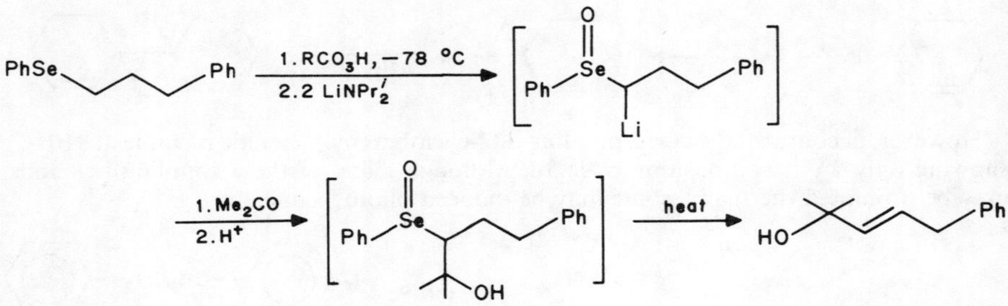

3. *Removal of—ER (E = Se, Te)*

As explained earlier, often an organyl selenide or telluride is an intermediate produced by reaction of RSeX with an organic substrate. The removal of the RSe (or RTe) group is then required to give the final product. In the examples immediately above both thermolysis and oxidation have been effective. A variety of other methods are available, however. For example, Clive *et al.*[171] have used triphenyltin hydride to cleave R—SePh bonds:

$$\text{RSePh} \xrightarrow[120\ ^\circ\text{C}]{\text{Ph}_3\text{SnH}} \text{RH} + \text{Ph}_3\text{SnSePh}$$

The reaction with RTePh goes under even milder conditions (ca. 80 °C). The products, Ph_3SnEPh (E = Se, Te), have been the subject of a Mössbauer investigation[148]. Other reagents, e.g. KIO_4, may serve as alternative oxidants[172]:

$$\underset{\textbf{(9)}}{\text{PhCOCH(Me)SePh}} \xrightarrow{\text{KIO}_4} \text{PhCOCH} = \text{CH}_2$$

In the above example, the selenide is formed by reaction of the carbonyl compound with strong base ($LiNR_2$) which traps the enol form:

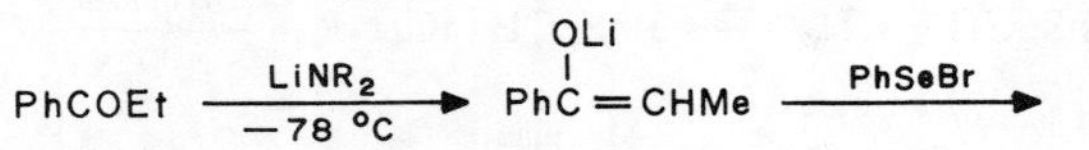

This reaction was dicovered simultaneously and independently by Reich *et al.*[173] and Sharpless *et al.*[174,175].

A further example of alkene synthesis involves the intermediate formation of a bis-telluride via a *vic*-dibromide. Recent examples include the reaction of 5α, 6β-dibromocholestan-3β-ol with $NaBH_4$–dithienyl ditelluride [i.e. $NaTe(C_4H_3S)$], which in ethanol gave a 90% yield of cholesterol[176]. Similarly, $NaTe(C_6H_4OEt\text{-}p)$ reacts with a range of 1, 2-dibromo compounds to give alkene via concerted elimination of ditelluride[72]. Similar effects may be obtained with the corresponding selenium reagents. Thus *vic*-dichlorides undergo a *syn* reductive elimination with NaSeMe or NaSePh and other *vic*-dihalides undergo *anti* elimination[177a].

It was recently reported that the 2-pyridylseleno group is a better leaving group than PhSe[177b] in selenoxide elimination to enones. An example is

$$\text{RCH}_2\text{C(O)R}' + \text{2-PyrSeBr} \longrightarrow \text{R}'\text{C(O)CH(R)SePyr-2} \xrightarrow{[\text{O}]} \text{R}'\text{C(O)CH} = \text{CHR}^2$$

$$\text{R}^2\text{CH} = \text{R}$$

Also, treatment of alkyl phenyl selenides and tellurides with *m*-chloroperbenzoic acid in alcoholic media leads to substitution of PhSe or PhTe by alkoxide, e.g.

$$\mathrm{Me(CH_2)_9SePh} \xrightarrow[\text{MeOH, 1 h}]{3\,m\text{-}\mathrm{ClC_6H_4C(O)OOH}} \mathrm{Me(CH_2)_9OMe}$$

The reaction appears to proceed via the selenone or tellurone[177c].

4. Carbon to oxygen bond formation

The reaction between $PhSe^-$ and a carbonyl compound, RCOR′, gives a bis-selenide, $RR'C(SePh)_2$, which may then react with butyllithium to give RR′C(Li)SePh[178]. This reagent may be exploited to afford a new route to epoxides[179,180], e.g.

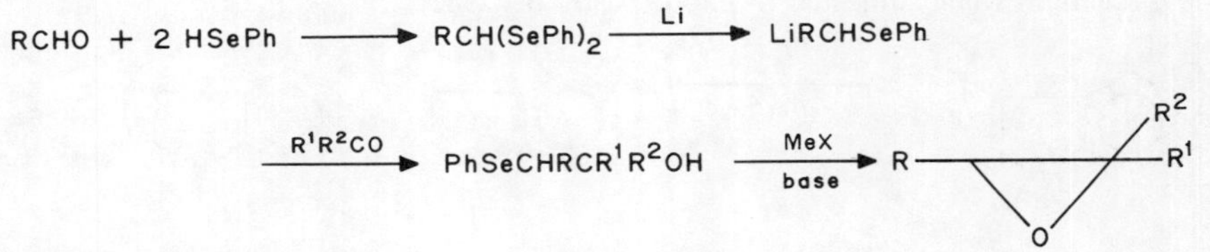

Uemura *et al.*[181] recently demonstrated a new ether synthesis by reaction of *m*-chloroperbenzoic acid with alkyl phenyl selenides, e.g.

$$\mathrm{Me(CH_2)_{11}CH(Me)SePh} \xrightarrow[\text{MeOH, r.t.}]{m\text{-}\mathrm{ClC_6H_4C(O)OOH}} \underset{100\%}{\mathrm{Me(CH_2)_{11}CH(Me)OMe}}$$

It now seems clear that both organoselenium and organotellurium reagents are well established within the synthetic armament of the organic chemist. Those readers requiring more detailed information and more comprehensive coverage of the subject are advised to consult the review literature referenced here.

D. Photochemical Reactions

Clive *et al.*[171] have warned that tellurides of satisfactory purity may only be obtained under red light (photographic dark-room conditions). This does not seem to reflect general experience in this field; however, it is the present writer's experience that tellurides of the (alkyl)(aryl)Te class tend to be less photolytically stable than $(alkyl)_2Te$ or $(aryl)_2Te$, and it is generally true that many tellurides encountered in organic synthesis are indeed of the (alkyl)(aryl)Te type. Hence Clive *et al.*'s warning should be heeded when such materials are studied.

Despite clear evidence that organyl selenium and tellurium compounds may be photosensitive, the number of papers devoted to studies of their photochemistry is not extensive. The topic was reviewed in 1980 by Martens and Praefcke[182].

1. Selenides and tellurides

The choice of solvent can have a profound influence on the products and/or product ratios when an organyl selenium (or tellurium) compound is photolysed, e.g.[183]

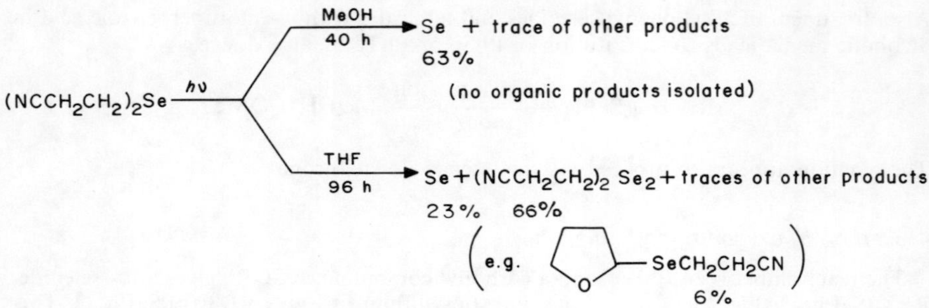

In some cases selenium will be more cleanly eliminated, e.g. diindolyl selenide gives almost 100% selenium when subjected to photolysis in benzene[184]:

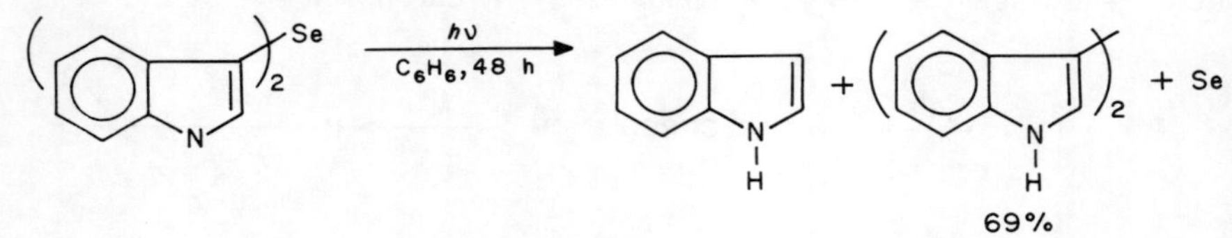

It is not generally true that irradiation of selenides and tellurides will produce elemental selenium and tellurium. For example, the following interesting cyclization reaction was observed[185]:

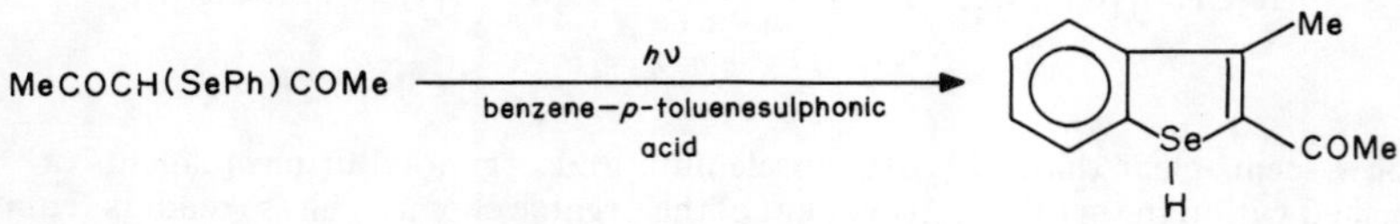

The cyclization is considered to go via an enol to a selenocarbonyl ylid which rearranges to **10**, which in turn undergoes acid-catalysed dehydration to the product.

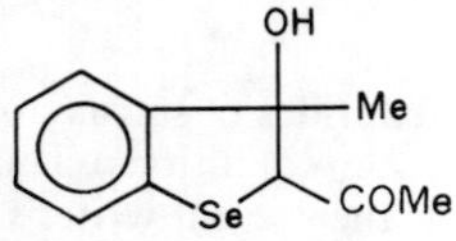

(10)

Another example involves aryl alkyl selenides, e.g. $PhSeCH_2Ph$:

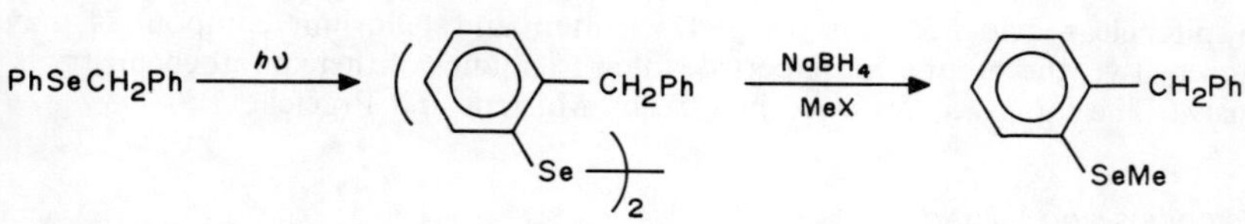

This reaction is general for PhSeR and has been exploited to form selenium-substituted *C*-nucleosides[186].

Azides may eliminate dinitrogen and rearrange. As shown in the following reaction[187] very different product ratios are obtained on thermolysis and photolysis:

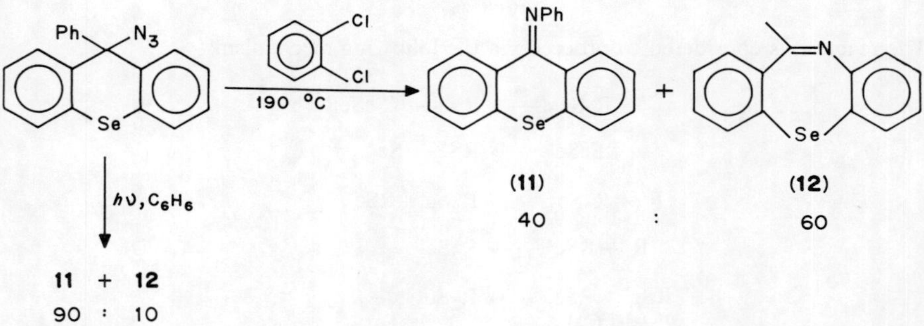

Tellurides also undergo interesting reactions, for example the telluride (**13**) forms a thioxanthone via a cyclization reaction[188]:

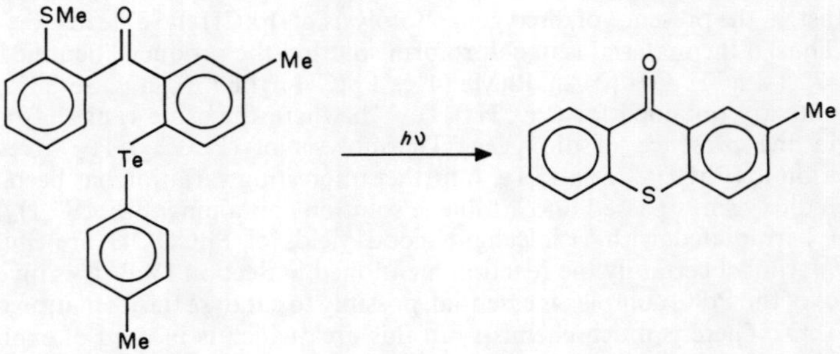

(13)

Photo-oxidation reactions may occur. For example, a new route to selenoxides is provided by the following reaction[189]:

$$\text{RSeR}' \xrightarrow[h\nu]{O_2} \text{RR}'\text{SeO}$$

$$28\text{–}95\%$$

$$R = Ph;\ R' = Me,\ CH_2Ph,\ CHPrCH(OH)Pr$$

Although not strictly organyl tellurium compounds, it is of interest that the following reaction was recently observed[190]:

$$\text{Te}(S_2CNR_2)_n \xrightarrow[h\nu]{O_2} TeO_2 + \frac{n}{2}R_2NC(S)SS(S)CNR_2$$

$(n = 2 \text{ or } 4)$

2. *Diselenides and ditellurides*

The photolysis pathway for diselenides and ditellurides is very dependent on the presence or absence of dioxygen in the reaction system. In the absence of dioxygen, photolysis of dibenzyl diselenide proceeds as follows[191]:

$$(PhCH_2)_2Se_2 \xrightarrow[\text{MeCN, 2h}]{h\nu(\lambda > 280\,\text{nm})} Se + (PhCH_2)_2Se$$

The reaction is considered to proceed via the following mechanism:

$$R_2Se_2 \xrightarrow{h\nu} R^{\bullet} + - {}^{\bullet}SeSeR$$

$$RSeSe^{\bullet} \longrightarrow RSe^{\bullet} + Se$$

$$R^{\bullet} + R_2Se_2 \longrightarrow R_2Se + RSe^{\bullet}$$

$$R^{\bullet} + RSe^{\bullet} \longrightarrow R_2Se$$

$$RSe^{\bullet} + RSe^{\bullet} \longrightarrow R_2Se_2$$

$$R = PhCH_2$$

The quantum yield for disappearance of the diselenide is 0.16. Although ESR evidence has been obtained for the formation of $PhSe^{\bullet}$ radicals on UV irradiation of Ph_2Se_2[192], it is believed that C—Se bond scission is the main mechanistic pathway in the above case.

In contrast, in the presence of dioxygen, photolysis of $(PhCH_2)_2Se_2$ leads to a variety of products. Thus, if the material is in chloroform solution the products identified included PhCHO (47%), $PhCH_2OH$ (24%), PhMe (4%), $Ph(CH_2)_2Ph$ (5%) and Se (53%)[193]. Very similar results are obtained for $(PhCH_2)_2Te_2$[193], where the same range of products is obtained in the presence of dioxygen. Thermolysis of $(PhCH_2)_2Te_2$ gives a clean decomposition to $(PhCH_2)_2Te$ and Te. A further interesting variation has been observed. Thus, when dioxygen is passed into a toluene solution containing $(p\text{-}EtOC_6H_4)_2Te_2$ and $[Pd(PPh_3)_4]$ irradiated with visible light, good yields of PhCOOH are obtained via PhCHO[194]. Almost certainly the reaction mentioned in Section IV.B[149] is involved, but the presence of the Pd(0) complex is essential, possibly to catalyse the formation of radicals such as $RTeO^{\bullet}$. There is much chemistry in this area which is in need of exploitation.

Cross and coworkers[195,196] have observed that elimination of selenium on photolysis of diselenides under oxygen-free conditions may be suppressed in the presence of triorganylphosphines. Thus:

$$R_3P + R'_2Se_2 \xrightarrow[\text{350 nm, MeCN}]{h\nu} R_3PSe + R'_2Se$$

The rates of the reactions decreased in the orders: $R_3 = Me_2Ph > MePh_2 > Ph_3$ and $R' = CH_2Ph > Et > Me > Ph$. A mechanism involving photocleavage of the Se—Se bond was favoured. Presumably this is readily reversible, but the intermediate $RSe^{\bullet}$ radicals may be captured by R_3P:

$$R'_2Se \xrightarrow{h\nu} R'Se^{\bullet} + {}^{\bullet}SeR'$$

$$R'Se^{\bullet} + PR_3 \longrightarrow R'^{\bullet} + R_3PSe$$

$$R'^{\bullet} + R'_2Se_2 \longrightarrow R'Se^{\bullet} + R'_2Se$$

$$2R'^{\bullet} \longrightarrow R'R'$$

Similar observations have been made by others[197] who estimate the quantum yield for the disappearance of $(PhCH_2)_2Se_2$ to be 1. Identical observations are again made for ditellurides, but in this case the triorganyl phosphine tellurides, R_3PTe, deposit tellurium after irradiation has ceased[198]. In the presence of dioxygen, elemental selenium and R_3PO are products in the diselenide reactions.

Another group exploiting the photolysis of diselenides and ditellurides in the presence of triphenylphosphine observed CIDNP effects[199].

UV irradiation of $(CF_3)_2Se_2$ gave $(CF_3)Se^{\bullet}$ radicals, which were trapped by metal carbonyls, e.g. $[Mn_2(CO)_{10}]$ to give $[Mn(CO)_4(SeCF_3)]_2$[200]. $RTe^{\bullet}$ radicals have been produced in pulse radiolysis experiments[201].

Although not strictly photochemistry of ditellurides, the following reaction is of interest:

$$Ph_2Te_2 \xrightarrow[NH_3(l)]{Na} PhTe^- \xrightarrow[h\nu]{RX} PhTeR$$

$$R = Ph,\ 1\text{-Naph}$$

A photo-$S_{RN}1$ mechanism has been proposed[202].

E. Lewis Base Reactions

Both diorganyl chalcogenides and diorganyl dichalcogenides have lone pairs of electrons which should confer the property of Lewis basicity. This is indeed the case and a considerable growth in the ligand chemistry of selenium and tellurium is detectable in the current literature. The topic has been most comprehensively reviewed by Gysling[203,204], who is also a contributor to this series, hence the topic will receive only brief treatment here.

Selenides and, even more particularly, tellurides show very pronounced class B[205] or 'soft'[206] character as ligands. Thus the majority of complexes reported are either with metals of strong class B character, e.g. Hg^{II}, with metals in low oxidation states or with those such as Pd^{II}, Pt^{II} and Rh^{III} which may be considered borderline A/B in their behaviour. Tellurium ligands, for example, show no affinity for copper(II) but readily form complexes with copper(I). It is generally advisable to carry out complex-forming reactions under dinitrogen since, in some cases, there is evidence that ions such as copper(II) may catalyse the oxidation of tellurides, e.g.[73]

$$RTeCH_2TeR \xrightarrow[\textit{but}\ O_2-Cu^{II},\ 15\,min]{O_2,\ 16\,h} RTe(O)CH_2Te(O)R$$

$$R = p\text{-}EtOC_6H_4$$

The telluroxide is isolated as a copper(II) complex.

1. Selenides and tellurides as ligands

Typical examples of the type of complex formed are presented here. Complexes with metals in low oxidation states are, for example, $[Mo(CO)_3(Phen)(SePh_2)]$[207] and *trans*-$[RhCl(CO)(EEt_2)_2]$ (E = Se, Te)[208] and $[RhCl(TePh_2)_3]$[209]. *trans*-$[PdCl_2(SeEt_2)]$ has been the subject of a crystallographic study[210]. *Fac* and *mer* isomers of $[RhCl_3(TePh_2)_3]$ have been prepared[211], as have similar iridium(III) complexes and compounds of the type $[RhCl_3(bipy)(TePh_2)]$[211]. A complex of $Pd(SCN)_2$ with $Te[(CH_2)_3SiMe_3]_2$ (= L) has been shown by X-ray crystallography to be a thiocyanato complex, *trans*-$[Pd(SCN)_2L_2]$[212].

As more crystallographic studies become available, it is notable that the telluride ligands often bridge metal ions. For example, $[CuCl(TeEt_2)]$ has a structure based on layers of CuCl bridged by $TeEt_2$ with Cu→Te bond lengths of 2.625 and 2.535 Å[213]. A further example is produced by the complex $[IAg(\mu\text{-}TePh_2)AgI]$[214].

In solution, fluxional behaviour is often shown by the chalcogen ligand. Thus, ^{77}Se NMR studies on $[PtMe_3X(MeSe(CH_2)_nSeMe)]$ ($n = 2$ or 3; X = Cl, Br, or I) clearly show the presence of invertomers[215]. For the case of $n = 1$ (dimeric complexes) metallotropic 1,3-shifts are observed[216,217].

2. Diselenides and ditellurides as ligands

Very frequently the diorganyl dichalcogenides undergo oxidative addition to a metal centre, resulting in complexes of RE^- (E = Se, Te) which are not relevant to this chapter. However, Mehdi and Miller[159] express the view that simple coordination of ditelluride may be a precursor step to rupture of the Te—Te bond. There are some examples in the literature in which the Se—Se or Te—Te bond of R_2E_2 (E = Se, Te) remains intact on coordination. Thus, for example, the rhenium complexes $[(OC)_3Re(\mu\text{-}Br)_2(\mu\text{-}E_2Ph_2)Re(CO)_3]$ (E = Se[218] or Te[219]) have been crystallographically characterized. Variable-temperature NMR studies of $[M(CO)_5(Me_3SiCH_2SeSeCH_2SiMe_3)]$ have established that not only does the coordinated selenium atom undergo pyramidal inversion but also, above ambient temperature, a novel 1,2-shift between adjacent selenium atoms occurs[220]. Copper(I) complexes $[CuCl{\cdot}R_2Te_2]$ (R = p-$EtOC_6H_4$) have been reported and are probably polymeric[221]. It has also been suggested that Lewis acids such as copper(I) may trap unsymmetrical ditellurides[222].

3. Bi- and poly-dentate ligands

$MeSe(CH_2)_2SeMe$ has been mentioned in another context[215] (Section 1 IV.E.1) and the syntheses of some Pt^{II}[223] and Pt^{IV}[224] complexes have been described. Compound **14** is a further example of a bidentate selenium ligand used to complex Pt^{IV}[224]. In this instance the carbon backbone is rigid. Compound **15** is related to **14** and is of interest in that it has been shown to form well defined complexes with nickel(II), e.g. $[Ni(\mathbf{15})X]ClO_4$ (X = Cl, Br, I, NCS), which are trigonal bipyramidal complexes. The bis-complex $[Ni(\mathbf{15})_2](ClO_4)_2$ contains octahedral nickel(II)[225]. A tetradentate selenoether, $MeSe(CH_2)_2Se(CH_2)_3Se(CH_2)_2SeMe$, has been reported, which gives binuclear complexes with $[PdCl_4]^{2-}$ in which the ligand is bidentate to each palladium atom[226]. Bis(2-pyridylethyl) selenide forms 1:1 complexes with copper(II) halides, nitrate and perchlorate[227]. It is often the case that the presence of a nitrogen donor atom will increase the affinity of the ligands for 'harder' acids.

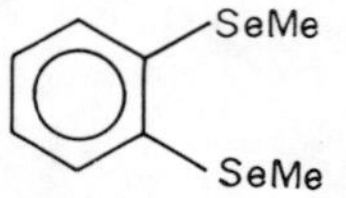

(14)

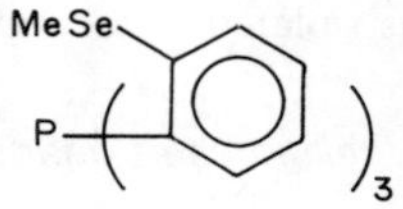

(15)

Very few bi- or poly-dentate tellurium ligands are known but $Ph_2P(o\text{-}C_6H_4TePh)$ (= L) is an exception, and the complex $[PtL_2][Pt(SCN)_4]{\cdot}2DMF$ has been the subject of an X-ray study[228]. The author's group has recently attempted to add more tellurium ligands to this list. Thus **6** (Section III.C) readily forms complexes with Pd^{II}, Pt^{II} and Rh^{I}. Of particular interest is the monomeric complex $[HgCl_2(\mathbf{6})]$, in which the two tellurium atoms, but not the nitrogen atoms, are coordinated to mercury thus forming a 13-membered chelate ring. In fact the nitrogen–tellurium interaction seen in uncoordinated **6** is retained in the complex; thus tellurium is simultaneously a Lewis acid and a Lewis base[108]. The ligands $RTe(CH_2)_nTeR$ (R = p-$EtOC_6H_4$; $n = 7, 9$ or 10) give soluble macrocyclic complexes with $PdCl_2$ and $PtCl_2$. In the solid state the palladium complexes appear to be *trans* and the platinium complexes *cis*, but in solution ^{125}Te NMR data reveal a *cis*$\rightleftharpoons$*trans* equilibrium[229].

4. Scales of Lewis acidity

Quantitative scales of Lewis acidity and basicity are very attractive concepts which, if realized, would greatly enhance the accuracy of chemical predictions. Great problems arise because the acid–base pair can rarely be considered free of their environment in practice, so lattice or solvation factors could become dominant. Drago[230] has given an interesting treatment of the problems involved. ^{125}Te Mössbauer parameters are rarely sensitive to effects more than one atom away from tellurium. When a telluride, R_2Te, coordinates to a Lewis acid, p electron density will be removed from tellurium, thus reducing the imbalance of occupancy of the tellurium p orbital set. Hence it is expected that the quadrupole splitting should decrease on coordination and, further, that this decrease will be independent of lattice effects and may be related directly to the coordinate bond strength. It was suggested that a scale of Lewis acidity might be set up relative to (p-$EtOC_6H_4)_2Te$, selected for its ease of preparation, handling and purification. The quadrupole splitting did indeed decrease in the order $Cu^I > Pd^{II} \approx Pt^{II} > Hg^{II}$[231]. However, as more crystallographic data become available it is clear that the model in its present form is too simplistic. Thus, tellurides often bridge metal centres (e.g. Cu^I[213]), bond angles about coordinated tellurium often depart significantly from 90°, bringing into question the initial assumption of p orbital bonding, and sometimes the metal may be in more than one coordination environment in the same complex, e.g. $[HgI_2(TePh_2)]_4$[232]. The original postulate is not without merit, but it requires refinement in a well designed Mössbauer/crystallographic study.

5. Conclusion

This brief, and to some extent superficial, survey of ligand chemistry claims no more than to place the topic in context in this chapter. It is safe to predict, however, that this will prove to be a growth area in organyl selenium and tellurium chemistry, not only because the new ligands will be of interest to coordination chemists but also because of the possibilities that some of the new complexes may possess interesting electrical or catalytic properties.

F. Charge-Transfer Complexes

In Section IV.A reference was made to charge-transfer complexes of tellurides with organic halides[72,146]. Even simple selenides when reacted with iodine do not undergo the oxidative addition reaction but form a complex $R_2Se \cdots I_2$[233]; indeed, the following may be general:

$$AE^{II}B + XY \rightleftharpoons (AB)E^{II} \cdots XY \rightleftharpoons (AB)E^{IV}(XY)$$
$$E = Se, Te$$

where the position of the equilibrium depends on the identity of groups A, B, X and Y and also on E.

It is to be expected that organylchalcogens will form charge-transfer (CT) complexes with acceptors such as TCNQ (7, 7, 8, 8-tetracyano-*p*-quinodimethane), but studies with simple organyl selenium or tellurium compounds are scarce. However, Heller *et al.*[234] have made 1:1 complexes of **16** (X = Se, Te) with the related tetracyanoethene. The same group more recently reported complexes of the same donors with TCNQ[235]. The interactions were considered to be of a π–π type.

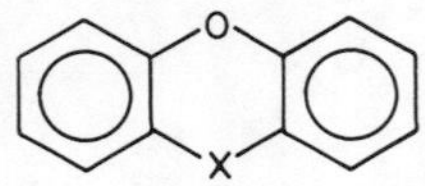

(16)

The range of TCNQ complexes of simple organyltellurium compounds was recently extended[236] to include donors such as diphenyltelluride (1:1), 1,3-dihydro-2-telluraindene (1:1), R_2Te_2 (1:2) (R = Ph or *p*-$EtOC_6H_4$) and dibenzotellurophene (1:1). The crystal structure of the last complex[236] showed interesting differences from that of the dibenzothiophene complex[237] in that significant Te···N interactions occur between the mixed stacks. It was suggested that ^{125}Te Mössbauer spectroscopy will prove a useful tool to determine the degree of charge transfer in tellurium complexes. Certainly the difference in quadrupole splitting for the 1,3-dihydro-2-telluraindene complexes of TCNQ ($8.72\,mm\,s^{-1}$) and chloranil ($10.24\,mm\,s^{-1}$) is significant. The former complex is considered to have an ionic ground state[236].

A major impetus for the study of CT complexes is the hope that some may exhibit enhanced solid-state conductivities. Much of the synthetic work involves heterocyclic compounds which are beyond the scope of this chapter, but some of the work does involve diselenides and ditellurides. For example, tetraselenatetracene (TSeT) forms a metallic compound $(TSeT)_2{}^+Cl^-$, but the $PF_6{}^-$ and $AsF_6{}^-$ salts are only semiconducting, a fact attributed to the greater volume expansion induced by the larger anions which is likely to weaken the essential Se···Se contacts[238,239]. Tetratelluratetracene (TTeT) will form CT complexes with reagents such as TCNQ, $CuCl_2$, $CuBr_2$ and I_2. The materials have compacted disc conductivities in the range 0.1–$1\,\Omega^{-1}\,cm^{-1}$[240].

Poly(*p*-phenylene selenide), a yellow powder which may have some diselenide linkages, gives a black solid on exposure to AsF_5 vapour and undergoes a 40% increase in weight. The compacted disc conductivity of the black material was 10^{-2}–$10^{-3}\,\Omega^{-1}\,cm^{-1}$[238]. The polymeric diselenides mentioned in Section II.B.1[44] seem to undergo only surface reaction with reagents such as AsF_5. Poly(methylene ditelluride) undergoes a semiconductor–metal transition at 280 K[106].

V. SPECTROSCOPY

Other chapters in both Volumes I and II of this book cover, in some detail, aspects of NMR and ESR spectroscopy, UV and vibrational spectroscopy, photoelectron spectroscopy and ^{125}Te Mössbauer spectroscopy. All that will be attempted here is to place these techniques in the context of this chapter. Since it is probable that NMR spectroscopy is likely to be the most important of these techniques in the future, it is given more prominence than the others.

A. Nuclear Magnetic Resonance Spectroscopy

In addition to the obvious possibility of studying organylselenium and tellurium compounds by ^{1}H and ^{13}C NMR spectroscopy, both elements contain isotopes which permit selenium and tellurium NMR spectra to be recorded directly. Details of the magnetic isotopes are given in Table 5.

The advent of FT NMR spectrometers with multinuclear facilities has greatly eased the experimental difficulties in determining selenium and tellurium NMR spectra, with the result that there is now a minor explosion of new data appearing in the literature, although

TABLE 5. Properties of NMR nuclei of selenium and tellurium

Species	Spin	Natural abundance(%)	Receptivity ($^{13}C = 1$)
^{77}Se	$\frac{1}{2}$	7.50	2.9
^{125}Te	$\frac{1}{2}$	6.99	12.5
^{123}Te	$\frac{1}{2}$	0.87	0.88

the data for ^{77}Se remain more numerous. The chemical shift range for ^{77}Se is about 2700 ppm and for ^{125}Te 4700 ppm. ^{125}Te chemical shifts show a marked temperature fluctuation during long accumulation times[241]. The nuclear Overhauser effect is rarely observed for ^{125}Te NMR, and when it is observed it leads to a reduction in signal intensity. Spin–lattice relaxation times are, in general, 4–6 times shorter for ^{125}Te than for ^{77}Se[242], which together with the greater receptivity of the tellurium nucleus does make it the more attractive to study. Three excellent reviews are available which relate particularly to organyl selenium[243] and organyl tellurium[244,245] compounds.

The relative electronegativities of selenium and the bonded group seem to be the major influence on selenium shielding in organylselenium compounds, high shielding being characteristic of negative selenium[250]. Data for dialkyl selenides show some surprises; thus the alkyl shielding effect is the inverse of expectation if it depended entirely on inductive effects, e.g. Me_2Se, $\delta = 0$ppm; Me(Et)Se, $\delta = 108$ppm[246]. On the other hand, replacement of Me by CF_3 produces an additive effect on the chemical shift: $Me(CF_3)Se$, $\delta = 370$ppm; $(CF_3)_2Se$, $\delta = 737$ppm[247]. The selenium chemical shift in alkyl aryl selenides is dependent on the alkyl group for constant aryl and is also sensitive to substituents in the aryl group for constant alkyl. Symmetrical diaryl selenides also show sensitivity to the substituent on the aryl group. The general trend is, as expected, that electron-with drawing groups are deshielding and electron-releasing groups are shielding, e.g. for p-XC_6H_4SeMe: X = OMe, $\delta = 189$ppm; X = H, $\delta = 202$ppm; X = NO_2, $\delta = 233$ppm[248].

The selenium chemical shifts for diselenides are sensitive to the organic group and also show substituent effects when aryl groups are present. It is interesting to consider the spectra of unsymmetrical diselenides, e.g. Ph_2Se_2, $\delta = 460$ppm; PhSeSeMe, $\delta = 445$, 294 ppm; Me_2Se_2, $\delta = 275$ppm. Not only is the presence of the unsymmetrical species readily detected by ^{77}Se NMR, but the data indicate that the effect of alkyl/aryl substitution is almost entirely on the shielding at the selenium atom bonded to the new group[243].

It appears that for heavier atoms such as ^{77}Se and ^{125}Te the paramagnetic contribution (σ_p) to the shielding is dominant[249]. Similar trends are observed in ^{125}Te data to those discussed above for ^{77}Se data. Thus, in particular, the shielding is dependent on the relative electronegativities and for diaryl tellurides the same influence of substituents is seen, e.g. Ph_2Te, $\delta = 688$ppm; (p-$MeOC_6H_4$)PhTe, $\delta = 668$ppm; (p-BrC_6H_4)PhTe, $\delta = 1079.6$ppm[249] [*N.B.* All shift data are relative to Me_2Se (^{77}Se) or Me_2Te (^{125}Te).]

Although no unsymmetrical ditellurides have been isolated, their existence in solution is readily demonstrated by ^{125}Te NMR spectroscopy[147,245]. An exchange reaction, slow on the NMR time scale, occurs when two symmetrical ditellurides are mixed:

$$R_2Te_2 + R'_2Te_2 \cdots 2RTeTeR'$$

The reaction is under entropy control since the equilibrium constant is independent of temperature and approaches a value of 4[147,148]. In the presence of dioxygen anomalous chemical shifts are observed and in some spectra CIDNP effects are seen; the radicals causing these effects result from a competing reaction of dioxygen with the ditellurides[149], which was discussed in Section IV.B. For the unsymmetrical species ^{125}Te–^{125}Te coupling constants of the order of 200 Hz are observed. As for the diselenides, the major effect of changing an organic group in a ditelluride is on the shielding at the tellurium atom to which the group is bonded. Thus, for the following equilibrium, chemical shifts are as indicated[147]:

$$\underset{448.5}{(p\text{-}ClC_6H_4)_2Te_2} + \underset{457.9}{(p\text{-}EtOC_6H_4)_2Te_2} \rightleftharpoons 2(\underset{438.5}{p\text{-}ClC_6H_4)Te}\text{—}\underset{469.6\ \text{ppm}}{Te(C_6H_4OEt\text{-}p)}$$

Other workers[245] have noted a much larger influence on the shielding of the second tellurium atom in alkyl aryl ditellurides, e.g. Pr_2Te_2, $\delta = 115$ ppm; PrTeTePh, $\delta = 223$, 305 ppm; Ph_2Te_2, $\delta = 420$ ppm. It is now certain that, for ditellurides, reliable chemical shift data can only be obtained in the absence of dioxygen. It has been suggested that tellurium–proton coupling patterns can be a valuable aid to spectral assignment in ^{125}Te NMR studies[250].

One area in which ^{77}Se and ^{125}Te NMR spectroscopy is likely to find increasing application is in the ligand chemistry of the elements. For example, in complexes of metals such as Pd^{II} and Pt^{II} differences in shielding are experienced by the ^{77}Se nucleus in *cis* and *trans* isomers of the complexes $[MCl_2(Me_2Se)_2]$[246]. The same is true for ^{125}Te NMR, which was used to demonstrate a solution equilibrium of the *cis* and *trans* isomers of $[\{(PhCH_2CH_2)_2Te\}_2PtCl_2]$[251]; similarly, ^{125}Te data for $[MCl_2RTe(CH_2)_nTeR]$ (M = Pd, Pt; $n = 7, 9$ or 10; R = p-$EtOC_6H_4$) show a solution equilibrium of the *cis* and *trans* isomers[252].

A number of ^{13}C NMR studies of organylselenium and tellurium compounds have been reported. Thus, for example, Chadha and Miller[253] have surveyed a range of organotellurium compounds which included a few tellurides and one ditelluride. For aryl compounds, the resonance of the carbon atom bond to tellurium can be assigned via the low intensity of the resonance or, better, by measurement of the spin–lattice relaxation time, T_1, since carbons bearing no hydrogen atoms generally relax on order of magnitude more slowly than those bonded to hydrogen. In this way the ^{13}C chemical shift of carbon bonded to chalcogen in the series Ph_2S_2 (137 ppm), Ph_2Se_2 (130.9 ppm) and Ph_2Te_2 (108 ppm) were assigned (all vs. TMS)[254]. Relative to benzene (128.8 ppm), there is a downfield shift for Ph_2S_2 and Ph_2Se_2, but an upfield shift for Ph_2Te_2. When related phosphorus and tellurium compounds are compared, e.g. $Ph_2P(CH_2)_nPPh_2$ and $RTe(CH_2)_nTeR$ (R = p-$EtOC_6H_4$), the resonance of the aliphatic carbon bonded to tellurium is ca. 14 ppm upfield compared with the similar resonance in the phosphorus compound. Interestingly, when the tellurium compound coordinates to a metal, the carbon resonance undergoes a large downfield shift[73]. A very interesting development will be the application of solid-state high-resolution ^{13}C NMR spectroscopy to organyl chalcogen compounds. In their study of dithiocarbamate and xanthate complexes of tellurium(IV) and tellurium(II), Zumbulyadis and Gysling illustrated the potential of the technique for probing crystal as opposed to molecular symmetry[255].

Other applications of NMR spectroscopy in this field are conventional. For example 2,11-diselena[3.3]metacyclophane (**17**) undergoes conformational flipping between *syn* and *anti* conformers at $-110\,^{\circ}C$ and the energy barrier was calculated to be 33 kJ mol^{-1}[256]. In the same way the dynamic stereochemistry of the related bis-diselenide **18** is illustrated by the fact that the methylene group shows an AB spectrum at low temperature and a singlet at high temperature. The coalescence temperature was 208 K and the free energy of activation is 46.5 kJ mol^{-1}. The data were interpreted in terms of the interconversion of *syn* conformers[257]. Russian workers[258] have demonstrated barriers to rotation around aromatic ring Se of the order of $42\,kJ\,mol^{-1}$.

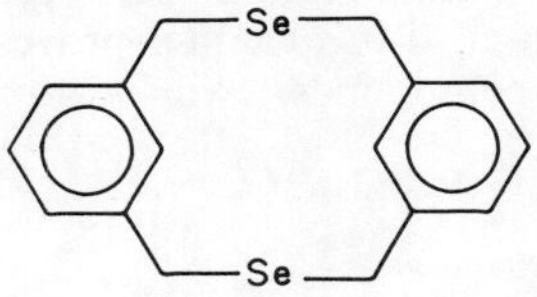

(17)

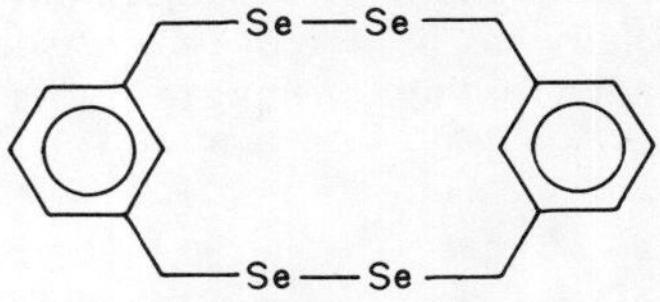

(18)

B. Vibrational Spectra

Only a few observations will be made. Carbon–selenium stretching modes in dialkyl selenides usually appear between 500 and $700\,mol^{-1}$. They are intense in the Raman spectrum[259]. In Me_2Te $\nu_s(TeC)$ and $\nu_{as}(TeC)$ are accidently coincident at $528\,cm^{-1}$[260], but they are seen as distinct vibrations when Me_2Te is coordinated to Pd^{II} or Pt^{II}[261]. For diaryl chalcogenides assignment of $\nu(EC)$ (E = Se, Te) is more difficult, although isotopic studies on diphenyltellurium(IV) dihalides suggest that $\nu(TeC)$ occurs at ca. $300\,cm^{-1}$ (ν_{as}) and ca. $240\,cm^{-1}$ (ν_s)[262]. Rotational isomers of C_1 and C_3 symmetry for $Pr^i(Me)Se$ have been shown to co-exist in the liquid state, whereas for $(Pr^i)_2Se$ three (C_2, C_3, C_1) have been shown to co-exist in the liquid phase[263].

A remarkable claim by Hamada and Morishita[264] that vibrational data for R_2E (E = O, S, Se, Te) suggested the molecules to be linear in structure has been shown to be based on a misinterpretation of the data[265].

C. ^{125}Te Mössbauer Spectra

This topic is covered comprehensively in this volume in Chapter 2 by Berry. The chemical isomer shift data for tellurides and ditellurides are relatively insensitive to the organic group. The shifts for ditellurides are, on average, slightly more positive (ca. 0.28–$0.37\,mm\,s^{-1}$) than for tellurides (ca. 0.06–$0.18\,mm\,s^{-1}$) (both with respect to I/Cu). Since the more positive the chemical isomer shift the greater is the s electron density at the nucleus, it follows that in both tellurides and ditellurides there are relatively small s electron densities at the tellurium nucleus[266,267].

Both tellurides and ditellurides have large (and similar) quadrupole splittings of the order of 10–$11\,mm\,s^{-1}$, which correspond to a large imbalance in the 5p orbital population on tellurium; this must, in turn, imply that the Te—C and Te—Te bonds have considerable covalent character. If the chemical isomer shift data, quadrupole splitting data and structural data (Section III) are considered together, the conclusion is that the tellurium must be using bonding orbitals which are between p^3 and sp^3 in character in tellurides and ditellurides[268].

The influence of coordination or charge-transfer interactions on, particularly, the quadrupole splitting of tellurides and ditellurides has been covered in Section IV. For the future, combined ^{125}Te NMR and ^{125}Te Mössbauer studies will be of considerable interest. One caveat in comparing Mössbauer data with other data is that, owing to the low recoil free fraction for organyltellurium compounds, it is necessary to record the spectra at 4 K. In some cases this might represent a different regime of stability for a particular material.

VI. CONCLUSION

This chapter has attempted to occupy the middle ground between coverage of selenium and tellurium analogues of ethers and peroxides that may be appropriate to a general text on organometallic chemistry and the comprehensive coverage that would be expected in a specialist monograph on the subject. Overlap with other chapters in the series is inevitable but hopefully has been kept to a reasonable minimum. If this chapter proves a useful starting point for those wishing to know more of these compounds, it will have achieved its objective.

VII. REFERENCES

1. R. A. Zingaro and W. C. Cooper (Eds.), *Selenium*, Van Nostrand Reinhold, New York, 1975, Chapt. 8, pp. 408–545.
2. K. J. Irgolic, *The Organic Chemistry of Tellurium*, Gordon and Breach, New York, 1974.
3. D. L. J. Clive, *Modern Organoselenium Chemistry*, Pergamon Press, Oxford, 1979.
4. P. D. Magnus, in *Comprehensive Organic Chemistry* (Ed. D. N. Jones), Vol. 3, Pergamon Press, Oxford, pp. 491–538.
5. D. L. J. Clive, *Tetrahedron*, **34**, 1049 (1978).
6. K. J. Irgolic, *J. Organomet. Chem.*, **103**, 91 (1975); **130**, 411 (1977); **158**, 235 (1978); **189**, 65 (1980); **203**, 367 (1980).
7. *Proceedings of the Second International Symposium on Organic Selenium and Tellurium Chemistry, Including Biochemistry, Chem. Scr.*, **8A**, 1–117 (1975).
8. D. Cagniant and G. Kirsch (Eds.), *Proceedings of the Third International Symposium on Organic Selenium and Tellurium Compounds*, Université de Metz, Metz, 1979, pp. 1–413.
9. F. J. Berry and W. R. McWhinnie (Eds.), *Proceedings of the Fourth International Conference on the Organic Chemistry of Selenium and Tellurium*, University of Aston in Birmingham, Birmingham, 1983, pp. 1–678.
10. J. P. Mila and J. F. Labarre, *C. R. Acad. Sci., Ser. C*, **263**, 1481 (1966).
11. R. Paetzold, U. Lindner, G. Bochmann and P. Reich, *Z. Anorg. Allg. Chem.*, **352**, 295 (1967).
12. D. J. Sandman, J. C. Stark and B. M. Foxman, *Organometallics*, **1**, 739 (1982).
13. D. J. Sandman, M. Rubner and L. A. Samuelson, *J. Chem. Soc., Chem. Commun.*, 1133 (1982).
14. L. A. Acampora, D. L. Dugger, T. Emma, J. Mohammed, M. Rubner, L. A. Samuelson, D. J. Sandman and S. Tripathy, in *Polymers in Electronics* (Ed. T. Davidson), American Chemical Society, Washington, DC, 1983, p. 461.
15. S. C. Cohen, M. L. N. Reddy and A. G. Massey, *J. Organometal. Chem.*, **11**, 563 (1968).
16. H. Reinboldt, in *Methoden der Organischen Chemie, Vol. IX, Schwefel–, Selen– und Tellurverbindungen* (Eds. E. Muller, O. Bayer, H. Meerwein and K. Ziegler), 4th ed., Georg Thieme, Stuttgart, 1955; see F. Tabourg, *C. R. Acad. Sci.*, **138**, 982 (1904).
17. G. B. Deacon and J. C. Parrott, *J. Organomet. Chem.*, **22**, 287 (1970).
18. R. C. Paul, K. K. Bhasin, and R. K. Chadra, *J. Inorg. Nucl. Chem.*, **37**, 2337 (1975).
19. H. M. Leicester, *J. Am.. Chem. Soc.*, **60**, 619 (1938).
20. A. Arase and Y. Masuda, *Chem. Lett.*, 419 (1975).
21. F. N. Alquist and R. E. Nelson, *J. Am. Chem. Soc.*, **53**, 4033 (1931).
22. R. F. Riley, J. Flats and D. Bengels, *J. Org. Chem.*, **27**, 2651 (1962).
23. H. J. Reich, in Ref. 9, pp. 258–272.
24. G. A. Krauss and M. J. Taschner, *J. Org. Chem.*, **45**, 1175 (1980).
25. G. Zima and D. Liotta, *Synth. Commun.*, **9**, 697 (1979); see also D. Liotta, *Acc. Chem. Res.*, **17**, 28 (1984).
26. A. E. Feiring, *J. Org. Chem.*, **45**, 1962 (1980).
27. S. Raucher, *J. Org. Chem.*, **42**, 2950 (1977).
28. N. Miyoshi, H. Ishii, S. Murai and N. Sonoda, *Chem. Lett.*, 873 (1979).
29. A. Osuka, N. Ohmassa and H. Suzuki, *Synthesis*, 857 (1982).
30. P. A. Grieco, S. Gilman and M. Nishizawa, *J. Org. Chem.*, **41**, 1485 (1976).
31. E. G. Kataev, T. G. Mannafov, A. B. Remizov and O. A. Komarouskaya, *Zh. Org. Khim.*, **11**, 2322 (1975).
32. E. G. Kataev, T. G. Mannafov and Yu. Yu. Samitov, *Zh. Org. Khim.*, **11**, 2324 (1975).
33. B. Lindgren, *Tetrahedron Lett.*, 4347 (1974).
34. S. -I. Murahashi and T. Yano, *J. Am. Chem. Soc.*, **102**, 2456 (1980).
35. H. Hauptmann and W. E. Walter, *J. Am. Chem. Soc.*, **77**, 4929 (1955).

36. Y. Okamoto and T. Yano, *J. Organomet. Chem.*, **29**, 99 (1971).
37. R. H. Mitchell, *Can. J. Chem.*, **54**, 238 (1976).
38. R. Weber, L. Christiaens, Ph. Thibaut, M. Renson, A. Croisy and P. Jacquignon, *Tetrahedron.*, **30**, 3865 (1974).
39. N. Petragnani, J. V. Comasseto, R. Rodrigues and T. J. Bracksem, *J. Organomet. Chem.*, **124**. 1 (1974).
40. E. Weissflog, *Phosphorus Sulfur*, **8**, 87 (1980).
41. R. Paetzold, H. D. Schumann and A. Simon, *Z. Anorg. Allg. Chem.*, **305**, 88 (1960).
42. Ref. 16, p. 1337.
43. J. C. Stark, R. Reed, L. A. Acampora, D. J. Sandman, S. Jansen, M. T. Jones and B. M. Foxman, *Organometallics*, **3**, 732 (1984); see also A. Yamahira, T. Nogami and H. Mikawa, *J. Chem. Soc., Chem. Commun.*, 904 (1983).
44. D. J. Sandman, J. C. Stark, M. Rubner, G. P. Hamill, L. A. Acampora, L. A. Samuelson, M. A. McGrath and G. W. Allen, *Proc. Sixth ICCOSS, Freiburg, Germany, 4–8 Oct., 1982.*
45. A. Runet and M. Renson, *Bull. Soc. Chim. Belg.*, **75**, 157 (1966).
46. W. Ried and G. Sell, *Synthesis*, 447 (1976).
47. W. H. H. Gunther and J. Y. C. Chu, *J. Org. Chem.*, **43**, 2672 (1978).
48. E. Rebane, *Ark. Kemi*, **20**, 205 (1962).
49. T. Naito, S. Ina, S. Inove, H. Hanai and S. Katagiri, *Nagoya Shiritsu Daigaku Yakagakubu Kiyo*, **5**, 43 (1957); *Chem. Abstr.*, **52**, 6248e (1958).
50. K. Y. Jen and M. P. Cava, *J. Org. Chem.*, **48**, 1449 (1983).
51. B. Kh. Strelets, M. Gel'mont, Yu. I. Akulin and L. S. Efros, *Khim. Geterotsikil. Soedin.*, 1205 (1979).
52. M. Nakazaki, *J. Chem. Soc. Jpn. Pure Chem. Sect.*, **75**, 338 (1954).
53. J. Meinwald, D. Dauphaise, F. Wudl and J. J. Hauser, *J. Am. Chem. Soc.*, **99**, 255 (1977).
54. K. A. Lestrup and D. O. Cowan, *J. Phys. (Paris), Colloq.*, 1247 (1983).
55. H. Reinboldt and E. Giesbrecht, *Chem. Ber.*, **88**, 1 (1955).
56. O. Foss and V. Janickis, *J. Chem. Soc., Dalton Trans.*, 628 (1980).
57. R. P. Shibaeva, V. F. Kaminskii, E. E. Kostynchenko and E. B. Yagubskii, *J. Organomet. Chem.*, **260**, 293 (1984).
58. R. E. Hollands, A. G. Osborne and I. Townsend, *Inorg. Chim. Acta.*, **37**, L541 (1979).
59. W. L. Dorn, A. Knoechel, P. Schulz and G. Klar, *Z. Naturforsch., Teil B*, **31**, 1043 (1976).
60. R. E. Cobbledick, F. W. B. Einstein, W. R. McWhinnie and F. H. Musa, *J. Chem. Res. (S)*, 145 (1979); *(M)* 1901 (1979).
61. F. Wohler, *Justus Liebigs Ann. Chem.*, **35**, 111 (1840).
62. G. Natta, *Giorn. Chim. Ind. Appl.*, **8**, 367 (1926).
63. L. Chugaev and W. Chlopin, *Chem. Ber.*, **47**, 1269 (1914).
64. B. A. Trofimov, N. K. Gurasova, A. A. Tatarinova, S. V. Amosova, L. M. Sinegovskaya, V. V. Keiko and V. A. Potapov, *Zh. Org. Khim.*, **20**, 1802 (1984).
65. B. A. Trofimov, S. V. Amosova, N. K. Gurasova, V. A. Potapov and A. A. Tatarinova, *Sulfur Lett.*, **1**, 151 (1983).
66. N. K. Gurasova, V. A. Potapov, S. V. Amosova and B. A. Trofimov, *Zh. Org. Khim.*, **19**, 2477 (1983).
67. L. Brandsma and H. E. Wijers, *Recl. Trav. Chim. Pays-Bas*, **82**, 68 (1963).
68. K. Lestrup, D. Talham, A. Bloch, T. Poeler and D. Cowan, *J. Chem. Soc., Chem. Commun.*, 336 (1982).
69. F. Wudl and E. Ahron-Shalom, *J. Am. Chem. Soc.*, **104**, 1154 (1982).
70. D. Seebach and A. K. Beck, *Chem. Ber.*, **108**, 314 (1975).
71. N. Dereu and J. L. Piette, *Bull. Soc. Chim. Fr.*, 623 (1979).
72. K. G. K. De Silva, Z. Monsef-Mirzai and W. R. McWhinnie, *J. Chem. Soc., Dalton Trans.*, 2143 (1983).
73. H. M. K. K. Pathirana and W. R. McWhinnie, *J. Chem. Soc., Dalton Trans.*, 2003 (1986).
74. S. R. Buzilova, I. D. Sadekov, T. V. Lipovich, T. M. Flippora and L. I. Vereshchagin, *Zh. Obshch. Khim.*, **47**, 1999 (1977).
75. N. V. Kondratenko, V. I. Popov, A. A. Kolomeitsev, I. D. Sadekov, V. I. Minkin and L. M. Yagupol'skii, *Zh. Org. Khim.*, **15**, 1561 (1979).
76. T. Kauffmann and H. Ahlers, *Chem. Ber.*, **116**, 1001 (1983).
77. J. F. Suttle and C. R. F. Smith, *Inorg. Synth.*, **3**, 140 (1950).

78. C. Elgy and W. R. McWhinnie, unpublished observation.
79. G. T. Morgan and F. H. Burstall, *J. Chem. Soc.*, 2599 (1930).
80. H. D. K. Drew, *J. Chem. Soc.*, 223 (1926).
81. W. H. H. Gunther, J. Nepywoda and J. Y. C. Chu, *J. Organomet. Chem.*, **74**, 79 (1974).
82. P. Thavornyntikarn and W. R. McWhinnie, *J. Organomet. Chem.*, **50**, 135 (1973).
83. H. Reinboldt and N. Petragnani, *Chem. Ber.*, **89**, 1270 (1956).
84. W. R. McWhinnie and M. G. Patel, *J. Chem. Soc., Dalton Trans.*, 199 (1972).
85. K. Lederer, *Chem. Ber.*, **49**, 2663 (1916).
86. H. Reinboldt and G. Vicentini, *Chem. Ber.*, **89**, 624 (1956).
87. B. C. Pant, *J. Organomet. Chem.*, **54**, 19 (1973).
88. R. F. Ziolo and W. H. H. Gunther, *J. Organomet. Chem.*, **146**, 245 (1978).
89. S. S. Abed-Ali and W. R. McWhinnie, *J. Organomet. Chem.*, **277**, 365 (1984).
90. W. Lohner and K. Praefcke, *J. Organomet. Chem.*, **194**, 173 (1980).
91. A. Luxen and L. Christiaens, *Tetrahedron Lett.*, **23**, 3905 (1982).
92. L. Engman, *J. Org. Chem.*, **48**, 2920 (1983).
93. P. Wiriyachitra, S. J. Falcone and M. P. Cava, *J. Org. Chem.*, **44**, 3957 (1979).
94. R. J. Lagow, L. L. Gerchman, R. A. Jacob and J. A. Morrison, *J. Am. Chem. Soc.*, **97**, 518 (1975).
95. T. J. Juhlke, R. W. Braun, T. R. Bierscheuk and R. J. Lagow, *J. Am. Chem. Soc.*, **101**, 3229 (1979).
96. J. Bergman and L. Engman, *Tetrahedron Lett.*, 1509 (1979).
97. J. Bergman and L. Engman, *Synthesis*, 569 (1980).
98. G. Bergson, *Acta Chem. Scand.*, **11**, 571 (1957).
99. M. T. Chen and J. W. George, *J. Organomet. Chem.*, **12**, 401 (1968).
100. W. S. Haller and K. J. Irgolic, *J. Organomet. Chem.*, **38**, 97 (1972).
101. L. Lange and W. W. Du Mont, *J. Organomet. Chem.*, **286**, C1 (1985).
102. N. Petragnani and M. de Moura Compos, *Tetrahedron*, **21**, 13 (1965).
103. M. A. K. Ahmed, W. R. McWhinnie and T. A. Hamor, *J. Organomet. Chem.*, **281**, 205 (1985).
104. W. V. Farrar, *Research*, **4**, 177 (1951).
105. G. T. Morgan and H. D. K. Drew, *J. Chem. Soc.*, 531 (1925).
106. C. W. Dirk, D. Nalewajeh, G. B. Blauchet, H. Sfchaffer, F. Moraes, R. M. Boysel and Wudl, *J. Am. Chem. Soc.*, **107**, 675 (1985).
107. A. Cisar and J. D. Corbett, *Inorg. Chem.*, **16**, 632 (1977).
108. W. R. McWhinnie and N. I. Al-Salim, (In Press).
109. N. Dereu and M. Renson, *J. Organomet. Chem.*, **258**, 163 (1983).
110. R. J. Gillespie and R. S. Nyholm, *Q. Rev. Chem. Soc.*, **11**, 339 (1957).
111. L. Pauling, *The Nature of the Chemical Bond*, 3rd ed., Cornell University Press, Ithaca, NY, 1960.
112. J. A. Van Vechten and J. C. Phillips, *Phys. Rev. B*, **2**, 2160 (1970).
113. J. F. Beecher, *J. Mol. Spectrosc.*, **21**, 414 (1966).
114. H. J. M. Bowen, *Trans. Faraday Soc.*, **50**, 452 (1954).
115. W. R. Blackmore and S. C. Abrahams, *Acta Crystallogr.*, **8**, 323 (1955).
116. L. Karle, *Angew. Chem., Int. Ed. Engl.*, **7**, 811 (1968).
117. R. Kirekas and T. Laitalaineu, *Acta Chem. Scand., Ser. B*, **37**, 61 (1983).
118. V. Ramgopalakrishnan and R. Walter, *J. Am. Chem. Soc.*, **106**, 3413 (1984).
119. Yu. E. Ovchinnikov, V. E. Shklover, Yu. T. Struchkov, V. A. Palyulin, V. I. Rokitskaya, M. Yu. Aismont and V. F. Traren, *J. Organomet. Chem.*, **290**, 25 (1985).
120. H. T. Palmer and R. A. Palmer, *Acta Crystallogr.*, **258**, 1090 (1969).
121. R. E. Marsh, *Acta Crystallogr.*, **5**, 458 (1952).
122. F. H. Kruse, R. E. Marsh and J. D. McCullough, *Acta Crystallogr.*, **10**, 201 (1957).
123. A. E. McCarthy and H. B. Singh, *J. Organomet. Chem.*, **275**, 57 (1984).
124. J. Lamotte, H. Campsteyn, L. Dupont and M. Vermeire, *Cryst. Struct. Commun.*, **6**, 749 (1977).
125. N. Al-Salim, T. A. Hamor and W. R. McWhinnie, *J. Chem. Soc., Chem. Commun.*, 453 (1986).
126. W. R. Blackmore and S. C. Abrahams, *Acta Crystallogr.*, **8**, 317 (1955).
127. C. L. Ralston, R. J. Secomb and A. H. White, *J. Chem. Soc., Dalton Trans.*, 2307 (1976).
128. J. C. Dewan and J. Silver, *Acta Crystallogr., Sect. B*, **33**, 1469 (1977).
129. J. C. Dewan and J. Silver, *J. Chem. Soc., Dalton Trans.*, 644 (1977).
130. J. C. Dewan and J. Silver, *J. Organomet. Chem.*, **125**, 125 (1977).
131. J. C. Dewan and J. Silver, *Aust. J. Chem.*, **30**, 487 (1977).

132. D. P. Rainville, R. A. Zingaro and E. A. Meyers, *J. Fluorine Chem.*, **16**, 245 (1980).
133. G. C. Pappalardo, S. Gruttadauria and K. J. Irgolic, *J. Organomet. Chem.*, **97**, 173 (1975).
134. D. J. Sandman, J. C. Stark, G. P. Hamill, W. A. Burke and B. M. Foxman, *Mol. Cryst. Liq. Cryst.*, **86**, 79 (1982).
135. G. Llabres, O. Dideberg and L. Dupont, *Acta Crystallogr., Sect. B*, **28**, 2438 (1972).
136. M. R. Spirlet, G. Van den Bossche, O. Dideberg and L. Dupont, *Acta Crystallogr., Sect. B*, **35**, 1727 (1979).
137. A. E. McCarthy and S. Ludlow, *J. Organomet. Chem.*, **219**, 169 (1981).
138. D. J. Sandman, J. C. Stark and B. M. Foxman, *Organometallics*, **1**, 739 (1982).
139. R. F. Galiallira, E. A. Pavlenko, O. N. Pruzhkov and T. K. Postnikova, *Tr. Khim. Tekhnol. Fak. Donetsk. Ind. Inst.*, 82 (1975).
140. J. C. Scaiano, P. Schmid and K. U. Ingold, *J. Organomet. Chem.*, **121**, C4 (1976).
141. R. A. McClelland and M. Leung, *J. Org. Chem.*, **45**, 187 (1980).
142. J. L. Piquard and L. Hevessi, *Tetrahedron Lett.*, **21**, 1901 (1980).
143. H. B. Singh and W. R. McWhinnie, *J. Organomet. Chem.*, **264**, 305 (1984).
144. S. Herberg and D. Naumaun, *Z. Anorg. Allg. Chem.*, **494**, 151 (1982).
145. T. N. Srivastava, R. C. Srivastava and V. K. Srivastava, *Indian J. Chem., Sect. A*, **22**, 503 (1983).
146. N. S. Dance, W. R. McWhinnie, J. Mallaki and Z. Monsef-Mirzai, *J. Organomet. Chem.*, **198**, 131 (1980).
147. P. Granger, S. Chapelle, W. R. McWhinnie and A. Z. Al Rubaie, *J. Organomet. Chem.*, **220**, 149 (1981).
148. N. S. Dance, W. R. McWhinnie and C. H. W. Jones, *J. Organomet. Chem.*, **125**, 291 (1977).
149. R. T. Mehdi, *PhD Thesis*, University of Aston in Birmingham, 1981.
150. J. V. Comasseto, J. T. B. Ferreira and J. A. Foutanillas Val, *J. Organomet. Chem.*, **277**, 261 (1984).
151. R. T. Mehdi and J. D. Miller, *J. Chem. Soc., Dalton Trans.*, 1071 (1983).
152. W. L. Dom, A. Knoechel, P. Schulz and G. Klar, *Z. Naturforsch., Teil B*, **31**, 1043 (1976).
153. F. Bigoli, E. Leporati, M. A. Pellinghelli, G. Crisponi, P. Delplano and E. F. Trogu, *J. Chem. Soc., Dalton Trans.*, 1763 (1983).
154. J. Passmore, E. K. Richardson and P. Taylor, *J. Chem. Soc., Dalton Trans.*, 1006 (1979).
155. I. P. Romm, I. V. Oliferenko, E. N. Gur'yanova, V. V. Troitsky, V. A. Chernoplekova, L. M. Kataeva and E. G. Kataev, *Zh. Obshch. Khim.*, **53**, 477 (1983).
156. I. V. Oliferenko, I. P. Romm, E. N. Gur'yanova, V. V. Troitsky, L. M. Kataeva, V. A. Chernoplekova and E. G. Kataev, *J. Organomet. Chem.*, **265**, 27 (1984).
157. R. Pellicciari, M. Curini, P. Ceccherelli and R. Fringuelli, *J. Chem. Soc., Chem. Commun.*, 440 (1979).
158. L. Y. Chia and W. R. McWhinnie, *J. Organomet. Chem.*, **148**, 165 (1978).
159. R. T. Mehdi and J. D. Miller, *J. Chem. Soc., Dalton Trans.*, 1065 (1984).
160. H. J. Reich, *Acc. Chem. Res.*, **12**, 22 (1979).
161. A. Krief, W. Dumont, A. Cravador, J. N. Dennis, S. Halozy, L. Hevesi, D. Labor, J. Lucchetti and J. Remion, *Bull. Soc. Chim. Fr.*, 519 (1980).
162. D. Liotta, in Ref. 9, pp. 278–296.
163. J. Bergman, *Tetrahedron*, **28**, 3323 (1972).
164. J. Bergman, in Ref. 9, pp. 215–241.
165. D. L. J. Clive, P. C. Anderson, N. Moss and A. Singh, *J. Org. Chem.*, **47**, 1641 (1982).
166. S. Halazy and A. Krief, *Tetrahedron Lett.*, 4233 (1979).
167. S. Rancher and G. A. Koolpe, *J. Org. Chem.*, **43**, 4252 (1978).
168. W. Dumont, P. Bayet and A. Krief, *Angew. Chem.*, **86**, 857 (1974).
169. H. J. Reich, S. K. Shah and F. Chow, *J. Am. Chem. Soc.*, **101**, 6648 (1979).
170. H. J. Reich and S. K. Shah, *J. Am. Chem. Soc.*, **97**, 3250 (1975).
171. D. L. J. Clive, G. J. Chittattu, V. Farina, W. A. Keil, S. M. Meuchen, C. G. Russell, A. Singh, Chi Kwong Wong and N. J. Curtis, *J. Am. Chem. Soc.*, **102**, 4438 (1980).
172. H. J. Reich, J. M. Renga and I. L. Reich, *J. Am. Chem. Soc.*, **97**, 5434 (1975).
173. H. J. Reich, I. L. Reich and J. M. Renga, *J. Am. Chem. Soc.*, **95**, 5813 (1973).
174. K. B. Sharpless, R. F. Laver and A. Y. Teranishi, *J. Am. Chem. Soc.*, **95**, 6137 (1973).
175. K. B. Sharpless, in Ref. 7, pp. 9–13.
176. L. Engman, *Tetrahedron Lett.*, **23**, 3601 (1982).

177. (a) M. Sevrin, J. N. Davis and A. Krief, *Tetrahedron Lett.*, **21**, 1877 (1980); (b) A. Toshimitsu, H. Owada, K. Terao and S. Uemura, *J. Org. Chem.*, **49**, 3796 (1984); (c) S. Uemura and S. I. Fukuzawa, *J. Chem. Soc., Perkin Trans. 1*, 471 (1985).
178. D. Seebach and A. K. Beck, *Angew. Chem.*, **86**, 859 (1974).
179. W. Dumont and A. Krief, *Angew. Chem.*, **87**, 347 (1975).
180. D. Van Erde and A. Krief, *Tetrahedron Lett.*, 457 (1976).
181. S. Uemura, S. Fukuzawa and A. Toshimitsu, *J. Chem. Soc., Chem. Commun.*, 1501 (1983).
182. J. Martens and K. Praefcke, *J. Organomet. Chem.*, **198**, 321 (1980).
183. G. Ahlgren, B. Åkermark and M. Nils, *J. Organomet. Chem.*, **30**, 303 (1971).
184. J. Bergman, *Acta Chem. Scand.*, **22**, 1883 (1968).
185. A. G. Schultz, *J. Org. Chem.*, **40**, 3466 (1975).
186. J. L. Fourrey, G. Henry and P. Jovin, *Tetrahedron Lett.*, 455 (1980).
187. J. P. Le Roux, P. L. Desbene and M. Seguin, *Tetrahedron Lett.*, 3141 (1976).
188. W. Lohner, J. Martens, K. Praefcke and H. Simon, *J. Organomet. Chem.*, **154**, 263 (1978).
189. L. Hevesi and A. Krief, *Angew. Chem.*, **88**, 413 (1976).
190. M. A. K. Ahmed, P. Granger and W. R. McWhinnie, *Polyhedron*, **5**, 859 (1986).
191. J. Y. C. Chu, D. G. Marsh and W. H. H. Gunther, *J. Am. Chem. Soc.*, **97**, 4905 (1975).
192. U. Schmidt, A. Muller and K. Markau, *Chem. Ber.*, **97**, 405 (1964).
193. K. H. Spencer and M. P. Cava, *J. Org. Chem.*, **42**, 2937 (1977).
194. Z. Monsef-Mirzai, *PhD Thesis*, University of Aston in Birmingham, 1982.
195. R. J. Cross and D. Millington, *J. Chem. Soc., Chem. Commun.*, 455 (1975).
196. D. H. Brown, R. J. Cross and D. Millington, *J. Chem. Soc., Dalton Trans.*, 159 (1977).
197. J. Y. C. Chu and D. G. Marsh, *J. Org. Chem.*,**41**, 3204 (1976).
198. D. H. Brown, R. J. Cross and D. Millington, *J. Organomet. Chem.*, **125**, 219 (1977).
199. G. Vermeersch, N. Febray-Garot, S. Caplain, A. Couture and A. Lablanche-Combier, *Proc. IUPAC Symp. Photochem. 7th*, 347 (1978); *Chem. Abstr.*, **90**, 120645u (1979).
200. M. K. Chaudhuri, A. Haas and A. Wensky, *J. Organomet. Chem.*, **116**, 323 (1976).
201. J. Bergman, N. Eklund, T. E. Eriksen and J. Lind, *Acta Chem. Scand., Ser. A*, **32**, 455 (1978).
202. A. B. Pierini and R. A. Rossi, *J. Organomet. Chem.*, **168**, 163 (1978).
203. H. J. Gysling, *Coord. Chem. Rev.*, **42**, 133 (1982).
204. H. J. Gysling, in Ref. 9, pp. 32–82.
205. S. Ahrland, J. Chatt and N. R. Davies, *Q. Rev. Chem. Soc.*, **12**, 265 (1958).
206. R. G. Pearson, *J. Am. Chem. Soc.*, **85**, 3533 (1963).
207. L. W. Houk and G. R. Dobson, *Inorg. Chem.*, **5**, 2119 (1966).
208. F. Faraone, R. Pietropaolo and S. Sergi, *J. Organomet. Chem.*, **24**, 797 (1970).
209. S. A. Gardner, *J. Organomet. Chem.*, **190**, 289 (1980).
210. P. E. Skakke and S. E. Rasmussen, *Acta Chem. Scand.*, **24**, 2634 (1970).
211. V. Rattanaphani, *PhD Thesis*, University of Aston in Birmingham, 1973.
212. H. J. Gysling, H. R. Luss and D. L. Smith, *Inorg. Chem.*, **18**, 2696 (1979).
213. R. K. Chadha and J. E. Drake, *J. Organomet. Chem.*, **286**, 121 (1985).
214. Y. Hayashi, T. Arahawa, T. Shiga, M. Ozaki, H. Kobayashi (Asahi Chem. Ind. Co. Ltd.), *Jpn. Kokai*, 78 65, 827 (1978); *Chem. Abstr.*, **89**, 146588 (1978).
215. E. W. Abel, K. G. Orrell and A. W. G. Platt, *J. Chem. Soc., Dalton Trans.*, 2435 (1983).
216. E. W. Abel, A. R. Khan, Kite, K. G. Orrell and V. Šik, *J. Chem. Soc., Dalton Trans.*, 2208 (1980).
217. E. W. Abel, K. Kite, K. G. Orrell, V. Šik and B. L. Williams, *J. Chem. Soc., Dalton Trans.*, 2439 (1981).
218. J. Korp, I. Bernal, J. L. Atwood, F. Calderazzo and D. Vitali, *J. Chem. Soc., Dalton Trans.*, 1492 (1979).
219. F. Calderazzo, D. Vitali, R. Poli, J. L. Atwood, R. D. Rodgers, J. M. Cummings and I. Bernal, *J. Chem. Soc., Dalton Trans.*, 1004 (1981).
220. E. W. Abel, S. K. Bhargava, P. K. Mittal, K. G. Orrell and V. Šik, *J. Chem. Soc., Dalton Trans.*, 1561 (1985).
221. I. Davies and W. R. McWhinnie, *Inorg. Nucl. Chem. Lett.*, **12**, 763 (1976).
222. A. Z. Al-Rubaie and W. R. McWhinnie, *J. Organomet. Chem.*, **235**, 59 (1982).
223. E. W. Abel, S. K. Bhargava, K. Kite, K. G. Orrell, V. Šik and B. L. Williams, *Polyhedron*, **1**, 289 (1982).
224. E. W. Abel, S. K. Bhargava, K. Kite, K. G. Orrell, V. Šik and B. L. Williams, *J. Chem. Soc., Dalton Trans.*, 583 (1982).

225. G. Dyer and D. W. Meek, *Inorg. Chem.*, **6**, 149 (1967).
226. W. Levason, C. A. McAuliffe and S. G. Murray, *J. Chem. Soc., Dalton Trans.*, 269 (1976).
227. E. Uhlig, B. Borek and H. Glänzer, *Z. Anorg. Allg. Chem.*, **348**, 189 (1966).
228. H. J. Gysling and H. R. Luss, *Organometallics*, **3**, 596 (1984).
229. H. M. K. K. Pathirana and W. R. McWhinnie, unpublished observations.
230. R. S. Drago, *J. Chem. Educ.*, **51**, 300 (1974).
231. I. Davies, W. R. McWhinnie, N. S. Dance and C. H. W. Jones, *Inorg. Chim. Acta*, **29**, L203 (1978).
232. F. W. B. Einstein, C. H. W. Jones, T. Jones and R. D. Sharma, *Inorg. Chem.*, **22**, 3925 (1983).
233. J. D. McCullough, G. Y. Chao and D. E. Zuccaro, *Acta Crystallogr.*, **12**, 815 (1959).
234. C. A. Heller, R. A. Zingaro and E. A. Meyers, *Can. J. Chem.*, **52**, 3814 (1974).
235. D. P. Rainville, R. A. Zingaro and J. P. Ferroris, *Can. J. Chem.*, **58**, 1133 (1980).
236. H. B. Singh, W. R. McWhinnie, R. F. Ziolo and C. H. W. Jones, *J. Chem. Soc., Dalton Trans.*, 1267 (184).
237. J. D. Wright and Z. A. Ahmad, *Acta Crystallogr., Sect. B*, **37**, 1848 (1981).
238. D. J. Sandaman, J. C. Stark, M. Rubner, G. P. Hamill, L. A. Acampora, L. A. Samuelson, M. A. McGrath and G. W. Allen, in Ref. 9, pp. 637–670.
239. A. V. Zvarykina, L. A. Kushch, V. N. Laukhin, O. L. Lependina, I. M. Yanovskaya and E. B. Yagubskii, *Izv. Akad. Nauk SSSR, Ser. Khim.*, 1625 (1983).
240. D. J. Sandman, J. C. Stark, M. Rubner, L. A. Acampora, L. A. Samuelson and B. M. Foxman, *Mol. Cryst. Liq. Cryst.*, **93**, 293 (1983).
241. P. Granger and S. Chapelle, *J. Magn. Reson.*, **39**, 329 (1980).
242. D. B. Denney, D. Z. Denney, P. J. Hammond and Y. F. Hsu, *J. Am. Chem. Soc.*, **103**, 2340 (1981).
243. M. Baiwir, in Ref. 9, pp. 406–467.
244. P. Granger, in Ref. 8, pp. 303–316.
245. D. H. O'Brien, K. J. Irgolic and C. K. Huang, in Ref. 9, pp. 468–491. *See also*: D. H. O'Brien, N. Dereu, R. A. Grigsby, K. J. Irgolic and F. F. Knapp, Jr., *Organometallics*, **1**, 513 (1982).
246. W. McFarlane and R. J. Wood, *J. Chem. Soc., Dalton Trans.*, 1397 (1972).
247. G. Llabres, M. Baiwir, L. Christiaens, J. Densel, L. Laitem and J. L. Piette, *Can. J. Chem.*, **56**, 2008 (1978).
248. G. A. Kalabin, D. F. Kushnarev, G. A. Chmutova and L. V. Kashurnikova, *Zh. Org. Khim.*, **15**, 24 (1979).
249. R. K. Chadha and J. M. Miller, *Can. J. Chem.*, **60**, 2256 (1982).
250. C. H. W. Jones and R. D. Sharma, *J. Organomet. Chem.*, **255**, 61 (1983).
251. H. J. Gysling, N. Zumbulyadis and J. A. Robertson, *J. Organomet. Chem.*, **209**, C41 (1981).
252. H. M. K. K. Pathirana, W. R. McWhinnie and P. Granger, unpublished observations.
253. R. K. Chadha and J. M. Miller, *J. Chem. Soc., Dalton Trans.*, 117 (1982).
254. A. Forchioni, V. Galasso, K. J. Irgolic and G. Pappalardo, *J. Organomet. Chem.*, **135**, 327 (1977).
255. N. Zumbulyadis and H. J. Gysling, *Inorg. Chem.*, **21**, 564 (1982).
256. R. H. Mitchell, *Tetrahedron Lett.*, 1363 (1975).
257. K. R. Dixon and R. H. Mitchell, *Can. J. Chem.*, **61**, 1598 (1983).
258. V. I. Naddaka, I. I. Logacheva and V. S. Yur'eva, *Zh. Org. Khim.*, **12**, 2606 (1976).
259. K. Ohno, T. Hirokawa, S. Aono and H. Murata, *Bull Chem. Soc. Jpn.*, **50**, 305 (1977).
260. J. R. Allkins and P. J. Hendra, *Spectrochim. Acta*, **22**, 2075 (1966).
261. J. R. Allkins and P. J. Hendra, *Spectrochim. Acta, Part A*, **24**, 1305 (1968).
262. N. S. Dance and W. R. McWhinnie, *J. Chem. Soc., Dalton Trans.*, 43 (1975).
263. K. Ohno, A. Mitsui and H. Murata, *Bull. Chem. Soc. Jpn.*, **52**, 2178 (1979).
264. K. Hamada and H. Morishita, *J. Mol. Struct.*, **44**, 119 (1975).
265. D. C. McKean and J. L. Duncan, *J. Mol. Struct.*, **51**, 149 (1979).
266. C. H. W. Jones, R. Schultz, W. R. McWhinnie and N. S. Dance, *Can. J. Chem.*, **54**, 3234 (1976).
267. F. J. Berry, E. H. Kustan and B. C. Smith, *J. Chem. Soc., Dalton Trans.*, 1323 (1975).
268. F. J. Berry, C. H. W. Jones and W. R. McWhinnie, in Ref. 9, pp. 492–506.

The Chemistry of Organic Selenium and Tellurium Compounds Volume 2
Edited by S. Patai
© 1987 John Wiley & Sons Ltd.

CHAPTER **14**

Organic selenocyanates and tellurocyanates and related compounds

AKIO TOSHIMITSU AND SAKAE UEMURA

Institute for Chemical Research, Kyoto University, Uji, Kyoto 611, Japan

I. INTRODUCTION . . . 541
II. ORGANIC SELENOCYANATES . . . 542
A. Preparation . . . 542
1. By exchange of halogen . . . 542
2. From diazonium salts . . . 545
3. From organometallic compounds . . . 545
4. By selenocyanation . . . 546
5. By formation of an Se—CN bond . . . 547
B. Physical Properties . . . 548
C. Reactions . . . 550
1. Additions to the C≡N bond . . . 550
2. Substitution of the CN group . . . 553
a. Substitution by heteroatoms . . . 553
b. Substitution by carbon atoms . . . 554
c. Conversion of alcohol to selenide . . . 557
d. Selenenylation of olefins and acetylenes . . . 569
3. Elimination of the CN group . . . 571
4. Formation of complexes . . . 573
III. ORGANIC ISOSELENOCYANATES . . . 574
IV. ORGANIC TELLUROCYANATES . . . 577
V. ORGANIC ISOTELLUROCYANATES . . . 585
VI. REFERENCES . . . 585

I. INTRODUCTION

The chemistry on selenocyanates and related compounds has been reviewed by Bulka[1], the literature coverage being up to 1973. Since then, this field has grown steadily and much progress has been achieved in the use of organic selenocyanates and in the chemistry of organic tellurocyanates. This chapter deals with recent advances in this field after 1973, focusing mainly on the preparation, properties and reactions of organic selenocyanates (RSeCN), isoselenocyanates (RN=C=Se) and tellurocyanates (RTeCN), where R is any organyl group. The literature coverage is nearly complete up to the end of 1985.

II. ORGANIC SELENOCYANATES

A. Preparation

1. By exchange of halogen

The substitution of halogen by a selenocyanato group is known as a general route for the preparation of alkyl selenocyanates. The reaction has been utilized in the identification of alkyl iodide[2] produced by the ring opening of a selenetane as shown in equation 1. Potassium and sodium selenocyanates are normally used as a source of selenocyanate. They are conveniently prepared by the reaction of potassium and sodium cyanides with selenium in dimethylformamide[3] or *N, N*-dimethylacetamide[4].

The preparation and fungicidal effect of the furfuryl selenocyanate derivative **1** has been reported[5,6]. This reaction also proceeds when the alkyl halide bears a hydroxy[7] or isothiocyanato[8] group on the β-carbon atom. In the case of 1, 3-dibromopropan-2-ol, however, only one bromine atom was substituted by an excess of selenocyanate to afford 1-bromo-3-selenocyanatopropan-2-ol (**2**) (equation 2)[9]. When the alkyl halide has an alkyl-[10] or aryl-seleno[11] substituent on the β-carbon atom, the selenocyanate ion attacks the selenium atom of an episelenonium ion intermediate (**3**) to afford an alkylselenenyl selenocyanate which decomposes to elemental selenium and 2-bromoethyl selenocyanate, and the latter reacts with potassium selenocyanate to produce bis(selenocyanato)ethane (equation 3). The substitution of halogen α to carbonyl has been utilized in the preparation of α-selenocyanatocarbonyl compounds[12–14]. A kinetic study of the substitution of bromine of 2-bromo-1-phenylethanone by selenocyanate in acetonitrile revealed that the reaction was almost 20 times faster than that of methyl iodide, owing to acceleration by the presence of the α-carbonyl group[15]. The rate and equilibrium constants of similar reactions were compared (equation 4), and the nucleophilicity order was found to be $NCSe^- \geqslant NCS^- > Cl^- \gg Br^-$ and the leaving group ability was $Br^- \gg Cl^- > NCSe^- > NCS^-$[16]. A similar reactivity order was also observed in the substitution reactions of benzyl halides by pseudohalide anions[17,18].

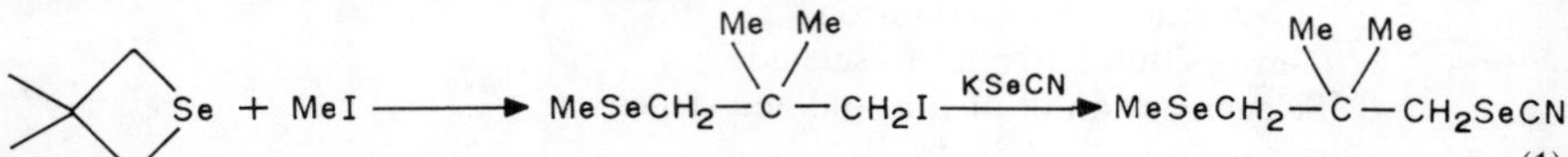

(1)

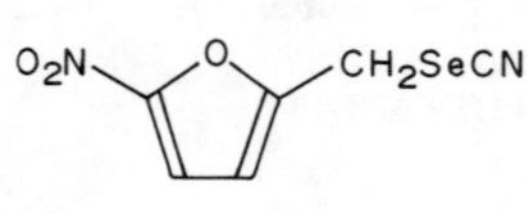

(**1**)

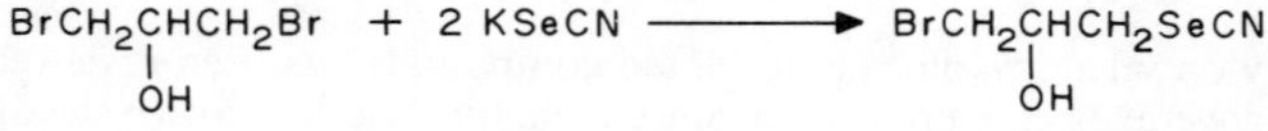

(2)

(**2**)

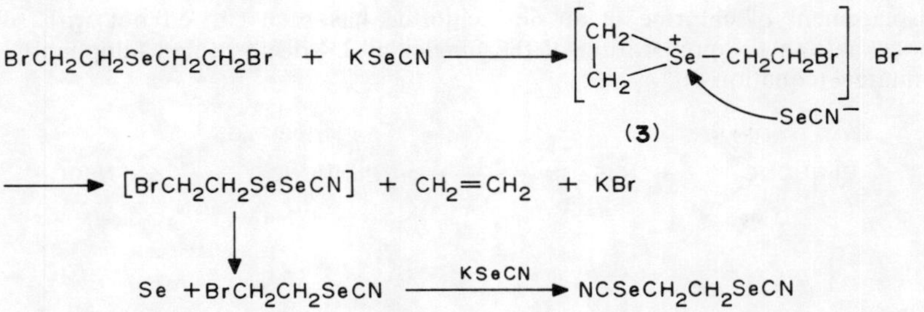

(3)

$$PhCOCH_2X + Y^- \rightleftharpoons^{MeCN} PhCOCH_2Y + X^-$$ (4)

X = Cl, Br
Y = SeCN, SCN

The reaction of iodonium ion with potassium selenocyanate has been utilized in the preparation of perfluoroalkyl selenocyanate[19], the leaving group being iodobenzene in this case (equation 5). The reaction of an epiiodonium ion with a selenocyanate ion has also been reported (equation 6)[20].

$$Ph\overset{+}{I}C_3F_7\ Cl^- + KSeCN \xrightarrow{DMF} C_3F_7SeCN + PhI + KCl$$ (5)

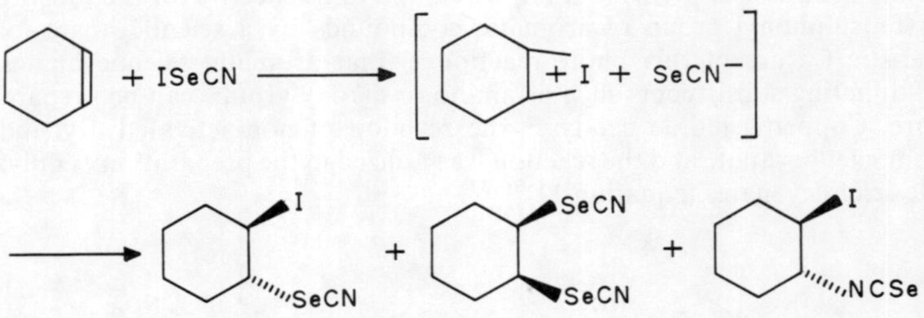

(6)

There are several examples of the substitution of an oxygen functional group by a selenocyanato group. Thus, the conversion of alcohols to alkyl selenocyanates by reaction with triphenylphosphine and selenocyanogen has been reported and the substitution of oxygen by SeCN was postulated as shown in equation (7)[21]. A tosyl group has also been utilized as a leaving group in the preparation of alkyl selenocyanates[22,23].

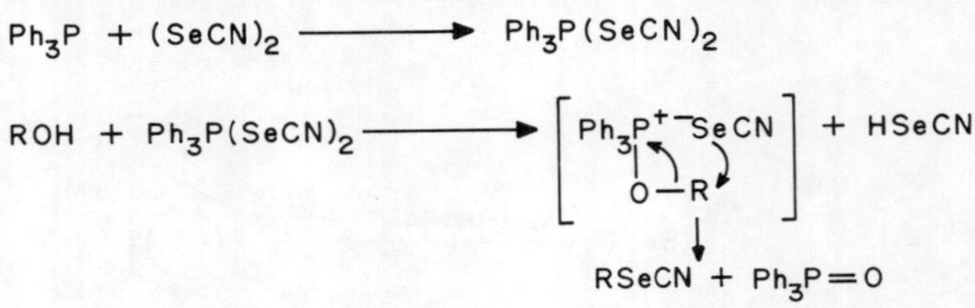

(7)

Displacement of chlorine of an acyl chloride has been carried out with sodium selenocyanate in the preparation of the fungicidal 2,4-disubstituted allophanoyl selenocyanate **4** (equation 8)[24].

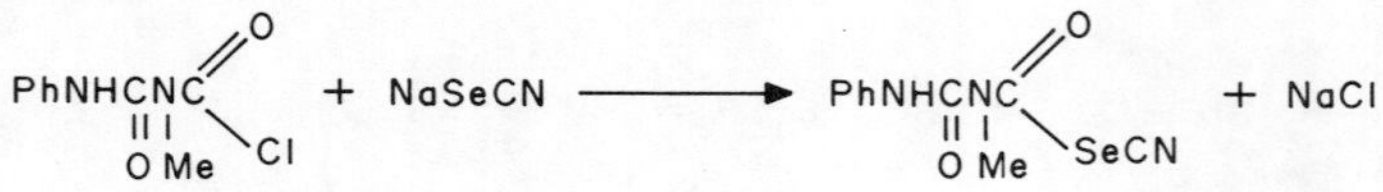

(8)

(**4**)

A bromine bound to the vinyl carbon of a cyclobutenedione derivative was substituted by a SeCN to afford the corresponding selenocyanate (**5**) (equation 9)[25].

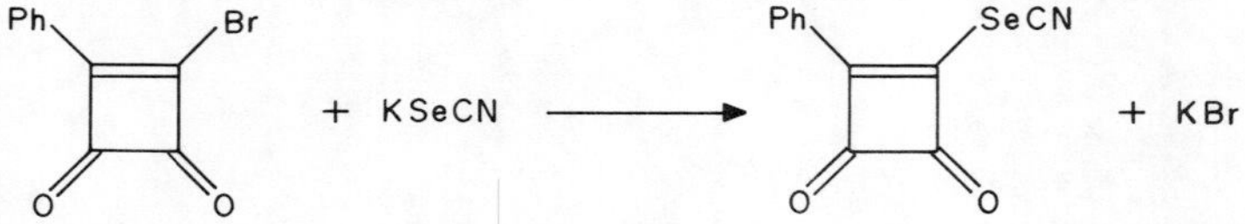

(9)

(**5**)

It is known that the halogen in an aromatic halide is substituted by a selenocyanate ion if the aromatic halide is activated by electron-withdrawing substituents. It was reported that halomethylsulphonyl groups enhanced the displacement of aromatic halogen (X), whereas the a halogen (X′) of the halomethyl group itself resisted the displacement as shown in equation (10)[26]. The reaction of potassium selenocyanate with 2-nitro-3-bromo-thiophene or -selenophene has also been reported to afford the substitution products[27]. UV irradiation was found to be effective for the substitution of a halogen or sulphonyl group of aromatic compounds by a selenocyanate ion[28,29]. A characteristic feature of this photo-reaction is that aromatic selenocyanates bearing electron-donating substituents such as amino or alkoxy groups can be prepared by this procedure. Copper(I) iodide catalyses the reaction of non-activated aryl iodide with potassium selenocyanate and the reaction was utilized in the preparation of polyalkylated aromatic selenocyanates (equation 11)[30,31].

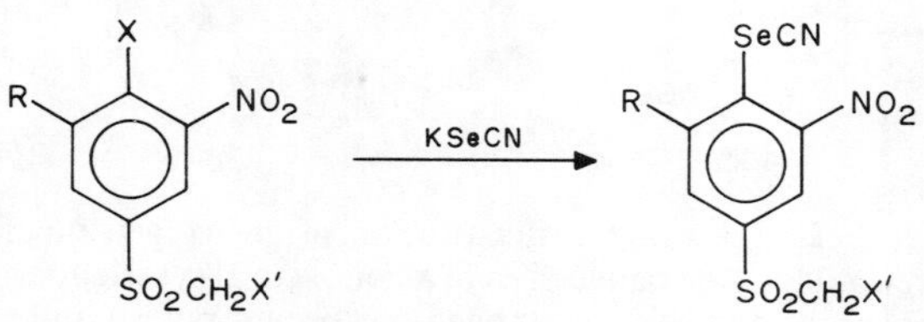

(10)

$X = Cl, Br, I$
$X' = Cl, Br$
$R = NO_2, H$

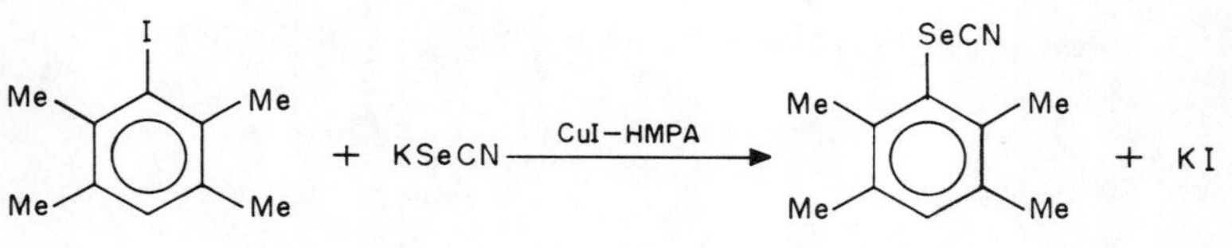

(11)

The substitution of a halogen of disubstituted azulenes by a selenocyanato group of $Cu(SeCN)_2$ was examined (equation 12)[32]. The reaction rates decreased in the order $R = MeO > Ph > COPh > SO_2Ph > NO_2$. This reaction was assumed to proceed through one-electron transfer to form a radical anion, which then reacted with $[Cu(SeCN)_2]^{+\bullet}$.

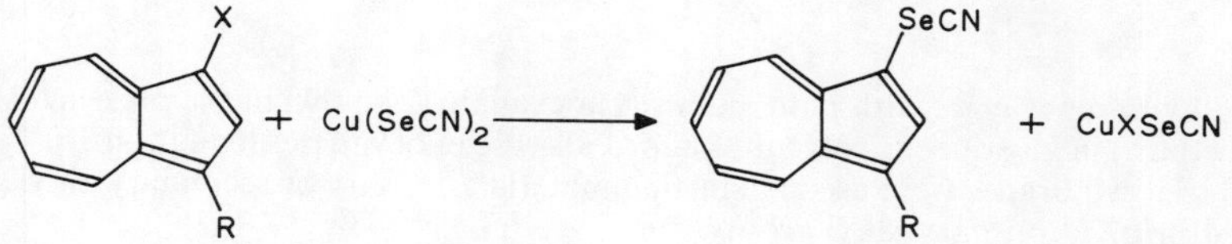

(12)

X = I, Br, Cl
$R = OMe, Ph, COPh, SO_2Ph, NO_2$

2. From diazonium salts

The reaction of diazonium salts with potassium (or sodium) selenocyanate is the most reliable procedure for the preparation of aromatic selenocyanates. Not only phenyl selenocyanate derivatives bearing various substituents[4,33–37] but also selenocyanatothiophene and -selenophene derivatives[38] (equation 13) were prepared by this procedure. This reaction is feasible with diazotized poly(*p*-aminostyrene), providing a route to polymer-supported selenium reagents[39].

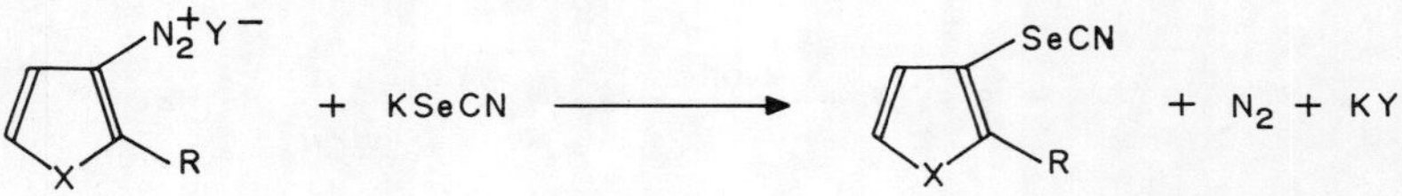

(13)

X = S, Se
$R = CHO, COMe, NO_2$

3. From organometallic compounds

The reaction of arylthallium(III) compounds (**6**) with a mixture of copper(II) sulphate and potassium selenocyanate affords aryl selenocyanates in good to excellent yields (equation 14)[40,41]. In the absence of a copper(II) salt, the yields of arylselenocyanates were poor. When arylmercury(II) compounds were used, the results were unsatisfactory owing to side reactions of aryl selenocyanate with the starting materials[42]. In the case of alkylthallium(III) compounds (7), a thallium(III) group was replaced by a selenocyanato group by treatment with only potassium selenocyanate. This reaction was utilized in a one-pot conversion of olefins to β-alkoxyalkyl selenocyanates (equation 15)[40,43].

$$p\text{-}RC_6H_4TlXY + Cu(SO_4)_2 + KSeCN \longrightarrow p\text{-}RC_6H_4SeCN \qquad (14)$$

(**6**)

R = H, Me, MeO
$X,Y = ClO_4, OAc$
$(CF_3COO)_2$
$(CCl_3COO)_2$

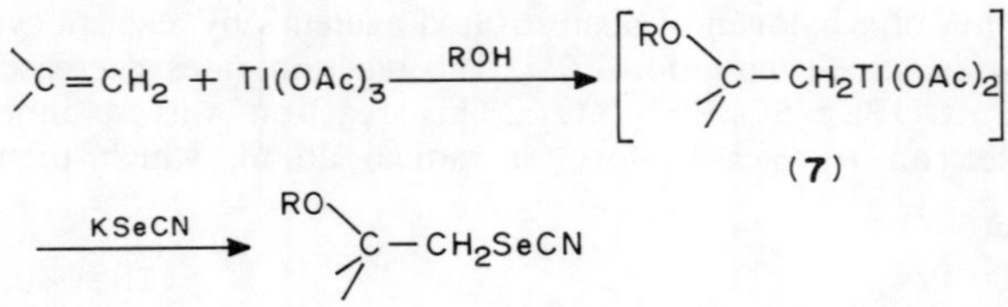

(15)

Trialkylboranes react with potassium selenocyanate (2 equiv.) in the presence of Fe^{3+} ion (2 equiv.) in an aqueous solvent to afford alkyl selenocyanates (equation 16)[44]. When mixed trialkylboranes were used as starting materials, tertiary or secondary alkyl groups reacted prior to primary alkyl groups[45].

$$R_3B + 2\ KSeCN + Fe_2(NH_4)_2(SO_4)_2 . 24H_2O \xrightarrow{H_2O} RSeCN \qquad (16)$$

4. By selenocyanation

Electrophilic substitution of 3-substituted thiophenes or selenophenes has been reported to produce 2-selenocyanato-thiophene or -selenophene derivatives (equation 17)[46,47]. Selenocyanogen bromide was prepared *in situ* by the reaction of potassium selenocyanate and bromine. Dicyanotriselenide was used in the selenocyanation of polystyrene, the selenium content in the product being estimated to be 1.6 mmol g^{-1} (equation 18)[48].

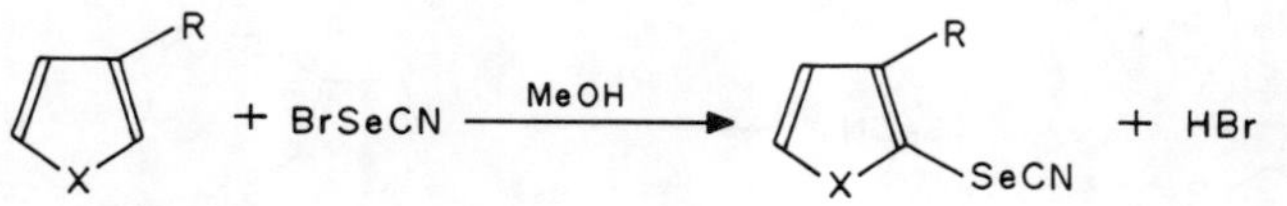

(17)

X = S, Se

R = OMe, NH_2, NHCOMe, NHCHO

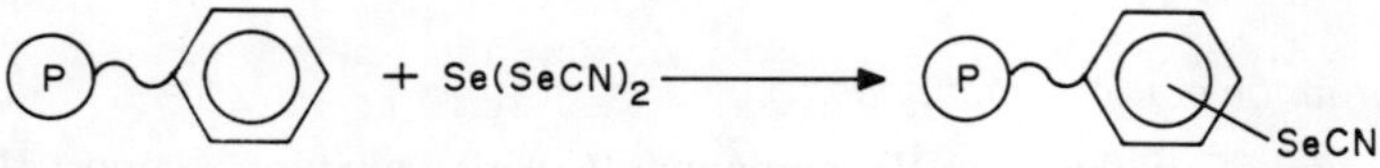

(18)

Selenocyanogen was generated electrochemically by anodic oxidation and it reacted with electron-rich olefins to afford bis(selenocyanato) adducts (equation 19)[49], probably by electrophilic addition of an SeCN group. An electrophilic SeCN species was also generated by the reaction of potassium selenocyanate with copper(II) halides. It reacted with olefins to afford β-alkoxyalkyl selenocyanates (**8**) in alcoholic solvents and β-halogenoalkyl selenocyanates (**9**) in acetonitrile (equation 20)[50]. The conversion of (**9**) to (**8**) was catalysed by copper(I) or -(II) salts and proceeded with retention of configuration, suggesting the intermediacy of an episelenonium ion.

$$SeCN^- - e^- \longrightarrow (SeCN)^{\bullet}$$

$$2\,(SeCN)^{\bullet} \longrightarrow (SeCN)_2 \qquad (19)$$

$$ROCH{=}CH_2 + (SeCN)_2 \longrightarrow ROCH(SeCN)CH_2SeCN$$

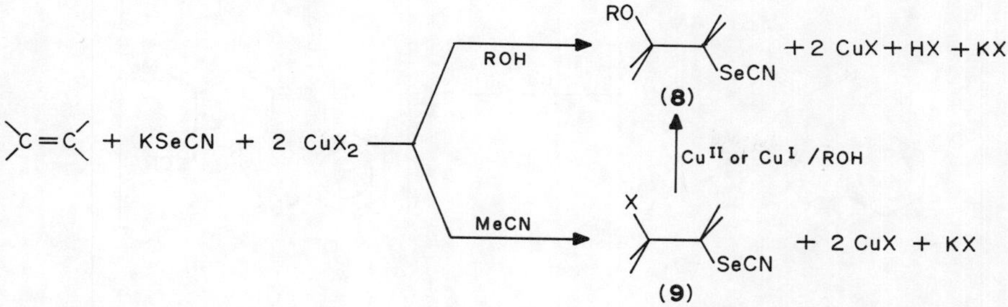

(20)

Nucleophilic addition to carbon—carbon triple bonds has been utilized in the preparation of vinyl selenocyanates. By the reaction of acetylene carboxylic acid with potassium selenocyanate, *cis*-*β*-selenocyanatoacrylic acid was produced through *trans*-addition of a selenocyanato group and a hydrogen (equation 21)[51]. The attack of selenocyanate on a propargyl bromide derivative was accompanied by loss of bromine to form allenyl selenocyanates (equation 22)[52].

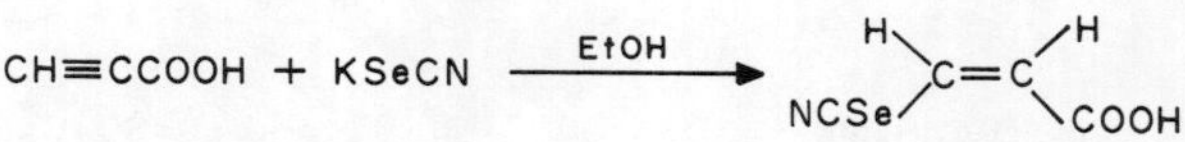

(21)

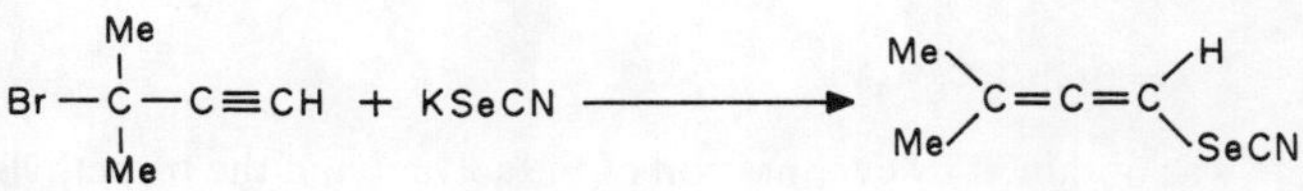

(22)

5. By formation of an Se—CN bond

The preparation of selenocyanates by the reaction of certain selenenyl compounds with potassium or silver cyanide is a well known procedure and is often used for the trapping of reactive intermediates or the identification of the products.

Chlorodifluoromethylselenenyl chloride ($F_2CClSeCl$), one of the products of chlorine cleavage of polymerized Se=CF_2, when treated with silver cyanide affords chlorodifluoromethyl selenocyanate ($F_2CClSeCN$)[53]. Dichlorofluoromethylselenenyl chloride ($FCCl_2SeCl$), prepared from carbon diselenide, fluorine chloride and chlorine, also reacts with silver cyanide to give $FCCl_2SeCN$[54].

The episelenonium ion **3**, when treated with potassium cyanide, produces 2-bromoethyl selenocyanate as a final product[10,55].

UV irradiation or heating of a benzeneselenazole derivative (**10**) induced its rearrangement to a phenyl selenocyanate bearing the pyridine nucleus in the *ortho* position (**11**) (equation 23)[56]. The attack of a cyanide anion on selenium was proposed as the final step of this reaction, as silver cyanide was formed if the reaction was carried out in the presence of silver nitrate.

Both selenenyl bromide and sulphonate afforded the corresponding selenocyanates with potassium cyanide (equation **24**)[57].

One of the most convenient procedures for the preparation of phenyl selenocyanate is the reaction of phenylselenenyl chloride with trimethylsilyl cyanide (equation 25)[58,59] at room temperature in dichloromethane or tetrahydrofuran. Essentially pure phenyl

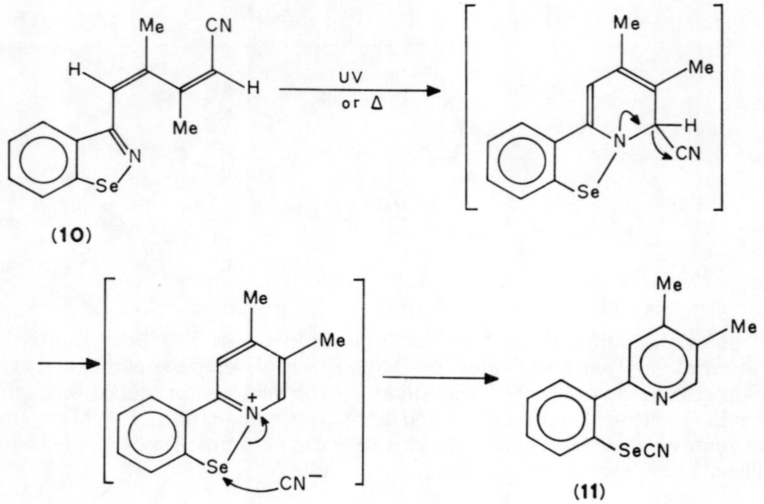

(10) (11) (23)

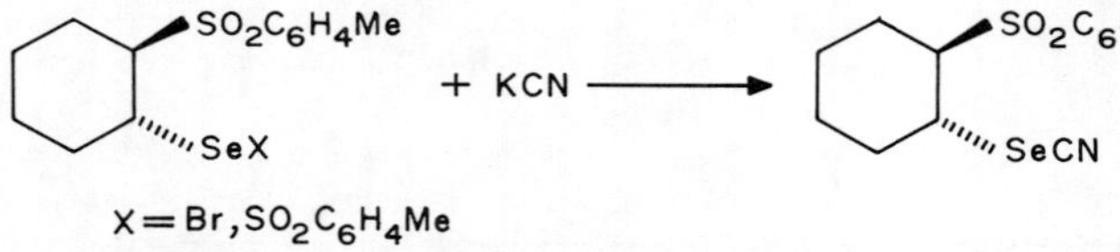

(24)

selenocyanate was obtained by evaporation of the solvent and the trimethylsilyl chloride produced.

$$PhSeCl + Me_3SiCN \longrightarrow PhSeCN + Me_3SiCl \qquad (25)$$

An example in which selenium acts as a nucleophile in the Se—CN bond-forming reaction (equation 26)[60] is the reaction of aryl methyl selenide with lithium methyl selenide when nucleophilic substitution on methyl carbon gives dimethyl selenide and aryl lithium selenide. The latter was trapped by cyanogen iodide to afford aryl selenocyanate, the arylseleno group reacting as a nucleophile.

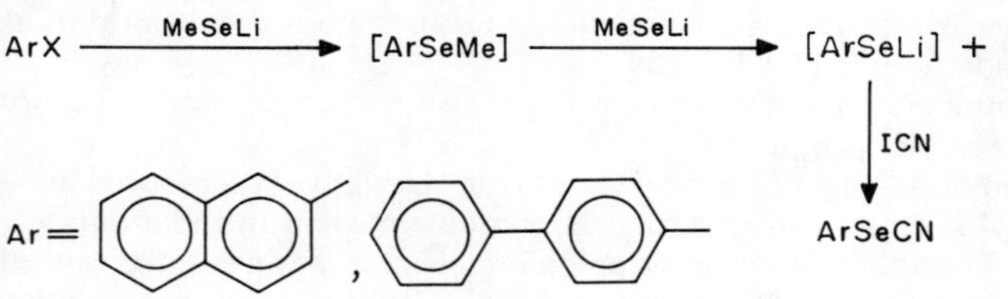

(26)

B. Physical Properties

The polarized IR spectra of 4-RC_6H_4SeCN (R = H, NO_2, NH_2) were recorded for a conformational study[61]. It was indicated that the CN group was in the plane of the aromatic ring. The NO_2 group was not coplanar with the ring; the protons of the NH_2

group were both above the ring plane. The IR spectra of XC_6H_4SeCN($X = H$, 4-F, 4-Cl, 4-Br, 3-F, 3-Cl, 3-Br) were recorded in the range 400–4000 cm^{-1}. The spectra were interpreted in terms of phenyl ring or SeCN group vibrations, with little interaction between the normal modes of the two frameworks[62]. IR and Raman spectra of benzyl selenocyanate was recorded and tentative assignments were proposed[63], based on force fields transferred from similar molecules such as ethyl selenocyanate[64] and ethyl thiocyanate after appropriate scalings.

The UV spectrum of phenyl selenocyanate was compared with those of the analogous cyanate and thiocyanate. All were found to agree well with the results of MO-LCAO calculations[65].

The He(I) photoelectron spectrum of CH_3SeCN was found to exhibit a series of five ionization bands below 15 eV, the first four being assigned to a sequence of n, π and σ orbitals of an SeCN group, based both on empirical correlations with simpler molecules and on semi-empirical and *ab initio* quantum mechanical calculations[66].

The structure of 4-nitrobenzyl selenocyanate and benzyl selenocyanate has been determined by X-ray crystallography[67]. They both exist in the *syn*-clinal (*gauche*) conformation in the solid state. More detailed results for the corresponding tellurocyanate are presented in Section IV.

Microwave spectral measurements were carried out using methyl and ethyl selenocyanates. For the main isotopic species of methyl selenocyanate (^{80}Se, ^{78}Se, ^{77}Se, ^{76}Se, ^{82}Se, ^{12}C and ^{13}C) a full centrifugal distortion analysis was reported[68]. Internal rotation splittings were analysed to give a barrier height to the methyl internal rotation of 1241 $\pm$ 50 and 1228 $\pm$ 50 cal mol^{-1} for the CH_3 and CD_3 species, respectively[69]. Stark measurements yielded dipole moments of $\mu = 4.42 \pm 0.05$ D for $CH_3{}^{80}SeCN$ and $\mu = 4.36 \pm 0.04$ D for $CD_3{}^{80}SeCN$[69]. Microwave spectroscopic studies of ethyl selenocyanate indicated that the preferred conformation of the linear SeCN group is *syn*-clinal[70].

A conformational study of *para*-substituted phenyl selenocyanates (RC_6H_4SeCN) was carried out by the determination of dipole moments and Kerr constants. The molecule is planar when R is an electron-donating group, while the coplanarity of the C_6H_4SeCN fragment is disputed when R is an electron-withdrawing group[71]. In solutions, *meta*-substituted selenocyanates (**12**) were a mixture of *syn* (**12b**) and *anti* (**12a**) forms; the proportion of the latter and the dihedral angle between the aromatic ring and the SeCN plane increased with increasing electron-withdrawing ability of R. *Ortho*-substituted phenyl selenocyanates (2-RC_6H_4SeCN; R = Me, Cl, Br, NO_2) existed as a planar or nearly planar *anti* conformers[72]. The coplanarity in the bis-selenocyanate [*p*-$(SeCN)_2C_6H_4$] was lower than in the corresponding monosubstituted compound[73].

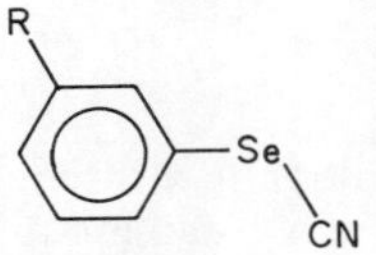

(**12a**)

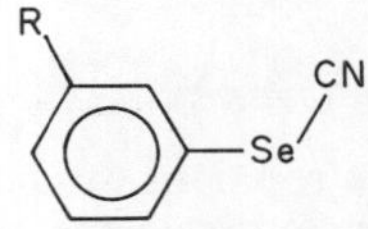

(**12b**)

R = Me, Cl, Br, NO_2

The mass spectrum of *m*-nitrophenyl selenocyanate was compared with that of the corresponding benzeneseleninic acid[74]. In the acid, a peak due to the diselenide was observed in addition to a very weak peak of a molecular ion. The selenocyanate showed an intense molecular ion peak and fragmentation peaks by loss of CN or NO_2, while the peak due to diselenide was of very low intensity. Mass spectral measurements on 2-(selenocyanatomethyl)-3-cyanobenzoselenophene were reported, together with the spectra of other 2,3-disubstituted benzoselenophenes[75].

Selenium-77 magnetic resonance is increasingly becoming a more popular spectroscopic technique. Technical developments of NMR spectroscopy, especially the use of Fourier transformation, have overcome the potential drawbacks such as the low natural abundance of the ^{77}Se isotope (7.58%) and an NMR sensitivity of 6.97×10^{-3} with respect to that of the proton at constant field. Hence ^{77}Se NMR spectra of a number of organoselenium compounds were determined, including organoselenocyanates[76–79]. A range of over 800 ppm was observed for the ^{77}Se chemical shift. The effect of the medium on the shielding of ^{77}Se chemical shifts was examined and it was found that the solvent exerted much less effect than the structure of the selenide[80]. It was reported that *ortho*-substitution by a carbonyl function had a highly deshielding effect on the ^{77}Se nucleus of aryl selenocyanates. In conjunction with the decrease in the carbonyl stretching frequencies in IR spectra, this was attributed to the intramolecular interaction between the carbonyl group and the selenium atom as depicted in **13**[81].

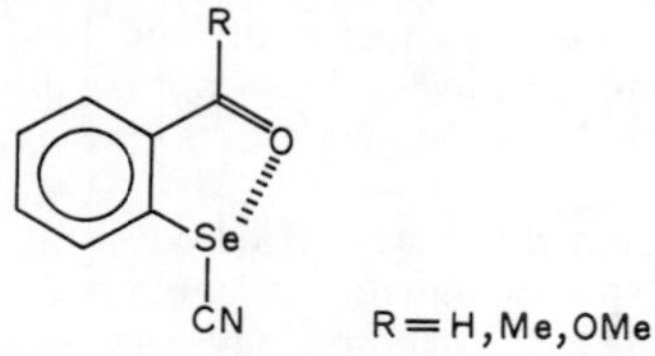

(13)

The coupling constants of ^{77}Se with α- and β-^{13}C atoms were determined in RC_6H_4SeCN (R = H, 2-MeO, 4-NH_2, 4-NO_2) and compared with those of selenides, diselenides and selenophene. The coupling constants were affected by the hybridization of the carbon atoms, the nature of the substituents and the orientation of the unshared electron pair of the selenium[82].

The ^{13}CNMR spectra of 2-$RCOC_6H_4SeCN$ (R = H, Me, MeO, EtO) were measured and compared with those of $RCOC_6H_5$ and 2-RC_6H_4SeX. In all cases except one (R = EtO), the phenyl carbon attached to a selenocyanato group was deshielded relative to the *ortho* carbon of $RCOC_6H_5$. The variation in the chemical shifts of the phenyl carbon with various X revealed the transmission of the inductive effect of X through the selenium atom[83].

C. Reactions

1. Additions to the C≡N bond

Some cyclization reactions can be regarded as addition to a carbon—nitrogen triple bond. Other cyclization reactions involving the loss of a cyano group are classfied into substitution reactions on a selenium atom and are described in Section II.C.2.

The reaction of hydrazoyl halides with potassium selenocyanate afforded the substitution products **14**, which cyclized spontaneously to 1,3,4-selenadiazole derivatives (**15**) (equation 27)[84]. Treatment of phenacyl selenocyanate with aryldiazonium salts afforded similar intermediates (**14**; R = PhCO) which underwent the same cyclization[85] as shown in equation 27. When this reaction was applied to the diazonium salts derived from anthranilic acid or its methyl ester, the 1,3,4-selenadiazole derivatives produced were susceptible to further cyclization to give 1,3,4-selenadiazolo[2,3-*b*]quinazoline derivatives (**16**) in good yields (equation 28)[86,87].

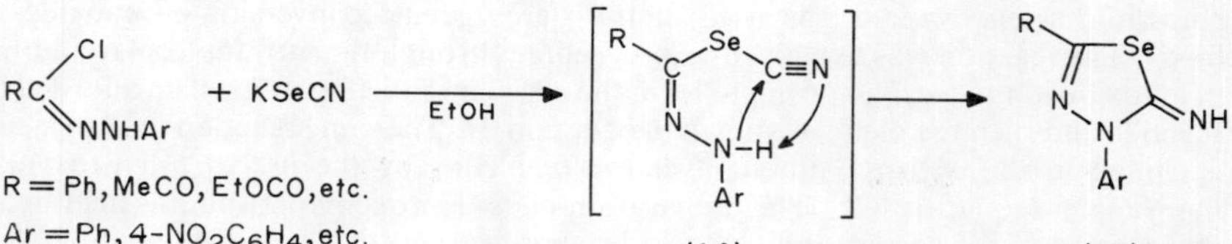

(14) (15)

(27)

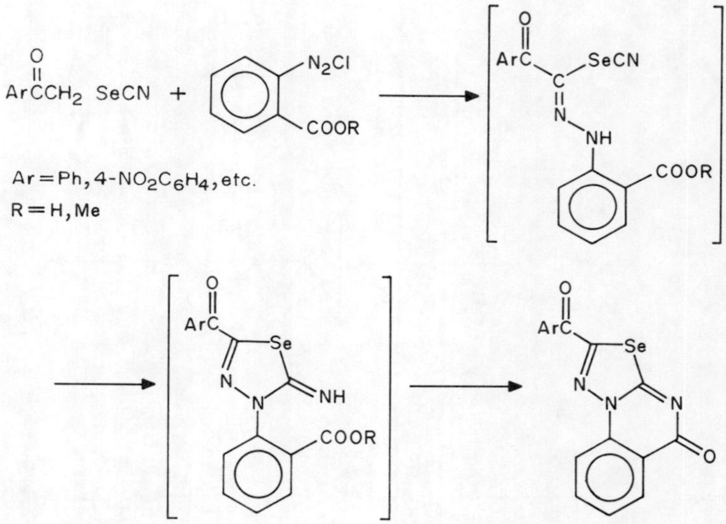

(16)

(28)

Intramolecular addition of an acyl carbon was also reported. The reaction of *cis-β*-selenocyanatoacrylic acid with phosphorus pentachloride produced the corresponding acid chloride, which cyclized to 2-chloro-1, 3-selenazin-4-one (equation 29)[51]. Cyclization of a thiazolidin-2-ylidene cyanoacetate derivative was also reported (equation 30)[88].

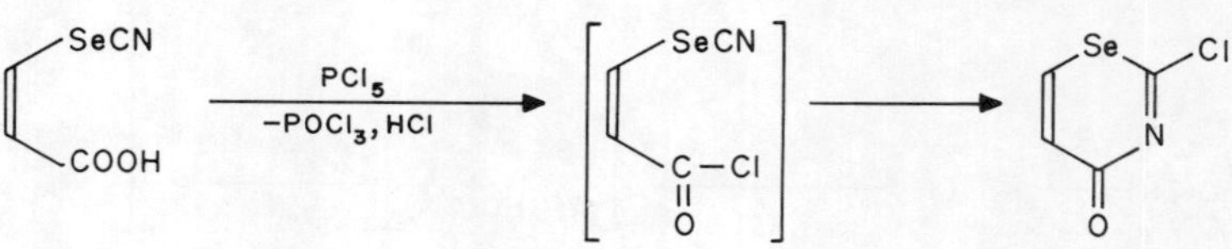

(29)

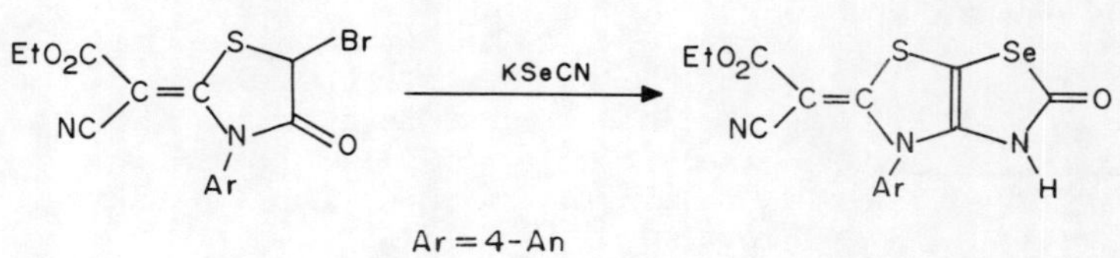

(30)

Ar = 4-An

Potassium selenocyanate was used in the stereospecific conversion of epoxides to olefins[89]. The reaction was explained as proceeding through the intramolecular addition of an alkoxide ion to a cyano group to form the cyclic intermediate **17**, which afforded the episelenide and then the olefin as shown in equation 31. The same reaction sequence may be included in the *cis–trans* interconversion of olefins by the use of β-hydroxyalkyl selenocyanates (equation 32)[7]. The preparation of *cis*-1-propenylphosphonic acid by the deoxygenation of the corresponding epoxide was also reported[90]. 2-*N*-Acetylimino-1, 3-oxaselenoles, which have closely related structures to the postulated intermediate **17**, were isolated by the reaction of *gem*-dicyano epoxides with potassium selenocyanate in acetic anhydride as solvent (equation 33)[91].

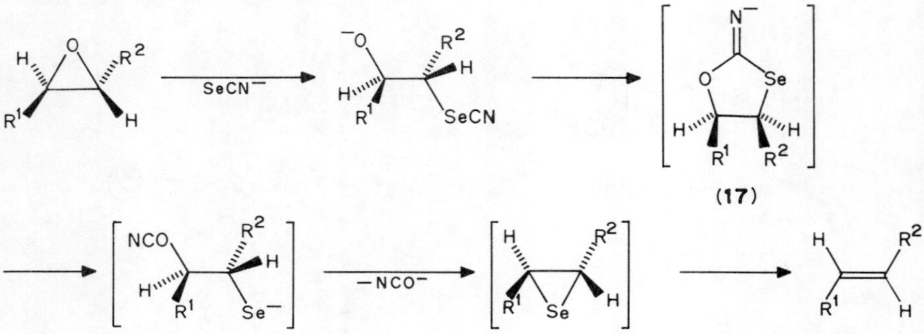

(31)

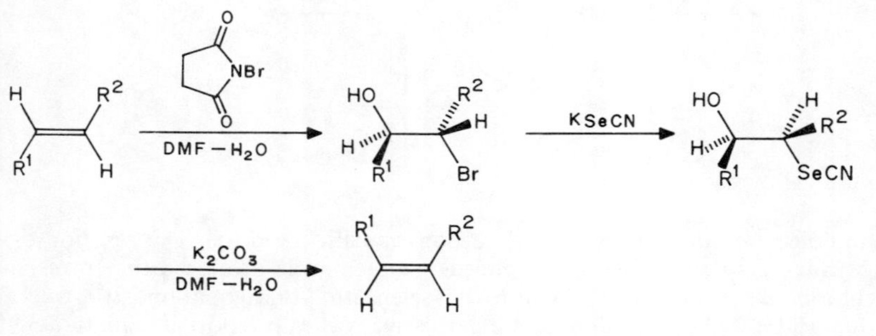

(32)

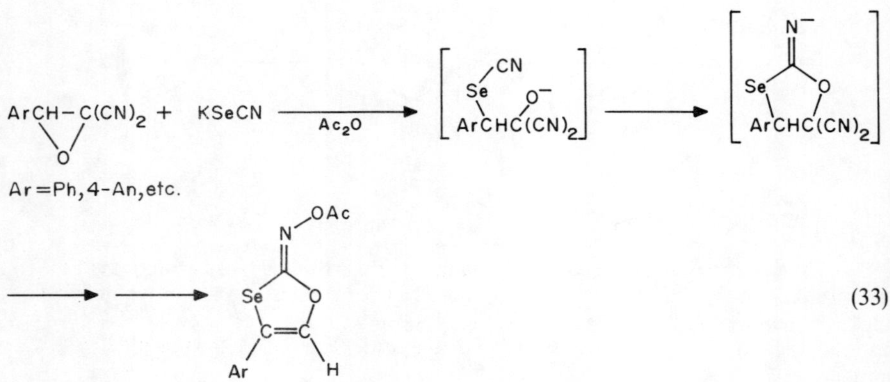

(33)

2. Substitution of the CN group

a. Substitution by heteroatoms. By the reaction of aryl selenocyanates with a halogen such as chlorine[37] or bromine[34,35], a cyano group was replaced by the halogen to afford arylselenenyl halides quantitatively, and the corresponding trihalides were formed when two equivalents of a halogen were employed (equation 34)[11].

$$\mathrm{ArSeCN} + \mathrm{X_2} \longrightarrow \mathrm{ArSeX} + \mathrm{XCN}$$

$$\mathrm{ArSeCN} + 2\,\mathrm{X_2} \longrightarrow \mathrm{ArSeX_3} + \mathrm{XCN} \qquad (34)$$

$$\mathrm{X} = \mathrm{Cl, Br}$$

The substitution of a cyano group by a nitrogen atom was also observed. An unsaturated *N, N*-dimethyliminium salt was treated successively with sodium selenocyanate and an arylamine to afford a 1, 2-selenazolinium salt (equation 35)[92]. The final step was intramolecular substitution of a CN group by a nitrogen atom of the imine. A nitrogen atom of the oxime was also reported to replace the cyano group of aryl selenocyanate to afford cyclization to a benzoselenazole derivative (equation 36)[36].

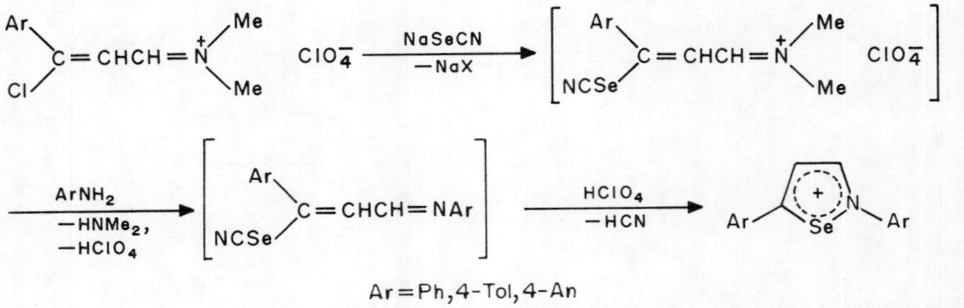

(35)

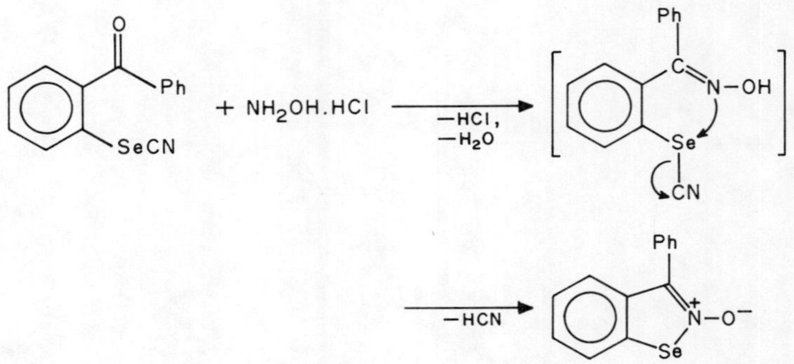

(36)

The reactions of *o*-nitrophenyl selenocyanate with arylthiols were investigated under anhydrous conditions in a nitrogen atmosphere[93]. The products were found to be selenide sulphides (equation 37). When *p*-nitrophenyl selenocyanate was used as the starting material, however, the product was diphenyl diselenide. The reason for this difference was not clarified.

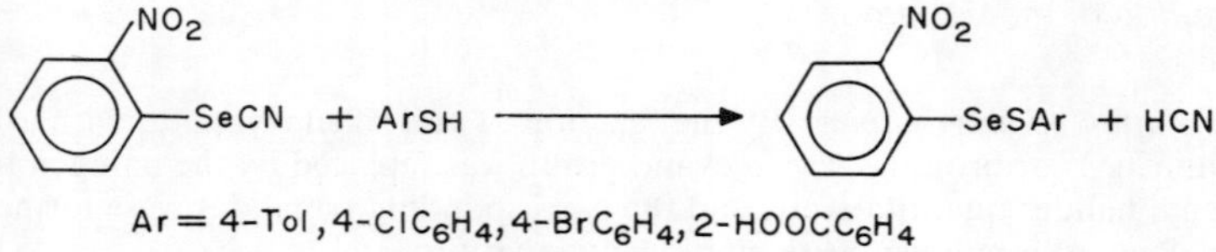

(37)

Selenocyanates are easily converted to diselenides in various procedures (equation 38). They include the reduction selenol and subsequent oxidation ($NaBH_4$–air[22], $Na_2S_2O_4$–H_2O_2[94], Na_2S–air[47]) and hydrolysis under acidic[56] or basic[3,21] conditions. Air oxidation might take part in the last reactions. This reaction was utilized in the preparation of diselenide polymers from 4,4′-bis(selenocyanato)azobenzene[95] or -benzene[33] for the study of photoreactions of the polymers. For the preparation of selenacyclophanes, hydrolysis of a selenocyanate shown in equation 39 was adopted, since the intermediate selenocyanates can easily be obtained and purified. The hydrolysis was carried out at high dilution to avoid the formation of linear polymeric diselenides[96].

$$\text{RSeCN} \longrightarrow \text{RSeSeR} \tag{38}$$

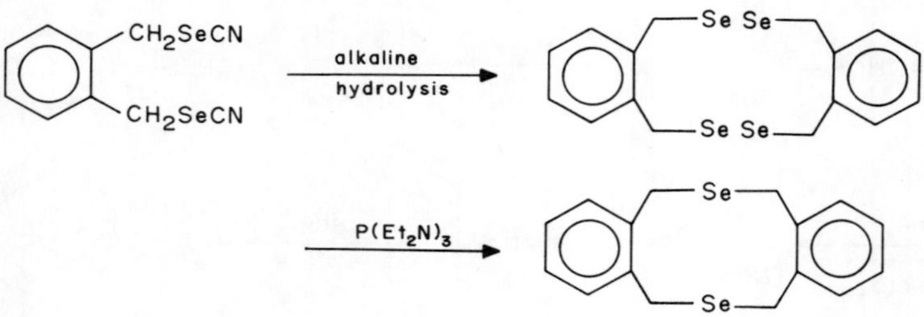

(39)

Synthesis of 1, 1, 4, 4-tetramethyl-1*H*, 4*H*-selenolo[3, 4-*c*]selenophene was reported using allenyl selenocyanate as a starting material[52]. The first step of this reaction was the formation of diselenide. Subsequent two isomerization reactions afforded the product as shown in equation 40.

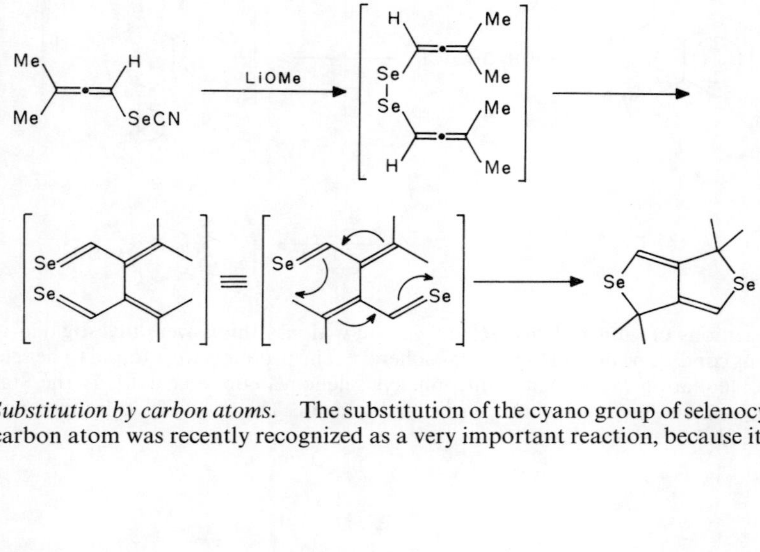

(40)

b. Substitution by carbon atoms. The substitution of the cyano group of selenocyanates by a carbon atom was recently recognized as a very important reaction, because it results in the introduction of an organoselenium group into the carbon atom. This reaction has usually been carried out through reduction of selenocyanates and subsequent reaction with alkyl halides or with their equivalents. Direct substitution by a nucleophilic carbon (a carbanion or its equivalents) is also possible. These reactions are summarized in this section. Two mechanistically different reactions are described in Sections II.C.2.c and d.

As an arylseleno group was usually used as a leaving group in selenoxide elimination reactions[97], many examples of the conversion of aryl selenocyanates to alkyl aryl selenides have been reported. By the reduction of aryl selenocyanates with sodium borohydride, sodium arylselenolates were formed. The subsequent reaction with alkyl halides[98–103] or epoxides[104] afforded alkyl aryl selenides in almost quantitative yields (equation 41). Polymer-bound phenyl selenocyanate was alkylated in a similar manner[39]. This reaction was utilized in the conversion of ethyl 2-bromobutyrate to ethyl crotonate by a combination with selenoxide elimination. After the elimination, the polymer-bound selenium reagent was easily removed by filtration. When benzoyl chloride was used instead of alkyl halide in reaction 41, aryl selenobenzoate was produced[105]. By the reaction of 4-methyl-2-nitrophenyl selenocyanate with sodium borohydride and then with propargyl bromide, aryl propargyl selenide was produced (equation 42)[106]. The difference between this result and that in equation (22) is interesting.

$$\mathrm{ArSeCN} \xrightarrow[\text{or } \mathrm{Na_2S}]{\mathrm{NaBH_4}} [\mathrm{ArSeNa}] \xrightarrow{\mathrm{RX}} \mathrm{ArSeR} \qquad (41)$$

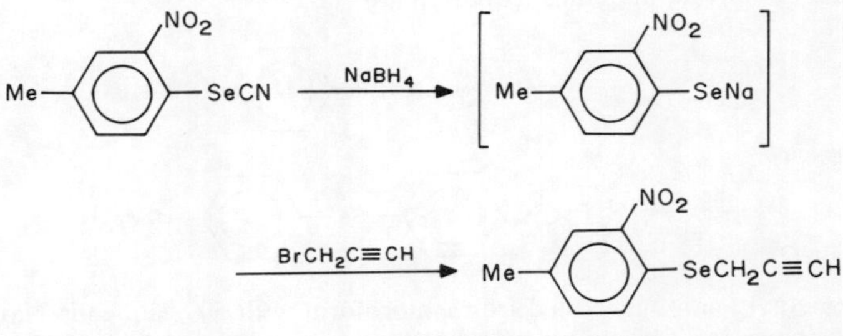

(42)

The order of reduction of selenocyanates and addition of halides can be altered. By the addition of a mixture of benzyl selenocyanate and benzyl halide to sodium borohydride in ethanol, dibenzyl selenide was produced almost quantitatively[107]. A similar reaction was carried out under high dilution with two doubly substituted reagents and diselena[3, 3]cyclophanes were prepared[108] in yields more than ten times higher than those in the previous procedure[96] (equation 43).

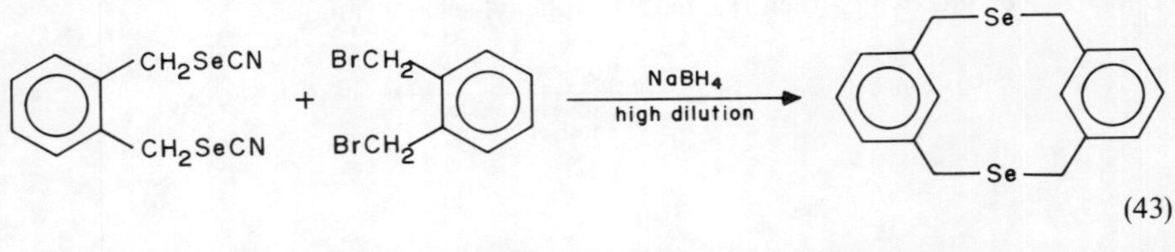

(43)

If necessary, it is possible to isolate the intermediate selenium species as a selenol. Thus, benzeneselenol and hexaneselenol were isolated by distillation under nitrogen after quenching by the addition of aqueous hydrogen chloride, the reaction mixture of the selenocyanates with sodium borohydride[109].

The use of reducing agents other than sodium borohydride was also reported. 2-Selenocyanatothiophene or selenophene derivatives (equation 17) were alkylated on the selenium atom after reduction with sodium sulphide[46,47]. 1, 2-Diselenocyanatoethane was reduced by hypophosphorous acid to 1, 2-ethanediselenol, which was then treated with thiocarbonyldiimidazole to afford 1, 3-diselenolane-2-thione (**18**) (equation 44)[3,110].

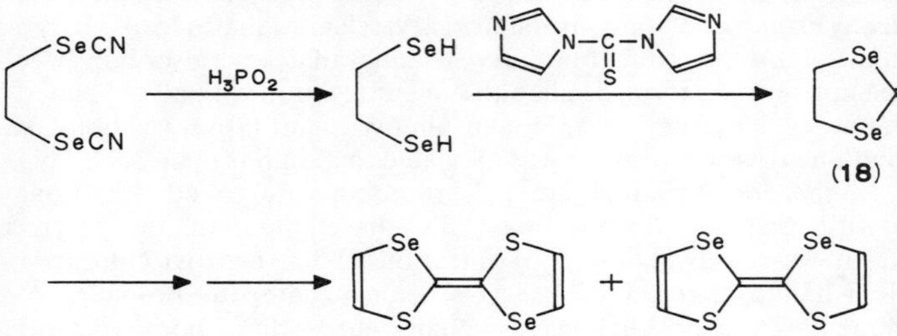

(44)

Compound **18** is an important precursor of *sym*-(*E*/*Z*)-diselenadithiafulvalene, the conductivity of whose TCNQ complex was shown to be $700 \pm 300\,\Omega^{-1}$ at room temperature.

Aryllithium or arylmagnesium compounds are known to react with aryl selenocyanates to afford diaryl selenides[1]. This reaction was applied to the preparation of polymer-bound diaryl selenide using 1% cross-linked polystyryllithium and *p*-methoxyphenyl selenocyanate as starting materials (equation 45)[111].

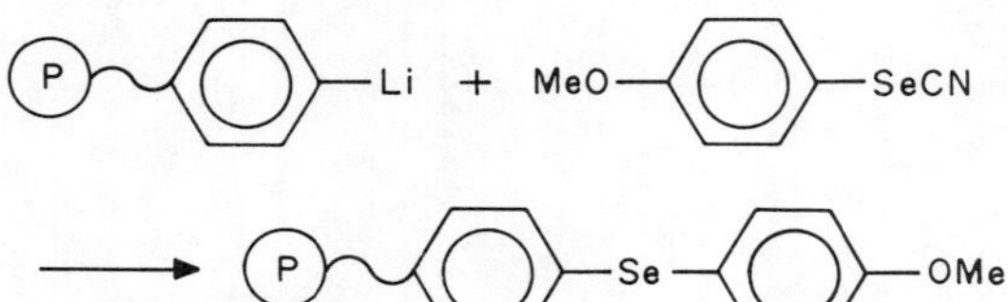

(45)

Trichloromethyl anion, generated from chloroform with 50% aqueous NaOH in the presence of a phase-transfer catalyst, was shown to react with benzyl selenocyanate to give the substitution product (equation 46)[112]. Benzyl trichloromethyl selenide is a good precursor of selenium-stabilized carbenes and was used in the preparation of the (5,10,15, 20-tetraphenylporphinato)iron(II)–selenocarbonyl complex (equation 47)[112]. The substitution product was also obtained by the reaction of phenyl selenocyanate with terminal acetylenes in the presence of triethylamine and copper(I) catalyst (equation 48)[113]. The role of the copper(I) catalyst was explained to be the formation of copper(I) acetylide, which then reacted with phenyl selenocyanate.

$$PhCH_2SeCN + CHCl_3 \xrightarrow[\text{phase-transfer catalyst}]{50\%\ \text{aq. NaOH}} PhCH_2SeCCl_3 \qquad (46)$$

$$Fe(TPP) + PhCH_2SeCCl_3 \xrightarrow[-Cl_2]{Na_2S_2O_4-H_2O} Fe(TPP)[C(Cl)SeCH_2Ph]$$

$$\xrightarrow[-PhCH_2Cl]{FeCl_2} Fe(TPP)(C{=}Se) \qquad (47)$$

$$PhSeCN + HC\equiv CR \xrightarrow[\text{CuBr cat.}]{\text{2 equiv. }Et_3N} PhC\equiv CR \qquad (48)$$

The substitution of a cyano group by an intramolecular carbon nucleophile has been reported. Treatment of 2-acetyl-3-selenocyanatothiophene with ammonia afforded a bicyclic compound (equation 49)[39]. The reaction of 2-acetylphenyl selenocyanate with phenylhydrazine produced benzoselenopheno[3, 2-*b*]indole through the Fischer indole synthesis and cyclization by carbon—selenium bond formation (equation 50)[36].

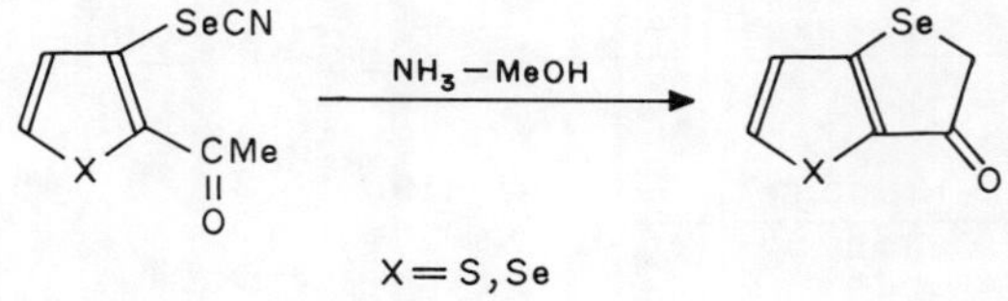

(49)

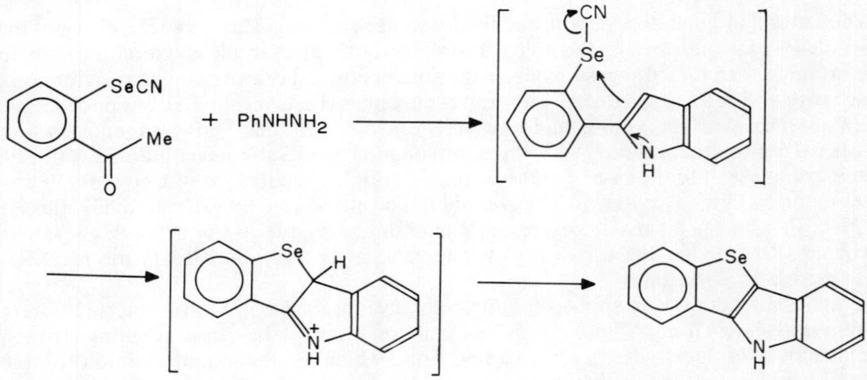

(50)

c. *Conversion of alcohol to selenide.* In 1976, Grieco *et al.*[114] reported the reaction of o-nitrophenyl selenocyanate with alcohols in the presence of tributylphosphine to afford the alkyl *o*-nitrophenyl selenides in almost quantitative yields (equation 51). Formally, this reaction involves the substitution of a cyano group by an alkyl carbon on the selenium atom, as already mentioned in the previous section. Actually, this reaction seems to proceed through the substitution by a phosphine atom as a first step as shown in Scheme 1, although the formation of Bu_3P^+ $CNArSe^-$ in the initial step cannot be ruled out.

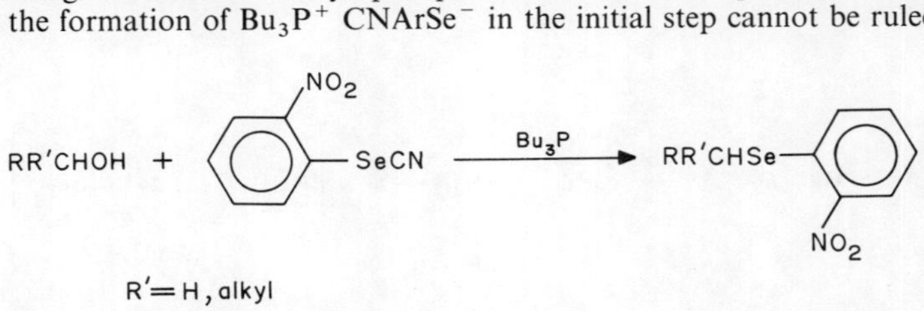

(51)

Later, the stereochemistry of this reaction was studied using secondary alcohols as substrates[115] (Scheme 2). This reaction proceeds with inversion of configuration on the

$$ArSeCN + Bu_3P \longrightarrow ArSe\overset{+}{P}Bu_3\ CN^-$$

$$ArSe\overset{+}{P}Bu_3\ CN^- + RCH_2OH \longrightarrow RCH_2O\overset{+}{P}Bu_3 + HCN + ArSe^-$$

$$RCH_2O\overset{+}{P}Bu_3 + ArSe^- \longrightarrow RCH_2SeAr + Bu_3P{=}O$$

SCHEME 1

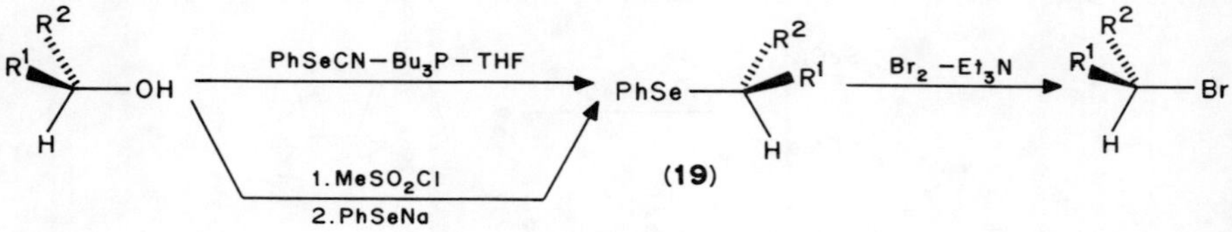

SCHEME 2

carbon atom, in good accord with the final step in Scheme 1. The stereochemistry of the selenide **19** was confirmed by the preparation of an authentic sample and its conversion to the bromide with the known stereochemistry (inversion in both steps). The reaction took place with aryl selenocyanates (phenyl and *o*-nitrophenyl selenocyanate), but poor results were obtained with diaryl diselenides or arylselenenyl chlorides[114]. It was reported that *N*-phenylselenophthalimide (N-PSP) gave analogous results to those obtained with aryl selenocyanates[116]. In the case of the secondary alcohol group in a steroid, however, it was later reported that an arylseleno group could not be introduced into the steroid by the use of N-PSP, whereas a satisfactory result was obtained by the use of aryl selenocyanate (Scheme 3)[117]. Hence, the use of aryl selenocyanates is recommended in the reactions described in this section.

When the reaction was applied to aldehydes, cyanoselenenylation products (**20**) were obtained as shown in equation (52)[118]. This reaction was explained as proceeding through the reaction of aldehydes with a cyano ion, which is the counter anion of the selenophosphonium salt, to afford the cyanohydrin intermediate (**21**) (equation 53). When ketones were used instead of aldehydes, cyanohydrins were isolated as products, presumably owing to the steric hindrance in the nucleophilic substitution by selenium in the final step of the reaction. Carboxylic acids also reacted with phenyl selenocyanate in the presence of tributylphosphine to give the benzeneselenol esters (equation 54)[119].

$$RCHO + ArSeCN \xrightarrow[-Bu_3PO]{Bu_3P} RCH(CN)(SeAr) \quad \textbf{(20)} \tag{52}$$

$$[ArSe\overset{+}{P}Bu_3\ CN^-] + RCHO \longrightarrow \left[RCH(CN)(O^-)\ .\ ArSe\overset{+}{P}Bu_3 \right] \quad \textbf{(21)} \tag{53}$$

$$RC(=O)OH + PhSeCN \xrightarrow{Bu_3P} RC(=O)SePh \tag{54}$$

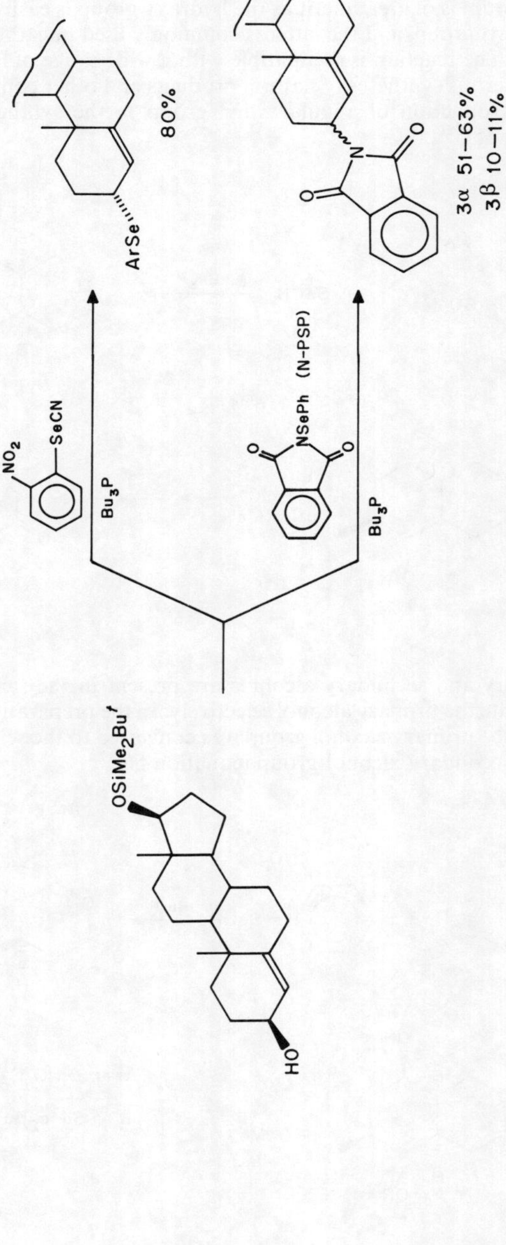

$OSiMe_2Bu^t$
HO
NO_2
SeCN
Bu_3P
NSePh (N-PSP)
Bu_3P
ArSe
80%
N
3α 51–63%
3β 10–11%

SCHEME 3

When the reaction 51 is combined with selenoxide elimination[97], these reactions constitute a very mild and elegant procedure for the introduction of a double bond into organic molecules. For faster rates and better yields e.g. of the terminal olefins *o*- or *p*-nitrophenyl selenocyanate was favourably used rather than phenyl selenocyanate. Considering the carbon atom, reaction 51 is the substitution of the hydroxy group by an arylseleno group. This direct substitution of the hydroxy group (without derivatization to a different leaving group) is of great merit as the hydroxy group is easily produced by the reduction of a carbonyl group, and the latter is commonly used in carbon—carbon bond formation reactions. The reaction is compatible with a wide range of functional groups, and was used in numerous syntheses of natural products and other compounds. The first example was the introduction of angular vinyl group in the synthesis of (±)-deoxyvernolepin (equation 55)[120].

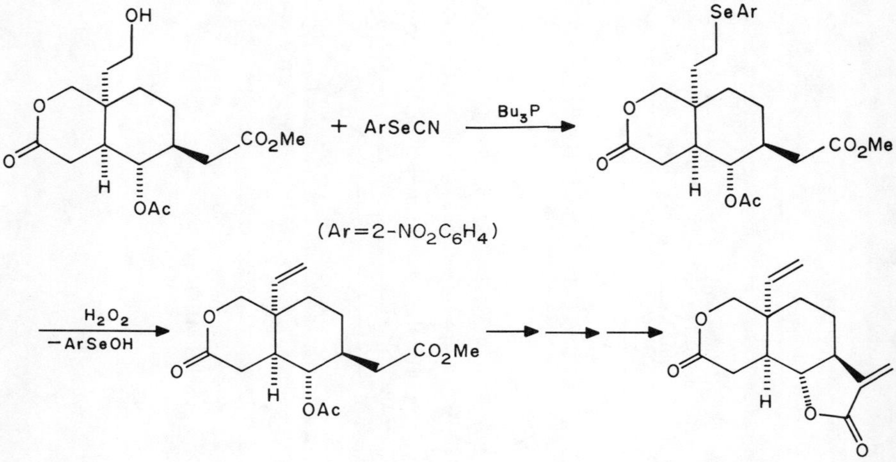

(55)

When both primary and secondary alcohols are present in the same molecule, it is possible to selenenylate the primary alcohol selectively. In the preparation of 8-deoxy-9*β*-hydroxyvernolepin, the primary alcohol group was converted to the selenide without the need to protect the secondary alcohol group (equation 56)[121].

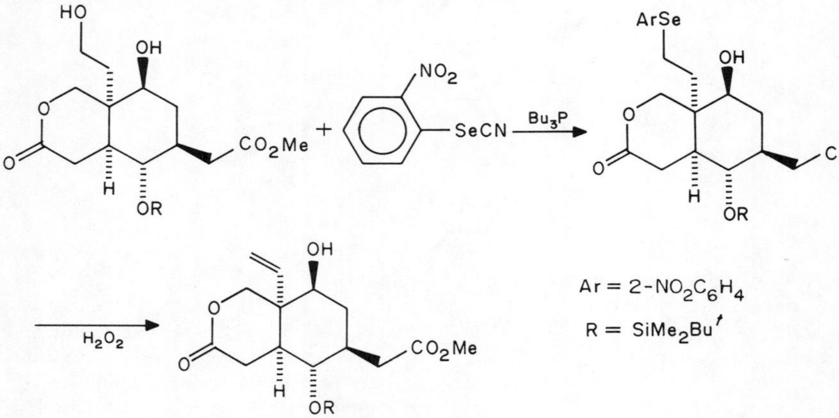

(56)

During the total synthesis of estrone, a diol (**22**) was treated with 1.5 equiv. of *o*-nitrophenyl selenocyanate to afford the selenide **23** selectively (equation 57)[122]. In diols, **24**[123], **25**[124] and **26**[125], again only the primary alcohol groups were converted to the selenides using the same procedure. In certain cases, discrimination between two primary alcohols is also possible, the more hindered alcohol being left intact. A good example was found in the synthesis of the alkaloid (±)-antirhine (equation 58)[126].

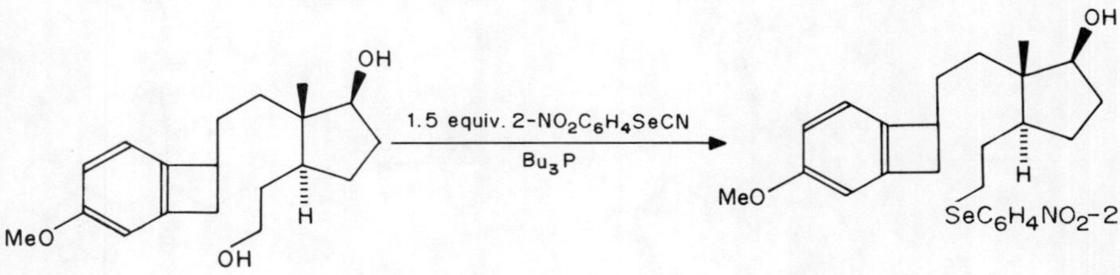

(**22**) (**23**)

(57)

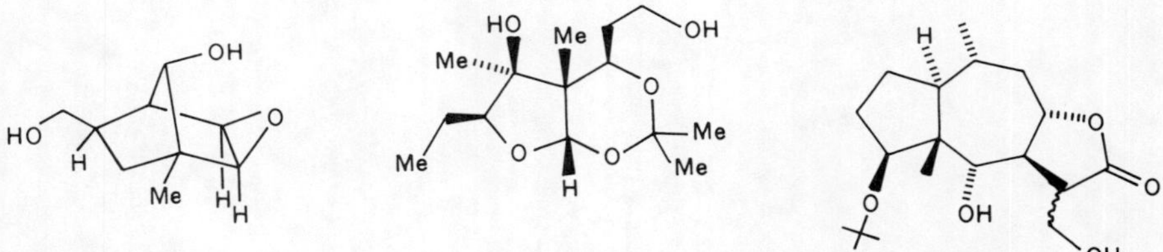

(**24**) (**25**) (**26**)

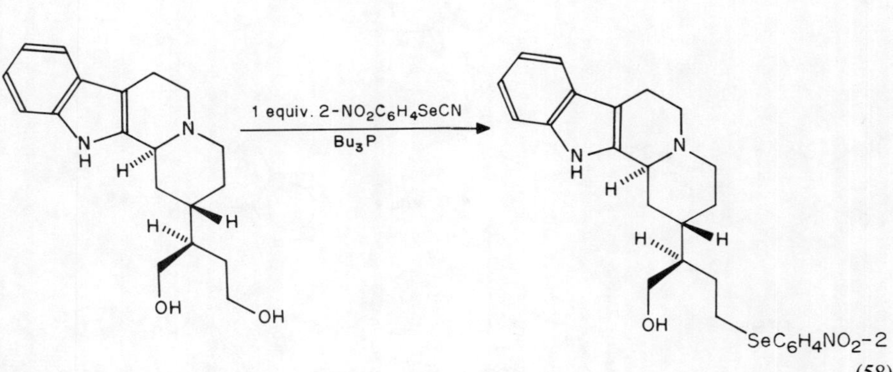

(58)

In the total synthesis of (+)-costunolide, it was reported that diselenenylation was not possible even by the use of excess of aryl selenocyanate. Although the difference was only in the relative position of a methyl substituent (β or γ), the hydroxy group with the methyl group in the γ-position was converted to the selenide selectively. Thus, the desired diene **27** was prepared by stepwise selenenylation and elimination (equation 59)[127]. A similar stepwise procedure was adopted in the conversion of the diol **28** to the diene in the total synthesis of temisin[128]. Later, a closely related diol (**29**) was treated with 4.8 equiv. of *o*-nitrophenyl selenocyanate for a much longer time to afford a mixture of di- and monoselenenylated products in 79% and 13% yields respectively (equation 60)[129]. A similar

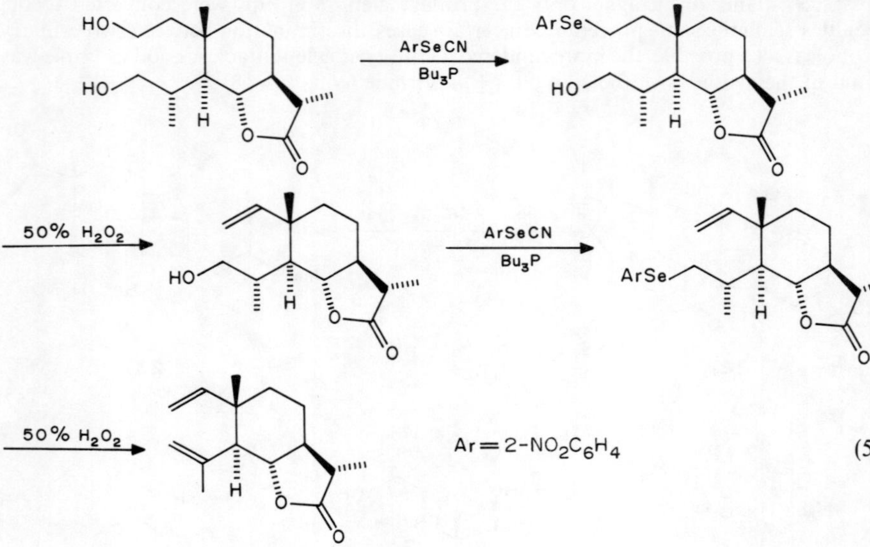

(27)

Ar = 2-$NO_2C_6H_4$ (59)

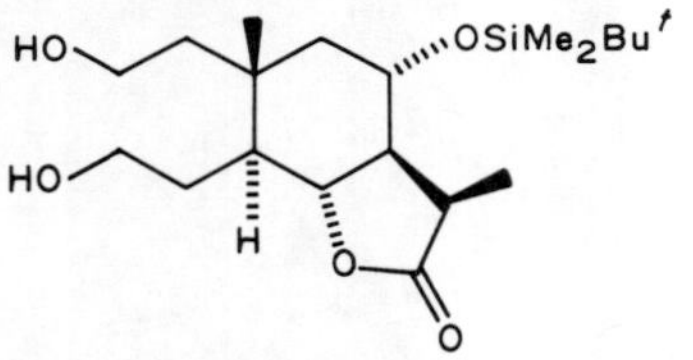

(28)

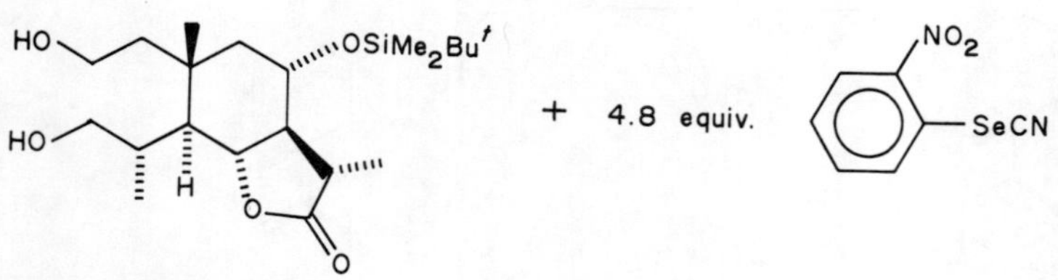

(29)

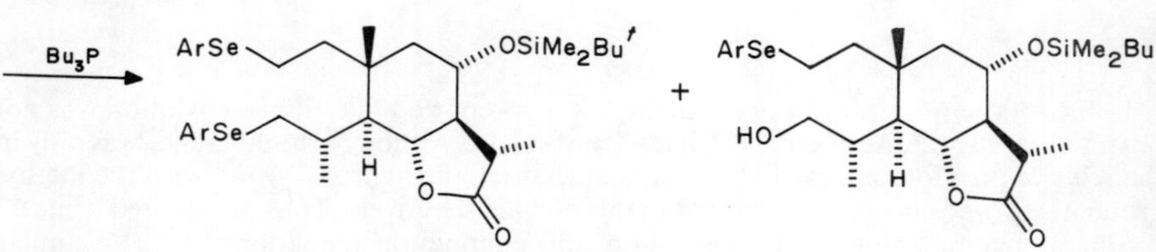

79% 13% (60)

mixture was obtained from the diol **30** (R = Me)[130], but in the cases of simpler diols such as **30** (R = H)[130], **31**[131] and **32**[132], it was not difficult to displace both OH groups in a one-step reaction. As expected, the tertiary OH group in **32** was not converted to the selenide.

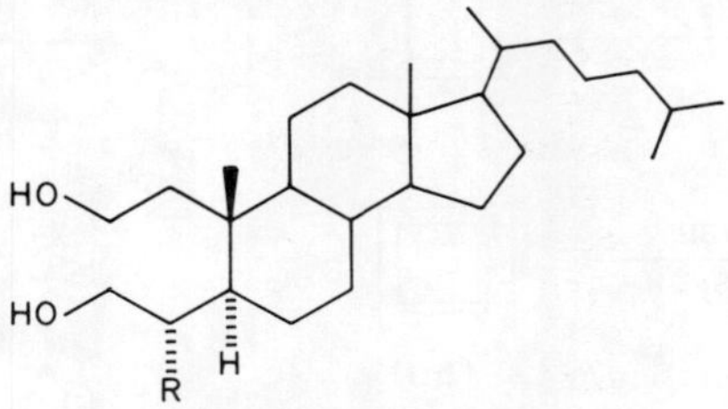

(**30**) R = H, Me

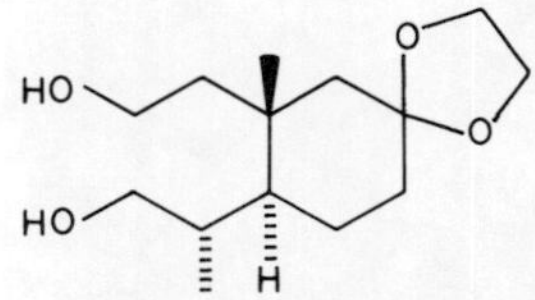

(**31**)

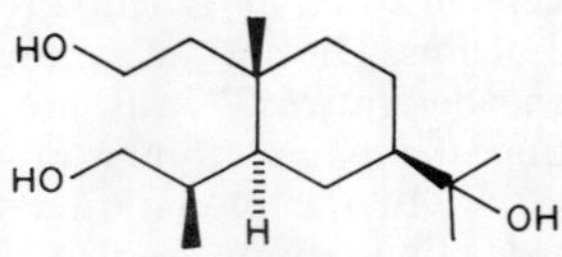

(**32**)

Although the reaction is slower, a secondary OH group can be replaced by an arylseleno group using this procedure, as already mentioned in Scheme 2 and 3. Another example was found in the conversion of benzyl alcohol derivatives to β-aminostyrene derivatives (**33**) (equation 61)[133]. This reaction was utilized in the synthesis of cyclic and acyclic peptide alkaloids, zizyphine A (**34**)[133] and celenamide A (**35**)[134], respectively (the double bonds introduced are shown by arrows).

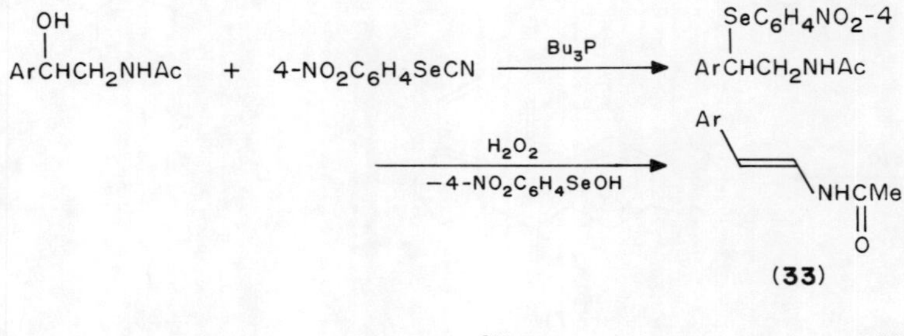

(61)

(**33**)

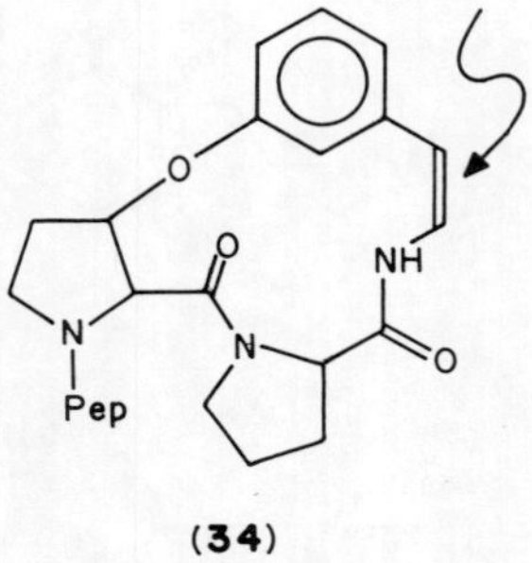

(**34**)

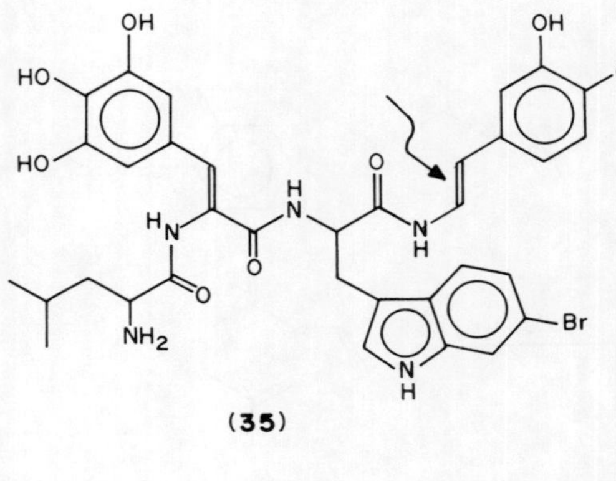

(**35**)

The preparation of thermally unstable oxete was also carried out using this procedure (equation 62)[135]. The formation of **36** was detected by NMR spectroscopy.

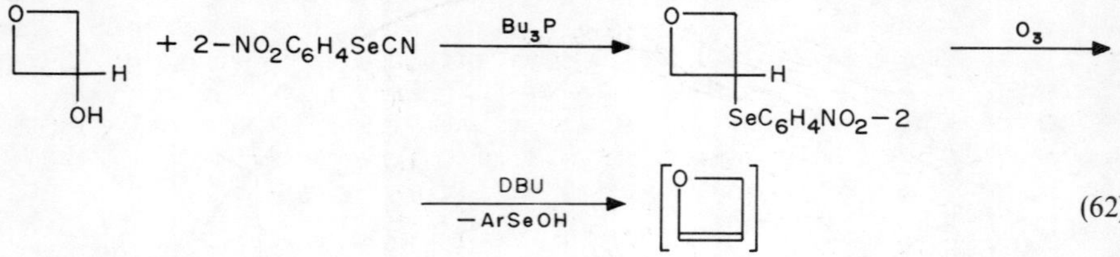

(62)

(**36**)

In other numerous examples of this double bond-forming reaction, primary alcohols were used as substrates. Replacement by an arylseleno group and subsequent selenoxide elimination produced terminal olefins. During the synthesis of the skeleton of lycorine alkaloids, the 1-aminobutadiene derivative **37** was prepared by this selenenylation–elimination sequence, indicating the compatibility of this procedure with reactive functional groups (equation 63)[136]. In the preparation of the irregular monoterpene (*R*)-santolinatriene, the selenoxide elimination was also carried out in the presence of a non-conjugated diene group in the molecule (equation 64)[137].

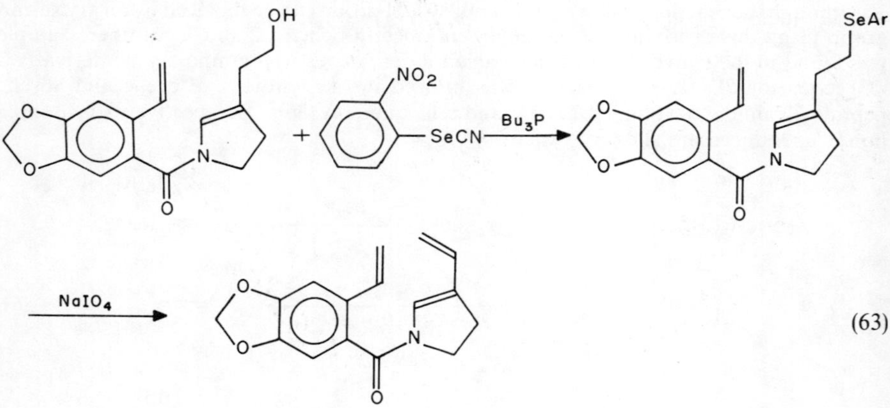

(63)

(**37**)

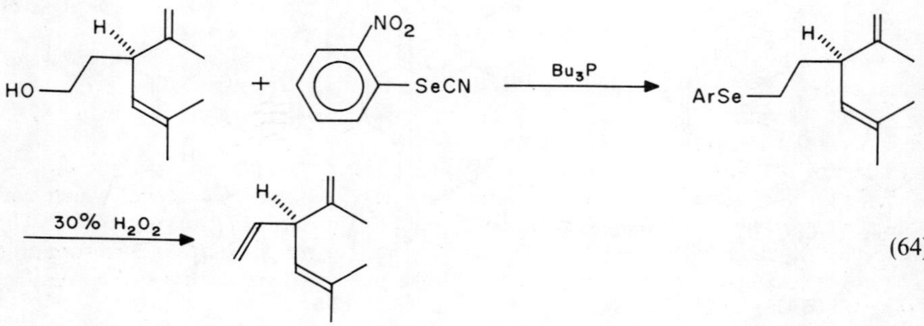

(64)

Terminal double bonds were introduced using this procedure in the syntheses of estradiol[138,139], the sex pheromone of the pine sawflies[140], *dl*-3β-bromo-8-epicaparrapi oxide[141], the ionophore calimycin (A-23187)[142], (+)-8-deoxyvernolepin[143,144], erythromycin[145], Prelog–Djerassi lactone[146], monoterpene, semburin and isosemburin[147], the wing gland pheromone of the African sugar-cane borer[148] and miyaginin, *p*-allylphenyl glycoside[149].

This procedure was also utilized in the preparation of the starting materials for the studies of the tandem Cope–Claisen rearrangement, namely **38** (equation 65)[150], and for the photochemical rearrangement, namely **39**[151].

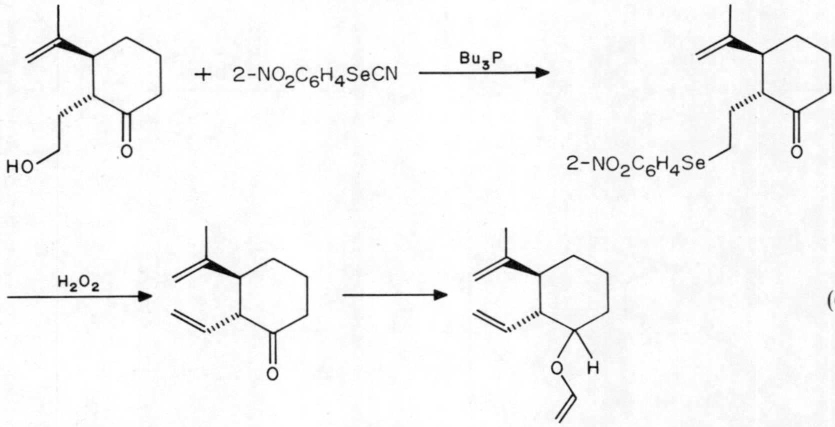

(65)

(38)

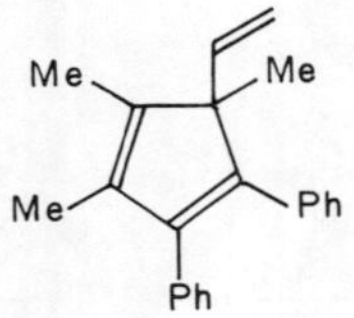

(39)

The procedure was also effective for the preparation of an exocyclic double bond. For example, in the total synthesis of ($\pm$)-β-chamigrene, the ester function was converted to an exocyclic methylene group in three steps, viz. reduction, selenenylation and elimination (equation 66)[152]. Similar procedures were used in the total syntheses of pederin[153,154], ($\pm$)-atractylon and ($\pm$)-lindestrene[155]. The exocyclic double bonds of **40**[156] and **41**[157] were also prepared using this procedure.

In the case of highly strained polycyclic compounds, the procedure was also shown to be effective for the introduction of exocyclic methylene groups. Isocomene, which was isolated from a toxic plant, was synthesized using this procedure (equation 67)[158]. The polycyclic alcohols **42**[159], **43**[160] and **44**[161] were converted to the corresponding exomethylene compounds by treatment with *o*-nitrophenyl selenocyanate and subsequent oxidation (ozone or 30% H_2O_2).

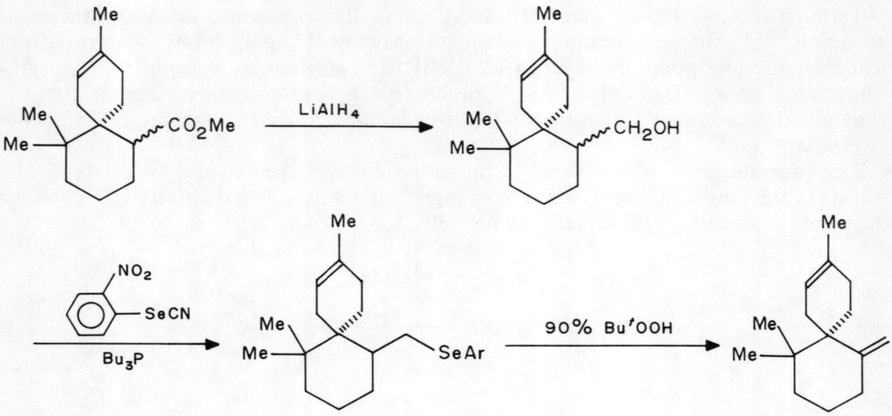

(66)

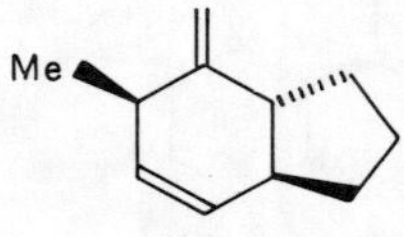

(40)

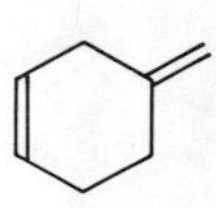

(41)

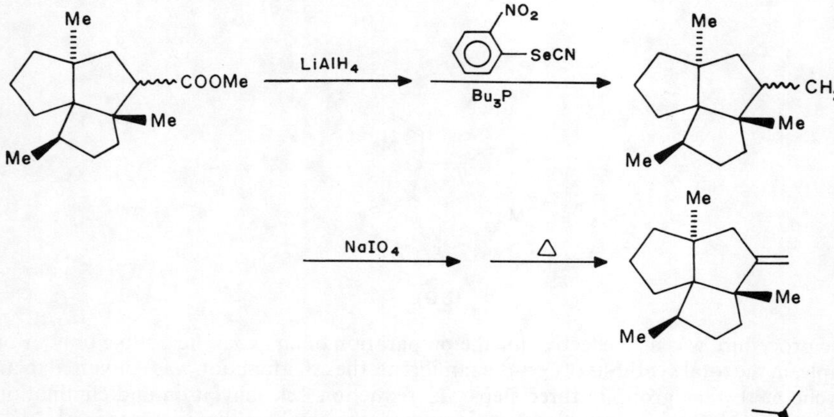

(67)

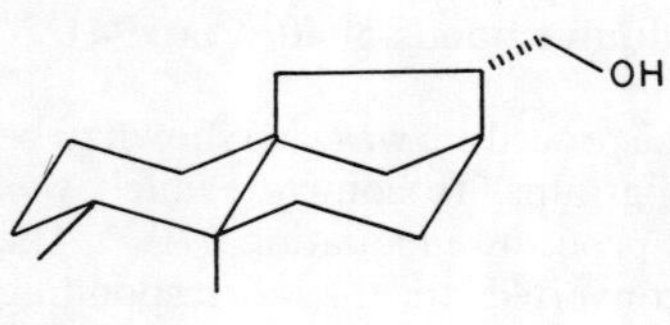

(42)

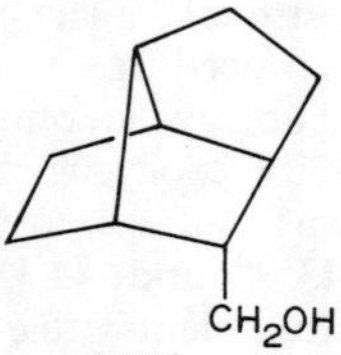

(43)

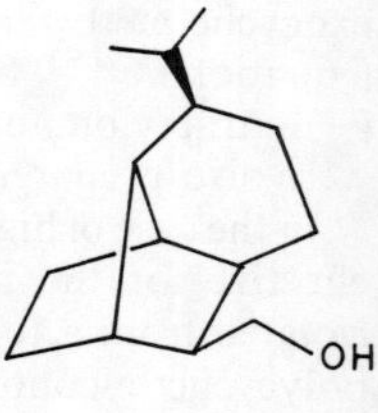

(44)

The formation of an exocyclic double bond conjugated with an endocyclic double bond was also achieved similarly. Examples are found in the total syntheses of (±)-periplanone-B[162], which is the sex excitant pheromone of the American cockroach (equation 68), and in the total synthesis of the sesquiterpene (±)-silphiperfol-6-ene (equation 69)[163].

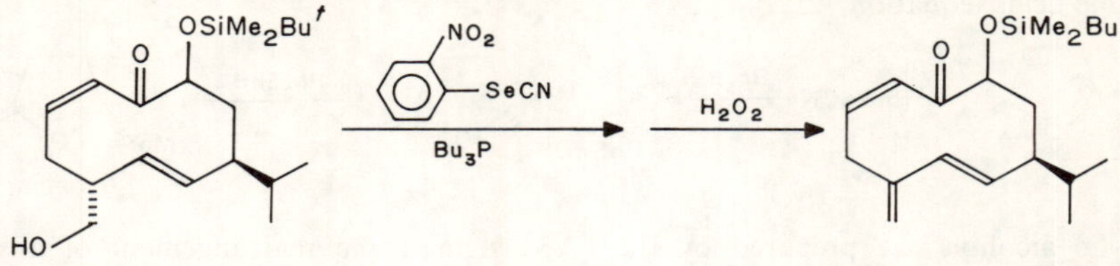

(68)

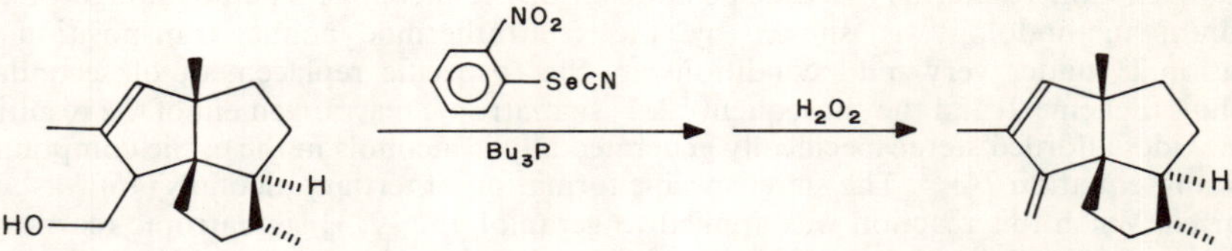

(69)

An elegant application is the preparation of α-(methylenecyclopropyl)glycine, which shows significant biological activity (equation 70)[164].

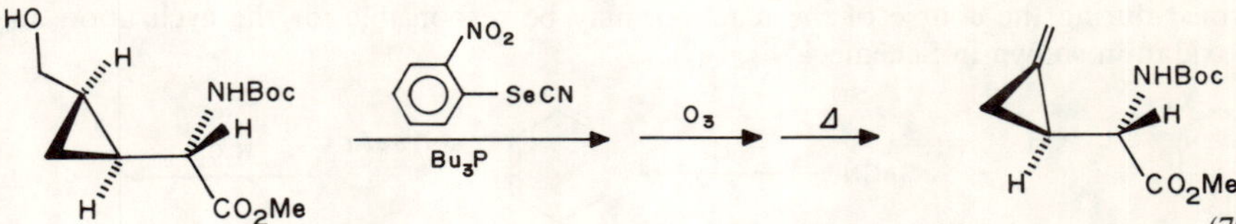

(70)

In carbohydrate chemistry the selenenylation–elimination sequence was used in the preparation of a key intermediate (**45**) for the synthesis of antibiotics related to angustmycin A and nucleocidine (equation 71)[165–168]. This reaction was also carried out without protection of the secondary OH groups.

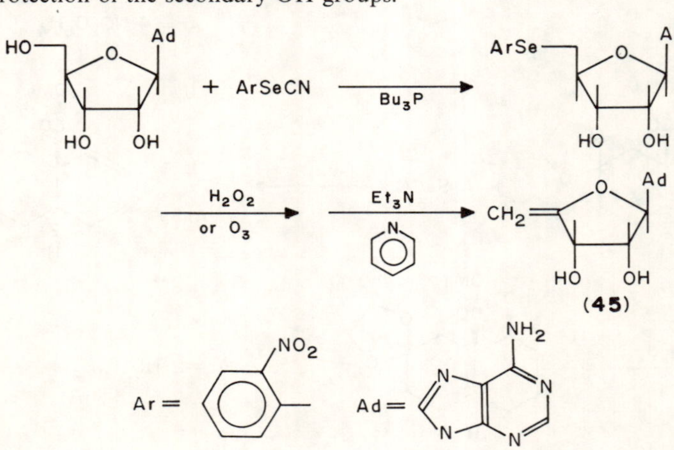

(71)

(**45**)

When conversion of an alcohol to a selenide is combined with reductive removal of the introduced arylseleno group, the overall reaction is the replacement of a hydroxy group by hydrogen[169]. This was carried out in the presence of a lactone group in the preparation of (*R*)- and (*S*)-α-benzylidene-γ-butyrolactone from the readily available (*S*)- and (*R*)-glutamic acids (equation 72)[170].

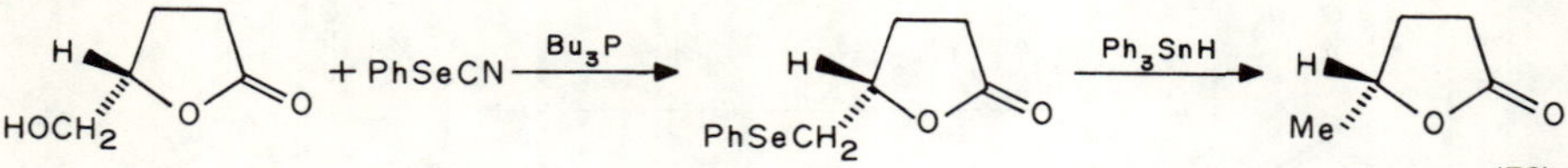

(72)

Allylic alcohols are prepared by the [2, 3] sigmatropic rearrangement of allylic selenoxides[97]. If allylic selenides are prepared using the so far described reaction, the overall reaction results in the transposition of allylic alcohols. The advantage of this selenium methodology is shown by the contrathermodynamic transposition in equation 73 under very mild conditions[171]. Stereospecific replacement of secondary alcohols (Scheme 2) and the subsequent [2, 3] sigmatropic rearrangement of the resulting selenoxides afforded stereospecifically generated allylic alcohols in the cyclic compounds shown in equation 74[172]. The stereospecific formation of tertiary alcohols (**46**) was also reported. When this reaction was applied to geraniol, the [2, 3] sigmatropic rearrangement was accompanied by epoxidation of the isolated double bond to afford **47** in the presence of pyridine. In the absence of pyridine, the product was found to be a tetrahydrofuran derivative (**48**), which was independently prepared from **47** by the action of *p*-toluenesulphonic acid (Scheme 4)[173,174]. Arylseleninic acid or its peracid, which were formed during the course of the reaction, may be responsible for the cyclization and epoxidation shown in Scheme 4.

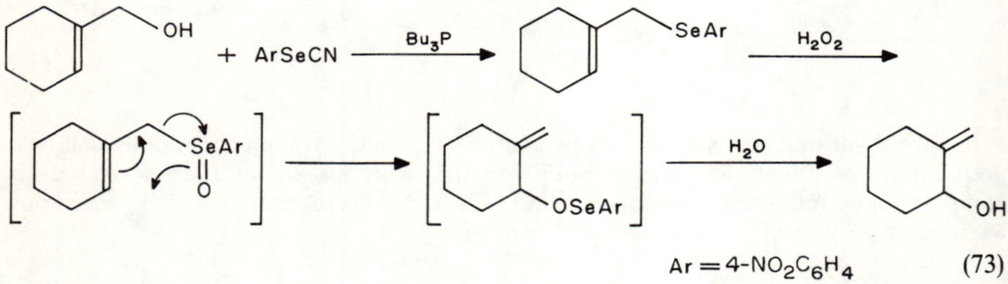

(73)

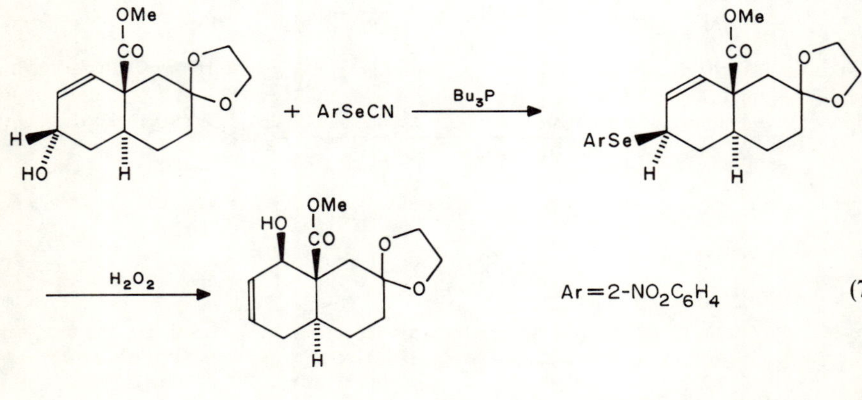

(74)

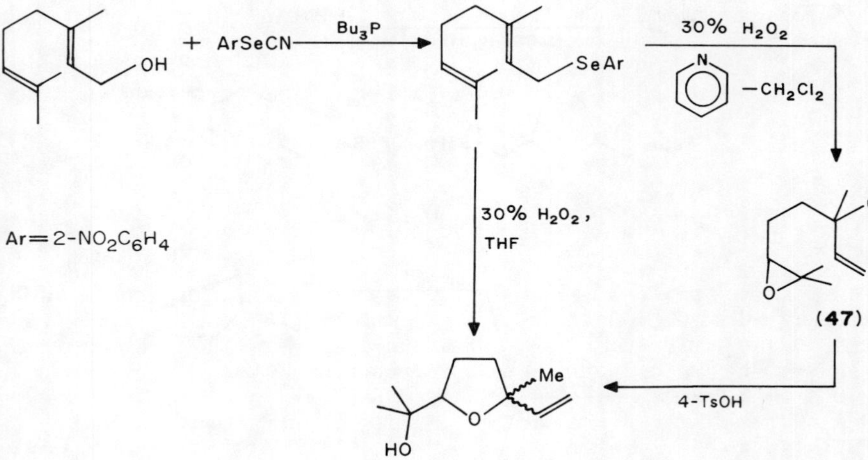

(48)

SCHEME 4

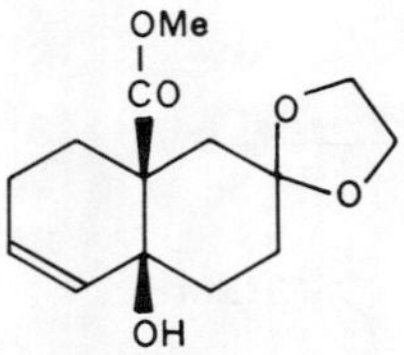

(46)

d. Selenenylation of olefins and acetylenes. The reactions described in this section are formally a substitution of the cyano group of the —SeCN moiety by a carbon atom similarly to the reactions in the previous sections (II.C.2.b and c). These reactions are, however, best described as an electrophilic addition of organoselenium reagents to carbon—carbon multiple bonds. This methodology constitutes one of the most important branches in the selenenylation reactions.

When the reactions are carried out in hydroxylic solvents in the presence of metal catalysts, phenylseleno and alkoxy groups are added to the olefins to form β-alkoxyalkyl phenyl selenides (**49**) (equation 75)[175,176]. Cu^{I}, Cu^{II} and Ni^{II} were found to be most effective when used in the form of their chlorides or bromides. The stereochemistry of the addition reaction was confirmed to be *trans*. Oxidation of **49** produces allylic ethers, acetates and alcohols[97]. When non-conjugated dienes were used as substrates, cyclic ethers bearing two phenylseleno groups were produced (Scheme 5)[177–179].

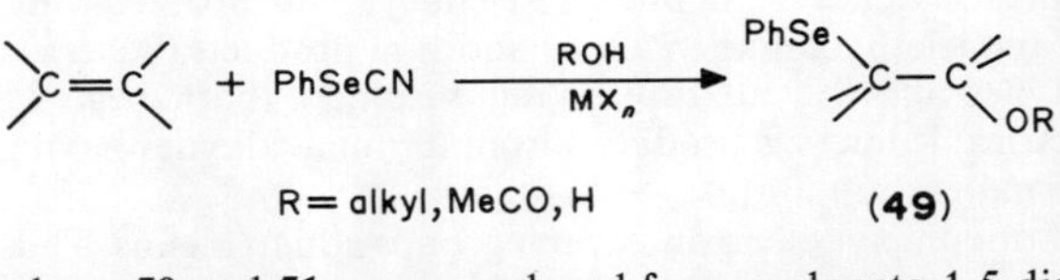

(75)

R= alkyl, MeCO, H **(49)**

Two bicyclic ethers, **50** and **51**, were produced from cycloocta-1,5-diene, the reaction being highly solvent-dependent; **50** was produced selectively (95:5) in methanol and **51** was

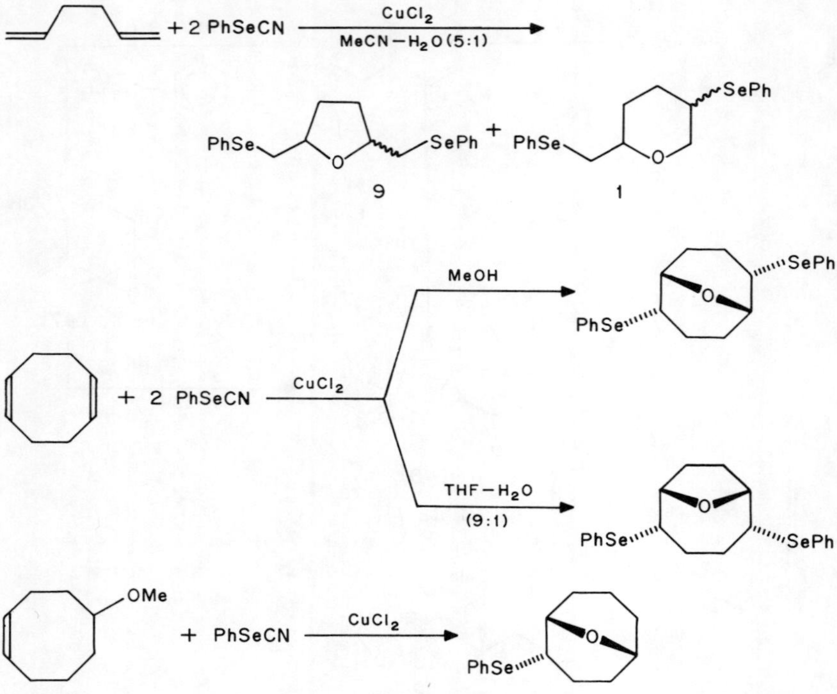

(50)

(51)

SCHEME 5

the sole product in acetonitrile–water (5:1). The second step of these reactions is an intramolecular oxyselenation and it was confirmed that a methoxy group reacted with an episelenonium ion to form cyclic ethers (Scheme 5). By the application of equation 20 to non-conjugated dienes, selenium-containing cyclic compounds were produced (equation 76)[43]. The second step of this reaction was also an oxyselenation as described above.

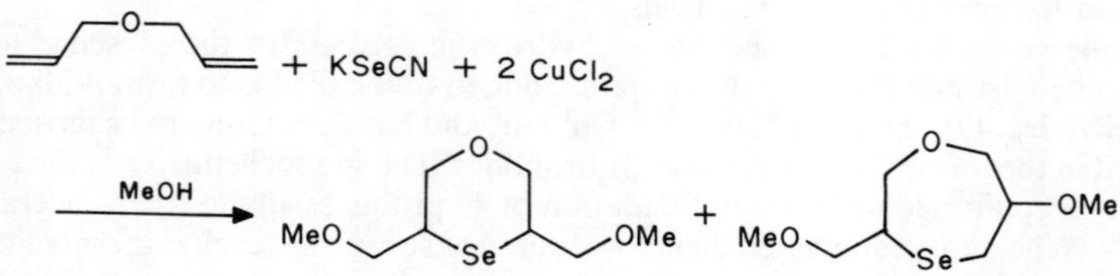

(76)

The reaction of acetylenes with phenyl selenocyanate proceeds in the presence of copper(II) halide and triethylamine to afford addition products (Scheme 6)[180]. Reactions with internal alkynes afford a mixture of regio-isomers (both *trans* adducts), while a mixture of geometrical isomers is produced from terminal alkynes having a phenylseleno group at the internal carbon atom.

In other cases, both phenylseleno and cyano groups add to olefins. Thus, in the reaction of phenyl selenocyanate with enamines, addition products were obtained almost quantitatively (equation 77)[181]. The ethanol solvent did not attach the double bond, but

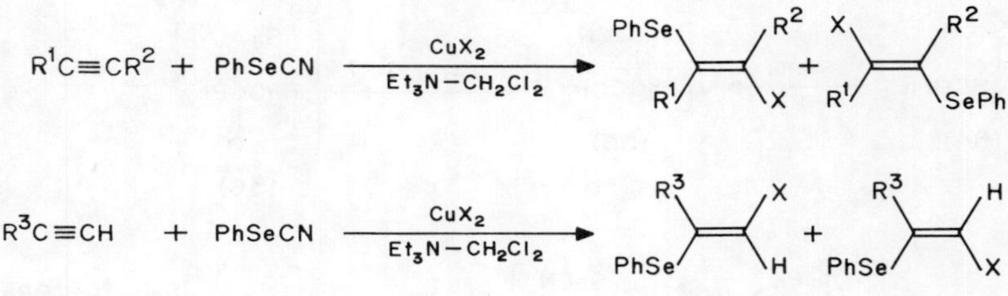

$R^1, R^2, R^3 = alkyl; X = Cl, Br$

SCHEME 6

the cyano group was incorporated in the products. As the products were found to consist of one isomer, the *trans* addition of the shown regioselectivity was strongly suggested. The addition also proceeded when ketene dialkyl acetals or ketene alkyl silyl acetals were used as olefins (equation 78)[182]. In this case, non-stereoselective addition was observed. Oxidation of **52** afforded β,γ-unsaturated-α,α-dioxycarbonitriles via selenoxide elimination. Alkenes bearing only alkyl substituents were less reactive than the olefins described above, and the addition of Lewis acid catalysts was necessary, tin(IV) chloride giving the best results (equation 79)[183]. Stereospecific addition was observed in this case (see **53**) and this was utilized in the stereoselective syntheses of α,β-unsaturated nitriles by selenoxide *syn* elimination. From unsymmetrical olefins, a mixture of regioisomers was produced.

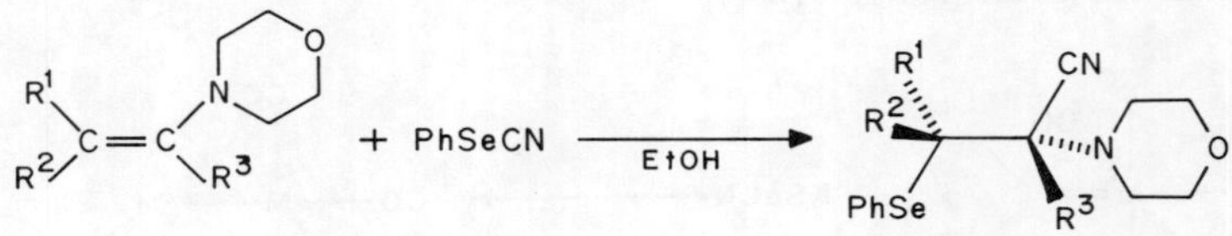

(77)

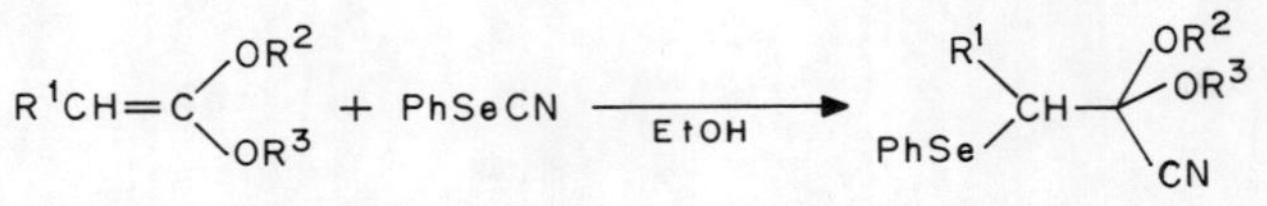

(78)

(52)

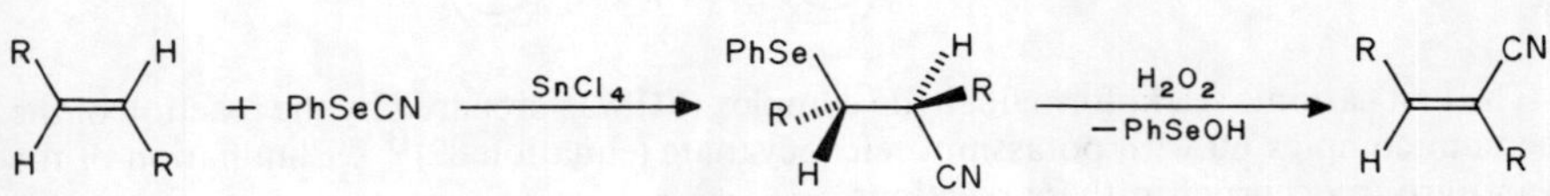

(53)

(79)

3. *Elimination of the CN group*

The oxaselenole **57** was produced by the reaction of the selenocyanate **54** with sodium hydride (equation 80)[13]. This reaction was explained to proceed through the selenoketone intermediate **56**, which was formed by base-induced elimination of HCN from **54**. The intermediate **56** then reacted with another mole of **55** to give the cyclization product.

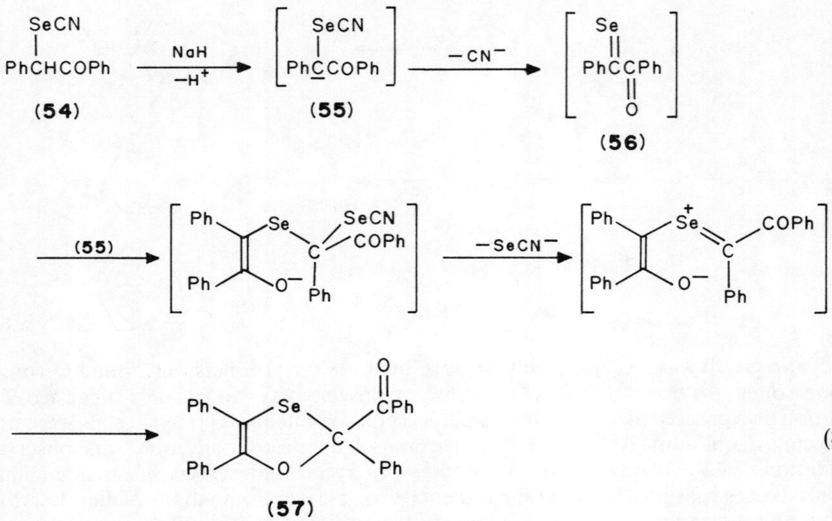

(80)

The formation of selenoketones was observed as the metal complex with chromium or tungsten carbonyls. By the treatment of [aryl(phenyl)carbene]pentacarbonyl complexes (**58**) with potassium selenocyanate, selenium was inserted to the metal—carbene bond to give the complex (**59**) (equation 81)[184].

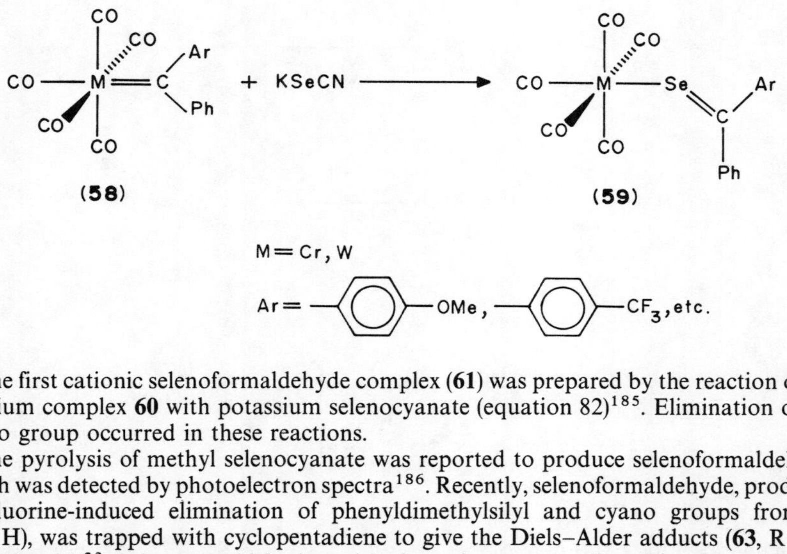

(81)

The first cationic selenoformaldehyde complex (**61**) was prepared by the reaction of the rhenium complex **60** with potassium selenocyanate (equation 82)[185]. Elimination of the cyano group occurred in these reactions.

The pyrolysis of methyl selenocyanate was reported to produce selenoformaldehyde, which was detected by photoelectron spectra[186]. Recently, selenoformaldehyde, produced by fluorine-induced elimination of phenyldimethylsilyl and cyano groups from **62** (R = H), was trapped with cyclopentadiene to give the Diels–Alder adducts (**63**, R = H) (equation 83)[23]. Selenoacetaldehyde and its homologues as well as selenobenzaldehyde were produced and trapped effectively by this procedure.

In the formation of alkyl isoselenocyanates by the reaction of carboimidoyl dichlorides with potassium selenocyanate, the elimination of cyanogen chloride from the intermediate (**64**) was postulated (equation 84)[187].

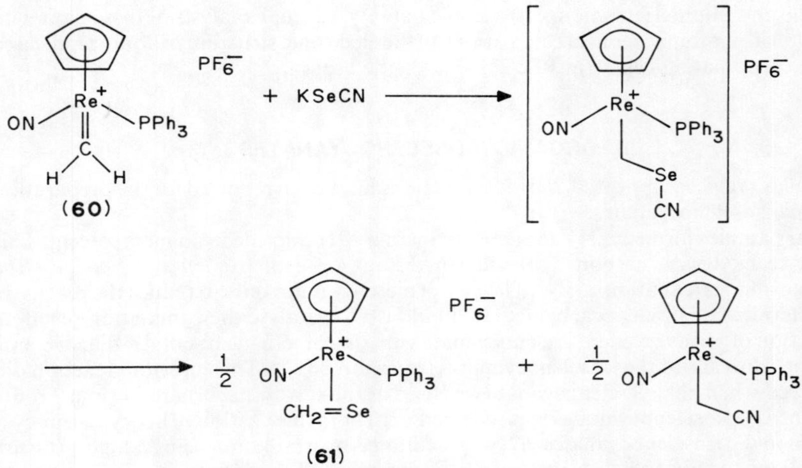

(82)

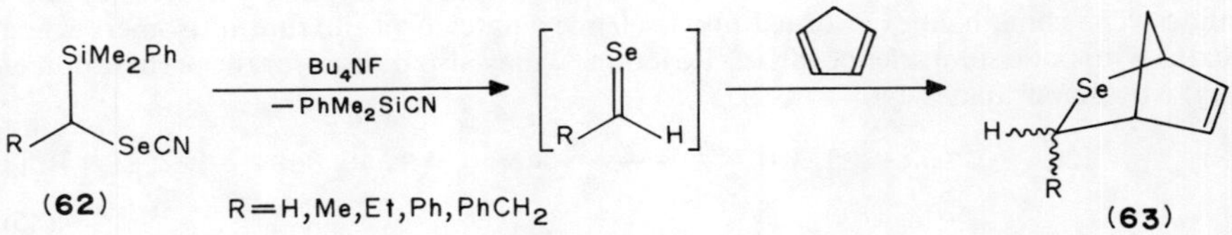

(83)

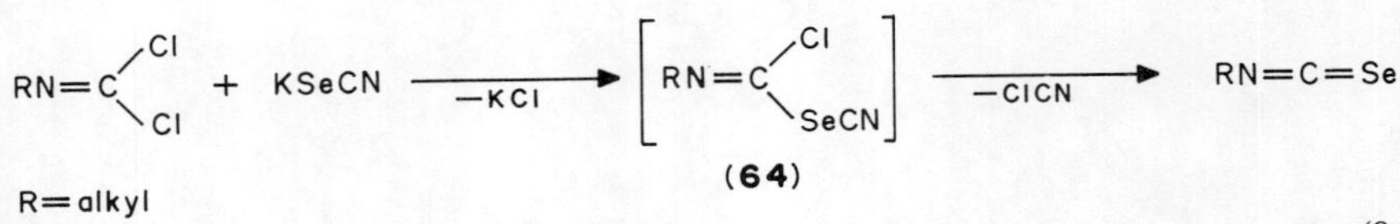

(84)

4. Formation of complexes.

Organic selenocyanate complexes, $M(CO)_5(RSeCN)$ (M = Cr, Mo), were prepared by abstraction of an iodide ion from the anion $[M(CO)_5I]^-$ by a silver ion, followed by the addition of RSeCN[188]. It was concluded from the ^1H and ^{13}C NMR and IR spectra that the ligands coordinated through the cyano-N rather than the Se atom. Palladium complexes of either the formula $PdCl_2 \cdot RSeCN$ or $PdCl_2 \cdot 2RSeCN$ were prepared, depending on the preparation procedure[189]. The IR and far IR spectra of these complexes were discussed.

The charge-transfer spectra of the 1:1 TCNE $[(NC)_2C{=}C(CN)_2]$ complex with RC_6H_4SeCN (R = H, *o*-, *m*- or *p*-Me, Cl, Br, MeO, *p*-NH_2, Me_2N) were examined[190,191]. The energies of the first charge-transfer transitions were linearly correlated with σ^+ constants. The formation of 1:1 $RSeCN{-}I_2$ complexes (R = Me, Ph) was established from electronic and photoelectronic spectroscopic data[192]. Mutual analysis of these data with the results of quantum chemical calculations of the electronic structure of donors indicated that the coordinating centre in RSeCN was the Se atom.

III. ORGANIC ISOSELENOCYANATES

Since Bulka's review[1] appeared, only five reactions have been reported for the preparation of organic isoselenocyanates.

Primary amines formed a 1:1 adduct with mercury(II) chloride in an inert solvent. This complex reacted with carbon diselenide in the presence of triethylamine to give the isoselenocyanates (equation 85)[193]. The role of mercury is not only to reduce the reactivity of the intermediate diselenocarbamate (**65**) but also to catalyse the elimination reaction. The reaction of phenylselenenyl selenocyanate with diazomethane or ethyl α-diazoacetate was reported to afford the isoselenocyanates (equation 86)[194]. This insertion reaction did not proceed when phenyl selenocyanate was used as the substrate. A photochemical route to some benzyl isoselenocyanates was reported[195]. The isomerization of benzyl selenocyanates (**66**) to the isoselenocyanates (**67**) was facilitated by irradiation using a high-pressure mercury lamp to afford a 9:1 mixture of **67** and **66** within 20 min (equation 87). Compound **67** was isolated by column chromatography. To investigate the ambident nature of the SeCN group, alkyl halides were treated with mercury(II) selenocyanate in hexane as solvent[196], when alkyl isoselenocyanates were obtained from secondary or tertiary alkyl halides. The same halides afforded alkyl selenocyanates or a mixture of isomers when treated with potassium selenocyanate. Reaction 84 may also be used for the preparation of alkyl isoselenocyanates[187].

$$RNH_2 + HgCl_2 + CSe_2 + 2\ Et_3N \longrightarrow RN{=}C{=}Se + HgSe + 2\ [Et_3N.HCl] \tag{85}$$

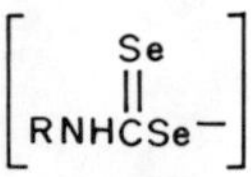

(65)

$$PhSeSeCN + RCHN_2 \longrightarrow PhSeCH(R)NCSe \tag{86}$$

R=H,COOEt

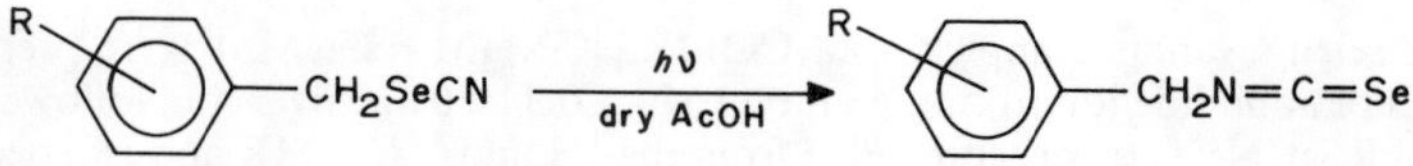

(87)

(66) **(67)**

The photoelectron spectra of $Ph_2CHNCSe$ and Ph_3CNCSe have been reported[66]. Although the spectra were dominated by the presence of strong benzenoid bands, some general trends related to structural effects can be recognized.

The reactions of isoselenocyanates with amines to form selenoureas are well known. By the reaction of benzoyl chloride with potassium selenocyanate in the presence of diisopropylamine, the selenourea **68** was produced through the addition of the amine to the intermediate isoselenocyanate (equation 88)[197]. In the case of acetylenic amines, the amine addition to a N=C bond was accompanied by the selenium addition to the triple bond to afford the selenazolines **69** (equation 89)[198]. If the solvent of the reaction described in equation 9 was changed from diethyl ether–tetrahydrofuran to dimethoxyethane–dioxane, the product became the isoselenocyanate, which was trapped by the addition of amine to give the selenourea **70**[25].

(68)

(88)

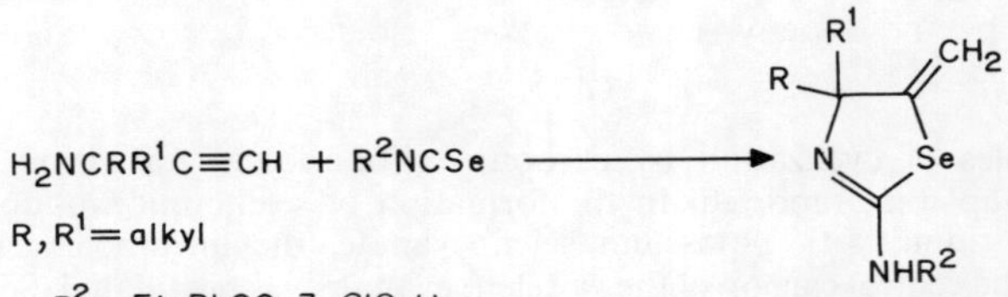

(69)

(89)

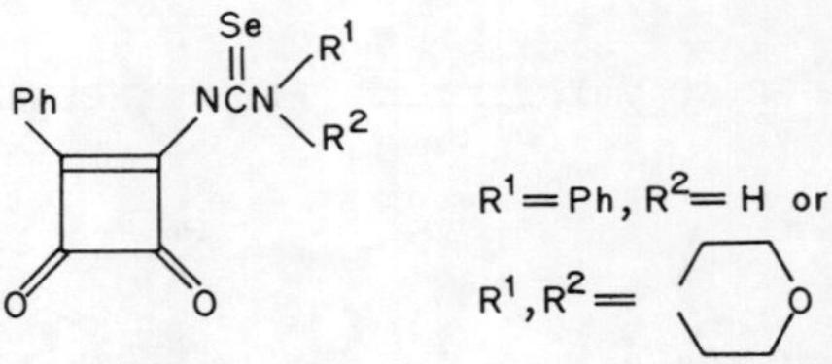

(70)

By the reaction of *N*-acylchloroformamidine (**71**) with potassium selenocyanate, a cyclic compound (**72**) was produced by the addition of oxygen to the carbon atom of the introduced isoselenocyanate group (equation 90)[199].

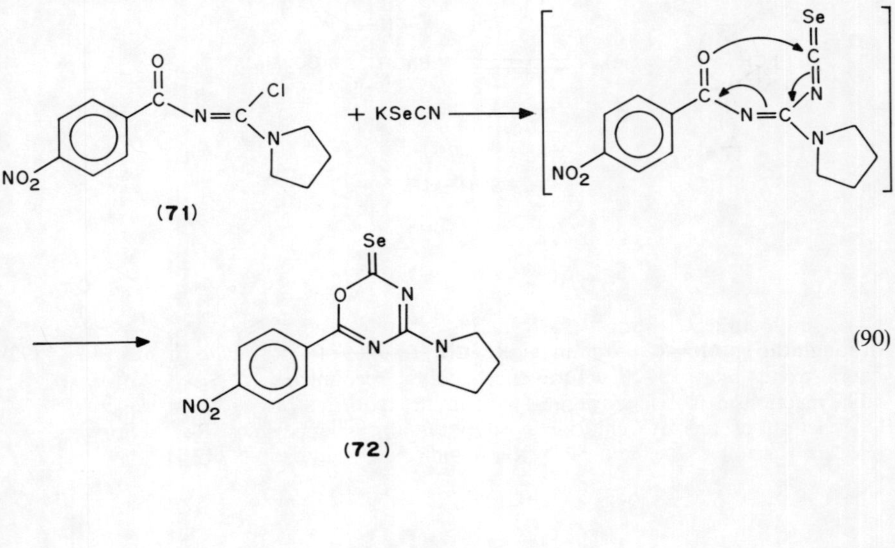

(71)

(90)

(72)

The addition of phosphorus to an isoselenocyanato group was reported. Addition products (**73**) were obtained by the reaction of trimethyl or triethyl phosphite with acetyl isoselenocyanate, which was, in turn, prepared from acetyl chloride and potassium selenocyanate and was used without isolation (equation 91)[200].

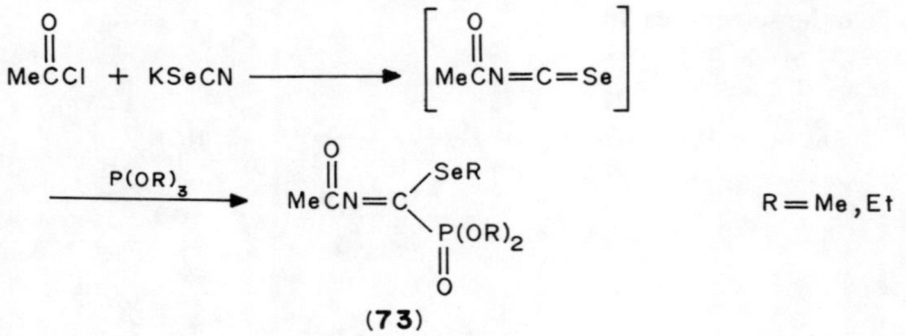

(91)

(**73**)

Two examples of cyclization by carbon—carbon bond formation in an isoselenocyanato group were reported. In the formation of selenoimidazolidone (**74**) by the reaction of aziridine with potassium selenocyanate, the intramolecular attack of a carbanion on the central carbon of the isoselenocyanate was postulated (Scheme 7)[201]. By the reaction of *N*-vinylchloroformamidines with potassium selenocyanate, a heterocyclic compound (**75**) was produced through carbon—carbon bond formation (equation 92)[202].

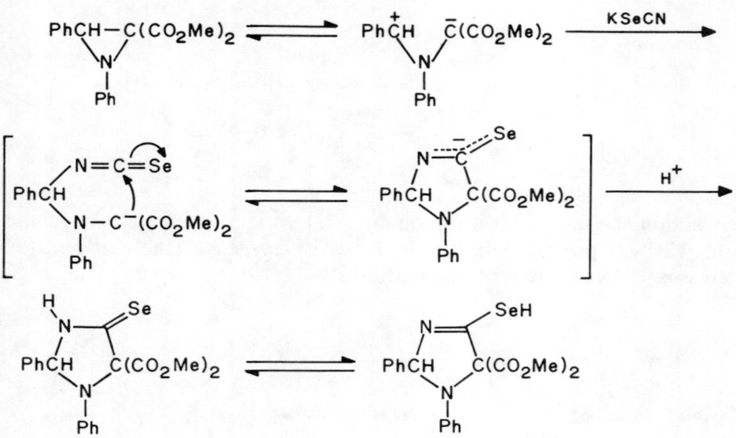

(**74**)

SCHEME 7

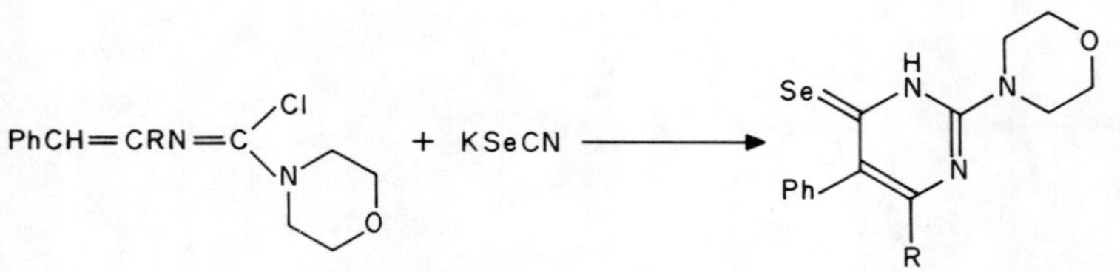

(92)

R=H, Ph

(**75**)

The complex **59** in equation (81) was also prepared by the reaction of **58** with PhNCSe instead of KSeCN. A selenium atom was transferred from the isoselenocyanate to the metal complex, leaving phenyl isocyanide[184].

IV. ORGANIC TELLUROCYANATES

Compared with selenocyanates, organic tellurocyanates (RTeCN) are little known. Although the formation of various such compounds has been claimed in many patents[203,204], none of them has been characterized. The first concrete example of an organic tellurocyanate seems to be 1-azulenyl tellurocyanate (**76**), prepared by Nefedov in 1968[205]. The compound (fine violet needles, m.p. 80–81 °C) was synthesized in 7% yield by treatment of azulene with a combination of copper(II) acetate and potassium tellurocyanate (KTeCN) in acetonitrile at reflux temperature (equation 93), KTeCN being produced separately by heating potassium cyanide and tellurium at 100–250 °C. The *in situ* formation of copper(II) tellurocyanate [$Cu(TeCN)_2$] and the direct replacement of hydrogen in azulene by this species were proposed. This substitution method, however, does not seem to have been applied to any other aromatic compound. Another method for aryl tellurocyanates was developed by Renson and coworkers, who treated stable aryltellurium(II) halides with silver cyanide to produce substitution of halide by cyanide (equation 94)[206,207]. Similar reactions also proceeded with potassium cyanide (equation 95)[208]. In these reactions a carbonyl-containing function or a nitro group was necessary at the position *ortho* to the tellurium atom to obtain remarkably stable aryl tellurocyanates, attributed to the interaction of the tellurium with the carbonyl or nitro oxygen atom, which is then coordinatively less unsaturated[209]. Another route to aryl tellurocyanates is a carbon—tellurium bond cleavage of 3-oxo-2,3-dihydrobenzotellurophene by aqueous hydrogen cyanide (equation 96)[210].

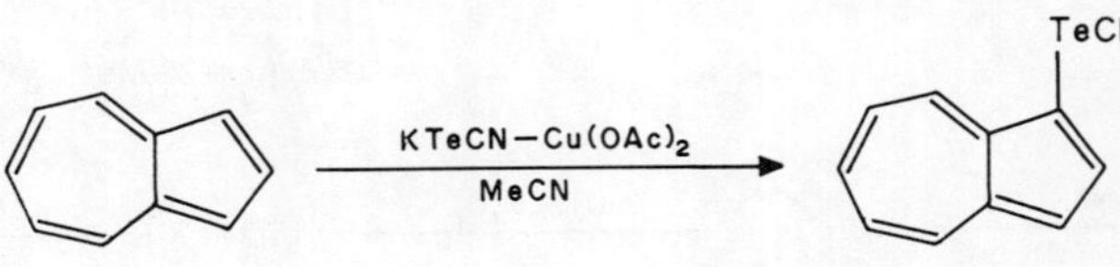

(**76**)

(93)

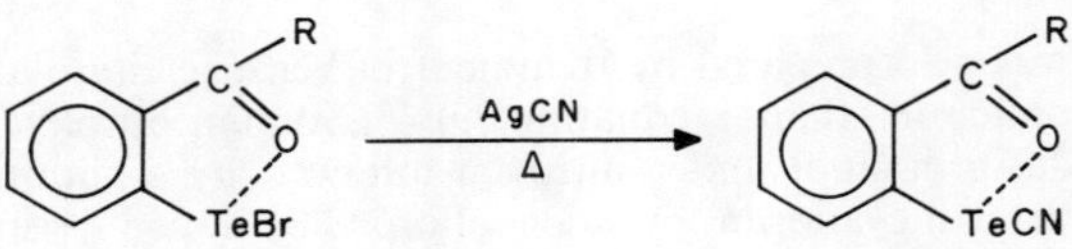

(**77**) R=H
(**78**) R=Ph

(94)

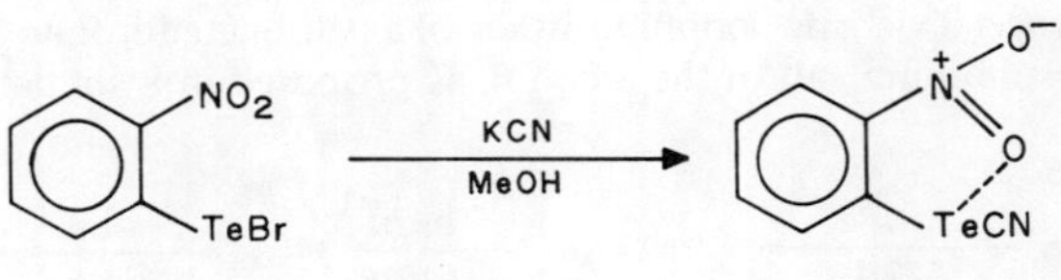

(**79**)

(95)

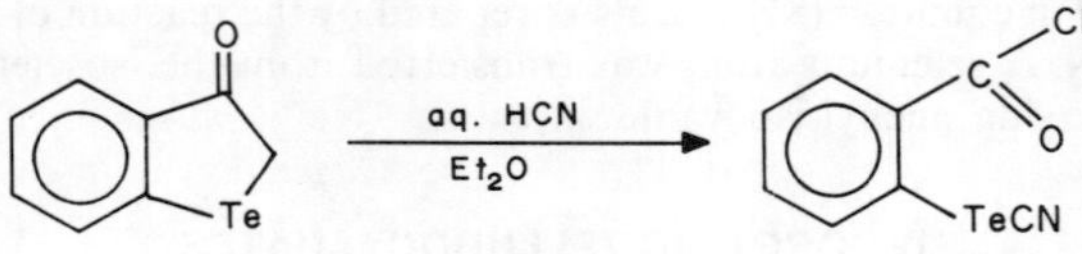

(80)

(96)

These methods, unfortunately, are of limited applicability for the synthesis of aryl tellurocyanates because aryltellurium(II) halides are normally unstable polymeric species, unless stabilized by a group such as *o*-formyl, *o*-acetyl, *o*-benzoyl or *o*-nitro, and also available tellurophenes are very limited.

It was later proved, however, that *ortho*-substituents are not necessarily required for the preparation of phenyl tellurocyanates. Thus, Falcone and Cava[208] succeeded in syntheses of tellurocyanates by two procedures. One is a direct nucleophilic attack by a cyanide ion on one of the tellurium atoms of a diaryl ditelluride, the reaction being carried out in DMSO at room temperature (equation 97). By this method phenyl tellurocyanate (**81**) and 4-methoxyphenyl tellurocyanate (**82**) were prepared, but the yield never surpassed 20%. The other method consists in partial reduction of aryltellurium trihalides with aqueous sodium metabisulphite ($Na_2S_2O_5$) to aryltellurium(II) halides (**84**), followed by the addition of potassium cyanide (equation 98). By this method **81** and **82** were prepared in 86% and 93% yields respectively. The reaction conditions are critical, and the best results were obtained when the trihalide was first stirred with 2 equiv. of sodium hydrogen carbonate to yield an aryltellurinyl halide (**83**), and then about 4 equiv. each of sodium metabisulphite and potassium cyanide were added.

$$\text{ArTeTeAr} + \text{CN}^- \xrightarrow{\text{DMSO}} \text{ArTe}^- + \text{ArTeCN} \tag{97}$$

(**81**) Ar = Ph

(**82**) Ar = 4-$MeOC_6H_4$

$$\underset{(X = Cl, Br)}{\text{ArTeX}_3} \xrightarrow{\text{aq. NaHCO}_3} \underset{(\mathbf{83})}{\text{ArTe(O)X}} \xrightarrow{\text{aq. Na}_2\text{S}_2\text{O}_5} \underset{(\mathbf{84})}{[\text{ArTeX}]} \xrightarrow{\text{KCN}} \mathbf{81, 82} \tag{98}$$

Compound **81** was also prepared by treatment of benzenetellurolate anion ($PhTe^-$) with cyanogen bromide in ethanol (equation 99)[211]. Sodium borohydride reduction of diphenyl ditelluride in ethanol under nitrogen followed by addition of an ethanolic solution of commercial cyanogen bromide afforded **81** in 58% isolated yield. The tellurocyanate thus prepared could be converted to various alkyl phenyl tellurides in moderate to good yields by treating it with alcohol in dichloroethane or tetrahydrofuran at room temperature in the presence of tributylphosphine (equation 99)[211]. Such a transformation did not occur by using other organophosphorus compounds such as triphenylphosphine or triethyl phosphite. Although details are not known, this telluride formation may proceed via a nucleophilic attack of a benzenetellurolate anion on an alkyl group of the phosphonium salt of the alcohol, as proposed in a similar reaction of aryl selenocyanate[114].

$$\text{PhTeTePh} \xrightarrow[\text{EtOH}]{\text{NaBH}_4} [\text{PhTeNa}] \xrightarrow[\text{EtOH}]{\text{BrCN}} \mathbf{81} \xrightarrow[\text{Bu}_3\text{P, CH}_2\text{Cl}_2]{\text{ROH}} \text{PhTeR} \tag{99}$$

Compound **81** has also been used as a source of introduction of phenyltellurium species to other organic compounds. For example, treatment of cyclohexene with **81** in methanol in the presence of copper(II) chloride afforded (2-methoxycyclohexyl)phenyltellurium dichloride in 62% yield (equation 100)[212,213]. The reaction is the so-called oxytelluriation of olefins, with the phenyltellurium species acting as an electrophile. A contrasting example is a nucleophilic substitution by the species for preparation of tellurium-containing amino acid[214]. Thus, treatment of **81** with β- or γ-halogenoamino acids in the presence of sodium borohydride in ethanol gave the corresponding phenyltelluroamino acids in ca. 30% yield (equation 101).

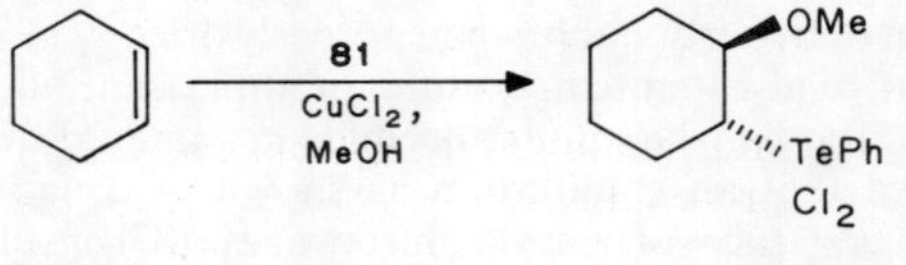

(100)

$$X(CH_2)_nCH(NH_3Cl)CO_2H \xrightarrow[\text{NaBH}_4,\ \text{EtOH}]{81} PhTe(CH_2)_nCH(NH_3Cl)CO_2H \qquad (101)$$

$X = Cl, Br$
$n = 1, 2$

Two aryl tellurocyanates were recently prepared as by-products in the synthesis of diaryl tellurides from aromatic amines[215]. Aryldiazonium tetrafluroroborates were converted to symmetrical diaryl tellurides in moderate yields by treatment with potassium tellurocyanate in DMSO (for KTeCN–DMSO, see below). Although the expected aryl tellurocyanates are seldom stable enough to be isolable under the reaction conditions, in two cases they were isolated by chromatography in addition to diaryl tellurides, viz., **79** and 2, 6-dimethylphenyl tellurocyanate (**85**), in 11% and 40% yields, respectively (equation 102). The main reaction course was the disproportionation of the intermediate aryl tellurocyanates to diaryl tellurides and tellurium dicyanide, the latter of which decomposed instantly to elemental tellurium and dicyanogen (equation 103).

$$ArN_2^+BF_4^- + KTeCN \xrightarrow{\text{DMSO}} ArTeCN + KBF_4 + N_2 \qquad (102)$$

(**79**) $Ar = 2\text{-}NO_2C_6H_4$

(**85**) $Ar = 2,6\text{-}Me_2C_6H_3$

$$2\,ArTeCN \xrightarrow{\text{DMSO}} Ar_2Te + Te(CN)_2 \qquad (103)$$

$$Te(CN)_2 \longrightarrow Te + (CN)_2$$

An attempt to prepare 2, 3, 5, 6-tetramethylphenyl tellurocyanate by the reaction of the corresponding aryl iodide with KTeCN in a mixture of hexamethylphosphoramide and DMSO in the presence of copper(I) iodide resulted in the formation of the corresponding aryl cyanide as the sole identifiable product[31]. The initially formed copper(I) tellurocyanate decomposed immediately to elemental tellurium and copper(I) cyanide, the latter being the reactive species for the aryl cyanide. In contrast, the corresponding aryl selenocyanate was obtained in 45–61% yields by a similar procedure.

Although a tellurocyanate ion has been shown to be a much stronger nucleophile than selenocyanate and thiocyanate ions by kinetic studies using benzylic halides[18,216] and dichloromethane[217] as substrates, the preparation of organic tellurocyanates by nucleophilic substitution by the tellurocyanate ion was not known until a decade ago, probably because of the instability of the ion and the products towards air and/or water. Thus, alkali metal tellurocyanates and their solutions were decomposed instantly by water with the liberation of elemental tellurium[218,219] and a sterically large counter cation was necessary to stabilize the tellurocyanate anion[18,216,219]. By using the stabilized tellurocyanate ion, Austad *et al.*[216] succeeded in isolating from benzyl bromide in acetonitrile a stable and almost colourless crystalline substance, which was suggested to be tetraphenylarsonium bromocyanobenzyl tellurate (**86**) (equation 104). Tetraphenylarsonium tellurocyanate was prepared by the reaction of tetraphenylarsonium cyanide with tellurium in acetonitrile and shown to be fairly stable when properly dried[219]. Freshly prepared **86** had no odour, but on storage at room temperature an unpleasant odour was produced. On crystallization of **86**, tetraphenylarsonium bromide crystallized out as a minor product. The authors suggested that an equilibrium as shown in equation 104 exists and the unpleasant odour is due to traces of benzyl tellurocyanate (**87**) or subsequent products, but they could not isolate and characterize it[216]. A complex similar to **86** was also prepared in high yield[18] from 4-nitrobenzyl chloride and bis(triphenylphosphine)iminium tellurocyanate [m.p. 190–193 °C (decomp.)][220] in acetone namely **88** [yellow–brownish needles, m.p. ca. 120 °C (decomp.)] (equation 105). Structural determination of the anionic part of **88** by X-ray crystallography indicated that the C—Te—C bond angle is 87.4° and the Te—Cl bond is approximately *trans* to the Te—C (cyano) bond. In contrast Spencer *et al.*[221] succeeded in isolating (61% yield) and characterizing **87** (white needles, m.p. 126–127 °C), the first alkyl tellurocyanate ever isolated, by the reaction of benzyl chloride with potassium or sodium tellurocyanate in DMSO at room temperature (equation 106). The key problem in the reaction was to prepare stable alkali metal tellurocyanates or their solutions, which were known to be very sensitive to air and water as described above. Although it was already known that KTeCN was produced in the solid state by heating potassium cyanide and tellurium at 100–250 °C and it could be dissolved in acetonitrile[205,222]*, Spencer *et al.*[221] found that stable solutions of potassium or sodium tellurocyanate were readily prepared by stirring together 1 equiv. each of powdered tellurium and dry, powdered alkali metal cyanide in dry DMSO at 100 °C under an inert atmosphere until all of the tellurium dissolved. The resulting pale yellow solution decomposed instantly if added to water, with the formation of elemental tellurium, and also attempts to precipitate the alkali metal tellurocyanates by addition of acetone or diethyl ether resulted in the formation of a tellurium mirror. A detailed IR spectrum of **87**

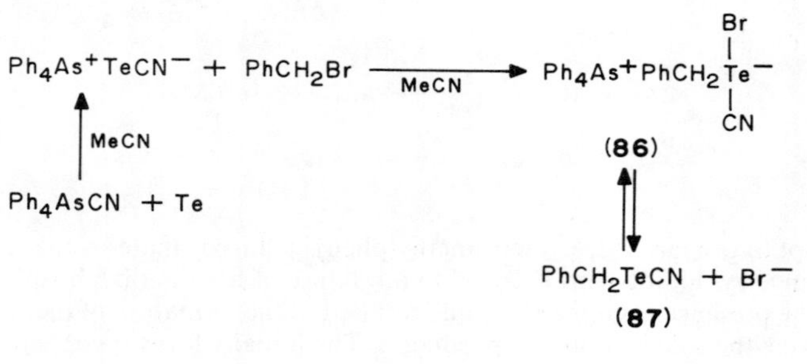

(104)

*Contrary to Ref. 205, Greenwood *et al.*[222] could not observe any reaction of powdered tellurium with fused NaCN–KCN eutectic, with aqueous or methanolic KCN under reflux or with a suspension of KCN in benzene.

in the solid state was recorded together with those of the corresponding thiocyanate and selenocyanate and tentative assignments, based on force fields transferred from similar molecules, were proposed[63].

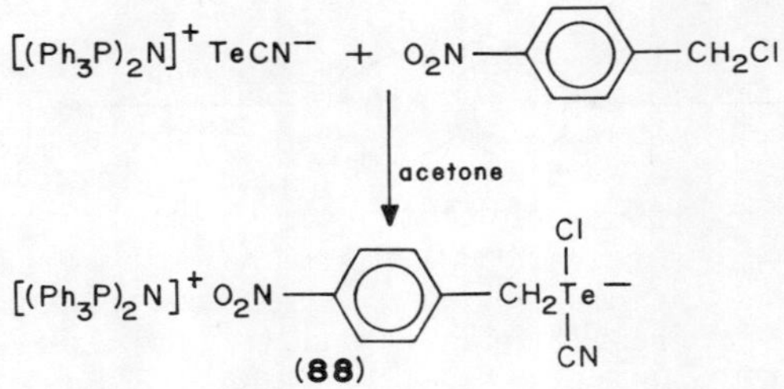

(88)

(105)

$$\mathrm{MCN + Te \xrightarrow[\text{DMSO}]{\Delta} [MTeCN] \xrightarrow{PhCH_2Cl} \mathbf{87}} \qquad (106)$$

$(M{=}K, Na)$

The reaction shown in equation 106 was also applied to 4-chloro-, 4-methoxy- and 1, 2, 3, 4, 5-pentamethyl-benzyl chlorides, and the corresponding ring-substituted benzyl tellurocyanates (**89**)–(**91**) were produced in 46–64% yields (equation 107)[223]. All these tellurocyanates are light sensitive and thermally unstable white to pale yellow solids. When 4-nitrobenzyl chloride was treated under similar conditions, 4-nitrobenzyl tellurocyanate (**92**) could not be isolated, but it changed to the corresponding ditelluride during the work-up procedure. However Maartmann-Moe *et al.*[224] succeeded in isolating **92** in 49% yield by a similar treatment and they claimed that the compound is a stable solid and can be stored for months even in daylight and moist air.

$$\mathrm{ArCH_2Cl \xrightarrow[\text{DMSO}]{[KTeCN]} ArCH_2TeCN}$$

(**89**) $Ar = 4\text{-}ClC_6H_4$

(**90**) $Ar = 4\text{-}MeOC_6H_4$ (107)

(**91**) $Ar = 1,2,3,4,5\text{-}Me_5C_6$

(**92**) $Ar = 4\text{-}O_2NC_6H_4$

The X-ray crystal structural determination of **92** revealed the following[67,224]. There are two distinctly different Te—C bond lengths in the molecule, Te—C_1 (benzylic) and Te—C_2 (cyano) being 2.167 and 2.06 Å, respectively. The C_1—Te—C_2 bond angle is 90.6°. The tellurium atom forms two strong intermolecular 'secondary bonds' to oxygen atoms of 2.949 and 3.182 Å, these two bonds and the two Te—C bonds being strictly coplanar. The compound may be considered as a distorted square-planar tellurium(II) complex and must owe its stability in the crystalline state to intermolecular contacts between the tellurium atom and the two oxygen atoms. The corresponding selenocyanate has been shown to have a similar structure[67].

Tellurenyl halides (**93**) obtained by halogenolysis of tellurophthalide reacted with ethanol and then potassium cyanide to give a 33–69% yield of 2-carbethoxybenzyl tellurocyanate (**94**) (equation 108), which was also synthesized independently from 2-

carbethoxybenzyl bromide in 85% yield by treatment with KTeCN in DMSO (equation 109)[225].

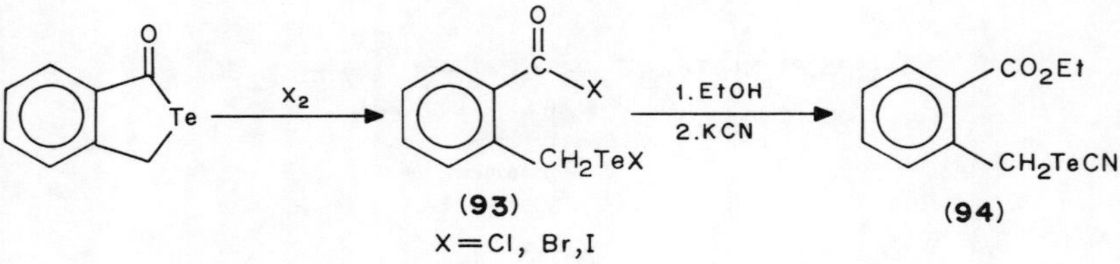

(108)

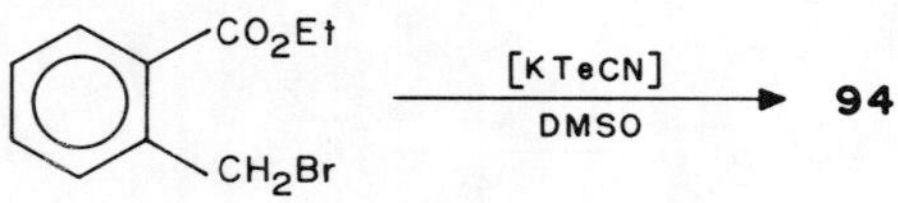

(109)

Organic tellurocyanates so far isolated and characterized are summarized in Table 1.

The tellurocyanate **87** was stable to light alone in the absence of oxygen, but darkened in the presence of air with the precipitation of a black solid. Organic products were revealed to be benzaldehyde (60%) and benzyl alcohol (40%). Several other reactions of **87** were also examined by Cava and coworkers, as follows[221,226]. Crystallization of an equimolar mixture of **87** and tetraphenylarsonium bromide afforded the complex **86**, but the properties of this substance suggested that it is probably a loose molecular complex of its two components, rather than a compound[216] containing an anionic tellurium in the form of **86** as previously suggested. Chromatography of the complex on silica gel in the dark afforded **87** with a recovery of 90%. The reaction of **87** with methanolic sodium hydroxide

TABLE 1. Organic tellurocyanates isolated and characterized

Compound	Equation[a]	Isolated yield (%)	M.p. (°C) (colour)	Spectral data: ^{1}HNMR δ(ppm), IR ν(cm^{-1}), mass m/e	Ref.
76	93	7	80–81 (violet)	IR 2148, 778, 743	205
77	94	—[b]	104		206
78	94	80	135	NMR 7.3–7.7 (m, 7H), 7.85–8.5 (m, 2H)	207
79	95	65	148–149 (red–brown)	IR 2400, 1600, 1580, 1500, 1320 Mass 278 (M^+), 250 ($M^+ - CN$)	208
79	102	11	160	NMR 7.62 (t, 1H), 7.74 (t, 1H), 8.46–8.52 (m, 2H) IR 2150	215
80	96	50	117		210
81	98	86	74–75 (white)	IR 2170, 1400, 1170, 730, 690 Mass 233 (M^+), 207 ($M^+ - CN$)	208

Table 1. (Contd.)

Compound	Equation[a]	Isolated yield(%)	M.p. (°C) (colour)	Spectral data: ^{1}HNMR δ(ppm), IR ν(cm^{-1}), mass m/e	Ref.
81	97	20	74–75		208
81	99	58	67–70 (decomp.) (colourless)		211
82	98	93	75–77 (white)	IR 2160, 1590, 1500, 1400, 1260, 1190 Mass 263 (M$^+$), 237 (M$^+$ – CN)	208
82	97	18	75–77		208
85	102	40	114–115	NMR 2.69 (s, 6H), 7.23–7.26 (m, 3H) IR 2140	215
87	106 (M = K)	61	126–127 (white)	NMR 4.52 (s, 2H), 7.38 (br, s, 5H) Mass 247 (M$^+$), 91	221
87	106 (M = Na)	57			221
87	107	72	126–127 (decomp.)	NMR 4.48 (s, 2H), 7.26 (s, 5H) IR 2150, 1600, 1495, 1455, 1140, 1060, 760, 695	223
89	107	46	75–80 (decomp.)	NMR 4.36 (s, 2H), 7.14 (s, 4H) IR 2150, 1485, 1410, 1150, 1140, 1090, 1015, 835, 805	223
90	107	59	76–79 (decomp.)	NMR 3.80 (s, 3H), 4.52 (s, 2H), 6.92 (d, 2H, J = 9 Hz), 7.22 (d, 2H, J = 9 Hz) IR 2140, 1605, 1500, 1460, 1300, 1240, 1175, 1030, 830–815, 795, 735	223
91	107	64	161–162 (decomp.) (white–pale yellow)	NMR 2.22 (s, 9H), 2.32 (s, 6H), 4.72 (s, 2H) IR 2130, 1430, 1140, 1050, 990, 840, 760	223
92	107	49	123 (decomp.) (green–yellow)	NMR 4.36 (s, 2H), 7.91 (m, 4H) IR 2157, 430 Mass 292 (13)[c], 290 (11), 288 (7), 136 (100) UV(MeCN) ~ 342 nm (log$\varepsilon \leqslant$ ~ 3.33)	224
94	109	85	72–73 (dark)	NMR 1.41 (t, 3H), 4.22 (s, 2H), 4.36 (q, 2H), 7.32–7.41 (m, 2H), 7.54 (t, 1H), 7.98 (d, 1H) IR 2110, 1660 Mass 391 (1)[c], 163 (27), 135 (100), 118 (15), 105 (11)	225
94	108 (X = Br)	69			225
94	108 (X = Cl)	33			225
94	108 (X = I)	44			225

[a]Equation number employed for synthesis.
[b]Not determined.
[c]The value in parentheses is a relative intensity.

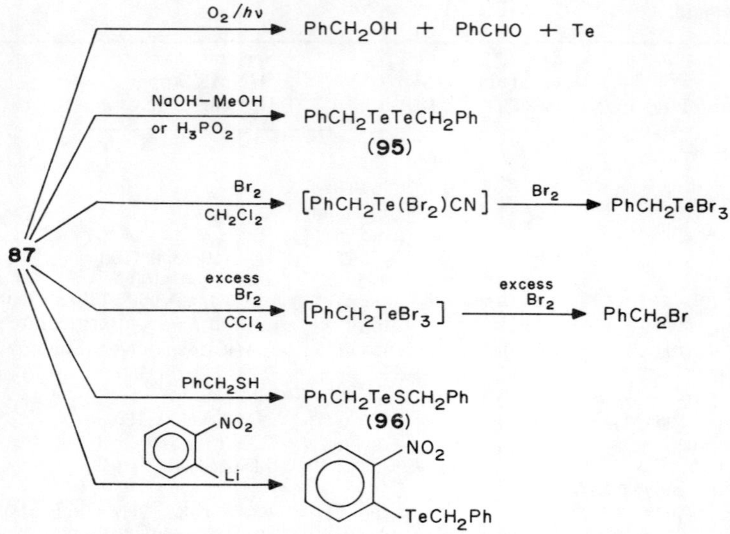

SCHEME 8

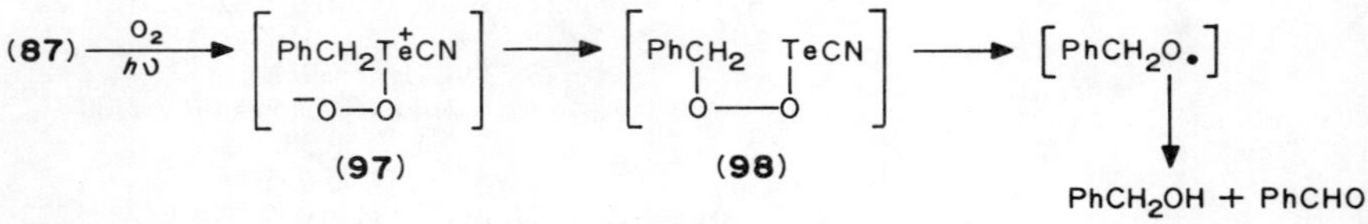

SCHEME 9

under argon gave a photosensitive dibenzyl ditelluride (**95**) in fair yield (29%), which was produced more effectively and almost quantitatively by treatment of **87** with warm hypophosphorous acid(H_3PO_2). Similar transformations were also observed with **89**–**91**[223]. Treatment of **87** with excess of bromine produced benzyl bromide via a benzyltellurium tribromide. The reaction of **87** with benzyl thiol in carbon tetrachloride afforded dibenzyl thiotelluride (**96**). The attack of a carbanion on the tellurium atom of **87** was also expected and, in fact, 2-nitrophenyllithium reacted smoothly with **87** to give the organe–red benzyl 2-nitrophenyl telluride in 72% yield at $-100\,°C$. All the results are summarized in Scheme 8. The proposed mechanism of the photooxidation of **87** is shown in Scheme 9[221]. The major reaction pathway may involve singlet oxygen, which could attack the tellurium atom to give a 1,3-dipolar peroxide (**97**). Rearrangement of **97** to an unstable benzylperoxytellurocyanate (**98**) might occur in which an oxygen—oxygen bond fission gives benzyloxy radicals, leading to the alcohol and the aldehyde. It was proposed that reduction of **87** with hypophosphorous acid may give a very unstable benzyl tellurol ($PhCH_2TeH$) rather than benzyltellurium radical ($PhCH_2Te^{\bullet}$) and the former reacts instantaneously with another mole of **87** to give **95**. This assumption is consistent with the facile formation of **96** when treated with benzyl thiol.

$$TeCN^- - e^- \rightleftharpoons TeCN^\bullet$$

$$2TeCN^\bullet \longrightarrow (TeCN)_2$$

$$(TeCN)_2 + TeCN^- \rightleftharpoons (TeCN)_3^-$$

$$2(TeCN)_3^- - 2e^- \rightleftharpoons 3(TeCN)_2$$

$$(TeCN)_2 \longrightarrow 2Te + (CN)_2$$

SCHEME 10

The electrochemical oxidation of tetraethylammonium tellurocyanate in acetonitrile produced tellurocyanogen, $(TeCN)_2$, via the $(TeCN)_3{}^-$ ion as an intermediate[227]. This compound was unstable and decomposed to tellurium and dicyanogen (Scheme 10). The instability of $(TeCN)_2$ will make it difficult to use this compound as a reagent to introduce the TeCN group into organic molecules and there seems to be no report of such attempts.

V. ORGANIC ISOTELLUROCYANATES

Organic isotellurocyanates (RN=C=Te) have not been prepared so far, in spite of several attempts at their synthesis from isonitriles or organocyanosilanes and tellurium[228–230] and also by photoinduced isomerization of organic tellurocyanates[223]. When organic tellurocyanates were subjected to photolysis, only tellurium precipitation was observed[223], as expected from the reaction shown in Scheme 8.

VI. REFERENCES

1. E. Bulka, in *The Chemistry of Cyanates and Their Thio Derivatives, Part 2* (Ed. S. Patai), Wiley, Chichester, 1977, Ch. 19.
2. B. Lindgren, *Chem. Scr.*, **16**, 24 (1980).
3. M. V. Lakshmikantham and M. P. Cava, *J. Org. Chem.*, **45**, 2632 (1980).
4. T. Hori and K. B. Sharpless, *J. Org. Chem.*, **43**, 1689 (1978).
5. K. Alzbeta, V. Konecny, M. Fandlova and J. Kovac, *Czech. Pat.*, 169352; *Chem. Abstr.*, **88**, P152599g (1978).
6. A. Krutosikova, V. Konecny, M. Fandlova and J. Kovac, *Czech. Pat.*, 170750; *Chem. Abstr.*, **89**, P101858d (1978).
7. D. V. Ende and A. Krief, *Tetrahedron Lett.*, 2709 (1975).
8. R. C. Cambie, P. S. Rutledge, V. Y. Y. Tan and P. D. Woodgate, *J. Chem. Res.* (*S*), 392 (1984).
9. A. P. Arnold and A. J. Canty, *Aust. J. Chem.*, **36**, 815 (1983).
10. B. Lindgren, *Acta Chem. Scand., Ser. B*, **30**, 941 (1976).
11. B. Lindgren, *Acta Chem. Scand., Ser. B*, **31**, 1 (1977).
12. F. Asinger and M. K. Schmitz, *Monatsh. Chem.*, **113**, 1191 (1982).
13. J. Gramza, R. B. Mitchell and D. C. Ditter, *J. Org. Chem.*, **49**, 2057 (1984).
14. V. N. Bodner, M. O. Lozinskii and P. S. Pel'kis, *Ukr. Khim. Zh.*, **47**, 424 (1981); *Chem. Abstr.*, **95**, 80775f (1981).
15. A. Halvorsen and J. Songstad, *J. Chem. Soc., Chem. Commun.*, 327 (1978).
16. T. Thorstenson and J. Songstad, *Acta Chem. Scand., Ser. A*, **32**, 133 (1978).
17. T. Thorstenson, R. Eliason and J. Songstad, *Acta Chem. Scand., Ser. A*, **31**, 276 (1977).
18. K. Maartmann-Moe, K. A. Sanderud and J. Songstad, *Acta Chem. Scand., Ser. B*, **36**, 211 (1982).
19. L. M. Yagupol'skii, I. I. Maletina, N. V. Kondratenko and V. V. Orda, *Synthesis*, 835 (1978).
20. P. D. Woodgate, H. H. Lee and P. S. Rutledge, *Synthesis*, 152 (1978).
21. Y. Tamura, M. Adachi, T. Kawasaki and Y. Kita, *Tetrahedron Lett.*, 2251 (1979).
22. C. H. Banks, J. R. Daniel and R. A. Zingaro, *J. Med. Chem.*, **22**, 572 (1979).

23. G. A. Krafft and P. T. Meinke, *J. Am. Chem. Soc.*, **108**, 1314 (1986).
24. H. G. Werchan, G. Dittrich and H. Lehmann, *East Ger. Pat*; DD 202699; *Chem. Abstr.*, **100**, P138785z (1984).
25. W. Ried and H. Dietschmann, *Justus Liebigs Ann. Chem.*, 1003 (1981).
26. Z. Ejmocki, *Pr. Nauk. Politech. Warz., Chem.*, **17**, 93 (1975); *Chem. Abstr.*, **86**, 42730u (1977).
27. C. Paulmier and F. Outurquin, *J. Heterocycl. Chem.*, **20**, 113 (1980).
28. A. V. El'tsov, A. N. Frolov, E. V. Smirnov and V. V. Yunnikov, *USSR Pat.*, 520354; *Chem. Abstr.*, **85**, P94065e (1976): A. N. Frolov, A. V. El'tsov, E. V. Smirnov and O. V. Kul'bitskaya, *Zh. Org. Khim.*, **13**, 2007 (1977); *Chem. Abstr.*, **88**, 6482x (1978).
29. A. N. Frolov, E. V. Smirnov, O. V. Kul'bitskaya and A. V. El'tsov, *Zh. Org. Khim.*, **16**, 2302 (1980); *Chem. Abstr.*, **94**, 174536k (1981).
30. H. Suzuki and M. Shinoda, *Synthesis*, 640 (1977).
31. H. Suzuki, K. Miyoshi and M. Shinoda, *Bull. Chem. Soc. Jpn.*, **53**, 1765 (1980).
32. V. A. Nafedov, L. K. Tarygina, L. V. Kryuchkova and Yu. S. Ryabokobylko, *Zh. Org. Khim.*, **17**, 570 (1981); *Chem. Abstr.*, **95**, 41864 (1981).
33. W. H. H. Gunther, *US Pat.*, 3905958; *Chem. Abstr.*, **84**, P37310m (1976).
34. D. Herrmann and R. Gollmick, *Z. Chem.*, **17**, 290 (1977).
35. H. J. Reich, *US Pat.*, 4242285; *Chem. Abstr.*, **94**, P139188b (1981).
36. D. E. Ames, A. G. Singh and W. F. Smith, *Tetrahedron*, **39**, 831 (1983).
37. G. H. Schmid and D. G. Garratt, *J. Org. Chem.*, **48**, 4169 (1983).
38. C. Paulmier, *Bull. Soc. Chim. Fr. II*, 237 (1979).
39. M. Kato, R. Michels and W. Heitz, *J. Polym. Sci., Polym. Lett.*, **14**, 413 (1976); M. Kato, *Ger. Offen.*, 2649163; *Chem. Abstr.*, **87**, 53851f (1977).
40. S. Uemura, A. Toshimitsu, M. Okano and K. Ichikawa, *Bull. Chem. Soc. Jpn.*, **48**, 1925 (1975).
41. S. Uemura, H. Miyoshi, M. Wakasugi, M. Okano, O. Itoh, T. Izumi and K. Ichikawa, *Bull. Chem. Soc. Jpn.*, **53**, 553 (1980).
42. S. Uemura, A. Toshimitsu, Y. Kozawa and M. Okano, *Bull. Inst. Chem. Res. Kyoto Univ.*, **59**, 354 (1981).
43. A. Toshimitsu, S. Uemura and M. Okano, *J. Chem. Soc., Perkin Trans. 1*, 1206 (1979).
44. A. Arase and Y. Masuda, *Chem. Lett.*, 785 (1976).
45. A. Arase and Y. Masuda, *Chem. Lett.*, 1115 (1976).
46. C. Paulmier, *Tetrahedron Lett.*, 1797 (1978).
47. C. Paulmier, *Bull. Soc. Chim. Fr. II*, 592 (1979).
48. R. Michels, M. Kato and W. Heitz, *Makromol. Chem.*, **177**, 2311 (1976).
49. G. Cauquis and G. Pierre, *Tetrahedron*, **34**, 1475 (1978).
50. A. Toshimitsu, S. Uemura and M. Okano, *J. Chem. Soc., Perkin Trans. 1*, 1273 (1978).
51. G. Simchen and G. Entenmann, *Justus Liebigs Ann. Chem.*, 1249 (1977).
52. S. Braverman, M. Freund and I. Goldberg, *Tetrahedron Lett.*, **21**, 3617 (1980).
53. A. Haas, B. Koch and N. Welcman, *Chem.-Ztg.*, **98**, 511 (1974).
54. A. Haas, B. Koch and N. Welcman, *Z. Anorg. Alg. Chem.*, **427**, 114 (1976).
55. B. Lindgren, *Tetrahedron Lett.*, 4347 (1979).
56. M. R. Bryce, C. D. Reynolds, P. Hanson and J. M. Vernan, *J. Chem. Soc., Perkin Trans. 1*, 607 (1981).
57. Y. H. Kang and J. L. Kice, *J. Org. Chem.*, **49**, 1507 (1984).
58. S. Tomoda, Y. Takeuchi and Y. Nomura, *Chem. Lett.*, 1069 (1981).
59. S. Tomoda, Y. Takeuchi and Y. Nomura, *Synthesis*, 212 (1985).
60. M. Tiecco, L. Testaferri, M. Tingoli, D. Chianelli and M. Montanucci, *J. Org. Chem.*, **48**, 4289 (1983).
61. E. G. Yarkova, N. R. Safiullina, G. A. Chmutova and I. V. Bautina, *Zh. Obshch. Khim.*, **49**, 2025 (1979); *Chem. Abstr.*, **92**, 75714 (1980).
62. S. Giorgianni, A. Passerini and R. Passerini, *Spectrosc. Lett.*, **16**, 775 (1983).
63. C. E. Sjøgren, *Acta Chem. Scand., Ser. A*, **38**, 657 (1984).
64. W. J. Franklin, R. L. Werner and R. A. Ashby, *Spectrochim. Acta, Part A*, **30**, 387 (1974).
65. F. Ritschl, *Z. Chem.*, **18**, 411 (1978); *Chem. Abstr.*, **90**, 120811v (1979).
66. M. V. Andreocci, M. Bossa, C. Furlani, M. N. Piancastelli and T. Tarantelli, *J. Chem. Soc., Faraday Trans. II*, **75**, 105 (1979).
67. K. Maartmann-Moe, K. A. Sanderud and J. Songstad, *Acta Chem. Scand., Ser. A*, **38**, 187 (1984).
68. B. M. Landsberg, *Z. Naturforsch., Teil A*, **33**, 1333 (1978).

69. T. Sakaizumi, Y. Kohri, O. Ohashi and I. Yamaguchi, *Bull. Chem. Soc. Jpn.*, **51**, 3411 (1978).
70. T. Sakaizumi, T. Itakura and O. Ohashi, *Koen Yoshishu-Bunshi Kozo Toronkai*, 524 (1979); *Chem. Abstr.*, **93**, 176230j (1980).
71. G. A. Chmutova, I. V. Bautina and A. N. Vereshchagin, *Izv. Akad. Nauk SSSR, Ser. Khim.*, 2469 (1976); *Chem. Abstr.*, **86**, 120659u (1977).
72. G. A. Chmutova, I. V. Bautina and A. N. Vereshchagin, *Izv. Akad. Nauk SSSR, Ser. Khim.*, 1554 (1981); *Chem. Abstr.*, **95**, 149826t (1981).
73. G. A. Chmutova, I. V. Bautina, R. U. Belyzlov and A. N. Vereshchagin, *Zh. Org. Khim.*, **17**, 2249 (1981); *Chem. Abstr.*, **96**, 34149c (1982).
74. A. Benedetti C. Preti, L. Tassi and G. Tosi, *Aust. J. Chem.*, **35**, 1365 (1982).
75. A. Croisy, P. Jacquignon and L. Christiaens, *Org. Mass. Spectrom.*, **10**, 1084 (1975).
76. L. Christiaens, J. L. Piette, L. Laitem, M. Baiwir, J. Denoel and G. Llabres, *Org. Magn. Reson.*, **8**, 354 (1976).
77. W. H. Pan and J. P. Fackler, Jr., *J. Am. Chem. Soc.*, **100**, 5783 (1978).
78. G. A. Kalabin, D. F. Kushnarev, G. A. Chmutova and L. V. Kashurnikova, *Zh. Org. Khim.*, **15**, 24 (1979); *Chem. Abstr.*, **90**, 167514r (1979).
79. P. Granger, S. Chapelle and C. Paulmier, *Org. Magn. Reson.*, **14**, 240 (1980).
80. R. B. Valeef, G.A. Kalabin and D. F. Kushnarev, *Zh. Org. Khim.*, **16**, 2482 (1980); *Chem. Abstr.*, **94**, 120695t (1981).
81. G. Llabres, M. Baiwir, J. L. Piette and L. Christiaens, *Org. Magn. Reson.*, **15**, 152 (1981).
82. G. A. Kalabin and D. F. Kushnarev, *Zh. Strukt. Khim.*, **20**, 617 (1979); *Chem. Abstr.*, **92**, 110126c (1980).
83. G. Llabres, M. Baiwir, L. Christiaens and J. L. Piette, *Org. Magn. Reson.*, **21**, 461 (1983).
84. H. M. Hassaneen, A. Shetta and A. S. Shawali, *J. Heterocycl. Chem.*, **17**, 1185 (1980).
85. M. Takahashi and M. Kurosawa, *Bull. Chem. Soc. Jpn.*, **53**, 1185 (1980).
86. H. M. Hassaneen, A. O. Abdelhamid, A. Shetta and A. S. Shawali, *Gazz. Chim. Ital.*, **112**, 545 (1982).
87. A. O. Abdelhamid, H. M. Hassaneen and A. S. Shawali, *J. Heterocycl. Chem.*, **20**, 719 (1983).
88. K. Peseke and N. C. Catanedo, *East Ger. Pat.*, 147361; *Chem. Abstr.*, **95**, P203939x (1981).
89. J. M. Bahan, R. A. W. Johnstone and M. J. Wright, *J. Chem. Soc., Perkin Trans. 1*, 1216 (1975).
90. Meiji Seika Kaisha, *Jpn. Kokai Tokkyo Koho*, JP 5852299 [8352299]; *Chem. Abstr.*, **99**, P53972u (1983).
91. A. Lobert and A. Le Marechal, *J. Chem. Soc., Chem. Commun.*, 447 (1978).
92. J. Liebscher and H. Hartmann, *Synthesis*, 273 (1976).
93. E. R. Clark and M. A. S. Al-Turahi, *J. Organomet. Chem.*, **134**, 181 (1977).
94. J. T. Groves, S. M. Lindenauer, B. J. Haywood, J. A. Knol and J. S. Schultz, *J. Med. Chem.*, **17**, 902 (1974).
95. D. Herrmann and M. Schlenzig, *Z. Chem.*, **18**, 94 (1978).
96. R. H. Mitchell, *Can. J. Chem.*, **54**, 238 (1976).
97. For reviews, see D. L. J. Clive, *Tetrahedron*, **34**, 1049 (1978); H. J. Reich, in *Oxidation of Organic Compounds, Part C* (Ed. W. Trahanovsky), Academic Press, New York, 1978, pp. 1–130.
98. K. B. Sharpless and M. W. Young, *J. Org. Chem.*, **40**, 947 (1975).
99. P. A. Grieco, J. A. Noguez and Y. Masaki, *Tetrahedron Lett.*, 4213 (1975).
100. H. J. Reich, S. Wollowitz, J. E. Trend, F. Chow and D. F. Wendelborn, *J. Org. Chem.*, **43**, 1697 (1978).
101. K. Mori, S. Masuda and T. Suguro, *Tetrahedron*, **37**, 1329 (1981).
102. H. J. Reich and S. Wollowitz, *J. Am. Chem. Soc.*, **104**, 7051 (1982).
103. T. Otsubo, F. Ogura, H. Yamaguchi, Y. Sakata and S. Misumi, *Bull. Chem. Soc. Jpn.*, **55**, 182 (1982).
104. K. Furuichi, S. Yogai and T. Miwa, *J. Chem. Soc., Chem. Commun.*, 66 (1980).
105. G. P. Mullen, N. P. Luthra, R. B. Dunlap and J. D. Odom, *J. Org. Chem.*, **50**, 811 (1985).
106. H. J. Reich, S. K. Shah, P. M. Gold and R. E. Olson, *J. Am. Chem. Soc.*, **103**, 3112 (1981).
107. T. Otsubo, F. Ogura, H. Yamaguchi and S. Misumi, *Synth. Commun.*, **10**, 595 (1980).
108. H. Higuchi and S. Misumi, *Tetrahedron Lett.*, **23**, 5571 (1982).
109. N. Esaki, H. Tanaka, S. Uemura, T. Suzuki and K. Soda, *Biochemistry*, **18**, 407 (1979).
110. M. V. Lakshmikantham, M. P. Cava and A. F. Garito, *J. Chem. Soc., Chem. Commun.*, 383 (1975).
111. N. X. Hu, Y. Aso, T. Otsubo and F. Ogura, *Chem. Lett.*, 603 (1985).

112. J. P. Battioni, D. Mansuy and J. C. Chottard, *Inorg. Chem.*, **19**, 791 (1980).
113. S. Tomoda, Y. Takeuchi and Y. Nomura, *Chem. Lett.*, 253 (1982).
114. P. A. Grieco, S. Gilman and M. Nishizawa, *J. Org. Chem.*, **41**, 1485 (1976).
115. M. Sevrin and A. Krief, *J. Chem. Soc., Chem. Commun.*, 656 (1980).
116. P. A. Grieco, J. Y. Jaw, D. A. Claremon and K. C. Nicolaou, *J. Org. Chem.*, **46**, 1215 (1981).
117. T. G. Back and D. J. McPhee, *J. Org. Chem.*, **49**, 3842 (1984).
118. P. A. Grieco and Y. Yokoyama, *J. Am. Chem. Soc.*, **99**, 5210 (1977).
119. P. A. Grieco, Y. Yokoyama and E. Williams, *J. Org. Chem.*, **43**, 1283 (1978).
120. P. A. Grieco, J. A. Noguez and Y. Masaki, *J. Org. Chem.*, **42**, 495 (1977).
121. J. A. Marshall and G. A. Flynn, *J. Org. Chem.*, **44**, 1391 (1979).
122. P. A. Grieco, T. Takigawa and W. Shillinger, *J. Org. Chem.*, **45**, 2247 (1980).
123. W. R. Roush and T. E. D'Ambra, *J. Org. Chem.*, **46**, 5045 (1981).
124. S. L. Schreiber and K. Satake, *J. Am. Chem. Soc.*, **105**, 6723 (1983).
125. K. Nagao, M. Chiba and S.-W. Kim, *Chem. Pharm. Bull.*, **31**, 414 (1983).
126. S. Takano, M. Takahashi and K. Ogasawara, *J. Am. Chem. Soc.*, **102**, 4282 (1980).
127. P. A. Grieco and M. Nishizawa, *J. Org. Chem.*, **42**, 1717 (1977).
128. M. Nishizawa, P. A. Grieco, S. D. Burke and W. Metz, *J. Chem. Soc., Chem. Commun.*, 76 (1978).
129. M. Arnó, B. Garciá, J. R. Pedro and E. Seoane, *Tetrahedron Lett.*, **24**, 1741 (1983).
130. M. Arnó, G. M. Begona, J. R. Pedro and E. Seoane, *Steroids*, **43**, 305 (1984).
131. G. Majetich, P. A. Grieco and M. Nishizawa, *J. Org. Chem.*, **42**, 2327 (1977).
132. J. P. Cutney and A. K. Singh, *Can. J. Chem.*, **61**, 1111 (1983).
133. U. Schmidt, A. Lieberknecht, H. Bökens and H. Griesser, *J. Org. Chem.*, **48**, 2680 (1983).
134. U. Schmidt and J. Wild, *Angew. Chem. Int. Ed. Engl.*, **23**, 991 (1984).
135. L. E. Friedrich and P. Y. -S. Lam, *J. Org. Chem.*, **46**, 306 (1981).
136. G. Stork and D. J. Morgans, Jr., *J. Am. Chem. Soc.*, **101**, 7110 (1979).
137. S. Takano, M. Tanaka, K. Seo, H. Hirama and K. Ogasawara, *J. Org. Chem.*, **50**, 931 (1985).
138. T. Kametani, H. Matsumoto, H. Nemoto and F. Fukumoto, *J. Am Chem. Soc.*, **100**, 6218 (1978).
139. T. Kametani, *Jpn. Kokai Tokkyo Koho* Jp 79125639; *Chem. Abstr.*, **92**, P128474 (1980).
140. K. Mori and S. Masuda, *Agric. Biol. Chem.*, **42**, 1015 (1978).
141. T. R. Hoye and M. J. Kourth, *J. Org. Chem.*, **44**, 3461 (1979).
142. P. A. Grieco, E. Williams, H. Tanaka and S. Gilman, *J. Org. Chem.*, **45**, 3537 (1980).
143. Y. Fujimoto, H. Miura, T. Shimidzu and T. Tatsuno, *Tetrahedron Lett.*, **21**, 3409 (1980).
144. Institute of Physical and Chemical Research, *Jpn. Kokai Tokkyo Koho,* JP 81139487; *Chem. Abstr.*, **96**, P104552 (1982).
145. R. B. Woodward, E. Logusch, K. P. Nambier, K. Sakan, D. E. Ward, B.-W. Au-Yeung, P. Balaram, L. J. Browne, P. J. Card, C. H. Chen, R. B. Chênevert, A. Fliri, K. Frobel, H.-J. Gais, D. G. Garratt, K. Hayakawa, W. Heggie, D. P. Hesson, D. Hoppe, J. A. Hyatt, D. Ikeda, P. A. Jacobi, K. S. Kim, Y. Kobuke, K. Kojima, K. Krowicki, V. J. Lee, T. Leutert, S. Malchenko, J. Martens, R. S. Matthews, B. S. Ong, J. B. Press, T. V. Rajan Babu, G. Rousseau, H. M. Sauter, M. Suzuki, K. Tatsuta, L. M. Tolbert, E. A. Truesdale, I. Uchida, Y. Ueda, T. Uyehara, A. T. Vasella, W. C. Vladuchick, P. A. Wade, R. M. Williams and H. N.-C. Wong, *J. Am. Chem. Soc.*, **103**, 3210 (1981).
146. D. J. Morgans, Jr., *Tetrahedron Lett.*, **22**, 3721 (1981).
147. S. Takano, M. Morimoto, K. Masuda and K. Ogasawara, *Chem. Pharam. Bull.*, **30**, 4238 (1982).
148. T. Uematsu, T. Umemura and K. Mori, *Agric. Biol. Chem.*, **47**, 597 (1983).
149. M. Ojika, H. Kuyama, H. Niwa and K. Yamada, *Bull. Chem. Soc. Jpn.*, **57**, 2893 (1984).
150. F. E. Ziegler and J. J. Piwinsky, *J. Am. Chem. Soc.*, **104**, 7181 (1982).
151. H. E. Zimmerman and W. D. Ramsden, *Can. J. Chem.*, **62**, 2592 (1984).
152. R. E. Ireland, W. C. Dow, J. D. Godfrey and S. Thaisrivongs, *J. Org. Chem.*, **49**, 1001 (1984).
153. M. A. Adams, A. J. Duggan, J. Smolanoff and J. Meinwald, *J. Am. Chem. Soc.*, **101**, 5364 (1979).
154. T. Nakata, S. Nagao and T. Oishi, *Tetrahedron Lett.*, **26**, 6465 (1985).
155. M. C. Hanan, *Tetrahedron Lett.*, **26**, 6393 (1985).
156. W. R. Roush, H. R. Gillis and A. I. Ko, *J. Am. Chem. Soc.*, **104**, 2269 (1982).
157. P. G. Gassman and S. M. Bonser, *Tetrahedron Lett.*, **24**, 3431 (1983).
158. W. Oppolzer, K. Baettig and T. Hudlicky, *Helv. Chim. Acta*, **62**, 1493 (1979).
159. R. L. Funk, L. H. M. Horcher, II, J. U. Daggett and M. M. Hansen, *J. Org. Chem.*, **48**, 2632 (1983).
160. A. Nickon and A. G. Stern, *Tetrahedron Lett.*, **26**, 5915 (1985).

162. W. C. Still, *J. Am. Chem. Soc.*, **101**, 2493 (1979).
163. P. A. Wender and S. K. Singh, *Tetrahedron Lett.*, **26**, 5987 (1985).
164. N. Kurokawa and Y. Ohfune, *Tetrahedron Lett.*, **26**, 83 (1985).
165. H. Takaku, T. Nomoto and K. Kimura, *Chem. Lett.*, 1221 (1981).
166. Sankyo Chemical Industries, *Jpn. Kokai Tokkyo Koho*, JP 58 04780 [83 04780]; *Chem. Abstr.*, **98**, P161120c (1983).
167. Sankyo Kagaku K. K., *Jpn. Kokai Tokkyo Koho*, JP 58 04797 [83 04797]; *Chem. Abstr.*, **98**, P161121d (1983).
168. C. Boullais, N. Zylber, J. Zylber, J. Guilhem and A. Gaudemer, *Tetrahedron*, **39**, 759 (1983).
169. D. L. J. Clive, G. J. Chittattu, V. Farina, W. A. Kiel, S. M. Menchen, C. G. Russell, A. Singh, C. K. Wong and N. J. Curtis, *J. Am Chem. Soc.*, **102**, 4483 (1980).
170. T. Uematsu, N. Matsuo and Y. Sanemitsu, *Agric. Biol. Chem.*, **48**, 2477 (1984).
171. D. L. J. Clive, G. Chittattu, N. J. Curtis and S. M. Menchen, *J. Chem. Soc., Chem. Commun.*, 770 (1978).
172. P. A. Zoretie, R. J. Chambers, G. D. Marbury and A. A. Riebiro, *J. Org. Chem.*, **50**, 2981 (1985).
173. T. Kametani, H. Nemoto and K. Fukumoto, *Heterocycles*, **6**, 1365 (1977).
174. T. Kametani, H. Nemoto and K. Fukumoto, *Bioorg. Chem.*, **7**, 215 (1978).
175. A. Toshimitsu, S. Uemura and M. Okano, *J. Chem. Soc., Chem. Commun.*, 166 (1977).
176. A. Toshimitsu, T. Aoai, S. Uemura and M. Okano, *J. Org. Chem.*, **45**, 1953 (1980).
177. S. Uemura, A. Toshimitsu, T. Aoai and M. Okano, *J. Chem. Soc., Chem. Commun.*, 610 (1979).
178. S. Uemura, A. Toshimitsu, T. Aoai and M. Okano, *Chem. Lett.*, 1359 (1979).
179. A. Toshimitsu, T. Aoai, S. Uemura and M. Okano, *J. Org. Chem.*, **46**, 3021 (1981).
180. S. Tomoda, Y. Takeuchi and Y. Nomura, *Chem. Lett.*, 1715 (1981).
181. S. Tomoda, Y. Takeuchi and Y. Nomura, *Tetrahedron Lett.*, **23**, 1361 (1982).
182. S. Tomoda, Y. Takeuchi and Y. Nomura, *Chem. Lett.*, 1733 (1982).
183. S. Tomoda, Y. Takeuchi and Y. Nomura, *J. Chem. Soc., Chem. Commun.*, 871 (1982).
184. H. Fischer and S. Zeuner, *Z. Naturforsch., Teil B*, **38**, 1365 (1983).
185. F. B. McCormick, *Organometallics*, **3**, 1924 (1984).
186. H. Bock, S. Augen, P. Rasmus, B. Solouki and E. Weissflog, *Chem. Ber.*, **117**, 187 (1984).
187. S. Tanaka, S. Uemura and M. Okano, *Bull. Chem. Soc. Jpn.*, **50**, 722 (1977).
188. M. H. Quick and R. J. Angelici, *Inorg. Chem.*, **15**, 160 (1976).
189. S. Uemura, A. Toshimitsu, Y. Kozawa and M. Okano, *Bull. Inst. Chem. Res. Kyoto Univ.*, **59**, 350 (1981).
190. G. A. Chmutova and I. V. Bautina, *Zh. Obshch. Khim.*, **49**, 465 (1979); *Chem. Abstr.*, **90**, 186237u (1979).
191. G. A. Chmutova and I. V. Bautina, *Zh. Obshch. Khim.*, **51**, 1657 (1981); *Chem. Abstr.*, **95**, 167972s (1981).
192. G. A. Chmutova, M. A. Kazymova and L. V. Ermolaeva, *Zh. Obschch. Khim.*, **55**, 422 (1985); *Chem. Abstr.*, **102**, 211792x (1985).
193. L. Henrikson and U. Ehrbar, *Synthesis*, 519 (1976).
194. T. G. Back and R. G. Kerr, *J. Organomet. Chem.*, **286**, 171 (1985).
195. H. Suzuki, M. Usuki and T. Hanafusa, *Synthesis*, 705 (1979).
196. S. Uemura, N. Watanabe, A. Toshimitsu and M. Okano, *Bull. Chem. Soc. Jpn.*, **51**, 1818 (1978).
197. R. Herzschuh, B. Birner, L. Beyer, F. Dietze and E. Hoyer, *Z. Anorg. Allg. Chem.*, **464**, 159 (1980).
198. I. N. Azerbaev, L. A. Tsoi, A. D. Salimbaeva, S. T. Cholpankulova, G. A. Ryskieva, L. T. Kalkabaeva and M. Zh. Aitkhozhaeva, *Tr. Inst. Khim. Nauk Akad. Nauk Kaz. SSR*, **52**, 128 (1980); *Chem. Abstr.*, **94**, 208766c (1981).
199. W. E. Ried and E. Hanns, *Justus Liebigs Ann. Chem.*, 201 (1982).
200. W. J. Stec, K. Lesiak and M. Sudol, *Synthesis*, 785 (1975).
201. M. Vaultier and R. Carrie, *Tetrahedron*, **32**, 2525 (1976).
202. W. Reid and H. E. Erle, *Chem. Ber.*, **112**, 640 (1979).
203. A. von Wasserman, *Br. Pat.*, 978; *Chem. Abstr.*, **7**, 2286 (1913).
204. J. N. Barglin, *US Pat.*, 2209184; *Chem. Abstr.*, **35**, 272 (1941); *US Pat.*, 2217611, 2217612, 2217613, 2217614, 2217615; *Chem. Abstr.*, **35**, 1156 (1941); *US Pat.*, 2227058, 2227059, 2227061; *Chem. Abstr.*, **35**, 2646 (1941); *US Pat.*, 2239495; *Chem. Abstr.*, **35**, 4781 (1941); *US Pat.*, 2263716; *Chem. Abstr.*, **36**, 1730 (1942); *US Pat.*, 2275606; *Chem. Abstr.*, **36**, 4240 (1942); *US Pat.*, 2400095; *Chem. Abstr.*, **40**, 4747 (1946).
205. V. A. Nefedov, *Zh. Obshch. Khim.*, **38**, 2191 (1968); *Chem. Abstr.*, **70**, 28556 (1969).

206. J.-L. Piette, R. Lysy and M. Renson, *Bull. Soc. Chim. Fr.*, 3559 (1972).
207. J.-L. Piette, P. Thibaut and M. Renson, *Tetrahedron*, **34**, 655 (1978).
208. S. J. Falcone and M. P. Cava, *J. Org. Chem.*, **45**, 1044 (1980).
209. M. Baiwir, G. Llabres, J. Denoel and J.-L. Piette, *Mol. Phys.*, **25**, 1 (1973).
210. J.-M. Talbot, J.-L. Piette and M. Renson, *Bull. Soc. Chim. Fr.*, 294 (1976).
211. F. Ogura, H. Yamaguchi, T. Otsubo and K. Chikamatsu, *Synth. Commun.*, **12**, 131 (1982).
212. S. Uemura, S. Fukuzawa, A. Toshimitsu and M. Okano, *Tetrahedron Lett.*, **23**, 1177 (1982).
213. S. Uemura, S. Fukuzawa and A. Toshimitsu, *J. Organomet. Chem.*, **250**, 203 (1985).
214. Ajinomoto Co., *Jpn. Kokai Tokkyo Koho*, JP 59 216868 [84216868]; *Chem. Abstr.*, **102**, P185497h (1985).
215. L. Engman, *J. Org. Chem.*, **48**, 2920 (1983).
216. T. Austad, S. Esperås and J. Songstad, *Acta Chem. Scand.*, **27**, 3594 (1973).
217. S. Bekkevoll, I. Svorstøl, H. Høiland and J. Songstad, *Acta Chem. Scand., Ser. B*, **37**, 935 (1983).
218. A. W. Downs, *Chem. Commun.*, 1290 (1968).
219. T. Austad, J. Songstad and K. Åse, *Acta Chem. Scand.*, **25**, 331 (1971).
220. A. Martinsen and J. Songstad, *Acta Chem. Scand., Ser. A*, **31**, 645 (1977).
221. H. K. Spencer, M. V. Lakshmikantham and M. P. Cava, *J. Am. Chem. Soc.*, **99**, 1470 (1977).
222. N. N. Greenwood, R. Little and M. J. Spraque, *J. Chem. Soc.*, 1292 (1964).
223. H. Suzuki, K. Miyoshi and A. Osuka, *Nippon Kagaku Kaishi*, 472 (1981).
224. K. Maartmann-Moe, K. A. Sanderud and J. Songstad, *Acta Chem. Scand., Ser. A*, **35**, 151 (1981).
225. L. Engman and M. P. Cava, *J. Org. Chem.*, **46**, 4194 (1981).
226. P. Wiriyachitra, J. S. Falcone and M. P. Cava, *J. Org. Chem.*, **44**, 3957 (1979).
227. G. Cauquis and G. Pierre, *Bull. Soc. Chim. Fr.*, 736 (1976).
228. M. Lipp, F. Dallacker and I. Meier zu Köcker, *Monatsh. Chem.*, **90**, 41 (1959).
229. W. J. Franklin and R. L. Werner, *Tetrahedron Lett.*, 3003 (1965).
230. J. S. Thayer, *Inorg. Chem.*, **7**, 2599 (1968).

The Chemistry of Organic Selenium and Tellurium Compounds Volume 2
Edited by S. Patai
© 1987 John Wiley & Sons Ltd.

CHAPTER **15**

Organic compounds containing bonds between Se or Te with P, As, Sb and Bi

WOLF-WALTHER DU MONT, RALPH HENSEL, SILVIA KUBINIOK AND LUTZ LANGE

Fachbereich Chemie, Universität Oldenburg, Carl-von-Ossietzky-Strasse 9–11, D-2900 Oldenburg, FRG

I. INTRODUCTION 593
II. PREPARATION OF ORGANYL-SELENO- AND -TELLURO-PHOSPHANES, -ARSANES, -STIBANES AND -BISMUTHANES . . 595
A. Organyl-seleno- and -telluro-phosphinous Acid Esters and Related Arsenic, Antimony and Bismuth Compounds 595
1. Phosphinous acid esters, $R_2{}^1PER^2$ (E = Se, Te). 595
a. Organylseleno(diorganyl)phosphanes 595
b. Organyltelluro(diorganyl)phosphanes 599
2. Arsinous acid esters, $R_2{}^1AsER^2$ (E = Se, Te) 602
a. Organylseleno(diorganyl)arsanes 602
b. Organyltelluro(diorganyl)arsanes. 603
3. Organyl-seleno- and -telluro-(diorganyl)stibanes, $R_2{}^1SbER^2$ (E = Se, Te) 604
4. Organyl-seleno- and -telluro-(diorganyl)bismuthanes, $R_2{}^1BiER^2$ (E = Se, Te) 604
B. Organyl-seleno and -telluro Esters of Phosphonous Acids and Related Arsenic, Antimony and Bismuth Compounds 605
1. Seleno- and telluro-phosphonous esters, $R^1P(ER^2)_2$ (E = Se, Te) . . 605
a. Organylbis(organylseleno)phosphanes 605
b. Organylbis(organyltelluro)phosphanes 606
2. Seleno- and telluro-arsonous esters and related Sb and Bi compounds 606
a. Reaction of dichloro(phenyl)arsane with trimethyl-(methylseleno)silane 606
b. Organylbis(organyl-seleno or -telluro)stibanes 607
c. Methylbis(phenyl-seleno and -telluro)bismuthane 607
C. Organyl-seleno and -telluro Esters of Phosphorous Acid and Related Arsenic, Antimony and Bismuth Compounds 607

1. Phosphorous acid esters, $P(ER)_3$ (E = Se, Te) 607
   a. Tri(organylseleno)phosphanes and halogeno(organylseleno)-phosphanes 607
   b. Tri(*p*-anisyltelluro)phosphane 609
2. Arsenous acid esters, $As(SeR)_3$ [tri(organylseleno)arsanes] . . . 609

III. ORGANYL-SELENO- AND -TELLURO-DIPHOSPHANES . . . 610

IV. SELENO- OR TELLURO-BIS-PHOSPHANES, -ARSANES, -STIBANES AND -BISMUTHANES 611

A. Seleno- or Telluro-phosphinous Anhydrides and Related Arsenic, Antimony and Bismuth Compounds 611
1. Preparation of Seleno- and Telluro-phosphinous anhydrides, R_2PEPR_2 (E = Se, Te) 611
   a. Selenobisphosphanes 611
   b. tellurobisphosphanes 613
2. Synthesis and structures of μ-9,10-chalcogeno-9,10-diarsa-anthracenes 614
3. Preparation of seleno- and telluro-bis(diorganylstibanes) 615
4. Seleno- and telluro-bis(diorganylbismuthanes) 616

B. Cyclic Seleno- or Telluro-bisphosphanes and -bisarsanes 616
1. Chalcogenocyclophosphanes, $(RP)_nE_m$ (E = Se, Te). 616
   a. Selenocyclophosphanes 616
   b. Tellurocyclophosphanes. 617
2. Selenocycloarsanes. 619
3. Selenocyclostibanes 621

V. ORGANYL-SELENO- AND -TELLURO-PHOSPHONIUM IONS . . 621

A. Preparation of Selenophosphonium Ions 621
B. Preparation of Tellurophosphonium Ions. 623

VI. ORGANYL-SELENO- AND -TELLURO-PHOSPHINATES, -PHOSPHONATES AND -PHOSPHATES 624

A. Organylseleno Derivatives of Pentavalent Phosphorus 624
1. Selenophosphinic acid *Se*-organyl esters, $R_2{}^1P(=E)SeR^2$ (E = O, S, Se) 624
2. Selenophosphonic acid *Se*-esters, $R^1P(=E)(XR^2)(SeR^3)$ 625
3. Selenophosphoric acid *Se*-esters 627
   a. Oxidation of phosphites with arylselenenyl halides or with diselenides 627
   b. Oxidation of Selenophosphites with oxygen or sulphur 627
   c. Rearrangement of selenophosphoric acid *O*-esters containing at least one n-alkyl or 2-alkenyl group R^2 628
   d. Alkylation and acylation of anionic selenophosphates 628
   e. Hydrogen chloride elimination from phosphoric acid chlorides with phenyl- or butyl-selenols in the presence of triethylamine . 629
   f. Chloroselenation reactions. 630

B. Organyltelluro Derivatives of Pentavalent Phosphorus 630

VII. PENTAVALENT THREE-COORDINATE PHOSPHORUS COMPOUNDS WITH P=Se DOUBLE BONDS 631

A. Methylene(selenoxo)phosphoranes, $R^1P(=R^2)(=Se)$ 631
B. metaselenophosphonates, $R^1P(=E)(=Se)$ 633
C. metaselenophosphates, $R^1R^2NP(=NR^1)(=Se)$ 633

VIII. FOUR-COORDINATE P^V, As^V AND Sb^V COMPOUNDS WITH DOUBLE BONDS TO SELENIUM OR TELLURIUM 634

A. Four-coordinate Phosphorus Compounds with P=Se or P=Te Bonds 634

1. Chalcogenophosphoranes, $R^1P_2{}^2P{=}E$ ($E = Se, Te$; R^1, R^2 = alkyl, aryl or H) ('phosphane chalcogenides') 634
a. Phosphane selenides 634
b. Phosphane tellurides 635
2. Chalcogenophosphinic esters, $R_2{}^1(R_n{}^2X)P{=}E$ ($E = Se, Te$). . . . 636
a. Phosphinic acid derivatives with P=Se bonds 636
b. Tellurophosphinic acid esters 637
3. Chalcogenophosphonic esters, $R^1(R_n{}^2X)_2P{=}E$ ($E = Se, Te$) . . . 637
a. Selenophosphonic acid derivatives with P=Se bonds 637
b. Tellurophosphonic acid derivatives with P=Te bonds 638
4. Phosphoric acid derivatives with P=Se or P=Te bonds 639
a. Selenophosphoric acid esters, amides and other derivatives . . 639
b. Tellurophosphoric acid amides and esters 639
B. Arsenic Compounds with As=Se or As=Te Double Bonds 641
1. Tertiary arsane selenides (selenoarsoranes) 641
2. Selenoarsinic acid derivatives containing a terminal As—Se bond . 641
C. Stibane Selenides 642
IX. STRUCTURE, BONDING AND REACTIONS OF ORGANIC SELENIUM AND TELLURIUM COMPOUNDS WITH BONDS BETWEEN Se OR Te AND P, As, Sb OR Bi 642
A. ^{31}P, ^{77}Se and ^{125}Te NMR Spectra 642
B. Vibrational, UV–Visible and He(I)–PE Spectra. 644
C. Bond Lengths and Angles 645
D. Selected Reactions Involving Bonds between Se or Te and P, As, Sb or Bi 647
1. Reactions of seleno- and telluro-phosphanes and -arsanes with cationic and anionic hydrides 647
2. Types of coordination of ligands with bonds between Se or Te and P, As, Sb or Bi 648
a. P^{III} and As^{III} ligands 648
b. η^2P-, Se-coordination 649
c. Se- or Te-coordination 650
3. Alkylation, oxidation, deoxygenation and other reactions 650
X. ACKNOWLEDGEMENT 652
XI. REFERENCES 652

I. INTRODUCTION

Recently, considerable interest has been focused on the properties of compounds containing single, double or triple bonds or secondary bonding involving the heavier main group elements. As far as bonds between chalcogens and Group VA (Group 15) elements (pnictogens) are concerned, the basic chemistry of compounds with P—S and P—Se bonding is well covered in books and recent reviews[1–5], whereas only scattered information is available on compounds with P—Te bonds and on organic compounds containing bonds between Se or Te and As, Sb or Bi. The main concern of this chapter will be the presentation of recent work on P^{III}, As^{III}, Sb^{III} and Bi^{III} compounds with organylseleno and organyltelluro substituents or generally with biscoordinate Se or Te (Sections II–IV). Organylseleno- and organyltelluro-substituted compounds with four-coordinate P and As are treated in Sections V (organylseleno- and -telluro-phosphonium ions) and VI (phosphinic, phosphonic or phosphoric acid *Se*- or *Te*-esters and anhydrides). As far as compounds with (at least formally) double bonds between Se or Te and P, As and Bi are concerned, which do not contain Se—C or Te—C bonds, only very selected results on structures, bonding and reactivity will be presented. Among these are the formation of novel seleno-metaphosphinates, -metaphosphonates and -metaphosphates (Section VII, selenoxo-phosphoranes of P^V with coordination number 3) and novel chalcogen transfer reactions between four-coordinate P^V, As^V and Sb^V compounds with double (or ylidic) bonds to Se or Te and the corresponding phosphanes, arsanes and stibanes (Section VIII). The mainly preparative Sections II–VIII are followed by a discussion of spectroscopic and structural features of compounds of Se or Te with bonds to P, As, Sb or Bi and of chemical reactions involving these bonds (Section IX). Finally, all compounds of Se or Te with bonds to P, As, Sb or Bi contain lone pairs of electrons that give them the ability to coordinate with Lewis acids via lone pairs at Se/Te or the main Group V element or both. Recent results on these Lewis base to Lewis acid interactions will also be presented in Section VIII; for general coordination chemistry of organic Se and Te compounds, the review[6] by Gysling and also his chapter in this volume can be consulted.

TABLE 1. Classification and naming of organic compounds of Se or Te with bonds to P, As, Sb and Bi

Structure	Section	Compounds
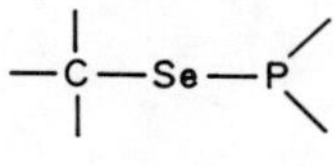	II	Organylselenophosphanes (selenophosphinous, -phosphonous and -phosphorus esters) and related Te and As, Sb, Bi compounds)
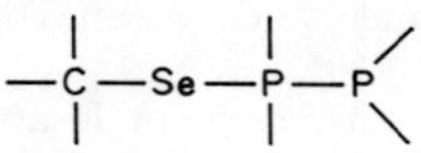	III	Organylselenodiphosphanes (and related Te and As compounds)
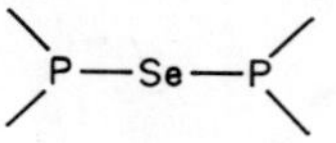	IV	Selenobisphosphanes (selenophosphinous, -phosphonous and -phosphorous anhydrides) (and related compounds, including seleno- or telluro-cyclophosphanes)
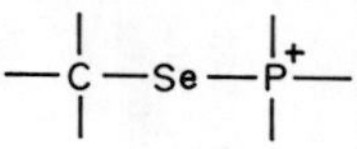	V	Organylselenophosphonium ions (and related Te and As compounds)
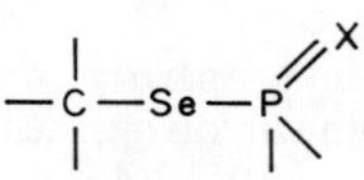	VI	Seleno-phosphinic, -phosphonic and -phosphoric acid and *Se*-esters
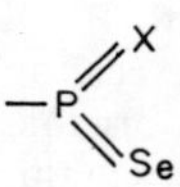	VII	Selenometa-phosphinates, -phosphonates and -phosphates (imino- or methylene-selenoxophosphoranes)
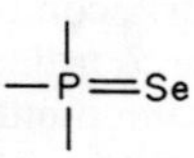	VIII	Selenophosphinic, selenophosphonic and selenophosphoric acids, esters and amides; selenophosphoranes ('phosphane selenides') with P=Se bonds and related compounds

A note on nomenclature may be useful as the nomenclature of P, As, Sb and Bi compounds is somewhat confused[1,7]. To find comparable names for structurally comparable P, As, Sb and Bi compounds, the trivalent derivatives will preferably be named phosphanes, arsanes, stibanes or bismuthanes, but classification as derivatives of phosphinous, phosphonous or phosphorous acid derivatives (arsinous, etc) will also be mentioned. Few selenium and no tellurium compounds of pentavalent As and Sb are known. P^V compounds will preferably be named as derivatives of phosphinic, phosphonic and phosphoric acids.

The classification and naming of the compounds concerned are outlined in Table 1.

It should be mentioned that various inorganic phosphorus selenides are known as molecular compounds, but a phosphorus telluride has not yet been well characterized. Arsenic gives both molecular and polymeric chalcogenides, and antimony and bismuth chalcogenides are known to have lattice structures with interesting solid-state properties[8]. Electronegativity differences between Se or Te heavier main Group V elements are generally low (close to zero in the case of P/Te). Owing to increasing atomic radii, the single bond strengths decrease from P to Bi, and this is even more true for the tendency to form double bonds; π-overlap seems to be very poor between Te and As, Sb and Bi and generally in all cases where Bi would be involved in (formal) double bonding (see Section IX).

II. PREPARATION OF ORGANYL-SELENO- AND -TELLURO-PHOSPHANES, -ARSANES, -STIBANES AND -BISMUTHANES

A. Organyl-seleno- and -telluro-phosphinous Acid Esters and Related Arsenic, Antimony and Bismuth Compounds.

1. Phosphinous acid esters, $R_2^1PER^2$ ($E = Se, Te$)

a. Organylseleno(diorganyl)phosphanes. Early work on the formation of selenophosphanes and -arsanes was based on the reaction of bis(trifluoromethylseleno)mercury with iodophosphanes and iodoarsanes[9]. Bis(trifluoromethyl)-trifluoromethylselenophosphane(**1**) was isolated in quantitative yield by trap to trap condensation as colourless liquid by Emeleus *et al.*[9].

$$2\,(CF_3)_2PI + Hg(SeCF_3)_2 \longrightarrow 2\,(CF_3)_2PSeCF_3 + HgI_2 \qquad (1)$$

(1)

$$Ph_2PI + MSeCF_3 \longrightarrow Ph_2PSeCF_3 + MI \qquad (2)$$

$M = Cs, Tl$ **(2)**

In a similar manner, trifluoromethylselenolates of Cs and Tl^I react with iododiphenylphosphane to give diphenyl(trifluoromethylseleno)phosphane (**2**)[10] as a reddish–yellow oil that is fairly soluble in pentane. Various reactions which lead to diphenyl(phenylseleno)phosphane (**3**) as an intermediate have been described by Petragnani *et al.*[11]. Air oxidation and hydrolysis led in all cases to the oxidized species **4**. The basic types of P—Se bond formation were the reactions of chlorodiphenylphosphane with phenylselenomagnesium bromide, with phenylselenol or with diphenyl diselenide. Diphenyl diselenide is also cleaved by diphenylphosphinomagnesium bromide, and finally phenylselenyl bromide reacts with diphenylphosphane or with the phosphorus Grignard reagent[11].

$$Ph_2PCl + PhSeMgBr \xrightarrow[-\text{'MgBrCl'}]{} [Ph_2PSePh] \qquad (3)$$

(3)

$$Ph_2PCl + PhSeH \xrightarrow[-HCl]{} [3] \qquad (4)$$

$$Ph_2PMgBr + PhSeBr \xrightarrow[-MgBr_2]{} [3] \qquad (5)$$

$$Ph_2PH + PhSeBr \xrightarrow[-HBr]{} [3] \qquad (6)$$

$$Ph_2PCl + PhSeSePh \longrightarrow [3] + PhSeCl \qquad (7)$$

$$Ph_2PMgBr + PhSeSePh \longrightarrow [3] + PhSeMgBr \qquad (8)$$

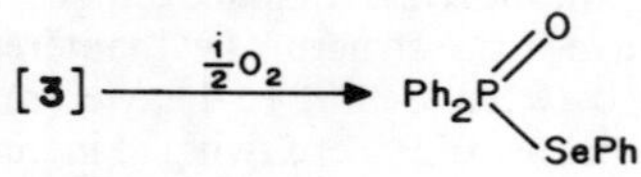

(9)

(4)

Compound [3] is halogenated by phenylselenyl halides, but hydrolysis of the postulated halogenophosphoranes led also to **4**, which was isolated in all cases as colourless crystals (see Section VI)[11]. Pure [3] was obtained by adding chlorodiphenylphosphane to an equimolar quantity of phenylselenol and a slight excess of triethylamine in diethyl ether under inert gas at $-10\,°C$. Distillation led to the pure, pale yellow product. The compound is oxidized in air within a few hours. It was reported to isomerize to $Ph_3P{=}Se$ within 3–4 h at 100 °C, but no details were given on the evidence for this isomerization[12]. The attempted preparation of [3] by reaction of sodium phenylselenolate with chlorodiphenylphosphane lead mainly to Ph_3PSe, even at lower temperatures. It was proposed that the selenophosphane to selenophosphorane (phosphaneselenide) rearrangement is due to the weaker Se—C bond in [3] compared with the S—C bond of the corresponding sulphur compound, which did not rearrange under comparable conditions[12]. A versatile method for the formation of methylseleno derivatives of phosphorus and arsenic is the reaction of trimethyl(methylseleno)silane with chlorophosphanes (and chloroarsanes) which gives quantitative yields of chloro(trimethyl)silane[13].

$$Ph_2PCl + PhSeH + Et_3N \longrightarrow Ph_2PSePh + Et_3NH^+Cl^- \qquad (10)$$

$$Ph_2PCl + PhSeNa \longrightarrow Ph_3P=Se + NaCl \qquad (11)$$

$$Ph_2PCl + MeSeSiMe_3 \longrightarrow \underset{(5)}{Ph_2PSeMe} + Me_3SiCl \qquad (12)$$

Compound **5** was characterized by performing the reaction in an NMR tube. Removal of the volatile material containing chlorotrimethylsilane and excess of trimethyl(methylseleno)silane led to **5** as viscous yellow oil that is thermally stable. In a similar manner, the reaction of disilylselane with bromodifluorophosphane lead to P—Se bond formation; an NMR study of the reaction showed that with excess of bromodifluorophosphane, disilylselane was completely consumed in the equilibrium, but digermylselane was not[14] (see also Section IV).

$$F_2PBr + (H_3Si)_2Se \rightarrow \underset{(6)}{H_3SiSePF_2} + H_3SiBr \qquad (13)$$

$$2H_3SiSePF_2 \rightleftharpoons (H_3Si)_2Se + \underset{(7)}{Se(PF_2)_2} \qquad (14)$$

A detailed study on reaction paths to thio-, seleno- and telluro-phosphinous esters and related arsenic compounds by Dehnert, Grobe *et al.*[15] showed that so-called (2 + 2) dismutation reactions of diorganyldichalcogenides with tetraorganyl-diphosphanes or -diarsanes[16–19] are very effective for the formation of bonds between sulphur, selenium or tellurium and phosphorus or arsenic.

$$R_2{}^1PPR_2{}^1 + R^2SeSeR^2 \longrightarrow 2\,R_2{}^1PSeR^2 \qquad (15)$$

R^1	R^2	Compound	Ref.
Me	Ph	Me_2PSePh **(8)**	16
Me	Me	Me_2PSeMe **(9)**	15
CF_3	Me	$(CF_3)_2PSeMe$ **(10)**	15
CF_3	CF_3	$(CF_3)_2PSeCF_3$ **(1)**	15
Me	CF_3	Me_2PSeCF_3 **(11)** (unstable above − 30 °C)	15

All these P—P/Se—Se dismutation reactions give quantitatively crude yields of the P—Se-bonded species (as do most diphosphane/disulphide reactions). The dismutation can even occur when one phosphorus atom of the diphosphane is coordinated to pentacarbonyl chromium, molybdenum or tungsten[20,21]. Bis (trifluoromethyl) diselenide is more reactive in dismutation reactions than dimethyl diselenide.

$$(CO)_5M{-}PMe_2{-}PMe_2 + MeSeSeMe \xrightarrow[40-50\,^{\circ}C\,(C_6D_6)]{>5\,d} (CO)_5M{-}PMe_2SeMe + \underset{(\mathbf{9})}{Me_2PSeMe} \qquad (16)$$

M = Cr (**12**)
M = Mo (**13**)
M = W (**14**)

$$(CO)_5Mo{-}PMe_2PMe_2 + CF_3SeSeCF_3 \xrightarrow[(C_6D_5CD_3)]{-10\,^{\circ}C} \underset{(\mathbf{15})}{(CO)_5Mo{-}PMe_2SeCF_3} + \underset{(\mathbf{11})}{Me_2PSeCF_3} \qquad (17)$$

The molybdenum complex **15** is significantly more thermally stable than non-coordinated dimethyl(trifluoromethylseleno)phosphane (**11**). The complexes **12**–**15** were easily separated from the volatile ligands **9** and **11**[21]. Dismutation reactions of coordinated tetramethyldiphosphane are generally slower than with tetramethyldiphosphane itself; double coordination of the diphosphane (as bridging bidentate ligand between pentacarbonyl-chromium, -molybdenum and -tungsten moieties) leads to complete loss of the ability to undergo (2 + 2) dismutations even with very reactive bis(trifluoromethyl) disulphide[21].

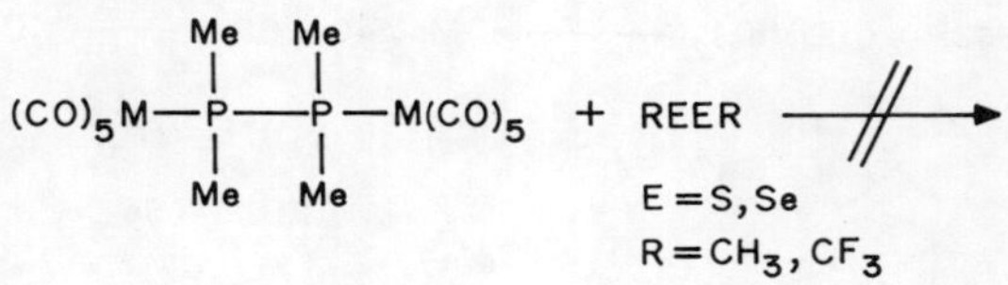

(18)

A novel way to make P—Se bonds is the selenation or hydroselenation of phosphenes (alkylidenephosphanes). Addition of methyselenol to difluoromethylidene-(trifluoromethyl)phosphane leads to chiral difluoromethyl(trifluoromethyl)-(methylseleno)phosphane (**16**), i.e. hydroselenation across the P=C double bond occurs via nucleophilic attack at phosphorus in this (first) case[22].

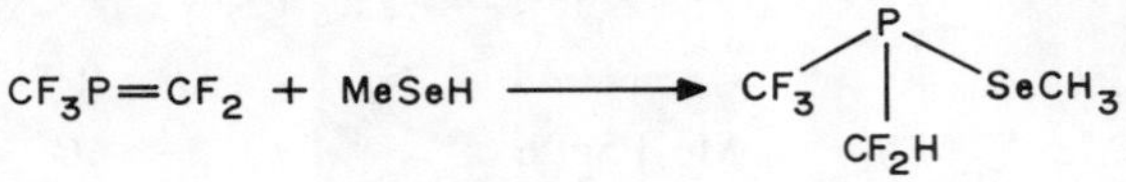

(19)

(16)

With elemental selenium (as with sulphur), alkylidenephosphanes react to give either 1,2-λ^3-selenophosphiranes or the isomeric alkylideneselenoxophosphoranes (see Section VII).

A decision between the two types of structures A and B can easily be made with help of ^{31}P NMR spectroscopy. Typically, ^{31}P NMR shifts of phosphiranes A are upfield compared with open-chain phosphanes (effect of small ring size on ^{31}P resonance), and the NMR coupling constant $^{n}J(^{77}Se^{31}P)$ is much less than 200 Hz in all known phosphiranes and diphosphiranes. Imino- or alkylidene selenoxophosphoranes B also give ^{31}P resonances at comparatively high field in most cases, but $^{1}J(^{77}Se^{31}P)$ is very large (available values are all more than 800 Hz) (for details, see Sections VII and IX).

A B

1,2-di-*t*-butylmethylenephosphane reacts with elemental sulphur or selenium at room temperature within 2 weeks (S) or 3 days (Se). Only phosphirane formation was observed in each case[23].

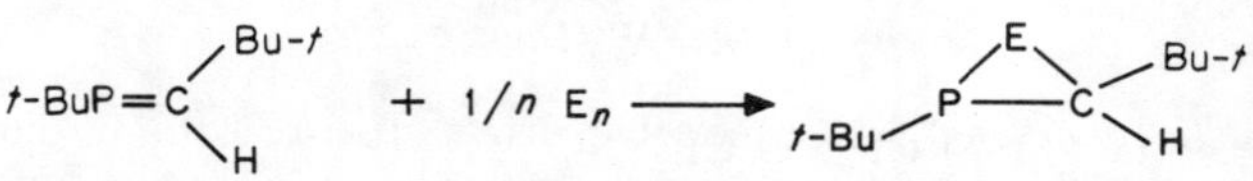

(20)

(17) E=S

(18) E=Se

The preference for the three-membered rings **17** and **18** over the isomeric methylene(thioxo and selenoxo)phosphoranes (type B) was rationalized by molecular orbital calculations by Schoeller and Niecke[24], which predict a non-polar π-system with a low-energy LUMO for the parent 1,2-di-*t*-butylmethylenephosphane (see Section VII). Similarly, bis(trimethylsilyl)methylene(trimethylsilylethinyl)phosphane gives '(2 + 1) cycloaddition' with S and Se leading to 1,2-λ^3-thia- and -selena-phosphiranes[25].

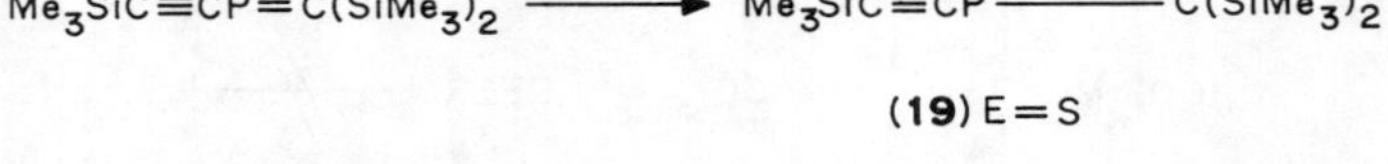

(21)

(19) E=S

(20) E=Se

Compounds **19** and **20** are yellow liquids that partially decompose to the starting materials at room temperature, but purification has been achieved by distillation *in vacuo*. Both compounds can be desulphurated or deselenated with triphenylphosphane (regeneration of the alkylidenephosphane). With 2,3-dimethylbutadiene, cycloaddition occurs with initial loss of sulphur or selenium; NMR evidence for the cycloreversion of **19** was provided by NMR identification of the tetrahydrophosphabenzene intermediate that is sulphurated to give the final product **21**.

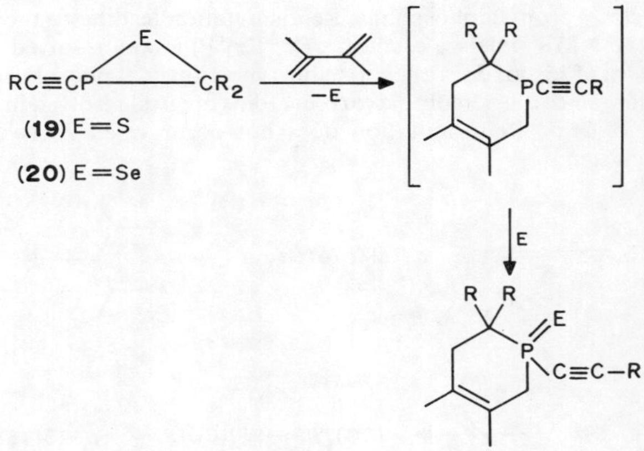

(**19**) E=S

(**20**) E=Se

(**21**) E=S

(**22**) E=Se

(22)

These findings indicate that thia- and selena-phosphiranes may be useful precursors of species with doubly bonded phosphorus that are easily formed by desulphuration and deselenation reactions. Selenaphosphiranes **20** and **22** are apparently more stable than seleniranes, which decompose thermally with alkene formation (see Section IX).

b. Organyltelluro(diorganyl)phosphanes. The dismutation reaction of dimethylditelluride with tetramethyldiphosphane leads only to an equilibrium mixture of dimethyl(methyltelluro)phosphane (**23**) with the starting materials; **23** cannot be separated from Me_2Te_2 and Me_4P_2. Quite differently, the dismutation of dimethylditelluride with tetrakis(trifluoromethyl)diphosphane is quantitative after 4 days at 50 °C and methyltellurobis(trifluoromethyl)phosphane (**24**) was the first tellurophosphane that could be obtained in pure state as an orange–yellow liquid that boils at 118 °C[15].

$$Me_2PPMe_2 + MeTeTeMe \rightleftharpoons 2\ Me_2PTeMe \qquad (23)$$

(23)

$$(CF_3)_2PP(CF_3)_2 + MeTeTeMe \rightarrow 2\ (CF_3)_2PTeMe \qquad (24)$$

(24)

Bulky substituents at the diphosphane appear to favour dismutations with di-*p*-tolylditelluride. Dismutation with tetra-*t*-butyldiphosphane is complete after heating the educts for 1 h in boiling toluene, but with tetra-*i*-propyldiphosphane the (2 + 2) dismutations with the same ditelluride leads to a reaction mixture that contains the tellurophosphane **25** as the main product with a few percent of the educts still present in the equilibrium. Compounds **25** and **26** were obtained as pure compounds by distillation *in vacuo*[26] as yellow liquids.

$$R_2PPR_2 + p\text{-TolTeTeTol-}p \longrightarrow 2\ R_2PTeTol\text{-}p \qquad (25)$$

(25) R = *i*-Pr
(26) R = *t*-Bu

Older (originally distilled) samples of **25** contain, again, small amounts of di-*p*-tolylditelluride and tetra-*i*-propyldiphosphane. Re-dismutation leading to the educts is not a fast process on the NMR time-scale, since $^1J(^{125}Te^{31}P)$ is well resolved in the ^{31}P and ^{125}Te NMR spectra of **25**, nor does one tolytelluro substituent at phosphorus lead to a low barrier of inversion, since diastereotopic carbon atoms are well resolved in the (25 °C) ^{13}C NMR spectrum of **25**[26]. Re-dismutation does not occur when **25** is coordinated to pentacarbonyltungsten[27].

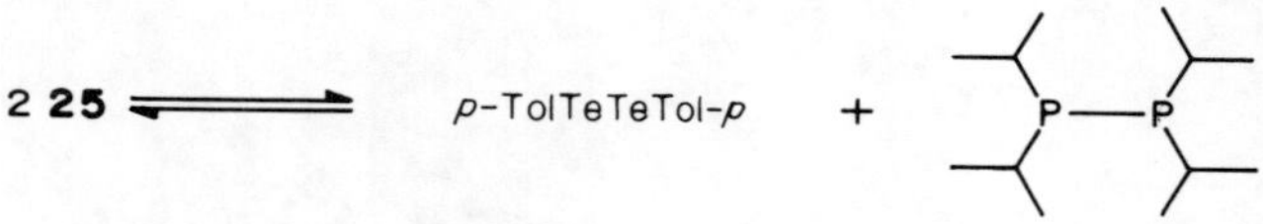

(26)

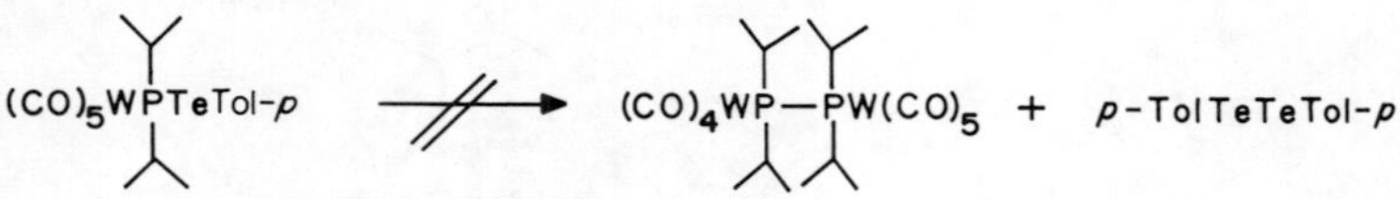

(27)

This finding is consistent with Grobe and Le Van's observation[21] that the reverse reaction (of coordinated tetramethyldiphosphane as bridging bidentate ligand with disulphides and diselenides) does not occur. Di-*t*-butyl(trimethyl-silyl, -germyl and -stannyl)tellurophosphanes are formed by insertion of elemental tellurium into the P—Si, P—Ge and P—Sn bonds of di-*t*-butyl(trimethyl-silyl, -germyl and -stannyl)phosphane[28]. n-$Bu_3P{=}Te$ may also serve as source of tellurium atoms; in each case, the short-lived intermediates $R_2P(=Te)MMe_3$ rearrange with migration of the trimethyl-silyl, -germyl and -stannyl groups from phosphorus to tellurium (a kind of retro-Arbuzov rearrangement). Subsequently, a rearrangement centred at tellurium leads to equilibrium mixtures of **27–29** with the symmetrically substituted tellurium compounds (equation 29). In this case, no significant retro-(2 + 2) dismutation (which would give the ditellurides $Me_3MTeTeMMe_3$, M = Si, Ge, Sn)[28] occurs.

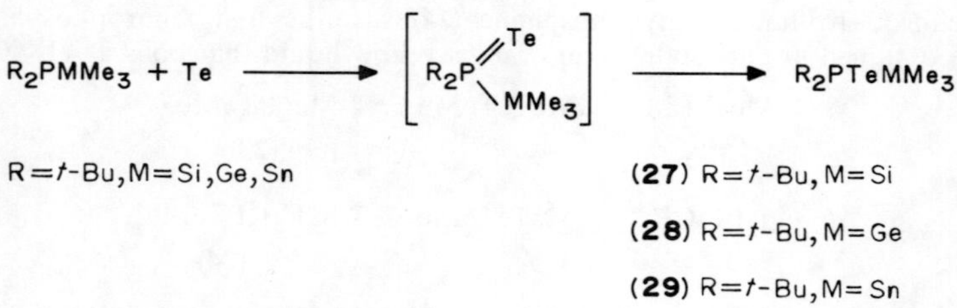

(28)

$$2\ R_2PTeMMe_3 \rightleftharpoons (Me_3M)_2Te + Te(PR_2)_2 \qquad (29)$$

(27–29) M = Si, Ge, Sn **(30)** R = *t*-Bu

Compounds **27**–**29** are not further oxidized by excess of tellurium; in contrast, excess of sulphur or selenium oxidizes R_2PSiMe_3 to the pentavalent phosphorus compounds $t\text{-}Bu_2P({=}E)ESiMe_3$ (E=S, Se)[29]. NMR spectra of compounds related to **27**–**29** have also been recorded after reacting $(H_3M)_2Te$ (M = Si, Ge) with bromodifluorophosphane[14]. Compounds **31**–**33** decomposed thermally by precipitation of elemental tellurium[14].

$$(H_3M)_2Te \xrightarrow[-H_3MBr]{F_2PBr} F_2PTeMH_3 \xrightarrow[-H_3MBr]{F_2PBr} F_2PTePF_2 \qquad (30)$$

$$\textbf{(31)}\ M = Si \qquad \textbf{(33)}$$
$$\textbf{(32)}\ M = Ge$$

P—Te bonds may also be formed by metal halide elimination from a telluromagnesium or tellurolithium reagent with di-*t*-butylchlorophosphane. Thus **26** is obtained conveniently in high yield with *p*-tolyltelluromagnesium bromide[30,31]:

$$t\text{-}Bu_2PCl + p\text{-}TolTeMgBr \xrightarrow{(THF)} t\text{-}Bu_2PTeTol\text{-}p + \text{‘MgBrCl’} \qquad (31)$$
$$\textbf{(26)}$$

$$(t\text{-}Bu)_2PCl + (Me_3Si)_3CTeLi \xrightarrow[-LiCl]{(THF)} (Me_3Si)_3CTeP(Bu\text{-}t)_2 \qquad (32)$$
$$\textbf{(34)}$$

In contrast to the thermally stable **26**, the bulky **34** rearranges at room temperature with formation of the corresponding symmetrically substituted tellurium compounds[32]:

$$2\ \textbf{34} \longrightarrow (Me_3Si)_3CTeC(SiMe_3)_3 + (t\text{-}Bu)_2PTeP(Bu\text{-}t)_2 \qquad (33)$$
$$\textbf{(30)}$$

Starting with the tris(trimethylsilyl)methyl ditelluride anion, the first ditellurophosphane (**35**) was characterized by NMR spectroscopy as a thermally labile compound that disproportionates into an organic tritelluride and **34**[32].

$$(t\text{-}Bu)_2PCl + (Me_3Si)_3CTeTeLi \xrightarrow[-LiCl]{(THF)} (Me_3Si)_3CTeTeP(Bu\text{-}t)_2$$
$$\textbf{(35)}$$

$$2\ \textbf{35} \longrightarrow \textbf{34} + (Me_3Si)_3CTeTeTeC(SiMe_3)_3 \qquad (34)$$

Extremely bulky substituents at tellurium disfavour (2 + 2) dismutation reactions with tetra-*t*-butyldiphosphane and tetra-*i*-propyldiphosphane, but P—Te bonds can be made with 2, 4, 6-tri-*t*-butylphenyltellurolithium[33]:

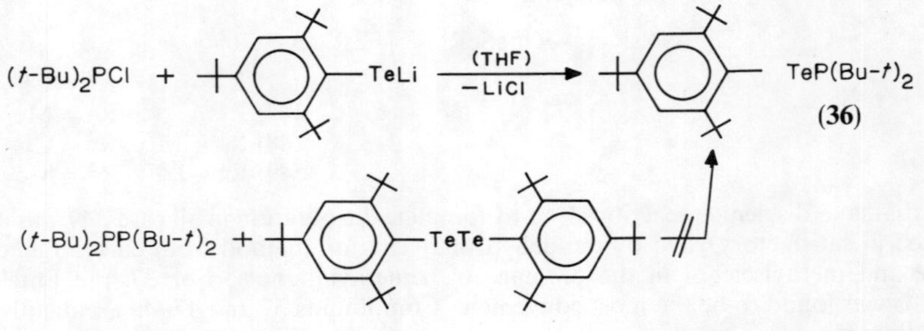

(35)

(36)

(36)

2. Arsinous acid esters, $R_2^1AsER^2$ (E = Se, Te)

a. Organylseleno(diorganyl)arsanes. As mentioned previously (Section II.A.1.a), the first perfluoroalkylselenoarsanes were prepared by Emeleus *et al.*[9] from mercury-selenium compounds with two equivalents of iodobis(trifluoromethyl)arsane. Vacuum fractionation of the volatile products gave the pure colourless compounds **37** and **38**.

$$2\,(CF_3)_2AsI + Hg(SeR)_2 \longrightarrow 2\,(CF_3)_2AsSeR + HgI_2 \tag{37}$$

(37) $R = CF_3$
(38) $R = n\text{-}C_3F_7$

Sagan *et al.*[34] found that protolysis of dialkyl(diethylamino)arsanes with alcohols, thiols and selenols is a convenient method for the synthesis of compounds that contain As—O, As—S and As—Se bonds. Qualitative observation of the progress of amine formation showed the selenols to be most reactive and alcohols to be least reactive, suggesting that the reaction rate depends on the acidity of the protic agents.

$$R_2^1AsNEt_2 + R^2SeH \longrightarrow R_2^1AsSeR^2 + Et_2NH \tag{38}$$

(39) $R^1 = Me$, $R^2 = Et$
(40) $R^1 = Me$, $R^2 = Ph$
(41) $R^1 = Me$, $R^2 = CH_2Ph$
(42) $R^1 = Me$, $R^2 = 1\text{-}C_{10}H_7$
(43) $R^1 = Et$, $R^2 = Ph$
(44) $R^1 = Et$, $R^2 = CH_2Ph$
(45) $R^1 = Pr$, $R^2 = Ph$

After expelling the amine by refluxing the reaction mixture and removing excess of the selenol by distillation, **39**–**45** were obtained in nearly quantitative yield (excess of the selenol leads to precipitation of diethylammonium alkaneselenolates)[34]. Compounds **39**–**45** are fairly inert towards air and moisture, but on extended exposure to air dialkylarsinic acids are formed. Dialkyl (alkoxy)arsanes are much more sensitive and the reactivity of dialkyl(alkylthio)arsanes towards moist air is intermediate. Infrared spectra confirmed that Arbuzov-like rearrangements did not occur with **39**–**45**.

Trimethyl(methylseleno)silane reacts with chlorodiphenylarsane to give methylselenodiphenylarsane (**46**) under mild conditions. The preparation and characterization of **46** are identical with that of the corresponding phosphane **5**[13].

$$Ph_2AsCl + MeSeSiMe_3 \longrightarrow Ph_2AsSeMe + Me_3SiCl \tag{39}$$

(46)

Dehnert, Grobe *et al.*[15] used both substitution and (2 + 2) dismutation reactions for the preparation of dialkyl(alkylseleno)arsanes:

$$R_2^1AsAsR_2^1 + R^2SeSeR^2 \longrightarrow 2R_2^1AsSeR^2 \tag{40}$$

(37) $R^1 =$ $R^2 = CF_3$
(47) $R^1 =$ $R^2 = Me$
(48) $R^1 = Me$, $R^2 = CF_3$
(49) $R^1 = CF_3$, $R^2 = Me$

The diarsane/diselenide reactions lead to complete dismutation in all cases; **47** was also obtained in satisfactory yield by the dehydrohalgenation method from chlorodimethylarsane and methylselenol in the presence of trimethylamine[15]. For **37**, the Emeleus synthesis was found to be the most convenient. Compounds **37** and **47**–**49** are distillable yellow liquids.

In the course of a synthetic study on biologically active glucose esters of dimethylarsinous acid, Zingaro and coworkers[35,36] used the (2 + 2) dismutation of tetramethyldiarsane with 6,6-diselenobis(1,2,3,4-tetra-*O*-acetyl-α- or -β-D-glucopyranose) to give 1,2,3,4-tetra-*O*-acetyl-6-*Se*-dimethylarsino-6-seleno-α- and -β-D-glucopyranose (**50** and **51**). 6-*Se*-Dimethylarsino-6-seleno-β-D-glucopyranose (**52**) was prepared as a yellow solid in the same manner. Under alkaline conditions, it is possible to deacetylate thio sugar esters of dimethylarsinous acid, but under similar conditions the C—Se—$AsMe_2$ linkage is also subject to hydrolysis (in these cases, hydrolysis of the As—Se compound under aerobic conditions is about six times faster than hydrolysis of the corresponding sulphur compound).

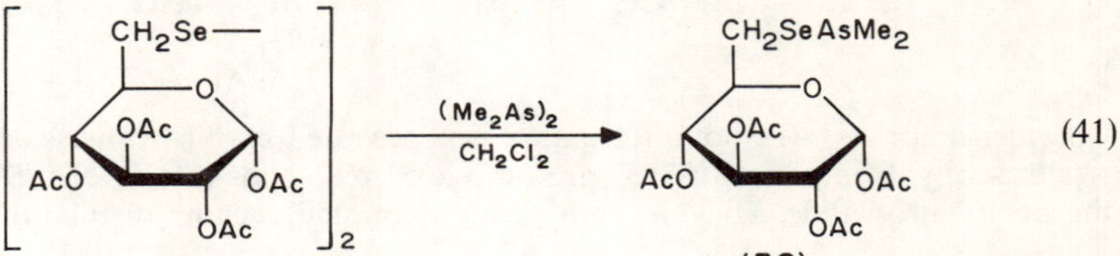

(**50**) (41)

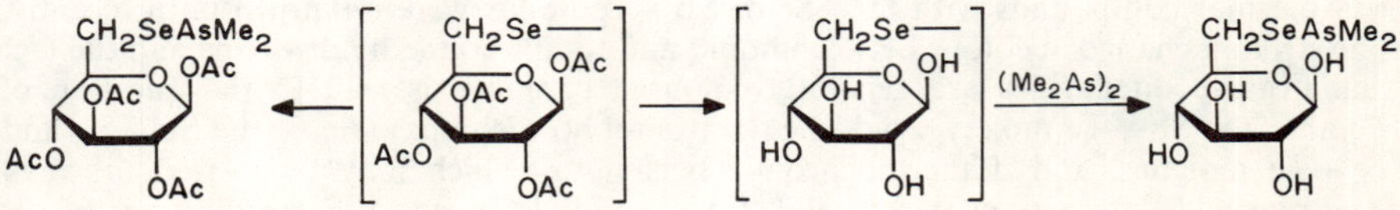

(**51**) (**52**)

(42)

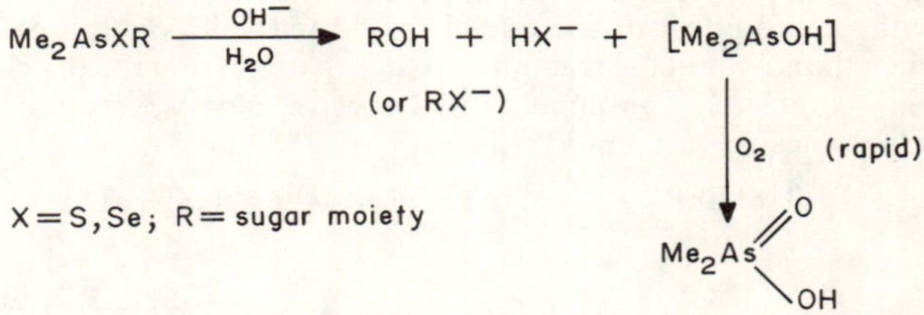

(43)

A further remarkable difference in the reactivity of 6-seleno and 6-thio sugars is that unlike the 6,6-diselenobis(β-D-glucopyranose), the corresponding 6,6-dithiobis(β-D-glucopyranose) did not react with tetramethyldiarsane.

b. Organyltelluro(diorganyl)arsanes. The (2 + 2) dismutations of diarsanes with dimethylditelluride are very similar to those with diphosphanes: with tetramethyldiarsane only an equilibrium mixture is obtained, but with tetrakis(trifluoromethyl)diarsane, methyltellurobis-(trifluoromethyl)arsane (**54**) is formed as a pure orange-yellow compound[15].

$$Me_2AsAsMe_2 + MeTeTeMe \rightleftharpoons 2Me_2AsTeMe \quad (44)$$

(**53**)

$$(CF_3)_2AsAs(CF_3)_2 + MeTeTeMe \longrightarrow 2(CF_3)_2AsTeMe \quad (45)$$
(54)

Aryltelluro-dimethyl- and -diethyl-arsanes (**55**–**57**) are formed in about 90% yield from diphenyl ditelluride or di-*p*-tolyl ditelluride with the corresponding diarsanes (equation 46) or with excess of dimethylarsane [37](equation 47).

$$R_2AsAsR_2 + ArTeTeAr \longrightarrow 2R_2AsTeAr \quad (46)$$
(**55**) R = Me, Ar = Ph
(**56**) R = Me, Ar = *p*-Tol
(**57**) R = Et, Ar = Ph

$$Me_2AsH + ArTeTeAr \longrightarrow Me_2AsTeAr + ArTeH \quad (47)$$
(**55**) Ar = Ph
(**56**) Ar = *p*-Tol

Evidence for As—Te bond formation was provided by NMR, mass and UV–visible spectroscopy. Compounds **55**–**57** are orange to red solids (**56**) or oils (**55**, **57**) that are thermally fairly labile. They decompose on attempted vacuum distillation.

3. *Organyl-seleno- and -telluro-(diorganyl)stibanes, $R_2^1SbER^2$ (E = Se, Te)*

Molecular compounds with Sb—Se or Sb—Te bonds were unknown until recently. The striking change in colour between liquid and solid tetramethyldistibane and the fact that organic ditellurides are generally coloured compounds lead to the question of whether the Sb—Te moiety would be a chromophore comparable to the Sb—Sb and Te—Te moieties, and if thermochromic behaviour, which may be related to weak intermolecular interactions in the solid state, would also occur in heterodiatomic heavy non-metal moieties. Surprisingly, (2 + 2) dismutations of di-*p*-tolyl ditelluride with tetramethyldistibane[38] and tetraethyldistibane[39] are fast and quantitative, whereas equilibrium mixtures are formed from non-strained diphosphanes with ditellurides[15]. The reaction between di-*p*-tolyl ditelluride and tetraalkyldistibanes can be carried out like a titration, with the liquid distibane added to the red solid ditelluride. The reaction is complete when the colour of the reaction mixture changes from red to yellow. At this stage the educts are completely consumed, NMR spectra, mass spectra and analytical data confirm the presence of pure liquid tellurostibanes **58** and **59**.

$$R_2SbSbR_2 + p\text{-TolTeTeTol-}p \longrightarrow 2R_2SbTeTol\text{-}p \quad (48)$$
(**58**) R = Me[38]
(**59**) R = Et[39]

Yellow **58** and orange **59** are less coloured than both educts and they do not show significant temperature-dependent colour changes. Both compounds are sensitive to air; among the oxidation products is red di-*p*-tolyl ditelluride. The (2 + 2) dismutation of diphenyldiselenide with tetramethyldistibane is also quantitative, leading to dimethyl(phenylseleno)stibane (**60**)[40], which is a yellowish compound and non-thermochromic.

4. *Organyl-seleno- and -telluro-(diorganyl)bismuthanes, $R_2^1BiER^2$ (E = Se, Te)*

Tetraorganyldibismuthanes were regarded as being unstable until recently. After the characterization of the first pure dibismuthanes, it became clear that the colour of such compounds in the solid state can be completely different from that in the liquid[41–44]. Secondary bonding in the solid state is at present a point of discussion when coloured dibismuthanes, distibanes and ditellurides are concerned and has led to the search for heteronuclear bonds between Bi, Sb and Te.

Tetramethyldibismuthane reacts with diphenyl disulphide, diphenyl diselenide and diphenyl ditelluride in diethyl ether solution below 0 °C (equation 49)[45]. Similarly, tetra-*n*-propyldibismuthane reacts quantitatively with di-*p*-tolyl ditelluride to give *p*-tolyltellurodi-*n*-propylbismuthane (equation 50)[46].

$$Me_2BiBiMe_2 + PhEEPh \longrightarrow 2\,Me_2BiEPh \quad (49)$$

yellow crystals
(**61**) E = S
(**62**) E = Se
(**63**) E = Te

$$Pr_2BiBiPr_2 + p\text{-TolTeTeTol-}p \longrightarrow 2\,Pr_2BiTeTol\text{-}p \quad (50)$$

(**64**)

The thermal stability of **61**–**63** depends strongly on the chalcogen: the tellurobismuth ane **63** decomposes (slowly) even at −30 °C, whereas dimethyl(phenyl-thio and -seleno)bismuthane appears to be stable at this temperature. Thermal decomposition of **61**–**63** leads in the first step to methylbis(phenylchalcogeno)bismuthanes[45].

$$2\,Me_2BiEPh \longrightarrow Me_3Bi + MeBi(EPh)_2 \quad (51)$$

(**61**–**63**) (**65**) E = S
(**66**) E = Se
(**67**) E = Te

For **65** and **66**, this disproportionation can be used as a preparative method (see Section II.B.2.c); **67** cannot be isolated owing to further decomposition. In solution, only PhTeMe could be characterized, and the unsoluble black residue seemed to contain mainly polymeric $(BiCh_3)_x$. Formation of PhTeMe is due to a dismutation of trimethylbismuth with diphenyl ditelluride[45].

$$BiMe_3 + PhTeTePh \longrightarrow PhTeMe + Me_2BiTePh \quad (52)$$

(**63**)

Red–brown liquid **64** (orange in solution) appears to be thermally more stable than **63** neither of them showing typical thermochromic behaviour. Compound **64** was stable enough in C_6D_6 solution to allow the recording of a ^{125}Te NMR spectrum (acquisition time several hours)[46]. Alkyl scrambling occurred when mass spectra of the tellurobismuthanes were run. The (2 + 2)dismutations with dichalcogenides belong to the first reactions of dibismuthanes that lead selectively to Bi—Bi bond cleavage.

B. Organyl-seleno and -telluro Esters of Phosphonous Acids and Related Arsenic, Antimony and Bismuth Compounds

1. *Seleno- and telluro-phosphonous esters, $R^1P(ER^2)_2$ (E = Se, Te)*

a. Organylbis(organylseleno)phosphanes. Phenylbis(methylseleno)phosphane (**69**) and chloro(phenyl) (methylseleno)phosphane (**68**) are formed from dichlorophenylphosphane, depending on the amount of added trimethyl(methylseleno)silane, shown by ^{1}H NMR spectra in sealed tubes[13]. Diiodo(phenyl)phosphane reacts with bis(trifluoromethylseleno)mercury in benzene to give phenylbis(trifluoromethylseleno)phosphane (**70**) in 87% yield as the pure compound[47]. Alternatively, **70** is prepared by reaction of pentaphenylcyclopentaphosphane with bis(trifluoromethyl)-diselenide for 10h at 40 °C in a sealed vessel; **70** is also formed when 1,2-diiodo-1,2-diphenyldiphosphane reacts with $Hg(SeCF_3)_2$ in C_6H_6 for 1h at 0 °C. One of the products of these reactions is CF_3SeHgI[47].

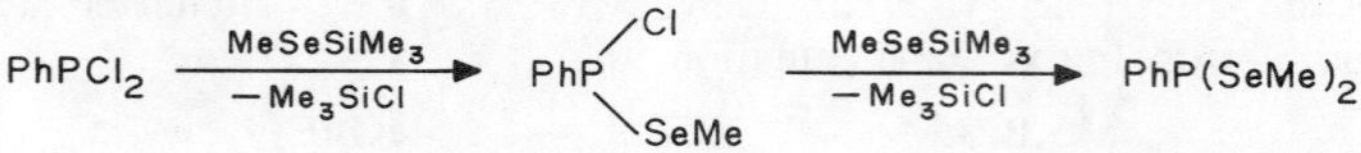

(68) (69) (53)

$$PhPI_2 + Hg(SeCF_3)_2 \xrightarrow{-HgI_2} PhP(SeCF_3)_2 \quad (54)$$

(70)

$$1/5\ (PhP)_5 + CF_3SeSeCF_3 \longrightarrow \mathbf{(70)} \quad (55)$$

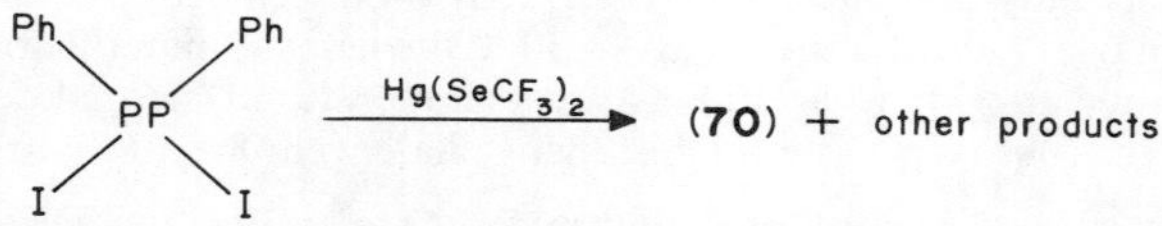

(56)

Compound **70** is fairly soluble in pentane and benzene and is thermally stable up to 80 °C.

b. Organylbis(organyltelluro)phosphanes. The cleavage of tri-*t*-butylcyclotriphosphane with ditolyl ditelluride did not lead to *t*-butylbis(*p*-tolyltelluro) phosphane (**71**), but a ditolyldiphosphane was formed (see Section III). 1,2-Di-*t*-butyl-1,2-dichlorodiphosphane did not react with di-*p*-tolyl ditelluride to give a simple (2 + 2) dismutation that should lead to *t*-butyl(chloro)(*p*-tolyltelluro)phosphane (**72**), but disproportionation occurred and diphosphanes were obtained (Section III) with only traces of **72** (NMR evidence)[30,31]. Finally, reaction of *t*-butyldichlorophosphane with two equivalents of *p*-tolyltelluromagnesium bromide in THF led again to disproportionation and **71** was only observed as a byproduct [$\delta(^{31}P = +51.7$ ppm]; which was enriched in pentane solution and characterized by MS (molecular ion at *m/e* 526 with correct isotopic distribution)[30,31].

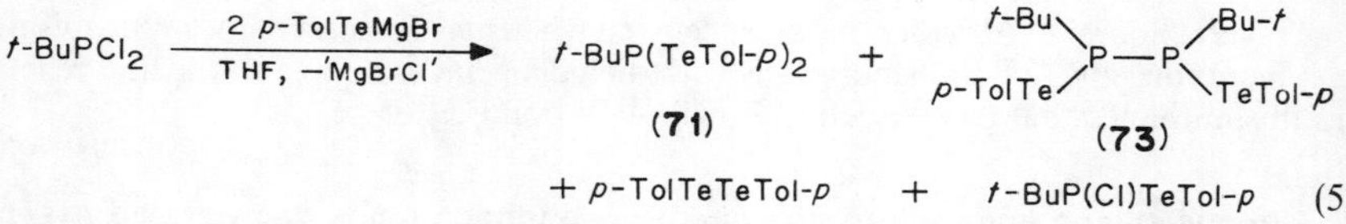

(71) (73) (72) (57)

2. *Seleno- and telluro-arsonous esters and related Sb and Bi compounds*

a. Reaction of dichloro(phenyl)arsane with trimethyl(methylseleno)silane[13].

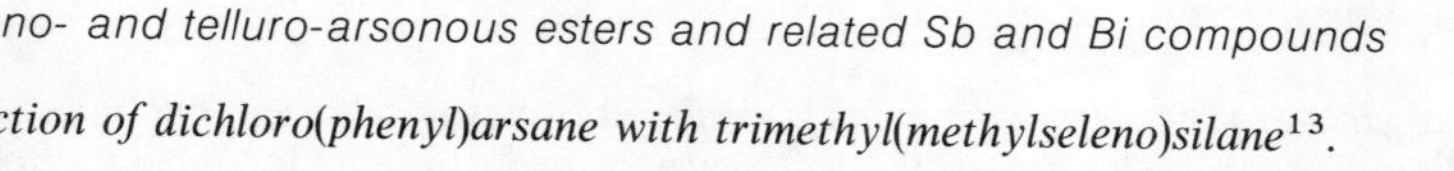

(74) (75) (58)

The formation and characterization of **74** and **75** were similar to those of the related phosphorus compounds **68** and **69**.

b. Organylbis(organyl-seleno or -telluro)stibanes. These appear to be unknown at present (see also Section IV.B.3).

c. Methylbis(phenyl-seleno- and -telluro)bismuthane. Wieber and Sauer[45] used the thermal decomposition of dimethyl(phenylseleno)bismuthane (**62**) in a stream of nitrogen at 80 °C for the preparation of methylbis(phenylseleno)bismuthane (**66**) (equation 51); **66** is an orange powder and the monomeric compound (mass spectrometry) decomposes with melting at 120 °C. The corresponding bismuth compound decomposes at room temperature[45] (see Section II.A.4).

C. Organyl-seleno and -telluro Esters of Phosphorous Acid and Related Arsenic, Antimony and Bismuth Compounds

1. Phosphorous acid esters, $P(ER)_3$ ($E = Se, Te$)

a. Tri(organylseleno)phosphanes and halogeno(organylseleno)phosphanes. Phenylselenol reacts with phosphorus trichloride with evolution of hydrogen chloride in cold benzene to give tri(phenylseleno)phosphane (**76**); spontaneous oxidation leads to the isolation of triselenophosphoric acid *Se,Se,Se*-triphenyl ester (**77**)[11]. Pure **76** was obtained as yellow crystals from finely divided white phosphorus suspended in acetone with diphenyl diselenide in the presence of few drops of concentrated KOH. Pure tri(methylseleno)phosphane (**78**) was prepared similarly as a yellow oil with a very repulsive odour (b.p. 45–50 °C/0.1 Torr)[48].

$$PCl_3 + 3\,PhSeH \xrightarrow{-3\,HCl} \underset{(\mathbf{76})}{[P(SePh)_3]} \xrightarrow{1/2\,O_2} \underset{(\mathbf{77})}{(PhSe)_3P{=}O} \qquad (59)$$

$$P_4 + 6\,RSeSeR \xrightarrow[\text{acetone}]{(KOH)} 4\,P(SeR)_3 \qquad (60)$$

$$(\mathbf{76})\ R = Ph$$
$$(\mathbf{78})\ R = Me$$

Compound **76** is only slightly soluble in organic solvents, but **78** is fairly soluble in inert solvents and was studied by ^{31}P NMR (Section IX). NMR and Raman spectra served to confirm the formation of **78** when phosphorus trichloride reacted with excess of trimethyl(methylseleno)silane[13]. With a deficit of trimethyl(methylseleno)silane, mixed species were detected in equilibrium by ^{1}H NMR[13]:

$$PCl_3 + 3\,MeSeSiMe_3 \longrightarrow \underset{(\mathbf{78})}{P(SeMe)_3} + 3\,Me_3SiCl \qquad (61)$$

$$PCl_3 \xrightarrow[-Me_3SiCl]{MeSeSiMe_3\ (\text{deficit})} \underset{(\mathbf{79})}{Cl_2PSeMe} + \underset{(\mathbf{80})}{ClP(SeMe)_2} \qquad (62)$$

$$2\,Cl_2PSeMe \rightleftharpoons ClP(SeMe)_2 + PCl_3 \qquad (63)$$
$$\textbf{(79)} \qquad\qquad \textbf{(80)}$$

$$2\,ClP(SeMe)_2 \rightleftharpoons P(SeMe)_3 + Cl_2PSeMe \qquad (64)$$
$$\textbf{(80)} \qquad\quad \textbf{(78)} \qquad\quad \textbf{(79)}$$

$$(EtO)_2PCl \xrightarrow{MeSeSiMe_3} (EtO)_2PSeMe + EtO(Cl)PSeMe \qquad (65)$$
$$\textbf{(81)} \qquad\qquad \textbf{(82)}$$

The mixed species **83** was prepared by a more selective route[49]:

$$2\,PF_2I + Hg(SeCF_3)_2 \longrightarrow 2\,F_2PSeCF_3 + HgI_2 \qquad (66)$$
$$\textbf{(83)}$$

Similarly, the mixed species (**85**) or tris(trifluoromethylseleno)phosphane (**84**) were formed from the mercury compound with phosphorus tribromide[47]. Compound **84** was isolated as pure yellow liquid at $-30\,°C$. Interestingly, **84** is thermally unstable at room temperature, when a redox disproportionation leads to bis(trifluoromethyl) diselenide, but this decomposition remains uncomplete. After 24 h at 60 °C, diselenide formation is accompanied by precipitation of an orange–red insoluble solid that was suggested to have the approximate composition $(PSeCF_3)_x$[47]. Alkoxibis(butylseleno)phosphanes were prepared from the corresponding alkoxydichlorophosphanes with butylselenol in the presence of triethylamine in benzene solution[50].

$$2\,PBr_3 + 3\,Hg(SeCF_3)_2 \longrightarrow 2\,P(SeCF_3)_3 + 3\,HgBr_2 \qquad (67)$$
$$\textbf{(84)}$$

$$PBr_3 + Hg(SeCF_3)_2 \longrightarrow BrP(SeCF_3)_2 + HgBr_2 \qquad (68)$$
$$\textbf{(85)}$$

$$ROPCl_2 + 2\,BuSeH + 2\,NEt_3 \longrightarrow ROP(SeBu)_2 + 2\,HNEt_3Cl \qquad (69)$$
$$\textbf{(86)}\ R = Et$$
$$\textbf{(87)}\ R = Pr$$
$$\textbf{(88)}\ R = Bu$$

$$EtOP(SeBu)_2 + \tfrac{1}{2}O_2 \longrightarrow EtOP({=}O)\,(SeBu)_2 \qquad (70)$$

Compounds **86–88** are colourless, distillable liquids that are easily oxidized by oxygen or sulphur (see Section VI). The first derivatives of triselenophosphorous acid were tris(diselenoxanthogenato)phosphanes (orange–yellow solids) that were prepared straightforwardly from the potassium diselenoxanthogenates with PCl_3[51]. Reaction of a potassium diselenoxanthogenate with $POCl_3$ leads also to the tris(diselenoxanthogenato)phosphane and a diselenide $[i\text{-PropOC(Se)Se}]_2$ as oxidation product (equation 72)[52]. Separation of the products was achieved by crystallization of **90**, which is less soluble in diethyl ether[52].

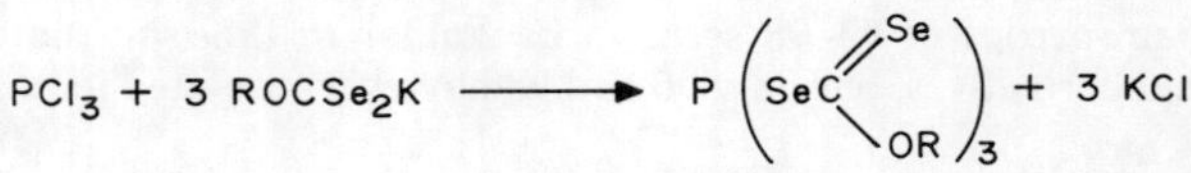 (71)

(**89**) R=Et

(**90**) R=*i*-Pr

(**91**) R=*s*-Bu

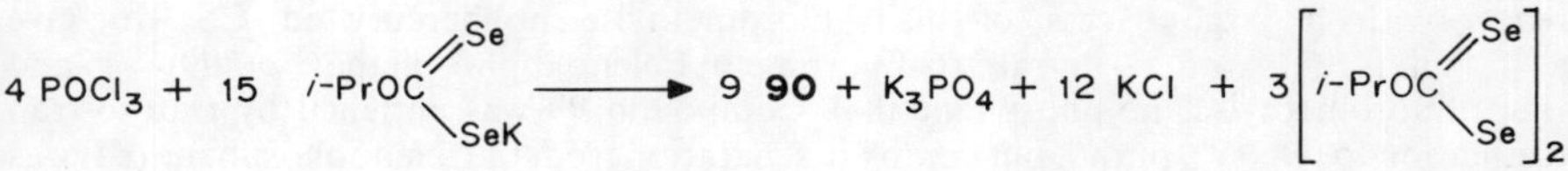

(72)

b. Tri(p-anisyltelluro)phosphane. This compound has been prepared from finely divided white phosphorus with di-*p*-anisylditelluride in the presence of KOH[48]. At −50°C, the compound separated in moderate yield as rust-coloured crystals from the dark-red acetone solution. When dried under vacuum **92** gave satisfactory analytical data, but when an inert gas was admitted to the flask containing the compounds the crystals became beige–white and evolved white fumes. Under the mother liquor (acetone solution), crystalline **92** appears to be perfectly stable at −20 °C.

$$P_4 + 6\,p\text{-AnTeTeAn-}p \xrightarrow[\text{acetone}]{\text{(KOH)}} 4\,P(\text{TeAn-}p)_3 \quad (73)$$

(**92**)

2. *Arsenous acid esters, As(SeR)$_3$ [tri(organylseleno)arsanes]*

The preparation of these compounds is similar to that of the corresponding phosphorus compounds, i.e. arsenous halides were treated with trimethyl(methylseleno)silane[13], with bis(trifluoromethylseleno)mercury[47] and with potassium isopropyl diselenoxanthogenate[52].

$$AsCl_3 \xrightarrow[-3\,Me_3SiCl]{3\,MeSeSiMe_3} As(SeMe)_3 \quad (74)$$

(**93**)

$$AsCl_3 \xrightarrow[-2\,Me_3SiCl]{2\,MeSeSiMe_3} ClAs(SeMe)_2 \rightleftharpoons \tfrac{1}{2}Cl_2AsSeMe + \tfrac{1}{2}\mathbf{93} \quad (75)$$

(**94**) (**95**)

$$AsCl_3 \xrightarrow[-HgCl_2]{Hg(SeCF_3)_2} As(SeCF_3)_3 + ClAs(SeCF_3)_2 \quad (76)$$

(**96**) (**97**)

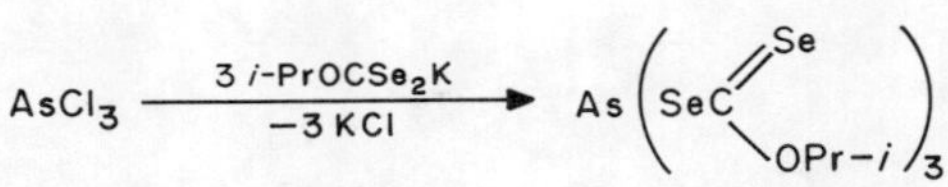 (77)

(**98**)

The physical properties of **93–98** seem to be similar to those of the corresponding phosphorous acid derivatives, but only **96** and **98** have been isolated in analytically pure states[47,52]*.

III. ORGANYL-SELENO- AND -TELLURO-DIPHOSPHANES

Only two compounds that contain an organyl-seleno or -telluro substituent bonded to a diphosphorus moiety have appeared in literature. Diphosphorus tetraiodide reacts with a large excess of bis(trifluoromethylseleno)mercury in CS_2 to give a mixture of tetrakis(trifluoromethylseleno)diphosphane (**99**) and tris(trifluoromethylseleno)phosphane (**84**). Compound **99** was enriched by trap-to-trap distillation to 90–95% purity, but complete separation from the selenophosphane (**84**) was not achieved; **99** was characterized by ^{19}F and ^{31}P NMR and mass spectroscopy[47]. Thermal decomposition of the diphosphane leads to the trisselenophosphane (**84**) and a solid orange material which was not further characterized.

$$P_2I_4 + \xrightarrow{4\,Hg(SeCF_3)_2} \underset{(\mathbf{99})}{(CF_3Se)_2PP(SeCF_3)_2} + \underset{(\mathbf{84})}{P(SeCF_3)_3} + HgI_2 + CF_3SeHgI \qquad (78)$$

$$\underset{(\mathbf{99})}{(CF_3Se)_2PP(SeCF_3)_2} \xrightarrow{\Delta} \underset{(\mathbf{84})}{P(SeCF_3)_3} + \text{orange solid} \qquad (79)$$

It should be recalled (Section II.C.1.a) that the monophosphane **84** itself undergoes partial thermal decomposition leading to bis(trifluoromethyl) diselenide and an orange solid that has the approximate composition of $(CF_3SeP)_x$[47].

This decomposition pathway resembles the reverse reaction of the degradation of P_4 or $(PhP)_5$[48,53] with dialkyl disulphides or dimethyl diselenide[48], which may be regarded as a sequence of P—P bond cleavages by (2 + 2) dismutation with the dichalcogenide. The appearance of the diselenide and the (probably P—P-bond) polymers $(CF_3SeP)_x$ from **99** and **84**, respectively, could be explained by (2 + 2) dismutation/retrodismutation (2 >PSe— ⇌ >PP< + —SeSe—) equilibria or by α-elimination at P^{III} leading to phosphinidene (P^I) intermediates ($—\ddot{P}(SeR)_2 \rightarrow —\ddot{P}: + R_2Se_2$). In the case of Ph(I)PP(I)Ph, $Hg(SeCF_3)_2$ leads to complete oxidation of the diphosphane to give the fairly stable $PhP(SeCF_3)_2$ (**70**). In the aryl-*t*-butyltellurium–phosphorus systems, the equilibrium of this redox [or (2 + 2) dismutation] reaction seems to be more on the side of the P—P-bonded species. Tetra-*t*-butylcyclotetraphosphane does not react at all with di-*p*-tolyl ditelluride[54]. Strained tri-*t*-butylcyclotriphosphane does react with the diaryl ditelluride, but even with a large excess of the ditelluride only 1, 2-di-*t*-butyl-1, 2-di-*p*-tolytellurodiphosphane (**73**) is formed, which does not react further with the ditelluride[30,31].

$$(t\text{-BuP})_4 + p\text{-TolTeTeTol-}p \nrightarrow \qquad (80)$$

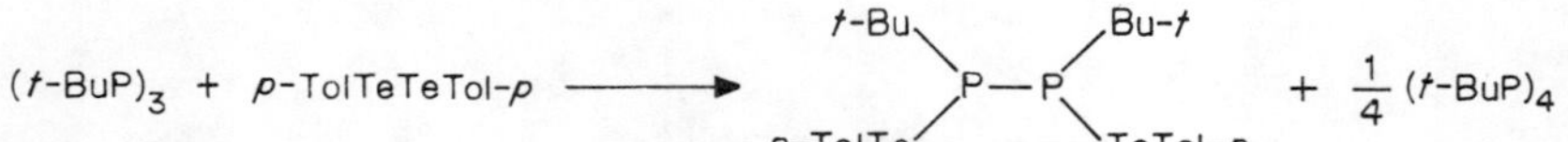

(**73**) (81)

*For diselenophosphato complexes $M[Se_2P(OR)_2]_3$ (M = As, Sb, Bi), made by Zingaro's group, see Ref. 3 and references cited therein.

On the other hand, the attempted synthesis of *t*-butylbis(*p*-tolyltelluro)phosphane (**71**) (equation 57) led to the diphosphane (**73**) and di-*p*-tolyl ditelluride. Compound **71** appears to decompose by re-dismutation of one P—Te bond to give P—P-bonded **73** and the ditelluride. Compound **73** is thermally stable at room temperature. In the melt, irreversible decomposition leads to di-*p*-tolyl ditelluride and tetra-*t*-butylcyclotetraphosphane (see equation 80)[30,31]. The fair thermal stability of the diphosphane **73** is obviously due to kinetic stabilization from bulky *t*-butyl groups. A high-yield synthesis of **73** is the reaction of 1,2-di-*t*-butyl-1,2-dichlorodiphosphane with *p*-tolyltelluromagnesium bromide. The pure compound (yellow crystals from pentane) was obtained in 75% yield from this reaction[30,31].

$$t\text{-Bu(Cl)PP(Cl)Bu-}t + 2p\text{-TolTeMgBr} \xrightarrow[-2\text{'MgBrCl'}]{\text{(THF)}} (\mathbf{73}) \qquad (82)$$

Surprisingly, **73** was also one of the main products when 1,2-di-*t*-butyl-1,2-dichlorodiphosphane was stirred with di-*p*-tolyl ditelluride at 50–55 °C in toluene for 24 h[30,54].

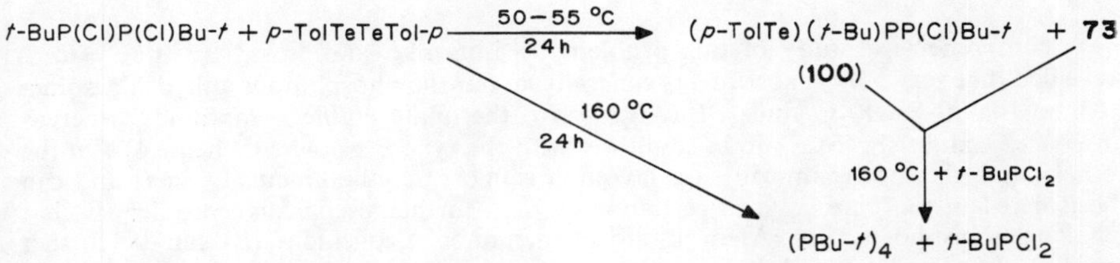

(83)

Compound **100** was detected by its AX-type ^{31}P NMR spectra [$\delta(P_A) + 124.8$ ppm, $\delta(P_X) + 34.3$ ppm; $J(AX) \pm 322.3$ Hz]; chiral discrimination leads obviously to the predominance of one isomer of **100**. At 160 °C, irreversible decomposition of both diphosphanes occurs and only $(t\text{-BuP})_4$ and $t\text{-BuPCl}_2$ were detected by ^{31}P NMR after 24 h at this temperature. In solutions of **73**, two isomers are present that give ^{31}P NMR signals at +39.6 and +29.2 ppm. The isomer with ^{31}P NMR shift to higher field predominates (about 88%, probably *meso*-**73**). ^{31}P NMR 'tellurium satellites' and the ^{125}Te NMR spectrum of the predominant isomer are AA′ and X parts of an AA′X spin system; analysis of the spectra led to $^1J(^{125}\text{Te}^{31}\text{P}) = -520$ Hz, $^2J(^{125}\text{Te}^{31}\text{P}) = -22.5$ Hz and $^1J(^{31}\text{P}^{31}\text{P}) = -319$ Hz. A diphosphanditelluride structure with P=Te bonds can therefore be ruled out[30,54].

IV. SELENO- OR TELLURO-BIS-PHOSPHANES, -ARSANES, -STIBANES AND -BISMUTHANES

A. Seleno- or Tellurophosphinous Anhydrides and Related Arsenic, Antimony and Bismuth Compounds

1. *Preparation of seleno- and telluro-phosphinous anhydrides, R_2PEPR_2 (E = Se, Te)*

a. Selenobisphosphanes. These compounds are very rare in the literature. The inorganic selenobisphosphane F_2PSePF_2 (**101**) was prepared from bromodifluorophosphane with disilylselane[14]. The NMR spectra of **101** are strongly temperature dependent: ^2J(PP) and

^4J(FF) increase by about 50% when the temperature of the sample is lowered from room temperature to 173K. This should be due to conformational changes (changes in the thermal population of conformations) that lead to differences in lone pair–lone pair interactions which influence ^1J(PP) and long-range couplings[14].

Generally, chalcogenobisphosphanes are in equilibria with diphosphane chalcogenides. The latter may rearrange to give diphosphane dichalcogenides and diphosphanes:

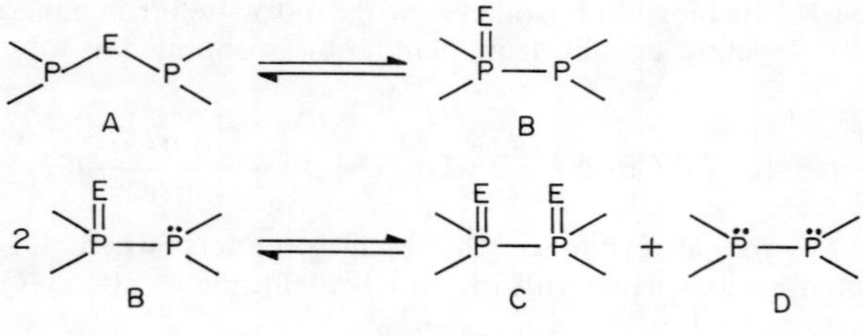

SCHEME 1

A comprehensive study of this problem by Lutsenko and Foss[55] (with E = O, S) revealed that very bulky substituents generally favour the phosphinous anhydride isomer (A), so that $(t\text{-Bu}_2P)_2O$ and $(t\text{-Bu}_2P)_2S$[56] are thermally stable compounds, whereas $(i\text{-Pr}_2P)_2O$ and $(i\text{-Pr}_2P)_2S$ are in equilibria with the type B isomers (Scheme 1). On the other hand, thermodynamically less favoured isomers may be kinetically inert and can be isolated in the pure state in certain cases[55]. A further preparative problem is that chalcogenobisphosphanes (A) or diphosphanemonochalcogenides (B) can be further oxidized by elemental chalcogen.

Tetra-*t*-butyldiphosphane reacts with equimolar amounts of selenium to give two products, the selenobis(di-*t*-butylphosphane) (**102**) and the diseleno species (**103**) (NMR spectroscopic evidence). Compound **103** is the only product if the ratio of tetra-*t*-butyldiphosphane to selenium is 1:2. The diseleno species (**103**) was isolated in the pure state as colourless crystals from diethyl ether–pentane[57].

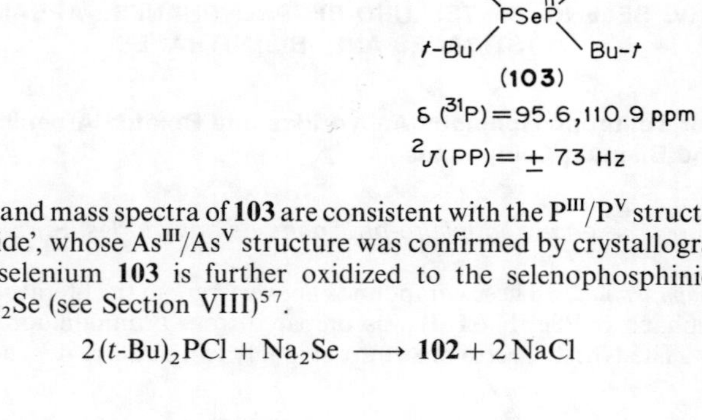

(84)

The NMR, IR and mass spectra of **103** are consistent with the P^{III}/P^V structure related to 'cacodyl disulphide', whose As^{III}/As^V structure was confirmed by crystallographic data[58]. With excess of selenium **103** is further oxidized to the selenophosphinic anhydride, $[(t\text{-Bu})_2P({=}Se)]_2Se$ (see Section VIII)[57].

$$2\,(t\text{-Bu})_2PCl + Na_2Se \longrightarrow \mathbf{102} + 2\,NaCl \qquad (85)$$

From di-*t*-butylchlorophosphane with sodium selenide, **102** was obtained as thermally stable colourless needles from toluene–pentane[57].

b. Tellurobisphosphanes. Tellurobis(di-*t*-butylphosphane) (**30**) was first obtained in a pure state when di-*t*-butylchlorophosphane reacted with sodium telluride[28,31]; **30** is also formed in equilibria with di-*t*-butyl(trimethylsilyl-, -germyl- and -stannyl-telluro)phosphanes (**27–29**) (equation 28) or when di-*t*-butyl-[tris(trimethylsilyl)methyltelluro]phosphane (**34**) decomposes to give the symmetrically substituted tellurium compounds[32]. A very straightforward preparation of **30** is tellurium insertion into the P—P bond of tetra-*t*-butyldiphosphane. Heating the diphosphane for 3 days with powdered tellurium in boiling toluene yields the tellurobisphosphane in quantitative crude yield; from pentane pure yellowish needles are obtained, which melt at 70–80 °C without decomposition[31,57].

$$2\,(t\text{-Bu})_2PCl + Na_2Te \longrightarrow \underset{\textbf{(30)}}{(t\text{-Bu})_2PTeP(Bu\text{-}t)_2} + 2\,NaCl \qquad (86)$$

$$(t\text{-Bu})_2PP(Bu\text{-}t)_2 + Te \xrightarrow[3\ d]{110\ ^\circ C} \textbf{30} \qquad (87)$$

The latter preparation is much easier than the synthesis of the corresponding selenium compound **102** because neither further oxidation of **30** to give a species related to **103** nor significant di-insertion (to give a ditellurobisphosphane) occur with excess of tellurium. In contrast, reaction of sodium ditelluride with di-*t*-butylchlorophosphane leads to the formation of **30** with precipitation of elemental tellurium[30,54]. Photochemically, **30** decomposes[59] with extrusion of elemental tellurium (reverse reaction of the thermal synthesis from equation 87) and formation of tetra-*t*-butyldiphosphane. By-products are $(t\text{-Bu})_2PH$ [from $(t\text{-Bu})_2P$ radicals] and the tellurocyclophosphane $(t\text{-BuP})_3Te$ (from P—C cleavage; see Section IV.B).

$$\underset{\textbf{(30)}}{(t\text{-Bu})_2PTeP(Bu\text{-}t)_2} \xrightarrow[\substack{C_6H_6/C_6D_6\\ \text{(sealed tube)}\\ 30\ ^\circ C}]{h\nu\ (UV)} (t\text{-Bu})_2PP(Bu\text{-}t)_2 + Te$$

$$[\,+ (t\text{-Bu})_2PH + (t\text{-BuP})_3Te + H_2C{=}CMe_2] \qquad (88)$$

Tetraisopropyldiphosphane reacts with elemental tellurium under milder conditions than tetra-*t*-butyldiphosphane, but in this case tellurium insertion into the P—P bond is far from quantitative; equilibrium mixtures are obtained as yellow solutions that do not contain more than 60% of the tellurobisphosphane **104**[57]. Similar observations had been made when tri-*n*-alkylphosphanes were found to react 'incompletely' with elemental tellurium to give mixtures of phosphane tellurides and unreacted phosphane[60], but the formation of **104** according to equation 89 is the first case of an incomplete tellurium insertion.

$$(i\text{-Pr})_2PP(Pr\text{-}i)_2 + Te \rightleftharpoons \underset{\textbf{(104)}}{(i\text{-Pr})_2PTeP(Pr\text{-}i)_2} \qquad (89)$$

The sodium telluride–chlorophosphane reaction also does not lead to pure **104**, but some tellurium precipitation with formation of tetraisopropyldiphosphane again, occurs.

$$2\,(i\text{-Pr})_2PCl + Na_2Te \longrightarrow \textbf{104}\ [\,+ Te + (i\text{-Pr})_4P_2] \qquad (90)$$

Crystallization from pentane afforded small amounts of yellow crystalline **104** that gave correct analytical data. In contrast to the *t*-butyl compound **30**, **104** is extremely sensitive to air, moisture and thermal decomposition[31,57]. From equilibrium mixtures with tetraisopropyldiphosphane, **104** could be trapped by selective coordination with norbornadienetetracarbonylchromium(0).

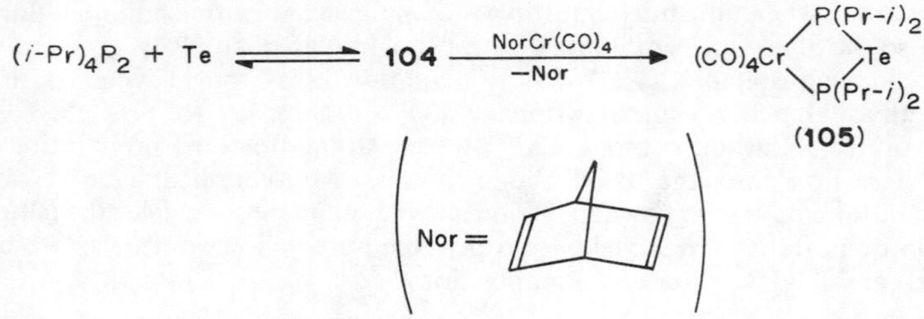

(91)

Compared with the ligand **104**, the four-membered chelate complex **105** is remarkably stable to thermal detelluration or hydrolytic decomposition. Pure red crystals of **105** were isolated after several recrystallizations from pentane–toluene[57]. Tetramethyldiphosphane reacts with elemental tellurium at room temperature to give a yellow solution, but ^{31}P NMR spectra reveal that again there is a large excess of the diphosphane present in solution. Attempts to isolate tellurobis(dimethylphosphane) (**106**) in a pure state (yellow crystals) are in progress. Reactions that should lead to tellurobis(diphenylphosphane) gave only tetraphenyldiphosphane and elemental tellurium[59].

$$Me_2PPMe_2 + Te \rightleftharpoons \underset{\textbf{(106)}}{Me_2PTePMe_2} \qquad (92)$$

$$2\,Ph_2PSiMe_3 + Te \longrightarrow Ph_4P_2 + (Me_3Si)_2Te \qquad (93)$$

2. Synthesis and structures of μ-9,10-chalcogeno-9,10-diarsaanthracenes

The so-called 'arsanthrene oxide' (**107**) has been known since 1921; with hydrogen sulphide, selenide and telluride the corresponding μ-9,10-chalcogeno-9,10-diarsaanthracenes **108–110** are formed as stable crystalline compounds[61].

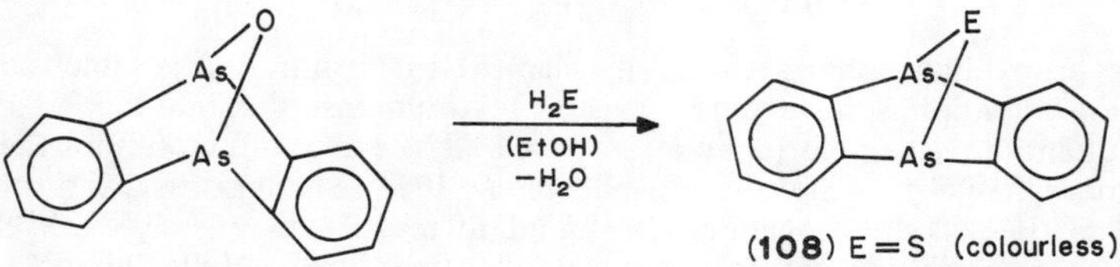

(94)

(**107**)

(**108**) E = S (colourless)
(**109**) E = Se (yellow)
(**110**) E = Te (red)

X-ray crystal structure analyses have been carried out on the chalcogenobisarsanes **107–110**[62]. All four compounds are in the 'butterfly' conformation about the plane of the As—X—As moiety and the molecules of all four compounds show non-crystallographic symmetry of 2*mm* (corresponding to molecular symmetry point group C_{2v}) about an axis bisecting the As—E—As bond angle. The As···As distances are determined by the increase in covalent radii from O to Te and by their preferred valency angles. In non-bridged 9,10-dihydro-9,10-dimethyl-9,10-diarsaanthracene, the As···As distance is 3.27 Å. The As···As distances in the selenium and tellurium compounds do not adapt sufficiently to allow the Se and Te atoms their preferred valency angle (Table 2).

TABLE 2. Comparison of structural features of μ-9, 10-chalcogeno-9, 10-diarsaanthracenes **107–110**[62] with $C_2AsEAsC_2$ moieties

E	Compound	d(AsE) (Å)[a]	AsÊAs	d(As ··· As) (Å)	CÂsC
O	**107**	1.84 1.82	108.8	2.98	95.0
S	**108**	2.23 2.26	91.5	3.23	92.7
Se	**109**	2.39 2.40	86.2	3.28	93.4
Te	**110**	2.57	80.8	3.35	95.7

[a] Where two values are given, this is due to a slight distortion of the molecules in the crystals (deviation from ideal point group C_{2v}).

Compounds **108–110** are fairly inert towards electrophiles such as methyl iodide, but they are attacked by oxidizing agents and by nucleophiles such as methylmagnesium iodide[61].

3. Preparation of seleno- and telluro-bis(diorganylstibanes)

The first seleno- and telluro-bisstibanes, $R_2SbESbR_2$ (E = Se, Te), were prepared recently by Breunig and Jawad[39].

$$R_2SbSbR_2 + Se \longrightarrow R_2SbSeSbR_2 \tag{95}$$

(**111**) R = Me (orange liquid)
(**112**) R = Et (yellow crystals)
(**113**) R = Ph (pale yellow crystals)

$$R_2SbSbR_2 + Te \longrightarrow R_2SbTeSbR_2 \tag{96}$$

(**114**) R = Me (brown liquid)
(**115**) R = Et (red liquid)

Compounds **111–115** are formed under mild conditions by slightly exothermic reactions of the distibanes with elemental selenium or tellurium; the pure compounds have been isolated in high yields. Contrary to the reactions of non-strained diphosphanes, the distibanes are quantitatively mono-selenated and mono-tellurated. Compounds **111–115** are sufficiently thermally stable to be handled at room temperature in the pure state under inert gas and in inert solvents. A remarkable feature of the dimethylantimony selenium and tellurium derivatives **111** and **114** is their thermochromic behaviour. At 5 °C the orange–red liquid of **111** becomes a red crystalline solid, but below −20 °C the colour of the solid is yellow. The properties of the tellurium compound **114** are even more striking: at room temperature **114** is a brown liquid and just below the melting point (8 °C) the crystals are blue–violet and look very much like solid iodine. The finding that only the methyl derivatives (and not the bulkier ethyl substituted compounds **112** and **115**) show this thermochromic behaviour led to the prediction that unusual short intermolecular contacts in the solids (a kind of secondary bonding) might be involved[39]. Such contacts have been confirmed to exist in solid tetramethyldistibane, which is also strongly thermochromic[63,64]. Spectroscopic investigations confirmed the chalcogenobisstibane structure A for **111–115** and the distibane chalcogenide structure B could be ruled out.

4. *Seleno- and telluro-bis(diorganylbismuthanes)*

Recently, the formation of the first seleno- and telluro-bisbismuthanes was reported at a conference[65]. These compounds are also strongly coloured and they appear to be thermally less stable than the corresponding antimony compounds. The decomposition pathways seem to involve alkyl scrambling (compare Section II.A.4).

$$R_2BiBiR_2 + E \longrightarrow R_2BiEBiR_2 \qquad (97)$$

(116) E = S, R = Me (Ref. 41)
(117) E = Se, R = Pr (Ref. 65)
(118) E = Se, R = Pr (Ref. 65)
(119) E = Te, R = Pr (Ref. 65)

B. Cyclic Seleno- or Telluro-bisphosphanes and -bisarsanes

1. *Chalcogenocyclophosphanes, $(RP)_nE_m$ (E = Se, Te)*

a. Selenocyclophosphanes. Within the scope of their work on phosphorus ring systems, Baudler *et al.*[66] prepared the first selenadiphosphirane (**120**) by reaction of 1,2-di-*t*-butyl-1,2-dichlorodiphosphane with bis(trimethylstannyl)selenide.

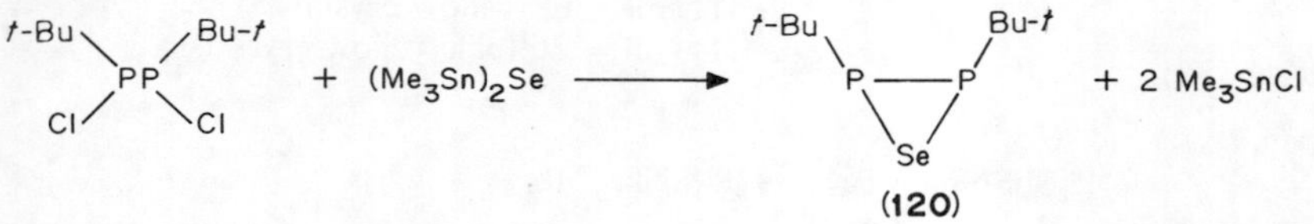

(98)

Selenadiphosphirane **120** tends to give dimerization and subsequent elimination reactions leading to other cyclic compounds containing Se atoms and *t*-butylphosphosphandiyl groups as ring members which were detected by ^{31}P NMR spectroscopy. Typically, **120** shows an upfield ^{31}P shift and a small coupling constant $J(^{77}Se^{31}P)$.

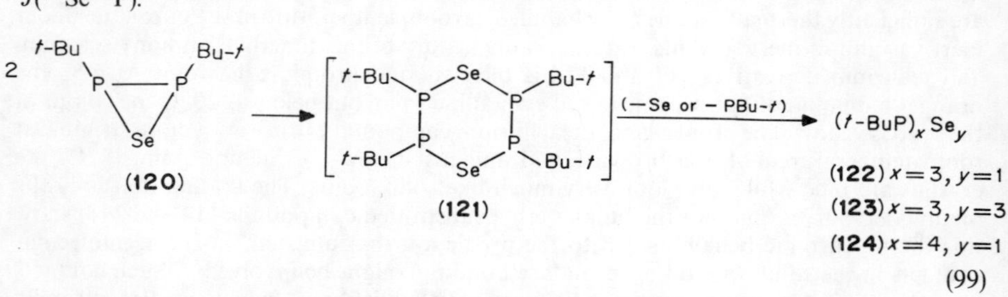

(99)

The thiadiphosphirane corresponding to **120** was prepared similarly[66].

An alternative route to selenadiphosphiranes is the reaction of a diphosphene with elemental selenium. Formation of the selenadiphosphirane competes with di(seleno)metaphosphonate formation[67].

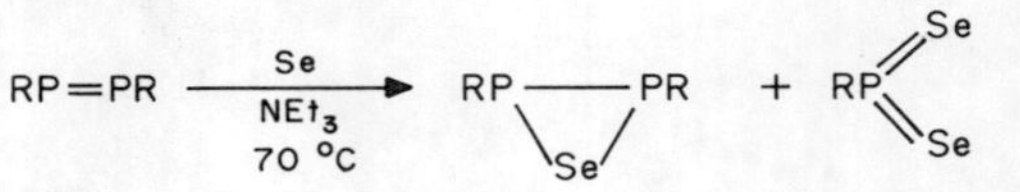

17% (125) 4% (126) (100)

R=2,4,6-tri-*t*-butylphenyl

Reaction of the diphosphene with sulphur proceeds thermally with formation of a diphosphene sulphide (**127**), which can be rearranged photochemically to give the thiadiphosphirane that corresponds to **125**[68].

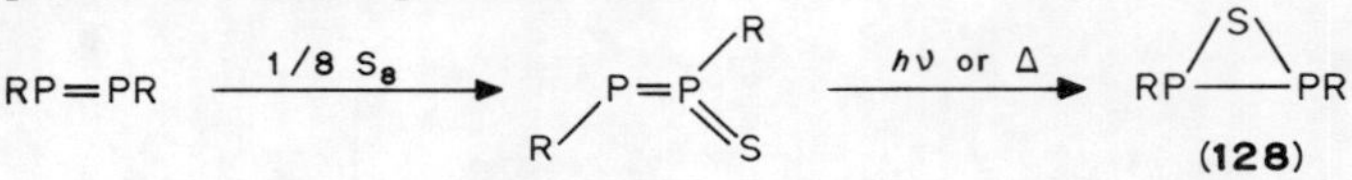

(127) (128) (101)

R=2,4,6-tri-*t*-butylphenyl

Much earlier, Maier found that phenylphosphane reacts with red selenium with evolution of hydrogen selenide and the formation of a material that had previously been assigned a tetraphenylcyclotetraphosphane–tetraselenide structure[69,70]. Similar arsenic compounds have also been reported[70a].

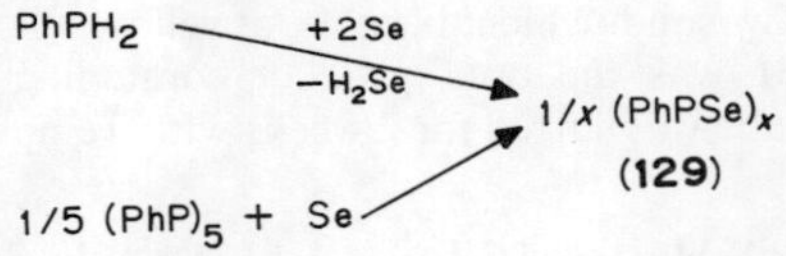

(129) (102)

b. Tellurocyclophosphanes. Tellurocyclophosphanes are formed in a substitution and redox reaction between *t*-butyldichlorophosphane and sodium telluride. Essentially the same products are formed (in a different distribution) when the chlorophosphane reacts with bis(trimethylsilyl)telluride[30,54,59].

$$t\text{-BuPCl}_2 \xrightarrow[\text{or } (\text{Me}_3\text{Si})_2\text{Te},\ -2\text{Me}_3\text{SiCl}]{\text{Na}_2\text{Te},\ -2\text{NaCl}} 1/x\ (t\text{-BuP})_x\text{Te} + (1-\tfrac{1}{x})\text{Te} \qquad (103)$$

(130) $x = 2$
(131) $x = 3$
(132) $x = 4$.

The main products of this reaction are the telluradiphosphirane **130** and the telluratriphosphetane **131**, but several other species are also present in the red oil that was obtained as the pentane-soluble fraction of the reaction mixture. The telluradiphosphirane **130** is the only main product when sodium telluride is treated with *t*-butyldichlorophosphane in pentane solution. Application of a more polar solvent (such as diethyl ether) leads to dimerization of **130** to give the six-membered heterocycle **133**, which decomposes further either with loss of a Te atom (to give **132**) or with loss of a *t*-butylphosphandiyl unit to give the ditelluratriphospholan $(t\text{-BuP})_3Te_2$ (**134**). Further extrusion of tellurium leads from **134** to **131** and from **132** to tetra-*t*-butylcyclotetraphosphane.

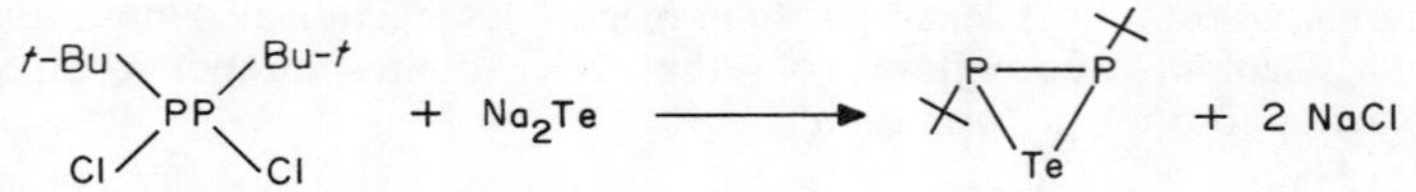

(130)

(104)

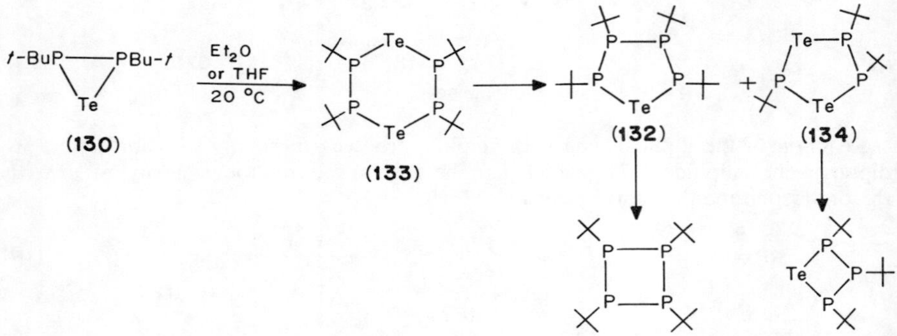

(130) (133) (132) (134) (131)

(105)

When the reaction according to equation 104 is carried out in pentane at 35–40 °C (5h), evaporation of the solvent yields a yellow residue that contains mainly **130**(^{31}P NMR δ −69 ppm). Pure **130** (26%) was isolated by distillation into a flask that was kept at 78 K. It condensed as a yellow solid which became a yellow oil at about −40 °C. The telluratriphosphetane **131** was the only product containing P atoms when *t*-butylbis(trimethylsilyl)phosphane was stirred for 2 weeks with Te powder in the dark(without solvent)[59].

$$t\text{-BuP(SiMe}_3)_2 + 4/3\,\text{Te} \longrightarrow \tfrac{1}{3}\,\mathbf{131} + (\text{Me}_3\text{Si})_2\text{Te} \qquad (106)$$

Distillation of bis(trimethylsilyl)telluride into a 78 K trap within several days gave **131** as a red oil that gave correct analytical data. Compound **131** is thermally stable at room temperature. Evidence for the telluratriphosphetane structure is provided by EI and FI MS and by ^{31}P and ^{125}Te NMR data[30,54,59]. Selected data of the various tellurocyclophosphanes are compiled in Table 3. The ^{125}Te NMR shift of **131** given in Ref. 59 was incorrect owing to 'signal folding'.

TABLE 3. Properties of tellurocyclophosphanes[30,54,59] $(t\text{-BuP})_n\text{Te}_m$

Method	$n=4, m=2$ (**133**)	$n=3, m=2$ (**134**)	$n=4, m=1$ (**132**)	$n=3, m=1$ (**131**)	$n=2, m=1$ (**130**)
MS	—	520(M^+)	482(M^+)	394(M^+)	306(M^+)
	520(M^+ − C_4H_9P)	463(M^+ − C_4H_9)	425(M^+ − C_4H_9)	337(M^+ − C_4H_9)	250(M^+ − C_4H_8)
^{31}NMR	+46.9(s)	+99.8(d)	+89.9	−4.1(t)	−70.0
		+153.7(t)	+94.8	−60.5(d)	
		($J = \pm 26$ Hz)	(AA′BB′)	($J = \pm 172.6$)	
^{125}NMR[a]	δ −33.7(m)	δ ca. +160(m)	δ ca. +50(m)	δ −789(d, t)	δ −1123.7(t)

[a] ^{125}Te NMR: ditolyl ditelluride standard, Bruker WP80 instrument (25.27 MHz, ^{125}Te).

Mixtures containing some or all of the tellurocyclophosphanes **130–134** are formed from various reactions that lead to bond formation between Te and t-BuP groups[30,54].

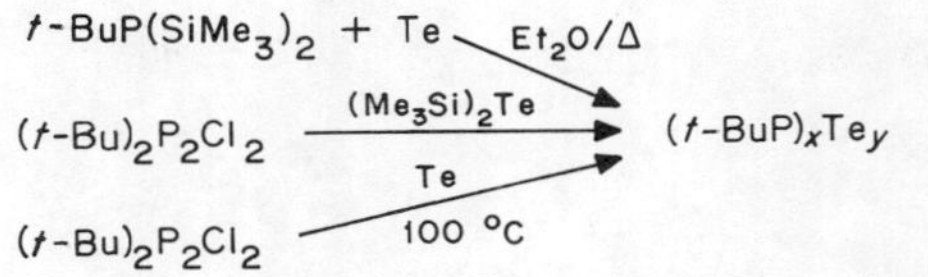

(107)

(108)

(109)

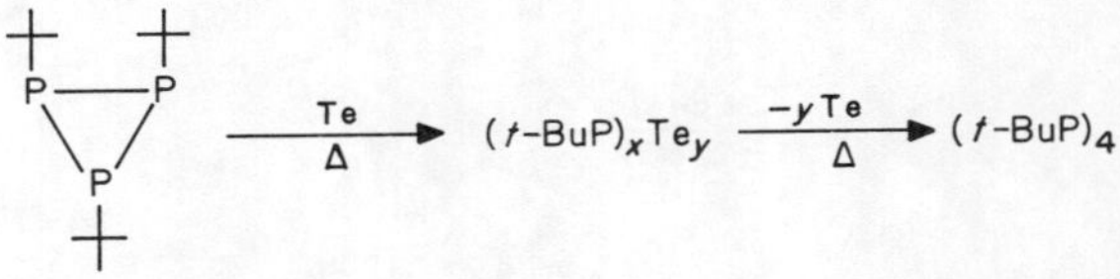

(110)

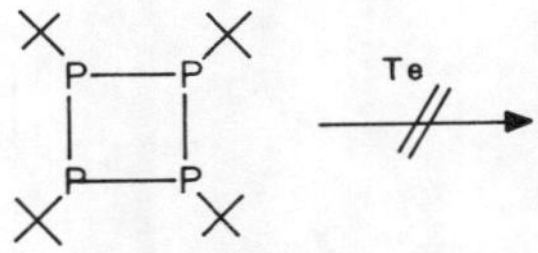

(111)

Tri-t-butylcyclotriphosphane reacts with elemental tellurium, but tetra-t-butylcyclotetraphosphane does not. Thermal decomposition reactions of tellurocyclophosphanes lead finally to tetra-t-butylcyclotetraphosphane (equation 110). Surprisingly, reactions under very mild conditions such as that of t-butyldichlorophosphane and bis(trimethylsilyl) telluride do not lead to 1, 3-ditelluradiphosphetane but yield preferentially the three-membered telluradiphosphirane and products of its thermal decomposition (equation 103 and 107). The 1, 2- or 1, 3-ditelluradiphosphetane (A, B) are either unstable with respect to loss of tellurium with ring contraction, or the intermediate telluroxophosphane (t-BuP═Te), having a highly ylidic character, dimerizes not with $(2 + 2)$ cycloaddition (to give a 1, 3- or 1, 2-ditelluradiphosphetane) but in a kind of $(2 + 1)$ addition leading to the short-lived 'dimer' C, which undergoes spontaneous detelluration to give **130** (Scheme 2).

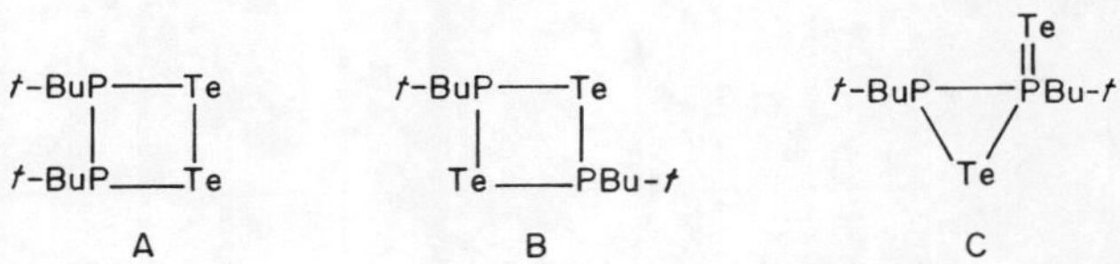

SCHEME 2. Molecules that have *not* been detected in tellurocyclophosphane mixtures.

2. *Selenocycloarsanes*

The reaction of 1, 1, 1-tris(diiodoarsinomethyl)ethane with water or with hydrogen sulphide gives cage compounds of the adamantane type, but with sodium hydridoselenide a noradamantane compound (**135**) containing one As—As bond is formed. Sodium telluride leads to complete reduction and the cyclotriarsane (**136**) is formed[71–73].

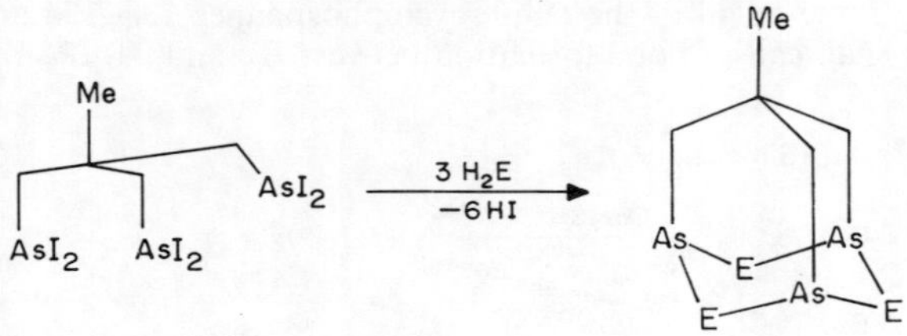

E=O,S (112)

(135)

(113)

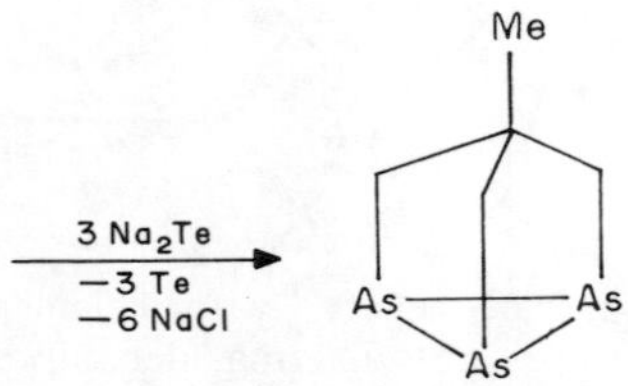

(136)

(114)

Reaction of the nortricyclene-type triarsane (**136**) with elemental selenium also led to the diseleno derivative **135** (sulphur reacts similarly); excess of sulphur or selenium did not lead to any further oxidation (i.e. S or Se insertion). With elemental tellurium, triarsane **136** did not react at all[71–73]. The structure of **135** was confirmed by an X-ray crystal structure analysis[71]. Compound **135** is less strained than a comparable compound containing $C(COOEt)_2$ groups in place of the selenium atoms. The As_3Se_2 fragment is a non-planar five-membered ring. The SeAsSe and AsSeAs angles (97–99°) are smaller than the AsAsSe angle (103°); the As—As bond is 0.05 Å longer than the sum of the arsenic covalent radii, the dihedral angle at the SeAsAsSe fragment being close to zero. The conformation of **135** is shown in Fig. 1.

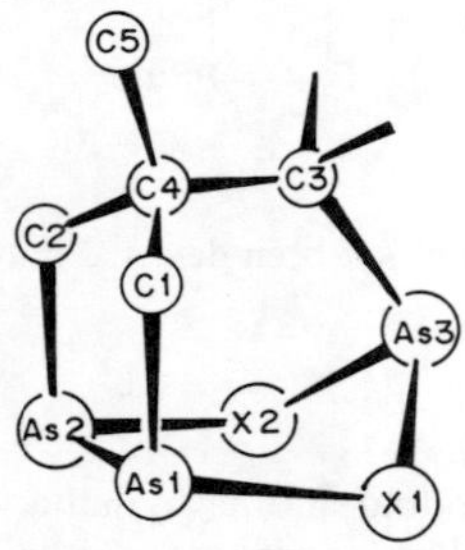

FIGURE 1. Conformation of $MeC(CH_2As)_3Se_2$ (**135**); X1, X2 = selenium atoms. Reproduced by permission of Verlag der Zeitschrift für Naturforschung from Reference 71

3. Selenocyclostibanes[74]

1, 1, 1-Tris(dichlorostibinomethyl)ethane reacts with NaSeH in THF to give the cage compounds **137** (60%) and **138** (20%).

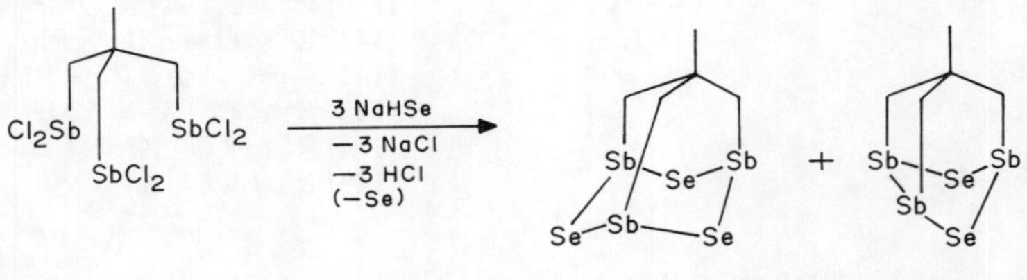

(115)

(137) **(138)**

V. ORGANYL-SELENO- AND -TELLURO-PHOSPHONIUM IONS

A. Preparation of Selenophosphonium ions

Alkoxyphosphonium ions are key intermediates in the course of the Arbuzov rearrangement. They can be stable when nucleophilic attack of the counter ion (generally leading to an alkyl halide and a P=O bond) is precluded by the use of bulky alkyl groups or/and by the use of non-nucleophilic anions. Thiophosphonium ions may react with nucleophiles not only at carbon (leading to a P=S bond) or at phosphorus (leading to substitution reactions at P), but also at the sulphur atom, leading to reduction to give the phosphane; this type of reaction can be used for the synthesis of certain chiral phosphanes[75]. Organylseleno- and organyltelluro-phosphonium salts are even more susceptible to nucleophilic attack at Se or Te, but at present very little is known about their chemistry. Three basic routes lead to organylselenophosphonium ions: (1) *P* alkylation of a selenophosphane; (2) *Se* alkylation of a selenophosphorane (phosphane selenide); (3) heteroatom (e.g. *O*) alkylation of an organylseleno-phosphinate, -phosphonate or -phosphate (se Section VI).

$$R_2{}^1PSeR^2 + R^3X \longrightarrow (R_2{}^1R^3PSeR^2)^+X^-$$

$$R_3{}^1P = Se + R^2X \longrightarrow (R_3{}^1PSeR^2)^+X^-$$

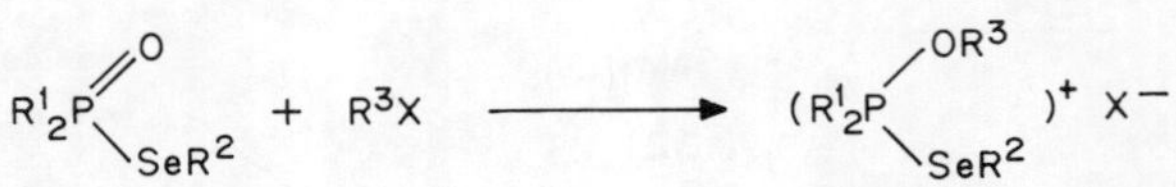

Little is known about the alkylation of selenophosphanes (route 1); alkylation of selenophosphoranes (which are generally more easily accessible than selenophosphanes) is the favourite method of preparation in all published studies dealing with selenophosphonium ions (route 2). A complete series of methoxy-, methylthio- and methylselenophosphonium ions have been prepared and studied with the aid of NMR and IR spectroscopy by Schmidpeter and Brecht[76].

Only the trimethoxy(methylseleno)phosphonium ion was too labile to be studied by NMR (instantaneous decomposition of $(MeO)_3P{=}Se$ and formation of the rearranged product $(MeO)_2(MeSe)P{=}O$ occurred). The alkylation of aminophosphaneselenides proceeds even with methyl iodide[77].

$$R^1R_2{}^2P{=}Se + Me_3O^+\ SbCl_6{}^- \xrightarrow{-Me_2O} R^1R_2{}^2PSeMe^+\ SbCl_6{}^- \qquad (115)$$

$[\text{or } (MeO)_2SO_2/HSbCl_6]$

(**139**) $R^1 = R^2 = Ph$
(**140**) $R^1 = Ph,\ R^2 = OMe$
(**141**) $R^1 = OMe,\ R^2 = Ph$
(**142**) $R^1 = NMe_2,\ R^2 = Ph$
(**143**) $R^1 = Ph,\ R^2 = NMe_2$
(**144**) $R^1 = R^2 = NMe_2$
(**145**) $R^1 = R^2 = Bu$

$$R^1R_2{}^2P{=}Se + MeI \longrightarrow R^1R_2{}^2PSeMe^+I^- \qquad (116)$$

(**144a**) $R^1 = R^2 = NMe_2$
(**146**) $R^1 = R^2 = NEt_2$
(**147**) $R^1 = Et,\ R^2 = NMe_2$

Similarly, ethylation of triphenylphosphane selenide worked with triethyloxonium tetrafluoroborate, but with $(RO)_3P{=}Se$ loss of selenium or rearrangement to $(RO)_2(EtSe)P{=}O$ occurred[78].

$$Ph_3P{=}Se + (Et_3O)BF_4 \longrightarrow \underset{(\mathbf{148})}{(Ph_3PSeEt)^+}\ BF_4{}^- + Et_2O \qquad (117)$$

Methyl triflate is also a very efficient methylating agent for phosphane selenides or for the *O*-methylation of phosphinic acid *Se*-esters[79]. The methylselenium cation moiety can be transferred to more nucleophilic phosphanes (equation 118), or[80] removed with EtS^- or SH^- with retention of configuration at phosphorus[75], i.e. soft nucleophiles tend to attack selenophosphonium ions at the selenium atoms.

$$(Me_2N)_3P + \underset{(\mathbf{149})}{R^1R^2R^3P^*SeMe^+CF_3SO_3{}^-} \rightarrow \underset{(\text{retention})}{R^1R^2R^3P^*} + \underset{(\mathbf{150})}{(Me_2N)_3PSeMe^+CF_3SO_3{}^-} \qquad (118)$$

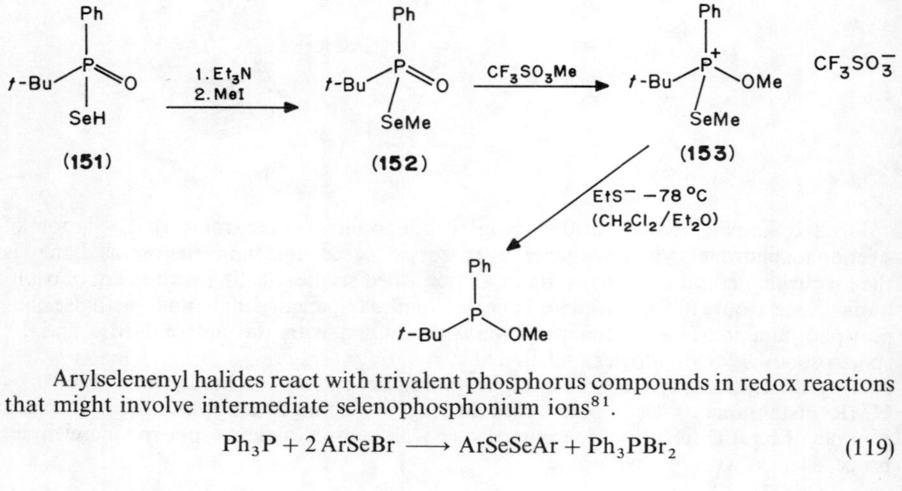

Arylselenenyl halides react with trivalent phosphorus compounds in redox reactions that might involve intermediate selenophosphonium ions[81].

$$Ph_3P + 2\,ArSeBr \longrightarrow ArSeSeAr + Ph_3PBr_2 \qquad (119)$$

Ar = 2,4-$(NO_2)_2C_6H_3$ (120)

A kind of zwitterionic thio- and seleno-phosphonium aluminates are formed when thio- of seleno-metaphosphates react with various aluminium compounds with organoalumination of a P=N bond. The four-membered heterocycles contain two-coordinated Se bonded to four-coordinated P, but no Se—C bonds are present in the molecules[82].

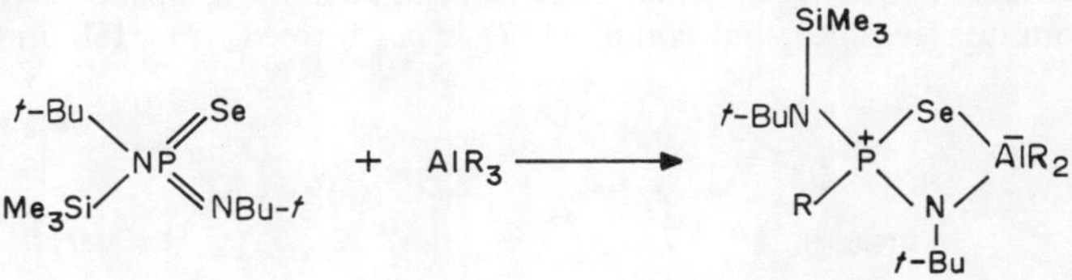

(121)

R = alkyl, phenyl, Cl

B. Preparation of Tellurophosphonium Ions

When triphenylphosphane reacted with 2-naphthyltellurenyl iodide (prepared *in situ* from the ditelluride with iodine), a yellow intermediate was isolated, which might be an aryltelluro(triphenyl)phosphonium iodide, but the structure of **154** is unknown[81]. Decomposition of the yellow adduct yielded dinaphthyl ditelluride and hydrolysis of the P-containing product (probably Ph_3PI_2) gave triphenylphosphane oxide[81].

$$\text{2-NaphTeI} + Ph_3P \rightarrow \underset{\textbf{(154)}}{\text{(1:1 adduct)}} \xrightarrow{H_2O} \text{2-NaphTeTeNaph-2} + Ph_3P{=}O \quad (122)$$

The first well characterized alkyl tellurophosphonium ions were prepared by Kuhn and Schumann[83] by addition of methyl iodide to a series of tellurophosphoranes (phosphane tellurides). The methyltellurophosphonium salts precipitate instantaneously from benzene solution.

$$R_3P{=}Te + MeI \longrightarrow R_3PTeMe^+\ I^- \quad (123)$$

(155) R = Me
(156) R = *i*-Pr
(157) R = Bu
(158) R = *t*-Bu
(159) R = NMe_2

Compounds **156–159** are moderately soluble in nitromethane, allowing ^{31}P NMR spectra to be run, but within 1 h in solution about 50% of the tellurophosphonium salts decompose[83]. The reaction of dialkyl(*p*-tolyltelluro)phosphanes **25** and **26** with methyl iodide proceeds with P-methylation, and ^{31}P NMR spectra provide evidence for the formation of aryltellurophosphonium iodides, but subsequently nucleophilic attack at the P—Te bond leads to redox reactions and a part of the tolyltelluro groups is lost as *p*-tolylditelluride[84].

Cleavage of tellurophosphonium ions with nucleophiles leads preferentially to attack at Te, i.e. they react like phosphane-stabilized tellurenyl cations. With methyllithium,

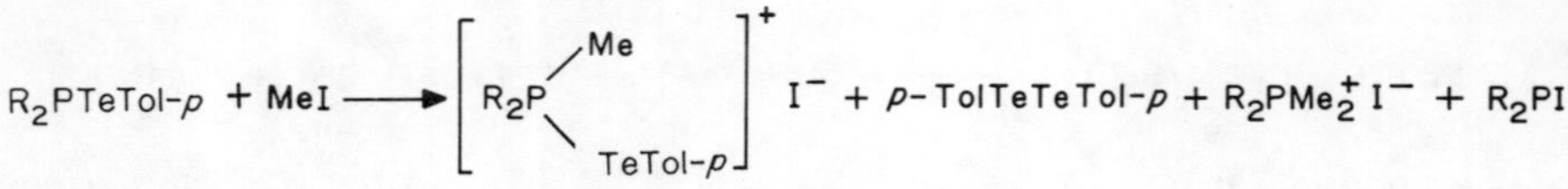

(**160**) R = *i*-Pr
(**161**) R = *t*-Bu (124)

dimethyl telluride is formed from methyltellurophosphonium iodide with reduction leading back to the tertiary phosphanes[83]. Methylation of tellurotri-*tert*-butylphosphorane occurs even when it is coordinated as a ligand with the pentacarbonyltungsten acceptor, but complex **163** is *not* formed from **158** and $(CO)_5W$–THF[85].

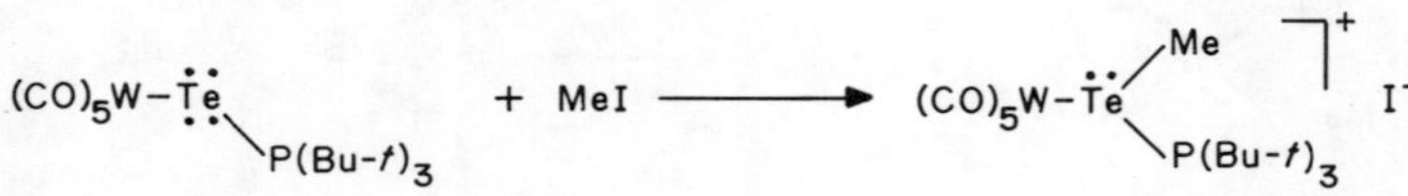 (125)

(**162**) (**163**)

Protonation of seleno- or telluro-phosphoranes with strong acids leads to species of the type R_3PEH^+ (E = Se, Te), which have been the topic of NMR spectroscopic studies (see Section IX), but none of these cations has been isolated in the pure state[78]. Similarly, complexes such as **162**, which may be regarded as zwitterions containing a chalcogenophosphonium centre, are not included in this section.

VI. ORGANYL-SELENO- AND -TELLURO-PHOSPHINATES, -PHOSPHONATES AND -PHOSPHATES

Numerous organoselenium derivatives of pentavalent phosphorus have been described, but very few corresponding tellurium compounds are known. Organo-selenium and -tellurium derivatives of pentavalent arsenic, antimony and bismuth are in general very rare.

A. Organylseleno Derivatives of Pentavalent Phosphorus

1. *Selenophosphinic acid Se-organyl esters, $R_2^1P(=E)SeR^2$ (E = O, S, Se)*

The attempted synthesis of diphenyl(phenylseleno)phosphane (**3**) with work-up in the presence of air led to diphenylselenophosphinic acid *Se*-phenyl ester (**4**); **4** is also obtained when tetraphenyldiphosphane dioxide reacts with phenylselenenyl bromide or when diphenylphosphinic chloride reacts with phenylselenomagnesium bromide[11]. Compound **4** is stable in water and the colourless compound melts at 78–80 °C.

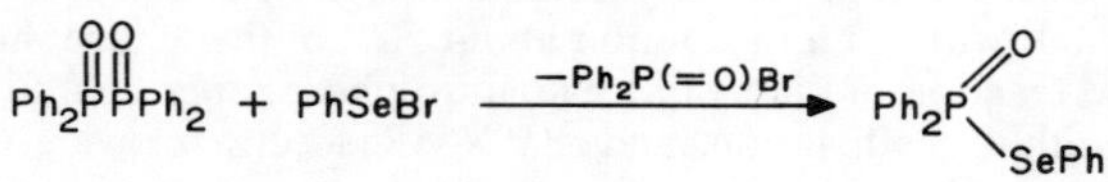 (126)

(**4**)

 (127)

Seleno(thio)- and diseleno-phosphinic acid derivatives have been prepared by Kuchen and Knop[86].

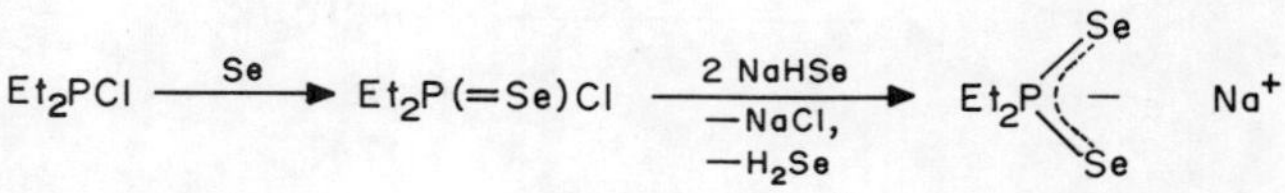

(164)

(128)

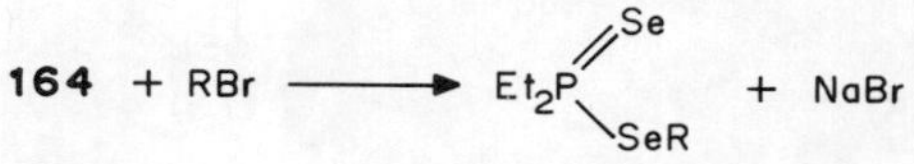

(165) R=Et

(166) R=Pr

(129)

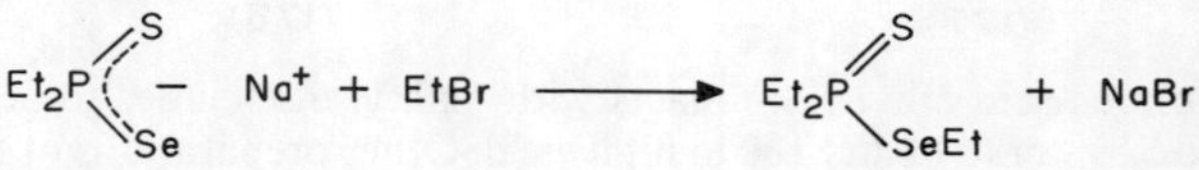

(167) (168)

(130)

Compounds **165**, **166** and **168** are colourless liquids that can be distilled under reduced pressure. *Se*-alkylation of **167** is preferred to *S*-alkylation; bromination of diethylselenothiophosphinic acid *Se*-ethyl ester (**168**) leads to diethylthiophosphinic bromide by P—Se bond cleavage[86]. *Se*-alkylation is preferred to *O*-alkylation when *tert*-butyl(phenyl)selenophosphinic acid (**151**) reacts with methyl iodide[75,87] (see Section V.A).

Selenophosphinic acid *O*-alkyl esters give the rearranged *Se*-esters by a thermal reaction[88]; the *Se*-esters are also formed in the course of transesterification reactions from P=Se bonded compounds with alkyl bromides (see also Section V)[89].

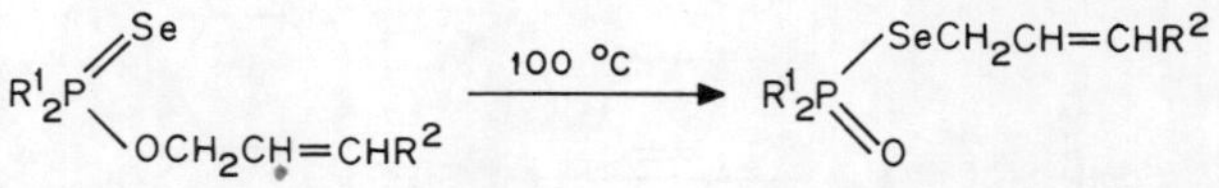

(169) (170)

R^1=Et

R^2=Me

(131)

A decision between *Se*- and *O*-ester structures is easily made by ^{31}PNMR[88,89].

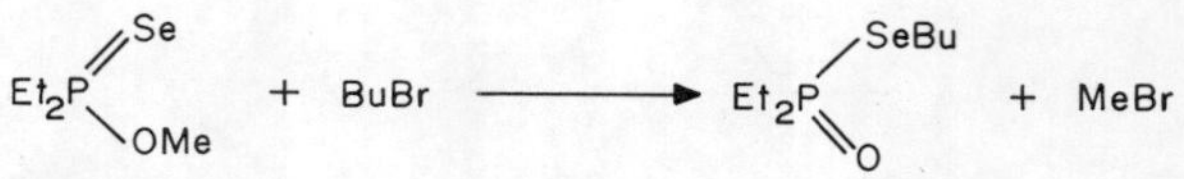

(171) (172)

(132)

2. *Selenophosphonic acid Se-esters, $R^1P(=E)(XR^2)(SeR^3)$*

Ethyl(seleno)phosphonic acid *O*-ethyl ester reacts with octyl iodide with *Se*-alkylation[90] (equation 133). The selenophosphonate **175** reacts with 2-diethylaminoethyl chloride with *Se*-alkylation to give highly toxic **176**[91], and ethyl(seleno)phosphonic acid *O*,*O*-diethyl ester (**177**) reacts with butyl bromide to the rearranged product **178**[89].

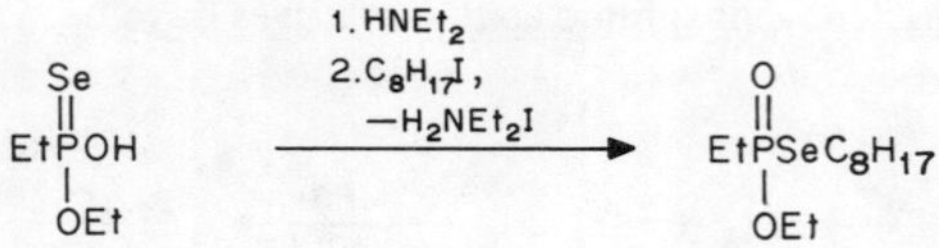

(173) (174) (133)

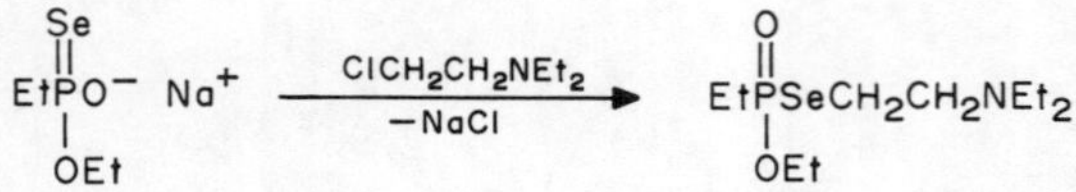

(175) (176) (134)

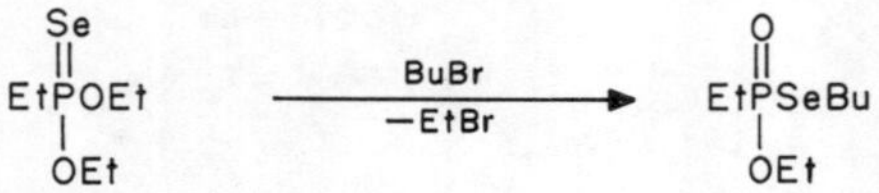

(177) (178) (135)

Ethylphosphonous acid ethyl ester reacts with dialkyl diselenides to give the selenophosphonic acid *Se*-esters **179** and **180** in high yield. Other preparations of selenophosphonic acid *O*, *Se*-diesters are the chloroselenation of alkenes with EtP(=O)(OEt)SeCl[92] or the reaction of ethyldi(ethoxy)phosphane with phenylselenenyl chloride or bromide[93].

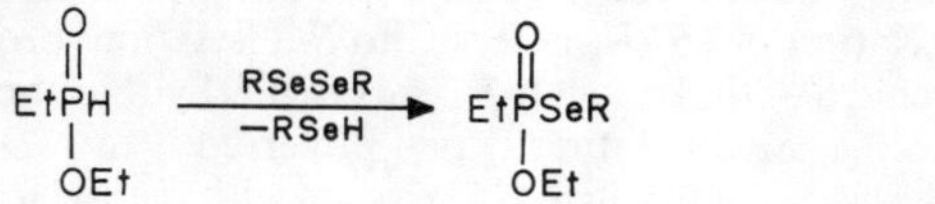

(136)

(**179**) R = Me (Ref. 94)
(**180**) R = Et (Ref. 94)

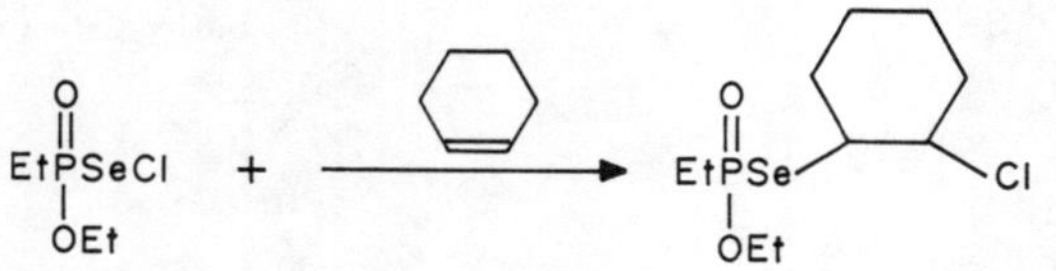

(181) (182) (137)

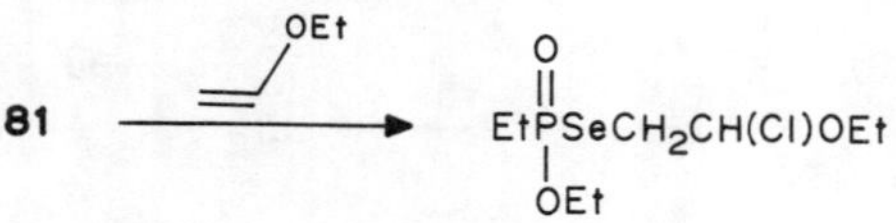

(183) (138)

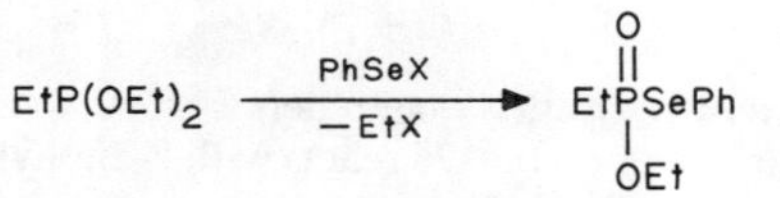

(184) (139)

X = Cl or Br

Selenophosphonic acid *O,Se*-diesters in most cases are thermally stable colourless liquids that can be distilled *in vacuo*. UV spectra of these compounds show a bathochromic shift and a significant decrease in the extinctions of λ_{max} compared with the isomeric *O,O*-diesters containing P=Se chromophores[95].

3. Selenophosphoric acid Se-esters

An excellent review covering (also) selenophosphoric acid *Se*-esters up to 1980 appeared in Houben-Weyl[2]. Here only basic types of the formation of P—Se bonds leading to this class of compounds will be considered.

a. Oxidation of phosphites with arylselenenyl halides or with diselenides[96]. See Table 4.

TABLE 4. Oxidation of phosphorous acid derivatives with arylselenenyl halides or with organic diselenides[2]. In most cases R^1 is Me or Et and R^2 is aryl or CF_3

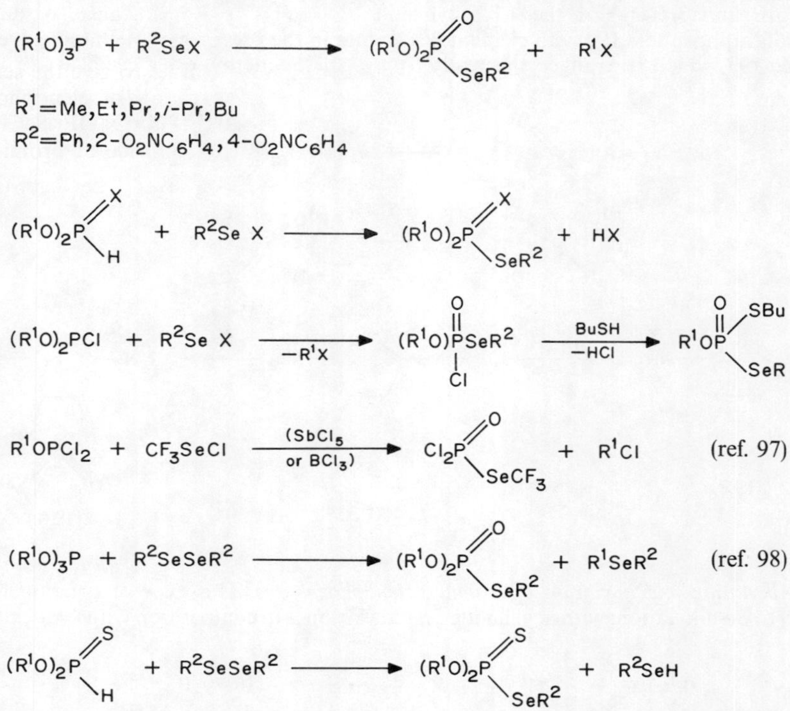

b. Oxidation of selenophosphites with oxygen or sulphur[11,48]

$$(SePh)_3P + Y \longrightarrow (PhSe)_3P{=}Y \qquad (140)$$

$Y = 1/2\ O_2,\ 1/8\ S_8$

(77) $Y = O$

(185) $Y = S$

$$ROP(SeBu)_2 + Y \longrightarrow RO(BuSe)_2P{=}Y \quad \text{(ref. 50)} \qquad (141)$$

(86–88)

$Y = 1/2\ O_2,\ 1/8\ S_8$

(186) R = Et, Y = O
(187) R = Et, Y = S
(188) R = Pr, Y = O
(189) R = Pr, Y = S
(190) R = Bu, Y = O
(191) R = Bu, Y = S

For comparison, the ethoxy derivatives **186** and **187** were also prepared from EtOP(=Y)Cl_2 (Y = O, S) with butylselenol in the presence or triethylamine[50] (see Section VI.A.3.e).

c. Rearrangement of selenophosphoric acid O-esters containing at least one n-alkyl or 2-alkenyl group R^2 [2,3]. These rearrangements are catalysed by Lewis acids such as BF_3 and many other metal or metalloid halides, but there are cases known where Lewis acids are unnecessary for the 1,3-alkyl shift (R^2 = allyl[88]). For instance, oxidation of the diazaphosphole **192** with elemental selenium in the presence of methanol gave the selenide **193**, which rearranged thermally to give the *Se*-methyl ester **194**[99].

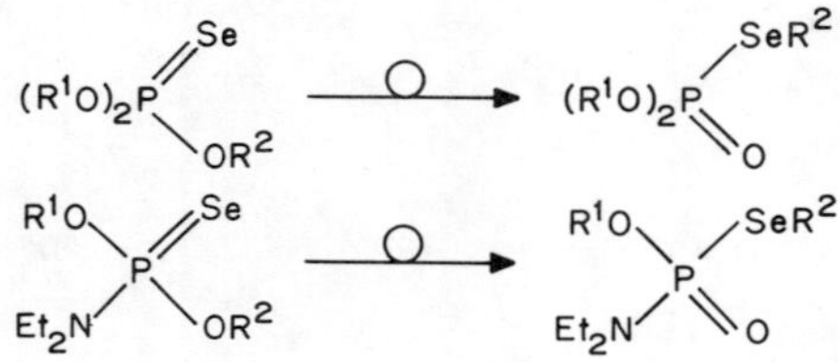

(142)

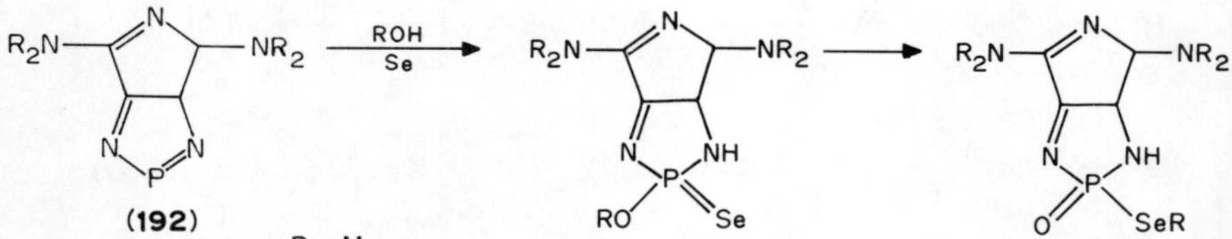

(192) R=Me **(193)** **(194)**

(143)

d. Alkylation and acylation of anionic selenophosphates. This type of reaction leads mainly to *Se*-alkylation; with acyl halides, *Se*-acylation is in competition with *O*-acylation.

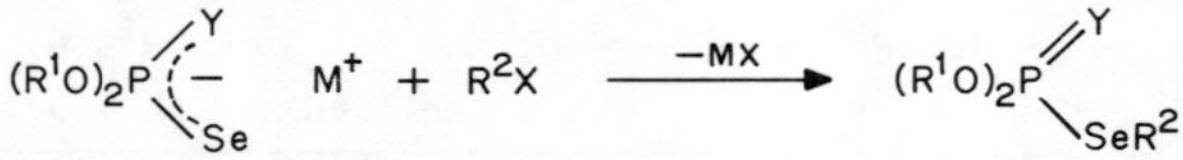

R^1 is Et, *i*-Pr or Ph in most cases, Y may be O, S or Se and various alkyl and 2-alkenyl halides were found to give this reaction. In a similar manner, aziridine is cleaved to give the highly toxic ester **196**[91].

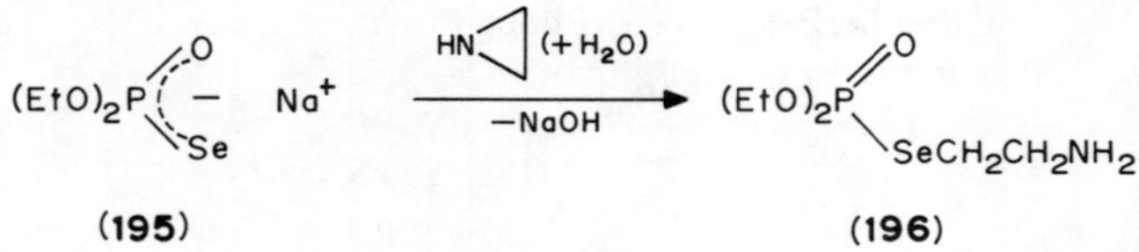

(144)

(195) **(196)**

The anionic selenophosphates may be prepared *in situ*[100]. Acylation of seleno(thio)phosphoric acid *O,O*-diethyl and -diphenyl esters led to a series of 'mixed anhydrides'[101].

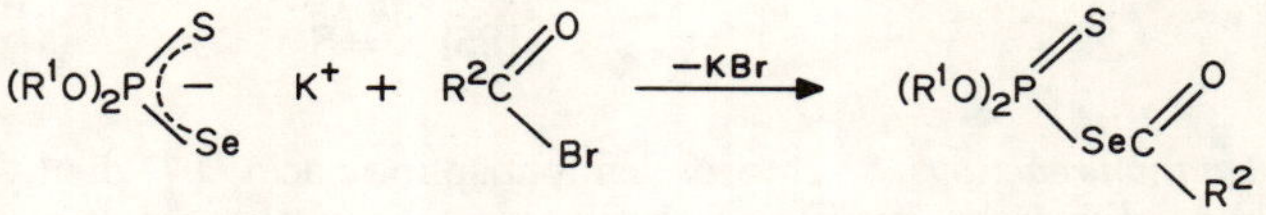

(145)

(**197**) R^1=Et, R^2=Me
(**198**) R^1=Et, R^2=Et
(**199**) R^1=Et, R^2=Pr
(**200**) R^1=Et, R^2=Ph
(**201**) R^1=Et, R^2=*i*-Pr
(**202**) R^1=Et, R^2=OEt
(**203**) R^1=Ph, R^2=Me
(**204**) R^1=Ph, R^2=Et
(**205**) R^1=Ph, R^2=Ph
(**206**) R^1=Ph, R^2=OEt

Se-acylation was predominant, but some *S*-acylation also occurred, as shown by thin-layer chromatography[101]. The reaction of Na[SeOP(OPr-*i*)$_2$] with various main group halides was followed by NMR and IR spectroscopy. *Se*-alkylation and *O*-acylation occurred[78], with organotin halides both types of reactions were observed (see equation 146).

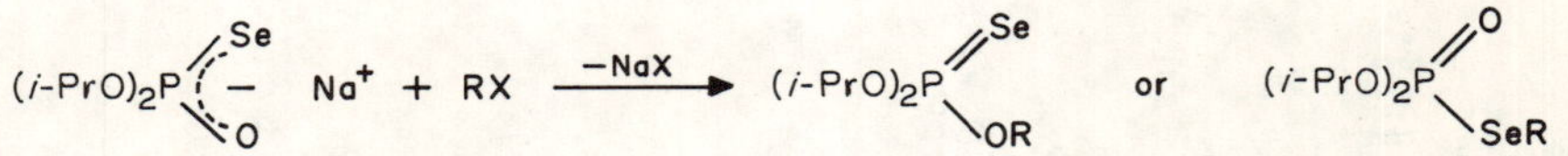

(**207**) A B

(**208**) R=COPh
(**209**) R=$SnMe_3$
(**210**) R=1/2 $SnPh_2$
(**211**) R=1/3 SnPh
(**212**) R=1/3 P

(**213**) R=Et
(**214**) R=$SiMe_3$
(**215**) R=$GePh_3$
(**216**) R=SO_2Ph
(**217**) R=1/3 As

(146)

e. Hydrogen chloride elimination from phosphoric acid chlorides with phenyl- or butylselenols in the presence of triethylamine[2,50].

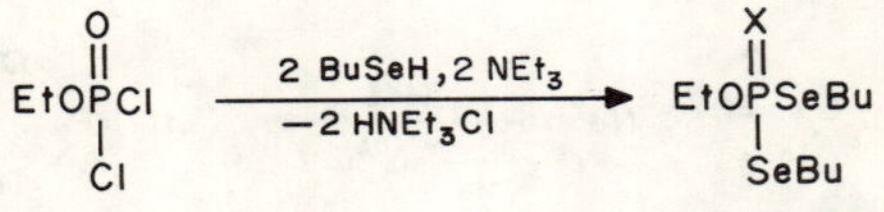

(147)

(**218**) X=O
(**219**) X=Se

$$P(=X)Cl_3 \xrightarrow[-3\ HNEt_3Cl]{3\ PhSeH,\ 3\ NEt_3} (PhSe)_3P{=}X \tag{148}$$

(77) X = O
(185) X = S

f. Chloroselenation reactions. Se-chloroselenophosphoric acid *O,O*-diethyl ester reacts with *N,N*-dialkylanilines by aromatic substitution to give the corresponding *Se*-aryl esters[102].

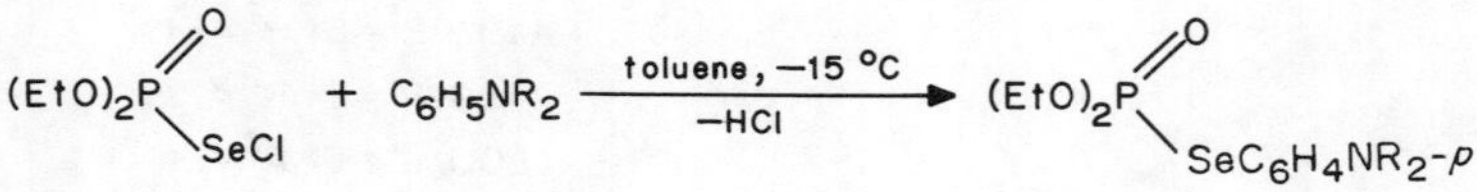

(149)

(220)

(221) R = Me
(222) R = Et

Compound **220** adds also to cyclohexane and to ethyl vinyl ether by chloroselenation [**220** is prepared *in situ* from $(EtO)_3P{=}Se$ with SO_2Cl_2[92,93,102]].

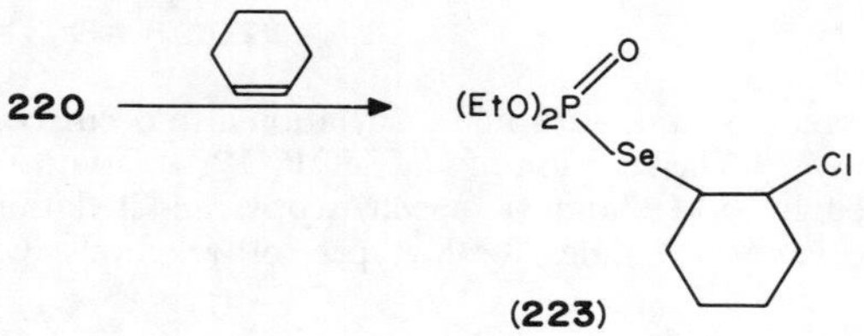

(150)

(223)

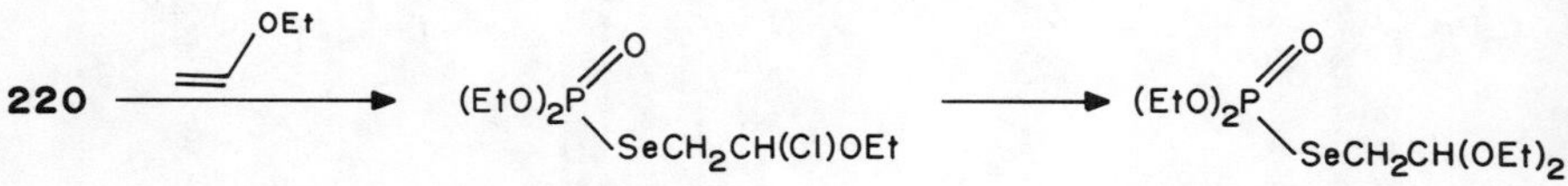

(224) **(225)**

(151)

B. Organyltelluro Derivatives of Pentavalent Phosphorus

Sodium or potassium diethyl phosphite react with elemental tellurium in ethanol to give colourless crystalline tellurophosphates which are sensitive to air and moisture[103] (corresponding selenophosphates were formed in a similar manner[104]). With methyl iodide, methylation occurs at tellurium[105] (equation 152). Tellurophosphoric acid *O,O*-diethyl-*Te*-methyl ester **227** has been characterized in solution by ^{1}H and ^{31}P NMR. *Te*-alkylation has been proposed as the key step of the oxirane deoxygenation with **226** (equation 153).

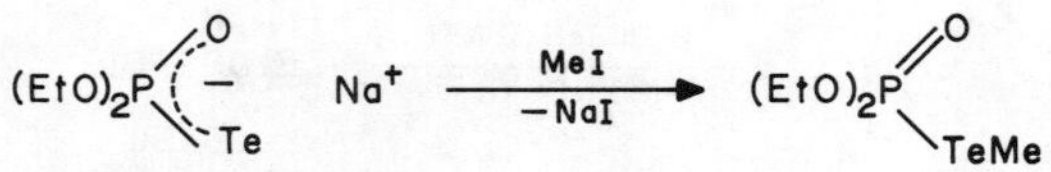

(152)

(226) **(227)**

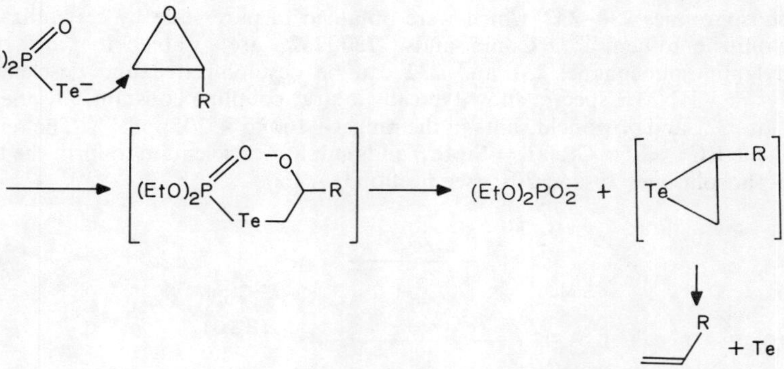

(153)

The first tellurocarbohydrates were prepared by *Te*-alkylation of a tellurophosphoric acid *O*, *O*-diester anion with 2, 3, 4, 6-tetra-*O*-acetyl-α-D-glucopyranosyl bromide[106]. The reaction proceeds with inversion at the anomeric carbon atom. Spectroscopic evidence for *Te*-alkylation comes from the magnitude of the NMR coupling constant, $^{1}J(^{125}Te^{31}P) = 1245$ Hz[106]. Oxidation of β-D-glucosyl tellurophosphate (**229**) led to novel di(β-D-glucosyl) ditelluride.

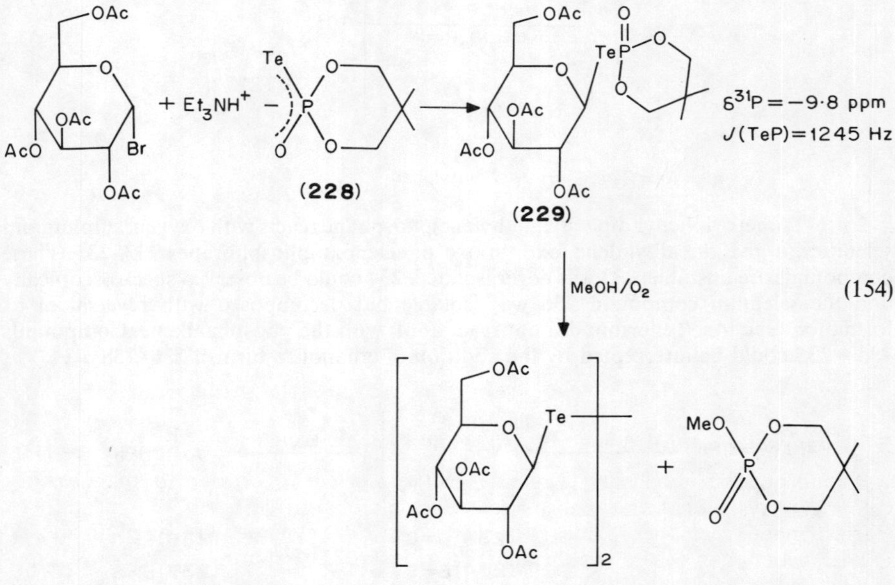

(154)

VII. PENTAVALENT THREE-COORDINATE PHOSPHORUS COMPOUNDS WITH P=Se DOUBLE BONDS

A. Methylene(selenoxo)phosphoranes, R¹P(=R²) (=Se)

2, 4, 6-Tri-*tert*-butylphenyl[bis(trimethylsilyl)methylene]phosphanes reacts with ozone (−78 °C), sulphur[107] or selenium to give the alkylidene(oxo, thioxo or selenoxo)phosphoranes **230–232**, which were obtained in pure state by crystallization from acetonitrile–toluene[108]. Compounds **230–232** are stable to air; with tris(dimethylamino)phosphane, **231** and **232** can be desulphurated and deselenated, respectively[108]. ^{31}P NMR spectra show typically a large coupling constant, $^1J(^{77}Se^{31}P)$ ($\pm$ 846 Hz for **232**), and downfield shifts in the range + 160 to + 205 ppm[108]. The related phosphaalkene RC≡CP=CR$_2$ (R = SiMe$_3$) adds sulphur or selenium to form the thio- and seleno-phosphiranes **19** and **20**[25] (see Section II.A.1.a).

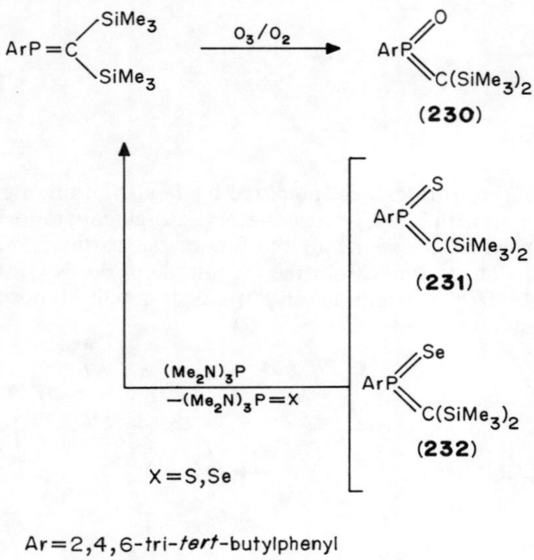

(155)

2, 4, 6-Trimethylphenyl(diphenylmethylene)phosphane reacts with oxygen, sulphur and selenium to give the alkylidene (oxo, thioxo or selenoxo)phosphoranes **233–235**. These compounds are unstable: **233** was never isolated, **234** could be observed spectroscopically and the selenium compound **235** was isolable but decomposed with reversal of its formation reaction. Tellurium did not react at all with the phosphaalkene. Compounds **233—235** could be intercepted by the addition of ethanol to furnish **236–238**[109].

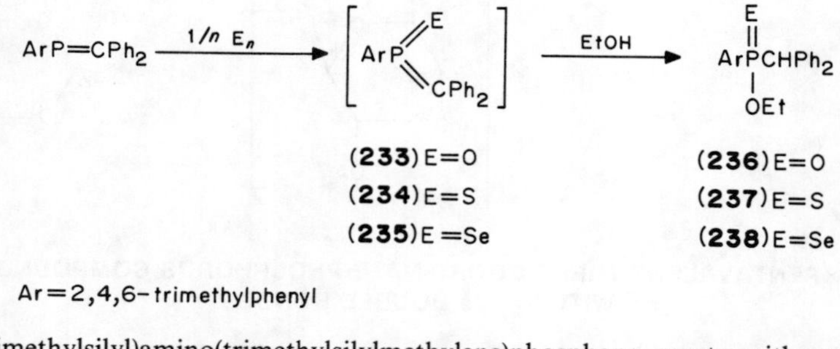

(156)

Bis(trimethylsilyl)amino(trimethylsilylmethylene)phosphane reacts with sulphur or selenium to give the corresponding thioxo- and selenoxo-phosphoranes **239** and **240**. The latter was not isolated in the pure state, but NMR data [δ(P) = 172.4 ppm, $J(^{77}Se^{31}P)$ 882.5 Hz] confirm that the alternative selenaphosphirane structure can be ruled out[110].

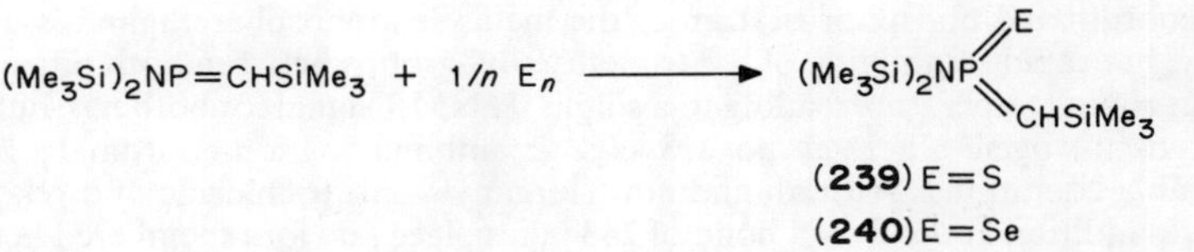

(157)

B. Metaselenophosphonates, $R^1P(=E)(=Se)$

Yoshifuji showed that bis(2, 4, 6-tri-*tert*-butylphenyl)diphosphene gives various types of oxidation products with sulphur and selenium[67,68]. With sulphur, isomers **127** and **128** have been characterized; with selenium, a thermal reaction in the presence of triethylamine afforded the selenadiphosphirane (**125**) and a true metadiselenophosphonate (**126**). The metaselenophosphonate structure was confirmed by NMR data.

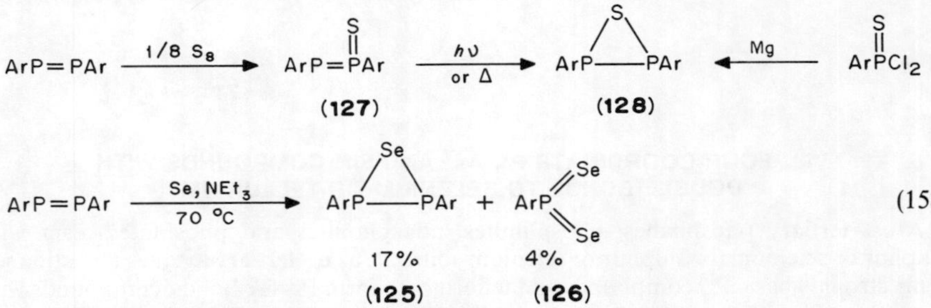

(158)

Ar=2,4,6-tri-*tert*-butylphenyl

C. Metaselenophosphates, $R^1R^2NP(=NR^1)(=Se)$

Monomeric amino(imino)phosphanes ['phospha(III)azenes'] with bulky substituents are oxidized by selenium to metaselenophosphoric acid derivatives[111,112].

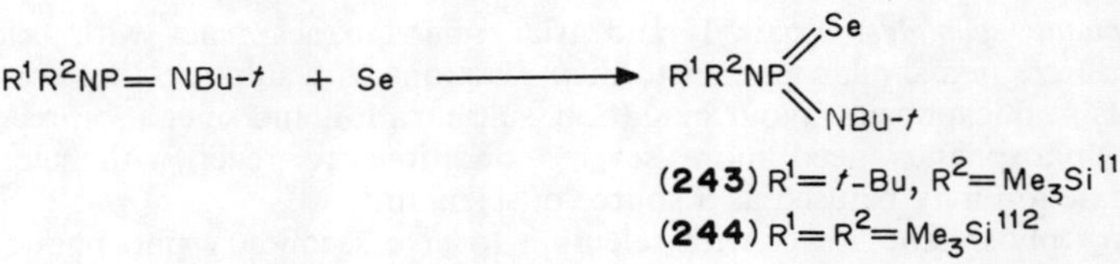

(159)

Compounds **243** and **244** are thermally stable; they are soluble in inert solvents, but sensitive to protic and oxidizing agents (as are the corresponding metathiophosphates); **243** is an orange and **244** a greenish yellow liquid. With ethylenebis-(triphenylphosphane)platinum(0), the metaselenophosphate **243** reacts with η_2-coordination of the P=Se bond with the bis(triphenylphosphane)platinum acceptor moiety[113].

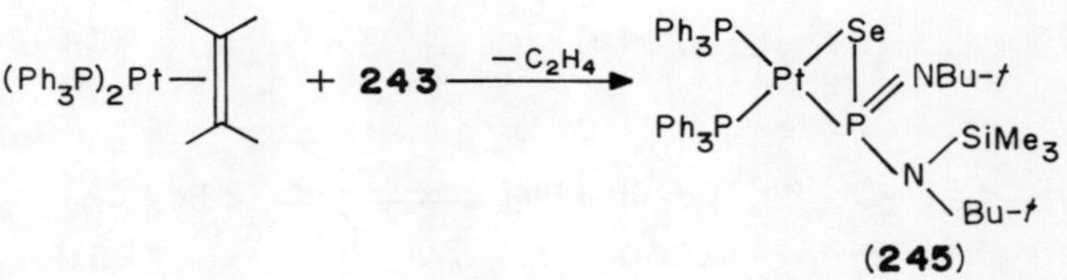

(160)

The coordinated phosphorus atom of the metaselenophosphate ligand is a centre of chirality, but rapid scrambling of the trimethylsilyl groups between both nitrogen atoms occurs at room temperature leading to a single ^{1}H NMR signal for both *tert*-butyl groups bonded to nitrogen. There is no 1,3-silyl scrambling in non-coordinate **243** under comparable conditions. With aluminium triorganyls and trichloride and related Lewis acids, 1,2-addition at the P—N bond of **243** takes place and four membered heterocycles such as **246** and **247** are formed. The corresponding metathiophosphate behaves similarly[82].

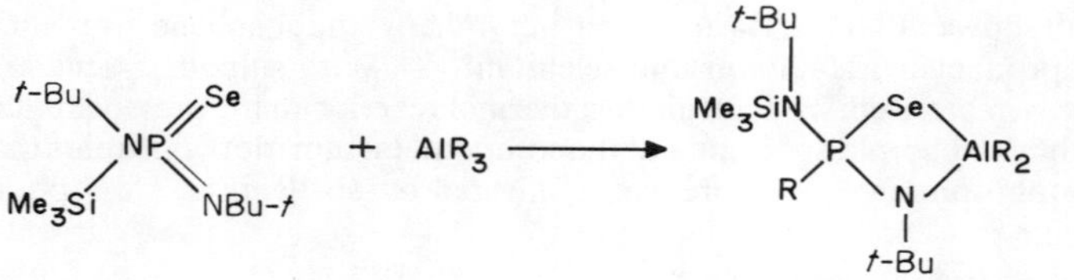
(161)

(246) R=Me
(247) R=Ph

VIII. FOUR-COORDINATE P^V, As^V AND Sb^V COMPOUNDS WITH DOUBLE BONDS TO SELENIUM OR TELLURIUM

Most tertiary phosphanes, phosphinites, phosphonites and phosphites react with sulphur or selenium (or sulphur or selenium sources), as do tertiary arsanes and stibanes. Only strongly basic P^{III} compounds add tellurium to form P=Te bonds; compounds with As=Te or Sb=Te and any compounds with Bi=X (X = S, Se, Te) bonds are unknown.

A. Four-coordinate Phosphorus Compounds with P=Se or P=Te Bonds

1. Chalcogenophosphoranes, $R^1P_2{}^2P$=E (E = Se, Te; R^1, R^2 = alkyl, aryl or H) (Phosphane chalcogenides)

a. Phosphane selenides. Trialkyl- and triaryl-phosphanes react with selenium or selenium sources in a similar manner to their reactions with sulphur. Electron-donating substituents at phosphorus favour oxidation, sulphuration and selenation reactions; di- and tri-arylphosphanes need more severe conditions to react with selenium; the selenocyanate ion may be used as a source of selenium[1–3,5].

Secondary phosphanes react with selenium to give secondary phosphane selenides (selenophosphinous acids), $R_2P(=Se)H$; primary $PhPH_2$ gave evolution of H_2Se and a cyclic oligomer of (PhPSe) was formed (Section IV). Bis(diphenylphosphino)methane gives a mono- and a di-selenide[114]. Mixing the 'diselenide' **248** with the diphosphane leads to immediate selenium transfer to give the 'monoselenide' **249**.

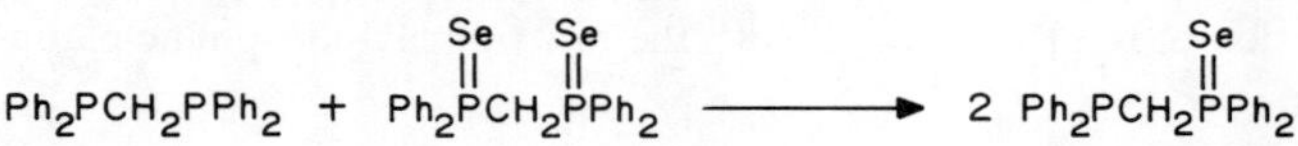
(162)

(248) **(249)**

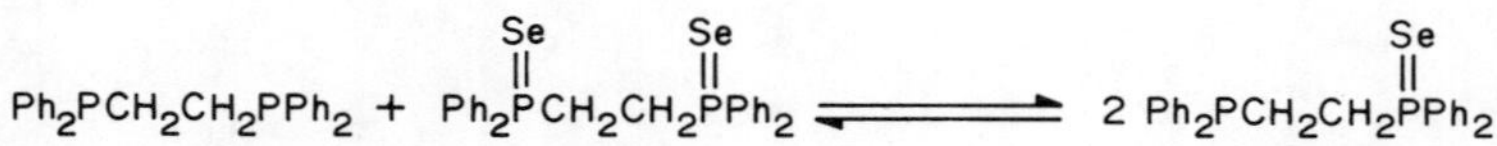
(163)

(250) **(251)**

With 1, 2-bisdiphenylphosphinoethane a statistical distribution of selenium take place (equation 163) and pure 'monoselenide' **251** is not obtained[114]. Compound **248** is less favoured in the equilibrium because the electron-deficient P^V atom deactivates the close P^{III} atom in the molecule, so that monoselenide **249** is formed according to equation 162. Temperature-dependent NMR spectra have shown that coalescence temperatures for the above and various related systems are in the range 90–160°C in most cases; their concentration dependence indicates that even with chelating phosphanes, selenium transfer is at least partly intermolecular.

b. Phosphane tellurides. Early work by Zingaro and coworkers showed that trialkylphosphanes tend to react incompletely with elemental tellurium. Pure phosphane tellurides could be obtained by crystallization of the compounds from mother liquors that contained an excess of unreacted phosphane[60,115]. Aryl substituents at phosphorus lead to unfavourable equilibria and triphenylphosphane does not react with elemental tellurium. Telluration of triphenylphosphane was achieved with tetraphenylarsonium tellurocyanate in the presence of lithium perchlorate. Under these conditions, a remarkable compound of composition $(Ph_3P)_2Te$ separated as yellow crystals from acetone solution that contained a large excess of triphenylphosphane. X-ray diffraction analysis of a disordered crystal led to the conclusion that a linear P=Te... P arrangement with very different P—Te bond distances is present in the crystal (P=Te 2.27–2.42 Å; P... Te 3.37–3.95 Å)[116].

$$R_3P + 1/x\,Te_x \rightleftharpoons R_3P=Te \qquad (164)$$

$$2\,Ph_3P + Ph_4As^+\,TeCN^- \underset{}{\overset{(LiClO_4)\ \text{acetone}}{\rightleftharpoons}} \underset{\textbf{(252)}}{Ph_3PTe\cdots PPh_3} + Ph_4As^+\,CN^- \qquad (165)$$

^1H, ^{31}P and ^{125}Te NMR spectra of phosphane telluride–phosphane mixtures revealed that rapid transfer of tellurium atoms occurs at room temperature in solutions in all known cases[117–121]. Transition states related to **252** have been proposed, i.e. nucleophilic attack of phosphane at coordinated tellurium leads to the tellurium transfer reaction (S_N-type of associative reaction).

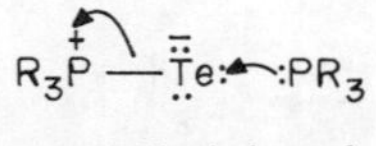

The course of the reaction of a trialkylphosphane with tellurium is sensitive to branching at the α-C atom. With tributylphosphane, 1:1 mixtures of phosphane and phosphane telluride are formed in toluene solution, but with tri-*tert*-butylphosphane telluration proceeds quantitatively. Tributylphosphane telluride (**253**) suffers from decomposition by loss of tellurium, but tri-*tert*-butylphosphane telluride (**254**) is remarkably thermally stable in the pure state and in inert solvents[117].

$$Bu_3P + Te \rightleftharpoons \underset{\textbf{(253)}}{Bu_3P=Te} \qquad (166)$$

$$t\text{-}Bu_3P + Te \rightarrow \underset{\textbf{(254)}}{t\text{-}Bu_3P=Te} \qquad (167)$$

2. *Chalcogenophosphinic esters and amides, $R_2^1(R_n^2X)P{=}E$ (E = Se, Te)*

a. Phosphinic acid derivatives with P=Se bonds. Oxidation of secondary phosphane selenides (selenophosphinous acids) in the presence of triethylamine leads to the corresponding selenophosphinic acid, e.g. equation 168[87].

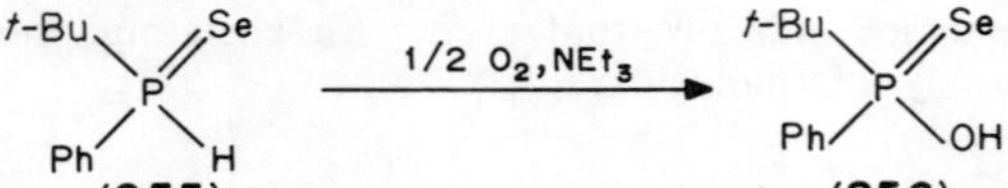

(255) (256) (168)

Dialkyl(chloro)phosphanes are easily selenated[1,2]. Selenation in the presence of alcohols and triethylamine leads to selenophosphinic acid *O*-esters (equation 169) (For reactions leading to the corresponding *Se*-esters, see Section VI).

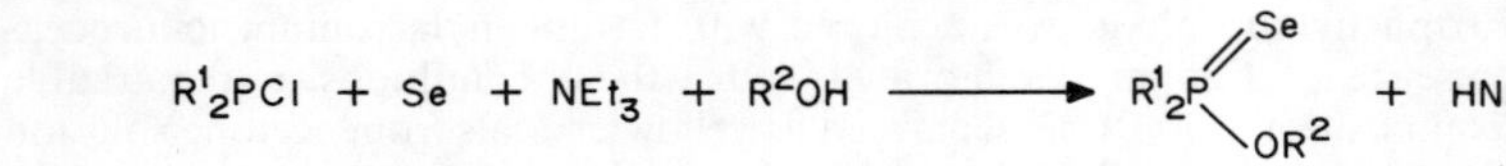

(169)

Selenophosphinic acid amides are generally obtained by selenation of the parent aminophosphanes. Seleno(thio)phosphinic acid *S*-esters are correspondingly prepared by selenation of organylthiophosphanes or from $R_2P({=}Se)Cl$ with a thiol in the presence of triethylamine. Diselenophosphinates and corresponding esters were prepared according to equations 128, 129 and 169)[86].

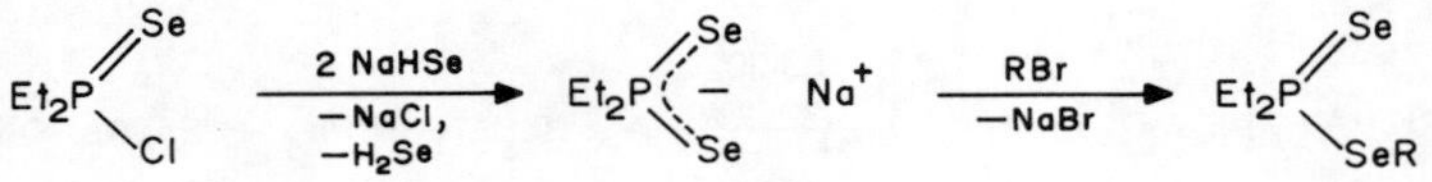

(257) (164) (165) R = Et (166) R = Pr

Dithio-, seleno(thio)- and diseleno-phosphinates are soft ligands that can act as bidentate chelating or bridging ligands with many metal acceptors. Protonation of **164** leads to bis(selenophosphinyl)-selenide, -diselenide and -triselenide[86].

$$\mathbf{164} \xrightarrow[\text{dioxane}]{\text{HCl}} (Et_2P({=}Se))_2Se_n \qquad (170)$$

(258) $n = 1$
(259) $n = 2$ (unstable)
(260) $n = 3$

$$\mathbf{164} \xrightarrow[H_2O]{KI_3} \mathbf{258} + \mathbf{260} + \text{other products} \qquad (171)$$

Oxidation or protonation of sodium diethylseleno(thio)phosphinate leads to Se—Se bond formation as in **259**–**261**; the P=S bonds remain in terminal positions [X-ray crystallographic evidence on $Et_2P({=}S)SeSeP({=}S)Et_2$[122]].

$$\mathbf{164} + \mathbf{257} \longrightarrow Et_2P({=}Se)SeP({=}Se)Et_2 + NaCl \qquad (172)$$

(258)

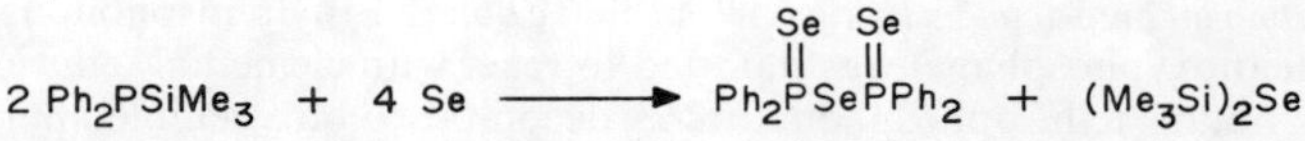

(261) (173)

Bis(selenophosphinyl) selenide (**258**) was formed according to equation 170 or 172. A bis(selenophosphinyl)selenide (**261**) related to **258** was formed when diphenyl(trimethylsilyl)phosphane reacted with elemental selenium[123].

With sulphur, the silylphosphane gave diphenyldithiophosphinic acid trimethylsilyl ester (**262**)[123]. In contrast, di-*tert*-butyl(trimethylsilyl)phosphane reacts with excess of sulphur or selenium to give di-*tert*-butyldichalcogenophosphinic acid silyl esters in both cases[29].

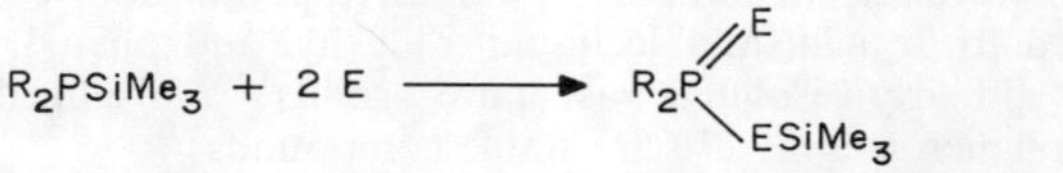

(174)

(**262**) R=Ph, E=S
(**263**) R=*t*-Bu, E=S
(**264**) R=*t*-Bu, E=Se

b. Tellurophosphinic acid esters. A compound of this type was prepared by telluration of the corresponding phosphinous acid ester with elemental tellurium[124].

3. Chalcogenophosphonic esters and amides $R^1(R_n^2X)_2P{=}E$ (E = Se, Te)

a. Selenophosphonic acid derivatives with P=Se bonds. These compounds are generally prepared by methods related to those described for selenophosphinic halides, esters and amides[1–3,5].

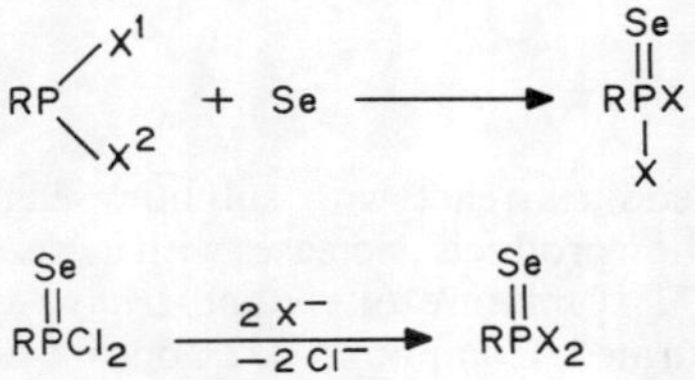

$X = F^-, EtO^-, R^2CO_2^-, R^2S^-, R^2_2N^-$

Alkylphosphonous dihalides do not react very straightforwardly with elemental selenium; red selenium and catalytic amounts of aluminium trichloride enhance the selenation. Electron-donating substituents such as dialkylamino groups favour the tendency of phosphonites to react with selenium (as with other oxidizing agents). Cl/F exchange occurs with SbF_3 and catalytic amounts of $SbCl_5$[125]; monofluoridation [to give RP(=Se) (Cl)F] has also been achieved. Selenophosphonic acid *O*-monoesters, $R^1P(Se)(OR^2)OH$, may be split into their optical antipodes with help from chiral bases[126]. For *O*-ester to *Se*-ester rearrangements of selenophosphonates, see Section VI.A.2.

Selenophosphonates are also formed when amino(imino)phosphanes (**243**) react with triorganylaluminium compounds with organoalumination of the P=N double bond (leading, finally, to **246** and **247**)[82].

Reaction of *tert*-butylbis(trimethylsilyl)phosphane with excess of selenium to give bis(trimethylsilyl)telluride and a high yield of dimeric diselenometaphosphonic anhydride $[t\text{-}BuPSe_2]_2$ containing a diselenadiphosphetane structure[57] (compare equation 106).

b. Tellurophosphonic acid derivatives with P=Te bonds. Ethylphosphonous acid diethyl ester (ethyldiethoxyphosphane) was reported to react with elemental tellurium to give a red solution, but the tellurophosphonate (**265**) decomposed on attempted purification[126].

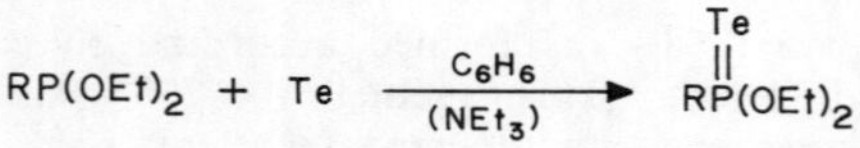

(**265**) R=Et
(**266**) R=$CH_2CH{=}CH_2$

(175)

Measurement of density and refraction index confirmed that the reaction had occurred. Decomposition led to 'tellurium mirrors'[127]. The corresponding alkyl tellurophosphonate (**266**) was reported to be a distillable liquid[128]. Ethyl and phenyl tellurophosphonic diamides were similarly isolated as pure solids[129]. Tellurides of bidentate bis(diamino)phosphanes such as **270** are stable compounds[130].

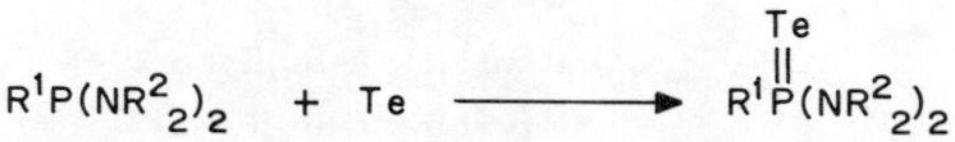

(**267**) R^1=Et, R^2=Me (ref.129)
(**268**) R^1=Ph, R^2=Me (ref.129)
(**269**) R^1=Me, R^2=Et (ref.130)

(176)

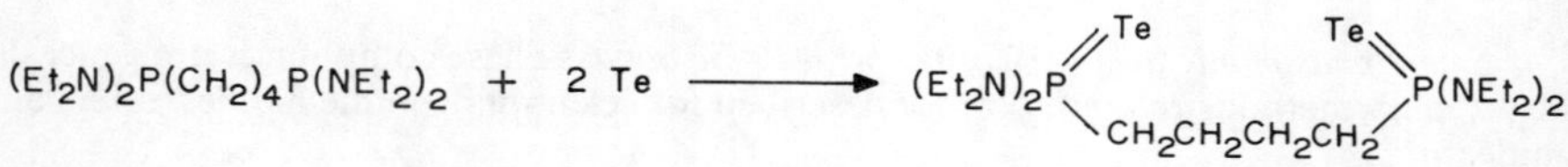

(**270**)

(177)

N-, *P*-Alkyldiazadiphosphetidines react with sulphur, selenium and tellurium. The thermal stability of the oxidation products decreases with increasing atomic number of the chalcogen[131]. Compound **271** is mixture of *cis* and *trans* isomers; the Cl-substituted phosphorus atom is not tellurated. Compound **272** appears to be only one isomer.

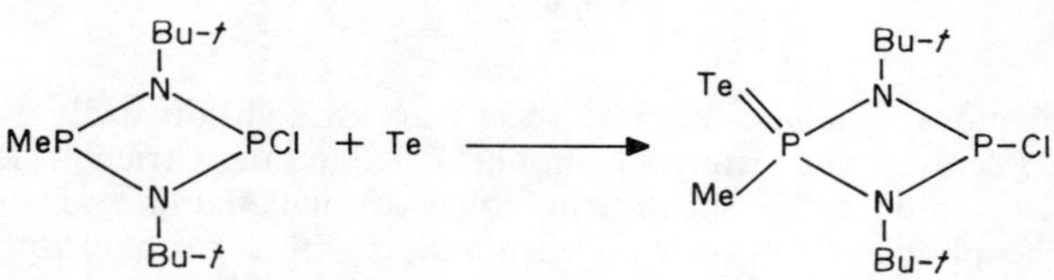

(**271**)

(178)

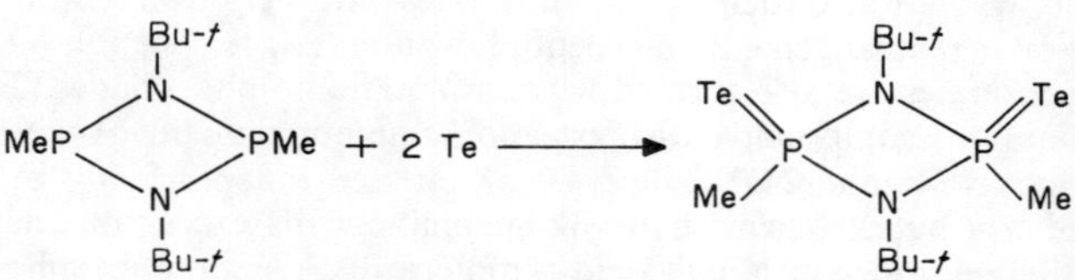

(**272**)

(179)

The yellow crystalline compounds are stable in the pure state, but **271** suffers from loss of tellurium in solution[131]. Reaction of a *cis*-1, 3, 2λ^3, 4λ^3-diazadiphosphetidine (equation 180) with an approximately equimolar amount of tellurium leads to a monotelluride (**273**), which undergoes rapid tellurium transfer in solution[120].

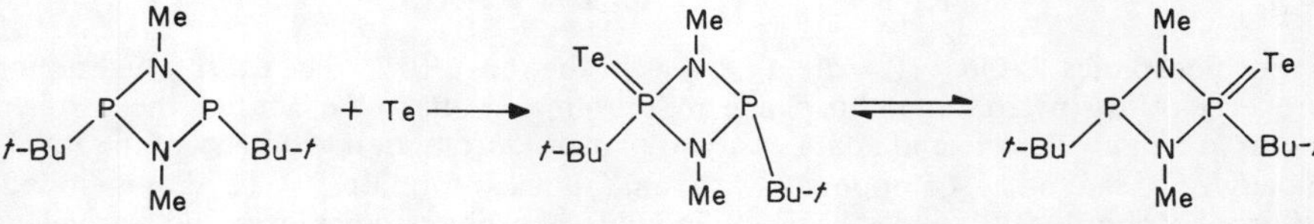

(273)

(180)

The concentration dependence of the temperature of coalescence indicates that tellurium transfer is not strictly intramolecular. At $-60\,°C$, separate signals for λ^3-P [δ(P) + 200.4 ppm] and λ^5-P [δ(P) + 119.5 ppm] and also ^2J(PP) (18.0 Hz) are well resolved[120]. An X-ray crystal structure analysis confirmed that Te is tightly bonded to *one* P atom, and there is no indication of any type of secondary bonding between Te and λ^3-P. The deviation from planarity of the P_2N_2 ring of **273** is about 30% larger than that in the related ditelluride or disulphide[132,133]. Stronger folding of the P_2N_2 ring favours an intramolecular tellurium atom transfer. The *trans*-diazadiphosphetidine ditelluride contains a planar P_2N_2 core[133].

4. *Phosphoric acid derivatives with P=Se or P=Te bonds*

a. Selenophosphoric acid esters, amides and other derivatives. Synthetic aspects of this well studied class of compounds are covered up to 1980 in Houben-Weyl[2]. Selenophosphates with various substitution patterns are preparatively fairly readily available (see also Section VI.A.3.a). The main synthetic pathways are the direct selenation of P^{III} compounds or substitution reactions at P—Cl bonds, for instance with alcohols or thiols in the presence of triethylamine or with excess of a secondary amine. Diselenophosphates can be obtained from diphosphorus pentaselenide with various alcohols[2,3]. Transesterification reactions are also known. Oxidation of diselenophosphoric acid *O,O*-diesters leads to di(selenophosphoryl) selenides and diselenides[134,135].

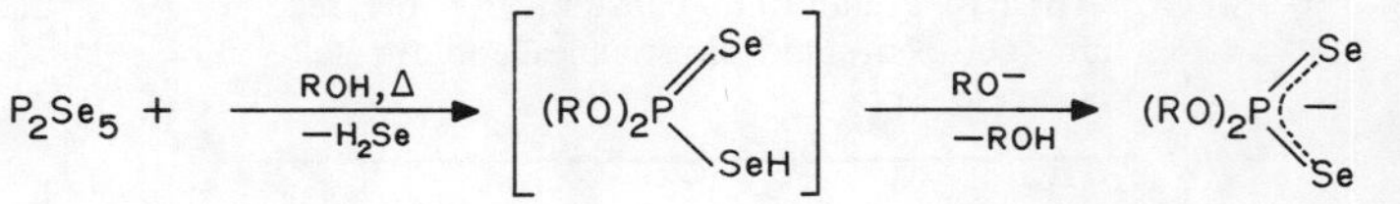

(181)

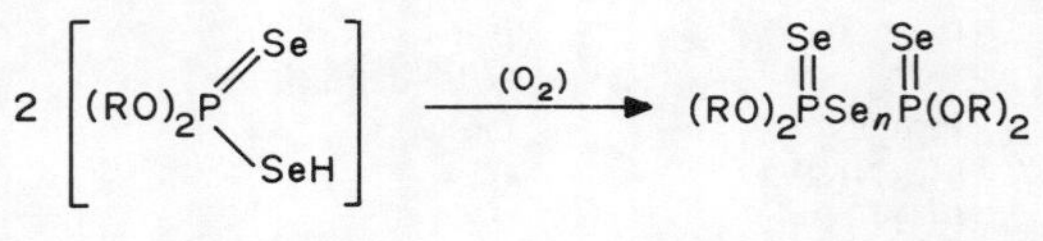

(182)

b. Tellurophosphoric acid amides and esters. The first tellurophosphoric acid triamides [tris(dialkylamino)phosphanetellurides] were prepared independently by two different groups[129,136]. The reaction of the aminophosphanes was reported to proceed 'with difficulty' compared with addition of sulphur or selenium to amides of trivalent phosphorus, but the reaction is much closer to being quantitative than the reaction of most trialkylphosphanes with elemental tellurium[129] (Section VIII.A.1.b). A careful study of the 'tellurium basicity' of trivalent phosphorus species revealed an exceptional tellurophilicity of tri(pyrrolidino)phosphane, the only phosphane being superior to the cyanide ion (see Table 5)[137].

$$R_3{}^1P + TeCN^- \rightleftharpoons R_3{}^1P=Te + CN^- \qquad (183)$$

For phosphites, $P(OR)_3$ (R = alkyl), K was in the range 10^{-3}. The exceptional donor properties of tris(pyrrolidino)phosphane are presumably due to the fact that the nitrogen atoms are highly flexible and adapt easily to planar coordination geometry of the PNC_2 moiety[137]. Like other compounds with an (at least formal) P=Te double bond, tellurophosphoric acid amides behave like labile phosphane complexes and zerovalent tellurium; in the presence of non-coordinated p^{III} compounds, tellurium atom transfer takes place as a reaction that is fast on the ^1H or ^{31}P NMR time-scales[114,117–120,138,139].

$$\underset{\textbf{(254)}}{(t\text{-Bu})_3PTe} + (Me_2N)_3P \rightleftharpoons (t\text{-Bu})_3P + \underset{\textbf{(274)}}{(Me_2N)_3PTe} \qquad (184)$$

$$274 + (Me_3N)_3P^* \rightleftharpoons (Me_2N)_3P + (Me_2N)_3P^*Te \qquad (185)$$

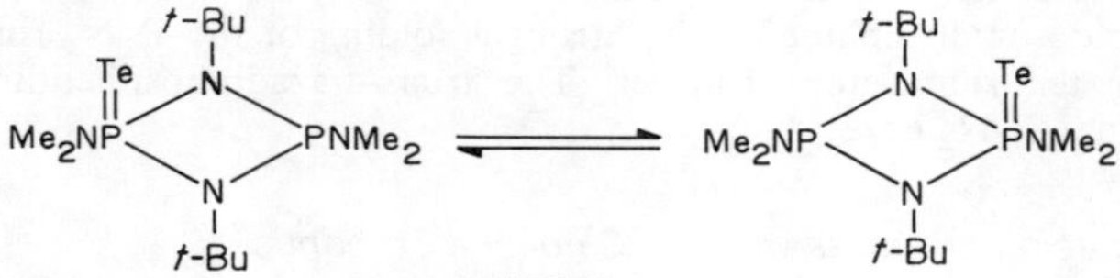

(186)

(275)

In 1950, Foss[103] showed that potassium diethylphosphite reacts with tellurium to give telluro-*O*,*O*-diethylphosphate (**226**) (Section VI.B) as colourless hygroscopic needles. Recently, this compound has been used for the deoxygenation of expoxides[105]. In contrast to tellurophosphoric acid *O*,*O*-diester anions, the neutral species (like tellurophosphoric acid trialkyl esters) appear to be unstable. Interestingly, tellurophosphorous acid bis(trimethylsilyl) ester, $(Me_3SiO)_2P(H)Te$ (**276**), was reported to be a distillable liquid[140].

TABLE 5. Equilibrium constants for the reactions between some phosphanes and the tellurocyanate ion[137]

$R_3{}^1P^a$	K
$(Pyrr)_3P$	80
CN^-	1 (by definition)
$(Pip)_3P$	0.6
$(Me_2N)_3P$	0.54
$(Et_2N)_3P$	0.27
$(n\text{-}Bu)_3P$	0.17
$(Mor)_3P$	0.09
$[(n\text{-}Pr)_2N]_3P$	0.048
Ph_3P	0

[a] Pyrr = pyrrolidino; Pip = piperidino; Mor = morpholino.

B. Arsenic Compounds with As=Se or As=Te Double Bonds

1. Tertiary arsane selenides (selenoarsoranes)

Trialkylarsane selenides are easily prepared by heating tertiary arsanes with a solvent and powdered grey selenium[141]. Triphenylarsane selenide can be prepared from dichlorotriphenylarsorane with ammonium selenide[142]; the thermolabile compound decomposes readily with precipitation of selenium. Two infrared absorptions in the range 330–360 cm^{-1} were assigned to the As=Se stretching vibration[143]. Ditertiary arsanes give diselenides that are in equilibrium with the monoselenides and elemental selenium[144,145]. Formation of the monoselenide **277** is favoured in the methylenebis(dimethylarsane)/selenium system. The reason for this behaviour should be deactivation of As^{III} by a neighbouring electron-withdrawing selenoarsinyl group (as in the case of phosphane selenides **248** and **249**)[114]. Additionally, monoselenide **277** gives a low temperature of coalescence (60 °C) in ^{1}H NMR, which is due to rapid selenium transfer.

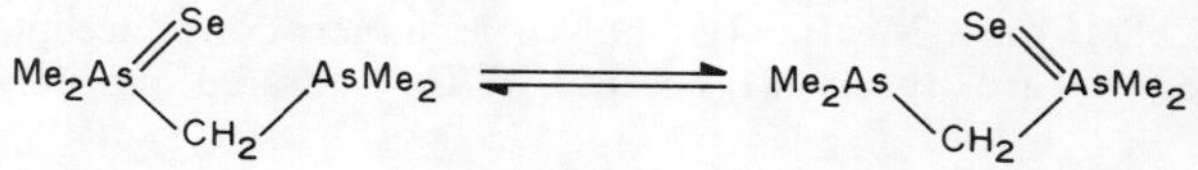
(187)

(277)

In contrast to ^{1}H NMR line broadening of the Me_2As proton resonances of **277** in solution, ditertiary arsane (mono)selenides, in which the As atoms were separated by three or four methylene groups, gave sharp separated signals for Me_2As and Me_2As (=Se) groups[145]. Line broadening due to selenium transfer was also observed in the ^{13}C NMR spectra of tri-*tert*-butylarsane selenide/tri-*tert*-butylarsane mixtures in inert solvents[117,118].

$$\underset{\textbf{(278)}}{(t\text{-Bu})_3As=Se} + (t\text{-Bu})_3As^* \rightleftharpoons (t\text{-Bu})_3As + (t\text{-Bu})_3As^*=Se \qquad (188)$$

Tetra-tertiary diphenylarsinomethanes such as $C(CH_2AsPh_2)_4$ are reluctant to react with sulphur or selenium, but the dimethylarsinomethanes are easily oxidized to give selenides such as $C[CH_2As(=Se)Me_2]_4$[146]. A tertiary arsane telluride has not yet been isolated. Even tri-*tert*-butylarsane (first ionization potential only 7.8 eV[147]) is not able to coordinate significantly with tellurium. Spectroscopic evidence for $(t\text{-Bu})_3As \rightarrow Te$ coordination was recently provided by the NMR spectra of the (kinetically) labile tri-*tert*-butylphosphane telluride in presence of a three-fold excess of tri-*tert*-butylarsane.

$$\underset{\textbf{(254)}}{(t\text{-Bu})_3PTe} + (t\text{-Bu})_3As \xrightleftharpoons{(K\approx 10^{-3})} (t\text{-Bu})_3P + \underset{\textbf{(279)}}{(t\text{-Bu})_3AsTe} \qquad (189)$$

Tri-*tert*-butylphosphane telluride and (de-tellurated) tri-*tert*-butylphosphane give a single ^{1}H NMR doublet signal; the magnitude of the averaged $^{3}J(^{31}P^{1}H)$ of the reaction mixture is smaller than that of pure **254**[117,118].

2. Selenoarsinic acid derivatives containing a terminal As—Se bond

The formation of sodium dimethyldiselenoarsinate (**281**) is very similar to that of the corresponding sulphur compound **280**[148].

Addition of water to an ethanolic solution of **280** or **281** leads to crystallization as colourless dihydrates $[Me_2As(E)ENa.2H_2O]$. Metal complexes based on **281** were much less stable than those with the dithioarsinate as anionic ligand[148].

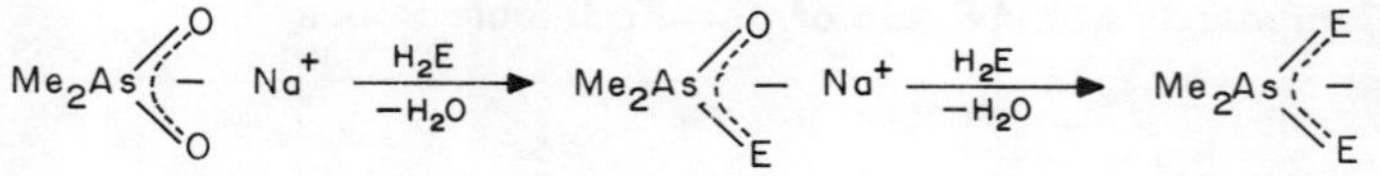 (190)

(280) E=S

(281) E=Se

C. Stibane Selenides

Trialkylstibanes react with powdered selenium to yield stable trialkylstibane selenides[149]. Only with trimethylstibane is a labile compound of the composition Me_3SbSe_2 formed[150]. Acyclic dimer structure was proposed for this stibane diselenide; attempts to trap the compound by coordination with cyclopentadienyl(dicarbonyl)iron(I) as a two-electron acceptor led to loss of selenium and formation of the stibane selenide complex **284**[151]. Among all $Me_3M{=}E$ molecules (M = P, E = S, Se, Te, M = As, E = S, Se; M = Sb, E = S, Se), only $Me_3P{=}Te$ showed a larger donor/acceptor ratio than $Me_3Sb{=}Se$ (**283**) towards the iron(I) acceptor [ν(CO) infrared and $\delta^{13}C({=}O)$ NMR evidence)[151].

$$Me_3Sb + 2\,Se \longrightarrow \tfrac{1}{2}(Me_3SbSe_2)_2 \qquad (191)$$

(282)

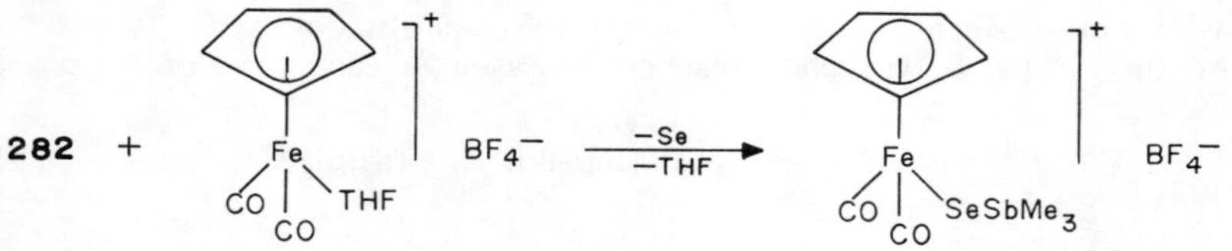

(284)

(192)

IX. STRUCTURE, BONDING AND REACTIONS OF ORGANIC SELENIUM AND TELLURIUM COMPOUNDS WITH BONDS BETWEEN Se OR Te AND P, As, Sb OR Bi

A. ^{31}P, ^{77}Se and ^{125}Te NMR Spectra

Owing to the occurrence of satellite spectra from $J(^{77}Se^{31}P)$ and $J(^{125}Te^{31}P)$, ^{31}P NMR is a very useful tool for the assignment of structures when P—Se or P—Te bonds are involved. δ(P) follows empirically known trends as far as typical substituent effects, steric crowding influence, small ring effects or coordination shifts are concerned. More specific information comes from one-bonded coupling constants ^1J(SeP) and ^1J(TeP). The magnitude of these coupling constants is large in all cases of terminal $P^{III}{=}Se$, $P^{V}{=}Se$ and $P^{V}{=}Te$ bonds. Electron-withdrawing substituents such as halogen atoms or CF_3 groups lead to increased ^1J(SeP), and ^1J(SeP) steric effects are also of some importance. The influence of conformational effects on ^1J(SeP) was demonstrated by the different P—Se couplings in the enantiomers of **285** (see Table 6). J(TeP) generally follows the trends of J(SeP), but its magnitude is larger in all comparable cases. An excellent review appeared a few years ago[152]; some typical data are compiled in Table 6.

TABLE 6. NMR data for PSe compounds: ^{31}P and ^{77}Se NMR[a]

Compound	$\delta^{31}P$ (ppm)	$J(^{77}Se^{31}P)$ (Hz)	$\delta^{77}Se$(ppm)	Ref.
$(MeO)_3P=Se$	+ 77.5	− 963	− 396	153
$(Me_2N)_3P=Se$	+ 81.8	− 805	− 366	153
$Me_3P=Se$	+ 8	− 648	− 235	153
$(t\text{-}Bu)_3P=Se$	+ 92.5	± 711	− 428	85, 117, 118
$Ph_3P=Se$	+ 34	− 736	− 275	78, 154
$(p\text{-}Tol)_3P=Se$	+ 32	− 726		154
$(o\text{-}Tol)_3P=Se$	+ 26	− 708		154
Me_2PSeMe	+ 0.8	− 218	+ 58	15, 155, 156
$Me_2P(=S)SeMe$		− 341	+ 196	156
$Me_2(MeSe)PW(CO)_5$	− 16.4	± 309		155
$(CF_3)_2PSeMe^2$	+ 27.9	± 294		15, 155
$(CF_3)_2(MeSe)PW(CO)_5$	+ 45.1	± 432		155
$(CF_3)_2PSeP(CF_3)_2$	+ 246.9		+ 700.8	14
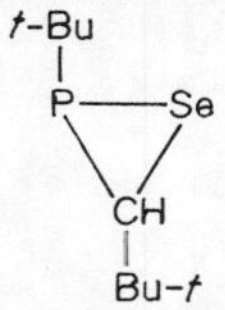	− 62	± 53.7		23
ArP(Se) = C(SiMe$_3$)$_2$	+ 195.2	± 846		108
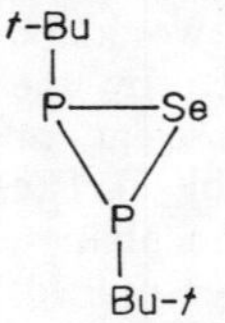	− 76.9	± 135.2		66
 (**285**)	+ 79.7 + 78.0	− 346.2, − 760.4 − 390.6, − 751.1		157

[a] ^{31}P shifts to low field from 85% H_3PO_4; ^{77}Se shifts to low field from Me_2Se.[152]

Coordination of seleno- or telluro-phosphanes leads in most cases (non-chelating P ligands) to an increase in ^1J(SeP) or ^1J(TeP). Coordination of phosphane selenides or phosphane tellurides leads to decreased (in some cases fairly constant)[158,85] Se–P or Te–P coupling constants[78]. Three-membered heterocycles containing phosphorus and selenium or tellurium give very small magnitudes of J(SeP) or J(TeP) and upfield ^{31}P shifts. However, the isomeric selenoxophosphoranes containing P^V with coordination number three have deshielded ^{31}P nuclei that couple strongly with ^{77}Se. Several early papers on the preparation of phosphane tellurides reported that the compounds were so unstable in solution that ^{31}P NMR spectroscopic data could not be collected. In fact, difficulties arise from rapid tellurium transfer reactions, especially in the presence of larger amounts of P^{III} compounds (when the reaction with elemental tellurium was not complete). Concentration-dependent line broadening up to complete coalescence occurs in many cases at temperatures close to 25 °C. To resolve ^{125}Te satellites, samples have to be very pure and/or cooling to temperatures below 0 °C is helpful. Similarly, ^{125}Te NMR signals of phosphane tellurides appear as singlets due to tellurium transfer when a non-coordinated phosphane is present in solution. ^{77}Se and ^{125}Te shifts of phosphane selenides

TABLE 7. Hetero-NMR data for tellurophosphanes

	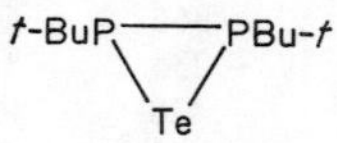 (130)	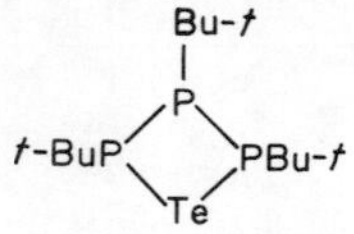 (131)	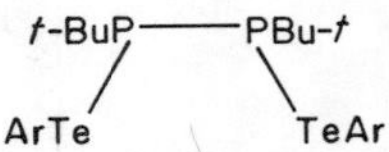 (73)
$\delta^{31}P$ (ppm)	$-$ 70.9[a]	$-4.1-60.5$ ($J = \pm 172.6$ Hz)	$+29$ (88%) $+39$ (12%)
$\delta^{125}Te$ (ppm)[b]	-1123.7[b](t)	-789.2 (d,t)	-23 (X-part of ABX) (main isomer)
J(TeP) (Hz)	$\pm$ 226.4[c]	± 84 (^2J) ± 10.7 (^1J,^3J)	$^1J = -519.9$ $^2J = -22.5$

[a]At $-30\,°C$.
[b]δ relative to di-*p*-tolyl ditelluride[152].
[c]at $-30\,°C$ $J - 235.4$ Hz.

and tellurides are at the upfield edge of ^{77}Se or ^{125}Te shift scales, which would be consistent with a predominance of an ylidic type of bonding (described by the phosphoniochalcogenate structure $R_3\overset{+}{P}—\overset{-}{E}$ (E = Se, Te), but apparently certain seleno- or telluro-phosphanes $R_2{}^1PER^2$ are in a similar ^{77}Se and ^{125}Te shift range (Table 7). The ^{125}Te shifts of the first tellurostibanes are upfield from comparable tellurophosphanes, and the ^{125}Te NMR signal of tellurobismuthane (**64**)[46] appears even further upfield. Telluradiphosphirane (**130**) appears in ^{125}Te NMR far upfield of larger Te—P heterocycles and from acyclic tellurodiphosphanes such as **73** (Table 7)[30,31,54].

B. Vibrational, UV–Visible and He(I)–PE Spectra

Normal coordinate analyses have been carried out on basic molecules with P–Se and P=Se (and related main Group 5/main Group 6) (group 15/group 16) subunits. In the case of trimethylphosphane chalcogenides, P=E stretching force constants were determined to decrease in the following order (all values $\times\ 10^2$ N/m)[159,160]:

Ref.	Me_3PO	Me_3PS	Me_3PSe	Me_3PTe
159	7.04	3.73	2.84	2.20
160	8.03	3.37	2.35	

The assignment of strong infrared absorptions (and corresponding Raman emissions) in the region of 500 cm^{-1} to the P=Te stretching vibration led to the calculation of force constants that were significantly larger than f(P=S)[136]. This would suggest unusually strong P=Te bonds, but crystallographic bond length determinations do not support this idea. Possibly, strong coupling of ν(P=E) with symmetric stretching modes (like $\nu s(PC_3)$ in tri-*tert*-butylphosphane chalcogenides)[117] leads to sets of IR and Raman bands that cannot be unequivocally assigned to ν(P = Se), ν(P = Te) or $\nu(PC_3)$ vibrations. Similarly, intensity criteria were not sufficient for the assignment of P—Se and As—Se stretching vibrations in methylthio- and methylseleno-bis(trifluoromethyl)-phosphanes and -arsanes (**10, 49** and related sulphur compounds)[161]. Owing to the mass of the CF_3 groups, $\nu_s(PC_2)$

is close to ν(PSe), and coupling of the vibrational modes leads to difficulties in determining precise force constants in these cases. Nevertheless, ν(PSe) may be used empirically for the determination of the abundance of conformational isomers[162] or for the decision between P(=Se)OR and P(=O)SeR type structural isomers and related problems.

In UV-visible spectroscopy, differences in the absorption between P=Se- and P—Se-bonded isomers can also be used for the prediction of structures. As in the case of compounds with P=S bonds, substituents lead to bathochromic shifts of λ_{max} of the P=Se chromophore in the order $OEt < NEt_2 < Et < Cl \approx SEt$ (210–260nm)[95]. Most phosphane tellurides are yellow owing to an absorption in the region of 280cm^{-1} that stretches into the visible region.

The dipolar character of P=X bonds increases in the order P=S < P=Se < P=Te; arsane chalcogenides are more polar than the corresponding phosphane chalcogenides[163]. Electron-withdrawing substituents in phosphane or arsane chalcogenides lower the polarity of the compounds.

The first ionization potential (IP_1) of phosphane oxides, sulphides and selenides is assigned to orbitals exhibiting predominant lone pair character on O, S or Se. IP_1 may be correlated with the sum of substituent electronegativities and 'coordination' of a phosphane with O, S or Se is paralleled by an overall stabilization of the corresponding phosphane energies. As the atomic ionization potentials decrease from O to Se, a related decrease of IP_1 was observed on going from phosphane oxides to phosphane selenides[164].

The first two He(I)—PE bands of dimethyl(methylthio and seleno)-phosphanes and -arsanes, Me_2MEMe (M = P, As; E = S, Se), are due to orbitals that arise from linear combinations of the lone pairs at P or As and S or Se. The large splitting between the two linear combinations of the lone pair-type orbitals indicates gauche conformations in these molecules (torsion angle $\tau \neq 0°$, $\neq 180°$)[165].

C. Bond Lengths and Angles

Most structure determinations that have been carried out on organic selenium and tellurium compounds with bonds to P, As, Sb or Bi are from phosphane selenides and tellurides, $R_3P{=}E$ (E = Se, Te). P=Se bond lengths are in the region of 210 ± 2 pm in most tertiary phosphane selenides and related dialkylamino- and alkoxy-phosphane selenides (seleno-phosphinic, -phosphonic and -phosphoric esters and amides), compared with about 207 pm estimated from P and Se 'double bond radii'[166]. P=Te bond lengths determined in tellurophosphonic and tellurophosphoric amides range from 231 to 235pm; these values are significantly longer than the sum of the predicted double bond radii[167]. The deviation of determined P=Se and P=Te bond lengths from the estimated double bond distances was interpreted as evidence for increasing ylidic character of the P=Se and P=Te bonds compared with P=O or P=S bonds[168].

A B

Increased participation of the mesomeric structure B would also explain the high polarity of phosphane selenides and tellurides and their strong nucleophilic character. In a series of papers on studies of the P—N bond, Rømming and coworkers compared the structures of tris(dialkylamino)phosphanes with those of the corresponding phosphane selenides and tellurides[137,168–171]. The structures of the pentacovalent phosphorus compounds $(R_2N)_3P{=}E$ (R = dimethylamino, morpholino, piperidino or pyrrolidino) were found to be closely related to the structures of the parent aminophosphanes. In all cases, the molecular skeleton $P(NC_2)_3$ does not possess C_3 symmetry (in the solid). At least two different nitrogen atoms (essentially sp^3 hybridized/pyramidal and sp^2 hybridized/planar, respectively) with different P—N bond distances are present in the molecules. Two nitrogen lone pairs are nearly orthogonal to the phosphorus lone pair or to the P=Se bond and the lone pair of the third nitrogen atom is close to being antiparallel to the phosphorus lone pair or to the P=Se or P=Te bond, respectively. In these conformations the sum of the repulsive lone pair–lone pair interactions is obviously a minimum. P=Se or P=Te bonds seem to exert the same conformational influence as the phosphorus lone pair. Compared with the aminophosphanes, the sums of NPN angles are 11–12° larger in the phosphane selenides and tellurides (average about 4° per NPN angle), the individual angles varying from 102° to 114°. Hence phosphorus is far from having a regular tetrahedral coordination geometry in compounds with the $N_3P{=}E$ (E = Se, Te) skeleton. Correspondingly, the Se=P—C angles in tertiary phosphane selenides (like the Se=P—N angles in aminophosphane selenides) are in the range 112–114°.

Coordination of phosphane selenides with Lewis acids weakens the P=Se bonds. Complex formation of triphenylphosphane selenide [d(PSe) = 210.6 pm][172] with $HgCl_2$ leads to elongation of the PSe bond $(Ph_3PSeHgCl_2)_2$:d(PSe = 217 pm][173] half the way to P^V—Se single bond distances [d(PSe) = 224–228 pm][122,166,174]. Similarly, d(PSe) of the complexes **286** and **287** is close to 216 pm[175].

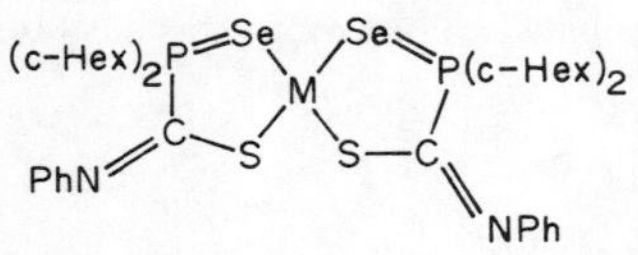

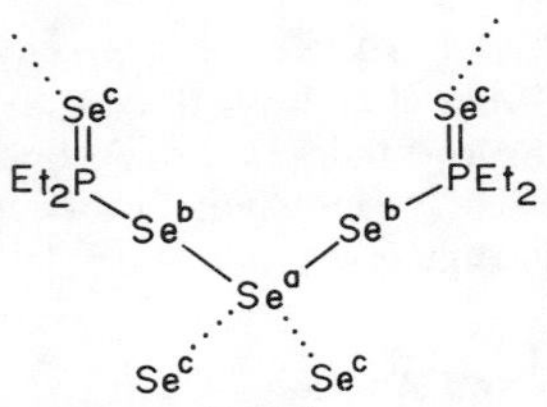

(260)

(**286**) M = Zn, d(P—Se) = 215.9 pm
∠ZnSeP = 96°
(**287**) M = Cd, d(P—Se) = 215.6 pm
∠CdSeP = 89.2°

d(P=Se) = 209.5 pm, d(Se—Se) = 235.2 pm
d(P—Se) = 223.9 pm, d(Se···Se) = 367.9 pm
$\angle Se^bSe^aSe^b = 103.9°$
$\angle Se^cSe^aSe^c = 89.9°$ (intermolecular)

Weak intermolecular secondary bonding (Se···Se 367.9 pm) between double-bonded selenium atoms and the central Se atom of the triselenium chain of **260** was confirmed by an X-ray crystal structure analysis[166]. In $(CO)_5WTeP(Bu\text{-}t)_3$ (**162**), d(TeP) (243.9 pm) is close to a Te—P single bond length[85].

Coordination of tellurobis(di-*tert*-butylphosphane) (**30**) with tetracarbonylchromium led to a chelate complex, $(CO)_4Cr(\mu\text{-}PBu\text{-}t_2)_2Te$. In **288**, the X-ray crystal structure determination gave 248 pm for the P—Te bond distance, fitting well the value expected for a single bond[176]. The P—Te—P bond angle is only 78°; tellurium seems to be the softest atom of the CrP_2Te heterocycle as far as the ability to adapt to small bond angles is concerned. The small P—Te—P bond angle in the four-membered chelate is paralleled by μ-9, 10-telluro-9, 10-diarsaanthracene (**110**), where the As—Te—As angle (80.8°) adapts to the preferred geometry of the 9, 10-diarsaanthrancene system, which demands As···As distances of about 330 pm[62]. The small angles at tellurium do not lead to any special reactivity which could be expected if significant ring strain were present in both compounds.

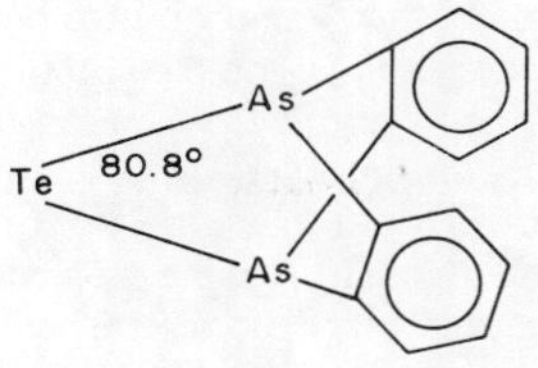

(110)

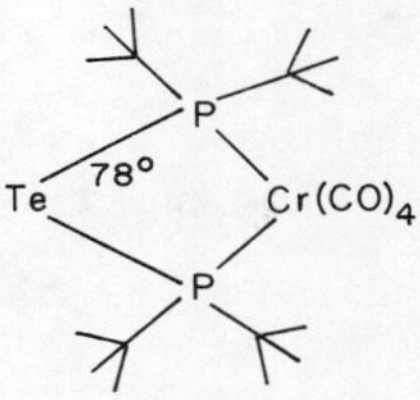

(288)

1. Selected Reactions Involving Bonds between Se or Te and P, As, Sb or Bi

1. Reactions of seleno- and telluro-phosphanes and -arsanes with cationic and anionic hydrides. As highly polar compounds, phosphane oxides, selenides and tellurides can give hydrogen bonds with protic agents such as phenol or carboxylic acids[78,177]. In contrast, seleno- and telluro-phosphanes are of very low polarity and fairly inert to water and methanol. Owing to the small differences in the electronegativities of P, As, Se and Te, electron-withdrawing substituents might lead to 'umpolung' of the P—Se, P—Te or As—Te bond polarity[28,178]. With hydrogen iodide, $Me_2AsSeCF_3$ is cleaved to give CF_3SeH as a protic reaction product, but $(CF_3)_2PSeMe$ and $(CF_3)_2TeMe$ lead to $(CF_3)_2PH$[178]. Cleavage of the same educts with trimethyltin hydride leads in all cases to Me_2AsH or $(CF_3)_2PH$ and organotin-selenium or -tellurium compounds. Bond polarity is obviously not sufficient to explain the nature of these reaction products. Primary products (according to P—Se or As—Se bond polarity) can rearrange to give the observed final product[177].

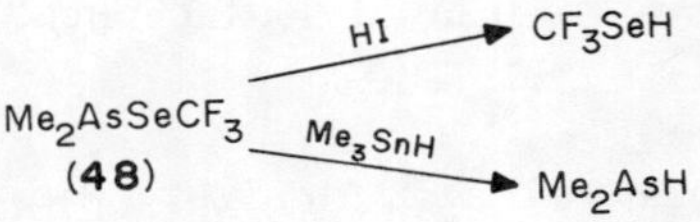

(193)

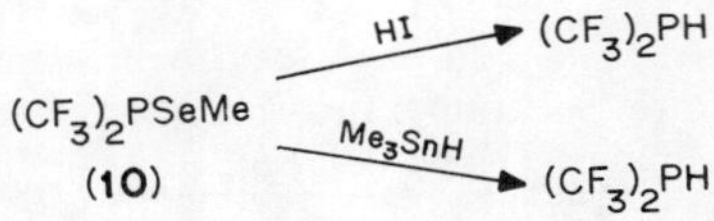

(194)

A careful study of the cleavage of **10** with trimethyltin hydride revealed that at $-40\,°C$, 20 min after mixing the educts significant amounts of $(CF_3)_2PSnMe_3$ are present, and in a separate experiment it was shown that $(CF_3)_2PSnMe_3$ is rapidly cleaved by MeSeH to give the final products of equation 194, $(CF_3)_2PH$ and $Me_3SnSeMe$ (i.e. P appears to be negatively charged and Se should be positively charged in **10**)[178]. Similarly, $(CF_3)_2AsSeMe$ (**49**) reacts stepwise with cyclopentadienyltricarbonyl-chromium, -molybdenum and -tungsten hydride[179].

Cleavage of the tellurophosphane $(t\text{-Bu})_2PTeTol\text{-}p$ (**26**) with gaseous HCl leads to di-*p*-tolyl ditelluride and products that are related to products of the cleavage of tetra-*tert*-butyldiphosphane with HCl[29].

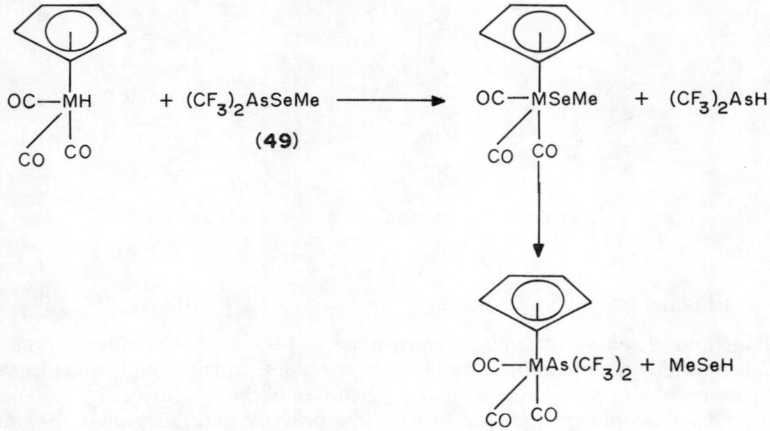

(195)

2. *Types of coordination of ligands with bonds between Se or Te and P, As, Sb or Bi*

a. P^{III} and As^{III} ligands. Thio-, seleno- and telluro-phosphanes coordinate with transition metal acceptors preferentally via the lone pair at phosphorus and the corresponding arsanes via As. With monodentate (P) ligands, mono- and di-substituted complexes have been characterized; bidentate $[(t\text{-Bu})_2P]_2Te$ gives four-membered chelate complexes with metal carbonyls and platinum or palladium dichloride.

$$(CO)_5MTHF + R_2{}^1PER^2 \xrightarrow{-THF} (CO)_5MPR_2{}^1ER^2$$ (refs. 20, 155, 180, 181, 183) (196)

$$(CO)_4M(\eta^4\text{-}C_7H_8) + 2R_2{}^1PER^2 \xrightarrow{-C_7H_8} (CO)_4M(PR_2{}^1ER^2)_2$$ (ref. 182) (197)

(*cis* ⟶ *trans* rearrangement)

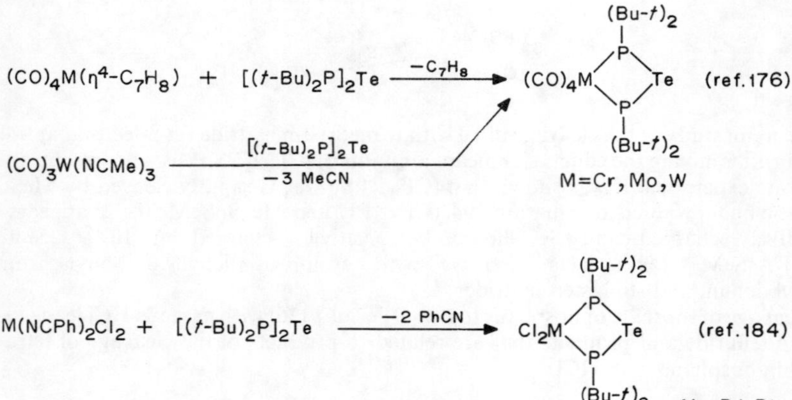

(198)

(199)

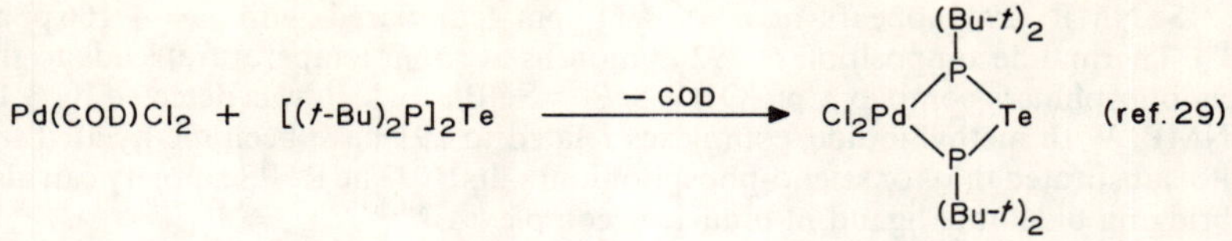

(200)

b. η^2-P-, Se-coordination. The metaselenophosphate **243** (Section VII.C) coordinates with the bis(triphenylphosphane)platinum(0) acceptor by η^2-orientation of the P=Se double bond to platinum[185].

Addition of sulphur or selenium to W=P double bonds leads to three-membered ring systems (compare Section II.A.1.a) that contain $(t\text{-Bu})_2$PS or $(t\text{-Bu})_2$PSe moieties as η^2-coordinated three-electron ligands[186].

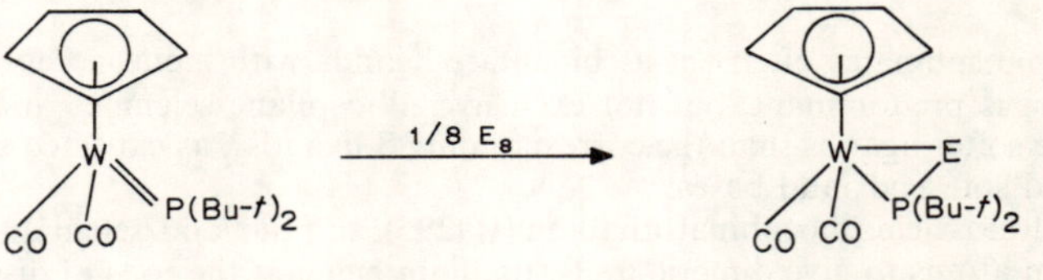

(201)

(289) E=S

(290) E=Se

In solid **290**, the P—Se bond length determined to be 217.1 pm [d(WP) = 243.1 pm, d(WSe) = 268.9 pm] (compare Section IX.C).

Alternatively, cyclopentadienyltricarbonyl(diphenylphosphino)tungsten reacts with sulphur or red selenium to give the corresponding diphenyl-thio- or -seleno-phosphinyl complexes (**291**, **292**) that are decarbonylated to give the π^2-diphenylchalcogenophosphinite complexes (**293**, **294**)[187].

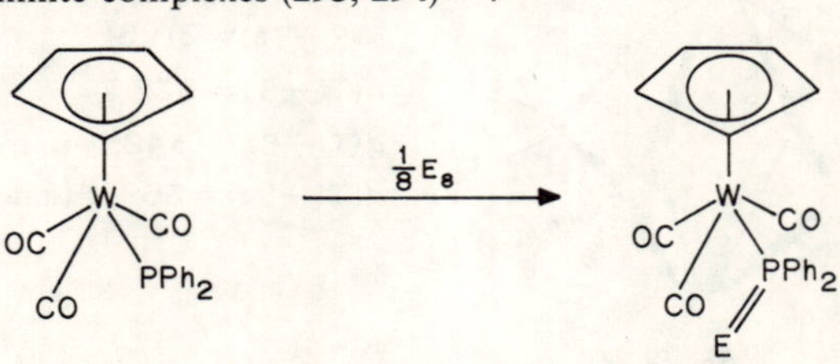

(291) E=S

(292) E=Se

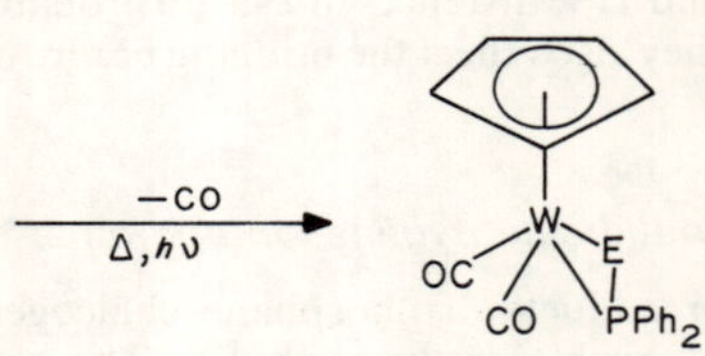

(202)

(293) E=S

(294) E=Se

In ^{77}Se NMR, **294** appears at $\delta = -911$ ppm (compared with $\delta = +166$ ppm for **292**[186]). Thermal decomposition of **292** (3 months at room temperature) leads to the η^1-diselenophosphinate complex $Cp(CO)_3WSeP({=}Se)Ph_2$, which was detected by ^{31}P and ^{77}Se NMR. With methyl iodide, complexes related to **291** have been methylated to give metallo-substituted thio- or seleno-phosphonium salts[186]. The R_2PSe moiety can also act as a bridging bidentate ligand in dinuclear complexes[188,189].

c. Se- or Te-coordination. Tertiary phosphane selenides give complexes with many main group and transition metal acceptors[3]. Examples of coordination compounds with R_3PTe, R_3AsSe and R_3SbSe ligands are comparatively rare[151]. Triphenylphosphane sulphide is a better donor than triphenylphosphane selenide towards boron trichloride and boron tribromide as shown by ligand exchange reactions[190]. The geometry of phosphane selenide complexes is strongly related to that of phosphane sulphide complexes (angular M—Se—P arrangements due to stereochemically active lone pairs of electrons at Se)[3,173].

Diselenophosphinates often act as bidentate ligands with metal acceptors; chelating coordination is predominant, but not exclusive. Phosphane selenides and selenophosphinates are softer ligands than the corresponding S ligands[3], as expected from the rules for hard and soft acids and bases.

In diethylthioselenophosphinatothallium(I) (**295**), sulphur and selenium are bridging two thallium atoms to give dimeric units, thallium being at the tops of distorted square pyramids. The Tl—Se distances within the dimer are long (342.4 pm), but there are further intermolecular contacts between selenium and the thallium atoms of two different dimers. These intermolecular Tl—Se bonds link the dimeric units together to give a two-dimensional polymeric layer structure[191]. The relatively strong P—S and P—Se bonds in **295** are consistent with findings from IR spectroscopy[192].

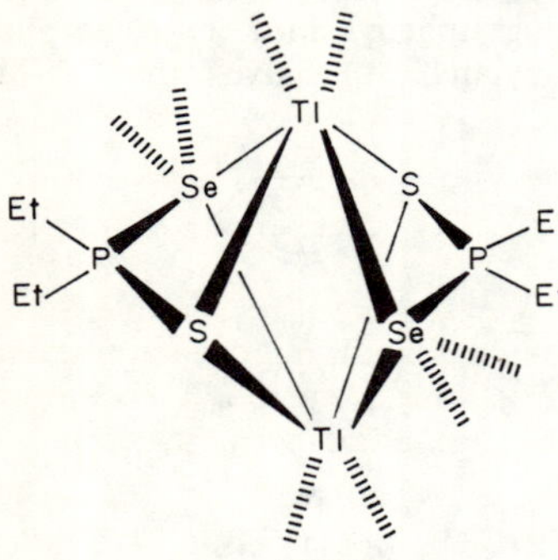

(**295**)

	pm
d(P—S)=	204.2
d(P—Se)=	211.5
d(Tl—S)=	323.7
d(Tl—Se)=	342.4 (intramolecular)
d(Tl—Se)=	359.4 (intermolecular)

The large TlS and TlSe distances in **295** are probably partly due to a high degree of polar bond character, they also reflect the bridging nature of the [η^2-]sulphur and [η^4-]selenium atoms.

3. *Alkylation, oxidation, deoxygenation and other reactions*

Charge-transfer adducts of phosphane chalcogenides with iodine have been investigated using UV–visible spectroscopy[193]. The equilibrium constants increased from E = O to E = Se in each series of compounds[193,194].

$$R_3P{=}E + I_2 \rightleftharpoons R_3P{=}E \cdots I_2$$

Structural evidence is not yet available. Protonation of phosphane selenides occurs with strong acids and may be followed by ^{31}P NMR[78]. Similarly, strong solvation of phosphane selenides by liquid SO_2 is indicated by significant downfield shifts in ^{77}Se NMR and by decreasing coupling constants $^{1}J(^{77}Se^{31}P)$ in SO_2 solution[195]. Tris(dimethylamino)phosphane telluride, $(Me_2N)_3P{=}Te$ (**274**), gives a deep red adduct with liquid SO_2; an upfield shift in ^{31}P and lowering of $^{1}J(^{125}Te^{31}P)$ is also observed[195]. With methyl iodide, seleno- and telluro-phosphanes are primarily subject to *P*-alkylation (subsequently P—Se or P—Te bond breaking occurs, Section V); phosphane selenides need strong alkylating agents such as dimethyl sulphate or trialkyloxonium salts, but compounds that are activated by electron-donating dialkylamino groups can even react with methyl iodide (for rearrangements under the action of alkylating halides, see Section VI).

Recent experiments have shown that phosphane tellurides are also methylated with methyl iodide under mild conditions[83].

Triphenylphosphane selenide gives a reaction related to the Wittig alkene synthesis with hexafluoroacetone or bis(trisfluoromethyl)ketene to give 1, 3-diselenetanes[196]. The latter synthesis is less successful than the reaction of triphenylphosphane sulphide with bis(trifluoromethyl)ketene, which gives the 1, 3-dithietane in 60% yield.

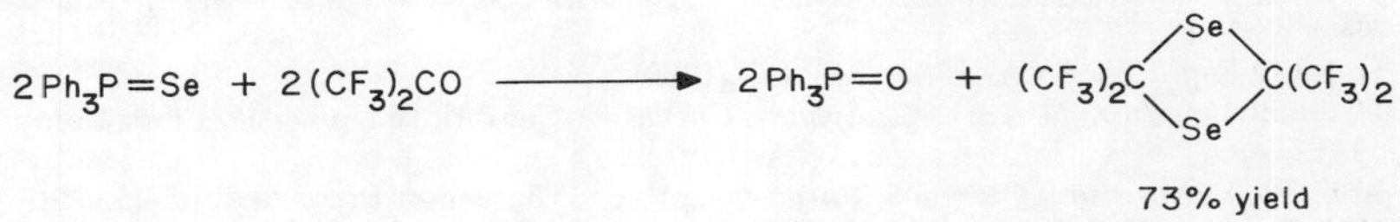

(203)

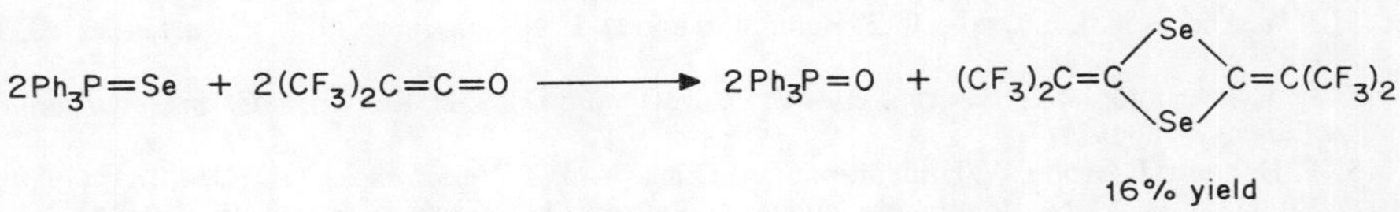

(204)

Reactions of epoxides with tertiary phosphane selenides in the presence of trifluoroacetic acid lead to deoxygenation of the epoxide and to precipitation of elemental selenium. In the $Ph_3P{=}S$–epoxide reaction, episulphides are formed under mild conditions, but episelenides (seleniranes), being less stable, suffer spontaneous deselenation, leading stereospecifically to alkenes[197,198].

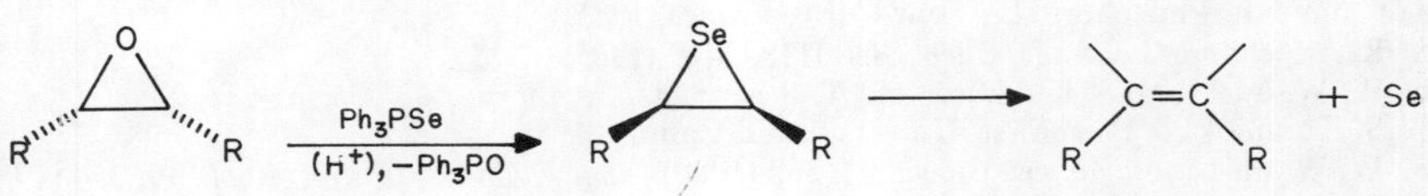

(205)

Sodium tellurodiethyl phosphate leads in a similar manner to deoxygenation of epoxides, possibly via extremely labile telluriranes[105]. The reaction may even be carried out starting from sodium diethyl phosphite with catalytic amounts of elemental tellurium[105]. The reducing properties of selenodiethyl phosphate have been used for the deoxygenation of various sulphoxides[199]. Selenophosphonic and selenophosphoric acid *Se*-esters can act as phosphorylating cholinesterase inhibitors; some of them are extremely toxic[91]. Appropriate care should be taken in handling these compounds.

X. ACKNOWLEDGEMENT

We thank all our colleagues who contributed to this chapter by providing information about their very recent work by sending us reprints, preprints and unpublished manuscripts. Special thanks are due to K. J. Irgolic for card files on Se(Te) compounds.

XI. REFERENCES

1. K. Sasse, 'Organische Phosphorverbindungen', in *Houben-Weyl, Methoden der Organischen Chemie*, 4. Aufl., Vols. XII/1 and XII/2, Georg Thieme, Stuttgart, 1963.
2. M. Regitz (Ed.), 'Organische Phosphorverbindungen', in *Houben-Weyl, Methoden der Organischen Chemie*, Vols. E1 and E2, Georg Thieme, Stuttgart, 1982.
3. R. A. Zingaro and W. C. Cooper (Eds.), *Selenium*, Van Nostrand Reinhold, New York, 1974.
4. K. J. Irgolic, *The Organic Chemistry of Tellurium*, Gordon and Breach, New York, 1974.
5. G. M. Kosolapoff and L. Maier, *Organic Phosphorus Compounds*, Vol. 1, Wiley–Interscience, New York, 1972.
6. H. J. Gysling, *Coord. Chem. Rev.*, **42**, 133 (1982).
7. H. Goldwhite, *Introduction to Phosphorus Chemistry*, Cambridge University Press, Cambridge, 1981.
8. A. F. Wells, *Structural Inorganic Chemistry*, 5th ed., Clarendon Press, Oxford, 1984.
9. H. J. Emeleus, K. J. Parker and N. Weleman, *J. Chem. Soc.*, 2529 (1962).
10. A. Darmadi, A. Haas and B. Koch, *Z. Naturforsch., Teil B*, **35**, 526 (1980).
11. N. Petragnani, V. G. Toscano and M. de Moura Campos *Chem. Ber.*, **101**, 3070 (1968).
12. R. A. N. McLean, *Inorg. Nucl. Chem. Lett.*, **5**, 745 (1969).
13. J. W. Anderson, J. E. Drake, R. T. Hemmings and D. L. Nelson, *Inorg. Nucl. Chem. Lett.*, **11**, 233 (1975).
14. D. E. J. Arnold, J. S. Dryburgh, E. A. V. Ebsworth and D. W. H. Rankin, *J. Chem. Soc., Dalton. Trans.*, 2518 (1972).
15. P. Dehnert, J. Grobe, W. Hildebrandt and Duc le Van, *Z. Naturforsch., Teil B*, **34**, 1646 (1979).
16. Yu. N. Shlyk, G. M. Bogolyubov and A. A. Petrov, *Zh. Obshch. Khim.*, **38**, 193 (1968).
17. W. R. Cullen, D. S. Dhaliwal and W. B. Fox, *Inorg. Chem.*, **3**, 1332 (1964).
18. W. R. Cullen and D. S. Dhaliwal, *Can. J. Chem.*, **45**, 379 (1967).
19. I. B. Mishra and A. B. Burg, *Inorg. Chem.*, **11**, 664 (1972).
20. J. Grobe and Duc le Van, *Z. Naturforsch., Teil B*, **35**, 694 (1980).
21. J. Grobe and D. le Van, *J. Fluor. Chem.*, **19**, 279 (1982).
22. J. Grobe, D. le Van, J. Nientiedt, *Z. Naturforsch., Teil B*, **41**, 149 (1986).
23. E. Niecke and E. Symalla, *Chimia*, **39**, 320 (1985).
24. W. W. Schoeller and E. Niecke, unpublished work, presented at the Colloque Nationale sur la Chimie du Phosphore, La Tour Carol, France, 1985.
25. R. Appel and C. Casser, *Chem. Ber.*, **118**, 3419 (1985).
26. W.-W. du Mont, S. Kubiniok and T. Severengiz, *Z. Anorg. Allg. Chem.*, **521**, 21 (1985).
27. S. Kubiniok, *Diplomarbeit*, Universität Oldenburg, 1984.
28. W.-W. du Mont, *Angew. Chem.*, **92**, 562 (1980); *Angew. Chem., Int. Ed. Engl.*, **19**, 554 (1980).
29. W.-W. du Mont and R. Hensel, unpublished results.
30. T. Severengiz, *PhD Thesis*, Oldenburg, 1985.
31. W.-W. du Mont, R. Hensel and T. Severengiz, *Phosphorus Sulfur*, **18**, 73 (1983).
32. F. Sladky, B. Bildstein and D. Obendorf, *J. Organomet. Chem.*, **295**, C1 (1985).
33. L. Lange and W.-W. du Mont, unpublished results.
34. L. S. Sagan, R. A. Zingaro and K. J. Irgolic, *J. Organomet. Chem.*, **39**, 301 (1972).
35. G. C. Chen, J. R. Daniel and R. A. Zingaro, *Carbohydr. Res.*, **50**, 53 (1976).
36. G. C. Chen, R. A. Zingaro and C. R. Thompson, *Carbohydr. Res.*, **39**, 61 (1975).
37. R. A. Pyles, K. J. Irgolic and G. C. Pappalardo, unpublished work, 1983.

38. W.-W. du Mont, T. Severengiz and H. J. Breunig, *Z. Naturforsch., Teil B*, **38**, 1306 (1983).
39. H. J. Breunig and H. Jawad, *J. Organomet. Chem.*, **277**, 257 (1984).
40. H. J. Breunig and S. Gülec, *Z. Naturforsch., Teil B*, in press.
41. M. Wieber and I. Sauer, *Z. Naturforsch., Teil B*, **39**, 887 (1984).
42. H. J. Breunig and D. Müller, *Angew. Chem.*, **94**, 448 (1982); *Angew. Chem., Int. Ed. Engl.*, **21**, 439 (1982).
43. G. Becker and M. Rössler, *Z. Naturforsch., Teil B*, **37**, 91 (1982).
44. A. J. Ashe, III, E. G. Ludwig and J. Oleksyszyn, *Organometallics*, **2**, 1859 (1983).
45. M. Wieber and I. Sauer, *Z. Naturforsch., Teil B*, **39**, 1668 (1984).
46. W.-W. du Mont, T. Severengiz, H. J. Breunig and D. Müller, *Z. Naturforsch., Teil B*, **40**, 848 (1985).
47. A. Darmadi, A. Haas and M. Kaschani-Motlagh, *Z. Anorg. Allg. Chem.*, **448**, 35 (1979); see also A. G. Osborne, R. E. Hollands, R. F. Bryan and S. Lockhart, *J. Organomet. Chem.*, **288**, 207 (1985).
48. L. Maier, *Helv. Chim. Acta*, **59**, 252 (1976).
49. C. J. Marsden, *J. Fluor. Chem.*, **5**, 423 (1975).
50. Ya. I. Kolodii, Ya. I. Mel'nik, V. V. Turkevich and A. P. Vas'kiv, *Zh. Obshch. Khim.*, **48**, 331 (1978).
51. A. Rosenbaum, H. Kirchberg and E. Leibnitz, *J. Prakt. Chem.*, **19**, 1 (1963).
52. A. Rosenbaum, *J. Prakt. Chem.*, **37**, 200 (1968).
53. M. Schmidt, I. Boie, C. Osterroth, R. Schröer and H. F. Grützmacher, *Chem. Ber.*, **101**, 1381 (1968).
54. W.-W. du Mont, S. Kubiniok and T. Severengiz, 4th International Symposium on Inorganic Ring Systems, Orsay, France, 1985, Abstract 37.
55. I. F. Lutsenko and V. L. Foss, *Pure Appl. Chem.*, **52**, 917 (1980).
56. K. Issleib and M. Hoffmann, *Chem. Ber.*, **99**, 1320 (1966).
57. R. Hensel, *PhD Thesis*, Universität Oldenburg (FRG), 1987; W.-W. du Mont, R. Hensel, S. Kubiniok, L. Lange and T. Severengiz, *X Int. Conf. on Phosphorus Chemistry, Bonn, 1986*, Abstract B42.
58. N. Camerman and J. Trotter, *J. Chem. Soc.*, 219 (1964).
59. W.-W. du Mont, T. Severengiz and B. Meyer, *Angew. Chem.*, **95**, 1025 (1983); *Angew. Chem., Int. Ed. Engl.*, **22**, 983 (1983).
60. R. A. Zingaro, B. H. Steeves and K. Irgolic, *J. Organomet. Chem.*, **4**, 320 (1965).
61. D. W. Allen, J. C. Coppola, O. Kennard, F. G. Mann, W. D. S. Motherwell and D. G. Watson, *J. Chem. Soc. C*, 810 (1970).
62. O. Kennard, D. L. Wampler, J. C. Coppola, W. D. S. Motherwell, F. G. Mann, D. G. Watson, C. H. MacGillavry, C. H. Stam and P. Benci, *J. Chem. Soc. C*, 1511 (1971).
63. O. Mundt, M. Riffel, G. Becker and A. Simon, *Z. Naturforsch., Teil B*, **39**, 317 (1984).
64. A.-J. Ashe, III, E. G. Ludwig, J. Oleksyszyn and J. C. Huffmann, *Organometallics*, **3**, 337 (1984).
65. H. J. Breunig and D. Müller, VI FECHEM Conference on Organometallic Chemistry, Riga/Yurmala, USSR, 1985, Abstract 136; H. J. Breunig and D. Müller, *Z. Naturforsch., Teil B*, **41**, (1986).
66. M. Baudler, H. Suchomel, G. Fürstenberg and U. Schings, *Angew. Chem.*, **93**, 1087 (1981); *Angew. Chem., Int. Ed. Engl.*, **20**, 1044 (1981).
67. M. Yoshifuji, K. Shibayama and N. Inamoto, *Chem. Lett.*, 6036 (1984).
68. M. Yoshifuji, K. Shibayama, K. Ando and M. Inamoto, *Heterocycles*, **21**, 475 (1984).
69. H. L. Krauss and H. Jung, *Z. Naturforsch., Teil B*, **15**, 545 (1960).
70. L. Maier, *Helv. Chim. Acta*, **48**, 1190 (1965).
70a. D. Herrmann, *Z. Anorg. Allg. Chem.*, **416**, 50 (1975).
71. G. Thiele, H. W. Rotter and M. Lietz and J. Ellermann, *Z. Naturforsch., Teil B*, **39**, 1344 (1984).
72. J. Ellermann and L. Brehm, *Chem. Ber.*, **117**, 2675 (1984).
73. J. Ellermann and M. Lietz, *Z. Naturforsch., Teil B*, **35**, 1514 (1980).
74. J. Ellermann and A. Veit, *Z. Naturforsch., Teil B*, **40**, 948 (1985).
75. M. Mikolajczyk, *Pure Appl. Chem.*, **52**, 959 (1980).
76. A. Schmidpeter and H. Brecht, *Z. Naturforsch., Teil B*, **24**, 179 (1969).
77. I. A. Nuretdinov, N. A. Buina and N. P. Grechkin, *Izv. Akad. Nauk SSSR, Ser. Khim.*, 169 (1969) (Engl. transl., 163).
78. C. Glidewell and E. J. Leslie, *J. Chem. Soc., Dalton Trans.*, 527 (1977).

79. J. Omelańczyk, W. Perliskowa and M. Mikoloajczyk, *Phosphorus Sulfur*, **18**, 444 (1983).
80. J. Omelańczyk and M. Mikolajczyk, *Tetrahedron Lett.*, **25**, 2493 (1984).
81. N. Petragnani and M. de Moura Campos, *Tetrahedron*, **21**, 13 (1965).
82. V. D. Romanenko, V. F. Shulgin, V. V. Skopenko, A. N. Chernega, M. Yu. Antipin, Yu. T. Struchkov, I. E. Boldeskul and L. N. Markovski, *Zh. Obshch. Khim.*, **55**, 282 (1985).
83. N. Kuhn and H. Schumann, *Phosphorus Sulfur*, **26**, 199 (1986).
84. S. Kubiniok and W.-W. du Mont, unpublished results.
85. N. Kuhn, H. Schumann and G. Wolmershäuser, *J. Chem. Soc., Chem. Commun.*, 1595 (1985); N. Kuhn, unpublished work.
86. W. Kuchen and B. Knop, *Angew. Chem.*, **77**, 259 (1965); *Chem. Ber.*, **99**, 1663 (1966).
87. B. Krawiecka, Z. Skrzypczynski and J. Michalski, *Phosphorus*, **3**, 177 (1973).
88. I. A. Nuretdinov, N. A. Buina and E. V. Bayandika, *Zh. Obshch. Khim.*, **48**, 1073 (1978) (Engl. transl., 977).
89. E. V. Bayandina, I. A. Nuretdinov and E. I. Loginova, *Izv. Akad. Nauk SSSR, Ser. Khim.*, 1627 (1976) (Engl. transl. 1543).
90. I. A. Nuretdinov, E. V. Bayandina and D. N. Sadkova, *Dokl. Akad. Nauk SSSR*, **239**, 1110 (1978) (Engl. transl., 142).
91. S. Akerfeldt and L. Fagerlind, *J. Med. Chem.*, **10**, 115 (1967).
92. A. Markovska, *Bull. Acad. Pol. Sci., Ser. Sci. Chem.*, **15**, 153 (1967); *Chem. Abstr.*, **67**, 90891B (1967).
93. E. G. Kataev, T. G. Mannafov and G. I. Kostina, *Zh. Obshch. Khim.*, **38**, 363 (1968).
94. I. A. Nuretdinov, D. N. Sadkova and E. V. Bayandina, *Izv. Akad. Nauk SSSR. Ser. Khim.*, 2635 (1977) (Engl. transl., 2441).
95. R. R. Shagidullin, A. V. Chernova, I. A. Nuretdinov, G. M. Doroshkina and E. V. Bayandina, *Izv. Akad. Nauk SSSR, Ser. Khim.*, **1**, 184 (1976) (Engl. transl. 174).
96. W. Lorenz and B. Schrader, to Bayer AG, *US Pat.*, 3 082 240 (1960/63); *Chem. Abstr.*, **59**, 5077 (1963).
97. A. Haas and J. Mikolajczak, *Chem. Ber.*, **114**, 829 (1981).
98. A. Haas and A. Lopusiński, *Chem. Ber.*, **114**, 3176 (1981).
99. A. Schmidpeter, personal communication, 1985.
100. W. J. Stec, A. Okrusecek, K. Lesiak, B. Uznanski and J. Michalski, *J. Org. Chem.*, **41**, 227 (1976).
101. N. I. Zemlyanski, N. M. Chernaya and V. V. Turkevich, *Zh. Obshch. Khim.*, **37**, 495 (1967) (Engl. transl., 464).
102. A. Markowska and W. Buchowiecki, *Bull. Acad. Pol. Sci, Ser. Sci. Chim.*, **21**, 455 (1973).
103. O. Foss, *Acta Chem. Scand.*, **4**, 1241 (1950).
104. O. Foss, *Acta Chem. Scand.*, **1**, 8 (1947).
105. D. L. J. Clive and S. M. Menchen, *J. Chem. Soc. Chem. Commun.*, 658 (1977).
106. J. Czyzewska-Chlebny and M. Michalska, *J. Chem. Soc., Chem. Commun.*, 693 (1985).
107. M. Caira, R. H. Neilson, W. H. Watson, P. Wisian-Neilson and Z.-M. Xie, *J. Chem. Soc., Chem. Commun.*, 698 (1984).
108. R. Appel and C. Casser, *Tetrahedron Lett.*, **25**, 4109 (1984).
109. T. A. van der Knaap and F. Bickelhaupt, *Phosphorus Sulfur*, **18**, 47 (1983).
110. E. Niecke and D.-A. Wildbredt, *J. Chem. Soc., Chem. Commun.*, 72 (1981).
111. O. J. Scherer and N. Kuhn, *J. Organomet. Chem.*, **78**, C17 (1974).
112. O. J. Scherer and N. Kuhn, *J. Organomet. Chem.*, **82**, C3 (1974).
113. G. J. Scherer, H. Jungmann, C. Krüger and G. Wolmershäuser, *Chem. Ber.*, **117**, 2382 (1984).
114. D. H. Brown, R. J. Cross and R. Keat, *J. Chem. Soc., Dalton Trans.*, 871 (1980).
115. R. A. Zingaro, *J. Organomet. Chem.*, **1**, 200 (1963).
116. T. Austad, T. Rød, K. Åse, J. Songstad and A. H. Norbury, *Acta Chem. Scand.*, **27**, 1939 (1973).
117. W.-W. du Mont and H.-J. Kroth, *J. Organomet. Chem.*, **113**, C35 (1976).
118. W.-W. du Mont, *Z. Naturforsch., Teil B*, **40**, 1453 (1985).
119. W.-W. du Mont and H.-J. Kroth, *Z. Naturforsch., Teil B*, **36**, 332 (1981) [the δ^{125}Te values in this paper have been withdrawn (see ref. 26)].
120. O. J. Scherer and G. Schnabl, *Angew. Chem.*, **89**, 500 (1977).
121. W.-W. du Mont, paper presented at the DFG Symposium and EUCHEM Conference on 'Shortlived Species in Inorganic Chemistry, Their Reactivity and Their Stabilization', Burghausen, F.R.G., 1984.
122. S. Husebye, *Acta Chem. Scand.*, **20**, 51 (1966).

123. H.-G. Horn and H.-G. Lindner, *Chem. Ztg.*, **109**, 77 (1985).
124. I. A. Nuretdinov and E. I. Loginova, *Izv. Akad. Nauk SSSR, Ser. Khim.*, 2827 (1973).
125. H. W. Roesky and W. Kloker, *Z. Naturforsch., Teil B*, **28**, 697 (1973).
126. I. A. Nuretdinov, N. A. Buina, E. V. Bayandina and E. G. Sibgatulind, *Izv. Akad. Nauk SSSR, Ser. Khim.*, 483 (1974).
127. E. Gryszkiewicz-Trochimowski, J. Quinchon and O. Gryszkiewicz-Trochimowski, *Mém. Présentés Soc. Chim.*, **299**, 1794 (1960).
128. A. I. Razumov, B. G. Liovbev, M. B. Gazizov and Z. M. Khammatova, *Zh. Obshch. Khim.*, **34**, 1851 (1964).
129. N. P. Grechkin, I. A. Nuretdinov and N. A. Buina, *Izv. Akad. Nauk SSSR, Ser. Khim.*, **1**, 168 (1969).
130. K. Diemert, W. Kuchen and J. Kutter, *Phosphorus Sulfur*, **15**, 155 (1983).
131. O. J. Scherer and G. Schnabl, *Chem. Ber.*, **109**, 2996 (1976).
132. S. Pohl, *Z. Naturforsch., Teil B*, **33**, 610 (1978).
133. S. Pohl, *Z. Naturforsch., Teil B*, **34**, 256 (1979).
134. N. I. Zemlyanskii and R. D. Gorak, *Zh. Obshch. Khim.*, **41**, 2446 (1971).
135. R. D. Gorak and N. I. Zemlyanskii, *Zh. Obshch. Khim.*, **41**, 1994 (1971).
136. F. Räuchle, W. Pohl, B. Blaich and J. Boubean, *Ber. Bunsenges. Phys. Chem.*, **75**, 66 (1971).
137. C. Rømming, A. J. Iversen and J. Songstad, *Acta Chem. Scand., Ser. A*, **34**, 333 (1980).
138. D. H. Brown, R. J. Cross and D. Millington, *J. Organomet. Chem.*, **125**, 219 (1977).
139. R. Keat and D. G. Thompson, *J. Organomet. Chem.*, **141**, C13 (1977).
140. K. Issleib, W. Mögelin and A. Balszuweit, *Z. Anorg. Allg. Chem.*, **530**, 16 (1986).
141. R. A. Zingaro and A. Merijanian, *Inorg. Chem.*, **3**, 580 (1964).
142. K. A. Jensen and P. H. Nielsen, *Acta Chem. Scand.*, **17**, 1875 (1963).
143. F. L. Kolar, R. A. Zingaro and J. Laane, *J. Mol. Struct.*, **18**, 319 (1973).
144. P. B. Chi and F. Kober, *Z. Anorg. Allg. Chem.*, **466**, 183 (1980).
145. F. Kober and P. B. Chi, *Z. Anorg. Allg. Chem.*, **454**, 24 (1979).
146. F. Kober and P. B. Chi, *Z. Anorg. Allg. Chem.*, **477**, 139 (1981).
147. H. Schumann, U. Franck, W.-W. du Mont and F. Marschner, *J. Organomet. Chem.*, **222**, 217 (1981).
148. W. Kuchen, M. Förster, H. Hertel and B. Höhn, *Chem. Ber.*, **105**, 3310 (1972).
149. G. O. Doak and L. D. Freedman, *Organometallic Compounds of Arsenic, Antimony and Bismuth*, Wiley–Interscience, New York, London, Sydney, Toronto, 1970.
150. R. A. Zingaro and A. Merijanian, *J. Organomet. Chem.*, **1**, 369 (1964).
151. N. Kuhn and H. Schumann, *J. Organomet. Chem.*, **288**, C51 (1985); *J. Organomet. Chem.*, **304**, 181 (1986).
152. C. Rodger, N. Sheppard, C. McFarlane and W. McFarlane, in *NMR and the Periodic Table* (Eds. R. K. Harris and B. E. Mann), Academic Press, New York, 1978, p. 402.
153. W. McFarlane and D. S. Rycroft, *J. Chem. Soc., Chem. Commun.*, 902 (1972); *J. Chem. Soc., Dalton Trans.*, 2162 (1973).
154. R. P. Pinell, C. A. Megerle, S. L. Manatt and P. A. Kroon, *J. Am. Chem. Soc.*, **95**, 977 (1973).
155. J. Grobe and D. le Van, *J. Fluor. Chem.*, **24**, 25 (1984).
156. W. McFarlane and J. A. Nash, *J. Chem. Soc., Chem. Commun.*, 913 (1969).
157. I. J. Colquhoun, H. C. E. McFarlane, W. McFarlane, J. A. Nash, R. Keat, D. S. Rycroft and D. G. Thompson, *Org. Magn. Reson.*, **12**, 473 (1979).
158. I. J. Colquhoun and W. McFarlane, *J. Chem. Soc., Dalton Trans.*, 658 (1981).
159. F. Watari, *Inorg. Chem.*, **20**, 1776 (1981).
160. E. J. Jacob and S. Samdal, *J. Am. Chem. Soc.*, **99**, 5656 (1977).
161. P. Dehnert, R. Demuth and J. Grobe, *Spectrochim. Acta Part A*, **36**, 3 (1979).
162. I. I. Vandyukova, R. R. Shagidullin and I. A. Nuretdinov, *Izv. Akad. Nauk SSSR, Ser. Khim.*, 1390 (1976).
163. R. R. Carlson and D. W. Meek, *Inorg. Chem.*, **13**, 1741 (1974).
164. S. Elbel and H. tom Dieck, *J. Chem. Soc., Dalton Trans.*, 1762 (1976).
165. M. C. Böhm, M. Eckert-Maksić, R. Gleiter, J. Grobe and D. le Van, *Chem. Ber.*, **114**, 2300 (1981).
166. S. Husebye and G. Helland-Madsen, *Acta Chem. Scand.*, **23**, 1398 (1969).
167. A. Bondi, *J. Phys. Chem.*, **68**, 441 (1964).
168. C. Rømming and J. Songstad, *Acta Chem. Scand., Ser. A*, **32**, 689 (1979).
169. C. Rømming, K. Maartmann-Moe and J. Songstad, *Acta Chem. Scand., Ser. A*, **38**, 349 (1984).

170. C. Rømming and J. Songstad, *Acta Chem. Scand., Ser. A*, **33**, 187 (1979).
171. C. Rømming and J. Songstad, *Acta Chem. Scand., Ser. A*, **32**, 689 (1978).
172. P. W. Codding and K. A. Kerr, *Acta Crystallogr., Sect. B*, **35**, 1261 (1979).
173. L. S. Dent Glasser, L. Ingram, M. G. King and G. P. McQuillan, *J. Chem. Soc. A*, 2501 (1969).
174. S. Husebye, *Acta Chem. Scand.*, **23**, 1389 (1969).
175. R. Kramolowsky, J. Sawluk, unpublished work; J. Sawluk, *PhD Thesis*, Universität Hamburg, 1985.
176. R. Hensel, W.-W. du Mont, R. Boese, D. Wewers and L. Weber, *Chem. Ber.*, **118**, 1580 (1985).
177. R. R. Shagidullin, I. P. Lipatova, I. A. Nuretdinov and S. A. Samartseva, *Dokl. Akad. Nauk SSSR*, **211**, 1363 (1973).
178. P. Dehnert, J. Grobe and D. le Van, *Z. Naturforsch., Teil B*, **36**, 48 (1981).
179. J. Grobe and R. Haubold, *Z. Anorg. Allg. Chem.*, **522**, 159 (1985).
180. J. Grobe and D. le Van, *Z. Naturforsch., Teil B*, **34**, 1653 (1979).
181. J. Grobe and D. le Van, *Z. Allg. Anorg. Chem.*, **518**, 36 (1984).
182. J. Grobe, M. Köhne-Wächter and D. le Van, *Z. Anorg. Allg. Chem.*, **519**, 67 (1984).
183. W.-W. du Mont, S. Kubiniok, T. Severengiz, W. Saak and S. Ponl, unpublished results (X-ray crystal structure analysis).
184. R. Hensel and W.-W. du Mont, *Chem. Ber.*, **118**, 5016 (1985).
185. O. J. Scherer, H. Jungmann, C. Krüger and G. Wolmershäuser, *Chem. Ber.*, **117**, 2382 (1984).
186. W. Malisch, unpublished work (1985).
187. W. Malisch, R. Maisch, I. J. Colquhoun and W. McFarlane, *J. Organomet. Chem.*, **220**, C1 (1981).
188. B. Walther, B. Messbauer and H. Meyer, *Inorg. Chim. Acta*, **37**, L525 (1979).
189. J. Grobe, J. Vetter and D. Rehder, *Z. Naturforsch., Teils B*, **40**, 975 (1985).
190. M. Schmidt and H. D. Block, *Chem. Ber.*, **103**, 3705 (1970).
191. S. Esperås and S. Husebye, *Acta Chem. Scand.*, **27**, 3355 (1973).
192. W. Kuchen and H. Hertel, *Angew. Chem.*, **81**, 127 (1969); *Angew. Chem. Int. Ed. Engl.*, **8**, 89 (1969).
193. P. Bruno, M. Caselli, C. Fragale and S. Magrino, *J. Inorg. Nucl. Chem.*, **39**, 1757 (1977).
194. F. L. Kolar, R. A. Zingaro and K. Irgolic, *J. Inorg. Nucl. Chem.*, **28**, 2981 (1966).
195. P. A. W. Dean, *Can. J. Chem.*, **57**, 754 (1979).
196. M. S. Raasch, *J. Org. Chem.*, **45**, 3517 (1980).
197. T. H. Chan and J. R. Finkenbine, *Tetrahedron Lett.*, 2091 (1974).
198. D. L. J. Clive and C. V. Denyer, *J. Chem. Soc., Chem. Commun.*, 253 (1973).
199. D. L. J. Clive, W. A. Kiel, S. M. Menchen and C. K. Wong, *J. Chem. Soc., Chem. Commun.*, 657 (1977).

The Chemistry of Organic Selenium and Tellurium Compounds Volume 2
Edited by S. Patai
© 1987 John Wiley & Sons Ltd.

CHAPTER **16**

Free radical reactions of organoselenium and organotellurium compounds

LAURENCE CASTLE and M. JOHN PERKINS

Department of Chemistry, Royal Holloway and Bedford New College, University of London, Egham, Surrey TW20 0EX, UK

I. INTRODUCTION 657
II. SELENYL RADICALS (RSe•) 658
A. From Diselenides 658
B. From Selenols 662
III. S_H2 DISPLACEMENTS AT SELENIUM 664
A. Reactions of Monoselenides 665
B. Reactions with Diselenides 667
IV. SELENIUM RADICALS IN S_{RN} SUBSTITUTION PROCESSES . . . 669
V. α-SELENOALKYL RADICALS 670
VI. OXYSELENIUM AND OXYTELLURIUM SPECIES 670
VII. REFERENCES 671

I. INTRODUCTION

The review[1] by Shine in Klayman and Gunther's 1973 book on organic selenium compounds is punctuated by statements such as '…very little is known about the preparation and chemistry of organoselenium radicals'. It is also clear from that review that there is mechanistic uncertainty about many of the reactions of organoselenium compounds. In many respects little has changed, although analogy with the extensive and relatively well understood organic free-radical chemistry of sulphur has led recently to a small number of key papers in which quantitative kinetic data, or clearly defined synthetic applications, have begun to appear.

The synthetic work depends largely on qualitative ideas. For example, it seems reasonable to assume that Se—H, Se—C and Se—Se bonds are weaker than their counterparts in sulphur chemistry and that one-electron oxidation of organoselenium compounds occurs more readily than does that of the corresponding organosulphur species. Many of these ideas are borne out by qualitative experience, such as the facile air oxidation of the evil-smelling selenols to give diselenides. Some also appear to be backed by quantitative experimental data, such as the first ionization potentials of the dimethyl chalcogenides[2]. On the other hand, values for the strength of the Se—Se bond in organic compounds that differ by 20 kcal mol^{-1} can be found among the pages of these volumes. Even the comment on oxidation of selenols must be viewed in the light of the different pK_a values of selenols; much greater concentrations of the readily oxidized conjugate base are likely to be present in solutions of the more acidic selenols than in those of corresponding thiols.

Despite these words of caution, the qualitative picture seems reasonable and can no doubt be extrapolated to tellurium, for which even less quantitative information is available.

Analogy with sulphur chemistry suggests that the major emphasis in this chapter will be on selenyl radicals, RSe$^{\bullet}$, and on their formation (e.g. from diselenides and selenols) and this is reflected in Section II. Almost equally important, however, is Section III on S_H2 reactions[3] at selenium, in which radical attack proceeds (at least in some cases) via a discrete selenuranyl radical (**1**) which then dissociates.

$$Y^{\bullet} + R_2Se \rightarrow R_2\dot{Se}Y \rightarrow R^{\bullet} + RSeY$$

(**1**)

Section IV describes S_{RN} processes involving selenium radicals, and Section V briefly considers α-selenoalkyl radicals. There is special interest here in examples in which the radical centre also carries an electron-withdrawing group, for which 'capto-dative' stabilization appears to be important.

The emphasis throughout this chapter is on organic reactions involving selenium- or tellurium-containing radicals. Structure and spectroscopic properties (ESR) are formally addressed elsewhere. Therefore, relatively scant attention is given to radicals $RSeO_2^{\bullet}$ and RSeO$^{\bullet}$ in the final section (Section VI), since there seems to be little documented 'organic chemistry' for these species.

II. SELENYL RADICALS (RSe$^{\bullet}$)

A. From Diselenides

Studies of the thermolysis and photolyis of diselenides have been largely mechanistic in intent, with a central question being the relative importance of C—Se versus Se—Se bond fission. Little chemistry of a preparative nature has been described.

Whilst the pyrolytic equilibration of monomeric and polymeric trimethylene diselenide[4], $(CH_2)_3Se_2$ (equation 1), is suggestive of a weak Se—Se bond, the bulk of the reaction may well involve an S_H2 chain process rather than diradical intermediates. On the other hand, there seems little doubt that photolysis of diselenides does produce selenyl radicals. Experiments[5] in which solutions of dimethyl diselenide were directly photolysed in the cavity of an ESR spectrometer failed to reveal methylselenyl radicals but, in the presence of a nitrone spin trap (for a review of spin trapping, see Ref. 6), or of di-*tert*-butyl selenoketone, spin adducts **2** and **3**, respectively, were easily detected. The spectrum of the spin adduct **2** showed no ^{77}Se satellites, but had an unusually high g-factor attributable to the presence of selenium.

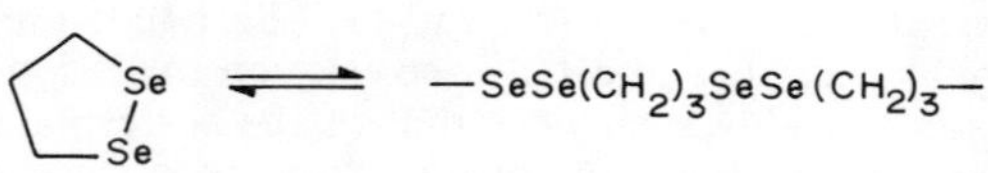

(1)

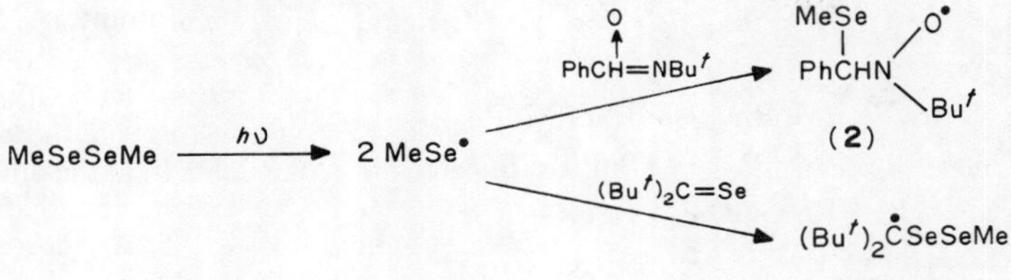

Spin trapping has also been employed to examine the photolytic scission of diphenyl and dibenzyl diselenides[7]. Using nitrosodurene, compelling evidence for scission of the former to give PhSe• was presented; in this case the spectrum of the spin adduct (**4**) ($a_N = 18.5$ G) shows not only a very high g-value (2.0099) but also well resolved ^{77}Se satellite lines.

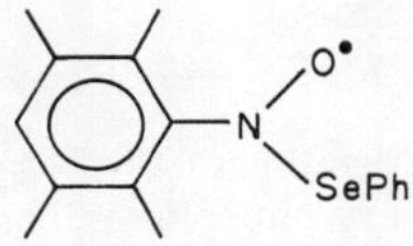

(**4**)

In contrast, dibenzyl diselenide gave only spin adducts of the benzyl radical. Earlier workers had already suggested[8] that both C—Se and Se—Se scission occur competitively when this diselenide is exposed to ultraviolet light. Photoreaction in an inert atmosphere gives quantitative yields of dibenzyl selenide and red selenium, while benzyl chloride and benzaldehyde are among the products of reaction in the presence of CCl_4 and O_2, respectively. Analogous products have been identified from the photolysis of dibenzyl ditelluride in the presence and absence of oxygen[9]. The products from diselenide photolysis have been accounted for in terms of C—Se scission alone; scission of the weaker Se—Se bond is considered to occur but judged to be a non-productive process.

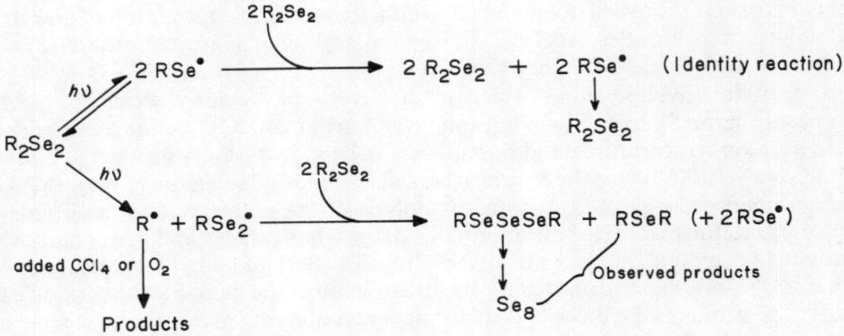

This suggested unimportance of Se—Se scission in the pathway to product formation contrasts with the results of experiments where a trap for the RSe• radical is present. Thus Millington's group[10,11] and also Chu and Marsh[12] studied the photolysis of diselenides (RSeSeR) in the presence of tertiary phosphines (equation 2; Ch = Se) and found similar overall results when R = Me, Ph or $PhCH_2$. They interpreted their findings in terms of the chain reaction shown in Scheme 1 involving a selenophosphoranyl intermediate, and initiated by Se—Se bond fission. Of course, initiation by C—Se fission could, by addition

$$\mathrm{RSeSeR} \xrightarrow{h\nu} \mathrm{2RSe^{\bullet}} \qquad \text{initiation}$$

$$\left.\begin{aligned} \mathrm{RSe^{\bullet} + PMePh_2} &\longrightarrow \mathrm{RSe\dot{P}MePh_2} \\ \mathrm{RSe\dot{P}MePh_2} &\longrightarrow \mathrm{SePMePh_2 + R^{\bullet}} \\ \mathrm{R^{\bullet} + RSeSeR} &\longrightarrow \mathrm{RSeR + RSe^{\bullet}} \end{aligned}\right\} \qquad \text{propagation}$$

SCHEME 1

of reaction 3, give stoichiometric formation of the phosphine selenide and RSeR by the same propagation sequence.

$$\mathrm{RChChR + R'_3P} \xrightarrow{h\nu} \mathrm{R'_3PCh + RChR} \qquad \mathrm{Ch = S\ or\ Se} \qquad (2)$$

$$\mathrm{RSeSe^{\bullet} + PMePh_2} \longrightarrow \mathrm{RSe^{\bullet} + SePMePh_2} \qquad (3)$$

Subsequently, Brown *et al.*[13] demonstrated a similar photochemical transfer of tellurium from ditellurides to tertiary phosphines (equation 2; Ch = Te).

A CIDNP study[14] of the dibenzyl diselenide/tertiary phosphine system appears to indicate both C—Se and Se—Se fission.

In view of the general similarity between allylic and benzylic systems, the thermal and photochemical stability of dibenzyl telluride[9] contrasts with the instability of bis-3-cyclohexenyl telluride (**5**); this compound has been observed by NMR but decomposes rapidly in solution to bicyclohexenyl and tellurium[15].

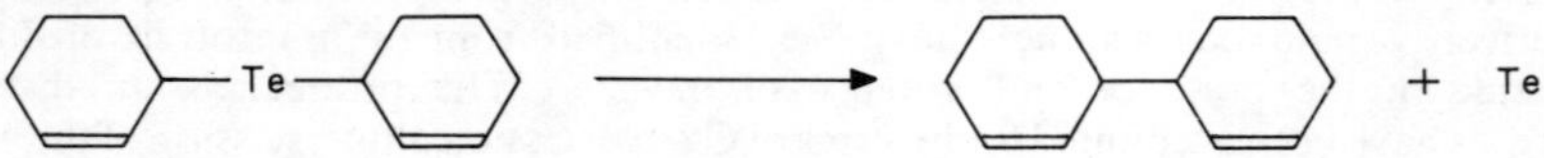

(**5**)

Other extrusion processes involving tellurium are seen in the formation of diaryls from diaryl tellurium dichlorides, Ar_2TeCl_2[16], or from aryl tellurium trichlorides, $ArTeCl_3$[17], on treatment with degassed Raney nickel.

Bis(diphenylmethyl)diselenide, which must have a particularly weak Se—C bond, decomposes thermally to give *sym*-tetraphenylethane at 200 °C[18]. Some monoselenide is isolated at lower temperatures, and this appears to be a key primary product; it undergoes thermolysis at 140 °C to give the hydrocarbon, diselenide and selenium as major products.

Trifluoromethylselenyl radicals from photolysis of the corresponding diselenide have been found to displace carbon monoxide from manganese and iron carbonyls[19]. Contrasting behaviour is found with $PhSe^{\bullet}$ (from Ph_2Se_2) and vinyl mercurials, in which attack by $PhSe^{\bullet}$ is apparently not at the metal atom but at the mercury-substituted carbon centre[20]. The products are those of displacement reactions at vinyl carbon (Scheme 2). An alternative mechanism, involving S_H2 displacement of vinyl radicals from mercury, was discounted.

When Ph_2Se_2 is photolysed in the presence of allyltin derivatives an addition–elimination sequence occurs with high efficiency. A chain reaction is propagated by S_H2 displacement of $PhSe^{\bullet}$ from diselenide by the displaced tin radical (Scheme 3)[21].

Discussions of the addition of selenyl radicals to alkenes and alkynes have most frequently drawn examples from reactions of selenols, for which there is in fact usually some mechanistic uncertainty. The position is much clearer where the photolysis of

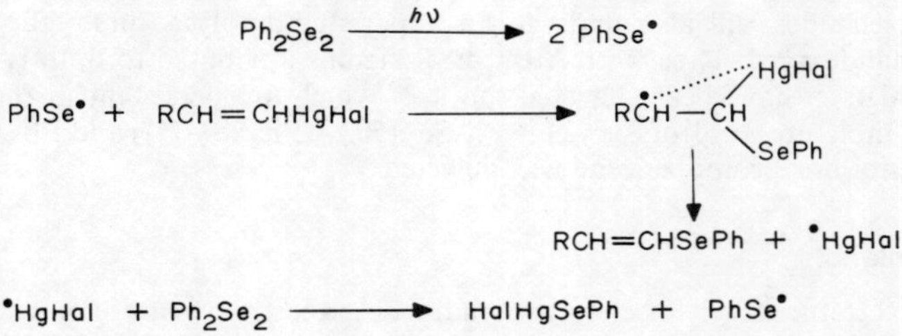

SCHEME 2

$$PhSeSePh \xrightarrow{h\nu} 2PhSe^{\cdot}$$
$$PhSe^{\cdot} + RCH{=}CHCH_2SnBu_3 \longrightarrow PhSeCH(R)CH{=}CH_2 + Bu_3Sn^{\cdot}$$
$$Bu_3Sn^{\cdot} + Ph_2Se_2 \longrightarrow Bu_3SnSePh + PhSe^{\cdot}$$

SCHEME 3

diselenides is used as a source of authentic selenyl radicals. Although radical adducts of $MeSe^{\cdot}$ to 1,1-di-*tert*-butylethene could not be observed by ESR[5], a flash photolysis study[22] of diphenyl diselenide in the presence of various vinyl monomers has yielded not only rate data for addition of $PhSe^{\cdot}$ to these substrates but also for the facile reverse of this reaction, i.e. fragmentation of the adducts. The experiments involved direct observation of the transient absorption of $PhSe^{\cdot}$ at 490 nm. The self-reaction of this species in CCl_4 was found to have a second-order rate constant of $7 \times 10^9\,l\,mol^{-1}\,s^{-1}$ at 23 °C. Although this bimolecular decay is unaffected by the presence either of oxygen or of an alkene, its rate is reduced when both are present together; the (reversibly formed) adducts with the vinyl monomers may be intercepted by oxygen (Scheme 4). Kinetic analysis yields values for k_1 and the ratio k_{-1}/k_2. Selected values are given in Table 1. The rates of $PhSe^{\cdot}$ addition were consistently between one and two orders of magnitude lower than those for the corresponding addition reactions of $PhS^{\cdot}$. Hexane, like CCl_4, was found to be inert towards $PhSe^{\cdot}$.

$$PhSe^{\cdot} + CH_2{=}CHX \underset{k_{-1}}{\overset{k_1}{\rightleftarrows}} PhSeCH_2C^{\cdot}HX$$
$$PhSeCH_2C^{\cdot}HX + O_2 \xrightarrow{k_2} PhSeCH_2CHX(O{-}O^{\cdot}) \longrightarrow \text{products}$$

SCHEME 4

TABLE 1. Rate data for Scheme 4

X	k_1 ($l\,mol^{-1}\,s^{-1}$)	k_{-1}/k_2 ($mol\,l^{-1}$)
Ph	2.2×10^6	$3.1 + 10^{-2}$
CN	1.4×10^4	1.8×10^{-2}
EtO	1.5×10^4	5.9×10^{-3}
AcO	$\sim 10^3$	–

Diaryl diselenides will also undergo an apparent homolysis under the influence of Group III halides[23,24]. Thus, with $AlBr_3$ ESR signals attributed to both $[ArSe_2AlBr_3]^{\bullet}$ and $[ArSeAlBr_3]^{\bullet}$ have been detected, both of which are persistent in the absence of oxygen and moisture. Hydrolysis yields Ph_2Se_2 (50%), Ph_2Se (43%) and Ph_2 (7%); neither benzeneselenol nor bromobenzene was detected.

B. From Selenols

In contrast to the largely mechanistic studies described above for diselenides, examples of the use of selenols in radical-mediated chemistry are generally of synthetic utility, and the intermediacy of selenyl radicals is presumed (albeit with sound chemical reasoning on occasions) rather than proved.

The mechanistic ambiguity of selenol additions to carbon—carbon multiple bonds was addressed by Shine[1]. Several more recent examples seem to proceed by unambiguous radical mechanisms. Benzeneselenol undergoes nucleophilic (Michael) addition to $PhC{\equiv}CCO_2Et$ under basic conditions to give **6** (predominantly the *Z*-isomer), but reacts by radical addition under neutral conditions to give **7** (stereochemistry not assigned)[25]. Earlier workers had apparently made an incorrect structural assignment to the product of the neutral reaction[26].

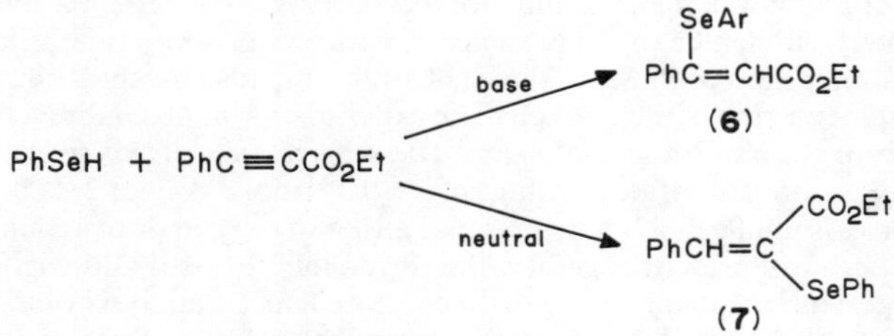

A clear distinction has also been reported between base-promoted and neutral attack of benzeneselenol on alkene **8**[27]. Addition to other alkynes has been found to be complicated by reversibility[28]; for example, neat PhSeH with 1-phenylbutyne gives nearly pure (*Z*)-**9** on reaction at 50 °C in the absence of base, but at 120 °C the *Z*/*E* ratio approaches unity and the monomeric alkene **10** is an additional product. Under these latter reaction conditions (sealed tube, 120 °C), pure (*Z*)-**9** was found to isomerize to a *Z*–*E* mixture, but no comment was made on whether or not alkene **10** also formed in this control experiment. There is also no discussion of how it might arise in the selenol reactions. It seems likely that **9** is indeed an intermediate in the formation of **10**.

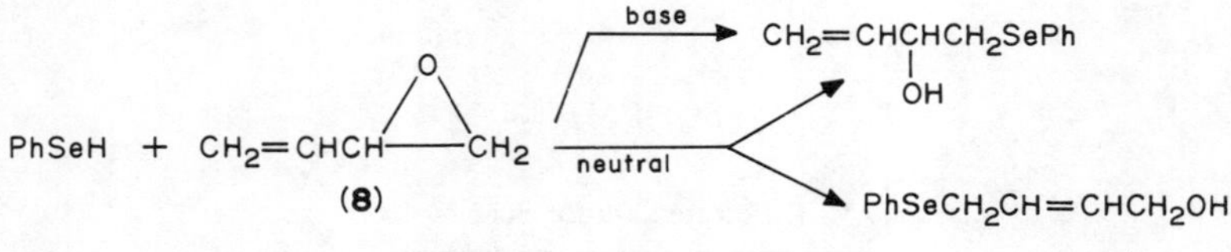

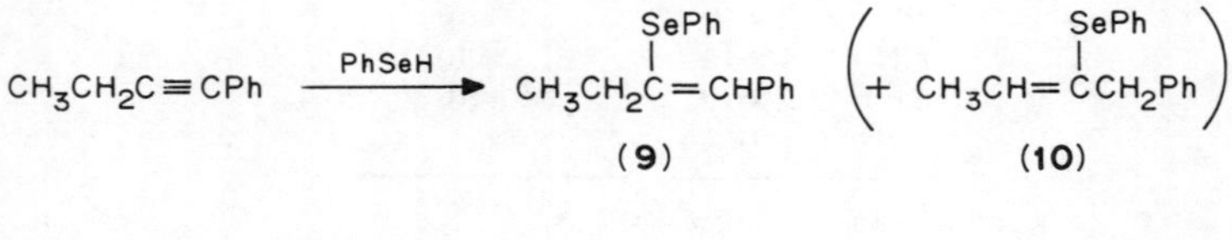

Photo-initiated additions of Et_3M—SeH, where M = Si or Ge, to a variety of alkenes have also been reported[29].

The ease of oxidation of selenols has prompted investigation of the use of these compounds as reducing agents. In various successful examples diselenides are produced but, as is so often the case, there is no compelling evidence for radical involvement. One of the oldest examples is reduction of disulphides to thiols (equation 4). This may be effected with stoichiometric amounts of selenol, or by using catalytic amounts of selenol or diselenide and a suitable reagent which will effect the reduction of diselenide to selenol, such as hypophosphorous acid[30]. The catalytic system is also effective with sulphoxides[30]. Neither disulphides nor sulphoxides are reduced by hypophosphorous acid alone. The efficient deoxygenation of a cephalosporin *S*-oxide by stoichiometric amounts of PhSeH is indicative of the mildness of this procedure[31]. Sulphoxides have also been reduced by *O*, *O*-diethyl hydrogen phosphoroselenate[32], and by the Se—B heterocycle **11**[33], but there has been little speculation on the mechanisms of these reactions[32,34].

$$R_2S_2 + 2R'SeH \longrightarrow 2RSH + R'_2Se_2 \tag{4}$$

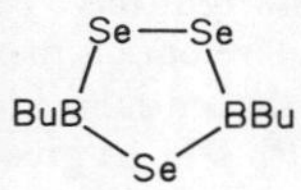

(11)

The reaction of diazonium salts with iodide ion is generally considered to be initiated by electron transfer (equation 5); this is followed by reactions of the derived aryl radicals. When diazonium decomposition is promoted by hydrogen iodide, reduction products can be isolated[35]. Diazonium decomposition may also be brought about by $ArSe^-$ and this affords a recognized route to unsymmetrical diaryl selenides (equation 6)[36]. When a suspension of *p*-$ClC_6H_4N_2{}^+BF_4{}^-$ in dichloromethane–acetone (10:1) is treated with PhSeH there is rapid evolution of nitrogen and formation of chlorobenzene and *p*-chlorophenyl phenyl selenide[37]. The mechanism outlined in Scheme 5 was suggested. Very different chemistry results when the solvent is dichloromethane alone. In this case there is no evolution of nitrogen, and a solid remains in suspension. Diphenyl diselenide is formed, however, and the suspended solid is transformed quantitatively to hydrazinium fluoroborate (equation 7). Whilst the reaction in the presence of acetone seems to be a straightforward free-radical process, the reaction in its absence, although synthetically useful, is less easily understood. Possibly the influence of the acetone is to change the redox potentials of species present in the system; in its absence a species ArN═NSePh may be sufficiently long-lived to be reduced by hydrogen atom transfer from the selenol. It is known that azobenzene is easily reduced to hydrazobenzene[30].

$$ArN_2{}^+ + I^- \longrightarrow I^\bullet + [ArN_2{}^\bullet] \longrightarrow Ar^\bullet + N_2 \tag{5}$$

$$ArN_2{}^+ + PhSe^- \longrightarrow ArSePh + N_2 \tag{6}$$

$$\begin{aligned} ArN_2{}^+ + PhSeH &\longrightarrow Ar^\bullet + N_2 + PhSe^\bullet + H^+ \\ Ar^\bullet + PhSeH &\longrightarrow ArH + PhSe^\bullet \\ 2PhSe^\bullet &\longrightarrow Ph_2Se_2 \\ Ar^\bullet + PhSeSePh &\longrightarrow ArSePh + PhSe^\bullet \end{aligned}$$

SCHEME 5

$$4PhSeH + PhN_2^+BF_4^- \longrightarrow 2Ph_2Se_2 + PhNHNH_3^+BF_4^- \qquad (7)$$

Other reductions by benzeneselenol include the conversion of nitro, nitroso and hydroxylamino groups to amino[38]. These transformations may also be achieved by the use of hydrogen telluride, H_2Te[39], which will also reduce aromatic and aliphatic aldehydes to alcohols[40,41]. Benzenetellurol has recently been employed for these reactions[42] and, for aromatic nitro compounds, reduction has been made catalytic in tellurol by the use of sodium borohydride as the stoichiometric reducing agent[43].

It has long been known that certain carbon—carbon double bonds may be reduced by selenols, and that hydrogen transfer reductions, e.g. of cinnamic acid by tetralin, may be promoted by diphenyl diselenide[44]. Comparison of this type of behaviour with the use of elemental selenium to dehydrogenate hydroaromatic compounds has been drawn by Orchin and coworkers[45,46], who have had success in using diselenides and other organoselenium compounds for dehydrogenations, and have concluded that ArSe$^\bullet$ is likely to be a key hydrogen acceptor in these reactions.

In related work, simple alkyl radicals generated in the gas phase have been adsorbed on to a freshly deposited film of amorphous selenium; on heating to ca. 150 °C the film releases the alkyl radicals and not, as might have been expected, alkylselenyl radicals[47].

The reduction of chalcone by benzeneselenol is promoted by ultraviolet irradiation, and this probably involves photolysis of the initial adduct **12**[48]. By analogy, benzyl phenyl selenide is rapidly photoreduced by PhSeH to give toluene.

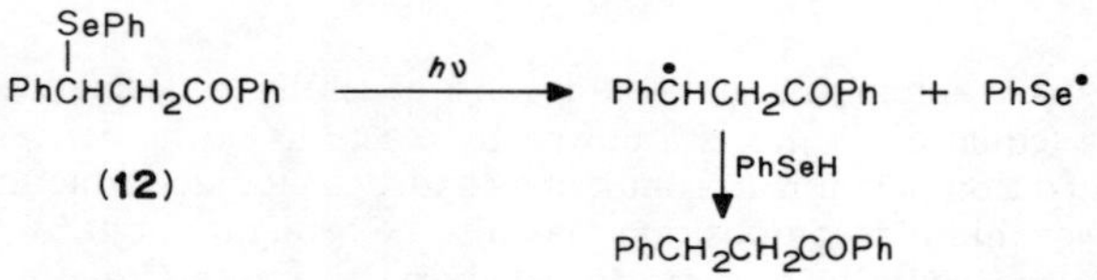

(12)

Photoreduction of several compounds containing the C=N group in the presence of benzeneselenol has also been demonstrated[48]. Benzaldehyde oxime gives benzylhydroxylamine, and benzylideneaniline gives *N*-benzylaniline. Benzaldehyde phenylhydrazone yields benzylphenylhydrazine, not necessarily via the azo tautomer, since the *N*-methyl derivative (**13**) is also reduced.

$$\underset{(\mathbf{13})}{PhCH=NNMePh} \xrightarrow[PhSeH]{h\nu} PhCH_2NHNMePh$$

It seems possible that hydrogen atom transfer is involved in most, if not all, of the above reactions, but it must again be stressed that there is little firm evidence for this.

III. S_H2 DISPLACEMENTS AT SELENIUM

In a 1971 book on the S_H2 reaction (substitution, homolytic, bimolecular), Ingold and Roberts[3] were able to report only one example of such a displacement reaction in Group VIA (Se, Te, Po) organometallic compounds. This situation has altered dramatically in recent years and the S_H2 reaction at selenium is now firmly established in a number of synthetic strategies.

Radical attack at selenium proceeds formally via a selenuranyl radical and, in certain cases, this species may be observed directly by ESR. Examples of this include **14** and **15**, which were observed in liquid solution and are believed to have the structures indicated[49]. Selenuranyl species have also been detected by ESR in solid matrices at low temperatures[50,51]. The precise fates of **14** and **15** are not clear, but where one of the substituents is weakly bonded it can be expected to break away in a fragmentation step. As we shall see, control of the direction of this fragmentation is the key to the use of the S_H2 reaction in synthesis.

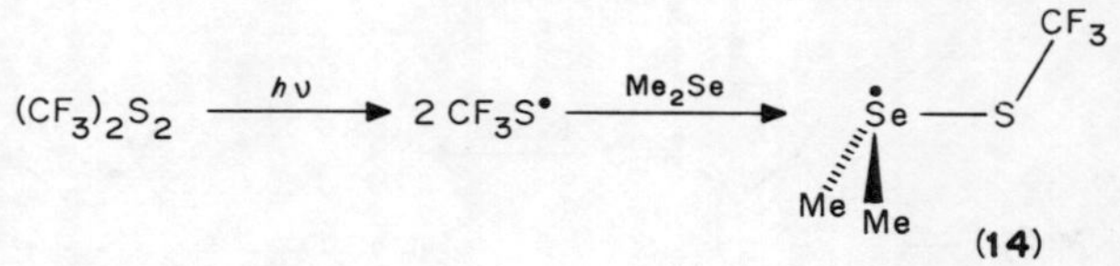

(14)

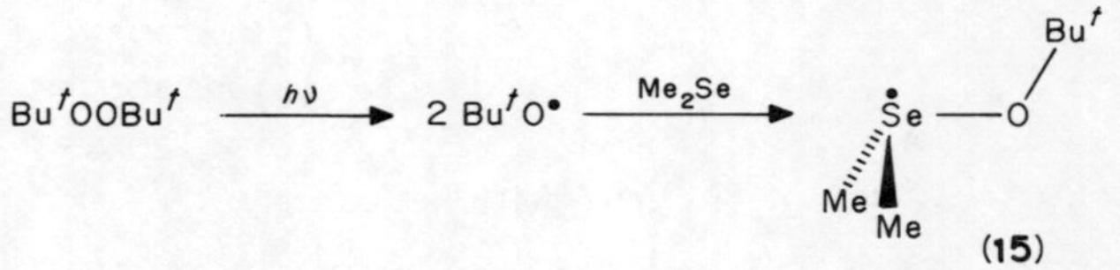

(15)

A. Reactions of Monoselenides

There are many methods for elaboration of complex organic structures which depend on the use of selenium-containing reagents. Intermediates are produced incorporating the selenium, which must then be removed. Commonly this is achieved oxidatively via selenoxide elimination, but increasingly a reductive procedure using tin hydrides is being adopted. In this reaction, alkyl phenyl selenides are cleaved to give alkane, presumably by the chain propagation sequence shown in Scheme 6[52]. An application of this procedure is given in Scheme 7[53]; similar chemistry has been deployed by Nicolaou and coworkers[54,55]

$$R'SePh + R_3Sn^{\bullet} \xrightarrow{S_H2} R'^{\bullet} + PhSeSnR_3$$

$$R'^{\bullet} + R_3SnH \longrightarrow R'H + R_3Sn^{\bullet}$$

SCHEME 6

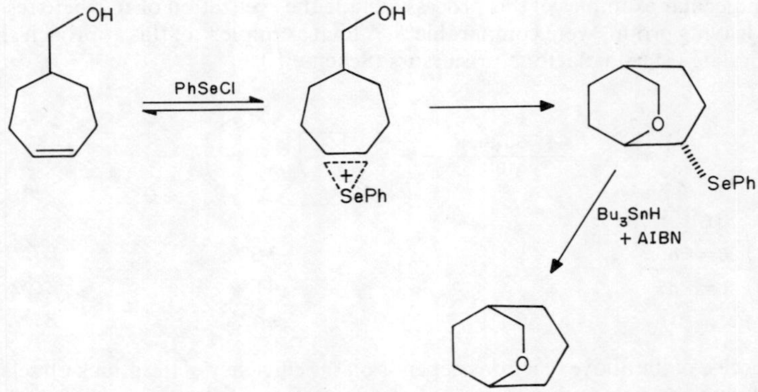

SCHEME 7

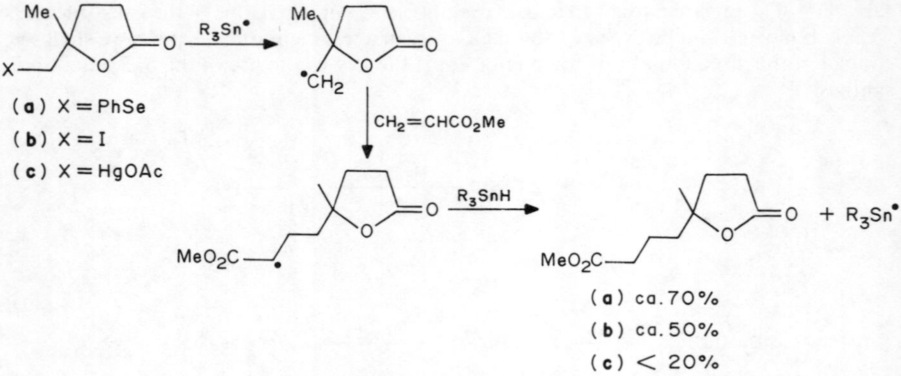

SCHEME 8

and Corey *et al.*[56,49] in syntheses of various lactones, including prostacyclin, in which carboxyl instead of hydroxyl is used as the neighbouring group in the phenylselenation step.

Clive and coworkers[52,57] have extended the reaction to the reduction of selenoketals; this affords a new alternative to the Wolff–Kishner reduction of carbonyl to methylene They have also demonstrated that the phenyltelluryl group may be removed in the same way. In the absence of a radical initiator, such as 2,2′-azobis(2-methylpropionitrile) (AIBN), the reactions with tellurium compounds generally proceed at lower temperatures (ca. 50 °C) than do those of the selenium analogues (ca. 100 °C), but the organotellurium compounds are very sensitive to photolysis, extending the precautions that have to be taken to include working in red light.

Modifications of this hydride procedure whereby the intermediate alkyl radical is intercepted either inter-[58,59] or intra-molecularly[60,61] by addition to an alkene have been developed. The phenylselenyl leaving group in reductive alkylation with methyl acrylate (Scheme 8)[58] proved superior to either iodine or acetoxymercury. In a variation of this intramolecular trapping procedure[59], the alkene group to be appended is present originally in the tin reagent and a simplified chain process involving only two reagents leads to the product (Scheme 9).

Intramolecular examples of this process include the cyclization of **16** where results with different leaving groups were comparable[60]. A final example[61] of this approach shows the use of acrylate esters as lactone precursors (Scheme 10).

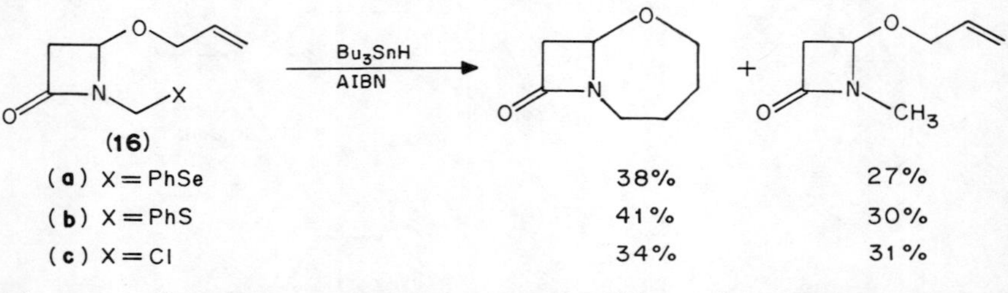

The success of the above reactions depends on the efficiency of the attack on selenium by the trialkyltin radicals and the mode of fragmentation of the intermediate selenuranyl radical. Some quantitative estimates of the rates of such S_H2 displacements by $(n\text{-Bu})_3Sn^•$

$$\mathrm{RSePh} \xrightarrow[\text{or AIBN}]{h\nu} \mathrm{R^{\bullet}}$$

$$\mathrm{R^{\bullet} + CH_2{=}CHCH_2SnBu_3 \longrightarrow RCH_2CH{=}CH_2 + Bu_3Sn^{\bullet}}$$

$$\mathrm{Bu_3Sn^{\bullet} + RSePh \longrightarrow Bu_3SnSePh + R^{\bullet}}$$

SCHEME 9

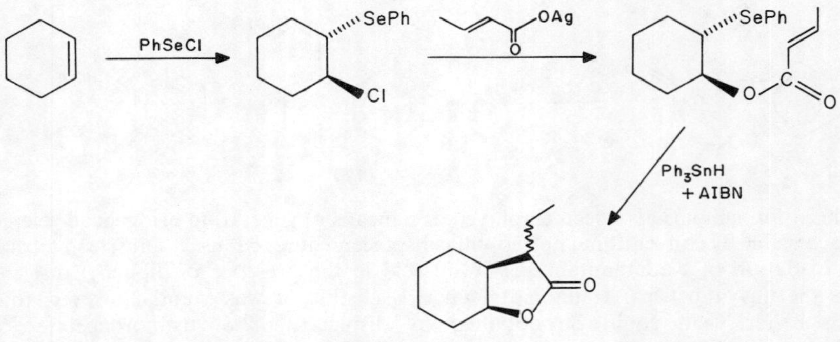

SCHEME 10

are available from the work of Scaiano *et al.*[62], who found the expected order of reactivity $Me_2Te > Me_2Se > Me_2S$. For $(n\text{-Bu})_3Sn^{\bullet} + Me_2Se$ at 25 °C in a hydrocarbon solvent, a rate constant of $3.7 \times 10^5\ \mathrm{l\,mol^{-1}\,s^{-1}}$ was determined. Concerning the fragmentation of the $R^1R^2R^3Se^{\bullet}$ intermediate, the substituents are invariably $R^1 = Ph$, $R^2 =$ tin reagent and $R^3 =$ alkyl (sp^3) carbon of the compound to be modified. Preferential scission of the Se—R^3 bond ensures synthetic utility.

B. Reactions with Diselenides

S_H2 reactions of diselenides lead to selenium-containing products (equation 8). In many cases ionic routes to the selenium-containing products are more convenient, but the radical process is often efficient and may permit unusual precursors to be used. A probable early example of this reaction, not recognized as such, involved mercury dialkyls and diselenides which, when heated together in dioxane, gave good yields of unsymmetrical selenide and mercury metal (equation 9)[63].

$$\mathrm{R'^{\bullet} + RSeSeR \longrightarrow R'SeR + RSe^{\bullet}} \tag{8}$$

$$\mathrm{R_2Hg + R'_2Se_2 \longrightarrow 2R'SeR + Hg^0} \tag{9}$$

R = alkyl
R′ = alkyl or aryl

The first example[69] of a thermal S_H2 reaction involving diselenide precursors to be recognized is that sole example reported in the book by Ingold and Roberts[3] and involves the redistribution reactions of dichlorodiselenide with either dimethyl selenide or dimethyl diselenide to give first CH_3Se_4Cl and thence methyl-terminated selenium chains and elemental selenium by a sequence of condensation plus scrambling reactions. A further early example is the pyrolytic equilibration of monomeric and polymeric trimethylene diselenide, discussed in Section II.A (equation 1).

The final step of Scheme 5 proposes S_H2 displacement on diselenide by an aryl radical. This has been adapted to a convenient and efficient synthesis of unsymmetrical diaryl selenides using copper-promoted decomposition of the diazonium salt in the presence of diphenyl diselenide[65]. When a neighbouring double bond is present, the aryl radical intermediate may be intercepted to give a cyclic product, e.g. **17**.

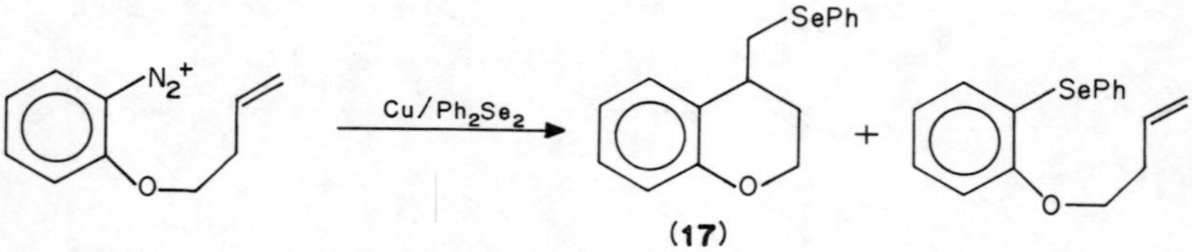

(17)

This approach has also been employed as a means of generating bridgehead selenides not accessible by conventional nucleophilic displacement reactions[66]. Thus, lead tetraacetate oxidation of 1-adamantanecarboxylic acid in the presence of diphenyl diselenide gives the unsymmetrical 1-adamantyl phenyl selenide. It was found, however, that a much better yield could be obtained by displacement of trifluoroacetate from 1-adamantyl trifluoroacetate using benzeneselenol. Other leaving groups were ineffective, and an electron-transfer mechanism is suspected.

More recently, a general strategy for synthesizing unsymmetrical disubstituted chalcogenides (**19**) has been developed by Barton's group[67]. In this, chain decomposition of the readily prepared carboxylate (**18**) occurs in the presence of the appropriate dichalcogenide. Among the products obtained this way was a bridgehead adamantyl telluride.

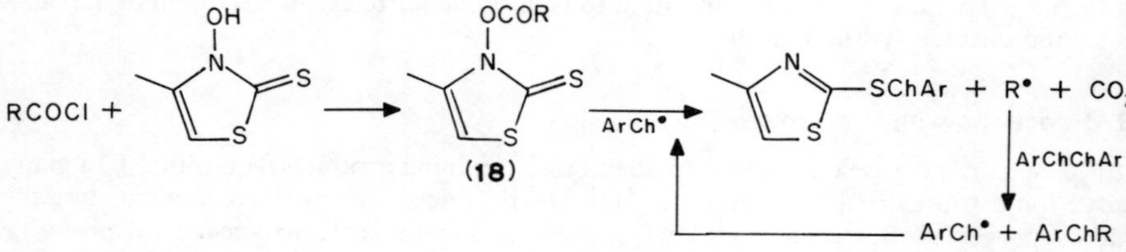

(18)

(19)

The rate of reaction of Ph_2Se_2 with primary alkyl radicals has been the subject of two independent studies[66,68], both of which used the 5-hexenyl radical 'clock', in which the bimolecular reaction of 5-hexenyl radical (**20**) with diselenide competes with cyclization, the rate constant (k_1) for which is known. In one investigation, the hexenyl radicals were generated by thermolysis of the corresponding heptenoyl peroxide[66]; the other employed a photoinitiated chain reaction with hexenylmercury(II) halides in which the selenyl radical displaces hexenyl from mercury[68]. In both cases, the rate constant for the bimolecular process (k_2) can then be calculated from a knowledge of the ratio of selenides **21** and **22** and of the concentration of Ph_2Se_2. The values of k_2 are ca. 1×10^7 and $5 \times 10^7\,\mathrm{l\,mol^{-1}\,s^{-1}}$ for reaction in benzene at 45 and 80 °C, respectively. A value for the corresponding reaction with diphenyl ditelluride at 45 °C is $5 \times 10^7\,\mathrm{l\,mol^{-1}\,s^{-1}}$[68]. These reactions are preparatively more efficient with the diselenides and ditellurides; reactions of primary alkyl radicals with diphenyl disulphide are significantly slower ($k \approx 10^5\,\mathrm{l\,mol^{-1}\,s^{-1}}$ at 80 °C)[65,68].

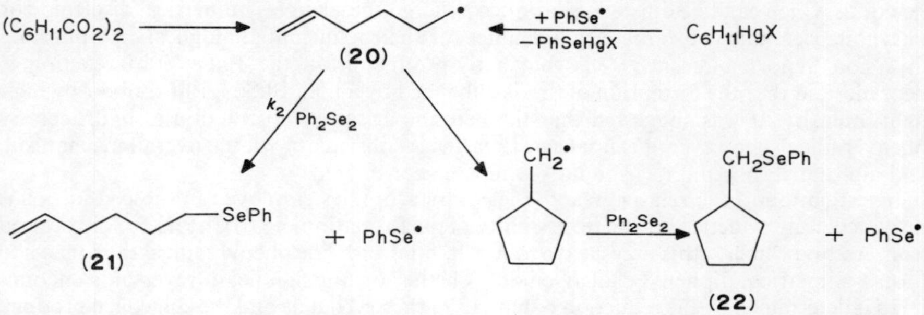

Alkyl cobaloximes will also react with diphenyl diselenide to give the mixed alkyl phenyl selenide[69]. If the alkyl group is attached to cobalt via a chiral carbon then racemization is observed in the product, consistent with a mechanism involving free alkyl radicals.

IV. SELENIUM RADICALS IN S_{RN} SUBSTITUTION PROCESSES

Radical nucleophilic substitution reactions (S_{RN}) of both aliphatic and aromatic systems are now well documented, and many such reactions have been effected with selenium and tellurium nucleophiles. Rossi's group has been particularly active in this field[70,71]. A typical reaction sequence is illustrated in Scheme 11[72], in which the reactions, in liquid ammonia, are stimulated by ultraviolet light. Table 2 shows some representative results.

This is a procedure that is generally satisfactory for the preparation of unsymmetrical diaryl chalcogenides[73,74]; it has also been extended to the use of Se^{2-} and Te^{2-} in

$$\mathrm{ArHal} + \mathrm{PhSe}^- \xrightarrow[\mathrm{NH_3}]{h\nu} [\mathrm{ArHal}]^{-\bullet} + \mathrm{PhSe}^\bullet \qquad \text{(initiation)}$$

$$\left.\begin{aligned} [\mathrm{ArHal}]^{-\bullet} &\longrightarrow \mathrm{Ar}^\bullet + \mathrm{Hal}^- \\ \mathrm{Ar}^\bullet + \mathrm{PhSe}^- &\longrightarrow [\mathrm{PhSeAr}]^{-\bullet} \\ [\mathrm{PhSeAr}]^{-\bullet} + \mathrm{ArHal} &\longrightarrow \mathrm{PhSeAr} + [\mathrm{ArHal}]^{-\bullet} \\ \text{Overall: } \mathrm{ArHal} + \mathrm{PhSe}^- &\longrightarrow \mathrm{PhSeAr} + \mathrm{Hal}^- \end{aligned}\right\} \text{propagation}$$

SCHEME 11

TABLE 2. Yields (%) of diaryl chalcogenides (ArChPh; Ch = Se, Te) from the photostimulated reactions $ArX + PhCh^- \longrightarrow ArChPh + X^-$

	$PhCh^-$	
ArX	$PhSe^-$	$PhTe^-$
PhBr	0	0
PhI	–	90
1-Chloronaphthalene	73	37
4-Chlorobiphenyl	52	–
9-Bromophenanthrene	72	–

reactions which can be adjusted to produce either areneselenol, symmetrical diselenide or unsymmetrical diaryl or alkyl aryl selenides (or their tellurium analogues)[75]. A puzzling diversion, reported recently[76], is that a by-product from the $PhSe^-$/PhI reaction is benzene, and that the formation of this is enhanced by a high $PhSe^-$/PhI ratio or by high light intensity. It was suggested that the benzene arises by abstraction of hydrogen by phenyl radicals; however, it is not clear that this is consistent with the overall mechanism, and alternative possibilities should perhaps be considered.

Investigation of the relative reactivities of nucleophiles employed in these studies has provided clear evidence for the reversibility of the formation of $[ArChPh]^{-\bullet}$[77]. Produced from $Ar^\bullet$ and $PhCh^-$, this may cleave to $ArCh^-$ and $Ph^\bullet$. The phenyl radical then reacts to form symmetrical diphenyl chalcogenide. Whether or not significant reversal is encountered is determined by the reduction potentials of the aryl halide and the chalcogen product. Reversal has not been observed when Ch = S and is most important for Ch = Te.

A similar photostimulated displacement of iodide from trifluoromethyl iodide by $PhSe^-$ has also been reported[78].

V. α-SELENOALKYL RADICALS

Hydrogen abstraction by *tert*-butoxy radicals from dimethyl selenide is more facile than that from saturated alkanes[5], yet whilst in the latter reactions alkyl radicals are detectable by ESR the methylselenomethyl radicals are not. This has been attributed to strong electron interaction with the selenium lone pair, resulting in line broadening by both spin–rotation and spin–orbit mechanisms. In contrast, hindered selenoalkyl radicals formed by radical addition to the selenium atom of di-*tert*-butyl selenoketone are readily observable[5,79]. They are forced by steric effects to adopt conformations in which there is little interaction of the unpaired electron with the selenium electrons; at low temperatures they exist in equilibrium with diamagnetic dimers. The butoxyl adduct of the selenoketone, t-BuOSeC$^\bullet$(Bu-t)$_2$, decays unimolecularly by fission of the C—O bond to give initially t-Bu$^\bullet$ and the selone oxide. In an interesting extension of this work, Scaiano[80] found that di-*tert*-butyl selenoketone could be used to intercept triplet Norrish Type II biradicals generated by ultraviolet irradiation of aryl ketone (equation 10).

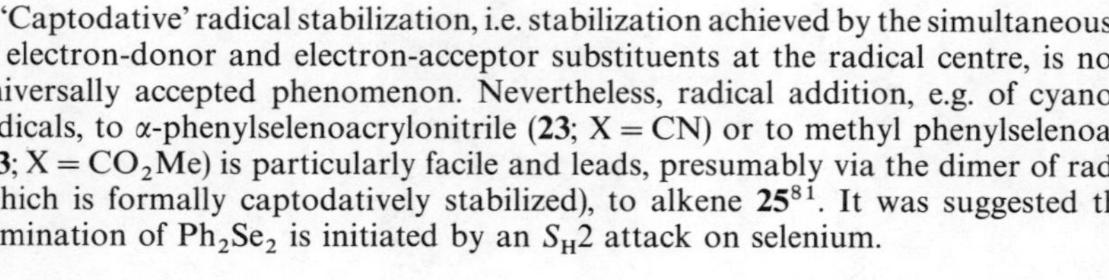

(10)

'Captodative' radical stabilization, i.e. stabilization achieved by the simultaneous action of electron-donor and electron-acceptor substituents at the radical centre, is not yet a universally accepted phenomenon. Nevertheless, radical addition, e.g. of cyanopropyl radicals, to α-phenylselenoacrylonitrile (**23**; X = CN) or to methyl phenylselenoacrylate (**23**; X = CO_2Me) is particularly facile and leads, presumably via the dimer of radical **24** (which is formally captodatively stabilized), to alkene **25**[81]. It was suggested that the elimination of Ph_2Se_2 is initiated by an S_H2 attack on selenium.

VI. OXYSELENIUM AND OXYTELLURIUM SPECIES

We have been unable to locate any organic chemistry to report in this section. Studies by ESR of, for example, $PhSeO^\bullet$ and $PhSeO_2^\bullet$ have been described, but not their organic

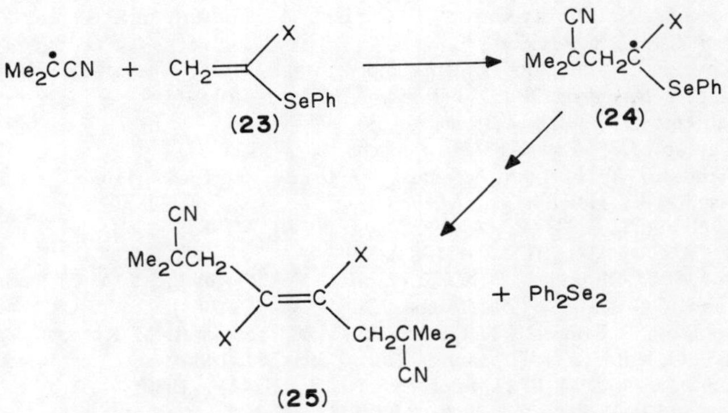

reactions. Benzeneselenenic anhydride has been used extensively as an oxidizing agent, but these reactions are generally written as two-electron processes. Such was the case for the interesting oxidation of *N*-phenylbenzohydroxamic acid[82], which gives, *inter alia*, the oxygenated product (**26**), but in unpublished work this system has been found to give intense ESR signals attributable to benzoyl phenyl nitroxide (**27**). The possibility of one-electron processes in the chemistry of this important oxidant may therefore merit further study.

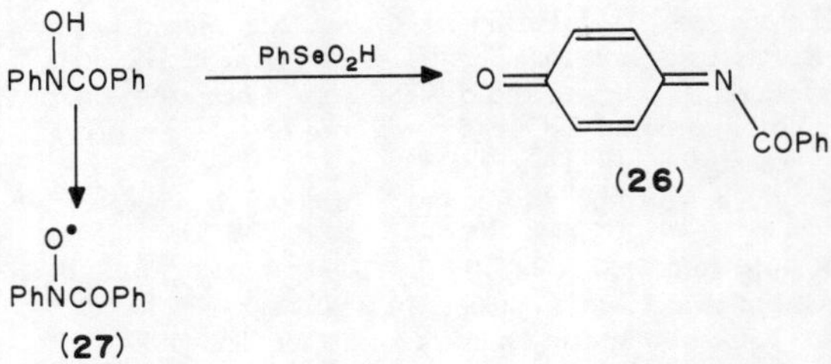

Finally, we are unaware of any selenium or tellurium analogues of the release of methyl radical from dimethyl sulphoxide by HO$^{\bullet}$.

VII. REFERENCES

1. H. J. Shine, in *Organic Selenium Compounds: Their Chemistry and Biology* (Eds. D. L. Klayman and W. H. H. Gunther), Wiley-Interscience, New York, 1973.
2. G. D. Sturgeon and M. L. Gross, in *The Chemistry of Organic Selenium and Tellurium Compounds* (Eds. S. Patai and Z. Rappoport), Vol. 1, Chapter 7, Section 1.A.
3. K. U. Ingold and B. P. Roberts, *Free Radical Substitution Reactions*, Wiley–Interscience, New York, 1971.
4. J. R. Brown, G. P. Gillman and M. H. George, *J. Polym. Sci., Part A*, **5**, 903 (1967).
5. J. C. Scaiano and K. U. Ingold, *J. Am. Chem. Soc.*, **99**, 2079 (1977).
6. M. J. Perkins, *Adv. Phys. Org. Chem.*, **17**, 1 (1980).
7. R. Franzi and M. Geoffroy, *J. Organomet. Chem.*, **218**, 321 (1981).
8. J. Y. C. Chu, D. G. Marsh and W. H. H. Gunther, *J. Am. Chem. Soc.*, **97**, 4905 (1975).
9. H. K. Spencer and M. P. Cava, *J. Org. Chem.*, **42**, 2937 (1977).
10. D. H. Brown, R. J. Cross and D. Millington, *J. Chem. Soc., Dalton Trans.*, 159 (1977).
11. R. J. Cross and D. Millington, *J. Chem. Soc., Chem. Commun.*, 455 (1975).
12. J. Y. C. Chu and D. G. Marsh, *J. Org. Chem.*, **41**, 204 (1976).
13. D. H. Brown, R. J. Cross and D. Millington, *J. Organomet. Chem.*, **125**, 219 (1977).

14. G. Vermeersch, N. Febvay-Garot, S. Caplain, A. Couture and A. Lablanche-Combier, *Tetrahedron Lett.*, **7**, 609 (1979).
15. D. L. J. Clive, P. C. Anderson, N. Moss and A. Singh, *J. Org. Chem.*, **47**, 1641 (1982).
16. G. Bergman, *Tetrahedron*, **28**, 3323 (1972).
17. J. Bergman and L. Engman, *Tetrahedron*, **36**, 1275 (1980).
18. J. Y. C. Chu and J. W. Lewicki, *J. Org. Chem.*, **42**, 2491 (1977).
19. M. K. Chaudhuri, A. Haas and A. Wensky, *J. Organomet. Chem.*, **116**, 323 (1976).
20. G. A. Russell and J. Hershberger, *J. Am. Chem. Soc.*, **102**, 7603 (1980).
21. G. A. Russell and L. L. Herold, *J. Org. Chem.*, **50**, 1037 (1985).
22. O. Ito, *J. Am. Chem. Soc.*, **105**, 850 (1983).
23. I. P. Romm, I. V. Olifirenko, E. N. Guryanova, V. V. Troitskii, V. A. Chernoplekova, L. M. Kataeva and E. G. Kataeva, *Zh. Obshch. Khim.*, **53**, 477 (1983).
24. I. V. Olifirenko, I. P. Romm, E. N. Guryanova, V. V. Troitskii, L. M. Kataeva, V. A. Chernoplekova and E. G. Kataeva, *J. Organomet. Chem.*, **265**, 272 (1984).
25. D. H. Wadsworth and M. R. Detty, *J. Org. Chem.*, **45**, 4611 (1980).
26. J. Gosselck and E. Wolters, *Chem. Ber.*, **95**, 1237 (1962).
27. G. I. Zaitseva and V. M. Al'bitskaya, *J. Org. Chem. USSR*, **4**, 1987 (1968).
28. J. V. Comasseto and J. T. B. Ferreira, *J. Organomet. Chem.*, **216**, 287 (1981).
29. N. S. Vyazankin, M. N. Bochkarev and L. P. Maiorova, *Zh. Obshch. Khim.*, **39**, 468 (1969).
30. W. H. H. Gunther, *J. Org. Chem.*, **31**, 1202 (1966).
31. M. J. Perkins, B. V. Smith, B. Terem and E. S. Turner, *J. Chem. Res.* (*S*), 341 (1979).
32. D. L. J. Clive, W. A. Kiel, S. M. Menchen and C. K. Wong, *J. Chem. Soc., Chem. Commun.*, 657 (1977).
33. D. L. J. Clive and S. M. Menchen, *J. Chem. Soc., Chem. Commun.*, 168 (1979).
34. M. Mikolajczyk, *Chem. Ind.* (*London*), 2059 (1966).
35. D. H. Hey, G. H. Jones and M. J. Perkins, *J. Chem. Soc., Chem. Commun.*, 1375 (1969).
36. L. Chieriu and R. Passerini, *Boll. Sci. Fac. Chim. Ind. Bologna*, **12**, 56 (1954).
37. F. G. James, M. J. Perkins, O. Porta and B. V. Smith, *J. Chem. Soc., Chem. Commun.*, 131 (1977).
38. K. Fujimori, H. Yoshimoto and S. Oae, *Tetrahedron Lett.*, 4397 (1979).
39. L. Engman, *Acc. Chem. Res.*, **18**, 274 (1985).
40. N. Kambe, K. Kondo, S. Morita, S. Murai and N. Sonoda, *Angew. Chem., Int. Ed. Engl.*, **19**, 1009 (1980).
41. N. Kambe, K. Kondo and N. Sonoda, *Angew. Chem., Int. Ed. Engl.*, **19**, 1009 (1980).
42. M. Akiba and M. P. Cava, *Synth. Commun.*, **14**, 1119 (1984).
43. N. Ohira, Y. Aso, T. Otsubo and F. Ogura, *Chem. Lett.*, 853 (1984).
44. S. Morimoto, *J. Chem. Soc. Jpn.*, **75**, 557 (1954).
45. H. A. Silverwood and M. Orchin, *J. Org. Chem.*, **27**, 3401 (1962).
46. E. J. Levi and M. Orchin, *J. Org. Chem.*, **31**, 4302 (1966).
47. S. A. Zavyalov, I. A. Myasnikov and E. E. Gutman, *Zh. Fiz. Khim.*, **53**, 1303 (1979).
48. M. J. Perkins, B. V. Smith and E. S. Turner, *J. Chem. Soc., Chem. Commun.*, 977 (1980).
49. J. R. M. Giles, B. P. Roberts, M. J. Perkins and E. S. Turner, *J. Chem. Soc., Chem. Commun.*, 504 (1980).
50. M. Geoffrey, *J. Chem. Phys.*, **70**, 1497 (1979).
51. R. Franzi, M. Geoffrey, L. Givet and V. Leray, *J. Phys. Chem.*, **83**, 2898 (1979).
52. D. L. J. Clive, G. Chittattu and C. K. Wong, *J. Chem. Soc., Chem. Commun.*, 41 (1978).
53. K. C. Nicolaou, R. L. Magolda, W. J. Spiro, W. E. Barnette, Z. Lysenko and M. M. Joullie, *J. Am. Chem. Soc.*, **102**, 3784 (1980).
54. K. C. Nicolaou, G. P. Gasic and W. E. Barnette, *Angew. Chem., Int. Ed. Engl.*, **17**, 293 (1978).
55. K. C. Nicolaou, S. P. Seitz, W. J. Spiro and J. F. Blount, *J. Am. Chem. Soc.*, **101**, 3884 (1979).
56. E. J. Corey, H. L. Pearce, I. Szekely and M. Ishiguro, *Tetrahedron Lett.*, 1023 (1978).
57. D. L. J. Clive, G. J. Chittattu, V. Farina, W. A. Kiel, S. M. Menchen, C. G. Russell, A. Singh, C. K. Wong and N. J. Curtis, *J. Am. Chem. Soc.*, **102**, 4438 (1980).
58. S. D. Burke, W. F. Fobare and D. M. Armistead, *J. Org. Chem.*, **47**, 3348 (1982).
59. G. E. Keck and J. B. Yates, *J. Am. Chem. Soc.*, **104**, 5829 (1982).
60. M. D. Bachi and C. Hoornaert, *Tetrahedron Lett.*, **22**, 2693 (1981).
61. D. L. J. Clive and P. L. Beaulien, *J. Chem. Soc., Chem. Commun.*, 307 (1983).
62. J. C. Scaiano, P. Schmid and K. U. Ingold, *J. Organomet. Chem.*, **121**, C4 (1976).
63. Y. Okamoto and T Yano, *J. Organomet. Chem.*, **29**, 99 (1971).

64. D. Grant and J. R. Van Wazer, *J. Am. Chem. Soc.*, **86**, 3012 (1964).
65. P. Bishop, *PhD Thesis*, London, 1985.
66. M. J. Perkins and E. S. Turner, *J. Chem. Soc., Chem. Commun.*, 139 (1981).
67. D. H. R. Barton, D. Bridon and S. Z. Zard, *Tetrahedron Lett.*, **25**, 5777 (1984).
68. G. A. Russell and H. Tashtoush, *J. Am. Chem. Soc.*, **105**, 1398 (1983).
69. J. Deniau, K. N. V. Duong, A. Gaudener, P. Bougeard and M. D. Johnson, *J. Chem. Soc., Perkin Trans. 2*, 393 (1981).
70. R. A. Rossi, *Acc. Chem. Res.*, **15**, 164 (1982).
71. R. A. Rossi, *J. Chem. Educ.*, **59**, 310 (1982).
72. A. B. Pierini and R. A. Rossi *J. Org. Chem.*, **44**, 4667 (1979).
73. A. B. Pierini and R. A. Rossi, *J. Organomet. Chem.*, **144**, C12 (1978).
74. A. B. Pierini and R. A. Rossi, *J. Organomet. Chem.*, **168**, 163 (1979).
75. R. A. Rossi and A. B. Penenory, *J. Org. Chem.*, **46**, 4580 (1981).
76. A. B. Penenory, A. B. Pierini and R. A. Rossi, *J. Org. Chem.*, **49**, 3834 (1984).
77. A. B. Pierini, A. B. Penenory and R. A. Rossi, *J. Org. Chem.*, **49**, 486 (1984).
78. V. G. Voloshocnuk, V. N. Bioko and L. M. Yagupolskii, *Zh. Org. Khim.*, **13**, 2008 (1977).
79. J. C. Scaiano and K. U. Ingold, *J. Chem. Soc., Chem. Commun.*, 205 (1976).
80. J. C. Scaiano, *J. Am. Chem. Soc.*, **99**, 1494 (1977).
81. Z. Janousek, S. Piettre, F. Gorissen-Hervens and H. G. Viehe, *J. Organomet. Chem.*, **250**, 197 (1983).
82. T. Frejd and K. B. Sharpless, *Tetrahedron Lett.*, 2239 (1978).

The Chemistry of Organic Selenium and Tellurium Compounds Volume 2
Edited by S. Patai
© 1987 John Wiley & Sons Ltd.

CHAPTER **17**

Synthesis of selenium and tellurium ylides and carbanions: application to organic synthesis*

A. KRIEF

Universitaires Notre-Dame Facultés de la Paix, Namur, Belgium

I. α-SELENOORGANOMETALLICS: SYNTHESIS AND SYNTHETIC USEFULNESS . . . 677
A. General Aspects . . . 677
1. Reactivity of selenides . . . 677
2. Synthesis of selenides . . . 681
B. Synthesis and Reactivity of α-Selenoorganometallics . . . 682
1. Synthesis of α-selenoalkylmetals in which the carbanionic centre is sp - hybridized and the selenyl moiety is uncharged . . . 682
a. General . . . 682
b. Theoretical and practical aspects . . . 682
i. Synthesis of α-selenoalkylmetals by metalation . . . 682
ii. Synthesis of α-selenoalkyllithiums by cleavage of the C—Se bond of selenoacetals . . . 686
iii. Synthesis of α-selenoalkyllithiums via halogen–metal exchange . 693
iv. Synthesis of α-selenoalkyllithiums by addition of alkyllithiums to phenyl vinyl selenides . . . 695
2. Reactivity of α-selenoalkylmetals in which the carbanionic centre is sp^3-hybridized and the selenyl moiety is uncharged . . . 695
a. Reactivity of α-selenoalkylmetals bearing hydrogens or one or two alkyl groups on the carbanionic centre . . . 695
b. Reactivity of α-functionalized α-selenoalkylmetals. . . . 695
c. Further comments on the reactivity of α-selenoalkylmetals . . . 696
i. Reaction with alkyl and allyl halides, epoxides and oxetanes. . 696
ii. Reaction with carbonyl compounds . . . 696
iii. Reaction with enals, enones and enoates . . . 700

*This contribution involves the work published prior to 1985.

3. Synthesis and reactivity of sp^3-hybridized α-selenoalkylmetals bearing a charged selenyl moiety . . . 703
a. Synthesis of α-metalloselenoxides . . . 705
b. Reactivity of α-metalloselenoxides . . . 705
c. Synthesis and reactivity of selenonium ylides . . . 705
4. Synthesis and reactivity of 1-metallo-1-selenoalkenes . . . 706
a. Synthesis . . . 706
i. By metalation of vinyl selenides . . . 706
ii. By cleavage of the C—Se bond of 1,1-bis(seleno)alkenes . . . 708
b. Reactivity . . . 709
5. Synthesis of allyllithiums, benzyllithiums, α-thioalkyllithiums and α-silylalkyllithiums from the corresponding selenides by selenium–metal exchange . . . 709
a. Synthesis of allyl- and benzyllithiums . . . 709
b. Synthesis of α-thioalkyllithiums . . . 710
c. Synthesis of α-silylalkyllithiums . . . 710
d. Miscellaneous syntheses of organometallics by selenium–metal exchange . . . 711
C. Reactions of Selenides and Functionalized Selenides . . . 711
1. Reactions involving the reduction of the C—Se bond . . . 711
2. Reactions involving the transformation of the C—Se bond to a C—halogen bond . . . 712
3. Selenoxide synthesis and reactivity . . . 713
a. Synthesis . . . 713
b. Reactions involving the selenoxide elimination . . . 715
c. Regiochemistry of the selenoxide elimination . . . 715
d. Structure–reactivity relationships in the selenoxide elimination . . . 715
e. α-Silyl selenoxides . . . 716
f. Allylic, allenic and propargylic selenoxides . . . 716
g. β-Alkyloxyselenoxides . . . 716
h. Vinyl selenoxides . . . 716
i. Reduction of selenoxides to selenides . . . 716
4. Synthesis and reactivity of selenonium salts . . . 718
a. Synthesis . . . 718
b. Reactivity . . . 719
5. Synthesis using functionalized selenides . . . 720
a. Synthesis of olefins from β-hydroxyselenides . . . 720
b. Synthesis of ketones from β-hydroxyselenides . . . 723
D. Use of Selenoalkylmetals in Organic Synthesis . . . 726
1. Alkanes or deuteriated alkanes . . . 726
a. From aldehydes and ketones . . . 726
b. From alkyl halides . . . 730
2. Non-functionalized olefins . . . 730
a. From aldehydes and ketones . . . 730
b. From alkyl halides . . . 730
c. From alkyl halides via alkyllithiums . . . 737
3. Functionalized olefins . . . 737
a. α-Heterosubstituted olefins . . . 737
b. 1,3-Dienes . . . 739
c. Allyl alcohols . . . 739
d. α,β-Unsaturated aldehydes and ketones . . . 739
e. Ene-4-ones . . . 744
f. α,β-Unsaturated esters, silyl esters and acids . . . 744

g. α-Alkilydene lactones 744
h. γ-Functionalized olefins such as homoallyl alcohols 744
4. Non-functionalized alcohols 744
5. Functionalized alcohols. 744
6. Heterocycles. 746
a. Epoxides 746
b. Oxetanes 746
c. Tetrahydrofurans 746
7. Alkyl halides 746
8. Non-functionalized aldehydes and ketones 747
a. Synthesis from carbonyl compounds 747
b. Synthesis from alkyl halides 749
c. Synthesis from alkylmetals 752
9. Functionalized aldehydes and ketones. 752
a. α-Hydroxy-carbonyl compounds 752
b. 1,4-Diketones 752
10. Carboxylic acids and esters 752
II. α-TELLUROORGANOMETALLICS: SYNTHESIS AND SYNTHETIC USEFULNESS 753
A. Synthesis of α-Telluroorganometallics 753
B. Reactivity of α-Telluroorganometallics 754
C. Reactivity of Tellurides and Functionalized Tellurides 754
III. REFERENCES 757

I. α-SELENOORGANOMETALLICS: SYNTHESIS AND SYNTHETIC USEFULNESS

A. General Aspects

When present in organic molecules, the Se atom confers on them unique properties[1–8]. These properties result from the location of Se in the Periodic Table in the neighbourhood of C and Br, two atoms which play an important role in organic chemistry.

The C—Se bond (243 kJ mol^{-1}) is weaker than the C—S bond (272 kJ mol^{-1}) and the C—Br bond (285 kJ mol^{-1}). On the other hand the electronegativity* of Se (2.55) is very close to that of C (2.55) (S 2.58, Br 2.96).

1. Reactivity of selenides

Selenides are usually stable compounds. They are however transformed selectively to several other functional groups by use of the proper reagents.

The high nucleophilicity and oxidizability of the Se atom plays an important role in the reactivity of selenides and functionalized selenides. These properties are usually not observed with alkyl bromides and are much more pronounced than those found in sulphides[7,9], two types of molecules possessing a heteroatom located in the Periodic Table of elements in the neighbourhood of the Se atom. The electrophilicity of the Se atom in selenides allows valuable synthetic transformations. The same type of electrophilicity is often observed in the case of alkyl bromides (i.e. halogen–metal exchange) but is unusual for sulphides[8].

The electrophilic properties of the Se atom in selenides is probably due to the presence of low-energy, unoccupied 4d orbitals. These are responsible for the unusual reactivity of alkyllithiums which cleave the C—Se bond of selenides rather than metalate them[8,10–12]

*These values refer to Pauling's method [A. L. Albred, *J. Inorg. Nucl. Chem.*, **17**, 215 (1961)]. Other methods give C 250, Se 248 [Albred–Rochow values; *J. Inorg. Nucl. Chem.*, **5**, 264 (1958)] and C 263, Se 233 [Mulliken-type values H. O. Pritchard and H. A. Skinner, Journal, **55**, 745 (1955)].

$$PhSeMe + BuLi \rightarrow [PhLi] + BuSeMe$$

n-BuLi: ether, reflux/19 h	30%	(as PhCOOH after carbonation)[10]
n-BuLi: THF/hexane, 20°C/0.5 h	93%[a]	(as Ph_2CHOH after reaction with PhCH=O)[11]
n-BuLi: THF/hexane, − 78°C/1.3 h	20%[a]	(as Ph_2CHOH after reaction with PhCH=O)[12]
n-BuLi: THF/hexane, − 50°C/1.75 h	89%	(as Ph_2CHOH after reaction with PhCH=O)[12]
s-BuLi: ether/hexane, − 50°C/1.5 h	57%[b]	(as Ph_2CHOH after reaction with PhCH=O)[12]
t-BuLi: ether/hexane, − 78°C/1.5 h	81%[a]	(as Ph_2CHOH after reaction with PhCH=O)[12]
n-BuLi: ether/hexane, − 50°C/1.5 h	10%[c]	(as Ph_2CHOH after reaction with PhCH=O)[12]

[a] These reactions have also been followed by ^{77}Se-NMR: the time given is the minimum required for complete reaction at that temperature.
[b] 41% recovery of the starting selenide.
[c] 88% recovery of the starting selenide.

SCHEME 1

(Scheme 1) as observed in sulphides[8,10]. Since metalloamides are not strong enough to metalate dialkyl or phenylalkyl selenides (α-seleno carbanions are less stabilized than their thio analogues)[13], these are not valuable starting materials for the synthesis of α-selenoalkyllithiums[8]. The Se–metal exchange is less easy than the Br–metal exchange but several orders of magnitude faster than the S–metal exchange in sulphides[8].

Selenides can be reduced to alkanes with Raney nickel[14], lithium in ethylamine[14] or tin hydrides[15,16,221] more efficiently than sulphides[16] (Scheme 2).

$$C_{11}H_{23}CH(SeR)Me \rightarrow C_{11}H_{23}CH_2Me \qquad (a)$$

R = Ph, Me: Raney Ni/ethanol, 80°C/24 h 65%[14]
R = Ph, Me: $Li/EtNH_2$, − 10°C/2 h 80%[14]

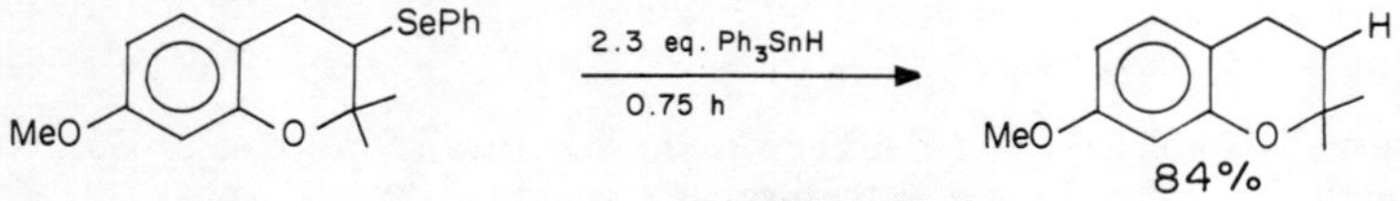

(Ref. 15) (b)

SCHEME 2

$Ph_2Se + MeI \xrightarrow[20\,°C/2\,h]{AgBF_4} Ph_2\overset{+}{Se}Me\ BF_4^-$

→ (PhCH=O; *t*-BuOK, 20°C/2 h) → $PhCH\text{—}CH_2$ (O, epoxide) $+ Ph_2Se$ 90% (Ref. 19) (a)

→ (1. BuLi/THF, − 78°C; 2. PhCH=O) → $Ph_2CHOH + Ph_2CHOMe$ 40% 32% (Ref. 11) (b)

SCHEME 3

They are very good complexing agents for metal ions[3] and powerful nucleophiles. Selenonium salts are therefore readily available on reaction with halogens[17,18] or alkyl halides[3,17,19], dialkyl sulphates[19] or alkyl fluorosulphonates[20] (Schemes 3 and 4).

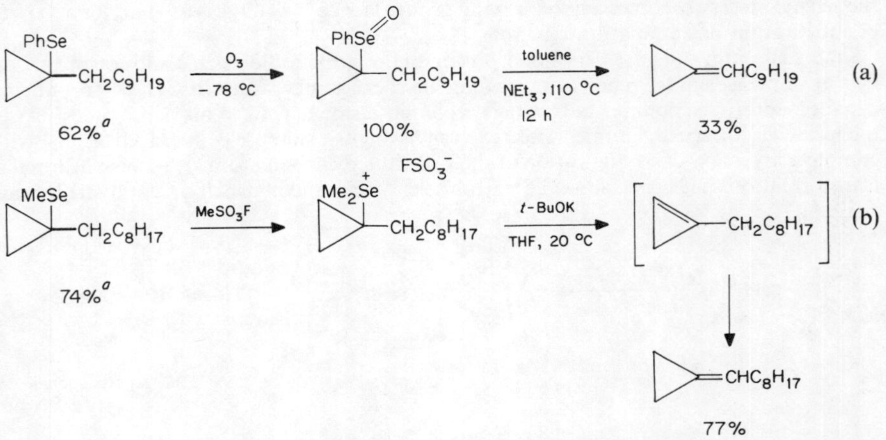

[a] This yield refers to the synthesis of the compound by formation of the bond shown by a heavy line.

SCHEME 4[20]

Selenides are easily oxidized to selenoxides[3,6,8] under a large variety of conditions (see Section i.e. 3a and Schemes 4a and 7). The overoxidation to selenone has been described as being particularly difficult[5,6]. It was however recently shown in the author's laboratory that under suitable conditions (*m*-chloroperbenzoic acid or $KMnO_4$, 20 °C) selenones can be obtained in good yield*.

Selenonium salts and selenoxides are more prone to metalation than selenides. *t*-BuOK in DMSO and KOH in water under phase-transfer catalysis are strong enough to allow the hydrogen–metal exchange in selenonium salts[11,19,20] (Scheme 3a). In the case of selenoxides metalation has been achieved[21,22] with LDA in THF even at − 78 °C (Scheme 5). Again alkyllithiums do not act as bases and react on the Se atom of these derivatives leading, for example, in the case of *n*-BuLi and diphenylmethylselenonium fluoroborate to an exchange of ligand on the Se atom producing phenyllithium[11]. This is further trapped

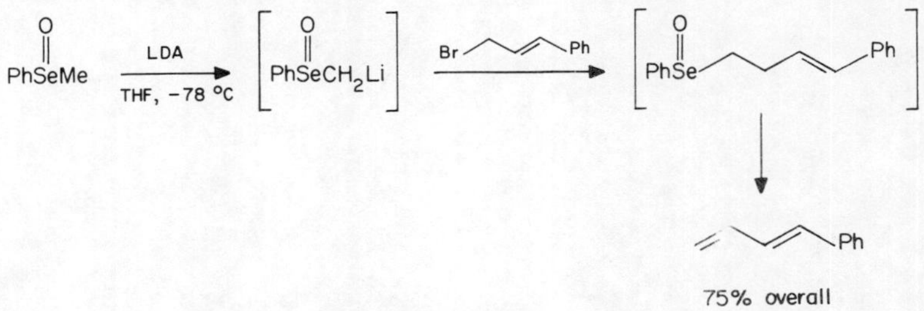

SCHEME 5[22]

*A. Krief and J. N. Denis, unpublished results. It was in several instances noticed[3–6,8] that selenides are much more easily oxidized to selenoxides than sulphides are to sulphoxides, but that the reverse was found in further oxidation to selenone and sulphone.

by addition of benzaldehyde to the medium and results in the formation of diphenylcarbinol (Scheme 3b).

Compounds possessing a positively charged Se atom are valuable derivatives in which the selenyl moiety becomes a good leaving group. Hence they have been used for a variety of substitution or elimination reactions:

(*a*) Selenonium salts are very good precursors of alkyl iodides, chlorides or bromides. For example, selenides react with bromine at room temperature in CH_2Cl_2 producing instantaneously selenonium salts which decompose slowly (2 h) to alkyl bromides[17,18] in the presence of triethylamine. The reaction is highly stereoselective and occurs with complete inversion of configuration at the substituted carbon atom[18] (Scheme 6). Similar transformations have been achieved[23] from selenoxides and hydrochloric or hydrobromic acids.

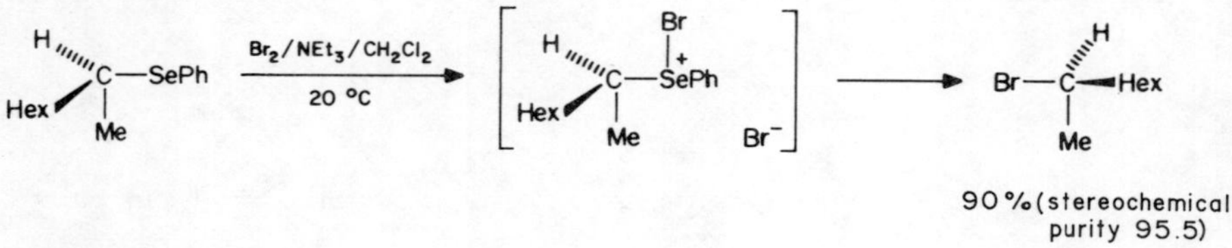

SCHEME 6[18]

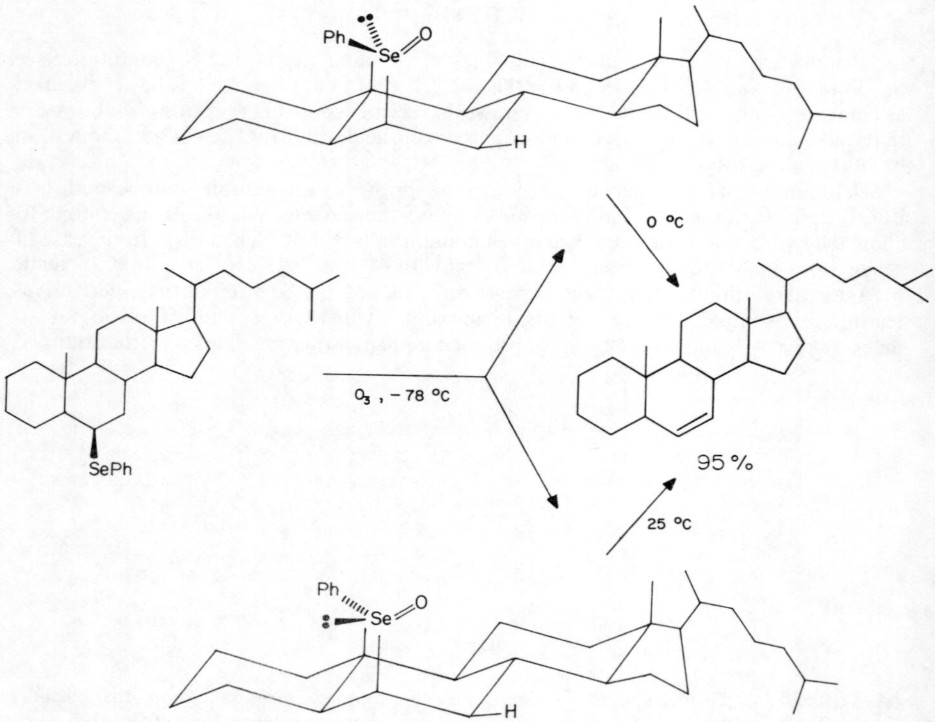

SCHEME 7[26]

(*b*) Elimination reactions leading to olefins have also been specifically achieved from selenoxides[2,4–6,8,27,28] and selenonium salts[8,20,24,25] (Schemes 4, 5, 7 and 8). Since the discovery, by Jones and coworkers[26] (Scheme 7), and development of the method simultaneously by Sharpless[2,27], Reich[5,6] and Clive[4], the selenoxide elimination leading to olefins has been frequently used for the total synthesis of natural products[28] (Scheme 8). This reaction has contributed much to the development of the chemistry of organoselenium compounds. Selenoxides decompose thermally to produce olefins in high yield. The reaction is analogous to the aminoxide and sulphoxide elimination reactions[29], but whereas these occur respectively at 260° and 160 °C, the selenoxide elimination takes place even at 20 °C. The stereochemistry of the reaction has been determined and the elimination of selenenic acid is found to occur in a *syn* fashion[27] (Scheme 8).

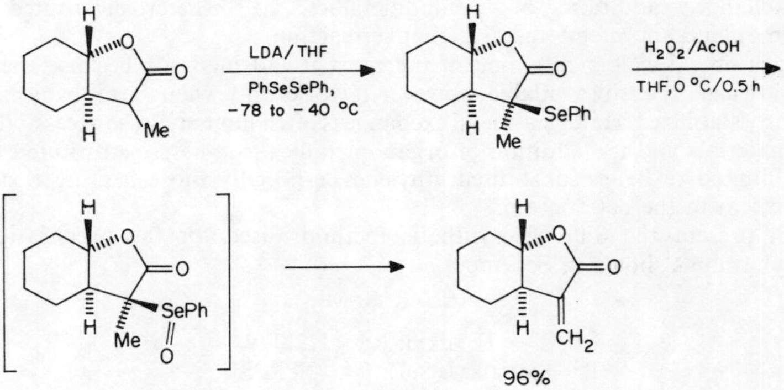

SCHEME 8[28]

(*c*) Olefins can be prepared[20] by reaction of selenonium salts with bases. The transformation probably involves the intermediary formation of an ylide which is thought to produce the olefin via a process related to selenoxide elimination (compare a and b, Scheme 4).

2. *Synthesis of selenides*

Selenides and functionalized selenides can be easily prepared from selenols or selenolates and compounds containing an electrophilic carbon atom. The high nucleophilicity of the Se atom in these derivatives and the high acidity of the selenols make them versatile compounds for the synthesis of Se-containing molecules[1–6,8].

The selenyl moiety can also be introduced as an electrophilic species into molecules possessing a nucleophilic carbon atom. Diselenides, selenenyl halides, selenenic acids and esters and seleninic halides and anhydrides are now the choice reagents[2,4–6]. These methods are very useful but they usually do not allow the introduction of the selenyl moiety with the concomitant formation of C—C bonds between two different entities. However, the reaction of α-selenoalkylmetals with electrophiles fulfills this requirement[2,4–68]. We shall discuss below:

(*a*) the synthesis of α-selenoalkylmetals,
(*b*) their reactivity towards electrophiles,
(*c*) some transformations which allow the selective removal of the selenyl moiety on the products prepared, and
(*d*) the description of functional group transformations using α-selenoalkylmetals.

B. Synthesis and Reactivity of α-Selenoorganometallics

1. Synthesis of α-selenoalkylmetals in which the carbanionic centre is sp^3-hybridized and the selenyl moiety is uncharged

a. General. α-Selenoalkylmetals belong to the well-known family of α-heterosubstituted organometallics[30–66]. Those bearing nitrogen, phosphorus or sulphur moieties (charged or not), alkoxy or silyl groups and halogens have been prepared during the last thirty years[8].

There are several synthetic routes to such organometallics which have all their advantages and inconveniences[8]. These include hydrogen–metal exchange, halogen–metal exchange, addition of organometallics to α-heterosubstituted olefins heteroatom–metal and metal–metal exchange reactions.

Hydrogen–metal exchange is one of the most useful methods because the starting materials are usually easily available; however, it works only when the carbanionic species is sufficiently stabilized. Halogen–metal exchange is often limited by the accessibility of the starting materials and the addition of organometallics to α-heterosubstituted olefins is generally limited to heterosubstituted ethylenes, especially those bearing a stabilizing group adjacent to the heteroatom.

We will present the different synthetic methods used for the preparation of α-selenoalkyllithiums shown in Scheme 9.

$$RSeCR^1R^2M$$

$$R^1 = H, \text{alkyl};\ R^2 = H, \text{alkyl}$$
$$R^1 = H, \text{alkyl}, SeR;\ R^2 = SeR, SiR'_3$$
$$R^1 = H, \text{alkyl};\ R^2 = RCO, CO_2R, CONR_2,$$

CN, Ph, Ar, 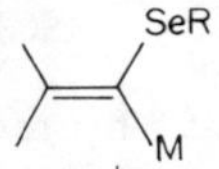

$RS(=O)CR^1R^2M$ $\quad R_2\overset{+}{Se}CR^1R^2M \; X^-$

SCHEME 9

b. Theoretical and practical aspects. The strategy used for the synthesis of α-selenoalkylmetals greatly depends upon the substituents present around the carbanionic centre expected to be formed.

(i) Synthesis of α-selenoalkylmetals by metalation. Dialkyl or aryl alkyl selenides cannot usually be metalated since metalloamides are not strong enough to perform the reaction and alkylmetals usually react on the Se atom of these compounds (Scheme 1).

It has been shown however[67] that (phenylselenomethyl) lithium can be prepared, although in modest yield, by metalation of methyl phenyl selenide with *n*-BuLi when tetramethylethylene diamine (TMEDA) is used as the cosolvent (Scheme 10).

$$PhSeCH_3 \xrightarrow[> 0^\circ C]{BuLi/TMEDA/THF} \underset{38\%}{PhSeCH_2Li}$$

SCHEME 10[67]

The presence of a substituent able to produce extra stabilization at the expected carbanionic centre often permits the synthesis of the α-selenoalkylmetals from the corresponding selenides and metalloamides:

(*a*) Aryl selenides bearing another heteroatomic moiety in the α-position, such as phenylselenoacetals[67–72] (Scheme 11), α-thiophenyl selenides[73] (Scheme 12), α-silylphenylselenomethane[22,72,74–76] (Scheme 13), α-methoxyarylselenomethane[72] as well as α-phosphono[101], α-phosphonatophenyl selenides[102], α-nitrophenyl selenides[103], α-stannylphenyl selenides[34] and α-germylphenyl selenides[300] have been successfully metalated. There is in fact a big difference of reactivity between the parent compounds which are usually metalated with lithium diisobutyramide (LiDBA) in THF[67] or lithium diisopropylamide (LDA) in THF[68,75,76], and the higher homologues which require stronger basic systems such as lithium tetramethylpiperidide (LiTMP) in THF–HMPT[68] or diglyme[73] or potassium diisopropylamide (KDA) in THF[69] (Schemes 11–13). Methylseleno derivatives of these types are however completely inert under all these conditions.

$$(RSe)_2CHR^1 \xrightarrow{\text{base}} (RSe)_2CMR^1 \xrightarrow{\text{HexBr}} (RSe)_2C(Hex)R^1$$

$R = Ph, R^1 = H$: $(i\text{-}Bu)_2NLi$/THF–hexane, – 78°C then 30°C/0.2 h	80%[67,68]
$R = Ph, R^1 = Me$: LDA/THF–HMPT, – 30°C then 30°C/0.2 h	74%[68]
$R = Ph, R^1 = Me$: LiTMP/THF/HMPT, – 30°C then 30°C/0.2 h	83%[68]
$R = Ph, R^1 = Hex$: LDA/THF/HMPT, – 30°C then 30°C/0.2 h	44%[68]
$R = Ph, R^1 = Hex$: LiTMP/THF/HMPT, – 30°C then 30°C/0.2 h	86%[68]
$R = Ph, R^1 = Me$: KDA/THF, – 78°C	95%[69]
$R = Me, R^1 = H$: LiTMP/THF/HMPT, – 30°C then 30°C/0.2 h	0%[68]
$R = m\text{-}CF_3C_6H_4, R^1 = H$: LDA/THF, – 78°C/0.1 h	85%[a] [70]

[a] This yield refers to the β-hydroxyselenoacetal derived from benzaldehyde.

SCHEME 11

$$\underset{(\mathbf{A})}{RSeCH(SR^2)R^1} \xrightarrow{\text{base}} RSeCLi(SR^2)R^1 \xrightarrow{\text{PhCHO}} RSeC(SR^2)R^1CH(OH)Ph$$

$R = R^2 = Ph, R^1 = H$: LiTMP/THF, – 24°C/1.5 h	91%	
$R = R^2 = Ph, R^1 = Me$: LiTMP/THF, – 24°C/1.5 h	68%	(**A**: 10%)
$R = R^2 = Ph, R^1 = Pen$: LiTMP/THF, – 24°C/1.5 h	0%	(**A**: 85%)
$R = R^2 = Ph, R^1 = Pen$: LiTMP/DME, – 24°C/0.5 h	21%	(**A**: 64%)
$R = R^2 = Ph, R^1 = Pen$: LiTMP/diglyme, – 24°C/0.5 h	65%	(**A**: 17%)
$R = Ph, R^2 = Me, R^1 = H$: LiTMP/THF, – 78°C/1 h	0%	(**A**: 80%)
$R = Ph, R^2 = Me, R^1 = H$: LiTMP/DME, – 78°C/1 h	45%	—
$R = Ph, R^2 = Me, R^1 = H$: LiTMP/diglyme, – 78°C/1 h	75%	—

SCHEME 12[73]

$$RSeCH_2SiMe_3 \xrightarrow[\text{THF/hexane}]{\text{LDA}} [RSeCHLiSiMe_3] \xrightarrow{C_{10}H_{21}Br} RSeCH(C_{10}H_{21})SiMe_3 \xrightarrow{H_2O_2} C_{10}H_{21}CHO$$

R = Ph	– 78°C/3 h	– 78°C	5%[74]
	– 30°C/0.5 h	– 30°C	75%[74,75]
R = Me	– 30°C/0.5 h	– 30°C	0%[74]

SCHEME 13

(*b*) Selenides bearing two other heteroatomic moieties are more acidic than those presented above and have been in both the phenylseleno[67] and the methylseleno[68,74] series readily metalated with LiDBA[67,71] or LDA[68,74] in THF (Scheme 14). This is for example the case for trisphenylseleno-[67] and trismethylseleno-[68,74] methane as well as α-silyldiselenomethane[71].

$$(PhSe)_3CH \xrightarrow{\text{LiDBA/THF, }-78\,^\circ\text{C}} (PhSe)_3CLi \xrightarrow{\text{MeI}} \underset{93\%}{(PhSe)_3CMe} \qquad \text{(Ref. 67)} \quad \text{(a)}$$

$$(MeSe)_3CH \xrightarrow{\text{LDA/THF, }-78\,^\circ\text{C}} (MeSe)_3CLi \xrightarrow{\text{MeI}} \underset{80\%}{(MeSe)_3CMe} \qquad \text{(Ref. 68)} \quad \text{(b)}$$

$$(PhSe)_2CHSiMe_3 \rightarrow [(PhSe)_2CLiSiMe_3]$$

$$\xrightarrow{\text{EtCHO}} [(PhSe)_2C(SiMe_3)\underset{\displaystyle O^-}{\underset{|}{C}}HEt] \rightarrow \underset{70\%}{(PhSe)_2C{=}CHEt} \qquad \text{(Ref. 71)} \quad \text{(c)}$$

SCHEME 14

(*c*) Primary allyl[72,77–79,321,326], benzyl[22,72,78,80,83] and ethynyl[72,81] aryl selenides as well as their methylseleno analogues[78,83] are readily metalated at −78 °C with LDA and LiTMP/THF, respectively (Scheme 15), but the reaction seems to be restricted to the parent compounds and does not work in the case of higher homologues[22,77].

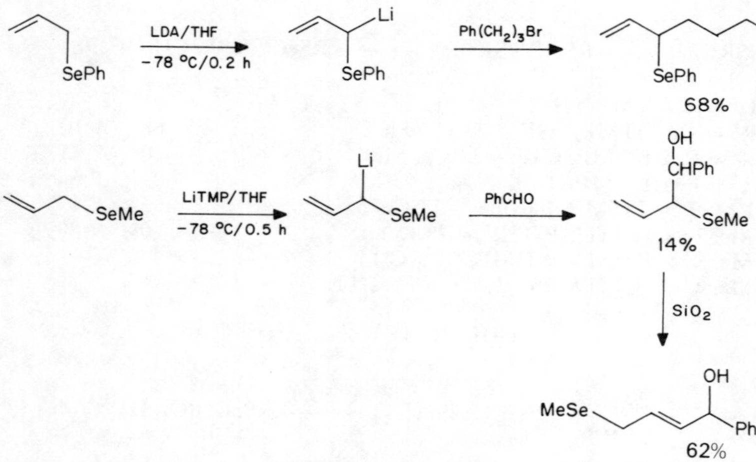

(Ref. 77) (a)

(Ref. 78) (b)

$$PhCH_2SePh \xrightarrow{\text{LDA/THF}} PhC(SePh)HLi \xrightarrow{\text{EtBr}} PhC(SePh)HEt$$

80% (Ref. 80) (c)

$$PhCH_2SeMe \xrightarrow[-78\,^\circ\text{C}/0.5\text{h}]{\text{LiTMP/THF}} PhC(SeMe)HLi \xrightarrow{\text{PhCHO}} PhC(SeMe)HC(OH)HPh$$

68% (Ref. 78) (d)

$$\mathrm{PhCHRSePh} \not\xrightarrow{\mathrm{LDA/THF}} \mathrm{PhCRLiSePh} \qquad \text{(Ref. 72)} \quad \text{(e)}$$

$R = Me, SiMe_3$

SCHEME 15

(*d*) Finally, α-seleno ketones[72,84–86,256b], esters[72,87–89], acids[72,90], lactones[88,91] and nitriles[92] produce the corresponding lithio derivatives on reaction with LDA in THF (Scheme 16 and 17). α-Phenylselenoacetonitrile has also been successfully metalated under phase-transfer catalysis and produces[93] 1-phenylseleno-1-cyanocyclopropane if the reaction is conducted in the presence of 1,2-dibromopropane.

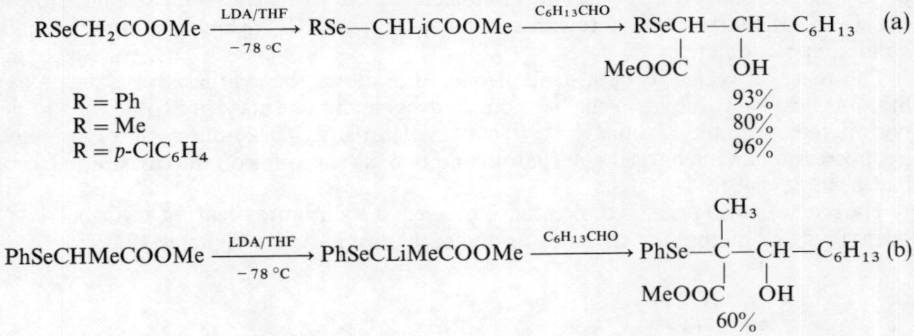

SCHEME 16[88]

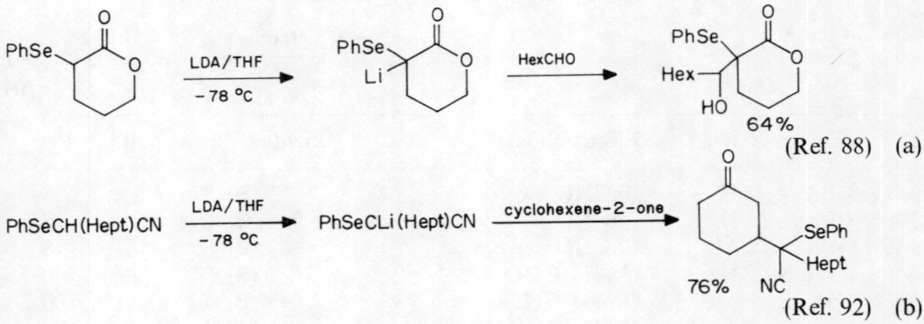

SCHEME 17

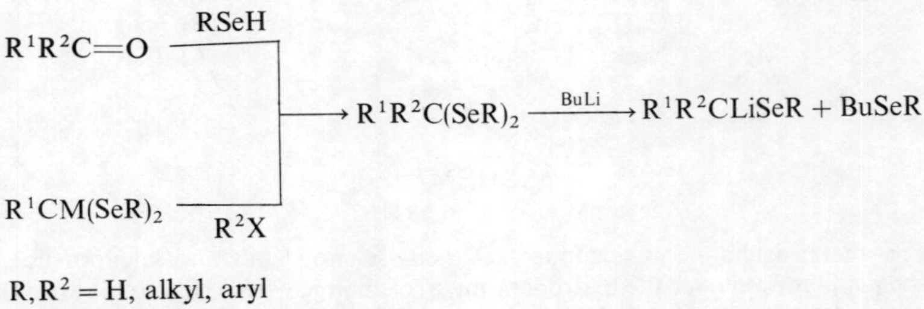

$R, R^2 = H$, alkyl, aryl
$R =$ alkyl, aryl

SCHEME 18

(ii) Synthesis of α-selenoalkyllithiums by cleavage of the C—Se bond of selenoacetals. Except in the case of (phenylseleno) methyllithium, the general route to α-selenoalkyllithiums is the reaction of selenoacetals with alkyllithiums[2,4–6,8,11,67,94,126,127,133] (Scheme 18). Most of the work has been carried out with phenyl- and methyl-selenoacetals themselves, readily available from carbonyl compounds[11,19,67,94–99,126b,127] or by alkylation of 1-metallo-1, 1-bisselenoalkanes[67–69,74,133].

The Se–metal exchange in selenoacetals is usually performed at − 78 °C with *n*-BuLi in THF/hexane (1:1) and leads to the quantitative formation of the selenoalkyllithiums and aryl or alkyl butyl selenides. Under these conditions the selenoalkyllithiums are stable for prolonged periods (~1 day), they are completely unreactive towards the selenides formed, but are particularly reactive towards electrophiles such as carbonyl compounds, alkyl halides, epoxides, etc.

The reaction seems to be a general one, and allows the synthesis of α-selenoalkyllithiums bearing two hydrogens[67,94], one hydrogen and one alkyl group[11,94,126] or even two alkyl or cycloalkyl groups[11,94,126,127,133]. Before this work none of the organometallics possessing an uncharged heteroatom and two alkyl groups on the carbanionic centre had been described.

The reaction also permits the synthesis of selenoalkyllithiums bearing aryl[78], silyl[74,100], selenyl[67,68] or methoxy[72] moieties on the carbanionic centre (Scheme 19).

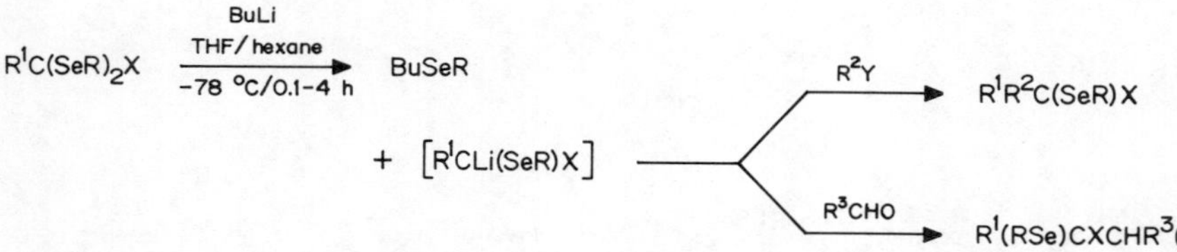

(A)

(B)

R	X	R¹	Electrophile	Yield (%) in **A** or **B**	Ref.
Me	Ph	H	PhCHO	(**B**) 87	78
Ph	Ph	H	PhCHO	(**B**) 83	78
Me	Ph	Me	PhCHO	(**B**) 85	78
Me	Me_3Si	H	$C_{10}H_{21}CHO$	(**B**) 54	100
Me	Me_3Si	Me	$C_{10}H_{21}CHO$	(**B**) 50	100
			$C_{10}H_{21}Br$	(**A**) 95	74
Me	MeSe	H	Me_3SiCl	(**A**) 72	68
Me	MeSe	Me	Me_3SiCl	(**A**) 83	68
			$C_6H_{13}CHO$	(**B**) 65	68
Ph	PhSe	Me	MeI	(**A**) 72	67
Ph	PhSe	PhSe	H_2O	(**A**) —	67

SCHEME 19

The Se–metal exchange in selenoacetals, α-silylselenoacetals and selenoorthoesters is dramatically favoured over the hydrogen–metal exchange when the latter is formally also possible, and only in very special cases such as trisphenylselenoorthoformate does competitive metalation occur[67] (Scheme 20).

$$(PhSe)_3CH \xrightarrow[n\text{-BuSePh}]{\text{BuLi/THF/hexane}} (PhSe)_2CHLi + (PhSe)_3CLi$$

SCHEME 20[67]

α-Selenoalkyllithiums are stable at − 78 °C in THF/hexane and isomerization to the ring-metalated derivatives or to the more stable selenomethyllithium is usually not observed under these conditions but can occur at higher temperatures[104] (Scheme 21). This isomerization has been reported once in a very special case[105] (Scheme 22).

$$R^1R^2CLi(SePh) \xrightarrow{\text{ether or THF, } -78^\circ C} \!\!\!\!\!\!\!\!\times\;\; R^1R^2CH(O\text{-}C_6H_4Li) \qquad \text{(Refs. 8, 78)} \quad \text{(a)}$$

$$R^1R^2CLi(SeCH_3) \xrightarrow{\text{ether or THF, } -78^\circ C} \!\!\!\!\!\!\!\!\times\;\; R^1R^2CH(SeCH_2Li) \qquad \text{(Refs. 8, 78)} \quad \text{(b)}$$

$$BuCH_2CH(SePh)_2 \xrightarrow[-78^\circ C]{t\text{-BuLi/ether}} BuCH_2CHLi(SePh) \xrightarrow[\text{2. heptanal}]{\text{1. } 20^\circ C/1\,h} \underset{16\%}{HexSePh}$$

$$+ \underset{26\%}{HexSeC_6H_4CH(OH)Hept} + \underset{16\%}{BuCH{=}CHSePh} \qquad \text{(Ref. 106b)} \quad \text{(c)}$$

SCHEME 21

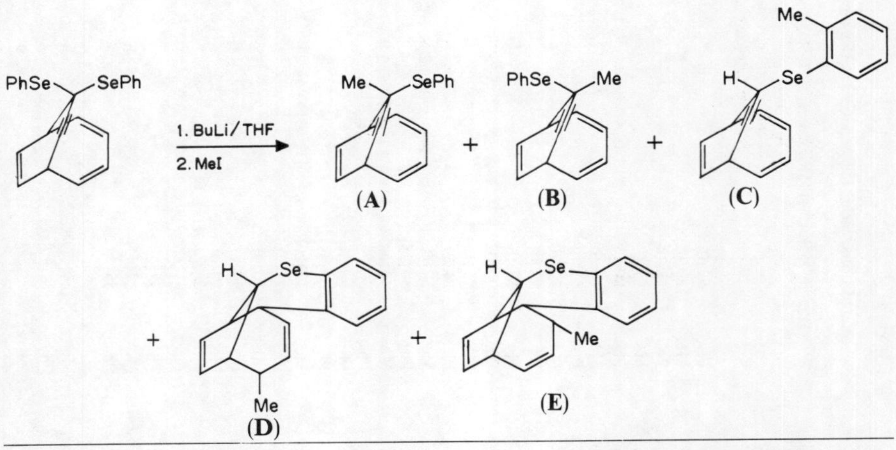

			Individual yields (%)				
Temp (°C)	Time (h)	Overall yield (%)	A	B	C	D	E
− 78	0.1	96	—	15	85	—	—
+ 20	1	84	—	—	—	27	73

SCHEME 22[105]

Unusual products resulting from the decomposition of the selenoalkyllithiums have been reported in the reaction of α-selenomethyllithiums with epoxides[106] or of triphenylselenomethyllithium with triphenyl phosphine[67].

$$R^1R^2C(SeR)_2 \xrightarrow{R^3Li} R^1R^2CLi(SeR) \xrightarrow{PhCHO} R^1R^2C(SeR)CHPh(OH)$$

			Conditions; yield, 1% (recovery of the starting material, 1%)			
			n-BuLi–THF/hexane, – 78°C			
Entries	Acetal	R	0.1 h	Minimum time required for complete cleavage	s-BuLi–THF/hexane, – 78 °C Minimum time required for complete cleavage	n-BuLi-ether/hexane, – 78 °C/1 h
1a	$CH_2C(SeR)_2$	Ph	82(0)			88 (0)
1b		Me	80(0)			
2a	$CH_3CH(SeR)_2$	Ph	70(0)			73 (16)
2b		Me	81(0)			
3a	$C_{10}H_{21}CH(SeR)_2$	Ph	80(0)			40 (44)
3b		Me	83(0)			
4a	t-$BuCH(SeR)_2$	Ph	92(0)			
4b		Me	82(10)			
5a	$PhCH(SeR)_2$	Ph	83(0)			81 (0)
5b		Me	87(0)			89 (0)
6a	$MeC(Me)(SeR)_2$	Ph	94(0)			
6b		Me	75(16)			
7a	$MeC(C_6H_{13})(SeR)_2$	Ph	81(2)			
7b		Me	50(40)	90 (0)/2 h		
8a	Me(t-Bu)$C(SeR)_2$	Ph	59(25)			
8b		Me	32(60)	76 (6)/2 h		
9a	Me(p-Tol)$C(SeR)_2$	Ph				
9b		Me	90(0)			80 (0)
10a	$(C_6H_{13})_2C(SeR)_2$	Ph	47(48)			
10b		Me	17(80)	93 (0)/4 h	83 (0)/0.1 h	
11a	$(CH_2)_nC(SeR)_2 (n = 2)$	Ph	91(0)			35 (64)
11b		Me	80(0)			
12a	$(n = 3)$	Ph	90(0)			
12b		Me	80(0)			
13a	$(n = 4)$	Ph	90(0)			
13b		Me	81(0)			
14a	$(n = 5)$	Ph	88(0)			
14b		Me	62(35)	87 (0)/2 h		
15a	1,1-Di(RSe)-2-methylcyclohexane	Ph	55(43)			
15b		Me	25(72)	82 (0)/4 h	87 (0)/0.1 h	
16a	2,2-Di(RSe)adamantane	Ph	0(89)			
16b		Me	0(97)	0 (98)/2 h	81 (0)/0.5 h	

SCHEME 23[78]

As general trends, the following comments on the synthesis of α-selenoalkyllithiums by Se–metal exchange can be made:

(*a*) Phenylselenoacetals are more easily cleaved than their methylseleno analogues[8,78] (Scheme 23, compare entries a and b).

(*b*) Selenoacetals derived from aldehydes are already cleaved after 0.1 h at − 78 °C by *n*-BuLi in THF/hexane whereas those derived from ketones required longer times[78] (0.5 to 4 h: Scheme 23, compare entries 1–4 to entries 6–8).

(*c*) In the series of acetals derived from methyl ketones with more bulky R^2 groups the reaction is more difficult (Scheme 23, examples 6–10 and 15–17). It is usually better in these difficult cases to use a longer period of time at − 78 °C rather than to increase the reaction temperature since the organometallics are stable at − 78 °C, but readily decompose at higher temperatures.

(*d*) The presence of a stabilizing group such as aryl, selenyl or silyl enhances the speed of the reaction (Scheme 23, compare entry 9b to 6b and 7b; see also Scheme 18).

(*e*) Selenoacetals derived from cyclopropanone[133], cyclobutanone[127] or cyclopentanone[78] are more easily cleaved than those of cyclohexanone (Scheme 23, entries 11–16). The conditions just reported are not suitable for the synthesis of the α-seleno-alkyllithiums derived from adamantanone (Scheme 23, entry 16).

As to the reactivity of selenoacetals under various experimental conditions:

(*a*) Several other sets of experimental conditions have been successfully tried for the cleavage of selenoacetals, and, e.g., *s*-BuLi in THF/hexane has proved to be particularly powerful[78]. Thus 7,7-bis(methylseleno)tridecane, which required 4 h to be completely cleaved by *n*-BuLi in THF/hexane at − 78 °C (Scheme 23, entry 10b), has been completely transformed to the corresponding α-selenoalkyllithium after 0.1 h reaction at the same temperature using *s*-BuLi. Interestingly these conditions allow the quantitative cleavage of methylselenoacetal derived from 2-methylcyclohexanone (0.1 h instead of 4 h for *n*-BuLi at − 78 °C)[78] and of the one derived from adamantanone[78] (100% after 0.5 h instead of 0% after 2 h for *n*-BuLi at − 78 °C). Closely related results have been obtained[78] with *t*-BuLi used instead of *s*-BuLi.

(*b*) *s*-BuLi also allows the cleavage of several selenoacetals when ether/hexane[8,78,107–112] is used instead of THF/hexane[8,11,67,78,94,107–112,126,127,133]. In this solvent system *s*-BuLi is less reactive than in ether (see Scheme 23) but the medium offers in several instances advantages over THF/hexane[107–112]. Thus, the α-selenoalkyllithiums are more stable in this medium and do not decompose at − 50 °C. They also have a higher propensity to add across the carbonyl group of particularly hindered[108] or enolizable[107,109,110] ketones. Moreover, the ratio of axial/equatorial attack on rigid cyclohexanones is often very different if the reactions are conducted in ether rather than in THF[108].

(*c*) *n*-BuLi in ether has also been used for the cleavage of the C—Se bond in selenoacetals. At − 78 °C, only phenylselenoacetals derived from formaldehyde, and phenyl- and methyl-selenoacetals derived from aromatic aldehydes and ketones are quantitatively cleaved. The reaction is very slow and requires at least 1 h for completion (Scheme 23). Under similar conditions, phenylselenoacetals of aliphatic aldehydes and cyclopropanone are only partially cleaved (Scheme 23). Phenylselenoacetals derived from aliphatic ketones and all methylselenoacetals except those of aromatic carbonyl compounds remain unchanged. If the reaction is performed at − 50 °C, however, all the phenylselenoacetals derived from aldehydes are cleaved in 0.5 h, but those derived from ketones are cleaved very slowly[78].

(*d*) Methyl- and phenyl-lithiums have also been reacted[8,78] with selenoacetals in THF/hexane but, with the exclusion of selenoacetals of aromatic compounds[12], they are not suitable for the synthesis of α-selenoalkyllithiums.

Although only preliminary work has been done in this field, it seems that some chemoselectivity can be expected. In the case of molecules bearing two selenoacetal groups derived from differently substituted carbonyl compounds, it is found[111] that phenyl- and methyl-selenoacetals behave differently.

In the case of phenylselenoacetals it is possible to cleave the selenoacetal group derived from an aldehyde without affecting the one derived from a ketone by taking advantage of the difference of reactivity of *n*-BuLi in ether towards these two groups (see above)[78,111]. At -50 °C the discrimination can be achieved under kinetically controlled conditions and a specific example is shown in Scheme 24.

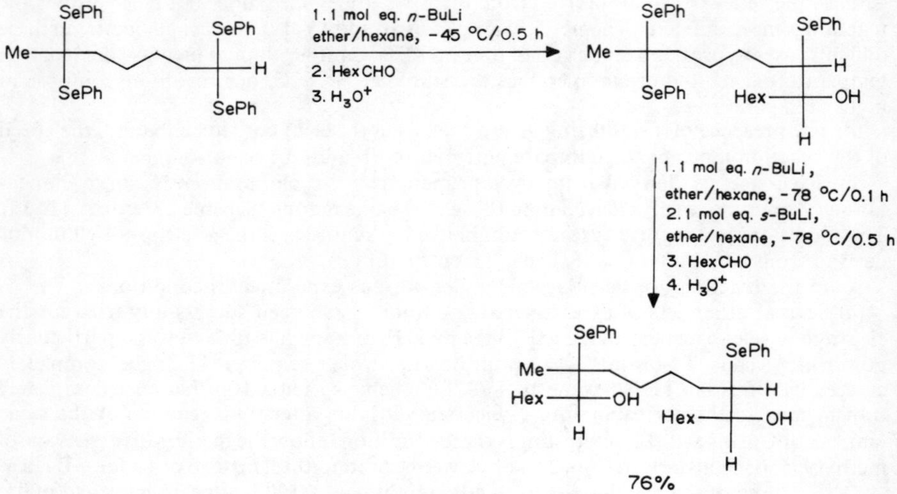

SCHEME 24[78,111]

In the case of methylselenoacetals the reactions have not been performed on one molecule bearing both different acetal functions but on a mixture of two different selenoacetals[111] (Scheme 25). The difference in reactivity of these acetals towards *n*-BuLi in THF or *s*-BuLi in ether is not strong enough to permit a high selectivity favouring the cleavage of the acetal derived from the aldehyde (Scheme 25a–d) (75/25). It was found however[111] that α-selenoalkyllithiums derived from aldehydes are thermodynamically more stable than those derived from ketones. It is therefore possible to increase the

$$\text{HexCH(SeMe)}_2 + \text{HexCMe(SeMe)}_2 \xrightarrow[\text{2. PhCHO}]{\text{1. RLi}} \text{HexCH(SeMe)CHPh(OH)} \quad \textbf{(A)}$$

$$+ \text{HexCMe(SeMe)CHPh(OH)} \quad \textbf{(B)}$$

Reaction–solvent	Temp (°C)	Time (h)	**A** (%)	**B** (%)	**A**/(**A** + **B**)
(a) *s*-BuLi–ether/hexane	– 78	0.5	55	20	73
(b) *n*-BuLi–THF/hexane	– 100	0.5	15	5	75
(c) *n*-BuLi–THF/hexane	– 78	0.05	37	11	77
(d) *n*-BuLi–THF/hexane	– 78	0.1	67	23	74
(e) *n*-BuLi–THF/hexane	– 78 then – 40	0.1 1	78	8	90

SCHEME 25[78,111]

formation of the former derivatives by carrying out the reaction under thermodynamically controlled experimental conditions[111]. If the reaction is performed in the more polar THF at − 78 °C and then stirred for two hours at − 4 °C, a net increase of the percentage of the organometallic derived from aldehyde is observed as compared with the results obtained by quenching the mixture after 0.1 h reaction at − 78 °C (Scheme 25e).

Preliminary results show that phenylselenoacetals are much more easily cleaved than their methylseleno analogues. Experiments have been performed on a 1:1 mixture of

TABLE 1. ^{77}Se-NMR data for selenoacetals, $R^1R^2C(SeR)_2$, and their corresponding α-selenoalkyllithiums, $R^1R^2CLi(SeR)$[78 a,b]

			$R^1R^2C(SeR)_2$	$R^1R^2CLi(SeR)$
R^1	R^2	R	^{77}Se-NMR (ppm)	^{77}Se-NMR (ppm)
H	H	Ph	324	278
		Me	104	61
H	Me	Ph	422	389
		Me	194	180
H	Et	Ph	378	335
		Me	155	126
H	$C_{10}H_{21}$	Ph	388	347
		Me	163	138
H	*t*-Bu	Ph	382	307
		Me	155	95
H	Ph	Me	237	54
Me	Me	Ph	530	475
		Me	277	249
Me	Et	Ph	479	408
		Me	241	187
Me	$C_{10}H_{21}$	Ph	490	425
		Me	249	202
Me	*t*-Bu	Ph	479	397
		Me	247	179
Hex	Hex	Ph	448	405
		Me	207	182
Me	Ph	Me	359	168
$CH_2—CH_2$		Ph	527	492
		Me	297	184
$CH_2—CH_2—CH_2$		Ph	479	452
		Me	253	213
$CH_2—(CH_2)_2—CH_2$		Ph	439	401
		Me	235	186
$CH_2—(CH_2)_3—CH_2$		Ph	558–402	476
		Me	313–166	246

[a] ^{77}Se-NMR Jeol FX 90Q, solvent THF/hexane, standard conditions for 10^{-3} mol of the Se compound (Me_2Se as external standard), temperature − 78 °C.

[b] ^{77}Se-NMR shift (ppm) of some selenides in THF: Me_2Se (0), MeSeBu (82), MeSePh (196), PhSeBu (288), Ph_2Se (408).

phenylseleno- and methylseleno-acetals derived from acetone, using *n*-BuLi in THF. After 0.1 h at $-78\,^{\circ}C$ a 86:14 ratio of 2-phenylseleno-: 2-methylseleno-2-propyllithiums is observed.

Most of the work we have done for synthetic purposes has been recently carefully reinvestigated by monitoring the reactions by ^{77}Se-NMR. In the case of reactions between selenoacetals and alkyllithiums, the peaks attributed to the Se atom in the acetals, α-selenoalkyllithiums and butyl selenides are well differentiated[78] (Table 1). Hence one can follow the changes depending upon the experimental conditions. From the temperature at which the reaction starts [78] (Scheme 23), information related to the speed of the reaction[78] and the temperature at which it occurs[8,78] have been derived.

Little information is available on the intimate mechanism of the Se bond cleavage of selenoacetals by alkyllithiums. If the reaction is followed by ^{77}Se-NMR, even at $-85\,^{\circ}C$ only peaks attributed to the organometallic and to the butyl selenide are concomitantly formed at the expense of the one corresponding to the selenoacetals, and in no case has an intermediate, such as for example an ate complex, been detected[8,78].

The metal plays an important role in the further reaction of organometallics, but only a little is known in the case of α-selenoalkyl derivatives. For example, up to now, alkyllithiums are the only organometallics which have been used to cleave the C—Se bond of selenoacetals. No reports have appeared concerning the reaction of alkylsodiums, Grignard or organozinc reagents or cuprates with selenoacetals.

The metal–metal interchange on α-selenoalkyllithiums has been successfully achieved by adding $MgBr_2$[78] at $-78\,^{\circ}C$ or CuI[112–117] at $-110\,^{\circ}C$. In both cases the ^{77}Se-NMR of the solution of the selenoalkyllithiums before and after the addition of the metal salts were different. Specific examples concerning the addition of CuI are presented in Scheme 26.

	$C_6H_{13}CHLiSeMe$	$C_9H_{19}CLi(Me)(SeMe)$	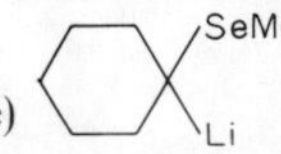
THF/hexane, $-78^{\circ}C$, $SeMe_2$ as external standard	136 ppm	202 ppm	245 ppm
As above, but after addition of 0.5 mol eq. $CuI–SMe_2$	102 ppm 103 ppm	178 ppm 179 ppm	208 ppm 209 ppm

SCHEME 26[78]

As expected, the reactivity of the new species is somewhat different from the one of the α-selenoalkyllithiums. For example, in the presence of $MgBr_2$ the percentage of the C(3) additions of the selenoalkyllithiums to cyclohexenone increases[78] (Scheme 27) and in the presence of 0.5 eq. of CuI the yield of allylation of α-selenoalkyllithiums is greatly enhanced[113–116] (Scheme 28).

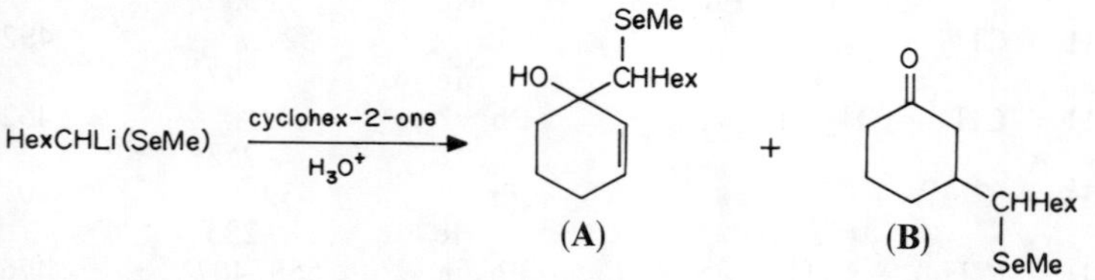

	A(%)	B(%)	A/(A + B)
Without additive	90	0	100
In the presence of $MgBr_2$	17	17	50

SCHEME 27[78]

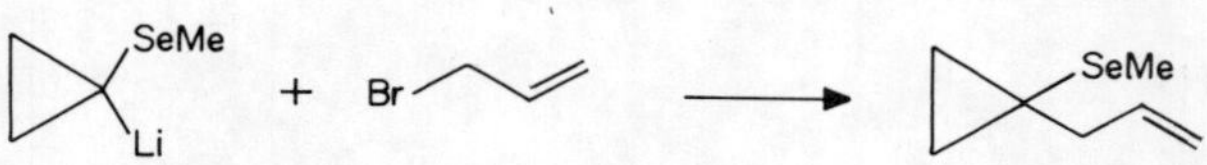

(Ref. 115) (a)

THF, −78°C	0%
THF/HMPT, −78°C	0%
THF, −78°C in the presence of 0.5 eq. CuI	70%

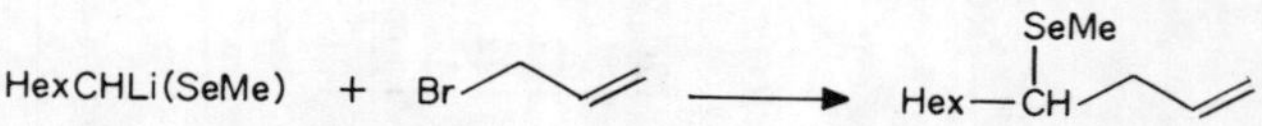

(Ref. 113) (b)

THF, −100°C/0.5 h	50%
THF, −100°C in the presence of CuI/Me_2S	83%
Ether, −100°C in the presence of CuI/Me_2S	85%

SCHEME 28

Finally it was found[112,117] that oxidative coupling of the α-selenoalkyllithiums takes place around −40 °C in the presence of CuI, and produces olefins in high yield (Scheme 29). The mechanism of the process still remains unknown.

$$\text{HexCHLi(SeMe)} + 0.5\text{ mol eq. CuI/SMe}_2 \longrightarrow \text{HexCH=CHHex}$$

THF, −78 to +20°C/0.7 h	80%
Ether, −78 to +20°C/0.7 h	85%

SCHEME 29[112,117]

(*iii*) *Synthesis of α-selenoalkyllithiums* via *halogen–metal exchange*. The whole series of α-phenylselenoalkyllithiums, even the dialkyl-substituted ones, have been successfully synthesized[118,119] from the corresponding α-bromophenyl selenides and *n*-BuLi in THF/hexane (Scheme 31). The cleavage of the C—Br bond is highly selective and occurs quite instantaneously at −78 °C, but the yields of α-selenoalkyllithiums (60–75%) are lower than those obtained by the route involving the cleavage of the C—Se bond in selenoacetals. Moreover, the α-bromoselenides are not in all cases stable and they must often be prepared just before they are reacted.

$R^1CH{=}CR^2(SePh) + HBr$

$R^1R^2C{=}O + PhSeH + HBr$

$R^1R^2C(SePh)OSiMe_3 + HBr$

$$\longrightarrow R^1R^2CH(SePh)Br \xrightarrow[\text{THF, } -78\,^\circ\text{C}]{n\text{-BuLi}} R^1R^2CH(SePh)Li \xrightarrow{\text{HexCHO}} R^1R^2CH(SePh)CHHex(OH)$$

$R^1 = H, R^2 = H$:	70%
$R^1 = Me, R^2 = H$:	69%
$R^1 = Hex, R^2 = H$:	63%
$R^1 = Me, R^2 = H$:	62%

SCHEME 30[118]

This synthetic route to α-selenoalkyllithiums is not general and does not work for methylseleno derivatives. Moreover, C—Se rather than C—Cl bond cleavage is observed if α-chlorophenyl selenides are reacted[118] instead of α-bromophenyl selenides.

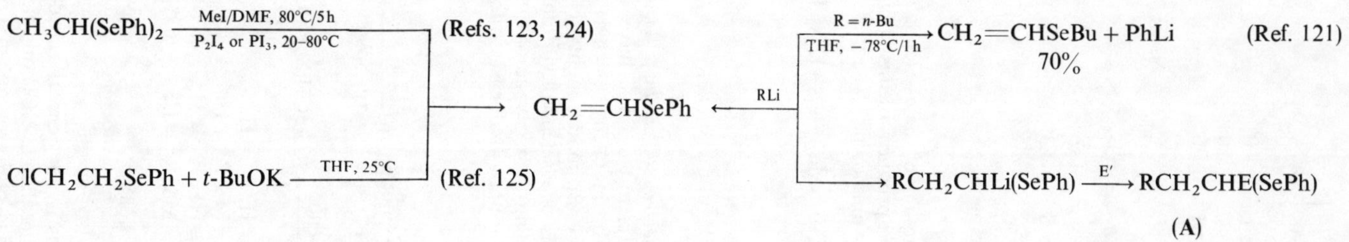

R	Conditions	E′	E	**A** (%)	Ref.
n-Bu	Ether, 20°C/0.5–1 h	H_2O	H	75	120
n-Bu	Ether, 20°C/0.2 h	HexCHO	HexCHOH	40	121
n-Bu	DME, 0°C	PhCHO	PhCHOH	71	122
i-Pr	Ether, 0°C	Me_2CO	Me_2CHOH	72	122
s-Bu	Ether, 20°C/0.5–1 h	H_2O	H	25	121
t-Bu	Ether, 0°C	D_2O	D	85	122
t-Bu	Ether, 20°C/0.5–1 h	H_2O	H	25	121

SCHEME 31

(*iv*) *Synthesis of α-selenoalkyllithiums by addition of alkyllithiums to phenyl vinyl selenides.* Three research groups[120–122] have simultaneously but independently described the addition of alkyllithiums to phenyl vinyl selenide. The reaction proceeds readily in DME or ether between 0 and 20 °C producing α-lithioalkyl phenyl selenides which may be trapped with various electrophiles (Scheme 31). The reaction must be performed as rapidly as possible since the organometallic formed is unstable at this temperature and isomerizes easily to the corresponding aryllithium[106b,121] (Scheme 32, compare also to Scheme 21c).

$$BuLi + CH_2{=}CHSePh \xrightarrow[20°C/0.2\,h]{ether} BuCH_2CHLi(SePh) \xrightarrow[HexCHO]{20°C/0.8\,h} \underset{30\%}{HexSePh} + \underset{18\%}{HexSeC_6H_4CHOH}$$

SCHEME 32[106b,121]

The proper choice of reaction conditions is essential for the success of the addition since α-deprotonation leading to 1-lithio-1-phenylselenoethylene or cleavage of the C—Se bond producing butyl phenyl selenides may also occur[121,122]. The last reaction is in fact the predominant one if THF is used instead of ether or DME as the solvent and is particularly favoured if the reactions are performed at low temperature[121,122] (Scheme 31).

2. *Reactivity of α-selenoalkylmetals in which the carbanionic centre is sp^3-hybridized and the selenyl moiety is uncharged*

a. Reactivity of α-selenoalkylmetals bearing hydrogens or one or two alkyl groups on the carbanionic centre. These derivatives have proved to be particularly reactive. They have been protonated[14,120,121], deuteriated[14,67b,122,126a], reacted with primary alkyl halides[14,20,106b,122,127], allyl halides[14,113,114], trialkyl and triallyl boranes[128], terminal epoxides[106,127,129], oxetanes[106], aldehydes[11,21,94,107,118,126,127,130–133] and ketones[11,21,67,72,94,107,118,122,126,127,130,131,133–135,137], as well as α-seleno aldehydes[139], α,β-unsaturated aldehydes[138,154] and ketones[72,138,154], esters[136], acid chlorides[136,140], anhydrides[136], chlorocarbonates[136] and carbonic anhydride[115,136], dimethylformamide[136,139], diphenyl disulphide[126,141] and related derivatives[126b], silyl chlorides[122,141–143], phenylselenenyl halides[122] and nitriles[122].

b. Reactivity of α-functionalized α-selenoalkylmetals. The following functional groups attached to α-selenoalkyllithiums were investigated:

(*i*) Allyl moieties react with primary[77,321] and secondary[72,78,321] alkyl halides, epoxides[77,321], silyl chlorides[77,78,321], aldehydes[78], ketones[77,78,321] and enones[79].

(*ii*) Benzyl groups react with primary and secondary alkyl halides[72,78,80], epoxides[72], allyl bromides[72] and aldehydes[72,78,145] and ketones[145] including hindered ones.

(*iii*) Ethynyl groups react with primary and secondary alkyl halides[82] and benzyl halides[82].

(*iv*) Ketone groups react with allyl halides[85].

(*v*) Carboxyl groups react with secondary allyl bromides[90], terminal epoxides[90] and aromatic aldehydes[72].

(*vi*) Ester groups react with primary alkyl halides[87,88a], aldehydes and ketones[88a] and α,β-unsaturated carbonyl compounds[88a,b,89].

(*vii*) Lactone rings react with primary alkyl halides[91], aldehydes and ketones[88a].

(*viii*) Nitrile groups react with enones[92] and alkyl halides[93].

(*ix*) Silyl groups react with primary alkyl halides[74,75], benzyl halides[76], epoxides[147] and aldehydes and ketones[74].

(*x*) The α-thio moiety reacts with aldehydes[73].

(*xi*) α-Selenyl moieties react with trimethyl silyl chloride[68,71,74], deuteriated water[67], primary alkyl halides[67–69,126,133], secondary alkyl halides[69], tosylates[133], benzyl halides[69], epoxides[68,69], aldehydes[68,69,72,187] and ketones[21,67–69], α,β-unsaturated ketones[69,148,149], α,β-unsaturated aldehydes[150] and α,β-unsaturated esters[150].

(*xii*) α-Methoxy groups react with aldehydes[72] and ketones[72].

(*xiii*) The α-nitro group reacts with formaldehyde[103].

(*xiv*) An α-phosphonium salt reacts with aldehydes[67,87], but it does not react with ketones.

(*xv*) An α-phosphonato group reacts with aldehydes[67,187], but it does not react with ketones.

(*xvi*) α,α-Diseleno groups (orthoesters) react with D_2O[67], primary alkyl halides[67,74], trimethylsilyl chloride[74], epoxides[133] and aldehydes[67].

c. Further comments on the reactivity of α-selenoalkylmetals. Some of the reactions presented require further comments which are given below.

(*i*) *Reaction with alkyl and allyl halides, epoxides and oxetanes.* (*a*) Selenoalkyllithiums react in THF at −78 °C exclusively with terminal alkylhalides, epoxides or oxetanes to form, respectively, selenides, γ- or δ-hydroxyselenides. In the last two cases the reaction occurs exclusively at the less hindered side of the heterocycle.

(*b*) Selenomethyllithiums are less reactive than their higher homologues. The reaction with primary alkyl halides and terminal epoxides or oxetanes occurs around −30 °C instead of around −60 °C for the higher homologues. At that temperature however part of the organometallic decomposes and produces selenolates which further react with the electrophiles present. The side-reactions can be avoided if the substitution is performed in THF at −78 °C but in the presence of HMPT[106].

(*c*) When secondary or tertiary alkyl halides are used instead of primary ones, no alkylation is observed in THF at from −100 °C to −20 °C and not even in the presence of CuI.

(*d*) Selenoalkyllithiums are not efficiently allylated[113,115] in THF at −78 °C or −100 °C. However, at −100 °C in the presence of 0.5 mol eq. of CuI the reaction works particularly well[113–115] (Scheme 28). In the case of geranyl and neryl chlorides the reaction is both regio- and also highly stereo-selective[113,114]. It occurs in an S_N instead of an $S_{N'}$ manner and with retention of the stereochemistry of the C=C bonds. When CuI is used the reaction must be performed around −100 °C, since at higher temperatures a side-reaction occurs[112,117] (Scheme 29) which leads to an olefin resulting from the coupling of two selenoalkyl moieties followed formally by the elimination of diselenide.

(*ii*) *Reaction with carbonyl compounds.* α-Selenoalkyllithiums react with a large variety of aldehydes and ketones including α,β-unsaturated, highly enolizable or hindered ones.

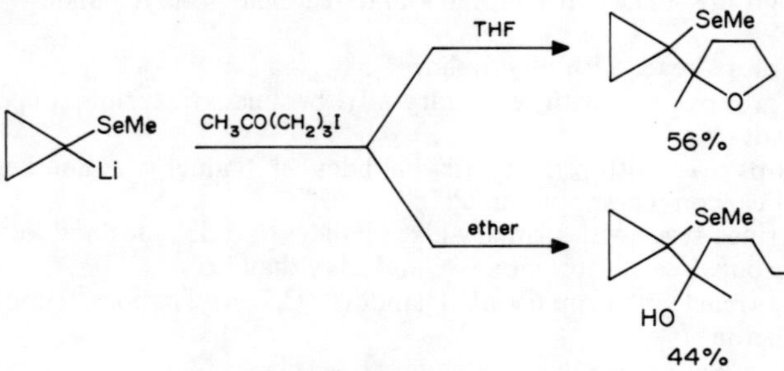

SCHEME 33[115]

(*a*) The reaction of α-selenoalkyllithiums, even of those possessing two alkyl groups on the carbanionic centre is readily (few minutes) achieved in THF/ or ether/hexane usually at − 78 °C. The reaction proceeds much more readily with carbonyl compounds than, for example, with alkyl halides[115] (Scheme 33).

(*b*) The reaction is not usually stereoselective with aliphatic or aromatic aldehydes and ketones since a mixture (1:1 to 7:3) of the two possible stereoisomers around the newly formed C—C bond is usually obtained without any regard to the solvent or the conditions used.

(*c*) In the case of rigid cyclohexanones high stereoselectivity is observed and products resulting from an axial or equatorial attack are found depending upon the solvent used (THF/ or ether/hexane, − 78 °C)[108,144] (Scheme 34). With α-seleno aldehydes, β-hydroxyselenides are formed with a very high stereoselectivity. The Cram[291] and Felkin[292] rules can account for the results observed[139] (Scheme 35).

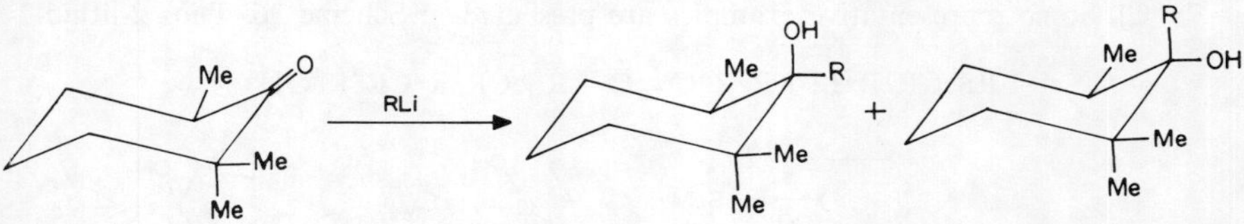

(A) **(B)** (a)

R	Conditions	Total yield (%)	**A** (%)	**B** (%)	Ref.
$MeSeCH_2$	Ether/hexane, − 78°C	86	31	69	144
	THF/hexane, − 78°C	92	18	82	
MeSeCHMe	Ether/hexane, − 78°C	84	95	5	144
	THF/hexane, − 78°C	85	95	5	
$MeSeCMe_2$	Ether/hexane, − 78°C	85	95	5	108
	THF/hexane, − 45°C	48	33	67	
$PhSeCMe_2$	Ether/hexane, − 78°C	84	95	5	108
	THF/hexane, − 78°C	0	—	—	

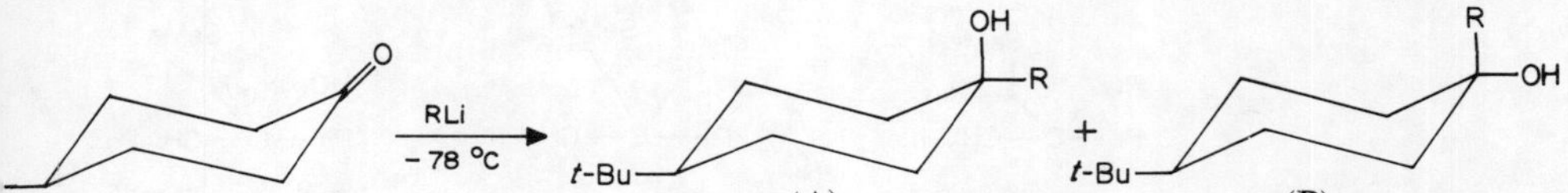

(A) **(B)**

(Ref. 144) (b)

R	Conditions	Total yield (%)	**A** (%)	**B** (%)
$MeSeCH_2$	THF	85	75	25
	Ether	75	69	31
MeSeCHMe	THF	70	90	10
	Ether	79	90	10
$MeSeCMe_2$	THF	75	100	00
	Ether	75	100	00

SCHEME 34

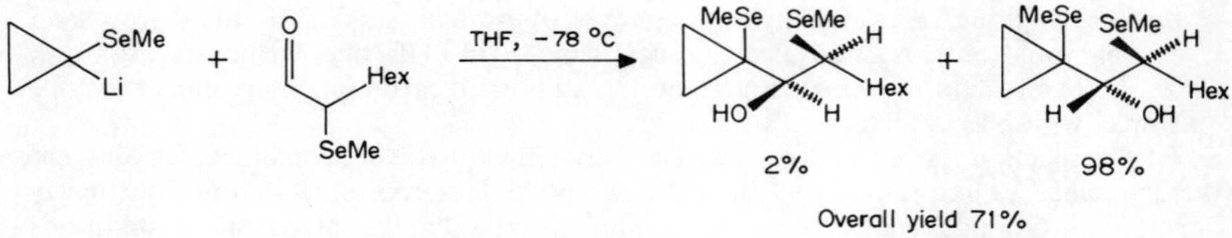

SCHEME 35[139]

(*d*) α-Selenoalkyllithiums exhibit particularly high nucleophilicity towards aldehydes and ketones even those bearing a highly hindered or enolizable carbonyl group. Best results are obtained with methylseleno derivatives, especially if the reactions are performed in ether/hexane rather than in THF/hexane and at low temperatures (usually − 78 °C). Some representative examples are presented in Scheme 36. Thus 2-lithio-2-

$$RSeCR^1R^2Li + R^3R^4C{=}O \rightarrow RSeCR^1R^2CR^3R^4OH$$

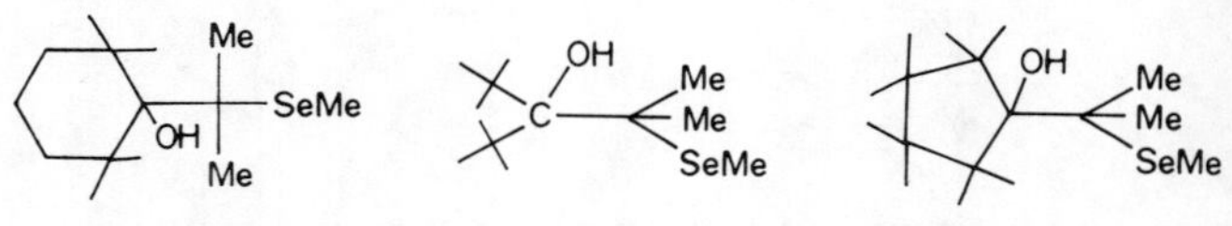

THF/hexane	77%[94]	72%[144]	—
Ether/hexane	83%[144]	—	23%[151]

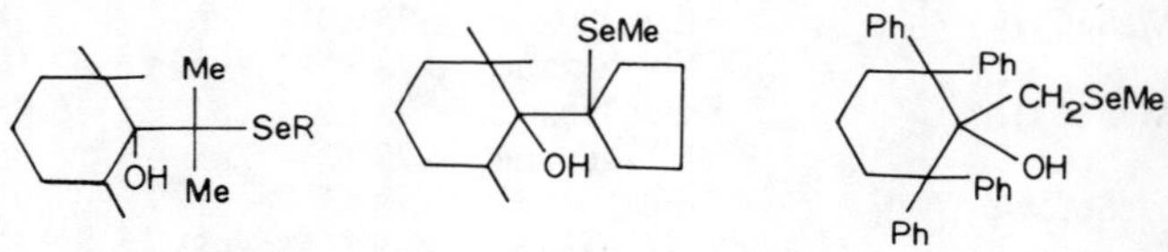

	R = Me	R = Ph		
THF/hexane	59%[108]	0%[108]	63%[115]	0%[155]
Ether/hexane	83%[108]	88%[108]	—	0%[115]

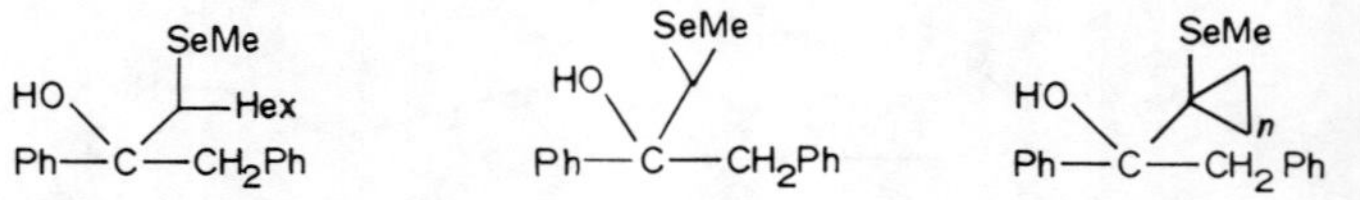

			$n = 1$	$n = 2$
THF/hexane	75%[144]	80%[94]	56%[115]	68%[115]
Ether/hexane	80%[144]	80%[144]	70%[115]	—
Ether/hexane/ HMPT(1 eq.)	12%[144]	—	—	—

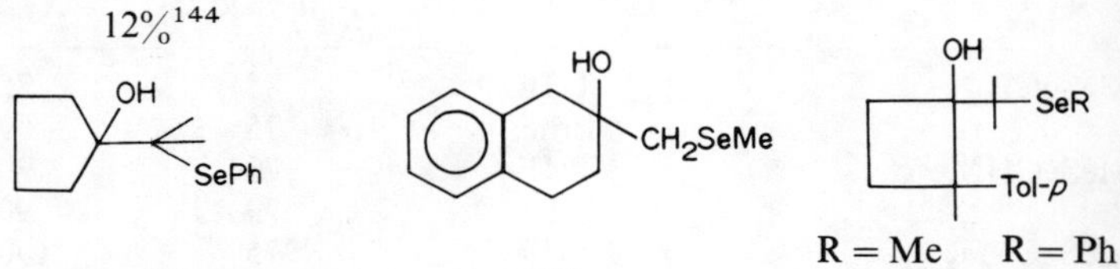

			R = Me	R = Ph
THF/hexane	45%[115]	44%[115]	0%	0%
Ether/hexane	—	—	66%[109]	49%[109]

SCHEME 36

methylselenopropane and α-lithiocyclopropyl selenides lead to a high yield of β-hydroxyselenides on reaction with deoxybenzoin[94, 130,131], 2,2,6,6-tetramethylcyclohexanone[94,108,144], 2,2,6-trimethylcyclohexanone[108,115] and di(*t*-butyl) ketone[144] (Scheme 36). Rather modest yields are found in the case of cyclopentanone[144], tetralone[144] and permethylcyclopentanone[151]. The last reaction is exceptional since the permethylcyclopentanone carbonyl is so hindered that it is not reduced by $AlLiH_4$ (Scheme 36). 2,2,6,6-Tetraphenylcyclohexanone[152] however was reluctant to form a new carbon–carbon bond, even with the less hindered (methylseleno) methyllithium.

(*e*) α-Heterosubstituted α-selenoalkyllithiums are much less nucleophilic towards carbonyl compounds than polyalkylated α-selenoalkyllithiums. Thus α-selenyl α-lithioselenides lead to modest to good yields of the corresponding alcohol on reaction with carbonyl compounds[68]. A comparison between the yield of C—C bond formation of various organometallics bearing an α-Se atom and some carbonyl compounds is presented in Scheme 37.

$$R^1R^2CXLi + HexCH{=}O \xrightarrow{THF} R^1R^2CXCH(OH)Hex \qquad (a)$$

X	R^1	R^2	Yield (%)	Ref.
PhSe	Me	Me	70	11
MeSe	Me	Et	82	94
MeSe	MeSe	Me	65	68
MeSe	Me_3Si	Me	40	74
MeSe	MeSe	MeSe	75	68

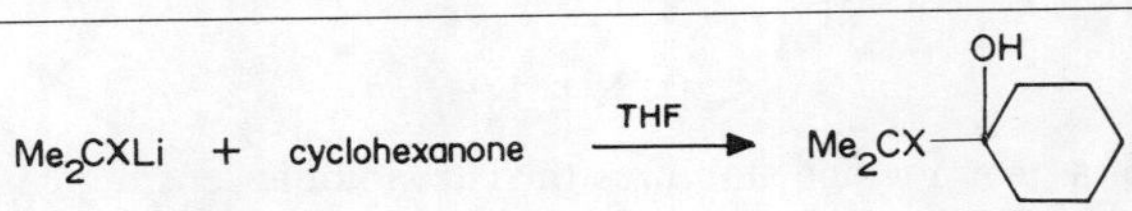

(b)

X	Yield (%)	Ref.
PhSe	78	11
	85	126a
MeSe	82	135
PhSeO	55	22

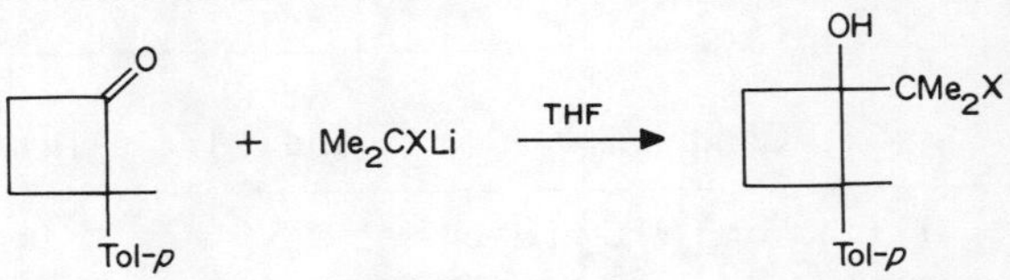

(c)

	THF/hexane		Ether/hexane	
X	Yield (%)	Ref.	Yield (%)	Ref.
SeMe	0	109	66	109
SePh	0	109	49	109
Se(O)Ph	39	153	—	—

SCHEME 37

(*iii*) *Reaction with enals, enones and enoates.* α-Selenoalkyllithiums bearing hydrogen or alkyl groups on the carbanionic centre have a very high tendency to add at the C(1) site of enals[138,154] and enones[138,154]. Chalcone[138,154] however produces a significant percentage of C(3) addition (Scheme 38). Due to side-reactions addition of CuI[112,117] (Scheme 29) or of HMPT[8,155], known for favouring the C(3) addition of organometallics at the C(3) site of α, β-unsaturated carbonyl compounds, has proved to be successful for favouring the C(3) addition of α-selenoalkyllithiums.

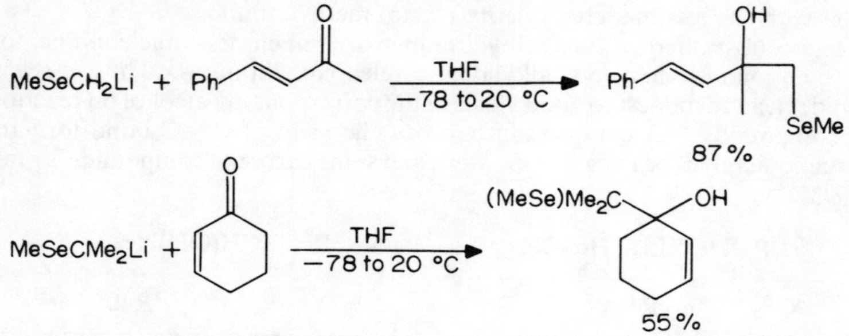

$$\mathrm{MeSeCR^1R^2Li + PhCH{=}CHCOPh \xrightarrow[\text{or} -78\,\text{to} +20\,°\text{C}]{\text{THF}, -78\,°\text{C}} PhCH{=}CH} \qquad \text{(c)}$$

$$\mathrm{—CPh(OH)CR^1R^2(SeMe) + MeSeCR^1R^2CHPhCH_2COPh}$$

$R^1 = R^2 = H$:	35%	35%
$R^1 = R^2 = Me$:	21%	47%

SCHEME 38[154]

The presence of a group which stabilizes the carbanionic centre greatly modifies the course of the reaction[8].

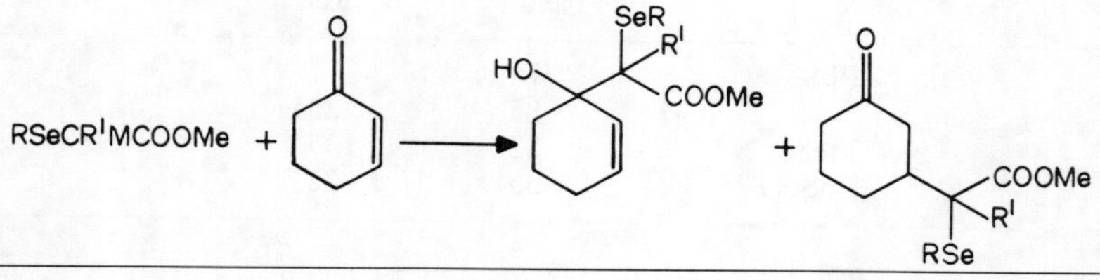

R	R^1	M	Conditions	Overall yield (%)	C(1)(%)	C(3)(%)
Me	Me	K	1. Ether or THF, − 110 or − 78 °C, 0.2 or 1 h 2. H_3O^+ at − 110 or − 78 °C	75	0	100
Me	Me	Li	1. Ether, − 110 or − 78 °C, 0.2 or 1 h 2. H_3O^+ at − 110 or − 78 °C	60	100	0
Me	Me	Li	THF, − 110 or − 78 °C, 0.2 or 1 h	75	85	15

SCHEME 39[89]

Exclusive addition at the C(3) site of cyclohexanone is observed with α-lithiophenylselenonitrile[92] and α-potassioseleno esters[89] (Scheme 39) and in the case of chalcone with α-lithioseleno esters[88b,89] (Scheme 40).

$$\text{MeSeCMMeCOOMe} + \text{PhCH}{=}\text{CHCOPh} \longrightarrow \text{(MeSe)MeC(COOMe)CHPhCH}_2\text{COPh}$$

M = Li:THF, − 78 or − 110°C, 0.2 h or 1 h	83%
M = Li:ether, − 78 or − 110°C	67%
M = K:THF or ether, − 78 or − 110°C	75%

SCHEME 40[89]

In the case of α-lithioseleno esters and enones (excluding chalcone) it is possible to control the regiochemistry of the addition: C(1) adducts are formed under kinetically controlled conditions especially if the reactions are performed[88b,89] in THF or, better, in ether at low temperature (− 78 °C) (Scheme 39). At higher temperature or when HMPT is added C(3) adducts are exclusively obtained[88b,89] and have been shown to be formed under thermodynamically controlled conditions[88b,89]. Whatever the conditions used, phenylseleno derivatives have a higher tendency to add at C(3) to enones than their methylseleno analogues and higher substitution at the carbanionic centre enhances this tendency[89] (Scheme 41).

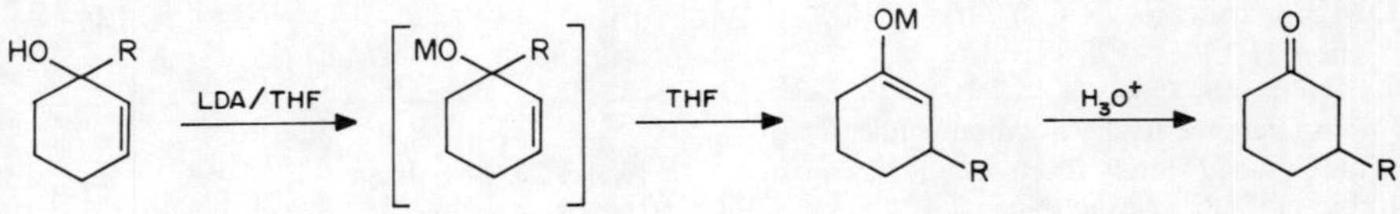

R	R¹	Temp. (°C)	t(h)[a]
MeSeCR¹(COOMe)	Me	− 50	1
PhSeCR¹(COOMe)	H	− 25	1.5
	Me	− 25	inst.
		− 40	0.1
		− 60	0.2

[a]Minimum time required for complete C (1) to C (3) isomerization. If potassium replaces lithium, the reaction produces the C (3) adduct instantaneously.

SCHEME 41[89]

In the case of 1-metallo-1,1-bis(seleno) alkanes[8,117] the temperature does not seem to have a significant effect on the ratio of product resulting from the C(1)/C(3) addition to enones. The structure of the organometallic, the nature of the metal and especially the nature of the solvent used controls the regiochemistry of the reactions (Schemes 42 and 43):

(*a*) This it is found that C(1) additions to enones[8,69,148,149] and enals[8,150] usually occur regioselectively if ether is used whereas C(3) adducts are formed[148,149] in THF in the presence of HMPT. Interestingly, both reactions were found to occur under kinetic control[18,148–150] (Scheme 42).

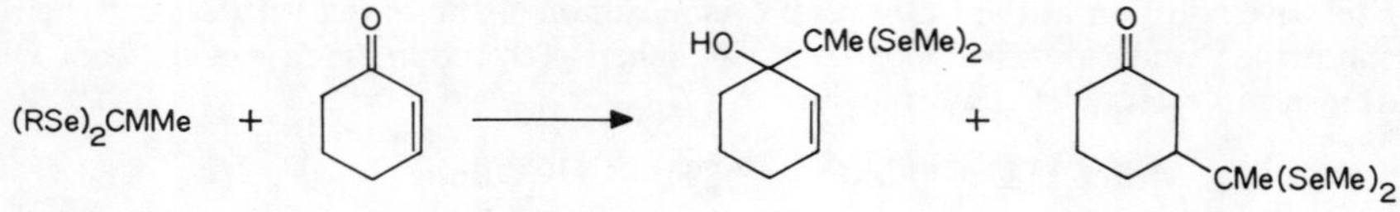

Conditions	R	M	Overall yield (%)	C(1) (%)	C(3) (%)	Ref.
Ether/hexane, $-70°C/0.2$ h,	Me	Li	78	97	03	117
then H_3O^+, $-78°C$	Ph	Li	71	93	07	117
THF/hexane, $-78°C/0.2$ h,	Me	Li	63	73	27	117
then H_3O^+, $-78°C$	Ph	Li	80	00	100	117
	Ph	K	76	100	0	69
THF/hexane + HMPT(1.1 eq.)[a],	Me	Li	78	00	100	117, 149
$-78°C/0.2$ h then H_3O^+, $-78°C$	Ph	Li	80	00	100	117, 149
THF/hexane + Ph_3PO(1.1 eq.)[a],	Me	Li	68	23	77	117
$-78°C/0.2$ h, then H_3O^+, $-78°C$						
Ether/hexane + HMPT(0.5 eq.)[a], $-78°C/0.2$ h, then H_3O^+, $-78°C$	Me	Li	74	75	25	117
Ether/hexane + kryptofix 2.2.2 (1 eq.), $-78°C/0.2$ h, then H_3O^+, $-78°C$	Me	Li	60	28	72	149
DME/hexane, $-78°C/0.2$ h, then H_3O^+, $-78°C$	Me	Li	61	13	87	149

[a] With reference to the organometallic.
[b] Trade name (Merck, Dormstat) for hexaoxa—4, 7, 13, 16, 21, 24—diaza—1, 10-bicyclo[8.8.8]-hexacosane.

SCHEME 42

$(RSe)_2CMMe + PhCH{=}CHCOPh \longrightarrow PhCH{=}CH(OH)PhC(SeR)_2Me$
$+ PhCHC(SeR)_2MeCH_2COPh$

Conditions	R	Overall yield (%)	C(1) (%)	C(3) (%)
THF/hexane, $-78°C/0.2$ h,	Me	91	25	75
then H_3O^+, $-78°C$	Ph	60	55	45
THF/hexane + HMPT(1.1 eq.)[a],	Me	77	95	5
$-78°C/0.2$ h, then H_3O^+, $-78°C$	Ph	63	98	2

[a] With reference to the organometallic.

SCHEME 43[156]

(*b*) In the case of enoates again the C(1) adduct is found in ether or THF and the C(3) adduct is obtained if the reactions are performed in THF–HMPT or in dimethoxyethane[150] (Scheme 42). Surprisingly, chalcone has, as the only one among various enones used, the reverse tendency[156] (Scheme 43).

(*c*) As a general trend phenylseleno derivatives have a higher tendency to add C(3) to enones and enoates[150] (Schemes 42 and 44) than their methylseleno analogues and surprisingly the reverse is found with enals[117,150] (Scheme 45).

$$(MeSe)_2CMeLi + PenCH{=}CHCO_2Me \longrightarrow (MeSe)_2CMeCHPenCH_2CO_2Me$$

THF/hexane + HMPT(1.1 eq.), $-78°C/0.2\,h$, then H_3O^+, $-78°C$	48%
DME, $-78°C/0.2\,h$, then H_3O^+, $-78°C$	40%

SCHEME 44[150]

$$(RSe)_2CMeLi + PrCH{=}CHCH{=}O \longrightarrow PrCH{=}CHCH(OH)C(SeR)_2Me + PrCHC(SeR)_2MeCH_2CHO$$

Conditions	R	Overall yield (%)	C(1) (%)	C(3) (%)	Ref.
THF/hexane, $-78°C/0.2\,h$,	Me	74	100	0	150
then H_3O^+, $-78°C$	Ph	80	100	0	150
THF/hexane + HMPT(1.1 eq.)[a],	Me	68	28	72	150
$-78°C/0.2\,h$, then H_3O^+, $-78°C$	Ph	60	100	0	150
THF/hexane + HMPT(2.2 eq.)	Me	51	21	79	117
THF/hexane + HMPT(4.4 eq.)	Me	27	18	82	117
THF/hexane + kryptofix 2.2.1	Me	[b]	68	32	117
DMF/hexane + TPPT[c]	Me	58	24	76	117
DME	Me	[b]	83	17	117

[a] With reference to the organometallic.
[b] *GC*[2] ratio.
[c] Trimethyl pyrrolidinophosphotriamide.

SCHEME 45

(*d*) The addition of organometallics to enones produces enolates which have been trapped with various electrophiles[117,148]. Silylated enolates are obtained[117] on reaction with Me_3SiCl and 2-alkylated ketones are formed[117] on reaction with alkyl halides. The stereochemistry of the C(3) addition of 1-lithio-1,1-bis(methylseleno)ethane with methylcyclohexenones has been established[117,148] as well as that of products resulting from the methylation of the adduct between the same organometallics and cyclohexenone[117,148] (Scheme 46).

(*e*) Finally the cleavage of the C—Se bond of highly functionalized selenoacetals (Scheme 47) has been successfully achieved[148] and the resulting α-selenoalkyllithium has been subjected to further reaction (Scheme 47).

3. Synthesis and reactivity of sp³-hybridized α-selenoalkylmetals bearing a charged selenyl moiety

The presence of a positive charge on the selenyl moiety as in selenoxides[21,22] and selenonium salts[19] allows, unlike most of the selenides, easy metalation of such compounds by metalloamides in the first case (-78 °C, 0.3 h) or by *t*-BuOK in the second one.

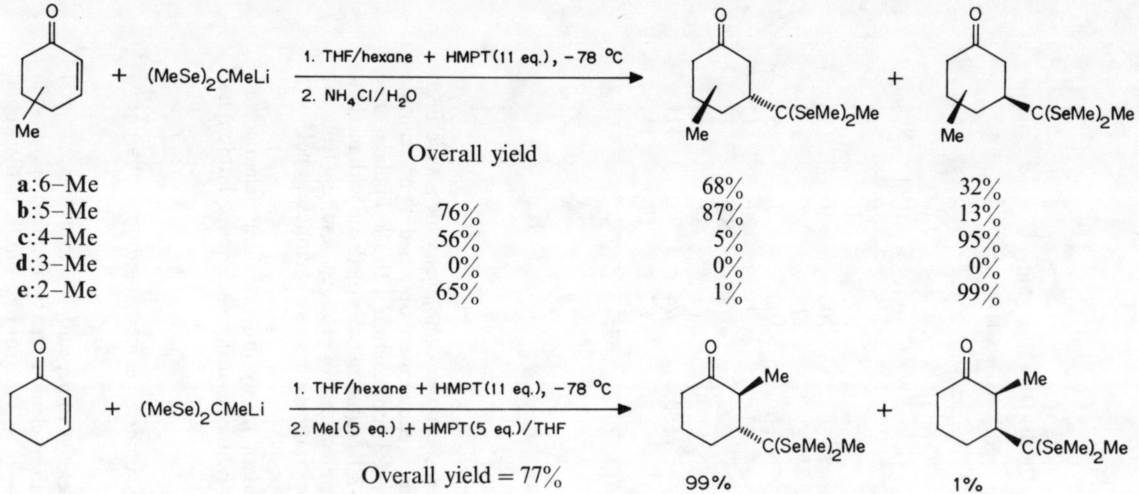

SCHEME 46[148]

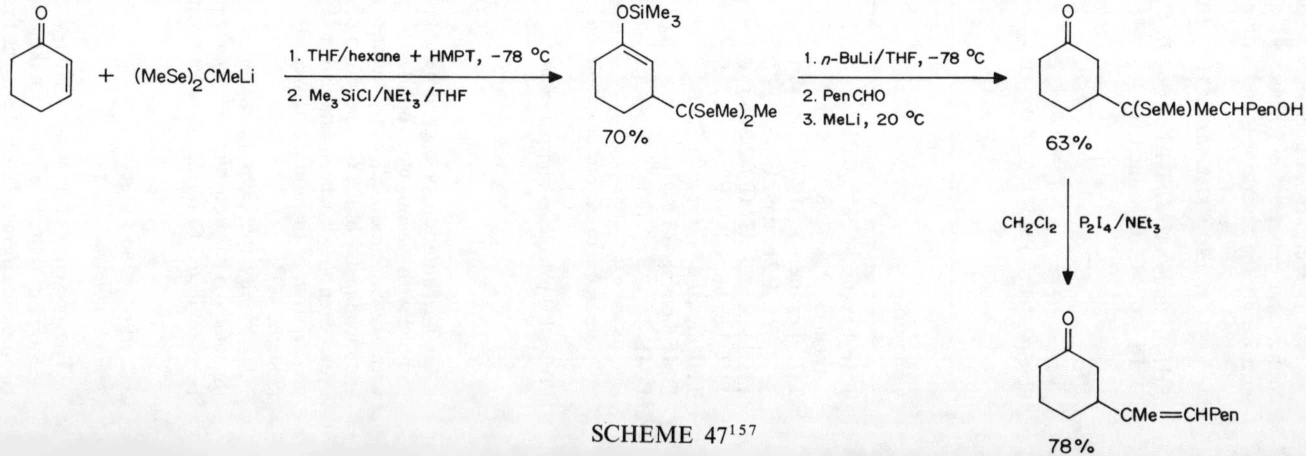

SCHEME 47[157]

a. Synthesis of α-metalloselenoxides. Selenoxides must in most cases be handled below 0 °C to avoid olefin formation[22,153,158]. A procedure has been therefore developed by Reich and Shah[22], in which the phenyl alkyl selenide is oxidized by *m*-chloroperbenzoic acid in THF at 0 °C[21,158] or at – 10 °C[22,158] or by ozone in ether[22] at – 78 °C and then immediately deprotonated by LDA/THF at – 78 °C (2 eq. of LDA, if perbenzoic acid is used) (Scheme 48).

$$\mathrm{PhSe(O)CHR^1R^2} \xrightarrow[-78\,^\circ\mathrm{C}]{\mathrm{LDA,\,THF}} \mathrm{PhSe(O)CR^1R^2Li} \xrightarrow{\mathrm{E'}} \mathrm{PhSe(O)CR^1R^2E}$$

R^1	R^2	E′	E	Yield (%)
H	H	$PhCH{=}CHCH_2Br$	$PhCH{=}CHCH_2{-}$	75
H	$(CH_2)_2Ph$	$Me_2C{=}CHCH_2Br$	$Me_2C{=}CHCH_2{-}$	64
H	$(CH_2)_2Ph$	EtCHO	EtCHOH	87
Me	Me	PhCHO	PhCHOH	81
Me	Me	$PhCO_2Me$	PhC=O	81

SCHEME 48[22]

b. Reactivity of α-metalloselenoxides. These organometallics react[21,22] with primary alkyl, allyl and benzyl halides[21,22,158], aldehydes[21,22,72,158], ketones[21,22,153,154] and aromatic esters[21,22,158]. The corresponding functionalized selenoxides are then allowed to fragment to olefins and selenenic acids[21,22] or are reduced to the corresponding selenides by bisulphite solution[21,22], by P_2I_4[159] or by PI_3[160].

c. Synthesis and reactivity of selenonium ylides. Stable selenonium ylides have been prepared by the action of active methylene compounds on selenide dichlorides in basic methanol[161–164,302,303] or on selenoxides in the presence of dicyclohexylcarbodiimide[302]. A further method involves the thermal or photolytic decomposition of diazo compounds in the presence of selenides[164a,302].

Selenonium ylides stabilized by groups in which the negative charge can delocalize are quite uncreative. Some of them have been isolated[302,303] and their structure determined by X-ray crystallography[162]. Thus, $Me_2Se{=}CH{-}C({=}O)Ph$ is stable at room temperature[165], although it trimerizes to tribenzoylcyclopropane when heated[165]; if the reaction is conducted in the presence of chalcone, 1,2-dibenzoyl-3-phenylcyclopropane is formed[165].

Non-stabilized selenonium ylides which bear H or R groups on the carbanionic centre are much more reactive. They have been generated[19] at 20 °C on reaction of *t*-BuOK in DMSO with selenonium salts (the latter being readily available from selenides and alkyl halides[19]), and have been trapped *in situ* with non-enolizable aldehydes and ketones producing epoxides in moderate to good yields (Scheme 49).

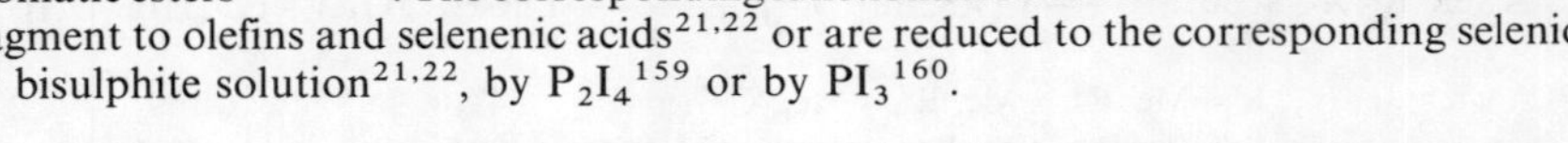

R	R^1	R^2	X	Yield (%)
Me	H	H	1	90
Ph	H	H	BF_4	90
Ph	H	Me	BF_4	69

SCHEME 49[19]

Under similar[19] or closely related[166] conditions enolizable carbonyl compounds such as heptanal, cyclohexanone and acetophenone do not form significant amounts of the corresponding oxiranes. In the case of acetophenone and trimethylselenonium fluoroborate, polymethylation of the carbonyl compound occurs. These results can be explained by enolate formation which is further alkylated by the selenonium salts[19] (Scheme 50).

$$\mathrm{I^-, R_2\overset{+}{S}eMe + PhC(O)Me \xrightarrow{\mathit{t}\text{-}BuOK/DMSO} [PhCOCH_2K] \xrightarrow{R_2{}^+SeMe, I^-} \underset{40\%}{PhCOMe}}$$
$$\mathrm{R = Me, Ph}$$
$$\mathrm{+ \underset{13\%}{PhCOCH_2Me} + \underset{28\%}{PhCOCHMe_2}}$$

SCHEME 50[19]

In one case the ylide has been generated[19] from diphenylmethylselenonium fluoroborate and $LiCH_2Cl$ in DME at $-78\,°C$. Further addition of acetophenone leads to the formation of 1-phenyl-1-methyloxirane in 30% yield.

Selenoxides[22] and selenonium ylides[19] also react with alkyllithiums but no metalation is observed. In the case of selenonium salts[11] and BuLi $-78\,°C$ 0.2 h), selenanes are probably formed. On reaction with aldehydes and ketones selenanes release one group producing good yield of alcohols[11] (Scheme 51). Ph groups migrate in preference to Me groups or to other alkyl groups[11].

$$\mathrm{R\overset{+}{S}eR^1R^2, X^- + BuLi \xrightarrow{THF} [RSeBuR^1R^2] \xrightarrow{PhCHO} PhCHR^1(OH) + PhCHBu(OH)}$$

	PhCHR¹(OH)	PhCHBu(OH)
$R = Me, R^1 = Me, R^2 = Me, X = I$:	54%	16%
$R = Ph, R^1 = Ph, R^2 = Me, X = MeSO_4$:	40%	0%
$R = Ph, R^1 = Ph, R^2 = Ph, X = BF_4$:	56%	0%

SCHEME 51[11]

4. *Synthesis and reactivity of 1-metallo-1-selenoalkenes*

a. Synthesis. The synthesis of these organometallics has been successfully achieved by metalation of vinyl selenides or by cleavage of the C—Se bond of 1,1-bis(seleno)alkanes.

(i) By metalation of vinyl selenides. Vinyl aryl selenides are valuable precursors of metallo-1-selenoalkenes. They have been prepared by several routes[324] including addition of selenols to terminal acetylenes[101,167], dehydrohalogenation of β-halo-[125,169–173,180] and α-halo-selenides[174,175,325] reaction of selenoacetals[123,124,176] (Scheme 31), reaction of alkynyl trialkylborate[177], alkenylborane[180], alkenyllithium[178], magnesium[179] and alkenylmercurial derivatives[180] with selenyl halides, and the reduction of 1-(phenylseleno)alkynes[180]. The reaction of Wittig[101] or Wittig–Horner[101,102] reagents with aldehydes and α-lithiosilylselenides with aldehydes[74,181] as well as the addition of selenamides to acetylene[183] and base-catalysed isomerization of allyl selenides[184] have also been used for the synthesis of vinyl selenides.

Deprotonation of the parent compound phenyl vinyl selenide has been successfully achieved by LDA at $-78\,°C$ in THF[70,121] by KDA[69] under similar conditions (Scheme 52) or with LDA in THF in the presence of HMPT[121]. Deprotonation with LDA is shown to be reversible[70] and during competitive deprotonation studies (LDA, THF) aryl vinyl sulphides are found to be thermodynamically less acidic than aryl vinyl selenides[70] [$K_{S/Se} = 0.21$ for phenyl vinyl and 0.3 for *m*-(trifluoromethyl)phenyl vinyl]. Metalation with LiTMP is irreversible and competitive deprotonation studies show vinyl

$$R^1R^2C{=}CH(SeAr) \xrightarrow{\text{base}} R^1R^2C{=}CM(SeAr) \xrightarrow{E'} R^1R^2C{=}CE(SeAr)$$

Ar	R[1]	R[2]	Conditions	E′	E	Yield (%)	Ref.
Ph	H	H	1.5 eq. LDA/THF, – 78°C/1 h	D_2O	D	80	70,121
				HexCHO	HexCHOH	40	121
			LDA/THF–HMPT(20:1)	DecBr	Dec	80	121
	H	H	KDA/THF, – 78°C	DecBr	Dec	94	69
m-$CF_3C_6H_4$	H	H	LDA/THF–HMPT, – 78°C	BuI	Bu	85	70
Ph	H	Me[a]	LDA/THF, – 78°C/5.8 h	MeI	Me	3[b]	70
m-$CF_3C_6H_4$	H	Me[a]	LDA/THF, – 78°C/5 h	MeI	Me	17[b]	70
m-$CH_3C_6H_4$	H	Et	1.5 eq. LiTMP/THF, – 50°C	MeI	Me	61	70
Ph	Bu	H	KDA/THF, – 78°C	MeI	Me	85	69
	H	nBu	KDA/THF, – 78°C	MeI	Me	85	69
Ph	Me	Me	KDA/THF or LiTMP	—	—	—	69

[a] Mixture of isomers used.
[b] Deprotonation at the allylic position is also observed.

SCHEME 52

selenides to be kinetically more acidic than their thio analogues [$K_{S/Se} = 0.37$ for phenyl vinyl and 0.42 for *m*-(trifluoromethylphenyl) vinyl].

Phenyl vinyl selenides with β-monoalkyl substituents are not completely metalated with LDA in THF. Deprotonation proceeds quite readily until half of the LDA is used up and then slows down[70]. Both vinyl and allyl protons are removed under these conditions[70]. In these cases, LiTMP is not satisfactory at − 78 °C and the reaction requires an excess of reagent (1.5 mol eq.) and high temperatures to go to completion[70]. KDA in THF[69] works much better and metalation is complete at − 78 °C. The metalation is found to be highly regio- and stereo-selective and occurs with retention of configuration[69].

β,β-Dialkyl-substituted phenyl vinyl selenides are metalated at the allylic positions with LiTMP[70] or KDA[69] but not at the vinylic one. The other substituent present on the Se atom has a big influence on the feasibility of the reaction. *m*-(Trifluoromethyl)phenylseleno derivatives are much more acidic than the phenyl ones[70] and methylseleno derivatives have not yet been metalated[185].

(*ii*) *By cleavage of the C—Se bond of* 1,1-*bis*(*seleno*)*alkenes.* 1-Metallo-1-selenoalkenes can be conveniently prepared from 1,1-bis(seleno)alkenes and BuLi in THF[185,186], taking advantage of the cleavage of the C—Se bond (Scheme 53). 1,1-Bis(selenoalkenes, are readily available from alkyl[124] or β-hydroxyalkyl[187] orthoesters which in turn have been prepared from α-metallo orthoesters[67,68] and alkyl halides[67,68], aldehydes or ketones, respectively. The alkenes have also been directly prepared from carbonyl compounds and 1,1-bis(seleno)-1-silylalkanes[186] taking advantage of the Peterson elimination of the silyloxy group.

$$R^1R^2C{=}C(SeR)_2 \xrightarrow[-78^\circ C/1\,h]{BuLi/THF} R^1R^2C{=}C(SeR)Li \xrightarrow{E'} R^1R^2C{=}C(SeR)E$$

R	R¹	R²	E′	E	Yield (%)	Ref.
Ph	H	Ph	H_2O	H	61(*E*)	186
			PhCOBr	PhCO	55(*E*)	186
Me	H	Dec	MeI	Me	80(*Z* + *E*)	185
			HexCHO	HexCHOH	69(*E* + *Z*)	185
			DMF	CHO	79(*E*)	185
			CO_2	CO_2H	80(*E* + *Z*)	185

SCHEME 53

The synthesis of 1-metalloseleno alkenes from the corresponding ketene selenoacetal and BuLi is an easy reaction which takes place rapidly (0.5 h) at − 78 °C in THF/hexane. Both ketene phenylselenoacetals[186] and their methylseleno analogues[185], are cleaved under these conditions and the organometallic is obtained in both cases in quantitative yield. The stereochemistry of the reaction is not well defined. The only available results are those concerning the stereochemistry of the compounds resulting from further reactions of the organometallic with electrophiles[185,186].

Both stereoisomers are obtained if the methylseleno organometallic[185] is hydrolysed or reacted with aldehydes and ketones whereas only one isomer, of α-seleno-α, β-unsaturated carbonyl compounds, whose stereochemistry has not been determined, has been found on reaction of α-methylseleno derivatives with DMF[185] or of a phenylseleno derivative with phenacyl bromide[186].

b. Reactivity. α-Metalloselenoalkenes react with alkyl halides[69,70,121,185], unhindered aldehydes and ketones[69,70,121,185], ethylene oxide[69], DMF[185], chlorocarbonates[185], phenacyl bromide[186], carbonic anhydride[185], trimethylsilyl chloride[70], dimethyl disulphite[70] and diphenyl diselenide[70]. 1,2-Addition is exclusively observed with cyclohexenone[69] with either Li^{+}[185] or K^{+}[69] used as a counterions.

5. Synthesis of allyllithiums, benzyllithiums, α-thioalkyllithiums and α-silylalkyllithiums from the corresponding selenides by selenium–metal exchange

a. Synthesis of allyl- and benzyl-lithiums. The cleavage of the C—Se bond of selenides and functionalized selenides by alkyllithiums is a powerful synthetic route to α-selenoalkyllithiums and bis(selenoalkyl)lithiums and permits the synthesis of other organometallics often difficult to prepare by other routes, such as allyllithiums,

$$CH_2{=}CH{-}CH_2Br \xrightarrow[EtOH]{RSeH/NaOH} CH_2{=}CHCH_2SeR \xrightarrow[78°C/0.1\,h]{BuLi/THF} CH_2{=}CHCH_2Li \xrightarrow{PhCHO} CH_2{=}CHCH(OH)Ph$$

R = Ph	81%	86%
R = Me	80%	95%

SCHEME 54[12]

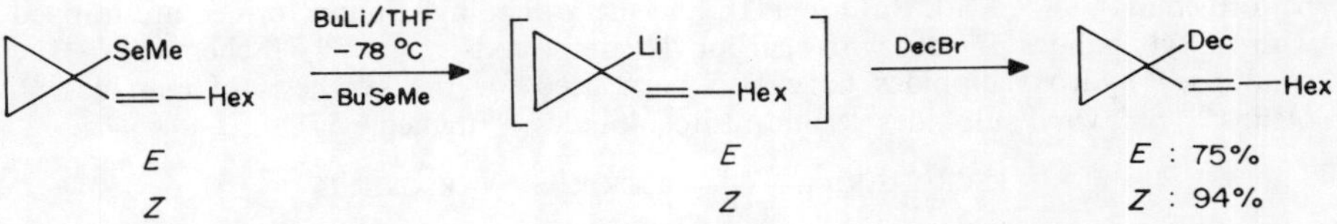

SCHEME 55[188]

$$ArC(SeR)R^1R^2 \Rightarrow ArCH_2X + RSeNa \quad X = Cl, Br \quad 62–90\%$$

$$ArC(SeR)R^1R^2 \Rightarrow ArCR^1R^2OH + RSeH + H_2SO_4 \quad R^1 = alkyl;\ R^2 = H, alkyl \quad 55–78\%$$

$$ArC(SeR)R^1R^2 \Rightarrow ArCR^1(SeR)Li + R^2X \Rightarrow ArCR^1(SeR)_2 \text{ or } ArCHR^1(SeR)$$

A

$$\mathbf{A} \xrightarrow[-78\,°C/0.1\,h]{BuLi/THF} ArCR^1R^2Li \xrightarrow{PhCHO} ArCR^1R^2CHPh(OH)\ \mathbf{B}$$

Ar	R^1	R^2	R		**B**(%)
Ph	H	H	Ph	(reverse addition of **A** to BuLi)	86
Ph	H	H	Me		98
Ph	H	Me	Me	(reverse addition)	71
Ph	Me	Me	Me		80
p-ClC_6H_4	H	H	Me		91
p-FC_6H_4	H	H	Me		83
o-ClC_6H_4	H	H	Me		94

SCHEME 56[12]

benzyllithiums and α-thio- and α-silyl-alkyllithiums. Phenyl methyl selenide is quantitatively cleaved[78] by BuLi in THF/hexane after 1.8 h at $-50\,^{\circ}C$. Usually the more stabilized of the two possible organometallics is formed and the rate of the cleavage of a selenide is faster if the carbanionic species formed is more stabilized[78]. Allyl and benzyl phenyl selenides as well as their methylseleno analogues are instantaneously cleaved by BuLi in THF at $-78\,^{\circ}C$, producing quantitatively allyl-[326] or benzyl-lithiums[12]. The latter have been alkylated or hydroxyalkylated with alkyl halides[12] or carbonyl compounds[12,326] (Schemes 54–56). The method is particularly useful since these organometallics cannot be easily prepared by simple metalation or by halogen–metal exchange[189–193,201] and often require the reaction of lithium on the corresponding ether[194–196] or thio ether[198] or the reaction of alkyllithiums on vinylic or benzylic tin derivatives[197–200].

b. Synthesis of α-thioalkyllithiums. Methylthio- and phenylthio-methyllithiums can be prepared[8] by metalation of the corresponding sulphide either with BuLi/TMEDA[202] or BuLi/DABCO[203]. The method proved unsuitable for higher homologues. These monoalkylated phenylthioalkyllithiums have been prepared by addition of alkylmetals to vinyl phenyl sulphide[204,205] or by metalation of an alkyl phenyl sulphide[206].

The discovery that all the α-phenylthioalkyllithiums (even the most substituted ones), and all the α-methylthio analogues can be prepared by C—Se bond cleavage of mixed S,Se acetals is therefore an important observation[126,207–209]. The reactions are usually performed at $-78\,^{\circ}C$ with BuLi in THF and the α-thioalkyllithiums formed are trapped with alkyl halides[126] or with carbonyl compounds[126,207,209] (Scheme 57). The resulting β-hydroxysulphides have also been used for the synthesis of epoxides[208], olefins[209,210], vinyl sulphides[210] and azidosulphides[210] (Scheme 57).

$$RSCR^2R^3SeR^1 \xrightarrow[-78\ C]{BuLi/THF} RSCR^2R^3Li \xrightarrow{E'} RSCR^2R^3E$$

R	R^1	R^2	R^3	E′	E	Yield (%)	Ref.
Ph	Ph	H	H	PhCHO	PhCHOH	94	207, 208
Ph	Ph	H	Me	$PhC(=O)CH_2Ph$	$PhC(OH)CH_2Ph$	92	208
Ph	Ph	H	$CHMe_2$	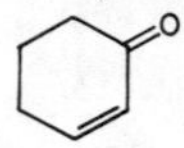	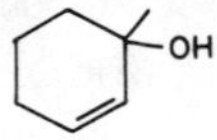	60	135
Ph	Ph	$-CH_2(CH_2)_3CH_2-$		PhCHO	PhCHOH	85	126, 207
Ph	Ph	Me	Me	HexCHO	HexCHOH	57	126, 208
Me	Me	H	Me	DecCHO	DecCHOH	73	209
Ph	Me	H	Me	DecCHO	DecCHOH	71	209
Me	Me	Me	Me	DecCHO	DecCHOH	70	209
Me	Ph	H	Me	DecCHO	DecCHOH	80	207

SCHEME 57

c. Synthesis of α-silylalkyllithiums. The synthesis of α-silylalkyllithiums by cleavage of the C—Se bond in α-silyl selenides is not as facile as that of α-seleno- and α-thio-alkyllithiums which we have just presented. Thus α-silyl phenyl selenides do not produce α-silylalkyllithiums but exclusively α-silyl butyl selenides and phenyllithium[211] (Scheme 58).

$$Et_3SiCH(Hex)SePh + BuLi \xrightarrow{THF,\ -78^{\circ}C} \underset{66\%}{Et_3SiCH(Hex)SeBu} + PhLi$$

SCHEME 58[211]

If methylseleno derivatives are used, however, the α-lithioselenides are obtained. The reaction proceeds at 0 °C for monosubstituted alkyl derivatives and produces moderate (~ 50%) yields of the desired organometallics[211] (Scheme 59). Under similar conditions secondary alkyl silyl selenides do not react except for the cyclopropyl derivative which leads to α-silylcyclopropyllithium in very good yield[143] at − 20 °C (last entry in Scheme 59).

$$R^3_3SiCR^1R^2SeMe + BuLi \xrightarrow[BuSeMe]{THF} R^3_3SiCR^1R^2Li \xrightarrow{R^4CHO} R^3_3SiCR^1R^2CHR^4OH$$

R^3	R^1	R^2	Conditions	R^4	Yield (%)	Ref.
Et	H	Me	0°C/1 h	Hex	45	211
Me	H	Hex	0°C/1 h		50	211
Et	H	Hex	0°C/1 h	Hex	30	211
Me	—CH_2CH_2—		− 20°C/1 h	Dec	85	143

SCHEME 59

α-Silylalkylmetals are rather difficult to prepare. The parent compound forms in low yield on metalation of tetramethylsilane[8,212,213] or by halogen–metal exchange in α-bromotriphenylsilylmethane[8,214]. The last reaction has been used for the preparation of trimethylsilyl/ethyllithium and more recently for the synthesis of α-silylcyclopropyllithium[143]. These compounds remain until now the only two examples of α-silylalkyllithiums bearing two alkyl groups on the carbanionic centre[143,215,216]. Compounds bearing only one alkyl group on the carbanionic centre (at the exclusion of α-silylethyllithiums) have also been obtained by addition of organometallics across the C═C bond of vinylsilanes[8,217–219].

d. Miscellaneous syntheses of organometallics by selenium–metal exchange. C—Se bond cleavage in selenides has also been used for the synthesis of a few aryllithiums[8,10–12], some metallo heteroaromatic compounds and α-lithiotriphenylethylgermane[34].

C. Reactions of Selenides and Functionalized Selenides

1. Reactions involving the reduction of the C—Se bond

Reactions involving the reduction of a C—Se bond to a C—H bond have been achieved using Raney nickel in ethanol[14,220], lithium in ethylamine[14] and trialkyl- or triaryl-tin hydrides[15,108,148,221] with or without azobisisobutyronitrile AIBN (Scheme 2).

Thus substituted selenides[222] and β- and γ-hydroxyselenides[108,221] have been reduced to the corresponding alkanes or hydroxyalkanes with $Li/EtNH_2$ or with tin hydrides.

Selenoacetals, including functionalized ones, have been reduced selectively[15,148,221] to selenides or to the corresponding alkanes depending upon the number of equivalents (1 or 2) of tin hydride used. Direct reduction to alkanes has also been performed on reaction with Raney nickel[14] or lithium in ethylamine[14].

The reduction of vinyl selenides is more complex. Thus whereas methyl vinyl selenides are transformed to the corresponding olefins on reaction with tin hydride[185], involving the reduction of the Se—sp² carbon Bond, Se—methyl bond cleavage is observed[185] if lithium in ethylamine is used (Scheme 60).

$$OctCH{=}C(SeMe)Me \xrightarrow[\text{toluene, 90°C/3h}]{\text{Bu}_3\text{SnH/AIBN}} OctCH{=}CHMe \quad 51\%$$

$$OctCH{=}C(SeMe)Me \xrightarrow[-15\text{°C/1 h}]{\text{Li/EtNH}_2} OctCH{=}C(Me)Se{-}SeC(Me){=}CHOct \quad 68\%$$

SCHEME 60[185]

Aluminium amalgam[153], P_2I_4[223] and PI_3[223] efficiently reduce α-selenoketones to ketones (Scheme 61). P_2I_4 surprisingly allows the direct transformation of α-seleno-α-cyclopropyl ketones to γ-iodoketones[223] (Scheme 61b).

$$OctCH(SeR)COMe \xrightarrow[\text{R = Ph or Me}]{0.5 \text{ mol eq. } P_2I_4,\ 20\text{°C/0.4 h}} OctCH_2COMe \quad (>80\%) \qquad \text{(Ref. 223)} \quad \text{(a)}$$

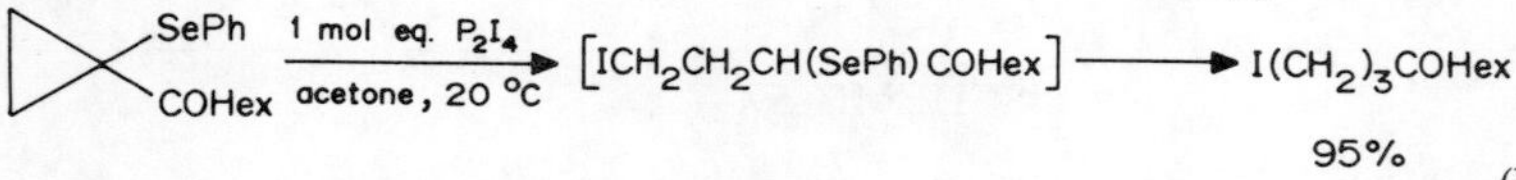

(Ref. 223) (b)

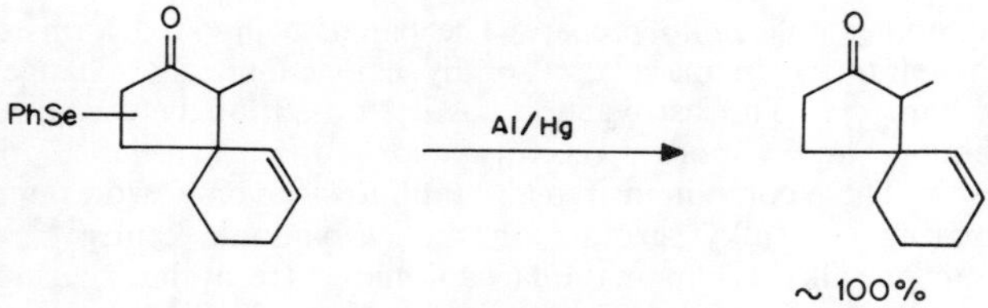

(Ref. 153) (c)

SCHEME 61

As already mentioned, the reduction of the C—Se bond can be performed with alkyllithiums. Thus α-selenoalkyllithiums and α-lithioselenoacetals (prepared from selenoacetals or selenoortho esters) hydrolyse or deuterolyse to selenides and selenoacetals or deuteriated analogues.

Selenides are also reduced with alkyllithiums and useful reactions have been performed on allyl or benzyl selenides which quantitatively produce allyl- and benzyl-lithiums[12] on reaction with BuLi in THF.

2. *Reactions involving the transformation of the C—Se bond to a C—halogen bond*

Selenides are valuable precursors of alkyl halides. The reactions can be directly achieved from dialkyl or aryl alkyl selenides and bromine (in CH_2Cl_2/NEt_3[17,18] or in ethanol[18]) or alkyl halides (CH_3I[17] or $BrCH_2COOMe$[106b] in DMF).

These reactions proceed through selenonium salts which can be isolated if the reactions are performed at low temperature but which decompose slowly in the presence of triethylamine or by refluxing in DMF to produce alkyl halides. The Br_2/NEt_3 method is particularly valuable for the synthesis of secondary or tertiary alkyl bromides from the corresponding selenides and occurs with complete inversion of configuration at the substituted carbon atom in case of the secondary alkyl selenides[18].

Coupled with the Grieco reaction[224] which allows the synthesis of selenides from alcohols with inversion of configuration[18], these reactions permit two-step synthesis of alkyl bromides from alcohols with overall retention of configuration[18] (Scheme 62).

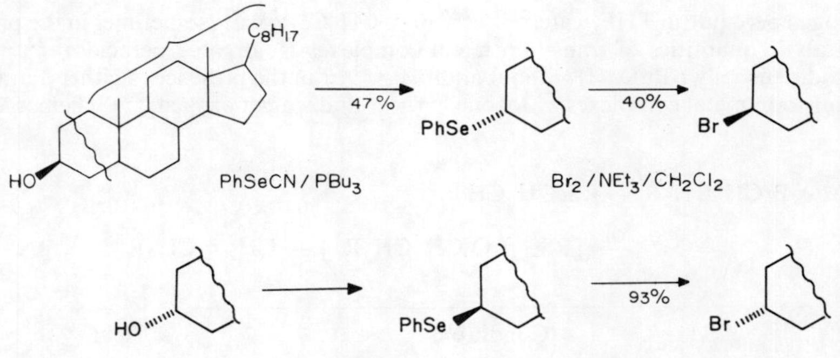

SCHEME 62[18]

The RX/DMF method proceeds efficiently with selenides possessing a primary alkyl chain, but with secondary alkyl selenides it mainly leads to olefins.

Both methods have been successfully applied[106] to the synthesis of γ-halogenoselenides from γ-hydroxyselenides (Scheme 63a) but exclusively produce tetrahydrofurans[106] if applied to the higher homologues, namely δ-hydroxyselenides (Scheme 63b).

$$\text{HexCH(SePh)CH}_2\text{CH(OH)Me} \xrightarrow[80^\circ\text{C/3 h}]{\text{Br}_2\text{/EtOH–H}_2\text{O}} \text{HexCHBrCH}_2\text{CH(OH)Me} \quad \text{(a)}$$

80%

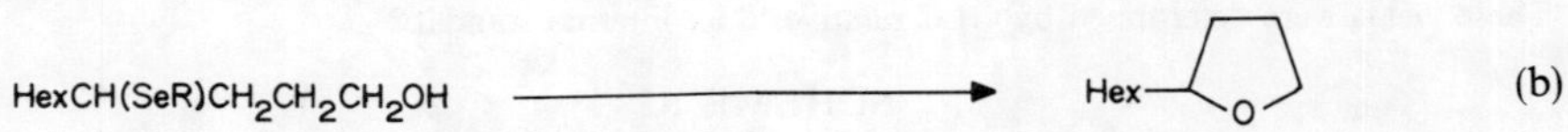

R = Me	Br_2/EtOH–H_2O, 80°C/3 h	77%
R = Ph	Br_2/EtOH–H_2O, 80°C/3 h	70%
R = Me	Br_2/NEt_3–CH_2Cl_2, 20°C/3 h	80%
R = Ph	Br_2/NEt_3–CH_2Cl_2, 20°C/3 h	74%

SCHEME 63[106a]

S—Br exchange has also been performed on vinyl selenides[34,120,185,225], but the addition of bromine to the C=C bond as well as to the Se atom can occur. In benzene or chlorobenzene vinylbromide was formed[34,120,185], whereas a mixture of vinyl bromide and α-bromoketone, arising presumably from the addition of bromine across the C=C bond, was observed[185] in CH_2Cl_2 or CH_2Cl_2/NEt_3 (Scheme 64).

$$\text{OctCH=CMe(SeMe)} \longrightarrow \text{OctCH=CBrMe} + \text{OctCHBrCOMe}$$

Br_2/PhH or PhCl, 20°C/15 h	51%	0%
Br_2/CH_2Cl_2	43%	24%
Br_2/NEt_3–CH_2Cl_2	60%	24%

SCHEME 64[185]

3. *Selenoxide synthesis and reactivity*

a. Synthesis. Selenides are readily oxidized to selenoxides[3–6,8]. Among the various reagents used for selenoxide formation the following have been the most frequently used:

hydrogen peroxide in THF/water[7,226,227] or in CH_2Cl_2[125,227], sometimes in the presence of catalytic quantities of transition metal complexes[228], organic peracids[22,227,229,230], peroxides (usually t-BuO_2H) without additives[231] or in the presence of either alumina[232] or transition metal complexes[228], ozone[9,227,232] and singlet oxygen[320] (Schemes 4a, 7, 8, 65).

$$RSeM + BrCH_2CH_2R^1 \rightarrow RSeCH_2CH_2R^1$$

$$\xrightarrow{(O)} [RSe({=}O)CH_2CH_2R^1] \rightarrow CH_2{=}CH_2R^1$$

R	R^1	Conditions	Yield (%)	Ref.
Ph	$C_{10}H_{21}$	10 eq. H_2O_2/THF, 25°C/20 h	77	233
o-$NO_2C_6H_4$	$C_{10}H_{21}$	10 eq. H_2O_2/THF, 0°C/0.5 h	62, 91[a]	233
m-$CF_3C_6H_4$	$C_{10}H_{21}$	10 eq. H_2O_2/THF, 25°C/2 h	70[a]	233
p-$MeOC_6H_4$	$C_{10}H_{21}$	10 eq. H_2O_2/THF, 25°C/7 h	60[a]	233
Ph	C_8H_{17}	O_3/CH_2Cl_2, – 78°C then 45°C/24 h	30	232
		O_3/CH_2Cl_2, 3 eq. NEt_3, 45°C/24 h	75	232
		t-BuO_2H/Al_2O_3/THF, 65°C/3 h	86	232
Me	C_8H_{17}	O_3/CH_2Cl_2, – 78°C then 45°C/24 h	< 1	232
		O_3/CH_2Cl_2, 3 eq. NEt_3	23	232
		t-BuO_2H/Al_2O_3/THF, 65°C/3 h	83	232

[a]These yields were determined by GLC relative to an internal standard.

SCHEME 65

The reaction with ozone can be performed at low temperature (– 78 °C) and seems to be highly selective, selenides being oxidized selectively even in the presence of an olefin which remains unchanged. Selenoxides are stable at this low temperature and do not eliminate.

Hydrogen peroxide in THF is one of the best methods if one wishes to oxidize the selenide, followed by its decomposition to olefin. The reaction is often conducted in the presence of an excess of hydrogen peroxide which further oxidizes the selenenic acid, concomitantly formed, to selenenic acid[3–6,8,26,27], avoiding the addition of selenenic acid across the C=C bond of the olefin. The hydrogen peroxide method does not always permit selenoxide formation. In the case of vinyl selenides for example the C=C bond is cleaved and carbonyl compounds are formed[123] (Scheme 66).

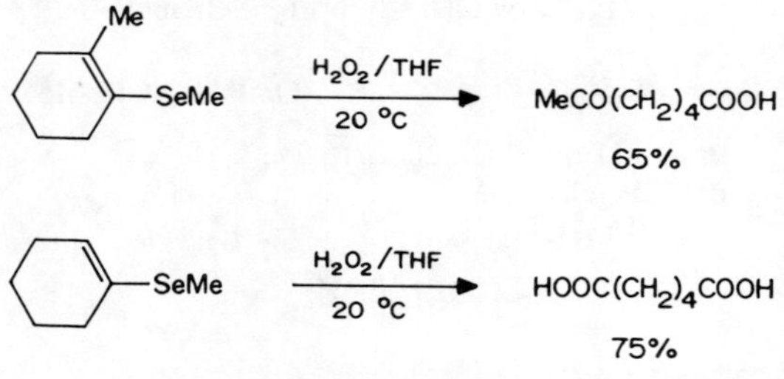

(a)

(b)

SCHEME 66[123]

If the selenoxide is used as an intermediate for olefin synthesis it is often important to remove the selenenic acid from the medium[3–6,8,26], in order to avoid its addition to the olefin. This can be done as above with an excess of oxidant, or if the reactions are carried out in the presence of an amine[227,232,234] or basic alumina[232] which trap the selenenic acid.

b. Reactions involving the selenoxide elimination. The stereochemistry of the selenoxide elimination has been investigated[27]. It proceeds, as in the case of sulphoxide elimination, in a *syn* fashion[27,91]. Further insight into its intimate mechanism has been described[235]. The reaction is not usually regio- and stereo-selective if several hydrogens are available on the C atoms β to the one bearing the selenoxy group. In the product mixture of different olefins the less substituted ones predominate, in which the formation of the *E* isomer is greatly favoured over the *Z*. The *syn* elimination of selenenic acid is an imperative requirement and therefore olefin formation does not occur when this stereochemical requirement cannot be achieved[28,91] (Scheme 67).

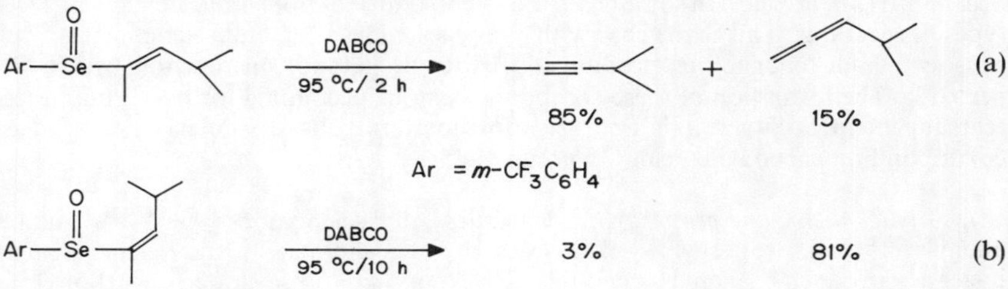

Ar = m-$CF_3C_6H_4$—

SCHEME 67[264]

c. Regiochemistry of the selenoxide elimination. In several instances the reaction is regioselective. Thus in the case of β-hydroxyselenides[11,22,108,126,127,132,233,241,252] the elimination occurs away from the hydroxyl group and produces allyl alcohols except in the case where the only hydrogen available is the one α to the hydroxyl group.

Similar observations have been made with β-acetoxy-[226] or trifluoromethylacetoxy-[242] selenoxides[220,226,243–246], β-alkoxyselenoxides[220,226,243–247], β-selenoxyacetals[248], β-halogenoselenoxides[226,249], at the exclusion of the chloro compounds, β-selenoxyalkylamides[250] and β-selenoxyamines[168] which lead, respectively, to allyl acetates or trifluoromethyl acetates, allyl ethers, allylic halides, enamides and amines.

β-Azidoselenoxides[251] for their part eliminate in both directions producing a mixture of vinyl and allyl azides. Finally, a series of functionalized selenoxides have been successfully transformed to functionalized olefins. That is the case among others of α-selenoxy aldehydes[14,136], α-selenoxyketones[2,22,136,227,229,253–255], α-selenoxy esters[136,229,256,257], α-selenoxylactones[91,258,259], α-selenoxy acids[2,136], α-selenoxyamides[260], α-selenoxynitriles[92,261], α-selenoxynitroalkanes[103,262] and α-selenoxyphosphonium salts[230], which produce respectively α,β-unsaturated aldehydes, ketones, esters, lactones, acids, amides, nitriles and nitroalkanes and vinyl phosphonium salts. In most cases the synthesis of these compounds has been directly achieved at 20 °C or below from the corresponding selenide and an oxidant, without isolation of the intermediary selenoxide.

d. Structure–reactivity relationships in the selenoxide elimination. The nature of the selenoxy group has a great influence on the elimination. The *o*-nitrophenylselenoxy[233], *m*-chlorophenylselenoxy[233], pyridylselenoxy[263] and *m*-(trifluoromethyl)phenylselenoxy[264] groups eliminate much faster than their phenylselenoxy analogues, which themselves react at a temperature below 20 °C. Methylselenoxy derivatives[232] react much more slowly in CH_2Cl_2 or THF. In the two latter cases valuable improvements are observed if the reactions are carried out in the presence of an amine[227,232,234] or at reflux temperature in THF in the presence of *t*-butyl hydroperoxide and basic alumina[232].

All things being equal, secondary alkyl selenoxides react faster than primary ones and the presence of a carbonyl group on the carbon bearing the selenoxy group favours the reaction, whereas a cyclopropyl[20] or a vinyl[123,264] group disfavours it. In the latter cases the selenoxide often releases the oxygen and reverts to the starting selenide, but the presence of a group with a strong inductive effect linked to the Se atom can restore the reactivity. Thus, whereas vinyl phenyl selenoxides do not fragment[123,264] their *m*(trifluoromethyl)phenyl analogues afford allenic and/or acetylenic compounds by selenenic acid elimination[264] (Scheme 67).

e. α-Silyl selenoxides. In some cases the selenoxide does not eliminate but leads to other types of reactions. This is the case with α-silyl selenides[74–76] and selenoacetals[266] which possess a high tendency to produce aldehydes or ketones on reaction with hydrogen peroxide. The formation of these compounds can be accounted for by a Pummerer-type rearrangement. α-Silylacetals and selenoorthoesters behave similarly and lead to the corresponding carboxylic acids[74] in high yields.

f. Allylic, allenic and propargylic selenoxides. Allyl selenoxides[27,77,267,269] and related allenyl[5,268] and propargyl[81] selenoxides do not eliminate[82] the selenenic acid but instead produce allenyl alcohols[78,267] or α,β-unsaturated carbonyl compounds[5,81,268], respectively. The reaction therefore implies an allyl selenoxide → allyl seleninate rearrangement rather than the selenoxide elimination reaction reported previously. Arylselenoxy derivatives are much more prone to rearrangement than their methylselenoxy[264,269] or arylsulphoxy[81,269] analogues. For example the cyclopropyl allyl derivatives shown in Scheme 68 rearrange to alkylidene cyclopropane in the phenylseleno series[269] but not in the methylseleno or phenylthio series.

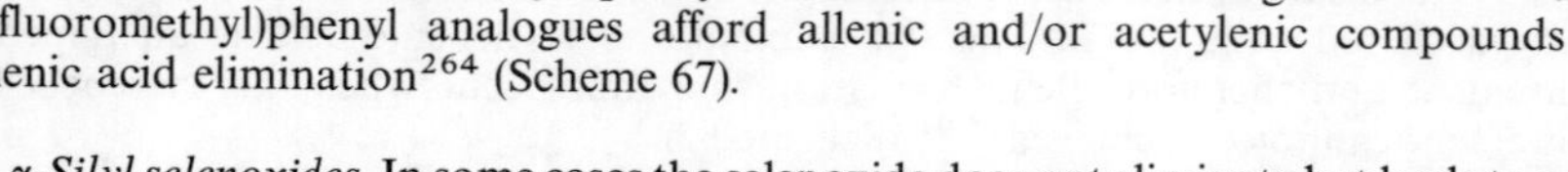

X = SePh	H_2O_2/THF–H_2O	0%
	H_2O_2/MeOH	0%
	$NaIO_4$/MeOH, extraction then piperidine/cyclohexane	77%
X = SeMe	As above	16%
X = SPh	As above	0%

SCHEME 68[269]

g. β-Alkyloxyselenoxides. As already mentioned β-hydroxyselenides fragment to allyl alcohols. A ring-enlargement leading to cyclopentanones takes place[153] when β-alkoxyselenoxides, derived from cyclobutanones, are heated in THF (Scheme 69).

h. Vinyl selenoxides. A series of vinyl selenoxides have been reacted with enolates, and have led via α-lithioselenoxides, to α-cyclopropyl ketones and esters in fair to good yield. A specific example of this kind of reaction[271] is shown in Scheme 70.

i. Reduction of selenoxides to selenides. Selenoxides have been reduced to selenides by a large variety of reagents which include P_2I_4 and PI_3[159,160,270], phosphine, tin

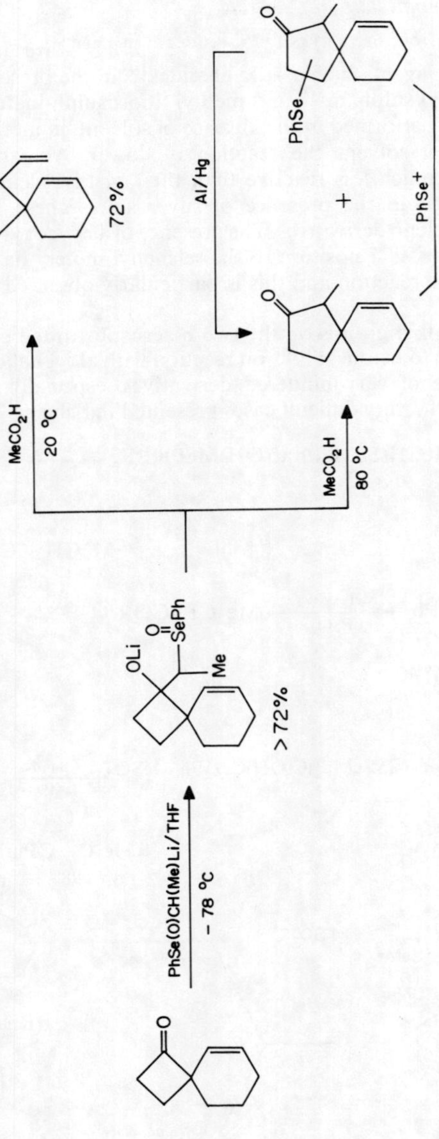

SCHEME 69[153]

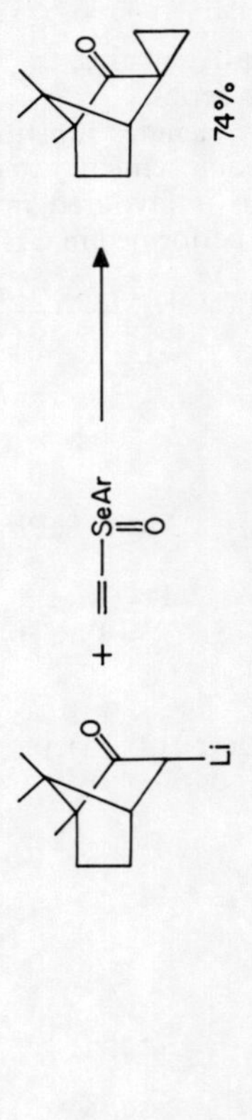

Ar = p-ClC_6H_4

SCHEME 70[271]

chloride[21] and selenolate[271]. The selenoxides have also been used as mild and selective oxidizing agents[6,272–279]. Their transformation to alkyl chlorides and bromides has been successfully performed[23] with HCl or HBr.

4. Synthesis and reactivity of selenonium salts

a. Synthesis. Selenonium salts[19,94,106,108,129,135,137,138,140,251,269,280] are usually prepared from selenides and alkylating agents such as alkyl halides[19], in the presence or not of silver tetrafluoroborate[19], dialkyl sulphates[19] and methyl fluorosulphonate[20,108] (magic methyl). The alkylation is often performed in the absence of solvent, in methylene dichloride or in ether but in the last solvent the reaction is slower. As expected, phenylseleno derivatives[19,137,140] are much less reactive than their methylseleno analogues[94,138], and are usually alkylated in the presence of silver salts. These strong conditions are not required for methylseleno derivatives. The presence of a hydroxyl group in the β-[94,135,137,138,281], γ-[106b,129] or δ-[106b] position to the selenenyl moiety has been found to greatly enhance the rate of the reaction and this is particularly observed in hte methylseleno series.

β-Hydroxyselenides substituted by alkyl groups on the two heterosubstituted carbon atoms have some tendency to rearrange to ketones[19,108] on reaction with alkyl halides[137]. This reaction is favoured in the case of very hindered derivatives, especially if the alkylation is performed in CH_2Cl_2. In the very difficult cases presented in Scheme 71, the

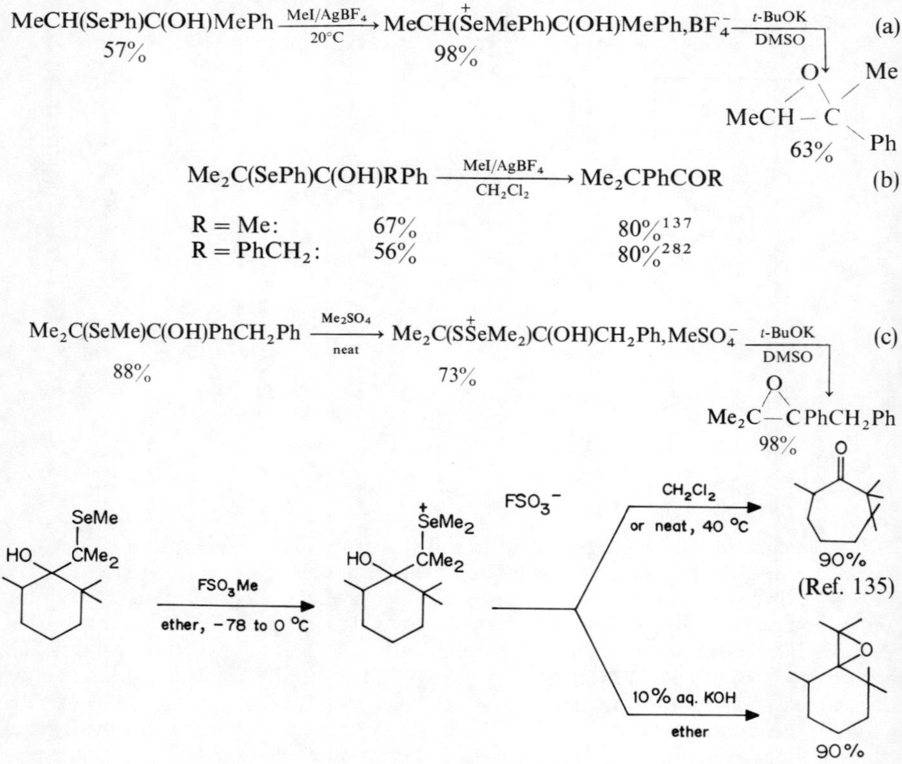

SCHEME 71

solvent was found to have a crucial role. Thus the hindered β-hydroxyselenonium salts can be quantitatively prepared if ether[108] is used as the solvent and rearrange immediately to ketones when they are dissolved in CH_2Cl_2 (Scheme 71).

b. Reactivity. Selenonium salts are reactive compounds. They react very specifically but in very different ways with nucleophilic/basic reagents. Substitution is observed with enolates[19] and alcoholates[19] (Schemes 3 and 50) whereas attack on the Se atom takes place with alkyllithiums[11] (Schemes 3 and 51).

Metalation has been found to occur on reaction with a base such as *t*-BuOK in DMSO[20] or in THF[20], with KOH in the same solvents[20] or with metalloamides[11]. The resulting ylides have been trapped *in situ* with carbonyl compounds to produce epoxides[11,166].

In the absence of carbonyl compounds an internal elimination involving the selenide and a β-hydrogen, if available, takes place and produces a very good yield of olefin. This reaction is related to the selenoxide elimination[20] but is far superior to it for the synthesis of alkylidene cyclopropanes[20,269] (Schemes 4b and 72) and allylidene cyclopropanes[269] from cyclopropyl selenides. If several hydrogens are available in the β-position, the elimination is not regiospecific and leads to a mixture of regioisomeric olefins in which the less substituted ones predominate[20]. Moreover, in the case of α,β-disubstituted olefins the *E* isomer is exclusively formed[20].

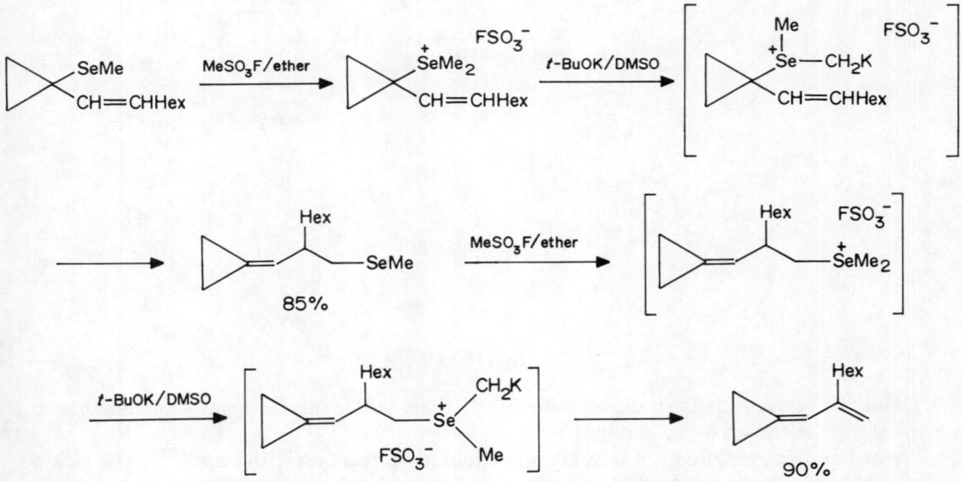

SCHEME 72[269]

The reaction of selenonium salts with bases requires further comments. Thus the presence of a hydroxyl group in the molecule often changes the course of the reaction:

(*i*) β-Hydroxyselenonium salts do not usually produce allyl alcohols[108] by the elimination reaction just reported but instead lead to epoxides by an internal substitution reaction. The reaction proceeds stereospecifically[2,140] and allows the synthesis of terminal[94,137] and polyalkylated derivatives[94,137,140] and hindered[108] and α,β-unsaturated compounds[138] (Scheme 73).

(*ii*) γ-Hydroxyselenonium salts in which the selenyl moiety is attached to a methylene group produce oxetanes[106b,307] (Scheme 74a), whereas homoallyl alcohols are exclusively observed[129] if a hydrogen is present on the carbon δ to the hydroxyl group (Scheme 74b).

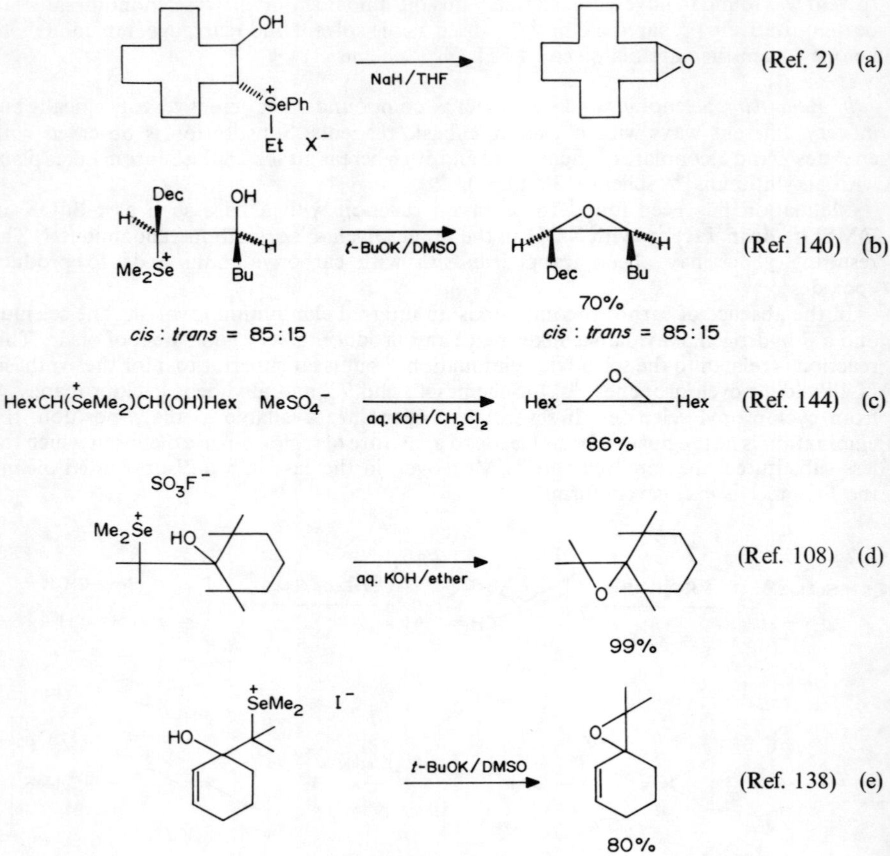

SCHEME 73

The high regioselectivity of the reaction (no allyl alcohol formed) suggests an anchimeric assistance of the alkoxide group[127,129].

(*iii*) δ-Hydroxyselenonium salts are also cyclized on reaction with bases[106b]. The yields of tetrahydrofurans are modest and other unidentified products are formed (Scheme 75). Similar results are observed if the selenonium salts are heated at 80 °C in chloroform[106b] in the absence of base.

5. Synthesis using functionalized selenides

a. Synthesis of olefins from β-hydroxyselenides. As already shown selenides are valuable precursors of olefins taking advantage of the easy thermal elimination involving the corresponding selenoxides or selenonium ylides. Other reactions producing olefins are in some cases more flexible than the just reported ones. While oxidation of selenoacetals[266], orthoseleno esters[128] and α-silyl selenides[74–76] leads to the formation of carbonyl compounds, vinyl selenides[123,124,324,325] and ketene selenoacetals[124] are however prepared from selenoacetals[266] and orthoesters[74], respectively, on reaction with MeI in DMF at 80 °C[123] or with PI_3 or P_2I_4 in chlorinated solvents[124].

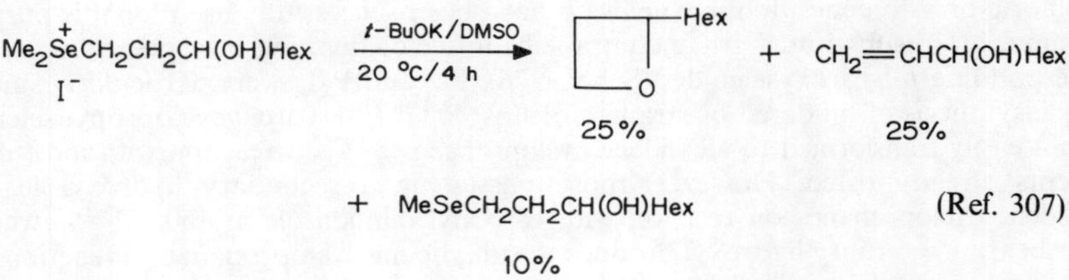

(Ref. 307) (a)

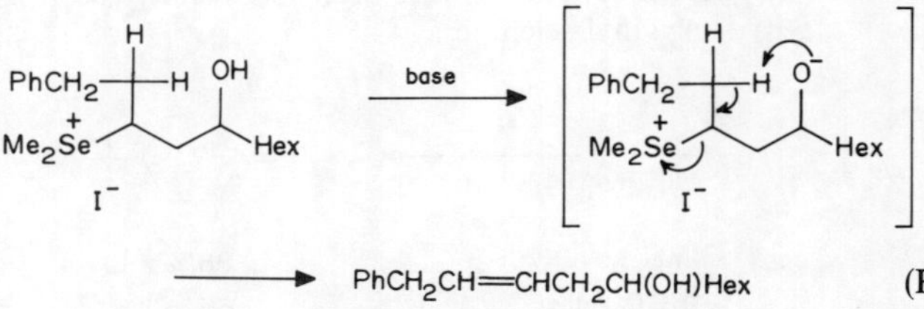

(Ref. 129) (b)

1 mol eq. *t*-BuOK/DMSO, 20°C/4 h	91%
aq. 50% KOH, Adogen 464[a], CH_2Cl_2, 20°C/4 h	73%

[a] Trade name (Ashland Chemical Co.) for methyltrialkyl (C_8–C_{10}) ammonium chlorides.

SCHEME 74

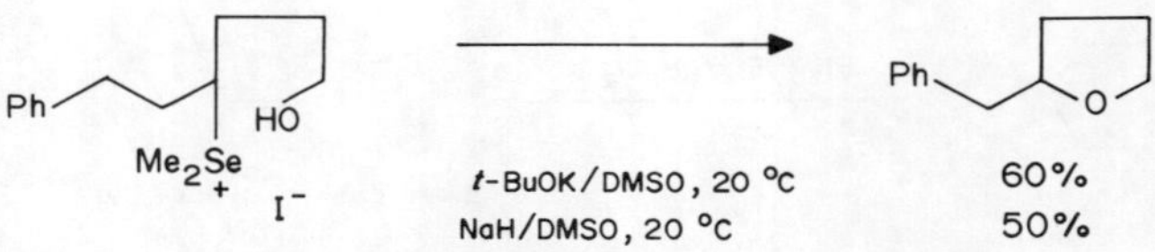

SCHEME 75[106b]

As already pointed out elimination of a hydrogen and a selenyl moiety leads to regio- and stereo-isomeric mixtures of olefins when several topologically different hydrogens are available.

Concomitant elimination of a selenenyl and another heteroatomic moiety in β-heterosubstituted selenides overcomes these problems. Thus β-halogenoselenides are regio- and stereo-selectively transformed to olefins on reaction with selenolate ions[283] or in the presence of an amine[21,131]. The reaction was found to occur by formal *trans* elimination[72,100,130,131,140] of the two heteroatomic moieties. Episeleniiranium salts have been proposed as intermediates in these reactions[72,131]. Attack of the selenolate in the first case and the amine in the second on the charged Se atom of the seleniiranium ion can explain both the reaction and its stereochemical course.

β-Hydroxyselenides are readily available by reaction of α-selenoalkyllithiums with carbonyl compounds[2,4–8,11,67,100,108,126,130,139], by reduction of β-hydroxyselenoxides[21], by reaction of organometallics on α-selenocarbonyl compounds[39,40,143], by ring-opening of epoxides with selenolates[130,131,284], by reaction of olefins with selenenic acids[231,281,327,337] or related reagents or by hydrolysis of β-halogenoselenides[251]. β-Acetoxy-[226] and trifluoroacetoxy-[285,286] selenides are also found to be valuable precursors of olefins: the reaction takes advantage of the easy transformation of the hydroxyl group to a better leaving group.

Mesyl chloride[21,72,287], thionyl chloride[130], trifluoroacetic anhydride[131,288], phosphorus oxychloride[88,100], diphosphorus tetraiodide[108] or phosphorus triiodide[107,108,269] in the presence of an amine, trimethylsilyl chloride and NaI in MeCN[289] as well as perchloric or *p*-toluenesulphonic acids[131] have been successfully used for this purpose. Terminal, α, β-disubstituted, tri- and tetra-substituted olefins have been prepared from the corresponding β-hydroxyselenides (Scheme 76). PI_3 and P_2I_4 were particularly suitable for the synthesis of hindered or strained olefins[107,108] β-Hydroxycyclopropyl selenides are not easily transformed to alkylidene cyclopropanes[8b,107] on reaction with most of the reagents already listed. However, those possessing a secondary hydroxyl lead to alkylidene cyclopropanes on reaction with carbonyl diimidazole at 160 °C[8b,107] whereas those bearing a tertiary hydroxyl produce the alkylidene cyclopropanes on reaction with PI_3 already at 20 °C[107]. The last conditions have been successfully used for the synthesis of an allene from a β-hydroxyvinyl selenide[185].

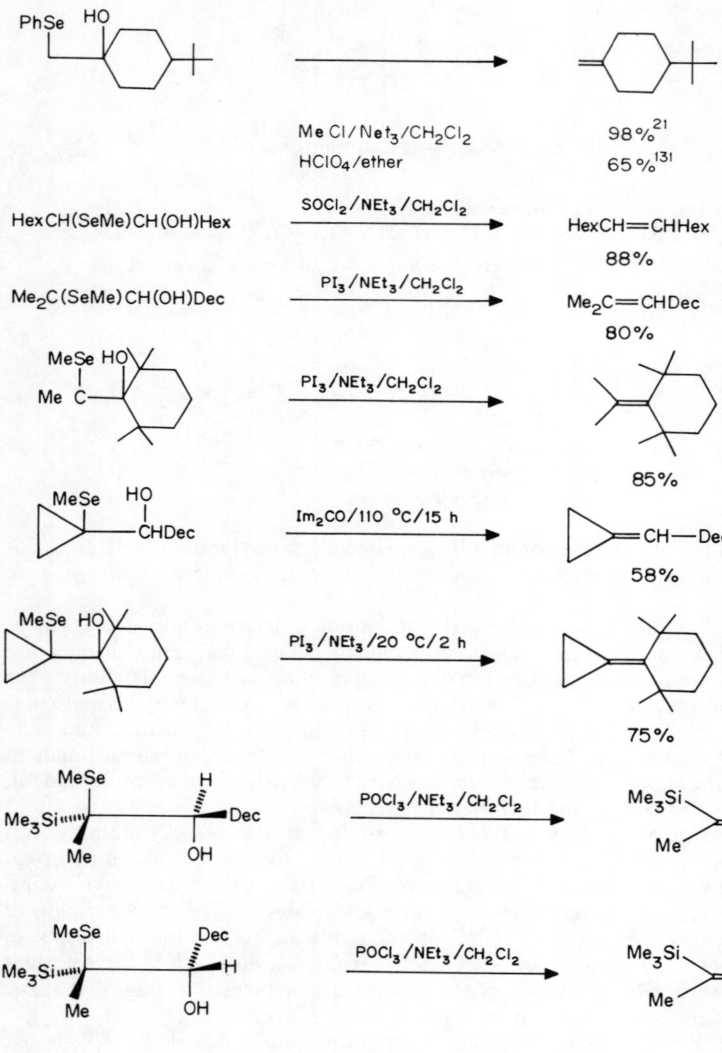

(a)

(Ref. 130) (b)

(Ref. 144) (c)

(Ref. 108) (d)

(Ref. 107) (e)

(Ref. 107) (f)

(Ref. 100) (g)

(Ref. 100) (h)

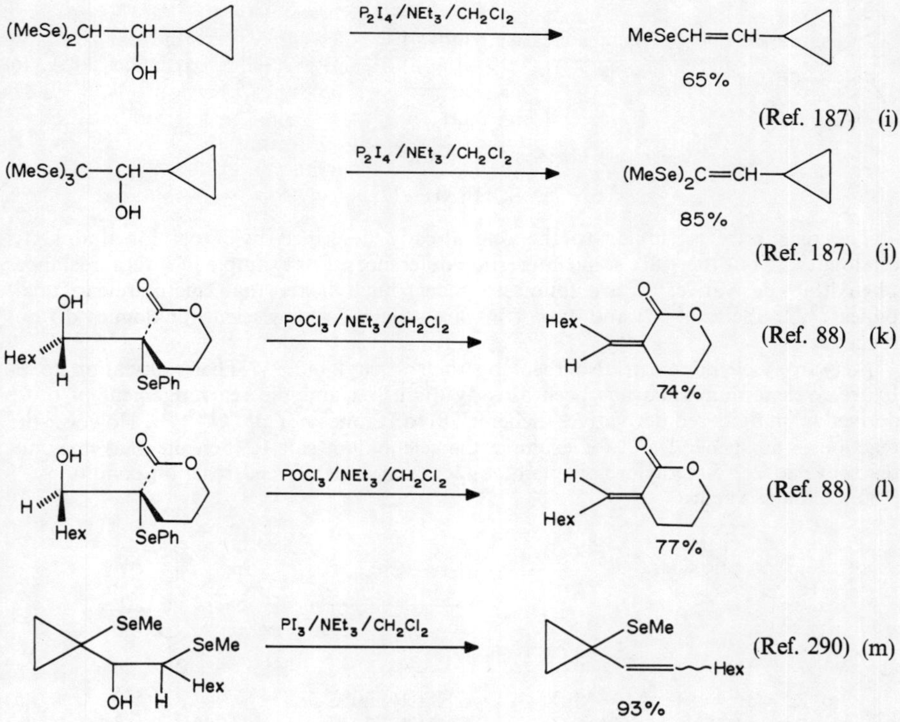

SCHEME 76

b. Synthesis of ketones from β-hydroxyselenides. The presence of two potential leaving groups β to each other makes β-hydroxyselenides valuable candidates for pinacolic-type rearrangements. The selective transformation of each of the two groups to leaving groups has already been described. Usually, transformation of the hydroxyl group to a better leaving group does not lead to carbonyl compounds but to olefins (see above). However, β-hydroxycyclopropyl selenides produce cyclobutanones on reaction with acids[8b,107,109,188] (Scheme 77).

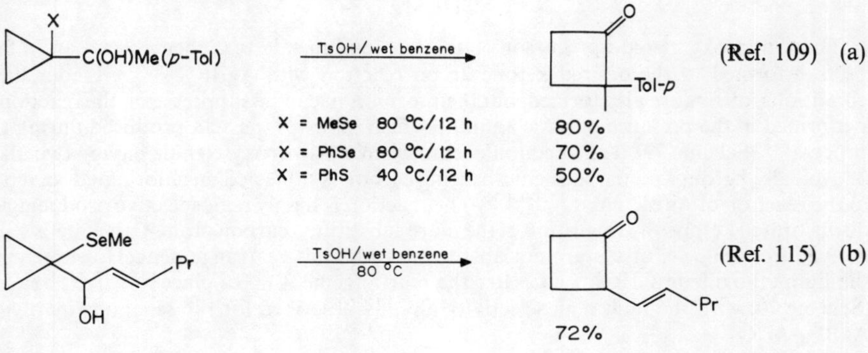

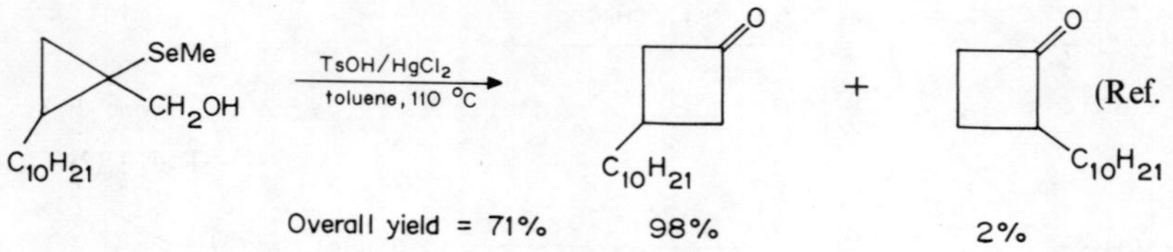

(Ref. 115) (c)

SCHEME 77

The reaction[8b] is similar to the one already described by Trost for thiophenyl analogues [328] but there are some interesting differences. For example in several instances phenylthio derivatives[109] are found to react much faster than methylseleno analogues[109,115] (Scheme 77) and except in rare cases[109] phenylseleno analogues do not react[115].

β-Hydroxyselenides can also react by their selenyl moiety. Their alkylation to β-hydroxyselenonium salts has been already discussed and the rearrangement of those derived from hindered derivatives (Scheme 78) to ketones noted[108,109,135]. However the reaction is not general and for example the selenonium salt **1c** (Scheme 78a) does not rearrange to 2,2-dimethylcycloheptanone (**2c**) even under forced reaction conditions[135] (180 °C, under vacuo).

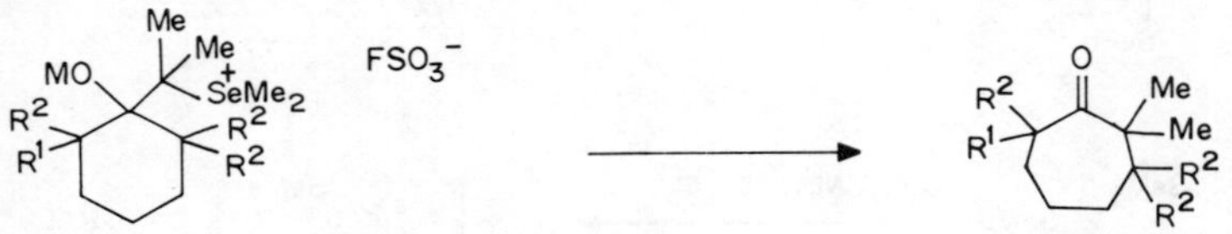

(1) **(2)**

a) $R^1 = R^2 = Me$; $M = MgX$	CH_2Cl_2, 20°C	94%[108,135]	(a)
b) $R^1 = H, R^2 = Me$; $M = H$	CH_2Cl_2, 20°C	90%[108,135]	(b)
c) $R^1 = R^2 = H$; $M = H$	neat, 180°C	trace[135]	(c)

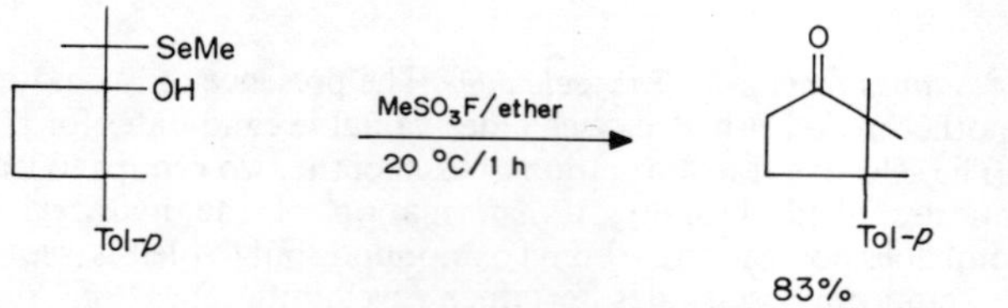

(Ref. 109) (d)

83%

(±)-Cuparenone

SCHEME 78

The β-hydroxyselenide precursor of the selenonium salt **1c** (Scheme 78a) can in fact be transformed to the desired ketone **2c** on reaction with $AgBF_4$[109,135]. Under these conditions, olefins are also formed, but their formation can be suppressed if the reaction is performed in the presence of basic alumina which removes the acid produced during the process[135] (Scheme 79). This reaction occurs when a β-hydroxyselenide having two alkyl or cycloalkyl groups on the selenenyl-bearing carbon is subjected, in chlorinated solvents, to the reaction of $AgBF_4/Al_2O_3$[109,135]. The reaction is highly regioselective producing α-disubstituted ketones by migration of the more substituted carbon atom. Other metal salts derived from Hg, Ni or Cu are not able to promote this rearrangement. However with thallium ethoxide in $CHCl_3$ or $CHBr_3$ the rearrangement takes place[282,339,340] readily (Scheme 79) with the high regioselectivity already described for the reactions involving $AgBF_4/Al_2O_3$.

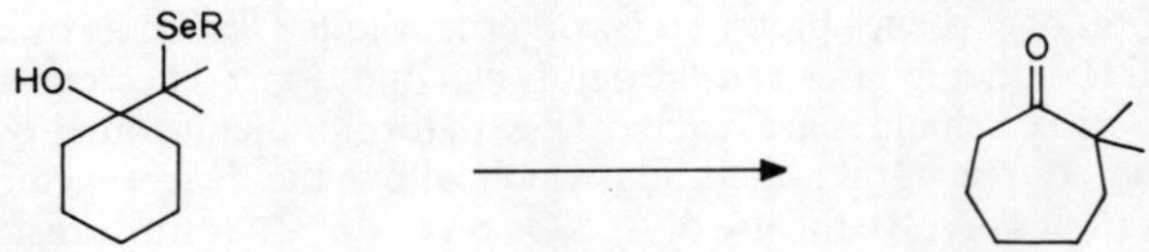

(Ref. 135)

R = Me, Ph	$AgBF_4/CHCl_3/40°/0.2$ h	40%	(a)
R = Me	$AgBF_4/Al_2O_3/CHCl_3$	58%	(b)
R = Ph	$AgBF_4/Al_2O_3/CHCl_3$	85%	(c)

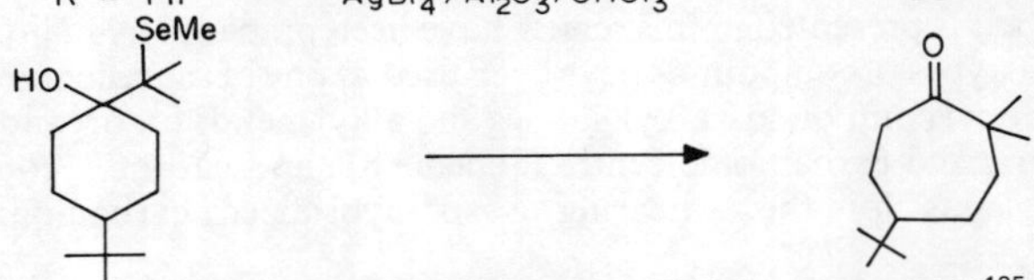

	$AgBF_4/CHCl_3$, 20 °C/8 h	61% [135]	(d)
	$AgBF_4/Al_2O_3/CHCl_3$, 20 °C/4 h	62% [135]	(e)
	$TlOEt/CHCl_3$, 20 °C/8 h	96% [282]	(f)

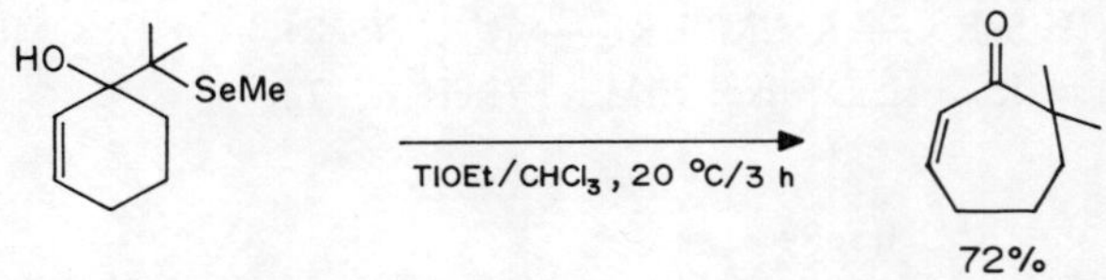

(Ref. 282) (g)

$$\text{HeptCH(OH)C(SePh)Me}_2 \xrightarrow{\text{TlOEt/CHCl}_3,\ 20°\text{C}/15\text{h}} \text{HCOCHeptMe}_2 + \text{HeptCOCHMe}_2$$

Overall yield = 70% 60% 40%

(Ref. 282) (h)

SCHEME 79

Thallium ethoxide also reacts with other β-hydroxyselenides[282]. Those which bear one or two hydrogens on the carbon bearing the selenyl moiety do not lead to rearranged ketones but to epoxides with high stereoselectivity[282] (Scheme 80).

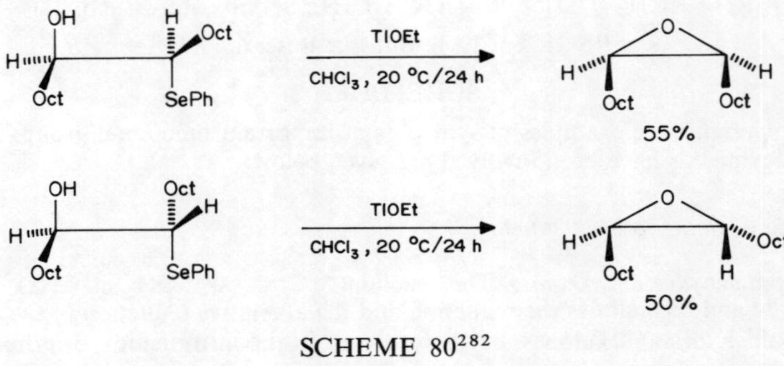

(a)

(b)

SCHEME 80[282]

Preliminary results seem to indicate the intermediary formation of a carbene by reaction of TlOEt with $CHCl_3$ or $CHBr_3$, one of these being required for the success of the reactions[282]. In fact a similar rearrangement is observed if β-hydroxyselenides are subjected to the reaction of carbenes prepared by other methods[340] such as by $CHCl_3$/*t*-BuOK or $CHCl_3$/KOH under phase-transfer catalysis conditions. The rearrangement takes place also if β-hydroxyselenides are reacted[340] with (bromodichloromethyl)phenylmercury as a carbene source. Other methods which allow the rearrangement of β-hydroxyselenides to ketones involve the use of $AgNO_3$ on celite[329], or the transformation of the selenyl moiety to a selenoxy[153,241] or selenonyl[330] one.

D. Use of Selenoalkylmetals in Organic Synthesis

The α-selenoalkylmetals presented in this review have been prepared by many different routes in which several types of compounds have been used as one of the reaction partners. Among these are carbonyl compounds, alkyl halides and alkyl metals for organometallics possessing a sp^3-hybridized carbanionic centre (Scheme 81) and carbonyl compounds, alkyl halides and olefins for those bearing a sp^2-hybridized carbanionic centre (Schemes 82 and 83).

(a) Carbonyl compounds

$$R^1R^2C{=}O \xrightarrow{a^*} R^1R^2C(SeR)_2 \xrightarrow{b} R^1R^2CM(SeR) \quad \text{(Refs. 8, 11, 67, 94, 126)}$$

$$R^1CHO \xrightarrow{a} R^1CH(SeR)_2 \xrightarrow{c} R^1CM(SeR)_2 \xrightarrow{d} R^1R^2C(SeR)_2 \xrightarrow{b} R^1R^2CM(SeR)$$

$$R^2 = H, D, \text{alkyl}, SiMe_3 \quad \text{(Refs. 67–72)}$$

(b) Alkyl halides

$$R^1R^2CHX + MSeR \rightarrow R^1R^2CH(SeR)$$

$$\rightarrow R^1R^2CHSe({=}O)R \xrightarrow{c} R^1R^2CMSe({=}O)R \quad \text{(Refs. 21, 22)}$$

$$\rightarrow R^1R^2CH(\overset{+}{S}eR_2)\,X^- \xrightarrow{c} R^1R^2CM(\overset{+}{S}eR_2)\,X^- \quad \text{(Ref. 19)}$$

$$R^1R^2CHX + SeR_2 \rightarrow R^1R^2CH(\overset{+}{S}eR_2)\,X^- \xrightarrow{c} R^1R^2CM(\overset{+}{S}eR_2)\,X^- \quad \text{(Ref. 19)}$$

$$R^2X + R^1CM(SeR)_2 \rightarrow R^1R^2C(SeR)_2 \xrightarrow{b} R^1R^2CM(SeR) \quad \text{(Refs. 11, 67, 72, 94, 126)}$$

$$R^2X + MC(SeR)_3 \rightarrow R^2C(SeR)_3 \xrightarrow{b} R^2CM(SeR)_2 \quad \text{(Refs. 67, 68)}$$

(c) Alkyl metals

$$R^1Li + CH_2{=}CHSePh \rightarrow R^1CH_2CHLi(SePh) \quad \text{(Refs. 106b, 21)}$$

∗ a: RSeH, $ZnCl_2$; b: BuM; c: Base; d: $[R^{2+}]$

SCHEME 81

Some representative examples of syntheses of important functional groups in which α-selenoalkylmetals have been involved are given below.

1. Alkanes or deuteriated alkanes

a. From aldehydes and ketones. The reaction($\gt C{=}O \rightarrow \gt CH_2, \gt CHD, \gt CHR$) (Schemes 84 and 85a) allows the reduction and the alkylative reduction of the carbonyl group of aldehydes and ketones, but is restricted to the introduction of primary alkyl groups.

The reaction ($\gt C{=}O \rightarrow \gt C{<}^{R^1}_{R^2}$) (Scheme 86a) is restricted to aromatic carbonyl compounds. R^1 and R^2 can be primary or secondary alkyl groups. In the case of aromatic aldehydes the following process is even possible:

$$ArCHO \rightarrow ArCR^1R^2R^3 \quad \text{(Scheme 86a)}$$

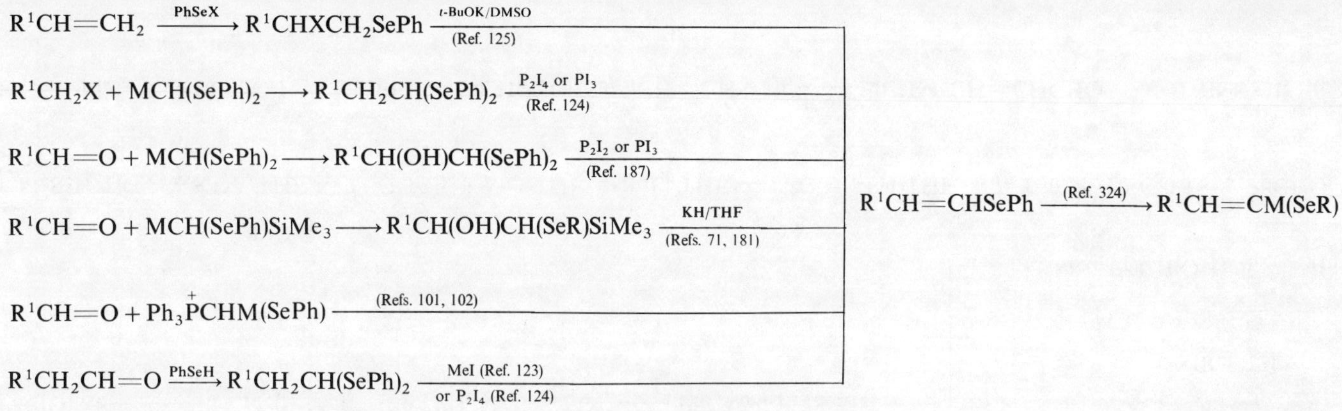

SCHEME 82

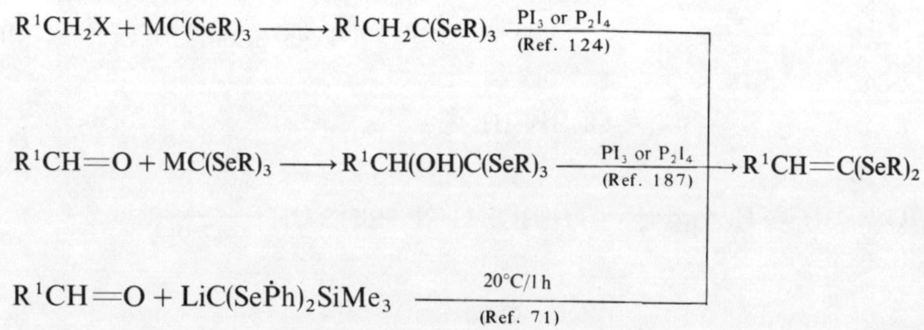

SCHEME 83

$$Dec_2C{=}O \rightarrow Dec_2CLi(SeMe) \xrightarrow{D_2O} \underset{75\%}{Dec_2CD(SeMe)} \xrightarrow{Li/EtNH_2} \underset{87\%}{Dec_2CHD}$$

SCHEME 84[14]

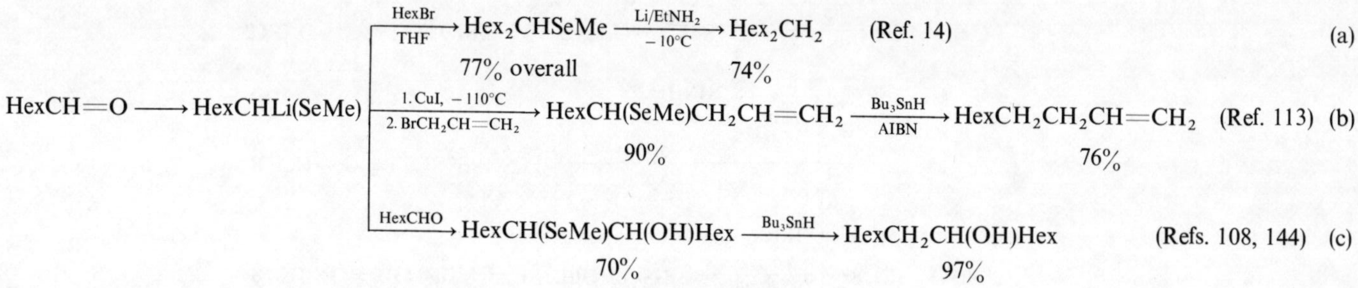

SCHEME 85

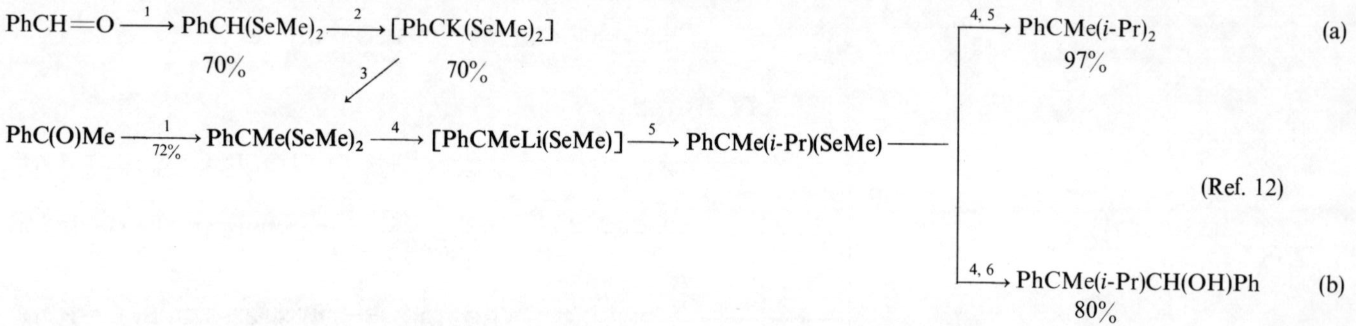

1: $MeSeH/TiCl_4$; 2: KDA/THF, − 78°C/0.2 h; 3: MeI/THF; 4: BuLi/THF, − 78°C; 5: *i*-PrI/THF, − 78°C/0.3 h, 20°C/0.5 h; 6: PhCHO, − 78°C

EtCHBrMe + PhSeCHLiCH=CHSePh ⟶ EtCHMeCHSePhCH=CHSePh $\xrightarrow{H_2O_2}$ EtCHMeCH=CHCHO (75%) (Ref. 321) (c)

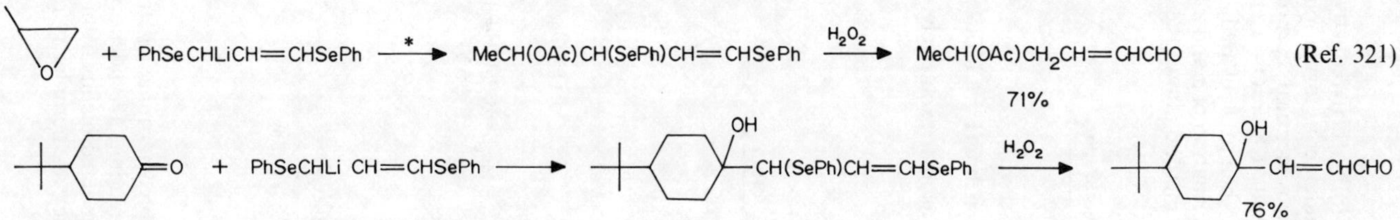

(Ref. 321) (d)

(Ref. 321) (e)

$$Me_2CHBr + PhSeCHLiCH{=}C(SePh)SiMe_3 \rightarrow Me_2CHCH(SePh)CH{=}C(SePh)SiMe_3 \xrightarrow[2.\ Et_2NH]{1.\ MeCO_3H} Me_2CHCH{=}CHC(O)SiMe_3$$

75% (Ref. 321) (f)

$$Me_2C{=}O + PhSeCHLiCH{=}C(SePh)SiMe_3 \rightarrow Me_2C(OH)C(SePh)CH{=}CHSePh$$

$$Me_2C(OH)C(SePh)CH{=}CHSePh \xrightarrow[2.\ Et_2NH]{1.\ MeCO_3H} Me_2C(OH)CH{=}CHC(O)SiMe_3 \quad 60\%$$

$$Me_2C(OH)C(SePh)CH{=}CHSePh \xrightarrow{MeSO_2Cl/NEt_3} Me_2C{=}CHCH{=}CHSePh \quad 80\%$$

(Ref. 321)

*then quenching with acetic anhydride

SCHEME 86

b. From alkyl halides. Schemes 85a and 86a show (C—X→C—R) reactions.

2. Non-functionalized olefins

a. From aldehydes and ketones. This can be achieved (*i*) by their reductive allylation (C(1)=O→C(1)H—C—C=C) (Scheme 85b), (*ii*) by simultaneous formation of σ and π bonds (>C(1)=O→>C(1)=C<) by coupling of two identical carbonyl groups (Scheme 87), (*iii*) by simultaneous formation of σ and π bonds (C(1)=O→C(1)=C) from two different carbonyl compounds, one of them being activated as an α-selenoalkyllithium. (Scheme 88) or (*iv*) By simultaneous formation of σ and π bonds from carbonyl compounds and cyclopropane derivatives; this method allows the synthesis of strained alkylidenecycloalkanes (Schemes 89 and 90).

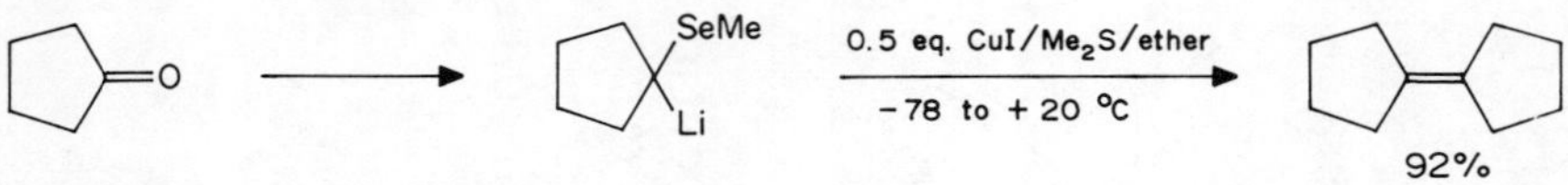

(a)

$$t\text{-BuCHO} \rightarrow t\text{-BuCHLi(SeMe)} \xrightarrow[-78\text{ to }+20^{\circ}\text{C}]{0.5\text{eq. CuI/Me}_2\text{S/ether}} t\text{-BuCH=CHBu-}t$$

E isomer exclusively (b)

SCHEME 87[117]

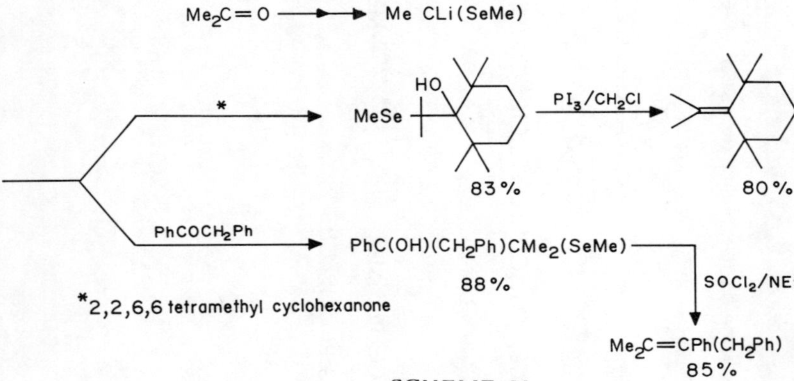

(Ref. 108)

(Ref. 130)

SCHEME 88

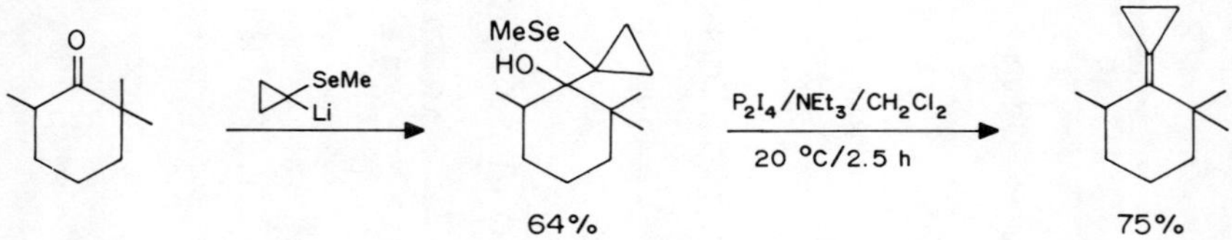

SCHEME 89[107]

Reactions (*iii*) and (*iv*) permit the regioselective formation of terminal, di-, tri-, and tetra-substituted olefins as well as alkylidenecycloalkanes including alkylidenecyclobutanes. They have been successfully used for the synthesis of hindered olefins from hindered or easily enolizable carbonyl compounds and although they are completely regioselective they are not usually stereoselective.

b. From alkyl halides. Primary alkyl halides can be used for the preparation of olefins. (C(1)—X→C(1)=C) (Scheme 91). Since the reactions involve at one stage the

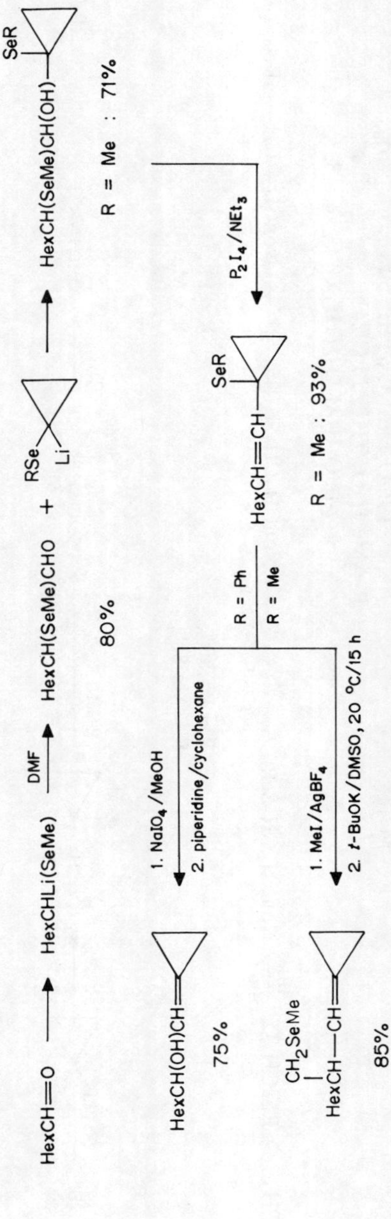

SCHEME 90[139,269]

$$CH_3(CH_2)_3I + PhC(Se{=}O)LiPh \rightarrow CH_3(CH_2)_3CPh[\overset{+}{S}e(Ph)O^-] \rightarrow CH_3(CH_2)_2CH{=}CHPh$$ (Ref. 22) (a)

81% overall

$$OctCH_2Br + RSeCH_2Li \xrightarrow[HMPT]{THF} OctCH_2CH_2SeR \xrightarrow{[OctCH_2CH_2\overset{+}{S}e(R)O^-]} OctCH{=}CH_2$$ (Ref. 232) (b)

R = Ph:80%
R = Me:88%

R = Ph: H_2O_2/THF, 20°C/24 h 30%
t-BuO_2H/Al_2O_3/THF, 60°C/3 h 86%
R = Me: H_2O_2/THF, 20°C/24 h 1%
t-BuO_2H/Al_2O_3/THF, 60°C/3 h 83%

$$NonCH_2Br + RSeCH_2Li \xrightarrow[HMPT]{THF} NonCH_2CH_2SeR \xrightarrow[2.\ t\text{-}BuOK/DMSO,\ 20^\circ C]{1.\ MeI/AgBF_4} NonCH_2CH_2Se(R)CH_2^- \longrightarrow NonCH{=}CH_2$$ (Refs. 20, 127) (c)

R = Ph:80%
R = Me:88%

R = Ph:71%
R = Me:90%

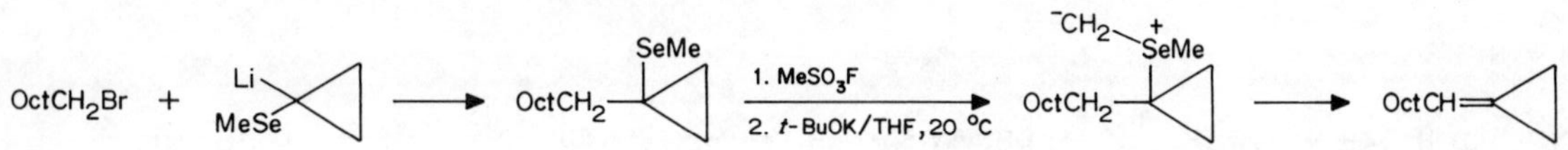

(Ref. 20) (d)

74% 77%

$$PhCH_2Br + PhSeCHLiPh \rightarrow PhCH_2CH(SePh)Ph \xrightarrow{H_2O_2} PhCH{=}CHPh$$ (Ref. 80) (e)

82% overall

SCHEME 91

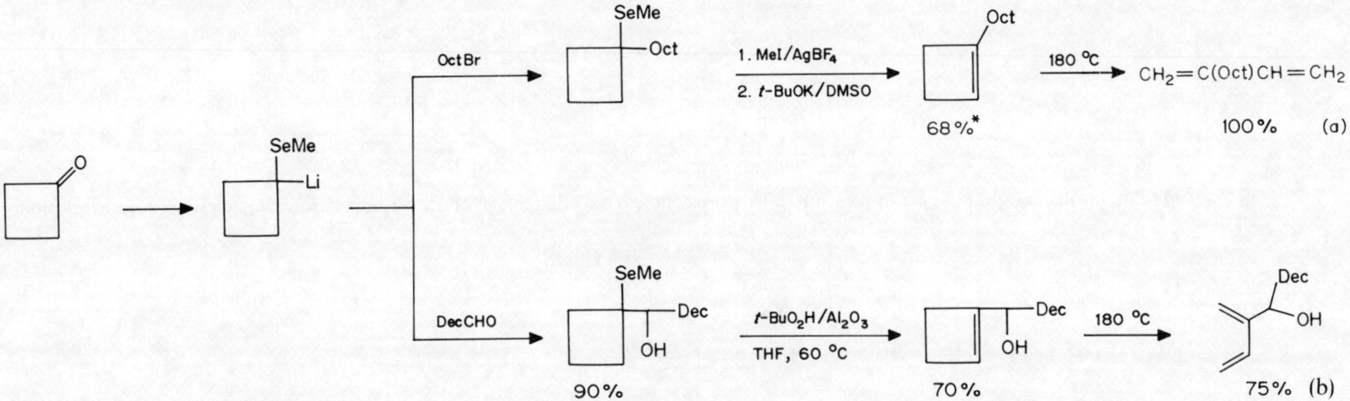

*Contains 7% octylidene cyclobutane

SCHEME 92[127]

$$BuLi + CH_2{=}CHSePh \xrightarrow[\text{then } H_3O^+]{\text{ether or DME}} \underset{97\%}{BuCH_2CH_2SePh} \xrightarrow[\text{THF}]{t\text{-}BuO_2H/Al_2O_3} BuCH_2CH_2\overset{+}{S}e(Ph)O^- \rightarrow \underset{74\%}{BuCH{=}CH_2}$$ (Ref. 121) (a)

$$i\text{-}PrLi + CH_2{=}CHSePh \rightarrow [i\text{-}PrCH_2CHLiSePh] \xrightarrow{Me_2C{=}O} i\text{-}PrCH_2CH(SePh)C(OH)Me_2 \xrightarrow[2.\ i\text{-}Pr_2NH/CCl_4,\ 80°C/0.5\,h]{1.\ O_3,\ -78°C/0.1\,h}$$ (Ref. 122) (b)

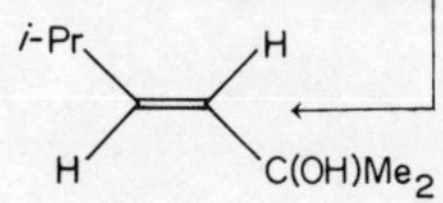

SCHEME 93

$$DecCH{=}O + Me_3SiCLiMe(SeMe) \rightarrow \underset{\mathbf{A}\ \ 50\%}{DecCH(OH)CMe(SeMe)SiMe_3}$$

(a) $\mathbf{A} \xrightarrow{POCl_3/NEt_3/CH_2Cl_2} \underset{83\%\ \ *}{DecCH{=}CMeSiMe_3} \xrightarrow[20^\circ C/2\,h]{Br_2/CCl_4} \underset{80\%}{DecCH{=}CBrMe}$ (Ref. 100)

(b) $\mathbf{A} \xrightarrow[THF/55^\circ C]{t\text{-}BuOK} \underset{82\%^{100}}{DecCH{=}CMe(SeMe)} \xrightarrow[aq.\ MeCN]{HgCl_2} \underset{80\%^{308}}{DecCH_2C(O)Me}$

SCHEME 94

*Pure *Z* and pure *E* isomers are available on reaction with one of the two stereoisomers of **A** easily purified by SiO_2/thin layer chromatography.

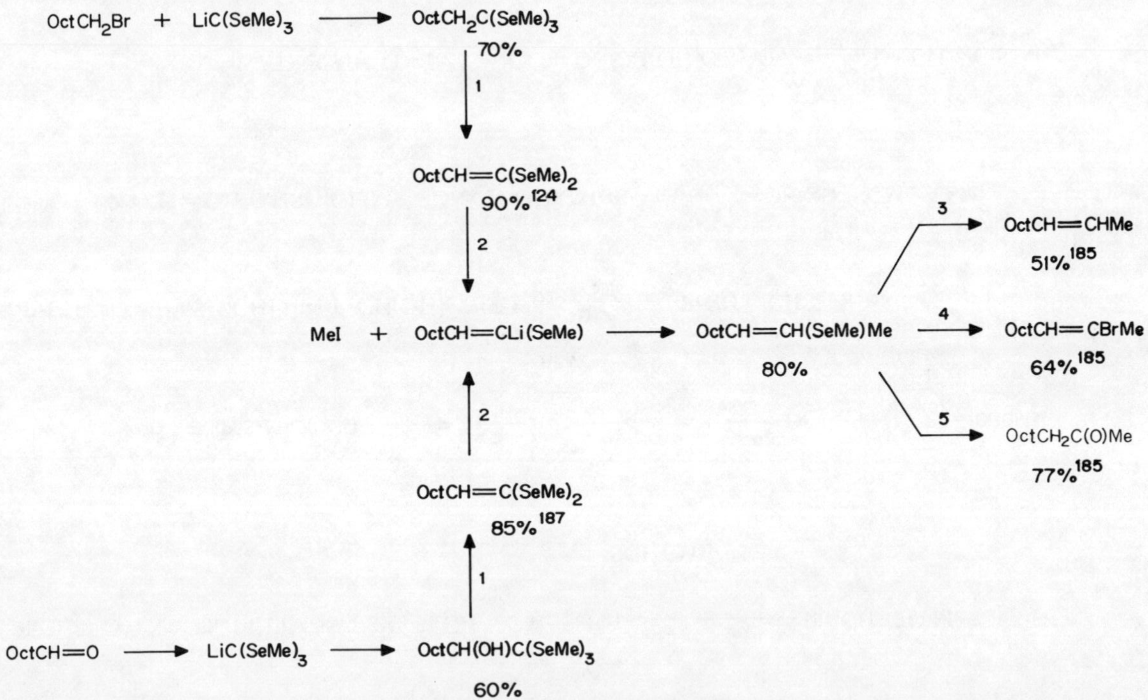

1: $P_2I_4/NEt_3/CH_2Cl_2$, 20°C/1 h; 2: BuLi/THF, − 78°C; 3: Bu_3SnH/AIBN, benzene, 90°C/3 h; 4: Br_2/benzene, 20°C/15 h; 5: Br_2/EtOH, 90°C/6 h

SCHEME 95

$$\mathrm{DecCHO + MeCLi(SeMe)SiMe_3 \rightarrow \underset{50\%}{DecCH(OH)CMe(SeMe)SiMe_3}} \xrightarrow{H_2O_2/THF} \underset{50\%}{\mathrm{DecCH(OH)COMe}}$$ (Ref. 100) (a)

$$\mathrm{DecCH(OH)CMe(SeMe)SiMe_3} \xrightarrow[20^\circ C/12\,h]{HgCl_2/aq.\ MeCN} \underset{75\%}{\mathrm{DecC(O)CH_2Me}}$$ (Ref. 100) (b)

$$\mathrm{PhCH{=}O \rightarrow X^- \ Ph_3\overset{+}{P}CLiMe(SePh) \rightarrow \underset{90\%}{PhCH{=}CMe(SePh)}} \xrightarrow{HgCl_2/aq.\ MeCN} \underset{90\%}{\mathrm{Ph_2CH_2C(O)Me}}$$ (Ref. 101) (c)

SCHEME 96

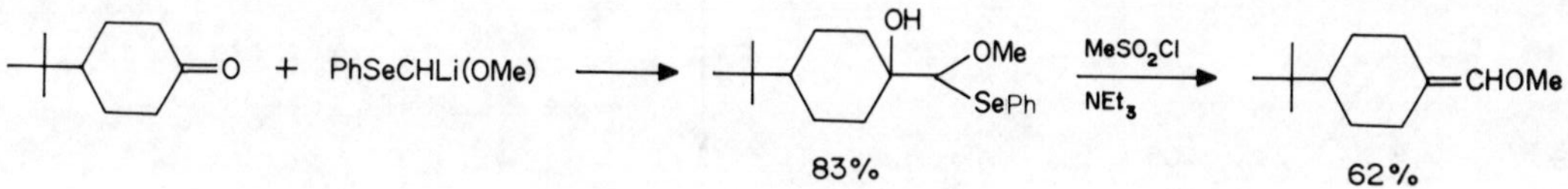

(Ref. 5) (a)

$$i\text{-}\mathrm{PrCH{=}O + (PhSe)_2CHLi \rightarrow } i\text{-}\mathrm{PrCH(OH)CH(SePh)_2} \xrightarrow[NEt_3]{MeSO_2Cl} \underset{65\%}{i\text{-}\mathrm{PrCH{=}CHSePh}}$$ (Ref. 21) (b)

$$\mathrm{DecCH{=}O + (MeSe)_2CLiMe \rightarrow \underset{35\%}{DecCH(OH)CMe(SeMe)_2}} \xrightarrow[NEt_3]{P_2I_4} \underset{90\%}{\mathrm{DecCH{=}CMe(SeMe)}} \xrightarrow[aq.\ MeCN]{HgCl} \underset{80\%}{\mathrm{DecCH_2COMe}}$$

(Refs. 187, 308) (c)

$$\mathrm{DecCH{=}O + (MeSe)_3CLi \rightarrow \underset{63\%}{DecCH(OH)C(SeMe)_3}} \xrightarrow[NEt_3]{P_2I_4} \underset{85\%}{\mathrm{DecCH{=}C(SeMe)_2}}$$ (Ref. 187) (d)

SCHEME 97

elimination of the selenyl moiety and one hydrogen they are regioselective, but only in those cases where topologically identical hydrogens are available for that purpose (Scheme 92).

c. From alkyl halides via alkyllithiums. Scheme 93 show (C(1)—X→C(1)—M→C(1)—C=C) reactions. Primary, secondary and tertiary alkyllithiums can be used, but at present only terminal monoalkylated olefins can be prepared by this method, which implies formal vinylation of an organometallic compound.

3. Functionalized olefins

a. α-Heterosubstituted olefins. The reactions have been performed according to (C(1)=O→C(1)=C—X). Vinyl silanes (Scheme 94a), vinyl selenides (Schemes 94c, 95, 96c and 97b, c) and functionalized vinyl selenides (Scheme 98), vinyl ethers (Scheme 97a), vinyl bromides (Scheme 94a and 95) and ketene selenoacetals (Schemes 97d and 98) are obtained from carbonyl compounds. The transformations are completely regioselective and allow in several instances the synthesis of stereoisomerically pure compounds.

Vinyl selenides (Scheme 95), vinyl bromides (Scheme 95), and ketene selenoacetals (Schemes 95 and 98) can also be obtained from alkyl halides according to the following description (C(1)—X→C(1)=C—Y).

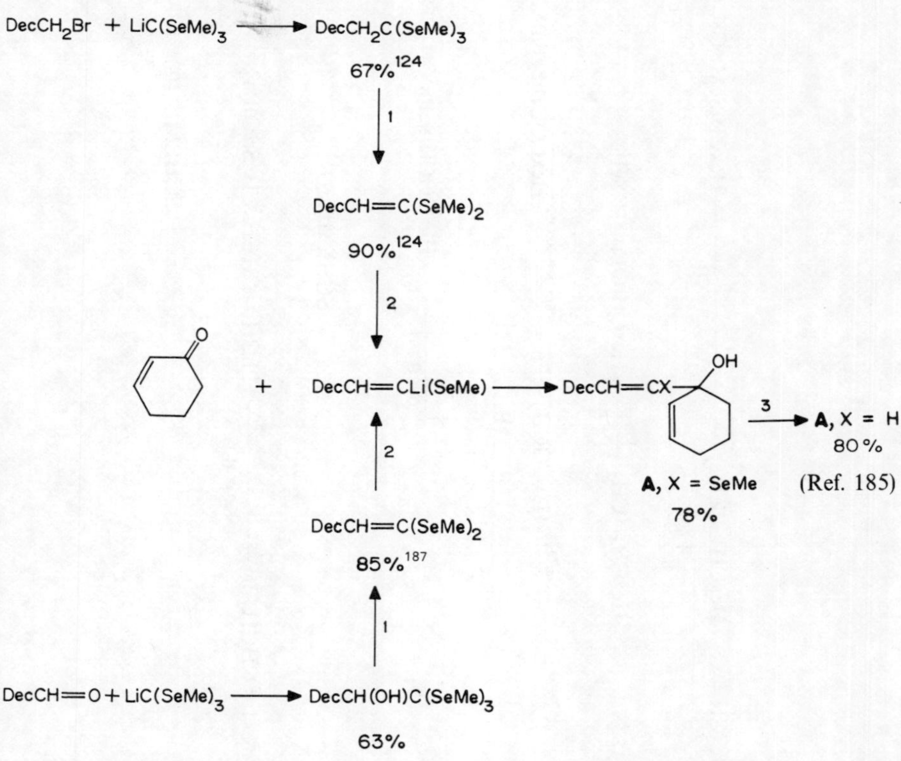

1: $P_2I_4/NEt_3/CH_2Cl_2$, 20°C/1 h; 2: BuLi/THF, −78°C; 3: Bu_3SnH/AIBN, benzene, 80°C/3 h

SCHEME 98

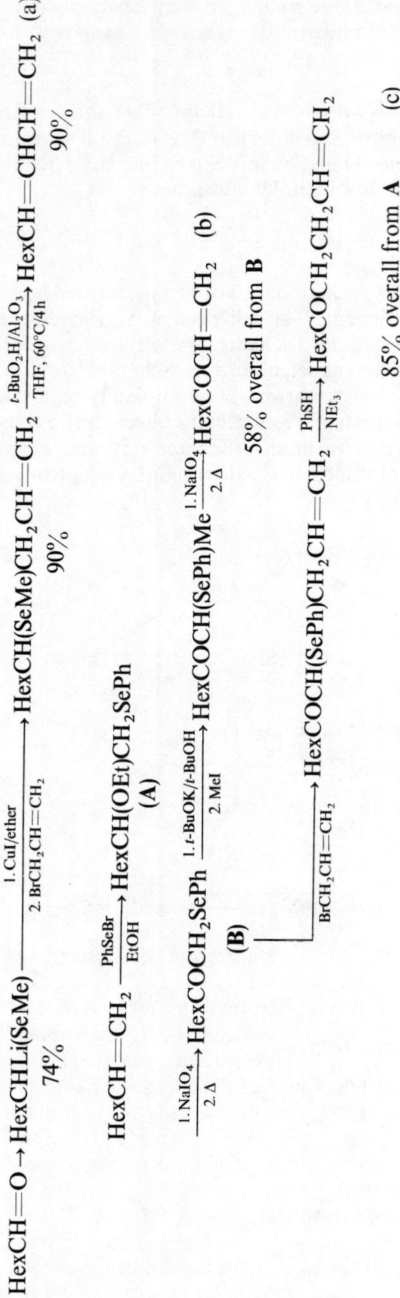

SCHEME 99[12]

SCHEME 100[22]

b. 1, 3-*Dienes.* 1, 3-Dienes can be synthesized:

(*i*) from aldehydes or ketones and allyl halides (C(1)=O → C(1)=C—C=C) (Scheme 99a); interestingly the stereochemistry of the C=C bond of the allyl halide is retained during the process.

(*ii*) from allyl halides (C(1)=C(2)—C(3)—X → C(1)=C(2)—C(3)=C) (Schemes 99a and 100). The last reaction permits an easy synthesis of allylidene cyclopropanes.

(*iii*) from α-enones (C(1)=C(2)—C(3)=O → C(1)=C(2)—C(3)=C) as shown in Scheme 101 or from alkyl halides and cyclobutanone (C(1)—X → C=C—C—(C(1))=C) (Scheme 93). The latter reaction has been extended to the synthesis of functionalized dienes when using epoxides (Scheme 102) or carbonyl compounds (Scheme 93b).

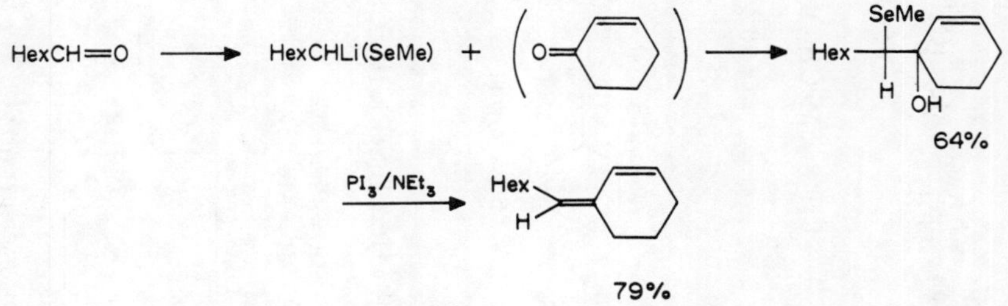

SCHEME 101[309]

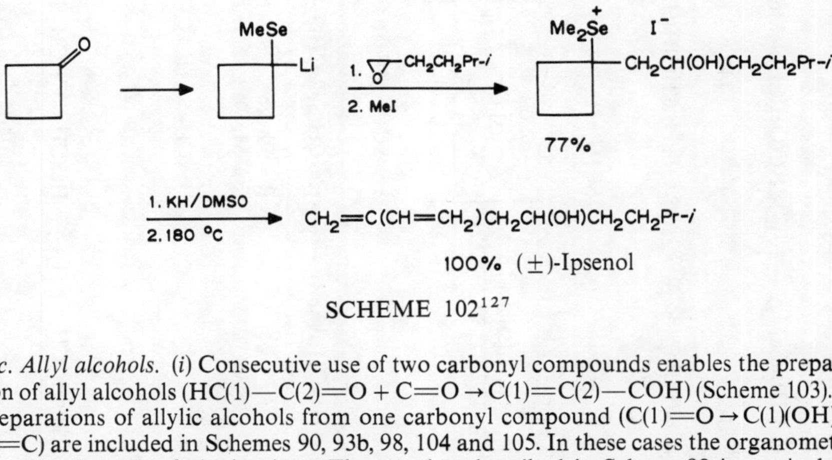

100% (±)-Ipsenol

SCHEME 102[127]

c. Allyl alcohols. (*i*) Consecutive use of two carbonyl compounds enables the preparation of allyl alcohols (HC(1)—C(2)=O + C=O → C(1)=C(2)—COH) (Scheme 103). (*ii*) Preparations of allylic alcohols from one carbonyl compound (C(1)=O → C(1)(OH)—C=C) are included in Schemes 90, 93b, 98, 104 and 105. In these cases the organometallics play the role of vinyl anions. The reaction described in Scheme 98 is particularly suitable for the synthesis of biallylic alcohols. (*iii*) The reaction using alkyllithiums (C(1)—Li → C(1)—C=C—COH) (Scheme 93b) involves the introduction of a vinylic moiety using addition of the alkyllithium across the C=C bond of vinyl selenide which therefore plays the role of a + C=C – synthon. (*iv*) Allyl alcohols have also been prepared from alkyl halides and α-metalloallylselenide and the reaction involves an allylselenoxide → allylseleninate rearrangement (C(1)—X → C(1)—C=C—COH) (Scheme 106).

d. α, β-Unsaturated aldehydes and ketones. α, β-Unsaturated aldehydes and ketones are both available by the following methods: (*i*) from two carbonyl compounds, one of them being an aldehyde (C(1)—C(2)=O + CH=O → C(1)=C(2)—C=O) (Scheme 107);

$$Me_2C{=}O \rightarrow Me_2CLi(SeMe) \xrightarrow{NonCH_2CHO} Me_2C(SeMe)CH(OH)CH_2Non \xrightarrow[THF,\ 80°C]{t\text{-}BuO_2H/Al_2O_3} CH_2{=}CMeCH(OH)CH_2Non$$

92% 65% (Ref. 144) (a)

$$NonCH_2CH{=}O \rightarrow NonCH_2CHLi(SeMe) \xrightarrow{Me_2CO} Me_2C(OH)CH(SeMe)CH_2Non \xrightarrow[THF,\ 80°C]{t\text{-}BuO_2H/Al_2O_3} Me_2C(OH)CH{=}CHNon$$

74% 78% (Ref. 144) (b)

$$Me_2C{=}O \rightarrow Me_2CLi(SePh) \xrightarrow{HexCHO} Me_2C(SePh)CH(OH)Hex \xrightarrow[20°C]{H_2O_2/THF} CH_2{=}CMeCH(OH)Hex$$

70% 80% (Ref. 11) (c)

SCHEME 103

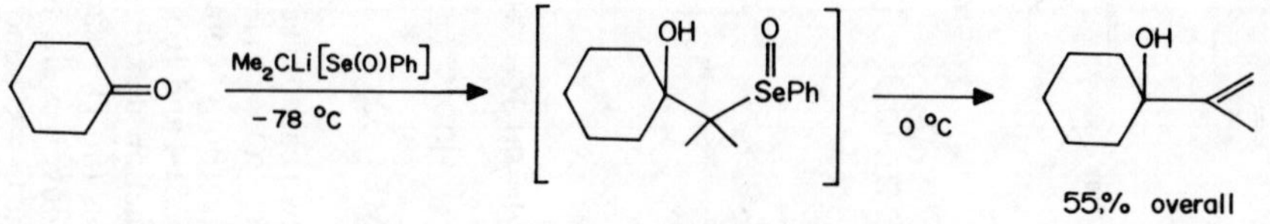

SCHEME 104[108]

$$HexCH{=}O \rightarrow HexCHLi(SeR) \xrightarrow{DMF} HexCH(SeR)CHO \xrightarrow{CH_2{=}PPh_3} HexCH(SeR)CH{=}CH_2$$

R = Ph: 75% R = Ph: 78%
R = Me: 91% R = Me: 85%

$$HexCH{=}CHCH_2OH \xleftarrow[R\,=\,Ph]{H_2O_2/CH_2Cl_2\text{-pyridine}} HexCH(SeR)CH{=}CH_2$$

70% *E* only

$$HexCH(OH)CH{=}CH_2 \xleftarrow{(O)} HexCH{=}CHCH_2SeR \leftarrow HexCH(SeR)CH{=}CH_2$$

100% R = Ph: $h\nu$, 20°C/0.7 h
100% R = Me: TsOH, 20 °C/3 h

SCHEME 105[290]

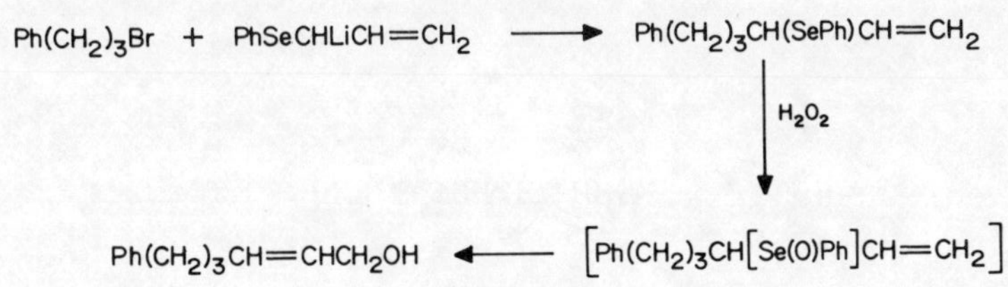

SCHEME 106[77]

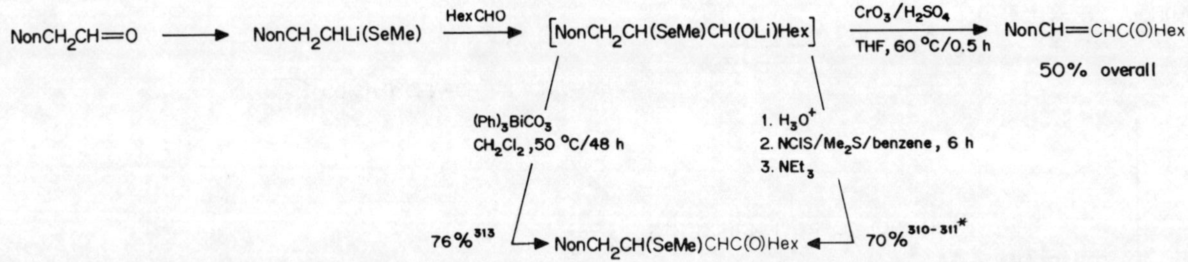

*See Ref. 312 for a related transformation.

SCHEME 107

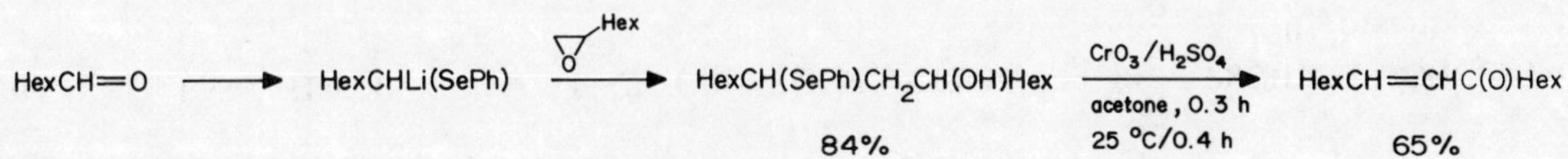

SCHEME 108[174]

$$PhCH_2Br + PhSeCHLiCH{=}CClMe \rightarrow PhCH_2CH(SePh)CH{=}CClMe \xrightarrow[20°C/0.2\,h]{H_2O_2/CH_2Cl_2} PhCH_2CH{=}CHCOMe$$

70% overall

SCHEME 109[77]

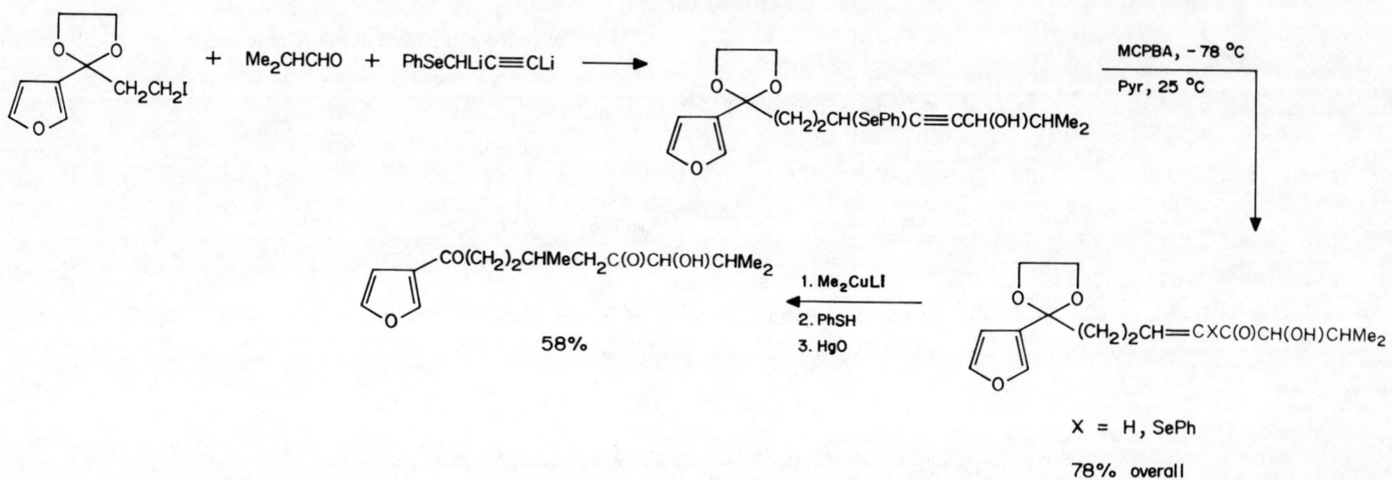

SCHEME 110[314]

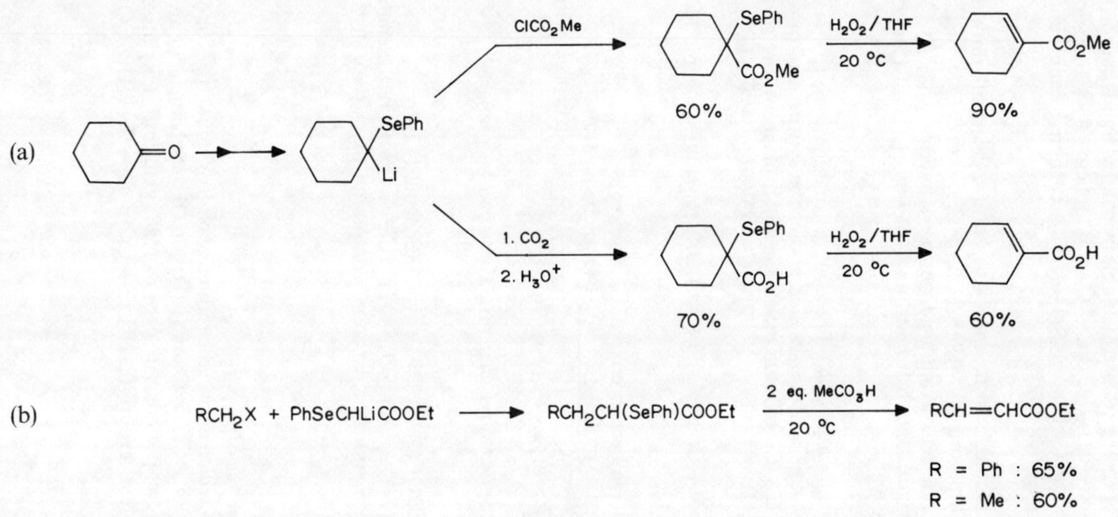

SCHEME 111[135]

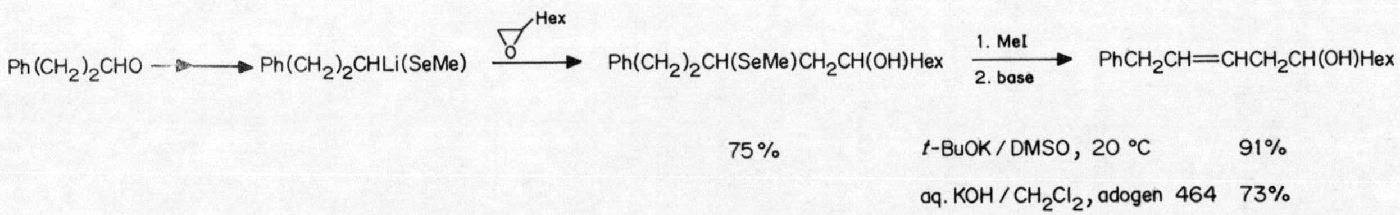

SCHEME 112[129]

(*ii*) from one carbonyl compound and an epoxide (C=O + HC(1)—C(2) → C=C(1)—
O
C(2)=O) (Scheme 108); (*iii*) from alkyl halides (C(1)—X → C(1)—C=C—C=O) (Schemes 109 and 110) and an Se containing reagent which plays the role of an unsaturated homoenolate —C=C—C=O; (*iv*) from olefins (C(1)=C(2) → C=C—C(2)—C(1)=O) (Scheme 99c).

α, β-Unsaturated aldehydes are for their part also available from alkyl halides according to the following equation (C(1)—X → C(1)—C=C—C=O) (Scheme 86c) and γ or δ hydroxy analogues have been prepared when an epoxide or a carbonyl compound is reacted instead of an alkyl halide (Scheme 86d).

e. Ene-4-ones. Ene-4-ones are available from ene-2-ones and α-selenoalkyllithiums which play the role of masked vinyllithiums (O=C(1)—C(2)=C(3) → O=C(1)—C(2)—C(3)—C=C) (Scheme 47). They are also available from allyl halides (C(1)=C(2)—C(3)—X → C(1)=C(2)—C(3)—C=O) and from terminal olefins (C(1)=C(2) → C=C—C—C(2)—C(1)=O) (Scheme 99c).

f. α, β-Unsaturated esters, silyl esters and acids. These are available:

(*i*) from carbonyl compounds through a process which allows the synthesis of pure *Z* and *E* enoates (C(1)=O → C(1)=C—C(=O)(OR) (Scheme 76k,l) or according to the following strategy (HC(1)—C(2)=O → C(1)=C(2)—C(=O)OR; R = Me, H) (Scheme 111a).

(*ii*) from alkyl halides (C(1)H—X → C(1)=C—C(=O)OR; R = alky (Scheme 111b) or R = SiMe (Scheme 86f).

g. α-Alkylidene lactones. These have been prepared:

(*i*) from alkyl halides (C(1)—X → C(1)=C—C(=O)—O) (Scheme 8) or

(*ii*) from carbonyl compounds (C(1)=O → C(1)=C—C(=O)—O) in a process which allows the synthesis of pure *Z* or *E* isomers (Scheme 76k,l).

h. γ-Functionalized olefins such as homoallyl alcohols. These can be prepared:

O
(*i*) from epoxides (C(1)—C(2) → HOC(1)—C(2)—C=C) (Scheme 112); or

(*ii*) from carbonyl compounds (C(1)—C(2)=O → C(1)=C(2)—C—COH) (Schemes 102 and 112). In the latter transformation the carbonyl compound plays the role of a masked vinyl anion.

4. Non-functionalized alcohols

Non-functionalized alcohols have been prepared:

(*i*) from two ketones in a reaction which allows the reductive hydroxyalkylation of one of the two carbonyl groups (C(1)=O + C(2)=O → C(1)(H)—COH) (Scheme 85c) or, in the case of aromatic carbonyl compounds, it permits the gem-*ipso* alkylation–hydroxyalkylation of their carbonyl group (C(1)=O + C(2)=O → C(1)(H)—COH) (Scheme 86a).

5. Functionalized alcohols

γ-Halogeno alcohols have been obtained from carbonyl compounds (C(1)=O → Br—
O
C(1)—C—C—OH) and from epoxides (C(1)—C(2) → HOC(1)—C(2)—CBr) (Scheme 113).

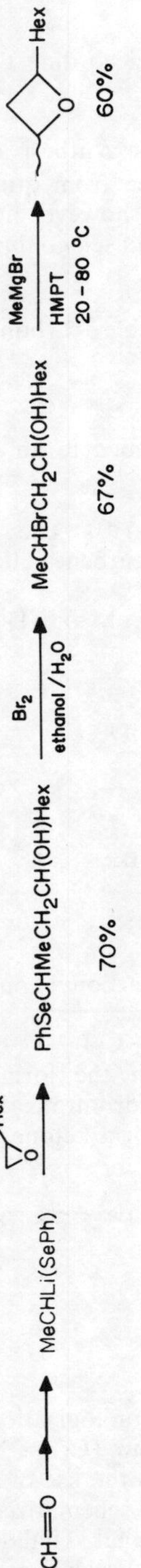

SCHEME 113[106]

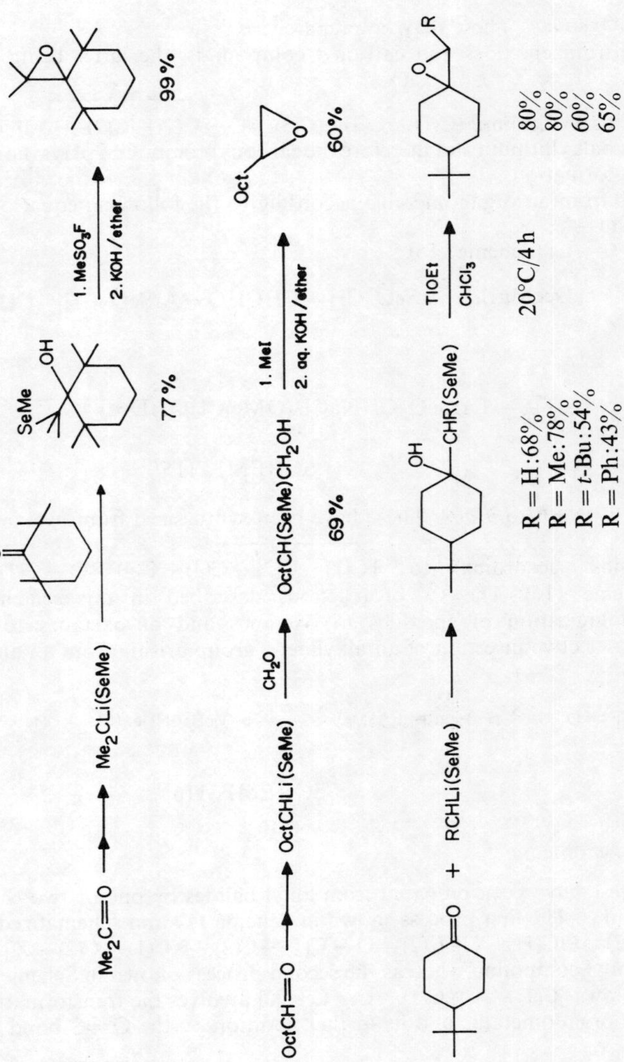

SCHEME 114

6. Heterocycles

a. Epoxides. These may be prepared from two carbonyl compounds according to:

O

(C(1)=O + C(2)=O → C(1)—C(2) (Scheme 114). The reaction permits the synthesis of terminal, disubstituted, trisubstituted and tetrasubstituted epoxides even from quite hindered and/or enolizable carbonyl compounds. Oxaspiropentanes are however not available by this method. The method involving a carbonyl compound and selenonium

O

ylides (C(1)=O → C(1)—C) (Scheme 3) is restricted to non-enolizable carbonyl compounds.

b. Oxetanes. These may be prepared:

(*i*) from epoxides and carbonyl compounds, the latter being transformed to an α-

O ┌——O——┐

selenoalkyllithium (C(1)—C(2) + C(3)=O → C(1)—C(2)—C(3) (Scheme 113). The α-selenoalkyllithium and therefore the carbonyl compound plays the role of a carbene in this transformation.

(*ii*) from an organometallic according to the following equation: (C(1)—M → C(1)—

┌—O—┐

C—C—C) (Scheme 115).

$$\text{DecMgBr} + \text{ArSeO}_2\text{CH}{=}\text{CHCHO} \rightarrow \text{ArSeO}_2\text{CH}{=}\text{CHCH(OH)Dec}$$

$$\xrightarrow[\text{MeONa}]{\text{MeOH}} [\text{ArSeO}_2\text{CHNaCH(OMe)CHOHDec}] \longrightarrow$$

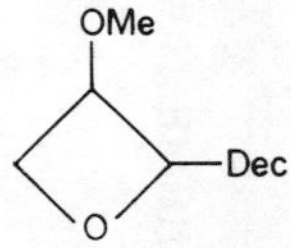

SCHEME 115[293] 78% overall

c. Tetrahydrofurans. These have been synthesized from oxetanes and carbonyl compounds according to [C(1)—C(2)—C(3) + C(4)=O → C(1)—C(2)—C(3)—C(4)] (Scheme 116). The set of reactions described in this scheme permit the formal homologization of epoxides to oxetanes and of oxetanes to tetrahydrofurans by regioselective insertion of an alkylidene group arising from an aldehyde or a ketone.

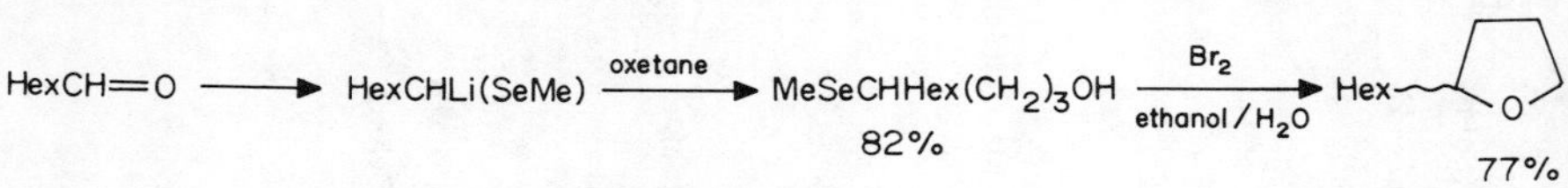

SCHEME 116[106]

7. Alkyl halides

These have been prepared from alkyl halides by one or two carbon homologization reactions. The first process shown in Scheme 117 and schematized as follows [C(1)—X + C(2)=OC(1)—X + C(2)=O → C(1)—C(2)—XC(1)—C(2)—X] requires the use of a carbonyl compound whereas the second process, shown in Scheme 118 and schematized as follows [C(1)—X → C(1)—C—C—X] involves the transformation of the alkyl halide to an organometallic and its further addition on the C=C bond of a vinyl selenide.

$$\text{NonI} + \text{LiCH}_2\text{SeR} \xrightarrow[\text{1 eq. HMPT}]{\text{THF}} \text{NonCH}_2\text{SeR} \xrightarrow[\text{DMF/80°C}]{\text{MeI}} \text{NonCH}_2\text{I}$$

$$\begin{array}{ll} \text{R} = \text{Me}: 88\% & 90\% \\ \text{R} = \text{Ph}: 80\% & 90\% \end{array}$$

SCHEME 117[17]

$$\text{BuX} \xrightarrow[65\%]{\text{Li/hexane}} \text{BuLi} + \text{CH}_2{=}\text{CHSePh} \rightarrow \text{BuCH}_2\text{CH}_2\text{SePh} \xrightarrow[\text{MeI/DMF}]{\text{Br}_2\text{/EtOH}} \text{BuCH}_2\text{CH}_2\text{X}$$

X = Br: 40%[120]
X = I: 80%[121]

SCHEME 118

8. Non-functionalized aldehydes and ketones

a. Synthesis from carbonyl compounds. Ketones have been synthesized

(*i*) from aldehydes and α-silyl-α-selenoalkyllithiums according to the following [HC(1)=O → C—C(1)(=O)] (Scheme 96b). These, however, are not particularly good nucleophiles toward carbonyl compounds.

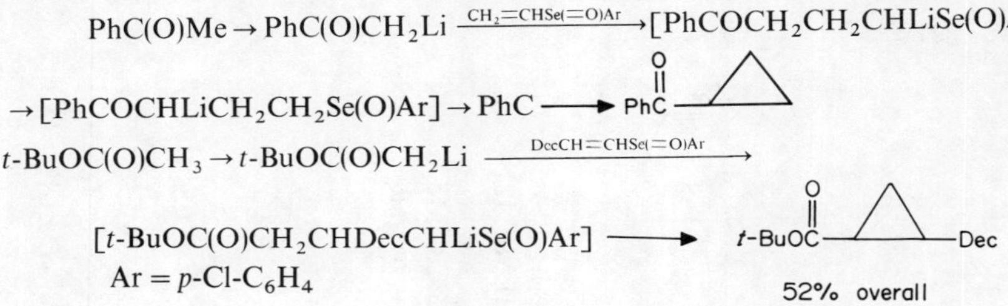

SCHEME 119[271]

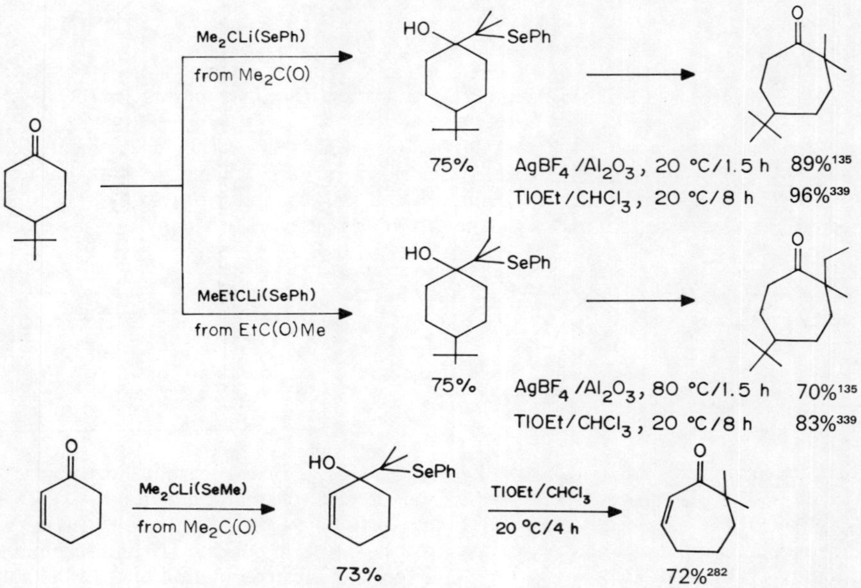

SCHEME 120

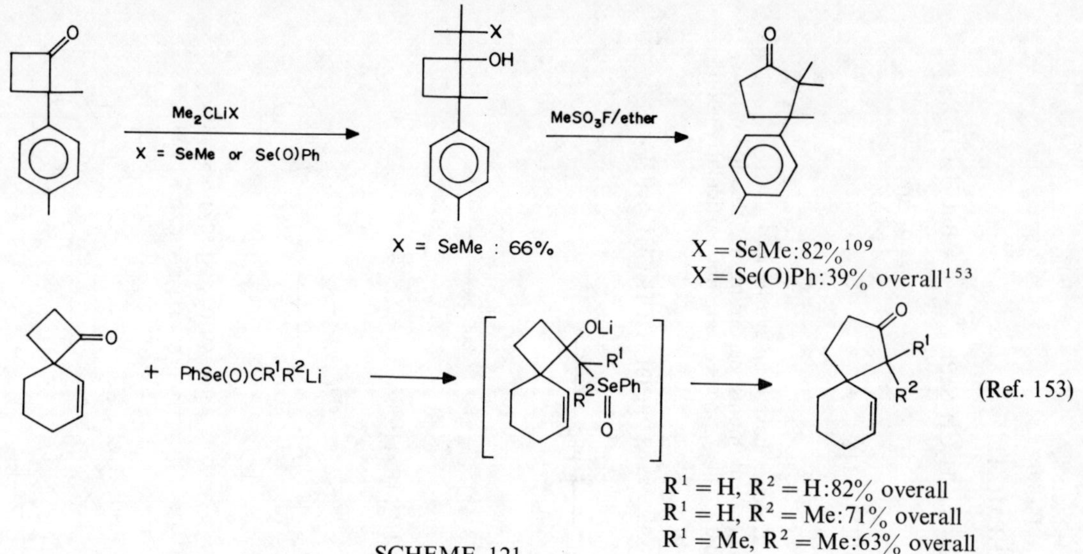

SCHEME 121

DecBr —

(a) $\xrightarrow{(PhSe)_2CKMe}$ DecCMe(SePh)$_2$ $\xrightarrow[0.1\ h]{CuCl_2/CuO}$ DecC(O)Me, 85% overall (Ref. 69)

(b) $\xrightarrow{(MeSe)_2CLiMe}$ DecCMe(SeMe)$_2$ $\xrightarrow[0.1\ h]{CuCl_2/CuO}$ DecC(O)Me, 72% overall (Refs. 68, 266)

(c) $\xrightarrow[-78°C,\ THF]{Me_3SiCHLi(SeMe)}$ DecCH(SeMe)SiMe$_3$ $\xrightarrow[NaIO_4/EtOH]{H_2O_2/THF\ or}$ DecCH=O, 77% (Ref. 74)

(d) $\xrightarrow[-78°C,\ THF]{Me_3SiCHLi(SePh)}$ DecCH(SePh)SiMe$_3$, 80% $\xrightarrow{H_2O_2/THF}$ DecCH=O, 77% (Refs. 74, 75)

SCHEME 122

(*ii*) according to [HC(1)=O → C(1)—C=O] (Schemes 94c, 95, 96c and 97c). The organometallic used in this transformation allows the homologization of the carbonyl compounds.

(*iii*) in the preparation of a cyclopropyl ketone [HC(1)—C(2)=O → $\begin{matrix}\text{C}\\|\\\text{C}\end{matrix}\!>$C(1)—C=O) Scheme 119). The carbonyl compound plays the role of an α-carbonyl carbene and the alkyl phenyl selenoxide plays the role of an olefin.

(*iv*) from two ketones according to the following equation [C(1)—C(2)=O + C(3)=O → C(1)—C(3)—C(2)=O]. Schemes 120 and 121 show methods which allow the introduction of a carbon atom bearing two alkyl groups (often prepared from a ketone) between the carbonyl group and usually its more substituted α-carbon atom. The reaction applies also to hindered or strongly enolizable carbonyl compounds.

b. Synthesis from alkyl halides. Ketones and aldehydes may be prepared from alkyl halides and α-selenoalkylmetals which play the role of masked acyl anions [C(1)—X + C(2)=O → C(1)—C(2)=O] (Schemes 95, 122 and 123). Among these are the α-selenoalkylmetals bearing an sp^3 (Scheme 122) or an sp^2 (Schemes 95 and 123) carbanionic centre.

$$\text{DecBr} + \text{(PhSe)CLi}{=}\text{CH}_2 \xrightarrow{\text{THF/HMPT}} \underset{70\%}{\text{DecC(SePh)}{=}\text{CH}_2} \xrightarrow{\text{H}_2\text{SO}_4/\text{H}_2\text{O}} \underset{80\%}{\text{DecC(O)Me}}$$

SCHEME 123[121]

$$\text{BuLi} + \text{CH}_2{=}\text{CHSePh} \rightarrow [\text{BuCH}_2\text{CH(SePh)Li}] \xrightarrow{\text{PhSeBr}} \underset{80\%^{120}}{\text{BuCH}_2\text{CH(SePh)}_2}$$

$$\xrightarrow[\text{acetone 1\%}]{\text{CuCl}_2/\text{CuO}} \underset{70\%^{266}}{\text{BuCH}_2\text{CH}{=}\text{O}} \qquad \text{(a)}$$

$$i\text{-PrLi} + \text{CH}_2 = \text{CHSePh} \rightarrow [i\text{-PrCH}_2\text{CH(SePh)Li}] \xrightarrow{\text{Me}_3\text{SiCl}} \underset{86\%^{122}}{i\text{-PrCH}_2\text{CH(SePh)SiMe}_3}$$

$$\xrightarrow{\text{H}_2\text{O}_2/\text{THF}} \underset{70\text{–}80\%^{75}}{i\text{-PrCH}_2\text{CH}{=}\text{O}} \qquad \text{(b)}$$

SCHEME 124

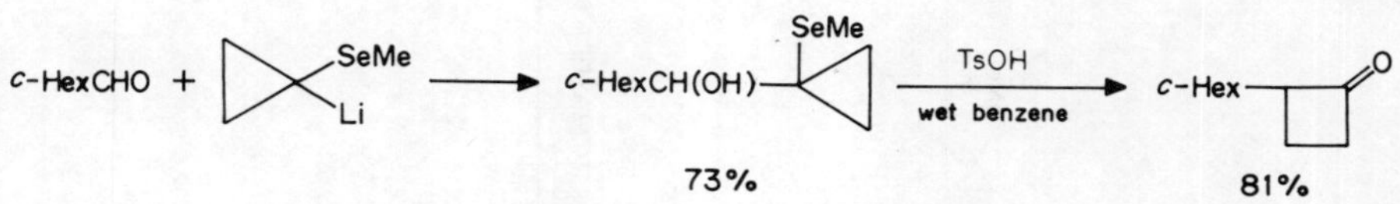

(Ref. 115) (a)

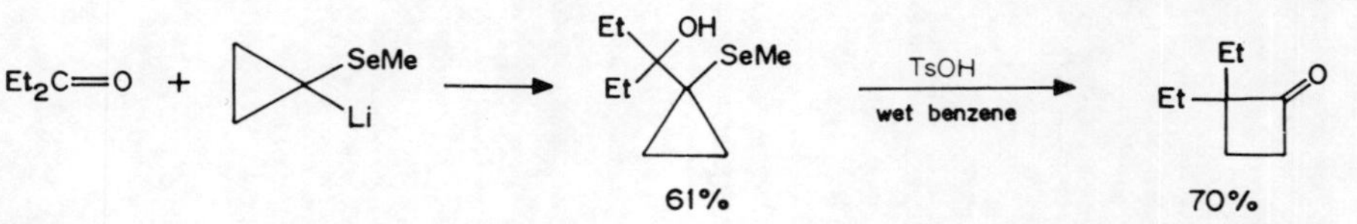

(Refs. 107, 115) (b)

SCHEME 125

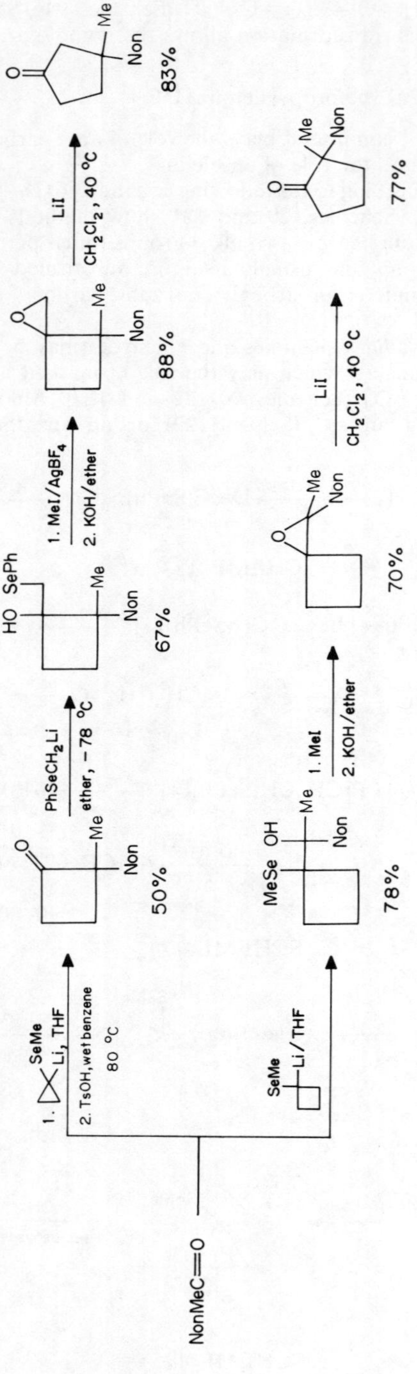

SCHEME 126[110]

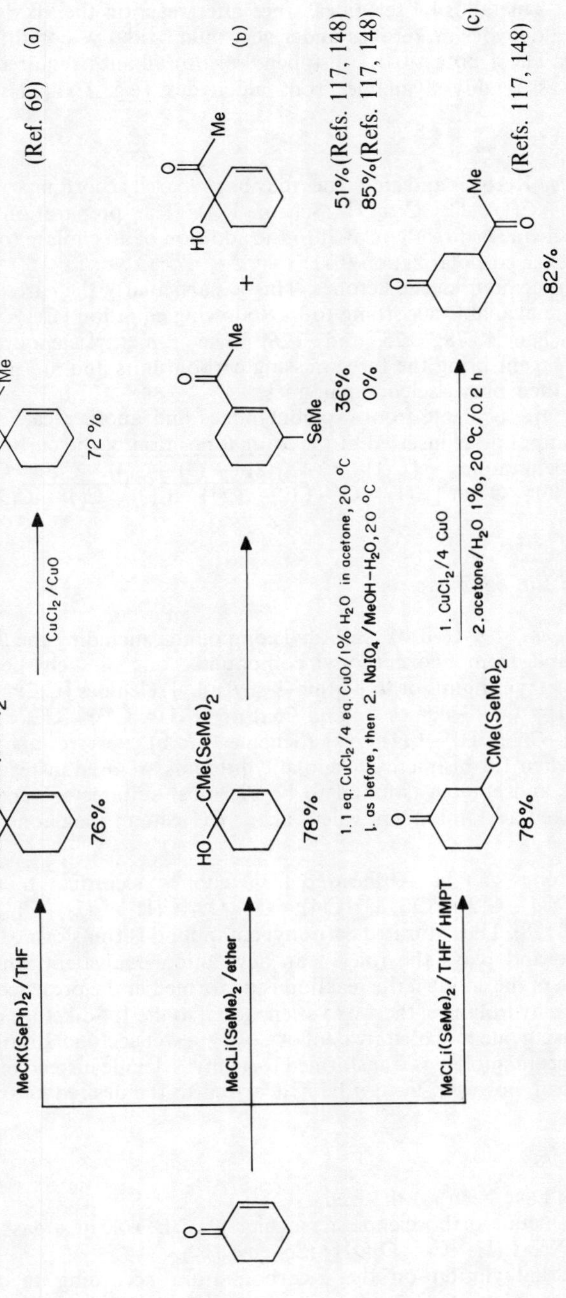

SCHEME 127[117,148,157]

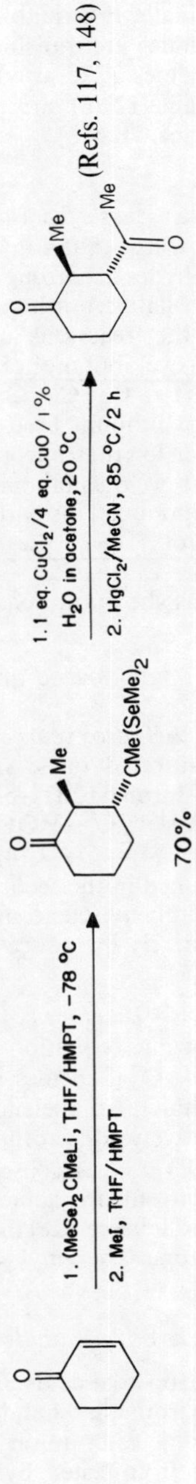

SCHEME 128[117]

Particularly valuable are the α-metallosilyl selenides, since after reaction the alkylsilyl selenides are transformed to aldehydes or ketones under very mild oxidative conditions (Schemes 122c, d and 124b). The 1-potassio-1,1-bis(phenylseleno) alkanes required in Scheme 122(a) are themselves readily available from aldehydes [e.g. CH_3CHO → $(PhSe)_2CKMe$].

c. Synthesis from alkylmetals. Ketones and aldehydes may be prepared according to the following equation [C(1)M → C(1)—C—C=O] (Scheme 124). The preparation of aldehydes according to this scheme is directly related to the addition of an enolate to an alkyl halide but involves a reversed polarization ($C(1)^-$ + C^+—C=O).

Other reactions are more specific to cyclic ketones. This is particularly the case of:

(*i*) cyclobutanones which are available according to the following equation [C(1)=O → C(1)—C—C—C=O) (Schemes 78, 125 and 126). The α-methylselenocyclopropyllithiums used as a coreagent bring the three missing carbon units and are much better for that purpose than their phenylseleno analogues.

(*ii*) cyclopentanones which are available from cyclobutanones and another carbonyl compound. This carbonyl compound is inserted at the α- or β-position of the carbonyl group of the cyclopentanone [C(1)=O + C(2)—C(3)—C(4)—C(5)=O → C(1)—C(2)—C(3)—C(4)—C(5)=O] or [C(1)=O → C(2)—C(1)—C(3)—C(4)—C(5) = O] as shown in Scheme 126.

9. Functionalized aldehydes and ketones

a. α-Hydroxycarbonyl compounds. α-Hydroxycarbonyl compounds, including the β, γ-unsaturated ones, are available from two carbonyl compounds, one of them being transformed to 1, 1-bis(seleno)alkyllithiums or to 1-lithio-1-silyl alkyl selenides [C(2)=O + HC(1)=O → HOC(2)—C(1)=O] (Schemes 42 and 96a) or [C(1)=C(3)—C(2)=O + $H_2C(4)$=O → C(4)=C(3)—C(2)(OH)—C(1)=O] (Scheme 127a,b). Best results are obtained in the second case when 1, 1-bis(methylseleno)alkyllithiums are used instead of their phenylseleno analogues and ether was found to be the most suitable solvent to selectively introduce the masked acyl anion equivalent at the C(1) site of the enone.

b. 1, 4-Diketones. 1, 4-Diketones can be synthesized from enones according to the following equation [O=C(1)—C(2)=C(3) + HC(4)=O → O=C(1)—C(2)—C(3)—C(4)=O] (Schemes 127c and 128). The saturated carbonyl compound is transformed to 1-lithio-1, 1-bis(seleno) alkane and plays the role of an acyl anion equivalent which selectively adds at the C(3) site of the enone if the reaction is performed in the presence of HMPT used as a cosolvent. The hydrolysis of the 4-oxo-selenoacetal to the 1, 4-diketone can be conveniently achieved by using one molar equivalent of $CuCl_2$ in wet acetone. In some cases (Scheme 128) the selenoacetal moiety is transformed to a vinyl selenide instead of to a ketone. The vinyl selenide can, however, in turn be hydrolysed to the desired ketone.

10. Carboxylic acids and esters

Carboxylic acids and esters have been synthesized

(*a*) from alkyl halides and α-metalloorthoseleno esters which play the role of a masked carbonic ester anion [C(1)—X → C(1)—C(=O)OH] (Scheme 129).

(*b*) from ester by formal dialkylation on the α-carbon atom according to the following equation [$H_2C(1)$—C(2)(=O)OC → C—C—C(1)—C(2)(=O)—O—C(3)] (Scheme 119b).

$$\text{HexBr} + (\text{MeSe})_3\text{CLi} \xrightarrow{-78\,^\circ\text{C/THF}} \underset{80\%}{\text{HexC(SeMe)}_3} \xrightarrow{\text{H}_2\text{O}_2\text{/THF}} \underset{82\%}{\text{HexCO}_2\text{H}}$$

SCHEME 129[74]

II. α-TELLUROORGANOMETALLICS: SYNTHESIS AND SYNTHETIC USEFULNESS

A. Synthesis of α-Telluroorganometallics

Telluroalkylmetals are much less well known that their seleno analogues. The first description of such compounds appeared in 1970 when Lloyd and coworkers described the first telluronium ylide[295] (Scheme 130) as a rather unstable compound[295,302].

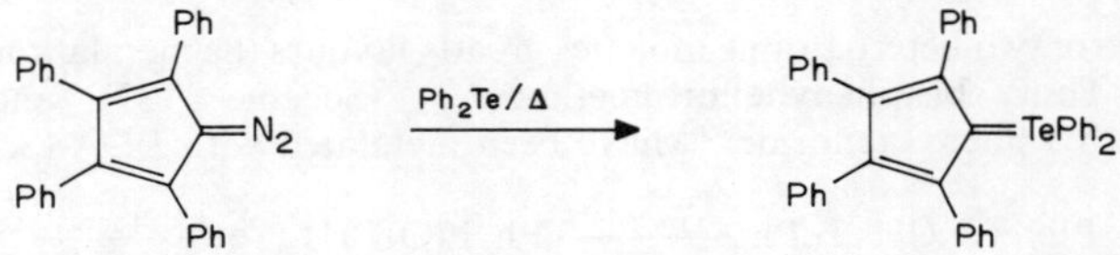

SCHEME 130[295]

Since that date other stabilized telluronium ylides have been synthesized[304,306,316–317] and some of them have been used in synthesis[316,317]. For example dialkyltelluronium carbethoxymethylides are reported[316] to produce α,β-unsaturated esters on reaction with carbonyl compounds whereas dialkyl telluronium allylides[317] lead to the formation of allyl epoxides on reaction with the same derivatives (Scheme 131).

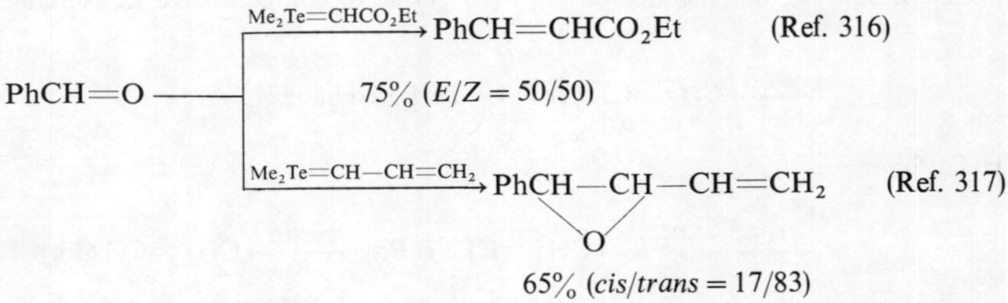

SCHEME 131

The synthetic methods used for the synthesis of these ylides[295,302,304–306] are similar to those used for the Se analogues and imply the reaction of diazo compounds with tellurides[295,302], or the reaction of active methylene derivatives with dihalogenotellurides.

Dialkyltelluronium carbethoxymethylides and dialkyl telluronium allylides are prepared[316,317] on reaction of the corresponding telluronium salts with *t*-BuOK in THF at − 20 ° and − 78 °C, respectively. Whereas dialkyltelluronium allylides are quantitatively formed by reaction with *t*-BuOK in THF on the corresponding telluronium salts, diphenyltelluronium analogues do not produce the ylide but instead produce diphenyl telluride and allyl bromides.

On the other hand telluromethyllithium has been synthesized by Seebach[294] from 1,1-bis(phenyltelluro)methane and methyllithium, butyllithium or *t*-butyllithium. The reaction takes advantage of the cleavage of the C—Te bond (Scheme 132). The reaction seems to be easier than the related C—Se bond cleavage in selenoacetals already described in this review. Phenyltelluromethyllithium is surprisingly found to be thermally stable[34,294] and does not rearrange even after 12 h at 20 °C.

$$(PhTe)_2CH_2 \xrightarrow[-78^\circ C]{LDA/THF} (PhTe)_2CHLi \xrightarrow{PhCH_2Br} (PhTe)_2CHCH_2Ph \quad 100\%$$

$$(PhTe)_2CH_2 \xrightarrow[-\,PhTeMe]{MeLi} PhTeCH_2Li \xrightarrow{PhCHO} PhTeCH_2CH(OH)Ph \quad 60\%$$

SCHEME 132[294]

The presence of two heteroatomic moieties greatly favours the metalation of the telluro compounds. Thus bis(phenyltelluro)methane[297] (Scheme 132) and (diphenylphosphino)methyl phenyl telluride[299] have been metalated with LDA (Scheme 133).

$$Ph_2P(O)CH_2TePh \xrightarrow[-78^\circ C]{LDA/THF} Ph_2P(O)CHLiTePh \xrightarrow[-78^\circ C]{PhCHO} Ph_2P(O)CH{=}CHPh(\textit{trans}) \quad 41\%$$

SCHEME 133[299]

Vinyl phenyl telluride has also been metalated [34,297] (Scheme 134). The best results are obtained when lithium dicyclohexylamide is used. BuLi[297] also permits the metalation of vinyl phenyl telluride, but the yield is poor (10%) due to competing Te/Li exchange.

$$CH_2{=}CH(TePh) \xrightarrow{BuLi} CH_2{=}CLi(TePh) + \text{Te/Li exchange} \quad 10\%$$

$$CH_2{=}CH(TePh) \xrightarrow{LiN(C{-}C_6H_{11})_2} CH_2{=}CLi(TePh)\ (51\%) \xrightarrow{PhCHO} CH_2{=}C(TePh)CH(OH)Ph \quad 88\%$$

SCHEME 134[297]

It has been noticed[34] that the telluryl moiety stabilizes an α-carbanion better than a selenyl moiety. Such results have been obtained from competitive metalation experiments between differently substituted tellurides and selenides[34].

B. Reactivity of α-Telluroorganometallics

Little is known about the reactivity of these organometallics towards electrophiles.

C. Reactivity of Tellurides and Functionalized Tellurides[319]

Similarly to selenides[221] alkyl phenyl and alkyl methyl tellurides are reduced to the corresponding alkane on reaction with triphenyltin hydride[221]. As expected the reaction is usually faster with tellurides than with selenides (Scheme 135). The reduction is even faster if dichlorotelluronium salts (Scheme 135) are used in place of the telluride[221].

$$C_{12}H_{25}X \longrightarrow C_{12}H_{26}$$

X = SePh	2.5 mol eq. Ph_3SnH, 120°C/2.5 h	73%
X = TePh	2.4 mol eq. Ph_3SnH, 20°C/5 h	87%
X = $TeCl_2$	3.5 mol eq. Ph_3SnH, 20°C/2 h	70%

$$C_8H_{17}-CH(OR)-CH_2TePh \longrightarrow C_8H_{17}-CH(OR)-CH_3$$

R = H	2.2 mol eq. Ph_3SnH, 80°C/1.5 h	88%
R = Ac	2 mol eq. Ph_3SnH, 80°C/0.7 h	71%

SCHEME 135[221]

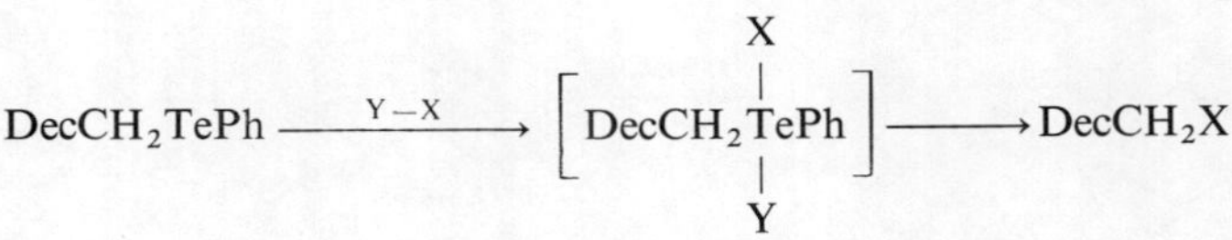

Y—X	Conditions	Yields
Cl—Cl	DMF, 160°C	89%
Br—Br	DMF or ethanol, 70°C	93%
I—I	CH_3CN or THF, 85°C	71%
Me—I	DMF or THF, 55°C	88%

SCHEME 136[322]

Phenyltelluroalkanes react with sulphuryl chloride[322], bromine[217,322] and iodine[322] and produce the corresponding telluronium dihalides in quantitative yield. These telluronium derivatives subsequently heated at 70–100 °C afford alkyl halides smoothly (Scheme 136). Although the pyrolysis is expected to occur via a 1, 2-halogen shift, the yields are much improved[322] by the addition of alkali metal halides or ammonium halides[322,297].

1, 1-Bis(phenyltelluro)alkanes also react[322] with Br_2 and lead to 1, 1-dibromoalkanes (Scheme 137); on the other hand the same treatment with NaI and I_2 produces aldehydes (Scheme 137). Closely related reactions have already been described for selenoanalogues[18].

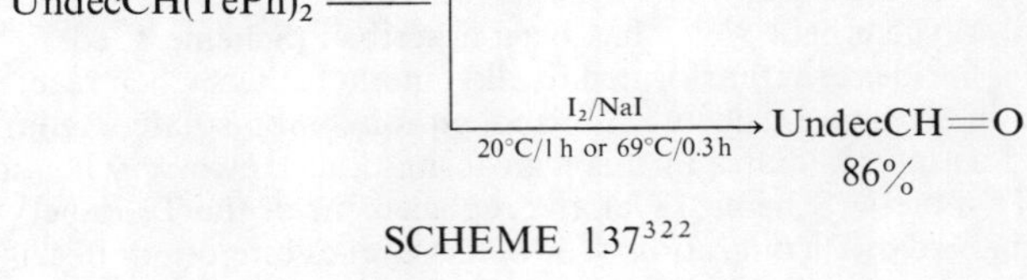

SCHEME 137[322]

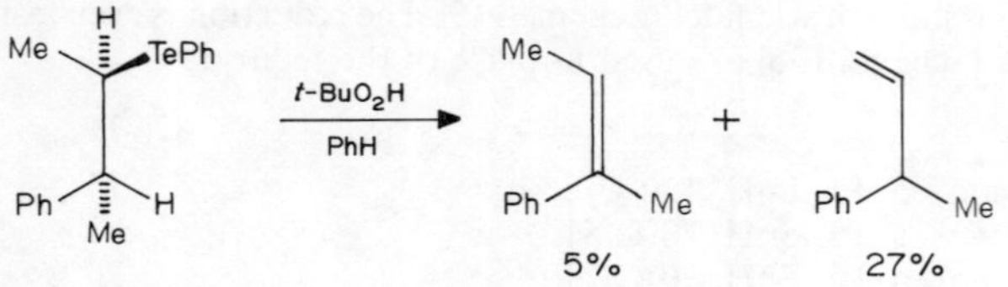

(Ref. 2) (a)

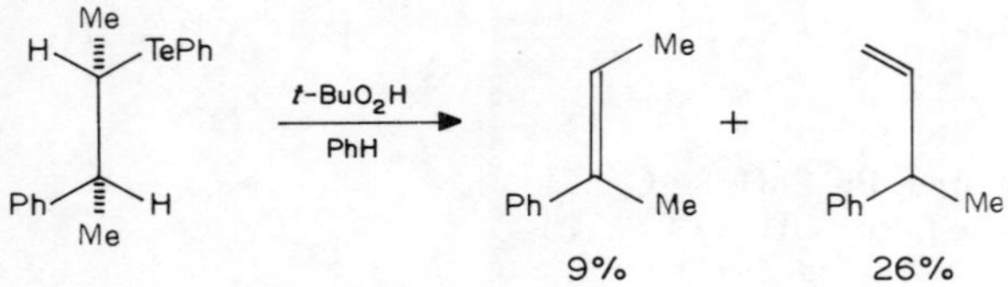

(Ref. 2) (b)

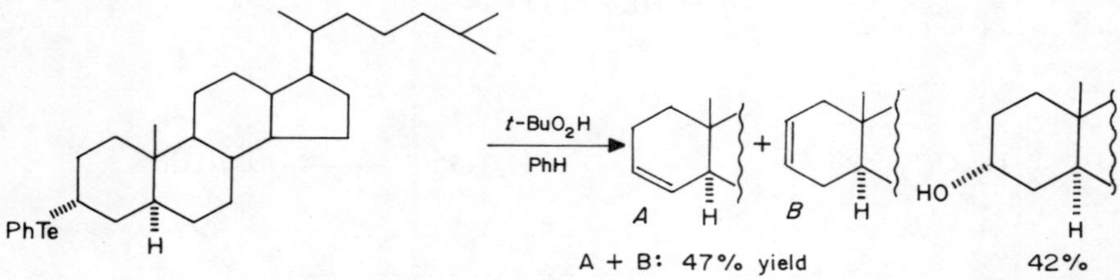

($\alpha:\beta = 1:1$)

(Ref. 2) (c)

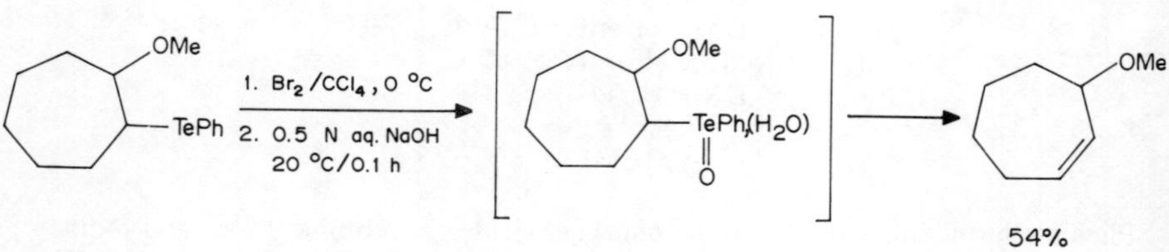

(Refs. 331, 332) (d)

SCHEME 138

Tellurides lead to olefins on reaction with *t*-butyl hydroperoxide in benzene (Scheme 138): a telluroxide is proposed as a n intermediate[2]. The yield of olefin is rather low (35%) probably due to a side-reaction involving a rearrangement of the telluroxide which produces an alcohol (Scheme 138c)[2,318]. The mechanism of the reaction has been described by Sharpless[2] and involves, as in the case of sulphoxides and selenoxides, a *syn* elimination reaction. Finally if different hydrogens are available for the elimination reaction, those leading to the less substituted olefins are the more prone to be eliminated. More recently a facile elimination of some telluroxides, leading to olefins[332], allylic alcohols[332] and allylic ethers[331,332] has been described (Scheme 138d).

Phenyl alkyl tellurides are transformed to alkyl methyl ethers when reacted with excess perbenzoic acid in methanol (Scheme 139) and, although not isolated, tellurones[333] have been postulated as intermediates in this transformation. However when a Ph group is vicinal to the Te moiety (Scheme 139c), the replacement of the Te moiety by the MeO group is accompanied by Ph migration[331], whereas cyclic compounds bearing a MeO and a telluryl group in the β-position readily afford the dimethyl acetals of the ring-contracted cyclic aldehydes[331].

$$\mathrm{DecCH_2CH_2TePh} \xrightarrow[20^\circ\mathrm{C/1\,h,\ 50\%\ yield}]{4\ \mathrm{mCPBA/MeOH}} \mathrm{DecCH_2CH_2OMe} \qquad \text{(a)}$$

1. Br_2/CCl_4, 0°C
2. 0.5 N aq. NaOH, 20°C/0.1 h

$$\longrightarrow \mathrm{DecCH_2CH_2Te(O)Ph} \xrightarrow[20^\circ\mathrm{C/1\,h,\ 95\%\ yield}]{2\,m\text{-}\mathrm{CPBA/MeOH}} \mathrm{DecCH_2CH_2OMe}$$

$$\mathrm{PhCH(Me)CH_2TePh} \xrightarrow[20^\circ\mathrm{C/1\,h}]{3\,m\text{-}\mathrm{CPBA/MeOH}} \underset{90\%}{\mathrm{MeOCH(Me)CH_2Ph}} \qquad \text{(b)}$$

$$\mathrm{PhCH(OMe)CH_2TePh} \xrightarrow[20^\circ\mathrm{C/1\,h}]{2\,m\text{-}\mathrm{CPBA/MeOH}} \underset{90\%}{\mathrm{(MeO)_2CHCH_2Ph}} \qquad \text{(c)}$$

SCHEME 139[331]

On the other hand, β-hydroxytellurides are not prone to elimination of the two heteroatomic moieties[301] and under conditions where β-hydroxyselenides are transformed in high yield to olefin[4–6,8], the telluro analogues only lead to a very low yield[301] of the olefins (Scheme 140).

$$\mathrm{PhTeCH_2CH(OH)Ph} \xrightarrow[110^\circ\mathrm{C}]{\text{toluene}} \underset{15\%}{\mathrm{CH_2{=}CHPh}}$$

SCHEME 140[301]

Despite the results just reported, Kauffman has found[299] that 1-diphenyl-phosphino-1-phenyltelluromethyllithium reacts with benzaldehyde and immediately loses the telluryl and hydroxyl moieties (Scheme 133).

III. REFERENCES

1. H. Rheinboldt (Ed.), *Methoden zur Herstellung und Umwandlung organischer Selenium und Tellur Verbindungen: Methoden der Organischen Chemie IX*, Georg Thieme Verlag, Stuttgart, 1955, p. 916
2. K. B. Sharpless, K. M. Gordon, R. F. Lauer, D. W. Patrick, S. P. Singer and M. W. Young, *Chem. Scripta*, **8A**, 9 (1975).
3. D. L. Klayman and W. H. H. Gunther (Eds.), *Organic Selenium Compounds: Their Chemistry and Biology*, John Wiley and Sons, New York, 1973.
4. D. L. J. Clive, *Tetrahedron*, **34**, 1049 (1978).
5. H. J. Reich, *Acc. Chem. Res.*, 22 (1979).
6. H. J. Reich, in *Oxidation in Organic Chemistry, Part* C (Ed. W. S. Trahanovsky), *Organic Chemistry: A series of Monographs*, Vol. 5C, Academic Press, New York, 1978.
7. Ref. 6, p. 11.
8. (a) A. Krief, *Tetrahedron*, **36**, 2531 (1980).
8. (b) A. Krief, *Synthesis and Synthetic Applications of 1-Metallid-1-selenocyclopropanes and Cyclobutanes and related 1-Metallo-1-silylcyclopropanes. Topics Curr. Chem.*, Vol. 135, 1 (1986).
9. G. Avrey, D. Barnard and D. T. Woodbridge, *J. Chem. Soc.* 2089 (1962).
10. H. Gilman and F. J. Webb, *J. Amer. Chem. Soc.*, **71**, 4062 (1949).
11. W. Dumont, P. Bayet and A. Krief. *Angew. Chem. Int. Ed.*, **13**, 804 (1974).
12. (a) M. Clarembeau and A. Krief, *Tetrahedron Letters*, **26**, 1093 (1985); (b) M. Clarembeau and

A. Krief, unpublished results; (c) M. Clarembeau and A. Krief, *Tetrahedron Letters*, **27**, 1719 (1986); (d) M. Clarembeau and A. Krief, *Tetrahedron Letters*, **27**, 1723 (1986); (e) A. Krief. M. Clarembeau, *J.C.S. Chem. Commun*, 688 (1986).
13. Ref. 8, p. 2608.
14. M. Sevrin, D. Van Ende and A. Krief. *Tetrahedron Letters*, 2643 (1976).
15. D. L. J. Clive, G. Chittattu and C. K. Wong, *J. Chem. Soc. Chem. Commun.*, 41 (1978).
16. D. L. J. Clive, G. J. Chittattu, V. Farina, W. A. Kiel, S. M. Menchen, C. G. Russell, A. Singh, C. K. Wong and N. J. Curtis, *Proceedings of the Third International Symposium on Organic Selenium and Tellurium Compounds* (Metz, 9–12 July 1979).
17. M. Sevrin, W. Dumont, L. Hevesi and A. Krief, *Tetrahedron Letters*, 2647 (1976).
18. M. Sevrin and A. Krief, *J. Chem. Soc., Chem. Commun.*, 656 (1980).
19. W. Dumont, P. Bayet and A. Krief, *Angew. Chem. Int. Ed.*, **13**, 274 (1974).
20. S. Halazy and A. Krief. *Tetrahedron Letters*, 4233 (1979).
21. H. J. Reich and F. Chow, *J. Chem. Soc., Chem. Commun.*, 790 (1975).
22. H. J. Reich and S. K. Shah, *J. Amer. Chem. Soc.*, **97**, 3250 (1975).
23. L. Hevesi, M. Sevrin and A. Krief, *Tetrahedron Letters*, 2651 (1976).
24. A. Krief, *Proceedings of the Third International Symposium on Organic Selenium and Tellurium Compounds*, (Metz, 9–12 July 1979).
25. A. Krief, W. Dumont, A. Cravador, J. N. Denis, S. Halazy, L. Hevesi, D. Labar, J. Lucchetti, J. Rémion, M. Sevrin and D. Van Ende, *Bull. Soc. Chim. Fr. II*, 519 (1980).
26. D. N. Jones, D. Mundy and R. D. Whitehouse, *J. Chem. Soc., Chem. Commun.*, 86 (1970).
27. K. B. Sharpless, M. W. Young and R. F. Lauer, *Tetrahedron Letters*, 1979 (1973).
28. P. A. Grieco and M. Miyashita, *J. Org. Chem.*, **39**, 120 (1974).
29. C. A. Kingsbury and D. J. Cram, *J. Amer. Chem. Soc.*, **82**, 1810 (1960).
30. B. J. Wakefield, *The Chemistry of Organolithium Compounds*, Pergamon Press, Oxford, 1974.
31. D. Seebach and K. H. Geiss, in *New Applications of Organometallic Reagents in Organic Synthesis* (Ed. D. Seyferth), Elsevier, Amsterdam, 1976.
32. D. J. Cram, *Fundamentals of Carbanion Chemistry—Organic Chemistry: A Series of Monographs*, Vol. 4, Academic Press, New York, 1965.
33. D. J. Peterson, *Organomet. Chem. Rev.*, **A7**, 295 (1972).
34. (a) T. Kauffmann, *Top. Curr. Chem.*, **92**, 109 (1980).
(b) T. Kauffmann, *Angew. Chem. Int. Ed.*, **21**, 410 (1982).
35. U. Schöllkopf, *Angew. Chem. Int. Ed.*, **10**, 763 (1970).
36. (a) U. Schöllkopf, in *Methoden der Organischen Chemie XIII/1*, Georg Thieme Verlag, Stuttgart, 1970, p. 87.
(b) K. Nutzel, H. Gilman and G. F. Wright, in *Methoden der Organischen Chemie XIII/2a*, Georg Thieme Verlag, Stuttgart, 1973, p. 53.
37. D. Seebach, *Angew. Chem. Int. Ed.*, **8**, 639 (1969).
38. O. W. Lever, *Tetrahedron*, **32**, 1943 (1976).
39. A. W. Johnson, *Ylid Chemistry*, Academic Press, New York (1966).
40. P. Beak and D. B. Reitz, *Chem. Rev.*, **78**, 275 (1978).
41. D. Seebach and D. Enders, *Angew. Chem. Int. Ed.*, **14**, 15 (1975).
42. (a) D. Hoppe, *Angew. Chem. Int. Ed.*, **13**, 789 (1974).
(b) U. Schöllkopf, *Angew. Chem. Int. Ed.*, **16**, 339 (1977).
43. H. Ahlbrecht, *Chimia*, **31**, 391 (1977).
44. T. H. Chan, *Acc. Chem. Res.*, **10**, 442 (1977).
45. E. W. Colvin, *Chem. Soc. Rev.*, **7**, 15, (1978).
46. W. S. Wadsworth, *Organic Reactions*, **25**, 73 (1977)
47. J. Boutagy and R. Thomas, *Chem. Rev.* **74**, 87 (1974) and references cited therein.
48. A. Maercker, *Organic Reactions*, **14**, 270 (1965).
49. U. Schöllkopf, *Newer Methods of Preparative Organic Chemistry*, Vol. 3, Academic Press, New York, 1964, p. 111; *Angew. Chem.*, **71**, 260 (1959).
50. L. D. Bergelson and M. M. Shemyakin, *Angew. Chem. Int. Ed.*, **3**, 250 (1964).
51. H. J. Bestmann, Newer Methods of Preparative Organic Chemistry, Vol 5, Academic Press, New York, 1968, p. 1; *Angew. Chem. Int. Ed.*, **4**, 583, 645, 830 (1965).
52. H. Pommer, *Angew. Chem. Int. Ed.*, **16**; 423 (1977) and references cited therein.
53. E. Block, *Reaction of Organosulfur Compounds—Organic Chemistry: A Series of Monographs*, Vol. 37, Academic Press, New York, 1978.

54. D. Seebach, *Synthesis*, 17 (1969).
55. B. T. Gröbel and D. Seebach, *Synthesis*, 357 (1977).
56. L. A. Paquette, *Acc. Chem. Res.*, **1**, 209 (1968).
57. P. D. Magnus, *Tetrahedron*, **33**, 2019 (1977).
58. B. M. Trost and L. S. Melvin, *Sulfur Ylides Emerging Synthetic Intermediates—Organic Chemistry: A Series of Monographs*, Vol. 31, Academic Press, New York, 1975, and references cited therein.
59. G. Köbrich, *Angew. Chem. Int. Ed.*, **6**, 41 (1967).
60. G. Köbrich, *Bull. Soc. Chim. Fr.*, 2712 (1969).
61. J. Villieras, *Organomet. Chem. Rev.*, **A7**, 81, (1971).
62. M. S. Newman and B. J. Magerlein, *Organic Reactions*, **5**, 413 (1949).
63. M. Ballester, *Chem. Rev.*, **55**, 283 (1955).
64. J. Hine, *Divalent Carbon*, Ronald Press, New York, 1964.
65. W. Kirmse, *Carbene Chemistry—Organic Chemistry: A Series of Monographs*, Vol. 1, Academic Press, New York, 1971.
66. M. Jones and R. A. Moss, *Carbenes* (Ed. R. A. Moss) Vols I and II, Wiley–Interscience, New York, 1973.
67. (a) D. Seebach and N. Peleties, *Angew. Chem. Int. Ed.*, **8**, 450 (1969).
(b) D. Seebach and N. Peleties, *Chem. Ber.*, **105**, 511 (1972).
68. D. Van Ende, A. Cravador and A. Krief, *J. Organometal. Chem.*, **1**, 177 (1979).
69. S. Raucher and G. A. Koolpe, *J. Org. Chem.*, **43**, 3794 (1978).
70. H. J. Reich, W. W. Willis and P. D. Clark, *J. Org. Chem.*, **46**, 2775 (1981).
71. B. T. Gröbel and D. Seebach, *Chem. Ber.*, **110**, 852 (1977).
72. H. J. Reich, F. Chow and S. K. Shah, *J. Amer. Chem. Soc.*, **101**, 6638 (1979).
73. P. Bouhy, *Mémoire de Licence*, Namur, June 1980.
74. D. Van Ende, W. Dumont and A. Krief, *J. Organometal Chem.*, **C10**, 149, (1978).
75. K. Sachdev and H. S. Sachdev, *Tetrahedron Letters*, 4223 (1976).
76. H. J. Reich and S. K. Shah, *J. Org. Chem.*, **42**, 1773 (1977).
77. H. J. Reich, *J. Org. Chem.*, **40**, 2570 (1975).
78. M. Clarembeau, unpublished results.
79. M. R. Binns, R. K. Haynes, T. L. Houston and W. R. Jackson, *Tetrahedron Letters*, 573 (1980).
80. R. H. Mitchell, *J. Chem. Soc., Chem. Commun.*, 990 (1974).
81. H. J. Reich and S. K. Shah, *J. Amer. Chem. Soc.*, **99**, 263 (1977).
82. In one case however selenoxide *syn* elimination exclusively occurred, see Ref. 4 in our Ref. 81.
83. F. Zutterman and A. Krief, unpublished results.
84. D. Liotta, M. Saindane and D. Brothers, *J. Org. Chem.*, **47**, 1598 (1982).
85. H. J. Reich and M. L. Cohen, *J. Amer. Chem. Soc.*, **101**, 1307 (1979).
86. F. G. Bordwell, J. E. Bares, J. E. Bartmess, G. E. Drucker, J. Gerhold, G. J. Mc Collum, M. Van Der Puy, N. R. Vanier and W. S. Matthews, *J. Org. Chem.*, **42**, 326 (1977).
87. K. B. Sharpless, R. F. Lauer and A. Y. Teranishi, *J. Amer. Chem. Soc.* **95**, 6137 (1973).
88. (a) J. Lucchetti and A. Krief, *Tetrahedron Letters*, 2693 (1978).
(b) J. Lucchetti and A. Krief, *Tetrahedron Letters*, 2697 (1978).
89. J. Lucchetti and A. Krief, *J. Chem. Soc., Chem. Commun.*, 127 (1982).
90. N. Petragnani and H. M. C. Ferraz, *Synthesis*, 476 (1978).
91. P. A. Grieco, *Synthesis*, 67 (1975) and references cited therein.
92. P. A. Grieco and Y. Yokoyama, *J. Amer. Chem. Soc.*, **99**, 5210 (1977).
93. Y. Masuyama, Y. Ueno and M. Okawara, *Chem. Letters*, 835 (1977).
94. D. Van Ende, W. Dumont and A. Krief, *Angew. Chem. Int. Ed.*, **14**, 700 (1975).
95. W. Dumont and A. Krief, *Angew. Chem. Int. Ed.*, **16**, 540 (1977).
96. A. Cravador, A. Krief and L. Hevesi, *J. Chem. Soc., Chem. Commun.*, 451 (1980).
97. J. Lucchetti and A. Krief, *Compt. Rend., Ser. C*, **289**, 287 (1979).
98. (a) D. L. J. Clive and S. M. Menchen, *J. Chem. Soc., Chem. Commun.*, 356 (1978).
(b) D. L. J. Clive and S. M. Menchen, *J. Org. Chem.*, **44**, 1883 (1979).
99. A. Cravador and A. Krief, *Compt. Rend., Ser. C*, **289**, 267 (1979).
100. W. Dumont, D. Van Ende and A. Krief, *Tetrahedron Letters*, 485 (1979).
101. N. Petragnani, R. Rodrigues and J. V. Comassetto, *J. Organomet. Chem.*, **114**, 281 (1976).
102. J. V. Comassetto and C. A. Brandt, *J. Chem. Res. (S)*, 56 (1982).
103. T. Sakakibara, S. Ikuta and R. Sudoh, *Synthesis*, 261 (1982).

104. Ref. 8, p. 2605.
105. H. M. J. Gillissen, P. Schipper, P. J. J. M. van Ool and H. M. Buck, *J. Org. Chem.*, **45**, 319 (1980).
106. (a) M. Sevrin and A. Krief, *Tetrahedron Letters*, 585 (1980).
(b) M. Sevrin, Ph.D. *Thesis*, Namur, October 1980.
107. S. Halazy and A. Krief, *J. Chem. Soc., Chem. Commun.*, 1136 (1979).
108. D. Labar and A. Krief, *J. Chem. Soc., Chem. Commun.*, 564 (1982).
109. S. Halazy, F. Zutterman and A. Krief, *Tetrahedron Letters*, 4385 (1982).
110. S. Halazy and A. Krief, *J. Chem. Soc., Chem. Commun.*, 1200 (1982).
111. W. Dumont and A. Krief, *Société Chimique de France, Journée d'automne, Palaiseau*, September 1981.
112. J. Lucchetti, J. Remion and A. Krief, *Compt. Rend., Ser. C*, **288**, 553 (1979).
113. J. L. Bertrand, *Mémoire de Licence*, Namur, June 1983.
114. M. Clarembeau and J. L. Bertrand and A. Krief, *Israel J. Chem.*, **24**, 125 (1984).
115. S. Halazy, *Ph.D. Thesis*, Namur, December 1982.
116. F. Zutterman and A. Krief, *J. Org. Chem.*, **48**, 1135 (1983).
117. J. Lucchetti, *Ph.D. Thesis*, Namur, March 1983.
118. W. Dumont, M. Sevrin and A. Krief, *Angew. Chem. Int. Ed.*, **16**, 541 (1977).
119. Ref. 22, footnote 13.
120. Th. Kauffmann, H. Ahlers, H.J. Tilhard and A. Woltermann, *Angew. Chem. Int. Ed.*, **16**, 710 (1977).
121. M. Sevrin, J. N. Denis and A. Krief, *Angew. Chem. Int. Ed.*, **17**, 526 (1978).
122. S. Raucher and G. A. Koolpe, *J. Org. Chem.*, **43**, 4252 (1978).
123. M. Sevrin, W. Dumont and A. Krief, *Tetrahedron Letters*, 3835 (1977) and references cited therein.
124. J. N. Denis and A. Krief, *Tetrahedron Letters*, 3407 (1982) and references cited therein.
125. S. Raucher, *J. Org. Chem.*, **43**, 2950 (1977).
126. (a) D. Seebach and A. K. beck, *Angew. Chem. Int. Ed.*, **13**, 806 (1974).
(b) D. Seebach, N. Meyer and A. K. Beck, *Liebigs Ann. Chem.*, 846 (1977).
127. S. Halazy and A. Krief, *Tetrahedron Letters*, 1997 (1980).
128. W. Dumont, A. Cravador and A. Krief, unpublished results
129. M. Sevrin and A. Krief, *Tetrahedron Letters*, 187 (1978).
130. J. Remion and A. Krief, *Tetrahedron Letters*, 3743 (1976).
131. J. Remion, W. Dumont and A. Krief, *Tetrahedron Letters*, 1385 (1976).
132. D. Labar, W. Dumont, L. Hevesi and A. Krief, *Tetrahedron Letters*, 1145 (1978).
133. S. Halazy, J. Lucchetti and A. Krief, *Tetrahedron Letters*, 3971 (1978).
134. S. Halazy and A. Krief, *Tetrahedron Letters*, 1829 (1981).
135. D. Labar, J. L. Laboureur and A. Krief, *Tetrahedron Letters*, 983 (1982).
136. J. N. Denis, W. Dumont and A. Krief, *Tetrahedron Letters*, 453 (1976).
137. W. Dumont and A. Krief, *Angew. Chem. Int. Ed.*, **14**, 350 (1975).
138. D. Van Ende and A. Krief, *Tetrahedron Letters*, 457 (1976).
139. S. Halazy and A. Krief, *Tetrahedron Letters*, 1833 (1981).
140. A. M. Leonard-Coppens and A. Krief, *Tetrahedron Letters*, 3227 (1976).
141. A. Anciaux, A. Eman, W. Dumont, D. Van Ende and A. Krief, *Tetrahedron Letters*, 1613 (1975).
142. W. Dumont and A. Krief, *Angew. Chem. Int. Ed.*, **15**, 161 (1976).
143. S. Halazy, W. Dumont and A. Krief, *Tetrahedron Letters*, 4737 (1981).
144. D. Labar and A. Krief, unpublished results.
145. F. Zutterman, unpublished results.
146. J. Lucchetti, W. Dumont and A. Krief, *Tetrahedron Letters*, 2695 (1979).
147. J. D. White, M. C. Kang and B. G. Sheldon, *Tetrahedron Letters*, 4539 (1983).
148. J. Lucchetti and A. Krief, *Tetrahedron Letters*, 1623 (1981).
149. J. Lucchetti, W. Dumont and A. Krief, *Tetrahedron Letters*, 2695 (1979).
150. J. Lucchetti and A. Krief, *J. Organometal. Chem.*, **C49**, 194 (1980).
151. L. Fitjer, H. J. Shauerman and D. Wehle, *Tetrahedron Letters*, **25**, 2329 (1984).
152. 2,2,6,6-Tetraphenylcyclohexanone was kindly provided by prof. D. H. R. Barton, Institut des Substances Naturelles, Gif-sur-Yvette, France.
153. R. C. Gadwood, *J. Org. Chem.*, **48**, 2098 (1983).

154. D. Van Ende, *Ph.D. Thesis*, Namur, 1976.
155. L. Wartski, M. El Bouz, J. Seyden-Penne, W. Dumont and A. Krief, *Tetrahedron Letters*, 1543 (1979).
156. W. Dumont, J. Lucchetti and A. Krief, *J. Chem. Soc., Chem. Commun.*, 66 (1983).
157. J. Lucchetti and A. Krief, *Synth. Commun.*, **14**, 1153 (1983).
158. H. J. Reich, S. K. Shah and F. Chow, *J. Amer. Chem. Soc.*, **101**, 6648 (1979).
159. J. N. Denis and A. Krief, *Tetrahedron Letters*, 3995 (1979).
160. J. N. Denis and A. Krief, *J. Chem. Soc., Chem. Commun.*, 544 (1980).
161. N. N. Magdesieva, R. A. Kandgeteyan and A. A. Ibragimov, *J. Organometal. Chem.*, **42**, 399 (1972) and references cited therein.
162. K. H. Wei, I. C. Paul, M.-M. Y. Chang and J. I. Musher, *J. Amer. Chem. Soc.*, **96**, 4099 (1974).
163. E. Ernstbrunner and D. Lloyd, *Liebigs Ann. Chem.*, **753**, 196 (1971).
164. (a) D. Lloyd and M. I. C. Singer, *J. Chem. Soc., Chem. Commun.*, 390 (1967).
(b) E. D. Hughes and K. I. Kuriyan, *J. Chem. Soc.*, 1609 (1935).
165. W. W. Lotz and J. Gosselck, *Tetrahedron*, **29**, 917 (1973).
166. (a) Recently a report from Negoro appeared (Ref. 166b), 8 years after our own paper and seems to be incorrect. We have tried without success to repeat the reactions therein described. Moreover the ^{1}H-NMR published in this paper do not refer to epoxides as proposed but to the starting carbonyl compound.
(b) K. Negoro, K. Takaki and M. Yasumura, *Angew. Chem. Int. Ed.*, **20**, 671 (1981).
167. E. G. Kataev and V. N. Petrov, *J. Gen. Chem. USSR*, **32**, 3626 (1962).
168. H. J. Reich and J. M. Renga, *J. Org. Chem.*, **40**, 3313 (1975).
169. W. Hölzle and W. Jenny, *Helv. Chim. Acta*, **41**, 593 (1958).
170. E. G. Kataev, T. G. Maunafov, E. A. Berdnikov and O. A. Komarovskaya, *J. Org. Chem. USSR*, **9**, 1998 (1973).
171. T. A. Hase and P. Kukk'ola, *Synth. Commun.*, **10**, 451 (1980).
172. A. E. Feiring, *J. Org. Chem.*, **45**, 1958, 1962 (1980).
173. I. Kuwajima and M. Shimizu, *Tetrahedron Letters*, 1277 (1978).
174. M. Sevrin, W. Dumont and A. Krief, *Tetrahedron Letters*, 183 (1978).
175. D. J. Buckley, S. Kulkowit and A. McKervey, *J. Chem. Soc., Chem. Commun.*, 506 (1980).
176. H. J. Reich and M. L. Cohen, *J. Amer. Chem. Soc.*, **101**, 1307 (1979).
177. J. Hooz and R. Mortimer, *Tetrahedron Letters*, 805 (1976).
178. B. Harirchian and P. Magnus, *J. Chem. Soc., Chem. Commun.*, 522 (1977).
179. Y. Okamoto, R. Homsany and T. Yano, *Tetrahedron Letters*, 2529 (1972).
180. S. Raucher, *J. Org. Chem.*, **43**, 4885 (1978).
181. W. Dumont, D. Van Ende and A. Krief, *Tetrahedron Letters*, 485 (1979).
182. H. J. Reich and J. M. Renga, *J. Org. Chem.*, **40**, 3313 (1975).
183. H. J. Reich, J. M. Renga and J. E. Trend, *Tetrahedron Letters*, 2217 (1976).
184. E. G. Kataev, T. G. Mannafov and Y. Y. Samitov, *Zh. Org. Khim.*, **11**, 2324 (1975).
185. J. N. Denis and A. Krief, *Tetrahedron Letters*, 3411 (1982).
186. B. T. H. Gröbel and D. Seebach, *Chem. Ber.*, **110**, 867 (1977).
187. J. N. Denis, S. Desauvage, L. Hevesi and A. Krief, *Tetrahedron Letters*, 4009 (1981).
188. S. Halazy and A. Krief, *Tetrahedron Letters*, 4341 (1981).
189. D. Seyferth (Ed.) *New Applications of Organometallic Reagents in Organic Synthesis* Elsevier, Amsterdam, 1976.
190. D. Seyferth and M. A. Weiner, *J. Org. Chem.*, **26**, 4797 (1961).
191. R. M. Magid and J. G. Welch, *J. Amer. Chem. Soc.*, **90**, 5211 (1968).
192. D. E. Applequist and E. G. Saurborn, *J. Org. Chem.*, **37**, 1676 (1972).
193. A. H. Haubein and H. Gilman, *J. Amer. Chem. Soc.*, **66**, 1515 (1944).
194. P. Miginiac and C. Bouchoule, *Bull. Soc. Chim. Fr.*, 4156 (1968).
195. J. J. Eisch and A. M. Jacobs, *J. Org. Chem.*, **28**, 2145 (1963).
196. H. A. McNinch and H. Gilman, *J. Org. Chem.*, **26**, 3723 (1961).
197. D. Seyferth and M. A. Weiner, *J. Org. Chem.*, **24**, 1395 (1959).
198. M. Micha-Screttas and C. G. Screttas, *J. Org. Chem.*, **44**, 713 (1979).
199. M. A. Weiner and D. Seyferth, *Organic Synthesis*, **41**, 30 (1961).
200. D. Seyferth, R. Suzuki, C. J. Murphy and C. R. Sabet, *J. Organometal. Chem.*, 431 (1964).
201. D. F. Hoeg and D. I. Lusk, *J. Amer. Chem. Soc.*, **86**, 928 (1964).
202. D. J. Peterson, *J. Org. Chem.*, **32**, 1717 (1967) and references cited therein.

203. E. J. Corey and D. Seebach, *J. Org. Chem.*, **31**, 4097 (1966).
204. W. E. Parham and R. F. Motter, *J. Amer. Chem. Soc.*, **81**, 2146 (1959).
205. W. E. Parham, M. A. Kalnins and D. R. Theissen, *J. Org. Chem.*, **27**, 2698 (1962).
206. T. M. Dolak and T. A. Bryson, *Tetrahedron Letters*, 1961 (1977).
207. W. Dumont, A. Anciaux, A. Eman, J. N. Denis and A. Krief, to be published.
208. A. Anciaux, A. Eman, W. Dumont and A. Krief, *Tetrahedron Letters*, 1617 (1975).
209. J. N. Denis, W. Dumont and A. Krief, *Tetrahedron Letters*, 4111 (1979).
210. W. Dumont and A. Krief, *J. Chem. Soc., Chem. Commun.*, 673 (1980).
211. W. Dumont and A. Krief, *Angew. Chem. Int. Ed.*, **15**, 161 (1976).
212. D. J. Peterson, *J. Organometal. Chem.*, **9**, 373 (1967).
213. G. A. Gornowicz and R. West, *J. Amer. Chem. Soc.*, **90**, 4478 (1968).
214. A. G. Brook, J. M. Duff and D. G. Anderson, *Can. J. Chem.*, **48**, 561 (1970).
215. T. Hiyama, A. Kanakura, Y. Morizawa and H. Nozaki, *Tetrahedron Letters*, 1279 (1982).
216. (a) L. A. Paquette, K. A. Horn and G. J. Wells, *Tetrahedron Letters*, 259 (1982).
(b) T. J. Cohen, personal communication, 1981.
217. L. F. Cason and H. G. Brooks, *J. Org. Chem.*, **19**, 1278 (1954).
218. G. R. Buell, R. Corriu, G. Guerin and L. Spialter, *J. Amer. Chem. Soc.*, **92**, 7424 (1970).
219. T. H. Chan and E. Chang, *J. Org. Chem.*, **39**, 3264 (1974).
220. K. C. Nicolaou and Z. Lysenko, *J. Amer. Chem. Soc.*, **99**, 3185 (1977).
221. D. L. J. Clive, G. J. Chittattu, V. Farina, W. A. Kiel, S. M. Menchen, C. G. Russell, A. Singh, C. K. Wong and N. J. Curtis, *J. Amer. Chem. Soc.*, **102**, 4438 (1980).
222. S. Halazy, unpublished results.
223. J. N. Denis and A. Krief, *J. Chem. Soc., Chem. Commun.*, 229 (1983).
224. P. A. Grieco, S. Gilman and M. Nishizawa, *J. Org. Chem.*, **41**, 1485 (1976).
225. G. Hölzle and W. Jenny, *Helv. Chim. Acta*, **89**, 712 (1958).
226. K. B. Sharpless and R. F. Lauer, *J. Org. Chem.*, **39**, 429 (1974).
227. H. J. Reich, J. M. Renga and I. L. Reich, *J. Amer. Chem. Soc.*, **97**, 5434 (1975).
228. W. G. Salmond, The Upjohn Company. *U. S. Patent*, No. 3,970,676 (1976)
229. K. B. Sharpless, R. F. Lauer and A. Y. Teranishi, *J. Amer. Chem. Soc.*, **95**, 6137 (1973).
230. T. Minami, H. Sako, T. Ikehira, T. Hanamoto and I. Hirao, *J. Org. Chem.*, **48**, 2569 (1983).
231. T. Hori and K. B. Sharpless, *J. Org. Chem.*, **43**, 1689 (1978).
232. D. Labar, L. Hevesi, W. Dumont and A. Krief, *Tetrahedron Letters*, 1141 (1978).
233. K. B. Sharpless and M. W. Young, *J. Org. Chem.*, **40**, 947 (1975).
234. R. D. Clark and C. H. Heathcock, *J. Org. Chem.*, **41**, 1396 (1976).
235. L. D. Kwart, A. G. Horgan and H. Kwart, *J. Amer. Chem. Soc.*, **103**, 1232 (1981).
236. K. Torssell, *Tetrahedron Letters*, 623 (1974).
237. Y. Tsuda, T. Sano, J. Taga, K. Isobe, J. Toda, H. Irie, H. Tanaka, S. Takagi, M. Yamaki and M. Murata, *J. Chem. Soc., Chem. Commun.*, 933 (1975).
238. S. David, A. Lubineau and J. M. Vatele, *J. Chem. Soc., Chem. Commun.*, 701 (1975); *J. Chem. Soc. Perkin Trans. 1*, 1831 (1976).
239. L. A. Paquette, C. C. Liao, R. L. Burson, R. E. Wingard, Jr., C. N. Shih, J. Fayos and J. Clardy, *J. Amer. Chem. Soc.*, **99**, 6935 (1977).
240. L. A. Paquette, C. C. Liao, D. C. Liotta and W. E. Fristad, *J. Amer. Chem. Soc.*, **98**, 6412 (1976).
241. B. M. Trost and P. H. Scudder, *J. Amer. Chem. Soc.*, **99**, 7601 (1977).
242. H. J. Reich, *J. Org. Chem.*, **39**, 428 (1974).
243. K. Isobe, J. Taga and Y. Tsuda, *Tetrahedron Letters*, 2331 (1976).
244. K. C. Nicolaou and W. E. Barnette, *J. Chem. Soc., Chem. Commun.*, 331 (1977).
245. K. C. Nicolaou and Z. Lysenko, *Tetrahedron Letters*, 1257 (1977).
246. E. J. Corey, G. E. Keck and I. Szekely, *J. Amer. Chem. Soc.*, **99**, 2006 (1977).
247. D. L. J. Clive, G. Chittattu, N. J. Curtis, W. A. Kiel and C. K. Wong, *J. Chem. Soc., Chem. Commun.*, 725 (1977).
248. A. G. Gonzalez, C. Betancor, C. G. Francisco, R. Hernandez, J. A. Salazar and E. Suarez, *Tetrahedron Letters*, 2959 (1977).
249. D. Liotta and G. Zima, *J. Org. Chem.*, **45**, 2551 (1980).
250. A. Toshimitsu, H. Owada, T. Aoai, S. Uemura and M. Okano, *J. Chem. Soc., Chem. Commun.*, 546 (1981).
251. J. N. Denis, J. Vicens and A. Krief, *Tetrahedron Letters*, 2697 (1979).

252. P. S. Liu, V. E. Marquez, J. A. Kelley, and J. S. Driscoll, *J. Org. Chem.*, **45**, 5225 (1980).
253. F. M. Dean and B. K. Park, *J. Chem. Soc., Chem. Commun.*, 142 (1975).
254. S. H. Korzeniowski, D. P. Vanderbilt and L. B. Hendry, *Org. Prep. Proc. Int.*, **8**, 81 (1976).
255. J. Schwartz and Y. Hayasi, *Tetrahedron Letters*, 1497 (1980).
256. J. Tsuji, K. Masaoka and T. Takahashi, *Tetrahedron Letters*, 2267 (1977).
257. J. Tsuji, K. Masaoka, T. Takahashi, A. Suzuki and N. Miyaura, *Bull. Chem. Soc. Japan*, **50**, 2507 (1977).
258. P. A. Grieco and M. Miyashita, *J. Org. Chem.*, **39**, 120 (1974).
259. P. A. Grieco and M. Nishizawa, *J. Chem. Soc., Chem. Commun.*, 582 (1976).
260. P. A. Zoretic and R. Soja, *J. Org. Chem.*, **41**, 3587 (1976); *J. Heterocycl. Chem.*, **14**, 681 (1977).
261. D. N. Brattesani and C. H. Heathcock, *Tetrahedron Letters*, 2279 (1974); *J. Org. Chem.*, **40**, 2165 (1975).
262. T. Sakakibara, I. Takai, E. Omara and R. Sudoh, *J. Chem. Commun.*, 261 (1981).
263. A. Toshimitsu, H. Owada, S. Uemura and M. Okano, *Tetrahedron Letters*, 2105 (1982).
264. H. J. Reich and W. W. Willis, *J. Amer. Chem. Soc.*, **102**, 5967 (1980).
265. T. G. Back and S. Collins, *Tetrahedron Letters*, 5111 (1981).
266. A. Burton, L. Hevesi, W. Dumont, A. Cravador and A. Krief, *Synthesis*, 877 (1979).
267. W. G. Salmond, M. A. Barta, A. M. Cain and M. C. Sobala, *Tetrahedron Letters*, 1683 (1977).
268. S. K. Shah, *Ph. D. Thesis*, University of Wisconsin, Madison, 1977.
269. S. Halazy and A. Krief, *Tetrahedron Letters*, 2135 (1981).
270. Diphosphorus tetraiodide (P_2I_4) a powerful deoxygenator, Aldrich, Technical Information 191, 1980, Product No. 21865–0.
271. M. Shimizu and I. Kuwajima, *J. Org. Chem.*, **45**, 2921 (1980).
272. F. Ogura, H. Yamaguchi, T. Otsubo and H. Tanaka, *Bull. Chem. Soc. Japan*, **55**, 641 (1982).
273. D. Barnard and D. T. Woodbridge, *Chem. Ind. (London)*, 1603 (1959).
274. K. Balenovic, R. Lazic, V. Polak and P. Stern, *Bull. Sci. Cons. Acad. Sci. Arts RSF Yougosl. Sect. A17*, 147 (1972); *Chem. Abstr.*, **77**, 139499j (1972).
275. N. Bregant, I. Perina and K. Balenovic, *Bull. Sci. Cons. Acad. Sci. Arts RSF Yougosl. Sect. A17*, 148 (1972); *Chem. Abstr.*, **77**, 152092j (1972).
276. I. Perina, N. Bregant and K. Balenovic, *Bull. Sci. Cons. Acad. Sci. Arts RSF Yougosl. Sect. A18*, 4 (1973); *Chem. Abstr.*, **78**, 160007w (1973).
277. M. Poje and K. Balenovic, *Bull. Sci. Cons. Acad. Sci. Arts RSF Yougosl. Sect. A20*, 1 (1975); *Chem. Abstr.*, **83**, 43558u (1975).
278. I. Perina, N. Bregant and K. Balenovic, *Bull. Sci. Cons. Acad. Sci. Arts RSF Yougosl. Sect. A18*, 3 (1973); *Chem. Abstr.*, **79**, 5096b (1973).
279. K. Balenovic, N. Bregant and I. Perina, *Synthesis*, 172 (1973).
280. A. Cravador and A. Krief, *Tetrahedron Letters*, 2491 (1981).
281. D. Labar, A. Krief and L. Hevesi, *Tetrahedron Letters*, 3967 (1978).
282. J. L. Laboureur and A. Krief, *Tetrahedron Letters*, **25**, 4569 (1984).
283. M. Sevrin, J. N. Denis and A. Krief, *Tetrahedron Letters*, 1877 (1980).
284. K. B. Sharpless and R. F. Lauer, *J. Amer. Chem. Soc.*, **95**, 2697 (1973).
285. D. L. J. Clive, *J. Chem. Soc., Chem. Commun.*, 100 (1974).
286. K. B. Sharpless and R. F. Lauer, *J. Org. Chem.*, **39**, 428 (1974).
287. D. L. J. Clive, C. G. Russell and S. C. Suri, *J. Org. Chem.*, **47**, 1632 (1982).
288. C. J. Kowalski and J. S. Dung, *J. Amer. Chem. Soc.*, **102**, 7951 (1980).
289. D. L. J. Clive and V. N. Kale, *J. Org. Chem.*, **46**, 231 (1981).
290. T. Di Giamberardino, S. Halazy, W. Dumont and A. Krief, *Tetrahedron Letters*, 3413 (1983).
291. (a) D. J. Cram and F. A. Abd. Elhafez, *J. Amer. Chem. Soc.*, **74**, 5828 (1952).
(b) T. J. Leitereg and D. J. Cram, *J. Amer. Chem. Soc.*, **90**, 4011, 4019 (1968).
292. (a) M. Cherest, H. Felkin and N. Prudent, *Tetrahedron Letters*, 2199 (1968).
(b) M. Cherest and H. Felkin, *Tetrahedron Letters*, 2205 (1968).
293. M. Shimizu and I. Kuwajima, *J. Org. Chem.*, **45**, 4063 (1980).
294. D. Seebach and A. K. Beck, *Chem. Ber.*, **108**, 314 (1975).
295. (a) B. H. Freeman and D. Lloyd, *J. Chem. Soc., Chem. Commun.*, 924 (1970).
(b) B. H. Freeman, D. Lloyd and M. I. C. Singer, *Tetrahedron*, **28**, 343 (1972).
296. E. J. Corey and M. Jautelat, *Tetrahedron Letters*, 5787 (1968).
297. Ref. 34b, p. 417.
298. Ref. 34b, p. 426.

299. Ref. 34b, p. 427.
300. Ref. 34b, p. 427.
301. T. Kauffmann, H. Ahlers, R. Joussen, R. Kriegsmann, A. Vahrenhorst and R. Woltermann, *Tetrahedron Letters*, 4399 (1978).
302. D. Lloyd, *Chem. Scripta*, **8a**, 14 (1975).
303. N. S. Zefirov, *Chem. Scripta*, **8a**, 20 (1975).
304. N. N. Magdesieva, R. A. Kyandzhetsian and O. A. Rakitin, *Zh. Org. Khim.*, **11**, 2562 (1975).
305. N. N. Magdesieva, R. A. Kyandzhetsian and O. A. Rakitin, *Zh. Org. Khim.*, **12**, 36 (1976).
306. I. D. Sadekov, A. I. Usachev, A. A. Maksimenko and V. I. Minkin, *Zh. Obshch. Khim.*, **45**, 2562 (1975).
307. M. Sevrin, unpublished results.
308. (a) A. Burton, *Mémoire de Licence*, Namur, June 1978.
(b) A. Burton, L. Hevesi and A. Krief, unpublished results.
309. S. Spreutel, *Mémoire de Licence*, Namur, June 1979.
310. J. Lucchetti and A. Krief, *Compt. Rend., Ser. C*, **288**, 537 (1979).
311. R. Baudat and M. Petrzilka, *Helv. Chim. Acta*, **62**, 1406 (1979).
312. G. H. Posner and M. J. Chapdelaine, *Tetrahedron Letters*, 3227 (1977).
313. D. H. R. Barton, D. J. Lester, W. B. Motherwell and M. T. Barros Papoula, *J. Chem. Soc., Chem. Commun.*, 246 (1980).
314. H. J. Reich, P. M. Gold and F. Chow, *Tetrahedron Letters*, 4433 (1979).
315. I. D. Sadekov, A. I. Usachev, A. A. Maksimenko and A. I. Minkin, *Zh. Obsch. Khim.*, **48**, 934 (1974).
316. A. Osuka, Y. Mori, H. Shimizu and H. Suzuki, *Tetrahedron Letters*, 2599 (1983).
317. A. Osuka and H. Suzuki, *Tetrahedron Letters*, 5109 (1983).
318. S. Uemura and S. Fukuzawa, *Tetrahedron Letters*, 4347 (1983).
319. For some synthetic methods based on tellurium chemistry see the references cited in our Ref. 318.
320. L. Hevesi and A. Krief, *Angew. Chem. Int. Ed.*, **15**, 381 (1976).
321. H. J. Reich, M. C. Clark and W. W. Willis, *J. Org. Chem.*, **47**, 1618 (1982).
322. K. Chikamatsu, T. Otsubo, F. Ogura, H. Yamaguchi, *Chem. Letters*, 1081 (1982).
323. T. Otsubo, F. Ogura, H. Yamaguchi, H. Higuchi, Y. Sakata and S. Misumi, *Chem. Letters*, 447 (1981).
324. J. V. Comasseto, *J. Organometal. Chem.*, **253**, 131 (1983).
325. D. H. R. Barton, G. Bashiardes and J. L. Fourrey, *Tetrahedron Letters*, **25**, 1287 (1984).
326. M. Clarembeau and A. Krief, *Tetrahedron Letters*, **25**, 3629 (1984).
327. H. J. Reich, S. Wollowitz, J. E. Trend, F. Chow and D. F. Wandelborn, *J. Org. Chem.*, **43**, 1697 (1978).
328. B. M. Trost, D. E. Keeley, H. C. Ardnt and M. J. Bogdanowicz, *J. Amer. Chem. Soc.*, **99**, 3088 (1977).
329. H. Nishuyama, T. Kitajima, A. Yamamoto and K. Itho, *J. Chem. Soc., Chem. Commun.*, 1232 (1982).
330. J. N. Denis, J. L. Laboureur and A. Krief, unpublished results.
331. S. Uemura and S. Fukuzawa, *Tetrahedron Letters*, 4347 (1983).
332. S. Uemura and S. Fukuzawa, *J. Amer. Chem. Soc.*, **105**, 2748 (1983).
333. H. Lee and M. P. Cava, *J. Chem. Soc., Chem. Commun.*, 277 (1981).
334. A. Krief, W. Dumont, J. N. Denis, G. Evrard and B. Norberg, *J. Chem. Soc. Chem. Commun.*, 569 (1985).
335. A. Krief, W. Dumont and J. N. Denis, *J. Chem. Soc. Chem. Commun.*, 570 (1985).
336. M. Clarembeau, A. Cravadoi, W. Dumont, L. Hevesi, A. Krief, J. Lucchetti and D. Van Ende, *Tetrahedron*, in press.
337. D. Labar and A. Krief, *Bull. Soc. Chem. Belg.*, **93**, 1005 (1984) and references cited.
338. D. Labar, PhD thesis, Namur, March 1985.
339. (a) J. L. Laboureur and A. Krief, *Tetrahedron Letters*, **25**, 2713 (1984).
(b) A. Krief and J. L. Laboureur, J. C. S., *Chem. Commun.*, 702 (1986).
(c) Schmit, S. Sahroui-Taleb, E. Differding, G. G. Dehasse, S. De Lombaert and L. Ghosez, *Tetrahedron Letters*, **25**, 5043 (1984).
340. A. Krief, J. L. Laboureur and W. Dumont, *Tetrahedron Letters*, **28**, 1549 (1987).
(b) A. Krief and J. L. Laboureur, *Tetrahedron Letters*, **28**, 1545 (1987).

The Chemistry of Organic Selenium and Tellurium Compounds Volume 2
Edited by S. Patai
© 1987 John Wiley & Sons Ltd.

CHAPTER **18**

Selenium and tellurium derivatives of carbohydrates and nucleoside analogs

ZBIGNIEW J. WITCZAK

Whistler Center For Carbohydrate Research, Purdue University, Smith Hall, West Lafayette, Indiana 47907, USA

I. INTRODUCTION 765
II. SELENOGLYCOSIDES AND OTHER SELENIUM DERIVATIVES OF CARBOHYDRATES: SYNTHESIS AND REACTIONS . . . 766
III. SUGARS WITH SELENIUM IN THE SUGAR RING 775
IV. SELENONUCLEOSIDE ANALOGS: SYNTHESIS AND REACTIONS 778
V. TELLUROCARBOHYDRATES AND ORGANOTELLURIUM REAGENTS IN CARBOHYDRATE CHEMISTRY 788
VI. BIOLOGICAL ACTIVITY OF SELENONUCLEOSIDE ANALOGS AND SELENO SUGARS 789
VII. CONCLUSIONS 790
VIII. REFERENCES 790

I. INTRODUCTION

The chemistry of organic selenium compounds has developed rapidly over the past 15 years and has acquired new dimensions[1–8] especially in the area of the synthesis of natural products[9], selenocarbohydrates[10–16], selenonucleosides[17] and other selenium-containing molecules of biological interest[18,19]. Various aspects of biologically important organic selenium compounds have been extensively studied and reviewed[20]. The biomedical and biochemical aspects are treated in three books[21–23].

In carbohydrate chemistry, the first article[24] reporting on the introduction of selenium into a sugar moiety was published in 1917, since when a number of reports dealing with the synthesis of many natural products[9] through organoselenium intermediates have been published. This chapter collates information on the synthesis and reactivity of organoselenium and organotellurium intermediates and illustrates some of the chemical properties that have contributed to the synthesis of selenocarbohydrates and selenonucleosides.

Knowledge of selenium intermediates, seleno sugars and selenonucleosides is growing steadily and they now constitute key intermediates in routes to various heterocyclic derivatives of sugars and to free sugars.

II. SELENOGLYCOSIDES AND OTHER SELENIUM DERIVATIVES OF CARBOHYDRATES: SYNTHESIS AND REACTIONS

Bonner and Robinson[13] synthesized several selenoglycosides using the approach previously adopted by Schneider and Wrede[24]. The reaction sequence starts with α-D-glycopyranosyl bromide (equation 1) and proceeds with Walden inversion at $C_{(1)}$. The same methodology was used by Wagner and Lehmann[14] for the synthesis of the various substituted selenoglycosides (equation 2).

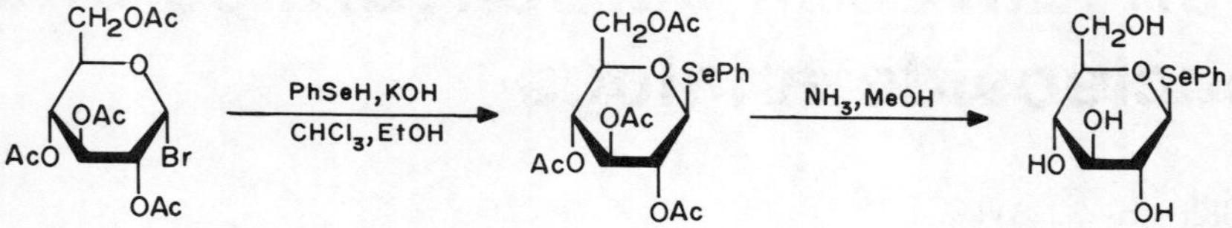

(1)

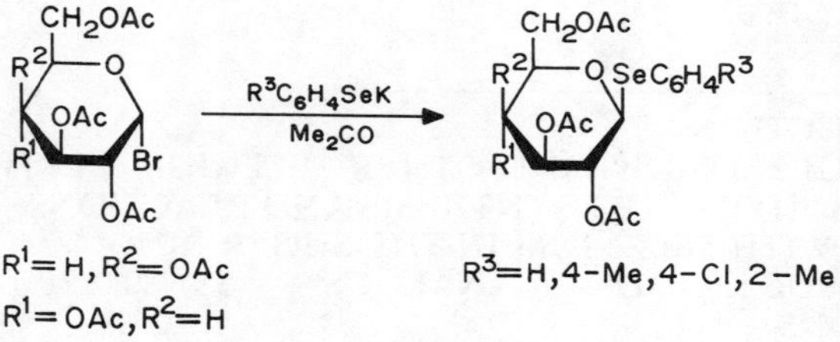

(2)

A similar approach was employed for the preparation of the sodium salt of 1-seleno-D-glucose[16]. This sequence proceeded by treatment of α-D-glucosyl bromide with PhCOSeK, followed by methanolysis (equation 3). This sodium salt of 1-seleno-D-glucose was an excellent precursor in the synthesis of selenoglycosides and sugar diselenides, as reported by Wagner and Nühn[25]. However, they proceeded by treatment of α-D-glucosyl bromide with selenourea, followed by basic hydrolysis of the intermediate 2, 3, 4, 6-tetra-*O*-acetyl-1-selenopseudourea-β-D-glucopyranose hydrobromide, with formation of the sodium salt of 1-seleno-D-glucose (equation 4).

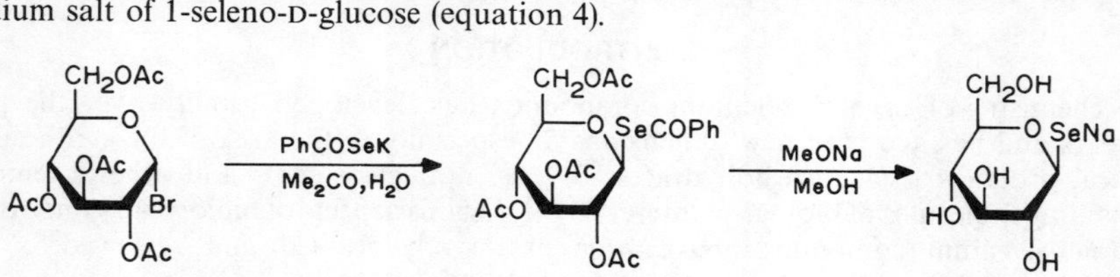

(3)

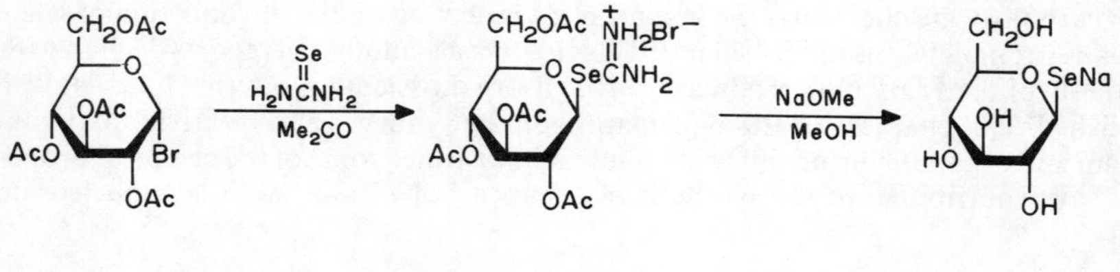

(4)

Other interesting examples of the application of similar procedures have appeared in the literature[26].

All of the above-mentioned methods of synthesis of selenoglycosides lead preferentially to the formation of β-anomers, which are probably thermodynamically more stable than the α-anomers. However, Frenzel and coworkers[26] found that the formation of both anomers of selenoglycosides occurs during opening of the epoxide ring of Brigl's anhydride (3, 4, 6-tri-*O*-acetyl-1, 2-anhydro-α-D-glucopyranose) with benzeneselenol (equation 5). The anomeric products were separated by thin-layer chromatography and the purity, configuration and conformation of the glycosides were confirmed by ^{1}H NMR spectroscopy. It is noteworthy that the protons *cis* to the aglycone are shifted to lower field in comparison with *O*- and *S*-series of glycosides.

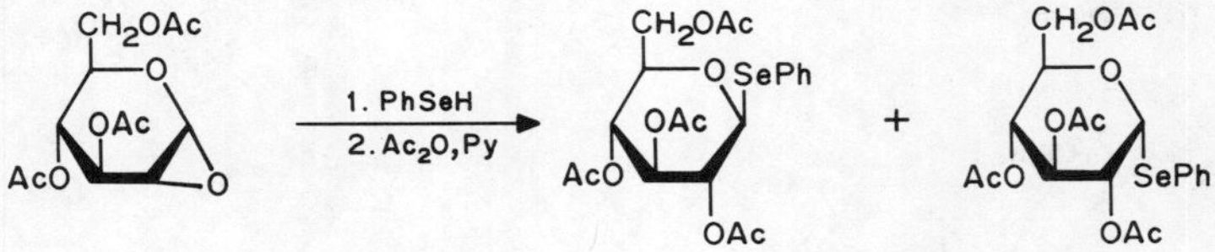

(5)

Another interesting example of the application of Brigl's anhydride in selenoorganic chemistry is its reaction with the triethylammonium salt of 2-oxo-2-seleno-5, 5-dimethyl-1, 3, 2-dioxaphosphorinane, with the formation of 3, 4, 6-tri-*O*-acetyl-D-glucal and a diselenide[27] (equation 6).

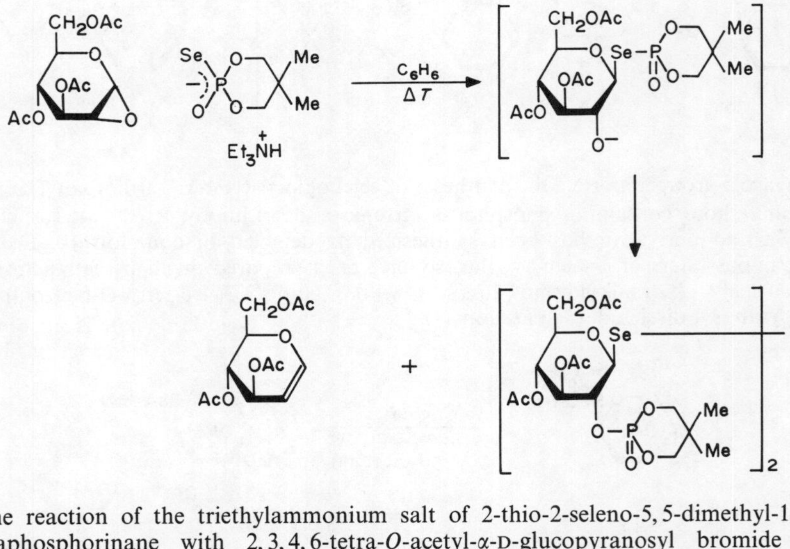

(6)

The reaction of the triethylammonium salt of 2-thio-2-seleno-5, 5-dimethyl-1, 3, 2-dioxaphosphorinane with 2, 3, 4, 6-tetra-*O*-acetyl-α-D-glucopyranosyl bromide and 2, 3, 4, 6-tetra-*O*-acetyl-α-D-galactopyranosyl bromide gave both seleno- and thioglycosides (equation 7). Analogous reactions of α-D-glucosyl bromides with triethylammonium salts of seleno acids of phosphorus have been reported[28]. Interesting transformations involving selenoglycosides have also been reported[29]. The first is anomerization of a selenoglucoside (equation 8) on heating in boiling xylene for 12 h, and the second is a selenono–selenolo rearrangement of a selenous ester to yield a selenoglycoside (equation 9).

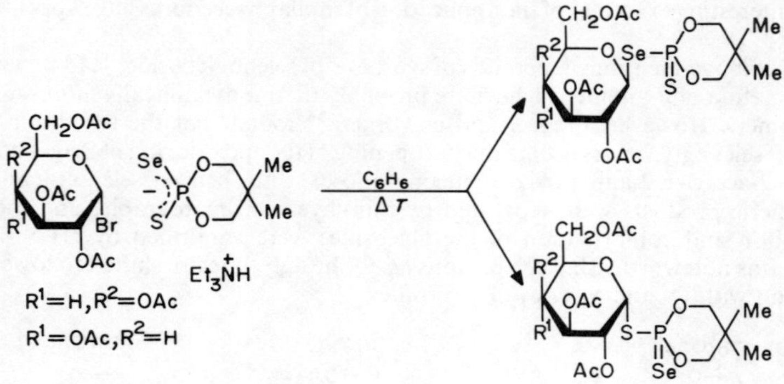

(7)

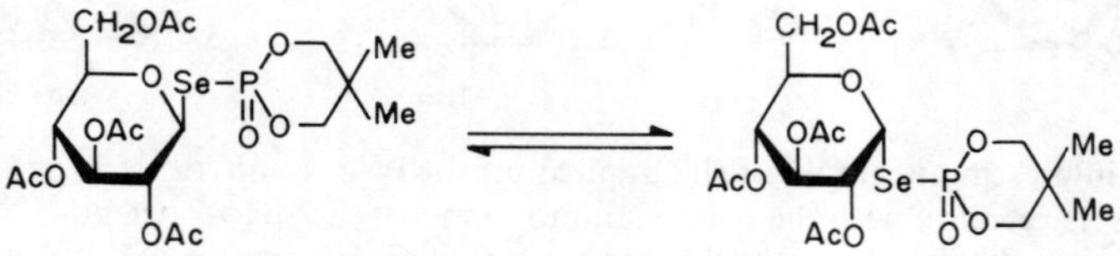

(8)

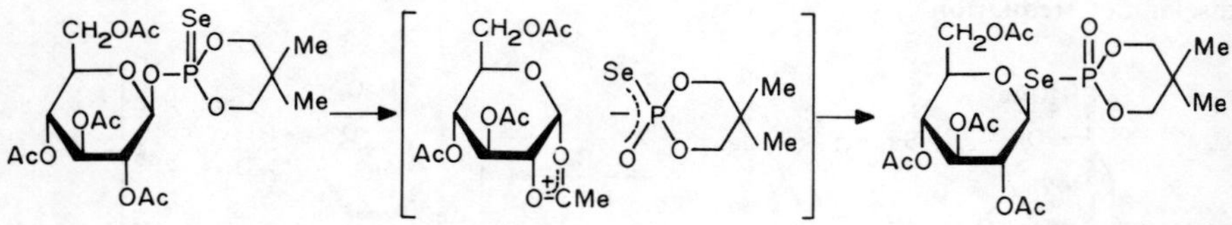

(9)

Zingaro's group reported the synthesis of selenoglucose esters of diorganyl groups, including those containing phosphorus, arsenic and antimony[12,30,31]. A 1-selenodimethylarsino derivative has been synthesized as depicted in equation 10. Dialkylphosphinous esters of *β*-seleno-D-glucose have been prepared through the reactions of symmetrical tetraalkyldiphosphines with 6-bis(2, 3, 4, 6-tetra-*O*-benzoyl)-*β*-D-glucopyranosyl diselenide[32] (equation 11).

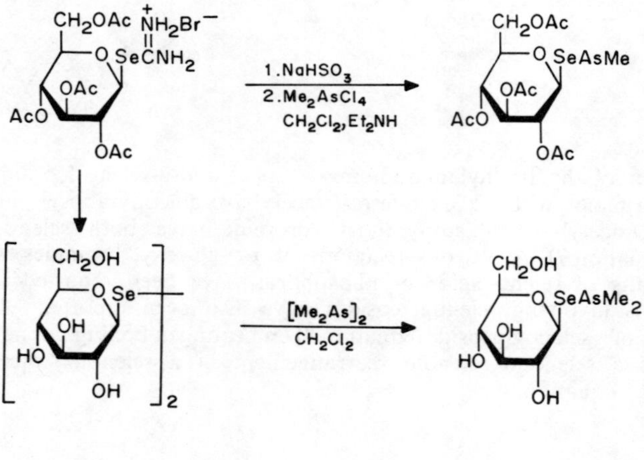

(10)

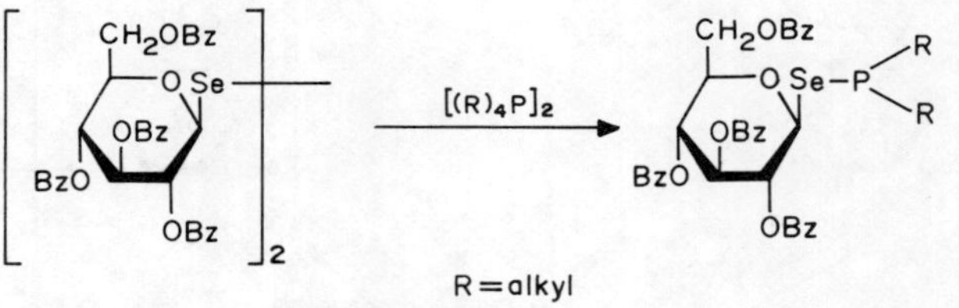

(11)

The antimony analogues are prepared by the use of an analogous series of reactions. However, the use of dimethylstibine halides in the two-phase water–dichloromethane reaction is obviated by the extreme hydrolytic instability of the antimony—halogen bond. Therefore, these derivatives were prepared exclusively by the addition of tetramethylstibine to the diselenide[33]. Zingaro[12] and Daniel and Zingaro[34] also reported the synthesis of dimethylarsinous acid esters of 1-seleno-D-galactose, as shown in equation 12.

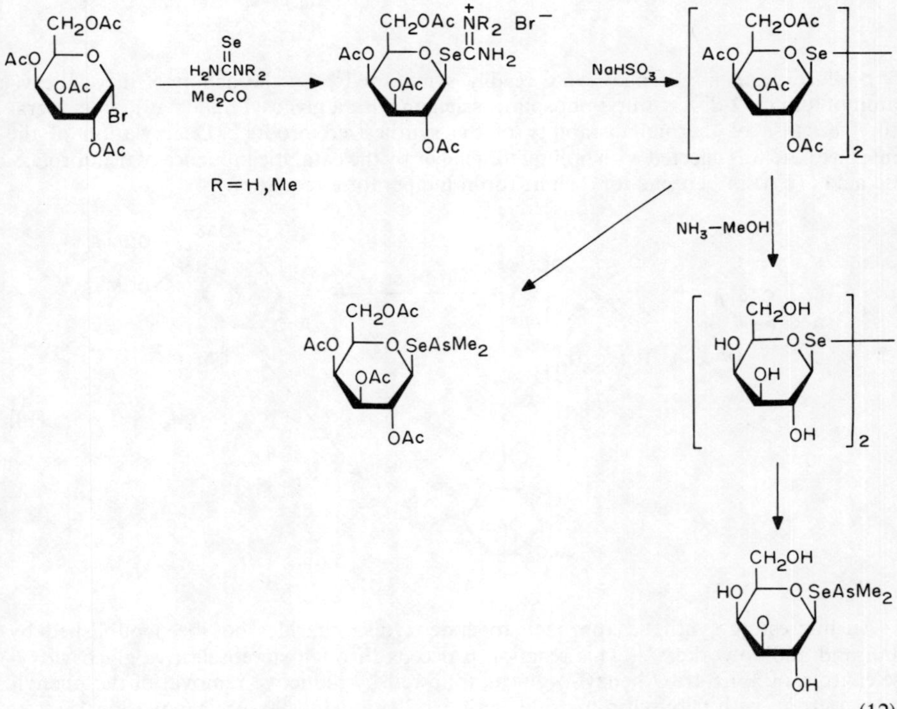

(12)

The synthesis of 1-selenodimethylarsin-2-acetamido-2-deoxy-α-D-glucopyranose has been reported by Chen and coworkers[35]. Reaction with dimethylselenourea is followed by reduction of the intermediate selenoureide with sodium hydrogen sulfite to the diselenide and subsequent condensation with dimethylarsine (equation 13). Interestingly, the coupling constant and chemical shift of the anomeric proton ($J_{1,2} = 4$ Hz, $\delta = 5.38$ ppm) of the intermediate diselenide is indicative of α-D-anomeric configuration, in contrast to a previous report of a similar reaction of 1-halogeno sugars with selenourea[16] and with the selenophenolate ion[25].

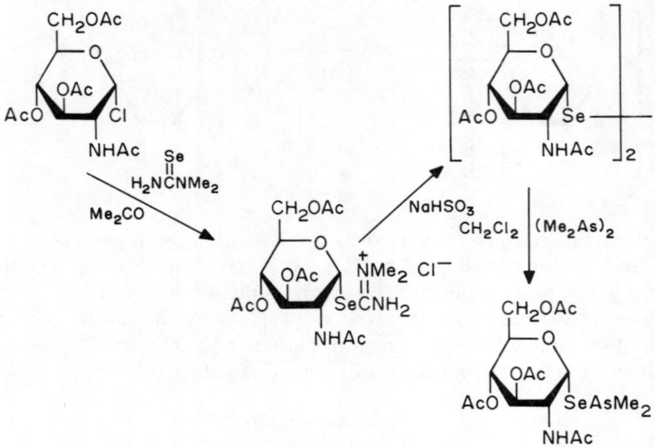

(13)

D-Glucosyl selenophosphates are readily synthesized[36] by the action of dicyclohexylammonium *O*, *O*-di-*tert*-butyl phosphoroselenate with a glycosyl halide at low temperature, because of thermal instability of the synthesized product. Dealkylation of the intermediate was effected with boiling toluene or by the catalytic influence of trifluoroacetic acid (TFA) in benzene for 24 h at room temperature (equation 14).

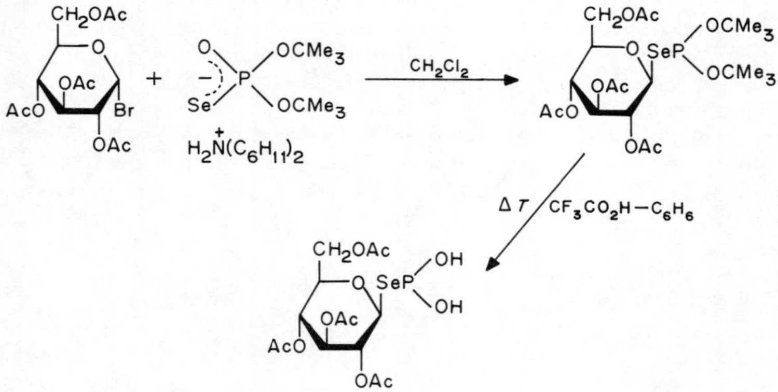

(14)

An interesting synthetic approach to 2′-deoxydisaccharides has been published by Jaurand and coworkers[37]. This reaction proceeds through stereoselective glycosyloxyselenation of 3, 4, 6-tri-*O*-benzyl-D-glucal, followed by reductive removal of the phenylseleno group with tributyltin hydride, with excellent yields (90–95%) (equation 15).

Introduction of the arylseleno residue into the sugar moiety has been accomplished with phenylselenyl halides[38,39], using methods based on the Current and Sharpless[39] procedure of addition of ArSeX to olefins. Examples of ArSeX addition to olefins involving cyclization through reactions with internal nucleophiles have also been reported[40–45]. Another method of introduction of the arylseleno residue is the opening of the oxirane ring of sugars with PhSeH, which has been reported by many workers[46–52]. These methods of converting epoxides into phenyl selenides which were intermediates to unsaturated derivatives of sugars are based on the Sharpless and Lauer method of synthesis of allylic alcohols from epoxides[53–57].

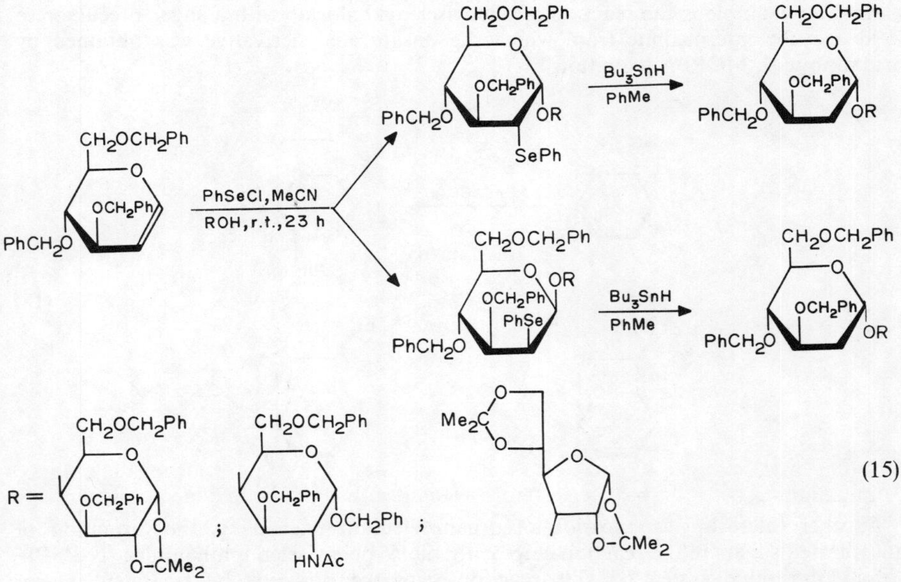

(15)

Paulsen and coworkers[58] successfully employed 2-methyl-2-selenoxo-1,3-benzothiazole (MSBT) the conversion of various anhydro sugars into unsaturated derivatives. The conversion proceeds in 5–10 h at room temperature with high yields and, usually, a high degree of purity of the product (equation 16). Under these conditions the 1,6-anhydro ring of 1,6:3,4- or 1,6:2,3-dianhydro sugars is unreactive and cleavage of only the 2,3- or 3,4-anhydro ring is observed[58]. This selectivity of the reaction constitutes a new entry to the rare class of unsaturated 1,6-anhydro sugars (equation 17).

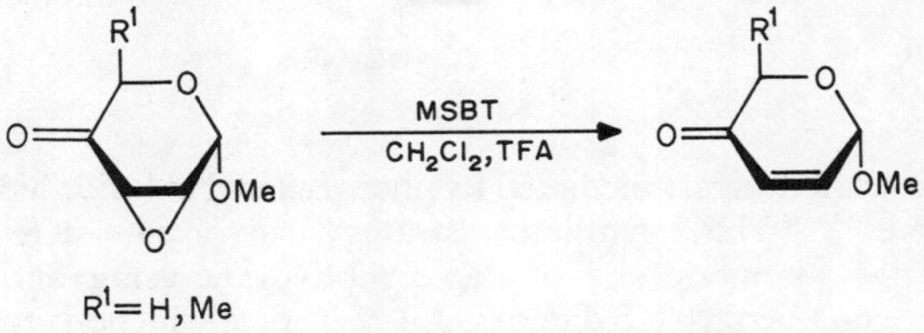

(16)

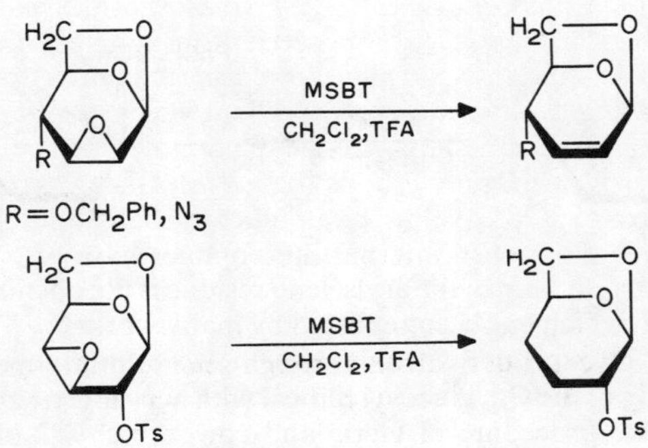

(17)

Another example is the reaction of phenylselenyl chloride with a sugar precursor to yield a cyclic intermediate from which the unsaturated derivative was obtained by oxidation with MCPBA (equation 18)[59].

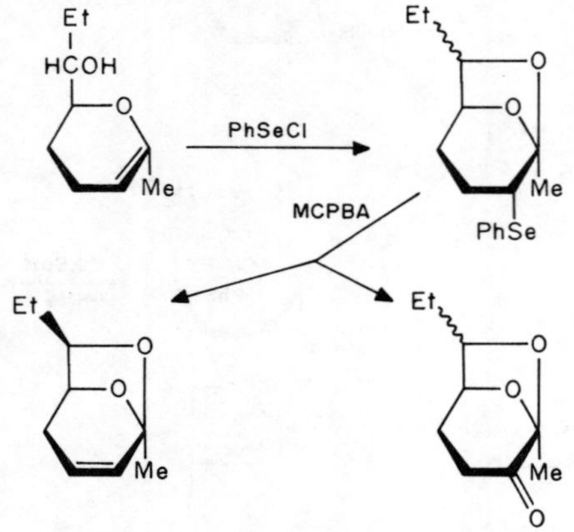

(18)

Another interesting application of organoselenium reagents is the formation of unsaturated sugars in the reaction employing the *N*-phenylselenophthalimide (N-PSP)–tributylphosphine system[60]. This reaction sequence proceeds by treatment of an appropriately protected methyl α-D-glucoside with N-PSP–Bu_3P in oxolane, followed by oxyselenation with hydrogen peroxide (equation 19).

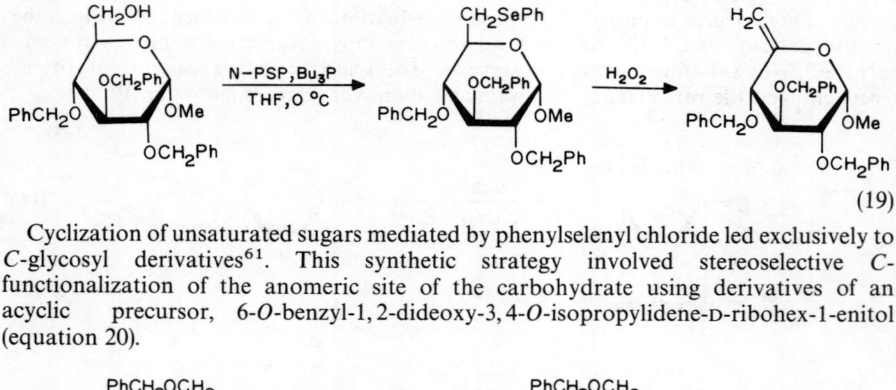

(19)

Cyclization of unsaturated sugars mediated by phenylselenyl chloride led exclusively to *C*-glycosyl derivatives[61]. This synthetic strategy involved stereoselective *C*-functionalization of the anomeric site of the carbohydrate using derivatives of an acyclic precursor, 6-*O*-benzyl-1, 2-dideoxy-3, 4-*O*-isopropylidene-D-ribohex-1-enitol (equation 20).

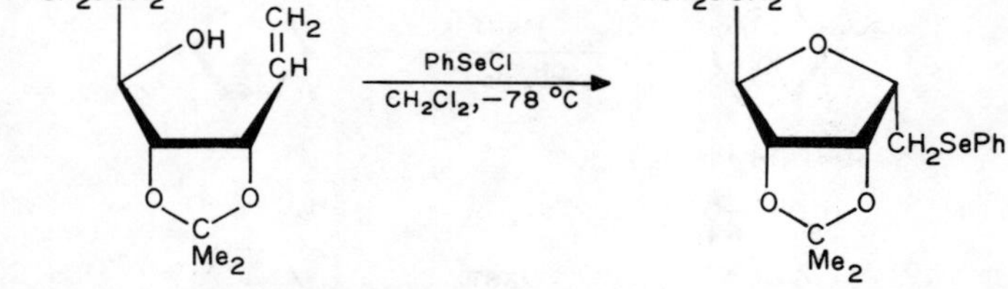

(20)

Another route to *C*-glycosyl derivatives through selenium intermediates involves the generation and trapping of the $C_{(1)}$ glucosyl radical with acrylic ester[62] (equation 21). This concept is based on the procedure of Geise and coworkers[63–66] of coupling glucosyl bromide with acrylonitrile or with methyl acrylate with axial arrangement of the substituents in $C_{(1)}$.

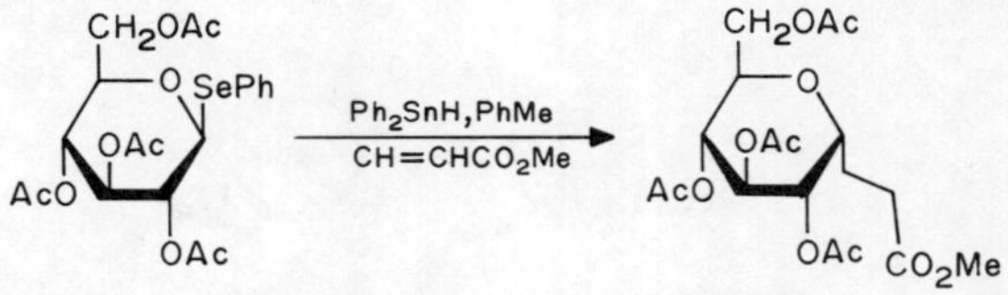

(21)

An interesting synthetic approach to 3-*C*-substituted glycals has been reported by Rollin and coworkers[67]. The reaction sequence starts from 3,4,6-tri-*O*-acetyl-D-glucal and proceeds by formation of a vinyl glycoside, which by thermal rearrangement forms a 3-*C*-substituted glycal (equation 22).

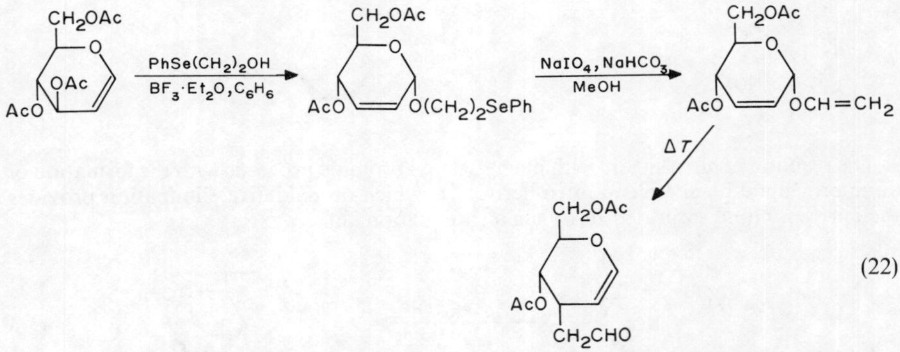

(22)

The previously mentioned phenylselenyl chloride-mediated cyclization of unsaturated sugars (equation 20) developed by Nicolaou and coworkers[41,68] and commonly called phenylselenolactonization has also been employed in the synthesis of butenolides[69] and is illustrated in equation 23.

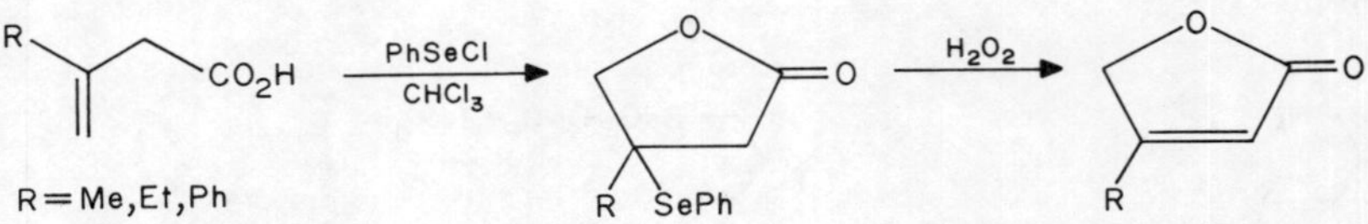

(23)

The phenylselenyl group in a sugar moiety may undergo the interesting photorearrangement[70] depicted in equation 24.

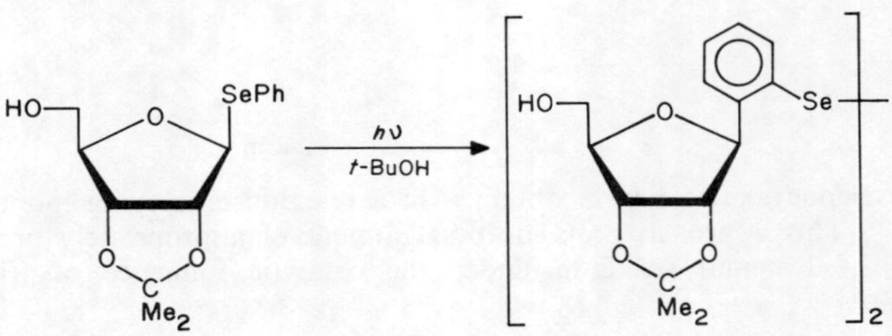

(24)

The very reactive *o*-nitrophenylselenyl group undergoes eliminative oxidation by treatment with hydrogen peroxide much more easily than other arylselenyl groups and is therefore employed in the synthesis of unsaturated sugars[71] (equation 25).

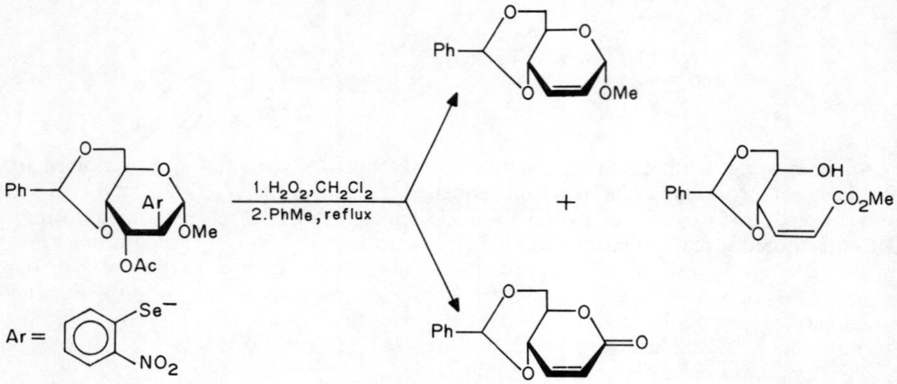

(25)

The reaction of nitro sugars with phenylselenyl bromide proceeds with the formation of an intermediate phenylseleno nitro derivative, which on oxidative elimination provides the corresponding saturated nitro sugar[72] (equation 26).

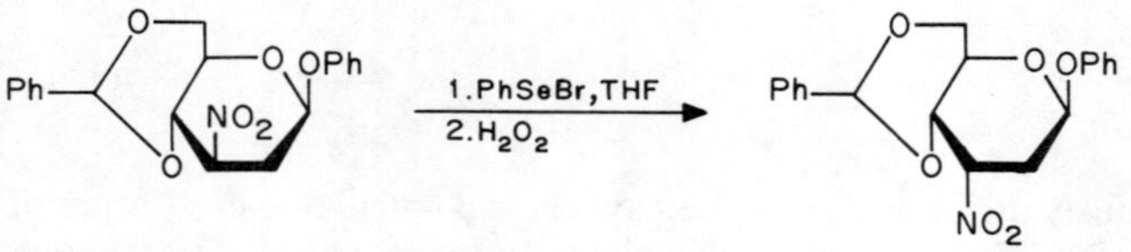

(26)

The *o*-nitrophenyl selenocyanate–tributylphosphine system developed by Grieco and coworkers[73,74] has been employed in the synthesis of the antibiotic calcimycin[74] (equation 27).

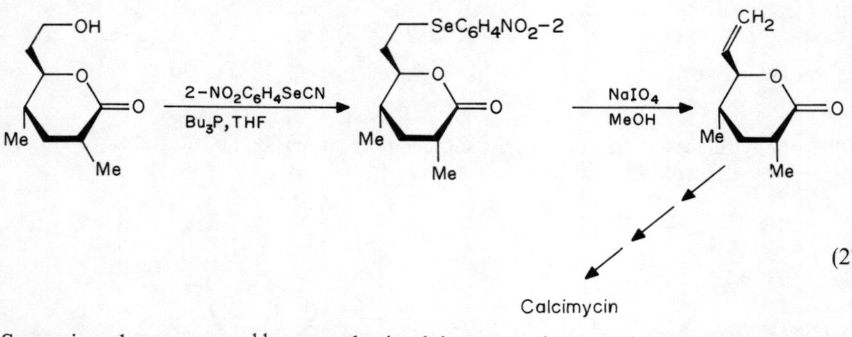

(27)

Sugar isoselenocyanates[11] were obtained in a reaction employing monosaccharide isocyanides[75]. This approach proceeds by treatment of appropriately protected isocyanides with elemental selenium under the catalytic influence of triethylamine (equation 28).

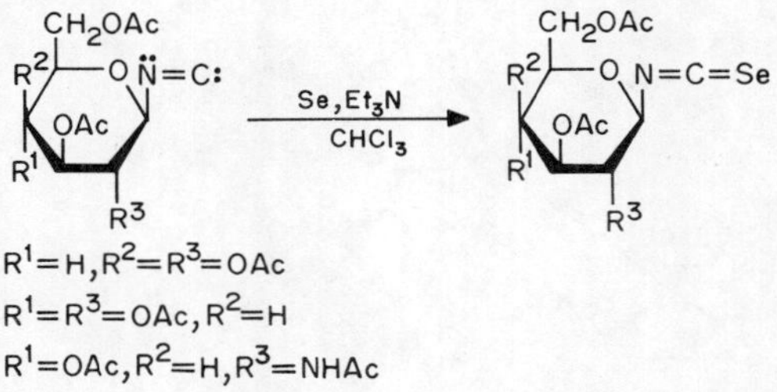

(28)

III. SUGARS WITH SELENIUM IN THE SUGAR RING

The introduction of sulfur[76], nitrogen[77,78] and phosphorus atoms[79–81] into furanose and pyranose rings of simple monosaccharides is successful and well known, but the introduction of selenium has usually failed or is very difficult[82–86]. Blumberg and coworkers[82] reported the first successful introduction of a selenium atom as the sugar ring heteroatom into L-arabinose, D-ribose, and D-xylose. The synthesis of benzyl 1, 5-diseleno-L-arabinopyranoside is illustrated in equation 29.

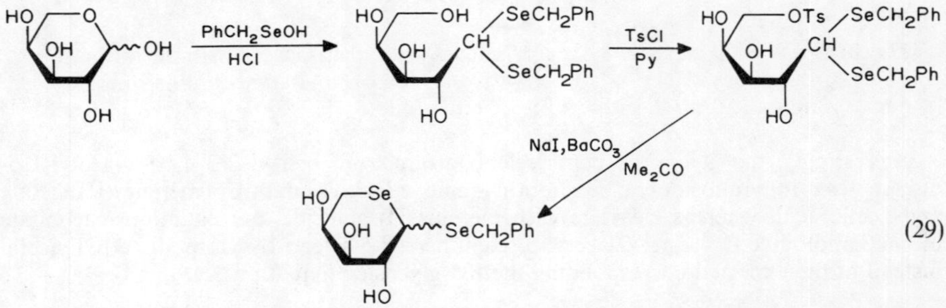

(29)

A similar approach has been employed for the synthesis of the benzyl 2, 3, 4-tri-*O*-acetyl-1, 5-diseleno-D-xylopyranoside and benzyl 2, 3, 4-tri-*O*-benzyl-1, 5-diseleno-D-xylopyranoside, using as the starting material 1, 2-*O*-isopropylidene-5-*O*-*p*-tolylsulfonyl-α-D-xylofuranose (equation 30). Analogously, the benzyl 2, 3, 4-tri-*O*-acetyl-1, 5-diseleno-D-ribopyranoside has been prepared from methyl 2, 3-*O*-isopropylidene-5-*O*-*p*-toluenesulfonyl-β-D-ribofuranoside (equation 31).

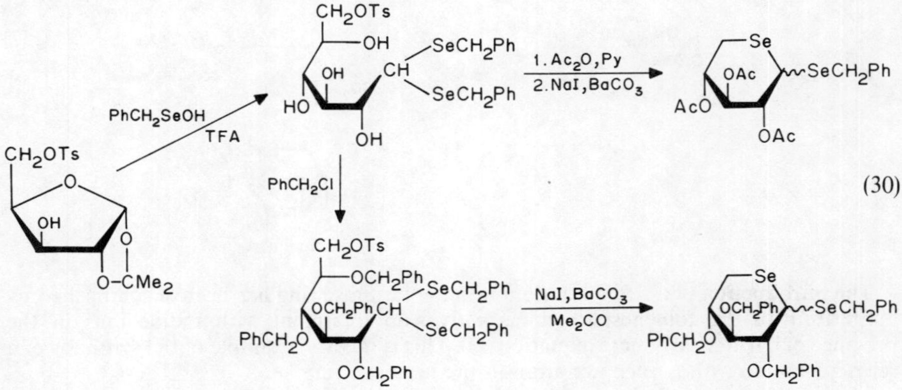

(30)

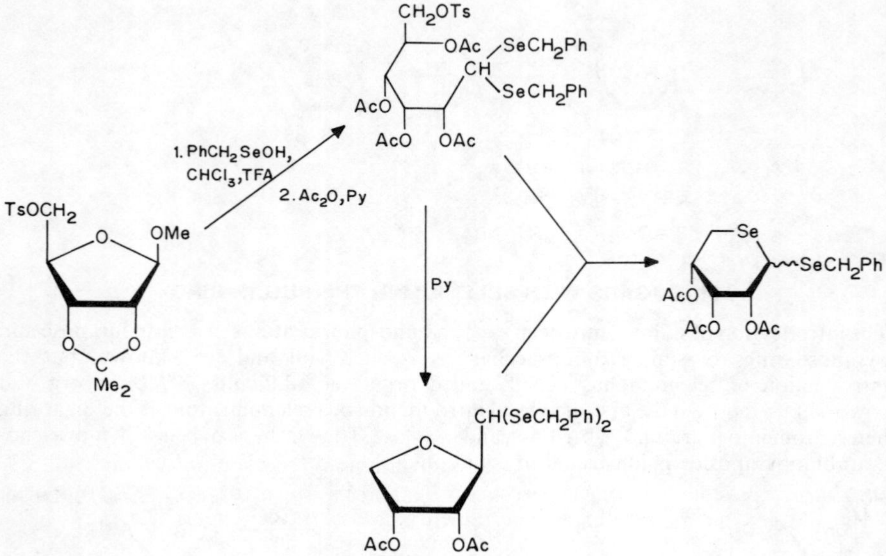

(31)

Interestingly, the glycosidic benzylseleno group from benzyl 2, 3, 4-tri-*O*-acetyl-1, 5-diseleno-D-xylopyranoside can be smoothly removed with a mixture of mercury(II) acetate and acetic acid, whereas cleavage with mercury(II) chloride and cadmium carbonate in methanol give 2, 3, 4-tri-*O*-acetyl-5-selenobenzyl-5-seleno-D-xylose dimethyl acetal instead of the expected corresponding methyl glycoside (equation 32).

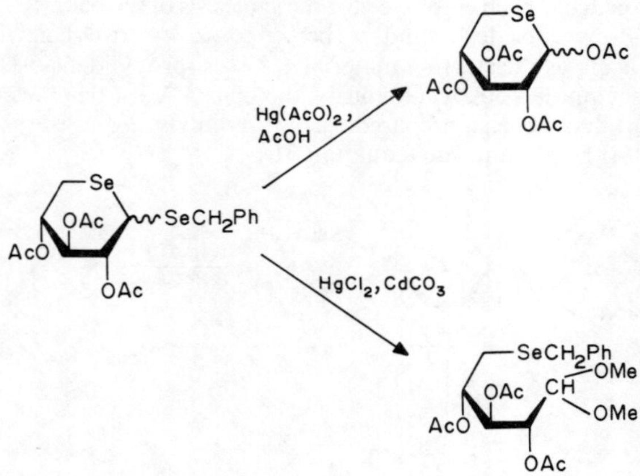

(32)

The introduction of a selenium atom into the furanose ring has been accomplished by application of the toluenesulfonyl derivative, on treatment with iodide ions in the presence of barium carbonate (equation 33). This is the first example of the synthesis of a selenoglycoside with a selenium atom in the furanose ring.

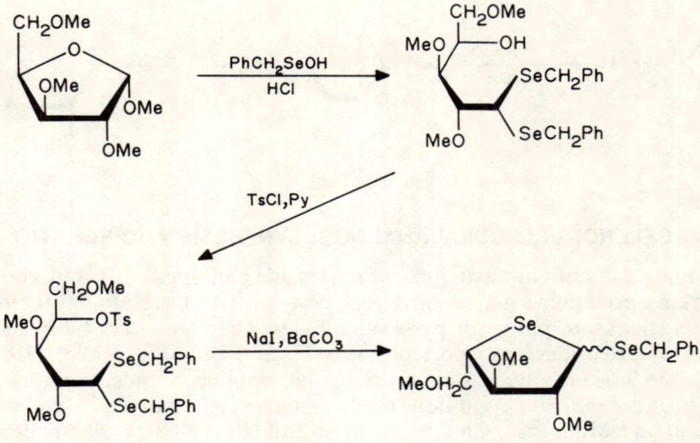

(33)

Van Es and Whistler[83] also reported a synthesis of the sodium salt of 5-deoxy-5-seleno-D-xylose, a convenient precursor to the 3,4-dihydroxy-2,3,4,5-tetrahydro-D-*threo*-2-selenophene-2-carbaldehyde dimethyl acetal (equation 34). The same acetal can also be obtained by removing the benzylseleno group from benzyl 2,3,4-tri-*O*-acetyl-1,5-diseleno-D-xylopyranose with mercury(II) acetate in acetic acid, followed by methanolysis with sodium methoxide, as reported by Blumberg and coworkers[82] (equation 35).

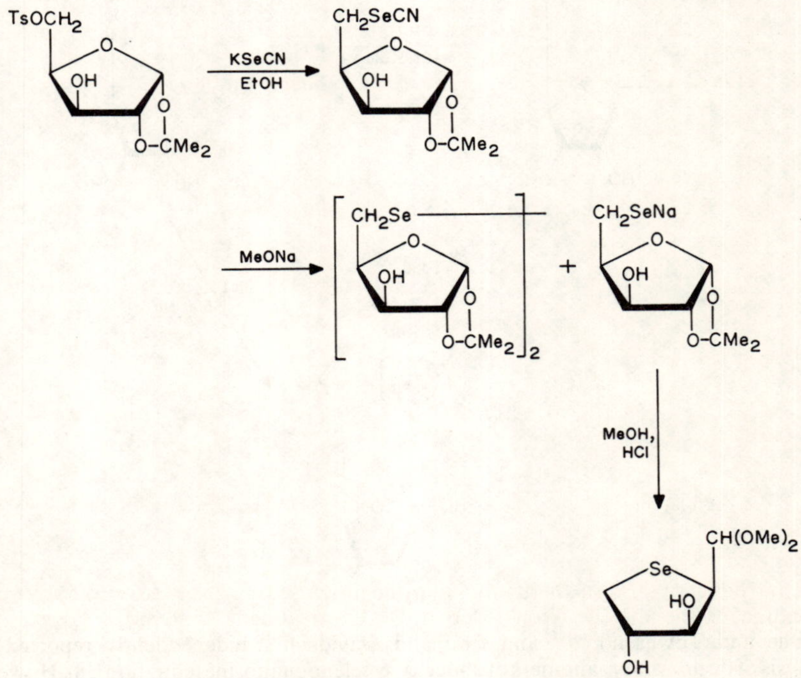

(34)

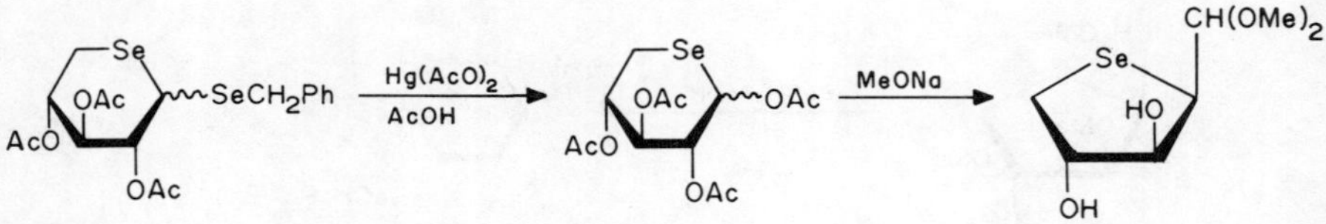

(35)

IV. SELENONUCLEOSIDE ANALOGS: SYNTHESIS AND REACTIONS

The antitumor and antiviral activity[87] of nucleoside analogs[88,89] is well known. These properties have prompted a major search for a new synthetic methodology for introducing sulfur or selenium into purine or pyrimidine bases. The first successful introduction of selenium into nucleosides was reported in 1960 by Jaffe and Mautner[90,91]. However, the 6-selenopurine-9-β-D-ribonucleoside was highly unstable[90] under neutral and basic conditions. Under the same conditions 6-selenoguanosine[92] has been found to be stable.

Townsend and Milne[92,93], Chu[94] and Shine and Chu[95] independently described the synthesis of 6-selenoguanosine (equation 36), which with methyl iodide under basic conditions afforded 6-methylselenoguanosine (equation 37). Similarly, 6-methylselenoinosine has been synthesized.

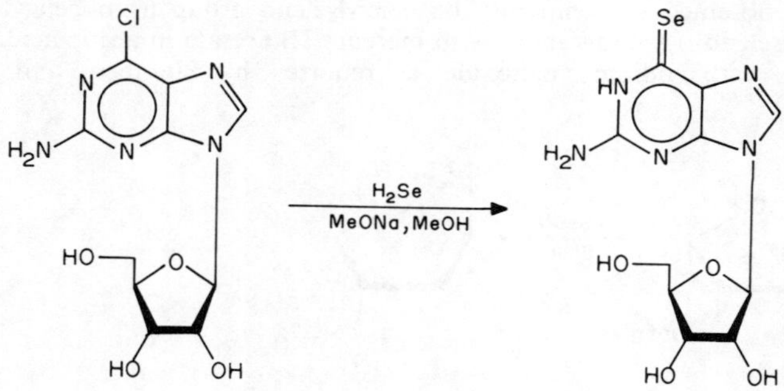

(36)

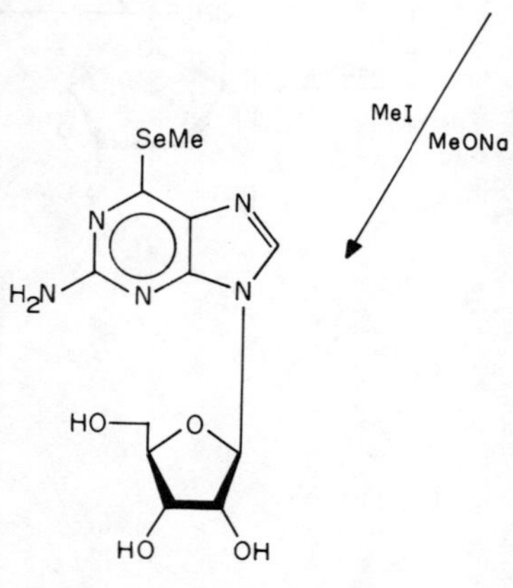

(37)

Milne and Townsend[96,97] and Chu and Davidson[98] independently reported the synthesis of both α-and β-anomers of 2′-deoxy-6-selenoguanosine (equation 38). However, these derivatives are unstable in aqueous solution and decompose at room temperature after 24 h.

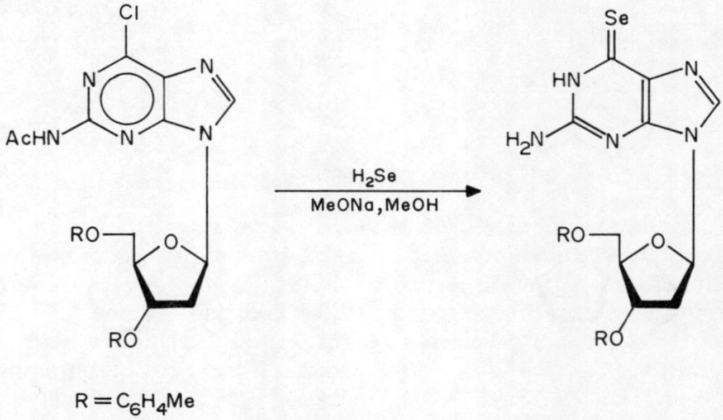

(38)

The synthesis of 7-selenoxo-3-(*β*-D-ribofuranosyl)pyrazolo[4, 3-*d*]pyrimidine (selenoformicin B) has also been reported by Milne and Townsend[99]. The synthesis involves a simple nucleophilic displacement of the chlorine with selenourea in ethanol at reflux temperature (equation 39).

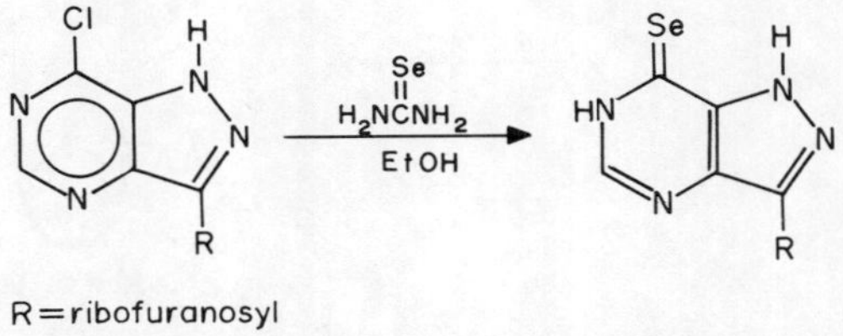

(39)

Wise and Townsend[100,101] also reported the synthesis of 2- and 4-selenouridine. The synthetic sequence begins with silylation of 2-selenouracil, followed by condensation with 2, 3, 5-tri-*O*-benzoyl-1-*O*-acetyl-α-D-ribofuranose. Deprotection of the condensation product with sodium methoxide in methanol produced 2-selenouridine in 30% yield (equation 40). The synthetic approach to 4-selenouridine starts from 4-chloro-1-(2, 3, 5-tri-*O*-benzoyl-*β*-D-ribofuranosyl)pyrimidin-2-one, which on treatment with selenourea in methanol, and subsequent debenzoylation, affords 4-selenouridine. However, the seleno group in both 2-seleno- and 4-selenouridine is very labile[100,101].

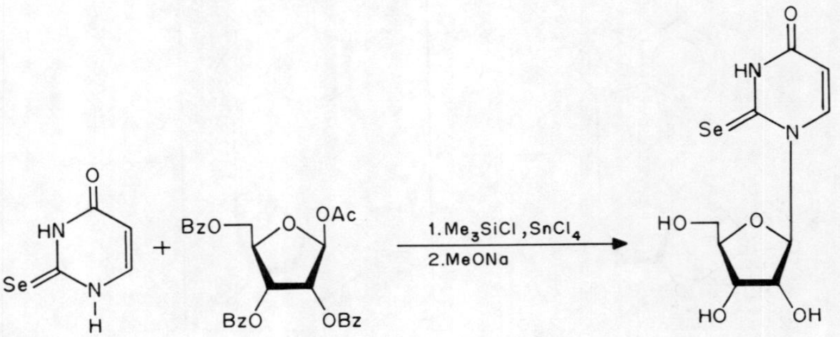

(40)

Chu and coworkers[102] reported the synthesis of some 8-substituted cyclic selenonucleotides by treatment of 8-bromoadenosine 3′,5′-cyclic monophosphate with sodium hydrogen selenide in refluxing methanol and subsequent alkylation (equation 41).

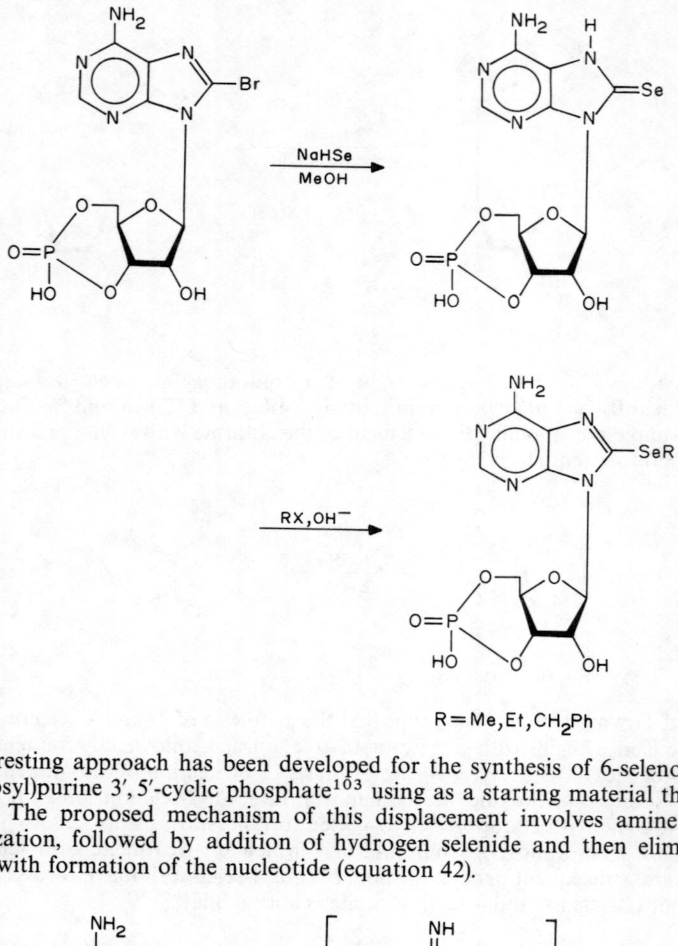

(41)

An interesting approach has been developed for the synthesis of 6-selenoxo-9-(β-D-ribofuranosyl)purine 3′,5′-cyclic phosphate[103] using as a starting material the 6-amino precursor. The proposed mechanism of this displacement involves amine to imine, tautomerization, followed by addition of hydrogen selenide and then elimination of ammonia with formation of the nucleotide (equation 42).

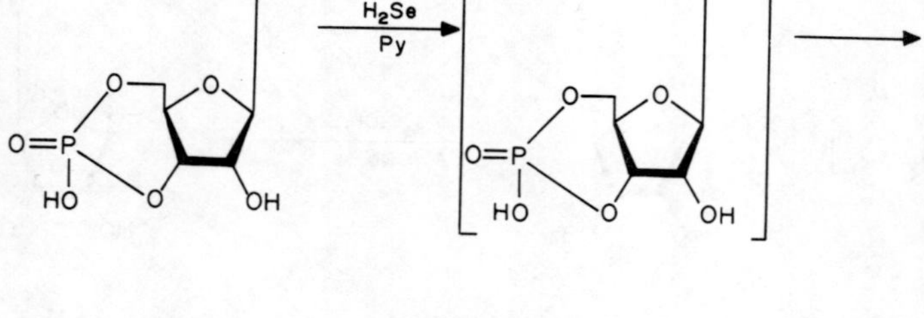

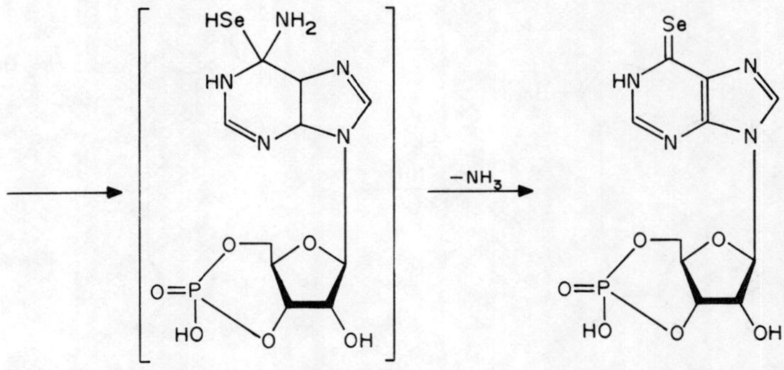

(42)

Alkylation, chlorination and seleno–thiono conversion have also been reported (equation 43).

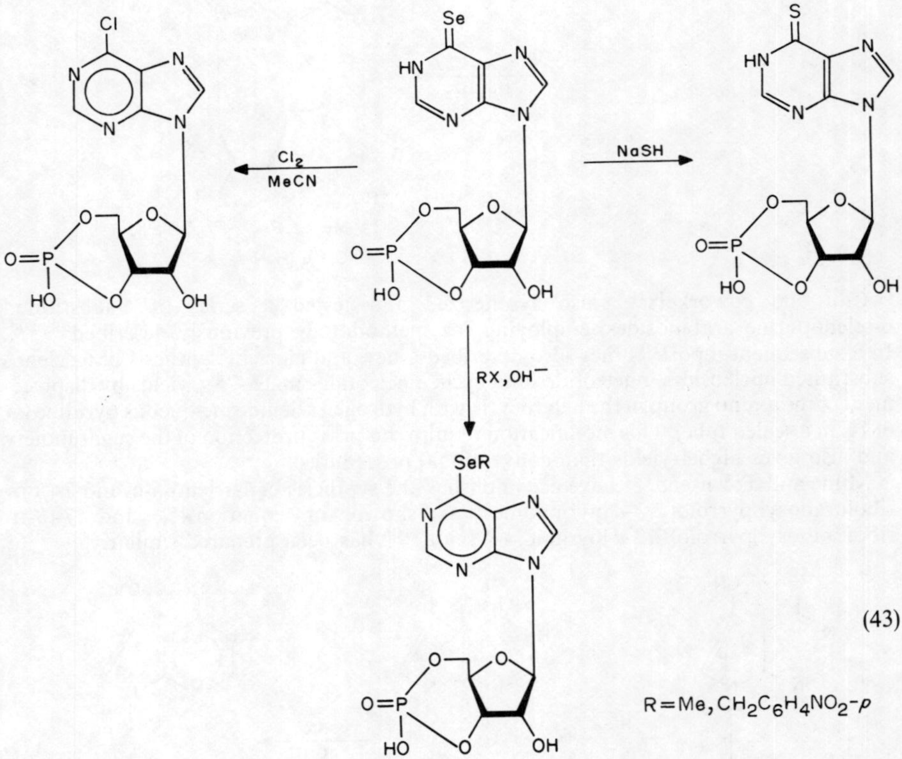

(43)

R = Me, $CH_2C_6H_4NO_2$-*p*

8-Seleno derivatives of guanosine 3′,5′-cyclic phosphate (cGMP) have also been prepared[104] using as a starting material cGMP, which by direct bromination gave 8-bromo-cGMP. Treatment with selenourea gave the isoselenouronium hydrobromide intermediate, which on alkylation yielded the required nucleotides (equation 44). 8-Substituted selenoguanosine 5′-monophosphates and selenoguanosines were prepared similarly[104].

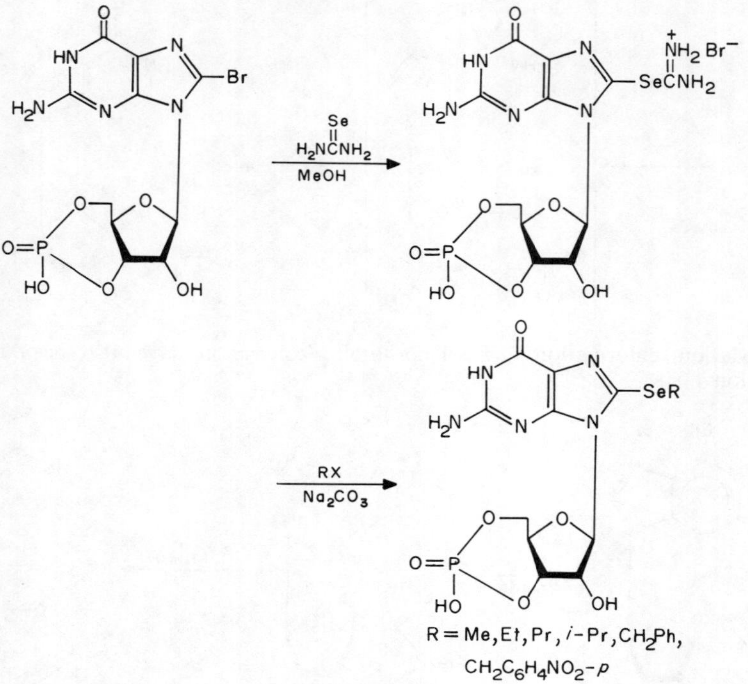

(44)

Chu and coworkers[105] also synthesized and tested a series of 6-substituted 6-selenopurine arabinosides, employing the methodology previously described[99–104]. In a subsequent report[106] they also described a new and efficient synthesis of 6-seleno-substituted nucleosides, nucleotides and cyclic nucleotides in 22–75% yields by displacement of the amino group in the heterocycle with hydrogen selenide in aqueous pyridine (at 68 °C in a sealed tube). This modification requires no prior protection of the sugar moiety and also gives higher yields than conventional procedures.

Milne and Townsend[107] have accomplished the synthesis of 4-seleno-5-cyano-7-(β-D-ribofuranosyl)pyrrolo[2, 3-*d*]pyrimidines as shown in equation 45, and 7-(β-D-ribofuranosyl)pyrrolo[2, 3-*d*]pyrimidin-4-selone[108] has been prepared similarly.

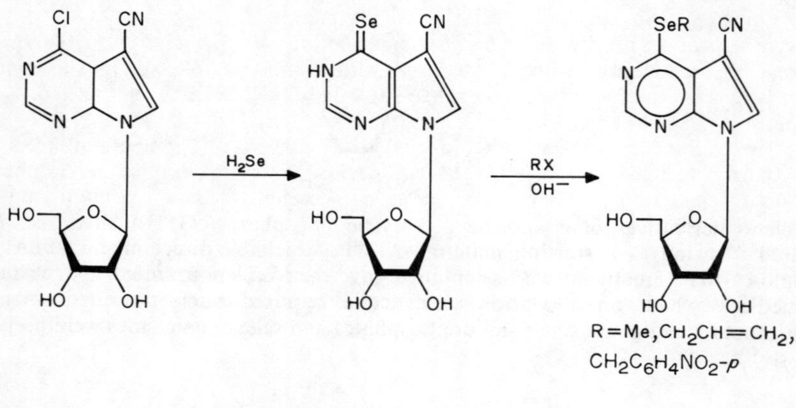

(45)

Interesting results on the reactivity of the 4-substituted seleno group and the 5-cyano group in both nucleosides towards nucleophilic reagents such as hydrazine and hydroxylamine have been reported[107] (equation 46). The configuration (*Z* or *E*) of the products has not been established.

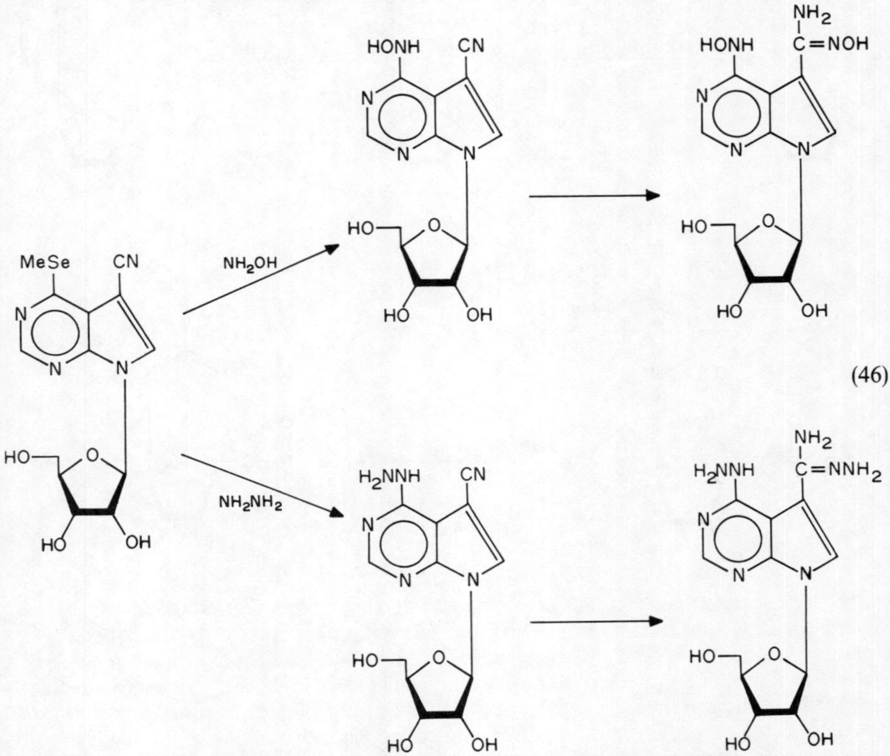

(46)

An interesting report on the effects of exocyclic atoms (O, S and Se) in nucleosides on the chemical shifts of the anomeric proton in the sugar moieties and protons at $C_{(5)}$ and $C_{(6)}$ positions of the hetero ring has been published by Wise and Townsend[109]. They also reported[110] the first synthesis of a selenium-bridged cyclonucleoside, Se,2′-cyclo-2-selenocytidine, using 2-selenocytosine as a starting material.

The synthesis of the first cyclic selenopurine uncleotide was also reported by Wise and coworkers[111]. Treatment of 8-bromoadenosine with selenourea, followed by subsequent reaction with 2-acetoxyisobutyryl chloride(AIBC), furnished Se^8, 2′-cyclo-8-seleno-β-D-arabinofuranosyladenine in 54% yield (equation 47).

Tributylphosphine combined with diphenyl diselenide in acetonitrile has been reported[112] to yield new types of selenonucleotides, e.g. 3′-*O*-acetylthymidine-*Se*-phenyl 5′-phosphoroselenoate is obtained from the monopyridinium salt of 3′-acetylthymidine 5′-phosphate (equation 48).

Treatment of thymidine 5′-phosphate with bis(trimethylsilyl)acetamide (BSA) in dimethylformamide solution with selenium powder afforded a mixture of two products (equation 49), thymidine 5′-phosphoroselenoate and its autooxidized product having an Se—Se bond. Alkylation of this mixture afforded *Se*-ethylthymidine 5′-phosphoroselenoate.

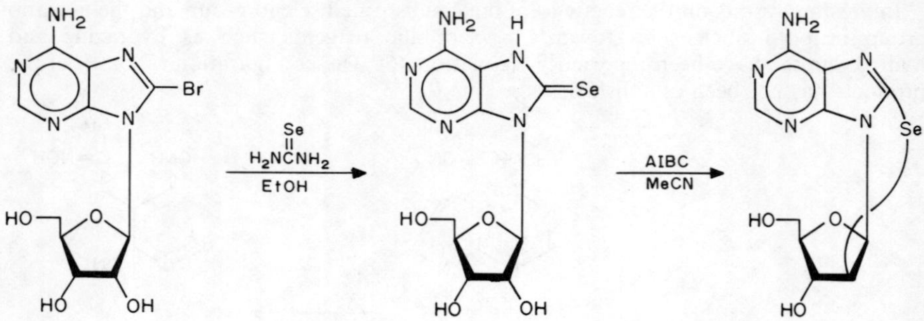
NH2
Br
Se
H2NCNH2
EtOH
H
Se
AIBC
MeCN
HO
OH

(47)

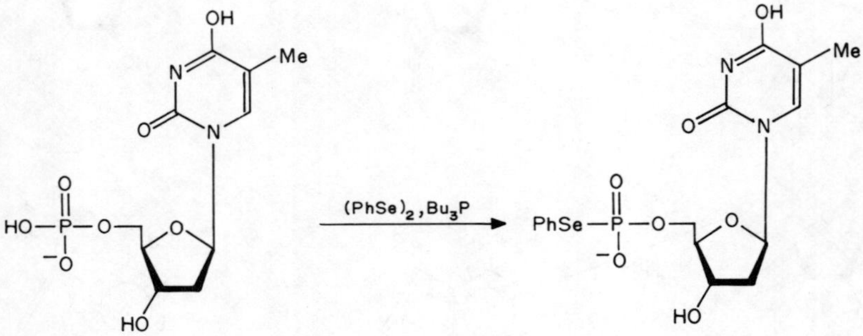
OH
Me
(PhSe)2,Bu3P
PhSe

(48)

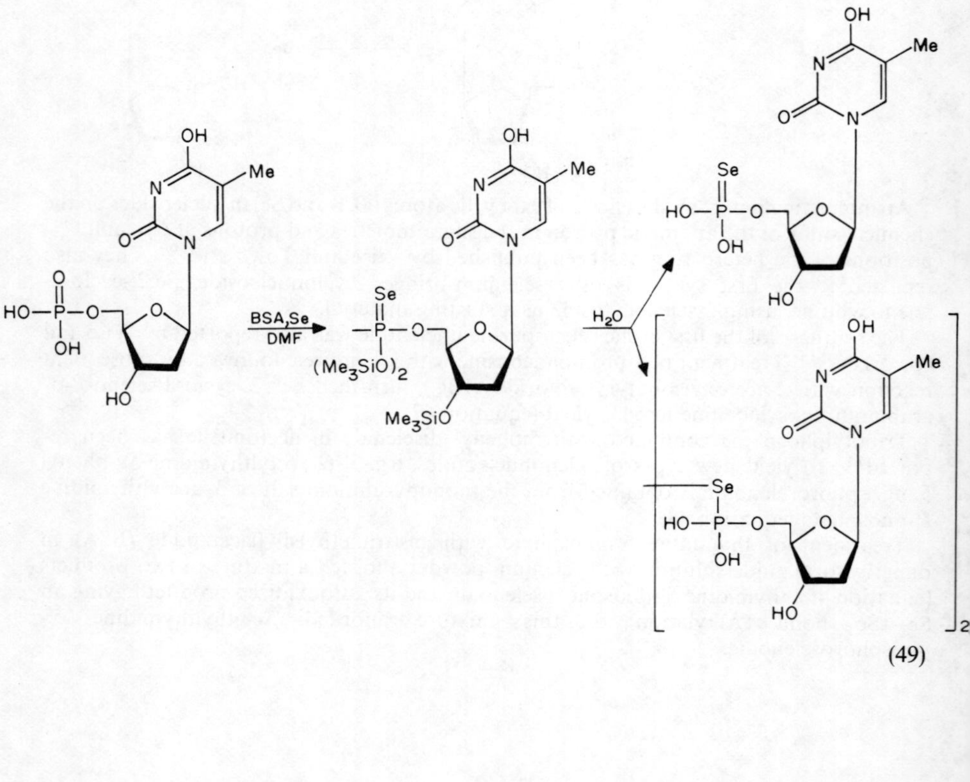
OH
Me
BSA,Se
DMF
(Me3SiO)2
Me3SiO
H2O
Se
HO
OH

(49)

Takaku and coworkers[113] reported the synthesis of 5′-*Se*-(2-nitrophenyl)-5′-selenoxyadenosine as an excellent precursor to 9-(5′-deoxy-β-D-*erythro*-pent-4-enofuranosyl)adenine (equation 50).

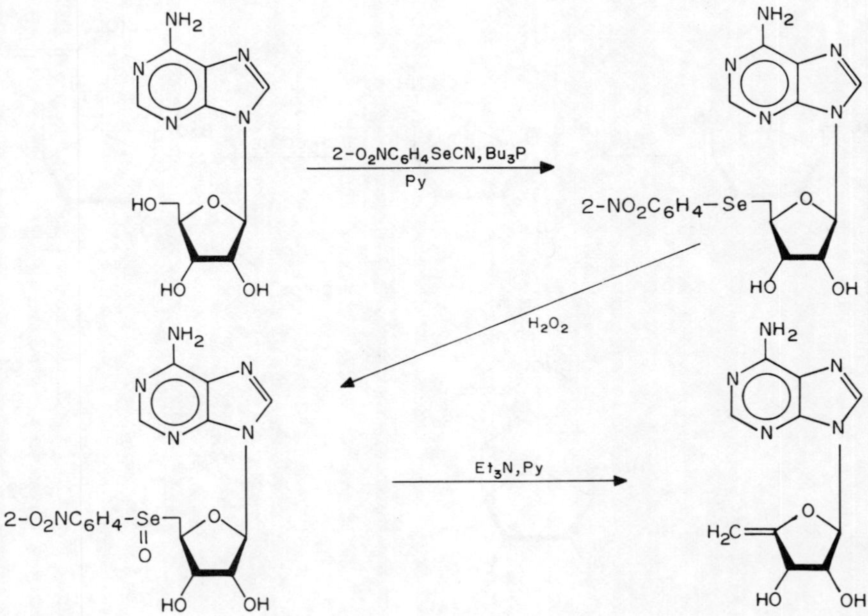

(50)

Treatment of adenosine with 2-nitrophenyl selenocyanate and tributylphosphine gave an intermediate selenide, which on oxidation, followed by treatment with triethylamine in pyridine, yielded an unsaturated nucleoside. Triethylamine promotes *syn* elimination of the selenoxide group and hydrogen at the 4′-position, similarly to the results previously reported by Zylber and coworkers[114]. However, a recent report by Boullais and coworkers[115] showed that triethylamine in dimethyl sulfoxide promotes elimination effectively and with higher yields. Alternatively, the elimination of the selenoxide group with 1, 8-diazabicyclo[5.4.0]undec-7-ene (DBU) in toluene solution is also very effective (equation 51).

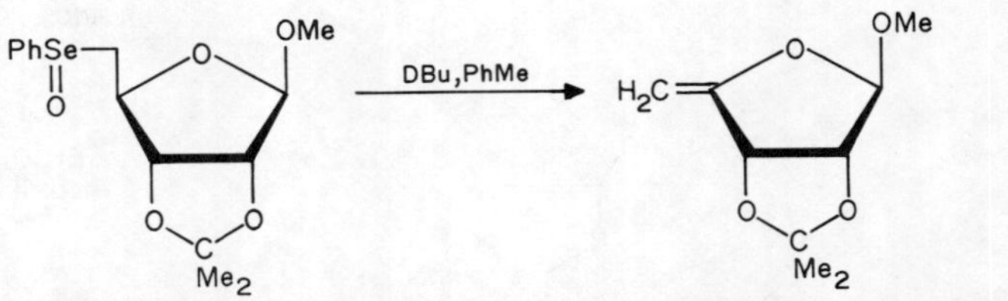

(51)

The first introduction of selenazole as a base in nucleosides was reported in 1983 by Srivastava and Robins[116]. The synthetic approach started with 2, 3, 5-tri-*O*-benzoyl-β-D-ribofuranosyl-1-carbonitrile, which on treatment with hydrogen selenide in the presence of 4-(dimethylamino)pyridine as catalyst afforded the intermediate selenoamide. Subsequent condensation with ethyl bromopyruvate provided an anomeric mixture of ethyl 2-(2, 3, 5-tri-*O*-benzoyl-D-ribofuranosyl)selenazole-4-carboxylate, which after separation and deprotection afforded selenazofurin in moderate yield (equation 52). Two novel convenient syntheses[117,118] of selenazofurin have been published recently (equation 53).

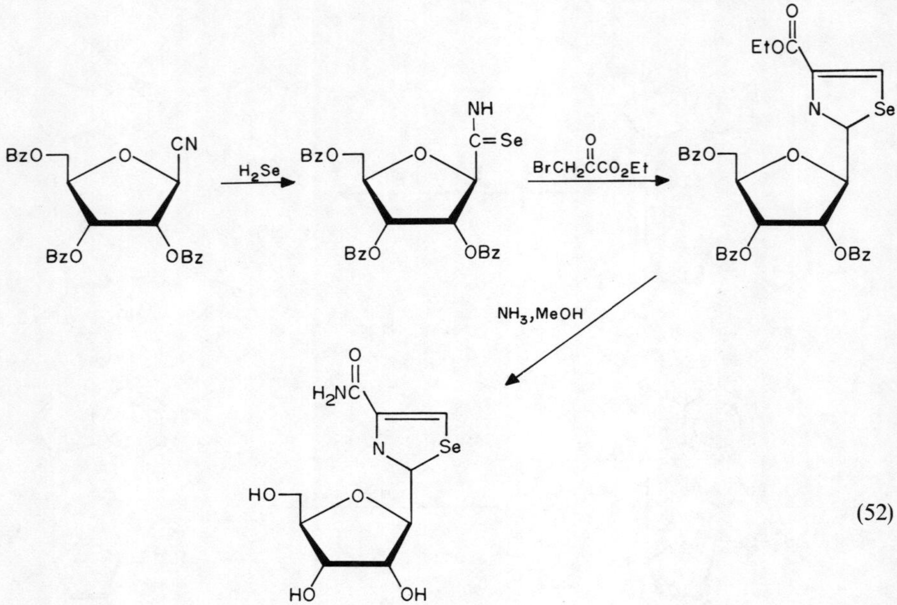

(52)

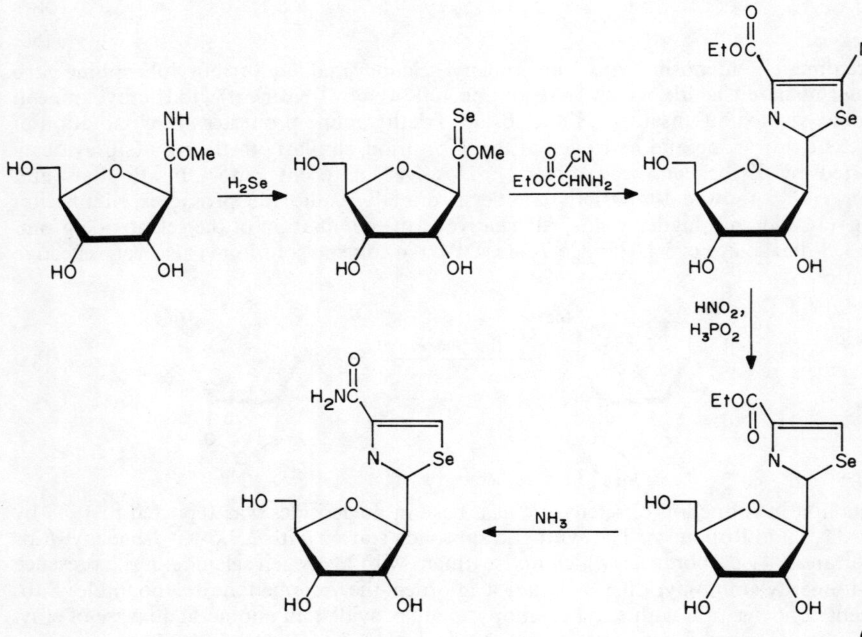

(53)

Pyrazolo[4, 3-*d*]pyrimidine nucleosides continue to be of considerable interest from both chemical and biological points of view. Two representatives of this class have recently been synthesized, 1-methyl-*β*-D-ribofuranosylpyrazolo[4, 3-*d*]pyrimidine-7(6*H*)-selone[119] and 1-*β*-D-ribofuranosylpyrazolo[3, 4-*d*]pyrimidine-4(5*H*)-selone[120]. The synthetic approaches are illustrated in equation 54.

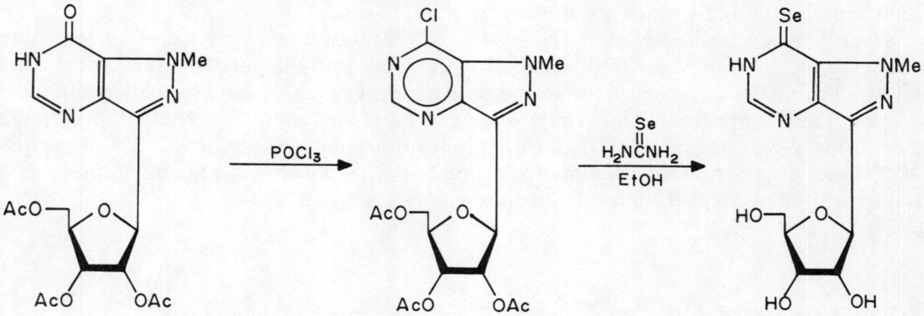

(54)

Schinazi and coworkers[121] reported the synthesis of several 5-phenylselenyl derivatives of pyrimidine nucleosides, by electrophilic addition of phenylselenyl chloride to the nucleoside under basic conditions (equation 55).

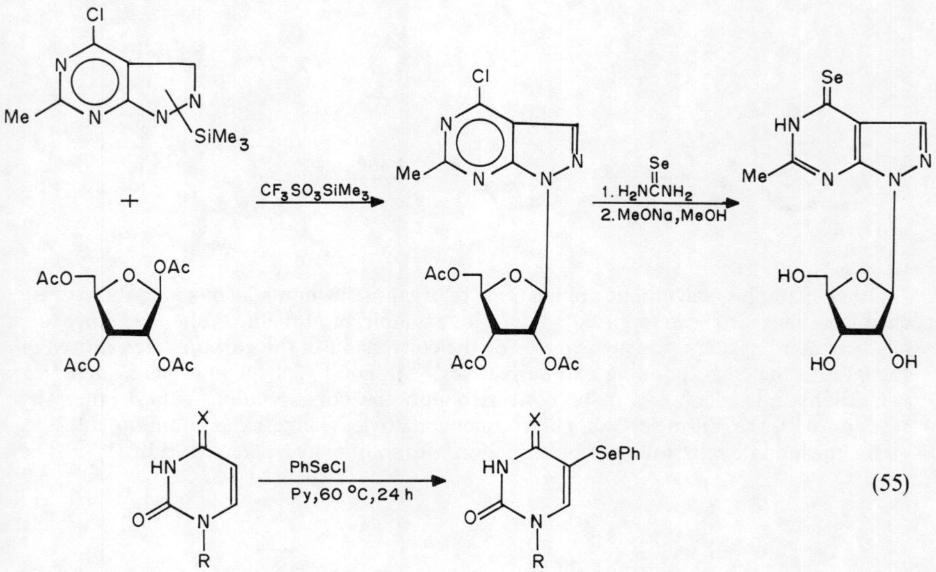

(55)

X=O, R=ribosyl
X=NH, R=ribosyl
X=O, R=2-deoxyribosyl
X=O, R=arabinosyl
X=NH, R=arabinosyl

V. TELLUROCARBOHYDRATES AND ORGANOTELLURIUM REAGENTS IN CARBOHYDRATE CHEMISTRY

Chemical transformations involving tellurium were, until recently, very rare. The explosive development of organoselenium chemistry has called the attention to the potential of tellurium reagents and a number of interesting transformations based on tellurium-containing species are now known.

Several review articles[122,123] and books[124–126] about many concepts of tellurium chemistry[127,128] have been published. However, no study devoted to the synthesis of tellurocarbohydrates has been published until a recent paper by Czyżewska-Chlebny and Michalska[129] appeared. This route involves condensation of a triethylammonium 2-tellurido-2-oxo-5, 5-dimethyl-1, 3, 2-dioxaphosphorinane salt with 2, 3, 4, 6-tetra-*O*-acetyl-α-D-glucopyranosyl bromide (equation 56). This is the first known synthesis of a tellurocarbohydrate, through a tellurophosphorus reagent.

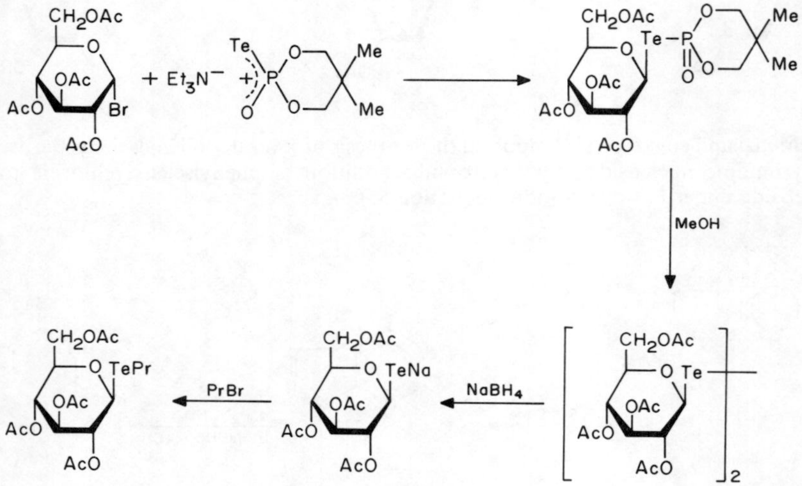

(56)

Other examples of synthetic applications of organotellurium reagents in carbohydrate chemistry have also been reported[130–132]. For example, bis(4-methoxyphenyl) telluroxide has been used as a convenient reagent for the conversion of thiocarbonyl derivatives of sugars into the corresponding oxo derivatives[130] in good yields (equations 57 and 58).

Carbohydrate alcohols can be converted into the corresponding benzyl ethers by reaction with the Vilsmeier salt, chloro(phenylmethylene)dimethylammonium chloride, giving the imidate salts, followed by reduction with sodium hydrogen telluride[131,132]. The

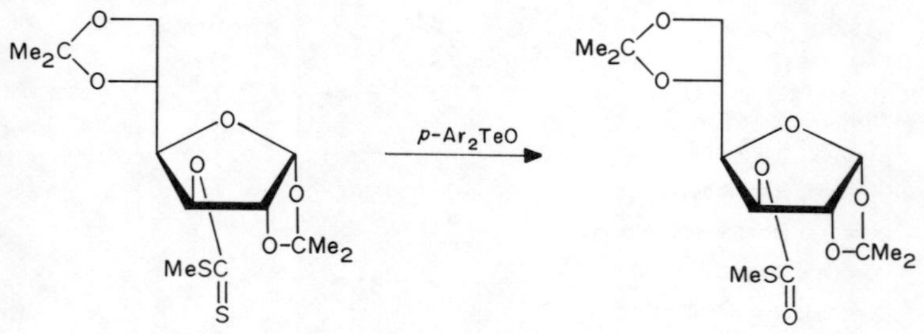

(57)

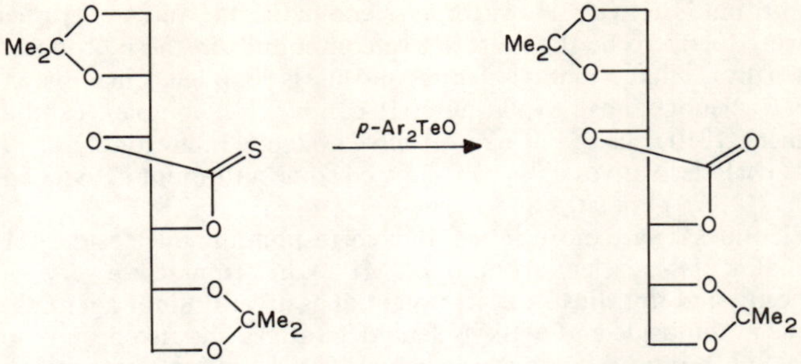

(58)

reaction proceeds under mild conditions, probably through the tellurobenzoate and hydrogen atom transfer (equation 59).

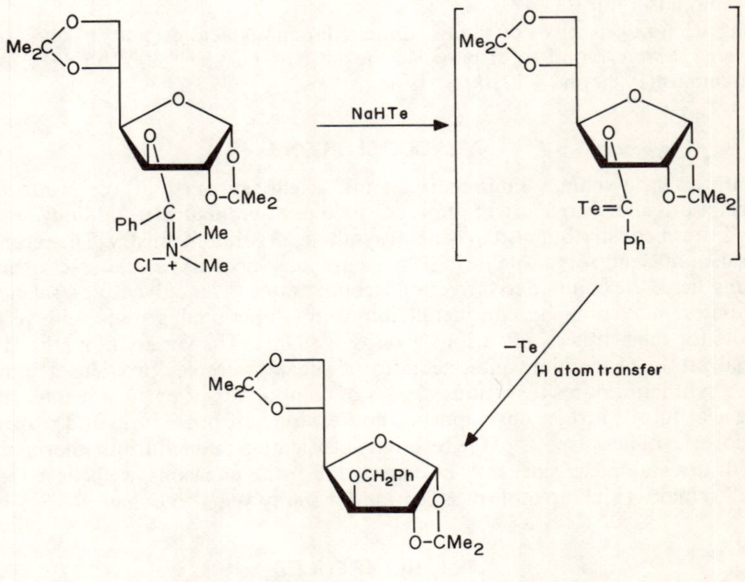

(59)

VI. BIOLOGICAL ACTIVITY OF SELENONUCLEOSIDE ANALOGS AND SELENO SUGARS

The biological activity of seleno-organic derivatives[90] in bioorganic chemistry has been known for a long time, especially the antitumor activity of selenonucleoside analogs[90,91]. For example, 6-selenoguanine and 6-thioguanine, which are good antitumor agents, prompted the synthesis of 6-selenoguanosine[92] as a potentially much more effective antitumor agent. Further, a comparative investigation showed that selenoguanine and selenoguanosine inhibit the growth of Sarcoma 180 ascites cells more effectively than the corresponding thionucleosides. Interestingly, β-2′-deoxy-6-selenoguanosine was found to have an activity approximately equal to that of the 6-thio congener, whereas the α-anomer was much less active than the corresponding α-2′-deoxy-6-thioguanosine. This important observation was confirmed by Milne and Townsend[96,97], who also observed that alkylation of both anomers at the exocyclic selenium atom appeared to cause a marked decrease in antitumor activity. However, 6-selenoguanosine and 6-alkylselenoguanosine derivatives were found to be the most active compounds in these groups.

It is noteworthy[133] that 6-methylselenoguanosine is completely inactive as an inhibitor, whereas the 6-selenoguanosine–platinum(II) complex[134] complex exhibits antitumor activity against L1210 cells in mice and *in vitro* systems. Interestingly, the 8-substituted seleno cyclic GMP derivatives[102,103,105] showed some antitumor activity against murine leukemic cells (LS178Y) *in vitro* and *in vivo*.

Cyclic nucleotides[102] are more active than corresponding nucleosides. The previously mentioned analog of 8-isoselenouronium-cGMP hydrobromide is a very active inhibitor (98% inhibition)[5], and this illustrates its potential as an antitumor agent. Cytotoxicity of 6-selenopurine arabinoside and 6-alkylseleno derivatives has also been reported, but the cytotoxic effect of the above analogs was below 50% inhibition. Interestingly, the *in vitro* antitumor activity of 2-β-D-ribofuranosylselenazole 4-carboxamide[116,117] (selenazofurin) and its 5′-phosphate in comparison with the corresponding thiazole congeners were found to be more active towards P388 and L1210 cells in culture and also effective against Lewis lung carcinoma in mice.

Among seleno derivatives of sugars, dimethylarsinous acid esters[12] of glucose[31] and galactose[30–34] were found to display carcinostatic activity[12,30,31,34,135] *in vivo* against mice leukemias (P388 and L1210 test systems).

VII. CONCLUSIONS

The synthesis of selenium intermediates and seleno sugars and the study of their transformations afford an heuristic approach for a new synthetic methodology not only in carbohydrate chemistry but also in general synthetic organic chemistry. The recent use of organoselenium and organotellurium reagents has become a key factor in these fascinating fields. Selenium derivatives of carbohydrates (selenoglycosides, selenophenyl intermediates) may now be considered important functional groups and very good precursors for the synthesis of various groups of sugars. The variety of methods for the functionalization of selenium intermediates of sugar molecules provides a number of attractive synthetic routes to various classes of compounds of particular interest.

In the near future, further developments concerning new procedures and reagents and also discoveries of new aspects of the reactivity of selenium and tellurium intermediates of carbohydrates and nucleosides may be expected. For these reasons, we believe that these fields will remain a rich area of investigation for many years to come.

VIII. REFERENCES

1. K. J. Irgolic and M. V. Kudchadker, '*Organic Chemistry of Selenium*', in *Selenium* (Eds. R. A. Zingaro and W. C. Cooper), Van Nostrand Reinhold, New York, 1974, Chap. 8, p. 408.
2. D. L. Klayman and W. H. S. Gunther (Eds.), *Organic Selenium Compounds: Their Chemistry and Biology*, Wiley, New York, 1973.
3. H. J. Reich, *Organoselenium Oxidations*, in *Oxidation in Organic Chemistry*: Part C (Ed. W. S. Trachanowsky), Academic Press, New York, 1978, Chap. 1, p. 1.
4. D. L. J. Clive, *Aldrichim. Acta*, **11**, 43 (1978).
5. D. L. J. Clive, *Tetrahedron*, **34**, 1049 (1978).
6. H. J. Reich, *Acc. Chem. Res.*, **12**, 22 (1978).
7. S. V. Ley, *Chem. Ind.* (*London*), 101 (1985).
8. D. Liotta and R. Monaham, III, *Science*, **231**, 356 (1968).
9. K. C. Nicolaou and N. A. Petasis, *Selenium in Natural Product Synthesis*, CIS, Philadelphia, 1984.
10. Z. J. Witczak and R. L. Whistler, *Heterocycles*, **19**, 1719 (1982).
11. Z. J. Witczak, *Tetrahedron*, **41**, 4781 (1985).

12. R. A. Zingaro, *Chem. Scr.*, **8A**, 51 (1975).
13. W. A. Bonner and A. Robinson, *J. Am. Chem. Soc.*, **72**, 354 (1950).
14. G. Wagner and G. Lehmann, *Pharm. Zentralhalle*, **100**, 160 (1961).
15. G. Wagner and P. Nün, *Pharmazle*, **22**, 548 (1967).
16. J. Kocourek, J. Klenha and V. Jiracek, *Chem. Ind.* (*London*), 1937 (1963).
17. Z. J. Witczak, *Nucleosides Nucleotides*, **2**, 295 (1983).
18. R. Walter, in *Peptides*: *Chem. Biochem., Proc. Am. Peptide Symp., 1st*, Marcel Dekker, New York, 1968, p. 467; *Chem. Abstr.*, **73**, 56379e (1970).
19. L. Laitem, P. Thibaut and L. Christiaens, *J. Heterocycl. Chem.*, **13**, 469 (1976).
20. J. Jaurequi-Adeu, *Adv. Protein Chem.*, **21**, 387 (1966).
21. O. H. Muth (Ed.), *Selenium in Biomedicine*, Avi Publishing, Westport, 1966.
22. I. Rosenfold and O. Beath, *Selenium Biobotany, Biochemistry, Toxicity and Nutrition*, Academic Press, New York, 1964.
23. R. J. Shamberger, *Biochemistry of Selenium*, Plenum Press, New York, 1983.
24. W. Schneider and F. Wrede, *Chem. Ber.*, **50**, 793 (1917).
25. G. Wagner and P. Nühn, *Arch. Pharm.*, **297**, 461 (1964).
26. H. Frenzel, P. Nühn and G. Wagner, *Arch. Pharm.*, **302**, 63 (1969).
27. W. Kudelska and M. Michalska, *Tetrahedron*, **37**, 2994 (1981).
28. M. Michalska, J. Michalski and I. Orlich-Krężel, *Tetrahedron*, **34**, 2821 (1978).
29. M. Michalska, J. Michalski and I. Orlich-Krężel, *Pol. J. Chem.*, **53**, 253 (1979).
30. J. R. Daniel and R. A. Zingaro, *Phosphorus Sulfur*, **4**, 179 (1978).
31. R. A. Zingaro and J. K. Thompson, *Carbohydr. Res.*, **29**, 147 (1973).
32. C. D. Mickey, P. H. Javora and R. A. Zingaro, *J. Carbohydr. Nucleos. Nucleot.*, **1**, 291, (1974).
33. C. L. Baimbridge, C. D. Mickey and R. A. Zingaro, *J. Chem. Soc. Perkin Trans. 1*, 1395 (1975).
34. J. R. Daniel and R. A. Zingaro, *Carbohydr. Res.*, **64**, 69 (1978).
35. G. C. Chen, C. H. Banks, K. J. Irgolic and R. A. Zingaro, *J. Chem. Soc. Perkin Trans. 1*, 2288 (1980).
36. R. Lipka and M. Michalska, *Carbohydr. Res.*, **113**, 317 (1983).
37. G. Jaurand, J. M. Beau and P. Sinaÿ, *J. Chem. Soc., Chem. Commun.*, 572, 701 (1981).
38. K. Mori, *Tetrahedron Lett.*, 1609 (1976).
39. S. Current and K. B. Sharpless, *Tetrahedron Lett.*, 5078 (1978).
40. K. C. Nicolaou and Z. Lysenko, *Tetrahedron Lett.*, 1259 (1977).
41. K. C. Nicolaou and Z. Lysenko, *J. Am. Chem. Soc.*, **99**, 3186 (1977).
42. D. L. J. Clive, G. Chittattu, N. J. Curtis, W. A. Kiel and C. K. Wong, *J. Chem. Soc., Chem. Commun.*, 725 (1978).
43. D. L. J. Clive, G. Chittattu and C. K. Wong, *Can. J. Chem.*, **55**, 3894 (1977).
44. D. L. J. Clive, C. K. Wong, W. A. Kiel and S. M. Menchen, *J. Chem. Soc., Chem. Commun.*, 379 (1978).
45. D. L. J. Clive, G. Chittattu and C. K. Wong, *J. Chem. Soc, Chem. Commun.*, 441 (1978).
46. S. David, A. Lubineau and J. M. Vatele, *J. Chem. Soc., Chem. Commun.*, 701 (1975).
47. S. David, A. Lubineau and J. M. Vatele, *J. Chem. Soc., Perkin Trans. 1*, 1831 (1976).
48. K. Torssel and M. P. Tyagi, *Acta Chem. Scand., Ser. B*, **31**, 297 (1977).
49. S. David and A. Lubineau, *Nouv. J. Chim.*, **1**, 375 (1977).
50. Y. Nishimura, H. Umezawa and S. Umezawa, *Tetrahedron Lett.*, **22**, 77 (1981).
51. T. Van Es, *Carbohydr. Res.*, **5**, 282 (1967).
52. J. Mieczkowski, A. Konowal and A. Zamojski, *Pol. J. Chem.*, **57**, 75 (1983),
53. A. P. Kozikowski, K. L. Sorgi and R. J. Schmiesing, *J. Chem. Soc., Chem. Commun.*, 477 (1980); A. P. Kozikowski, R. J. Schmiesing and K. L. Sorgi, *J. Am. Chem. Soc.*, **102**, 6580 (1980).
54. K. B. Sharpless and R. F. Lauer, *J. Am. Chem.*, **94**, 7154 (1972).
55. K. B. Sharpless and R. F. Lauer, *J. Org. Chem.*, **37**, 3973 (1972).
56. K. B. Sharpless, M. W. Young and F. R. Lauer, *J. Am. Chem. Soc.*, **94**, 154 (1972); **95**, 2697 (1973).
57. K. B. Sharpless, M. W. Young and F. R. Lauer, *Tetrahedron Lett.*, 1979 (1973).
58. H. Paulsen, F. R. Heiker, J. Feldmann and K. Heyns, *Synthesis*, 636 (1980).
59. B. P. Mundy and W. G. Bornmann, *J. Org. Chem.*, **49**, 5264 (1984).
60. J. M. Lancelin, J. R. Pougny and P. Sinaÿ, *Carbohydr. Res.*, **136**, 369 (1985).
61. F. Freeman and K. D. Robarge, *Carbohydr. Res.*, **137**, 89 (1985).
62. R. M. Adlington, J. E. Baldwin, A. Basak and R. P. Kozyrod, *J. Chem. Soc., Chem. Commun.*, 944 (1983).

63. B. Geise and J. Dupuis, *Angew. Chem., Int. Ed. Engl.*, **22**, 622 (1983).
64. B. Geise and J. Dupuis, *Tetrahedron Lett.*, **25**, 1369 (1984).
65. G. Geise, J. Dupuis, T. Hasskerl and J. Meixner, *Tetrahedron Lett.*, **24**, 703 (1983).
66. J. Dupuis, B. Geise, D. Rüegge, H. Fischer, H. G. Korth and R. Sustmann, *Angew. Chem., Int. Ed. Engl.*, **23**, 896 (1984).
67. P. Rollin, V. Verez Bencomo and P. Sinaÿ, *Synthesis*, 134 (1984).
68. K. C. Nicolaou, S. P. Seitz, W. J. Sipio and J. F. Blount, *J. Am. Chem. Soc.*, **101**, 3884 (1979); **102**, 3784 (1980).
69. D. Goldsmith, D. Liotta, C. Lee and G. Zima, *Tetrahedron Lett.*, 4801 (1979).
70. J. L. Fourney, G. Henry and P. Jouin, *Tetrahedron Lett.*, **21**, 455 (1980).
71. K. Furuichi, S. Yogai and T. Miwa, *J. Chem. Soc., Chem. Commun.*, 66 (1980).
72. T. Sakakibara, I. Takei, E. Ohara and R. Sudoh, *J. Chem. Soc., Chem. Commun.*, 261 (1981).
73. P. A. Grieco, Y. Yokoyama and E. Williams, *J. Org. Chem.*, **43**, 1283 (1978).
74. P. A. Grieco, E. Williams, H. Tanaka and S. Gilman, *J. Org. Chem.*, **45**, 3537 (1980).
75. Z. J. Witczak, *J. Carbohydr. Chem.*, **3**, 359 (1984).
76. H. Paulsen and K. Todt, *Adv. Carbohydr. Chem.*, **23**, 115 (1968).
77. H. Paulsen, *Angew. Chem., Int. Ed. Engl.*, **5**, 495 (1966).
78. D. Horton and D. H. Hutson, *Adv. Carbohydr. Chem.*, **18**, 123 (1963).
79. S. Inokawa, *Kagaku* (*Kyoto*), **24**, 901 (1969); *Chem. Abstr.*, **72**, 90757r (1970).
80. Z. J. Witczak and R. L. Whistler, *J. Carbohydr. Chem.*, **2**, 351 (1983).
81. H. Yamamoto and S. Inokawa, *Adv. Carbohydr. Chem. Biochem.*, **42**, 135 (1984).
82. K. Blumberg, A. Fuccello and T. Van Es, *Carbohydr. Res.*, **59**, 351 (1977).
83. T. Van Es and R. L. Whistler, *Tetrahedron*, **23**, 2849 (1967).
84. T. Van Es and J. J. Rabelo, *Carbohydr. Res.*, **29**, 252 (1973).
85. J. J. Rabelo and T. Van Es, *Carbohydr. Res.*, **30**, 202 (1973); **30**, 318 (1973); **32**, 175 (1974).
86. T. Van Es and J. J. Rabelo, *Carbohydr. Res.*, **36**, 408 (1974).
87. R. J. Suhadolnik, *Nucleoside Antibiotics*, Wiley–Interscience, New York, 1970; R. J. Suhadolnik, *Nucleosides as Biological Probes*, Wiley–Interscience, New York, 1979.
88. S. Hanessian and A. G. Pernet, *Adv. Carbohydr. Chem. Biochem.*, **33**, 111 (1976).
89. G. D. Daves and C. C. Cheng, *Prog. Med. Chem.*, **13**, 303 (1976).
90. J. J. Jaffe and H. G. Mautner, *Cancer Res.*, **20**, 381 (1960).
91. J. J. Jaffe and H. G. Mautner, *Cancer Res.*, **18**, 938 (1958).
92. L. B. Townsend and G. H. Milne, *J. Heterocycl. Chem.*, **7**, 753 (1970).
93. G. H. Milne and L. B. Townsend, *J. Heterocycl. Chem.*, **8**, 379 (1971).
94. S. M. Chu, *J. Med. Chem.*, **14**, 254 (1971).
95. C. Y. Shine and S. H. Chu, in *Nucleic Acid Chemistry: Improved and New Synthetic Procedures, Methods and Techniques* (Eds. L. B. Townsend and R. S. Tipson), Wiley–Interscience, New York, 1978, Vol. 2, p. 673.
96. G. H. Milne and L. B. Townsend, *Biochim. Biophys. Acta*, **269**, 344 (1972).
97. G. H. Milne and L. B. Townsend, *J. Med. Chem.*, **17**, 263 (1974).
98. S. H. Chu and D. D. Davidson, *J. Med. Chem.*, **15**, 1088 (1972).
99. G. H. Milne and L. B. Townsend, *J. Chem. Soc., Perkin Trans. 1*, 2677 (1972).
100. D. S. Wise and L. B. Townsend, *J. Heterocycl. Chem.*, **9**, 1461 (1972).
101. D. S. Wise and L. B. Townsend, in *Nucleic Acid Chemistry: Improved and New Synthetic Procedures, Methods and Techniques* (Eds. L. B. Townsend and R. S. Tipson), Wiley–Interscience, New York, 1978, Vol. 1, p. 413.
102. S. M. Chu, C. Y. Shine and M. Y. Chu, *J. Med. Chem.*, **17**, 406 (1974).
103. C. Y. Shine and S. H. Chu, *J. Heterocycl. Chem.*, **12**, 493 (1975).
104. C. Y. Shine and S. H. Chu, *J. Chem. Soc., Chem. Commun.*, 319 (1975).
105. S. H. Chu, C. Y. Shine and M. Y. Chu, *J. Med. Chem.*, **18**, 559 (1975).
106. S. H. Chu, C. Y. Shine and M. Y. Chu, *J. Pharm. Sci.*, **64**, 1343 (1975).
107. G. H. Milne and L. B. Townsend, *J. Heterocycl. Chem.*, **13**, 745 (1976).
108. G. H. Milne and L. B. Townsend, *J. Carbohydr. Nucleosides Nucleotides*, **3**, 177 (1976).
109. D. S. Wise and L. B. Townsend, *Tetrahedron Lett.*, 755 (1977).
110. D. S. Wise and L. B. Townsend, *J. Chem. Soc., Chem. Commun.*, 271 (1979).
111. D. S. Wise, G.H. Milne and L. B. Townsend, *Heterocycles*, **15**, 345 (1981).
112. M. Sekine and T. Hata, *Chem. Lett.*, 801 (1979).
113. H. Takaku, T. Namoto and K. Kimura, *Chem. Lett.*, 1221 (1981).

114. N. Zylber, J. Zylber and A. Gaudemer, *J. Chem. Soc., Chem. Commun.*, 1084 (1978).
115. C. Boullais, N. Zylber, J. Zylber, J. Guilhem and A. Gaudemer, *Tetrahedron*, **39**, 759 (1983).
116. P. C. Srivastava and R. K. Robins, *J. Med. Chem.*, **26**, 445 (1981).
117. W. J. Hennen, B. C. Hinshaw, T. A. Riley, S. G. Wood and R. K. Robins, *J. Org. Chem.*, 1741 (1985).
118. P. D. Cook and D. J. McNamara, *J. Heterocycl. Chem.*, **23**, 155 (1986).
119. B. G. Ugarkar, R. K. Robins and G. R. Revankar, *Nucleosides, Nucleotides*, **3**, 233 (1984).
120. B. G. Ugarkar, H. B. Cottam, P. A. McKernan, R. K. Robins and G. R. Revankar, *J. Med. Chem.*, **27**, 1026 (1984).
121. R. F. Schinazi, J. Arbiser, J. J. S. Lee, T. I. Kalman and N. H. Prusoff, *J. Med. Chem.*, **29**, 1293 (1986).
122. N. Petragnani and M. Moura Campos, *Organomet. Chem. Rev.*, **2**, 61 (1967).
123. K. J. Irgolic and R. Zingaro, in *Organometallic Reactions* (Eds. E. Becker and M. Tsutsui), Wiley, New York, London, 1971.
124. W. C. Cooper (Ed.), *Tellurium*, Van Nostrand Reinhold, New York, 1971.
125. K. J. Irgolic, in *Houben-Weyl, Methoden der Organischen Chemie*, 4th ed., Vol. E12, Georg Thieme, Stuttgart, New York, in press.
126. K. J. Irgolic, *The Organic Chemistry of Tellurium*, Gordon and Breach, New York, 1974.
127. K. J. Irgolic, *J. Organomet. Chem.*, **103**, 91 (1975); **130**, 411 (1978); **158**, 235 (1978); **189**, 65 (1980); **203**, 368 (1980).
128. N. Petragnani and J. V. Comasseto, *Synthesis*, 1 (1986).
129. J. Czyżewska-Chlebny and M. Michalska, *J. Chem. Soc., Chem. Commun.*, 693 (1985).
130. S. V. Ley, C. A. Meerholz and D. H. R. Barton, *Tetrahedron, Suppl. 1*, **37**, 213 (1981).
131. D. H. R. Barton and S. W. McCombie, *J. Chem. Soc., Perkin Trans. 1*, 1574 (1975).
132. A. G. M. Barrett, R. W. Read and D. H. R. Barton, *J. Chem. Soc., Perkin Trans. 1*, 2184 (1980).
133. A. I. Ross, K. G. Agrawal, S. H. Chu and R. E. Parks, Jr., *Biochem. Pharmacol.*, **22**, 141 (1973).
134. M. Maeda, N. Abiko and T. Sasaki, *J. Med. Chem.*, **24**, 167 (1981).
135. C. H. Banks, J. R. Daniel and R. A. Zingaro, *J. Med. Chem.*, **22**, 572 (1979).

Author Index

This author index is designed to enable the reader to locate an author's name and work with the aid of the reference numbers appearing in the text. The page numbers are printed in normal type in ascending numerical order, followed by the reference numbers in parentheses. The numbers in *italics* refer to the pages on which the references are actually listed.

Abatjoglou, A. G., 124 (376), *201*
Abdelhamid, A. O., 550 (86, 87), *587*
Abed-Ali, S. S., 509 (89), *536*
Abel, E. W., 528 (215–217, 220, 223, 224), *538*
Abell, P. I., 310 (101), *336*
Abiko, N., 386 (117), *391*, 790 (134), *793*
Abrahams, S. C., 62, 65, 81 (31), *88*, 513 (115), 515 (126), *536*
Abramovitch, R. A., 326 (148), *337*
Aburaki, S., 98, 129 (64), *195*
Acampora, L. A., 341 (16), *346*, 485 (116), *494*, 497 (14), 501 (43), 502 (44), 506 (14), 511 (44), 514 (43), 530 (238, 240), *534, 535, 539*
Achiba, Y., 39, 40 (157), *50*
Adachi, K., 108, 169, 172 (204), *198*
Adachi, M., 543, 554 (21), *585*
Adam, F., 102 (139), *196*
Adams, M. A., 565 (153), *588*
Adlington, R. M., 772 (62), *791*
Agawa, T., 114 (265), 186 (905), *199, 211*
Agnew, W. F., 369 (23, 26), *375*
Agrawal, K. C., 263 (257), *272*
Agrawal, K. G., 790 (133), *793*
Agris, P. F., 362 (55), *365*
Aharon-Shalom, E., 341, 345 (18), *346*, 474 (69), 480, 485 (96), *493, 494*
Ahern, D., 131, 139 (422), *202*
Ahlbrecht, H., 681, 682 (43), *758*
Ahlers, H., 132 (434), 159 (698), *202, 207*, 507 (76), *535*, 694, 695, 713, 747, 749 (120), 757 (301), *760, 764*
Ahlers, K. D., 245 (141), 247, 248 (159, 162), 255 (159), *270*
Ahlgren, G., 284 (35), *335*, 523 (183), *538*
Ahluwalia, G. S., 386 (123), *391*
Ahmad, R., 172 (786), *209*
Ahmad, Z. A., 530 (237), *539*
Ahmed, M. A. K., 53, 62, 85 (13), *87*, 511 (103), 525 (190), *536, 538*
Ahmed, Z., 172 (787), *209*
Ahran-Shalom, E., 245 (139), *270*
Ahrland, S., 527 (205), *538*
Ahron-Shalom, E., 506 (69), *535*
Aime, S., 345 (49), *347*
Aismont, M. Yu. 513, 514 (119), *536*
Aitkhozhaeva, M. Zh. 575 (198), *589*
Aizawa, M., 177 (846), *210*
Akamatu, H., (130), *494*
Akerfeldt, S., 625, 628, 652 (91), *654*
Åkermark, B., 523 (183), *538*
Akermark, B., 284 (35), *335*
Akiba, M., 193 (970), *213*, 664 (42), *672*
Akulin, Y., 453 (153), *460*
Akulin, Yu. I., 503 (51), *535*
Akulinin, O. B., 25 (91), *48*
Alaghamandan, H., 374 (98), *376*
Alaydros, E., 266 (277), *273*
Albeck, M., 133 (442, 443), 154 (442), 164 (730), *202, 208*, 245 (139, 140), *270*, 341 (17), *346*, 408 (92), *418*
Al'bitskaya, V. M., 662 (27), *672*
Albizati, K. F., 144 (526), *204*
Alcacer, L., 467, 485, 489 (54), *493*
Alcock, N. W., 343 (35, 36a), 344 (35), *347*
Alderdice, M., 143 (517), *204*
Alekseevskii, V. A., 260 (236), *272*
Aleshine, K. P., 53 (14), *87*
Alexander, J. D. Jr., 383 (60), *389*
Alexander, M., 369 (16), *374*
Alexander, R. P., 167 (755), *208*

Alfthan, G., 379 (10), *388*
Allen, D. W., 614, 615 (61), *653*
Allen, G. W., 502, 511 (44), 530 (238), *535, 539*
Allinger, N. L., 32 (113), *49*
Allkins, J. R., 533 (260, 261), *539*
Alnajar, M. S., 312 (108), *336*
Alonso, R. A., 296, 298, 299 (66), *335*
Alpert, J. S., 371 (56), *375*
Al Rubaie, A. Z., 518, 531, 532 (147), *537*
Al-Rubaie, A. Z., 528 (222), *538*
Al-Salim, N. 515, 516 (125), *536*
Al-Salim, N. I., 512, 528 (108), *536*
Al-Turahi, M. A. S., 553, 555 (93), *587*
Alzbeta, K., 542 (5), *585*
Alzner-DeWeerd, B., 362, 363 (56), *365*
Ambrose, K. R., 371 (63), 372 (72–74, 77), *375, 376*, 388 (164), *392*
Ambrosius, K., 445 (78), *458*
Amdur, M. L., 368 (6), *374*
Ames, A., 106 (186), 163 (717–719), 169, 172, 187 (718, 719), *197, 208*, 410 (125), *419*
Ames, D., 446 (83), *458*
Ames, D. E., 545, 553, 557 (36), *586*
Aminabhavi, T. M., 373 (86–88, 90, 91), *376* 384 (78–80), *390*
Amos, R. A., 141, 161 (497), *203*
Amosova, S. V., 407 (100, 101), *419*, 505 (64–66), *535*
Amzil, J., 487 (125), *494*
Anciaux, A. 154 (642, 662), 155 (642), 156 (662), *206, 207*, 695 (141), 710 (207, 208), *760, 762*
Andersen, J. R., 474 (77), 478 (77, 87), *493*
Andersen, K. K., 219, 263 (18), 264 (18, 268), 265 (268), *268, 273*
Anderson, D. G., 711 (214), *762*
Anderson, J. N., 388 (168), *392*
Anderson, J. R., 388 (170), *392*
Anderson, J. W., 596, 602, 605–607, 609, 648 (13), *652*
Anderson, P. C., 165 (733), *208*, 397 (22), *417*, 520 (165), *537*, 660 (15), *672*
Ando, K., 617, 633 (68), *653*
Ando, R., 146 (543, 544), 163 (720–722), *204, 208*
Andre, J.-M., 456 (203), *461*
Andreocci, M. V., 5, 18, 20, 21, (18), *47*, 327 (152), *337*, 549, 574 (66), *586*
Andres, K., 467, 469 (39, 43), *492*
Andrews, G. D., 99 (95), *195*
Andrews, L., 297 (69), *335*
Angelici, R. J., 573 (188), *589*
Angelov, C. M., 143 (509, 510), *204*
Angoh, A. G., 145 (540, 541), 184, 186 (899), *204, 211*
Anjaneyulu, V., 123 (370), *201*
Anthoni, U., 249 (172), 253 (197, 198), 254, 255 (205), *271*
Antipin, M. Yu., 623, 634, 637 (82), *653*
Antsyshkina, A. S., 267 (288), *273*
Anzai, H., 467 (15), *492*
Aoai, T., 97 (52, 53), 108 (208), 116 (208, 291), 126 (52, 53, 393), 144 (520, 521a, 521b, 523), *195, 198, 199, 201, 204*, 569 (176–179), *589*, 715 (250), *762*
Aono, S., 533 (259), *539*
Apostolescu, M., 455 (199), *461*
Appel, R., 431 (40), *457*, 598 (25), 632 (25, 108), 643 (108), *652, 654*
Applequist, D. E., 710 (192), *761*
Arad-Yellin, R., 450 (132), *459*
Arase, A., 498 (20), *534*, 546 (44, 45), *586*
Arbelot, M., 40 (158), *50*
Arbiser, J., 787 (121), *793*
Ardnt, H. C., 724 (328), *764*
Arduser, F., 382 (49), *389*
Aria, F., 151 (588), *205*, 450 (121), *459*
Arigoni, D., 118 (328), *200*
Arita, M., 108, 169, 172 (204), *198*
Ariyoshi, K., 147 (548), *204*
Armen, G. H., 327 (150), *337*
Armistead, D. M., 145 (535), *204*, 666 (58), *672*
Armstrong, J. B., 383 (60), *389*
Arnó, M., 562 (129), 563 (130), *588*
Arnold, A. P., 542 (9), *585*
Arnold, D. E. J., 596, 601, 611, 612, 643 (14), *652*
Arnold, D. R., 314 (114), *336*
Arnott, M. S., 385, 386 (104), *390*
Arshadi, M., 453 (171), *460*
Arthur, J. R., 384 (85), *390*
Arunachalam, T., 97, 108, 116, 172 (49), *195*
Asa, K., 400 (51), *418*
Asakura, T., 370 (49), *375*
Åsbrink, L., 6 (33, 38), 7 (38), 22, 25 (75, 76), 40 (159), *47, 48, 50*
Ascherl, A., 446 (80), *458*
Åse, K., 580 (219), *590*, 635 (116), *654*
Ashby, R. A., 549 (64), *586*
Ashdown, D. H. J., 187 (923), *212*
Ashe, A.-J., III 615 (64), *653*
Ashe, A. J., III 604 (44), *653*
Asinger, F., 238 (109), *269*, 406 (86), *418*, 542 (12), *585*
Aso, Y., 169, 172 (774), 178 (853), *209, 210*, 556 (III), *587*, 664 (43), *672*
Assadi, F., (219), *272*
Attarwala, S. T., 476, 477 (80), *493*
Atwood, J. L., 528 (218, 219), *538*
Augen, S., 572 (186), *589*
Aust, S. D., 380 (26), *389*
Austad, T., 400 (51), *418*, 579 (216), 580 (216, 219), 582 (216), *590*, 635 (116), *654*
Avenarius, I. A., 53 (17), *87*
Aviado, D. M., 378 (2), *388*
Avrey, G., 677, 681 (9), *757*

Aygen, S., 26 (100), 37 (140, 141), 38 (140, 141, 147), 39 (147, 151), 40 (151), *48–50*, 228, 266 (80), *269*, 450 (118, 133), 453 (158), *459, 460*
Ayorinde, F. O., 133 (441), *202*
Azerbaev, I. N., 575 (198), *589*
Azman, A., 266 (280), *273*
Azvedo, L. J., 483 (108), *494*

Baba, H., 369 (32), *375*
Babeshkim, A. M., 53 (18), *87*
Babushikina, T. A., 86 (88), *89*
Bachi, M. D., 145 (538a, 538b), *204*, 666 (60), *672*
Back, T. G., 98 (70, 71), 100 (110–112), 109 (222, 223), 111 (240, 249, 250), 114 (271), 124 (223, 382), 126 (111), 136 (70, 71, 463), 137 (70, 71), 138 (70, 110–112, 222), 146 (240, 250), 149 (249, 567, 574, 575), 150 (249, 567, 577, 578), 151 (250), 152 (223, 382), 162 (714), 167 (746, 747), 173 (808), 174 (112, 814), 179 (808), 180 (112, 222), 184 (893), 189 (271, 463), 190 (463), 191 (893, 942), *195, 196, 198, 199, 201, 203, 205, 208–212*, 219 (15), 222 (15, 52), 223–225, 245 (52), *268*, 329 (161), *337*, 398 (28), *417*, 427 (24), *457*, 558 (117), 574 (194), *588, 589*, (265), *763*
Bäckvall, J.-E., 129 (412), 130 (416), *202*
Badet, B., 356 (29), *364*
Baerends, E. J., 18 (68), *48*
Baettig, K., 565 (158), *588*
Bagnall, K. W., 220, 221 (23), *268*
Bahan, J. M., 552 (89), *587*
Bailey, A., 467 (24), 485 (111), *492, 494*
Bailey, J. M., 379 (12), *388*
Bailey,T. R., 113 (258), *199*
Baimbridge, C. L., 769, 790 (33), *791*
Baird, N. C., 6 (28, 29), *47*
Baiwir, M., 342 (26), *347*, 448 (101), *459*, 531 (243, 247), *539*, 550 (76, 81, 83), 577 (209), *587, 590*
Bajwa, G. S., 480 (95), *494*
Bak, B., 25 (89, 90), *48*, 243 (128–130), 267 (128), *270*, 449 (112), 450 (117), *459*
Baker, A. D., 2 (1, 3, 8), 10 (49), 13 (54), 32, 33 (116), 35 (129), 36 (116), *46, 47, 49*, 327 (150), *337*
Baker, C., 2 (8), 10 (49), 13 (54), *46, 47*
Baker, F. J., 369 (11), *374*
Baker, S. S., 379 (16), *389*
Balachandran, S. 388 (165), *392*
Balakrishnan, P., 107 (199), *198*
Balassa, J. J., 384 (91), *390*
Baldwin, J. E., 99, 141 (104), 142 (507), *196, 204*, 772 (62), *791*
Baldwin, R. L., 358 (37), *364*
Balenovic, K., 128 (405), 147 (554, 555), 148 (562), 149 (573, 576), *202, 204, 205*, 718 (274–279), *763*
Balkovec, J. M., 99, 122, 183 (94), *195*
Ballard, R. E., 2, 30 (11), *47*
Ballester, M., 681, 682 (63), *759*
Balodis, K. A., 485 (117), *494*
Balogh, V., 102, 122 (143), *196*
Balton, Ya. G., 439, 440 (56), *458*
Balzani, V., 306 (90), *336*
Bancroft, G. M., 52 (5), 53, 64 (12), *87*
Banks, C. H., 386 (120), *391*, 543, 554 (22), *585*, 769 (35), 790 (135), *791, 793*
Banks, R. H., 467 (13, 22), 469 (22), *492*
Bannou, T., 108, 116, 141 (206), *198*
Barashenkov, G., 455 (194–196), *461*
Barashenkov, G. G., 438 (52, 53), 439 (55), 440 (61), *458*
Barattesani, D. N., 113 (260), *199*
Bares, J. E., 108 (207), *198*, 685 (86), *759*
Barglin, J. N., 577 (204), *589*
Barili, P., 453 (162), *460*
Barker, G. K., 42, 44 (168), *50*
Barltrop, J. A., 317 (126), *336*
Barnarc, P. W. C., 87 (92), *89*
Barnard, D., 253 (192), *271*, 404 (70), *418*, 677, 681 (9), 718 (273), *757, 763*
Barner, B. A., 109 (220), *198*
Barnes, R. G., 73 (55), *88*
Barnette, W. E., 98 (77, 78), 116 (285), 139 (78, 479–481), 141 (285, 479), 142 (77, 78, 508), 144 (285, 524), 152 (78, 508, 610), 174 (78, 285, 480), *195, 199, 203, 204, 206*, 427 (25a), *457*, 665 (53, 54), *672*, 715 (244), *762*
Barnikov, G., 246 (149), *270*
Barnum, C., 32, 33, 36 (116), *49*, 103 (151), 106 (192), 107 (192, 199), 184 (892), 187 (892, 934), *197, 198, 211, 212*, 327 (150), *337*, 406 (84), *418*
Barnum, C. S., 184, 187 (889), *211*
Baroni, A., 438, 440 (50), *458*
Barraclough, C. G., 260 (235), *272*
Barrett, A. G., 219 (16), 233 (16, 92, 93), 234 (93), 264 (16), *268, 269*
Barrett, A. G. M., 153 (628), 175 (628, 822), 178 (860), *206, 210*, 327 (155), *337*, 788 (132), *793*
Barrett, J., 82 (71), *88*
Barrie, A., 26 (102), *48*
Barrière, J.-C., 153 (620), *206*
Barros Papoula, M. T., 741 (313), *764*
Barta, M. A., 96, 179 (37), *194*, 716 (267), *763*
Bartlett, P. A., 97 (48), 172 (48, 782), *195, 209*
Bartmess, J. E., 108 (207), *198*, 685 (86), *759*
Barton, D. H. R., 99, 102 (102), 111 (238–242, 245, 246), 112 (251), 120 (245, 345, 350), 121 (354–357), 122 (246), 127 (395), 128 (403), 132 (102), 146 (240–242), 147 (551, 557, 558), 148 (558, 564–566), 149 (569–

571), 150 (551, 569–571, 579, 580), 152 (551, 571), 153 (571, 617–619, 623, 624, 627, 628), 154 (571), 163 (102), 164 (731), 167 (746, 747), 168 (356), 174 (395, 566), 175 (628, 821, 822), 178 (821, 855, 859, 860), 186 (902), 191 (102, 966), *196, 198–202, 204–206, 208, 210, 211, 213*, 219 (15, 16), 221 (38), 222 (15, 52), 223–225 (52), 227 (72), 229 (38), 231 (90), 232 (38, 90), 233 (16, 38, 92, 93), 234 (93), 240 (38), 245 (52), 246 (154), 264 (16), *268–270*, 322 (138), 323 (139), 327 (155), 329 (161), *337*, 398 (28), *417*, 429, 430, 436 (32), *457*, 668 (67), *673*, 699 (152), 706, 720 (325), 741 (313), *760, 764*, 788 (130–132), *793*
Barton, T. A., 286, 319 (38), *335*
Barton, T. J., 314, 317 (116), *336*
Basak, A., 772 (62), *791*
Basch, H., 40 (152), *50*
Bashiardes, G., 150 (580), *205*, 706, 720 (325), *764*
Basmadjian, G. P., 388 (165, 167), *392*
Bates, G. S., 173 (811), *210*
Bats, J., 456 (204), *461*
Battersby, A. R., 104 (166), *197*
Bättig, K., 187 (929), *212*
Battioni, J. P., 556 (112), *588*
Batzel, N., 451 (134), *459*
Baudat, R., 118, 184, 186 (321), *200*, 741 (311), *764*
Baudler, M., 616, 643 (66), *653*
Bauer, C., 405 (81), *418*
Baumann, C. A., 384, 385 (92), *390*
Baumann, H., 453 (166), *460*
Bautina, I. V., 548 (61), 549 (71–73), 573 (190, 191), *586, 587, 589*
Bayandika, E. V., 625, 628 (88), *654*
Bayandina, E. V., 413 (139), *419*, 625 (89, 90), 626 (94), 627 (95), 637, 638 (126), 645 (95), *654*
Bayer, C., 106, 107 (192), *197*
Bayet, P., 154, 155 (640), 161 (708), *206, 207*, 521 (168), *537*, 677 (11), 678, 679, 681, 685 (11, 19), 686, 689, 695, 699 (11), 703, 705 (19), 706 (11, 19), 711, 715 (11), 718 (19), 719 (11, 19), 721 (11), 726 (11, 19), 740 (11), *757, 758*
Beak, P., 681, 682 (40), *758*
Beams, F. A., 385, 386 (103), *390*
Beath, O., 765 (22), *791*
Beau, J.-M., 97 (51), 98 (64), 99 (51), 117 (51, 320), 129 (64), *195, 200*
Beau, J. M., 770 (37), *791*
Beaulien, P. L., 666 (61), *672*
Beaulieu, P. L., 131 (423), 139 (423, 486), 145 (537), 176 (831), *202–204, 210*
Becci, P. J., 385 (134, 135), *391*
Becher, J., 237 (105), 266 (278), *269, 273*
Bechgaard, K., 41 (163), *50*, 249 (175, 178), 251 (183), *271*, 413 (143), *419*, 467 (21–23, 26, 27, 37, 38, 41, 44, 45), 469 (22, 23, 37, 38, 41, 44–46), 473, 474 (66), 475 (73, 75), 478 (87, 88), 489 (134), *492–494*
Beck, A. K., 154 (639, 653, 664), 155, 156 (639, 653), 157 (653), 189 (664), *206, 207*, 408 (105), *419*, 506, 509, 519 (70), 523 (178), *535, 538*, 685, 686, 689, 695, 696 (126a, 126b), 699 (126a), 710, 715, 721, 726 (126a, 126b), 753, 754 (294), *760, 763*
Becker, G., 604 (43), 615 (63), *653*
Becker, W., 234, 238 (95), *269*, 451 (139a), *460*
Bednarz, K., 263 (259), *272*
Beecher, J. F., 513 (113), *536*
Beelitz, K., 299 (79), 302 (79, 81), *335*
Beens, W., 62 (20), *87*
Begona, G. M., 563 (130), *588*
Behan, J. M., 176 (842), *210*, 398 (33), *417*
Beierwaltes, W. H., 388 (165), *392*
Bekkevoll, S., 580 (217), *590*
Belin, C., 346 (54), *347*
Bell, R. R., 382 (52), *389*
Bellus, D., 321 (134), *337*
Beloeil, J.-C., 102, 122 (143), *196*
Belyzlov, R. U., 549 (73), *587*
Benci, P., 614, 615, 646 (62), *653*
Bencomo, V. V., 97 (42), *194*
Bender, S. L., 245 (138), *270*, 341 (17), *346*, 408 (93, 94), *418*
Benedek, G., 467, 469 (38), *492*
Benedetti, A., 549 (74), *587*
Benedetti, E., 454 (191), *461*
Benedict, C. R., 371 (54), *375*
Benforemo, N., 407 (110), *419*
Bengels, D., 498 (22), *534*
Bennett, F. C., Jr., 255 (210, 211), *271*
Beno, M., 467, 483 (51), *493*
Beno, M. A., 341 (15), *346*, 467 (50), 483 (50, 108), *493, 494*
Berding, H., 238 (109), *269*
Berdnikov, E. A., 134 (450), *202*, 706 (170), *761*
Berezov, T. T., 356, 357 (31), *364*
Berg, C., 243 (130), *270*, 450 (117), *459*, 478 (87), *493*
Bergelson, L. D., 681, 682 (50), *758*
Berger, B., 102 (137), *196*
Berger, H., 471 (62), *493*
Bergman, G., 660 (16), *672*
Bergman, J., 33 (121), *49*, 116 (305), 120 (352, 353), 122 (353), 124 (305), 129 (412), 139 (483), 149, 150 (572), 164 (725–727, 729), 174 (483), 191 (964), *200–203, 205, 208, 213*, 245 (140), *270*, 279 (18), 287 (40), *334, 335*, 341 (17), *346*, 397 (23–25), *417*, 510 (96, 97), 520 (163, 164), 521 (164), 524 (184), 527 (201), *536–538*, 660 (17), *672*
Bergmann, F., 263 (253), *272*

Bergson, G., 266 (279), *273*, 511 (98), *536*
Berkowitz, J., 7 (39, 41), 8 (41), *47*
Berlin, K. D., 480 (95), *494*
Berlinsky, A. J., 466 (5), *491*
Berman, E. M., 97, 142 (56), *195*
Bernal, I., 528 (218, 219), *538*
Bernardi, F., 33 (119), *49*
Berry, D. A., 386 (124), *391*
Berry, F. J., 52 (8, 9), 53 (13), 60 (8), 62 (13, 23, 26–29), 63, 65 (23), 66 (38), 67 (23, 26, 43), 70 (26, 43), 71 (23, 26, 43), 72, 73 (26), 74 (23, 26), 76, 79 (26), 80 (27, 62), 81 (62), 85 (13), *87*, *88*, 497 (9), 533 (267, 268), *534*, *539*
Bert, G., 37, 38 (142), *49*
Bertini, V., 400 (48), *417*, 446 (81), 448, 452 (100), 453 (161), 454 (81, 187b, 190–192), *458–461*
Bertrand, J. L., 100, 154, 155, 157 (124), *196*, 692 (113, 114), 693 (113), 695, 696 (113, 114), 728 (113), *760*
Bertz, S. H., 474 (69), *493*
Bérubé, G., 117 (319), *200*
Besserer, J. A., 386 (124), *391*
Bestmann, H.-J., 118, 119 (342), *200*
Bestmann, H. J., 118 (322), *200*, 681, 682 (51), *758*
Betancor, C., 109 (227), *198*, 715 (248), *762*
Betteridge, D., 31, 32 (112), *49*, 327 (151), *337*
Betz, H., 453 (178), *460*
Beveridge, D. L., 6 (23, 27), *47*
Bewick, A., 126 (394), *201*
Beyer, L., 575 (197), *589*
Bezzi, S., 261 (240), *272*
Bhacca, N., 310 (102), *336*
Bhargava, S. K., 528 (220, 223, 224), *538*
Bhasin, K. K., 498, 508 (18), *534*
Bianchini, C., 411 (130), *419*
Bianco, J. A., 371 (56), *375*
Bianco, R., 255 (208), *271*
Bickelhaupt, F., 632 (109), *654*
Bielska, M. T., 153 (619), *206*
Bieri, G., 39 (149), *50*
Bierscheuk, T. R., 510 (95), *536*
Bigoli, F., 519 (153), *537*
Bigotto, A., 33 (121), *49*
Bildstein, B., 601, 613 (32), *652*
Billion, A., 147 (557), *205*
Billmers, J. M., 99, 180 (88), *195*
Binns, M. R., 159 (696), *207*, 684, 695 (79), *759*
Bioko, V. N., 297 (74), *335*, 670 (78), *673*
Biradar, N. S., 373 (86–91), *376*, 384 (78–80), *390*
Birchall, J. M., 297 (72), *335*
Bird, C., 423 (10), *457*
Bird, P. H., 343 (33), *347*
Bird, S. R. A., 82 (71), *88*
Birner, B., 575 (197), *589*
Birt, D. F., 385 (138), *391*
Bishop, P., 668 (65), *673*
Bjerre, C., 243 (130), *270*, 450 (117), *459*
Black, D., 445 (76), *458*
Blackadder, E. S., 368 (8), *374*
Blackmore, W. R., 62, 65, 81 (31), *88*, 513 (115), 515 (126), *536*
Blackwell, D. S. L., 322 (137), *337*
Blaha, J., 453 (185, 186a), *461*
Blaich, B., 639, 644 (136), *655*
Blair, J., 266 (277), *273*
Blair-West, J. R., 383 (60), *389*
Blanc, J., 326 (149), *337*
Blanck, T. J. J., 356 (26), *364*
Blau, M., 388 (169), *392*
Blauchet, G. B., 512, 530 (106), *536*
Blecher, A., 414 (149), *420*
Blechschmitt, K., 259 (231), *272*
Bloch, A., 407 (117), *419*, 466 (1, 3), 467 (1), 480 (97), 483 (1), 485 (97), 489 (3), *491*, *494*, 506 (68), *535*
Bloch, A. N., 41 (163), *50*, 249 (175), *271*, 466 (2), 467 (2, 9, 13, 14, 21–23, 47), 469 (22, 23, 47), 473 (66), 474 (66, 70, 72, 74), 483 (103), 489 (132), *491–494*
Bloch, R., 106 (197), *197*
Block, A. N., 250 (179, 181), *271*
Block, E., 40 (155), *50*, 681,, 682 (53), *758*
Block, H. D., 650 (190), *656*
Blount, J. F., 139, 141 (474a), *203*, 665 (55), *672*, 773 (68), *792*
Blumbach, J., 102 (136), *196*
Blumberg, K., 775, 777 (82), *792*
Boatman, R. J., 120 (346), *200*
Bochkarev, L. N., 398, 399 (35, 36), 414 (36), *417*
Bochkarev, M. N., 281, 309 (23), *334*, 398 (35), 399 (35, 41, 42, 44), 414 (44, 147), *417*, *420*, 663 (29), *672*
Bochmann, G., 497 (11), *534*
Bock, H., 15 (59), 16 (64), 26 (100), 33 (117, 118, 120), 35 (130, 131, 134, 135), 36 (117, 134, 138), 37 (138, 140–143), 38 (140–143, 147), 39 (147, 151, 154), 40 (151, 153, 155), *47–50*, 228, 266 (80), *269*, 412 (135), *419*, 450 (118), 453 (158), *459*, *460*, 572 (186), *589*
Bocker, K., 379 (13), *388*
Bodner, V. N., 542 (14), *585*
Bodor, N., 6 (31), *47*
Boeckman, R. K., Jr., 116 (284), *199*
Boek, H., 450 (133), *459*
Boese, R., 451 (134), *459*, 646 (176), *656*
Boettger, S. D., 104, 106 (165), *197*
Bogan, L. E. Jr., 343 (37), 345 (37, 45, 46), *347*
Bogdanowicz, M. J., 724 (328), *764*
Boger, D. L., 172 (789), *209*
Bogolyubov, G. M., 597 (16), *652*

Bohm, B. A., 310 (101), *336*
Böhm, M. C., 44, 45 (169), *50*, 645 (165), *655*
Boie, I., 610 (53), *653*
Boivin, J., 147 (557), *205*
Boje, L., 240, 241 (119), *270*
Bökens, H., 97 (54), *195*, 563, (133), *588*
Boldeskul, I. E., 623, 634, 637 (82), *653*
Bolivar, R. A., 314 (115), 317 (115, 124, 125, 128), *336*
Bonamico, M., 267 (287), *273*
Bondi, A., 645 (167), *655*
Bonner, W. A., 765, 766 (13), *791*
Bonser, S. M., 565 (157), *588*
Bontempelli, G., 483 (101), *494*
Boon, G. D., 369, 370 (22), *375*
Booth, B. A., 263 (257), *272*
Booth, R., 53 (19), *87*
Borbe, H., 447 (92), *459*
Borch, G., 40, 41 (160), *50*
Bordeleau, L., 143 (519), *204*
Bordwell, F. G., 108 (207), *198*, 312 (107), *336*, 685 (86), *759*
Borek, B., 528 (227), *538*
Boritzki, T. J., 386 (124), *391*
Bornmann, W. G., 124, 140 (378), *201*, 772 (59), *791*
Bossa, M. 5, 18, 20, 21 (18), *47*, 327 (152), *337*, 549, 574 (66), *586*
Bottger, J., 455 (198), *461*
Bottino, N. R., 369 (15), *374*
Botto, R., 453 (165), *460*
Boubean, J., 639, 644 (136), *655*
Boucher, C. A., 371 (66), *376*
Bouchoule, C., 710 (194), *761*
Bougeard, P., 304, 305 (86), *336*, 669 (69), *673*
Bouhy, P., 683, 695 (73), *759*
Boullais, C., 567 (168), *589*, 785 (115), *793*
Bourguignon, J., 325 (142), *337*
Boutagy, J., 681, 682 (47), *758*
Boutique, J.-P., 456 (203), *461*
Bowen, H. J. M., 513, 514 (114), *536*
Bowen, L. H., 71 (46), 73 (54), *88*
Boxler, D., 99, 173 (99), *195*
Boyne, R., 384 (85), *390*
Boysel, R. M., 512, 530 (106), *536*
Boyson, R. A., 297 (72), *335*
Bracksem, T. J., 501 (39), *535*
Bradley, D. E., 370 (40), *375*
Bradt, W. E., 221 (42), *268*
Braga, A. L., 114 (264), *199*
Brähler, G., 36, 37 (138), *49*
Bramwell, F., 456 (202), *461*
Branca, S. J., 116 (289), *199*
Brandsma, L., 235 (99), 239 (114), 243 (99), 244 (135), *269, 270*, 408 (95, 97), *418*, 505 (67), *535*
Brandt, C. A., 154 (665), 169 (760), 189 (665), *207, 209*, 683, 696, 706, 727 (102), *759*
Brasted, R. C., 340 (1), *346*
Brattesani, D. N., 113 (259), *199*, 715 (261), *763*
Braun, J., 453 (157), *460*
Braun, R. W., 510 (95), *536*
Braunstein, A. E., 352 (12), *364*
Braverman, S., 547, 554 (52), *586*
Brecht, H., 621 (76), *653*
Bregadze, V. I., 416· (155, 156), *420*
Bregant, N., 128 (405), 147 (554, 555), 149 (576), *202, 204, 205*, 718 (275, 276, 278, 279), *763*
Brehm, L., 401 (54), *418*, 619, 620 (72), *653*
Bremmer, M. L., 118 (337), *200*
Breunig, H. J., 401 (57), *418*, 604 (38–40, 42), 605 (46), 615 (39), 616 (65), 644 (46), *653*
Brewster, A. G., 111 (240), 121 (354, 355), 128 (403), 146 (240), *198, 201, 202*, 429, 430, 436 (32), *457*
Bridge, H. J., 221 (46), *268*
Bridges, A. J., 131 (419–421), *202*
Bridon, D., 191 (966), *213*, 668 (67), *673*
Brill, A. B., 371 (69), *376*
Bringmann, G., 246 (154), *270*
Brisk, M., 35 (129), *49*
Brittain, W., 453 (185, 186a), *461*
Britten-Kelly, M. R., 127 (395), 167 (746, 747), 174 (395), *201, 208*, 219 (15), 222 (15, 52), 223–225, 245 (52), *268*, 329 (161), *337*, 398 (28), *417*
Britton, W. E., 266 (277), *273*
Brocksom, T. J., 103, 169 (152), *197*
Brogli, F., 39 (150), *50*
Brook, A. G., 711 (214), *762*
Brooks, H. G., 711, 755 (217), *762*
Brothers, D., 184 (890), *211*, 685 (84), *759*
Brown, C. K., 385 (136), *391*
Brown, D. H., 283 (31), *335*, 412 (138), 413 (138, 142), *419*, 526 (196), 527 (198), *538*, 634, 635 (114), 640 (114, 138), 641 (114), *654, 655*, 659 (10), 660 (13), *671*
Brown, D. L., 118 (337), *200*
Brown, J. H., 383 (71, 72), *390*
Brown, J. R., 658 (4), *671*
Brown, R. D., 25 (92), *48*
Brown, R. F. C., 106 (198), 187 (916), *197, 212*
Brown, R. S., 129 (414), *202*
Brundle, C. R., 2 (3, 8), 10 (49), 13 (54), 40 (152), *46, 47, 50*
Brunner, H., 405 (79), *418*
Bruno, P., 650 (193), *656*
Brutus, M., 487 (125), *494*
Bryan, R. F.,416 (154), *420*
Bryant, D. R., 124 (376), *201*
Bryant, R. W., 379 (12), *388*
Bryce, M. R., 547, 554 (56), *586*
Bryden, W. A., 489 (132), *494*
Bryson, T. A., 104 (178), *197*, 710 (206), *762*
Buchardt, O., 334 (169), *337*

Buchler, G., 431 (40), *457*
Buchowiecki, W., 630 (102), *654*
Buchwald, H., 452 (142), *460*
Buck, H. M., 687 (105), *759*
Buckley, D. J., 189 (938, 939), *212*, 706 (175), *761*
Buckmaster, G. W., 381 (42), *389*
Buell, G. R., 711 (218), *762*
Buina, N. A., 621 (77), 625, 628 (88), 637 (126), 638 (126, 129), 639, 640 (129), *653–655*
Bulka, E., 245 (141), 247 (159, 162), 248 (159, 162, 166), 255 (159), 257 (166), *270*, 541, 574 (1), *585*
Bunnelle, W. H., 99, 106 (106), *196*
Bunnenberg, E., 151 (604), *205*, 264, 265 (269), *273*
Bunnet, J. F., 294, 296 (57), *335*
Bunnett, J. F., 294 (62, 63), *335*
Buravov, L. I., 467, 483 (48), *492*
Bureneva, M., 453 (186b), *461*
Burg, A. B., 597 (19), *652*
Burger, F., 39 (149), *50*
Burger, H., 246 (148), *270*
Burger, K., 413 (144), *419*
Burk, R. F., 381 (36–38, 41, 43), *389*
Burke, S., 101 (130), 104 (176), *196, 197*
Burke, S. D., 145 (535), 173, 184 (799, 802, 803), *204, 209*, 561 (128), *588*, 666 (58), *672*
Burke, W. A., 516, 517 (134), *537*
Burnett, D. A., 145 (539), *204*
Burrow, P. D., 25 (85), 45 (172), *48, 50*
Burson, R. L., 721 (239), *762*
Burstall, F. H., 507 (79), *536*
Burton, A., 153–155 (622), *206*, 716, 720 (266), 734, 736 (308a, 308b), 748, 749 (266), *763, 764*
Bus, J. S., 380 (26), *389*
Butler, T. A., 388 (166), *392*
Buzilova, S. R., 507 (74), *535*
Byström, S. E., 175 (830), *210*

Cadenas, E., 374 (92), *376*, 384 (75), *390*, 447 (88), *459*
Cagen, S. Z., 381 (35), *389*
Cagniant, D., 497 (8), *534*
Cain, A. M., 96, 179 (37), *194*, 716 (267), *763*
Cain, M., 110, 118 (236), 119, 148 (236, 343), *198, 200*
Caira, M., 631 (107), *654*
Calderari, G., 116 (304), *200*
Calderazzo, F., 528 (218, 219), *538*
Call, E. W., 374 (98), *376*
Callahan, A. P., 371 (58, 59, 61, 63–65, 67), 372 (58, 61, 64, 67, 72–74, 77), 373 (81), *375, 376*, 388 (164, 166), *392*
Callear, A. B., 281, 318 (24, 25), *334*
Calò, V., 176 (840), *210*
Calogero, S., 87 (92), *89*
Calvery, H. O., 384 (89), *390*
Calvo, C., 266 (285), *273*
Camasseto, J. V., 169 (760), *209*
Cambie, R. C., 542 (8), *585*
Camerman, N., 612 (58), *653*
Campbell, T., 423 (3), *456*
Campos, O., 110 (236), 118 (236, 334), 119, 148 (236, 334, 343), *198, 200*
Camps, F., 118 (340), *200*
Campsteyn, H., 454 (189), *461*, 515, 516 (124), *536*
Candela, G. A., 467 (7), *491*
Canonico, P. G., 374 (96), *376*
Cantoni, G. L., 351 (7), *364*
Canty, A. J., 542 (9), *585*
Capeller, L., 446 (80), *458*
Caplain, S., 284 (33), *335*, 527 (199), *538*, 660 (14), *672*
Caplin, A., 450 (131), *459*
Carassiti, V., 306 (90), *336*
Cardoso, J. M., 153 (619), *206*
Carlsen, J. B., 242 (124), *270*
Carlson, K. D., 467 (50), 483 (50, 108), *493, 494*
Carlson, R., 397 (24), *417*
Carlson, R. R., 645 (163), *655*
Carlsson, R., 164 (726), *208*
Carlton, W. W., 369 (28), *375*
Carmack, M., 452 (150), *460*
Carneiro, K., 467 (27, 45), 469 (45, 46), *492*
Carrie, R., 576 (201), *589*
Carrol, P., 408 (92), *418*
Carroll, P., 245 (140), *270*, 341 (17), *346*
Carroll, P. J., 341, 345 (18), *346*
Carruthers, T., 466, 467, 483 (1), *491*
Carruthers, T. F., 250 (179), *271*, 466, 467 (2), 474 (70), *491, 493*
Caruso, A. J., 104, 140 (177), 169 (177, 770), *197, 209*
Caselli, M., 650 (193), *656*
Cason, L. F., 711, 755 (217), *762*
Caspi, E., 97, 108, 116, 172 (49), *195*
Casser, C., 598 (25), 632 (25, 108), 643 (108), *652, 654*
Castro, G., 467 (16), *492*
Catanedo, N. C., 551 (88), *587*
Catani, V., 167 (754), *208*
Cauletti, C., 5, 18, 20, 21 (18), *47*, 327 (152), *337*
Cauquis, G., 546 (49), 585 (227), *586, 590*
Cava, M., 449, 454 (114), *459*
Cava, M. C., 328 (157), *337*
Cava, M. P., 100 (113), 147, 152 (549), 172 (786, 787), 176 (832), 187 (916), 193 (969, 970), *196, 204, 209, 210, 212, 213*, 236, 238 (102), 245 (139, 140), *269, 270*, 283, 328 (30), *334*, 341 (17, 18), 342 (23), 345 (18), *346, 347*, 396 (18), 407 (108, 110, 111), 408

(92), 409 (122), 412 (18), *417–419*, 477 (84, 85), 478 (86), 483 (104), *493, 494*, 503 (50), 510 (93), 526 (193), *535, 536, 538*, 542, 554 (3), 556 (3, 110), 577, 578 (208), 580 (221), 582 (208, 221, 225, 226), 583 (208, 221, 225), 584 (221), *585, 587, 589, 590*, 659, 660 (9), 664 (42), *671, 672*, 756 (333), *764*
Cavallini, D., 351 (11), *364*
Ceasar, G. P., 36, 37 (137), *49*
Ceccherelli, P., 190 (940), *212*, 519 (157), *537*
Cederbaum, L. S., 25 (78), 39, 40 (156), *48, 50*
Chabaud, B., 118 (330), 119 (344), *200*
Chadha, R. K., 527, 529 (213), 531 (249), 532 (253), *538, 539*
Chadra, R. K., 498, 508 (18), *534*
Chaikin, P. M., 466 (6), 467 (16), *491, 492*
Challenger, F., 369 (13), *374*
Chambers, J. Q., 483 (102, 109), *494*
Chambers, R. J., 179 (868), *211*, 568 (172), *589*
Chan, D. M. T., 97, 99 (41), *194*
Chan, L. Y. Y., 67 (42), *88*
Chan, T. H., 176 (837), *210*, 651 (197), *656*, 681, 682 (44), 711 (219), *758, 762*
Chan, W. K., 173 (811), *210*
Chandhuri, M. K., 306 (91), *336*
Chandler, C., 453 (177), *460*
Chang, C. C., 356 (25), *364*
Chang, E., 711 (219), *762*
Chang, J.-H., 361 (51), *364*
Chang, M.-M. Y., 705 (162), *761*
Chanon, M., 40 (158), *50*
Chao, G. Y., 67 (41), *88*, 529 (233), *539*
Chapdelaine, M. J., 186 (903), *211*, 741 (312), *764*
Chapelle, S., 448 (102), *459*, 518 (147), 531 (147, 241), 532 (147), *537, 539*, 550 (79), *587*
Chapman, H. L. Jr., 383 (60), *389*
Chapman, O., 450 (119), *459*
Chapman, O. L., 228, 244 (81), *269*, 314 (113), 325 (145), *336, 337*
Chapmann, O. L., 289, 326 (44), *335*
Chappell, J. S., 489 (132), *494*
Chaslus-Dancla, E., 370 (39), *375*
Chasseau, D., 467, 485, 489 (54), *493*
Chasseur, M.-L., 370 (38), *375*
Chatt, J. 527 (205), *538*
Chau, Y. K., 369 (14), *374*
Chaudhuri, M. K., 527 (200), *538*, 660 (19), *672*
Chavdarian, C. G., 97, 105 (57), *195*
Cheeke, P. R., 381 (42), *389*
Cheeseman, G., 423 (10), 453 (164), *457, 460*
Chen, C.-S., 360 (44), 361 (44, 47, 51), *364*
Chen, D., 383 (68), *390*
Chen, G. C., 603 (35, 36), *652*, 769 (35), *791*
Chen, H. W., 220 (32), *268*
Chen, J., 370 (45), *375*
Chen, M. T., 72 (52), *88*, 511 (99), *536*
Chen, X., 370 (44, 45), *375*
Cheng, C. C., 120 (348, 349), *200*, 778 (89), *792*
Cheng, J. T., 369 (23), *375*
Cherayil, J. D., 263 (249), *272*, 362 (54), *365*
Cherest, M., 697 (292a, 292b), *763*
Chernaya, N. M., 629 (101), *654*
Chernega, A. N., 623, 634, 637 (82), *653*
Chernoplekova, V. A., 519 (155, 156), *537*, 662 (23, 24), *672*
Chernova, A. V., 627, 645 (95), *654*
Cheshire, D. R., 145, 154, 155 (536), *204*
Cheyne, R. M., 68 (34), 73 (56), 78, 79 (34), 86 (90, 91), 87 (34, 56), *88, 89*
Chhabra, B. R., 118 (339), *200*
Chi, P. B., 641 (144–146), *655*
Chia, L. Y., 519 (158), *537*
Chianelli, D., 169 (763), *209*, 409 (121), *419*, 548 (60), *586*
Chiang, L.-Y., 250 (181), *271*, 407 (115), *419*, 466 (3), 467 (24, 47), 469 (47), 474 (72, 74), 480 (90), 489 (3), *491–493*
Chiang, T. C., 489 (129), *494*
Chiaroni, A., 153 (620), *206*
Chiba, M., 387 (151), *391*, 561 (125), *588*
Chidichimo, G., 453 (161, 162), *460*
Chieriu, L., 663 (36), *672*
Chiesi Villa, A., 258 (224), *272*
Chikamatsu, K., 133, 154, 155 (435), 173 (809), *202, 209*, 578, 583 (211), *590*, 755 (322), *764*
Chi Kwong Wong 522, 523, (171), *537*
Ching, W.-M., 262 (247), *272*, 361 (45, 46, 49, 50, 52), 362 (45, 46, 49, 50, 52, 56, 57), 363 (46, 56), *364, 365*
Chittattu, G., 139 (471, 472, 477), 140 (471), 141 (472, 494), 144 (525), 149 (472), 174 (477, 525, 817), 179 (867), *203, 204, 210, 211*, 568 (171), *589*, 665, 666 (52), *672*, 678, 681, 711 (15), 715 (247), *758, 762*, 770 (42, 43, 45), *791*
Chittattu, G. J., 172, 174 (780), *209*, 522, 523 (171), *537*, 568 (169), *589*, 666 (57), *672*, 678 (16, 221), 681 (16), 711 (221), 716 (16), 754, 755 (221), *758, 762*
Chiu, M., 453 (169), *460*
Chlopin, W., 505 (63), *535*
Chmutova, G. A., 531 (248), *539*, 548 (61), 549 (71–73), 550 (78), 573 (190, 191), 574 (192), *586, 587, 589*
Cho, G. J., 382 (48), *389*
Cho, G. L., 382 (46), *389*
Chocat, P., 353 (14, 15), 355 (15), 356 (15, 41), 359 (41), *364*
Choi, J.-K., 145 (539), *204*
Cholpankulova, S. T., 575 (198), *589*
Chottard, J. C., 556 (112), *588*
Chouinard, P. M., 97, 172 (48), *195*

Chow, F., 96 (34, 38), 98, 99, 115, 116 (38), 123 (38, 373), 152 (373), 154, 155 (34, 656), 157 (34, 656, 679), 158 (694), 160 (34, 656, 679), 184, 187 (694), *194, 201, 207*, 521 (169), *537*, 555 (100), *587*, 679, 681 (21), 683–686 (72), 695, 696 (21, 72), 703 (21), 705 (21, 72, 158), 718 (21), 721 (21, 72, 327), 726 (21, 72), 736 (21), 742 (314), *758, 759, 761, 764*
Christensen, B. G., 97 (55), *195*
Christensen, D., 25 (89, 90), *48*
Christiaens, L., 172 (788), *209*, 315, 316 (119), *336*, 372 (79), *376*, 423 (14), 447 (14, 93, 97), 448 (101), 450 (115), *457, 459*, 481, 483 (99), *494*, 500 (38), 509 (91), 531 (247), *535, 536, 539*, 549 (75), 550 (76, 81, 83), *587*, 765 (19), *791*
Christie, R., 456 (111), *459*
Christofferson, G. D., 67 (39, 40), *88*
Chu, J. Y., 228 (78), *269*
Chu, J. Y. C., 276, 282 (6), 283 (29), 285, 286 (36), 328 (6), *334, 335*, 344 (40), *347*, 396, 403 (13), 412 (13, 137), 413 (137, 140), *417, 419*, 503 (47), 508 (81), 526 (191, 197), *535, 536, 538*, 659 (8, 12), 660 (18), *671, 672*
Chu, M.-Y., 386 (115, 116), *391*
Chu, M. Y., 263 (256), *272*, 780 (102), 781 (105), 782 (102, 105, 106), 790 (102), *792*
Chu, S.-H., 230 (87), 263 (250, 256), *269, 272*, 386 (112–116), 388 (162), *390, 391*
Chu, S. H., 778 (95, 98), 780 (103), 781 (104, 105), 782 (103–106), 790 (103, 133), *792, 793*
Chu, S. M., 778 (94), 780, 782, 790 (102), *792*
Chu, S.-W., 263 (252), *272*
Chugaev, L., 505 (63), *535*
Chung, C., 383 (59), *389*
Cisar, A., 512 (107), *536*
Ciufolini, M., 97, 142 (56), *195*
Clardy, J., 108 (217), *198*, 407 (113, 114), *419*, 721 (239), *762*
Clarembeau, M., 100 (124), 154 (124, 631, 659, 660), 155 (124, 659, 660), 157 (124, 659), 173 (797), *196, 206, 207, 209*, 677, 678, 681 (12a, 12b), 684 (78, 326), 685 (336), 686–688 (78), 689 (12a, 12b, 78), 690, 691 (78), 692, 695 (78, 114), 696 (114), 709 (12a, 12b), 710 (12a, 12b, 78, 326), 712 (12a, 12b, 326), 728, 738 (12a, 12b), *757–760, 764*
Claremon, D. A., 116 (285, 286), 129 (286), 139 (474b), 141 (285, 286, 474b), 144 (285, 286), 152 (611, 612), 173 (286, 804), 174 (285), *199, 203, 206, 209*, 427 (25a), 428 (26), *457*, 558 (116), *588*
Clark, C. W., 110 (228), *198*
Clark, E. R., 553, 555 (93), *587*
Clark, G. R., 345 (46), *347*
Clark, L. C., 385 (96), *390*
Clark, M. C., 157, 158 (680), *207*, 684, 695, 729 (321), *764*
Clark, P. A., 26 (95), *48*
Clark, P. D., 154, 155, 157, 159 (636), *206*, 683, 706–709, 726 (70), *759*
Clark, P. S., 408 (104), *419*
Clark, R. D., 99 (103), *196*, 715 (234), *762*
Clausen, J., 387 (150), *391*
Clayton, C. C., 384, 385 (92), *390*
Clayton, E. M., 265 (220), *272*
Clayton, J., 452 (141), *460*
Clementi, E., 6 (22), *47*
Cléophax, J., 153 (620), *206*
Clive, D. L., 398 (32), *417*
Clive, D. L. J., 94 (6, 7), 95, 108 (30), 115, 116 (275), 139 (471, 472, 475, 477), 140 (471), 141 (472, 494), 142 (502, 503), 144 (525), 145 (536, 537, 540, 541), 149 (472), 154, 155 (536), 157 (475), 165 (733), 172 (780), 174 (477, 525, 780, 817), 176 (831, 836, 841a, 841b), 177 (862–864), 179 (867), 184, 186 (899–901), *194, 199, 203, 204, 208–211*, 310 (100), *336*, 356 (24), *364*, 397 (22), *417*, 497, 498, 516 (3, 5), 520 (3, 5, 165), 522, 523 (171), *534, 537*, 555, 560 (97), 568 (97, 169, 171), *587, 589*, 630, 640 (105), 651 (105, 198), 652 (199), *654, 656*, 660 (15), 663 (32, 33), 665 (52), 666 (52, 57, 61), *672*, 677 (4), 678 (15, 16, 221), 679 (4), 681 (4, 15, 16), 685 (4, 98a, 98b), 711 (15, 221), 713, 714 (4), 715 (247), 716 (16), 721 (285, 287, 289), 754, 755 (221), 757 (4), *757–759, 762, 763*, 765 (4, 5), 770 (42–45), 790 (5), *790, 791*
Cobbledick, R., 455 (193), *461*
Cobbledick, R. E., 342 (24), *347*, 505, 511, 512 (60), *535*
Coch, E. H., 351 (9), *364*
Codding, P. W., 646 (172), *655*
Cody, R., 453 (187a), *461*
Coe, D. E., 126 (394), *201*
Coffey, J. L., 372 (74), *376*, 388 (164, 166), *392*
Coghlan, M. J., 109, 184, 186 (221), *198*
Cohen, H. J., 379 (16), *389*
Cohen, J. M., 467 (10), *491*
Cohen, M. L., 172 (792), 182, 184 (880), *209, 211*, 340 (5), *346*, 408 (104), *419*, 685, 695 (85), 706 (176), *759, 761*
Cohen, S. C., 498 (15), *534*
Cohen, T. J., 711 (216b), *762*
Cohen, V., 451 (139b), *460*
Cohen, V. I., 228 (77), 229, 233 (83), 234 (96), 235 (77), 238 (77, 83), 255 (207), (219), *269, 271, 272*
Cole, G., 221 (40), *268*
Coleman, L. B., 467 (10), *491*
Coll, J., 118 (340), *200*

Collard-Charon, C., 229 (82), 245 (142), 246 (142, 151), 247, 248 (158), 255 (206), *269–271*
Collienne, R., 230 (86), *269*
Collin, J. E., 4, 27, 28 (13), *47*
Collins, S., 98 (70, 71), 100 (110–112), 109 (222), 126 (111), 136, 137 (70, 71), 138 (70, 110–112, 222), 149 (567, 574, 575), 150 (567), 162 (714), 174 (112), 180 (112, 222), *195, 196, 198, 205, 208*, (265), *763*
Colona, F. P., 36 (136), *49*, 327 (154), *337*
Colonna, F. P., 21–25 (74), 30 (110), 33 (74), 34, 35 (124), *48, 49*
Colquhoun, I., 612–614 (57), *653*
Colquhoun, I. J., 643 (158), 649 (187), (157), *655, 656*
Colter, M. A., 134 (448), *202*
Colton, R. J., 15 (60), *47*
Colvin, E. W., 681, 682 (45), *758*
Colwell, R., 369 (12), *374*
Coman, M., 121 (363), *201*
Comasseto, J. V., 103 (152), 114 (263, 264, 269), 126 (269, 388), 141 (499), 154 (665), 167 (754), 169 (152, 263, 269), 174 (499), 189 (263, 665), *197, 199, 201, 203, 207, 208*, 406, 414, 416 (83), *418*, 501 (39), 518 (150), *535, 537*, 662 (28), *672*, 706, 720, 727 (324), *764*, 788 (128), *793*
Comassetto, J. V., 683, 696, 706, 727 (101, 102), 736 (101), *759*
Combrink, W., 151 (594, 596), *205*
Combs, G. F., 380 (28, 29), 381 (45), *389*
Comrie, A. M., 248 (167), *270*
Congeri, M., 453 (161), *460*
Conner, M. W., 385 (128), *391*
Connington, P. H., 5 (17), *47*
Connor, B. R. O., 477 (83), *493*
Connor, J., 281, 318 (27), *334*
Cook, J. M., 110 (236), 118 (236, 334), 119, 148 (236, 334, 343), *198, 200*
Cook, P. D., 374 (98), *376*, 386 (124), *391*, 786 (118), *793*
Cooke, F., 181 (878), *211*
Coon, J. B., 11 (51), *47*
Cooney, D. A., 386 (123), *391*
Cooper, J. R., 467, 469 (23), *492*
Cooper, W., 217 (9), *267*, 423 (5, 7), *456*
Cooper, W. C., 217 (2), *267*, 394 (2), *416*, 497, 498, 501, 516 (1), *534*, 593, 610, 628, 634, 637, 648, 650 (3), *652*, 788 (124), *793*
Coppinger, G., 423 (3), *456*
Coppola, J. C., 614, 615 (61, 62), 646 (62), *653*
Copps, T., 467, 483 (50), *493*
Corbett, J. D., 512 (107), *536*
Corbio, B., 370 (38), *375*
Cordes, A. W., 266 (282), *273*
Corey, E. J., 139 (476), 174 (816), 184 (888), *203, 210, 211*, 666 (56), *672*, 710 (203), (296), *761, 763*
Corey, E. R., 40 (155), *50*
Corey, E. Y., 715 (246), *762*
Cornford, A. B., 15 (58), *47*
Correia, M. A., 381 (43), 382 (55), *389*
Corriu, R., 711 (218), *762*
Cossey, A. L., 104 (155), *197*
Costello, S. M., 388 (163), *392*
Cottam, H. B., 787 (120), *793*
Cotton, F. A., 356 (23), *364*
Couch, D. A., 72 (51), *88*
Coulon, C., 467 (31), *492*
Coulston, K. J., 106 (198), *197*
Couture, A., 165 (740), *208*, 284 (32, 33), *335*, 396 (14), *417*, 527 (199), *538*, 660 (14), *672*
Cowan, D., 407 (115, 117), *419*, 466, 467 (1), 480 (97), 483 (1), 485 (97, 111), *491, 494*, 506 (68), *535*
Cowan, D. O., 41 (161, 163), *50*, 249 (175), 250 (179, 181, 182), 251 (182), *271*, 341 (19), *347*, 407 (116), *419*, 466 (2, 3), 467 (2, 7–9, 11, 13, 14, 21–24, 47), 469 (22, 23, 47), 471 (93), 473 (66), 474 (66, 70, 72, 74), 480 (89–91, 93, 98), 483 (103), 485 (98), 489 (3, 132), *491–494*, 504 (54), *535*
Cowley, A. H., 42 (164, 165), *50*
Cox, E. R., 369 (15), *374*
Cox, S. D., 341, 345 (18), *346*
Coyle, J. D., 317 (126), *336*
Crabtree, G. W., 467 (50), 483 (50, 108), *493, 494*
Cradock, S., 10, 12 (47, 48), 21 (70, 71), (107), *47–49*
Cram, D. J., 681 (29, 32), 682 (32), 697 (291a, 291b), *758, 763*
Cravadoi, A., 685 (336), *764*
Cravador, A., 126 (392), 153 (621, 622), 154 (621, 622, 631), 155 (621, 622), 172 (783), *201, 206, 209*, 520 (161), *537*, 681 (25, 68), 683, 684 (68), 685 (68, 96, 99), 686 (68), 695 (128), 696, 699, 708 (68), 716 (266), 718 (280), 720 (128, 266), 721 (25), 726 (68), 748, 749 (266), *758–760, 763*
Crawford, I. P., 358 (38), *364*
Creary, X., 294 (62, 63), *335*
Credali, L., 221 (47–50), *268*
Crich, D., 99, 102 (102), 120 (345), 132, 163, 191 (102), *196, 200*, 322 (138), 323 (139), *337*
Criddle, R. S., 355 (22), *364*
Crimmin, M. J., 135 (455), *203*
Cripe, W. S., 383 (60), *389*
Crisponi, G., 519 (153), *537*
Cristofferson, G. D., 344 (39a), *347*
Cristol, S. J., 312 (109–111), *336*
Crofts, J. C., 25 (92), *48*

Croisy, A., 500 (38), *535*, 549 (75), *587*
Cross, C. E., 380 (23, 27), *389*
Cross, R. J., 283 (31), *335*, 412 (138), 413 (138, 141, 142), *419*, 526 (195, 196), 527 (198), *538*, 634, 635 (114), 640 (114, 138), 641 (114), *654, 655*, 659 (10, 11), 660 (13), *671*
Crouse, G. D., 98 (75, 76), 108 (214), 137 (75, 76), 157, 184, 186 (214), *195, 198*
Crowley, C., 370 (47), *375*
Culbertson, E., 395 (6, 7), *417*
Culen, E. R., 219, 263, 264 (18), *268*
Cullen, E. R., 167 (749–752), *208*, 219 (21), 222 (21, 54, 55), 223 (21), 225 (21, 61), 263, 264 (260), *268, 269, 272*, 293 (55), *335*, 398 (29, 31), *417*
Cullen, W. R., 597 (17, 18), *652*
Cummings, J. M., 528 (219), *538*
Cunningham, E. B., 371 (65), *376*
Curini, M., 190 (940), *212*, 519 (157), *537*
Curran, D. P., 141 (501), *203*
Current, S., 140 (489), *203*, 770 (39), *791*
Curtis, N. J., 139, 140 (471), 172, 174 (780), 179 (867), *203, 209, 211*, 522, 523 (171), *537*, 568 (169, 171), *589*, 666 (57), *672*, 678 (16, 221), 681 (16), 711 (221), 715 (247), 716 (16), 754, 755 (221), *758, 762*, 770 (42), *791*
Cussans, N. J., 153 (617–619, 623, 624), *206*
Cuthbertson, E., 165 (735, 741, 742), *208*, 396 (21), *417*
Cutney, J. P., 563 (132), *588*
Czarny, M. R., 147 (556, 560), *205*
Czauderna, M., 403 (66), *418*
Czyzewska-Chlebny, J., 371 (53), *375*, 631 (106), *654*, 788 (129), *793*

Daggett, J. U., 565 (159), *588*
Dahl, B. M., 253 (198), *271*
Dahl, L. F., 345 (50), *347*
Dahn, D. J., 469 (58), *493*
Dailey, B. P., 59, 60, 64 (22), *88*
Dallacker, F., 247 (163), *270*, 585 (228), *590*
D'Ambra, T. E., 561 (123), *588*
Danby, C. J., 10, 13 (50), *47*
Dance, N. S., 62–65, 67, 70, 71, 74 (25), 75 (57, 58), 76 (57, 59), 77–79 (25), 80, 81 (61), 83 (61, 78, 79, 81), 84 (79, 81, 82), *88, 89*, 517 (146), 518, 522 (148), 529 (146, 231), 531 (148), 533 (262, 266), *537, 539*
d'Angelo, J., 106 (185), *197*
Daniel, J. R., 386 (120), *391*, 543, 554 (22), *585*, 603 (35), *652*, 768 (30), 769 (34), 790 (30, 34, 135), *791, 793*
Danieli, B., 174 (813), *210*
Danishefsky, S., 97 (56), 108 (210, 211, 215–217), 142 (56, 504), 168 (504), *195, 198, 204*, 428 (25d), *457*
D'Antonio, P., 35 (133), *49*
D'Antuono, J., 118 (337), *200*
Dapporto, P., 454 (192), *461*
Darmadi, A., 595 (10), 605, 606, 608–610 (47), *652, 653*
Das, R. P., 387 (148), *391*
Daupbaise, D., 504 (53), *535*
Dauplaise, D., 407 (113, 114), *419*
Dauplase, D., 36, 37 (138), *49*
Davalian, D., 450 (116), *459*
Davalion, D., 243 (134), *270*
Daves, G. D., 778 (89), *792*
David, S., (238), *762*, 770 (46, 47, 49), *791*
Davidson, D. D., 386 (114), *391*, 778 (98), *792*
Davies, D. I., 312 (110), *336*
Davies, I., 83 (78, 81), 84 (81), *89*, 528 (221), 529 (231), *538, 539*
Davies, M. H., 381 (39), *389*
Davies, N. R., 527 (205), *538*
Davis, F. A., 99, 180 (88, 89), *195*, 426 (18), 429, 430, 436 (31), *457*
Davis, J. N., 522 (177a), *537*
Davis, L., 357–359 (34), *364*
Davis, M., 449 (108, 109), 453 (160), *459, 460*
Davis, M. A., 388 (163), *392*
Davison, A., 260 (235), *272*
Davoud, M., 486 (123, 124), 487 (124), *494*
Dawson, J., 453 (173), *460*
Dawson, W. H., 340 (9), *346*
Deacon, G. B., 498 (17), *534*
Deady, L., 453 (160), *460*
De Alti, G., 24 (83), *48*
Dean, F. M., 715 (253), *762*
Dean, P. A. W., 651 (195), *656*
De Checchi, C., 221 (48–50), *268*
Dechend, F., von 234 (94), *269*
Decleva, P., 24 (83), *48*
Dehe, D., 467, 485, 489 (53), *493*
Dehnert, P., 597, 602–604, 643 (15), 644 (161), 647 (178), *652, 655, 656*
Deicha, C., 453 (172), *460*
DeJong, R., 235, 243 (99), *269*
Delaini, F., 379 (20), *389*
Delhaes, P., 467 (31), *492*
Delhalle, J., 456 (203), *461*
Delplano, P., 519 (153), *537*
Delwiche, J., 4, 27, 28 (13), *47*
DeMarco, C., 351 (11), *364*
De Mark, M. R., Van 412 (136), *419*
De Meio, R. H., 368 (3–5), 370 (48, 49), *374, 375*
De Munno, A., 446 (81), 448, 452 (100), 453 (161), 454 (81, 190–192), *458–461*
DeMunno, A., 400 (48), *417*
Demuth, R., 644 (161), *655*
Dendel, J., 448 (101), *459*
Deniau, J., 304, 305 (86), *336*, 669 (69), *673*
Denis, J. N., 97 (59), 100 (126), 116 (59), 126

(391), 132, (391, 431), 154, 155 (644), 157 (391), 685, 686), 159 (391, 431), 160 (431), 163 (724), 175 (824), 186 (904), *195, 196, 201, 202, 206–208, 210, 211*, 679 (334, 335), 681 (25), 694 (121, 124), 695 (121, 136), 696 (187), 705 (159, 160), 706, 707 (121), 708 (124, 185, 187), 709 (121, 185), 710 ((207, 209), 711 (185), 712 (185, 223), 713 (185), 715 (136, 251), 716 (159, 160), 718 (251), 720 (124), 721 (25, 251, 283), 722 (185), 723 (187), 726 (330), 727 (124, 187), 733 (121), 735 (124, 185, 187), 736 (187), 737 (124, 185, 187), 747, 749 (121), *758, 760–764*
Dennes, G., 87 (92), *89*
Denney, D. B., 531 (242), *539*
Denney, D. Z., 531 (242), *539*
Dennis, J. N., 520 (161), *537*
Dennis, R., 453 (187a), *461*
Denniston, M. L., 297 (73), *335*
Denoel, J., 550 (76), 577 (209), *587, 590*
Densel, J., 531 (247), *539*
Dent Glasser, L. S., 646, 650 (173), *655*
Denton, D. A., 383 (60), *389*
Denyer, C. A., 176 (836), *210*
Denyer, C. V., 398 (32), *417*, 651 (198), *656*
Dereu, N., 384 (74), *390*, 506 (71), 512 (109), 531, 532 (245), *535, 536, 539*
Dereu, N. L. M., 396 (11), *417*
Derkach, N., 455 (194–197), *461*
Derkach, N. Ya., 428 (28), 430 (34), 431 (38), 432 (34, 38), 433 (34), 435, 436 (43–45), 437 (47–49), 438 (52, 53), 439 (55, 57), 440 (59, 61), 441 (49), 443 (57), (58), *457, 458*
Derrick, P. J., 22, 25 (75, 76), *48*
Deryabina, L. A., 260 (236), *272*
Desai, S. R., 118 (324), *200*
Desauvage, S., 157 (685), *207*, 696, 708, 723, 727, 735–737 (187), *761*
Desbene, P. L., 325 (146, 147), 326 (147), *337*, 525 (187), *538*
De Silva, K. G. K., 85 (85), *89*, 506, 518, 522, 529 (72), *535*
Dessy, G., 267 (287), *273*
DeTitta, G. T., 101 (130), *196*
Detty, M. R., 99, 116 (84), 134 (446, 447), 144 (84), 169, 172 (772), 177 (865, 866), 187 (446, 772, 914), *195, 202, 209, 211, 212*, 221 (37), 262 (243), *268, 272*, 341 (17), *346*, 408 (94), *418*, 487 (126–128), 488 (127), *494*, 662 (25), *672*
De Wames, R. E., 11 (51), *47*
Dewan, J. C., 81 (66–70), 82 (70), *88*, 515 (128, 131), *536*
Dewar, M. J. S., 6 (28–32), *47*
Dewar, P. S., 32, 33, 36 (115), *49*
Déziel, R., 143 (519), *204*
Dhaliwal, D. S., 597 (17, 18), *652*
Dhawan, S., 449 (110), *459*
Dhawan, S. N., 151 (601), *205*
Dickson, D. P. E., 52, 60 (8), *87*
Dideberg, O., 62, 65 (33), *88*, 342 (26), 345 (44), *347*, 517 (135, 136), *537*
Diebeler, V. N., 27 (106), *49*
Dieck, H. tom, 42 (166, 167), 43 (166), 44 (166, 167), *50*, 645 (164), *655*
Diehl, D. R., 113 (262), *199*
Diemert, K., 638 (130), *655*
Dietrich, C. O., 127 (396), *201*, 438 (51), *458*
Dietschmann, H., 544, 575 (25), *586*
Dietze, F., 575 (197), *589*
Di Giamberardino, T., 723, 740 (290), *763*
DiGiamberardino, T., 154–156 (652), *206*
Dingwall, D., 248 (167), *270*
Dini, G., 378 (3), *388*
Dion, R. L., 386 (123), *391*
DiPerro, M., 119, 148 (343), *200*
Dirk, C. W., 512, 530 (106), *536*
DiSalvo, F. J., 467, 469 (43), *492*
Distefano, G., 6 (36), 21, 22 (72, 74), 23 (72, 74, 81), 24 (72, 74, 80, 81), 25 (72, 74, 81, 85), 28 (108), 29 (80, 108), 30 (80, 110), 33 (74, 119, 121), 34, 35 (124–126), 36 (136), *47–49*, 327 (154), *337*
Ditter, D. C., 542, 571 (13), *585*
Dittrich, G., 544 (24), *586*
Divakar, M. C., 373 (89), *376*
Dixon, K. R., 532 (257), *539*
Dixon, R. N., 6 (37), *47*
Dixon, W. B., 25 (89), *48*
Djerassi, C., 151 (604), *205*
Doak, G. O., 642 (149), *655*
Dobosh, P. A., 6 (27), *47*
Dobrowolski, P. J., 161 (705), *207*
Dobson, G. R., 527 (207), *538*
Dobud, P., 75, 76 (57), *88*
Doherty, A. M., 140 (491), *203*
Dolak, T. M., 710 (206), *762*
Dolenko, G. N., 416 (157), *420*
Dolle, R. E., 152 (612), *206*
Dom, W. L., 519 (152), *537*
Dombsky, M., 87 (93), *89*
Domcke, W., 39, 40 (156), *50*
Donaldson, J. D., 82 (71–76), 87 (92), *88, 89*
Donati, M. B., 379 (20), *389*
Donohue, J., 345 (51), *347*
Dopper, J. H., 315 (120), 316 (120–122), *336*
Dorfman, J. R., 341 (15), *346*
Dorn, W. L., 505, 507 (59), *535*
Doroshkina, G. M., 627, 645 (95), *654*
Dostal, K., 424 (17), 429 (29, 30), 431 (29), 450 (17), *457*
Doughty, P. F., 370 (48), *375*
Douglass, I. B., 246 (143), *270*
Dow, W. C., 565 (152), *588*

Dowbenko, R., 312 (105), *336*
Dowd, P., 169 (769), *209*
Downs, A. W., 580 (218), *590*
Drabowicz, J., 153 (615, 616), *206*
Dräger, M., 248 (169, 170), *271*
Drago, R. S., 529 (230), *539*
Drake, J. E., 220 (28), *268*, 340 (2c), *346*, 527, 529 (213), *538*, 596, 602, 605–607, 609, 648 (13), *652*
Draper, H. H., 382 (52), *389*
Dreier, C., 237 (105), 266 (278), *269, 273*
Drew, H. D. K., 66 (37), *88*, 508 (80), 512 (105), *536*
Drimal, J., 378 (2), *388*
Driscoll, J. S., 116 (287), *199*, 715 (252), *762*
Drofenik, M., 266 (280), *273*
Droste, D., 453 (178), *460*
Droste-Tran-Viet, D., 452 (147), *460*
Dru, D., 369 (26), *375*
Drucker, G. E., 108 (207), *198*, 685 (86), *759*
Dryburgh, J. S., 596, 601, 611, 612, 643 (14), *652*
DuBois, K. P., 387 (141), *391*
Duckett, S., 369 (24, 25, 27), 370 (35, 36), *375*
Due, M., 254 (203), *271*
Duff, J. M., 711 (214), *762*
Duggan, A. J., 565 (153), *588*
Dugger, D. L., 497, 506 (14), *534*
Duke, E., 370 (43), *375*
Dumont, W., 99 (83), 100 (125), 125 (384, 386), 126 (390), 132 (428), 153 (622), 154 (384, 386, 622, 631, 640–644, 647, 649, 650, 652, 662, 663, 673, 677), 155 (384, 386, 622, 640–644, 647, 649, 650, 652), 156 (384, 386, 641, 643, 647, 650, 652, 662, 678), 157 (384, 650, 681, 686), 161 (708), 163 (724), 171 (681), *195, 196, 201, 202, 206–208*, 520 (161), 521 (168), 523 (179), *537, 538*, 677 (11), 678 (11, 19), 679 (11, 17, 19, 334, 335), 680 (17), 681 (11, 17, 19, 25), 683, 684 (74), 685 (11, 19, 74, 94, 95, 336), 686 (11, 74, 94, 100), 689 (11, 94, 111), 690, 691 (111), 693 (118), 694 (123), 695 (11, 74, 94, 118, 128, 131, 132, 136, 137, 141–143), 696 (74, 149), 698 (94, 155), 699 (11, 74, 94, 131), 700 (155), 701 (149), 702 (149, 156), 703, 705 (19), 706 (11, 19, 74, 174, 181), 710 (207–211), 711 (11, 143, 211), 712 (17), 714 (123, 232), 715 (11, 123, 132, 136, 232), 716 (74, 266), 718 (19, 94, 137), 719 (11, 19, 94, 137), 720 (74, 123, 128, 266), 721 (11, 25, 100, 131), 722 (100, 131), 723 (290), 725 (340), 726 (11, 19, 94), 727 (123, 181), 732 (232), 734, 736 (100), 740 (11, 290), 741 (174), 747 (17), 748 (74, 266), 749 (266), 753 (74), (146), *757–764*
Du Mont, W. W., 511 (101), *536*
duMont, W.-W., 400 (49, 50, 52), *417, 418*
Dunathan, H. C., 356 (28), *364*
Dunbar, F. X., 228, 229 (75), *269*
Duncan, J. L., 533 (265), *539*
Duncan, W., 10, 12 (47, 48), *47*
Dung, J.-S., 99 (107), 184, 186 (898), *196, 211*
Dung, J. S., 721 (288), *763*
Dunkin, I., 291, 292 (51), *335*, 453 (159), *460*
Dunlap, R. B., 555 (105), *587*
Duong, K. N. V., 304, 305 (86), *336*, 669 (69), *673*
Dupois, J., 772 (63–66), *792*
DuPont, L., 342 (26), *347*
Dupont, L., 62, 65 (33), *88*, 345 (44), *347*, 454 (189), *461*, 515, 516 (124), 517 (135, 136), *536, 537*
Dürner, G., 102 (139), *196*
Dürr, H., 276 (3), *334*
Durrant, M. L., 106 (195), *197*
Dutta, S. R., 123 (367–369), *201*
D'yachenko, S., 453 (186b), *461*
Dyer, G., 528 (225), *538*
Dyke, J. M., 7 (40), 15 (57), *47*
Dzhunaev, I. A., 407 (112), *419*

Easton, D. B. J., 477 (82), *493*
Eastwood, F. W., 106 (198), *197*
Eastwood, M. A., 388 (170), *392*
Eaton, P. E., 99 (95, 106), 106 (106), *195, 196*
Ebsworth, E. A., 246 (146), *270*
Ebsworth, E. A. V., 21 (70), *48*, 596, 601, 611, 612, 643 (14), *652*
Echter, T., 151 (597), *205*, 450 (124), *459*
Eck, V., 38 (148), *50*
Eckert-Maksić, M., 645 (165), *655*
Eckert-Maksic, M., 44, 45 (169), *50*
Edqvist, O., 6, 7 (38), 22, 25 (75, 76), *47, 48*
Edwards, A. J., 67, 70, 71 (43), *88*
Edwards, M. P., 152 (613, 614), 153 (614), *206*
Edwards, P. D., 113 (258), *199*
Efros, L., 453 (153), *460*
Efros, L. S., 503 (51), *535*
Egdell, R. G., 2 (7), *46*
Eggert, H., 402 (62), 403, 404 (68), 415 (62), *418*
Egli, H., 118 (332), *200*
Ehrbar, U., 574 (193), *589*
Ehrbor, J., 246 (150), *270*
Eian, G. L., 325 (145), *337*
Eid, A., 451 (138), *460*
Einbrodt, H. J., 368, 369 (9), *374*
Einstein, F. W. B., 67 (42), 83 (80), *88, 89*, 342 (24), *347*, 505, 511, 512 (60), 529 (232), *535, 539*
Eisch, J. J., 710 (195), *761*
Eismont, M. Yu., 35 (127, 128), *49*
Ejmocki, Z., 544 (26), *586*
Ekelund-Price, J., 371 (54), *375*
Eklund, N., 279 (18), *334*, 527 (201), *538*

Eland, J. H., 26 (94), *48*
Eland, J. H. D., 2 (9), 10, 13 (50), *46, 47*
Elbel, S., 42 (166, 167), 43 (166), 44 (166, 167), *50*, 645 (164), *655*
El Bouz, M., 698, 700 (155), *760*
Elgy, C., 507 (78), *535*
Elhafez, F. A., Abd. 697 (291a), *763*
Eliason, R., 542 (17), *585*
Ellermann, J., 401 (54, 56), *418*, 619, 620, (71–73), 621 (74), *653*
Ellis, P. D., 340 (9), *346*
Elmaleh, D., 371, 372 (61), *375*
Elmaleh, D. R., 371 (57, 66, 68, 69), *375, 376*
Elmes, P. S., 72 (51), *88*
Elsayed, N. M., 380 (24), *389*
El'tsov, A. V., 297 (67), *335*, 544 (28, 29), *586*
El'tsov, V., 258 (226), *272*
Eman, A., 154 (642, 662), 155 (642), 156 (662), *206, 207*, 695 (141), 710 (207, 208), *760, 762*
Emeleus, H. J., 595, 602 (9), *652*
Emge, T., 467, 483 (51), *493*
Emge, T. J., 467, 483 (50), *493*
Emma, T., 497, 506 (14), *534*
Ende, D. V., 542, 552 (7), *585*
Enders, D., 113 (256), *199*, 681, 682 (41), *758*
Endres, H., 485, 489 (113), *494*
Engel, P., 99, 106 (106), *196*
Engler, E. M., 41 (161, 162), *50*, 249 (177), 251 (177, 184), 252 (188, 189), *271*, 408 (88, 89), *418*, 467 (17, 18, 20, 40), 469 (40, 59, 110), 472 (63, 65), 473 (67), 474, 478 (77), 483 (107, 109, 110), *492–494*
Engman, J. L., 95, 129 (22), *194*
Engman, L., 33 (121), *49*, 116 (305), 120 (352, 353), 122 (353), 124 (305), 129 (412), 130 (415–417), 139 (483), 147 (549), 149, 150 (572), 152 (549), 164 (727), 165 (417), 174 (415, 483), 175 (829, 830), 176 (832), 191 (964), 193 (969), *200–205, 208, 210, 213*, 245 (139, 140), *270*, 341 (17), *346*, 407 (111), 408 (92), 409 (122), *418, 419*, 510 (92, 96, 97), 522 (176), *536, 537*, 579 (215), 582, 583 (215, 225), *590*, 660 (17), 664 (39), *672*
Enkelman, V., 485, 489 (114), *494*
Enoki, T., 467 (25), 483 (106), *492, 494*
Ensley, H., 107 (199), *198*
Enstein, F., 455 (193), *461*
Entenmann, G., 547, 551 (51), *586*
Epp, D., 350 (1), *363*
Epstein, A. J., 36, 37 (137), *49*, 251, 252 (186), *271*, 486 (121), *494*
Erashko, V. I., 113 (252), *199*
Eremeeva, G., 453 (153), *460*
Erickson, N. E., 53, 54, 62 (16), *87*
Eriksen, T. E., 279 (18), *334*, 527 (201), *538*
Erlanger, M. W., 379 (8), *388*
Erle, H. E., 576 (202), *589*
Ermann, P., 118 (322), *200*
Ermolaeva, L. V., 574 (192), *589*
Ermolenko, M. S., 163 (715, 716), *208*
Ernstbrunner, E., 32, 33, 36 (115), *49*, 705 (163), *761*
Esaki, N., 351, 352 (10), 353 (10, 13–15), 355 (15, 21), 356 (15, 21, 31–33, 39–41), 357 (31–33, 36), 358 (39), 359 (41), *364*, 555 (109), *587*
Esperås, S., 404 (72), *418*, 579, 580, 582 (216), *590*, 650 (191), *656*
Esperas, S., 343 (38), *347*
Es-Seddiki, S., 486 (123, 124), 487 (124), *494*
Etemad, S., 249, 251 (177), *271*, 467 (16, 18–20), *492*
Etheredge, S. J., 97 (56), 108 (210, 211), 142 (56), *195, 198*
Etschenberg, E., 447 (94, 95), *459*
Evanega, G. R., 317 (123), *336*
Evans, S., 2 (4), *46*
Evers, M., 172 (788), *209*, 447 (97), *459*
Evrard, G., 679 (334), *764*
Exon, J. H., 384, 387 (88), *390*
Eyley, S. C., 129 (414), *202*

Fabvey-Garot, N., 284 (33), *335*
Facchnetti, T., 379 (20), *389*
Fackler, J. P., 220 (32), *268*
Fackler, J. P. Jr., 550 (77), *587*
Fadley, C. S., 5, 8 (15), *47*
Faehl, L. G., 152, 154 (609), *206*
Fagerlind, L., 625, 628, 652 (91), *654*
Fahrbach, G., 430, 432 (33), 441, 442 (63), *457, 458*
Fainzil'berg, A. A., 113 (252), *199*
Falcone, J. S., 582 (226), *590*
Falcone, S. J., 184 (891), *211*, 342 (23), *347*, 510 (93), *536*, 577, 578, 582, 583 (208), *589*
Fallon, J. T., 379 (16), *389*
Fanaii, G., 450 (128), *459*
Fanconi, F., 378 (3), *388*
Fandlova, M., 542 (5, 6), *585*
Fanghänel, E., 472, 477 (64), *493*
Fanghångel, E., 408 (90), *418*
Fankhauser, J. E., 127 (398–400, 402), *201*
Faraone, F., 527 (208), *538*
Farina, V., 142 (503), 172, 174 (780), *204, 209*, 522, 523 (171), *537*, 568 (169), *589*, 666 (57), *672*, 678 (16, 221), 681 (16), 711 (221), 716 (16), 754, 755 (221), *758, 762*
Farnier, M., 26, 27 (104), *49*
Farrar, W. V., 512, 519 (104), *536*
Fateley, W., 453 (186a), *461*
Fauconnier, A., 446 (82), *458*
Fausel, M., 374 (93), *376*, 384 (76), *390*
Fayos, J., 721 (239), *762*
Fazio, R., 131, 139 (422), *202*

Febray-Garot, N., 527 (199), *538*
Febvay-Garot, N., 660 (14), *672*
Fehler, F., 220 (24, 27), *268*
Fehlner, T. P., 9, 10 (45), *47*
Feinendegen, L. E., 379 (13), *388*
Feiring, A. E., 135 (451, 452), *202*, 499 (26), *534*, 706 (172), *761*
Fekih, A., 178 (855, 859), *210*
Felbert, G., 247 (161), 256 (214), *270, 271*
Feldmann, J., 771 (58), *791*
Felix 165 (740), *208*
Felix, G., 284 (32), *335*, 396 (14), *417*
Felkin, H., 697 (292a, 292b), *763*
Ferguson, J. E., 72 (51), *88*
Ferrans, V. J., 369 (22), 370 (22, 37), *375*, 379 (18, 19), *389*
Ferraris, J., 467 (8), *491*
Ferraris, J. P., 41 (163), *50*, 467 (7, 9, 11, 14), *491, 492*
Ferraz, H. M. C., 160 (702, 703), *207*, 685, 695 (90), *759*
Ferreira, D., 127, 174 (395), *201*
Ferreira, J. T. B., 518 (150), *537*, 662 (28), *672*
Ferreira, T. W., 162 (710), *207*
Ferren, L. A., 371 (58, 59, 63, 64), *372* (58, 64), 373 (81), *375, 376*, 388 (166), *392*
Ferroris, J. P., 529 (235), *539*
Fetizon, M., 102, 122 (143), *196*
Feuerstein, S., 379 (20), *389*
Feustel, M., 131 (424), *202*
Fichtner, M. W., 341 (17), *346*
Ficini, J., 106 (185), *197*
Fidler, J. W., 382 (51), *389*
Filer, C. N., 131, 139 (422), *202*
Findlay, R. H., 24 (82), 26 (102), *48*
Finkenbine, J. R., 176 (837), *210*, 651 (197), *656*
Firestone, M. A., 483 (108), *494*
Fischer, E. O., 410 (128), *419*
Fischer, H., 259 (232–234), 260 (232), *272*, 410 (127), *419*, 447 (92), *459*, 572, 577 (184), *589*, 772 (66), *792*
Fischer, J. W., 131 (419–421), *202*
Fishbein, L., 368 (2), *374*
Fisher, A. P., III 251, 252 (186), *271*
Fisher, E. O., 229 (85), *269*
Fissi, A., 400 (48), *417*
Fitjer, L., 698, 699 (151), *760*
Fitzhugh, O. G., 384 (89), *390*
Fitzner, J. N., 127 (399–402), *201*
Flandrois, S., 467 (31), *492*
Flannagan, N., 32, 33, 36 (116), *49*, 327 (150), *337*
Flats, J., 498 (22), *534*
Fleming, R. W., 369 (16), *374*
Fleming, S. A., 187 (928), *212*
Fleming, W. P., 106 (194), *197*
Flippora, T. M., 507 (74), *535*
Flood, L. A., 121, 123, 124, 147 (361), *201*
Floss, H. G., 356 (25), *364*
Flynn, G. A., 560 (121), *588*
Fobare, W. F., 145 (535), *204*, 666 (58), *672*
Foltz, C. M., 380 (31), *389*
Fookes, C. J. R., 104 (166), *197*
Forchioni, A., 532 (254), *539*
Förster, M., 641 (148), *655*
Fort, R. C., 298 (75), *335*
Fortier, S., 101 (130), *196*
Foss, O., 342, 343 (27a-c, 27g), *347*, 404 (76), 415 (150, 151), *418, 420*, (62), *458*, 504, 514 (56), *535*, 630 (103, 104), 640 (103), *654*
Foss, V. L., 612 (55), *653*
Foster, D. C., 408 (103), *419*
Foster, D. G., 340 (7), *346*
Foster, S. J., 111 (244), *199*
Fourney, J. L., 773 (70), *792*
Fourrey, J.-L., 150 (580), *205*, 320 (132), *336*
Fourrey, J. L., 524 (186), *538*, 706, 720 (325), *764*
Foutanillas Val, J. A., 518 (150), *537*
Fox, W. B., 597 (17), *652*
Foxman, B. M., 341 (16), *346*, 485 (116, 118), *494*, 497 (12), 501 (12, 43), 506 (12), 514 (43), 516 (134), 517 (134, 138), 530 (240), *534, 535, 537, 539*
Fragale, C., 650 (193), *656*
Francisco, C. G., 109 (227), 183 (885), *198, 211*, 715 (248), *762*
Franck, U., 641 (147), *655*
Frandsen, E. G., 237 (105), 266 (278), *269, 273*
Franghanel, F., 455 (198), *461*
Franklin, W. J., 246 (153), *270*, 549 (64), 585 (229), *586, 590*
Franzi, R., 313 (112), *336*, 659 (7), 665 (51), 671, 672
Fraser, F. J., 378 (5), *388*
Fredericksen, E., 247, 255 (156), *270*
Fredericksen, P. A., 249 (173), *271*
Frederiksen, P. A., 254 (201), *271*
Freedman, L. D., 642 (149), *655*
Freeman, B. H., 753 (295a, 295b), *763*
Freeman, F., 772 (61), *791*
Frei, B., 187 (925), *212*
Freire, R., 183 (885), *211*
Frejd, T., 184 (886), *211*, 671 (82), *673*
French, K., 23–25 (81), *48*
Frenzel, H., 767 (26), *791*
Freund, M., 547, 554 (52), *586*
Fridh, C., 6 (33), *47*
Friedman, L., 312 (106), *336*
Friedrich, L. E., 564 (135), *588*
Friend, R. H., 467 (27), *492*
Frimer, A. A., 328 (159), *337*
Fringuelli, F., 21–23 (72–74), 24 (72, 74), 25 (72–74), 33 (74), *48*, 265 (273), *273*
Fringuelli, R., 190 (940), *212*, 519 (157), *537*

Fripiat, J., 456 (203), *461*
Fristad, W. E., (240), *762*
Fritsch, N., 108 (217), *198*
Frolov, A. N., 297 (67), *335*, 544 (28, 29), *586*
Frolow, F., 145 (538b), *204*, 372 (80), *376*
Frost, C. F., 385, 386 (103), *390*
Frost, D. C., 10–12 (46), 15 (58), 17, 18 (67), *47, 48*
Frost, D. V., 384 (91), 385 (93), *390*
Fry, D. W., 386 (124), *391*
Fuccello, A., 775, 777 (82), *792*
Fujibayashi, S., 139 (484), *203*
Fujimori, K., 178 (854, 858), *210*, 664 (38), *672*
Fujimoto, K., 154, 155 (638), *206*, 382 (53), *389*
Fujimoto, N., 369 (29, 30), *375*, 387 (151), *391*
Fujimoto, Y., 565 (143), *588*
Fujita, E., 99 (86), *195*
Fujita, S., 141 (498), *203*, 427 (25c), *457*
Fukuda, N., 223 (58), *269*
Fukumoto, F., 565 (138), *588*
Fukumoto, K., 123 (374, 375), 144, 172 (532, 534), 177 (846), 179 (374, 375), *201, 204, 210*, 568 (173, 174), *589*
Fukumura, M., 430–434 (37), *457*
Fukushima, A. A., 474, 478 (77), *493*
Fukuyama, T., 124, 140 (379), *201*
Fukuzawa, S., 100 (116–120), 102 (438), 116 (116, 118, 119), 124 (377a, 377b), 125 (117, 283, 387), 133 (436–439), 162 (712, 713), 180 (869), 183 (117, 283, 387), 184 (887), *196, 199, 201, 202, 208, 211*, 523 (181), *538*, 579 (212, 213), *590*, 756 (318, 331, 332), 757 (331), *764*
Fukuzawa, S. I., 397 (26), *417*, 523 (177c), *537*
Fuller, G. B., 126 (394), *201*
Fuller, W. W., 467, 469 (47), *492*
Füllgrabe, H.-J., 399 (43), *417*
Fung, N. Y. M., 222 (57), 227 (71), *268, 269*, 322, 329, 330 (136), *337*
Funk, R. L., 565 (159), *588*
Furin, G. G., 416 (157), *420*
Furlani, C., 5, 18, 20, 21 (18), *47*, 327 (152), *337*, 549, 574 (66), *586*
Fürstenberg, G., 616, 643 (66), *653*
Furui, S., 116 (301), *200*
Furuichi, K., 97 (43), *194*, 555 (104), *587*, 774 (71), *792*
Furukawa, N., 430–434 (37), *457*
Fuwa, K., 453 (184), *461*

Gabriel, H., 374 (95), *376*, 447 (91), *459*
Gabrio, T., 246 (149), *270*
Gadwood, R. C., 160, 183 (701), *207*, 699, 705, 712, 716, 717, 726, 748 (153), *760*
Galasso, V., 30 (110), 33 (121), 34, 35 (124–126), 36 (136), *49*, 327 (153), *337*, 532 (254), *539*
Gale, L. H., 311 (104), *336*
Galiallira, R. F., 517 (139), *537*
Gall, J. H., 165 (742), *208*, 395 (6), *417*
Gallagher, C. H., 380 (33), *389*
Gammill, R. B., 108 (217), *198*
Gancarz, R. A., 98 (72, 73), 115 (279), 137 (72, 73), 145 (73), 153 (73, 629), *195, 199, 206*, 280, 310, 311 (22), *334*
Ganem, B., 104, 124 (167), *197*
Ganguly, R., 187 (908), *211*
Ganonico, P. G., 386 (125), *391*
Ganther, H. E., 382 (57, 58), 384 (84), 387 (146, 149), *389 – 391*, 403 (64), *418*
Gaprindashvili, V., 453 (175, 176), *460*
Garbe, J. E., 220 (29), *268*
Garbuglio, C., 261 (241), *272*
Garciá, B., 562 (129), *588*
Gardner, S. A., 344 (42), *347*, 527 (209), *538*
Garito, A., 449, 454 (114), *459*
Garito, A. F., 467 (10), 477 (84), 478 (86), 483 (104), *491, 493, 494*, 556 (110), *587*
Gar'kin, V. P., 409 (119), *419*, 432 (41), 441, 442 (41, 64–66), 443 (68–71), *457, 458*
Garkin, V. P., 148 (561), *205*
Garner, B. J., 486 (119), *494*
Garratt, D. G., 116 (299), 129 (410, 411), 131 (423), 136 (461, 462), 139 (423, 486), 144 (299), (280), *199, 200, 202, 203*, 545, 553 (37), *586*
Garvey, D. S., 172 (795), *209*
Gash, D. M., 264, 265 (268), *273*
Gasic, G. P., 665 (54), *672*
Gasiewicz, T. A., 387 (155), *391*
Gassman, P. G., 182 (883), *211*, 565 (157), *588*
Gathehouse, B. M., 106 (198), *197*
Gattermayer, R., 399 (45), *417*
Gattow, G., 248 (169, 170), *271*
Gaudemer, A., 96, 97 (36), *194*, 304, 305 (86), *336*, 567 (168), *589*, 785 (114, 115), *793*
Gaudener, A., 669 (69), *673*
Gaultier, J., 467, 485, 489 (54), *493*
Gautheron, B., 34 (122, 123), *49*, 305 (88, 89), *336*
Gawish, A., 119, 148 (343), *200*
Gaydoul, K.-R., 159 (697), *207*
Gaynor, T., 386 (118), *391*
Gazizov, I. G., 33 (120), *49*
Gazizov, M. B., 638 (128), *655*
Ge, K., 370 (45), *375*, 379 (16), *389*
Gebeyehu, G., 386 (123), *391*
Geise, B., 772 (63–66), *792*
Geiser, U., 483 (108), *494*
Geiss, K. H., 681, 682 (31), *758*
Gellender, M., 35 (129), *49*
Gel'mont, M., 503 (51), *535*
Gemmer, R. V., 467 (13), *492*
Gemroth, T. C., 102 (141), *196*

Geoffrey, M., 665 (50, 51), *672*
Geoffroy, M., 313 (112), *336*, 659 (7), *671*
George, C., 35 (133), *49*
George, J. W., 72 (50, 52), *88*, 511 (99), *536*
George, M. H., 658 (4), *671*
Gerchman, L. L., 510 (94), *536*
Gerhold, J., 108 (207), *198*, 685 (86), *759*
Gerlach, M. L., 384 (81–83), *390*
Géro, S. D., 153 (620), *206*
Geserich, H., 452 (147), *460*
Gewald, K., 247 (155), (217), *270, 272*
Ghandehari, M., 450 (116), *459*
Ghandehari, M. H., 243 (134), *270*
Ghysels, G., 370 (38), *375*
Giannotti, C., 305 (87), *336*
Gibb, T. C., 52 (3, 7), 59 (3), *87*
Gibson, J. E., 380 (26), 381 (35), *389*
Giddings, P. J., 182 (881, 882), 191 (881, 882, 941), *211, 212*
Gieren, A., 452 (147), *453* (178), *460*
Giersch, W., 104 (158), *197*
Giesbrecht, E., 504 (55), *535*
Giese, R. W., 388 (163), *392*
Gilbert, B., 453 (169), *460*
Gilbert, E. E., 123 (371), *201*
Gilchrist, T., 451 (136), 455 (200), *460, 461*
Gilchrist, T. L., 262 (244), *272*
Giles, J. R. M., 664, 666 (49), *672*
Gill, D. S., 411 (129), *419*
Gill, G. B., 141 (500), *203*
Gill, S. P., 388 (165), *392*
Gillespie, R. J., 345, 346 (47), *347*, 513 (110), *536*
Gillis, H. R., 565 (156), *588*
Gillissen, H. M. J., 687 (105), *759*
Gillman, G. P., 658 (4), *671*
Gilman, H., 677, 678 (10), 681 (10, 36b), 682 (36b), 710 (193, 196), 711 (10), *757, 758, 761*
Gilman, S., 96, 99, (39), 122 (366), 123 (372), 173 (39), *194, 201*, 499 (30), *534*, 557, 558 (114), 565 (142), 578 (114), *588*, 712 (224), *762*, 774 (74), *792*
Gilmor, J. R., 32, 33, 36 (115), *49*
Ginodman, V. B., 467 (35), *492*
Giorgianni, S., 549 (62), *586*
Giotti, A., 378 (3), *388*
Giovannini, E., 39 (150), *50*
Gipstein, E., 104, 114 (169), *197*
Girshovich, M. Z., 258 (226), *272*
Giua, M., 255 (208), *271*
Givet, L., 665 (51), *672*
Gladysz, J. A., 220 (29), *268*
Glänzer, H., 528 (227), *538*
Glasebrook, A. L., 228, 229 (73), *269*
Glasso, V., 24, 25 (84), *48*
Glavincevski, B. M., 220 (28), *268*
Gleason, R. W., 312 (109), *336*
Gledel, J., 370 (38), *375*
Gleiter, R., 26 (95, 104), 27 (104), 41 (161, 163), 44, 45 (169), *48–50*, 645 (165), *655*
Glidewell, C., 622, 624, 629, 643, 647, 651 (78), *653*
Glover, S. A., 164 (731, 732), *208*, 402 (60), *418*
Gnonlonfoun, N., 144 (529), *204*
Godfrey, C. R. A., 111, 146 (242), *198*
Godfrey, J. D., 565 (152), *588*
Godfrey, M. 32, 33, 36 (115), *49*
Godovikov, N. N., 416 (156), *420*
Godwin, K. O., 378 (4, 5), *388*
Goerdeler, J., 451 (139c), *460*
Goetze, U., 246 (148), *270*
Gogoz, F., 373 (82), *376*
Goh, L. Y., 405 (78), *418*
Gokhale, U., 100, 138, 174, 180 (112), *196*
Gold, P. M., 158, 184, 187 (694, 695), *207*, 555 (106), *587*, 742 (314), *764*
Goldberg, I., 547, 554 (52), *586*
Goldberg, M. E., 358 (37), *364*
Goldberg, S. M., 5, 8 (15), *47*
Goldsmith, D., 141 (495), *203*, 773 (69), *792*
Goldsmith, D. J., 106 (193), 154, 155 (658), 169, 170 (768), *197, 207, 209*
Goldwhite, H., 595 (7), *652*
Golebiewski, A., 6 (21), *47*
Golgolab, H., 151 (585, 586), *205*
Gollmick, R., 545, 553 (34), *586*
Golob, L., 7 (40), *47*
Gombler, W., 263 (261, 262), *272*
Gondeau, D., 40 (159), *50*
Gonser, U., 52 (6), *87*
González, A. G., 109 (227), *198*
Gonzalez, A. G., 715 (248), *762*
Goodman, M. M., 371 (56, 58–62, 64, 69), 372 (58, 60–62, 64), 374 (62), *375, 376*
Goodrich-Smith, M., 385 (136), *391*
Gorak, R. D., 639 (134, 135), *655*
Gordon, F., 411 (129), *419*
Gordon, K. M., 94 (3), 100, 102 (114), 121 (362), 122 (114), *194, 196, 201*, 677, 681, 685, 715, 719, 720, 756 (2), *757*
Gordon, R. D., 453 (167), *460*
Gorissen-Hervens, F., 135 (453), *203*, 670 (81), *673*
Gornowicz, G. A., 711 (213), *762*
Goryachenkova, E. V., 352 (12), *364*
Gosselck, J., 161 (709), *207*, 278 (17), *334*, 662 (26), *672*, 705 (165), *761*
Goto, J., 379 (11), *388*
Goto, T., 114 (270), 117 (317, 318), *199, 200*
Goudie, C., 387 (146), *391*
Gould, R. O., 404 (73), *418*
Graber, D. R., 187 (911), *212*
Grabley, F.-F., 408 (96), *418*

Grabley, F. F., 244 (136), *270*
Graf, E., 384 (74), *390*
Graf, P., 374 (92), *376*, 384 (75), *390*, 447 (88), *459*
Graf, W., 174 (818, 819), *210*, 330, 332 (165), *337*
Gramza, J., 542, 571 (13), *585*
Grandclaudon, P., 315, 316 (119), *336*
Granger, P., 448 (102), *459*, 518 (147), 525 (190), 531 (147, 241, 244), 532 (147, 252), *537–539*, 550 (79), *587*
Grant, D., 667 (64), *673*
Grant, K. E., 385 (128), *391*
Gravel, D., 143 (519), *204*
Grechkin, N. P., 621 (77), 638–640 (129), *653, 655*
Greeder, G. A., 385, 386 (108), *390*
Green, D. C., 483 (109), *494*
Green, M., 411 (129), *419*
Greene, E., 369 (26), *375*
Greene, H. D., 385 (136), *391*
Greene, R. C., 351 (8, 9), *364*
Greene, R. L., 466 (6), 467 (16, 40), 469 (40), 483 (107), *491, 492, 494*
Greenfield, M. C., 72 (51), *88*
Greenwood, L., 379 (18), *389*
Greenwood, N. N., 52, 59 (3), *87*, 580 (222), *590*
Grieco, P. A., 96 (32, 39), 99 (39, 98, 99), 101 (130), 104 (32, 172, 173, 176), 108 (213), 122 (366), 123 (372), 139 (480), 161 (704), 173 (39, 98, 99, 173, 798–804, 810), 174 (480), 184 (799, 802, 803), 187 (704, 909), *194–198, 201, 203, 207, 209–211*, 427 (23), 428 (26), *457*, 499 (30), *534*, 555 (99), 557 (114), 558 (114, 116, 118, 119), 560 (120), 561 (122, 127, 128), 563 (131), 565 (142), 578 (114), *587, 588*, 681 (28), 685, 695 (91, 92), 701 (92), 712 (224), 715 (28, 91, 92, 258, 259), 721 (28), *758, 759, 762, 763*, 774 (73, 74), *792*
Grieg, G., 281, 318 (27), *334*
Griesser, H., 97 (54), *195*, 563 (133), *588*
Griffin 256 (215), *272*
Griffin, A. C., 385 (104, 137), 386 (104), *390, 391*
Griffin, T. S., 220 (26), *268*
Grigsby, R. A., 371 (63), 373 (81), *375, 376*, 531, 532 (245), *539*
Grimm, H. G., 256 (213), *271*
Grimsey, R. M. A., 87 (92), *89*
Grobe, J., 597 (15, 20, 21), 598 (22), 600 (21), 602–604 (15), 628 (22), 643 (15, 155), 644 (161), 645 (165), 647 (178, 179), 648 (182), 650 (189), (180, 181, 183), *652, 655, 656*
Gröbel, B.-T., 154–156 (635), 159 (699), *206, 207*
Gröbel, B. T., 681, 682 (55), 683, 684, 695, 696, 726, 727 (71), *758, 759*
Gröbel, B. T. H., 708, 709 (186), *761*
Grober, J., 44, 45 (169), *50*
Gronowitz, S., 21–23, 25 (73), 26, 27 (104), *48, 49*, 299 (79), 302 (79, 81), *335*, 407 (109), *419*, 480 (92), *493*
Gross, D., 451 (139c), *460*
Gross, M. L., 658 (2), *671*
Gross, R., 379 (13), *388*
Grossman, J., 32, 33, 36 (116), *49*, 327 (150), *337*
Grossoni, G., 258 (224), *272*
Groutas, W. C., 386 (118), *391*
Groves, J. T., 554, 555 (94), *587*
Gruber, J. M., 386 (119), *391*
Gruner, G., 467, 485 (28), *492*
Gruttadauria, S., 516 (133), *536*
Grützmacher, H. F., 610 (53), *653*
Gryskiewicz-Trochimowski, E., 638 (127), *655*
Gryskiewicz-Trochimowski, O., 638 (127), *655*
Grzejszczak, S., 114 (268), *199*
Gschwend, H., 447 (96), *459*
Guang-di, Y., 32, 33, 36 (116), *49*, 327 (150), *337*
Gubser, D. U., 467, 469 (47), *492*
Guddatt, L. W., 106 (198), *197*
Guerin, G., 711 (218), *762*
Guerra, M., 23–25 (81), *48*
Gugel, H., 151 (595), *205*
Guilhem, J., 567 (168), *589*, 785 (115), *793*
Guillaume, M., 372 (79), *376*
Guillemonat, A., 118 (326), *200*
Guimon, C., 26 (101), 34 (122), 35 (132), 40 (158, 159), *48–50*
Guimon, M. F., 35 (132), *49*
Guingant, A., 106 (185), *197*
Gunning, H. E., 281, 318 (26), *334*
Günther, H., 104 (166), *197*
Günther, W. H., 220 (22), *268*
Gunther, W. H., 221 (33), 228 (76, 78), *268, 269*
Günther, W. H. H., 394 (1), 396 (10), 403, 406, 407 (1), 412, 413 (137), 415 (1), *416, 417, 419*
Gunther, W. H. H., 94 (1), *194*, 217 (1), *267*, 276 (6, 7), 282, 328 (6), *334*, 423–425, 428 (4), *456*, 503 (47), 508 (81), 509 (88), 526 (191), *535, 536, 538*, 545, 554 (33), *586*, 659 (8), 663 (30), *671, 672*, 677, 679, 681, 713, 714 (3), *757*
Gunther, W. H. S., 765 (2), *790*
Gurskii, M. E., 279 (19), *334*
Gur'yanova, E. N., 519 (155, 156), *537*
Guryanova, E. N., 662 (23, 24), *672*
Gusarova, N. K., 407 (100, 101), *419*, 505 (64–66), *535*
Gutman, A. L., 104 (166), *197*

Gutman, E. E., 664 (47), *672*
Guyer, C. E., 388 (166), *392*
Guy Orpen, A., 411 (129), *419*
Guziec, F. S., 329 (161), *337*
Guziec, F. S. Jr., 167 (746–753), *208*, 219 (15, 17, 18, 21), 221 (34), 222 (15, 21, 52–55), 223 (21, 34, 52), 224 (52), 225 (21, 52, 61), 226 (62, 63, 65), 227 (72), 245 (52), 249, 251, 252 (176), 263 (17, 18, 260), 264 (18, 260, 268), 265 (268, 272), 266 (277), *268, 269, 271–273*, 293 (55), *335*, 398 (28, 29, 31), *417*
Guzman, F., 110, 118, 119, 148 (236), *198*
Gyles, G. L., 370 (41), *375*
Gyllenberg, H. G., 369 (20), *375*
Gysling, H. J., 248 (168), *271*, 342 (20), 344 (42), *347*, 527 (203, 204, 212), 528 (228), 532 (251, 255), *538, 539*, 594 (6), *652*
Gyul'maliev, A., 453 (156), *460*

Haas, A., 306 (91), *336*, 527 (200), *538*, 547 (53, 54), *586*, 595 (10), 605, 606, 608–610 (47), 627· (97, 98), *652–654*, 660 (19), *672*
Haba, A., 424 (17), 429 (30), 450 (17), *457*
Haber, S. B., 142 (507), *204*
Haces, A., 180 (874), *211*
Hacker, A. D., 380 (24), *389*
Hacker, N., 291, 292 (52), *335*, 400 (46), *417*, 453 (160), *460*
Haddon, R., 456 (202, 203), *461*
Hadjimarkos, D. M., 382 (50), *389*
Hadjiminolis, S., 82 (74), *89*
Hadzi, D., 266 (280), *273*
Hafeman, D. F., 380 (32), *389*
Hafeman, D. G., 380 (34), 382 (58), *389*
Hagan, C. P., 106 (194), *197*
Hakiki, M., 487 (125), *494*
Halazy, S., 100 (121–123), 154 (121, 122, 648, 650, 652, 661, 666, 668, 670, 671), 155 (122, 648, 650, 652, 661), 156 (121, 122, 650, 652, 666, 671), 157 (122, 650, 666, 668), 167 (757), 179 (123), 182 (661, 668, 670, 671), 183 (670, 671), *196, 206–208*, 521 (166), *537*, 679 (20), 681 (20, 25), 685, 686 (127, 133), 689 (107, 109, 110, 127, 133), 692, 693 (115), 695 (20, 107, 115, 127, 133, 134, 139, 143), 696 (115, 133), 697 (115, 139), 698 (109, 115, 139), 699 (109, 115), 709, 710 (188), 711 (143, 222), 715 (20, 127), 716 (269), 718 (20, 109, 269), 719 (20, 269), 720 (127), 721 (25, 107, 139, 269), 722 (107), 723 (107, 109, 115, 188, 290), 724 (109, 115), 730 (107), 731 (139, 269), 732 (20, 127), 733, 739 (127), 740 (290), 749 (107, 115), 750 (110), *758, 760–763*
Haley, N. F., 245 (138), *270*, 341 (17), *346*, 408 (93, 94), *418*
Hall, D., 37, 38 (144), *50*
Hall, G. G., 6 (20), *47*
Hall, L. N., 467, 483 (50), *493*
Hall, T. W., 193 (971), *213*
Haller, W. S., 408 (107), *419*, 511 (100), *536*
Halozy, S., 520 (161), *537*
Halpern, A., 86 (89), *89*
Halverson, A. W., 382 (56), *389*
Halvorsen, A., 542 (15), *585*
Ham, N. S., 5 (17), *47*
Hamada, K., 340 (2a), *346*, 533 (264), *539*
Hamada, Y., 97, 116 (50), *195*
Hamel, E. J., 385 (136), *391*
Hamill, G. P., 502, 511 (44), 516, 517 (134), 530 (238), *535, 537, 539*
Hammar, G. W., 255 (210), *271*
Hammett, A., 2 (2, 5), *46*
Hammond, D. A., 102 (136), *196*
Hammond, P. J., 531 (242), *539*
Hamor, T. A., 511 (103) 515, 516, (125), *536*
Hampson, P., 264 (265), *273*
Hanafusa, T., 246 (145), *270*, 320 (133), *336*, 574 (195), *589*
Hanai, H., 503 (49), *535*
Hanamoto, T., 114 (266), *199*, 714, 715 (230), *762*
Hanan, M. C., 565 (155), *588*
Hanazaki, Y., 443 (67), *458*
Handa, K., 103, 107 (154), *197*
Hanessian, S., 778 (88), *792*
Hanns, E., 575 (199), *589*
Hanocq, M., 453 (168, 180, 181), *460*
Hanold, N., 151 (599), *205*, 450 (123), *459*
Hansen, M., M., 565 (159), *588*
Hansen, M. R., 134 (448), *202*
Hansen, P.-E., 231 (91), *269*
Hansen, P. E., 102 (140), *196*, 231, 232 (90), *269*
Hansen-Nygaard, L., 25 (89, 90), *48*
Hanson, P., 547, 554 (56), *586*
Hanson, R., 449 (108, 109), *459*
Harada, Y., 25 (79), *48*
Hardte, I., 398 (30), *417*
Harfield, J., 382 (47), *389*
Harget, A., 6 (31), *47*
Hari, T., 99, 115, 116, 123 (82), *195*
Harirchian, B., 114 (272), *199*, 706 (178), *761*
Harms, R. H., 467, 485, 489 (53), *493*
Harnett, N. M., 370 (41), *375*
Harrington, J. K., 312 (111), *336*
Harris, J. C., 340 (10), *346*
Harrison, W. D., 343, (35, 36a), 344 (35), *347*
Harrit, H., 453 (159), *460*
Harrit, N., 288 (42), 291, 292 (51), *335*
Hart, D. J., 145 (539), *204*
Hart, H., 489 (131), *494*

Hartmann, H. 236 (103, 104), 238 (104, 112), *269*, 446 (84), *458*, 553, 555 (92), *587*
Hartzler, H. D., 477 (81), *493*
Haruna, M., 105 (184), 106 (190), 118 (335), *197, 200*
Harvey, A. B., 340 (2b), *346*
Harvey, R. G., 121 (358, 359), *201*
Hase, T. A., 187 (933), *212*, 706 (171), *761*
Haseda, N., 369 (31), *375*
Hasegawa, G., 380 (23), *389*
Haselbach, E., 6 (30, 31), *47*
Hashimoto, C., 148 (565, 566), 174 (566), *205*
Hassaneen, H. M., 550 (84, 86, 87), *587*
Hasskerl, T., 772 (65), *792*
Hata, T., 783 (112), *792*
Hatanaka, I., 153 (625), *206*
Hatanaka, M., 358 (38), *364*
Hattori, K., 184 (887), 193 (968), *211, 213*
Hau, S., 383 (67), *390*
Haubein, A. H., 710 (193), *761*
Haubold, R., 647 (179), *656*
Hauge, S., 342 (27d, 27f, 27g, 29b), 343 (27d, 27f, 27g), *347*
Haupt, E., 102 (139), *196*
Hauptmann, H., 500 (35), *534*
Hauser, F. M., 139 (485), *203*
Hauser, J. J., 504 (53), *535*
Hausmann, H., 340 (4), *346*
Hauson, R. N., 388 (163), *392*
Hawkes, W. C., 355 (18), *364*
Hayama, T., 98 (79, 80), 137 (79, 80, 466), *195, 203*
Hayano, K., 118 (339), *200*
Hayase, Y., 173 (811), *210*
Hayashi M., 387 (159, 160), *391*
Hayashi, Y., 187 (924), *212*
Hayasi, Y., 187 (931), *212*, 715 (255), *763*
Hayes, P. C., 109, 184, 186 (221), *198*
Haynes, R. K., 159 (696), 173 (812), *207, 210*, 684, 695 (79), *759*
Hayward, G. C., 344 (39b), *347*
Haywood, B. J., 554, 555 (94), *587*
Hazell, A. C., 343 (36b), *347*
Headford, C. E. 259 (230), *272*
Heathcock, C. H., 99 (103), 102 (141), 104 (170), 113 (259, 260), 114 (170), *196, 197, 199*, 715 (234, 261), *762, 763*
Hebold, M., 398 (30), *417*
Hecht, C., 308 (98), *336*
Hecht, K., 379 (9), *388*
Heck, J. V., 97 (55), *195*
Heeger, A. J., 467 (10), *491*
Heiker, F. R., 771 (58), *791*
Heilbronner, E., 26 (95), 29, 30 (109), 39 (149, 150), *48–50*
Heilmann, S., 483 (100), *494*
Heinrich, M. A., Jr. 378 (1), 380 (22), *388, 389*
Heinzerling, R. H., 384 (81–83), *390*
Heitz, W., 103, 124 (150), *197*, 545 (39), 546 (48), 555, 557 (39), *586*
Helland-Madsen, G., 645, 646 (166), *655*
Heller, C. A., 529 (234), *539*
Hellwinckel, D., 430, 432 (33), 441, 442 (63), *457, 458*
Hemidy, J. F., 487 (125), *494*
Hemmings, R. T., 340 (2c), *346*, 596, 602, 605–607, 609, 648 (13), *652*
Henberger, C., 330, 332 (165), *337*
Hende, J. H., van den 262 (242), *272*
Hendra, P. J., 344 (39b), *347*, 533 (260, 261), *539*
Hendriksen, L., 246 (150), *270*, 327 (151), *337*
Hendry, L. B., 715 (254), *763*
Hengge, E., 398 (38, 39), *417*
Hennen, W. J., 786, 790 (117), *793*
Henriksen, L., 31, 32 (112), *49*, 220, 221 (31), 231 (89), 237 (105), 238 (113), 240 (116, 119), 241 (119), 242, 243 (126), 247 (164), 248 (171), 249 (171, 173, 175, 178), 253 (191, 199), 254 (116, 191, 201), 256 (126, 191, 199), 264 (116, 267), 265 (116), 266 (278), *268–271, 273*, 402 (62), 403 (68), 404 (68, 69, 71, 74), 405 (77), 407, 409 (102), 413 (102, 143), 414 (102, 145), 415 (62), *418, 419*, 451 (140), *460*, 473 (66, 68), 474 (66), 475 (75), 476 (76), *493*
Henrikson, L., 574 (193), *589*
Henriques, F. C. Jr., 368 (4), *374*
Henry, G., 320 (132), *336*, 524 (186), *538*, 773 (70), *792*
Hensel, R., 400 (49), *417*, 601 (29, 31), 606, 610, 611 (31), 612 (57), 613, 614 (31, 57), 637 (29), 644 (31), 646 (176), 647 (29), 648 (184), 649 (29), *652, 653, 656*
Herberg, S., 517 (144), *537*
Herberhold, M., 308 (98), *336*, 416 (158), *420*
Hermann, W. A., 259 (231), *272*
Hernández, R., 109 (227), 183 (885), *198, 211*
Hernandez, R., 715 (248), *762*
Herold, L. L., 303, 305 (85), *336*, 660 (21), *672*
Herring, F., G., 15 (58), *47*
Herrmann, D., 401 (55), *418*, 545, 553 (34), 554, 555 (95), *586, 587*
Herrmann, W. A., 308 (98), *336*, 405 (81), *418*
Herro, G., 386 (118), *391*
Hershberger, J., 303, 305 (83, 84), *336*, 660 (20), *672*
Hertel, H., 641 (148), 650 (192), *655, 656*
Herve, Y., 99, 102, 132, 163, 191 (102), *196*, 322 (138), *337*
Herzschuh, R., 575 (197), *589*
Hessel Andersen, N., 467 (26), *492*
Heuberger, C., 174 (818), *210*
Hevesi, L., 98 (69), 99 (83, 90), 100 (127), 115, 116 (273), 122 (69), 132 (428, 432), 153 (622), 154 (622, 631, 649), 155 (622, 649),

157 (685), 167 (69, 756, 757), *195, 196, 199, 202, 206–208*, 520 (161), 525 (189), *537, 538*, 679 (17), 680 (17, 23), 681 (17, 23, 25), 685 (96, 336), 695 (132), 696, 708 (187), 712 (17), 714 (232), 715 (132, 232), 716 (266), 718 (23, 281), 720 (266), 721 (23, 25, 281), 723, 727 (187), 732 (232), 734 (308b), 735 (187), 736 (187, 308b), 737 (187), 747 (17), 748, 749 (266), (320), *758–764*
Hevessi, L., 517 (142), *537*
Hewett, W. A., 312 (107), *336*
Hey, D. H., 663 (35), *672*
Heydt, H., 293 (56), *335*
Heyns, K., 771 (58), *791*
Hickey, M. J., 32 (113), *49*
Hielsen, P. H., 40, 41 (160), *50*
Higa, T., 369 (29, 30), *375*
Higuchi, H., 100, 154, 155 (115), 165 (734, 737, 738), 166 (743, 744), *196, 208*, 396 (15–17, 19, 20), 397 (16), *417*, 555 (108), *587*, (323), *764*
Hildebrandt, W., 597, 602–604, 643 (15), *652*
Hill, C. H., 387 (143, 144, 156), *391*
Hill, R. K., 187 (930), *212*
Hillen, L. W., 403 (67), *418*
Hillier, I. H., 13 (56), *47*
Hillman, G. R., 388 (162), *391*
Hilse, H., 379 (9), *388*
Hilti, B., 467, 485 (30), *492*
Hilti, H., 467 (31), *492*
Himbert, G., 131 (424), *202*
Himmelsbach, R. J., 102 (135), *196*
Hindenlae, D. M., 356 (26), *364*
Hine, J., 681, 682 (64), *759*
Hinsberg, O., 452 (145), *460*
Hinshaw, B. C., 786, 790 (117), *793*
Hinze, J., 219 (20), *268*
Hirabayashi, T., 35 (135), 37, 38 (141), 39 (154), 40 (155), *49, 50*, 412 (135), *419*, 453 (158), *460*
Hirabayashi, Y., 166 (745), *208*, 241 (120, 121), *270*, 397, 398 (27), *417*
Hirai, K., 154, 155 (638), *206*
Hirama, H., 564 (137), *588*
Hirao, I., 114 (266), *199*, 714, 715 (230), *762*
Hiraoka, T., 148 (563), *205*
Hirashima, T., 178 (849), *210*
Hiroi, K., 99 (98), 102 (148), 173 (98), *195, 196*
Hirokawa, T., 533 (259), *539*
Hiyama, T., 711 (215), *762*
Hlavacek, R. J., 255 (212), *271*
Hnatowich, D., 371 (56), *375*
Ho, P.-T., 98, 129 (63), *195*
Ho, T.-L., 193 (971), *213*
Hoeg, D. F., 710 (201), *761*
Hoekstra, W., 154, 155 (658), *207*
Hoekstra, W. A., 382 (57, 58), *389*
Hoekstra, W. G., 355 (17), *364*, 380 (30, 32, 33), 385 (129), 387 (149), *389, 391*
Hoekstra, W. H., 387 (146), *391*
Hoffman, J. L., 360 (42), 362 (53), *364, 365*
Hoffman, J. W., 263 (248), *272*
Hoffmann, M., 612 (56), *653*
Hoffmann, R., 6 (35), *47*, 292 (54), *335*
Höfler, M., 340 (4), *346*
Hofmann, H., 452 (143), *460*
Hofmann, W., 405 (80), *418*
Hoggard, R., 262 (246), *272*, 453 (173), *460*
Höhn, B., 641 (148), *655*
Hohne, G., 278, 301, 321 (16), *334*
Høiland, H., 580 (217), *590*
Holker, J. S. E., 128 (404), *202*
Hollander, M. I., 167 (749), *208*, 222 (54), *268*
Hollands, R. E., 407 (118), 416 (118, 154), *419, 420*, 504 (58), *535*
Hollo, Z. M., 387 (152), *391*
Holm, A., 242 (124), 243 (128–130), 254 (200, 203), 267 (128), *270, 271*, 288 (42), *335*, 449 (112), 450 (117), 451 (140), *459, 460*
Holmes, T. J., 36, 37 (137), *49*, 251, 252 (186), *271*, 486 (121), *494*
Holt, D., 407 (115), *419*, 480 (90), *493*
Holtan, R. C., 135, 180 (460), *203*
Holtzman, J. L., 380 (29), *389*
Hölzle, G., 713 (225), *762*
Holzle, W., 706 (169), *761*
Homfeld, E., 453 (160), *460*
Hommes, H., 408 (95), *418*
Homsany, R., 706 (179), *761*
Hoornaert, C., 145 (538a, 538b), *204*, 666 (60), *672*
Hooz, J., 109 (224), 135 (457, 458), *198, 203*, 706 (177), *761*
Hope, H., 255 (209), 266 (283), *271, 273*
Hopf, G., (60), *458*
Hopkins, P. B., 127 (398–402), *201*
Hoppe, D., 681, 682 (42a), *758*
Horcher, L. H. M., II 565 (159), *588*
Horgan, A. G., 715 (235), *762*
Hori, M., 430, 432–434 (36), *457*
Hori, T., 127 (396), 131 (426, 427), *201, 202*, 438 (51), *458*, 542, 545 (4), *585*, 714, 721 (231), *762*
Horie, S., 369 (33), *375*
Horn, H.-G., 637 (123), *654*
Horn, K. A., 711 (216a), *762*
Horn, V., 437 (46), *457*
Hornby, J. L., 220 (29), *268*
Hőrnfeldt, A.-B., 407 (109), *419*
Hornfeldt, A. B., 480 (92), *493*
Horozoglu Armen, G., 32, 33, 36 (116), *49*
Horton, D., 775 (78), *792*
Hoshino, O., 97, 116 (50), *195*
Hoshino, S., 154, 155, 169 (654), *206*
Hosoi, S., 101, 125 (128, 129), *196*

Houk, L. W., 527 (207), *538*
House, H. O., 99 (105), *196*
Houston, T. L., 684, 695 (79), *759*
Hovey, M. C., 187 (927), *212*, 328 (158), *337*
Howard, J. A. K., 416 (154), *420*
Howell, G. O., 387 (143), *391*
Hoye, T. R., 104, 140 (177), 169 (177, 770), *197, 209*, 565 (141), *588*
Hoyer, E., 253 (195), *271*, 575 (197), *589*
Hsu, C. Y., 385, 386 (109), *390*
Hsu, Y. F., 531 (242), *539*
Hu, C., 466, 467, 483 (1), *491*
Hu, M.-L., 383 (59), *389*
Hu, N. X., 556 (111), *587*
Huang, C. K., 531, 532 (245), *539*
Huang, W.-S., 407 (110), *419*
Huber, R. E., 355 (22), *364*
Hubner, T., 453 (178), *460*
Hudlicky, T., 565 (158), *588*
Huestis, L., 453 (177), *460*
Huey, J. L., 374 (97), *376*
Huffman, J. H., 374 (98), *376*
Huffmann, J. C., 615 (64), *653*
Huggins, J. W., 374 (96), *376*, 386 (125), *391*
Hughes, E. D., 705 (164b), *761*
Huguet, J. L., 95 (23), *194*
Hui, R. A. H. F., 111 (240, 245, 246), 120 (245, 350), 122 (246), 146 (240), *198–200*
Hull, S. E., 6 (37), *47*
Huls, R., 247 (157, 158), 248 (158), *270*
Hundt, R., 399 (45), *417*
Hunger, B., 408 (91), *418*
Hunig, S., 486 (119), *494*
Hurst, K. M., 104, 106 (165), *197*
Hurtley, W. R. H., 480 (94), *493*
Husebye, S., 68, 78, 79, 87 (34), *88*, 342 (27a, 27c), 343 (27a, 27c, 38), *347*, 404 (72, 75), *418*, 636 (122), 645 (166), 646 (122, 166, 174), 650 (191), *654–656*
Hussong, R., 293 (56), *335*
Hutchins, R. O., 174 (815), *210*
Hutson, D. H., 775 (78), *792*
Huttunen, J. K., 379 (10), *388*
Huu, P. M., 388 (167), *392*
Hwang, K.-J., 182 (884), *211*
Hyono, T., 187 (924), *212*

Ibata, T., 290 (50), *335*
Ibragimov, A. A., 705 (161), *761*
Ibrahim, N., 109 (223), 124, 152 (223, 382), *198, 201*
Ice, R. D., 388 (165), *392*
Ichikawa, K., 117 (314), *200*, 545 (40, 41), *586*
Ichikawa, Y., 114 (270), 117 (318), *199, 200*
Ichinoe, T., 380 (25), *389*
Idris, M. S. H., 141 (500), *203*
Iijima, I., 101 (131), *196*
Iio, H., 110 (237), *198*
Iitaka, Y., 97, 116 (50), *195*
Ikeda, T., 192 (943), *212*
Ikehira, T., 114 (266), *199*, 714, 715 (230), *762*
Ikota, N., 104, 124 (167), *197*
Ikuta, S., 113 (254), *199*, 683, 696, 715 (103), *759*
Imoto, T., 118 (325), *200*
Imura, N., 387 (153, 154, 158), *391*
Imura, T., 452 (148), *460*
Ina, S., 503 (49), *535*
Inaba, M., 171 (775, 776), *209*, 469 (55), *493*
Inamoto, M., 617, 633 (68), *653*
Inamoto, N., 223 (58), 226 (64), *269*, 400 (53), *418*, 617, 633 (67), *653*
Inanaga, J., 108 (213), *198*
Indorato, C., 173 (812), *210*
Indue, M., 449 (104), *459*
Ingold, K. U., 227 (66–68), *269*, 517 (140), *537*, 658 (3, 5), 661 (5), 667 (3, 62), 670 (5, 79), *671–673*
Ingram, L., 646, 650 (173), *655*
Innorta, G., 21–25 (72), *48*
Inokawa, S., 775 (79, 81), *792*
Inokuchi, H., 467 (25), 483 (106), *492, 494*
Inokucki, H., (130), *494*
Inoue, K., 371 (51), *375*, 383 (62), *390*
Inouye, M., 175 (825), *210*
Inove, S., 503 (49), *535*
Ip, C., 385 (100, 102, 133), 386 (102), *390, 391*
Ireland, R. E., 174 (820), *210*, 565 (152), *588*
Irgolic, K., 423 (6, 8), *456*, 613, 635 (60), 650 (194), *653, 656*
Irgolic, K. J., 23–25 (81), 34, 35 (124–126), 36 (136), *48, 49*, 94 (2), 95 (14–20), *194*, 217 (7, 8), *267*, 276 (8), *334*, 340 (3), *346*, 368 (1), 369 (15), 371 (63), 373 (81, 83), *374–376*, 394, 406, 407 (3), 408 (107), 414, 416 (3), *417, 419*, 497 (2, 6), 511 (100), 516 (2, 6, 133), 531 (245), 532 (245, 254), *534, 536, 539*, 593 (4), 602 (34), 604 (37), *652*, 765 (1), 769 (35), 788 (123, 125–127), *790, 791, 793*
Irie, H., (237), *762*
Iroshnikova, N. G., 86 (88), *89*
Irvin, G. P., 297 (72), *335*
Iset, L. C., 486 (120), *494*
Isett, L. C., 467, 485 (29), *492*
Ishida, M., 186 (905), *211*
Ishida, T., 107 (201, 202), *198*, 449 (104), *459*
Ishido, Y., 113 (255), *199*
Ishiguro, M., 174 (816), *210*, 666 (56), *672*
Ishiguro, T., 467 (15), *492*
Ishihara, H., 166 (745), *208*, 241 (120–122), 242 (123), *270*, 397, 398 (27), *417*
Ishii, A., 223 (58), 226 (64), *269*
Ishii, H., 169 (773), 187 (919), 191 (962, 963), *209, 212, 213*, 499 (28), *534*
Ishii, K., 187 (925), *212*

Ishmael, D., 385, 386 (106), *390*
Isobe, K., 116 (295), *199*, 715 (243), (237), *762*
Isobe, M., 114 (270), 117 (317, 318), *199, 200*
Israel, M., 176 (834), *210*
Issleib, K., 612 (56), 640 (140), *653, 655*
Itagaki, K., 181 (876), 184 (896), *211*
Itakura, T., 549 (70), *587*
Itho, K., 726 (329), *764*
Ito, H., 449 (105a), *459*
Ito, K., 105 (184), 106 (190), 118 (335), *197, 200*
Ito, O., 661 (22), *672*
Ito, Y., 108, 169, 172 (204), *198*
Itoh, A., 187 (920, 921), *212*
Itoh, K., 181 (876, 877), 184 (896, 897), *211*
Itoh, O., 545 (41), *586*
Iversen, A. J., 640, 645 (137), *655*
Iwano, Y., 154, 155 (638), *206*
Iwata, H., 371 (50, 51), *375*, 379 (11), 382 (54), 383 (61–63), 387 (147, 159, 160), *388–391*
Iwata, S., 39, 40 (157), *50*
Iyenger, G. V., 379 (13), *388*
Izawa, T., 172 (779), *209*
Izumi, T., 545 (41), *586*

Jackson, C. L., 276 (5), *334*
Jackson, R. C., 386 (124), *391*
Jackson, W. P., 143 (511, 512, 518), *204*
Jackson, W. R., 684, 695 (79), *759*
Jacob, E. J., 644 (160), *655*
Jacob, R. A., 510 (94), *536*
Jacobs, A. M., 710 (195), *761*
Jacobs, M. M., 385 (103–105, 137), 386 (103–105), *390, 391*
Jacobsen, C., 251 (183), *271*, 478 (88), *493*
Jacobsen, C. S., 467 (26, 44, 45), 469 (44–46), 478 (87), *492, 493*
Jacobsen, M., 451 (140), *460*
Jacquignon, P., 500 (38), *535*, 549 (75), *587*
Jaffe, H. H., 219 (20), *268*
Jaffe, J. J., 263 (254), *272*, 386 (112), *390*, 778, 789 (90, 91), *792*
Jagdmann, G. E. Jr., 113 (258), *199*
Jaitner, P., 306 (92, 93), 307 (94), *336*
Jakovac, I. J., 169, 170 (771), *209*
Jamall, I. S., 379 (21), *389*
James, F. G., 178 (857), *210*, 663 (37), *672*
James, M. N., 266 (286), *273*
Janickis, V., 415 (150, 151), *420*, (62), *458*, 504, 514 (56), *535*
Jankowski, K., 117 (319), *200*
Janossy, A., 467, 485 (28), *492*
Janousek, Z., 135 (453, 454), *203*, 670 (81), *673*
Jansen, S., 501, 514 (43), *535*
Jansson, B., 385 (137), *391*
Jaurand, G., 97, 99 (51), 117 (51, 320), *195, 200*, 770 (37), *791*
Jaurequi-Adeu, J., 765 (20), *791*
Javora, P. H., 768, 790 (32), *791*
Jaw, D. Y., 428 (26), *457*
Jaw, J. Y., 173 (804), *209*, 558 (116), *588*
Jawad, H., 401 (57), *418*, 604, 615 (39), *653*
Jayaram, H. N., 386 (123), *391*
Jeffs, P. W., 121 (360), *201*
Jefson, M., 102 (144), *196*, 426 (19a), *457*
Jeger, O., 187 (925), *212*
Jehanno, N., 325, 326 (147), *337*
Jen, K.-Y., 407 (108, 110), *419*
Jen, K. Y., 193 (970), *213*, 503 (50), *535*
Jenkins, K. L., 403 (63), *418*
Jenny, W., 706 (169), 713 (225), *761, 762*
Jensen, H. B., 118 (328, 329), *200*
Jensen, K. A., 217 (4), 230 (88), 238 (110, 113), 239 (88), 240 (88, 119), 241 (119), 242 (124–126), 243 (126), 247 (156, 161), 248 (168), 249 (172, 173), 253 (88, 194), 254 (200, 201, 203), 255 (156), 256 (126, 214), 264 (263, 267), *267, 269–273*, 641 (142), *655*
Jensen, L. S., 387 (142), *391*
Jensen, S., 485 (116), *494*
Jephcote, V. J., 98 (67), *195*
Jernssi, R. A., 110 (232), *198*
Jerome, D., 466 (4), 467 (4, 23, 27, 37, 38, 41, 42), 469 (23, 37, 38, 41, 42), *491, 492*
Jerris, P. J., 102, 179 (138), *196*
Jetter, W. W., 368 (5), *374*
Jiracek, V., 765, 766, 769 (16), *791*
Joesten, M. D., 340 (11), *346*
Johannesen, O., 342, 343 (27f), *347*
Johannsen, I., 249 (178), 251 (183), *271*, 413 (143), *419*, 475 (73, 75), 478 (88), *493*
John, D. I., 182 (881, 882), 191 (881, 882, 941), *211, 212*
Johns, D. G., 386 (123), *391*
Johnson, A. W., 681, 682 (39), *758*
Johnson, B. F. G., 405 (82), *418*
Johnson, G., 327 (155), *337*
Johnson, K. H., 25 (86), *48*
Johnson, K. L., 381 (39), *389*
Johnson, M., 107 (203), *198*
Johnson, M. D., 304, 305 (86), *336*, 669 (69), *673*
Johnson, R. A., 379 (16), *389*
Johnston, J., 187 (918), *212*
Johnston, R. A. W., 26 (103), *48*
Johnstone, J. J., 71 (45), 86 (90), *88, 89*
Johnstone, R. A. W., 176 (842), *210*, 398 (33), *417*, 552 (89), *587*
Jolocam, M., 371 (52), *375*
Jonathan, N., 7 (40), 15 (57), *47*
Jones, C., 453 (179), *460*
Jones, C. H. W., 62 (25, 26, 28, 29), 63 (25, 30), 64 (25), 65 (25, 30, 35), 66 (35, 36, 38), 67 (25, 26), 68 (34), 70 (25, 26), 71 (25, 26, 45), 72 (26), 73 (26, 36, 56), 74 (25, 26, 35, 36), 75 (57, 58), 76 (26, 57), 77 (25), 78 (25, 34),

79 (25, 26, 34), 80, 81 (61), 83 (61, 78–81), 84 (79, 81–84), 85 (86), 86 (87, 90, 91), 87 (34, 56, 93), *88, 89*, 518, 522 (148), 529 (231, 232), 530 (236), 531 (148, 250), 532 (250), 533 (266, 268), *537, 539*
Jones, D., 23–25 (81), *48*
Jones, D. N., 95, 96 (24), *194*, 680, 681, 714, 721 (26), *758*
Jones, F. N., 477 (83), *493*
Jones, G. H., 663 (35), *672*
Jones, J. B., 169, 170 (771), *209*
Jones, M., 681, 682 (66), *759*
Jones, M. T., 485 (116), *494*, 501, 514 (43), *535*
Jones, T., 83 (80), *89*, 529 (232), *539*
Jonkers, G., 16 (65), 17 (65, 66), 18 (66, 68), 19 (66), *48*
Jonsson, B.-Ö., 22, 25 (75, 76), *48*
Jordan, K. D., 45 (172), *50*
Jorgensen, C. K., 248 (168), *271*
Jouin, P., 320 (132), *336*, 773 (70), *792*
Joullie, M. M., 665 (53), *672*
Joussen, R., 757 (301), *764*
Jovin, P., 524 (186), *538*
Judge, R. H., 228 (79), *269*
Juhlke, T. J., 510 (95), *536*
Julius, A. D., 385 (138), *391*
Jung, H., 617 (69), *653*
Jungmann, H., 633 (113), 649 (185), *654, 656*

Kabalka, G. W., 371 (64, 65, 67), 372 (64, 67), *376*
Kabo, A., 116, 144 (299), (280), *199, 200*
Kagan, J., 151 (601), *205*, 325 (141), *337*, 449 (110), *459*
Kägi, B., 256 (214), *271*
Kagoshima, S., 467 (15), *492*
Kahn, M., 108 (210, 211), *198*
Kai, Y., 165 (739), *208*
Kaiho, T., 172 (795), *209*
Kaiser, I. I., 360–362 (43), *364*
Kaji, A., 333 (168), *337*
Kajimura, K., 467 (15), *492*
Kalabin, G. A., 531 (248), *539*, 550 (78, 80, 82), *587*
Kalè, V. N., 139, 157 (475), *203*
Kale, V. N., 721 (289), *763*
Kalkabaeva, L. T., 575 (198), *589*
Kallen, R. G., 356 (26), *364*
Kalman, T. I., 787 (121), *793*
Kalnins, M. A., 710 (205), *761*
Kalyanasundaram, S. K., 175 (826, 827), *210*
Kambe, N., 103, 107 (153), 177 (844, 845), 178 (851), 191 (962, 963), 192 (958), *197, 210, 212, 213*, 220 (25), 221 (36), *268*, 664 (40, 41), *672*
Kametani, T., 116 (297, 298), 123 (374, 375), 144 (297, 298, 531–534), 172 (531–534), 177 (846), 179 (374, 375), *200, 201, 204, 210*, 565 (138, 139), 568 (173, 174), *588, 589*
Kaminskii, V. F., 467 (32, 49), 483 (49), *492*, 504 (57), *535*
Kampel, V. Ts., 416 (155, 156), *420*
Kanakura, A., 711 (215), *762*
Kandgeteyan, R. A., 705 (161), *761*
Kaneda, T., 165 (739), *208*, 382 (53), *389*
Kanefusa, T., 428 (27), *457*
Kaneko, K., 99 (86), *195*
Kaneko, Y., 172 (779), *209*
Kang, M. C., (147), *760*
Kang, Y.-H., 137 (467), 138 (469), 145 (467), (470), *203*, 311–313 (103), *336*
Kang, Y. H., 547 (57), *586*
Kano, T., 369 (31), *375*
Kaplan, M., 456 (202, 203), *461*
Kappas, A., 381 (40), *389*
Kar, A. B., 387 (148), *391*
Karajagi, G. V., 373 (89), *376*
Karasaki, Y., 234, 238 (97), *269*
Karatsovnik, M. V., 467, 483 (48), *492*
Karimov, Y. S., 467 (34), *492*
Karle, I. J., 266 (281), *273*
Karle, I. L., 340 (14), *346*
Karle, J., 35 (133), *49*, 266 (281), *273*, 340 (14), *346*
Karle, L., 513 (116), *536*
Karsch, H. H., 612–614 (57), *653*
Kasai, N., 165 (739), *208*
Kasamatsu, S., 371 (51), *375*, 383 (62), *390*
Kaschani-Motlagh, M., 605, 606, 608–610 (47), *653*
Kashurnikova, L. V., 531 (248), *539*, 550 (78), *587*
Kashyap, R., 266 (277), *273*
Kataev, E. G., 134 (449, 450), *202*, 499 (31, 32), 519 (155, 156), *534, 537*, 626, 630 (93), *654*, 706 (167, 170, 184), *761*
Kataeva, E. G., 662 (23, 24), *672*
Kataeva, L. M., 519 (155, 156), *537*, 662 (23, 24), *672*
Katagiri, S., 503 (49), *535*
Kataoka, T., 430, 432–434 (36), *457*
Kato, H., 118 (325), *200*
Kato, M., 103, 124 (150), *197*, 545 (39), 546 (48), 555, 557 (39), *586*
Kato, S., 241 (122), 242 (123), *270*
Kato, Y., 177 (847), *210*
Katov, A. I., 467 (33), *492*
Katritzky, A., 453 (159), *460*
Katsumata, S., 39, 40 (157), *50*
Katsuura, K., 116 (288), *199*
Katzenellenbogen, J. A., 141, 161 (497), *203*
Kauffmann, T., 132 (434), 159 (697, 698), *202, 207*, 507 (76), *535*, 681, 682 (34a, 34b). 683

(34a, 34b, 300), 711, 713 (34a, 34b), 754 (34a, 34b, 297, 299), 755 (297), 757 (299, 301), (298), *758, 763, 764*
Kauffmann, Th., 694, 695, 713, 747, 749 (120), *760*
Kaupp, G., 446 (80), *458*
Kawada, M., 193 (972), *213*
Kawagishi, M., 193 (972), *213*
Kawagishi, T., 98, 187 (68), *195*
Kawamura, T., 168 (759), *209*
Kawasaki, T., 543, 554 (21), *585*
Kawase, T., 469 (55, 56), *493*
Kazymova, M. A., 574 (192), *589*
KcKinnon, D. M., 471 (61), *493*
Keall, J. H. H., 368 (7), *374*
Keat, R., 634, 635 (114), 640 (114, 139), 641 (114), (157), *654, 655*
Keay, B. A., 151 (600), *205*
Kebarle, P., 281, 318 (26), *334*
Keck, G. E., 139 (476), 168 (758), *203, 208*, 666 (59), *672*, 715 (246), *762*
Keeley, D. E., 724 (328), *764*
Keil, W. A., 522, 523 (171), *537*
Keller, H. J., 467 (52, 53), 485, 489 (52, 53, 113), *493, 494*
Kelley, J. A., 116 (287), *199*, 386 (123), *391*, 715 (252), *762*
Kellogg, R. M., 319 (130), *336*
Kelly, M. J., 135 (460), 180 (460, 873), *203, 211*
Kelly, W. A., 369 (28), *375*
Kendall, R. V., 237 (106), *269*
Kende, A. S., 120 (346), *200*
Kennard, O., 614, 615 (61, 62), 646 (62), *653*
Kennedy, M. S. F., 26 (97, 98, 103), *48*
Kennedy, P., 169 (769), *209*
Kerekes, I., 133 (444), *202*
Kerr, K. A., 646 (172), *655*
Kerr, R. G., 100, 126 (111), 136 (463), 138 (111), 149 (567), 150 (567, 578), 184 (893), 189, 190, (463), 191 (893, 942), *196, 203, 205, 211, 212*, 574 (194), *589*
Kessler, W. V., 388 (168), *392*
Ketcham, R., 480 (92), *493*
Kezar, H. S., III 106 (192, 193), 107 (192), *197*
Khammatova, Z. M., 638 (128), *655*
Khan, A. R., 528 (216), *538*
Khavtasi, N., 453 (176), *460*
Khayat, A. I. Y., 370 (46), *375*
Khetrapal, C., 453 (163), *460*
Khidekel, M. L., 467 (32, 33), *492*
Khoi, N., 111, 122 (247), *199*
Khorrami, J., 151 (587), *205*, 290 (48), *335*
Khrapov, V. V., 86 (88), *89*
Kice, J. L., 98 (72, 73), 115 (278, 279), 137 (72, 73, 467), 138 (469), 145 (73, 467), 152 (609), 153 (73, 629), 154 (609), (470), *195, 199, 203, 206*, 280, 310 (22), 311 (22, 103), 312, 313 (103), *334, 336*, 403 (65), *418*, 547 (57), *586*
Kiedrowski, G. V., 102 (137), *196*
Kiel, W. A., 139, 140 (471), 142 (502, 503), 172, 174 (780), 177 (864), *203, 204, 209, 211*, 568 (169), *589*, 652 (199), *656*, 663 (32), 666 (57), *672*, 678 (16, 221), 681 (16), 711 (221), 715 (247), 716 (16), 754, 755 (221), *758, 762*, 770 (42, 44), *791*
Kiem, J., 379 (13), *388*
Kiguchi, T., 148 (565, 566), 174 (566), *205*
Kikuchi, M., 387 (151), *391*
Kikukawa, K., 168 (759), *209*
Kim, H., 108 (213), *198*
Kim, J.-R., 191 (965), *213*
Kim, S.-W., 561 (125), *588*
Kimura, K., 39, 40 (157), *50*, 567 (165), *588*, 785 (113), *792*
Kindt, S., 374 (94), *376*, 384 (77), *390*, 447 (90), *459*
King, C. E., 34, 35 (124), *49*
King, L. C., 255 (212), *271*
King, M. G., 646, 650 (173), *655*
Kingsbury, C. A., 681 (29), *758*
Kini, A., 466, 489 (3), *491*
Kinney, W. A., 98, 137 (74, 76), *195*
Kinoshita, M., 193 (967, 968), *213*
Kirchberg, H., 608 (51), *653*
Kirekas, R., 513 (117), *536*
Kirkien, A. M., 265 (271), *273*
Kirmse, W., 681, 682 (65), *759*
Kirsch, G., 371, 372, 374 (62), *375*, 497 (8), *534*
Kirsi, J. J., 374 (96), *376*, 386 (125), *391*
Kirspuu, H., 253 (193), *271*
Kisch, H., 451 (137), *460*
Kisch, H. J., 262 (245), *272*
Kishi, Y., 110 (237), *198*
Kisilenko, A. A., 443, 444 (72), *458*
Kistenmacher, T., 467 (24), *492*
Kistenmacher, T. J., 250, 251 (182), *271*, 341 (19), *347*, 407 (116), *419*, 480 (89, 98), 485 (98), *493, 494*
Kita, Y., 543, 554 (21), *585*
Kitajima, T., 184 (897), *211*, 726 (329), *764*
Kitamura, M., 117 (317, 318), *200*
Kitchin, J., 142 (507), *204*
Kite, K., 528 (216, 217, 223, 224), *538*
Kito, H., 387 (159, 160), *391*
Kiuchi, F., 101, 125 (128), *196*
Klaeboe, P., 40, 41 (160), *50*, 342 (27h, 29a), 343 (27h), *347*
Klages, C.-P., 224, 233, 240, 263 (59), 266 (276), *269, 273*
Klar, G., 152 (608), *206*, 402 (61), *418*, 505, 507 (59), 519 (152), *535, 537*
Klásek, A., 172 (790), *209*

Klayman, D. L., 94 (1), *194*, 217 (1), 220 (26), 254, 255 (204), 256 (204, 215), *267, 268, 271, 272*, 276 (7), *334*, 394 (1), 396 (10), 403, 406, 407, 415 (1), *416, 417*, 677, 679, 681, 713, 714 (3), *757*, 765 (2), *790*
Klaymann, D., 423–425, 428 (4), *456*
Kleiner, E., 102 (139), *196*
Klenha, J., 765, 766, 769 (16), *791*
Kleschick, W. A., 99 (105), 104, 114 (170), *196, 197*
Klingberg, E., 262 (242), *272*
Klingbserg, E., 471 (60), *493*
Klinke, H., 451 (139c), *460*
Kloker, W., 637 (125), *654*
Klopman, G., 6 (32), *47*
Kluger, E. W., 426 (18), 429, 430, 436 (31), *457*
Knaap, T. A., van der 632 (109), *654*
Knapp, F. F. Jr., 371 (55–69), 372 (55, 58, 60–62, 64, 67, 71–75, 77), 373 (81), 374 (62), *375, 376*, 388 (164, 166), *392*, 531, 532 (245), *539*
Knjazschanski, M. I., 319 (131), *336*
Knochel, P., 116 (304), *200*
Knoechel, A., 505, 507 (59), 519 (152), *535, 537*
Knol, J. A., 554, 555 (94), *587*
Knop, B., 625, 636 (86), *654*
Ko, A. I., 565 (156), *588*
Ko, S. S., 116 (284), *199*
Kobayashi, M., 26, 27 (104), 41 (161, 163) *49, 50*, 100, 138 (109), *196*
Kobayashi, T., 148 (563), *205*
Kobelt, D., 343 (34), *347*
Kober, F., 641 (144–146), *655*
Köbrich, G., 681, 682 (59, 60), *759*
Koch, B., 547 (53, 54), *586*, 595 (10), *652*
Koch, P., 192 (949, 955), *212*, 467, 485, 489 (52, 53), *493*
Kochetkov, N. K., 163 (715, 716), *208*
Kocienski, P., 172 (796), *209*
Kocourek, J., 765, 766, 769 (16), *791*
Koehler, W. H., 415 (152), *420*
Koft, E. R., 110 (234), *198*
Koga, T., 118 (325), *200*
Köhne-Wächter, M., 648 (182), *656*
Kohri, Y., 549 (69), *587*
Koishi, M., 192 (947), *212*
Kokkinidis, M., 452 (147), *460*
Kokoska, S., 385 (134), 386 (110), *390, 391*
Kolar, F. L., 641 (143), 650 (194), *655, 656*
Koller, L. D., 384, 387 (88), *390*
Kolodii, Ya. I., 608, 628, 629 (50), *653*
Kolomeitsev. A. A., 507 (75), *535*
Kolshorn, H., 450 (123), *459*
Kolt, R. J., 98, 129 (63), *195*
Komarouskaya, O. A., 499 (31), *534*
Komarovskaya, O. A., 134 (449, 450), *202*, 706 (170), *761*
Kometani, T., 169, 170 (765), *209*
Komin, A., 452 (150), *460*
Komina, T. V., 33 (120), *49*
Konar, A., 26, 27 (104), *49*
Kondo, A., 175 (828), *210*
Kondo, K., 116 (294), 149 (568), 172 (781), 177 (844, 845), 178 (849–851), 187 (919), 191 (961–963), 192 (943–948, 950–954, 956, 957, 959, 960), *199, 205, 209, 210, 212, 213*, 220 (25), 221 (36), *268*, 409, 410 (124), *419*, 664 (40, 41), *672*
Kondratenko, N. V., 507 (75), *535*, 543 (19), *585*
Konecny, V., 542 (5, 6), *585*
Konetzka, W. A., 369 (10), *374*
Konishi, H., (216), *272*
Kono, S., 5, 8 (15), *47*
Kononovich, P. A., 467, 483 (48), *492*
Konor, A., 407 (109), *419*
Konowal, A., 770 (52), *791*
Koolpe, G. A., 154, 155, 159 (637), 160 (700), *206, 207*, 521 (167), *537*, 683, 685 (69), 694, 695 (122), 696, 701, 702, 706–709, 726 (69), 733 (122), 748 (69), 749 (122), 751 (69), *759, 760*
Koopmans, T., 3 (12), *47*
Kopecky, M. J., 387 (146), *391*
Kopitsya, N.I., 253 (193), *271*
Kopiwoda, S., 371 (57, 68), *375, 376*
Korbacz, K., 114 (268), *199*
Kornis, G., 411 (132), *419*
Korp, J., 528 (218), *538*
Korth, H. G., 772 (66), *792*
Korzeniowski, S. H., 715 (254), *763*
Kosolapoff, G. M., 593, 634, 637, 648 (5), *652*
Kostina, G.I., 626, 630 (93), *654*
Kostynchenko, E. E., 504 (57), *535*
Kosugi, K., 162 (711), *208*
Kotera, K., 97, 116 (50), *195*
Kotov, A. I., 467 (32), *492*
Kourth, M. J., 565 (141), *588*
Kovac, J., 542 (5, 6), *585*
Kowalski, C. J., 184, 186 (898), *211, 721 (288), 763*
Kozawa, Y., 117 (313), *200*, 545 (42), 573 (189), *586, 589*
Kozikowski, A. P., 106 (186), 117 (315, 316), 163 (717–719), 169, 172, 187 (718, 719), *197, 200, 208*, 410 (125), *419*, 770 (53), *791*
Kozuka, S., 153 (625), *206*
Kozyrod, R. P., 772 (62), *791*
Krackov, M. H., 230 (87), *269*
Kraemer, W. P., 25 (78), *48*
Krafft, F., 412 (133), *419*
Krafft, G. A., 543, 572 (23), *586*
Krainova, N. Yu., 409 (123), *419*
Kramolowsky, R., 646, 648 (175), *656*

Krantz, A., 288 (41, 43), *335*
Krasnov, V. P., 409 (119), *419*, 432 (39, 41), 441, 442 (41), (75), *457, 458*
Krasnyanskaya, T., 455 (197), *461*
Kraus, H.-J., 114 (267), *199*
Krause, W., 379 (9), *388*
Krauss, G. A., 498 (24), *534*
Krauss, H. L., 617 (69), *653*
Krawiecka, B., 625, 636 (87), *654*
Krebs, A., 398 (30), *417*
Krebs, E.-P., 99 (95),*195*
Kresze, G., 439, 441 (54), *458*
Kretzschmar, G., 323 (139), *337*
Krief, A., 97 (59), 99 (83, 90–92), 100 (91, 121–126), 115 (273, 274), 116 (59, 273, 274), 125 (384, 386), 126 (390–392), 132 (391, 428–432), 153 (621, 622), 154 (91, 121, 122, 124, 384, 386, 429, 621, 622, 630, 631, 634, 640–652, 659–663, 666–677), 155 (91, 122, 124, 384, 386, 621, 622, 634, 640–652, 659–661), 156 (91, 121, 122, 384, 386, 429, 641, 643, 646, 647, 650–652, 662, 666, 671, 678), 157 (122, 124, 384, 391, 650, 651, 659, 666, 668, 672, 681, 683–687), 159 (391, 431), 160 (431), 161 (683, 708), 163 (724), 171 (681, 684), 172 (783), 173 (430, 797), 174 (634, 645), 175 (824), 177 (843), 179 (123), 182 (661, 668, 670, 671), 183 (667, 669–671), 184 (895), 186 (91, 92, 895, 904), 187 (91), *195, 196, 199, 201, 202, 206–211*, 520 (161), 521 (166, 168), 522 (177a), 523 (179, 180), 525 (189), *537, 538*, 542, 552 (7), 557 (115), *585, 588*, 677 (8, 11, 12a, 12b), 678 (8, 11, 12a, 12b, 13, 14, 19), 679 (8, 11, 17–20, 334, 335), 680 (17, 18, 23), 681 (8, 11, 12a, 12b, 13, 14, 17–20, 23–25, 68), 682 (8), 683 (68, 74), 684 (68, 74, 83, 326), 685 (8, 11, 19, 68, 74, 88a, 88b, 89, 94–97, 99, 127, 133, 336), 686 (11, 68, 74, 94, 100, 127, 133), 687 (8, 104, 106a) 689 (8, 11, 12a, 12b, 94, 107–112, 127, 133), 690, 691 (111), 692 (8, 112, 114, 116), 693 (112, 118), 694 (121, 123, 124), 695 (11, 14, 20, 74, 88a, 88b, 89, 94, 106a, 107, 114, 118, 121, 127–143), 696 (68, 74, 106a, 108, 112, 114, 133, 148–150, 187), 697 (108, 139, 144), 698 (94, 108, 109, 139, 144, 155), 699 (11, 68, 74, 94, 108, 109, 130, 131, 135, 144), 700 (8, 89, 112, 138, 155), 701 (8, 18, 88b, 89, 148–150), 702 (149, 150, 156), 703 (19, 148, 150, 157), 704 (157), 705 (19, 159, 160), 706 (11, 19, 74, 121, 174, 181), 707 (121), 708 (68, 124, 185, 187), 709 (12a, 12b, 121, 185, 188), 710 (8, 12a, 12b, 135, 188, 207–211, 326), 711 (8, 11, 14, 108, 143, 148, 185, 211), 712 (12a, 12b, 17, 18, 185, 223, 326), 713 (8, 18, 106a, 185), 714 (8, 123, 232, 337), 715 (11, 14, 20, 108, 123, 127, 132, 136, 232, 251), 716 (74, 159, 160, 266, 269), 718 (19, 20, 23, 94, 106a, 108, 109, 129, 135, 137, 138, 140, 251, 269, 280, 281), 719 (11, 19, 20, 94, 108, 129, 137, 140, 269), 720 (74, 108, 123, 124, 127–129, 138, 140, 144, 266), 721 (11, 23–25, 88a, 88b, 100, 107, 108, 129–131, 139, 140, 251, 269, 281, 283, 337), 722 (100, 107, 108, 130, 131, 144, 185), 723 (88a, 88b, 107, 109, 187, 188, 290), 724 (108, 109, 135, 339), 725 (135, 340), 726 (8, 11, 19, 68, 94, 330), 727 (123, 124, 181, 187), 728 (12a, 12b, 14, 108, 144), 730 (107, 108, 130), 731 (139, 269), 732 (20, 127, 232), 733 (121, 127), 734 (100, 308b), 735 (124, 185, 187), 736 (100, 187, 308b), 737 (124, 185, 187), 738 (12a, 12b), 739 (127), 740 (11, 108, 144, 290), 741 (174, 310), 743 (129, 135), 745 (106a, 108, 144), 746 (106a), 747 (17, 121, 135), 748 (74, 266), 749 (107, 121, 266), 750 (110), 751 (148), 753 (74), 754 (310), 755 (18), 757 (8), (146, 320), *757–764*
Kriegsmann, R., 757 (301), *764*
Krishnamachari, S. L. N. G., 286 (37), *335*
Krishnan, V., 248 (168), 253 (194), *271*
Kristiansen, E. S., 220, 221 (31), 253, 254, 256 (191), *268, 271*, 473 (68), *493*
Krogh, J. A., 151 (602, 603, 606, 607), *205, 206*
Krohnke, C., 485, 489 (114), *494*
Kroon, S. L., 643 (154), *655*
Kroth, H.-J., 400 (52), *418*, 635, 640 (117, 119), 641, 643, 644 (117), *654*
Krow, G. R., 126 (389), *201*
Kruchten, E. M. G. A., van 180 (874), *211*
Kruck, T., 340 (4), *346*
Krug, W., 466, 467 (1), 483 (1, 105), *491, 494*
Krug, W. P., 483 (103), *494*
Krüger, C., 114 (267), *199*, 633 (113), 649 (185), *654, 656*
Krumbein, W. E., 370 (42), *375*
Krupoder, S. A., 416 (157), *420*
Kruse, F. H., 62, 65 (32), *88*, 514, 517 (122), *536*
Krutosikova, A., 542 (6), *585*
Kryuchkova, L. V., 545 (32), *586*
Ksander, G. M., 107 (203), *198*
Kubiniok, S., 600 (26, 27), 623 (84), *652, 654*
Kubiniol, S., 611, 613, 617, 618, 644 (54), *653*
Kuchadker, M. V., 217 (8), *267*
Kuchen, W., 625, 636 (86), 638 (130), 641 (148), 650 (192), *654*–656
Kuchitsu, K., 25 (79), *48*
Kudchadker, M. V., 94 (2), *194*, 765 (1), *790*
Kudelska, W., 767 (27), *791*

Kuder, J. E., 265 (270), *273*
Kudo, K., 191 (965), *213*
Kuebler, N. A., 40 (152), *50*
Kuehn, K., 380 (24), *389*
Kugimiya, M., 165 (738), *208*, 396 (19), *417*
Kuhl, P., 447 (92), *459*
Kuhn, N., 308 (95, 97), *336*, 623 (83), 624 (83, 85), 633 (111, 112), 642 (151), 643, 646 (85), 650 (151), 651 (83), *654, 655*
Kuhn, R., 221 (44), *268*
Kuivila, H. G., 312 (108), *336*
Kukk'ola, P., 706 (171), *761*
Kukkola, P., 187 (933), *212*
Kul'bitskaya, O. V., 544 (28, 29), *586*
Kul'bitzkaya, O. V., 297 (67), *335*
Kulikova, M. F., 369 (17), *374*
Kulkarni, Y. D., 373 (85), *376*
Kulkarni, Y. S., 118 (338), *200*
Kulkowit, S., 189 (938), *212*, 706 (175), *761*
Küllmer, V., 340 (4), *346*
Kumagawa, T., 172 (779), *209*
Kumar, R., 373 (84), *376*
Kumar, V., 343 (33), *347*
Kumari, S., 452 (149), *460*
Kumler, P. L., 282, 327, 328 (28), *334*, 412 (136), *419*
Kumler, W. D., 247 (165), 263 (258), 265 (165), *270, 272*
Kumogai, H., 467 (25), *492*
Kunai, A., 99 (95), *195*
Kung, H. F., 388 (169), *392*
Kunitskaya, G. P., 444 (73, 74), 445 (74), *458*
Kunwar, A., 453 (163), *460*
Kunz, D., 266 (275), *273*
Kupper, R., 113 (257), *199*
Kuppermann, K., 22, 25 (77), *48*
Kurbanova, N. Z., 369 (17), *374*
Kuriyan, K. I., 705 (164b), *761*
Kurobe, H., 116 (297, 298), 144 (297, 298, 531–534), 172 (531–534), 177 (846), *200, 204, 210*
Kurokawa, N., 567 (164), *588*
Kurosawa, M., 550 (85), *587*
Kurz, J. H., 340 (10), *346*
Kushch, L. A., 530 (239), *539*
Kushnarev, D. F., 531 (248), *539*, 550 (78, 80, 82), *587*
Kustan, E. H., 62, 63, 65, 67, 71, 74 (23), *88*, 533 (267), *539*
Kutter, J., 638 (130), *655*
Kuwajima, I., 115 (306–309), 116 (306–310), 146 (542–547), 154, 155 (654), 163 (720–723), 169 (654), 184 (894), *200, 204, 206, 208, 211*, 706 (173), 716–718 (271), 746 (293), 747 (271), *761, 763*
Kuyama, H., 565 (149), *588*
Kuzmin, R. N., 53 (17), *87*
Kwak, J. F., 483 (108), *494*
Kwart, H., 715 (235), *762*
Kwart, L. D., 715 (235), *762*
Kyandzhetsian, R. A., 753 (304, 305), *764*
Kyba, E. P., 326 (148), *337*

Laane, J., 641 (143), *655*
Laarif, A., 34 (123), *49*
Labar, D., 99 (83), 115, 116 (273, 274), 154 (649), 651, 669), 155 (649, 651), 156, 157 (651), 183 (669), *195, 199, 206, 207*, 681 (25), 689 (108), 695 (132, 135), 696 (108), 697, 698 (108, 144), 699 (108, 135, 144), 710 (135), 711 (108), 714 (232, 337), 715 (108, 132, 232), 718 (108, 135, 281), 719 (108, 338), 720 (108, 144), 721 (25, 108, 281, 337), 722 (108, 144), 724 (108, 135), 725 (135), 728 (108, 144), 730 (108), 732 (232), 740 (108, 144), 743 (135), 745 (108, 144), 747 (135), *758, 760, 762–764*
Labarre, J. F., 497 (10), *534*
Lablache-Combier, A., 165 (740), *208*, 284 (32, 33), 315, 316 (119), *335, 336*
Lablanche-Combier, A., 396 (14), *417*, 527 (199), *538*, 660 (14), *672*
Labor, D., 520 (161), *537*
Laboureur, J. L., 154 (667, 669), 156 (678), 183 (667, 669), *207*, 695, 699, 710 (135), 718 (135, 282), 724 (135, 282, 339), 725 (135, 282, 340), 726, (282, 330), 743 (135), 745 (282), 747 (135, 282), *760, 763, 764*
Ladenstein, R., 350 (1), *363*
Lafont, J.-P., 370 (39), *375*
Laghai, A., 356 (25), *364*
Lagier, R., 483 (107), *494*
Lagow, R. J., 510 (94, 95), *536*
Laishes, B. A., 385 (129), *391*
Laishev, V., 450 (129), *459*
Laishev, V. J., 408 (99), *418*
Laishev, V. Z., 245 (137), *270*
Laitalainen, T., 110 (233), *198*
Laitalaineu T., 513 (117), *536*
Laitem, L., 448 (101), *459*, 531 (247), *539*, 550 (76), *587*, 765 (19), *791*
Lakatos, T., 369 (18), *374*
Lakshimikantham, M. V., 580, 582–584 (221), *590*
Lakshmikantam, M. V., 477 (84, 85), 478 (86), *493*
Lakshmikantham, M. V., 193 (970), *213*, 328 (157), *337*, 341 (17, 18), 345 (18), *346*, 542, 554 (3), 556 (3, 110), *585, 587*
Lakshmikanthan, M. V., 245 (139, 140), *270*, 408 (92), *418*
Lakshmilkantham, M., 449, 454 (114), *459*
Lalezari, I., 151 (581–587, 589), *205*, 243 (131, 133), (218), *270, 272*, 290 (47, 48), *335*, 340

(12), *346*, 446 (79), 449 (79, 106), 450 (127), *458, 459*
Lam, P. Y.-S., 564 (135), *588*
Lambert, C., 447 (98), *459*, 481, 483 (99), *494*
Lambert, S., 453 (174), *460*
Lamm, V., 453 (178), *460*
Lamotte, G., 246 (154), *270*
Lamotte, J., 515, 516 (124), *536*
Lamotte-Brasseur, J., 454 (189), *461*
Lancelin, J. M., 772 (60), *791*
Landsberg, B. M., 549 (68), *586*
Lane, A. G., 310 (102), *336*
Lane, H. W., 385 (131), 387 (127), *391*
Lane, J. M., 381 (36–38), *389*
Lange, C. A., de 14, 15 (61), 16 (65), 17 (65, 66), 18 (66, 68, 69), 19 (66), 20, 42 (69), *47, 48*
Lange, L., 511 (101), *536*, 601 (33), *652*
LaPlaca, S. J., 467 (16), *492*
Lapouyade, R., 165 (740), *208*, 396 (14), *417*
Lappert, M. F., 42, 44 (168), *50*, 258–260, 267 (227), *272*, 303, 305 (82), *335*, 396 (8), *417*
Lardon, M. A., 396, 403, 412 (12), *417*
Larin, G. M., 253 (193), *271*
Larsen, B. D., 288 (42), *335*
Larsen, C., 253 (198), 254, 255 (205), *271*
Larsen, R. D. Jr., 147, 148 (559), *205*
Larsen, S., 104 (179), *197*, 404 (69), 414 (145), *418, 419*
Laszlo, S. E., de 140 (487), *203*
Latshaw, J. D., 382 (51), *389*
Lattman, M., 42 (164, 165), *50*
Lauer, R. F., 94 (3), 95, 96 (26–28), 97 (40), 98 (28, 40), 99 (28), 100 (114), 101 (27), 102 (114), 103 (27), 115, 116 (40), 118 (327), 122 (114), 171 (26), 182 (879), *194, 196, 200, 211*, 677 (2), 681 (2, 27), 685 (2, 87), 695 (87), 714 (27, 226, 229), 715 (2, 27, 226, 229), 716 (27), 719, 720 (2), 721 (27, 226, 284, 286), 756 (2), *757–759, 762, 763*, 770 (54–57), *791*
Laughlin, D., 82 (72), *88*
Laughlin, D. R., 82 (76), *89*
Laukhin, V. N., 467 (33, 36, 48), 483 (48), *492*, 530 (239), *539*
Laureni, J., 288 (41, 43), *335*
Laver, R. F., 522 (174), *537*
Law, K.-W., 100 (112), 109 (222), 138 (112, 222), 162 (714), 174 (112), 180 (112, 222), *196, 198, 208*
Lawrence, R. A., 381 (36), *389*
Lawson, T. A., 385 (138), *391*
Layer, M., 151 (593, 596), *205*, 450 (122), *459*
Laypouyade, R., 284 (32), *335*
Lazic, R., 149 (573), *205*, 718 (274), *763*
Learn, K., 174 (815), *210*
Leaver, D., 471 (61), 477 (82), *493*
Lebedev, R. A., 53 (18), *87*
Lebedev, V. A., 53 (18), *87*
Leber, J. D., 102, 122 (142), *196*
LeBoeuf, R. A., 380 (30), 385 (129), *389, 391*
LeCostumer, G., 487 (125), *494*
Le Coustumer, G., 486 (123, 124), 487 (124), *494*
Lederev, K., 508 (85), *536*
Lee, C., 141 (495), *203*, 773 (69), *792*
Lee, E., 118 (341), *200*
Lee, H., 100 (113), *196*, 756 (333), *764*
Lee, H. H., 543 (20), *585*
Lee, H. S., 476, 477 (80), *493*
Lee, J. J. S., 787 (121), *793*
Lee, J.-S., 486 (121), *494*
Lee, K. H., 322 (137), *337*
Lee, M., 250, 251 (182), *271*, 467 (24, 47), 469 (47), 480 (89), *492, 493*
Lee, S. T., 9 (44), 10 (44, 46), 11, 12 (46), *47*
Lee, V. Y., 251 (184), *271*, 469 (110), 483 (107, 110), *494*
Lee, W.-J., 356, 357 (32), *364*
Lee, Y.-W., 118 (341), *200*
Leeuw, D. M., 14, 15, (61), *47*
Lefur, D., 467, 469 (23), *492*
Lehmann, G., 765, 766 (14), *791*
Lehmann, H., 544 (24), *586*
Leibfritz, D., 102 (139), *196*
Leibnitz, E., 608 (51), *653*
Leicester H. M., 498 (19), *534*
Leitereg, T. J., 697 (291b), *763*
Lelj, F., 453 (162), *460*
Le Marechal, A., 552 (91), *587*
Le Minor, L., 370 (38), *375*
Lendor, P. W., 303, 305 (82), *335*
Leonard, W. R., 187 (922), *212*
Léonard-Coppens, A. M., 184, 186 (895), *211*
Leonard-Coppens, A. M., 695, 718–721 (140), *760*
Leopold, W. R., 386 (124), *391*
Lependina, O. L., 530 (239), *539*
Leporati, E., 519 (153), *537*
Leray, V., 665 (51), *672*
Lerouge, P., 102 (145, 146), 128, 139, 180 (406), *196, 202*, 426 (19b), *457*
Le Roux, J.-P., 325 (146), *337*
Le Roux, J. P., 525 (187), *538*
Lerstrup, K., 249 (178), 250, 251 (182), *271*, 341 (19), *347*, 407 (116, 117), 413 (143), *419*, 466 (3), 467 (24), 475 (75), 480 (89, 91, 97, 98), 485 (97, 98, 111), 489 (3), *491–494*
Lerstrup, K. A., 240, 254, 264, 265 (116), *270*
Lesch, D. A., 345 (48), 346 (52), *347*
Lesiak, K., 576 (200), *589*, 629 (100), *654*
Leslie, E. J., 622, 624, 629, 643, 647, 651 (78), *653*
Lesma, G., 99 (87), 174 (813), *195, 210*
Lesser, R., 446 (87), *459*

Lester, D. J., 111 (238–240, 245), 120 (245), 146 (240), 149 (569, 570), 150 (569, 570, 579), 186 (902), *198, 199, 205, 211*, 741 (313), *764*

Lestrup, K., 506 (68), *535*

Lestrup, K. A., 504 (54), *535*

Leung, M., 517 (141), *537*

Leung, P., 467, 483 (51), *493*

Le Van, D., 44, 45 (169), *50*

Levason, W., 528 (226), *538*

Levchenko, E., 455 (196, 197), *461*

Levchenko, E. S., 428 (28), 435, 436 (43–45), 437 (47, 48), 438 (53), 439 (56), 440 (56, 59), (58), *457, 458*

Levenberg, P. A., 102, 179 (138), *196*

Lever, O. W., 681, 682 (38), *758*

Levi, E. J., 664 (46), *672*

Lewicki, J. W., 228 (78), *269*, 285, 286 (36), *335*, 344 (40), *347*, 396, 403, 412 (13), *417*, 660 (18), *672*

Lewis, D. T., 240 (118), *270*

Lewis, J., 405 (82), *418*

Ley, S. V., 94 (9), 111 (238–242, 245, 246), 120 (245, 350), 121 (354, 355, 357), 122 (246), 128 (403), 129 (413), 140 (490, 491), 143 (511–516, 518), 146 (240–242), 147 (551), 149 (569–571), 150 (551, 569–571, 579), 152 (551, 571, 613, 614), 153 (571, 614, 617–619, 623, 624, 627), 154 (571), 164 (731), 173 (805–807), 174 (513), *194, 198–206, 208, 209*, 427 (25b), 429, 430, 436 (32), *457*, 765 (7), 788 (130), *790, 793*

Leyck, S., 383 (73), 384 (73, 74), *390*, 447 (94, 95), *459*

Lginova, E. I., 625 (89), *654*

Li, J., 370 (44), *375*, 383 (68), *390*

Liao, C. C., 330 (164), *337*, 721 (239), (240), *762*

Liao, T. K., 120 (348, 349), *200*

Lieberknecht, A., 97 (54), *195*, 563 (133), *588*

Liebscher, J., 236 (104), 238 (104, 112), *269*, 446 (84), *458*, 553, 555 (92), *587*

Lietz, M., 401 (56), *418*, 619, 620 (71, 73), *653*

Lightner, D. A., 264, 265 (269), *273*

Lin, T., 453 (157), *460*

Lind, J., 279 (18), *334*, 527 (201), *538*

Lindenauer, S. M., 554, 555 (94), *587*

Lindgren, B., 500 (33), *534*, 542 (2, 10, 11), 547 (10, 55), 553 (11), *585, 586*

Lindholm, E., 6 (33, 38), 7 (38), 22, 25 (75, 76), *47, 48*

Lindner, H.-G., 637 (123), *654*

Lindner, U., 497 (11), *534*

Lintermans, P., 370 (38), *375*

Liotta, D., 32, 33, 36 (116), *49*, 94 (10–12), 103 (151), 106 (192), 107 (192, 199), 129 (409), 141 (495), 169 (761, 762, 767), 184 (889, 890, 892), 187 (889, 892, 918, 932, 934–936), *194, 197, 198, 202, 203, 209, 211, 212*, 327 (150), *337*, 406 (84), *418*, 499 (25), 520 (25, 162), *534, 537*, 685 (84), 715 (249), *759, 762*, 765 (8), 773 (69), *790, 792*

Liotta, D. C., 154, 155 (658), *207*, (240), *762*

Liovbev, B. G., 638 (128), *655*

Lipatova, I. P., 647 (177), *656*

Lipka, R., 770 (36), *791*

Lipovich, T. V., 507 (74), *535*

Lipp, M., 247 (163), *270*, 585 (228), *590*

Lish, P. M., 378 (2), *388*

Lister, S. G., 152 (613, 614), 153 (614), *206*

Liston, S. R., 27 (106), *49*

Little, R., 580 (222), *590*

Litvinov, V. P., 407 (112), *419*

Liu, P. S., 116 (287), *199*, 715 (252), *762*

Livdane, A. D., 485 (117), *494*

Livinghouse, T., 187 (922), *212*

Livingston, R., 284 (34), *335*

Llabres, G., 62, 65 (33), *88*, 342 (26), 345 (44), *347*, 448 (101), *459*, 517 (135), 531 (247), *537, 539*, 550 (76, 81, 83), 577 (209), *587, 590*

Lloyd, C. M., 11 (51), *47*

Lloyd, D., 705 (163, 164a, 302), 753 (295a, 295b, 302), *761, 763, 764*

Lloyd, D. R., 13 (55), *47*

Lobert, A., 552 (91), *587*

Lodder, G., 297 (70), *335*

Lodge, P. G., 405 (82), *418*

Logacheva, I. I., 532 (258), *539*

Loginova, E. I., 637 (124), *654*

Logusch, E., 565 (145), *588*

Lohner, W., 278 (15, 16), 290 (49), 301 (16), 321 (15, 16), *334, 335*, 509 (90), 525 (188), *536, 538*

Lok, K. P., 169, 170 (771), *209*

Lombardo, L., 104 (155–157), *197*

Long, C. G., 249, 251, 252 (176), *271*

Long, G. G., 71 (46), *88*

Loogen, F., 379 (13), *388*

Lopez, A. F., 294 (64), *335*

Lopez, L., 176 (840), *210*

Lopez-Castro, A., 266 (284), *273*

Lopusiński, A., 627 (98), *654*

Lorah, D. P., 120 (346), *200*

Loreh, M., 450 (124), *459*

Lorenz, B., 253 (195), *271*

Lorenz, M. G., 370 (42), *375*

Lorenz, W., 627 (96), *654*

Lorenzon, G., 370 (43), *375*

Loss, H. R., 341 (17), *346*

Lotz, W. W., 161 (709), *207*, 278 (17), *334*, 705 (165), *761*

Lowerey A. H., 35 (133), *49*

Lozinskii, M. O., 542 (14), *585*

Lubineau, A., (238), *762*, 770 (46, 47, 49), *791*

Lucchesini, F., 446 (81), 448, 452 (100), 454 (81, 187b, 190–192), *458, 459, 461*
Lucchetti, J., 99 (92), 154 (631, 634, 648, 672–677), 155 (634, 648), 157 (672, 683, 687), 161 (683), 174 (634), 186 (92), *195, 206 207*, 520 (161), *537*, 681 (25), 685 (88a, 88b, 89, 97, 133, 336), 686 (133), 689 (112, 133), 692, 693 (112, 117), 695 (88a, 88b, 89, 133), 696 (112, 117, 133, 148–150), 700 (89, 112, 117), 701 (88b, 89, 117, 148–150), 702 (117, 149, 150, 156), 703 (117, 148, 150, 157), 704 (157), 711 (148), 721 (25, 88a, 88b), 723 (88a, 88b), 730 (117), 741 (310), 751 (117, 148), 754 (310), (146), *758–761, 764*
Lucci, R. P., 325 (143), *337*
Luce, E., 117 (319), *200*
Luczac, J., 153 (626), *206*
Luczak, J., 258 (225), *272*
Ludersdorf, R., 277, 300 (12), *334*
Ludlow, S., 517 (137), *537*
Ludwig, E. G., 604 (44), 615 (64), *653*
Lukjanow, B. S., 319 (131), *336*
Lukman, B., 266 (280), *273*
Luppold, E., 329 (160), *337*
Lusinchi, X., 148 (564–566), 174 (566), 178 (855, 859), *205, 210*
Lusk, D. I., 710 (201), *761*
Luss, H. R., 245 (138), 262 (243), *270, 272*, 344 (42), *347*, 408 (93), *418*, 527 (212), 528 (228), *538*
Luthra, N. P., 555 (105), *587*
Lutsenko, I. F., 612 (55), *653*
Luttinghaus, A., 471 (62), *493*
Luxen, A., 447 (97), *459*, 509 (91), *536*
Luzinchi, X., 147, 148 (558), *205*
Lyalikova, N. N., 369 (17), *374*
Lyapina, F., 455 (194, 195), *461*
Lyapina, T. V., 428 (28), 430, 432, 433 (34), 435, 436 (44, 45), 437 (47, 49), 440 (59),441 (49), (58), *457, 458*
Lygo, B., 140 (490, 491), 143 (514, 515), *203, 204*, 427 (25b), *457*
Lynn, D. G., 121 (360), *201*
Lyons, D. E., 355 (18), *364*
Lyons, R. E., 221 (42, 43), 227 (43), *268*, 412 (133), *419*
Lysenko, Z., 139 (478), 141, 174 (493), *203*, 665 (53), *672*, 711 (220), 715 (220, 245), *762*, 768, 790 (30, 31), *791*
Lysy, R., 342 (22), *347*, 577, 582 (206), *589*
Lyubovskii, R. B., 467 (32), *492*
Lyzwa, P., 153 (616), *206*

Maartman-Moe, K., 581, 583 (224), *590*
Maartmann-Moe, K., 342 (29c), *347*, 542 (18), 549 (67), 579, 580 (18), 581 (67), *585, 586*, 645 (169), *655*
MacCanon, D. M., 378 (1), 380 (22), *388, 389*
MacConnell, K. P., 382 (46), *389*
MacDiarmid, A. G., 407 (110), *419*
MacDonald, C. B., 17, 18 (67), *48*
Macdonald, J. E., 182 (884), *211*
MacGillavry, C. H., 614, 615, 646 (62), *653*
Machado, R., 317 (124), *336*
MacNicol, D. D., 165 (735, 741, 742), *208*, 395 (6, 7), 396 (21), *417*
Maddock, A. G., 53, 54, 62 (16), *87*
Maeda, A., 99 (86), *195*
Maeda, M., 386 (117), *391*, 790 (134), *793*
Maeda, S., 369 (15), *374*
Maekawa, E., 137 (468), 141 (498), *203*, 427 (25c), 428 (27), *457*
Maercker, A., 681, 682 (48), *758*
Maese, C. O., (111), *269*
Magdesieva, N. N., 25 (91), *48*, 705 (161), 753 (304, 305), *761, 764*
Magerlein, B. J., 681, 682 (62), *759*
Magid, R. M., 710 (191), *761*
Magnolda, R. L., 139 (480, 481), 174 (480), *203*, 665 (53), *672*
Magnus, P., 114 (272), 181 (878), *199, 211*, 706 (178), *761*
Magnus, P. D., 94, 95 (8), 121 (356, 357), 168 (356), *194, 201*, 217 (11), 258 (222), *267, 272*, 497, 516 (4), *534*, 681, 682 (57), *759*
Magolda, R. L., 98 (77, 78), 109 (225), 139 (78, 474b), 141 (474b), 142 (77, 78, 508), 152 (78, 508, 610–612), 174 (78), *195, 198, 203, 204, 206*
Magrino, S., 650 (193), *656*
Mahalanabis, K. K., 106 (187), *197*
Maier, J. P., 26 (96), 37, 38 (144), 39 (149), *48, 50*
Maier, L., 593 (5), 607, 609, 610 (48), 617 (70), 627, 629 (48), 634, 637, 648 (5), *652, 653*
Mailbard, Ph., 305 (87), *336*
Maines, M. D., 381 (40), *389*
Maiorova, L. P., 281, 309 (23), *334*, 663 (29), *672*
Mairova, L. P., 399 (41), *417*
Maisch, R., 649 (187), *656*
Majetich, G., 173 (801), *209*, 563 (131), *588*
Majewski, M., 106 (189), *197*
Makani, T., 346 (54), *347*
Maksimenko, A. A., 441, 442 (65), *458*, 753 (306), (315), *764*
Malashkhiya, M., 453 (176), *460*
Malazow, L. N., 416 (157), *420*
Malek-Yazdi, F., 229 (84), 235 (98), 264 (84), *269*, (46), *335*, 450 (126), *459*
Maletina, I. I., 543 (19), *585*
Malisch, W., 649 (186, 187), 650 (186), *656*
Mallaki, J., 517, 529 (146), *537*

Mammi, M., 221 (47, 48), 261 (241), *268, 272*
Manandhar, M. D., 113 (255), *199*
Manapov, R. A., 53 (14), *87*
Manatt, S. L., 643 (154), *655*
Mandai, T., 193 (972), *213*
Mandel, G. S., 174 (820), *210*
Mandel, N. S., 174 (820), *210*
Mander, L. N., 104 (155–157), *197*
Manderson, W. G., 368 (8), *374*
Manghi, N., 378 (3), *388*
Mangini, A., 33 (119), *49*
Mangion, M. M., 81, 83 (64), *88*
Mann, F. G., 614, 615 (61, 62), 646 (62), *653*
Mannafov, T. G., 134 (449, 450), *202*, 499 (31, 32), *534*, 626, 630 (93), *654*, 706 (184), *761*
Manson, S. T., 5, 8 (16), *47*
Mansuy, D., 556 (112), *588*
Mantei, R., 119, 148 (343), *200*
Mao, M. K.-T., 99, 122, 183 (94), *195*
Marafante, E., 370 (43), *375*
Marbury, G. D., 179 (868), *211*, 568 (172), *589*
Margolis, D. S., 221, 222 (45), *268*
Marino, G., 21–25 (72, 74), 33 (74), *48*
Marino, J. P., 147 (553, 559), 148 (559), 168 (553), *204, 205*
Marinovic, N., 173, 184 (799, 802), *209*
Markau, K., 279 (20), *334*, 526 (192), *538*
Markiewicz, W., 169 (761, 762), *209*
Markovska, A., 626, 630 (92), *654*
Markovski, L. N., 444 (73, 74), 445 (74), *458*, 623, 634, 637 (82), *653*
Markowska, A., 630 (102), *654*
Markowski, L. N., 443, 444 (72), *458*
Maroy, K., 342, 343 (27a, 27b), *347*
Marquez, V. E., 116 (287), *199*, 386 (123), *391*, 715 (252), *762*
Marschner, F., 641 (147), *655*
Marsden, C. J., 608 (49), *653*
Marsden, K., 411 (129), *419*
Marsh, D. G., 276, 282 (6), 283 (29), 285, 286 (36), 328 (6), *334, 335*, 344 (40), *347*, 412 (137), 413 (137, 140), *419*, 526 (191, 197), *538*, 659 (8, 12), *671*
Marsh, P. F., 72 (49), *88*
Marsh, R. E., 62, 65 (32), *88*, 514 (121, 122), 517 (122), *536*
Marshall, J., 456 (202), *461*
Marshall, J. A., 124 (380), *201*, 560 (121), *588*
Marshall, M. V., 385, 386 (104), *390*
Martens, J., 276 (9), 277 (11, 12), 278 (15), 300 (12, 80), 321 (11, 15, 135), 333 (167), *334, 335, 337*, 523 (182), 525 (188), *538*
Martin, B., 59, 62 (21), *88*
Martin, D. R., 297 (73), *335*
Martin, J. E., 384 (81–83), *390*
Martin, J. L., 353 (16), *364*
Martin, N. H., 368 (7), *374*
Martin, R. L., 260 (235), *272*
Martin, S. F., 118 (324), *200*
Martin, T. R., 258–260, 267 (227), *272*, 396 (8), *417*
Martini, F., 378 (3), *388*
Martinsen, A., 580 (220), *590*
Marusic, N., 379 (17), *389*
Maruyama, K., 139 (482), 157, 158 (682, 688), *203, 207*
Marx, J. N., 161 (705), *207*
Masaki, Y., 99 (99), 173 (99, 798, 800), *195, 209*, 555 (99), 560 (120), *587, 588*
Masamune, S., 172 (795), 173 (811), *209, 210*
Masaoka, K., 104 (159, 180), *197*, 685 (256), 715 (256, 257), *763*
Masey, A. G., 498 (15), *534*
Mason, R. P., 380 (29), *389*
Massot, J. C., 305 (87), *336*
Masters, B. S. S., 381 (41), *389*
Masuda, K., 565 (147), *588*
Masuda, S., 555 (101), 565 (140), *587, 588*
Masuda, Y., 498 (20), *534*, 546 (44, 45), *586*
Masukawa, T., 371 (50, 51), *375*, 379 (11), 382 (54), 383 (61–63), 387 (159, 160), *388–391*
Masuyama, Y., 113, 161 (261), *199*, 685, 695 (93), *759*
Mataka, S., 452 (148), *460*
Mathews, W. S., 108 (207), *198*
Mathey, F., 176 (838), *210*
Mathias, A., 264 (265), *273*
Mathiasch, B., 414 (148, 149), *420*
Mathies, P., 187 (925), *212*
Maticoli, F. J., 326 (149), *337*
Matsuda, T., 168 (759), *209*
Matsui, T., 111 (248), *199*
Matsumoto, H., 565 (138), *588*
Matsumoto, T., 118 (339), 187 (924), *200, 212*
Matsumoto, Y., 369 (29–32), *375*
Matsunaga, Y., (130), *494*
Matsuo, N., 568 (170), *589*
Matsuura, T., 177 (846), *210*
Mattes, R., 242, 266 (127), *270*
Matthews, W. S., 685 (86), *759*
Matz, C., 242, 266 (127), *270*
Maul, J. J., 32 (113), *49*
Maunafov, T. G., 706 (170), *761*
Mausner, L. F., 371, 372 (61), *375*
Mautner, H., 265 (220), *272*
Mautner, H. G., 217 (6), 230 (87), 237 (108), 247 (165), 263 (108, 254, 255, 258), 265 (108, 165, 274), 266 (279, 286), *267, 269, 270, 272, 273*, 386 (112), 388 (161, 162), *390, 391*, 778, 789 (90, 91), *792*
Maxfield, M., 489 (132), *494*
May, L., 52 (4), *87*
Mayer, C. W., 467 (30, 31), 485 (30), *492*
Mayer, R., 247 (155), 266 (275), (217), *270, 272, 273*, 408 (87), *418*
Mayerle, J. J., 467 (16), *492*

Maynard, E. P., 379 (16), *389*
Mayo, P., de 222 (56), 227 (70, 71), *268, 269*, 276 (4), 322 (136, 137), 329 (4, 136), 330 (4, 136, 164), *334, 337*
Mayr, A., 453 (173), *460*
Mayr, A. J., 262 (246), *272*
Mays, M., 467 (24), *492*
Mays, M. J., 246 (146), *270*
Mazaud, A., 467, 469 (37, 41), *492*
Mazzochin, G., 483 (101), *494*
M'Buyi, M., 446 (86), 447 (97), 448 (86), *459*
McAfee, F., 115 (278), *199*
McAulifee, C. A., 528 (226), *538*
McCarthy, A. E., 53, 62, 85 (13), *87*, 515 (123), 517 (137), *536, 537*
McCauley, J. P., Jr. 99, 180 (89), *195*
McClelland, R. A., 517 (141), *537*
McCollum, G. J., 108 (207), *198*, 685 (86), *759*
McCombie, S. W., 175, 178 (821), *210*, 221, 229, 232, 233, 240 (38), *268*, 788 (131), *793*
McConnell, K. P., 263 (248), *272*, 360 (42), 362 (53), *364, 365*, 382 (48), *389*
McConnell, M. L., 21 (70), *48*
McCormick, F. B., 572 (185), *589*
McCullough, J. D., 62, 65 (32), 67 (39–41), 72 (49), 80 (60), 81 (63), 82 (60), *88*, 344 (39a), *347*, 514, 517 (122), 529 (233), *536, 539*
McCullough, R., 467 (24), *492*
McCullough, R. D., 471, 480 (93), *493*
McCurry, P. M. Jr., 108 (216, 217), *198*
McDowell, C. A., 10–12 (46), 15 (58), 17, 18 (67), *47, 48*
McFarlane, C., 642–644 (152), *655*
McFarlane, H. C. E., (157), *655*
McFarlane, W., 340 (13), *346*, 531, 532 (246), *539*, 612–614 (57), 642 (152), 643 (152, 153, 156, 158), 644 (152), 649 (187), (157), *653, 655, 656*
McGrath, M. A., 502, 511 (44), 530 (238), *535, 539*
McKean, D. C., 533 (265), *539*
McKennis, J., 453 (173), *460*
McKenzie, S., 235 (101), *269*
McKernan, P. A., 374 (96), *376*, 386 (125), *391*, 787 (120), *793*
McKervey, A., 189 (938), *212*, 706 (175), *761*
McKervey, M. A., 189 (939), *212*
McKinnon, B. J., 227 (70), *269*
McKusick, K. A., 371 (57, 68), *375, 376*
McLaughlin, G. M., 258–260, 267 (227), *272*, 396 (8), *417*
McLean, R. A. N., 596 (12), *652*
McMurry, J. E., 107 (203), *198*
McNamara, D. J., 786 (118), *793*
McNinch, H. A., 710 (196), *761*
McPhee, D. J., 109, 124, 152 (223), 173, 179 (808), *198, 209*, 427 (24), *457*, 558 (117), *588*
McQuaid, L. A., 172 (782), *209*
McQuillan, G.P., 646, 650 (173), *655*
McShane, W. J., 369 (15), *374*
McWhan, D. B., 467, 469 (39, 43), *492*
McWhinnie, W., 453 (179), 455 (193), *460, 461*
McWhinnie, W. R., 53 (13), 62 (13, 25, 28, 29), 63–65, 67 (25), 68 (44), 70, 71 (25), 72 (48), 74 (25), 76 (59), 77–79 (25), 83 (78, 81), 84 (81, 82), 85 (13, 85, 86), *87–89*, 342 (24), *347*, 497 (9), 505 (60), 506 (72, 73), 507 (78), 508 (82, 84), 509 (89), 511 (60, 103), 512 (60, 108), 515, 516 (125), 517 (143, 146), 518 (72, 147, 148), 519 (158), 522 (72, 148), 525 (190), 527 (73), 528 (108, 221, 222), 529 (72, 146, 229, 231), 530 (236), 531 (147, 148), 532 (73, 147, 252), 533 (262, 266, 268), *534–539*
Medina, D., 385 (107, 131, 132), 386 (107), 387 (126, 127), *390–391*
Medne, R. S., 485 (117), *494*
Meek, D. K., 528 (225), *538*
Meek, D. W., 645 (163) *655*
Meeker, L. D., 385 (134), 386 (110), *390, 391*
Meerholz, C. A., 147 (551), 149 (571), 150, 152 (551, 571), 153 (571, 627), 154 (571), *204–206*, 788 (130), *793*
Meese, C. O., 425 (21), 427 (22), 429 (21, 22), 430 (21), *457*
Megerle, C. A., 643 (154), *655*
Mehdi, R. T., 280 (21), *334*, 518 (149), 519 (151, 159), 526 (149), 528 (159), 532 (149), *537*
Mehrotra, M. M., 184 (888), *211*
Meier, H., 151 (590–599), *205*, 243 (132), *270*, 289 (45), *335*, 449 (107), 450 (122–124, 130), *459*
Meier zu Köcker, I., 585 (228), *590*
Meiji Seika Kaisha 552 (90), *587*
Meineken, G. E., 371 (69), *376*
Meinke, P. T., 543 572 (23), *586*
Meinwald, J., 36, 37 (138), *49*, 102 (144), *196*, 407 (113, 114), *419*, 426 (19a), *457*, 504 (53), *535*, 565 (153), *588*
Meir zu Köchen, I., 247 (163), *270*
Meister, A., 355 (19), *364*
Meixner, J., 772 (65), *792*
Mekle, U., 450 (124), *459*
Meli, A., 411 (130), *419*
Meller, A., 399 (43), *417*
Mellon, F. A., 26 (103), *48*
Mellor, J. M., 32, 33, 36 (115), *49*, 126 (394), *201*
Mel'nik, Ya. I., 608, 628, 629 (50), *653*
Meloan, C., 453 (185, 186a), *461*
Melvin, L. S., 681, 682 (58), *759*
Menassen, J., 372 (80), *376*
Menchen, S. M., 142 (502, 503), 172, 174 (780), 176 (841a, 841b), 177 (862–864), 179 (867), *203, 204, 209–211*, 568 (169, 171), *589*,

630, 640, 651 (105), 652 (199), *654, 656,* 663 (32, 33), 666 (57), *672,* 678 (16, 221), 681 (16), 685 (98a, 98b), 711 (221), 716 (16), 754, 755 (221), *758, 759, 762,* 770 (44), *791*
Menconi, A., 400 (48), *417*
Mengers, E. A., 81, 83 (64), *88*
Mente, P., 451 (135, 136), *460*
Mente, P. G., 262 (244), *272*
Menzel, I., 151 (590, 592), *205,* 243 (132), *270,* 289 (45), *335*
Merenyi, R., 135 (454), *203*
Merijanian, A., 641 (141), 642 (150), *655*
Merrick, B. A., 381 (39), *389*
Merrick, M. V., 388 (170), *392*
Messbauer, B., 650 (188), *656*
Meter, H., 449 (113), *459*
Meth-Cohn, O., 423 (11, 13), *457*
Metz, P., 97, 117 (47), *194*
Metz, W., 173, 184 (803), *209,* 561 (128), *588*
Metzger, H., 256 (213), *271*
Metzler, D. E., 357–359 (34), *364*
Meuchen, S. M., 522, 523 (171), *537*
Meunier, Ph., 34 (122), *49*
Meyer,B., 400 (50), *418,* 613, 614, 617, 618 (59), *653*
Meyer, H., 650 (188), *656*
Meyer, J., 234, 238 (95), *269,* 451 (139a), *460*
Meyer, N., 154–157 (653), *206,* 685, 686, 689, 695, 696, 710, 715, 721, 726 (126b), *760*
Meyers, A.I., 113 (258), *199*
Meyers, E. A., 515 (132), 529 (234), *536, 539*
Michalska, M., 371 (53), *375,* 631 (106), *654,* 767 (27–29), 770 (36), 788 (129), *791, 793*
Michalski, J., 625 (87), 629 (100), 636 (87), *654,* 767 (28, 29), *791*
Michand, R. L., 263 (257), *272*
Micha-Screttas, M., 710 (198), 761
Michejda, C. J., 113 (257), 199
Michelotti, E. L., 162 (710), *207*
Michels, R., 103, 124 (150), *197,* 545 (39), 546 (48), 555, 557 (39), *586*
Michels, S., 368, 369 (9), *374*
Mickey, C. D., 768 (32), 769 (33), 790 (32, 33), *791*
Mieczkowski, J., 770 (52), *791*
Miginiac, P., 710 (194), *761*
Mihaly, G., 467, 485 (28), *492*
Mikawa, H., 485 (115), *494,* 501, 514 (43), *535*
Miki, K., 165 (739), *208*
Mikolajczak, J., 627 (97), *654*
Mikolajczyk, M., 114 (268), 153 (615, 616, 626), *199, 206,* 258 (225), *272,* 621 (75), 622 (75, 80), 625 (75), *653,* 663 (34), *672*
Mikoloajczyk, M., 622 (79), *653*
Mila, J. P., 497 (10), *534*
Miles, E. W., 356 (39, 40), 358 (38, 39), *364*
Miles, M. G., 469 (58), *493*
Miller, A. C., 118 (324), *200*
Miller, B., 485 (112), *494*
Miller, J. D., 280 (21), *334,* 519 (151, 159), 528 (159), *537*
Miller, J. M., 531 (249), 532 (253), *539*
Milliet, P., 147 (558), 148 (558, 564–566), 174 (566), *205*
Millington, D., 283 (31), *335,* 412 (138), 413 (138, 141, 142), *419,* 526 (195, 196), 527 (198), *538,* 640 (138), *655,* 659 (10, 11), 660 (13), *671*
Milne, G. H., 263 (251), *272,* 778 (92, 93, 96, 97), 779 (99), 782 (99, 107), 783 (107, 108, 111), 789 (92, 96, 97), *792*
Milner, J. A., 385, 386 (101, 108, 109), *390*
Milone, L., 345 (49), *347*
Minami, T., 114 (265, 266), *199,* 714, 715 (230), *762*
Mincuzzi, A., 176 (840), *210*
Mingoia, Q., 240 (117), *270*
Minh, T. Q., 315, 316 (119), *336*
Minki, V. I., 148 (561), *205*
Minkin, A. I, (315), *764*
Minkin, V. I., 409 (119), *419,* 432 (39, 41), 435, 439 (42), 441, 442 (41, 64), 443 (68–70), (75), *457, 458,* 753 (306), *764*
Minkin, W. I., 319 (131), *336*
Minkiu, V. I., 507 (75), *535*
Miranda, R., 383 (60), *389*
Mirzai, H., (111), *269*
Mishra, I. B., 597 (19), *652*
Miskey, M., 369 (18), *374*
Misumi, S., 100, 154, 155 (115), 165 (734, 737–739), 166 (743, 744), *196, 208,* 396 (15–17, 19, 20), 397 (16), *417,* 555 (103, 107, 108), *587,* (323), *764*
Mitchell, R. B., 542, 571 (13), *585*
Mitchell, R. H., 96 (35), 154, 155 (35, 657), 165 (736), *194, 207, 208,* 500 (37), 532 (256, 257), *535,* 539, 554, 555 (96), *587,* 684, 695, 732 (80), *759*
Mitsui, A., 533 (263), *539*
Mittal, P. K., 528 (220), *538*
Miura, H., 565 (143), *588*
Miura, M., 162 (711), *208*
Miura, T., 100, 138 (109), 182 (883), *196, 211*
Miwa, T., 97 (43), *194,* 555 (104), *587,* 774 (71), *792*
Miyake, J., 176 (839), *210,* 398 (34), *417*
Miyake, J.-I., 234, 238 (97), 239 (115), *269, 270*
Miyamoto, I., 118 (325), *200*
Miyano, M., 98, 129 (65), *195*
Miyashita, M., 96, 104 (32), 187 (906, 907, 909), *194, 211,* 681 (28), 715 (28, 258), 721 (28), *758,* 763
Miyata, T., 178 (849), *210*

Miyaura, N., 104 (159), *197*, 715 (257), *763*
Miyazaki, H., 333 (166), *337*
Miyoshi, H., 133 (440), *202*, 545 (41), *586*
Miyoshi, K., 544 (31), 581, 583–585 (223), *586, 590*
Miyoshi, N., 103, 107, (153), 116 (294, 300–303), 169 (773), 172 (781), 187 (919), 191 (961), 192 (945, 946), *197, 199, 200, 209, 212*, 499 (28), *534*
Mizuno, M., 483 (104), *494*
Mizutaki, S., 184 (887), *211*
Mlochowski, J., 120, 147 (351), *201*, 415 (153), *420*
Mobilio, D., 118, 157, 158 (336), *200*
Modelli, A., 6 (36), 23, 24 (81), 25 (81, 85), *47, 48*
Moeller, J., 453 (170), *460*
Mogensen, B., 467 (27), *492*
Mohamand, S., 35 (135), 37, 38 (141), 39 (154), 40 (155), *49, 50*
Mohammed, J., 497, 506 (14), *534*
Mohmada, S., 453 (158), *460*
Mohmand, S., 412 (135), *419*
Molines, H., 143 (514, 515), *204*
Molle, L., 383 (66), *390*, 453 (168, 180, 181), *460*
Moller, J., 456 (111), *459*
Mollier, Y., 486 (123, 124), 487 (124, 125), *494*
Molz, T., 450 (124), *459*
Monaham, R., III 765 (8), *790*
Mondovi, B., 351 (11), *364*
Monsef-Mirzai, Z., 85 (85), *89*, 506 (72), 517 (146), 518, 522 (72), 526 (194), 529 (72, 146), *535, 537, 538*
Mont, W.-W., du 600 (26, 28), 601 (29, 31, 33), 604 (38, 40), 605 (46), 606, 610 (31), 611 (31, 54), 612 (57), 613 (28, 31, 54, 57, 59), 614 (31, 57, 59), 617, 618 (54, 59), 623 (84), 635 (117–119, 121), 637 (29), 640 (117–119), 641 (117, 147), 643 (117), 644 (31, 46, 54, 117), 646 (176), 647 (28, 29), 648 (184), 649 (29), *652–656*
Montag, R. A., 42 (164, 165), *50*
Montanucci, M., 169 (763), *209*, 409 (121), *419*, 548 (60), *586*
Montero, L., 317 (124), *336*
Montiel, A., 453 (183), *461*
Moore, E. C., 263 (257), *272*
Moore, I., 411 (129), *419*
Mooyman, R., 14, 15 (61), 16 (65), 17 (65, 66), 18, 19 (66), *47, 48*
Moradpour, A., 475 (73), *493*
Moraes, F., 512, 530 (106), *536*
Morel, J., 325 (142), *337*
Morella, A. M., 132 (433), *202*
Morgan, G. T., 66 (37), *88*, 507 (79), 512 (105), *536*
Morgans, D. J., Jr. 564 (136), 565 (146), *588*
Mori, K., 97, 116 (50), 187 (910, 926), *195, 212*, 555 (101), 565 (140, 148), *587, 588*, 770 (38), *791*
Mori, Y., 161 (706), *207*, 753 (316), *764*
Morii, H., 165 (739), *208*
Morimoto, F., 220 (25), *268*
Morimoto, M., 565 (147), *588*
Morimoto, S., 664 (44), *672*
Morino, Y., 351–353 (10), *364*
Morishita, H., 533 (264), *539*
Morishita, M., 340 (2a), *346*
Morisset, V. M., 131 (423), 139 (423, 486), *202, 203*
Morita, H., 118 (325), *200*
Morita, M., 453 (184), *461*
Morita, S., 177 (844), *210*, 221 (36), *268*, 664 (40), *672*
Moriwake, T., 171 (775, 776), *209*
Morizawa, Y., 711 (215), *762*
Morris, A., 7 (40), 15 (57), *47*
Morris, J. G., 383 (60), *389*
Morrison, J. A., 510 (94), *536*
Mortensen, K., 251 (183), *271*, 467, 469 (44), 478 (88), *492, 493*
Mortezaei-Zandjani, G., 151 (588), *205*
Mortezali-Zandjani, G., 450 (121), *459*
Mortillaro, L., 221 (47–50), *268*
Mortimer, R., 135 (457), *203*, 706 (177), *761*
Mortimer, R. D., 135 (458), *203*
Morton, J. A., 143 (511, 512, 514), *204*
Morzycki, J. W., 111, 146 (241, 242), *198*
Moses, P. R., 483 (102), *494*
Moss, N., 165 (733), *208*, 397 (22), *417*, 520 (165), *537*, 660 (15), *672*
Moss, R. A., 681, 682 (66), *759*
Mössbauer, R. L., 52 (1), *87*
Mossman, A., 182 (883), *211*
Motherwell, R. S., 246 (154), *270*
Motherwell, W. B., 111 (241, 242), 112 (251), 146 (241, 242), 186 (902), *198, 199, 211*, 246 (154), *270*, 741 (313), *764*
Motherwell, W. D S., 614, 615 (61, 62), 646 (62), *653*
Motter, R. F., 710 (204), *761*
Moule, D. C., 228 (79), *269*
Moura Campos, M., 788 (122), *793*
Moura Campos, M., de 141 (492), *203*, 342 (21), *347*, 595, 596 (11), 622, 623 (81), 624, 627, 629, 648 (11), *652, 653*
Moura Compos, M., de 408 (106), *419*, 511 (102), *536*
Moustakis, C. A., 167 (752, 753), *208*, 219 (17, 21), 221 (34), 222 (21), 223 (21, 34), 225 (21), 226 (62, 63, 65), 263 (17), 265 (272), *268, 269, 273*
Moxon, A. L., 387 (140, 141), *391*

Mpango, G., B., 106 (189), *197*
Mrani, M., 221, 223 (35), *268*
Mrotsek, H., (111), *269*
Msezane, A., 5, 8 (16), *47*
Mudd, S. H., 351 (7), *364*
Mueller, A., 374 (92, 95), *376*
Mukaiyama, T., 140 (488), *203*
Mukerji, B., 387 (148), *391*
Mullen, G. P., 555 (105), *587*
Muller, A., 279 (20), *334*, 384 (75), *390*, 447 (88, 91), *459*, 526 (192), *538*
Müller, A. K., 408 (87, 91), *418*
Müller, C., 38 (148), *50*
Müller, D., 604 (42), 605 (46), 616 (65), 644 (46), *653*
Müller, E., 411 (131), *419*
Muller, E., 329 (160), *337*, 450 (125), *459*
Muller, G., 176 (838), *210*
Muller, J.-F., 26 (93), *48*
Müller, P., 172 (793), *209*
Müller, R., 398 (30), *417*
Mullins, M. J., 99 (85), *195*
Mumtaz, M., 106 (187), *197*
Munakata, T., 25 (79), *48*
Mundt, O., 615 (63), *653*
Mundy, B. P., 124, 140 (378), *201*, 772 (59), *791*
Mundy, D., 95, 96 (24), *194*, 680, 681, 714, 721 (26), *758*
Muniz-Miranda, M., 454 (191), *461*
Munk, M. E., 184 (891), *211*
Murahashi, S., 172 (791), *209*
Murahashi, S.-I., 500 (34), *534*
Murai, S., 103, 107 (153), 108 (209), 116 (294, 300–303), 149 (568), 169 (773), 172 (781), 176 (839), 177 (844, 845), 178 (849), 187 (919), 191 (961, 962), 192 (945, 946, 950, 956–958), *197–200, 205, 209, 210, 212, 213*, 221 (36), 234, 238 (97), 239 (115), *268–270*, 398 (34), *417*, 499 (28), *534*, 664 (40), *672*
Murata, H., 533 (259, 263), *539*
Murata, K., 192 (946), *212*
Murata, M., (237), *762*
Mureson, A., 121 (363), *201*
Murphy, C. J., 167 (748–752), *208*, 219 (18, 21), 222 (21, 53–55), 223 (21), 225 (21, 61), 263, 264 (18, 260), *268, 269, 272*, 293 (55), *335*, 398 (31), *417*, 710 (200), *761*
Murphy, C. T., 398 (29), *417*
Murray, B. J., 487 (126–128), 488 (127), *494*
Murray, B. K., 374 (96), *376*, 386 (125), *391*
Murray, P. J., 143 (513, 516), 174 (513), *204*
Murray, S. G., 528 (226), *538*
Musa, F., 455 (193), *461*
Musa, F. H., 342 (24), *347*, 505, 511, 512 (60), *535*
Musher, J. I., 705 (162), *761*
Musorin, G. K., 407 (100), *419*
Mustafa, M. G., 380 (24), *389*
Muth, O. H., 765 (21), *791*
Muto, S., 241 (122), *270*
Myasnikov, I. A., 664 (47), *672*
Myer, R., 408 (91), *418*

Naber, E. C., 382 (51), *389*
Naddaka, V. I., 148 (561), *205*, 409 (119), *419*, 432 (39, 41), 435, 439 (42), 441, 442 (41, 64–66), 443 (68–71), (75), *457, 458*, 532 (258), *539*
Nafedov, V. A., 545 (32), *586*
Nagano, K., 220 (25), *268*
Naganuma, A., 387 (153, 154, 158), *391*
Nagao, K., 561 (125), *588*
Nagao, S., 565 (154), *588*
Nagao, Y., 99 (86), *195*
Nagaoka, H., 104, 106 (165), 110 (237), *197, 198*
Nagashima, H., 117, 184 (311), *200*
Nago, F., 483 (101), *494*
Nagubandi, S., 327 (155), *337*
Nagy-Felsobuki, E., 15 (62, 63), 16 (62), *48*
Naito, T., 503 (49), *535*
Najima, M., 133 (444), *202*
Nakai, A., 101 (128), 105 (184), 125 (128), *196, 197*
Nakajima, T., 225 (60), *269*
Nakamura, K., 193 (967), *213*, 330 (163), *337*
Nakamura, N., 340 (6), *346*
Nakamura, T., 351, 352 (10), 353 (10, 13–15), 355, 356 (15, 21), *364*
Nakamura, W., 154, 189 (665), *207*
Nakanishi, K., 118 (341), *200*
Nakata, T., 565 (154), *588*
Nakaya, S., 383 (61), *389*
Nakayama, M., 111 (248), *199*
Nakayama, T., 356, 357 (31, 32), *364*
Nakayasu, E., 369 (31), *375*
Nakazaki, M., 503 (52), *535*
Nakhdjavan, B., 152 (608), *206*, 402 (61), *418*
Nalewajeh, D., 512, 530 (106), *536*
Nalewajek, D., 250 (180), *271*, 409 (120), *419*, 467, 469 (39, 43), 474 (71), *492, 493*
Nambier, K. P., 565 (145), *588*
Namoto, T., 785 (113), *792*
Narayanan, V. L., 386 (121), *391*
Nardelli, M., 258 (224), *272*
Narimatsu, S., 181 (877), *211*
Narita, M., 249 (174), *271*
Naruta, Y., 139 (482), *203*
Nash, J. A., 643 (156), (157), *655*
Natalis, P., 4, 27, 28 (13), *47*
Nath, A., 86 (89), *89*
Natsukawa, K., 192 (956), *212*

Natta, G., 505 (62), *535*
Naumaun, D., 517 (144), *537*
Nayak, U. R., 118 (323), *200*
Neal, R. A., 381 (44), *389*
Neckers, D. C., 315 (120), 316 (120–122), *336*
Nefedov, V. A., 577, 580, 582 (205), *589*
Negishi, H., 382 (53), *389*
Negoro, K., 147 (550), *204*, 706, 719 (166b), *761*
Neidlein, R., 452 (147), *460*
Neijzen, B. J. M., 18, 20, 42 (69), *48*
Neil, J., 26 (102), *48*
Neiland, O. V., 485 (117), *494*
Neillein, R., 453 (178), *460*
Neilson, R. H., 631 (107), *654*
Nelson, A. A., 384 (89), *390*
Nelson, A. J., 314, 317 (116), *336*
Nelson, D. L., 596, 602, 605–607, 609, 648 (13), *652*
Nemoto, H., 116 (297, 298), 123 (374, 375), 144 (297, 298, 531–534), 172 (531–534), 177 (846), 179 (374, 375), *200, 201, 204, 210*, 565 (138), 568 (173, 174), *588, 589*
Nepywoda, J., 508 (81), *536*
Nesmeyanov, A. N., 53 (18), *87*
Neuhaus, D., 173 (806), *209*
Neve, J., 453 (168, 180, 181), *460*
Newberne, P. M., 385 (128), *391*
Newman, M. S., 681, 682 (62), *759*
Nguyen, B. T., 374 (97), *376*
Nickolson, R. C., 187 (913), *212*
Nickon, A., 187 (908), *211*, 565 (160), *588*
Nicolaou, K. C., 94 (13), 98 (77, 78), 109 (225), 116 (285, 286), 129 (286), 139 (78, 473, 474a, 474b, 478–481), 141 (285, 286, 474a, 474b, 479, 493), 142 (77, 78, 508), 144 (285, 286, 524), 152 (78, 508, 610–612), 173 (286, 804), 174 (78, 285, 480, 493), *194, 195, 198, 199, 203, 204, 206, 209*, 427 (25a), 428 (26), *457*, 558 (116), *588*, 665 (53–55), *672*, 711 (220), 715 (220, 244, 245), *762*, 765 (9), 770 (40, 41), 773 (41, 68), *790–792*
Niecke, E., 598 (23, 24), 632 (110), 643 (23), *652, 654*
Nielsen, O., 402 (62), 403, 404 (68), 415 (62), *418*, 449 (112), *459*
Nielsen, O. J., 243 (128, 129), 267 (128), *270*
Nielsen, P., 36, 37 (137), *49*
Nielsen, P. H., 238 (110), 242 (124), 253 (198), 254, 255 (205), 264 (267), *269–271, 273*, 641 (142), *655*
Nielson, C. J., 342 (29a), *347*
Niemala, S. I., 369 (20), *375*
Niessen, W., von 25 (78), *48*
Nigro, N. D., 385 (139), *391*
Nils, M., 523 (183), *538*
Nilsson, M., 284 (35), *335*
Ninomiya, I., 148 (565, 566), 174 (566), *205*
Nischigaki, M., 428 (27), *457*
Nishikawa, N., 187 (924), *212*
Nishimura, S., 362 (58), *365*
Nishimura, T., 382 (54), *389*
Nishimura, Y., 770 (50), *791*
Nishio, T., 187 (925), *212*
Nishitani, K., 104 (174, 175), 109 (219, 220), *197, 198*
Nishiyama, H., 181 (876, 877), 184 (896, 897), *211*
Nishizawa, M., 96, 99 (39), 104 (172, 173), 123 (372), 173 (39, 173, 799, 801–803), 184 (799, 802, 803), 187 (924), *194, 197, 201, 209, 212*, 499 (30), *534*, 557, 558 (114), 561 (127, 128), 563 (131), 578 (114), *588*, 712 (224), 715 (259), *762, 763*
Nishuyama, H., 726 (329), *764*
Nitz, T. J., 116 (296), *199*
Niwa, H., 565 (149), *588*
Niwa, I., 108 (209), *198*
Niworozschkin, L. E., 319 (131), *336*
Nogami, T., 501, 514 (43), *535*
Nogami, Y., 118 (325), *200*
Noguez, J. A., 99 (98), 173 (98, 798, 800), *195, 209*, 555 (99), 560 (120), *587, 588*
Nohami, T., 485 (115), *494*
Nolan, C., 370 (43), *375*
Noltemeyer, M., 399 (43), *417*
Nomoto, T., 567 (165), *588*
Nomura, Y., 98 (79–81), 102 (147), 131 (147, 425), 136 (81, 464, 465), 137 (79, 80, 466), 173 (147), *195, 196, 202, 203*, 547 (58, 59), 556 (113), 570 (180, 181), 571 (182, 183), *586, 588, 589*
Noodleman, L., 18 (68), *48*
Norbeck, D. W., 174 (820), *210*
Norberg, B., 679 (334), *764*
Norburg, A. H., 400 (51), *418*
Norbury, A. H., 635 (116), *654*
North, J. A., 374 (96), *376*, 386 (125), *391*
Nothe, D., 467, 485, 489 (52, 53), *493*
Notikova, N., 452 (151), *460*
Novacek, E. J., 351, 353 (6), *364*
Novak, I., 7, 8 (42), *47*
Noyori, R., 98, 187 (68), *195*
Nozaki, H., 187 (920, 921), *212*, 711 (215), *762*
Nsunda, K. M., 98 (69), 100 (127), 122, 167 (69), *195, 196*
Nühn, P., 766 (25), 767 (26), 769 (25), *791*
Nün, P., 765 (15), *791*
Nunez, L., 483 (108), *494*
Nuretdinov, I. A., 413 (139), *419*, 621 (77), 625 (88–90), 626 (94), 627 (95), 628 (88), 637 (124, 126), 638 (126, 129), 639, 640 (129), 645 (95, 162), 647 (177), *653–656*

Nutzel, K., 681, 682 (36b), *758*
Nyholm, R. S., 513 (110), *536*

Oae, S., 178 (854, 858), *210*, 430–432 (35, 37), 433, 434 (37), 445 (77), *457, 458*, 664 (38), *672*
Oakleaf, J. A., 106 (188), *197*
Obayan, K. V., 443 (68, 70, 71), *458*
Obendorf, D., 601, 613 (32), *652*
Oborn, C. J., 387 (126), *391*
O'Brien, D. H., 531, 532 (245), *539*
O'Brien, E., 128 (404), *202*
O'Brien, M., 369 (12), *374*
O'Callaghan, W. B., 281, 318 (26), *334*
Ochiai, M., 99 (86), *195*
Odenigbo, G., 450 (125), *459*
Odom, J. D., 340 (9), *346*, 555 (105), *587*
Oehme, P., 379 (9), *388*
Offermanns, H., 238 (109), *269*
Ogara, F., 396 (15, 16), 397 (16), *417*
Ogasawara 561 (126), *588*
Ogasawara, K., 169 (766), *209*, 564 (137), 565 (147), *588*
Ogawa, A., 176 (839), 192 (957, 958), *210, 212*, 234, 238 (97), 239 (115), *269, 270*, 398 (34), *417*
Ogawa, H., 118 (325), *200*
Ogoshi, H., (216), *272*
Ogura, F., 100 (115), 133 (435), 147 (548, 552), 152, 153 (552), 154 (115, 435, 552), 155 (115, 435), 165 (734), 166 (743, 744), 169, 172 (774), 173 (809), 178 (853), *196, 202, 204, 208–210*, 396 (20), *417*, 555 (103, 107), 556 (111), 578, 583 (211), *587, 590*, 664 (43), *672*, 718 (272), 755 (322), (323), *763, 764*
Oguri, T., 101 (130), 173, 184 (802), *196, 209*
Oh, S. H., 387 (146), *391*
O'Hanlon, P. J., 135 (455), *203*
Ohara, E., 113 (253), *199*, 774 (72), *792*
Ohashi, O., 549 (69, 70), *587*
Ohba, N., 111 (243), 122 (243, 364), *198, 201*
Ohe, K., 100 (120), 180 (869), 191 (965), *196, 211, 213*
Ohfune, Y., 105, 109 (183), 122 (366), *197, 201*, 567 (164), *588*
Ohira, N., 178 (853), *210*, 664 (43), *672*
Ohlendorf, D., 187 (923), *212*
Ohloff, G., 104 (158), *197*, 327 (156), *337*
Ohmassa, N., 499 (29), *534*
Ohno, A., 219 (14), *268*, 330 (163), *337*
Ohno, K., 533 (259, 263), *539*
Ohno, M., 108, 169, 172 (204), *198*
Ohno, Y., 116 (294), *199*
Ohsawa, Y., 387 (147), *391*
Ohshima, T., 101, 125 (128), *196*
Ohshiro, Y., 114 (265), 186 (905), *199, 211*
Ohtsuka, T., 118 (339), *200*
Oishi, T., 565 (154), *588*
Ojika, M., 565 (149), *588*
Oka, S., 330 (163), *337*
Okabe, H., 297 (68), *335*
Okada, R. D., 371 (66), *376*
Okamoto, H., 371 (51), *375*, 383 (62), 387 (147), *390, 391*
Okamoto, Y., 476, 477 (78–80), *493*, 500 (36), *535*, 667 (63), *672*, 706 (179), *761*
Okamura, H., 162 (711), *208*
Okamura, W. H., 180 (874), *211*
Okano, M., 97 (52, 53, 60), 99 (60, 96, 100, 101), 100 (116), 103 (96, 149), 108 (96, 208), 116 (101, 116, 208, 290, 291), 117 (313, 314), 124 (377a, 377b), 126 (52, 53, 393), 133 (439, 440), 136 (60), 144 (520), 521a, 521b, 522, 523, 527), 159 (101), 164, 168 (728), (282), *195, 196, 198–202, 204, 208*, 545 (40–43), 546 (50), 569 (175–179), 570 (43), 572 (187), 573 (189), 574 (187, 196), 579 (212), *586, 589, 590*, 715 (250, 263), *762, 763*
Okawara, M., 113, 161 (261), *199*, 685, 695 (93), *759*
Okazaki, R., 223 (58), 226 (64), *269*
Okono, Y., 263 (258), *272*
Okrusecek, A., 629 (100), *654*
Olah, G. A., 133 (444), *202*, 225 (60), *269*
Oldfield, J. E., 383 (69), *390*
O'Leary, M. H., 356 (25), *364*
Oleksyszyn, J., 604 (44), 615 (64), *653*
Oliferenko, I. V., 519 (155, 156), *537*
Olifirenko, I. V., 662 (23, 24), *672*
Olliver, J., 108, 184 (212), *198*
Olofson, R. A., 237 (106), *269*
Olsen, M., 467 (45), 469 (45, 46), *492*
Olson, O. E., 351, 353 (6), *364*
Olson, R. E., 135 (459, 460), 158 (695), 180 (460), 184, 187 (695), *203, 207*, 555 (106), *587*
Omara, E., 715 (262), *763*
Omaye, S. T., 380 (23, 27), *389*
Omelańczyk, J., 622 (79, 80), *653*
Ono, M., 108 (205, 206), 116 (205, 206, 292, 293), 141 (206), *198, 199*
Onyamboko, N., 446 (82), 448 (102), *458, 459*
Ool, P. J. J. M., van 687 (105), *759*
Opalenko, A. A., 53 (17), *87*
Oppolzer, W., 104 (161), 187 (929), *197, 212*, 565 (158), *588*
Orchard, A. F., 2 (2, 4–6), *46*
Orchin, M., 412 (134), *419*, 664 (45, 46), *672*
Orda, V. V., 543 (19), *585*
Orlich-Krezel, I., 767 (28, 29), *791*
Orrell, K. C., 528 (220), *538*
Orrell, K. G., 528 (215–217, 223, 224), *538*

Orvane, P., 106 (197), *197*
Osaka, K., 193 (972), *213*
Osaka, N., 184 (896), *211*
Osawa, Y., 369 (33), *375*
Osborn, M. E., 106, 145 (196), *197*
Osborne, A. G., 407 (118), 416 (118, 154), *419, 420*, 504 (58), *535*
Oshima, K., 187 (920, 921), *212*
Ostadalova, I., 387 (145), *391*
Ostashkova, N., 453 (186b), *461*
Osterroth, C., 610 (53), *653*
Osuka, A., 161 (706, 707), 172 (785), 175 (828), 176 (833, 835), 178 (852), *207, 209, 210*, 499 (29), *534*, 581, 583–585 (223), *590*, 753 (316, 317), *764*
Otera, J., 193 (972), *213*
Otsubo, T., 100 (115), 133 (435), 147 (548, 552), 152, 153 (552), 154 (115, 435, 552), 155 (115, 435), 165 (734, 738), 166 (743, 744), 169, 172 (774), 173 (809), 178 (853), *196, 202, 204, 208–210*, 396 (15, 16, 19, 20), 397 (16), *417*, 555 (103, 107), 556 (111), 578, 583 (211), *587, 590*, 664 (43), *672*, 718 (272), 755 (322), (323), *763, 764*
Otter, R., 374 (93), *376*, 384 (76), *390*
Ottlinger, R., 413 (144), *419*
Oudenes, J., 109 (224), *198*
Outurquin, F., 544 (27), *586*
Ovchinnikov, Yu. E., 513, 514 (119), *536*
Owada, H., 97 (52, 53), 99 (96, 100, 101), 103 (96, 149), 108 (96, 208), 116 (101, 208), 126 (52, 53), 159 (101), *195, 196, 198*, 522 (177b), *537*, 715 (250, 263), *762, 763*
Owen, B. A., 371, 372 (67), *376*
Owens, W., 139, 174 (480), *203*
Oyama, H., 223 (58), *269*
Ozawa, S., 187 (920, 921), *212*

Pacheco, D., 314 (117, 118), 317 (127), *336*
Paetzold, R., 437 (46), (60), *457, 458*, 497 (11), 501 (41), *534, 535*
Painter, E., 423 (2), *456*
Pakzad, B., 299 (78), *335*
Pakzad, P., 277, 300 (12), *334*
Palacios, S. M., 296 (66), 298 (66, 76, 77), 299 (66, 77), *335*
Palmer, B. D., 143 (516), 152 (613, 614), 153 (614), *204, 206*
Palmer, H. T., 514 (120), *536*
Palmer, I. S., 351, 353 (6), *364*
Palmer, M. H., 26 (97, 98, 102, 103), *48*
Palmer, R. A., 514 (120), *536*
Palmisano, G., 99 (87), 174 (813), *195, 210*
Palyulin, V. A., 513, 514 (119), *536*
Pan, W.-H., 220 (32), *268*
Pan, W. H., 550 (77), *587*
Panek, J. S., 172 (789), *209*
Pannell, K., 453 (173), *460*
Pannell, K. H., 262 (246), *272*
Pant, B. C., 76 (59), *88*, 508 (87), *536*
Papahatjis, D. P., 152 (611, 612), *206*
Pape, L. A., 371 (56), *375*
Papirnik, M., 453 (186b), *461*
Papoula, M. T. B., 186 (902), *211*
Pappalardo, G., 532 (254), *539*
Pappalardo, G. C., 23–25 (81), 33 (121), 36 (136), *48, 49*, 516 (133), *536*, 604 (37), *652*
Pappolardo, G. C., 34, 35 (124–126), *49*
Paquette, L. A., 98 (74–76), 106 (196), 108 (214), 109 (221), 116 (296), 137 (74–76), 145 (196), 154, 155 (655), 157 (214, 655), 171, 172 (655), 184, 186 (214, 221), *195, 197–199, 207*, 681, 682 (56), 711 (216a), 721 (239), (240), *758, 762*
Parente, A., 118 (340), *200*
Parham, W. E., 710 (204, 205), *761*
Parish, R. V., 83 (77), *89*
Parizek, J., 387 (145), *391*
Park, B. K., 128 (404), *202*, 715 (253), *762*
Parker, K. J., 595, 602 (9), *652*
Parkin, S. S., 251 (184), *271*
Parkin, S. S. P., 469 (110), 483 (107, 110), *494*
Parks, R. E. Jr., 790 (133), *793*
Parnham, M., 447 (90, 92), *459*
Parnham, M. J., 374 (94), *376*, 383 (73), 384 (73, 74, 77), *390*
Parrott, J. C., 498 (17), *534*
Parsons, P. J., 129 (414), *202*
Partovi, M., 450 (116), *459*
Partovi, M. H., 243 (134), *270*
Pascoe, G. A., 382 (55), *389*
Pasmurtseva, N. A., 430, 432, 433 (34), 435, 436 (43, 44), 437 (48, 49), 439 (57), 441 (49), 443 (57), *457, 458*
Passerini, A., 549 (62), *586*
Passerini, R., 549 (62), *586*, 663 (36), *672*
Passmore, J., 519 (154), *537*
Pasternak, M., 53 (15), *87*
Pastour, P., 325 (142), *337*
Patel, M. G., 68 (44), *88*, 508 (84), *536*
Patel, V. V., 249, 251 (177), 252 (188, 189), *271*, 408 (88, 89), *418*, 467 (17, 20), 469 (59), 472 (63, 65), 473 (67), 474, 478 (77), *492, 493*
Paterson, I., 167 (755), *208*
Pathirana, H. M. K. K., 506, 527 (73), 529 (229), 532 (73, 252), *535, 539*
Patil, C. S., 373 (86–88, 90, 91), *376*, 384 (78–80), *390*
Patil, S. R., 124 (377b), 162 (713), *201, 208*, 397 (26), *417*
Patrick, D. W., 94 (3), 100, 102, 122 (114), *194, 196*, 677, 681, 685, 715, 719, 720, 756 (2), *757*

Patsaev, A. K., 187 (915), *212*
Pattenden, G., 423 (9), *457*
Paty, P. B., 187 (918), *212*
Paul, B. C., 343 (33), *347*
Paul, I. C., 705 (162), *761*
Paul, R. C., 498, 508 (18), *534*
Paul, W., 259 (228, 229), *272*
Pauling, L., 219 (19), *268*, 513, 515, 516 (111), *536*
Paull, K. D., 386 (121), *391*
Paulmier, C., 102 (145, 146), 128, 139, 180 (406), *196, 202*, 325 (142), *337*, 426 (19b), *457*, 544 (27), 545 (38), 546 (46, 47), 550 (79), 556 (46, 47), *586, 587*
Paulsen, H., 771 (58), 775 (76, 77), *791, 792*
Paulus, E. F., 343 (34), *347*
Pavlenko, E. A., 517 (139), *537*
Pawlak, K., 374 (97), *376*
Payne, N. C., 227 (70), *269*
Payo, E., 317 (123), *336*
Peake, S. L., 123, 152 (373), *201*
Pearce, H. L., 174 (816), *210*, 666 (56), *672*
Pearson, M. J., 104 (168), *197*
Pearson, P. S., 72 (53), *88*, 342 (28, 31), 343 (28), *347*
Pearson, R. G., 406 (85), *418*, 527 (206), *538*
Pearson, T. G., 228, 229 (74), *269*
Pecile, C., 246 (152), *270*
Pedersen, C., 452 (152), 453 (159, 160, 170), 456 (111), *459, 460*
Pedersen, C. L., 151 (605), *206*, 287 (39), 291 (51–53), 292 (51, 52), 323 (39, 140), 324, 333 (53), 334 (140), *335, 337*, 400 (46, 47), *417*
Pedersen, C. T., 246 (144), 247 (160, 161), *270*
Pedersen, H. J., 469 (46), *492*
Pedersen, L. M., 40, 41 (160), *50*
Pedersen, N. D., 382 (47), *389*
Pedersen, S., 453 (177), *460*
Pedley, J. B., 42, 44 (168), *50*
Pedro, J. R., 562 (129), 563 (130), *588*
Peel, J. B., 15 (62, 63), 16 (62), *48*
Peevey, R. M., 127 (398), *201*
Pegg, W. J., 176 (834), *210*
Peierls,R. E., 489 (135), *494*
Peleties, N., 154, 155, 157, (632, 633), *206*, 681–687, 689, 695, 696, 708, 721, 726, (67a, 67b,), 759
Pel'kis, P. S., 542 (14), *585*
Pellicciari, R., 190 (940), *212*, 519 (157), *537*
Pellinghelli, M. A., 519 (153), *537*
Pellizer, G., 33 (121), *49*
Pelter, A., 187 (923), *212*
Penenory, A. B., 296 (65), *335*, 670 (75–77), *673*
Penn, R. E., 40 (155), *50*
Pennanen, S. I., 181 (875), *211*
Penney, T., 249, 251 (177), *271*, 467 (18, 20), *492*
Perez-Albuerne, E. A., 467, 485 (29), *492*
Perez-Rodriguez, M., 266 (284), *273*
Perina, I., 128 (405), 147 (554, 555), 149 (576), *202, 204, 205*, 718 (275, 276, 278, 279), *763*
Perkins, M. J., 177 (848, 861), 178 (857), *210, 211*, 658 (6), 663 (31, 35, 37), 664 (48, 49), 666 (49), 668 (66), *671–673*
Perliskowa, W., 622 (79), *653*
Perlstein, J., 467 (8), *491*
Perlstein, J. H., 467 (7), 486 (120), 487 (126), *491, 494*
Pernet, A. G., 778 (88), *792*
Perrier, M., 144 (281), *199*, 456 (205), *461*
Perrotti, E., 192 (955), *212*
Perry, E. F., 379 (8), *388*
Perry, H. M. Jr., 378 (7), 379 (8), *388*
Pesce, G., 176 (840), *210*
Peseke, K., 551 (88), *587*
Pesin, V., 452 (151), 453 (186b), *460, 461*
Petasis, N. A., 94 (13), 116, 129, 141, 144, 173 (286), *194, 199*, 765 (9), *790*
Peters, P., 453 (177), *460*
Petersen, H., 151 (597), *205*, 449 (113), *459*
Peterson, D. J., 681, 682 (33), 710 (202), 711 (212), *758, 761, 762*
Peterson, F. J., 380 (28, 29), *389*
Petragnani, N., 103 (152), 114 (263, 264, 269), 126 (269), 141 (492, 499), 154 (665), 160 (702, 703), 169 (152, 263, 269), 174 (499), 189 (263, 665, 937), *197, 199, 203, 207, 212*, 317 (129), *336*, 342 (21), 343 (32), *347*, 406 (83), 408 (106), 414, 416 (83), *418, 419*, 501 (39), 508 (83), 511 (102), *535, 536*, 595, 596 (11), 622, 623 (81), 624, 627, 629, 648 (11), *652, 653*, 683 (101), 685, 695 (90), 696, 706, 727, 736 (101), *759*, 788 (122, 128), *793*
Petrasiunas, G. L. R., 222 (56), *268*, 322 (137), *337*
Petrov, A., 450 (129), *459*
Petrov, A. A., 245 (137), *270*, 408 (99), *418*, 597 (16), *652*
Petrov, M., 450 (129), *459*
Petrov, M. L., 245 (137), *270*, 408 (99), *418*
Petrov, V. N., 706 (167), *761*
Petrzilka, M., 97 (44–46), 117 (44), 118, 184, 186 (321), *194, 200*, 741 (311), *764*
Pettersen, R. C., 262 (246), *272*
Peyrussan, V., 475 (73), *493*
Pfaff, E., 104 (162), *197*
Pfenninger, J., 174 (818, 819), *210*, 330, 332 (165), *337*
Pfenninger, M., 106 (198), *197*
Pfisterer, H., 259 (231), *272*
Pfister-Guillouzo, G., 26 (101), 34 (122), 35 (132), 40 (158, 159), *48–50*

Pham Zuy Hein 53 (10), *87*
Phillips, G. B., 104, 114 (171), *197*
Phillips, G. W., 118 (324), *200*
Phillips, J. C., 513 (112), *536*
Phillips, J. G., 109 (220), *198*
Phillips, N. H., 342, 343 (30), *347*
Piancastelli, M. N., 5, 18, 20, 21 (18), *47*, 327 (152), *337*, 549, 574 (66), *586*
Picker, K., 231, 232 (90), *269*
Pierini, A. B., 294 (59–61), 295 (59), 296 (65), *335*, 527 (202), *538*, 669 (72–74), 670 (76, 77), *673*
Pierre, G., 546 (49), 585 (227), *586, 590*
Pietra, S., 261 (241), *272*
Pietropaolo, R., 527 (208), *538*
Piette, J.-L., 448 (101), 454 (188), *459, 461*, 577 (206, 207, 209, 210), 582 (206, 207, 210), *589, 590*
Piette, J. L., 342 (22, 23, 25, 26), *347*, 372 (79), 376, 506 (71), 531 (247), *535, 539*, 550 (76, 81, 83), *587*
Piettre, S., 135 (453, 454), *203*, 670 (81), *673*
Pignataro, S., 21–23 (72, 74), 24 (72, 74, 80), 25 (72, 74), 28 (108), 29 (80, 108), 30 (80), 33 (74, 119), *48, 49*
Pikkarainen, J., 379 (10), *388*
Pilgram, K., 453 (174), *460*
Pinell, R. P., 643 (154), *655*
Piquard, J. L., 517 (142), *537*
Pisareva, I. V., 409 (123), *419*
Pitteloud, R., 97 (46), *194*
Pittman, C. U. Jr., 249 (174), *271*
Pittman, R. W., 221 (45, 46), 222 (45), *268*
Piwinsky, J. J., 565 (150), *588*
Platt, A. W. G., 528 (215), *538*
Plieninger, H., 104 (162), *197*
Plimmer, J. R., 265 (271), *273*
Plowman, J., 386 (121), *391*
Pluim, H., 187 (917), *212*
Poehler, T., 407 (117), *419*, 466 (1, 3), 467 (1, 24), 480 (97), 483 (1), 485 (97), 489 (3), *491, 492, 494*
Poehler, T. O., 250 (181), *271*, 466 (2), 467 (2, 9, 13, 14, 22, 47), 469 (22, 47), 474 (72, 74), 483 (103), 489 (132), *491–494*
Poeler, T., 506 (68), *535*
Poelshcuk, O. K., 416 (157), *420*
Pogonowski, C. S., 104 (176), *197*
Pohl, H. A., 480 (95), *494*
Pohl, P., 370 (38), *375*
Pohl, S, 639 (132, 133), *655*
Pohl, W., 639, 644 (136), *655*
Poje, M., 148 (562), *205*, 718 (277), *763*
Polak, V., 149 (573), *205*, 718 (274), *763*
Poleschner, H., 408 (90), *418*, 455 (198), *461*, 472, 477 (64), *493*
Poli, R., 528 (219), *538*
Poliakoff, M., 291, 292 (51), *335*, 453 (159), *460*
Pollock, S. H., 383 (72), *390*
Polonsky, J., 111, 122, (247), *199*
Pommer, H., 681, 682 (52), *758*
Ponomareva, O. B., 416 (155, 156), *420*
Pope, A. L., 382 (57, 58), *389*
Pope, L., 36 (136), *49*
Pople, J. A., 6 (23–25, 27), *47*
Popov, V. I., 507 (75), *535*
Popova, L. L., 435, 439 (42), *457*
Porai-Koshits, M. A., 267 (288), *273*
Porta, O., 178 (857), *210*, 663 (37), *672*
Porter, A. E., 246 (154), *270*
Posner, G H., 172 (784), 186 (903), *209, 211*, 741 (312), *764*
Postnikova, T. K., 517 (139), *537*
Potapor, V. A., 407 (101), *419*
Potapov, V. A., 505 (64–66), *535*
Potember, R. S., 467, 469 (47), *492*
Potier, P., 99, 102, 132, 163, 191 (102), *196*, 322 (138), *337*
Potts, A. W., 2 (7), 4 (14), 7 (42), 8 (42, 43), 9 (43), 13, 15 (53), 27, 28 (14), *46, 47*
Potts, K., 453 (187a), *461*
Pougny, J.-R., 98, 129 (64), *195*
Pougny, J. R., 772 (60), *791*
Pouly, S., 34 (122, 123), *49*, 305 (88, 89), *336*
Pozeev, N. M., 25 (91), *48*
Pradhan, B. P., 123 (367–369), *201*
Praefcke, K., 276 (9), 277 (11–14), 278 (15, 16), 290 (49), 299 (14, 78, 79), 300 (12, 14, 80), 301 (13, 16), 302 (79, 81), 321 (11, 15, 16, 135), 333 (167), *334, 335, 337*, 509 (90), 523 (182), 525 (188), *536, 538*
Prasada Rao, Y. S., 362 (54), *365*
Prasado Rao, Y. S., 263 (249), *272*
Pratt, D. V., 127 (401), *201*
Pratt, R., 263 (258), *272*
Preisler, P. W., 258 (223), *272*
Preti, C., 549 (74), *587*
Price, W. C., 4, 27, 28 (14), *47*
Prinzbach, H., 471 (62), *493*
Pritzkow, H., 71 (47), *88*, 467, 485, 489 (52, 53), 493
Prokhorova, T. G., 467, 483 (49), *492*
Prousa, R., 408 (91), *418*
Prudent, N., 697 (292a), *763*
Prusoff, N. H., 787 (121), *793*
Pruzhkov, O. N., 517 (139), *537*
Ptitsina, O. A., 279 (19), *334*
Puff, H., 399 (45), *417*
Puleo, R., 106, 107 (192), *197*
Purcell, R. H., 228, 229 (74), *269*
Puska, P., 379 (10), *388*
Pustoslemsek, P., 260 (238), *272*
Pyle, R. E., 467 (13, 22), 469 (22), *492*
Pyles R. A., 604 (37), *652*

Quekborner, J., 485, 489 (113), *494*
Quick, M. H., 573 (188), *589*
Quinchon, J., 638 (127), *655*
Quinkert, G., 102 (139), *196*

Raasch, M. S., 651 (196), *656*
Rabalais, J. W., 2 (10), 15 (60), *46, 47*
Rabelo, J. J., 775 (84–86), *792*
Rabet, F., 151 (584), *205*
Rabinovich, D., 372 (80), *376*
Rabjohn,N., 110 (229, 230), *198*
Radchenko, S. I., 408 (98), *418*
Rainbow, I., 106 (198), *197*
Rainville, D. P., 515 (132), 529 (235), *536, 539*
Raithby, P. R., 405 (82), *418*
Raja, T. K., 175 (823), *210*
Rakiewicz, D. M., 141 (501), *203*
Rakitin, O. A., 753 (304, 305), *764*
Ralston, C. L., 515 (127), *536*
Ramasamy, K., 175 (826, 827), *210*
Ramazanova, O. A., 260 (236), *272*
Ramgopalakrishnan, V., 513 (118), *536*
Ramsden, W. D., 565 (151), *588*
Rancher, S., 521 (167), *537*
Rando, R. R., 357 (36), *364*
Rankin, D. W. H., 21 (70, 71), *48*, 596, 601, 611, 612, 643 (14), *652*
Rao, G. S., 123 (370), *201*
Rao, S. N., 184, 186 (899), *211*
Rapoport, H., 101 (132), *196*
Rashi, M., 263 (253), *272*
Rasmus, P., 572 (186), *589*
Rasmussen, F., 467 (45), 469 (45, 46), *492*
Rasmussen, S. E., 527 (210), *538*
Rastogi, S. C., 387 (150), *391*
Raston, C. L., 81, 82 (65), *88*, 344 (41), *347*
Rastrup-Andersen, J. 25 (89, 90), *48*
Rattanaphani, V., 527 (211), *538*
Raucher, S., 98 (62), 117 (312), 129 (62, 408), 134 (408, 448), 154, 155, 159 (637), 160 (700), 182 (884), *195, 200, 202, 206, 207, 211*, 499 (27), *534*, 683, 685 (69), 694 (122, 125), 695 (122), 696, 701, 702 (69), 706 (69, 125, 180), 707–709 (69), 724 (125), 726 (69), 727 (125), 733 (122), 748 (69), 749 (122), 751 (69), *759–761*
Rauchfuss, T. B., 343 (37), 345 (37, 45, 46, 48), 346 (52), *347*
Räuchle, F., 639, 644 (136), *655*
Razavi, A., 308 (98), *336*
Razumov, A. I., 638 (128), *655*
Razuvaev, G. A., 398 (35), 399 (35, 41), *417*
Razuvayev, G. A., 399 (42), *417*
Read, C. M., 121 (355), *201*
Read, R. W., 153 (628), 175 (628, 822), 178 (860), *206, 210*, 219 (16), 233 (16, 92, 93), 234 (93), 264 (16), *268, 269*, 788 (132), *793*

Reap, J. J., 99, 173 (98), *195*
Rebane, E., 503 (48), *535*
Rebar, A. H., 379 (18), *389*
Records, R., 264, 265 (269), *273*
Redchenko, V. V., 35 (127), *49*
Reddock, A. H., 489 (129), *494*
Reddy, K. A., 380 (23, 27), *389*
Reddy, M. L. N., 498 (15), *534*
Reed, N. V., 99, 141 (104), *196*
Reed, R., 485 (116), *494*, 501, 514 (43), *535*
Rees, C., 451 (135, 136), 455 (200), *460, 461*
Rees, C. W., 111 (244), *199*, 262 (244), *272*
Regitz, M., 293 (56), *335*, 593, 627, 634, 636, 637, 639 (2), *652*
Rehder, O., 650 (189), *656*
Reich, H. J., 94 (4, 5), 95 (29), 96 (33, 34, 38), 97 (58), 98, (38, 66), 99 (38, 93), 100 (108), 102 (29, 133, 134), 103, 104 (134), 106 (133, 134, 191), 107 (58, 133), 108 (133, 134), 112, 114 (134), 115 (38, 276, 277), 116 (38, 58, 276), 118 (134, 276), 123 (38, 373), 125 (383), 128 (58, 407), 129 (66), 135 (459, 460), 152 (373), 154 (34, 636, 656), 155 (33, 34, 636, 656), 157 (34, 636, 656, 679, 680, 690–692), 158 (680, 691–695), 159 (636, 690), 160 (33, 34, 656, 679), 162 (691), 172 (792), 180 (66, 460, 693, 870–873), 182 (692, 880), 184 (694, 695, 880), 187 (133, 134, 694, 695), *194–197, 199, 201–203, 206, 207, 209, 211*, 340 (5), 342, 343 (30), *346, 347*, 408 (104), 410 (126), *419*, 424 (16a), 425, 426 (16a, 16b), 427 (20), *457*, 498 (23), 520 (23, 160), 521 (23, 169, 170), 522 (172, 173), *534, 537*, 545, 553 (35), 555 (97, 100, 102, 106), 560, 568 (97), *586, 587*, 677 (5–7), 679 (5, 6, 21, 22), 681 (5–7, 21, 22), 683 (22, 70, 72, 76), 684 (22, 72, 77, 81, 321), 685 (5, 6, 72, 85), 686 (72), 693 (119), 695 (21, 72, 77, 85, 321), 696 (21, 72), 699 (22), 703 (21, 22), 705 (21, 22, 72, 158), 706 (22, 70, 176, 182, 183), 707–709 (70), 713 (5, 6), 714 (5–7, 22, 227), 715 (22, 168, 227, 242, 264), 716 (5, 76, 77, 81, 264), 718 (6, 21), 720 (76), 721 (21, 22, 72, 327), 726 (21, 22, 70, 72), 729 (321), 732 (22), 736 (5, 21), 738 (22), 741 (77), 742 (77, 314), 757 (5, 6), *757–764*, 765 (3, 6), *790*
Reich, I. L., 95 (29), 98 (66), 102 (29, 133, 134), 103, 104 (134), 106 (133, 134), 107 (133), 108 (133, 134), 112, 114, 118 (134), 129 (66), 157, 158 (692), 180 (66, 870), 182 (692), 187 (133, 134), *194–196, 207, 211*, 342, 343 (30), *347*, 522 (172, 173), *537*, 714, 715 (227), *762*
Reich, P., 497 (11), *534*
Reichelt, I., 108 (218), *198*

Reid, D., 217 (10), *267*, 423 (12), 456 (111), *457, 459*
Reid, D. H., 235 (101), 251 262 (185), *269, 271*
Reid, W., 576 (202), *589*
Reilly, M. P., 370 (49), *375*
Reinboldt, H., 498 (16, 21), 501, 503 (42), 504 (42, 55), 508 (83, 86), *534–536*
Reiner, D., 416 (158), *420*
Reissig, H.-U., 108 (218), *198*
Reiter, R., 447 (89), *459*
Reitz, D. B., 681, 682 (40), *758*
Rémion, J., 157 (681, 684, 687), 171 (681, 684), *207*, 681, 721 (25), *758*
Remion, J., 520 (161), *537*, 689, 692. 693 (112), 695 (130, 131), 696 (112), 699 (130, 131), 700 (112), 721, 722 (130, 131), 730 (130), *760*
Remizov, A. B., 134 (449), *202*, 499 (31), *534*
Ren, S., 383 (68), *390*
Renard, M., 167 (756), *208*
Renbaum, L. A., 187 (930), *212*
Renga, J. M., 95 (29), 97 (58), 99 (85), 102 (29, 133, 134), 103, 104 (134), 106 (133, 134, 191), 107 (58, 133), 108 (133, 134), 112, 114 (134), 116 (58), 118 (134), 128 (58, 407), 187 (133, 134), *194–197, 202*, 424–426 (16a), 427 (20), *457*, 522 (172, 173), *537*, 706 (182, 183), 714 (227), 715 (168, 227), *761, 762*
Renken, T. L., 40 (155), *50*
Renson, M., 217 (12), 229 (82), 230 (86), 245 (142), 246 (142, 151), 247 (158), 248 (12, 158), 255 (206), 257 (12), *268–271*, 342 (22, 23, 25), *347*, 423, 445 (15), 446 (15, 82, 85), 448 (15, 102, 103), 449–452 (15), 454 (188), 456 (206), *457–459, 461*, 500 (38), 502 (45), 512 (109), *535, 536*, 577, 582 (206, 207, 210), *589, 590*
Renson, M. R., 247 (157), *270*
Reuter, H., 379 (13), *388*
Reutov, O. A., 279 (19), *334*
Revankar, G. R., 787 (119, 120), *793*
Reverdy, G., 322 (137), *337*
Revis, N. W., 379 (17), *389*
Rewinski, J. W., 319 (131), *336*
Reynolds, C. D., 547, 554 (56), *586*
Reynolds, G. A., 486 (120), *494*
Rhee, R. P., 139 (485), *203*
Rheinboldt, H., 677 681, 718 (1), *757*
Rheingold, A. L., 343 (37), 345 (37, 45), *347*
Rheinholt, H., 423–426, 428 (1), *456*
Ribault, M., 467, 469 (37, 38, 41), *492*
Rice, F. O., 228, 229 (73), *269*
Rice, J. E., 476, 477 (79), *493*
Rice, K. C., 101 (131), *196*
Richards, P., 371, 372 (61), *375*
Richardson, E. K., 519 (154), *537*
Riche, C., 153 (620), *206*
Richmond, R. E., 104 (163), 116 (289), *197, 199*
Rickards, R. W., 97, 172 (61), *195*
Ridyard, A., 26 (102), *48*
Riebiro, A. A., 179 (868), *211*, 568 (172), *589*
Riech, H. J., 310 (99), *336*
Rieche, M., 86 (89), *89*
Ried, W., 260 (238), *272*, 502 (46), *535*, 544, 575 (25), *586*
Ried, W. E., 575 (199), *589*
Riede, J., 259, 260 (232), *272*
Riedmüller, S., 229 (85), *269*, 410 (128), *419*
Riegel, F., 414, 416 (416), *419*
Rieger, J. A., 388 (167), *392*
Riffel, M., 615 (63), *653*
Riga, J., 456 (203), *461*
Rihs, G., 467 (30, 31), 485 (30), *492*
Riley, R. F., 498 (22), *534*
Riley, T. A., 786, 790 (117), *793*
Rindorf, G., 469 (46), *492*
Ritschl, F., 549 (65), *586*
Rivas, C., 314 (115, 117, 118), 317 (115, 124, 125, 127, 128), *336*
Rivory, J., 486 (122), *494*
Robarge, K. D., 772 (61), *791*
Roberge, R., 25 (88), *48*
Roberts, B. P., 658 (3), 664, 666 (49), 667 (3), *671, 672*
Roberts, J., 453 (165), *460*
Roberts, M. E., 383 (70), *390*
Roberts, M., R., 107 (200), *198*
Roberts, P. J., 13 (55), *47*
Robertson, J. A., 532 (251), *539*
Robertson, J. D., 264, 265 (268), *273*
Robertson, W. A. H., 471 (61), *493*
Robin, M. B., 40 (152), *50*
Robins, B. D., 124, 140 (379), *201*
Robins, R. K., 374 (96–99), *376*, 386 (111, 121–123, 125), *390, 391*, 785 (116), 786 (117), 787 (119, 120), 790 (116, 117), *793*
Robinson, A., 765, 766 (13), *791*
Robinson, M. F., 385 (97), *390*
Rφd, T., 400 (51), *418*, 635 (116), *654*
Rodger, C., 642–644 (152), *655*
Rodgers, R. D., 528 (219), *538*
Rodin, O. G., 35 (127, 128), *49*
Rodrigo, R., 151 (600), *205*
Rodrigues, R., 103 (152), 114
263), 189 (263), *197, 19*
683, 696, 706, 727, 7
Rodriguez, A. D., 187 (9
Rodriguez, E., 101 (13
Rodriguez, H., 447 (
Roesky, H., 445 (7
460, 461
Roesky, H. W.,
Rogers, D. Z.,

Rogers, N. H., 135 (455), *203*
Roise, D., 356 (29, 30), *364*
Rokitskaya, V. I., 513, 514 (119), *536*
Rolandson, Å., 404 (72), *418*
Rollin, P., 97 (42), *194*, 773 (67), *792*
Rollinson, S. W., 141, 161 (497), *203*
Romanenko, V. D., 623, 634, 637 (82), *653*
Romer, A., 447 (92), *459*
Romm, I. P., 519 (155, 156), *537*, 662 (23, 24), *672*
Rømming, C., 640 (137), 645 (137, 168–171), *655*
Romming, C., 102 (140), *196*
Roothaan, C. C. J., 6 (19), *47*
Roper, W. A., 259 (230), *272*
Rosan, A. M., 346 (53), *347*
Rose, H., 264 (266), *273*
Rosenbaum, A., 608 (51, 52), 609, 610 (52), *653*
Rosenberg, R. A., 9, 10 (44), *47*
Rosenfeld, M. N., 121 (354–357), 128 (403), 168 (356), *201, 202*, 429, 430, 436 (32), *457*
Rosenfold, I., 765 (22), *791*
Rosenkilde S., 288 (42), *335*
Rosmus, P., 15 (59), 26 (100), 37, 38 (140–144), 39 (151), 40 (151, 153), *47–50*, 228, 266 (80), *269*, 450 (133), 453 (158), *459, 460*
Ross, A. I., 790 (133), *793*
Ross, D. L., 326 (149), *337*
Ross, H. M., 388 (170), *392*
Ross, R. J., 116 (296), *199*
Ross, S. D., 82 (72, 74), *88, 89*
Rossetti, R., 345 (49), *347*
Rossi, R. A., 294 (58–61, 64), 295 (59), 296 (65, 66), 298 (66, 76, 77), 299 (66, 77), *335*, 527 (202), *538*, 669 (70–74), 670 (75–77), 673
Rossi, R. H., de 294 (64), *335*
Rössler, M., 604 (43), *653*
Roth, R. W., 286, 319 (38), *335*
Rotruck, J. T., 382 (57, 58), *389*
Rotter, H. W., 401 (56), *418*, 619, 620 (71), *653*
Röttinger, E., 340 (4), *346*
Rouessac, A., 141 (496), 144 (530), *203, 204*
Rouessac, F., 141 (496), 144 (528–530), *203, 204*
Rouessac, R., 144 (281), *199*
Roush, W. R., 561 (123), 565 (156), *588*
Roy, J., 95, 97, 99 (25), *194*
Royce, R. D. Jr., 124 (380), *201*
Roziere, J., 346 (54), *347*
Rubner, M., 341 (16), *346*, 497 (13, 14), 502 (44), 506 (14), 511 (44), 530 (238, 240), *534, 535, 539*
Rudzinski, W. E., 373 (86–88, 90, 91), *376*, 384 (78–80), *390*
Rüegge, D., 772 (66), *792*
[illegible]uge, B., 227 (70, 71), *269*, 322, 329, 330 (136), [illegible]
[illegible] 398 (30), *417*
Ruhlmann, K., 452 (142), *460*
Ruider, G., 486 (119), *494*
Runet, A., 502 (45), *535*
Runice, C. E., 385 (138), *391*
Rupp, L. W. Jr., 467, 469 (43), *492*
Ruppel, W., 452 (147), *460*
Rusek, J. J., 135 (459), *203*
Ruskin, J. N., 379 (16), *389*
Russell, C. G., 139, 141, 149 (472), 172, 174 (780), 184, 186 (899–901), *203, 209, 211*, 522, 523 (171), *537*, 568 (169), *589*, 666 (57), *672*, 678 (16, 221), 681 (16), 711 (221), 716 (16), 721 (287), 754, 755 (221), *758, 762, 763*
Russell, G. A., 303, 305 (83–85), *336*, 660 (20, 21), 668 (68), *672, 673*
Russo, J. M., 249, 251, 252 (176), *271*
Russo, M., 221 (48–50), *268*
Russo, U., 87 (92), *89*
Ruther, F., 118 (332), *200*
Rutherford, J. S., 266 (285), *273*
Rutledge, P. S., 542 (8), 543 (20), *585*
Ryabokobylko, Yu. S., 545 (32), *586*
Ryan, M. D., 116 (299), 136 (461), 144 (299), *200, 203*
Ryan, R. C., 345 (50), *347*
Rycroft, D. S., 643 (153), (157), *655*
Ryskieva, G. A., 575 (198), *589*
Ryu, I., 108 (209), *198*

Saá, J. M., 172 (786), *209*
Sabbioni, E., 370 (43), *375*
Sabet, C. R., 710 (200), *761*
Sachdev, H. S., 104, 114 (169), 125, 154, 155 (385), *197, 201*, 683, 695, 716, 720, 748, 749 (75), *759*
Sachdev, K., 125, 154, 155 (385), *201*, 683, 695, 716, 720, 748, 749 (75), *759*
Sachleben, R. A., 124, 140 (379), *201*
Sachs, F., 456 (201), *461*
Sadee, W., 374 (97), *376*
Sadeh, T., 372 (79), *376*
Sadek, S. A., 388 (167, 168), *392*
Sadekov, I. D., 432 (41), 441, 442 (41, 64, 65), *457, 458*, 507 (74, 75), *535*, 753 (306), (315), *764*
Sadkova, D. N., 625 (90), 626 (94), *654*
Saelinger, D. A., 360 (42), *364*
Safayhi, H., 374 (93), *376*, 384 (76), *390*
Safiullina, N. R., 548 (61), *586*
Sagai, M., 380 (25), *389*
Sagan, L. S., 602 (34), *652*
Said, G., 369 (24), 370 (36), *375*
Saigh, G. S., 228, 229 (74), *269*
Saindane, M. 32, 33, 36 (116), *49*, 103 (151), 107 (199), 184 (889, 890), 187 (889, 934, 936), *197, 198, 211, 212*, 327 (150), *337*, 406 (84), *418*, 685 (84), *759*

Saito, G., 467 (25), 483 (106), *492, 494*
Saito, S., 171 (775, 776), *209*
Saito, Y., 157, 158 (682, 688), *207*
Sakaguchi, R., 111 (243), 122 (243, 364), *198, 201*
Sakai, Y., 101, 125 (128), *196*
Sakai-Wong, J., 382 (55), *389*
Sakaizumi, T., 549 (69, 70), *587*
Sakaki, K., 430–432 (35), 445 (77), *457, 458*
Sakakibara, J., 449 (104, 105a, 105b), *459*
Sakakibara, T., 113 (253–255), *199*, 683, 696 (103), 715 (103, 262), *759, 763*, 774 (72), *792*
Sakamoto, T., 120 (347), *200*
Sakan, K., 565 (145), *588*
Sakan, T., 187 (924), *212*
Sakasai, T., 120 (347), *200*
Sakata, K., 172 (779), *209*
Sakata, Y., 100, 154, 155 (115), 165 (738), 166 (743, 744), *196, 208*, 396 (16, 19, 20), 397 (16), *417*, 555 (103), *587*, (323), *764*
Sako, H., 114 (266), *199*, 714, 715 (230), *762*
Sakurai, H., 178 (850), 192 (948, 952, 954, 959, 960), *210, 212*, 340 (8), *346*, 409, 410 (124), *419*
Sakurgai, Y., 118 (325), *200*
Sakuta, K., 181 (876), *211*
Salahub, D. R., 25 (88), *48*
Salama, A., 451 (138), *460*
Salaün, J., 108, 184 (212), *198*
Salazar, J. A., 109 (227), 183 (885), *198, 211*, 715 (248), *762*
Saleh, G., 114 (265), *199*
Salimbaeva, A. D., 575 (198), *589*
Salmasi, S., 385 (138), *391*
Salmona, G., 26 (101), *48*
Salmona, M., 379 (20), *389*
Salmond, W. G., 96, 179 (37), *194*, 714 (228), 716 (267), *762, 763*
Salonen, J. T., 379 (10), *388*
Samartseva, S. A., 647 (177), *656*
Samdal, S., 644 (160), *655*
Samitov, Yu. Yu., 499 (32), *534*
Samitov, Y. Y., 706 (184), *761*
Sammes, P. G., 297 (71), *335*
Samochoka, K., 403 (66), *418*
Samoilau, B. N., 53 (14), *87*
Samuelson, L. A., 341 (16), *346*, 497 (13, 14), 502 (44), 506 (14), 511 (44), 530 (238, 240), *534, 535, 539*
Sanche, L., 45 (170), *50*
Sanchez, A., 383 (60), *389*
Sandaman, D. J., 530 (238), *539*
Sander, W., 450 (119), *459*
Sander, W. W., 228, 244 (81), *269*, 289, 326 (44), *335*
Sanderud, K. A., 542 (18), 549 (67), 579, 580 (18), 581 (67, 224), 583 (224), *585, 586, 590*
Sandholm, M. 383 (64), *390*
Sandman, D. J., 36, 37 (137), *49*, 251, 252 (186), *271*, 341 (16), *346*, 467 (10), 485 (116, 118), 486 (121), *491, 494*, 497 (12–14), 501 (12, 43), 502 (44), 506 (12, 14), 511 (44), 514 (43), 516 (134), 517 (134, 138), 530 (240), *534, 535, 537, 539*
Sandstroem, J., 31, 32 (112), 40 (159), *49, 50*
Sandstrom, J., 264 (263), *272*, 327 (151), *337*
Sanemitsu, Y., 568 (170), *589*
San Filippo, L. J., 167 (752), *208*, 219, 222, 223, 225 (21), 226 (65), *268, 269*
Sanina, L. P., 398 (36), 399, 414 (36, 44), *417*
Sankyo Kagaku, K. K., 567 (167), *589*
Sano, T., (237), *762*
Santiago, A. N., 298 (76, 77), 299 (77), *335*
Santiesteban, H., 169 (761, 762, 767), *209*
Santini, C., 124 (381), *201*
Santorelli, A. C., 263 (257), *272*
Santry, D. P., 6 (24, 26), *47*
Saris, L. E., 236, 238 (102), *269*
Sarkar, S. D., 388 (165), *392*
Sarkar, T., 181 (878), *211*
Sartorelli, A. C., 386 (112), *390*
Saruta, T., 369 (29–31), *375*
Sasaki, K., 169, 172 (774), *209*
Sasaki, T., 386 (117), *391*, 790 (134), *793*
Sasaoka, M., 489 (131), *494*
Sashida, H., 97, 116 (50), *195*
Sasse, K., 593, 595, 634, 636, 637 (1), *652*
Sastry, K. A. R., 371 (64, 65, 67), 372 (64, 67), *376*
Satake, K., 561 (124), *588*
Sato, S., 102 (148), *196*, 241 (121), *270*
Satoh, M., 387 (153), *391*
Satoh, T., 111 (243), 122 (243, 364, 365), 172 (779), *198, 201, 209*
Satyanarayana, N., 118 (323), *200*
Sauer, Ch., 86 (89), *89*
Sauer, I., 604 (41), 605, 607, 619 (45), *653*
Saunders, V. R., 13 (56), *47*
Saurborn, E. G., 710 (192), *761*
Sauve, J. P., 487 (125), *494*
Savolainen, H., 370 (34), *375*
Sawai, H., 108, 169, 172 (204), *198*
Sawaki, S., 97, 116 (50), *195*
Sawicki, R. A., 142 (506), *204*
Sawluk, J., 646, 648 (175), *656*
Sazonova, O. M., 113 (252), *199*
Sbrana, G., 454 (191), *461*
Scaiano, J. C., 227 (66–69), *269*, 517 (140), *537*, 658, 661 (5), 667 (62), 670 (5, 79, 80), *671–673*
Scarborough, R. M. Jr., 102 (138), 144 (524), 169 (764a, 764b), 172 (764a, 764b, 794), 179 (138), *196, 204, 209*
Schaad, L. J., 340 (11), *346*
Schafer, A., 398 (40), *417*

Schäfer, H.-J., 97, 117 (47), *194*
Schäffer, W., 21–23, 25 (73), *48*
Schafter, D. E., 467 (11), *491*
Schapkin, A. A., 25 (91), *48*
Scharrer, E., 382 (49), *389*
Schaumann, E., 244 (136), 264 (266), *270, 273*, 408 (96), *418*
Scheithauer, S., 266 (275), *273*
Scheller, A., 411 (131), *419*
Schenk, W., 486 (119), *494*
Scherer, G. J., 633 (113), *654*
Scherer, O. J., 633 (111, 112), 635 (120), 638 (131), 639 (120, 131), 640 (120), 649 (185), *654–656*
Scherowsky, G., 252 (187), *271*, 469 (57), *493*
Schill, G., 189 (937), *212*, 317 (129), *336*
Schilling, F., 456 (202), *461*
Schilling, W., 173 (811), *210*
Schinazi, R. F., 787 (121), *793*
Schings, U., 616, 643 (66), *653*
Schinner, A., 221 (41), *268*
Schipper, P., 687 (105), *759*
Schirber, J. E., 483 (108), *494*
Schlenzig, M., 554, 555 (95), *587*
Schlessinger, R. H., 107 (200), *198*
Schleyer, P. v. R., 298 (75), *335*
Schmelzer, A., 29, 30 (109), *49*
Schmid, G., 110 (237), *198*
Schmid, G. H., 129 (410, 411), 131 (418), *202*, 545, 553 (37), *586*
Schmid, P., 517 (140), *537*, 667 (62), *672*
Schmidbaur, H., 114 (267), *199*
Schmidkonz, B., 308 (98), *336*
Schmidpeter, A., 621 (76), 628 (99), *653, 654*
Schmidt, A. H., 260 (238), *272*
Schmidt, D., 277, 301 (13), *334*
Schmidt, H., 425, 429, 430 (21), *457*
Schmidt, M., 610 (53), 650 (190), *653, 656*
Schmidt, U., 97 (54), *195*, 279 (20), *334*, 526 (192), *538*, 563 (133, 134), *588*
Schmiesing, R. J., 117 (315, 316), *200*, 770 (53), *791*
Schmitz, M. K., 406 (86), *418*, 542 (12), *585*
Schnabl, G., 635 (120), 638 (131), 639 (120, 131), 640 (120), *654, 655*
Schneider, C. J., 385 (95), *390*
Schneider, W., 765, 766 (24), *791*
Schnell, R. C., 381 (39), *389*
Schniepp, S., 151 (594, 596), *205*
Schobert, R., 118, 119 (342), *200*
Schoeller, W. W., 598 (24), *652*
Schollhorn, H., 308 (98), *336*
Schöllkopf, U., 681, 682 (35, 36a, 42b, 49), *758*
Schönberg, A., 396 (9), *417*
Schönberger, A., 276, 329 (2), *334*
Schrader, B., 627 (96), *654*
Schrauzer, G., 451 (137), *460*
Schrauzer, G. N., 262 (245), *272*, 380 (24), 385 (95, 99, 106), 386 (106), *389, 390*
Schreiber, S. L., 561 (124), *588*
Schröder, A., 221 (39), *268*
Schroeder, H. A., 384 (91), *390*
Schroeder, M. C., 157, 158, 182 (692), *207*
Schröer, R., 610 (53), *653*
Schubert, U., 416 (158), *420*
Schule, R., 59, 62 (21), *88*
Schulert, A. R., 381 (44), *389*
Schulte, K. H., 327 (156), *337*
Schulte-Elte, K. H., 104 (158), *197*
Schultz, A. G., 277 (10), 325 (10, 143, 144), *334, 337*, 524 (185), *538*
Schultz, J. S., 554, 555 (94), *587*
Schultz, R., 62–65, 67, 70, 71, 74, 77–79 (25), *88*, 533 (266), *539*
Schultz, R. D., 384 (87), *390*
Schulz, G. J., 45 (170, 171), *50*
Schulz, H. J., 466 (4), 467 (4, 41), 469 (41), *491, 492*
Schulz, P., 505, 507 (59), 519 (152), *535, 537*
Schulz, R., 26, 37–39 (99), *48*, 450 (120a, 102b), *459*
Schulze, U., 277 (14), 278 (16), 299 (14), 300 (14, 80), 301, 321 (16), 333 (167), *334, 335, 337*
Schumaker, R. R., 251 (184), *271*, 467 (16), 469 (59, 110), 474, 478 (77), 483 (107, 110), *492–494*
Schumann, H., 308 (95–97), *336*, 623 (83), 624 (83, 85), 641 (147), 642 (151), 643, 646 (85), 650 (151), 651 (83), *654, 655*
Schumann, H. D., 501 (41), *535*
Schuster, H. C., 398 (38, 39), *417*
Schuster, R., 39 (150), *50*
Schütz, M., 398 (30), *417*
Schwartz, A., 147, 168 (553), *204*
Schwartz, J., 187 (931), *212*, 715 (255), *763*
Schwartz, K., 380 (31), *389*
Schwarz, H., 333 (167), *337*
Schwedi, G., 453 (182), *460*
Schweig, A., 21–23, 25 (73), 26 (99), 32, 36 (114), 37 (99, 145), 38 (99, 146, 148), 39 (99), 41 (162), *48–50*, 450 (120a, 120b), *459*
Schweitzer, D., 467 (52, 53), 485, 489 (52, 53, 113), *493, 494*
Sciscia-Santoro, S., 351 (11), *364*
Scortia, T. N., 258 (223), *272*
Scott, B. A., 249, 251 (177), *271*, 467 (18, 20), *492*
Scott, J. C., 407 (113), *419*, 469 (46), 483 (107), *492, 494*
Scott, K. A., 388 (161), *391*
Scott, M. L., 381 (45), *389*
Scottlander, M., 25 (89), *48*
Screttas, C. G., 710 (198), *761*

Scudder, E. D., 221, 227 (43), *268*
Scudder, P. H., 172, 183 (777), *209*, 715, 726 (241), *762*
Secomb, R. J., 81, 82 (65), *88*, 344 (41), *347*, 515 (127), *536*
Sedov, Y. A., 133 (445), *202*
Seebach, D., 113 (256), 116 (304), 154 (632, 633, 635, 639, 653, 664), 155 (632, 633, 635, 639, 653), 156 (635, 639, 653), 157 (632, 633, 653), 159 (699), 189 (664), *199, 200, 206, 207*, 408 (105), *419*, 506, 509, 519 (70), 523 (178), *535, 538*, 681, 682 (31, 37, 41, 54, 55, 67a, 67b), 683, 684 (67a, 67b, 71), 685, 686 (67a, 67b, 126a, 126b), 687 (67a, 67b), 689 (67a, 67b, 126a, 127b), 695, 696 (67a, 67b, 71, 126a, 126b), 699 (126a), 708 (67a, 67b, 186), 709 (186), 710 (126a, 126b, 203), 715 (126a, 126b), 721 (67a, 67b, 126a, 126b), 726 (67a, 67b, 71, 126a, 126b), 727 (71), 753, 754 (294), *758–761, 763*
Seeber, R., 483 (101), *494*
Sega, A., 454 (192), *461*
Segal, G. A., 6 (24–26), *47*
Segmuller, B. E., 97, 142 (56), *195*
Seguin, M., 325 (146), *337*, 525 (187), *538*
Seiden, P. E., 467 (18), *492*
Seidler, M. D., 134 (447), 177 (866), *202, 211*, 221 (37), *268*, 487 (128), *494*
Seitz, S. P., 116 (285), 139 (474a), 141 (285, 474a), 144, 174 (285), *199, 203*, 427 (25a), *457*, 665 (55), *672*, 773 (68), *792*
Sekido, E., 340 (6), *346*
Sekine, M., 783 (112), *792*
Seko, T., 137 (468), *203*, 428 (27), *457*
Selin, L. E., 6, 7 (38), *47*
Sell, G., 502 (46), *535*
Sell, J. A., 22, 25 (77), *48*
Semmelhack, M. F., 104 (164, 165), 106 (165), 109 (226), *197, 198*
Sens, T., 362 (55), *365*
Seo, K., 564 (137), *588*
Seoane, E., 562 (129), 563 (130), *588*
Serfass, R. E., 384 (84), *390*
Sergeev, V. 452 (151), *460*
Sergi, S., 527 (208), *538*
Serrano, R., 259 (231), *272*
Seshadri, R., 176 (834), *210*
Set, L., 145, 154, 155 (536), *204*
Severengis, T., 400 (49, 50), *417, 418*
Severengiz, T., 600 (26), 601 (30, 31), 604 (38), 605 (46), 606, 610 (30, 31), 611 (30, 31, 54), 613 (30, 31, 54, 59), 614 (31, 59), 617, 618 (30, 54, 59), 619 (30), 644 (30, 31, 46, 54), *652, 653*
Sevrin, M., 99 (91), 100 (91, 125), 126 (390), 132 (428–432), 154 (91, 429, 645, 663), 155 (91, 645), 156 (91, 429), 159, 160 (431), 173 (430), 174 (645), 175 (824), 186, 187 (91), *195, 196, 201, 202, 206, 207, 210*, 522 (177a), *537*, 557 (115), *588*, 678 (14), 679 (17, 18), 680 (17, 18, 23), 681 (14, 17, 18, 23, 25), 687 (106a, 106b), 693 (118), 694 (121, 123), 695 (14, 106a, 106b, 118, 121, 129), 696 (106a, 106b), 701 (18), 706 (121, 174), 707, 709 (121), 711 (14), 712 (17, 18, 106b), 713 (18, 106a, 106b), 714 (123), 715 (14, 123), 718 (23, 106a, 106b, 129), 719 (129, 307), 720 (106b, 123, 129), 721 (23, 25, 106b, 129, 283, 307), 726 (106b), 727 (123), 728 (14), 733 (121), 741 (174), 743 (129), 745, 746 (106a, 106b), 747 (17, 121), 749 (121), 755 (18), *758, 760, 761, 763, 764*
Seyden-Penne, J., 698, 700 (155), *760*
Seyferth, D., 710 (189, 190, 197, 199, 200), *761*
Sfchaffer, H., 512, 530 (106), *536*
Shafer, D. E., 485 (112), *494*
Shafiee, A., 151 (581–585, 587, 588), *205*, 243 (131, 133), (218, 219), *270, 272*, 290 (47, 48), *335*, 446 (79), 449 (79, 106), 450 (121, 128), *458, 459*
Shagidullin, R. R., 627 (95), 645 (95, 162), 647 (177), *654–656*
Shah, S. K., 96 (33, 34), 125 (383), 154 (34, 656), 155 (33, 34, 656), 157 (34, 656), 158 (693, 695), 160 (33, 34, 656), 180 (693), 184, 187 (695), *194, 201, 207*, 521 (169, 170), *537*, 555 (106), *587*, 679, 681 (22), 683 (22, 72, 76), 684 (22, 72, 81), 685, 686 (72), 693 (119), 695, 696 (72), 699, 703 (22), 705 (22, 72, 158), 706, 714, 715 (22), 716 (76, 81, 268), 720 (76), 721, 726 (22, 72), 732, 738 (22), *758–761, 763*
Shahabi, S., 5, 8 (16), *47*
Shaik, S., 133, 154 (442), *202*
Sham, H. L., 108 (213), *198*
Shamberger, R. J., 369 (21), *375*, 378 (6), 385 (93, 94, 130), *388, 390, 391*, 765 (23), *791*
Shams, N. A., 110 (235), *198*
Shankaranarayana, M. L., 253 (196), *271*
Shanmugam, P., 175 (826, 827), *210*
Shapiro, V. G., 53 (10), *87*
Sharghi, N., 340 (12), *346*
Sharma, K., 452 (149), *460*
Sharma, R. D., 63 (30), 65 (30, 35), 66 (35, 36), 73 (36), 74 (35, 36), 83 (80), 84 (83, 84), *88, 89*, 529 (232), 531, 532 (250), *539*
Sharp, G. J., 42, 44 (168), *50*
Sharp, K. W., 415 (152), *420*
Sharpless, K. B., 94 (3), 95, 96 (26–28), 97 (40), 98 (28, 40), 99 (28, 82, 97), 100 (114), 101 (27), 102 (114), 103 (27), 104 (160), 115, 116 (40, 82), 118 (327–330, 333), 119 (344), 121 (362), 122 (114), 123 (82), 127 (160,

396, 397), 131 (426, 427), 140 (489), 171 (26), 182 (879), 184 (886), *194–197, 200–203, 211*, 438 (51), *458*, 522 (174, 175), *537*, 542, 545 (4), 555 (98), *585, 587*, 671 (82), *673*, 677 (2), 681 (2, 27), 685 (2, 87), 695 (87), 714 (27, 226, 229, 231, 233), 715 (2, 27, 226, 229, 233), 716 (27), 719, 720 (2), 721 (27, 226, 231, 284, 286), 756 (2), *757–759, 762, 763*, 770 (39, 54–57), *791*
Shauerman, H. J., 698, 699 (151), *760*
Shaw, D. A., 126 (389), *201*
Shaw, S. M., 388 (168), *392*
Shawali, A. S., 550 (84, 86, 87), *587*
Shawl, E. T., 260 (235), *272*
Shchegolev, I. F., 467 (32–34, 36, 48), 483 (48), *492*
Shea, R. G., 127 (399, 400, 402), *201*
Shealy, Y., 452 (141), *460*
Sheffy, B. E., 384 (87), *390*
Shefter, E., 266 (286), *273*
Sheldon, B. G., (147), *760*
Shelton, E. J., 131, 139 (422), *202*
Shemyakin, M. M., 681, 682, (50), *758*
Shepherd, F., 385 (107, 131, 132), 386 (107), *390, 391*
Sheppard, N., 642–644 (152), *655*
Sheraga, H. A., 37 (139), *49*
Shermann, E., 453 (174), *460*
Sherwin, P. F., 40 (155), *50*
Shetta, A., 550 (84, 86), *587*
Shibaeva, R. P., 467 (32, 49), 483 (49), *492*, 504 (57), *535*
Shibata, Y., 449 (104, 105b), 453 (184), *459, 461*
Shibayama, K., 400 (53), *418*, 617, 633 (67, 68), *653*
Shibutani, Y., 370 (49), *375*
Shih, C. N., 721 (239), *762*
Shikazono, N., 53 (11), *87*
Shillinger, W., 561 (122), *588*
Shimidzu, T., 565 (143), *588*
Shimizu, H., 161 (706), 178 (852), *207, 210*, 430, 432–434 (36), *457*, 753 (316), *764*
Shimizu, M., 115 (306–309), 116 (306–310), 146 (542–547), 154, 155 (654), 163 (722, 723), 169 (654), *200, 204, 206, 208*, 706 (173), 716–718 (271), 746 (293), 747 (271), *761, 763*
Shine, C. Y., 778 (95), 780 (102, 103), 781 (104, 105), 782 (102–106), 790 (102, 103), *792*
Shine, H. J., 657, 662 (1), *671*
Shine, R. J., 217, 248 (5), 254–256 (204), 257 (5), 265 (271), *267, 271, 273*
Shinmon, N., 187 (916), *212*
Shinoda, M., 544 (30, 31), *586*
Shirahama, H., 118 (339), *200*
Shirhatti, V., 187 (908), *211*
Shirley, D. A., 9, 10 (44), *47*
Shiue, C.-Y., 263 (250, 256), *272*, 386 (115, 116), *391*
Shklover, V. E., 513, 514 (119), *536*
Shlyk, Yu. N., 597 (16), *652*
Shomaker, V., (111), *49*
Shorji, T., 53 (11), *87*
Shorobogatova, V. I., 505 (64), *535*
Shpinel, V. S., 53 (10, 17), *87*
Shreiber, S. I., 124 (381), *201*
Shrift, A., 351 (5), *364*
Shtyrkov, G. L., 86 (88), *89*
Shu, A. Y. L., 151 (604), *205*
Shu, P., 250 (179), *271*, 407 (115), *419*, 466 467 (1), 474 (70), 480 (90), 483 (1), *491, 493*
Shulgin, V. F., 623, 634, 637 (82), *653*
Shull, L. R., 381 (42), *389*
Shultz, A., 467, 483 (51), *493*
Shulz, R., 37 (145), 38 (146, 148), *50*
Siami, G., 381 (44), *389*
Sibgatulind, E. G., 637, 638 (126), *654*
Sidwell, R. W., 374 (98), *376*
Siebert, W., 414, 416 (146), *419*
Sies, H., 374 (92, 95), *376*, 384 (75), *390*, 447 (88, 91), *459*
Sih, J. C., 187 (911), *212*
Šik, V., 528 (216, 217, 220, 223, 224), *538*
Silveira, C. C., 167 (754), *208*
Silver, J., 62 (27), 80 (27, 62), 81 (62, 66–70), 82 (70–75), *88, 89*, 515 (128–131), *536*
Silverman, R. B., 217, 222 (3), *267*
Silverton, J. V., 101 (131), *196*
Silverwood, A., 412 (134), *419*
Silverwood, H. A., 664 (45), *672*
Šimánek, V., 172 (790), *209*
Simchen, G., 547, 551 (51), *586*
Simiti, I., 121 (363), *201*
Simon, A., 501 (41), *535*, 615 (63), *653*
Simon, H., 104 (166), *197*, 277 (11), 278 (15, 16), 299 (78), 301 (16), 321 (11, 15, 16, 135), 333 (167), *334, 335, 337*, 525 (188), *538*
Simon, T. C., 379 (12), *388*
Simpkins, N. S., 173 (805–807), *209*
Sinaÿ, P., 97 (42, 51), 98 (64), 99 (51), 117 (51, 320), 129 (64), *194, 195, 200*, 770 (37), 772 (60), 773 (67), *791, 792*
Singer, E., 396 (9), *417*
Singer, H. S., 312 (111), *336*
Singer, M. I. C., 705 (164a), 753 (295a), *761, 763*
Singer, S. P., 94 (3), 99 (85), 100, 102 (114), 104 (160), 122 (114), 127 (160, 397), *194–197, 201*, 677, 681, 685, 715, 719, 720, 756 (2), *757*
Singh, A., 139, 141 (472), 142 (503), 149 (472), 165 (733), 172, 174 (780), *203, 204, 208,*

209, 397 (22), *417*, 446 (83), *458*, 520 (165), 522, 523 (171), *537*, 568 (169), *589*, 660 (15), 666 (57), *672*, 678 (16, 221), 681 (16), 711 (221), 716 (16), 754, 755 (221), *758, 762*
Singh, A. G., 545, 553, 557 (36), *586*
Singh, A. K., 563 (132), *588*
Singh, H., 453 (179), *460*
Singh, H. B., 85 (86), *89*, 515 (123), 517 (143), 530 (236), *536, 537, 539*
Singh, R., 452 (149), *460*
Singh, R. K., 108 (217), *198*
Singh, S. K., 567 (163), *588*
Sink, C. W., 340 (2b), *346*
Sinn, E., 405 (78), *418*
Sipio, W. J., 98 (78), 109 (225), 139 (78, 474a, 474b), 141 (474a, 474b), 142, 152, 174 (78), *195, 198, 203*, 773 (68), *792*
Sirat, H. M., 172 (778), *209*
Sjøberg, B., 164 (726), *208*, 397 (24), *417*
Sjøgren, C. E., 549, 581 (63), *586*
Sju, P., 41 (161), *50*
Skakke, P. E., 527 (210), *538*
Skinner, A. F., 86 (87), *89*
Skopenko, V. V., 623, 634, 637 (82), *653*
Skrzypczynski, Z., 625, 636 (87), *654*
Skyarevskii, V. V., 53 (14), *87*
Sladky, F., 601, 613 (32), *652*
Slatarov, S., 387 (152), *391*
Slater, J. C., 25 (87), *48*
Slebocka-Tilk, H., 115 (278), *199*, 403 (65), *418*
Sliwa, W., 452 (144, 146), *460*
Sliwkowski, M., 351 (2), *363*
Slyusarenko, E. I., 439 (55), *458*
Smiles, S., 480 (94), *493*
Smirnov, E. V., 297 (67), *335*, 544 (28, 29), *586*
Smith, A. B. III, 102 (138), 104 (163), 110 (234), 116 (289), 144 (524), 169 (764a, 764b), 172 (764a, 764b, 794), 179 (138), *196–199, 204, 209*
Smith, B. C., 62, 63, 65 (23), 66 (38), 67, 71, 74 (23), *88*, 533 (267), *539*
Smith, B. V., 177 (848, 861), 178 (857), *210, 211*, 663 (31, 37), 664 (48), *672*
Smith, C. R. F., 507 (77), *535*
Smith, D. L., 527 (212), *538*
Smith, J. C., 379 (21), 387 (155), *389, 391*
Smith, K. V., 62, 74 (24), *88*
Smith, M. R., 81, 83 (64), *88*
Smith, W. F., 545, 553, 557 (36), *586*
Smith, W. V., 73 (55), *88*
Smolanoff, J., 565 (158), *588*
Smyth, W., 446 (83), *458*
Snider, B. B., 104, 114 (171), 118 (388), *197, 200*
Snieckus, V., 106 (187–189), 116 (288), *197, 199*
Snitman, D. L., 102 (135), *196*
Snowden, R. L., 565 (161), *588*
Sobala, M. C., 96, 179 (37), *194*, 716 (267), *763*
Sobczak, R. L., 106, 145 (196), *197*
Sochorkowa, M., 429, 431 (29), *457*
Soda, K., 351, 352 (10), 353 (10, 13–15), 355 (15, 20, 21), 356 (15, 20, 21, 27, 30–33, 39–41), 357 (31–33, 36), 358 (39), 359 (41), *364*, 372 (79), *376*, 555 (109), *587*
Soja, P., 105 (181, 182), *197*
Soja, R., 715 (260), *763*
Soliven, E., 382 (55), *389*
Soll, D., 362 (55), *365*
Solouki, B., 15 (59), 16 (64), 26 (100), 37 (140–143), 38 (140–143, 147), 39 (147, 151), 40 (151, 153, 155), *47–50*, 228, 266 (80), *269*, 450 (118, 133), 453 (158), *459, 460*, 572 (186), *589*
Solozhenkin, P. M., 253 (193), *271*
Soltani, A., 151 (587), *205*, 290 (48), *335*
Som, P., 371 (69), *376*
Sonada, N., 234, 238 (97), 239 (115), *269, 270*
Songstad, J., 342 (29c), *347*, 400 (51), *418*, 542 (15–18), 549 (67), 579 (18, 216), 580 (18, 216, 217, 219, 220), 581 (67, 224), 582 (216), 583 (224), *585, 586, 590*, 635 (116), 640 (137), 645 (137, 168–171), *654, 655*
Sonoda, N., 103, 107 (153), 108 (209), 116 (294, 300–303), 149 (568), 169 (773), 172 (781), 176 (839), 177 (844, 845), 178 (849–851), 187 (919), 191 (961–963), 192 (943–948, 950–954, 956–960), *197–200, 205, 209, 210, 212, 213*, 220 (25), 221 (36), 254 (202), *268, 271*, 398 (34), 409, 410 (124), *417, 419*, 499 (28), *534*, 664 (40, 41), *672*
Sorgi, K. L., 117 (315, 316), *200*, 770 (53), *791*
Sosnovsky, G., 151 (602, 603, 606, 607), *205, 206*
Soullier, B. K., 385 (139), *391*
Sowa, L., 467, 483 (51), *493*
Spallholz, J. E., 383 (59), 384 (81–83, 86), *389, 390*
Spanget-Larsen, J., 26, 27 (104), 41 (161, 163), *49, 50*
Sparks, R. A., 67 (40), *88*, 344 (39a), *347*
Spence, G. G., 334 (169), *337*
Spencer, H., 449, 454 (114), *459*
Spencer, H. K., 283 (30), 328 (30, 157), *334, 337*, 396, 412 (18), *417*, 478 (86), *493*, 580, 582–584 (221), *590*, 659, 660 (9), *671*
Spencer, K. H., 526 (193), *538*
Speth, D. R., 118 (331), *200*
Spialter, L., 711 (218), *762*
Spies, H., 247 (155), (217), *270, 272*
Spijkervat, A. L., 53 (15), *87*
Spirlet, M. R., 517 (136), *537*
Spiro, W. J., 665 (53, 55), *672*

Spisak, J. F., 369 (19), *375*
Spraque, M. J., 580 (222), *590*
Sprecher, M., 133 (443), *202*
Spreutel, S., 739 (309), *764*
Springer, J. P., 116 (296), *199*
Spunta, G., 33 (119), *49*
Srivastava, K. C., 387 (150), *391*
Srivastava, P. C., 371 (65, 67), 372 (67), 373 (84), 374 (96), *376*, 386 (111, 121, 125), *390, 391*, 785, 790 (116), *793*
Srivastava, R. C., 517 (145), *537*
Srivastava, S., 373 (85), *376*
Srivastava, S. C., 371 (69), *376*
Srivastava, T. N., 373 (84), *376*, 517 (145), *537*
Srivastava, V. K., 517 (145), *537*
Stadtman, T., 262 (247), *272*
Stadtman, T. C., 351 (2–4), 360 (44), 361 (44, 45, 49, 52), 362 (45, 49, 52, 56), 363 (56), *363–365*
Stam, C. H., 614, 615, 646 (62), *653*
Stanghellini, P. L., 345 (49), *347*
Stankevich, V. G., 53 (14), *87*
Stanko, V. I., 86 (88), *89*
Stanley, W., 282, 327, 328 (28), *334*, 412 (136), *419*
Starace, A. F., 5, 8 (16), *47*
Stark, J. C., 341 (16), *346*, 485 (116, 118), *494*, 497 (12), 510 (12, 43), 502 (44), 506 (12), 511 (44), 514 (43), 516 (134), 517 (134, 138), 530 (238, 240), *534, 535, 537, 539*
Stavaux, M. 487 (125), *494*
Stec, W. J., 576 (200), *589*, 629 (100), *654*
Steeves, B. H., 613, 635 (60), *653*
Steliou, K., 221, 223 (35), *268*
Stenkevich, I., 453 (156), *460*
Stenlake, J. B., 248 (167), *270*
Stepanov, B. I., 35 (127, 128), *49*
Stepanov, E. P., 53 (14), *87*
Stephan, W., 396 (9), *417*
Stephenson, L. M., 118 (331), *200*
Stern, A. G., 565 (160), *588*
Stern, P., 149 (573), *205*, 718 (274), *763*
Stetsenko, A., 453 (186b), *461*
Steudel, R., 402 (58, 59), *418*
Stevens, A. I., 467, 469 (43), *492*
Stevens, A. L., 467, 469 (39), *492*
Stevens, J. G., 71 (46), *88*
Stevens, R. V., 144 (526), *204*
Stevenson, D. P., (111), *49*
Stewart, I. M., 260 (235), *272*
Still, W. C., 118, 157, 158 (336), *200*, 567 (162), *588*
Stock, G., 564 (136), *588*
Stockton, R. A., 369 (15), *374*
Stone, A., 411 (129), *419*
Strauss, E.-M., 402 (58, 59), *418*
Strauss, H. W., 371 (57, 66, 68, 69), *375, 376*
Strauss, M. D., 371, 372 (61), *375*
Strausz, O. P., 281, 318 (26, 27), *334*
Streeter, D. G., 386 (122), *391*
Streets, D. G., 7, 8 (41), *47*
Streitwiesser, A. Jr., 6 (34), *47*
Strelets, B. Kh., 503 (51), *535*
Strelets, T., 453 (153), *460*
Strezelecka, H., 486 (122), *494*
Stringer, O. D., 99, 180, (88, 89), *195*
Stropnik, C., 235 (100), *269*
Struchkov, Yu. T., 513, 514 (119), *536*, 623, 634, 637 (82), *653*
Stukalo, E. A. 443 (72), 444 (72–74), 445 (74), *458*
Sturgeon, G. D., 658 (2), *671*
Su, W., 184 (888), *211*
Suárez, E., 109 (227), 183 (885), *198, 211*
Suarez, E., 715 (248), *762*
Suchi, R., 342 (29a), *347*
Suchomel, H., 616, 643 (66), *653*
Sudoh, R., 113 (253, 254), *199*, 683, 696 (103), 715 (103, 262), *759, 763*, 774 (72), *792*
Sudol, M., 576 (200), *589*
Suemitsu, R. 177 (847), *210*
Suemura, S., 118 (325), *200*
Sugawara, T., 163 (720–722), *208*
Sugie, K., 356, 357 (31), *364*
Sugita, N., 191 (965), *213*
Suguro, T., 555 (101), *587*
Suhadolnik, R. J., 778 (87), *792*
Sukhai, R. S., 235 (99), 239 (114), 243 (99), 244 (135), *269, 270*, 408 (97), *418*
Sukumaran, K. B., 121 (358, 359), 201
Sumi, H., 467 (15), *492*
Sunay, U., 169 (762), *209*
Sunde, M. L., 387 (146), *391*
Sunde, R. A., 355 (17), *364*
Suri, S. C., 184, 186 (899, 901), *211*, 721 (287), *763*
Surya Prakash, G. K., 225 (60), *269*
Suschitzky, H., 423 (11, 13), *457*
Sustmann, R., 6 (29), *47*, 772 (66), *792*
Suttle, J. F., 507 (77), *535*
Suzuki, A., 104 (159), *197*, 715 (257), *763*
Suzuki, F., 105 (184), 106 (190), *197*
Suzuki, H., 161 (706, 707), 172 (785), 175 (825, 828), 176 (833, 835), 178 (852, 856), *207, 209, 210*, 246 (145), *270*, 320 (133), 333 (166), *336, 337*, 443 (67), *458*, 499 (29), *534*, 544 (30, 31), 574 (195), 581, 583–585 (223), *586, 589, 590*, 753 (316, 317), *764*
Suzuki, K., 116 (297, 298), 140 (488), 144 (297, 298), *200, 203*
Suzuki, M., 98, 187 (68), *195*
Suzuki, R., 710 (200), *761*
Suzuki, T., 187 (907), *211*, 351–353 (10), 356 (33), 357 (33, 36), *364*, 555 (109), *587*

Svanholm, U., 247 (161), 264 (264), *270, 273*
Svanholt, H., 243 (128–130), 267 (128), *270*, 449 (112), 450 (117), *459*
Sviridov, A. F., 163 (715, 716), *208*
Svorstøl, I., 580 (217), *590*
Swanson, A. B., 382 (58), *389*
Swift, P., 26 (102), *48*
Symalla, E., 598, 643 (23), *652*
Syper, L., 120, 147 (351), *201*, 415 (153), *420*
Székely, I., 139 (476), 174 (816), *203, 210*
Szekely, I., 666 (56), *672*, 715 (246), *762*
Szperl, L., 221 (51), *268*

Taga, J., 116 (295), *199*, 715 (243), (237), *762*
Tahara, S., 111 (248), *199*
Tahir, T. A., 280 (21), *334*
Tainturier, G., 34 (122, 123), *49*, 305 (88, 89), *336*
Takagi, M., 168 (759), *209*
Takagi, S., (237), *762*
Takahashi, K., 452 (148), *460*
Takahashi, M., 550 (85), 561 (126), *587, 588*
Takahashi, T., 104 (159, 180), 117, 184 (311), *197, 200*, 685 (256), 715 (256, 257), *763*
Takai, I., 113 (253), *199*, 715 (262), *763*
Takai, Y., 116 (300), *200*
Takaki, K., 147 (550), *204*, 706, 719 (166b), *761*
Takaku, H., 567 (165), *588*, 785 (113), *792*
Takano, K. 135 (456), *203*
Takano, S., 169 (766), *209*, 561 (126), 564 (137), 565 (147), *588*
Taka-Oka, K., 172 (785), *209*
Takaoka, K., 178 (856), *210*
Takarada, M., 192 (950), *212*
Takebayashi, M., 290 (50), *335*
Takechi, K., 176 (835), *210*
Takechi, S., 187 (926), *212*
Takeda, R., 115, 116 (306, 307, 309), 184 (894), *200, 211*
Takeda, S., 443 (67), *458*
Takei, H., 162 (711), *208*
Takei, I., 774 (72), *792*
Takekoshi, H., 53 (11), *87*
Takeuchi, Y., 98 (79–81), 102 (147), 131 (147, 425), 136 (81, 464, 465), 137 (79, 80, 466), 169, 170 (765), 173 (147), *195, 196, 202, 203, 209*, 453 (159), *460*, 547 (58, 59), 556 (113), 570 (180, 181), 571 (182, 183), *586, 588, 589*
Takigawa, T., 561 (122), *588*
Takita, S., 111, 122 (243), *198*
Talbot, J.-M., 577, 582 (210), *590*
Talcott, P. A., 384, 387 (88), *390*
Talham, 466, 489 (3), *491*
Talham, D., 407 (117), *419*, 480, 485 (97), *494*, 506 (68), *535*
Tamagaki, S., 153 (625), *206*, 430–432 (35), 445 (77), *457, 458*
Tamai, H., 171 (775), *209*
Tamaki, K., 184 (887), *211*
Tamari, T., 133 (443), 164 (730), *202, 208*
Tamura, N., 111, 122 (243), 169 (766), *198, 209*
Tamura, S., 369 (29, 31), *375*
Tamura, Y., 543, 554 (21), *585*
Tan, V. Y. Y., 542 (8), *585*
Tanaka, H., 147, 152–154 (552), *204*, 340 (8), *346*, 351, 352 (10), 353 (10, 13–15), 355 (15, 21), 356 (15, 21, 31–33, 39–41), 357 (31–33, 36), 358 (39), 359 (41), *364*, 555 (109), 565 (142), *587, 588*, 718 (272), (237), *762, 763*, 774 (74), *792*
Tanaka, I., 356, 357 (32), *364*
Tanaka, J., 467 (25), *492*
Tanaka, K., 395 (5), *417*
Tanaka, M., 564 (137), *588*
Tanaka, S., 572, 574 (187), *589*
Tanaka, T., 154, 155, 169 (654), *206*
Tancheva, C., 143 (510), *204*
Taneja, S. P., 84 (83), *89*
Tang, S., 370 (44), *375*
Tanigawa, H., 165 (739), *208*
Taniguchi, Y., 118 (325), *200*
Tanikaga, R., 333 (168), *337*
Tanizawa, K., 353 (15), 355, 356 (15, 20), *364*
Tanner, D. B., 467, 469 (44), *492*
Tanovnik, B. S., 235 (100), *269*
Tappel, A. L., 355 (18), *364*
Tarantelli, T., 5, 18, 20, 21 (18), *47*, 246 (152), *270*, 327 (152), *337*, 549, 574 (66), *586*
Tarygina, L. K., 545 (32), *586*
Taschner, M. J., 498 (24), *534*
Tashiro, M., 452 (148), *460*
Tashtoush, H., 668 (68), *673*
Tassi, L., 549 (74), *587*
Tatarinova, A. A., 407 (101), *419*, 505 (64, 65), *535*
Tate, S. S., 355 (19), *364*
Taticchi, A., 21–23 (72–74), 24 (72, 74), 25 (72–74), 33 (74), *48*, 265 (273), *273*
Tatsuno, T., 565 (143), *588*
Taylor, E. C., 334 (169), *337*
Taylor, H. S., 6 (21), *47*
Taylor, J. L., 383 (71), *390*
Taylor, P., 519 (154), *537*
Taylor, R. T., 121, 123, 124, 147 (361), *201*
Temkin, H., 407 (113), *419*
Teranishi, A. Y., 95, 96, 101, 103 (27), *194*, 522 (174), *537*, 685, 695 (87), 714, 715 (229), *759, 762*
Terao, K., 99 (96, 101), 103, 108 (96), 116 (101), 142 (505), 159 (101), *195, 196, 204*, 522 (177b), *537*
Terem, B., 177 (861), *211*, 663 (31), *672*
Terrier, F., 453 (172), *460*

Testaferri, L., 169 (763), *209*, 409 (121), *419*, 548 (60), *586*
Tezuka, T., 333 (166), *337*
Thaisrivongs, S., 565 (152), *588*
Thavornyntikarn, P., 508 (82), *536*
Thavornyutikarn, P., 72 (48), *88*
Thayer, J. S., 62, 74 (24), *88*, 246 (147), *270*, 585 (230), *590*
Theissen, D. R., 710 (205), *761*
Theobald, F., 34 (123), *49*
Theodorakis, M. C., 386 (118), *391*
Thewalts, U., 308 (98), *336*
Thi, M. P. N., 172 (793), *209*
Thibaut, P., 342 (23), *347*, 577, 582 (207), *589*, 765 (19), *791*
Thibaut, Ph., 447 (97), *459*, 500 (38), *535*
Thiele, G., 401 (56), *418*, 619, 620 (71), *653*
Thierry, J., 99, 102, 132, 163, 191 (102), *196*, 322 (138), *337*
Thirring, K., 104 (161), *197*
Thomas, A., 452 (144, 146), *460*
Thomas, E. J., 98 (67), 99 (104), 106 (195), 141 (104), 172 (778), 182 (881, 882), 191 (881, 882, 941), *195–197, 209, 211, 212*
Thomas, G. A., 467 (11, 12, 39, 43), 469 (39, 43), *491, 492*
Thomas, M. J. K., 82 (76), *89*
Thomas, M. T., 106 (188, 189), *197*
Thomas, R., 681, 682 (47), *758*
Thompson, C. R., 603 (36), *652*
Thompson, D. G., 640 (139), (157), *655*
Thompson, H. J., 385 (134, 135), 386 (110), *390, 391*
Thompson, J. K., 768, 790 (31), *791*
Thompson, M., 2 (3), *46*
Thomson, C. D., 385 (97), *390*
Thon, N., 32, 36 (114), 41 (162), *49, 50*
Thorstenson, T., 542 (16, 17), *585*
Thorup, N., 467, 469 (44), *492*
Thottathil, J. K., 169, 170 (768), *209*
Tice, C. M., 102 (141), *196*
Tiecco, M., 169 (763), *209*, 409 (121), *419*, 548 (60), *586*
Tiegs, G., 374 (93), *376*, 384 (76), *390*
Tietze, L. F., 102 (137), *196*
Tilhard, H.-J., 132 (434), *202*
Tilhard, H. J., 694, 695, 713, 747, 749 (120), *760*
Timofeeva, T., 453 (153), *460*
Tingoli, M., 169 (763), *209*, 409 (121), *419*, 548 (60), *586*
Tischenko, N. P., 431, 432 (38), *457*
Tishaninova, A. A., 113 (252), *199*
Tisler, M., 235 (100), *269*
Titus, D. D., 486 (121), *494*
Tobiason, F., 453 (177), *460*
Toda, F., 395 (5), *417*
Toda, J., (237), *762*
Todd, M. R., 21 (70), *48*
Toder, B. H., 116 (289), 169, 172 (764b), *199, 209*
Todres, Z., 453 (156), *460*
Todt, K., 775 (76), *792*
Togami, M., 187 (924), *212*
Toghraie, S., 151 (588), *205*, 450 (121), *459*
Tollari, S., 99 (87), 174 (813), *195, 210*
Tolosa, E. A., 352 (12), *364*
Tomimatsu, K., 430, 432–434 (36), *457*
Tominaga, T., 104 (174, 175), *197*
Tomita, M., 105, 109 (183), *197*
Tomkins, W. A. F., 386 (118), *391*
Tomods, S., 98 (79–81), 102 (147), 104 (164, 165), 106 (165), 131 (147, 425), 136 (81, 464, 465), 137 (79, 80, 466), 173 (147), *195–197, 202, 203*, 547 (58, 59), 556 (113), 570 (180, 181), 571 (182, 183), *586, 588, 589*
Torii, S., 103, 107 (154), 108 (205, 206), 116 (205, 206, 292, 293), 135 (456), 139 (484), 141 (206), *197–199, 203*
Toriumi, K., 483 (106), *494*
Torrance, J. B., 489 (133), *494*
Torssel, K. 770 (48), *791*
Torssell, K., (236), *762*
Toru, T., 137 (468), 141 (498), *203*, 427 (25c), 428 (27), *457*
Toscano, V. G., 595, 596, 624, 627, 629, 648 (11), *652*
Toshimitsu, A., 95 (31), 97 (52, 53, 60), 99 (60, 96, 100, 101), 100 (116, 119), 103 (96, 149), 108 (96, 208), 116 (101, 116, 119, 208, 290, 291), 117 (313, 314), 125 (387), 126 (52, 53, 393), 136 (60), 142 (505), 144 (520, 521a, 521b, 522, 523, 527), 159 (101), 183 (387), (282), *194–196, 198–201, 204*, 522 (177b), 523 (181), *537, 538*, 545 (40, 42, 43), 546 (50), 569 (175–179), 570 (43), 573 (189), 574 (196), 579 (212, 213), *586, 589, 590*, 715 (250, 263), *762, 763*
Tosi, G., 549 (74), *587*
Townes, C. H., 59, 60, 64 (22), *88*
Townsend, I., 407, 416 (118), *419*, 504 (58), *535*
Townsend, L. B., 263 (251), *272*, 778 (92, 93, 96, 97), 779 (99–101), 782 (99–101, 107), 783 (107–111), 789 (92, 96, 97), *792*
Tracey, C. M., 387 (127), *391*
Trachtenberg, E. N., 110 (231), *198*
Traren, V. F., 513, 514 (119), *536*
Traven, V. F., 35 (127, 128), *49*
Traverso, G., 252 (190), 261 (239), *271, 272*
Traynham, J. L., 310 (102), *336*
Trend, J. E., 96, 98 (38), 99 (38, 93), 115, 116, 123 (38), 128 (407), *194, 195, 202*, 427 (20), *457*, 555 (100), *587*, 706 (183), 721 (327), *761, 764*
Triebwasser, K. C., 382 (56), *389*

Tripathy, S., 497, 506 (14), *534*
Trofimov, B. A., 407 (100, 101), *419*
Trogu, E. F., 519 (153), *537*
Troitskii, V. V., 662 (23, 24), *672*
Troitsky, V. V., 519 (155, 156), *537*
Tromfimov, B. A., 505 (64–66), *535*
Trost, B. M., 97 (41), 99 (41, 94), 122 (94), 172 (777), 183 (94, 777), *194, 195, 209*, 681, 682 (58), 715 (241), 724 (328), 726 (241), *759, 762, 764*
Trotter, J., 612 (58), *653*
Trotter, J. W., 106 (194), *197*
Troup, J., 345 (43), *347*
Truesdale, L. K., 127 (396), *201*, 438 (51), *458*
Tsai, L., 262 (247), *272*, 361 (49, 52), 362 (49, 52, 57), *364, 365*
Tsai, Y.-M., 145 (539), *204*
Tsay, D.-T., 382 (56), *389*
Tscherkes, C. A., 384 (90), *390*
Tschmutova, C. A., 33 (118, 120), *49*
Tschmutova, G., 33, 36 (117), *49*
Tschmutowa, G., 35, 36 (134), *49*
Tseng, P., 53 (11), *87*
Tsintsadze, G. U., 267 (288), *273*
Tskalobadze, L., 453 (175), *460*
Tsoi, L. A., 187 (915), *212*, 575 (198), *589*
Tsuda, Y., 101 (128, 129), 105 (184), 116 (295), 125 (128, 129), *196, 197, 199*, 715 (243), (237), *762*
Tsuji, J., 104 (159, 180), 117, 184 (311), *197, 200*, 685 (256), 715 (256, 257), *763*
Tsutsumi, S., 192 (943, 944, 947, 951, 953), *212*, 254 (202), *271*
Tsveniashvili, V., 453 (175, 176), *460*
Tuan, H. M., 361 (47), *364*
Tuan, N. W., 361 (51), *364*
Tucek, E., 247, 248 (162), *270*
Tucker, E. M., 370 (47), *375*
Tully, C. R., 286, 319 (38), *335*
Tunbridge, R. E., 368 (7), *374*
Tuovinen, O. H., 369 (20), *375*
Turkevich, V. V., 608, 628 (50), 629 (50, 101), *653, 654*
Turner, C., 453 (164), *460*
Turner, D. W., 2 (8), 10 (49), 13 (54), 26 (96), *46–48*
Turner, E. S., 177 (848, 861), *210, 211*, 663 (31), 664 (48, 49), 666 (49), 668 (66), *672, 673*
Turner, J. V., 104 (156), *197*
Turro, N. J., 330 (162), *337*
Tuthill, P. A., 108 (213), *198*
Tyagi, M. P., 770 (48), *791*
Tyerman, W. J. R., 281, 318 (24–26), *334*

Uchida, A., 453 (166), *460*
Ueda, K., 290 (50), *335*
Ueda, T., 449 (104), *459*
Uematsu, T., 565 (148), 568 (170), *588, 589*
Uemura, M., 187 (924), *212*
Uemura, S., 95 (21, 31), 97 (52, 53, 60), 99 (60, 96, 100, 101), 100 (116–120), 102 (438), 103 (96, 149), 108 (96, 208), 116 (101, 116, 118, 119, 208, 290, 291), 117 (313, 314), 124 (377a, 377b), 125 (117, 283, 387), 126 (52, 53, 393), 133 (436–440), 136 (60), 142 (505), 144 (520, 521a, 521b, 522, 523, 527), 159 (101), 162 (712, 713), 164, 168 (728), 180 (869), 183 (117, 283, 387), 184 (887), 191 (965), (282), *194–196, 198–202, 204, 208, 211, 213*, 356, 357 (33), *364*, 397 (26), *417*, 522 (177b), 523 (177c, 181), *537, 538*, 545 (40–43), 546 (50), 555 (109), 569 (175–179), 570 (43), 572 (187), 573 (189), 574 (187, 196), 579 (212, 213), *586, 587, 589, 590*, 715 (250, 263), 756 (318, 331, 332), 757 (331), *762–764*
Ueno, Y., 113, 161 (261), *199*, 685, 695 (93), *759*
Ugarkar, B. G., 787 (119, 120), *793*
Uhlemann, E., 258 (221), 260 (237), *272*
Uhlig, E., 528 (227), *538*
Ujjainwalla, M., 136 (461), *203*
Ulevitch, R. J., 356 (26), *364*
Ulmann, A., 372 (80), *376*
Umbreit, M. A., 118 (333), *200*
Umeda, S., 165 (739), *208*
Umemura, T., 565 (148), *588*
Umezawa, B., 97, 116 (50), *195*
Umezawa, H., 770 (50), *791*
Umezawa, S., 770 (50), *791*
Umhoefer, S. G., 151 (603), *205*
Uneyama, K., 103, 107 (154), 108 (205, 206), 116 (205, 206, 292, 293), 135 (456), 139 (484), 141 (206), *197–199, 203*
Uno, H., 139 (482), *203*
Uohama, M., 330 (163), *337*
Urabe, H., 146 (546, 547), *204*
Usachev, A. I., 753 (306), (315), *764*
Ushanov, V. Z., 187 (915), *212*
Usiatinsky, A. Ya., 416 (155, 156), *420*
Usui, Y., 171 (775), *209*
Usuki, M., 246 (145), *270*, 320 (133), *336*, 574 (195), *589*
Uznanski, B., 629 (100), *654*

Vahrenhorst, A., 757 (301), *764*
Vahrenkamp, H., 340 (4), *346*
Valeef, R. B., 550 (80), *587*
Valkonen, S., 370 (34), *375*
Valle, G., 221 (47, 48), *268*
Van, D. le, 643 (155), 645 (165), 647 (178), 648 (182), (180, 181, 183), *655, 656*
Van, Duc le, 597 (15, 20, 21), 600 (21), 602–604, 643 (15), *652*
Van Allen, J. A., 486 (120), *494*
Van Bleet, J. F., 379 (18, 19), *389*

VanDeMark, M. R., 282, 327, 328 (28), *334*
Van den Bossche, G., 517 (136), *537*
Vanderbilt, D. P., 715 (254), *763*
Van Der Puy, M., 108 (207), *198*, 685 (86), *759*
Vandyukova, I. I., 645 (162), *655*
Van Ende, D., 125 (384, 386), 154 (384, 386, 631, 642, 643, 645, 646), 155 (384, 386, 642, 643, 645, 646), 156 (384, 386, 643, 646), 157 (384), 174 (645), 177 (843), *201, 206, 210*, 678 (14), 681 (14, 25, 68), 683, 684 (68, 74), 685 (68, 74, 94, 336), 686 (68, 74, 94, 100), 689 (94), 695 (14, 74, 94, 138, 141, 154), 696 (68, 74), 698 (94), 699 (68, 74, 94), 700 (138, 154), 705 (154), 706 (74, 181), 708 (68), 711, 715 (14), 716 (74), 718 (94, 138), 719 (94), 720 (74, 138), 721 (25, 100), 722 (100), 726 (68, 94), 727 (181), 728 (14), 734, 736 (100), 748, 753 (74), *758–761, 764*
VanEnde, D., 153–155 (621), *206*
Van Erde, D., 523 (180), *538*
Van Es. T., 770 (51), 775 (82–86), 777 (82, 83), *791, 792*
Vanier, N. R., 108 (207), *198*, 685 (86), *759*
Vanino, L., 221 (41), *268*
Van Vechten, J. A., 513 (112), *536*
Van Vleet, J. F., 369 (22), 370 (22, 37), *375*
Van Wazer, J. R., 667 (64), *673*
Vargas. F., 314 (117, 118), 317 (124, 127), *336*
Vasella, A., 118 (328), *200*
Vas'kiv, A. P., 608, 628, 629 (50), *653*
Vasudev, P., 71 (45), 73, 87 (56), *88*
Vatele, J. M., (238), *762*, 770 (46, 47), *791*
Vaughan, D., 455 (200), *461*
Vaultier, M., 576 (201), *589*
Veda, T., 449 (105a, 105b), *459*
Vedejs, E., 99 (85), 104 (179), *195, 197*
Veigel, J., 485, 489 (113), *494*
Veit, A., 621 (74), *653*
Veracini, C., 453 (161, 162), *460*
Verbist, J., 456 (203), *461*
Vereshchagin, A. N., 549 (71–73), *587*
Vereshchagin, L. I., 507 (74), *535*
Verez Bencomo, V., 773 (67), *792*
Verheul, H., 62 (20), *87*
Verkade, J. C., 42 (164, 165), *50*
Verkruijsse, H. D., 408 (95), *418*
Vermeersch, G., 284 (33), *335*, 527 (199), *538*, 660 (14), *672*
Vermeire, M., 454 (189), *461*, 515, 516 (124), *536*
Vernan, J. M., 547, 554 (56), *586*
Verner, H., 38 (148), *50*
Vertongen, F., 383 (66), *390*
Vetter, J., 650 (189), *656*
Veve, J., 383 (66), *390*
Vial, C., 104 (158), *197*
Vialle, J., 456 (205), *461*
Vicens, J., 97, 116 (59), *195*, 715, 718, 721 (251), *762*
Vidoni Toni, M. E., 258 (224), *272*
Viehe, H. G., 135 (453, 454), *203*, 670 (81), *673*
Vikane, O., 342 (27d-f, 27h, 29a, 29b), 343 (27d-f, 27h), *347*
Vilas, N. N., 382 (52), *389*
Villieras, J., 681, 682 (61), *759*
Vincent, E. J., 26 (101), *48*
Vinkitachalam, T. V., 286 (37), *335*
Vinokurova, G. M., 413 (139), *419*
Violet, C. E., 53 (19), *87*
Vitali, D., 528 (218, 219), *538*
Vivier, H., 34 (123), *49*
Voet, J. G., 356 (26, 28), *364*
Voigt, E., 151 (591), *205*
Volgarev, M. N., 384 (90), *390*
Volmer, M., 154, 155 (658), *207*
Volonchuk, V. G., 431, 432 (38), *457*
Voloshchuk, V. G., 297 (74), *335*
Voloshocnuk, V. G., 670 (78), *673*
Voorthius, H., 62 (20), *87*
Vorbrüggen, H., 187 (913), *212*
Voss, J., 224, 233, 240, 263 (59), 266 (276), *269, 273*
Vtyurina, N. N., 33 (118, 120), *49*
Vuilhorgne, M., 153 (620), *206*
Vyazankin, N. S., 281, 309, (23), *334*, 398 (35, 36), 399 (35, 36, 41, 42, 44), 414 (36, 44, 147), *417, 420*, 663 (29), *672*
Vyaznikovtsev, L. V., 187 (915), *212*

Wachter, J., 405 (79), *418*
Wada, K., 107 (201, 202), *198*
Wadsworth, D. H., 187 (914), *212*, 662 (25), *672*
Wadsworth, W. S., 681, 682 (46), *758*
Wagenknecht, J. H., 469 (58), *493*
Wagner, G., 35 (130, 131), *49*, 765 (14, 15), 766 (14, 25), 767 (26), 769 (25), *791*
Wagner, P., 387 (146), *391*
Wagner, P. A., 387 (149), *391*
Wagner, P. J., 276, 329 (1), *334*
Waitkins, G. R., 110 (228), *198*
Wakasugi, M., 133 (439), 164, 168 (728), *202, 208*, 545 (41), *586*
Wakefield, B. J., 681, 682 (30), *758*
Walatka, V., 467 (8), *491*
Walatka, V. W., 467 (7), *491*
Walker, D. F., 383 (60), *389*
Walker, H., 423 (3), *456*
Wallis, J. D., 172 (778), *209*
Wallmark, I., 230 (87), *269*
Walsh, C., 357 (35), *364*
Walsh, C. T., 356 (29, 30), *364*
Walsh, W. M. Jr., 467, 469 (43), *492*

Walter, R., 95, 97, 99 (25), *194*, 513 (118), *536*, 765 (18), *791*
Walter, W., 264 (266), (111), *269, 273*, 425, 429, 430 (21), *457*
Walter, W. E., 500 (35), *534*
Walther, B., 650 (188), *656*
Wampler, D. L., 614, 615, 646 (62), *653*
Wandelborn, D. F., 721 (327), *764*
Wang, A., 467, 483 (51), *493*
Wang, H. G., 483 (108), *494*
Wang, H. H., 467, 483 (50), *493*
Ward, A. D., 132 (433), *202*
Ward, D. E., 565 (145), *588*
Ward, R. S., 187 (923), *212*
Warner, H., 405 (80), *418*
Warner, J. S., 220 (30), *268*
Warpehoski, M. A., 118 (330), *200*
Wart, H. E. van, 37 (139), *49*
Wartski, L., 698, 700 (155), *760*
Wasserman, A. von, 577 (203), *589*
Watanabe, K., 99 (86), *195*
Watanabe, N., 97, 99, 136 (60), *195*, 574 (196), *589*
Watanabe, T., 369 (33), *375*, 378 (2), *388*
Watari, F., 644 (159), *655*
Watson, D. G., 614, 615, (61, 62), 646 (62), *653*
Watson, K., 445 (76), *458*
Watson, W. H., 631 (107), *654*
Watson, W. P., 97, 172 (61), *195*
Watt, D. S., 102 (135), *196*
Wawzonek, S., 483 (100), *494*
Waykole, L., 154, 155 (658), *207*
Wazczak, J. V., 467, 469 (43), *492*
Weast, R. C., 27, 30 (105), *49*
Weaver, J. L., 285, 286 (36), *335*, 344 (40), *347*
Webb, F. J., 677, 678, 681, 711 (10), *757*
Webb, K. S., 483 (108), *494*
Webb, R. P. II, 142, 168 (504), *204*
Webb, R. R. II, 428 (25b), *457*
Weber, K., 456 (204), *461*
Weber, L., 646 (176), *656*
Weber, R., 446 (82, 85, 99), 447 (99), 448 (103), 449 (99), 454 (188), *458, 459, 461*, 500 (38), *535*
Webster, R. G., 235 (101), *269*
Wee, A. G. H., 151 (604), *205*
Weedon, A. C., 222 (56), 227 (71), *268, 269*, 322, 329, 330 (136), *337*
Weger, M., 467, 469 (23, 44), *492*
Wegner, G., 485, 489 (114), *494*
Wehle, D., 698, 699 (151), *760*
Wehne, D., 467, 485, 489 (52), *493*
Wei, C., 405 (78), *418*
Wei, K. H., 705 (162), *761*
Wei, K. S., 284 (34), *335*
Weichmann, J., 259 (231), *272*, 405 (81), *418*
Weidenbruch, H., 398 (40), *417*
Weiland, J., 252 (187), *271*, 469 (57), *493*
Weiler, L., 143 (517), *204*
Weiner, M. A., 710 (190, 197, 199), *761*
Weirich, W. E., 379 (19), *389*
Weiss, R., 446 (87), *459*
Weissflog, E., 39, 40 (151), *50*, 228, 266 (80), *269*, 501 (40), *535*, 572 (186), *589*
Welch, J. G., 710 (191), *761*
Welch, S. C., 106 (194), *197*, 386 (119), *391*
Welcman, N., 547 (53, 54), *586*
Weleman, N., 595, 602 (9), *652*
Weller, D. D., 101 (132), *196*
Wells, A. F., 595 (8), *652*
Wells, G. J., 154, 155, 157, 171, 172 (655), *207*, 711 (216a), *762*
Welsch, C. W., 385 (136), *391*
Welter, A., 447 (93–95, 97), *459*
Wen, T. N., 361 (47), *364*
Wen, Z., 370 (45), *375*, 379 (16), *389*
Wendel, A., 350 (1), *363*, 374 (93), *376*, 379 (20), 384 (76), *389, 390*, 447 (89), *459*
Wendelborn, D. F., 96, 98, 99, 115, 116, 123 (38), *194*, 555 (100), *587*
Wender, P. A., 567 (163), *588*
Wenkert, E., 162 (710), *207*
Wennerbeck, I., 31, 32 (112), *49*, 327 (151), *337*
Wensky, A., 306 (91), *336*, 527 (200), *538*, 660 (19), *672*
Wentink, T., 12 (52), *47*
Wentrup, C., 254 (203), *271*
Werchan, H. C., 544 (24), *586*
Werner, H., 259 (228, 229), *272*
Werner, M., 467, 485, 489 (52, 53), *493*
Werner, R. L., 246 (153), *270*, 403 (67), *418*, 549 (64), 585 (229), *586, 590*
Wertheim, G. K., 52 (2), *87*
West, F. G., 104 (179), *197*
West, R., 398 (37), *417*, 711 (213), *762*
Westermark, T., 383 (64, 65), *390*
Westwood, N. P. C., 17, 18 (67), 42, 44 (168), *48, 50*
Weswig, P. H., 382 (47), 383 (69), 387 (157), *389–391*
Wetter, H., 157, 158 (689), *207*
Wewers, D., 646 (176), *656*
Whangbo, M.-H., 341 (15), *346*
Whanger, P. D., 382 (47), 383 (69), 385, 386 (98), 387 (157), *389–391*
Whistler, R. L., 765 (10), 775 (80, 83), 777 (83), *790, 792*
White, A. H., 81, 82 (65), *88*, 344 (41), *347*, 515 (127), *536*
White, D. A., 385 (95), *390*
White, D. H., 106 (194), *197*
White, J. D., (147), *760*
White, M. G., 9, 10 (44), *47*
White, R., 369 (25), *375*

Whiteford, R. A., (107), *49*
Whitehead, E. I., 351, 353 (6), *364*, 382 (56), *389*
Whitehouse, R. D., 95, 96 (24), *194*, 680, 681, 714, 721 (26), *758*
Whiting, D. A., 102 (136), *196*
Whittle, A. J., 129 (413), 143 (518), 173 (805–807), *202, 204, 209*
Wiberg, K. B., 394 (4), *417*
Wieber, M., 604 (41), 605, 607, 619 (45), *653*
Wietzke, M., 398 (30), *417*
Wijekoon, W. M., 264, 265 (269), *273*
Wijers, H. E., 505 (67), *535*
Wilckens, R., 452 (147), *460*
Wild, J., 563 (134), *588*
Wildbredt, D.-A., 632 (110), *654*
Wilke, G., 258 (221), *272*
Wilke, M., 398 (30), *417*
Wilkens, C. J., 72 (51), *88*
Wilkinson, G., 356 (23), *364*
Wille, F., 446 (80), *458*
Willhardt, I. H., 352 (12), *364*
Williams, B. L., 528 (217, 223, 224), *538*
Williams, D. F., 489 (129), *494*
Williams, D. J., 111 (244, 246), 122 (246), 140 (491), 152, 153 (614), 182, 191 (882), *199, 203, 206, 211*
Williams, D. R., 109 (219, 220), 118 (337), *198, 200*
Williams, E., 173 (810), *210*, 427 (23), *457*, 558 (119), 565 (142), *588*, 774 (73, 74), *792*
Williams, F. D., 228, 229 (75), *269*
Williams, J., 467, 483 (50, 51), *493*
Williams, J. M., 341 (15), *346*, 483 (108), *494*
Williams, J. R., 102, 122 (142), *196*
Williams, M. A., 31, 32 (112), *49*, 327 (151), *337*
Williams, R. M., 99 (107), *196*
Williams, T. A., 8, 9 (43), *47*
Williard, P. G., 140 (487), *203*
Willis, C. E., 385 (94), *390*
Willis, W. W., 683 (70), 684, 695 (321), 706–709 (70), 715, 716 (264), 726 (70), 729 (321), *759, 763, 764*
Willis, W. W. Jr., 100 (108), 115 (277), 154, 155 (636), 157 (636, 680, 690), 158 (680), 159 (636, 690), *196, 199, 206, 207*
Willson, C. G., 104, 114 (169), *197*
Wilson, C. A. II, 104 (178), *197*
Wilson, J. D., 469 (58), *493*
Wilson, P. S., 385 (139), *391*
Wilson, S. R., 142 (506), *204*
Wingard, R. E. Jr., 721 (239), *762*
Winkelmann, J., 383 (73), 384 (73, 74), *390*
Winter, W., 329 (160), *337*
Wintergerst, H., 405 (79), *418*
Winther, W., 411 (131), *419*
Wiorogorski, W., 221 (51), *268*
Wiriyachitra, P., 342 (23), *347*, 510 (93), *536*, 582 (226), *590*
Wirtz, P., 447 (93), *459*
Wise, D. S., 779, 782 (100, 101), 783 (109–111), *792*
Wisian-Neilson, P., 631 (107), *654*
Witczak, Z. J., 765 (10, 11, 17), 774 (11, 75), 775 (80), *790–792*
Wittek, P. J., 120 (348, 349), *200*
Wittmann, J., 37, 38 (141), *49*, 453 (158), *460*
Wittwer, A. J., 262 (247), *272*, 361 (48, 49, 52), 362 (49, 52, 57), *364, 365*
Wiygul, F., 467 (24), *492*
Wiygul, F. M., 250, 251 (182), *271*, 480 (89), *493*
Woggon, W.-D., 118 (332), *200*
Wohler, F., 505 (61), *535*
Wojciechowski, P. S., 476, 477 (78, 79), *493*
Wojnowska, M., 398 (37), 399 (43), *417*
Wojnowski, W., 398 (37), *417*
Wolf, G. C., 388 (168), *392*
Wolf, H. R., 187 (925), *212*
Wolff, M. E., 372 (70), *376*
Wolffram, S., 382 (49), *389*
Wollenberg, R. H., 187 (912), *212*
Wollowitz, S., 96, 98, 99 (38), 115 (38, 277), 116, 123 (38), 180 (870–872), *194, 199, 211*, 555 (100, 102), *587*, 721 (327), *764*
Wolmershäuser, G., 624 (85), 633 (113), 643, 646 (85), 649 (185), *654, 656*
Wolmershauser, G., 308 (97), *336*
Woltermann, A., 132 (434), *202*, 694, 695, 713, 747, 749 (120), *760*
Woltermann, R., 757 (301), *764*
Wolters, E., 662 (26), *672*
Wong, C. K., 139 (471, 477), 140 (471), 142 (502, 503), 144 (525), 172 (780), 174 (477, 525, 780, 817), 177 (864), *203, 204, 209–211*, 568 (169), *589*, 652 (199), *656*, 663 (32), 665 (52), 666 (52, 57), *672*, 678 (15, 16, 221), 681 (15, 16), 711 (15, 221), 715 (247), 716 (16), 754, 755 (221), *758, 762*, 770 (42–45), *791*
Wong, P. T. S., 369 (14), *374*
Wong, S. K., 227 (71), *269*, 322, 329, 330 (136), *337*
Wong, T. C., 219 (17, 18), 263 (17, 18, 260), 264 (18, 260), *268, 272*
Wong, Y. X., 379 (13), *388*
Woo, D. V., 372 (74), *376*, 388 (164), *392*
Wood, R. J., 340 (13), *346*, 531, 532 (246), *539*
Wood, S. G., 786, 790 (117), *793*
Woodbridge, D. T., 253 (192), *271*, 404 (70), *418*, 677, 681 (9), 718 (273), *757, 763*
Woodgate, P. D., 542 (8), 543 (20), *585*
Woodward, P., 411 (129), *419*

Woodward, R. B., 292 (54), *335*, 565 (145), *588*
Wooten, F., 53 (19), *87*
Wovkulich, P. M., 102, 179 (138), *196*
Wozniak, J., 112 (251), *199*
Wrede, F., 765, 766 (24), *791*
Wright, G. F., 681, 682 (36b), *758*
Wright, J. D., 530 (237), *539*
Wright, M. J., 176 (842), *210*, 398 (33), *417*, 552 (89), *587*
Wu, A., 106 (188, 189), *197*
Wu, M., 9, 10 (45), *47*
Wu, S. H., 383 (69), *390*
Wucherpfennig, W., 439, 441 (54), *458*
Wudl 512, 530 (106), *536*
Wudl, F., 245 (139, 140), 250 (180), *270, 271*, 341 (17, 18), 345 (18), *346*, 409 (120), *419*, 450 (132), *459*, 467 (11, 39, 43), 469 (39, 43), 474 (69, 71), 480 (96), 485 (96, 112), *491–494*, 504 (53), 506 (69), *535*
Wynberg, H., 187 (917), *212*, 315 (120), 316 (120, 121), *336*
Wynne, K. J., 72 (50, 53), *88*, 342 (28, 31), 343 (28), *347*

Xie, Z.-M., 631 (107), *654*

Yagi, T., 356 (30), *364*
Yagubskii, E. B., 467 (32, 33, 48, 49), 483 (48, 49), *492*, 504 (57), 530 (239), *535, 539*
Yagupol'skii, L. M., 297 (74), *335*, 507 (75), *535*, 543 (19), *585*
Yagupolskii, L. M., 670 (78), *673*
Yakobson, G. G., 416 (157), *420*
Yalpani, M., 151 (581–584), *205*, 229 (84), 235 (98), 243 (131, 133, 134), 264 (84), *269, 270*, 290 (47), (46), *335*, 446 (79), 449 (79, 106), 450 (116, 126), *458, 459*
Yamada, K., 565 (149), *588*
Yamagishi, F. G., 467 (10), *491*
Yamaguchi, H., 100 (115), 133 (435), 147 (548, 552), 152, 153 (552), 154 (115, 435, 552), 155 (115, 435), 165 (734), 166 (743, 744), 173 (809), *196, 202, 204, 208, 209*, 396 (15, 16, 20), 397 (16), *417*, 453 (166), *460*, 555 (103, 107), 578, 583 (211), *587, 590*, 718 (272), 755 (322), (323), *763, 764*
Yamaguchi, I., 549 (69), *587*
Yamahira, A., 485 (115), *494*, 501, 514 (43), *535*
Yamakawa, K., 104 (174, 175), 111 (243), 122 (243, 364, 365), 172 (779), *197, 198, 201, 209*
Yamaki, M., (237), *762*
Yamamoto, A., 184 (897), *211*, 726 (329), *764*
Yamamoto, G., 192 (956), *212*, 254 (202), *271*
Yamamoto, H., 775 (81), *792*
Yamamoto, K., 371 (69), *376*
Yamamoto, T., 103, 107 (153), *197*
Yamamoto, Y., 157, 158 (682, 688), *207*
Yamanaka, H., 120 (347), *200*
Yamashita, M., 177 (847), *210*
Yamauchi, K., 193 (967, 968), *213*
Yamauchi, T., 184 (887), *211*
Yamazaki, T., 39, 40 (157), *50*
Yan, C. F., 108 (215–217), *198*
Yan, T.-H., 154, 155, 157, 171, 172 (655), *207*
Yang, G., 370 (45), *375*
Yano, T., 172 (791), *209*, 500 (34, 36), *534, 535*, 667 (63), *672*, 706 (179), *761*
Yanovskaya, I. M., 530 (239), *539*
Yarkova, E. G., 548 (61), *586*
Yashunsky, D. V., 163 (715, 716), *208*
Yasuda, D., 371 (66), *376*
Yasuda, T., 371 (57, 68), *375, 376*
Yasuhara, T., 192 (943), *212*
Yasumura, M., 147 (550), *204*, 706, 719 (166b), *761*
Yates, J. B., 168 (758), *208*, 666 (59), *672*
Yavari, I., 453 (165), *460*
Yeates, C., 172 (796), *209*
Yefremova, L. L., 352 (12), *364*
Yelm, K. E., 180 (872), *211*
Yogai, S., 97 (43), *194*, 555 (104), *587*, 774 (71), *792*
Yokohama, A., 340 (8), *346*
Yokoyama, M., 387 (153), *391*
Yokoyama, S., 191 (961), 192 (945), *212*
Yokoyama, Y., 122 (366), 123 (372), 161 (704), 173 (810), 187 (704), *201, 207, 210*, 558 (118, 119), *588*, 685, 695, 701, 715 (92), *759*, 774 (73), *792*
Yokuyama, Y., 427 (23), *457*
Yonashiro, M., 160 (703), *207*
Yondea, S., 469 (56), *493*
Yoneda, F., 453 (166), *460*
Yoneda, S., 469 (55), *493*
Yonekura, Y., 371 (69), *376*
Yorifuji, T., 356 (27), *364*
Yoshida, Y., 192 (947), *212*
Yoshida, Z., 469 (55, 56), *493*
Yoshida, Z. I., (216), *272*
Yoshifiyi, M., 400 (53), *418*
Yoshifuji, M., 617, 633 (67, 68), *653*
Yoshii, E., 169, 170 (765), *209*
Yoshikoshi, A., 187 (906, 907), *211*
Yoshimoto, H., 178 (854, 858), *210*, 664 (38), *672*
Young, J. D., 370 (47), *375*
Young, M. W., 94 (3), 95, 96, 98 (28), 99 (28, 97), 100, 102, 122 (114), *194–196*, 555 (98), *587*, 677 (2), 681 (2, 27), 685 (2), 714 (27, 233), 715, (2, 27, 233), 716 (27), 719, 720 (2), 721 (27), 756 (2), *757, 758, 762*, 770 (56, 57), *791*
Young, P. A., 360–362 (43), *364*

Yunnikov, V. V., 544 (28), *586*
Yurchakevich, E. E., 467 (35), *492*
Yur'eva, V. S., 532 (258), *539*

Zabolotnaya, T., 455 (197), *461*
Zaiko, E. J., 99 (105), *196*
Zaitseva, G. I., 662 (27), *672*
Zak, Z., 424 (17), 429 (30), 450 (17), *457*
Zakharin, L. I., 409 (123), *419*
Zamarlik, H., 141 (496), 144 (528, 529), *203, 204*
Zamboni, R., 108 (210, 211), *198*
Zambransky, B. J., 62, 74 (24), *88*
Zamojski, A., 770 (52), *791*
Zanati, G., 372 (70), *376*
Zard, S. Z., 112 (251), 191 (966), *199, 213*, 668 (67), *673*
Zask, A., 109 (226), *198*
Zavyalov, S. A., 664 (47), *672*
Zdansky, G., 372 (76), *376*
Zefirov, N. S., 705 (303), *764*
Zemlyanski, N. I., 629 (101), *654*
Zemlyanskii, N. I., 639 (134, 135), *655*
Zetzeke, A., 450 (122), *459*
Zetzsche, A., 151 (593), *205*
Zeuner, S., 259 (232–234), 260 (232), *272*, 410 (127), *419*, 572, 577 (184), *589*
Zheng, M., 371 (56), *375*
Zherikhina, L. N., 467 (35), *492*
Zhigarev, G. G., 409 (123), *419*
Zhu, Z., 383 (67), *390*
Zibarev, A. V., 416 (157), *420*
Ziegenhagen, B., 398 (30), *417*
Ziegler, F. E., 565 (150), *588*
Ziegler, M. L., 259 (231), *272*
Zima, G., 103 (151), 106, 107 (192), 129 (409), 141 (495), 184 (892), 187 (892, 918, 932, 934–936), *197, 202, 203, 211, 212*, 406 (84), *418*, 499, 520 (25), *534*, 715 (249), *762*, 773 (69), *792*
Zima, G. C., 32, 33, 36 (116), *49*, 327 (150), *337*
Zimmer, O., 449 (107), 450 (130), *459*
Zimmer, R., 399 (45), *417*
Zimmer-Gasser, B., 416 (158), *420*
Zimmerman, H. E., 113 (262), 187 (927, 928), *199, 212*, 565 (151), *588*
Zimmermann, H. W., 340 (4), *346*
Zingaro, R., 788 (123), *793*
Zingaro, R. A., 81, 83 (64), *88*, 217 (2), 248 (168), 255 (210, 211), *267, 271*, 369 (15), 371 (54), *374, 375*, 386 (120), *391*, 396 (11), *417*, 423 (5), *456*, 497, 498, 501 (1), 515 (132), 516 (1), 529 (234, 235), *534, 536, 539*, 543, 554 (22), *585*, 593 (3), 602 (34), 603 (35, 36), 610 (3), 613 (60), 628, 634 (3), 635 (60, 115), 637 (3), 641 (141, 143), 642 (150), 648 (3), 650 (3, 194), *652–656*, 765 (12), 768 (12, 30–32), 769 (12, 33–35), 790 (12, 30–34, 135), *791, 793*
Ziolo, R., 453 (179), *460*
Ziolo, R. F., 85 (86), *89*, 345 (43), *347*, 509 (88), 530 (236), *536, 539*
Zolotukhin, S. P., 467 (32, 34), *492*
Zoretic, P. A., 105 (181, 182), 179 (868), *197, 211*, 715 (260), *763*
Zoretie, P. A., 568 (172), *589*
Zountsas, J., 151 (598), *205*, 450 (130), *459*
Zrimsek, Z., 235 (100), *269*
Zuccaro, D. E., 529 (233), *539*
Zumbulyadis, N., 532 (251, 255), *539*
Zutterman, F., 154, 182, 183 (670), *207*, 684 (83), 689 (109), 692 (116), 695 (145), 698, 699, 718, 723, 724 (109), *759, 760*
Zvarykina, A. V., 467, 483 (48), *492*, 530 (239), *539*
Zybill, C., 114 (267), *199*
Zylber, J., 96, 97 (36), *194*, 567 (168), *589*, 785 (114, 115), *793*
Zylber, N., 96, 97 (36), *194*, 567 (168), *589*, 785 (114, 115), *793*

Subject Index

Ab initio self-consistent field (SCF) calculations 6
Acetals—*see also* Diselenoacetals, Ditelluroacetals, Selenoacetals
  dihydropyran—*see* Dihydropyran acetals
  ketene—*see* Ketene acetals
  synthesis of 140
Acetoxyketones, synthesis of 122
Acetoxymethylation 120
Acetylenic sulfones, synthesis of 100, 138
Acidity 339, 340
Acyl halides, synthesis of 133
N-Acylimines 151
Alcohols,
  allenic—*see* Allenic alcohols
  allylic—*see* Allylic alcohols
  dehydration of 173
  homoallylic—*see* Homoallylic alcohols
  propargylic—*see* Propargylic alcohols
  protection of 193
  synthesis of 177, 744
Aldehydes—*see also* Arylselenoaldehydes, Selenoaldehydes, Telluroaldehydes
  regeneration of 150, 153
  synthesis of 120, 749–752
Alkanes, synthesis of 133, 726–730, 754
Alkenes—*see also* Metalloselenoalkenes, Nitroalkenes
  arylation of 168
  cyclization of 145
  dihalogenation of 132, 133
  hindered 167
  isomerization of 130, 177
  reaction with chalcogen halides 129
  synthesis of 720–723, 730–744, 756, 757
Alkylaryltellurium(IV) halides, Mössbauer spectra of 72, 73
Alkyl chalcogenides, photoelectron spectra of 27–30
Alkyl halides, synthesis of 132, 133, 173, 746, 747
α-Alkylidene lactones, synthesis of 744
Alkyl phenyl chalcogenides, photoelectron spectra of 32, 33
Alkynes,
  reactions of 131, 145, 163
  synthesis of 100, 151
Allenes,
  cyclization of 145
  synthesis of 100
Allenic alcohols, synthesis of 180
Allenic sulfones, synthesis of 100, 138
Allyl halides,
  dimerization of 165
  synthesis of 98, 129, 131
Allylic acetates, synthesis of 115
Allylic alcohols,
  synthesis of 115, 118, 155, 156, 158–160, 171, 180, 181, 739
  transposition of 179
Allylic ethers, synthesis of 115
Allylic selenides 133, 154, 157, 158, 162
Allylic tosylamides 127
Amides—*see* Carbinolamides, Carboxamides, Cyclic amides, Selenenamides, Seleninamides, Selenoamides, Selenonamides, Telluroamides, Tellurophosphonic acid amides, Tellurophosphoric acid amides, Tosylamides
β-Amidoselenides 126
Amidoselenious esters 440
β-Amidoselenoxides 97
Amine *N*-oxides, synthesis of 148
Amines—*see also* α-Cyanoamines
  cleavage of 172
Amino acids—*see also* Selenoamino acids, Telluroamino acids
  synthesis of 127
Amino ethers, synthesis of 140
Aminoglycosides 233
Aminoselanes 441

Ammonium salts, cleavage of 172
Anilines, synthesis of 128
Anisoles, photoelectron spectra of 32, 33
Antibodies, tellurium-containing analogues of 373
Arsenous acid esters, synthesis of 609, 610
Arsinous acid esters, synthesis of 604
Arsonous acid esters, synthesis of 606, 607
Aryl halides, synthesis of 133
Aryl ketones, synthesis of 120
Arylselenoaldehydes 118
Atomic orbitals 5
Azaselenapentalenes 456
Azides 226
β-Azidoselenides 127
β-Azidoselenoxides 97
Azines 225
  hindered 167
Azo compounds, synthesis of 148, 150

Baeyer–Villiger reactions 102, 122, 123
Benzeneselenenic acid derivatives 115
Benzeneselenenyl compounds 101, 102, 104, 106–109, 115, 118, 126, 129, 131, 133, 135–137, 139–144, 150–152, 173, 181, 187
Benzeneseleninic acid derivatives 111, 112, 116, 120–122, 124, 126, 128, 146–151, 153, 154, 168, 429
Benzeneseleninyl halides 112, 147, 151
Benzisoselenazoles 446
Benzisoselenazolinones 446–448
Benzocyclobutanes, synthesis of 165
Benzodichalcogenazolium salts 449
Benzofurans, synthesis of 140
Benzopyrans, synthesis of 140
Benzoselenadiazoles, photoelectron spectra of 26, 39
Benzoselenazepines 448
Benzoselenazines 448
Benzoselenazoles, photoelectron spectra of 26
Benzoselenophenoindoles 557
Benzoselenophens, photoelectron spectra of 26
Benzotellurophens, photoelectron spectra of 26
Benzyl halides, synthesis of 133
Benzyl selenides 154
Bibenzyls, synthesis of 165
Bidentate ligands 528
Biphenyls, synthesis of 164
Bis(arylseleno)propenes 158
Bis(organyltelluro)methane halides. Mössbauer spectra of 73, 74
Bis(organyltelluro)methanes, Mössbauer spectra of 66
Bis(pentafluorophenyl)tellurium(IV) dihalides, Mössbauer spectra of 74, 75
Bisselenenylation 102, 105
Bis(trifluoromethyl)tellurium(IV) dihalides, Mössbauer spectra of 74, 75
Bond angles 645–647
Bond lengths 645, 646
α-Bromophenyl selenides, in synthesis of α-selenoalkyllithiums 693

Cannizzaro disproportionation, photolytic 324
Captodative stabilization 670
Carbinolamides, synthesis of 152
Carbocyclic ring closure 143, 144
Carbon—chalcogen multiple bonds 218–220
Carbonylation 191, 192
1,2-Carbonyl transposition 125
Carboselenoic acid esters, photochemistry of 330
Carboxamides, synthesis of 163, 171, 173
Carboxylic acids—*see also* Selenocarboxylic acids, Tellurocarboxylic acids
  synthesis of 752, 753
Chalcanthrenes, photoelectron spectra of 34, 35
Chalcogen dihalides, photoelectron spectra of 14–17
Chalcogen dioxides 103, 107, 110, 118–121, 123, 133, 139, 147, 148, 150, 151, 153, 510
  photoelectron spectra of 13
Charge-transfer complexes 529, 530
  Mössbauer spectra of 85
Cheletropic reactions 292, 293
Chiroptical properties 265
Chlorohydrin esters, synthesis of 130
Chloroselenation 630
Te-Chloro-*N*-tosyltellurimides 443, 444
Cholic acids, side-chain degradation of 112
Claisen rearrangements 117, 170, 182
Cleavage reactions 106, 108, 137, 169, 171, 172
Cobaloximes 304, 669
Complex formation 342–345, 573, 574
Configuration interaction 3
Cope–Claisen rearrangements 565
Cope rearrangements 186, 244
α-Cyanoamines, synthesis of 147
Cyclic amides, synthesis of 142
Cyclic diselenides 503, 504
Cyclic ethers, synthesis of 139, 143, 144
Cyclic phosphonates, synthesis of 143
Cyclic sulfides, synthesis of 142
Cyclic tellurium compounds, Mössbauer spectra of 66, 67, 80–83
Cyclization 139–146, 666–669
Cycloaddition 244, 247, 317
Cyclobutanes—*see also* Benzocyclobutanes
  synthesis of 182
Cyclofunctionalization 139

Cyclophanes—*see also* Diselenacyclophanes, Selenacyclophanes
  synthesis of 165
Cyclopropanes,
  cleavage of 106, 108, 172
  synthesis of 154, 157, 161, 163
Cyclopropyl selenides 182
Cycloreversion 290–292

Decarbonylation 174
Decarboxylation 104, 141, 170, 174, 191
Dehalogenation 175, 176
Dehydration 173
Deoxygenation 99, 171, 174, 176, 177
Deselenation, photochemical 286
Detelluration 259
  photooxidative 329
1, 2-Diacetoxy compounds, synthesis of 124
1, 2-Diacylhydrazines, synthesis of 149
Se-Diamido-*Se*-dichloroselenium compounds 440
Diamino-*Te*-tellurium imides 445
Dianions,
  from β-keto lactones 107
  from propargyl selenide 158
  from α-selenocarboxylic acids 160
Diaryl diselenides 103, 104, 107, 108, 111, 115, 116, 120, 126, 127, 131, 144, 146, 169, 173, 190, 240, 553
  photochemistry of 661
  reactions of 604, 605, 668, 669
  synthesis of 408
Diaryl ditellurides,
  reactions of 578
  synthesis of 408, 584
Diarylselenium compounds 147, 148
Diaryltellurium (IV) dicarboxylates, Mössbauer spectra of 76, 77
Diaryltellurium dihalides 168
Diatomic chalcogen molecules, photoelectron spectra of 6–8
Diazo compounds, reactions of 189–191, 222, 225
Diazonium salts 509
  reactions of 545, 663, 668
Dibenzofurans, synthesis of 164
Diboranyl diselenides 414
Dicarbonyl selenides 426
Dichalcogenides, photoelectron spectra of 35–37
Dichalcogenopyranylidenes 485–488
Diels–Alder reactions 108, 131, 137, 187, 193
Dienes,
  regeneration of 175
  synthesis of 180, 739
Dienones, synthesis of 102, 111
Dihalo-*Te*-diaminotelluriums 445
Se-Dihaloselenium imides 439, 440
Dihalotellurium imides 444
2, 5-Dihydrofurans, synthesis of 139
Dihydropyran acetals, synthesis of 117
Diimides, synthesis of 149
Diketones, synthesis of 111, 121, 124, 166
cis-Diols, synthesis of 124
Diorganotellurium(IV) halides, Mössbauer spectra of 67–71
Dipole moment studies 265
Diselenacyclophanes 555
Diselenadihydrofulvenes 243
Diselenafulvalenes 450
Diselenametacyclophanes 532
Diselenetanes 243
Diselenides 220–222, 227, 228, 240–242, 246–251, 253
  as ligands 528
  cyclic—*see* Cyclic diselenides
  dehydrogenation by 664
  diaryl—*see* Diaryl diselenides
  diboranyl—*see* Diboranyl diselenides
  photochemistry of 279, 282, 306, 307, 317, 328, 525–527, 658–662
  pyrolysis 165
  reactions of 412, 413, 416, 518–520, 667–669
  structure of 513, 514
  sugar—*see* Sugar diselenides
  synthesis of 501–504
  thermolysis of 658–662
Diselenins 450
Diselenoacetals 154, 409
  ketene—*see* Ketene diselenoacetals
Diselenocarbamates 249, 250, 253, 413
Diselenocarboxylato complexes 410
Diselenoesters 230, 239
Diselenoketals 154
Diselenolates 251
Diselenols 556
Diselenourethanes 250
Dismutation reactions 597, 599–605
Disulfides 152
  reduction of 663
Ditellurates 245
Ditelluretanes 341
Ditellurides 255
  as ligands 528
  diaryl—*see* Diaryl tellurides
  Mössbauer spectra of 62–66
  photochemistry of 279, 283, 525–527
  reactions of 412, 518–520, 603, 604
  structure of 516
  synthesis of 510–512
Ditelluroacetals 154, 155
Ditellurodihydrofulvenes 245
Dithiaselanes, reactions of 403
Dithiins, photoelectron spectra of 30, 31
Dithiolium ions 471

Electrical conductivity 463–466
Electrochemical reduction 266
of carbon diselenide 483
Electrochemical selenenylation 103, 107, 108, 126, 135, 139, 141
Electron transmission spectroscopy 45, 46
Enamides, synthesis of 128
Enamine sulfones, synthesis of 138
Ene-4-ones, synthesis of 744
Enones—*see also* α-Haloenones, α-Selenoenones
reactions of 118, 177, 187
synthesis of 101–103, 107, 110, 112, 113, 118, 135, 158, 184, 189
Episelenides 167, 225
Episelenonium ions 542, 546
Epitellurides 130
Epoxidation 99, 123, 156, 161, 163, 186
Epoxides 171, 176
cleavage of 171
deoxygenation of 171, 176
synthesis of 746
Esters—*see also* Diselenoesters, Selenoesters, Telluroesters
amidoselenious—*see* Amidoselenious esters
arsenous acid—*see* Arsenous acid esters
arsinous acid—*see* Arsinous acid esters
arsonous acid—*see* Arsonous acid esters
carboselenoic acid—*see* Carboselenoic acid esters
chlorohydrin—*see* Chlorohydrin esters
cleavage of 169
β-keto—*see* β-Keto esters
selenohydroxamic—*see* Selenohydroxamic esters
phosphinic acid—*see* Phosphinic acid esters
phosphinous acid—*see* Phosphinous acid esters
phosphonic acid—*see* Phosphonic acid esters
phosphonous acid—*see* Phosphonous acid esters
phosphoric acid—*see* Phosphoric acid esters
phosphoroselenoic acid—*see* Phosphoroselenoic acid esters
phosphorous acid—*see* Phosphorous acid esters
selenol—*see* Selenol esters
α, β-unsaturated—*see* α, β-Unsaturated esters
Ethers,
allylic—*see* Allylic ethers
amino—*see* Amino ethers
cleavage of 172
cyclic—*see* Cyclic ethers
methyl—*see* Methyl ethers
synthesis of 231, 233

Flash photolysis 281, 286, 323
Flash vacuum pyrolysis 165, 166
Fluoroalkanes, synthesis of 133
Franck–Condon factors 11, 12
Free-radical reactions 657–671
Friedel–Crafts reactions 163, 167
photolytic 277, 301, 321
Fries rearrangements, photolytic 277, 278, 300, 301, 321
Furans,
photoelectron spectra of 21–25
synthesis of 104

Gelius model 5
Germaneselenols, photochemistry of 309
Germanium chalcogenides, photoelectron spectra of 9
Germylselenols, photochemistry of 281

α-Haloenones 189
synthesis of 129
Haloselenides 129, 131, 154, 167, 175, 193, 713
β-Haloselenoxides 98
Hartree–Fock calculations 6
Hartree–Fock–Slater calculations 17
Heterocycles, as organic conductors 466–490
Heterocyclic tellurium compounds, Mössbauer spectra of 80–83
Heterocyclopentadienes, photoelectron spectra of 21–27
Heterofulvalenes 467–485
Homoallylic alcohols, synthesis of 156, 744
Hückel molecular orbital (HMO) method 6
Hunsdiecker photodecarboxylative rearrangement 322
Hydrazones 222, 223, 227
Hydrogen bonding 339, 340
Hydroselenation 598
Hydroselenohydrins 221
Hydroxamic acids, synthesis of 151
Hydroxydienones, synthesis of 121
Hydroxyketones, synthesis of 122
Hydroxyselenides 427
β-Hydroxyselenides 144, 154, 156, 171, 183, 184, 186, 718
reactions of 720–726
synthesis of 699
δ-Hydroxyselenides 713
β-Hydroxyselenoxides 160

Imides—*see also* Selenimides, Selenoimides, Selenoximides, Tellurimides
diamino-*Te*-tellurium—*see* Diamino-*Te*-tellurium imides
Se-dihaloselenium—*see* *Se*-Dihaloselenium imides

dihalotellurium—*see* Dihalotellurium imides
synthesis of 151
Imines—*see also* *N*-Acylimines, Selenenimines, Seleninimines 226
hindered 167
synthesis of 147
Indoles, synthesis of 148
Infrared spectroscopy 264, 580
Insulators 464
Iodine-125 emission Mössbauer studies 85, 86
Ionization energies 2, 3
Isonitriles 245
Isoselenazoles 446–449
Isoselenazolium salts 446
Isoselenocyanates 237, 254–256, 320, 574–577, 774, 775
photoelectron spectra of 21
reactions of 246, 247
synthesis of 245, 246
Isotellurazoles 454
Isotellurocyanates 585
Isothiazoles, photoelectron spectra of 25
β-Isothiocyanoselenoxides 97

Jahn–Teller effect 3

Ketals—*see also* Diselenoketals
synthesis of 140
Ketene acetals, synthesis of 117
Ketene diselenoacetals 159
Ketene selenoacetals 100
synthesis of 737
β-Keto esters, synthesis of 166
Ketones—*see also* Acetoxyketones, Hydroxyketones, α-(2-Pyridylseleno)ketones, Selenoketones, Telluroketones
aryl—*see* Aryl ketones
methyl—*see* Methyl ketones
propargylic—*see* Propargylic ketones
regeneration of 150, 153
α-selenocyclopropyl—*see* α-Selenocyclopropyl ketones
silyl—*see* Silyl ketones
synthesis of 723–726, 747–752
β-Keto sulfones, synthesis of 138
Koopman's theorem 3

Lactams—*see* α-Methylene-β-lactams, Selenolactams, α, β-Unsaturated lactams
Lactones—*see also* Selenolactones
α-alkylidene—*see* α-Alkylidene lactones
cleavage of 169
α-methylene—*see* α-Methylene lactones
synthesis of 141, 149
α, β-unsaturated—*see* α, β-Unsaturated lactones
Lead chalcogenides, photoelectron spectra of 9, 10
Lewis base reactions 527–529
Ligand-exchange reactions 306
Linear combination of bond orbitals (LCBO) 6

Macrolides, synthesis of 141
Mannich reaction 158
Mass spectroscopy 265
Metalloselenoalkenes 706–709
α-Metalloselenoxides 705
Metal selenides 220, 221, 234, 251, 414
Metal tellurides 221, 414
Mössbauer spectra of 84, 85
Metaselenophosphates 633, 634, 649
Metaselenophosphonates 633
Methylation 193
α-Methylene-β-lactams, synthesis of 186
α-Methylene lactones,
protection of 187
synthesis of 104, 160, 182
Methylene(selenoxo)phosphoranes 631–633
Methyl ethers, synthesis of 125
Methyl ketones, synthesis of 191
Microwave studies 266, 267
Molecular ionic states 3
Molecular orbitals 2–6
highest occupied (HOMOs) 5, 19, 464, 465
lowest unoccupied (LUMOs) 33, 464, 645
theoretical calculations for 5,6
Mössbauer effect 52, 53
Mössbauer quadrupole splittings, additivity model for 77–79
Mössbauer spectroscopy 51–87, 533, 534

Naphthalene-1, 8-dichalcogenides, photoelectron spectra of 36, 37
Natural products, synthesis of 560–567
Nitriles 229, 234
synthesis of 147, 151
α, β-unsaturated—*see* α, β-Unsaturated nitriles
Nitroalkenes, synthesis of 113, 137
Nitrogen compounds, reduction by selenols and tellurols 663, 664
Nitroso compounds,
synthesis of 150
α, β-unsaturated—*see* α, β-Unsaturated nitroso compounds
Nuclear magnetic resonance spectroscopy 263, 264, 340, 415, 518, 530–533, 642–644, 692

O_2, photoelectron spectra of 6, 7
Organocuprates, reactions of 138, 162, 163

Organometallic compounds—*see also* Organocuprates, α-Selenoalkylmetals, α-Telluroalkylmetals
containing tellurium 345, 346
insertion of chalcogen atoms into 399, 405–411, 414–416
reactions of 226, 242, 302–309, 498, 503, 506–509, 545, 546, 660, 661
synthesis of 154, 709–711
Organometallic substitutions 302–309
Organotellurium(IV) halides, Mössbauer spectra of 67–80
Organotellurium ligands, Mössbauer spectra of 83, 84
Oxaselenazines 455
Oxaselenoles 552, 571
2-Oxazines, cleavage of 171
2-Oxazolines, cleavage of 171
Oxetanes, synthesis of 146, 156, 746
Oxyselenenylation, reagents for 116
Oxyselenium radicals 670, 671
Oxytellurenylation, reagents for 116
Oxytelluriation 579

Paterno–Buchi oxetane cycloaddition 317
Peierls' distortion 489
Perseleninic acids 122, 152
Peterson reactions 156
Phenolic coupling 121, 168
Phenols, synthesis of 102
Phenoxachalcogenins, photoelectron spectra of 34
Phenyl chalcogenides, photoelectron spectra of 32–35
Phenylselenolactonization 773
β-Phenylselenosulfones 137
Phosphane selenides 634, 635
complexes of 646, 650
reactions of 650, 651
Phosphane tellurides 635, 651
Phosphine derivatives, photoelectron spectra of 42–44
Phosphine oxides, synthesis of 153
Phosphinic acid esters 624, 625, 636, 637
Phosphinous acid anhydrides, synthesis of 611–614
Phosphinous acid esters, synthesis of 595–601
Phosphonates—*see* Cyclic phosphonates, Selenophosphonates
Phosphonic acid esters 625–627, 637, 638, 652
Phosphonous acid esters, synthesis of 605, 606
Phosphoric acid esters 627–630, 639, 640, 652
Phosphoroselenoic acid esters, photoelectron spectra of 42
Phosphorous acid esters, synthesis of 607–609
Phosphorus compounds,
reactions of 114
α,β-unsaturated—*see* α,β-Unsaturated phosphorus compounds
Photoaddition 309–314
Photocyclization 300
Photocycloaddition 314–317
Photoelectron spectroscopy 1–46, 266, 644, 645
Photoelimination 281–293
Photofragmentation 276–281
Photoionization cross-sections 4
Photoisomerization 326, 327
Photooxidation 327–329
Photorearrangement 318–326, 773
Photoreduction 329–334
Photosubstitution 294–309
Polydentate ligands 528
Polyselanes 403, 412, 415
Polyselenides 415, 497
structure of 514
Polytellurides 512
Porphyrins, tellurium-containing analogues of 372
Propargylic alcohols, synthesis of 119
Propargylic ketones, synthesis of 119, 180
Pseudohalide derivatives, photoelectron spectra of 17–21
Pulse radiolysis 279
Pummerer reactions 102, 124
'Push–pull' interactions 31
2-Pyridylselenenyl halides 108
α-(2-Pyridylseleno)ketones 103, 108
Pyrroles, synthesis of 104

Quinones, synthesis of 121, 147

Reductive deselenization 148, 171, 174, 187
Reductive elimination 156, 160, 161, 175, 176, 186
Resonance stabilization 259–262
Retrocyclization 225

S_2, photoelectron spectra of 7
Schrödinger equation 25
S_H2 displacements 664–669
Se_2, photoelectron spectra of 7, 8
Selenacyclophanes 554
Selenadiazines 455, 456
Selenadiazoles—*see also* Benzoselenadiazoles 237, 242–244, 251 449–453, 477, 478
photochemistry of 287–293, 323, 333
photoelectron spectra of 26, 37
Selenadiazolines 167, 225, 229
Se extrusion reactions of 398

Selenadiazoloquinazolines 550
Selenafulvalenes 469–474, 476, 556
Selenals 228, 244, 326
Selenatriazoles 451, 453
Selenazinones 551
Selenaziridines 445
Selenazoles—*see also* Benzoselenazoles 446–449, 547, 553
  photoelectron spectra of 26
Selenazolines 575
Selenenamides 102, 107, 128, 148, 187
  reactions of 425–428
  synthesis of 424, 425
Selenenic acid anhydrides 671
Selenenic acids 99, 115, 122, 144
Selenenimines 427, 429, 430
Selenenylation 426, 569
trans-Selenenylation 184
Selenenyl halides—*see also* 2-Pyridylselenenyl halides 127, 129, 547
  polymer-supported 103
  reactions of 114, 627
Selenenyl sulfides 428
Selenetanes, ring opening of 542
Selenides—*see also* β-Amidoselenides, β-Azidoselenides, Haloselenides, Hydroxyselenides
  allylic—*see* Allylic selenides
  as ligands 527, 528
  benzyl—*see* Benzyl selenides
  bridgehead 668
  α-bromophenyl—*see* α-Bromophenyl selenides
  cyclopropyl—*see* Cyclopropyl selenides
  dicarbonyl—*see* Dicarbonyl selenides
  from alcohols 173
  metal—*see* Metal selenides
  metalation of 684, 685, 709, 710
  phosphane—*see* Phosphane selenides
  photochemistry of 279, 284, 523–525
  pyrolysis of 165, 166
  reactions of 98, 125, 168, 169, 172, 176, 187, 228, 229, 234–237, 241, 244, 246, 247, 249–251, 253–256, 262, 396, 397, 516–518, 556, 665–667, 677–681, 712, 713
  silyl—*see* Silyl selenides
  structure of 513
  synthesis of 220, 221, 303, 497–501, 555, 569, 681
  trialkylarsane—*see* Trialkylarsane selenides
  trialkylstibane—*see* Trialkylstibane selenides
  trimethylsilyl—*see* Trimethylsilyl selenides
  vinyl—*see* Vinyl selenides
  with β-oxygen substituents 115
Selenimides
  reactions of 432–437
  synthesis of 430–432, 435
Seleninamides 428
Selenines 224
Seleninic acids 122, 149, 150, 152–154
  polymer-supported 121, 124, 147
Seleninimines 429, 430
Seleniranium ions 115, 129, 144
Selenium,
  as an essential trace element 368–370, 374, 378–384
  elemental 103, 107, 114, 149, 191, 220, 245, 247, 249, 253, 497
  extrusion of 225, 246, 284, 287, 396–398, 403, 404, 411–413, 416
  insertion of 317, 318, 395, 396, 402, 413–416, 478–483
    into As—As bonds 400, 401
    into carbon—metal bonds 406–411
    into Co—Co bonds 405
    into Ge—Ge bonds 399
    into Mo—Mo bonds 405
    into P—P bonds 400
    into Rh—Rh bonds 405
    into Sb—Sb bonds 401
    into Si—Si bonds 398
    into Sn—Sn bonds 399
    into W—W bonds 405
  transfer of 635, 641
Selenium compounds,
  anticarcinogenic properties of 384, 385
  biological effects of 378–384
  interaction with heavy metals 387
  medical applications of 387, 388
  toxicity of 377–388
  use in organic synthesis 520–523
Selenium diimides 127, 438, 439
Selenium halides 129, 133, 151, 223, 497, 498
Selenium nucleic acids 360–363
Selenium ylides 161, 182
Selenoacetals 100, 167, 232, 241
  ketene—*see* Ketene selenoacetals
  metalation of 685–693
  reduction of 711
Selenoaldehydes 227, 228, 259, 426, 427, 572
  photoelectron spectra of 39, 40, 266
α-Selenoaldehydes 101, 102, 109, 116, 184, 186
α-Selenoalkylmetals,
  NMR spectra of 692
  reactions of 692, 695–703
  synthesis of 682–695
  use in organic synthesis 726–753
Selenoalkyl radicals 227, 670
Selenoamides 105, 176, 229, 233, 243, 244, 253
  electrochemical studies of 266
  mass spectra of 265
  NMR spectra of 264
  reactions of 238–240

synthesis of 234–237
Selenoamino acids 351–360
Selenoarsanes,
reactions of 647
synthesis of 602, 603, 606, 607, 609, 610
Selenoarsinic acid esters 641, 642
Selenoazepines, photochemistry of 325
Selenobis(diorganylbismuthanes), synthesis of 616
Selenobis(diorganylstibanes), synthesis of 615, 616
Selenobismuthanes, synthesis of 604, 605, 607
Selenobisphosphanes, synthesis of 611–613
Selenoboranes 169, 172, 177, 187
Selenocarbamates 253, 254, 404, 407, 473–476
α-Seleno carbanions 154, 161
Selenocarbohydrates 766–775
Selenocarbonates 174, 253, 254, 256, 477
photochemistry of 332
reactions of 251–253
synthesis of 248–251
Selenocarbonyl compounds,
biological 262, 263
nomenclature of 217, 218
spectroscopic studies of 263–267
synthesis of 220, 221
Selenocarbonyl phthalides 229
Selenocarboxylate salts 241, 242
Selenocarboxylic acids 240–242
metalation of 685
α-Seleno cations 167
Selenochromenes, photochemistry of 319
Selenocyanates 117, 131, 136, 144, 173, 176, 177, 179, 187, 190, 246, 247, 255, 259, 296
complexes of 573, 574
photochemistry of 320
photoelectron spectra of 18–21
physical properties of 548–550
reactions of 550–574
synthesis of 542–548
Selenocyanation 546
Selenocycloarsanes 619, 620
Selenocyclophosphanes 616, 617
α-Selenocyclopropyl ketones 186
Selenocyclostibanes 621
Selenodiarsaanthracenes 614, 615
α-Seleno-β-dicarbonyl compounds 106
Selenodiphosphanes 610, 611
α-Selenoenones 102, 107, 128, 135, 158, 187, 189
Selenoenzymes 350, 351, 367, 368
Selenoesters 243, 428
electrochemical studies of 266
metalation of 685
photochemistry of 299, 300, 321
reactions of 104, 163, 166, 174, 175, 191, 231–233, 235
synthesis of 149, 169, 173, 229–231
UV and visible spectra of 264
X-ray studies of 266
Selenofulvalenes, photoelectron spectra of 41, 42
Selenoglycosides 766, 767
Selenohydrazides 253, 254
Selenohydroxamic esters 191
Selenoimidazolidones 576
Selenoimides, synthesis of 128
Selenoisocyanates—*see* Isoselenocyanates
Selenoketenes 229, 235, 242–245, 288, 289
microwave studies of 267
photoelectron spectra of 37–39
Selenoketones 167, 426, 469, 572
chiroptical properties of 265
dipole moments of 265
electrochemical studies of 266
mass spectra of 265
metalation of 685
NMR spectra of 263
photochemistry of 322, 323, 326, 329, 330
photoelectron spectra of 40, 41
reactions of 224–227, 397, 658, 670, 712
synthesis of 222–224
UV and visible spectra of 264
α-Selenoketones 101, 107–109, 116, 118, 124–126, 135, 184, 191, 685
β-Selenoketones 108, 187
Selenolactams 105
IR spectra of 264
Selenolactones 104, 427
metalation of 685
Selenolates 169, 171, 172, 175–177, 187, 232, 243, 249, 253, 260, 450
Selenoles 469
Selenol esters, photochemistry of 277
Selenoloselenophenes 554
Selenols—*see also* Germaneselenols, Germylselenols 172, 176, 177, 187, 224, 235, 261, 555
acidity and hydrogen bonding in 339, 340
as reducing agents 662–664
Selenonamides 428, 429
Selenones 95, 125, 146, 163, 183
Selenonic acids 152, 154
Selenonium salts 100, 132, 155, 156, 186, 193
photochemistry of 279
reactions of 719, 720
synthesis of 718, 719
Selenonium ylides 705, 706
Selenonucleosides 262, 263, 778–787
biological activity of 789, 790
Selenoorthoesters 100
Selenophenes—*see also* Selenoloselenophenes
photochemistry of 286, 314, 315, 319, 325
Selenophenones 259

Selenophens—*see also* Benzoselenophens
  photoelectron spectra of 21–27
Selenophilic addition 226
Selenophosphanes,
  reactions of 647–648
  synthesis of 595–599, 605–610
Selenophosphates 177, 770
Selenophosphinic acid esters 624, 625, 636, 637
Selenophosphinous acid anhydrides, synthesis of 611–613
Selenophosphinous acid esters, synthesis of 595–599
Selenophosphonates 169, 293
Selenophosphonic acid esters 625–627, 637
  toxicity of 652
Selenophosphonium salts 558
  synthesis of 621–623
Selenophosphonous acid esters, synthesis of 605, 606
Selenophosphoranes 103, 169
Selenophosphoric acid esters 627–630, 639
  toxicity of 628, 652
Selenophosphorous acid esters, synthesis of 607–609
Selenophthalimides 129, 139, 141–144, 173, 190, 427, 558, 772
Selenophthenes, photochemistry of 314
α-Selenopropanoyl halides 186
Selenopyridinones 237
Selenopyridones, photoelectron spectra of 266
Selenopyrones 251, 252
Selenosemicarbazides,
  NMR spectra of 264
  reactions of 247, 248
  synthesis of 246, 247
Selenosemicarbazones 247, 248, 263
α-Selenosilanes 154
Selenostibanes, synthesis of 604, 607
Selenosuccinimides 141, 144, 427
Seleno sugars 775–778
  biological activity of 789, 790
Selenosulfides 154
  synthesis of 142
Selenosulfonates 136, 137, 149, 153, 189
  photochemistry of 280, 310–312
Selenosulfonation 137, 138, 145, 162, 310–313
Selenothiocarbazates 254
Selenothiocarboxylates 137
Selenothiophthenes 261, 262
Selenoureas 246, 250, 474, 574, 575, 766, 769
  mass spectra of 265
  NMR spectra of 264
  reactions of 257, 258
  synthesis of 255–257
  X-ray studies of 266
Selenourethanes 254
Selenoxanthenes, photochemistry of 325
Selenoxanthones 277, 299
Selenoxide elimination 715, 716
Selenoxide hydrates 99
Selenoxides—*see also* β-Amidoselenoxides, β-Azidoselenoxides, β-Haloselenoxides, β-Hydroxyselenoxides, β-Isothiocyanoselenoxides, α-Metalloselenoxides 120, 124, 128, 147–149, 152–154, 168
  α-anions of 160
  diastereomeric 96
  elimination from 95–100, 715–718
  photochemistry of 333
  polymer-supported 124
  rearrangement of 178–181
  reduction of 99, 177
  α-silyl 125
  γ-stannyl 98
  substituent effects in 96–99
  synthesis of 713–715
  vinyl—*see* Vinyl selenoxides
  with β-oxygen substituents 97
Selenoximides 437
Selenoxophosphans 293
Selenoxophospholes 293
Selenuranyl radicals 658, 664–669
Selenylcobaloximes 304
Selenyl radicals 658–664
  reactions of 659–661
  spin trapping of 658, 659
Selenylzirconocenes 305
Self-association 341, 342
Selones—*see* Selenoketones
Selone thiocarbamates 247
Selonethiol esters 230
Semiconductors 464
Semi-empirical calculations 6
Semimetals 465
Se...Se interactions 345, 346
[1, 3] Sigmatropic rearrangements 182, 320
[1, 5] Sigmatropic rearrangements 159, 182
[2, 3] Sigmatropic rearrangements 118, 127, 156, 158, 178–182, 568
Silyl ketones, synthesis of 125, 135, 158, 180, 184
Silyl selenides 155, 156, 177
Spectral fine structure 3
Spin-orbit coupling 3
Spiroketals, synthesis of 140
S_{RN} substitution processes 669, 670
Sugar diselenides 766
Sulfides—*see* Cyclic sulfides, Selenenyl sulfides, Selenosulfides
Sulfones—*see also* α-Phenylselenosulfones
  acetylenic—*see* Acetylenic sulfones
  allenic—*see* Allenic sulfones

enamine—*see* Enamine sulfones
β-keto—*see* β-Keto sulfones
synthesis of 152, 153
vinyl—*see* Vinyl sulfones
β-Sulfonyl allylic alcohols 138
Sulfoxides,
reduction of 177, 663
synthesis of 152, 153
Sulfur compounds,
reactions of 114
α, β-unsaturated—*see* α, β-Unsaturated sulfur compounds
Superconductors 469, 483

Te_2, photoelectron spectra of 7, 8
Telluracyclohexanes, Mössbauer spectra of 66, 67
Telluradiazoles 450, 454, 455
Tellura fatty acids 371, 372
Tellurasteroids 372
Tellurates—*see also* Tetrahaloaryltellurates 580
reduction of 369
Tellurazoles 454
Tellurenic acids 100
Tellurenyl halides 342
Tellurides 303, 369
as ligands 527, 528
from alcohols 173
metal—*see* Metal tellurides
Mössbauer spectra of 62–66
phosphane—*see* Phosphane tellurides
photochemistry of 279, 281, 523–525
pyrolysis of 165
reactions of 125, 133, 154, 162, 168, 169, 233, 234, 240, 254, 255, 396, 397, 516–518, 754–757
structure of 515, 516
synthesis of 164, 504–510
vinyl—*see* Vinyl tellurides
Tellurimides—*see also* *Te*-Chloro-*N*-tosyltellurimides
reactions of 443
synthesis of 441, 442
Tellurites, reduction of 369
Tellurium,
as an essential trace element 374
elemental 165, 245, 254, 258, 505, 506, 510, 511, 600
extrusion of 258, 402, 617
insertion of 317, 318, 396, 416, 600, 613
into C—I bonds 414
into Co—Co bonds 405
into Ge—Ge bonds 399
into P—P bonds 400
into Sb—Sb bonds 401
into Si—Si bonds 398
into Sn—Sn bonds 399
transfer of 635, 639, 640
Tellurium compounds,
biological interactions with 369–371
cyclic—*see* Cyclic tellurium compounds
effect on nervous system 369, 370, 373
heterocyclic—*see* Heterocyclic tellurium compounds
synthetic analogues of naturally occurring 371–373
toxicity of 368
uses in organic synthesis 520–523
Tellurium diimides 444
Tellurium halides 129, 130, 133, 141, 152, 154, 164, 369, 370, 507–509, 511
Tellurium-125 Mössbauer spectroscopy 51–87, 533, 534
general principles of 53–58
parameters for 58–61
Tellurium oxides 120, 122, 124, 139, 149, 150
Tellurium ylides 161
Telluroaldehydes 229, 259
α-Telluroalkylmetals, synthesis of 753, 754
Telluroamides 240
IR spectra of 264
UV and visible spectra of 265
Telluroamino acids 372, 579
Telluroarsanes, synthesis of 603, 604
Tellurobis(diorganylbismuthanes), synthesis of 616
Tellurobis(diorganylstibanes), synthesis of 615, 616
Tellurobismuthanes, synthesis of 604, 605, 607
Tellurobisphosphanes, synthesis of 613, 614
Tellurocarbamates 409
α-Telluro carbanions 154
Tellurocarbazates 254
Tellurocarbohydrates 371, 631, 788, 789
Tellurocarbonyl compounds,
nomenclature of 217, 218
spectroscopic studies of 263–267
synthesis of 220, 221
Tellurocarboxylic acids 242
Tellurocoumaranones 290
Tellurocyanates 173, 259, 510, 577–585
photochemistry of 328
Tellurocyclophosphanes 617–619
Tellurodiarsaanthracenes 614, 615
Tellurodiphosphanes 610, 611
Telluroesters 153, 175, 240
NMR spectra of 264
photochemistry of 278, 301, 321
reactions of 234
synthesis of 233
UV and visible spectra of 265
Tellurohydrazides 254
IR spectra of 264
UV and visible spectra of 265

Telluroketenes 245
Telluroketones 227
Tellurolates 172, 175, 176, 245
Tellurols 584
 acidity and hydrogen bonding in 339, 340
 as reducing agents 664
Tellurones 125, 147, 152, 183
Telluronium salts 133, 755
Tellurophenes 577, 578
 photochemistry of 286
Tellurophenones 259
Tellurophenopyridazines, photochemistry of 328
Tellurophens—*see also* Benzotellurophens
 photoelectron spectra of 21–27
Tellurophosphanes,
 reactions of 647, 648
 synthesis of 599–601, 606, 609, 610
Tellurophosphates 176
Tellurophosphinic acid esters 637
Tellurophosphinous acid anhydrides, synthesis of 613, 614
Tellurophosphinous acid esters, synthesis of 599–601
Tellurophosphonic acid amides 638, 639
Tellurophosphonic acid esters 638
Tellurophosphonium ions, synthesis of 623, 624
Tellurophosphonous acid esters, synthesis of 606
Tellurophosphoranes 308
Tellurophosphoric acid amides 639, 640
Tellurophosphoric acid esters 630, 639, 640
Tellurophosphorous acid esters, synthesis of 609
Tellurostibanes, synthesis of 604, 607
Telluroureas 258–260
 X-ray studies of 267
Telluroxide hydrates 100
Telluroxides 147, 149, 150, 152–154, 193, 756
 elimination from 100
 reduction of 177
TeO, photoelectron spectra of 8
Teratogenicity 373
Te...Te interactions 345, 346
Tetrahaloaryltellurates, Mössbauer spectra of 75, 76
Tetrahydrofurans 29
 synthesis of 139, 146, 156, 746
Tetrahydropyrans, synthesis of 139
Tetraselenafulvalenes 469–473, 476
Tetraselenametacyclophanes 500
Tetraselenides 504, 514
Tetraselenopolyacenes 485
Tetratellurafulvalenes 341, 485
Tetratelluropolyacenes 485
Tetratellurotetracenes 341
Tetrazenes, synthesis of 150
Thermochemistry 657, 658
Thiaselenazines 455
Thiaselenins, photoelectron spectra of 30, 31
Thiatellurins, photoelectron spectra of 30
Thiazoles, photoelectron spectra of 25
Thienothiophens, photoelectron spectra of 26, 27
Thiocarbonyl compounds 153
Thiones 224–226
Thiophenes, synthesis of 104
Thiophens, photoelectron spectra of 21–27
Thiourea–tellurium complexes, Mössbauer spectra of 86, 87
Through-bond interactions 30
Tin chalcogenides, photoelectron spectra of 9, 10
Tosylamides,
 allylic—*see* Allylic tosylamides
 synthesis of 127
Townes and Dailey theory 64
Trialkylarsane selenides 641
Trialkylstibane selenides 642
Triatomic molecules, photoelectron spectra of 10–17
Trimethylsilyl halides, synthesis of 134
Trimethylsilyl selenides 134
Triselanes 404
Triselenides 253, 256, 504, 514
Triselenocarbonates 231, 246
Triselenoorthoesters 154, 157, 167

Ultraviolet and visible spectroscopy 264, 265, 644, 645
α, β-Unsaturated carbonyl compounds, synthesis of 739–744
α, β-Unsaturated esters, synthesis of 103, 744
α, β-Unsaturated lactams, synthesis of 105, 112
α, β-Unsaturated lactones, synthesis of 103, 112
α, β-Unsaturated nitriles, synthesis of 113, 137, 187
α, β-Unsaturated nitroso compounds 113
α, β-Unsaturated phosphorus compounds, synthesis of 114
α, β-Unsaturated sulfur compounds, synthesis of 114

Valence-electron-only-model-potential (VEOMP) calculations 16, 17
Vibrational spectroscopy 533, 644, 645
Vilsmeier–Haack formylation 235
Vinyl compounds, synthesis of 737
Vinyl halides, synthesis of 98, 129, 132, 133
Vinyl isothiocyanates, synthesis of 137
Vinyl selenides 100, 126, 132, 134, 135, 143, 150, 154, 159, 160, 162, 169, 190

in synthesis of α-selenoalkylmetals 695, 706
reactions of 711, 713
synthesis of 737
Vinyl selenoxides 100, 180
Vinyl sulfones, synthesis of 137
Vinyl tellurides 159, 162

Wittig reactions 156, 169, 501
Wolff rearrangements 289

X-ray studies 266, 267, 581

Ylides 161, 182, 705, 706

Zwitterions 340